Ready for student success?

Set your goals high with *Algebra 2* from Holt, Rinehart and Winston.

READY

Holt mathematics goes the extra distance to help you and your students achieve more. ***Algebra 2*** provides traditional math instruction with frequent practice while including options for students to communicate and explore content in ways that illuminate the transitions between concrete and abstract thinking.

With traditional learning in focus, and flexible support on demand, an important range of educational needs can be mastered with confidence:

- **Reaching all learners**
- **Teaching for understanding**
- **Integrating with technology**

Need more than just an assortment of teaching tools?

You'll find clear and accessible goals for support in *Algebra 2.*

PREPARE

Even when teaching with traditional tools, it's important to find clarification and support for alternative resources. Whether you use activity notes, teaching suggestions, content connections, or technology tips for your lessons, Holt's ***Algebra 2*** carefully incorporates your options so that you can find them and use them most efficiently when and if you need them.

The *Annotated Teacher's Edition,* along with Holt's outstanding assortment of print and technology components, ensure well-defined and comprehensive support throughout all phases of your lessons:

- **Preparing with ease**
- **Teaching with options**
- **Assessing with confidence**

TABLE OF CONTENTS

READY...

Each chapter in ***Algebra 2*** includes important features that get students ready to learn. Familiar contexts and clear entry points help students to focus and gain motivation as they proceed from basic skills towards understanding of new concepts.

Features that get your students ready to learn:

Quick Warm-Up helps students review previously taught and prerequisite skills.

NAME ______ CLASS ______ DATE ______

Quick Warm-Up: Assessing Prior Knowledge
2.6 Special Functions

Find the greatest integer that is *less than* or *equal to* the given number.

1. 3.97 ______ 2. −3.97 ______ 3. 0.0051 ______ 4. −7 ______

Evaluate.

5. |25.1| ______ 6. |−0.7| ______

Each manageable **Objective** helps students focus by breaking down the skills and concepts to be learned in each lesson.

Objective

- Write, graph, and apply special functions: piecewise, step, and absolute value.

Special Functions

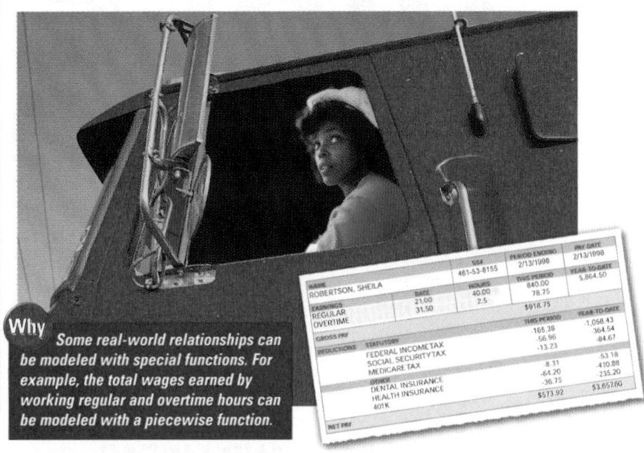

Objective

- Write, graph, and apply special functions: piecewise, step, and absolute value.

Why *Some real-world relationships can be modeled with special functions. For example, the total wages earned by working regular and overtime hours can be modeled with a piecewise function.*

Piecewise Functions

APPLICATION
INCOME

A truck driver earns $21.00 per hour for the first 40 hours worked in one week. The driver earns time-and-a-half, or $31.50, for each hour worked in excess of 40. The pair of function rules below represent the driver's wage, $w(h)$, as a function of the hours worked in one week, h.

$$w(h) = \begin{cases} 21h & \text{if } 0 < h \le 40 \\ 31.5h - 420 & \text{if } h > 40 \end{cases}$$

This function is an example of a *piecewise function*. A **piecewise function** consists of different function rules for different parts of the domain.

CRITICAL THINKING Where does the number −420 in the piecewise function w come from?

Exploring Piecewise Functions

You will need: graph paper

1. Copy and complete the table below, using the function for the truck driver's wages given above.

Hours worked, *h*	10.0	30.0	35.0	40.0	52.5
Wage, *w*(*h*)					

2. Extend your table by choosing other values of h in the interval $0 < h \le 60$.
3. Plot the ordered pairs from your table and make a scatter plot.

CHECKPOINT ✔ 4. If you connect the consecutive points in the scatter plot, what observations can you make about the graph?

124 CHAPTER 2

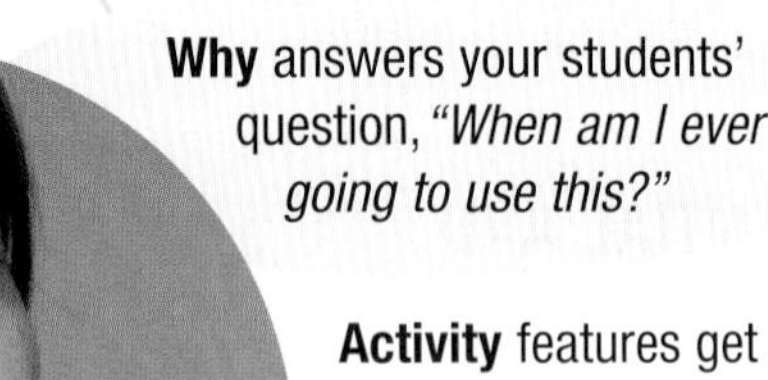

Why answers your students' question, *"When am I ever going to use this?"*

WHY?

Activity features get students practicing, thinking, and making connections.

Every **Application** motivates and helps your students connect with math by demonstrating how concepts and principles are used in the real world.

For Students

SET...

Example 1 shows you how to use the appropriate function rule for each domain of a piecewise function when graphing it.

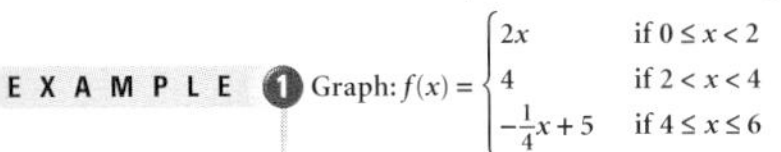

EXAMPLE 1 Graph: $f(x) = \begin{cases} 2x & \text{if } 0 \le x < 2 \\ 4 & \text{if } 2 < x < 4 \\ -\frac{1}{4}x + 5 & \text{if } 4 \le x \le 6 \end{cases}$

Notice that an open circle is graphed at (2, 4) because 2 is not included in the domain.

SOLUTION

1. Graph $y = 2x$ for $0 \le x < 2$.
   $0 = 2(0)$ and $4 = 2(2)$

TRY THIS Grap

St

The ... of fu... same...

A st... rang... basic...

On many graphics calculators, this function is denoted **int**.

Example 2 shows how first-class postage can be modeled with a rounding-up function.

EXAMPLE 2 The cost of mailing a first-class letter (up to 11 ounces inclusive) in 1998 is given by the function $p(w) = 32 + 23\lceil w - 1 \rceil$, where w represents the weight in ounces and p represents the postage in cents.

APPLICATION
CONSUMER ECONOMICS

Find the cost of mailing a first-class letter given each weight.
a. 2.2 ounces **b.** 4.8 ounces

SOLUTION

a. $p(2.2) = 32 + 23\lceil 2.2 - 1 \rceil$
$= 32 + 23\lceil 1.2 \rceil$
$= 32 + 23(2)$
$= 78$
The cost is \$0.78.

b. $p(4.8) = 32 + 23\lceil 4.8 - 1 \rceil$
$= 32 + 23\lceil 3.8 \rceil$
$= 32 + 23(4)$
$= 124$
The cost is \$1.24.

TRY THIS Find the cost of mailing an 0.8-ounce and a 2.9-ounce first-class letter.

Example 3 shows you how to graph a variation, or transformation, of a basic step function.

EXAMPLE 3 Graph $g(x) = 2\lceil x \rceil$.

SOLUTION

PROBLEM SOLVING **Make a table** to compare values of $f(x) = \lceil x \rceil$ with values of $g(x)$.

Interval of x	$f(x) = \lceil x \rceil$	$g(x) = 2\lceil x \rceil$
$-4 < x \le -3$	-3	-6
$-3 < x \le -2$	-2	-4
$-2 < x \le -1$	-1	-2
$-1 < x \le 0$	0	0
$0 < x \le 1$	1	2
$1 < x \le 2$	2	4
$2 < x \le 3$	3	6

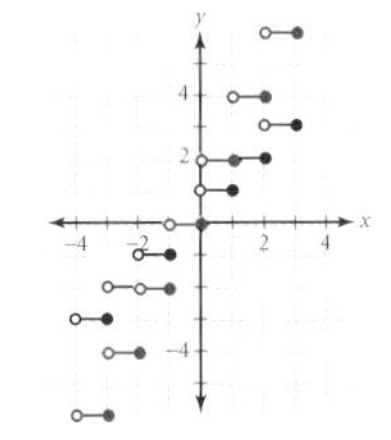

TRY THIS Graph $g(x) = \lceil x \rceil - 1$.

CRITICAL THINKING Let $f(x) = [x]$ and $g(x) = \lceil x \rceil$. Compare and contrast $f - g$ with $g - f$.

126 CHAPTER 2

Each chapter in *Algebra 2* supports both skills-based and concept-building instruction—targeting traditional lesson requirements and including options for changing demands in your classroom:

Example-Solution-Try This format helps to segment lessons into manageable bites.

Skills practice abounds, beginning with **Communicate** and **Guided Skills Practice** in the chapter and extending to **Extra Practice** in the **Info Bank.**

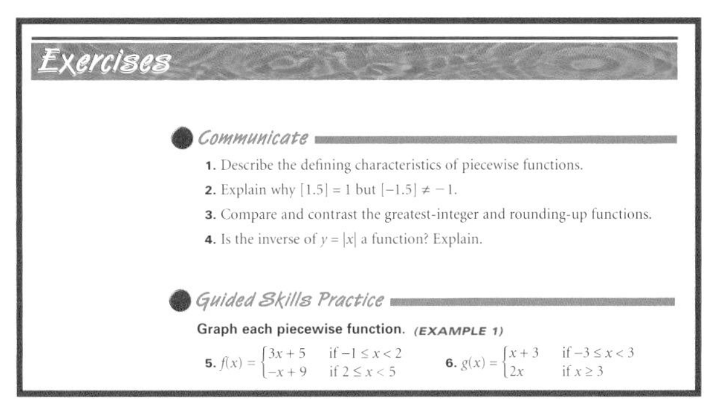

Exercises

Communicate

1. Describe the defining characteristics of piecewise functions.
2. Explain why $[1.5] = 1$ but $[-1.5] \ne -1$.
3. Compare and contrast the greatest-integer and rounding-up functions.
4. Is the inverse of $y = |x|$ a function? Explain.

Guided Skills Practice

Graph each piecewise function. *(EXAMPLE 1)*

5. $f(x) = \begin{cases} 3x + 5 & \text{if } -1 \le x < 2 \\ -x + 9 & \text{if } 2 \le x < 5 \end{cases}$
6. $g(x) = \begin{cases} x + 3 & \text{if } -3 \le x < 3 \\ 2x & \text{if } x \ge 3 \end{cases}$

Extra Practice

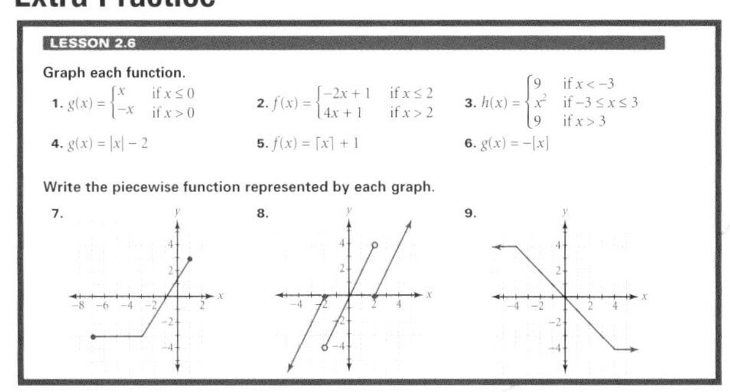

LESSON 2.6

Graph each function.

1. $g(x) = \begin{cases} x & \text{if } x \le 0 \\ -x & \text{if } x > 0 \end{cases}$
2. $f(x) = \begin{cases} -2x + 1 & \text{if } x \le 2 \\ 4x + 1 & \text{if } x > 2 \end{cases}$
3. $h(x) = \begin{cases} 9 & \text{if } x < -3 \\ x^2 & \text{if } -3 \le x \le 3 \\ 9 & \text{if } x > 3 \end{cases}$
4. $g(x) = |x| - 2$
5. $f(x) = \lceil x \rceil + 1$
6. $g(x) = -[x]$

Write the piecewise function represented by each graph.

7. 8. 9.

Look Back

77. CONSUMER ECONOMICS A valet parking lot charges a fixed fee of \$2.00 to park a car plus \$1.50 per hour for covered parking. *(LESSON 2.3)*
a. Write a linear function, c, to model the valet parking charge for h hours of covered parking.
b. If a car is parked in covered parking for 3.5 hours, what is the charge?

Let $f(x) = x + 2$ and $g(x) = 3x$. Perform each function operation. State any domain restrictions. *(LESSON 2.4)*

78. Find $f \circ g$. **79.** Find $g \circ f$.
80. Find $f \circ f$. **81.** Find $g \circ g$.
82. Find $f - g$. **83.** Find $f + g$.
84. Find $f \cdot g$. **85.** Find $\frac{f}{g}$.

Find an equation for the inverse of each function. Then use composition to verify that the equation you wrote is the inverse. *(LESSON 2.5)*

86. $f(x) = 3x - \frac{1}{2}$ **87.** $a(x) = \frac{3}{4}(x - 2)$
88. $h(x) = -2(x - 4) + 1$ **89.** $g(x) = -x + 8$

Look Beyond

90 TRANSFORMATIONS Graph $f(x) = \frac{|x|}{x}$, $g(x) = \frac{2|x|}{x}$, and $h(x) = \frac{-2|x|}{x}$. State the domain and range for each function.

Critical Thinking questions develop problem-solving and higher-order thinking skills.

Look Back and **Look Beyond** exercises offer practice, reinforce learning, and foreshadow upcoming topics or challenge students to stretch their knowledge.

For Students

GO FARTHER!

Algebra 2 offers students ample opportunities to review, extend, and apply what they've learned. Holt's outstanding integration fosters lasting comprehension—with quick access to an assortment of unique and optional tools.

Engaging and practical options for ensuring comprehension:

Real-world integration includes the compelling stories and actual news events of **Eyewitness Math.**

REAL-WORLD

More than 3600 space missions since 1957 have left thousands of large and millions of smaller debris objects in near-Earth space. Information about the orbital debris is needed to determine the current and future hazards that this debris may pose to space operations. Only the largest objects can be repeatedly tracked and cataloged.

The table at right shows the estimated number of cataloged rocket bodies and fragmentation debris from 1965 to 1990 in five-year periods.

Space Debris Table

Year	Rocket bodies	Fragmentation debris
1965	175	900
1970	350	1850
1975	525	2250
1980	700	2600
1985	875	3200
1990	1050	2900

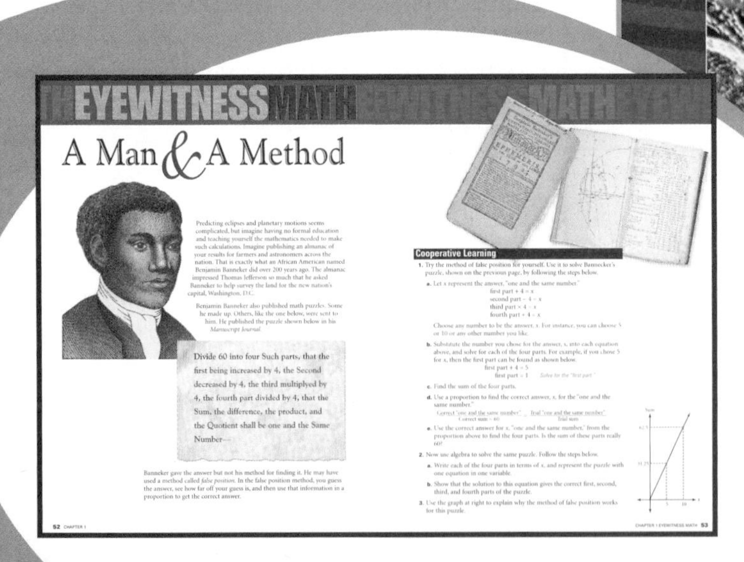
EYEWITNESS MATH

A Man & A Method

Divide 60 into four Such parts, that the first being increased by 4, the Second decreased by 4, the third multiplyed by 4, the fourth part divided by 4, that the Sum, the difference, the product, and the Quotient shall be one and the Same Number—

Cooperative Learning

For Students

Skylab, *the United States' first space station*

Activity 1

1. Use the data from the Space Debris Table on the previous page, which shows the estimated number of cataloged rocket bodies and fragmentation debris for the years 1965 to 1990. Find the average annual rate of change in the number of rocket bodies and debris from 1965 to 1990. Then find the rate of change for each five-year interval. Compare the individual five-year rates with the average rate.
2. Find a linear model for all of the rocket body data from Step 1. Let the years be represented by x (where $x = 0$ represents 1965 and $x = 5$ represents 1970). Verify your equation by using data from the table.

Activity 2

1. Create a scatter plot for the number of other cataloged debris objects for the given years from 1965 to 1990 from the Space Debris Table on the previous page. Plot the years on the horizontal axis (where $x = 0$ represents 1965) and number of other debris objects on the vertical axis.
2. Sketch the curve on your scatter plot that you think best models your data. Describe the trend you see in the scatter plot. Do you think you can make reliable predictions by using the model you sketched? Explain.
3. Using a graphics calculator, find linear, quadratic, and exponential regression equations for this data. Discuss which model best approximates the data.

Cataloged Space Debris

Number of debris objects: 200, 400, 600, 800, 1000, 1200

Altitude (km): 400, 800, 1200, 1600, 2000, 2400, 2800

Activity 3

1. Using the Cataloged Space Debris graph at left, which shows the distribution of cataloged debris objects by altitude, describe what happens to the number of debris objects as the altitude increases. Write an approximate step function that models the data in the graph.
2. Using the step function that you wrote in Step 1, estimate the number of objects at an altitude of 725 kilometers, 1450 kilometers, and 1900 kilometers. Discuss the usefulness of your model.

CHAPTER 2 PROJECT **143**

Integration of **Multiple Representations** includes verbal, symbolic, tabular, and graphical representations.

Performance assessment in ***Algebra 2*** encourages your students' independence and confidence in math and helps to increase the probability that they will remember what they learn.

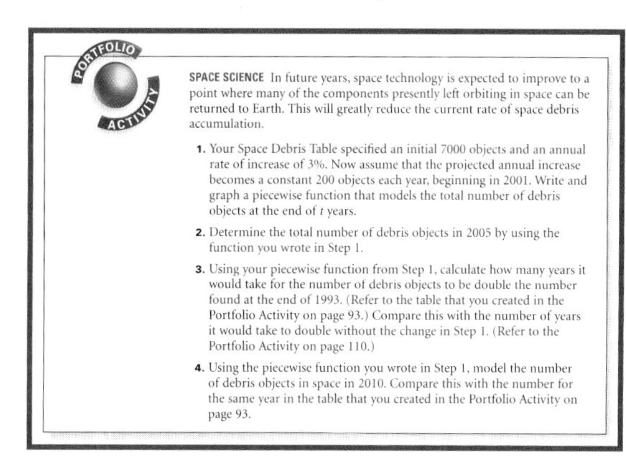

PORTFOLIO ACTIVITY

SPACE SCIENCE In future years, space technology is expected to improve to a point where many of the components presently left orbiting in space can be returned to Earth. This will greatly reduce the current rate of space debris accumulation.

1. Your Space Debris Table specified an initial 7000 objects and an annual rate of increase of 3%. Now assume that the projected annual increase becomes a constant 200 objects each year, beginning in 2001. Write and graph a piecewise function that models the total number of debris objects at the end of t years.
2. Determine the total number of debris objects in 2005 by using the function you wrote in Step 1.
3. Using your piecewise function from Step 1, calculate how many years it would take for the number of debris objects to be double the number found at the end of 1993. (Refer to the table that you created in the Portfolio Activity on page 93.) Compare this with the number of years it would take to double without the change in Step 1. (Refer to the Portfolio Activity on page 110.)
4. Using the piecewise function you wrote in Step 1, model the number of debris objects in space in 2010. Compare this with the number for the same year in the table that you created in the Portfolio Activity on page 93.

Assessment options include:

- Portfolio Activity
- Chapter Project
- College Entrance Exam Practice

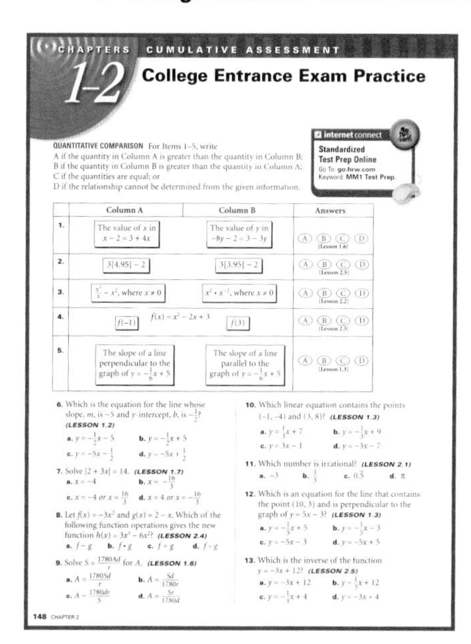

CHAPTERS 1–2 CUMULATIVE ASSESSMENT

College Entrance Exam Practice

internet connect — Standardized Test Prep Online — Go To: go.hrw.com Keyword: MM1 Test Prep

QUANTITATIVE COMPARISON For Items 1–5, write
A if the quantity in Column A is greater than the quantity in Column B;
B if the quantity in Column B is greater than the quantity in Column A;
C if the quantities are equal; or
D if the relationship cannot be determined from the given information.

	Column A	Column B	Answers
1.	The value of x in $x - 2 = 3 + 4x$	The value of y in $-8y - 2 = 3 - 3y$	A B C D [Lesson 1.6]
2.	$3[4.95] - 2$	$3[3.95] - 2$	A B C D [Lesson 2.5]
3.	$\frac{x^3}{x} - x^2$, where $x \neq 0$	$x^2 \cdot x^{-2}$, where $x \neq 0$	A B C D [Lesson 2.2]
4.	$f(-1)$ $f(x) = x^2 - 2x + 3$	$f(3)$	A B C D [Lesson 2.3]
5.	The slope of a line perpendicular to the graph of $y = -\frac{1}{6}x + 5$	The slope of a line parallel to the graph of $y = -\frac{1}{6}x + 5$	A B C D [Lesson 1.3]

6. Which is the equation for the line whose slope, m, is -5 and y-intercept, b, is $-\frac{1}{2}$? *(LESSON 1.2)*
   a. $y = -\frac{1}{2}x - 5$ b. $y = -\frac{1}{2}x + 5$ c. $y = -5x - \frac{1}{2}$ d. $y = -5x + \frac{1}{2}$
7. Solve $|2 + 3x| = 14$. *(LESSON 1.7)*
   a. $x = -4$ b. $x = -\frac{16}{3}$ c. $x = -4$ *or* $x = \frac{16}{3}$ d. $x = 4$ *or* $x = -\frac{16}{3}$
8. Let $f(x) = -3x^2$ and $g(x) = 2 - x$. Which of the following function operations gives the new function $h(x) = 3x^3 - 6x^2$? *(LESSON 2.4)*
   a. $f - g$ b. $f \cdot g$ c. $f + g$ d. $f \div g$
9. Solve $S = \frac{1780Ad}{r}$ for A. *(LESSON 1.6)*
   a. $A = \frac{1780Sd}{r}$ b. $A = \frac{Sd}{1780r}$ c. $A = \frac{1780dr}{S}$ d. $A = \frac{Sr}{1780d}$
10. Which linear equation contains the points $(-1, -4)$ and $(3, 8)$? *(LESSON 1.3)*
    a. $y = \frac{1}{3}x + 7$ b. $y = -\frac{1}{3}x + 9$ c. $y = 3x - 1$ d. $y = -3x - 7$
11. Which number is irrational? *(LESSON 2.1)*
    a. -3 b. $\frac{1}{3}$ c. $0.\overline{5}$ d. π
12. Which is an equation for the line that contains the point $(10, 3)$ and is perpendicular to the graph of $y = 5x - 3$? *(LESSON 1.3)*
    a. $y = -\frac{1}{5}x + 5$ b. $y = -\frac{1}{5}x - 3$ c. $y = -5x - 3$ d. $y = -5x + 5$
13. Which is the inverse of the function $y = -3x + 12$? *(LESSON 2.5)*
    a. $y = -3x + 12$ b. $y = \frac{1}{3}x + 12$ c. $y = -\frac{1}{3}x + 4$ d. $y = -3x + 4$

148 CHAPTER 2

For Students

PREPARE WITH EASE

No matter what your teaching preference may be, the *Annotated Teacher's Edition* for ***Algebra 2*** saves you time and offers you superior guidance in lesson preparation.

Chapter and lesson support includes a variety of well-organized traditional tools and strategies, including:

- Background Information
- Chapter Resources
- Chapter Objectives

About Chapter 2

Background Information

Chapter 2 begins with a review of real numbers, their properties, the order of operations, and the properties of exponents. At the heart of the chapter is a study of basic function concepts: functions as relations, operations with functions, composition of functions, and inverses of functions. After a brief look at "special" functions—piecewise, step, and absolute-value functions—the chapter concludes with an overview of basic transformations of the graphs of functions.

CHAPTER RESOURCES

- Block-Scheduling Handbook
- Writing Activities for Your Portfolio
- Tech Prep Masters
- Long-Term Project
- Assessment Resources: Mid-Chapter Assessment, Chapter Assessments, Alternative Assessments
- Test and Practice Generator
- Technology Handbook

Chapter Objectives

- Identify and use properties of real numbers. [2.1]
- Evaluate expressions by using the order of operations. [2.1]
- Evaluate expressions involving exponents. [2.2]
- Simplify expressions involving exponents. [2.2]
- Graph a relation, state its domain and range, and tell whether it is a function. [2.3]
- Write a function in function notation and evaluate it. [2.3]

Numbers and Functions

FUNCTIONS ARE USED IN THE REAL WORLD to quantify trends and relationships between two variables. For example, the relationship between the *speed* at which an amusement park ride rotates and the *force* that holds riders in their seat can be described by a function.

Lessons

About the Photos

Any object that is moving in a circular path experiences *centripetal force.* Centripetal force is always directed toward the center of the circular path and depends on the object's mass, m, its velocity, v, and the radius of the circular path, r. The formula used to calculate centripetal force is $F = \frac{mv^2}{r}$.

Suppose that a person is on an amusement park ride that moves in a circular path with a radius of 10 meters (≈30 feet) at about 4.5 meters per second (≈10 miles per hour). If the person's mass is 55 kilograms (≈120 pounds), then the force is $\frac{55(4.5)^2}{10}$, or approximately 111 newtons. [A] newton is the amount of force required to ac[celer]ate a 1-kilogram mass at a rate of 1 meter pe[r sec]ond per second.)

The centripetal force is said to be a fu[nction] of each rider's mass. For this particular[...] $\frac{v^2}{r} = \frac{4.5^2}{10} = 2.025$, so the relationship betwe[en the] mass of a rider and the centripetal force f[or] that rider can be written as $F(m) = 2.025m$.

Prepare

NCTM PRINCIPLES & STANDARDS 1, 2, 5–10

QUICK WARM-UP

Find the greatest integer that is *less than or equal to* the given number.

1. 3.97 3
2. −3.97 −4
3. 0.0051 0
4. −7 −7

Evaluate.

5. |25.1| 25.1
6. |−0.7| 0.7

Also on Quiz Transparency 2.6

Teach

Why Ask students to imagine that a telephone company's charge for a long-distance call is 15¢ per minute or any fraction of a minute. This means that if you talk for $4\frac{1}{2}$ minutes, you are charged for 5 minutes. Tell them that the function relating the cost of a call, y, to the number of minutes, x, is called a *step function.* Elicit their conjectures about the reason that the function is given this name.

Special Functions

Objective
- Write, graph, and apply special functions: piecewise, step, and absolute value.

Why Some real-world relationships can be modeled with special functions. For example, the total wages earned by working regular and overtime hours can be modeled with a piecewise function.

Piecewise Functions

APPLICATION INCOME

A truck driver earns $21.00 per hour for the first 40 hours worked in one week. The driver earns time-and-a-half, or $31.50, for each hour worked in excess of 40. The pair of function rules below represent the driver's wage, $w(h)$, as a function of the hours worked in one week, h.

$$w(h) = \begin{cases} 21h & \text{if } 0 < h \le 40 \\ 31.5h - 420 & \text{if } h > 40 \end{cases}$$

This function is an example of a *piecewise function.* A **piecewise function** consists of different function rules for different parts of the domain.

CRITICAL THINKING Where does the number −420 in the piecewise function w come from?

Activity

Exploring Piecewise Functions

You will need: graph paper

1. Copy and complete the table below, using the function for the truck driver's wages given above.

Hours worked, h	10.0	30.0	35.0	40.0	52.5
Wage, w(h)					

2. Extend your table by choosing other values of h in the interval $0 < h \le 60$.
3. Plot the ordered pairs from your table and make a scatter plot.

CHECKPOINT ✓ 4. If you connect the consecutive points in the scatter plot, what observations can you make about the graph?

Alternative Teaching Strategy

USING TECHNOLOGY Have students analyze the **int** function of their graphics calculators. Use their observations to initiate a discussion about the rounding-down and rounding-up functions.

If students are using TI-82 or TI-83 graphics calculators, tell them to press Y= and clear any existing equations. They should then place the cursor next to **Y1** and enter the following:

TI-83: MATH ▶ NUM 5:int X,T,θ,n)

TI-82: MATH ▶ NUM 4:int X,T,θ)

To see the distinct "steps" of the graph, students must have their calculators in **Dot** mode. Have them press MODE, use the arrow keys to highlight **Dot**, and press ENTER. To view the graph, they should press ZOOM and select **4:ZDecimal**.

To have the cursor trace the graph, instruct students to press TRACE and use the arrow keys. Tell them to observe the display at the bottom of the screen as they do this. Now elicit their ideas about the meaning of the **int** function.

124 LESSON 2.6

- Lesson Resources
- Lesson Objectives
- Quick Warm-Up

For Teachers

About the Chapter Project

Real-world situations are often very complex, with changing or unknown factors. Mathematical models can be used to represent such real-world situations and to predict probable outcomes. In the Chapter Project, *Space Trash*, you will use functions to model data related to the growing problem of space debris orbiting the Earth.

After completing the Chapter Project, you will be able to do the following:

- Use a table to represent the relationship between time in years and the number of space debris objects, and show that an appropriate function models this relationship.
- Find and discuss models for the accumulation of space debris.
- Determine the piecewise function that describes the relationship between altitude and number of orbital debris objects.

About the Portfolio Activities

Throughout the chapter, you will be given opportunities to complete Portfolio Activities that are designed to support your work on the Chapter Project.

- Finding the projected number of space debris objects at the end of each year through the year 2010 is included in the Portfolio Activity on page 93.
- Using exponents to project the number of space debris objects in a given year is included in the Portfolio Activity on page 101.
- Comparing regression models for the space debris data is included in the Portfolio Activity on page 110.
- Operating on function models is included in the Portfolio Activity on page 117.
- Using piecewise functions to model trends that change over time is included in the Portfolio Activity on page 132.

- Perform operations with functions to write new functions. [2.4]
- Find the composite of two functions. [2.4]
- Find the inverse of a relation or function. [2.5]
- Determine whether the inverse of a function is a function. [2.5]
- Write, graph, and apply special functions: piecewise, step, and absolute value. [2.6]
- Identify the transformation(s) from one function to another. [2.7]

Portfolio Activities appear at the end of Lessons 2.1, 2.2, 2.3, 2.4, and 2.6. Each serves as preparation for the Chapter Project. The Portfolio Activities as well as the Chapter Project Activities are appropriate for inclusion in the student's portfolio. Students should be encouraged to include in their portfolios any other work in which they feel a sense of pride or a sense of accomplishment.

internet connect go.hrw.com

Chapter Internet Features and Online Activities

Lesson	Keyword	Page	Lesson	Keyword	Page
2.1	MB1 Homework Help	90	2.4	MB1 Homework Help	115
2.2	MB1 Homework Help	99		MB1 UV	114
	MB1 Athletes	101	2.5	MB1 Homework Help	122
2.3	MB1 Homework Help	108	2.6	MB1 Homework Help	130
	MB1 Overpopulation	109		MB1 Piecewise	132
	MB1 Exponential	110	2.7	MB1 Homework Help	140

CHAPTER 2 **85**

Alternatives for teaching preparation are conveniently located when you need them, including:

- About the Chapter Project
- About the Portfolio Activities
- Internet Connect (links to go.hrw.com)
- In-text references to Lesson Keywords
- Online Technology Updates
- *One-Stop Planner® CD-ROM*

ONE-STOP PLANNER: THE ULTIMATE PLANNING TOOL

ALGEBRA 2 LESSON PLANNER

Lesson 2.1 Operations With Numbers pp. 86-93

NCTM Principles and Standards

Standard 1: Number and Operation
Standard 7: Reasoning and Proof
Standard 9: Connections
Standard 2: Patterns, Functions, and Algebra
Standard 8: Communications
Standard 10: Representation

PREPARE

Objectives
- Identify and use properties of Real Numbers.
- Evaluate expressions by using the order of operations.

Pacing
Traditional: 1 day   Block Schedule: 1/2 day   Two-Year: 2 days

Vocabulary and Key Concepts
- intercepts • irrational numbers • natural numbers • order of operations
- Properties of Real Numbers • rational numbers • Real Numbers • whole numbers

Quick Warm-up
Reviews evaluating expressions containing negative numbers, squares, and square roots, p. 86

For Teachers

TEACH WITH OPTIONS

With ***Algebra 2*** you can expect full teaching support organized for maximum convenience throughout the wrap. Separation between traditional and alternative options makes for easy selection.

Options include:

- Additional Examples
- Teaching Tips
- Critical Thinking references

Teaching Tip

Use some numerical examples to help clarify the piecewise definition of the absolute-value function. For example, because −3 is less than 0, use the second part of the definition to find its absolute value: $|-3| = -(-3) = 3$.

Additional teaching support includes:

- Applications
- Math Connections
- Alternative Teaching Strategies
- Inclusion Strategies
- Enrichment
- Reteaching the Lesson
- Reduced ancillaries at point-of-use

CHECKPOINT ✔
$\lfloor 5 \rfloor = 5$; $\lfloor 3.2 \rfloor = 3$; $\lfloor -4.4 \rfloor = -5$;
$\lceil 5 \rceil = 5$; $\lceil 3.2 \rceil = 4$; $\lceil -4.4 \rceil = -4$

ADDITIONAL EXAMPLE 2

The cost of parking at a local garage is given by the function $c(h) = 5 + 1.25\lceil h - 1 \rceil$, where h is the number of hours that you park and c is the cost in dollars. **Find the cost of parking at this garage for each period of time.**

a. $\frac{3}{4}$ hour **b.** $8\frac{1}{4}$ hours

$5.00 $15.00

TRY THIS
$0.32; $0.78

ADDITIONAL EXAMPLE 3

Graph $g(x) = 0.5\lceil x \rceil$.

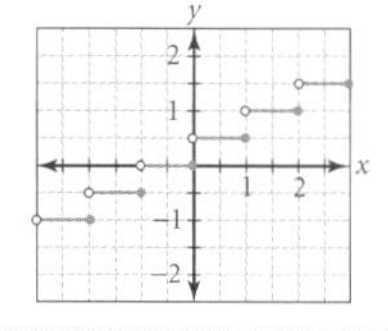

Math CONNECTION

TRANSFORMATIONS The transformation in Example 3 is a *vertical stretch* of the parent step function by a factor of 2. The Try This exercise that follows demonstrates a *translation* of 1 unit down. Students will learn more about transformations like these in Lesson 2.7.

The domain of both $f(x) = \lfloor x \rfloor$ and $f(x) = \lceil x \rceil$ is the set of all real numbers the range of both functions is the set of all integers.

CHECKPOINT ✔ Evaluate $\lfloor 5 \rfloor$, $\lfloor 3.2 \rfloor$, and $\lfloor -4.4 \rfloor$. Evaluate $\lceil 5 \rceil$, $\lceil 3.2 \rceil$, and $\lceil -4.4 \rceil$.

Example 2 shows how first-class postage can be modeled with a rounding function.

EXAMPLE 2 The cost of mailing a first-class letter (up to 11 ounces inclusive) in 1998 is given by the function $p(w) = 32 + 23\lceil w - 1 \rceil$, where w represents the weight in ounces and p represents the postage in cents.

APPLICATION
CONSUMER ECONOMICS

Find the cost of mailing a first-class letter given each weight.
a. 2.2 ounces **b.** 4.8 ounces

SOLUTION

a. $p(2.2) = 32 + 23\lceil 2.2 - 1 \rceil$
$= 32 + 23\lceil 1.2 \rceil$
$= 32 + 23(2)$
$= 78$
The cost is $0.78.

b. $p(4.8) = 32 + 23\lceil 4.8 - 1 \rceil$
$= 32 + 23\lceil 3.8 \rceil$
$= 32 + 23(4)$
$= 124$
The cost is $1.24.

TRY THIS Find the cost of mailing an 0.8-ounce and a 2.9-ounce first-class letter.

Example 3 shows you how to graph a variation, or transformation, of a ba step function.

EXAMPLE 3 **Graph $g(x) = 2\lceil x \rceil$.**

SOLUTION

PROBLEM SOLVING **Make a table** to compare values of $f(x) = \lceil x \rceil$ with values of $g(x)$.

CONNECTION
TRANSFORMATIONS

Interval of x	$f(x) = \lceil x \rceil$	$g(x) = 2\lceil x \rceil$
$-4 < x \le -3$	−3	−6
$-3 < x \le -2$	−2	−4
$-2 < x \le -1$	−1	−2
$-1 < x \le 0$	0	0
$0 < x \le 1$	1	2
$1 < x \le 2$	2	4
$2 < x \le 3$	3	6

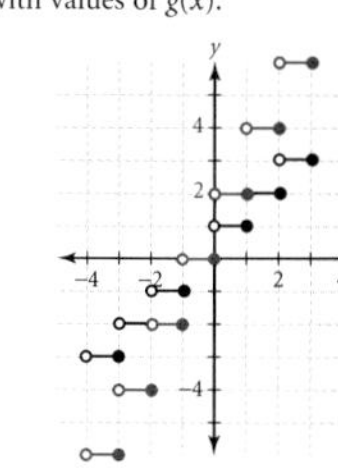

TRY THIS Graph $g(x) = \lceil x \rceil - 1$.

CRITICAL THINKING Let $f(x) = \lfloor x \rfloor$ and $g(x) = \lceil x \rceil$. Compare and contrast $f - g$ with $g - f$.

Inclusion Strategies

VISUAL LEARNERS Some students will find it helpful to know that the rounding-up and rounding-down functions also are called the *ceiling function* and the *floor function*, respectively. These names, together with the basic concept of a step function, provide some rich visual imagery. In fact, students with graphics ability may enjoy creating imaginative instructional posters for these functions. If you wish, you can hold a contest. Display students' posters in the classroom and have the class vote on the most original poster.

Enrichment

Consider the function $f(x) = x - \lfloor x \rfloor$, where x

1. Describe the function in words. **It gives decimal part of a positive real number.**
2. Graph the function. **See graph below.**
3. Write $f(x) = x - \lfloor x \rfloor$ as a piecewise functio over the interval $0 \le x < 6$. **$f(x) = x - (n -$ for $n - 1 \le x < n$, where n is a whole num**

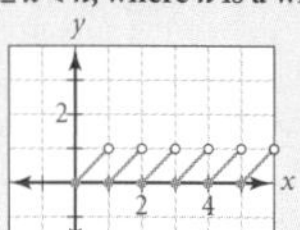

For Teachers

 HOLT, RINEHART AND WINSTON

Absolute-Value Functions

The **absolute-value function**, denoted by $f(x) = |x|$, can be defined as a piecewise function as follows:

$$f(x) = \begin{cases} |x| = x & \text{if } x \geq 0 \\ |x| = -x & \text{if } x < 0 \end{cases}$$

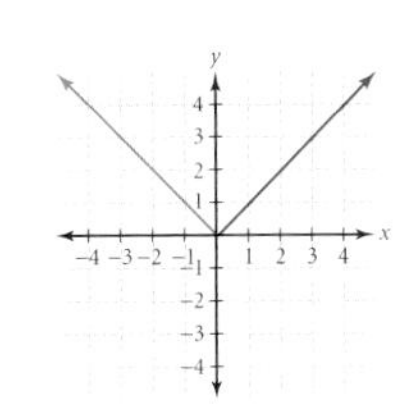

The graph of the absolute-value function has a characteristic V-shape, as shown. The domain of $f(x) = |x|$ is the set of all real numbers, and the range is the set of all nonnegative real numbers.

Example 4 involves a transformation of the absolute-value function.

EXAMPLE 4 **Graph $g(x) = 3|x| - 1$ by making a table of values. Then graph the inverse of g on the same coordinate plane.**

SOLUTION

PROBLEM SOLVING

Make a table of values to compare values of g with values of $f(x) = |x|$.

CONNECTION
TRANSFORMATIONS

x	$f(x) = \|x\|$	$g(x) = 3\|x\| - 1$
−2	2	$3\|-2\| - 1 = 5$
−1	1	$3\|-1\| - 1 = 2$
0	0	$3\|0\| - 1 = -1$
1	1	$3\|1\| - 1 = 2$
2	2	$3\|2\| - 1 = 5$

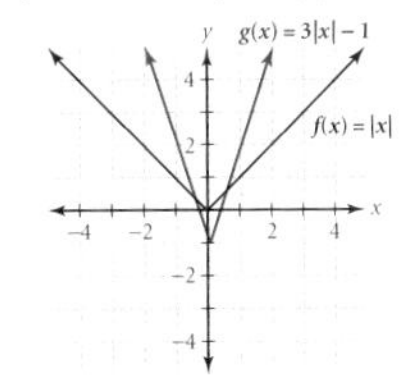

To graph the inverse of g, interchange the values of x and $g(x)$ given in the table above.

TECHNOLOGY
GRAPHICS CALCULATOR
Keystroke Guide, page 152

x	Inverse of g
5	−2
2	−1
−1	0
2	1
5	2

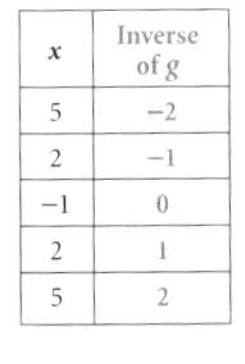

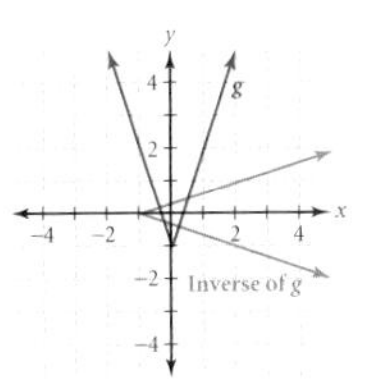

CHECK

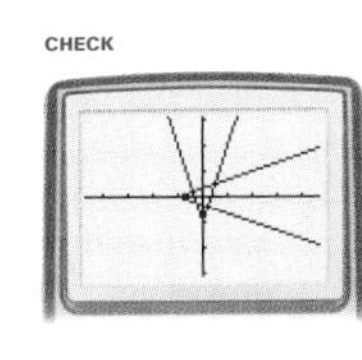

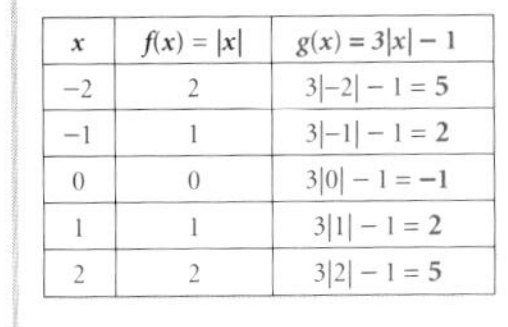

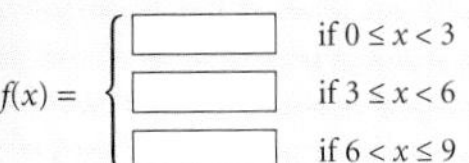

TRY THIS Graph $f(x) = \frac{1}{2}|x| + 1$ by making a table of values. Then graph the inverse of f on the same coordinate plane.

TRY THIS

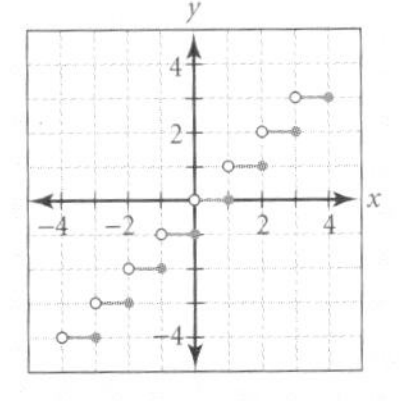

CRITICAL THINKING

If x is an integer, then $(f-g)(x) = 0$ and $(g-f)(x) = 0$.

If x is not an integer, then $(f-g)(x) = -1$ and $(g-f)(x) = 1$.

ADDITIONAL EXAMPLE 4

Graph $f(x) = 3|x| + 2$ by making a table of values. Then graph the inverse of f on the same coordinate plane.

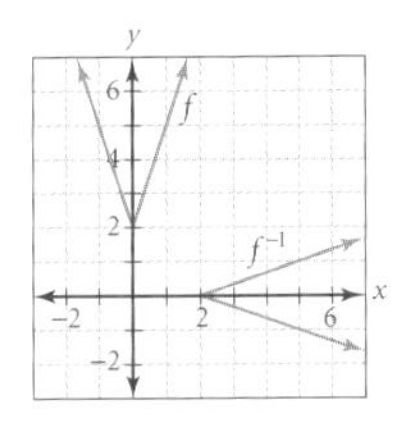

Math CONNECTION

TRANSFORMATIONS Example 4 involves a combination of transformations of the parent absolute-value function: a vertical stretch by a factor of 3 and a translation

Reteaching the Lesson

COOPERATIVE LEARNING Display the following on the board or overhead:

$$f(x) = \begin{cases} \square & \text{if } 0 \leq x < 3 \\ \square & \text{if } 3 \leq x < 6 \\ \square & \text{if } 6 < x \leq 9 \end{cases}$$

Have students work in groups of three or four. Each group should fill in the boxes with linear expressions to create a piecewise ... group should then graph the fun...

After all groups have completed ... volunteers to share their gra... Have each student in the cla... function that the graphs re... dents' answers, and discu... arise.

LESSON 2.6 127

Interactive teaching options feature:

- ***Mathepedia***®—an interactive mathematics encyclopedia on CD-ROM!
- **go.hrw.com**—our award-winning Web site that helps you "click" with technology

MATHEPEDIA

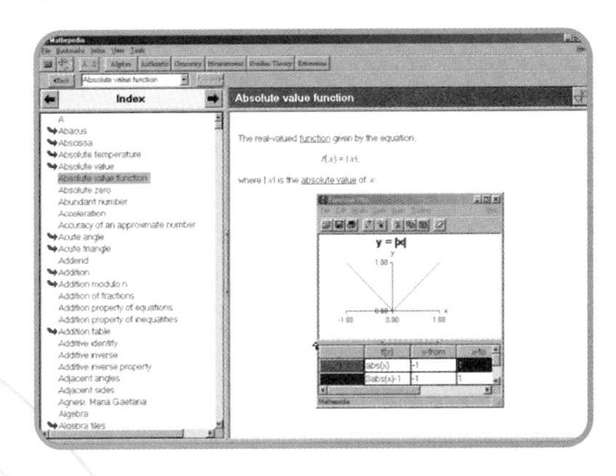

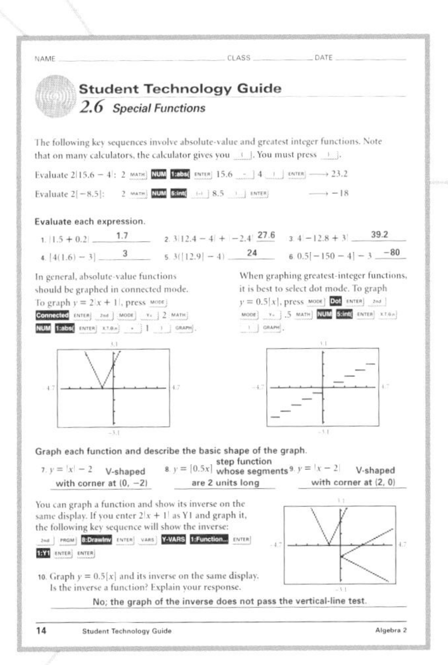

NAME _____ CLASS _____ DATE _____

Student Technology Guide
2.6 Special Functions

Evaluate each expression.

Graph each function and describe the basic shape of the graph.

10. Graph $y = 0.5[x]$ and its inverse on the same display. Is the inverse a function? Explain your response.
No; the graph of the inverse does not pass the vertical-line test.

14 Student Technology Guide Algebra 2

For Teachers

ASSESS WITH CONFIDENCE

Algebra 2 covers a wealth of potential assessment goals for your classroom. Tools address skill-oriented assessment goals and additional needs that challenge students to solve problems, reason critically, understand concepts, and better communicate mathematically.

Skill-oriented assessment includes:

- Practice and Apply for each lesson
- Leveled practice with Practice Masters, Levels A, B, C
- Chapter Test

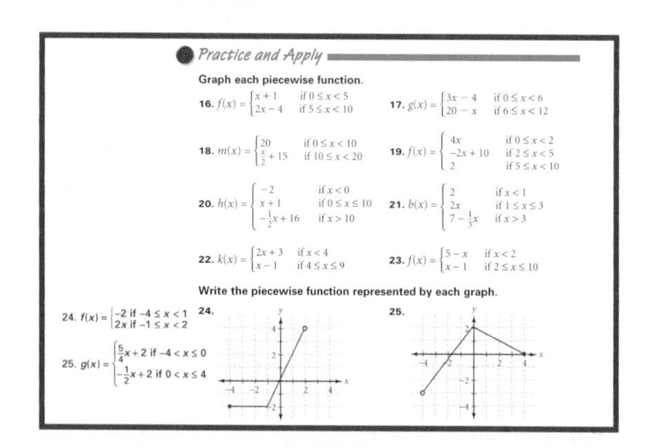

Practice and Apply

Graph each piecewise function.

16. $f(x) = \begin{cases} x+1 & \text{if } 0 \le x < 5 \\ 2x-4 & \text{if } 5 \le x < 10 \end{cases}$ **17.** $g(x) = \begin{cases} 3x-4 & \text{if } 0 \le x < 6 \\ 20-x & \text{if } 6 \le x < 12 \end{cases}$

18. $m(x) = \begin{cases} 20 & \text{if } 0 \le x < 10 \\ \frac{x}{2}+15 & \text{if } 10 \le x < 20 \end{cases}$ **19.** $f(x) = \begin{cases} 4x & \text{if } 0 \le x < 2 \\ -2x+10 & \text{if } 2 \le x < 5 \\ 2 & \text{if } 5 \le x < 10 \end{cases}$

20. $h(x) = \begin{cases} -2 & \text{if } x < 0 \\ x+1 & \text{if } 0 \le x \le 10 \\ -\frac{1}{2}x+16 & \text{if } x > 10 \end{cases}$ **21.** $b(x) = \begin{cases} 2 & \text{if } x < 1 \\ 2x & \text{if } 1 \le x \le 3 \\ 7-\frac{1}{3}x & \text{if } x > 3 \end{cases}$

22. $k(x) = \begin{cases} 2x+3 & \text{if } x < 4 \\ x-1 & \text{if } 4 \le x \le 9 \end{cases}$ **23.** $f(x) = \begin{cases} 5-x & \text{if } x < 2 \\ x-1 & \text{if } 2 \le x \le 10 \end{cases}$

Write the piecewise function represented by each graph.

24. $f(x) = \begin{cases} -2 \text{ if } -4 \le x < 1 \\ 2x \text{ if } -1 \le x < 2 \end{cases}$

25. $g(x) = \begin{cases} \frac{5}{4}x+2 \text{ if } -4 < x \le 0 \\ -\frac{1}{2}x+2 \text{ if } 0 < x \le 4 \end{cases}$

24. **25.**

2 Chapter Test

Evaluate each expression by using the order of operations.

1. $5 + 2(7 - 4)^2$ **2.** $12 - 9 \div 3 + 2 \cdot 5$

3. $\frac{4+6}{2} + 2 \cdot 5$ **4.** $5 \cdot 4 \div 2 + 3^{(4-1)}$

State the property that is used in each statement. All variables represent real numbers.

5. $5x \cdot 1 = 5x$ **6.** $7d - 14 = 7(d - 2)$

7. $\left(\frac{2}{r}\right)\left(\frac{r}{2}\right) = 1$ **8.** $4(yz) = (yz)4$

Simplify each expression. Assume that no variable equals zero.

9. $y^3(x^2y)$ **10.** $(9rt)^2(3rst)^{-3}$

11. $\frac{14r^2s^{-3}t^4}{35r^{-2}s^5t^3}$ **12.** $\left(\frac{3p^4q^{-1}}{8p^{-2}q^3}\right)^{-2}$

13. PHYSICS Kinetic energy can be measured in joules and is given by the formula $k = \frac{1}{2}mv^2$. What is the kinetic energy in joules of an object with mass $m = 100$ kg and velocity $v = 5$ meters per second?

State whether the following are functions, and if so give the domain and range.

14. $\{(5, 7), (7, 12), (9, 7), (11, 12), (13, 7)\}$

15. $\{(-2, 4), (0, 6), (-2, 8), (0, 10), (-2, 12)\}$

Evaluate each function for $x = -2$, $x = 0$, and $x = 2$

$= \frac{x^2}{2} + x - 4$

ost of

Evaluate each composite function.

22. $(g \circ f)(3)$ **23.** $(f \circ g)(2)$

24. BIOLOGY The number of times a cricket chirps per minute can be predicted using the formula $n = 4(F - 40)$, where n is the number of chirps per minute and F is the Fahrenheit temperature. The Celsius scale for temperature can be converted to Fahrenheit using $F = \frac{9}{5}C + 32$. Find the number of cricket chirps per minute as a function of Celsius temperature.

Find the inverse of each function. State whether the inverse is a function.

25. $\{(3, 4), (4, 3), (5, 6), (6, 5), (7, 8), (8, 7)\}$

26. $\{(1, 2), (5, 6), (2, 2), (6, 6), (3, 4)\}$

Find an equation for the inverse and state whether the inverse is a function.

27. $f(x) = \frac{3x - 8}{4}$ **28.** $f(x) = x^4 - 3$

Graph each function.

29. $f(x) = \begin{cases} 2x + 5 & \text{if } x \le 0 \\ -\frac{1}{3}x + 3 & \text{if } x > 0 \end{cases}$

30. $f(x) = 2[x] - 1$

Evaluate.

31. $[4.5] + 3.5$ **32.** $[-1.7] + [1.7]$

Identify each transformation from the parent function f to g.

$\sqrt{x}$, $g(x) = \sqrt{x} - 3$

x^2, $g(x) = (x + 3)^2 - 4$

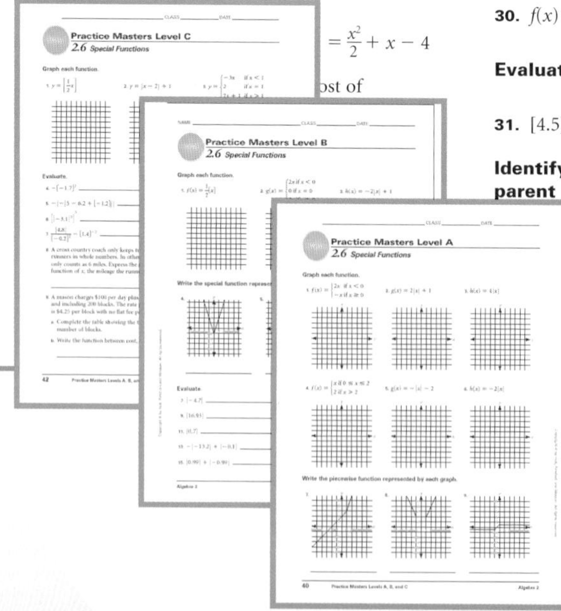

For Teachers

CHAPTERS 1–2 CUMULATIVE ASSESSMENT

College Entrance Exam Practice

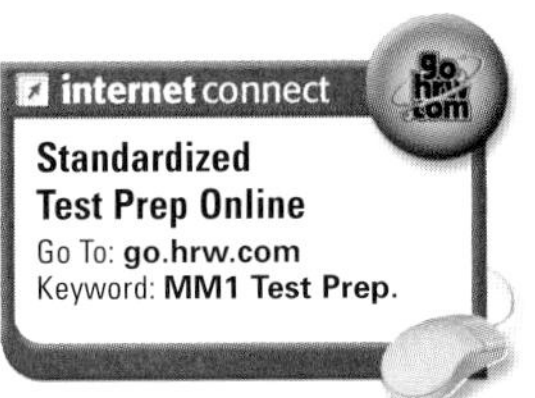

QUANTITATIVE COMPARISON For Items 1–5, write
A if the quantity in Column A is greater than the quantity in Column B;
B if the quantity in Column B is greater than the quantity in Column A;
C if the quantities are equal; or
D if the relationship cannot be determined from the given information.

	Column A	Column B	Answers
1.	The value of x in $x - 2 = 3 + 4x$	The value of y in $-8y - 2 = 3 - 3y$	(A) (B) (C) (D) [Lesson 1.6]
2.	$3[4.95] - 2$	$3[3.95] - 2$	(A) (B) (C) (D) [Lesson 2.5]
3.	$\frac{x^3}{x} - x^2$, where $x \neq 0$	$x^2 \cdot x^{-2}$, where $x \neq 0$	(A) (B) (C) (D) [Lesson 2.2]
4.	$f(-1)$ $f(x) = x^2 - 2x + 3$	$f(3)$	(A) (B) (C) (D) [Lesson 2.3]
5.	The slope of a line perpendicular to the graph of $y = -\frac{1}{6}x + 5$	The slope of a line parallel to the graph of $y = -\frac{1}{6}x + 5$	(A) (B) (C) (D) [Lesson 1.3]

6. Which is the equation for the line whose slope, m, is -5 and y-intercept, b, is $-\frac{1}{2}$? **(LESSON 1.2)**
a. $y = -\frac{1}{2}x - 5$ **b.** $y = -\frac{1}{2}x + 5$
c. $y = -5x - \frac{1}{2}$ **d.** $y = -5x + \frac{1}{2}$

7. Solve $|2 + 3x| = 14$. **(LESSON 1.7)**
a. $x = -4$ **b.** $x = -\frac{16}{3}$
c. $x = -4$ or $x = \frac{16}{3}$ **d.** $x = 4$ or $x = -\frac{16}{3}$

8. Let $f(x) = -3x^2$ and $g(x) = 2 - x$. Which of the following function operations gives the new function $h(x) = 3x^3 - 6x^2$? **(LESSON 2.4)**
a. $f - g$ **b.** $f \cdot g$ **c.** $f \div g$ **d.** $f \circ g$

9. Solve $S = \frac{1780Ad}{r}$ for A. **(LESSON 1.6)**
a. $A = \frac{1780Sd}{r}$ **b.** $A = \frac{Sd}{1780r}$
c. $A = \frac{1780dr}{S}$ **d.** $A = \frac{Sr}{1780d}$

10. Which linear equation contains the points $(-1, -4)$ and $(3, 8)$? **(LESSON 1.3)**
a. $y = \frac{1}{3}x + 7$ **b.** $y = -\frac{1}{3}x + 9$
c. $y = 3x - 1$ **d.** $y = -3x - 7$

11. Which number is irrational? **(LESSON 2.1)**
a. -3 **b.** $\frac{1}{3}$ **c.** $0.\overline{5}$ **d.** π

12. Which is an equation for the line that contains the point $(10, 3)$ and is perpendicular to the graph of $y = 5x - 3$? **(LESSON 1.3)**
a. $y = -\frac{1}{5}x + 5$ **b.** $y = -\frac{1}{5}x - 3$
c. $y = -5x - 3$ **d.** $y = -5x + 5$

13. Which is the inverse of the function $y = -3x + 12$? **(LESSON 2.5)**
a. $y = -3x + 12$ **b.** $y = \frac{1}{3}x + 12$
c. $y = -\frac{1}{3}x + 4$ **d.** $y = -3x + 4$

Alternative Assessment

The following suggest alternative assessments for students who may benefit from a different type of assessment than the regular chapter quizzes and the mid-chapter/end-of-chapter test. Visit the HRW web site to get additional Alternative Assessment material.

internet connect
Alternative Assessment
Go To: **go.hrw.com**
Keyword: **MB1 Alt Assess**

Performance Assessment

1. a. Simplify $\left[\frac{(2a^{-4}b^3)(3b^5)}{(ab^2)^2}\right]^{-2}$, justifying each step with one of the properties of exponents.
b. Evaluate the expression in part **a** for $a = 6$ and $b = 2$. Give an exact answer.
c. How does the simplified expression from part **a** make it easier to find the solution to part **b**?

2. Let $f(x) = -2x + 5$ and $g(x) = 3x - 4$.
a. Find $f + g$, $f - g$, fg, and $f \circ g$.
b. Show that $f \circ g$ is not the same as $g \circ f$.
c. List the inverses of f and g. Explain how you can determine whether these inverses are functions.

Portfolio Project

Suggest that students choose one of the following projects for inclusion in their portfolios.

1. a. Graph $f(x) = [x]$, $h(x) = [0.5x + 1]$. Compare the graphs of these three functions. Then generalize the behavior of the graph of $g(x) = [mx + b]$ over the intervals $0 < m < 1$ and $m > 1$.
b. Describe the graph of $f(x) = [x^2]$. Graph the function to confirm or deny your conjecture.
c. Explore the effect of applying the greatest-integer function to other types of functions. To graph a greatest-integer function, set **Y1=int(X)** by pressing MATH, selecting NUM, choosing 5:int(, and pressing X,T,θ,n) ENTER.

internet connect

The table below identifies the pages in this chapter that contain internet and technology information.

Content Links

Activities Online	pages 101, 109, 114
Portfolio Extensions	pages 110, 132
Homework Help Online	pages 90, 99, 108, 115, 122, 130, 140
Graphic Calculator Support	page 150

Resource Links

Parents can go online and find concepts that students are learning—lesson by lesson—and questions that pertain to each lesson, which facilitate parent-student discussion.

Go To: **go.hrw.com**
Keyword: **MB1 Parent Guide**

Technical Support

The following may be used to obtain technical support for any HRW software product.
Online Help: **www.hrwtechsupport.com**
e-mail: **tschrw@hrwtechsupport.com**
HRW Technical Support Center: **(800)323-9239**
7 AM to 10 PM Monday through Friday Central Time

Visit the HRW math web site at: **www.hrw.com/math**

84C CHAPTER 2 INTERLEAF

Alternative assessment includes:

- College Entrance Exam Practice
- Performance Assessment
- Portfolio Project
- Internet Connect with Portfolio Links
- *One-Stop Planner CD-ROM with Dynamic Test Generator*

TEST GENERATOR

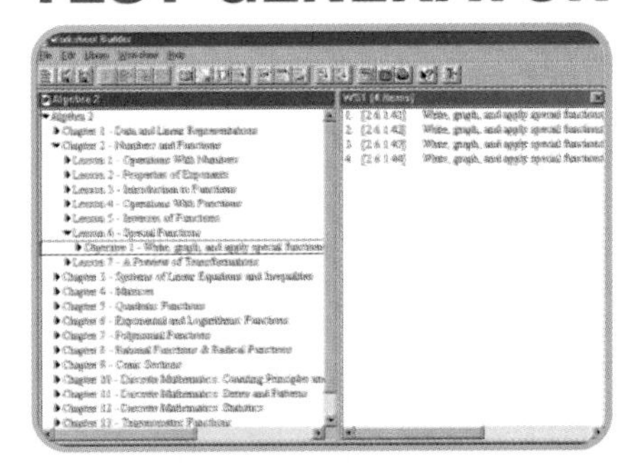

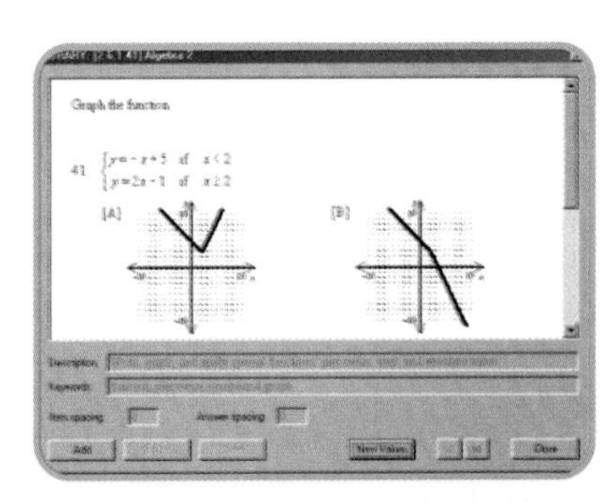

For Teachers

HOLT, RINEHART AND WINSTON

TEACHING RESOURCES

OVERHEAD TRANSPARENCIES

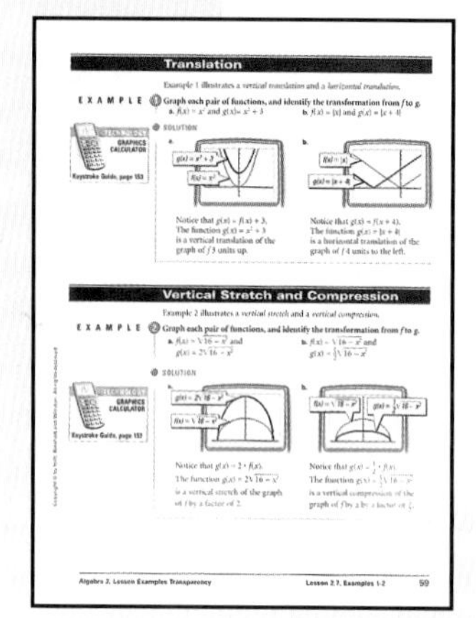
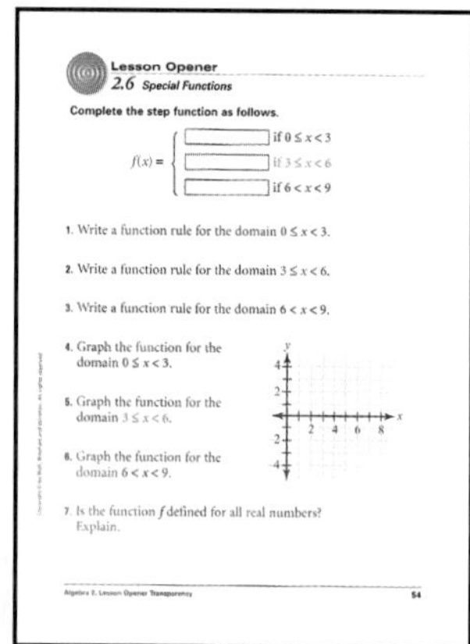

Lesson Presentation Transparencies

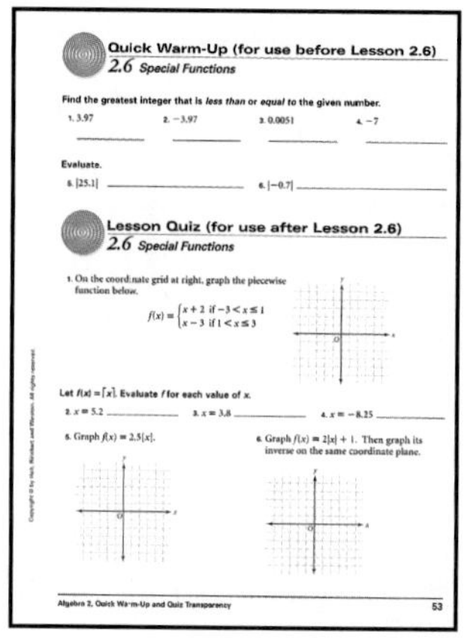

Quiz Transparencies

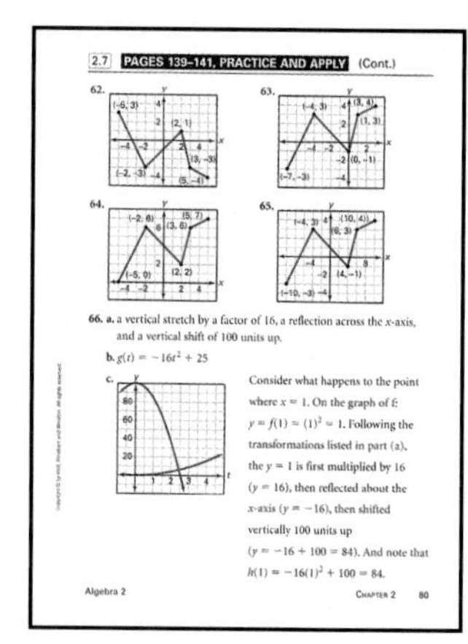

Answer Key Transparencies

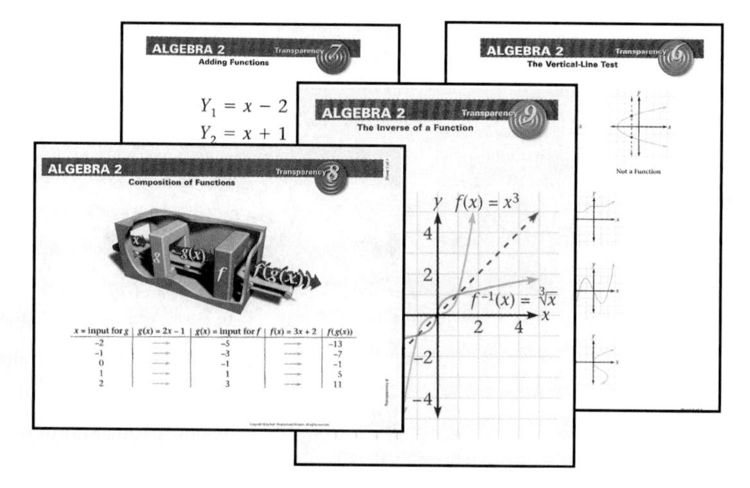

Teaching Transparencies

PRACTICE & ASSESSMENT

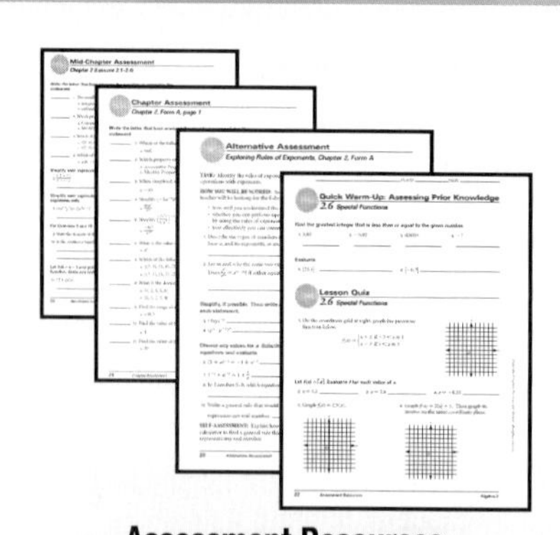

Assessment Resources

Quick Warm-Up, Lesson Quiz, Mid-Chapter Assessment, Chapter Assessments A & B, Alternative Assessments A & B

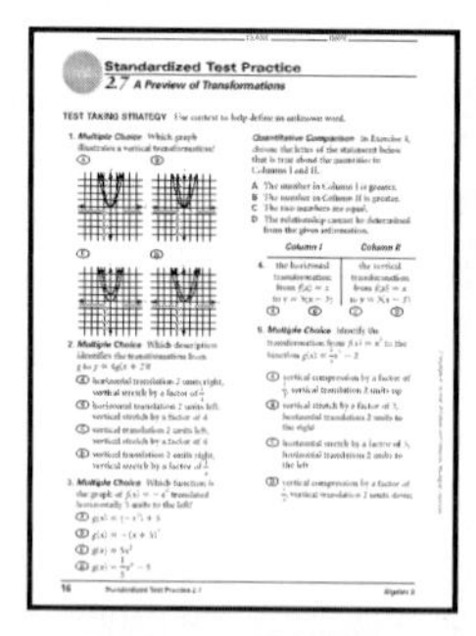

Test and Practice Generator Item Listing

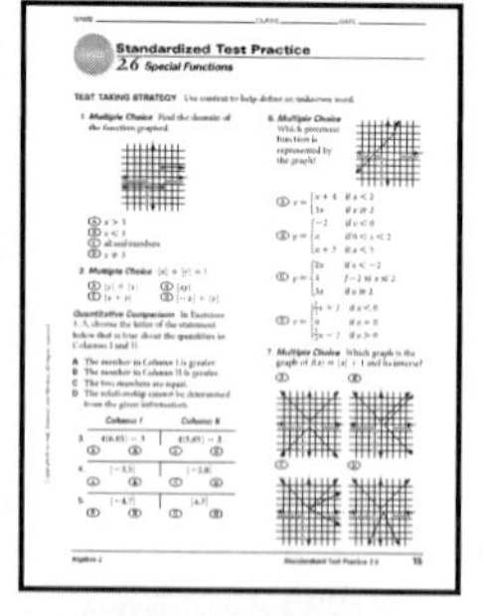

Standardized Test Practice Masters

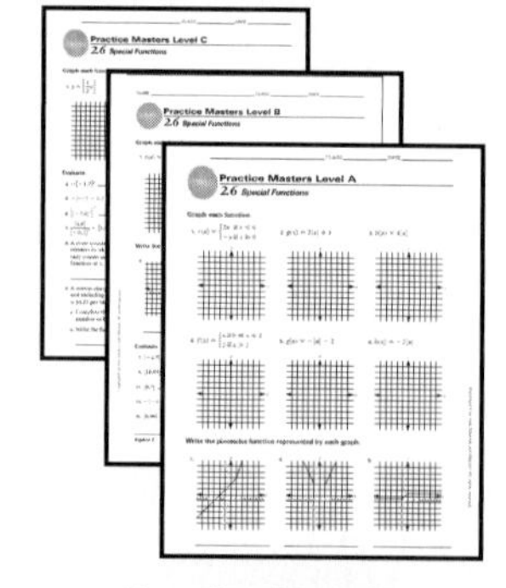

Practice Masters Levels A, B, C

STUDENT HELP

- Homework Help Online
- Standardized Test Prep Online

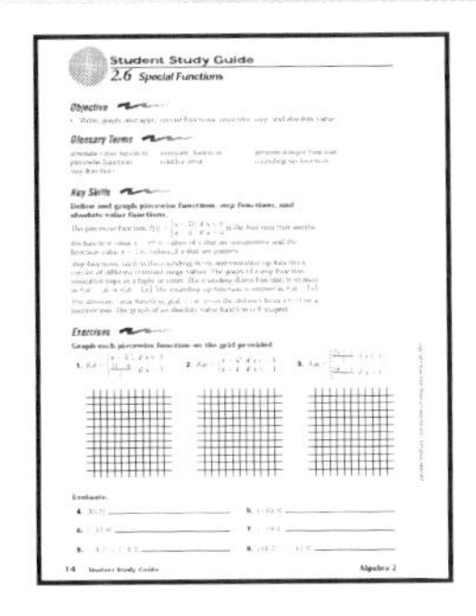

Student Study Guide

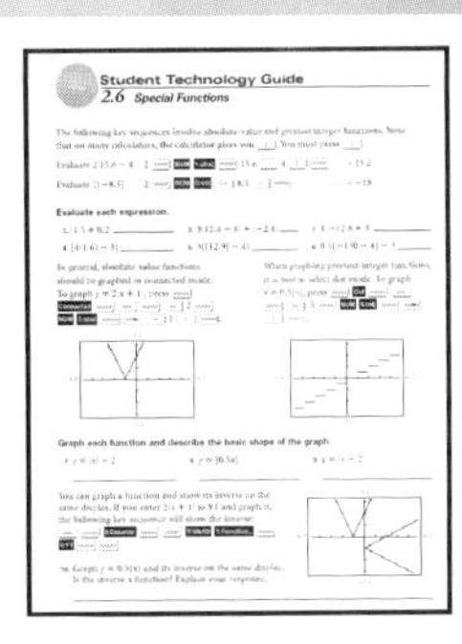

Student Technology Guide

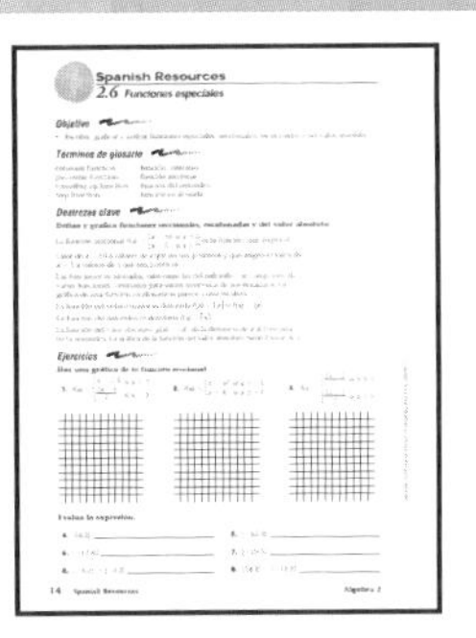

Spanish Resources

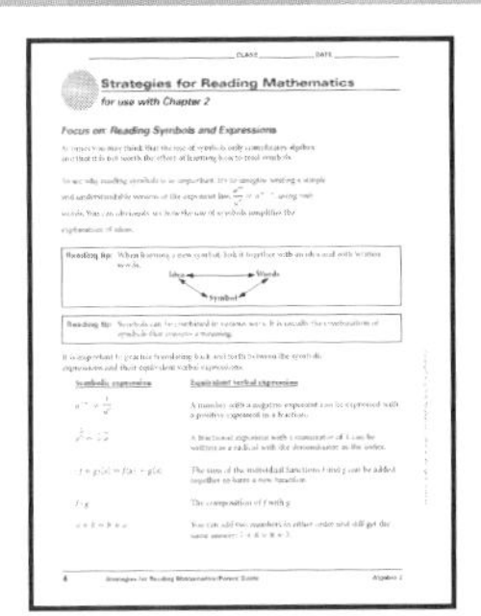

Building Success in Mathematics

ACTIVITIES & EXTENSIONS

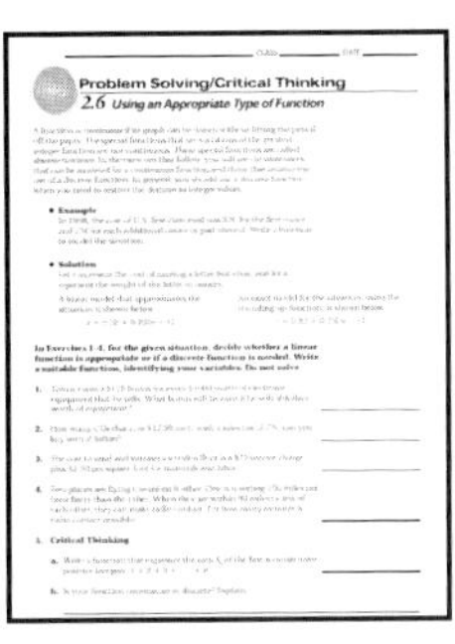

Problem Solving/ Critical Thinking Masters

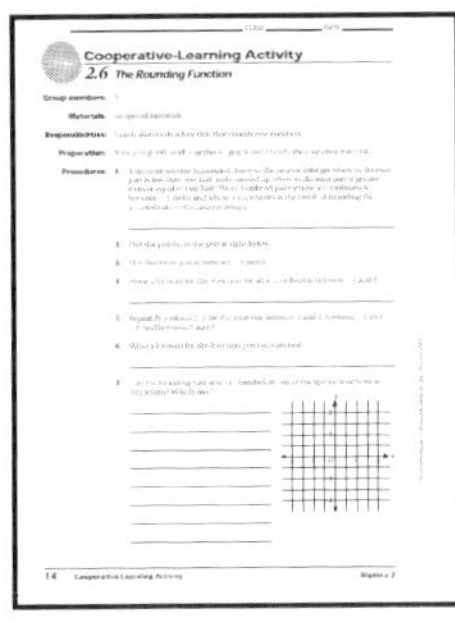

Cooperative-Learning Activities

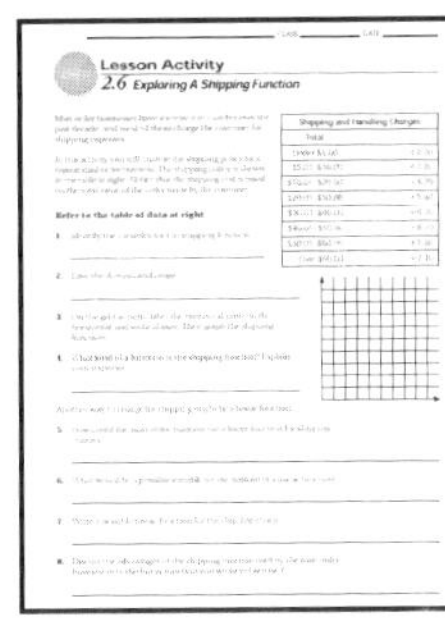

Lesson Activities

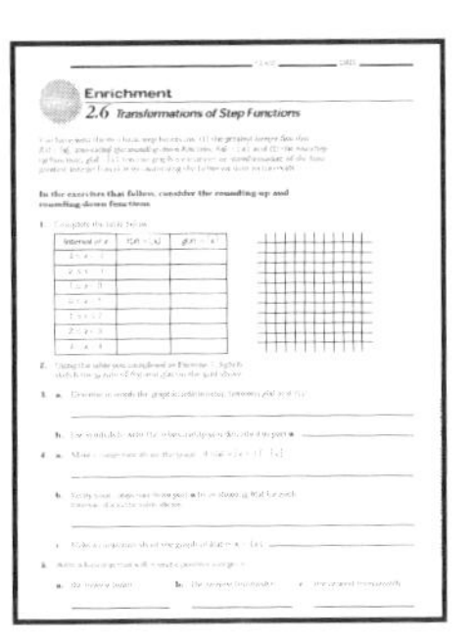

Enrichment Masters

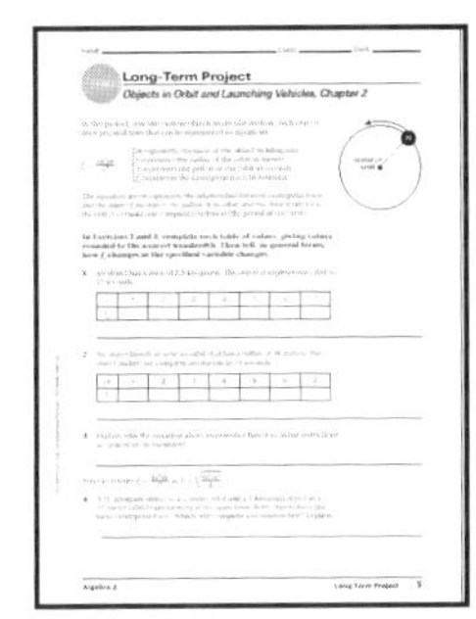

Long-Term Projects

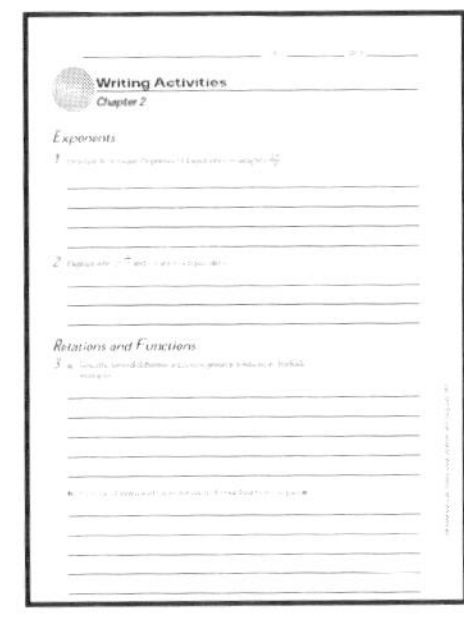

Writing Activities for Your Portfolio

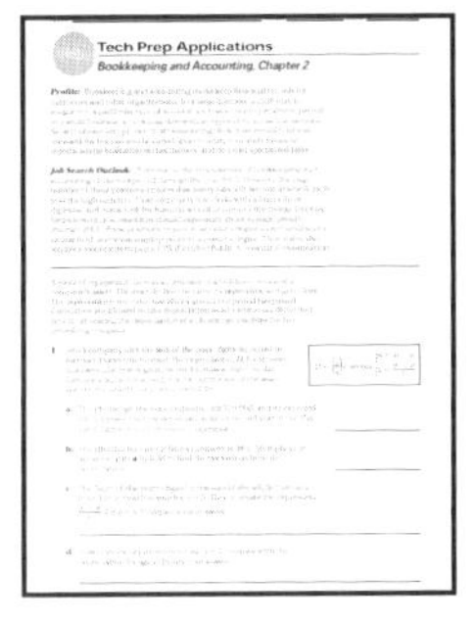

Tech Prep Masters

TEACHER'S TOOLS

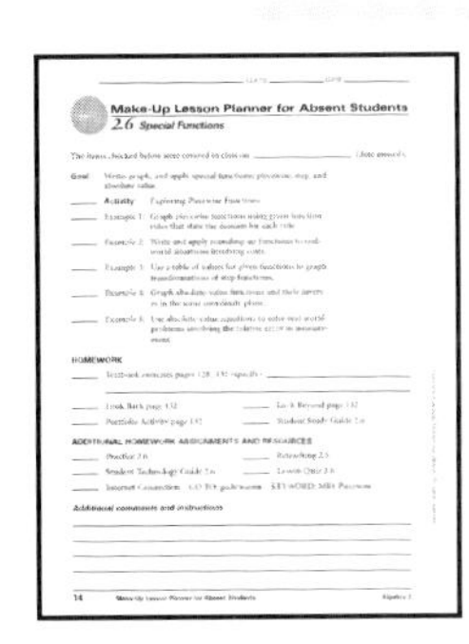

Make-Up Lesson Planner for Absent Students

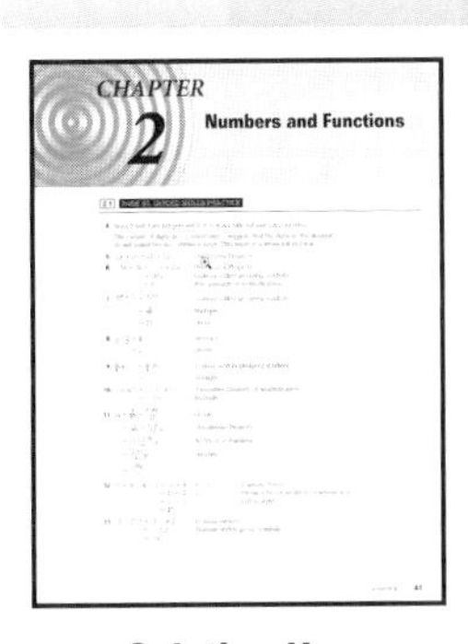

Solution Key

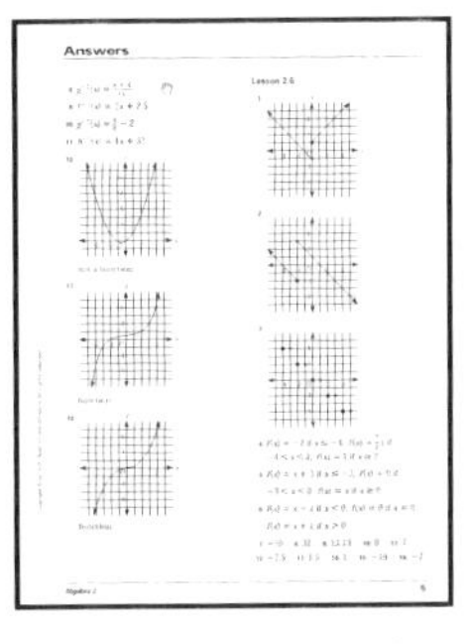

Practice Workbook Answer Key

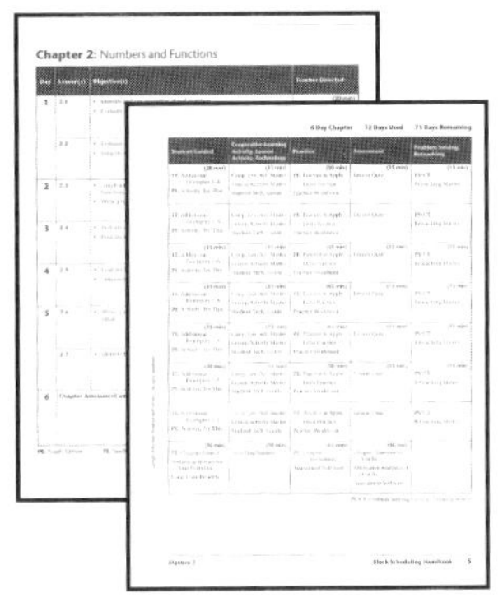

Block-Scheduling Handbook

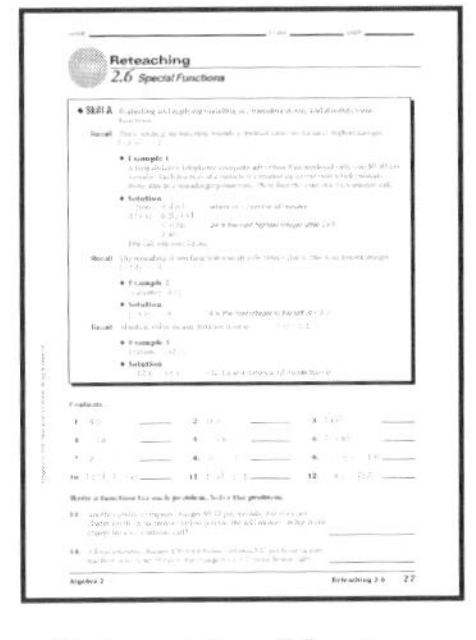

Reteaching Masters

TECHNOLOGY

ALGEBRA 2 ONE-STOP PLANNER CD-ROM WITH DYNAMIC TEST GENERATOR

The *One-Stop Planner CD-ROM with Dynamic Test Generator* is an all-in-one, comprehensive management tool designed to make planning, management, and assessment easier and more efficient. This unique CD-ROM allows you to conveniently view and print teaching resources from your computer.

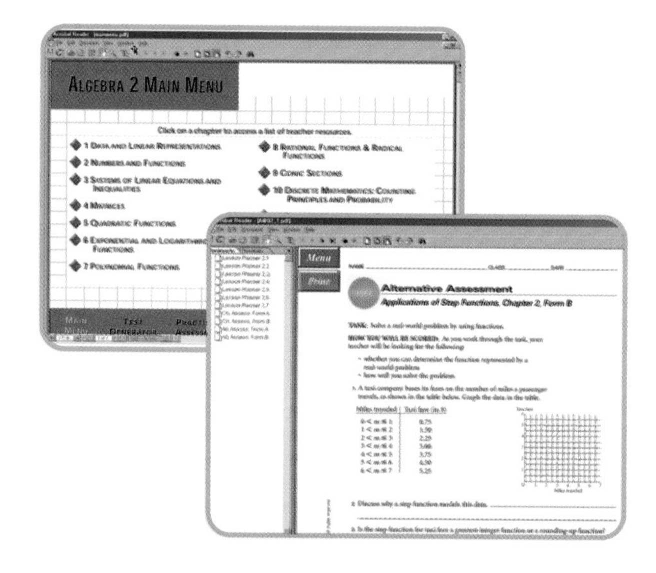

The *Dynamic Test Generator* is also available separately. It includes these new and exciting capabilities:

- Worksheet Builder—a flexible test generator
- Management Module—where all student, teacher, and gradebook functions are performed
- Student Module—where students can take tests and review their performance on screen

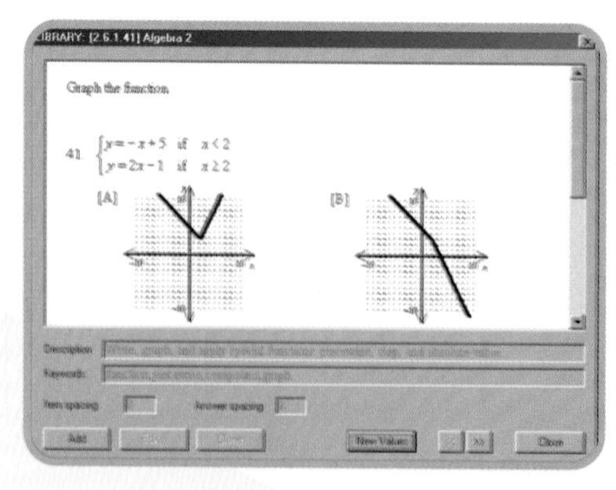

LESSON PRESENTATIONS ON CD-ROM

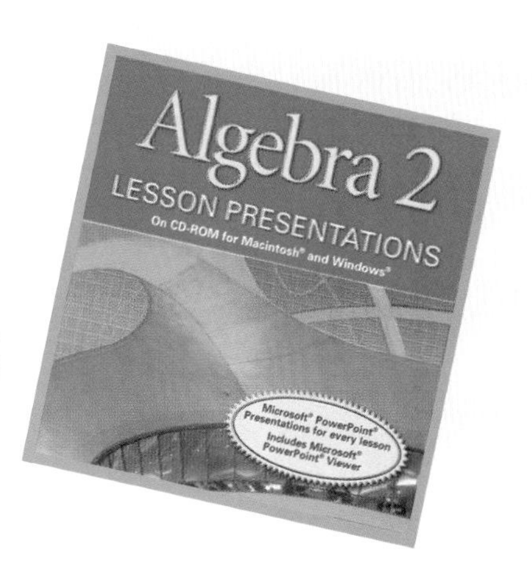

Looking for some interactive teaching resource options? ***Lesson Presentations on CD-ROM*** contains interactive presentations for each lesson in the *Pupil's Edition.* Useful for direct presentation of lesson content to a class, the program can also be used by individual students as a tutorial.

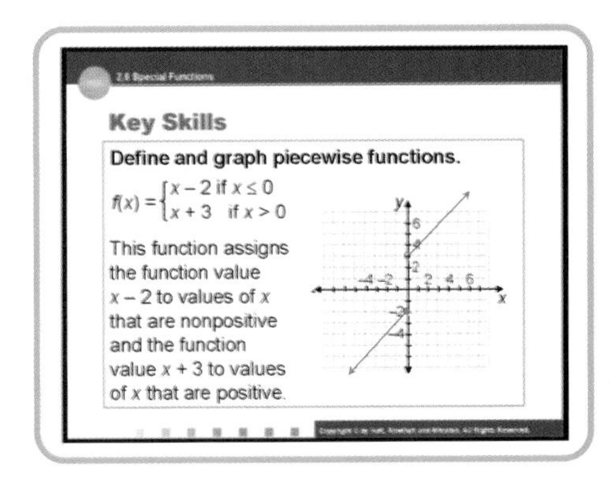

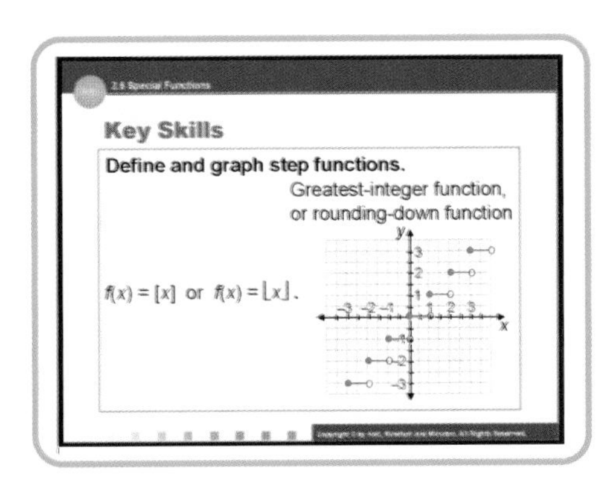

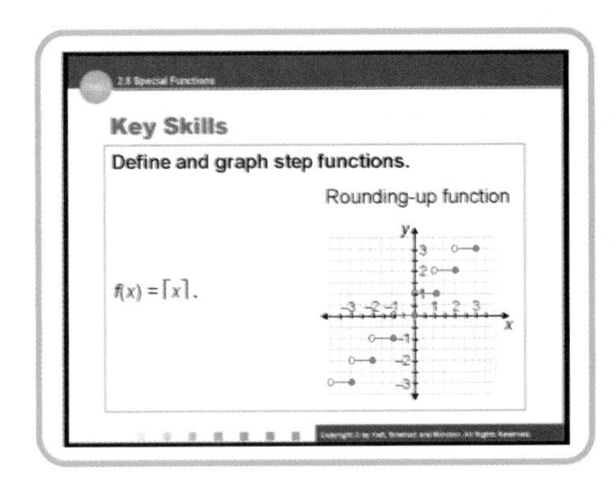

TECHNOLOGY

MATHEPEDIA CD-ROM

The ***Mathepedia CD-ROM*** features over 1,000 middle school and high school math topics. Organized like a dictionary or encyclopedia and enhanced with graphics and interactive math objects, this program includes the following:

- Exercises to reinforce understanding
- A fully-functioning graphing calculator
- Plotting tools available within the graphing button in the toolbar
- Alphabetical list of all terms available
- Subsets of terms organized into mini-dictionaries

Mathepedia offers convenient help for:

- Teachers illustrating and enhancing lessons
- Parents learning or relearning math along with your students
- Students reviewing and practicing concepts

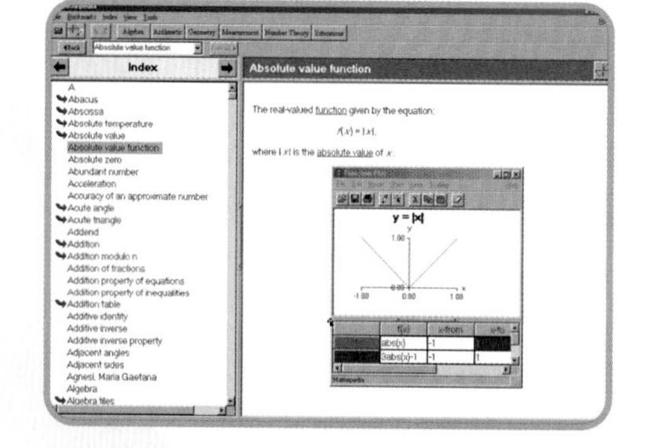

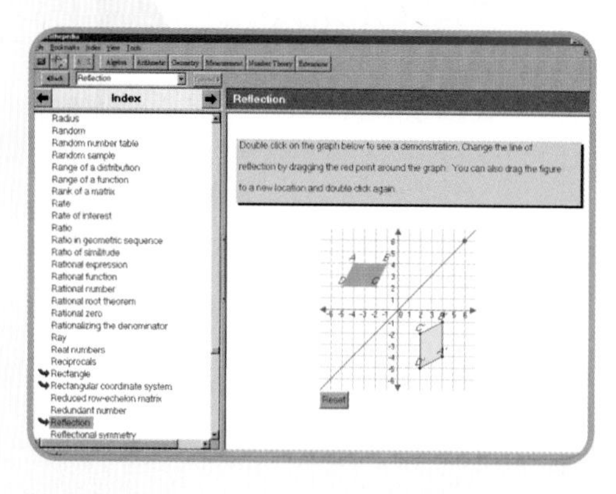

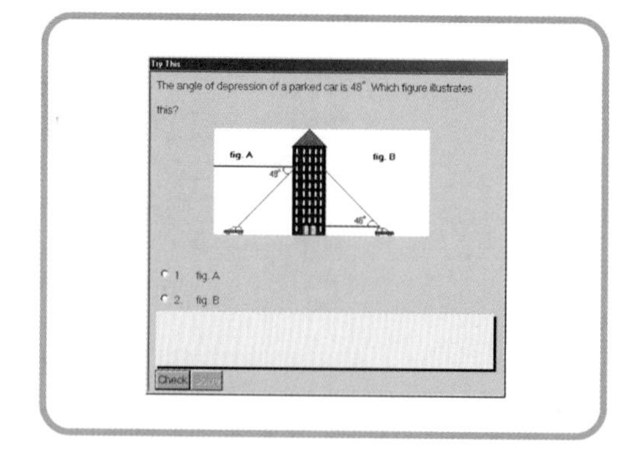

TECHNOLOGY GUIDES

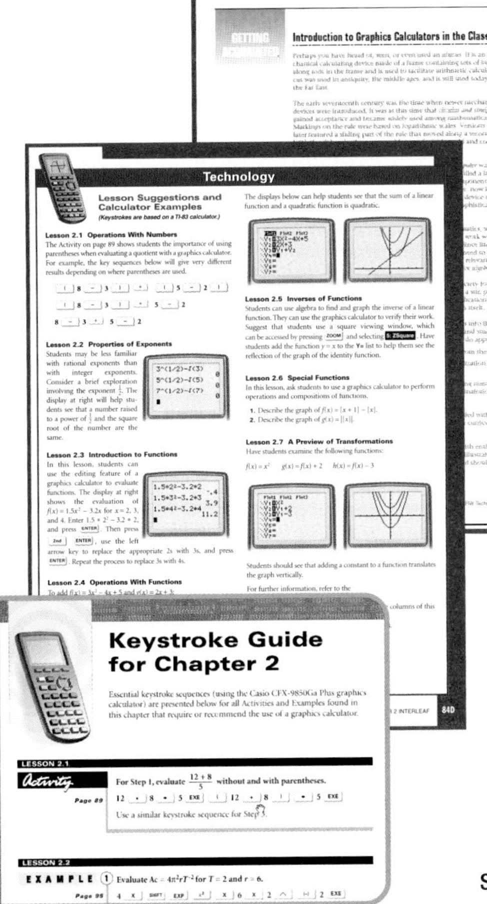

You can feel comfortable and confident implementing technology in your classroom with the following in-text and out-of-text support:

Technology is fully supported in the *Pupil's Edition* through **Graphics Calculator Keystroke Guides** at the end of each chapter, providing specific, step-by-step keystrokes for each operation presented in lesson Examples and Activities. Additional support is found in the *Annotated Teacher's Edition* and online.

Our ***Student Technology Guide*** provides blackline masters with computer and calculator activities that offer additional practice and alternative technology which helps build students' skills further and support the material presented in the book.

Printed in the United States of America

GO
FARTHER

WITH

Integrate technology with **Internet Connect**—a unique resource that helps you to enrich and expand lessons and strengthen students' computer skills. In-text references lead you and your students to **go.hrw.com** where resources help you review, extend, apply, and assess learning.

Teachers, students, and parents can benefit from these exceptional resources—all available online:

- **Portfolio Extension**
- **Alternative Assessment**
- **Homework Help Online**
- **Activities Online**
- **Standardized Test Prep Online**

go.hrw.com also features:

- **Calculator Keystroke Guides**
- **Parent Guides**
- **State-specific Resources**

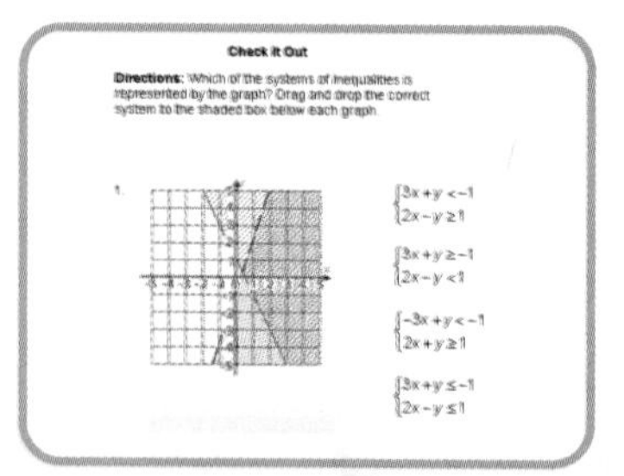

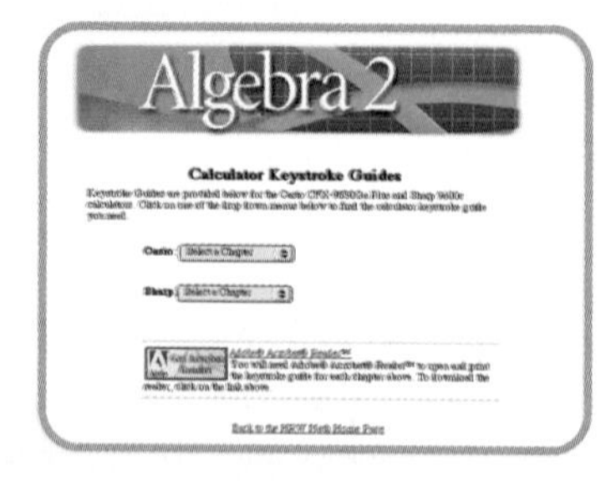

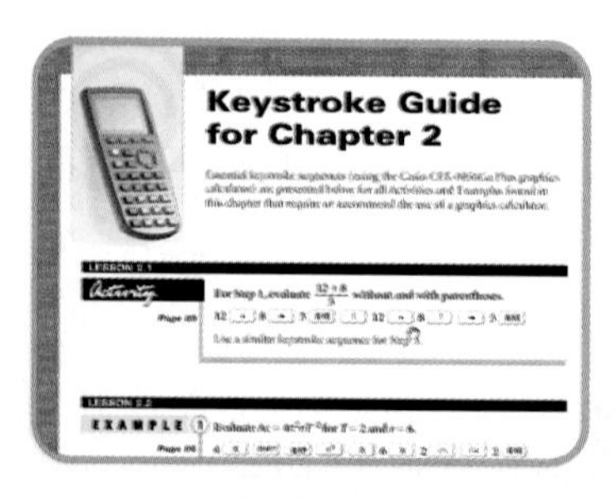

ANNOTATED TEACHER'S EDITION

Algebra 2

HOLT, RINEHART AND WINSTON

A Harcourt Classroom Education Company

Austin • New York • Orlando • Atlanta • San Francisco • Boston • Dallas • Toronto • London

Acknowledgments appear on pages 1095-1096, which are extensions of the copyright page.

Printed in the United States of America

ISBN: 0-03-066056-4

1 2 3 4 5 6 7 048 05 04 03 02 01

AUTHORS

James E. Schultz, *Senior Series Author*
Dr. Schultz has over 30 years of experience teaching at the high school and college levels and is the Robert L. Morton Professor of Mathematics Education at Ohio University. He helped to establish standards for mathematics instruction as a co-author of the NCTM *Curriculum and Evaluation Standards for School Mathematics* and *A Core Curriculum: Making Mathematics Count for Everyone.*

Wade Ellis, Jr., *Senior Author*
Professor Ellis has co-authored numerous books and articles on how to integrate technology realistically and meaningfully into the mathematics curriculum. He was a key contributor to the landmark study *Everybody Counts: A Report to the Nation on the Future of Mathematics Education.*

Kathleen A. Hollowell
Dr. Hollowell is an experienced high school mathematics and computer science teacher who currently serves as Director of the Mathematics & Science Education Resource Center, University of Delaware. Dr. Hollowell is particularly well versed in the special challenge of motivating students and making the classroom a more dynamic place to learn.

Paul A. Kennedy
Dr. Kennedy is a professor in the Department of Mathematics at Southwest Texas State University. His research focuses on developing algebraic reasoning, the uses of technology in mathematics, and the recruitment and retention of mathematics teachers. He is a recognized state and national leader in mathematics education with expertise in developing staff development materials. He currently serves on the Teachers Teaching with Technology Academic Council and is part of the national faculty at the University of Notre Dame's ACE program.

CONTRIBUTING AUTHOR

Martin Engelbrecht
A mathematics teacher at Culver Academies, Culver, Indiana, Mr. Engelbrecht also teaches statistics at Purdue University, North Central. An innovative teacher and writer, he integrates applied mathematics with technology to make mathematics accessible to all students.

TECHNOLOGY WRITER AND CONSULTANT

Betty Mayberry
Ms. Mayberry is the mathematics department chair at Gallatin High School, Gallatin, Tennessee. She has received the Presidential Award for Excellence in Teaching Mathematics and the Tandy Technology Scholar award. She is a Teachers Teaching with Technology instructor and a popular speaker for the effective use of technology in mathematics instruction.

REVIEWERS

Reba W. Allen
Gallatin High School
Gallatin, TN

Judy B. Basara
St. Hubert High School
Philadelphia, PA

Audrey M. Beres
Bassick High School
Bridgeport, CT

Mark Budahl
Mitchell High School
Mitchell, SD

Michael Chronister
J. K. Mullen High School
Denver, CO

Suellyn Chronister
Littleton High School
Littleton, CO

Craig Duncan
The Chinquapin School
Highlands, TX

Robert S. Formentelli
Uniontown Area High School
Uniontown, PA

Richard Frankenberger
Hazelwood School District
Florissant, MO

Mary Hutchinson
South Mecklenburg High School
Charlotte, NC

James A. Kohr
Eldorado High School
Albuquerque, NM

David Landreth
Carlsbad Senior High School
Carlsbad, NM

Pam Mason
Mathematics Advisor
Austin, TX

Gilbert Melendez
Gasden High School
Las Cruces, NM

Cheryl Mockel
Mt. Spokane High School
Mead, WA

Steve Murray
Cerritos High School
Huntington Beach, CA

Stacey Pearson
North Gaston High School
Dallas, NC

James R. Ringstrom
La Costa Canyon High School
Carlsbad, CA

Douglas E. Roberts
Franklin Heights High School
Columbus, OH

Alice B. Robertson
Central High School
Carrolton, GA

James F. Rybolt
Mt. Diablo USA
Antioch, CA

Jeffrey C. Schmook
Yough Senior High School
Herminie, PA

Harry Sirockman
Central Catholic High School
Pittsburgh, PA

Richard Soendlin
Arsenal Technical High School
Indianapolis, IN

Charleen Strain
Roswell High School
Roswell, NM

Harry D. Stratigos
Pennsylvania Department of Education
Harrisburg, PA

Jeff Vollmer
Myers Park High School
Charlotte, NC

Table of Contents

DATA AND LINEAR REPRESENTATIONS 2

MATH CONNECTIONS

APPLICATIONS

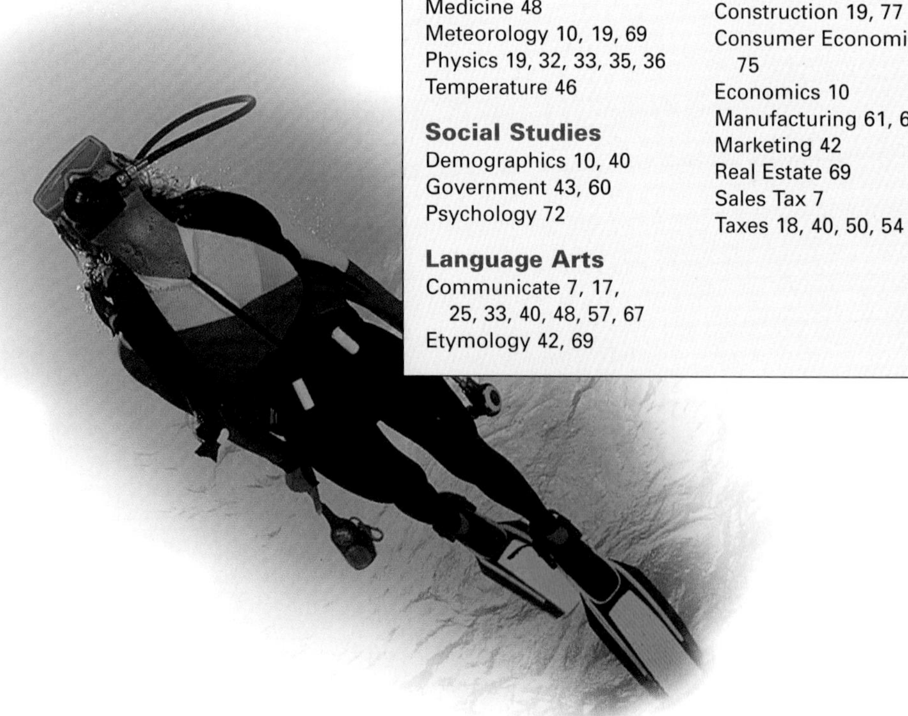

2 NUMBERS AND FUNCTIONS 84

SYSTEMS OF LINEAR EQUATIONS & INEQUALITIES 154

MATH CONNECTIONS

APPLICATIONS

MATRICES 214

MATH CONNECTIONS

APPLICATIONS

Ancient clay tablet believed to contain Pythagorean triples

6 EXPONENTIAL AND LOGARITHMIC FUNCTIONS 352

MATH CONNECTIONS

APPLICATIONS

POLYNOMIAL FUNCTIONS

MATH CONNECTIONS

APPLICATIONS

RATIONAL FUNCTIONS & RADICAL FUNCTIONS 478

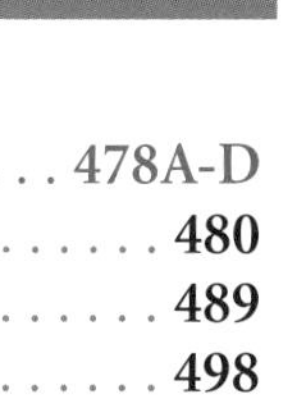

MATH CONNECTIONS

APPLICATIONS

CONIC SECTIONS 560

MATH CONNECTIONS

APPLICATIONS

DISCRETE MATHEMATICS: COUNTING PRINCIPLES AND PROBABILITY 626

MATH CONNECTIONS

APPLICATIONS

11 DISCRETE MATHEMATICS: SERIES AND PATTERNS 688

MATH CONNECTIONS

APPLICATIONS

12 DISCRETE MATHEMATICS: STATISTICS 762

MATH CONNECTIONS

APPLICATIONS

TRIGONOMETRIC FUNCTIONS 826

MATH CONNECTIONS

APPLICATIONS

14 FURTHER TOPICS IN TRIGONOMETRY 884

MATH CONNECTIONS

APPLICATIONS

INFO BANK 940

Bringing Math *to Life*

Our lives are touched daily by the application of mathematics. Mathematics is the main tool of science, providing tools to quantify, verify, and describe results and theories. Computers, with circuitry and software that applies mathematical logic, control many of the dimensions of the infrastructure around us. Traffic lights, power grids, and automated teller machines are among the many things that depend on computer-based mathematical logic structures to work properly.

In ancient Egypt, problem sets were written specifically to drill a particular mathematical skill. The drill-and-practice approach may be valid at certain points in the instructional sequence, but when used as the primary means of instruction, it decontextualizes the mathematics and reinforces the impression that mathematics is an abstraction that has no practical application. Most teachers agree that a combination of drill and practice and an instructional context consisting of applications, connections, and problem solving is the best way to teach mathematics.

APPLICATIONS

Throughout Holt, Rinehart and Winston's *Algebra 2*, students are introduced to concepts through real-world applications. These applications provide connections to domains of human activity such as business, consumer economics, science, life skills, vocational skills, and leisure activities. These applications anchor the student in a real-world context that is familiar and provides a springboard to the discovery of the underlying mathematical concepts.

Applications provide evidence that mathematics is important in our world and that the algebra learned in the classroom today does have practical uses. Students may also recognize the power of mathematics to advance their opportunities in life.

Appropriate use of technology to support the exploration of applications helps students to prepare for real-world problem-solving situations. With the use of technology, students can explore realistic, rather than contrived, data. Students who learn to tackle algebraic problems by actively exploring various approaches, collaborating with others, and comparing results to look for alternative strategies will be equipped to transfer these approaches to their personal and professional challenges later in life.

CONNECTIONS

Mathematics provides the conceptual basis for the structure of many things around us. For example, the architecture of a building is influenced by formulas that help determine stress and weight factors. Mathematics is the main tool of science, providing statistical modeling to support experimental research. Even so-called pure mathematics sometimes finds practical and direct use in helping to describe the nature of things (for example, consider fractal geometry). Our insurance rates, tax rates, and interest rates are all determined by mathematical relationships. Mathematics is itself a connection between domains of knowledge.

Mathematics is also connected within itself. A simple geometric construction of a 45-45-90 right triangle can be used to discover irrational numbers such as $\sqrt{2}$. The distance formula is an algebraic representation of the length of a line segment.

Holt, Rinehart and Winston's *Algebra 2* emphasizes connections within mathematical disciplines and to other domains of knowledge. Instruction that emphasizes curricular connections is more likely to motivate students' interest in mathematically oriented subjects such as chemistry, physics, logic, economics, statistics, and computer science.

Technology

TECHNOLOGY GOALS

To meet changing curriculum requirements, algebra teachers are continually searching for the best way to integrate technology into their classroom. They know that technology allows their students to become active participants in the learning process. Through the use of technology, today's mathematics classrooms are being transformed into laboratories where students explore and experiment with mathematical concepts rather than just memorize isolated facts. Students make generalizations and reach conclusions about mathematical concepts and relationships and apply them to real-world situations. Holt, Rinehart and Winston's *Algebra 2* supports instruction that utilizes technology with activities and examples that encourage students to make and test conjectures and to confirm mathematical ideas for themselves.

Teachers also know that technology encourages cooperative learning. Cooperative-learning groups allow students to compare results, brainstorm, and reach conclusions based on group results. As in real life, where scientists and financial analysts often consult each other, students in the technology-oriented classroom learn to communicate and consult with each other. Students learn that such consultations are not "cheating," but are rather a method of sharing information that will be used to solve a problem.

Furthermore, the use of technology provides students with an avenue to explore real-world applications that would not be practical to venture into without the use of such technology. Students are able to study realistic data instead of data that simply works out nicely. For instance, when calculating by hand, most students study only simple interest. The use of technology allows students to explore compound interest, which is how interest is calculated by today's financial institutions.

Real data from students' local town or neighborhood can be used to find the average price or rental cost of a home. Students can study data that are more closely related to their own lives, such as comparing the cost of phone plans and determining when one plan costs less than another.

Many lessons in Holt, Rinehart and Winston's *Algebra 2* utilize the power of technology as an exploration tool. For example, in Lesson 5.1, students investigate the graphs of quadratic functions. Students use a graphics calculator to compare the graphs of two linear functions, f and g, with the graph of the product of the linear functions, $f \cdot g$.

TYPES OF TECHNOLOGY

The most prominent use of technology in Holt, Rinehart and Winston's *Algebra 2* is the graphics calculator. Computer spreadsheet software and Internet resources are also utilized.

Most educators agree that mathematics instruction can be enhanced by the effective use of technology. Graphics calculators are commonly used in second-year algebra courses because they can be powerful tools for visualizing mathematical concepts. Holt, Rinehart and Winston's *Algebra 2* carefully incorporates graphics calculators in its instruction so that their use is appropriate and promotes mathematical reasoning.

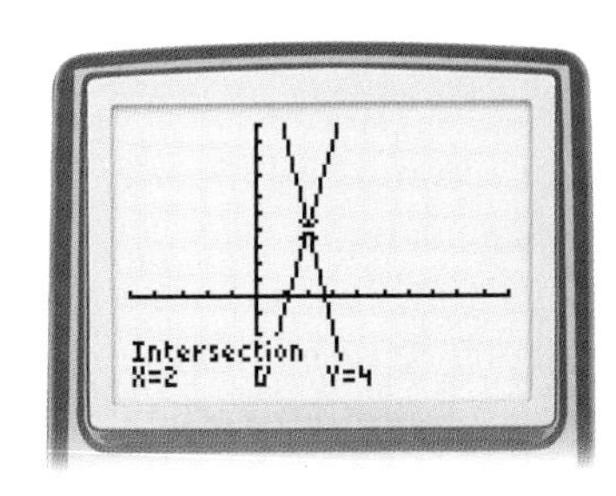

Graphics calculators perform many specialized and complex mathematical actions, including graphing. These calculators use certain keys to access full-screen menus that show additional operations. The scope of capabilities is wide, including traditional arithmetic calculations, matrix operations, statistics, function graphing, parametric graphing, polar graphing, sequence graphing, and more. Graphics calculators are also capable of running user-defined programs. When using the graphing capabilities of the calculators, the use of a range-setting window adds greatly to the usefulness of the technology.

The power of technology is most apparent when students analyze the behavior of functions, which can be very tedious and time consuming to study without the use of a graphics calculator or a computer spreadsheet. For example, in Lesson 7.1, students explore the graphs of polynomial functions. By using graphics calculators, students can quickly and accurately graph third- and fourth-degree polynomials and look for patterns to make conjectures.

TEACHING AIDS

Holt, Rinehart and Winston's *Algebra 2* often recommends that students have access to graphics calculators. Some topics even require the use of a graphics calculator. A Keystroke Guide that provides the essential keystroke sequences for TI-82, TI-83, and TI-83 Plus graphics calculators follows each chapter. (Keystrokes for other models of graphics calculators are found on the HRW Web site.) These keystrokes are provided for all activities and examples found in *Algebra 2* that require or recommend the use of a graphics calculator.

Another teaching aid that is gaining wide acceptance is the Internet. Throughout both the Teacher's Edition and Pupil's Edition of *Algebra 2*, you

will find references to Holt, Rinehart and Winston's Web site. These "Internet Connects" provide math resources or educational links for tutorial assistance, references for student research, classroom activities, and teaching resources. The HRW Web site also provides updates to the technology used in the text.

An effective, powerful, yet easy to use tool in the classroom is *Mathepedia* computer software. This tool allows students to explore mathematics by utilizing a dictionary design and over 140 interactive mathematics tools. This comprehensive reference tool contains over 900 terms and definitions with extensive hypertext links to show the connections between math topics. Topics can be arranged alphabetically and according to the strands of the NCTM Principles and Standards. This powerful tool has outstanding graphing capabilities with nine types of available graphs and the option of displaying five different graphs at one time on the same axes. *Mathepedia* may be used by students in the classroom, the computer lab, or at home. Students can use it for reviewing concepts, exploring topics they are currently studying, and engaging in exciting enrichment activities.

Assessment

ASSESSMENT GOALS

An essential aspect of any learning environment, such as an algebra classroom, is the process of assessing or evaluating what students have learned. Informally, this is done with paper-and-pencil tests given by the teacher on a regular basis to measure students' knowledge of the material. Formal evaluations with standardized tests are generally conducted over a period of years to establish performance records for both individuals and groups of students within a school or school district. Both types of tests are very good at measuring the ability of a student to use a particular mathematical skill or to recall a specific fact. They fall short, however, in evaluating other key goals of learning mathematics, such as being able to solve problems, to reason critically, to understand concepts, and to communicate mathematically, both verbally and in writing. Other techniques, usually referred to as alternative assessment, are needed to evaluate students' performance on these goals of instruction.

The goals of an alternative-assessment program are to provide a means of evaluating students' progress in nonskill areas of mathematical learning. Thus, the design and structure of alternative assessment techniques must be quite different from those of the skill-oriented, paper-and-pencil tests of the past.

TYPES OF ALTERNATIVE ASSESSMENT

In the world outside school, a person's work is evaluated by what that person can do (that is, by the results the person achieves), not by taking a test. For example, a musician may demonstrate skill by making music, a pilot by flying an airplane, a writer by writing a book, and a surgeon by performing an operation. Students learn to think mathematically and to solve problems on a continuous basis over a long period of time as they study mathematics at many grade levels. Students, too, can demonstrate what they have learned by collecting a representative sample of their best work in a portfolio. A portfolio should illustrate achievements in problem solving, critical thinking, writing about mathematics, mathematical applications or connections, and any other activity that demonstrates an understanding of both concepts and skills.

Specific examples of the kinds of work that students can include in their portfolio are solutions to nonroutine problems, graphs, tables or charts, computer printouts, group reports or reports of individual research, simulations, and artwork or models. Each entry should be dated and should be chosen to show the student's growth in mathematical competence and maturity.

A portfolio is just one way for students to demonstrate their performance on a mathematical task. Performance assessment can also be achieved in other ways, such as by asking students questions and evaluating their answers, by observing their work in cooperative-learning groups, by having students give verbal presentations, by working on extended projects and investigations, and by keeping journals.

Peer assessment and self-evaluation are also valuable methods of assessing students' performance. Students should be able to critique their classmates' work and their own work against standards set by the teacher. In order to evaluate their work, students need to know the teacher's goals of instruction and the criteria (scoring rubrics) that have been established for evaluating performance against the goals. Students can help to design their own self-assessment forms that they then fill out on a regular basis and give to the teacher. They can also help to construct test items that are incorporated into tests given to their classmates. This work is ideally done in small groups of four students. The teacher can then choose items from each group to construct the test for the entire class. Another alternative testing technique is to have students work on take-home tests that pose more open-ended and nonroutine questions and problems. Students can devote more time to such tests and, in doing so, demonstrate their understanding of concepts and skills and their ability to do mathematics independently.

SCORING

The use of alternative assessment techniques implies the need to have a set of standards against which students' work is judged. Numerical grades are no longer sufficient because growth in understanding and problem solving cannot be measured by a single number or letter grade. Instead, scoring rubrics or criteria can allow the teacher more flexibility to recognize and comment on all aspects of a student's work, pointing out both strengths and weaknesses that need to be corrected.

A scoring rubric can be created for each major instructional goal, such as being able to solve problems or communicate mathematically. A rubric generally consists of four or five short descriptive paragraphs that can be used to evaluate a piece of work. For example, if a five-point paragraph scale is used, a rating of 5 may denote that the student has completed all aspects of the assignment and has a comprehensive understanding of problems. The content of the fifth paragraph specifies the details of what constitutes the rating of highly satisfactory. On the other hand, a rating of 1 designates an essentially unsatisfactory performance, and the first paragraph specifies the details of what constitutes an unsatisfactory rating. The other three paragraphs provide an opportunity for the teacher to recognize significant accomplishments by the student as well as aspects of the work that need improvement. Thus, scoring rubrics are a far more realistic and educationally substantive way to evaluate a student's performance than a single grade, which is usually determined by an answer being either right or wrong.

The Holistic Scoring Rubric illustrated on the next page is an excellent and effective guide for use by math teachers who are practicing performance assessment in their classrooms. The scorer gathers evidence about a student's mathematical ability. The rubric is then used to assign a single performance rating based on an overall view of the full contents of the student's portfolio. The Holistic Scoring Rubric lists the types of work and tools for entries that are appropriate for a student to place in his or her portfolio.

Holt, Rinehart and Winston's *Algebra 2* provides numerous opportunities for portfolio assessment in each chapter of the Pupil's Edition. Investigation and discovery are encouraged in all of the activities in the student lessons of the program. Interdisciplinary topics from astronomy to zoology and applications from the arts to transportation are designated throughout the lessons. Projects are found at the end of every chapter, and Portfolio Activities are found within the chapter. Communicate exercises offer excellent writing opportunities. Tools such as graphics calculators can be used to support the activities that students include in their portfolios.

ASSESSMENT AND *ALGEBRA 2*

Throughout the textbook, students are asked to explain their work; describe what they are doing; compare and contrast different approaches; analyze a problem; make sketches, graphs, tables, and other models; hypothesize, conjecture, and look for counterexamples; and make and prove generalizations.

All of these activities, including the more traditional responses to routine problems, provide the teacher with a wealth of assessment opportunities to see how well students are progressing in their understanding and knowledge of algebra. The assessment task can be aligned with the major process goals of instruction, with scoring rubrics established for each goal. For example, a teacher may decide to organize his or her assessment tasks based on the following general skills of mathematics: problem solving, reasoning, communicating, and connecting.

Within each of these areas, specific goals can be written and shared with students. In this way, the assessment process becomes an integral part not only of evaluating students' progress, but also of the instructional process itself. The results of assessment can be used to modify the instructional approach in order to enhance learning for all students.

In addition to the many opportunities for performance assessment found in the lessons and chapter tests of *Algebra 2*, a variety of assessment types are integrated into the chapter-end material. The Chapter Review and Assessment and the Cumulative Assessment include both traditional and alternative assessment. The Cumulative Assessments are formatted in the style of college entrance exams. In addition to multiple-choice and free-response items, each Cumulative Assessment contains quantitative-comparison questions that emphasize the concepts of equality, inequality, and estimation. Other types of college entrance exam questions found in the Cumulative Assessments are student-produced response questions with solutions recorded in answer grids.

Holistic Scoring Rubric

	PERFORMANCE GOALS AND CRITERIA				
	Mathematical Reasoning Selecting and using appropriate types of reasoning and methods of proof through inductive and deductive reasoning	**Problem Solving** Solving of problems through the use of exploration, appropriate strategies, and a systematic approach	**Communication** Communicating ideas, thoughts, and approaches through the use of everyday language, mathematical language and symbols, graphs, tables, charts, and diagrams	**Mathematical Connections** Recognizing connections between different mathematical ideas or between mathematics and other disciplines	**Use of Tools** Using technology (calculators, computers, etc.) and/or manipulatives
LEVEL FOUR: SUPERIOR	• Uses sophisticated mathematical reasoning • Provides strong supporting arguments • Includes examples and counterexamples	• Shows thorough understanding of the problems' mathematical ideas and processes • Uses and synthesizes multiple strategies that lead to correct solutions	• Contains a complete response with clear, precise, and appropriate language • Uses effective diagrams such as graphs, tables, or charts	• Demonstrates a comprehensive knowledge of connections to other mathematical topics or other disciplines	• Makes appropriate use of technology and manipulatives to demonstrate mathematical concepts
LEVEL THREE: COMPETENT	• Uses sound mathematical reasoning • Includes some supporting arguments	• Shows basic understanding of the problems' mathematical ideas and processes • Uses appropriate strategies that lead to correct solutions	• Contains a solid response but is expressed less elegantly and less completely • Uses accurate diagrams	• Demonstrates some knowledge of connections to other mathematical topics or other disciplines	• Uses some technology and manipulatives to demonstrate solutions to problems
LEVEL TWO: MARGINAL	• Uses somewhat appropriate mathematical reasoning • Includes incomplete or faulty arguments	• Shows a partial understanding of the problem • Uses some appropriate strategies that lead to partially correct solutions	• Contains a fairly complete response but uses unclear language • Uses inappropriate and/or unclear diagrams	• Demonstrates few connections	• Occasionally uses technology and manipulatives appropriately
LEVEL ONE: LIMITED	• Uses limited mathematical reasoning • Includes no arguments	• Shows little understanding of the problems • Uses poor or inappropriate strategies that lead to incorrect solutions	• Uses some appropriate mathematical language • Uses few, if any, diagrams	• Does not demonstrate or demonstrates inappropriate connections to other mathematical topics or other disciplines	• Rarely uses technology and manipulatives appropriately

NCTM Principles and Standards

In 1989, the National Council of Teachers of Mathematics published the Curriculum and Evaluation Standards for School Mathematics, generally referred to as the Standards, with the overall objective of improving students' mathematical education. Just as important, the Standards acknowledged the need for school mathematics to respond to and anticipate the changing needs of an increasingly technological and quantitative world. This document called for a focus across all grade levels in the areas of problem solving, communicating, reasoning, and making connections. Content changes for grades 5 through 8 called for increased attention in the areas of developing number and operation sense, identifying and using functional relationships, developing and using tables, graphs, and rules to describe situations, and developing an understanding of variables, expressions, and equations. Other changes called for increased attention in using statistical methods to describe, analyze, evaluate, and make decisions; creating experimental and theoretical models of situations involving probabilities; and developing an understanding of geometric objects and relationships. Changes in mathematical content for grades 9 through 12 called for increased attention in the use of real-world problems in algebra to motivate students and apply theory, integration of geometry topics at all grade levels, realistic trigonometry applications and modeling, and the integration of functions across topics at all grade levels.

An underlying precept of the original Standards was to provide an ongoing means for improving student performance in mathematical education. Revision of the original NCTM Standards was planned in the early development of the 1989 version. It was assumed that additional research and emerging technologies would necessitate the need for modification of the Standards' themes and messages. It is also important to articulate any language in the original Standards that caused confusion or lead to a misunderstanding of the intended messages. For example, "mathematical topics to receive decreased attention" was sometimes interpreted as areas for elimination. Since the creation of the original document, it has also become clear that certain mandates for change—such as basic skills and conceptual learning—needed adjustment and further clarification. Thus, the major purpose of the new set of standards is based on improvements in the field, research on learning and teaching, and technological advances that have occurred over the last 10 years.

Even though the new document, which will be referred to as Principles and Standards, includes a number of substantive and structural features that are different from the original Standards, a goal of the new Principles and Standards is for students to be able to achieve the goals established in the original Standards. Such goals as students learning to value mathematics, becoming confident in their own ability, becoming mathematical problem solvers, learning to communicate mathematically, and learning to reason mathematically remain important foundations of the new Principles and Standards.

The new document is built around ten standards organized across all grade levels. Five of these standards are made up of mathematical content, and five are made up of process standards. Each standard is elaborated at four groups of grade levels — pre-k–2, grades 3–5, grades 6–8, and grades 9–12. Within each standard, a number of focus areas are identified and main points are elaborated with detailed descriptions of the content at each group of grade levels.

Five standards describe the mathematical content that students should learn and the goals for students' understanding of concepts and procedures.

- Number and Operation
- Patterns, Functions, and Algebra
- Geometry and Spatial Sense
- Measurement
- Data Analysis, Statistics, and Probability

Five standards describe the mathematical processes through which students should acquire and use their mathematical knowledge. These are described in terms of students' mathematical learning outcomes.

- Problem Solving
- Reasoning and Proof
- Communications
- Connections
- Representation

The first four process standards listed above are present across all grade levels in the original Standards. The fifth process standard, Representation, has been added to the new Principles and Standards to highlight the importance of such processes as organizing, recording, and communicating mathematical ideas and using representation in modeling. Representations are used to create, organize, record, and communicate mathematical ideas; develop a repertoire of mathematical representations that can be used purposefully, flexibly, and appropriately; and model and interpret physical, social, and mathematical phenomena. The following table correlates the NCTM Principles and Standards to the lessons contained in Holt, Rinehart and Winston's *Algebra 2*. Standard 10, Representation, is expanded into six descriptive categories of representations.

HOLT, RINEHART AND WINSTON'S *ALGEBRA 2* and NCTM Principles and Standards Correlation

LESSON		NCTM STANDARDS	REPRESENTATIONS						
			CONCRETE	DRAWING	GRAPH	SYMBOL	TABLE	VERBAL	VISUAL
1.1	Tables and Graphs of Linear Equations	1, 2, 4–10			•	•	•	•	•
1.2	Slopes and Intercepts	1, 2, 6–10	•		•	•	•	•	•
1.3	Linear Equations in Two Variables	1, 2, 6–10			•	•		•	•
1.4	Direct Variation and Proportion	1–4, 6–10		•		•	•	•	•
1.5	Scatter Plots and Least-Squares Lines	1, 2, 5–10			•	•	•	•	•
1.6	Introduction to Solving Equations	1–4, 6, 8–10		•		•		•	•
	Eyewitness Math: *A Man and a Method*	1–3, 6–10			•	•		•	•
1.7	Introduction to Solving Inequalities	1, 2, 6, 8–10			•	•		•	•
1.8	Solving Absolute-Value Equations and Inequalities	1–4, 6–10			•	•		•	•
	Chapter Project: *Correlation Exploration*	1, 2, 4–6, 9, 10	•		•	•	•	•	•
2.1	Operations With Numbers	1, 2, 7–10			•	•		•	•
2.2	Properties of Exponents	1, 2, 6–10		•		•		•	•
2.3	Introduction to Functions	1, 2, 6–10		•	•	•	•	•	•
2.4	Operations With Functions	1–3, 5–10		•		•	•	•	•
2.5	Inverses of Functions	1–3, 6–10		•	•	•		•	•
2.6	Special Functions	1, 2, 5–10			•	•	•	•	•
2.7	A Preview of Transformations	1–10			•	•	•	•	•
	Chapter Project: *Space Trash*	1–6, 8–10			•	•	•	•	•
3.1	Solving Systems by Graphing or Substitution	1–4, 6, 8–10	•		•	•		•	•
3.2	Solving Systems by Elimination	1–4, 6–10	•			•	•	•	•
3.3	Linear Inequalities in Two Variables	1–4, 6–10			•	•	•	•	•
3.4	Systems of Linear Inequalities	1–4, 6, 8–10			•	•	•	•	•
3.5	Linear Programming	1–4, 6–10			•	•	•	•	•
3.6	Parametric Equations	1–4, 6–10	•	•	•	•	•	•	•
	Chapter Project: *Maximum Profit/Minimum Cost*	1–4, 6, 8–10			•	•		•	•
4.1	Using Matrices to Represent Data	1–10			•	•	•	•	•
4.2	Matrix Multiplication	1–10		•		•	•	•	•
4.3	The Inverse of a Matrix	1–10				•	•	•	•
	Eyewitness Math: *How Secret Is Secret?*	1, 2, 5, 6, 8–10		•		•		•	•
4.4	Solving Systems With Matrix Equations	1–4, 6–10		•	•	•		•	•
4.5	Using Matrix Row Operations	1–10			•	•	•	•	•
	Chapter Project: *Spell Check*	1–3, 6, 8–10		•		•		•	•
5.1	Introduction to Quadratic Functions	1–4, 6–10		•	•	•	•	•	•
5.2	Introduction to Solving Quadratic Equations	1–4, 6–10		•	•	•	•	•	•
5.3	Factoring Quadratic Expressions	1–10	•		•	•	•	•	•
5.4	Completing the Square	1–4, 6–10	•	•	•	•		•	•
5.5	The Quadratic Formula	1–10		•	•	•	•	•	•
5.6	Quadratic Equations and Complex Numbers	1–4, 6–10			•	•	•	•	•

(Continued)

LESSON		NCTM STANDARDS	REPRESENTATIONS						
			CONCRETE	DRAWING	GRAPH	SYMBOL	TABLE	VERBAL	VISUAL
5.7	Curve Fitting With Quadratic Models	1–10		•	•	•	•	•	•
5.8	Solving Quadratic Inequalities	1–10			•	•	•	•	•
	Chapter Project: *Out of This World*	1–3, 5, 6, 8–10				•	•	•	•
6.1	Exponential Growth and Decay	1, 2, 4, 6, 8–10	•	•		•	•	•	•
6.2	Exponential Functions	1–10			•	•	•	•	•
6.3	Logarithmic Functions	1–4, 6–10		•	•	•	•	•	•
6.4	Properties of Logarithmic Functions	1, 2–10	•	•		•	•	•	•
6.5	Applications of Common Logarithms	1–4, 6–10			•	•	•	•	•
6.6	The Natural Base, *e*	1–4, 6–10			•	•	•	•	•
	Eyewitness Math: *Meet e in St. Louis*	1–3, 6, 8–10			•	•	•	•	•
6.7	Further Applications and Modeling	1–4, 6, 8–10			•	•	•	•	•
	Chapter Project: *Warm Ups*	1, 2, 4–10	•			•	•	•	•
7.1	An Introduction to Polynomial Functions	1–4, 6–10		•	•	•	•	•	•
7.2	Polynomial Functions and Their Graphs	1–10			•	•	•	•	•
7.3	Products and Factors of Polynomials	1–4, 6–10		•	•	•	•	•	•
7.4	Solving Polynomial Equations	1–4, 6–10		•	•	•	•	•	•
	Eyewitness Math: *Scream Machine*	1, 2, 5, 6, 8–10				•		•	•
7.5	Zeros of Polynomial Functions	1–4, 6, 8–10		•	•	•		•	•
	Chapter Project: *Fill it up!*	1–4, 6–10	•			•	•	•	•
8.1	Inverse, Joint, and Combined Variation	1–4, 6–10		•	•	•	•	•	•
8.2	Rational Functions and Their Graphs	1–4, 6–10		•	•	•	•	•	•
8.3	Multiplying and Dividing Rational Expressions	1–3, 6–10		•	•	•	•	•	•
8.4	Adding and Subtracting Rational Expressions	1–4, 6–10		•	•	•		•	•
8.5	Solving Rational Equations and Inequalities	1–4, 6–10			•	•	•	•	•
8.6	Radical Expressions and Radical Functions	1–4, 6–10		•	•	•	•	•	•
8.7	Simplifying Radical Expressions	1–4, 6–10		•		•		•	•
8.8	Solving Radical Equations and Inequalities	1–4, 6–10		•	•	•		•	•
	Chapter Project: *Means to an End*	1, 2, 4, 6–10				•		•	•
9.1	Introduction to Conic Sections	1–4, 6–10		•	•	•		•	•
9.2	Parabolas	1–4, 6–10		•	•	•	•	•	•
9.3	Circles	1–4, 6–10			•	•		•	•
9.4	Ellipses	1–4, 6–10	•	•	•	•		•	•
9.5	Hyperbolas	1–4, 6–10		•	•	•	•	•	•
	Eyewitness Math: *What's So Fuzzy?*	1–10			•	•	•	•	•
9.6	Solving Nonlinear Systems of Equations	1–4, 6–10		•	•	•		•	•
	Chapter Project: *Focus on This!*	1–3, 6–10			•	•		•	•
10.1	Introduction to Probability	1–6, 8–10	•	•		•	•	•	•
10.2	Permutations	1, 2, 4–10		•		•		•	•
10.3	Combinations	1, 2, 4–10		•		•		•	•

(Continued)

LESSON		NCTM STANDARDS	REPRESENTATIONS						
			CONCRETE	DRAWING	GRAPH	SYMBOL	TABLE	VERBAL	VISUAL
	Eyewitness Math: *Let's Make a Deal*	1, 2, 4–10				•	•	•	•
10.4	Using Addition With Probability	1–10	•	•		•	•	•	•
10.5	Independent Events	1, 2, 4–10		•		•		•	•
10.6	Dependent Events and Conditional Probability	1–10		•		•	•	•	•
10.7	Experimental Probability and Simulation	1–10	•	•		•	•	•	•
	Chapter Project: *Next, please...*	1, 2, 4–10				•	•	•	•
11.1	Sequences and Series	1–4, 6–10		•		•	•	•	•
11.2	Arithmetic Sequences	1–3, 6, 8–10			•	•		•	•
11.3	Arithmetic Series	1–3, 5, 8–10		•		•		•	•
11.4	Geometric Sequences	1–4, 6–10			•	•	•	•	•
11.5	Geometric Series and Mathematical Induction	1, 2, 4, 6–10	•	•		•	•	•	•
11.6	Infinite Geometric Series	1–4, 6–10	•	•	•	•	•	•	•
11.7	Pascal's Triangle	1, 2, 4–6, 8–10		•		•	•	•	•
11.8	The Binomial Theorem	1–4, 6–10		•		•	•	•	•
	Chapter Project: *Over the Edge*	1–4, 6–10		•		•	•	•	•
12.1	Measures of Central Tendency	1, 2, 4–10				•	•	•	•
12.2	Stem-and-Leaf Plots, Histograms, and Circle Graphs	1–10			•	•	•	•	•
12.3	Percentiles and Box-and-Whisker Plots	1, 2, 4–10		•	•	•	•	•	•
	Eyewitness Math: *Is It Random?*	1, 2, 4–10			•	•	•	•	•
12.4	Measures of Dispersion	1, 2, 4–10			•	•	•	•	•
12.5	Binomial Distributions	1–10	•	•	•	•	•	•	•
12.6	Normal Distributions	1–10		•	•	•		•	•
	Chapter Project: *That's Not Fair!*	1, 2, 6–10				•	•	•	•
13.1	Right-Triangle Trigonometry	1–4, 6–10		•		•	•	•	•
13.2	Angles of Rotation	1–4, 6–10		•	•	•	•	•	•
13.3	Trigonometric Functions of Any Angle	1–4, 6–10		•	•	•		•	•
13.4	Radian Measure and Arc Length	1–4, 6–10		•	•	•		•	•
13.5	Graphing Trigonometric Functions	1–4, 6–10		•	•	•	•	•	•
13.6	Inverses of Trigonometric Functions	1–4, 6–10		•	•	•		•	•
	Chapter Project: *Reinventing the Wheel*	1–4, 6–10			•	•		•	•
14.1	The Law of Sines	1–4, 6–10		•		•		•	•
14.2	The Law of Cosines	1–4, 6–10		•	•	•		•	•
14.3	Fundamental Trigonometric Identities	1, 2, 4, 6–10		•	•	•		•	•
14.4	Sum and Difference Identities	1–4, 6–10			•	•		•	•
14.5	Double- and Half-Angle Identities	1–4, 6–10		•	•	•		•	•
14.6	Solving Trigonometric Equations	1, 2, 4, 6–10			•	•	•	•	•
	Chapter Project: *Gearing Up*	1–4, 6–10	•			•		•	•

Data and Linear Representations

Lesson Presentation CD-ROM
PowerPoint® presentations for each lesson 1.1–1.8

CHAPTER PLANNING GUIDE

Lesson	1.1	1.2	1.3	1.4	1.5	1.6	1.7	1.8	Project and Review
Pupil's Edition Pages	4–11	12–20	21–28	29–36	37–44	45–51	54–60	61–69	52–53, 70–79
Practice and Assessment									
Extra Practice (Pupil's Edition)	942	942	943	943	944	944	945	945	
Practice Workbook	1	2	3	4	5	6	7	8	
Practice Masters Levels A, B, and C	1–3	4–6	7–9	10–12	13–15	16–18	19–21	22–24	
Standardized Test Practice Masters	1	2	3	4	5	6	7	8	
Assessment Resources	1	2	3	4	5	6	7	8	5, 10–15
Visual Resources									
Lesson Presentation Transparencies Vol. 1	1–4	5–8	9–12	13–16	17–20	21–24	25–28	29–32	
Teaching Transparencies					1, 2, 3	4			
Answer Key Transparencies	1–7	8–14	15–17	18–20	21–22	23–25	26–30	31–34	35–40
Quiz Transparencies	1.1	1.2	1.3	1.4	1.5	1.6	1.7	1.8	
Teacher's Tools									
Reteaching Masters	1–2	3–4	5–6	7–8	9–10	11–12	13–14	15–16	
Make-Up Lesson Planner for Absent Students	1	2	3	4	5	6	7	8	
Student Study Guide	1	2	3	4	5	6	7	8	
Spanish Resources	1	2	3	4	5	6	7	8	
Block Scheduling Handbook									2–3
Activities and Extensions									
Lesson Activities	1	2	3	4	5	6	7	8	
Enrichment Masters	1	2	3	4	5	6	7	8	
Cooperative-Learning Activities	1	2	3	4	5	6	7	8	
Problem Solving/ Critical Thinking	1	2	3	4	5	6	7	8	
Student Technology Guide	1	2	3	4	5	6	7	8	
Long Term Projects									1–4
Writing Activities for Your Portfolio									1–3
Tech Prep Masters									1–4
Building Success in Mathematics									1–3

LESSON PACING GUIDE

Lesson	1.1	1.2	1.3	1.4	1.5	1.6	1.7	1.8	Project and Review
Traditional	1 day	1 day	2 days	2 days	1 day	1 day	1 day	1 day	2 days
Block	$\frac{1}{2}$ day	$\frac{1}{2}$ day	1 day	1 day	$\frac{1}{2}$ day	$\frac{1}{2}$ day	$\frac{1}{2}$ day	$\frac{1}{2}$ day	1 day
Two-Year	2 days	2 days	4 days	4 days	2 days	2 days	2 days	2 days	4 days

CONNECTIONS AND APPLICATIONS

Lesson	1.1	1.2	1.3	1.4	1.5	1.6	1.7	1.8	Review
Algebra	4–11	12–20	21–28	29–36	37–44	45–51	54–60	61–69	70–79
Geometry			23, 27	31, 34, 35		47, 49			75, 76, 79
Business and Economics	7, 10, 12	18, 19			40, 42	50	54, 60	61, 66, 69	74, 75, 77
Statistics					38, 39		60		74
Life Skills	4, 5, 8, 9		28	33		50, 51	56, 58, 59		74, 76
Science and Technology	10	19		32, 33, 35, 36	37, 42, 43, 44	46, 48	59	68, 69	72, 79
Social Studies	10				40, 43		60		72
Sports and Leisure			21, 22, 25, 27	29, 30	39, 41	50, 51	60	67, 68, 69	77
Cultural Connection: Asia				35					
Cultural Connection: Africa									77
Other	10	12, 20	28		42		60	69	

BLOCK SCHEDULING GUIDE

Day	Lesson	Teacher Directed: Lesson Examples, Teaching Transparencies	Student Guided Activity, Try This	Cooperative-Learning Activity, Lesson Activity, Student Technology Guide	Practice: Practice & Apply, Extra Practice, Practice Workbook	Assessment: Quiz, Mid-Chapter Assessment	Problem Solving, Reteaching
1	1.1	10 min	10 min	8 min	25 min	8 min	8 min
	1.2	10 min	10 min	7 min	25 min	7 min	7 min
2	1.3	10 min	15 min	15 min	65 min	15 min	15 min
3	1.4	10 min	15 min	15 min	65 min	15 min	15 min
4	1.5	10 min	10 min	8 min	25 min	8 min	8 min
	1.6	10 min	10 min	7 min	25 min	7 min	7 min
5	1.7	10 min	10 min	8 min	25 min	8 min	8 min
	1.8	10 min	10 min	7 min	25 min	7 min	7 min
6	Assess.	50 min **PE:** Chapter Review	90 min **PE:** Chapter Project, Writing Activities	90 min Tech Prep Masters	65 min **PE:** Chapter Assessment, Test Generator	30 min Chap. Assess. (A or B), Alt. Assess. (A or B), Test Generator	

PE: Pupil's Edition

Alternative Assessment

The following suggest alternative assessments for students who may benefit from a different type of assessment than the regular chapter quizzes and the mid-chapter/end-of-chapter test. Visit the HRW web site to get additional Alternative Assessment material.

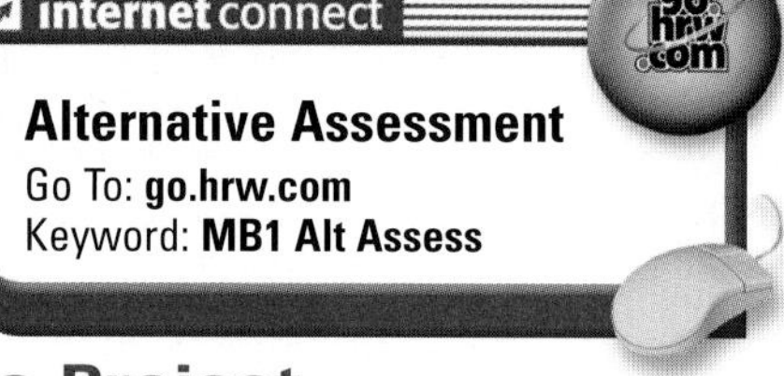

Performance Assessment

1. You are given three sets of specifications for lines.
- Line l_1 contains $(-3, 2)$ and $(5, 7)$.
- Line l_2 has a slope of -0.5 and contains $(3, 5.5)$.
- Line l_3 has a slope of 2 and a y-intercept of 6.

a. Write an equation for each line.
b. Are any of the lines parallel? perpendicular?

2. Consider the equations $2x = 4$, $2x + 5 = 12$, $2(x + 5) = 3x - 7$, $2|x + 5| = 12$, and $\frac{x+3}{4} = \frac{x}{5}$.
a. Solve each equation.
b. Which equations cannot be solved by using only the properties of equality?

3. The table below relates the variables x and y.

x	−3	−2	−1	0	1	2
y	−6.0	−3.5	−1.0	1.5	4.0	6.5

a. Does the table represent a linear relationship? a direct variation equation?
b. Find an equation to represent the data. Find y when $x = 10$.

Portfolio Project

Suggest that students perform the following project for inclusion in their portfolios.

1. Suppose that a, b, c, d, r, and s are numbers.
a. Let $ax + b = cx + d$. What values of the constants will make the expressions on each side of the equation equivalent?
b. Let $\frac{ax+b}{cx+d} = \frac{r}{s}$. Solve for x. Under what conditions will your formula work?
c. Experiment with other types of equations.

2. a. On a number line, graph $|x - n| \le \frac{1}{4}$, where n is an integer. Describe the graph.
b. What inequality would have a solution with the left endpoint at an integer and the right endpoint halfway between two consecutive integers?
c. Experiment with other solution sets and inequalities.

internet connect

The table below identifies the pages in this chapter that contain internet and technology information.

Content Links

Activities Online	**pages 8, 36, 41**
Portfolio Extensions	**pages 11, 20, 44**
Homework Help Online	**pages 8, 17, 26, 34, 41, 49, 58, 68**
Graphic Calculator Support	**page 80**

Resource Links

Parents can go online and find concepts that students are learning—lesson by lesson—and questions that pertain to each lesson, which facilitate parent-student discussion.

Go To: **go.hrw.com**
Keyword: **MB1 Parent Guide**

Technical Support

The following may be used to obtain technical support for any HRW software product.

Online Help: **www.hrwtechsupport.com**
e-mail: **tschrw@hrwtechsupport.com**
HRW Technical Support Center: **(800)323-9239**
7 AM to 10 PM Monday through Friday Central Time

Visit the HRW math web site at: **www.hrw.com/math**

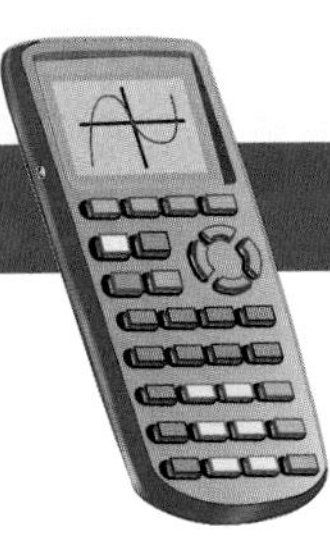

Technology

Lesson Suggestions and Calculator Examples

(Keystrokes are based on a TI-83 calculator.)

Lesson 1.1 Tables and Graphs of Linear Equations
To find out how much experience students have had with graphics calculators, have students enter the equation $y = 2x + 1$, select appropriate window settings, graph the equation, and use the TRACE key to read the coordinates of a point on the line.

Lesson 1.2 Slopes and Intercepts
In this lesson, students will explore how different values of m and b affect the graph of the equation $y = mx + b$. The displays below show graphs of equations of the form $y = mx$ and $y = 2x + b$, where m is $\frac{1}{2}$, 1, 2, and 3 and b is -3, -1, 0, and 3.

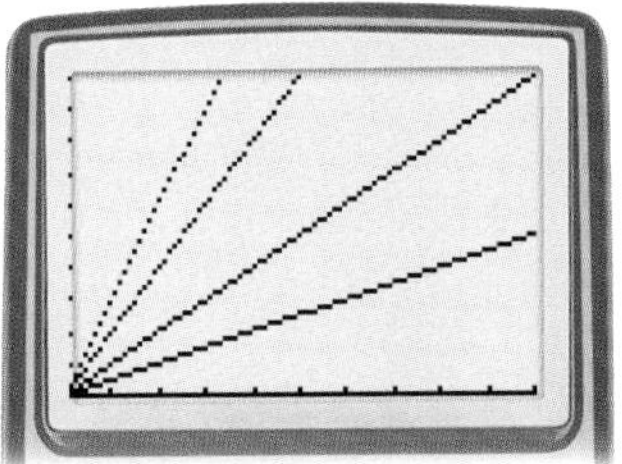

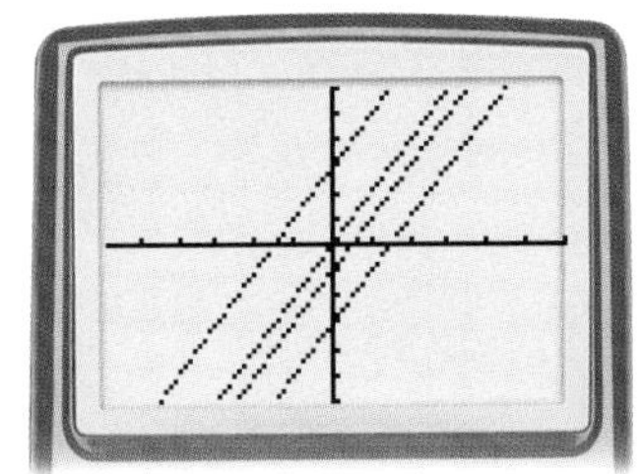

Have students graph these and other equations, testing different values of m and b. Ask students to draw conclusions and make conjectures.

Lesson 1.3 Linear Equations in Two Variables
Students can use a graphics calculator to determine whether a pair of equations derived algebraically are parallel or perpendicular. For the check to be valid, students must use a square window. The displays below show the graphs of the perpendicular lines $y = 2x + 1$ and $y = -0.5x + 2$ in square and non-square windows, respectively.

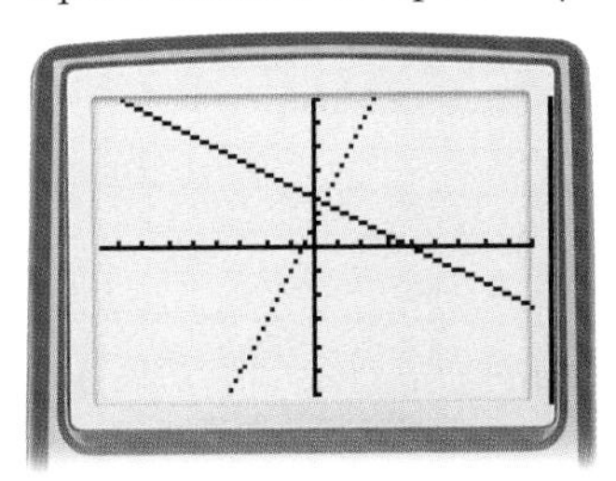

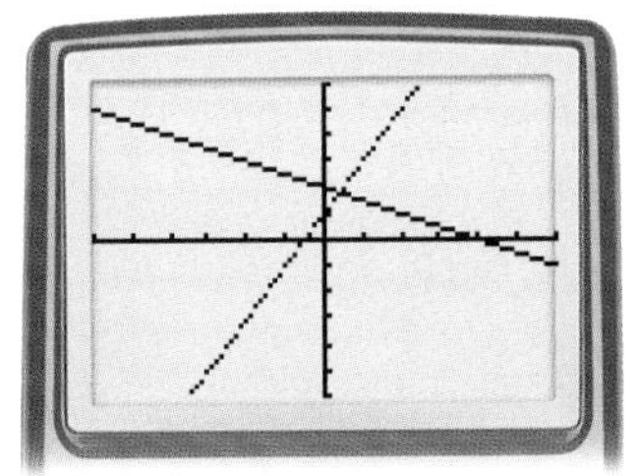

Lesson 1.4 Direct Variation and Proportion
Explain to students that the direct-variation relationship $y = kx$, where k is a nonzero constant, is a *graph compatible* form. That is, students can enter the expression for kx into the **Y=** menu, graph it, and then use the TRACE key to read coordinates from the graph.

Lesson 1.5 Scatter Plots and Least-Squares Lines
To create a scatter plot of data and find an equation of best fit, guide students through the following steps:

1. Enter the data into lists **L1** and **L2**.
2. Create the scatter plot to verify that the pattern is linear.
3. Obtain an equation for the least-squares line.

(See Lesson 1.5 of the Keystroke Guide in the textbook for the appropriate key sequences.)

Lesson 1.6 Introduction to Solving Equations
Example 3 on page 47 shows a graphical solution to a linear equation. This approach is helpful when the coefficients are decimals. It also illustrates the use of the intersection feature, which can be accessed as follows:

2nd TRACE **CALCULATE** **5:intersect**. Have students graph the equation $y = 3.24x - 4.09 - (-0.72x + 3.65)$ and find the x-value where the line crosses the x-axis.

Lesson 1.7 Introduction to Solving Inequalities
Have students graph $x \geq 3$ and $x \leq 5$. To access $\geq$ and $\leq$, press 2nd MATH and select **4:≥** or **6:≤**. When using the **LOGIC** menu, select **1:and** or **2:or**.

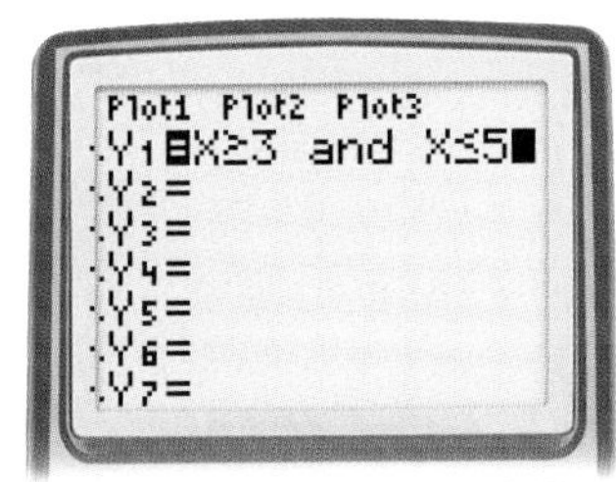

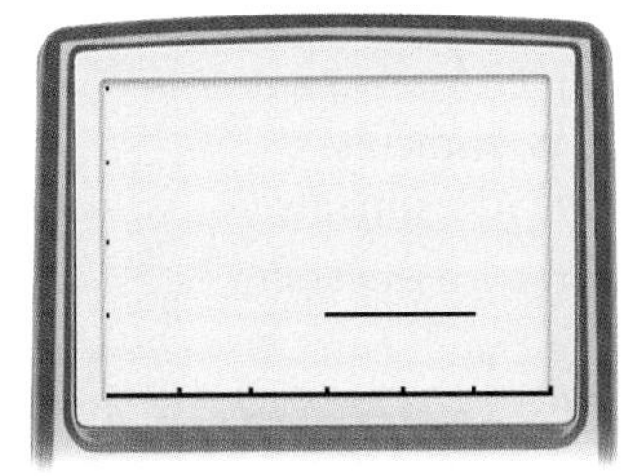

Students can use the TRACE key to locate and identify the endpoints of the solution.

Lesson 1.8 Solving Absolute-Value Equations and Inequalities
The discussion on page 62 paves the way for a theorem about $|x| = a$. Students can explore absolute-value equations with two solutions, one solution, or no solution, depending on the value substituted for a. Have students explore $|x| + 2 = a$ for $a = 4$, 2, and 1.

For further information, refer to the

- technology discussions in the lessons.
- lesson-related teacher's commentary in the side columns of this *Teacher's Edition*.
- lesson-related *Student Technology Guide* masters.
- *HRW Technology Handbook*.

internet connect

For keystrokes of other graphing calculators models, visit the HRW web site at **go.hrw.com** and enter the keyword **MB1 CALC**.

About Chapter 1

Background Information

An equation whose graph is a line is called a linear equation. In Chapter 1, students first graph and analyze linear equations in two variables and then use the equations to solve direct-variation problems and to model trends in data. In the second part of the chapter, students review the properties of equality and inequality and use them to solve linear equations and inequalities in one variable. The chapter concludes with a study of absolute-value equations and inequalities.

CHAPTER RESOURCES

- Block-Scheduling Handbook
- Writing Activities for Your Portfolio
- Tech Prep Masters
- Long-Term Project
- Assessment Resources:
  Mid-Chapter Assessment
  Chapter Assessments
  Alternative Assessments
- Test and Practice Generator
- Technology Handbook

Chapter Objectives

- Represent a real-world linear relationship in a table, graph, or equation. [**1.1**]
- Identify linear equations and linear relationships between variables in a table. [**1.1**]
- Graph a linear equation. [**1.2**]
- Write a linear equation for a given line in the coordinate plane. [**1.2**]
- Write a linear equation in two variables given sufficient information. [**1.3**]

1 Data and Linear Representations

COLLECTING, ORGANIZING, AND REPRESENTING data are important skills in the real world. One example is the research of the annual migration of millions of monarch butterflies. Observers across North America collect data and share their information over the World Wide Web.

Tagging a monarch butterfly

Lessons

About the Photos

Each year in North America, millions of monarch butterflies migrate an average of 50 kilometers, or 31 miles, per day. Researchers have found that many of the butterflies travel a total distance of approximately 4500 kilometers, or 2796 miles, on their journey from eastern Canada to their winter sites in Mexico. The migration may take up to 90 days. If the average length of a monarch butterfly is 3 centimeters, or 1.2 inches, each butterfly will have flown its body length approximately 150,000,000 times. In order for a person who is 6 feet tall to make an equivalent journey, he or she would need to travel around the earth about 11 times.

Data about the migration of monarch butterflies has been gathered for years by observers at different locations along the migration path. Although the data suggests several hypotheses regarding the migration, most of these hypotheses have not yet been thoroughly tested, so any conclusions may have to wait until well into the next century.

About the Chapter Project

Algebra provides the power to model real-world data. In the Chapter Project, *Correlation Exploration,* you will investigate the correlations between many common variables. Throughout this book, you will learn about different types of algebraic models that can be used to investigate trends and make predictions. In this chapter, you will focus on linear models.

After completing the Chapter Project, you will be able to do the following:

- Represent real-world data by using scatter plots.
- Find and use linear models to predict other possible data values.

About the Portfolio Activities

Throughout the chapter, you will be given opportunities to complete Portfolio Activities that are designed to support your work on the Chapter Project.

- Creating a graph to represent your data is included in the Portfolio Activity on page 11.
- Estimating a linear model for your data is included in the Portfolio Activity on page 20.
- Using your linear model to predict other possible data values is included in the Portfolio Activity on page 36.
- Using technology to find a least-squares regression line for your data set is included in the Portfolio Activity on page 44.
- Using the least-squares regression line to predict other data values is included in the Portfolio Activity on page 51.

internet connect go.hrw.com

Chapter Internet Features and Online Activities

Lesson	Keyword	Page	Lesson	Keyword	Page
1.1	MB1 Homework Help	8	1.5	MB1 Homework Help	41
	MB1 Paycheck	8		MB1 Bicycle	41
	MB1 Linear	11		MB1 Regression	41
1.2	MB1 Homework Help	17	1.6	MB1 Homework Help	49
	MB1 Rate	20	1.7	MB1 Homework Help	58
1.3	MB1 Homework Help	26	1.8	MB1 Homework Help	68
1.4	MB1 Homework Help	34			
	MB1 Metrics	36			

- Write an equation for a line that is parallel or perpendicular to a given line. [**1.3**]
- Write and apply direct-variation equations. [**1.4**]
- Write and solve proportions. [**1.4**]
- Create a scatter plot and draw an informal inference about any correlation between the variables. [**1.5**]
- Use a graphics calculator to find an equation for the least-squares line and use it to make predictions or estimates. [**1.5**]
- Write and solve a linear equation in one variable. [**1.6**]
- Solve a literal equation for a specified variable. [**1.6**]
- Write, solve, and graph linear inequalities in one variable. [**1.7**]
- Solve and graph compound linear inequalities in one variable. [**1.7**]
- Write, solve, and graph absolute-value equations and inequalities in mathematical and real-world situations. [**1.8**]

Portfolio Activities appear at the end of Lessons 1.1, 1.2, 1.4, 1.5, and 1.6. Each serves as preparation for the Chapter Project. The Portfolio Activities as well as the Chapter Project Activities are appropriate for inclusion in the student's portfolio. Students should be encouraged to include in their portfolios any other work in which they feel a sense of pride or a sense of accomplishment.

Prepare

NCTM PRINCIPLES & STANDARDS 1, 2, 4, 6–10

QUICK WARM-UP

Graph each point on the same coordinate plane.

1. $A(-1, 2)$ **2.** $B(4, 7)$

3. $C(-3, 0)$ **4.** $D(0, 3)$

Check students' graphs.

5. Refer to Exercises 1–4. Which equation below gives the relationship between each pair of x- and y-coordinates?

a. $y = 3x$ **b.** $y = -3x$

c. $y = x + 3$ **d.** $y = x - 3$

Choice c is correct.

Also on Quiz Transparency 1.1

Teach

Why A linear relationship between two variables permits "what if" reasoning. Have students compare a compensation plan that offers high base pay and low commission with a compensation plan offering low base pay and high commission. Discuss under what conditions each plan is favorable to an employee.

Tables and Graphs of Linear Equations

Objectives

- Represent a real-world linear relationship in a table, graph, or equation.
- Identify linear equations and linear relationships between variables in a table.

Why *Linear relationships between two variables occur in a wide variety of situations. For example, the wages of a salesperson who earns a commission are often linearly related to the dollar amount of his or her sales.*

A salesperson in a music store usually earns commission.

Activity
Investigating Commission

APPLICATION INCOME

You will need: graph paper and a straightedge

Suppose that you work part-time in a music store. You earn $40 per week plus a 10% commission on all of the sales you make.

1. Discuss what it means to earn commission.

PROBLEM SOLVING

2. You can represent this relationship between weekly sales and weekly wages by **making a table.** Copy and complete the table below.

Weekly sales, s (in dollars)	Weekly wages, w (in dollars)
100	$40 + 0.10(100) = 50$
200	?
300	?
400	?
500	?

3. What observations can you make about successive entries in the weekly sales column? in the weekly wages column?
4. You can represent each row in the table as an ordered pair (s, w). Plot each ordered pair, and connect the first and last points with a straightedge. Does each point that you plotted appear to be contained in this line segment?

CHECKPOINT ✔

5. Write an equation to represent the relationship between s and w.

Alternative Teaching Strategy

USING TECHNOLOGY Students can use the table feature of a graphics calculator to generate many of the tables in this lesson. For instance, if they are using a TI-82 or TI-83 graphics calculator, they can follow these steps to create the table of sales and wages in the Activity above.

First press [Y=] to display the **Y=** editor. Place the cursor to the right of **Y1=**. If there already is an expression there, press [CLEAR] to erase it. Then enter **40** [+] **.1** [X,T,θ,n].

To display the **TABLE SETUP** screen, press [2nd] [WINDOW]. Then make these choices.

Indpnt: Auto Ask
Depend: Auto Ask

The values for **TblStart** (or **TblMin**) do not matter.

Now press [2nd] [GRAPH] to display the table. To generate values for the five rows of the table in the Activity, use the following key sequence:

100 200 300 400 500

In the Activity on page 4, a linear relationship that can be modeled by a linear equation is described. Example 1 provides another instance of a linear relationship.

EXAMPLE 1 **An attorney charges a fixed fee of $250 for an initial meeting and $150 per hour for all hours worked after that.**

APPLICATION
INCOME

A lawyer, or attorney, in court

a. Make a table of the total charge for 1, 2, 3, and 4 hours worked.
b. Graph the points represented by your table and connect them.
c. Write a linear equation to model this situation.
d. Find the charge for 25 hours of work.

SOLUTION

a.

Hours worked	1	2	3	4
Total charge	$400	$550	$700	$850

b.

c. Translate the verbal description into an equation involving c and h.

$$\text{total charge} = \text{variable charge} + \text{fixed fee}$$
$$c = 150h + 250$$

Thus, $c = 150h + 250$.

d. Use the equation. Substitute 25 for h in the equation.

$$c = 150h + 250$$
$$c = 150(25) + 250$$
$$c = 4000$$

TECHNOLOGY
GRAPHICS CALCULATOR
Keystroke Guide, page 80

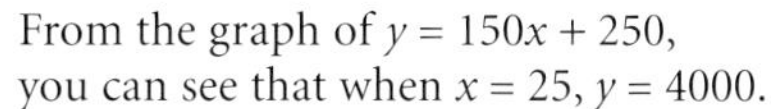

CHECK
From the graph of $y = 150x + 250$, you can see that when $x = 25$, $y = 4000$.

Thus, the attorney charges $4000 for 25 hours of work.

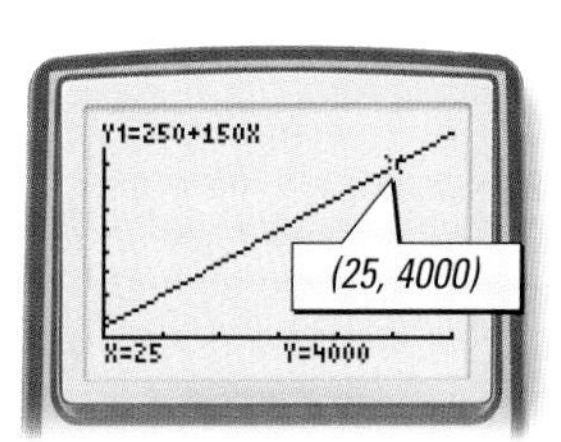

TRY THIS A water tank already contains 55 gallons of water when Darius begins to fill it. Water flows into the tank at a rate of 9 gallons per minute.
a. Make a table for the volume of water in the tank after 1, 2, 3, and 4 minutes.
b. Graph the points represented by your table and connect them.
c. Write a linear equation to model this situation.
d. Find the volume of water in the tank 20 minutes after Darius begins filling the tank.

The equations in Example 1 and the Try This exercise above have a characteristic in common. Each has the form shown below.

total amount = variable amount + fixed amount

In general, if a relationship between x and y can be written as $y = mx + b$, where m and b are real numbers, then x and y are **linearly related**. The equation $y = mx + b$ is called a **linear equation**. The graph of a linear equation is a straight line.

Interdisciplinary Connection

PHYSICS *Acceleration* is the rate at which the velocity of an object changes. Suppose that an airplane is traveling in a straight line at a speed of 60 meters per second. The plane then accelerates, and its speed increases at a constant rate of 0.5 meters per second.

a. Write a linear equation relating the speed of the airplane, v, to the number of seconds, t, of constant acceleration. $v = 60 + 0.5t$

b. What is the speed of the airplane after 15 seconds? 67.5 m/s

Inclusion Strategies

VISUAL LEARNERS Use diagrams such as the one at the right to give students a means of visualizing constant differences between x-values and between y-values.

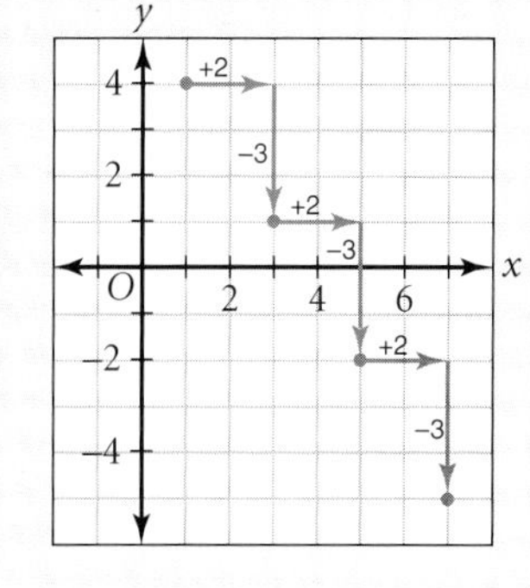

Answers to Activity steps can be found in Additional Answers of the Teacher's Edition.

Activity Notes

Students should observe that a constant difference of $100 in the sales column results in a constant difference of $10 in the wage column. The relationship between s and w is linear.

Checkpoint questions provide an opportunity for ongoing assessment. The answers are provided in the Teacher's Edition side copy.

CHECKPOINT 
5. $w = 0.10s + 40$
The weekly wage is $40 plus 10 percent of weekly sales.

ADDITIONAL EXAMPLE 1

A plumber charges a base fee of $55 for a service call plus $35 per hour for all hours worked during the call.

a. **Make a table of the total charge for 1, 2, 3, and 4 hours worked.**
Check students' tables.

b. **Graph the points represented by your table.**
Check students' graphs.

c. **Write a linear equation to model this situation.**
$c = 35h + 55$, where c is the total charge and h is the number of hours worked

d. **Find the charge for 5 hours of work.** $230

TRY THIS
a. Check students' tables.
b. Check students' graphs.
c. $V = 9t + 55$, where V is volume in gallons and t is time in minutes
d. 235 gal

CHECKPOINT ✔
$m = 150, b = 250$

ADDITIONAL
EXAMPLE 2

Graph $y = 3x - 5$.

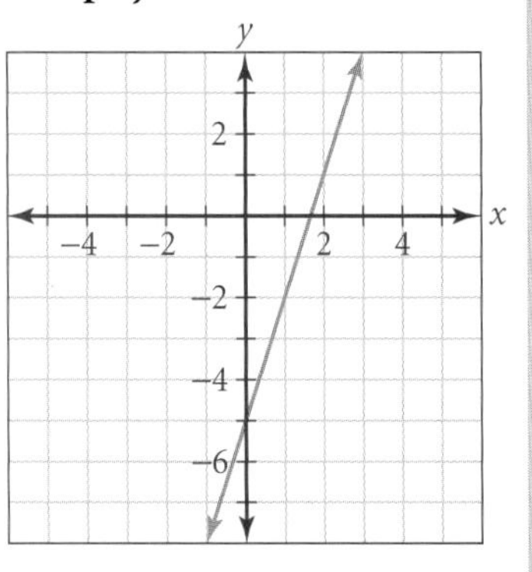

Teaching Tips about technology refer to keystrokes for the TI-82, TI-83, or TI-83 Plus.

Teaching Tip

TECHNOLOGY To check Example 2, press WINDOW and set the viewing window as follows:

Xmin=–4 **Ymin=–3**
Xmax=4 **Ymax=3**
Xscl=1 **Yscl=1**

Press Y= . Place the cursor next to **Y1** and enter (**2** ÷ **3**) X,T,θ,n – **1.** Be sure to clear all other equations. Then press GRAPH .

A keystroke guide is provided at the end of each chapter for examples given in the Pupil's Edition. If you are using a different graphics calculator, such as a Casio or Sharp, go to the HRW Internet site at go.hrw.com and enter the keyword MB1 CALC to find keystroke guides.

TRY THIS
Check students' graphs.

CHECKPOINT ✔ What are the values of m and b in the equation $c = 150h + 250$ from Example 1?

EXAMPLE 2 **Graph $y = \frac{2}{3}x - 1$.**

SOLUTION

Because $y = \frac{2}{3}x - 1$ is of the form $y = mx + b$, where $m = \frac{2}{3}$ and $b = -1$, its graph is a straight line. A line is determined by two points, so you need to plot only two ordered pairs that satisfy $y = \frac{2}{3}x - 1$ and draw the line through them.

x	0	3
y	$y = \frac{2}{3}(0) - 1 = -1$	$y = \frac{2}{3}(3) - 1 = 1$

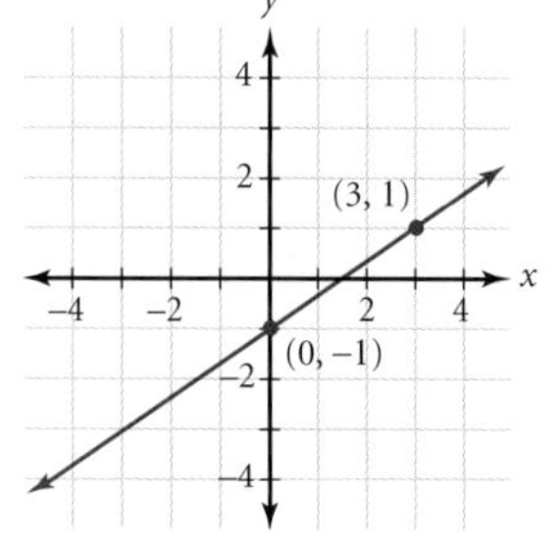

Plot (0, –1) and (3, 1), and draw a line through them.

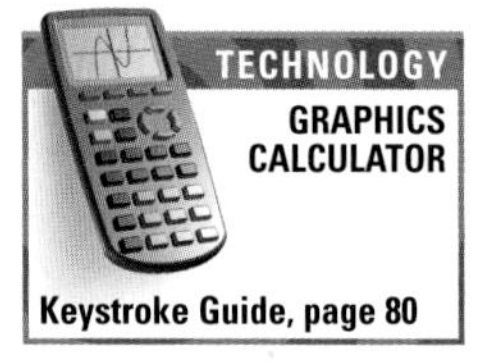

CHECK

Graph $y = \frac{2}{3}x - 1$ on a graphics calculator, and verify that the points (0, –1) and (3, 1) are on the line.

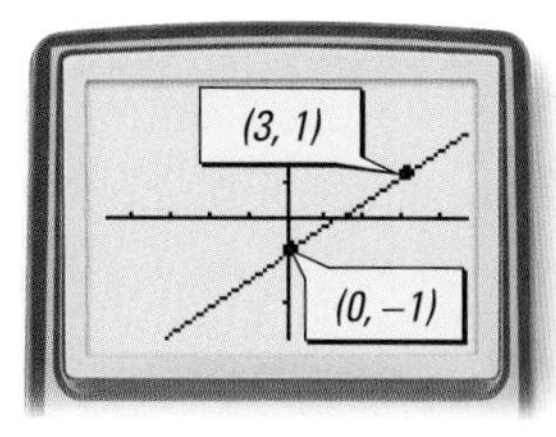

TRY THIS Graph $y = \frac{5}{4}x + 3$.

When variables represented in a table of values are linearly related and there is a constant difference in the x-values, there is also a constant difference in the y-values. For example, consider the linear equation $y = -2x + 5$.

Make a table of values by choosing x-values that have a constant difference, such as 1, 2, 3, 4, and so on.

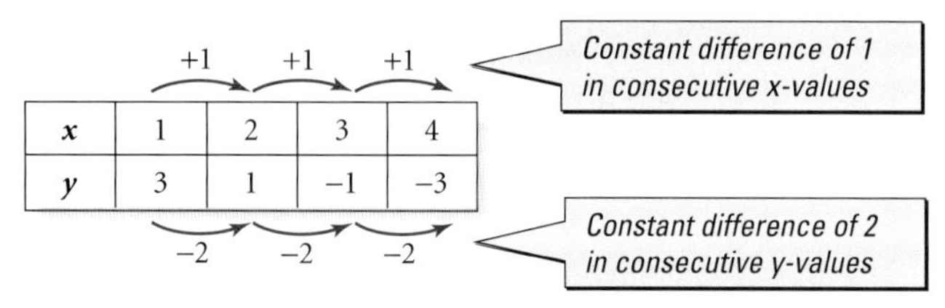

x	1	2	3	4
y	3	1	–1	–3

In a linear relationship, a constant difference in consecutive x-values results in a constant difference in consecutive y-values.

CHECKPOINT ✔ Suppose that you are making a table of values to determine whether the variables are linearly related. Describe the relationship. What must exist between the x-values that you choose?

Enrichment

Give students this table of values for $y = x^2$.

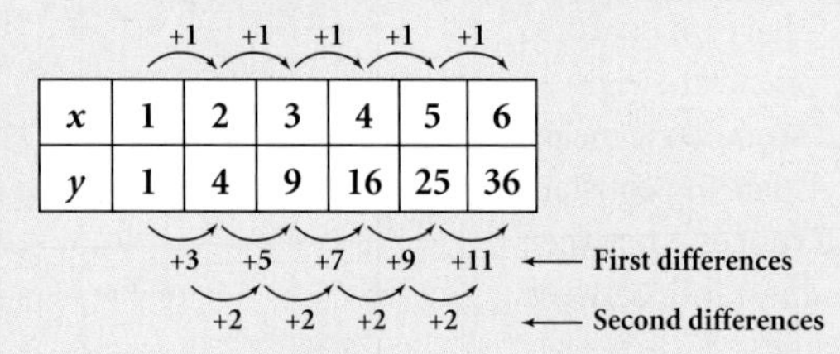

x	1	2	3	4	5	6
y	1	4	9	16	25	36

Point out that the *first differences* are not constant, but the *second differences* are.

Give students the equations in the next column. Have them make a table of values for each equation and find the level at which the differences are constant.

1. $y = 2x^2$ second **2.** $y = x^2 + 2$ second
3. $y = x^2 + x$ second **4.** $y = 2x^3$ third
5. $y = x^3 + 2$ third **6.** $y = x^3 + x^2 + x$ third

Have students use their results to make and test a conjecture about constant differences. **Students' conjectures will take on various forms. In general, if $y = a_nx^n + a_{n-1}x^{n-1} + \cdots + a_1x^1 + a_0x^0$, there will be constant nth differences.**

EXAMPLE 3 **Does the table of values at right represent a linear relationship between x and y? Explain. If the relationship is linear, write the next ordered pair that would appear in the table.**

x	7	12	17	22	27	32
y	11	8	5	2	−1	−4

SOLUTION

Find differences in consecutive x-values and consecutive y-values.

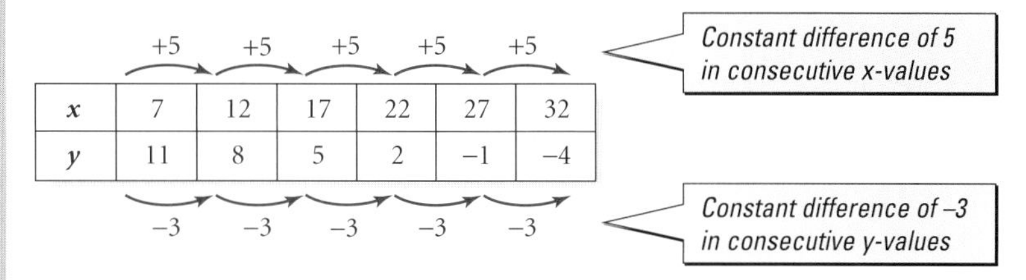

x	7	12	17	22	27	32
y	11	8	5	2	−1	−4

PROBLEM SOLVING **Look for a pattern.** Because there is a constant difference in the x-values and a constant difference in the y-values, the relationship between x and y is linear.

The next table entry for x is $32 + 5$, or 7.
The next table entry for y is $-4 + (-3)$, or -7.

TRY THIS Does the table of values at right represent a linear relationship between x and y? Explain. If the relationship is linear, write the next ordered pair that would appear in the table.

x	−2	2	6	10	14	18
y	1	2	4	8	16	32

CRITICAL THINKING Does the table at right represent a linear relationship between x and y? Explain.

x	9	6	3	0	−3	−6
y	5	5	5	5	5	5

CHECKPOINT ✔
You must choose x-values that have a constant difference.

ADDITIONAL EXAMPLE 3

Does the table of values below represent a linear relationship between x and y? Explain. If the relationship is linear, write the next ordered pair that would appear in the table.

x	5	13	21	29	37	45
y	11	9	7	5	3	1

There is a constant difference between consecutive x-values (8) and a constant difference between consecutive y-values (−2), so the relationship between x and y is linear. The next table entry for x is 53, and the next table entry for y is −1.

Exercises

Communicate

1. Discuss the relationships among the table, equation, and graph shown here.

x	−4	−2	0	2	4
y	−7	−4	−1	2	5

$$y = \frac{3}{2}x - 1$$

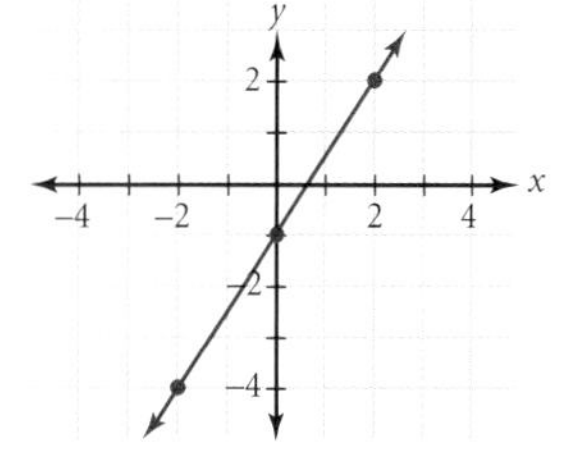

APPLICATION

2. **SALES TAX** Suppose that a state sales tax is 7%. Make a table showing the amount of tax on items with prices of \$6, \$8, \$10, and \$12. How can you find whether the price and amount of sales tax are linearly related?

3. Explain how to verify that the points $(-1, 7)$, $(0, 4)$, and $(2, -2)$ are all on the same line.

Try This questions provide an opportunity for ongoing assessment. The answers are provided in the Teacher's Edition side copy as well as in the Pupil's Edition Selected Answers.

TRY THIS
The relationship is not linear because there is not a constant difference in the y-values.

Critical Thinking questions provide students an opportunity to analyze, assimilate, and expand their understanding of lesson concepts. The answers are provided in the Teacher's Edition side copy.

CRITICAL THINKING
Yes, the relationship is linear because there is a constant difference in the x-values, −3, and a constant difference in the y-values, 0.

Reteaching the Lesson

WORKING BACKWARD Display the graph at right on the board or overhead. Tell students to make a table of six sets of coordinates on the line, choosing the coordinates in such a way that the table demonstrates constant differences between x-values and between y-values. Ask several volunteers to write their table on the board. Discuss the results with the class, taking care to note similarities and differences among the tables. Then repeat the activity for the line given by the equation $y = -1.5x$.

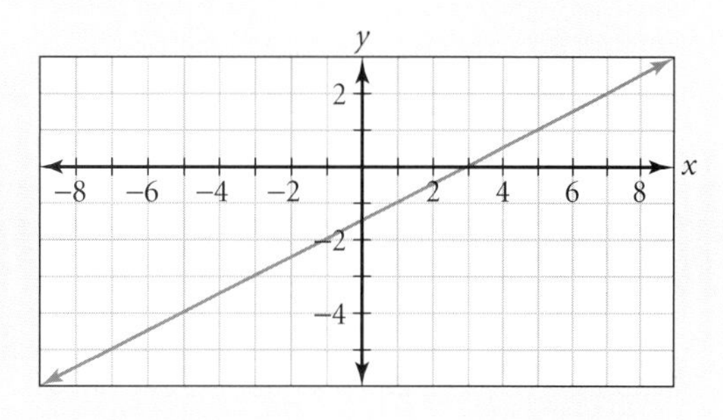

sample table:

x	−5	−3	−1	1	3	5
y	−4	−3	−2	−1	0	1

Assess

Communicate exercises provide an opportunity for students to discuss their discoveries, conjectures, and the lesson concepts. Throughout the text, the answers to all Communicate exercises can be found in the Teacher's Edition Additional Answers.

Guided Skills Practice provides an opportunity for students to practice the skills and concepts discovered in each example before working independently on the Practice and Apply exercises. The answers to all Guided Skills Practice can be found in the Pupil's Edition Selected Answers as well as in the Teacher's Edition side copy or Additional Answers.

Selected Answers

Exercises 4–6, 7–69 odd

ASSIGNMENT GUIDE

In Class	1–6
Core	7–18, 19–29 odd, 31–47, 49–53 odd
Core Plus	8–30 even, 31–54
Review	55–70
Preview	71

Extra Practice can be found beginning on page 940.

Answers to odd-numbered Extra Practice exercises can be found immediately after Selected Answers in the Pupil's Edition.

Guided Skills Practice

APPLICATION

internet connect Activities Online Go To: go.hrw.com Keyword: MB1 Paycheck

4. INCOME Suppose that you work part-time at a department store, earning a base salary of \$50 per week plus a 15% commission on all sales that you make. ***(EXAMPLE 1)***

Weekly sales, x	Weekly income, y
100	$50 + (0.15)(100) = 65$
200	? **80**
300	? **95**
400	? **110**
x	**$0.15x + 50$** ?

a. Copy and complete the table.

b. Graph the points represented in the table and connect them.

c. Write a linear equation to represent the relationship between the weekly sales, x, and the weekly income y. **$y = 0.15x + 50$**

d. Find the weekly income, y, for weekly sales of \$1200. **\$230**

5. Graph $y = 3x - 2$. ***(EXAMPLE 2)***

6. Does the table below represent a linear relationship between x and y? If the relationship is linear, write the next ordered pair that would appear in the table. ***(EXAMPLE 3)*** **yes; (26, 49)**

x	−4	1	6	11	16	21
y	13	19	25	31	37	43

Practice and Apply

State whether each equation is a linear equation.

7. $y = -3x$ **linear**
8. $y = -x$ **linear**
9. $y = 12 + 2x$ **linear**
10. $y = 5 - 4x$ **linear**
11. $y = \frac{1}{2}x - 3$ **linear**
12. $y = -\frac{2}{3}x$ **linear**
13. $y = \frac{1}{x}$ **not linear**
14. $y = \frac{-4}{x}$ **not linear**
15. $y = -x^2 + 1$ **not linear**
16. $y = 2 + 5x^2$ **not linear**
17. $y = 5.5x - 2$ **linear**
18. $y = 11 - 1.2x$ **linear**

internet connect Homework Help Online Go To: go.hrw.com Keyword: MB1 Homework Help for Exercises 19–30

Graph each linear equation.

19. $y = 2x + 1$
20. $y = 4x + 3$
21. $y = 3x - 6$
22. $y = 6x - 3$
23. $y = 5 - 2x$
24. $y = 3 - 5x$
25. $y = -x + 5$
26. $y = -x - 2$
27. $y = \frac{2}{3}x + 4$
28. $y = \frac{1}{3}x - 5$
29. $y + 3 = x + 6$
30. $y + 4 = x - 3$

For Exercises 31–38, determine whether each table represents a linear relationship between x and y. If the relationship is linear, write the next ordered pair that would appear in the table.

31. **linear (4, 58)**

x	y
0	10
1	22
2	34
3	46

32. **linear (12, 11)**

x	y
0	−5
3	−1
6	3
9	7

33. **not linear**

x	y
3	−5
4	1
5	6
6	11

34. **not linear**

x	y
−2	1
−3	2
−4	4
−5	8

4b.

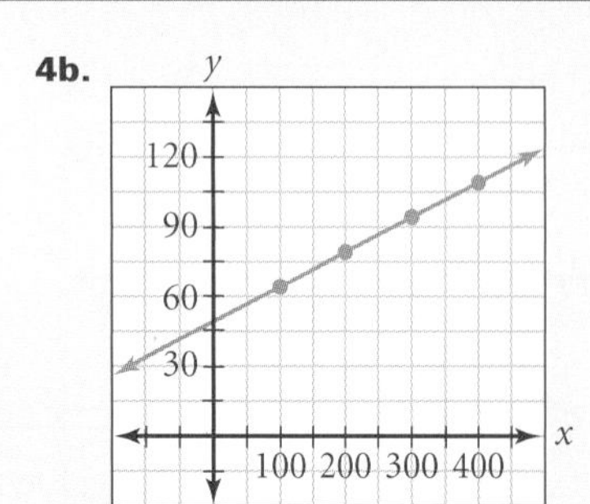

5.

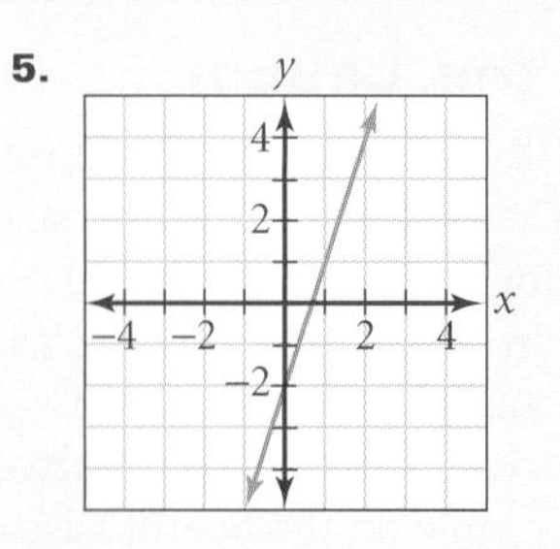

The answers to Exercises 19–30 can be found in Additional Answers beginning on page 1002.

35. linear (0, 4)

x	y
8	28
6	22
4	16
2	10

36. linear (0, −23)

x	y
12	−3
9	−8
6	−13
3	−18

37. not linear

x	y
6	115
9	100
12	85
15	75

38. not linear

x	y
−6	58
−9	44
−12	32
−15	20

For each graph, make a table of values to represent the points. Does the table represent a linear relationship? Explain.

39.

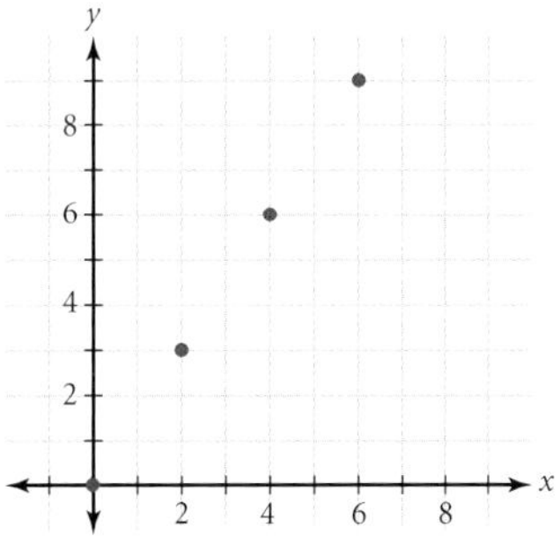

40.

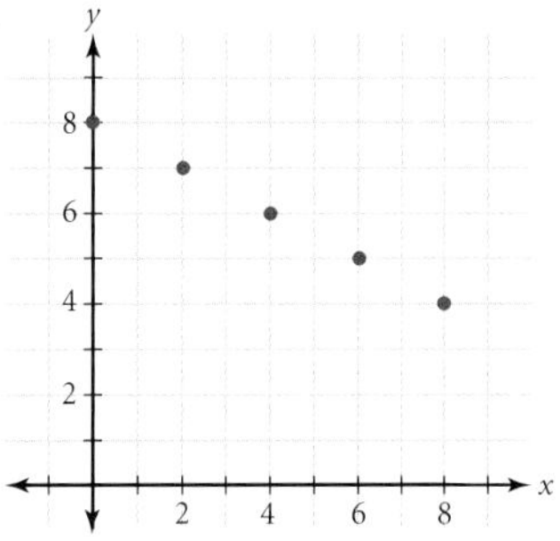

41. Make a table of values for the equation $y = 4x - 1$, and graph the line. Is the point (2, 6) on this line? Explain how to answer this question by using the table, the graph, and the equation.

Calculator button indicates that a graphics calculator is recommended.

Use a graphics calculator to graph each equation. Then sketch the graph on graph paper.

42 $y = -3x + 1.5$ **43** $y = -x - 2.5$ **44** $y = 12 - 2.5x$

45 $y = 4 - 0.5x$ **46** $y = \frac{1}{2}x - \frac{3}{5}$ **47** $y = -\frac{2}{3}x + \frac{1}{2}$

CHALLENGE

48. What can you determine about the graph of $y = mx + b$ when $x = 0$? What can you determine about the graph of $y = mx + b$ when $y = 0$?

APPLICATION

49. INCOME A video rental store charges a $6 membership fee and $3 for each video rented. In the graph at right, the x-axis represents the number of videos rented by a customer and the y-axis represents the store's revenue from that customer.

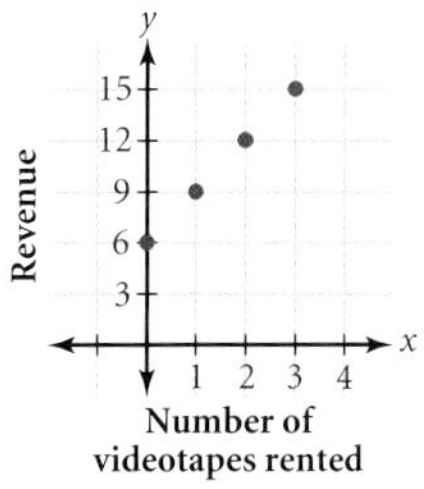

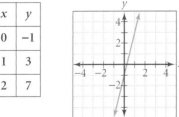

a. Make a table for the data points on the graph.

b. If 15 videos are rented, what is the revenue? **$51**

c. If a new member paid the store a total of $27, how many videos were rented? **7**

d. Explain how to find answers to parts **b** and **c** by using an extended table and an extended graph.

Error Analysis

Students often make simple arithmetic errors when using a given equation to calculate the coordinates of a point. Calculating coordinates for just two points may result in an incorrect graph. Encourage students to determine coordinates for at least three points. If all three points do not lie on one line, they will know that at least one set of coordinates is incorrect.

39.

x	y
0	0
2	3
4	6
6	9

This is a linear relationship because there is a constant difference of 2 in consecutive x-values and a constant difference of 3 in consecutive y-values.

40.

x	y
0	8
2	7
4	6
6	5
8	4

This is a linear relationship because there is a constant difference of 2 in consecutive x-values and a constant difference of −1 in consecutive y-values.

41.

x	y
0	−1
1	3
2	7

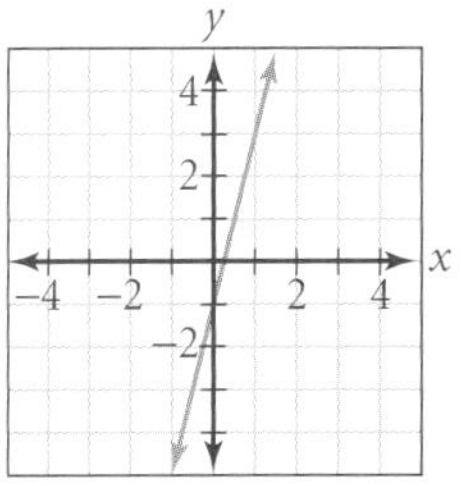

The point (2, 6) is not on the line since in the table, $x = 2$ is paired with $y = 7$. The point (2, 6) also does not make the equation true. Finally, the point (2, 6) is not a point on the graph of $y = 4x - 1$.

42. $y = -3x + 1.5$

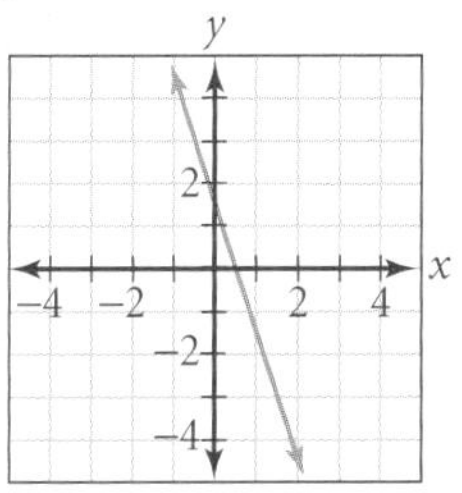

43. $y = -x - 2.5$

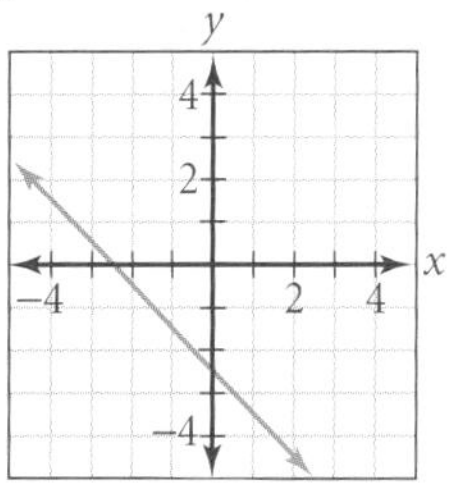

The answers to Exercises 44–49 can be found in Additional Answers beginning on page 1002.

Look Back exercises regularly review important concepts and skills. It is recommended that they be assigned to all students. Answers to the odd-numbered Look Back exercises can be found in the Selected Answers of the Pupil's Edition. Answers to all of the Look Back exercises can be found in the side copy or Additional Answers of the Teacher's Edition.

Look Beyond exercises are of three types: those that foreshadow upcoming concepts, those that extend concepts introduced in the lesson, and those that involve nonroutine problem solving. Answers to Look Beyond exercises can be found in the side copy or Additional Answers of the Teacher's Edition.

50a.

x	y
0	2200
1	2270
2	2340
3	2410
4	2480

Practice

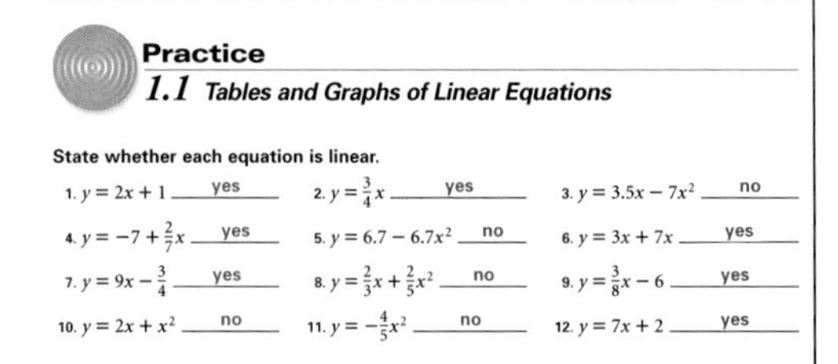

NAME ______ CLASS ______ DATE ______

Practice

1.1 ***Tables and Graphs of Linear Equations***

State whether each equation is linear.

1. $y = 2x + 1$ yes
2. $y = \frac{3}{4}x$ yes
3. $y = 3.5x - 7x^2$ no
4. $y = -7 + \frac{2}{7}x$ yes
5. $y = 6.7 - 6.7x^2$ no
6. $y = 3x + 7x$ yes
7. $y = 9x - \frac{3}{4}$ yes
8. $y = \frac{2}{3}x + \frac{2}{5}x^2$ no
9. $y = \frac{3}{8}x - 6$ yes
10. $y = 2x + x^2$ no
11. $y = -\frac{4}{5}x^2$ no
12. $y = 7x + 2$ yes

Graph each linear equation.

13. $y = \frac{2}{3}x + 4$
14. $y = \frac{4}{5}x$
15. $y = 3x + 2$

Determine whether each table represents a linear relationship between *x* and *y*. If so, write the next ordered pair that would appear in the table.

16.

x	y
2	11
3	13
4	15
5	17
6	19

17.

x	y
0	4
1	8
2	12
3	16
4	20

18.

x	y
−1	5
−2	11
−3	21
−4	35
not	linear

19.

x	y
0	−3
−1	−8
−2	−13
−3	−18
−4	23

20.

x	y
0	3
2	−5
4	−13
6	−21
8	−29

21.

x	y
1	−6
2	−8
3	−10
4	−12
5	−14

APPLICATIONS

Registration day at a community college

50. DEMOGRAPHICS City Community College plans to increase its enrollment capacity to keep up with an increasing number of student applicants. The college currently has an enrollment capacity of 2200 students and plans to increase its capacity by 70 students each year.

a. Let x represent the number of years from now, and let y represent the enrollment capacity. Make a table of values for x and y with x-values of 0, 1, 2, 3, and 4.

b. What will the enrollment capacity be 3 years from now? **2410**

c. Write a linear equation that could be used to find the enrollment capacity, y, after x years. $\mathbf{y = 2200 + 70x}$

51. BUSINESS An airport parking lot charges a basic fee of $2 plus $1 per half-hour parked.

Half-hours	Total charge ($)
0	2 + (1)(0) = 2
1	? **3**
2	? **4**
3	? **5**
? **10**	12

a. Copy and complete the table.

b. Graph the points represented in the table. Label each axis, and indicate your scale.

c. Write an equation for the total charge, c, in terms of the number of half-hours parked, h. $\mathbf{c = 2 + h}$

d. How many half-hours is 72 hours? What is the total charge for parking in the lot for 72 hours? **144; $146**

52. METEOROLOGY At 6:00 A.M., the temperature was 67°F. As a cold front passed, the temperature began to drop at a steady rate of 4°F per hour.

a. Write a linear equation relating the temperature in degrees Fahrenheit, t, to the number of hours, h, after the initial temperature reading.

b. Estimate, to the nearest 15 minutes, how long it would take to reach freezing (32°F) if the drop in temperature continued at the same rate.

52a. $\mathbf{t = 67 - 4h}$

b. 8 hr, 45 min

Supply *is the amount that manufacturers are willing to produce at a certain price.* ***Demand*** *is the amount that consumers are willing to buy at a certain price.*

53. ECONOMICS This table gives the price, the *supply*, and the *demand* for a video game.

Price ($)	Supply	Demand
20	150	500
30	250	400
50	450	200

a. Graph the points representing price and supply and the points representing price and demand on the same coordinate plane.

b. Estimate the price at which the supply of video games meets the demand. Estimate the supply and demand at this price.

c. What happens to the supply and to the demand when the price of the video game is higher than the price in part **b**? lower than the price in part **b**?

54. BUSINESS Casey has a small business making dessert baskets. She estimates that her fixed weekly costs for rent, electricity, and salaries is $200. The ingredients for one dessert basket cost $2.50.

a. If Casey makes 40 dessert baskets in a given week, what will her total weekly costs be? **$300**

b. Casey's total costs for last week were $500. How many dessert baskets did she make? **120**

51b.

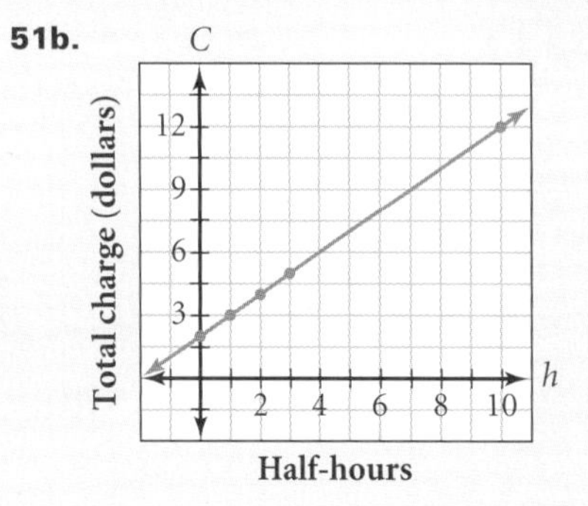

53a.

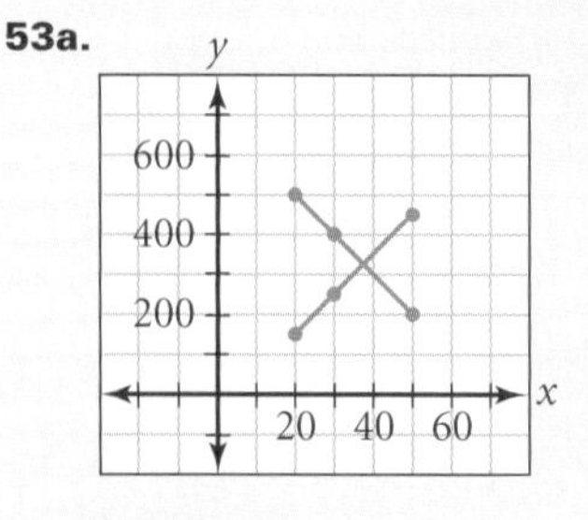

b. $37.50; 325

c. When the price is higher, supply is greater than demand. When the price is lower, supply is less than demand.

Look Back

The following *Rules of Divisibility* are useful in finding factors of numbers.
If a number is

- divisible by 3, the sum of the digits of the number is divisible by 3.
- divisible by 6, the number is divisible by 3 and is even.
- divisible by 9, the sum of the digits of the number is divisible by 9.
- divisible by 4, the number formed by the last two digits is divisible by 4.

Find numbers a and b that meet the following conditions:

55. $ab = 36$ and $a + b = 13$ **9, 4**

56. $ab = 51$ and $a + b = 20$ **17, 3**

57. $ab = 82$ and $a + b = 43$ **41, 2**

58. $ab = 72$ and $a + b = 22$ **18, 4**

59. $ab = 128$ and $a + b = 24$ **8, 16**

60. $ab = 56$ and $a + b = 15$ **8, 7**

61. $ab = 48$ and $a + b = 19$ **16, 3**

62. $ab = 52$ and $a + b = 17$ **13, 4**

Evaluate each expression. Write your answer in simplest form.

63. $\frac{1}{2} \times \frac{4}{7}$ $\mathbf{\frac{2}{7}}$

64. $\frac{2}{3} \times \frac{6}{11}$ $\mathbf{\frac{4}{11}}$

65. $3 \times \frac{2}{3}$ **2**

66. $\frac{13}{14} \times 7$ $\mathbf{6\frac{1}{2}}$

67. $7 \div \frac{7}{8}$ **8**

68. $5 \div \frac{5}{6}$ **6**

69. $21 \div \frac{7}{8}$ **24**

70. $10 \div \frac{5}{6}$ **12**

Portfolio Extension
Go To: **go.hrw.com**
Keyword:
MB1 Linear

Look Beyond

71. Graph the equations $y = 2x$, $y = 2x + 3$, and $y = 2x - 4$ on the same coordinate plane. How are the graphs alike? How are they different?

1. Describe three real-world situations in which a distance changes at a fairly constant rate over time. For example, the distance driven on an interstate highway or the distance walked in a walkathon changes at a fairly constant rate.

2. Choose one of your three real-world situations from Step 1, and determine a suitable way to collect some time and distance data. Collect and record a minimum of seven data values. This data will become your **portfolio data set.**

3. Organize the data from your portfolio data set in a table of values.

4. Use graph paper to graph your portfolio data set. Label the x-axis with units of time, and label the y-axis with units of distance. The point that represents the first distance measure that you took should have an x-coordinate of 0.

WORKING ON THE CHAPTER PROJECT

You should now be able to complete Activity 1 of the Chapter Project.

Look Beyond

In Exercise 71, students graph three equations on the same coordinate plane and then analyze the graphs. They should observe that the three lines have the same slant, or *slope*, but intersect the axes at different points. Students will study slopes and intercepts in greater depth in Lesson 1.2.

ALTERNATIVE

Assessment

Portfolio Activity

The Portfolio Activity can be used as preparation for the Chapter Project or as a separate activity. In the Portfolio Activity on this page, students use tables and graphs to make a visual representation of a linear relationship between time and distance.

71.

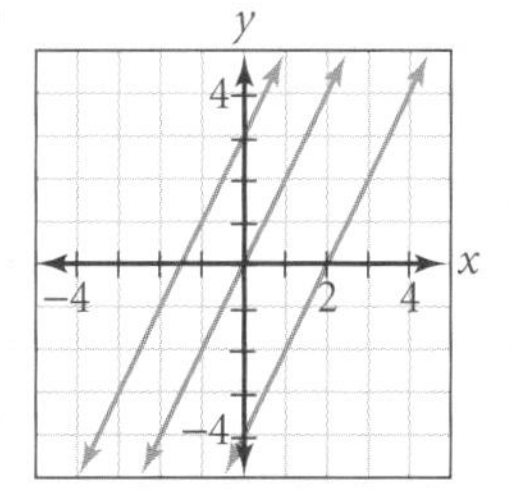

All the lines have the same slope and are parallel but have different y-intercepts.

Answers to Portfolio Activities can be found in Additional Answers of the Teacher's Edition.

Student Technology Guide

NAME ______ CLASS ______ DATE ______

Student Technology Guide
***1.1* Tables and Graphs of Linear Equations**

You can use a graphics calculator to graph a linear equation in x and y. To graph $y + 3 = \frac{2}{3}x + 4$ follow the steps below.

Step 1: Solve for y.

$$y + 3 = \frac{2}{3}x + 4$$
$$y = \frac{2}{3}x + 1 \longrightarrow \frac{2x}{3} + 1$$

Step 2: Press [Y=]. Enter the expression as Y1 in the function list.

2 [X,T,θ,n] [÷] 3 [+] 1

Step 3: Press [WINDOW]. Enter ranges for x and for y.

Step 4: Press [GRAPH].

```
Plot1 Plot2 Plot3
\Y1=2X/3+1
\Y2=
\Y3=
\Y4=
\Y5=
\Y6=
\Y7=
```

```
WINDOW
 Xmin=-10
 Xmax=10
 Xscl=1
 Ymin=-10
 Ymax=10
 Yscl=1
 Xres=1
```

You can also use a graphics calculator to confirm or deny that a point is on the graph of a linear equation. To determine if (4, 3) is on the graph of $y = \frac{2}{3}x + 1$, follow the procedure below.

With the graph displayed, press [TRACE]. Use [◄] and [►] to move the cursor to the point where x is about 4. The display shows that the point (4, 3) is below the line, not on it.

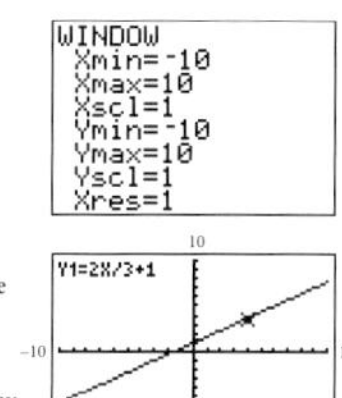

Use a graphics calculator to graph each linear equation. State the viewing window that you used. Settings may vary. Sample settings:

1. $y = -2x + 1.8$ — [−4.7, 4.7] × [−3.1, 3.1]
2. $y = -x - 62$ — [−100, 100] × [−100, 50]
3. $y = 30 + 4.5x$ — [−30, 30] × [−20, 80]
4. $y = 3.5x + 5.1$ — [−10, 10] × [−10, 10]
5. $y = -\frac{1}{4}x - \frac{3}{8}$ — [−3, 3] × [−2, 2]
6. $y = -\frac{2}{3}x - \frac{1}{2}$ — [−3, 3] × [−2, 2]

Is the indicated point on, above, or below the given line?

7. $(2, 7.2)$; $y - 3.4 = 0.5x + 2.8$ — on the line
8. $(1, 3)$; $y - \frac{1}{2} = \frac{1}{5}x + \frac{4}{5}$ — above the line
9. $(-1, -9)$; $y = 3.2x - 5.2$ — below the line

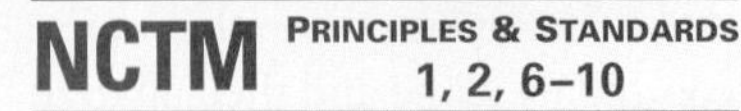

Prepare

NCTM PRINCIPLES & STANDARDS 1, 2, 6–10

QUICK WARM-UP

In Exercises 1–3, solve for y.

1. $4x + y = 3$ $\quad y = -4x + 3$
2. $x + 2y = 10$ $\quad y = -\frac{1}{2}x + 5$
3. $-3 + 6y = 2x$ $\quad y = \frac{1}{3}x + \frac{1}{2}$
4. Graph the points $A(2, 5)$, $B(3, 7)$, and $C(5, 11)$. Connect the points with a line. Estimate where the line crosses the y-axis.

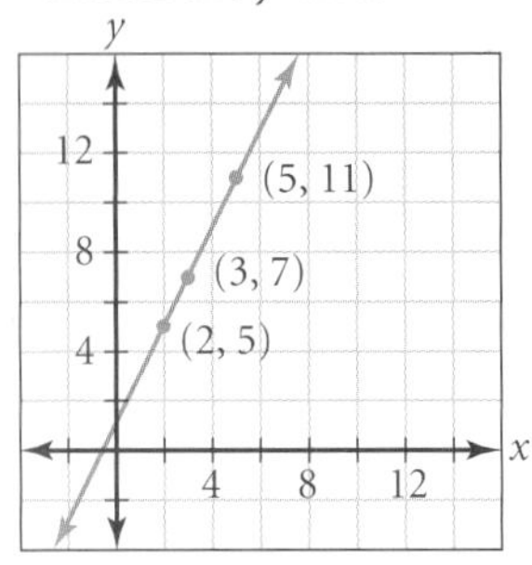

The line crosses the y-axis at $(0, 1)$.

Also on Quiz Transparency 1.2

Teach

Why Discuss with students some everyday uses of the word *slope*. Examples might include the slope of a road, the slope of a roof, and a ski slope. Ask them how they think the idea of slope relates to the position of a line in a coordinate plane.

Slopes and Intercepts

Why *Trends in the real world, such as the increase in cellular phone use, can often be modeled by a linear equation, in which the slope indicates a rate of change.*

Objectives

- Graph a linear equation.
- Write a linear equation for a given line in the coordinate plane.

APPLICATION
COMMUNICATIONS

Every year more and more people in the United States become cellular phone subscribers. The table below represents the recent trend in cellular phone subscriptions. The third column in the table gives the change in the number of subscribers from one year to the next.

Year	Number of subscribers (in millions, rounded to the nearest 100,000)	Yearly change (in millions, rounded to the nearest 100,000)	
1990	5.3		
1991	7.6	1990–1991	2.3
1992	11.0	1991–1992	3.4
1993	16.0	1992–1993	5.0
1994	24.0	1993–1994	8.0
1995	33.8	1994–1995	9.8

U.S. Cellular Subscribers

Number of subscribers (to the nearest 100,000): 0, 5, 10, 15, 20, 25, 30, 35

Years after 1990: 0, 1, 2, 3, 4, 5

Notice in the highlighted row above that the number of new subscribers from 1990 to 1991 is 7.6 – 5.3, or 2.3. This difference can also be represented by the ratio shown below.

$$\frac{\text{Change in subscribers}}{\text{Change in years}} = \frac{7.6 - 5.3}{1991 - 1990} = \frac{2.3}{1} = 2.3$$

All entries in the third column can be found by using a ratio. Each of the ratios gives the *rate of change* from one year to the next.

You can also find the *average rate of change* from 1990 to 1995. On average, from 1990 to 1995, there were about 5.7 million new subscribers per year. This average rate of change is indicated by the red dashed line on the graph at left.

$$\frac{\text{Total change in subscribers}}{\text{Total change in years}} = \frac{33.8 - 5.3}{1995 - 1990} = \frac{28.5}{5} = 5.7$$

CHECKPOINT ✔ Estimate the average rate of change from 1990 to 1993 and from 1993 to 1995.

Alternative Teaching Strategy

USING TABLES Have students complete the table at right. Have them make similar tables for $y = 5x - 1$ and $y = -0.5x + 4$. Then ask them to compare the ratio in the fifth column of each table with the equation. Elicit the observation that the ratio is equal to the coefficient of x. Use this as an introduction to a discussion of slope.

To introduce slope-intercept form, have students make a similar table for $4x + 2y = 6$. Discuss what must be done to the equation so that the slope ratio can be read from the equation.

Equation: $y = x + 5$

x	y	Change from previous x	Change from previous y	$\frac{\text{Change in } y}{\text{Change in } x}$
–7	–2	—	—	—
–4	1	+3	+3	1
–1	4	+3	+3	1
2	7	+3	+3	1
5	10	+3	+3	1
8	13	+3	+3	1

In a graph, the *slope* of a line is the change in vertical units divided by the corresponding change in horizontal units.

The slope of a line is sometimes referred to as $m = \frac{rise}{run}$.

Slope of a Line

If points (x_1, y_1) and (x_2, y_2) lie on a line, then the slope, m, of the line is given by the ratio below.

$$m = \frac{\text{change in } y}{\text{corresponding change in } x} = \frac{y_2 - y_1}{x_2 - x_1}$$

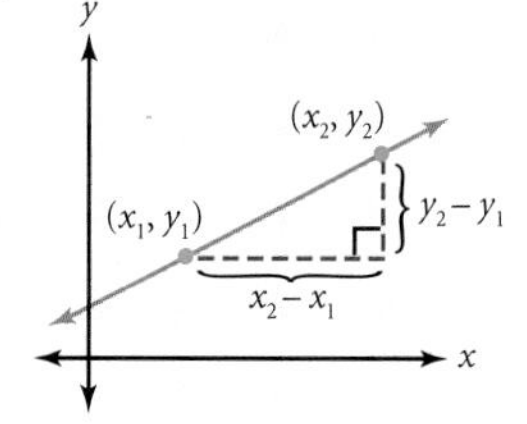

You can find the slope of a line if you know the coordinates of two points on the line. This is shown in Example 1.

EXAMPLE 1 **Find the slope of the line containing the points (0, 4) and (3, 1).**

SOLUTION

PROBLEM SOLVING

Use a formula. Let $(x_1, y_1) = (0, 4)$ and $(x_2, y_2) = (3, 1)$. Apply the definition of slope.

$$m = \frac{y_2 - y_1}{x_2 - x_1} = \frac{1 - 4}{3 - 0} = \frac{-3}{3} = -1$$

The line containing (0, 4) and (3, 1) has a slope of –1.

TRY THIS Find the slope of the line containing the points (–5, 3) and (3, –4).

Activity: Exploring Slopes

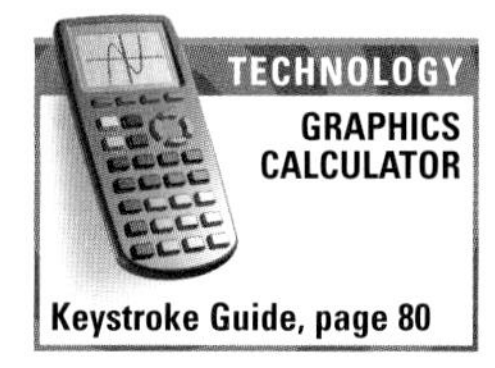

You will need: a graphics calculator

1. Each equation below has the form $y = mx$, where m is the slope. Graph each pair of equations. Describe how the slopes for each pair of lines are alike and how they are different.
   a. $y = \frac{1}{5}x$ and $y = -\frac{1}{5}x$
   b. $y = \frac{2}{3}x$ and $y = -\frac{2}{3}x$
   c. $y = x$ and $y = -x$
   d. $y = \frac{3}{2}x$ and $y = -\frac{3}{2}x$
   e. $y = 5x$ and $y = -5x$
2. Make a conjecture about the slopes of $y = mx$ when $m < 0$ and when $m > 0$. Explain how the graphs of these lines are related.
3. **CHECKPOINT ✔** Verify your conjecture from Step 2 by writing and graphing another pair of equations with the relationship that you described in Step 2.

CHECKPOINT ✔
1990 to 1993: ≈3.57 million new subscribers per year

1993 to 1995: 8.9 million new subscribers per year

ADDITIONAL EXAMPLE 1

Find the slope of the line containing the points (3, 5) and (–2, –6).

The slope is $\frac{11}{5}$.

TRY THIS

$-\frac{7}{8}$

Activity Notes

In this Activity, students graph pairs of equations of the form $y = mx$ and $y = -mx$. They should observe that the graphs in each pair are reflection images of each other across the y-axis.

Teaching Tips about technology refer to keystrokes for the TI-82, TI-83, or TI-83 Plus.

Teaching Tip

TECHNOLOGY To view the graphs in the Activity, you may wish to use a window in which the scales for the x-axis and y-axis are identical. A convenient way to do this is to press ZOOM and then select 4:ZDecimal or 6:ZStandard.

A keystroke guide is provided at the end of each chapter for examples given in the Pupil's Edition.

CHECKPOINT ✔
3. Answers may vary. sample answer: $y = 7x$, $y = -7x$

Interdisciplinary Connection

HEALTH The table at right gives temperature readings for a patient admitted to a hospital.

a. Display the data in a line graph with *Hours after admission* on the horizontal axis and *Temperature (°F)* on the vertical axis. **Check students' graphs.**

b. Between which two hours did the temperature increase most? What was the average rate of change per hour? **16, 20; +0.6°F per hour**

c. Between which two hours did the temperature decrease most? What was the average rate of change per hour? **24, 28; –0.575°F per hour**

Hour	Temp. (°F)	Hour	Temp. (°F)
0	99.0	28	100.2
4	99.8	32	100.0
16	98.6	36	100.2
20	101.0	40	100.5
24	102.5	48	99.2

Intercepts

The y-coordinate of the point where the graph of a linear equation crosses the y-axis is called the **y-intercept** of the line. In the graph below, the y-intercept of **line r** is 3 and the y-intercept of **line s** is –2.

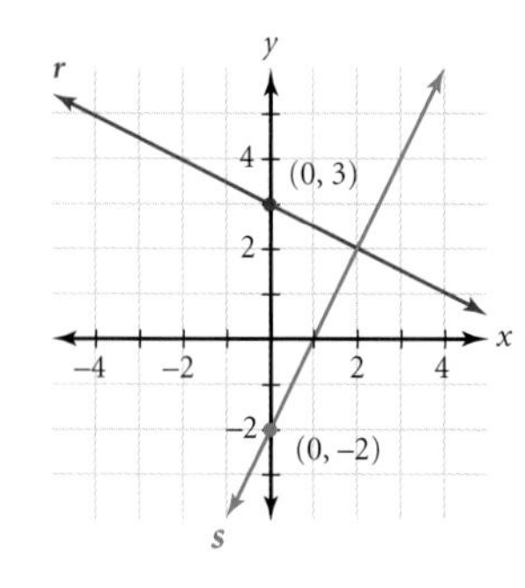

To find the y-intercept of a line, substitute 0 for x in an equation for the line.

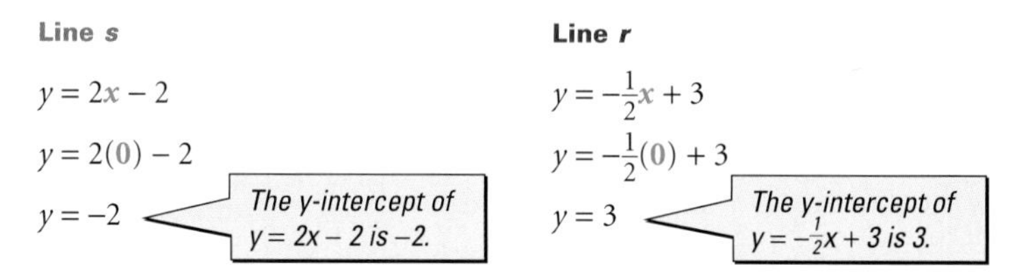

Slope-Intercept Form

The **slope-intercept form** of a line is $y = mx + b$, where m is the slope and b is the y-intercept.

The slope of a line tells you about the steepness and direction of the line. The y-intercept tells you where the line crosses the y-axis.

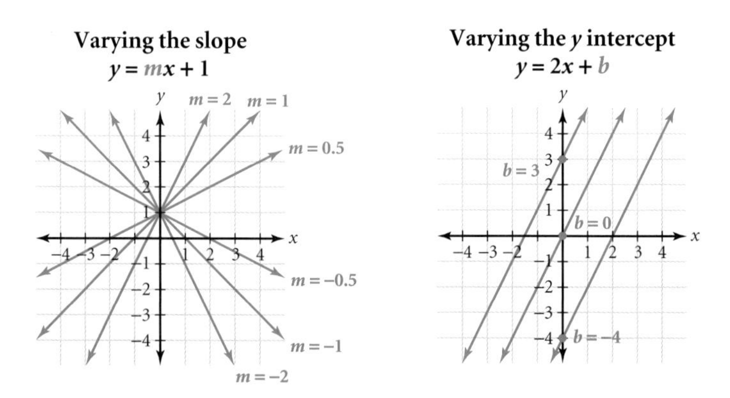

Teaching Tip

Check that students know how to read the subscripted variables that are used in this lesson. Tell them that y_1 is read as "y sub one," y_2 as "y sub two," and so on. Point out that this is simply an easier way to say "the first y-coordinate," "the second y-coordinate," and so on.

Teaching Tip

Remind students that the slope of a horizontal line is 0. As the line is rotated counterclockwise, its slope increases through positive values until the line is vertical, at which point the slope is undefined. A line just to the left of vertical has large negative slope, and as the line is rotated farther counterclockwise, the slope increases through large negative values until the line is horizontal again, at which point the slope is again 0.

Inclusion Strategies

ENGLISH LANGUAGE DEVELOPMENT Ask students to describe some nonmathematical uses of the word *intercept*. For example, they might mention passes in a football game being *intercepted* by a quarterback or newspaper headlines proclaiming enemy planes being *intercepted* by army fighters. Have students explain the meaning of *intercept* in each situation. Then ask them to relate these uses of *intercept* to the mathematical use in this lesson. Tell them to write a sentence using *intercept* in a mathematical context.

EXAMPLE 2 **Use the slope and y-intercept to graph the equation $-2x + y = -3$.**

SOLUTION

1. Write the equation in slope-intercept form, $y = mx + b$.

$$-2x + y = -3$$
$$y = 2x - 3$$

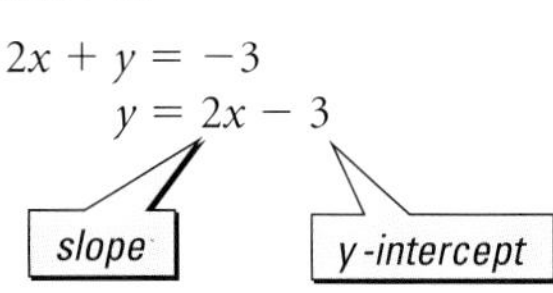

The slope is 2 and the y-intercept is -3.

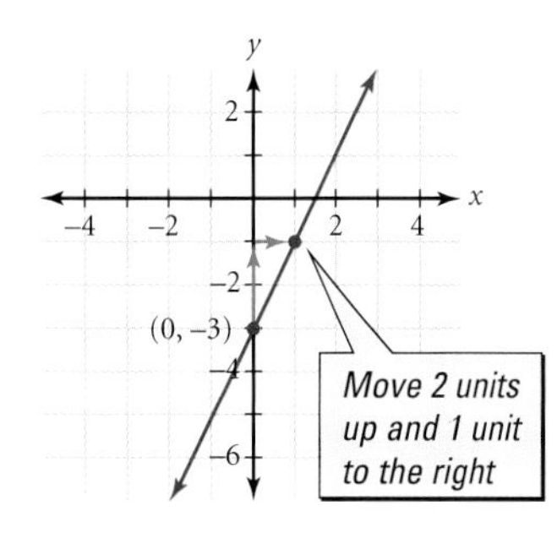

2. Plot the point $(0, -3)$, and use the slope to find a second point.

$$m = \frac{\text{change in } y}{\text{corresponding change in } x} = \frac{2}{1}$$

3. Connect the two points to graph the line.

TRY THIS Use the slope and y-intercept to graph the equation $2x + y = 3$.

Example 3 shows you how to write an equation for a line that is graphed.

EXAMPLE 3 **Write the equation, in slope-intercept form, for the line graphed.**

SOLUTION

Because the point $(0, 2)$ is on the graph, the y-intercept is 2. Use another convenient point on the line, such as $(3, 0)$, to find the slope.

Let $(x_1, y_1) = (0, 2)$ and $(x_2, y_2) = (3, 0)$.

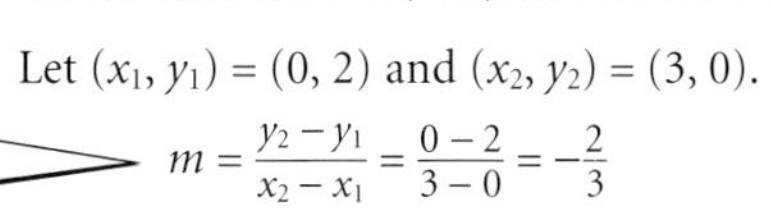

$$m = \frac{y_2 - y_1}{x_2 - x_1} = \frac{0 - 2}{3 - 0} = -\frac{2}{3}$$

The same result occurs when $(x_1, y_1) = (3, 0)$ and when $(x_2, y_2) = (0, 2)$.

The equation in slope-intercept form is $y = -\frac{2}{3}x + 2$.

TRY THIS Write the equation, in slope-intercept form, for the line that passes through $(1, 4)$ and has a y-intercept of 3.

Standard Form

The **standard form** of a linear equation is $Ax + By = C$, where A, B, and C are real numbers and A and B are not *both* 0.

Example 4 on page 16 shows you how to use the x- and y-intercepts to sketch the graph of a linear equation. The **x-intercept** of a graph is the x-coordinate of the point where the graph crosses the x-axis.

ADDITIONAL EXAMPLE 2

Use the slope and y-intercept to graph $-3x + y = -6$.

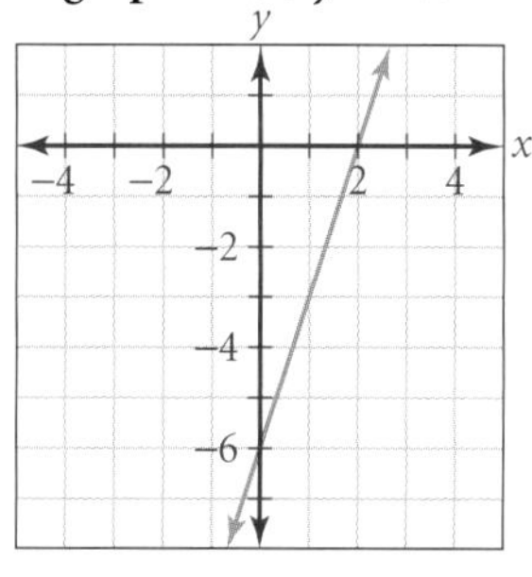

TRY THIS

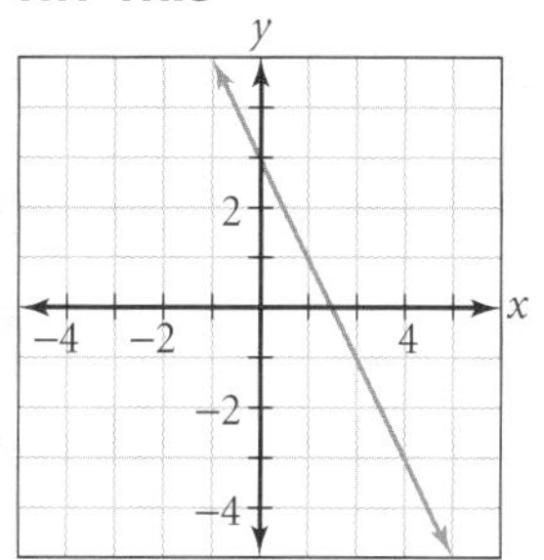

ADDITIONAL EXAMPLE 3

Write the equation, in slope-intercept form, for the line graphed.

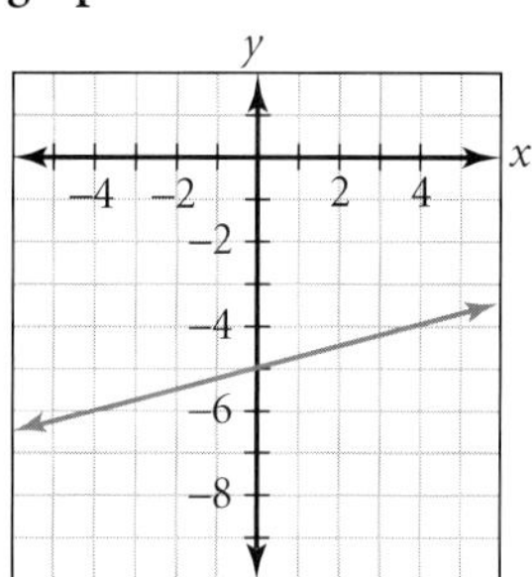

$y = \frac{1}{4}x - 5$

TRY THIS

$y = x + 3$

Enrichment

Give students the graph at right. Have them work in pairs to identify two different situations that the graph might represent. Then each partner should choose one of the situations and write a paragraph describing it. Students should be sure to identify what is represented by the scale on each axis. They also should interpret the slopes of the segments. After all students have written their paragraphs, have a class discussion of their interpretations.

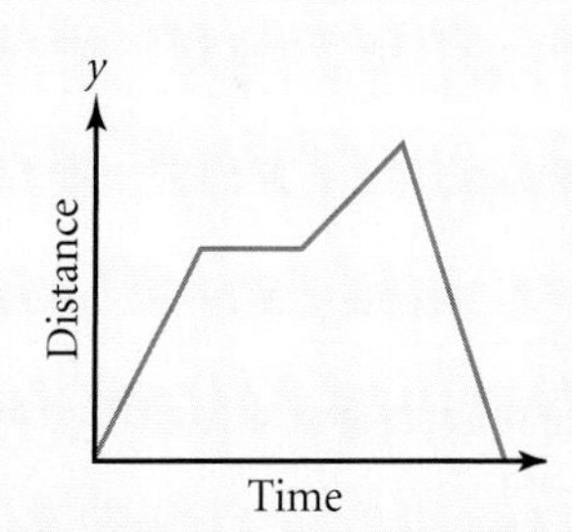

Sample situation: A person leaves home, jogs for a while at a constant rate, rests for a while, walks farther away from home at a constant rate for a while longer, and then jogs back home at a constant rate.

ADDITIONAL
EXAMPLE 4

Use intercepts to graph the equation $3x + 6y = -18$.

x-intercept: $(-6, 0)$
y-intercept: $(0, -3)$

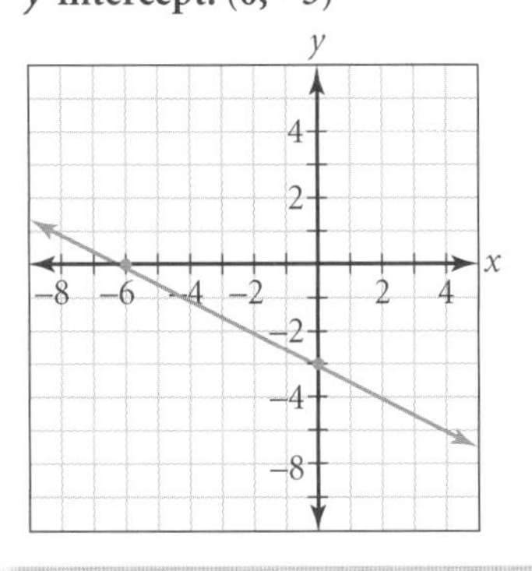

TRY THIS

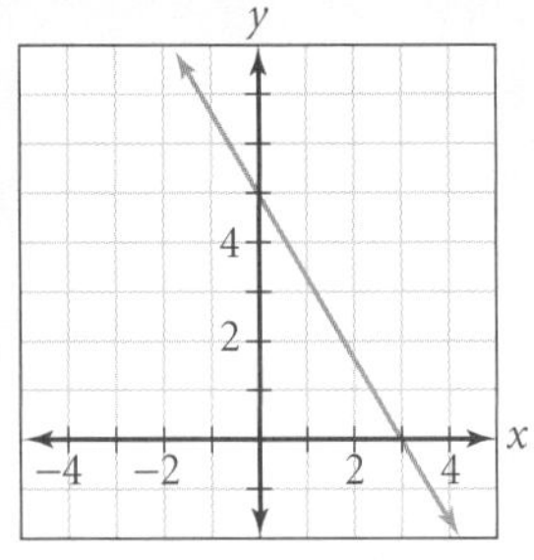

CHECKPOINT ✔
horizontal; vertical

CRITICAL THINKING
$y = \frac{C}{B}$ is the horizontal line (zero slope) through $\left(0, \frac{C}{B}\right)$; $x = \frac{C}{A}$ is the vertical line (undefined slope) through $\left(\frac{C}{A}, 0\right)$.

EXAMPLE 4 **Use intercepts to graph the equation $2x - 3y = 6$.**

SOLUTION

To find the x-intercept, let $y = 0$.

$$2x - 3y = 6$$
$$2x - 3(0) = 6$$
$$x = 3$$

To find the y-intercept, let $x = 0$.

$$2x - 3y = 6$$
$$2(0) - 3y = 6$$
$$y = -2$$

Graph the points $(3, 0)$ and $(0, -2)$. Draw the line through these two points.

TRY THIS Use intercepts to graph the equation $5x + 3y = 15$.

Horizontal and Vertical Lines

For the **horizontal line** at right, the formula for slope gives $\frac{3-3}{2-(-3)} = \frac{0}{5}$, which is 0.

A **horizontal line** is a line that has a slope of 0.

For the **vertical line** at right, the formula for slope gives $\frac{1-(-3)}{-2-(-2)} = \frac{4}{0}$, which is undefined.

A **vertical line** is a line that has an undefined slope.

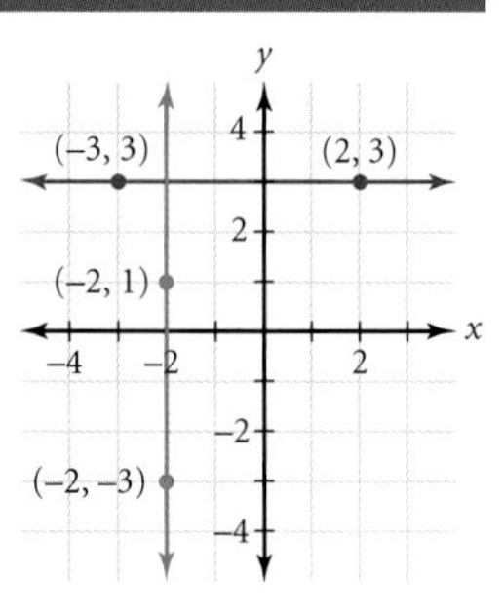

CHECKPOINT ✔ Which type of line has a y-intercept but no x-intercept? Which type of line has an x-intercept but no y-intercept?

EXAMPLE 5 **Graph each equation.**

a. $y = -2$ **b.** $x = -3$

SOLUTION

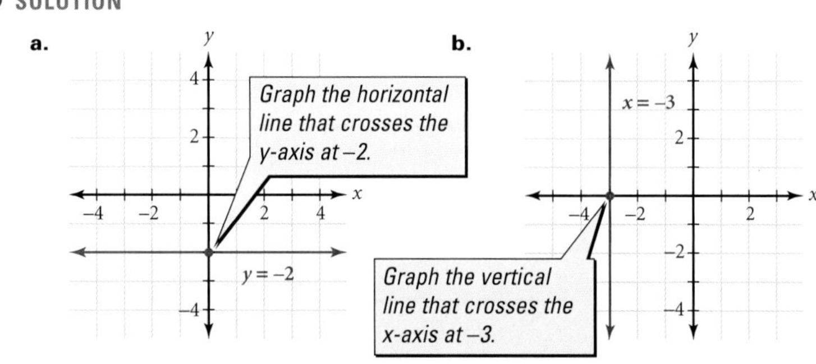
a.
Graph the horizontal line that crosses the y-axis at −2.
y = −2
b.
x = −3
Graph the vertical line that crosses the x-axis at −3.

CRITICAL THINKING Verify that every line of the form $By = C$ is horizontal and that every line of the form $Ax = C$ is vertical.

Reteaching the Lesson

COOPERATIVE LEARNING Have students work in groups of three or four. Give them the descriptions of the lines at right, one at a time. Each student in the group should write an equation in slope-intercept form for a line that matches the description. If possible, each student should choose a line that is different from those of the other students in the group. If the group decides that only one line matches the description, they should be prepared to defend that decision. Discuss the results with the class after each exercise.

- a slope of -5 $\mathbf{y = -5x + b}$
- a y-intercept of 1.5 $\mathbf{y = mx + 1.5}$
- a slope of 1 and a y-intercept of -3 $\mathbf{y = x - 3}$
- through $(0, -2)$ $\mathbf{y = mx - 2}$
- through $(4, 0)$ $\mathbf{y = mx - 4m}$
- through $(-2, 0)$ and $(0, -4)$ $\mathbf{y = -2x - 4}$

Exercises

Communicate

1. Explain how to arrange the linear equations given below in ascending order of steepness, that is, from least steep to most steep.
   a. $y = 3x - 5$ **b.** $y = \frac{1}{3}x + 5$ **c.** $y = \frac{1}{2}x + 5$
   d. $y = 5x + 5$ **e.** $y = 5$ **f.** $y = 3x + 5$
2. Describe how to sketch the graph of the line $3x + 2y = 4$.
3. Explain how to find the y-intercept for the graph of $2x - 5y = 10$.
4. Describe how to write the equation $x - y = 2$ in slope-intercept form.

Guided Skills Practice

5. Find the slope of the line containing the points $(-2, 4)$ and $(8, -3)$. $-\frac{7}{10}$ *(EXAMPLE 1)*
6. Use the slope and y-intercept to graph the equation $\frac{1}{2}x + y = -4$. *(EXAMPLE 2)*
7. Write the equation in slope-intercept form for the line graphed at right. *(EXAMPLE 3)* $y = \frac{1}{2}x - 3$

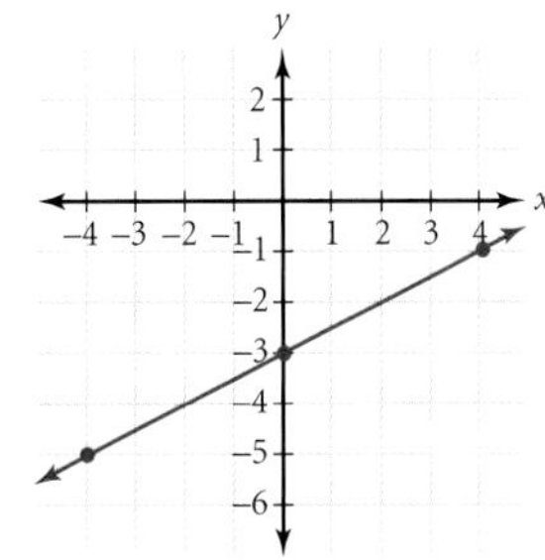

8. Use intercepts to graph $-2x - 4y = 8$. *(EXAMPLE 4)*

Graph each equation. *(EXAMPLE 5)*

9. $x = \frac{1}{2}$ 10. $y = \frac{3}{2}$

Practice and Apply

Write the equation in slope-intercept form for the line that has the indicated slope, m, and y-intercept, b.

11. $m = 2, b = 0.75$ $y = 2x + 0.75$
12. $m = -5, b = 0$ $y = -5x$
13. $m = 0, b = -3$ $y = -3$
14. $m = -\frac{1}{8}, b = 2$ $y = -\frac{1}{8}x + 2$
15. $m = -3, b = 7$ $y = -3x + 7$
16. $m = -\frac{2}{3}, b = -1$ $y = -\frac{2}{3}x - 1$
17. $m = \frac{1}{4}, b = -\frac{3}{4}$ $y = \frac{1}{4}x - \frac{3}{4}$
18. $m = 0.08, b = -2.91$ $y = 0.08x - 2.91$

Find the slope of the line containing the indicated points.

19. $(0, 0)$ and $(3, 30)$ **10**
20. $(1, -3)$ and $(3, -5)$ **−1**
21. $(3, -2)$ and $(4, 5)$ **7**
22. $(-10, -4)$ and $(-3, -3)$ $\frac{1}{7}$
23. $(-6, -6)$ and $(-3, 1)$ $\frac{7}{3}$
24. $(-2, 8)$ and $(-2, -1)$ **undefined**
25. $\left(\frac{1}{2}, -3\right)$ and $\left(3, -\frac{1}{2}\right)$ **1**
26. $(-4, 8)$ and $(-3, -6)$ **−14**

internet connect — Homework Help Online. Go To: go.hrw.com Keyword: MB1 Homework Help for Exercises 19–26

6. $\frac{1}{2}x + y = -4$

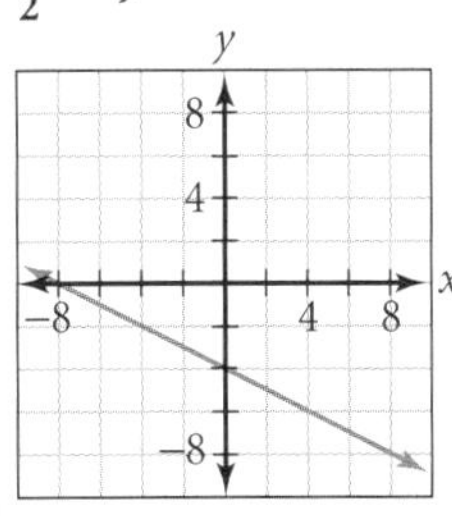

8. $-2x - 4y = 8$

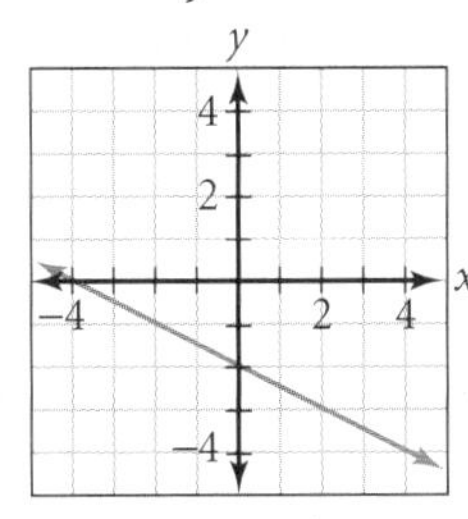

9. $x = \frac{1}{2}$

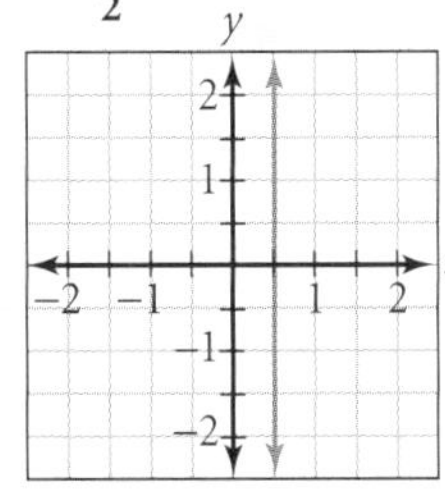

10. $y = \frac{3}{2}$

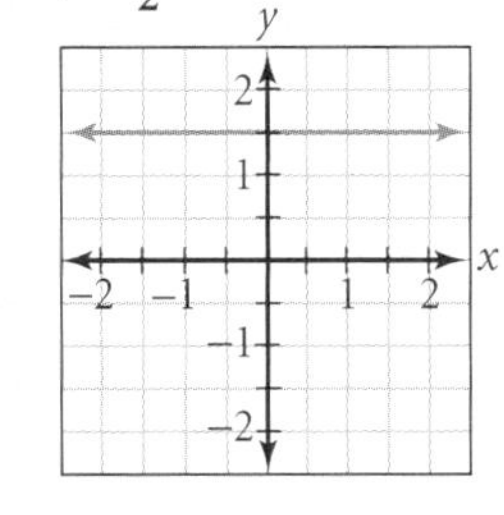

ADDITIONAL EXAMPLE 5

Graph each equation.

a. $y = -7$

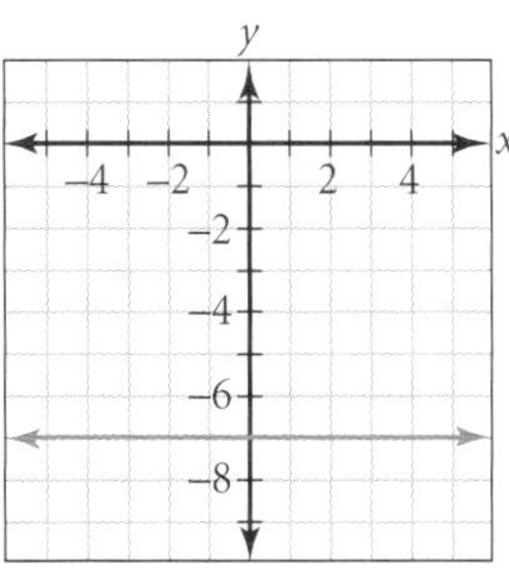

b. $x = -5$

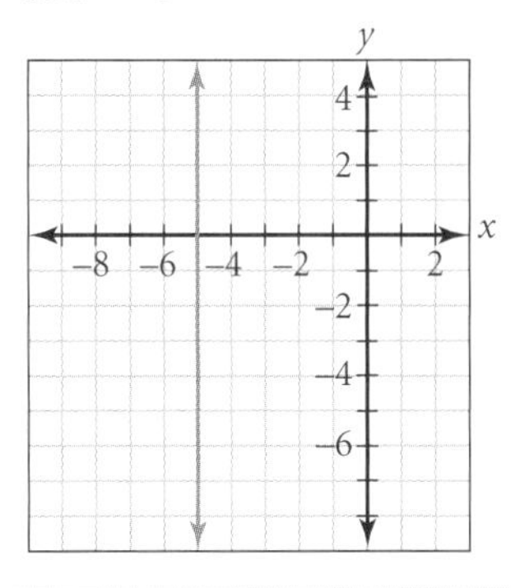

Assess

Selected Answers

Exercises 5–10, 11–73 odd

ASSIGNMENT GUIDE

In Class	1–10
Core	11–18, 19–35 odd, 36–39, 41–69 odd
Core Plus	12–34 even, 36–39, 40–62 even, 64–69
Review	70–74
Preview	75

Extra Practice can be found beginning on page 940.

Answers to odd-numbered Extra Practice exercises can be found immediately after Selected Answers in the Pupil's Edition.

Error Analysis

Students may calculate a slope incorrectly because they have subtracted the x-coordinates and y-coordinates in different orders. Suggest that they write each set of coordinates in a different color and then use the same colors when applying the slope formula. In this way, students can make a visual check that the colors appear in the same order in each subtraction. Remind students that the difference of the y-coordinates is the numerator of the slope and that the difference of the x-coordinates is the denominator.

27. $m = -2, b = 0$

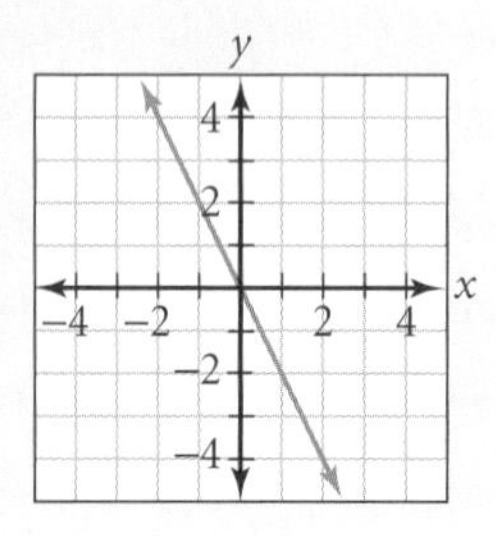

28. $m = 0, b = 2$

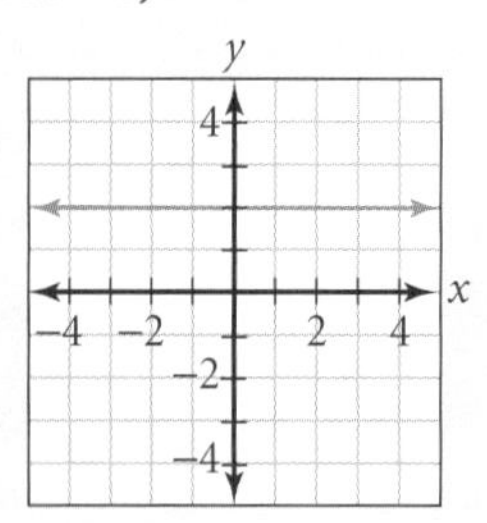

Practice

NAME ______ CLASS ______ DATE ______

Practice
1.2 Slopes and Intercepts

Write the equation in slope-intercept form for the line that has the indicated slope, m, and y-intercept, b.

1. $m = 2, b = -5$ $y = 2x - 5$
2. $m = 3, b = 1$ $y = 3x + 1$
3. $m = -4, b = 3$ $y = -4x + 3$
4. $m = \frac{4}{5}, b = -\frac{2}{5}$ $y = \frac{4}{5}x - \frac{2}{5}$
5. $m = \frac{1}{6}, b = 3$ $y = \frac{1}{6}x + 3$
6. $m = \frac{1}{4}, b = 4$ $y = \frac{1}{4}x + 4$

Find the slope of the line containing the indicated points.

7. (3, 0) and (−3,4) $-\frac{2}{3}$
8. $\left(-1, -\frac{1}{5}\right)$ and $\left(\frac{2}{3}, \frac{3}{4}\right)$ $\frac{57}{100}$
9. (2, 6) and (1, 5) 1
10. (−1, −5) and (2, 4) 3

Identify the slope, m, and the y-intercept, b, for each line.

11. $3x + 4y = 6$ $m = -\frac{3}{4}, b = \frac{3}{2}$
12. $\frac{3}{4}x + 2y = -3$ $m = -\frac{3}{8}, b = -\frac{3}{2}$
13. $-2x - y = 4$ $m = -2, b = -4$
14. $15x + 5y = -35$ $m = -3, b = -7$

Write an equation in slope-intercept form for each line.

15.

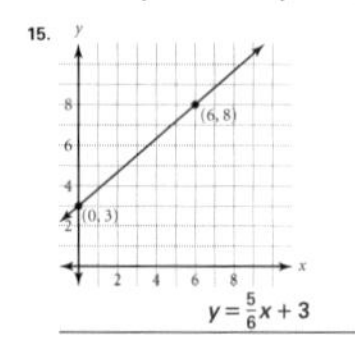

$y = \frac{5}{6}x + 3$

16.

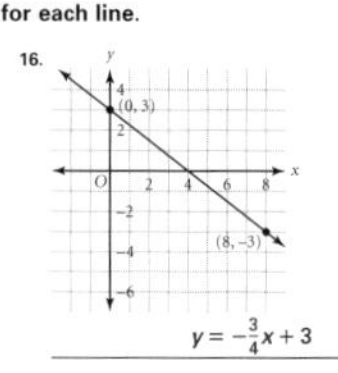

$y = -\frac{3}{4}x + 3$

Identify the slope, m, and the y-intercept, b, for each line. Then graph.

27. $y + 2x = 0$ **28.** $y = 2$ **29.** $-\frac{1}{3}x + y = -7$

30. $x + y = 6$ **31.** $y = x$ **32.** $-2x = 8 + 4y$

33. $-0.6x + y = -4$ **34.** $2x + y = 1$ **35.** $x = -3$

Write an equation in slope-intercept form for each line.

36. $y = 3x + 2$

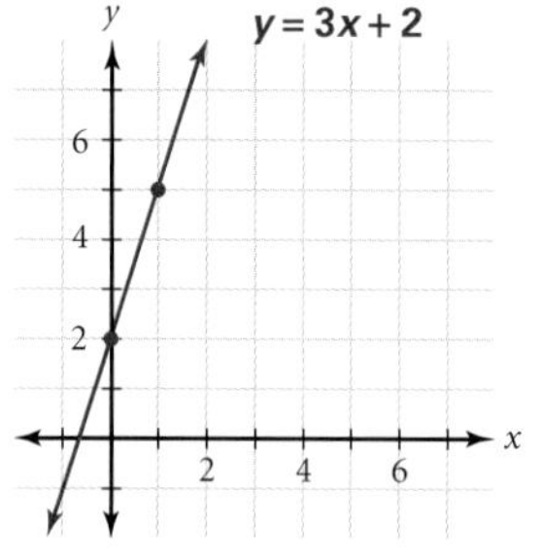

37. $y = \frac{3}{2}x - 1$

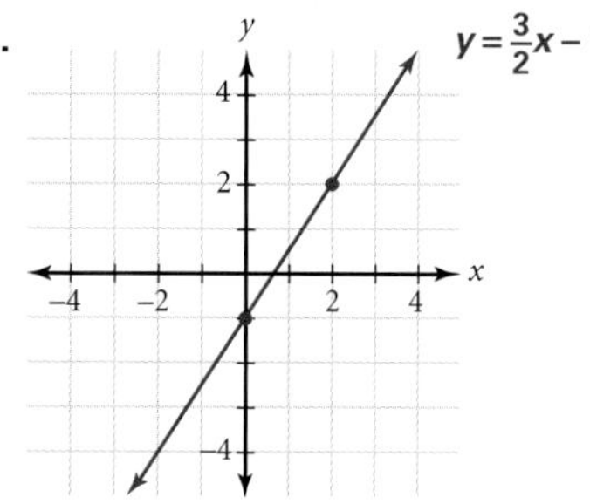

38. $y = -\frac{1}{2}x + 4$

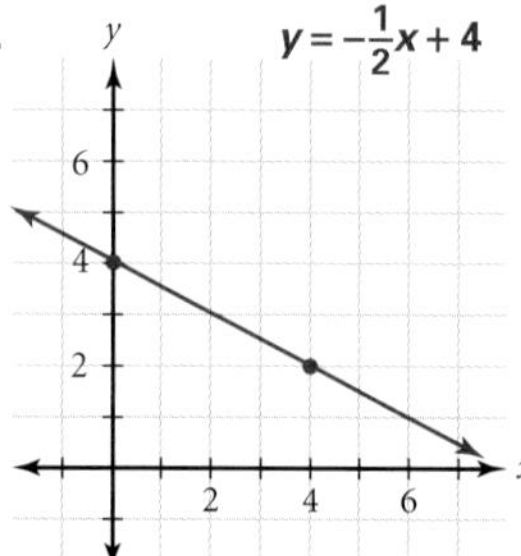

39. $y = -3x + 2$

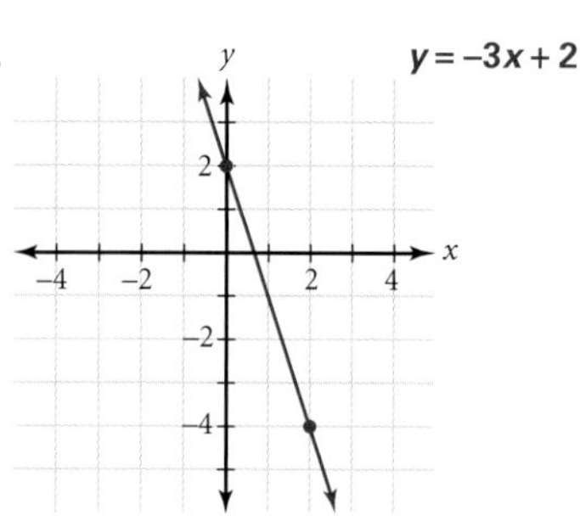

Use intercepts to graph each equation.

40. $4x + y = -4$ **41.** $x + 3y = 12$ **42.** $2x - y = 8$

43. $-x + 2y = 5$ **44.** $7x + 3y = 2$ **45.** $-x + 8y = -6$

46. $-3x + y = -9$ **47.** $-x - 7y = 3$ **48.** $x - y = -1$

49. $-\frac{1}{2}x + 3y = 7$ **50.** $5x - 8y = 16$ **51.** $x + \frac{1}{2}y = -2$

Find the slope of each line. Then graph.

52. $x = 5$ **undefined** **53.** $x = -2$ **undefined** **54.** $y = 8$ **0**

55. $y = -5$ **0** **56.** $x = -1$ **undefined** **57.** $x = 9$ **undefined**

58. $y = -8$ **0** **59.** $y = 7$ **0** **60.** $x = -\frac{1}{3}$ **undefined**

61. $x = -\frac{1}{4}$ **undefined** **62.** $y = \frac{3}{4}$ **0** **63.** $y = \frac{2}{3}$ **0**

CHALLENGE

64. The points (−2, 4), (0, 2), and (3, a − 1) are on one line. Find a. **$a = 0$**

APPLICATION

65. TAXES Tristan buys a computer for \$3600. For tax purposes, he declares a linear depreciation (loss of value) of \$600 per year. Let y be the declared value of the computer after x years.

a. What is the slope of the line that models this depreciation? **−600**

b. Find the y-intercept of the line. **3600**

c. Write a linear equation in slope-intercept form to model the value of the computer over time. **$y = -600x + 3600$ $(0 \leq x \leq 6)$**

d. Find the value of Tristan's computer after 4.5 years. **\$900**

29. $m = \frac{1}{3}, b = -7$

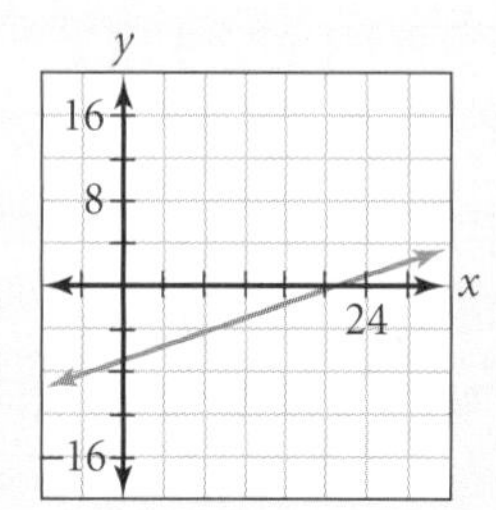

30. $m = -1, b = 6$

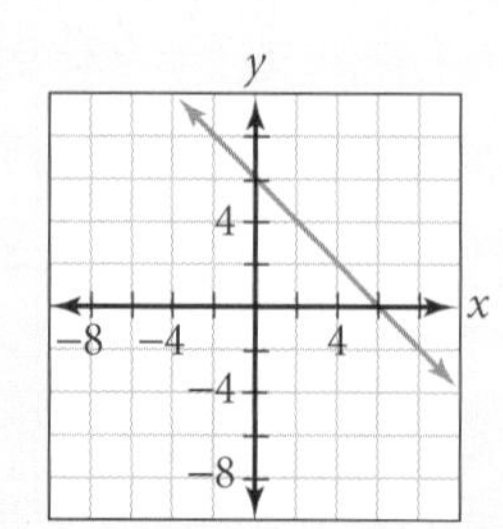

31. $m = 1, b = 0$

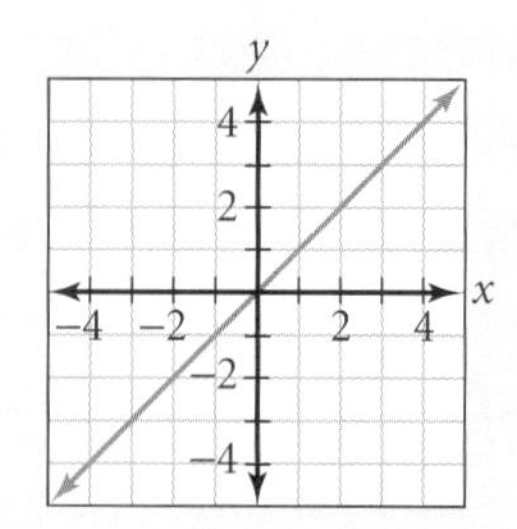

The answers to Exercises 32–35 and 40–63 can be found in Additional Answers beginning on page 1002.

APPLICATIONS

66. PHYSICS The graph at right shows data for the distance of an object from a motion detector.

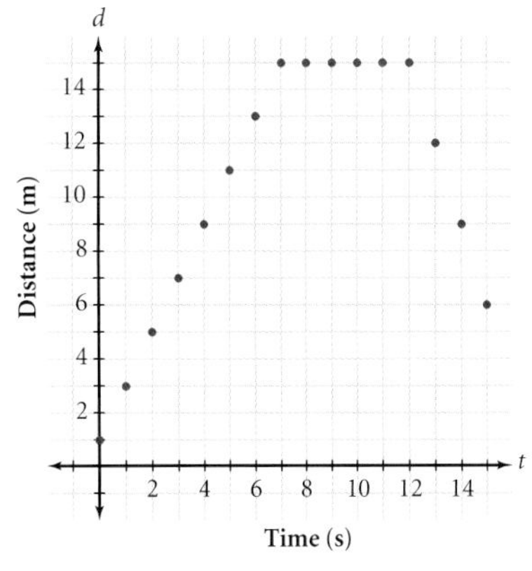

a. Describe how the motion of the object changes with time in the graph.

b. Model the distance of the object from the detector by using three linear equations.

c. For what times is each equation valid?

d. During which time interval is the object moving the fastest?

e. Is the slope of the segment representing the fastest movement positive or negative? Explain what the sign of the slope indicates about the motion of the object.

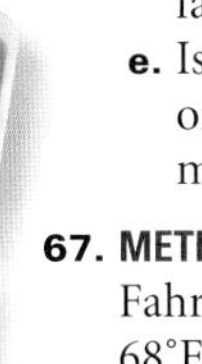

67. METEOROLOGY The thermometer at left shows temperatures in degrees Fahrenheit, F, and in degrees Celsius, C. Room temperature is 20°C, or 68°F.

a. Choose two other Fahrenheit temperatures, and use the thermometer at left to estimate the Celsius equivalents.

b. Let the temperature equivalents from part **a** represent points of the form (F, C). Graph the three points and draw a line through the points.

c. Find the slope and y-intercept of the line.

d. Write the equation of the line in slope-intercept form.

68. CONSTRUCTION The slope of a roof is called the pitch and is defined as follows:

$$\text{pitch} = \frac{\text{rise of roof}}{\frac{1}{2} \times \text{span of roof}}$$

a. Find the pitch of a roof if the rise is 12 feet and the span is 30 feet. $\frac{4}{5}$

b. Find the pitch of a roof if the rise is 18 feet and the span is 60 feet. $\frac{3}{5}$

c. Find the pitch of a roof if the rise is 4 feet and the span is 50 feet. $\frac{4}{25}$

d. If the pitch is constant, is the relationship between the rise and span linear? Explain. **yes; for a constant pitch, rise = $\frac{1}{2}$(span)(pitch)**

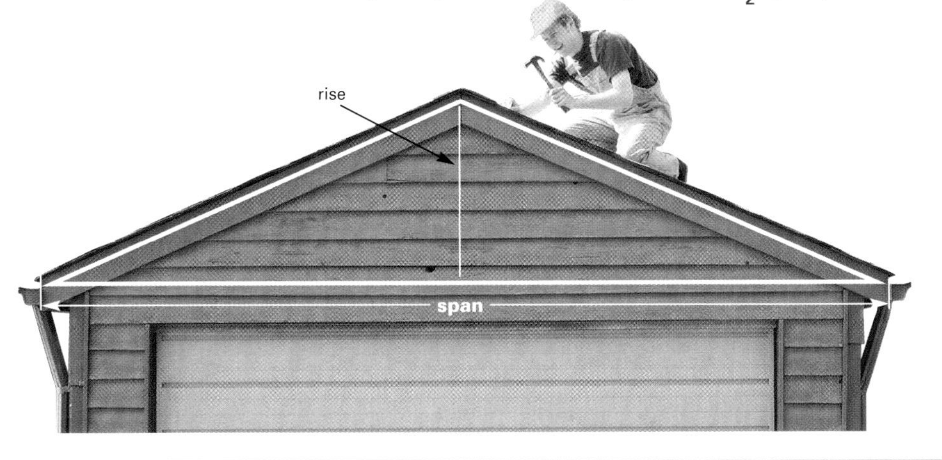

66a. The object is sensed 1 m from the detector and moves away from the detector at a constant rate of 2 m/s for 7 s. The object then stops at 15 m away for 5 s. Finally, the object moves towards the detector at a constant rate of 3 m/s for 3 s.

b. $y = 2x + 1$, $y = 15$, $y = -3x + 51$

c. $y = 2x + 1$ is valid from 0 to 7 s; $y = 15$ is valid from 7 s to 12 s; $y = -3x + 51$ is valid from 12 s to 15 s.

d. 3rd time interval from 12 s to 15 s

e. Negative; if the sign of the slope is negative, then the object is moving toward the detector. If the sign of the slope is positive, then the object is moving away from the detector.

Student Technology Guide

NAME ________ CLASS ________ DATE ________

Student Technology Guide
1.2 Slopes and Intercepts

A graphics calculator can give you the intercepts of the graph of a function. First you must make sure that your ranges for x and y are such that the points where the intercepts occur are visible on the display.

To find the intercepts of the graph of $3x + 2y = 12$, first write an equation for y in terms of x.

$$3x + 2y = 12 \longrightarrow y = \frac{12 - 3x}{2}$$

Enter your expression into the calculator's function list. Press Y= (12 − 3 X,T,θ,n) ÷ 2.

Press WINDOW. Choose $-10 \le x \le 10$ and $-10 \le y \le 10$.

Press GRAPH. Then you will see the display shown at right. Notice that the points where the intercepts occur are visible on the display.

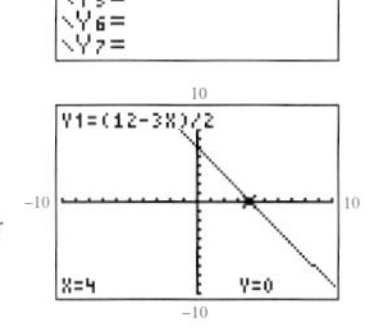

Finally, press TRACE. Using the arrow keys, move the cursor to each point where the line crosses an axis. You will find that the x-intercept equals 4 and the y-intercept equals 6.

Rewrite each equation for y in terms of x. Then use a graphics calculator to find the intercepts of the graph of each equation.

1. $4x - 2y = 8$
   $y =$ (8 − 4x)/(−2)
   x-intercept: 2
   y-intercept: −4

2. $x + 3y = 6$
   $y =$ (6 − x)/3
   x-intercept: 6
   y-intercept: 2

3. $-5x + 3y = 15$
   $y =$ (15 + 5x)/3
   x-intercept: −3
   y-intercept: 5

4. $2x + 4y = 12$
   $y =$ (12 − 2x)/4
   x-intercept: 6
   y-intercept: 3

5. $-\frac{x}{4} + \frac{y}{5} = 1$
   $y =$ 5(1 + x/4)
   x-intercept: −4
   y-intercept: 5

6. $\frac{2}{3}x + \frac{1}{8}y = 4$
   $y =$ 8(4 − 2x/3)
   x-intercept: 6
   y-intercept: 32

7. Graph $7x - 60y = 1260$ for $-10 \le x \le 10$ and $-10 \le y \le 10$. Are the intercepts of the graph visible on the display? What window settings should you use to see the intercepts?
   Answers may vary. No; the range for x must include 180, and the range for y must include −61.

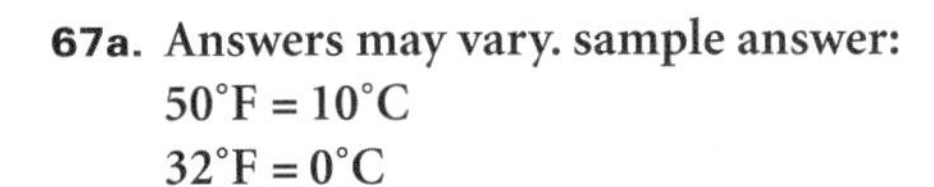

67a. Answers may vary. sample answer:
50°F = 10°C
32°F = 0°C

b.

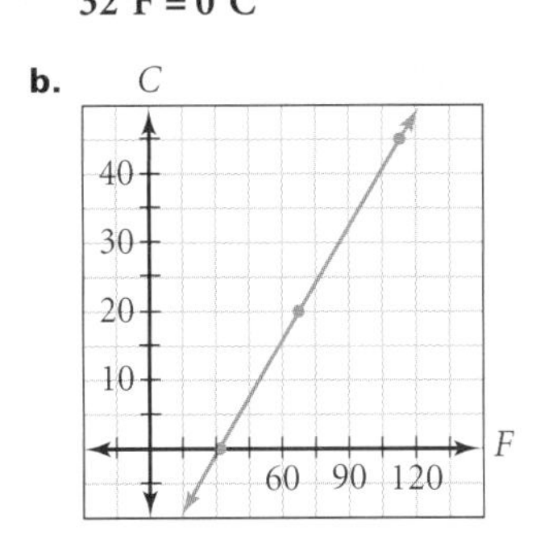

c. Accept approximate answers. Exact values are given here. $m = \frac{5}{9}$; $b = -\frac{160}{9}$

d. Accept approximate answers. Exact values are given here. $C = \frac{5}{9}F - \frac{160}{9}$

In Exercise 75, students graph two linear equations simultaneously and identify the point of intersection. This previews the concept of a *system of equations*. Students will study systems of linear equations in depth in Chapter 3.

Teaching Tips about technology refer to keystrokes for the TI-82, TI-83, or TI-83 Plus.

Teaching Tip

TECHNOLOGY To obtain the graph for Exercise 75, press Y=, clear any existing equations, and enter the given equations as **Y1** and **Y2**. Press WINDOW and use these window settings:

Xmin=–10	Ymin=–10
Xmax=20	Ymax=20
Xscl=1	Yscl=1

Then press GRAPH. To find the point of intersection, press 2nd TRACE to access the **CALCULATE** menu. Choose **5:intersect** and then press .

A keystroke guide is provided at the end of each chapter for examples given in the Pupil's Edition.

ALTERNATIVE Assessment

Portfolio Activity

The Portfolio Activity can be used as preparation for the Chapter Project or as a separate activity. In the Portfolio Activity on this page, students visually estimate a line of best fit for a set of data. Then they find an equation for the line and use it to identify a rate of change.

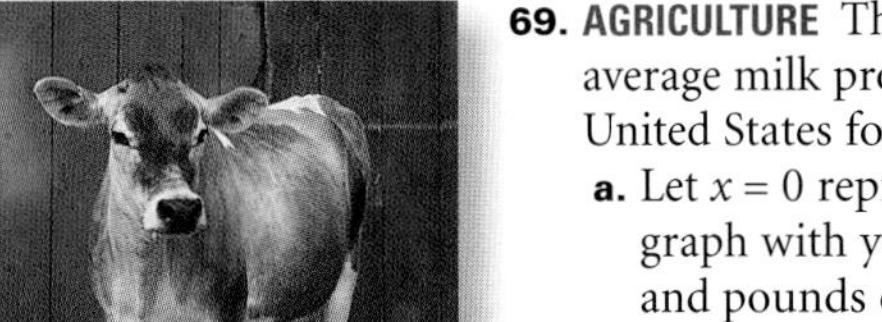

Shown above is a Jersey cow, one type of dairy cow.

69. AGRICULTURE The table at right shows the average milk production of dairy cows in the United States for the years from 1993 to 1996.

Year	Pounds of milk per day
1993	42.7
1994	44.3
1995	45.2
1996	45.2

[*Source: U.S. Dept. of Agriculture*]

a. Let $x = 0$ represent 1990. Make a line graph with years on the horizontal axis and pounds of milk on the vertical axis.

b. In what year did the average milk production increase the most? What is the slope of the graph for that year? **1993–1994; 1.6 lb/day**

c. In what year did the average milk production increase the least? What is the slope of the graph for that year? **1995–1996; 0 lb/day**

Look Back

30 70. Find the value of V in the equation $V = lwh$ when $l = 2$, $w = 3$, and $h = 5$.

$160 71. Use the formula $I = prt$ to find the interest, I, in dollars when the principal, p, is $1000; the annual interest rate, r, is 8%; and the time, t, is 2 years.

64 ft 72. Use the formula $P = 4s$ to find the perimeter of a square, P, in feet when the length of a side, s, is 16 feet.

internet connect Portfolio Extension — Go To: **go.hrw.com** Keyword: **MB1 Rate**

73. Does the table of values at right represent a linear relationship between x and y? Explain. If the relationship is linear, write the next ordered pair that would appear in the table. ***(LESSON 1.1)***

x	–8	–5	–2	1	4	7
y	9	7	5	3	1	–1

74. Make a table of values for the equation $y = -3x + 7$, and graph the line. Is the point $(4, -4)$ on this line? Explain how to answer this question by using the table, the graph, and the equation. ***(LESSON 1.1)***

Calculator button indicates that a graphics calculator is recommended.

Look Beyond

75 Graph the equations $y = 2.12x - 3.7$ and $y = x + 5.4$ on the same screen. Find the coordinates of any points of intersection. **(8.125, 13.525)**

PORTFOLIO ACTIVITY

Refer to your portfolio data set from the Portfolio Activity on page 11.

1. Choose two data points from your portfolio data set. Choose points that seem to fit the overall trend of your portfolio data set, not data points that contains values which may be far higher or lower than the others.
2. Use a straightedge to draw a line through the points on your graph. This line will be your linear model for your portfolio data set. Find the equation of your linear model.
3. What rate of change is indicated by your linear model? Include the appropriate units of measurement.

WORKING ON THE CHAPTER PROJECT

You should now be able to complete Activity 2 of the Chapter Project.

69a.

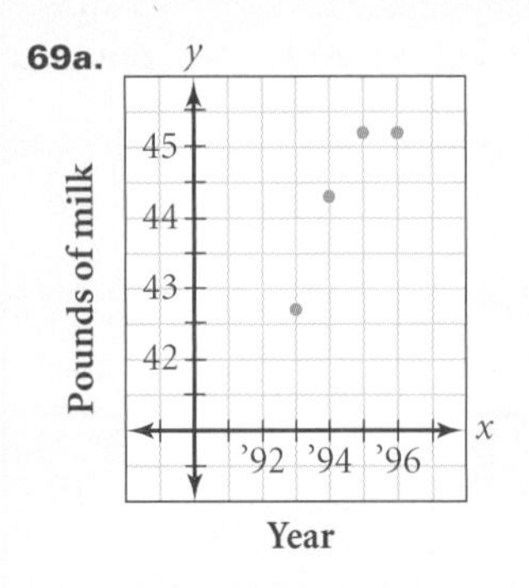

73. Yes; there is a constant difference of 3 in the x-values and a constant difference of –2 in the y-values; (10, –3).

The answers to Exercise 74 and Portfolio Activity 1–3 can be found in Additional Answers beginning on page 1002.

Linear Equations in Two Variables

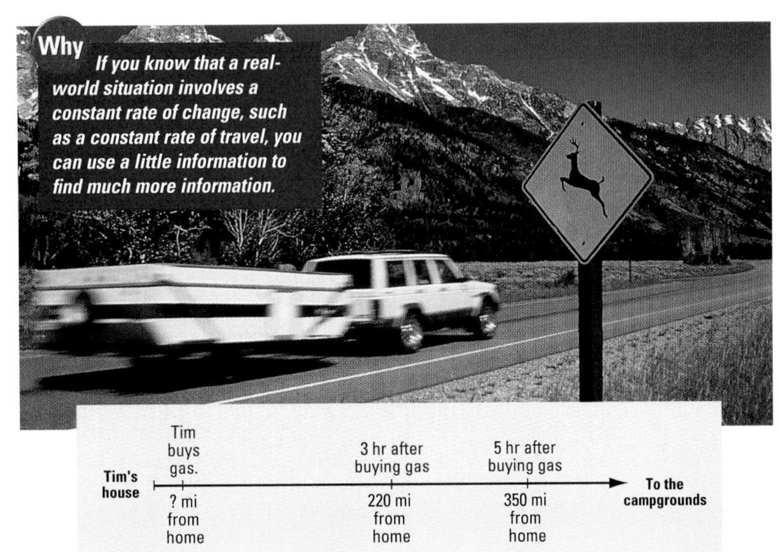

Objectives

- Write a linear equation in two variables given sufficient information.
- Write an equation for a line that is parallel or perpendicular to a given line.

APPLICATION
TRAVEL

Tim leaves his house and drives at a constant speed to go camping. On his way to the campgrounds, he stops to buy gasoline. Three hours after buying gas, Tim has traveled 220 miles from home, and 5 hours after buying gas he has traveled 350 miles from home. How far from home was Tim when he bought gas? To answer this question, you can use the information given to write a linear equation in two variables. *You will solve this problem in Example 3.*

Example 1 shows you how to write a linear equation given two points on the line.

EXAMPLE 1 **Write an equation in slope-intercept form for the line containing the points (4, −3) and (2, 1).**

SOLUTION

1. Find the slope of the line.

$$m = \frac{1-(-3)}{2-4} = \frac{4}{-2} = -2$$

2. Find the y-intercept of the line.

Substitute −2 for m and the coordinates of either point into $y = mx + b$.

$$y = mx + b$$
$$1 = -2(2) + b$$
$$1 = -4 + b$$
$$5 = b$$

The point (2, 1) is used.

3. Write an equation.

$$y = mx + b$$
$$y = -2x + 5$$

Prepare

NCTM PRINCIPLES & STANDARDS 1, 2, 6–10

QUICK WARM-UP

1. Find the slope of the line containing the points (−1, 12) and (5, −6). −3

2. Identify the slope and y-intercept for the line $y = -5x + 7$.
slope: −5; y-intercept: 7

3. Write the equation in slope-intercept form for the line that has a y-intercept of 0 and a slope of −1. $y = -x$

Also on Quiz Transparency 1.3

Teach

Why By using a linear equation in two variables to model an event, students can see how a change in one variable causes change in another. Students know many such relationships, but they probably have not considered using a linear equation to represent them. Ask them to suggest several linear relationships, such as distance traveled over time and the price and demand of a product.

ADDITIONAL EXAMPLE 1

Write an equation in slope-intercept form for the line containing the points (6, 9) and (5, −6).
$y = 15x - 81$

Alternative Teaching Strategy

INVITING PARTICIPATION On the board or overhead, display the graph of $y = -2x + 3$. Have students copy the line onto graph paper. Tell them to draw a line that they believe is parallel to it. Then instruct them to write the equation of the parallel line in slope-intercept form. Ask several volunteers to share their equations with the class, and make a list of the equations as they are given.

Discuss the list with the class. Ask if everyone agrees that all the equations identify lines parallel to the given line. Resolve any disagreements. Then ask what all the equations have in common. Students should note that in each equation, the coefficient of x is −2, so the slope of the line is −2. Lead them to see that if two lines have the same slope, then they are parallel.

Repeat the activity for perpendicular lines.

You can use the *point-slope form* to write an equation of a line if you are given the slope and the coordinates of any point on the line.

Point-Slope Form

If a line has a slope of m and contains the point (x_1, y_1), then the **point-slope form** of its equation is $y - y_1 = m(x - x_1)$.

ADDITIONAL EXAMPLE 2

Write an equation in slope-intercept form for the line that has a slope of $\frac{3}{4}$ and contains the point (−12, 2). $y = \frac{3}{4}x + 11$

EXAMPLE 2 **Write an equation in slope-intercept form for the line that has a slope of $\frac{1}{2}$ and contains the point (−8, 3).**

SOLUTION

$y - y_1 = m(x - x_1)$ *Begin with point-slope form.*

$y - 3 = \frac{1}{2}[x - (-8)]$ *Substitute $\frac{1}{2}$ for m, 3 for y_1, and −8 for x_1.*

$y - 3 = \frac{1}{2}x + 4$

$y = \frac{1}{2}x + 7$ *Write the equation in slope-intercept form.*

The distance traveled by a motorist driving at a constant speed can be modeled by a linear equation.

ADDITIONAL EXAMPLE 3

Marva left her house and drove at a constant speed to a conference in another state. She picked up Delia along the way. Two hours after picking up Delia, they were 140 miles from Marva's house, and 5 hours after picking up Delia, they were 344 miles from Marva's house. **How far from her house was Marva when she picked up Delia?** 4 mi

EXAMPLE 3 Refer to the travel problem described at the beginning of the lesson.

How far from home was Tim when he bought gas?

APPLICATION
TRAVEL

SOLUTION

Write a linear equation to model Tim's distance, y, in terms of time, x. Three hours after buying gas, Tim has traveled 220 miles, and 5 hours after buying gas, he has traveled 350 miles.

The line contains (3, 220) and (5, 350).

Find the slope. $m = \frac{350 - 220}{5 - 3} = 65$

Write an equation. Begin with point-slope form.

$y - y_1 = m(x - x_1)$

$y - 220 = 65(x - 3)$ *Use either point. The point (3, 220) is used here.*

$y - 220 = 65x - 195$

$y = 65x + 25$ *Write the equation in slope-intercept form.*

Thus, $y = 65x + 25$ models Tim's distance from home with respect to time. Since x represents the number of hours he traveled *after* he bought gas, he bought gas when $x = 0$. Thus, he bought gas when he was 25 miles from home.

CHECKPOINT ✔ What does the slope of the line $y = 65x + 25$ in Example 3 represent?

CHECKPOINT ✔ Tim's speed (65 mph)

Interdisciplinary Connection

BUSINESS Many salespeople earn an amount of money that is based on the amount of the sales they make. The amount they earn, which is usually calculated as a percent of sales, is called a *commission*. The percent is called the *commission rate*. In addition, some salespeople are paid a regular salary, called *base pay*, regardless of the amount of sales. Write a formula that can be used to calculate a salesperson's earnings, E, for one month in terms of monthly base pay, b, the commission rate, r, and the total sales for the month, t. $E = b + rt$

Suppose that, in a certain month, a salesperson makes sales of $15,000 and has total earnings of $2700. In the next month, the sales are $22,500 and the total earnings are $2925. Find this salesperson's monthly base pay and commission rate. **monthly base pay: $2250; commission rate: 3%**

Parallel and Perpendicular Lines

The graph at right shows line ℓ_2 parallel to ℓ_1 and ℓ_3 perpendicular to ℓ_1 at point P.

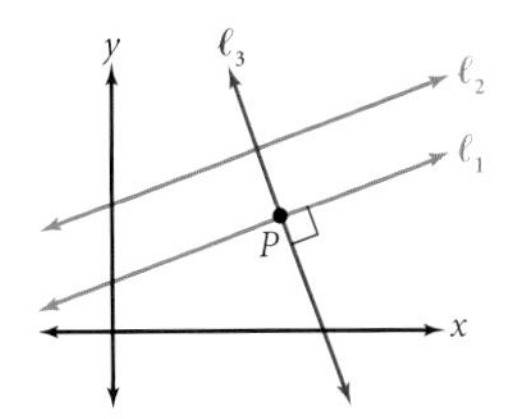

In the Activity below, you can explore how parallel lines and perpendicular lines are related.

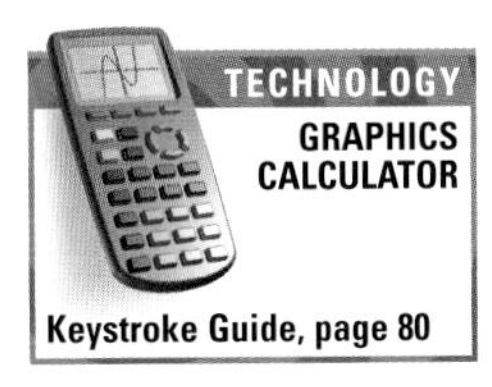

Activity Exploring Parallel and Perpendicular Lines

You will need: a graphics calculator

1. Graph $y = 2x + 1$. On the same screen, graph $y = 2x$, $y = 2x - 2.5$, and $y = 3x + 1$. Which equations have graphs that appear to be parallel to that of $y = 2x + 1$? What do these equations have in common?

CHECKPOINT ✔ 2. Write an equation in slope-intercept form for a line whose graph you think will be parallel to that of $y = 2x + 1$. Verify by graphing.

3. Graph $y = 2x + 1$. On the same screen, graph $y = -\frac{1}{2}x + 2$, $y = \frac{1}{2}x + 2$, and $y = -\frac{1}{2}x + 3$. Which equations have graphs that appear to be perpendicular to that of $y = 2x + 1$? What do these equations have in common?

CHECKPOINT ✔ 4. Write an equation in slope-intercept form whose graph you think will be perpendicular to that of $y = 2x + 1$. Verify by graphing.

The relationships between the slopes of parallel lines are stated below.

Parallel Lines

If two lines have the same slope, they are parallel.
If two lines are parallel, they have the same slope.

All vertical lines have an undefined slope and are parallel to one another.
All horizontal lines have a slope of 0 and are parallel to one another.

The graphs of three parallel lines are shown at right.

$y = 2x + 3$
$y = 2x - 1$
$y = 2x - 4$

Notice that the lines do not intersect.

Because they are parallel, the lines will *never* intersect.

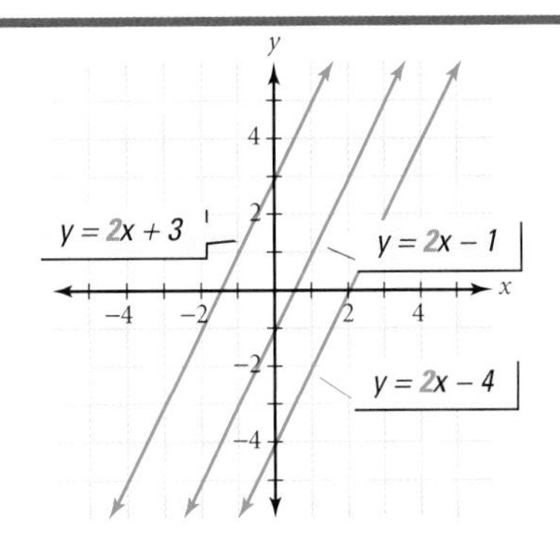

Math CONNECTION

COORDINATE GEOMETRY Review with students the definitions of parallel and perpendicular lines and right angles, which they should recall from their work in previous courses. *Parallel lines* are lines in the same plane that do not intersect. *Perpendicular lines* are lines that intersect to form an angle with a measure of 90°, which is called a *right angle*.

Activity Notes

In this Activity, students use a graphics calculator to explore the relationship between slopes of parallel lines and the relationship between slopes of perpendicular lines. They should discover that two lines with the same slope are parallel. Two lines with slopes that are negative reciprocals are perpendicular.

Teaching Tip

TECHNOLOGY To view the graphs in the Activity, be sure that students are using a square viewing window so that the right angles appear "true."

A keystroke guide is provided at the end of each chapter for examples given in the Pupil's Edition.

CHECKPOINT ✔

2. Answers may vary. sample answer: $y = 2x + 3$

CHECKPOINT ✔

4. Answers may vary. sample answer: $y = -\frac{1}{2}x - 3$

Enrichment

Have students demonstrate that each of the following is a true statement:

1. An equation of the line that contains the points (x_1, y_1) and (x_2, y_2), where $x_1 \neq x_2$, is $y - y_1 = \frac{y_2 - y_1}{x_2 - x_1}(x - x_1)$. (This is called the *two-point form* of the equation.)
The slope of the line containing the points (x_1, y_1) and (x_2, y_2) is $\frac{y_2 - y_1}{x_2 - x_1}$. Substitute this expression for m in the point-slope form of the equation.

2. An equation of the line with an x-intercept of a and a y-intercept of b is $\frac{x}{a} - \frac{y}{b} = 1$, where $a \neq 0$ and $b \neq 0$. (This is called the *intercept form* of the equation.)
The line contains $(a, 0)$ and $(0, b)$, so the slope is $\frac{b - 0}{0 - a}$, or $-\frac{b}{a}$. Substitute this expression for m in the slope-intercept form to obtain $y = -\frac{b}{a}x + b$. Dividing each side by b yields $\frac{y}{b} = -\frac{x}{a} + 1$, or $\frac{x}{a} + \frac{y}{b} = 1$.

ADDITIONAL
EXAMPLE 4

Write an equation in slope-intercept form for the line that contains the point (−2, 7) and is parallel to the graph of $y = 3x - 4$.
$y = 3x + 13$

Teaching Tips about technology refer to keystrokes for the TI-82, TI-83, or TI-83 Plus.

Teaching Tip

TECHNOLOGY When checking the solutions to Examples 4 and 5, you may want the calculator to display the graphs simultaneously rather than sequentially. To do this on a TI-82 or TI-83 graphics calculator, press MODE. Move the cursor to highlight Simul in the sixth line of the display. Then press ENTER GRAPH.

A keystroke guide is provided at the end of each chapter for examples given in the Pupil's Edition. If you are using a different graphics calculator, such as a Casio or Sharp, go to the HRW Internet site at go.hrw.com and enter the keyword MB1 CALC to find keystroke guides.

TRY THIS
$y = -4x - 16$

ADDITIONAL
EXAMPLE 5

Write an equation in slope-intercept form for the line that contains the point (8, −1) and is perpendicular to the graph of $y = 7x + 3$.

$y = -\frac{1}{7}x + \frac{1}{7}$

EXAMPLE 4 **Write an equation in slope-intercept form for the line that contains the point (−1, 3) and is parallel to the graph of $y = -2x + 4$.**

SOLUTION

Because the line is parallel to the graph of $y = -2x + 4$, the slope is also −2.

$$y - y_1 = m(x - x_1)$$
$$y - 3 = -2[x - (-1)]$$
$$y - 3 = -2x - 2$$
$$y = -2x + 1$$

CHECK

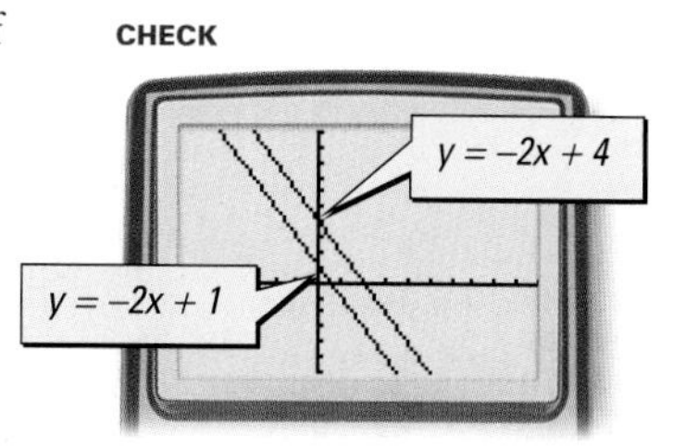

TRY THIS Write an equation in slope-intercept form for the line that contains the point (−3, −4) and is parallel to the graph of $y = -4x - 2$.

The relationships between the slopes of perpendicular lines are stated below.

Perpendicular Lines

If a nonvertical line is perpendicular to another line, the slopes of the lines are negative reciprocals of one another.

All vertical lines are perpendicular to all horizontal lines.

All horizontal lines are perpendicular to all vertical lines.

The graphs of two perpendicular lines are shown at right.

$$y = 2x + 1 \text{ and } y = -\frac{1}{2}x - 3$$

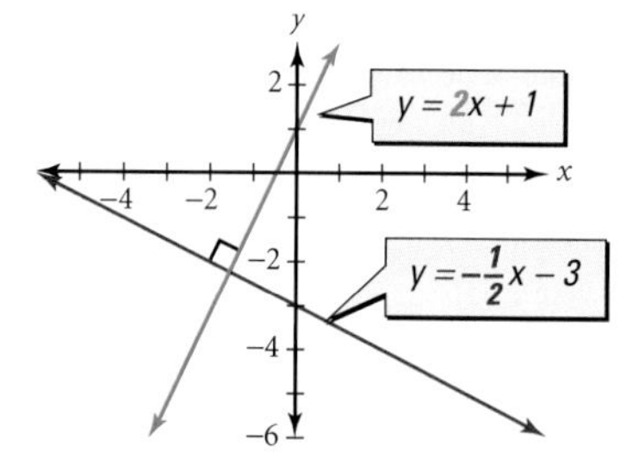

EXAMPLE 5 **Write an equation in slope-intercept form for the line that contains the point (4, −3) and is perpendicular to the graph of $y = 4x + 5$.**

SOLUTION

Because the line is perpendicular to the graph of $y = 4x + 5$, the slope is $-\frac{1}{4}$.

$$y - y_1 = m(x - x_1)$$
$$y - (-3) = -\frac{1}{4}(x - 4)$$
$$y + 3 = -\frac{1}{4}x + 1$$
$$y = -\frac{1}{4}x - 2$$

CHECK

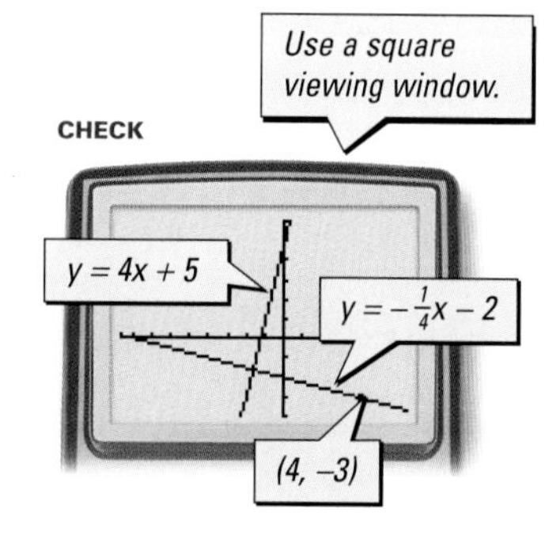

Inclusion Strategies

TACTILE LEARNERS Have students use geoboards with pegs arranged in a square array to model lines in a coordinate plane. The figure at right shows part of the line containing (4, −3) and (2, 1) from Example 1. Students can confirm that the slope is −2 and can stretch the rubber band to see that the line intercepts the y-axis at (0, 5).

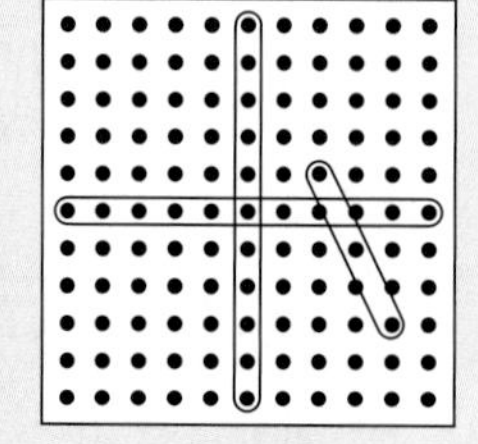

Reteaching the Lesson

COOPERATIVE LEARNING Review with students these three sets of conditions from which they have learned to identify the equation of a line.

1. contains the points (□, □) and (□, □)
2. contains the point (□, □) and has a slope of □
3. contains the point (□, □) and is parallel (or perpendicular) to □

Have each student create one exercise for each set of conditions, write the answers on a separate sheet of paper, and then trade exercises with a partner. Partners should answer each other's exercises and work together to resolve any disagreements.

TRY THIS Write an equation in slope-intercept form for the line that contains the point $(-1, 5)$ and is perpendicular to the graph of $y = -4x - 2$.

CRITICAL THINKING The graph of $3x - y = 4$ is perpendicular to the graph of $Ax + 2y = 8$ for some value of A. Find A.

Exercises

Communicate

1. Describe how to write an equation in slope-intercept form for the line containing two given points, such as $(1, 3)$ and $(4, -2)$.
2. Explain how to use the different forms of a linear equation to write the equation of the line graphed at right.
3. Describe how to determine whether the lines $5x + 6y = 12$ and $6x - 5y = 15$ are parallel, perpendicular, or neither.
4. Explain how to write the equation for the line that contains the point $(3, -1)$ and is perpendicular to the line $x + 2y = 4$.

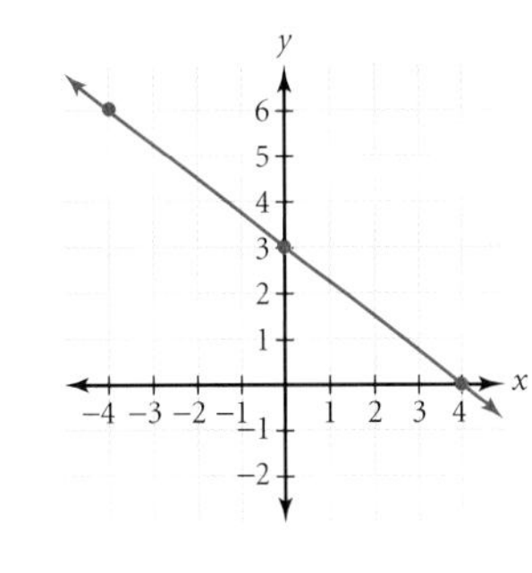

Guided Skills Practice

5. Write an equation in slope-intercept form for the line containing the points $(3, 3)$ and $(-5, -1)$. ***(EXAMPLE 1)*** $y = \frac{1}{2}x + \frac{3}{2}$
6. Write an equation in slope-intercept form for the line that has a slope of 3 and contains the point $(4, 7)$. ***(EXAMPLE 2)*** $y = 3x - 5$

APPLICATION

7. **TRAVEL** Tina leaves home and drives at a constant speed to college. On her way to the campus, she stops at a restaurant to have lunch. Two hours after leaving the restaurant, Tina has traveled 130 miles, and 4 hours after leaving the restaurant, she has traveled 240 miles. How far from home was Tina when she had lunch? ***(EXAMPLE 3)*** 20 mi

8. Write an equation in slope-intercept form for the line that contains the point $(6, -5)$ and is parallel to the line $2x - 5y = -3$. ***(EXAMPLE 4)*** $y = \frac{2}{5}x - \frac{37}{5}$
9. Write an equation in slope-intercept form for the line that contains the point $(-7, 3)$ and is perpendicular to the line $2x + 5y = 3$. ***(EXAMPLE 5)*** $y = \frac{5}{2}x + \frac{41}{2}$

TRY THIS

$y = \frac{1}{4}x + \frac{21}{4}$

CRITICAL THINKING

$A = \frac{2}{3}$

Assess

Selected Answers

Exercises 5–9, 11–77 odd

ASSIGNMENT GUIDE

In Class	1–9
Core	11–59 odd, 63
Core Plus	10–30 even, 32–34, 36–50 even, 51–64
Review	65–78
Preview	79, 80

Extra Practice can be found beginning on page 940.

Answers to odd-numbered Extra Practice exercises can be found immediately after Selected Answers in the Pupil's Edition.

Error Analysis

When finding an equation of a line given two points on the line, students sometimes calculate the slope correctly and then inadvertently switch the values of x and y when making the substitution to find the y-intercept. The coordinates of both given points must satisfy the equation, so encourage them to substitute the coordinates of the second point into their equation to check their work.

internet connect

Homework Help Online
Go To: **go.hrw.com**
Keyword:
MB1 Homework Help for Exercises 10–21

Practice and Apply

Write an equation for the line containing the indicated points.

10. (0, 0) and (3, 30) $y = 10x$
11. (1, −3) and (3, −5) $y = -x - 2$
12. (−4, −4) and (−3, −3) $y = x$
13. (−10, −4) and (−3, −3) $y = \frac{1}{7}x - \frac{18}{7}$
14. (−6, −6) and (−3, 1) $y = \frac{7}{3}x + 8$
15. (−2, 8) and (−2, −1) $x = -2$
16. (4, −8) and (3, −6) $y = -2x$
17. (8, −3) and (−8, 3) $y = -\frac{3}{8}x$
18. $\left(-\frac{1}{2}, 7\right)$ and $\left(-4, \frac{1}{2}\right)$ $y = \frac{13}{7}x + \frac{111}{14}$
19. $\left(\frac{1}{2}, -3\right)$ and $\left(3, -\frac{1}{2}\right)$ $y = x - \frac{7}{2}$
20. (−9, 1) and $\left(-\frac{1}{2}, 1\right)$ $y = 1$
21. (−5, 4) and $\left(-5, -\frac{2}{3}\right)$ $x = -5$

Write an equation in slope-intercept form for the line that has the indicated slope, *m*, and contains the given point.

22. $m = -\frac{1}{2}$, (8, 1) $y = -\frac{1}{2}x + 5$
23. $m = -\frac{2}{3}$, (6, −5) $y = -\frac{2}{3}x - 1$
24. $m = -4$, (5, −3) $y = -4x + 17$
25. $m = 5$, (−1, −3) $y = 5x + 2$
26. $m = 0$, (2, 3) $y = 3$
27. $m = 0$, (−7, 8) $y = 8$
28. $m = 4$, (9, −3) $y = 4x - 39$
29. $m = 3$, (−4, 9) $y = 3x + 21$
30. $m = -\frac{1}{5}$, (8, −2) $y = -\frac{1}{5}x - \frac{2}{5}$
31. $m = -\frac{2}{3}$, (5, −4) $y = -\frac{2}{3}x - \frac{2}{3}$

Write a linear equation to model each table of values. For each equation, state what the slope represents.

32.

Hours	Miles
3	135
5	225

$y = 45x$; avg. speed (mph)

33.

Items	Cost ($)
4	14.00
7	21.50

$y = 2.5x + 4$; avg. cost per item

34. $y = 1.5x + 2$; avg. cost per hr

Hours	Parking fee ($)
3	6.50
7	12.50

Write an equation in slope-intercept form for the line that contains the given point and is parallel to the given line.

35. (−2, 3), $y = -3x + 2$ $y = -3x - 3$
36. (5, −3), $y = 4x + 2$ $y = 4x - 23$
37. (0, −4), $y = \frac{1}{2}x - 1$ $y = \frac{1}{2}x - 4$
38. (−6, 2), $y = -\frac{2}{3}x - 3$ $y = -\frac{2}{3}x - 2$
39. (−1, −3), $2x + 5y = 15$ $y = -\frac{2}{5}x - \frac{17}{5}$
40. (4, −3), $3x + 4y = 8$ $y = -\frac{3}{4}x$
41. (3, 0), $-x + 2y = 17$ $y = \frac{1}{2}x - \frac{3}{2}$
42. (4, −3), $-4x + y = -7$ $y = 4x - 19$

Write an equation in slope-intercept form for the line that contains the given point and is perpendicular to the given line.

43. (−2, 5), $y = -2x + 4$ $y = \frac{1}{2}x + 6$
44. (1, −4), $y = 3x - 2$ $y = -\frac{1}{3}x - \frac{11}{3}$
45. (8, 5), $y = -x + 2$ $y = x - 3$
46. (0, −5), $y = x - 5$ $y = -x - 5$
47. (2, 5), $6x + 2y = 24$ $y = \frac{1}{3}x + \frac{13}{3}$
48. (3, −1), $12x + 4y = 8$ $y = \frac{1}{3}x - 2$
49. (−2, 4), $x - 6y = 15$ $y = -6x - 8$
50. (5, −2), $2x - 5y = 15$ $y = -\frac{5}{2}x + \frac{21}{2}$

51. Write an equation for the line that is perpendicular to the line $2x + 5y = 15$ at the y-intercept. $y = \frac{5}{2}x + 3$

52. Write an equation for the line that is perpendicular to the line $x - 3y = 9$ at the x-intercept. $y = -3x + 27$

Practice

NAME ______ CLASS ______ DATE ______

Practice
1.3 Linear Equations in Two Variables

Write an equation for the line containing the indicated points.

1. (2, 4) and (3, 5) $y = x + 2$
2. (−1, 3) and (3, −1) $y = -x + 2$
3. (3, 1) and $\left(\frac{1}{2}, \frac{3}{2}\right)$ $y = -\frac{1}{5}x + \frac{8}{5}$
4. (2, 0) and (−6, 4) $y = -\frac{1}{2}x + 1$
5. (−1, −4) and (−2, 5) $y = -9x - 13$
6. $\left(\frac{1}{2}, \frac{3}{2}\right)$ and $\left(-2, -\frac{1}{2}\right)$ $y = \frac{4}{5}x + \frac{11}{10}$

Write an equation in slope-intercept form for the line that has the indicated slope, *m*, and contains the given point.

7. $m = 1$ and (3, 3) $y = x$
8. $m = -\frac{1}{2}$ and (4 ,6) $y = -\frac{1}{2}x + 8$
9. $m = \frac{3}{4}$ and (4, −2) $y = \frac{3}{4}x - 5$
10. $m = 4$ and (4, 3) $y = 4x - 13$
11. $m = -2$ and (−2, 3) $y = -2x - 1$
12. $m = -\frac{1}{4}$ and (8, 6) $y = -\frac{1}{4}x + 8$

Write an equation in slope-intercept form for the line that contains the given point and is parallel to the given line.

13. (1, 4); $y = -3x + 2$ $y = -3x + 7$
14. (−2, 3); $y = -4x + 2$ $y = -4x - 5$
15. (4, −2); $y = \frac{3}{4}x + \frac{1}{4}$ $y = \frac{3}{4}x - 5$
16. (−6, 3); $y = 2x + 2$ $y = 2x + 15$
17. (2, −1); $y = -3x - 6$ $y = -3x + 5$
18. (3, −4); $y = 4x - 3$ $y = 4x - 16$
19. (2, −2); $y = -\frac{1}{2}x - 3$ $y = -\frac{1}{2}x - 1$
20. (1, −1); $y = 3x - 2$ $y = 3x - 4$
21. (2, −2); $y = \frac{1}{2}x + 3$ $y = \frac{1}{2}x - 3$
22. (1, 0); $y = -3x - 2$ $y = -3x + 3$

Write an equation in slope-intercept form for the line that contains the given point and is perpendicular to the given line.

23. (2, 4); $y = \frac{1}{2}x + 3$ $y = -2x + 8$
24. (6, −4); $y = 3x - \frac{3}{4}$ $y = -\frac{1}{3}x - 2$
25. (6, −7); $y = -2x - 5$ $y = \frac{1}{2}x - 10$
26. (2, −5); $y = 2x - 4$ $y = -\frac{1}{2}x - 4$
27. $\left(3, \frac{11}{4}\right)$; $y = 4x + 6$ $y = -\frac{1}{4}x + \frac{7}{2}$
28. (3, 5); $y = -x - 1$ $y = x + 2$
29. $\left(1, \frac{2}{3}\right)$; $y = \frac{3}{4}x + 3$ $y = -\frac{4}{3}x + 2$
30. (1, 4); $y = -\frac{3}{4}x - 4$ $y = \frac{4}{3}x + \frac{8}{3}$
31. (3, −1); $y = 3x + \frac{3}{4}$ $y = -\frac{1}{3}x$
32. $\left(-1, -\frac{7}{2}\right)$; $y = 4x - 3$ $y = -\frac{1}{4}x - \frac{15}{4}$

CONNECTIONS

COORDINATE GEOMETRY For Exercises 53–58, refer to the lines graphed on the coordinate plane below.

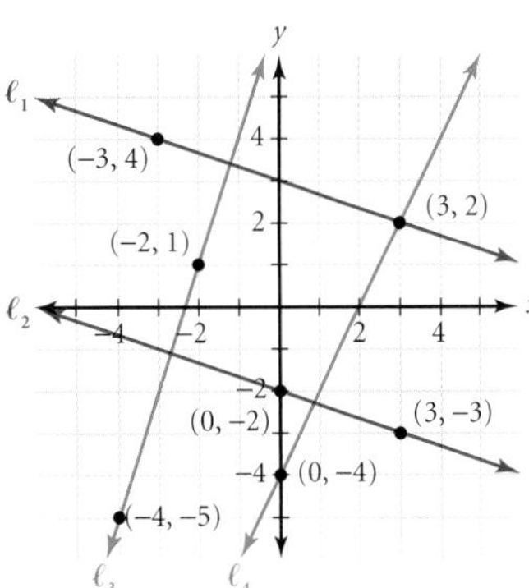

53. Use slopes to determine whether ℓ_1 is parallel to ℓ_2. **yes**

54. Use slopes to determine whether ℓ_3 is parallel to ℓ_4. **no**

55. Use slopes to determine whether ℓ_1 is perpendicular to ℓ_3. **yes**

56. Use slopes to determine whether ℓ_2 is perpendicular to ℓ_3. **yes**

57. Use slopes to determine whether ℓ_2 is perpendicular to ℓ_4. **no**

58. Use slopes to determine whether ℓ_1 is perpendicular to ℓ_4. **no**

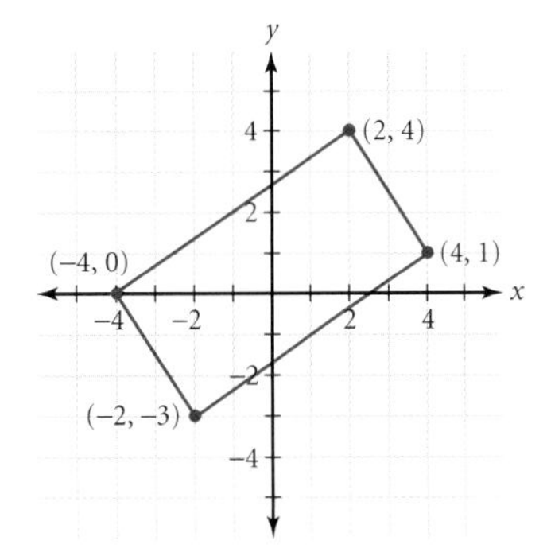

59. **COORDINATE GEOMETRY** Opposite sides of a parallelogram are parallel. Use slopes to determine whether the quadrilateral graphed in the coordinate plane at right is a parallelogram.

60. **COORDINATE GEOMETRY** A rectangle has opposite sides that are parallel and four right angles. Use slopes to determine whether the quadrilateral graphed in the coordinate plane at right is a rectangle.

CHALLENGE

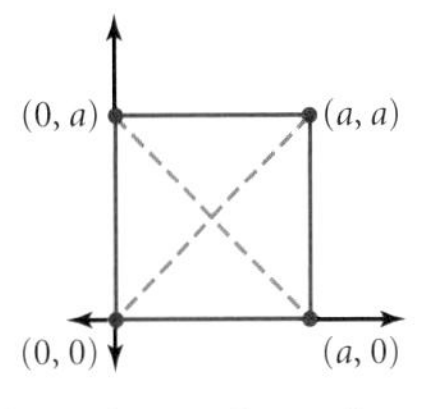

61. Use the diagram at right to prove that the diagonals of any square are perpendicular.

APPLICATIONS

62. **TRAVEL** Mac bikes at a nonconstant rate of speed from home through town. He then begins his training ride, at a constant speed of 25 miles per hour. After 3 hours of biking at a constant speed, his odometer shows that he has traveled 83 miles since he left home.

a. Write a linear equation in slope-intercept form for the distance, d, in miles that Mac has traveled in terms of the time, t, in hours since he began his training ride. **$d = 25t + 8$**

b. When Mac began his training ride, how far from home was he? **8 miles**

59. $m_1 = \frac{4-1}{2-4} = \frac{3}{-2}$

$m_2 = \frac{0-(-3)}{-4-(-2)} = \frac{3}{-2}$

$m_3 = \frac{4-0}{2-(-4)} = \frac{2}{3}$

$m_4 = \frac{1-(-3)}{4-(-2)} = \frac{2}{3}$

Since there are two pairs of parallel sides, the quadrilateral is a parallelogram.

60. The slopes of opposite sides are equal, and the slopes of adjacent sides are negative reciprocals. Therefore, opposite sides are parallel and there are four right angles. The quadrilateral is a rectangle.

61. slope of $D_1 = \frac{a-0}{0-a} = -1$

slope of $D_2 = \frac{a-0}{a-0} = 1$

Since the slopes are negative reciprocals of each other, the diagonals are perpendicular.

Student Technology Guide

NAME ____________ CLASS ________ DATE ________

Student Technology Guide
1.3 *Linear Equations in Two Variables*

Suppose you want to graph the equation of the line that passes through the point $(-4, -1)$ and is perpendicular to the line $y = 2x + 3$. The new line has a slope of $-\frac{1}{2}$ and contains the point $(-4, -1)$. Using the point-slope form, you can easily find an equation for the line, as shown at right.

$$y - y_1 = m(x - x_1)$$
$$y - (-1) = -\tfrac{1}{2}(x - (-4))$$
$$y = -\tfrac{1}{2}(x + 4) + (-1)$$
$$y = -\tfrac{1}{2}x - 3$$

To graph $y = 2x + 3$ and $y = -\frac{1}{2}x - 3$, enter the equations into the function list, choose window settings such as those shown at right, and press GRAPH. You might be tempted to conclude that the lines do not look perpendicular. However, these lines *are* perpendicular, because the slopes 2 and $-\frac{1}{2}$ are negative reciprocals of one another.

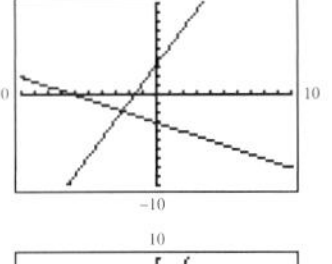

When viewing perpendicular lines, it is important to use a square viewing window, that is, a viewing window in which one unit on the x-axis is the same length as one unit on the y-axis. You can convert any viewing window to a square window by pressing ZOOM, selecting 5:ZSquare, and pressing ENTER. This converts the window above to the one at right.

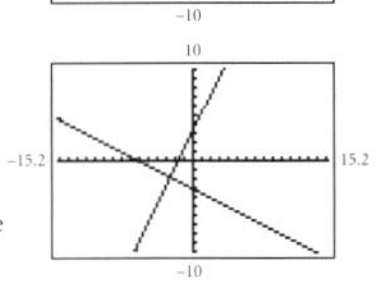

Write an equation in slope-intercept form for the line that contains the given point and is perpendicular to the given line. Graph the equations in a square viewing window, and state your window settings.

1. $(2, 2); y = -\frac{1}{3}x + 1$ **$y = 3x - 4$; sample settings: $[-4.7, 4.7] \times [-3.1, 3.1]$**
2. $(3, 4); y = \frac{3}{2}x - 2$ **$y = -\frac{2}{3}x + 6$; sample settings: $[-9.4, 9.4] \times [-6.2, 6.2]$**
3. $(-4, -2); y = -4x + 3$ **$y = \frac{1}{4}x - 1$; sample settings: $[-4.7, 4.7] \times [-3.1, 3.1]$**
4. $(12, -6); y = \frac{6}{5}x + 4$ **$y = -\frac{5}{6}x + 4$; sample settings: $[-14.1, 14.1] \times [-9.3, 9.3]$**
5. $(9, 27); y = -\frac{3}{4}x + 28$ **$y = \frac{4}{3}x + 15$; sample settings: $[-47, 47] \times [-31, 31]$**
6. $(-5, 2); y = -\frac{3}{4}x + 5$ **$y = \frac{4}{3}x + \frac{26}{3}$; sample settings: $[-15.2, 15.2] \times [-10, 10]$**

Exercises 79 and 80 are a simple investigation that can serve as a transition from the topic of linear equations to the concept of direct variation. Students will study direct variation in depth in Lesson 1.4.

APPLICATIONS

63. ACADEMICS A professor gives a test, and the scores range from 40 to 80. The professor decides to *scale* the test in order to make the scores range from 60 to 90. Let x represent an original score, and let y represent a converted score.

a. Use the ordered pairs (40, 60) and (80, 90) to write the equation that the professor will use to scale the test scores. $y=\frac{3}{4}x+30$

b. What will an original score of 45 become? **64**

c. If a converted score is 84, what was the original score? **72**

64. INCOME Trevor is a salesperson who earns a weekly salary and a commission that is 7% of his weekly sales. In one week Trevor's sales were $952.00 and his weekly income was $466.64. In another week his sales were $2515.00 and his weekly income was $576.05.

a. Write a linear equation in slope-intercept form for Trevor's weekly income, y, in terms of his weekly sales, x. $y=0.07x+400$

b. What is Trevor's weekly salary? **$400**

Look Back

Copy and complete the table. Write the fractions in simplest form.

	Fraction	Decimal	Percent
65.	$\frac{1}{3}$	$0.\overline{33}$	$33\frac{1}{3}\%$
66.	$\frac{7}{8}$	0.875	$87\frac{1}{2}\%$
67.	$\frac{1}{50}$	0.02	2%
68.	$\frac{1}{20}$	0.05	5%
69.	$\frac{1}{8}$	0.125	$12\frac{1}{2}\%$
70.	$\frac{2}{3}$	$0.\overline{6}$	$66\frac{2}{3}\%$
71.	$\frac{1}{6}$	$0.1\overline{6}$	$16\frac{2}{3}\%$
72.	$\frac{1}{10{,}000}$	0.0001	0.01%
73.	$\frac{4}{5}$	0.80	80%
74.	$\frac{2}{5}$	0.4	40%
75.	$\frac{9}{20}$	0.45	45%
76.	$\frac{5}{6}$	$0.8\overline{3}$	$83\frac{1}{3}\%$

77. Use the formula $d = rt$ to find the distance, d, in meters when the rate, r, is 50 meters per second and the time, t, is 4 seconds. **200 m**

78. Use the formula $C = \pi d$ to find the circumference, C, in centimeters when the diameter, d, is 8 centimeters. Use 3.14 for π. **25.12 cm**

Look Beyond

79. Let $y = 4x$.

a. $\frac{y}{x} =$ __?__ **4**

b. If $y = 3$, then $x =$ __?__. $\frac{3}{4}$

80. Let $y = mx$. If $y = 4$ and $x = 2$, then $m =$ __?__. **2**

Direct Variation and Proportion

Why *Many events in the real world have a direct-variation relationship. For example, the distance you travel when bicycling can have a direct-variation relationship with time.*

Objectives

- Write and apply direct-variation equations.
- Write and solve proportions.

APPLICATION
RECREATION

Each day Johnathon rides his bicycle for exercise. When traveling at a constant rate, he rides 4 miles in about 20 minutes. At this rate, how long would it take Johnathon to travel 7 miles? To answer this question, you can use a *direct-variation equation* or a *proportion*. You will solve this problem in Example 2.

Recall that distance, d, rate, r, and elapsed time, t, are related by the equation $d = rt$. You can say that *d varies directly as t* because as time increases, the distance traveled increases proportionally.

Direct Variation

The variable y varies directly as x if there is a nonzero constant k such that $y = kx$. The equation $y = kx$ is called a **direct-variation equation** and the number k is called the **constant of variation.**

EXAMPLE 1 **Find the constant of variation, k, and the direct-variation equation if y varies directly as x and $y = -24$ when $x = 4$.**

SOLUTION

$y = kx$ — Use the direct-variation equation.
$-24 = k \cdot 4$ — Substitute 24 for y and 4 for x.
$\frac{-24}{4} = k$ — Solve for k.
$-6 = k$

The direct-variation equation is $y = -6x$.

TRY THIS Find the constant of variation, k, and the direct-variation equation if y varies directly as x and $y = 15$ when $x = 3$.

Prepare

NCTM PRINCIPLES & STANDARDS 1–4, 6–10

QUICK WARM-UP

Solve each equation.

1. $27 = c(-4)$ — $c = -6.75$
2. $-8 = -\frac{3}{4}z$ — $z = \frac{32}{3}$
3. $\frac{k}{4} = -20$ — $k = -80$
4. $\frac{n}{2} = \frac{5}{8}$ — $n = 1.25$
5. $2(4y + 1) = 3y$ — $y = -0.4$

Also on Quiz Transparency 1.4

Teach

Why There are many situations in which one quantity varies directly as another. For example, in many jobs an employee's wages vary directly as the number of hours worked. If you buy taxable merchandise, the amount of sales tax varies directly as the total price of the merchandise. Have students brainstorm a list of other direct-variation situations.

ADDITIONAL EXAMPLE 1

Find the constant of variation, k, and the direct-variation equation if y varies directly as x and $y = -72$ when $x = -18$.

$k = 4$; $y = 4x$

TRY THIS

$k = 5$; $y = 5x$

Alternative Teaching Strategy

HANDS-ON STRATEGIES Conduct a lab in which students take measurements that demonstrate a direct-variation relationship.

For example, provide rulers, string, and several circular objects. Have students measure the diameter and circumference of each object and then set up and complete a table with these headings.

Diameter (x)	Circumference (y)	$\frac{\text{Circumference}}{\text{Diameter}}$ $\left(\frac{y}{x}\right)$

Some measurement error will be introduced, but the ratio in the third column should be nearly constant. Students should recognize it as π. Elicit from them the direct-variation equation $y = \pi x$. Lead them to see that the familiar formula for circumference, $C = \pi d$, is really a direct-variation equation, with π being the constant of variation.

Another direct-variation relationship that students can measure in a lab activity is the height of the water in a cylindrical container and the number of fluid ounces of water.

ADDITIONAL
EXAMPLE 2

When traveling at a constant rate, Adia drives her car 12 miles in about 15 minutes. **At this rate, how long would it take Adia to drive 30 miles?** 37.5 min

Teaching Tips about technology refer to keystrokes for the TI-82, TI-83, or TI-83 Plus.

Teaching Tip

TECHNOLOGY To obtain the graph in Example 2, press Y=, clear any existing equations, and enter **(1/5)X** next to **Y1.** Then press WINDOW and use these settings.

Xmin=0	Ymin=−10
Xmax=50	Ymax=25
Xscl=5	Yscl=5

To view the graph, press GRAPH.

To check that (35, 7) is on the line, press 2nd TRACE to access the **CALCULATE** menu. Select **1:value**. Enter **35** next to **X=**, press ENTER, and **Y=7** will appear.

A keystroke guide is provided at the end of each chapter for examples given in the Pupil's Edition. If you are using a different graphics calculator, such as a Casio or Sharp, go to the HRW Internet site at go.hrw.com and enter the keyword MB1 CALC to find keystroke guides.

TRY THIS
72 min

EXAMPLE 2 Refer to the problem described at the beginning of the lesson.

APPLICATION
RECREATION

At the constant rate that Johnathon bikes, how long would it take him to travel 7 miles?

SOLUTION

1. Write a direct-variation equation, $d = rt$, that models Johnathon's distance as it varies with time.

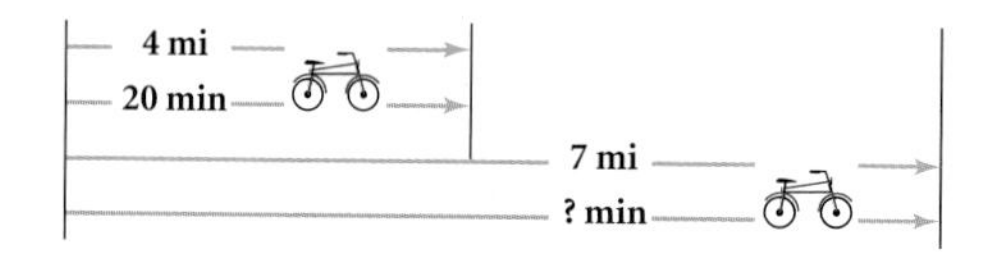

Find the constant of variation, r.

$$r = \frac{4 \text{ mi}}{20 \text{ min}} = \frac{1}{5} \text{ mile per minute}$$

Write the direct-variation equation.

distance in miles ↘ *time in minutes* ↙

$$d = \frac{1}{5}t$$

2. Use the direct-variation equation to solve the problem.

$d = \frac{1}{5}t$

$7 = \frac{1}{5}t$ *Substitute 7 for d.*

$35 = t$ *Solve for t.*

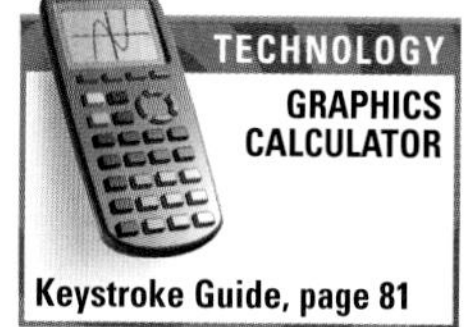

TECHNOLOGY
GRAPHICS CALCULATOR
Keystroke Guide, page 81

CHECK

Graph the equation $y = \frac{1}{5}x$, and check to see that the point (35, 7) is on the line.

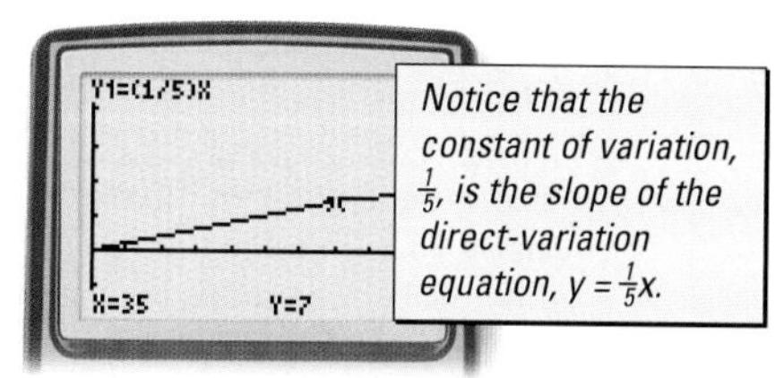

Notice that the constant of variation, $\frac{1}{5}$, is the slope of the direct-variation equation, $y = \frac{1}{5}x$.

Thus, at the rate given it will take Johnathon 35 minutes to travel 7 miles.

TRY THIS Suppose that when Johnathon is riding, he travels 5 miles in about 30 minutes. At this rate, how long would it take Johnathon to travel 12 miles?

The *Proportion Property* given below applies to all direct-variation relationships.

Proportion Property of Direct Variation

For $x_1 \neq 0$ and $x_2 \neq 0$:

If (x_1, y_1) and (x_2, y_2) satisfy $y = kx$, then $\frac{y_1}{x_1} = k = \frac{y_2}{x_2}$.

Interdisciplinary Connection

CHEMISTRY Absolute temperature is measured in *kelvins (K)*. The size of one kelvin on the Kelvin scale is equal to the size of one degree on the Celsius scale, but 0 *K* is equal to approximately −273°C.

When a fixed amount of gas is maintained at a constant pressure, its volume varies directly as its absolute temperature. This principle is known as *Charles' law* and is represented by the equation $\frac{V_1}{T_2} = \frac{V_2}{T_2}$, where V_1 is the initial volume, V_2 is the final volume, T_1 is the initial temperature, and T_2 is the final temperature. Suppose that 0.25 cubic meter of a gas is heated at a constant pressure from a temperature of 13°C to a new temperature of 65°C. What is the new volume of the gas? (Hint: Be sure to convert the temperatures to kelvins by using the formula $K = 273.15 + C$.) **≈0.3 m³**

In the Activity below, you can see a connection between the concepts of geometric similarity, proportion, and direct variation.

Exploring Similarity and Direct Variation

You will need: a calculator

Recall from geometry that *similar* figures have the same shape. This means that the corresponding angles of similar polygons are congruent, and their corresponding sides are proportional.

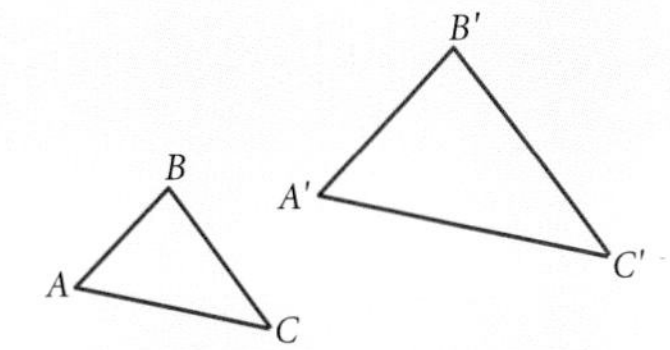

1. Copy and complete the table below to compare the lengths of the sides in $\triangle A'B'C'$ with the corresponding lengths in $\triangle ABC$.

Length in $\triangle ABC$	Length in $\triangle A'B'C'$	Ratio of $\triangle A'B'C'$ to $\triangle ABC$
$AB = 16$	$A'B' = 24$	$\frac{A'B'}{AB} = ?$
$BC = 20$	$B'C' = 30$	$\frac{B'C'}{BC} = ?$
$AC = 24$	$A'C' = 36$	$\frac{A'C'}{AC} = ?$

CHECKPOINT ✔ 2. Do your calculations in the third column indicate a direct-variation relationship between the lengths of the sides of $\triangle A'B'C'$ and those of $\triangle ABC$? Explain your response.

It is said that if y varies directly as x, then y is *proportional* to x.

A **proportion** is a statement that two *ratios* are equal. A ratio is the comparison of two quantities by division. A proportion of the form $\frac{a}{b} = \frac{c}{d}$ can be rearranged as follows:

$$\frac{a}{b} = \frac{c}{d}$$
$$\frac{a}{b} \cdot bd = \frac{c}{d} \cdot bd$$
$$ad = bc$$

The result is called the *Cross-Product Property of Proportions.*

Cross-Product Property of Proportions

For $b \neq 0$ and $d \neq 0$:
If $\frac{a}{b} = \frac{c}{d}$, then $ad = bc$.

In a proportion of the form $\frac{a}{b} = \frac{c}{d}$, a and d are the *extremes* and b and c are the *means*. By the Cross-Product Property, the product of the extremes equals the product of the means.

Activity Notes

From their work in previous courses, students should recall that corresponding sides of similar polygons are proportional. In this Activity, students investigate this relationship from the perspective of direct variation. They should discover that the constant ratio between pairs of corresponding sides is a constant of variation.

Math CONNECTION

GEOMETRY The constant ratio between corresponding sides of similar polygons is sometimes called the *scale factor of the similarity* or the *ratio of similitude.*

CHECKPOINT ✔

2. There is a direct-variation relationship. If the side lengths of $\triangle ABC$ are represented by x and the side lengths of $\triangle A'B'C'$ are represented by y, then the direct-variation equation is $y = \frac{3}{2}x$.

Inclusion Strategies

LINGUISTIC LEARNERS Encourage students to read a proportion like $\frac{a}{b} = \frac{c}{d}$ as "a is to b as c is to d." They may find this practice especially helpful in translating the language of a real-world problem into mathematical symbols. For example, another way of looking at the problem in Example 3 is to ask this question: A weight of 38 pounds on Mars is to a weight of 100 pounds on Earth as a weight of how many pounds on Mars is to a weight of 24.3 pounds on Earth?

Enrichment

Give students the table below. Have them use it to calculate their own weight on each planet. (The equivalent weight on Pluto remains unknown.)

Equivalent Weights for a 100-Pound Person

Planet	Weight (pounds)	Planet	Weight (pounds)
Mercury	38	Jupiter	253
Venus	91	Saturn	107
Earth	100	Uranus	91
Mars	38	Neptune	113

ADDITIONAL EXAMPLE 3

The ratio of weight on Jupiter, W_J, to weight on Earth, W_E, is given by $\frac{W_J}{W_E} \approx \frac{254}{100}$. The rover *Sojourner* weighed 24.3 pounds on Earth. Suppose that *Sojourner* was sent to Jupiter instead of Mars.

a. Find *Sojourner's* weight on Jupiter to the nearest tenth of a pound.
≈61.7 lb

b. Write a direct-variation equation that gives the weight of an object on Jupiter, W_J, in terms of its weight on Earth, W_E.
$W_J \approx 2.54W_E$

Teaching Tips about technology refer to keystrokes for the TI-82, TI-83, or TI-83 Plus.

Teaching Tip

TECHNOLOGY Set the viewing window as follows for Example 3:

Xmin=0	Ymin=0
Xmax=30	Ymax=15
Xscl=5	Yscl=1

Press GRAPH. Then press 2nd TRACE, enter **24.3** next to **X=**, and press ENTER.

ADDITIONAL EXAMPLE 4

Solve $\frac{2a-1}{5} = \frac{3a}{7}$. Check your answer. $a = -7$

TRY THIS
$x = 4$

CRITICAL THINKING
Using the cross-product property, $x^2 = a^2$, so $x = a$ or $x = -a$.

APPLICATION PHYSICS

Using Newton's law of universal gravitation, ratios that compare the weight of an object on Earth with its weight on another planet can be calculated. For Mars and Earth, the ratio is shown below.

$$\frac{W_M}{W_E} \approx \frac{38}{100} \quad \begin{matrix} \leftarrow \text{weight on Mars} \\ \leftarrow \text{weight on Earth} \end{matrix}$$

EXAMPLE 3 ***Sojourner* is the name of the first rover (robotic roving vehicle) that was sent to Mars. *Sojourner* weighs 24.3 pounds on Earth and is about the size of a child's small wagon.**

a. Find the weight of *Sojourner* on Mars to the nearest tenth of a pound.

b. Write a direct-variation equation that gives the weight of an object on Mars, W_M, in terms of its weight on Earth, W_E.

The Sojourner

SOLUTION

a. Solve the proportion for the weight of *Sojourner* on Mars.

$$\frac{W_M}{24.3} \approx \frac{38}{100}$$

$$(W_M)(100) \approx (24.3)(38) \quad \text{Use the Cross-Product Property.}$$

$$W_M \approx \frac{(24.3)(38)}{100} \quad \begin{matrix} \leftarrow \text{weight on Mars} \\ \leftarrow \text{weight on Earth} \end{matrix}$$

$$W_M \approx 9.2$$

On Mars, Sojourner would weigh about 9.2 pounds.

b.

$$\frac{W_M}{W_E} \approx \frac{38}{100}$$

$$W_M \approx \frac{38W_E}{100} \quad \begin{matrix} \leftarrow \text{weight on Mars} \\ \leftarrow \text{weight on Earth} \end{matrix}$$

$$W_M \approx 0.38W_E$$

TECHNOLOGY GRAPHICS CALCULATOR
Keystroke Guide, page 81

CHECK
Graph $y = 0.38x$, and confirm that a weight of 24.3 pounds on Earth, x, corresponds to a weight of about 9.2 pounds on Mars, y.

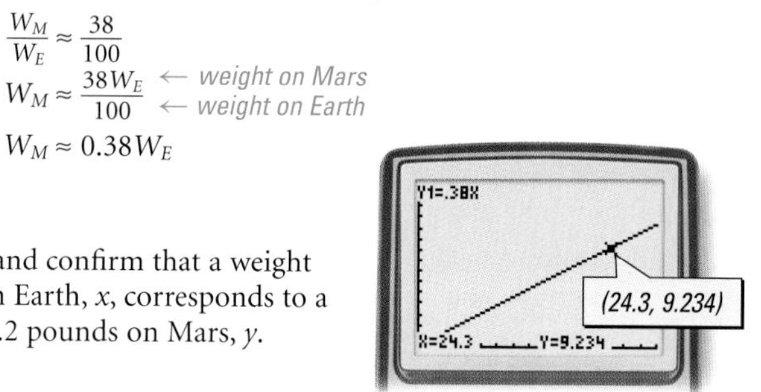

EXAMPLE 4 **Solve $\frac{3x-1}{5} = \frac{x}{2}$. Check your answer.**

SOLUTION

$$\frac{3x-1}{5} = \frac{x}{2}$$

$$(3x-1)(2) = (5)(x) \quad \text{Use the Cross-Product Property.}$$

$$6x - 2 = 5x$$

$$x - 2 = 0$$

$$x = 2$$

CHECK

$$\frac{3x-1}{5} = \frac{x}{2}$$

$$\frac{3(2)-1}{5} \stackrel{?}{=} \frac{2}{2}$$

$$1 = 1 \quad \textbf{True}$$

TRY THIS Solve $\frac{3x+2}{7} = \frac{x}{2}$. Check your answer.

CRITICAL THINKING Let $a > 0$. How many solutions does $\frac{x}{a} = \frac{a}{x}$ have? Find the solutions. Justify your answer.

Reteaching the Lesson

USING TABLES Have students complete this table.

x	3	4	5	6	7	8
y	4.5	6	7.5	9	10.5	12
$\frac{y}{x}$	1.5	1.5	1.5	1.5	1.5	1.5

Write $\frac{y}{x} = 1.5$ on the board or overhead. Multiply each side by x to obtain $y = 1.5x$. Lead students to see that this is a direct-variation equation. The ratio is the constant of variation.

Some students believe that a pattern of constant *differences* always leads to a direct-variation relationship. To demonstrate that this is not true, have students complete this table.

x	3	4	5	6	7	8
y	4.5	5.5	6.5	7.5	8.5	9.5
$\frac{y}{x}$	1.5	1.375	1.3	1.25	≈1.2	1.1875

Point out that the *ratio* is not constant, so it is not possible to write an equation of the form $\frac{y}{x} = k$ that applies to every entry in the table.

Exercises

Communicate

1. Suppose that y varies directly as x and that $y = 18$ when $x = 9$. Describe how you would find an equation of direct variation that relates these two variables.
2. When are linear equations *not* direct variations? How do their graphs differ from those of direct variations?
3. Describe two methods for solving the following problem:

   If y varies directly as x and $y = 8$ when $x = -2$, what is the value of x when $y = 12$?

Determine whether each equation describes a direct variation. Explain your reasoning.

4. $y = x + 5$
5. $y = x - 5$
6. $y = 5x$
7. $y = \frac{x}{5}$

Guided Skills Practice

8. Find the constant of variation, k, and the direct-variation equation if y varies directly as x and $y = 1000$ when $x = 200$. ***(EXAMPLE 1)*** **5; $y = 5x$**

APPLICATIONS

9. **PHYSICS** The speed of sound in air is about 335 feet per second. At this rate, how far would sound travel in 25 seconds? ***(EXAMPLE 2)*** **8375 ft**
10. **INCOME** Workers at a particular store earn hourly wages. A person who worked 18 hours earned \$114.30. ***(EXAMPLE 3)***
    a. How many hours must this person work to earn \$127? **20**
    b. Write a direct-variation equation that gives the income of this person in terms of the hours worked. What does the constant of variation represent? **$y = 6.35x$; hourly wage**

Solve each equation for x. Check your answers. ***(EXAMPLE 4)***

11. $\frac{4x - 1}{21} = \frac{x}{6}$ **2**
12. $\frac{x + 4}{-4} = \frac{3x}{36}$ **−3**
13. $\frac{2x}{8} = \frac{x + 3}{7}$ **4**

Practice and Apply

In Exercises 14–29, y varies directly as x. Find the constant of variation, and write an equation of direct variation that relates the two variables.

14. $y = 21$ when $x = 7$ **3; $y = 3x$**
15. $y = 2$ when $x = 1$ **2; $y = 2x$**
16. $y = -16$ when $x = 2$ **−8; $y = -8x$**
17. $y = 1$ when $x = \frac{1}{3}$ **3; $y = 3x$**
18. $y = \frac{4}{5}$ when $x = \frac{1}{5}$ **4; $y = 4x$**
19. $y = -\frac{6}{7}$ when $x = -\frac{18}{35}$ **$\frac{5}{3}$; $y = \frac{5}{3}x$**
20. $y = -2$ when $x = 9$ **$-\frac{2}{9}$; $y = -\frac{2}{9}x$**
21. $y = 5$ when $x = -0.1$ **−50; $y = -50x$**
22. $y = 1.8$ when $x = 30$ **0.06; $y = 0.06x$**
23. $y = 0.4$ when $x = -1$ **−0.4; $y = -0.4x$**
24. $y = 24$ when $x = 8$ **3; $y = 3x$**
25. $y = 12$ when $x = \frac{1}{4}$ **48; $y = 48x$**

Assess

Selected Answers

Exercises 8–13, 15–81 odd

ASSIGNMENT GUIDE	
In Class	1–13
Core	15–35 odd, 37–51, 53–65 odd
Core Plus	14–36 even, 37–51, 52–56 even, 59–71
Review	72–81
Preview	82

Extra Practice can be found beginning on page 940.

Answers to odd-numbered Extra Practice exercises can be found immediately after Selected Answers in the Pupil's Edition.

Mid-Chapter Assessment for Lessons 1.1 through 1.4 can be found on page 5 of the *Assessment Resources.*

Error Analysis

When solving equations such as $\frac{3x-1}{5} = \frac{x}{2}$ in Example 4, students sometimes forget to apply the distributive property after they have written the cross products. Suggest that they use arrows like those shown below as a reminder to multiply each term inside the parentheses by the factor outside.

$$(3x-1)(2) = (5)(x)$$
$$6x - 2 = 5x$$

26. $y = -\frac{5}{8}$ when $x = -1$ **$\frac{5}{8}$; $y = \frac{5}{8}x$**
27. $y = 4$ when $x = 0.2$ **20; $y = 20x$**
28. $y = 0.6$ when $x = -3$ **−0.2; $y = -0.2x$**
29. $y = -1.2$ when $x = 4$ **−0.3; $y = -0.3x$**

Write an equation that describes each direct variation.

30. p varies directly as q. **$p = kq$**
31. a is directly proportional to b. **$a = kb$**

For Exercises 32–36, a varies directly as b.

32. If a is 2.8 when b is 7, find a when b is −4. **−1.6**
33. If a is 6.3 when b is 70, find b when a is 5.4. **60**
34. If a is −5 when b is 2.5, find b when a is 6. **−3**
35. If b is $-\frac{3}{5}$ when a is $-\frac{9}{10}$, find a when b is $\frac{1}{3}$. **$\frac{1}{2}$**
36. If b is $-\frac{1}{2}$ when a is $-\frac{3}{10}$, find a when b is $-\frac{5}{9}$. **$-\frac{1}{3}$**

internet connect
Homework Help Online
Go To: **go.hrw.com**
Keyword:
MB1 Homework Help
for Exercises 37–51

Solve each proportion for the variable. Check your answers.

37. $\frac{w}{4} = \frac{10}{12}$ **$3\frac{1}{3}$**
38. $\frac{5}{q} = \frac{7}{8}$ **$5\frac{5}{7}$**
39. $\frac{1}{8} = \frac{x}{100}$ **$12\frac{1}{2}$**
40. $\frac{9}{10} = \frac{6}{r}$ **$6\frac{2}{3}$**
41. $\frac{x}{3} = \frac{-7}{10}$ **$-2\frac{1}{10}$**
42. $\frac{3}{5} = \frac{x}{2}$ **$1\frac{1}{5}$**
43. $\frac{7}{x} = \frac{3}{4}$ **$9\frac{1}{3}$**
44. $\frac{x+5}{2} = \frac{4}{3}$ **$-2\frac{1}{3}$**
45. $\frac{x}{-5} = x - 6$ **5**
46. $\frac{x-1}{56} = \frac{x}{64}$ **8**
47. $\frac{3x+1}{5} = \frac{x}{2}$ **−2**
48. $\frac{-4x}{-7} = x - 3$ **7**
49. $\frac{x+1}{9} = \frac{5x}{40}$ **8**
50. $\frac{6x-3}{9} = \frac{8x}{8}$ **−1**
51. $\frac{5x}{-30} = \frac{x-5}{4}$ **3**

Determine whether the values in each table represent a direct variation. If so, write an equation for the variation. If not, explain.

52.

x	2	3	4	5	6
y	−4	−9	−16	−25	−36

53.

x	5	6	7	8	9
y	0.10	0.12	0.14	0.16	0.18

54.

x	−1	0	1	2	3
y	8	10	12	14	16

55.

x	−2	−1	0	1	2
y	1	0.5	0	−0.5	−1

56.

x	−7	−3	1	5	9
y	133	57	−19	−95	−171

57.

x	1	−5	−11	−17	−23
y	−6	42	90	138	186

53. yes; $y = 0.02x$
55. yes; $y = -\frac{1}{2}x$
56. yes; $y = -19x$

52. No; there is no k such that $y = kx$ for every x in the table.
54. No; there is no k such that $y = kx$ for every x in the table.
57. No; there is no k such that $y = kx$ for every x in the table.
58. If x varies directly as y, then $x = ky$ for some constant $k \neq 0$. Therefore, $y = \frac{1}{k}x$. Since k is a nonzero constant, $\frac{1}{k}$ is also a nonzero constant. Hence, y varies directly as x.

58. Show that if x varies directly as y, then y varies directly as x.

CHALLENGE

59. If a varies directly as c and b varies directly as c, show that $a + b$ varies directly as c.

CONNECTION

60. **COORDINATE GEOMETRY** Which of the lines shown in the graph at right represents a direct variation? Explain your reasoning.

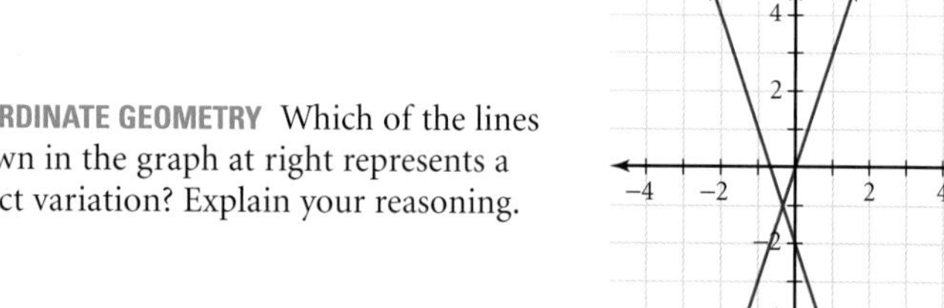

59. If a varies directly as c, then $a = k_1c$ for some nonzero constant k_1. If b varies directly as c, then $b = k_2c$ for some nonzero constant k_2. Add each side of these equations together to get $a + b = (k_1 + k_2)c$. Since $k_1 + k_2$ is a nonzero constant, $a + b$ varies directly as c.

60. l_2 represents a direct variation since the line passes through the origin and the equation of the line is in the form $y = kx$. l_1 does not pass through the origin and is not in the form $y = kx$, so it does not represent a direct variation.

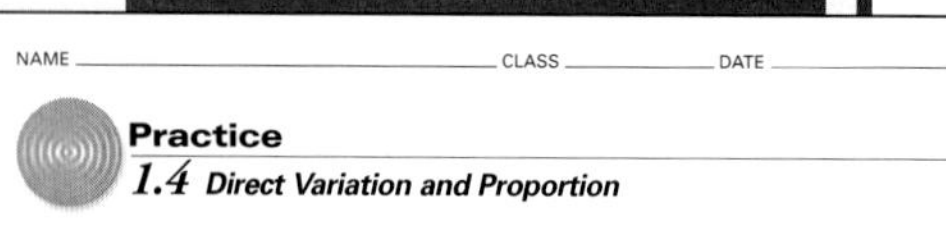

Practice

NAME ______ CLASS ______ DATE ______

Practice
***1.4* Direct Variation and Proportion**

In Exercises 1–8, y varies directly as x. Find the constant of variation, and write an equation of direct variation that relates the two variables.

1. $y = -10$, for $x = 2$ $k = -5$; $y = -5x$
2. $y = 7$, for $x = 3$ $k = \frac{7}{3}$; $y = \frac{7}{3}x$
3. $y = 4$, for $x = -3$ $k = -\frac{4}{3}$; $y = -\frac{4}{3}x$
4. $y = 3.2$, for $x = 12.8$ $k = \frac{1}{4}$; $y = \frac{1}{4}x$
5. $y = -2$, for $x = -7$ $k = \frac{2}{7}$; $y = \frac{2}{7}x$
6. $y = 5$, for $x = 6$ $k = \frac{5}{6}$; $y = \frac{5}{6}x$
7. $y = \frac{2}{3}$, for $x = \frac{1}{3}$ $k = 2$; $y = 2x$
8. $y = -\frac{3}{5}$, for $x = \frac{1}{5}$ $k = -3$; $y = -3x$

Solve each proportion for the variable. Check your answers.

9. $\frac{x}{4} = \frac{9}{12}$ $x = 3$
10. $\frac{x+2}{3} = \frac{3x}{6}$ $x = 4$
11. $\frac{y}{6} = \frac{18}{24}$ $y = \frac{9}{2}$
12. $\frac{2x-5}{10} = \frac{3x}{20}$ $x = 10$
13. $\frac{z}{10} = \frac{60}{240}$ $z = \frac{5}{2}$
14. $\frac{3y}{10} = \frac{y-1}{6}$ $y = -\frac{5}{4}$
15. $\frac{x+2}{5} = \frac{5}{25}$ $x = -1$
16. $\frac{5z}{7} = \frac{z+3}{14}$ $z = \frac{1}{3}$

Determine whether the values in each table represent a direct variation. If so, write an equation for the variation. If not, explain why not.

17.

x	y
−2	−6
−1	−3
0	0
1	3
2	6

yes; $y = 3x$

18.

x	y
5	49
4	28
3	20
2	5
1	2

no; there is no constant, k, such that $y = kx$.

19.

x	y
1	2
3	6
5	10
7	14
9	18

yes; $y = 2x$

CULTURAL CONNECTION: ASIA The Harappan civilization flourished in an area near present-day Pakistan around 2500 B.C.E. They used balancing stones in their system of weights and measures. The Vedic civilization, which followed the Harappan civilization, used gunja seeds to weigh precious metals. The smallest Harappan stone has the same mass as 8 gunja seeds.

The scale is balanced with 16 gunja seeds on the left and the second smallest Harappan stone on the right.

61. The mass of a Harappan stone, m, varies directly as the number of gunja seeds, g. Find the constant of variation and the direct-variation equation for this relationship. **0.11; $m = 0.11g$**

62. How many gunja seeds are equivalent to a Harappan stone whose mass is 3.52 grams? **32**

63. The largest Harappan stone is equivalent to 320 gunja seeds. What is the mass of this stone? **35.2 grams**

Smallest Harappan stone

0.88 gram

1.76 grams

3.52 grams

320 gunja seeds
Largest Harappan stone

CONNECTIONS

64. GEOMETRY In the figure at right, the height of each object is directly proportional to the length of its shadow. The person is $5\frac{1}{2}$ feet tall and casts an 8-foot shadow, while the tree casts a 33-foot shadow. How tall is the tree? **22.6875 ft, or 22 ft $8\frac{1}{4}$ in.**

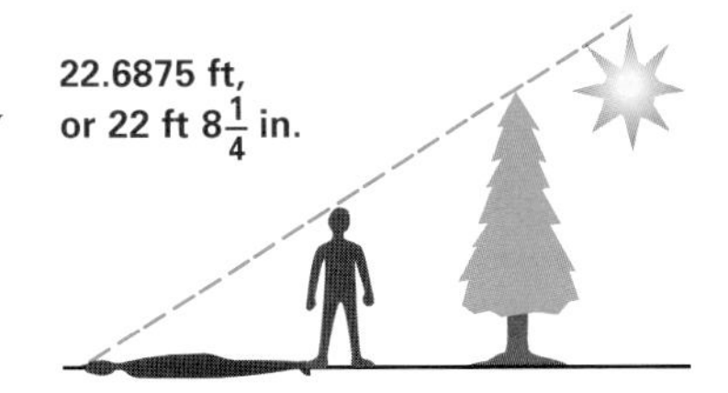

65. GEOMETRY In an aerial photograph, a triangular plot of land has the dimensions given in the figure at right. If the actual length of the longest side of the plot is 50 kilometers, find the actual lengths of the two shorter sides. **35 km, 30 km**

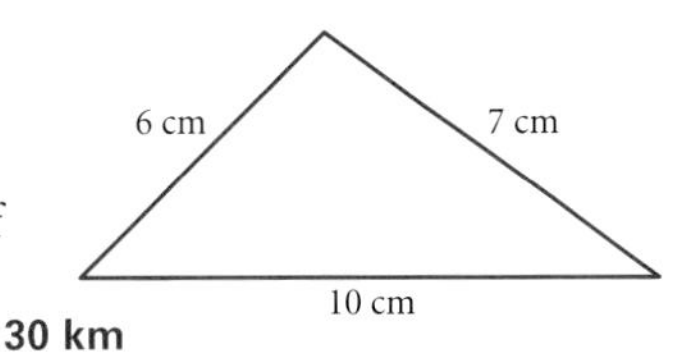

APPLICATION

PHYSICS In an electric circuit, Ohm's law states that the voltage, V, measured in volts varies directly as the electric current, I, measured in amperes according to the equation $V = IR$. The constant of variation is the electrical resistance of the circuit, R, measured in ohms.

66. An iron is plugged into a 110-volt electrical outlet, creating a current of 5.5 amperes in the iron. Find the electrical resistance of the iron. **20 ohms**

67. A heater is plugged into a 110-volt outlet. If the resistance of the heater is 11 ohms, find the current in the heater. **10 amps**

68. Find the current, to the nearest hundredth of an ampere, in a night light with a resistance of 300 ohms that is plugged into a 110-volt outlet. **0.37 amp**

69. Find the current, to the nearest hundredth of an ampere, in a lamp that has a resistance of 385 ohms and is plugged into a 110-volt outlet. **0.29 amp**

Cultural Connections appear throughout the textbook in the instructional sections as well as in the exercises. Because they reflect the contributions from many of the diverse world's cultures to contemporary mathematics, students from different cultures are able to identify with the contributions of their culture.

Student Technology Guide

NAME ______ CLASS ______ DATE ______

Student Technology Guide
1.4 Direct Variation and Proportion

In Lesson 1.4, you used algebraic methods to solve the proportion $\frac{3x-1}{5} = \frac{x}{2}$. Here is another way to solve this equation by using a graphics calculator.

Step 1: Press Y=. Enter the left side of the equation, $\frac{3x-1}{5}$, for Y1. Enter the right side, $\frac{x}{2}$, for Y2.

Y1: (3 X,T,θ,n − 1) ÷ 5

Y2: X,T,θ,n ÷ 2

Plot1 Plot2 Plot3
\Y1 = (3X−1)/5
\Y2 = X/2
\Y3 =
\Y4 =
\Y5 =
\Y6 =
\Y7 =

Step 2: Press WINDOW. Enter $-5 \le x \le 5$ and $-2 \le y \le 3$.

Step 3: Press TRACE. Use ◄ or ► to move to the point of intersection. While tracing, you can use ▲ or ▼ to move back and forth between curves. The solution appears to be near $x = 2$.

Y1=(3X−1)/5
X=2.0212766 Y=1.012766

Step 4: To determine if $x = 2$ is the *exact* solution, press 2nd TRACE 1:value ENTER 2 ENTER. Then use ▲ or ▼ to move back and forth between curves. This shows that the point (2, 1) is on both curves.

Use a graphics calculator to solve each proportion.

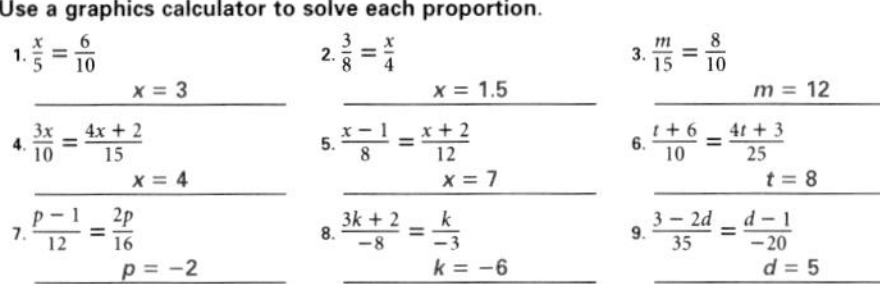

1. $\frac{x}{5} = \frac{6}{10}$ $x = 3$
2. $\frac{3}{8} = \frac{x}{4}$ $x = 1.5$
3. $\frac{m}{15} = \frac{8}{10}$ $m = 12$
4. $\frac{3x}{10} = \frac{4x+2}{15}$ $x = 4$
5. $\frac{x-1}{8} = \frac{x+2}{12}$ $x = 7$
6. $\frac{t+6}{10} = \frac{4t+3}{25}$ $t = 8$
7. $\frac{p-1}{12} = \frac{2p}{16}$ $p = -2$
8. $\frac{3k+2}{-8} = \frac{k}{-3}$ $k = -6$
9. $\frac{3-2d}{35} = \frac{d-1}{-20}$ $d = 5$

10. Use a graphics calculator to solve $\frac{x}{2} = \frac{x+3}{3} = \frac{x+24}{10}$. Explain what you did.
$x = 6$; Explanations may vary. Enter the three expressions into the function list, graph them, and then use tracing to get an exact answer.

In Exercise 82, students investigate an inverse-variation equation and its graph. Students may recall from previous courses that the shape of this type of graph is called a *hyperbola.* Students will study inverse-variation relationships in greater detail in Lesson 8.1.

ALTERNATIVE
Assessment

Portfolio Activity

The Portfolio Activity can be used as preparation for the Chapter Project or as a separate activity. In the Portfolio Activity on this page, students analyze both a table of data and its graph from the perspective of direct variation and proportion.

Answers to Portfolio Activities can be found in Additional Answers of the Teacher's Edition.

70a. $k = 0.433$; $p = 0.433d$

b. 43.3 lb/in.^{2}

c. $k = 0.445$; $p = 0.445d$

d. 44.5 lb/in.^{2}

71a. 0.1875; $s = 0.1875w$; the constant represents the "stretchiness" of the spring as compared to that of other springs.

b. 16 lb

c. 7.5 in.

82. Answers may vary depending on the choice of k but should show a vertical stretch or compression of the graph of $y = \frac{1}{x}$.

APPLICATIONS

70. PHYSICS As a scuba diver descends, the increase in water pressure varies directly as the increase in depth below the water's surface. However, the constant of variation is smaller in fresh water than in salt water. For example, at 80 feet below the surface of a typical freshwater lake, the pressure is 34.64 pounds per square inch greater than the pressure at the surface. In a typical ocean, where the water is salty, the pressure at 80 feet is 35.6 pounds per square inch greater than the pressure at the surface.

- **a.** Find the constant of variation and the equation of direct variation for the increase in pressure in a typical freshwater lake.
- **b.** Find the increase in pressure at 100 feet below the surface of a lake.
- **c.** Find the constant of variation and the equation of direct variation for the increase in pressure in a typical ocean.
- **d.** Find the increase in pressure at 100 feet below the surface of an ocean.

71. PHYSICS The distance a spring stretches varies directly as the amount of weight that is hanging on it. A weight of 32 pounds stretches the spring 6 inches, and a weight of 48 pounds stretches it 9 inches.

- **a.** Find the constant of variation and the equation of direct variation for the stretch of the spring. What does the constant of variation represent?
- **b.** How heavy is the weight hanging on the spring when it is stretched 3 inches?
- **c.** Find the stretch of the spring when a weight of 40 pounds is hanging on it.

Look Back

Write the prime factorization for each number.

72. 261 $3^2 \cdot 29$ **73.** 860 $2^2 \cdot 5 \cdot 43$ **74.** 315 $3^2 \cdot 5 \cdot 7$ **75.** 180 $2^2 \cdot 3^2 \cdot 5$ **76.** 154 $2 \cdot 7 \cdot 11$ **77.** 490 $2 \cdot 5 \cdot 7^2$

Evaluate.

78. $\dfrac{\frac{11}{13}}{\frac{11}{26}}$ 2 **79.** $\dfrac{\frac{5}{6}}{\frac{15}{12}}$ $\frac{2}{3}$ **80.** $\dfrac{-\frac{1}{3}}{-\frac{4}{21}}$ $1\frac{3}{4}$ **81.** $\dfrac{-\frac{2}{5}}{\frac{28}{25}}$ $-\frac{5}{14}$

Look Beyond

Calculator button indicates that a graphics calculator is recommended.

82 An equation of the form $xy = k$, where k is a constant greater than zero, is called an *inverse-variation equation.* Choose a positive value for k, and graph the equation. Describe the graph.

Refer to your portfolio data set from the Portfolio Activity on page 11.

1. In your portfolio data set, does one variable vary directly as the other variable? Explain.
2. Using data values from your portfolio data set, write a proportion of the form $\frac{y_1}{x_1} = \frac{y_2}{x_2}$. Is the proportion true for some values? Is the proportion true for all values? Explain.
3. Using points on your linear model from the Portfolio Activity on page 20, write a proportion of the form $\frac{y_1}{x_1} = \frac{y_2}{x_2}$. Is the proportion true for some values? Is the proportion true for all values? Explain.

Scatter Plots and Least-Squares Lines

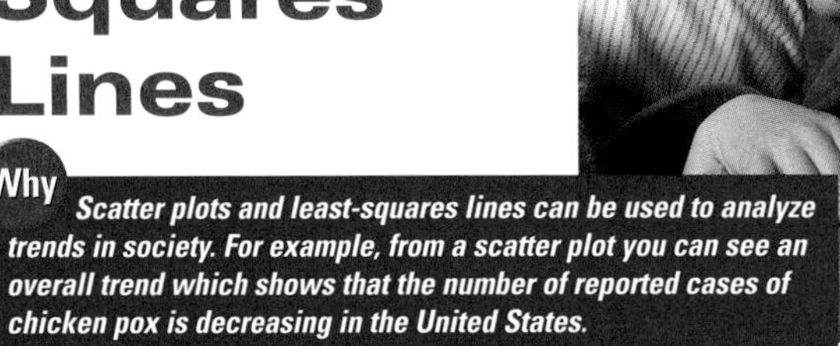

Why *Scatter plots and least-squares lines can be used to analyze trends in society. For example, from a scatter plot you can see an overall trend which shows that the number of reported cases of chicken pox is decreasing in the United States.*

Objectives

- Create a scatter plot and draw an informal inference about any correlation between the variables.
- Use a graphics calculator to find an equation for the least-squares line and use it to make predictions or estimates.

APPLICATION
HEALTH

In many real-world problems, you will find data that relate two variables such as time and distance or age and height. You can view the relationship between two variables with a **scatter plot**.

The following data on the number of reported cases of chicken pox in thousands in the United States is graphed in a scatter plot. The variable x represents the number of years after 1988 ($x = 0$ represents 1988) and y represents the number of cases in thousands.

Chicken Pox in the United States

Year	Reported cases (in the thousands)
1989	185.4
1990	173.1
1991	147.1
1992	158.4
1993	134.7
1994	151.2

[*Source: Centers for Disease Control and Prevention*]

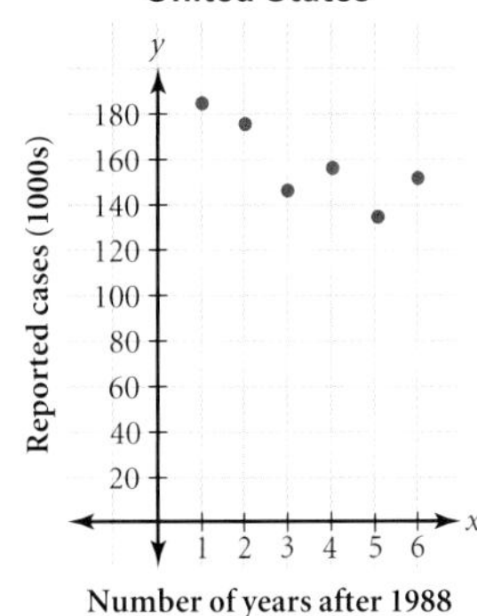

Activity Investigating a Scatter Plot

TECHNOLOGY
GRAPHICS CALCULATOR
Keystroke Guide, page 81

You will need: a graphics calculator

1. Create a scatter plot for the data on reported cases of chicken pox in the United States from 1989 to 1994. Let $x = 1$ represent the year 1989.
2. Write a linear equation in slope-intercept form that closely fits the data points. Graph your equation along with the data points.

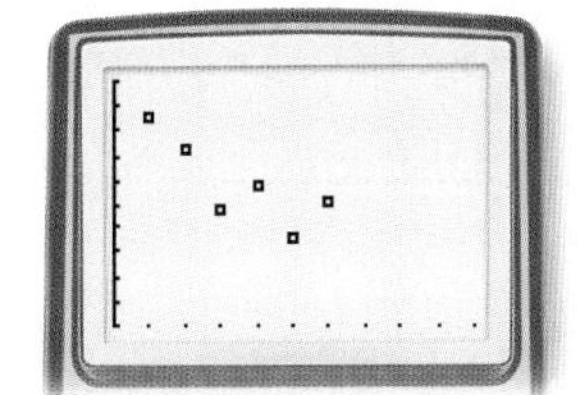

Prepare

NCTM PRINCIPLES & STANDARDS 1, 2, 5–10

QUICK WARM-UP

Graph each point in the same coordinate plane.

1. $A(1, 400)$ 2. $B(3, 300)$
3. $C(6, 150)$ 4. $D(8, 50)$

1–4. Check students' graphs.

5. Refer to Exercises 1–4. Suppose that a line were drawn to connect the four points. Would its slope be positive or negative? **negative**

Also on Quiz Transparency 1.5

Teach

Why In previous lessons, students studied two-variable relationships that could be modeled precisely by a linear equation. However, real-world situations often are more complex. Many relationships are *nearly* linear but do not conform exactly to a linear model. Have students suggest situations in which this might be true.

Use Teaching Transparency 1.

Activity Notes

In this Activity, students use the statistics feature of a graphics calculator to create a scatter plot. However, they make and test their own estimates for a line of best fit rather than use the regression equation option of the calculator. They should observe that the slope of the line of best fit gives a rate whose units are taken from the scales of the axes.

Alternative Teaching Strategy

USING VISUAL MODELS Discuss with students the idea of fitting a line to a set of data. Give them the first graph at right. Have them find the sum of the squares of the vertical distances from the points to the line. $1^2 + 3^2 + 2^2 + 1^2 = 15$ Repeat with the second graph. $1^2 + 2^2 + 2^2 + 0^2 = 9$ Point out that the sum of the squares is less for the graph of $y = 3$, so it is a *better* line of fit for the data than $y = 0.5x + 1$. The *line of best fit* will be the line for which this sum of squares is the least; thus, it is called the *least-squares* line. Have students find the equation of the least-squares line for the data on the graphs by using a calculator. $y \approx -0.05x + 2.9$ The correlation coefficient is approximately -0.08, so the data are very weakly correlated. A linear model is not well suited to the data.

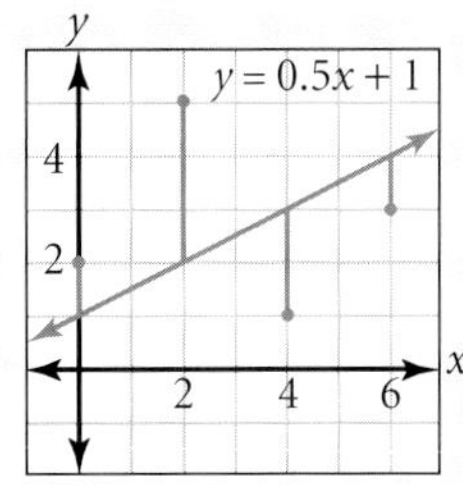

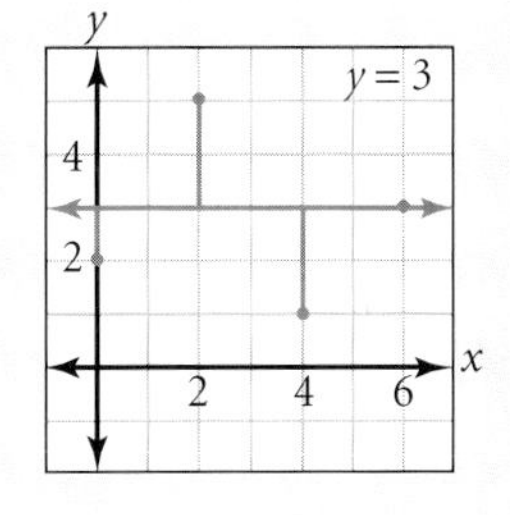

☞ For keystrokes to create a scatter-plot and calculate the regression equation, see pages 81 and 82.

CHECKPOINT ✔

4. Slope ≈ −8; the slope indicates that 8000 fewer cases of chicken pox are reported each year.

ADDITIONAL EXAMPLE 1

Create a scatter plot for the data shown below. Describe the correlation. Then find and graph an equation for the least-squares line.

x	0	2	3	6	7	9	12
y	14	19	22	26	26	32	38

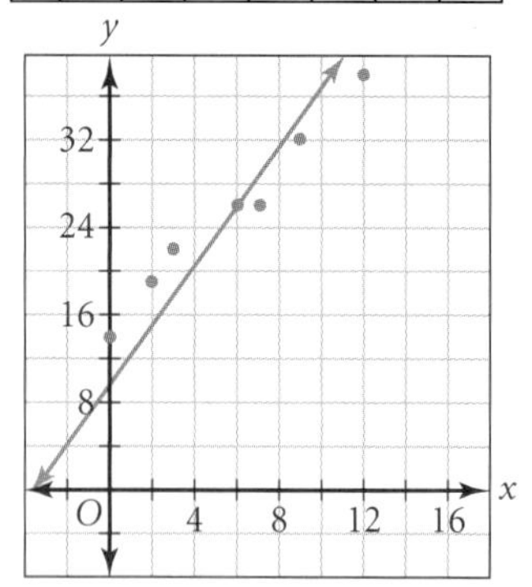

positive; $y \approx 1.89x + 14.75$

Use Teaching Transparency 2.

Math CONNECTION

STATISTICS The actual values in a set of data are called *observed values*. The points generated by the linear model are called *predicted values*. The vertical distance between an observed value and the corresponding predicted value is called the *error* or *deviation* in the prediction.

3. Adjust the slope and y-intercept of your equation until you think the graph best fits the data points. Record your *best-fit* equation.

CHECKPOINT ✔ 4. What rate of change is indicated by the slope of your linear equation? Write a sentence that states what the slope indicates about the reported cases of chicken pox.

The chicken-pox data in the Activity involves a two-variable data set that has a *negative correlation*. In general, there is a *correlation* between two variables when there appears to be a line about which the data points cluster. The diagrams below show the three possible correlations.

Positive correlation

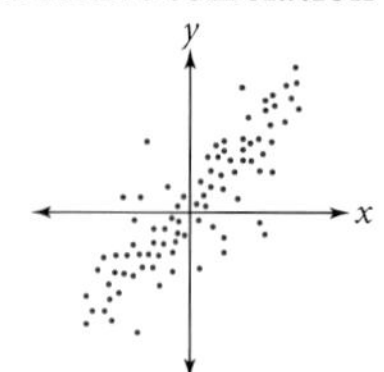

Negative correlation

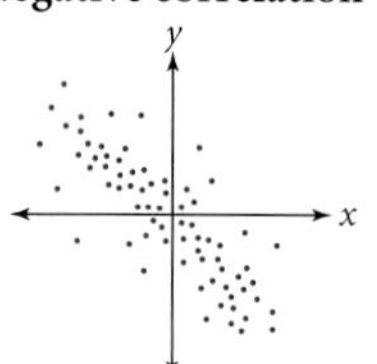

No reliable correlation

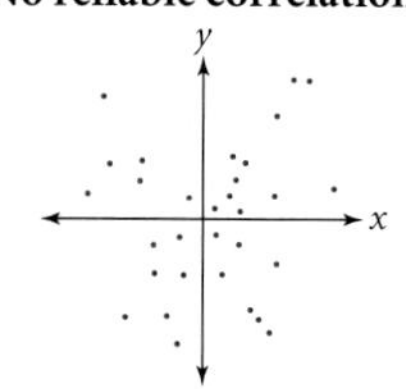

Finding the Least-Squares Line

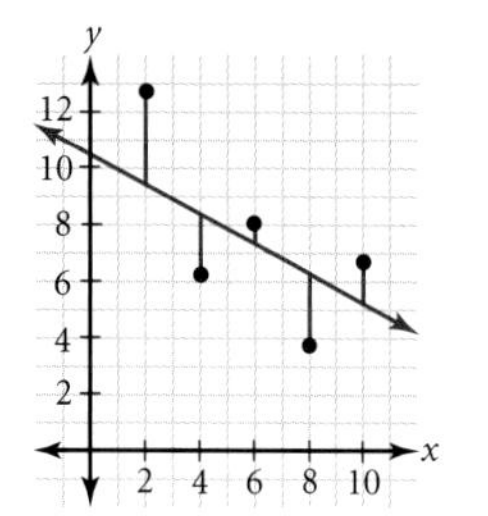

A scatter plot can help you see patterns in data involving two variables. If you think there may be a linear correlation between the variables, you can use a calculator to find a *linear-regression line*, also called the *least-squares line*, that best fits the data.

The graph at left shows the vertical distance from each point in a scatter plot to a fitted line. The fit of a **least-squares line** is based on *minimizing* these vertical distances for a data set. A least-squares line is one type of linear model for a data set.

EXAMPLE 1 **Create a scatter plot for the data shown at right. Describe the correlation. Then find and graph an equation for the least-squares line.**

Keystroke Guide, page 82

x	y
0	−3.2
2	1.2
4	5.0
6	8.8
7	11.6
8	13.0
10	17.5

SOLUTION

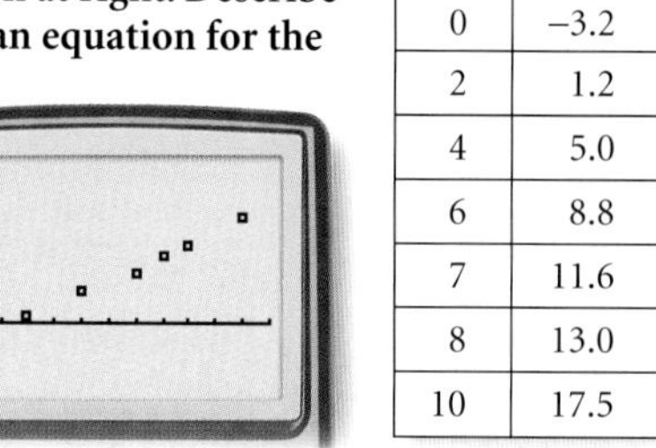

1. Create the scatter plot.
2. Describe the correlation.

   Because the points rise from left to right, the correlation is positive.

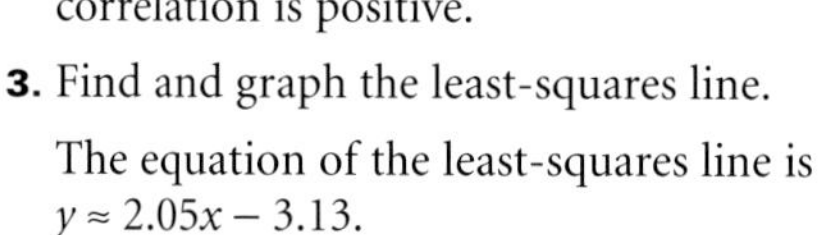

3. Find and graph the least-squares line.

   The equation of the least-squares line is $y \approx 2.05x - 3.13$.

   Graph the least-squares line on the scatter plot with the data points.

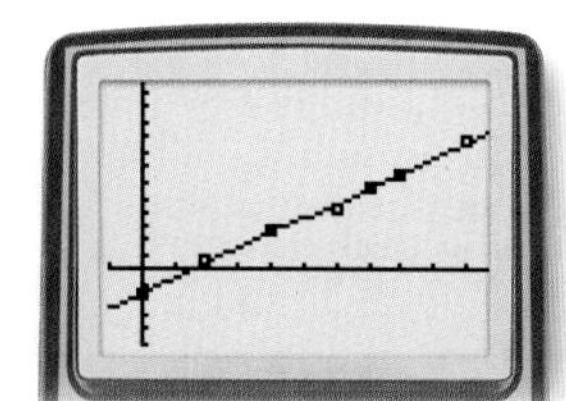

Interdisciplinary Connection

ECONOMICS The *consumer price index* (CPI) is a measure of the average change in price for a fixed *market basket* of goods and services. The table below gives the CPI for several recent years.

Year (19—)	80	82	84	86	88	90	92	94
CPI	82	97	104	110	118	131	140	148

Let x represent the year after 1980 and y represent the CPI. Find an equation of the least-squares line for the data and estimate the CPI in 1985 and in 2000. $y \approx 4.56x + 84.33$; ≈107; ≈176$y \approx 4.56x + 84.33$; ≈107; ≈176

Inclusion Strategies

COOPERATIVE LEARNING In the Activity on pages 37 and 38, students are asked to make a visual estimate of a line of best fit. Some students may be uncomfortable because there is no precise procedure for doing this. It may be helpful to have students do the Activity in groups of three or four. Each student in the group can make an independent estimate, and then the group can share and discuss their results. Students should feel reassured that several lines appear to fit the data, so there is no single "right answer" for the visual estimate.

Correlation and Prediction

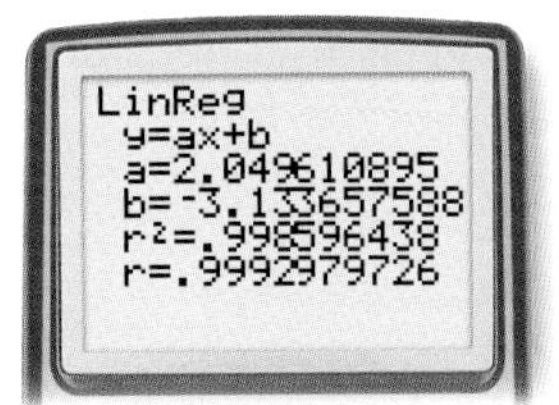

Examine the graphics calculator display at left, which shows the linear-regression equation for Example 1. Notice that the display also shows a value of about 0.9993 for r. The **correlation coefficient,** denoted by r, indicates how closely the data points cluster around the least-squares line.

The correlation coefficient can vary from −1, which is a perfect fit for a negative correlation, to +1, which is a perfect fit for a positive correlation.

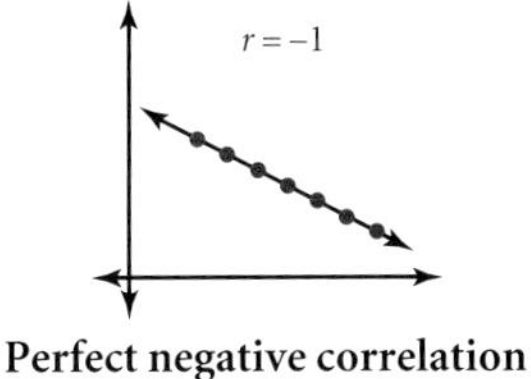

Perfect negative correlation

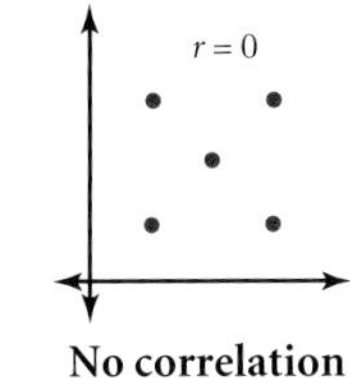

No correlation

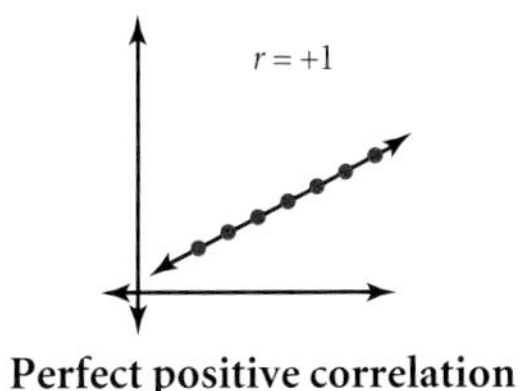

Perfect positive correlation

The closer the correlation coefficient is to −1 or +1, the better the least-squares line fits the data.

CHECKPOINT ✔ Refer to the data and the least-squares line found in Example 1. What is the correlation coefficient for this least-squares line? Is the correlation strong?

EXAMPLE 2

APPLICATION
SPORTS

The winning times for the men's Olympic 1500-meter freestyle swimming event are given in the table. Notice that there is not a winning time recorded for the year 1940 (the Olympic games were not held during World War II).

Estimate what the winning time for this event could have been in 1940.

Year	Time (min:sec)	Time (min)
1908	22:48.4	22.81
1912	22:00.0	22.00
1920	22:23.2	22.39
1924	20:06.6	20.11
1928	19:51.8	19.86
1932	19:12.4	19.21
1936	19:13.7	19.23
1948	19:18.5	19.31
1952	18:30.3	18.51
1956	17:58.9	17.98
1960	17:19.6	17.33
1964	17:01.7	17.03
1968	16:38.9	16.65
1972	15:52.58	15.88
1976	15:02.40	15.04
1980	14:58.27	14.97
1984	15:05.20	15.09
1988	15:00.40	15.00
1992	14:43.48	14.72
1996	14:56.40	14.94

SOLUTION

Let x represent the number of years after 1900. Let y represent the winning time in minutes. Enter the data into your calculator, and make a scatter plot.

TECHNOLOGY
GRAPHICS CALCULATOR
Keystroke Guide, page 82

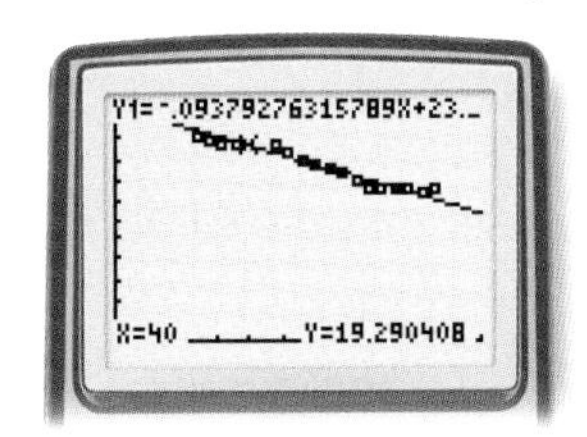

Using the equation for the least-squares line calculated from columns 1 and 2, the y-value that corresponds to $x = 40$ is about 19.29. Thus, for the men's Olympic 1500-meter freestyle in 1940, the winning time might have been about 19.29 minutes, or about 19:17.42.

Math CONNECTION

STATISTICS The correlation coefficient, r, is defined mathematically by the formula $r = \frac{\overline{xy} - \overline{x} \cdot \overline{y}}{s_x s_y}$, where $\overline{x}$ is the mean of the x-values, $\overline{y}$ is the mean of the y-values, $\overline{xy}$ is the mean of the products of corresponding x-values and y-values, s_x is the standard deviation of the x-values, and s_y is the standard deviation of the y-values.

CHECKPOINT ✔

$r = 0.9993$; very strong positive correlation

ADDITIONAL EXAMPLE 2

The table below shows the number of juniors enrolled in advanced algebra at West High School in six of the last seven years. The number for the third year has been misplaced. **Estimate the number of students enrolled in advanced algebra in the third year.**

Year	1	2	4	5	6	7
No.	33	29	45	43	43	61

≈37 students

Use Teaching Transparency 3.

Teaching Tips about technology refer to keystrokes for the TI-82, TI-83, or TI-83 Plus.

Enrichment

Anthropometry is the study of human body measurements. One conjecture made in anthropometry is that there is a linear relationship between the circumference of a person's neck and the circumference of the wrist. Have students collect data on neck and wrist circumferences from their classmates. Have them make a scatter plot of the data, find a regression equation and correlation coefficient, and use their results to assess the conjecture. **Exact results will vary. Students will probably find a strong positive correlation.**

Reteaching the Lesson

WORKING BACKWARD Have students work in pairs. All students should graph eight points that they think represent a set of data with a strong positive correlation, but not a perfect positive correlation. Partners should exchange graphs, and each should use a graphics calculator to find a linear-regression equation and correlation coefficient for the other's data. Partners should then discuss the results. Do they agree that there is a strong positive correlation? If not, they should work together to adjust the data. Repeat the activity for a strong negative correlation.

Teaching Tip

TECHNOLOGY You also can estimate the time in Example 2 without graphing the equation. First enter the data and find the linear-regression equation. Then, on the home screen, enter **40** STO▸ ALPHA X,T,θ,n ENTER. This stores 40 in memory as the value of x. Now press VARS and choose **5:Statistics...**. From the **EQ** menu, choose **1:RegEQ** (or **7:RegEQ**) and press ENTER. The predicted value for 1940 will appear on the home screen.

Teaching Tip

Data values that involve years are generally converted to a number of years *after* a selected year. This is done primarily to reduce the number of significant digits in the data values, and it often generates a regression equation that is slightly more accurate than if the years themselves were used. It also yields an equation that is easier to graph because the y-intercept is closer to the origin.

TRY THIS

about 13.66 minutes, or about 13:39.72

CRITICAL THINKING

The assumption is that the winning time in the year 2000 fits the correlation of the rest of the data.

Teaching Tips about technology refer to keystrokes for the TI-82, TI-83, or TI-83 Plus.

Teaching Tip

TECHNOLOGY To show the coefficient of determination, r^2, and the correlation coefficient, r, on the TI-83 graphics calculator, press 2nd **0** to show the catalog listing. Use the down arrow key, ▼, to scroll down to **DiagnosticOn** and then press ENTER twice. Now the coefficient of determination and the correlation coefficient will show under the coefficients when you calculate a regression model. To turn off this feature, select **DiagnosticOff** from the catalog listing and press ENTER twice. When using the TI-82, the correlation coefficient, r, is shown under the coefficients of linear, logarithmic, exponential and power regression models only.

A keystroke guide is provided at the end of each chapter for examples given in the Pupil's Edition. If you are using a different graphics calculator, such as a Casio or Sharp, go to the HRW Internet site at go.hrw.com and enter the keyword MB1 CALC to find keystroke guides.

TRY THIS Use the least-squares line in Example 2 to estimate the winning time in this Olympic event in the year 2000.

CRITICAL THINKING What assumption is made by using the least-squares line in the Try This exercise above?

Exercises

Communicate

For Exercises 1–3, decide whether each statement is true or false. If it is false, explain why.

1. A correlation coefficient can be equal to 3.
2. For a given data set, if the slope of the least-squares line is positive, then the correlation coefficient is positive.
3. A data set with a correlation coefficient of 0.2 has a stronger linear relationship than a data set with a correlation coefficient of –0.9.

APPLICATIONS

4. **DEMOGRAPHICS** As a population increases, the area available per person decreases. Give another example of a situation that you would expect to have a strong negative correlation with population.
5. **TAXES** As a population increases, the government revenue from taxes tends to increase. Give another example of a situation that you would expect to have a strong positive correlation with population.

Describe the correlation among data that have the given correlation coefficient.

6. $r = 0.02$
7. $r = -0.61$
8. $r = 0.96$

Guided Skills Practice

Create a scatter plot of the data in each table. Describe the correlation. Then find an equation for the least-squares line. ***(EXAMPLE 1)***

Calculator button indicates that a graphics calculator is recommended.

strong neg.; $y \approx -0.8x + 9.2$

9

x	5	0	2	6	9	4	5	3	6	4	2	6	1	7	5	2
y	6	8	7	5	2	5	7	8	3	6	8	4	9	3	5	8

fairly strong pos.; $y \approx 0.83x + 1.1$

10

x	4	4	0	5	2	9	7	6	1	8	2	7	8	3	9	5
y	2	6	1	6	1	7	5	7	2	7	2	9	9	5	8	7

no reliable correlation

11

x	1	6	9	8	2	7	4	9	1	3	6	5	0	5	8	3
y	2	5	8	2	8	1	7	9	9	1	9	4	9	5	1	2

APPLICATION

Ray Harroun, 1911

Jim Rathman, 1960

Arie Luyendyk, 1990

12 **SPORTS** The Indianapolis 500 auto race is held each year on Memorial Day. The table below gives the average speed, in miles per hour, of the winner for selected years from 1911 to 1996. In 1945, the race was not held. Estimate what could have been the average winning speed in 1945. Let $x = 0$ represent the year 1900. ***(EXAMPLE 2)*** **112.58 mph**

Year	Winner	Average speed	Year	Winner	Average speed
1911	Ray Harroun	74.602	1960	Jim Rathman	138.767
1915	Ralph DePalma	89.010	1965	Jimmy Clark	150.686
1920	Gaston Chevrolet	88.618	1970	Al Unser, Sr.	155.749
1925	Peter DePaolo	101.127	1975	Bobby Unser	149.213
1930	Billy Arnold	100.448	1980	Johnny Rutherford	142.862
1935	Kelly Petillo	106.240	1985	Danny Sullivan	152.982
1940	Wilbur Shaw	114.277	1990	Arie Luyendyk	185.984
1946	George Robson	114.820	1994	Al Unser, Jr.	160.872
1950	Johnnie Parsons	124.002	1995	Jacques Villenueve	153.616
1955	Bob Sweikert	128.209	1996	Buddy Lazier	147.956

[*Source: Sportsline USA, Inc., 1997*]

Practice and Apply

13. fairly strong pos.; $y \approx 0.8x + 1.36$

14. $y \approx -0.04 + .97x$
$x \approx 7$

13 Create a scatter plot of the data in the table below. Describe the correlation. Then find an equation for the least-squares line.

x	8	4	1	5	4	4	9	8	5	2	7	1	6	3	2	4
y	7	6	2	5	6	4	8	8	6	3	8	3	6	4	1	3

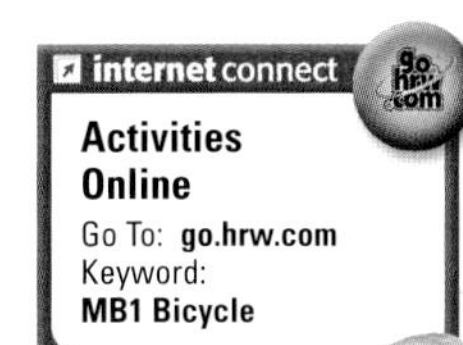

14 *Find an equation* for the least-squares line of the data below. Use the equation to predict the x-value that corresponds to a y-value of 7.

x	9	5	8	6	2	4	7	3	1	2	6	5	7	2	4	6
y	8	4	9	5	1	4	8	3	2	1	5	5	6	2	5	6

Match each correlation coefficient with one of the data sets graphed.

$r = 1$ $\quad r \approx 0.87$ $\quad r \approx 0.63$ $\quad r = -1$ $\quad r \approx -0.91$ $\quad r \approx -0.84$

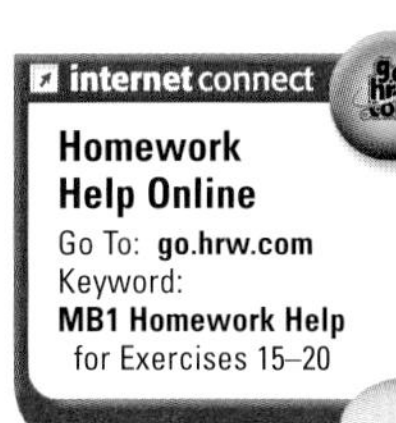

15. **$r = -1$**

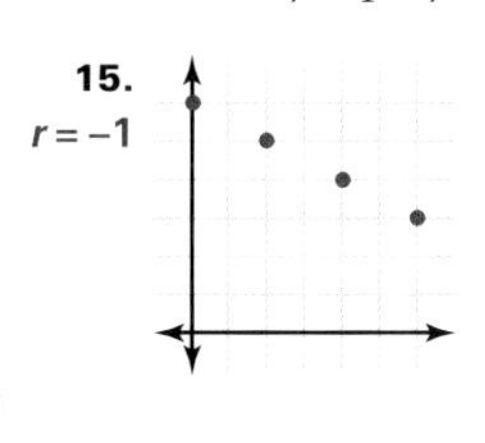

16. **$r \approx 0.87$**

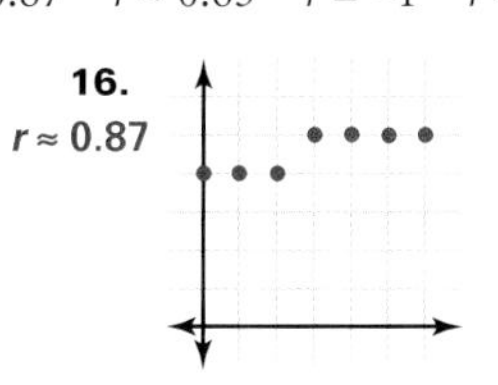

17. **$r = 1$**

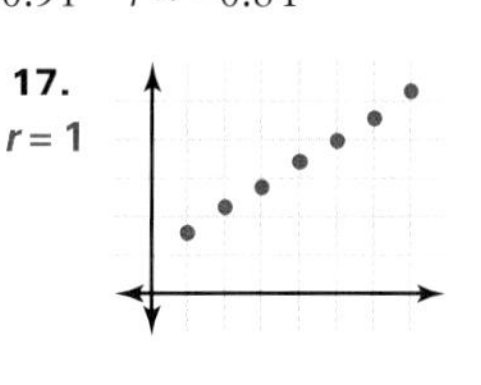

18. **$r \approx -0.84$**

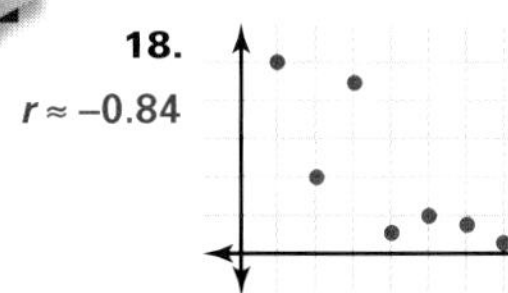

19. **$r \approx 0.63$**

20. **$r \approx -0.91$**

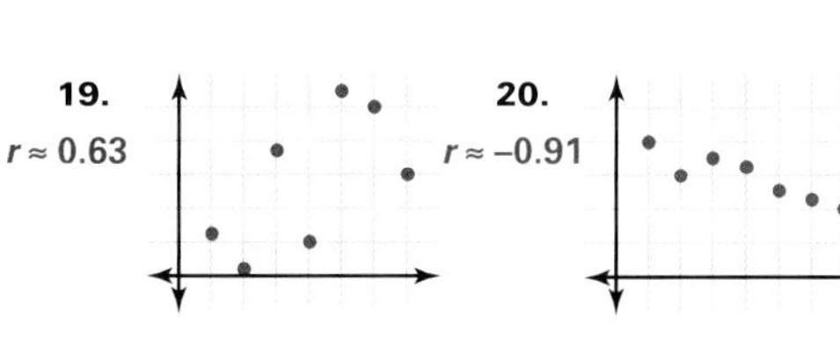

Assess

Selected Answers

Exercises 9–12, 13–35 odd

ASSIGNMENT GUIDE

In Class	1–12
Core	13–20, 23–25 odd
Core Plus	13–20, 21, 22–26 even
Review	27–35
Preview	36–39

Extra Practice can be found beginning on page 940.

Answers to odd-numbered Extra Practice exercises can be found immediately after Selected Answers in the Pupil's Edition.

Error Analysis

A graphics calculator will calculate a least-squares line for any set of data, and this leads some students to believe that a linear model is always appropriate. Remind them that the value of r is a critical piece of information. If r does not indicate a strong correlation, the linear model is not suitable. Tell students that they will learn about nonlinear models as the course progresses.

21. ***Interpolate*—to estimate values (of a function) between two known values.**

***Extrapolate*—to predict by projecting past known data.**

CHALLENGE

21. ETYMOLOGY Look for the words *interpolate* and *extrapolate* in a dictionary. For each word, write a definition that you think best applies to using a least-squares regression line to make predictions.

APPLICATIONS

22 MARKETING Sixteen people of various ages were polled and asked to estimate the number of CDs they had bought in the previous year. The following table contains the collected data:

Age	18	20	20	22	24	25	25	26	28	30	30	31	32	33	35	45
CDs	12	15	18	12	10	8	6	6	4	4	4	2	2	3	6	1

a. Let x represent age, and let y represent the number of CDs purchased. Enter the data in a graphics calculator, and find the equation of the least-squares line. **$y \approx -0.6x + 23.7$**

b. Find the correlation coefficient, r, to the nearest tenth. Explain how the value of r describes the data. **$r \approx -0.8$; fairly strong negative**

c. Use the least-squares line to predict the number of CDs purchased by a person who is 27 years old. **about 8 CDs**

d. Use the least-squares line to predict the age of a person who purchased 15 CDs in the previous year. **about 14 or 15 years old**

23 ANATOMY The following tables give the height and shoe size of some adults. Let x represent height, in inches, and let y represent shoe size.

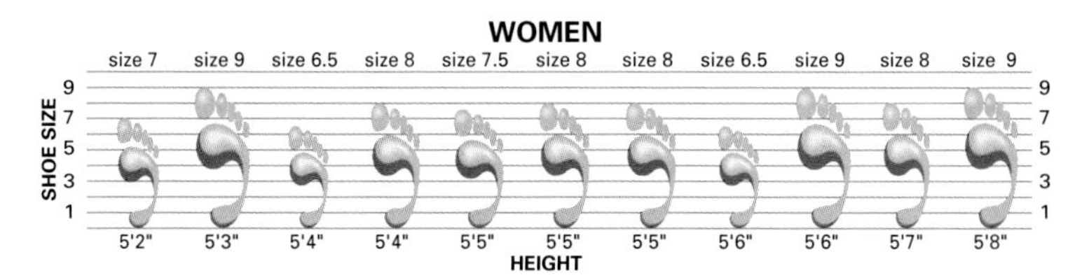

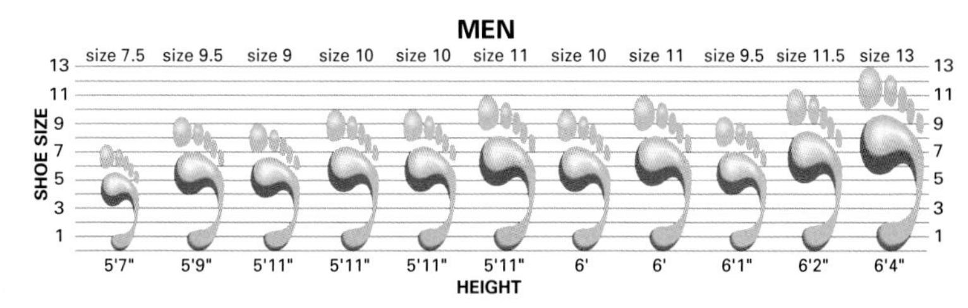

a. Enter the data for women, and find the equation of the least-squares line. **$y \approx 0.17x - 2.97$**

$y \approx 0.52x - 26.91$ **b.** Enter the data for men, and find the equation of the least-squares line.

c. Find the correlation coefficients for the women's data and for the men's data. Explain how the different correlation coefficients describe the two data sets. **women: $r \approx 0.3$, weak pos.; men: $r \approx 0.9$, strong pos.**

d. Use the appropriate least-squares line to predict the shoe size of a woman who is 5'1" tall. **about $7\frac{1}{2}$**

e. Use the appropriate least-squares line to predict the shoe size of a man who is 6'2" tall. **about $11\frac{1}{2}$**

f. Use the appropriate least-squares line to predict the height of a man who wears size 12 shoes. **about 6'3"**

g. Use the appropriate least-squares line to predict the height of a woman who wears size 8 shoes. **about 5'5"**

Practice

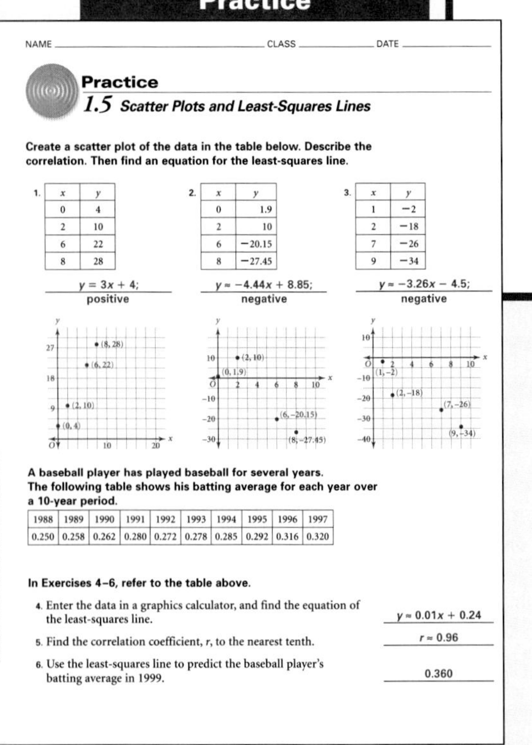

NAME ______ CLASS ______ DATE ______

Practice
***1.5* Scatter Plots and Least-Squares Lines**

Create a scatter plot of the data in the table below. Describe the correlation. Then find an equation for the least-squares line.

1.

x	y
0	4
2	10
6	22
8	28

$y = 3x + 4$; positive

2.

x	y
0	1.9
2	10
6	−20.15
8	−27.45

$y \approx -4.44x + 8.85$; negative

3.

x	y
1	−2
2	−18
7	−26
9	−34

$y \approx -3.26x - 4.5$; negative

A baseball player has played baseball for several years. The following table shows his batting average for each year over a 10-year period.

1988	1989	1990	1991	1992	1993	1994	1995	1996	1997
0.250	0.258	0.262	0.280	0.272	0.278	0.285	0.292	0.316	0.320

In Exercises 4–6, refer to the table above.

4. Enter the data in a graphics calculator, and find the equation of the least-squares line. $y \approx 0.01x + 0.24$

5. Find the correlation coefficient, r, to the nearest tenth. $r \approx 0.96$

6. Use the least-squares line to predict the baseball player's batting average in 1999. 0.360

L. Douglas Wilder served as governor of Virginia from 1990 to 1994. He was the first elected African American governor in U.S. history.

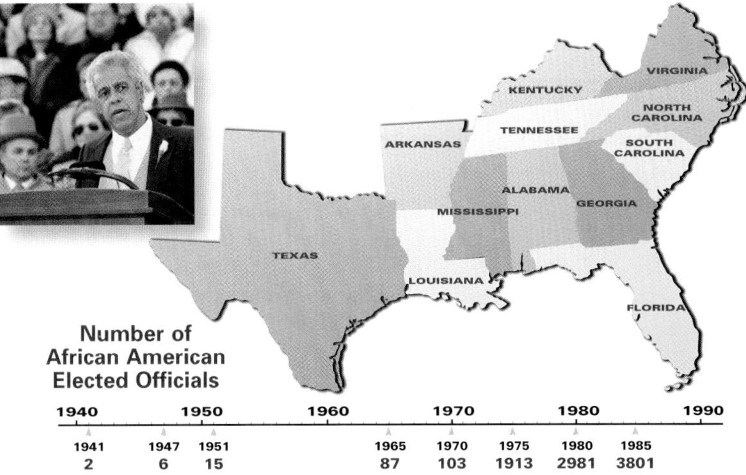

APPLICATIONS

24 **GOVERNMENT** The time line above shows the number of African American elected officials in Southern states from 1941 to 1985.

a. Create a scatter plot for this information, with the years on the x-axis. Let $x = 0$ represent 1900.

b. Find the least-squares line for the data from 1941 to 1965.

c. Find the least-squares line for the data from 1965 to 1985.

d. Explain how the slopes of the two lines are different. What happened in the 1960s that might explain the extreme change?

24a.

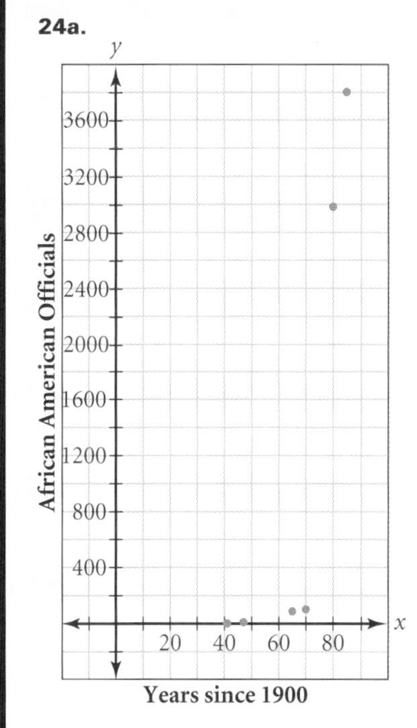

b. $y \approx 3.76x - 164.40$

c. $y \approx 206.12x - 13{,}682$

d. The slope from the data in part **c** is far steeper than the slope from the data in part **b.** The dramatic increase in African American officials elected beginning in the 1960s is due to the Civil Rights Act, which was passed in 1964.

25 **HEALTH** The table below gives information about cigarette smokers between the ages of 18 and 24 in the United States for selected years from 1965 to 1993.

a. Enter the data for females, and find the equation of the least-squares line. Let $x = 0$ represent 1900.

25a. $y \approx -0.57x + 77.12$

b. Enter the data for males, and find the equation of the least-squares line.

b. $y \approx -0.94x + 111.97$

c. Find the correlation coefficients for the female's data and for the male's data. Explain what the different correlation coefficients tell you about the two data sets.

c. female: $r \approx -0.89$
male: $r \approx -0.95$
Male's data set is more linear.

Percent of Population That Are Cigarette Smokers in the United States

Year	Female (18–24)	Male (18–24)
1965	38.1	54.1
1974	34.1	42.1
1979	33.8	35.0
1983	35.5	32.9
1985	30.4	28.0
1987	26.1	28.2
1990	22.5	26.6
1992	24.9	28.0
1993	22.9	28.8

[*Source: Statistical Abstract of the United States, 1996*]

d. Use the appropriate least-squares line to estimate the percent of females between the ages of 18 and 24 who were smokers in 1980.

d. about 32%

e. Use the appropriate least-squares line to estimate the percent of males between the ages of 18 and 24 who were smokers in 1970.

e. about 46%

f. Use the appropriate least-squares line to estimate the year in which 50% of males between the ages of 18 and 24 were smokers.

f. about 1966

g. Use the appropriate least-squares line to estimate the year in which 22% of males between the ages of 18 and 24 were smokers.

g. about 1996

Student Technology Guide

NAME ______ CLASS ______ DATE ______

Student Technology Guide
1.5 *Scatter Plots and Least-Squares Lines*

Actual and projected elementary and secondary school enrollment from 1996 to 2006 for the United States is given in the table below. Each data value is in thousands of students.

Year	1996	1997	1998	1999	2000	2006
Enrollment	51,484	52,217	52,725	53,132	53,465	54,388

You can use a graphics calculator to make a scatter plot of the data.

To clear all functions from the function list, press STAT 4: ClrList ENTER 2nd 1 , 2nd 2 ENTER. To enter the data, press STAT 1: Edit... ENTER. Using the arrow keys, highlight the entry in row 1 of list L1. To enter the x-values, press 6 ENTER 7 ENTER ... 10 ENTER 16 ENTER. Using the arrow keys, highlight the entry in row 1 of list L2. To enter the y-values, press 51484 ENTER 52217 ENTER ... 53465 ENTER 54388 ENTER.

To make the necessary range settings, press WINDOW. Choose $0 \le x \le 16$ and $50{,}000 \le y \le 55{,}000$.

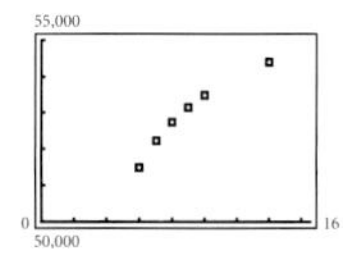

To make the scatter plot, press 2nd Y= 1: Plot1...On ENTER On Type: ENTER GRAPH.

You can also find an equation for the least-squares line. Press STAT CALC 4: LinReg(ax + b) ENTER ENTER.

For the enrollment data, $y \approx 264.73x + 50{,}431.01$.

Choose a base year for the data. Then use a graphics calculator to write an equation for the least-squares line of each data set. Answers may vary.

1. Energy consumption in the United States in quadrillion BTUs

Year	1980	1985	1990	1992	1994	1996
BTUs	75.96	73.98	81.27	82.14	85.59	89.81

sample: 1980; $y \approx 0.89x + 72.97$

2. College enrollment in the United States in thousands

Year	1985	1990	1995	1996
Number	12,247	13,189	14,262	13,917

sample: 1980; $y \approx 170.84x + 11{,}439.12$

In Exercises 36–39, students investigate whether the operations of addition, subtraction, multiplication, and division are commutative. They will study the commutative property and other properties of real numbers in Lesson 2.1.

ALTERNATIVE

Assessment

Portfolio Activity

The Portfolio Activity can be used as preparation for the Chapter Project or as a separate activity. In the Portfolio Activity on this page, students use a graphics calculator to find an equation of the least-squares line for their portfolio data set. Then they compare this equation to the equation of the line of best fit that they estimated visually.

Answers to Portfolio Activities can be found in Additional Answers of the Teacher's Edition.

APPLICATION

26 **AGRICULTURE** The table below gives the number of acres in an average farm in the United States from 1940 to 1995.

Year	1940	1950	1960	1970	1980	1995
Number of acres	174	213	297	374	426	469

[*Source: The World Almanac, 1997*]

a. Let $x = 0$ represent the year 1900, and let y represent the number of acres in an average farm in the United States. Enter the data, and find the equation of the least-squares line. **$y \approx 5.78x - 54.76$**

b. Find the correlation coefficient, r, to the nearest tenth. Explain what the value of r tells you about the data. **$r \approx 1.0$; extremely strong pos.**

c. Use the least-squares line to estimate the number of acres in an average farm in 1955. **≈263**

d. Use the least-squares line to predict the year in which there were about 325 acres in an average farm. **about 1966**

Without using a calculator, write an equivalent decimal for each fraction.

27. $\frac{1}{3}$ **$0.\overline{3}$** **28.** $-\frac{3}{5}$ **−0.6** **29.** $\frac{17}{4}$ **4.25** **30.** $\frac{5}{3}$ **$1.\overline{6}$** **31.** $\frac{7}{20}$ **0.35**

Without using a calculator, evaluate each expression. Write your answer as a decimal.

32. $5 \div 2$ **2.5** **33.** $5 \div 0.2$ **25** **34.** $5 \div 0.02$ **250**

35. The line whose equation is $y = -1.6x + 1$ is parallel to another line whose equation is $y = mx - 4$. Find m. **(LESSON 1.3)** **$m = -1.6$**

Look Beyond

Determine whether each equation is true when *a, b,* and *x* are real numbers.

36. $ax + bx = bx + ax$ **true** **37.** $ax - bx = bx - ax$ **false**

38. $ax \cdot bx = bx \cdot ax$ **true** **39.** $ax \div bx = bx \div ax$ **false**

1. Use a graphics calculator to find the equation of the least-squares line for your portfolio data set.
2. Plot the least-squares line on the same coordinate plane as your portfolio data points and with the linear model that you created in the Portfolio Activity on page 36.
3. Compare the slope of the least-squares line with the slope of your linear model.
4. What is the correlation coefficient for your least-squares line?

WORKING ON THE CHAPTER PROJECT

You should now be able to complete Activity 3 of the Chapter Project.

Introduction to Solving Equations

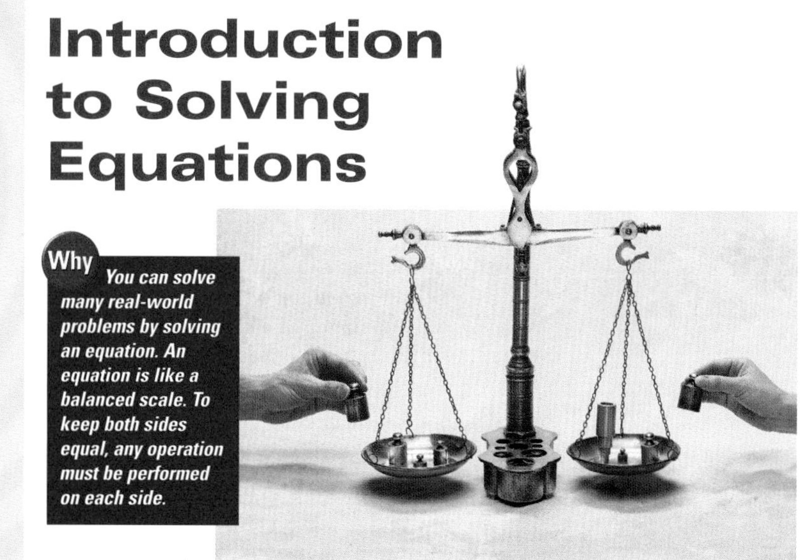

Why *You can solve many real-world problems by solving an equation. An equation is like a balanced scale. To keep both sides equal, any operation must be performed on each side.*

Objectives

- Write and solve a linear equation in one variable.
- Solve a literal equation for a specified variable.

An **equation** is a statement that two expressions are equal. An equation usually contains one or more variables. A **variable** is a symbol that represents many different numbers in a set of numbers.

Equation in *one* variable, w: $12w = 10$
Equation in *two* variables, x and y: $2x + 3y = 12$

Any value of a variable that makes an equation true is called a **solution of the equation.** For example, $12w = 10$ is an equation in one variable, w. Because $\frac{5}{6}$ satisfies the equation, $\frac{5}{6}$ is a solution.

$$12w = 10$$
$$12\left(\frac{5}{6}\right) \stackrel{?}{=} 10$$
$$10 = 10 \quad \textbf{True}$$

To solve equations, the *Properties of Equality,* shown below, or the *Substitution Property,* shown on page 46, may be used.

Properties of Equality

For real numbers a, b, and c:

Reflexive Property	$a = a$
Symmetric Property	If $a = b$, then $b = a$.
Transitive Property	If $a = b$ and $b = c$, then $a = c$.
Addition Property	If $a = b$, then $a + c = b + c$.
Subtraction Property	If $a = b$, then $a - c = b - c$.
Multiplication Property	If $a = b$, then $ac = bc$.
Division Property	If $a = b$, then $\frac{a}{c} = \frac{b}{c}$, where $c \neq 0$.

Alternative Teaching Strategy

USING MANIPULATIVES Have students use algebra tiles to model the equation $2x - 8 = 5x + 7$. They should place 2 x-tiles and 8 negative unit tiles to the left of the equal sign and 5 x-tiles and 7 positive unit tiles to the right.

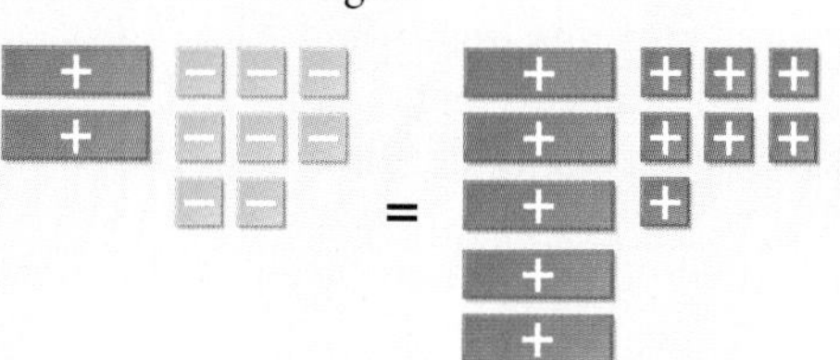

Now have them model the solution process as follows:

1. Add 2 negative x-tiles to each side.
2. Remove 2 *neutral pairs* of x-tiles from each side.
3. Add 7 negative unit tiles to each side.
4. Remove 7 neutral pairs of unit tiles from the right side.
5. Divide the tiles on each side of the equal sign into 3 groups of equal size.

Each x-tile can be matched with a group of 5 negative unit tiles, so the solution is $x = -5$.

Prepare

NCTM PRINCIPLES & STANDARDS 1–4, 6, 8–10

Quick Warm-Up

Solve each equation.

1. $-12 + r = 3$ $r = 15$
2. $-12b = 3$ $b = -0.25$
3. $\frac{k}{-12} = 3$ $k = -36$
4. $\frac{-12}{z} = 3$ $z = -4$
5. $-12 - m = 3$ $m = -15$
6. $-12 = 5x + 3$ $x = -3$

Also on Quiz Transparency 1.6

Teach

Why Ask students to give examples of equations that they can solve easily by using mental math. Have them describe some of the thought processes they use to arrive at their solutions. Point out that many equations which are used to model real-world situations are too difficult to solve by using mental math alone. The properties of equality make it possible to find solutions to more complicated equations.

Substitution Property

If $a = b$, you may replace a with b in any true statement containing a and the resulting statement will still be true.

In an expression, such as $5 + 3x - x - 1$, the parts that are added or subtracted are called **terms**. The terms $3x$ and x are called **like terms** because they contain the *same form of the variable* x. The constant terms, 5 and 1, are also like terms. An expression is **simplified** when all the like terms have been combined and all the parentheses have been removed.

$$5 + 3x - x - 1$$
$$= 2x + 4 \quad \textit{simplified}$$

EXAMPLE 1 The relationship between the Celsius temperature, C, and the Fahrenheit temperature, F, is given by $F = \frac{9}{5}C + 32$.

APPLICATION TEMPERATURE

Find the Celsius temperature that is equivalent to 86°F.

SOLUTION

$$F = \frac{9}{5}C + 32$$
$$86 = \frac{9}{5}C + 32 \quad \textit{Substitute 86 for F.}$$
$$86 - 32 = \frac{9}{5}C + 32 - 32 \quad \textit{Use the Subtraction Property.}$$
$$54 = \frac{9}{5}C \quad \textit{Simplify.}$$
$$\left(\frac{5}{9}\right)54 = \left(\frac{5}{9}\right)\left(\frac{9}{5}C\right) \quad \textit{Use the Multiplication Property.}$$
$$30 = C \quad \textit{Simplify.}$$

Thus, 30°C is equivalent to 86°F.

ADDITIONAL EXAMPLE 1

Using the equation given in Example 1, find the Celsius temperature that is equivalent to 122°F. 50°C

EXAMPLE 2 **Solve $2x + 7 = 5x - 9$. Check your solution by using substitution.**

SOLUTION

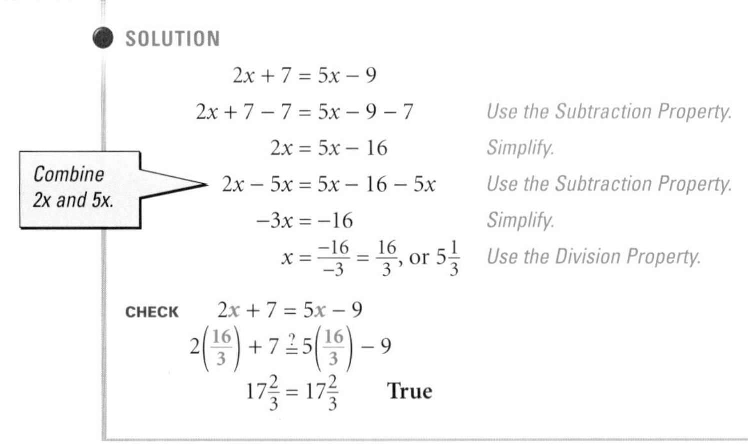

$$2x + 7 = 5x - 9$$
$$2x + 7 - 7 = 5x - 9 - 7 \quad \textit{Use the Subtraction Property.}$$
$$2x = 5x - 16 \quad \textit{Simplify.}$$
$$2x - 5x = 5x - 16 - 5x \quad \textit{Use the Subtraction Property.}$$
$$-3x = -16 \quad \textit{Simplify.}$$
$$x = \frac{-16}{-3} = \frac{16}{3}, \text{ or } 5\frac{1}{3} \quad \textit{Use the Division Property.}$$

CHECK
$$2x + 7 = 5x - 9$$
$$2\left(\frac{16}{3}\right) + 7 \stackrel{?}{=} 5\left(\frac{16}{3}\right) - 9$$
$$17\frac{2}{3} = 17\frac{2}{3} \quad \textbf{True}$$

ADDITIONAL EXAMPLE 2

Solve $3x - 8 = 5x - 20$. Check your solution by using substitution. $x = 6$

Interdisciplinary Connection

CONSUMER EDUCATION When buying an expensive item, many people pay just part of the purchase price as a *down payment* and take out a loan for the remainder. Then they pay off the loan in a number of equal monthly payments. Thus, the actual cost of the purchase, C, is given by the formula $C = d + np$, where d is the amount of the down payment, n is the number of monthly payments, and p is the amount of each monthly payment. Solve $C = d + np$ for p. $p = \frac{C - d}{n}$

Inclusion Strategies

ENGLISH LANGUAGE DEVELOPMENT Students often have difficulty translating the conditions of a real-world problem into the symbols of an equation. To provide them with a different perspective, reverse the process. That is, give students a linear equation in one variable and have them create real-world problems that can be solved by using it. Ask volunteers to share their problems with the class. In this way, students can appreciate the fact that a single equation may be a model for several different situations.

TRY THIS Solve $3x + 12 = -5x + 24$. Check your solution by using substitution.

An algebraic solution method was shown in Example 2. In the Activity below, you can explore a graphic solution method for solving equations.

TECHNOLOGY GRAPHICS CALCULATOR
Keystroke Guide, page 82

Exploring Graphic Solution Methods

You will need: a graphics calculator

1. In the equation $x + 3 = 9 - 2x$, what two expressions are equal?
2. Use a graphics calculator to graph $y = x + 3$ and $y = 9 - 2x$ on the same screen. For what value of x do $x + 3$ and $9 - 2x$ have the same value?
3. Check to see if this value is the solution to the original equation.

CHECKPOINT ✔ 4. Describe how to solve $2x - 1 = 2 - x$ by using a graphics calculator.

TRY THIS
$x = \frac{3}{2}$

Activity Notes

In this Activity, students investigate a method for finding the solution to a one-variable equation by graphing two linear equations in two variables. Students should observe that the x-coordinate of the point of intersection of the two graphs is the solution of the given equation.

CHECKPOINT ✔
4. Graph $y = 2x - 1$ and $y = 2 - x$ on the same screen, and find the x-coordinate of the point where the graphs intersect: $x = 1$.

Use Teaching Transparency 4.

EXAMPLE 3 **Solve $3.24x - 4.09 = -0.72x + 3.65$ by graphing.**

SOLUTION

Write the original equation as the pair of equations below.

$$3.24x - 4.09 = -0.72x + 3.65$$
$$y = 3.24x - 4.09 \quad \text{and} \quad y = -0.72x + 3.65$$

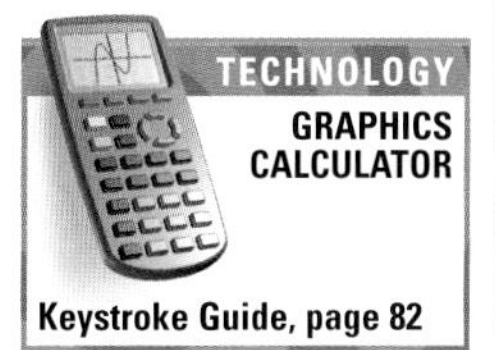
TECHNOLOGY GRAPHICS CALCULATOR
Keystroke Guide, page 82

Graph the two equations on the same screen, and find the point of intersection.

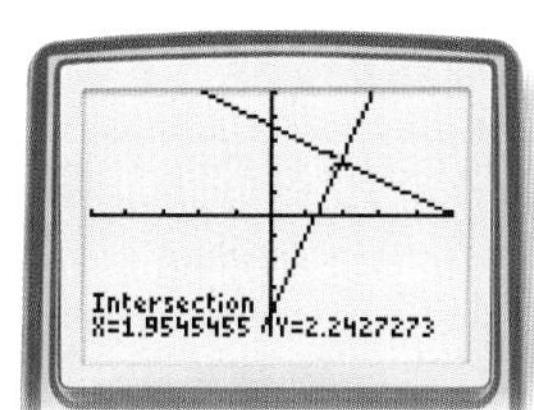

Read the x-coordinate of the point where the graphs intersect.

From the calculator display, the solution is $x \approx 1.95$.

TRY THIS Solve $2.24x - 6.24 = 4.26x - 8.76$ by graphing.

ADDITIONAL EXAMPLE 3

Solve the equation $6.25x - 3.36 = -0.8x + 8.9$ by graphing.

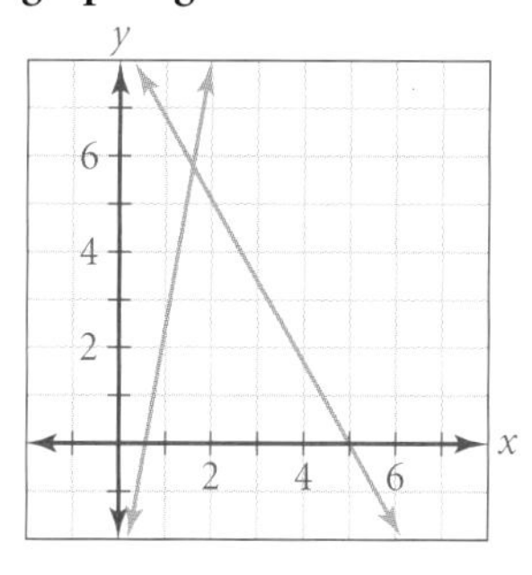

$x \approx 1.74$

TRY THIS
$x \approx 1.25$

Literal Equations

A **literal equation** is an equation that contains two or more variables.

CONNECTION GEOMETRY

Formulas are examples of literal equations. The following examples of literal equations are from geometry:

Volume of a cube, V:
$$V = s^3, \text{ where } s \text{ is the side length}$$
Area of a circle, A:
$$A = \pi r^2, \text{ where } r \text{ is the radius}$$
Volume of a square pyramid, V:
$$V = \frac{1}{3}s^2h, \text{ where } s \text{ is the side length and } h \text{ is the altitude}$$

Enrichment

Replace each __?__ in the equations below with a number or expression so that the solution to the resulting equation is

a. −1. **b.** 0.
c. any real number. **d.** no real number.

1. $5c =$ __?__
 a. −5 b. 0 c. $5c$ d. $5c + 1$
2. $3t + 8 =$ __?__ $+ 1$
 a. 4 b. 7 c. $3t + 7$ d. $3t$
3. $4y +$ __?__ $= -2y - 9$
 a. −3 b. −9 c. $-6y - 9$ d. $-6y$

Reteaching the Lesson

USING DISCUSSION Have students compare different ways of solving the same equation. For instance, the equation in Example 2 can be solved as shown at right. Some students may find this solution more "friendly" because it results in a positive coefficient for x.

$$\begin{aligned} 2x + 7 &= 5x - 9 \\ 2x + 7 - 2x &= 5x - 9 - 2x \\ 7 &= 3x - 9 \\ 7 + 9 &= 3x - 9 + 9 \\ 16 &= 3x \\ \frac{16}{3} &= x \\ x &= \frac{16}{3} \end{aligned}$$

Math CONNECTION

GEOMETRY This may be the first time that students think of formulas from geometry as literal equations. They will be asked to solve some geometric formulas for an indicated variable in Exercises 47, 48, 50, 51, and 63.

ADDITIONAL EXAMPLE 4

Solve $a = \frac{V - V_0}{t}$ for V_0.

$V_0 = V - at$

CRITICAL THINKING

$a = \frac{12c}{d - c}$

CHECKPOINT ✓

Substitute 30 for C.

$86 = \frac{9}{5}C + 32$ $\quad 54 = \frac{9}{5}C$

$86 = \frac{9}{5}(30) + 32$ $\quad 54 = \frac{9}{5}(30)$

$86 = 86$ $\quad 54 = 54$

$C = 30$

$30 = 30$

Assess

Selected Answers

Exercises 6–11, 13–83 odd

ASSIGNMENT GUIDE	
In Class	1–11
Core	13–39 odd, 41–46, 47–61 odd, 65, 69
Core Plus	12–40 even, 41–46, 48–62 even, 64–71
Review	72–84
Preview	85–88

Extra Practice can be found beginning on page 940.

Answers to odd-numbered Extra Practice exercises can be found immediately after Selected Answers in the Pupil's Edition.

APPLICATION MEDICINE

Young's formula is used to relate a child's dose of a medication to an adult's dose of the same medication. The formula applies to children from 1 to 12 years old.

$$\frac{a}{a+12} \times d = c, \text{ where } \begin{cases} a \text{ represents the child's age} \\ d \text{ represents the adult's dose} \\ c \text{ represents the child's dose} \end{cases}$$

EXAMPLE 4 **Solve $\frac{a}{a+12} \times d = c$ for d.**

SOLUTION

$$\frac{a}{a+12} \times d = c$$

$(a+12)\frac{a}{a+12} \times d = (a+12)c$ — *Use the Multiplication Property.*

$ad = c(a+12)$ — *Simplify and use the Commutative Property.*

$d = \frac{c(a+12)}{a}$ — *Use the Division Property.*

CRITICAL THINKING Solve $\frac{a}{a+12} \times d = c$ for a.

CHECKPOINT ✓ Two equations are **equivalent** if they have the same solution.

Use substitution to verify that the following equations are equivalent:

$86 = \frac{9}{5}C + 32$ $\qquad 54 = \frac{9}{5}C$ $\qquad C = 30$

Exercises

Communicate

Tell which Properties of Equality you would use to solve each equation.

1. $52 = -2.7x - 3$ **2.** $\frac{x}{5} = x + 2.2$ **3.** $x - 5 = -2x - 2$

4. Describe one way to obtain an equation that is equivalent to $4x - 7 = 14$.

5. Describe how to solve $\frac{2(x+3)}{7} = \frac{9(x-3)}{5}$ by graphing.

Guided Skills Practice

Solve each equation. Check your solution. ***(EXAMPLES 1 AND 2)***

6. $4x + 12 = 20$ **2** **7.** $\frac{x}{5} + 3 = 4$ **5**

8. $-\frac{5}{2}x + \frac{5}{2} = 2 - 3x$ **−1** **9.** $7 - 6x = 2x - 9$ **2**

Calculator button indicates that a graphics calculator is recommended.

10 Solve $\frac{4(x+5)}{3} = \frac{-3(x-7)}{5}$ by graphing. ***(EXAMPLE 3)*** **$x \approx -1.28$**

11. Solve $Ax + By = C$ for y. ***(EXAMPLE 4)*** **$y = -\frac{A}{B}x + \frac{C}{B}$**

Practice and Apply

Solve each equation.

12. $1 = 2x - 5$ **3** **13.** $-2x - 7 = 9$ **-8** **14.** $2x - 1 = -5$ **-2**

15. $3x - 3 = 5$ $\mathbf{\frac{8}{3}}$ **16.** $2x - 5 = 19$ **12** **17.** $5x - 3 = 12$ **3**

18. $20 = 6x - 10$ **5** **19.** $4 - 5x = 19$ **-3** **20.** $3x + 1 = \frac{1}{2}$ $\mathbf{-\frac{1}{6}}$

21. $4x + 80 = -6x$ **-8** **22.** $5x + 15 = 2x$ **-5** **23.** $7x = -2x + 5$ $\mathbf{\frac{5}{9}}$

24. $5x + 3 = 2x + 18$ **5** **25.** $-4x - 3 = x + 7$ **-2** **26.** $3x - 8 = 2x + 2$ **10**

27. $\frac{1}{5}x + 3 = 2$ **-5** **28.** $\frac{1}{4}x - \frac{5}{2} = -2$ **2** **29.** $\frac{1}{6}x + \frac{3}{2} = 2$ **3**

30. $0 = \frac{1}{2}x + 2$ **-4** **31.** $-\frac{3}{5}x + 12 = 4$ $\mathbf{\frac{40}{3}}$ **32.** $-5 = \frac{3}{2}x - 2$ **-2**

33. $\frac{1}{3}x = -x + 4$ **3** **34.** $x - 5 = -\frac{3}{2}x + \frac{5}{2}$ **3**

35. $-\frac{1}{3}x + 1 = \frac{3}{2}x - 1$ $\mathbf{\frac{12}{11}}$ **36.** $-2x + 5 = -\frac{1}{3}x - 6$ $\mathbf{\frac{33}{5}}$

37. $\frac{2}{3}x - 9 = -\frac{1}{2}x + 4$ $\mathbf{\frac{78}{7}}$ **38.** $\frac{1}{4}x - 3 = 6x$ $\mathbf{-\frac{12}{23}}$

39. $\frac{1}{3}x - \frac{4}{3} = -\frac{1}{6}x - 1$ $\mathbf{\frac{2}{3}}$ **40.** $\frac{2}{5}x + \frac{6}{5} = x - 3$ **7**

Solve each equation by graphing. Give your answers to the nearest hundredth.

internet connect 

Homework Help Online
Go To: **go.hrw.com**
Keyword:
MB1 Homework Help for Exercises 41–46

41 $0.24x + 1.1 = 2.56x - 1.5$ **1.12** **42** $1.05x - 4.28 = -2.65x + 4.1$ **2.26**

43 $-0.75x + 12.42 = 4.36$ **10.75** **44** $0.35x - 2.72 = 5.83x$ **-0.50**

45 $0.67x - 8.75 = -0.48x + 3.99$ **11.08** **46** $5.9(0.33x - 1.33) = -1.03x - 5.72$ **0.71**

Solve each literal equation for the indicated variable.

47. $\frac{1}{2}bh = A$ for b **48.** $P = 2l + 2w$ for w

49. $\frac{1}{R} = \frac{1}{r_1} + \frac{1}{r_2}$ for r_2 **50.** $A = \frac{1}{2}h(b_1 + b_2)$ for b_2

51. $A = \frac{1}{2}h(b_1 + b_2)$ for h **52.** $y = \frac{u+1}{u+2}$ for u

53. $ax + b = cx + d$ for x **54.** $ax + b = cx + d$ for d

55. $I = P(1 + rt)$ for r **56.** $I = P(1 + rt)$ for t

47. $b = \frac{2A}{h}$
48. $w = \frac{P - 2l}{2}$
49. $r_2 = -\frac{r_1R}{R - r_1}$ *or* $\frac{Rr_1}{r_1 - R}$
50. $b_2 = \frac{2A}{h} - b_1$
51. $h = \frac{2A}{b_1 + b_2}$
52. $u = -\frac{2y - 1}{y - 1}$ *or* $\frac{1 - 2y}{y - 1}$
53. $x = -\frac{b - d}{a - c}$ *or* $\frac{d - b}{a - c}$

Solve each literal equation for v.

57. $x = vt$ $\mathbf{v = \frac{x}{t}}$ **58.** $x = vt + \frac{1}{2}at^2$ **59.** $y = \frac{1}{2}xv$ $\mathbf{v = \frac{2y}{x}}$

60. Given the equation $y = 4x + 7$, use substitution to solve $-2x + y = 19$ for x. **6**

61. Given the equation $x = -y + 9$, use substitution to solve $3x - 5y = 59$ for x. **13**

CONNECTIONS

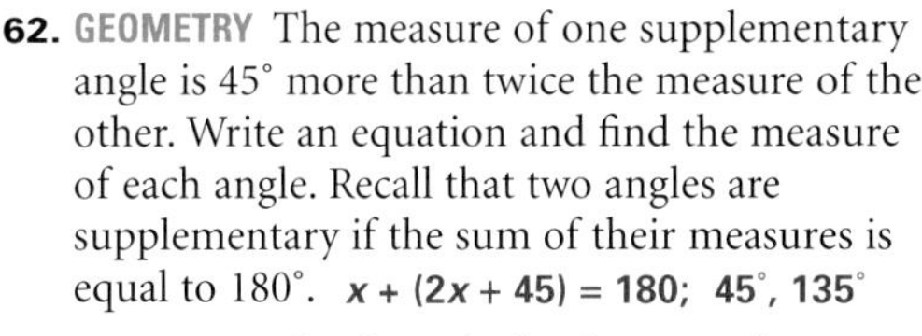

62. GEOMETRY The measure of one supplementary angle is 45° more than twice the measure of the other. Write an equation and find the measure of each angle. Recall that two angles are supplementary if the sum of their measures is equal to 180°. **$x + (2x + 45) = 180$; 45°, 135°**

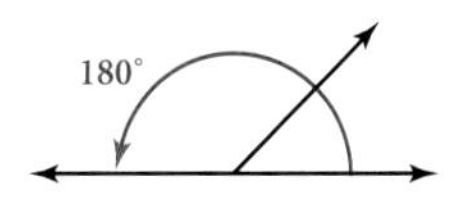

CHALLENGE

63. GEOMETRY The formula for the area of a cone in terms of the slant height, s, and the radius of the base, r, is $A = \pi rs + \pi r^2$. Write a formula for the slant height of a cone in terms of its area and the radius of its base. $\mathbf{s = \frac{A - \pi r^2}{\pi r}}$

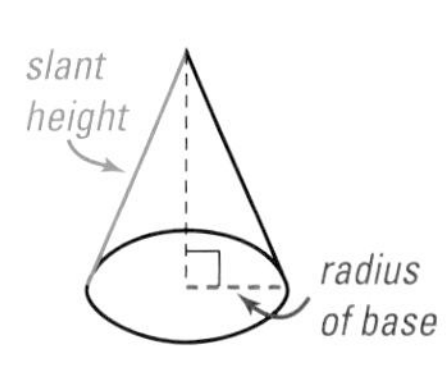

54. $d = (a - c)x + b$

55. $r = \frac{I - P}{Pt}$

56. $t = \frac{I - P}{Pr}$

58. $v = \frac{2x - at^2}{2t}$ *or* $\frac{x - \frac{1}{2}at^2}{t}$

Error Analysis

When using a graphics calculator to solve an equation in one variable, students may become confused and select the y-coordinate of the point of intersection as the solution to the given equation. Stress the importance of using substitution to check that the proposed solution satisfies the original equation.

APPLICATIONS **Write and solve an appropriate equation for each situation.**

64. RECREATION A summer carnival charges a \$2 admission fee and \$0.50 for each ride. If Tamara has \$10 to spend, how many rides can she go on? **16**

65. TAXES Aaron's mother purchases a new computer for \$1750. If she claims a linear depreciation (loss of value) on the computer at a rate of \$250 per year, how long will it take for the value of the computer to be \$0? **7 yr**

66. CONSUMER ECONOMICS The receipt for repairs on Victor's car is shown at right.

a. Write an equation to model the total bill in terms of parts and labor. $\mathbf{3.5x + 72 = 272}$

b. What hourly rate does the repair shop charge for labor? **about \$57.14/hr**

AUTO REPAIR
#111 Auto Lane
Lana, TX 78787

INVOICE

Date: July 7

ITEM	AMOUNT
Parts:	
Brake Fluid	\$ 6.00
Wheel Cylinder	\$28.50
Rear Brake Shoes	\$20.00
Front Brake Pads	\$15.00
Shop Supplies	\$2.50
Labor	3.5 hours
TOTAL	\$ 272.00

67. INCOME Louis has two different job offers for a position in shoe sales. One pays \$25 per week plus a \$2 commission for each pair of shoes sold. The second job pays \$40 per week plus a \$1.50 commission for each pair of shoes sold. How many shoes would Louis have to sell to make the same total salary in either job? **30 pairs**

68. BANKING Carmen has taken out a loan for \$800 to buy a car. She plans to pay back the loan at a rate of \$40 per month. Ramona has borrowed \$500 to buy a car, which she plans to pay back at a rate of \$20 per month.

a. How long will it take Carmen to pay back her loan? **20 months**

b. How long will it take Ramona to pay back her loan? **25 months**

c. If Carmen and Ramona take out their loans at the same time, how long will it take for their remaining balances to be equal? What are their remaining balances after this amount of time? **15 months; \$200**

69. INCOME Amelia has a job baby-sitting for a neighbor. She is paid \$20 per week plus \$2.50 for each hour on the job. If Amelia wants to earn \$40 to buy a new sweater, how many hours would she need to work? **8 hr**

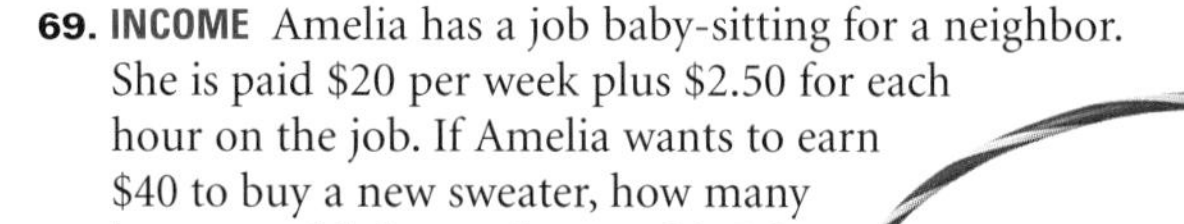

Practice

NAME ______________ CLASS ________ DATE ________

Practice
1.6 ***Introduction to Solving Equations***

Solve each equation.

1. $4x + 4(2x - 1) = 20$ $\underline{x = 2}$
2. $4x + 20 = 5(x + 3)$ $\underline{x = 5}$
3. $5x + 15 = 10(x - 3)$ $\underline{x = 9}$
4. $2x + 5 = 17$ $\underline{x = 6}$
5. $3x - 4 = 4(3x - 19)$ $\underline{x = 8}$
6. $3(2x - 4) = 3x - 5(x + 1)$ $\underline{x = \frac{7}{8}}$
7. $-0.4x - 6(3x - 2) = 48.8$ $\underline{x = -2}$
8. $2(x + 3) = 5x + 15$ $\underline{x = -3}$
9. $2x + 1 = 7 - 10x$ $\underline{x = \frac{1}{2}}$
10. $5x - 3 = 15 - 4x$ $\underline{x = 2}$
11. $4x - 10 = 3(x + 2)$ $\underline{x = 16}$
12. $6(x + 2) = 5x - 9$ $\underline{x = -21}$
13. $5x + 10(4x + 3) = 15$ $\underline{x = -\frac{1}{3}}$
14. $2(x + 3) = 5(x - 3)$ $\underline{x = 7}$
15. $-4x + 7 = 5(x + 2)$ $\underline{x = -\frac{1}{3}}$
16. $7x = 2(x - 3)$ $\underline{x = -\frac{6}{5}}$
17. $5x - 15 = 4x + 3$ $\underline{x = 18}$
18. $5(x + 0.5) = -1.5(x + 3x)$ $\underline{x = -\frac{5}{22}}$
19. $2(2x + 2) + x = 3x - 4$ $\underline{x = -4}$
20. $2x = 3(x + 2)$ $\underline{x = -6}$
21. $2x + 4(3x + 6) = 12$ $\underline{x = -\frac{6}{7}}$
22. $2x + 2(2x - 3) = -3$ $\underline{x = \frac{1}{2}}$

Solve each literal equation for the indicated variable.

23. $L \times W \times D = V$, for W $\underline{W = \frac{V}{LD}}$
24. $C = 2\pi r$, for r $\underline{r = \frac{C}{2\pi}}$
25. $V_1P_1 = V_2P_2$, for P_1 $\underline{P_1 = \frac{V_2P_2}{V_1}}$
26. $q = q_p \times D \times Q$, for q_p $\underline{q_p = \frac{q}{D \times Q}}$
27. $T = T_o - a(z - z_0)$, for a $\underline{a = \frac{T_0 - T}{z - z_0}}$
28. $A = (a + b)h$, for h $\underline{h = \frac{A}{a + b}}$

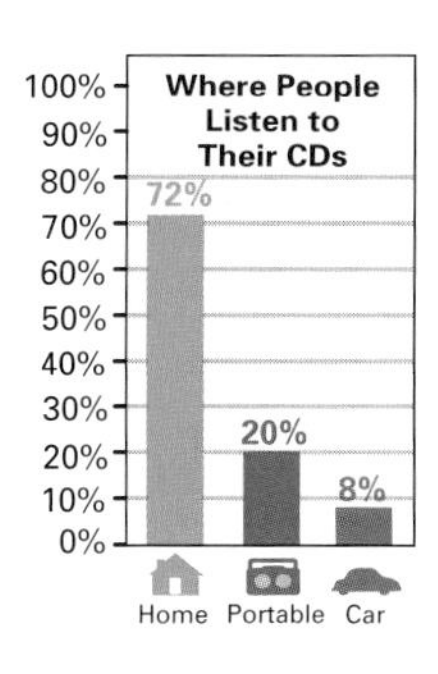

70. RECREATION The results of a survey of CD listeners in 1993 show that 72% usually listen to CDs at home, 20% usually listen to CDs on a portable player, and 8% usually listen to CDs in a car. If 180 of the respondents say that they usually listen to CDs on a portable player, how many people were surveyed? **900**

71. INCOME Anthony wants to buy a used car that will cost \$185.00 per month. If Anthony earns \$5.35 per hour, how many hours must Anthony work each month in order to pay for the car? **about 35 hr**

Look Back

Identify the slope, *m*, and *y*-intercept, *b*, for each line. Then graph the equation. ***(LESSON 1.2)***

72. $y = 2x - 6$ **$m = 2$, $b = -6$**

73. $3x + 4y = 9$ **$m = -\frac{3}{4}$, $b = \frac{9}{4}$**

74. $y = 2$ **$m = 0$, $b = 2$**

Write each number in decimal notation.

75. 5.736×10^4 **57,360**

76. 7.4609×10^3 **7460.9**

77. 46.72×10^6 **46,720,000**

78. 6.72×10^{-6} **0.00000672**

Write each number in scientific notation.

79. 25,000 **2.5×10^4**

80. 720,000 **7.2×10^5**

81. 260.07 **2.6007×10^2**

82. 5.7002 **5.7002×10^0**

83. 0.05 **5×10^{-2}**

84. 0.0002046 **2.046×10^{-4}**

Look Beyond

Explain what each expression means.

85. $y > -5$

86. $-3 < x < 3$

87. $-1 \leq y \leq 1$

88. $x \leq -3$

1. Choose a *y*-value (distance) that is different from those in your portfolio data set. Substitute this *y*-value into the equation for the least-squares line, and make a prediction about the corresponding time.
2. Show your results from Step 1 on your graph.

WORKING ON THE CHAPTER PROJECT

You should now be able to complete the Chapter Project.

Look Beyond

In Exercises 85–88, students are asked to interpret some simple inequalities. They should be familiar with sentences of this type from their work in previous courses. Students will solve inequalities in Lesson 1.7, so you may wish to use these exercises as a transition to that lesson.

ALTERNATIVE Assessment

Portfolio Activity

The Portfolio Activity can be used as preparation for the Chapter Project or as a separate activity. In the Portfolio Activity on this page, students use the equation of the least-squares line to predict the corresponding time for a distance that is not included in their portfolio data sets.

Answers to Portfolio Activities can be found in Additional Answers of the Teacher's Edition.

Student Technology Guide

NAME ______ CLASS ______ DATE ______

Student Technology Guide
1.6 Introduction to Solving Equations

You can use a graphics calculator to solve linear equations in one variable. For example, to solve the equation $3.1x + 6.2 = -4.5x - 1.8$, you can use the graphical approach explained below.

The graphical method involves the transformation of an equation in one variable into a pair of equations that are each in two variables. The pair of equations needed to solve $3.1x + 6.2 = -4.5x - 1.8$ is shown at right.

$$\begin{cases} y = 3.1x + 6.2 \\ y = -4.5x - 1.8 \end{cases}$$

Press Y=. Then enter the expressions. Press WINDOW. Enter $-10 \leq x \leq 10$ and $-10 \leq y \leq 10$.

Be sure to use (-) for a negative sign, and - for subtraction.

Now you have a choice as to what to do next.

- Method 1: Press TRACE. Then use ◄ or ► to move to the point of intersection.
- Method 2: Press 2nd TRACE 5:intersect ENTER ENTER ENTER ENTER.

Method 1

Y1=3.1X+6.2
X=-1.06383 Y=2.9021277

Method 2

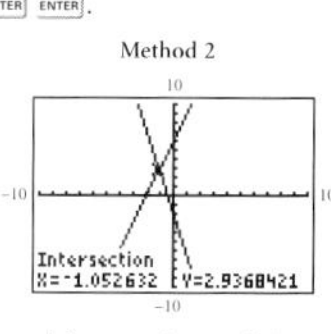

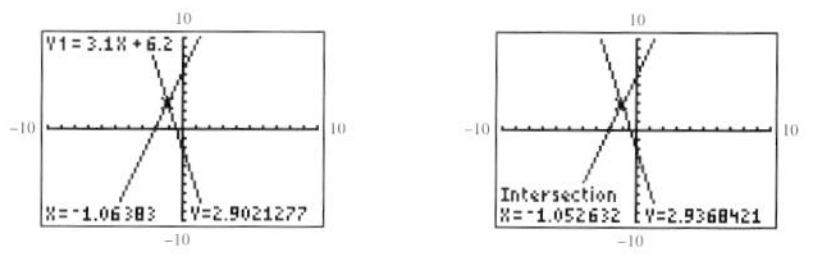

Calculator displays may vary somewhat in their reports of the approximate solution. According to the screens shown above, $x \approx -1.05$.

Solve each equation by graphing. Round to the nearest hundredth, if necessary.

1. $6.4x - 3.5 = -0.5x + 4.2$ $x \approx 1.12$
2. $45x + 59.5 = 35x - 8$ $x \approx -6.75$
3. $3.1(0.43x - 1.8) = -2.8x - 5.9$ $x \approx -0.08$
4. $\frac{3.5(x - 6)}{9.2} = x - 4.8$ $x \approx 4.06$
5. $8.3x - 9.1 = \frac{x}{6.5}$ $x \approx 1.12$
6. $-0.15x + 1.1 = 3.8x - 6.2$ $x \approx 1.85$
7. $0.1x - 0.5x = 2.8x + 0.9$ $x \approx -0.52$
8. $10.5x + 7.2 - 0.3x = 3.9$ $x \approx -0.32$
9. $25x + 3.25 = 42.8 + 8.9x$ $x \approx 2.46$

72. slope = 2
y-intercept = −6

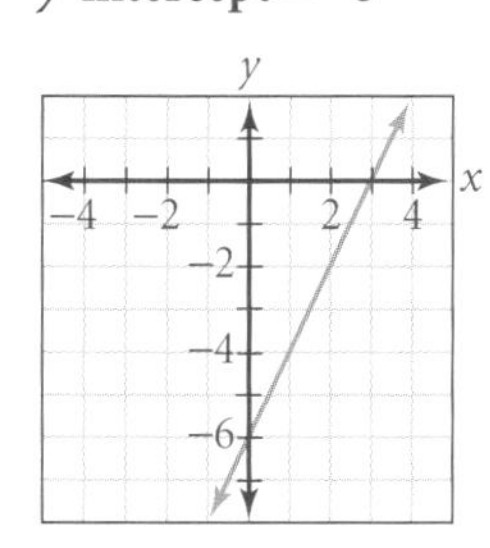

73. slope = $-\frac{3}{4}$
y-intercept = $\frac{9}{4}$

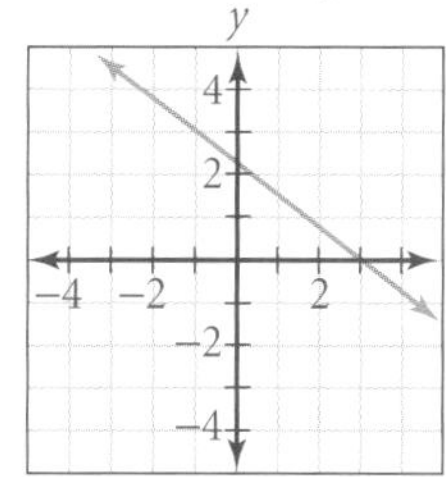

74. slope = 0
y-intercept = 2

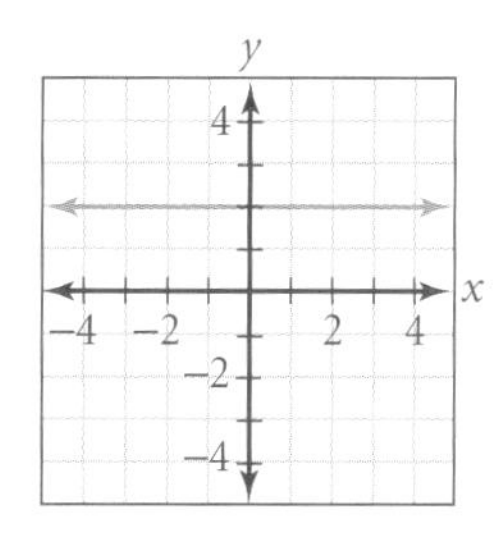

The answers to Exercises 85–88 can be found in Additional Answers beginning on page 1002.

Focus

The text on this page recalls the genius of a man who, without the benefit of formal education, became a published authority on astronomy, a respected surveyor, and a creator of popular mathematical puzzles. Students will compare two methods of solving a problem that he posed: the method of *false position* and a more familiar algebraic method. They should not only gain insight into the logical process of problem solving, but also develop some insight into the contributions of a truly remarkable man.

Motivate

Have students read the first paragraph of the text. Ask them to share other facts that they may know about the life and accomplishments of Benjamin Banneker.

Students may be interested to learn that Banneker was 57 years old when he borrowed some astronomy books and equipment and taught himself enough to publish his own almanac of astronomy. The woodcut of Banneker shown on this page is a self-portrait that appeared on the title page of his almanac.

Tell students to read the puzzle. Be sure that they are able to follow the unfamiliar style of writing. You may want to discuss how the English language has changed over the last two centuries. Ask one or two volunteers to restate the puzzle in their own words.

A Man & A Method

Predicting eclipses and planetary motions seems complicated, but imagine having no formal education and teaching yourself the mathematics needed to make such calculations. Imagine publishing an almanac of your results for farmers and astronomers across the nation. That is exactly what an African American named Benjamin Banneker did over 200 years ago. The almanac impressed Thomas Jefferson so much that he asked Banneker to help survey the land for the new nation's capital, Washington, D.C.

Benjamin Banneker also published math puzzles. Some he made up. Others, like the one below, were sent to him. He published the puzzle shown below in his *Manuscript Journal.*

> Divide 60 into four Such parts, that the first being increased by 4, the Second decreased by 4, the third multiplyed by 4, the fourth part divided by 4, that the Sum, the difference, the product, and the Quotient shall be one and the Same Number—

Banneker gave the answer but not his method for finding it. He may have used a method called *false position.* In the false position method, you guess the answer, see how far off your guess is, and then use that information in a proportion to get the correct answer.

1a. Answers may vary. Sample choice of number: 10

b. Answers may vary. Sample answer:
first part + 4 = 10; first part = 6
second part −4 = 10; second part = 14
third part × 4 = 10; third part = $\frac{10}{4}$, or 2.5
fourth part ÷ 4 = 10; fourth part = 40

c. Answers may vary. Sample answer:
6 + 14 + 2.5 + 40 = 62.5

d. Answers may vary. Sample answer:
$\frac{\text{correct number}}{60} = \frac{10}{62.5}$; correct number = 9.6

e. first part + 4 = 9.6; first part = 5.6
second part − 4 = 9.6; second part = 13.6
third part × 4 = 9.6; third part = $\frac{9.6}{4}$, or 2.4
fourth part ÷ 4 = 9.6; fourth part = 38.4
Sum: 5.6 + 13.6 + 2.4 + 38.4 = 60

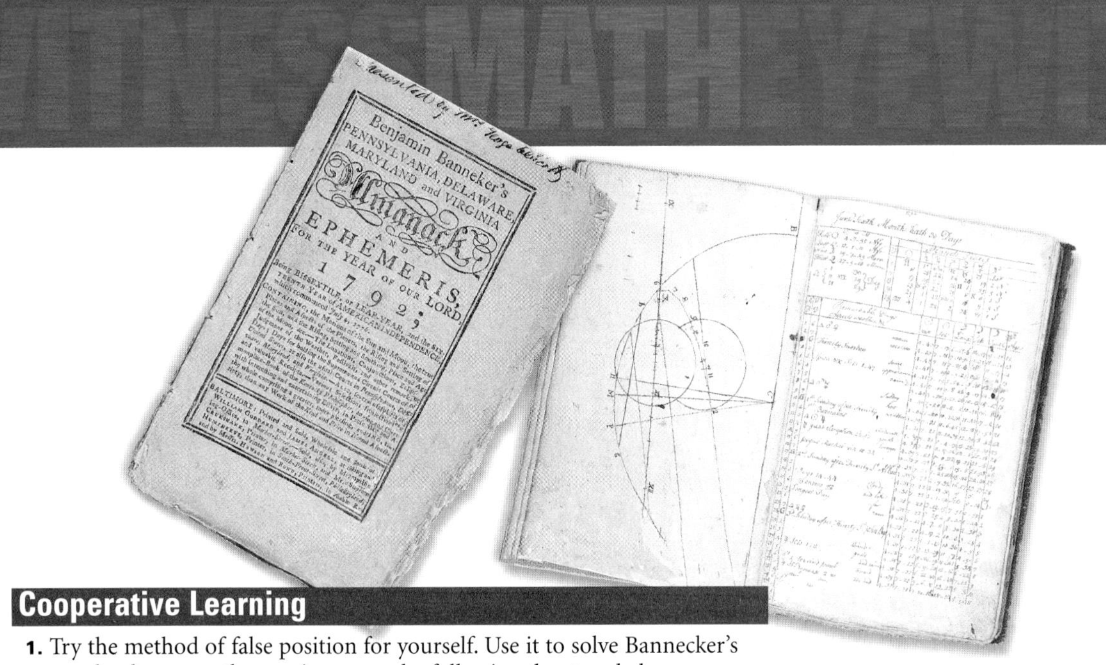

Cooperative Learning

1. Try the method of false position for yourself. Use it to solve Bannecker's puzzle, shown on the previous page, by following the steps below.
   a. Let x represent the answer, "one and the same number."

$$\begin{aligned} &\text{first part} + 4 = x \\ &\text{second part} - 4 = x \\ &\text{third part} \times 4 = x \\ &\text{fourth part} \div 4 = x \end{aligned}$$

   Choose any number to be the answer, x. For instance, you can choose 5 or 10 or any other number you like.

   b. Substitute the number you chose for the answer, x, into each equation above, and solve for each of the four parts. For example, if you chose 5 for x, then the first part can be found as shown below.

$$\begin{aligned} \text{first part} + 4 &= 5 \\ \text{first part} &= 1 \quad \textit{Solve for the "first part."} \end{aligned}$$

   c. Find the sum of the four parts.

   d. Use a proportion to find the correct answer, x, for the "one and the same number."

$$\frac{\text{Correct "one and the same number"}}{\text{Correct sum} = 60} = \frac{\text{Trial "one and the same number"}}{\text{Trial sum}}$$

   e. Use the correct answer for x, "one and the same number," from the proportion above to find the four parts. Is the sum of these parts really 60?

2. Now use algebra to solve the same puzzle. Follow the steps below.
   a. Write each of the four parts in terms of x, and represent the puzzle with one equation in one variable.
   b. Show that the solution to this equation gives the correct first, second, third, and fourth parts of the puzzle.

3. Use the graph at right to explain why the method of false position works for this puzzle.

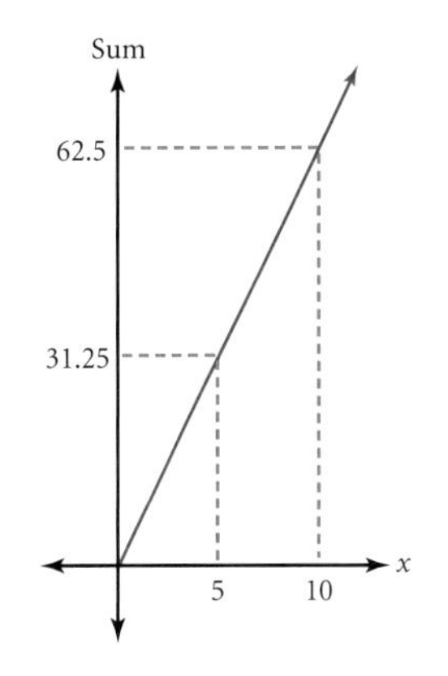

2a. $(x-4)+(x+4)+\left(\frac{x}{4}\right)+(4x)=60$

b. $(x-4)+(x+4)+\left(\frac{x}{4}\right)+(4x)=60$

$$x-4+x+4+\frac{1}{4}x+4x=60$$

$$6\frac{1}{4}x=60$$

$$x=9\frac{3}{5}, \text{ or } 9.6$$

3. **Answers may vary. Sample answer: The graph represents a direct variation, so the ratio of the trail number to the resulting sum is proportional to the ratio of the correct number and the correct sum.**

Cooperative Learning

You may wish to have students work in groups of three or four. In Step 1, each student can solve the puzzle independently, with each group member choosing a different value for x. After all group members have completed part **e**, the group should compare results and work together to resolve any difficulties.

Begin Step 2 with each group member again working independently to complete parts **a** and **b**. The group can then come together to discuss their findings.

Each group should discuss the question in Step 3 together. One group member should be chosen to summarize the group's answer.

Discuss

Have the designated member of each group present the group's explanation to the class. Discuss any differences.

Have students compare the false-position method with the algebraic method. What do they see as the advantages and disadvantages of each? Ask them to speculate why the false-position method is not commonly known. Discuss why this method is not used in other situations.

Prepare

NCTM PRINCIPLES & STANDARDS 1, 2, 6, 8–10

QUICK WARM-UP

1. Graph and label the following points on the same number line:

a. -1 **b.** 3.5

c. $-\frac{9}{4}$ **d.** 0

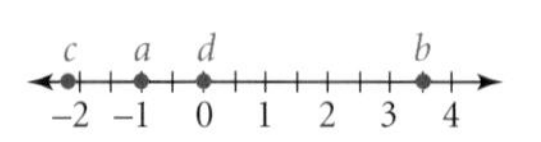

2. Graph each inequality on a separate number line.

a. $x < 2$ **b.** $x \geq -3$

a. −2 −1 0 1 2 3 4

b. −6 −4 −2 0 2 4 6

Also on Quiz Transparency 1.7

Teach

Why The income-tax schedule described on this page is just one of many real-world situations that can be modeled by an inequality. Have students brainstorm a list of other such situations.

Introduction to Solving Inequalities

Why *There are many situations in life that involve inequalities. For example, an income-tax bracket can be described by using an inequality.*

Objectives

- Write, solve, and graph linear inequalities in one variable.
- Solve and graph compound linear inequalities in one variable.

APPLICATION
TAXES

The federal tax calculation table above applies to single people filing a tax return in 1998 for the tax year 1997.

Suppose, for example, that your taxable income is x dollars and the tax due is t dollars. You can read the second row as follows:

$$\text{If } x > 24{,}650 \text{ and } x \leq 59{,}750, \text{ then } t = 3697.50 + 0.28(x - 24{,}650).$$

The statements $x > 24{,}650$ *and* $x \leq 59{,}750$ are examples of *linear inequalities in one variable.* In general, an **inequality** is a mathematical statement involving $<, >, \leq, \geq,$ or $\neq$.

Just as there are Properties of Equality that you can use to solve equations, there are *Properties of Inequality* that you can use to solve inequalities.

Properties of Inequality

For all real numbers a, b, and c, where $a \leq b$:

Addition Property	$a + c \leq b + c$
Subtraction Property	$a - c \leq b - c$
Multiplication Property	If $c \geq 0$, then $ac \leq bc$. If $c \leq 0$, then $ac \geq bc$.
Division Property	If $c > 0$, then $\frac{a}{c} \leq \frac{b}{c}$. If $c < 0$, then $\frac{a}{c} \geq \frac{b}{c}$.

Similar statements can be written for $a < b$, $a \geq b$, and $a > b$.

Alternative Teaching Strategy

USING TABLES Have students use the table feature of a graphics calculator to explore the inequality in Example 2: $4 - 3p > 16 - p$. If they are using a TI-82 or TI-83, have them press [Y=], clear any existing equations, and then enter **4–3X** next to **Y1=** and **16–X** next to **Y2=**.

Next instruct students to press [2nd] [WINDOW] and make these choices in the **TABLE SETUP** screen.

TblStart=0 (or **TblMin=0**) **Indpnt: Auto Ask**
ΔTbl=.5 **Depend: Auto Ask**

If they now press [2nd] [GRAPH], they will see a table.

Ask students to describe the relationship between **Y1** and **Y2** in the portion of the table that they see. **Y1 < Y2** Have them use the [▲] key to scroll up the table until they arrive at the place where **Y1 = Y2**. Ask them to name the corresponding value of **X**. **–6** Point out that –6 is the solution of the equation $4 - 3p = 16 - p$.

Now have students examine the table and describe all values of **X** for which **Y1 > Y2. X < –6** This means that the solution of $4 - 3p > 16 - p$ is $p < -6$.

Any value of a variable that makes an inequality true is called a **solution of the inequality**. For example, $6x + 1 < 13$ is an inequality in one variable, x. Values such as $\frac{1}{2}$ and -1 are solutions of the inequality, as shown below.

$$6x + 1 < 13 \qquad\qquad 6x + 1 < 13$$
$$6\left(\frac{1}{2}\right) + 1 \overset{?}{<} 13 \qquad\qquad 6(-1) + 1 \overset{?}{<} 13$$
$$4 < 13 \quad \textbf{True} \qquad\qquad -5 < 13 \quad \textbf{True}$$

CHECKPOINT ✔ Find two or more solutions of $6x + 1 < 13$, and show that they are solutions.

CHECKPOINT ✔
Answers may vary. sample answer: $x = 0$ *or* $x = 1$

EXAMPLE 1 Solve $4x - 5 \geq 13$.

SOLUTION

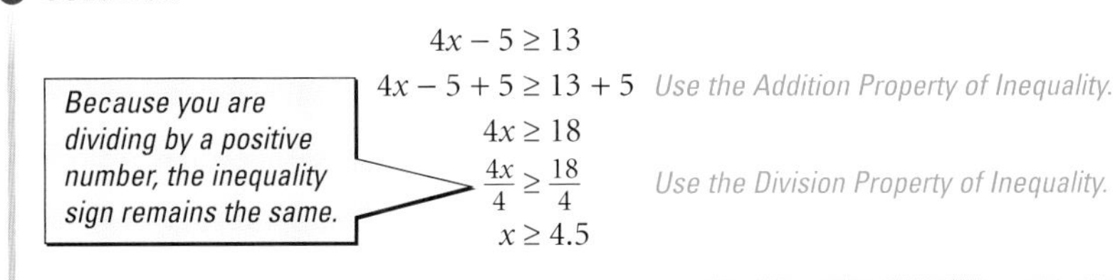

$4x - 5 \geq 13$
$4x - 5 + 5 \geq 13 + 5$ Use the Addition Property of Inequality.
$4x \geq 18$
$\frac{4x}{4} \geq \frac{18}{4}$ Use the Division Property of Inequality.
$x \geq 4.5$
Because you are dividing by a positive number, the inequality sign remains the same.

TRY THIS Solve $-4 < 7 - 3x$.

ADDITIONAL EXAMPLE 1

Solve $8d - 3 \leq 9$. $d \leq 1.5$

TRY THIS

$x < \frac{11}{3}$

You can represent the solution of an inequality in one variable on a number line. The number line shown below is the graph of $x \geq 4.5$.

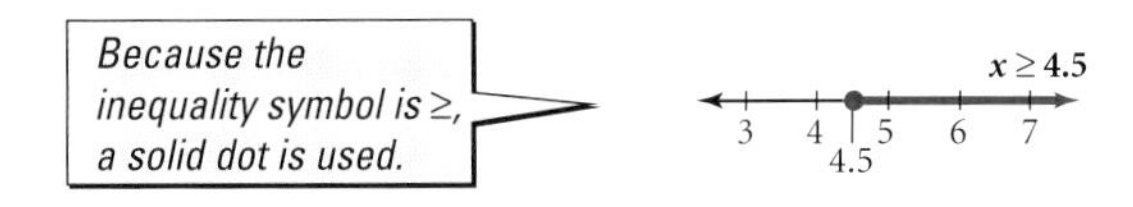

Because the inequality symbol is ≥, a solid dot is used.
$x \geq 4.5$
3 4 5 6 7
4.5

EXAMPLE 2 Solve $4 - 3p > 16 - p$. Graph the solution on a number line.

SOLUTION

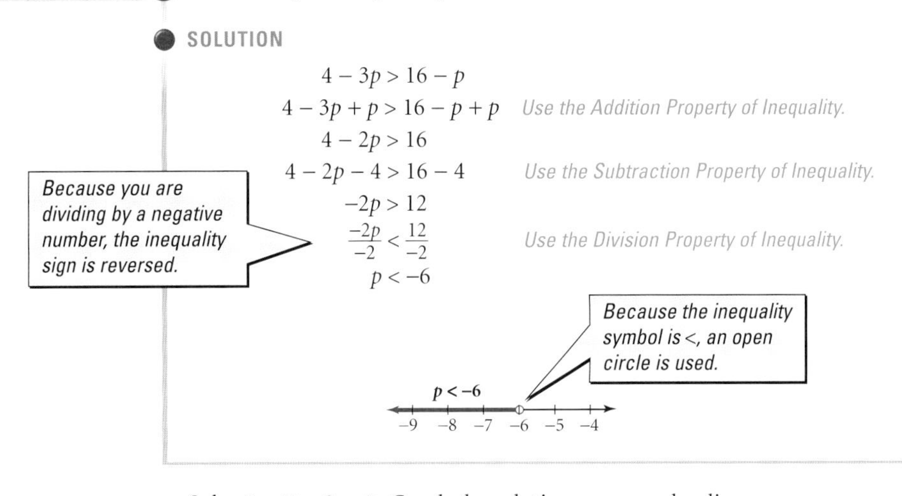

TRY THIS Solve $5 - 7t > 8 - 4t$. Graph the solution on a number line.

ADDITIONAL EXAMPLE 2

Solve $2y + 9 < 5y + 15$. Graph the solution on a number line.

$y > -2$

Teaching Tip

Emphasize the fact that multiplying or dividing each side of an inequality by a negative number produces an equivalent inequality *only* if the inequality sign is reversed.

TRY THIS

$t < -1$

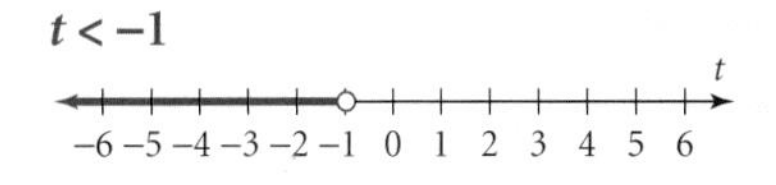

Interdisciplinary Connection

CHEMISTRY The pH of a substance is a measure of how *acidic* or *alkaline* it is. A pH is measured on a scale from 0 to 14, with 0 being highly acidic, 14 being highly alkaline, and 7 being neutral. According to most guidelines, the water in a swimming pool should have a pH between 7.2 and 7.6, inclusive. Suppose that two samples of water from a swimming pool have a pH of 7.5 and 7.8. What pH would be required for a third sample in order for the average of the three samples to fall within the acceptable range?
between 6.3 and 7.5, inclusive

Inclusion Strategies

ENGLISH LANGUAGE DEVELOPMENT Give students inequalities like the ones below, and have them list as many verbal interpretations as they can. **Sample interpretations are given.**

$x \geq 1$	**x is greater than or equal to 1; x is at least 1; x is no less than 1.**
$x \leq 7$	**x is less than or equal to 7; x is at most 7; x is no greater than 7.**
$x > 2$	**x is greater than 2; x exceeds 2.**
$x < 4$	**x is less than 4; x is up to but not including 4.**
$0 \leq x \leq 8$	**x is between 0 and 8, inclusive.**

ADDITIONAL EXAMPLE 3

Michael's test average in his mathematics class is 92, and his homework average is 81. The test average is 60% of the final grade, the quiz average is 15% of the final grade, and the homework average is 25% of the final grade. **What quiz average does Michael need in order to have a final grade of at least 90?** Michael's quiz average must be at least 97.

Activity Notes

In this Activity, students find the solution of a one-variable inequality by using the graphs of two linear equations. Students should recognize that this as an extension of the technique of using graphs to solve a one-variable linear equation. They should observe that the x-coordinate of the point of intersection of the two graphs is the boundary of the given inequality.

CHECKPOINT ✔

5. Graph $y = 3x + 2$ and $y = 5$ on the same screen. Find the values of x for which the graph of $y = 3x + 2$ is above the graph of $y = 5$.

CRITICAL THINKING

Yes, the method applies. Graph the related equation for each side of the inequality, find the intersection point of the two lines, and determine the values of x for which one line is above or below the other, depending on the sense of the inequality.

EXAMPLE 3

APPLICATION ACADEMICS

Claire's test average in her world history class is 90. The test average is $\frac{2}{3}$ of the final grade and the homework average is $\frac{1}{3}$ of the final grade.

What homework average does Claire need in order to have a final grade of at least 93?

SOLUTION

PROBLEM SOLVING

Write an equation.

$$\text{Final grade} = \frac{2}{3}\left(\text{Test average}\right) + \frac{1}{3}\left(\text{Homework average}\right)$$

$$f = \frac{2}{3}(90) + \frac{1}{3}h$$

Claire wants a final grade of at least 93, or $f \geq 93$.

$f \geq 93$	
$\frac{2}{3}(90) + \frac{1}{3}h \geq 93$	*Substitute $\frac{2}{3}(90) + \frac{1}{3}h$ for f.*
$60 + \frac{1}{3}h \geq 93$	*Simplify.*
$\frac{1}{3}h \geq 33$	*Use the Subtraction Property of Inequality.*
$h \geq 99$	*Use the Multiplication Property of Inequality.*

Claire's homework average must be at least 99 in order for her to have a final grade of at least 93.

Activity Exploring Inequalities Graphically

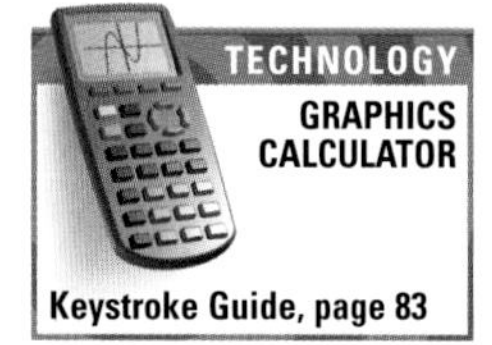

You will need: a graphics calculator

1. Solve $2x - 3 < 3$ for x.
2. Use a graphics calculator to graph $y = 2x - 3$ and $y = 3$ on the same screen.
3. For what values of x is the graph of $y = 2x - 3$ below the graph of $y = 3$?
4. Explain how the answer to Step 3 helps you to solve $2x - 3 < 3$.

CHECKPOINT ✔

5. How would you use graphs to solve $3x + 2 > 5$? List and explain your steps.

CRITICAL THINKING

Does the method you explored in the Activity above apply to solving inequalities such as $2x - 3 < x + 4$ and $4 \geq 3x + 1$? Justify your response.

The inequalities $x > 24{,}650$ *and* $x \leq 59{,}750$, which describe the income-tax bracket stated at the beginning of this lesson, form a compound inequality. A **compound inequality** is a pair of inequalities joined by *and* or *or*.

Enrichment

Tell whether each statement is true or false for all real numbers a, b, c, and d. If a statement is false, give a *counterexample* to show why it is false.

1. If $a < b$ and $c < d$, then $a + c < b + d$. **true**
2. If $a < b$ and $c < d$, then $a - b < c - d$. **false; $0 < 1$ and $2 < 5$, but $0 - 1 > 2 - 5$**
3. If $a < b$, then $a < b^2$. **false; $0.25 < 0.5$, but $0.25 = (0.5)^2$**
4. If $a < b$, then $a^2 < b^2$. **false; $-1 < 0$, but $(-1)^2 > 0$**
5. If $a < b$, then $a^3 < b^3$. **true**

Reteaching the Lesson

WORKING BACKWARD Tell students to fill in the boxes with real numbers so that $n > -3$ is the solution of the resulting inequality. **Samples given:**

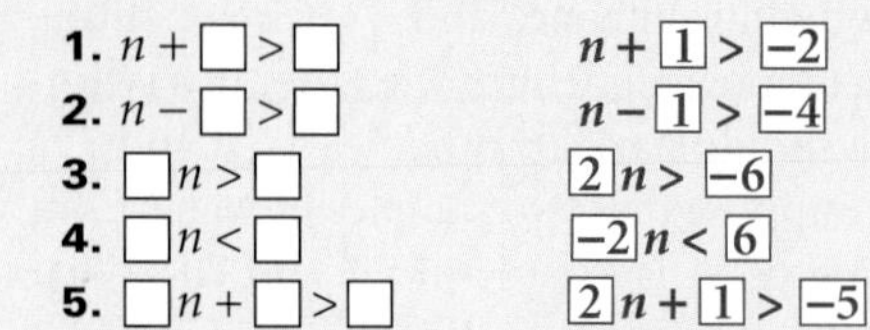

1.	$n + \square > \square$	$n + \boxed{1} > \boxed{-2}$
2.	$n - \square > \square$	$n - \boxed{1} > \boxed{-4}$
3.	$\square n > \square$	$\boxed{2}n > \boxed{-6}$
4.	$\square n < \square$	$\boxed{-2}n < \boxed{6}$
5.	$\square n + \square > \square$	$\boxed{2}n + \boxed{1} > \boxed{-5}$
6.	$\square n - \square < \square$	$\boxed{-2}n - \boxed{1} < \boxed{5}$

For each exercise, ask students to share their inequalities with the class. Discuss the results.

To solve an inequality involving *and*, find the values of the variable that satisfy *both* inequalities. This is shown in Example 4.

EXAMPLE 4 **Solve $2x + 1 \geq 3$ *and* $3x - 4 \leq 17$. Graph the solution.**

SOLUTION

$$\begin{aligned} 2x + 1 &\geq 3 & \quad \textit{and} \quad 3x - 4 &\leq 17 \\ 2x &\geq 2 & 3x &\leq 21 \\ x &\geq 1 & x &\leq 7 \end{aligned}$$

The solution is all values of x between 1 and 7 inclusive.

$x \geq 1$ *and* $x \leq 7$

0 1 2 3 4 5 6 7 8

ADDITIONAL EXAMPLE 4

Solve $2x + 3 > 1$ *and* $5x - 9 < 6$. Graph the solution.

$-1 < x < 3$

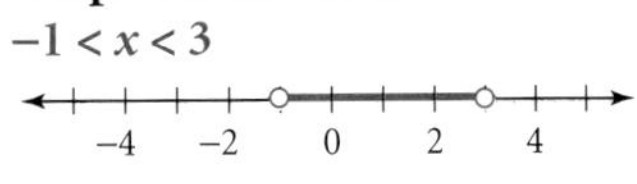

TRY THIS Solve $-2x + 5 \geq 3$ *and* $x - 5 > -12$. Graph the solution.

TRY THIS

$x \leq 1$ *and* $x > -7$

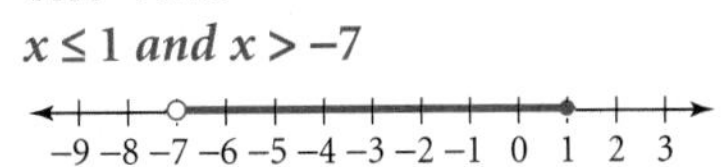

The solution $x \geq 1$ *and* $x \leq 7$ in Example 4 can also be written as $1 \leq x \leq 7$. In general, the compound statement $x > a$ *and* $x < b$, where $a < b$, can be written as $a < x < b$.

CHECKPOINT ✔ What is another way to express the statement $x < 3$ *and* $x > -4$?

CHECKPOINT ✔

$-4 < x < 3$

When you solve a compound inequality involving *or*, find those values of the variable that satisfy *at least one* of the inequalities. This is shown in Example 5.

EXAMPLE 5 **Solve $5x + 1 > 21$ *or* $3x + 2 < -1$. Graph the solution.**

SOLUTION

$$\begin{aligned} 5x + 1 &> 21 & \quad \textit{or} \quad 3x + 2 &< -1 \\ 5x &> 20 & 3x &< -3 \\ x &> 4 & x &< -1 \end{aligned}$$

The solution is all values of x less than −1 or greater than 4.

$x < -1$ *or* $x > 4$

−2 −1 0 1 2 3 4 5

ADDITIONAL EXAMPLE 5

Solve $2b - 3 \geq 1$ *or* $3b + 7 \leq 1$. Graph the solution.

$b \leq -2$ *or* $b \geq 2$

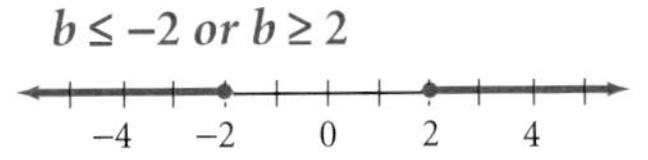

TRY THIS Solve $2x \leq 5$ *or* $7x + 1 > 36$. Graph the solution.

TRY THIS

$x \leq \frac{5}{2}$ *or* $x > 5$

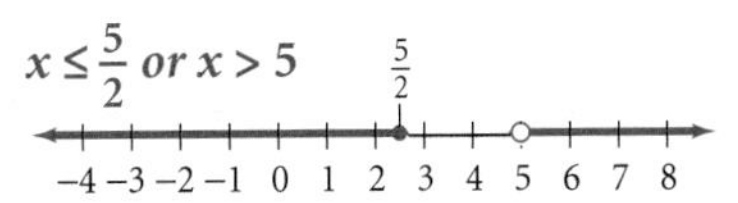

Exercises

Communicate

1. Describe the steps you would take to graph $7x - 7 > 0$ on a number line.
2. How does the graph of $7x - 7 > 0$ differ from the graph of $7x - 7 \geq 0$? from the graph of $7x - 7 < 0$?
3. Is $x < 16$ equivalent to $-x < -16$? Explain.
4. How can you express "x is nonnegative" by using an inequality?

Assess

Selected Answers

Exercises 5–9, 11–87 odd

ASSIGNMENT GUIDE	
In Class	1–9
Core	10–19, 21–49 odd, 50, 51, 53–75 odd
Core Plus	10–19, 20–48 even, 50, 51, 52–70 even, 72–76
Review	77–88
Preview	89

Extra Practice can be found beginning on page 940.

Answers to odd-numbered Extra Practice exercises can be found immediately after Selected Answers in the Pupil's Edition.

6. $q > 5$

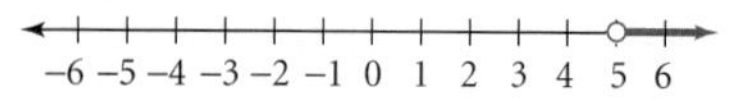

8. $x \geq -2$ *and* $x < 6$

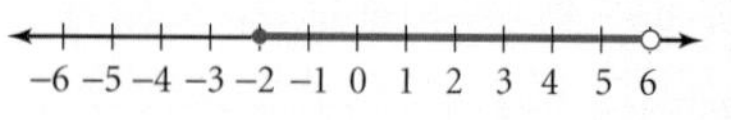

Guided Skills Practice

5. Solve $3x + 1 < 13$. ***(EXAMPLE 1)*** $x < 4$

6. Solve $q + 4 < 4q - 11$. Graph the solution on a number line. ***(EXAMPLE 2)***

APPLICATION

7. ACADEMICS Connor's homework average in English class is 92. The test average is $\frac{3}{4}$ of the final grade, and the homework average is $\frac{1}{4}$ of the final grade. What test average does Connor need in order to have a final grade of at least 80? ***(EXAMPLE 3)*** $t \geq 76$

8. Solve $3x - 7 \geq -13$ *and* $2x + 3 < 15$. Graph the solution. ***(EXAMPLE 4)***

9. Solve $2x + 4 \geq -10$ *or* $4x - 6 > 14$. Graph the solution. ***(EXAMPLE 5)***

Practice and Apply

Write an inequality that describes each graph.

10. (number line −6 to 6) $x \leq 1$ **11.** (number line −6 to 6) $x > -1$

12. (number line −6 to 6) $x < -3$ **13.** (number line −6 to 6) $x \geq -1$

14. (number line −6 to 6) $x > -6$ **15.** (number line −6 to 6) $x \leq 5$

16. (number line −6 to 6) $x > 1$ **17.** (number line −6 to 6) $x \geq 2$

18. (number line −6 to 6) $x > -5$ **19.** (number line −6 to 6) $x < 3$

internet connect — Homework Help Online. Go To: go.hrw.com Keyword: MB1 Homework Help for Exercises 20–49

Solve each inequality, and graph the solution on a number line.

20. $5x > 10$ $x > 2$ **21.** $35x > 70$ $x > 2$ **22.** $-5x > 10$ $x < -2$

23. $-35x > 70$ $x < -2$ **24.** $-5x > -10$ $x < 2$ **25.** $-35x > -70$ $x < 2$

26. $s - 2 > 10$ $s > 12$ **27.** $y + 5 < -3$ $y < -8$ **28.** $3x + 7 < 31$ $x < 8$

29. $2x - 3 \geq 19$ $x \geq 11$ **30.** $\frac{1}{2}d - 1 \geq -15$ $d \geq -28$ **31.** $\frac{1}{5}x - 2 \leq 28$ $x \leq 150$

32. $-2x > 14$ $x < -7$ **33.** $-5x \leq 30$ $x \geq -6$ **34.** $-x + 8 < 41$ $x > -33$

35. $-5x - 15 > 60$ $x < -15$ **36.** $-10 < -5x$ $x < 2$ **37.** $-81 \leq -9x$ $x \leq 9$

38. $\frac{-x}{3} \geq 10$ $x \leq -30$ **39.** $\frac{-t}{32} < 2$ $t > -64$ **40.** $-6(p + 4) < 12$ $p > -6$

41. $6 - (4x - 3) \geq 8$ $x \leq \frac{1}{4}$ **42.** $4y - 12 > 7y - 15$ **43.** $8a - 11 < 4a + 9$ $a < 5$

44. $3(4x - 5) < 8x + 3$ **45.** $6(x - 9) \geq 21 + x$ **46.** $-4x - 3 < -6x - 17$

47. $-x + 5 \geq -4x - 7$ **48.** $2(x - 5) < -4(3x + 2)$ **49.** $-5(3x + 2) \geq 4(x - 1)$

42. $y < 1$
44. $x < 4.5$
45. $x \geq 15$
46. $x < -7$
47. $x \geq -4$
48. $x < \frac{1}{7}$
49. $x \leq -\frac{6}{19}$

50. Graph each compound inequality on a number line.

a. $x > -4$ *and* $x < 2$ $-4 < x < 2$ **b.** $x > -4$ *and* $x > 2$ $x > 2$

c. $x > -4$ *or* $x < 2$ all real numbers **d.** $x > -4$ *or* $x > 2$ $x > -4$

51. Graph each compound inequality on a number line.

a. $x < -4$ *and* $x < 2$ $x < -4$ **b.** $x < -4$ *and* $x > 2$ no solution

c. $x < -4$ *or* $x < 2$ $x < 2$ **d.** $x < -4$ *or* $x > 2$ $x < -4$ *or* $x > 2$

Practice

NAME ______ CLASS ______ DATE ______

Practice
1.7 Introduction to Solving Inequalities

Write an inequality that describes each graph.

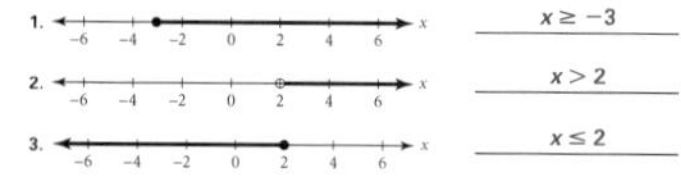

1. $x \geq -3$
2. $x > 2$
3. $x \leq 2$

Solve each inequality, and graph the solution on a number line.

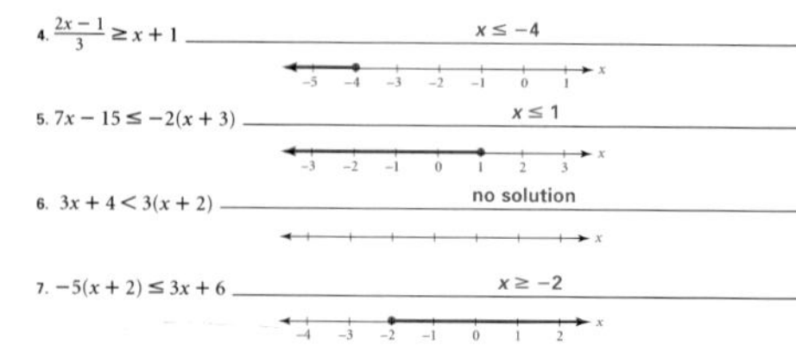

4. $\frac{2x-1}{3} \geq x + 1$ $x \leq -4$
5. $7x - 15 \leq -2(x + 3)$ $x \leq 1$
6. $3x + 4 < 3(x + 2)$ no solution
7. $-5(x + 2) \leq 3x + 6$ $x \geq -2$

Graph the solution of each compound inequality on a number line.

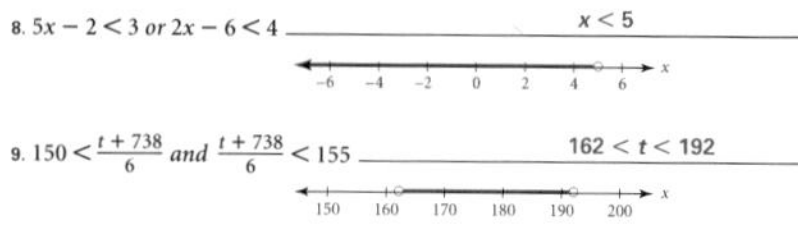

8. $5x - 2 < 3$ *or* $2x - 6 < 4$ $x < 5$
9. $150 < \frac{t + 738}{6}$ *and* $\frac{t + 738}{6} < 155$ $162 < t < 192$

9. $x \geq -7$ *or* $x > 5$

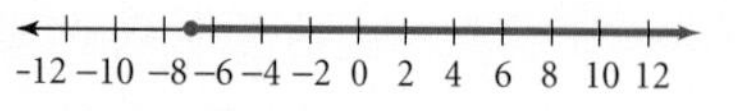

20. and 21. $x > 2$

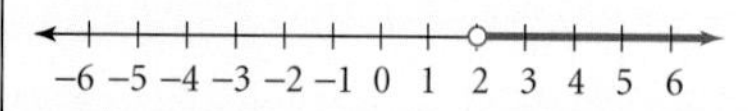

22. $x < -2$

−6 −5 −4 −3 −2 −1 0 1 2 3 4 5 6

23. $x < -2$

−6 −5 −4 −3 −2 −1 0 1 2 3 4 5 6

24. $x < 2$

−6 −5 −4 −3 −2 −1 0 1 2 3 4 5 6

25. $x < 2$

26. $s > 12$

0 2 4 6 8 10 12 14 16 18 20 22 24

The answers to Exercises 27–51 can be found in Additional answers beginning on page 1002.

Graph the solution of each compound inequality on a number line.

52. $n + 4 < 16$ *and* $n - 3 > 12$

53. $y - 2 < 4$ *and* $y + 4 > 7$

54. $s + 7 > 4$ *or* $s - 2 < 2$

55. $x + 8 < 5$ *or* $x - 1 > 3$

56. $x + 9 \leq 5$ *and* $4x \geq 12$

57. $5y \geq 15$ *and* $y + 8 \geq 2$

58. $c - 8 \leq 2$ *or* $6c \geq -18$

59. $x + 9 \leq 5$ *or* $4x \geq 12$

60. $5a + 12 < 2$ *and* $5a - 12 < 3$

61. $3t + 5 > 11$ *and* $4t - 1 < 15$

62. $-9x > -81$ *and* $2(x + 6) > -4$

63. $-5d < 40$ *and* $4(d - 3) < -8$

64. $20 - 3x \geq 11$ *or* $-4x \leq -20$

65. $14 - 3x \leq 2$ *or* $5 - 4x \geq 17$

66. $5 - 2b > -3$ *or* $-3(b - 3) < -6$

67. $-6x - 11 < 13$ *or* $3(x + 2) \leq -9$

68. $\frac{1}{2}(x + 9) \leq -3$ *and* $-10 < -5x$

69. $\frac{4m}{3} + 5 > 2$ *and* $4 \leq -2(m - 3) - 7$

70. $2x < 7x - 10$ *or* $8x \leq 3x - 15$

71. $2x - 7 < 5x + 8$ *or* $\frac{1}{2}(16 - 4x) \geq 0$

CHALLENGE

72. Solve $-2a \leq 3x + a < 10a$ for x. **$-a \leq x < 3a$**

APPLICATIONS

73. FUND-RAISING A charity is planning to raffle off a new car donated by a local car dealer. The charity wants to raise at least \$70,000. It expects to sell at least 1250 tickets and to spend \$5000 promoting the raffle. Find the possible ticket prices, p, by solving the inequality below. **\$60 or greater**

$$1250p - 5000 \geq 70{,}000$$

HEALTH One study has found that people who reduced their fat intake to less than 20% of their total calories suffered fewer headaches. [*Source: Loma Linda University School of Public Health, CA*]

74. Write and solve an inequality to find the total number of calories consumed by people in this study before they reduced their fat intake to 324 fat calories. **more than 1620 calories**

75. Write and solve an inequality to find the number of fat calories consumed by someone in this study who consumed a total of 1850 calories before reducing the fat intake. **less than 370 fat calories**

Hamburger with french fries: about 46 grams of fat

Grilled chicken with rice and carrots: about 10 grams of fat

Error Analysis

Students often forget to reverse the inequality sign when multiplying or dividing by a negative number. Encourage them to check values at and near the boundary of the solution. For instance, if the proposed solution is $a < -2$, they should check that

- -2 is the solution of the related *equation*,
- -3 is a solution of the given inequality, and
- -1 is *not* a solution of the given inequality.

52. $n < 12$ *and* $n > 15$; no solution

53. $y < 6$ *and* $y > 3$

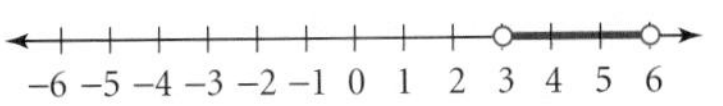

54. $s > -3$ *or* $s < 4$; all real numbers

55. $x < -3$ *or* $x > 4$

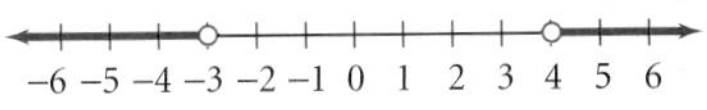

56. $x \leq -4$ *and* $x \geq 3$; no solution

57. $y \geq 3$ *and* $y \geq -6$

−6 −5 −4 −3 −2 −1 0 1 2 3 4 5 6

58. $c \leq 10$ *or* $c \geq -3$; all real numbers

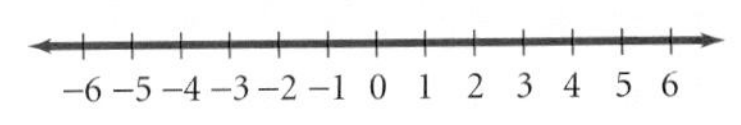

Student Technology Guide

NAME ______ CLASS ______ DATE ______

Student Technology Guide
***1.7* Introduction to Solving Inequalities**

Certain graphics calculators can be used to find the approximate solutions to compound inequalities. For example, to solve $2x - 1 > 1.2$ *and* $3x - 4 \leq 5.8$, use the logical "and" statement located under the MATH LOGIC menu. The steps are shown below.

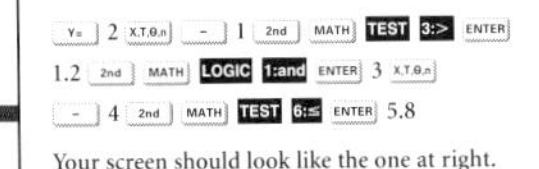

Y= 2 X,T,θ,n − 1 2nd MATH TEST 3:> ENTER
1.2 2nd MATH LOGIC 1:and ENTER 3 X,T,θ,n
− 4 2nd MATH TEST 6:≤ ENTER 5.8

Plot1 Plot2 Plot3
\Y1 ■ 2X−1>1.2 and 3X−4≤5.8
\Y2=
\Y3=
\Y4=
\Y5=
\Y6=

Your screen should look like the one at right.

Press GRAPH to see the solution of $2x - 1 > 1.2$ *and* $3x - 4 \leq 5.8$. Press TRACE to approximate the endpoints of the line segment. If greater accuracy is needed, use ZOOM.

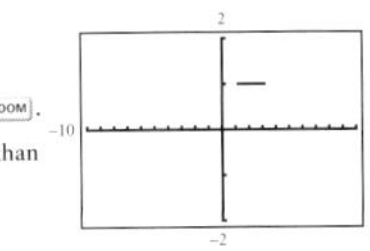

The solution is all values of x greater than 1.1 and less than or (approximately) equal to 3.27. You can write this as $x > 1.1$ *and* $x \leq 3.27$.

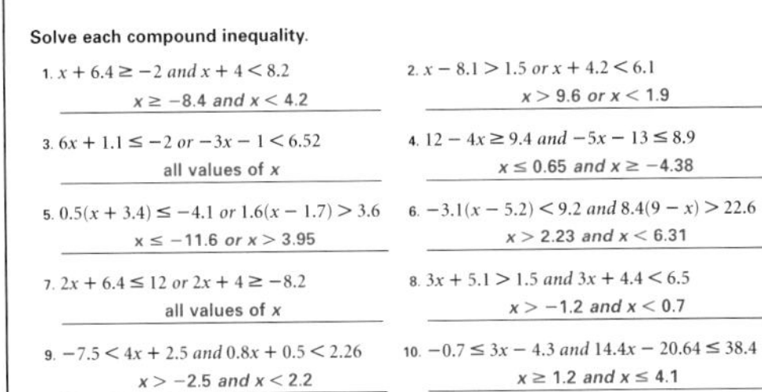

Solve each compound inequality.

1. $x + 6.4 \geq -2$ *and* $x + 4 < 8.2$ — **$x \geq -8.4$ and $x < 4.2$**
2. $x - 8.1 > 1.5$ *or* $x + 4.2 < 6.1$ — **$x > 9.6$ or $x < 1.9$**
3. $6x + 1.1 \leq -2$ *or* $-3x - 1 < 6.52$ — **all values of x**
4. $12 - 4x \geq 9.4$ *and* $-5x - 13 \leq 8.9$ — **$x \leq 0.65$ and $x \geq -4.38$**
5. $0.5(x + 3.4) \leq -4.1$ *or* $1.6(x - 1.7) > 3.6$ — **$x \leq -11.6$ or $x > 3.95$**
6. $-3.1(x - 5.2) < 9.2$ *and* $8.4(9 - x) > 22.6$ — **$x > 2.23$ and $x < 6.31$**
7. $2x + 6.4 \leq 12$ *or* $2x + 4 \geq -8.2$ — **all values of x**
8. $3x + 5.1 > 1.5$ *and* $3x + 4.4 < 6.5$ — **$x > -1.2$ and $x < 0.7$**
9. $-7.5 < 4x + 2.5$ *and* $0.8x + 0.5 < 2.26$ — **$x > -2.5$ and $x < 2.2$**
10. $-0.7 \leq 3x - 4.3$ *and* $14.4x - 20.64 \leq 38.4$ — **$x \geq 1.2$ and $x \leq 4.1$**

59. $x \leq -4$ *or* $x \geq 3$

−6 −5 −4 −3 −2 −1 0 1 2 3 4 5 6

60. $a < -2$ *and* $a < 3$

−6 −5 −4 −3 −2 −1 0 1 2 3 4 5 6

61. $t > 2$ *and* $t < 4$

−6 −5 −4 −3 −2 −1 0 1 2 3 4 5 6

62. $x < 9$ *and* $x > -8$

−12 −10 −8 −6 −4 −2 0 2 4 6 8 10 12

63. $d > -8$ *and* $d < 1$

−12 −10 −8 −6 −4 −2 0 2 4 6 8 10 12

64. $x \leq 3$ *or* $x \geq 5$

−6 −5 −4 −3 −2 −1 0 1 2 3 4 5 6

65. $x \geq 4$ *or* $x \leq -3$

−6 −5 −4 −3 −2 −1 0 1 2 3 4 5 6

The answers to Exercises 66–71 can be found in Additional Answers beginning on page 1002.

Look Beyond

In Exercise 89, students find two real numbers that have an absolute value of 4. Students will learn to solve absolute-value equations and inequalities in Lesson 1.8.

76c.

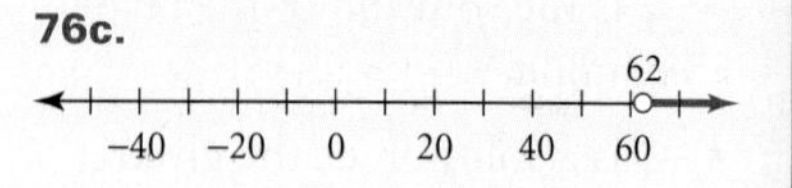

APPLICATION

76a. $54x > 40x + 868$; $x > 62$
b. 63

76. BUSINESS The money earned, or *revenue R*, from selling *x* units of a product is $R = 54x$. The cost of producing *x* units is $C = 40x + 868$. In order to make a profit, the revenue must be greater than the cost.
a. Write and solve an inequality in one variable that describes this relationship between revenue and cost.
b. How many units of the product must be sold in order to make a profit?
c. Graph the solution on a number line.

Look Back

Find the slope of each line. ***(LESSON 1.2)***

77. $y + 2x = 3$ **−2**

78. $3x - y = 6$ **3**

79. $x - 3y = -8$ $\frac{1}{3}$

80. $2x - 4y = 3(x - y) + 7$ **−1**

81. $y = -\frac{3}{2}x + \frac{7}{2}$
82. $y = \frac{7}{9}x - \frac{53}{9}$
83. $y = -\frac{8}{3}x - \frac{26}{3}$

Write the equation in slope-intercept form of a line that passes through the given points. ***(LESSON 1.3)***

81. (1, 2) and (3, −1) **82.** (5, −2) and (−4, −9) **83.** (8, −30) and (−1, −6)

APPLICATIONS

84. $d = 55t$

84. TRAVEL Michelle finds that after 4 hours of driving at a constant speed, she is 220 miles from her starting point. After 6 more hours, she is 550 miles from her starting point. Write an equation in slope-intercept form for the distance traveled, *d*, in miles in terms of the elapsed time, *t*, in hours. ***(LESSON 1.3)***

85. 60,000

85. GOVERNMENT In order to determine how people feel about a school-bond proposal, a public opinion poll is taken. Of a sample of 300 registered voters, 240 favor the bond proposal. If the number of people who favor the bond proposal is directly proportional to the number of registered voters, how many of the 75,000 registered voters favor the bond proposal? ***(LESSON 1.4)***

CONNECTION

Calculator button indicates that a graphics calculator is recommended.

86a. negative; $y \approx -15.50x + 69.15$
b. $y \approx 25.75$

86 STATISTICS Enter the data from the table below in a graphics calculator. ***(LESSON 1.5)***

x	1.0	1.3	1.5	1.6	1.8	1.9	2.0	2.2	2.3	2.5
y	58	47	50	39	40	35	41	31	34	36

a. Create a scatter plot, identify the correlation as positive or negative, and find the equation of the least-squares line.
b. Use the equation of the least-squares line that you found in part **a** to predict the value of *y* when *x* is 2.8.

Solve each literal equation for the indicated variable. ***(LESSON 1.6)***

87. $A = p + prt$ for t $t = \frac{A - p}{pr}$

88. $SA = 2ab + 2ac + 2bc$ for a $a = \frac{SA - 2bc}{2(b + c)}$

Look Beyond

89. What two real numbers have an absolute value of 4? **+4, −4**

Solving Absolute-Value Equations and Inequalities

Objective

- Write, solve, and graph absolute-value equations and inequalities in mathematical and real-world situations.

Why *Measurement usually involves an allowable amount of error, called measurement tolerance, which can be expressed by using absolute-value notation. Measurement tolerance is important in many fields, including manufacturing.*

APPLICATION
MANUFACTURING

A company manufactures a small gear for a car according to design specifications. If the gear is made too large, it will not fit. If it is made too small, the car will not run properly. What measurement tolerance is close enough for this gear? *You will solve this problem in Example 5.*

Definition of Absolute Value

$\lvert -3 \rvert = 3$	$\lvert 3 \rvert = 3$
The absolute value of a negative number is its opposite.	The absolute value of a nonnegative number is itself.

Notice that both 3 and −3 are 3 units from 0 on the number line.

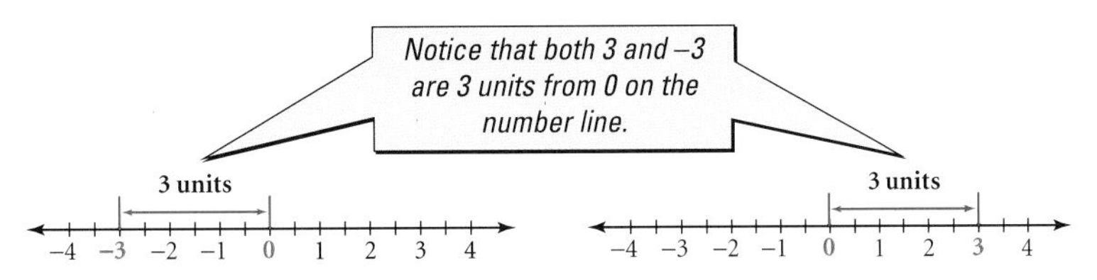

Prepare

NCTM PRINCIPLES & STANDARDS 1–4, 6–10

QUICK WARM-UP

Solve each equation.

1. $m + 2 = -2$ $\quad m = -4$
2. $m + 2 = 3m - 2$ $\quad m = 2$
3. $m + 2 = -(3m - 2)$ $\quad m = 0$

Solve each inequality.

4. $k - 3 > 5$ $\quad k > 8$
5. $k - 3 > 2k + 9$ $\quad k < -12$
6. $k - 3 > -(2k + 9)$ $\quad k > -2$

Also on Quiz Transparency 1.8

Teach

Why Ask students whether they have noticed differences between items that are supposed to be identical, such as two pairs of jeans of the same size. Explain that manufacturers allow size variations within certain limits. Ask students why this is done and how it can affect them. Point out that absolute value can be used to describe *allowances*, or *tolerances*, such as this.

Alternative Teaching Strategy

USING VISUAL MODELS On a number line, the distance between two numbers a and b is $|a - b|$. Use the steps demonstrated to show how this fact can be used to solve $|x - 6| = 9$.

$\lvert x - 6 \rvert$	$=$	9
↓	↓	↓
The distance between x and 6	is	9 units.

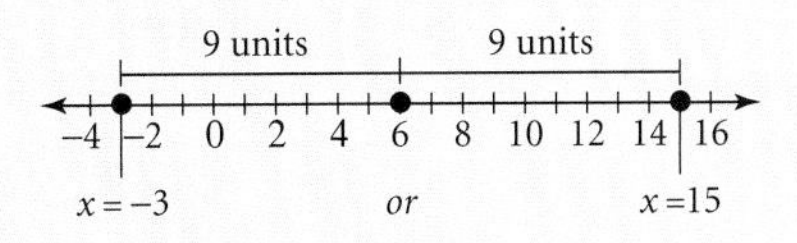

Extend the technique to solve the equation in Example 1:

$$|2x + 3| = 4 \rightarrow |2x - (-3)| = 4$$

The distance between $2x$ and -3 is 4.

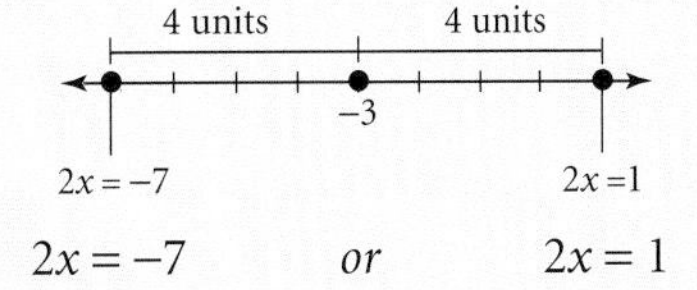

$2x = -7$ *or* $2x = 1$

$x = -3.5$ $\quad$ $x = 0.5$

The technique can also be applied to absolute-value inequalities.

CHECKPOINT ✔
$|-2| = -(-2) = 2$ and -2 is 2 units away from 0; $|-1| = -(-1) = 1$ and -1 is 1 unit away from 0; $|0| = 0$ and 0 is 0 units away from 0; $|1| = 1$ and 1 is 1 unit away from 0; $|2| = 2$ and 2 is 2 units away from 0.

Teaching Tips about technology refer to keystrokes for the TI-82, TI-83, or TI-83 Plus.

Teaching Tip

TECHNOLOGY To create the graphs on this page, first press WINDOW and use these settings.

Xmin=–5	**Ymin=–2**
Xmax=5	**Ymax=4**
Xscl=1	**Yscl=1**

Press Y=, clear the equations, place the cursor next to **Y1** and enter the following:

TI-83: MATH ► NUM 1:abs(ENTER X,T,θ,n) ▼ **2** GRAPH

TI-82: 2nd x^{-1} X,T,θ,n ▼ **2** GRAPH

This gives the first graph. To obtain the second graph, press Y= ▼ **(Y2=) 3** GRAPH. For the third graph, set **Ymin=–1** and **Ymax=5**, and then press Y= ▼ **(Y2=) 4** GRAPH.

A keystroke guide is provided at the end of each chapter for examples given in the Pupil's Edition. If you are using a different graphics calculator such as a Casio or Sharp, go to the HRW Internet site at go.hrw.com and enter the keyword MB1 CALC to find keystroke guides.

CHECKPOINT ✔
$x = 5$ *or* $x = -5$

The algebraic and geometric definitions of absolute value are given below.

Absolute Value

Let x be any real number.

Algebraic definition:
The **absolute value** of x, denoted by $|x|$, is given by the following:
If $x \geq 0$, then $|x| = x$.
If $x < 0$, then $|x| = -x$.

Geometric definition:
The **absolute value** of x is the distance from x to 0 on the number line.

CHECKPOINT ✔ Verify for the following values of x that the algebraic definition and the geometric definition of absolute value give the same result:

$$\{-2, -1, 0, 1, 2\}$$

Absolute-Value Equations

You can use a graph to better understand absolute-value equations. Examine the graphs below.

$y = |x|$ and $y = 2$ | $y = |x|$ and $y = 3$ | $y = |x|$ and $y = 4$

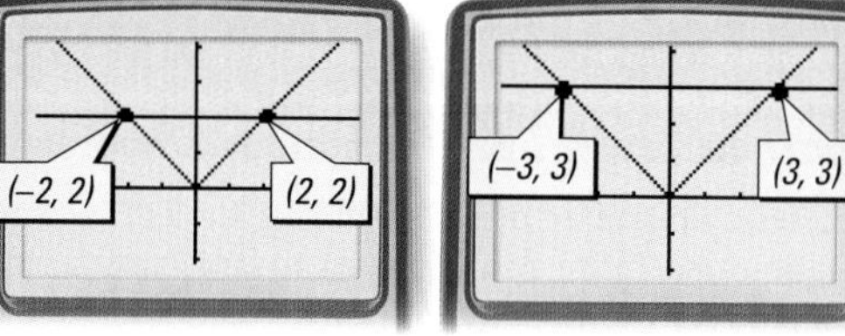

If $|x| = 2$, then $x = -2$ *or* $x = 2$. | If $|x| = 3$, then $x = -3$ *or* $x = 3$. | If $|x| = 4$, then $x = -4$ *or* $x = 4$.

The graphs suggest the following fact:

Absolute-Value Equations

If $a > 0$ and $|x| = a$, then $x = a$ *or* $x = -a$.

By definition, $|x|$ is a distance and is therefore always nonnegative. Notice, however, that the solution to an absolute-value equation can be negative.

CHECKPOINT ✔ Solve the absolute-value equation $|x| = 5$.

Interdisciplinary Connection

MANUFACTURING A quality-control inspector uses a *control chart* to record information about changes over time. The control chart at right shows an inspector's measurements of the lengths of a manufactured part at one-hour intervals. Write an absolute-value inequality to represent the measurement tolerance of the part. What appears to be happening to the lengths?

$|l - 6.200| \leq 0.150$, where l is the length of the part; the lengths appear to be approaching the upper control limit.

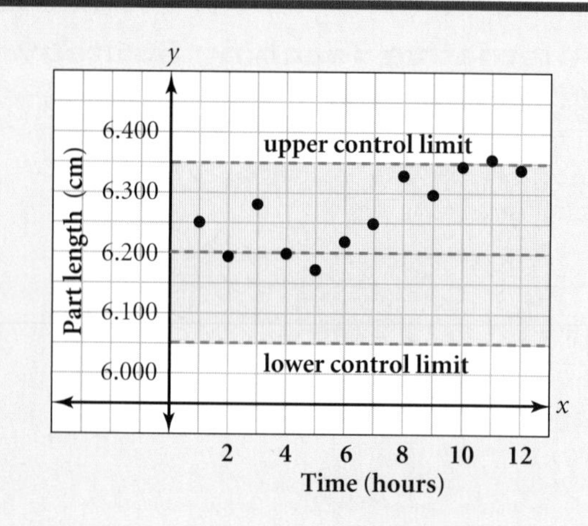

EXAMPLE 1 **Solve $|2x + 3| = 4$. Graph the solution on a number line.**

SOLUTION

Solve the two related equations.

$2x + 3 = 4$ *or* $2x + 3 = -4$

$2x = 1$ $\quad$ $2x = -7$

$x = 0.5$ $\quad$ $x = -\frac{7}{2}$, or -3.5

CHECK

Let $x = 0.5$.

$|2x + 3| = 4$

$|2(0.5) + 3| \stackrel{?}{=} 4$

$|4| = 4$ **True**

Let $x = -3.5$.

$|2x + 3| = 4$

$|2(-3.5) + 3| \stackrel{?}{=} 4$

$|-4| = 4$ **True**

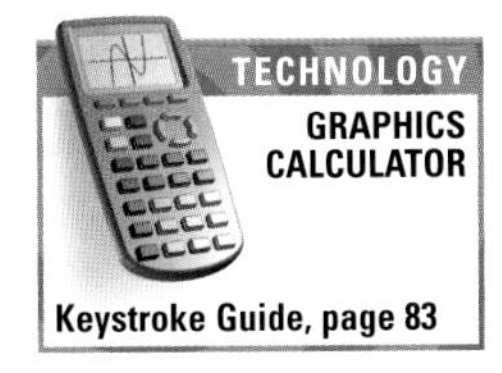

You can also check your solutions by graphing $y = |2x + 3|$ and $y = 4$ on the same screen.

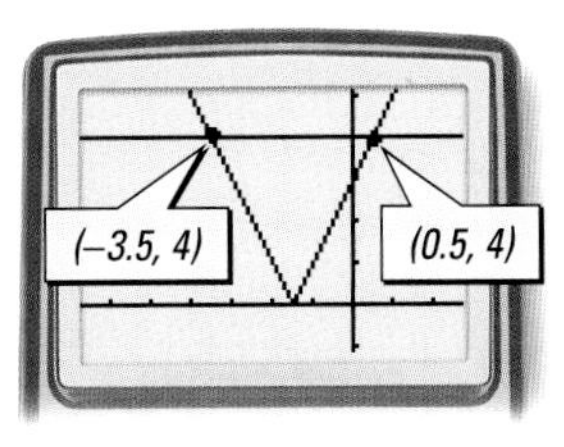

The graph shows that if $|2x + 3| = 4$, then $x = -3.5$ *or* $x = 0.5$.

The solution is graphed on the number line at right.

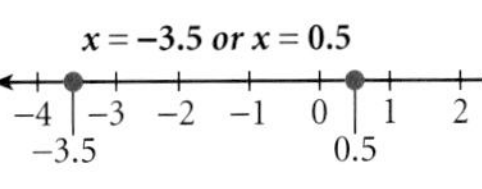

TRY THIS Solve $|3x + 5| = 7$. Graph the solution on a number line.

Activity Exploring Solution Possibilities

TECHNOLOGY GRAPHICS CALCULATOR Keystroke Guide, page 83

You will need: a graphics calculator

The display at right shows the graphs of $y = |x|$ and $y = 2x - 2$. These equations model $|x| = mx + b$, where $m = 2$ and $b = -2$.

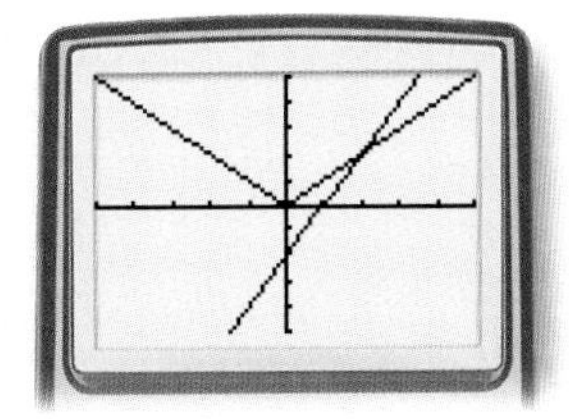

1. Graph these equations on the same screen. At how many points do the graphs intersect? From the graph, find the solution of $|x| = 2x - 2$.

PROBLEM SOLVING

2. **Guess and check.** Modify $y = 2x - 2$ so that the graph of the modified equation intersects the graph of $y = |x|$ at two points. Graph the equations to check. What are the solutions of $|x| = mx + b$ for your modified values of m and b?
3. Find values of m and b such that the graphs of $y = |x|$ and $y = mx + b$ have no points in common. Graph the equations to check.
4. Find values of m and b such that the graphs of $y = |x|$ and $y = mx + b$ have infinitely many points in common. Graph the equations to check.

CHECKPOINT ✔

5. Summarize the possible solutions of $|x| = mx + b$.

Inclusion Strategies

KINESTHETIC LEARNERS Draw a number line on a roll of paper, with the integers from −8 to 8 spaced about 1 foot apart. Place it on the floor. Ask for volunteers to act out absolute-value equations.

For instance, to act out $|x| = 3$, have one student stand at 0, holding a card marked with a large 0. Write a large X on two cards and give one to each of two other students. Each of these students should stand 3 units from 0, but in opposite directions. Elicit from the class that the solution is $x = 3$ *or* $x = -3$.

Repeat the demonstration for $|x - 4| = 3$, with the first student standing at 4 and the others standing at 1 and 7. Contrast this with $|x + 4| = 3$, or $|x - (-4)| = 3$. Now the first student should stand at −4 and the others at −1 and −7.

Then have students act out $|2x + 4| = 3$, giving cards labeled $2x$ to two students. Point out that, this time, the students holding the $2x$ cards should be standing at −1 and −7. If $2x = -1$ *or* $2x = -7$, it follows that $x = -0.5$ *or* $x = -3.5$.

ADDITIONAL EXAMPLE 1

Solve $|2x - 1| = 3$. Graph the solution on a number line.

$x = -1$ *or* $x = 2$

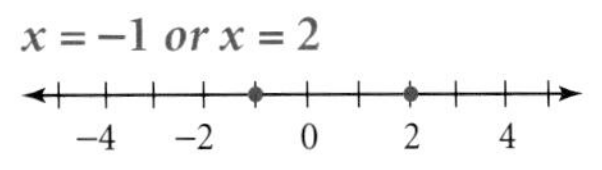

TRY THIS

$x = -4$ *or* $x = \frac{2}{3}$

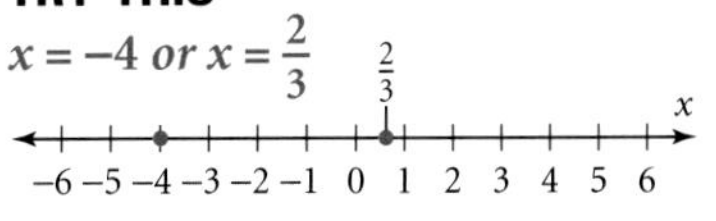

Activity Notes

In this Activity, students investigate the solutions of equations of the form $|x| = mx + b$ by examining the intersection of the graphs of $y = |x|$ and $y = mx + b$. Students should observe that the nature of the intersection, and thus the nature of the solution, depends on the slope of the line as well as its y-intercept.

Cooperative Learning

You may wish to have students do the Activity in groups of three. The groups can begin by doing Step 1 together. Then one group member can do Step 2, one can do Step 3, and the third can do Step 4. After each member has completed the assigned step, the group can come together to discuss the results and to create the summary for Step 5.

CHECKPOINT ✔

5. **There are either 0, 1, 2, or infinitely many solutions.**

ADDITIONAL

EXAMPLE 2

Solve $|2x + 1| = x + 5$. Check your solution(s).
$x = -2$ *or* $x = 4$

Teaching Tip

TECHNOLOGY When using a TI-83 or TI-82 graphics calculator to check the solution in Example 2, set the viewing window as follows:

Xmin=–3	Ymin=–3
Xmax=7	Ymax=7
Xscl=1	Yscl=1

The absolute-value function on a TI-83 graphics calculator is accessed by pressing MATH, selecting the **NUM** menu, and choosing **1:abs(**.

On a TI-82 graphics calculator, the absolute-value function is accessed by pressing 2nd x^{-1}.

To graph both sides of the equation, enter the following into the **Y=** editor:

Y1=abs(X–3)
Y2=3X+5

Press GRAPH to view the graphs.

To find the intersection point, press 2nd TRACE to access the **CALCULATE** menu. Choose **5:intersect**, and then press ENTER ENTER ENTER.

A keystroke guide is provided at the end of each chapter for examples given in the Pupil's Edition. If you are using a different graphics calculator, such as a Casio or Sharp, go to the HRW Internet site at go.hrw.com and enter the keyword MB1 CALC to find keystroke guides.

TRY THIS

$x = \frac{3}{2}$

EXAMPLE 2 **Solve $|x - 3| = 3x + 5$. Check your solution.**

SOLUTION

Solve the two related equations.

$$x - 3 = 3x + 5 \qquad \text{or} \qquad x - 3 = -(3x + 5)$$
$$-2x = 8 \qquad\qquad x - 3 = -3x - 5$$
$$x = -4 \qquad\qquad 4x = -2$$
$$x = -\frac{1}{2}, \text{ or } -0.5$$

CHECK

Let $x = -4$.
$$|x - 3| = 3x + 5$$
$$|(-4) - 3| \stackrel{?}{=} 3(-4) + 5$$
$$|-7| = -7 \quad \textbf{False}$$

Let $x = -0.5$.
$$|x - 3| = 3x + 5$$
$$|(-0.5) - 3| \stackrel{?}{=} 3(-0.5) + 5$$
$$|-3.5| = 3.5 \quad \textbf{True}$$

Since −4 does not satisfy the given equation and −0.5 does satisfy the given equation, the only solution is −0.5.

TECHNOLOGY GRAPHICS CALCULATOR
Keystroke Guide, page 83

You can also check your solution by graphing $y = |x - 3|$ and $y = 3x + 5$ on the same screen and looking for any points of intersection.

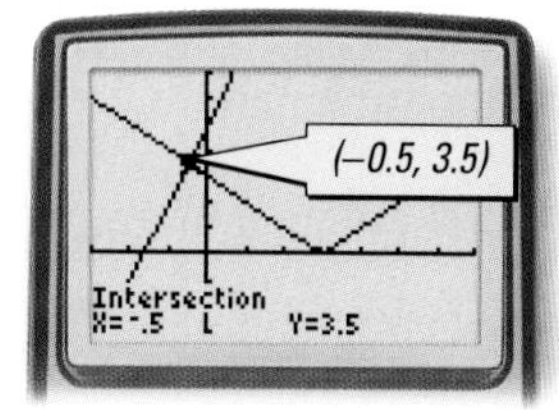

The graph shows that $|x - 3| = 3x + 5$ is true only when $x = -0.5$.

TRY THIS Solve $|x - 4| = x + 1$. Check your solution.

Absolute-Value Inequalities

The red dots on the number line below illustrate the solution of $|x| = 3$. Notice that these solutions divide the number line into three distinct regions.

This suggests a fact that will help you solve absolute-value inequalities.

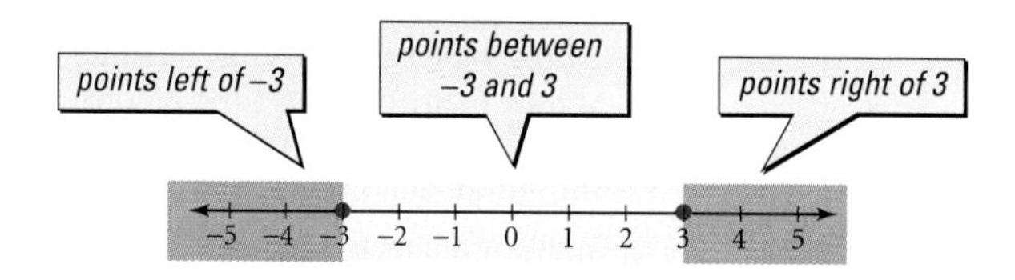

Absolute-Value Inequalities

If $a > 0$ and $|x| < a$, then $x > -a$ *and* $x < a$.
If $a > 0$ and $|x| > a$, then $x < -a$ *or* $x > a$.
You can write similar statements for $|x| \leq a$ and $|x| \geq a$.

Enrichment

Have students work in pairs to solve each of these absolute-value inequalities.

1. $1 < |x| < 5$ — $-5 < x < -1$ *or* $1 < x < 5$
2. $0 < |x| < 12$ — $-12 < x < 12$
3. $3 \leq |x - 2| \leq 9$ — $-7 \leq x \leq -1$ *or* $5 \leq x \leq 11$
4. $7 \leq |4x + 3| < 15$ — $-4.5 < x \leq -2.5$ *or* $1 \leq x < 3$
5. $-5 \leq |2x - 3| \leq -3$ — no real solutions

Now have each student write three inequalities similar to these, with the solutions, on a separate sheet of paper. Partners should exchange inequalities and solve them. Students should check each other's solutions, and they should work together to resolve any disagreements.

EXAMPLE 3 **Solve $|5 - 3x| > 9$. Graph the solution on a number line.**

SOLUTION

The inequality is of the form $|x| > a$, so solve an *or* statement.

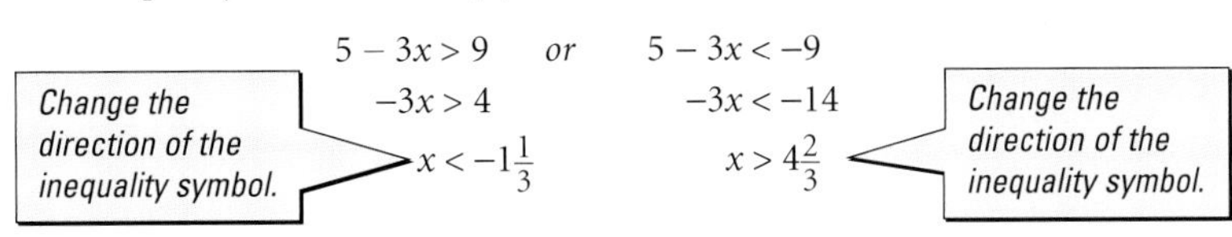
Change the direction of the inequality symbol.
Change the direction of the inequality symbol.

$$5 - 3x > 9 \quad or \quad 5 - 3x < -9$$
$$-3x > 4 \qquad\qquad -3x < -14$$
$$x < -1\tfrac{1}{3} \qquad\qquad x > 4\tfrac{2}{3}$$

TECHNOLOGY
GRAPHICS CALCULATOR
Keystroke Guide, page 83

CHECK

Check your solution by graphing $y = |5 - 3x|$ and $y = 9$ on the same screen.

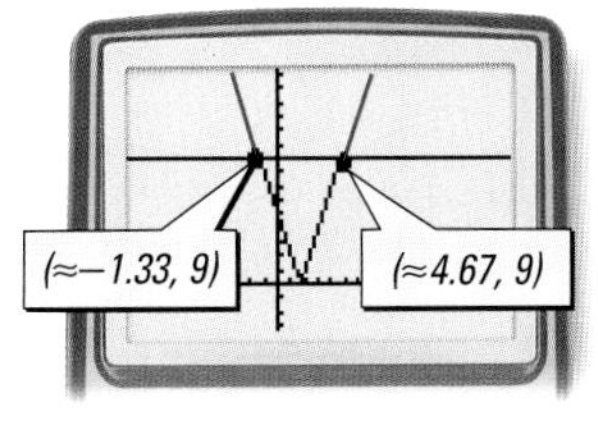
(≈–1.33, 9)
(≈4.67, 9)

The graph shows that if $|5 - 3x| > 9$, then $x < -1\frac{1}{3}$ *or* $x > 4\frac{2}{3}$.

The solution is graphed on the number line at right.

$x < -1\frac{1}{3}$ *or* $x > 4\frac{2}{3}$

TRY THIS Solve $|3x - 7| > 1$. Graph the solution on a number line.

EXAMPLE 4 **Solve $|5 - 3x| < 9$. Graph the solution on a number line.**

SOLUTION

The inequality is of the form $|x| < a$, so solve an *and* statement.

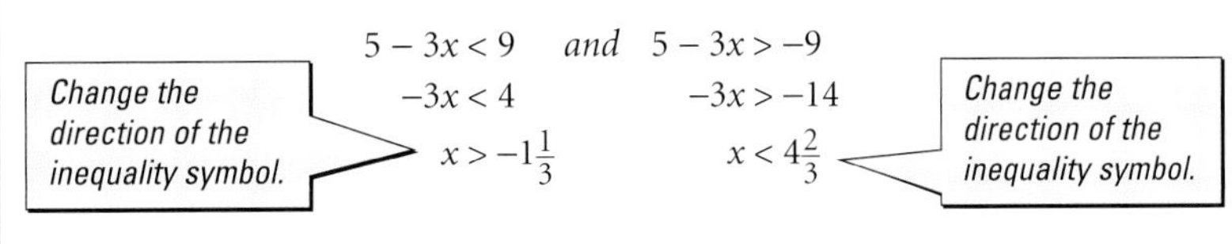
Change the direction of the inequality symbol.
Change the direction of the inequality symbol.

$$5 - 3x < 9 \quad and \quad 5 - 3x > -9$$
$$-3x < 4 \qquad\qquad -3x > -14$$
$$x > -1\tfrac{1}{3} \qquad\qquad x < 4\tfrac{2}{3}$$

TECHNOLOGY
GRAPHICS CALCULATOR
Keystroke Guide, page 83

CHECK

Check your solution by graphing $y = |5 - 3x|$ and $y = 9$ on the same screen.

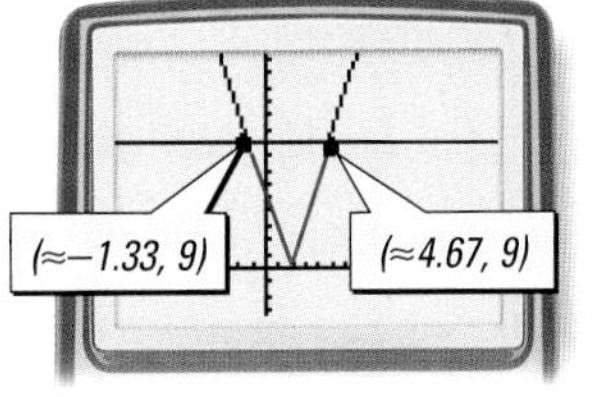
(≈–1.33, 9)
(≈4.67, 9)

The graph shows that if $|5 - 3x| < 9$, then $x > -1\frac{1}{3}$ *and* $x < 4\frac{2}{3}$, or $-1\frac{1}{3} < x < 4\frac{2}{3}$.

The solution is graphed on the number line at right.

$x > -1\frac{1}{3}$ *and* $x < 4\frac{2}{3}$

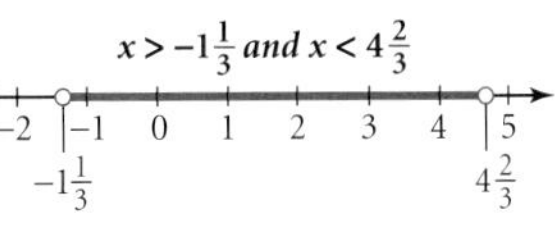
–2
–1
0
1
2
3
4
5
$-1\frac{1}{3}$
$4\frac{2}{3}$

TRY THIS Solve $|5x - 3| < 7$. Graph the solution on a number line.

CHECKPOINT ✔ Compare and contrast the problems in Examples 3 and 4.

Reteaching the Lesson

COOPERATIVE LEARNING Review with students the number-line interpretation of absolute value. Then have students work in pairs. Display graphs of the solutions to absolute-value equations or inequalities on the board or overhead, one at a time. (Four samples are shown at right.) Tell students that they are to write two different absolute-value equations or inequalities whose solutions are given by the graph. Discuss the results for each graph with the class.

Sample answers are given.

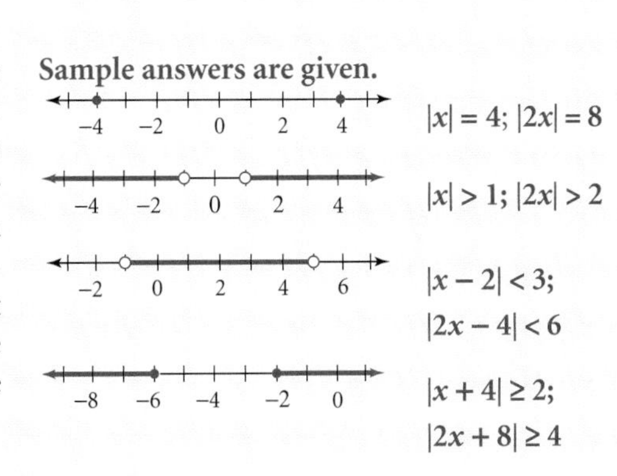

$|x| = 4$; $|2x| = 8$

$|x| > 1$; $|2x| > 2$

$|x - 2| < 3$; $|2x - 4| < 6$

$|x + 4| \geq 2$; $|2x + 8| \geq 4$

ADDITIONAL EXAMPLE 3

Solve $|3x + 2| > 4$. Graph the solution on a number line.

$x < -2$ *or* $x > \frac{2}{3}$

TRY THIS

$x < 2$ *or* $x > \frac{8}{3}$

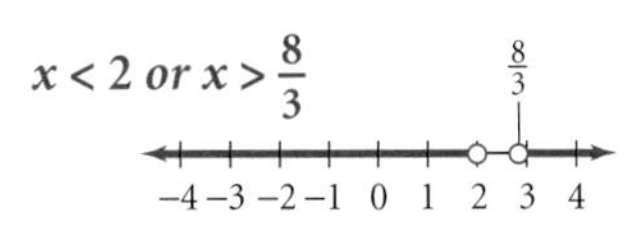

ADDITIONAL EXAMPLE 4

Solve $|5x + 2| \leq 8$. Graph the solution on a number line.

$-2 \leq x \leq 1\frac{1}{5}$

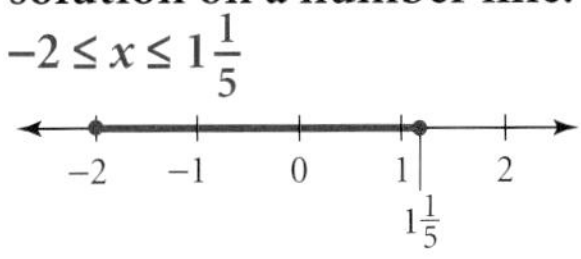

TRY THIS

$-\frac{4}{5} < x < 2$

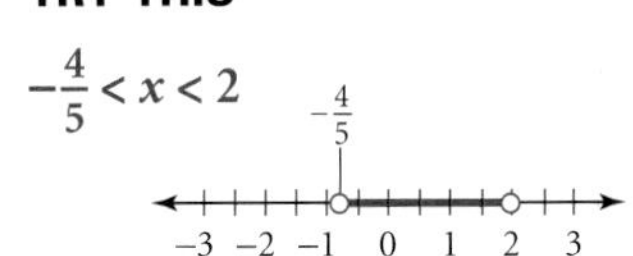

CHECKPOINT ✔

The expressions are the same, but the inequality sign is reversed.

ADDITIONAL
EXAMPLE 5

A connecting rod is designed with a specification of 54.0 centimeters for its length. The rod will work if it is within 0.0004 centimeter of the specified length. **Write an absolute-value inequality to represent the measurement tolerance for the length of this connecting rod.**

$|l - 54.0| \leq 0.0004$, where l represents the length of the connecting rod

TRY THIS
$|t - 12.00| \leq 0.01$

CHECKPOINT ✓
Answers may vary. sample answer: $|x + 2| \leq -3$; $|x + 2| \geq 0$

CRITICAL THINKING
The value of a must be less than 0, and the solution to $|x| > a$ is all real numbers.

You can write an absolute-value inequality to describe the measurement tolerance for a machine part.

EXAMPLE 5

APPLICATION
MANUFACTURING

A gear is designed with a specification of 3.50 centimeters for the diameter. It will work if it is no more than ±0.01 centimeter of the specified measurement.

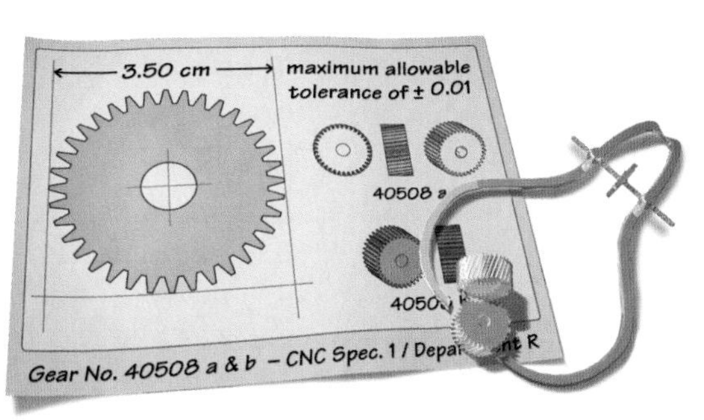

Write an absolute-value inequality to represent the measurement tolerance for the diameter of this gear.

SOLUTION

Let d represent an acceptable gear diameter.

$$d \geq 3.50 - 0.01 \textit{ and } d \leq 3.50 + 0.01$$

Write this compound inequality as follows:

$$3.50 - 0.01 \leq d \leq 3.50 + 0.01$$

Solve the compound inequality.

$$3.50 - 0.01 \leq \quad d \quad \leq 3.50 + 0.01$$
$$3.50 - 0.01 - 3.50 \leq d - 3.50 \leq 3.50 + 0.01 - 3.50$$
$$-0.01 \leq d - 3.50 \leq 0.01$$

Thus, $|d - 3.50| \leq 0.01$ represents the tolerance for the diameter of this gear.

TRY THIS Write $12.00 - 0.01 \leq t \leq 12.00 + 0.01$ as an absolute-value inequality.

An absolute-value inequality may have no solution, or any real number may be a solution. Examine the graphs of $y = |2x - 1|$ and $y = -3$ below.

Notice that the graph of $y = |2x - 1|$ is never below the graph of $y = -3$. Thus, $|2x - 1| < -3$ has no solution. There is no value of x for which the absolute value, $|2x - 1|$, is less than -3.

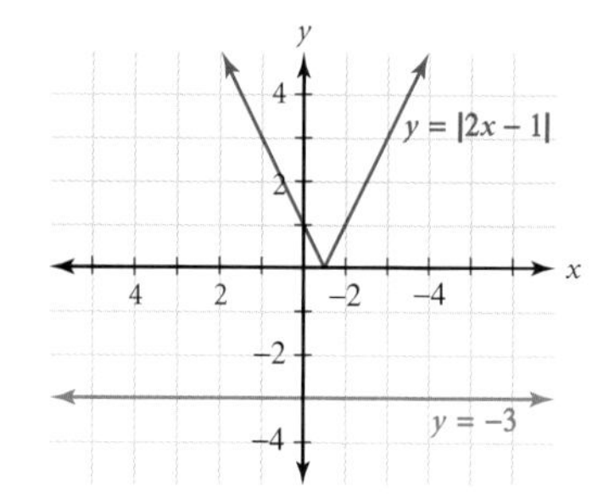

Notice also that the graph of $y = |2x - 1|$ is always above the graph of $y = -3$. Thus, any real number is a solution to $|2x - 1| > -3$. The absolute value of any real number is greater than -3.

CHECKPOINT ✓ Write an absolute-value inequality that contains ≤ and has no solution. Write an absolute-value inequality that contains ≥ and has all real numbers as its solution.

CRITICAL THINKING If $|x| \leq a$ has no solution, what can you conclude about the possible values of a and about the solution to $|x| > a$?

Exercises

Communicate

1. Explain why the equation $|3x - 5| + 4 = 3$ has no solution.
2. Discuss why it is necessary to always check your solution when solving absolute-value equations.
3. Explain why an absolute-value equation can have two solutions.
4. Discuss the meanings of the words *and* and *or*. Compare the mathematical meanings of these words with their common meanings.
5. Use a graph to describe the type of absolute-value inequality whose solution is any real number.

Guided Skills Practice

Solve each equation. Check your solution. ***(EXAMPLES 1 AND 2)***

6. $|x - 10| = 4$ **14, 6**
7. $|2x - 5| = 3$ **4, 1**
8. $10 = |7 - 3x|$ **$-1, \frac{17}{3}$**
9. $x + 4 = |x - 2|$ **-1**
10. $\frac{1}{2}x + 1 = |x - 2| - 1$ **8, 0**
11. $\frac{1}{2}x + 1 = |x + 3|$ **no solution**

Solve each inequality. Graph the solution on a number line. ***(EXAMPLES 3 AND 4)***

12. $2 < |4 - x|$
13. $|2x + 1| \geq 5$
14. $|5 + x| < \frac{1}{2}$
15. $\frac{1}{2}|2x + 1| \geq 2$
16. $3|x + 1| \leq 2$
17. $3|x + 1| + 3 > 2$

APPLICATION

18. **RECREATION** Ashley tosses a horseshoe at a stake 25 feet away. The horseshoe lands no more than 2 feet from the stake. ***(EXAMPLE 5)***
   a. Write an absolute-value inequality that represents the range of distances that the horseshoe traveled.
   b. Solve this inequality and graph it on a number line.

18a. $|x - 25| \leq 2$ or $|25 - x| \leq 2$

b. $23 \leq x \leq 27$

Practice and Apply

Match each statement on the left with a statement or sentence on the right.

19. $|x + 2| = 4$ **b**
20. $|x + 2| < 4$ **a**
21. $|x + 2| < -4$ **d**
22. $|x + 2| > -4$ **e**
23. $|x + 2| > 4$ **c**
24. $|x + 2| = -4$ **d**

a. $x < 2$ *and* $x > -6$
b. $x = 2$ *or* $x = -6$
c. $x > 2$ *or* $x < -6$
d. There is no solution.
e. The solution is all real numbers.
f. none of the above

Assess

Selected Answers

Exercises 6–18, 19–77 odd

ASSIGNMENT GUIDE	
In Class	1–18
Core	19–24, 25–57 odd, 58–60, 63, 65
Core Plus	19–24, 26–56 even, 58–65
Review	66–78
Preview	79

Extra Practice can be found beginning on page 940.

Answers to odd-numbered Extra Practice exercises can be found immediately after Selected Answers in the Pupil's Edition.

12. $x < 2$ *or* $x > 6$

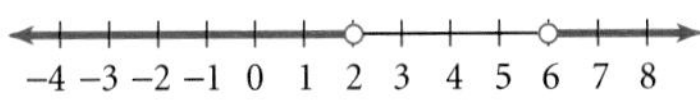

13. $x \leq -3$ *or* $x \geq 2$

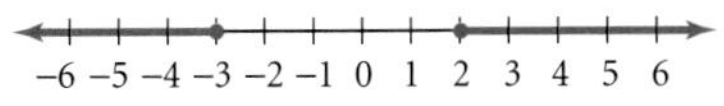

14. $-\frac{11}{2} < x < -\frac{9}{2}$

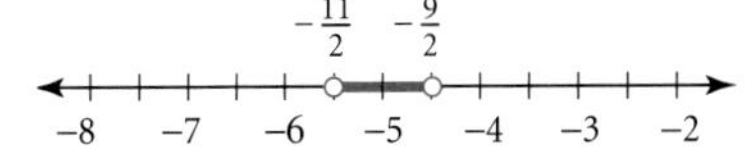

15. $x \leq -1$ *or* $x \geq 1$

16. $x \geq -\frac{3}{2}$

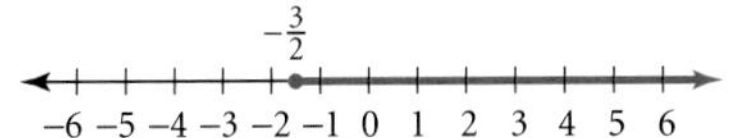

17. all real numbers

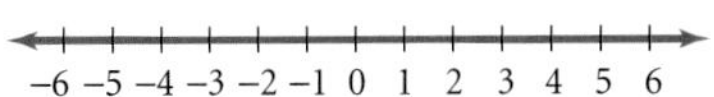

18b. $23 \leq x \leq 27$

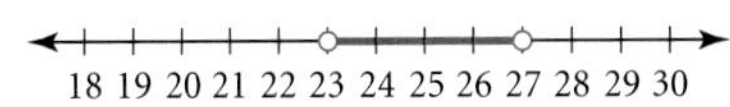

Error Analysis

When solving an absolute-value equation or inequality, some students solve only one of the related equations or inequalities. For instance, given $|3x - 5| = 9$, they may solve only $3x - 5 = 9$. Since absolute-value equations and inequalities generally have two related parts, suggest that students begin each exercise by writing the following on their papers:

☐ *and/or* ☐

This can serve as a visual reminder that they are not finished until they have filled in both boxes and have chosen the correct conjunction.

25. $x = 4$ *or* $x = -12$

26. $x = -7$ *or* $x = 17$

27. $x = 8$ *or* $x = -12$

28. $x = 7$ *or* $x = 9$

29. $x = -7$ *or* $x = 11$

30. $x = 6$ *or* $x = -16$

31. $x = 2$ *or* $x = 13$

32. $x = -10$ *or* $x = 2$

33. $x = -4\frac{1}{2}$ *or* $x = 9\frac{1}{2}$

34. $x = -5\frac{1}{2}$ *or* $x = 3$

Practice

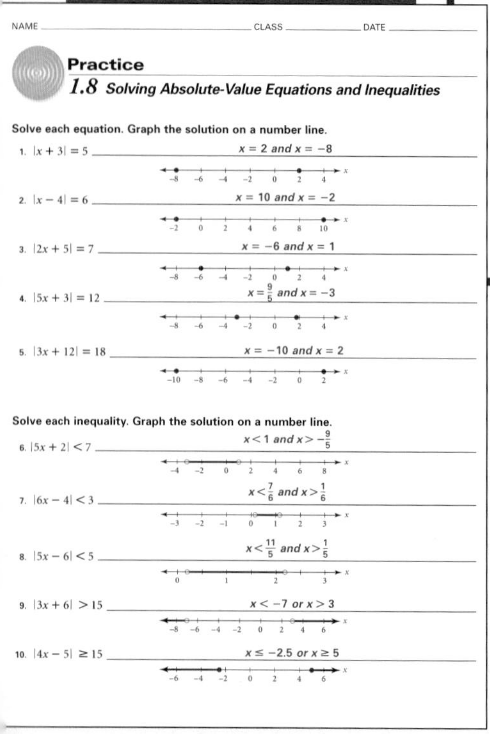

NAME ______ CLASS ______ DATE ______

Practice
1.8 Solving Absolute-Value Equations and Inequalities

Solve each equation. Graph the solution on a number line.

1. $|x + 3| = 5$ — *$x = 2$ and $x = -8$*
2. $|x - 4| = 6$ — *$x = 10$ and $x = -2$*
3. $|2x + 5| = 7$ — *$x = -6$ and $x = 1$*
4. $|5x + 3| = 12$ — *$x = \frac{9}{5}$ and $x = -3$*
5. $|3x + 12| = 18$ — *$x = -10$ and $x = 2$*

Solve each inequality. Graph the solution on a number line.

6. $|5x + 2| < 7$ — *$x < 1$ and $x > -\frac{9}{5}$*
7. $|6x - 4| < 3$ — *$x < \frac{7}{6}$ and $x > \frac{1}{6}$*
8. $|5x - 6| < 5$ — *$x < \frac{11}{5}$ and $x > \frac{1}{5}$*
9. $|3x + 6| > 15$ — *$x < -7$ or $x > 3$*
10. $|4x - 5| \geq 15$ — *$x \leq -2.5$ or $x \geq 5$*

internet connect go.hrw.com
Homework Help Online
Go To: **go.hrw.com**
Keyword:
MB1 Homework Help
for Exercises 25–57

Solve each equation.

25. $|x + 4| = 8$
26. $|x - 5| = 12$
27. $|2 + x| = 10$
28. $|8 - x| = 1$
29. $|x - 2| = 9$
30. $|x + 5| = 11$
31. $|2x - 15| = 11$
32. $|3x + 12| = 18$
33. $|10 - 4x| = 28$
34. $|5 + 4x| = 17$
35. $|5x - 6| = 2$
36. $|10 - 3x| + 5 = 2$
37. $|10x + 2| - 18 = -12$
38. $|4 - 3x| - 9 = 3$
39. $|2x - 8| + 2 = 1$

Solve each inequality. Graph the solution on a number line. If the equation has no solution, write *no solution*.

40. $|x - 4| > 1$
41. $|x + 5| \leq 7$
42. $|3x| > 15$
43. $|-2x| \leq 12$
44. $|4x| \leq -8$
45. $|3 - x| \geq -5$
46. $|2 + 5x| \leq 3$
47. $|2x - 3| < 11$
48. $|4x + 6| \leq 14$
49. $\left|\frac{2x + 3}{-5}\right| < 3$
50. $|4x - 5| \geq 15$
51. $|2x - 1| \geq -5$
52. $|5x + 3| > -2$
53. $|7 - 6x| < -4$
54. $|9x + 4| \leq -11$
55. $-2|4x + 1| \leq -4$
56. $-2|4x + 1| \geq -4$
57. $\left|\frac{3}{2} - \frac{5}{2}x\right| < -\frac{7}{2}$

SUMMARY
Three different representations of an inequality are given below.

Verbal	Algebraic	Graphic
The distance between x and 3 is less than 5.	$\lvert x - 3\rvert < 5$	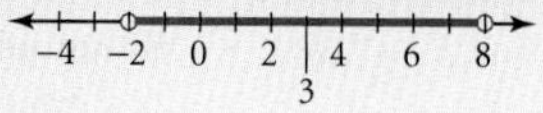

Examine the summary box above. For Exercises 58–60, write the two missing representations for each inequality.

58. The distance between x and 7 is less than 4.

59. $|x - 4| < 1$

60. 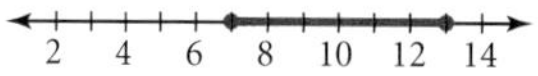

CHALLENGE

61. Solve the inequality $\left|\frac{4x}{3}\right| \leq 2x + 5$. $x \geq -\frac{3}{2}$

For Exercises 62–65, write and solve an absolute-value inequality.

APPLICATIONS

62. between 114 lb and 126 lb

62. HEALTH Antonio weighs 120 pounds, and his doctor said that his weight differs from his ideal body weight by less than 5 percent. What are the possible values, to the nearest pound, for Antonio's ideal body weight?

63. ENTERTAINMENT A tightrope walker is 10 feet from one end of the rope. If he then takes 3 steps and each step is 11 inches long, how far is he now from the same end of the rope? Give both possible answers. **7.25 ft or 12.75 ft**

35. $x = \frac{4}{5}$ *or* $x = \frac{8}{5}$

36. no solution

37. $x = -\frac{4}{5}$ *or* $x = \frac{2}{5}$

38. $x = -\frac{8}{3}$ *or* $x = \frac{16}{3}$

39. no solution

40. $x < 3$ *or* $x > 5$

41. $x \geq -12$ *and* $x \leq 2$

42. $x < -5$ *or* $x > 5$

43. $x \geq -6$ *and* $x \leq 6$

44. no solution

45. all real numbers

46. no solution

47. $x > -4$ *and* $x < 7$

48. $x \geq -5$ *and* $x \leq 2$

49. $x < \frac{12}{7}$

50. $x \leq -\frac{5}{2}$ or $x \geq 5$

51. all real numbers

52. all real numbers

53. no solution

54. no solution

55. $x < 2$ *or* $x > \frac{8}{3}$

56. $x > -\frac{4}{5}$ *and* $x < 2$

57. no solution

58. $x > 3$ *and* $x < 11$

59. The distance between x and 4 is less than 1.

60. The distance between x and 10 is less than or equal to 3; $|x - 10| \leq 3$.

APPLICATIONS

64. from 28 ft/s to 32 ft/s

64. METEOROLOGY An instrument called an *anemometer* measures a wind speed of 30 feet per second. The true wind speed is within 2 feet per second (inclusive) of the measured wind speed. What range is possible for the true wind speed?

An anemometer is used to measure wind speed.

65. RECREATION A recent poll reported that 68 percent of moviegoers eat popcorn during the movie. The margin of error for the poll was 3%. What are the minimum and maximum possible percents according to this poll? **65%, 71%**

Look Back

APPLICATIONS

66b. $v = 90{,}000 - 4500t$
c. $22,500

66. REAL ESTATE A rental property is purchased for $90,000. For tax purposes, a depreciation of 5% of property's initial value, or $4500, is assumed per year. ***(LESSON 1.1)***

a. Make a table of values for the value of the property, v, after t years.
b. Write a linear equation for the value of the property, v, after t years.
c. What is the value of the property after 15 years?

67. SPORTS A baseball pitcher allows 10 runs in 15 innings. At this rate, how many runs would you expect the pitcher to allow in a 9-inning game? **6** ***(LESSON 1.4)***

Solve each proportion for *x*. *(LESSON 1.4)*

68. $\frac{2}{x} = \frac{5}{8}$ **$\frac{16}{5}$**

69. $\frac{x-3}{4} = \frac{2x}{16}$ **6**

70. $\frac{10x}{-60} = \frac{2x-10}{8}$ **3**

71. Solve $\frac{1}{2}bh = A$ for h. ***(LESSON 1.6)*** $h = \frac{2A}{b}$

72. Solve $P = 2l + 2w$ for l. ***(LESSON 1.6)*** $l = \frac{P-2w}{2}$

Solve each inequality, and graph the solution on a number line. *(LESSON 1.7)*

73. $4x - 5 < \frac{1}{3}(8x + 3)$

74. $x - 9 \geq \frac{1}{6}(21 + x)$

Graph each compound inequality on a number line. *(LESSON 1.7)*

75. $x > -1$ *and* $x < 5$

76. $x < 3$ *and* $x > -3$

77. $x \leq -2$ *or* $x > 4$

78. $x > 2$ *or* $x \leq -1$

Look Beyond

APPLICATION

79. ETYMOLOGY Look up the word *rational* in the dictionary. Write the definition that is related to math. What is the meaning of the root word, *ratio,* that relates to the math usage?

79. *Rational*—a number or quantity that can be expressed as the quotient of two integers, where the divisor is not zero. *Ratio*—the indicated quotient of two mathematical expressions.

The answer to Exercise 66 can be found in Additional Answers beginning on page 1002.

In Exercise 79, students are asked to research the meaning and derivation of the word *rational* as it is used in mathematics. Students will study rational numbers when they study the structure of the real-number system in Lesson 2.1. You may wish to use this exercise as a transition to Chapter 2.

73. $x < \frac{9}{2}$

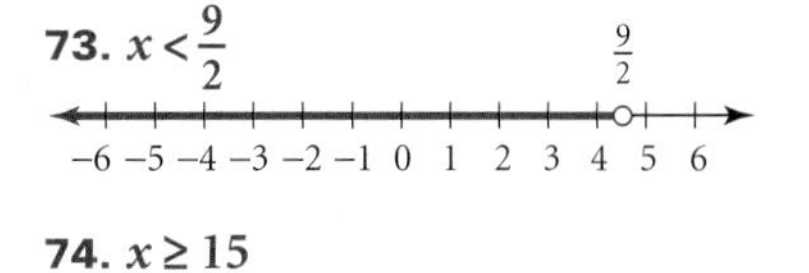

74. $x \geq 15$

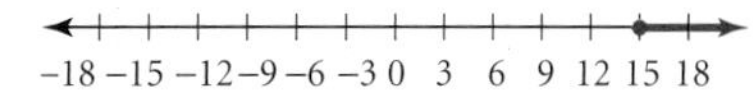

75. $-1 < x < 5$

–6 –5 –4 –3 –2 –1 0 1 2 3 4 5 6

76. $-3 < x < 3$

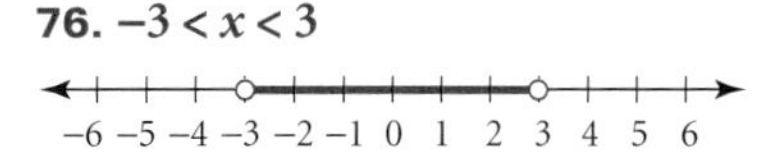

77. $x \leq -2$ *or* $x > 4$

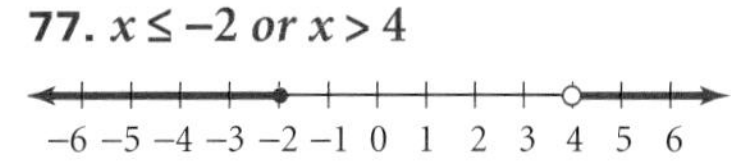

78. $x \leq -1$ *or* $x > 2$

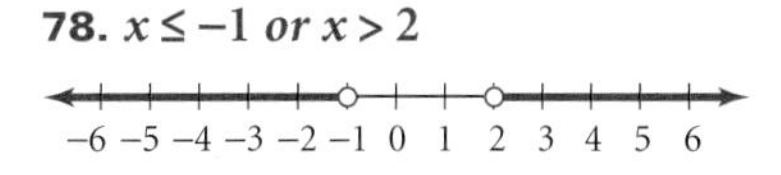

Student Technology Guide

NAME ______ CLASS ______ DATE ______

Student Technology Guide
***1.8* Absolute-Value Equations and Inequalities**

The logic feature of a graphics calculator can be used to solve an absolute-value inequality. For example, to solve $|x - 4| > 3$, enter the following key strokes.

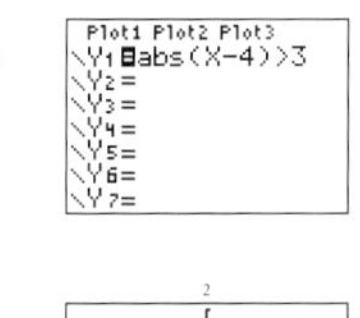

You will see the screen at the right.

Press WINDOW. Choose $-10 \leq x < 10$ and $-2 \leq y \leq 2$.

Press GRAPH. The endpoints of the two horizontal line segments on the display give you a clue to the solution. To locate the endpoints, press TRACE. The endpoints have x-values of 1 and 7. Since the inequality involves $>$, the solution has two parts, $x < 1$ *or* $x > 7$.

Use a graphics calculator to solve each absolute-value inequality.

1. $|x - 3.4| > 2.2$ — $x < 1.2$ *or* $x > 5.6$
2. $|-3.9x| \geq 12.5$ — $x \leq -3.21$ *or* $x \geq 3.21$
3. $|3.8x - 1.8| > 3.6$ — $x \leq -0.47$ *or* $x \geq 1.42$
4. $|2.4x + 7.1| \geq 3.4$ — $x \leq -4.38$ *or* $x \geq -1.54$
5. $|5.8x - 4.3| > 8.2$ — $x \leq -0.67$ *or* $x \geq 2.16$
6. $|1.5x - 3.2| \geq 12.6$ — $x \leq -6.27$ *or* $x \geq 10.53$

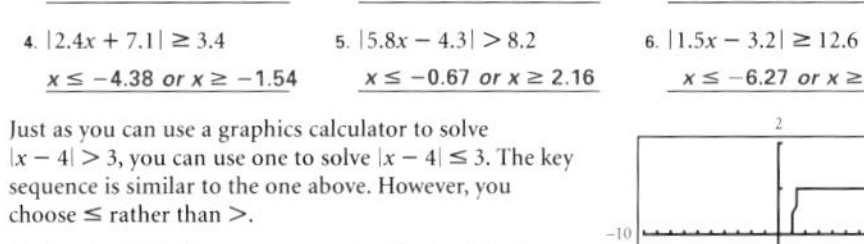

Just as you can use a graphics calculator to solve $|x - 4| > 3$, you can use one to solve $|x - 4| \leq 3$. The key sequence is similar to the one above. However, you choose $\leq$ rather than $>$.

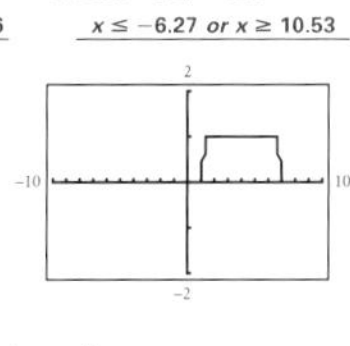

Notice the single line segment on the display. As before, locate the endpoints to help find the solution to the given inequality. Because the inequality involves $\leq$, the endpoints of the segment are part of the solution. The solution to $|x - 4| \leq 3$ is $x \geq 1$ *and* $x \leq 7$.

Use a graphics calculator to solve each absolute-value inequality.

7. $|1.5 - 0.25x| \leq 6$ — $x \geq -18$ *and* $x \leq 30$
8. $|6.4 - 3.2x| + 5.1 < 10.4$ — $x > -0.34$ *and* $x < 3.66$
9. $|x + 4.3| < 1.8$ — $x > -6.1$ *and* $x < -2.5$
10. $|1.4 - 2.4x| \leq 2.8$ — $x \geq -0.58$ *and* $x \leq 1.75$
11. $|2(x - 4.5)| \leq 5.6$ — $x \geq 1.7$ *and* $x \leq 7.3$
12. $|3 - 2.5(x + 1)| \leq 10$ — $x \geq -3.8$ *and* $x \leq 4.2$

Focus

To most people, it will seem natural that there are relationships among body measurements. For instance, it makes sense that the taller of two people will also have the longer arm span. In this Chapter Project, students take measurements and investigate whether some specific relationships among them can be modeled by linear equations.

Motivate

Before starting the project, ask students to estimate the hand span, the arm span, and the distance from head to ceiling for a basketball player who is 7 feet tall. Have them explain their estimates. Ask if they think that there is a relationship between the player's height and each of these measurements. If they believe that there is a relationship, ask them to describe it.

Activity 1

1. Answers will vary. Sample answer: Yes, it seems reasonable that taller people would have longer arm spans. If you consider the arm span of a young child, it is much shorter than that of a teenager.
2. Answers will vary. Sample answer: Yes, it seems reasonable that taller people would have bigger hand spans. For example, children's and adult's gloves are different sizes.
3. Answers will vary. Sample answer: Yes, the distance from the top of a person's head to the ceiling is equal to the height of the ceiling minus the person's height.
4. Answers will vary. Sample answer: No, the amount of money a person has in his or her pocket is not in any way related to the person's height.

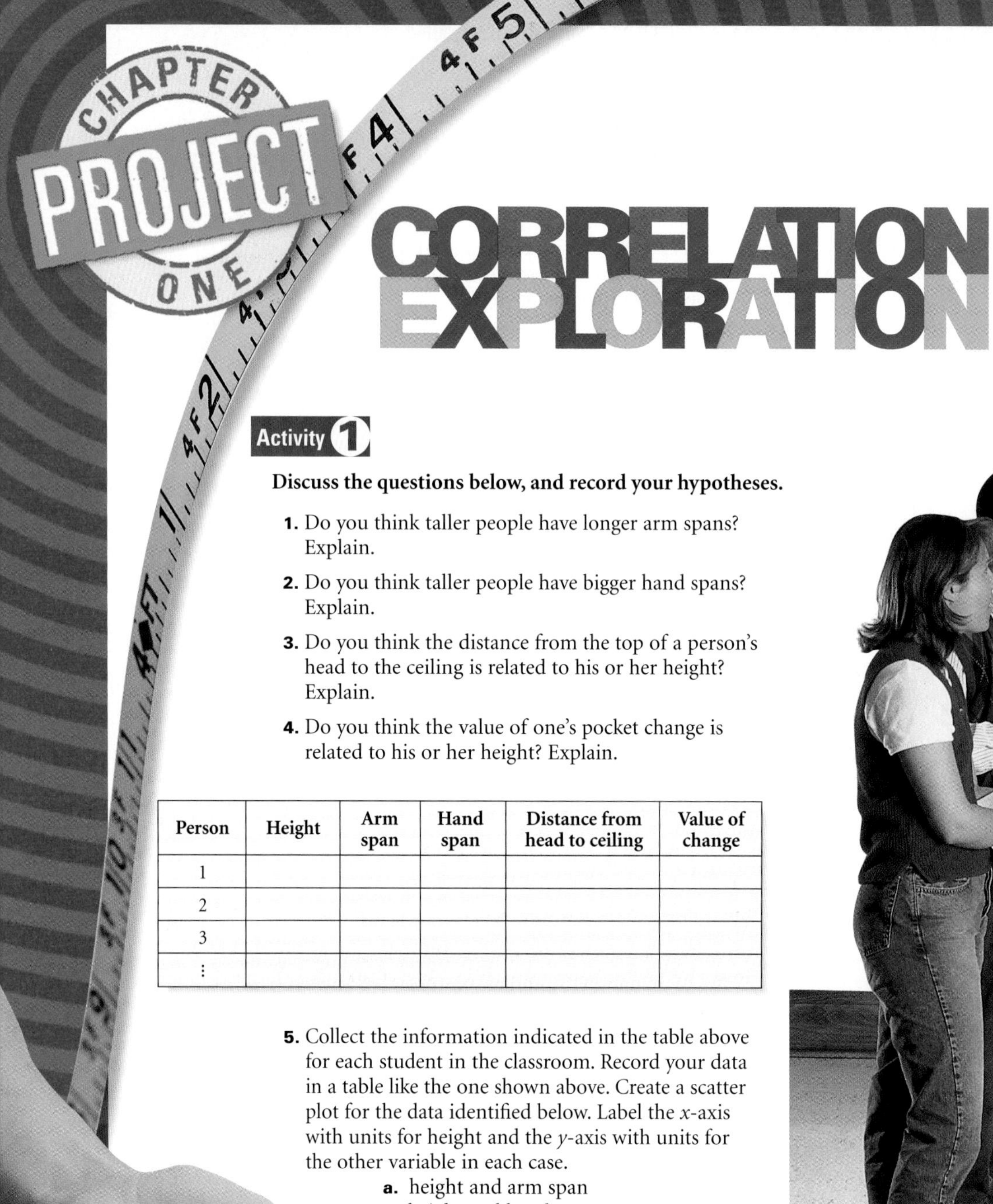

CORRELATION EXPLORATION

Activity 1

Discuss the questions below, and record your hypotheses.

1. Do you think taller people have longer arm spans? Explain.
2. Do you think taller people have bigger hand spans? Explain.
3. Do you think the distance from the top of a person's head to the ceiling is related to his or her height? Explain.
4. Do you think the value of one's pocket change is related to his or her height? Explain.

Person	Height	Arm span	Hand span	Distance from head to ceiling	Value of change
1					
2					
3					
⋮					

5. Collect the information indicated in the table above for each student in the classroom. Record your data in a table like the one shown above. Create a scatter plot for the data identified below. Label the x-axis with units for height and the y-axis with units for the other variable in each case.
   a. height and arm span
   b. height and hand span
   c. height and distance from the top of one's head to the ceiling
   d. height and value of pocket change

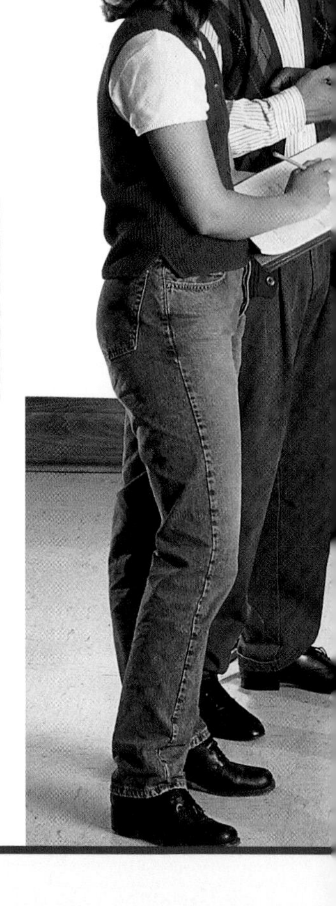

Activity 2

Answers will vary. Sample answers:

1–2a. height and arm span: $y = x$

b. height and hand span: $y = 0.07x + 3.24$

c. height and distance from head to ceiling: $y = 120 - x$

d. height and coin value: $y = -13.9x + 1061.8$ (Note that the data points do not seem to be related, so a linear model may not be useful in finding further data points.)

3a. The linear model is $y = x$, with a slope of 1 and a y-intercept of 0.

b. The linear model is $y = 0.07x + 3.24$, with a slope of 0.07 and a y-intercept of 3.24.

c. The linear model is $y = 120 - x$, with a slope of -1 and a y-intercept of 120.

d. The linear model is $y = -13.9x + 1061.8$, with a slope of -13.9 and a y-intercept of 1061.8.

Activity 2

1. Use a straightedge to estimate a linear model for each scatter plot.
2. Write an equation for each of your linear models.
3. Identify the slope and y-intercept for each of your linear models.
4. For each linear model, what does the slope tell you about the relationship between the variables?
5. For each linear model, what does the y-intercept tell you about the relationship between the variables?

Activity 3

1. Describe the correlation between the variables represented in each scatter plot.
2. Find the correlation coefficient for each scatter plot.
3. For each scatter plot, find and graph an equation for the least-squares line.
4. For each scatter plot, was your linear model reasonably close to the least-squares line? Explain.
5. Use the equations of the least-squares lines to make each prediction below.
   a. the arm span of a person who is 5 feet tall
   b. the hand span of a person who is 5 feet tall
   c. the distance to the ceiling from the head of a person who is 5 feet tall
   d. the value of the pocket change of a person who is 5 feet tall

Activity 4

Solve the equation of the appropriate least-squares line to make each prediction below. Name the Properties of Equality that you use to solve each equation.

1. the height of a person with an arm span of 62 inches
2. the height of a person with a hand span of $8\frac{1}{2}$ inches
3. the height of a person whose head is 28 inches from the ceiling
4. the height of a person with \$1.26 in pocket change

4a. **The slope tells you that there is a positive relationship between height and arm span.**

b. **The slope tells you that there is a positive relationship between height and hand span.**

c. **The slope tells you that there is a negative relationship between height and distance from head to ceiling.**

d. **The slope tells you that there is a negative relationship between height and coin value but that the model is not appropriate since there is really no relationship between these two variables. Also, the points do not tend to cluster around the line for the linear model.**

The answers to Step 5 of Activity 2 and all steps of Activities 3 and 4 can be found in Additional Answers beginning on page 1002.

Cooperative Learning

Have students work in groups of four. For **Activity 1,** begin by having each group discuss the questions posed in Steps 1–4. Then each group member should be assigned one of the measurements that is being compared with height—arm span, hand span, distance from head to ceiling, and coin value.

Each group member should now complete Activity 1 from the perspective of the assigned measurement. That is, each student should record the group's prediction for the measurement, take the actual measurement for each group member, and collect and compile the class data.

Continue in this manner for **Activities 2–4,** with each group member doing the steps of the activity as they relate to his or her assigned measurement. Conclude each activity by having the group come together to compare and discuss the results.

Discuss

After all groups have completed all of the activities, bring the class together for a general discussion. Ask questions such as the following:

- How well did the least-squares lines model the data?
- Which relationship was best modeled by a linear equation?

Have the class reconsider the 7-foot tall basketball player. Have them use their regression equations to estimate arm span, hand span, and distance from head to ceiling for this player. Compare the results with the estimates made at the beginning of the Chapter Project.

1 Chapter Review and Assessment

VOCABULARY

Chapter Test, Form A

NAME ____________ CLASS ________ DATE ________

Chapter Assessment
Chapter 1, Form A, page 1

Write the letter that best answers the question or completes the statement.

__c__ 1. Which of the following equations is a linear equation?
a. $y = 3x^2 + 7$ b. $y = \frac{8}{3x} + 4$ c. $y = \frac{8}{3}x + 4$ d. $y = 2 + \sqrt{x}$

__a__ 2. What is the y-intercept of the line $2x - 3y = 15$?
a. -5 b. -3 c. 2 d. 15

__d__ 3. What is the slope of the line containing the points $(5, -2)$ and $(2, 10)$?
a. $\frac{1}{4}$ b. $-\frac{1}{4}$ c. 4 d. -4

__d__ 4. What is the equation of the line whose slope is -2 and whose y-intercept is 8?
a. $y = 8x + 2$ b. $y = 8x - 2$ c. $y = 2x - 8$ d. $y = -2x + 8$

__b__ 5. The correlation coefficient for the data graphed at right is
a. close to -1.
b. close to 0.
c. close to 1.
d. close to 2.

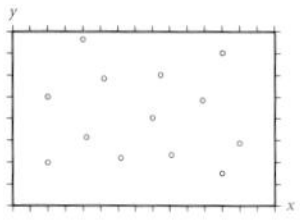

__d__ 6. Find the equation for the least-squares line of the data below.

x	2	5	9	15	22	30	40
y	5	10	15	25	35	50	70

a. $y \approx 1.7x + 0.99$ b. $y \approx 0.4x + 1.7$
c. $y \approx 0.4x + 0.99$ d. $y \approx 1.7x + 0.4$

__d__ 7. What is the equation of the line that contains the point $(6, 8)$ and is parallel to the line $y = \frac{2}{3}x + 2$?
a. $y = \frac{3}{2}x - 1$ b. $y = -\frac{3}{2}x + 17$ c. $y = -\frac{2}{3}x + 12$ d. $y = \frac{2}{3}x + 4$

__a__ 8. A tree service charges a basic fee of $50 to make a house call plus $20 per hour to trim trees. The equation that represents the total cost, c, of a house call and tree trimming in terms of the number of hours spent, h, is
a. $c = 20h + 50$ b. $c = 50h + 20$ c. $c = 20h - 50$ d. $c = 20h$

__a__ 9. If y varies directly as x and $y = 25$ when $x = 4$, find the constant of variation.
a. $k = 6.25$ b. $k = -6.25$ c. $k = 0.16$ d. $k = 4$

NAME ____________ CLASS ________ DATE ________

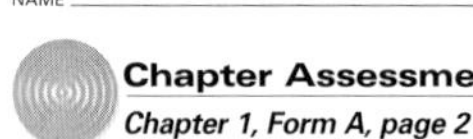

Chapter Assessment
Chapter 1, Form A, page 2

__b__ 10. Find the solution to the equation $2x + 3(x - 7) = -2(x - 21)$.
a. $x = 2$ b. $x = 9$ c. $x = 5$ d. $x = -3$

__c__ 11. Find the solution to the equation $|5x - 10| = 20$.
a. $x = -6$ b. $x = -2$ c. $x = 6$ *or* $x = -2$ d. $x = -6$ *or* $x = 2$

__c__ 12. Find the solution to the equation $\frac{6x + 5}{3} = \frac{5 - 2x}{4}$.
a. $x = -\frac{1}{3}$ b. $x = \frac{1}{3}$ c. $x = -\frac{1}{6}$ d. $x = \frac{1}{6}$

__c__ 13. Find the solution to the inequality $\frac{6 - 2x}{3} > -4$.
a. $x < -15$ b. $x > -9$ c. $x < 9$ d. $x > 15$

__a__ 14. Find the solution to the inequality $3x - 15 < 9 + 7x$.
a. $x > -6$ b. $x < -6$ c. $x > 6$ d. $x < 6$

__c__ 15. Which of the following is the graph of the solution set of the compound inequality $3x + 4 \leq 13$ *and* $6 + 2x \geq -2$?
a. (number line: −4, 0, 3) b. (number line: −3, 0, 4)
c. (number line: −4, 0, 3) d. (number line: −3, 0, 4)

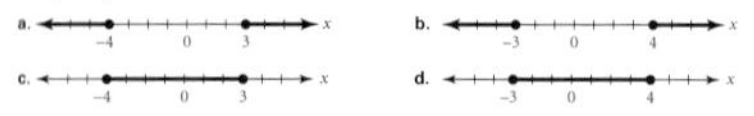

__d__ 16. Find the solution to the inequality $|4x - 6| > 14$.
a. $x > 5$ b. $x < -2$ c. $x > 5$ *and* $x < -2$ d. $x > 5$ *or* $x < -2$

__b__ 17. Which property is exemplified by $9(x - 1) = 9x - 9$?
a. Addition Property of Equality b. Distributive Property
c. Division Property of Equality d. Substitution Property

__a__ 18. The cost of catering a banquet varies directly as the number of people who attend the banquet. If it costs $3875 to cater a banquet that is attended by 250 people, how much will it cost to cater a banquet that is attended by 400 people?
a. $6200 b. $7000 c. $7750 d. $8500

__c__ 19. Solve $V = \frac{ax + b}{r^2}$ for b.
a. $Vr^2 + ax$ b. $Vax + r^2$ c. $Vr^2 - ax$ d. $Vax - r^2$

__a__ 20. Find the solution to the equation $|2x - 3| = 3x + 8$.
a. $x = 5$ b. $x = -11$ c. $x = 5$ *or* $x = -11$ d. $x = 5$ *and* $x = -11$

Key Skills & Exercises

LESSON 1.1

Key Skills

Identify linear equations and linear relationships between variables in a table.

$y = 8x + 4$ is a linear equation in the form $y = mx + b$, where $m = 8$ and $b = 4$.

x	-2	0	2	4
$y = 8x + 4$	-12	4	20	36

(x increases by +2, +2, +2; y increases by +8, +8, +8)

Represent a real-world linear relationship in a table, graph, or equation.

Projected college enrollments are given below.

Year	Enrollment (in millions)
1998	14.3
2000	14.8
2002	15.3
2004	15.8

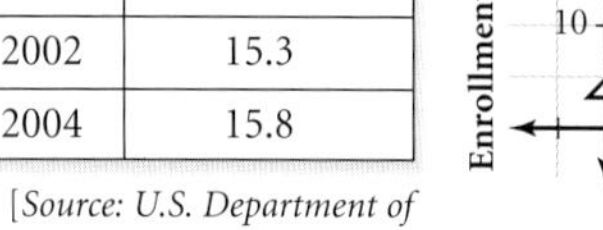

[*Source*: U.S. Department of Education]

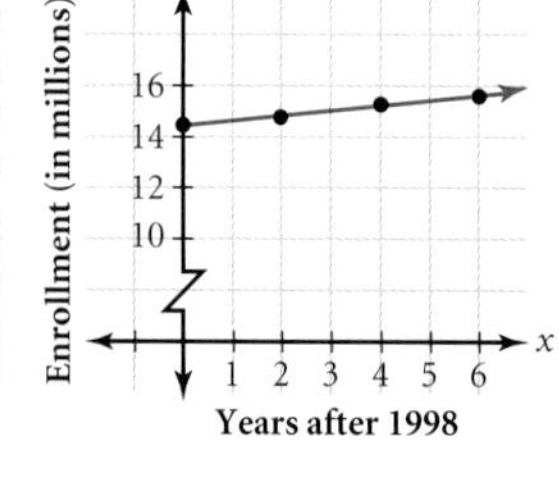

The equation is $y = 0.5x + 14.3$, where $x = 0$ represents 1998 and y is the projected enrollment.

Exercises

State whether each relationship is linear; if so, write the next ordered pair that would appear in the table.

1.

x	3	6	9	12
y	3	6	12	24

not linear

2.

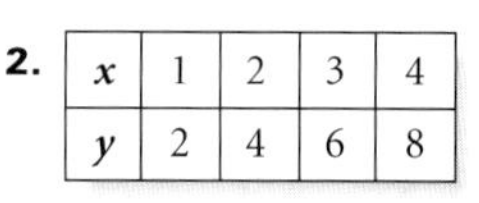

x	1	2	3	4
y	2	4	6	8

linear; (5, 10)

3.

x	7	14	21	28
y	5	10	15	20

linear; (35, 25)

PSYCHOLOGY Psychologists define intelligence quotient (IQ) as 100 times a person's mental age divided by his or her chronological age. The result is rounded to the nearest integer.

4. Find the IQ for an individual whose chronological age is 15 and whose mental age is the following: 10, 14, 15, 19, and 25. **67, 93, 100, 127, 167**

5. Represent the linear relationship from Exercise 4 in a table, a graph, and an equation.

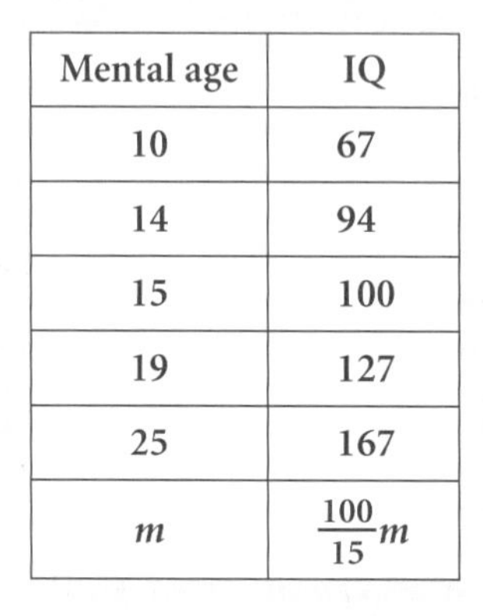

Mental age	IQ
10	67
14	94
15	100
19	127
25	167
m	$\frac{100}{15}m$

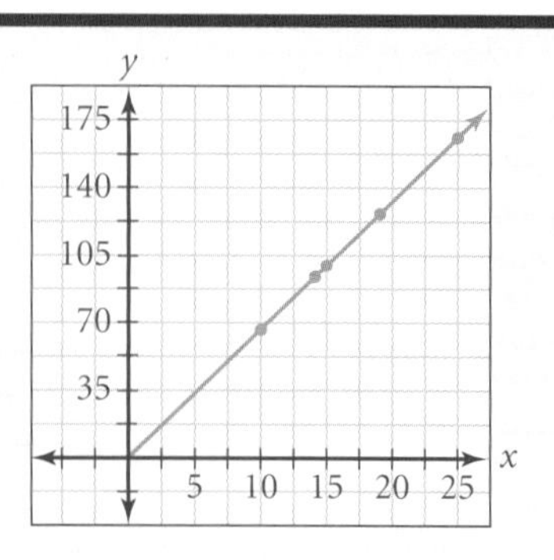

$y = \frac{100}{15}m$

LESSON 1.2

Key Skills

Graph a linear equation by using the slope and y-intercept.

Graph $y = -3x + 4$.

The y-intercept is 4.

The slope is -3.

$m = -3 = \frac{-3}{1}$ or $\frac{3}{-1}$

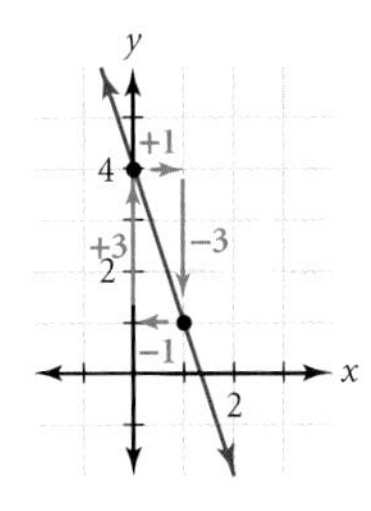

Exercises

Graph each equation.

6. $y = -\frac{1}{2}x$ **7.** $y = \frac{4}{5}x + 1$

8. $y = 3x - 1$ **9.** $y = -2x$

10. $2x + y = 3$ **11.** $3y - x = 1$

12. $y = 1$ **13.** $x = -2$

LESSON 1.3

Key Skills

Write a linear equation in two variables given sufficient information.

slope of $\frac{2}{3}$ and contains the point $(6, 9)$

$$y - y_1 = m(x - x_1)$$
$$y - 9 = \frac{2}{3}(x - 6)$$
$$y = \frac{2}{3}x - 5$$

contains the points $(2, 500)$ and $(3, 1500)$

Find m.

$$m = \frac{1500 - 500}{3 - 2} = 1000$$

Find b.

$$y = 1000x + b$$
$$500 = 1000(2) + b$$
$$b = -1500$$

Thus, $y = 1000x - 1500$.

Write an equation in slope-intercept form for the line that contains a given point and is perpendicular or parallel to a given line.

perpendicular to $y = 4x + 10$ and contains $(60, 40)$

$$y - y_1 = m(x - x_1)$$
$$y - 40 = -\frac{1}{4}(x - 60)$$
$$y = -\frac{1}{4}x + 55$$

Substitute the negative reciprocal of 4 for m.

parallel to $y = 4x + 10$ and contains $(60, 40)$

$$y - y_1 = m(x - x_1)$$

$$y - 40 = 4(x - 60)$$
$$y = 4x - 200$$

Substitute 4 for m.

Exercises

Write an equation in slope-intercept form for the line that has the indicated slope, m, and contains the given point.

14. $m = -3$, $(5, 8)$ $\mathbf{y = -3x + 23}$

15. $m = 100$, $(-2, 198)$ $\mathbf{y = 100x + 398}$

16. $m = \frac{1}{20}$, $(0, 5)$ $\mathbf{y = \frac{1}{20}x + 5}$

17. $m = 0$, $(-5, 4)$ $\mathbf{y = 4}$

Write an equation for the line that contains the indicated points.

18. $(3, 4)$ and $(5, 4)$ $\mathbf{y = 4}$

19. $(-2, 8)$ and $(-2, -1)$ $\mathbf{x = -2}$

20. $\left(-5\frac{3}{4}, 2\right)$ and $\left(-3\frac{1}{4}, 3\frac{1}{2}\right)$ $\mathbf{y = \frac{3}{5}x + \frac{109}{20}}$

21. $(6.8, 2)$ and $(3.6, 6)$ $\mathbf{y = -1.25x + 10.5}$

Write an equation for the line that contains the given point and is perpendicular or parallel to the given line.

22. $(3, 0)$, $y = -2x - 5$, perpendicular $\mathbf{y = \frac{1}{2}x - \frac{3}{2}}$

23. $(-3, -2)$, $y = \frac{1}{3}x - 5$, parallel $\mathbf{y = \frac{1}{3}x - 1}$

24. $(4, -1)$, $y = -\frac{2}{3}x + 7$, parallel $\mathbf{y = -\frac{2}{3}x + \frac{5}{3}}$

25. $(4, -1)$, $y = -8$, perpendicular $\mathbf{x = 4}$

7. $y = \frac{4}{5}x + 1$

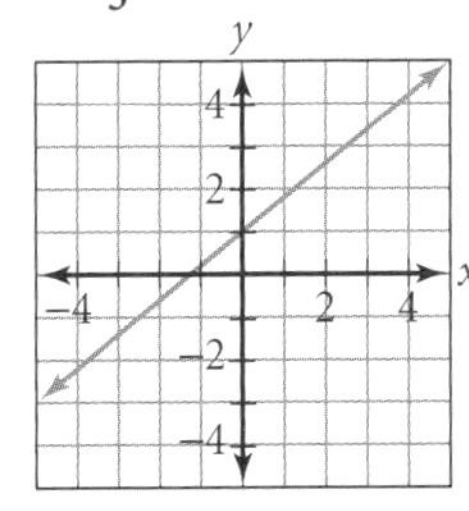

8. $y = 3x - 1$

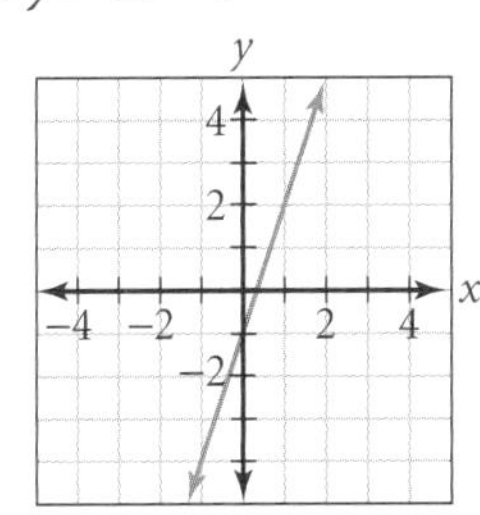

9. $y = -2x$

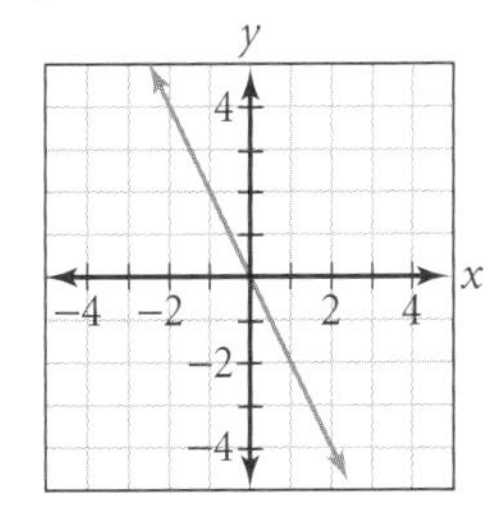

The answers to Exercises 10–13 can be found in Additional Answers beginning on page 1002.

6. $y = -\frac{1}{2}x$

y

x

4, 2, −2, −4; −4, −2, 4

Chapter Test, Form B

NAME ________ CLASS ________ DATE ________

Chapter Assessment

Chapter 1, Form B, page 1

1. Does the table at right represent a linear relationship between x and y? If the relationship is linear, write the next ordered pair that would appear in the table. **yes; (19, 63)**

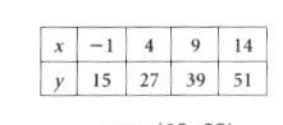

x	−1	4	9	14
y	15	27	39	51

2. Find the slope of the line containing the points $(-3, 5)$ and $(-7, 8)$. $m = -\frac{3}{4}$

3. Graph $y = -2x + 3$.

4. Graph $x = -2$.

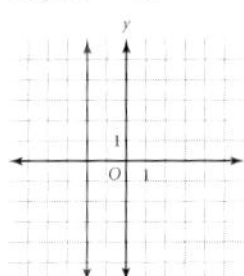

5. Write an equation in slope-intercept form for the line that has a y-intercept of 7 and a slope of -4. $y = -4x + 7$

6. Write the equation of the line that passes through the point $(6, -4)$ and is perpendicular to the line $-2x + 3y = 8$. $y = -\frac{3}{2}x + 5$

The table below contains data on the number of car accidents within a 24-hour period compared with the amount of snow, in inches, accumulated in the town of Culver.

Inches of snow, x	0	1.5	4	5	0.5	7	12	5.5
No. of accidents, y	2	9	6	5	1	10	15	7

7. Find the correlation coefficient, r, to the nearest hundredth for the data. $r \approx 0.86$

8. Find an equation for the least-squares line of the data. $y \approx 0.98x + 2.51$

9. Predict the number of accidents within a 24-hour period in the town of Culver if 9 inches of snow has accumulated. 11 or 12 accidents

Solve.

10. $\frac{3x - 2}{5} = 8$ $x = 14$

11. $\frac{2x - 6}{4} = \frac{3x + 5}{8}$ $x = 17$

NAME ________ CLASS ________ DATE ________

Chapter Assessment

Chapter 1, Form B, page 2

Solve each equation.

12. $|8x - 12| - 4 = 14$ $x = 3.75$ or $x = -0.75$

13. $|2x - 6| = 4 - 3x$ $x = -2$

Solve each inequality.

14. $8(5 - 3x) \leq -24$ $x \geq \frac{8}{3}$

15. $\frac{5}{3}x + 15 > \frac{3}{4}x - 18$ $x > -36$

Solve each inequality. Graph your solution on the number line.

16. $|x + 5| \leq 2$ $x \leq -3$ and $x \geq -7$

−7, −3, 0

17. $|4x - 5| > 7$ $x > 3$ or $x < -0.5$

−3, 0, 3

Solve each literal equation for the indicated variable.

18. $C = hr^2 + mr$, for h $h = \frac{C - mr}{r^2}$

19. $\frac{A + B}{dt} = \frac{F}{mn}$, for t $t = \frac{mn(A + B)}{Fd}$

20. Find the constant of variation, k, and the direct variation equation if y varies directly as x and $y = 20$ when $x = 80$. $k = 0.25$; $y = 0.25x$

21. The distance that Amy can drive in her car varies directly as the amount of gas that she has in her car. If Amy can drive 330 miles on 12 gallons of gas, how many gallons of gas will she need to drive 1100 miles? 40 gallons

22. Sam is a salesperson at a department store. He earns a weekly salary and a commission on his weekly sales. In one week, Sam's sales were \$3500, and his income for the week was \$475. The next week his sales were \$5000, and his income for the week was \$625.
    a. Write a linear equation in slope-intercept form for Sam's weekly income, y, in terms of his weekly sales, x. $y = 0.1x + 125$
    b. What were Sam's sales in a week in which his income was \$325? \$2000

23. In a factory, the time to assemble a VCR should vary no more than 5 minutes from the desired time of 40 minutes. Write an absolute-value inequality to represent the acceptable time that it takes to assemble a VCR in the factory. $|t - 40| \leq 5$

LESSON 1.4

Key Skills

Write a direct-variation equation for given variables.

Find the constant of variation and the direct-variation equation if y varies directly as x and $y = 2.25$ when $x = 9$.

$$k = \frac{2.25}{9} = 0.25$$

$$y = 0.25x$$

Write and solve proportions.

$$\frac{10}{12} = \frac{x}{24}$$

$$12x = 240$$

$$x = 20$$

Exercises

For Exercises 26–28, y varies directly as x. Find the constant of variation and the direct-variation equation.

26. $y = 750$ when $x = 25$ **30; $y = 30x$**

27. $y = 0.05$ when $x = 10$ **0.005; $y = 0.005x$**

28. $y = -2$ when $x = 14$ **$-\frac{1}{7}$; $y = -\frac{1}{7}x$**

PHYSICS The bending of a beam varies directly as the mass of the load it supports. Suppose that a beam is bent 20 millimeters by a mass of 40 kilograms.

29. How much will the beam bend when it supports a mass of 100 kilograms? **50 mm**

30. If the beam bends 25 millimeters, what load is the beam supporting? **50 kg**

LESSON 1.5

Key Skills

Make a scatter plot, and describe the correlation.

x	y
1	−6
2	−3
3	−4
4	0
5	−6
6	3
7	5
8	4
9	12
10	7

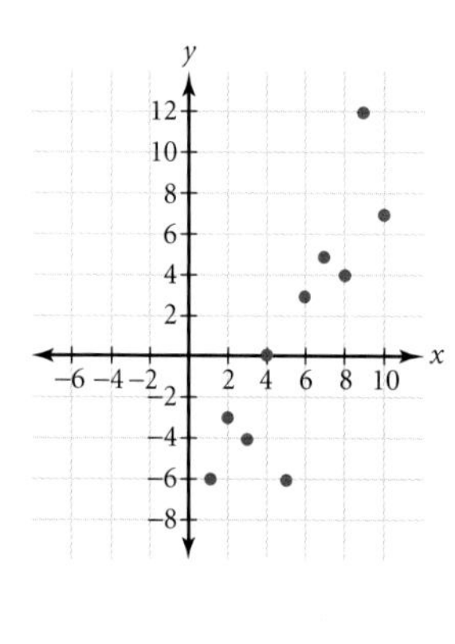

Because the data points go upward from left to right, the correlation is positive.

Find an equation for the least-squares line, and use it to make predictions.

The least-squares line for the data above is $y \approx 1.73x - 83$.

Exercises

31. STATISTICS Make a scatter plot of the data below. Describe the correlation. **neg.**

x	−4	−3	−2	−1	0	1	2	3	4
y	30	25	24	25	17	10	13	5	−1

32. FREIGHT CHARGES A statistician for the Civil Aeronautics Board wants to be able to predict the freight charge for a standard-sized crate. The statistician takes a sample of 10 freight invoices from different companies. Let d represent the distance in miles and let c represent the freight charge in dollars. Make a scatter plot of the data below. Describe the correlation. Then find an equation for the least-squares line.

d	500	600	900	1000	1200	1400	1600	1700	2200
c	41	55	50	70	60	78	75	89	105

pos.; $y \approx 0.035x + 26.6$

LESSON 1.6

Key Skills

Write and solve a linear equation in one variable.

INCOME Johanna works as a waitress. Her wages are \$2.50 per hour plus tips, which average \$50 for each 4-hour shift. How many 4-hour shifts does she have to work to earn \$300?

Let n represent the number of shifts.

$$(\text{hourly rate} \times \text{no. of hours} + \text{tips})n = 300$$
$$[2.5(4) + 50]n = 300$$
$$n = 5$$

Johanna must work 5 shifts to earn \$300.

Solve each literal equation for the indicated variable.

Solve $A = \pi rs + \pi r^2$ for s.

$$A = \pi rs + \pi r^2$$
$$A - \pi r^2 = \pi rs$$
$$\frac{A - \pi r^2}{\pi r} = s$$
$$\frac{A}{\pi r} - r = s$$

Exercises

CONSUMER ECONOMICS Jorge's monthly bill from his Internet service provider was \$25. The service provider charges a base rate of \$15 per month plus \$1 for each hour that the service is used.

33. Write a linear equation to represent the total monthly charges. **$t = h + 15$**

34. Find the number of hours that Jorge was charged for that month. **10 hr**

Solve each literal equation for the indicated variable.

35. $F = \frac{9}{5}C + 32$ for C **$C = \frac{5}{9}F - \frac{160}{9}$**

36. $S = s_0 + v_0 + \frac{1}{2}gt^2$ for v_0 **$v_0 = S - s_0 - \frac{1}{2}gt^2$**

37. $\frac{1}{f} = \frac{1}{f_1} + \frac{1}{f_2}$ for f **$f = \frac{f_1 f_2}{f_1 + f_2}$**

38. $A = \frac{h}{2}(b_1 + b_2)$ for b_2 **$b_2 = \frac{2A}{h} - b_1$**

Lesson 1.7

Key Skills

Write, solve and graph linear inequalities in one variable.

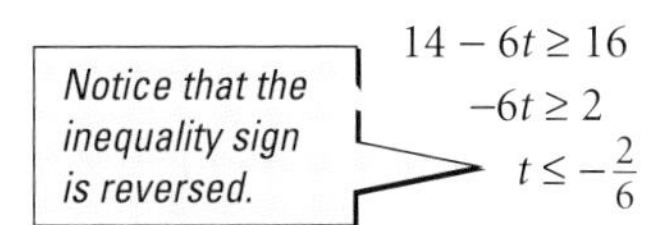

$$14 - 6t \geq 16$$
$$-6t \geq 2$$
$$t \leq -\frac{2}{6}$$

Solve and graph compound linear equalities in one variable.

$$5x + 7(x - 2) > 6 \quad or \quad 4x + 6 \leq -6$$
$$x > \frac{5}{3} \qquad\qquad x \leq -3$$

$x > \frac{5}{3}$ or $x \leq -3$

−4 −3 −2 −1 0 1 $\frac{5}{3}$ 2 3 4

$$5x + 7(x - 2) < 6 \quad and \quad 4x + 6 \geq -6$$
$$x < \frac{5}{3} \qquad\qquad x \geq -3$$

$-3 \leq x < \frac{5}{3}$

−4 −3 −2 −1 0 1 $\frac{5}{3}$ 2 3 4

Exercises

Write an inequality for each situation. Graph the solution.

39. GEOMETRY If the width of a rectangle is 10 meters and the perimeter is not to exceed 140 meters, how long can the length be?

40. BUSINESS An organization wants to sell tickets to a concert. It plans on selling 300 reserved-seat tickets and 150 general-admission tickets. The price of a reserved-seat ticket is \$2 more than a general-admission ticket. If the organization wants to collect at least \$3750, what is the minimum price it can charge for a reserved-seat ticket?

Solve and graph.

41. $4x - 3 < 29$ *and* $4 - 3x < -5$ **$x < 8$ and $x > 3$**

42. $-3x - 8 \leq 7$ *and* $-4x > -18$ **$x \geq -5$ and $x < \frac{9}{2}$**

43. $4x - 3 < 29$ *or* $4 - 3x < -5$ **$x < 8$ or $x > 3$**

44. $-3x - 8 \leq 7$ *or* $-4x > -18$ **$x \geq -5$ or $x < \frac{9}{2}$**

39. $2(10) + 2l \leq 140$
$l \leq 60$ m

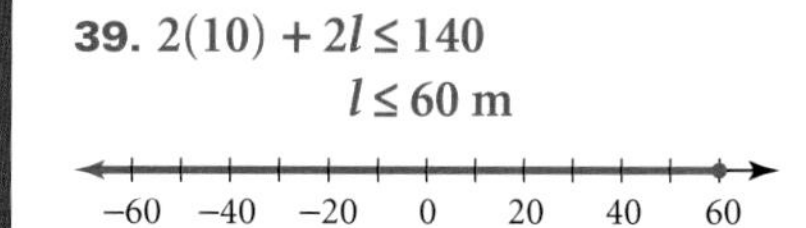

40. $150g + 300(g + 2) \geq 3750$
At least \$9 must be charged.

3 4 5 6 7 8 9 10 11 12 13 14 15

41. $x < 8$ *and* $x > 3$

−12 −10 −8 −6 −4 −2 0 2 4 6 8 10 12

42. $x \geq -5$ *and* $x < \frac{9}{2}$

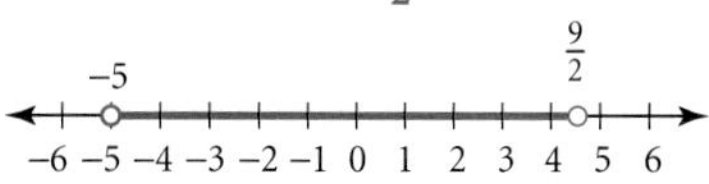

43. all real numbers

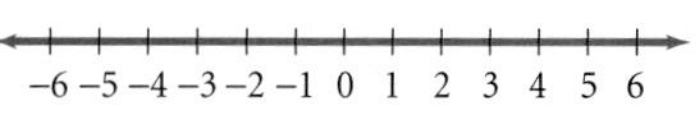

44. all real numbers

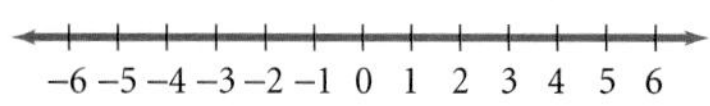

45. $x = -40$ *or* $x = 40$

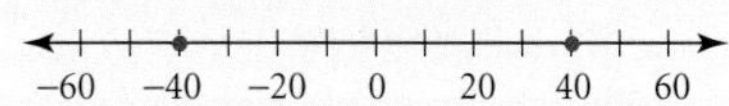

46. $x = -20$ *or* $x = 20$

−60 −40 −20 0 20 40 60

47. $x = -4.5$ *or* $x = 4.5$

−6 −5 −4 −3 −2 −1 0 1 2 3 4 5 6

48. $x = -22$ *or* $x = 14$

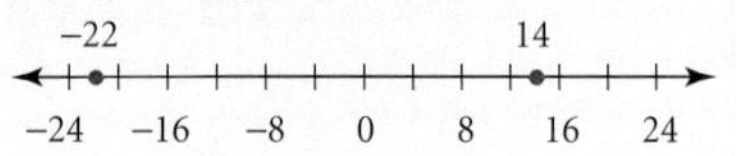

49. $x < -40$ *or* $x > 40$

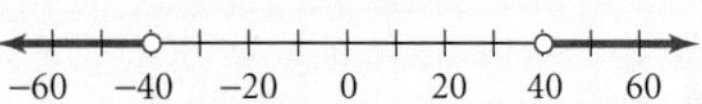

50. no solution

51. all real numbers

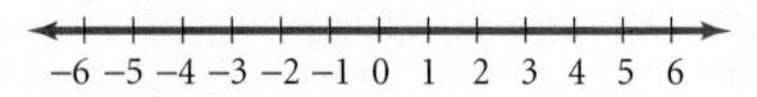

52. no solution

LESSON 1.8

Key Skills

Solve and graph absolute-value equations.

a.

$$|5x| = 45$$
$$5x = 45 \quad or \quad 5x = -45$$
$$x = 9 \qquad\quad x = -9$$

−9 −6 −3 0 3 6 9

b.

$$\tfrac{1}{2}|5x| + 5 = 45$$
$$\tfrac{1}{2}|5x| = 40$$
$$5x = 80 \quad or \quad x = -80$$
$$x = 16 \qquad\quad x = -16$$

−20 −16 −12 −8 −4 0 4 8 12 16 20

Solve and graph absolute-value inequalities.

a.

$$|x + 5| \ge 45$$
$$x + 5 \ge 45 \quad or \quad x + 5 \le -45$$
$$x \ge 40 \qquad\quad x \le -50$$

$x \le -50$ *or* $x \ge 40$

−50 −20 0 20 40

b.

$$|x + 5| \le 45$$
$$x + 5 \le 45 \quad and \quad x + 5 \ge -45$$
$$x \le 40 \qquad\quad x \ge -50$$

$-50 \le x \le 40$

−50 −20 0 20 40

Exercises

Solve and graph on a number line.

45. $\left|\frac{1}{2}x\right| = 20$ **$x = -40$ *or* $x = 40$**

46. $\left|\frac{4}{5}x\right| = 16$ **$x = -20$ *or* $x = 20$**

47. $12|2x| = 108$ **$x = -4.5$ *or* $x = 4.5$**

48. $\frac{2}{3}\left|x + 4\right| - 5 = 7$ **$x = -22$ *or* $x = 14$**

Solve and graph on a number line.

49. $\left|\frac{1}{2}x\right| > 20$ **$x < -40$ *or* $x > 40$**

50. $-\frac{1}{2}\left|\frac{1}{2}x\right| - 4 > 20$ **no solution**

51. $-5|6x - 7| \le 35$ **all real numbers**

52. $|6x - 7| \le -35$ **no solution**

Applications

GEOMETRY The length of the hypotenuse of a 30-60-90 triangle varies directly as the length of the side opposite the 30° angle. The length of the hypotenuse, h, is 45 units when the side opposite the 30° angle, s, is 22.5 units.

53. Find the constant of variation and the direct-variation equation. **2; $h = 2s$**

54. Find the length of the side opposite the 30° angle when the hypotenuse is 13 inches long. **6.5 in.**

PERSONAL FINANCE The majority of American families save less than $\frac{1}{10}$ of their income. Suppose that an average American family saves \$2850.

55. Write and solve an inequality that models this situation. **$2850 < \frac{1}{10}I$, for $I > 0$; $I > 28{,}500$**

56. Graph the solution on a number line. \$28,500 \$20,000 \$30,000

Chapter Test

State whether each relationship is linear. If so, write the next ordered pair that would appear in the table. not linear

1.

x	2	4	6	8	10
y	4	16	36	64	100

2.

x	1	4	7	10	13
y	5	12	19	26	33

linear; (16, 40)

3. BUSINESS For automobile repairs Wayne charges \$50 plus \$30 per hour. Write a linear equation to model this situation and find the charge for 8.5 hours of work. $c = 30h + 50$; \$305

Graph each equation.

4. $y = 2x$ **5.** $3x - 5y = 15$

6. $2x + 5 = y$ **7.** $x = 2$

Write an equation in slope-intercept form for the line that

8. has slope = 2 and contains point $(-1, 4)$
$y = 2x + 6$

9. has slope 0 and contains point $(-5, 7)$
$y = 7$

10. contains points (3, 5) and (4, −7)
$y = -12x + 41$

11. contains (1, 2) and is parallel to $y = 4x + 3$. $y = 4x - 2$

12. contains (−5, 9) and is perpendicular to $2x + 3y = 4$. $y = \frac{3}{2}x + \frac{33}{2}$

Solve each variation problem.

13. If y varies directly as x and $y = 3$ when $x = 7$, find y when $x = 63$. $y = 27$

14. If $y = 25$ when $x = 2125$, find the constant of variation and the direct variation equation. $\frac{1}{85}$, $y = \frac{x}{85}$

15. GEOGRAPHY A California map shows a scale indicating that 1 inch = 50 miles. The distance on the map between Sacramento and Los Angeles is $8\frac{1}{2}$ inches. What is the actual distance between the two cities? 425 mi

Create a scatter plot for the data in the table below. Describe the correlation, and then find an equation for the least squares line.

16.

x	2	5	3	8	6
y	4	6	5	7	6

fairly strong positive correlation; $y = 0.46x + 3.37$

Solve each equation.

17. $4x - 3 = 17$ $x = 5$ **18.** $\frac{x}{3} - 2 = 16$ $x = 54$

19. $2x - 0.8 = -2.4$
$x = -0.8$

20. $8x + 4 = 2x - 32$
$x = -6$

21. GEOMETRY Complementary angles equal 90°. If the measure of one complementary angle is 30° more than twice the other angle measure, write an equation and find the measure of each angle. $x + (2x + 30) = 90$; 20°, 70°

22. Solve the equation $A = \frac{1}{2}r^2\,\theta$ for θ. $\theta = \frac{2A}{r^2}$

Solve and graph each inequality on a number line.

23. $-3x - 6 > 15$
$x < -7$

24. $2(4x - 5) < 6x - 6$
$x < 2$

Solve and graph each compound inequality on a number line.

25. $3x + 4 > 7$ *and* $2x - 3 < 5$
$x > 1$ and $x < 4$

26. $5 - x \geq 3$ *or* $-2 + 4x \geq 10$
$x \leq 2$ or $x \geq 3$

27. ENTERTAINMENT Three adults and 5 children want to see a movie. They have \$40 between them. Write and solve an inequality to find the price a child's ticket cannot exceed, if an adult ticket costs \$7. $3(7) + 5(x) \leq 40$; $x \leq 3.80$

Solve and graph on a number line.

28. $\left|\frac{1}{2}x - 4\right| = 3$
$x = 2$ or $x = 14$

29. $|5x + 3| \geq -2$
all real numbers

30. $|2x + 13| \leq -3$
no solution

31. $\left|\frac{3}{5}x + 6\right| \geq 9$

$x \geq 5$ or $x \leq -25$

Chapter Test

16.

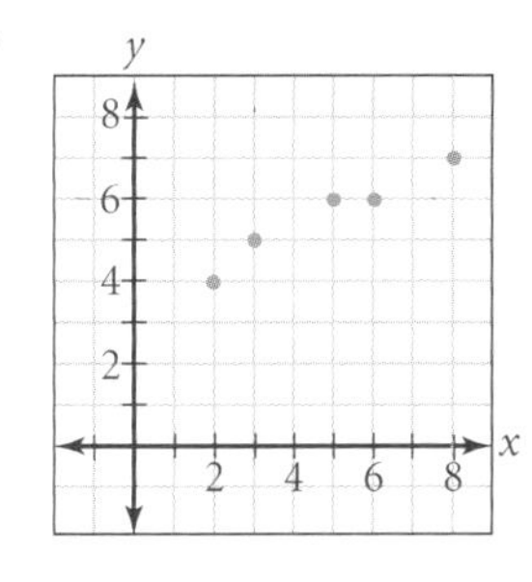

23.

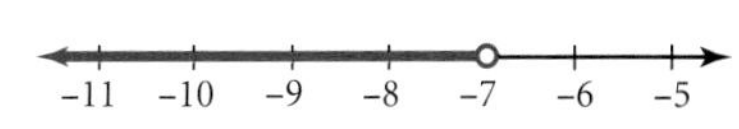

24.

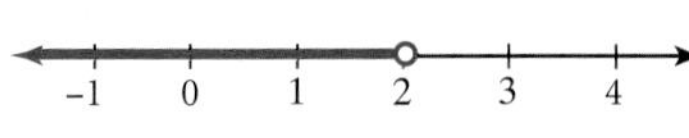

25.

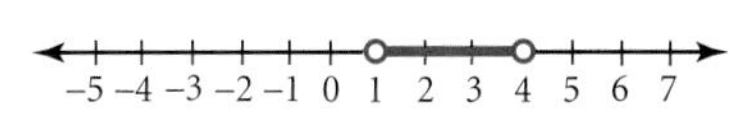

26.

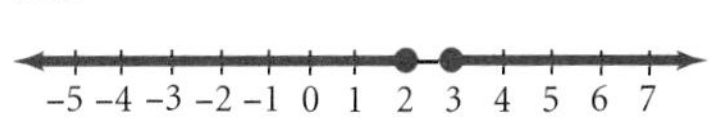

28.

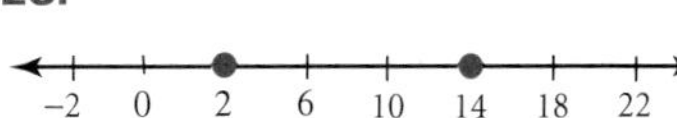

29.

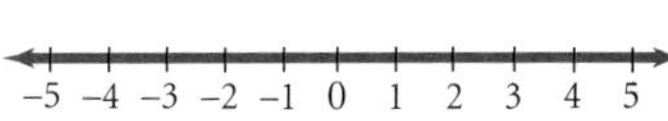

30. no solution

31.

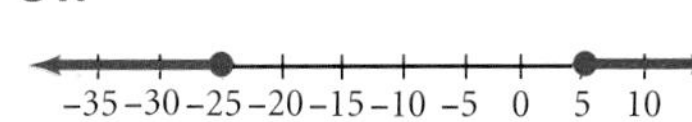

4.

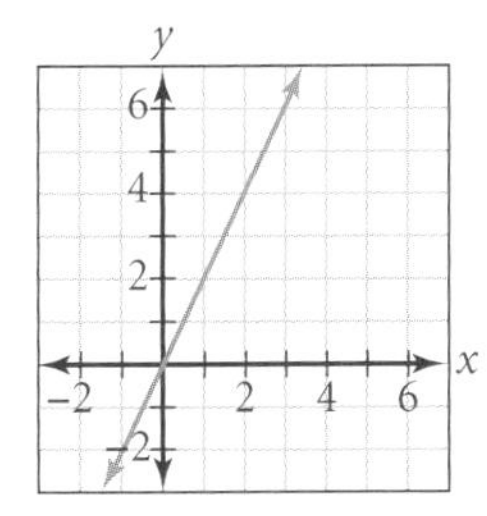

5.

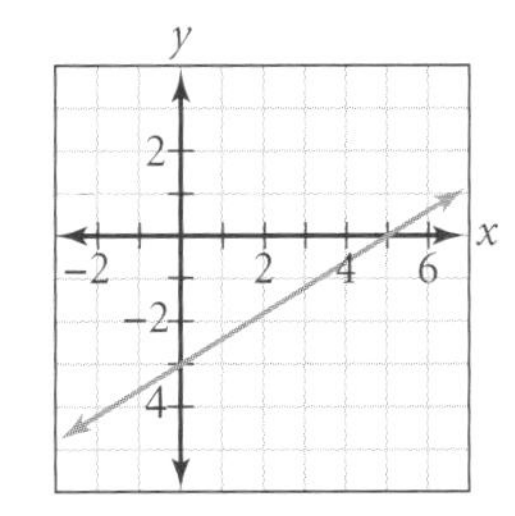

6.

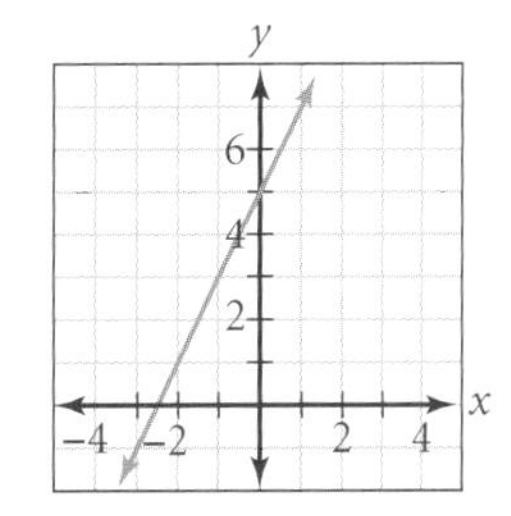

7.

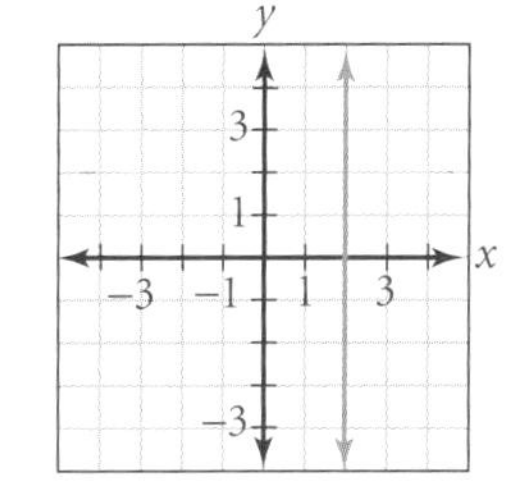

College Entrance Exam Practice

Multiple-Choice and Quantitative-Comparison Samples

The first half of the Cumulative Assessment contains two types of items found on standardized tests—multiple-choice questions and quantitative-comparison questions. Quantitative-comparison items emphasize the concepts of equality, inequality, and estimation.

Free-Response Grid Samples

The second half of the Cumulative Assessment is a free-response section. This part of the Cumulative Assessment requires student-produced response items like those commonly found on college entrance exams. These questions require the use of machine-scored answer grids. You may wish to have students practice answering these items in preparation for standardized tests.

College Entrance Exam Practice

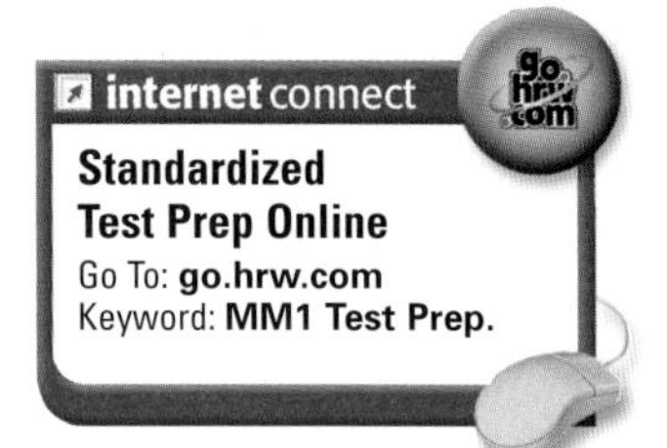

QUANTITATIVE COMPARISON For Items 1–5, write
A if the quantity in Column A is greater than the quantity in Column B;
B if the quantity in Column B is greater than the quantity in Column A;
C if the quantities are equal; or
D if the relationship cannot be determined from the given information.

	Column A	Column B	Answers
1. B	$2y - 3x = 4$ slope of the graph of the equation	y-intercept of the graph of the equation	(A) (B) (C) (D) [Lesson 1.2]
2. B	slope of the line graphed	y-intercept of the graph	(A) (B) (C) (D) [Lesson 1.2]
3. C	slope of the line containing (4, 3) and (−4, −3)	(−3, −4) and (1, −1)	(A) (B) (C) (D) [Lesson 1.2]
4. B	the value of x $\frac{1-2x}{5} = \frac{x}{6}$	$\frac{2x-4}{3} = \frac{x}{5}$	(A) (B) (C) (D) [Lesson 1.4]
5. D	the value of x $\lvert 3x - 1 \rvert = 2$	$\lvert 5x \rvert - 3 = 7$	(A) (B) (C) (D) [Lesson 1.8]

6. Which one of the following equations is not a linear equation? ***(LESSON 1.1)*** c

a. $2x + 3y = 11$ **b.** $y = \frac{3-4x}{7}$
c. $y = \frac{7}{3-4x}$ **d.** $x = 3 - y$

7. Which gives the slope and y-intercept for the graph of $2x + 3y = 2$? ***(LESSON 1.2)*** a

a. $m = -\frac{2}{3}, b = \frac{2}{3}$ **b.** $m = \frac{2}{3}, b = -\frac{2}{3}$
c. $m = 2, b = -2$ **d.** $m = -2, b = 2$

8. Find the slope of the line containing the points (2, −1) and (−5, 0). ***(LESSON 1.2)*** d

a. −7 **b.** −3
c. $\frac{1}{7}$ **d.** $-\frac{1}{7}$

9. Which equation represents a line with no y-intercept? ***(LESSON 1.2)*** b

a. $y = 4$ **b.** $x = -\frac{1}{4}$
c. $x + y = 2$ **d.** $y = 3x$

10. Which statement is *not* true? ***(LESSON 1.2)***
b
a. All horizontal lines are perpendicular to vertical lines.
b. The slopes of two nonvertical lines that are perpendicular to each other are reciprocals.
c. The slope of any horizontal line is 0.
d. The slope of any vertical line is undefined.

11. Which equation contains the point $(-4, 6)$ and is parallel to the graph of $y = -2x - \frac{1}{4}$? ***(LESSON 1.3)***
d
a. $y = -2x + 2$ **b.** $y = 2x - 2$
c. $y = 2x + 2$ **d.** $y = -2x - 2$

12. If x varies directly as y and $x = 4$ when $y = -5$, what is x when $y = 2.5$? ***(LESSON 1.4)***
d
a. $-\frac{1}{2}$ **b.** $\frac{1}{2}$ **c.** 2 **d.** -2

13. Solve $T = \frac{24I}{B(n+1)}$ for I. ***(LESSON 1.6)***
d
a. $I = \frac{24T}{B(n+1)}$ **b.** $I = \frac{24B}{T(n+1)}$
c. $I = \frac{T(n+1)}{24B}$ **d.** $I = \frac{TB(n+1)}{24}$

14. Graph the equation $y = -5x - 12$. ***(LESSON 1.1)***

15. Does the information in the table below represent a linear relationship between x and y? Explain your response. ***(LESSON 1.1)***

x	0	1	2	3
y	13	17	21	25

16. Find the equation of the line containing the points $(-3, 6)$ and $(-5, 8)$. ***(LESSON 1.3)*** $y = -x + 3$

17. Write the equation in slope-intercept form for the line that contains the point $(5, -8)$ and is perpendicular to the graph of $y = \frac{1}{3}x - 4$. ***(LESSON 1.3)*** $y = -3x + 7$

18. Write an equation in slope-intercept form for the line that contains the points $(2, -4)$ and $(3, -1)$. ***(LESSON 1.3)*** $y = 3x - 10$

ANATOMY The following table lists the heights and weights for 10 randomly selected young adult males with medium frames. The heights are in inches, and the weights are in pounds. ***(LESSON 1.5)***

Heights, x	67	66	70	67	67	68	69	66	70	71
Weights, y	146	145	157	148	145	149	151	141	154	159

19 Create a scatter plot for this data. Is the correlation positive, negative, or none? **pos.**

20 Find the correlation coefficient, r. $r \approx 0.97$

21 Graph the least-squares line with the scatter plot. Predict the weight (to the nearest pound) of a young adult male selected at random from this group who is 63 inches tall. **134 lb**

22. Solve and graph the inequality $5x - 6(x + 9) < 1$. ***(LESSON 1.7)*** $x > -55$

23. Solve $|2 + 3x| \geq 14$. ***(LESSON 1.8)*** $x \leq -\frac{16}{3}$ or $x \geq 4$

FREE RESPONSE GRID The following questions may be answered by using a free-response grid such as that commonly used by standardized-test services.

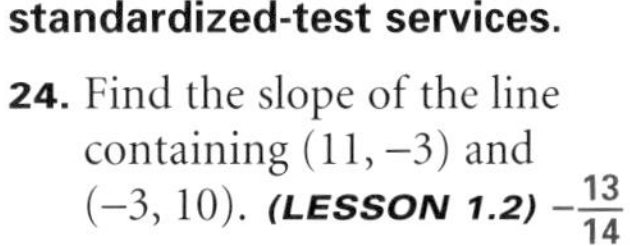

24. Find the slope of the line containing $(11, -3)$ and $(-3, 10)$. ***(LESSON 1.2)*** $-\frac{13}{14}$

25. Find the y-intercept of the graph of $3x - 5y = 2$. ***(LESSON 1.2)*** $-\frac{2}{5}$

26. Find the slope of the graph of $3x - 5y = 2$. ***(LESSON 1.2)*** $\frac{3}{5}$

GEOMETRY The perimeter, p, of a square varies directly as the length, l, of a side. ***(LESSON 1.4)***

27. What is the constant of variation that relates the perimeter to the length of a side? **4**

28. If a square has a side length of 3.5 centimeters, how many centimeters long is its perimeter? **14**

14. $y = -5x - 12$

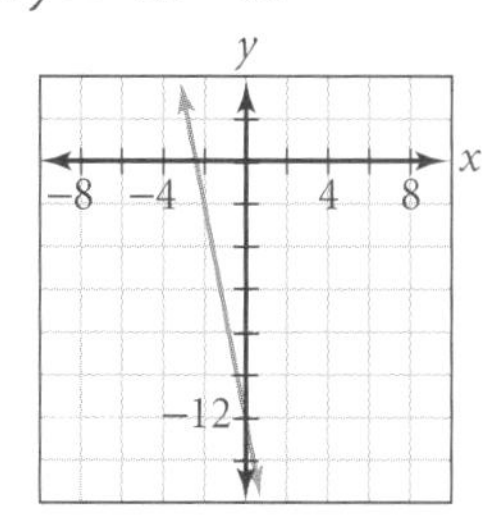

15. The difference in x-values is a constant 1. The difference in y-values is a constant 4. This is a linear relationship.

22.

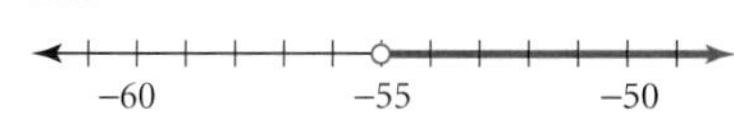

Keystroke Guide for Chapter 1

Essential keystroke sequences (using the model TI-82 or TI-83 graphics calculator) are presented below for all Activities and Examples found in this chapter that require or recommend the use of a graphics calculator.

internet connect

For Keystrokes of other graphing calculator models, visit the HRW web site at **go.hrw.com** and enter the keyword **MB1 CALC**.

LESSON 1.1

EXAMPLE 1 Page 5

Graph $y = 150x + 250$, and find the value of y when $x = 25$.

Use viewing window [0, 30] by [–500, 5000].

Graph the equation:

Y= 150 X,T,θ,n + 250 GRAPH

Evaluate for $x = 25$:

2nd TRACE (CALC) 1:value ENTER ($x =$) 25 ENTER

EXAMPLE 2 Page 6

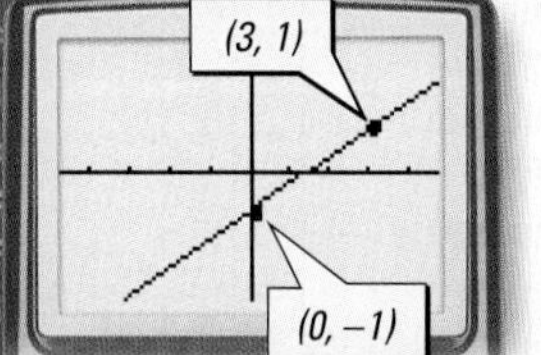

Graph $y = \frac{2}{3}x - 1$, and verify that (0, –1) and (3, 1) are on the line.

Use viewing window [–4, 4] by [–3, 3].

Graph the equation:

Y= (2 ÷ 3) X,T,θ,n – 1 GRAPH

Evaluate for $x = 0$ and $x = 3$:
Use a keystroke sequence similar to that used in Example 1.

LESSON 1.2

Page 13

Graph each pair of equations in Step 1.

Use the standard viewing window [–10, 10] by [–10, 10]. Use a keystroke sequence similar to that used in Example 2 of Lesson 1.1.

LESSON 1.3

Page 23

For Steps 1 and 3, graph the given equations on the same screen.

Use viewing window [–3, 3] by [–3, 3].

Use a keystroke sequence similar to that used in Example 1 of Lesson 1.1.

Use keystrokes ZOOM 5:ZSquare ENTER to obtain a square viewing window.

EXAMPLES 4 and 5 For Example 4, graph $y = -2x + 4$ and $y = -2x + 1$ on the same screen.

Page 24

Use viewing window [–6, 10] by [–6, 10].

Y= (–) 2 X,T,θ,n + 4 ENTER (Y2=) (–) 2 X,T,θ,n + 1 GRAPH

For Example 5, use viewing window [–6, 10] by [–6, 10].

Use a keystroke sequence similar to that used in Example 2 of Lesson 1.1.
Use keystrokes ZOOM 5:ZSquare ENTER to obtain a square viewing window.

LESSON 1.4

EXAMPLE 2 Graph $y = \frac{1}{5}x$, and verify that the point (35, 7) is on the line.

Page 30

Y1=(1/5)X

X=35 Y=7

Use viewing window [0, 50] by [–4, 25].

Graph the equation:

Y= (1 ÷ 5) X,T,θ,n GRAPH

Evaluate for $x = 35$:

Use a keystroke sequence similar to that used in Example 1 of Lesson 1.1.

EXAMPLE 3 Graph the equation $y = 0.38x$, and confirm that an x-value of 24.3 corresponds to a y-value of about 9.2 kilograms.

Page 32

Use viewing window [0, 30] by [0, 15].

To graph the equation and evaluate for $x = 24.3$, use a keystroke sequence similar to that used in Example 1 of Lesson 1.1.

LESSON 1.5

Activity

Page 37

First clear old data and equations.

For Step 1, create a scatter plot of the data.

Use viewing window [0, 10] by [100, 200].

Enter the data:

STAT EDIT 1:EDIT ENTER L1 1 ENTER 2 ENTER 3 ENTER 4 ENTER 5 ENTER 6 ENTER ► L2 185.4 ENTER 173.1 ENTER 147.1 ENTER 158.4 ENTER 134.7 ENTER 151.2 ENTER

Create a scatter plot:

2nd Y= (STAT PLOT) STAT PLOTS 1:Plot 1 ENTER On ENTER ▼ (Type:) [scatter] ENTER

▼ (Xlist:) 2nd 1 (L1) ▼ (Ylist:) 2nd 2 (L2) ▼ (Mark:) [□]

⇑ TI-82: L1 ENTER ⇑ TI-82: L2 ENTER

ENTER GRAPH

EXAMPLE 1 Page 38

Create a scatter plot for the data. Then graph the least-squares line.

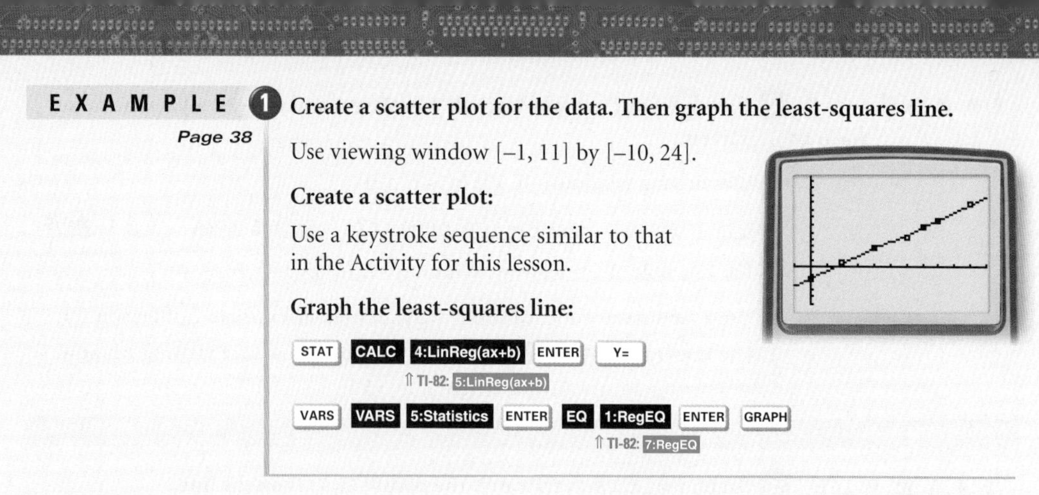

Use viewing window [–1, 11] by [–10, 24].

Create a scatter plot:

Use a keystroke sequence similar to that in the Activity for this lesson.

Graph the least-squares line:

STAT CALC 4:LinReg(ax+b) ENTER Y=

⇑ TI-82: 5:LinReg(ax+b)

VARS VARS 5:Statistics ENTER EQ 1:RegEQ ENTER GRAPH

⇑ TI-82: 7:RegEQ

EXAMPLE 2 Page 39

Create a scatter plot for the data, graph the least-squares line, and find the y-value that corresponds to $x = 40$.

Use viewing window [0, 100] by [0, 24].

Create a scatter plot, and graph the least-squares line:
Use a keystroke sequence similar to that in the Activity for this lesson and in Example 1 above.

Evaluate for $x = 40$:

2nd TRACE (CALC) 1: value ENTER (X=) 40 ENTER

LESSON 1.6

Activity Page 47

For Step 2, graph $y = x + 3$ and $y = 9 - 2x$ on the same screen. Find the value of x for which $y = x + 3$ and $y = 9 - 2x$ are equal.

Use the friendly viewing window [–4.7, 4.7] by [–10, 10].

Y= X,T,θ,n + 3 ENTER (Y2=) 9 – 2 X,T,θ,n GRAPH

Press TRACE and move the cursor toward the point of intersection.
Use ▲ and ▼ to move the cursor from one line to the other.

EXAMPLE 3 Page 47

Graph $y = 3.24x - 4.09$ and $y = -0.72x + 3.65$ on the same screen, and find any points of intersection.

Use viewing window [–5, 5] by [–5, 5].

Graph the equations:
Use a keystroke sequence similar to that in the Activity for this lesson.

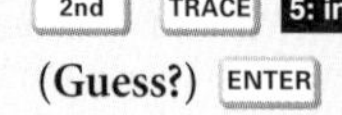

Find any points of intersection:

Move your cursor as indicated.

2nd TRACE (CALC) 5: intersect ENTER (First curve?) ENTER (Second curve?) ENTER (Guess?) ENTER

LESSON 1.7

Activity
Page 56

For Step 2, graph $y = 2x - 3$ and $y = 3$ on the same screen.

Use the friendly viewing window [–4.7, 4.7] by [–5, 5]. Use a keystroke sequence similar to that in the Activity for Lesson 1.6.

LESSON 1.8

Activity
Page 63

For Step 1, graph $y = |x|$ and $y = 2x - 2$, and find any points of intersection.

Use viewing window [–5, 5] by [–5, 5].

Graph the equations:

ENTER (Y2=) 2 X,T,θ,n – 2 GRAPH

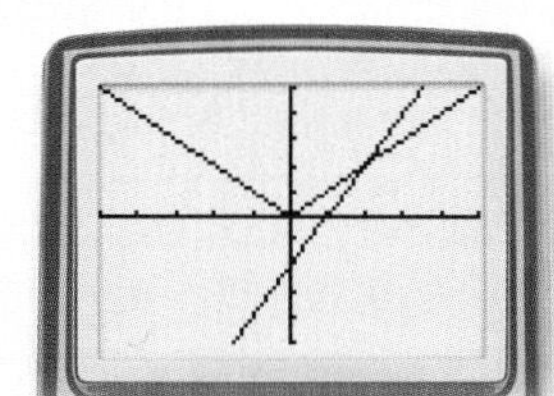

Find the point of intersection:
Use a keystroke sequence similar to that in Example 3 of Lesson 1.6.

EXAMPLES 1 and 2 Pages 63 and 64

Graph the two equations on the same screen, and find any points of intersection.

For Example 1, use viewing window [–7, 3] by [–1, 5].
For Example 2, use viewing window [–3, 7] by [–3, 7].

Use a keystroke sequence similar to that in the Activity for this lesson.

EXAMPLES 3 and 4 Page 65

For Examples 3 and 4, graph $y = |5 - 3x|$ and $y = 9$ on the same screen, and find any points of intersection.

Use viewing window [–11, 17] by [–3, 15].

Use a keystroke sequence similar to that in the Activity for this lesson.

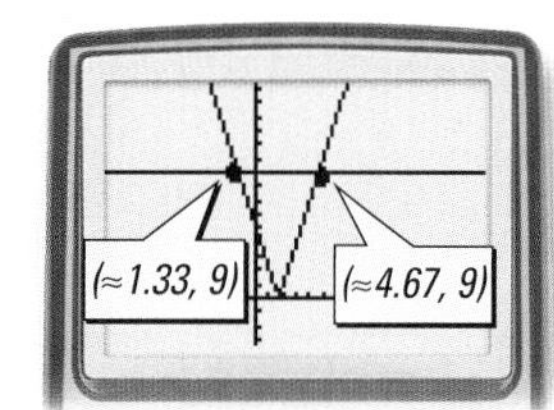

2 Numbers and Functions

Lesson Presentation CD-ROM
PowerPoint® presentations for each lesson 2.1–2.7

CHAPTER PLANNING GUIDE

Lesson	2.1	2.2	2.3	2.4	2.5	2.6	2.7	Project and Review
Pupil's Edition Pages	86–93	94–101	102–110	111–117	118–123	124–132	133–141	142–149
Practice and Assessment								
Extra Practice (Pupil's Edition)	946	946	947	947	948	948	949	
Practice Workbook	9	10	11	12	13	14	15	
Practice Masters Levels A, B, and C	25–27	28–30	31–33	34–36	37–39	40–42	43–45	
Standardized Test Practice Masters	10	11	12	13	14	15	16	17
Assessment Resources	16	17	18	19	21	22	23	20, 24–29
Visual Resources								
Lesson Presentation Transparencies Vol. 1	33–36	37–40	41–44	45–48	49–52	53–56	57–60	
Teaching Transparencies	5		6	7, 8	9, 10	11		
Answer Key Transparencies	41–46	47–50	51–55	56–61	62–66	67–75	76–81	82–86
Quiz Transparencies	2.1	2.2	2.3	2.4	2.5	2.6	2.7	
Teacher's Tools								
Reteaching Masters	17–18	19–20	21–22	23–24	25–26	27–28	29–30	
Make-Up Lesson Planner for Absent Students	9	10	11	12	13	14	15	
Student Study Guide	9	10	11	12	13	14	15	
Spanish Resources	9	10	11	12	13	14	15	
Block Scheduling Handbook								4–5
Activities and Extensions								
Lesson Activities	9	10	11	12	13	14	15	
Enrichment Masters	9	10	11	12	13	14	15	
Cooperative-Learning Activities	9	10	11	12	13	14	15	
Problem Solving/ Critical Thinking	9	10	11	12	13	14	15	
Student Technology Guide	9	10	11	12	13	14	15	
Long Term Projects								5–8
Writing Activities for Your Portfolio								4–6
Tech Prep Masters								5–8
Building Success in Mathematics								4–6

LESSON PACING GUIDE

Lesson	2.1	2.2	2.3	2.4	2.5	2.6	2.7	Project and Review
Traditional	1 day	1 day	2 days	2 days	2 days	1 day	1 day	2 days
Block	$\frac{1}{2}$ day	$\frac{1}{2}$ day	1 day	1 day	1 day	$\frac{1}{2}$ day	$\frac{1}{2}$ day	1 day
Two-Year	2 days	2 days	4 days	4 days	4 days	2 days	2 days	4 days

CONNECTIONS AND APPLICATIONS

Lesson	2.1	2.2	2.3	2.4	2.5	2.6	2.7	Review
Algebra	86–93	94–101	102–110	111–117	118–123	124–132	133–141	142–149
Geometry		99	102, 104, 109					
Statistics	91							149
Transformations						126, 127, 130, 132		
Business and Economics	88, 90, 92			116	123	128, 129, 130		
Life Skills			106, 107, 108, 109	111, 114, 115, 117	123	124, 126 129, 131, 132		146, 149
Science and Technology	86, 93	94, 97, 98, 100, 101	110	116, 117	118, 121	132	133, 141	147
Social Studies	92			116				
Sports and Leisure			102		123			146
Cultural Connection: Asia	91							
Other	90	101	107					

BLOCK SCHEDULING GUIDE

Day	Lesson	Teacher Directed: Lesson Examples, Teaching Transparencies	Student Guided Activity, Try This	Cooperative-Learning Activity, Lesson Activity, Student Technology Guide	Practice: Practice & Apply, Extra Practice, Practice Workbook	Assessment: Quiz, Mid-Chapter Assessment	Problem Solving, Reteaching
1	2.1	10 min	10 min	8 min	25 min	8 min	8 min
	2.2	10 min	10 min	7 min	25 min	7 min	7 min
2	2.3	10 min	15 min	15 min	65 min	15 min	15 min
3	2.4	10 min	15 min	15 min	65 min	15 min	15 min
4	2.5	10 min	15 min	15 min	65 min	15 min	15 min
5	2.6	10 min	10 min	8 min	25 min	8 min	8 min
	2.7	10 min	10 min	7 min	25 min	7 min	7 min
6	Assess.	50 min **PE:** Chapter Review	90 min **PE:** Chapter Project, Writing Activities	90 min Tech Prep Masters	65 min **PE:** Chapter Assessment, Test Generator	30 min Chap. Assess. (A or B), Alt. Assess. (A or B), Test Generator	

PE: Pupil's Edition

Alternative Assessment

The following suggest alternative assessments for students who may benefit from a different type of assessment than the regular chapter quizzes and the mid-chapter/end-of-chapter test. Visit the HRW web site to get additional Alternative Assessment material.

internet connect

Alternative Assessment
Go To: **go.hrw.com**
Keyword: **MB1 Alt Assess**

Performance Assessment

1. a. Simplify $\left[\frac{(2a^{-4}b^3)(3b^5)}{(ab^2)^2}\right]^{-2}$, justifying each step with one of the properties of exponents.

b. Evaluate the expression in part **a** for $a = 6$ and $b = 2$. Give an exact answer.

c. How does the simplified expression from part **a** make it easier to find the solution to part **b**?

2. Let $f(x) = -2x + 5$ and $g(x) = 3x - 4$.

a. Find $f + g$, $f - g$, fg, and $f \circ g$.

b. Show that $f \circ g$ is not the same as $g \circ f$.

c. List the inverses of f and g. Explain how you can determine whether these inverses are functions.

Portfolio Project

Suggest that students choose one of the following projects for inclusion in their portfolios.

1. a. Graph $f(x) = [x]$, $h(x) = [0.5x + 1]$. Compare the graphs of these three functions. Then generalize the behavior of the graph of $g(x) = [mx + b]$ over the intervals $0 < m < 1$ and $m > 1$.

b. Describe the graph of $f(x) = [x^2]$. Graph the function to confirm or deny your conjecture.

c. Explore the effect of applying the greatest-integer function to other types of functions. To graph a greatest-integer function, set **Y1=int(X)** by pressing MATH, selecting NUM, choosing 5:int(, and pressing X,T,θ,n) ENTER.

internet connect

The table below identifies the pages in this chapter that contain internet and technology information.

Content Links

Activities Online	**pages 101, 109, 114**
Portfolio Extensions	**pages 110, 132**
Homework Help Online	**pages 90, 99, 108, 115, 122, 130, 140**
Graphic Calculator Support	**page 150**

Resource Links

Parents can go online and find concepts that students are learning—lesson by lesson—and questions that pertain to each lesson, which facilitate parent-student discussion.

Go To: **go.hrw.com**
Keyword: **MB1 Parent Guide**

Technical Support

The following may be used to obtain technical support for any HRW software product.

Online Help: **www.hrwtechsupport.com**
e-mail: **tschrw@hrwtechsupport.com**
HRW Technical Support Center: **(800)323-9239**
7 AM to 10 PM Monday through Friday Central Time

Visit the HRW math web site at: **www.hrw.com/math**

Technology

Lesson Suggestions and Calculator Examples

(Keystrokes are based on a TI-83 calculator.)

Lesson 2.1 Operations With Numbers

The Activity on page 89 shows students the importance of using parentheses when evaluating a quotient with a graphics calculator. For example, the key sequences below will give very different results depending on where parentheses are used.

(8 − 3) ÷ (5 − 2)

(8 − 3) ÷ 5 − 2

8 − 3 ÷ 5 − 2

Lesson 2.2 Properties of Exponents

Students may be less familiar with rational exponents than with integer exponents. Consider a brief exploration involving the exponent $\frac{1}{2}$. The display at right will help students see that a number raised to a power of $\frac{1}{2}$ and the square root of the number are the same.

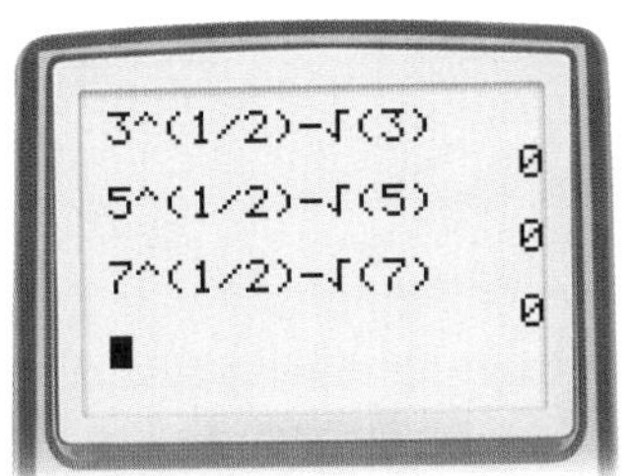

Lesson 2.3 Introduction to Functions

In this lesson, students can use the editing feature of a graphics calculator to evaluate functions. The display at right shows the evaluation of $f(x) = 1.5x^2 - 3.2x$ for $x = 2$, 3, and 4. Enter $1.5 * 2^2 - 3.2 * 2$, and press ENTER. Then press 2nd ENTER, use the left arrow key to replace the appropriate 2s with 3s, and press ENTER. Repeat the process to replace 3s with 4s.

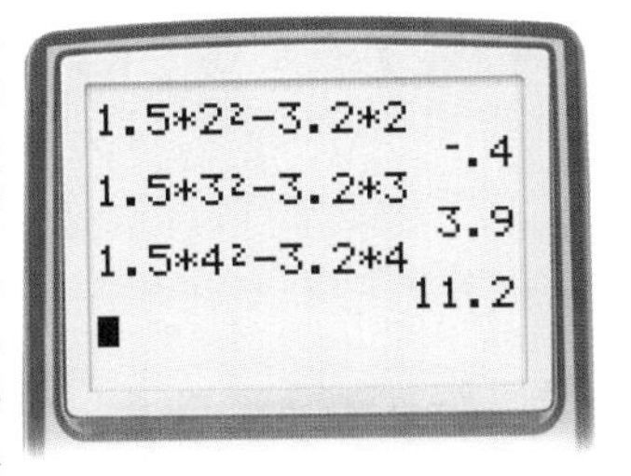

Lesson 2.4 Operations With Functions

To add $f(x) = 3x^2 - 4x + 5$ and $g(x) = 2x + 3$:

1. Enter functions *f* and *g* into **Y1** and **Y2**.
2. Let **Y3=** $f + g$ by pressing VARS, selecting **Y-VARS** **1:Function...** **1:Y1**, pressing +, and then repeating the process for **Y2**.

The displays below can help students see that the sum of a linear function and a quadratic function is quadratic.

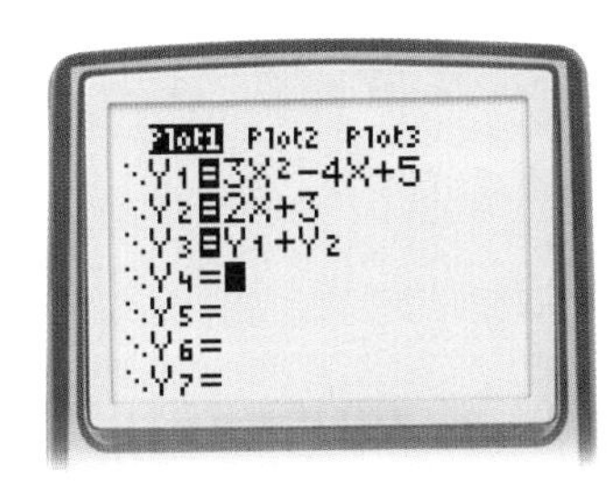

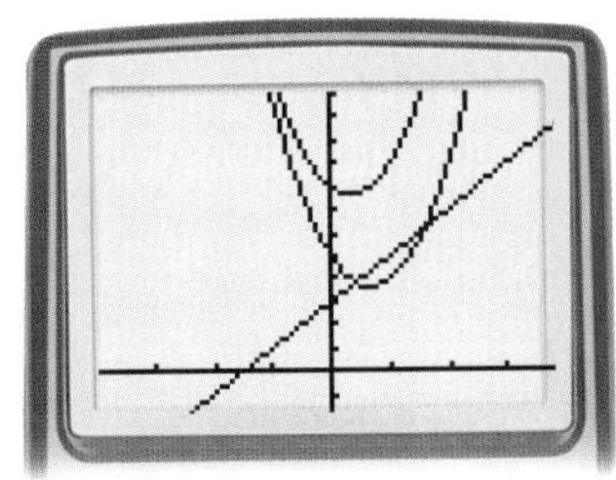

Lesson 2.5 Inverses of Functions

Students can use algebra to find and graph the inverse of a linear function. They can use the graphics calculator to verify their work. Suggest that students use a square viewing window, which can be accessed by pressing ZOOM and selecting **5: ZSquare**. Have students add the function $y = x$ to the **Y=** list to help them see the reflection of the graph of the identity function.

Lesson 2.6 Special Functions

In this lesson, ask students to use a graphics calculator to perform operations and compositions of functions.

1. Describe the graph of $f(x) = [x + 1] - [x]$.
2. Describe the graph of $g(x) = |[x]|$.

Lesson 2.7 A Preview of Transformations

Have students examine the following functions:

$f(x) = x^2$ $\quad g(x) = f(x) + 2$ $\quad h(x) = f(x) - 3$

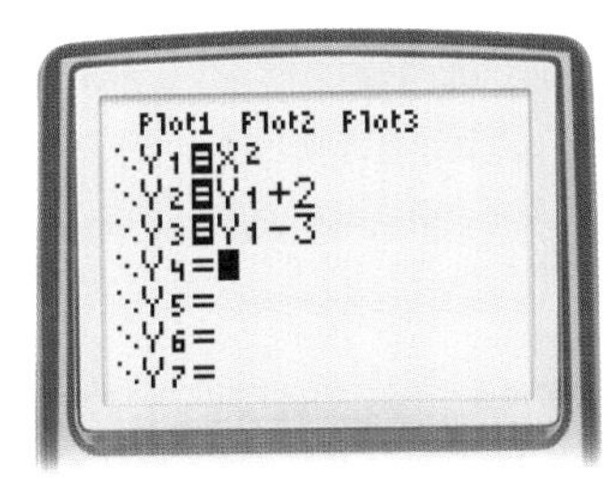

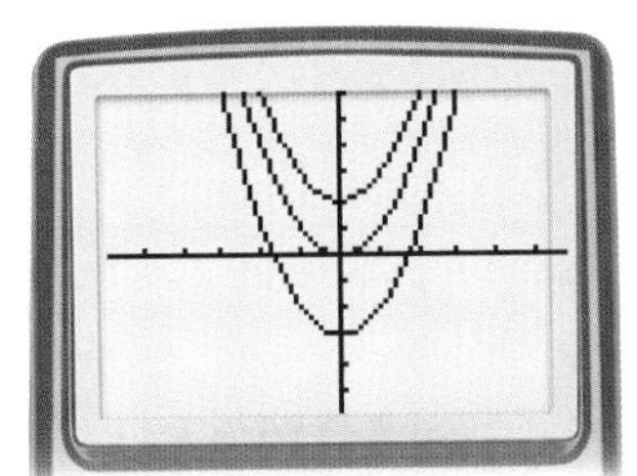

Students should see that adding a constant to a function translates the graph vertically.

For further information, refer to the

- technology discussions in the lessons.
- lesson-related teacher's commentary in the side columns of this *Teacher's Edition*.
- lesson-related *Student Technology Guide* masters.
- *HRW Technology Handbook*.

internet connect

For keystrokes of other graphing calculators models, visit the HRW web site at **go.hrw.com** and enter the keyword **MB1 CALC**.

About Chapter 2

Background Information

Chapter 2 begins with a review of real numbers, their properties, the order of operations, and the properties of exponents. At the heart of the chapter is a study of basic function concepts: functions as relations, operations with functions, composition of functions, and inverses of functions. After a brief look at "special" functions—piecewise, step, and absolute-value functions—the chapter concludes with an overview of basic transformations of the graphs of functions.

CHAPTER RESOURCES

- Block-Scheduling Handbook
- Writing Activities for Your Portfolio
- Tech Prep Masters
- Long-Term Project
- Assessment Resources:
  - Mid-Chapter Assessment
  - Chapter Assessments
  - Alternative Assessments
- Test and Practice Generator
- Technology Handbook

Chapter Objectives

- Identify and use properties of real numbers. [2.1]
- Evaluate expressions by using the order of operations. [2.1]
- Evaluate expressions involving exponents. [2.2]
- Simplify expressions involving exponents. [2.2]
- Graph a relation, state its domain and range, and tell whether it is a function. [2.3]
- Write a function in function notation and evaluate it. [2.3]

Numbers and Functions

FUNCTIONS ARE USED IN THE REAL WORLD to quantify trends and relationships between two variables. For example, the relationship between the *speed* at which an amusement park ride rotates and the *force* that holds riders in their seat can be described by a function.

Lessons

About the Photos

Any object that is moving in a circular path experiences *centripetal force*. Centripetal force is always directed toward the center of the circular path and depends on the object's mass, m, its velocity, v, and the radius of the circular path, r. The formula used to calculate centripetal force is $F = \frac{mv^2}{r}$.

Suppose that a person is on an amusement park ride that moves in a circular path with a radius of 10 meters (≈30 feet) at about 4.5 meters per second (≈10 miles per hour). If the person's mass is 55 kilograms (≈120 pounds), then the force felt is $\frac{55(4.5)^2}{10}$, or approximately 111 newtons. (One newton is the amount of force required to accelerate a 1-kilogram mass at a rate of 1 meter per second per second.)

The centripetal force is said to be a *function* of each rider's mass. For this particular ride $\frac{v^2}{r} = \frac{4.5^2}{10} = 2.025$, so the relationship between the mass of a rider and the centripetal force felt by that rider can be written as $F(m) = 2.025m$.

About the Chapter Project

Real-world situations are often very complex, with changing or unknown factors. Mathematical models can be used to represent such real-world situations and to predict probable outcomes. In the Chapter Project, *Space Trash,* you will use functions to model data related to the growing problem of space debris orbiting the Earth.

After completing the Chapter Project, you will be able to do the following:

- Use a table to represent the relationship between time in years and the number of space debris objects, and show that an appropriate function models this relationship.
- Find and discuss models for the accumulation of space debris.
- Determine the piecewise function that describes the relationship between altitude and number of orbital debris objects.

About the Portfolio Activities

Throughout the chapter, you will be given opportunities to complete Portfolio Activities that are designed to support your work on the Chapter Project.

- Finding the projected number of space debris objects at the end of each year through the year 2010 is included in the Portfolio Activity on page 93.
- Using exponents to project the number of space debris objects in a given year is included in the Portfolio Activity on page 101.
- Comparing regression models for the space debris data is included in the Portfolio Activity on page 110.
- Operating on function models is included in the Portfolio Activity on page 117.
- Using piecewise functions to model trends that change over time is included in the Portfolio Activity on page 132.

internet connect

Chapter Internet Features and Online Activities

Lesson	Keyword	Page
2.1	MB1 Homework Help	90
2.2	MB1 Homework Help	99
	MB1 Athletes	101
2.3	MB1 Homework Help	108
	MB1 Overpopulation	109
	MB1 Exponential	110
2.4	MB1 Homework Help	115
	MB1 UV	114
2.5	MB1 Homework Help	122
2.6	MB1 Homework Help	130
	MB1 Piecewise	132
2.7	MB1 Homework Help	140

- Perform operations with functions to write new functions. [2.4]
- Find the composite of two functions. [2.4]
- Find the inverse of a relation or function. [2.5]
- Determine whether the inverse of a function is a function. [2.5]
- Write, graph, and apply special functions: piecewise, step, and absolute value. [2.6]
- Identify the transformation(s) from one function to another. [2.7]

Portfolio Activities appear at the end of Lessons 2.1, 2.2, 2.3, 2.4, and 2.6. Each serves as preparation for the Chapter Project. The Portfolio Activities as well as the Chapter Project Activities are appropriate for inclusion in the student's portfolio. Students should be encouraged to include in their portfolios any other work in which they feel a sense of pride or a sense of accomplishment.

Prepare

NCTM PRINCIPLES & STANDARDS 1, 2, 7–10

QUICK WARM-UP

Evaluate each expression.

1. $\frac{1}{5} - \frac{1}{3}$ $\quad -\frac{2}{15}$
2. $-3 + \frac{1}{3}$ $\quad -2\frac{2}{3}$
3. $(-3)\left(\frac{1}{3}\right)$ $\quad -1$
4. $(-3) \div \left(-\frac{1}{3}\right)$ $\quad 9$
5. 25^2 $\quad 625$
6. $\sqrt{25}$ $\quad 5$

Also on Quiz Transparency 2.1

Teach

Why People in virtually every walk of life use numbers for some part of their work. Ask students to name various occupations, the types of numbers that are used in those occupations, and how the numbers are used. **Samples: Carpenters use fractions for measurement. Retailers use decimals for weights and volumes.**

Use Teaching Transparency 5.

Operations With Numbers

Why *In everyday life, people use many different types of numbers. A weather reporter, for example, will use integers to give temperature measurements, and rational numbers to give barometric pressure readings.*

Objectives

- Identify and use Properties of Real Numbers.
- Evaluate expressions by using the order of operations.

APPLICATION METEOROLOGY

A typical weather report might say that the temperature is 82°F, which is an integer, and that the barometric pressure is 29.98 inches of mercury, which is a positive rational number written as a decimal. These, as well as other types of numbers, typically belong to more than one *number set.*

Number Sets

Natural numbers	1, 2, 3, . . .
Whole numbers	0, 1, 2, 3, . . .
Integers	. . . , −3, −2, −1, 0, 1, 2, 3, . . .
Rational numbers	$\frac{p}{q}$, where p and q are integers and $q \neq 0$
Irrational numbers	numbers whose decimal part does not terminate or repeat
Real numbers	all rational and all irrational numbers

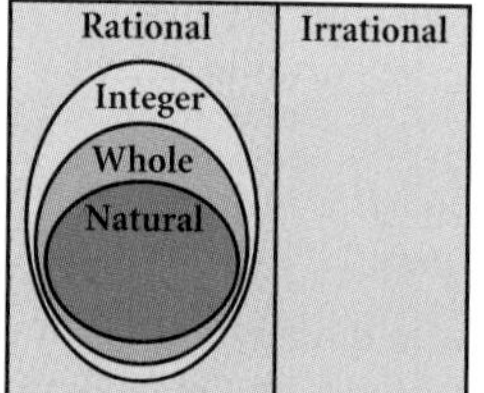

The *Venn diagram* at left shows the relationship between the various number sets. An important fact about rational numbers is stated in the following theorem, which will help you differentiate between rational and irrational numbers:

Every rational number can be written as a terminating or repeating decimal. Every terminating or repeating decimal represents a rational number.

When a repeating decimal is written, a bar is used to indicate the digit or digits that repeat. For example, $0.\overline{3}$ represents the repeating rational number 0.333333 . . .

Alternative Teaching Strategy

HANDS-ON STRATEGIES Have students work in groups of two or three. Give each group a set of boxes that have been labeled and nested to make a three-dimensional model of the Venn diagram for the real-number system that is shown on the pupil's page.

Prepare sets of index cards that display various types of numbers, writing one number on each card. Give each group a different set of cards. Have the group work together to place each card in the appropriate location inside the set of boxes.

Bring the class together to discuss the results. Starting with *natural numbers,* ask groups to report the numbers that they placed in each location. Have one student make a list of the numbers on the board or overhead. Ask students to describe the numbers in each location. For example, you might elicit an observation such as the following: In the whole-number box, zero is the only number that is not inside the natural-number box.

EXAMPLE 1 Classify each number in as many ways as possible.

a. −2.77 **b.** 178,000 **c.** 12.020002000002 . . .

SOLUTION

a. Since −2.77 is a terminating decimal, it is a rational and a real number.

b. Since 178,000 is positive and has no decimal part, it is a natural number, a whole number, an integer, a rational number, and a real number.

c. The pattern of digits in the decimal part of 12.020002000002 . . . suggests that the decimal part does not terminate and does not repeat. This number is irrational and real.

Every real number corresponds to a point on the real-number line. Conversely, for every point on the number line, you can assign a real-number coordinate.

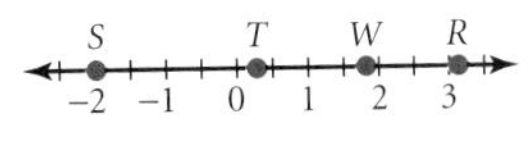

On the number line at left, point S corresponds to −2, point T corresponds to 0.5, point W corresponds to $1\frac{2}{3}$, and point R corresponds to the irrational number π, which is about 3.14.

Properties of Real Numbers

Properties for the fundamental operations of addition and multiplication with real numbers are listed below. Addition and multiplication are linked by the *Distributive Property.*

Properties of Addition and Multiplication

For all real numbers a, b, and c:

	Addition	Multiplication
Closure	$a + b$ is a real number.	ab is a real number.
Commutative	$a + b = b + a$	$ab = ba$
Associative	$(a + b) + c = a + (b + c)$	$(ab)c = a(bc)$
Identity	There is a number 0 such that $a + 0 = a$ and $0 + a = a$.	There is a number 1 such that $1 \cdot a = a$ and $a \cdot 1 = a$.
Inverse	For every real number a, there is a real number $-a$ such that $a + (-a) = 0$.	For every nonzero real number a, there is a real number $\frac{1}{a}$ such that $a\left(\frac{1}{a}\right) = 1$.

The Distributive Property

For all real numbers a, b, and c:

$$a(b + c) = ab + ac \text{ and } (b + c)a = ba + ca$$

Interdisciplinary Connection

COMPUTER APPLICATIONS In a computer spreadsheet, a *formula* is a sequence of numbers, operation symbols, cell references, and spreadsheet functions that instructs the computer to calculate a new value from existing values. In many spreadsheet programs, you use an equal sign, = , to signal the beginning of a formula. For instance, the formula **=A1+A2** instructs the computer to add the values in cells **A1** and **A2.** Write a spreadsheet formula for each calculation at right. (Note: Use the symbols +, −, *, and / for the four basic operations, and use ^ to indicate an exponent. Use parentheses as grouping symbols.)

a. the average of the values in cells **B1**, **B2**, and **B3**
sample: =(B1+B2+B3)/3

b. one-fourth of the average of the values in cells **C1, C2, C3,** and **C4**
sample: =.25*((C1+C2+C3+C4)/4)

c. the square of the difference when the value in cell **D1** is subtracted from the value in cell **D2,** divided by 7
sample: =((D2−D1)^2)/7

ADDITIONAL EXAMPLE 1

Classify each number in as many ways as possible.

a. 0.12122122212222…
irrational number, real number

b. 56
natural number, integer, rational number, real number

c. $4.\overline{77}$
rational number, real number

Teaching Tip

To help students understand the meaning and validity of the properties of the real numbers, suggest that they substitute numbers for a, b, and c and simplify the resulting expressions.

Teaching Tip

Some students may need practice with the properties of real numbers. Have them state a property to justify the following statements:

1. $6 + (-3) = (-3) + 6$
   Commutative, Add.
2. $2(4 - 5) = (4 - 5)2$
   Commutative, Mult.
3. $(-10)(-7) = (-7)(-10)$
   Commutative, Mult.
4. $-2 + (x - 5) = (-2 + x) - 5$
   Associative, Add.
5. $x[(-w) + y] = x(-w) + x(y)$
   Distributive
6. $(m - n) + [-(m - n)] = 0$
   Add. Inverse
7. $(-2) \cdot \frac{1}{-2} = 1$
   Mult. Inverse
8. $c = 1 \cdot c$
   Mult. Identity
9. If $7 + x = 7 + x^2$, then $x = x^2$
   Add. Prop. of Equality
10. $\frac{1}{2}(-3) + \pi$ is a real number
    Closure of real numbers

ADDITIONAL
EXAMPLE 2

Write and justify each step in the simplification of $(r+s)(5+t)$.

$= r(5+t) + s(5+t)$ Dist.
$= r \cdot 5 + rt + s \cdot 5 + st$ Dist.
$= 5r + rt + 5s + st$ Comm.

TRY THIS

$(a+b)(c-d)$
$= a(c-d) + b(c-d)$ Dist.
$= ac - ad + bc - bd$ Dist.

ADDITIONAL
EXAMPLE 3

A bank pays 2.7% simple interest on a savings account quarterly. This means that every three months 2.7% of the amount in the savings account is added to the account. The amount, A, in the account at the end of a quarter is given by $A = p + 0.027p$, where p is the amount at the beginning of the quarter. **Show that $A = 1.027p$. Justify each step.**

$A = p + 0.027p$
$A = 1p + 0.027p$ Identity
$A = (1 + 0.027)p$ Distributive
$A = 1.027p$ Addition

TRY THIS

$T = c + 0.06c$
$= 1c + 0.06c$ Identity
$= (1 + 0.06)c$ Distributive
$= 1.06c$ Addition

CRITICAL THINKING

$r\% = \frac{r}{100}$, so the total cost, T, is given by $T = c + \frac{r}{100}c$.

$T = c + \frac{r}{100}c$
$= 1 \cdot c + \frac{r}{100}c$ Identity
$= \left(1 + \frac{r}{100}\right)c$ Distributive

Example 2 shows you how to use Properties of Real Numbers to rewrite an expression.

EXAMPLE 2 **Write and justify each step in the simplification of $(z + x)(2 + w)$.**

SOLUTION

$(z + x)(2 + w) = (z + x)(2) + (z + x)(w)$ *Use the Distributive Property.*
$= z \cdot 2 + x \cdot 2 + zw + xw$ *Use the Distributive Property.*
$= 2z + 2x + wz + wx$ *Use the Commutative Property.*

TRY THIS Write and justify each step in the simplification of $(a + b)(c - d)$.

EXAMPLE 3 **APPLICATION** **TAXES**

When you purchase an item costing c dollars in a state that has a sales tax of 5%, the total cost, T, is given by $T = c + 0.05c$.

Show that $T = 1.05c$. Justify each step.

SOLUTION

$T = c + 0.05c$
$T = 1c + 0.05c$ *Use the Identity Property.*
$T = (1 + 0.05)c$ *Use the Distributive Property.*
$T = 1.05c$ *Add.*

TRY THIS If the sales tax is 6%, then $T = c + 0.06c$. Show that $T = 1.06c$. Justify each step.

CRITICAL THINKING When you purchase an item costing c dollars in a state that has a sales tax of r%, show that $T = \left(1 + \frac{r}{100}\right)c$. Justify each step.

Order of Operations

If an expression involves only numbers and operations, you can evaluate the expression by using the *order of operations.*

Order of Operations

1. Perform operations within the innermost grouping symbols according to Steps 2–4 below.
2. Perform operations indicated by exponents (powers).
3. Perform multiplication and division in order from left to right.
4. Perform addition and subtraction in order from left to right.

Inclusion Strategies

LINGUISTIC LEARNERS A mnemonic device can help some students remember the order of operations. Students can make up their own mnemonics or use *PEMDAS*. The letters represent the order of operations—Parentheses, Exponents, Multiplication, Division, Addition, Subtraction—and students can remember it with a phrase such as *Please Excuse My Dear Aunt Sally.*

Enrichment

Have students work in pairs. Give them the expression $3 + 1 - 2 \cdot 4 - 3^2 + 2$. Challenge them to add grouping symbols to the expression so that it has each value below.

1. −8 **sample:** $3 + (1 - 2) \cdot 4 - 3^2 + 2$
2. 4 **sample:** $3 + 1 - 2 \cdot (4 - 3)^2 + 2$
3. −11 **sample:** $3 + (1 - 2 \cdot 4) - 3^2 + 2$
4. −3 **sample:** $(3 + 1 - 2) \cdot 4 - (3^2 + 2)$

Example 4 shows you how to evaluate an expression by using the order of operations.

EXAMPLE 4 **Evaluate $\frac{2^2(12+8)}{5}$ by using the order of operations.**

SOLUTION

$$\frac{2^2(12+8)}{5} = \frac{2^2(20)}{5}$$ *Evaluate within grouping symbols.*

$$= \frac{4(20)}{5}$$ *Evaluate powers.*

$$= \frac{80}{5}$$ *Multiply within grouping symbols.*

$$= 16$$

A fraction bar is a grouping symbol.

TRY THIS Evaluate $\frac{18-2\cdot 5}{15+3(-3)}$ by using the order of operations.

In the Activity below, you can explore how calculators use the order of operations.

Activity: Exploring the Order of Operations

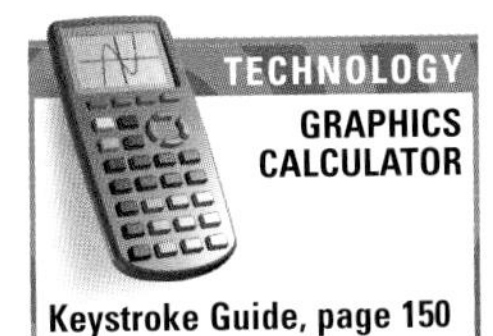

TECHNOLOGY GRAPHICS CALCULATOR

Keystroke Guide, page 150

You will need: a scientific or graphics calculator

1. Evaluate $\frac{12+8}{5}$ on a calculator as follows:
   a. without any parentheses for grouping
      12 [+] 8 [÷] 5
   b. with parentheses to group the numerator
      [(] 12 [+] 8 [)] [÷] 5

CHECKPOINT ✔

2. Explain why your results in parts **a** and **b** in Step 1 are not equal. Which result is correct?
3. The slope of the line containing the points (2, 3) and (5, 8) is $\frac{8-3}{5-2}$. Evaluate this expression on a calculator as follows:
   a. without any parentheses for grouping
      8 [-] 3 [÷] 5 [-] 2
   b. with parentheses to group the numerator but not the denominator
      [(] 8 [-] 3 [)] [÷] 5 [-] 2
   c. with parentheses to group the numerator and the denominator
      [(] 8 [-] 3 [)] [÷] [(] 5 [-] 2 [)]

CHECKPOINT ✔

4. Discuss why there are three different results to Step 3. Which result is correct?

When you use a calculator to evaluate expressions such as $\frac{9-4}{7-5}$, you need to enclose the numerator and the denominator in separate sets of parentheses.

CHECKPOINT ✔ What keystrokes will give the correct answer for $\frac{9-4}{7-5}$?

ADDITIONAL EXAMPLE 4

Evaluate $\frac{6(11+3^2)}{8}$ by using the order of operations. 15

TRY THIS

$\frac{4}{3}$

Activity Notes

In this Activity, students use their calculators to evaluate a fraction whose numerator or denominator contains an addition expression. They should discover that the fraction bar is another type of grouping symbol. Therefore, when evaluating fractions of this type on a calculator, they must take care to enclose the addition expressions within parentheses.

Teaching Tip

TECHNOLOGY If you omit an open parenthesis, [(], when using a TI-82 or TI-83 graphics calculator, you will receive an error message. You are allowed to omit the closing parenthesis, [)], but alert students to the potential for error that arises as a result. For instance, if a student carelessly enters (12 + 8) ÷ 5 as [(] **12** [+] **8** [÷] **5** [ENTER], the calculator will proceed to give the result 13.6 rather than the intended result of 4.

☞ For answers to Checkpoints, see page 90.

Reteaching the Lesson

COOPERATIVE LEARNING Have students work in groups of three or four. Direct their attention to the chart of properties on page 87. For each property, the group should work together to write a sentence that states the property in a different way. Whenever possible, they should write the statement without using variables.

To get them started, you might suggest the following statements for the closure property:

The sum or product of two real numbers is another real number.

When you add or multiply two real numbers, the result is always a real number.

When all the groups have completed their work, bring the class together for a discussion of the results.

CHECKPOINT ✓

2. In part **a**, division is performed before addition.

In part **b**, the parentheses change the order of operations, so addition is performed before division. The result from part **b** is correct.

CHECKPOINT ✓

4. In part **a**, the division, $3 \div 5 = 0.6$, is performed first, yielding $8 - 0.6 - 2$. The result is 5.4.

In part **b**, the subtraction within the parentheses, $8 - 3$, is performed first and then the division is done before the last subtraction. The result is -1.

In part **c**, both subtractions, $8 - 3$ and $5 - 2$, should be performed before division. The result is $1.\overline{6}$. The result from part **c** is correct.

CHECKPOINT ✓

$(9 - 4) \div (7 - 5)$

Assess

Selected Answers

Exercises 4–15, 17–93 odd

ASSIGNMENT GUIDE	
In Class	1–15
Core	16–31, 33–37 odd, 38–55, 57–77 odd
Core Plus	16–31, 32–36 even, 38–55, 56–70 even, 71–80
Review	81–94
Preview	95

Extra Practice can be found beginning on page 940.

Exercises

Communicate

1. Discuss two times in the past week when you added numbers. What type of numbers did you add? Can you think of occasions when you used any of the other types of numbers?

APPLICATIONS

2. **ETYMOLOGY** Explain what the Commutative Properties of Addition and Multiplication are. Why is the word *commutative* appropriate for these properties?
3. **ETYMOLOGY** Explain what the Associative Properties of Addition and Multiplication are. Why is the word *associative* appropriate for this property?

Guided Skills Practice

4. Classify $\frac{3}{2}$ and $-2.101001000\ldots$ in as many ways as possible. ***(EXAMPLE 1)*** **rational and real; irrational and real**

Write and justify each step in the simplification of each expression. ***(EXAMPLE 2)***

5. $2(b + d)$ **$2b + 2d$**
6. $-3a + 3a$ **0**
7. $\frac{3(8 + 2)}{2}$ **15**
8. $\frac{7 - 1}{5 - 2}$ **2**
9. $\frac{1}{4}(4 \cdot 5)$ **5**
10. $-5(4t^2)$ **$-20t^2$**

APPLICATION

11. **BUSINESS** A monthly automotive insurance payment, m, is given by $m = \frac{p}{12} + \frac{0.06p}{12}$, where p is the yearly premium and $0.06p$ represents the annual processing fee. Show that $m = \frac{1.06p}{12}$. Justify each step. ***(EXAMPLE 3)***

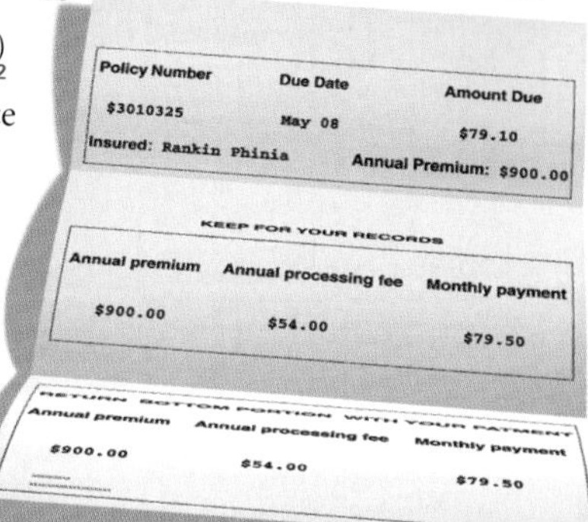

An insurance adjuster evaluates the damage to a car after an accident.

Evaluate each expression by using the order of operations. ***(EXAMPLE 4)***

12. $5^2 + 8 \div 4 - 2$ **25**
13. $(7 - 3^2)2$ **-4**
14. $\frac{5 \cdot 6 \div 3 \cdot 7}{12}$ **$\frac{35}{6}$**
15. $2[14 - 3(6 - 1)^2]$ **-122**

Practice and Apply

internet connect
Homework Help Online
Go To: **go.hrw.com**
Keyword:
MB1 Homework Help for Exercises 16–31

Classify each number in as many ways as possible.

16. -23
17. -5.1
18. $\sqrt{3}$
19. $\sqrt{2}$
20. $\frac{2}{3}$
21. $\frac{3}{9}$
22. $-0.\overline{85}$
23. $-1.\overline{063}$
24. $-\frac{5}{7}$
25. $\sqrt{25}$
26. $\frac{\sqrt{36}}{2}$
27. 1
28. 0
29. $-\pi$
30. $5.010010001\ldots$
31. $\sqrt{28}$

Graph each pair of numbers on a number line.

32. -3 and -2.5
33. -1.5 and -4
34. $\frac{13}{2}$ and 7
35. $4\frac{3}{8}$ and 2
36. $-3.\overline{6}$ and -4
37. $\sqrt{7}$ and 3

11. $m = \frac{p}{12} + \frac{0.06p}{12} = \frac{p + 0.06p}{12}$ (Add fractions.)

$= \frac{(1 + 0.06)p}{12}$ (Distributive)

$= \frac{1.06p}{12}$ (Add.)

16. -23 is an integer, a rational number, and a real number.

17. -5.1 is a rational and real number.

18. $\sqrt{3}$ is an irrational and real number.

19. $\sqrt{2}$ is an irrational and real number.

20. $\frac{2}{3}$ is a rational and real number.

21. $\frac{3}{9}$ is a rational and real number.

22. $-0.\overline{85}$ is a rational and real number.

23. $-1.\overline{063}$ is a rational and real number.

24. $-\frac{5}{7}$ is a rational and real number.

25. $\sqrt{25}$, or 5, is a natural number, a whole number, an integer, a rational number, and a real number.

38. Comm. Prop. of Mult.

42. Assoc. Add.

44. Inverse Mult.

State the property that is illustrated in each statement. All variables represent real numbers.

38. $v(3t) = (3t)v$

39. Assoc. Prop. of Mult.
39. $(25x)y = 25(xy)$

40. $4x + 13y = 13y + 4x$ **Comm. Add.**

41. $2.3 + x = x + 2.3$ **Comm. Add.**

42. $(2 + 3) + 5 = 2 + (3 + 5)$

43. Assoc. Add.
43. $(3 + a) + b = 3 + (a + b)$

44. $x\left(\frac{1}{x}\right) = 1$, where $x \neq 0$

45. $\frac{x}{3} \cdot \frac{3}{x} = 1$, where $x \neq 0$ **Inverse Mult.**

46. $-7 + 7 = 0$ **Inverse Add.**

47. $0 = 2x + (-2x)$ **Inverse Add.**

48. $1 \cdot (3x) = 3x$ **Ident. Mult.**

49. $63 \cdot 1 = 63$ **Identity Mult.**

50. $-5x + 0 = -5x$ **Ident. Add.**

51. $x + y = 0 + x + y$ **Ident. Add.**

52. $m(x^2 + x) = mx^2 + mx$ **Dist.**

53. $2(3 - y) = 2 \cdot 3 - 2y$ **Dist.**

54. $4yw = 4wy$ **Comm. Mult.**

55. $5(127) = 127(5)$ **Comm. Mult.**

Evaluate each expression by using the order of operations.

56. $3 \cdot 2^2 + 3$ **15**

57. $6 \div 3 \cdot 2$ **4**

58. $2^2(2 + 3) + 5$ **25**

59. $6 \div (3 - 1) \cdot 5$ **15**

60. $-3 \cdot 5^2 + 16$ **−59**

61. $5(2 - 3)^2$ **5**

62. $(3 - 2) + (5 - 4) - 2$ **0**

63. $30 - 3 \times 2 + 6 \div 3$ **26**

64. $16 \div 2 \times 6 - 1$ **47**

65. $(2^2 + 1) + 4 \div 2$ **7**

66. $6 \div 3 - (10 - 3^2)$ **1**

67. $2^{(3-1)} + (3 - 1)$ **6**

68. $3 \cdot 4 - 2^{(4-1)}$ **4**

69. $\frac{8-2}{3} + (2 + 1)$ **5**

70. $2 \cdot 4 + \frac{14}{5+2}$ **10**

71. Complete the following investigation:

a. Count the number of items in your home that display numbers.

b. What types of numbers are represented?

c. Name two examples of integers and two examples of rational numbers that you found.

CHALLENGE

72. Can a number be both rational and irrational? Explain your reasoning.

CONNECTION

73. STATISTICS While trying to find the average of 8, 10, 14, and 16, Ron entered 8 [+] 10 [+] 14 [+] 16 [÷] 4 [=] into a calculator and got 36 for an answer.

a. Did Ron get the correct average of 8, 10, 14, and 16? Explain.

b. What keystrokes should Ron have used?

74. CULTURAL CONNECTION: ASIA Ancient Babylonians used rational numbers as approximations of irrational numbers. For example, the Babylonians knew that the diagonal of a square was $\sqrt{2}$ times the length of a side. For the value of $\sqrt{2}$, the Babylonians used 1.4142. They thought this value was close enough for their practical purposes.

A Babylonian cuneiform tablet showing the calculation of areas

Use a calculator for the following exercises:

a. Show that $\sqrt{2}$ does not equal exactly 1.4142.

b. Find $\sqrt{2}$ on your calculator. Write down the result. Enter this number in your calculator, and square it. Is the result equal to 2? Explain why or why not.

Error Analysis

Students may classify some numbers incorrectly simply because they have forgotten how to work with radicals. Review some examples of simplified radicals, such as $\sqrt{49} = 7$. Also review the process of approximating radicals, such as locating $\sqrt{50}$ between 7 and 8, but closer to 7. Remind students how to use the square-root key on their calculators.

26. $\frac{\sqrt{36}}{2} = \frac{6}{2} = 3$ **is a natural number, a whole number, an integer, a rational number, and a real number.**

27. 1 is a natural number, a whole number, an integer, a rational number, and a real number.

28. 0 is a whole number, an integer, a rational number, and a real number.

29. $-\pi$ **is an irrational and real number.**

30. 5.01001001 . . . is irrational and real.

31. $\sqrt{28}$ **is an irrational and real number.**

32.

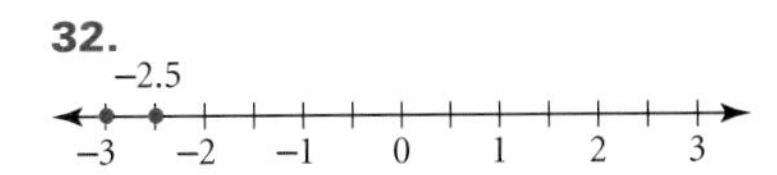

33.

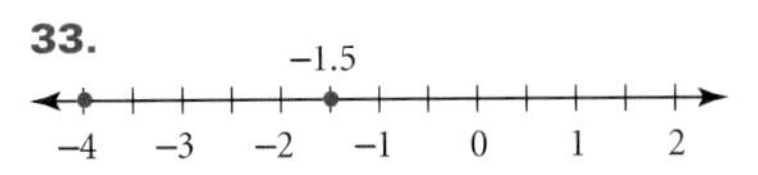

34.

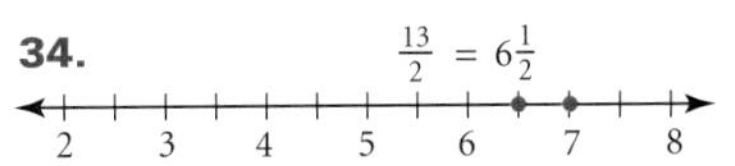

35.

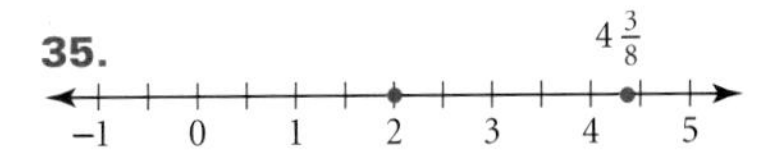

36.

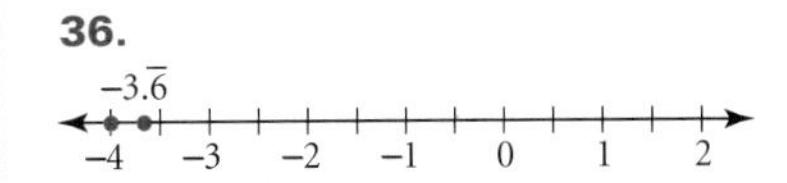

37.

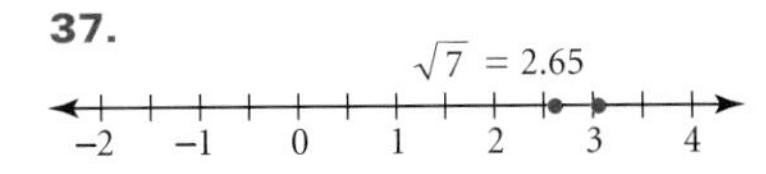

71a. Answers may vary.

b. sample: whole, rational, real, and negative integers

c. sample: integers: 0 through 10 on the phone, 0 on a scale; rational: 3.5-oz bag of microwave popcorn, $\frac{1}{4}$ on a measuring cup

72. No; a number written as a decimal can terminate or repeat, making it rational, or cannot terminate or repeat, making it irrational.

73a. No; Ron calculated $8 + 10 + 14 + \frac{16}{4}$, not $\frac{8 + 10 + 14 + 16}{4}$.

b. Ron should have entered the expression $(8 + 10 + 14 + 16) \div 4$, which would have given an answer of 12.

74a. $(1.4142)^2 = 1.99996164 \neq 2$

b. $\sqrt{2} = 1.41421352$; no, the result is 1.999999999; because $\sqrt{2}$ is irrational, it cannot be represented by a terminating decimal.

APPLICATIONS

75. BUSINESS A small business contributes $224 per month per worker for health insurance. This amount is half of the worker's monthly insurance premium and can be represented by $\frac{0.5p}{12} = 224$, where p is the yearly premium. Solve for p and justify each step. **$p = 5376$**

76. CURRENT EVENTS Use your local newspaper to answer the following questions: **Answers may vary. Check students' work.**

a. Find three articles that include numerical data.

b. Summarize each article. Include the reason that the article depends on numbers and the possible effect that each article may have on daily life.

c. What types of numbers were included in each article? What units, if any, were used with the numbers?

DEMOGRAPHICS Read the article below and answer the questions that follow.

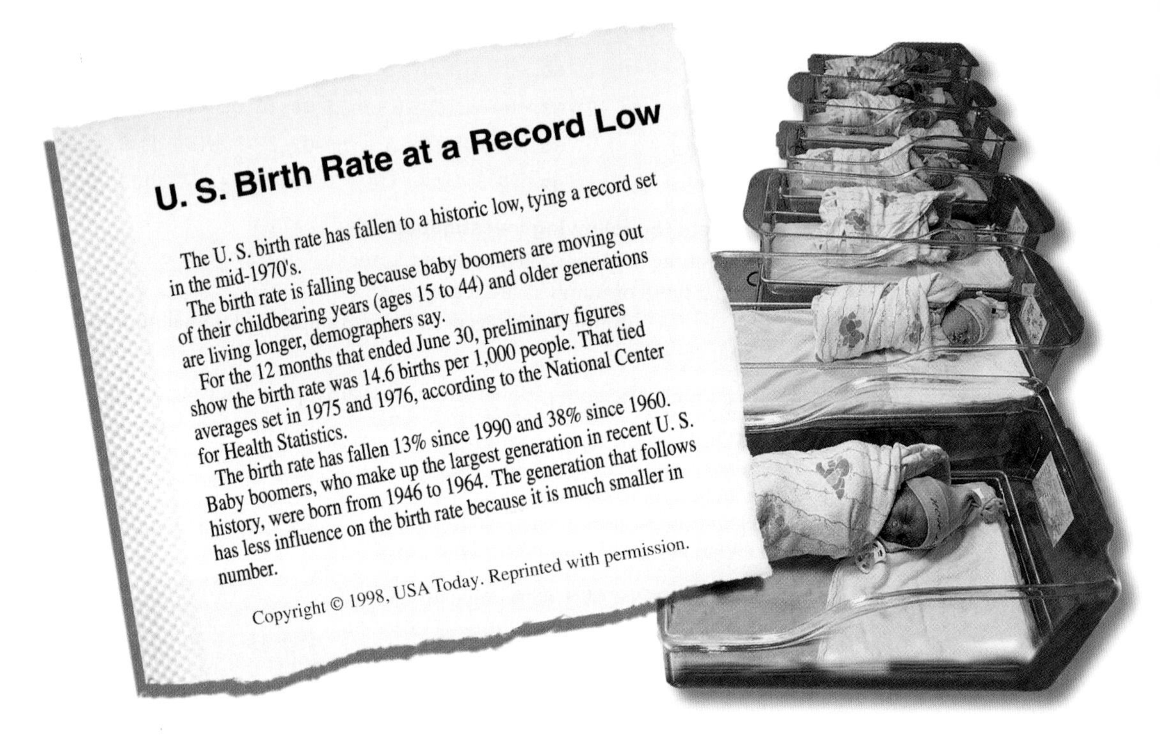

U. S. Birth Rate at a Record Low

The U. S. birth rate has fallen to a historic low, tying a record set in the mid-1970's.

The birth rate is falling because baby boomers are moving out of their childbearing years (ages 15 to 44) and older generations are living longer, demographers say.

For the 12 months that ended June 30, preliminary figures show the birth rate was 14.6 births per 1,000 people. That tied averages set in 1975 and 1976, according to the National Center for Health Statistics.

The birth rate has fallen 13% since 1990 and 38% since 1960. Baby boomers, who make up the largest generation in recent U. S. history, were born from 1946 to 1964. The generation that follows has less influence on the birth rate because it is much smaller in number.

77. Explain how rates such as 14.6 per 1000 can be considered rational numbers. **$\frac{14.6}{1000} = \frac{146}{10,000}$**

78. Explain how percents such as 13% can be considered rational numbers.

79. How much had the birth rate fallen since 1990 at the time of this article? Write this percent as a fraction, and classify it in as many ways as possible.

80. For the 12 months that ended June 30, 1998, what was the birth rate? Write this rate as a fraction, and classify it in as many ways as possible.

78. $13\% = \frac{13}{100}$

79. 13%; $\frac{13}{100}$; rational, real

80. 14.6 per 1000; $\frac{146}{10,000}$; rational, real

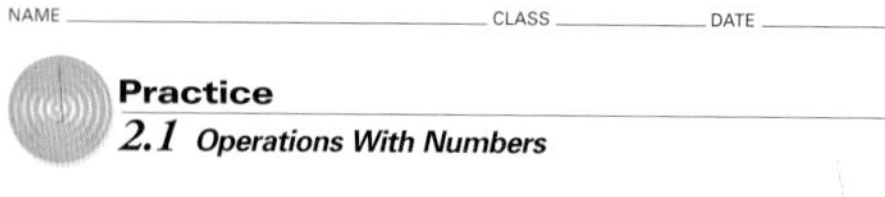

Practice

NAME ____ CLASS ____ DATE ____

Practice 2.1 Operations With Numbers

Classify each number in as many ways as possible.

1. $\frac{13}{17}$ — rational, real
2. $\sqrt{91}$ — irrational, real
3. 3.12112111211112... — irrational, real
4. 801.35 — rational, real
5. $-\sqrt{900}$ — integer, rational, real
6. $501.\overline{07}$ — rational, real

State the property that is illustrated in each statement. Assume that all variables represent real numbers.

7. $75 + (-75) = 0$ — Inverse Property of Addition
8. $181 \cdot 1 = 181$ — Identity Property of Multiplication
9. $-2 + (33 + 18) = (-2 + 33) + 18$ — Associative Property of Addition
10. $\frac{54}{k} \cdot \frac{k}{54} = 1$, where $k \neq 0$ — Inverse Property of Multiplication
11. $47y \cdot 3x = 3x \cdot 47y$ — Commutative Property of Multiplication
12. $14(x + 91) = 14x + 14(91)$ — Distributive Property
13. $\frac{7}{8} + 0 = \frac{7}{8}$ — Identity Property of Addition

Evaluate each expression by using the order of operations.

14. $-2 \cdot 4^2 - 1$ — -33
15. $52 \div (2 + 11)$ — 4
16. $27 + 8 \cdot 2$ — 43
17. $45 - 16 \div 8$ — 43
18. $13 \times 3 + 2 \times 5$ — 49
19. $12 + 8^2 \div 4$ — 28
20. $\frac{150 - 38}{4} - 4 + 2$ — 26
21. $(13 - 7)^2 \div 5$ — 7.2
22. $(77 - 50) - (13 - 42)$ — 56
23. $7 \cdot 12 + 30 \div 5$ — 90

Look Back

Find the slope of each line. ***(LESSON 1.2)***

81. $y = -3x$ **−3**
82. $y = 2x - 1$ **2**
83. $y = \frac{3x-1}{4}$ $\frac{3}{4}$
84. $y = -\frac{x}{2} + 1$ $-\frac{1}{2}$
85. $4y = 3x + 21$ $\frac{3}{4}$
86. $5x - 2y = 6$ $\frac{5}{2}$

Solve each equation. ***(LESSON 1.6)***

87. $3(x-5) = 4$ $\frac{19}{3}$
88. $-\frac{x}{3} + 9 = 2$ **21**
89. $-3x - 5 = x + 12$ $-\frac{17}{4}$
90. $\frac{1}{5}x - 4 = 3(x-5)$ $\frac{55}{14}$

Solve each absolute-value equation or inequality. Graph the solution on a number line. ***(LESSON 1.8)***

91. $|5x| = 12$ $x = \frac{12}{5}$ *or* $x = -\frac{12}{5}$
92. $\left|\frac{4}{5}x\right| = 12$ $x = 15$ *or* $x = -15$
93. $|4x + 2| > x + 3$ $x < -1$ *or* $x > \frac{1}{3}$
94. $|4x + 2| > 5x + 5$ $x < -\frac{7}{9}$

Look Beyond

95. The solutions to equations illustrate the need for different kinds of numbers. The equation $x + 7 = 5$ has the solution $x = -2$, which is a negative number, even though only nonnegative numbers appear in the equation. Similarly, only integers appear in the equation $2x = 5$, but the solution is $x = 2.5$, which is not an integer.

a. Find another example of an equation that includes only nonnegative numbers and that has a negative-number solution.

b. Find another example of an equation that includes only integers and that has a non-integer solution.

SPACE SCIENCE Scientists catalog only the space debris objects large enough to be repeatedly tracked by ground-based radar. At the end of 1993, the Air Force Space Command cataloged a total of 7000 debris objects in Earth orbit. At that time, it was projected that the number of cataloged space debris objects would grow at a rate of 3% per year.

Start with 7000 objects, the total number of space debris objects tracked at the end of 1993.

Using the annual growth rate of 3% (1.03 × the previous year's total), find the projected number of space debris objects at the end of each successive year through the year 2010. Round all results to the nearest whole number. Record your results in a table. Throughout the Portfolio Activities in this chapter, this will be your **Space Debris Table.**

Look Beyond

In Exercise 95, students see that the solution of an equation may be a type of number other than the types that appear in the equation. You may wish to use this exercise as a transition to Lesson 2.2. In that lesson, students will learn that although a fraction does not appear in an expression like 3^{-2}, the expression in fact represents a fraction.

ALTERNATIVE
Assessment

Portfolio Activity

The Portfolio Activity can be used as preparation for the Chapter Project or as a separate activity. In the Portfolio Activity on this page, students use given data and a simple equation to estimate the number of debris objects in Earth orbit during given years. They organize these estimates into a space debris table that they will use in subsequent Portfolio Activities in this chapter.

Answers to Portfolio Activities can be found in Additional Answers of the Teacher's Edition.

91. $x = \frac{12}{5}$ *or* $x = -\frac{12}{5}$

(Number line from −3 to 3 with closed points at $-\frac{12}{5}$ and $\frac{12}{5}$.)

92. $x = 15$ *or* $x = -15$

(Number line from −30 to 30 with closed points at −15 and 15.)

93. $x < -1$ *or* $x > \frac{1}{3}$

(Number line from −3 to 3 with open points at −1 and $\frac{1}{3}$.)

94. $x < -\frac{7}{9}$

(Number line from −3 to 3 with open point at $-\frac{7}{9}$.)

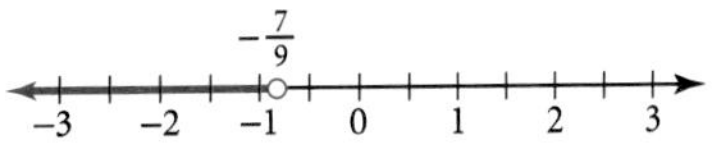

95a. Sample: $x + 8 = 4$ has the solution $x = -4$.

b. Sample: $3x = 4$ has the solution $x = \frac{4}{3}$.

Student Technology Guide

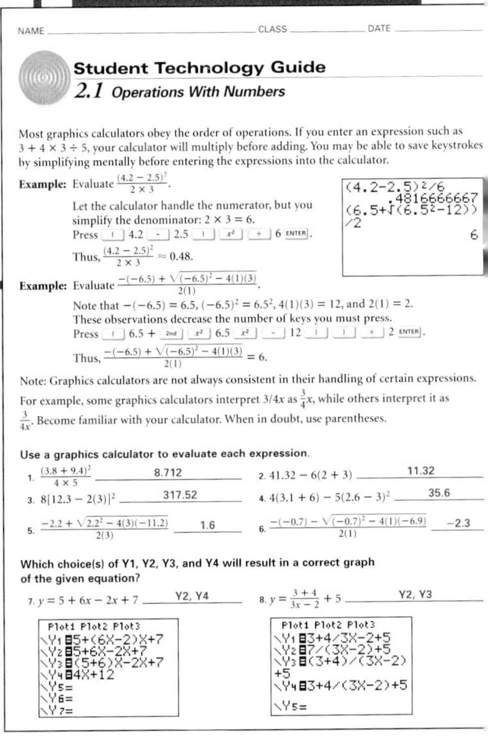

NAME ______ CLASS ______ DATE ______

Most graphics calculators obey the order of operations. If you enter an expression such as $3 + 4 \times 3 \div 5$, your calculator will multiply before adding. You may be able to save keystrokes by simplifying mentally before entering the expressions into the calculator.

Example: Evaluate $\frac{(4.2-2.5)^2}{2 \times 3}$.

Let the calculator handle the numerator, but you simplify the denominator: $2 \times 3 = 6$.
Press (4.2 - 2.5) x² ÷ 6 ENTER.

Thus, $\frac{(4.2-2.5)^2}{2 \times 3} \approx 0.48$.

```
(4.2-2.5)²/6
      .4816666667
(6.5+√(6.5²-12))
/2
                6
```

Example: Evaluate $\frac{-(-6.5) + \sqrt{(-6.5)^2 - 4(1)(3)}}{2(1)}$.

Note that $-(-6.5) = 6.5$, $(-6.5)^2 = 6.5^2$, $4(1)(3) = 12$, and $2(1) = 2$. These observations decrease the number of keys you must press.
Press (6.5 + 2nd x² 6.5 x² - 12)) ÷ 2 ENTER.

Thus, $\frac{-(-6.5) + \sqrt{(-6.5)^2 - 4(1)(3)}}{2(1)} = 6$.

Note: Graphics calculators are not always consistent in their handling of certain expressions. For example, some graphics calculators interpret $3/4x$ as $\frac{3}{4}x$, while others interpret it as $\frac{3}{4x}$. Become familiar with your calculator. When in doubt, use parentheses.

Use a graphics calculator to evaluate each expression.

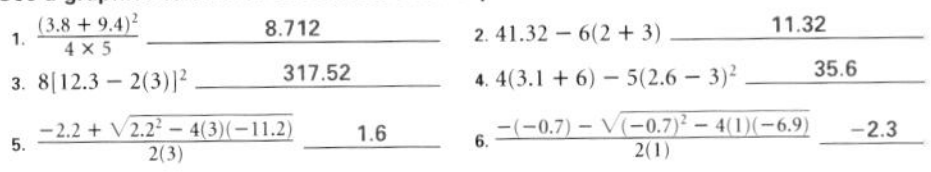

1. $\frac{(3.8+9.4)^2}{4 \times 5}$ **8.712**
2. $41.32 - 6(2+3)$ **11.32**
3. $8[12.3 - 2(3)]^2$ **317.52**
4. $4(3.1+6) - 5(2.6-3)^2$ **35.6**
5. $\frac{-2.2 + \sqrt{2.2^2 - 4(3)(-11.2)}}{2(3)}$ **1.6**
6. $\frac{-(-0.7) - \sqrt{(-0.7)^2 - 4(1)(-6.9)}}{2(1)}$ **−2.3**

Which choice(s) of Y1, Y2, Y3, and Y4 will result in a correct graph of the given equation?

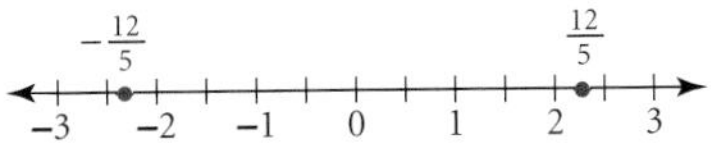

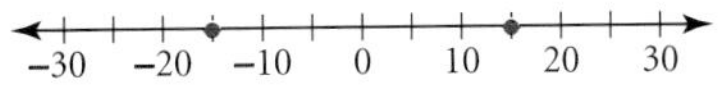

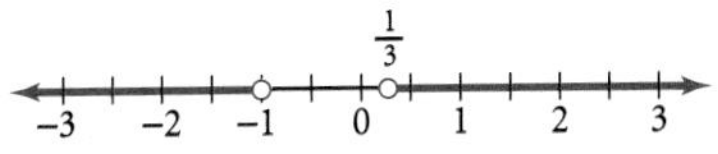

7. $y = 5 + 6x - 2x + 7$ **Y2, Y4**
8. $y = \frac{3+4}{3x-2} + 5$ **Y2, Y3**

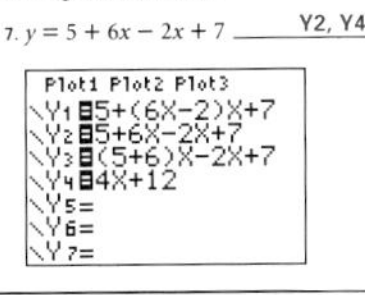

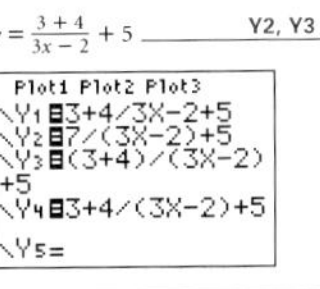

Prepare

NCTM PRINCIPLES & STANDARDS 1, 2, 6–10

QUICK WARM-UP

Evaluate each expression.

1. 3^3	27	**2.** 7^2	49
3. $2^5 \cdot 3^2$	288	**4.** $2^3 + 2^4$	24
5. $\frac{4^2}{2^3}$	2	**6.** $\left(\frac{4}{5}\right)^2$	$\frac{16}{25}$

Also on Quiz Transparency 2.2

Teach

Why Have students brainstorm ways that the properties of addition and multiplication can make calculations easier. As an example, show this use of the distributive property:

$$3(18) + 7(18) = 10(18) = 180$$

In this lesson, they will see how *properties of exponents* can be used to simplify calculations involving exponents.

2.2 Properties of Exponents

Objectives

- Evaluate expressions involving exponents.
- Simplify expressions involving exponents.

Why *Exponential expressions can be found in a wide variety of fields, including physics. For example, the centripetal acceleration of a rider who is traveling in a circle at a high speed on an amusement park ride can be described by using exponents.*

APPLICATION
PHYSICS

In an amusement park ride, a compartment with a rider travels in a circle at a high speed. The centripetal acceleration acting on the rider can be calculated by using the equation below. *You will use this equation in Example 1.*

$A_c = 4\pi^2 r T^{-2}$, where
- A_c represents the centripetal acceleration in feet per second squared
- r represents the radius of the circle in feet
- T represents the time for a full rotation in seconds

Recall that the expression a^n is called a **power** of a. In the expression, a is called the **base** and n is called the **exponent**.

Definition of Integer Exponents

Let a be a real number.

If n is a natural number, then $a^n = a \times a \times a \times \cdots \times a$, n times.

If a is nonzero, then $a^0 = 1$.

If n is a natural number, then $a^{-n} = \frac{1}{a^n}$.

In the expression a^0, a must be nonzero because 0^0 is undefined.

Alternative Teaching Strategy

USING TECHNOLOGY Have students use graphics calculators to create a graph and a table for $y = 2^x$. For TI-82 or TI-83 calculators, use the following procedure:

First press Y= and enter **2^X** to the right of **Y1=**. Then press WINDOW and enter these values.

Xmin=–4	**Xmax=4**	**Xscl=1**
Ymin=–5	**Ymax=10**	**Yscl=1**

Next press 2nd WINDOW and make these choices.

TblStart=1 (or TblMin=1)	**Indpnt: Auto Ask**
ΔTbl=1	**Depend: Auto Ask**

Now press 2nd GRAPH to see the table.

The integers 1, 2, 3, 4, 5, 6, and 7 are in the **X** column, and the corresponding powers of 2 are in the **Y1** column. Note the pattern 2, 4, 8, 16, and so on. Ask students to guess **Y1** for **X=0**. They will probably guess 1. Have them press GRAPH, and point out that the y-intercept of $y = 2^x$ is 1. Tell them to return to the table and press ▲. They will find that for **X=0, Y1=1,** so $2^0 = 1$.

Continue this investigation for negative integer powers of 2. Then repeat the analysis for other equations, such as $y = 3^x$ and $y = 5^x$.

Example 1 shows how you can use the definition $a^{-n} = \frac{1}{a^n}$.

EXAMPLE 1 Refer to the equation for centripetal acceleration given at the beginning of the lesson.

Find the centripetal acceleration in feet per second squared of a rider who makes one rotation in 2 seconds and whose radius of rotation is 6 feet.

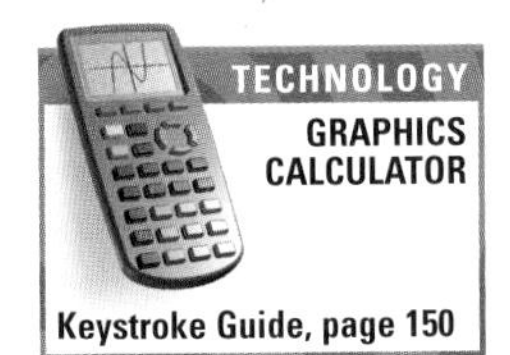

SOLUTION

Evaluate $A_c = 4\pi^2 r T^{-2}$ for $T = 2$ and $r = 6$.

$$A_c = 4\pi^2 r T^{-2}$$
$$= 4\pi^2(6)(2)^{-2}$$
$$= \frac{24\pi^2}{2^2} \qquad \text{Use } a^{-n} = \frac{1}{a^n}.$$
$$= 6\pi^2 \approx 59.2$$

The centripetal acceleration is about 59 feet per second squared.

CHECK

```
4π²*6*(2^-2)
59.21762641
```

TRY THIS Find the centripetal acceleration of a rider that makes one rotation in 5 seconds and whose radius of rotation is 6 feet.

Activity Exploring Properties of Exponents

You will need: no special materials

1. Rewrite $a^3 \cdot a^5$ by writing out all of the factors of a, counting them, and simplifying them as a power with a single exponent. What operation could you perform on the exponents in $a^3 \cdot a^5$ to obtain an equivalent expression with a single exponent?
2. Rewrite $(a^3)^5$ by writing out five sets of three factors of a, counting the factors of a, and simplifying them as a power with a single exponent. What operation could you perform on the exponents in $(a^3)^5$ to obtain an equivalent expression with a single exponent?

CHECKPOINT ✓ 3. Explain how to simplify $(a^7 \cdot a^3)^2$ by using addition and multiplication.

The results of the Activity suggest some *Properties of Exponents.*

Properties of Exponents

Let a and b be nonzero real numbers. Let m and n be integers.

Product of Powers	$(a)^m(a)^n = a^{m+n}$
Quotient of Powers	$\frac{a^m}{a^n} = a^{m-n}$
Power of a Power	$(a^m)^n = a^{mn}$
Power of a Product	$(ab)^n = a^n b^n$
Power of a Quotient	$\left(\frac{a}{b}\right)^n = \frac{a^n}{b^n}$

Interdisciplinary Connection

CONSUMER MATH When you borrow money from a financial institution, you generally agree to repay the loan in a specified number of equal monthly payments. Some people make several of these monthly payments and then decide to pay off the rest of the loan in one lump sum. This is called *prepaying* the loan. To determine the amount of the prepayment, A, you can use the formula $A = \frac{p[1-(1+r)^{-n}]}{r}$, where p is the amount of each monthly payment, r is the monthly interest rate, and n is the number of payments that remain.

Suppose that you take out a $10,000 loan for three years at a monthly interest rate of 1%. The amount of each monthly payment is $332.14. After making 24 payments, you decide to prepay the loan. What is the amount of the prepayment? How much will you save by prepaying the loan?
$3738.26; $247.42

ADDITIONAL EXAMPLE 1

Refer to the equation for centripetal acceleration given on page 94. **Find the centripetal acceleration in feet per second squared of a rider who makes one rotation in 4.5 seconds and whose radius of rotation is 8 feet.**
$\approx$15.6 ft/s^2

TRY THIS
$\approx$9.5 ft/s^2

Activity Notes

In this Activity, students use the definition of integer exponents to explore the product-of-powers property and the power-of-a-power property. You might suggest that they substitute numbers for the variable a to check their work. They should discover that when powers with the same base are multiplied, the exponents are added. When a power is raised to a power, the exponents are multiplied.

Cooperative Learning

You may wish to have students do the Activity in pairs. One partner can do Step 1 while the other does Step 2. Partners can then compare results, resolve any disagreements, and do Step 3 together.

CHECKPOINT ✓

3. **First add the exponents inside the parentheses: $a^7 \cdot a^3 = a^{7+3} = a^{10}$. Then multiply the resulting exponent by 2: $(a^{10})^2 = a^{10 \cdot 2} = a^{20}$.**

ADDITIONAL EXAMPLE 2

Simplify $9a^2b^3(-2a^5b^{-3})^2$. Write your answer with positive exponents only.

$\frac{36a^{12}}{b^3}$

TRY THIS

$\frac{30x^2}{z^2}$

Teaching Tip

Remind students that when multiplying or dividing powers with the same base, it is the exponent that changes, not the base.

ADDITIONAL EXAMPLE 3

Simplify $\left(\frac{-m^9n}{5m^{10}n^6}\right)^3$. Write your answer with positive exponents only.

$\frac{-1}{125m^3n^{15}}$

TRY THIS

$\frac{-27c^9}{b^{15}}$

CRITICAL THINKING

$a = 1, b = -2, c = -1$

In this lesson, you may assume that variables with negative exponents represent nonzero numbers.

EXAMPLE 2 **Simplify $3x^2y^{-2}(-2x^3y^{-4})$. Write your answer with positive exponents only.**

SOLUTION

$3x^2y^{-2}(-2x^3y^{-4}) = (3)(-2)x^2x^3y^{-2}y^{-4}$ *Use the Commutative Property.*

$= (3)(-2)x^{(2+3)}y^{[-2+(-4)]}$ *Use the Product of Powers Property.*

$= -6x^5y^{-6}$ *Simplify.*

$= -\frac{6x^5}{y^6}$ *Use $a^{-n} = \frac{1}{a^n}$.*

TRY THIS Simplify $2z(3x^2)(5z^{-3})$. Write your answer with positive exponents only.

PROBLEM SOLVING **Look for a pattern.** Examine what happens to powers with a *negative* base:

$$(-2)^2 = (-2)(-2) = 4$$
$$(-2)^3 = (-2)(-2)(-2) = -8$$
$$(-2)^4 = (-2)(-2)(-2)(-2) = 16$$
$$(-2)^5 = (-2)(-2)(-2)(-2)(-2) = -32$$

When the exponent of a negative base is *even*, the result is positive. When the exponent of a negative base is *odd*, the result is **negative**.

When simplifying an expression, be careful not to confuse the results of a negative base with the results of a negative exponent.

Even Exponent

$$(-2x)^{-2} = \frac{1}{(-2x)^2} = \frac{1}{(-2)^2x^2} = \frac{1}{4x^2}$$

Odd Exponent

$$(-2x)^{-3} = \frac{1}{(-2x)^3} = \frac{1}{(-2)^3x^3} = \frac{1}{-8x^3}, \text{ or } -\frac{1}{8x^3}$$

EXAMPLE 3 **Simplify $\left(\frac{-y^7}{2z^{12}y^3}\right)^4$. Write your answer with positive exponents only.**

SOLUTION

$\left(\frac{-y^7}{2z^{12}y^3}\right)^4 = \frac{(-y^7)^4}{(2z^{12}y^3)^4}$ *Use the Power of a Quotient Property.*

$= \frac{y^{28}}{16z^{48}y^{12}}$ *Use the Power of a Power Property.*

$= \frac{y^{28-12}}{16z^{48}}$ *Use the Quotient of Powers Property.*

$= \frac{y^{16}}{16z^{48}}$

TRY THIS Simplify $\left(\frac{-3b^2c^5}{c^2b^7}\right)^3$. Write your answer with positive exponents only.

CRITICAL THINKING Find a, b, and c such that $(x^{-2}y^3z^2)(y^az^bx^c) = x^{-3}y^4$ for all nonzero values of x, y, and z.

Inclusion Strategies

VISUAL LEARNERS Students with a more visual learning style may perceive expressions with exponents as a maze of symbols. Try modeling some expressions for them by using geometric shapes rather than letters as the variables. For example, you can model $x^3y^2(x^2)^3$ as follows:

x^3 y^2 $(x^2)^3$

(○○○)(□□)(○○)(○○)(○○)

↓

(○○○○○○○○○)(□□)

x^9 y^2

Enrichment

Have students work in pairs to find the value(s) of n for which each statement is true.

1. $(5^n)(5^4) = 125$ $n = -1$
2. $2^n = 4^3$ $n = 6$
3. $3^{2n} = 9^8$ $n = 8$
4. $(7^n)(7^4) = 49^{n+3}$ $n = -2$
5. $(2^n)^n = 4^8$ $n = 4$ *or* $n = -4$

Now have each student create three similar statements. Partners should exchange papers, and each should find the missing value(s) of n for the other's statements.

Rational Exponents

An expression with rational exponents can be represented in an equivalent form that involves the radical symbol, $\sqrt{\ }$.

For example, $a^{\frac{1}{3}}$ equals $\sqrt[3]{a}$ because when $a^{\frac{1}{3}}$ is cubed, the result is a, as shown at right. This is the definition of $\sqrt[3]{a}$.

$$\left(a^{\frac{1}{3}}\right)^3 = a^{\frac{1}{3} \cdot 3} = a$$

This relationship is true for all rational exponents. An expression with an exponent of $\frac{2}{3}$ is rewritten at right.

$$a^{\frac{2}{3}} = a^{\frac{1}{3} \cdot 2} = \left(a^{\frac{1}{3}}\right)^2 = \left(\sqrt[3]{a}\right)^2$$

Definition of Rational Exponents

For all positive real numbers a:

If n is a nonzero integer, then $a^{\frac{1}{n}} = \sqrt[n]{a}$.

If m and n are integers and $n \neq 0$, then $a^{\frac{m}{n}} = \left(a^{\frac{1}{n}}\right)^m = \left(\sqrt[n]{a}\right)^m = \sqrt[n]{a^m}$.

Example 4 shows how you can use the definition of rational exponents.

EXAMPLE 4 **Evaluate each expression.**

a. $16^{\frac{1}{4}}$

b. $27^{\frac{4}{3}}$

SOLUTION

a. $16^{\frac{1}{4}} = (2^4)^{\frac{1}{4}}$
$= 2^{4 \cdot \frac{1}{4}}$
$= 2^1 = 2$

b. $27^{\frac{4}{3}} = (3^3)^{\frac{4}{3}}$
$= 3^{3 \cdot \frac{4}{3}}$
$= 3^4 = 81$

CHECK

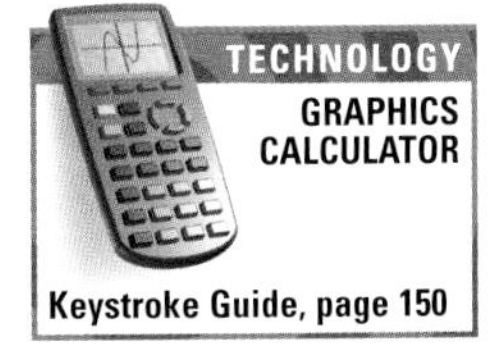

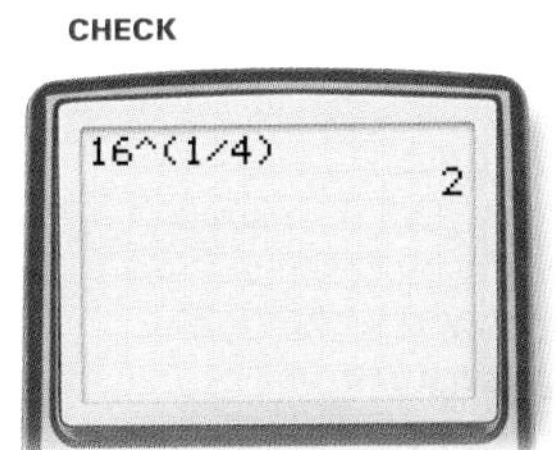

CHECK

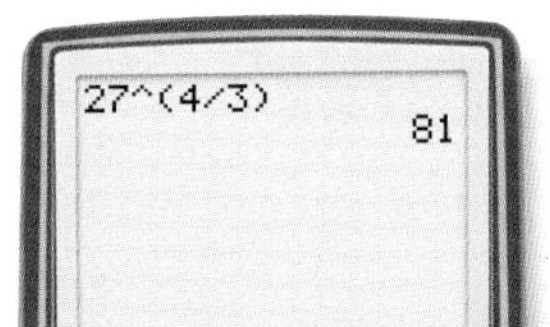

TRY THIS Evaluate $64^{\frac{1}{3}}$ and $36^{\frac{3}{2}}$.

The formula below is used to estimate a person's surface area based on his or her weight and height. This formula is used to calculate dosages for certain medications.

$$S = 0.007184 \times W^{0.425} \times H^{0.725},$$

where $\begin{cases} S \text{ is the surface area in square meters} \\ W \text{ is the weight in kilograms} \\ H \text{ is the height in centimeters} \end{cases}$

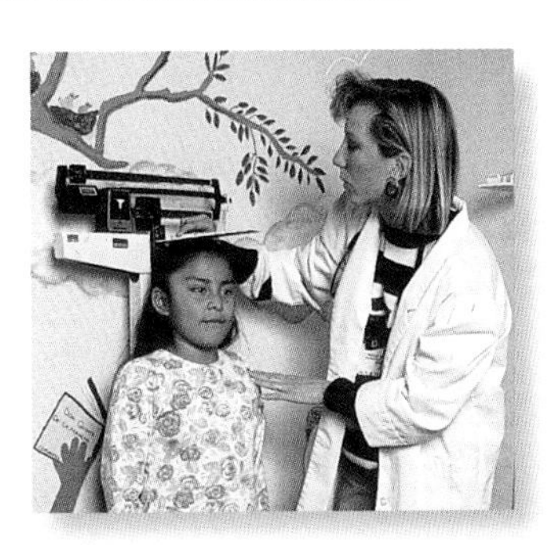

Teaching Tip

Terms, such as *index* and *radicand*, are introduced here and covered more thoroughly in Chapter 8.

ADDITIONAL EXAMPLE 4

Evaluate each expression.

a. $125^{\frac{1}{3}}$ 5

b. $8^{\frac{2}{3}}$ 4

Teaching Tip

TECHNOLOGY In part **a** of Example 4, point out to students that another way of evaluating the expression is to enter **16^.25.**

TRY THIS

4; 216

Reteaching the Lesson

COOPERATIVE LEARNING Have students work in groups of two or three. Display the following on the board or overhead:

$$x^{\square} \cdot x^{\square} = x^2$$

Tell students that they should find three different ways to fill the boxes with integers and make a true statement.

samples: $x^1 \cdot x^1 = x^2$; $x^0 \cdot x^2 = x^2$; $x^{-1} \cdot x^3 = x^2$

After all groups have completed their work, discuss the results with the entire class. Have each group share at least one of their answers. Use the results to summarize the product-of-powers property.

Repeat the activity to review the other properties of exponents, using expressions such as the ones below.

$$\frac{x^{\square}}{x^{\square}} = x^2 \qquad (x^{\square})^{\square} = x^2 \qquad (x^{\square}y^{\square})^{\square} = x^2y^4$$

ADDITIONAL
EXAMPLE 5

Refer to the formula for surface area given on page 97. **Estimate, to the nearest tenth of a square meter, the surface area of a person who stands 136.7 centimeters tall and weighs 34.2 kilograms.**
≈1.1 m²

TRY THIS
≈1.8 m²

Assess

Selected Answers
Exercises 5–18, 19–99 odd

ASSIGNMENT GUIDE	
In Class	1–18
Core	19–69 odd, 71–76, 79, 81, 82–86
Core Plus	20–70 even, 71–88
Review	89–100
Preview	101

Extra Practice can be found beginning on page 940.

EXAMPLE 5 **Estimate to the nearest tenth of a square meter the surface area of a person who stands 152.5 centimeters tall and weighs 57.2 kilograms.**

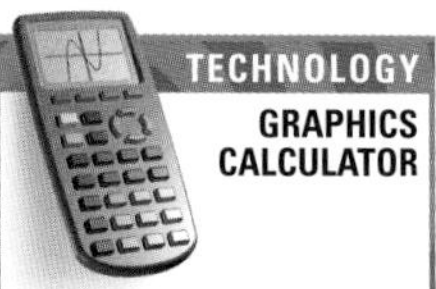
TECHNOLOGY
GRAPHICS CALCULATOR
Keystroke Guide, page 150

SOLUTION

Evaluate $S = 0.007184 \times W^{0.425} \times H^{0.725}$ for $W = 57.2$ and $H = 152.5$.

$$S = 0.007184 \times W^{0.425} \times H^{0.725}$$
$$= 0.007184 \times (57.2)^{0.425} \times (152.5)^{0.725}$$
$$\approx 1.54$$

The estimated surface area is about 1.5 square meters.

TRY THIS Estimate to the nearest tenth of a square meter the surface area of a person who stands 180 centimeters tall and weighs 62.3 kilograms.

Exercises

Communicate

1. Explain why x^5x^3 and $(x^5)^3$ are not equivalent expressions.
2. Explain why ax^2 and $(ax)^2$ are not equivalent expressions.
3. Describe how to evaluate 5^{-2}.
4. Describe how to evaluate $4^{\frac{3}{2}}$ by using the definition of rational exponents.

Guided Skills Practice

APPLICATION

5. **PHYSICS** Find the centripetal acceleration in feet per second squared of a model airplane that makes one rotation in 1.5 seconds and whose radius of rotation is 8 feet. ***(EXAMPLE 1)***
140 ft/s²

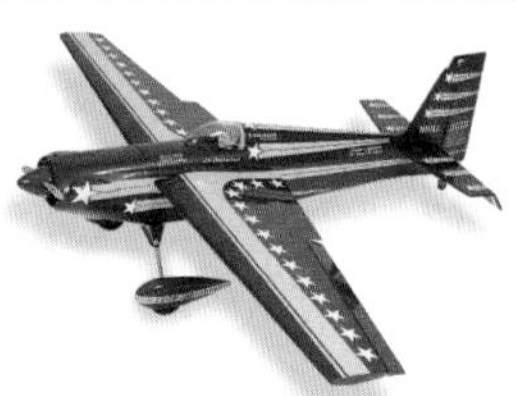

Simplify each expression, assuming that no variable equals zero. Write your answers with positive exponents only. ***(EXAMPLES 2 AND 3)***

6. x^4x^2 $\mathbf{x^6}$
7. $\frac{z^9}{z^3}$ $\mathbf{z^6}$
8. $(y^3)^6$ $\mathbf{y^{18}}$
9. $(a^3b^7)^4$ $\mathbf{a^{12}b^{28}}$
10. $(y^5y^{-2})^4$ $\mathbf{y^{12}}$
11. $\left(\frac{-2x^3y}{5x^7}\right)^2$ $\mathbf{\frac{4y^2}{25x^8}}$
12. $\left(\frac{a^3b^{-1}}{a^{-2}b^2}\right)^{-2}$ $\mathbf{\frac{b^6}{a^{10}}}$
13. $\left(\frac{1}{x^{-1}y^3z^0}\right)^{-1}$ $\mathbf{\frac{y^3}{x}}$

Evaluate each expression. ***(EXAMPLE 4)***

14. $100^{\frac{1}{2}}$ **10**
15. $9^{\frac{3}{2}}$ **27**
16. $27^{\frac{1}{3}}$ **3**
17. $64^{\frac{2}{3}}$ **16**

APPLICATION

18. **MEDICINE** Estimate to the nearest hundredth of a square meter the surface area of a person who stands 167.64 centimeters tall and weighs 53.64 kilograms. ***(EXAMPLE 5)*** **1.60 m²**

Practice and Apply

Evaluate each expression.

19. 3^0 **1** **20.** 9^0 **1** **21.** $(5a)^0$ **1** **22.** $(2^5 \cdot 2^3)^0$ **1**

23. 6^{-1} $\frac{1}{6}$ **24.** 4^{-2} $\frac{1}{16}$ **25.** $\left(\frac{3}{5}\right)^4$ $\frac{81}{625}$ **26.** $\left(\frac{4}{5}\right)^2$ $\frac{16}{25}$

27. $\left(\frac{1}{4}\right)^{-1}$ **4** **28.** $\left(\frac{2}{5}\right)^{-2}$ $\frac{25}{4}$ **29.** $\left(-\frac{1}{3}\right)^{-3}$ **−27** **30.** $\left(-\frac{2}{3}\right)^{-3}$ $-\frac{27}{8}$

31. $49^{\frac{1}{2}}$ **7** **32.** $27^{\frac{2}{3}}$ **9** **33.** $64^{\frac{4}{3}}$ **256** **34.** $25^{\frac{3}{2}}$ **125**

35. $36^{\frac{6}{4}}$ **216** **36.** $8^{\frac{2}{6}}$ **2** **37.** $-64^{\frac{2}{3}}$ **−16** **38.** $81^{-\frac{3}{2}}$ $\frac{1}{729}$

internet connect
Homework Help Online
Go To: **go.hrw.com**
Keyword:
MB1 Homework Help for Exercises 39–70

Simplify each expression, assuming that no variable equals zero. Write your answer with positive exponents only.

39. y^5y^2 y^7 **40.** $-2z^3z^5$ $-2z^8$ **41.** $-2y^3(5xy^4)$ $-10xy^7$ **42.** $6x^5 \cdot 3x^5 \cdot x^0$ $18x^{10}$

43. $\frac{m^9}{m^5}$ m^4 **44.** $\frac{bb^4}{b^2}$ b^3 **45.** $\frac{x^2x^{-5}}{x^4}$ $\frac{1}{x^7}$ **46.** $\frac{s^5t^2}{st^{-4}}$ s^4t^6

47. $(2x^4y)^3$ $8x^{12}y^3$ **48.** $(3st^{12})^3$ $27s^3t^{36}$ **49.** $(-5w^4v^5)^2$ **50.** $(-3x^2y^7)^3$

51. $\left(\frac{-2z^2}{x^3}\right)^7$ $-\frac{128z^{14}}{x^{21}}$ **52.** $\left(\frac{2b^4}{-a^2}\right)^3$ $-\frac{8b^{12}}{a^6}$ **53.** $\left(\frac{-2p^5q^{-4}}{q^3}\right)^3$ $-\frac{8p^{15}}{q^{21}}$ **54.** $\left(\frac{3m^2n^3}{m^{-1}}\right)^5$

55. $\left(\frac{3x^4}{y^{-2}}\right)^5$ **56.** $\left(\frac{-7y^{-2}}{-x^5}\right)^6$ $\frac{117{,}649}{x^{30}y^{12}}$ **57.** $\left(\frac{5r^2s^{-2}}{s^{-3}}\right)^{-1}$ $\frac{1}{5r^2s}$ **58.** $\left(\frac{x^{-2}y}{y^{-1}}\right)^{-3}$ $\frac{x^6}{y^6}$

49. $25w^8v^{10}$

50. $-27x^6y^{21}$

54. $243m^{15}n^{15}$

55. $243x^{20}y^{10}$

64. $-\frac{1}{8a^{21}b^{13}}$

67. $\frac{8x^{47}y^{18}z^{11}}{3}$

70. $16s^{18x}t^{8y-8}$

Simplify each expression, assuming that no variable equals zero. Write your answer with positive exponents only.

59. $\left(\frac{15xy^3}{3y^2}\right)^{-1}$ $\frac{1}{5xy}$ **60.** $\left[\frac{2x^{-3}}{(2x)^3}\right]^{-1}$ $4x^6$ **61.** $\left(\frac{4a^3b^{-3}}{a^{-1}b^2}\right)^{-2}$ $\frac{b^{10}}{16a^8}$

62. $\left(\frac{15a^2b^{-2}}{-3ab^{-3}}\right)^{-2}$ $\frac{1}{25a^2b^2}$ **63.** $(x^{-3}y^{-1})^{-1}(x^{-3}y^0)^2$ $\frac{y}{x^3}$ **64.** $(a^{-3}b^2)^4(-2a^3b^7)^{-3}$

65. $\left[\frac{(a^3b^5)^2}{a^5b^2}\right]^{-1}$ $\frac{1}{ab^8}$ **66.** $\left(\frac{s^{-3}}{4t}\right)^{-3}\left(\frac{5t}{s^{-7}}\right)^{-2}$ $\frac{64t}{25s^5}$ **67.** $\left(\frac{3z}{x^{-4}}\right)^2\left(\frac{3x^{-12}yz^{-3}}{2xy^7}\right)^{-3}$

68. $\left[\left(\frac{x^5y^2}{x^{-3}y}\right)^{-2}\left(\frac{y^{-3}}{2x^5}\right)^3\right]^{-1}$ $8x^{31}y^{11}$ **69.** $\left[\frac{(a^{-5}b^2)^{-1}}{(-a^1b^4c^{-1})^2}\right]^{-3}$ $\frac{b^{30}}{a^9c^6}$ **70.** $\left[\frac{(2s^{3x}t^{2y})^2}{(s^{3x}t^{-4})^{-1}}\right]^2$

Use a calculator to evaluate each expression to the nearest tenth.

71. $12^{6.05} + 8.8^{3.24}$ **3,382,159.1** **72.** $3.3^{2.7} - 5^{1.9} + 0.63^{0.95}$ **4.5**

73. $0.005^{21.53} + 9.05^{0.034}$ **1.1** **74.** $71.33^{0.44} + 478.2^{0.4}$ **18.3**

75. $11.7^{0.6} + 29.3^{1.23} - 6^{-2.2}$ **68.1** **76.** $89^{3.5} - 5.25^{9.25} + 324^{0.05}$ **2,064,248.4**

CHALLENGES

77. Show that if $y \neq 0$, then $y^{a-b} = \frac{1}{y^{b-a}}$.

78. Show that $\frac{x^{-1} - y^{-1}}{x - y} = -\frac{1}{xy}$.

CONNECTION

79. GEOMETRY The height, h, of a right circular cone can be calculated from the equation $h = \frac{3}{\pi}Vr^{-2}$, where V is the volume of the cone and r is the radius of the circular base.

a. Find the height to the nearest tenth of a right circular cone whose volume is 200 cubic centimeters and whose radius is 4 centimeters. **11.9 cm**

b. Write the equation for the height of a right circular cone with positive exponents only. $h = \frac{3V}{\pi r^2}$

h
r

Error Analysis

When multiplying powers with the same base, students often multiply the exponents. In Exercise 39, for example, they may give y^{10} as the answer. Suggest that they begin exercises like these by writing the powers in their expanded form, as shown below.

$$\overbrace{y \cdot y \cdot y \cdot y \cdot y}^{y^5} \cdot \overbrace{y \cdot y}^{y^2} = y^7$$

77. For all values of a and b, $-(a - b) = b - a$; therefore, $y^{a-b} = y^{-(b-a)} = \frac{1}{y^{b-a}}$.

78. $\frac{x^{-1} - y^{-1}}{x - y} = \frac{\frac{1}{x} - \frac{1}{y}}{x - y}$

$= \left(\frac{1}{x} - \frac{1}{y}\right)\left(\frac{1}{x - y}\right)$

$= \left(\frac{y - x}{xy}\right)\left(\frac{1}{x - y}\right)$

$= -1\left(\frac{x - y}{xy}\right)\left(\frac{1}{x - y}\right) = -\frac{1}{xy}$

APPLICATIONS

80. ENGINEERING The maximum load in tons that a foundation column can withstand is represented by the equation $F_{max} = \frac{9}{4}d^4l^{-2}$, where d is the diameter of the column in inches and l is the length of the column in feet.

a. Find the maximum load for a column that is 5 feet long and 10 inches in diameter. **900 tons**

b. Write the equation $F_{max} = \frac{9}{4}d^4l^{-2}$ with positive exponents only. $\mathbf{\frac{9d^4}{4l^2}}$

Foundation columns are cylindrical weight-bearing supports for structures such as the bridge shown above.

81. PHYSICS The resistance caused by friction between blood and the vessels that carry it can be modeled by the equation $R = \frac{2}{3}lr^{-4}$, where R is the resistance, l is the length of the blood vessel, and r is the radius of the blood vessel.

a. Find the resistance to the nearest tenth for a 0.2-meter long vessel with a 0.015-meter radius.

b. Write the resistance equation, $R = \frac{2}{3}lr^{-4}$, with positive exponents only.

81. a. 2,633,744.9
b. $\mathbf{\frac{2l}{3r^4}}$

82. PHYSICS The rate at which an object emits radiant energy can be expressed as $P = 5 \times 10^{-8}\left(\frac{A}{T^{-4}}\right)$, where P is the power radiated by the object in watts, A is the surface area of the object in square meters, and T is the temperature of the object in kelvins.

a. Write the equation $P = 5 \times 10^{-8}\left(\frac{A}{T^{-4}}\right)$ with positive exponents only.

b. Find the power radiated by a heater with a surface area of 0.25 square meters and a temperature of 300 K.

82. a. $\mathbf{P = \frac{5AT^4}{10^8}}$
b. 101.25 watts

CHEMISTRY Radioactive plutonium decays very slowly. The percent of plutonium remaining after x years can be represented by $A = 100\left(0.5^{\frac{x}{24,360}}\right)$. Find the percent of plutonium remaining after each number of years.

83. $x = 100$ **99.7%** **84.** $x = 500$ **98.6%** **85.** $x = 1000$ **97.2%** **86.** $x = 5000$ **86.7%**

PHYSICS Air pressure decreases with altitude according to the formula $P = 14.7(10)^{-0.000014a}$, where P is the air pressure in pounds per square inch and a is the altitude measured in feet above sea level.

87. Find the air pressure for Denver, Colorado, where the altitude is 1 mile (5280 feet) above sea level. **12.4 psi**

88. Find the air pressure at the top of Mount Everest, where the altitude is 29,028 feet above sea level. **5.8 psi**

Practice

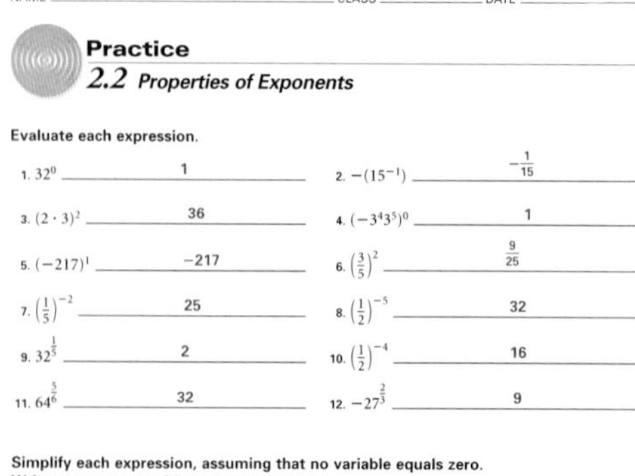

NAME ________ CLASS ________ DATE ________

Practice
2.2 Properties of Exponents

Evaluate each expression.

1. 32^0 **1**
2. $-(15^{-1})$ $-\frac{1}{15}$
3. $(2 \cdot 3)^2$ **36**
4. $(-3^4 3^5)^0$ **1**
5. $(-217)^1$ **−217**
6. $\left(\frac{3}{5}\right)^2$ $\frac{9}{25}$
7. $\left(\frac{1}{5}\right)^{-2}$ **25**
8. $\left(\frac{1}{2}\right)^{-5}$ **32**
9. $32^{\frac{1}{5}}$ **2**
10. $\left(\frac{1}{2}\right)^{-4}$ **16**
11. $64^{\frac{5}{6}}$ **32**
12. $-27^{\frac{2}{3}}$ **9**

Simplify each expression, assuming that no variable equals zero. Write your answer with positive exponents.

13. d^3d^{-4} $\frac{1}{d}$
14. $w^3y^4z \cdot wy^{-2}z$ $w^4y^2z^2$
15. $k^{-11}k^3$ $\frac{1}{k^8}$
16. $(x^7)^2$ x^{14}
17. $\frac{z^{15}}{z^{-2}}$ z^{17}
18. $\left(\frac{1}{x^{-7}}\right)^{-5}$ $\frac{1}{x^{35}}$
19. $\frac{y^{14}z^5}{y^9z^4}$ y^5z
20. $\frac{w^{21}w^{-12}}{w^9}$ **1**
21. $(3x^3y^5)^4$ $81x^{12}y^{20}$
22. $\left(-2a^3bc^6\right)^4$ $16a^{12}b^4c^{24}$
23. $(5a^2b^3)^3$ $125a^6b^9$
24. $\left(\frac{wz^4}{x^2}\right)^{-2}$ $\frac{x^4}{w^2z^8}$
25. $\left(\frac{a^{-2}}{b^{-3}}\right)^{-2}$ $\frac{a^4}{b^6}$
26. $\left(\frac{w^6}{k}\right)^3$ $\frac{w^{18}}{k^3}$
27. $\left(\frac{m^{-2}p^2}{2mp^3}\right)^{-4}$ $16m^{12}p^4$
28. $\left(\frac{xy^3z^2}{z^{-2}}\right)^{-1}$ $\frac{1}{xy^3z^4}$
29. $\left[\frac{(x^2y^2)^3}{x^5}\right]^{-1}$ $\frac{1}{xy^6}$
30. $\left(\frac{3x}{y}\right)^4\left[\frac{x^{-8}}{(xy)^3}\right]^{-2}$ $81x^{26}y^2$

Look Back

internet connect
Activities Online
Go To: **go.hrw.com**
Keyword:
MB1 Athletes

Graph the solution to each compound inequality on a number line. ***(LESSON 1.7)***

89. $x > -3$ *and* $x < 1$

90. $x > -\frac{1}{4}$ *and* $x > \frac{1}{2}$

91. $x > -3$ *or* $x < 1$

92. $x > -\frac{1}{4}$ *or* $x > \frac{1}{2}$

Classify each number in as many ways as possible. ***(LESSON 2.1)***

93. 9.373737... **rational, real**

94. $13\frac{1}{2}$ **rational, real**

95. 5.38388388838888 . . . **irrational, real**

96. −7.9 **rational, real**

Evaluate each expression by using the order of operations. ***(LESSON 2.1)***

97. $2(3 - 1) + 6 \div 3 \div 2$ **5**

98. $-3(9 - 12) - 2(7 - 3) - 1$ **0**

99. $3 \cdot 5^2 - 4(5 - 8)^2 \div 3$ **63**

100. $(5 - 3)^{\frac{(10-8)}{(13-12)}}$ **4**

Look Beyond

APPLICATION

101. GENEALOGY Your two parents are your first-generation ancestors, your four grandparents are your second-generation ancestors, and your eight great-grandparents are your third-generation ancestors. Make a table that represents these generations and the corresponding number of ancestors in each generation. Write an expression that represents the number of ancestors in the nth generation. (Hint: Use exponents in your expression.)

On January 22, 1997, the second-stage propellant tank of a Delta II launch vehicle landed near Georgetown, Texas, after spending nine months in Earth orbit.

SPACE SCIENCE Refer to the Space Debris Table from the Portfolio Activity on page 93.

1. For data in the table, let the years be represented by t (where $t = 0$ represents 1993) and let the total projected number of debris objects in orbit be represented by d.

Show that the data in the table can be modeled by the equation $d = 7000(1.03)^t$.

2. Use the equation in Step 1 to calculate the total projected number of debris objects in space at the end of the year 2020.

Look Beyond

In Exercise 101, students investigate an exponential-growth situation and write an expression to model it. They will study exponential growth in more depth in Lesson 6.1.

ALTERNATIVE Assessment

Portfolio Activity

The Portfolio Activity can be used as preparation for the Chapter Project or as a separate activity. In the Portfolio Activity on this page, students investigate an exponential equation that models the data from the Portfolio Activity on page 93. Then they use the equation to make a prediction.

Answers to Portfolio Activities can be found in Additional Answers of the Teacher's Edition.

Student Technology Guide

NAME ________ CLASS ________ DATE ________

Student Technology Guide
2.2 Properties of Exponents

Suppose that an object makes one revolution every T seconds around a circle with a radius of r feet. The object's centripetal acceleration, in feet per second squared, is given by $a_c = 4\pi^2 r T^{-2}$. If $T = 2$ and $r = 6$, you can find a_c by using the key sequence below.

4 [2nd] [^] [x²] [×] 6 [×] 2 [^] [(-)] 2 [ENTER]

```
4π²*6*2^-2
        59.21762641
```

Notice that you access π by pressing [2nd] [^] and that you enter a negative exponent by pressing [(-)].

Evaluating an expression with a rational exponent is just as easy. This process is shown in the examples below.

Example: Evaluate $3.5^{2.2}$: 3.5 [^] 2.2 [ENTER]

Example: Evaluate $2.3^{\frac{3}{4}}$.
Use a fraction: 2.3 [^] [(] 3 [÷] 4 [)] [ENTER]
Use a decimal: 2.3 [^] .75 [ENTER]

```
3.5^2.2
        15.73800567
2.3^(3/4)
        1.867651655
2.3^.75
        1.867651655
```

The exponent key is also useful when you need to deal with a radical expression.

Example: Evaluate $\sqrt[3]{100}$.

Change the expression to one with a rational exponent. Recall that taking the cube root of a number is the same as raising that number to the one-third power.

Press 100 [^] [(] 1 [÷] 3 [)] [ENTER].

```
100^(1/3)
        4.641588834
```

Use a graphics calculator to evaluate each expression. Round answers to the nearest tenth.

1. $12^{3.8}$ 12,615.0
2. $4.6^{5.1} - 5.9^{1.5}$ 2384.9
3. $0.75^{0.5} + 1.24^{1.2}$ 2.2
4. $9.1(7.2 - 4.1)^{5.8}$ 6440.8
5. $(0.58)^{-3.2}$ 5.7
6. $72.4^{1.38} - 6.53^{-4.2}$ 368.5
7. $10.4 - 8.4^{-3.2}$ 10.4
8. $6.2(7.4^{1.6} - 6.1^{0.5})$ 137.1
9. $8.2(5.2 - 1.8)^{-1.5}$ 1.3
10. $\sqrt[3]{120}$ 4.9
11. $\sqrt[4]{5^3}$ 3.3
12. $\sqrt[5]{(-3)^8}$ 5.8

89.

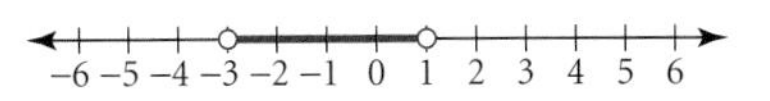

90.

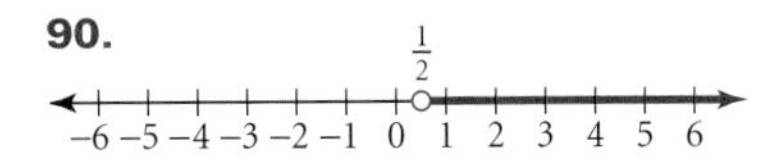

91. all real numbers

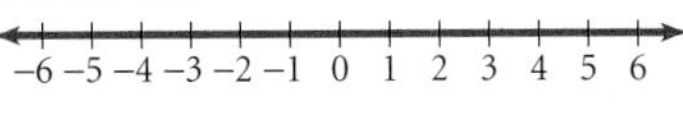

92.

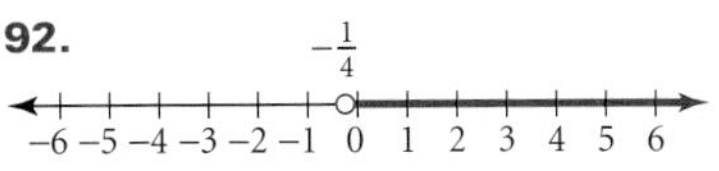

101.

Generation n ancestors	1	2	3	n
Numbers of ancestors	$2 = 2^1$	$4 = 2^2$	$8 = 2^3$	2^n

Prepare

NCTM PRINCIPLES & STANDARDS 1, 2, 6–10

QUICK WARM-UP

Evaluate each expression.	
1. $x - 2$, for $x = -5$	-7
2. $-2x$, for $x = -1.5$	3
3. x^2, for $x = -4$	16
4. $-x^2$, for $x = -1.2$	-1.44
5. x^3, for $x = -2$	-8
6. $-x^3$, for $x = -0.1$	0.001

Also on Quiz Transparency 2.3

Teach

Why The amount of change that students receive for a \$1.50 purchase depends on the amount given to the salesperson. The amount of change is said to be *a function of* the amount given. Many real-world situations can be represented by functions.

Math CONNECTION

GEOMETRY The perimeter and area of a square are both functions of the length of a side of the square.

2.3 Introduction to Functions

Objectives

- State the domain and range of a relation, and tell whether it is a function.
- Write a function in function notation and evaluate it.

Why *Functions and relations are commonly used to represent a variety of real-world relationships.*

APPLICATION ENTERTAINMENT

The table at right gives the number of theater screens in operation during each of the years from 1992 to 1996. You can represent this data as a collection of ordered pairs.

Theater Screens in Operation	
Year	**Total number of screens**
1992	25,105
1993	25,737
1994	26,586
1995	27,805
1996	29,690

[*Source: Motion Picture Association of America, Inc.*]

$\{(1992, 25{,}105), (1993, 25{,}737), (1994, 26{,}586), (1995, 27{,}805), (1996, 29{,}690)\}$

Notice that for each year, there is exactly one total number of screens. This relationship between years and the total number of screens is one example of a *function*.

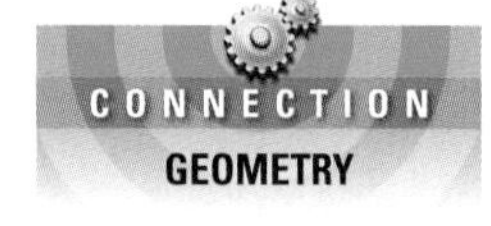

CONNECTION GEOMETRY

Recall from geometry the formulas for the perimeter and the area of a square.

$$\text{Perimeter: } P = 4s \quad \text{Area: } A = s^2$$

Each value for s defines exactly one perimeter and exactly one area. These relationships are also examples of *functions*.

Definition of Function

A **function** is a relationship between two variables such that each value of the first variable is paired with exactly one value of the second variable.

The **domain** of a function is the set of all possible values of the first variable. The **range** of a function is the set of all possible values of the second variable.

Alternative Teaching Strategy

INVITING PARTICIPATION Present students with the following situation: As a fund-raiser, your class plans to sell pizza at the snack bar during lunch for one week. After calling several pizza restaurants, you find one that charges \$8 for a large, two-topping pizza. Each large pizza has 12 slices. There also is a \$5 delivery charge.

Have the class estimate the number of pizzas they will need to buy. Call this number n. Have them make a table showing the cost of buying from $n - 5$ to $n + 5$ pizzas. Then the class should decide how much profit they want to make per pizza and use that amount to decide on the price they will charge for each slice of pizza.

Now have students write equations for the total cost of the pizzas, for the revenue they will take in each day, and for the profit they will make each day. Point out that each amount—the total cost, the revenue, and the profit—depends on the number of pizzas. This means that each amount *is a function of* the number of pizzas.

A function may also be represented by data in a table, as shown in Example 1.

EXAMPLE 1 **State whether the data in each table represents a function. Explain.**

a.

Domain, x	Range, y
1	−3.6
2	−3.6
3	4.2
4	4.2
5	10.7
6	12.1
52	52

b.

Domain, x	Range, y
3	7
3	8
3	10
4	42
10	34
11	18
52	52

SOLUTION

a. For each value of x in the table, there is exactly one value of y. The data set represents a function.

b. The data set does not represent a function because three different y-values 7, 8, and 10, are paired with one x-value, 3.

You can use the *vertical-line test* to determine if a graph represents a function.

Vertical-Line Test

If every vertical line intersects a given graph at no more than one point, then the graph represents a function.

Example 2 shows how you can use the vertical-line test to determine whether a graph represents a function.

EXAMPLE 2 **State whether each graph represents a function. Explain.**

a.

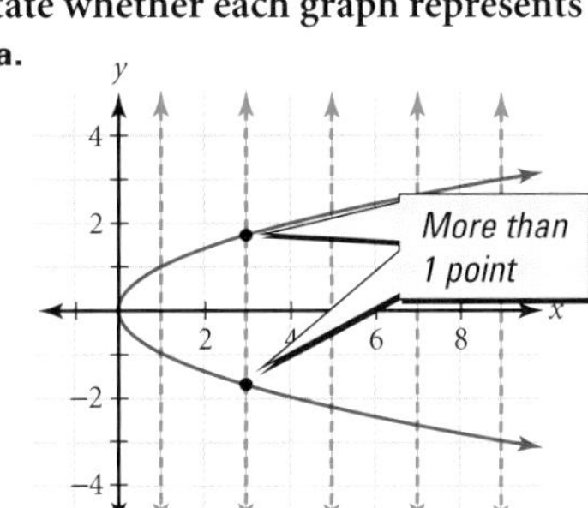

b.

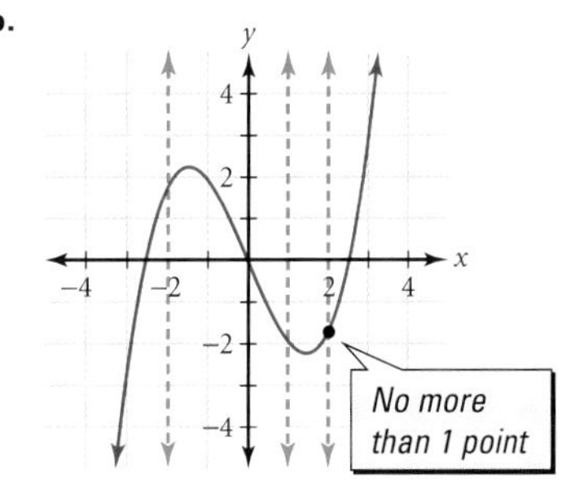

SOLUTION

a. The graph does not represent a function. There are many vertical lines that intersect the graph at two points.

b. The graph represents a function. Every vertical line intersects the graph at no more than one point.

Interdisciplinary Connection

HEALTH The number of times per minute that your heart beats is called your *heart rate*. When you exercise, your heart rate increases. To provide the greatest benefit to your cardiovascular system, your *maximum heart rate* should not exceed a certain value. It is generally agreed that this value is a function of age. Some fitness experts estimate the maximum heart rate for the average female by taking 70% of her age and subtracting it from 209. For the average male, they take 80% of his age and subtract it from 214.

a. Write a linear function to model the maximum heart rate for a female, f, as a function of age in years, a. What is the maximum heart rate for a 20-year-old female?
$f(a) = 209 - 0.7a$; about 195 beats per min

b. Write a linear function to model the maximum heart rate for a male, m, as a function of age in years, a. What is the maximum heart rate for a 20-year-old male?
$m(a) = 214 - 0.8a$; about 198 beats per min

ADDITIONAL EXAMPLE 1

State whether the data in each table represents y as a function of x. Explain.

a.

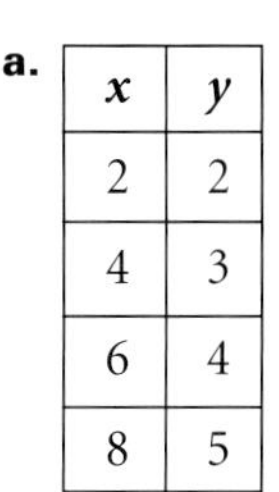

x	y
2	2
4	3
6	4
8	5

b.

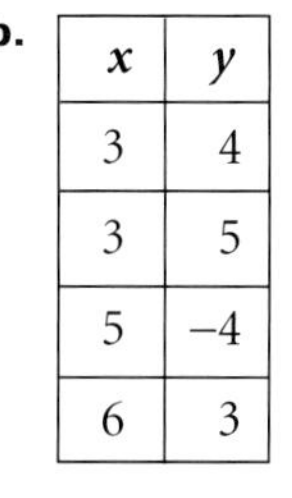

x	y
3	4
3	5
5	−4
6	3

a. The data set represents a function. For each value of x in the table, there is exactly one value of y.

b. The data set does not represent a function. One x-value (3) is paired with two different y-values (4 and 5).

Use Teaching Transparency 6.

ADDITIONAL EXAMPLE 2

State whether each graph represents a function. Explain.

a.

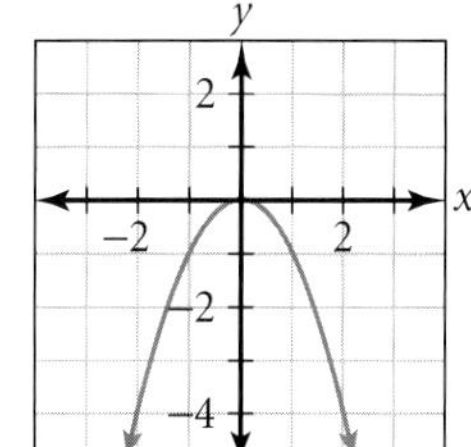

The graph represents a function. Any vertical line will intersect the graph at no more than one point.

b.

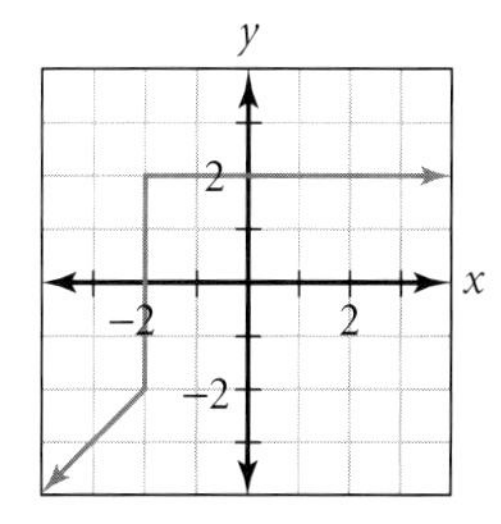

The graph does not represent a function. The line $x = -2$ intersects the graph at infinitely many points.

Activity Notes

In this Activity, students explore the concept of function in three different contexts. Although the Checkpoint question is open-ended, students' responses should indicate some understanding of a function as a relationship in which one quantity depends on another.

Math CONNECTION

GEOMETRY The volume, V, of a cylinder is given by the formula $V = \pi r^2 h$, where r is the radius of the cylinder and h is its height.

CHECKPOINT ✔

4. Sample: H is the height in feet of a ball thrown in the air as a function of time, t, in seconds.

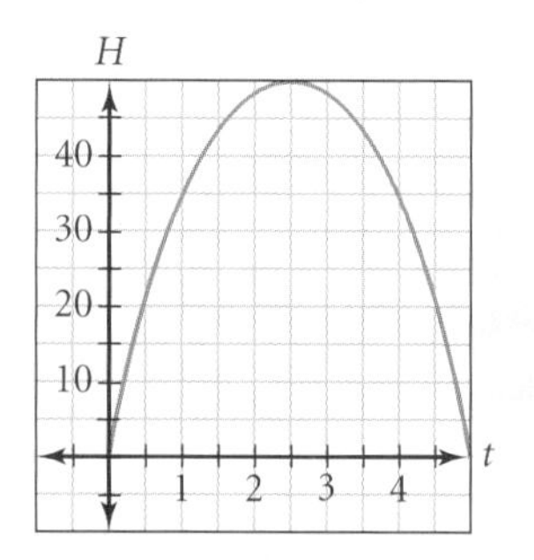

ADDITIONAL EXAMPLE 3

Let M be the set of the owners of motor vehicles that are registered in Ohio. Let R be the set of all motor vehicle registration numbers. **Is the correspondence between M and R a relation? a function? Explain.**

Every person who owns a registered motor vehicle has at least one registration number, so the correspondence is a relation. Some motor vehicle owners have two or more registered vehicles, so the relation is not a function.

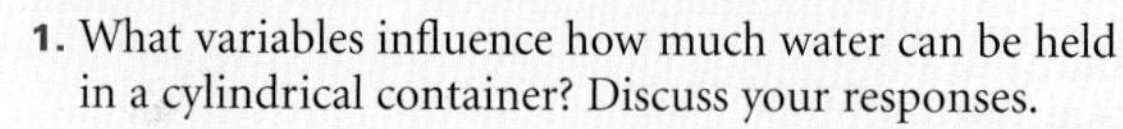

Exploring Functions

You will need: no special tools

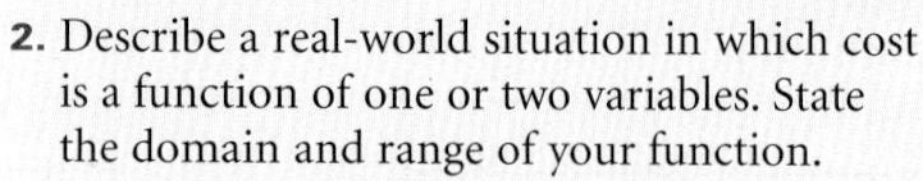

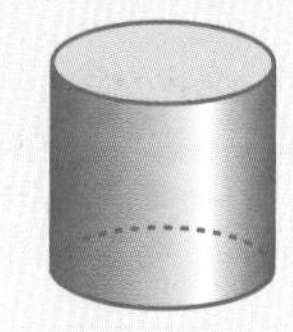

1. What variables influence how much water can be held in a cylindrical container? Discuss your responses.
2. Describe a real-world situation in which cost is a function of one or two variables. State the domain and range of your function.
3. The graph at right represents the volume, V, of water in a bathtub in gallons as a function of time, t, in minutes. Describe how V changes as t varies from 0 minutes to 16 minutes.

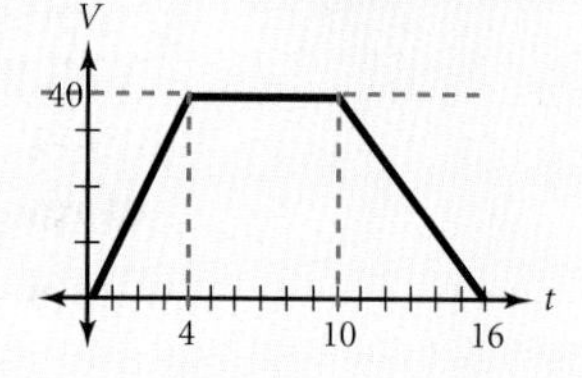

CHECKPOINT ✔ 4. Make up your own function like the one in Step 3. Describe your variables, and sketch a reasonable graph.

A function is a special type of *relation.*

Definition of Relation

A relationship between two variables such that each value of the first variable is paired with one or more values of the second variable is called a **relation.**

The **domain** is the set of all possible values of the first variable. The **range** is the set of all possible values of the second variable.

Example 3 gives an example of a real-world relation.

EXAMPLE 3 Let the first variable, R, represent residents in Brownsville, Texas, who have telephone service. Let the second variable, T, represent telephone numbers of residents in Brownsville, Texas.

Is the relationship between R and T a relation? a function? Explain.

SOLUTION

A resident with telephone service has at least one telephone number and may have more. Thus, the relationship is a relation.

Because some residents may have more than one telephone number, the relation is not a function.

TRY THIS Let the first variable, R, represent checking and savings account customers at a local bank. Let the second variable, N, represent checking and savings account numbers. Is the relationship between R and N a relation? a function? Explain.

CHECKPOINT ✔ Give two sets of ordered pairs that form a relation but not a function.

Inclusion Strategies

KINESTHETIC LEARNERS Use masking tape to create a large set of coordinate axes on the floor. Separate students into x- and y-coordinates. Assign one student to represent each x-value and several students to represent each y-value. Have them line up on their respective axes at one-unit intervals.

Call out a list of ordered pairs that do not form a function, one at a time. Have the appropriate students move to the position of the point. For example, if the ordered pair is (1, 2), the student at 1 on the x-axis and the student at 2 on the y-axis should move to the point (1, 2). When you call out a point for which the x-coordinate has already been used, no student will be available to meet the y-coordinate. Students should see that the list of ordered pairs does not form a function.

Now call out a list of ordered pairs that do form a function. After all students are in the correct positions, ask those who represent the domain to raise their hand. Then do the same for the range.

Example 4 shows you how to determine the domain and range of a function or relation from a graph.

EXAMPLE 4 **State the domain and range of each function graphed.**

a.

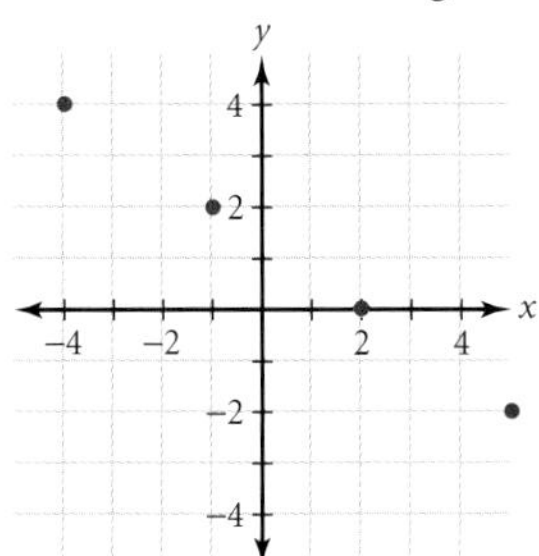

b.

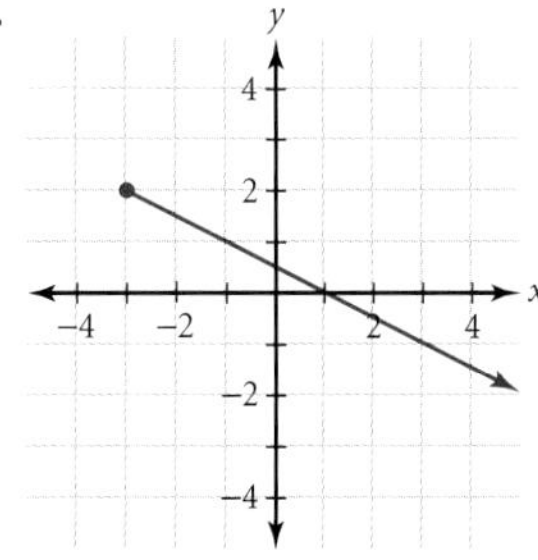

SOLUTION

The domain consists of the x-values, and the range consists of the y-values.

a. domain: $\{-4, -1, 2, 5\}$
range: $\{-2, 0, 2, 4\}$

b. domain: $x \geq -3$
range: $y \leq 2$

TRY THIS State the domain and range of each function graphed.

a.

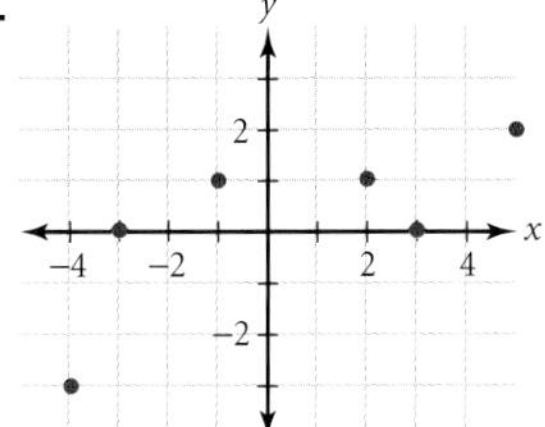

b.

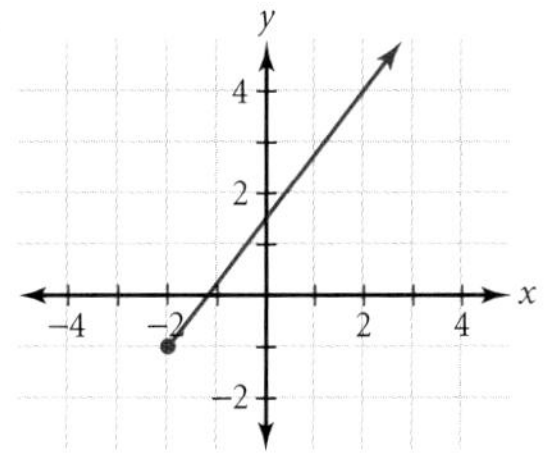

You can tell from a graph whether the function is *discrete* or *continuous.* The graph in part **a** of Example 4 illustrates a **discrete function**, whose graph is a set of individual points. The graph in part **b** of Example 4 illustrates a **continuous function**, whose graph is a line, ray, segment, or smooth curve.

CHECKPOINT ✔ Suppose that x is any real number. Is $y = 3x + 2$ a continuous function or a discrete function? Explain.

Functions and Function Notation

An equation can represent a function. The equation $y = 2x + 5$ represents a function. To express this equation as a function, use *function notation* and write $y = 2x + 5$ as $f(x) = 2x + 5$.

x	$f(x) = 2x + 5$	$f(x)$
-2	$f(-2) = 2(-2) + 5 = 1$	1
0	$f(0) = 2(0) + 5 = 5$	5
6	$f(6) = 2(6) + 5 = 17$	17
100	$f(100) = 2(100) + 5 = 205$	205

Enrichment

Have students write equations for two different functions that satisfy the given condition. **Sample answers are given.**

1. $f(2) = -3$ **$f(x) = x - 5$; $f(x) = -1.5x$**
2. $g(-6) = 0$ **$g(x) = x + 6$; $g(x) = 0.5x + 3$**
3. $h(0) = -6$ **$h(x) = x - 6$; $h(x) = -2(x + 3)$**
4. $j(2) = 2j(0)$ **$j(x) = x + 2$; $j(x) = -0.5x - 1$**
5. $k(3) = k(-3)$ **$k(x) = x^2$; $k(x) = |x|$**

TRY THIS
The correspondence between R and N is a relation because each customer has at least one number. However, some customers may have both savings and checking account numbers, so the relation is not a function.

CHECKPOINT ✔
Sample: Both of the sets $\{(1, 0), (1, 1)\}$ and $\{(4, 2), (4, -2)\}$ are relations but not functions.

ADDITIONAL EXAMPLE 4

State the domain and range of each function graphed below.

a.

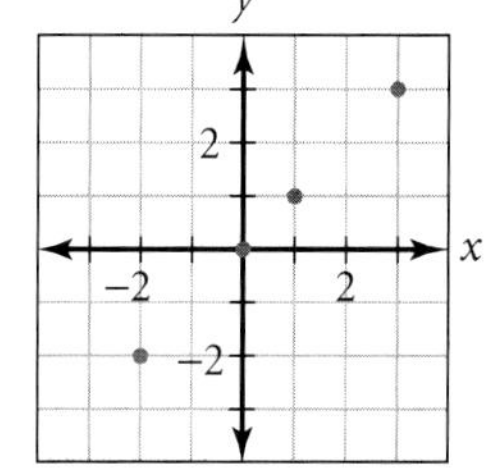

domain: $\{-2, 0, 1, 3\}$
range: $\{-2, 0, 1, 3\}$

b.

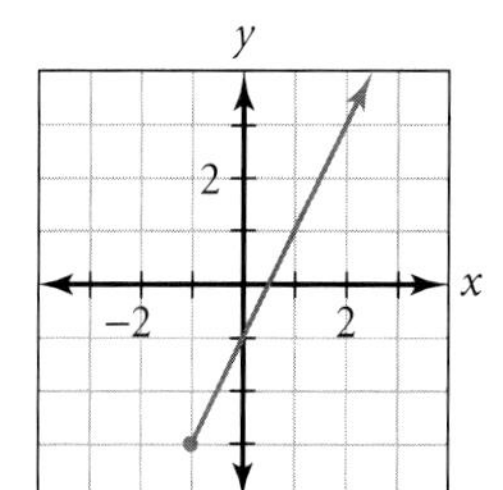

domain: $x \geq -1$
range: $y \geq -3$

TRY THIS
a. domain: $\{-4, -3, -1, 2, 3, 5\}$
range: $\{-3, 0, 1, 2\}$

b. domain: $x \geq -2$
range: $y \geq -1$

CHECKPOINT ✔
The function $y = 3x + 2$ is continuous. Its graph is a line, not a set of individual points.

ADDITIONAL EXAMPLE 5

Evaluate the function $g(x) = -1.2x^2 + 4x - 3$ for $x = 1$ and $x = 5$.
$g(1) = -0.2; g(5) = -13$

Teaching Tip

TECHNOLOGY When using a TI-83 graphics calculator, here is an alternative way to evaluate the function in Example 5. First press Y= and enter **.5X^2–3X+2** next to **Y1.** Press 2nd MODE and enter **4** STO▶ ALPHA STO▶ ENTER. This stores 4 in memory as the value of x. Now press VARS ▶ and choose **1:Function...**. From the **Function** menu, choose **1:Y1** and press ENTER. The value **–2** appears on the home screen.

The procedure for a TI-82 calculator is the same, except that you press 2nd VARS instead of VARS ▶.

ADDITIONAL EXAMPLE 6

A gift shop sells a specialty fruit-and-nut mix at a cost of $2.99 per pound. During the holiday season, you can buy as much of the mix as you like and have it packaged in a decorative tin that costs $4.95.

a. Write a linear function to model the total cost in dollars, c, of the tin containing the fruit-and-nut mix as a function of the number of pounds of the mix, n.
$c(n) = 4.95 + 2.99n$

b. Find the total cost of a tin that contains 1.5 pounds of the mix. $9.44

Function Notation

If there is a correspondence between values of the domain, x, and values of the range, y, that is a function, then $y = f(x)$, and (x, y) can be written as $(x, f(x))$. The notation $f(x)$ is read "f of x." The number represented by $f(x)$ is the value of the function f at x.

The variable x is called the **independent variable.**

The variable y, or $f(x)$, is called the **dependent variable.**

EXAMPLE 5 **Evaluate $f(x) = 0.5x^2 - 3x + 2$ for $x = 4$ and $x = 2.5$.**

SOLUTION

$f(x) = 0.5x^2 - 3x + 2$
$f(4) = 0.5(4)^2 - 3(4) + 2$
$f(4) = -2$

$f(x) = 0.5x^2 - 3x + 2$
$f(2.5) = 0.5(2.5)^2 - 3(2.5) + 2$
$f(2.5) = -2.375$

TECHNOLOGY GRAPHICS CALCULATOR
Keystroke Guide, page 150

CHECK

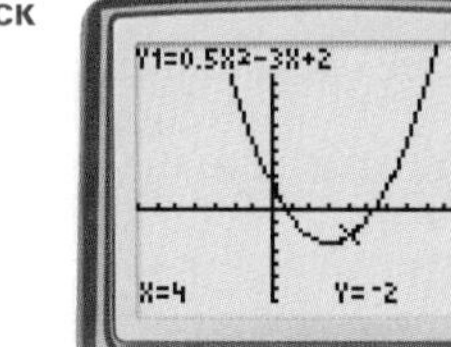

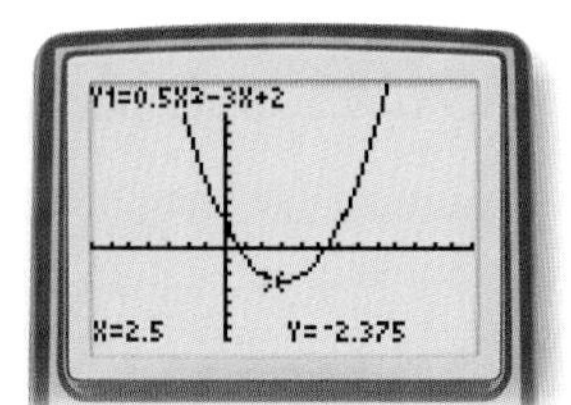

EXAMPLE 6 **Monthly residential electric charges, c, are determined by adding a fixed fee of $6.00 to the product of the amount of electricity consumed each month, x, in kilowatt-hours and a rate factor of 0.035 cents per kilowatt-hour.**

APPLICATION CONSUMER ECONOMICS

a. Write a linear function to model the monthly electric charge, c, as a function of the amount of electricity consumed each month, x.
b. If a household uses 712 kilowatt-hours of electricity in a given month, how much is the monthly electric charge?

SOLUTION

a. $c(x) =$ **electricity used + fixed fee**
$c(x) = 0.035x + 6.00$

The linear function is $c(x) = 0.035x + 6.00$.

b. Evaluate the function for $x = 712$.
$c(x) = 0.035x + 6.00$
$c(712) = 0.035(712) + 6.00$
$c(712) = 30.92$

For 712 kilowatt-hours of electricity, the monthly charge is $30.92.

CHECKPOINT ✓ State the independent and dependent variables in Example 6. Explain your response.

Reteaching the Lesson

WORKING BACKWARD Display the following sets on the board or overhead:

domain: $\{-2, -1, 0\}$ range: $\{-1, 0, 5\}$

Tell students to use this domain and range to create a relation that is a function and a relation that is not a function. Ask for volunteers to share their answers with the class, and discuss the results. **Sample answers are given below.**
function: $\{(-2, -1), (-1, 0), (0, 5)\}$
not a function: $\{(-2, -1), (-1, 0), (0, -1), (0, 5)\}$

Repeat the activity for the following domain and range:

domain: $x \geq -3$
range: $y \leq -1$

In this case, ask students to give their answers as graphs. **Sample graphs are given at right.**

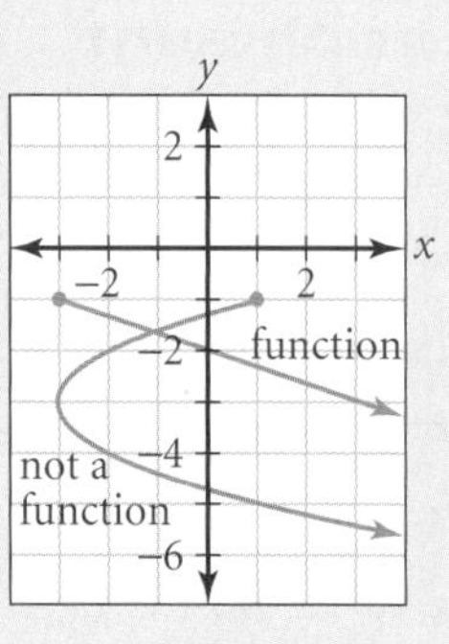

Exercises

Communicate

1. Explain how functions are different from relations. Sketch a graph of a relation that is not a function to illustrate your explanation.
2. Describe three ways to represent a function.
3. Explain how to find the domain and range of a set of ordered pairs such as $\{(4, 2), (3, 5), (-2, 0), (2, 5)\}$.
4. **INCOME** Cleo wants to graph the relationship between the dollar value of the meals she served and the amount of tips she received. Identify a real-world domain and range for this relationship and draw a sample graph for Cleo.

Guided Skills Practice

State whether the data in each table represents y as a function of x. Explain. ***(EXAMPLE 1)***

5. no; (5, 3) and (5, 7)

x	y
5	3
8	4
5	7
9	2

6. yes

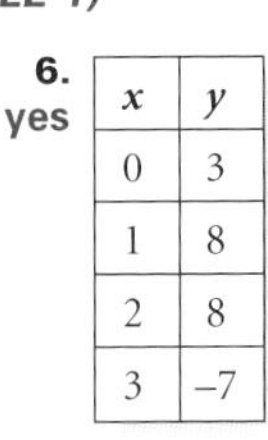

x	y
0	3
1	8
2	8
3	−7

7. yes

x	y
10	7
20	11
30	9
40	7

8. no; (2, 2) and (2, 1)

x	y
3	9
2	2
8	−3
2	1

State whether each graph represents a function. Explain. ***(EXAMPLE 2)***

9.

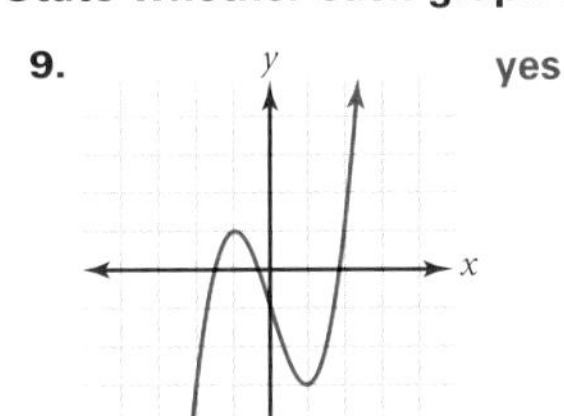

yes

10.

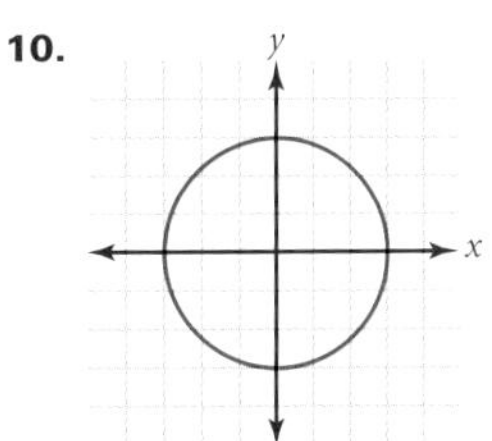

no; for example, the y-axis crosses graph at two points.

APPLICATION

11. **TRANSPORTATION** If the first variable, R, represents registered automobiles in your state that may be legally driven and the second variable, L, represents license plate numbers for these automobiles, is the relationship between R and L a relation? a function? Explain. ***(EXAMPLE 3)***

State the domain and range of each function graphed. ***(EXAMPLE 4)***

12.

domain: $x \geq -3$

range: $y \leq 4$

13.

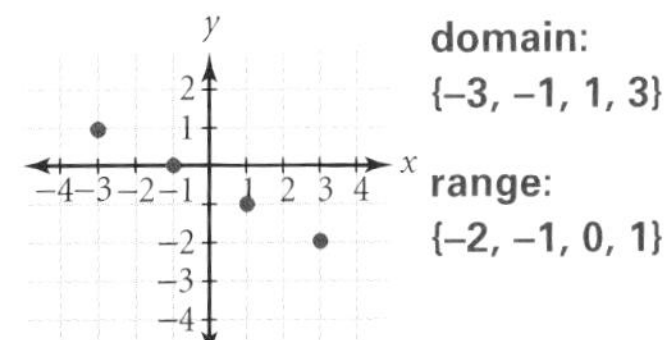

domain: $\{-3, -1, 1, 3\}$

range: $\{-2, -1, 0, 1\}$

11. A registered automobile has at least one license plate number. Thus, the correspondence is a relation. Since each registered automobile may not have more than one license plate number, the relation is also a function.

CHECKPOINT ✔

The independent variable is the number of kilowatt-hours and the dependent variable is the monthly electric charge. The monthly electric charge depends on the number of kilowatt-hours used.

Teaching Tip

Mathematicians often use the letter f to represent general functions and x to represent the independent variable. However, students should understand that any letters can be used. Point out that it is usually helpful to choose letters that are representative of the specific situation, as in Example 6.

Assess

Selected Answers

Exercises 5–15, 17–81 odd

ASSIGNMENT GUIDE

In Class	1–15
Core	17–49 odd, 51–60, 65–69 odd
Core Plus	16–50 even, 51–60, 62, 64–69
Review	70–81
Preview	82, 83

Extra Practice can be found beginning on page 940.

Error Analysis

When given a table of values such as the one in Exercise 6, students may think that it does not represent a function because two values of x correspond to a single value of y. Suggest that they plot the points represented by the ordered pairs on a set of coordinate axes. Although there is a *horizontal* line that contains two of the points, no *vertical* line passes through more than one point, so the relation represented by the table of values is a function.

14. Evaluate $f(x) = x^2 + 2x - 1$ for $x = 3$ and $x = 1.5$. ***(EXAMPLE 5)*** **14; 4.25**

APPLICATION

15. **INCOME** A plumber charges \$24 per hour of work plus \$20.00 to make a service call. ***(EXAMPLE 6)***
   a. Write a linear function to model the plumber's wages, w, for the number of hours worked, h. $\mathbf{w(h) = 24h + 20}$
   b. Find the plumber's wages for 5.5 hours of work. **\$152**

Practice and Apply

internet connect
Homework Help Online
Go To: **go.hrw.com**
Keyword:
MB1 Homework Help
for Exercises 16–30

State whether each relation represents a function.

16. $\{(0, 0), (1, 1)\}$ **yes**
17. $\{(1, 2), (2, 2), (3, 2)\}$ **yes**
18. $\{(1, -1), (1, -2), (1, -3)\}$ **no**
19. $\{(4, 1), (5, 2), (6, 3)\}$ **yes**
20. $\left\{\left(\frac{1}{2}, 1\right), \left(\frac{2}{5}, 2\right), \left(\frac{1}{3}, 1\right)\right\}$ **yes**
21. $\left\{\left(\frac{1}{3}, \frac{1}{4}\right), \left(\frac{1}{5}, \frac{1}{5}\right), \left(\frac{1}{4}, \frac{3}{4}\right)\right\}$ **yes**
22. $\{(11, 0), (12, -1), (21, -2)\}$ **yes**
23. $\{(0, 0), (2, 5), (3, 3)\}$ **yes**
24. $\{(1, 7), (-1, 7), (1, -7)\}$ **no**
25. $\{(-1, 8), (-1, 7), (0, 9)\}$ **no**

26. **no**

x	y
0	3
2	−5
2	1
4	7

27. **yes**

x	y
1	6
2	6
3	9
4	9

28. **no**

x	y
4	−2
4	2
6	−3
6	3

29. **yes**

x	y
−5	8
−3	8
−1	−2
1	−2

30. **no**

x	y
−2	−5
−2	−3
0	4
2	6

State whether each relation graphed below is a function. Explain.

31. **yes**

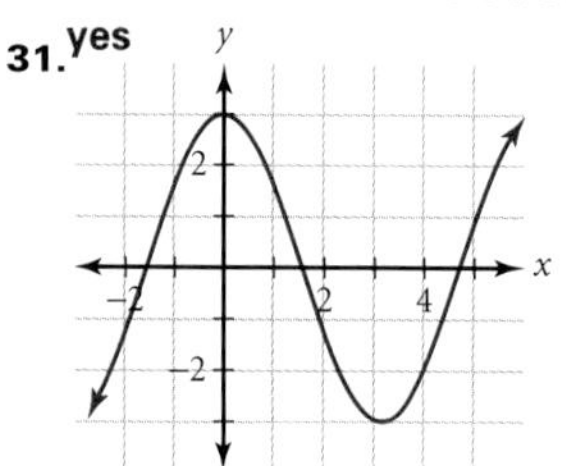

32. **no**

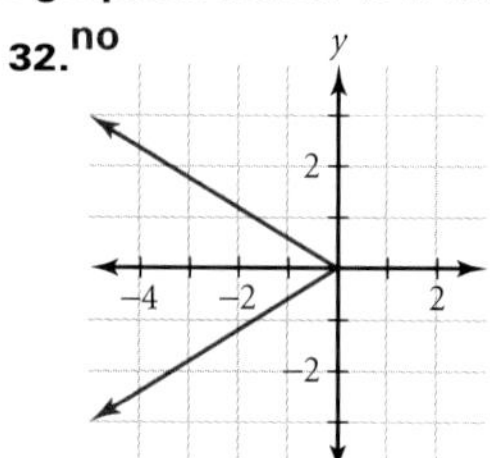

33. **no**

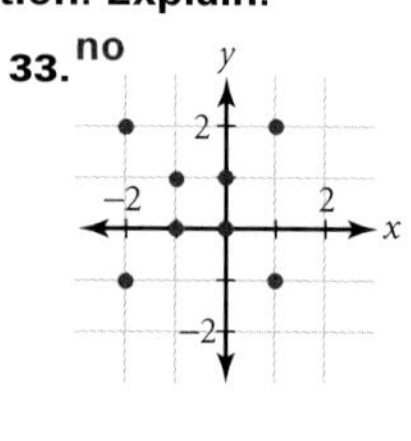

34. **yes**

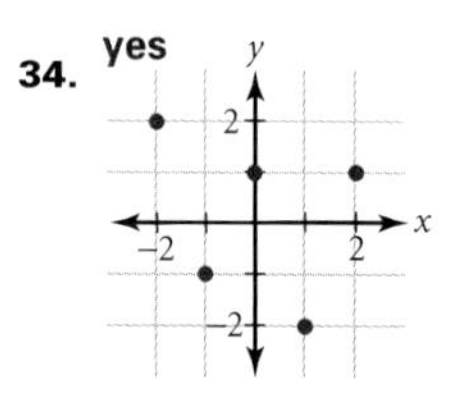

35. **yes**

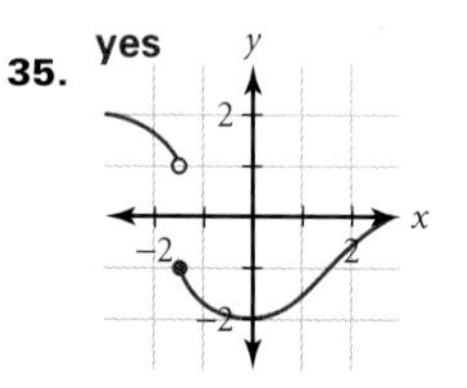

36. **no**

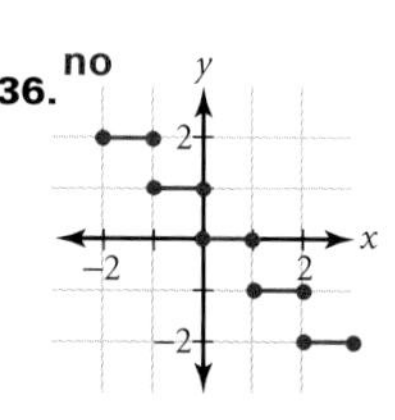

State the domain and range of each function.

37. $\{(0, 2), (3, 4)\}$
38. $\{(1, 5), (2, 5), (3, 5)\}$
39. $\{(9, -1), (8, -2), (7, -3)\}$
40. $\{(4, 1), (5, 2), (6, 3)\}$
41. $\{(6, -6), (5, -5), (4, -4)\}$
42. $\{(0, 0), (1.5, 0), (2.5, 0)\}$

37. domain: {0, 3}; range: {2, 4}

38. domain: {1, 2, 3}; range: {5}

39. domain: {7, 8, 9}; range: {−1, −2, −3}

40. domain: {4, 5, 6}; range: {1, 2, 3}

41. domain: {4, 5, 6}; range: {−6, −5, −4}

42. domain: {0, 1.5, 2.5}; range: {0}

Practice

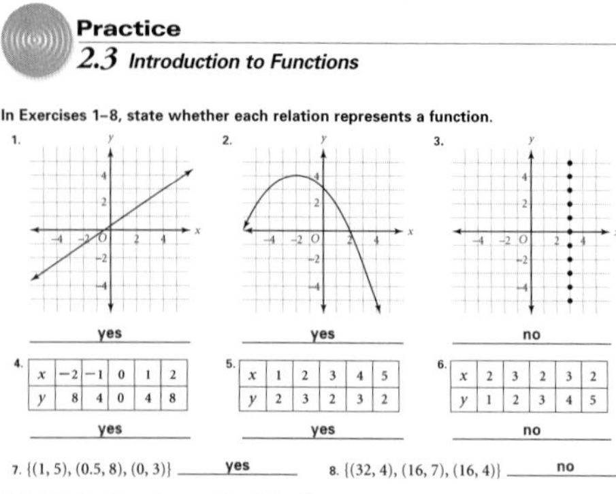

NAME ______ CLASS ______ DATE ______

Practice
2.3 Introduction to Functions

In Exercises 1–8, state whether each relation represents a function.

1. yes
2. yes
3. no

4.

x	−2	−1	0	1	2
y	8	4	0	4	8

yes

5.

x	1	2	3	4	5
y	2	3	2	3	2

yes

6.

x	2	3	2	3	2
y	1	2	3	4	5

no

7. $\{(1, 5), (0.5, 8), (0, 3)\}$ yes
8. $\{(32, 4), (16, 7), (16, 4)\}$ no

State the domain and range of each function.

9. $\left\{(-1, -3), (0, 1), \left(\frac{1}{2}, 3\right), \left(\frac{3}{2}, 7\right)\right\}$
domain: $\left\{-1, 0, \frac{1}{2}, \frac{3}{2}\right\}$; range: $\{-3, 1, 3, 7\}$

10. domain: $x \geq -5$; range: $y \geq -2$

11. $\{(-4.5, 6), (3, -1.5), (6.5, -5), (12, -10.5)\}$
domain: $\{-4.5, 3, 6.5, 12\}$; range: $\{6, -1.5, -5, -10.5\}$

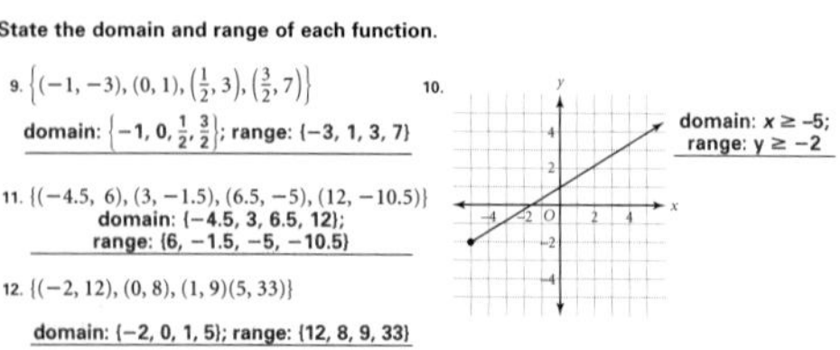

12. $\{(-2, 12), (0, 8), (1, 9)(5, 33)\}$
domain: $\{-2, 0, 1, 5\}$; range: $\{12, 8, 9, 33\}$

Evaluate each function for the given values of x.

13. $f(x) = 20x - 4$, for $x = -2$ and $x = 8$ — $f(-2) = -44$; $f(8) = 156$
14. $f(x) = 5x^2$, for $x = -3$ and $x = 5$ — $f(-3) = 45$; $f(5) = 125$
15. $f(x) = 12 - 3x$, for $x = 7$ and $x = -5$ — $f(7) = -9$; $f(-5) = 27$
16. $f(x) = 3x^2 - 2$, for $x = 11$ and $x = -4$ — $f(11) = 361$; $f(-4) = 46$
17. $f(x) = 3x - x^2$, for $x = 0.5$ and $x = 0$ — $f(0.5) = 1.25$; $f(0) = 0$

Evaluate each function for the given values of x.

43. $f(x) = 2x - 6$ for $x = 1$ and $x = 3$ **−4; 0**

44. $f(x) = 5 - 3x$ for $x = 1$ and $x = 3$ **2; −4**

45. $g(x) = \frac{2x - 1}{3}$ for $x = -1$ and $x = 1$ **$-1; \frac{1}{3}$**

46. $g(x) = \frac{x - 4}{5}$ for $x = -9$ and $x = 9$ **$-\frac{13}{5}; 1$**

47. $f(x) = 2x^2 - 3x$ for $x = 3$ and $x = -2.5$ **9; 20**

48. $f(x) = -x^2 + 4x - 1$ for $x = 2$ and $x = 1.5$ **3; 2.75**

49. $f(x) = \frac{1}{3}x^2$ for $x = -1$ and $x = \frac{3}{4}$ **$\frac{1}{3}; \frac{3}{16}$**

50. $f(x) = -4x^2$ for $x = \frac{3}{2}$ and $x = -2$ **−9; −16**

internet connect go.hrw.com
Activities Online
Go To: **go.hrw.com**
Keyword:
MB1 Overpopulation

Graph each function, and state the domain and range.

51 $y = -\frac{x}{2}$	**52** $y = \frac{2}{3}x - 5$	**53** $y = -2x^2$	**54** $y = x^2 + 2$
55 $y = 4$	**56** $y = -6$	**57** $y = x^3$	**58** $y = \left(\frac{x}{2}\right)^3$

59. Graph a function with a domain of $-3 \le x \le 3$ and a range of $-5 \le y \le 5$.

60. Graph a function with a domain of $-2 \le x \le 5$ and a range of $0 \le y \le 4$.

CHALLENGE

Given $f(t) = t^2 - 3$, find the indicated function value.

61. $f(\sqrt{2})$ **−1** **62.** $f(\sqrt{2} - 1)$ **$-2\sqrt{2}$** **63.** $f(a + \sqrt{2})$ **$a^2 + (2\sqrt{2})a - 1$**

CONNECTION

GEOMETRY The cube shown has volume V.

64. Express the volume of the cube as a function of s, the length of each side. **$V(s) = s^3$**

65. If the volume of the cube is 27 cubic meters, find the area of one face of the cube. **9 m²**

APPLICATIONS

66. CONSUMER ECONOMICS Computer equipment can depreciate very rapidly. Suppose a business assumes that a computer will depreciate linearly at a rate of 15% of its original price each year.

a. If a computer is purchased for $3200, write an equation in which the value for the computer is a function of its age in years.

b. Find the value of a computer after 3 years.

66a. **$V(t) = (1 - 0.15t)3200$**
b. $1760

CONSUMER ECONOMICS A clothing store is selling all out-of-season clothing at 30% off the original price.

67. Write a function that gives the discounted price as a function of the original price.

67. **$S(p) = 0.7p$**

68. Jason spent $47.25 on out-of-season items. Find the original cost of the items. **$67.50**

69. Helena purchased out-of-season items that originally cost $52. Find the sale price of these items. **$36.40**

51. domain: all real numbers; range: all real numbers

52. domain: all real numbers; range: all real numbers

53. domain: all real numbers; range: $y \ge 0$

54. domain: all real numbers; range: $y \ge 2$

55. domain: all real numbers; range: $y = 4$

56. domain: all real numbers; range: $y = -6$

57. domain: all real numbers; range: all real numbers

58. domain: all real numbers; range: all real numbers

59. Answers may vary. sample answer:

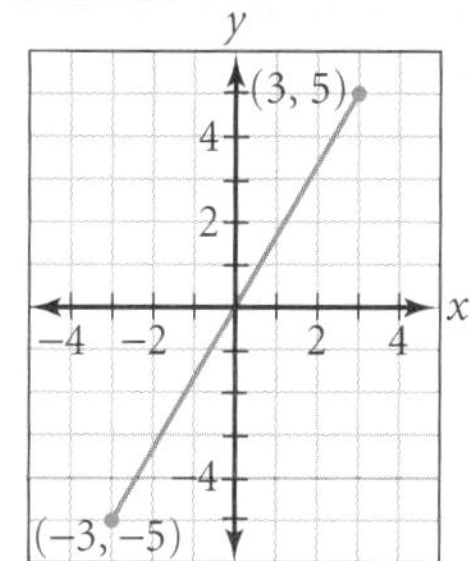

60. Answers may vary: sample answer:

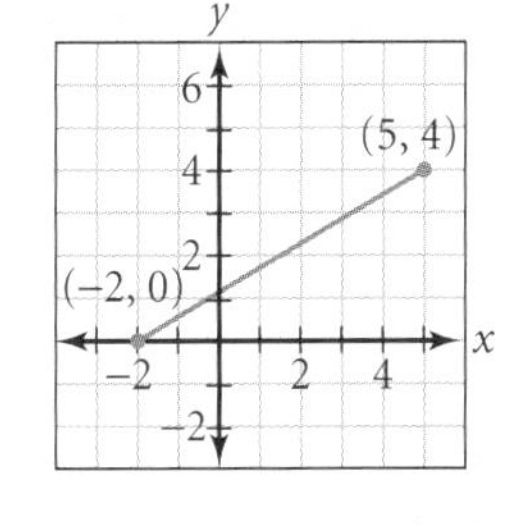

Student Technology Guide

NAME ____________ CLASS ________ DATE ________

Student Technology Guide
2.3 Introduction to Functions

To evaluate a function for more than one value of the variable, you can use the calculator's editing feature or use a graphical approach, which may give an approximate value.

Example: Evaluate $f(x) = 0.5x^2 - 3x + 2$ for $x = 4$ and $x = 2.5$. Because 2.5 contains a digit in the tenths place, enter 4.0 instead of just 4.

Press .5 [×] 4.0 [x^2] [−] 3 [×] 4.0 [+] 2 [ENTER]. Then press [2nd] [ENTER]. Use [◄] and [►] to change 4.0 to 2.5 in each place that it occurs in the expression. Press [ENTER] again.

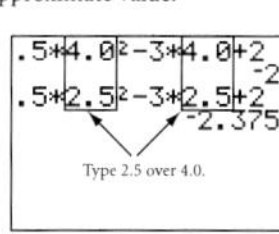

To verify your evaluations, graph $f(x) = 0.5x^2 - 3x + 2$. Then press [TRACE] and [►] and [◄] to move to the points with x-coordinates as close to 4.0 and 2.5 as possible.

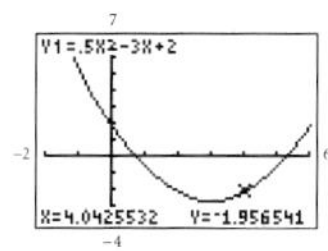

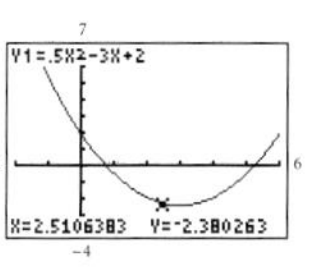

Use a calculator to evaluate each function for the given values of x. Round answers to the nearest hundredth. Then use a graph to verify your evaluations.

1. $f(x) = 0.5x - 2.5$ for $x = 1$ and $x = 3.5$ ____ −2.00; −0.75
2. $f(x) = 0.75x^2$ for $x = -4$ and $x = 0.5$ ____ 12.00; 0.19
3. $f(x) = 1.2x^3$ for $x = 1.74$ and $x = 6.25$ ____ 6.32; 292.97
4. $f(x) = 0.5x^4 - 1$ for $x = 1.5$ and $x = 2.2$ ____ 1.53; 10.71
5. $f(x) = -2.8x^2 + 3.2x + 4$ for $x = 0.5$ and $x = 3$ ____ 4.90; −11.60
6. Suppose that you want to evaluate $g(x) = -0.4x^2 + 3x + 2$ for $x = 4.0$. Explain how to edit the key sequence at the top of the page in order to evaluate g rather than f.
   Answers may vary. Press [2nd] [ENTER]. Use the left arrow key to replace 0.5 with −0.4 and replace + with − before the $3x$ term.

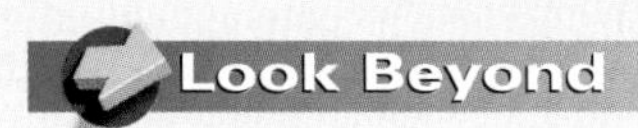

Look Beyond

In Exercises 82 and 83, students analyze the graphs of a quadratic equation and an exponential equation from the perspective of functions. Quadratic functions are the focus of Chapter 5, and exponential functions are studied in Chapter 6.

ALTERNATIVE
Assessment

Portfolio Activity

The Portfolio Activity can be used as preparation for the Chapter Project or as a separate activity. In the Portfolio Activity on this page, students graph the data from the Portfolio Activity on page 93 and analyze the function that is represented. They use the regression-equation capability of their graphics calculator to find a linear model and an exponential model for the data.

Answers to Portfolio Activities can be found in Additional Answers of the Teacher's Edition.

Look Back

Write the equation in slope-intercept form for the line that has the indicated slope, *m*, and contains the given point. *(LESSON 1.3)*

70. $m = 5, (2, 3)$ **$y = 5x - 7$**

71. $m = -3, (4, 1)$ **$y = -3x + 13$**

72. $m = \frac{1}{5}, (4, -11)$ **$y = \frac{1}{5}x - \frac{59}{5}$**

73. $m = -\frac{2}{3}, (-8, -3)$ **$y = -\frac{2}{3}x - \frac{25}{3}$**

Write an equation in slope-intercept form of the line containing the indicated points. *(LESSON 1.3)*

74. $(1, 4)$ and $(-3, 0)$ **$y = x + 3$**

75. $(0, 2)$ and $(-1, 1)$ **$y = x + 2$**

76. $(2, 3)$ and $(0, 0)$ **$y = \frac{3}{2}x$**

77. $(-2, -5)$ and $(5, -1)$ **$y = \frac{4}{7}x - \frac{27}{7}$**

Write an equation in slope-intercept form of the line that contains the given point and is perpendicular to a line with the given slope. *(LESSON 1.3)*

78. $P(6, -1), m = \frac{4}{3}$ **$y = -\frac{3}{4}x + \frac{7}{2}$**

79. $P(-5, 3), m = -\frac{1}{2}$ **$y = 2x + 13$**

Evaluate each expression by using the order of operations. *(LESSON 2.1)*

80. $3[2 - (5 - 3) - 7] \div 2$ **−10.5**

81. $-(-5^2)^3$ **15,625**

internet connect
Portfolio Extension
Go To: **go.hrw.com**
Keyword:
MB1 Exponential

Look Beyond

82 Graph $y = x^2 - 3x - 10$. Explain why this is a function. Give the domain and range of this function.

83 Graph $y = 2^x$. Explain why this is a function. Give the domain and range of this function.

SPACE SCIENCE Refer to the Space Debris Table from the Portfolio Activity on page 93.

1. Graph the ordered pairs in the Space Debris Table that you created in the Portfolio Activity on page 93. Let the years be represented by x (where $x = 0$ represents 1993) and the total number of debris objects by y. Describe how your graph illustrates that the total number of debris objects in space, y, is a function of time, x.
2. Give the domain and range of the function you described in Step 1. Then state the independent and dependent variables for this function.
3. Using the linear regression feature on your calculator, find a linear function that models this data. Next, using the exponential regression feature, find an exponential function that models this data.
4. Graph your linear model and your exponential model, and compare how well or how poorly the models appear to fit the data.

WORKING ON THE CHAPTER PROJECT

You should now be able to complete Activities 1 and 2 of the Chapter Project.

82. $y = x^2 - 3x - 10$ is a function because each vertical line drawn through the graph intersects it in at most one point; domain: all real numbers; range: $y \geq -16$.

83. $y = x^2$ is a function because each vertical line drawn through the graph intersects it in at most one point; domain: all real numbers; range: $y > 0$.

Operations With Functions

Why *Many real-world relationships can be described by operations with functions. For example, the distance traveled by a vehicle before coming to a complete stop is the sum of the distance that it travels while the driver reacts and the distance that it travels while the brakes are applied.*

Objectives

- Perform operations with functions to write new functions.
- Find the composition of two functions.

APPLICATION
TRANSPORTATION

Common sense tells you that the faster you drive, the farther you will travel while reacting to an obstacle and while braking to a complete stop. In other words, the reaction distance and braking distance depend on the speed of the car at the moment the obstacle is observed.

Stopping Distance Data

Speed (mph)	Reaction distance (ft)	Braking distance (ft)	Stopping distance (ft)
10	11	5	16
20	22	21	43
30	33	47	80
40	44	84	128
50	55	132	187
60	66	189	255
70	77	258	335

Activity: Investigating Braking Distance

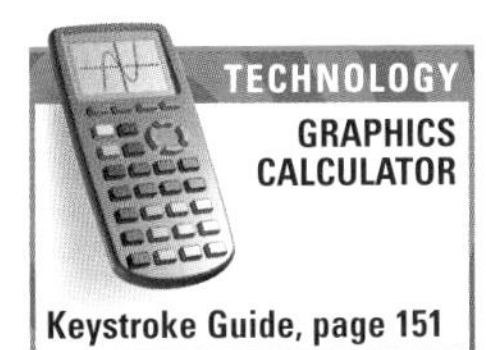
TECHNOLOGY
GRAPHICS CALCULATOR
Keystroke Guide, page 151

You will need: a graphics calculator

1. Using a graphics calculator, make a scatter plot of the data for speed and reaction distance. What type of function is represented by this data?
2. Using a graphics calculator, make a scatter plot of the data for speed and braking distance. What trend do you see in the graph?
3. Using the table above, identify a relationship between the stopping distance and the reaction and braking distances. Use this relationship to make predictions about the scatter plot of the data for speed and stopping distance.

CHECKPOINT ✔

4. Using a graphics calculator, make a scatter plot of the data for speed and stopping distance. What trend do you see in the graph?

Prepare

NCTM PRINCIPLES & STANDARDS 1–3, 5–10

QUICK WARM-UP

Simplify each expression.

1. $(x-2)+(x+5)$ $\quad 2x+3$
2. $(3x+4)-(2x-6)$ $\quad x+10$
3. $(x+2)(x-3)$ $\quad x^2-x-6$
4. $\frac{10x^2}{-2x}$ $\quad -5x$
5. $(-3x)^2-3x$ $\quad 9x^2-3x$

Also on Quiz Transparency 2.4

Teach

Why The cost of gasoline for an automobile trip depends on the number of gallons that you buy. The number of gallons that you buy, in turn, depends on the number of miles that you drive. A chain of relationships like this is an example of a *composition* of functions. Ask students to describe other situations that illustrate composition.

Activity Notes

In this Activity, students investigate the graphs of three sets of data. One set conforms to a linear model, one conforms to a quadratic model, and the third is the sum of the other two. Students should observe that the sum conforms to a quadratic model.

Alternative Teaching Strategy

VISUAL STRATEGIES Give students functions f and g defined as follows:

$f = \{(-3, -2), (0, 1), (4, 5)\}$

$g = \{(-2, 4), (1, 1), (5, 25)\}$

The range of f is the domain of g, so you can use this function diagram to show the relationships.

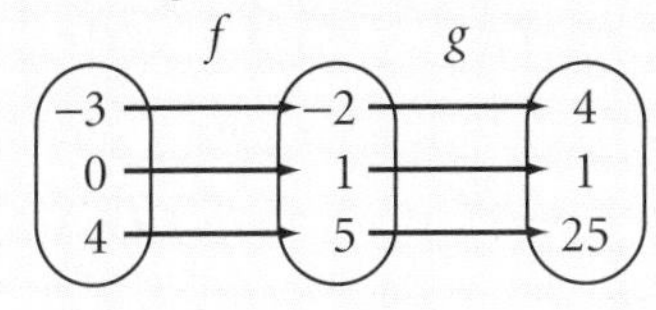

Point out that this leads to a new relationship: the *composite* function that is denoted $g \circ f$.

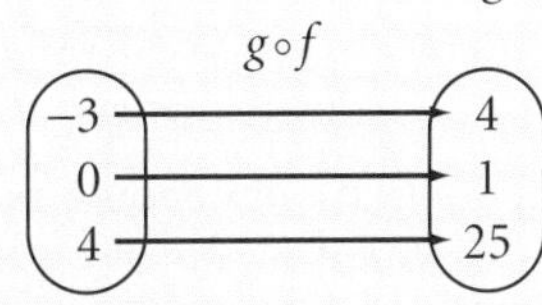

Discuss the following equations for f, g, and $g \circ f$:

$f(x) = x + 1 \quad g(x) = x^2 \quad (g \circ f)(x) = (x+1)^2$

Now redefine f and g, extending their domains to include all real numbers, and have students find an equation for $f \circ g$. $\;(f \circ g)(x) = x^2 + 1$

Cooperative Learning

You may wish to have students do the Activity in pairs. One partner can do Step 1 while the other does Step 2. Then they can do Step 3 together. For Step 4, the partners can first work individually and then compare and discuss their results.

CHECKPOINT ✔

4.

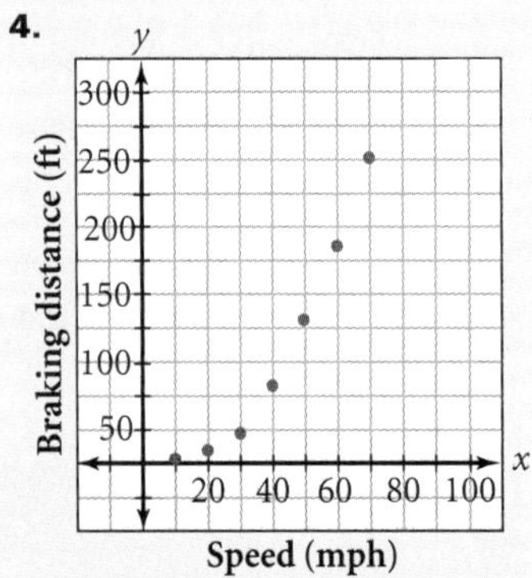

As speed increases, the stopping distance increases and the rate of change of the stopping distance increases.

Use Teaching Transparency 7.

ADDITIONAL EXAMPLE 1

Let $f(x) = 4x^2 + 6x - 9$ and $g(x) = 6x^2 - x + 2$.

a. **Find $f + g$.**
$(f + g)(x) = 10x^2 + 5x - 7$

b. **Find $f - g$.**
$(f - g)(x) = -2x^2 + 7x - 11$

TRY THIS

$(f + g)(x) = 12x - 2.5$

$(f - g)(x) = -14x^2 + 12x + 7.5$

The functions at right relate speed, x, in miles per hour to the reaction distance, r, and to braking distance, b, both in feet.

$r(x) = \frac{11}{10}x \qquad b(x) = \frac{1}{19}x^2$

$s(x) = r(x) + b(x)$

You can relate speed, x, to the total stopping distance, s, as a *sum* of r and b.

$s(x) = \frac{11}{10}x + \frac{1}{19}x^2$

Functions can also be combined by subtraction, multiplication, and division.

Operations With Functions

For all functions f and g:

Sum	$(f + g)(x) = f(x) + g(x)$
Difference	$(f - g)(x) = f(x) - g(x)$
Product	$(f \cdot g)(x) = f(x) \cdot g(x)$
Quotient	$\left(\frac{f}{g}\right)(x) = \frac{f(x)}{g(x)}$, where $g(x) \neq 0$

EXAMPLE 1 **Let $f(x) = 5x^2 - 2x + 3$ and $g(x) = 4x^2 + 7x - 5$.**

a. Find $f + g$. **b.** Find $f - g$.

SOLUTION

a. $(f + g)(x) = f(x) + g(x)$
$= (5x^2 - 2x + 3) + (4x^2 + 7x - 5)$
$= 5x^2 + 4x^2 - 2x + 7x + 3 - 5$ *Use the Commutative Property.*
$= 9x^2 + 5x - 2$ *Combine like terms.*

b. $(f - g)(x) = f(x) - g(x)$
$= (5x^2 - 2x + 3) - (4x^2 + 7x - 5)$
$= 5x^2 - 4x^2 - 2x - 7x + 3 - (-5)$ *Use the Commutative Property.*
$= x^2 - 9x + 8$ *Combine like terms.*

TRY THIS Let $f(x) = -7x^2 + 12x + 2.5$ and $g(x) = 7x^2 - 5$. Find $f + g$ and $f - g$.

EXAMPLE 2 **Let $f(x) = 5x^2$ and $g(x) = 3x - 1$.**

a. Find $f \cdot g$. **b.** Find $\frac{f}{g}$, and state any domain restrictions.

SOLUTION

a. $(f \cdot g)(x) = f(x) \cdot g(x)$
$= 5x^2(3x - 1)$
$= 5x^2(3x) - 5x^2(1)$
$= 15x^3 - 5x^2$

b. $\left(\frac{f}{g}\right)(x) = \frac{f(x)}{g(x)}$, where $g(x) \neq 0$
$= \frac{5x^2}{3x - 1}$, where $x \neq \frac{1}{3}$

$3x - 1 \neq 0$
$3x \neq 1$
$x \neq \frac{1}{3}$

CHECKPOINT ✔ Which property of real numbers is used in part **a** of Example 2?

TRY THIS Let $f(x) = 3x^2 + 1$ and $g(x) = 5x - 2$. Find $f \cdot g$ and $\frac{f}{g}$.

Interdisciplinary Connection

BUSINESS On a certain day, one British pound was worth 1.6603 U.S. dollars. On that same day, one U.S. dollar was worth 1.5325 Canadian dollars. Write a rule for the following functions:

a. B, representing a conversion from x British pounds to U.S. dollars **$B(x) = 1.6603x$**

b. C, representing a conversion from x U.S. dollars to Canadian dollars **$C(x) = 1.5325x$**

c. $C \circ B$ **$(C \circ B)(x) = 2.5444x$**

What does $C \circ B$ represent? **a conversion from x British pounds to Canadian dollars**

Inclusion Strategies

VISUAL LEARNERS Students may be overwhelmed by the complexity of the expressions used in compositions. It may help to use a visual device like a box to represent the variable in the "outer" function, then fill in the box with the expression for the "inner" function. For example, this is how you might show the compositions in Example 3.

$f(x) = x^2 - 1$	$g(x) = 3x$
$f(\square) = \square^2 - 1$	$g(\square) = 3\square$
$(f \circ g)(x) = \boxed{3x}^2 - 1$	$(g \circ f)(x) = 3\boxed{x^2 - 1}$

Composition of Functions

When you apply a function rule on the result of another function rule, you *compose* the functions. The illustration below shows how the composition $f \circ g$, or $f(g(x))$, read "f of g of x," works.

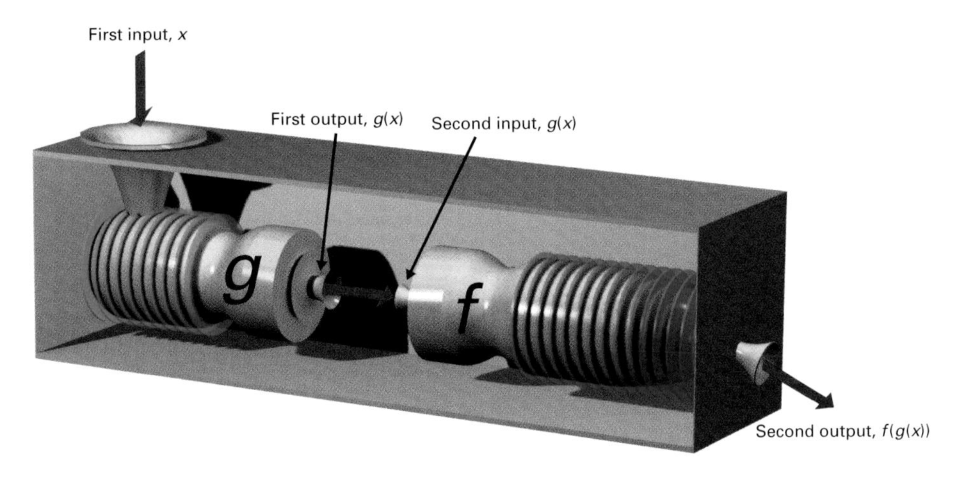

Composition of Functions

Let f and g be functions of x.

The composition of f with g, denoted $f \circ g$, is defined by $f(g(x))$.

The domain of $y = f(g(x))$ is the set of domain values of g whose range values are in the domain of f. The function $f \circ g$ is called the **composite** function of f with g.

Example 3 shows you how to form the composition of two functions.

EXAMPLE 3 Let $f(x) = x^2 - 1$ and $g(x) = 3x$.

a. Find $f \circ g$. **b.** Find $g \circ f$.

SOLUTION

a. $(f \circ g)(x) = f(g(x))$
$= f(3x)$
$= (3x)^2 - 1$
$= 9x^2 - 1$

b. $(g \circ f)(x) = g(f(x))$
$= g(x^2 - 1)$
$= 3(x^2 - 1)$
$= 3x^2 - 3$ *Distributive Property*

TRY THIS Let $f(x) = -2x^2 + 3$ and $g(x) = -2x$. Find $f \circ g$ and $g \circ f$.

In Example 3 and in most cases, $f \circ g$ and $g \circ f$ are not equivalent. That is, the composition of functions is not commutative.

CRITICAL THINKING Write two functions, f and g, whose composite functions $f \circ g$ and $g \circ f$ are equivalent. Justify your response.

ADDITIONAL EXAMPLE 2

Let $f(x) = 9x^2$ and let $g(x) = 4x + 3$.

a. Find $f \cdot g$.
$(f \cdot g)(x) = 9x^2(4x + 3)$
$= 36x^3 + 27x^2$

b. Find $\frac{f}{g}$, and state any domain restrictions.
$\left(\frac{f}{g}\right)(x) = \frac{9x^2}{4x+3}$, where $x \neq -\frac{3}{4}$

TRY THIS
$(f \cdot g) = 15x^3 - 6x^2 + 5x - 2$

$\left(\frac{f}{g}\right)(x) = \frac{3x^2+1}{5x-2}$, where $x \neq -\frac{2}{5}$

Use Teaching Transparency 8.

ADDITIONAL EXAMPLE 3

Let $f(x) = x^2 + 4$ and let $g(x) = 2x$.

a. Find $f \circ g$.
$(f \circ g)(x) = 4x^2 + 4$

b. Find $g \circ f$.
$(g \circ f)(x) = 2x^2 + 8$

TRY THIS
$(f \circ g)(x) = -8x^2 + 3$
$(g \circ f)(x) = 4x^2 - 6$

CRITICAL THINKING
samples:
$f(x) = x + 6$ and $g(x) = x - 6$;
$(f \circ g)(x) = x = (g \circ f)(x)$

Enrichment

Let $f(x) = x^2$, $g(x) = 3x$, and $h(x) = x + 5$. Have students work in pairs to find the composite functions below.

1. $[f \circ (g \circ h)](x)$ **$9x^2 + 90x + 225$**
2. $[g \circ (f \circ h)](x)$ **$3x^2 + 30x + 75$**
3. $[h \circ (g \circ f)](x)$ **$3x^2 + 5$**

Have students determine the three other possible composite functions and write equations for them.

$[f \circ (h \circ g)](x) = 9x^2 + 30x + 25$
$[g \circ (h \circ f)](x) = 3x^2 + 15$
$[h \circ (f \circ g)](x) = 9x^2 + 5$

Reteaching the Lesson

COOPERATIVE LEARNING Have students work in pairs. Write the expression $5x^2$ on the board or overhead. Tell them to find two functions, f and g, such that $(f + g)(x) = 5x^2$.
sample: $f(x) = 2x^2$ and $g(x) = 3x^2$

Now repeat the activity for $(f - g)(x)$, $(f \cdot g)(x)$, $\left(\frac{f}{g}\right)(x)$, and $(f \circ g)(x)$.

samples:
$(f - g)(x) = 5x^2$: $f(x) = 6x^2$ and $g(x) = x^2$
$(f \cdot g)(x) = 5x^2$: $f(x) = 5x$ and $g(x) = x$
$\left(\frac{f}{g}\right)(x) = 5x^2$: $f(x) = 5x^3$ and $g(x) = x$, $x \neq 0$
$(f \circ g)(x) = 5x^2$: $f(x) = 5x$ and $g(x) = x^2$

ADDITIONAL

EXAMPLE 4

A computer store is offering a \$40 rebate along with a 20% discount. Let x represent the original price in dollars of an item in the store.

a. Write a function, D, that represents the sale price after a 20% discount and a function, R, that represents the sale price after a \$40 rebate.

$D(x) = x - 0.2x = 0.8x$

$R(x) = x - 40$

b. Find the composite functions $R \circ D$ and $D \circ R$, and explain what they represent.

$(R \circ D)(x) = 0.8x - 40$ represents the price of the item if the discount is calculated before the rebate is given.

$(D \circ R)(x) = 0.8x - 32$ represents the price of the item if the rebate is given before the discount is calculated.

Assess

Selected Answers

Exercises 4–10, 11–71 odd

ASSIGNMENT GUIDE

In Class	1–10
Core	11–23 odd, 25–57 odd, 61, 63
Core Plus	12–58 even, 59–64
Review	65–71
Preview	72

Extra Practice can be found beginning on page 940.

Mid-Chapter Assessment for Lessons 2.1 through 2.4 can be found on page 20 of the *Assessment Resources.*

Example 4 shows you how composite functions can model problems that involve a series of actions.

EXAMPLE 4

APPLICATION
CONSUMER ECONOMICS

A local electronics store is offering a \$100.00 rebate along with a 10% discount. Let x represent the original price of an item in the store.

a. Write the function D that represents the sale price after a 10% discount and the function R that represents the sale price after a \$100 rebate.

b. Find the composite functions $(R \circ D)(x)$ and $(D \circ R)(x)$, and explain what they represent.

SOLUTION

a. Since a 10% discount on the original price is the same as paying 90% of the original price, $D(x) = x - 0.1x = 0.9x$.

The rebate function is $R(x) = x - 100$.

b. **10% discount first**

$(R \circ D)(x) = R(D(x)) = R(0.9x)$
$= (0.9x) - 100$
$= 0.9x - 100$

\$100 rebate first

$(D \circ R)(x) = D(R(x)) = D(x - 100)$
$= 0.9(x - 100)$
$= 0.9x - 90$

Notice that taking the 10% discount first results in a lower sale price.

Exercises

Communicate

internet connect
Activities Online
Go To: go.hrw.com
Keyword: MB1 UV

1. Explain how to write a function for reaction distance, r, in terms of the function for braking distance, b, and the function for stopping distance, s.
2. Explain how to find $f \circ g$ given $f(x) = 4x - 7$ and $g(x) = 2x^2 + 4$.
3. Show that $f \circ g$ and $g \circ f$ are not equivalent functions given $f(x) = 3x + 1$ and $g(x) = 2x$.

Guided Skills Practice

Let $f(x) = \frac{x}{2}$ and $g(x) = 3x + 1$. Perform each function operation. State any domain restrictions. *(EXAMPLES 1, 2, AND 3)*

4. $f + g$ $= \frac{7}{2}x + 1$ **5.** $f - g$ $= -\frac{5}{2}x - 1$ **6.** $f \cdot g$ $= \frac{3}{2}x^2 + \frac{x}{2}$

7. $\frac{f}{g}$ $= \frac{x}{2(3x+1)}, x \neq -\frac{1}{3}$ **8.** $f \circ g$ $= \frac{3x+1}{2}$ **9.** $g \circ f$ $= \frac{3}{2}x + 1$

10. Let x be the total cost of the meal.

$R(x) = x - 5$
$T(x) = x + 0.15x = 1.15x$
$(R \circ T)(x) = R(1.15x) = 1.15x - 5$
$(T \circ R)(x) = T(x - 5)$
$= 1.15(x - 5)$
$= 1.15x - 5.75$

$R \circ T$ represents the total cost of the meal plus the tip minus \$5. $T \circ R$ represents the total cost of the meal plus the tip minus \$5.75. $R \circ T$ represents the conditions of the coupon.

11. $(f + g)(x) = 4x + 8$; $(f - g)(x) = 4x - 2$

12. $(f + g)(x) = 17x + 7$; $(f - g)(x) = 23x + 7$

13. $(f + g)(x) = x^2 + 5x - 6$; $(f - g)(x) = x^2 - x + 4$

14. $(f + g)(x) = 5x^2 - 3x - 3$; $(f - g)(x) = -7x^2 - 3x + 3$

15. $(f + g)(x) = -3x^2 + x + 2$; $(f - g)(x) = 3x^2 + x - 6$

16. $(f + g)(x) = 2x^2 - x + 6$; $(f - g)(x) = 4x^2 - x - 6$

APPLICATION

10. CONSUMER ECONOMICS A coupon for \$5 off any meal states that a 15% tip will be added to the total check before the \$5 is subtracted. Let x represent the total check amount. Write a function, R, for the price reduction and a function, T, for the tip. Find the composite functions $(R \circ T)(x)$ and $(T \circ R)(x)$, and explain which composite function represents the conditions of the coupon. ***(EXAMPLE 4)***

Practice and Apply

Find $f + g$ and $f - g$.

11. $f(x) = 4x + 3$; $g(x) = 5$
12. $f(x) = 20x + 7$; $g(x) = -3x$
13. $f(x) = x^2 + 2x - 1$; $g(x) = 3x - 5$
14. $f(x) = -x^2 - 3x$; $g(x) = 6x^2 - 3$
15. $f(x) = x - 2$; $g(x) = 4 - 3x^2$
16. $f(x) = 3x^2 - x$; $g(x) = 6 - x^2$

internet connect — go.hrw.com
Homework Help Online
Go To: **go.hrw.com**
Keyword:
MB1 Homework Help
for Exercises 17–24

Find $f \cdot g$ and $\frac{f}{g}$. State any domain restrictions.

17. $f(x) = 3x^2$; $g(x) = x - 8$
18. $f(x) = 2x^2$; $g(x) = 7 - x$
19. $f(x) = -x^2$; $g(x) = 3x$
20. $f(x) = 4x^2 - 2x + 1$; $g(x) = -5x$
21. $f(x) = 2x^2$; $g(x) = 7 - x$
22. $f(x) = 7x^2$; $g(x) = 2x - 6$
23. $f(x) = -3x^2 - x$; $g(x) = 5 - x$
24. $f(x) = 4x^2 + 3$; $g(x) = 3x - 9$

Let $f(x) = x^2 - 1$ and $g(x) = 2x - 3$. Find each new function and write it in simplest form. Justify each step in the simplification, and state any domain restrictions.

25. $f + g$ **26.** $f - g$ **27.** $g - f$ **28.** $f \cdot g$ **29.** $\frac{f}{g}$

Let $f(x) = x - 3$ and $g(x) = x^2 - 9$. Find each new function and write it in simplest form. Justify each step in the simplification, and state any domain restrictions.

30. $f + g$ **31.** $f - g$ **32.** $g - f$ **33.** $f \cdot g$ **34.** $\frac{g}{f}$

Find $f \circ g$ and $g \circ f$.

35. $f(x) = x + 1$; $g(x) = 2x$
36. $f(x) = 3x$; $g(x) = 2x + 3$
37. $f(x) = 3x - 2$; $g(x) = x + 2$
38. $f(x) = 2x - 3$; $g(x) = x + 4$
39. $f(x) = -3x^2 - 1$; $g(x) = -5x$
40. $f(x) = -2x$; $g(x) = -2x^2 + 3$
41. $f(x) = -4x^2 + 3x - 1$; $g(x) = 3$
42. $f(x) = 2x^2 + 3x - 5$; $g(x) = 4$

Let $f(x) = 3x - 4$ and $g(x) = -x^2$. Evaluate each composite function.

43. $(f \circ g)(2)$ **−16** **44.** $(f \circ g)(-2)$ **−16** **45.** $(f \circ f)(2)$ **2** **46.** $(f \circ f)(-2)$ **−34**
47. $(g \circ g)(2)$ **−16** **48.** $(g \circ g)(-2)$ **−16** **49.** $(g \circ f)(0)$ **−16** **50.** $(f \circ g)(0)$ **−4**

Let $f(x) = 2x$, $g(x) = x^2 + 2$, and $h(x) = -4x + 3$. Find each composite function.

51. $f \circ g$ **52.** $g \circ f$ **53.** $f \circ h$ **54.** $h \circ g$
55. $f \circ f$ **56.** $h \circ f$ **57.** $h \circ (h \circ g)$ **58.** $(h \circ f) \circ g$

CHALLENGE

59. Let $h(x) = x^2 - 9$. Find two functions f and g such that $f \circ g = h$.

Teaching Tip

Some students may need practice using the properties of real numbers. Give them two functions and have them calculate the sum, difference, product, quotient, and composition of the two functions, justifying each step. An example is shown below.

For $f(x) = 2x - 3$ and $g(x) = x + 1$, find $f + g$, $f - g$, $f \cdot g$, $\frac{f}{g}$, and $f \circ g$.

$(f + g)(x)$
$= f(x) + g(x)$ Def. of $f + g$
$= (2x - 3) + (x + 1)$ Subst.
$= 2x + x - 3 + 1$ Comm., Add.
$= 3x - 2$ Simplify

$(f - g)(x)$
$= f(x) - g(x)$ Def. of $f - g$
$= (2x - 3) - (x + 1)$ Subst.
$= (2x - 3) - 1(x + 1)$ Mult. Ident.
$= 2x - 3 - x - 1$ Dist.
$= 2x - x - 3 - 1$ Comm., Add.
$= x - 4$ Simplify

$(f \cdot g)(x)$
$= f(x) \cdot g(x)$ Def. of $f \cdot g$
$= (2x - 3)(x + 1)$ Subst.
$= 2x(x + 1) - 3(x + 1)$ Dist.
$= 2x^2 + 2x - 3x - 3$ Dist.
$= 2x^2 - x - 3$ Simplify

$\left(\frac{f}{g}\right)(x)$
$= \frac{f(x)}{g(x)}, g(x) \neq 0$ Def. of $\frac{f}{g}$
$= \frac{2x - 3}{x + 1}, x \neq -1$ Subst.

$(f \circ g)(x)$
$= f[g(x)]$ Def. of comp.
$= f(x + 1)$ Subst.
$= 2(x + 1) - 3$ Def. of f
$= 2x + 2 - 3$ Dist.
$= 2x - 1$ Simplify

17. $(f \cdot g)(x) = 3x^3 - 24x^2$; $\left(\frac{f}{g}\right)(x) = \frac{3x^2}{x - 8}, x \neq 8$
18. $(f \cdot g)(x) = 14x^2 - 2x^3$; $\left(\frac{f}{g}\right)(x) = \frac{2x^2}{7 - x}, x \neq 7$
19. $(f \cdot g)(x) = -3x^3$; $\left(\frac{f}{g}\right)(x) = -\frac{x}{3}, x \neq 0$
20. $(f \cdot g)(x) = -20x^3 + 10x^2 - 5x$; $\left(\frac{f}{g}\right)(x) = -\frac{4x^2 - 2x + 1}{5x}, x \neq 0$
21. $(f \cdot g)(x) = 14x^2 - 2x^3$; $\left(\frac{f}{g}\right)(x) = \frac{2x^2}{7 - x}, x \neq 7$
22. $(f \cdot g)(x) = 14x^3 - 42x^2$; $\left(\frac{f}{g}\right)(x) = \frac{7x^2}{2x - 6}, x \neq 3$
23. $(f \cdot g)(x) = 3x^3 - 14x^2 - 5x$; $\left(\frac{f}{g}\right)(x) = \frac{-3x^2 - x}{5 - x}, x \neq 5$
24. $(f \cdot g)(x) = 12x^3 - 36x^2 + 9x - 27$; $\left(\frac{f}{g}\right)(x) = \frac{4x^2 + 3}{3x - 9}, x \neq 3$
25. $(f + g)(x) = x^2 + 2x - 4$
26. $(f - g)(x) = x^2 - 2x + 2$
27. $(g - f)(x) = -x^2 + 2x - 2$
28. $(f \cdot g)(x) = 2x^3 - 3x^2 - 2x + 3$
29. $\left(\frac{f}{g}\right)(x) = \frac{x^2 - 1}{2x - 3}, x \neq \frac{3}{2}$
30. $(f + g)(x) = x^2 + x - 12$

The answers to Exercises 31–42 and 51–59 can be found beginning on page 116.

Error Analysis

When stating restrictions on the domain of a quotient of functions, students may forget that the restrictions apply to the original quotient, not just the simplified quotient.

APPLICATIONS

60. DEMOGRAPHICS College enrollments in the United States are projected to increase from 1995 to 2005 for public and private schools. The models below give the total projected college enrollments in thousands, where $t = 0$ represents the year 1995. [*Source: U.S. National Center for Education Statistics*]

Public: $f(t) = 160t + 11{,}157$
Private: $g(t) = 50t + 3114$

a. Find $f + g$.
b. Evaluate $f + g$ for the year 2000.

60a. $(f + g)(t) = 210t + 14{,}271$
b. 15,321

61. MANUFACTURING The function $C(x) = 4x + 850$ closely approximates the cost of a daily production run of x picture frames. The number of picture frames produced is represented by the function $x(t) = 90t$, where t is the time in hours since the beginning of the production run.

a. Give the cost of a daily production run, C, as a function of time, t.
b. Find the cost of the production run that lasts 5 hours.
c. How many picture frames are produced in 5 hours?

61a. $C(t) = 360t + 850$
b. \$2650
c. 450

62. TEMPERATURE A temperature given in kelvins is equivalent to the temperature given in degrees Celsius plus 273. A temperature given in degrees Celsius is equivalent to $\frac{9}{5}$ times the quantity given by the temperature in degrees Fahrenheit minus 32.

a. Write a function expressing kelvins in terms of degrees Celsius. $K = C + 273$
b. Write a function expressing degrees Celsius in terms of degrees Fahrenheit. $C = \frac{5}{9}(F - 32)$
c. Write a function expressing kelvins in terms of degrees Fahrenheit. $K = \frac{5}{9}(F - 32) + 273$
d. Water freezes at 32°F. At what Kelvin temperature does water freeze? **273 kelvins**
e. Are kelvins and degrees Fahrenheit linearly related? Explain. **yes;** $K = \frac{5}{9}F + \frac{2297}{9}$

63. BUSINESS Jovante decides to start a business by making buttons and selling them for 25 cents each. The button machine costs \$125 dollars, and the materials for each button cost 10 cents.

a. Write a cost function, C, that represents the cost in dollars of making n buttons.
b. Write an income function, I, that represents the income in dollars generated by selling n buttons.
c. Write a profit function, P, that represents the income, I, minus the cost, C.
d. Find the number of buttons that Jovante must sell in order to make a \$100 profit.

63a. $C(n) = 0.10n + 125.00$
b. $I(n) = 0.25n$
c. $P(n) = 0.15n - 125.00$
d. 1500

Practice

NAME ______ CLASS ______ DATE ______

Practice
2.4 Operations With Functions

Find $f + g$ and $f - g$.

1. $f(x) = 7x^2 + 5x$; $g(x) = x^2 - 13$ $(f + g)(x) = 8x^2 + 5x - 13$; $(f - g)(x) = 6x^2 + 5x + 13$
2. $f(x) = 41 - 5x$; $g(x) = 13x^2$ $(f + g)(x) = 13x^2 - 5x + 41$; $(f - g)(x) = -13x^2 - 5x + 41$
3. $f(x) = x^2 + \frac{1}{3}x + 9$; $g(x) = -7x - 7$ $(f + g)(x) = x^2 - \frac{20}{3}x + 2$; $(f - g)(x) = x^2 + \frac{22}{3}x + 16$
4. $f(x) = -9x^2 + 6$; $g(x) = 12x^2$ $(f + g)(x) = 3x^2 + 6$; $(f - g)(x) = -21x^2 + 6$

Find $f \cdot g$ and $\frac{f}{g}$. State any domain restrictions.

5. $f(x) = 35x + 5$; $g(x) = 5$ $(f \cdot g)(x) = 175x + 25$; $\left(\frac{f}{g}\right)(x) = 7x + 1$
6. $f(x) = x^2 + 25$; $g(x) = 3x + 17$ $(f \cdot g)(x) = 3x^3 + 17x^2 + 75x + 425$; $\left(\frac{f}{g}\right)(x) = \frac{x^2 + 25}{3x + 17}, x \neq -5\frac{2}{3}$
7. $f(x) = x^2 + 16$; $g(x) = x^2 - 16$ $(f \cdot g)(x) = x^4 - 256$; $\left(\frac{f}{g}\right)(x) = \frac{x^2 + 16}{x^2 - 16}, x \neq -4$ and $x \neq 4$

Let $f(x) = -2x - 2$ and $g(x) = x + 10$. Find each new function, and state any domain restrictions.

8. $f + g$ $-x + 8$
9. $f - g$ $-3x - 12$
10. $g - f$ $3x + 12$
11. $f \cdot g$ $-2x^2 - 22x - 20$
12. $\frac{f}{g}$ $-\frac{2x + 2}{x + 10}, x \neq -10$
13. $\frac{g}{f}$ $-\frac{x + 10}{2x + 2}, x \neq -1$

Find $f \circ g$ and $g \circ f$.

14. $f(x) = 3x - 2$; $g(x) = \frac{1}{3}(x + 2)$ $(f \circ g)(x) = x$; $(g \circ f)(x) = x$
15. $f(x) = 4x$; $g(x) = x^2 - 1$ $(f \circ g)(x) = 4x^2 - 4$; $(g \circ f)(x) = 16x^2 - 1$
16. $f(x) = -x^2 + 1$; $g(x) = x$ $(f \circ g)(x) = -x^2 + 1$; $(g \circ f)(x) = -x^2 + 1$

Let $f(x) = 11x$, $g(x) = x^2 - 5$, and $h(x) = 2(x-4)$. Evaluate each composite function.

17. $(f \circ g)(-1)$ -44
18. $(h \circ f)(-2)$ -52
19. $(h \circ g)(2)$ -10
20. $(g \circ h)(4)$ -5
21. $(g \circ f)(0)$ -5
22. $(f \circ h)(5)$ 22
23. $(f \circ g)(0)$ -55
24. $(h \circ h)(-1)$ -28
25. $(f \circ f)(2)$ 44

31. $(f - g)(x) = -x^2 + x + 6$

32. $(g - f)(x) = x^2 - x - 6$

33. $(f \cdot g)(x) = x^3 - 3x^2 - 9x + 27$

34. $\left(\frac{g}{f}\right)(x) = \frac{x^2 - 9}{x - 3}$, or $\left(\frac{g}{f}\right)(x) = x + 3, x \neq 3$

35. $(f \circ g)(x) = 2x + 1$; $(g + f)(x) = 2x + 2$

36. $(f \circ g)(x) = 6x + 9$; $(g + f)(x) = 6x + 3$

37. $(f \circ g)(x) = 3x + 4$; $(g + f)(x) = 3x$

38. $(f \circ g)(x) = 2x + 5$; $(g + f)(x) = 2x + 1$

39. $(f \circ g)(x) = -75x^2 - 1$; $(g \circ f)(x) = 15x^2 + 5$

40. $(f \circ g)(x) = 4x^2 - 6$; $(g \circ f)(x) = -8x^2 + 3$

41. $(f \circ g)(x) = -28$; $(g + f)(x) = 3$

42. $(f \circ g)(x) = 39$; $(g + f)(x) = 4$

APPLICATION

T is total with tax.
D is amount after discount.
$T(x) = 1.06x$; $D(x) = 0.70x$
a. $T(D(x)) = 0.742x$
b. $D(T(x)) = 0.742x$
c. No, same either way; multiplication is commutative.

64. CONSUMER ECONOMICS A store is offering a discount of 30% on a suit. There is a sales tax of 6%.

a. Using a composition of functions, represent the situation in which the discount is taken before the sales tax is applied.

b. Using a composition of functions, represent the situation in which the sales tax is applied before the discount is taken.

c. Compare the composite functions from parts **a** and **b.** Does one of them result in a lower final cost? Explain why or why not.

Look Back

Write an equation in slope-intercept form for each line described. ***(LESSON 1.2)***

65. contains $(3, -2)$ and has a slope of $-\frac{1}{2}$ $\quad y = -\frac{1}{2}x - \frac{1}{2}$

66. contains $(-1, 4)$ and $(-3, 8)$ $\quad y = -2x + 2$

Describe the correlation among data that have the given correlation coefficient. ***(LESSON 1.5)***

67. $r = 0.09$ almost none **68.** $r = -0.95$ strong negative **69.** $r = 0.52$ weak positive

State the domain and range of each function. ***(LESSON 2.3)***

70. $\{(1, 3), (2, 5), (3, 7)\}$
domain: {1, 2, 3}
range: {3, 5, 7}

71. $y = -4x + 2$
domain: all real
range: all real

Look Beyond

72. Given $f(x) = 2x + 5$ and $g(x) = \frac{x-5}{2}$, find $f \circ g$ and $g \circ f$. Compare your results.

SPACE SCIENCE One source of space debris is nonfunctional spacecraft components. At the end of 1993, scientists tracked 1550 nonfunctional components in orbit and projected that 42 such objects are added each year. Suppose that an average of 15 of these are brought back to Earth each year.

1. Let *t* represent time in years. Write the function *D* that models the total number of cataloged nonfunctional components in orbit at the end of the year if 42 such objects are added each year to the initial 1550 objects in orbit.
2. Write the function *R* that models the total number of components that are brought back to Earth each year.
3. Write a function relating time, *t*, in years and the total number of nonfunctional components in orbit as a difference of *D* and *R*.
4. Graph the function from Step 3 on a graphics calculator. Examine the table of values for the function. Are there constant first differences in the *x*-values and in the *y*-values?
5. Identify the type of function that you graphed in Step 4 and justify your answer.

Look Beyond

In Exercise 72, students investigate functions *f* and *g* for which $f \circ g = g \circ f$. In Lesson 2.5, students will see how this result leads to the conclusion that *f* and *g* are *inverse functions*. You may wish to use this exercise as a transition to that lesson.

ALTERNATIVE
Assessment

Portfolio Activity

The Portfolio Activity can be used as preparation for the Chapter Project or as a separate activity. In the Portfolio Activity on this page, students investigate an aspect of the space-debris investigation that can be modeled as a difference of two functions.

Answers to Portfolio Activities can be found in Additional Answers of the Teacher's Edition.

Student Technology Guide

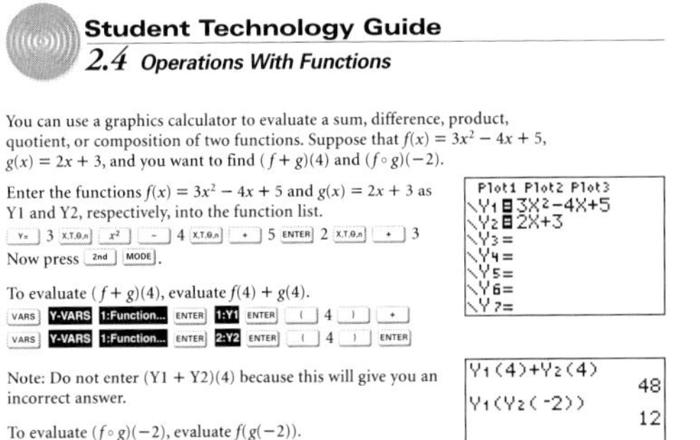

NAME ______________ CLASS ______________ DATE ______________

Student Technology Guide
2.4 Operations With Functions

You can use a graphics calculator to evaluate a sum, difference, product, quotient, or composition of two functions. Suppose that $f(x) = 3x^2 - 4x + 5$, $g(x) = 2x + 3$, and you want to find $(f + g)(4)$ and $(f \circ g)(-2)$.

Enter the functions $f(x) = 3x^2 - 4x + 5$ and $g(x) = 2x + 3$ as Y1 and Y2, respectively, into the function list.
Y= 3 X,T,θ,n x² - 4 X,T,θ,n + 5 ENTER 2 X,T,θ,n + 3
Now press 2nd MODE.

Plot1 Plot2 Plot3
\Y1 ■ 3X²-4X+5
\Y2 ■ 2X+3
\Y3 =
\Y4 =
\Y5 =
\Y6 =
\Y7 =

To evaluate $(f + g)(4)$, evaluate $f(4) + g(4)$.
VARS Y-VARS 1:Function... ENTER 1:Y1 ENTER (4) +
VARS Y-VARS 1:Function... ENTER 2:Y2 ENTER (4) ENTER

Note: Do not enter (Y1 + Y2)(4) because this will give you an incorrect answer.

To evaluate $(f \circ g)(-2)$, evaluate $f(g(-2))$.
VARS Y-VARS 1:Function... ENTER 1:Y1 ENTER (VARS Y-VARS 1:Function... ENTER 2:Y2 ENTER ((-) 2)) ENTER

Y1(4)+Y2(4) 48
Y1(Y2(-2)) 12

Let $f(x) = -2x + 5$ and $g(x) = -3x^2 + 4x + 3$. Use a calculator to evaluate each expression.

1. $(f + g)(5)$ **−57**
2. $(f \circ g)(3)$ **29**
3. $(g - f)(7.6)$ **−129.68**
4. $(f \cdot g)(3.5)$ **39.5**
5. $(g \circ f)(-4.1)$ **−466.92**
6. $\left(\frac{f}{g}\right)(-1)$ **−1.75**

You can use a graphics calculator to check your work when you perform operations on functions or find composite functions.

For example, suppose you find that $(f - g)(x) = 3x^2 - 6x + 2$. Choose several values of *x* and make sure that Y1(*x*) − Y2(*x*) and $3x^2 - 6x + 2$ give the same results for each value of *x*.

Y1(2)−Y2(2) 2
3(2)²−6(2)+2 2

Let $f(x) = 3x - 4$ and $g(x) = x^2 - 1$. Find each new function, and state any domain restrictions. Use a calculator to check your work.

7. $(f - g)(x)$ $\quad -x^2 + 3x - 3$
8. $\left(\frac{g}{f}\right)(x)$ $\quad \frac{x^2 - 1}{3x - 4}, x \neq \frac{4}{3}$
9. $(f \circ g)(x)$ $\quad 3x^2 - 7$

51. $(f \circ g)(x) = 2x^2 + 4$

52. $(g \circ f)(x) = 4x^2 + 2$

53. $(f \circ h)(x) = -8x + 6$

54. $(h \circ g)(x) = -4x^2 - 5$

55. $(f \circ f)(x) = 4x$

56. $(h \circ f)(x) = -8x + 3$

57. $(h \circ (h \circ g))(x) = 16x^2 + 23$

58. $((h \circ f) \circ g)(x) = -8x^2 - 13$

59. Answers may vary. Sample answer: Let $f(x) = x - 9$ and $g(x) = x^2$. Then $f \circ g = h$.

72. $(f + g)(x) = x$; $(g + f)(x) = x$; the functions *f* and *g* "undo" each other in the sense that whatever *g* does to *x*, *f* undoes it and returns *x*, and vice versa.

Prepare

NCTM PRINCIPLES & STANDARDS 1–3, 6–10

QUICK WARM-UP

Solve each equation for y.

1. $x = -4y$ $\quad y = -0.25x$
2. $x = 2y + 3$ $\quad y = 0.5x - 1.5$
3. $x = \frac{y+3}{5}$ $\quad y = 5x - 3$
4. $x = -\frac{1}{3}(y + 1)$ $\quad y = -3x - 1$
5. Let $f(x) = 2x - 4$ and let $g(x) = 0.5x + 2$. Find $(f \circ g)(x)$ and $(g \circ f)(x)$. $(f \circ g)(x) = (g \circ f)(x) = x$

Also on Quiz Transparency 2.5

Teach

Why You can use the function $f(x) = 12x$ to convert a number of feet to a number of inches. Someone else can use the function $g(x) = \frac{x}{12}$ to convert that number of inches back to a number of feet. In effect, one function "undoes" the other. Pairs of functions like these are called *inverse functions*. Ask students to describe other situations that illustrate inverse functions.

2.5 Inverses of Functions

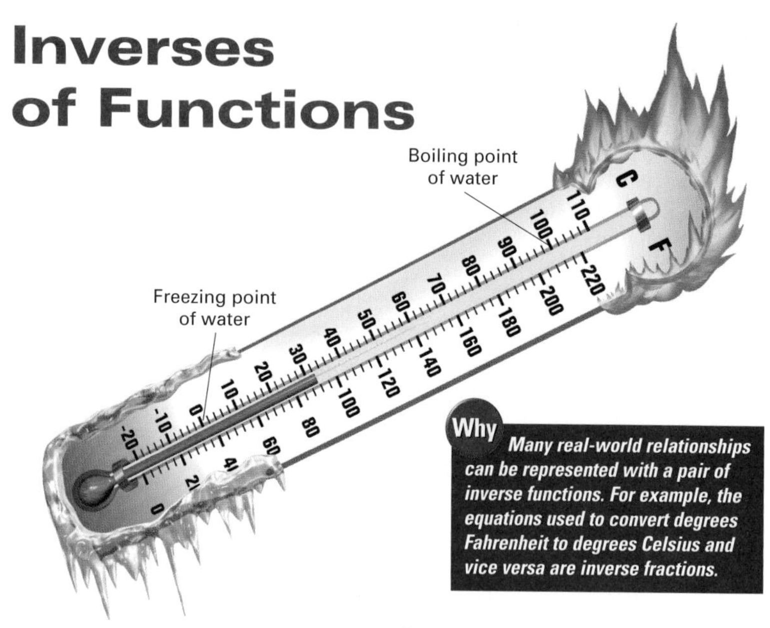

Why *Many real-world relationships can be represented with a pair of inverse functions. For example, the equations used to convert degrees Fahrenheit to degrees Celsius and vice versa are inverse fractions.*

Objectives

- Find the inverse of a relation or function.
- Determine whether the inverse of a function is a function.

APPLICATION
TEMPERATURE

In Lesson 1.6, you used the equation $F = \frac{9}{5}C + 32$ to find the Celsius temperature corresponding to 86°F. How can you find the Celsius temperatures that correspond to Fahrenheit temperatures? To answer this question, you need a function that gives C in terms of F. You will need to find the *inverse of a function*.

EXAMPLE 1 **Solve $F = \frac{9}{5}C + 32$ for C.**

SOLUTION

$$F = \frac{9}{5}C + 32$$
$$F - 32 = \frac{9}{5}C + 32 - 32$$
$$\frac{5}{9}(F - 32) = \frac{5}{9}\left(\frac{9}{5}C\right)$$
$$\frac{5}{9}(F - 32) = C$$

Thus, $C = \frac{5}{9}(F - 32)$.

Notice that subtraction of 32 and multiplication by $\frac{5}{9}$ are the inverse operations of those in $F = \frac{9}{5}C + 32$.

Example 1 indicates that finding the inverse of a function involves changing an ordered pair of the form (C, F) to an ordered pair of the form (F, C).

Inverse of a Relation

The **inverse of a relation** consisting of the ordered pairs (x, y) is the set of all ordered pairs (y, x).
The domain of the inverse is the range of the original relation.
The range of the inverse is the domain of the original relation.

Alternative Teaching Strategy

COOPERATIVE LEARNING Have students work in groups of two or three. Each student should draw a set of coordinate axes. On these axes, the student should graph the functions defined by $f(x) = x + 4$ and $g(x) = x - 4$. Now pose this question: How can you make one fold so that the graph of f lines up exactly with the graph of g? Each group should work together to answer this question. Be sure they all arrive at the correct conclusion: The equation of the desired "fold line" is $y = x$.

Repeat the activity for $f(x) = 2x$ and $g(x) = 0.5x$ and then again for $f(x) = 2x + 4$ and $g(x) = 0.5x - 2$.

Now have each group write equations for two different linear functions, f and g, whose graphs line up exactly when folded along the line $y = x$. Ask each group to share their equations with the class, and record them on the board or overhead. Use the pairs of equations to initiate a discussion of inverse functions and the relationship between their equations.

EXAMPLE 2 **Find the inverse of each relation. State whether the relation is a function. State whether the inverse is a function.**

a. $\{(1, 2), (2, 4), (3, 6), (4, 8)\}$ **b.** $\{(1, 5), (1, 6), (3, 6), (4, 9)\}$

SOLUTION

a. relation: $\{(1, 2), (2, 4), (3, 6), (4, 8)\}$
inverse: $\{(2, 1), (4, 2), (6, 3), (8, 4)\}$

The given relation is a function because each domain value is paired with exactly one range value. The inverse is also a function because each domain value is paired with exactly one range value.

b. relation: $\{(1, 5), (1, 6), (3, 6), (4, 9)\}$
inverse: $\{(6, 1), (5, 1), (6, 3), (9, 4)\}$

The given relation is not a function because the domain value 1 is paired with two range values, 5 and 6. The inverse is not a function because the domain value 6 is paired with two range values, 1 and 3.

Example 3 shows you how to find the inverse of a function by interchanging x and y and then solving for y.

EXAMPLE 3 **Find an equation for the inverse of $y = 3x - 2$.**

SOLUTION

In $y = 3x - 2$, interchange x and y. Then solve for y.

$$x = 3y - 2$$
$$x + 2 = 3y$$
$$\frac{x + 2}{3} = y$$
$$y = \frac{1}{3}x + \frac{2}{3}$$

TRY THIS Find an equation for the inverse of $y = 4x - 5$.

In the Activity below, you can explore the relationship between the graph of a function and the graph of its inverse.

Activity
Exploring Functions and Their Inverses

TECHNOLOGY GRAPHICS CALCULATOR
Keystroke Guide, page 151

You will need: a graphics calculator

1. Graph $y = 2x - 1$, its inverse, and $y = x$ in a square viewing window. Use the inverse feature of the calculator. How do the graphs of these functions relate to one another? Consider symmetry in your response.
2. Repeat Step 1 for each function listed at right.

CHECKPOINT ✔ 3. Write a generalization about the relationship between the graph of a function and the graph of its inverse.

Function
$y = 3x - 2$
$y = 3x + 2$
$y = -2x + 5$
$y = x^2$

Interdisciplinary Connection

BUSINESS To find the number of years it takes to double the amount of an investment, you can divide 72 by 100 times the annual interest rate. Financial analysts call this the *rule of 72*.

a. Write an equation for a function that gives the number of years, n, needed to double an investment at an annual interest rate of r.
$n = \frac{72}{100r}$

b. Write an equation for the inverse of this function. Describe what the inverse function represents. **Inverse: $r = \frac{72}{100n}$; the inverse gives the annual interest rate, r, needed to double an investment in n years.**

ADDITIONAL **EXAMPLE 1**

An object moves in a straight line at a constant speed of 50 meters per second. Then the object begins to accelerate at a constant rate of 3 meters per second squared. The speed, v, is related to the number of seconds of acceleration, t, according to the equation $v = 50 + 3t$. **Solve this equation for t.**

$t = \frac{1}{3}v - \frac{50}{3}$

ADDITIONAL **EXAMPLE 2**

Find the inverse of each relation. State whether the relation is a function. State whether the inverse is a function.

a. $\{(1, 2), (4, -2), (3, 2)\}$
Inverse: $\{(2, 1), (-2, 4), (2, 3)\}$; the relation is a function; the inverse is not a function because one domain value, 2, is paired with two range values, 1 and 3.

b. $\{(-2, 4), (3, -4), (-8, -5)\}$
Inverse: $\{(4, -2), (-4, 3), (-5, -8)\}$; both the relation and its inverse are functions.

ADDITIONAL **EXAMPLE 3**

Find an equation for the inverse of $y = \frac{2x + 3}{5}$.

$y = \frac{5}{2}x - \frac{3}{2}$

TRY THIS

$y = \frac{1}{4}x + \frac{5}{4}$

Use Teaching Transparency 9.

☞ For Activity Notes and the answer to the Checkpoint, see page 120.

Activity Notes

In this Activity, students graph four different functions on a graphics calculator and then use the calculator's "draw inverse" feature to graph the inverse of each function. They should observe that the graphs of a function and its inverse are reflections of each other across the line $y = x$.

Teaching Tip

TECHNOLOGY To get a more accurate picture of the symmetry of the graphs in the Activity on a TI-83 or TI-82 graphics calculator, use a square viewing window so that the line $y = x$ forms a 45° angle with each axis. You can do this by pressing ZOOM and selecting **4:ZSquare** (or **5:ZSquare**).

CHECKPOINT ✔

3. **The graph of the inverse of a function is the reflection of the graph of the function across the line $y = x$.**

Use Teaching Transparency 10.

Teaching Tip

When dealing with a function and its inverse, encourage students to write the function as $f(x)$ and its inverse as $f^{-1}(x)$ instead of letting y represent both.

If a function f and its inverse are both functions, the inverse of f is denoted by f^{-1}.

The graph of $f(x) = 3x - 2$ and its inverse, $f^{-1}(x) = \frac{1}{3}x + \frac{2}{3}$, from Example 3 are shown at right. Notice that the ordered pairs $(-2, 0)$ and $(0, -2)$ are "mirror images" or *reflections* of one another across the line $y = x$. This means every point (a, b) on the graph of f corresponds to a point (b, a) on the graph of f^{-1}. This is true for any two relations that are inverses of each other.

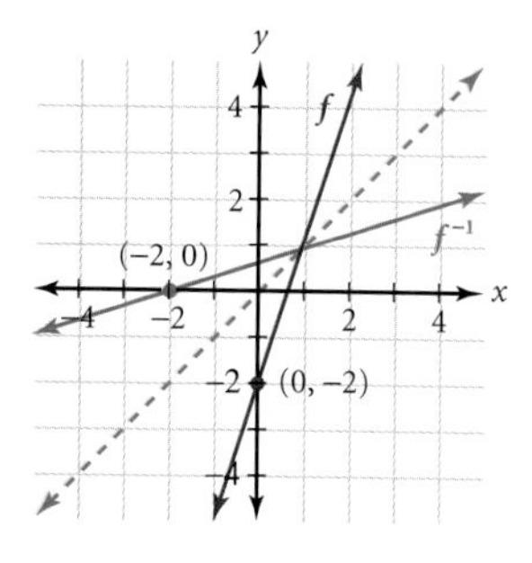

You can use the *horizontal-line test* to determine from a graph whether the inverse of a given function is a function.

Horizontal-Line Test

The inverse of a function is a function if and only if every horizontal line intersects the graph of the given function at no more than one point.

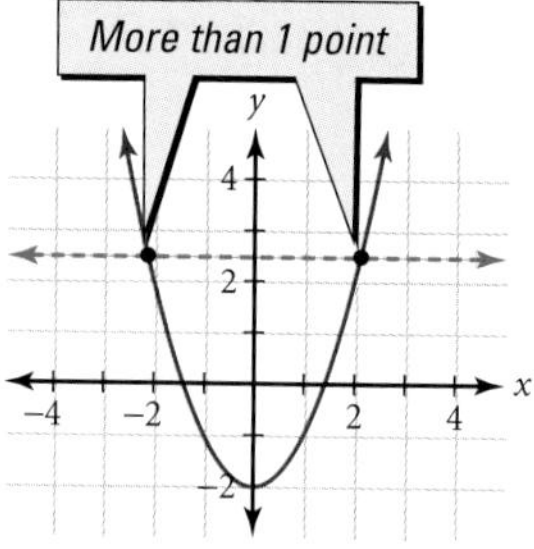

The horizontal-line test above shows that the inverse of this function is not a function.

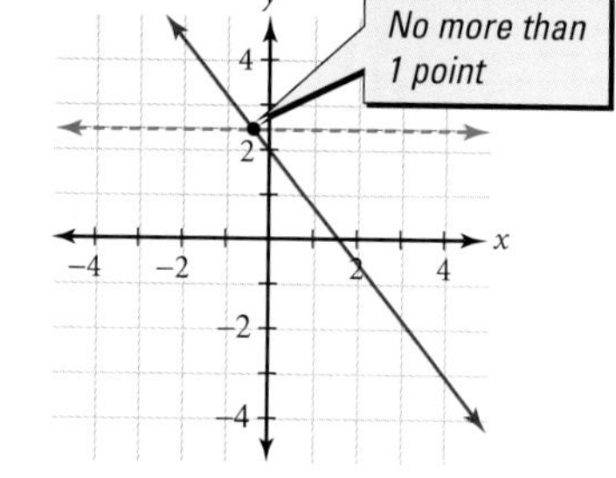

The horizontal-line test above shows that the inverse of this function is a function.

If a function has an inverse that is also a function, then the function is **one-to-one**. Every one-to-one function passes the horizontal-line test and has an inverse that is a function.

Just as the graphs of f and f^{-1} are reflections of one another across the line $y = x$, the composition of a function and its inverse are related to the *identity function*. The **identity function**, I, is defined as $I(x) = x$.

Composition and Inverses

If f and g are functions and $(f \circ g)(x) = (g \circ f)(x) = I(x) = x$, then f and g are inverses of one another.

Enrichment

1. Write an equation for the inverse of the linear parent function $f(x) = mx + b$.
   $f^{-1}(x) = \frac{x}{m} - \frac{b}{m}$**, where $m \neq 0$**
2. Is the inverse of every linear function also a linear function? Explain. **No; the inverse of a constant linear function, $f(x) = b$, is not a function.**
3. Is there any linear function that is its own inverse? Explain. **Yes; $f(x) = x$ is its own inverse, as is any function of the form $f(x) = -x + b$.**

Inclusion Strategies

TACTILE LEARNERS Give students a page that contains several graphs of linear functions. Have students place the edge of a transparent reflection tool, such as a geomirror, along the line $y = x$ and trace the reflection of each graph. Have them write an equation for each graph and for its reflected image. Then have students use the method described in Example 4 to verify that each graph and its reflected image are the graphs of inverse functions.

EXAMPLE 4 Show that $f(x) = 7x - 2$ and $g(x) = \frac{1}{7}x + \frac{2}{7}$ are inverses of each other.

SOLUTION

$$(f \circ g)(x) = f(g(x)) = f\left(\frac{1}{7}x + \frac{2}{7}\right) = 7\left(\frac{1}{7}x + \frac{2}{7}\right) - 2 = x + 2 - 2 = x$$

$$(g \circ f)(x) = g(f(x)) = g(7x - 2) = \frac{1}{7}(7x - 2) + \frac{2}{7} = x - \frac{2}{7} + \frac{2}{7} = x$$

CHECK

Graph $f \circ g$ and $g \circ f$ to check that the two functions are inverses.

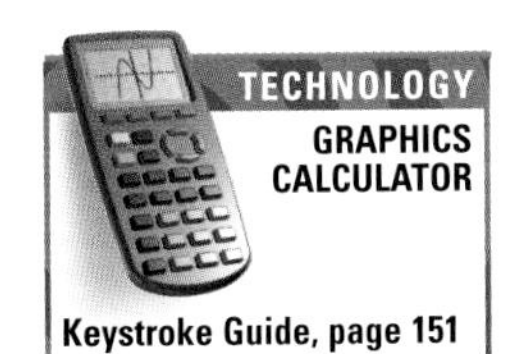

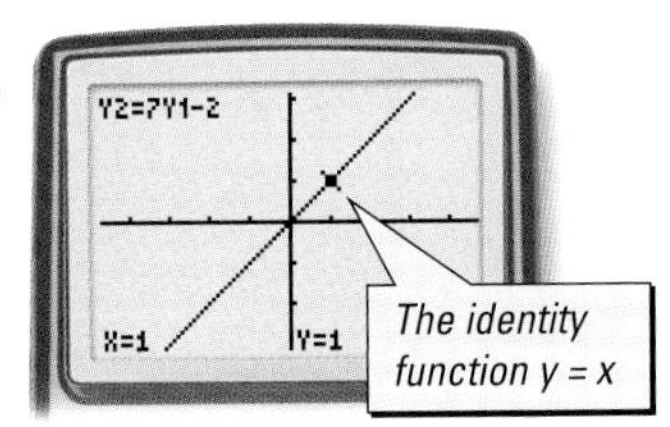

Because $(f \circ g)(x) = x$ and $(g \circ f)(x) = x$, the two functions are inverses of each other.

TRY THIS Show that $f(x) = -5x + 7$ and $g(x) = -\frac{1}{5}x + \frac{7}{5}$ are inverses of each another.

CRITICAL THINKING Let $f(x) = mx + b$, where $m \neq 0$. Find f^{-1}. Using composition, verify that f and f^{-1} are inverses of one another.

Exercises

Communicate

1. Explain what it means for a function to be a one-to-one function.
2. Describe when and why you would use the vertical-line test and the horizontal-line test.
3. Describe the procedure for finding the inverse of $y = 4x - 1$.
4. Explain how the graphs of a function and its inverse are related.

Guided Skills Practice

APPLICATION

5. **CHEMISTRY** A chemical reaction takes place at temperatures between 290 K and 300 K. The Fahrenheit and Kelvin temperature scales are related by the formula $K = \frac{5}{9}(F - 32) + 273$. Solve this equation for F. *(EXAMPLE 1)* $F = \frac{9}{5}(K - 273) + 32$

Find the inverse of each relation. State whether the relation is a function. State whether the inverse is a function. *(EXAMPLE 2)*

6. $\{(8, 3), (2, 2), (4, 3)\}$
7. $\{(3, 2), (9, 5), (2, 3), (4, 7)\}$

ADDITIONAL EXAMPLE 4

Show that $f(x) = 4x - 3$ and $g(x) = \frac{1}{4}x + \frac{3}{4}$ are inverses of each other.

$$(f \circ g)(x) = 4\left(\frac{1}{4}x + \frac{3}{4}\right) - 3 = (x + 3) - 3 = x$$

$$(g \circ f)(x) = \frac{1}{4}(4x - 3) + \frac{3}{4} = x - \frac{3}{4} + \frac{3}{4} = x$$

Since $(f \circ g)(x) = (g \circ f)(x) = x$, the two functions are inverses of each other.

TRY THIS

$$(f \circ g)(x) = f\left(-\frac{1}{5}x + \frac{7}{5}\right) = -5\left(-\frac{1}{5}x + \frac{7}{5}\right) + 7 = x - 7 + 7 = x$$

$$(g \circ f)(x) = g(-5x + 7) = -\frac{1}{5}(-5x + 7) + \frac{7}{5} = x - \frac{7}{5} + \frac{7}{5} = x$$

CRITICAL THINKING

$$f^{-1}(x) = \frac{1}{m}x - \frac{b}{m}$$

$$(f^{-1} \circ f)(x) = f^{-1}(mx + b) = \frac{1}{m}(mx + b) - \frac{b}{m} = x + \frac{b}{m} - \frac{b}{m} = x$$

$$(f \circ f^{-1})(x) = f\left(\frac{1}{m}x + \frac{b}{m}\right) = m\left(\frac{1}{m}x + \frac{b}{m}\right) + b = x - b + b = x$$

Reteaching the Lesson

USING VERBAL STRATEGIES Show students how to use *inverse operations* in reverse order to find an inverse function. Demonstrate this method with $f(x) = 7x - 2$.

		given function
Start with x.	x	
Multiply by 7.	$7x$	
Subtract 2.	$7x - 2$	$\rightarrow f(x) = 7x - 2$

		inverse function
Start with x.	x	
Add 2.	$x + 2$	
Divide by 7.	$\frac{x+2}{7}$	$\rightarrow f^{-1}(x) = \frac{1}{7}x + \frac{2}{7}$

6. Inverse: $\{(3, 8), (2, 2), (3, 4)\}$; relation is a function; inverse is not a function.
7. Inverse: $\{(2, 3), (5, 9), (3, 2)$ $(7, 4)\}$; relation is a function; inverse is a function.

Assess

Selected Answers

Exercises 5–10, 11–61 odd

ASSIGNMENT GUIDE	
In Class	1–10
Core	11–25 odd, 27, 28, 29–39 odd, 41–49, 51, 53
Core Plus	12–26 even, 27, 28, 30–40 even, 41–53
Review	54–61
Preview	62

Extra Practice can be found beginning on page 940.

Error Analysis

When a set of ordered pairs represents a function, some students automatically assume that its inverse also represents a function. Suggest that they graph the points represented by the ordered pairs for the inverse and apply the vertical-line test to the graph.

Find an equation for the inverse of each function. ***(EXAMPLE 3)***

8. $y = 3x + 9$ $\quad y = \frac{1}{3}x - 3$

9. $y = 5 - 3x$ $\quad y = -\frac{1}{3}x + \frac{5}{3}$

10. Verify that $f(x) = 6x - 5$ and $g(x) = \frac{1}{6}x + \frac{5}{6}$ are inverses of each other. ***(EXAMPLE 4)*** $(f \circ g)(x) = x$ and $(g \circ f)(x) = x$

Practice and Apply

internet connect

Homework Help Online
Go To: **go.hrw.com**
Keyword:
MB1 Homework Help
for Exercises 11–26

Find the inverse of each relation. State whether the relation is a function. State whether the inverse is a function.

11. $\{(3, 5), (6, 10), (9, 15)\}$

12. $\{(2, -3), (3, -4), (4, -2)\}$

13. $\{(5, 2), (4, 3), (3, 4), (2, 5)\}$

14. $\{(-1, -6), (0, 2), (1, 2), (3, 6)\}$

15. $\{(-3, -6), (-1, 2), (1, 2), (3, 6)\}$

16. $\{(2, 1), (4, 2), (2, 3), (8, 4)\}$

17. $\{(1, 2), (3, 4), (-3, 4), (-1, 2)\}$

18. $\{(9, -2), (4, -1), (1, 0), (3, 1), (7, 2)\}$

11. yes; yes
12. yes; yes
13. yes; yes
14. yes; no
15. yes; no
16. no; yes
17. yes; no
18. yes; yes

Find the inverse of each function. State whether the inverse is a function.

19. $\{(-1, 0), (-2, 1), (4, 3), (3, 4)\}$ **yes**

20. $\{(1, 4), (2, 3), (3, 2), (4, 1)\}$ **yes**

21. $\{(1, 2), (2, 3), (3, 2), (4, 1)\}$ **no**

22. $\{(3, -2), (2, -3), (1, -2), (0, -1)\}$ **no**

23. $\{(5, 2), (4, 3), (3, 5), (2, 3)\}$ **no**

24. $\{(3, 0), (2, -1), (1, 2), (0, 1), (-1, 2)\}$ **no**

25. $\{(0, 2), (2, 3), (3, 4), (1, 1)\}$ **yes**

26. $\{(-1, 2), (-2, 3), (-3, 4), (0, 0)\}$ **yes**

Determine whether the inverse of each function graphed is also a function.

27. **yes**

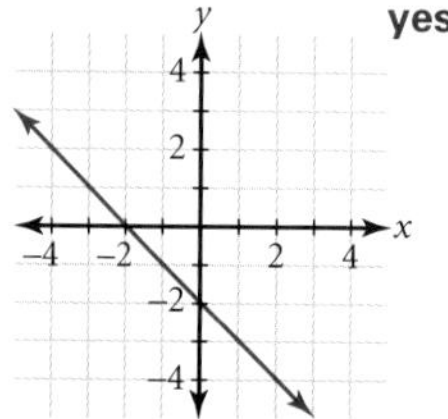

28. **no**

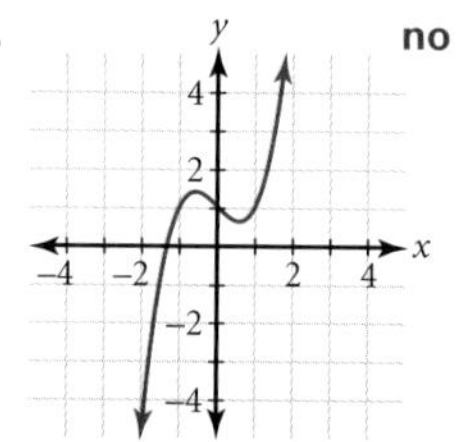

For each function, find an equation for the inverse. Then use composition to verify that the equation you wrote is the inverse.

29. $f(x) = 5x + 1$

30. $g(x) = -2x - 7$

31. $h(x) = -\frac{1}{2}x + 3$

32. $g(x) = \frac{x - 1}{4}$

33. $h(x) = \frac{x + 8}{3}$

34. $f(x) = \frac{x - 3}{2}$

35. $f(x) = \frac{2x - 3}{4}$

36. $f(x) = \frac{1}{3}x - 1$

37. $g(x) = 2x - \frac{3x}{4}$

38. $h(x) = \frac{1}{4}(x - 1)$

39. $g(x) = \frac{1}{2}(x + 2) - 3$

40. $h(x) = \frac{3}{2}(x - 3) + 2$

Graph each function, and use the horizontal-line test to determine whether the inverse is a function.

41 $f(x) = 1 - x^2$ **no**

42 $h(x) = -\frac{1}{3}x + 5$ **yes**

43 $g(x) = \frac{7 - 2x}{5}$ **yes**

44 $f(x) = 2x^3$ **yes**

45 $g(x) = 3$ **no**

46 $h(x) = \frac{3}{x}$ **yes**

47 $g(x) = x^4$ **no**

48 $f(x) = x^2 - 2x$ **no**

49 $f(x) = x^5$ **yes**

Practice

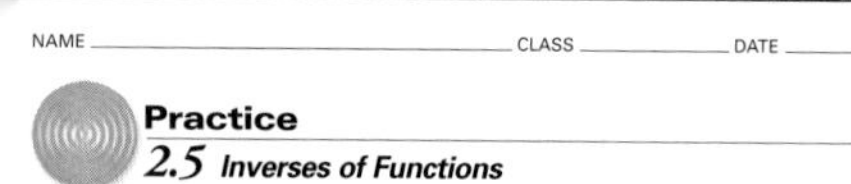

NAME ______ CLASS ______ DATE ______

Practice
2.5 ***Inverses of Functions***

Find the inverse of each relation. State whether the relation is a function and whether its inverse is a function.

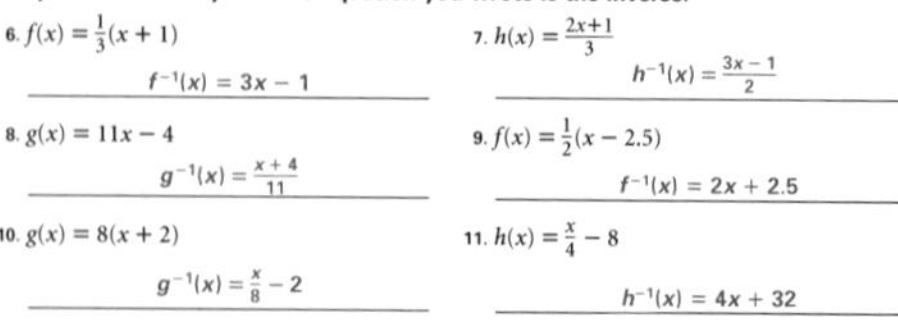

1. $\{(-1, -16), (0, -6), (2, 14)\}$ $\{(-16, -1), (-6, 0), (14, 2)\}$; yes; yes
2. $\{(7, 2), (6, 3), (7, 4), (6, 5)\}$ $\{(2, 7), (3, 6), (4, 7), (5, 6)\}$; no; yes
3. $\{(-2, 16), (-1, 1), (1, 1), (2, 16)\}$ $\{(16, -2), (1, -1), (1, 1), (16, 2)\}$; yes; no
4. $\{(-5, 7), (-3, 7), (-1, 7), (1, 7)\}$ $\{(7, -5), (7, -3), (7, -1), (7, 1)\}$; yes; no
5. $\{(-5, 4), (-3, 9), (1, 12), (7, 13)\}$ $\{(4, -5), (9, -3), (12, 1), (13, 7)\}$; yes; yes

For each function, find the equation of its inverse. Then use composition to verify that the equation you wrote is the inverse.

6. $f(x) = \frac{1}{3}(x + 1)$ $\quad f^{-1}(x) = 3x - 1$
7. $h(x) = \frac{2x+1}{3}$ $\quad h^{-1}(x) = \frac{3x - 1}{2}$
8. $g(x) = 11x - 4$ $\quad g^{-1}(x) = \frac{x + 4}{11}$
9. $f(x) = \frac{1}{2}(x - 2.5)$ $\quad f^{-1}(x) = 2x + 2.5$
10. $g(x) = 8(x + 2)$ $\quad g^{-1}(x) = \frac{x}{8} - 2$
11. $h(x) = \frac{x}{4} - 8$ $\quad h^{-1}(x) = 4x + 32$

Graph each function, and use the horizontal-line test to determine whether the inverse is a function.

12. $f(x) = x^2 + x$
13. $g(x) = x^3 + 1$
14. $h(x) = x^3 + x$

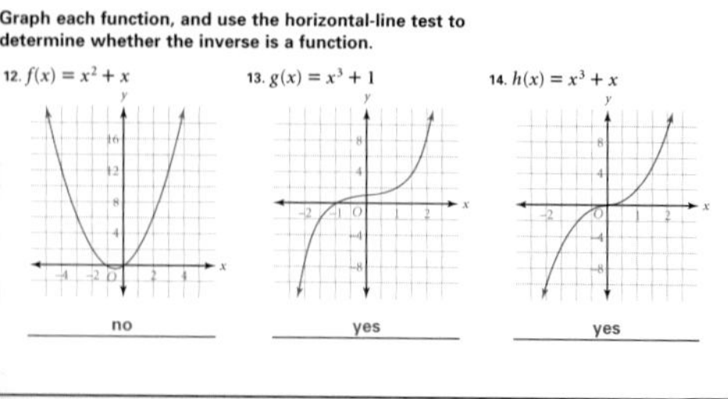

11. $\{(5, 3), (10, 6), (15, 9)\}$; yes; yes

12. $\{(-3, 2), (-4, 3), (-2, 4)\}$; yes; yes

13. $\{(2, 5), (3, 4), (4, 3), (5, 2)\}$; yes; yes

14. $\{(-6, -1), (2, 0), (2, 1), (6, 3)\}$; yes; no

15. $\{(-6, -3), (2, -1), (2, 1), (6, 3)\}$; yes; no

16. $\{(1, 2), (2, 4), (3, 2), (4, 8)\}$; no; yes

17. $\{(2, 1), (4, 3), (4, -3), (2, -1)\}$; yes; no

18. $\{(-2, 9), (-1, 4), (0, 1), (1, 3), (2, 7)\}$; yes; yes

19. $\{(0, -1), (1, 2), (3, 4), (4, 3)\}$; yes

20. $\{(4, 1), (3, 2), (2, 3), (1, 4)\}$; yes

21. $\{(2, 1), (3, 2), (2, 3), (1, 4)\}$; no

22. $\{(-2, 3), (-3, 2), (-2, 1), (-1, 0)\}$; no

23. $\{(2, 5), (3, 4), (5, 3), (3, 2)\}$; no

24. $\{(0, 3), (-1, 2), (2, 1), (1, 0), (2, -1)\}$; no

25. $\{(2, 0), (3, 2), (4, 3), (1, 1)\}$; yes

26. $\{(2, -1), (3, -2), (4, -3), (0, 0)\}$; yes

CHALLENGE

50. If a relation is not a function, can its inverse be a function? Explain and give examples.

APPLICATIONS

51. CONSUMER ECONOMICS New carpeting can be purchased and installed for \$17.50 per square yard plus a \$50 delivery fee.

a. Write an equation that gives the cost, c, of carpeting s square yards of a house. $c = 17.50s + 50$

b. Find the inverse of the cost function. $s = \frac{c - 50}{17.50}$

c. How many square yards can be carpeted for \$1485? **82 yd²**

52. REAL ESTATE New house prices depend on the cost of the property, or lot, and the size of the house. Suppose that a lot costs \$60,000 and a builder charges \$84 per square foot of living area in a house.

a. Write the function p that represents the price of a house with x square feet of living area.

b. Find the inverse of the price function, and discuss what it represents.

c. How big, to the nearest square foot, is a house that can be purchased for \$180,000?

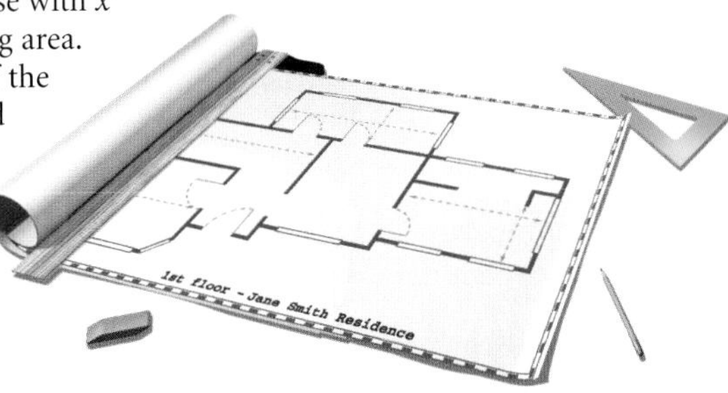

52a. $p = 84x + 60{,}000$

b. $x = \frac{p - 60{,}000}{84}$; x is number of ft², and p is number of dollars.

c. 1429 ft²

53. PUZZLES In a number puzzle, you are told to add 4 to your age, then multiply by 2, subtract 6, and finally divide by 2. You give the result, and you are immediately told your age. Use an inverse function to explain how the puzzle works.

Look Back

APPLICATION

54. BUSINESS The total revenue, R, is directly proportional to the number of cameras, x, sold. When 500 cameras are sold, the revenue is \$3800. *(LESSON 1.4)*

a. Find the revenue when 600 cameras are sold. **\$4560**

a. What is the constant of variation? $k = 7.6$

c. In this situation, what does the constant of variation represent? **cost per camera**

Evaluate each expression. *(LESSON 2.2)*

55. $2^3 \cdot 2^5$ **256**   **56.** $25^{-\frac{1}{2}}$ $\frac{1}{5}$ **57.** $(3^2)^3$ **729**

Let $f(x) = 3x - 2$ and $g(x) = 5x + 2$. *(LESSON 2.4)*

58. Find $f + g$. $(f + g)(x) = 8x$

59. Find $f \cdot g$. $(f \cdot g)(x) = 15x^2 - 4x - 4$

60. Evaluate $(f + g)(-1)$. **−8**

61. Evaluate $(f \cdot g)(-3)$. **143**

Look Beyond

62 Graph $f(x) = 2^x$. Why would you expect this function to have an inverse that is a function? Graph the inverse and state its domain.

Look Beyond

In Exercise 62, students analyze the graphs of an exponential function and its inverse, which is a logarithmic function. Students will study exponential and logarithmic functions in detail in Chapter 6.

29. $f^{-1}(x) = \frac{1}{5}x - \frac{1}{5}$

30. $g^{-1}(x) = -\frac{1}{2}x - \frac{7}{2}$

31. $h^{-1}(x) = -2x + 6$

32. $g^{-1}(x) = 4x + 1$

33. $h^{-1}(x) = 3x - 8$

34. $f^{-1}(x) = 2x + 3$

35. $f^{-1}(x) = \frac{4x + 3}{2}$

36. $f^{-1}(x) = 3x + 3$

37. $g^{-1}(x) = \frac{4}{5}x$

38. $h^{-1}(x) = 4x + 1$

39. $g^{-1}(x) = 2(x + 3) - 2$, or $g^{-1}(x) = 2x + 4$

40. $h^{-1}(x) = \frac{2}{3}(x - 2) + 3$, or $h^{-1}(x) = \frac{2}{3}x + \frac{5}{3}$

Student Technology Guide

NAME ______ CLASS ______ DATE ______

Student Technology Guide
2.5 *Inverses of Functions*

To find the inverse of a function, interchange x and y, and solve for y. You can use a graphics calculator to verify your answer.

Suppose that you find that $y = \frac{x+2}{7}$ is the inverse of $y = 7x - 2$.

To verify your work:

- Enter $7x - 2$ as Y1, $\frac{x+2}{7}$ as Y2, and x as Y3.
- Choose a square viewing window such as $-9 \le x \le 9$ and $-6 \le y \le 6$. You can make a square window by choosing a vertical range that is two-thirds that of the window's horizontal range or by pressing ZOOM ZSquare ENTER. Press GRAPH.

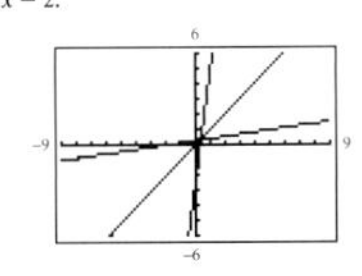

From the display at right above, you can see that the graphs of the given function and its inverse are reflections of one another across the line $y = x$.

Find the inverse of each function. Use a graphics calculator to verify your algebraic work.

1. $f(x) = 0.5x - 3.57$ $y = 2x + 7.14$
2. $f(x) = 6x - 2$ $y = \frac{1}{6}x + \frac{1}{3}$
3. $f(x) = 5x + 1$ $y = \frac{1}{5}x - \frac{1}{5}$
4. $f(x) = \frac{1}{3}x - 2$ $y = 3x - 6$
5. $f(x) = 4x - 0.5$ $y = 0.25x + 0.125$
6. $f(x) = 0.5x + 4$ $y = 2x - 8$

Another way to check your work is to write the composition of the two given functions. The graph of the composition should be the same as the graph of $y = x$. Enter the given functions as Y1 and Y2. Then press Y= Y3= VARS Y-VARS 1:Function... ENTER 1:Y1 ENTER (VARS Y-VARS 1:Function... ENTER 2:Y2 ENTER (X,T,θ,n)) GRAPH.

Plot1 Plot2 Plot3
\Y1=7X−2
\Y2=(X+2)/7
\Y3=Y1(Y2(X))
\Y4=
\Y5=
\Y6=
\Y7=

7. Find the inverse of $f(x) = -3x + 1$. Use the key sequence above to verify your algebraic work. $y = -\frac{1}{3}x + \frac{1}{3}$

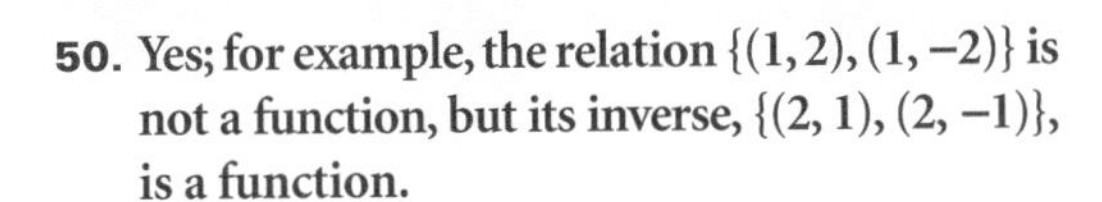

50. Yes; for example, the relation $\{(1, 2), (1, -2)\}$ is not a function, but its inverse, $\{(2, 1), (2, -1)\}$, is a function.

53. Let a be your age and g be the guess. Then the guess, g, can be represented by the function $g = \frac{2(a + 4) - 6}{2}$.

Now solve for a:

$$g = \frac{2(a + 4) - 6}{2}$$
$$2g = 2(a + 4) - 6$$
$$2g + 6 = 2(a + 4)$$
$$g + 3 = a + 4$$
$$g - 1 = a$$

In other words, your age is found by subtracting 1 from the guess.

62. The function has an inverse because it passes the horizontal-line test; domain of f^{-1}: $x > 0$.

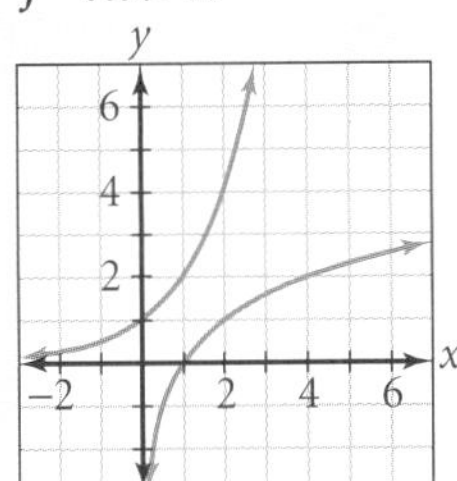

Prepare

NCTM PRINCIPLES & STANDARDS 1, 2, 5–10

QUICK WARM-UP

Find the greatest integer that is *less than or equal to* the given number.

1. 3.97   3   **2.** −3.97   −4

3. 0.0051   0   **4.** −7   −7

Evaluate.

5. |25.1| 25.1   **6.** |−0.7| 0.7

Also on Quiz Transparency 2.6

Teach

Why Ask students to imagine that a telephone company's charge for a long-distance call is 15¢ per minute or any fraction of a minute. This means that if you talk for $4\frac{1}{2}$ minutes, you are charged for 5 minutes. Tell them that the function relating the cost of a call, y, to the number of minutes, x, is called a *step function*. Elicit their conjectures about the reason that the function is given this name.

Special Functions

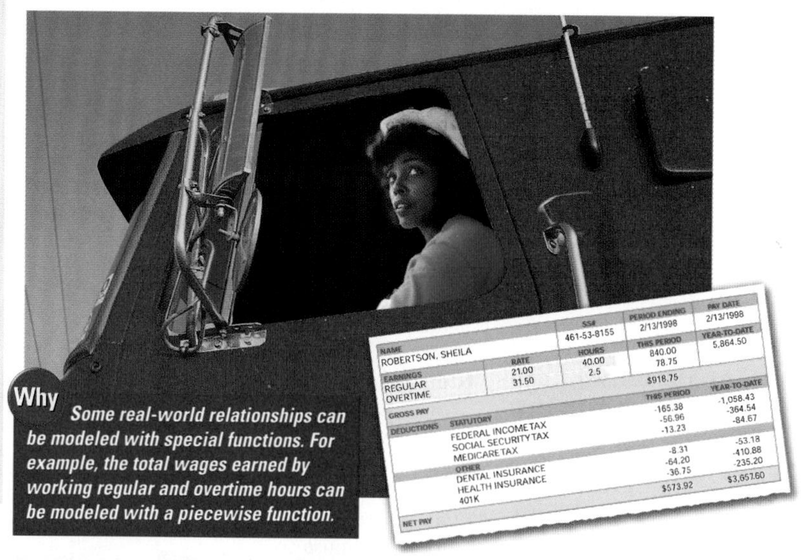

Objective

- Write, graph, and apply special functions: piecewise, step, and absolute value.

Why *Some real-world relationships can be modeled with special functions. For example, the total wages earned by working regular and overtime hours can be modeled with a piecewise function.*

Piecewise Functions

APPLICATION
INCOME

A truck driver earns $21.00 per hour for the first 40 hours worked in one week. The driver earns time-and-a-half, or $31.50, for each hour worked in excess of 40. The pair of function rules below represent the driver's wage, $w(h)$, as a function of the hours worked in one week, h.

$$w(h) = \begin{cases} 21h & \text{if } 0 < h \leq 40 \\ 31.5h - 420 & \text{if } h > 40 \end{cases}$$

This function is an example of a *piecewise function*. A **piecewise function** consists of different function rules for different parts of the domain.

CRITICAL THINKING Where does the number −420 in the piecewise function w come from?

Activity: Exploring Piecewise Functions

You will need: graph paper

1. Copy and complete the table below, using the function for the truck driver's wages given above.

Hours worked, h	10.0	30.0	35.0	40.0	52.5
Wage, $w(h)$					

2. Extend your table by choosing other values of h in the interval $0 < h \leq 60$.
3. Plot the ordered pairs from your table and make a scatter plot.

CHECKPOINT ✔

4. If you connect the consecutive points in the scatter plot, what observations can you make about the graph?

Alternative Teaching Strategy

USING TECHNOLOGY Have students analyze the **int** function of their graphics calculators. Use their observations to initiate a discussion about the rounding-down and rounding-up functions.

If students are using TI-82 or TI-83 graphics calculators, tell them to press [Y=] and clear any existing equations. They should then place the cursor next to **Y1** and enter the following:

TI-83: [MATH] [►] **NUM** **5:int(** [X,T,θ,n] [)]

TI-82: [MATH] [►] **NUM** **4:int** [(] [X,T,θ] [)]

To see the distinct "steps" of the graph, students must have their calculators in **Dot** mode. Have them press [MODE], use the arrow keys to highlight **Dot**, and press [ENTER]. To view the graph, they should press [ZOOM] and select **4:ZDecimal**.

To have the cursor trace the graph, instruct students to press [TRACE] and use the arrow keys. Tell them to observe the display at the bottom of the screen as they do this. Now elicit their ideas about the meaning of the **int** function.

Example 1 shows you how to use the appropriate function rule for each domain of a piecewise function when graphing it.

EXAMPLE 1 Graph: $f(x) = \begin{cases} 2x & \text{if } 0 \le x < 2 \\ 4 & \text{if } 2 < x < 4 \\ -\frac{1}{4}x + 5 & \text{if } 4 \le x \le 6 \end{cases}$

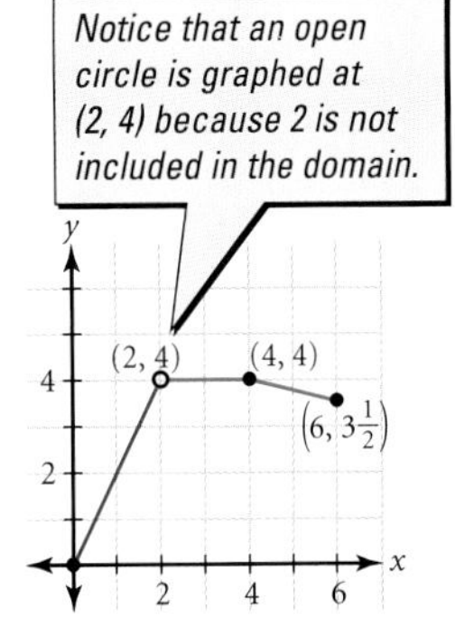

Notice that an open circle is graphed at (2, 4) because 2 is not included in the domain.
y
x
(2, 4)
(4, 4)
(6, 3½)

SOLUTION

1. Graph $y = 2x$ for $0 \le x < 2$.
   $0 = 2(0)$ and $4 = 2(2)$
   Connect (0, 0) and (2, 4). Use an open circle at (2, 4).
2. Graph $y = 4$ for $2 < x < 4$.
   Connect (2, 4) and (4, 4). Use an open circle at (2, 4).
3. Graph $y = -\frac{1}{4}x + 5$ for $4 \le x \le 6$.
   $4 = -\frac{1}{4}(4) + 5$ and $3\frac{1}{2} = -\frac{1}{4}(6) + 5$ Connect (4, 4) and $\left(6, 3\frac{1}{2}\right)$.

TRY THIS Graph: $f(x) = \begin{cases} -2x + 3 & \text{if } 0 \le x < 5 \\ -3x + 8 & \text{if } 5 \le x \le 10 \end{cases}$

Step Functions

The graph of a linear function with a slope of 0 is a horizontal line. This type of function is called a **constant function** because every function value is the same number.

A **step function** is a piecewise function that consists of different constant range values for different intervals of the domain of the function. The two basic step functions are shown below.

Greatest-integer function, or rounding-down function

$f(x) = [x]$, or $f(x) = \lfloor x \rfloor$

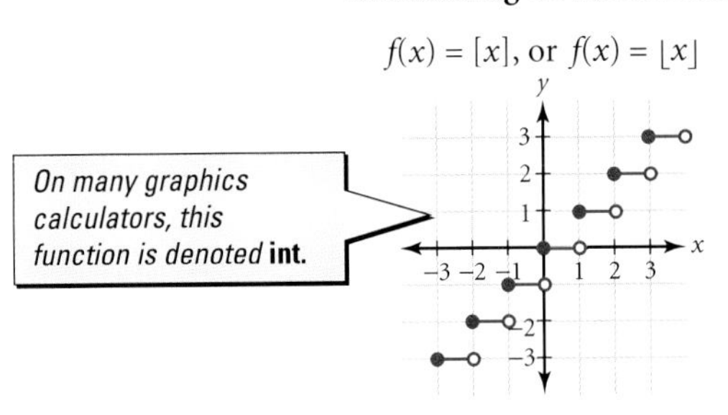

On many graphics calculators, this function is denoted **int**.
y
x

x	−3	−1.5	0	2.8
$f(x) = [x]$	−3	−2	0	2

Rounding-up function

$f(x) = \lceil x \rceil$

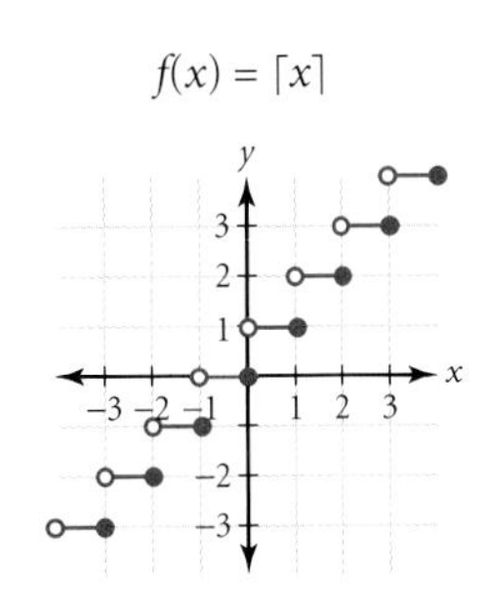

x	−3	−1.5	0	2.8
$f(x) = \lceil x \rceil$	−3	−1	0	3

Interdisciplinary Connection

SOCIAL STUDIES If an event occurred after the year 1752, you can use the following formula to find the day of the week on which it occurred:

$$w = d + 2m + \left[\frac{3(m+1)}{5}\right] + y + \left[\frac{y}{4}\right] - \left[\frac{y}{100}\right] + \left[\frac{y}{400}\right] + 2$$

In this formula, d is the day of the month, m is the month of the year, and y is the year. Note that March through December are months 3 through 12, but that January and February must be considered as months 13 and 14 *of the preceding year.*

After you calculate w, divide by 7. The remainder gives the day of the week: 0 represents Saturday, 1 represents Sunday, and so on.

On what day of the week did the events below occur?

1. The Declaration of Independence was adopted on July 4, 1776. **Thursday**
2. The first telephone call between New York and San Francisco occurred on January 25, 1915. **Monday**

CRITICAL THINKING

The driver's wage for working 40 hours is $840, not $1260. For more than 40 hours, $420 must be subtracted from 31.5h to calculate the correct wage.

Activity Notes

In this Activity, students make a table of values for a given piecewise function and graph the resulting points. When they connect the points that they graphed, they should see two segments joined at a common endpoint. Point out that the x-coordinate of this common endpoint is 40 and that this is the number of hours that signals a change in the driver's wage.

CHECKPOINT ✔

4. The graph appears to be made up of two straight line segments.

ADDITIONAL EXAMPLE 1

Graph the following function:

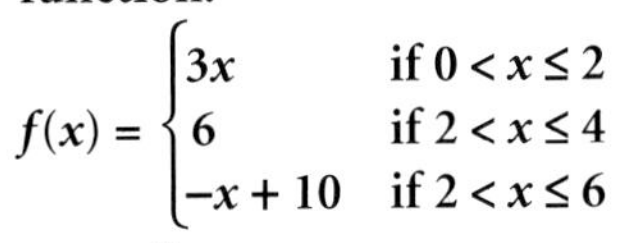

$f(x) = \begin{cases} 3x & \text{if } 0 < x \le 2 \\ 6 & \text{if } 2 < x \le 4 \\ -x + 10 & \text{if } 2 < x \le 6 \end{cases}$

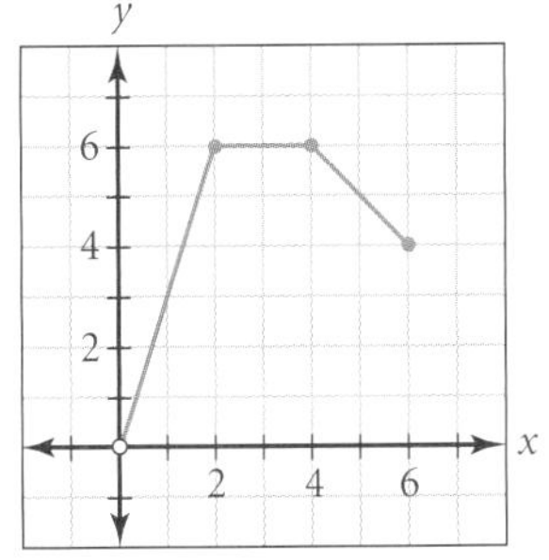

TRY THIS

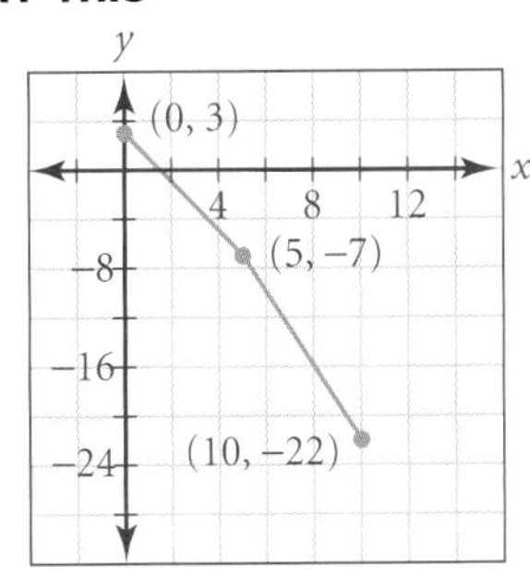

(0, 3)
(5, −7)
(10, −22)

Use Teaching Transparency 11.

The domain of both $f(x) = [x]$ and $f(x) = \lceil x \rceil$ is the set of all real numbers, and the range of both functions is the set of all integers.

CHECKPOINT ✔ Evaluate $[5]$, $[3.2]$, and $[-4.4]$. Evaluate $\lceil 5 \rceil$, $\lceil 3.2 \rceil$, and $\lceil -4.4 \rceil$.

Example 2 shows how first-class postage can be modeled with a rounding-up function.

EXAMPLE 2 The cost of mailing a first-class letter (up to 11 ounces inclusive) in 1998 is given by the function $p(w) = 32 + 23\lceil w - 1 \rceil$, where w represents the weight in ounces and p represents the postage in cents.

APPLICATION
CONSUMER ECONOMICS

Find the cost of mailing a first-class letter given each weight.

a. 2.2 ounces **b.** 4.8 ounces

SOLUTION

a. $p(2.2) = 32 + 23\lceil 2.2 - 1 \rceil$
$= 32 + 23\lceil 1.2 \rceil$
$= 32 + 23(2)$
$= 78$

The cost is \$0.78.

b. $p(4.8) = 32 + 23\lceil 4.8 - 1 \rceil$
$= 32 + 23\lceil 3.8 \rceil$
$= 32 + 23(4)$
$= 124$

The cost is \$1.24.

TRY THIS Find the cost of mailing an 0.8-ounce and a 2.9-ounce first-class letter.

Example 3 shows you how to graph a variation, or transformation, of a basic step function.

EXAMPLE 3 **Graph $g(x) = 2\lceil x \rceil$.**

SOLUTION

PROBLEM SOLVING **Make a table** to compare values of $f(x) = \lceil x \rceil$ with values of $g(x)$.

CONNECTION
TRANSFORMATIONS

Interval of x	$f(x) = \lceil x \rceil$	$g(x) = 2\lceil x \rceil$
$-4 < x \le -3$	-3	-6
$-3 < x \le -2$	-2	-4
$-2 < x \le -1$	-1	-2
$-1 < x \le 0$	0	0
$0 < x \le 1$	1	2
$1 < x \le 2$	2	4
$2 < x \le 3$	3	6

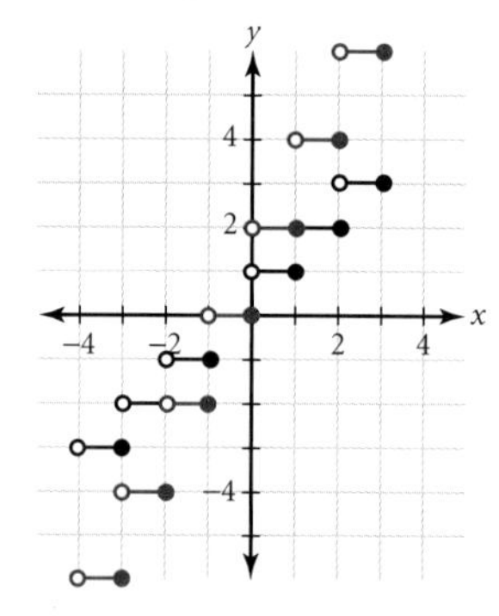

TRY THIS Graph $g(x) = \lceil x \rceil - 1$.

CRITICAL THINKING Let $f(x) = [x]$ and $g(x) = \lceil x \rceil$. Compare and contrast $f - g$ with $g - f$.

CHECKPOINT ✔
$[5] = 5$; $[3.2] = 3$; $[-4.4] = -5$;
$\lceil 5 \rceil = 5$; $\lceil 3.2 \rceil = 4$; $\lceil -4.4 \rceil = -4$

ADDITIONAL EXAMPLE 2

The cost of parking at a local garage is given by the function $c(h) = 5 + 1.25\lceil h - 1 \rceil$, where h is the number of hours that you park and c is the cost in dollars. **Find the cost of parking at this garage for each period of time.**

a. $\frac{3}{4}$ hour **b.** $8\frac{1}{4}$ hours

\$5.00 \$15.00

TRY THIS
\$0.32; \$0.78

ADDITIONAL EXAMPLE 3

Graph $g(x) = 0.5\lceil x \rceil$.

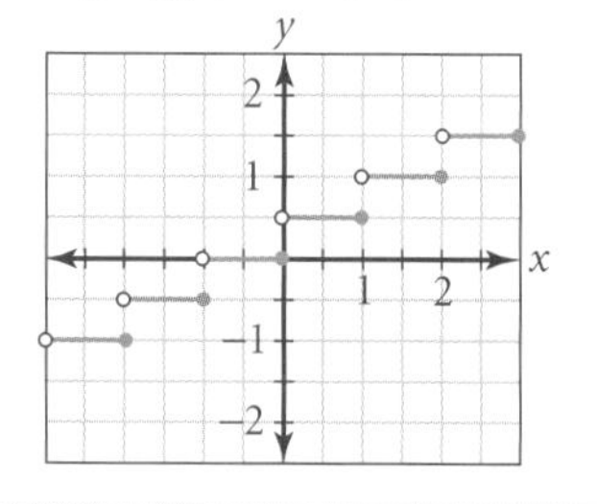

Math CONNECTION

TRANSFORMATIONS The transformation in Example 3 is a *vertical stretch* of the parent step function by a factor of 2. The Try This exercise that follows demonstrates a *translation* of 1 unit down. Students will learn more about transformations like these in Lesson 2.7.

Inclusion Strategies

VISUAL LEARNERS Some students will find it helpful to know that the rounding-up and rounding-down functions also are called the *ceiling function* and the *floor function*, respectively. These names, together with the basic concept of a step function, provide some rich visual imagery. In fact, students with graphics ability may enjoy creating imaginative instructional posters for these functions. If you wish, you can hold a contest. Display students' posters in the classroom and have the class vote on the most original poster.

Enrichment

Consider the function $f(x) = x - [x]$, where $x \ge 0$.

1. Describe the function in words. **It gives the decimal part of a positive real number.**
2. Graph the function. **See graph below.**
3. Write $f(x) = x - [x]$ as a piecewise function over the interval $0 \le x < 6$. **$f(x) = x - (n - 1)$ for $n - 1 \le x < n$, where n is a whole number**

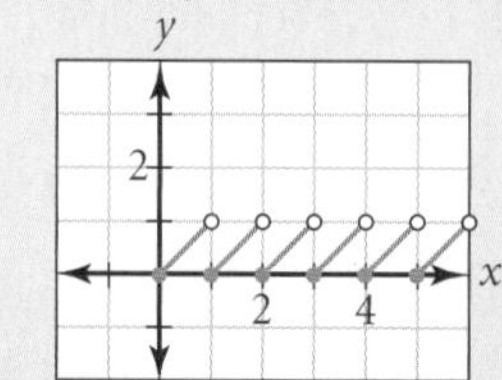

Absolute-Value Functions

The **absolute-value function**, denoted by $f(x) = |x|$, can be defined as a piecewise function as follows:

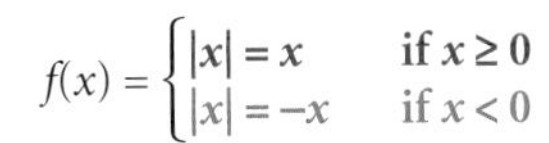

$$f(x) = \begin{cases} |x| = x & \text{if } x \geq 0 \\ |x| = -x & \text{if } x < 0 \end{cases}$$

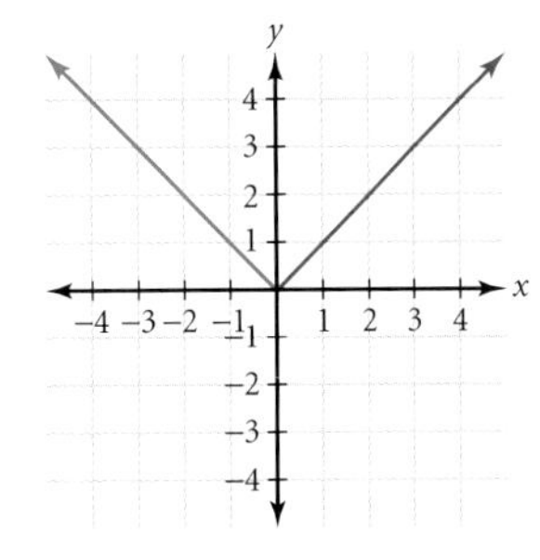

The graph of the absolute-value function has a characteristic V-shape, as shown. The domain of $f(x) = |x|$ is the set of all real numbers, and the range is the set of all nonnegative real numbers.

Example 4 involves a transformation of the absolute-value function.

EXAMPLE 4 **Graph $g(x) = 3|x| - 1$ by making a table of values. Then graph the inverse of g on the same coordinate plane.**

SOLUTION

PROBLEM SOLVING

Make a table of values to compare values of g with values of $f(x) = |x|$.

x	$f(x) = \|x\|$	$g(x) = 3\|x\| - 1$
−2	2	$3\|-2\| - 1 = 5$
−1	1	$3\|-1\| - 1 = 2$
0	0	$3\|0\| - 1 = -1$
1	1	$3\|1\| - 1 = 2$
2	2	$3\|2\| - 1 = 5$

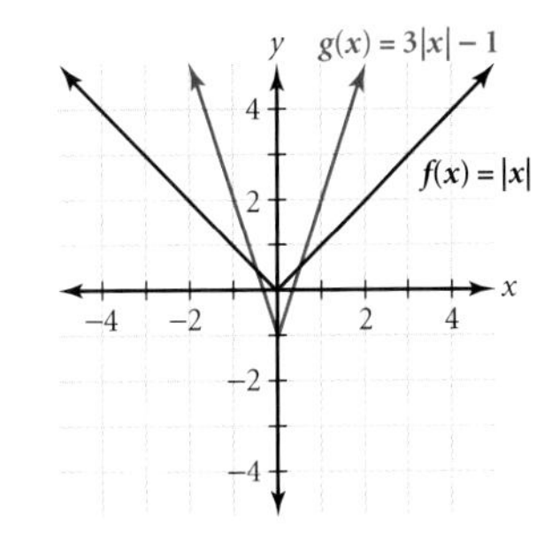

To graph the inverse of g, interchange the values of x and $g(x)$ given in the table above.

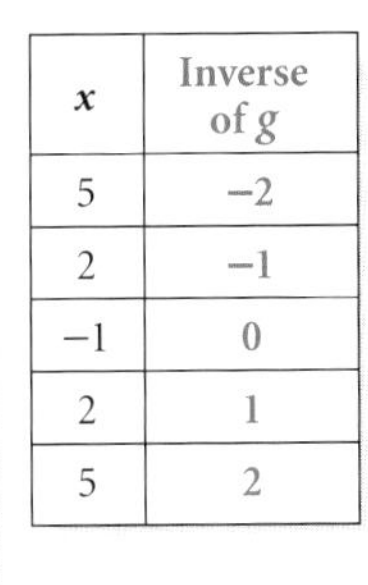

x	Inverse of g
5	−2
2	−1
−1	0
2	1
5	2

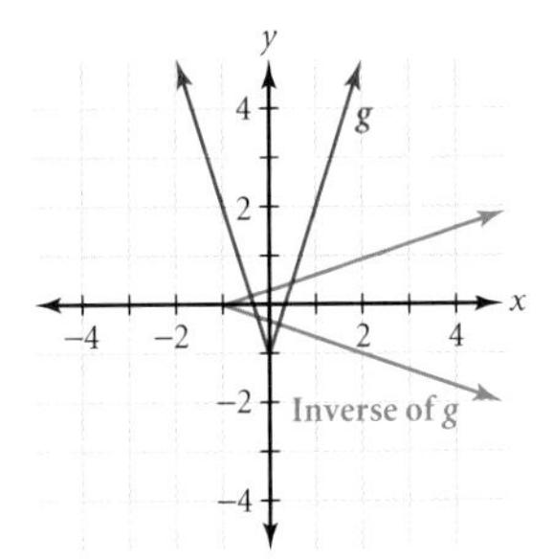

CHECK

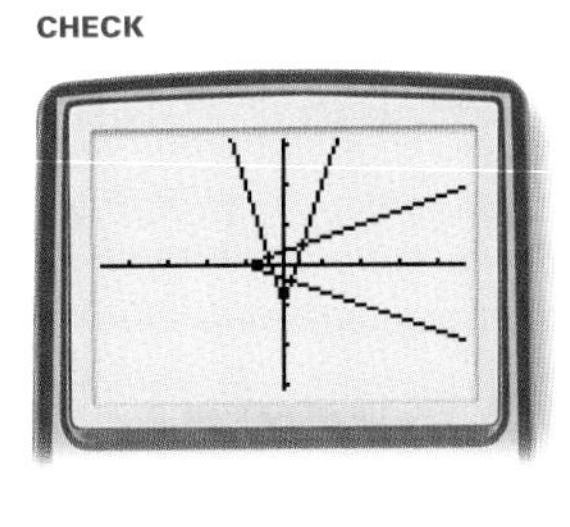

TRY THIS Graph $f(x) = \frac{1}{2}|x| + 1$ by making a table of values. Then graph the inverse of f on the same coordinate plane.

TRY THIS

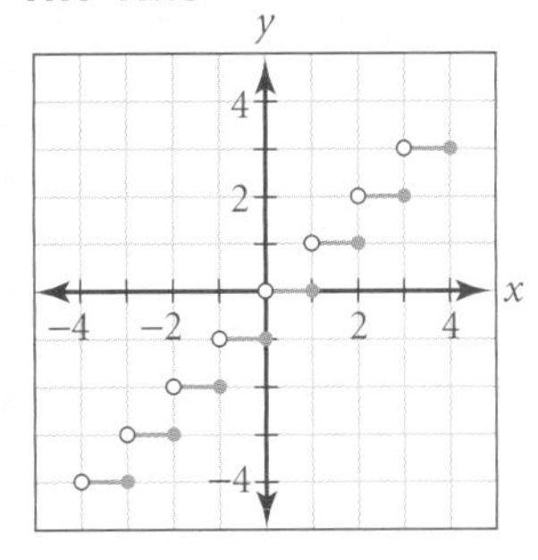

CRITICAL THINKING

If x is an integer, then $(f-g)(x) = 0$ and $(g-f)(x) = 0$.

If x is not an integer, then $(f-g)(x) = -1$ and $(g-f)(x) = 1$.

ADDITIONAL EXAMPLE 4

Graph $f(x) = 3|x| + 2$ by making a table of values. Then graph the inverse of f on the same coordinate plane.

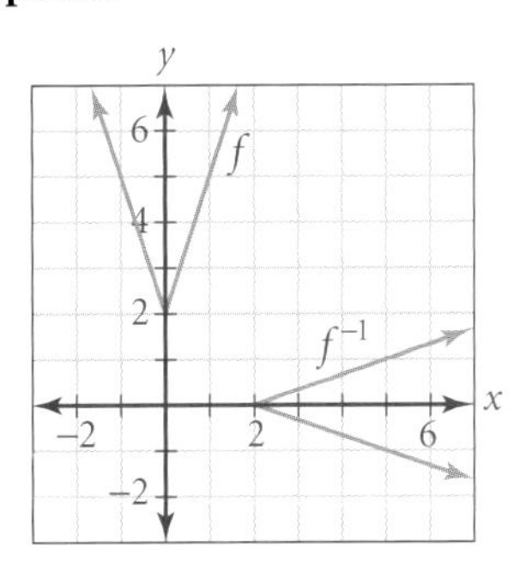

Math CONNECTION

TRANSFORMATIONS Example 4 involves a combination of transformations of the parent absolute-value function: a vertical stretch by a factor of 3 and a translation of 1 unit down.

TRY THIS

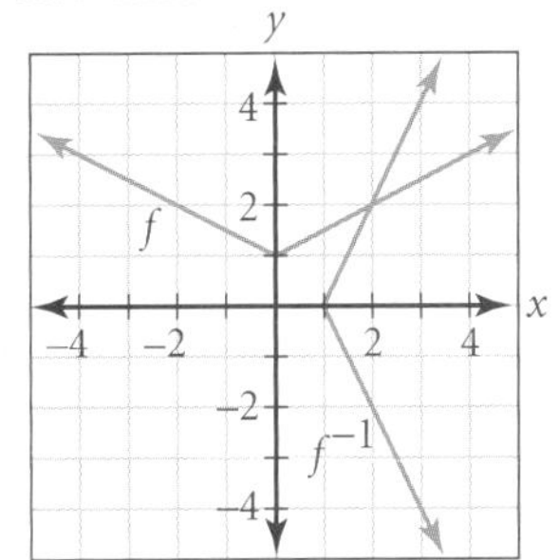

Reteaching the Lesson

COOPERATIVE LEARNING Display the following on the board or overhead:

$$f(x) = \begin{cases} \boxed{} & \text{if } 0 \leq x < 3 \\ \boxed{} & \text{if } 3 \leq x < 6 \\ \boxed{} & \text{if } 6 < x \leq 9 \end{cases}$$

Have students work in groups of three or four. Each group should fill in the boxes with linear is expressions to create a piecewise function. Each group should then graph the function.

After all groups have completed their work, ask for volunteers to share their graphs with the class. Have each student in the class write the piecewise function that the graphs represent. Check the students' answers, and discuss any difficulties that arise.

Teaching Tip

Use some numerical examples to help clarify the piecewise definition of the absolute-value function. For example, because −3 is less than 0, use the second part of the definition to find its absolute value: $|-3| = -(-3) = 3$.

ADDITIONAL EXAMPLE 5

A certain machine part's true length is 38.90 centimeters. Its length is measured as 39.25 centimeters. **Find the relative error in the measurement.**

about 0.009, or about 0.9%

TRY THIS

The relative error is approximately 0.00125, or 0.125%.

Assess

Selected Answers

Exercises 5–15, 17–89 odd

ASSIGNMENT GUIDE

In Class	1–15
Core	17–27 odd, 28–51, 53–63 odd, 69, 75
Core Plus	16–26 even, 28–51, 52–76 even
Review	77–89
Preview	90

Extra Practice can be found beginning on page 940.

When a measurement is performed on an object, the measured value and the true value may be slightly different. This difference may be due to factors such as the quality of the measuring device or the skill of the individual using the measuring device.

The *relative error* in a measurement of an object gives the amount of error in the measurement relative to the *true measure* of the object. If an object's true measure is x_t units, then the **relative error**, r, in the measurement, x, is given by $r(x) = \left|\frac{x_t - x}{x_t}\right|$.

EXAMPLE 5 A certain machine part's true length is 25.50 centimeters. Its length is measured as 25.55 centimeters.

APPLICATION MANUFACTURING

Find the relative error in the measurement.

SOLUTION

$$r(x) = \left|\frac{x_t - x}{x_t}\right|$$

$$r(25.55) = \left|\frac{25.50 - 25.55}{25.50}\right|$$

$$= \left|\frac{-0.05}{25.50}\right|$$

$$\approx 0.002$$

The relative error in the measurement, 25.55 centimeters, is about 0.002, or 0.2%.

TRY THIS The true length of a second machine part is 40.00 centimeters. Its length is measured as 40.05 centimeters. Find the relative error in this measurement.

Exercises

Communicate

1. Describe the defining characteristics of piecewise functions.
2. Explain why $[1.5] = 1$ but $[-1.5] \neq -1$.
3. Compare and contrast the greatest-integer and rounding-up functions.
4. Is the inverse of $y = |x|$ a function? Explain.

Guided Skills Practice

Graph each piecewise function. ***(EXAMPLE 1)***

5. $f(x) = \begin{cases} 3x + 5 & \text{if } -1 \le x < 2 \\ -x + 9 & \text{if } 2 \le x < 5 \end{cases}$

6. $g(x) = \begin{cases} x + 3 & \text{if } -3 \le x < 3 \\ 2x & \text{if } x \ge 3 \end{cases}$

5.

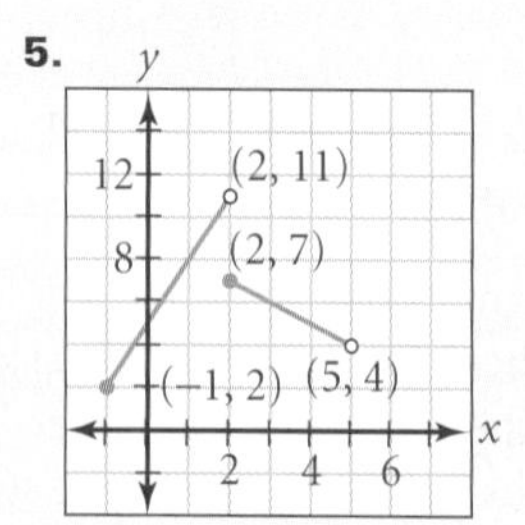

6.

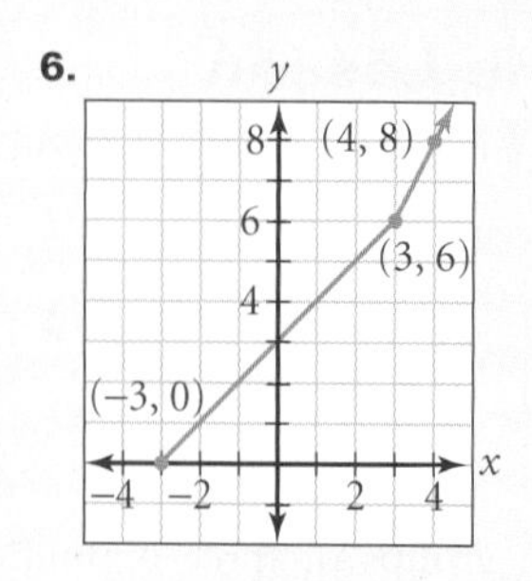

APPLICATION

CONSUMER ECONOMICS Darren and Joel called their grandmother during winter break. In 1998, this daytime call from Austin, Texas, to Denver, Colorado, was charged at a rate of \$0.14 per minute or fraction of a minute. The cost of this phone call can be modeled by $c(x) = 0.14\lceil x \rceil$, where c is the cost in dollars and x is the length of the call in minutes. Find the cost for each call. ***(EXAMPLE 2)***

7. \$3.08
9. \$4.76

7. 21.25 minutes **8.** 10 minutes **\$1.40**
9. 33.5 minutes **10.** 0.1 minutes **\$0.14**

Graph each step function by making a table of values. ***(EXAMPLE 3)***

11. $f(x) = 3\lceil x \rceil$ **12.** $g(x) = -4[x]$

Graph each absolute-value function by making a table of values. Then graph the inverse on the same coordinate plane. ***(EXAMPLE 4)***

13. $f(x) = |x| + 2$ **14.** $g(x) = 2|x| - 1$

APPLICATION

15. MANUFACTURING A certain machine part's true length is 16.000 centimeters. Its length is measured as 16.035 centimeters. Find the relative error in this measurement. ***(EXAMPLE 5)*** **0.002, or 0.2%**

Practice and Apply

Graph each piecewise function.

16. $f(x) = \begin{cases} x+1 & \text{if } 0 \le x < 5 \\ 2x-4 & \text{if } 5 \le x < 10 \end{cases}$ **17.** $g(x) = \begin{cases} 3x-4 & \text{if } 0 \le x < 6 \\ 20-x & \text{if } 6 \le x < 12 \end{cases}$

18. $m(x) = \begin{cases} 20 & \text{if } 0 \le x < 10 \\ \frac{x}{2}+15 & \text{if } 10 \le x < 20 \end{cases}$ **19.** $f(x) = \begin{cases} 4x & \text{if } 0 \le x < 2 \\ -2x+10 & \text{if } 2 \le x < 5 \\ 2 & \text{if } 5 \le x < 10 \end{cases}$

20. $h(x) = \begin{cases} -2 & \text{if } x < 0 \\ x+1 & \text{if } 0 \le x \le 10 \\ -\frac{1}{2}x+16 & \text{if } x > 10 \end{cases}$ **21.** $b(x) = \begin{cases} 2 & \text{if } x < 1 \\ 2x & \text{if } 1 \le x \le 3 \\ 7-\frac{1}{3}x & \text{if } x > 3 \end{cases}$

22. $k(x) = \begin{cases} 2x+3 & \text{if } x < 4 \\ x-1 & \text{if } 4 \le x \le 9 \end{cases}$ **23.** $f(x) = \begin{cases} 5-x & \text{if } x < 2 \\ x-1 & \text{if } 2 \le x \le 10 \end{cases}$

Write the piecewise function represented by each graph.

24. $f(x) = \begin{cases} -2 \text{ if } -4 \le x < 1 \\ 2x \text{ if } -1 \le x < 2 \end{cases}$

25. $g(x) = \begin{cases} \frac{5}{4}x+2 \text{ if } -4 < x \le 0 \\ -\frac{1}{2}x+2 \text{ if } 0 < x \le 4 \end{cases}$

24.

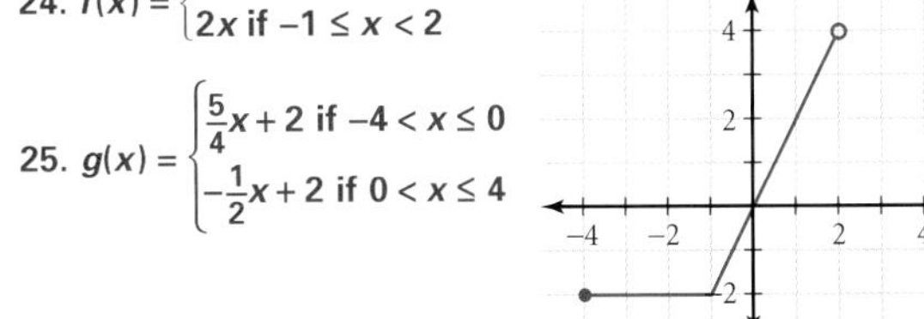

25.

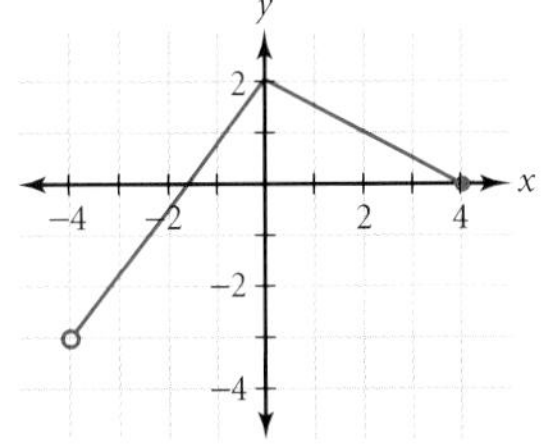

13.

x	$f(x) = \lvert x \rvert + 2$
−3	$\lvert -3 \rvert + 2 = 5$
−2	$\lvert -2 \rvert + 2 = 4$
−1	$\lvert -1 \rvert + 2 = 3$
0	$\lvert 0 \rvert + 2 = 2$
1	$\lvert 1 \rvert + 2 = 3$
2	$\lvert 2 \rvert + 2 = 4$
3	$\lvert 3 \rvert + 2 = 5$

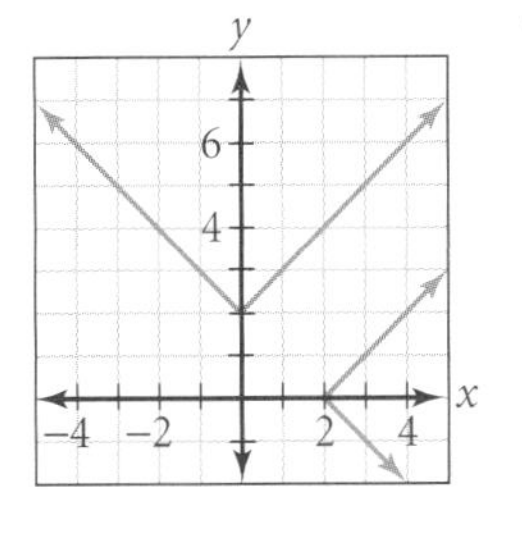

14.

x	$g(x) = 2\lvert x \rvert - 1$
−3	$2\lvert -3 \rvert - 1 = 5$
−2	$2\lvert -2 \rvert - 1 = 3$
−1	$2\lvert -1 \rvert - 1 = 1$
0	$2\lvert 0 \rvert - 1 = -1$
1	$2\lvert 1 \rvert - 1 = 1$
2	$2\lvert 2 \rvert - 1 = 3$
3	$2\lvert 3 \rvert - 1 = 5$

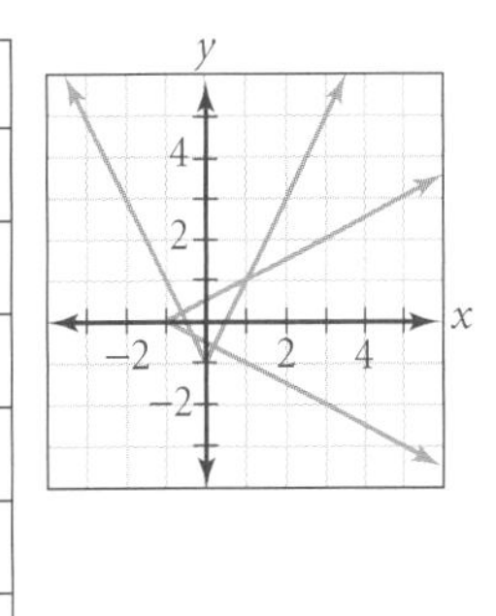

Error Analysis

When evaluating $[x]$ for negative values of x, students often just delete the decimal part of the number. For instance, they may evaluate $[-4.8]$ as -4 rather than -5. Encourage them to sketch a number line and locate the given number on it. Thus, *rounding down* means that they must locate the nearest integer *to the left*.

Interval of x	$f(x) = -3\lceil x \rceil$
$-4 < x \le -3$	$3(-3) = -9$
$-3 < x \le -2$	$3(-2) = -6$
$-2 < x \le -1$	$3(-1) = -3$
$-1 < x \le 0$	$3(0) = 0$
$0 < x \le 1$	$3(1) = 3$
$1 < x \le 2$	$3(2) = 6$
$2 < x \le 3$	$3(3) = 0$

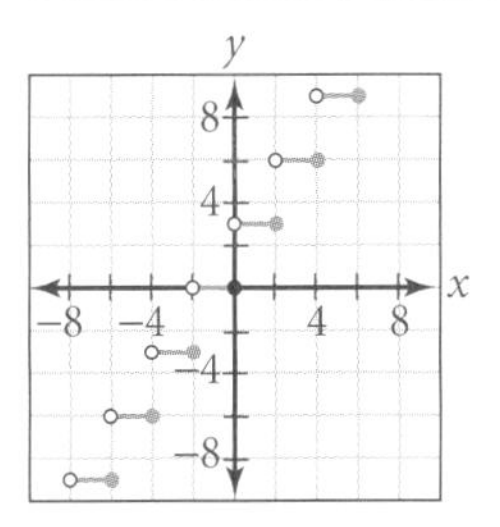

Interval of x	$g(x) = -4[x]$
$-3 \le x < -2$	$-4(-3) = 12$
$-2 \le x < -1$	$-4(-2) = 8$
$-1 \le x < 0$	$-4(-1) = 4$
$0 \le x < 1$	$-4(0) = 0$
$1 \le x < 2$	$-4(1) = -4$
$2 \le x < 3$	$-4(2) = -8$
$3 \le x < 4$	$-4(3) = -12$

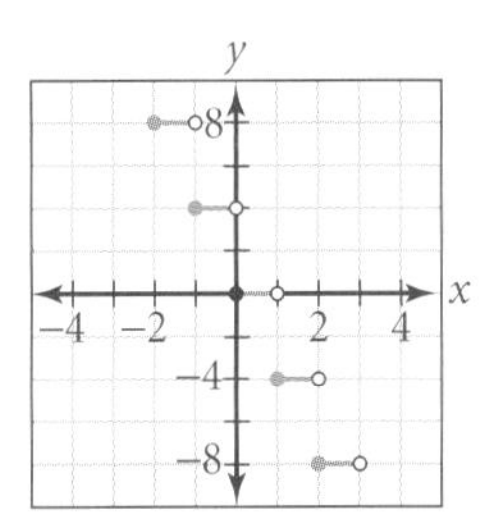

The answers to Exercises 16–23 can be found in Additional Answers beginning on page 1002.

52.

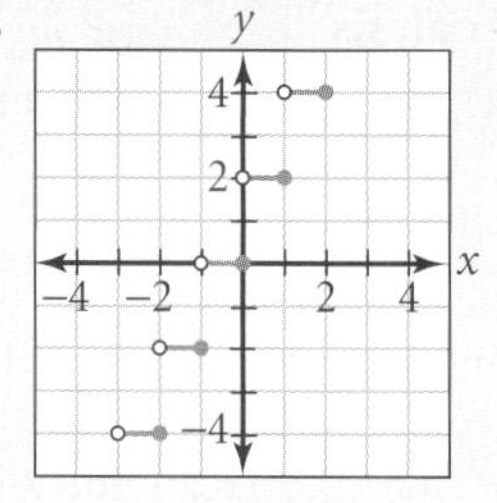

53.

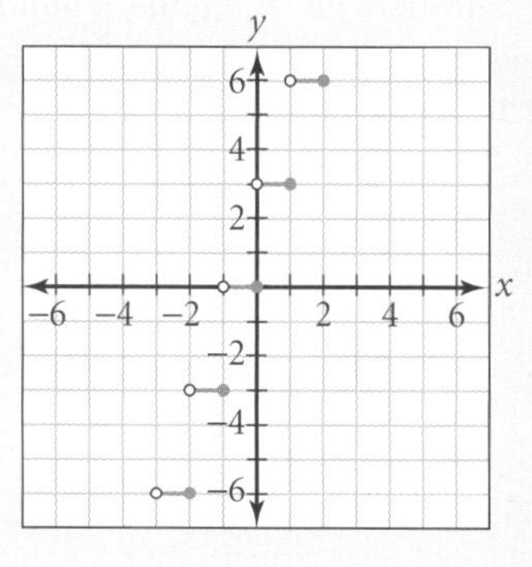

54.

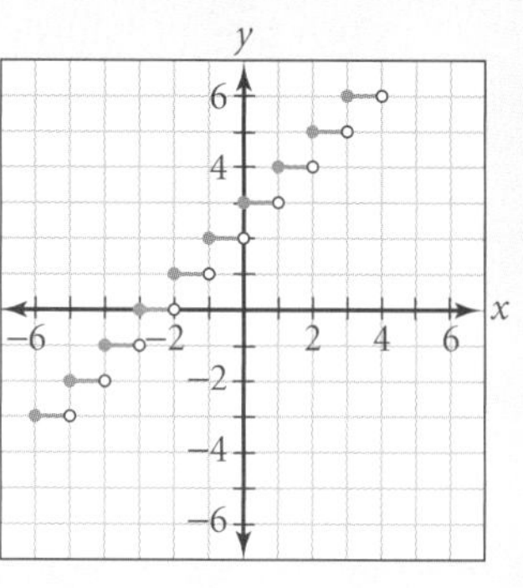

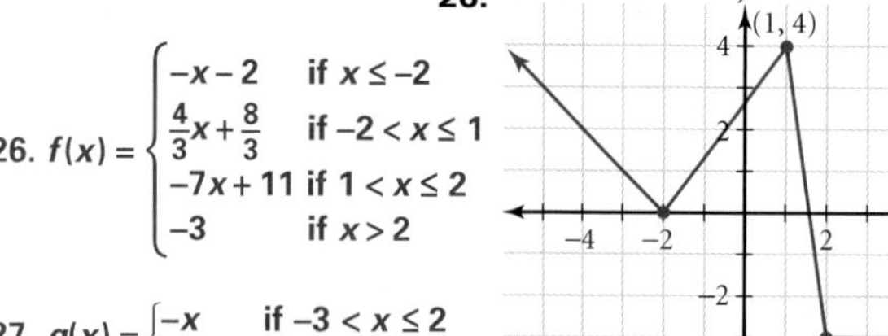

26. $f(x) = \begin{cases} -x-2 & \text{if } x \le -2 \\ \frac{4}{3}x + \frac{8}{3} & \text{if } -2 < x \le 1 \\ -7x + 11 & \text{if } 1 < x \le 2 \\ -3 & \text{if } x > 2 \end{cases}$

27. $g(x) = \begin{cases} -x & \text{if } -3 < x \le 2 \\ -x + 5 & \text{if } x > 2 \end{cases}$

26.

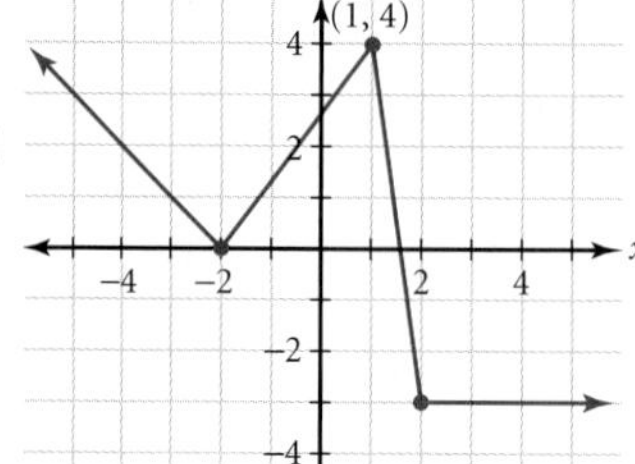

27.

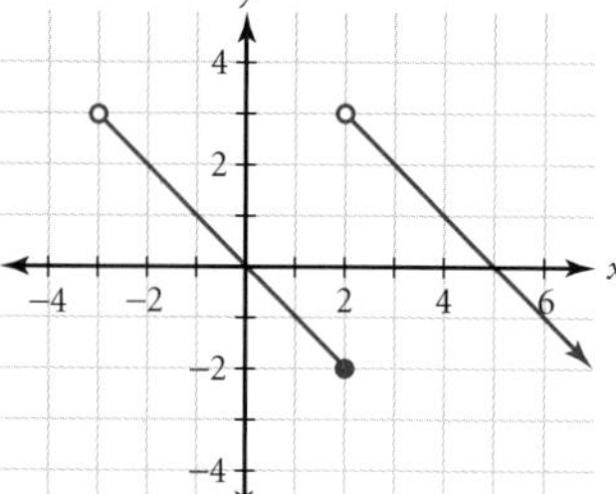

internet connect

Homework Help Online
Go To: **go.hrw.com**
Keyword: **MB1 Homework Help** for Exercises 28–51

Evaluate.

28. $[3.9]$ **3** **29.** $[-6.105]$ **−7** **30.** $\lceil 3.9 \rceil$ **4**
31. $\lceil -6.105 \rceil$ **−6** **32.** $[-4.1] - [-3.25]$ **−1** **33.** $[5.1] + [-2.01]$ **2**
34. $\lceil -4.1 \rceil - \lceil -3.25 \rceil$ **−1** **35.** $\lceil 5.1 \rceil + \lceil -2.01 \rceil$ **4** **36.** $[-2.5] + [1.999]$ **−2**
37. $[-2.99] + [2.99]$ **−1** **38.** $\lceil -5.1 \rceil - \lceil 5.1 \rceil$ **−11** **39.** $\lceil -2.5 \rceil - \lceil 1.999 \rceil$ **−4**
40. $[-2.3] - \lceil 5.6 \rceil$ **−9** **41.** $-\lceil 0.9 \rceil + [-8.7]$ **−10** **42.** $[6.2] - \lceil -4.7 \rceil$ **10**
43. $\lceil -8.99 \rceil - [-5.1]$ **−2** **44.** $-[0.25] - \lceil 0.25 \rceil$ **−1** **45.** $\lceil 2.75 \rceil - [2.75]$ **1**
46. $|4| - |-5|$ **−1** **47.** $|-7| - |2.2|$ **4.8** **48.** $|3| + |-4|$ **7**
49. $-|-1| - |1|$ **−2** **50.** $|-1| - |3|$ **−2** **51.** $|6| - |-2.67|$ **3.33**

Graph each function.

52. $f(x) = 2\lceil x \rceil$ **53.** $f(x) = 3\lceil x \rceil$ **54.** $g(x) = [x] + 3$
55. $g(x) = [x] - 3$ **56.** $g(x) = 2 - \lceil x \rceil$ **57.** $h(x) = 4 + \lceil x \rceil$

Graph each function and its inverse together on a coordinate plane.

58. $f(x) = 4|x| - 1$ **59.** $g(x) = \frac{1}{2}|x|$ **60.** $h(x) = 5|x| + 10$

Determine whether each statement is true or false. Explain.

61. $|x + y| = |x| + |y|$ **false** **62.** $|x - y| = |x| - |y|$ **false**
63. $|xy| = |x| \cdot |y|$ **true** **64.** $\left|\frac{x}{y}\right| = \frac{|x|}{|y|}$ **true**

65 Let $S(x) = \left|x - [x] - \frac{1}{2}\right|$ for $x \ge 0$. Graph S for $x \ge 0$. Describe the shape of the graph of S.

CHALLENGES

66. Compare and contrast $y = |[x]|$ with $y = [|x|]$. Include a discussion of the graphs of these functions.

67. Write a function f that rounds x up to the nearest tenth. Write another function g that rounds x down to the nearest hundredth.

CONNECTION

68. TRANSFORMATIONS The absolute-value function can be used to reflect a portion of the graph of g across the x-axis by using composition. Given $f(x) = |x|$ and $g(x) = x^2 - 2$, find and graph $f \circ g$. Compare the graphs of g and $f \circ g$.

APPLICATION

MANUFACTURING Find the relative error for each measurement of a machine part whose true length is 6.000 centimeters.

69. 6.035 centimeters **0.6%** **70.** 6.025 centimeters **0.4%**
71. 6.150 centimeters **2.5%** **72.** 6.300 centimeters **5%**

55.

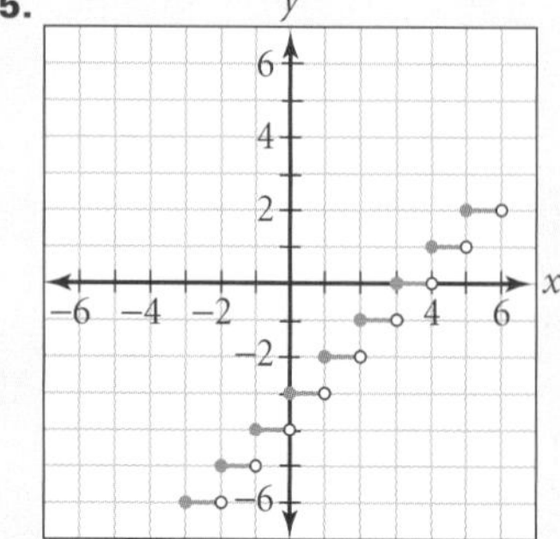

56.

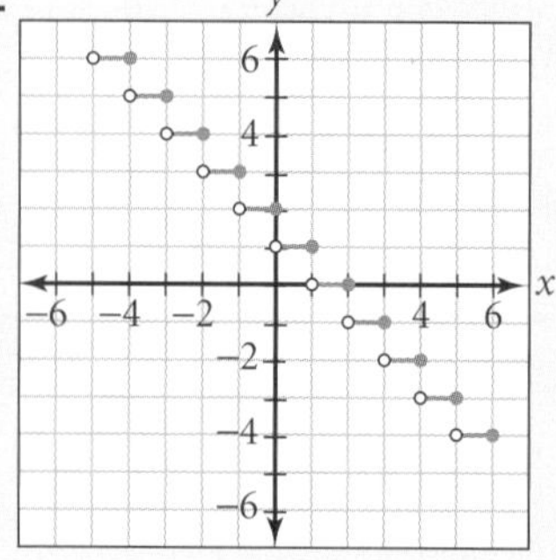

Practice

NAME ________ CLASS ________ DATE ________

Practice
2.6 Special Functions

Graph each function.

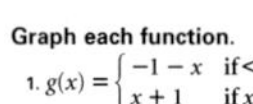

1. $g(x) = \begin{cases} -1 - x & \text{if} < 0 \\ x + 1 & \text{if } x \ge 0 \end{cases}$ 2. $f(x) = \begin{cases} -x - 4 & \text{if } x \le -2 \\ -x + 1 & \text{if } x > -2 \end{cases}$ 3. $h(x) = -2[x]$

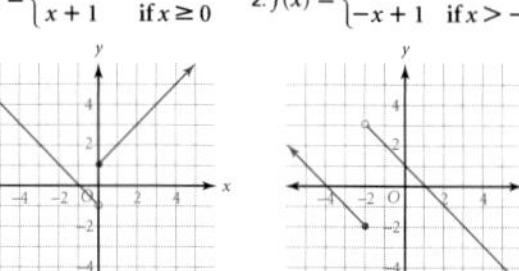

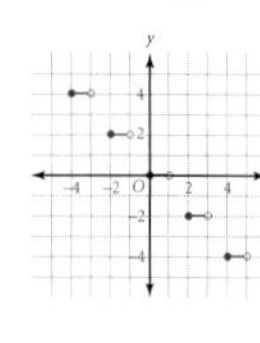

Write the piecewise function represented by each graph.

4.

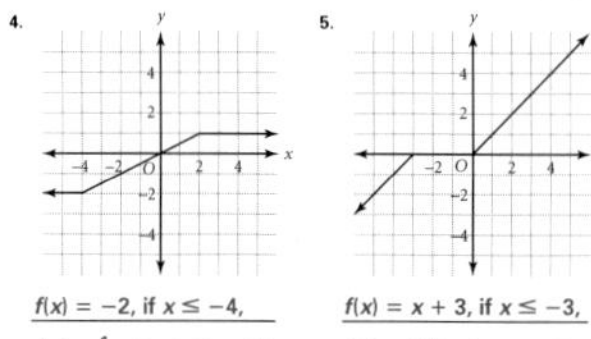

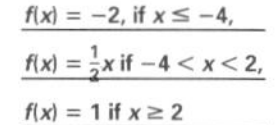

$f(x) = -2$, if $x \le -4$,
$f(x) = \frac{1}{2}x$ if $-4 < x < 2$,
$f(x) = 1$ if $x \ge 2$

5. $f(x) = x + 3$, if $x \le -3$,
$f(x) = 0$ if $-3 < x < 0$,
$f(x) = x$ if $x \ge 0$

6. $f(x) = x - 2$, if $x < 0$,
$f(x) = 0$ if $x = 0$,
$f(x) = x + 2$ if $x > 0$

Evaluate.

7. $[-9.23]$ −10 8. $\lceil 31.7 \rceil$ 32
9. $|13.13|$ 13.13 10. $\lceil -0.9 \rceil$ 0
11. $\lceil 7.8 \rceil + \lceil -1.88 \rceil$ 7 12. $[-2.22] - |-4.5|$ −7.5
13. $|5.25| - |-3.75|$ 1.5 14. $\lceil 2.5 \rceil - [2.5]$ 1
15. $-\lceil 12.95 \rceil - [6.3]$ −19 16. $-|-3| - [4.9]$ −7

APPLICATIONS

73. CONSUMER ECONOMICS A gourmet coffee store sells the house blend of coffee beans at \$9.89 per pound for quantities up to and including 5 pounds. For each additional pound, the price is \$7.98 per pound.

a. Construct a table to represent the cost for the domain $0 < x \le 10$, where x is the number of pounds of coffee beans purchased.

b. Graph the data in your table.

c. Write a function that relates the cost and the number of pounds of coffee beans purchased.

d. Determine the cost for 5.5 pounds of coffee beans.

73c. $c(x) = \begin{cases} 9.89x \text{ if } x \le 5 \\ 49.45 + 7.98(x - 5) \text{ if } x > 5 \end{cases}$

d. \$53.44

74. INCOME An air-conditioning salesperson receives a base salary of \$2850 per month plus a commission. The commission is 2% of the sales up to and including \$25,000 for the month and 5% of the sales over \$25,000 for the month.

a. Construct a table to represent the salesperson's total monthly income for the domain $0 < s \le 50{,}000$, where s is the sales for the month.

b. Graph the data in your table.

c. Write a function that relates the salesperson's total monthly income with his or her sales for the month.

d. Determine the salesperson's total monthly income if his or her sales were \$43,000 for the month.

74c. $I(s) = \begin{cases} 2850 + 0.02s \text{ if } 0 \le s \le 25{,}000 \\ 3350 + 0.05(s - 25{,}000) \text{ if } s > 25{,}000 \end{cases}$

d. \$4250

75. CONSUMER ECONOMICS The Break-n-Fix Repair Store charges \$45 for a service call that involves up to and including one hour of labor. For each additional half-hour of labor or fraction thereof, the store charges \$20.

a. Construct a table to represent the charges for the domain $0 < t \le 5$, where t is the labor measured in hours.

b. Graph the data in your table.

c. Write a function that relates the cost and the amount of labor involved in the service call.

d. Determine the cost of a service call that takes 3.75 hours.

75c. $c(t) = \begin{cases} 45 \text{ if } 0 < t \le 1 \\ 45 + 20\lceil 2(t - 1)\rceil \text{ if } 1 < t \le 5 \end{cases}$

d. \$165

76. CONSUMER ECONOMICS Residential water charges, c, are based on the monthly consumption of water, x, in thousands of gallons. Graph the piecewise function given below, which represents the rate schedule for monthly charges.

$$c(x) = \begin{cases} 1.25x + 4.50 & \text{if } 0 < x \le 3 \\ 2x + 2.25 & \text{if } 3 < x \le 7 \\ 2.60x - 7.25 & \text{if } 7 < x \le 15 \\ 3.80x - 22.75 & \text{if } x > 15 \end{cases}$$

57.

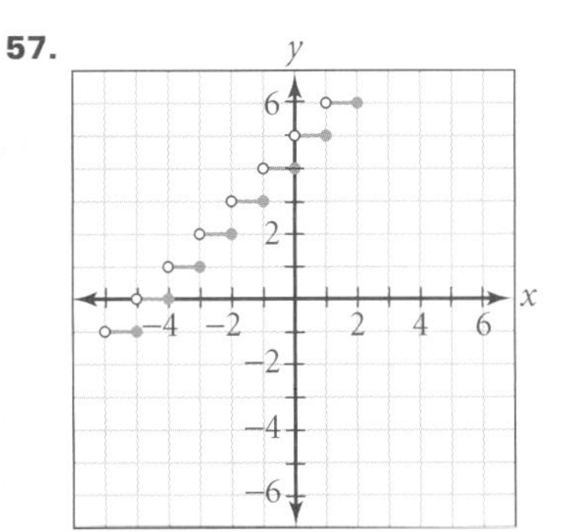

58.

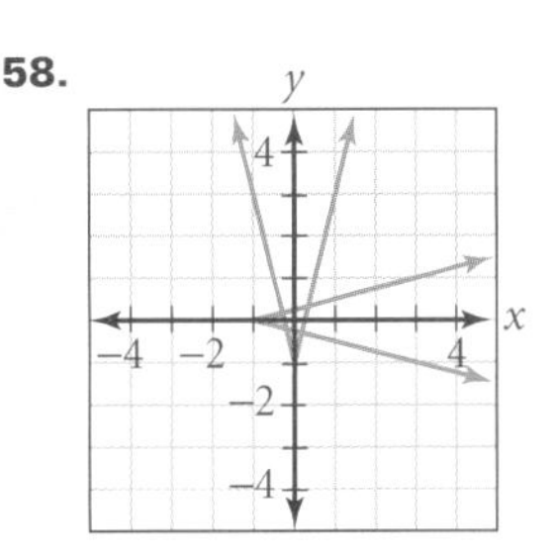

59.

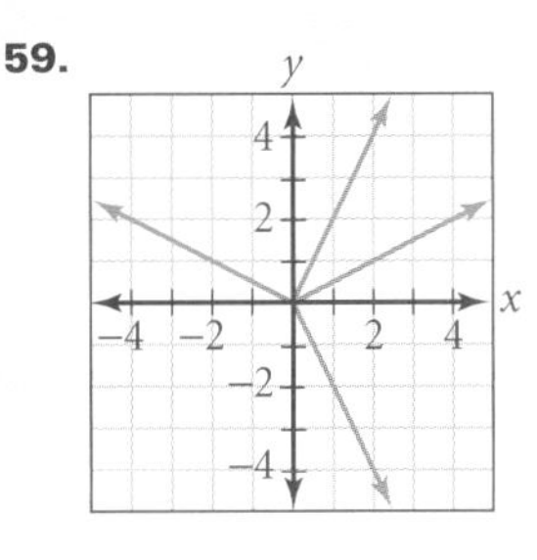

60.

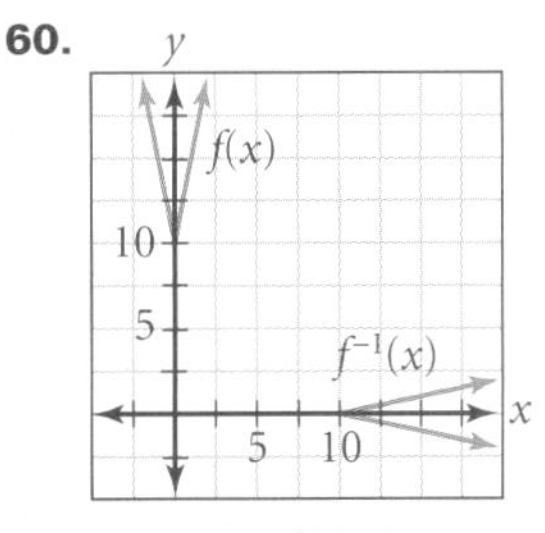

61. $|x + y| = |x| + |y|$
false; for example:
$|(-1) + 1| = |0| = 0$
$|-1| + |1| = 1 + 1 = 2$

62. $|x - y| = |x| - |y|$
false; for example:
$|1 - (-1)| = |2| = 2$
$|1| - |-1| = 1 - 1 = 0$

63. $|xy| = |x| \cdot |y|$ true for all values of x and y

64. $\left|\frac{x}{y}\right| = \frac{|x|}{|y|}$ true for all values of x and $y \ne 0$

65.

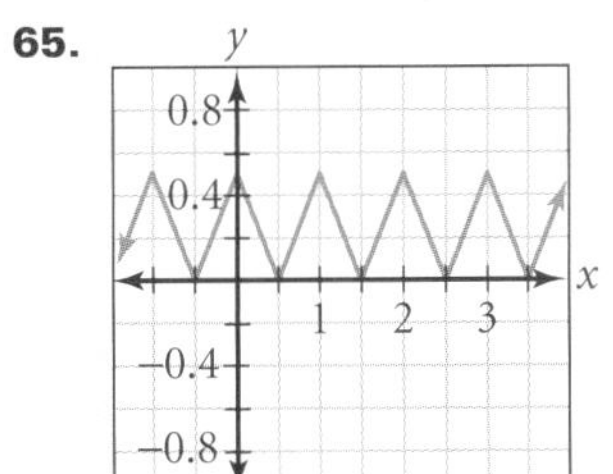

The shape of the graph is a sawtooth.

The answers to Exercises 66–68, and 73–76 can be found in Additional Answers beginning on page 1002.

Student Technology Guide

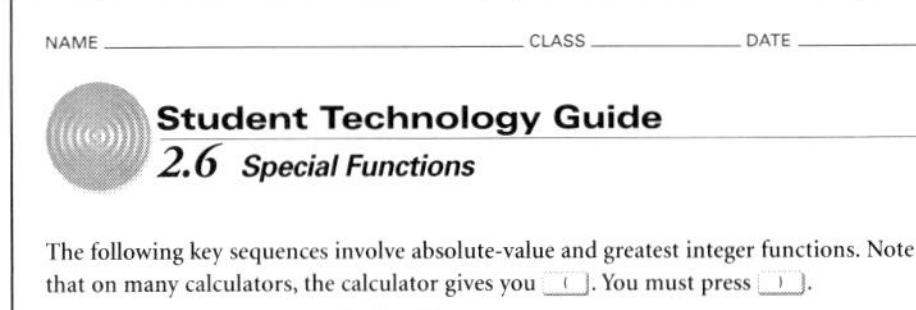

NAME ________ CLASS ________ DATE ________

Student Technology Guide
2.6 Special Functions

The following key sequences involve absolute-value and greatest integer functions. Note that on many calculators, the calculator gives you (. You must press) .

Evaluate 2|15.6 − 4|: 2 MATH NUM 1:abs(ENTER 15.6 − 4) ENTER ⟶ 23.2

Evaluate 2⟦−8.5⟧: 2 MATH NUM 5:int((-) 8.5) ENTER ⟶ −18

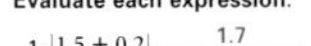

Evaluate each expression.

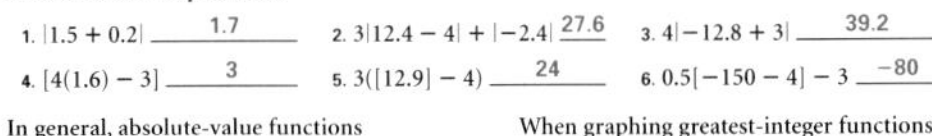

1. |1.5 + 0.2| 1.7
2. 3|12.4 − 4| + |−2.4| 27.6
3. 4|−12.8 + 3| 39.2
4. |4(1.6) − 3| 3
5. 3(⟦12.9⟧ − 4) 24
6. 0.5|−150 − 4| − 3 −80

In general, absolute-value functions should be graphed in connected mode. To graph $y = 2|x + 1|$, press MODE Connected ENTER 2nd MODE Y= 2 MATH NUM 1:abs(ENTER X,T,θ,n + 1) GRAPH.

When graphing greatest-integer functions, it is best to select dot mode. To graph $y = 0.5⟦x⟧$, press MODE Dot ENTER 2nd MODE Y= .5 MATH NUM 5:int(ENTER X,T,θ,n) GRAPH.

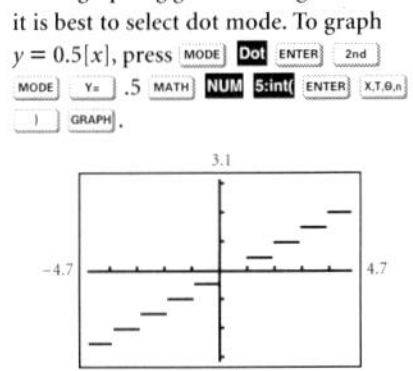

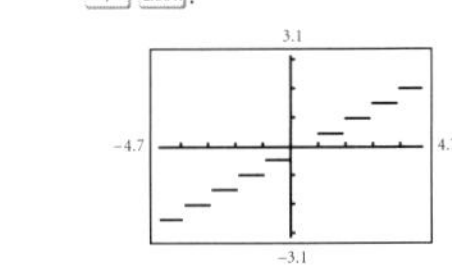

Graph each function and describe the basic shape of the graph.

7. $y = |x| - 2$ V-shaped with corner at (0, −2)
8. $y = ⟦0.5x⟧$ step function whose segments are 2 units long
9. $y = |x - 2|$ V-shaped with corner at (2, 0)

You can graph a function and show its inverse on the same display. If you enter $2|x + 1|$ as Y1 and graph it, the following key sequence will show the inverse: 2nd PRGM 8:DrawInv ENTER VARS Y-VARS 1:Function... ENTER 1:Y1 ENTER ENTER

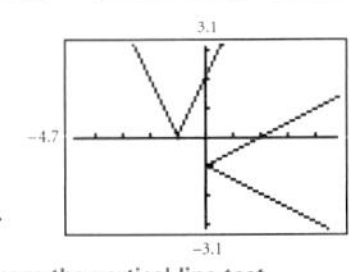

10. Graph $y = 0.5|x|$ and its inverse on the same display. Is the inverse a function? Explain your response.
No; the graph of the inverse does not pass the vertical-line test.

Look Beyond

In Exercise 90, students investigate three equations that contain an absolute-value expression but represents a step function. Then they look at two transformations of the function. Students will study transformations of the graphs of functions in Lesson 2.7.

ALTERNATIVE

Assessment

Portfolio Activity

The Portfolio Activity can be used as preparation for the Chapter Project or as a separate activity. In the Portfolio Activity on this page, students are given a new projection for the amount of space debris beginning in the year 2001 and use it to modify the model from the Portfolio Activity on page 101. Then they analyze the piecewise function that results.

Answers to Portfolio Activities can be found in Additional Answers of the Teacher's Edition.

Look Back

APPLICATION

77. CONSUMER ECONOMICS A valet parking lot charges a fixed fee of \$2.00 to park a car plus \$1.50 per hour for covered parking. ***(LESSON 2.3)***

a. Write a linear function, c, to model the valet parking charge for h hours of covered parking. $c(h) = 2.00 + 1.50h$

b. If a car is parked in covered parking for 3.5 hours, what is the charge? \$7.25

Let $f(x) = x + 2$ and $g(x) = 3x$. Perform each function operation. State any domain restrictions. ***(LESSON 2.4)***

78. Find $f \circ g$. $3x + 2$

79. Find $g \circ f$. $3x + 6$

80. Find $f \circ f$. $x + 4$

81. Find $g \circ g$. $9x$

82. Find $f - g$. $-2x + 2$

83. Find $f + g$. $4x + 2$

84. Find $f \cdot g$. $3x^2 + 6x$

85. Find $\frac{f}{g}$. $\frac{x+2}{3x}, x \neq 0$

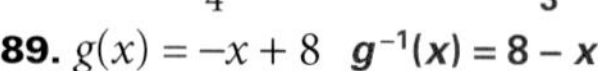

Find an equation for the inverse of each function. Then use composition to verify that the equation you wrote is the inverse. ***(LESSON 2.5)***

86. $f(x) = 3x - \frac{1}{2}$ $f^{-1}(x) = \frac{1}{3}x + \frac{1}{6}$

87. $a(x) = \frac{3}{4}(x - 2)$ $a^{-1}(x) = \frac{4}{3}x + 2$

88. $h(x) = -2(x - 4) + 1$ $h^{-1}(x) = -\frac{1}{2}x + \frac{9}{2}$

89. $g(x) = -x + 8$ $g^{-1}(x) = 8 - x$

internet connect

Portfolio Extension

Go To: go.hrw.com
Keyword:
MB1 Piecewise

Look Beyond

90 **TRANSFORMATIONS** Graph $f(x) = \frac{|x|}{x}$, $g(x) = \frac{2|x|}{x}$, and $h(x) = \frac{-2|x|}{x}$. State the domain and range for each function.

SPACE SCIENCE In future years, space technology is expected to improve to a point where many of the components presently left orbiting in space can be returned to Earth. This will greatly reduce the current rate of space debris accumulation.

1. Your Space Debris Table specified an initial 7000 objects and an annual rate of increase of 3%. Now assume that the projected annual increase becomes a constant 200 objects each year, beginning in 2001. Write and graph a piecewise function that models the total number of debris objects at the end of t years.
2. Determine the total number of debris objects in 2005 by using the function you wrote in Step 1.
3. Using your piecewise function from Step 1, calculate how many years it would take for the number of debris objects to be double the number found at the end of 1993. (Refer to the table that you created in the Portfolio Activity on page 93.) Compare this with the number of years it would take to double without the change in Step 1. (Refer to the Portfolio Activity on page 110.)
4. Using the piecewise function you wrote in Step 1, model the number of debris objects in space in 2010. Compare this with the number for the same year in the table that you created in the Portfolio Activity on page 93.

90. for $f(x) = \frac{|x|}{x}$, domain: all real numbers except 0; range: −1, 1

for $f(x) = \frac{2|x|}{x}$, domain: all real numbers except 0; range: −2, 2

for $f(x) = \frac{-2|x|}{x}$, domain: all real numbers except 0; range: −2, 2

A Preview of Transformations

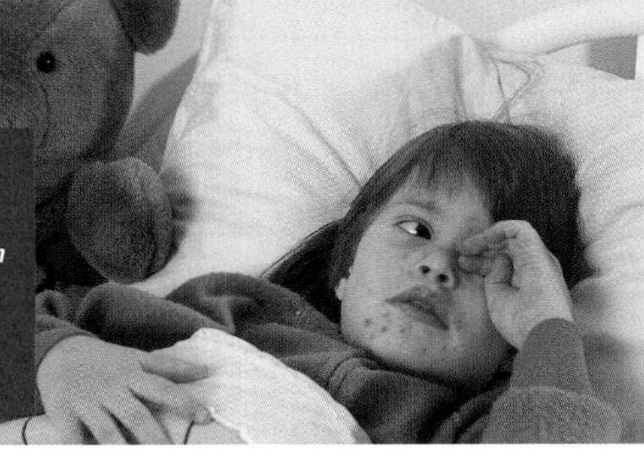

Why *Many real-world situations are represented by using transformations. For example, translations are often used to graph data involving trends such as the decreasing number of reported cases of chickenpox over time.*

Objective

- Identify the transformation(s) from one function to another.

APPLICATION HEALTH

The table of data below gives the number of reported cases of chickenpox in the United States in thousands from 1989 to 1994. In Lesson 1.5, this data was used to create a scatter plot by *translating* the x-values from the actual year to the number of years after 1988. Columns 2 and 3 in the table below show two other possible translations.

Year	Number of years after 1960	Number of years after 1970	Cases reported (in thousands)
1989	29	19	185.4
1990	30	20	173.1
1991	31	21	147.1
1992	32	22	158.4
1993	33	23	134.7
1994	34	24	151.2

[*Source: U.S. Centers for Disease Control and Prevention*]

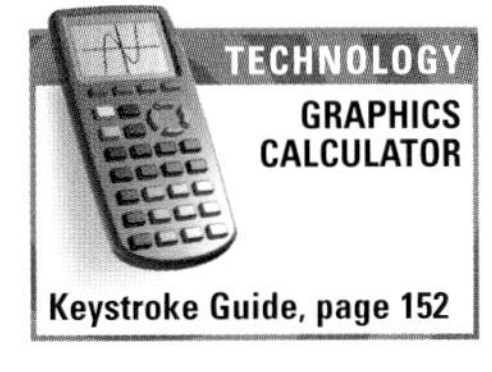

TECHNOLOGY GRAPHICS CALCULATOR

Keystroke Guide, page 152

Activity Exploring Translations of Data

You will need: a graphics calculator

1. Enter the data from columns 2 and 4 into your graphics calculator. Make a scatter plot. Then use linear regression to find an equation for the least-squares line.
2. Enter the data from columns 3 and 4 into your graphics calculator. Make a scatter plot. Then find an equation for the least-squares line.

CHECKPOINT ✔

3. How are the equations for the least-squares lines different? How are the graphs of the least-squares lines similar?

Just as you translate data, you can translate a function. A translation is one type of *transformation*. A **transformation** of a function is an alteration of the function rule that results in an alteration of its graph.

Prepare

NCTM PRINCIPLES & STANDARDS 1–10

QUICK WARM-UP

Evaluate each expression for $x = -2$.

1. $(x-6)^2$ 64
2. $x^2 - 6$ -2
3. $7x^2$ 28
4. $(7x)^2$ 196
5. $-x^2$ -4
6. $(-x)^2$ 4
7. $-3x^2 - 1$ -13
8. $-(3x-1)^2$ -49

Also on Quiz Transparency 2.7

Teach

Why Students most likely studied translations and reflections of geometric figures in previous courses, but they may recognize only informal names such as *slide* and *flip*. Elicit from them their understanding of these concepts. Tell them that graphs of functions are geometric figures and, as such, can be translated and reflected. Such *transformations* can be related directly to an equation for the function.

Activity Notes

In this Activity, students make two scatter plots of the same data, using a different "reference year" for each. Students should discover that the least-squares lines for the two scatter plots have the same slope, but that the second line is 10 units to the left of the first. This sets the stage for the study of translations of graphs.

Alternative Teaching Strategy

HANDS-ON STRATEGIES Have students graph the line $y = 2x$. Tell them to trace the line on a sheet of paper and translate the traced line 5 units up. Ask them to give an equation for the translated line. $y = 2x + 5$

Have them move the traced line back over the line $y = 2x$ and then translate it 5 units down. Ask for an equation of the translated line. $y = 2x - 5$

From this line, have students move the tracing 5 units to the right, back to its original position, and then 5 units to the left. Ask for the equations for the translated lines. $y = 2x - 10, y = 2x + 10$

Point out that alternative forms of these equations are $y = 2(x - 5)$ and $y = 2(x + 5)$.

Now have the students work in pairs. Each pair should experiment by translating other lines that pass through the origin. Tell each pair to make a conjecture about the relationship between the type of translation and the equation. Then have them apply their conjectures to translations of the graph of $y = x^2$.

Bring the class together to compare and discuss the results. Lead students to the generalizations about translations that are given on page 134.

CHECKPOINT ✔

3. The equations for the two regression lines have different y-intercepts but have the same slope.

ADDITIONAL

EXAMPLE 1

Graph each pair of functions, and identify the transformation from f to g.

a. $f(x) = |x|$ and $g(x) = |x| - 3$
vertical translation 3 units down; check students' graphs

b. $f(x) = x^2$ and $g(x) = (x + 2)^2$
horizontal translation 2 units to the left; check students' graphs

TRY THIS

a. **vertical translation 2 units down**

b. **horizontal translation 3 units to the right**

CHECKPOINT ✔

Writing the function as $j(x) = |x - (-4)|$ identifies h as -4, so the translation is to the left.

Translation

Example 1 illustrates a *vertical translation* and a *horizontal translation.*

EXAMPLE 1 **Graph each pair of functions, and identify the transformation from f to g.**

a. $f(x) = x^2$ and $g(x) = x^2 + 3$ **b.** $f(x) = |x|$ and $g(x) = |x + 4|$

SOLUTION

a.

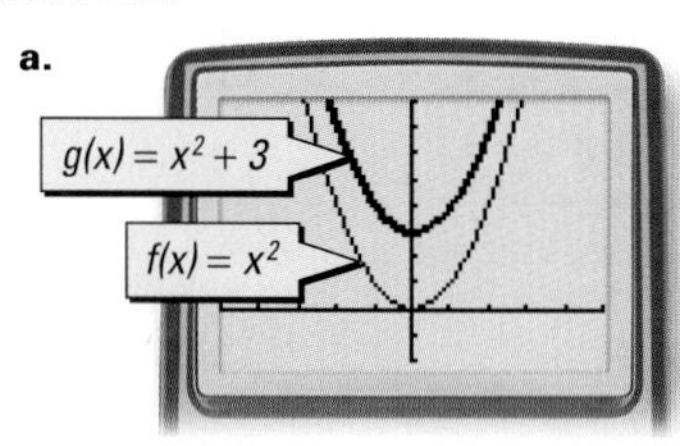

Notice that $g(x) = f(x) + 3$. The function $g(x) = x^2 + 3$ is a vertical translation of the graph of f 3 units up.

b.

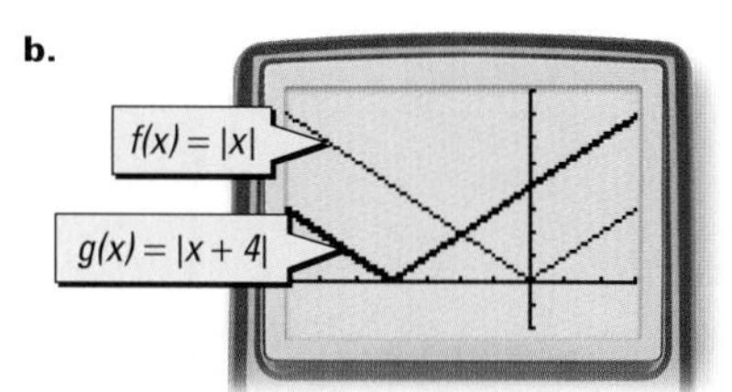

Notice that $g(x) = f(x + 4)$. The function $g(x) = |x + 4|$ is a horizontal translation of the graph of f 4 units to the left.

TRY THIS Graph each pair of functions, and identify the transformation from f to g.

a. $f(x) = x^2$ and $g(x) = x^2 - 2$ **b.** $f(x) = |x|$ and $g(x) = |x - 3|$

Notice that the shape of the transformed graphs in Example 1 do not change. Only the position of the graphs with respect to the coordinate axes changes.

Vertical and horizontal translations are generalized as follows:

Vertical and Horizontal Translations

If $y = f(x)$, then $y = f(x) + k$ gives a **vertical translation** of the graph of f. The translation is k units up for $k > 0$ and $|k|$ units down for $k < 0$.

If $y = f(x)$, then $y = f(x - h)$ gives a **horizontal translation** of the graph of f. The translation is h units to the right for $h > 0$ and $|h|$ units to the left for $h < 0$.

Vertical Translation

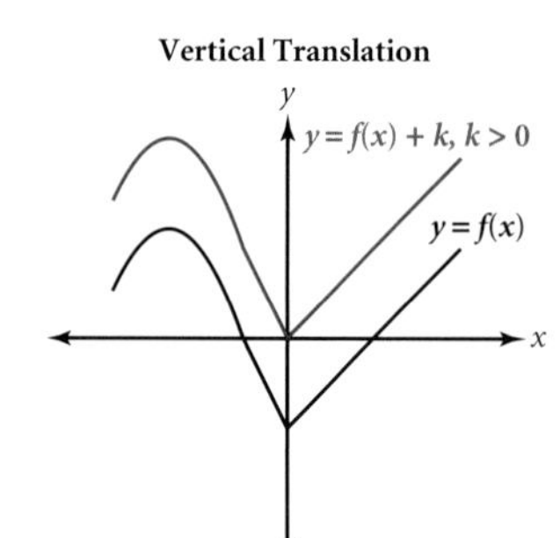

Horizontal Translation

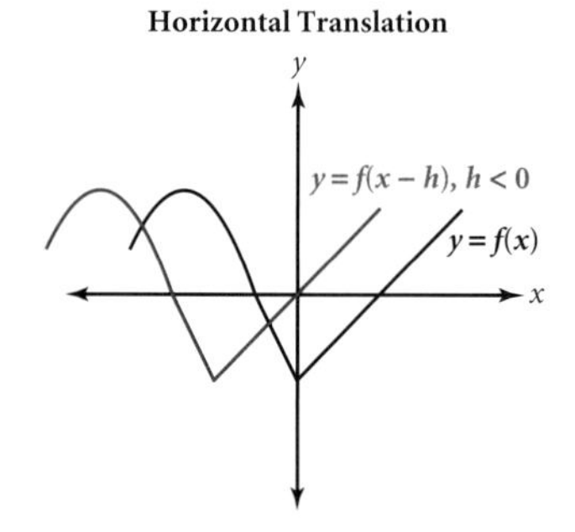

CHECKPOINT ✔ How does writing the function $j(x) = |x + 4|$ as $j(x) = |x - (-4)|$ help you to recognize that the translation is to the left?

Interdisciplinary Connection

ART The design at right utilizes several transformations of the parent function $f(x) = x^2$.

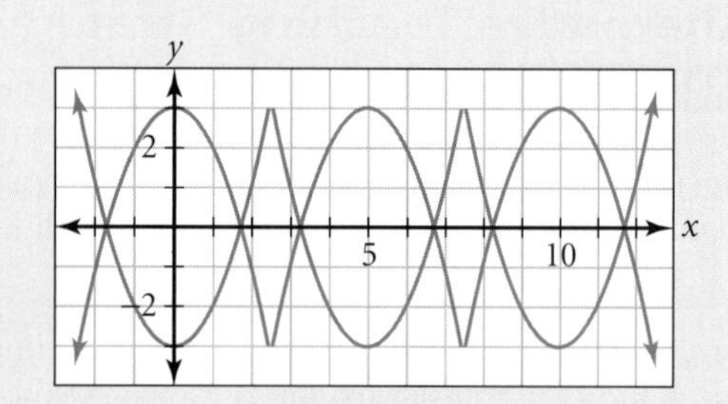

a. Give equations for all of the transformations shown. **$g(x) = x^2 - 3$, $h(x) = -x^2 + 3$, $j(x) = (x - 5)^2 - 3$, $k(x) = -(x - 5)^2 + 3$, $p(x) = (x - 10)^2 - 3$, $q(x) = -(x - 10)^2 + 3$**

b. Suppose that the design were continued indefinitely. Write general forms for all of the transformations. **for nonnegative integers a, $r(x) = (x - 5a)^2 - 3$, $s(x) = -(x - 5a)^2 + 3$**

c. Create an original design by using transformations of a parent function. Write equations for all the transformations involved. **Answers will vary**

Vertical Stretch and Compression

Example 2 illustrates a *vertical stretch* and a *vertical compression*.

EXAMPLE 2 **Graph each pair of functions, and identify the transformation from *f* to *g*.**

a. $f(x) = \sqrt{16 - x^2}$ and $g(x) = 2\sqrt{16 - x^2}$

b. $f(x) = \sqrt{16 - x^2}$ and $g(x) = \frac{1}{2}\sqrt{16 - x^2}$

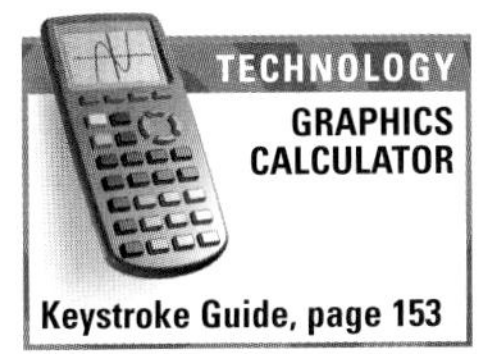

SOLUTION

a.

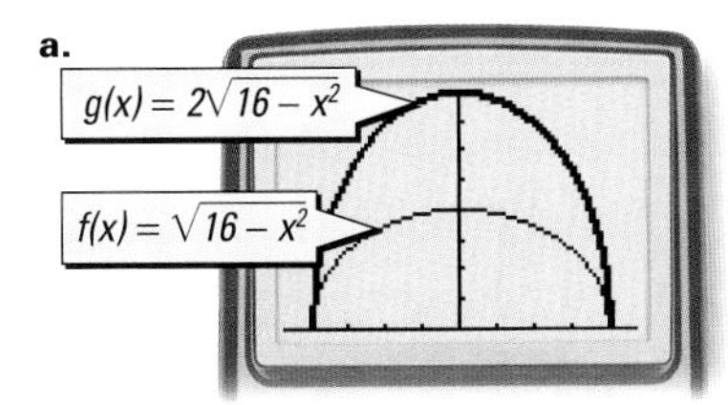

Notice that $g(x) = 2 \cdot f(x)$.
The function $g(x) = 2\sqrt{16 - x^2}$ is a vertical stretch of the graph of *f* by a factor of 2.

b.

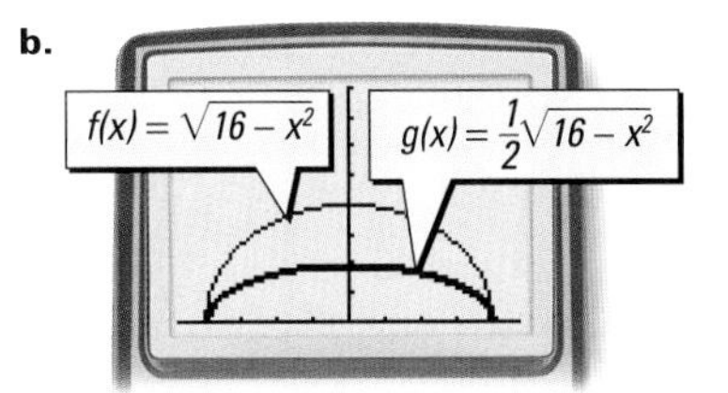

Notice that $g(x) = \frac{1}{2} \cdot f(x)$.
The function $g(x) = \frac{1}{2}\sqrt{16 - x^2}$ is a vertical compression of the graph of *f* by a by a factor of $\frac{1}{2}$.

TRY THIS Graph each pair of functions, and identify the transformation from *f* to *g*.

a. $f(x) = \sqrt{25 - x^2}$ and $g(x) = 3\sqrt{25 - x^2}$

b. $f(x) = \sqrt{25 - x^2}$ and $g(x) = \frac{1}{3}\sqrt{25 - x^2}$

A vertical stretch or compression moves the graph away from or toward the *x*-axis, respectively. It can be generalized as follows:

Vertical Stretch and Vertical Compression

If $y = f(x)$, then $y = af(x)$ gives a **vertical stretch** or **vertical compression** of the graph of *f*.

- If $a > 1$, the graph is stretched vertically by a factor of *a*.
- If $0 < a < 1$, the graph is compressed vertically by a factor of *a*.

Vertical Stretch

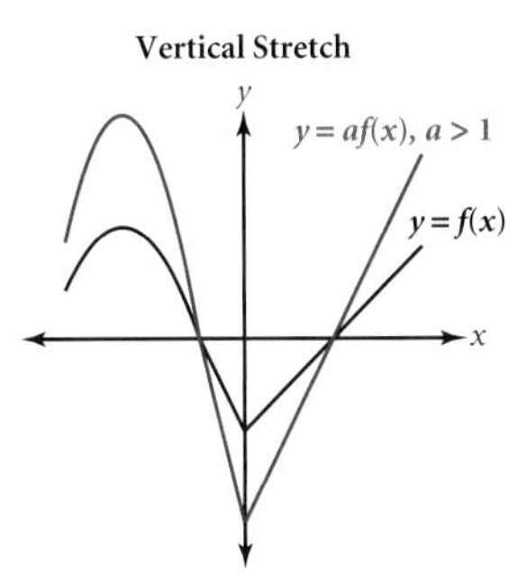

Vertical Compression

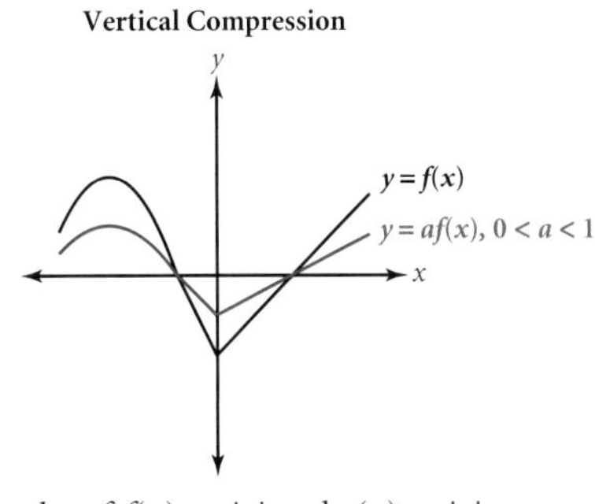

CRITICAL THINKING Let $0 < r < 1$ and $s > 1$. Compare the graphs of $f(x) = r|x|$ and $g(x) = s|x|$.

Inclusion Strategies

VISUAL LEARNERS Given a function such as $f(x) = -4(x + 1)^2 + 3$, some students will see only a maze of symbols. It may be helpful for them to look at just one transformation at a time, as shown in the graphs. Point out that any reflection across an axis should be done first.

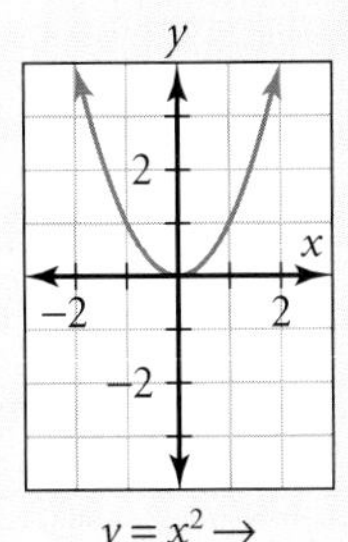

$y = x^2 \rightarrow$

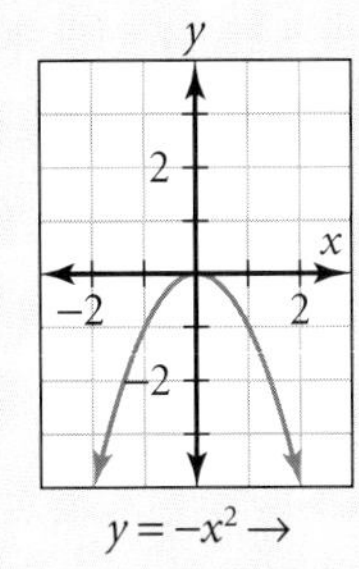

$y = -x^2 \rightarrow$

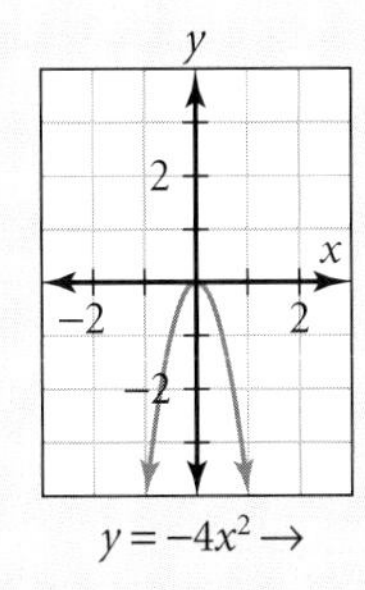

$y = -4x^2 \rightarrow$

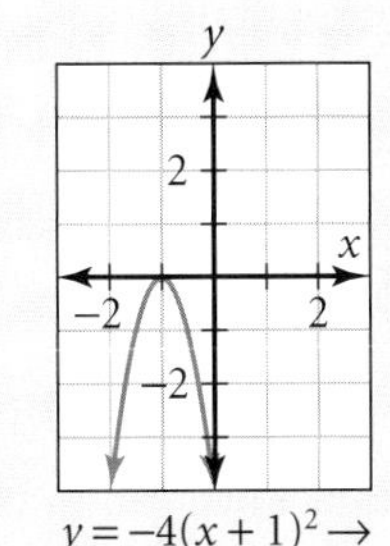

$y = -4(x + 1)^2 \rightarrow$

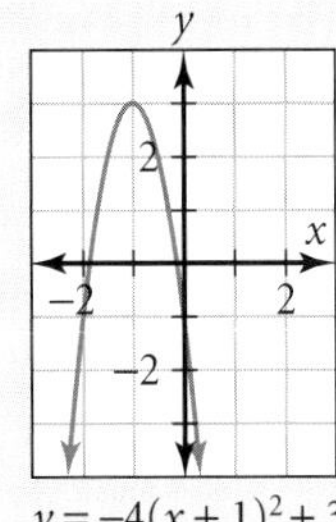

$y = -4(x + 1)^2 + 3$

ADDITIONAL EXAMPLE 2

Graph each pair of functions, and identify the transformation from *f* to *g*.

a. $f(x) = \sqrt{36 - x^2}$ and $g(x) = \frac{3}{2}\sqrt{36 - x^2}$
vertical stretch by a factor of $\frac{3}{2}$
Check students' graphs.

b. $f(x) = \sqrt{36 - x^2}$ and $g(x) = \frac{2}{3}\sqrt{36 - x^2}$
vertical compression by a factor of $\frac{2}{3}$
Check students' graphs.

Teaching Tip

TECHNOLOGY In Example 2, the graph of $f(x) = \sqrt{16 - x^2}$ is a semicircle. You may want students to use a square viewing window to get a better picture of the stretch and the compression.

TRY THIS

a. vertical stretch by a factor of 3

b. vertical compression by a factor of $\frac{1}{3}$

CRITICAL THINKING

If $0 < r < 1$, then the graph of $f(x) = r|x|$ is a vertical compression of the graph of $|x|$ by a factor of *r*. If $s > 1$, the graph of $g(x) = s|x|$ is a vertical stretch of the graph of $|x|$ by a factor of *s*. The graph of *f* does not rise as quickly as the graph of *g*.

ADDITIONAL
EXAMPLE 3

Graph each pair of functions, and identify the transformation from f to g.

a. $f(x) = \sqrt{36 - x^2}$ and $g(x) = \sqrt{36 - \left(\frac{3}{2}x\right)^2}$

horizontal compression by a factor of $\frac{2}{3}$

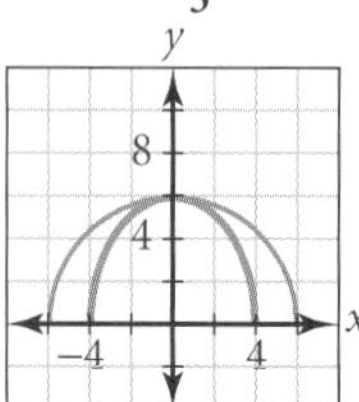

b. $f(x) = \sqrt{36 - x^2}$ and $g(x) = \sqrt{36 - \left(\frac{2}{3}x\right)^2}$

horizontal stretch by a factor of $\frac{3}{2}$

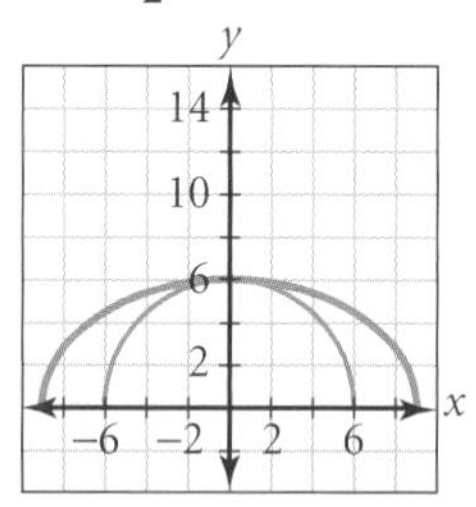

TRY THIS

a. horizontal compression by a factor of $\frac{1}{3}$

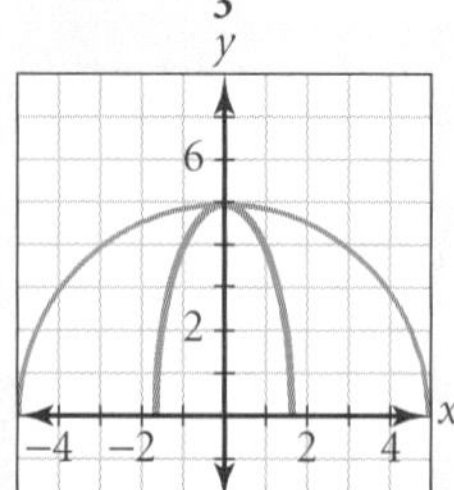

b. horizontal stretch by a factor of 4

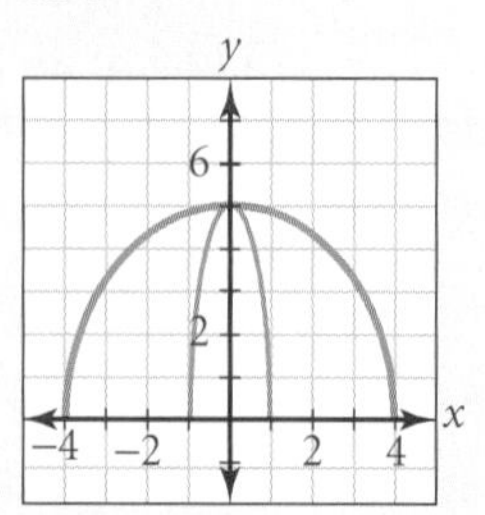

Horizontal Stretch and Compression

Example 3 illustrates a *horizontal stretch* and a *horizontal compression.*

EXAMPLE 3 **Graph each pair of functions, and identify the transformation from f to g.**

a. $f(x) = \sqrt{16 - x^2}$ and $g(x) = \sqrt{16 - (2x)^2}$

b. $f(x) = \sqrt{16 - x^2}$ and $g(x) = \sqrt{16 - \left(\frac{1}{2}x\right)^2}$

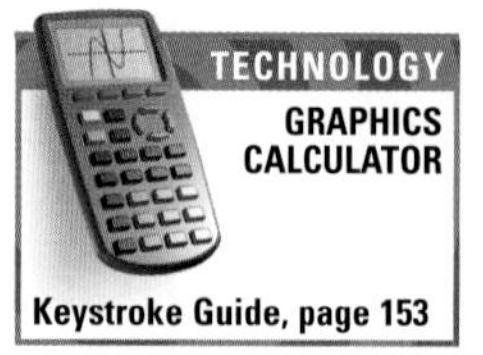

SOLUTION

a.

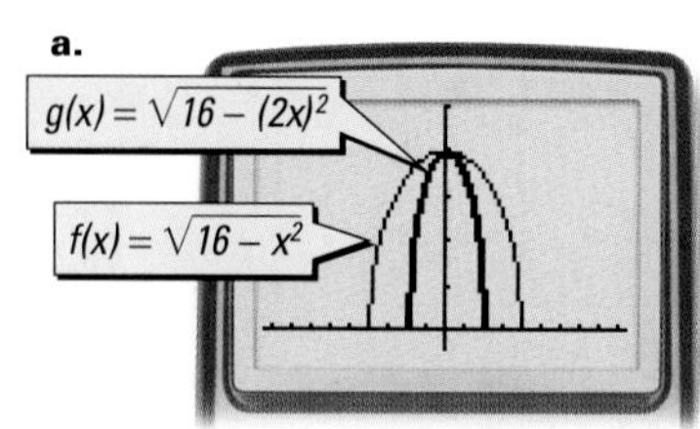

Notice that $g(x) = f(2x)$.
The function $g(x) = \sqrt{16 - (2x)^2}$ is a horizontal compression of the graph of f by a factor of $\frac{1}{2}$.

b.

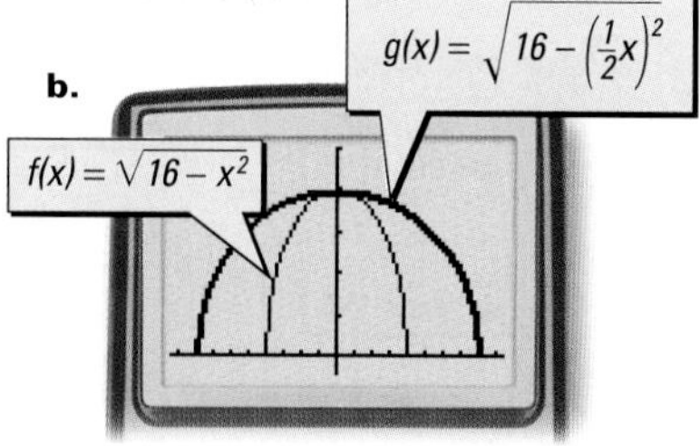

Notice that $g(x) = f\left(\frac{1}{2}x\right)$.

The function $g(x) = \sqrt{16 - \left(\frac{1}{2}x\right)^2}$ is a horizontal stretch of the graph of f by a factor of 2.

TRY THIS Graph each pair of functions, and identify the transformation from f to g.

a. $f(x) = \sqrt{25 - x^2}$ and $g(x) = \sqrt{25 - (3x)^2}$

b. $f(x) = \sqrt{25 - x^2}$ and $g(x) = \sqrt{25 - \left(\frac{1}{4}x\right)^2}$

A horizontal stretch or compression moves the graph away from or toward the y-axis, respectively. It can be generalized as follows:

Horizontal Stretch and Horizontal Compression

If $y = f(x)$, then $y = f(bx)$ gives a **horizontal stretch** or **horizontal compression** of the graph of f.

- If $b > 1$, the graph is compressed horizontally by a factor of $\frac{1}{b}$.
- If $0 < b < 1$, the graph is stretched horizontally by a factor of $\frac{1}{b}$.

Horizontal Stretch

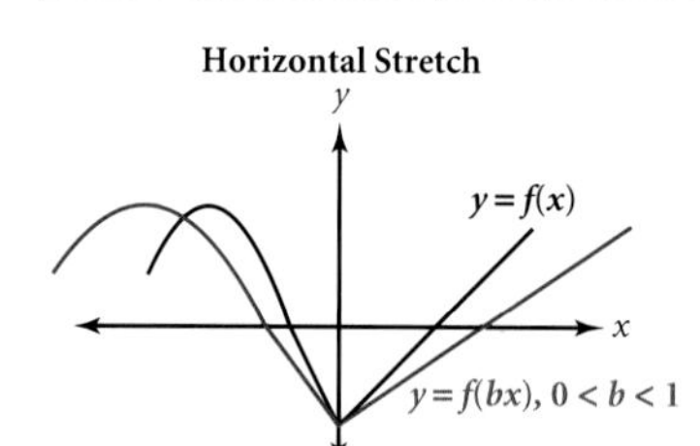

Horizontal Compression

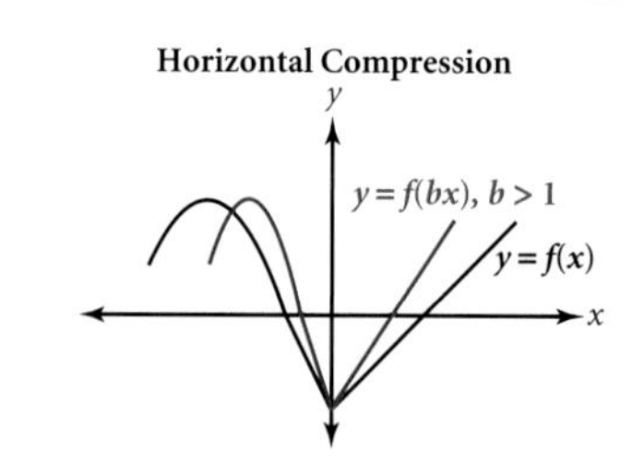

Enrichment

Have students work in pairs. Ask them to describe the transformation of a graph that results when the following changes are made to its equation:

Replace x with y, and replace y with $-x$.

The graph is rotated one-quarter turn, or 90°, counterclockwise about the origin.

After students have arrived at the correct conclusion, ask them to determine what changes in an equation will result in one-quarter turn *clockwise* about the origin.

Replace x with $-y$, and replace y with x.

Reflection

Example 4 illustrates reflections across the x-axis and across the y-axis.

EXAMPLE 4 **Graph each pair of functions, and identify the transformation from f to g.**

a. $f(x) = x^2$ and $g(x) = -(x^2)$ **b.** $f(x) = 2x + 3$ and $g(x) = 2(-x) + 3$

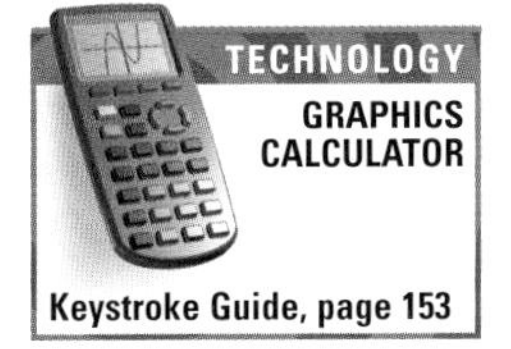

SOLUTION

a.

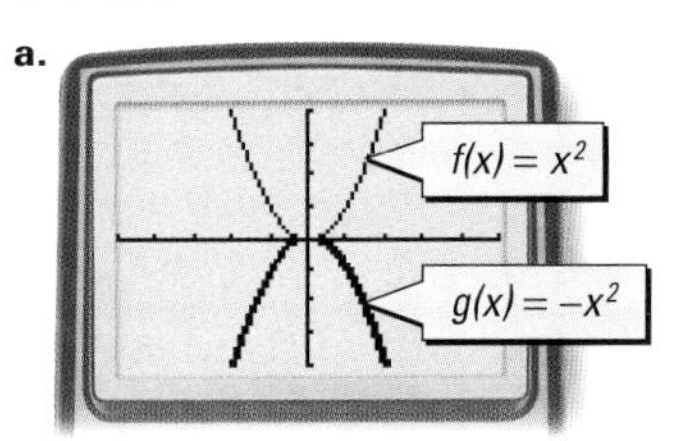

Notice that $g(x) = -f(x)$. The function $g(x) = -(x^2)$ is a reflection of the graph of f across the x-axis.

b.

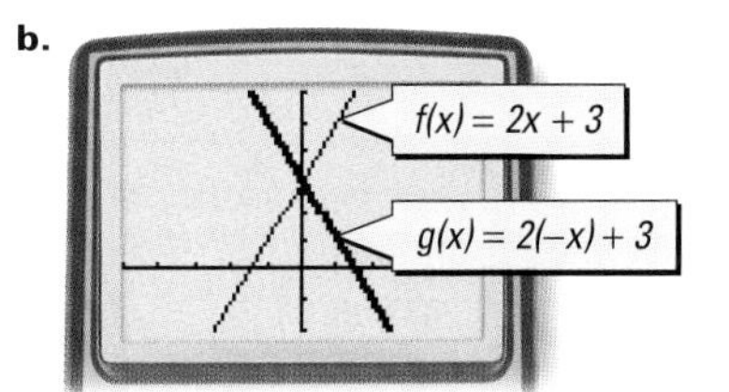

Notice that $g(x) = f(-x)$. The function $g(x) = 2(-x) + 3$, or $g(x) = -2x + 3$, is a reflection of the graph of f across the y-axis.

TRY THIS Graph each pair of functions, and identify the transformation from f to g.

a. $f(x) = |x|$ and $g(x) = -|x|$ **b.** $f(x) = 2x - 1$ and $g(x) = 2(-x) - 1$

CRITICAL THINKING Can the graph of a function be reflected across both the x- and y-axes simultaneously? Explain and give an example.

Reflections can be generalized as follows:

Reflections

If $y = f(x)$, then $y = -f(x)$ gives a **reflection** of the graph of f across the x-axis.

If $y = f(x)$, then $y = f(-x)$ gives a **reflection** of the graph of f across the y-axis.

Reflection Across the x-axis

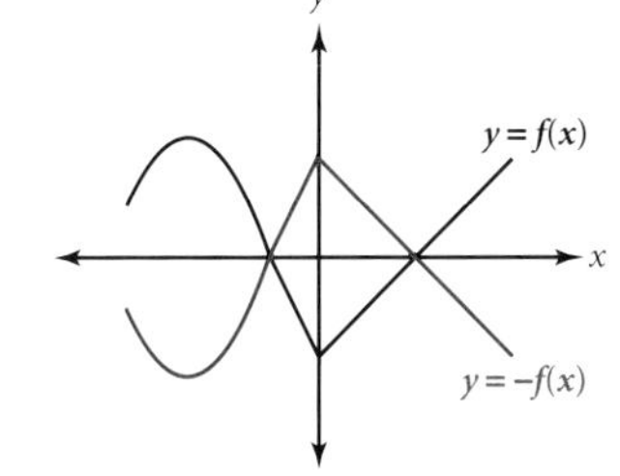

Reflection Across the y-axis

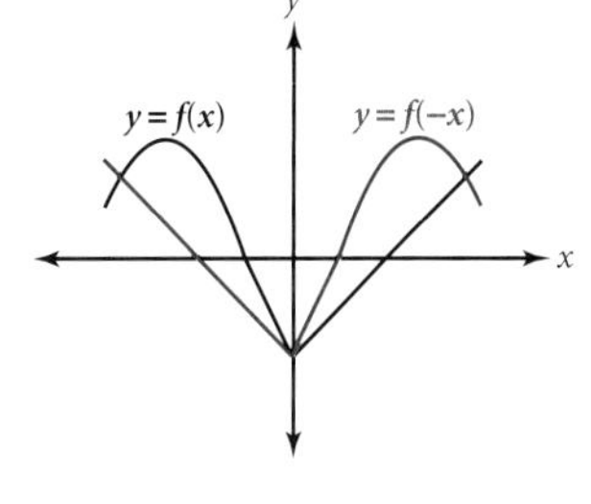

CRITICAL THINKING Let $f(x) = x^2$. What does $f(-x) = (-x)^2 = x^2$ tell you about the graph of $f(-x)$? Justify your response.

Reteaching the Lesson

GUIDED ANALYSIS Tell students to consider the parent function $f(x) = x^2$. Ask them to write an equation for a function, g, whose graph is a horizontal translation of the graph of f. Have volunteers share their equations, and display a list of the equations for the class to see. Discuss with the class what all of the correct equations have in common. **All are of the form $g(x) = (x \pm h)^2$.** Ask what characteristics all of the graphs have in common. **All touch the x-axis at exactly one point. The rest of the graphs lie entirely within the first and second quadrants.**

Repeat the activity for other transformations and combinations of transformations. In each case, make a list of students' equations and then lead the class through an analysis of common characteristics of the equations and of the graphs.

ADDITIONAL EXAMPLE 4

Graph each pair of functions, and identify the transformation from f to g.

a. $f(x) = \sqrt{36 - x^2}$ and $g(x) = -\sqrt{36 - x^2}$

vertical reflection across the x-axis

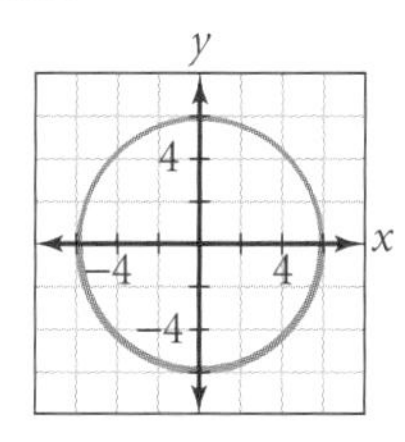

b. $f(x) = 3x + 5$ and $g(x) = 3(-x) + 5$

horizontal reflection across the y-axis

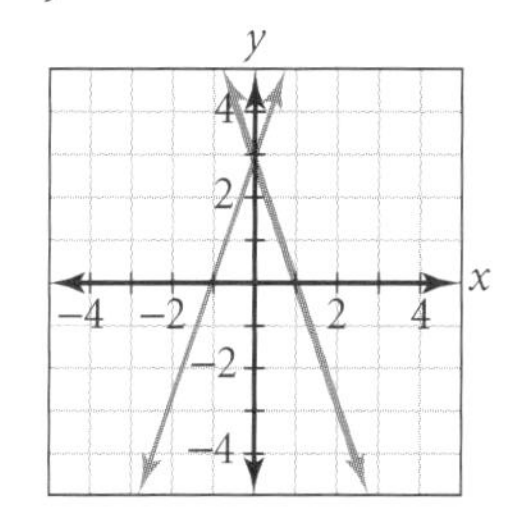

TRY THIS

a. reflection across the x-axis

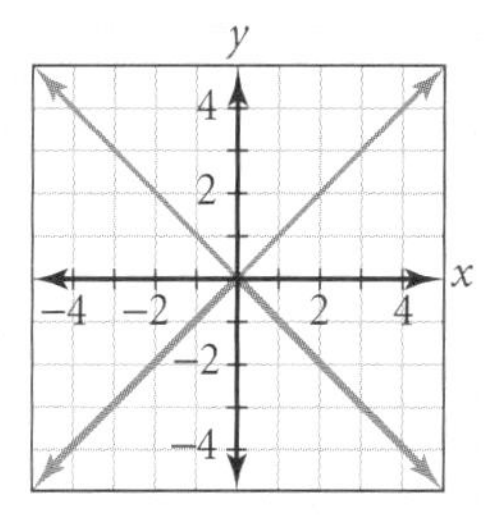

b. reflection across the y-axis

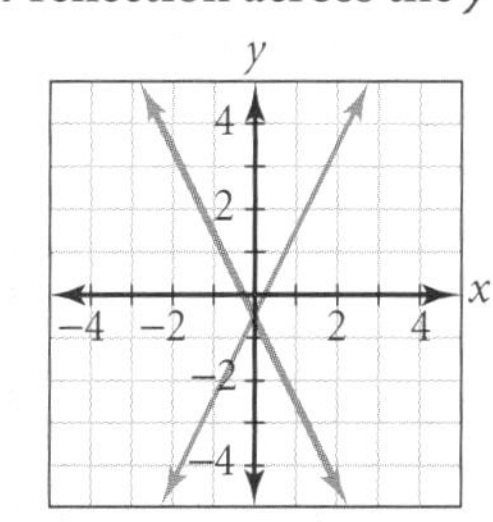

CRITICAL THINKING

Yes; for $f(x) = 2x - 1$, the function $-f(-x) = -[2(-x) - 1]$ would be a reflection of f across both the x- and y-axes.

CRITICAL THINKING

Because $f(-x) = (-x)^2 = x^2$ and $f(x) = x^2$, the graph of $f(-x)$ is symmetric with respect to the

Combining Transformations

The functions $f(x) = x^2$ and $h(x) = |x|$ are examples of *parent functions* because other functions are related to them by one or more transformations. Example 5 illustrates a combination of transformations.

EXAMPLE 5 **Graph each pair of functions, and identify the transformations from f to g.**

a. $f(x) = x^2$ and $g(x) = (x + 2)^2 - 1$ **b.** $f(x) = |x|$ and $g(x) = -4|x| + 3$

SOLUTION

a.

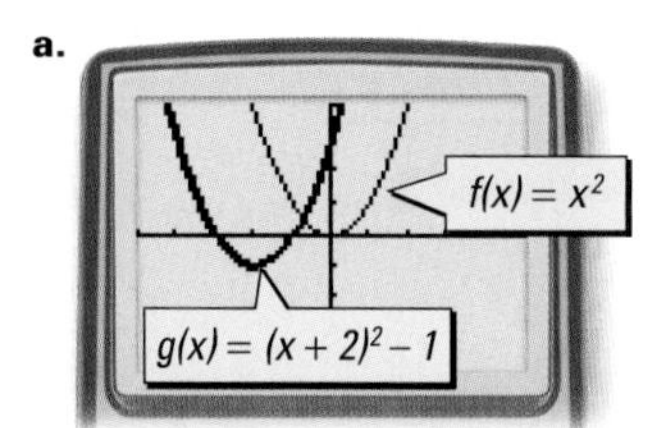

Notice that $g(x) = f(x + 2) - 1$, or $g(x) = f(x - (-2)) - 1$. There are two transformations of the parent function $f(x) = x^2$: a translation of 2 units to the left and a translation of 1 unit down.

$$g(x) = [x - (-2)]^2 - 1$$

horizontal translation (-2); vertical translation (-1)

b.

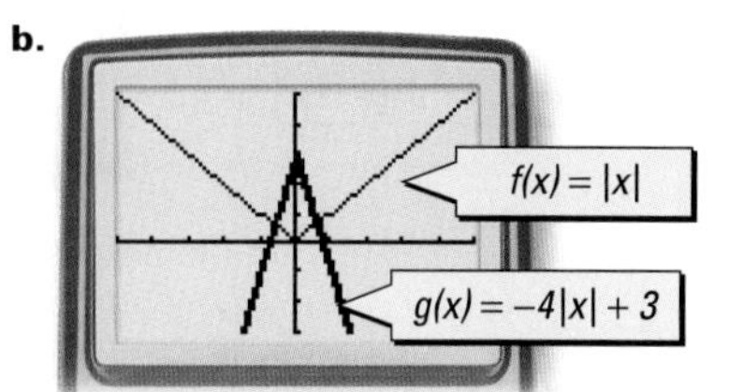

Notice that $g(x) = -4 \cdot f(x) + 3$. There are three transformations of the parent function $f(x) = |x|$: a vertical stretch by a factor of 4, a reflection across the x-axis, and a vertical translation of 3 units up.

$$g(x) = -4|x| + 3$$

reflection ($-$); vertical stretch (4); vertical translation ($+3$)

TRY THIS Graph each pair of functions, and identify the transformations from f to g.

a. $f(x) = x^2$ and $g(x) = 3(x - 1)^2$ **b.** $f(x) = |x|$ and $g(x) = \frac{1}{2}|x| - 2$

A summary of the transformations in this lesson is given below.

SUMMARY OF TRANSFORMATIONS	
Transformations of $y = f(x)$	**Transformed function**
Vertical translation of k units up	$y = f(x) + k$, where $k > 0$
Vertical translation of $\lvert k\rvert$ units down	$y = f(x) + k$, where $k < 0$
Horizontal translation of h units to the right	$y = f(x - h)$, where $h > 0$
Horizontal translation of $\lvert h\rvert$ units to the left	$y = f(x - h)$, where $h < 0$
Vertical stretch by a factor of a	$y = af(x)$, where $a > 1$
Vertical compression by a factor of a	$y = af(x)$, where $0 < a < 1$
Horizontal stretch by a factor of $\frac{1}{b}$	$y = f(bx)$, where $0 < b < 1$
Horizontal compression by a factor of $\frac{1}{b}$	$y = f(bx)$, where $b > 1$
Reflection across the x-axis	$y = -f(x)$
Reflection across the y-axis	$y = f(-x)$

ADDITIONAL EXAMPLE 5

Graph each pair of functions, and identify the transformations from f to g.

a. $f(x) = x^2$ and $g(x) = 3x^2 + 4$

vertical stretch by a factor of 3; vertical translation 4 units up

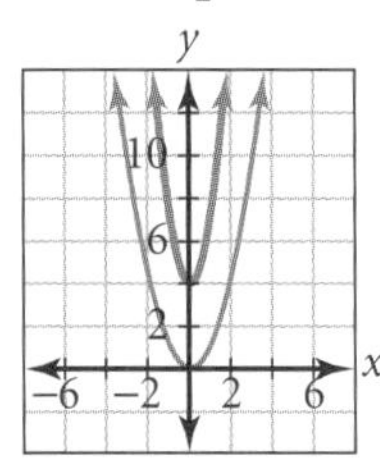

b. $f(x) = |x|$ and $g(x) = -|x - 2| - 1$

reflection across the x-axis; horizontal translation 2 units to the right; vertical translation 1 unit down

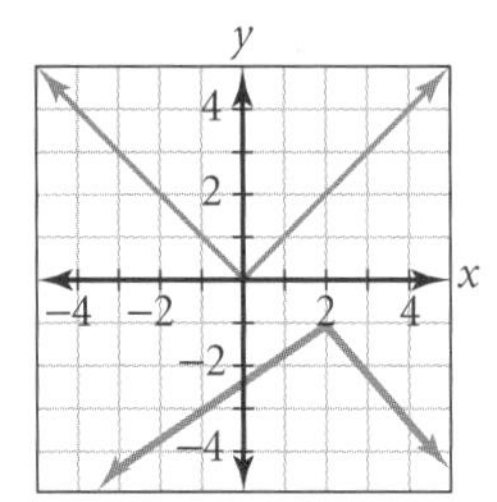

TRY THIS

a. translation 1 unit to the right and vertical stretch by a factor of 3

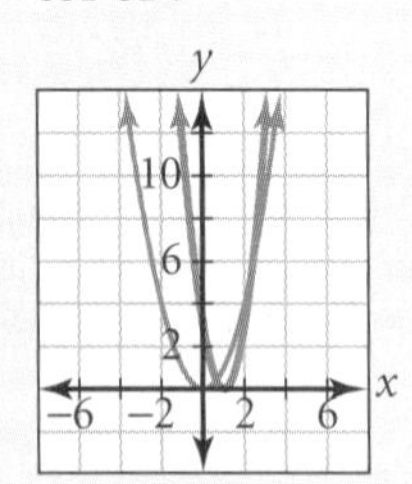

b. vertical compression by a factor of $\frac{1}{2}$ and vertical translation 2 units down

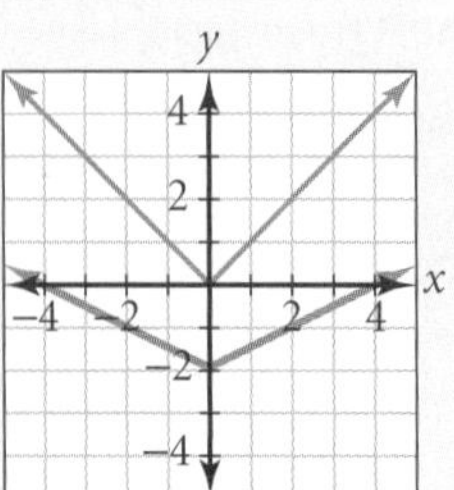

Exercises

Communicate

1. Describe each transformation of $f(x) = x^2$.

a. 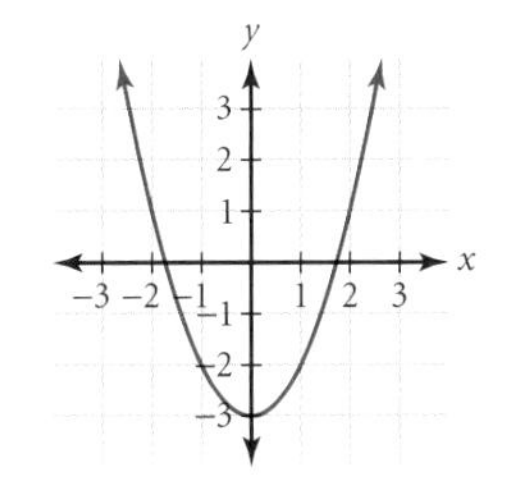

b. 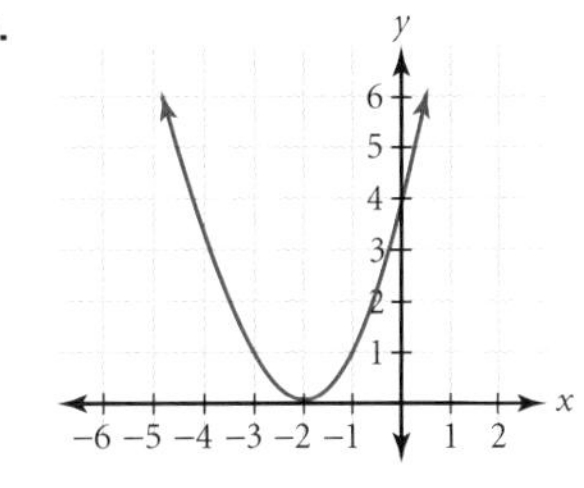

2. Describe the range of possible values of $\frac{1}{b}$ when $0 < b < 1$.
3. Compare and contrast reflections and translations.
4. For what values of h will the graph of $f(x - h)$ be translated to the right? to the left? Explain.
5. Differentiate between the effects of a vertical stretch and a horizontal stretch on the graph of a function.

Guided Skills Practice

Identify the transformations from f to g.

6. $f(x) = x^2$ and $g(x) = x^2 - 3$ ***(EXAMPLE 1)***
7. $f(x) = \sqrt{9 - x^2}$ and $g(x) = \frac{4}{3}\sqrt{9 - x^2}$ ***(EXAMPLE 2)***
8. $f(x) = \sqrt{36 - x^2}$ and $g(x) = \sqrt{36 - (2x)^2}$ ***(EXAMPLE 3)***
9. $f(x) = -3x + 1$ and $g(x) = -3(-x) + 1$ ***(EXAMPLE 4)***
10. $f(x) = x^2$ and $g(x) = (x - 3)^2 + 1$ ***(EXAMPLE 5)***

Practice and Apply

Identify each transformation from the parent function $f(x) = x^2$ to g.

11. $g(x) = 4x^2$
12. $g(x) = 5x^2$
13. $g(x) = (4x)^2$
14. $g(x) = (-5x)^2$
15. $g(x) = -\frac{1}{2}x^2$
16. $g(x) = -\frac{1}{5}x^2$
17. $g(x) = x^2 - 2$
18. $g(x) = x^2 + 3$
19. $g(x) = (x - 2)^2$
20. $g(x) = (x + 3)^2$
21. $g(x) = (-5x)^2 + 2$
22. $g(x) = 3(x - 1)^2$
23. $g(x) = \frac{1}{3}x^2 - 1$
24. $g(x) = -\frac{1}{4}x^2 + 3$
25. $g(x) = -2(x + 4)^2 + 1$
26. $g(x) = -5(x - 2)^2 - 4$

Assess

Selected Answers

Exercises 6–10, 11–75 odd

ASSIGNMENT GUIDE	
In Class	1–10
Core	11–57 odd, 60–65
Core Plus	12–56 even, 58–66
Review	67–75
Preview	76

Extra Practice can be found beginning on page 940.

6. vertical translation 3 units down

7. vertical stretch by a factor of $\frac{4}{3}$

8. horizontal compression by a factor of 2

9. reflection across the y-axis

10. two transformations: shifted 3 units to the right and shifted 1 unit up

11. vertical stretch by a factor of 4

12. vertical stretch by a factor of 5

13. horizontal compression by a factor of $\frac{1}{4}$

14. horizontal compression by a factor of $\frac{1}{5}$ and a reflection across the y-axis

15. vertical compression by a factor of $\frac{1}{2}$ and a reflection across the x-axis

16. vertical compression by a factor of $\frac{1}{5}$ and a reflection across the x-axis

17. vertical translation 2 units down

18. vertical translation 3 units up

19. horizontal translation 2 units to the right

20. horizontal translation 3 units to the left

21. horizontal compression by a factor of $\frac{1}{5}$, a vertical translation 2 units up, and a reflection across the y-axis

22. horizontal translation 1 unit to the right and a vertical stretch by a factor of 3

23. vertical compression by a factor of $\frac{1}{3}$ and a vertical translation 1 unit down

24. vertical compression by a factor of $\frac{1}{4}$, a reflection across the x-axis, and a vertical translation 3 units up

25. horizontal translation 4 units to the left, a vertical stretch by a factor of 2, a reflection across the x-axis, and a vertical translation 1 unit up

26. horizontal translation 2 units to the right, a vertical stretch by a factor of 5, a reflection across the x-axis, and a vertical translation 4 units down

Error Analysis

Students commonly associate positive numbers with right and negative numbers with left, so they often use the incorrect sign when writing an equation for a translation to the right or left. For instance, given $y = |x|$ and a translation of 5 units to the left, they may write $y = |x - 5|$. Encourage them to sketch the translated graph, identify the coordinates of two points on the graph, and verify that the coordinates satisfy their equation.

Identify each transformation from the parent function $f(x) = \sqrt{x}$ to g.

27. $g(x) = 4\sqrt{x}$ **28.** $g(x) = 3\sqrt{x}$ **29.** $g(x) = -\frac{1}{4}\sqrt{x}$

30. $g(x) = -\frac{1}{3}\sqrt{x}$ **31.** $g(x) = \sqrt{-4x}$ **32.** $g(x) = \sqrt{-3x}$

33. $g(x) = \sqrt{x} + 4$ **34.** $g(x) = \sqrt{x} - 3$ **35.** $g(x) = \sqrt{x + 4}$

36. $g(x) = \sqrt{x - 3}$ **37.** $g(x) = \sqrt{-2x} + 1$ **38.** $g(x) = -\sqrt{x + 3}$

39. $g(x) = -\sqrt{x - 4} + 3$ **40.** $g(x) = -\sqrt{3x} - 1$ **41.** $g(x) = -\sqrt{-x}$

Write the function for each graph described below.

42. the graph of $f(x) = |x|$ translated 4 units to the left **$g(x) = |x + 4|$**

43. the graph of $f(x) = x^2$ translated 2 units to the right **$g(x) = (x - 2)^2$**

44. the graph of $f(x) = |x|$ translated 5 units up **$g(x) = |x| + 5$**

45. the graph of $f(x) = x^2$ translated 6 units down **$g(x) = x^2 - 6$**

46. the graph of $f(x) = x^2$ vertically stretched by a factor of 3 **$g(x) = 3x^2$**

47. the graph of $f(x) = \sqrt{x}$ vertically compressed by a factor of $\frac{1}{3}$ **$g(x) = \frac{1}{3}\sqrt{x}$**

48. the graph of $f(x) = x^2$ horizontally compressed by a factor of $\frac{1}{5}$ **$g(x) = (5x)^2$**

49. the graph of $f(x) = \sqrt{x}$ horizontally stretched by a factor of 4

50. the graph of $f(x) = 3x + 1$ reflected across the x-axis **$g(x) = -(3x + 1)$**

51. the graph of $f(x) = 2x - 1$ reflected across the y-axis **$g(x) = 2(-x) - 1$**

52. the graph of $f(x) = x^2$ vertically stretched by a factor of 2 and translated 1 unit to the right **$g(x) = 2(x - 1)^2$**

53. the graph of $f(x) = |x|$ horizontally compressed by a factor of $\frac{1}{3}$, reflected across the x-axis, and translated 3 units down

54. the graph of $f(x) = x^2$ translated 7 units to the left **$g(x) = (x + 7)^2$**

55. the graph of $f(x) = x^2$ translated 5 units up **$g(x) = x^2 + 5$**

56. the graph of $f(x) = x^2$ stretched vertically by a factor of 2 **$g(x) = 2x^2$**

57. the graph of $f(x) = x^2$ reflected across the y-axis and stretched horizontally by a factor of 2 **$g(x) = \left(-\frac{1}{2}x\right)^2$**

49. $g(x) = \sqrt{\frac{1}{4}x}$

53. $g(x) = -|3x| - 3$

CHALLENGES

58. How are the domain and range of a function affected by a reflection across the y-axis? across the x-axis? Include examples in your explanation.

59. Show that a vertical compression can have the same effect on a graph as a horizontal stretch.

At right is the graph of the function f. Draw a careful sketch of each transformation of f.

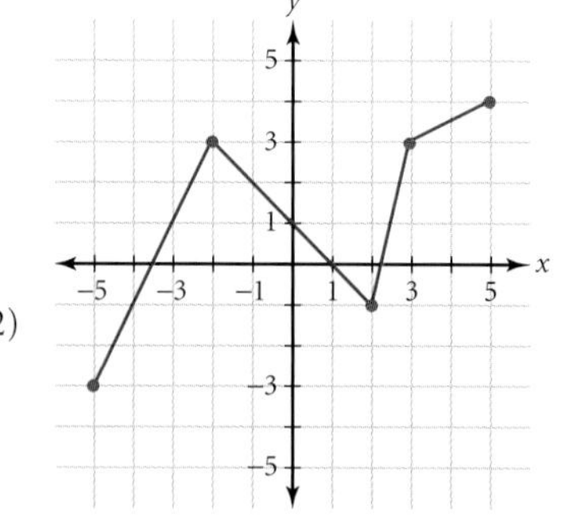

60. $g(x) = f(2x)$ **61.** $g(x) = 2f(x)$

62. $g(x) = -f(x)$ **63.** $g(x) = f(x + 2)$

64. $g(x) = f(x) + 3$ **65.** $g(x) = f\left(\frac{1}{2}x\right)$

27. vertical stretch by a factor of 4

28. vertical stretch by a factor of 3

29. vertical compression by a factor of $\frac{1}{4}$ and a reflection across the x-axis

30. vertical compression by a factor of $\frac{1}{3}$ and a reflection across the x-axis

31. horizontal compression by a factor of $\frac{1}{4}$ and a reflection across the y-axis

32. horizontal compression by a factor of $\frac{1}{3}$ and a reflection across the y-axis

33. vertical translation 4 units up

34. vertical translation 3 units down

35. horizontal translation 4 units to the left

36. horizontal translation 3 units to the right

37. horizontal compression by a factor of $\frac{1}{2}$, a reflection across the y-axis, and a vertical translation 1 unit up

38. horizontal translation 3 units to the left and a reflection across the x-axis

39. horizontal translation 4 units to the right, a reflection across the x-axis, and a vertical translation 3 units up

40. horizontal compression by a factor of $\frac{1}{3}$, a reflection across the x-axis, and a vertical translation 1 unit down

41. reflection across the x-axis and a reflection across the y-axis

Practice

NAME ______ CLASS ______ DATE ______

Practice
2.7 A Preview of Transformations

Identify each transformation from the parent function $f(x) = x^2$ to g.

1. $g(x) = (x + 7.5)^2$ a horizontal translation 7.5 units to the left
2. $g(x) = x^2 + 7.5$ a vertical translation 7.5 units up
3. $g(x) + (52x)^2$ a horizontal compression by a factor of 52
4. $g(x) = -2x^2$ a vertical stretch by a factor of 2 and a reflection across the x-axis
5. $g(x) = 14x^2 + 6$ a vertical stretch by a factor of 14 and a vertical translation 6 units up
6. $g(x) = 12(x - 7)^2$ a vertical stretch by a factor of 12 and a horizontal translation 7 units to the right

Identify each transformation from the parent function $f(x) = \sqrt{x}$ to g.

7. $g(x) = \sqrt{x + 21}$ a horizontal translation 21 units to the left
8. $g(x) = 17\sqrt{x}$ a vertical stretch by a factor of 17
9. $g(x) = \sqrt{\frac{1}{2}x}$ a horizontal stretch by a factor of 2
10. $g(x) = \sqrt{x} + 13.7$ a vertical translation 13.7 units up
11. $g(x) = -3\sqrt{x}$ a vertical stretch by a factor of 3 and a reflection across the x-axis
12. $g(x) = 41\sqrt{x - 8}$ a vertical stretch by a factor of 41 and a horizontal translation 8 units to the right

Write the function for each graph described below.

13. the graph of $f(x) = x^3$ reflected across the x-axis $g(x) = -x^3$
14. the graph of $f(x) = x^5$ translated 7 units to the left $g(x) = (x + 7)^5$
15. the graph of $f(x) = x^4$ stretched horizontally by a factor of 26 $g(x) = (26x)^4$
16. the graph of $f(x) = |x|$ compressed vertically by a factor of $\frac{1}{12}$ $g(x) = \left|\frac{x}{12}\right|$
17. the graph of $f(x) = \frac{2}{3}x + 9$ reflected across the y-axis $g(x) = -\frac{2}{3}x + 9$
18. the graph of $f(x) = x^3$ translated 33 units down $g(x) = x^3 - 33$

APPLICATION

66. PHYSICS Let the function $h(t) = -16t^2 + 100$ model the altitude of an eagle in free fall when it dives from an initial altitude of 100 feet to catch a fish.

a. Describe the transformations from the graph of $f(t) = t^2$ to the graph of h.

b. Write a transformed function, h, that represents the altitude of an eagle diving from an initial altitude of 25 feet.

c. Graph f and h on the same coordinate plane, and describe the transformation from f to h by following the transformation of a point.

A bald eagle, the national bird of the United States

Look Back

67. $y = 2x + 10$

67. Write an equation in slope-intercept form for the line that contains $(-3, 4)$ and is perpendicular to the graph of $y = -\frac{1}{2}x + 10$. ***(LESSON 1.3)***

68. Solve $\frac{3}{5} = \frac{5}{x}$. ***(LESSON 1.4)*** $x = \frac{25}{3}$

69. Solve the literal equation $D = \pi\left(\frac{b}{2}\right)^2 sn$ for s. ***(LESSON 1.6)*** $s = \frac{4D}{\pi b^2 n}$

Classify each number in as many ways as possible. ***(LESSON 2.1)***

70. 15.3849 rational, real

71. 14.393939 . . . rational, real

72. 17.1012012301234 . . . irrational, real

Find the inverse of each relation. State whether the relation is a function. State whether the inverse is a function. ***(LESSON 2.5)***

73. $\{(1, 100), (2, 200), (3, 300), (4, 400)\}$ yes; yes

74. $\{(1, 5), (2, 10), (3, 10), (4, 15)\}$ yes; no

75. Graph the piecewise function below. ***(LESSON 2.6)***

$$f(x) = \begin{cases} x & \text{if } -4 \le x \le -1 \\ x + 2 & \text{if } -1 < x < 3 \\ -3x + 18 & \text{if } 3 \le x \le 5 \end{cases}$$

Look Beyond

76 Graph the linear functions $x - 2y = -4$ and $3x + 2y = -4$ together. Find the coordinates of the point of intersection. Then add the corresponding sides of the two equations and solve the resulting equation for x. Is this x-value close to the x-value at the intersection of the two graphs?

Look Beyond

In Exercise 76, students investigate a system of linear equations and their graphs. Systems of linear equations are studied in detail in Chapter 3. You may wish to use this exercise as a transition to that chapter.

58. After a reflection across the y-axis, the domain of a function is also "reflected": if the domain was $x > a$, it becomes $x < -a$; if the domain was $x < a$, it becomes $x > -a$. The range remains unchanged.

After a reflection across the x-axis, the range of a function is "reflected": if the range was $y > b$, it becomes $y < -b$; if the range was $y < b$, it becomes $y > -b$. The domain remains unchanged.

Student Technology Guide

NAME ______ CLASS ______ DATE ______

Student Technology Guide
2.7 ***A Preview of Transformations***

You can see how a transformation is related to its parent function by viewing both graphs. Consider $y = x^2$ and $y = -(x - 4)^2$. Their graphs are shown at right. Notice the two graphs have the same shape and size. However, the second graph is a shift 4 units to the right and then a vertical reflection of the first.

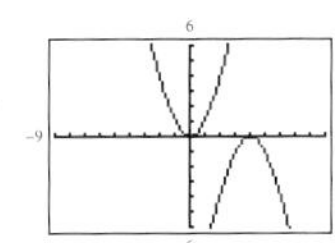

Graph each pair of functions. Identify the transformation(s) by which the second function is obtained from the first function.

1. $f(x) = |x|$ and $g(x) = 0.5|x|$ vertical compression by a factor of 0.5

2. $f(x) = x^2$ and $g(x) = x^2 - 0.5$ vertical translation 0.5 unit down

3. $g(x) = [x]$ and $h(x) = [x] + 4$ vertical translation 4 units up

4. $f(x) = x^2$ and $g(x) = 1.5(x - 2)^2$ horizontal translation 2 units to the right and vertical stretch by a factor of 1.5

5. $f(x) = \sqrt{x}$ and $g(x) = 1 - \sqrt{x}$ reflection across the x-axis and vertical translation 1 unit up

6. $h(x) = |x|$ and $g(x) = 3.2|x + 3|$ vertical stretch by a factor of 3.2 and horizontal translation 3 units to the left

You can also use a graphics calculator to graph given transformations of a parent function. Consider $y = x^2$ and its horizontal translation 4 units left followed by its vertical translation 2 units up.

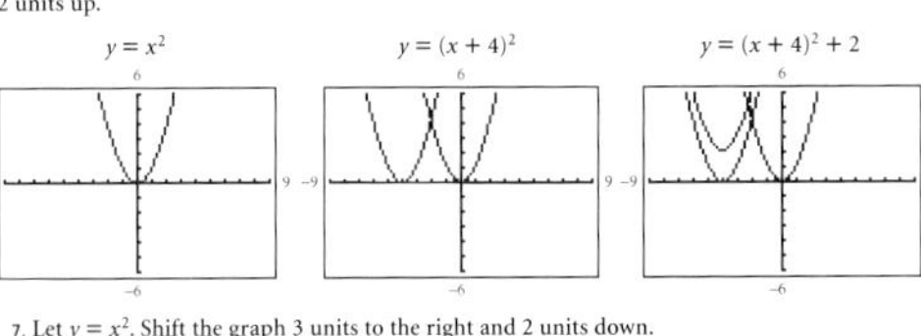

7. Let $y = x^2$. Shift the graph 3 units to the right and 2 units down. Write an equation for the graph that results.
$y = (x - 3)^2 - 2$

59. Answers may vary. Consider the vertical compression by a factor of $\frac{1}{4}$ to the graph of $f(x) = x^2$, which gives $f(x) = \frac{1}{4}x^2$, and the horizontal stretch given by $f\left(\frac{1}{2}x\right) = \left(\frac{1}{2}x\right)^2$. Since $\left(\frac{1}{2}x\right)^2 = \left(\frac{1}{2}\right)^2 x^2 = \frac{1}{4}x^2$, the effect of the vertical compression by a factor of $\frac{1}{4}$ and the horizontal stretch by a factor of 2 are identical.

60.

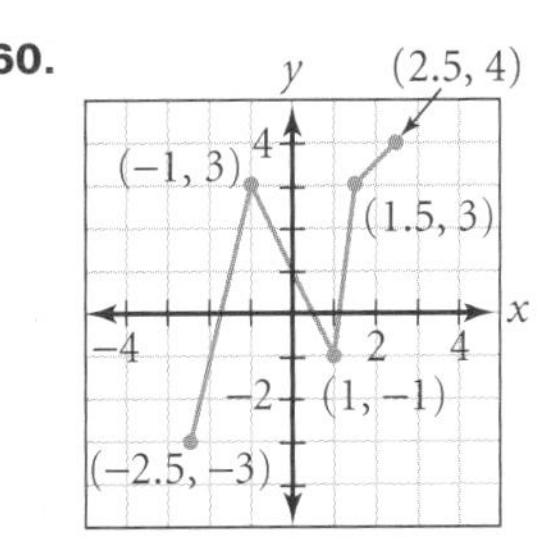

The answers to Exercises 61–66 and 73–76 can be found in Additional Answers beginning on page 1002.

Focus

As the era of space exploration progresses, the debris left behind in space is becoming a subject of increasing concern. In this Chapter Project, students are given raw data about the amount of this debris over several years. Using their knowledge of functions and regression equations, they analyze the data and make predictions.

Motivate

Discuss the problem of space debris with the class. Ask students to share their thoughts about the source of the debris. Encourage students who have a special interest in space exploration to relate any knowledge they have of the problem.

Have students examine the table on this page. Ask them for their perceptions of any trends in the data. Elicit their predictions about the amount of debris in the year 2005. Point out that the amount of debris is a function of the year. Tell them that as part of this Project, they will be searching for an equation—or equations—that might model this function.

Ask students why it might be important to reduce the rate at which space debris is accumulating. Point out that some objects are eliminated naturally as they are pulled toward Earth by gravity and burn up in the atmosphere. Elicit students' ideas about other actions that might be taken to eliminate the debris.

Space Trash

More than 3600 space missions since 1957 have left thousands of large and millions of smaller debris objects in near-Earth space. Information about the orbital debris is needed to determine the current and future hazards that this debris may pose to space operations. Only the largest objects can be repeatedly tracked and cataloged.

The table at right shows the estimated number of cataloged rocket bodies and fragmentation debris from 1965 to 1990 in five-year periods.

Space Debris Table

Year	Rocket bodies	Fragmentation debris
1965	175	900
1970	350	1850
1975	525	2250
1980	700	2600
1985	875	3200
1990	1050	2900

Activity 1

1. The average annual rate of change for each five-year period is 35. The overall average from 1965 to 1990 is $(1050 - 175) \div 125 = 35$. Because the rates are all the same, the function that represents the data is linear.

2. Since the data is modeled with a linear function, the slope will be the rate of change, 35, and the y-intercept will be the initial value, the value of y for $x = 0$, or 175: $y = 35x + 175$. To check, let $x = 15$—i.e., the year $1965 + 15 = 1980$: $y = 35(15) + 175 = 525 + 175 = 700$. This agrees with the value for 1980 on the table.

Activity 2

1.

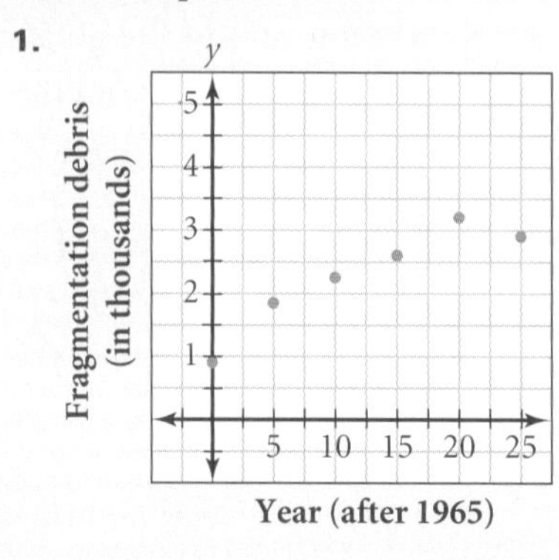

Activity 1

1. Use the data from the Space Debris Table on the previous page, which shows the estimated number of cataloged rocket bodies and fragmentation debris for the years 1965 to 1990. Find the average annual rate of change in the number of rocket bodies and debris from 1965 to 1990. Then find the rate of change for each five-year interval. Compare the individual five-year rates with the average rate.
2. Find a linear model for all of the rocket body data from Step 1. Let the years be represented by x (where $x = 0$ represents 1965 and $x = 5$ represents 1970). Verify your equation by using data from the table.

Skylab, *the United States' first space station*

Activity 2

1. Create a scatter plot for the number of other cataloged debris objects for the given years from 1965 to 1990 from the Space Debris Table on the previous page. Plot the years on the horizontal axis (where $x = 0$ represents 1965) and number of other debris objects on the vertical axis.
2. Sketch the curve on your scatter plot that you think best models your data. Describe the trend you see in the scatter plot. Do you think you can make reliable predictions by using the model you sketched? Explain.
3. Using a graphics calculator, find linear, quadratic, and exponential regression equations for this data. Discuss which model best approximates the data.

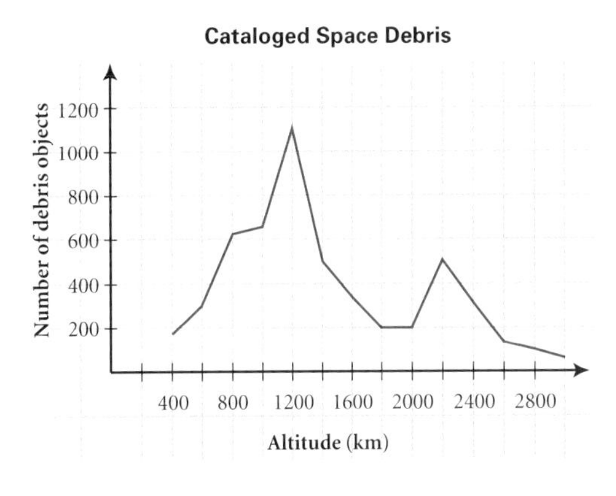

Activity 3

1. Using the Cataloged Space Debris graph at left, which shows the distribution of cataloged debris objects by altitude, describe what happens to the number of debris objects as the altitude increases. Write an approximate step function that models the data in the graph.
2. Using the step function that you wrote in Step 1, estimate the number of objects at an altitude of 725 kilometers, 1450 kilometers, and 1900 kilometers. Discuss the usefulness of your model.

2. Answers will vary. Sample: The number of objects increases fairly steadily until 1985 but then decreases in 1990. It would be difficult to make an accurate prediction because the last data point shows a decrease. You need more data to predict a trend.

3. Linear regression: $y = 82.3x + 1254.8$

Quadratic regression: $y = -3.9x^2 + 179.6x + 930.4$

Exponential regression: $y = 1227.6 \cdot 1.045^x$

Answers may vary. Sample: None of the models fits the data points very well. None of these equations takes into account the point where the data decrease. The quadratic model seems to have points that are closer than those of the other two models to the actual data points.

The answers to Activity 3 of the Chapter Project can be found in Additional Answers beginning on page 1002.

Cooperative Learning

Have students work in pairs. In **Activity 1**, the partners should first do Step 1 independently. They can then compare results and resolve any differences. For Step 2, one partner can read the data aloud while the other enters it into a graphics calculator.

Begin **Activity 2** with the partners again working independently. After each has completed Steps 1 and 2, they should compare and discuss their results. They can then do Step 3 together, this time switching roles in reading the data and entering it into the calculator.

Have the partners do Step 1 of **Activity 3** together. They should arrive at the conclusion that a piecewise function is the best model for the data. Suggest that they arrive at an agreement about appropriate intervals for the "pieces" of the function and then work independently to write the equations and make the required estimates. They can them compare their results.

Discuss

Compare students' predictions in Activity 2 with the predictions that they made at the beginning of the Project.

Discuss with students the limitations of their debris models. For instance, note that a sudden advance in technology might result in a dramatic increase in the amount of space exploration, which in turn might cause a sharp increase in the amount of debris. Elicit students' ideas about other occurrences that might limit the effectiveness of their models.

2 Chapter Review and Assessment

VOCABULARY

Key Skills & Exercises

LESSON 2.1

Key Skills

Identify and use Properties of Real Numbers.

The real numbers have the Closure, Commutative, Associative, Identity, Inverse, and Distributive Properties for addition and multiplication.

Evaluate expressions by using the order of operations.

$$\frac{-3 \times (6-4)^2}{6} = \frac{-3 \times 2^2}{6} = \frac{-3 \times 4}{6} = \frac{-12}{6} = -2$$

Exercises

State the property that is illustrated in each statement. All variables represent real numbers.

1. $a(2b) = (2b)a$ **Comm.**
2. $2 \times 1 = 2$ **Identity**
3. $a\left(\frac{1}{a}\right) = 1$ **Inverse**
4. $(2s)t = 2(st)$ **Assoc.**
5. $4x + 0 = 4x$ **Identity**
6. $-2a + 2a = 0$ **Inverse**
7. $(x + y) + 5 = 5 + (x + y)$ **Commutative**
8. $5(2 - x) = 5(2) + 5(-x)$ **Distributive**

Evaluate each expression.

9. $-1(5 + 3)^2 - 11$ **−75**
10. $\frac{(11-5)^2}{3 \times 2}$ **6**
11. $\frac{(6-12)5}{3^3}$ $-\frac{10}{9}$
12. $\frac{2^3 - (13+4)}{(-3)^2}$ **−1**

LESSON 2.2

Key Skills

Simplify and evaluate expressions by using the Properties of Exponents.

$$\left(\frac{(5^3)(5^{-2})}{5^2}\right)^2 = \left(\frac{5^{3-2}}{5^2}\right)^2 = \left(\frac{5^1}{5^2}\right)^2 = (5^{1-2})^2 = (5^{-1})^2 = 5^{-2} = \frac{1}{5^2} = \frac{1}{25}$$

Exercises

Evaluate or simplify each expression. Assume that no variable equals zero.

13. $x^4(3x)^2$ $9x^6$
14. $2a(5a^3b^5)^2$ $50a^7b^{10}$
15. $\frac{(u^2v)^3}{v^2}$ u^6v
16. $\left(\frac{p^{-1}q^2}{p^{-1}}\right)^{-4}\left(\frac{-p^5q^{-3}}{p^{-3}q^{-1}}\right)^{-3}$ $-\frac{1}{p^{24}q^2}$
17. $\frac{(2x)^3}{y^2}\left(\frac{x^2}{3}\right)^3$ $\frac{8x^9}{27y^2}$

Chapter Test, Form A

NAME ______ CLASS ______ DATE ______

Chapter Assessment
Chapter 2, Form A, page 1

Write the letter that best answers the question or completes the statement.

c 1. Which of the following sets of numbers does not contain $3.\overline{6}$?
a. real b. rational c. irrational d. all of these

d 2. Which property of addition is illustrated by the statement $a + (-a) = 0$?
a. Associative Property b. Commutative Property
c. Identity Property d. Inverse Property

c 3. When simplified, what is the value of $\frac{6 + 3 \cdot 2^3}{2(5-3) - 2 \cdot 7} + (-3)^2$?
a. −10 b. −6 c. 6 d. 10

a 4. Simplify $(-5a^{-4}b^3)(-2a^2b^{-3})^3$.
a. $\frac{40a^2}{b^6}$ b. $\frac{-40a^2}{b^6}$ c. $30ab^3$ d. $-30ab^3$

b 5. Simplify $\left(\frac{3x^7y}{x^2y^5}\right)^{-2}$.
a. $\frac{-6y^8}{x^{10}}$ b. $\frac{y^8}{9x^{10}}$ c. $\frac{-9y^{10}}{x^{10}}$ d. $\frac{y^{10}}{9x^{10}}$

b 6. What is the value of $27^{\frac{5}{3}}$?
a. 45 b. 243 c. 643,729 d. 4,782,969

d 7. Which of the following relations is not a function?
a. $\{(3, 7), (5, 8), (7, 9), (15, 10)\}$ b. $\{(7, 0), (10, 5), (13, 10), (16, 5)\}$
c. $\{(3, 5), (5, 5), (7, 5), (9, 5)\}$ d. $\{(9, 3), (16, 4), (9, -3), (16, -4)\}$

a 8. What is the domain of the relation $\{(0, 1), (2, 3), (4, 5), (6, 7), (8, 9)\}$?
a. $\{0, 2, 4, 6, 8\}$ b. $\{1, 3, 5, 7, 9\}$
c. $\{0, 1, 2, 3, 4\}$ d. $\{0, 1, 2, 3, 4, 5, 6, 7, 8, 9\}$

c 9. Find the range of the function $f(x) = -x^2 + 3$.
a. $x \le 3$ b. $x \ge 3$ c. $y \le 3$ d. $y \ge 3$

b 10. Find the value of the function $g(x) = \frac{x^3 + 9}{3x}$ when $x = -3$.
a. 4 b. 2 c. −3 d. −28

c 11. Find the value of the function $h(x) = -10[x]$ when $x = 7.7$.
a. 70 b. 80 c. −70 d. −80

NAME ______ CLASS ______ DATE ______

Chapter Assessment
Chapter 2, Form A, page 2

c 12. The inverse of the relation shown is
a. not a relation.
b. not a function.
c. a function.
d. none of the above.

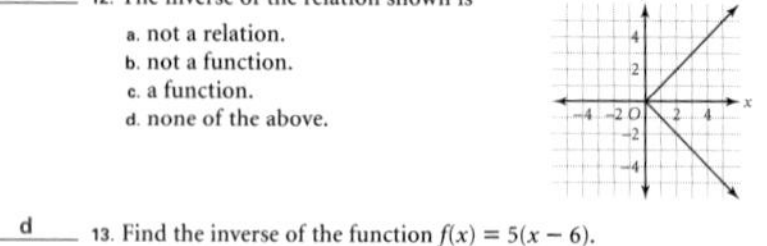

d 13. Find the inverse of the function $f(x) = 5(x - 6)$.
a. $y = 5(x + 6)$ b. $y = -5(x + 6)$ c. $y = \frac{x+6}{5}$ d. $y = \frac{x}{5} + 6$

a 14. If $f(x) = 6x + 3$ and $g(x) = 5x - 2$, what restrictions are on the domain of $\left(\frac{f}{g}\right)(x)$?
a. $x \ne \frac{2}{5}$ b. $x \ne -\frac{2}{5}$ c. $x \ne -\frac{1}{2}$ d. $x \ne -2$

a 15. If $f(x) = x^2 + 3$ and $g(x) = x + 1$, find $(g \circ f)(x)$.
a. $x^2 + 4$ b. $x^2 + 2x + 4$ c. $x^2 + 1$ d. $x^2 + 2x + 1$

b 16. Identify the transformation applied to $f(x) = x^3$ in order to obtain the function $g(x) = (-x)^3$.
a. reflection across the x-axis b. reflection across the y-axis
c. vertical translation d. horizontal translation

d 17. Identify the transformation applied to $f(x) = \sqrt{64 - x^2}$ in order to obtain the function $g(x) = \frac{3}{4}\sqrt{64 - x^2}$.
a. horizontal stretch by a factor of $\frac{4}{3}$ b. horizontal compression by a factor of $\frac{3}{4}$
c. vertical stretch by a factor of $\frac{4}{3}$ d. vertical compression by a factor of $\frac{3}{4}$

c 18. Find the function for the graph of $f(x) = \sqrt{x}$ after a translation of 4 units to the right and 3 units down.
a. $g(x) = 3 + \sqrt{x - 4}$ b. $g(x) = 3 + \sqrt{x + 4}$
c. $g(x) = -3 + \sqrt{x - 4}$ d. $g(x) = -3 + \sqrt{x + 4}$

b 19. The cost, c, to park in an airport parking lot can be modeled by the function $c(x) = 3 + \lceil x - 1 \rceil$, where x is the number of hours parked at the airport. How much will it cost to park at the airport for 2.7 hours?
a. \$6 b. \$5 c. \$4 d. \$3

LESSON 2.3

Key Skills

State the domain and range of a relation, and state whether it is a function.

The relation $\{(1, 2), (2, 4), (3, 6), (4, 8)\}$ is a function because each x-coordinate is paired with one and only one y-coordinate.

domain: $\{1, 2, 3, 4\}$ range: $\{2, 4, 6, 8\}$

Evaluate functions.

Evaluate $f(x) = 2x^2 - x + 3$ for $x = 5$.

$$f(5) = 2(5)^2 - 5 + 3$$
$$f(5) = 50 - 5 + 3$$
$$f(5) = 48$$

Exercises

State whether each relation is a function.

18. $\{(1, 2), (2, 3), (3, 5), (4, 7), (5, 11)\}$ **yes**

19. $\{(1, -1), (2, -2), (3, -3)\}$ **yes**

20. $\{(1, -1), (1, 1), (2, -3)\}$ **no**

21. $\{(1, 1), (1, -1), (2, 2), (2, -2)\}$ **no**

State the domain and range for each function.

22. $\{(-1, -6), (5, 8), (9, -1), (2, 3)\}$

23. $\{(-1, 2), (0, 6), (2, 7), (4, -7)\}$

Evaluate each function for $x = 1$ and $x = -1$.

24. $f(x) = 3x^2 - 2x + 1$ **2; 6**

25. $g(x) = 11x - 2$ **9; −13**

26. $f(x) = 3x^2 - 2$ **1; 1**

27. $h(x) = 2 - 3x$ **−1; 5**

LESSON 2.4

Key Skills

Add, subtract, multiply, and divide functions.

The sum, difference, product, and quotient of functions f and g are defined as follows:

$$(f + g)(x) = f(x) + g(x)$$
$$(f - g)(x) = f(x) - g(x)$$
$$(f \cdot g)(x) = f(x) \cdot g(x)$$
$$\left(\frac{f}{g}\right)(x) = \frac{f(x)}{g(x)}, \text{ where } g(x) \neq 0$$

Compose functions.

If f and g are functions with appropriate domains and ranges, then the composition of f with g, $f \circ g$, is defined by $f(g(x))$.

Exercises

Let $f(x) = 3x - 4$ and $g(x) = \frac{x}{2} - 5$. Find each new function, and state any domain restrictions.

28. $f + g$ **29.** $f - g$ **30.** $g - f$

31. $f \circ g$ **32.** $\frac{f}{g}$ **33.** $\frac{g}{f}$

Let $f(x) = 2x$ and $g(x) = -x + 2$. Find each composite function.

34. $f \circ g$ $= -2x + 4$

35. $g \circ f$ $= -2x + 2$

36. $f \circ f$ $= 4x$

37. $g \circ g$ $= x$

Let $f(x) = -3x - 5$ and $g(x) = 4x - 1$. Evaluate each composite function.

38. $(f \circ g)(3)$ **−38**

39. $(g \circ f)(2)$ **−45**

40. $(g \circ g)(-1)$ **−21**

LESSON 2.5

Key Skills

Find the inverse of a function.

To find the inverse of the function $f = \{(1, 2), (2, 4), (3, 8), (4, 16)\}$, reverse the order of the coordinates in each ordered pair. The inverse of f is $\{(2, 1), (4, 2), (8, 3), (16, 4)\}$.

To find the inverse of a function f defined by a function rule, replace $f(x)$ with y, interchange x and y, and solve for y.

Exercises

Find the inverse of each function.

41. $f = \{(-2, 0), (-1, 1), (0, 0), (1, 1), (2, 0)\}$

42. $f = \{(1, 2), (2, 3), (3, 4), (4, 5)\}$

Find an equation for the inverse of each function. Then use composition to verify that the equation you wrote is the inverse.

43. $f(x) = -\frac{2}{3}x + 4$

44. $f(x) = \frac{2 - x}{3}$

---

31. $(f \circ g)(x) = \frac{3x^2}{2} - 17x + 20$

32. $\left(\frac{f}{g}\right)(x) = \frac{3x - 4}{\frac{x}{2} - 5}, x \neq 10$

33. $\left(\frac{g}{f}\right)(x) = \frac{\frac{x}{2} - 5}{3x - 4}, x \neq \frac{4}{3}$

41. $\{(0, -2), (1, -1), (0, 0), (1, 1), (0, 2)\}$

42. $\{(2, 1), (3, 2), (4, 3), (5, 4)\}$

43. $f^{-1}(x) = -\frac{3}{2}x + 6$

44. $f^{-1}(x) = 2 - 3x$

---

22. domain: $\{-1, 5, 9, 2\}$; range: $\{-6, 8, -1, 3\}$

23. domain: $\{-1, 0, 2, 4\}$; range: $\{2, 6, 7, -7\}$

28. $(f + g)(x) = \frac{7x}{2} - 9$

29. $(f - g)(x) = \frac{5x}{2} + 1$

30. $(g - f)(x) = -\frac{5x}{2} - 1$

Chapter Test, Form B

NAME ______ CLASS ______ DATE ______

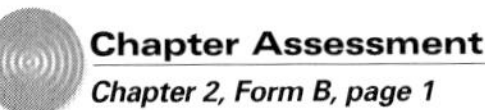

Chapter Assessment

Chapter 2, Form B, page 1

Classify each number in as many ways as possible.

1. -125 — real, rational, integer
2. $5.626226222\ldots$ — real, irrational
3. 12.345 — real, rational

Evaluate each expression.

4. $16^{\frac{3}{4}} + 15 \div 5 - 3^3(5 - 2)^{-2}$ — 8
5. $20 - \frac{15 + 3 \cdot 5^2}{5^2 - 4^2}$ — 10

Simplify each expression. Assume that no variable has a value of zero.

6. $(-2x^{-3}x^7)^4$ — $16x^{16}$
7. $\left(\frac{a^4b^{-3}}{b^{-5}}\right)^3\left(\frac{2ab^3}{a^3}\right)^{-2}$ — $\frac{a^{16}}{4}$

For Exercises 8–11, refer to the graph at right.

8. Is the relation a function? Explain your response.
Yes; no vertical line intersects the graph at more than one point.
9. Write the domain and range of the relation.
domain: $x \geq -2$; range: $y \geq 0$
10. Is the inverse of the relation a function? Explain your response.
No; there are horizontal lines that intersect the graph at more than one point.
11. Write the domain and range of the inverse.
domain: $x \geq 0$; range: $y \geq -2$

Find the value of each function for $x = -3$.

12. $f(x) = x^5 + 3x$ — $f(-3) = -252$
13. $g(x) = 5[x + 5.5]$ — $g(-3) = 10$

Let $f(x) = 2x + 1$ and $g(x) = x - 6$. Write an expression for each function. State any domain restrictions.

14. $(f + g)(x)$ — $3x - 5$
15. $(fg)(x)$ — $2x^2 - 11x - 6$
16. $(f \circ g)(x)$ — $2x - 11$
17. $\left(\frac{f}{g}\right)(x)$ — $\frac{2x + 1}{x - 6}, x \neq 6$

18. Find the inverse of the function $f(x) = \frac{x}{4} + 10$. — $f^{-1}(x) = 4(x - 10)$

NAME ______ CLASS ______ DATE ______

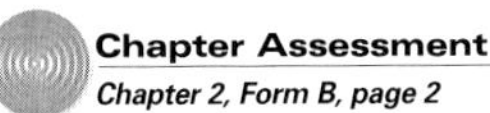

Chapter Assessment

Chapter 2, Form B, page 2

19. On the coordinate grid at right, graph the piecewise function below.
$$f(x) = \begin{cases} x & \text{if } -3 < x < 2 \\ 2x - 5 & \text{if } x \geq 2 \end{cases}$$
20. Identify the transformations applied to $f(x) = \sqrt{16 - x^2}$ in order to obtain the function $g(x) = \frac{3}{2}\sqrt{16 - (2x)^2}$.
horizontal compression by a factor of $\frac{1}{2}$ and vertical stretch by a factor of $\frac{3}{2}$
21. Find the function for the graph of $f(x) = [x]$ after a reflection across the x-axis and a horizontal stretch by a factor of 3. — $g(x) = -\left[\frac{1}{3}x\right]$
22. Justify each step in the simplification of $(3a + 3b) + (-3a)$.
   a. $(3a + 3b) + (-3a) = (3b + 3a) + (-3a)$ — Commutative Property of Addition
   b. $= 3b + [3a + (-3a)]$ — Associative Property of Addition
   c. $= 3b + 3[a + (-a)]$ — Distributive Property
   d. $= 3b + 3(0)$ — Inverse Property of Addition
   $= 3b + 0$
   e. $= 3b$ — Identity Property of Addition
23. The monthly rent for a store in a mall is \$200 plus \$1.50 per square foot of area covered by the store.
   a. Write a linear function to model the monthly rent, M, as a function of the number of square feet, n, that the store covers. — $M(n) = 200 + 1.50n$
   b. Find the inverse of the monthly rent function. — $n(M) = \frac{M - 200}{1.50}$
   c. What is the area of the largest store that can be rented for \$800 a month? — 400 square feet
24. A courier service charges a basic fee of \$5 to deliver packages. In addition, the company charges \$0.50 per pound or fraction of a pound that the package weighs.
   a. Write a function to model the cost, c, of sending a package that weighs n pounds. — $c(n) = 5 + 0.50\lceil n \rceil$
   b. Find the cost of sending a package that weighs 3.2 pounds. — \$7

Use the horizontal-line test to determine whether the inverse relation is a function.

If a function, f, has an inverse that is a function, then any horizontal line that intersects the graph of f will do so at only one point.

Graph each function, and use the horizontal-line test to determine whether the inverse is a function.

45 $f(x) = \frac{3x+3}{2}$ **yes** **46** $f(x) = x^2 - 1$ **no**

47 $f(x) = 2\left(\frac{2x}{3} + 1\right)$ **yes** **48** $f(x) = 3x + 5$ **yes**

LESSON 2.6

Key Skills

Define and graph piecewise functions, step functions, and absolute-value functions.

The piecewise function $f(x) = \begin{cases} 2x+1 & \text{if } x < 0 \\ 5x & \text{if } x \geq 0 \end{cases}$ is the function that assigns the function value $2x + 1$ to negative values of x and the function value $5x$ to nonnegative values of x.

A step function consists of different constant range values, and its graph resembles stair steps.

The absolute-value function $f(x) = |x|$ gives the distance from x to 0 on a number line and has a V-shaped graph.

Exercises

Graph each function.

49. $f(x) =1 \begin{cases} x & \text{if } x < 1 \\ -2x+3 & \text{if } x \geq 1 \end{cases}$

50. $f(x) = \begin{cases} x^2+1 & \text{if } x < 0 \\ 1 & \text{if } x \geq 0 \end{cases}$

51. $f(x) = 3\lfloor x \rfloor$ **52.** $f(x) = \frac{1}{2}|x| - 1$

Evaluate.

53. $\lfloor -77.99 \rfloor$ **−78** **54.** $\lfloor 4 \rfloor - \lfloor -7.1 \rfloor$ **12**

55. $\lceil 3.5 \rceil$ **4** **56.** $\lceil 4.0 \rceil - \lceil -7.001 \rceil$ **11**

57. $|-7| + \lfloor 1.09 \rfloor$ **8** **58.** $|3.5| - \lceil 2.9 \rceil$ **0.5**

LESSON 2.7

Key Skills

Identify transformations of functions.

The graph of $g(x) = 2[3(x-1)]^2 + 4$ is formed by the following transformations of the graph of $f(x) = x^2$:

- a horizontal translation of 1 unit to the right
- a horizontal compression by a factor of $\frac{1}{3}$
- a vertical stretch by a factor of 2
- a vertical translation of 4 units up

Exercises

Identify the transformations from f to g.

59. $f(x) = x^2$, $g(x) = (x-2)^2$

60. $f(x) = \sqrt{x}$, $g(x) = \sqrt{-x} + 3$

61. $f(x) = |x|$, $g(x) = -|x+3|$

Applications

62. CONSUMER ECONOMICS Hardwood flooring costs \$4.75 per square foot for amounts up to and including 500 square feet and \$4.50 for amounts of more than 500 square feet. Write and graph a function that describes the cost, c, of x square feet of hardwood flooring.

63. SPORTS The function $h(t) = -9.8t^2 + 1.5$ gives the height in meters of a basketball thrown from an initial height of 1.5 meters. Describe the transformations from $f(t) = t^2$ to h.

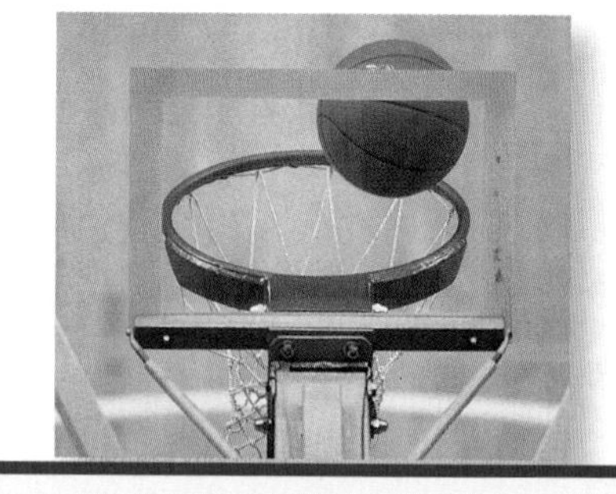

49.

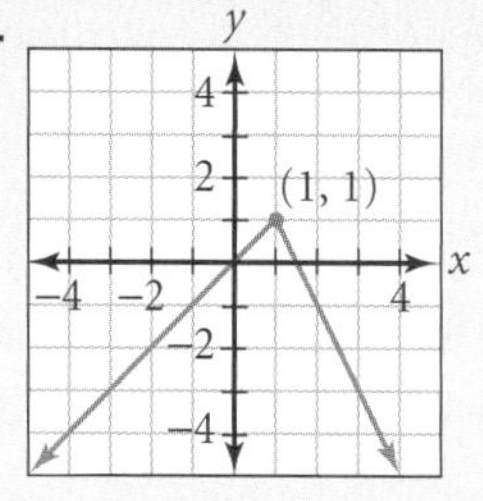

50.

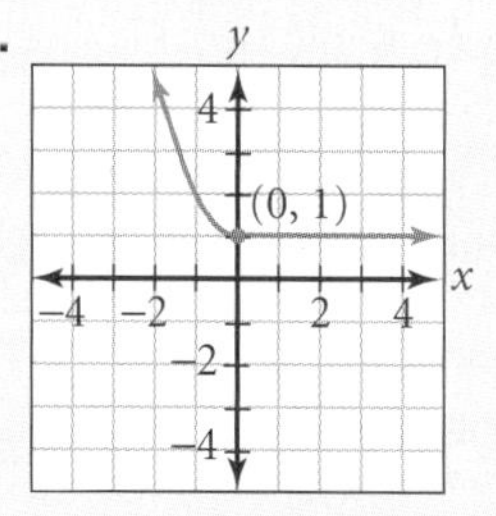

51.

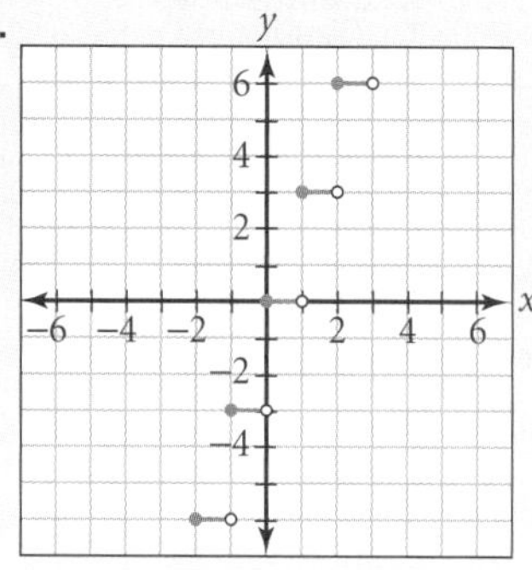

52.

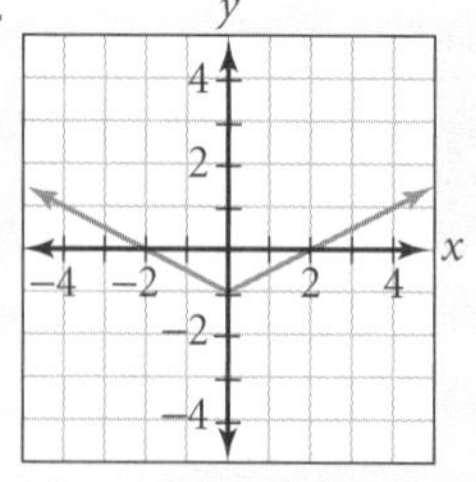

59. horizontal translation 2 units to the right

60. reflection across the y-axis and a vertical translation 3 units up

61. horizontal translation 3 units to the left and a reflection across the x-axis

62. $c(x) = \begin{cases} 4.75x, \text{ if } 0 \leq x \leq 500 \\ 2.375 + 4.50(x - 500), \text{ if } x > 500 \end{cases}$

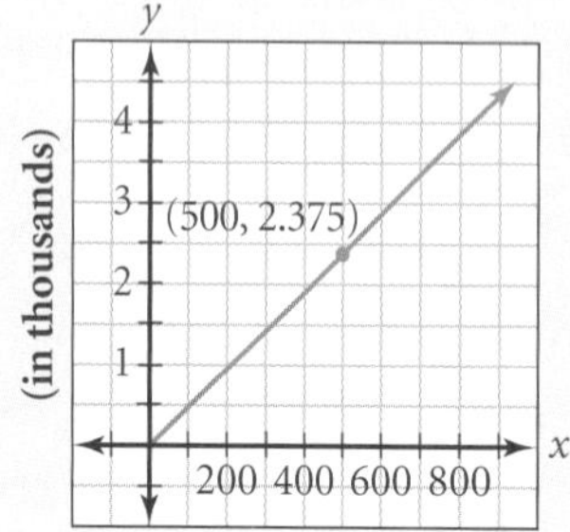

63. vertical stretch by a factor of 9.8, a reflection across the x-axis, and a vertical translation 1.5 units up

Chapter Test

Evaluate each expression by using the order of operations.

1. $5 + 2(7 - 4)^2$ 23
2. $12 - 9 \div 3 + 2 \cdot 5$ 19
3. $\frac{4+6}{2} + 2 \cdot 5$ 15
4. $5 \cdot 4 \div 2 + 3^{(4-1)}$ 37

State the property that is used in each statement. All variables represent real numbers.

5. $5x \cdot 1 = 5x$ Identity Mult.
6. $7d - 14 = 7(d - 2)$ Dist.
7. $\left(\frac{2}{r}\right)\left(\frac{r}{2}\right) = 1$ Inverse Mult.
8. $4(yz) = (yz)4$ Comm. Mult.

Simplify each expression. Assume that no variable equals zero.

9. $y^3(x^2y)$ x^2y^4
10. $(9rt)^2(3rst)^{-3}$ $\frac{3}{rs^3t}$
11. $\frac{14r^2s^{-3}t^4}{35r^{-2}s^5t^3}$ $\frac{2r^4t}{5s^8}$
12. $\left(\frac{3p^4q^{-1}}{8p^{-2}q^3}\right)^{-2}$ $\frac{64q^8}{9p^{12}}$

13. **PHYSICS** Kinetic energy can be measured in joules and is given by the formula $k = \frac{1}{2}mv^2$. What is the kinetic energy in joules of an object with mass $m = 100$ kg and velocity $v = 5$ meters per second? 1250 joules

State whether the following are functions, and if so give the domain and range.

14. $\{(5, 7), (7, 12), (9, 7), (11, 12), (13, 7)\}$ Yes
15. $\{(-2, 4), (0, 6), (-2, 8), (0, 10), (-2, 12)\}$ No

Evaluate each function for $x = -2$, $x = 0$, and $x = 2$.

16. $f(x) = 5x^2 - 4x + 7$ 35; 7; 19
17. $f(x) = \frac{x^2}{2} + x - 4$ −4; −4; 0

18. **CONSUMER ECONOMICS** The cost of enrollment at a community college is \$120 plus \$75 per credit, c, taken. Express the college costs as a function of c and find the cost when 12 credits are taken. $f(c) = 120 + 75c$; \$1020

For Items 19–23, let $f(x) = 2x + 7$ and $g(x) = x - 9$. Find new functions and state any domain restrictions, if appropriate.

19. $f + g$
20. $\frac{g}{f}$
21. $f \cdot g$

Evaluate each composite function.

22. $(g \circ f)(3)$
23. $(f \circ g)(2)$

24. **BIOLOGY** The number of times a cricket chirps per minute can be predicted using the formula $n = 4(F - 40)$, where n is the number of chirps per minute and F is the Fahrenheit temperature. The Celsius scale for temperature can be converted to Fahrenheit using $F = \frac{9}{5}C + 32$. Find the number of cricket chirps per minute as a function of Celsius temperature. $n = \frac{36}{5}c - 32$

Find the inverse of each function. State whether the inverse is a function.

25. $\{(3, 4), (4, 3), (5, 6), (6, 5), (7, 8), (8, 7)\}$
26. $\{(1, 2), (5, 6), (2, 2), (6, 6), (3, 4)\}$

Find an equation for the inverse and state whether the inverse is a function.

27. $f(x) = \frac{3x - 8}{4}$
28. $f(x) = x^4 - 3$

Graph each function.

29. $f(x) = \begin{cases} 2x + 5 & \text{if } x \leq 0 \\ -\frac{1}{3}x + 3 & \text{if } x > 0 \end{cases}$

30. $f(x) = 2[x] - 1$

Evaluate.

31. $[4.5] + 3.5$ 7.5
32. $[-1.7] + [1.7]$ −1

Identify each transformation from the parent function f to g.

33. $f(x) = \sqrt{x}$, $g(x) = \sqrt{x} - 3$ translated down 3 units
34. $f(x) = x^2$, $g(x) = (x + 3)^2 - 4$ translated left 3 units and down 4 units

Chapter Test

14. yes; domain: $\{5, 7, 9, 11, 13\}$; range: $\{7, 12\}$
19. $(f + g)(x) = 3x - 2$
20. $\left(\frac{g}{f}\right)(x) = \frac{x - 9}{2x + 7}, x \neq -\frac{7}{2}$
21. $(f \cdot g)(x) = 2x^2 - 11x - 63$
22. $(g \circ f)(3) = 4$
23. $(f \circ g)(2) = -7$
25. $\{(4, 3), (3, 4), (6, 5), (5, 6), (8, 7), (7, 8)\}$; yes
26. $\{(2, 1), (6, 5), (2, 2), (6, 6), (4, 3)\}$; no
27. $f^{-1}(x) = \frac{4x + 8}{3}$; yes
28. $f^{-1}(x) = \pm\sqrt[4]{x + 3}$; no
29.

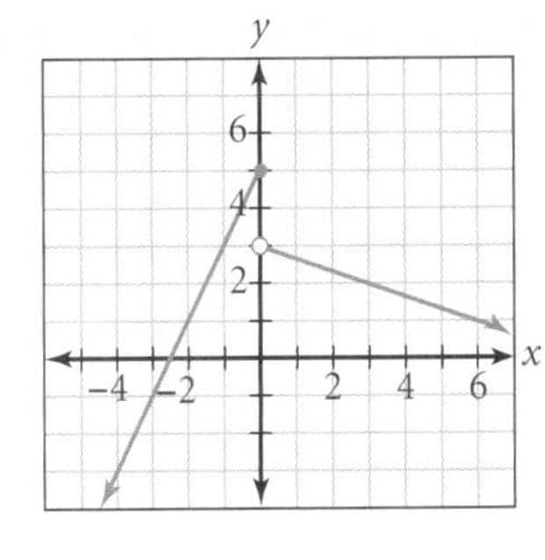

30.

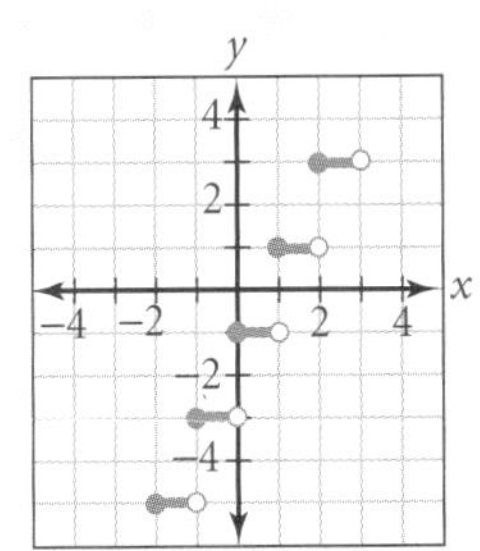

College Entrance Exam Practice

Multiple-Choice and Quantitative-Comparison Samples

The first half of the Cumulative Assessment contains two types of items found on standardized tests—multiple-choice questions and quantitative-comparison questions. Quantitative-comparison items emphasize the concepts of equality, inequality, and estimation.

Free-Response Grid Samples

The second half of the Cumulative Assessment is a free-response section. This part of the Cumulative Assessment requires student-produced response items like those commonly found on college entrance exams. These questions require the use of machine-scored answer grids. You may wish to have students practice answering these items in preparation for standardized tests.

CHAPTERS 1–2 CUMULATIVE ASSESSMENT

College Entrance Exam Practice

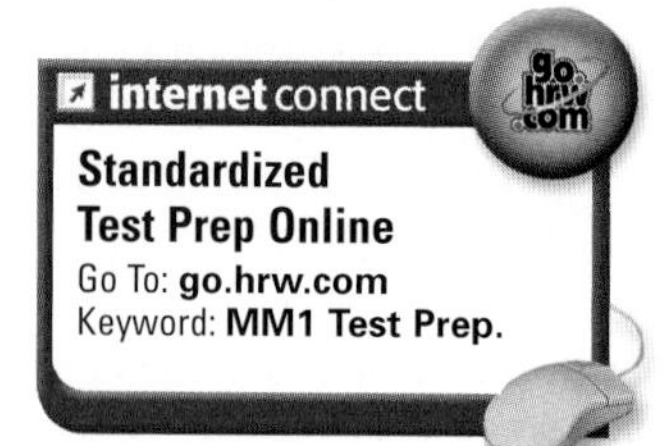

QUANTITATIVE COMPARISON For Items 1–5, write
A if the quantity in Column A is greater than the quantity in Column B;
B if the quantity in Column B is greater than the quantity in Column A;
C if the quantities are equal; or
D if the relationship cannot be determined from the given information.

	Column A	Column B	Answers
1. B	The value of x in $x - 2 = 3 + 4x$	The value of y in $-8y - 2 = 3 - 3y$	(A) (B) (C) (D) [Lesson 1.6]
2. A	$3[4.95] - 2$	$3[3.95] - 2$	(A) (B) (C) (D) [Lesson 2.5]
3. B	$\frac{x^3}{x} - x^2$, where $x \neq 0$	$x^2 \cdot x^{-2}$, where $x \neq 0$	(A) (B) (C) (D) [Lesson 2.2]
4. C	$f(x) = x^2 - 2x + 3$ $f(-1)$	$f(3)$	(A) (B) (C) (D) [Lesson 2.3]
5. A	The slope of a line perpendicular to the graph of $y = -\frac{1}{6}x + 5$	The slope of a line parallel to the graph of $y = -\frac{1}{6}x + 5$	(A) (B) (C) (D) [Lesson 1.3]

6. Which is the equation for the line whose slope, m, is -5 and y-intercept, b, is $-\frac{1}{2}$? **(LESSON 1.2)** c
   a. $y = -\frac{1}{2}x - 5$ **b.** $y = -\frac{1}{2}x + 5$
   c. $y = -5x - \frac{1}{2}$ **d.** $y = -5x + \frac{1}{2}$

7. Solve $|2 + 3x| = 14$. **(LESSON 1.7)** d
   a. $x = -4$ **b.** $x = -\frac{16}{3}$
   c. $x = -4$ *or* $x = \frac{16}{3}$ **d.** $x = 4$ *or* $x = -\frac{16}{3}$

8. Let $f(x) = -3x^2$ and $g(x) = 2 - x$. Which of the following function operations gives the new function $h(x) = 3x^3 - 6x^2$? **(LESSON 2.4)** b
   a. $f - g$ **b.** $f \cdot g$ **c.** $f \div g$ **d.** $f \circ g$

9. Solve $S = \frac{1780Ad}{r}$ for A. **(LESSON 1.6)** d
   a. $A = \frac{1780Sd}{r}$ **b.** $A = \frac{Sd}{1780r}$
   c. $A = \frac{1780dr}{S}$ **d.** $A = \frac{Sr}{1780d}$

10. Which linear equation contains the points $(-1, -4)$ and $(3, 8)$? **(LESSON 1.3)** c
    a. $y = \frac{1}{3}x + 7$ **b.** $y = -\frac{1}{3}x + 9$
    c. $y = 3x - 1$ **d.** $y = -3x - 7$

11. Which number is irrational? **(LESSON 2.1)** d
    a. -3 **b.** $\frac{1}{3}$ **c.** $0.\overline{5}$ **d.** π

12. Which is an equation for the line that contains the point $(10, 3)$ and is perpendicular to the graph of $y = 5x - 3$? **(LESSON 1.3)** a
    a. $y = -\frac{1}{5}x + 5$ **b.** $y = -\frac{1}{5}x - 3$
    c. $y = -5x - 3$ **d.** $y = -5x + 5$

13. Which is the inverse of the function $y = -3x + 12$? **(LESSON 2.5)** c
    a. $y = -3x + 12$ **b.** $y = \frac{1}{3}x + 12$
    c. $y = -\frac{1}{3}x + 4$ **d.** $y = -3x + 4$

14. Which description below identifies the transformations from $f(x)$ to $y = 3f(x - 2)$?
a *(LESSON 2.7)*
  a. a horizontal translation of 2 units to the right and a vertical stretch by a factor of 3
  b. a horizontal translation of 2 units to the left and a vertical stretch by a factor of 3
  c. a vertical translation of 2 units up and a vertical stretch by a factor of 3
  d. a vertical translation of 2 units down and a vertical stretch by a factor of 3

STATISTICS The following data set gives the resale price of a certain computer at monthly intervals. *(LESSON 1.5)*

Months	Price ($)	Months	Price ($)
1	3250	6	2700
2	3150	7	2700
3	3100	8	2450
4	2850	9	2350
5	2800	10	2300

15. Find the equation of the least-squares line. $y \approx -108.79x + 3363.3$

16. Use the least-squares line to predict the price after 12 months. **about $2058**

17. Find the correlation coefficient, and explain how it describes the data. $r = -0.99$; **strong negative**

Find the inverse of each function. State whether the inverse is a function. *(LESSON 2.5)*

18. $\{(3, 4)), (8, 4), (13, -4), (4, 0)\}$

19. $f(x) = \frac{x-2}{4}$ $f^{-1}(x) = 4x + 2$; **function**

20. Solve and graph the inequality $2(1 - 2x) \geq -x + 4$. *(LESSON 1.7)*

21. Write a relation that is not a function, and explain why it is not a function. *(LESSON 2.3)*

Let $f(x) = 2x - 2$ and $g(x) = 3x$. Perform each operation below, and write your answer in simplest form. *(LESSON 2.4)*

22. $f + g$ 23. $f - g$ 24. $f \cdot g$
25. $f \div g$ 26. $f \circ g$ 27. $g \circ f$

CONSUMER ECONOMICS Residential wastewater rates are based on a monthly customer charge of $4.00 plus $1.75 per 1000 gallons of water used up to and including 6900 gallons and $3.50 for each additional 1000 gallons of water. *(LESSON 2.6)*

28. Write a piecewise function to represent the monthly cost, c, in dollars for x gallons of water used.

29. Graph the function from Item 28. What is the monthly cost for using 8400 gallons of water in a month?

30. If the monthly wastewater bill is $25.00, how much water was used? **9450 gal**

FREE RESPONSE GRID
The following questions may be answered by using a free-response grid such as that commonly used by standardized-test services.

31. What value should replace the "?" in the table below in order for x and y to be linearly related? *(LESSON 1.1)* **−1**

x	3	5	7	9	11
y	2	?	−4	−7	−10

32. Evaluate $27^{\frac{2}{3}}$. *(LESSON 2.2)* **9**

33. If a varies directly as b and $a = -3$ when $b = 12$, what is b when $a = 1.5$? *(LESSON 1.4)* **−6**

34. Evaluate $-[-32.90]$. *(LESSON 2.6)* **33**

35. Evaluate $75 - \frac{3(4 + 12 \div 2)^2}{2 + 3}$. *(LESSON 2.1)* **15**

36. Evaluate $\frac{6^2 \cdot 6^{-3}}{6^{-1}}$. *(LESSON 2.2)* **1**

37. Find the slope of the line represented by $2x + 3y = 7$. *(LESSON 1.2)* $-\frac{2}{3}$

38. Find $f(0)$ for $f(x) = 5x^2 - x + 12$. *(LESSON 2.3)* **12**

39. What value of x is the solution to the equation $\frac{3x - 15}{2} = 9 - 4x$? *(LESSON 1.5)* **3**

40. Evaluate $\lceil -0.75 \rceil$. *(LESSON 2.6)* **0**

18. $\{(4, 3), (4, 8), (-4, 13), (0, 4)\}$ is not a function.

20. $x \leq -\frac{2}{3}$

21. Answers may vary. Sample answer: $\{(1, 1), (1, -1)\}$ is a relation but not a function because the domain element 1 is matched with more than one range element.

22. $5x - 2$

23. $-x - 2$

24. $6x^2 - 6x$

25. $\frac{2x - 2}{3x}$, if $x \neq 0$

26. $6x - 2$

27. $6x - 6$

28. For x in thousands of gallons, we have
$$c(x) = \begin{cases} 1.75x + 4, & \text{if } 0 \leq x \leq 6.9 \\ 3.50x - 8.08, & \text{if } x > 6.9 \end{cases}.$$

29.

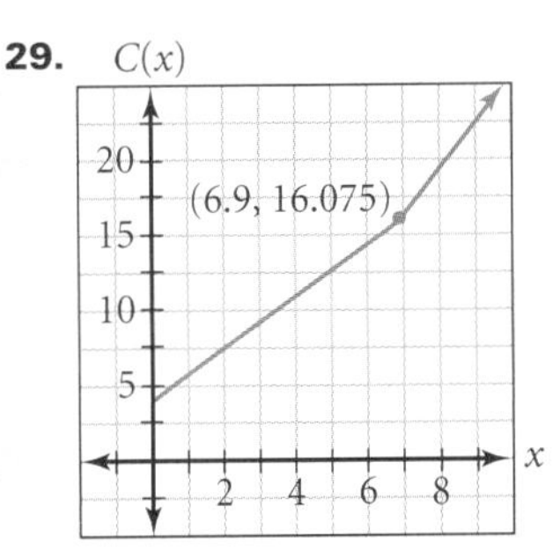

8400 gal will cost $21.33.

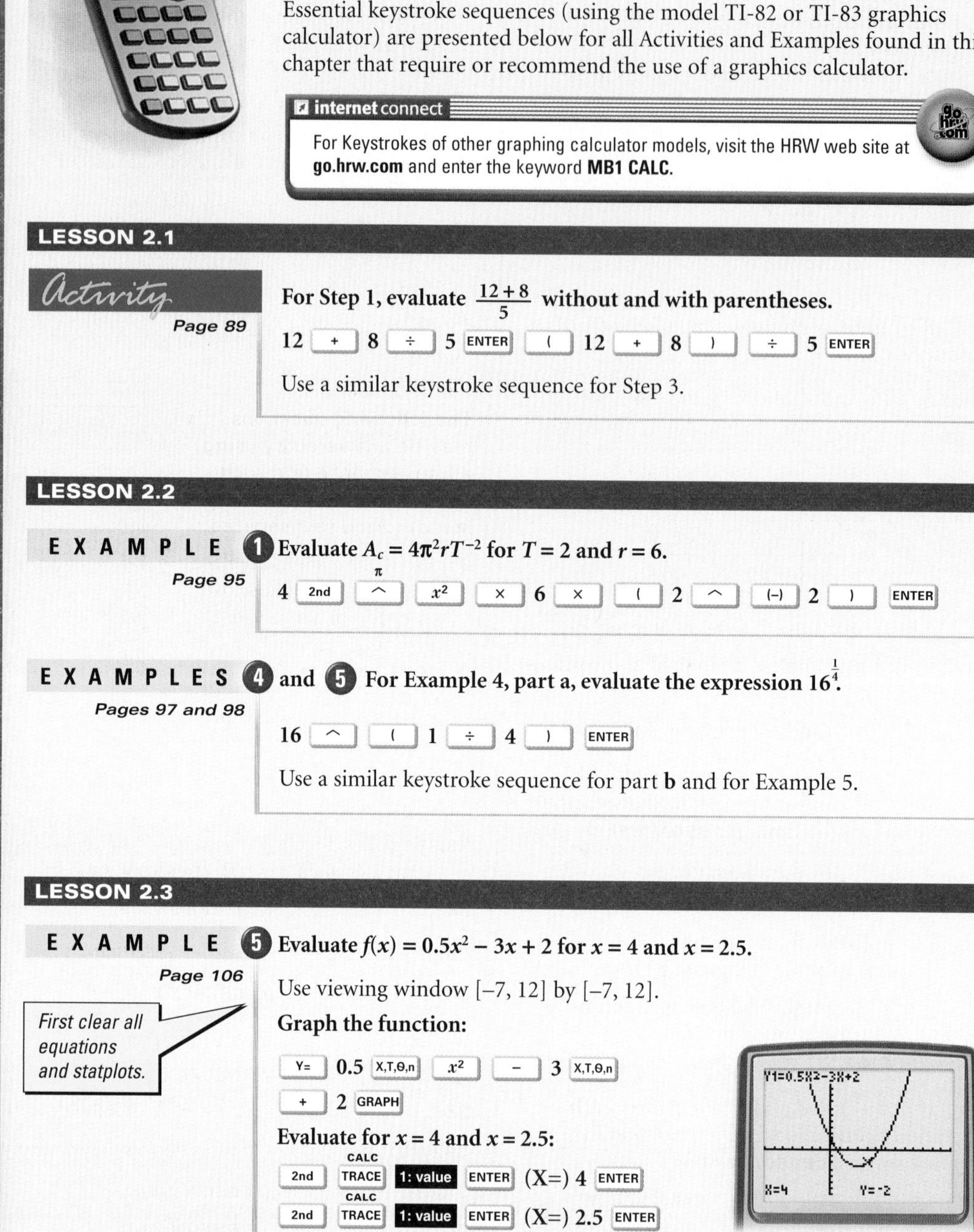

Keystroke Guide for Chapter 2

Essential keystroke sequences (using the model TI-82 or TI-83 graphics calculator) are presented below for all Activities and Examples found in this chapter that require or recommend the use of a graphics calculator.

internet connect

For Keystrokes of other graphing calculator models, visit the HRW web site at **go.hrw.com** and enter the keyword **MB1 CALC.**

LESSON 2.1

Activity
Page 89

For Step 1, evaluate $\frac{12+8}{5}$ without and with parentheses.

12 [+] 8 [÷] 5 [ENTER] [(] 12 [+] 8 [)] [÷] 5 [ENTER]

Use a similar keystroke sequence for Step 3.

LESSON 2.2

EXAMPLE 1 *Page 95*

Evaluate $A_c = 4\pi^2 rT^{-2}$ for $T = 2$ and $r = 6$.

4 [2nd] [^] (π) [x^2] [×] 6 [×] [(] 2 [^] [(-)] 2 [)] [ENTER]

EXAMPLES 4 and 5 *Pages 97 and 98*

For Example 4, part a, evaluate the expression $16^{\frac{1}{4}}$.

16 [^] [(] 1 [÷] 4 [)] [ENTER]

Use a similar keystroke sequence for part **b** and for Example 5.

LESSON 2.3

EXAMPLE 5 *Page 106*

First clear all equations and statplots.

Evaluate $f(x) = 0.5x^2 - 3x + 2$ for $x = 4$ and $x = 2.5$.

Use viewing window [–7, 12] by [–7, 12].

Graph the function:

[Y=] 0.5 [X,T,θ,n] [x^2] [–] 3 [X,T,θ,n] [+] 2 [GRAPH]

Evaluate for $x = 4$ and $x = 2.5$:

[2nd] [TRACE] (CALC) [1: value] [ENTER] (X=) 4 [ENTER]

[2nd] [TRACE] (CALC) [1: value] [ENTER] (X=) 2.5 [ENTER]

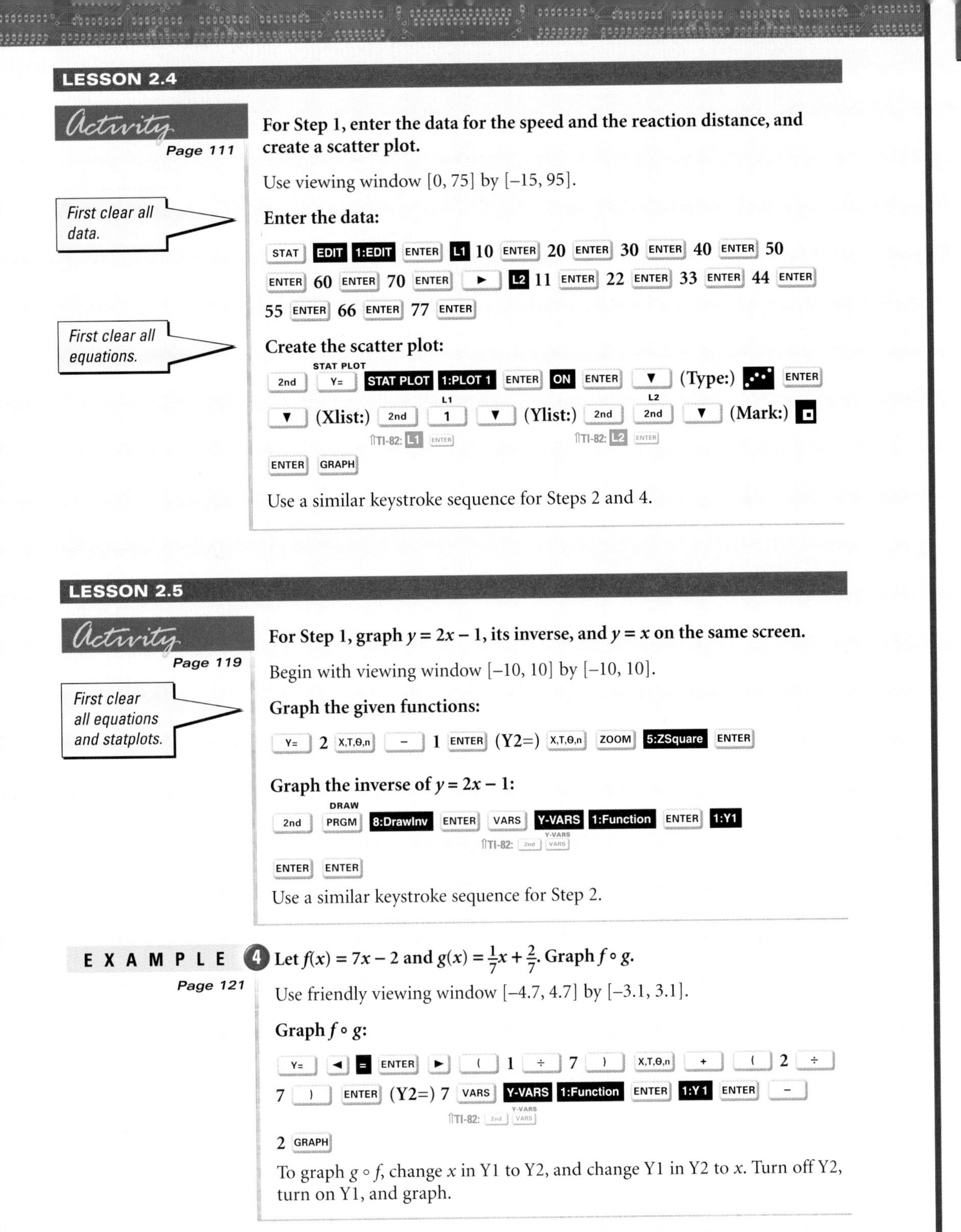

LESSON 2.4

Activity Page 111

For Step 1, enter the data for the speed and the reaction distance, and create a scatter plot.

Use viewing window [0, 75] by [–15, 95].

First clear all data.

Enter the data:

STAT | EDIT | 1:EDIT | ENTER | L1 | 10 | ENTER | 20 | ENTER | 30 | ENTER | 40 | ENTER | 50 | ENTER | 60 | ENTER | 70 | ENTER | ► | L2 | 11 | ENTER | 22 | ENTER | 33 | ENTER | 44 | ENTER | 55 | ENTER | 66 | ENTER | 77 | ENTER

First clear all equations.

Create the scatter plot:

2nd | Y= (STAT PLOT) | STAT PLOT | 1:PLOT 1 | ENTER | ON | ENTER | ▼ | (Type:) | ⁘ | ENTER | ▼ | (Xlist:) | 2nd | 1 (L1) | ▼ | (Ylist:) | 2nd | 2nd (L2) | ▼ | (Mark:) | ▫ | ENTER | GRAPH

⇑TI-82: L1 ENTER ⇑TI-82: L2 ENTER

Use a similar keystroke sequence for Steps 2 and 4.

LESSON 2.5

Activity Page 119

For Step 1, graph $y = 2x - 1$, its inverse, and $y = x$ on the same screen.

Begin with viewing window [–10, 10] by [–10, 10].

First clear all equations and statplots.

Graph the given functions:

Y= | 2 | X,T,θ,n | – | 1 | ENTER | (Y2=) | X,T,θ,n | ZOOM | 5:ZSquare | ENTER

Graph the inverse of $y = 2x - 1$:

2nd | PRGM (DRAW) | 8:DrawInv | ENTER | VARS | Y-VARS | 1:Function | ENTER | 1:Y1 | ENTER | ENTER

⇑TI-82: 2nd VARS (Y-VARS)

Use a similar keystroke sequence for Step 2.

EXAMPLE 4 Page 121

Let $f(x) = 7x - 2$ and $g(x) = \frac{1}{7}x + \frac{2}{7}$. Graph $f \circ g$.

Use friendly viewing window [–4.7, 4.7] by [–3.1, 3.1].

Graph $f \circ g$:

Y= | ◄ | ▪ | ENTER | ► | (| 1 | ÷ | 7 |) | X,T,θ,n | + | (| 2 | ÷ | 7 |) | ENTER | (Y2=) 7 | VARS | Y-VARS | 1:Function | ENTER | 1:Y1 | ENTER | – | 2 | GRAPH

⇑TI-82: 2nd VARS (Y-VARS)

To graph $g \circ f$, change x in Y1 to Y2, and change Y1 in Y2 to x. Turn off Y2, turn on Y1, and graph.

LESSON 2.6

EXAMPLE 4 Graph $y = 3|x| - 1$ and its inverse on the same screen.

Page 127

Use friendly viewing window [–4.7, 4.7] by [–3.1, 3.1].

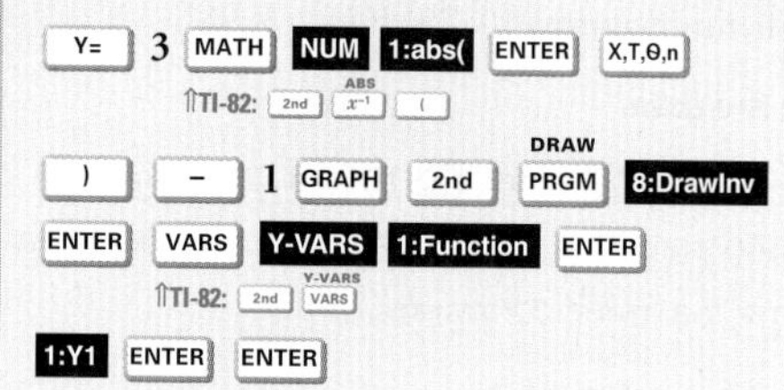

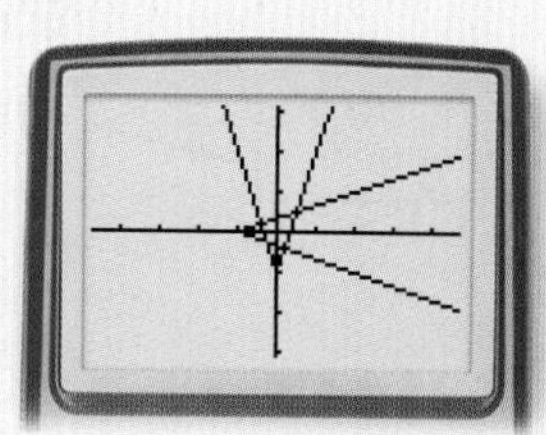

LESSON 2.7

Activity *Page 133*

For Step 1, enter the data from columns 2 and 4, create a scatter plot, and graph the least-squares line for the data.

Use viewing window [25, 35] by [120, 200].

Enter the data:

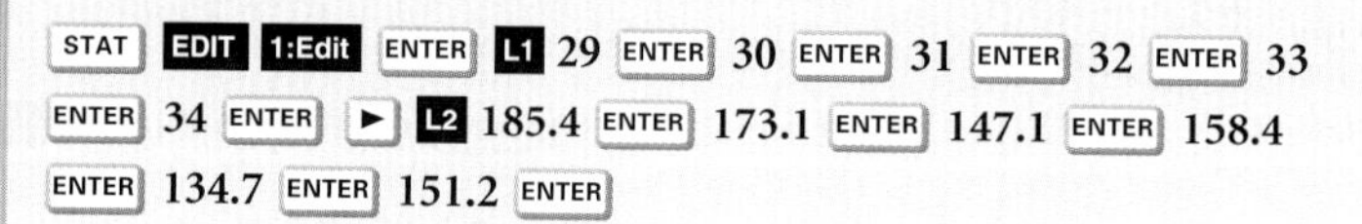

Create the scatter plot:

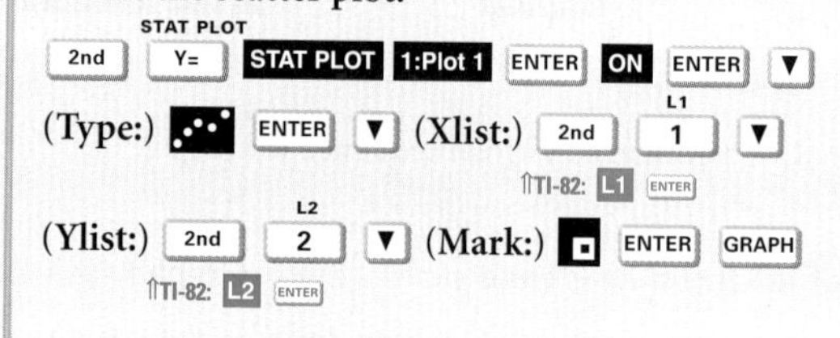

Graph the regression line for the data:

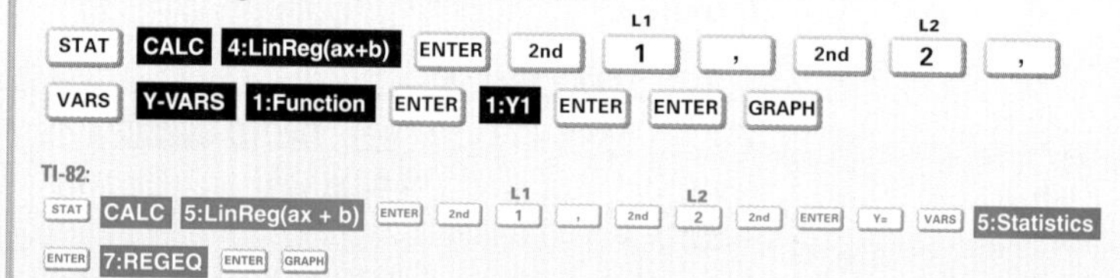

Use a similar keystroke sequence for Step 2.

EXAMPLE 1 Page 134

The TI-82 model does not draw thick lines. Omit cursor keys and ENTER key in Y2.

For part a, graph $y = x^2$ and $y = x^2 + 3$ on the same screen.

Use friendly viewing window [–4.7, 4.7] by [–2, 8].

Y= X,T,θ,n x^2 ENTER (Y2=) ◄ ◄ ENTER (⬊Y2=) ► ► X,T,θ,n x^2 + 3 GRAPH

For part b, graph $y = |x|$ and $y = |x + 4|$ on the same screen.

Use friendly viewing window [–4.7, 4.7] by [–2, 8].

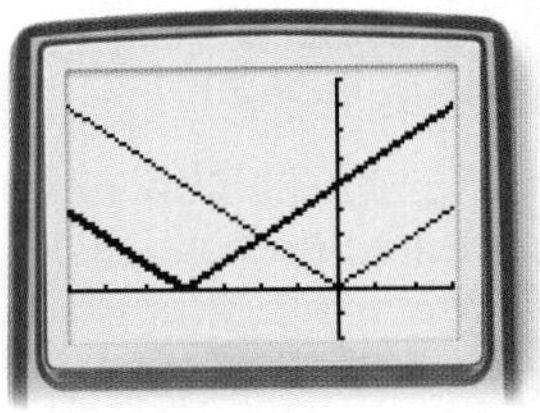

Y= MATH NUM 1:abs(ENTER
⇑TI-82: 2nd x^{-1} (
X,T,θ,n) ENTER
(Y2=) MATH NUM 1:abs(ENTER
⇑TI-82: 2nd x^{-1} (
X,T,θ,n + 3) GRAPH

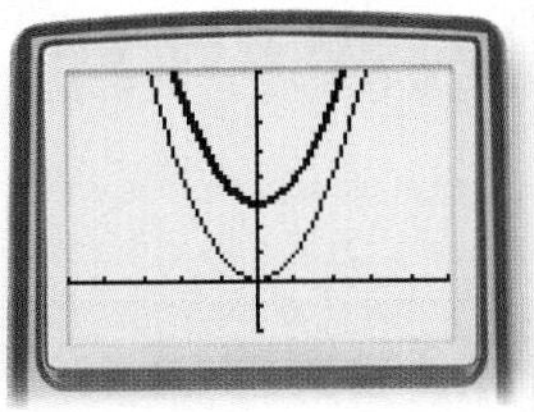

EXAMPLE 2 Page 135

For part a, graph $y = \sqrt{16 - x^2}$ and $y = 2\sqrt{16 - x^2}$ on the same screen.

Use friendly viewing window [–4.7, 4.7] by [0, 8].

Y= 2nd $\sqrt{\ }$ 16 – X,T,θ,n x^2
⇑TI-82: (
) ENTER (Y2=) 2 2nd $\sqrt{\ }$ 16 –
⇑TI-82: (
X,T,θ,n x^2) GRAPH

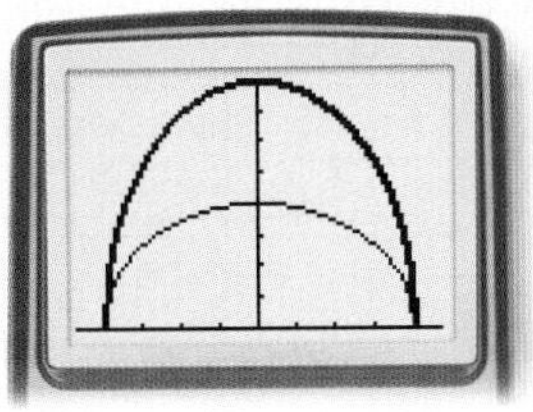

EXAMPLES 3, 4, and 5 Pages 136–138

For Example 3, part a, graph $y = \sqrt{(16 - x^2)}$ and $y = \sqrt{16 - (2x)^2}$ on the same screen.

Use friendly viewing window [–9.4, 9.4] by [0, 6.2].

Y= 2nd $\sqrt{\ }$ 16 – X,T,θ,n x^2
⇑TI-82: (
) ENTER (Y2=) 2nd $\sqrt{\ }$ 16 –
⇑TI-82: (
(2 X,T,θ,n) x^2) GRAPH

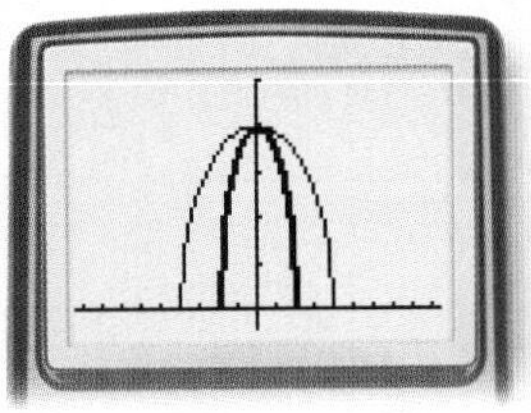

Use a similar keystroke sequence for part **b**.

Use a similar keystroke sequence for Examples 4 and 5.

3 Systems of Linear Equations and Inequalities

Lesson Presentation CD-ROM
PowerPoint® presentations for each lesson 3.1–3.6

CHAPTER PLANNING GUIDE

Lesson	3.1	3.2	3.3	3.4	3.5	3.6	Project and Review
Pupil's Edition Pages	156–163	164–171	172–178	179–186	187–194	195–201	202–209
Practice and Assessment							
Extra Practice (Pupil's Edition)	949	950	950	951	951	952	
Practice Workbook	16	17	18	19	20	21	
Practice Masters Levels A, B, and C	46–48	49–51	52–54	55–57	58–60	61–63	
Standardized Test Practice Masters	18	19	20	21	22	23	24
Assessment Resources	30	31	32	34	35	36	33, 37–42
Visual Resources							
Lesson Presentation Transparencies Vol. 1	61–64	65–68	69–72	73–76	77–80	81–84	
Teaching Transparencies			12		13		
Answer Key Transparencies	87–92	93–96	97–107	108–117	118–126	127–131	132–138
Quiz Transparencies	3.1	3.2	3.3	3.4	3.5	3.6	
Teacher's Tools							
Reteaching Masters	31–32	33–34	35–36	37–38	39–40	41–42	
Make-Up Lesson Planner for Absent Students	16	17	18	19	20	21	
Student Study Guide	16	17	18	19	20	21	
Spanish Resources	16	17	18	19	20	21	
Block Scheduling Handbook							6–7
Activities and Extensions							
Lesson Activities	16	17	18	19	20	21	
Enrichment Masters	16	17	18	19	20	21	
Cooperative-Learning Activities	16	17	18	19	20	21	
Problem Solving/ Critical Thinking	16	17	18	19	20	21	
Student Technology Guide	16	17	18	19	20	21	
Long Term Projects							9–12
Writing Activities for Your Portfolio							5–9
Tech Prep Masters							11–14
Building Success in Mathematics							7–9

LESSON PACING GUIDE

Lesson	3.1	3.2	3.3	3.4	3.5	3.6	Project and Review
Traditional	2 days	2 days	2 days	2 days	2 days	2 days	2 days
Block	1 day	1 day	1 day	1 day	1 day	1 day	1 day
Two-Year	4 days	4 days	4 days	4 days	4 days	4 days	4 days

CONNECTIONS AND APPLICATIONS

Lesson	3.1	3.2	3.3	3.4	3.5	3.6	Review
Algebra	156–163	164–171	172–178	179–186	187–194	195–201	202–209
Geometry	162	169	177	184	192		
Maximum/Minimum					188, 189, 190, 191, 192		
Transformations					197, 200, 201		
Business and Economics	161	164, 165, 168, 170, 171	172, 174, 176, 177, 178	184, 185, 186	192, 193, 194		206, 207, 209
Life Skills	162, 163	170, 171	172, 174, 176 177, 178	185	193		209
Science and Technology	156, 158, 163	170		182	187, 188, 190, 191, 193	195, 201	209
Sports and Leisure	163	170	177	179, 185		198, 199, 200	206
Cultural Connection: China		170					
Other				184-186			

BLOCK SCHEDULING GUIDE

Day	Lesson	Teacher Directed: Lesson Examples, Teaching Transparencies	Student Guided Activity, Try This	Cooperative-Learning Activity, Lesson Activity, Student Technology Guide	Practice: Practice & Apply, Extra Practice, Practice Workbook	Assessment: Quiz, Mid-Chapter Assessment	Problem Solving, Reteaching
1	3.1	10 min	15 min	15 min	65 min	15 min	15 min
2	3.2	10 min	15 min	15 min	65 min	15 min	15 min
3	3.3	10 min	15 min	15 min	65 min	15 min	15 min
4	3.4	10 min	15 min	15 min	65 min	15 min	15 min
5	3.5	10 min	15 min	15 min	65 min	15 min	15 min
6	3.6	10 min	15 min	15 min	65 min	15 min	15 min
7	Assess.	50 min **PE:** Chapter Review	90 min **PE:** Chapter Project, Writing Activities	90 min Tech Prep Masters	65 min **PE:** Chapter Assessment, Test Generator	30 min Chap. Assess. (A or B), Alt. Assess. (A or B), Test Generator	

PE: Pupil's Edition

Alternative Assessment

The following suggest alternative assessments for students who may benefit from a different type of assessment than the regular chapter quizzes and the mid-chapter/end-of-chapter test. Visit the HRW web site to get additional Alternative Assessment material.

internet connect

Alternative Assessment

Go To: **go.hrw.com**
Keyword: **MB1 Alt Assess**

Performance Assessment

1. Consider the system $\begin{cases} -2x + 3y = 7 \\ x + 5y = 2 \end{cases}$.

a. Solve this system by using an algebraic and a geometric method.

b. Compare and contrast algebraic and graphical methods of solution.

2. Consider the system $\begin{cases} -2x + 3y \le 7 \\ x + 5y \ge 2 \end{cases}$.

a. Graph the solution to this system of inequalities.

b. If you are also given $x \ge 0$ and $y \ge 0$, explain how to modify the solution.

c. Suppose that the solution region determined in parts **a** and **b** is a feasible region for a linear programming problem. Which solution maximizes $R = 5x - y$?

Portfolio Project

Suggest that students choose one of the following projects for inclusion in their portfolios.

1. You can use a system of equations and related inequalities to represent a polygon and its interior.

a. Represent region *ABCD* as a system of inequalities.

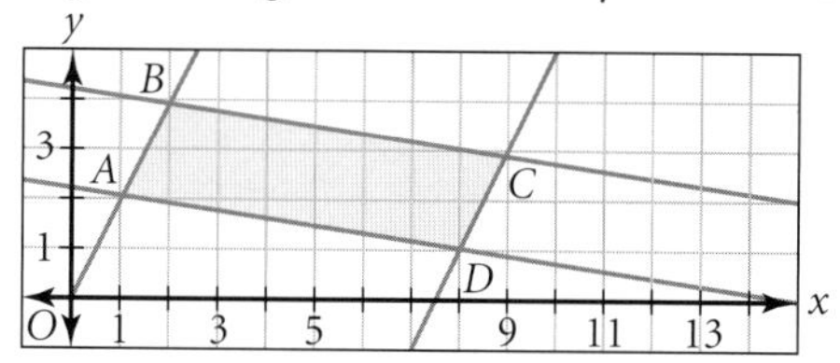

b. Investigate representations of other polygonal regions. Share your findings with the class.

2. Research the origins of linear programming. Find out how mathematicians used it to solve military supply problems during World War II.

internet connect

The table below identifies the pages in this chapter that contain internet and technology information.

Content Links

Activities Online	**pages 163, 183, 191**
Portfolio Extensions	**pages 186, 194**
Homework Help Online	**pages 161, 169, 176, 183, 192, 200**
Graphic Calculator Support	**page 210**

Resource Links

Parents can go online and find concepts that students are learning—lesson by lesson—and questions that pertain to each lesson, which facilitate parent-student discussion.

Go To: **go.hrw.com**
Keyword: **MB1 Parent Guide**

Technical Support

The following may be used to obtain technical support for any HRW software product.

Online Help: **www.hrwtechsupport.com**

e-mail: **tschrw@hrwtechsupport.com**

HRW Technical Support Center: **(800)323-9239**

7 AM to 10 PM Monday through Friday Central Time

Visit the HRW math web site at: **www.hrw.com/math**

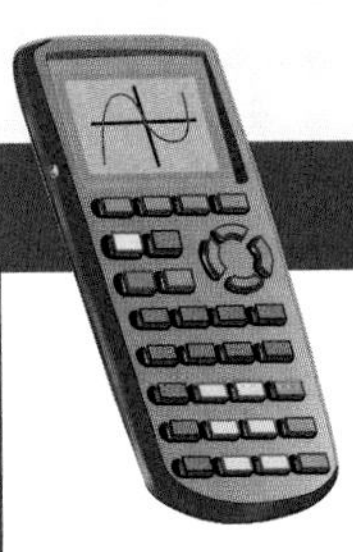

Technology

Lesson Suggestions and Calculator Examples

(Keystrokes are based on a TI-83 calculator.)

Lesson 3.1 Solving Systems by Graphing or Substitution

In the Activity on page 156, students will experiment with a pair of linear equations to determine whether they have a solution and, if so, how many solutions are possible. Consider having students enter a pair of linear equations, graph them, and find their point of intersection. Then have students edit the second equation so that the graphs reveal a different solution.

Lesson 3.2 Solving Systems by Elimination

Students learned the elimination method in Algebra 1 and will explore it in greater depth in this lesson. Stress the importance of performing steps one at a time.

Have students consider the system $\begin{cases} ax + by = c \\ dx + ey = f \end{cases}$, where students select values for a, b, c, d, e, and f. Ask students if they can find a relationship among a, b, c, and d that will show when the solution will be unique. The relationship should be $ad - bc \neq 0$.

Lesson 3.3 Linear Inequalities in Two Variables

In this lesson, students will learn how to use a graphics calculator to show a solution region. The displays below show how to graph the solution region for $y \geq 0.5x + 2$. Ask students to press **2nd** **PRGM** and select **DRAW** **7:Shade(** **ENTER**. Then have students enter the lower boundary and an upper boundary of their choice. Note that the upper boundary and **Ymax** are the same.

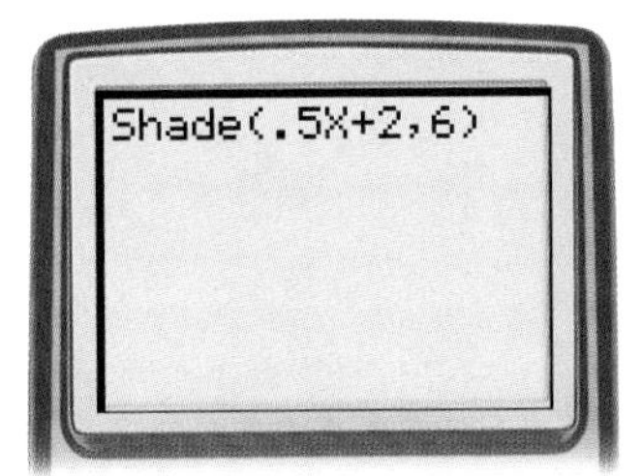

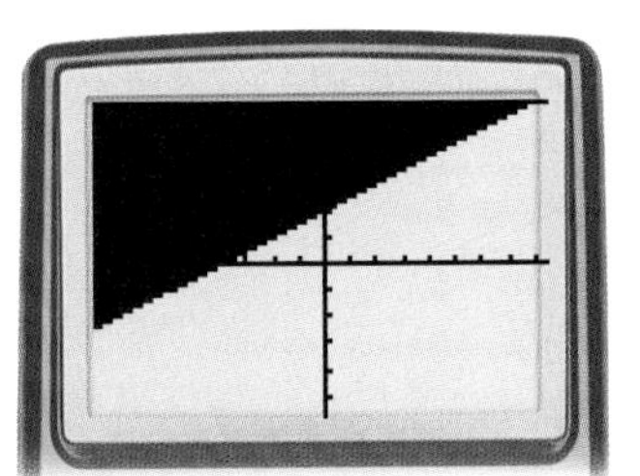

Lesson 3.4 Systems of Linear Inequalities

The calculator displays at top right show how to graph the solution of a pair of linear inequalities by using a method different from that described in the previous lesson.

Consider the system $\begin{cases} y \geq 0.5x + 2 \\ y < 3 \end{cases}$. Enter the equation into the **Y=** editor, ignoring the inequality signs. To shade the graphs, move the cursor to the left of **Y1** and press **ENTER** until the ◥ (shade above graph) symbol appears. **Y2** has a ≤ , so repeat the process and choose the ◣ (shade below graph) symbol. Then press **GRAPH**.

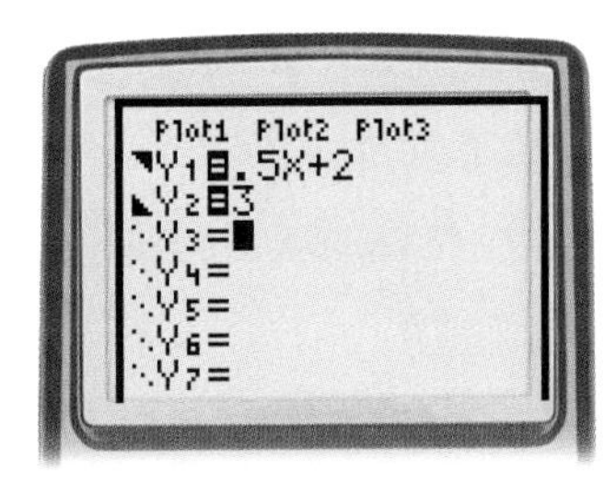

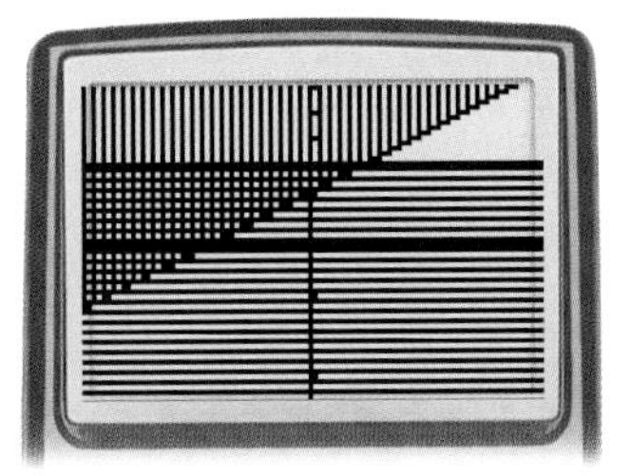

Lesson 3.5 Linear Programming

Linear programming problems are complex and require a great deal of work to solve, so computer programs are usually used to solve them. Discuss with students how the calculator skills they developed in earlier lessons in this chapter can help solve parts of a linear programming problem.

Lesson 3.6 Parametric Equations

Have students press **MODE** and select **Par** to graph parametric equations. To guide students, consider these steps.

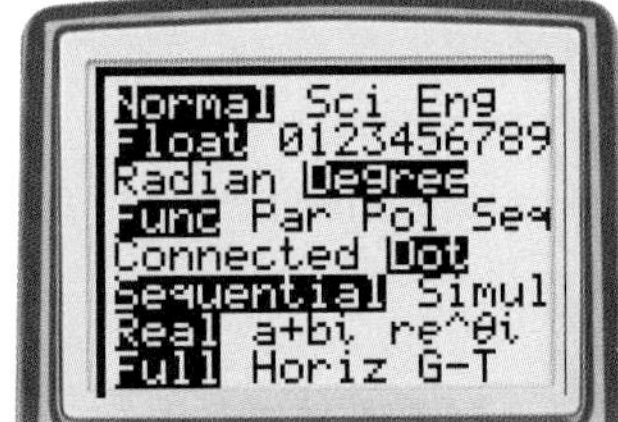

1. Choose the parametric mode setting.
2. Enter functions for x and y.
3. Choose **T** settings and window settings and then graph.

The displays below show $x(t) = 2t - 1$ and $y(t) = -t + 2$ over the intervals $0 \leq t \leq 2$ and $-2 \leq t \leq 4$.

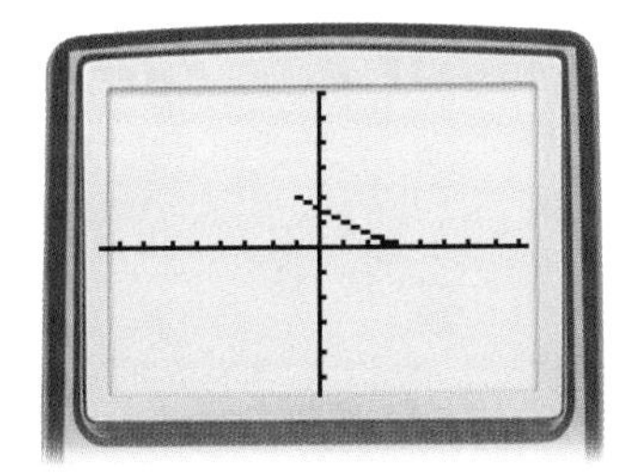

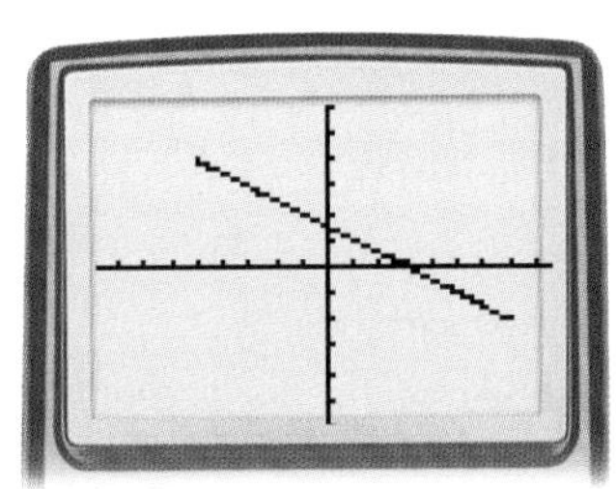

For further information, refer to the

- technology discussions in the lessons.
- lesson-related teacher's commentary in the side columns of this *Teacher's Edition.*
- lesson-related *Student Technology Guide* masters.
- *HRW Technology Handbook.*

internet connect

For keystrokes of other graphing calculators models, visit the HRW web site at **go.hrw.com** and enter the keyword **MB1 CALC**.

About Chapter 3

Background Information

In Chapter 3, students learn to solve systems of linear equations by using three basic techniques: graphing, substitution, and elimination. They also learn to graph linear inequalities and to solve systems of linear inequalities. This experience with systems is then used as the basis for a study of linear programming. In the final lesson of the chapter, students are introduced to parametric equations.

CHAPTER RESOURCES

- Block-Scheduling Handbook
- Writing Activities for Your Portfolio
- Tech Prep Masters
- Long-Term Project
- Assessment Resources:
  Mid-Chapter Assessment
  Chapter Assessments
  Alternative Assessments
- Test and Practice Generator
- Technology Handbook

Chapter Objectives

- Solve a system of linear equations in two variables by graphing. [**3.1**]
- Solve a system of linear equations by substitution. [**3.1**]
- Solve a system of two linear equations in two variables by elimination. [**3.2**]
- Solve and graph a linear inequality in two variables. [**3.3**]
- Use a linear inequality in two variables to solve real-world problems. [**3.3**]

Systems of Linear Equations and Inequalities

SYSTEMS OF LINEAR EQUATIONS AND INEQUALITIES are used to find optimal solutions to problems in business, finance, manufacturing, agriculture, and other fields such as photography.

For instance, the time required to develop a photograph involves both the *concentration* and the *temperature* of the developer fluid. Situations in which variables are related in more than one way can be represented with a system of equations.

Lessons

Supplies used to develop photographs.

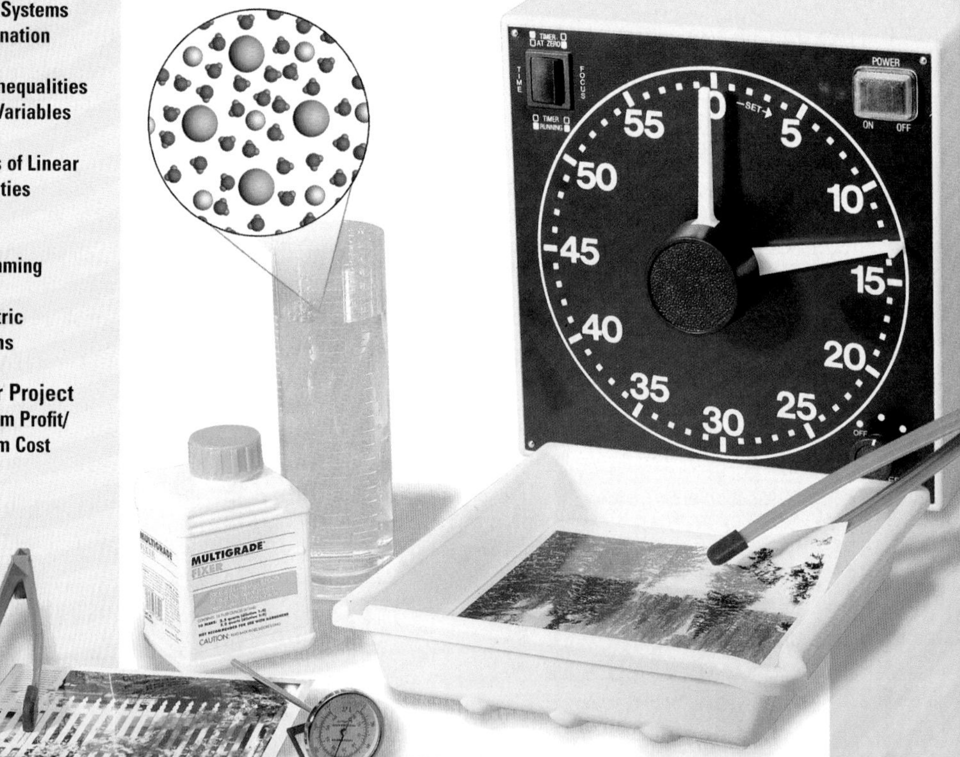

About the Photos

Linear programming is a new area of mathematics that came into being during the 1940s. During its brief history, linear programming has changed the way businesses make decisions from using intuition and experience to using an algorithm that produces an optimal solution based on available data.

The first businesses to use linear programming were the petroleum refineries, which used it for blending gasoline. The technique has since been adopted by almost every industry, including photography.

Some problems solved by linear programming today involve thousands of equations, sometimes with a million variables, and computers must be used to find the solutions. Linear programming accounts for 50% to 90% of nonroutine computer time used for management decisions.

About the Chapter Project

The Chapter Project, *Maximum Profit/Minimum Cost,* consists of two linear-programming problems. In one, you will be faced with a limited number of resources and will be asked to maximize profit. In the second, you will be asked to complete a job while minimizing the costs of labor. You will have an opportunity to use mathematics as it is often used in the real world to solve problems and make decisions.

After completing the Chapter Project, you will be able to do the following:

- Set up and solve linear-programming problems that involve finding maximums or minimums.
- Investigate how changes in the objective function or in the constraint inequalities affect the outcomes.

About the Portfolio Activities

Throughout the chapter, you will be given opportunities to complete Portfolio Activities that are designed to support your work on the Chapter Project.

- Writing and solving a system of equations to solve a production problem is included in the Portfolio Activity on page 171.
- Using a system of linear inequalities to find production limits and possibilities is included in the Portfolio Activity on page 186.
- Using linear programming to maximize the profits of a production company are included in the Portfolio Activity on page 194.

- Write and graph a system of linear inequalities in two variables. [**3.4**]
- Write a system of linear inequalities in two variables for a given solution region. [**3.4**]
- Write and graph a set of constraints for a linear-programming problem. [**3.5**]
- Use linear programming to find the maximum or minimum value of an objective function. [**3.5**]
- Graph a pair of parametric equations, and use them to model real-world applications. [**3.6**]
- Write the function represented by a pair of parametric equations. [**3.6**]

Portfolio Activities appear at the end of Lessons 3.2, 3.4, and 3.5. Each serves as preparation for the Chapter Project. The Portfolio Activities as well as the Chapter Project Activities are appropriate for inclusion in the student's portfolio. Students should be encouraged to include in their portfolios any other work in which they feel a sense of pride or a sense of accomplishment.

internet connect go.hrw.com

Chapter Internet Features and Online Activities

Lesson	Keyword	Page
3.1	MB1 Homework Help	161
	MB1 WBB	163
3.2	MB1 Homework Help	169
3.3	MB1 Homework Help	176
3.4	MB1 Homework Help	183
	MB1 Investments	183
	MB1 Diet	186
3.5	MB1 Homework Help	192
	MB1 Nutrition	191
	MB1 Simplex	194
3.6	MB1 Homework Help	200

Prepare

NCTM PRINCIPLES & STANDARDS 1–4, 6, 8–10

QUICK WARM-UP

Write each equation in slope-intercept form.

1. $x - y = 9$ $y = x - 9$
2. $2x = 5y$ $y = \frac{2}{5}x$, or $y = 0.4x$
3. $4x + 7y = 14$ $y = -\frac{4}{7}x + 2$
4. $3x - \frac{1}{5}y = 6$ $y = 15x - 30$
5. $2.5y + 8.1 = 7.5x$ $y = 3x - 3.24$

Also on Quiz Transparency 3.1

Teach

Why Systems of equations provide a means of solving a problem involving two variables when two or more conditions must be satisfied. Give students one or two examples of situations that can be modeled by a system of two linear equations in two variables. Ask them to suggest other such situations.

CHECKPOINT ✓

3. **If the slopes are the same, the system has no solution or infinitely many solutions. If the slopes are different, then the system has exactly one solution.**

Solving Systems by Graphing or Substitution

Why *Systems of equations are frequently used to model events that occur in daily life. A system of equations can be used to determine business profits or create exact mixtures.*

Objectives

- Solve a system of linear equations in two variables by graphing.
- Solve a system of linear equations by substitution.

APPLICATION CHEMISTRY

A laboratory technician is mixing a 10% saline solution with a 4% saline solution. How much of each solution is needed to make 500 milliliters of a 6% saline solution? *You will solve this problem in Example 3.*

A **system of equations** is a collection of equations in the same variables.

The solution of a system of two linear equations in x and y is any ordered pair, (x, y), that satisfies both equations. The solution (x, y) is also the point of intersection for the graphs of the lines in the system. For example, the ordered pair $(2, -1)$ is the solution of the system below.

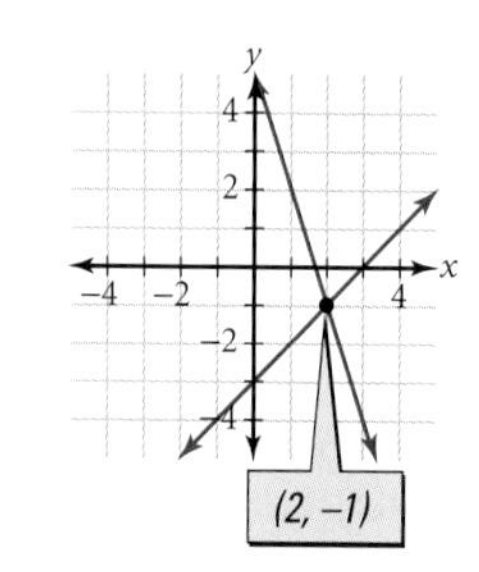

$$\begin{cases} y = x - 3 \\ y = 5 - 3x \end{cases} \rightarrow \begin{cases} (-1) = (2) - 3 & \text{True} \\ (-1) = 5 - 3(2) & \text{True} \end{cases}$$

Activity Exploring Graphs of Systems

TECHNOLOGY GRAPHICS CALCULATOR Keystroke Guide, page 210

You will need: graph paper or a graphics calculator

1. Graph system I at right.
   a. Are there any points of intersection?
   b. Can you find exactly one solution to the system? If so, what is it? If not, modify the system so that it has exactly one solution and state that solution.
2. Repeat Step 1 for systems II and III.
3. Describe the slopes of the lines whose equations form a system with no solution, with infinitely many solutions, and with exactly one solution. **CHECKPOINT ✓**

System
I. $\begin{cases} y = 2x - 1 \\ y = -x + 5 \end{cases}$
II. $\begin{cases} y = 2x - 1 \\ y = 2x + 1 \end{cases}$
III. $\begin{cases} y = \frac{8 - 3x}{4} \\ y = -\frac{3}{4}x + 2 \end{cases}$

Alternative Teaching Strategy

USING TECHNOLOGY/USING TABLES Refer to part **a** of Example 1. If students are using a TI-82 or TI-83 graphics calculators, have them press [Y=] and enter the equations as **Y1** and **Y2.** Then have them press [2nd] [WINDOW] and choose the following settings:

TblStart=0 (or **TblMin=0**) **Indpnt:** **Auto** **Ask**
ΔTbl=1 **Depend:** **Auto** **Ask**

If they press [2nd] [GRAPH], they will see a table. Point out that when the value in the **X** column of this table is 3, **Y1=Y2=2.** This means that a solution of the system is the ordered pair (3, 2). Discuss why this is the *only* solution.

Follow a similar procedure for part **b** of Example 1. This time, have students use [▲] and [▼] to search for a place in the table where **Y1=Y2.** They should conclude that this will never happen. To reinforce this fact, have them use the **VARS** menu to enter **Y3=Y1−Y2.** When they view the **Y3** column of the table, they will see a constant difference of −4. This confirms that the system has no solution.

You can graph a system of equations in two variables to find whether a solution for the system exists. The systems and graphs below illustrate the three possibilities for a system of two linear equations in two variables.

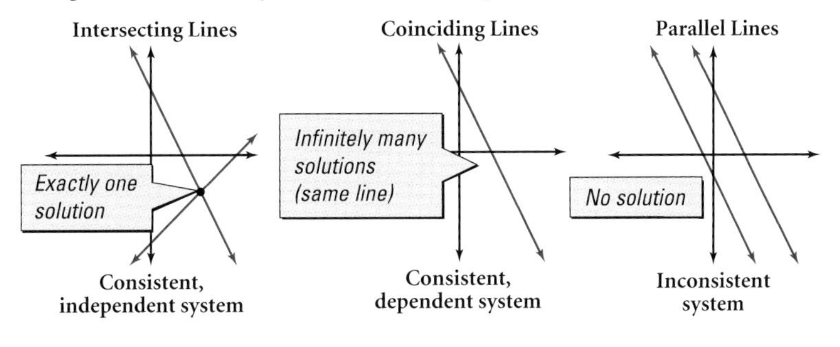

Classifying Systems of Equations

If a system of equations has at least one solution, it is called **consistent**.
- If a system has exactly one solution, it is called **independent**.
- If a system has infinitely many solutions, it is called **dependent**.

If a system does not have a solution, it is called **inconsistent**.

EXAMPLE 1 **Graph and classify each system. Then find the solution from the graph.**

a. $\begin{cases} x + y = 5 \\ x - 5y = -7 \end{cases}$ **b.** $\begin{cases} x - 2y = 3 \\ x + 5 = 2y \end{cases}$

TECHNOLOGY GRAPHICS CALCULATOR
Keystroke Guide, page 210

SOLUTION

Solve each equation for y.

a. $\begin{cases} x + y = 5 \\ x - 5y = -7 \end{cases} \rightarrow \begin{cases} y = -x + 5 \\ y = \frac{x + 7}{5} \end{cases}$

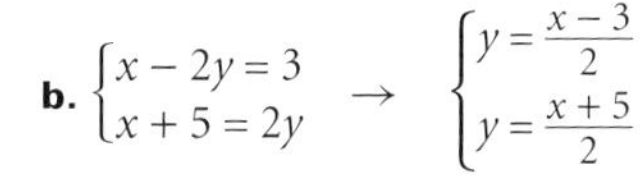

b. $\begin{cases} x - 2y = 3 \\ x + 5 = 2y \end{cases} \rightarrow \begin{cases} y = \frac{x - 3}{2} \\ y = \frac{x + 5}{2} \end{cases}$

Intersecting lines have different slopes.

Intersection X=3 Y=2

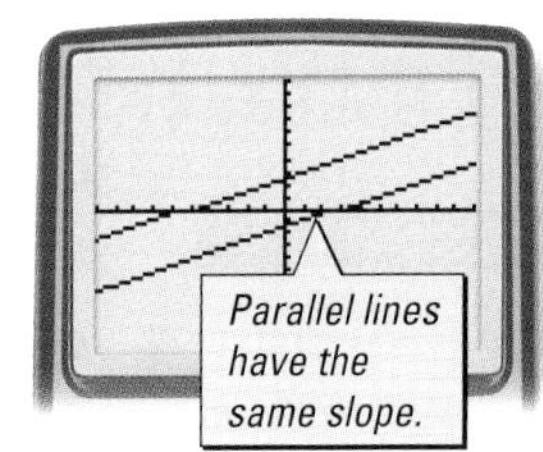

Because the lines intersect at exactly one point, the system is consistent and independent. The solution is (3, 2).

The lines appear to be parallel. You can verify this by examining the slopes of the lines—both are $\frac{1}{2}$. Thus, the system is inconsistent, and there is no solution.

TRY THIS Graph and classify $\begin{cases} y = 3x + 4 \\ y = -2x + 4 \end{cases}$. Then find the solution from the graph.

CRITICAL THINKING Classify $\begin{cases} y = mx \\ y = nx \end{cases}$, where $m \neq 0$ and $n \neq 0$, as thoroughly as possible.

Interdisciplinary Connection

CONSUMER EDUCATION Two local bookstores are each having a sale. At Book Bonanza, the sale price of any paperback book in stock is $2 off the list price. At Barrels of Books, the sale price of any paperback book in stock is 25% off the list price.

a. Write a system of equations that represents the sale prices at each store.
$\begin{cases} y = x - 2 \\ y = 0.75x \end{cases}$, **where x represents the list price in dollars and y represents the sale price in dollars**

b. What is the solution of the system, and what does it represent? **(8, 6); a paperback book with a list price of $8 will have a sale price of $6 at both stores.**

c. Which store has the better sale prices on paperback books? **For books with list prices less than $8, the sale prices at Book Bonanza are better. For books with list prices greater than $8, the sale prices at Barrels of Books are better. For a book with a list price of $8, the sale price is the same at both stores.**

Activity Notes

In this Activity, students investigate the slopes of lines that form the graph of a system of two linear equations in x and y. They should observe the following:

- If the slopes are not equal, then the system has exactly one solution.
- If the slopes are equal *and* the y-intercepts are different, then the system has no solution.
- If the slopes are equal *and* the y-intercepts are the same, then the system has infinitely many solutions.

ADDITIONAL EXAMPLE 1

Graph and classify each system. Then find the solution from the graph.

a. $\begin{cases} x - y = 2 \\ x + 2y = -6 \end{cases}$
Check students' graphs; consistent and independent; $\left(-\frac{2}{3}, -\frac{8}{3}\right)$

b. $\begin{cases} 9x - 3y = 3 \\ 21x + 4 = 7y \end{cases}$
Check students' graphs; inconsistent; no solution

TRY THIS
The system is consistent and independent. The solution is $x = 0$ and $y = 4$.

CRITICAL THINKING
Since both lines pass through the origin, the system has the solution (0, 0). If $m \neq n$, then the system is independent and consistent. If $m = n$, then the system is dependent.

ADDITIONAL

EXAMPLE 2

Use substitution to solve the system. Check your solution.

$$\begin{cases} 7x + y = 6 \\ 5x - y = -3 \end{cases}$$

(0.25, 4.25)

TRY THIS

(−1, 11)

ADDITIONAL

EXAMPLE 3

A laboratory technician is developing a cleaning solution that will remove stains from clothes. The technician mixes a 20% acid solution with a 5% acid solution. **How much of each solution, to the nearest milliliter, is needed to make 150 milliliters of a 15% acid solution?**

100 mL of the 20% acid solution and 50 mL of the 5% acid solution

Example 2 shows you how to use substitution to solve a system in which a variable has a coefficient of 1.

EXAMPLE 2 **Use substitution to solve the system. Check your solution.** $\begin{cases} 2x + y = 3 \\ 3x - 2y = 8 \end{cases}$

SOLUTION

Solve the first equation for y because it has a coefficient of 1.

$$2x + y = 3$$
$$y = 3 - 2x$$

Substitute $3 - 2x$ for y in the second equation.

$$3x - 2y = 8$$
$$3x - 2(3 - 2x) = 8 \quad \text{This equation has only one variable.}$$
$$3x - 6 + 4x = 8 \quad \text{Use the Distributive Property.}$$
$$7x = 14$$
$$x = 2$$

Substitute 2 for x in either original equation to find y.

First equation:	Second equation:
$2x + y = 3$	$3x - 2y = 8$
$2(2) + y = 3$	$3(2) - 2y = 8$
$y = -1$	$-2y = 8 - 6$
	$y = -1$

The solution is (2, −1).

CHECK

$$\begin{cases} 2x + y = 3 \\ 3x - 2y = 8 \end{cases} \rightarrow \begin{cases} 2(2) + (-1) = 3 & \textbf{True} \\ 3(2) - 2(-1) = 8 & \textbf{True} \end{cases}$$

Always check your answers by substituting them into both original equations.

TRY THIS Use substitution to solve the system. Check your solution. $\begin{cases} 3x + y = 8 \\ 18x + 2y = 4 \end{cases}$

EXAMPLE 3 Refer to the saline solution mixture described at the beginning of the lesson. **How much of each solution, to the nearest milliliter, is needed to make 500 milliliters of a 6% saline solution?**

APPLICATION
CHEMISTRY

PROBLEM SOLVING

SOLUTION

Write two equations in x and y. Let x and y represent the amounts of the 10% and 4% solutions, respectively.

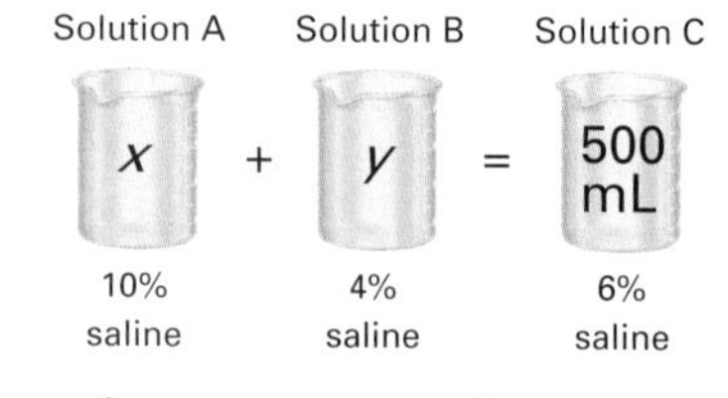

amount of 10% solution	+	amount of 4% solution	=	amount of 6% solution
x	+	y	=	500

saline in 10% solution	+	saline in 4% solution	=	saline in 6% solution
$0.10x$	+	$0.04y$	=	$0.06(500)$

Inclusion Strategies

ENGLISH LANGUAGE DEVELOPMENT The vocabulary of this lesson may be confusing to students who have reading difficulties, or for whom English is a second language. Discuss everyday uses of *intersection, consistent, inconsistent, independent, dependent,* and *substitution* (or *substitute*). Have students relate the mathematical interpretation to the everyday meaning. Pay special attention to the fact that within this lesson, the word *solution* refers not only to the ordered pair that satisfies a system of equations, but also to a chemical mixture.

Enrichment

Part of the graph of each system below is a triangle. Find the coordinates of the vertices of the triangle. Then calculate its area.

1. $\begin{cases} y = -x + 5 \\ y = -2 \\ x = 4 \end{cases}$ **(4, −2), (7, −2), (4, 1); 4.5 sq units**

2. $\begin{cases} y = x + 4 \\ y = -2x + 16 \\ y = 2 \end{cases}$ **(4, 8), (−2, 2), (7, 2); 27 sq units**

Method 1 Use the substitution method.

1. Solve the first equation for y.

$$x + y = 500$$
$$y = 500 - x$$

The first equation can be solved for either x or y.

2. Substitute $500 - x$ for y into the second equation, and solve for x.

$$0.10x + 0.04y = 0.06(500)$$
$$0.10x + 0.04(500 - x) = 0.06(500) \quad \text{Substitute.}$$
$$0.10x + 20 - 0.04x = 30 \quad \text{Simplify.}$$
$$0.06x = 10$$
$$x \approx 167$$

3. Substitute 167 for x into the first equation.

$$x + y = 500$$
$$167 + y \approx 500$$
$$y \approx 333$$

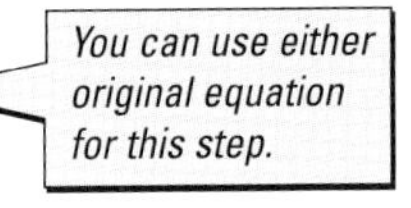

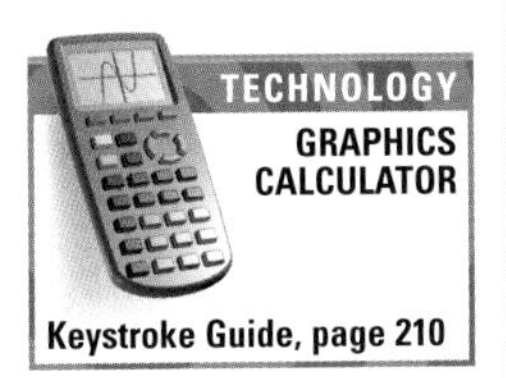

Method 2 Use the graphing method.

Graph the equations, and find any points of intersection.

To graph the system with a graphics calculator, solve for y and rewrite the system as shown below.

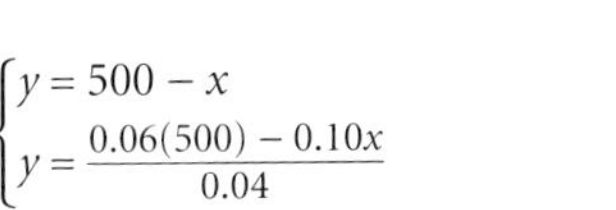

$$\begin{cases} y = 500 - x \\ y = \dfrac{0.06(500) - 0.10x}{0.04} \end{cases}$$

The technician needs to combine 167 milliliters of the 10% solution and 333 milliliters of the 4% solution.

TRY THIS If a 7% saline solution and a 4% saline solution are mixed to make 500 milliliters of a 5% saline solution, how much of each solution, to the nearest milliliter, is needed?

TRY THIS
≈167 mL of 7%;
≈333 mL of 4%

The solution of a system in three variables, such as in x, y, and z, is an *ordered triple* (x, y, z) that satisfies all three equations.

EXAMPLE 4 **Use substitution to solve the system. Check your solution.**

$$\begin{cases} x + y + z = 5 \\ 2x - 3y + z = -2 \\ 4z = 8 \end{cases}$$

SOLUTION

1. Solve the third equation for z.

$$4z = 8$$
$$z = 2$$

2. Substitute 2 for z in the first and second equations. Then simplify.

$$\begin{cases} x + y + z = 5 \\ 2x - 3y + z = -2 \\ z = 2 \end{cases} \rightarrow \begin{cases} x + y + 2 = 5 \\ 2x - 3y + 2 = -2 \end{cases} \rightarrow \begin{cases} x + y = 3 \\ 2x - 3y = -4 \end{cases}$$

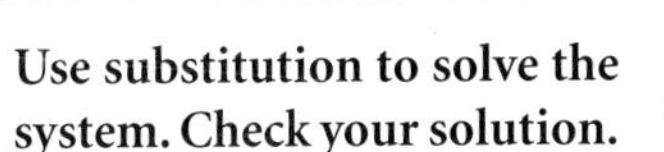

Use substitution to solve the system. Check your solution.

$$\begin{cases} x - y - z = -4 \\ 5x + 2y - 3z = 7 \\ 6z = -24 \end{cases}$$

(−3, 5, −4)

Reteaching the Lesson

COOPERATIVE LEARNING Display the partial system at right on the board or overhead. Have students work in pairs. Each pair must decide how to fill in the box with an equation so that the resulting system satisfies the conditions given below. If it is not possible to create such a system, the pair must explain why. Discuss results with the entire class after the groups have completed each exercise.

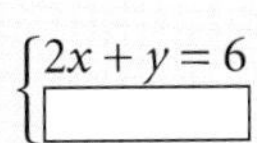

$$\begin{cases} 2x + y = 6 \\ \boxed{} \end{cases}$$

1. The only solution is (1, 4).
   sample: $y = x + 3$
2. (1, 4) is one of infinitely many solutions.
   sample: $4x + 2y = 12$
3. The only solution is (0, 0).
   Not possible; (0, 0) is not a solution of $2x + y = 6$, so it cannot be a solution of any system that includes $2x + y = 6$.
4. There is no solution.
   sample: $2x + y = 4$

TRY THIS
(5, 3, –3)

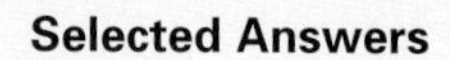

Assess

Selected Answers
Exercises 5–9, 11–77 odd

ASSIGNMENT GUIDE

In Class	1–9
Core	11–51 odd, 55, 57
Core Plus	10–46 even, 47–59
Review	60–78
Preview	79, 80

Extra Practice can be found beginning on page 940.

5.

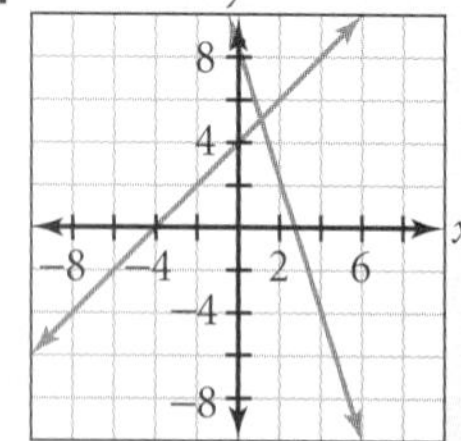

6.

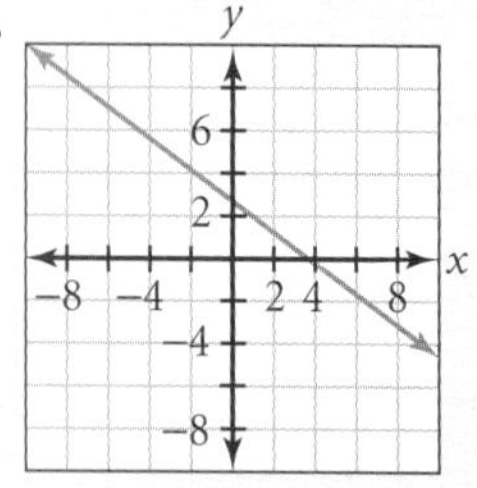

3. Use substitution to solve the resulting system. $\begin{cases} x + y = 3 \\ 2x - 3y = -4 \end{cases}$

$$\begin{aligned} x + y &= 3 \\ y &= 3 - x \end{aligned} \quad \rightarrow \quad \begin{aligned} 2x - 3y &= -4 \\ 2x - 3(3 - x) &= -4 \\ 2x - 9 + 3x &= -4 \\ 5x &= 5 \\ x &= 1 \end{aligned}$$

Substitute 3 – x for y in 2x – 3y = –4.

4. Now substitute 1 for x and 2 for z to find y.

$$\begin{aligned} x + y + z &= 5 \\ 1 + y + 2 &= 5 \\ y &= 2 \end{aligned}$$

You can use either the first or second equation of the original system.

Thus, the solution for the system is (1, 2, 2).

CHECK $\begin{cases} x + y + z = 5 \\ 2x - 3y + z = -2 \\ 4z = 8 \end{cases} \rightarrow \begin{cases} 1 + 2 + 2 = 5 & \textbf{True} \\ 2(1) - 3(2) + 2 = -2 & \textbf{True} \\ 4(2) = 8 & \textbf{True} \end{cases}$

TRY THIS Use substitution to solve the system. Check your solution. $\begin{cases} x + y + z = 5 \\ 2x - 3y + z = -2 \\ 4z = -12 \end{cases}$

Exercises

Communicate

1. Describe the graphs of the three types of systems of linear equations.

2. Create three systems of linear equations, one to demonstrate each type of system: inconsistent, dependent, and independent.

3. Explain how to solve the system $\begin{cases} y = 2x - 5 \\ x + y = 13 \end{cases}$ by graphing.

4. Explain how to use substitution to solve the system. $\begin{cases} p - q = -7 \\ 2p + q = 17 \end{cases}$

Guided Skills Practice

Graph and classify each system. Then find the solution from the graph. ***(EXAMPLE 1)***

5. $\begin{cases} x - y = -4 \\ 3x + y = 8 \end{cases}$ **independent; (1, 5)**

6. $\begin{cases} 3x + 4y = 12 \\ 4y - 12 = -3x \end{cases}$ **dependent; infinitely many solutions**

7. Use substitution to solve the system. Check your solution. ***(EXAMPLE 2)*** $\begin{cases} 2x + y = 8 \\ 6x + 2y = -8 \end{cases}$ **(–12, 32)**

13.

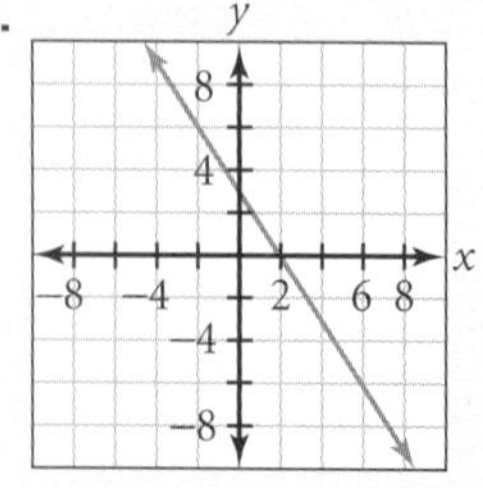

14.

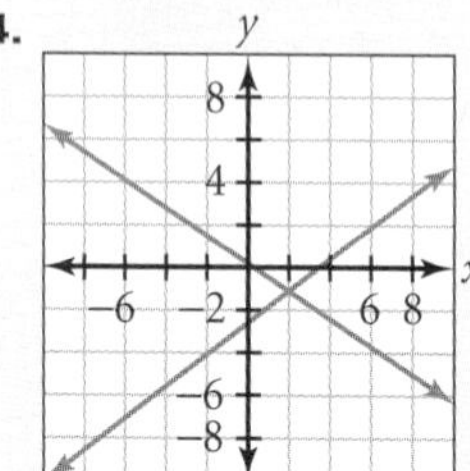

15.

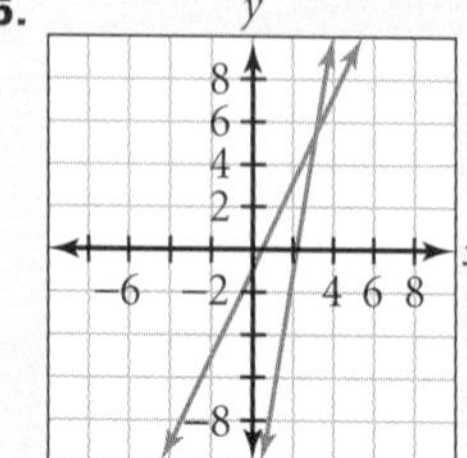

16.

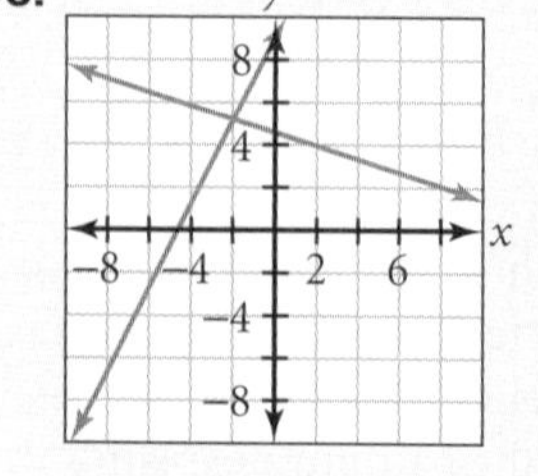

17.

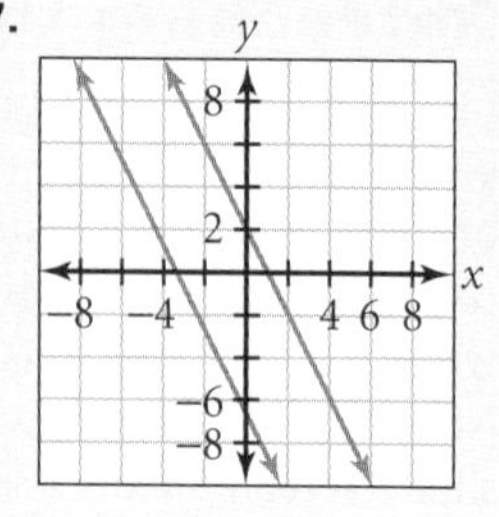

APPLICATION

8. SMALL BUSINESS A candy manufacturer wishes to mix two candies as a sales promotion. One candy sells for \$2.00 per pound and the other candy sells for \$0.75 per pound. The manufacturer wishes to have 1000 pounds of the mixture and to sell the mixture for \$1.35 per pound. How many pounds of each type of candy should be used in the mixture? ***(EXAMPLE 3)*** **480 lb of \$2.00 candy, 520 lb of \$0.75 candy**

9. Use substitution to solve the system. Check your solution. ***(EXAMPLE 4)*** **(1, 1, −3)** $\begin{cases} x + 6y + 2z = 1 \\ -2x + 3y - z = 4 \\ -1 = z + 2 \end{cases}$

Practice and Apply

Classify the type of system of equations represented by each graph below. If the system has exactly one solution, write it.

10.

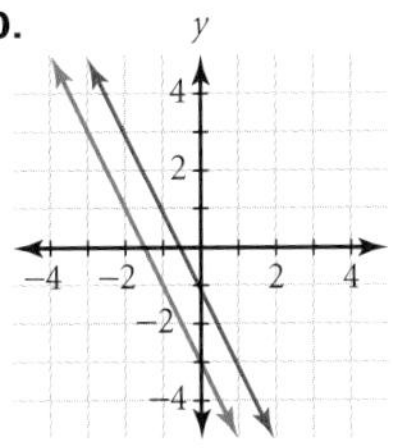

inconsistent

11.

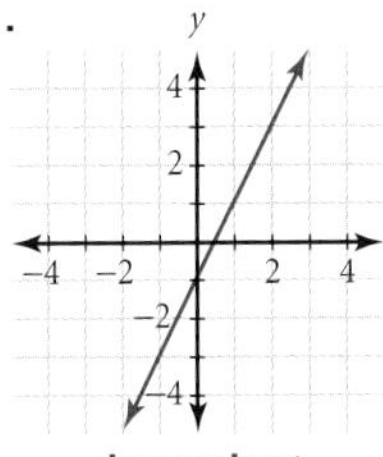

dependent

12.

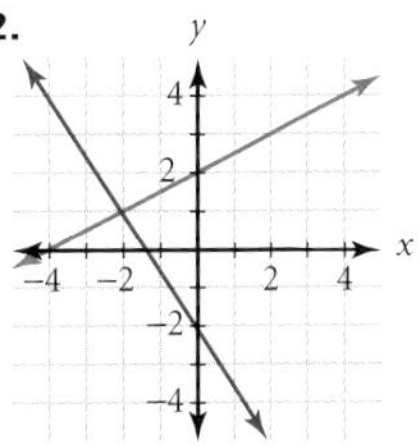

independent; (−2, 1)

Graph and classify each system. Then find the solution from the graph.

13. $\begin{cases} 6x + 4y = 12 \\ 2y = 6 - 3x \end{cases}$ **dep.**

14. $\begin{cases} 2x + 3y = 1 \\ -3x + 4y = -10 \end{cases}$ **indep. (2, −1)**

15. $\begin{cases} y = 2x - 1 \\ 6x - y = 13 \end{cases}$ **indep. (3, 5)**

16. $\begin{cases} x + 3y = 13 \\ 2x - y = -9 \end{cases}$ **indep. (−2, 5)**

17. $\begin{cases} y = -2x - 7 \\ 4x + 2y = 6 \end{cases}$ **incons.**

18. $\begin{cases} 2x + y = 5 \\ 4x + 2y = 6 \end{cases}$ **incons.**

19. $\begin{cases} -\frac{1}{2}x + y = 4 \\ x + 2y = 8 \end{cases}$ **indep. (0, 4)**

20. $\begin{cases} 3x - 6y = 9 \\ \frac{1}{2}x = y + \frac{3}{2} \end{cases}$ **dep.**

21. $\begin{cases} 4x + 5y = -7 \\ 3x - 6y = 24 \end{cases}$ **indep. (2, −3)**

22. $\begin{cases} -x + 2y = 3 \\ 2x - 4y = -6 \end{cases}$ **dep.**

23. $\begin{cases} 3x - y = 2 \\ -3x + y = 1 \end{cases}$ **incons.**

24. $\begin{cases} 6x - 3y = 9 \\ 3x + 7y = 47 \end{cases}$ **indep. (4, 5)**

internet connect
go.hrw.com
Homework Help Online
Go To: go.hrw.com
Keyword:
MB1 Homework Help for Exercises 25–40

Use substitution to solve each system of equations. Check your solution.

25. $\begin{cases} y = x + 3 \\ y = 2x - 4 \end{cases}$ **(7, 10)**

26. $\begin{cases} x = y + 4 \\ 2x + 3y = 43 \end{cases}$ **(11, 7)**

27. $\begin{cases} 4x + 2y = 20 \\ y = x - 2 \end{cases}$ **(4, 2)**

28. $\begin{cases} x - y = 3 \\ 2x + 2y = 2 \end{cases}$ **(2, −1)**

29. $\begin{cases} x - 2y = 0 \\ 2x - y = 6 \end{cases}$ **(4, 2)**

30. $\begin{cases} x + y = 0 \\ y + 2x = 4 \end{cases}$ **(4, −4)**

31. $\begin{cases} 2x + y = 8 \\ x - y = 3 \end{cases}$ $\left(\frac{11}{3}, \frac{2}{3}\right)$

32. $\begin{cases} x + 5y = 2 \\ x - 1 = 2y \end{cases}$ $\left(\frac{9}{7}, \frac{1}{7}\right)$

33. $\begin{cases} a = b + 2 \\ a = 5b - 6 \end{cases}$ **(4, 2)**

34. $\begin{cases} p = 3q - 3 \\ 2p + 5q = -17 \end{cases}$ **(−6, −1)**

35. $\begin{cases} 3t - r = 9 \\ r = 2t + 8 \end{cases}$ **(42, 17)**

36. $\begin{cases} 2x + y = -9 \\ 3x + y = 11 \end{cases}$ **(20, −49)**

37. $\begin{cases} x - 5y = 2 \\ 9x + 8 = 15y \end{cases}$ $\left(-\frac{7}{3}, -\frac{13}{15}\right)$

38. $\begin{cases} s = 10 + 5t \\ 2s = 40 + 4t \end{cases}$ $\left(\frac{80}{3}, \frac{10}{3}\right)$

39. $\begin{cases} -3x - y = 2 \\ -6x + 2y = -2 \end{cases}$ $\left(-\frac{1}{6}, -\frac{3}{2}\right)$

Error Analysis

Some students check that a solution satisfies just one equation in a system and then incorrectly identify it as the solution of the entire system. Tell them that a system of two equations is a special kind of compound sentence consisting of two equation "sentences" connected by the word *and*. The solution of the system must satisfy the first equation *and* the second equation, so it must be checked in both equations.

18.

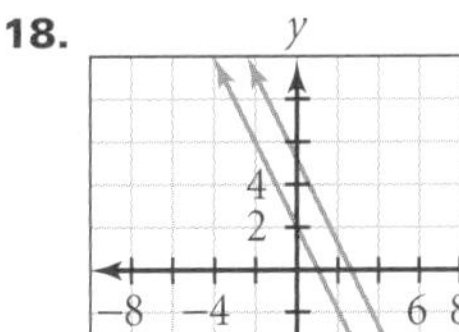

19.

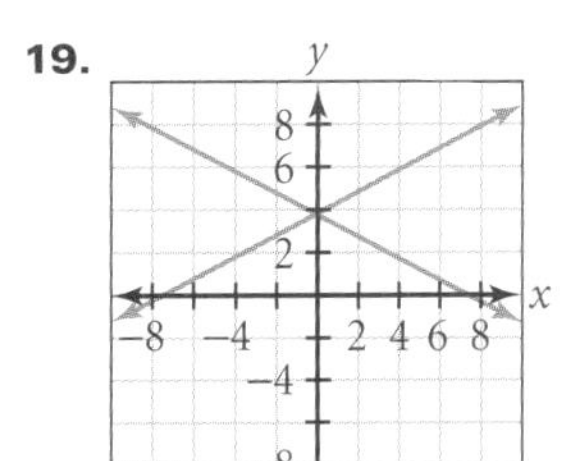

20.

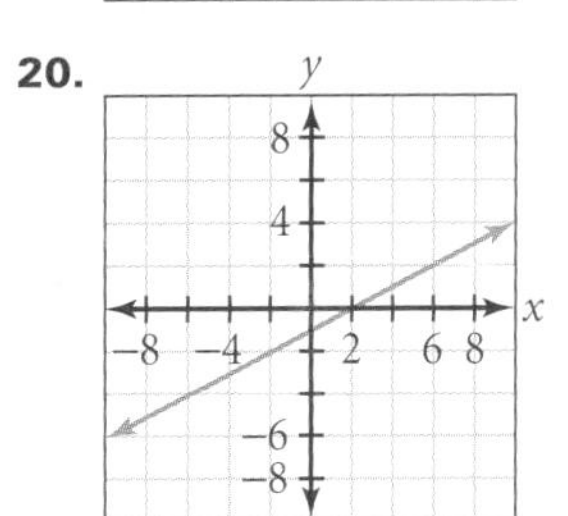

21.

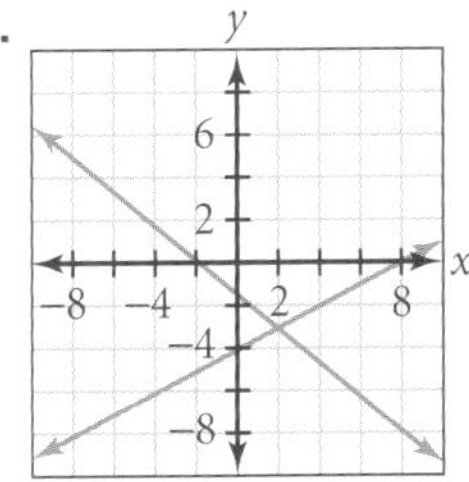

22.

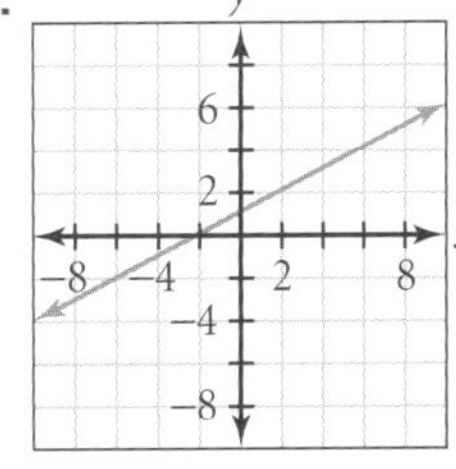

23.

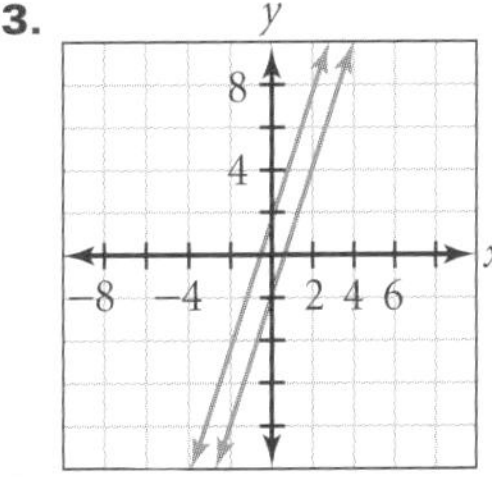

24.

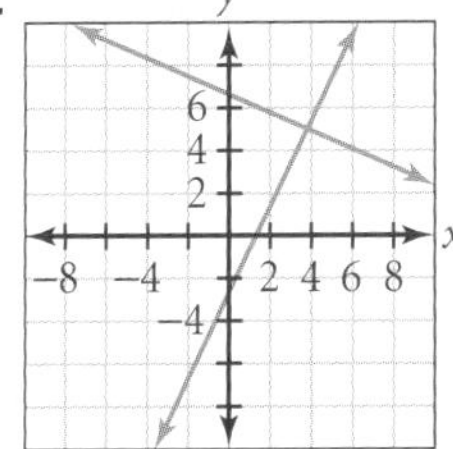

40. **Determine whether the given ordered pair is a solution of the given system.**

a. $(1, 3)$ $\begin{cases} 5x + 2y = 11 \\ x - y = 7 \end{cases}$ **no**

b. $(5, -2)$ $\begin{cases} 4x - 3y = 26 \\ 2x + y = 8 \end{cases}$ **yes**

c. $(2, 1)$ $\begin{cases} 2x - y = 8 \\ x + 3y = 5 \end{cases}$ **no**

d. $(5, 2)$ $\begin{cases} 4x - 2y = 16 \\ -8x + 4y = -32 \end{cases}$ **yes**

e. Of the four systems given above, one is dependent. Identify that system and give three additional ordered pairs that satisfy the system.
d is dependent; answers may vary; sample: (0, –8), (1, –6), and (2, –4)

Use substitution to solve each system of equations.

41. $\begin{cases} 2x - 3y - z = 12 \\ y + 3z = 10 \\ z = 4 \end{cases}$ **(5, –2, 4)**

42. $\begin{cases} 2x - 3y + 4z = 8 \\ 3x + 2y = 7 \\ x = 1 \end{cases}$ **(1, 2, 3)**

43. $\begin{cases} x + 3y - z = 8 \\ 2x - y + 2z = -9 \\ 3y = 9 \end{cases}$ **(–2, 3, –1)**

44. $\begin{cases} 2x + 3y - 2z = 4 \\ 3x - 3y + 2z = 16 \\ 2z = -5 \end{cases}$ $\left(\mathbf{4, -3, -\frac{5}{2}}\right)$

45. $\begin{cases} 3x + 5y = -3 \\ 10y - 2z = 2 \\ x = -z \end{cases}$ $\left(\mathbf{-2, \frac{3}{5}, 2}\right)$

46. $\begin{cases} a + b + c = 6 \\ 3a - b + c = 8 \\ 2b = c \end{cases}$ $\left(\mathbf{\frac{9}{4}, \frac{5}{4}, \frac{5}{2}}\right)$

Graph and classify each system. Then find the solution from the graph. Round your answers to the nearest hundredth when necessary.

47 $\begin{cases} y = 5x + 2.72 \\ y = 3.6x + 3.126 \end{cases}$

48 $\begin{cases} y = 4.3x - 0.44 \\ y = -2x + 4.6 \end{cases}$

49 $\begin{cases} -\frac{2}{5}x + y = -\frac{1}{10} \\ 3y - 2x = -\frac{5}{6} \end{cases}$

50 $\begin{cases} \frac{1}{7} = \frac{1}{14}x + 5\frac{1}{2}y \\ y = 4x + 14 \end{cases}$

51 $\begin{cases} 0.7y = 0.8x + 0.78 \\ -\frac{1}{5}x + \frac{1}{2}y = 2.1 \end{cases}$

52 $\begin{cases} 0.001y + \frac{4}{5}x = 0.2014 \\ 0.8x - 0.02y = 0.172 \end{cases}$

47. indep.; (0.29, 4.17)
48. indep.; (0.8, 3)
49. indep.; (0.67, 0.17)
50. indep.; (–3.48,0.07)
51. indep.; (4.15, 5.86)
52. indep.; (0.25, 1.4)

CHALLENGE

53. Solve $\begin{cases} ax + by = c \\ y = dx + e \end{cases}$ for x and y. Use the resulting expressions for x and y to solve a system in the same form, such as $\begin{cases} 2x + 3y = 21 \\ y = x + 2 \end{cases}$.

CONNECTION

GEOMETRY The perimeter of a rectangular swimming pool is 130 yards. Three times the length is equal to 10 times the width.

54. Find the length and the width of the pool. **50 yd, 15 yd**

55. Find the area of the pool. **750 sq yd**

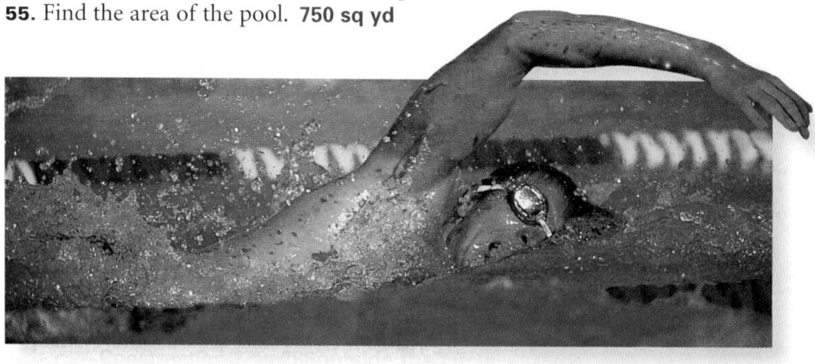

For Exercises 56–59, solve by writing a system of equations and using substitution. Check your answers.

APPLICATION

56. INCOME To earn money for college, Susan is making and selling earrings. Her weekly costs for advertising and phone calls are \$36, and each pair of earrings costs \$1.50 to produce. If Susan sells the earrings at \$6 per pair, how many pairs must she sell per week to break even? **8**

53. $x = \dfrac{c - be}{a + bd}$, $y = d\dfrac{c - be}{a + bd} + e$;

$(3, 5)$

Practice

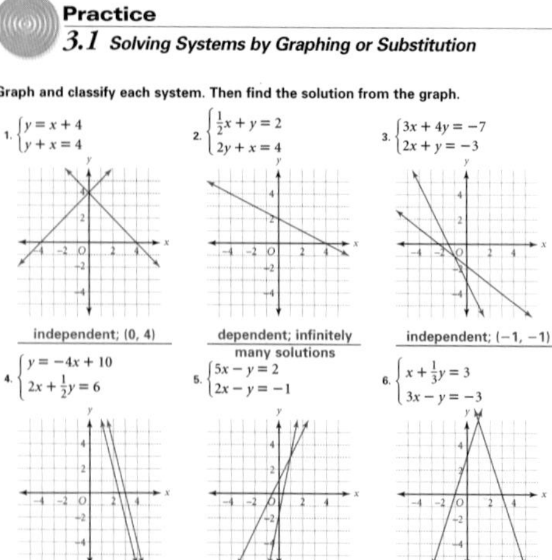

NAME ______ CLASS ______ DATE ______

Practice
3.1 *Solving Systems by Graphing or Substitution*

Graph and classify each system. Then find the solution from the graph.

1. $\begin{cases} y = x + 4 \\ y + x = 4 \end{cases}$ independent; (0, 4)

2. $\begin{cases} \frac{1}{2}x + y = 2 \\ 2y + x = 4 \end{cases}$ dependent; infinitely many solutions

3. $\begin{cases} 3x + 4y = -7 \\ 2x + y = -3 \end{cases}$ independent; (−1, −1)

4. $\begin{cases} y = -4x + 10 \\ 2x + \frac{1}{2}y = 6 \end{cases}$ inconsistent; no solution

5. $\begin{cases} 5x - y = 2 \\ 2x - y = -1 \end{cases}$ independent; (1, 3)

6. $\begin{cases} x + \frac{1}{3}y = 3 \\ 3x - y = -3 \end{cases}$ independent; (1, 6)

Use substitution to solve each system of equations. Check your solution.

7. $\begin{cases} y + 2x = 11 \\ x + y = 5 \end{cases}$ (6, −1)

8. $\begin{cases} x - 10y = 2 \\ x - 6y = 6 \end{cases}$ (12, 1)

9. $\begin{cases} 8x = y \\ 2x + y = 5 \end{cases}$ $\left(\frac{1}{2}, 4\right)$

10. $\begin{cases} 3x + y = -4 \\ \frac{1}{2}x + y = 6 \end{cases}$ (−4, 8)

11. $\begin{cases} x + 2y = 2 \\ 2x + 3y = -1 \end{cases}$ (−8, 5)

12. $\begin{cases} 3x + 4y = 11 \\ 2x + 4y = 8 \end{cases}$ $\left(3, \frac{1}{2}\right)$

APPLICATIONS

57. (2, 11); if < 2 hr, 2nd option is better, and if > 2 hr, 1st option is better.

58. 70 mL of 1% sol., 20 mL of 10% sol.

57. CONSUMER ECONOMICS Armando is comparing parking prices at a local concert. One option is a \$7 entry fee plus \$2 per hour. A second option is a \$5 entry fee plus \$3 per hour. What is the break-even point (intersection) for the two options? Which option do you think is better? Explain your reasoning.

58. CHEMISTRY To conduct a scientific experiment, students need to mix 90 milliliters of a 3% acid solution. They have a 1% and a 10% solution available. How many milliliters of the 1% solution and of the 10% solution should be combined to produce 90 milliliters of the 3% solution?

59. SPORTS Rebecca is the star forward on her high-school basketball team. In one game, her field-goal total was 23 points, made up of 2-point and 3-point baskets. If she made 4 more 2-point baskets than 3-point baskets, how many of each type of basket did she make?
7 2-point, 3 3-point

Solve each equation or inequality, and graph the solution on a number line. *(LESSON 1.8)*

60. $|x - 4| = 9$ **13, –5** **61.** $|x - 2| \leq 5$ **$-3 \leq x \leq 7$** **62.** $|x + 3| \leq -1$ **no solution**

State the property that is illustrated in each statement. All variables represent real numbers. *(LESSON 2.1)*

63. $\frac{2}{3}(6x + 12) = \frac{2}{3} \cdot 6x + \frac{2}{3} \cdot 12$ **Dist.** **64.** $5a + (3a + 6) = (5a + 3a) + 6$ **Assoc. of Add.**

65. $-4a + 4a = 0$ **Add. Inverse** **66.** $5x \cdot 9y = 9y \cdot 5x$ **Comm. of Mult.**

Evaluate each expression. *(LESSON 2.2)*

67. $5a^0$ **5** **68.** $36^{\frac{1}{2}}$ **6** **69.** $25^{-\frac{1}{2}}$ **$\frac{1}{5}$** **70.** $9^{\frac{3}{2}}$ **27**

Simplify each expression, assuming that no variable equals zero. *(LESSON 2.2)*

71. $\left(\frac{2x^3}{x^{-2}}\right)^2$ **$4x^{10}$** **72.** $\left(\frac{m^{-1}n^2}{n^{-3}}\right)^{-3}$ **$\frac{m^3}{n^{15}}$** **73.** $\left(\frac{2a^3b^{-2}}{-a^2b^{-3}}\right)^{-1}$ **$-\frac{1}{2ab}$** **74.** $\frac{(2y^2y)^{-2}}{3xy^{-4}}$ **$\frac{1}{12xy^2}$**

Find the inverse of each function and state whether the inverse is a function. *(LESSON 2.5)*

75. $\{(1, 4), (-3, 4), (2, 0)\}$ **76.** $\{(7, 8), (6, 4), (5, 0), (4, 4)\}$

77. $\{(3, 4), (4, 3), (3, -1), (11, -3)\}$ **78.** $\{(8, 6), (9, 4), (-8, 6), (0, 0)\}$

Use a graph to solve each nonlinear system of equations. Round your answers to the nearest hundredth when necessary.

79 $\begin{cases} y = x^2 + 3 \\ y = 4x \end{cases}$ **(1, 4), (3, 12)** **80** $\begin{cases} x + y = 19 \\ y = 2^x \end{cases}$ **(3.92, 15.08)**

internet connect
Activities Online
Go To: **go.hrw.com**
Keyword: **MB1 WBB**

In Exercises 79 and 80, students investigate nonlinear systems of equations. The system in Exercise 79 consists of a linear equation and a quadratic equation that can be solved by substitution or by using graphs, while the system in Exercise 80 consists of a linear equation and an exponential equation that can be solved *only* by using graphs. Students will study nonlinear systems in greater depth in Lesson 9.6.

75. The inverse is {(4, 1), (4, –3), (0, 2)}. The inverse is not a function.

76. The inverse is {(8, 7), (4, 6), (0, 5), (4, 4)}. The inverse is a function.

77. The inverse is {(4, 3), (3, 4), (–1, 3), (–3, 11)}. The inverse is a function.

78. The inverse is {(6, 8), (4, 9), (6, –8), (0, 0)}. The inverse is not a function.

Student Technology Guide

NAME ______ CLASS ______ DATE ______

Student Technology Guide
3.1 Solving Systems by Graphing or Substitution

When using a graphics calculator to solve a system of equations by graphing, first solve each equation for y. Then use TRACE or the intersect feature to find the solution.

Example: Solve the system $\begin{cases} 6x - 2y = 11 \\ 2x + 3y = 26 \end{cases}$.

- Solve $6x - 2y = 11$ for y. Because $-2y = 11 - 6x$, $y = \frac{11 - 6x}{-2}$, or $y = (11 - 6x) \div (-2)$. There is no need to simplify further.
- Solve $2x + 3y = 26$ for y. Because $3y = 26 - 2x$, $y = \frac{26 - 2x}{3}$, or $y = (26 - 2x) \div 3$. There is no need to simplify further.
- Enter the equations into your calculator. Then press GRAPH. If you obtain a graph like the one shown at left below, it may be tempting to assume that the system is dependent. But pressing TRACE ▼ reveals that the two equations do not have the same value at $x = 0$.

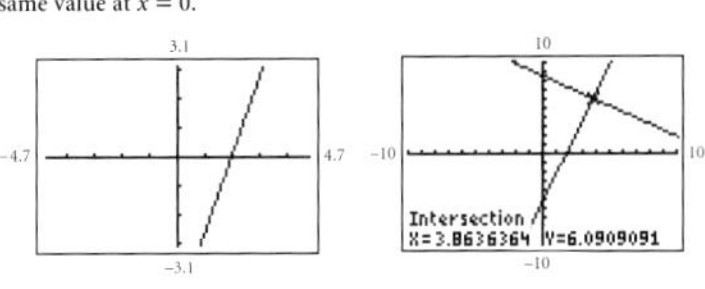

- Adjust the viewing window settings and press GRAPH 2nd TRACE 5:Intersect ENTER ENTER ENTER ENTER.
- Alternatively, use TRACE to find the approximate solution, which is (3.86, 6.09).

Solve each system of linear equations by graphing. If necessary, round answers to the nearest hundredth. Indicate whether each system is inconsistent, dependent, or independent.

1. $\begin{cases} 2x + 7y = -3 \\ 5x - 2y = 4 \end{cases}$ (0.56, –0.59); independent
2. $\begin{cases} 3x + 4y = 28 \\ 7x - y = 12 \end{cases}$ (2.45, 5.16); independent
3. $\begin{cases} 8x - 5y = -10.1 \\ 2x + 9y = 55.9 \end{cases}$ (2.3, 5.7); independent
4. $\begin{cases} 10.5x + 46.5y = 72 \\ 6.3x + 27.9y = 43.2 \end{cases}$ infinitely many solutions; dependent
5. $\begin{cases} 2.4x - 5.3y = 8.1 \\ 9.1x + 2.1y = -15 \end{cases}$ (–1.17, –2.06); independent
6. $\begin{cases} 6.4x - 5.6y = 7.2 \\ -54.4x + 47.6y = 12.3 \end{cases}$ no solution; inconsistent

Prepare

PRINCIPLES & STANDARDS 1–4, 6–10

QUICK WARM-UP

Simplify.

1. $(5x + 9y) + (-5x + 6y)$ **$15y$**
2. $(11y + 7x) + (4x - 11y)$ **$11x$**
3. $4y - (4x - y)$ **$-4x + 5y$**
4. $(8x + 2y) - (9x - 2y)$ **$-x + 4y$**
5. $(21x - 5y) - (15y + 4x)$ **$17x - 20y$**

Also on Quiz Transparency 3.2

Teach

Why Ask students what is meant by the term *process of elimination*. Have them suggest situations in which the process of elimination might be used as a problem-solving tool. Tell them that when solving systems of equations, elimination refers to the process of removing variables until just one variable remains.

Solving Systems by Elimination

Objective

- Solve a system of two linear equations in two variables by elimination.

Why *You can model real-world situations involving two variables, such as business cost and revenue, with a system of two equations in two variables.*

APPLICATION BUSINESS

In business, makers and sellers of goods must relate the cost of making goods to their selling price. They must also keep production costs within budget and maintain realistic expectations of revenue.

This table gives production costs and selling prices per frame for two sizes of picture frames. How many of each size should be made and sold if the production budget is \$930 and the expected revenue is \$1920? *You will solve this problem in Example 2.*

	Small	Large
Production cost	\$5.50	\$7.50
Selling price	\$12.00	\$15.00

You have learned to solve systems of linear equations by graphing and by using substitution. Some systems are more easily solved by another method called the *elimination method*. The **elimination method** involves multiplying and combining the equations in a system in order to eliminate a variable.

Independent Systems

EXAMPLE  **Use elimination to solve the system. Check your solution.** $\begin{cases} 2x + 5y = 15 \\ -4x + 7y = -13 \end{cases}$

SOLUTION

1. To eliminate x, multiply each side of the first equation by 2, and combine the resulting equations.

$$\begin{cases} 2x + 5y = 15 \\ -4x + 7y = -13 \end{cases} \rightarrow \begin{cases} 2(2x + 5y) = 2(15) \\ -4x + 7y = -13 \end{cases} \rightarrow \begin{cases} 4x + 10y = 30 \\ -4x + 7y = -13 \end{cases}$$

$$\begin{aligned} 4x + 10y &= 30 \\ -4x + 7y &= -13 \\ \hline 17y &= 17 \\ y &= 1 \end{aligned}$$

Use the Addition Property of Equality.

Solve for y.

Alternative Teaching Strategy

CONNECTING TO PRIOR KNOWLEDGE

Show students the system at right. $\begin{cases} 2x + 2y = 6 \\ 3x + 6y = 1 \end{cases}$ Elicit from them a set of steps for solving the system by substitution. The following is one possibility:

$$\begin{aligned} 2x + 2y = 6 &\rightarrow 2x = -2y + 6 \rightarrow x = -y + 3 \\ 3x + 6y = 12 &\rightarrow 3(-y + 3) + 6y = 12 \rightarrow y = 1 \\ 2x + 2y = 6 &\rightarrow 2x + 2(1) = 6 \rightarrow x = 2 \end{aligned}$$

Therefore, the solution is $(2, 1)$.

Then show students how the same solution can be obtained by the elimination method.

$$\begin{cases} 2x + 2y = 6 \\ 3x + 6y = 1 \end{cases} \rightarrow \begin{cases} -6x - 6y = -18 \\ 3x + 6y = 12 \end{cases}$$

$$\rightarrow -3x = -6$$

$$\rightarrow x = 2$$

$$2x + 2y = 6 \rightarrow 2(2) + 2y = 6 \rightarrow y = 1$$

Again, the solution is $(2, 1)$.

Now give students the system from Example 1. Lead them to see that substitution would result in decimal or fractional coefficients, so elimination would be considered a more efficient method for solving such a system.

2. Substitute 1 for y in either original equation to find x.

First equation:

$$2x + 5y = 15$$
$$2x + 5(1) = 15$$
$$x = 5$$

Second equation:

$$-4x + 7y = -13$$
$$-4x + 7(1) = -13$$
$$x = 5$$

The solution of the system is $x = 5$ and $y = 1$, or $(5, 1)$.

CHECK

$$\begin{cases} 2x + 5y = 15 \\ -4x + 7y = -13 \end{cases} \rightarrow \begin{cases} 2(5) + 5(1) = 15 & \textbf{True} \\ -4(5) + 7(1) = -13 & \textbf{True} \end{cases}$$

TRY THIS Use elimination to solve the system. Check your solution. $\begin{cases} 6r + 7s = -15 \\ -3r + s = -6 \end{cases}$

EXAMPLE 2 Refer to the picture-frame problem described at the beginning of the lesson.

APPLICATION
BUSINESS

How many small frames and how many large frames can be made and sold if the production budget is \$930 and the expected revenue is \$1920?

SOLUTION

1. Write a system of equations to represent the problem.

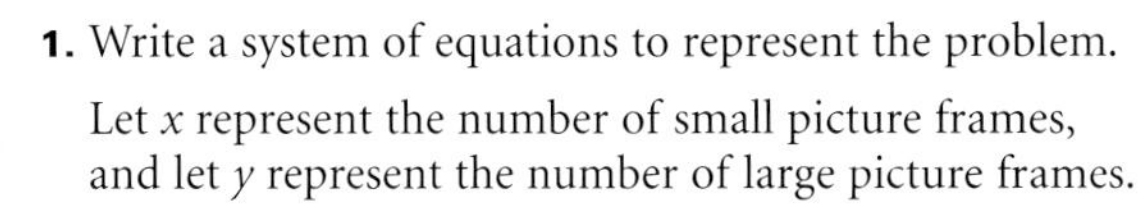

Let x represent the number of small picture frames, and let y represent the number of large picture frames.

	Small	Large	Total
Production cost	$5.5x$	$7.5y$	930
Selling price	$12x$	$15y$	1920

From the data in the table, write the system. $\begin{cases} 5.5x + 7.5y = 930 \\ 12x + 15y = 1920 \end{cases}$

2. To eliminate y, multiply each side of the first equation by -2, and combine the resulting equations.

$$\begin{cases} -2(5.5x + 7.5y) = -2(930) \\ 12x + 15y = 1920 \end{cases} \rightarrow \begin{cases} -11x - 15y = -1860 \\ 12x + 15y = 1920 \end{cases}$$

$$\begin{aligned} -11x - 15y &= -1860 \\ 12x + 15y &= 1920 \\ \hline x \qquad &= 60 \end{aligned}$$

Use the Addition Property of Equality.

3. Substitute 60 for x in either original equation to find y.

First equation:

$$5.5x + 7.5y = 930$$
$$5.5(60) + 7.5y = 930$$
$$7.5y = 600$$
$$y = 80$$

Second equation:

$$12x + 15y = 1920$$
$$12(60) + 15y = 1920$$
$$15y = 1200$$
$$y = 80$$

Thus, the solution is $(60, 80)$. To meet the required goals, 60 small picture frames and 80 large ones should be made.

Teaching Tip

Be sure that students understand the meaning of "combining the equations." Point out that the equations can be combined by using either the addition property of equality or the subtraction property of equality. You may wish to tell students that the elimination method is sometimes called the *linear-combination method.*

ADDITIONAL EXAMPLE 

Use elimination to solve the system below. Check your solution.

$$\begin{cases} 4x - 3y = 15 \\ 8x + 2y = -10 \end{cases}$$

$(0, -5)$

TRY THIS

$r = 1$ and $s = -3$, or $(1, -3)$

ADDITIONAL EXAMPLE 

To raise funds, the English Club is selling poetry cards and bouquets of flowers for Mother's Day. The club obtains the cards at a cost of \$2 each and the bouquets at a cost of \$3.50 each. They will sell the cards for \$6 and the bouquets for \$8. The club will spend \$360 from their account, and their expected revenue is \$900. **How many cards and bouquets can be obtained and sold?**

54 cards and 72 bouquets

Interdisciplinary Connection

CHEMISTRY A chemistry student recorded the following observations about an experiment:

When 3 milliliters of solution A are mixed with 3 milliliters of solution B, the result is a 40% salt solution.

When 4 milliliters of solution A are mixed with 2 milliliters of solution B, the result is a 35% salt solution.

a. Use a system of equations to find the concentration of salt in solution A and in solution B. **solution A: 25%; solution B: 55%**

b. Use a system of equations to find the number of milliliters of each solution that must be used to make 6 milliliters of a 50% salt solution. **solution A: 1 mL; solution B: 5 mL**

1. The only solution is $(1, 4)$.
sample: $y = x + 3$

2. $(1, 4)$ is one of infinitely many solutions.

CHECK

You can use a graph, a table, or substitution to verify the solution.

To use a graphics calculator, solve each equation for y.

$$\begin{cases} 5.5x + 7.5y = 930 \\ 12x + 15y = 1920 \end{cases} \rightarrow \begin{cases} y = \dfrac{930 - 5.5x}{7.5} \\ y = \dfrac{1920 - 12x}{15} \end{cases}$$

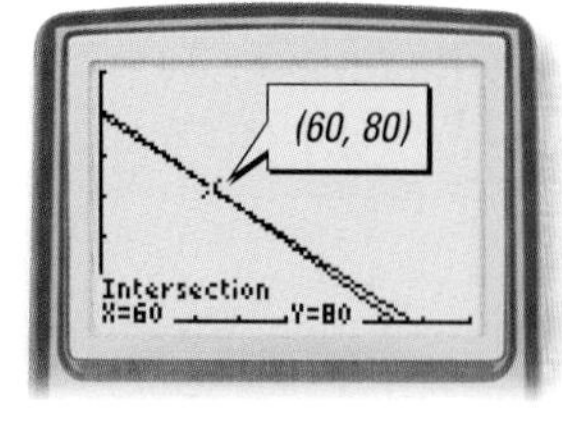

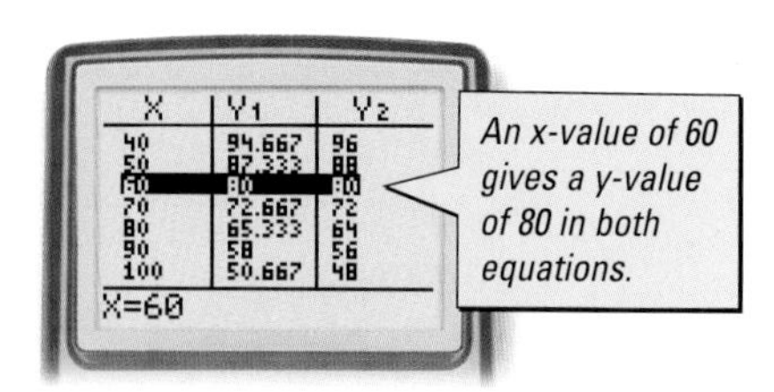

TRY THIS How does the solution to Example 2 change if the production budget is \$1245 and the revenue goal is \$2580?

CHECKPOINT ✔ Explain how you know that $5.5x + 7.5y$ and $12x + 15y$ are expressions that represent dollar amounts in Example 2.

CRITICAL THINKING The graph for the production budget and the graph for the revenue goal in Example 2 are quite similar. Describe what this means in terms of the number of small and large picture frames made and sold and in terms of the production budget and revenue goal.

Dependent and Inconsistent Systems

Investigating Systems

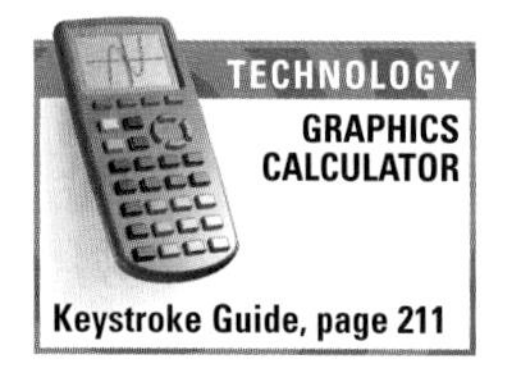

You will need: graph paper or a graphics calculator

System
I. $\begin{cases} x - y = -2 \\ -5x + 5y = 10 \end{cases}$
II. $\begin{cases} 3x + 3y = -5 \\ 2x + 2y = 7 \end{cases}$

1. Refer to system I in the first row.
   a. Graph the system and classify it as independent, dependent, or inconsistent.
   b. Solve the system by using elimination. Interpret the resulting mathematical statement.
2. Repeat parts **a** and **b** in Step 1 for system II.
3. **CHECKPOINT ✔** If an attempt to solve a system of equations results in the mathematical statement $0 = -3$, what can you say about this system?
4. **CHECKPOINT ✔** If an attempt to solve a system of equations results in the mathematical statement $0 = 0$, what can you say about this system?

You can use the elimination method to solve any system of two linear equations in two variables. Examples 3 and 4 show you how to interpret your results when the system is inconsistent or dependent.

TRY THIS
90 small and 100 large

CHECKPOINT ✔
$5.5x + 7.5y$ represents the total cost of production because \$5.50 and \$7.50 are the production costs of each small and each large frame, respectively.

CRITICAL THINKING
Any given combination of small and large picture frames produced is almost equal to the quantities sold. This means that a big loss or gain is unlikely to occur.

Activity Notes

In this Activity, students use the elimination method to solve two systems: one dependent and one inconsistent. They should observe that, in each case, both variables are eliminated. If the system is inconsistent, the resulting statement is false. If the system is dependent, the resulting statement is true.

Cooperative Learning

You may wish to have students do the Activity in pairs. One partner can do Step 1 while the other does Step 2. The partners can then compare their results and do Steps 3 and 4 together.

CHECKPOINT ✔

3. The statement is a false statement, so the system is inconsistent and there is no solution.

CHECKPOINT ✔

4. The statement is true, so the system is dependent and there are infinitely many solutions.

Inclusion Strategies

LINGUISTIC LEARNERS When using a system of equations in two variables to solve a real-world problem, some students will find it helpful to use variables other than x and y. In the solution of Example 2, for instance, you may want to suggest that they use s to represent the number of small picture frames and l to represent the number of large picture frames. If they then use a graphics calculator to check their solutions, caution them to keep track of which variable they translated into the independent variable x and which they translated into the dependent variable y.

Enrichment

Give students the system of three equations in three variables at right.

$$\begin{cases} 2x - y - 2z = 3 \\ 3x - 2y - 5z = 1 \\ x + 4y + 3z = 5 \end{cases}$$

Have them work in pairs to solve the system by using the elimination method. If they need a hint to get started, suggest that they choose any pair of the equations and eliminate one variable, and then choose a second pair and eliminate the same variable. **The solution is (3, –1, 2).**

EXAMPLE 3 **Use elimination to solve the system. Check your solution.** $\begin{cases} 2x + 5y = 12 \\ 2x + 5y = 15 \end{cases}$

SOLUTION

Combine the equations to eliminate a variable.

$$\begin{aligned} 2x + 5y &= 12 \\ 2x + 5y &= 15 \\ \hline 0 &= -3 \end{aligned}$$

This is a false statement. *Use the Subtraction Property of Equality.*

Because the result is a contradiction, the system is inconsistent. This means that the system has no solution.

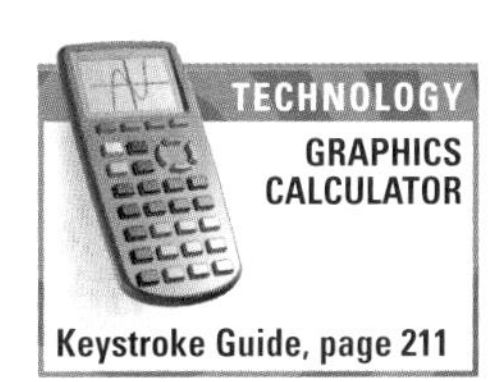

CHECK

Graph the system. $\begin{cases} y = \frac{12 - 2x}{5} \\ y = \frac{15 - 2x}{5} \end{cases}$

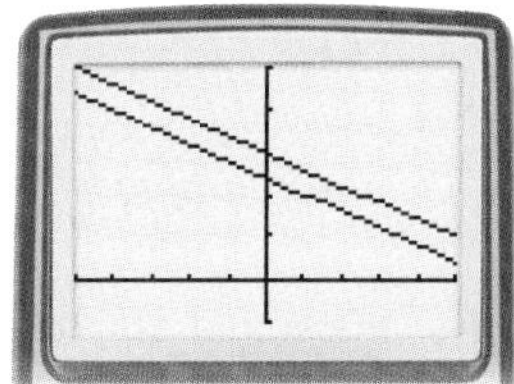

The graph indicates that the system is inconsistent. The slopes are equal.

TRY THIS Use elimination to solve the system. Check your solution. $\begin{cases} 6x - 2y = 9 \\ 6x - 2y = 7 \end{cases}$

CHECKPOINT ✔ Find the slope of each equation in the system in Example 3. How do the slopes of those equations confirm the conclusion reached in Example 3?

EXAMPLE 4 **Use elimination to solve the system. Check your solution.** $\begin{cases} 2x + 5y = 15 \\ -3x - 7.5y = -22.5 \end{cases}$

SOLUTION

Multiply each side of the first equation by 3, and multiply each side of the second equation by 2. Then combine the resulting equations to eliminate x.

$$\begin{cases} 3(2x + 5y) = 3(15) \\ 2(-3x - 7.5y) = 2(-22.5) \end{cases} \rightarrow \begin{cases} 6x + 15y = 45 \\ -6x - 15y = -45 \end{cases}$$

$$\begin{aligned} 6x + 15y &= 45 \\ -6x - 15y &= -45 \\ \hline 0 &= 0 \end{aligned}$$

This is a true statement. *Use the Addition Property of Equality.*

Because the result is true regardless of the values of the variables in the system, the system is consistent and dependent. This means that the solution of the system is all points on the graph of either equation.

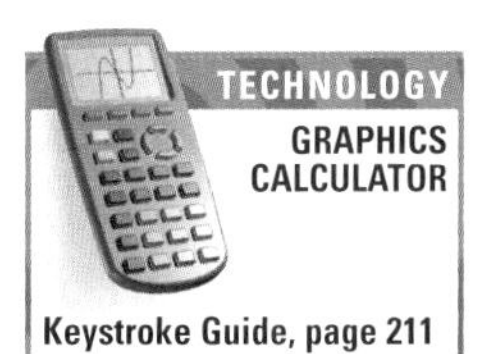

CHECK

Graph the system. $\begin{cases} y = \frac{15 - 2x}{5} \\ y = \frac{-22.5 + 3x}{-7.5} \end{cases}$

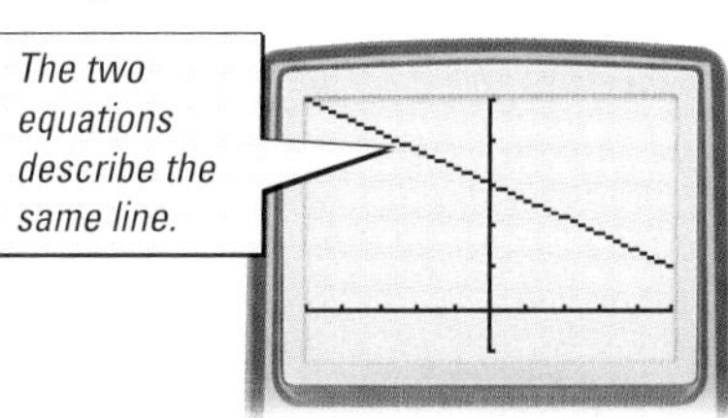

The graph indicates that the system is dependent.

ADDITIONAL EXAMPLE 3

Use elimination to solve the system. Check your solution.

$\begin{cases} 5x + 3y = 12 \\ 5x + 3y = 15 \end{cases}$

The system has no solution.

Teaching Tip

TECHNOLOGY When using a TI-82 or TI-83 graphics calculator to find the point of intersection of the graph in Example 3, you will get the error message **NO SIGN CHNG** (or **SIGN CHNG**). The calculator is evaluating **Y1–Y2** for corresponding values of **Y1** and **Y2**, and it is finding that the differences are constant. If the lines intersected, there would be a "sign change" in the differences at the point of intersection.

TRY THIS

There is no solution.

CHECKPOINT ✔

Both slopes are $-\frac{2}{5}$ and the y-intercepts are different, so the lines are parallel. Therefore, the system has no solution.

ADDITIONAL EXAMPLE 4

Use elimination to solve the system. Check your solution.

$\begin{cases} 16x - 9y = 8 \\ 4x - 2.25y = 12 \end{cases}$

The solution is all points on the graph of either equation.

Reteaching the Lesson

GUIDED ANALYSIS Display the systems at right on the board or overhead. Discuss with students how the solution process for system II will differ from that for system I. They may note, for example, that the solution for system I is begun by adding the two equations, while the solution of system II requires multiplication by –1. Then have them solve each system. **system I: (2, –1); system II: (–2, 5)**

I. $\begin{cases} 3x + 2y = 4 \\ 2x - 2y = 6 \end{cases}$

II. $\begin{cases} 3x + 2y = 4 \\ 2x + 2y = 6 \end{cases}$

Now display the systems at right. Discuss with students how the solution process for these systems will differ from the process for systems I and II. They most likely will note that the process must involve multiplication as well as addition. Have students solve each system and compare the two solution processes.

system III: (1.75, –0.625); system IV: (0, 2)

III. $\begin{cases} 3x + 2y = 4 \\ 2x - 4y = 6 \end{cases}$

IV. $\begin{cases} 3x + 2y = 4 \\ 2x + 3y = 6 \end{cases}$

TRY THIS
The system is dependent, so the solution is all points on the graph of either equation.

CHECKPOINT ✓
The slopes of both are $-\frac{2}{5}$ and the y-intercepts of both are 3, so the graphs of the equations coincide. The solution of the system is all points on the graph of either equation.

Assess

Selected Answers
Exercises 5–8, 9–61 odd

ASSIGNMENT GUIDE	
In Class	1–8
Core	9–35 odd, 41, 45
Core Plus	10–26 even, 27–46
Review	47–62
Preview	63, 64

Extra Practice can be found beginning on page 940.

TRY THIS Use elimination to solve the system. Check your solution. $\begin{cases} 8x + 4y = -16 \\ 2x + y = -4 \end{cases}$

CHECKPOINT ✓ Find the slope and y-intercept of each line represented in the system in Example 4. How do the slopes and y-intercepts of those lines confirm the conclusion reached in Example 4?

SUMMARY

THE ELIMINATION METHOD

1. Arrange each equation in standard form, $Ax + By = C$.
2. If the coefficients of x (or y) are the same number, use subtraction to eliminate the variable.
3. If the coefficients of x (or y) are opposites, use addition to eliminate the variable.
4. If the coefficients of x (or y) are different, multiply one or both equations by constants so that the coefficients of x (or y) are the same or opposite numbers. Then use Step 2 or 3 to eliminate the variable.
5. Use substitution to solve for the remaining variable.

Exercises

Communicate

1. Explain how to solve $\begin{cases} 3x - 4y = 3 \\ 2x + y = -5 \end{cases}$ by elimination.
2. When attempting to solve a system by elimination, what are the results for an inconsistent system? for a dependent system?
3. What property justifies adding the corresponding sides of two equations to create a new equation?
4. Compare the elimination method with the substitution method for solving systems of linear equations. What type of system is most easily solved by using substitution, and what type of system is most easily solved by using elimination?

Guided Skills Practice

5. Use elimination to solve the system. Check your solution. ***(EXAMPLE 1)*** $\begin{cases} 2x - 3y = 1 \\ 5x + 6y = 16 \end{cases}$ **(2, 1)**

APPLICATION

6. **BUSINESS** How does the solution to Example 2 change if the production costs for large and small picture frames are \$7 and \$6, respectively? ***(EXAMPLE 2)*** **85 small, 60 large**

Use elimination to solve each system of equations. Check your solution.
(EXAMPLES 3 AND 4)

7. $\begin{cases} 3x - y = -2 \\ 9x - 3y = 3 \end{cases}$ **no solution**

8. $\begin{cases} 3x + 2y = 5 \\ -6x - 4y = -10 \end{cases}$ **dependent; infinitely many solutions**

Practice and Apply

internet connect
Homework Help Online
Go To: **go.hrw.com**
Keyword:
MB1 Homework Help for Exercises 9–35

Use elimination to solve each system of equations. Check your solution.

9. $\begin{cases} 2x + y = 8 \\ x - y = 10 \end{cases}$ **(6, −4)**

10. $\begin{cases} 3x + 4y = 23 \\ -3x + y = 2 \end{cases}$ **(1, 5)**

11. $\begin{cases} -2x + 3y = -14 \\ 2x + 2y = 4 \end{cases}$ **(4, −2)**

12. $\begin{cases} 2p + 5q = 13 \\ p - q = -4 \end{cases}$ **(−1, 3)**

13. $\begin{cases} 2s - 5t = 22 \\ 2s - 3t = 6 \end{cases}$ **(−9, −8)**

14. $\begin{cases} 12y - 5z = 19 \\ 12y + 16z = 40 \end{cases}$ **(2, 1)**

15. $\begin{cases} x + y = 4 \\ 2x + 3y = 9 \end{cases}$ **(3, 1)**

16. $\begin{cases} 3a + 2b = 2 \\ a + 6b = 18 \end{cases}$ $\left(-\frac{3}{2}, \frac{13}{4}\right)$

17. $\begin{cases} 2x - 7y = 3 \\ 5x - 4y = -6 \end{cases}$ **(−2, −1)**

18. $\begin{cases} 5x + 3y = 2 \\ 2x + 20 = 4y \end{cases}$ **(−2, 4)**

19. $\begin{cases} 7b - 5c = 11 \\ -4c - 2b = -14 \end{cases}$ **(3, 2)**

20. $\begin{cases} 2y - 4x = 18 \\ -5x + 3y = 23 \end{cases}$ **(−4, 1)**

21. $\begin{cases} 5x - 8 = 3y \\ 10x - 6y = 18 \end{cases}$

22. $\begin{cases} 2x = 5 + 4y \\ 2y = 8 + x \end{cases}$ **no solution**

23. $\begin{cases} -8x + 4y = -2 \\ 4x - 2y = 1 \end{cases}$

24. $\begin{cases} 4y + 30 = 10x \\ 5x - 2y = 15 \end{cases}$

25. $\begin{cases} 3x - 4y = -1 \\ -10 + 8y = -6x \end{cases}$ $\left(\frac{2}{3}, \frac{3}{4}\right)$

26. $\begin{cases} 3x + 7y = 10 \\ 5x = 7 - 2y \end{cases}$ **(1, 1)**

21. no solution
23. infinitely many
24. infinitely many

Use any method to solve each system of linear equations. Check your solution.

27. $\begin{cases} -4.5x + 7.5y = -9 \\ 3x - 5y = 6 \end{cases}$ **infinitely many**

28. $\begin{cases} 5x - 3y = 8 \\ x + 0.6y = 1.8 \end{cases}$ $\left(\frac{17}{10}, \frac{1}{6}\right)$

29. $\begin{cases} 3y = 5 - x \\ x + 4y = 8 \end{cases}$ **(−4, 3)**

30. $\begin{cases} 7y - \frac{1}{2}x = 2 \\ -2y = x + 4 \end{cases}$ **(−4, 0)**

31. $\begin{cases} \frac{5}{2}x - y = 5 \\ 4y = 3x - 6 \end{cases}$ **(2, 0)**

32. $\begin{cases} 5x + 4y = 2 \\ 2x + 3y = 5 \end{cases}$ **(−2, 3)**

33. $\begin{cases} -2x + 5y = -23 \\ 24 + 4y = 3x \end{cases}$ **(4, −3)**

34. $\begin{cases} \frac{x+3}{2} = x + 4y \\ x = 5y \end{cases}$ $\left(\frac{15}{13}, \frac{3}{13}\right)$

35. $\begin{cases} 16x - 6y = -1 + 6y \\ \frac{2}{3}x = 4y + 3 \end{cases}$

35. $\left(-\frac{5}{7}, -\frac{73}{84}\right)$

CHALLENGE

Find the value(s) of k that satisfy the given condition for each system.

36. The system $\begin{cases} kx - 5y = 8 \\ 7x + 5y = 10 \end{cases}$ is consistent. $k \neq -7$

37. The system $\begin{cases} kx - 5y = 8 \\ 7x + 5y = 10 \end{cases}$ has infinitely many solutions. **no solution**

CONNECTION

38. COORDINATE GEOMETRY The ordered pairs (2, 9) and (6, 17) satisfy the linear equation $Ax + By = 5$, with coefficients A and B.

a. Substitute each ordered pair into the equation to obtain two new equations in two variables, A and B. $2A + 9B = 5,\ 6A + 17B = 5$

b. Solve the system of equations obtained in part **a** for A and B. **(−2, 1)**

c. Rewrite the linear equation $Ax + By = C$ with the values for A and B from part **b**. $-2x + y = 5$

d. Solve the linear equation for y and identify the slope and y-intercept.

e. Use the two ordered pairs and the slope formula to find the slope.

f. Do your slopes from parts **d** and **e** agree?

38d. $m = 2$, $b = 5$
e. $m = 2$
f. yes

Error Analysis

When solving systems by elimination, students often misinterpret the result 0 = 0. Some assign a value of 0 to one of the variables and solve for the other, creating an ordered pair that they identify as the solution of the system. Others believe there is something "wrong" because both variables have been eliminated, and they believe that the system has no solution. Emphasize the fact that when both variables are eliminated, it is important to focus on the truth of the remaining statement.

39. $\frac{25}{2}$, or 12.5, acres of good land and $\frac{175}{2}$, or 87.5, acres of bad land

39. CULTURAL CONNECTION: CHINA The *Jiuzhang*, or *Nine Chapters on the Mathematical Art*, was written around 250 B.C.E. in China. It includes problems with systems of linear equations, such as the one below. Solve the problem.

A family bought 100 acres of land and paid 10,000 pieces of gold. The price of good land is 300 pieces of gold per acre and the price of bad land is 500 pieces of gold for 7 acres. How many acres of good land and how many acres of bad land were purchased?

APPLICATIONS

40. 7 2-point, 2 3-point

40. SPORTS Greg is a star player on the basketball team. In one game, his field-goal total was 20 points, made up of 2-point and 3-point baskets. If Greg made a total of 9 baskets, how many of each type did he make?

INVESTMENTS Beth has saved $4500. She would like to earn $250 per year by investing her money. She received advice about two different investments: a low-risk investment that pays a 5% annual interest and a high-risk investment that pays a 9% annual interest.

41. How much should Beth invest in each type of investment to earn her annual goal? **$3875 at 5%, $625 at 9%**

42. How much should Beth invest in each type of investment if she wishes to earn $325 per year? **$2000 at 5%, $2500 at 9%**

43. BUSINESS A mail-order company charges for postage and handling according to the weight of the package. A package that weighs less than 3 pounds costs $2.00 for shipping and handling, and a package that weighs 3 pounds or more costs $3.00. An order of 12 packages had a total shipping and handling cost of $29.00. Find the number of packages that weighed less than 3 pounds and the number of packages that weighed 3 pounds or more. **7 pkgs. under 3 lb, 5 pkgs. at 3 lb or more**

NUTRITION One unit of whole-wheat flour has 13.6 grams of protein and 2.5 grams of fat. One unit of whole milk has 3.4 grams of protein and 3.7 grams of fat.

44. $\begin{cases} 13.6x + 3.4y = 75 \\ 2.5x + 3.7y = 15 \end{cases}$

44. Write a system of equations that can be used to find the number of units of whole-wheat flour and the number of units of whole milk that must be mixed to make a dough that has 75 grams of protein and 15 grams of fat.

45. 5.4 units of flour, 0.5 unit of milk

45. Solve the system and give the number of units, to the nearest tenth, of flour and of milk in the dough.

46. $5/hr baby-sitting, $6/hr restaurant

46. INCOME When Dale baby-sat for 8 hours and worked at a restaurant for 3 hours, he made a total of $58. When he baby-sat for 2 hours and worked at a restaurant for 5 hours, he made a total of $40. How much does Dale get paid for each type of work?

Practice

NAME ______________ CLASS ______________ DATE ______________

Practice 3.2 *Solving Systems by Elimination*

Use elimination to solve each system of equations. Check your solution.

1. $\begin{cases} -2x + 9y = -13 \\ 6x - 3y = 15 \end{cases}$ (2, −1)
2. $\begin{cases} \frac{2}{3}x - 3y = \frac{1}{5} \\ 2x - 9y = 4 \end{cases}$ no solution
3. $\begin{cases} 7y - x = 8 \\ x - y = 4 \end{cases}$ (6, 2)
4. $\begin{cases} 4x + y = 12 \\ 3x + \frac{1}{4}y = 9 \end{cases}$ (3, 0)
5. $\begin{cases} 5x + 9y = -7 \\ 2x + 3y = -1 \end{cases}$ (4, −3)
6. $\begin{cases} \frac{1}{2}x + y = 22 \\ 2x + 4y = 11 \end{cases}$ no solution
7. $\begin{cases} \frac{2}{3}x - y = -2 \\ 3x + 2y = -35 \end{cases}$ (−9, −4)
8. $\begin{cases} 6x - y = 26 \\ 3x - \frac{1}{2}y = 13 \end{cases}$ infinitely many solutions
9. $\begin{cases} \frac{1}{2}x + \frac{3}{4}y = 10 \\ 2x - y = 8 \end{cases}$ (8, 8)
10. $\begin{cases} x - 9y = -13 \\ 2x + y = -7 \end{cases}$ (−4, 1)
11. $\begin{cases} 13x + 7y = 19 \\ 9x - 2y = 20 \end{cases}$ (2, −1)
12. $\begin{cases} 5x + 2y = -9 \\ y - 3x = 12 \end{cases}$ (−3, 3)
13. $\begin{cases} 11x - 4y = 19 \\ 3x - 2y = 7 \end{cases}$ (1, −2)
14. $\begin{cases} 3x - 2y = 31 \\ 3x + 2y = -1 \end{cases}$ (5, −8)
15. $\begin{cases} 3x + 5y = 4 \\ 5x + 7y = 6 \end{cases}$ $\left(\frac{1}{2}, \frac{1}{2}\right)$

Use any method to solve each system of linear equations. Check your solution.

16. $\begin{cases} y = 5x + 2 \\ y = 2x + 5 \end{cases}$ (1, 7)
17. $\begin{cases} y = 6x \\ 2x + 5y = 16 \end{cases}$ $\left(\frac{1}{2}, 3\right)$
18. $\begin{cases} 4x + y = 9 \\ 2y = -8x + 18 \end{cases}$ infinitely many solutions
19. $\begin{cases} x + y = 1 \\ 2x + 5y = 4 \end{cases}$ $\left(\frac{1}{3}, \frac{2}{3}\right)$
20. $\begin{cases} y - 7 = x \\ x + y = 13 \end{cases}$ (3, 10)
21. $\begin{cases} 2x - 3y = 7 \\ \frac{2}{3}x - y = 9 \end{cases}$ no solution
22. $\begin{cases} 12x - y = 2 \\ 4x + 3y = 4 \end{cases}$ $\left(\frac{1}{4}, 1\right)$
23. $\begin{cases} 9x - 5y = -4 \\ 3x + 2y = 6 \end{cases}$ $\left(\frac{2}{3}, 2\right)$
24. $\begin{cases} 5x + 3y = 46 \\ 2x + 5y = 7 \end{cases}$ (11, −3)

Look Back

APPLICATION

47. LANDSCAPING The cost of plants is directly proportional to the area to be planted. Suppose that a nursery charges $78 for enough plants to cover an area of 50 square feet. How much would it cost to cover an area of 80 square feet? ***(LESSON 1.4)*** **$124.80**

Create a scatter plot of the data in each table. Then describe the correlation as positive or negative. ***(LESSON 1.5)***

48. **neg.**

x	3	5	6	0	8	5
y	9	4	3	11	1	3

49. **pos.**

x	1	3	5	2	9	4
y	2	5	5	1	12	4

Solve each equation. ***(LESSON 1.6)***

50. $3x - 12 = 24$ **12** **51.** $\frac{1}{5}x - 3 = x + 3$ $-\frac{15}{2}$ **52.** $5x - 2(3x - 1) = x - 10$ **6**

Solve each inequality, and graph the solution on a number line. ***(LESSON 1.7)***

53. $4x > 6$ $x > \frac{3}{2}$ **54.** $8x \le 24$ $x \le 3$ **55.** $-\frac{1}{2}x \le 5$ $x \ge -10$

Solve each compound inequality, and graph the solution on a number line. ***(LESSON 1.7)***

56. $n + 1 < 9$ *and* $n - 3 > 1$ $4 < n < 8$ **57.** $2y - 10 \le -6$ *and* $y + 8 \ge 2$ $-6 \le y \le 2$

58. $x + 11 \ge 7$ *or* $x - 4 \le 4$ **all real numbers** **59.** $-2a - 8 < 4$ *or* $3a - 9 < 21$ **all real numbers**

Evaluate each expression by using the order of operations. ***(LESSON 2.1)***

60. $7(6 - 5^2)$ **−133** **61.** $-(-3)^2 - 4^2$ **−25** **62.** $21 - 7 \times 3 + 8 \div 3$ $\frac{8}{3}$

Look Beyond

Solve each system of nonlinear equations by elimination.

63. $\begin{cases} x^2 + y^2 = 3 \\ x^2 + y^2 = 9 \end{cases}$ **no solution** **64.** $\begin{cases} y^2 = 1 - x^2 \\ x^2 + y^2 = 1 \end{cases}$ **infinitely many solutions**

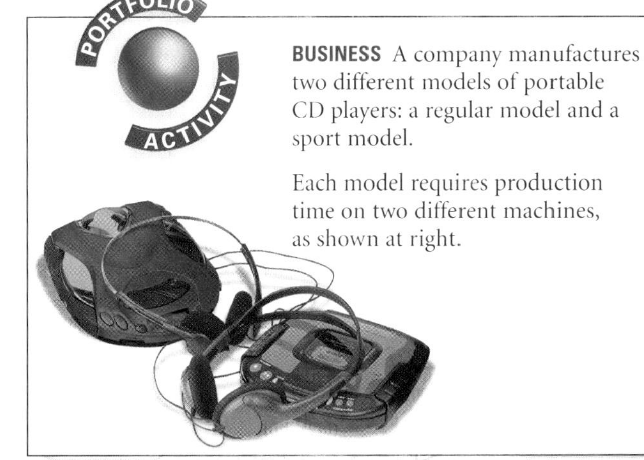

BUSINESS A company manufactures two different models of portable CD players: a regular model and a sport model.

Each model requires production time on two different machines, as shown at right.

	Regular model	Sport model
Machine A	2 minutes	1 minute
Machine B	1 minute	1 minute

Each machine is used for manufacturing many different items. In a given hour, machine A is used for CD-player production for 18 minutes and machine B is used for 10 minutes. How many CD players of each model are produced per hour?

Look Beyond

In Exercises 63 and 64, students investigate two nonlinear systems of equations. They should be able to adapt the techniques of this lesson to conclude that the systems are inconsistent and dependent, respectively. You may wish to tell them that the graph of the system in Exercise 63 is two concentric circles of different radii, while the graph of the system in Exercise 64 consists of identical circles. Students will learn more about nonlinear systems in Lesson 9.6.

ALTERNATIVE
Assessment

Portfolio Activity

The Portfolio Activity can be used as preparation for the Chapter Project or as a separate activity. In the Portfolio Activity on this page, students discover how a system of equations can be used to solve a problem about manufacturing.

Answers to Portfolio Activities can be found in Additional Answers of the Teacher's Edition.

Student Technology Guide

NAME ______ CLASS ______ DATE ______

Student Technology Guide
3.2 ***Solving Systems by Elimination***

A system of linear equations in x and y has a unique solution if the equations have different slopes. Consider $\begin{cases} ax + by = e \\ cx + dy = f \end{cases}$, where $b \ne 0$ and $d \ne 0$. Solving for y gives $\begin{cases} y = -\frac{a}{b}x + \frac{e}{b} \\ y = -\frac{c}{d}x + \frac{f}{d} \end{cases}$.

The system has a unique solution if $-\frac{a}{b} \ne -\frac{c}{d}$, or $ad - bc \ne 0$. Thus, a system of linear equations in standard form is independent if and only if:

(first x-coefficient)(second y-coefficient) − (first y-coefficient)(second x-coefficient) $\ne 0$

Example: Tell whether $\begin{cases} -2.45x + 4.25y = 12 \\ 4.64x + 1.16y = 16 \end{cases}$ has a unique solution.

Because $(-2.45)(1.16) - (4.25)(4.64) \ne 0$, the system has a unique solution.

Use a calculator to determine whether each system is independent.

1. $\begin{cases} 9.1x + 2.6y = 5.7 \\ 15.4x + 4.4y = 3.9 \end{cases}$ **dependent**
2. $\begin{cases} 2.9x + 5.1y = -2.7 \\ 5.3x - 6.1y = 9.8 \end{cases}$ **independent**
3. $\begin{cases} 9.3x - 14.1y = -6.3 \\ 6.2x - 9.4y = -4.2 \end{cases}$ **dependent**

Certain calculators, such as the TI-92, can be used to solve systems of equations algebraically by using substitution. Follow the steps outlined below to solve the system $\begin{cases} 3x + 5y = 11 \\ 4x - 7y = 6 \end{cases}$.

```
F1  F2▾ Algebra  F3▾ Calc  F4▾ Other  F5 PrgmIO  F6 Clear a-z...
■ solve(3·x+5·y=11,y)              y = ⁻(3·x−11)/5
■ solve(4·x−7·y=6,x) | y = ⁻(3·x−11)/5
                                    x = 107/41
■ y = ⁻(3·x−11)/5 | x = 107/41      y = 26/41
y=⁻(3*x−11)/5|x=107/41
MAIN        RAD AUTO        FUNC 3/30
```

- First, clear the entry line on the home screen.
- Using the first equation, solve for y.
  F2 1:Solve(ENTER 3 X + 5 Y = 11 , Y) ENTER
- Using the second original equation and the equation you found for y, solve for x.
  F2 1:Solve(ENTER 4 X − 7 Y = 6 , X) 2nd K ▲ ENTER ENTER
- Finally, obtain the value of y by pressing ▲ ▲ ▲ ENTER 2nd K ▲ ENTER ENTER.

The solution is $\left(\frac{107}{41}, \frac{26}{41}\right)$.

Use a calculator that can solve systems of equations algebraically to solve each system.

4. $\begin{cases} 4x - 3y = 17 \\ 3x + 5y = 23 \end{cases}$ $\left(\frac{154}{29}, \frac{41}{29}\right)$
5. $\begin{cases} 15x + 7y = 21 \\ -6x + 11y = 9 \end{cases}$ $\left(\frac{56}{69}, \frac{29}{23}\right)$
6. $\begin{cases} -5x + 8y = 12 \\ 4x + y = -7 \end{cases}$ $\left(-\frac{68}{37}, \frac{13}{37}\right)$

53. $x > \frac{3}{2}$

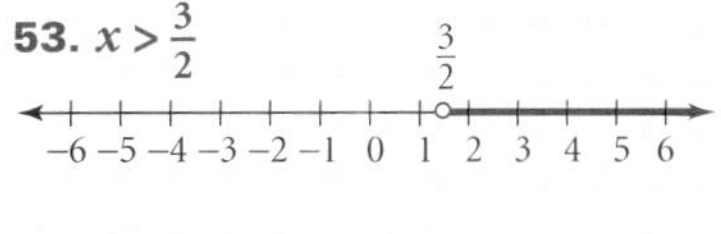

54. $x \le 3$

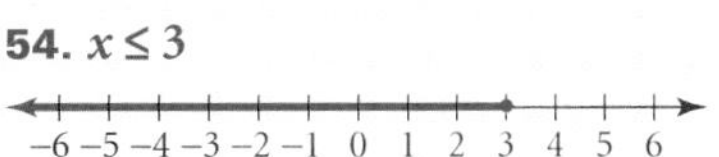

55. $x \ge -10$

−30 −20 −10 0 10 20 30

56. $4 < n < 8$

0 1 2 3 4 5 6 7 8 9 10 11 12

57. $-6 \le y \le 2$

−6 −5 −4 −3 −2 −1 0 1 2 3 4 5 6

58. all real numbers

−6 −5 −4 −3 −2 −1 0 1 2 3 4 5 6

59. all real numbers

−6 −5 −4 −3 −2 −1 0 1 2 3 4 5 6

Prepare

NCTM PRINCIPLES & STANDARDS 1–4, 6–10

QUICK WARM-UP

Solve each equation or inequality. Graph the solution on a number line.

1. $2c + 7 = 1$ $c = -3$;

2. $-12 \leq -6r$ $r \leq 2$;

3. $3m + 8 > 5$ $m > -1$;

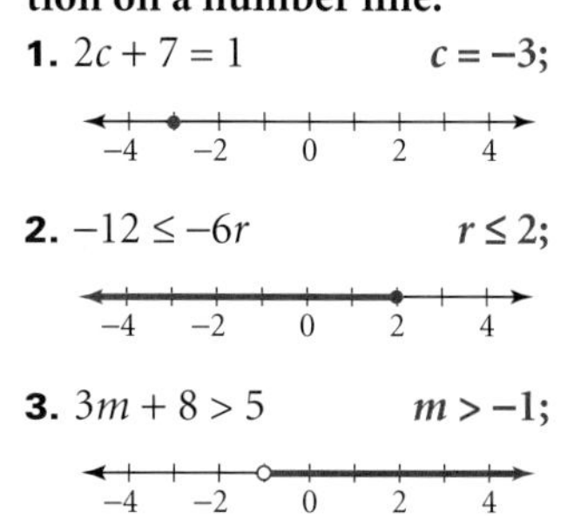

Also on Quiz Transparency 3.3

Teach

Why The *border* between two countries separates the land into three regions: the first country, the second country, and the border itself. In much the same way, a line separates a coordinate plane into three regions. In this lesson, students will learn how to use algebraic sentences to name regions.

Use Teaching Transparency 12.

Linear Inequalities in Two Variables

Why *Linear inequalities can be used to model real-world situations, such as the distance that a car with given fuel-economy ratio can be driven using no more than 20 gallons of gasoline.*

Objectives

- Solve and graph a linear inequality in two variables.
- Use a linear inequality in two variables to solve real-world problems.

APPLICATION
FUEL ECONOMY

The fuel-economy test results for three different automobiles are shown below.

EPA 1997 Fuel-Economy Program

Automobile	Fuel Economy (mpg) City	Fuel Economy (mpg) Highway
Automobile A	28.1	33.3
Automobile B	25.6	31.5
Automobile C	20.2	31.7

[*Source: Environmental Protection Agency*]

Consider automobile A. If you let x represent miles driven in the city and y represent the miles driven on the highway, then $\frac{x}{28.1} + \frac{y}{33.3}$ represents the number of gallons of gasoline consumed. Solve $\frac{x}{28.1} + \frac{y}{33.3} \leq 20$ to find how far you can drive using no more than 20 gallons of gasoline. You will solve this problem in Example 3.

The inequality $\frac{x}{28.1} + \frac{y}{33.3} \leq 20$ is an example of a *linear inequality in two variables.*

Linear Inequality in Two Variables

A **linear inequality in two variables**, x and y, is any inequality that can be written in one of the forms below, where $A \neq 0$ and $B \neq 0$.

$Ax + By \geq C$ $Ax + By > C$ $Ax + By \leq C$ $Ax + By < C$

Alternative Teaching Strategy

GUIDED ANALYSIS Display the graph below on the board or overhead. Have students list the coordinates of several points in the shaded region.

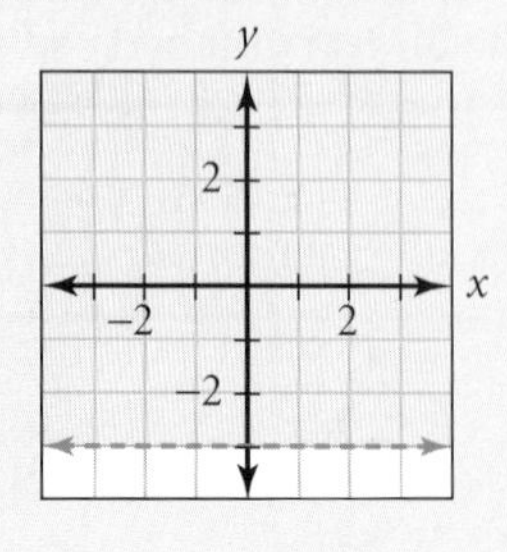

Ask students to examine the list of responses and name a characteristic common to all points in the shaded region. Elicit the fact that all the y-coordinates are greater than -3. Ask them to write an algebraic sentence that describes the shaded region. Lead them to the inequality $y > -3$.

Now make the line $y = -3$ solid, and ask students to adjust the algebraic sentence to the new graph. Lead them to the inequality $y \geq -3$.

Guide students through similar analyses for the inequalities $x < 3$, $y > x$, and $y < 2x + 4$.

A solution of a linear inequality in two variables, x and y, is an ordered pair (x, y) that satisfies the inequality. The solution to a linear inequality is a region of the coordinate plane and is called a *half-plane* bounded by a *boundary line.*

EXAMPLE 1 **Graph each linear inequality.**

a. $y < x + 2$ **b.** $y \geq -2x + 3$

SOLUTION

a. Graph the boundary line $y = x + 2$. Use a dashed line because the values on this line are not included in the solution.

Choose a point such as (0, 0) to test.

$$y \overset{?}{<} x + 2$$
$$0 < 0 + 2 \quad \textbf{True}$$

Since (0, 0) satisfies the inequality, shade the region that contains this point.

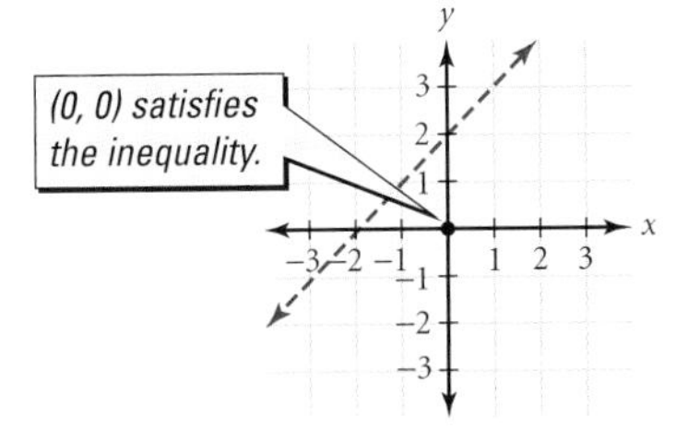

b. Graph the boundary line $y = -2x + 3$. Use a solid line because the values on this line are included in the solution.

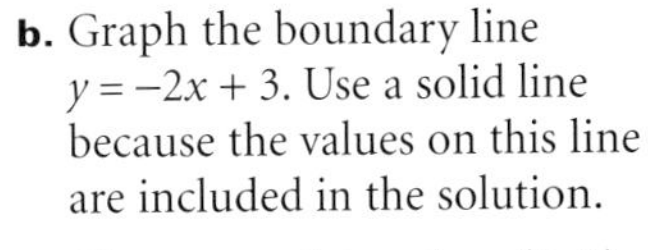

Choose a point such as (0, 0) to test.

$$y \overset{?}{\geq} -2x + 3$$
$$0 \geq -2(0) + 3 \quad \textbf{False}$$

Since (0, 0) does *not* satisfy the inequality, shade the region that does *not* contain this point.

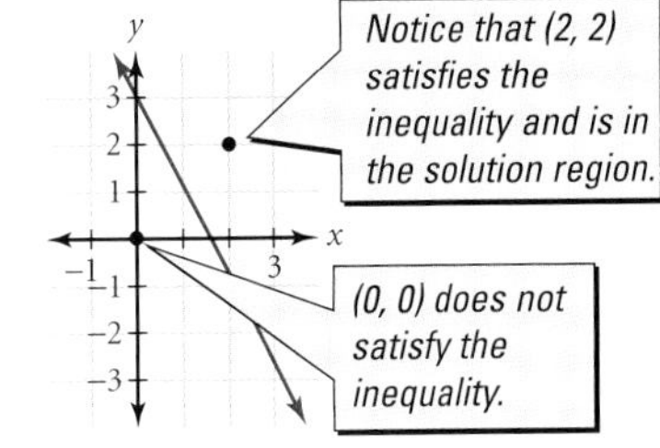

Notice that (2, 2) satisfies the inequality and is in the solution region.
(0, 0) does not satisfy the inequality.

TRY THIS Graph each linear inequality.

a. $y > -2x - 2$ **b.** $y \leq 2x + 5$

Sometimes you may need to solve for y before graphing a linear inequality, as shown in Example 2.

EXAMPLE 2 **Graph $-2x - 3y \leq 3$.**

SOLUTION

Solve for y in terms of x.

$$-2x - 3y \leq 3$$
$$-3y \leq 2x + 3$$
$$y \geq -\frac{2}{3}x - 1 \quad \textit{Change} \leq \textit{to} \geq.$$

Then graph $y \geq -\frac{2}{3}x - 1$.

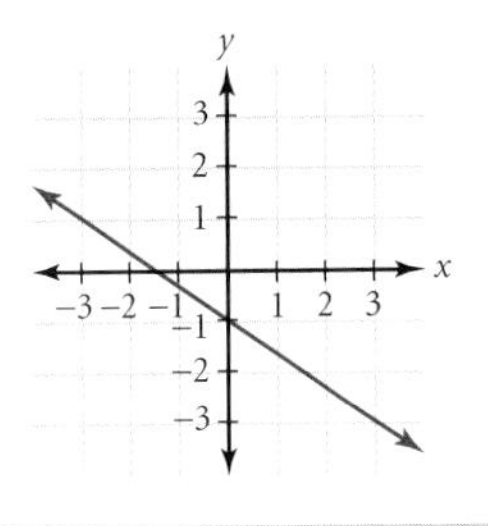

TRY THIS Graph $3x - 4y \geq 4$.

CHECKPOINT ✔ Is the solution to $y < mx + b$ above or below the boundary line? Is the solution to $y > mx + b$ above or below the boundary line?

Interdisciplinary Connection

BUSINESS A small business sells canvas backpacks and tote bags. The backpacks are sold for \$35 each, and the tote bags are sold for \$20 each. In order for the business to *break even,* the total amount of sales in one week must be \$2000.

a. Let b represent the number of backpacks sold in one week and t represent the number of tote bags sold. Write an algebraic sentence that describes the circumstances under which the business breaks even. **$35b + 20t = 2000$**

b. Write an algebraic sentence that describes the circumstances under which the business makes a profit. **$35b + 20t > 2000$**

c. Write an algebraic sentence that describes the circumstances under which the business incurs a loss. **$35b + 20t < 2000$**

d. How does the business fare in a week in which 40 backpacks and 62 tote bags are sold? Explain. **It makes a profit of \$640. The ordered pair (40, 62) satisfies $35b + 20t > 2000$.**

ADDITIONAL EXAMPLE 1

Graph each linear inequality.

a. $y > -x - 3$

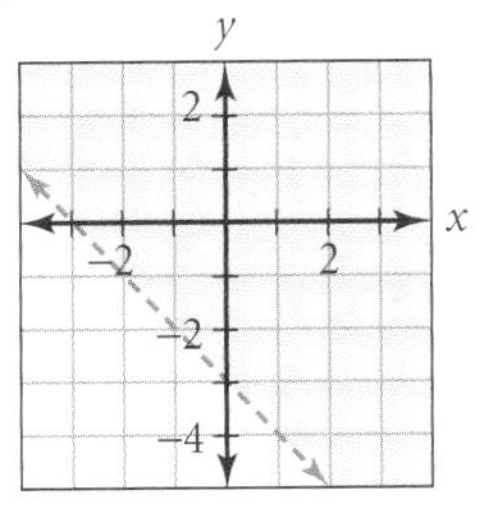

b. $y \leq 3x + 1$

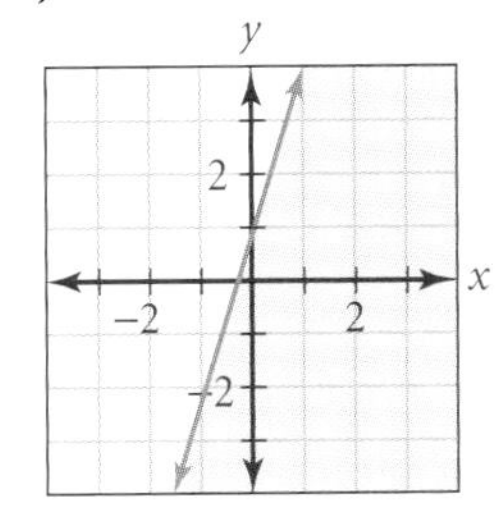

TRY THIS

a.

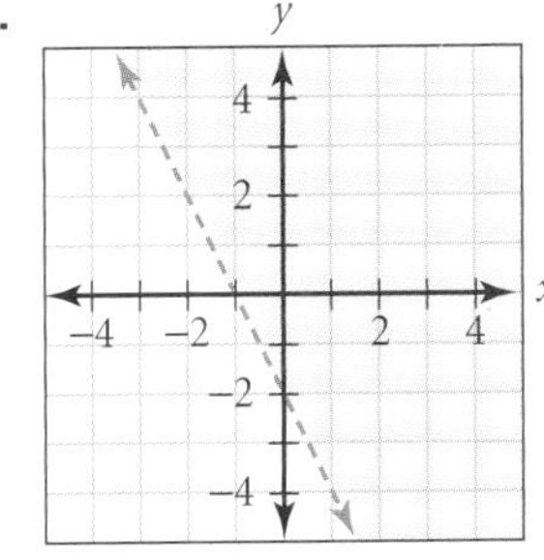

b.

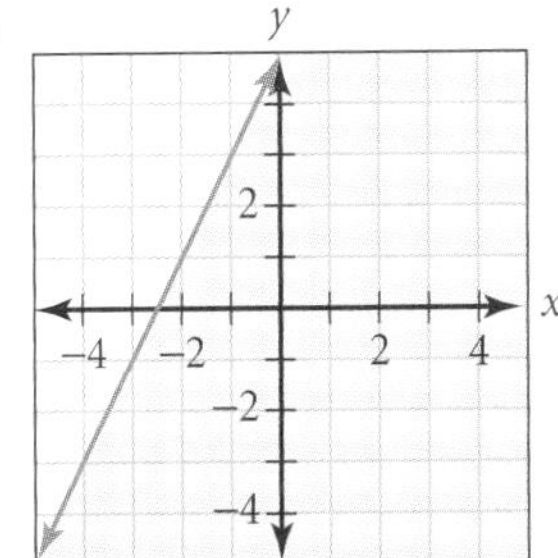

ADDITIONAL EXAMPLE 2

Graph $-5x - 2y > 4$.

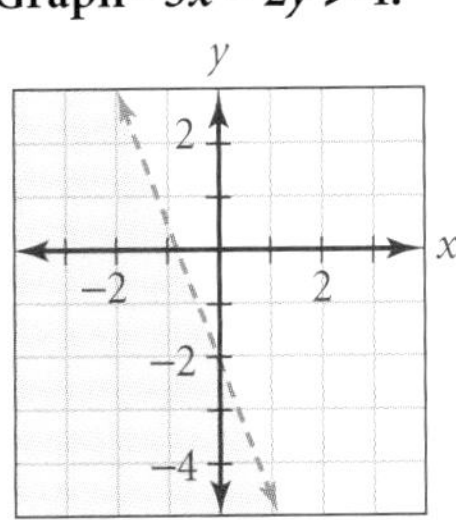

☞ For answers to Try This and Checkpoint, see page 174.

TRY THIS

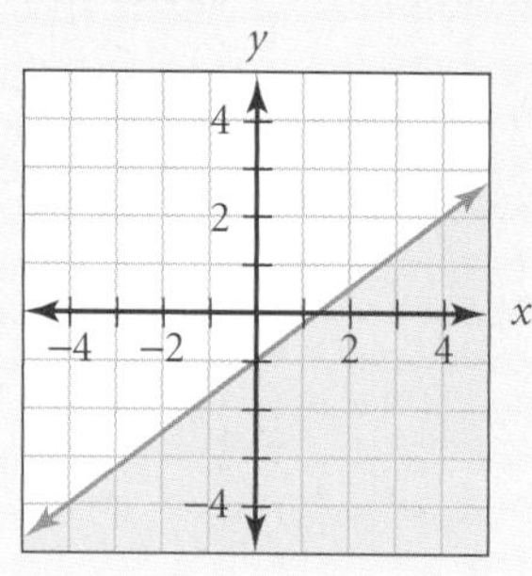

CHECKPOINT ✔
$y < mx + b$ is below the boundary line; $y > mx + b$ is above the boundary line.

ADDITIONAL EXAMPLE 3

Refer to the data for automobile B on page 172.

a. **Represent in a graph the possible distances automobile B can travel and use no more than 10 gallons of gasoline.**

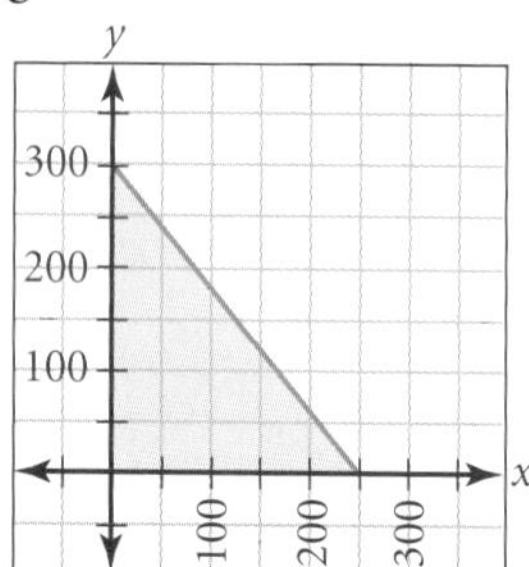

$\frac{x}{25.6} + \frac{y}{31.5} \leq 10$

b. **Can you drive 30 miles in the city and 200 miles on the highway in automobile B and use no more than 10 gallons of gasoline?** yes

EXAMPLE 3 **Refer to the fuel-economy test results given at the beginning of the lesson.**

APPLICATION FUEL ECONOMY

a. Represent in a graph the possible distances that automobile A can travel using no more than 20 gallons of gasoline.

b. Can you drive 25 miles in the city and 400 miles on the highway in automobile A and use no more than 20 gallons of gasoline?

TECHNOLOGY GRAPHICS CALCULATOR

Keystroke Guide, page 212

SOLUTION

a. Solve for y, and graph.

$$\frac{x}{28.1} + \frac{y}{33.3} \leq 20$$

$$y \leq -\frac{33.3}{28.1}x + 20(33.3)$$

Distance cannot be negative, so both $x \geq 0$ and $y \geq 0$.

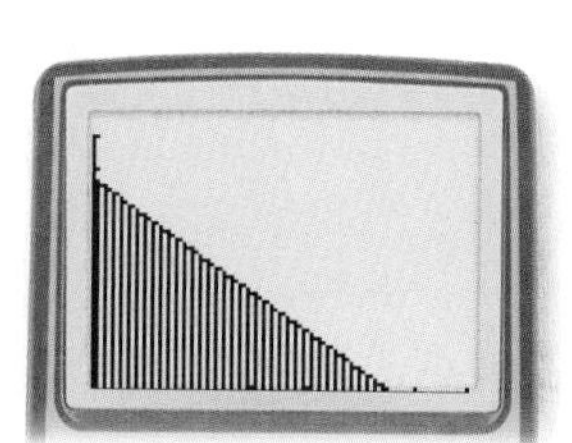

b. Test (25, 400) to see if the inequality is true.

$$\frac{25}{28.1} + \frac{400}{33.3} \stackrel{?}{\leq} 20 \quad \rightarrow \quad 12.9 \leq 20 \quad \textbf{True}$$

You can also plot the point (25, 400) to see whether the point is in the solution region.

Thus, you can drive 25 miles in the city and 400 miles on the highway and use no more than 20 gallons of gasoline in automobile A.

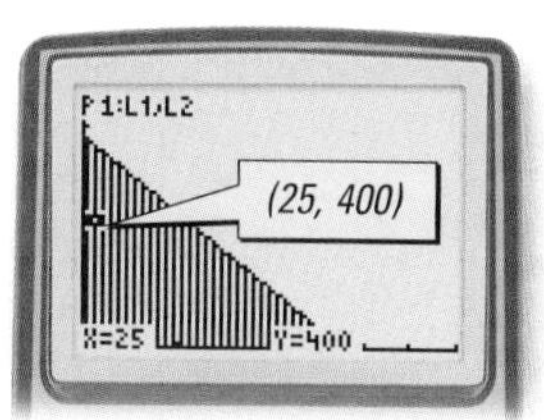

(25, 400)

The solution to a linear inequality in only one variable is also a half-plane when graphed in the coordinate plane.

EXAMPLE 4 **Graph each linear inequality.**

a. $x > -2$

b. $y \leq -1$

SOLUTION

a. Any ordered pair whose x-coordinate is greater than -2 is a solution. Graph the boundary line $x = -2$, using a dashed line, and shade the half-plane to the right of the line.

b. Any ordered pair whose y-coordinate is less than or equal to -1 is a solution. Graph the boundary line $y = -1$, using a solid line, and shade the half-plane below the line.

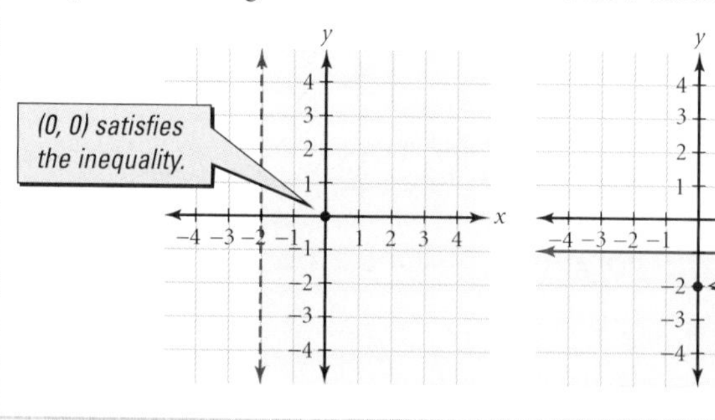

(0, 0) satisfies the inequality.
Notice that the point (0, −2) satisfies the inequality and is in the solution region.

TRY THIS Graph $x \leq 4$ and $y > 3$ separately.

Inclusion Strategies

KINESTHETIC LEARNERS Arrange the students' desks in rows if they are not already configured that way. Assign row and seat numbers so that the ordered pairs (row, seat) simulate the first quadrant of the coordinate plane. For example, the desk in the position "third row from the left and fifth seat from the front" would represent the ordered pair (3, 5). (Hint: The x-axis represents the front of the classroom.) Now have students act out several inequalities. For instance, ask students to stand if their y-coordinate is greater than or equal to their x-coordinate. Point out that the students who are standing represent the inequality $y \geq x$.

Enrichment

Have students work in pairs to graph the following inequalities:

1. $y \geq |x| - 1$

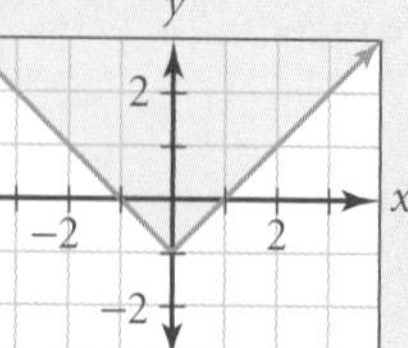

2. $|x| < 3$

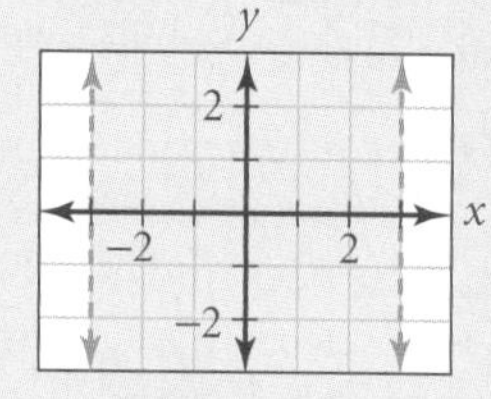

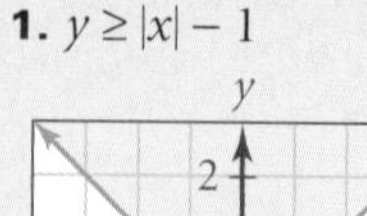

Now have each student write and graph two similar inequalities. Partners should exchange graphs and write the inequalities represented.

CHECKPOINT ✔ On which side of the boundary line is the solution region for $x > c$ (or $x \geq c$)? On which side of the boundary line is the solution region for $x < c$ (or $x \leq c$)?

As you will notice in the Activity below, sometimes the solution to a linear inequality is a set of *discrete* points in a region instead of the entire region.

Activity Exploring Discrete Solutions

You will need: no special materials

Daryll wants to buy some cassette tapes and CDs. A tape costs \$8 and a CD costs \$15. He can spend no more than \$90 on tapes and CDs.

PROBLEM SOLVING

1. Let x represent the number of tapes he can buy. Let y represent the number of CDs he can buy. **Write a linear inequality** in x and y to represent his possible purchases.

CHECKPOINT ✔

2. Explain why x and y must be whole numbers.
3. List four ordered pairs (x, y) that satisfy your inequality from Step 1.
4. How many tapes can Daryll buy if he chooses not to buy any CDs? How many CDs can Daryll buy if he chooses not to buy any tapes?
5. Graph all possible solutions to your inequality in Step 1 on a coordinate plane.

CHECKPOINT ✔

6. Describe other situations that can be modeled by a linear inequality that has only nonnegative integer solutions.

The procedure for graphing linear inequalities is summarized below.

SUMMARY
GRAPHING LINEAR INEQUALITIES

1. Given a linear inequality in two variables, graph its related linear equation.
   - For inequalities involving $\leq$ or $\geq$, use a solid boundary line.
   - For inequalities involving $<$ or $>$, use a dashed boundary line.
2. Shade the appropriate region.
   - For inequalities of the form $y \leq mx + b$ or $y < mx + b$, shade below boundary line.
   - For inequalities of the form $y \geq mx + b$ or $y > mx + b$, shade above boundary line.
   - For inequalities of the form $x \leq c$ or $x < c$, shade to the left of the boundary line.
   - For inequalities of the form $x \geq c$ or $x > c$, shade to the right of the boundary line.

CRITICAL THINKING

Let $Ax + By \geq C$, where $A \neq 0$ and $B \neq 0$. Under what conditions is the solution region of the inequality above the boundary line? Under what conditions is the solution region below the boundary line?

Reteaching the Lesson

COOPERATIVE LEARNING Have students work in groups of four. Each group member should choose a different one of the inequality symbols $>$, $<$, $\geq$, and $\leq$.

Now display the equation $y = 2x + 1$. Tell students to replace the equal sign with their chosen symbol and work *independently* to graph the resulting inequality. When all group members have completed their graphs, the group should come together to compare results. Point out that the boundaries of all four graphs should be in the same location, but the four graphs should each appear different. Each group should work together to arrive at the consensus that they have created a set of four appropriate graphs.

Repeat the activity with the equations $3x + 2y = 6$, $4y = -8$, and $-x = 5$. You may want to suggest that group members "rotate" the four inequality symbols among themselves so that each member has a chance to work with all four symbols.

ADDITIONAL EXAMPLE 4

Graph each linear inequality.

a. $x < -1$

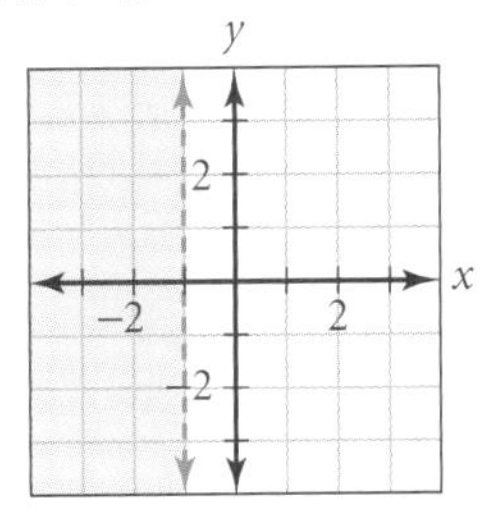

b. $y \geq 0$

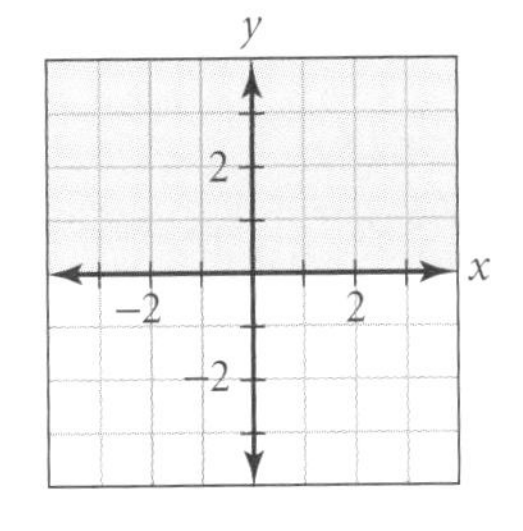

TRY THIS

$y \leq 4$

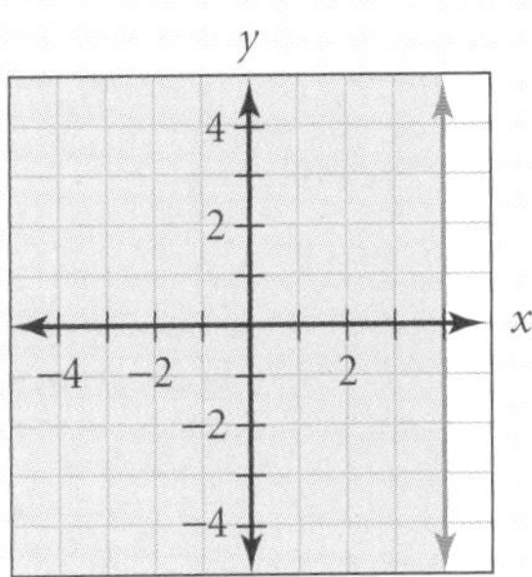

$y > 3$

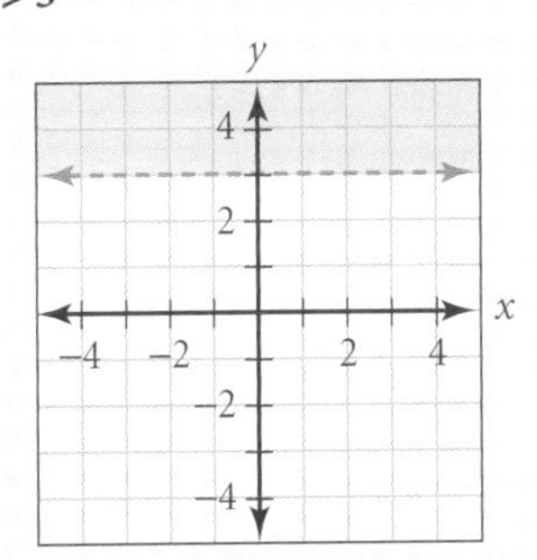

CHECKPOINT ✔

$x < c$ (or $x \leq c$) is on the left side of the boundary line; $x > c$ (or $x \geq c$) is on the right side of the boundary line.

☞ For Activity Notes and answers to Checkpoints and Critical Thinking, see page 176.

Activity Notes

In this Activity, students investigate a linear inequality whose solutions have only integral coordinates. Students should discover that the graph of these solutions is a set of discrete points rather than a shaded region. You may wish to tell them that points whose coordinates are integers are called *lattice points.*

CHECKPOINT ✔

2. Daryll cannot purchase a fraction of a tape or compact disc. He must purchase them in whole-number units.

CHECKPOINT ✔

6. any application in which x and y represent quantities measured in whole units

CRITICAL THINKING

The solution to $Ax + By \geq C$, where $A \neq 0$ and $B \neq 0$, is above the boundary line if $B > 0$ and is below the boundary line if $B < 0$.

Practice

NAME ____________ CLASS ________ DATE ________

Practice
3.3 *Linear Inequalities in Two Variables*

Graph each linear inequality.

1. $y < -x$
2. $y \geq -3x - 2$
3. $y > \frac{1}{4}x + 1$
4. $2x + y \leq -3$
5. $x - y > 4$
6. $x \geq -1.5$

7. Sheila earns a basic wage of \$8 per hour. Under certain conditions, she is paid \$12 per hour. The most that she can earn in one week is \$400.

a. Write an inequality that describes her total weekly wages for x hours at \$8 per hour and for y hours at \$12 per hour.
$8x + 12y \leq 400$

b. Graph the inequality on the grid at right.

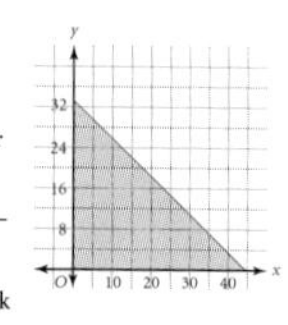

c. What is the maximum number of hours that Sheila can work for \$8 per hour? for \$12 per hour?
50 hours; $33\frac{1}{3}$ hours

Exercises

Communicate

1. Describe how to graph $7x - 5y > 0$ on graph paper.
2. When graphing a linear inequality, what determines whether the boundary line is dashed or solid?
3. Describe two ways to determine which region of the plane should be shaded for the different types of linear inequalities.

Guided Skills Practice

Graph each linear inequality. ***(EXAMPLES 1 AND 2)***

4. $y < 2x + 6$ **5.** $y \geq -\frac{1}{3}x + 6$ **6.** $2y + 3x \geq 12$ **7.** $4y - 5x < -8$

APPLICATION

8. FUEL ECONOMY Refer to the fuel-economy test results given on page 172. Can you drive 25 miles in the city and 400 miles on the highway in automobile B using no more than 20 gallons of gasoline? ***(EXAMPLE 3)***
Yes, (25, 400) is a solution.

Graph each linear inequality. ***(EXAMPLE 4)***

9. $x \leq 2$ **10.** $y > -3$

Practice and Apply

internet connect
Homework Help Online
Go To: **go.hrw.com**
Keyword:
MB1 Homework Help for Exercises 11–28, 32–44

Graph each linear inequality.

11. $y \geq 3x + 1$
12. $y > 5x + 2$
13. $y < 6x + 2$
14. $y \leq \frac{3}{2}x + 1$
15. $y \geq -\frac{1}{2}x + \frac{2}{3}$
16. $y > -3x - 4$
17. $y < -2x + \frac{1}{2}$
18. $y \leq -10x + 3$
19. $y - 5x \geq 2$
20. $2x + y > -2$
21. $x + 3y < 1$
22. $5x + 3y \leq 4$
23. $5x - y \geq 1$
24. $-2x - y > 0$
25. $6x - 4y > -2$
26. $3x - 2y > 5$
27. $\frac{3}{2}x - \frac{5}{4}y - 1 \leq 0$
28. $\frac{2}{3}x - \frac{1}{2}y \leq -2$

Write an inequality for each graph.

29. **$y < 3x + 2$**

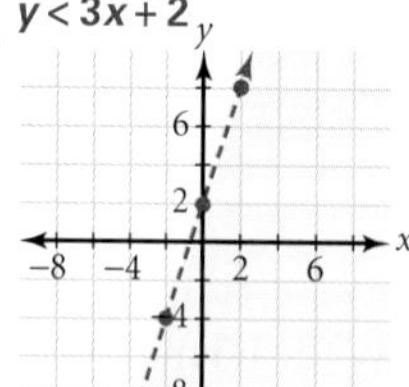

30. **$y \geq -2$**

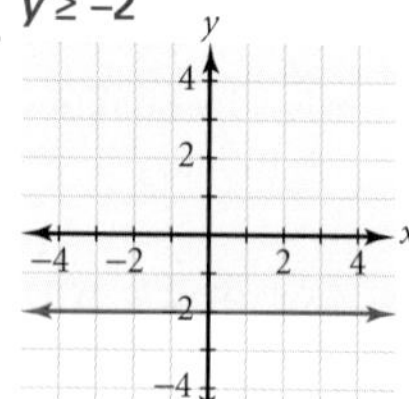

31. **$x \leq -1$**

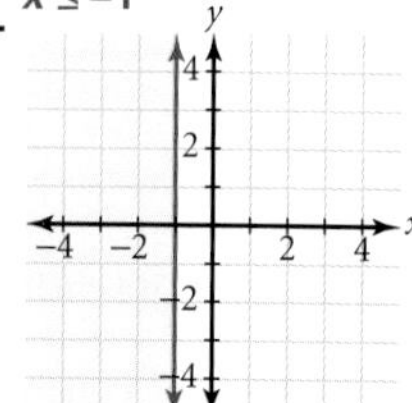

4.

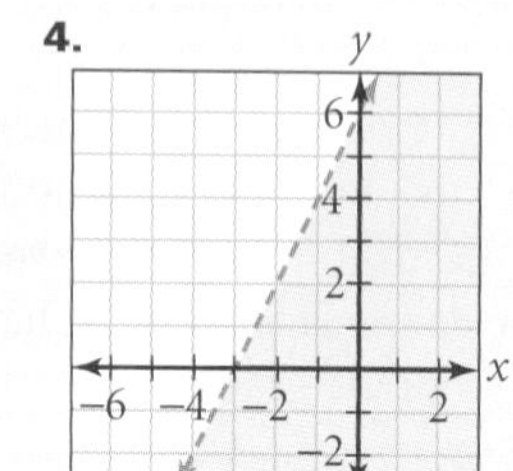

5.

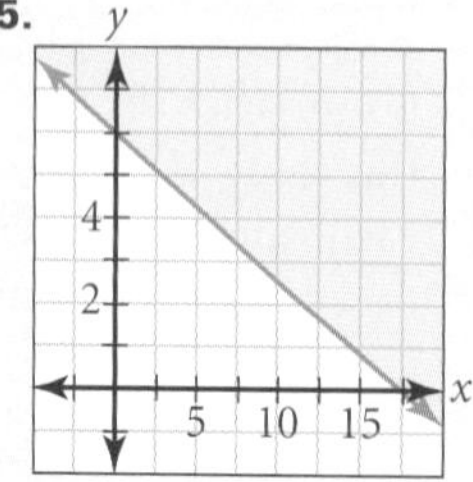

6.

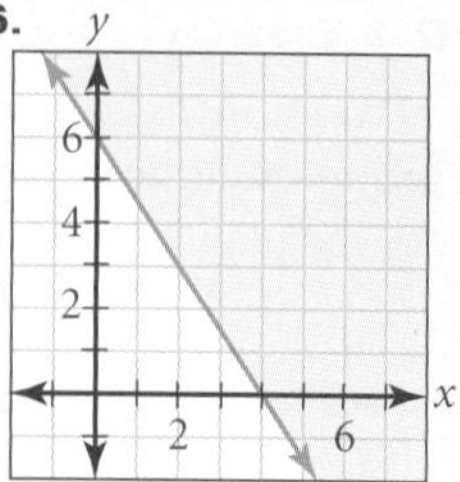

7.

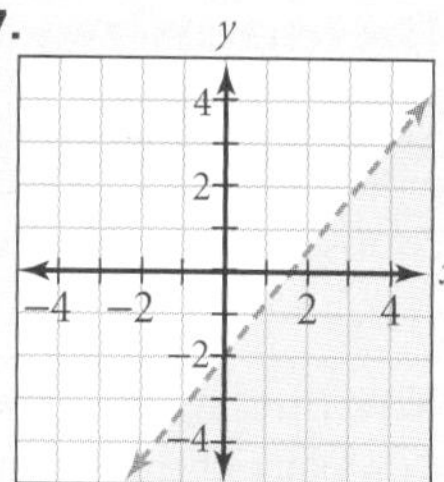

Graph each linear inequality.

32. $x < -1$ **33.** $x \leq 2$ **34.** $y \geq 3$ **35.** $y > -2$

36. $2y < 5$ **37.** $2y \leq -1$ **38.** $-x \leq 4$ **39.** $-\frac{5}{4}x < -2$

40. $-7y < 21$ **41.** $\frac{3 - 12y}{7} < 0$ **42.** $\frac{6x + 5}{3} \geq 8$ **43.** $3(4 - 2x) \leq -7$

44. Graph the inequality $y \geq \frac{1}{2}x + 5$.

a. Does the ordered pair (4, 1) satisfy the inequality? Support your answer by using both the inequality and the graph.

b. Identify three ordered pairs that have an x-coordinate of 6 and satisfy the inequality.

c. Identify three ordered pairs that have a y-coordinate of 8 and satisfy the inequality.

CHALLENGE

Describe the graphs in each *family* of inequalities for n-values of 1, 2, 3, and 4.

45. $x < n$ **46.** $y > (-1)^n x$ **47.** $ny \leq 2x$

CONNECTION

48a. $2x + 2y \leq 200$
$0 < x \leq 100$
$0 < y \leq 100$

48. GEOMETRY The perimeter of a rectangle with a length of x feet and a width of y feet cannot exceed 200 feet.

a. Write three linear inequalities to describe the restrictions on the values of the perimeter, of x, and of y.

b. Graph the solution region of the three inequalities from part **a**.

APPLICATIONS

SPORTS Michael is close to breaking his high-school record for field-goal points in a basketball game, needing 24 points to tie the record and 25 points to break the record. A field goal can be worth either 2 or 3 points. Michael will play in a game tonight. Write an equation or inequality for each situation below.

49. He fails to tie or break the record. $2x + 3y < 24$

50. He ties the record. $2x + 3y = 24$

51. He breaks the record. $2x + 3y > 24$ or $2x + 3y \geq 25$

52. **FUND-RAISING** The junior class is sponsoring a refreshment booth at home football games. They will earn a $0.25 profit on each soft drink sold and a $0.20 profit on each ice-cream bar sold. Their goal is to earn a profit of at least $50.

a. Write an inequality that describes the profit goal. $0.25x + 0.2y \geq 50$

b. Graph the inequality.

c. Give four ordered pairs that represent a profit of exactly $50.

d. Give three ordered pairs that represent a profit more than $50.

e. Give three ordered pairs that represent a profit less than $50.

53. **RECREATION** Amanda is planning a cookout. She has budgeted a maximum of $60 for hamburgers and turkey dogs. Hamburgers cost $3 per pound, and turkey dogs cost $2 per pound.

a. Write an inequality to describe the possible number of pounds of hamburgers and of turkey dogs that she can purchase. $3x + 2y \leq 60$

b. Graph the inequality.

c. What is the maximum number of pounds of hamburgers that she can purchase? **20 pounds**

d. What is the maximum number of pounds of turkey dogs that she can purchase? **30 pounds**

9.

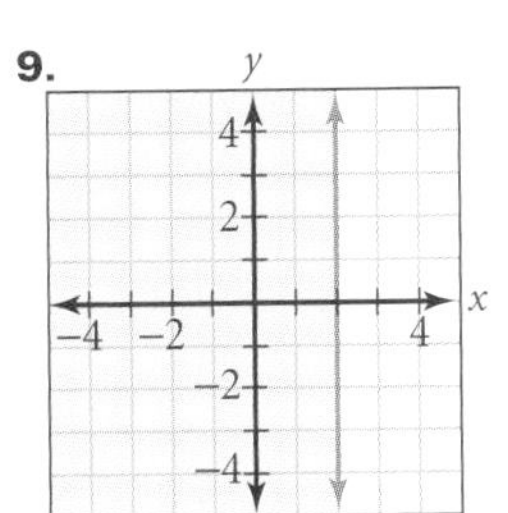

10.

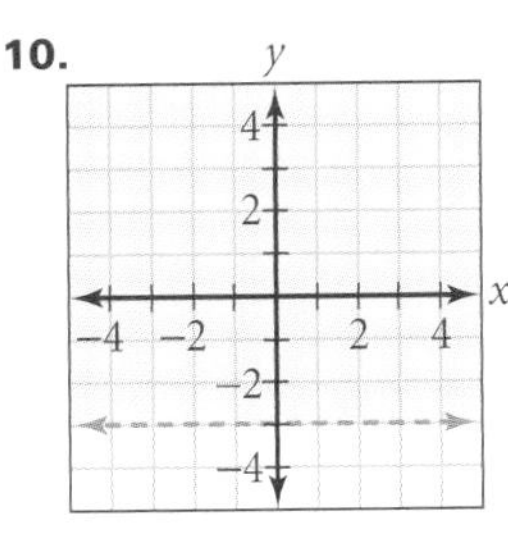

11.

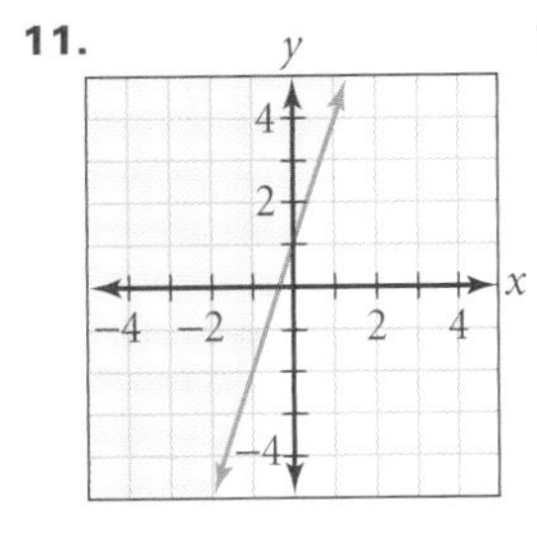

12.

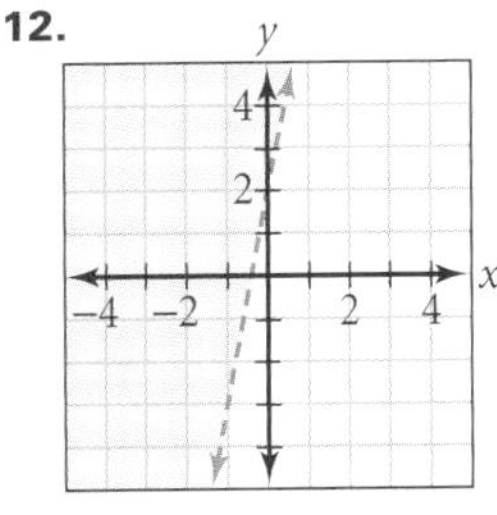

The answers to Exercises 13–28, 32–47, 52b–e, and 53b can be found in Additional Answers beginning on page 1002.

Assess

Selected Answers

Exercises 4–10, 11–71 odd

ASSIGNMENT GUIDE

In Class	1–10
Core	11–43 odd, 49–55 odd
Core Plus	12–28 even, 29–55
Review	56–71
Preview	72–75

Extra Practice can be found beginning on page 940.

Mid-Chapter Assessment for Lessons 3.1 through 3.3 can be found on page 33 of the *Assessment Resources.*

Student Technology Guide

NAME ______ CLASS ______ DATE ______

Student Technology Guide
3.3 Linear Inequalities in Two Variables

You can use the following procedure to graph inequalities using a graphics calculator.

Example: Graph $x - 2y \geq -2$.
First solve for y. $x - 2y \geq -2 \rightarrow -2y \geq -x - 2 \rightarrow y \leq \frac{1}{2}x + 1$
Note in the last step that $\geq$ becomes $\leq$.

Enter the expression $\frac{1}{2}x + 1$ as Y1: [Y=] .5 [X,T,θ,n] + 1

Use [◄] to move all the way to the left. The equation is of the form $y \leq mx + b$, so you should shade *below* the graph of the line. Press [ENTER] repeatedly until the symbol ◣ appears. Then press [GRAPH]. The calculator graphs the line and shades the region below it, as shown in the graph at right.

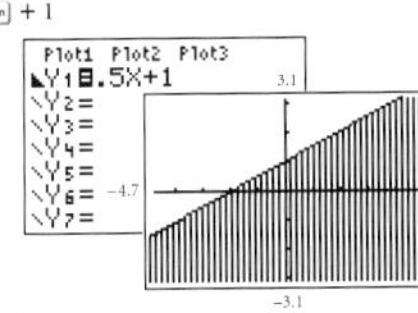

Graph each linear inequality and describe the graph.

1. $y \geq 2x - 1$ above and including $y = 2x - 1$
2. $y \leq \frac{1}{3}x + 2$ below and including $y = \frac{1}{3}x + 2$
3. $4x + y \leq 8$ below a[nd] including $y = -4x + 8$
4. $x - 2y \leq 6$ above and including $y = \frac{1}{2}x - 3$
5. $3x - 3y \geq 24$ below and including $y = x - 8$
6. $5x + 3y \geq 15$ above a[nd] including $y = -\frac{5}{3}x + 5$

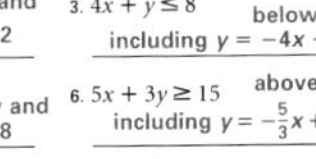

In many real-life situations, the values of the variables must be at least zero. In this case, you will want to choose the window settings Xmin = 0 and Ymin = 0.

A graph of $y \leq \frac{1}{2}x + 1$ for such a real-life situation is shown at right. Notice that the result is a shaded region in the first quadrant only.

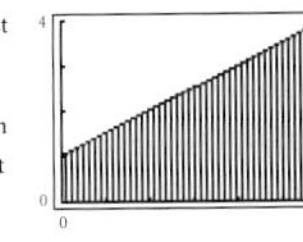

Assume x and y represent real-life quantities that must be at least zero. Graph each linear inequality and describe the graph.

7. $2x + 3y \leq 6$ triangular region with corners (0, 0), (3, 0), and (0, 2)
8. $4x - 5y \leq 20$ first quadrant above and including $y = \frac{4}{5}x - 4$
9. $6x + 3y \geq 15$ first quadrant, except the triangular region with corners (0, 0), (2.5, 0), and (0, 5)

Error Analysis

When graphing linear inequalities, students often shade the incorrect half-plane. Encourage them to check their work by substituting the coordinates of a point from *each* half-plane into the inequality. The coordinates of the point in the shaded half-plane must satisfy the inequality, while the coordinates of the point in the unshaded half-plane *must not* satisfy it.

In Lesson 2.6, students learned how to graph absolute-value equations in a coordinate plane. In Exercises 72–75, they apply their work with linear inequalities to an investigation of the graphs of absolute-value inequalities.

54a. $2x + y \le 20$, where x and y are nonnegative integers

b.

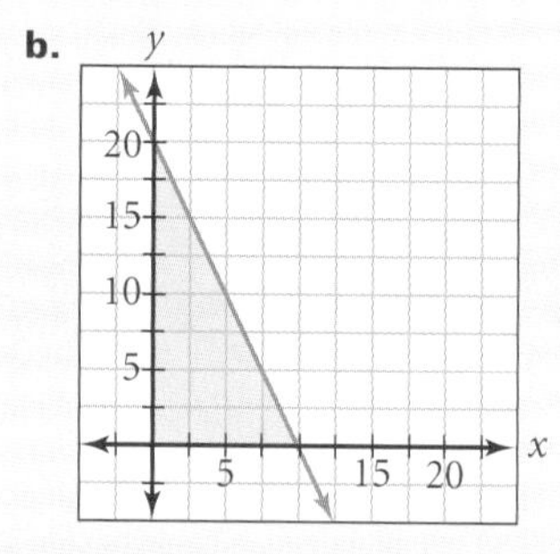

c. Answers may vary. sample answer: (0, 20), (0, 10), (5, 5)

55a. $0.95x + 1.25y \le 5.75$, where x and y are nonnegative integers

c.

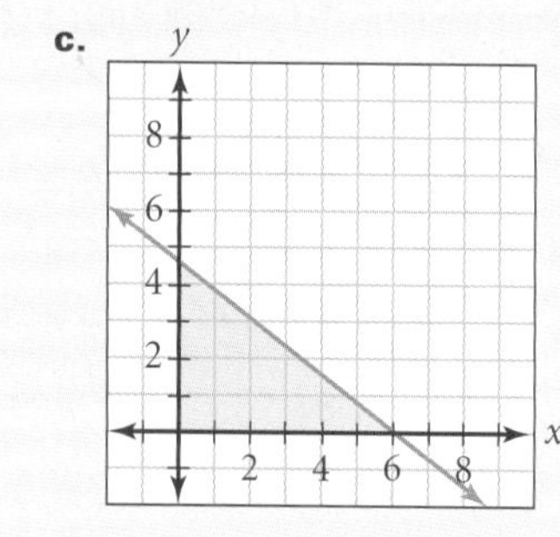

APPLICATIONS

54. SMALL BUSINESS A local bakery makes cakes for two special occasions: birthdays and holidays. A birthday cake requires 2 pounds of flour, and a holiday cake requires 1 pound of flour. The bakery currently has 20 pounds of flour available for making the two types of cakes.

a. Write an inequality to show the possible numbers of birthday and holiday cakes that the bakery can make. $\mathbf{2x + y \le 20}$

b. Graph the inequality.

c. Give three specific ordered pairs that satisfy the inequality.

55. CONSUMER ECONOMICS At the corner store, bags of popcorn cost \$0.95 and bags of peanuts cost \$1.25. Suppose that you want to buy x bags of popcorn and y bags of peanuts and that you have \$5.75.

a. Write an inequality to describe the number of bags of popcorn and the number of bags of peanuts you could buy. $\mathbf{0.95x + 1.25y \le 5.75}$

b. Solve the inequality for y. $\mathbf{y \le -0.76x + 4.6}$

c. Graph the inequality.

Look Back

For Exercises 56–59, y varies directly as x. Find the constant of variation and the direct-variation equation that relates the two variable. ***(LESSON 1.4)***

56. $x = 55$ when $y = 11$ $\mathbf{k = \frac{1}{5}; y = \frac{1}{5}x}$

57. $x = 4$ when $y = 32$ $\mathbf{k = 8; y = 8x}$

58. $x = -68$ when $y = \frac{1}{2}$ $\mathbf{k = -\frac{1}{136}; y = -\frac{1}{136}x}$

59. $x = \frac{1}{6}$ when $y = -45$ $\mathbf{k = -270; y = -270x}$

Solve each equation. ***(LESSON 1.6)***

60. $3x - 5 = 1 - 4x$ $\mathbf{\frac{6}{7}}$

61. $\frac{3}{2}x + 2 = 4 - 3x$ $\mathbf{\frac{4}{9}}$

62. $-3(x - 2) = -x$ **3**

Graph and classify each system as independent, dependent, or inconsistent. Then find the solution from the graph. ***(LESSON 3.1)***

63. $\begin{cases} x + 2y = -1 \\ 3x + 2y = 1 \end{cases}$ **indep. (1, −1)**

64. $\begin{cases} 2x - y = 3 \\ 3x + y = 7 \end{cases}$ **indep. (2, 1)**

65. $\begin{cases} 5x - 2y = 10 \\ x + 2y = 14 \end{cases}$ **indep. (4, 5)**

Use substitution to solve each system. ***(LESSON 3.1)***

66. $\begin{cases} y = -2x - 4 \\ y = 2x + 5 \end{cases}$ $\mathbf{\left(-\frac{9}{4}, \frac{1}{2}\right)}$

67. $\begin{cases} y = 8 - x \\ \frac{1}{2}y - x = \frac{5}{2} \end{cases}$ **(1, 7)**

68. $\begin{cases} y = 3x - 2 \\ 2y = 6x - 4 \end{cases}$ **infinitely many**

Use elimination to solve each system. ***(LESSON 3.2)***

69. $\begin{cases} 8x + 5y = 22 \\ 6x + 2y = 13 \end{cases}$ $\mathbf{\left(\frac{3}{2}, 2\right)}$

70. $\begin{cases} 2x + 2y = 12 \\ 7x - 4y = -13 \end{cases}$ **(1, 5)**

71. $\begin{cases} 2x - 3y = 3 \\ 4x + 2y = 14 \end{cases}$ **(3, 1)**

Look Beyond

Graph each inequality.

72. $|x + y| \le 5$ **73.** $|x - y| \le 5$ **74.** $|x| + |y| \le 5$ **75.** $2|x| + |y| \le 5$

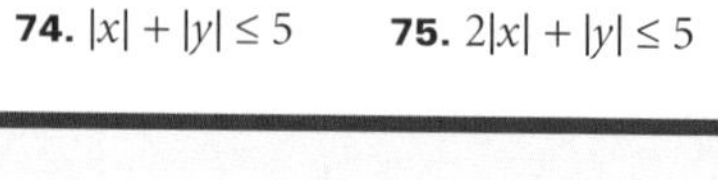

72.

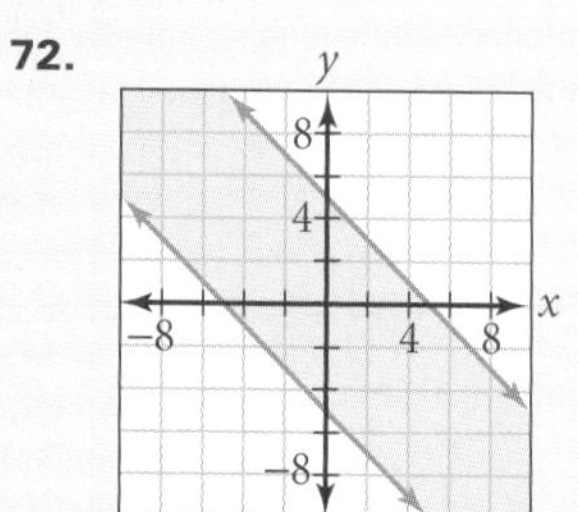

73.

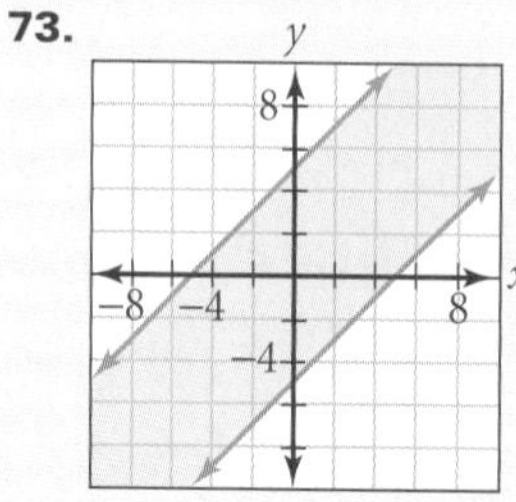

74.

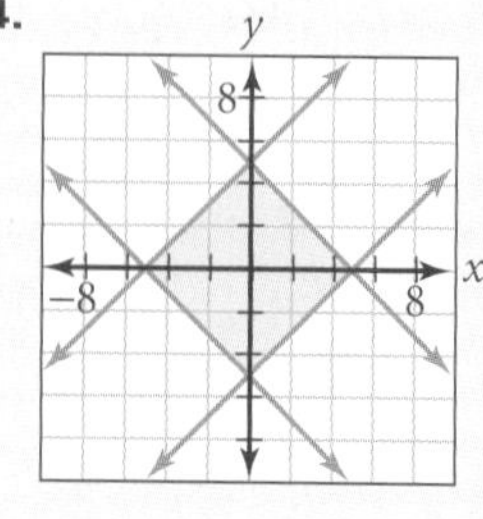

75.

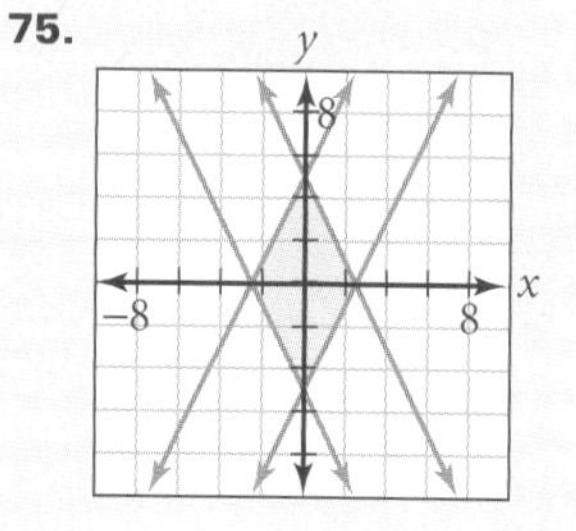

Systems of Linear Inequalities

Why *Systems of linear inequalities can be used to represent real-world situations such as the height and weight criteria that are needed for an actor.*

Objectives

- Write and graph a system of linear inequalities in two variables.
- Write a system of linear inequalities in two variables for a given solution region.

APPLICATION
DRAMA

Suppose that an actor is needed for the lead character in a play. The search is on for a male actor between the ages of 25 and 35, standing between 5' 8" and 5' 10" tall, with a medium build, and within the weight ranges given below for males with medium builds.

Height (in.)	Weight (lb)
66	139–151
67	142–154
68	145–157
69	148–160
70	151–163

[*Source: Taber's Cyclopedic Medical Dictionary, 15th Edition*]

You can represent the height and weight criteria for this actor with the *system of linear inequalities* below.

$$\begin{cases} 68 \le h \le 70 \\ w \ge 3h - 59 \\ w \le 3h - 47 \end{cases}$$

The system of linear inequalities can be graphed as shown at right. The dark blue shaded region indicates the range of acceptable heights and weights for the lead actor.

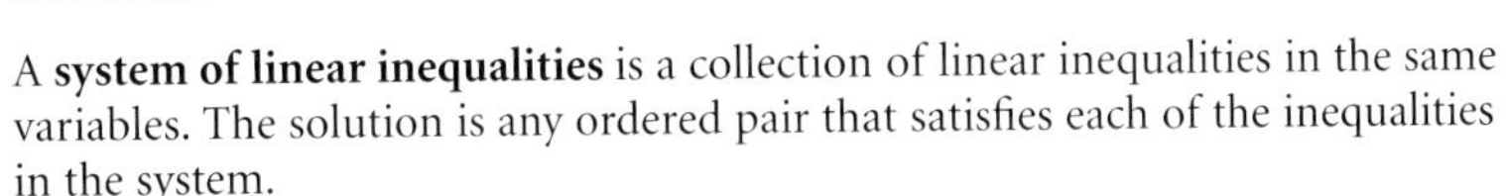

A **system of linear inequalities** is a collection of linear inequalities in the same variables. The solution is any ordered pair that satisfies each of the inequalities in the system.

To graph a system of linear inequalities, shade the part of the plane that is the intersection of all of the individual solution regions.

Prepare

NCTM PRINCIPLES & STANDARDS 1–4, 6, 8–10

QUICK WARM-UP

1. Graph $y < -x + 1$.

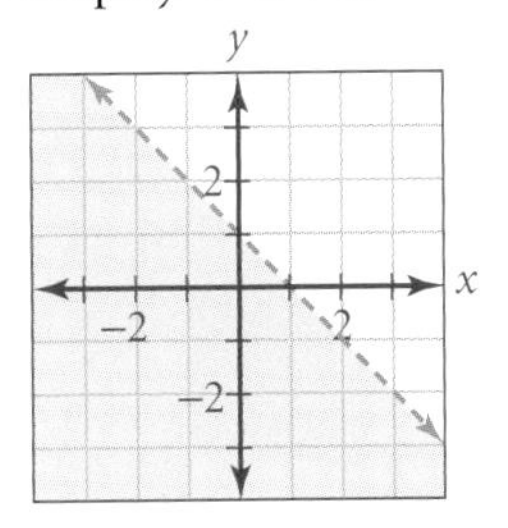

2. Graph the system.
$$\begin{cases} y = x + 2 \\ y = -2x - 1 \end{cases}$$

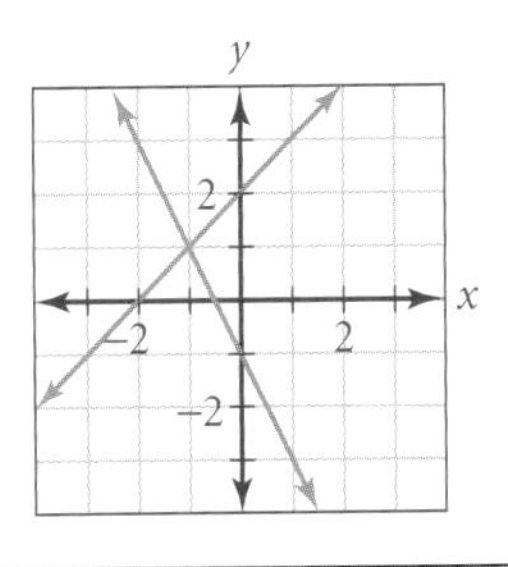

Also on Quiz Transparency 3.4

Teach

Why Ask students to compare the graph of $x > -2$ on a number line with its graph in a coordinate plane. Then show the number-line graph of $-2 < x < 3$. Ask them to make a conjecture about the graph of this inequality in a coordinate plane.

Alternative Teaching Strategy

GUIDED ANALYSIS Display the graph at right. Ask students to describe all points that lie either in the shaded region or on its boundary. Elicit the facts that the x- and y-coordinates are all nonnegative and that their sum is 4 or less. Lead students to see that these conditions are represented by the system at right.

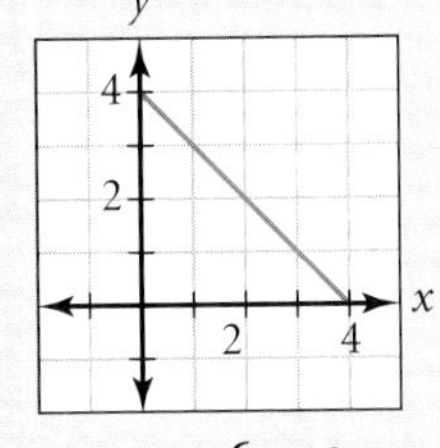

$$\begin{cases} x \ge 0 \\ x \ge 0 \\ x + y \le 4 \end{cases}$$

Inclusion Strategies

VISUAL LEARNERS Have students use a different color to graph each inequality in a system. Then they can see the graph of the solution as the region where a new color is formed. An example is shown at right.

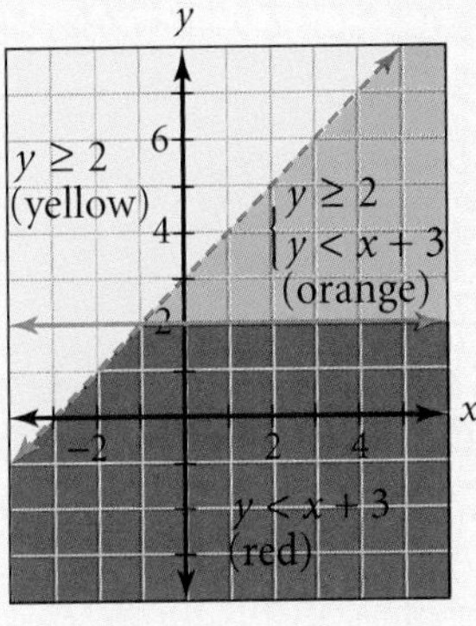

ADDITIONAL EXAMPLE 1

Graph the system.

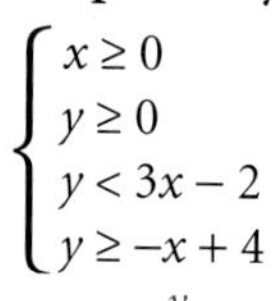

$$\begin{cases} x \geq 0 \\ y \geq 0 \\ y < 3x - 2 \\ y \geq -x + 4 \end{cases}$$

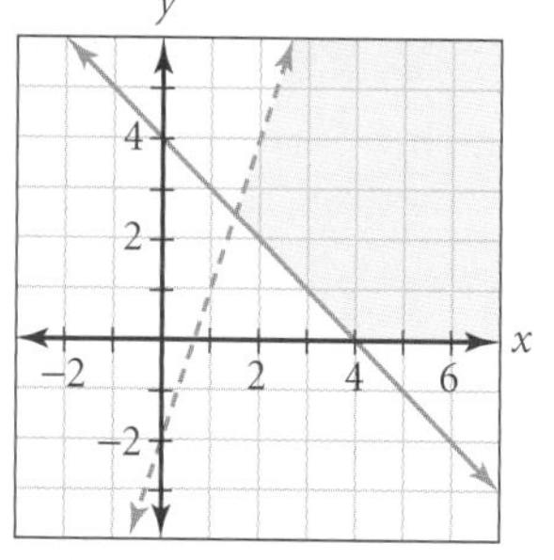

TRY THIS

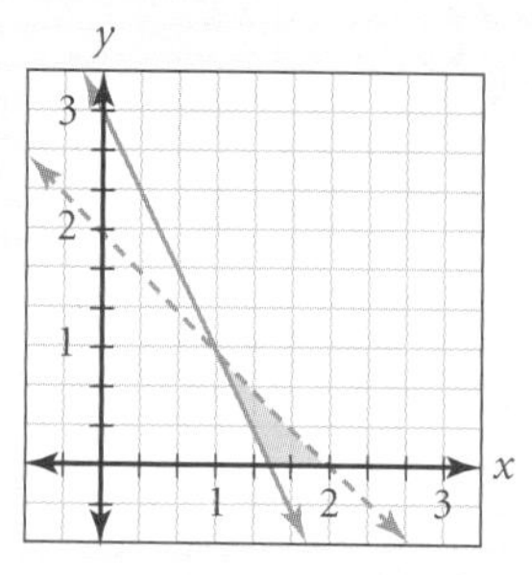

ADDITIONAL EXAMPLE 2

Graph the system.

$$\begin{cases} y \leq -2x - 2 \\ y > \frac{1}{2}x - 1 \\ x \geq -3 \end{cases}$$

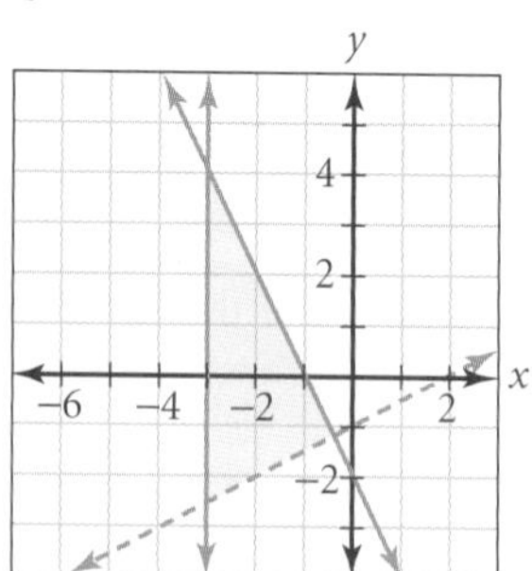

EXAMPLE 1 Graph the system. $\begin{cases} x \geq 0 \\ y \geq 0 \\ y > -2x + 5 \\ y \leq 3x + 1 \end{cases}$

SOLUTION

The inequalities $x \geq 0$ and $y \geq 0$ tell you that the final solution region is in the first quadrant of a coordinate plane and may include points on the positive axes.

Graph $y > -2x + 5$ in the first quadrant. Use a dashed boundary line.

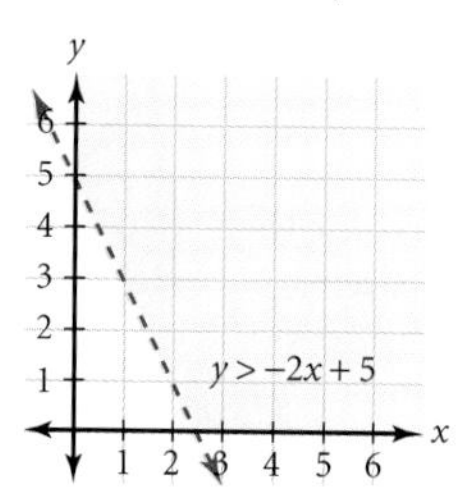

Graph $y \leq 3x + 1$ in the first quadrant. Use a solid boundary line.

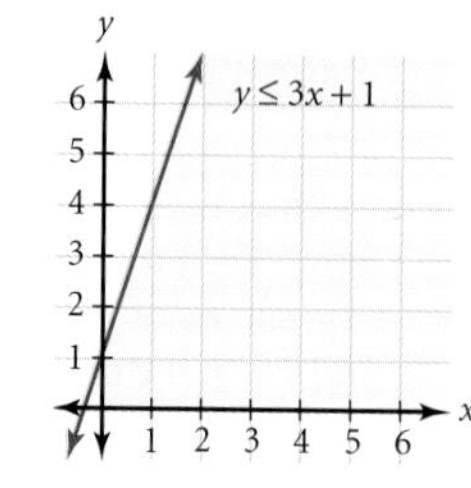

The solution is the intersection of these two regions.

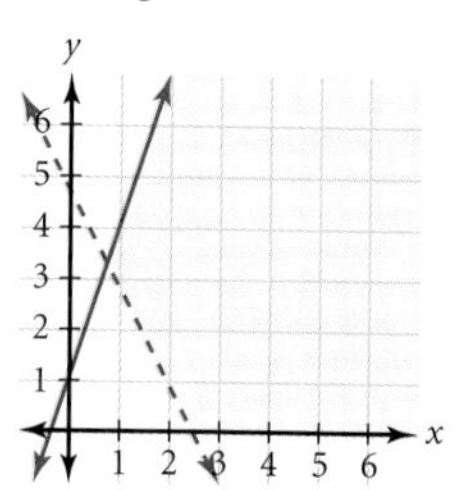

TRY THIS Graph the system. $\begin{cases} x \geq 0 \\ y \geq 0 \\ y < -x + 2 \\ y \geq -2x + 3 \end{cases}$

EXAMPLE 2 Graph the system. $\begin{cases} y \geq -x - 1 \\ y \leq 2x + 1 \\ x < 1 \end{cases}$

SOLUTION

Graph $y \geq -x - 1$. Use a solid boundary line.

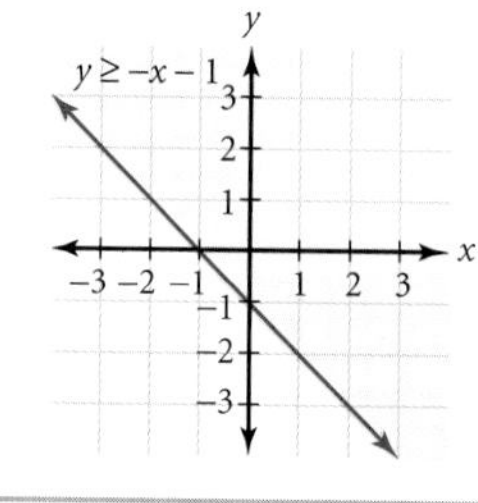

Graph $y \leq 2x + 1$. Use a solid boundary line.

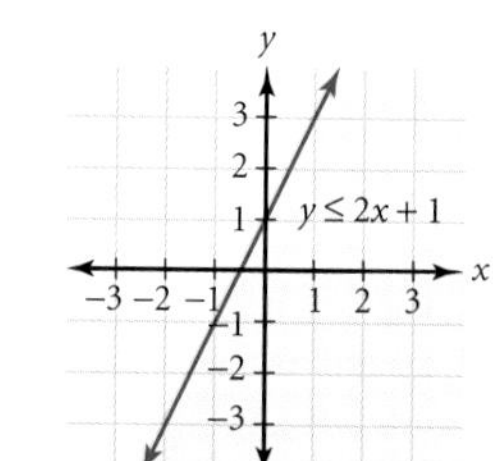

Graph the intersection of $y \geq -x - 1$ and $y \leq 2x + 1$ with $x < 1$.

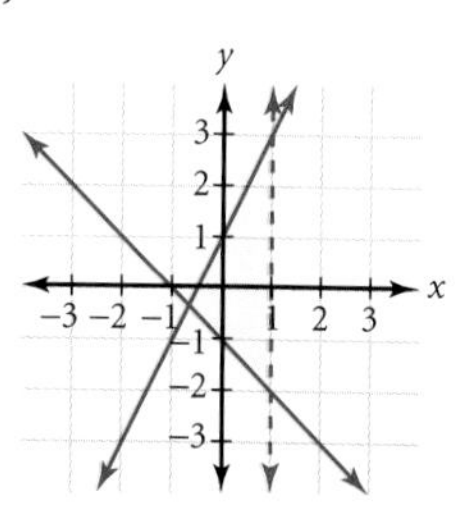

TRY THIS Graph the system. $\begin{cases} y > -x - 2 \\ y > x + 3 \\ y \leq 3 \end{cases}$

CRITICAL THINKING Find the coordinates of the vertices of the triangle graphed in Example 2. Do the coordinates of all three vertices satisfy the system? Justify your response.

Interdisciplinary Connection

CONSUMER EDUCATION Charts like the one at right appear on the packages of many articles of clothing. They show the range of heights and corresponding weights for which the given size is appropriate. This chart is for the medium size of a certain brand of women's clothing.

Let x represent height in inches and let y represent weight in pounds. Write a system of inequalities to represent the shaded region.

$$\begin{cases} y \leq 5x - 155 \\ y \leq 5x - 215 \\ y \leq -5x + 445 \\ y \leq -5x + 485 \end{cases}$$

Women's medium size

Weight (pounds): 115, 120, 125, 130, 135, 140, 145, 150, 155, 160, 165

Height (inches): 60, 61, 62, 63, 64, 65, 66, 67, 68, 69, 70

In a coordinate plane, the graph of a compound inequality in one variable can be a vertical or horizontal strip, as shown in Example 3.

EXAMPLE 3 **Graph each inequality in a coordinate plane.**

a. $-3 \le x \le 2$ **b.** $-1 < y < 3$

SOLUTION

a. The solution is the set of all ordered pairs whose x-coordinate is between -3 and 2 inclusive. The solution is a vertical strip.

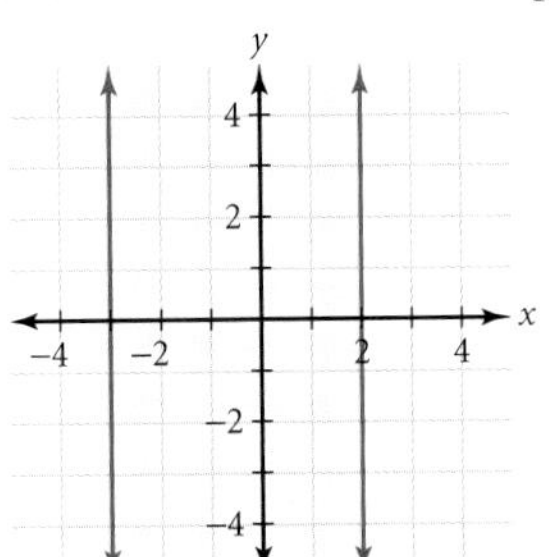

b. The solution is the set of all ordered pairs whose y-coordinate is between -1 and 3. The solution is a horizontal strip.

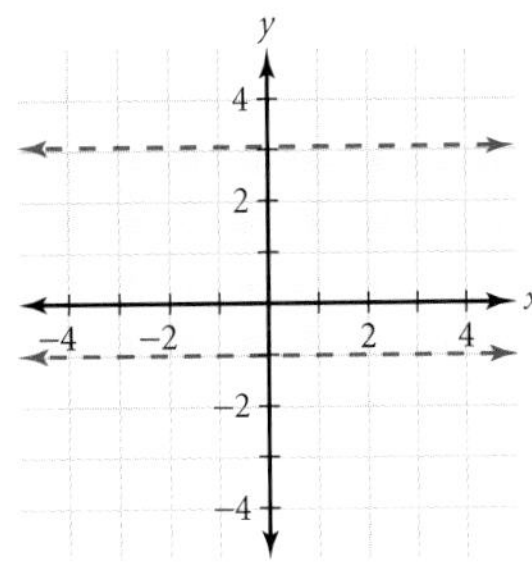

TRY THIS Graph each inequality in a coordinate plane.

a. $1 \le y < 4$ **b.** $0 < x \le 4$

CRITICAL THINKING In a coordinate plane, is the graph of a compound inequality in one variable always a vertical or horizontal strip? Explain.

EXAMPLE 4 **Write the system of inequalities graphed at right.**

SOLUTION

1. First find equations for the boundary lines.

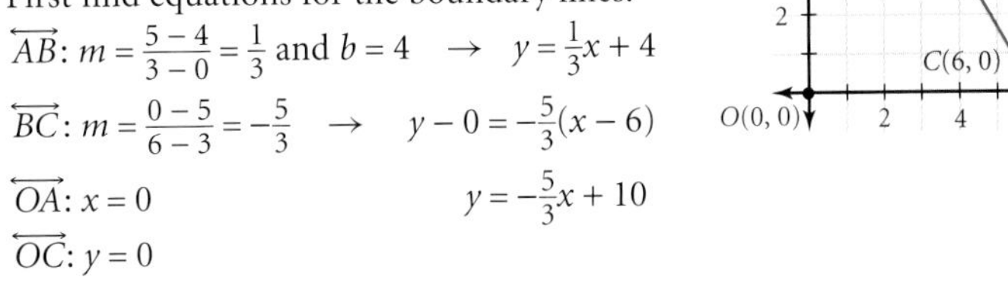

$\overleftrightarrow{AB}$: $m = \frac{5-4}{3-0} = \frac{1}{3}$ and $b = 4$ $\rightarrow$ $y = \frac{1}{3}x + 4$

$\overleftrightarrow{BC}$: $m = \frac{0-5}{6-3} = -\frac{5}{3}$ $\rightarrow$ $y - 0 = -\frac{5}{3}(x - 6)$

$y = -\frac{5}{3}x + 10$

$\overleftrightarrow{OA}$: $x = 0$

$\overleftrightarrow{OC}$: $y = 0$

2. Give each boundary line the appropriate inequality symbol. Because each boundary is solid, use $\ge$ or $\le$.

below $\overleftrightarrow{AB}$: $y \le \frac{1}{3}x + 4$ right of $\overleftrightarrow{OA}$: $x \ge 0$

below $\overleftrightarrow{BC}$: $y \le -\frac{5}{3}x + 10$ above $\overleftrightarrow{OC}$: $y \ge 0$

The system of inequalities is $\begin{cases} y \le \frac{1}{3}x + 4 \\ y \le -\frac{5}{3}x + 10 \\ x \le 0 \\ y \ge 0 \end{cases}$.

TRY THIS

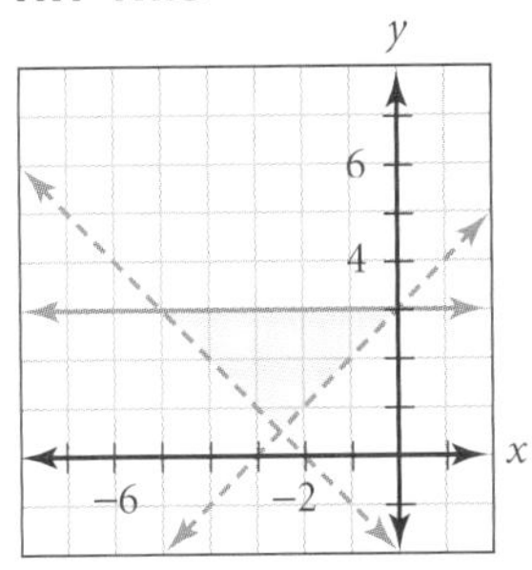

CRITICAL THINKING

(1, 3), (1, −2), and $\left(-\frac{2}{3}, -\frac{1}{3}\right)$; no; the x-coordinate of (1, 3) is not less than 1.

ADDITIONAL EXAMPLE 3

Graph each inequality in a coordinate plane.

a. $-4 \le x \le -1$

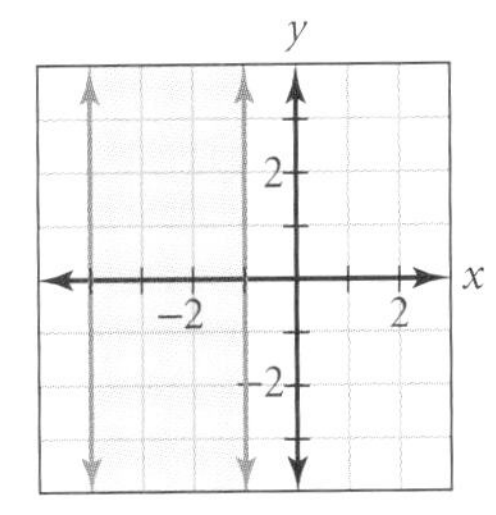

b. $-4 < y < 2$

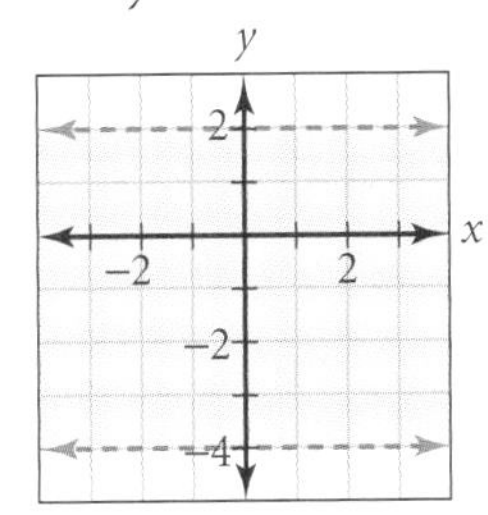

Enrichment

Have students work in pairs to graph the solution of this system.

$\begin{cases} |x| + |y| \ge 2 \\ |x| + |y| \le 4 \end{cases}$

See graph at right.

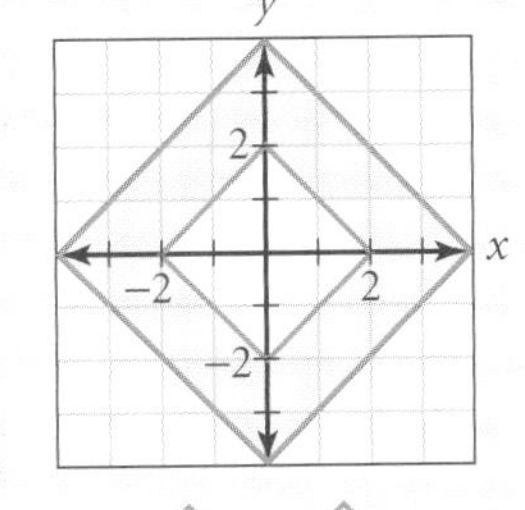

Then challenge them to create a system whose solution looks like the "figure 8" at right. **Answers will vary.**

Reteaching the Lesson

COOPERATIVE LEARNING

Give students the graph at right. Do not write any units on the axes. Have students work in pairs to create a system that might describe this graph. A sample response is given at right. Repeat the activity with several other graphs of systems.

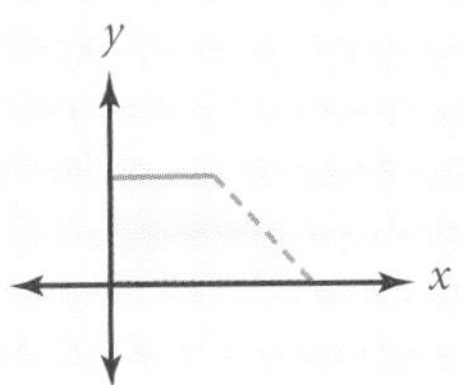

sample:

$\begin{cases} x \ge 0 \\ y \ge 0 \\ y \le 4 \\ y < -x + 8 \end{cases}$

☞ For answers to Try This and Checkpoint and for Additional Example 4, see page 182.

TRY THIS

a.

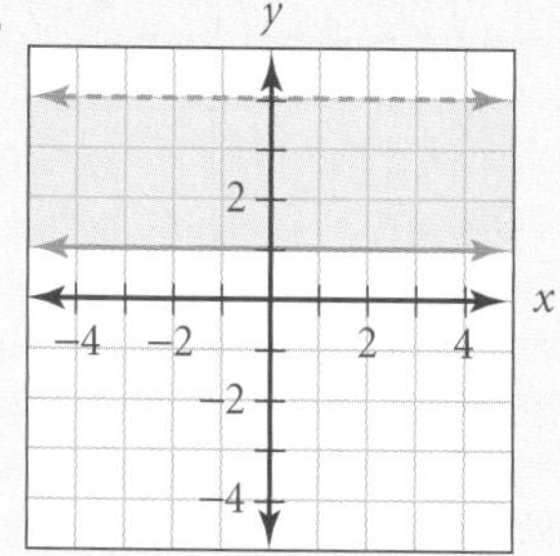

b.

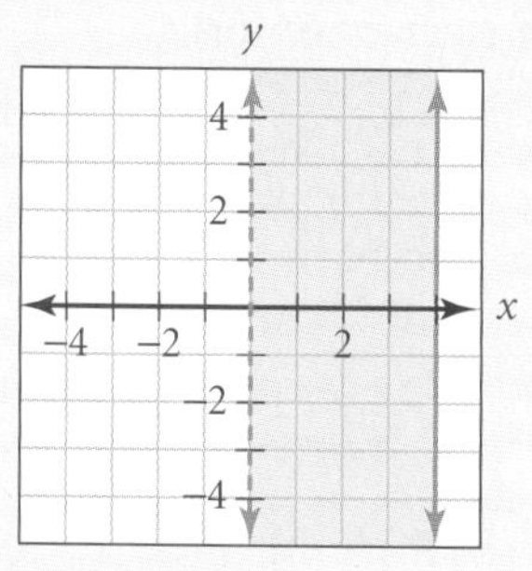

CRITICAL THINKING

No; for example, the graph of $1 \le |x| \le 2$ is 2 vertical strips.

ADDITIONAL EXAMPLE 4

Write a system of inequalities to represent the shaded region in the graph below.

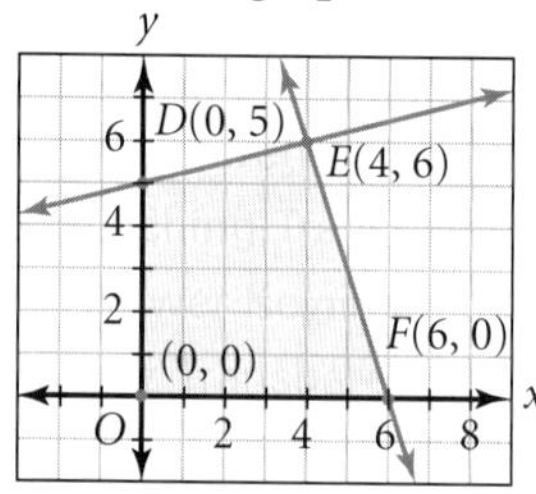

$$\begin{cases} y \le \frac{1}{4}x + 5 \\ y \le -3x + 18 \\ y \ge 0 \\ y \ge 0 \end{cases}$$

Activity Notes

In this Activity, students investigate a system of inequalities that defines the temperature/humidity "comfort zone" for the average individual. Students should observe that as the relative humidity increases, the temperatures at which a person feels comfortable decrease.

APPLICATION
HEALTH

Activity
Exploring the Comfort Zone

You will need: no special materials

The region labeled "Comfort Zone" in the graph at right shows the temperatures and relative humidity levels at which the average person feels comfortable.

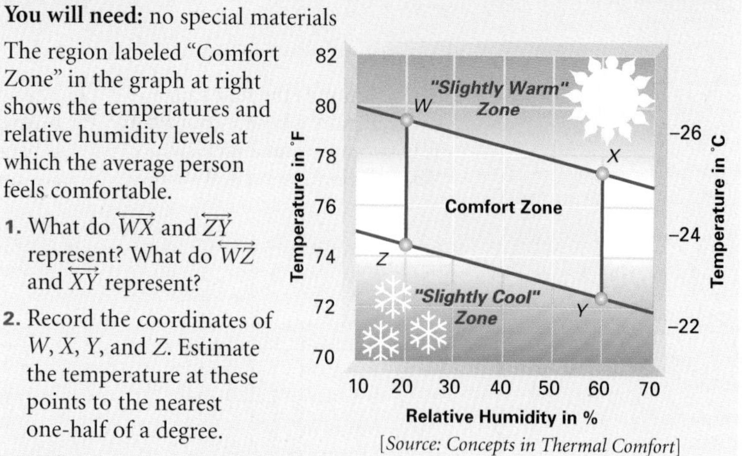

[Source: Concepts in Thermal Comfort]

1. What do $\overleftrightarrow{WX}$ and $\overleftrightarrow{ZY}$ represent? What do $\overleftrightarrow{WZ}$ and $\overleftrightarrow{XY}$ represent?
2. Record the coordinates of W, X, Y, and Z. Estimate the temperature at these points to the nearest one-half of a degree.
3. Write a system of linear inequalities that represents the comfort zone.

CHECKPOINT ✔

4. What can you say about the temperature in the comfort zone as the relative humidity increases? Explain how this relates to the slope of one of the boundary lines.

Exercises

Communicate

1. Describe when to use a solid line and when to use a dashed line to graph of a system of linear inequalities.
2. If the coordinates of a point above a boundary line make the inequality false, which side of the boundary line should be shaded?
3. Describe a system of linear inequalities that has no solution.

Guided Skills Practice

Graph each system. ***(EXAMPLES 1 AND 2)***

4. $\begin{cases} x \ge 0 \\ y \ge 0 \\ y \ge -x + 4 \\ y > x \end{cases}$

5. $\begin{cases} y \ge x \\ y \ge -x + 1 \\ y < 2 \end{cases}$

4.

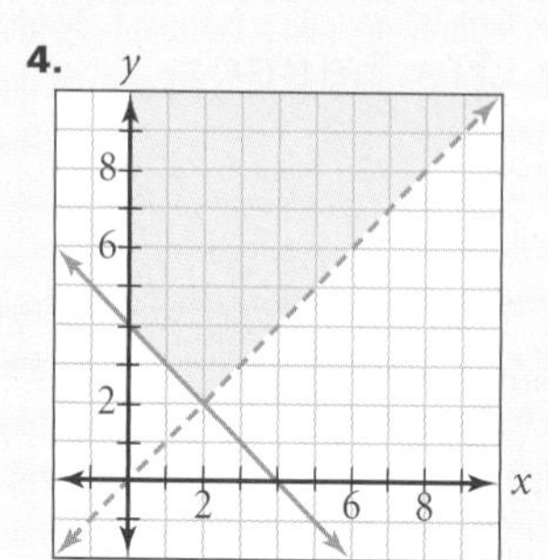

5.

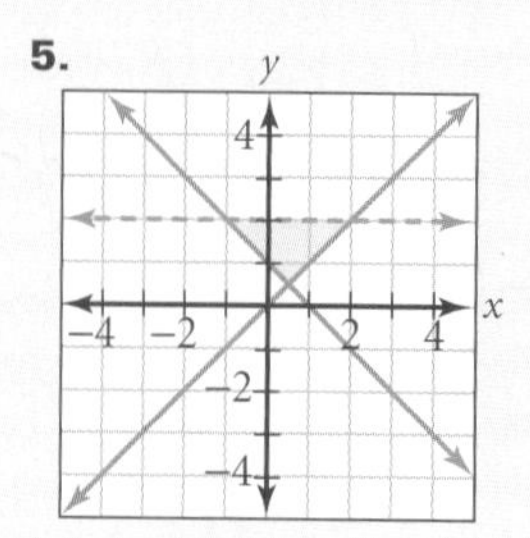

Graph each inequality in a coordinate plane. ***(EXAMPLE 3)***

6. $2 < x < 4$

7. $-1 \le y \le 5$

8. Write the system of inequalities graphed at right. ***(EXAMPLE 4)***

$$\begin{cases} x \ge 0,\ y \ge 0 \\ y \le \frac{2}{5}x + 2 \\ y \le -4x + 24 \end{cases}$$

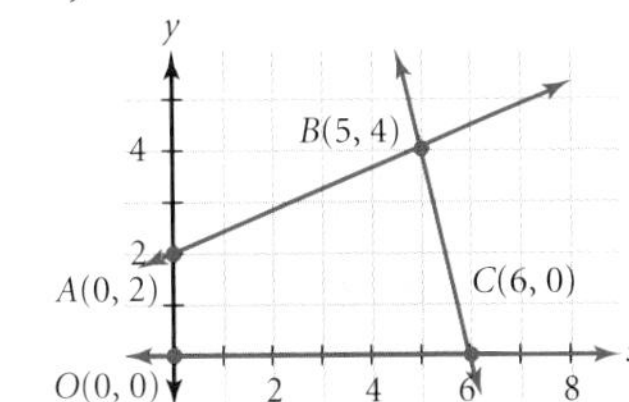

Practice and Apply

Graph each system of linear inequalities.

9. $\begin{cases} y \ge 2 \\ y < x + 1 \end{cases}$ **10.** $\begin{cases} x < 3 \\ y \le 2x + 2 \end{cases}$ **11.** $\begin{cases} y < 3x - 4 \\ y \ge 6 - x \end{cases}$

12. $\begin{cases} y \le 3 - x \\ y \ge x - 5 \end{cases}$ **13.** $\begin{cases} x \ge 0 \\ y \ge 0 \\ y > 2x + 1 \end{cases}$ **14.** $\begin{cases} x \ge 0 \\ y \le 0 \\ y < -x \end{cases}$

15. $\begin{cases} x \ge 0 \\ y \le 0 \\ y > 2x - 5 \end{cases}$ **16.** $\begin{cases} x \ge 0 \\ y \ge -1 \\ y < -2x + 3 \end{cases}$ **17.** $\begin{cases} y \ge 2x - 1 \\ x > 1 \\ y < 5 \end{cases}$

18. $\begin{cases} x \ge 0 \\ y \ge 1 \\ y \le 5 - x \end{cases}$ **19.** $\begin{cases} y < x - 1 \\ y + 2x < 3 \\ y \ge -1 \end{cases}$ **20.** $\begin{cases} y + 2x \ge 0 \\ y \ge 2x - 4 \\ y \le 3 \end{cases}$

Graph each compound inequality in a coordinate plane.

21. $-5 < y < 1$ **22.** $-1 \le y \le 3$ **23.** $2 \le x \le 8$

24. $-2 < x < 3$ **25.** $0 < y < 4$ **26.** $-6 \le y \le -2$

27. $-5 \le x < -1$ **28.** $-\frac{2}{3} < x \le \frac{1}{3}$ **29.** $-\frac{1}{4} \le y < -\frac{1}{5}$

30. $-4.4 < y \le -4$ **31.** $-1.5 \le x < -0.5$ **32.** $-5.5 < x \le -5.1$

Write the system of inequalities whose solution is graphed. Assume that each vertex has integer coordinates.

33.

34.

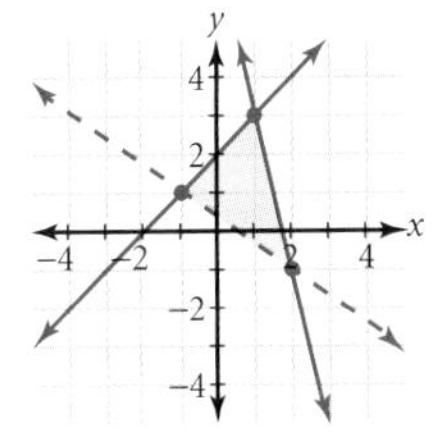

35.

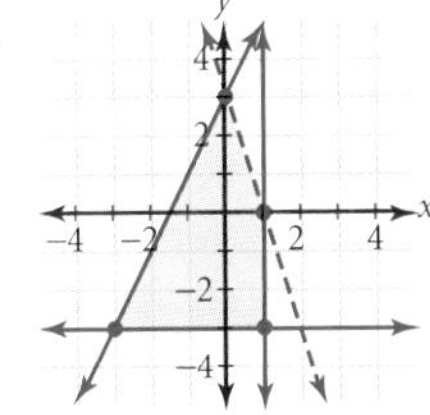

36.

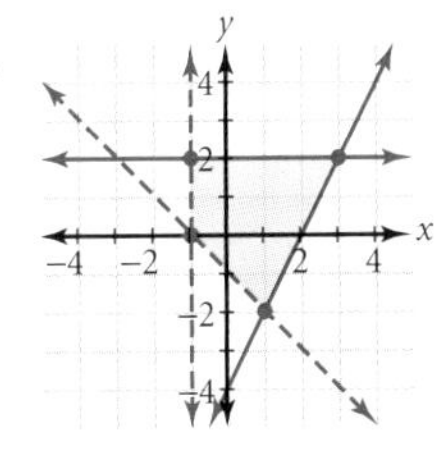

33. $\begin{cases} y > -2 \\ y \le 2x \\ y \le -2x + 4 \end{cases}$

34. $\begin{cases} y > -\frac{2}{3}x + \frac{1}{3} \\ y \le -4x + 7 \\ y \le x + 2 \end{cases}$

35. $\begin{cases} x \le 1 \\ y < -3x + 3 \\ y \le 2x + 3 \\ y \ge -3 \end{cases}$

36. $\begin{cases} x > -1 \\ y \le 2 \\ y > -x - 1 \\ y \ge 2x - 4 \end{cases}$

CHECKPOINT ✔

4. As relative humidity increases, the comfort zone temperature boundaries decrease. This inverse relationship is why the slopes are negative.

Assess

Selected Answers

Exercises 4–8, 9–67 odd

ASSIGNMENT GUIDE	
In Class	1–8
Core	9–45 odd, 47, 51
Core Plus	10–36 even, 37–52
Review	53–67
Preview	68–70

Extra Practice can be found beginning on page 940.

6.

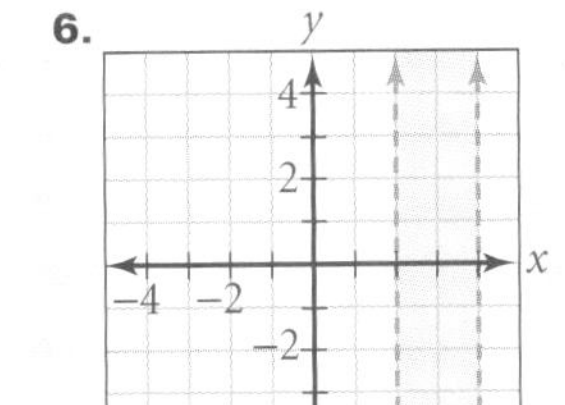

7.

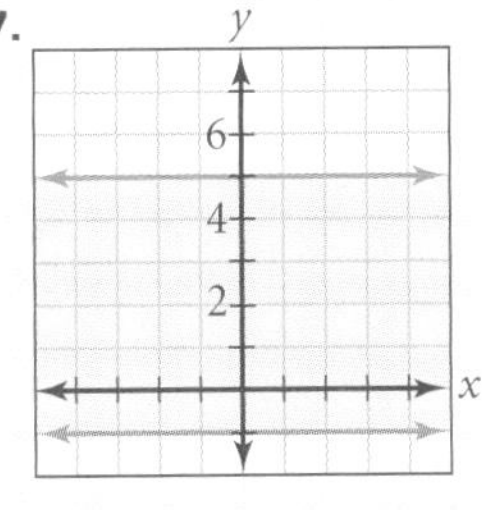

9.

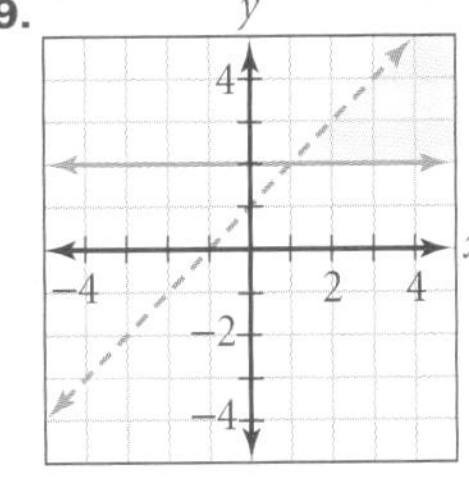

10.

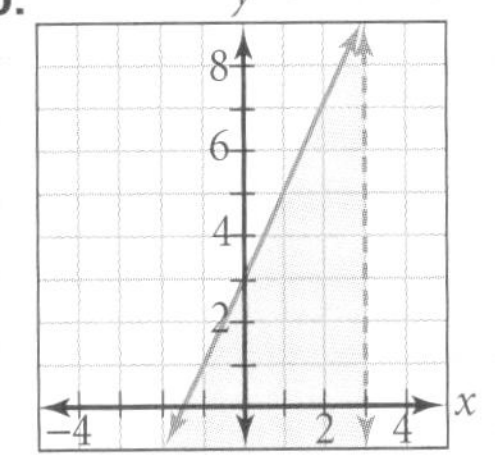

11.

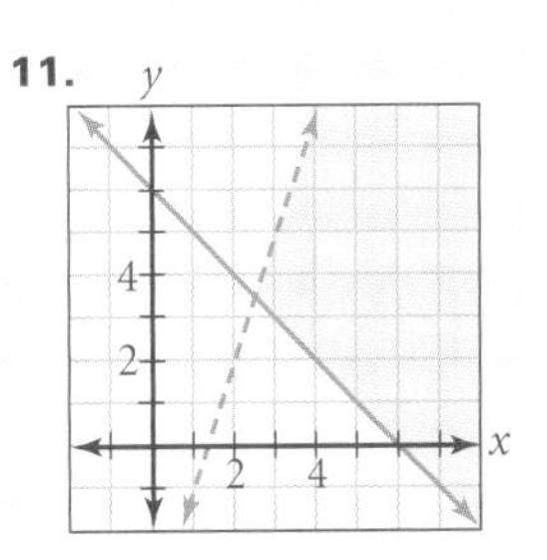

The answers to Exercises 12–32 can be found in Additional Answers beginning on page 1002.

Error Analysis

When graphing systems of inequalities, students often focus so intently on shading the correct region that they forget to consider whether the boundary lines should be dashed or solid. Suggest that they begin each exercise by writing the words "dashes? shading?" next to the exercise number. This simple note may help them remember that there are two types of decisions that affect the appearance of the graph.

37.

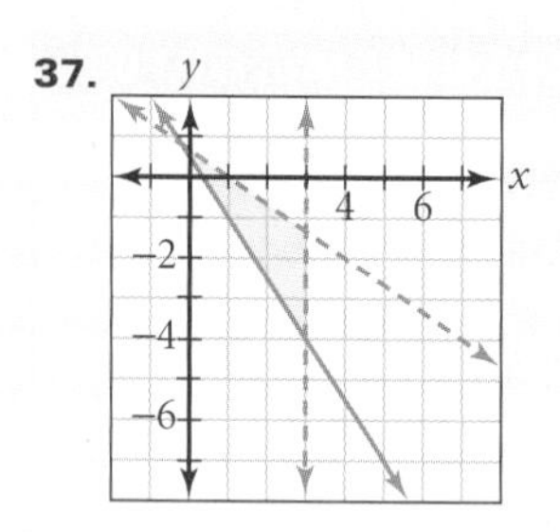

38.

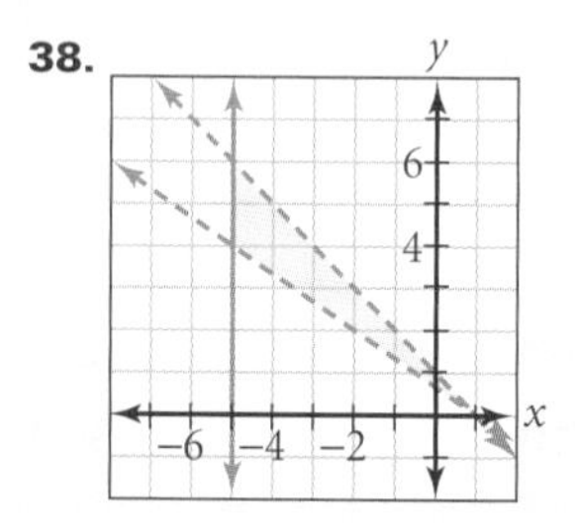

Graph each system of linear inequalities.

37. $\begin{cases} 3x + 2y \geq 1 \\ 2x + 3y < 2 \\ x < 3 \end{cases}$ **38.** $\begin{cases} x + y < 1 \\ 2x + 3y > 2 \\ x \geq -5 \end{cases}$ **39.** $\begin{cases} x + \frac{1}{2}y \leq 2 \\ 2x + 3y < 2 \end{cases}$

40. $\begin{cases} 2x + y \geq 2 \\ y \geq 3x + 2 \end{cases}$ **41.** $\begin{cases} x + y \leq 4 \\ 2x \leq y \end{cases}$ **42.** $\begin{cases} 2x - 2y < 1 \\ x + 2y \geq 2 \end{cases}$

43. $\begin{cases} 2x - y \leq 16 \\ x + y \leq 10 \\ x \geq 0 \\ y \geq 0 \end{cases}$ **44.** $\begin{cases} 3x - y \leq 15 \\ x + 2y \leq 10 \\ x \geq 0 \\ y \geq 0 \end{cases}$ **45.** $\begin{cases} 3x - 2y \geq 4 \\ x + y > 4 \\ x - y \leq 7 \\ x \geq 0 \\ y \geq 0 \end{cases}$

CONNECTIONS

46. GEOMETRY A parallelogram is a quadrilateral whose opposite sides are parallel. Create a graph of a parallelogram on a coordinate plane. Write a system of inequalities that represents the parallelogram and its interior.

CHALLENGE

47. GEOMETRY Classify the solution to $\begin{cases} y \leq a \\ y \geq |x| \end{cases}$, for $a > 0$, as a geometric figure.

APPLICATIONS

48. MANUFACTURING A small-appliance manufacturing company makes standard and deluxe models of a toaster oven. The company can make up to 200 ovens per week. The standard model costs \$20 to produce, and the deluxe model costs \$30 to produce. The company has budgeted no more than \$3600 per week to produce the ovens.

a. Let x represent the number of standard models, and let y represent the number of deluxe models. Write a system of linear inequalities to represent the possible combinations of standard and deluxe models that the company can make in one week.

b. Graph the system of inequalities.

c. Due to an increase in the rental costs for the company's factory, the company can budget no more that \$3000 per week to make the toaster ovens. Explain how this changes the possible combinations of standard and deluxe models that the company can make in one week.

48a. $\begin{cases} x + y \leq 200 \\ 20x + 30y \leq 3600 \\ x \geq 0 \\ y \geq 0 \end{cases}$

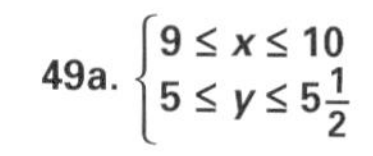

49a. $\begin{cases} 9 \leq x \leq 10 \\ 5 \leq y \leq 5\frac{1}{2} \end{cases}$

49. CRIMINOLOGY Officer Cheek is trying to solve a crime that was committed by a man with a shoe size from 9 to 10. According to witnesses, the man's height is between 5 feet and 5 feet 6 inches inclusive.

a. Let x represent the shoe size, and let y represent the height. Write a system of inequalities to represent the given information.

b. Graph the system of inequalities.

39.

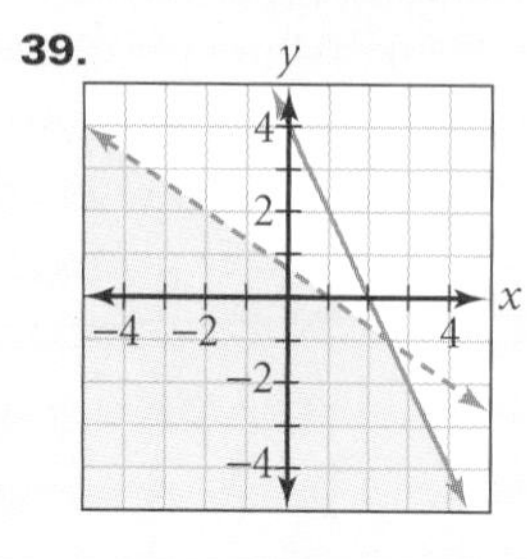

40.

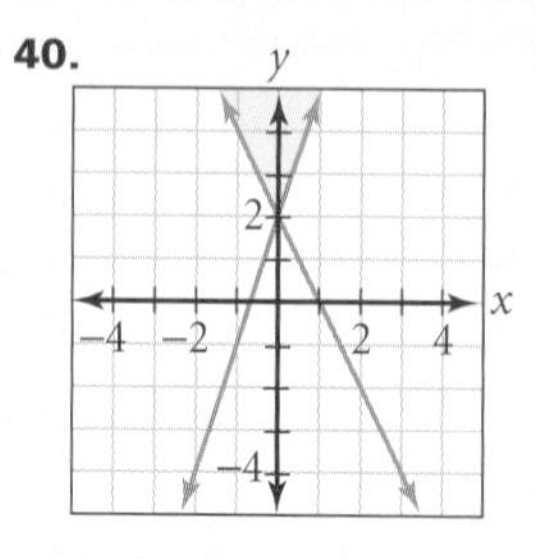

41.

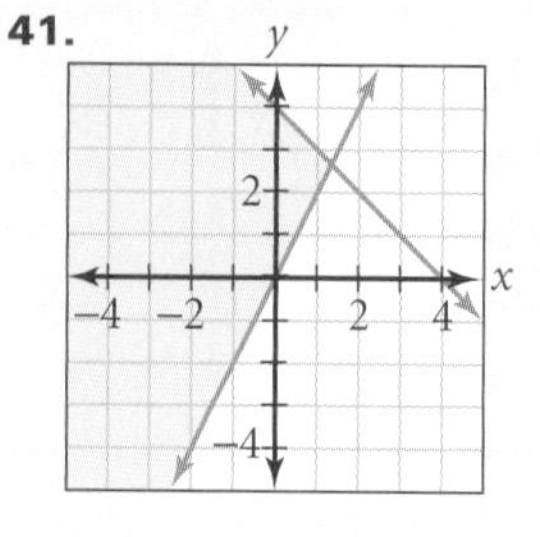

42.

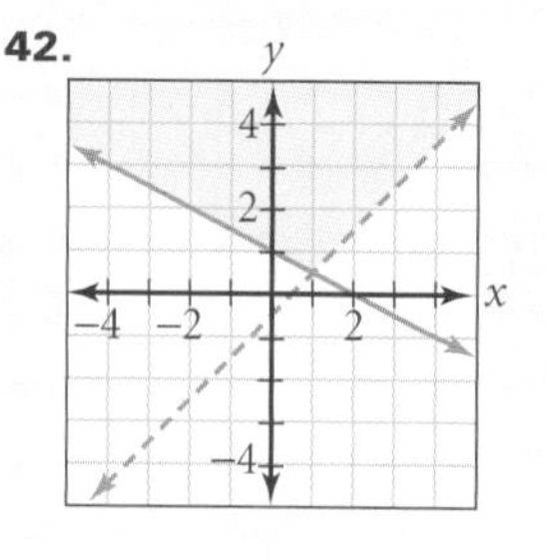

Practice

NAME ______ CLASS ______ DATE ______

Practice
3.4 Systems of Linear Inequalities

Graph each system of linear inequalities.

1. $\begin{cases} y \leq -x \\ y \geq 2x - 4 \end{cases}$ 2. $\begin{cases} y > -3x - 3 \\ y \leq \frac{1}{3}x - 1 \end{cases}$ 3. $\begin{cases} y \leq 4 \\ y \geq -3 \\ y > 3x - 2 \end{cases}$

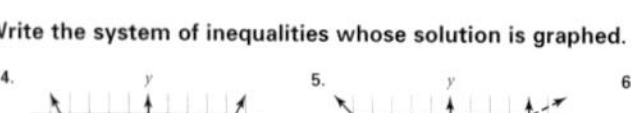

Write the system of inequalities whose solution is graphed.

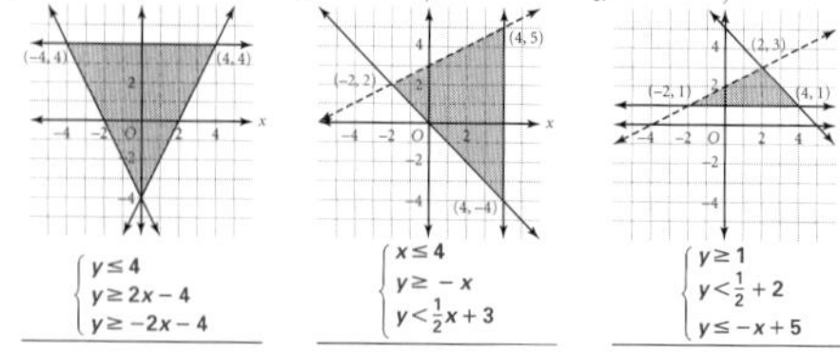

4. $\begin{cases} y \leq 4 \\ y \geq 2x - 4 \\ y \geq -2x - 4 \end{cases}$ 5. $\begin{cases} x \leq 4 \\ y \geq -x \\ y < \frac{1}{4}x + 3 \end{cases}$ 6. $\begin{cases} y \geq 1 \\ y < \frac{1}{2} + 2 \\ y \leq -x + 5 \end{cases}$

7. During the summer, Ryan works 30 hours or less per week mowing lawns and delivering newspapers. He earns \$6 per hour mowing lawns and \$7 per hour delivering papers. Ryan would like to earn at least \$126 per week. Let x be the number of hours mowing, and let y be the number of hours delivering papers. Write a system of inequalities to represent the possible hours and jobs that Ryan can work, and graph this system at right.

$x + y \leq 30$; $6x + 7y \geq 126$

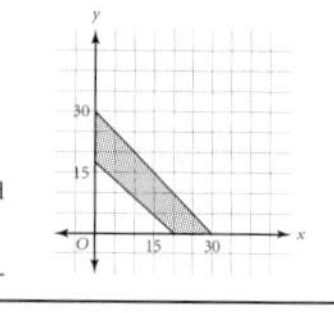

APPLICATIONS

50a. $\begin{cases} x + y \leq 40 \\ 20x + 10y \geq 500 \\ x \geq 0 \\ y \geq 0 \end{cases}$

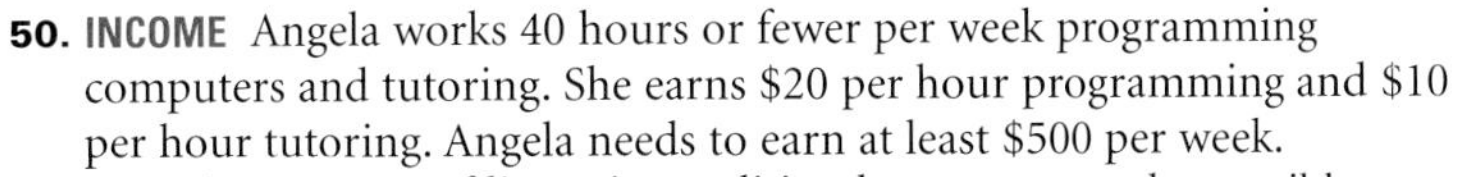

50. INCOME Angela works 40 hours or fewer per week programming computers and tutoring. She earns \$20 per hour programming and \$10 per hour tutoring. Angela needs to earn at least \$500 per week.

a. Write a system of linear inequalities that represents the possible combinations of hours spent tutoring and hours spent programming that will meet Angela's needs.

b. Graph the system of linear inequalities. Is the solution a polygon?

c. Find a point that is a solution to the system of linear inequalities. What are the coordinates of this point and what do the coordinates of this point represent?

d. Which point in the solution region represents the best way for Angela to spend her time? Explain why you think this is the best solution.

51a. $\begin{cases} x + y \leq 5000 \\ x \leq 3000 \\ y \leq 4000 \\ x \geq 0 \\ y \geq 0 \end{cases}$

51. ENTERTAINMENT A ticket office sells reserved tickets and general-admission tickets to a rock concert. The auditorium normally holds no more than 5000 people. There can be no more than 3000 reserved tickets and no more than 4000 general-admission tickets sold.

a. Let x represent the number of reserved tickets, and let y represent the number of general-admission tickets. Write a system of three linear inequalities to represent the possible combinations of reserved and general-admission tickets that can be sold. (Note that x and y must be non-negative integers.)

b. Graph the system of inequalities.

c. In order to increase the number of people that come to the concert, the auditorium increases its seating capacity to 5500. Explain how this addition changes the possible ticket combinations.

52a. $\begin{cases} x + y \geq 45 \\ 50x + 40y \geq 2000 \\ x \geq 0,\ y \geq 0 \end{cases}$

c. 30

52. BUSINESS A lawn and garden store sells gas-powered and electric hedge clippers. The store wants to sell at least 45 hedge clippers per month. The profit on each electric model is \$50, and the profit on each gas-powered model is \$40. The shop wants to earn at least \$2000 per month on the sale of hedge clippers.

a. Let x represent the number of electric hedge clippers, and let y represent the number of gas-powered hedge clippers. Write a system of linear inequalities to represent the possible combinations of each model sold.

b. Graph the system of inequalities.

c. If the shop sells 16 electric hedge clippers in one month, what is the minimum number of gas-powered hedge clippers that must be sold that month in order to meet their goals?

Look Back

53. Use intercepts to graph $5x - 3y = 7$. *(LESSON 1.2)*

54. Write the equation in slope-intercept form for the line that contains the point $(4, -3)$ and is parallel to $2x - 4y = 26$. *(LESSON 1.3)* $y = \frac{1}{2}x - 5$

Solve each proportion for the variable. Check your answers. *(LESSON 1.4)*

55. $\frac{3x}{2} = \frac{15}{12}$ $\frac{5}{6}$

56. $\frac{1 - 2x}{8} = \frac{7}{4}$ $-\frac{13}{2}$

57. $\frac{3x + 5}{6} = \frac{x - 2}{5}$ $-\frac{37}{9}$

Solve each equation. *(LESSON 1.6)*

58. $5x = 15 + 3x$ $\frac{15}{2}$

59. $7x - 13 = -2x$ $\frac{13}{9}$

43.

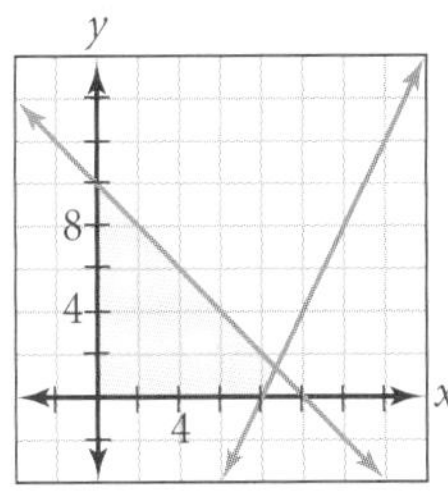

44.

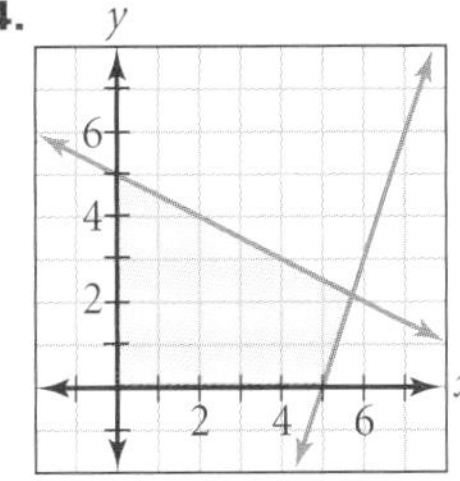

45.

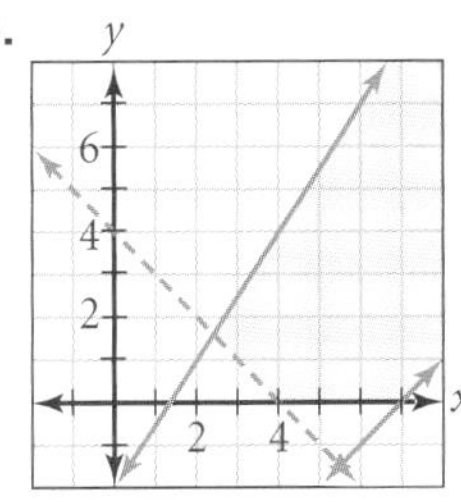

Student Technology Guide

NAME ________ CLASS ________ DATE ________

Student Technology Guide
3.4 Systems of Linear Inequalities

You can use a graphics calculator to graph a system of inequalities such as $\begin{cases} y \geq -\frac{3}{4}x \\ y \leq x - 1 \end{cases}$.

Method 1: Press Y=. Enter $-\frac{3x}{4}$ as Y1 and $x - 1$ as Y2. Use the arrow keys and ENTER to display ◥ to the left of Y1 and ◣ to the left of Y2. Press GRAPH. The calculator uses a different shading pattern for each inequality. The overlapped region indicates the solution.

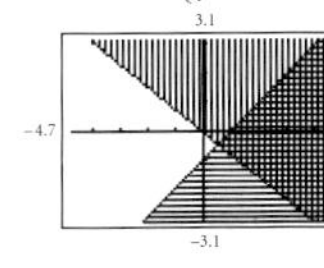

Method 2: Press Y=. Then use ▼ and CLEAR to remove any expressions. Press 2nd PRGM DRAW 7:Shade(ENTER (-) 3 X,T,θ,n ÷ 4 , X,T,θ,n - 1) ENTER. This tells the calculator to shade above the line $y = -\frac{3}{4}x$ and below the line $y = x - 1$.

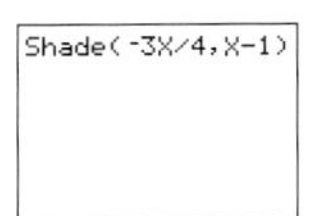

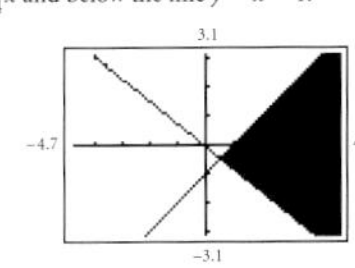

Note: If you use this method to solve several problems, you will need to press 2nd PRGM DRAW 1:Clr Draw ENTER after each problem.

Use a graphics calculator to graph each system. Describe the solution region.

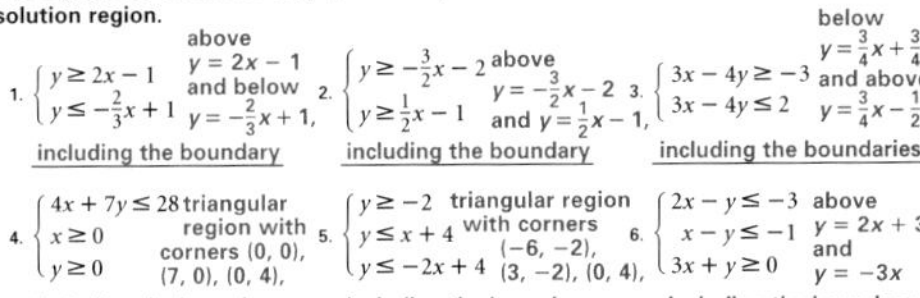

1. $\begin{cases} y \geq 2x - 1 \\ y \leq -\frac{2}{3}x + 1 \end{cases}$ above $y = 2x - 1$ and below $y = -\frac{2}{3}x + 1$, including the boundary

2. $\begin{cases} y \geq -\frac{3}{2}x - 2 \\ y \geq \frac{1}{2}x - 1 \end{cases}$ above $y = -\frac{3}{2}x - 2$ and $y = \frac{1}{2}x - 1$, including the boundary

3. $\begin{cases} 3x - 4y \geq -3 \\ 3x - 4y \leq 2 \end{cases}$ below $y = \frac{3}{4}x + \frac{3}{4}$ and above $y = \frac{3}{4}x - \frac{1}{2}$, including the boundaries

4. $\begin{cases} 4x + 7y \leq 28 \\ x \geq 0 \\ y \geq 0 \end{cases}$ triangular region with corners (0, 0), (7, 0), (0, 4), including the boundary

5. $\begin{cases} y \geq -2 \\ y \leq x + 4 \\ y \leq -2x + 4 \end{cases}$ triangular region with corners (−6, −2), (3, −2), (0, 4), including the boundary

6. $\begin{cases} 2x - y \leq -3 \\ x - y \leq -1 \\ 3x + y \geq 0 \end{cases}$ above $y = 2x + 3$ and $y = -3x$ including the boundary

46. Answers may vary.

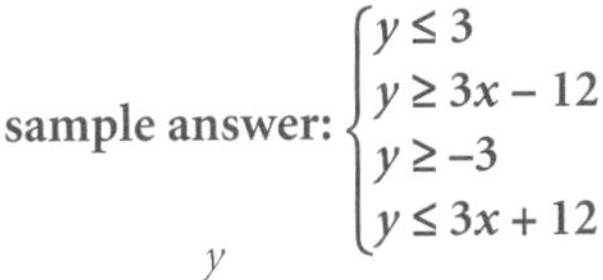

sample answer: $\begin{cases} y \leq 3 \\ y \geq 3x - 12 \\ y \geq -3 \\ y \leq 3x + 12 \end{cases}$

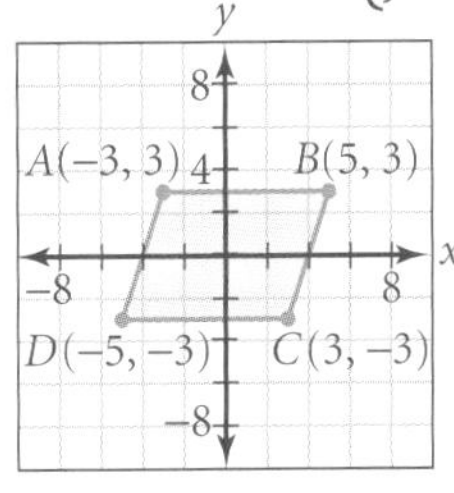

47. The figure is an isoceles right triangle.

☞ The answers to Exercises 48 and 49 can be found on page 186.

The answers to Exercises 50–53 can be found in Additional Answers beginning on page 1002.

In Exercises 68–70, students explore some ideas related to linear programming. In Exercises 68 and 69, they work with two equations of the type that typically represent an *objective function*. Exercise 70 lays the groundwork for the concept of the *feasible region*. Students will study linear programming in detail in Lesson 3.5. You may wish to use these exercises as a transition to that lesson.

ALTERNATIVE
Assessment

Portfolio Activity

The Portfolio Activity can be used as preparation for the Chapter Project or as a separate activity. In the Portfolio Activity on this page, students investigate a manufacturing problem that is similar to the problem in the Portfolio Activity on page 171. This time, however, they see how a system of linear *inequalities* can be used to solve the problem.

Answers to Portfolio Activities can be found in Additional Answers of the Teacher's Edition.

Let $f(x) = 3x + 2$ and $g(x) = 5 - x$. Find each new function, and state any domain restrictions. *(LESSON 2.4)*

60. $f + g = 2x + 7$ **61.** $f - g = 4x - 3$ **62.** $f \cdot g$ $-3x^2 + 13x + 10$ **63.** $\frac{f}{g} = \frac{3x+2}{5-x}$; $x \neq 5$

Use any method to solve each system of equations. Then classify each system as independent, dependent, or inconsistent. *(LESSONS 3.1 AND 3.2)*

64. $\begin{cases} 4x - 2y = 3 \\ 8y - 6x = 24 \end{cases}$ $\left(\frac{18}{5}, \frac{57}{10}\right)$; indep.

65. $\begin{cases} 5x + y = -1 \\ 10x - y = 3 \end{cases}$ $\left(\frac{2}{15}, -\frac{5}{13}\right)$; indep.

66. $\begin{cases} 2x + 5y = 10 \\ 2x - 5y = 0 \end{cases}$ $\left(\frac{5}{2}, 1\right)$; indep.

67. $\begin{cases} 3x + y = 6 \\ 6x + 2y = 12 \end{cases}$ infinitely many solutions; dep.

68. 210, 290, 340; increase

69. 1400, 3300, 5200; increase

Look Beyond

68. Evaluate $P = 20x + 30y$ for (3, 5), (4, 7), and (5, 8). As x- and y-values increase, does the value of P increase or decrease?

69. Evaluate $C = 400p + 500q$ for (1, 2), (2, 5), and (3, 8). As p- and q-values increase, does the value of C increase or decrease?

70. SYNONYMS Define the word *feasible*. Use a dictionary or write your own definition. What is a synonym for *feasible*?
Answers may vary. Sample: possible, reasonable, suitable

internet connect
Portfolio Extension
Go To: go.hrw.com
Keyword: MB1 Diet

PORTFOLIO ACTIVITY

BUSINESS Another company also manufactures two different models of portable CD players: a regular model and a sport model. Each model requires time on three different machines, as shown below.

	Regular model	Sport model
Machine A	2 minutes	1 minute
Machine B	1 minute	3 minutes
Machine C	1 minute	1 minute

Each machine is used for manufacturing many different items, so in a given hour, machine A is available for CD-player production a maximum of 18 minutes, machine B for a maximum of 24 minutes, and machine C for a maximum of 10 minutes.

1. Is it possible for the company to produce 4 regular models and 5 sport models in one hour? Explain your reasoning.
2. Is it possible for the company to produce 6 regular models and 5 sport models in one hour? Explain your reasoning.

WORKING ON THE CHAPTER PROJECT

You should now be able to complete Activities 1 and 3 of the Chapter Project.

48b.

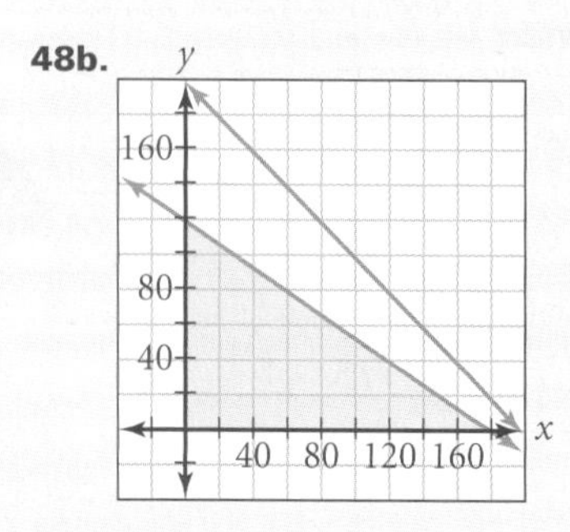

48c. The number of possible combinations is reduced. The maximum possible number of standard models changes from 180 to 150, and the maximum possible number of deluxe models changes from 120 to 100.

49b.

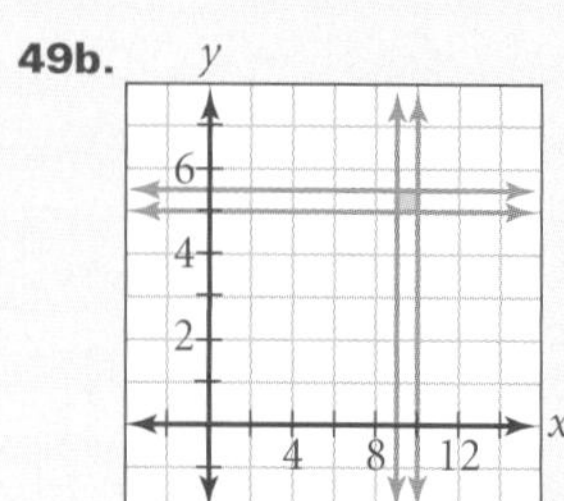

Linear Programming

Why *Linear programming is widely used in the management of business and agriculture to find optimal solutions to complex problems.*

Objectives

- Write and graph a set of constraints for a linear-programming problem.
- Use linear programming to find the maximum or minimum value of an objective function.

APPLICATION
AGRICULTURE

Max Desmond is a farmer who plants corn and wheat. In making planting decisions, he used the 1996 statistics at right from the United States Bureau of the Census.

Crop	Yield per acre	Average price
corn	113.5 bu	\$3.15/bu
soybeans	34.9 bu	\$6.80/bu
wheat	35.8 bu	\$4.45/bu
cotton	540 lb	\$0.759/lb
rice, rough	5621 lb	\$0.0865/lb

(The abbreviation for bushel is bu.)

Mr. Desmond wants to plant according to the following *constraints*:

- no more than 120 acres of corn and wheat
- at least 20 and no more than 80 acres of corn
- at least 30 acres of wheat

How many acres of each crop should Mr. Desmond plant to maximize the revenue from his harvest? You will answer this question in Example 2.

A method called **linear programming** is used to find optimal solutions such as the maximum revenue from Mr. Desmond's harvest. Linear-programming problems have the following characteristics:

- The inequalities contained in the problem are called **constraints**.
- The solution to the set of constraints is called the **feasible region**.
- The function to be maximized or minimized is called the **objective function**.

Prepare

NCTM PRINCIPLES & STANDARDS 1–4, 6–10

QUICK WARM-UP

1. Solve the system.

$\begin{cases} x + y = 7 \\ 2x - y = 8 \end{cases}$ $(5, 2)$

2. Graph the solution.

$\begin{cases} y < 2x + 3 \\ y + x < 5 \\ x \geq 0 \\ y > 1 \end{cases}$

Also on Quiz Transparency 3.5

Teach

Why Discuss with students the meanings of the words *constrain* and *constraint*. Ask them to suggest real-world constraints that might be represented by inequalities. For example, note that in many states, you can drive a motor vehicle under the constraint that your age in years, a, satisfies the inequality $a \geq 16$.

Alternative Teaching Strategy

GUIDED ANALYSIS Have students draw a set of coordinate axes on a sheet of graph paper. Tell them to graph all numbers x and y whose sum is less than or equal to 8. Point out that any point in the region satisfies the equation $x + y \leq 8$.

Now tell them that you are going to put these additional *constraints* on x and y: $2 \leq x \leq 5$ and $y \geq 2$. Tell them to shade the region of the graph that reflects all the constraints. Their graphs should look like the one at right. The darker shaded region contains all the *feasible* values of x and y.

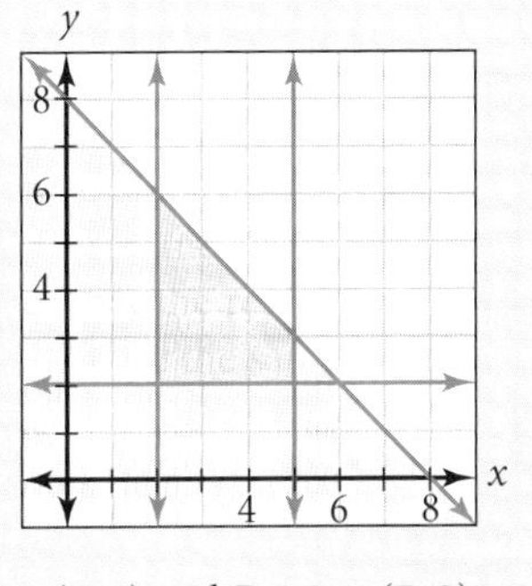

Now display the equation $D = x - y$. Tell students that their *objective* is to find the least and greatest possible values of D within or on the boundary of the darker region. They should arrive at $D = -4$ at $(2, 6)$ and $D = 3$ at $(5, 2)$.

Use this activity to introduce the concepts of *constraints*, *feasible region*, and *objective function*.

Use Teaching Transparency 13.

ADDITIONAL

EXAMPLE 1

A small company produces knitted afghans and sweaters and sells them through a chain of specialty stores. The company is to supply the stores with a total of no more than 100 afghans and sweaters per day. The stores guarantee that they will sell at least 10 and no more than 60 afghans per day and at least 20 sweaters per day. The company makes a profit of $10 on each afghan and a profit of $12 on each sweater.

a. Write a system of inequalities to represent the constraints. Let x represent the number of afghans. Let y represent the number of sweaters.

$$\begin{cases} 10 \le x \le 60 \\ y \ge 20 \\ x + y \le 100 \end{cases}$$

b. Graph the feasible region.

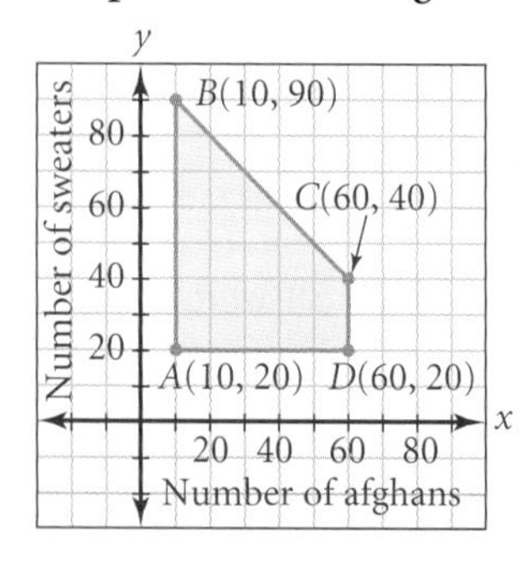

c. Write an objective function for the company's total profit, P, from the sales of afghans and sweaters.

$P = 10x + 12y$

Example 1 illustrates how to begin the method of linear programming.

EXAMPLE 1 **Refer to the planting problem described at the beginning of the lesson.**

APPLICATION
AGRICULTURE

a. Write a system of inequalities to represent the constraints.
b. Graph the feasible region.
c. Write an objective function for the revenue from Mr. Desmond's harvest.

SOLUTION

a. Let x represent the number of acres of corn. Let y represent the number of acres of wheat. Since x and y must be positive, $x \ge 0$ and $y \ge 0$.

Corn constraint: $20 \le x \le 80$
Wheat constraint: $y \ge 30$
Total acreage: $x + y \le 120$

The system is: $\begin{cases} 20 \le x \le 80 \\ y \ge 30 \\ x + y \le 120 \end{cases}$

b. The graph is shown below.

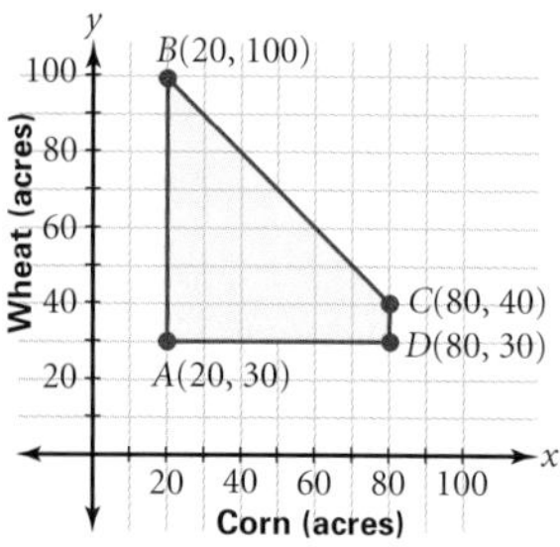

c. The objective function for the revenue is as follows:

$$R = \begin{pmatrix}\text{yield}\\\text{per}\\\text{acre}\end{pmatrix}\begin{pmatrix}\text{average}\\\text{price}\end{pmatrix}x + \begin{pmatrix}\text{yield}\\\text{per}\\\text{acre}\end{pmatrix}\begin{pmatrix}\text{average}\\\text{price}\end{pmatrix}y$$

$$R = (113.5)(3.15)x + (35.8)(4.45)y$$

$$R = 357.525x + 159.31y$$

For each point in the feasible region of a linear-programming problem, the objective function has a value. This value depends on both variables in the system that represents the feasible region.

In the Activity below, you will explore values in the feasible region for the objective function from Example 1.

Activity

Exploring the Objective Function

You will need: no special materials

1. Copy and complete the table to find the revenue at each of the four vertices of the feasible region from Example 1.

Vertex	Objective function
$A(20, 30)$	$R = 357.525(20) + 159.31(30) = 11{,}929.80$
$B(20, 100)$	?
$C(80, 40)$	?
$D(80, 30)$	?

2. Which vertex represents the greatest revenue? What do the coordinates of this vertex represent?

Interdisciplinary Connection

HOME ECONOMICS A home economics class has been asked to bake loaves of banana bread and some chocolate chip cookies for a charity bake sale. A local bakery has already donated 5 dozen eggs and 6 bags of flour. Each bag of flour contains $7\frac{1}{2}$ cups. All other ingredients will be donated as needed.

Each loaf of banana bread requires 1 egg and 2 cups of flour. Each batch of cookies requires 2 eggs and $1\frac{1}{2}$ cups of flour. The class has been asked to make at least 6 and at most 12 loaves of banana bread and at least 6 batches of cookies.

The loaves of bread will be sold for $5 each, and each batch of cookies will sell for $4. How many loaves of bread and how many batches of cookies should they bake with the supplies on hand in order to earn the maximum amount?

Let b be loaves of bread and c be batches of cookies.

$$\begin{cases} 6 \le b \le 12 & \textbf{bread} \\ c \ge 6 & \textbf{cookies} \\ b + 2c \le 60 & \textbf{eggs} \\ 2b + 1.5c \le 45 & \textbf{flour} \end{cases}$$

Objective function: $A = 5b + 4c$

max. amount: 6 loaves of bread and 22 batches of cookies

PROBLEM SOLVING

3. **Guess and check.** Choose points on the boundary lines of the feasible region. Find the corresponding revenues for these points. Can you find a point that gives a greater revenue than the vertex you chose in Step 2?

PROBLEM SOLVING

4. **Guess and check.** Choose points inside the feasible region. Find the corresponding revenues for these points. Can you find a point that gives a greater revenue than the vertex you chose in Step 2?

CHECKPOINT ✔

5. Do your investigations suggest that the maximum value of the objective function occurs at a vertex? Justify your response.

PROBLEM SOLVING

6. **Look for a pattern.** Repeat Steps 2–5 for the minimum revenue instead of the maximum revenue. Explain how the points that correspond to the maximum and minimum revenues are related.

In the Activity, you may have examined several points in the feasible region and found that the maximum and minimum revenues occur at vertices of the feasible region. The *Corner-Point Principle* confirms that you need to examine only the vertices of the feasible region to find the maximum or minimum value of the objective function.

Corner-Point Principle

In linear programming, the maximum and minimum values of the objective function each occur at one of the vertices of the feasible region.

EXAMPLE 2

APPLICATION
AGRICULTURE

Using the information in Example 1, maximize the objective function. Then graph the objective function that represents the maximum revenues along with the feasible region.

SOLUTION

PROBLEM SOLVING

Make a table containing the coordinates of the vertices of the feasible region. Evaluate $R = 357.525x + 159.31y$ for each ordered pair.

Vertex	Objective function
$A(20, 30)$	$R = 357.525(20) + 159.31(30) = 11{,}929.80$
$B(20, 100)$	$R = 357.525(20) + 159.31(100) = 23{,}081.50$
$C(80, 40)$	$R = 357.525(80) + 159.31(40) = 34{,}974.40$
$D(80, 30)$	$R = 357.525(80) + 159.31(30) = 33{,}381.30$

The maximum revenue of \$34,974.40 occurs at $C(80, 40)$. Thus, Mr. Desmond should plant 80 acres of corn and 40 acres of wheat.

Write $357.525x + 159.31y = 34{,}974.4$ as $y = \frac{34{,}974.4 - 357.525x}{159.31}$, and graph it along with the boundaries of the feasible region.

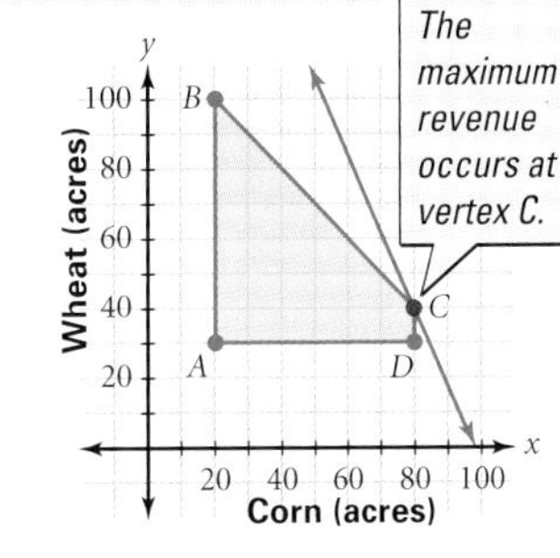

Activity Notes

In the Activity on page 188, students investigate the objective function from Example 1. They substitute the coordinates of several points from the feasible region of the graph into the equation for the objective function. They should discover that the maximum and minimum values of the function each occur at a vertex of the feasible region.

CHECKPOINT ✔

5. yes

Math CONNECTION

MAXIMUM/MINIMUM Using the coordinates of the vertices, students will discover the maximum and minimum revenues for the problem in Example 1.

ADDITIONAL EXAMPLE 2

Using the information in Additional Example 1 on page 188, maximize the objective function. Then graph the objective function, indicating the maximum profit and the feasible region.

The maximum profit occurs at $B(10, 90)$. The company makes the greatest profit when the stores sell 10 afghans and 90 sweaters in one day.

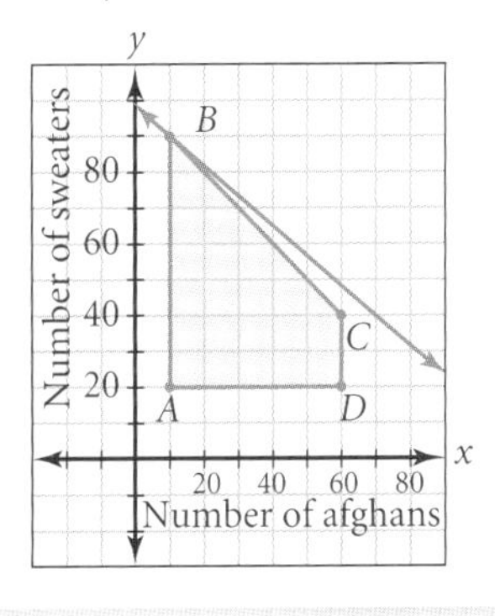

Inclusion Strategies

VISUAL LEARNERS Students may better grasp the concept of the objective function in Example 2 if they visualize all four lines that result when the coordinates of the vertices are substituted into the revenue equation. These lines are shown in the figure at right.

Point out that *any* line with an equation of the form $R = 357.525x + 159.31y$ will be parallel to these four lines. However, the one that intersects the feasible region *and produces the greatest possible value of R* is l_3.

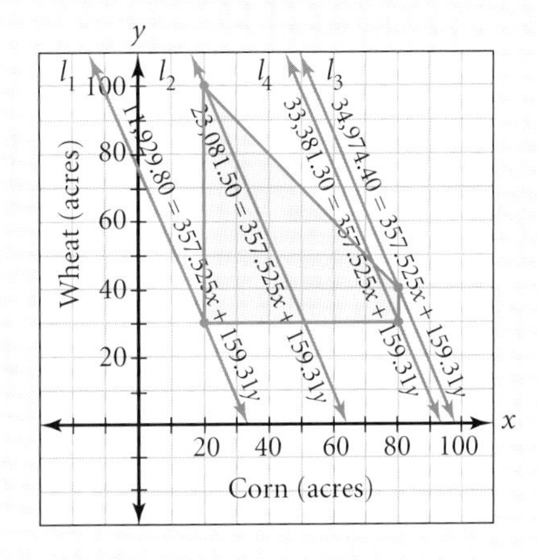

CHECKPOINT ✔
20 acres of corn and 30 acres of wheat

Math CONNECTION

MAXIMUM/MINIMUM Examples 1 and 2 illustrate a real-world situation in which linear programming is used to maximize a quantity. Be sure students understand that there are other situations in which linear programming is used to *minimize* a quantity. An example of such a situation can be found in Exercise 35 on page 193.

ADDITIONAL EXAMPLE 3

Find the maximum and minimum values, if they exist, of the objective function $T = 3x + 2y$ given the set of constraints below.

$$\begin{cases} x + y \le 10 \\ x + 2y \ge 12 \\ 4x + y \ge 13 \end{cases}$$

maximum: 28; minimum: 16

CHECKPOINT ✔
Answers may vary. Sample answer: A car dealer sells new and used cars. In order to maintain a balanced inventory, she needs to sell at least 2 new cars for every used car she sells.

CRITICAL THINKING
Probably not; a feasible region unbounded on the left would imply negative values of the variables.

CHECKPOINT ✔ How many acres of each crop give a minimum revenue?

EXAMPLE 3

Find the maximum and minimum values, if they exist, of the objective function $S = 2x + 3y$ given the set of constraints provided at right.

$$\begin{cases} x \ge 0, y \ge 0 \\ 5x + y \ge 20 \\ x + y \ge 12 \\ x + 2y \ge 16 \end{cases}$$

SOLUTION

1. Graph the feasible region as shown.
2. The objective function is $S = 2x + 3y$.
3. Find the coordinates of each vertex by solving the appropriate system.

$\begin{cases} 5x + y = 20 \\ x = 0 \end{cases}$ gives $A(0, 20)$

$\begin{cases} 5x + y = 20 \\ x + y = 12 \end{cases}$ gives $B(2, 10)$

$\begin{cases} x + y = 12 \\ x + 2y = 16 \end{cases}$ gives $C(8, 4)$

$\begin{cases} y = 0 \\ x + 2y = 16 \end{cases}$ gives $D(16, 0)$

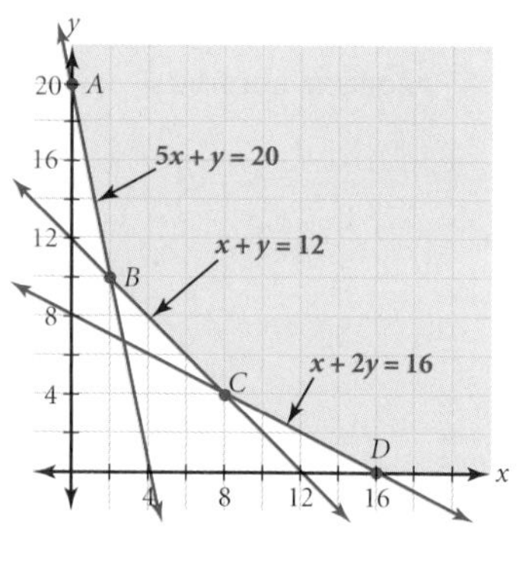

4. Evaluate $S = 2x + 3y$ for the coordinates of each vertex.

Vertex	Objective function
$A(0, 20)$	$S = 2(0) + 3(20) = 60$
$B(2, 10)$	$S = 2(2) + 3(10) = 34$
$C(8, 4)$	$S = 2(8) + 3(4) = 28$
$D(16, 0)$	$S = 2(16) + 3(0) = 32$

The feasible region is unbounded on the right of the vertices. Thus, there is no maximum value. The minimum value, 28, of $S = 2x + 3y$ occurs at $(8, 4)$.

CHECKPOINT ✔ Describe a real-world situation in which the feasible region would be unbounded on the right.

CRITICAL THINKING Can there be a real-world situation in which the feasible region would be unbounded on the left? Explain.

SUMMARY
LINEAR-PROGRAMMING PROCEDURE

Step 1. Write a system of inequalities, and graph the feasible region.
Step 2. Write the objective function to be maximized or minimized.
Step 3. Find the coordinates of the vertices of the feasible region.
Step 4. Evaluate the objective function for the coordinates of the vertices of the feasible region. Then identify the coordinates that give the required maximum or minimum.

Enrichment

The vertices of a feasible region are $P(4, 15)$, $Q(4, 6)$, and $R(16, 6)$. Write an equation of the form $A = Bx + Cy$ for an objective function that satisfies the following conditions:

1. P is the maximum, and Q is the minimum. sample: $A = 2x + 3y$
2. R is the maximum, and Q is the minimum. sample: $A = 3x + 2y$
3. P and R are both maximums, and Q is the minimum. sample: $A = 3x + 4y$

Reteaching the Lesson

COOPERATIVE LEARNING Some students will be feeling overwhelmed by the complexity of linear programming. Prepare three new problems. Divide the class into groups of three, and assign these roles: The *constrainer* identifies the constraints and writes the inequalities. The *grapher* graphs the feasible region and identifies its corner points. The *maximizer* (or *minimizer*) writes an equation for the objective function and maximizes (or minimizes) it. Have the members of each group work together to solve the three problems, with each member assuming a different role for each problem.

Exercises

Communicate

1. What is a *constraint* on a variable, such as x?
2. Discuss what the term *feasible* means when it is used to describe the possible solution region of a linear-programming problem.
3. In your own words, explain how to solve a linear-programming problem.

Guided Skills Practice

APPLICATION

AGRICULTURE Use the table of statistics at the beginning of the lesson to determine the constraints and to graph the feasible region for each situation below. Then write the corresponding objective function for the revenue. *(EXAMPLE 1)*

4. A farmer wants to plant corn and soybeans on 150 acres of land. The farmer wants to plant between 40 and 120 acres of corn and no more than 100 acres of soybeans.
5. A farmer wants to plant wheat and soybeans on 220 acres of land. The farmer wants to plant between 100 and 200 acres of wheat and no more than 75 acres of soybeans.

Find the number of acres of each crop that the farmer should plant to maximize the revenue for each situation. *(EXAMPLE 2)*

6. See Exercise 4. **120 acres of corn, 30 acres of soybean**
7. See Exercise 5. **145 acres of wheat, 75 acres of soybean**

CONNECTION

MAXIMUM/MINIMUM Find the maximum and minimum values, if they exist, of $C = 3x + 4y$ for each set of constraints. *(EXAMPLE 3)*

8. $\begin{cases} 3 \le x \le 8 \\ 2 \le y \le 6 \\ 2x + y \ge 12 \end{cases}$ **max = 48, min = 23**
9. $\begin{cases} 2 \le x \\ 4 \le y \le 8 \\ x + 2y \ge 16 \end{cases}$ **no max, min = 34**

Practice and Apply

Graph the feasible region for each set of constraints.

10. $\begin{cases} x + 2y \le 8 \\ 2x + y \ge 10 \\ x \ge 0, y \ge 0 \end{cases}$
11. $\begin{cases} 3x + 2y \le 12 \\ \frac{1}{2}x - y \le -2 \\ x \ge 0, y \ge 0 \end{cases}$
12. $\begin{cases} x + 2y \le 6 \\ 2x - y \le 7 \\ x \ge 2, y \ge 0 \end{cases}$
13. $\begin{cases} 3x + y \le 12 \\ 2x - 3y \ge -3 \\ x \le 0, y \le 6 \end{cases}$

Identify the vertices of the feasible region in

14. Exercise 10. **(5, 0), (4, 2), (8, 0)**
15. Exercise 11. **(0, 2), (0, 6), (2, 3)**
16. Exercise 12. **(2, 0), $\left(\frac{7}{2}, 0\right)$, (4, 1), (2, 2)**
17. Exercise 13. **(0, 1)**

Assess

Selected Answers

Exercises 4–9, 11–45 odd

ASSIGNMENT GUIDE	
In Class	1–9
Core	11–29 odd, 33, 35
Core Plus	10–30 even, 31, 32–36 even
Review	38–45
Preview	46–48

Extra Practice can be found beginning on page 940.

10.

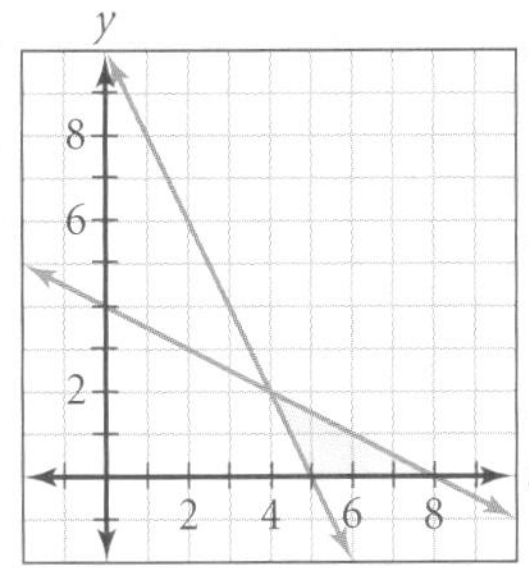

11.

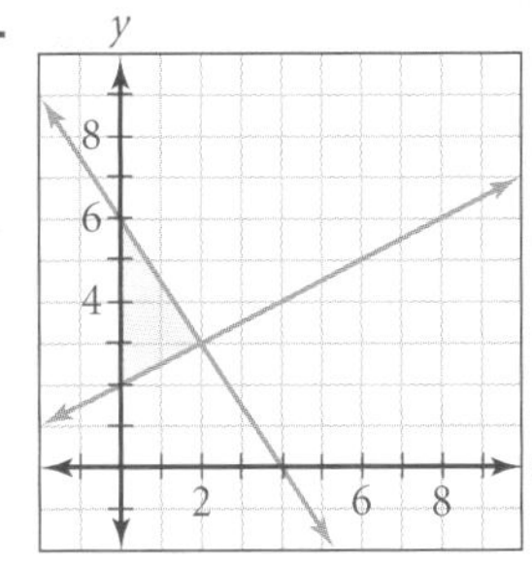

12.

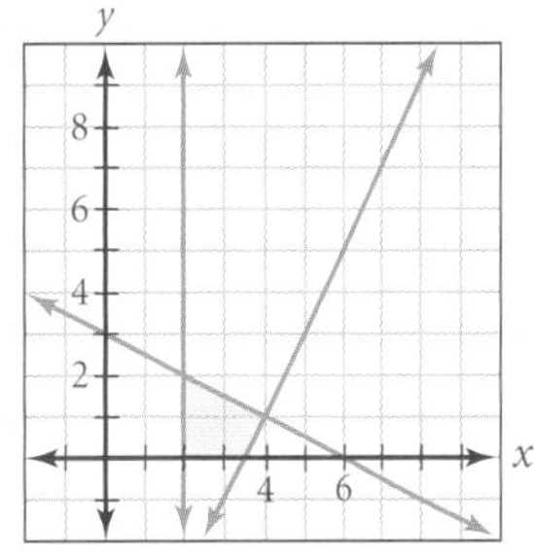

13.

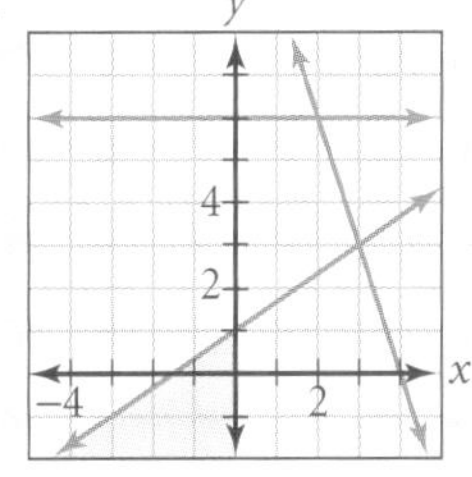

The answers to Exercises 4–5 can be found in Additional Answers beginning on page 1002.

Error Analysis

Some students may have difficulty separating the information that gives the constraints of the problem from the information that describes the objective function. Suggest that they first identify the objective function and then use the remaining information to list the constraints. It may help to point out that, in many cases, the objective function involves amounts of money.

30. Answers may vary. sample answer:

convex:

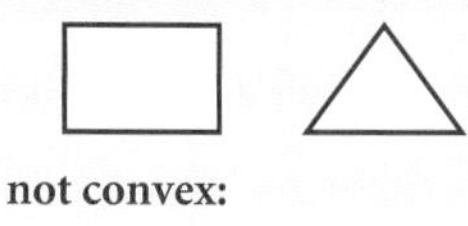

not convex:

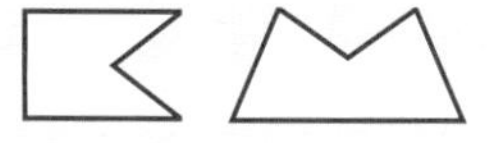

The feasible region for a set of constraints has vertices at (−2, 0), (3, 3), (6, 2), and (5, 1). Given this feasible region, find the maximum and minimum values of each objective function.

18. $C = 2x - y$ **max = 10, min = −4**

19. $M = 3y - x$ **max = 6, min = −2**

20. $I = 100x + 200y$ **max = 1000, min = −200**

21. $P = 3x + 2.5y$ **max = 23, min = −6**

Find the maximum and minimum values, if they exist, of each objective function for the given constraints.

22. $P = 5y + 3x$
Constraints:
$\begin{cases} x + y \le 6 \\ x - y \le 4 \\ x \ge 0 \\ y \ge 0 \end{cases}$ **30; 0**

23. $P = 3x + y$
Constraints:
$\begin{cases} x + y \ge 3 \\ 3x + 4y \le 12 \\ x \ge 0 \\ y \ge 0 \end{cases}$ **12; 3**

24. $P = 4x + 7y$
Constraints:
$\begin{cases} x + y \le 8 \\ y - x \le 2 \\ x \ge 0 \\ y \ge 0 \end{cases}$ **47; 0**

25. $P = 2x + 7y$
Constraints:
$\begin{cases} 4x - 2y \le 8 \\ x \ge 1 \\ 0 \le y \le 4 \end{cases}$ **36; 2**

26. $E = 2x + y$
Constraints:
$\begin{cases} x + y \ge 6 \\ x - y \le 4 \\ x \ge 0 \\ y \ge 0 \end{cases}$ **no max; min = 6**

27. $E = x + y$
Constraints:
$\begin{cases} x + 2y \ge 3 \\ 3x + 4y \ge 8 \\ x \ge 0 \\ y \ge 0 \end{cases}$ **no max; min = 2**

28. $E = 3x + 5y$
Constraints:
$\begin{cases} x - 2y \ge 0 \\ x + 2y \ge 8 \\ 1 \le x \le 6 \\ y \ge 0 \end{cases}$ **33; 22**

29. $E = 3x + 2y$
Constraints:
$\begin{cases} x + y \le 5 \\ y - x \ge 5 \\ 4x + y \ge -10 \end{cases}$ **10; −5**

CONNECTIONS

30. GEOMETRY If the feasible region for a linear-programming problem is bounded, it must form a convex polygon. Convex polygons cannot have "dents" and are defined as polygons in which any line segment connecting two points of the polygon has no part outside the polygon. Sketch two examples of convex polygons and two examples of polygons that are not convex (that is, concave).

CHALLENGE

31. MAXIMUM/MINIMUM An objective function can have a maximum (or minimum) value at two vertices if the graph of the objective function, equal to a constant function value, contains both vertices.

a. Draw the graph of a feasible region that has maximum values of 6 at two vertices for the objective function $P = 2x + 3y$.

b. Draw the graph of a feasible region that has minimum values of 6 at two vertices for the objective function $P = 2x + 3y$.

internet connect
Homework Help Online
Go To: go.hrw.com
Keyword: MB1 Homework Help for Exercises 32–37

32. MANUFACTURING A ski manufacturer makes two types of skis and has a fabricating department and a finishing department. A pair of downhill skis requires 6 hours to fabricate and 1 hour to finish. A pair of cross-country skis requires 4 hours to fabricate and 1 hour to finish. The fabricating department has 108 hours of labor available per day. The finishing department has 24 hours of labor available per day. The company makes a profit of $40 on each pair of downhill skis and a profit of $30 on each pair of cross-country skis.

a. Write a system of linear inequalities to represent the constraints.

b. Graph the feasible region.

c. Write the objective function for the profit, and find the maximum profit for the given constraints.

32a. $\begin{cases} 6x + 4y \le 108 \\ x + y \le 24 \\ x \ge 0 \\ y \ge 0 \end{cases}$

c. $P = 40x + 30y$; max = $780

Practice

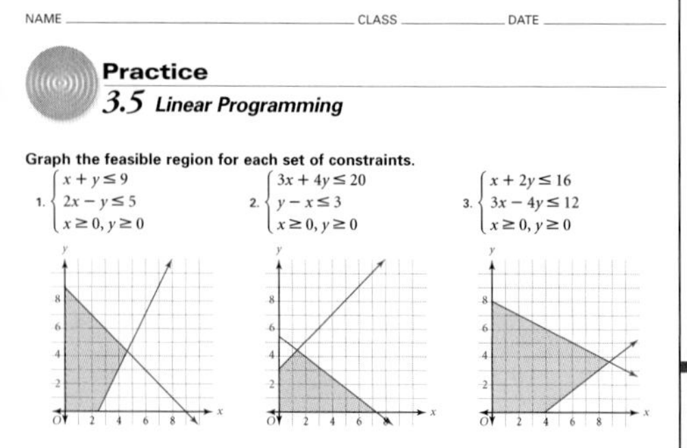

NAME ______ CLASS ______ DATE ______

Practice
3.5 Linear Programming

Graph the feasible region for each set of constraints.

1. $\begin{cases} x + y \le 9 \\ 2x - y \le 5 \\ x \ge 0, y \ge 0 \end{cases}$

2. $\begin{cases} 3x + 4y \le 20 \\ y - x \le 3 \\ x \ge 0, y \ge 0 \end{cases}$

3. $\begin{cases} x + 2y \le 16 \\ 3x - 4y \le 12 \\ x \ge 0, y \ge 0 \end{cases}$

The feasible region for a set of constraints has vertices at (2,0), (10, 1), (8, 5), and (0, 4). Given this feasible region, find the maximum and minimum values of each objective function.

4. $F = 4x + y$
maximum: 41
minimum: 4

5. $E = 2x - 3y$
maximum: 17
minimum: −12

6. $M = 3y - x$
maximum: 12
minimum: −7

Find the maximum and minimum values, if they exist, of each objective function for the given constraints.

7. $P = x + 5y$ maximum = 50, minimum = 0
Constraints:
$\begin{cases} 2x + y \le 10 \\ x - y \le 4 \\ x \ge 0 \\ y \ge 0 \end{cases}$

8. $E = 4x + 8y$ maximum = 68, minimum = 0
Constraints:
$\begin{cases} x + y \le 9 \\ y - x \le 7 \\ x \ge 0 \\ y \ge 0 \end{cases}$

9. $G = 20x + 10y$ maximum = 200, minimum = 20
Constraints:
$\begin{cases} 2x - y \ge 2 \\ x + y \le 10 \\ x \ge 0 \\ y \ge 0 \end{cases}$

10. $F = 12x - 5y$ maximum = 62, minimum = −25
Constraints:
$\begin{cases} 2x - y \le 10 \\ x + 2y \le 10 \\ x \ge 0 \\ y \ge 0 \end{cases}$

31. Answers may vary. Sample answers:

a. The objective function $P = 2x + 3y$ has its maximum, 6, at (0, 2) and $\left(\frac{6}{5}, \frac{6}{5}\right)$.

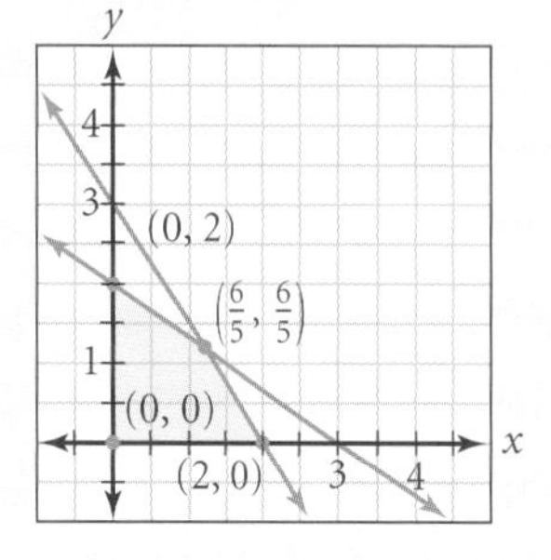

b.

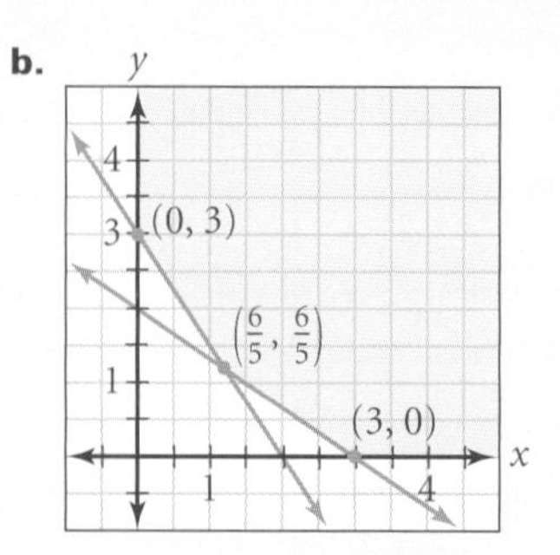

The objective function, $P = 2x + 3y$, has its minimum value, 6, at $\left(\frac{6}{5}, \frac{6}{5}\right)$ and (3, 0).

APPLICATIONS

33. TRANSPORTATION Trenton, Michigan, a small community, is trying to establish a public transportation system of large and small vans. It can spend no more than \$100,000 for both sizes of vehicles and no more than \$500 per month for maintenance. The community can purchase a small van for \$10,000 and maintain it for \$100 per month. The large vans cost \$20,000 each and can be maintained for \$75 per month. Each large van carries a maximum of 15 passengers, and each small van carries a maximum of 7 passengers.

a. Write a system of linear inequalities to represent the constraints.
b. Graph the feasible region.
c. Write the objective function for the number of passengers, and find the maximum number of passengers for the given constraints.

33a. $\begin{cases} 10{,}000x + 20{,}000y \le 100{,}000 \\ 100x + 75y \le 500 \\ x \ge 0 \\ y \ge 0 \end{cases}$

c. $P = 7x + 15y$; max = 75 passengers

34. BUSINESS A tourist agency can sell up to 1200 travel packages for a football game. The package includes airfare, weekend accommodations, and the choice of two types of flights: a nonstop flight or a two-stop flight. The nonstop flight can carry up to 150 passengers, and the two-stop flight can carry up to 100 passengers. The agency can locate no more than 10 planes for the travel packages. Each package with a nonstop flight sells for \$1200, and each package with a two-stop flight sells for \$900. Assume that each plane will carry the maximum number of passengers.

a. Write a system of linear inequalities to represent the constraints.
b. Graph the feasible region.
c. Write an objective function that maximizes the revenue for the tourist agency, and find the maximum revenue for the given constraints.

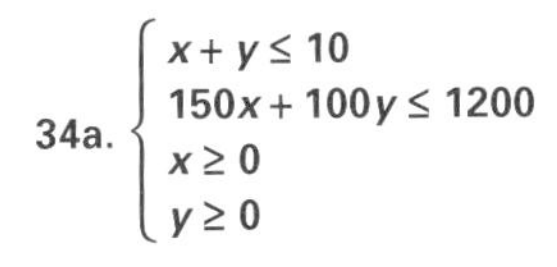

34a. $\begin{cases} x + y \le 10 \\ 150x + 100y \le 1200 \\ x \ge 0 \\ y \ge 0 \end{cases}$

c. $R = 1200x + 900y$; max = \$10,200

35. HEALTH A school dietitian wants to prepare a meal of meat and vegetables that has the lowest possible fat and that meets the Food and Drug Administration recommended daily allowances (RDA) of iron and protein. The RDA minimums are 20 milligrams of iron and 45 grams of protein. Each 3-ounce serving of meat contains 45 grams of protein, 10 milligrams of iron, and 4 grams of fat. Each 1-cup serving of vegetables contains 9 grams of protein, 6 milligrams of iron, and 2 grams of fat. Let x represent the number of 3-ounce servings of meat, and let y represent the number of 1-cup servings of vegetables.

a. Write a system of linear inequalities to represent the constraints.
b. Graph the feasible region.
c. Write the objective function for the number of grams of fat, and find the minimum number of grams of fat for the given constraints.

35a. $\begin{cases} 45x + 9y \ge 45 \\ 10x + 6y \ge 20 \\ x \ge 0 \\ y \ge 0 \end{cases}$

c. $M = 4x + 2y$; min = 7 grams

36a. $\begin{cases} x + y \le 90 \\ 4x + 6y \le 480 \\ 20x + 10y \le 1400 \\ x \ge 0 \\ y \ge 0 \end{cases}$

36. AGRICULTURE A farmer has 90 acres available for planting millet and alfalfa. Seed costs \$4 per acre for millet and \$6 per acre for alfalfa. Labor costs are \$20 per acre for millet and \$10 per acre for alfalfa. The expected income is \$110 per acre for millet and \$150 per acre for alfalfa. The farmer intends to spend no more than \$480 for seed and \$1400 for labor.

a. Write a system of linear inequalities to represent the constraints.
b. Graph the feasible region.
c. Write the objective function that maximizes the income, and find the maximum income for the given constraints. $I = 110x + 150y$; max = \$12,300

32b.

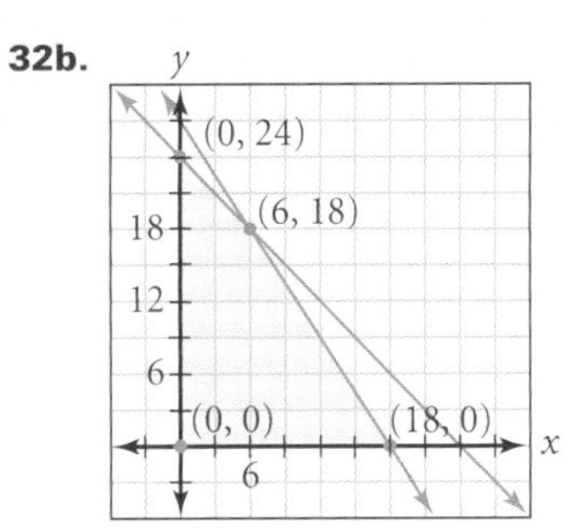

33b.

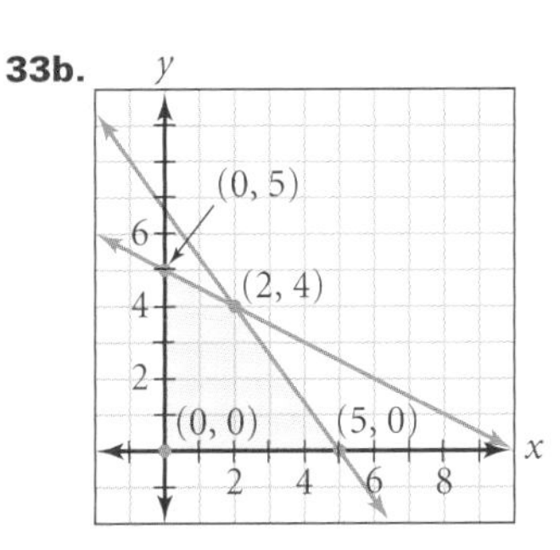

34b.

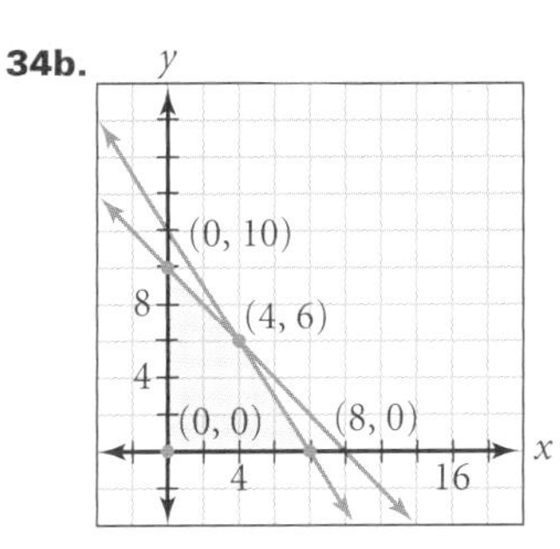

35b.

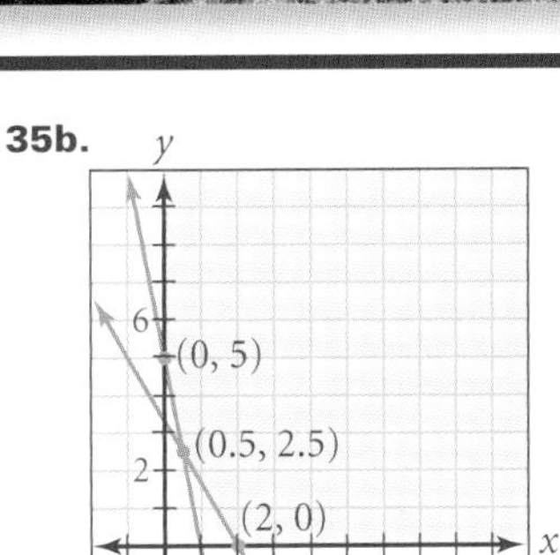

36b.

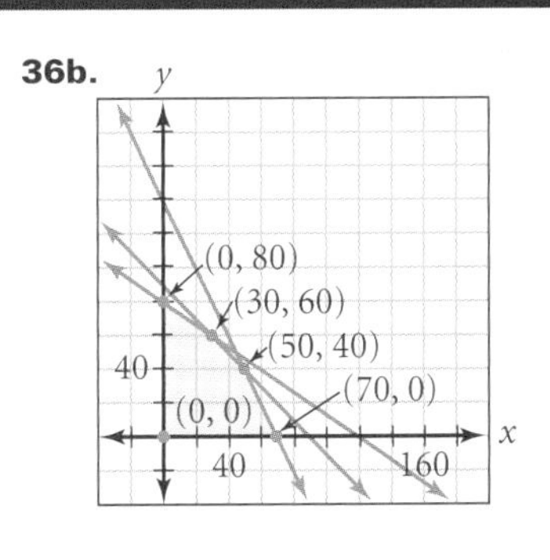

Student Technology Guide

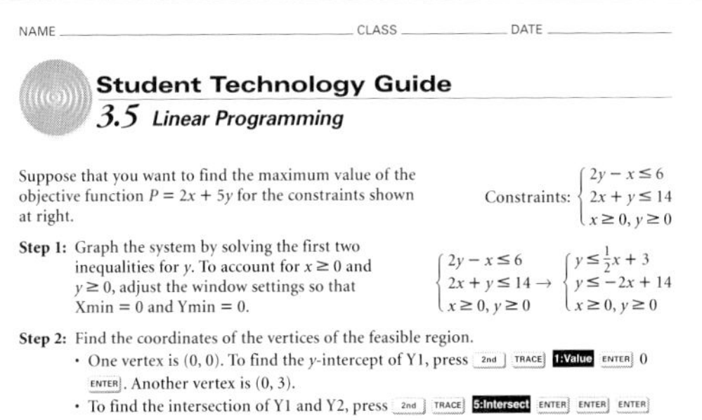

NAME ______ CLASS ______ DATE ______

Student Technology Guide
3.5 Linear Programming

Suppose that you want to find the maximum value of the objective function $P = 2x + 5y$ for the constraints shown at right. Constraints: $\begin{cases} 2y - x \le 6 \\ 2x + y \le 14 \\ x \ge 0, y \ge 0 \end{cases}$

Step 1: Graph the system by solving the first two inequalities for y. To account for $x \ge 0$ and $y \ge 0$, adjust the window settings so that Xmin = 0 and Ymin = 0. $\begin{cases} 2y - x \le 6 \\ 2x + y \le 14 \\ x \ge 0, y \ge 0 \end{cases} \rightarrow \begin{cases} y \le \frac{1}{2}x + 3 \\ y \le -2x + 14 \\ x \ge 0, y \ge 0 \end{cases}$

Step 2: Find the coordinates of the vertices of the feasible region.
- One vertex is (0, 0). To find the y-intercept of Y1, press 2nd TRACE 1:Value ENTER 0 ENTER. Another vertex is (0, 3).
- To find the intersection of Y1 and Y2, press 2nd TRACE 5:Intersect ENTER ENTER ENTER ENTER. A third vertex is (4.4, 5.2).
- To find the x-intercept of Y2, press 2nd TRACE 2:zero ENTER ▼ 0 ENTER 10 ENTER ENTER. The last vertex is (7, 0).

Step 3: Evaluate the objective function for the coordinates of the vertices of the feasible region by pressing STAT EDIT 1:Edit... ENTER. Enter the vertices as shown below. Then use the arrow keys to highlight L3 in the upper right corner. Because the objective function is $P = 2x + 5y$, enter 2 2nd 1 + 5 2nd 2 ENTER. The value of the objective function for each ordered pair appears in list L3. The maximum value is 34.8, which occurs at the point (4.4, 5.2).

L1	L2	L3
0	0	
0	3	
4.4	5.2	
7	0	

L3 =2L1+5L2

L1	L2	L3
0	0	0
0	3	15
4.4	5.2	34.8
7	0	14

L3(1)=0

Use a graphics calculator to maximize each objective function under the given constraints.

1. $P = 3x - 4y$ Constraints: $\begin{cases} 3x + 2y \le 16 \\ 5x + 3y \le 25 \\ x \ge 0, y \ge 0 \end{cases}$ 16 at $(5\frac{1}{3}, 0)$

2. $P = -2x + y$ Constraints: $\begin{cases} 3x - y \le 8 \\ 1 \le y \le 7 \\ x \ge 0 \end{cases}$ 7 at (0, 7)

3. $P = 5x + 8y$ Constraints: $\begin{cases} 6x + 5y \le 30 \\ 2x + 5y \le 20 \\ x \ge 0, y \ge 0 \end{cases}$ 36.5 at (2.5, 3)

In Exercises 46–48, students use a graphics calculator to investigate the relationship between the x-intercepts of the graph of a quadratic function and the factored form of the function. This relationship will be studied in depth in Lesson 5.3.

ALTERNATIVE
Assessment

Portfolio Activity

The Portfolio Activity can be used as preparation for the Chapter Project or as a separate activity. In the Portfolio Activity on this page, students apply linear programming techniques to the manufacturing problem that they began to investigate in the Portfolio Activity on page 186.

APPLICATION

37a. $\begin{cases} 6x + 2y \le 24 \\ x \ge 2 \\ y \ge 2 \end{cases}$

c. $P = 50x + 20y$; max = \$220

37. SMALL BUSINESS A carpenter makes bookcases in two sizes, large and small. It takes 6 hours to make a large bookcase and 2 hours to make a small one. The profit on a large bookcase is \$50, and the profit on a small bookcase is \$20. The carpenter can spend only 24 hours per week making bookcases and must make at least 2 of each size per week.

a. Write a system of linear inequalities to represent the constraints.
b. Graph the feasible region.
c. Write the objective function for the profit, and find the maximum profit for the given constraints.

Look Back

Determine whether each table represents a linear relationship between x and y. If the relationship is linear, write the next ordered pair that would appear in the table. *(LESSON 1.1)*

38.

x	0	2	4	6
y	5	10	20	40

nonlinear

39.

x	−1	−2	−3	−4
y	2	4	6	8

linear; (−5, 10)

Graph each function, and state the domain and the range. *(LESSON 2.3)*

40. $f(x) = 2x + 5$

41. $g(x) = 2x^2 - 3$

Find an equation for the inverse of each function. *(LESSON 2.5)*

42. $f(x) = 4x + 1$ $f^{-1}(x) = \frac{1}{4}x - \frac{1}{4}$

43. $g(x) = -2x + \frac{1}{4}$ $g^{-1}(x) = -\frac{1}{2}x + \frac{1}{8}$

44. $f(x) = 5 - 2x$ $f^{-1}(x) = -\frac{1}{2}x + \frac{5}{2}$

45. $g(x) = \frac{1}{4}x - 6$ $g^{-1}(x) = 4x + 24$

internet connect
Portfolio Extension
Go To: go.hrw.com
Keyword:
MB1 Simplex

Look Beyond

Find the x-intercepts for the graph of each function. Compare the factors of each function rule with the x-intercepts of the graph.

46 $f(x) = (x - 3)(x + 2)$

47 $f(x) = (x + 4)(x - 9)$

48 $f(x) = x(x + 12)$

BUSINESS Refer to the second company's CD-player production described in the Portfolio Activity on page 186.

1. Determine the coordinates of the vertices of the feasible region.
2. The second company estimates that it makes a \$20 profit for every regular model produced and a \$30 profit for every sport model produced. Write the objective function for the profit.
3. How many regular models and sport models should the second company produce per hour in order to maximize their profit? What is the maximum profit?

You should now be able to complete the Chapter Project.

37b.

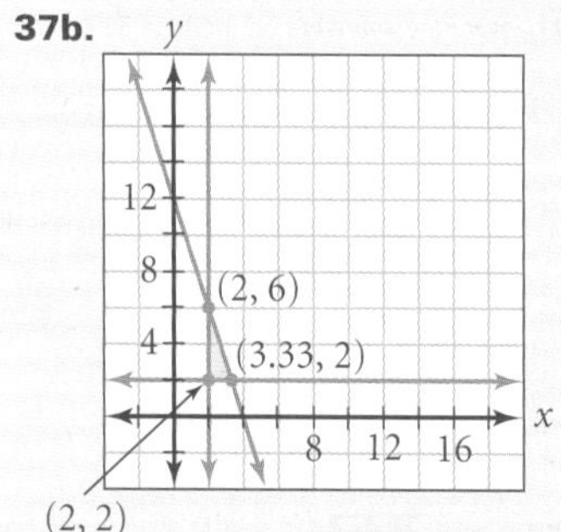

40.

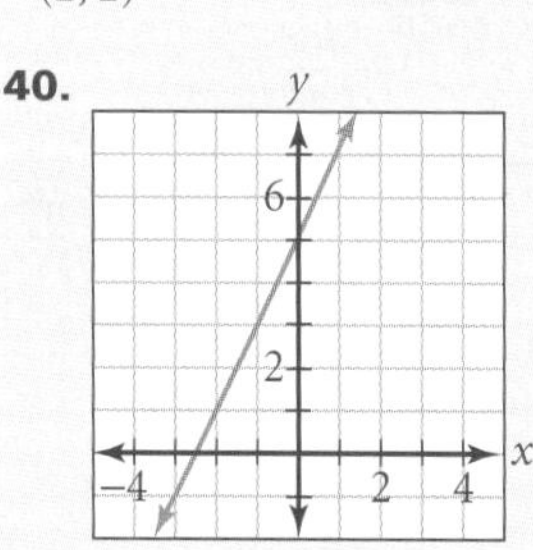

domain: all real numbers
range: all real numbers

41.

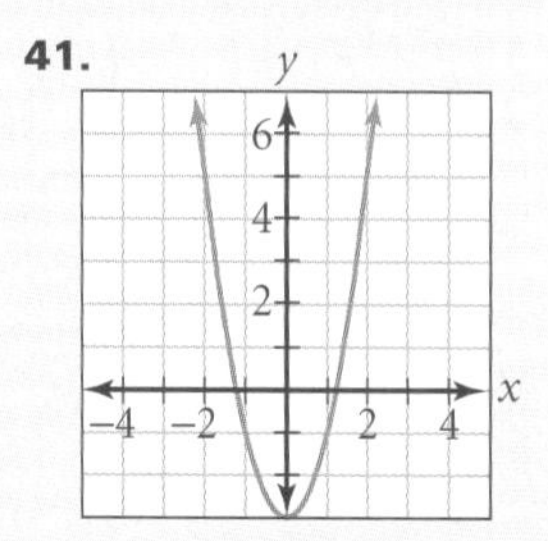

domain: all real numbers
range: $g(x) \ge -3$

46. The x-intercepts are 3 and −2. The factor $x - 3$ gives the x-intercept 3, and the factor $x + 2$ gives the x-intercept −2.

47. The x-intercepts are −4 and 9. The factor $x + 4$ gives the x-intercept −4, and the factor $x - 9$ gives the x-intercept 9.

48. The x-intercepts are 0 and −12. The factor x gives the x-intercept 0, and the factor $x + 12$ gives the x-intercept −12.

Parametric Equations

Why *Parametric equations can be used to describe real-world situations involving two independent variables, such as the linear and vertical positions of an airplane during takeoff.*

Objectives

- Graph a pair of parametric equations, and use them to model real-world applications.
- Write the function represented by a pair of parametric equations.

APPLICATION
AVIATION

A small airplane takes off from a field. One second after takeoff, the airplane is 120 feet down the runway and 13.4 feet above it. The airplane's ascent continues at a constant rate. To analyze the position of the airplane as a function of time, you can represent the horizontal and vertical distances traveled in terms of a third variable, the time after takeoff.

For the Activity below, use the information in the diagram above.

Activity Exploring the Position of an Airplane

You will need: a graphics calculator

Refer to the airplane described above. Let t represent the time in seconds after takeoff, let x represent the horizontal distance in feet traveled in t seconds, and let y represent the vertical distance, or altitude, in feet traveled in t seconds.

1. Copy and complete each table of values below.

t	0	1	2	3	4
x					

t	0	1	2	3	4
y					

2. Verify that the relationships between x and t and between y and t are linear. Write a pair of *parametric equations*: an equation for x in terms of t and an equation for y in terms of t. Then write a linear function for the altitude, y, in terms of horizontal distance traveled, x.

Alternative Teaching Strategy

VISUAL STRATEGIES Show students the graph at right. Tell them to imagine a point, P, that begins at location A and moves along the segment at a constant rate, arriving at location B after 1 second, at location C after 2 seconds, and so on. Ask them to give the coordinates of a point after t seconds. Elicit the response $(3t, 2t)$. Note that an alternative way of writing this is to use the pair of *parametric equations* $\begin{cases} x = 3t \\ y = 2t \end{cases}$. Repeat the activity with a segment that does not contain the origin.

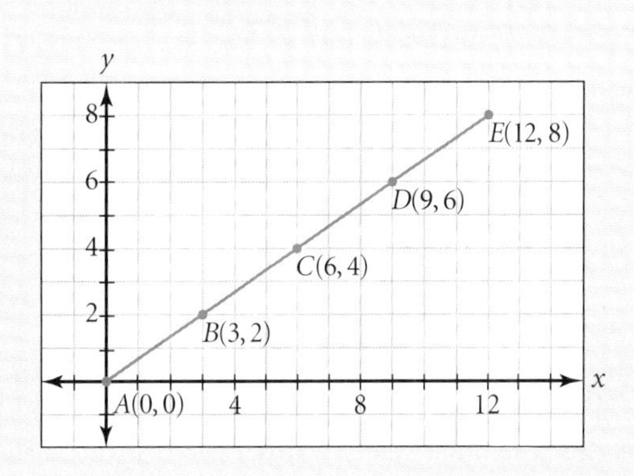

Prepare

NCTM PRINCIPLES & STANDARDS 1–4, 6–10

QUICK WARM-UP

Complete the table by evaluating each expression for the given values of x.

	x	$4x + 3$	$-6x - 5$
1.	−2	−5	7
2.	−1	−1	1
3.	0	3	−5
4.	1	7	−11
5.	2	11	−17

Also on Quiz Transparency 3.6

Teach

Why Consider a baseball t seconds after it has been hit. Ask students if the ball's position is determined only by its distance from home plate or only by its height above the ground. The distance from home plate and the height above the ground are two separate measures, but each depends on the amount of time that the ball has been in motion. In this lesson, students will learn how to describe the ball's position with a pair of *parametric equations.*

Activity Notes

In the Activity on page 195, students write a set of parametric equations to track the ascent of an airplane after takeoff. They should observe that the equations allow them to monitor relationships among three variables: horizontal distance from the point of takeoff, altitude, and elapsed time after takeoff.

Teaching Tip

TECHNOLOGY To enter the parametric equations from the Activity into a TI-82 or TI-83 graphics calculator, you must first press MODE, use the arrow keys to highlight **Par**, and then press ENTER. Now when you press Y=, you will access the parametric equation editor.

When you press WINDOW, you will see that you need to choose settings for the variable **T** as well as the usual settings for **X** and **Y**. If you enter **1** for the **Tstep** and press TRACE, you can use ▶ to track the progress of the airplane one second at a time.

CHECKPOINT ✔

4. No; the parametric equations give the time it takes to reach a certain altitude or horizontal distance.

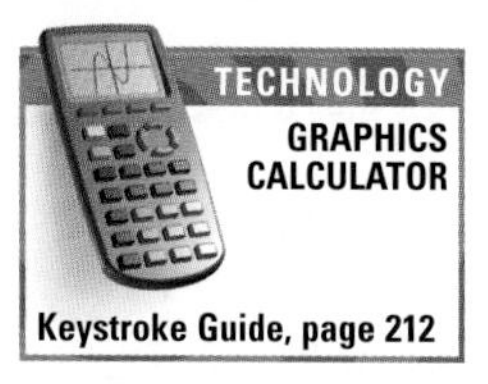

Keystroke Guide, page 212

3. Use a graphics calculator in parametric mode to find how many seconds it will take for the airplane to achieve an altitude of 1500 feet. What is the horizontal distance that the airplane will have traveled from the point of takeoff when it reaches this altitude (assuming that the airplane continues along a straight path)?

CHECKPOINT ✔

4. Can you use the linear function that you wrote in Step 2 to answer the questions in Step 3? What additional information does the pair of parametric equations give you that the linear function does not?

In general, a pair of **parametric equations** is a pair of continuous functions that define the *x*- and *y*-coordinates of a point in a coordinate plane in terms of a third variable, such as *t*, called the **parameter.**

To graph a pair of parametric equations, you can make a table of values or use a graphics calculator, as shown in Example 1.

EXAMPLE 1 **Graph the pair of parametric equations for $-3 \le t \le 3$.** $\begin{cases} x(t) = 2t - 1 \\ y(t) = -t + 2 \end{cases}$

SOLUTION

Method 1 Use graph paper.

PROBLEM SOLVING

Make a table of values, and graph each ordered pair (x, y) in the table. Draw a line segment through the points graphed.

t	x	y
−3	$2(-3) - 1 = -7$	$-(-3) + 2 = 5$
−2	$2(-2) - 1 = -5$	$-(-2) + 2 = 4$
−1	$2(-1) - 1 = -3$	$-(-1) + 2 = 3$
0	$2(0) - 1 = -1$	$-(0) + 2 = 2$
1	$2(1) - 1 = 1$	$-(1) + 2 = 1$
2	$2(2) - 1 = 3$	$-(2) + 2 = 0$
3	$2(3) - 1 = 5$	$-(3) + 2 = -1$

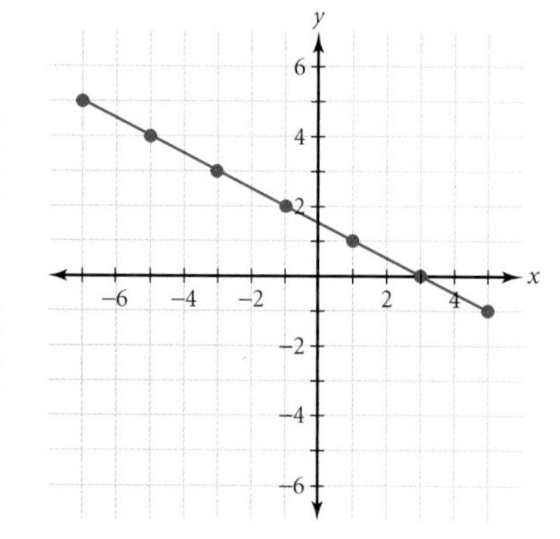

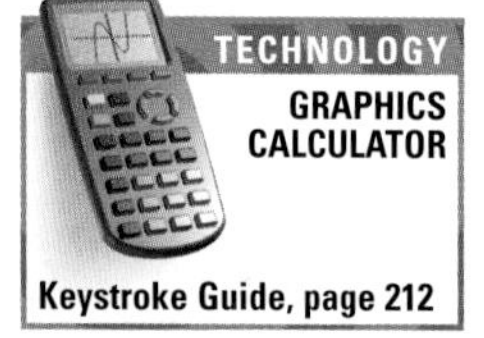

Keystroke Guide, page 212

Method 2 Use a graphics calculator.

In parametric mode, enter the functions for *x* and for *y* in terms of *t*.

Define your viewing window, including minimum and maximum values for *t*. Then graph.

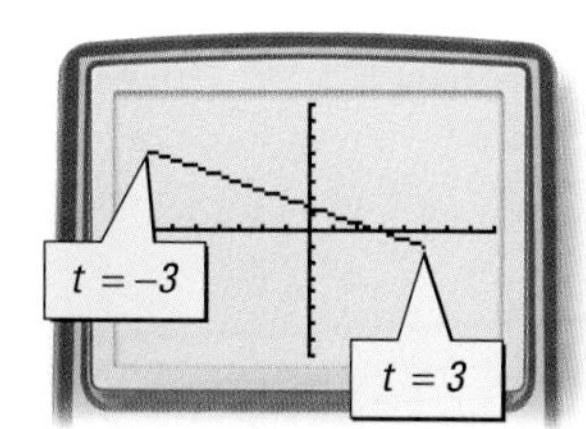

TRY THIS Graph the pair of parametric equations $\begin{cases} x(t) = -2t + 2 \\ y(t) = -t - 2 \end{cases}$ for $-4 \le t \le 4$.

Interdisciplinary Connection

PHYSICS The motion of the baseball described in Example 3 is an example of *projectile motion*. The following is a general set of parametric equations for any projectile motion:

$$\begin{cases} x(t) = d_0 + v_x t \\ y(t) = h_0 + v_y t - 0.5gt^2 \end{cases}$$

In these equations, d_0, and h_0 are the *x*- and *y*-coordinates at the time of release, respectively; v_x is the horizontal velocity at release; v_y is the vertical velocity at release; g is the acceleration due to gravity (32 ft/s^2, or 9.8 m/s^2); and t is the time in seconds after release.

Inclusion Strategies

TACTILE LEARNERS Have students graph functions and their inverses on graph paper, taking care to use the same scale on the *x*-axis and the *y*-axis. They can check their graphs by folding them along the line $y = x$ and holding the folded sheet up to a light source. If their work is correct, the graphs of the function and its inverse should coincide.

CRITICAL THINKING Let r and s be real numbers. Describe the line defined by each pair of parametric equations.

a. $\begin{cases} x(t) = r \\ y(t) = t \end{cases}$ **b.** $\begin{cases} x(t) = t \\ y(t) = s \end{cases}$

Example 2 shows how to describe a pair of parametric equations as an equation in two variables by eliminating the parameter.

EXAMPLE 2

Write the pair of parametric equations as a single equation in x and y. $\begin{cases} x(t) = 2t + 4 \\ y(t) = 5t - 2 \end{cases}$

SOLUTION

Method 1

1. Solve either equation for t.

$$x(t) = 2t + 4 \quad \rightarrow \quad x = 2t + 4$$
$$\frac{x-4}{2} = t$$

2. Substitute the expression for t in the other equation, and simplify.

$$y(t) = 5t - 2 \quad \rightarrow \quad y = 5t - 2$$
$$y = 5\left(\frac{x-4}{2}\right) - 2$$
$$y = \frac{5}{2}x - 10 - 2$$
$$y = \frac{5}{2}x - 12$$

Method 2

1. Solve both equations for t.

$$x(t) = 2t + 4 \quad \rightarrow \quad x = 2t + 4$$
$$\frac{x-4}{2} = t$$
$$y(t) = 5t - 2 \quad \rightarrow \quad y = 5t - 2$$
$$\frac{y+2}{5} = t$$

2. Set the resulting expressions for t equal to each other.

$$\frac{x-4}{2} = \frac{y+2}{5}$$
$$5(x-4) = 2(y+2)$$
$$5x - 20 = 2y + 4$$
$$y = \frac{5}{2}x - 12$$

TRY THIS Write the pair of parametric equations as a single equation in x and y. $\begin{cases} x(t) = -2t - 6 \\ y(t) = 3t - 1 \end{cases}$

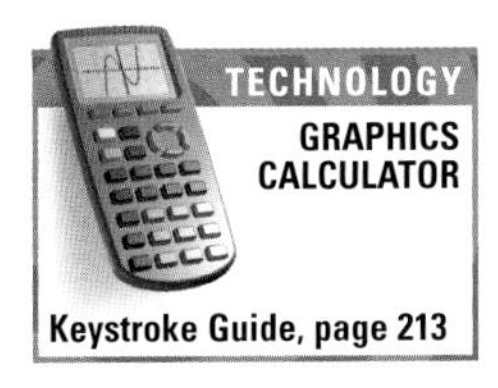

You can graph functions and their inverses by using parametric equations. If f is a function containing the point (x, y), then its inverse contains the point (y, x).

For example, $f(x) = x^2$ can be represented by $\begin{cases} x_1(t) = t \\ y_1(t) = t^2 \end{cases}$. The inverse of $f(x) = x^2$ can be represented by $\begin{cases} x_2(t) = t^2 \\ y_2(t) = t \end{cases}$.

Graph these two pairs of parametric equations for t-values from -5 to 5. Notice that the graphs are reflections of one another across the line $y = x$.

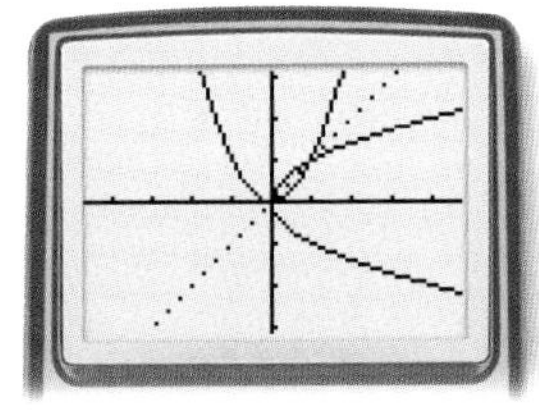

Enrichment

Parametric equations allow you to graph some interesting curves in a coordinate plane. Have students use a graphics calculator to graph the following pair of equations for $-1 \le t \le 1$:

$$\begin{cases} x(t) = (2 + t)(1 - t^2) \\ y(t) = (2 - t)(1 - t^2) \end{cases}$$

Direct them to use the following window settings:

Tmin=–1	Xmin=0	Ymin=0
Tmax=1	Xmax=4.7	Ymax=3.1
Tstep=.01	Xscl=.5	Yscl=.5

The graph is shown below.

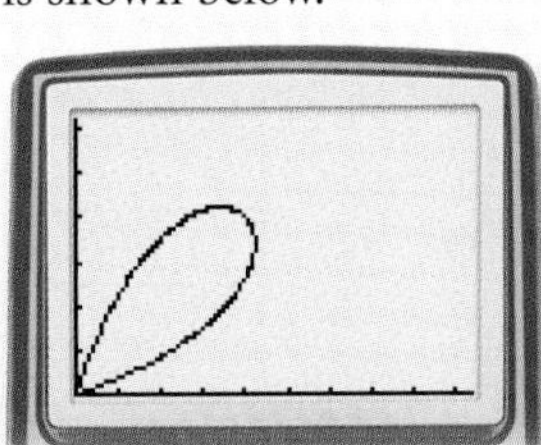

Have students work in pairs. Tell them to experiment with parametric equations to determine what other types of curves they might obtain. **Answers will vary.**

ADDITIONAL EXAMPLE 1

Graph the pair of parametric equations below for $-3 \le t \le 3$.

$$\begin{cases} x(t) = t - 1 \\ y(t) = -2t + 3 \end{cases}$$

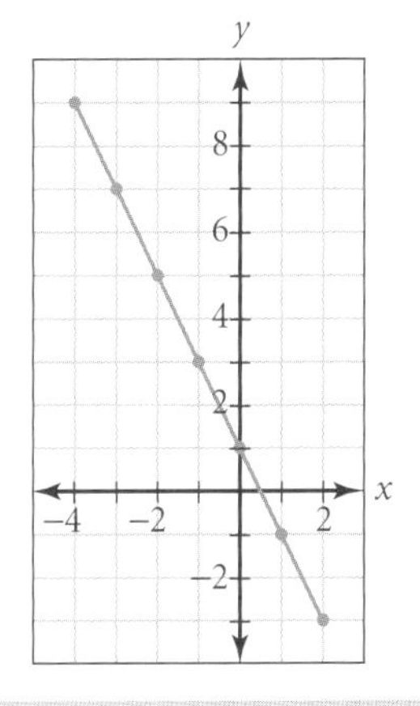

TRY THIS

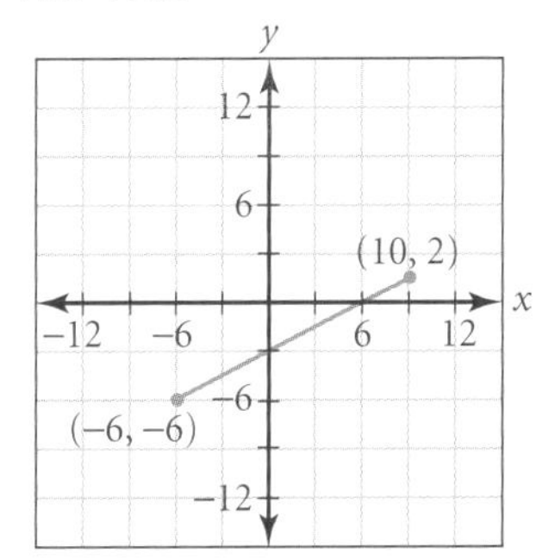

CRITICAL THINKING

a. the vertical line $x = r$

b. the horizontal line $y = s$

ADDITIONAL EXAMPLE 2

Write the pair of parametric equations as a single equation in x and y.

$$\begin{cases} x(t) = 3t - 5 \\ y(t) = -2t + 7 \end{cases}$$

$$y = -\frac{2}{3}x + \frac{11}{3}$$

TRY THIS

$$y = -\frac{3}{2}x - 10$$

Math CONNECTION

TRANSFORMATIONS When graphing inverse functions on a graphics calculator, urge students to use a square viewing window.

ADDITIONAL
EXAMPLE 3

An outfielder throws a softball to the third baseman 290 feet away to stop a runner. The ball is released 6 feet above the ground with a horizontal speed of 75 feet per second and a vertical speed of 44 feet per second. The third baseman's mitt is held 3 feet above the ground. The following parametric equations describe the path of the ball, where t is the number of seconds after the ball is released:

$$\begin{cases} x(t) = 75t \\ y(t) = 6 + 44t - 16t^2 \end{cases}$$

a. When does the ball reach its greatest altitude? The ball reaches its maximum altitude of about 36 ft after about 1.4 s.

b. Can the third baseman catch the ball? no

Teaching Tip

TECHNOLOGY When using parametric equations to create a nonlinear graph on a graphics calculator, it is important to choose a value for **Tstep** that will yield a sufficient number of points to define the curve correctly. In Example 3, for instance, if you use a value for **Tstep** that is too large, the graph will look like a broken line instead of a parabola.

TRY THIS

a. after about 0.95 s

b. no

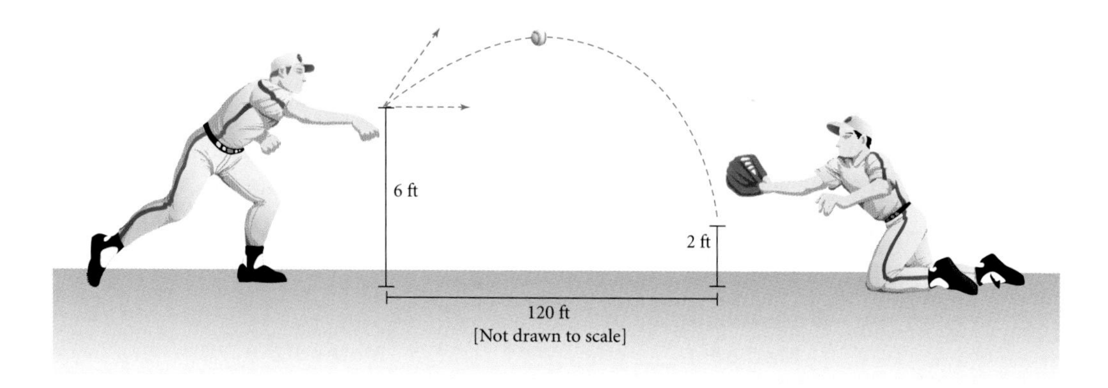

EXAMPLE 3

APPLICATION
SPORTS

An outfielder throws a baseball to the catcher 120 feet away to prevent a runner from scoring. The ball is released 6 feet above the ground with a horizontal speed of 70 feet per second and a vertical speed of 25 feet per second. The catcher holds his mitt 2 feet off the ground. The following parametric equations describe the path of the ball:

$$\begin{cases} x(t) = 70t & \text{x(t) gives the horizontal distance in feet after t seconds.} \\ y(t) = 6 + 25t - 16t^2 & \text{y(t) gives the vertical distance in feet after t seconds.} \end{cases}$$

a. When does the ball reach its greatest altitude?
b. Can the catcher catch the ball?

Keystroke Guide, page 213

SOLUTION

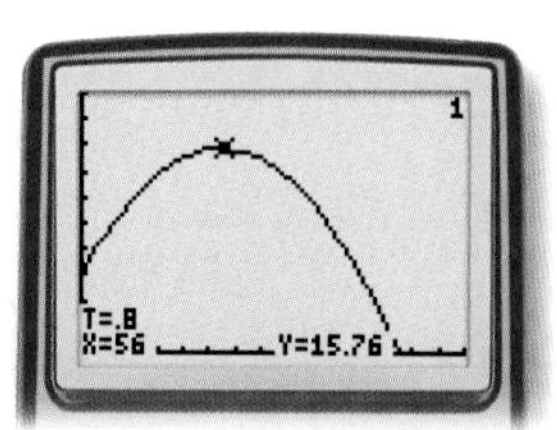

a. Graph the pair of parametric equations. Using the trace feature, you can find that the ball reaches its maximum altitude of about 16 feet after about 0.8 second.

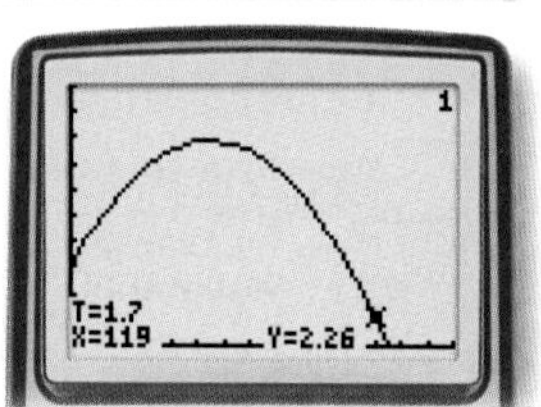

b. When the ball has traveled about 120 feet, the ball will be about 2.3 feet off the ground (and directly in front of the catcher). The catcher should be able to catch the ball.

TRY THIS Suppose that an outfielder throws the ball to the catcher 100 feet away. The ball is released 6 feet above the ground, and the catcher holds his mitt 2 feet off the ground. The following parametric equations describe the path of the ball:

$$\begin{cases} x(t) = 60t & \text{x(t) gives the horizontal distance in feet after t seconds.} \\ y(t) = 6 + 30t - 16t^2 & \text{y(t) gives the vertical distance in feet after t seconds.} \end{cases}$$

a. When does the ball reach its greatest altitude?
b. Can the catcher catch the ball?

Reteaching the Lesson

USING TABLES AND PATTERNS Display the tables on the board or overhead. Ask students to look for a pattern in each table and write a function rule for it. $x(t) = t + 3$; $y(t) = 2t - 1$

t	x
−2	1
−1	2
0	3
1	4
2	5

t	y
−2	−5
−1	−3
0	−1
1	1
2	3

Tell students that the two equations together form a pair of parametric equations.

Now have students make a table of just the x-and y-values. Stress the importance of keeping the values in the same order used in the original tables. Ask students to look for a pattern in the new table and write a function rule for it. See table at right; $y = 2x - 7$.

x	y
1	−5
2	−3
3	−1
4	1
5	3

Point out that they have written a single equation in x and y that is equivalent to their pair of parametric equations. Apply the methods used in Example 2 to verify this.

Exercises

Communicate

1. Describe the procedure for eliminating the parameter t in the pair of parametric equations $x(t) = 2t + 3$ and $y(t) = -t + 1$.
2. Explain what is lost when a pair of parametric equations are rewritten as a function in two variables.
3. Describe what each variable in the parametric equations in Example 3 represents.

Guided Skills Practice

Graph each pair of parametric equations for the given interval of t. ***(EXAMPLE 1)***

4. $\begin{cases} x(t) = t \\ y(t) = 3t + 2 \end{cases}$ for $-3 \le t \le 3$

5. $\begin{cases} x(t) = t + 7 \\ y(t) = 8 - t \end{cases}$ for $-4 \le t \le 4$

Write each pair of parametric equations as a single equation in x and y. ***(EXAMPLE 2)***

6. $\begin{cases} x(t) = 5t - 6 \\ y(t) = 3t + 1 \end{cases}$ $y = \frac{3}{5}x + \frac{23}{5}$

7. $\begin{cases} x(t) = 2t + 5 \\ y(t) = 3 - 2t \end{cases}$ $y = -x + 8$

APPLICATION

8. **SPORTS** A batter hits a ball 3 feet above the ground with a horizontal speed of 98 feet per second and a vertical speed of 45 feet per second toward the outfield fence. The fence is 250 feet from the batter and 10 feet high. If $x(t)$ gives the horizontal distance in feet after t seconds and $y(t)$ gives the vertical distance in feet after t seconds, the following parametric equations describe the path of the ball. ***(EXAMPLE 3)***

$$\begin{cases} x(t) = 98t \\ y(t) = 3 + 45t - 16t^2 \end{cases}$$

a. How long will it take the ball to reach the fence? approx. 2.6 seconds

b. Will the ball go over the fence? yes

Practice and Apply

Graph each pair of parametric equations for the given interval of t.

9. $\begin{cases} x(t) = 3t \\ y(t) = t - 2 \end{cases}$ for $-4 \le t \le 4$

10. $\begin{cases} x(t) = t + 5 \\ y(t) = 4t \end{cases}$ for $-3 \le t \le 3$

11. $\begin{cases} x(t) = 2t \\ y(t) = 6 - t \end{cases}$ for $-3 \le t \le 3$

12. $\begin{cases} x(t) = 5 - 2t \\ y(t) = \frac{t}{2} \end{cases}$ for $-4 \le t \le 4$

13. $\begin{cases} x(t) = 2t + 3 \\ y(t) = t^2 \end{cases}$ for $-3 \le t \le 3$

14. $\begin{cases} x(t) = t^2 \\ y(t) = 3t - 5 \end{cases}$ for $-4 \le t \le 4$

12.
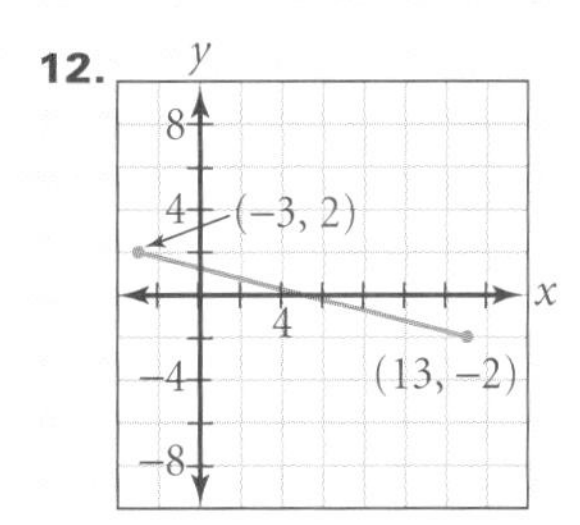

13.
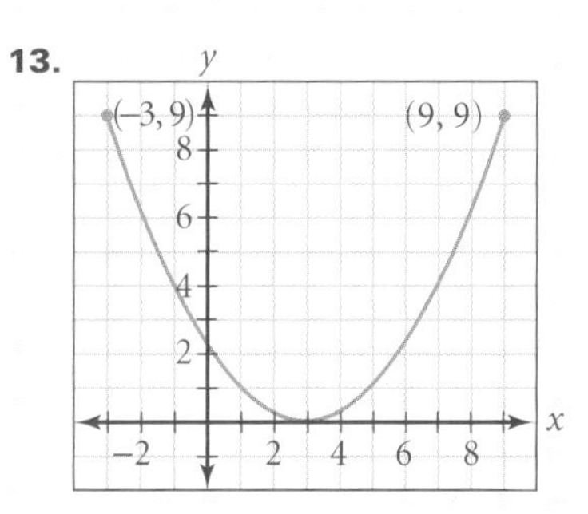

14.
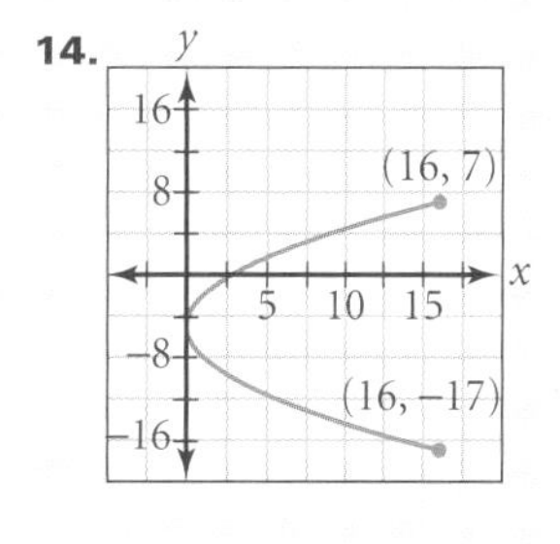

Assess

Selected Answers

Exercises 4–8, 9–49 odd

ASSIGNMENT GUIDE

In Class	1–8
Core	9–31 odd, 35
Core Plus	9–14, 16–26 even, 27–38
Review	39–50
Preview	51

Extra Practice can be found beginning on page 940.

4.
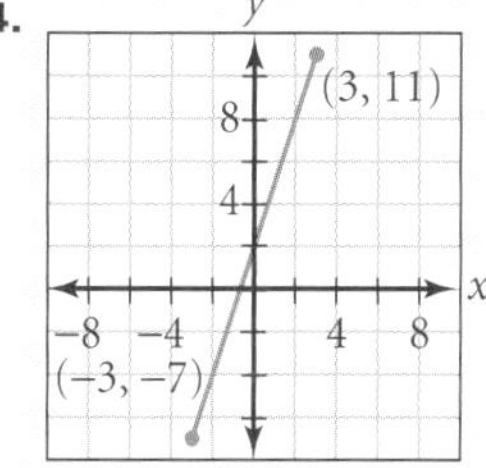

5.
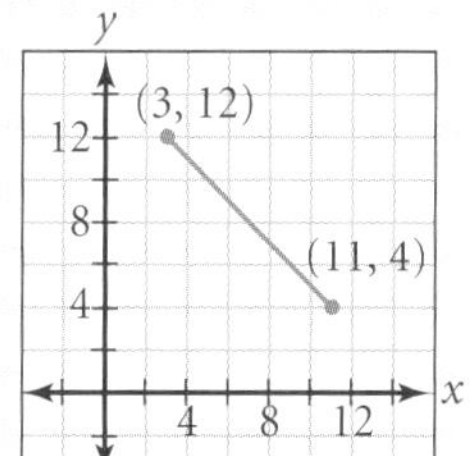

9.
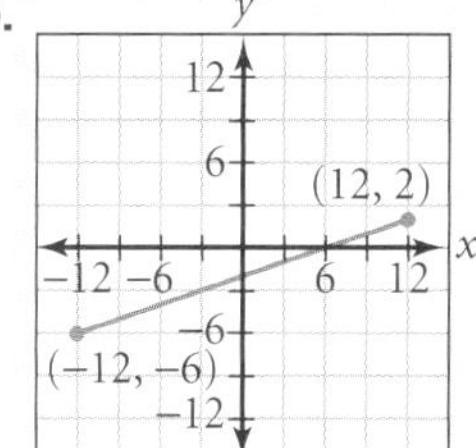

10.
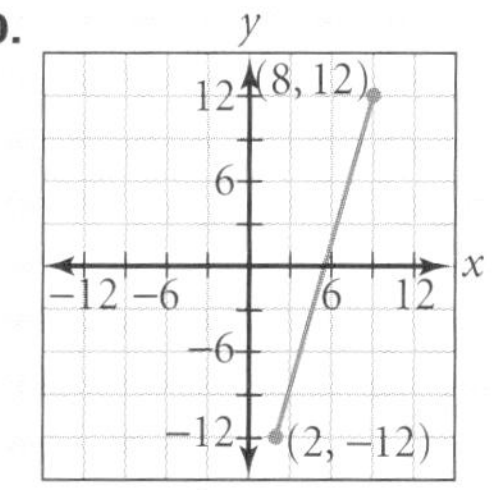

11.
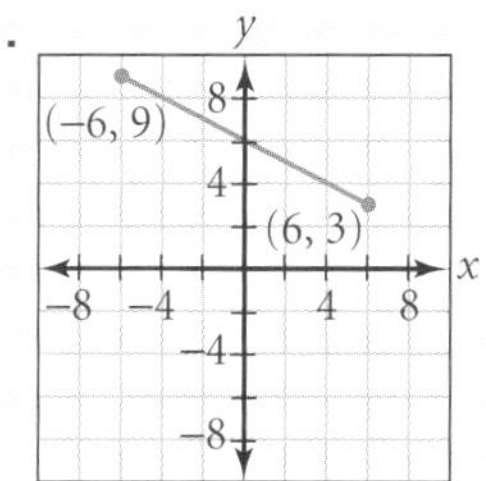

Error Analysis

Watch for students who choose a value for **Tmax** that is too large or too small. If a large value is chosen, a long time may elapse before the trace function can be used. If a small value is chosen, the graph will be incomplete. If either of these situations occurs, encourage students to adjust this value accordingly.

27. $\begin{cases} x(t) = t \\ y(t) = t^2 - 2 \end{cases}$

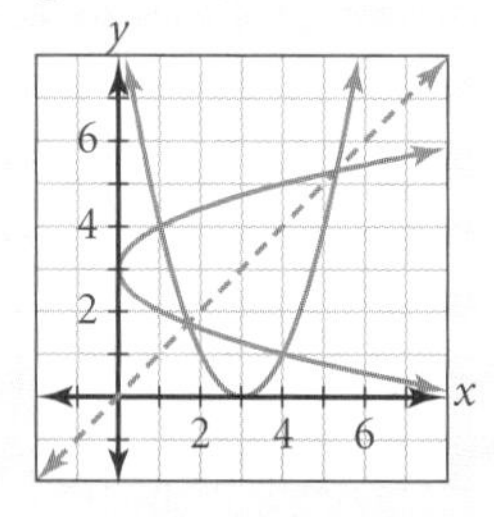

28. $\begin{cases} x(t) = t + 3 \\ y(t) = t^2 \end{cases}$

y
6
4
2
2
4
6
x

Practice

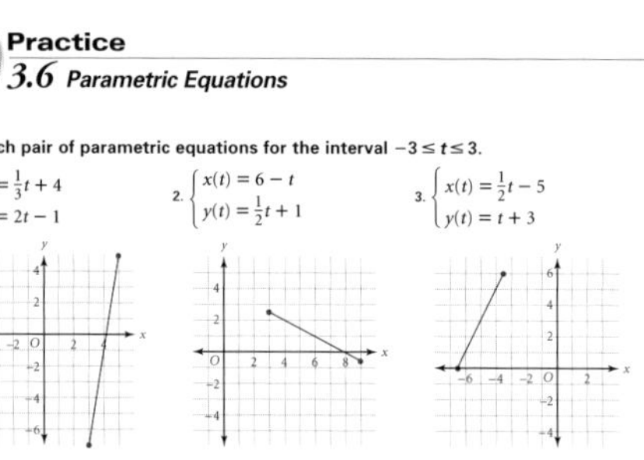

NAME ______ CLASS ______ DATE ______

Practice
3.6 Parametric Equations

Graph each pair of parametric equations for the interval $-3 \le t \le 3$.

1. $\begin{cases} x(t) = \frac{1}{3}t + 4 \\ y(t) = 2t - 1 \end{cases}$
2. $\begin{cases} x(t) = 6 - t \\ y(t) = \frac{1}{2}t + 1 \end{cases}$
3. $\begin{cases} x(t) = \frac{1}{2}t - 5 \\ y(t) = t + 3 \end{cases}$

Write each pair of parametric equations as a single equation in x and y.

4. $\begin{cases} x(t) = 2t + 12 \\ y(t) = t - 8 \end{cases}$ $y = \frac{1}{2}x - 14$
5. $\begin{cases} x(t) = \frac{1}{2}t \\ y(t) = 3t + 2 \end{cases}$ $y = 6x + 2$
6. $\begin{cases} x(t) = 4 + 2t \\ y(t) = 6 - t \end{cases}$ $y = -\frac{1}{2}x + 8$
7. $\begin{cases} x(t) = 10 - 2t \\ y(t) = t + 11 \end{cases}$ $y = -\frac{1}{2}x + 16$
8. $\begin{cases} x(t) = 3 - 12t \\ y(t) = 4t - 3 \end{cases}$ $y = \frac{1}{3}x - 2$
9. $\begin{cases} x(t) = 3t \\ y(t) = t^2 + 1 \end{cases}$ $y = \frac{x^2}{9} + 1$

An airplane is ascending at a constant rate. Its altitude changes at a rate of 12 feet per second. Its horizontal speed is 150 feet per second.

10. a. Write parametric equations that represent the plane's flight path. See right.
b. Graph the equations for the interval $0 \le t \le 30$. Use the grid at right.

10a. $\begin{cases} x(t) = 150t \\ y(t) = 12t \end{cases}$

11. a. How long will it take the plane to reach an altitude of 300 ft? 25 seconds
b. How far will the plane travel horizontally in that time? 3750 feet

Write each pair of parametric equations as a single equation in *x* and *y*.

15. $\begin{cases} x(t) = 2t \\ y(t) = t - 1 \end{cases}$ $y = \frac{1}{2}x - 1$

16. $\begin{cases} x(t) = t + 3 \\ y(t) = 3t \end{cases}$ $y = 3x - 9$

17. $\begin{cases} x(t) = 2t + 1 \\ y(t) = t + 5 \end{cases}$ $y = \frac{1}{2}x + \frac{9}{2}$

18. $\begin{cases} x(t) = t - 2 \\ y(t) = t + 7 \end{cases}$ $y = x + 9$

19. $\begin{cases} x(t) = 3t \\ y(t) = 1 - t \end{cases}$ $y = 1 - \frac{1}{3}x$

20. $\begin{cases} x(t) = t \\ y(t) = 3 - 2t \end{cases}$ $y = 3 - 2x$

21. $\begin{cases} x(t) = 2t \\ y(t) = t^2 - 1 \end{cases}$ $y = \frac{1}{4}x^2 - 1$

22. $\begin{cases} x(t) = t^2 \\ y(t) = \frac{t}{2} \end{cases}$ $x = 4y^2$

23. $\begin{cases} x(t) = \frac{1}{3}t \\ y(t) = t^2 \end{cases}$ $y = 9x^2$

24. $\begin{cases} x(t) = t^2 + 2t \\ y(t) = 2t \end{cases}$ $x = \frac{1}{4}y^2 + y$

25. $\begin{cases} x(t) = 5 - t^2 \\ y(t) = \frac{3}{2}t \end{cases}$ $x = 5 - \frac{4}{9}y^2$

26. $\begin{cases} x(t) = 2 - 3t^2 \\ y(t) = -\frac{1}{3}t \end{cases}$ $x = 2 - 27y^2$

internet connect go.hrw.com
Homework Help Online
Go To: **go.hrw.com**
Keyword:
MB1 Homework Help
for Exercises 27–32

Graph the function represented by each pair of parametric equations. Then graph its inverse on the same coordinate plane.

27 $\begin{cases} x(t) = t^2 - 2 \\ y(t) = t \end{cases}$

28 $\begin{cases} x(t) = t^2 \\ y(t) = t + 3 \end{cases}$

29 $\begin{cases} x(t) = t \\ y(t) = 6 - t^2 \end{cases}$

30 $\begin{cases} x(t) = 4 - t \\ y(t) = t^2 - 1 \end{cases}$

31 $\begin{cases} x(t) = t^2 + 5t - 1 \\ y(t) = t + 1 \end{cases}$

32 $\begin{cases} x(t) = 4 + 5t - t^2 \\ y(t) = t - 1 \end{cases}$

CHALLENGE

33. Write a pair of parametric equations to represent a line that has a slope of 3 and contains the point (4, –5). $x(t) = t,\ y(t) = 3t - 17$

CONNECTION

34. TRANSFORMATIONS Write the pair of parametric equations that represent a transformation of $\begin{cases} x(t) = t \\ y(t) = t^2 \end{cases}$ 1 unit down and 2 units to the right. $x(t) = t + 2,\ y(t) = t^2 - 1$

APPLICATION

35 SPORTS Frannie throws a softball from one end of a 200-foot field. The ball leaves her hand at a height of 6.5 feet with an initial velocity of 60 feet per second in the horizontal direction and 40 feet per second in the vertical direction. If $x(t)$ gives the horizontal distance in feet after t seconds and $y(t)$ gives the vertical distance in feet after t seconds, the following parametric equations describe the ball's path:

$$\begin{cases} x(t) = 60t \\ y(t) = 6.5 + 40t - 16t^2 \end{cases}$$

a. How high does the ball get? How long does it take for the ball to reach this height? **31.5 ft; 1.25 s**

b. What horizontal distance will the ball travel before it hits the ground? How long does it take for the ball to reach this point? **159.2 ft; 2.65 s**

36 SPORTS The server in a volleyball game serves the ball at an angle of 35° with the ground and from a height of 2 meters. The server is 9 meters from the 2.2-meter high net. The ball must not touch the net when served and must land within 9 meters of the other side of the net. If $x(t)$ gives the horizontal distance in meters after t seconds and $y(t)$ gives the vertical distance in meters after t seconds, the following parametric equations describe the ball's path:

$$\begin{cases} x(t) = 8.2t \\ y(t) = 2 + 5.7t - 4.9t^2 \end{cases}$$

a. How high above the net will the ball travel? **about 1.5 m**

b. According to the parametric equations, what horizontal distance will the ball travel before hitting the ground? **about 11.8 m**

c. Will the ball land within the area described? Explain. **yes**

29. $\begin{cases} x(t) = 6 - t^2 \\ y(t) = t \end{cases}$

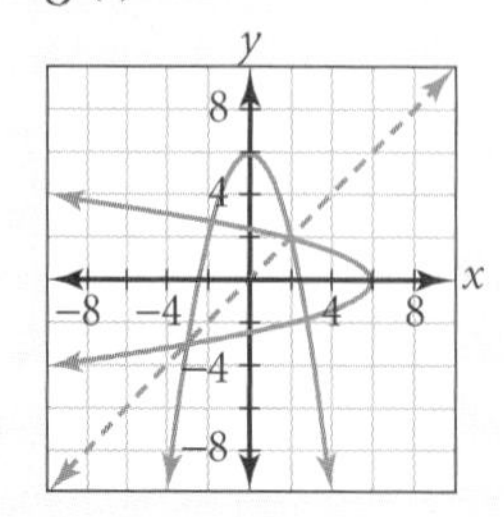

30. $\begin{cases} x(t) = t^2 - 1 \\ y(t) = 4 - t \end{cases}$

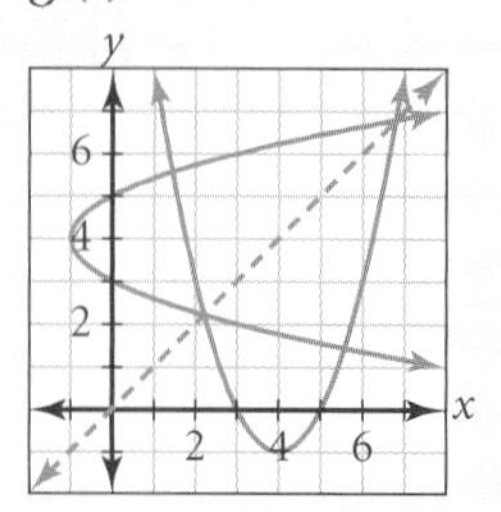

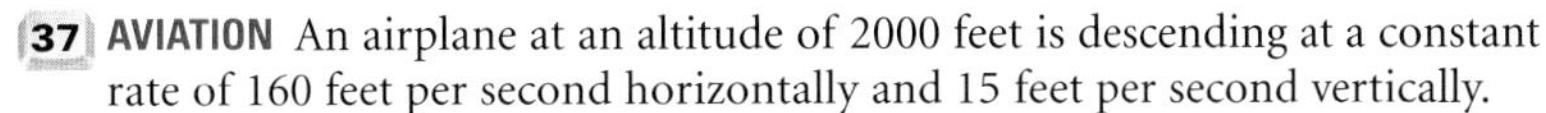

APPLICATION

37a. $\begin{cases} x(t) = 160t \\ y(t) = 2000 - 15t \end{cases}$

b. about 133 s

c. about 21,300 ft

37 **AVIATION** An airplane at an altitude of 2000 feet is descending at a constant rate of 160 feet per second horizontally and 15 feet per second vertically.

a. Write the pair of parametric equations that represent the airplane's flight path.

b. After how many seconds will the airplane touch down?

c. What horizontal distance will the airplane have traveled when it touches down on the runway?

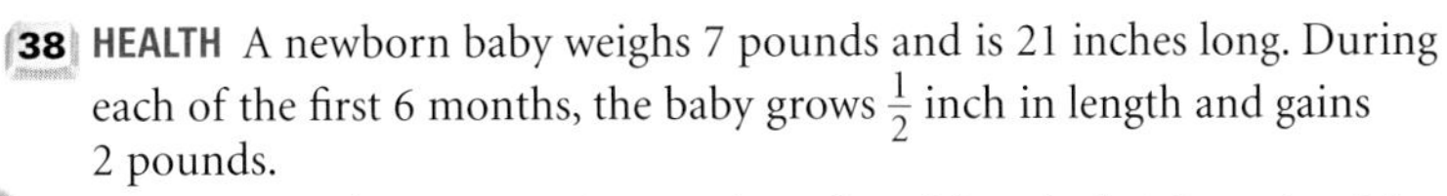

38 **HEALTH** A newborn baby weighs 7 pounds and is 21 inches long. During each of the first 6 months, the baby grows $\frac{1}{2}$ inch in length and gains 2 pounds.

a. Write the parametric equations describing the height and weight for t months, where $0 < t < 6$. $x(t) = 21 + \frac{1}{2}t,\ y(t) = 7 + 2t$

b. How many months will it take for the baby to weigh 14 pounds? How long is the baby at this time? $3\frac{1}{2}$ mo.; $22\frac{3}{4}$ in.

c. How many months will it take for the baby to reach 23 inches? How much does the baby weigh at this time? **4 mo.; 15 lb**

Look Back

Let $f(x) = x + 2$ and $g(x) = 1 - x$. ***(LESSON 2.4)***

39. Find $f \circ g$. $(f \circ g)(x) = 3 - x$

40. Find $g \circ f$. $(g \circ f)(x) = -1 - x$

41. Find $f \circ f$. $(f \circ f)(x) = x + 4$

42. Find $g \circ g$. $(g \circ g)(x) = x$

CONNECTIONS

TRANSFORMATIONS Identify each transformation from the parent function $f(x) = x^2$ to g. ***(LESSON 2.7)***

43. $g(x) = 12x^2 + 3$

44. $g(x) = -\left(\frac{1}{3}x\right)^2$

45. $g(x) = -\frac{1}{2}x^2 + 4$

46. $g(x) = -0.25(4x - 1)^2$

TRANSFORMATIONS Write the function for each graph described below. ***(LESSON 2.7)***

47. $g(x) = |x - 2|$

48. $g(x) = |x| + 1.5$

49. $g(x) = \frac{1}{5}x^2$

50. $g(x) = -\frac{1}{4}x^2$

47. the graph of $f(x) = |x|$ translated 2 units to the right

48. the graph of $f(x) = |x|$ translated 1.5 units up

49. the graph of $f(x) = x^2$ compressed vertically by a factor of $\frac{1}{5}$

50. the graph of $f(x) = x^2$ reflected across the x-axis and stretched horizontally by a factor of 4

Look Beyond

51 Graph the parametric equations $x(t) = \cos(t)$ and $y(t) = \sin(t)$ by using the SIN and COS keys on your calculator. Make sure your calculator is in radian and parametric modes. Use the following viewing window:

Tmin = 0	Tmax = 10	Tstep = 0.01
Xmin = −4.7	Xmax = 4.8	Xscl = 1
Ymin = −3.1	Ymax = 3.2	Yscl = 1

Did the calculator draw the figure in a clockwise or counterclockwise direction? What is the shape of the graph? **counterclockwise; circle**

Look Beyond

Exercise 50 is an informal introduction to basic concepts of trigonometric and analytical functions that are developed fully in Chapters 13 and 14.

31. $\begin{cases} x(t) = t + 1 \\ y(t) = t^2 + 5t - 1 \end{cases}$

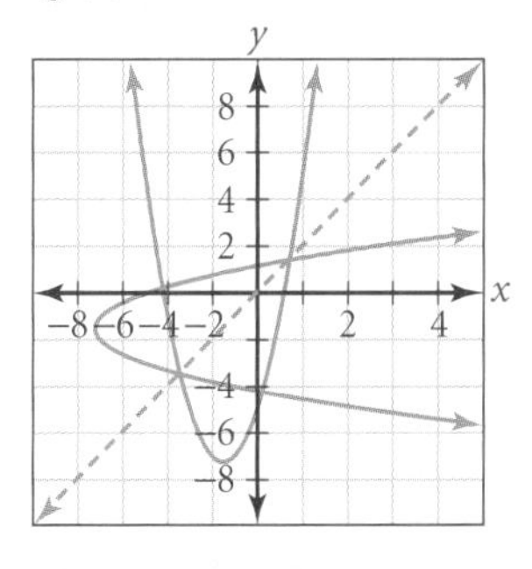

32. $\begin{cases} x(t) = t - 1 \\ y(t) = 4 + 5t - t^2 \end{cases}$

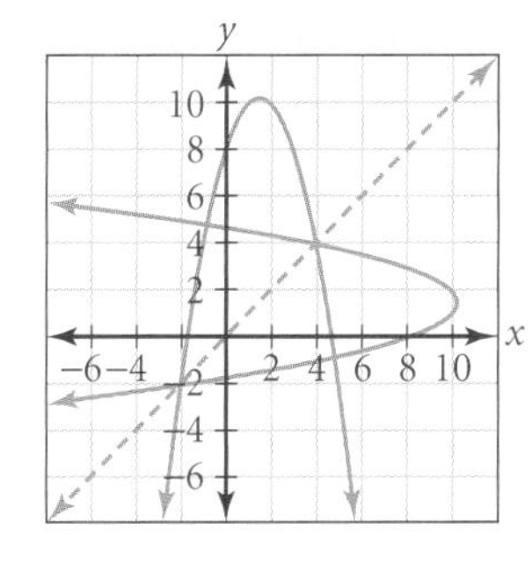

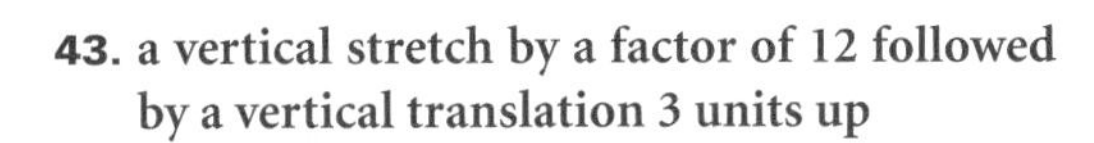

43. a vertical stretch by a factor of 12 followed by a vertical translation 3 units up

44. a horizontal stretch by a factor of 3 followed by a reflection across the x-axis

45. a horizontal stretch by the factor 2, a reflection across the x-axis, and then a vertical translation 4 units up

46. a horizontal compression by a factor of $\frac{1}{4}$, a horizontal translation 1 unit to the right, a vertical stretch by a factor of 0.25, and a reflection across the x-axis

Student Technology Guide

NAME ______ CLASS ______ DATE ______

Student Technology Guide
3.6 Parametric Equations

To graph the parametric equations $\begin{cases} x(t) = t^2 - 3 \\ y(t) = 2t + 3 \end{cases}$, for $-4 \le t \le 4$, follow these steps. Press MODE Par ENTER. Press WINDOW. Enter Tmin = −4, Tmax = 4, and Tstep = .1. Press Y= X,T,θ,n x^2 − 3 ENTER 2 X,T,θ,n + 3 GRAPH. Set Xmin = −9, Xmax = 18, Ymin = −6, and Ymax = 12.

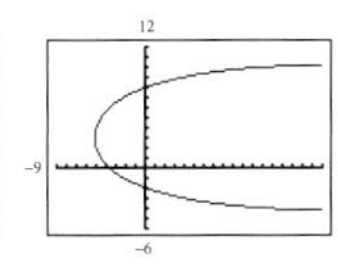

For each pair of parametric equations, give a viewing window that shows a characteristic graph for the given interval of t. Express your answer in the form $-9 \le x \le 18$ and $-6 \le y \le 12$. Answers may vary.

1. $\begin{cases} x(t) = 3t - 4 \\ y(t) = 0.6t \end{cases}$ $0 \le t \le 12$ — $-4 \le x \le 32,\ 0 \le y \le 7.2$

2. $\begin{cases} x(t) = -2t + 8 \\ y(t) = t^2 - 50 \end{cases}$ $-2 \le t \le 10$ — $-12 \le x \le 12,\ -50 \le y \le 50$

To graph the parametric equations in the example and its inverse, enter X1T and Y1T as above. Then follow the key sequence below.

X2T: VARS Y-VARS 2: Parametric... ENTER 2:Y1T ENTER

Y2T: VARS Y-VARS 2: Parametric... ENTER 1:X1T ENTER GRAPH

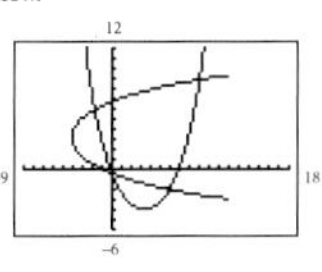

Find appropriate window settings to view each pair of equations and its inverse. Answers may vary.

3. $\begin{cases} x(t) = -2t + 5 \\ y(t) = 3t + 1 \end{cases}$ $-4 \le t \le 4$ — $-9 \le x \le 18,\ -6 \le y \le 12$

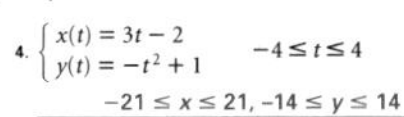

4. $\begin{cases} x(t) = 3t - 2 \\ y(t) = -t^2 + 1 \end{cases}$ $-4 \le t \le 4$ — $-21 \le x \le 21,\ -14 \le y \le 14$

Focus

When running a business, a major concern is attaining a specific goal while making the most efficient use of limited resources. In this Project, students will solve two business problems by using the linear programming techniques that they learned in this chapter. One problem focuses on the goal of maximizing profits, while the other focuses on minimizing labor costs.

Motivate

Discuss with students their perceptions of the way that businesses operate. Point out that the first two activities of the Project concern a florist. Ask students what types of expenses a florist might incur. Be sure they understand that to make a profit, these expenses must be offset by the prices affixed to the products sold. Discuss other factors that the florist must consider in setting prices, such as the need to remain competitive with other florists in the neighborhood.

When discussing the floral business, students should have noted that a major business expense is the cost of labor. In the second set of activities, students will shift their focus to the problem of minimizing labor costs. The situation they must consider is tile installation. Be sure they understand that a craftsman is highly skilled at the job but commands a high rate of pay. Apprentices receive less pay, but they generally cannot work as quickly or efficiently as a craftsman.

MAXIMUM PROFIT/ MINIMUM COST

FLORAL ARRANGEMENTS FOR MAXIMUM PROFITS

A local florist is making two types of floral arrangements for Thanksgiving: regular and special. Each regular arrangement requires 3 mums, 3 daisies, and 2 roses, and each special arrangement requires 4 mums, 2 daisies, and 4 roses. The florist has set aside 60 mums, 54 daisies, and 52 roses for the two types of arrangements.

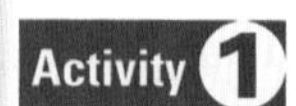

Let x represent the number of regular arrangements, and let y represent the number of special arrangements. Write a system of three equations to represent the florist's situation.

Activity 2

The florist will make a profit of \$2 on each regular arrangement and \$3 on each special arrangement.

1. Write an objective function for the profit.
2. Create a system of inequalities to represent the constraints. Graph the feasible region.
3. Identify the vertices of the feasible region.
4. How many of each type of arrangement should the florist make in order to maximize the profit? What is the maximum profit?
5. If the maximum profit is achieved, will there be any flowers left over? Explain your reasoning.

Activity 1

$3x + 4y = 60$

$3x + 2y = 54$

$2x + 4y = 52$

Activity 2

1. $P = 2x + 3y$

2. $\begin{cases} 3x + 4y \le 60 \\ 3x + 2y \le 54 \\ 2x + 4y \le 52 \\ x \ge 0, y \ge 0 \end{cases}$

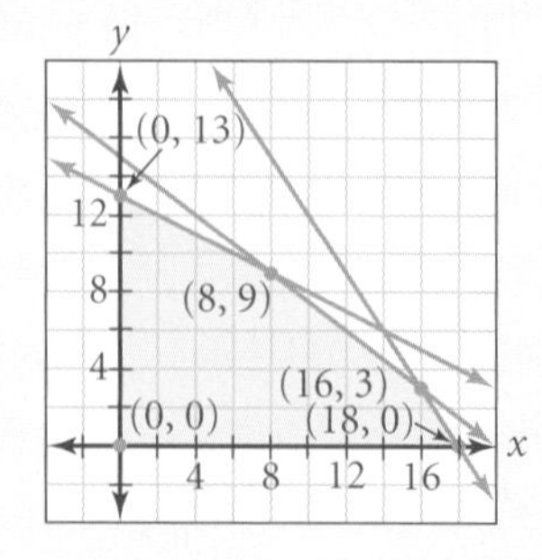

4. 8 regular and 9 special; \$43

5. flowers left: 0 mums, 12 daisies, and 0 roses

Activity 3

$500x + 100y = 2000$

$100x + 200y = 1600$

$100x + 100y = 1200$

The answer to Step 3 of Activity 2 can be found in Additional Answers beginning on page 1002.

TILING AT MINIMUM LABOR COST

A construction firm employs two levels of tile installers: craftsmen and apprentices. Craftsmen install 500 square feet of specialty tile, 100 square feet of plain tile, and 100 linear feet of trim in one day. An apprentice installs 100 square feet of specialty tile, 200 square feet of plain tile, and 100 linear feet of trim in one day. The firm has a one-day job that requires 2000 square feet of specialty tile, 1600 square feet of plain tile, and 1200 linear feet of trim.

Activity 3

Let x represent the number of craftsmen, and let y represent the number of apprentices. Write a system of three equations to represent the construction firm's situation with this job.

Activity 4

The construction firm pays craftsmen \$200 per day and pays apprentices \$120 per day.

1. Write an objective function for the labor costs.
2. Create a system of inequalities to represent the constraints. Graph the feasible region.
3. Identify the vertices of the feasible region.
4. How many craftsmen and how many apprentices should be assigned to this job so that it can be completed in one day with the minimum labor cost? What is the minimum labor cost?

Activity 5

Suppose that each apprentice's wages are increased to \$150 per day.

1. In this case, how many craftsmen and how many apprentices should be assigned to the job so that it can be completed in one day with the minimum labor cost? What is the minimum labor cost?
2. Do any points in the feasible region call for apprentices but no craftsmen? If so, is this a realistic scenario for the construction firm? What constraint could you add to ensure that every job has at least one craftsman assigned to it?

Activity 6

Suppose that union regulations require at least 1 craftsman for every 3 apprentices on a job.

1. Which of the vertices of the original feasible region (from Activity 4) satisfy this new constraint?
2. Write an inequality to represent the new constraint. Modify the feasible region on your graph by adding the boundary line for the new constraint.
3. Using the new constraints and the original labor cost of \$120 per day for an apprentice, how many craftsmen and how many apprentices should be assigned to the job so that it can be completed in one day with the minimum labor cost? What is the minimum labor cost?

Cooperative Learning

Have the students do the Project in pairs. Tell them that each pair is to act as a consulting partnership that has been hired to help the florist and the construction firm achieve their goals.

Have students begin **Activities 1 and 2**, working independently. After Step 3 of Activity 2, partners can compare results and resolve any discrepancies. They can then do Steps 4 and 5 together.

Conduct **Activities 3 and 4** in a similar fashion, with the partners working independently through Step 3 of Activity 4, and then coming together to do Step 4. The partners can then proceed to do **Activities 5 and 6** together.

Discuss

After all the pairs have completed the activities, bring the class together to compare and discuss the results. Encourage students to share any insights they have gained into the technique of linear programming.

Point out to students that the labor costs they have calculated for the tiling installation are only *projected* costs. Final costs of construction projects like this seldom are exactly equal to projected costs. Ask students what might happen to make the actual labor costs of the tiling installation less than or more than projected costs.

Activity 4

1. $C = 200x + 120y$
2. $\begin{cases} 500x + 100y \geq 2000 \\ 100x + 200y \geq 1600 \\ 100x + 100y \geq 1200 \\ x \geq 0, y \geq 0 \end{cases}$
3. The coordinates of the vertices are (0, 20), (2, 10), (8, 4), and (16, 0).
4. 2 craftsmen and
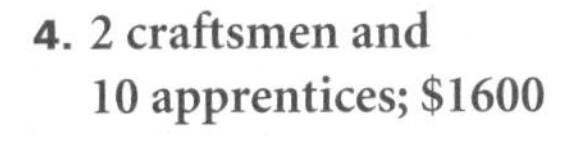
10 apprentices; \$1600

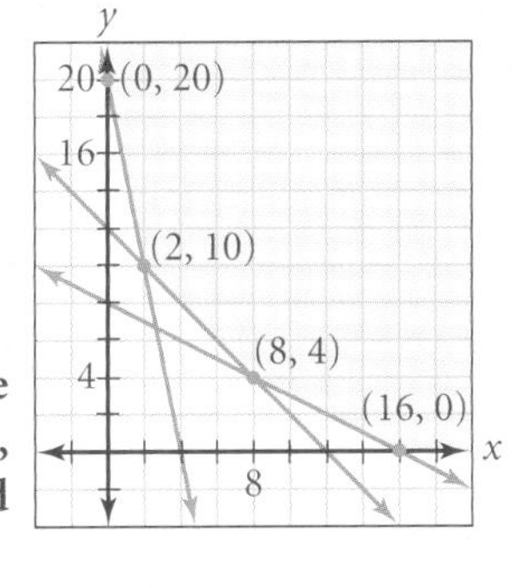

Activity 5

1. 2 craftsmen and 10 apprentices; \$1900
2. Yes, (0, 20); no, labor costs are much higher than the optimal solution; $x \geq 1$.

Activity 6

1. (8, 4) and (16, 0)
2. $3x - y \geq 0$;
3. 3 craftsmen and 9 apprentices; \$1680

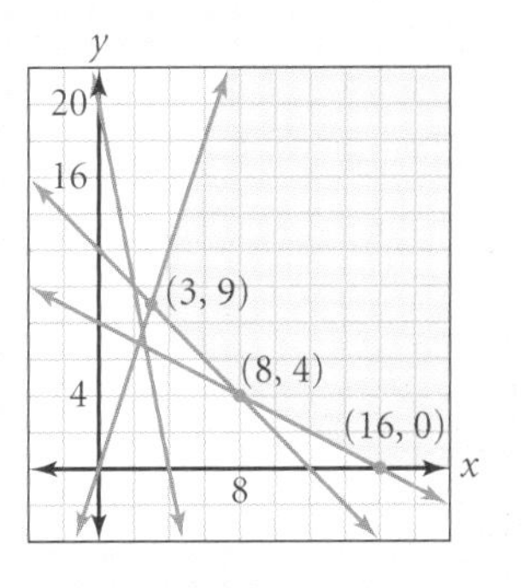

3 Chapter Review and Assessment

VOCABULARY

Key Skills & Exercises

LESSON 3.1

Key Skills

Solve a system of two linear equations in two variables graphically.

Classification of Systems of Equations	
Number of solutions	Type
0	inconsistent
1	consistent, independent
infinite	consistent, dependent

The solution to a consistent, independent system is given by the point of intersection of the graphs.

Use substitution to solve a system of linear equations.

Solve $\begin{cases} 3x - 5y = 28 \\ x + y = 4 \end{cases}$ by using substitution.

First solve $x + y = 4$ for one variable.

$$y = -x + 4$$

Then substitute $-x + 4$ for y in the other equation.

$$3x - 5y = 28$$
$$3x - 5(-x + 4) = 28$$
$$8x = 48$$
$$x = 6$$

Substitute 6 for x to find y.

$$x + y = 4$$
$$6 + y = 4$$
$$y = -2$$

Thus, the solution is $(6, -2)$.

Exercises

Graph and classify each system. Then find the solution from the graph.

1. $\begin{cases} x + y = 6 \\ 3x - 4y = 4 \end{cases}$ **indep. (4, 2)**
2. $\begin{cases} 3x + y = 11 \\ x - 2y = 6 \end{cases}$ **indep. (4, –1)**
3. $\begin{cases} y = 2x - 3 \\ -6x + 3y = -9 \end{cases}$ **dep.**
4. $\begin{cases} x + 2y = 4 \\ -3x - 6y = 12 \end{cases}$ **incons.**
5. $\begin{cases} 2x + 10y = -2 \\ 6x + 4y = 20 \end{cases}$ **indep. (4, –1)**
6. $\begin{cases} 5x + 6y = 14 \\ 3x + 5y = 7 \end{cases}$ **indep. (4, –1)**

Use substitution to solve each system. Check your solution.

7. $\begin{cases} y = 2x - 4 \\ 7x - 5y = 14 \end{cases}$ **(2, 0)**
8. $\begin{cases} y = 3x - 12 \\ 2x + 3y = -3 \end{cases}$ **(3, –3)**
9. $\begin{cases} 2x + 8y = 1 \\ x = 2y \end{cases}$ $\left(\frac{1}{6}, \frac{1}{12}\right)$
10. $\begin{cases} 4x + 3y = 13 \\ x + y = 4 \end{cases}$ **(1, 3)**
11. $\begin{cases} 6y = x + 18 \\ 2y - x = 6 \end{cases}$ **(0, 3)**
12. $\begin{cases} x + y = 7 \\ 2x + y = 5 \end{cases}$ **(–2, 9)**

Chapter Test, Form A

NAME ________ CLASS ________ DATE ________

Chapter Assessment
Chapter 3, Form A, page 1

Write the letter that best answers the question or completes the statement.

__b__ 1. How many solutions does the system $\begin{cases} 8x = 12y - 9 \\ 27y + 18x = 21 \end{cases}$ have?
a. 0 b. 1 c. 2 d. infinitely many

__c__ 2. The system $\begin{cases} 45x - 30y = -90 \\ -12x + 8y = 24 \end{cases}$ is
a. inconsistent. b. consistent and independent.
c. consistent and dependent. d. none of these.

__c__ 3. Which of the following is a solution to the system $\begin{cases} 3x + 2y - 3z = 15 \\ -2x + 4y + 2z = 6 \\ 4z + 12 = -8 \end{cases}$?
a. (2, –3, 5) b. (2, 3, –1) c. (–2, 3, –5) d. all of these

__d__ 4. Which inequality is graphed here?
a. $y > -2x - 2$
b. $y < -2x + 2$
c. $y \geq -2x + 2$
d. $y > -2x + 2$

__c__ 5. A vehicle manufacturer operates a plant that assembles and finishes both cars and trucks. It takes 5 person-days to assemble and 2 person-days to finish a truck. It takes 4 person-days to assemble and 3 person-days to finish a car. Assembly can take no more than 180 person-days per week, and finishing can take no more than 135 person-days per week. If x represents the number of trucks and y represents the number of cars, which of the following systems represent the weekly constraints on assembly and finishing?
a. $\begin{cases} 5x + 2y \leq 180 \\ 4x + 3y \leq 135 \end{cases}$ b. $\begin{cases} 5x + 4y \leq 180 \\ 3x + 2y \leq 135 \end{cases}$
c. $\begin{cases} 5x + 4y \leq 180 \\ 2x + 3y \leq 135 \end{cases}$ d. $\begin{cases} 4x + 5y \leq 180 \\ 2x + 3y \leq 135 \end{cases}$

__d__ 6. Which of the following is a point on the line defined by $\begin{cases} x(t) = 3t + 4 \\ y(t) = -5t + 6 \end{cases}$?
a. (3, –5) b. (10, 16) c. (–5, –9) d. (13, –9)

NAME ________ CLASS ________ DATE ________

Chapter Assessment
Chapter 3, Form A, page 2

__a__ 7. Which system is graphed at right?
a. $\begin{cases} x > -2 \\ y \leq -\frac{1}{2}x - 1 \end{cases}$ b. $\begin{cases} x < -2 \\ y \geq -\frac{1}{2}x - 1 \end{cases}$
c. $\begin{cases} x > -2 \\ y < -\frac{1}{2}x - 1 \end{cases}$ d. $\begin{cases} x \leq -2 \\ y > -\frac{1}{2}x - 1 \end{cases}$

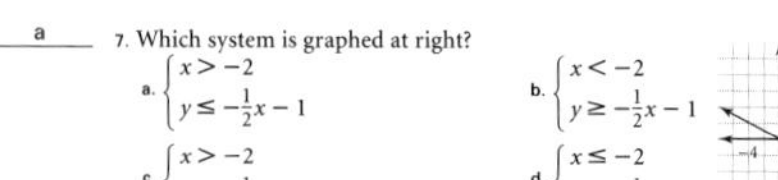

__b__ 8. Find the minimum value of the objective function $C = 12x + 20y$ on the feasible region shown at right.
a. –20 b. –44
c. –72 d. –104

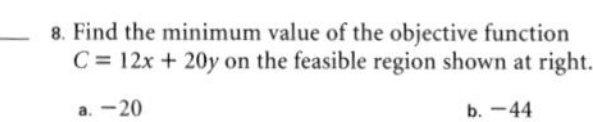
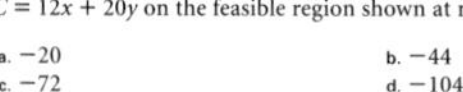
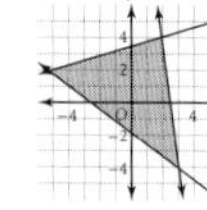

__b__ 9. Which of the following is equivalent to $\begin{cases} x(t) = 4t + 8 \\ y(t) = 6t - 1 \end{cases}$?
a. $y = -\frac{3}{2}x + 13$ b. $y = \frac{3}{2}x - 13$ c. $y = -\frac{2}{3}x - 13$ d. $y = \frac{3}{2}x + 13$

__d__ 10. A chemist needs to create a 38% saline solution by using a 20% saline solution and a 50% saline solution. If x is the amount of 20% solution and y is the amount of 50% solution, which system of equations must the chemist solve in order to find how much of each solution is needed to make 300 milliliters of a 38% solution?
a. $\begin{cases} x + y = 300 \\ 20x + 50y = 3800 \end{cases}$ b. $\begin{cases} x + y = 300 \\ 0.20x + 0.50y = 38 \end{cases}$
c. $\begin{cases} x + y = 38 \\ 0.20x + 0.50y = 300 \end{cases}$ d. $\begin{cases} x + y = 300 \\ 0.20x + 0.50y = 114 \end{cases}$

__c__ 11. A quarterback throws a football towards a receiver from a point 5 feet above the ground with a horizontal speed of 60 feet per second and a vertical speed of 33 feet per second. $\begin{cases} x(t) = 60t \\ y(t) = 5 + 33t - 16t^2 \end{cases}$
Parametric equations that describe the flight of the football are shown. If the receiver is 120 feet away from the quarterback, what is the height of the football when it reaches the receiver?
a. 3 feet b. 5 feet c. 7 feet d. 9 feet

21.

LESSON 3.2

Key Skills

Use elimination to solve a system of linear equations.

Solve $\begin{cases} 2x - 3y = -17 \\ 5x = 15 - 4y \end{cases}$ by using elimination.

Write each equation in standard form.

$$\begin{cases} 2x - 3y = -17 \\ 5x + 4y = 15 \end{cases}$$

Multiply the equations as needed, and then combine to eliminate one of the variables.

$$\begin{cases} 5(2x - 3y) = 5(-17) \\ -2(5x + 4y) = -2(15) \end{cases} \rightarrow \begin{aligned} 10x - 15y &= -85 \\ -10x - 8y &= -30 \\ \hline -23y &= -115 \\ y &= 5 \end{aligned}$$

Then use substitution to solve for the other variable.

$$5x = 15 - 4y$$
$$5x = 15 - 4(5)$$
$$x = -1$$

Thus, the solution is $(-1, 5)$.

Exercises

Use elimination to solve each system. Check your solution.

13. $\begin{cases} 2x - 5y = 1 \\ 3x - 4y = -2 \end{cases}$ **(−2, −1)**

14. $\begin{cases} 9x + 2y = 2 \\ 21x + 6y = 4 \end{cases}$ $\left(\frac{1}{3}, -\frac{1}{2}\right)$

15. $\begin{cases} -x + 2y = 12 \\ x + 6y = 20 \end{cases}$ **(−4, 4)**

16. $\begin{cases} 2x + 3y = 18 \\ 5x - y = 11 \end{cases}$ **(3, 4)**

17. $\begin{cases} 3y = 3x - 6 \\ y = x - 2 \end{cases}$ **infinitely many**

18. $\begin{cases} y = \frac{3}{2}x + 4 \\ 2y - 8 = 3x \end{cases}$ **infinitely many**

19. $\begin{cases} y = \frac{1}{2}x + 9 \\ 2y - x = 1 \end{cases}$ **no solution**

20. $\begin{cases} y = -2x - 4 \\ 2x + y = 6 \end{cases}$ **no solution**

LESSON 3.3

Key Skills

Graph a linear inequality in two variables.

Solve the inequality for y, reversing the inequality symbol if multiplying or dividing by a negative number. Graph the boundary line, using a dashed line for $<$ or $>$ and a solid line for $\leq$ or $\geq$. Substitute a point into the inequality to determine whether to shade above or below the boundary line.

Exercises

Graph each linear inequality.

21. $y > 2x - 3$

22. $y - 3x < 4$

23. $2x - y \leq 5$

24. $4x - 2y \leq -3$

25. $\frac{y}{4} \geq -\frac{x}{3} + 1$

26. $\frac{3}{2}x \leq \frac{1}{4}y - 3$

27. $y > -2$

28. $x \leq 7$

LESSON 3.4

Key Skills

Graph the system of linear inequalities.

Graph. $\begin{cases} y \geq -1 \\ y \leq \frac{4}{3}x + \frac{1}{3} \\ y < -4x + 11 \end{cases}$

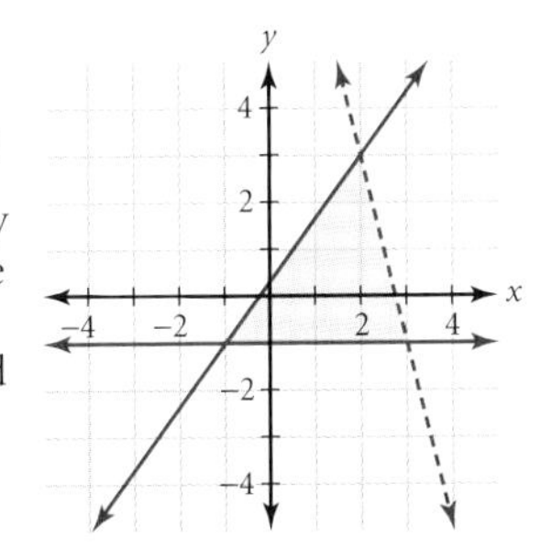

Graph each boundary line, using a solid line or a dashed line. The solution is the shaded region shown.

Exercises

Graph each system of linear inequalities.

29. $\begin{cases} y \geq 0 \\ x \geq 0 \\ y > -2x + 3 \\ y < 4x \end{cases}$

30. $\begin{cases} y \geq 0 \\ x \geq 0 \\ y \leq x + 8 \\ y \geq -2x + 3 \end{cases}$

31. $\begin{cases} y < 2x + 3 \\ y \geq 3x - 1 \\ x > 1 \end{cases}$

32. $\begin{cases} y \geq -x - 3 \\ y < 4x + 2 \\ x > -2 \end{cases}$

22.

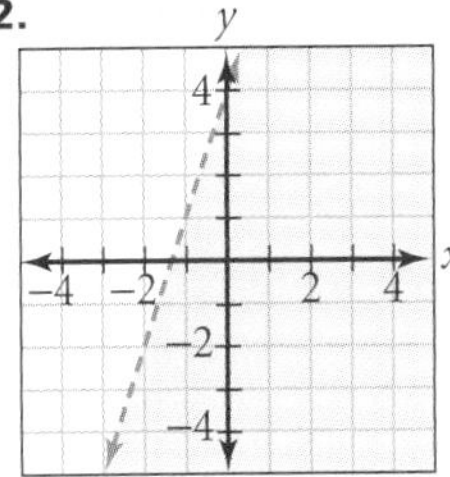

23.

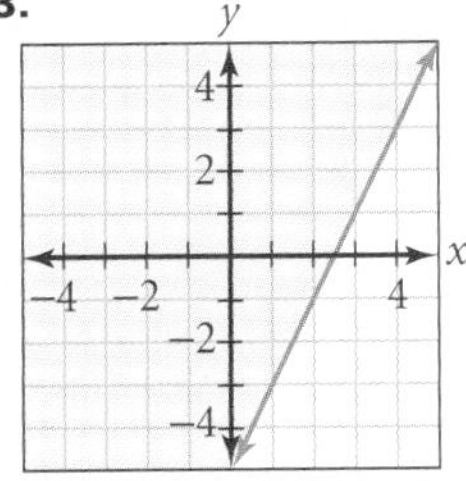

24.

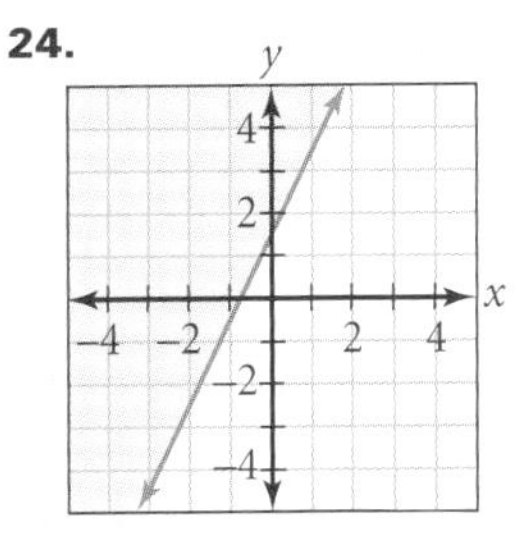

The answers to Exercises 25–32 can be found in Additional Answers beginning on page 1002.

Chapter Test, Form B

NAME ________ CLASS ________ DATE ________

Chapter Assessment

Chapter 3, Form B, page 1

1. a. Classify the system $\begin{cases} 24y - 8x = 40 \\ 3x = 9y - 21 \end{cases}$. **inconsistent**

   b. How many solutions does the system have? **0**

2. Solve $\begin{cases} 5x + 3y = 5 \\ 3x + y = -1 \end{cases}$ by substitution. **(−2, 5)**

3. Solve $\begin{cases} 2x - 3y = 24 \\ 6x + 2y = 6 \end{cases}$ by elimination. **(3, −6)**

4. Solve $\begin{cases} 5x - 3 = 7 \\ 3x + 5y - 4z = -13 \\ x - 3y + 5z = 16 \end{cases}$ by any method. **(2, −3, 1)**

5. Graph $2x + 3y \geq 9$ on the coordinate grid below.

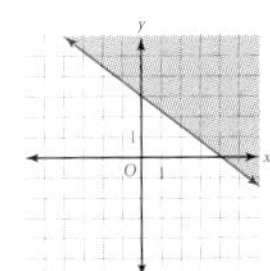

6. Graph the system $\begin{cases} -2x + y \leq -4 \\ x \leq 4 \\ y > -3 \end{cases}$ on the coordinate grid below.

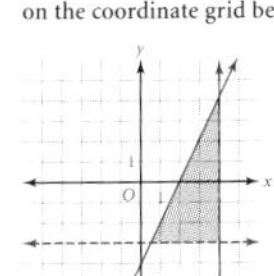

7. Some students want to start a business that cleans and polishes cars. It takes 1.5 hours of labor and costs $2.25 in supplies to clean a car. It takes 2 hours of labor and costs $4.50 in supplies to polish a car. The students can work a total of 120 hours in one week. They also decide that they want to spend no more than $160 per week on supplies. The students expect to make a profit of $7.75 for each car that they clean and a profit of $8.50 for each car that they polish.

   a. Write a system of inequalities to represent the constraints the students have set. **Let x be the number of cars cleaned and y be the number of cars polished.** $\begin{cases} x \geq 0,\ y \geq 0 \\ 1.5x + 2y \leq 120 \\ 2.25x + 4.50y \leq 160 \end{cases}$

   b. Write an objective function for the profit. $P = 7.75x + 8.50y$

NAME ________ CLASS ________ DATE ________

Chapter Assessment

Chapter 3, Form B, page 2

8. Find the maximum value of the objective function $P = 300x + 450y$ on the feasible area shown at right. **5100 at (5, 8)**

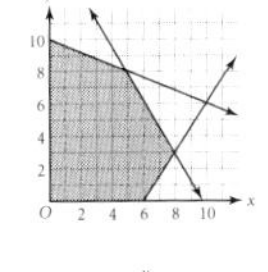

9. Consider the parametric equations $\begin{cases} x(t) = 3 - t \\ y(t) = 2t + 1 \end{cases}$.

   a. Graph the parametric equations for $-2 \leq t \leq 2$.

   b. Find a single equation in x and y that is equivalent to the parametric equations. $y = -2x + 7$

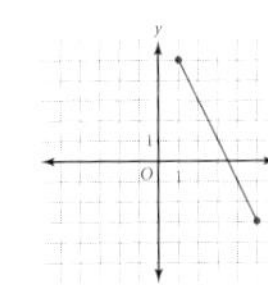

10. There are 120 seats on an airplane. If a passenger reserved a seat at least two weeks early, the passenger paid $150 for the flight. Other passengers paid $200. If all 120 seats were sold and the total amount paid by the passengers was $22,550, write and solve a system of equations to find how many passengers paid the lower fare. **Let x be number that paid lower fare and y be number that paid higher fare.** $\begin{cases} x + y = 120 \\ 150x + 200y = 22{,}550 \end{cases}$; **29 passengers paid the lower fare.**

11. A juice company wants to make a drink that is 60% fruit juice. The company already makes a drink that is 90% fruit juice and a drink that is 40% fruit juice. How much of each drink should they use to make 2 liters of a drink that is 60% fruit juice? **1.2 liters of 40% juice and 0.8 liters of 90% juice**

12. An arrow is shot from a point 20 feet above the ground with an initial horizontal velocity of 45 feet per second and an initial vertical velocity of 80 feet per second. The parametric equations that describe the arrow's path are shown at right. $\begin{cases} x(t) = 45t \\ y(t) = 20 + 80t - 16t^2 \end{cases}$

    a. The arrow is aimed at a target that is a horizontal distance of 90 feet from the point where the arrow is shot. If the arrow hits the target, how high is the target? **116 feet**

    b. How far does the arrow travel before it hits the ground? **≈ 235.7 feet**

LESSON 3.5

Key Skills

Use linear programming to find the maximum or minimum value of an objective function.

Step 1 Write a system of inequalities, and graph the feasible region.
Step 2 Write the objective function to be maximized or minimized.
Step 3 Find the coordinates of the vertices of the feasible region.
Step 4 Evaluate the objective function for the coordinates of the vertices of the feasible region. Then identify the coordinates that give the required maximum or minimum.

Exercises

33. **BROADCASTING** At a radio station, 6 minutes of each hour are devoted to news, and the remaining 54 minutes are devoted to music and commercials. Station policy requires at least 30 minutes of music per hour and at least 3 minutes of music for each minute of commercials. Use linear programming to find the maximum number of minutes available for commercials each hour. **13.5 min**

LESSON 3.6

Key Skills

Graph a pair of parametric equations for a given interval of *t*.

Make a table, plot the ordered pairs (x, y), and draw a line or curve through the points. Or use a graphics calculator in parametric mode.

Write a pair of parametric equations as a single equation in *x* and *y*.

Given the parametric equations $\begin{cases} x(t) = 3t + 1 \\ y(t) = t - 5 \end{cases}$, solve either equation for t and substitute the expression for t in the other equation.

$$y = t - 5 \rightarrow t = y + 5$$
$$x = 3t + 1$$
$$x = 3(y + 5) + 1$$
$$x = 3y + 16, \text{ or } y = \frac{x - 16}{3}$$

Exercises

Write each pair of parametric equations as a single equation in *x* and *y*.

34. $\begin{cases} x(t) = 2t \\ y(t) = t - 1 \end{cases}$ $y = \frac{1}{2}x - 1$

35. $\begin{cases} x(t) = t + 3 \\ y(t) = 3t - 1 \end{cases}$ $y = 3x - 10$

36. $\begin{cases} x(t) = t \\ y(t) = 3 - 2t \end{cases}$ $y = 3 - 2x$

37. $\begin{cases} x(t) = t + 5 \\ y(t) = 4t - 3 \end{cases}$ $y = 4x - 23$

38. **SPORTS** A goalie kicks a soccer ball from a height of 2 feet above the ground toward the center of the field. The ball is kicked with a horizontal speed of 40 feet per second and a vertical speed of 65 feet per second. If $x(t)$ gives the horizontal distance in feet after t seconds and $y(t)$ gives the vertical distance in feet after t seconds, then the ball's path can be described by the parametric equations

$\begin{cases} x(t) = 40t \\ y(t) = 2 + 65t - 16t^2 \end{cases}$. What horizontal distance does the ball travel before striking the ground? **about 164 ft**

Applications

39. **MANUFACTURING** A tire manufacturer has 1000 units of raw rubber to use for car and truck tires. Each car tire requires 5 units of rubber and each truck tire requires 12 units of rubber. Labor costs are \$8 for a car tire and \$12 for a truck tire. The manufacturer does not want to pay more than \$1500 in labor costs. Write and graph a system of inequalities to represent this situation.

39. $\begin{cases} 5x + 12y \le 1000 \\ 8x + 12y \le 1500 \end{cases}$

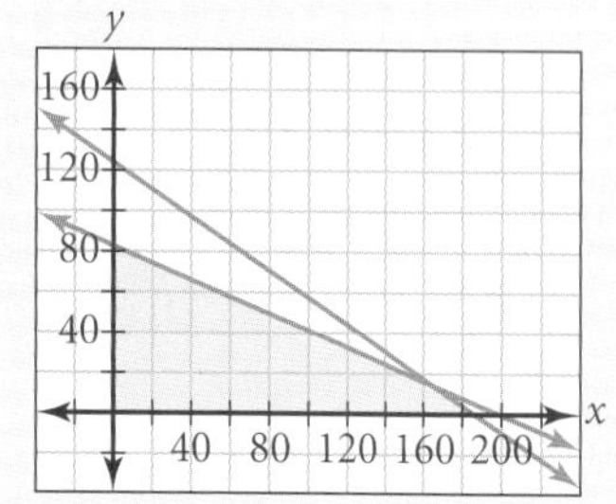

3 Chapter Test

Classify the type of system represented by each graph as independent, dependent, or inconsistent. If the system has exactly one solution, write it.

1.

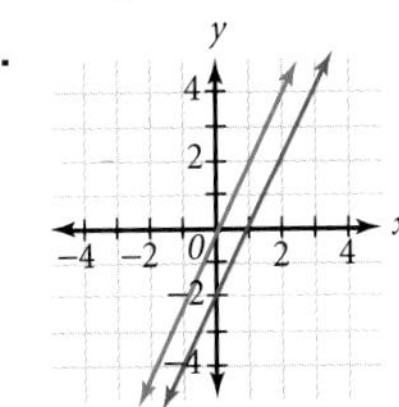

inconsistent

2.

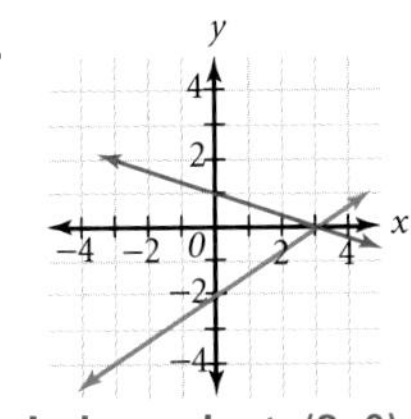

independent; (3, 0)

Use substitution to solve each system. Check your solution.

3. $\begin{cases} 2x - 3y = 1 \\ y = x - 2 \end{cases}$ (5, 3)

4. $\begin{cases} x + y + z = 7 \\ 2x + 3y = 3 \\ x = -2y \end{cases}$ (6, –3, 4)

5. NUMBER The sum of two numbers is 7. Four times the first number is one more than five times the second. Find the two numbers by setting up a system of equations and solving by substitution. **4 and 3**

Use elimination to solve each system. Check your solution.

6. $\begin{cases} x + y = 1 \\ x - 2y = -8 \end{cases}$ (–2, 3)

7. $\begin{cases} 5x + 2y = 24 \\ 2x - 12 = 4y \end{cases}$ $\left(5, -\frac{1}{2}\right)$

8. $\begin{cases} \frac{1}{3}x - y = 4 \\ 2x - 6y = 12 \end{cases}$ no solution

9. $\begin{cases} 4x + 3y = 0 \\ y - x = -7 \end{cases}$ (3, –4)

10. CONSUMER ECONOMICS Three blouses and four skirts are on sale for $72.50. Five blouses and two skirts would cost $66.00. Find the cost of one blouse and one skirt by setting up a system of equations and solving by elimination. **$8.50 blouse; $11.75 skirt**

Graph each linear inequality.

11. $y < 3x - 4$

12. $2x + 3y \geq 6$

13. $x \leq 3$

14. $\frac{2}{3}x - \frac{3}{4}y > -1$

15. RECREATION Misha has a 2500-meter spool of rope that he must cut into 50-meter and 75-meter lengths for his rock-climbing class. Write an inequality that will express the possible numbers of each length he can cut from this spool of rope. $50x + 75y \leq 2500$

Graph each system of linear inequalities.

16. $\begin{cases} x \geq 0 \\ y \geq 0 \\ x + y \leq 4 \\ x - y \leq 2 \end{cases}$

17. $\begin{cases} y > -2 \\ 2x - y > -2 \\ y \leq -2x + 6 \end{cases}$

18. MANUFACTURING A company produces windows and doors. A profit of $5 is realized on each window, and profit is $3 on each door. The company has 18 hours available for manufacturing in Plant A. Each window requires 3 hours, and each door 2 hours to manufacture. Plant B has 7.5 hours available for assembly. Each window requires 1.5 hours to assemble, and each door requires 0.75 hour. Use linear programming to determine how many windows and doors the manufacturer should produce to maximize profit.

Write each pair of parametric equations as a single equation in *x* and *y*.

19. $\begin{cases} x(t) = t - 4 \\ y(t) = 2t - 11 \end{cases}$ $y = 2x - 3$

20. $\begin{cases} x(t) = 5t \\ y(t) = -2t + 3 \end{cases}$ $y = -\frac{2}{5}x + 3$

21. AGRICULTURE A farmer recently planted a new crop of cherry trees in his orchard. The first year, the trees grow to 10 feet tall and produce $3\frac{1}{2}$ pounds of cherries each. The trees grow at the rate of $1\frac{1}{2}$ feet per year and increase their production of cherries by $\frac{3}{4}$ pounds each year. Write a system of parametric equations to determine the height $h(t)$ of a tree in t years and its cherry production $c(t)$ in t years. Then express the cherry production, c, as a function of height, h.

Chapter Test

11.

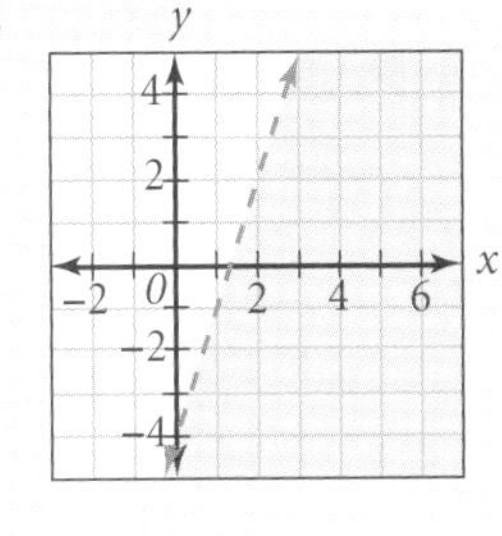

12.

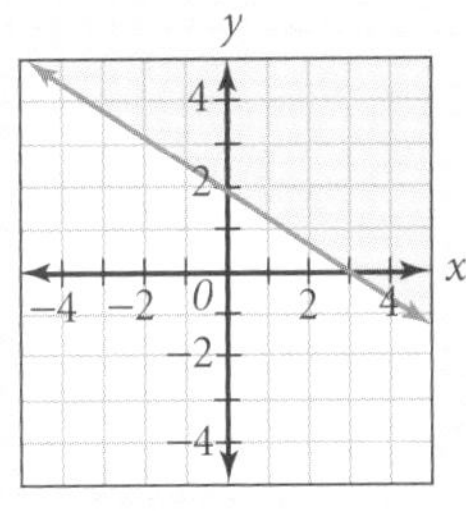

13.

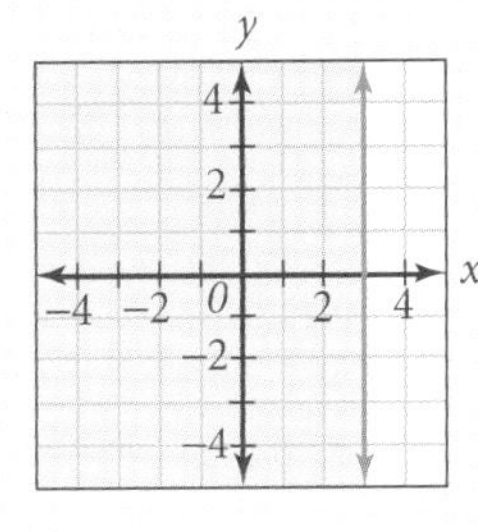

14.

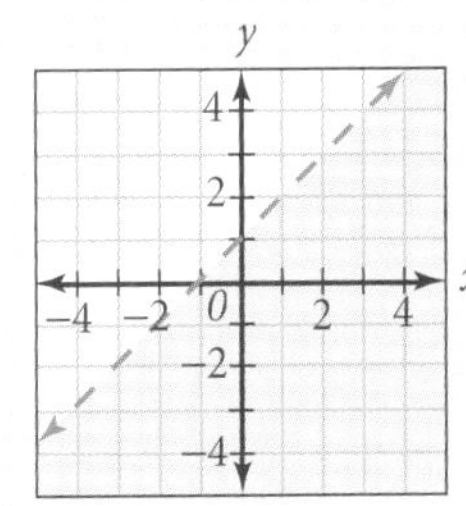

16.

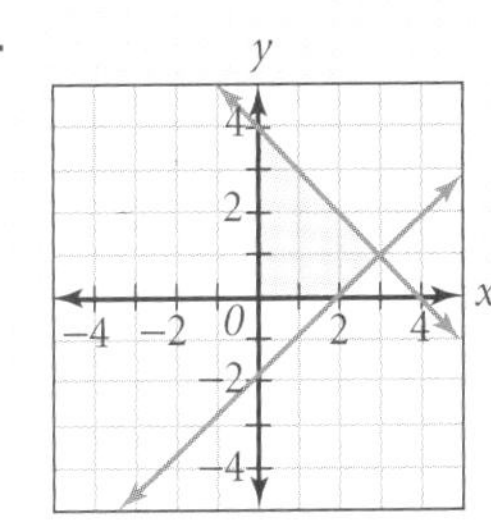

17.

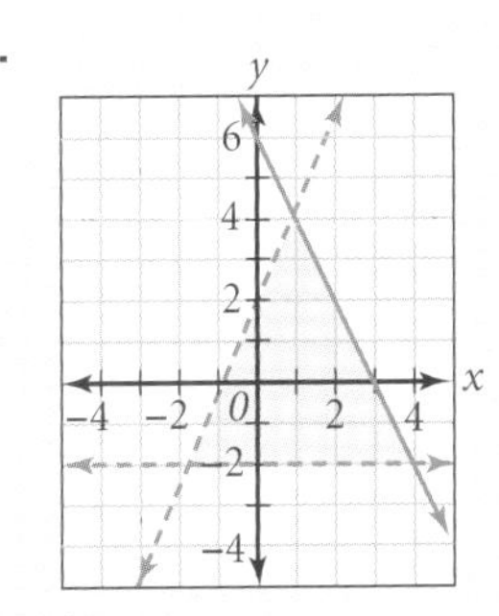

18.

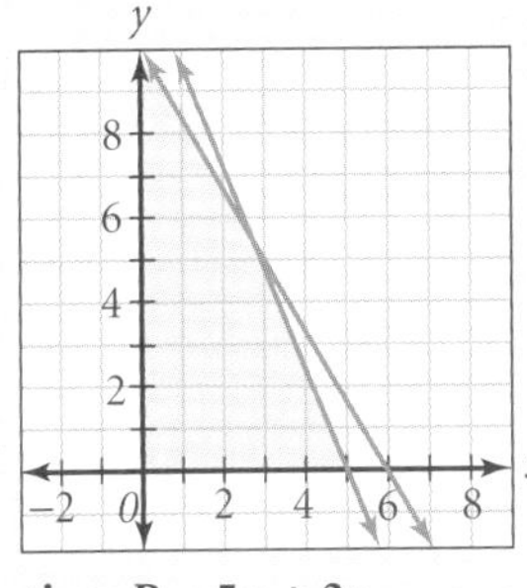

objective: $P = 5x + 3y$

$\begin{cases} x \geq 0 \\ y \geq 0 \\ 3x + 2y \leq 18 \\ 1.5x + 0.75y \leq 7.5 \end{cases}$

2 windows, 6 doors for a profit of $28

21. $\begin{cases} h(t) = 10 + 1\frac{1}{2}t \\ c(t) = 3\frac{1}{2} + \frac{3}{4}t \end{cases}$

$c(h) = \frac{1}{2}h - \frac{3}{2}$

College Entrance Exam Practice

Multiple-Choice and Quantitative-Comparison Samples

The first half of the Cumulative Assessment contains two types of items found on standardized tests—multiple-choice questions and quantitative-comparison questions. Quantitative-comparison items emphasize the concepts of equality, inequality, and estimation.

Free-Response Grid Samples

The second half of the Cumulative Assessment is a free-response section. This part of the Cumulative Assessment requires student-produced response items like those commonly found on college entrance exams. These questions require the use of machine-scored answer grids. You may wish to have students practice answering these items in preparation for standardized tests.

CHAPTERS 1–3 CUMULATIVE ASSESSMENT

College Entrance Exam Practice

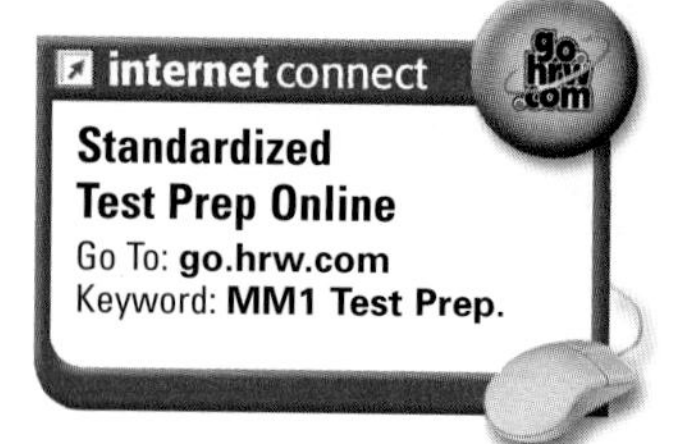

QUANTITATIVE COMPARISON For Items 1–5, write
A if the quantity in Column A is greater than the quantity in Column B;
B if the quantity in Column B is greater than the quantity in Column A;
C if the two quantities are equal; or
D if the relationship cannot be determined from the given information.

	Column A	Column B	Answers
1. A	$-0.\overline{72}$	$-\frac{4}{5}$	Ⓐ Ⓑ Ⓒ Ⓓ [Lesson 2.1]
2. D	The value of y when x is any real number $y = 4x$	$y = -4x$	Ⓐ Ⓑ Ⓒ Ⓓ [Lesson 2.3]
3. A	The value of x $\frac{x}{4} = \frac{x+2}{5}$	$\frac{x}{2} = \frac{21-x}{4}$	Ⓐ Ⓑ Ⓒ Ⓓ [Lesson 1.4]
4. D	$\|x+2\|$	$\|x-2\|$	Ⓐ Ⓑ Ⓒ Ⓓ [Lesson 1.8]
5. A	$f(x) = -3x$ and $g(x) = x - 4$ $(f \circ g)(-5)$	$(g \circ f)(-5)$	Ⓐ Ⓑ Ⓒ Ⓓ [Lesson 2.4]

6. Which is a solution of the system? d
$\begin{cases} -2x + y \le 8 \\ x - 3y > 9 \end{cases}$ ***(LESSON 3.4)***
 a. $(0, -3)$ b. $(0, 9)$
 c. both **a** and **b** d. neither **a** nor **b**

7. Which equation in x and y represents $\begin{cases} x(t) = 3t - 1 \\ y(t) = 2 - 3t \end{cases}$? ***(LESSON 3.6)*** c
 a. $y = x - 1$ b. $y = 3x - 1$
 c. $y = 1 - x$ d. $y = -3x - 1$

8. Evaluate $[3(2 + 1) + 3](2^2)$. ***(LESSON 2.1)*** d
 a. 36 b. 20
 c. 32 d. 48

9. How many solutions does the system have? d
$\begin{cases} 2x - 3y = 11 \\ 6x - 9y = 33 \end{cases}$ ***(LESSON 3.1)***
 a. 0 b. 1
 c. 2 d. infinite

10. Which function represents the graph of $f(x) = x^2$ translated 5 units down? ***(LESSON 2.7)*** b
 a. $f(x) = x^2 + 5$ b. $f(x) = x^2 - 5$
 c. $f(x) = (x - 5)^2$ d. $f(x) = (x + 5)^2$

11. Which is the range of $f(x) = -\left(\frac{x}{3}\right)^2$? ***(LESSON 2.3)*** d
 a. $f(x) \ge 0$ b. $f(x) \ge 3$
 c. $f(x) \le 3$ d. $f(x) \le 0$

12. Simplify $\left(\frac{2x^{-2}y^3}{x^2y^{-3}}\right)^{-1}$. ***(LESSON 2.2)*** d
 a. 2 b. $\frac{2x^4}{y^6}$
 c. $\frac{1}{2}$ d. $\frac{x^4}{2y^6}$

13. Which is the equation of the line that contains the point $(0, -2)$ and is parallel to the graph of $y = -\frac{1}{2}x - 1$? ***(LESSON 1.3)*** a
 a. $y = -\frac{1}{2}x - 2$ b. $y = -\frac{1}{2}x + 2$
 c. $y = 2x - 2$ d. $y = 2x + 2$

Match each statement on the left with a solution on the right. ***(LESSON 1.8)***

14. $|a - 5| < 3$ **c** — **a.** $a = 2$ *or* $a = 8$

15. $|a - 5| > 3$ **b** — **b.** $a < 2$ *or* $a > 8$

16. $|a - 5| = 3$ **a** — **c.** $a > 2$ *and* $a < 8$

17. $|a - 5| < -3$ **d** — **d.** There is no solution.

18. Graph $2y > -1$. ***(LESSON 3.3)***

19. Write the pair of parametric equations $\begin{cases} x(t) = 2 + t \\ y(t) = 3 + t \end{cases}$ as a single equation in x and y. ***(LESSON 3.6)*** $y = x + 1$

20. Classify $\begin{cases} 3y = 4x - 1 \\ x = \frac{4}{3}y \end{cases}$ as inconsistent, **indep.** dependent, or independent. ***(LESSON 3.1)***

21. Write the function for the graph of $f(x) = x^2$ translated 2 units up. ***(LESSON 2.7)*** $g(x) = x^2 + 2$

22. Solve. $\begin{cases} 3x - 3y = 1 \\ x + y = 4 \end{cases}$ ***(LESSON 3.1)*** $\left(\frac{13}{6}, \frac{11}{6}\right)$

23. Graph the system. $\begin{cases} y \le x + 1 \\ y \ge 2x - 1 \end{cases}$ ***(LESSON 3.4)***

24. Find the inverse of the function $f(x) = 3x - 2$. ***(LESSON 2.5)*** $f^{-1}(x) = \frac{1}{3}x + \frac{2}{3}$

25. Graph the ordered pairs below, and describe the correlation for the data as positive, negative, or none. ***(LESSON 1.5)*** **positive**

(10, 6), (20, 8), (20, 12), (30, 16), (40, 18), (50, 21), (60, 32), (80, 41), (100, 46), (110, 60), (120, 58)

26. Solve the literal equation $\frac{ax - b}{2} = c$ for x. ***(LESSON 1.6)*** $x = \frac{2c + b}{a}, a \neq 0$

27. Graph the parametric equations for $-3 \le t \le 3$. $\begin{cases} x(t) = t + 1 \\ y(t) = 2t - 1 \end{cases}$ ***(LESSON 3.6)***

28. Let $f(x) = 2x + 1$ and $g(x) = 3x^2$. Find $f \circ g$. ***(LESSON 2.4)*** $(f \circ g)(x) = 6x^2 + 1$

Solve each inequality, and graph the solution on a number line. ***(LESSON 1.7)***

29. $6(x - 4) \ge 6 + x$ $x \ge 6$

30. $\frac{-x}{4} > 3$ $x < -12$

Solve each equation. ***(LESSON 1.8)***

31. $|x - 4| = 9$ **13, −5**

32. $|3x + 12| = 18$ **2, −10**

FREE-RESPONSE GRID **The following questions may be answered by using a free-response grid such as that commonly used by standardized-test services.**

33. If a varies directly as b and $a = 16$ when $b = -8$, what is b when $a = 4$? ***(LESSON 1.4)*** **−2**

34. For what value of t does the graph of $\begin{cases} x(t) = t^2 + 1 \\ y(t) = t - 1 \end{cases}$ pass through the point (7.25, 1.5)? ***(LESSON 3.6)*** **2.5**

35. The graph of $g(x) = x + 1$ can be formed by translating the graph of $f(x) = x + 3$ how many units to the right? ***(LESSON 2.7)*** **2**

36. Evaluate $|-2| - |-2|$. ***(LESSON 2.6)*** **0**

37. Let $f(x) = |x| + 2x$. Find $f(-2)$. ***(LESSON 2.3)*** **−2**

38. Find the slope of a line that contains the points (−3, 4) and (−1, −8). ***(LESSON 1.2)*** **−6**

39. ENGINEERING The equation $F = \frac{9}{4}d^4l^{-2}$, where d is the diameter in inches and l is the length in feet, gives the maximum load in tons that a foundation column can support. Find the maximum load, in tons, for a column that is 6 feet long and 12 inches in diameter. ***(LESSON 2.2)*** **1296**

40. MANUFACTURING The true diameter of a pipe is 7.25 inches. Its diameter is measured as 7.29 inches. Find the relative error in the measurement to the nearest thousandth. ***(LESSON 2.6)*** **0.006**

41. INCOME Becky wants to buy a computer that costs \$1590. If Becky earns \$6.25 per hour, what is the minimum number of whole hours that she must work to earn enough money to buy the computer? ***(LESSON 1.7)*** **255**

18.

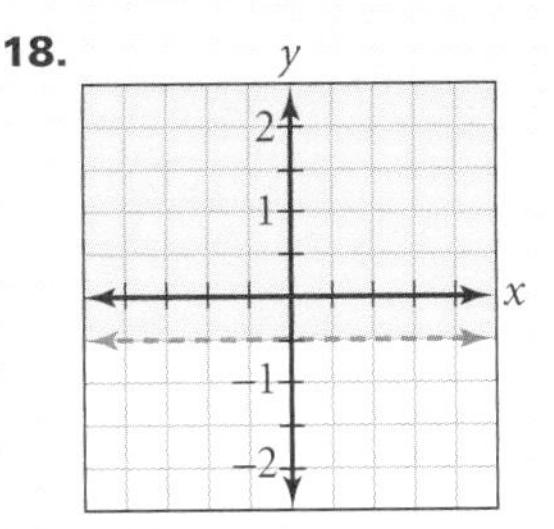

23.

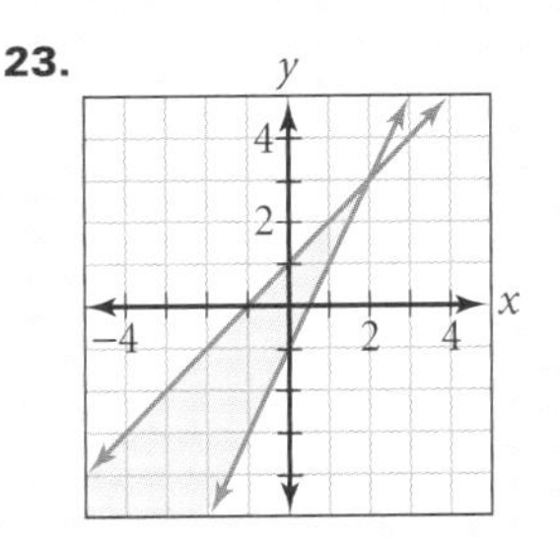

27.

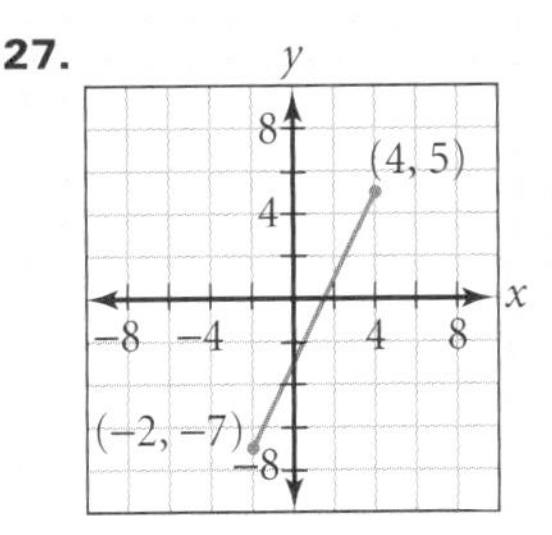

29. $x \ge 6$

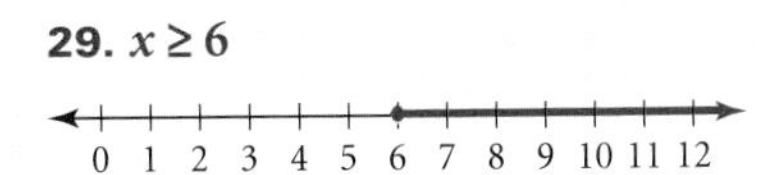

30. $x < -12$

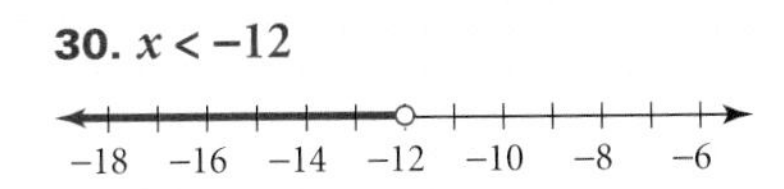

Keystroke Guide for Chapter 3

Essential keystroke sequences (using the model TI-82 or TI-83 graphics calculator) are presented below for all Activities and Examples found in this chapter that require or recommend the use of a graphics calculator.

internet connect

For Keystrokes of other graphing calculator models, visit the HRW web site at **go.hrw.com** and enter the keyword **MB1 CALC.**

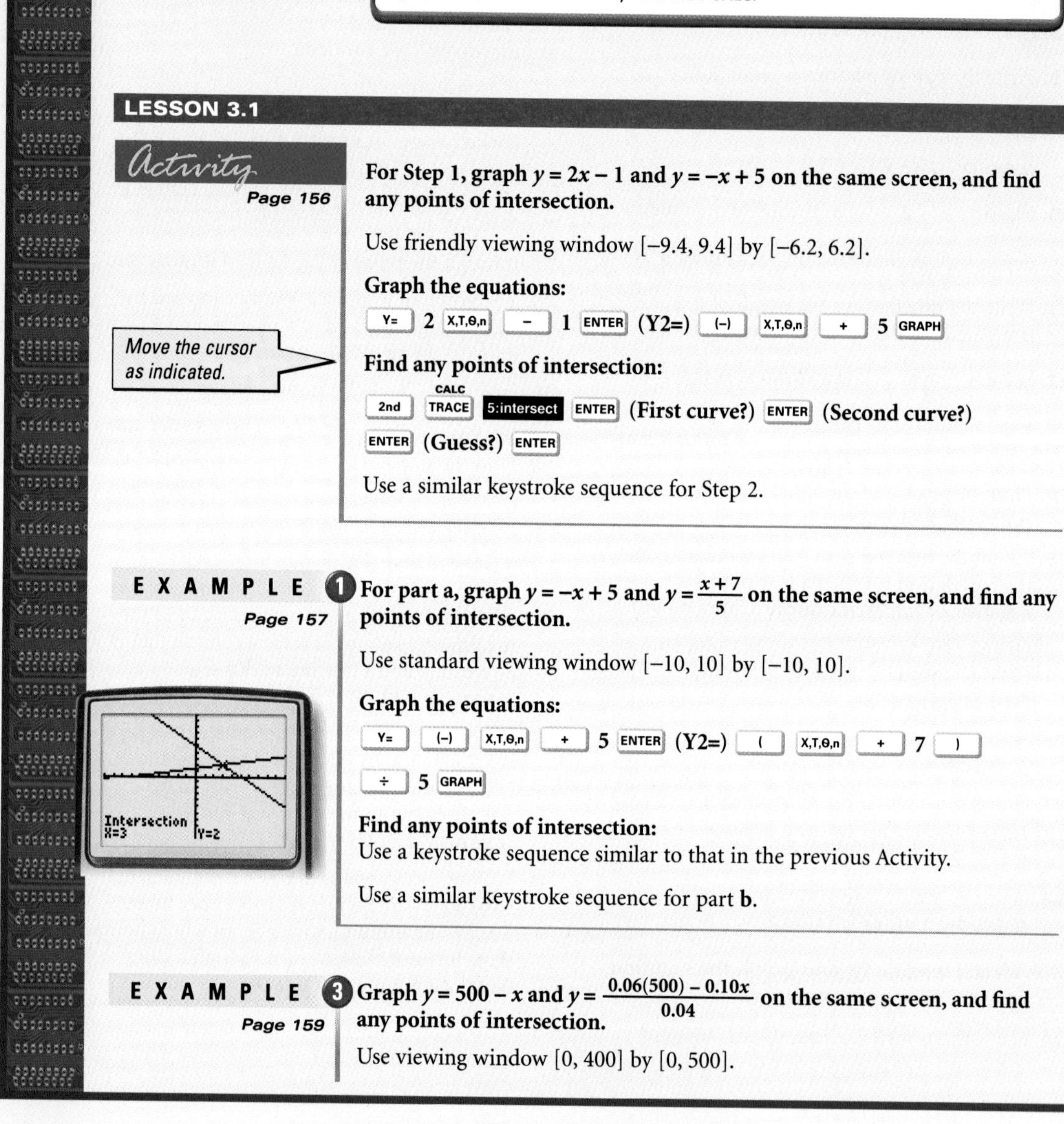

LESSON 3.1

Activity

Page 156

For Step 1, graph $y = 2x - 1$ and $y = -x + 5$ on the same screen, and find any points of intersection.

Use friendly viewing window [−9.4, 9.4] by [−6.2, 6.2].

Graph the equations:

Y= 2 X,T,θ,n − 1 ENTER (Y2=) (−) X,T,θ,n + 5 GRAPH

Move the cursor as indicated.

Find any points of intersection:

2nd TRACE (CALC) 5:intersect ENTER **(First curve?)** ENTER **(Second curve?)** ENTER **(Guess?)** ENTER

Use a similar keystroke sequence for Step 2.

EXAMPLE 1

Page 157

For part a, graph $y = -x + 5$ and $y = \frac{x+7}{5}$ on the same screen, and find any points of intersection.

Use standard viewing window [−10, 10] by [−10, 10].

Graph the equations:

Y= (−) X,T,θ,n + 5 ENTER (Y2=) (X,T,θ,n + 7) ÷ 5 GRAPH

Intersection X=3 Y=2

Find any points of intersection:

Use a keystroke sequence similar to that in the previous Activity.

Use a similar keystroke sequence for part **b**.

EXAMPLE 3

Page 159

Graph $y = 500 - x$ and $y = \frac{0.06(500) - 0.10x}{0.04}$ on the same screen, and find any points of intersection.

Use viewing window [0, 400] by [0, 500].

Graph the equations:

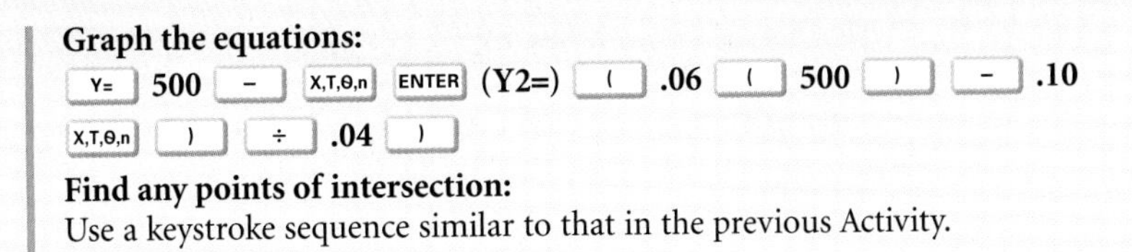

Y= 500 − X,T,θ,n ENTER (Y2=) (.06 (500) − .10
X,T,θ,n) ÷ .04)

Find any points of intersection:

Use a keystroke sequence similar to that in the previous Activity.

LESSON 3.2

EXAMPLE 2 Page 166

Graph $y = \frac{930 - 5.5x}{7.5}$ and $y = \frac{1920 - 12x}{15}$ on the same screen, and find any points of intersection. Then verify with a table.

Use viewing window [0, 200] by [0, 150].

Graph the equations, and find any points of intersection:

Use a keystroke sequence similar to that in the Activity and Examples 1 and 3 in Lesson 3.1.

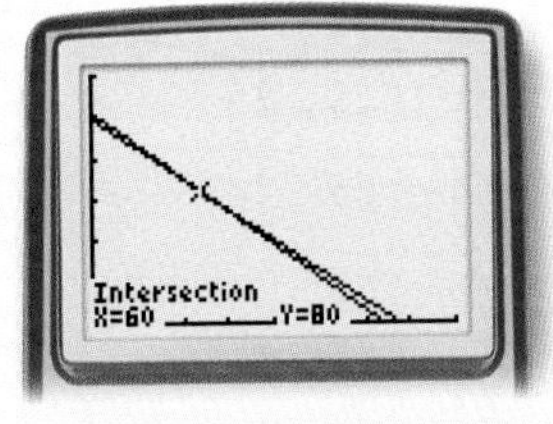

Verify the solution with a table:

2nd WINDOW (TBLSET) (TblStart =) 40 ENTER

⇑ TI-82: (Tbl Min =)

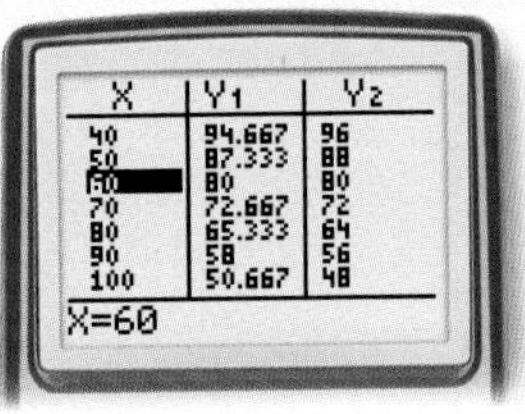

Enter the equations before using the table.

Activity Page 166

For Step 1, graph $y = x + 2$ and $y = \frac{10 + 5x}{5}$ on the same screen.

Use standard viewing window [−10, 10] by [−10, 10].

Use a keystroke sequence similar to that in the Activity in Lesson 3.1.

For Step 2, use a keystroke sequence similar to that in the Activity in Lesson 3.1. Use standard viewing window [−10, 10] by [−10, 10].

EXAMPLES 3 and 4 Page 167

For Example 3, graph $y = \frac{12 - 2x}{5}$ and $y = \frac{15 - 2x}{5}$ on the same screen.

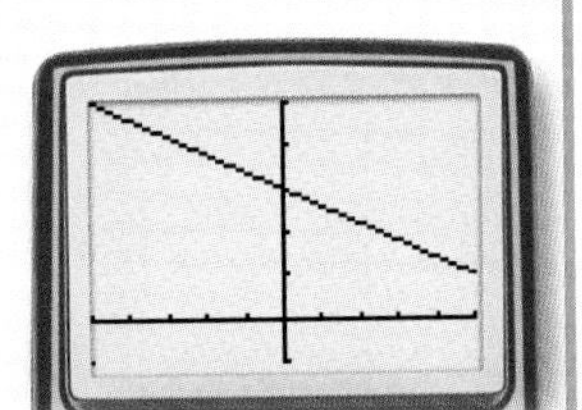

Use viewing window [−5, 5] by [−1, 5].

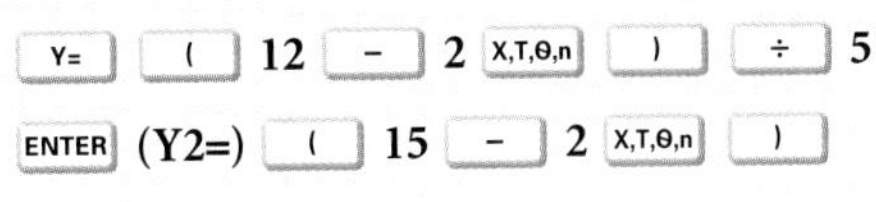

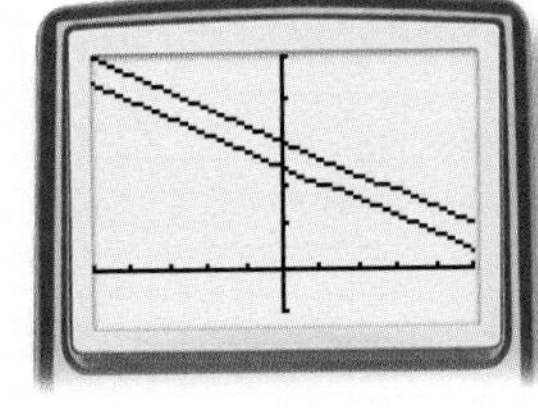

For Example 4, use a similar keystroke sequence. Use viewing window [−5, 5] by [−1, 5].

LESSON 3.3

EXAMPLE 3 Page 174

Graph $y \leq \frac{-33.3}{28.1}x + 20(33.3)$, and see whether (25, 400) satisfies the inequality.

Use viewing window [0, 800] by [0, 800].

Graph the inequality:

[Y=] [◄] [◄] [ENTER] [ENTER] [ENTER]
(◣Y1=) [►] [►] [(] [(-)]
33.3 [÷] 28.1 [)] [X,T,θ,n] [+]
20 [(] 33.3 [)] [GRAPH]

TI-82: Graph the line and use the shade feature.

First clear old data from L1 and L2.

Plot the point:

[STAT] EDIT 1:Edit [ENTER] L1 25 [ENTER] [►] L2
400 [ENTER] [2nd] [Y=] (STAT PLOT) 1:Plot 1 [ENTER] ON
[ENTER] [▼] (Type:) ⁛ [ENTER] [▼]
(Xlist:) [2nd] [1] (L1) [▼] (Ylist:) [2nd] [2] (L2)
[▼] (Mark:) + [ENTER] [GRAPH]

You can also move the cursor to find approximate coordinates.

P1:L1,L2
X=25 Y=400

LESSON 3.6

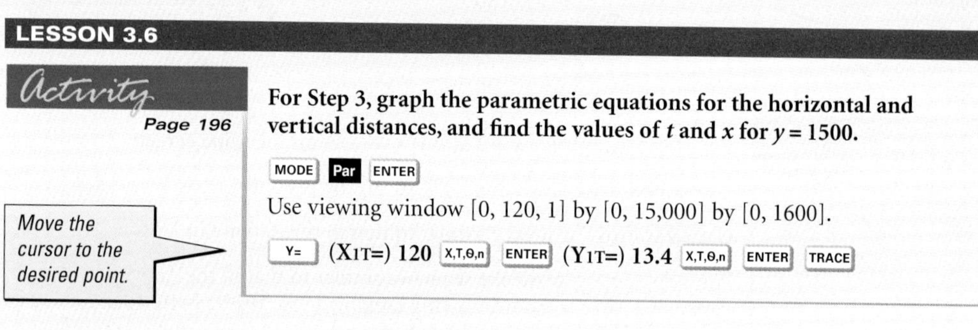

Activity Page 196

For Step 3, graph the parametric equations for the horizontal and vertical distances, and find the values of t and x for $y = 1500$.

[MODE] Par [ENTER]

Use viewing window [0, 120, 1] by [0, 15,000] by [0, 1600].

[Y=] (X1T=) 120 [X,T,θ,n] [ENTER] (Y1T=) 13.4 [X,T,θ,n] [ENTER] [TRACE]

Move the cursor to the desired point.

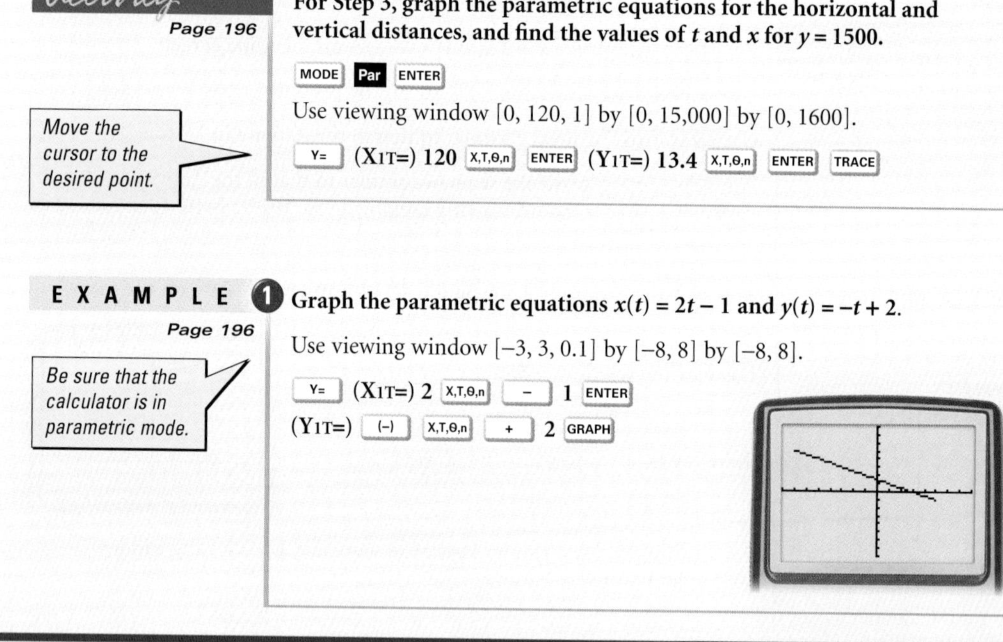

EXAMPLE 1 Page 196

Graph the parametric equations $x(t) = 2t - 1$ and $y(t) = -t + 2$.

Use viewing window [−3, 3, 0.1] by [−8, 8] by [−8, 8].

Be sure that the calculator is in parametric mode.

[Y=] (X1T=) 2 [X,T,θ,n] [−] 1 [ENTER]
(Y1T=) [(-)] [X,T,θ,n] [+] 2 [GRAPH]

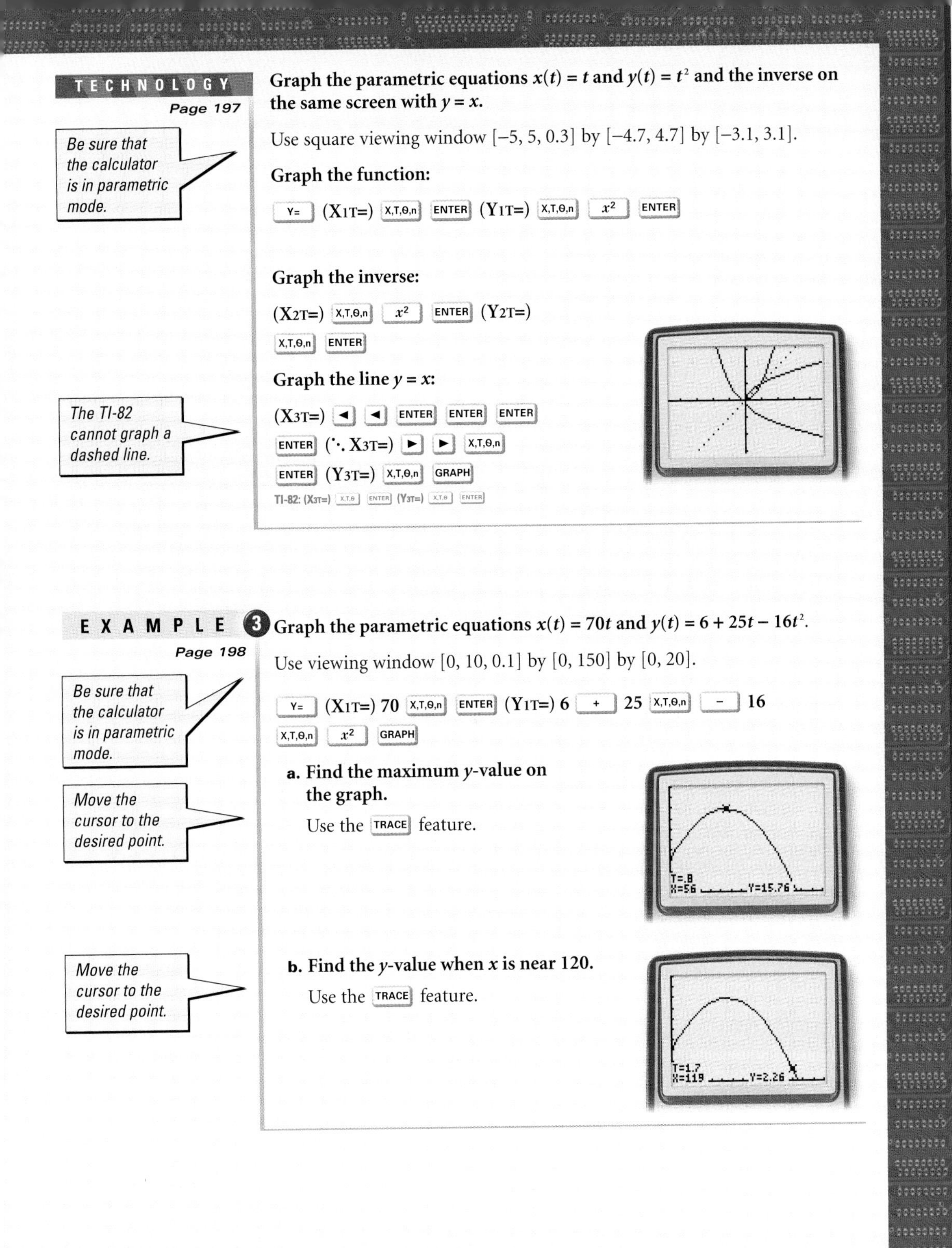

TECHNOLOGY
Page 197

Be sure that the calculator is in parametric mode.

Graph the parametric equations $x(t) = t$ and $y(t) = t^2$ and the inverse on the same screen with $y = x$.

Use square viewing window [–5, 5, 0.3] by [–4.7, 4.7] by [–3.1, 3.1].

Graph the function:

[Y=] (X1T=) [X,T,θ,n] [ENTER] (Y1T=) [X,T,θ,n] [x^2] [ENTER]

Graph the inverse:

(X2T=) [X,T,θ,n] [x^2] [ENTER] (Y2T=) [X,T,θ,n] [ENTER]

Graph the line $y = x$:

The TI-82 cannot graph a dashed line.

(X3T=) [◄] [◄] [ENTER] [ENTER] [ENTER] [ENTER] (∵. X3T=) [►] [►] [X,T,θ,n] [ENTER] (Y3T=) [X,T,θ,n] [GRAPH]

TI-82: (X3T=) [X,T,θ] [ENTER] (Y3T=) [X,T,θ] [ENTER]

EXAMPLE 3
Page 198

Be sure that the calculator is in parametric mode.

Graph the parametric equations $x(t) = 70t$ and $y(t) = 6 + 25t - 16t^2$.

Use viewing window [0, 10, 0.1] by [0, 150] by [0, 20].

[Y=] (X1T=) 70 [X,T,θ,n] [ENTER] (Y1T=) 6 [+] 25 [X,T,θ,n] [–] 16 [X,T,θ,n] [x^2] [GRAPH]

Move the cursor to the desired point.

a. Find the maximum y-value on the graph.

Use the [TRACE] feature.

Move the cursor to the desired point.

b. Find the y-value when x is near 120.

Use the [TRACE] feature.

4 Matrices

Lesson Presentation CD-ROM
PowerPoint® presentations for each lesson 4.1–4.5

CHAPTER PLANNING GUIDE

Lesson	4.1	4.2	4.3	4.4	4.5	Project and Review
Pupil's Edition Pages	216–224	225–233	234–241	244–250	251–259	242–243, 260–267
Practice and Assessment						
Extra Practice (Pupil's Edition)	952	953	953	954	954	
Practice Workbook	22	23	24	25	26	
Practice Masters Levels A, B, and C	64–66	67–69	70–72	73–75	76–78	
Standardized Test Practice Masters	25	26	27	28	29	30
Assessment Resources	43	44	45	47	48	46, 49–54
Visual Resources						
Lesson Presentation Transparencies Vol. 1	85–88	89–92	93–96	97–100	101–104	
Teaching Transparencies	14	15, 16				
Answer Key Transparencies	139–144	145–150	151–154	155–157	158–162	163–166
Quiz Transparencies	4.1	4.2	4.3	4.4	4.5	
Teacher's Tools						
Reteaching Masters	43–44	45–46	47–48	49–50	51–52	
Make-Up Lesson Planner for Absent Students	22	23	24	25	26	
Student Study Guide	22	23	24	25	26	
Spanish Resources	22	23	24	25	26	
Block Scheduling Handbook						8–9
Activities and Extensions						
Lesson Activities	22	23	24	25	26	
Enrichment Masters	22	23	24	25	26	
Cooperative-Learning Activities	22	23	24	25	26	
Problem Solving/ Critical Thinking	22	23	24	25	26	
Student Technology Guide	22	23	24	25	26	
Long Term Projects						13–16
Writing Activities for Your Portfolio						10–12
Tech Prep Masters						15–18
Building Success in Mathematics						10–11

LESSON PACING GUIDE

Lesson	4.1	4.2	4.3	4.4	4.5	Project and Review
Traditional	1 day	1 day	1 day	1 day	2 days	2 days
Block	$\frac{1}{2}$ day	$\frac{1}{2}$ day	$\frac{1}{2}$ day	$\frac{1}{2}$ day	1 day	1 day
Two–Year	2 days	2 days	2 days	2 days	4 days	4 days

CONNECTIONS AND APPLICATIONS

Lesson	4.1	4.2	4.3	4.4	4.5	Review
Algebra	216–224	225–233	234–241	244–250	251–259	260–267
Geometry	219, 221	227, 230	241	249	255	
Probablility					258	
Transformations	220, 221, 222	230				265
Business and Economics	216, 218, 221, 223	232, 233		244, 245, 248, 250	251, 253, 256, 258, 259	267
Life Skills	223					264
Science and Technology		227, 228, 229, 230, 231, 232, 233	234, 237, 239, 241	250		267
Social Studies	222					
Sports and Leisure		225, 231, 232		249	259	267
Cultural Connection: China					258	
Other						263

BLOCK SCHEDULING GUIDE

Day	Lesson	Teacher Directed: Lesson Examples, Teaching Transparencies	Student Guided Activity, Try This	Cooperative-Learning Activity, Lesson Activity, Student Technology Guide	Practice: Practice & Apply, Extra Practice, Practice Workbook	Assessment: Quiz, Mid-Chapter Assessment	Problem Solving, Reteaching
1	4.1	10 min	10 min	8 min	25 min	8 min	8 min
	4.2	10 min	10 min	7 min	25 min	7 min	7 min
2	4.3	10 min	10 min	8 min	25 min	8 min	8 min
	4.4	10 min	10 min	7 min	25 min	7 min	7 min
3	4.5	10 min	15 min	15 min	65 min	15 min	15 min
4	Assess	50 min **PE:** Chapter Review	90 min **PE:** Chapter Project, Writing Activities	90 min Tech Prep Masters	65 min **PE:** Chapter Assessment, Test Generator	30 min Chap. Assess. (A or B), Alt. Assess. (A or B), Test Generator	

PE: Pupil's Edition

Alternative Assessment

The following suggest alternative assessments for students who may benefit from a different type of assessment than the regular chapter quizzes and the mid-chapter/end-of-chapter test. Visit the HRW web site to get additional Alternative Assessment material.

internet connect

Alternative Assessment
Go To: **go.hrw.com**
Keyword: **MB1 Alt Assess**

Performance Assessment

1. Let $A = \begin{bmatrix} -1 & 2 & -3 \\ 0 & 3 & -3 \\ 0 & -3 & 1 \end{bmatrix}$ and $B = \begin{bmatrix} 0 & 1 & 2 \\ 4 & 5 & -2 \\ -2 & 8 & 1 \end{bmatrix}$.

a. Find $A + B$, $A - B$, $4A$, and AB.

b. Explain how to perform matrix operations, and state the conditions under which each operation can be performed.

2. Consider $\begin{cases} -x + 2y - 3z = 11 \\ 3y - 3z = -6 \\ -3y + z = 0 \end{cases}$.

a. Solve the system by using substitution, elimination, and row reduction.

b. The coefficient matrix for the system is matrix A in Exercise 1. Solve the system by using a matrix equation and an inverse matrix.

Portfolio Project

Suggest that students perform the following project for inclusion in their portfolios.

1. The vertices of $\triangle KLM$ are $K(0, 0)$, $L(4, 0)$, and $M(4, 3)$. Let $a > 1$ in the following matrices:

$$A = \begin{bmatrix} 1 & 0 \\ 0 & a \end{bmatrix} \quad B = \begin{bmatrix} 1 & a \\ 0 & 1 \end{bmatrix} \quad C = \begin{bmatrix} -1 & 0 \\ 0 & 1 \end{bmatrix}$$

$$D = \begin{bmatrix} 0 & 1 \\ 1 & 0 \end{bmatrix} \quad E = \begin{bmatrix} 0 & -1 \\ -1 & 0 \end{bmatrix} \quad F = \begin{bmatrix} 1 & 0 \\ 0 & -1 \end{bmatrix}$$

a. Using experimentation, find the geometric effect of various matrices on $\triangle KLM$, represented by $T = \begin{bmatrix} 0 & 4 & 4 \\ 0 & 0 & 3 \end{bmatrix}$.

b. Illustrate your discoveries by graphing.

internet connect

The table below identifies the pages in this chapter that contain internet and technology information.

Content Links

Activities Online	**pages 229, 240, 249**
Portfolio Extensions	**pages 224, 233**
Homework Help Online	**pages 221, 229, 239, 248, 256**
Graphic Calculator Support	**page 268**

Resource Links

Parents can go online and find concepts that students are learning—lesson by lesson—and questions that pertain to each lesson, which facilitate parent-student discussion.

Go To: **go.hrw.com**
Keyword: **MB1 Parent Guide**

Technical Support

The following may be used to obtain technical support for any HRW software product.

Online Help: **www.hrwtechsupport.com**
e-mail: **tschrw@hrwtechsupport.com**

HRW Technical Support Center: **(800)323-9239**
7 AM to 10 PM Monday through Friday Central Time

Visit the HRW math web site at: **www.hrw.com/math**

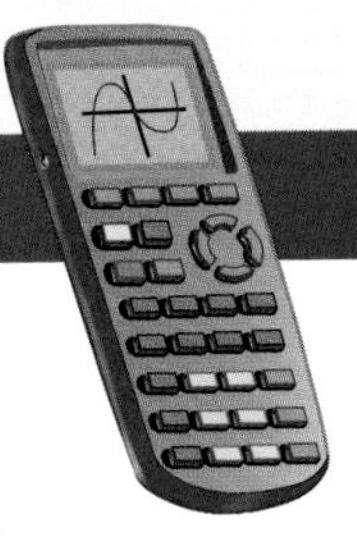

Technology

Lesson Suggestions and Calculator Examples

(Keystrokes are based on a TI-83 calculator.)

Lesson 4.1 Using Matrices to Represent Data

In this lesson, students will learn about matrix entries and operations. Let $A = \begin{bmatrix} -2 & -3 \\ 3 & 5 \end{bmatrix}$ and $B = \begin{bmatrix} 0 & 4 \\ -2 & 1 \end{bmatrix}$.

To enter matrices A and B, press MATRX, select **EDIT**, and choose a matrix name. (On a TI-83 Plus, press 2nd to access the matrix menu and proceed as described.) Next enter the dimensions and the entries in order row by row. To add matrix A and matrix B, students need to bring up matrix A by name, press +, bring matrix B up by name, and press ENTER.

Have students practice additional problems involving arithmetic of matrices.

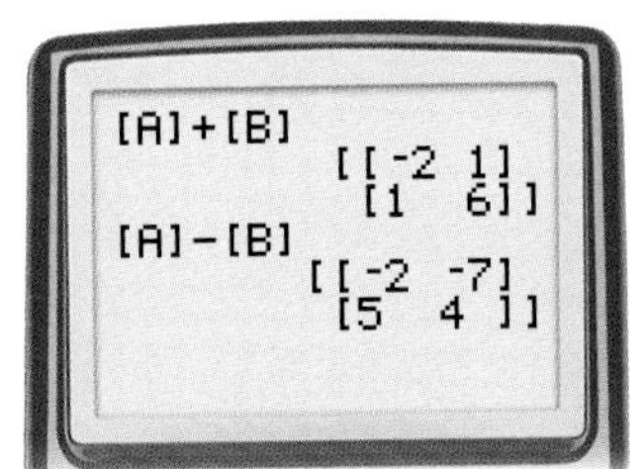

Lesson 4.2 Matrix Multiplication

In this lesson, students will learn how to perform matrix operations. Using matrices A and B above, have students calculate AB and BA. Ask students to compare the two products. Lesson 4.2 introduces a special kind of matrix multiplication, that is, multipication of a square matrix by itself. This operation is used in the study of networks, as discussed on page 228. Have students enter the adjacency

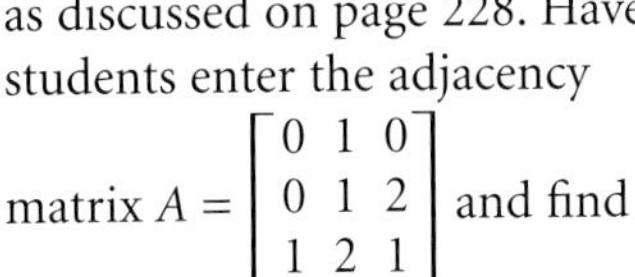

matrix $A = \begin{bmatrix} 0 & 1 & 0 \\ 0 & 1 & 2 \\ 1 & 2 & 1 \end{bmatrix}$ and find A^2, as shown at right. To help students appreciate the interplay between reason and technology, point out that the calculator will give A^2 but that students must interpret the display on their own.

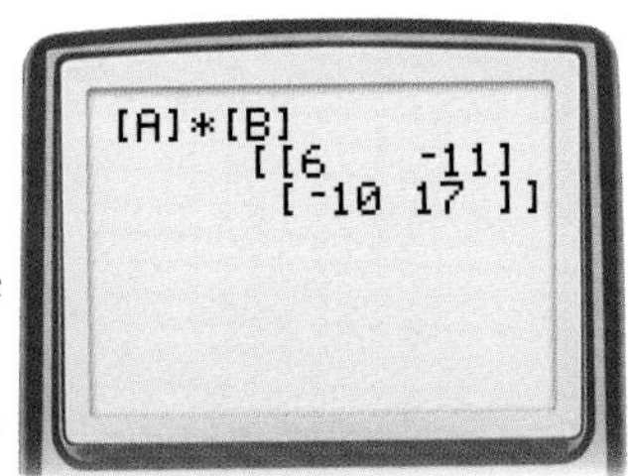

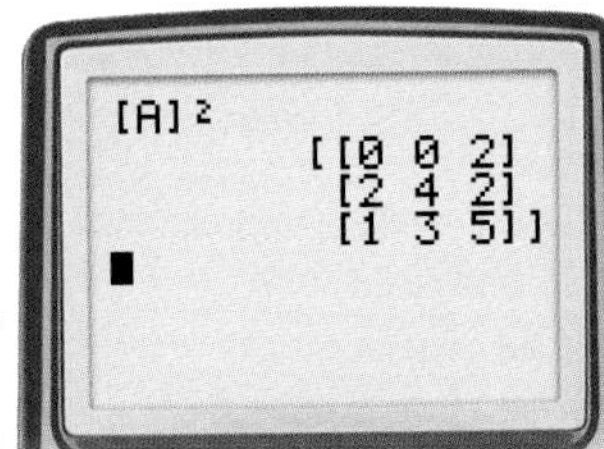

Lesson 4.3 The Inverse of a Matrix

There is a connection between arithmetic and matrix algebra. Point out that the multiplicative inverse of a matrix is similar to the multiplicative inverse of a real number but that there are also significant differences. Ask students to justify their answers to the following questions:

- Can you find the multiplicative inverse of a nonsquare matrix?
- Can you always find the multiplicative inverse of a square matrix?

After a brief discussion about the existence of inverses, you can have students go through the mechanics of finding the inverse, if it exists. For example, if $B = \begin{bmatrix} 2 & -1 & 1 \\ -1 & 3 & 4 \\ -2 & 1 & 0 \end{bmatrix}$, then the display at right shows its inverse. (This matrix is part **b** of Example 2 on page 236.)

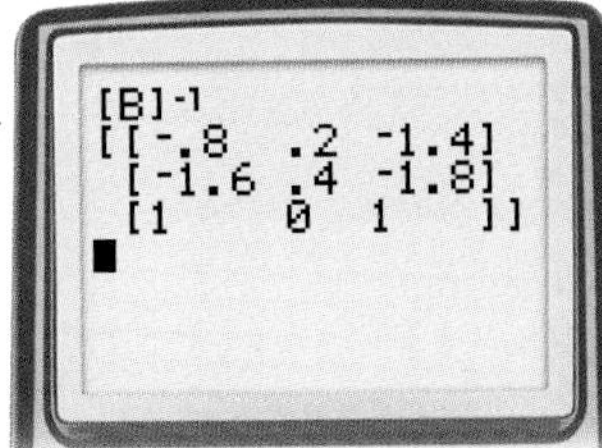

Lesson 4.4 Solving Systems With Matrix Equations

In a system of equations involving three equations in three variables a calculator will produce the solution quickly and with minimal risk of error.

The display at right shows the solution to the system

$$\begin{cases} 2y - z = -7 - 5x \\ x - 2y + 2z = 0 \\ 3y = 17 - z \end{cases}.$$

Note that, in this case, the equations must be rewritten before the matrices can be entered into the calculator.

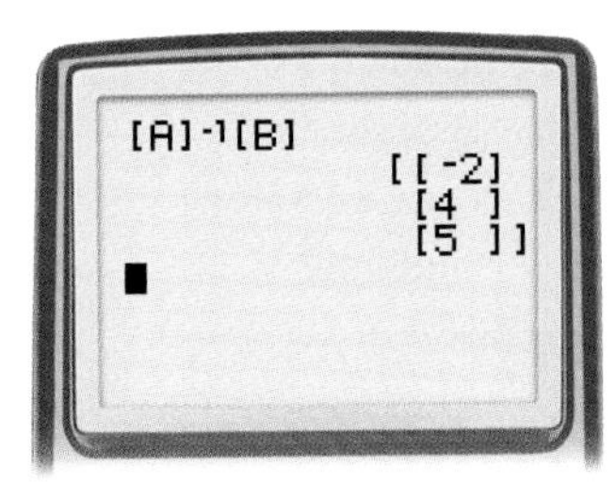

Lesson 4.5 Using Matrix Row Operations

On the TI-83, students can enter a matrix and execute a single command that will display the reduced row-echelon form of the matrix. Have students press MATRX, select **MATH** **B:rref(**, enter the matrix name, and press) ENTER.

For further information, refer to the

- technology discussions in the lessons.
- lesson-related teacher's commentary in the side columns of this *Teacher's Edition*.
- lesson-related *Student Technology Guide* masters.
- *HRW Technology Handbook.*

internet connect

For keystrokes of other graphing calculators models, visit the HRW web site at **go.hrw.com** and enter the keyword **MB1 CALC**.

About Chapter 4

Background Information

In Chapter 4, students study the structure and uses of matrices. They learn to find scalar multiples of a given matrix and to find sums, differences, and products of two or more matrices. They explore inverse matrices and use them to solve systems of linear equations. The final lesson of the chapter introduces matrix row operations, and students see how they can be used, in conjunction with augmented matrices, to solve systems of equations.

CHAPTER RESOURCES

- Block-Scheduling Handbook
- Writing Activities for Your Portfolio
- Tech Prep Masters
- Long-Term Project
- Assessment Resources:
  - Mid-Chapter Assessment
  - Chapter Assessments
  - Alternative Assessments
- Test and Practice Generator
- Technology Handbook

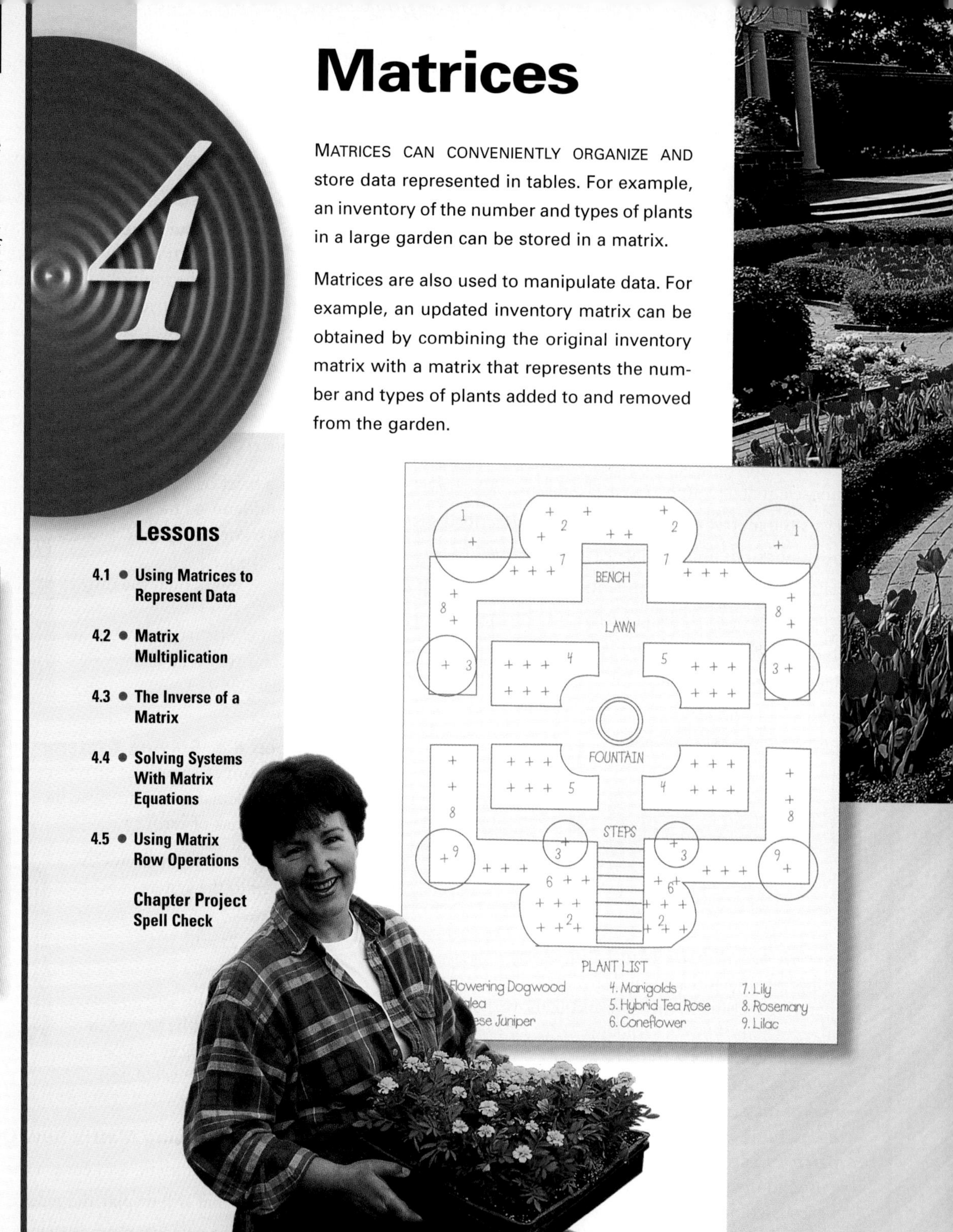

4 Matrices

MATRICES CAN CONVENIENTLY ORGANIZE AND store data represented in tables. For example, an inventory of the number and types of plants in a large garden can be stored in a matrix.

Matrices are also used to manipulate data. For example, an updated inventory matrix can be obtained by combining the original inventory matrix with a matrix that represents the number and types of plants added to and removed from the garden.

Lessons

About the Photos

Often an elaborate garden has several beds with a variety of plants. A gardener may use matrices to manage the number and cost of the plants incorporated into the garden. Suppose that a gardener wants to change the plants in three beds of the garden with a mixture of new plants which will be chosen from marigolds (M), juniper (J), coneflowers (C), azaleas (A), and rosemary (R). The number of new plants in each bed is shown in the matrix at right, along with a matrix showing the cost of each type of plant. The product of the two matrices gives the cost of the plants associated with each bed.

$$\text{Bed}\ \begin{matrix} \\ 1 \\ 2 \\ 3 \end{matrix} \begin{matrix} \begin{matrix} M & J & C & A & R \end{matrix} \\ \begin{bmatrix} 2 & 7 & 5 & 1 & 0 \\ 0 & 4 & 3 & 8 & 4 \\ 1 & 7 & 3 & 0 & 0 \end{bmatrix} \end{matrix} \quad \begin{matrix} \\ M \\ J \\ C \\ A \\ R \end{matrix} \begin{matrix} \text{Plant cost} \\ \begin{bmatrix} 1.95 \\ 5.34 \\ 1.57 \\ 8.37 \\ 2.51 \end{bmatrix} \end{matrix} = \text{Bed}\ \begin{matrix} \\ 1 \\ 2 \\ 3 \end{matrix} \begin{matrix} \text{Cost} \\ \begin{bmatrix} 57.50 \\ 103.07 \\ 44.04 \end{bmatrix} \end{matrix}$$

At a glance, the gardener can see the number of plants per bed, the price per plant, and the total cost for each of the three new beds.

About the Chapter Project

Mathematical models can apply to our everyday world in ways we rarely think about. By ordering seemingly random data into an organized format, models allow us to see underlying relations and to use them in fascinating ways.

In the Chapter Project, *Spell Check*, you will use the modeling process to examine spell-checking software.

After completing the Chapter Project, you will be able to do the following:

- Create a directed network and corresponding adjacency matrix to represent associated words in a spell-checking directory.
- Interpret powers of adjacency matrices.

About the Portfolio Activities

Throughout the chapter, you will be given opportunities to complete Portfolio Activities that are designed to support your work on the Chapter Project.

- Interpreting a directed network that represents paths between exhibits in the New York Museum of Natural History is included in the Portfolio Activity on page 224.
- Finding and interpreting powers of adjacency matrices that represent multistep paths between exhibits is included in the Portfolio Activity on page 233.

Chapter Objectives

- Represent mathematical and real-world data in a matrix. **[4.1]**
- Find sums and differences of matrices and the scalar product of a number and a matrix. **[4.1]**
- Multiply two matrices. **[4.2]**
- Use matrix multiplication to solve mathematical and real-world problems. **[4.2]**
- Find and use the inverse of a matrix, if it exists. **[4.3]**
- Find and use the determinant of a matrix. **[4.3]**
- Use matrices to solve systems of linear equations in mathematical and real-world situations. **[4.4]**
- Represent a system of equations as an augmented matrix. **[4.5]**
- Solve a system of linear equations by using elementary row operations. **[4.5]**

Portfolio Activities appear at the end of Lessons 4.1 and 4.2. Each serves as preparation for the Chapter Project. The Portfolio Activities as well as the Chapter Project Activities are appropriate for inclusion in the student's portfolio. Students should be encouraged to include in their portfolios any other work in which they feel a sense of pride or a sense of accomplishment.

internet connect go.hrw.com

Chapter Internet Features and Online Activities

Lesson	Keyword	Page	Lesson	Keyword	Page
4.1	MB1 Homework Help	221	4.3	MB1 Homework Help	239
	MB1 Matrix	224		MB1 Magic Squares	240
4.2	MB1 Homework Help	229	4.4	MB1 Homework Help	248
	MB1 Breakfast	229		MB1 Cramer	249
	MB1 Matrix Multiply	233	4.5	MB1 Homework Help	256

Prepare

NCTM PRINCIPLES & STANDARDS 1–10

QUICK WARM-UP

Evaluate each expression for $a = -5$, $b = 1.3$, and $c = -7$.

1. $a + b$ — -3.7
2. $b - c$ — 8.3
3. $a - b + c$ — -13.3
4. $-4b$ — -5.2

Solve each equation.

5. $16 = 2x - 5$ — $x = 10.5$
6. $3y + 8 = -y - 15$ — $y = -5.75$

Also on Quiz Transparency 4.1

Teach

Why Ask students to describe some methods that they have used in the past to organize information. Their answers may include structures such as tables, lists, and tree diagrams. Tell them that a *matrix* is a structure that makes it easy not only to organize information, but also to manipulate it.

Use Teaching Transparency 14.

4.1 Using Matrices to Represent Data

Why *Matrices can be used to organize data. For example, information about picnic tables and barbeque grills can be organized into matrices.*

Objectives

- Represent mathematical and real-world data in a matrix.
- Find sums and differences of matrices and the scalar product of a number and a matrix.

APPLICATION INVENTORY

The table below shows business activity for one month in a home-improvement store. The table shows stock (inventory on June 1), sales (during June), and receipt of new goods (deliveries in June).

	Inventory (June 1) Small	Inventory (June 1) Large	Sales (June) Small	Sales (June) Large	Deliveries (June) Small	Deliveries (June) Large
Picnic tables	8	10	7	9	15	20
Barbeque grills	15	12	15	12	18	24

You can represent the inventory data in a *matrix*.

Inventory matrix →

$$\begin{array}{r} \\ \textbf{Picnic tables} \\ \textbf{Barbeque grills} \end{array} \begin{array}{c} \textbf{Small}\ \ \textbf{Large} \\ \begin{bmatrix} 8 & 10 \\ 15 & 12 \end{bmatrix} \end{array} = M = \begin{bmatrix} m_{11} & m_{12} \\ m_{21} & m_{22} \end{bmatrix}$$

m_{21}: 2nd row, 1st column

A **matrix** (plural, *matrices*) is a rectangular array of numbers enclosed in a single set of brackets. The **dimensions** of a matrix are the number of horizontal rows and the number of vertical columns it has. For example, if a matrix has 2 rows and 3 columns, its dimensions are 2×3, read as "2 by 3." The inventory matrix above, M, is a matrix with dimensions of 2×2.

Each number in the matrix is called an **entry**, or element. You can denote the *address* of the entry in row 2 and column 1 of the inventory matrix, M, as m_{21} and state that $m_{21} = 15$. This entry represents 15 small barbeque grills in stock on June 1.

EXAMPLE 1 **Represent the June sales data in matrix S. Interpret the entry at s_{12}.**

SOLUTION

Sales matrix →

$$\begin{array}{r} \\ \textbf{Picnic tables} \\ \textbf{Barbeque grills} \end{array} \begin{array}{c} \textbf{Small}\ \ \textbf{Large} \\ \begin{bmatrix} 7 & 9 \\ 15 & 12 \end{bmatrix} \end{array} = S$$

In matrix S, $s_{12} = 9$. In June, 9 large picnic tables were sold.

TRY THIS Represent the delivery data in matrix D. Interpret the entry at d_{21}.

Alternative Teaching Strategy

HANDS-ON STRATEGIES Provide students with several bingo cards on which you have written row numbers along the left side. Ask them to describe the similarities and differences among the cards. Students should notice that all the cards have the same *dimensions*, that is, five rows and five columns. Point out that some cards may contain some of the same numbers, but no two cards have exactly the same numbers in every location.

Now name a specific row and column, such as row 4 of column N. Ask students to write the number that appears in that location. Point out that there is exactly one number in any given location on a single card. Use this discussion as an introduction to a definition of *matrix*.

Now have students work in pairs. Have them find the sums and the differences of the numbers in corresponding locations on their bingo cards. Use this activity to initiate a discussion of matrix addition and subtraction.

Two matrices are *equal* if they have the same dimensions and if corresponding entries are equivalent.

EXAMPLE 2 Solve $\begin{bmatrix} 2x+4 & 5 & 1 \\ -2 & -3y+5 & -4 \end{bmatrix} = \begin{bmatrix} 12 & 5 & 1 \\ -2 & 5y-3 & -4 \end{bmatrix}$ for x and y.

SOLUTION

Because the matrices are equal, $2x + 4 = 12$ and $-3y + 5 = 5y - 3$.

$2x + 4 = 12$	$-3y + 5 = 5y - 3$
$2x = 8$	$-8y = -8$
$x = 4$	$y = 1$

Thus, $x = 4$ and $y = 1$.

TRY THIS Solve $\begin{bmatrix} -3 & -2x-3 \\ -2 & 3y-12 \end{bmatrix} = \begin{bmatrix} -3 & -15 \\ -2 & -2y+13 \end{bmatrix}$ for x and y.

Addition and Scalar Multiplication

To find the sum (or difference) of matrices A and B with the same dimensions, find the sums (or differences) of *corresponding* entries in A and B.

EXAMPLE 3 Let $A = \begin{bmatrix} -2 & 0 & 1 \\ 5 & -7 & 8 \end{bmatrix}$ and $B = \begin{bmatrix} 5 & 7 & -1 \\ 0 & 2 & -8 \end{bmatrix}$.

a. Find $A + B$. **b.** Find $A - B$.

SOLUTION

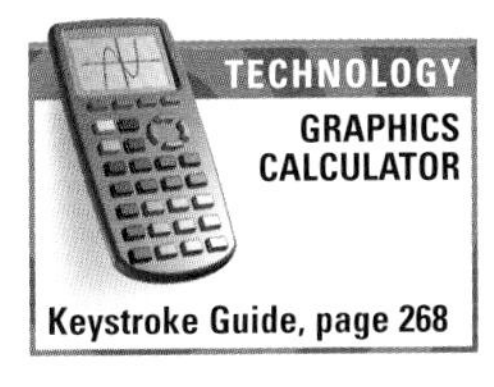

Keystroke Guide, page 268

a.

$$A + B = \begin{bmatrix} -2 & 0 & 1 \\ 5 & -7 & 8 \end{bmatrix} + \begin{bmatrix} 5 & 7 & -1 \\ 0 & 2 & -8 \end{bmatrix}$$
$$= \begin{bmatrix} -2+5 & 0+7 & 1+(-1) \\ 5+0 & -7+2 & 8+(-8) \end{bmatrix}$$
$$= \begin{bmatrix} 3 & 7 & 0 \\ 5 & -5 & 0 \end{bmatrix}$$

CHECK

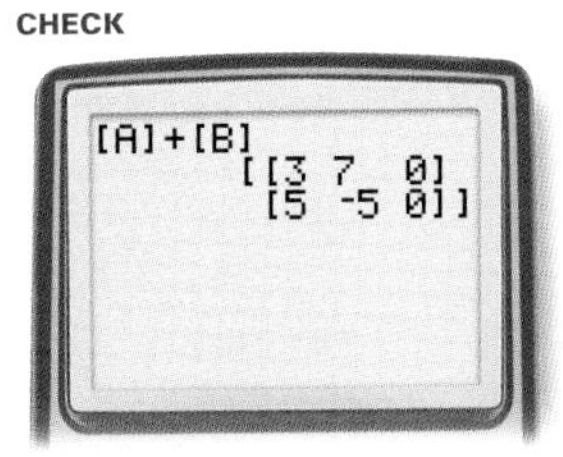

b.

$$A - B = \begin{bmatrix} -2 & 0 & 1 \\ 5 & -7 & 8 \end{bmatrix} - \begin{bmatrix} 5 & 7 & -1 \\ 0 & 2 & -8 \end{bmatrix}$$
$$= \begin{bmatrix} -2-5 & 0-7 & 1-(-1) \\ 5-0 & -7-2 & 8-(-8) \end{bmatrix}$$
$$= \begin{bmatrix} -7 & -7 & 2 \\ 5 & -9 & 16 \end{bmatrix}$$

CHECK

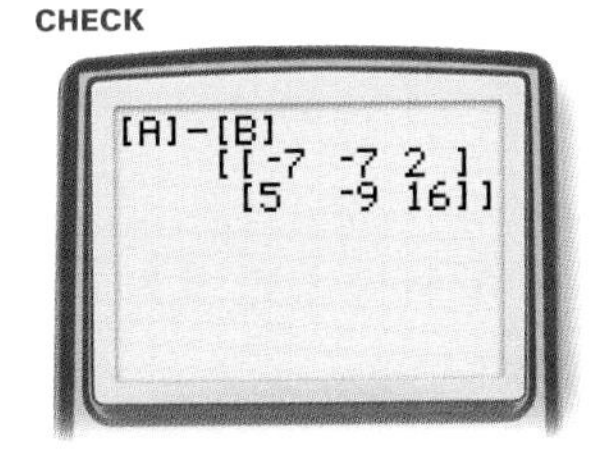

TRY THIS Let $A = \begin{bmatrix} 0 & 0 \\ 4 & 1 \\ -3 & -5 \end{bmatrix}$ and $B = \begin{bmatrix} -10 & 5 \\ 0 & 4 \\ -7 & 3 \end{bmatrix}$.

a. Find $A - B$. **b.** Find $A + B$.

Interdisciplinary Connection

BIOLOGY This table gives data about the females of a certain species of brown rats. Matrix L at right displays the same data in a *Leslie matrix.*

Age (months)	Birth rate	Survial rate
0–3	0	0.6
3–6	0.3	0.9
6–9	0.8	0.9
9–12	0.7	0.8
12–15	0.4	0.6

Explain how the data is arranged in the matrix.

$$L = \begin{bmatrix} 0 & 0.6 & 0 & 0 & 0 & 0 \\ 0.3 & 0 & 0.9 & 0 & 0 & 0 \\ 0.8 & 0 & 0 & 0.9 & 0 & 0 \\ 0.7 & 0 & 0 & 0 & 0.8 & 0 \\ 0.4 & 0 & 0 & 0 & 0 & 0.6 \end{bmatrix}$$

The birth rates are listed in the first column. The survival rates are listed on a diagonal that is immediately above the main diagonal.

ADDITIONAL EXAMPLE 1

The table shows the number of hours students studied math and science the week before midterm exams and the week before final exams.

Represent the number of hours studied in matrix T. Interpret the entry at t_{21}.

	Before midterms	Before finals
Math	8	15
Science	7	12

$T = $ (rows Math, Science; columns Midterms, Finals) $\begin{bmatrix} 8 & 15 \\ 7 & 12 \end{bmatrix}$

In matrix T, $t_{21} = 7$. The students spent 7 hours studying science before midterms.

TRY THIS

(rows Picnic tables, Barbecue grills; columns Small, Large) $\begin{bmatrix} 15 & 20 \\ 18 & 24 \end{bmatrix}$

$d_{21} = 18$; in June, 18 small barbecue grills were delivered.

ADDITIONAL EXAMPLE 2

Solve
$\begin{bmatrix} 4x+5 & 9 & 15 \\ 7 & -2y+3 & -1 \end{bmatrix} = \begin{bmatrix} 21 & 9 & 15 \\ 7 & y-12 & -1 \end{bmatrix}$ **for x and y.**

$x = 4$ and $y = 5$

TRY THIS

$x = 6$ and $y = 5$

☞ For Additional Example 3 and answer to Try This, see page 218.

ADDITIONAL EXAMPLE 3

Let $M = \begin{bmatrix} -5 & 9 & 2 \\ 11 & -1 & 3 \end{bmatrix}$ and $N = \begin{bmatrix} 2 & 6 & -31 \\ 22 & 10 & 0 \end{bmatrix}$.

a. Find $M + N$.

$$\begin{bmatrix} -3 & 15 & -29 \\ 33 & 9 & 3 \end{bmatrix}$$

b. Find $M - N$.

$$\begin{bmatrix} -7 & 3 & 33 \\ -11 & -11 & 3 \end{bmatrix}$$

TRY THIS

a. $\begin{bmatrix} 10 & -5 \\ 4 & -3 \\ 4 & -8 \end{bmatrix}$ **b.** $\begin{bmatrix} -10 & 5 \\ 4 & 5 \\ -10 & -2 \end{bmatrix}$

CHECKPOINT ✔

No; the matrices do not have the same dimensions.

ADDITIONAL EXAMPLE 4

Let $J = \begin{bmatrix} -3 & 2 \\ 0 & 5 \\ 9 & -8 \end{bmatrix}$,

$K = \begin{bmatrix} 0 & -1 \\ 7 & 6 \\ 2 & -4 \end{bmatrix}$, and

$L = \begin{bmatrix} 8 & 10 \\ -1 & -9 \\ 4 & 1 \end{bmatrix}$.

Find $J - K + L$.

$$\begin{bmatrix} 5 & 13 \\ -8 & -10 \\ 11 & -3 \end{bmatrix}$$

TRY THIS

$$\begin{bmatrix} -1 & 9 \\ 0 & -9 \end{bmatrix}$$

CHECKPOINT ✔ Is it possible to find the sum $\begin{bmatrix} -2 & 5 & 6 \\ 1 & -8 & 0 \end{bmatrix} + \begin{bmatrix} 1 & -5 \\ 8 & -6 \\ -3 & 0 \end{bmatrix}$? Explain.

Example 4 below shows how matrix addition and subtraction can be used in inventory calculations. You can perform matrix addition and subtraction in one step.

EXAMPLE 4

APPLICATION INVENTORY

Refer to the table of business activity at the beginning of the lesson. Let matrices M, S, and D represent the inventory, sales, and delivery data, respectively.

Find $M - S + D$. Interpret the final matrix.

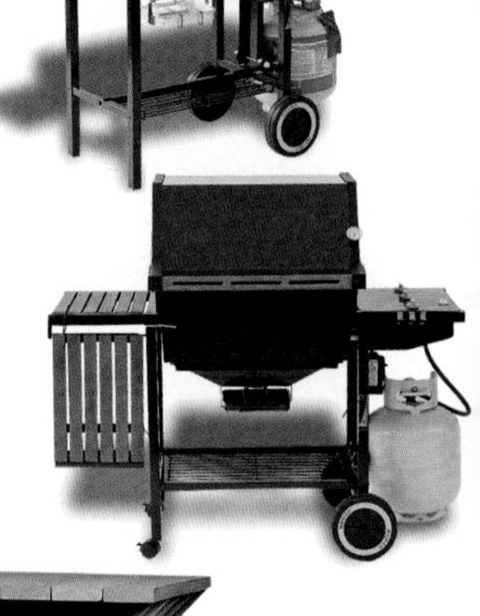

SOLUTION

$$M - S + D = \begin{bmatrix} 8 & 10 \\ 15 & 12 \end{bmatrix} - \begin{bmatrix} 7 & 9 \\ 15 & 12 \end{bmatrix} + \begin{bmatrix} 15 & 20 \\ 18 & 24 \end{bmatrix}$$

$$= \begin{bmatrix} 8 - 7 + 15 & 10 - 9 + 20 \\ 15 - 15 + 18 & 12 - 12 + 24 \end{bmatrix}$$

$$\begin{matrix} & \textbf{Small} & \textbf{Large} & \\ = & \begin{bmatrix} 16 & 21 \\ 18 & 24 \end{bmatrix} & & \begin{matrix} \textbf{Picnic tables} \\ \textbf{Barbeque grills} \end{matrix} \end{matrix}$$

At the end of June, the store has 16 small and 21 large picnic tables in stock. It also has 18 small and 24 large barbeque grills.

TRY THIS Find $\begin{bmatrix} 3 & 6 \\ -3 & -6 \end{bmatrix} + \begin{bmatrix} -4 & 5 \\ -7 & 8 \end{bmatrix} - \begin{bmatrix} 0 & 2 \\ -10 & 11 \end{bmatrix}$.

To multiply a matrix, A, by a real number, k, write a matrix whose entries are k times each of the entries in matrix A. This operation is called **scalar multiplication**.

EXAMPLE 5 **Let $A = \begin{bmatrix} 3 & 2 & 0 \\ -1 & -3 & 6 \\ 2 & 0 & -10 \end{bmatrix}$. Find $-2A$.**

SOLUTION

$$-2A = \begin{bmatrix} -2(3) & -2(2) & -2(0) \\ -2(-1) & -2(-3) & -2(6) \\ -2(2) & -2(0) & -2(-10) \end{bmatrix} = \begin{bmatrix} -6 & -4 & 0 \\ 2 & 6 & -12 \\ -4 & 0 & 20 \end{bmatrix}$$

CHECKPOINT ✔ What are the entries in matrix kA if A is a 2×3 matrix and $k = 0$?

Inclusion Strategies

AUDITORY LEARNERS Have students work in pairs. Give one student in each pair a matrix written on a sheet of paper. Be sure that the other students do not see the matrix. The students who have the matrix should now describe it to their partner so that the partner can re-create it on a clean sheet of paper. Encourage students to employ the language of matrices, taking care to use the terms *dimension*, *entry*, *row*, and *column* correctly. Have the partners switch roles and repeat the activity with a second matrix.

When $k = -1$, the scalar product kA is $-1A$, or simply $-A$, and is called the **additive inverse**, or *opposite*, of matrix A. For example,

if $A = \begin{bmatrix} 3 & -4 & 0 \\ 2 & -8 & 6 \\ 7 & 1 & -5 \end{bmatrix}$, then $-A = \begin{bmatrix} -3 & 4 & 0 \\ -2 & 8 & -6 \\ -7 & -1 & 5 \end{bmatrix}$ is the additive inverse of A.

CHECKPOINT ✔ Let $T = \begin{bmatrix} 3 & -5 \\ 0 & 2 \end{bmatrix}$. Write the sum of T and its additive inverse.

CRITICAL THINKING Let $k = 3$ and $A = \begin{bmatrix} 10 & -4 \\ -3 & 5 \end{bmatrix}$. Use matrix addition and scalar multiplication to show that $A + A + A = 3A$.

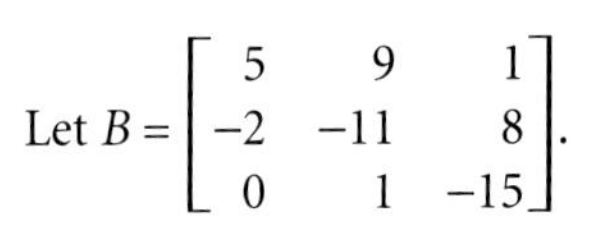

EXAMPLE 5

Let $B = \begin{bmatrix} 5 & 9 & 1 \\ -2 & -11 & 8 \\ 0 & 1 & -15 \end{bmatrix}$.

Find $-3B$.

$\begin{bmatrix} -15 & -27 & -3 \\ 6 & 33 & -24 \\ 0 & -3 & 45 \end{bmatrix}$

CHECKPOINT ✔

$\begin{bmatrix} 0 & 0 & 0 \\ 0 & 0 & 0 \end{bmatrix}$

CHECKPOINT ✔

$\begin{bmatrix} 0 & 0 \\ 0 & 0 \end{bmatrix}$

CRITICAL THINKING

$A + A + A = \begin{bmatrix} 30 & -12 \\ -9 & 15 \end{bmatrix}$

$3A = \begin{bmatrix} 30 & -12 \\ -9 & 15 \end{bmatrix}$

Properties of Matrix Addition

For matrices A, B, and C, each with dimensions of $m \times n$:

Commutative	$A + B = B + A$
Associative	$(A + B) + C = A + (B + C)$
Additive Identity	The $m \times n$ matrix having 0 as all of its entries is the $m \times n$ identity matrix for addition.
Additive Inverse	For every $m \times n$ matrix A, the matrix whose entries are the opposite of those in A is the additive inverse of A.

Geometric Transformations

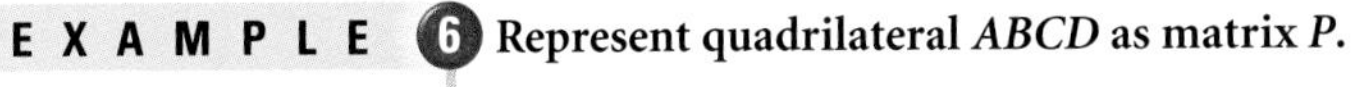

Example 6 shows you how to represent a polygon in the coordinate plane as a matrix.

EXAMPLE 6 Represent quadrilateral $ABCD$ as matrix P.

COORDINATE GEOMETRY

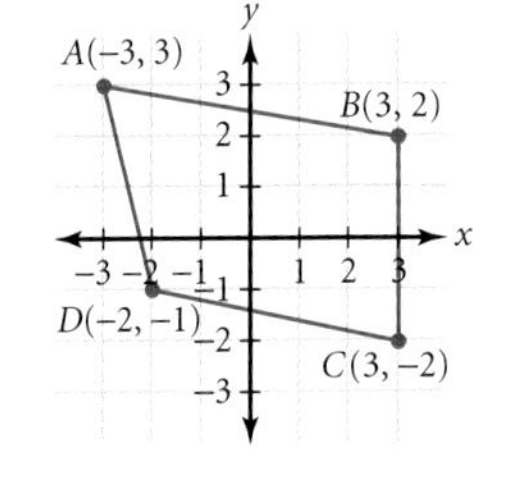

SOLUTION

Because each point has 2 coordinates and there are 4 points, create a 2×4 matrix.

$$P = \begin{bmatrix} -3 & 3 & 3 & -2 \\ 3 & 2 & -2 & -1 \end{bmatrix} \begin{matrix} x\text{-coordinates} \\ y\text{-coordinates} \end{matrix}$$

(columns A, B, C, D)

When you perform a *transformation* on one geometric figure to get another geometric figure, the original figure is called the **pre-image** and the resulting figure is the **image.** When you apply scalar multiplication to a matrix that represents a polygon, the product represents either an enlarged image or a reduced image of the pre-image polygon. This is shown in Example 7 on page 220.

ADDITIONAL

EXAMPLE 6

Represent the quadrilateral below as matrix R.

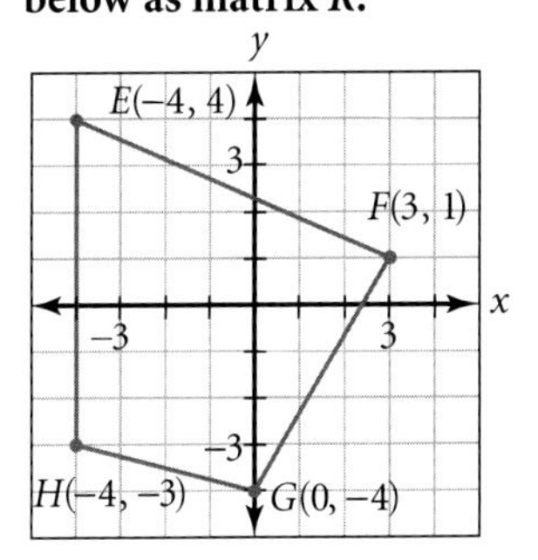

$R = \begin{bmatrix} -4 & 3 & 0 & -4 \\ 4 & 1 & -4 & -3 \end{bmatrix}$ (columns E, F, G, H)

Enrichment

Matrix A represents a triangle, and matrix B represents its image under some transformation. Have students determine the transformation and identify the matrix operation for it.

$A = \begin{bmatrix} 0 & 2 & 4 \\ 0 & 3 & 1 \end{bmatrix}$ (columns R, S, T)

$B = \begin{bmatrix} -4 & -2 & 0 \\ -1 & 2 & 0 \end{bmatrix}$ (columns R', S', T')

The translation is 4 units to the left and 1 unit down; matrix T at right is added to matrix A to produce matrix B.

$T = \begin{bmatrix} -4 & -4 & -4 \\ -1 & -1 & -1 \end{bmatrix}$

Students can check matrix operations by using the key sequences below. On a TI-82 or TI-83, press MATRX to access the **MATRX** menu, or on a TI-83 Plus, press 2nd x^{-1}. Select **EDIT** from the **MATRX** menu, choose a name for each matrix from the list shown, press ENTER, enter the dimensions of the matrix, and then enter the elements. After entering the matrices, access the matrix menu, select the name of the first matrix, press ENTER, press an operation key, such as +, return to the **MATRX** menu, select the name of the second matrix, and press ENTER.

Math CONNECTION

COORDINATE GEOMETRY In Example 6, remind students that a *quadrilateral* is a polygon that has four sides. The four points where the sides meet are the *vertices* of the quadrilateral. The coordinates of the vertices are used to create the matrix for the quadrilateral.

ADDITIONAL EXAMPLE 7

Refer to quadrilateral $EFGH$ and matrix R in Additional Example 6. **Graph the polygon that is represented by each matrix described below.**

a. $3R$

$$3R = \begin{bmatrix} -12 & 9 & 0 & -12 \\ 12 & 3 & -12 & -9 \end{bmatrix}$$

(columns E', F', G', H')

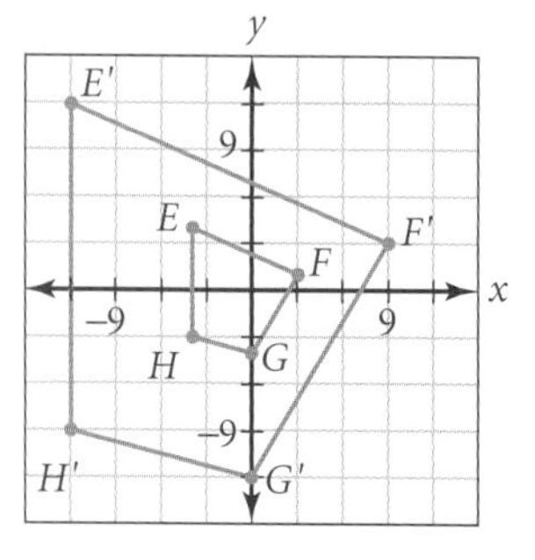

b. $\frac{1}{3}R$

$$\frac{1}{3}R = \begin{bmatrix} -\frac{4}{3} & 1 & 0 & -\frac{4}{3} \\ \frac{4}{3} & \frac{1}{3} & -\frac{4}{3} & -1 \end{bmatrix}$$

(columns E'', F'', G'', H'')

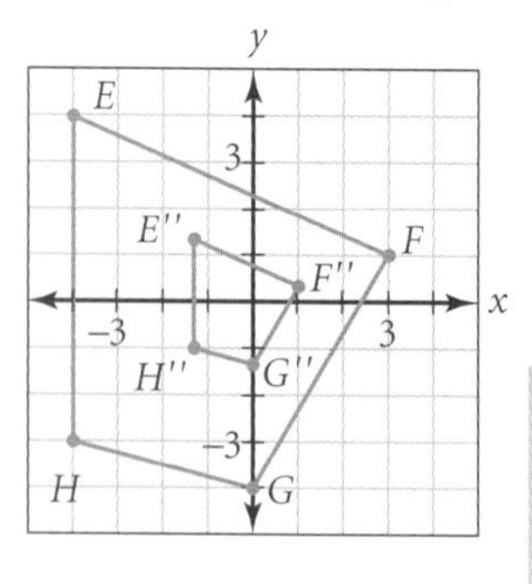

Math CONNECTION

TRANSFORMATIONS Remind students that the dilations in Example 7 result in quadrilaterals that are similar to the given quadrilateral.

EXAMPLE 7 Refer to quadrilateral $ABCD$ and matrix P in Example 6.

Graph the polygon that is represented by each matrix.

a. $2P$ **b.** $\frac{1}{2}P$

SOLUTION

a. Let quadrilateral $A'B'C'D'$ represent the image.

$$2P = \begin{bmatrix} -6 & 6 & 6 & -4 \\ 6 & 4 & -4 & -2 \end{bmatrix}$$

(columns A', B', C', D')

Graph $A'B'C'D'$.

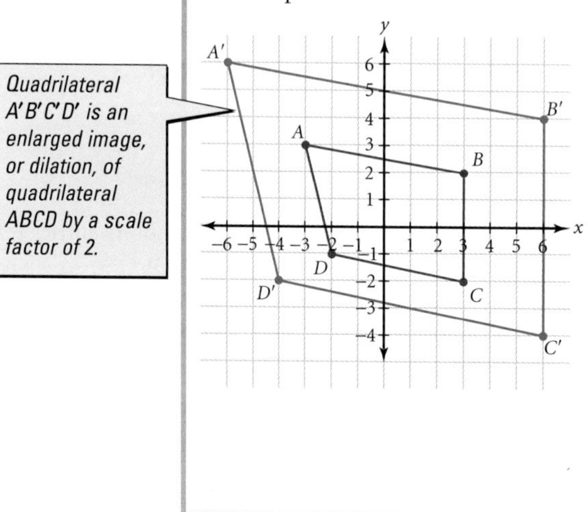

b. Let quadrilateral $A''B''C''D''$ represent the image.

$$\frac{1}{2}P = \begin{bmatrix} -\frac{3}{2} & \frac{3}{2} & \frac{3}{2} & -1 \\ \frac{3}{2} & 1 & -1 & -\frac{1}{2} \end{bmatrix}$$

(columns A'', B'', C'', D'')

Graph $A''B''C''D''$.

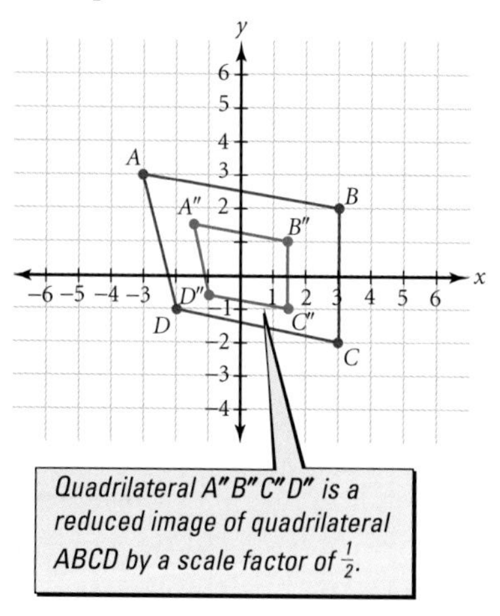

TRY THIS Refer to quadrilateral $ABCD$ and matrix P in Example 6. Sketch the polygon that is represented by each matrix.

a. $4P$ **b.** $\frac{1}{4}P$

Exercises

Communicate

1. Describe the location of entry m_{52} in a matrix called M.
2. Describe the location of entries b_{21}, b_{22}, b_{23}, and b_{24} in a matrix called B. What are the smallest possible dimensions of matrix B?
3. Explain how to represent a polygon in the coordinate plane as a matrix.
4. Explain how to use scalar multiplication to transform a polygon (pre-image) in the coordinate plane into another polygon (image).

Reteaching the Lesson

WORKING BACKWARD

Have students work in pairs. Give them matrix M at right. $M = \begin{bmatrix} 4 & 0 & -4 \\ 2 & 8 & -2 \end{bmatrix}$ Have them find matrices A and B and a real number k that meet the following conditions. Specify that no two entries in any one matrix can be equal. **Sample answers are given.**

1. $A + B = M$ $A = \begin{bmatrix} 1 & 3 & 5 \\ 2 & 4 & 6 \end{bmatrix}, B = \begin{bmatrix} 3 & -3 & -9 \\ 0 & 4 & -8 \end{bmatrix}$

2. $B - A = M$ $A = \begin{bmatrix} 6 & 5 & 2 \\ 5 & 3 & 1 \end{bmatrix}, B = \begin{bmatrix} 10 & 5 & -2 \\ 7 & 11 & -1 \end{bmatrix}$

3. $kA = M$ $k = 2, A = \begin{bmatrix} 2 & 0 & -2 \\ 1 & 4 & -1 \end{bmatrix}$

If students need additional practice, repeat the exercises for a different matrix M with different dimensions.

Guided Skills Practice

APPLICATION

5. $M = \begin{bmatrix} 12 & 28 & 17 \\ 15 & 32 & 45 \\ 6 & 20 & 30 \end{bmatrix}$

$m_{23} = 45$; 45 large T-shirts in inventory

6. $x = 3$, $y = -4$

7. a. $\begin{bmatrix} 11 & -7 \\ -7 & 5 \\ 0 & 6 \end{bmatrix}$

b. $\begin{bmatrix} -5 & 11 \\ -3 & -7 \\ -14 & 12 \end{bmatrix}$

8. $\begin{bmatrix} -3 & -9 & -9 \\ -12 & -1 & 15 \end{bmatrix}$

9. $\begin{bmatrix} \frac{1}{2} & 0 & -4 \\ -3 & 2 & -\frac{5}{2} \end{bmatrix}$

CONNECTIONS

10. $Q = \begin{bmatrix} -3 & 2 & 1 & -4 \\ 3 & 3 & -3 & -3 \end{bmatrix}$

5. INVENTORY Represent the inventory data at right in a matrix, M. Interpret the entry at m_{23}. ***(EXAMPLE 1)***

	Inventory		
	Small	Medium	Large
Jerseys	12	28	17
T-shirts	15	32	45
Sweatshirts	6	20	30

6. Solve $\begin{bmatrix} 6 & 5 \\ x+8 & 4 \\ 0 & 2y-1 \end{bmatrix} = \begin{bmatrix} 6 & 5 \\ 14-x & 4 \\ 0 & -13-y \end{bmatrix}$ for x and y. ***(EXAMPLE 2)***

7. Let $R = \begin{bmatrix} 3 & 2 \\ -5 & -1 \\ -7 & 9 \end{bmatrix}$ and $S = \begin{bmatrix} 8 & -9 \\ -2 & 6 \\ 7 & -3 \end{bmatrix}$. ***(EXAMPLE 3)***

a. Find $R + S$. **b.** Find $R - S$.

8. Find $\begin{bmatrix} 2 & -9 & -5 \\ -6 & 1 & 3 \end{bmatrix} - \begin{bmatrix} 5 & -2 & 12 \\ 2 & -4 & -5 \end{bmatrix} + \begin{bmatrix} 0 & -2 & 8 \\ -4 & -6 & 7 \end{bmatrix}$. ***(EXAMPLE 4)***

9. Let $A = \begin{bmatrix} -1 & 0 & 8 \\ 6 & -4 & 5 \end{bmatrix}$. Find $-\frac{1}{2}A$. ***(EXAMPLE 5)***

10. COORDINATE GEOMETRY Represent quadrilateral $ABCD$ at right as a matrix, Q. ***(EXAMPLE 6)***

11. TRANSFORMATIONS Refer to quadrilateral $ABCD$ at right and matrix Q in Exercise 10. Graph the polygon that is represented by each matrix below. ***(EXAMPLE 7)***

a. $3Q$ **b.** $\frac{1}{3}Q$

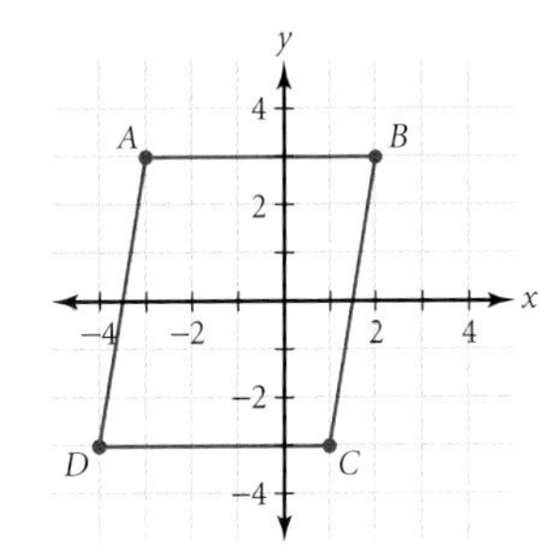

Practice and Apply

For Exercises 12–23,

let $A = \begin{bmatrix} 5 & 7 & -3 & 0 \\ -2 & 1 & 8 & 11 \end{bmatrix}$, $B = \begin{bmatrix} 8 & -5 & 2 \\ -1 & 4 & -2 \\ 0 & -5 & 3 \\ 5 & 7 & -6 \end{bmatrix}$, **and** $C = \begin{bmatrix} 7 \\ 2 \\ 6 \end{bmatrix}$.

internet connect
Homework Help Online
Go To: go.hrw.com
Keyword:
MB1 Homework Help for Exercises 12-17, 50-52, 55-59

Give the dimensions of each matrix.

12. A 2 × 4 **13.** B 4 × 3 **14.** C 3 × 1

Give the entry at the indicated address in matrix *A*, *B*, or *C*.

15. a_{23} 8 **16.** b_{12} −5 **17.** c_{31} 6

Find the indicated matrix.

18. $-A$ **19.** $-4C$ **20.** $-2B$

21. $-B$ **22.** $3A$ **23.** $\frac{1}{2}B$

TRY THIS

a.

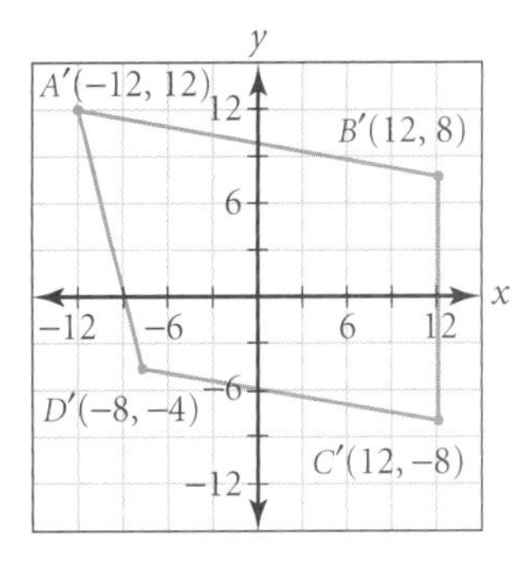

b.

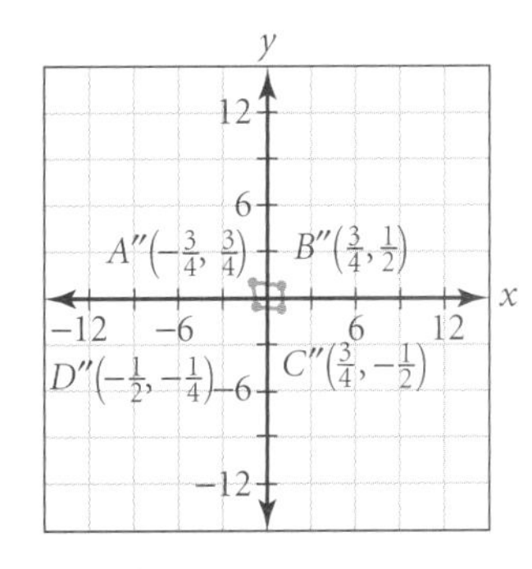

Assess

Selected Answers

Exercises 5–11, 13–71 odd

ASSIGNMENT GUIDE

In Class	1–11
Core	12–17, 19–49 odd, 50–54
Core Plus	12–17, 18–44 even, 46–65
Review	66–72
Preview	73

Extra Practice can be found beginning on page 940.

11a.

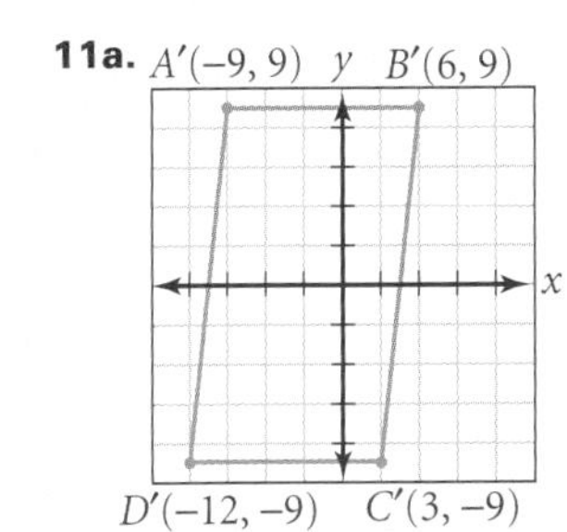

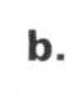

b.

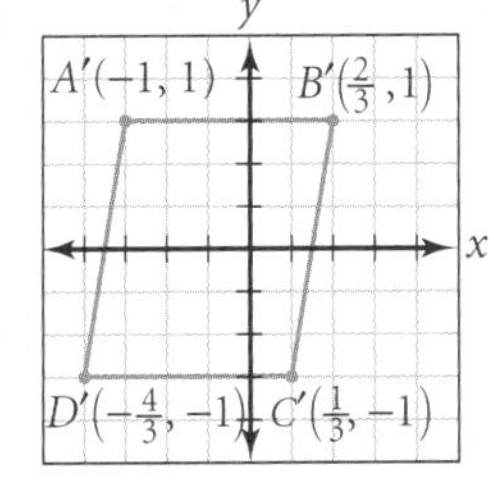

18. $\begin{bmatrix} -5 & -7 & 3 & 0 \\ 2 & -1 & -8 & -11 \end{bmatrix}$

19. $\begin{bmatrix} -28 \\ -8 \\ -24 \end{bmatrix}$

20. $\begin{bmatrix} -16 & 10 & -4 \\ 2 & -8 & 4 \\ 0 & 10 & -6 \\ -10 & -14 & 12 \end{bmatrix}$

21. $\begin{bmatrix} -8 & 5 & -2 \\ 1 & -4 & 2 \\ 0 & 5 & -3 \\ -5 & -7 & 6 \end{bmatrix}$

22. $\begin{bmatrix} 15 & 21 & -9 & 0 \\ -6 & 3 & 24 & 33 \end{bmatrix}$

23. $\begin{bmatrix} 4 & -\frac{5}{2} & 1 \\ -\frac{1}{2} & 2 & -1 \\ 0 & -\frac{5}{2} & \frac{3}{2} \\ \frac{5}{2} & \frac{7}{2} & -3 \end{bmatrix}$

Error Analysis

Students often reverse the order of the row and column when giving the dimensions of a matrix. Suggest that they think about locating a seat in a stadium or in a theater. That is, a person usually identifies the row first and then locates the seat within that row. Point out that the elements of a matrix are identified in a similar fashion: row first and then column.

30. $\begin{bmatrix} 13 & 3 & 10 & 2 \\ -7 & 10 & -8 & 5 \end{bmatrix}$

31. $\begin{bmatrix} 1 & 3 & -12 & 8 \\ 3 & 6 & 8 & -13 \end{bmatrix}$

32. $\begin{bmatrix} 14 & 6 & -2 & 10 \\ -4 & 16 & 0 & -8 \end{bmatrix}$

33. $\begin{bmatrix} -18 & 0 & -33 & 9 \\ 15 & -6 & 24 & -27 \end{bmatrix}$

34. $\begin{bmatrix} -1 & -3 & 12 & -8 \\ -3 & -6 & -8 & 13 \end{bmatrix}$

35. $\begin{bmatrix} 6 & 0 & 11 & -3 \\ -5 & 2 & -8 & 9 \end{bmatrix}$

36. $\begin{bmatrix} -4 & -12 & 48 & -32 \\ -12 & -24 & -32 & 52 \end{bmatrix}$

37. $\begin{bmatrix} 20 & 6 & 9 & 7 \\ -9 & 18 & -8 & 1 \end{bmatrix}$

38. $\begin{bmatrix} -1 & -3 & 12 & -8 \\ -3 & -6 & -8 & 13 \end{bmatrix}$

Practice

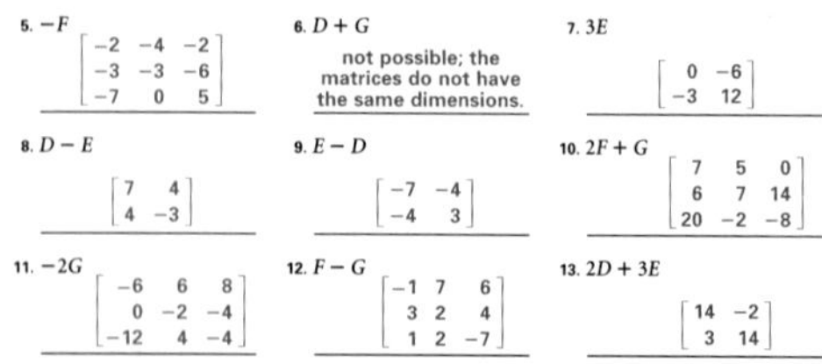

NAME ______________ CLASS ______ DATE ______

Practice
***4.1* Using Matrices to Represent Data**

In Exercises 1–13, let $D = \begin{bmatrix} 7 & 2 \\ 3 & 1 \end{bmatrix}$, $E = \begin{bmatrix} 0 & -2 \\ -1 & 4 \end{bmatrix}$, $F = \begin{bmatrix} 2 & 4 & 2 \\ 3 & 3 & 6 \\ 7 & 0 & -5 \end{bmatrix}$, **and** $G = \begin{bmatrix} 3 & -3 & -4 \\ 0 & 1 & 2 \\ 6 & -2 & 2 \end{bmatrix}$.

Give the dimensions of each matrix.

1. D 2 × 2 2. E 2 × 2 3. F 3 × 3 4. G 3 × 3

Find the indicated matrix.

5. $-F$ $\begin{bmatrix} -2 & -4 & -2 \\ -3 & -3 & -6 \\ -7 & 0 & 5 \end{bmatrix}$
6. $D + G$ not possible; the matrices do not have the same dimensions.
7. $3E$ $\begin{bmatrix} 0 & -6 \\ -3 & 12 \end{bmatrix}$
8. $D - E$ $\begin{bmatrix} 7 & 4 \\ 4 & -3 \end{bmatrix}$
9. $E - D$ $\begin{bmatrix} -7 & -4 \\ -4 & 3 \end{bmatrix}$
10. $2F + G$ $\begin{bmatrix} 7 & 5 & 0 \\ 6 & 7 & 14 \\ 20 & -2 & -8 \end{bmatrix}$
11. $-2G$ $\begin{bmatrix} -6 & 6 & 8 \\ 0 & -2 & -4 \\ -12 & 4 & -4 \end{bmatrix}$
12. $F - G$ $\begin{bmatrix} -1 & 7 & 6 \\ 3 & 2 & 4 \\ 1 & 2 & -7 \end{bmatrix}$
13. $2D + 3E$ $\begin{bmatrix} 14 & -2 \\ 3 & 14 \end{bmatrix}$

Matrix *M* at right represents the number of medals won by athletes from the United States, Germany, and Russia in the 1996 Summer Olympic Games.

	Gold	Silver	Bronze
United States	44	32	25
Germany	20	18	27
Russia	26	21	16

14. What are the dimensions of matrix *M*? 3 × 3
15. Find the total number of medals won by the United States. 101 medals
16. Find the total number of gold medals won by the three nations. 90 gold medals
17. Describe the data in location m_{23}. the number of bronze medals, 27, won by Germany
18. In the 1996 Summer Olympic Games, athletes from China won 16 gold medals, 22 silver medals, and 12 bronze medals. Write a new matrix, *M′*, that includes medals for all four countries. $\begin{bmatrix} 44 & 32 & 25 \\ 20 & 18 & 27 \\ 26 & 21 & 16 \\ 16 & 22 & 12 \end{bmatrix}$

24. $x = 3$, $y = \frac{1}{2}$
25. $x = -7$, $y = -18$
26. $x = 6$, $y = -3$
27. $x = 9$, $y = 8$
28. $x = -4$, $y = -1$
29. $x = 4$, $y = 3$

Solve for *x* and *y*.

24. $\begin{bmatrix} 3 & 4y \\ 5 & 8 \end{bmatrix} = \begin{bmatrix} 3 & 2 \\ 2x-1 & 8 \end{bmatrix}$

25. $\begin{bmatrix} -6 & 5 \\ -1 & 0 \end{bmatrix} = \begin{bmatrix} y+12 & 5 \\ -1 & x+7 \end{bmatrix}$

26. $\begin{bmatrix} 18 & \frac{1}{24}x \\ -\frac{2}{9}y & 15 \end{bmatrix} = \begin{bmatrix} 2x+6 & \frac{1}{4} \\ \frac{2}{3} & -5y \end{bmatrix}$

27. $\begin{bmatrix} \frac{2}{3}x & 12 \\ -4 & \frac{1}{2}y+5 \end{bmatrix} = \begin{bmatrix} 6 & x+3 \\ -4 & y+1 \end{bmatrix}$

28. $\begin{bmatrix} 2.5x & 3y+5 \\ 4 & y \end{bmatrix} = \begin{bmatrix} -10 & 2 \\ -x & y \end{bmatrix}$

29. $\begin{bmatrix} 4.1x & x \\ -100 & -3.7y \end{bmatrix} = \begin{bmatrix} 16.4 & x \\ -25x & -11.1 \end{bmatrix}$

For Exercises 30–45, let $A = \begin{bmatrix} 7 & 3 & -1 & 5 \\ -2 & 8 & 0 & -4 \end{bmatrix}$ **and** $B = \begin{bmatrix} 6 & 0 & 11 & -3 \\ -5 & 2 & -8 & 9 \end{bmatrix}$.

Perform the indicated operations.

30. $A + B$ **31.** $A - B$ **32.** $2A$ **33.** $-3B$
34. $B - A$ **35.** $A + B - A$ **36.** $4(B - A)$ **37.** $(B + A) - (-A)$
38. $-(A - B)$ **39.** $2A - (-B - A)$ **40.** $-\left(\frac{1}{2}B - A\right)$ **41.** $-3(B + A) - A$
42. $-\frac{1}{2}A + (B - A)$ **43.** $3B + 2A$ **44.** $\frac{1}{4}(B - 2A)$ **45.** $4\left(\frac{1}{2}A + \frac{2}{3}A\right)$

CHALLENGE

46. Construct a 3 × 3 square matrix, *A*, where $a_{ij} = i^2 + 2j - 3$.

CONNECTION

TRANSFORMATIONS For Exercises 47–49, refer to the coordinate plane at left.

47. a. Represent $\triangle MNO$ as matrix *A*.
b. Graph the polygon that is represented by $\frac{1}{2}A$.
c. Graph the polygon that is represented by $-\frac{1}{2}A$.
d. Graph the polygon that is represented by $4A$.

48. a. Represent $\triangle OPQ$ as matrix *B*.
b. Graph the polygon that is represented by $2B$.
c. Graph the polygon that is represented by $-4B$.
d. Graph the polygon that is represented by $\frac{1}{4}B$.

49. a. Represent $\triangle ORS$ as matrix *C*.
b. Graph the polygon that is represented by $2C$.
c. Graph the polygon that is represented by $-C$.
d. Graph the polygon that is represented by $-\frac{1}{2}C$.

APPLICATION

GEOGRAPHY Tracy and Renaldo both collect maps. Together they have a variety of maps from the 1960s to the 1990s. Matrix *M* shows the number of each type of map they have.

	'60s	'70s	'80s	'90s
Europe	3	1	4	2
Asia	5	3	6	3
North America	2	7	9	5
Africa	8	5	4	6

$= M$

50. What are the dimensions of matrix *M*? 4 × 4
51. Describe the entry at m_{42}. 5; 5 maps of Africa from the '70s
52. Describe the entry at m_{21}. 5; 5 maps of Asia from the '60s
53. What is the total number of maps of Africa that Renaldo and Tracy have? 23
54. What is the total number of maps from the 1960s that Tracy and Renaldo have? 18

39. $\begin{bmatrix} 27 & 9 & 8 & 12 \\ -11 & 26 & -8 & -3 \end{bmatrix}$

40. $\begin{bmatrix} 4 & 3 & -\frac{13}{2} & \frac{13}{2} \\ \frac{1}{2} & 7 & 4 & -\frac{17}{2} \end{bmatrix}$

41. $\begin{bmatrix} -46 & -12 & -29 & -11 \\ 23 & -38 & 24 & -11 \end{bmatrix}$

42. $\begin{bmatrix} -\frac{9}{2} & -\frac{9}{2} & \frac{25}{2} & -\frac{21}{2} \\ -2 & -10 & -8 & 15 \end{bmatrix}$

43. $\begin{bmatrix} 32 & 6 & 31 & 1 \\ -19 & 22 & -24 & 19 \end{bmatrix}$

44. $\begin{bmatrix} -2 & -\frac{3}{2} & \frac{13}{4} & -\frac{13}{4} \\ -\frac{1}{4} & -\frac{7}{2} & -2 & \frac{17}{4} \end{bmatrix}$

45. $\begin{bmatrix} \frac{98}{3} & 14 & -\frac{14}{3} & \frac{70}{3} \\ -\frac{28}{3} & \frac{112}{3} & 0 & -\frac{56}{3} \end{bmatrix}$

APPLICATIONS

55. $P = \begin{bmatrix} 27 & 31 & 24 & 18 \\ 48 & 72 & 61 & 25 \end{bmatrix}$

56. p_{13}

57. p_{21} represents the number of squash Jane sold.

CONSUMER ECONOMICS At a local farmer's market, Jane sold 27 squash, 31 tomatoes, 24 peppers, and 18 melons. Jose sold 48 squash, 72 tomatoes, 61 peppers, and 25 melons.

55. Create a 2×4 matrix of this data. Name this matrix P.

56. What is the address of the number of peppers that Jane sold?

57. What is the address of the data stored in the second row and first column. What does this entry represent?

58. Could you have created a matrix with different dimensions from the one you created in Exercise 55? **yes**

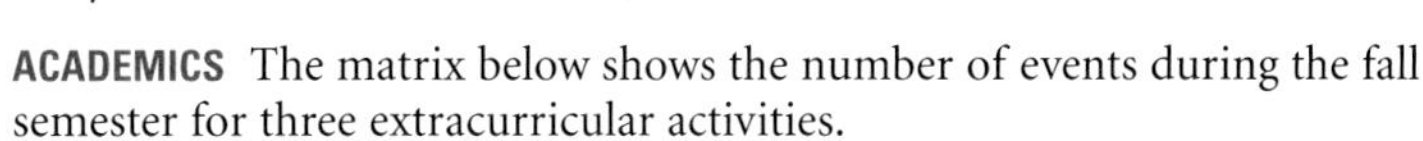

ACADEMICS The matrix below shows the number of events during the fall semester for three extracurricular activities.

	Aug.	Sept.	Oct.	Nov.	Dec.
Drama productions	0	1	2	1	2
Soccer games	1	4	3	3	0
Journalism publications	1	2	3	3	2

59. What are the dimensions of this matrix? **3 × 5**

60. Find the total number of events that occurred in September. **7**

61. Find the total number of drama productions during the fall semester. **6**

62. During which month did the most events occur? **October**

INVENTORY A music store manager wishes to organize information about his inventory. The store carries records, tapes, and compact discs of country, jazz, rock, blues, and classical music.

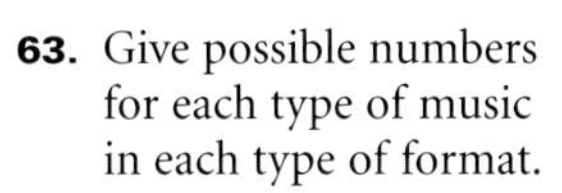

63. Give possible numbers for each type of music in each type of format.

64. Create a matrix to store this information. Name this matrix M.

65. Indicate what the entry at m_{23} of the matrix you created in Exercise 64 represents.

46. 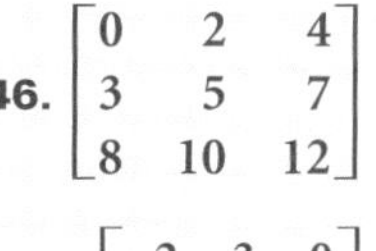

$\begin{bmatrix} 0 & 2 & 4 \\ 3 & 5 & 7 \\ 8 & 10 & 12 \end{bmatrix}$

47a. $\begin{bmatrix} -2 & 3 & 0 \\ 3 & 4 & 0 \end{bmatrix}$

b.

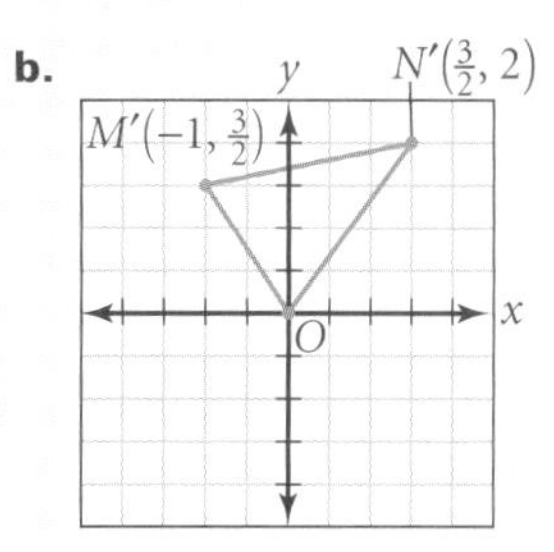

c.

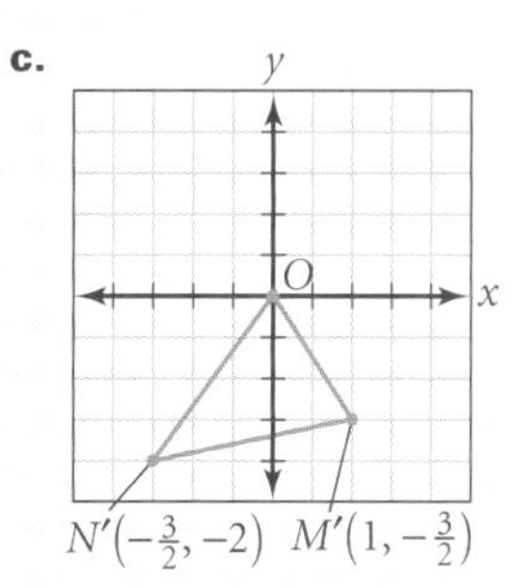

d.

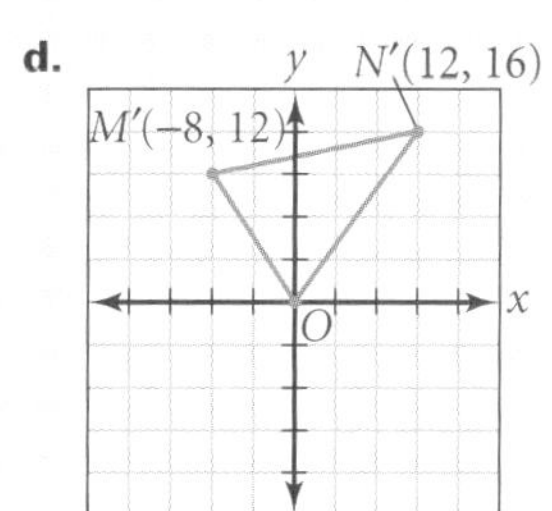

Student Technology Guide

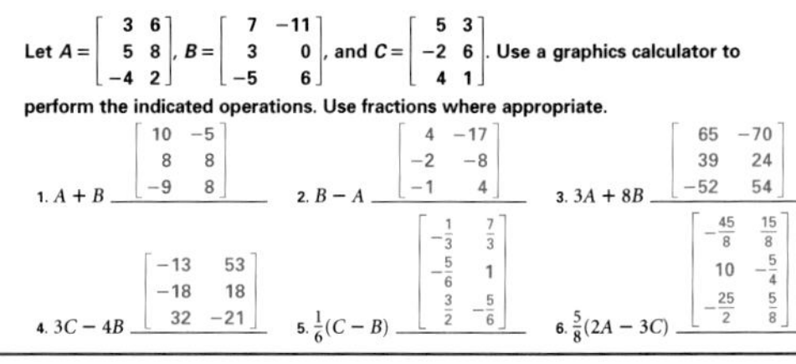

NAME ______ CLASS ______ DATE ______

Student Technology Guide

4.1 ***Using Matrices to Represent Data***

Graphics calculators include features for working with matrices.

Suppose you want to find $A + B$ and $3A - 2B$, where $A = \begin{bmatrix} 5 & 8 & 4 \\ -3 & 2 & 6 \end{bmatrix}$ and $B = \begin{bmatrix} 6 & -4 & -2 \\ 7 & 0 & 2 \end{bmatrix}$.

- To enter matrix A with dimensions 2 × 3, press MATRX EDIT 1:[A] ENTER 2 ENTER 3 ENTER 5 ENTER 8 ENTER 4 ENTER (-) 3 ENTER 2 ENTER 6 ENTER.
- To enter matrix B with dimensions 2 × 3, press MATRX EDIT 2:[B] ENTER 2 ENTER 3 ENTER 6 ENTER (-) 4 ENTER (-) 2 ENTER 7 ENTER 0 ENTER 2 ENTER.
- Press 2nd MODE to exit matrix edit mode.
- To find $A + B$ and $3A - 2B$, proceed as follows:
  $A + B$: MATRX NAMES 1:[A] ENTER + MATRX NAMES 2:[B] ENTER ENTER
  $3A - 2B$: 3 MATRX NAMES 1:[A] ENTER − 2 MATRX NAMES 2:[B] ENTER ENTER

To find $\frac{1}{2}A + \frac{1}{3}B$, you can follow a similar procedure, as shown. (Note that many calculators will give an error message if you try to enter the expression as [A]/2 + [B]/3.) The result does not fit on the calculator screen, but you can use ► and ◄ to see the complete result. You can also press MATH MATH 1:►Frac ENTER ENTER to see the final result displayed in fractional form.

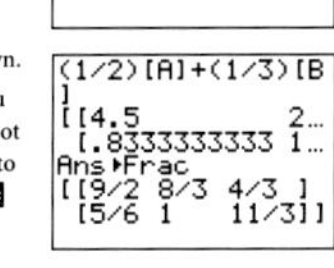

Let $A = \begin{bmatrix} 3 & 6 \\ 5 & 8 \\ -4 & 2 \end{bmatrix}$, $B = \begin{bmatrix} 7 & -11 \\ 3 & 0 \\ -5 & 6 \end{bmatrix}$, **and** $C = \begin{bmatrix} 5 & 3 \\ -2 & 6 \\ 4 & 1 \end{bmatrix}$. **Use a graphics calculator to perform the indicated operations. Use fractions where appropriate.**

1. $A + B$ $\begin{bmatrix} 10 & -5 \\ 8 & 8 \\ -9 & 8 \end{bmatrix}$
2. $B - A$ $\begin{bmatrix} 4 & -17 \\ -2 & -8 \\ -1 & 4 \end{bmatrix}$
3. $3A + 8B$ $\begin{bmatrix} 65 & -70 \\ 39 & 24 \\ -52 & 54 \end{bmatrix}$
4. $3C - 4B$ $\begin{bmatrix} -13 & 53 \\ -18 & 18 \\ 32 & -21 \end{bmatrix}$
5. $\frac{1}{6}(C - B)$ $\begin{bmatrix} -\frac{1}{3} & \frac{7}{3} \\ -\frac{5}{6} & 1 \\ \frac{3}{2} & -\frac{5}{6} \end{bmatrix}$
6. $\frac{5}{8}(2A - 3C)$ $\begin{bmatrix} -\frac{45}{8} & \frac{15}{8} \\ 10 & -\frac{5}{4} \\ -\frac{25}{2} & \frac{5}{8} \end{bmatrix}$

48a. 

$\begin{bmatrix} 0 & 2 & 0 \\ 0 & -2 & -3 \end{bmatrix}$

b.

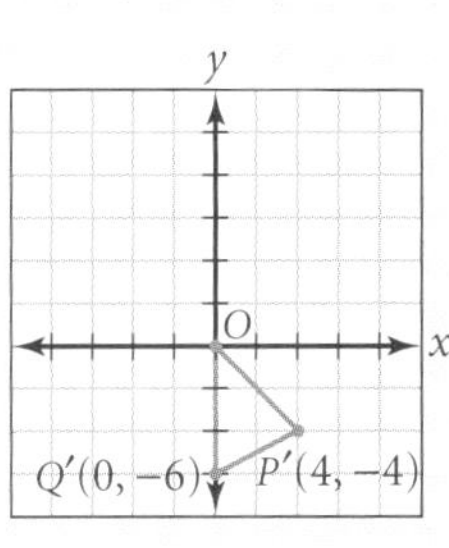

c.

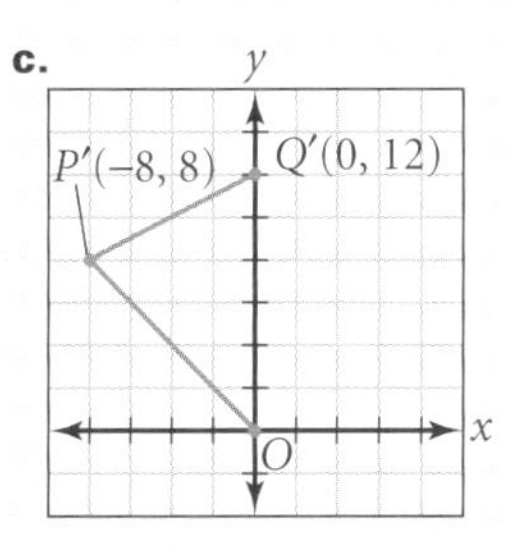

The answers to Exercises 48d and 49a–d can be found in Additional Answers beginning on page 1002.

Exercise 73 introduces students to the idea of associating a system of linear equations with a matrix. This serves as a preview of Lesson 4.4, where students will learn to write matrix equations in order to solve systems of linear equations.

ALTERNATIVE

Assessment

Portfolio Activity

The Portfolio Activity can be used as preparation for the Chapter Project or as a separate activity. In the Portfolio Activity on this page, students explore how matrices can be used to describe a *network* of pathways between a set of locations. Note that students will learn more about matrices and networks in Lesson 4.2.

Answers to Portfolio Activities can be found in Additional Answers of the Teacher's Edition.

66. $y = -\frac{11}{13}x + \frac{44}{13}$

67. $y = 4x - 37$

70. $y = \frac{2x+2}{3}$, or $x = \frac{3}{2}y - 1$

internet connect

Portfolio Extension

Go To: **go.hrw.com**
Keyword:
MB1 Matrix

Write an equation in slope-intercept form for the line containing the indicated points. ***(LESSON 1.3)***

66. (4, 0) and (−9, 11) **67.** (10, 3) and (8, −5)

68. **If y varies directly as x and y is 49 when x is 14, find x when y is 63.** ***(LESSON 1.4)*** **18**

69. Find the inverse of $f(x) = 2x - 1$. ***(LESSON 2.5)*** $f^{-1}(x) = \frac{x+1}{2}$

Write each pair of parametric equations as a single equation in x and y. ***(LESSON 3.6)***

70. $\begin{cases} x(t) = 3t - 1 \\ y(t) = 2t \end{cases}$ **71.** $\begin{cases} x(t) = 5 - t \\ y(t) = 3t \end{cases}$ **72.** $\begin{cases} x(t) = -6t \\ y(t) = t^2 \end{cases}$ $y = \frac{1}{36}x^2$

$y = -3x + 15$, or $x = 5 - \frac{1}{3}y$

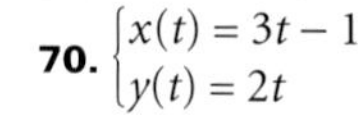

73. Write the system of linear equations represented by these equivalent matrices: $\begin{bmatrix} 5x - 2y \\ x + 4y \end{bmatrix} = \begin{bmatrix} 3 \\ 7 \end{bmatrix}$. $\begin{cases} 5x - 2y = 3 \\ x + 4y = 7 \end{cases}$

Finding your way around the exhibits in museums may appear to be a random process. However, to calculate the number of paths to and from exhibits requires a mathematical approach.

The exhibits on the fourth floor of the Museum of Natural History can be modeled by the network diagram below. The address n_{23} of matrix N below represents a path from the Evolution of Horses, H, to the Dinosaur Mummy, D.

Fourth Floor, NY Museum of Natural History

G: Glen Rose Trackway
T: Tyrannosaurus
D: Dinosaur Mummy
H: Evolution of Horses
M: Warren Mastodon

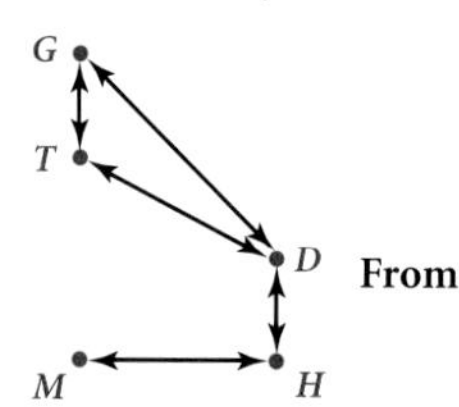

$$\text{From: } \begin{matrix} \\ M \\ H \\ D \\ T \\ G \end{matrix} \overset{\text{To:}}{\begin{matrix} M & H & D & T & G \\ \end{matrix}} \begin{bmatrix} 0 & 1 & 0 & 0 & 0 \\ 1 & 0 & 1 & 0 & 0 \\ 0 & 1 & 0 & 1 & 1 \\ 0 & 0 & 1 & 0 & 1 \\ 0 & 0 & 1 & 1 & 0 \end{bmatrix} = N$$

1. Describe the difference between locations n_{34} and n_{43} in matrix N.

2. Explain the meaning of the 1 at n_{54} and 0 at n_{14}.

3. Describe how to use matrix N to determine whether it is possible to reach Tyrannosaurus, T, directly from the Glen Rose Trackway, G.

4. Explain why each element of the main diagonal is 0.

5. Explain how to find the number of paths leading to each exhibit by using the matrix.

6. Explain how to find the number of paths going from each exhibit by using the matrix.

WORKING ON THE CHAPTER PROJECT

You should now be able to complete Activity 1 of the Chapter Project.

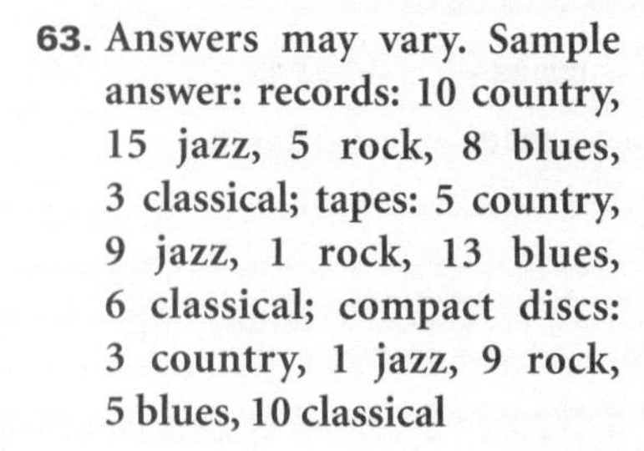

63. Answers may vary. Sample answer: records: 10 country, 15 jazz, 5 rock, 8 blues, 3 classical; tapes: 5 country, 9 jazz, 1 rock, 13 blues, 6 classical; compact discs: 3 country, 1 jazz, 9 rock, 5 blues, 10 classical

64. Answers may vary. Sample answer:

	Country	Jazz	Rock	Blues	Classical
Records	10	15	5	8	3
Tapes	5	9	1	13	6
CDs	3	1	9	5	10

65. Answers may vary. In the matrix at left, m_{23} = 1 rock tape.

Matrix Multiplication

Why *Many simple calculations, such as keeping track of the score of a football game, involve a process of calculation that is very similar to matrix multiplication. Matrix multiplication can also be used for more complicated calculations.*

Objectives

- Multiply two matrices.
- Use matrix multiplication to solve mathematical and real-world problems.

SPORTS

Matrix multiplication involves multiplication and addition. The process of matrix multiplication can be demonstrated by using football scores.

A football team scores 5 touchdowns, 4 extra points, and 2 field goals. A touchdown is worth 6 points, an extra point is 1 point, and a field goal is 3 points. The final score is evaluated as follows:

$$\begin{aligned}&(5 \text{ touchdowns})(6 \text{ pts}) + (4 \text{ extra points})(1 \text{ pt}) + (2 \text{ field goals})(3 \text{ pts})\\ &= \quad 30 \text{ points} \quad + \quad 4 \text{ points} \quad + \quad 6 \text{ points}\\ &= 40 \text{ points total}\end{aligned}$$

Matrix multiplication is performed in the same way.

Touch-downs, Extra points, Field goals × **Point values** = **Total score**

$$\begin{bmatrix}5 & 4 & 2\end{bmatrix} \times \begin{bmatrix}6\\1\\3\end{bmatrix} = [(5)(6) + (4)(1) + (2)(3)] = [40]$$

Notice that a 1×3 matrix multiplied by a 3×1 matrix results in a 1×1 matrix. To multiply any two matrices, the *inner dimensions* must be the same. Then the *outer dimensions* become the dimensions of the resulting product matrix.

Outer dimensions — Product dimensions

$$1 \times 3 \quad 3 \times 1 \quad = \quad 1 \times 1$$

Inner dimensions

Prepare

NCTM PRINCIPLES & STANDARDS 1–10

QUICK WARM-UP

Give the dimensions of each matrix.

1. $\begin{bmatrix}-5 & 0 & 7 & -11\\ 0 & 3 & 7 & -19\end{bmatrix}$ 2×4

2. $\begin{bmatrix}-15 & 7\\ 10 & 9\\ 6 & 6\end{bmatrix}$ 3×2

Identify the entry at each location of the matrix below.

$\begin{bmatrix}-7 & 0 & 13\\ -2 & 4 & -2\\ 12 & 1 & 10\end{bmatrix}$

3. b_{12} 0
4. b_{21} −2
5. b_{32} 1

Also on Quiz Transparency 4.2

Teach

Why Suppose that in a professional basketball game, a team makes 50 of its shots. Ask students how many points this team scored. Most will know that the answer depends on the type of shots that were made: 3-point baskets, 2-point baskets, or 1-point free throws. Tell students that *matrix multiplication* provides a way to record this information and calculate the score.

Alternative Teaching Strategy

USING TECHNOLOGY Have students enter these matrices into a graphics calculator and find AB.

$A = \begin{bmatrix}3 & 5 & -1\end{bmatrix} \quad B = \begin{bmatrix}1\\1\\1\end{bmatrix} \quad AB = [7]$

Then have them change A to a 1×3 matrix of their own choosing and find AB. Elicit their observations about the product. Lead them to see that each product is a 1×1 matrix, and that its single entry is the sum of the entries in matrix A.

Now have students re-enter $A = [3 \quad 5 \quad -1]$ and find AB for each of the following matrices:

$B = \begin{bmatrix}2\\2\\2\end{bmatrix}$ **[14]** $\quad B = \begin{bmatrix}3\\3\\3\end{bmatrix}$ **[21]** $\quad B = \begin{bmatrix}1\\2\\3\end{bmatrix}$ **[10]** $\quad B = \begin{bmatrix}1\\2\end{bmatrix}$ **does not exist**

Lead them in a discussion of the products. Use their insights to develop a rule for finding AB when A is a 1×3 matrix and B is a 3×1 matrix.

Now have students use their calculators to experiment with products of matrices with other dimensions. Lead them to the generalization for matrix multiplication that is given on page 226.

Use Teaching Transparency 16.

CHECKPOINT ✔
AB: No; the inner dimensions, 3 and 2, do not match.

BA: Yes; the inner dimensions, 2 and 2, match.

ADDITIONAL EXAMPLE 1

Let $J = \begin{bmatrix} 8 & 5 \\ 2 & 1 \end{bmatrix}$ and $K = \begin{bmatrix} 9 & -2 \\ 0 & 10 \end{bmatrix}$.

a. Find *JK*, if it exists.
$\begin{bmatrix} 72 & 34 \\ 18 & 6 \end{bmatrix}$

b. Find *KJ*, if it exists.
$\begin{bmatrix} 68 & 43 \\ 20 & 10 \end{bmatrix}$

TRY THIS

a. $\begin{bmatrix} -2 & -21 \\ 20 & 35 \\ -10 & 0 \end{bmatrix}$

b. does not exist

CRITICAL THINKING
sample answer:
$A = \begin{bmatrix} 1 & 1 \\ 1 & 2 \end{bmatrix}$; $B = \begin{bmatrix} 0 & 1 \\ 1 & 1 \end{bmatrix}$;
all such matrices would be square and have the same dimensions.

CHECKPOINT ✔ Let $A = \begin{bmatrix} 2 & -1 & 3 \\ 0 & 5 & 1 \end{bmatrix}$ and $B = \begin{bmatrix} 1 & 0 \\ 2 & -1 \end{bmatrix}$. Is it possible to find the product *AB*? Is it possible to find the product *BA*? Explain.

The procedure for finding the product of two matrices is given below.

Matrix Multiplication

If matrix *A* has dimensions $m \times n$ and matrix *B* has dimensions $n \times r$, then the product *AB* has dimensions $m \times r$.

Find the entry in row *i* and column *j* of *AB* by finding the sum of the products of the corresponding entries in row *i* of *A* and column *j* of *B*.

EXAMPLE 1 Let $H = \begin{bmatrix} 2 & -3 \\ 1 & 5 \end{bmatrix}$ and $G = \begin{bmatrix} 6 & 0 \\ 4 & 7 \end{bmatrix}$.

a. Find *HG*, if it exists. **b.** Find *GH*, if it exists.

SOLUTION

a. The product *HG* has dimensions of 2×2.

row 1 of H, column 1 of G; row 1 of H, column 2 of G

$$\underset{H}{\begin{bmatrix} 2 & -3 \\ 1 & 5 \end{bmatrix}} \underset{G}{\begin{bmatrix} 6 & 0 \\ 4 & 7 \end{bmatrix}} = \begin{bmatrix} (2)(6) + (-3)(4) & (2)(0) + (-3)(7) \\ (1)(6) + (5)(4) & (1)(0) + (5)(7) \end{bmatrix} = \underset{HG}{\begin{bmatrix} 0 & -21 \\ 26 & 35 \end{bmatrix}}$$

row 2 of H, column 1 of G; row 2 of H, column 2 of G

b. The product *GH* has dimensions of 2×2.

row 1 of G, column 1 of H; row 1 of G, column 2 of H

$$\underset{G}{\begin{bmatrix} 6 & 0 \\ 4 & 7 \end{bmatrix}} \underset{H}{\begin{bmatrix} 2 & -3 \\ 1 & 5 \end{bmatrix}} = \begin{bmatrix} (6)(2) + (0)(1) & (6)(-3) + (0)(5) \\ (4)(2) + (7)(1) & (4)(-3) + (7)(5) \end{bmatrix} = \underset{GH}{\begin{bmatrix} 12 & -18 \\ 15 & 23 \end{bmatrix}}$$

row 2 of G, column 1 of H; row 2 of G, column 2 of H

In Example 1, notice that although both *HG* and *GH* exist, the products are not equal; that is, $HG \neq GH$. Thus, matrix multiplication is *not* commutative.

TRY THIS Let $R = \begin{bmatrix} 2 & -3 \\ 0 & 5 \\ -2 & 0 \end{bmatrix}$ and $W = \begin{bmatrix} 5 & 0 \\ 4 & 7 \end{bmatrix}$.

a. Find *RW*, if it exists. **b.** Find *WR*, if it exists.

CRITICAL THINKING Find any matrices *A* and *B* such that $AB = A + B$. What can you say about the dimensions of any matrices *A* and *B* for which $AB = A + B$ is true?

Interdisciplinary Connection

BIOLOGY This table gives data about the number of females in an initial population of a species of brown rats. The data are displayed in matrix P_0 below. Matrix *L* at right contains data about the birth rates and survival rates for the females of this species.

Age (months)	Number
0–3	18
3–6	10
6–9	11
9–12	7
12–15	0
15–18	0

$$P_0 = [18 \quad 10 \quad 11 \quad 7 \quad 0 \quad 0]$$

$$L = \begin{bmatrix} 0 & 0.6 & 0 & 0 & 0 & 0 \\ 0.3 & 0 & 0.9 & 0 & 0 & 0 \\ 0.8 & 0 & 0 & 0.9 & 0 & 0 \\ 0.7 & 0 & 0 & 0 & 0.8 & 0 \\ 0.4 & 0 & 0 & 0 & 0 & 0.6 \\ 0 & 0 & 0 & 0 & 0 & 0 \end{bmatrix}$$

Biologists use the formula $P_n = P_0L^n$ to predict the population, P_n, after *n* three-month birth cycles. Use the formula to predict the population after four birth cycles. Round entries to the nearest whole number. $P_4 = [19 \quad 12 \quad 11 \quad 8 \quad 7 \quad 4]$

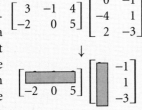

EXAMPLE 2

APPLICATION
NUTRITION

Karl and Kayla are making two snack mixes by mixing dried fruit and nuts. The amounts of protein, carbohydrates, and fat, in grams per serving, for the dried fruit and nuts are given in Table 1. The number of servings of dried fruit and nuts in each mix is given in Table 2.

Table 1

	Dried fruit	Nuts
Protein	3	20
Carbohydrates	65	21
Fat	1	52

Table 2

	Sports mix	Camp mix
Dried fruit	4	3
Nuts	2	3

a. Represent the information from Table 1 in a matrix called N. Represent the information from Table 2 in a matrix called G.
b. Find the product NG, and determine which mix has more protein and which mix has less fat.

SOLUTION

Notice that three distinct categories are represented: nutrients, ingredients, and type of mix. The categories that correspond to the *inner* dimensions of the matrices must be the same. The categories that correspond to the outer dimensions are different from one another and will become the labels of the product matrix.

a.
$$\begin{array}{r} \text{Protein} \\ \text{Carbohydrates} \\ \text{Fat} \end{array} \overset{\begin{array}{cc}\text{Dried fruit} & \text{Nuts}\end{array}}{\begin{bmatrix} 3 & 20 \\ 65 & 21 \\ 1 & 52 \end{bmatrix}} = N \qquad \begin{array}{r} \text{Dried fruit} \\ \text{Nuts} \end{array} \overset{\begin{array}{cc}\text{Sport mix} & \text{Camp mix}\end{array}}{\begin{bmatrix} 4 & 3 \\ 2 & 3 \end{bmatrix}} = G$$

b.
$$NG = \begin{bmatrix} 3 & 20 \\ 65 & 21 \\ 1 & 52 \end{bmatrix} \begin{bmatrix} 4 & 3 \\ 2 & 3 \end{bmatrix} = \begin{bmatrix} 3(4) + 20(2) & 3(3) + 20(3) \\ 65(4) + 21(2) & 65(3) + 21(3) \\ 1(4) + 52(2) & 1(3) + 52(3) \end{bmatrix}$$

$$= \overset{\begin{array}{cc}\text{Sport mix} & \text{Camp mix}\end{array}}{\begin{bmatrix} 52 & 69 \\ 302 & 258 \\ 108 & 159 \end{bmatrix}} \begin{array}{l} \text{Protein} \\ \text{Carbohydrates} \\ \text{Fat} \end{array}$$

The camp mix has more protein, 69 grams. The sport mix has less fat, 108 grams.

Activity Exploring Rotations in the Plane

CONNECTION
COORDINATE GEOMETRY

You will need: no special tools

You are given $\triangle KLM$ with vertices $K(0, 0)$, $L(4, 0)$, and $M(4, 3)$.

Let $A = \begin{bmatrix} 0 & -1 \\ 1 & 0 \end{bmatrix}$ and $B = \begin{bmatrix} 0 & 1 \\ -1 & 0 \end{bmatrix}$.

1. Sketch $\triangle KLM$ on graph paper. Then represent $\triangle KLM$ as matrix C.
2. Find AC. Graph the triangle represented by AC, $\triangle K'L'M'$, on the coordinate plane with $\triangle KLM$. How are $\triangle KLM$ and $\triangle K'L'M'$ related?

Inclusion Strategies

TACTILE LEARNERS

$$\begin{bmatrix} 3 & -1 & 4 \\ -2 & 0 & 5 \end{bmatrix} \begin{bmatrix} 0 & -1 \\ -4 & 1 \\ 2 & -3 \end{bmatrix}$$

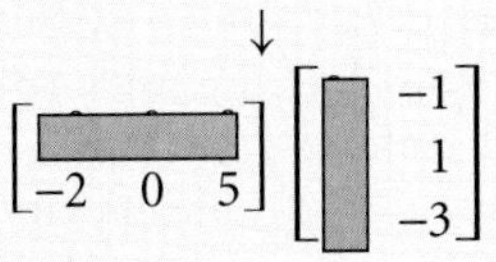

Write the matrix multiplication at right on a sheet of paper so that it covers most of one side of the paper. Give each student a copy. Have the students place two strips of paper in the positions shown. Now have them slide the strips to reveal the first pair of numbers to be multiplied. Tell them to write the product. **0**

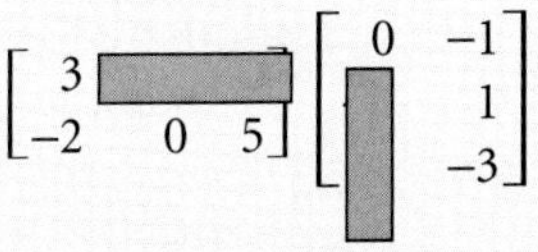

Have them continue in this way, revealing one pair of numbers at a time, until all the multiplications for this row and column are done. Now have them add the results to find the first entry of the product matrix. **12**

Lead them through a similar process to find the other three entries. **−16, 10, −13**

ADDITIONAL EXAMPLE 2

Refer to Example 2. Suppose that one serving of the dried fruit contains 2.9 grams of fiber, and one serving of the nuts contains 1.4 grams of fiber.

a. Rewrite matrix N to reflect this additional information.

$$N = \begin{array}{r} \text{Protein} \\ \text{Carb.} \\ \text{Fat} \\ \text{Fiber} \end{array} \overset{\begin{array}{cc}\text{Dried fruit} & \text{Nuts}\end{array}}{\begin{bmatrix} 3 & 20 \\ 65 & 21 \\ 1 & 52 \\ 2.9 & 1.4 \end{bmatrix}}$$

b. Find the new product NG.

$$NG = \begin{bmatrix} 52 & 69 \\ 302 & 258 \\ 108 & 159 \\ 14.4 & 12.9 \end{bmatrix}$$

c. Which mix has more fiber? The sports mix has more fiber.

Activity Notes

In this Activity, students investigate how certain rotations in a coordinate plane can be defined by matrix multiplication. Students should discover that when the matrix for $\triangle KLM$ is multiplied by matrix A, $\triangle KLM$ is rotated about the origin one quarter-turn *counterclockwise*, or 90°. When the matrix for $\triangle KLM$ is multiplied by matrix B, $\triangle KLM$ is rotated about the origin one quarter-turn *clockwise*, or −90°.

Math CONNECTION

GEOMETRY Remind students that it is important to specify the center of a rotation as well as its magnitude. For both rotations in the Activity, the center of rotation is the origin.

CHECKPOINT ✔

4. **Multiplying by matrix A rotates a geometric figure 90° counterclockwise about the origin.**

CHECKPOINT ✔

5. **Multiplying by matrix B rotates a geometric figure 90° clockwise about the origin.**

Teaching Tip

Remind students that an *adjacency* matrix shows the number of *one-step* paths between two vertices.

ADDITIONAL EXAMPLE 3

Refer to the directed network and adjacency matrix B below.

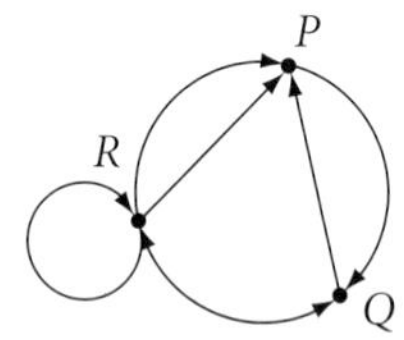

To:

$$B = \text{From: } \begin{matrix} P \\ Q \\ R \end{matrix} \overset{\begin{matrix} P & Q & R \end{matrix}}{\begin{bmatrix} 0 & 1 & 0 \\ 1 & 0 & 1 \\ 2 & 1 & 1 \end{bmatrix}}$$

a. **Find the matrix that gives the number of two-stage paths.**

To:

$$B^2 = \text{From: } \begin{matrix} P \\ Q \\ R \end{matrix} \overset{\begin{matrix} P & Q & R \end{matrix}}{\begin{bmatrix} 1 & 0 & 1 \\ 2 & 2 & 1 \\ 3 & 3 & 2 \end{bmatrix}}$$

b. **Interpret b_{32} in matrix B^2 and list the corresponding paths.**
The entry at b_{32} gives the number of two-stage paths from R to Q, 3. The paths are
$R \to P \to Q$ (using one path from R to P),
$R \to P \to Q$ (using the alternate path from R to P), and
$R \to R \to Q$.

3. Find BC. Graph triangle $\triangle K''L''M''$, represented by BC, on the coordinate plane with $\triangle KLM$. How are $\triangle KLM$ and $\triangle K''L''M''$ related?

CHECKPOINT ✔ 4. Make and verify a conjecture about the effect of matrix A on a geometric figure in the coordinate plane.

CHECKPOINT ✔ 5. Make and verify a conjecture about the effect of matrix B on a geometric figure in the coordinate plane.

APPLICATION
NETWORKS

A *network* is a finite set of connected points. Each point is called a **vertex** (plural, *vertices*). A **directed network** is a network in which permissible directions of travel between the vertices are indicated.

You can represent a network in an **adjacency matrix**, which indicates how many one-stage (direct) paths are possible from one vertex to another. A directed network and corresponding adjacency matrix are shown below.

Computer network connections

Directed network | **Adjacency matrix**

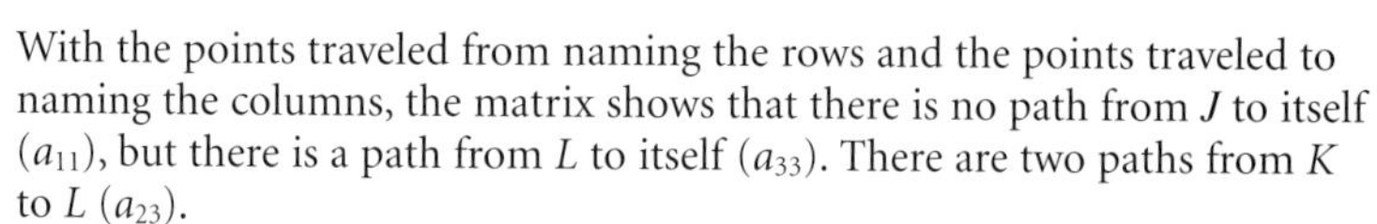

To:

$$\text{From: } \begin{matrix} J \\ K \\ L \end{matrix} \overset{\begin{matrix} J & K & L \end{matrix}}{\begin{bmatrix} 0 & 1 & 0 \\ 0 & 0 & 2 \\ 1 & 2 & 1 \end{bmatrix}} = A$$

With the points traveled from naming the rows and the points traveled to naming the columns, the matrix shows that there is no path from J to itself (a_{11}), but there is a path from L to itself (a_{33}). There are two paths from K to L (a_{23}).

If A is the adjacency matrix of a network, then the product $A \times A = A^2$ gives the number of two-stage paths from one vertex to another vertex by means of one intermediate vertex, such as from L to K by means of J.

EXAMPLE 3 **Refer to the directed network and adjacency matrix A above.**

a. Find the matrix that gives the number of two-stage paths.
b. Interpret a_{32} in matrix A^2. List the corresponding paths.

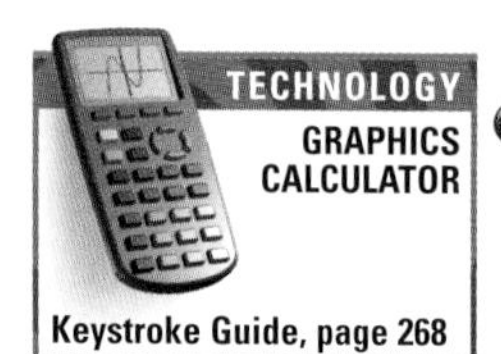

Keystroke Guide, page 268

SOLUTION

a. The matrix product A^2 gives the number of two-stage paths. Find A^2 or $A \times A$.
b. The entry in row 3 column 2 is 3.

To:

$$\text{From: } \begin{matrix} J \\ K \\ L \end{matrix} \overset{\begin{matrix} J & K & L \end{matrix}}{\begin{bmatrix} 0 & 0 & 2 \\ 2 & 4 & 2 \\ 1 & \textcircled{3} & 5 \end{bmatrix}} = A^2$$

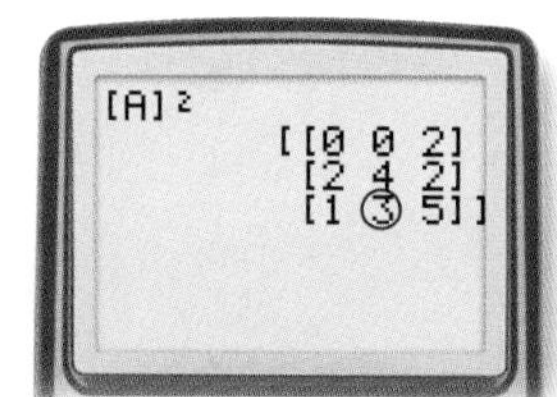

This number 3 represents the number of two-stage paths from L to K. From the directed network above you can see that these paths are as follows:

$L \to L \to K$ (using one path to K)
$L \to L \to K$ (using another path to K)
$L \to J \to K$

Enrichment

Give students the adjacency matrix below. Tell them to draw a directed network that might be represented by it.

To:

$$\text{From: } \begin{matrix} W \\ X \\ Y \\ Z \end{matrix} \overset{\begin{matrix} W & X & Y & Z \end{matrix}}{\begin{bmatrix} 0 & 0 & 1 & 0 \\ 2 & 1 & 2 & 0 \\ 0 & 1 & 0 & 1 \\ 1 & 0 & 1 & 1 \end{bmatrix}} = A$$

sample:

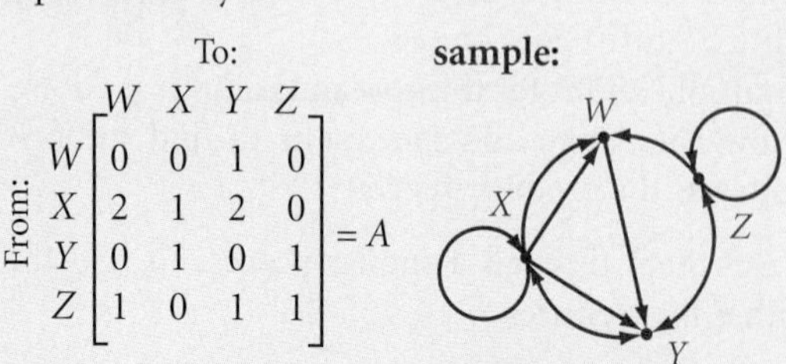

Reteaching the Lesson

INVITING PARTICIPATION Give each student a card with a different number on it. Arrange six students into a 3 × 2 matrix, named *A*, and four students into a 2 × 2 matrix, named *B*, at the front of the class. Have the students in the first row of matrix *A* and the first column of matrix *B* come forward and describe the operations for their numbers that will give the first entry of *AB*. All students should then return to their positions. Model the other three entries of *AB* in a similar manner. Repeat this activity with other matrices so that all students have a chance to participate.

Exercises

Communicate

1. What is necessary in order for two matrices to be multiplied?
2. Explain the steps you would use to multiply $\begin{bmatrix} 3 & -2 \\ 5 & 7 \end{bmatrix}$ and $\begin{bmatrix} 5 & -3 & 1 \\ -2 & -1 & 4 \end{bmatrix}$.
3. Explain how to represent a directed network with an adjacency matrix.

internet connect
Activities Online
Go To: **go.hrw.com**
Keyword:
MB1 Breakfast

Guided Skills Practice

4. Let $A = [-1 \quad 3 \quad 5]$ and $B = \begin{bmatrix} -4 \\ 2 \\ -5 \end{bmatrix}$. ***(EXAMPLE 1)***
   a. Find AB, if it exists. **[−15]**
   b. Find BA, if it exists. $\begin{bmatrix} 4 & -12 & -20 \\ -2 & 6 & 10 \\ 5 & -15 & -25 \end{bmatrix}$
5. **NUTRITION** Refer to Example 2 on page 227. Add a third kind of snack mix called trail mix, to matrix G. Let the trail mix contain 2 servings of dried fruit and 4 servings of nuts. ***(EXAMPLE 2)***
   a. What will be the entries in the new version of matrix G?
   b. Find the new product NG.
   c. Which of the three mixes has the greatest amount of protein? Which has the greatest amount of carbohydrates?
6. **NETWORKS** Represent the directed network at right in an adjacency matrix, M. ***(EXAMPLE 3)***
   a. Find the matrix that gives the number of two-stage paths.
   b. Interpret m_{22} in the resulting matrix, and list the corresponding paths.

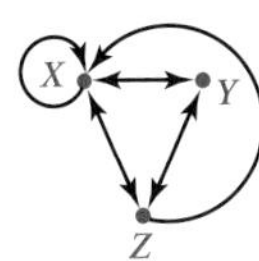

Practice and Apply

internet connect
Homework Help Online
Go To: **go.hrw.com**
Keyword:
MB1 Homework Help for Exercises 7–20

Find each product, if it exists.

7. $\begin{bmatrix} 2 \\ 0 \\ 6 \end{bmatrix}[1 \quad -3 \quad 4]$ $\begin{bmatrix} 2 & -6 & 8 \\ 0 & 0 & 0 \\ 6 & -18 & 24 \end{bmatrix}$

8. $[2 \quad 5 \quad 0]\begin{bmatrix} 8 & 1 \\ 0 & 4 \\ 2 & 5 \end{bmatrix}$ **[16 22]**

9. $\begin{bmatrix} 5 & 3 \\ 0 & 1 \end{bmatrix}\begin{bmatrix} 4 & 2 & -1 \\ 0 & 1 & 3 \end{bmatrix}$ $\begin{bmatrix} 20 & 13 & 4 \\ 0 & 1 & 3 \end{bmatrix}$

10. $\begin{bmatrix} 1 & 5 \\ -3 & 0 \end{bmatrix}\begin{bmatrix} 3 & -2 \\ -4 & 6 \end{bmatrix}$ $\begin{bmatrix} -17 & 28 \\ -9 & 6 \end{bmatrix}$

11. $\begin{bmatrix} 3 & 9 \\ 2 & -1 \end{bmatrix}\begin{bmatrix} 7 & 0 \\ 1 & 3 \end{bmatrix}$ $\begin{bmatrix} 30 & 27 \\ 13 & -3 \end{bmatrix}$

12. $\begin{bmatrix} -1 & 4 & 3 & 5 \\ 2 & 0 & -6 & 1 \end{bmatrix}\begin{bmatrix} 2 & 0 \\ 1 & 4 \\ 3 & -2 \\ -5 & 1 \end{bmatrix}$

13. $\begin{bmatrix} 4 & -6 & 5 \\ 2 & 4 & -1 \end{bmatrix}\begin{bmatrix} 2 & 7 & 1 \\ -1 & 5 & 2 \\ 1 & 4 & 1 \end{bmatrix}$

14. $\begin{bmatrix} 1 & 1 & -1 \\ 2 & 1 & 1 \\ 1 & -2 & 3 \end{bmatrix}\begin{bmatrix} 1 & -0.2 & 0.4 \\ -1 & 0.8 & -0.6 \\ -1 & 0.6 & -0.2 \end{bmatrix}$

12. $\begin{bmatrix} -14 & 15 \\ -19 & 13 \end{bmatrix}$

13. $\begin{bmatrix} 19 & 18 & -3 \\ -1 & 30 & 9 \end{bmatrix}$

14. $\begin{bmatrix} 1 & 0 & 0 \\ 0 & 1 & 0 \\ 0 & 0 & 1 \end{bmatrix}$

5a. $G = \begin{bmatrix} 4 & 3 & 2 \\ 2 & 3 & 4 \end{bmatrix}$

b. $NG = \begin{bmatrix} 52 & 69 & 86 \\ 302 & 258 & 214 \\ 108 & 159 & 210 \end{bmatrix}$

c. Trail mix has the most protein. Sport mix has the most carbohydrates.

6. $M = \begin{array}{c} \\ X \\ Y \\ Z \end{array}\begin{array}{c} X \quad Y \quad Z \\ \begin{bmatrix} 1 & 1 & 1 \\ 1 & 0 & 1 \\ 2 & 1 & 0 \end{bmatrix} \end{array}$

a. $M^2 = \begin{bmatrix} 4 & 2 & 2 \\ 3 & 2 & 1 \\ 3 & 2 & 3 \end{bmatrix}$

b. $m_{22} = 2$, the number of 2-stage paths from Y to Y: $Y \to Z \to Y$ and $Y \to X \to Y$.

Teaching Tip

TECHNOLOGY To access the **MATRX** menu on a TI-82 or TI-83, press [MATRX] and to access it on a TI-83 Plus, press [2nd] [x^{-1}]. Remind students that in order to create a matrix, they must access the **MATRX** menu, select **EDIT**, and then choose the matrix name. Stress to students that the dimensions of a matrix must be entered before the elements are entered and that the dimensions are always row × column. To compute a problem involving matrices, they must select **NAMES** and then choose each matrix name.

When performing matrix operations, [+], [−], and [×] can always be used, but [^] can be used with positive integers only. The multiplicative inverse of a matrix must be found by using [x^{-1}], not by using [^] [(−)] [1]. For matrices, the [^] key cannot be used with negative integers or with decimals, and division of matrices is not defined.

Assess

Selected Answers

Exercises 4–6, 9–41 odd

ASSIGNMENT GUIDE

In Class	1–6
Core	7–29 odd
Core Plus	7–14, 16–30 even
Review	31–42
Preview	43

Extra Practice can be found beginning on page 940.

Error Analysis

Students often make errors in matrix multiplications because they forget to perform some multiplications or they perform the same multiplication twice. Suggest that they begin each exercise by making an organized list of all multiplications that must be done. For instance, they might list the multiplications for Exercise 15 as follows:

$$\begin{bmatrix} \text{row B1}\times\text{col C1} & \text{row B1}\times\text{col C2} \\ \text{row B2}\times\text{col C1} & \text{row B2}\times\text{col C2} \\ \text{row B3}\times\text{col C1} & \text{row B3}\times\text{col C2} \\ \text{row B4}\times\text{col C1} & \text{row B4}\times\text{col C2} \end{bmatrix}$$

15. $\begin{bmatrix} -12 & 62 \\ 6 & 13 \\ -16 & 0 \\ -10 & 77 \end{bmatrix}$

19. $\begin{bmatrix} -188 & 222 \\ 12 & 794 \\ 60 & -142 \end{bmatrix}$

20. $\begin{bmatrix} -188 & 222 \\ 12 & 794 \\ 60 & -142 \end{bmatrix}$

21a. $A'(-1, 0), B'\left(\frac{3}{2}, 0\right), C'(0, -2)$

b.

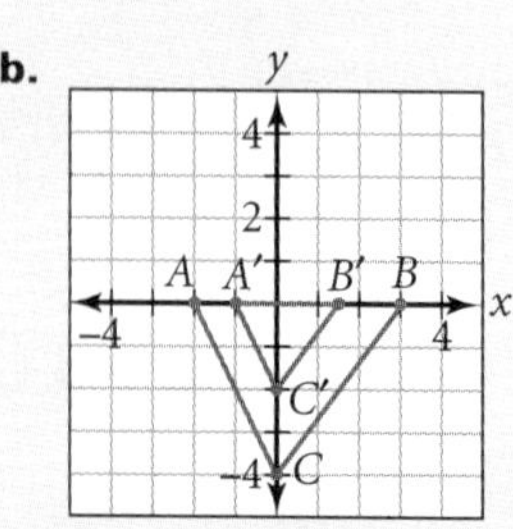

c. This transformation is the same as multiplying the original matrix by the scalar $\frac{1}{2}$.

22b.

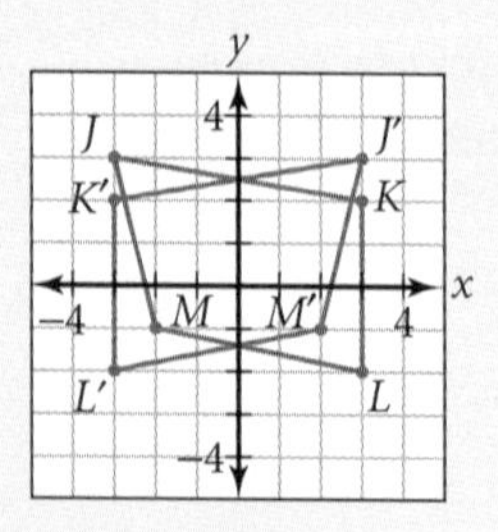

Let $A = \begin{bmatrix} 4 & -2 & 8 & 0 \\ 1 & 3 & -6 & 9 \\ -5 & 7 & 2 & 1 \end{bmatrix}$, $B = \begin{bmatrix} 7 & 6 \\ 2 & -3 \\ -1 & 8 \\ 9 & 5 \end{bmatrix}$, and $C = \begin{bmatrix} 0 & 8 \\ -2 & 1 \end{bmatrix}$. **Find each product, if it exists.**

15. BC **16.** CB **does not exist** **17.** BA **does not exist**

18. CA **does not exist** **19.** $A(BC)$ **20.** $(AB)C$

CONNECTIONS

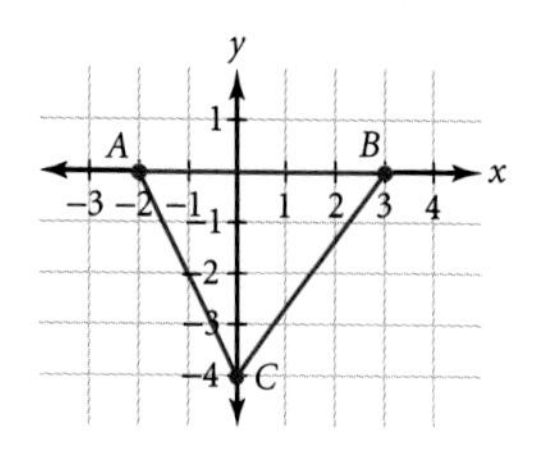

21. TRANSFORMATIONS Matrix $\begin{bmatrix} -2 & 3 & 0 \\ 0 & 0 & -4 \end{bmatrix}$ represents $\triangle ABC$ graphed at left.

a. Find the coordinates of the vertices of the image, $\triangle A'B'C'$, after multiplying the matrix above by the transformation matrix $\begin{bmatrix} \frac{1}{2} & 0 \\ 0 & \frac{1}{2} \end{bmatrix}$.

b. Sketch the image, $\triangle A'B'C'$, on the same plane as the pre-image, $\triangle ABC$.

c. Compare this transformation with the transformation resulting in an enlarged image or reduced image, which you learned in Lesson 4.1.

CHALLENGE

COORDINATE GEOMETRY For Exercises 22 and 23, let $A = \begin{bmatrix} -1 & 0 \\ 0 & 1 \end{bmatrix}$ **and** $B = \begin{bmatrix} 1 & 0 \\ 0 & -1 \end{bmatrix}$. **Matrix** $Q = \begin{bmatrix} -3 & 3 & 3 & -2 \\ 3 & 2 & -2 & -1 \end{bmatrix}$ **represents the vertices of quadrilateral** ***JKLM*** **at left.**

22. a. Find the product AQ. $AQ = \begin{bmatrix} 3 & -3 & -3 & 2 \\ 3 & 2 & -2 & -1 \end{bmatrix}$

b. Graph $J'K'L'M'$, the quadrilateral represented by AQ, on a coordinate plane with $JKLM$.

c. Make and verify a conjecture about the effect of matrix A on a geometric figure in the plane. (Hint: Describe the movement of the vertices from $JKLM$ to $J'K'L'M'$.) **Answers may vary. Sample: Reflection across *y*-axis.**

23. a. Find the product BQ.

b. Graph $J''K''L''M''$, the quadrilateral represented by BQ.

c. Make and verify a conjecture about the effect of matrix B on a geometric figure in the coordinate plane.

23. a. $\begin{bmatrix} -3 & 3 & 3 & -2 \\ -3 & -2 & 2 & 1 \end{bmatrix}$

c. Answers may vary. Sample: Reflection across *x*-axis.

APPLICATION

24. NUTRITION Jackson High School serves breakfast and lunch, each in two shifts. The cafeteria manager needs to estimate the number of meals needed during the first week of school. Table 1 gives the percentage of students who prefer meals with meat and of those who prefer meals with no meat at the first and second shifts. Table 2 shows the average number of students who come to first and second shifts of breakfast and of lunch.

Table 1

	1st shift	2nd shift
With meat	55%	62%
Without meat	45%	38%

Table 2

	Breakfast	Lunch
1st shift	72	102
2nd shift	85	130

a. Put the information from the tables into matrices.

b. Use matrix multiplication to find how many meals without meat are needed for breakfast and for lunch.

24a. $T_1 = \begin{bmatrix} 0.55 & 0.62 \\ 0.45 & 0.38 \end{bmatrix}$

$T_2 = \begin{bmatrix} 72 & 102 \\ 85 & 130 \end{bmatrix}$

b. 65 breakfasts, 96 lunches

23b.

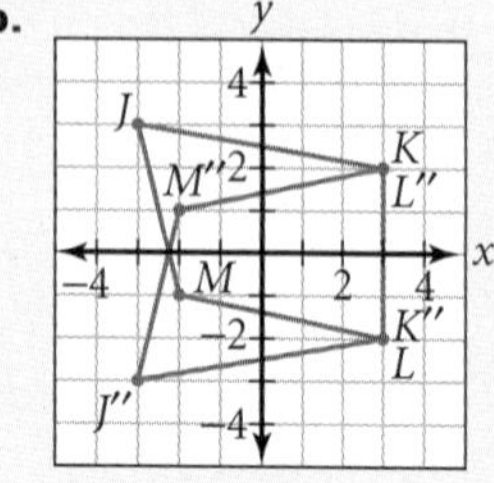

APPLICATIONS

25a. $\begin{bmatrix} 17 \\ 37 \\ 28 \\ 40 \end{bmatrix}$; 28; 122

b. 9 points; game 2

c. 9 rows, 1 column

25. SPORTS A high-school football team has played four games this season. Matrix S shows the number of touchdowns, extra points, and field goals scored in each game. Use matrix S and the point-values matrix, P, to answer parts **a–c**.

	Touchdowns	Extra points	Field goals
Game 1	2	2	1
Game 2	4	4	3
Game 3	3	1	3
Game 4	5	4	2

$$\begin{bmatrix} 2 & 2 & 1 \\ 4 & 4 & 3 \\ 3 & 1 & 3 \\ 5 & 4 & 2 \end{bmatrix} = S$$

	Point values
Touchdowns	6
Extra points	1
Field goals	3

$$\begin{bmatrix} 6 \\ 1 \\ 3 \end{bmatrix} = P$$

a. Find the product SP. What is the total number of points scored in game 3? in all four games?

b. Find the difference between the number of points scored in games 2 and 3. In which of these two games were the most points scored?

c. Suppose that matrix S included the information for all nine games of the season. How many rows and columns would the new product SP have?

26. NUTRITION A veterinarian has created formulas for producing her own mixtures of pet food. Using these formulas, she produces 3 mixtures from the 3 varieties of brand A food (regular, lite, and growth). She does the same for brand B. These mixtures are numbered 1, 2, and 3.

- The amounts of protein, fiber, and fat (in percent per serving) are given in matrix G for brand A and in matrix H for brand B.
- The three formulas that she is using, with the ingredients given in parts per serving, are stated in matrix J.

Brand A

	Regular	Lite	Growth
Protein	22	14	26
Fiber	3	15	3
Fat	13	4	17

$$\begin{bmatrix} 22 & 14 & 26 \\ 3 & 15 & 3 \\ 13 & 4 & 17 \end{bmatrix} = G$$

Brand B

	Regular	Lite	Growth
Protein	26	22	17
Fiber	5	5	4
Fat	15	12	28

$$\begin{bmatrix} 26 & 22 & 17 \\ 5 & 5 & 4 \\ 15 & 12 & 28 \end{bmatrix} = H$$

Formulas for mixtures

	1	2	3
Regular	1	2	1
Lite	2	1	1
Growth	1	1	2

$$\begin{bmatrix} 1 & 2 & 1 \\ 2 & 1 & 1 \\ 1 & 1 & 2 \end{bmatrix} = J$$

26a. $\begin{bmatrix} 76 & 84 & 88 \\ 36 & 24 & 24 \\ 38 & 47 & 51 \end{bmatrix}$

b. $\begin{bmatrix} 87 & 91 & 82 \\ 19 & 19 & 18 \\ 67 & 70 & 83 \end{bmatrix}$

c. brand A

d. brand A

a. Which two matrices must be multiplied to determine the nutritional content of mixtures 1, 2, and 3 from brand A? Find the product.

b. Which two matrices would be multiplied to determine the nutritional content of mixtures 1, 2, and 3 from brand B? Find the product.

c. The veterinarian wants mixture 3 to have the highest percentage of protein and fiber per serving. Should she use brand A or brand B for the mixture?

d. The veterinarian wants mixture 1 to have the lowest percentage of fat per serving. Determine whether brand A or brand B should be used in this mixture.

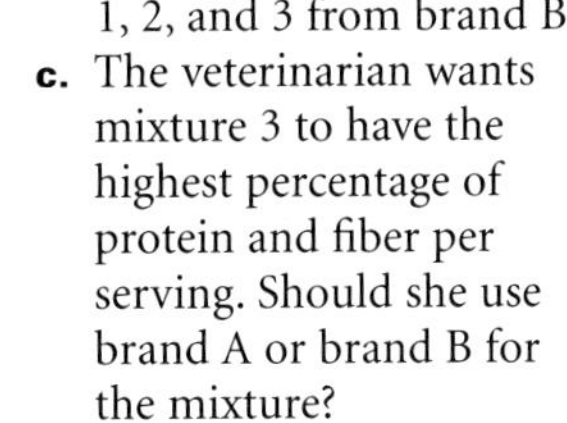

Look Beyond

In Exercise 43, students work backward from a given matrix equation to write the system of linear equations that it represents. This serves as a preview of Lesson 4.4, where students will learn to write matrix equations in order to solve systems of linear equations.

28a.
$$\begin{array}{c} \\ R \\ S \\ T \end{array}\begin{array}{c} R\quad S\quad T \\ \begin{bmatrix} 0 & 1 & 1 \\ 1 & 1 & 0 \\ 2 & 1 & 0 \end{bmatrix}\end{array}$$

b. 7 one-stage paths

c. R to S, R to T, S to R, S to S, T to R, T to R, T to S

29a.
$$\begin{array}{c} \\ R \\ S \\ T \end{array}\begin{array}{c} R\quad S\quad T \\ \begin{bmatrix} 3 & 2 & 0 \\ 1 & 2 & 1 \\ 1 & 3 & 2 \end{bmatrix}\end{array}$$

b. 15 two-stage paths

c. R to S to R, R to T to R, R to T to R, R to S to S, R to T to S, S to S to R, S to S to S, S to R to S, S to R to T, T to S to R, T to S to S, T to R to S, T to R to S, T to R to T, T to R to T

Practice

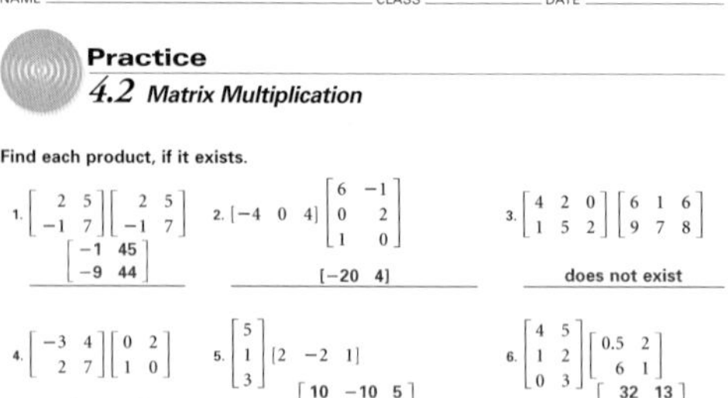

NAME ________ CLASS ________ DATE ________

Practice
4.2 Matrix Multiplication

Find each product, if it exists.

1. $\begin{bmatrix} 2 & 5 \\ -1 & 7 \end{bmatrix}\begin{bmatrix} 2 & 5 \\ -1 & 7 \end{bmatrix}$ $\begin{bmatrix} -1 & 45 \\ -9 & 44 \end{bmatrix}$

2. $[-4 \quad 0 \quad 4]\begin{bmatrix} 6 & -1 \\ 0 & 2 \\ 1 & 0 \end{bmatrix}$ $[-20 \quad 4]$

3. $\begin{bmatrix} 4 & 2 & 0 \\ 1 & 5 & 2 \end{bmatrix}\begin{bmatrix} 6 & 1 & 6 \\ 9 & 7 & 8 \end{bmatrix}$ does not exist

4. $\begin{bmatrix} -3 & 4 \\ 2 & 7 \end{bmatrix}\begin{bmatrix} 0 & 2 \\ 1 & 0 \end{bmatrix}$ $\begin{bmatrix} 4 & -6 \\ 7 & 4 \end{bmatrix}$

5. $\begin{bmatrix} 5 \\ 1 \\ 3 \end{bmatrix}[2 \quad -2 \quad 1]$ $\begin{bmatrix} 10 & -10 & 5 \\ 2 & -2 & 1 \\ 6 & -6 & 3 \end{bmatrix}$

6. $\begin{bmatrix} 4 & 5 \\ 1 & 2 \\ 0 & 3 \end{bmatrix}\begin{bmatrix} 0.5 & 2 \\ 6 & 1 \end{bmatrix}$ $\begin{bmatrix} 32 & 13 \\ 12.5 & 4 \\ 18 & 3 \end{bmatrix}$

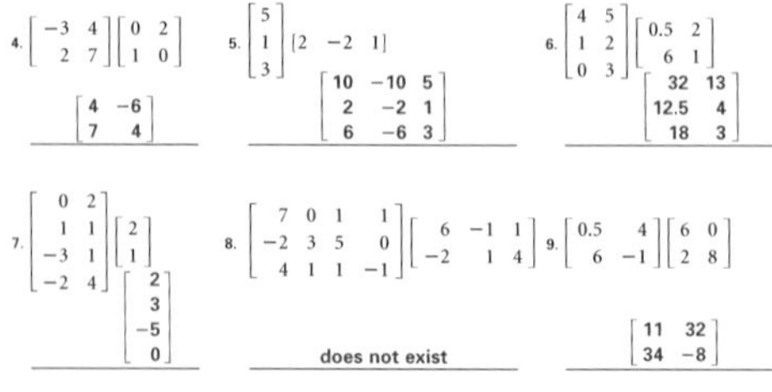

7. $\begin{bmatrix} 0 & 2 \\ 1 & 1 \\ -3 & 1 \\ -2 & 4 \end{bmatrix}\begin{bmatrix} 2 \\ 1 \end{bmatrix}$ $\begin{bmatrix} 2 \\ 3 \\ -5 \\ 0 \end{bmatrix}$

8. $\begin{bmatrix} 7 & 0 & 1 & 1 \\ -2 & 3 & 5 & 0 \\ 4 & 1 & 1 & -1 \end{bmatrix}\begin{bmatrix} 6 & -1 & 1 \\ -2 & 1 & 4 \end{bmatrix}$ does not exist

9. $\begin{bmatrix} 0.5 & 4 \\ 6 & -1 \end{bmatrix}\begin{bmatrix} 6 & 0 \\ 2 & 8 \end{bmatrix}$ $\begin{bmatrix} 11 & 32 \\ 34 & -8 \end{bmatrix}$

Matrix T, $\begin{bmatrix} -2 & 4 & 2 \\ 0 & 4 & -4 \end{bmatrix}$, represents triangle XYZ, which is graphed at right.

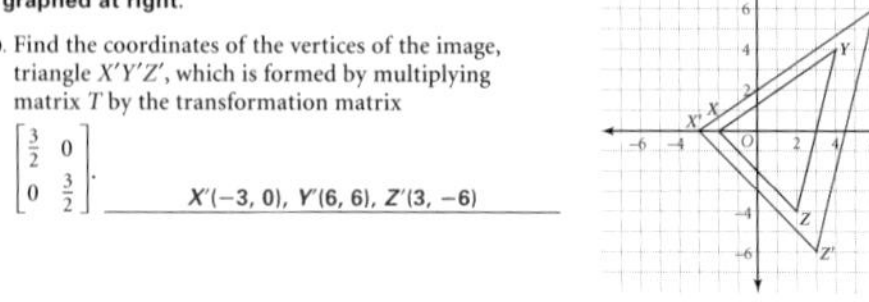

10. Find the coordinates of the vertices of the image, triangle $X'Y'Z'$, which is formed by multiplying matrix T by the transformation matrix $\begin{bmatrix} \frac{3}{2} & 0 \\ 0 & \frac{3}{2} \end{bmatrix}$. $X'(-3, 0)$, $Y'(6, 6)$, $Z'(3, -6)$

11. Sketch the image, triangle $X'Y'Z'$, on the grid at right above.

12. Describe the transformation. an enlargement by a factor of $\frac{3}{2}$

APPLICATIONS

27. INVENTORY A car rental agency has offices in New York and Los Angeles. Each month, $\frac{1}{2}$ of the cars in New York go to Los Angeles and $\frac{1}{3}$ of the cars in Los Angeles go to New York. If the company starts with 1000 cars at each office, how many cars are at each office n months later?

$$\text{Number of cars } \begin{array}{cc} \text{NY} & \text{LA} \\ [1000 & 1000] \end{array} = N \qquad \text{Origin: } \begin{array}{c} \\ \text{NY} \\ \text{LA} \end{array}\begin{array}{c} \text{Destination:} \\ \text{NY}\quad\text{LA} \\ \begin{bmatrix} \frac{1}{2} & \frac{1}{2} \\ \frac{1}{3} & \frac{2}{3} \end{bmatrix} \end{array} = P$$

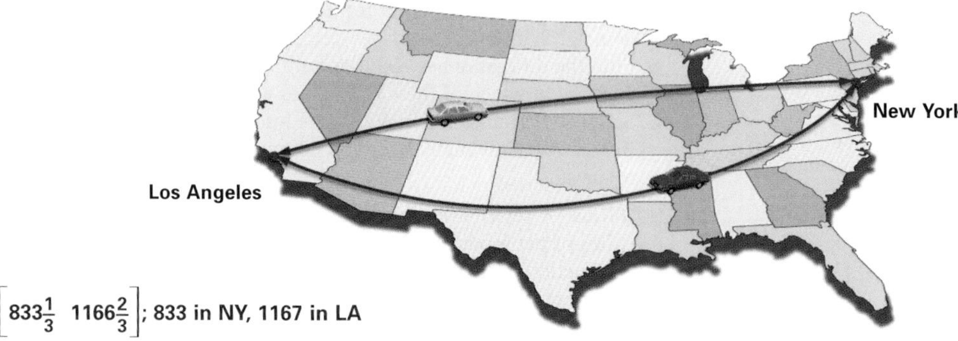

27a. $\begin{bmatrix} 833\frac{1}{3} & 1166\frac{2}{3} \end{bmatrix}$; 833 in NY, 1167 in LA

b. $\begin{bmatrix} 805\frac{5}{9} & 1194\frac{4}{9} \end{bmatrix}$; 806 in NY, 1194 in LA

c. 800 in NY, 1200 in LA; 4 months

a. The product NP represents the number of cars in each location after one month. Find NP. How many rental cars are in each location after 1 month?

b. Multiply matrix NP by P to find the number of cars in each location after 2 months. (Hint: Round up the entry in column one, and round down the entry in column two of the product.)

c. Continue to multiply each new product matrix by P until the product (with entries rounded to the nearest whole number) no longer changes. What is the final distribution of cars? How many months have passed?

NETWORKS For Exercises 28 and 29, refer to the directed network at left.

28. a. Create an adjacency matrix that represents the number of one-stage paths between the vertices.

b. How many one-stage paths does this directed network contain?

c. List the one-stage paths.

29. a. Create an adjacency matrix for the two-stage paths between the vertices.

b. How many two-stage paths does this directed network contain?

c. List the two-stage paths.

30a. $S = [5 \quad 4 \quad 2]$; $P = \begin{bmatrix} 6 \\ 1 \\ 3 \end{bmatrix}$

b. 40 points

30. SPORTS Suppose that a football team scores 5 touchdowns, 4 extra points, and 2 field goals in one game. A touchdown is worth 6 points, an extra point is 1 point, and a field goal is 3 points.

a. Construct a matrix, S, to represent the team's scoring events and a matrix, P, to represent the point values for each scoring event.

b. Multiply the two matrices to determine the team's total score for the game.

Look Back

31. undef. slope, no y-intercept

32. $-\frac{2}{3}$; -10

33. -18; 0

Find the slope and the y-intercept for each line. ***(LESSON 1.2)***

31. $4(2x - 7) = -6$ **32.** $y = -2\left(\frac{1}{3}x + 5\right)$ **33.** $6 - \frac{2}{3}(y + 9) = 12x$

Which of the following are true for all nonzero real numbers a and b? ***(LESSON 2.1)***

34. $a \div b = b \div a$ **false** **35.** $a + b = b + a$ **true**

36. $a - b = b - a$ **false** **37.** $a \cdot b = b \cdot a$ **true**

Let $f(x) = 2x + 3$ and $g(x) = 5x - 2$. ***(LESSON 2.4)***

38. Find $f \circ g$. **$= 10x - 1$** **39.** Find $g \circ f$. **$= 10x + 13$**

40. Is $f \circ g$ equal to $g \circ f$? **no** **41.** Find $(f \circ g)(6)$. **$= 59$**

APPLICATION

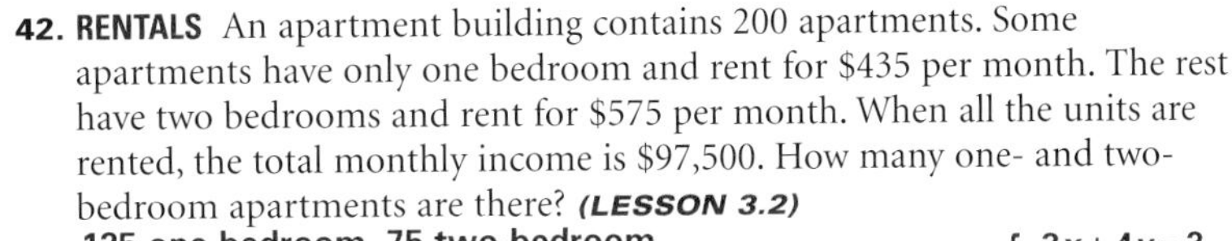

42. RENTALS An apartment building contains 200 apartments. Some apartments have only one bedroom and rent for \$435 per month. The rest have two bedrooms and rent for \$575 per month. When all the units are rented, the total monthly income is \$97,500. How many one- and two-bedroom apartments are there? ***(LESSON 3.2)***

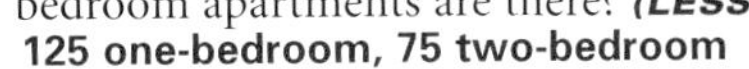

125 one-bedroom, 75 two-bedroom

Portfolio Extension
Go To: **go.hrw.com**
Keyword: **MB1 Matrix Multiply**

43. $\begin{cases} -3x + 4y = 3 \\ -6x + 8y = 6 \end{cases}$

Look Beyond

43. Write the system of two equations represented by this matrix equation.

$$\begin{bmatrix} -3 & 4 \\ -6 & 8 \end{bmatrix}\begin{bmatrix} x \\ y \end{bmatrix} = \begin{bmatrix} 3 \\ 6 \end{bmatrix}$$

NETWORKS Refer to the adjacency matrix N below, taken from the Portfolio Activity on page 224.

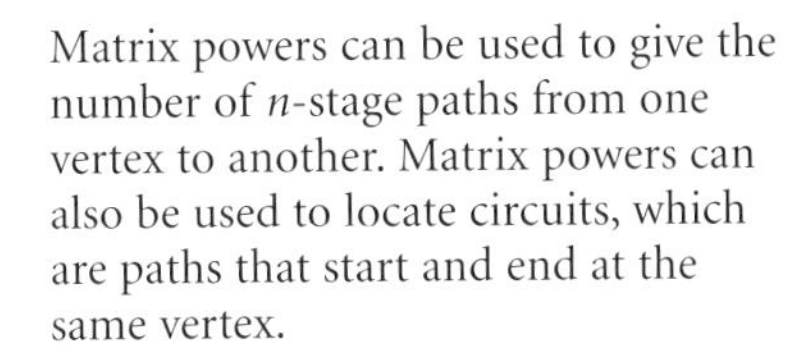

Matrix powers can be used to give the number of n-stage paths from one vertex to another. Matrix powers can also be used to locate circuits, which are paths that start and end at the same vertex.

$$\text{From: } \begin{matrix} & \text{To:} \\ & \begin{matrix} M & H & D & T & G \end{matrix} \\ \begin{matrix} M \\ H \\ D \\ T \\ G \end{matrix} & \begin{bmatrix} 0 & 1 & 0 & 0 & 0 \\ 1 & 0 & 1 & 0 & 0 \\ 0 & 1 & 0 & 1 & 1 \\ 0 & 0 & 1 & 0 & 1 \\ 0 & 0 & 1 & 1 & 0 \end{bmatrix} \end{matrix} = N$$

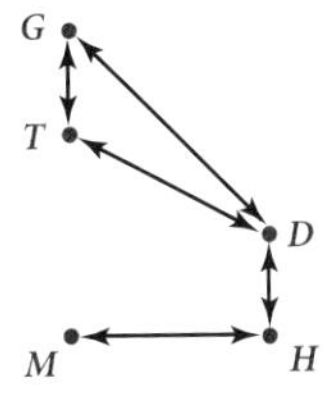

Matrix N^3 gives the number of three-stage paths from one vertex to another.

1. Find N^3. Interpret n_{54} in matrix N^3. List the corresponding paths.
2. Interpret n_{43} in matrix N^3. List the corresponding paths.
3. Find the sums of the rows of N^3. Use the sums to determine which exhibits have the greatest number of three-stage paths to themselves or other exhibits.
4. Find N^4. Interpret n_{13} in matrix N^4.
5. Find the number of four-stage paths from vertex D to itself. List the corresponding paths.

WORKING ON THE CHAPTER PROJECT

You should now be able to complete the Chapter Project.

ALTERNATIVE

Assessment

Portfolio Activity

The Portfolio Activity can be used as preparation for the Chapter Project or as a separate activity. In the Portfolio Activity on this page, students revisit the network and adjacency matrix from the Portfolio Activity on page 224. They apply and extend the techniques learned in this lesson, using the third and fourth powers of the adjacency matrix to find the number of three-stage and four-stage paths between the vertices of the network.

Answers to Portfolio Activities can be found in Additional Answers of the Teacher's Edition.

Student Technology Guide

NAME ______ CLASS ______ DATE ______

Student Technology Guide
4.2 Matrix Multiplication

Graphics calculators can perform matrix multiplication. If you try to multiply two matrices that cannot be multiplied, your calculator will give an error message. You can save time by checking the dimensions to make sure that the matrices can be multiplied *before* you enter them into your calculator.

Suppose you want to find the product AB, where $A = \begin{bmatrix} 3 & -2 \\ 6 & 9 \\ 4 & 1 \end{bmatrix}$ and $B = \begin{bmatrix} 3 & 4 & 0 & 7 \\ 6 & 1 & -2 & 8 \end{bmatrix}$.

- To enter matrix A, press MATRX EDIT 1:[A] ENTER 3 ENTER 2 ENTER 3 ENTER (-) 2 ENTER 6 ENTER 9 ENTER 4 ENTER 1 ENTER.
- To enter matrix B, press MATRX EDIT 2:[B] ENTER 2 ENTER 4 ENTER 3 ENTER 4 ENTER 0 ENTER 7 ENTER 6 ENTER 1 ENTER (-) 2 ENTER 8 ENTER.
- Press 2nd MODE to exit matrix edit mode.
- To find AB, press MATRX NAMES 1:[A] ENTER MATRX NAMES 2:[B] ENTER ENTER.

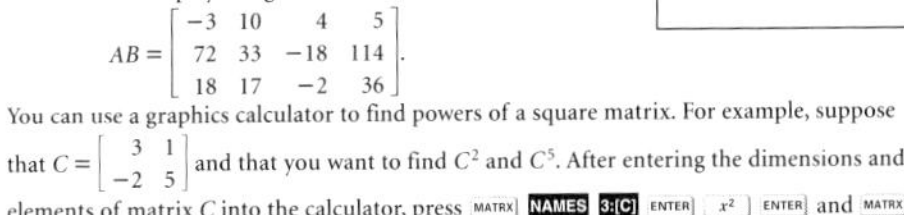

You can use ► and ◄ to see that there are no other columns to the right of what is shown. From the calculator display at right,

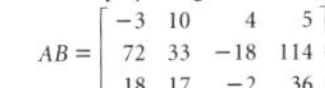

$$AB = \begin{bmatrix} -3 & 10 & 4 & 5 \\ 72 & 33 & -18 & 114 \\ 18 & 17 & -2 & 36 \end{bmatrix}.$$

You can use a graphics calculator to find powers of a square matrix. For example, suppose that $C = \begin{bmatrix} 3 & 1 \\ -2 & 5 \end{bmatrix}$ and that you want to find C^2 and C^5. After entering the dimensions and elements of matrix C into the calculator, press MATRX NAMES 3:[C] ENTER x^2 ENTER and MATRX NAMES 3:[C] ENTER ^ 5 ENTER.

Let $A = \begin{bmatrix} 6 & -4 & 2 \\ 3 & 1 & 9 \end{bmatrix}$, $B = \begin{bmatrix} 6 & -1 & 4 \\ 3 & 2 & -5 \\ 4 & 2 & 6 \end{bmatrix}$, $C = \begin{bmatrix} -5 & 3 \\ 2 & 7 \\ 1 & 6 \end{bmatrix}$, and $D = \begin{bmatrix} 2.3 & 6.2 \\ -1.4 & 0.6 \end{bmatrix}$.

Find each product, if it exists.

1. BC $\begin{bmatrix} -28 & 35 \\ -16 & -7 \\ -10 & 62 \end{bmatrix}$
2. AD does not exist
3. CD $\begin{bmatrix} -15.7 & -29.2 \\ -5.2 & 16.6 \\ -6.1 & 9.8 \end{bmatrix}$
4. $(AB)C$ $\begin{bmatrix} -124 & 362 \\ -190 & 656 \end{bmatrix}$
5. $(CD)A$ $\begin{bmatrix} -181.8 & 33.6 & -294.2 \\ 18.6 & 37.4 & 139 \\ -7.2 & 34.2 & 76 \end{bmatrix}$
6. $A(CB)$ does not exist
7. D^2 $\begin{bmatrix} -3.39 & 17.98 \\ -4.06 & -8.32 \end{bmatrix}$
8. B^2 $\begin{bmatrix} 49 & 0 & 53 \\ 4 & -9 & -28 \\ 54 & 12 & 42 \end{bmatrix}$
9. C^3 does not exist

Prepare

NCTM PRINCIPLES & STANDARDS 1–10

QUICK WARM-UP

Find the multiplicative inverse of each number.

1. -3 $-\frac{1}{3}$
2. 2.5 0.4
3. $-\frac{1}{7}$ -7
4. $1\frac{1}{5}$ $\frac{5}{6}$

Find each product.

5. $(-5)(-0.2)$ 1
6. $\left(\frac{2}{3}\right)\left(1\frac{1}{2}\right)$ 1

Also on Quiz Transparency 4.3

Teach

Why The inverse of an operation is said to "undo" that operation. Students have had experience with inverses of numerical operations, such as addition and multiplication, as well as inverses of functions. Lead students in a discussion of their experiences with inverses and identities.

The Inverse of a Matrix

Objectives

- Find and use the inverse of a matrix, if it exists.
- Find and use the determinant of a matrix.

Why *Just as inverse operations can be used to solve equations, inverse matrices can be used to decode messages.*

During World War II, Navaho code talkers, 29 members of the Navaho Nation, developed a code that was used by the United States Armed Forces.

APPLICATION
CRYPTOGRAPHY

The table at right is an assignment table for a code. Each letter of the alphabet is assigned a number. For example, the letter A is assigned the number 1 and Z is assigned the number 26. The dash, which represents the space between words in a message, is assigned the number 0. The question mark is assigned the number 27. For example, the phrase HELP ME would be encoded as 8 | 5 | 12 | 16 | 0 | 13 | 5.

– 0	G 7	N 14	U 21
A 1	H 8	O 15	V 22
B 2	I 9	P 16	W 23
C 3	J 10	Q 17	X 24
D 4	K 11	R 18	Y 25
E 5	L 12	S 19	Z 26
F 6	M 13	T 20	? 27

A matrix can be used to encode a message and another matrix, its inverse, is used to decode the message once it is received. *You will use a matrix to decode a message in Example 3.*

A **square matrix** is a matrix that has the same number of columns and rows. The following matrices are examples of square matrices:

$$\begin{bmatrix} 2 & -3 \\ \frac{1}{2} & 4 \end{bmatrix} \quad \begin{bmatrix} -1 & 5 & 3 \\ 4 & -8 & 1 \\ 3 & 7 & 0 \end{bmatrix} \quad \begin{bmatrix} -2 & -8 & 4 & 11 \\ 1 & -4 & 3 & -15 \\ 3 & 5 & 2 & 6 \\ 7 & 1 & 8 & 3 \end{bmatrix}$$

2×2 $\quad$ 3×3 $\quad$ 4×4

The *identity matrix for multiplication* for all 2×2 square matrices is $\begin{bmatrix} 1 & 0 \\ 0 & 1 \end{bmatrix}$.

An identity matrix, called I, has 1s on its *main diagonal* and 0s elsewhere.

$$I_{2\times 2} = \begin{bmatrix} 1 & 0 \\ 0 & 1 \end{bmatrix} \quad I_{3\times 3} = \begin{bmatrix} 1 & 0 & 0 \\ 0 & 1 & 0 \\ 0 & 0 & 1 \end{bmatrix} \quad I_{4\times 4} = \begin{bmatrix} 1 & 0 & 0 & 0 \\ 0 & 1 & 0 & 0 \\ 0 & 0 & 1 & 0 \\ 0 & 0 & 0 & 1 \end{bmatrix}$$

Alternative Teaching Strategy

USING TECHNOLOGY Give students matrix A at right. Have them experiment with their graphics calculators to find a matrix I such that $AI = IA = A$. They should arrive at matrix I at right.

$$A = \begin{bmatrix} -4 & 0 \\ 7 & 2 \end{bmatrix} \quad I = \begin{bmatrix} 1 & 0 \\ 0 & 1 \end{bmatrix}$$

Tell them that I is the *identity matrix* for multiplication for all 2×2 matrices. Have them enter a different 2×2 matrix of their own choosing, named B, and confirm that $BI = IB = B$.

Lead students to see that if matrix A has an *inverse* matrix A^{-1} under matrix multiplication, then $AA^{-1} = A^{-1}A = I$ must be true. Have them use their calculators to find A^{-1} for the given matrix A. They should arrive at matrix A^{-1} at right.

$$A^{-1} = \begin{bmatrix} -0.25 & 0 \\ 0.875 & 0.5 \end{bmatrix}$$

Have students investigate whether all 2×2 matrices have inverses. Direct them to look for a pattern among the 2×2 matrices that do not have inverses. Define the determinant of a matrix and discuss the value of $\det(A)$ when A is not invertible.

The Identity Matrix for Multiplication

Let A be a square matrix with n rows and n columns. Let I be a matrix with the same dimensions and with 1s on the main diagonal and 0s elsewhere. Then $AI = IA = A$.

The product of a real number and 1 is the same number. The product of a square matrix, A, and I is the same matrix, A.

$$\underset{A}{\begin{bmatrix} 3 & 1 \\ 2 & 1 \end{bmatrix}} \times \underset{I}{\begin{bmatrix} 1 & 0 \\ 0 & 1 \end{bmatrix}} = \begin{bmatrix} 3(1)+1(0) & 3(0)+1(1) \\ 2(1)+1(0) & 2(0)+1(1) \end{bmatrix} = \underset{A}{\begin{bmatrix} 3 & 1 \\ 2 & 1 \end{bmatrix}}$$

CHECKPOINT ✔ What is the product of $\begin{bmatrix} a & b \\ c & d \end{bmatrix}$ and $\begin{bmatrix} 1 & 0 \\ 0 & 1 \end{bmatrix}$?

The product of a real number and its multiplicative inverse is 1. The product of a square matrix and its *inverse* is the identity matrix I.

$$\underset{A}{\begin{bmatrix} 3 & 1 \\ 2 & 1 \end{bmatrix}} \times \underset{\text{inverse of } A}{\begin{bmatrix} 1 & -1 \\ -2 & 3 \end{bmatrix}} = \begin{bmatrix} 3(1)+1(-2) & 3(-1)+1(3) \\ 2(1)+1(-2) & 2(-1)+1(3) \end{bmatrix} = \underset{I}{\begin{bmatrix} 1 & 0 \\ 0 & 1 \end{bmatrix}}$$

The Inverse of a Matrix

Let A be a square matrix with n rows and n columns. If there is an $n \times n$ matrix B such that $AB = I$ and $BA = I$, then A and B are inverses of one another. The inverse of matrix A is denoted by A^{-1}. $\left(\text{Note: } A^{-1} \neq \frac{1}{A}\right)$

In general, to show that matrices are inverses of one another, you need to show that the multiplication of the matrices is commutative and results in the identity matrix.

EXAMPLE 1 Let $A = \begin{bmatrix} 2 & 3 \\ 3 & 5 \end{bmatrix}$ and $B = \begin{bmatrix} 5 & -3 \\ -3 & 2 \end{bmatrix}$.

Show that A and B are inverses of one another.

SOLUTION

$$AB = \begin{bmatrix} 2 & 3 \\ 3 & 5 \end{bmatrix}\begin{bmatrix} 5 & -3 \\ -3 & 2 \end{bmatrix} = \begin{bmatrix} 2(5)+3(-3) & 2(-3)+3(2) \\ 3(5)+5(-3) & 3(-3)+5(2) \end{bmatrix} = \begin{bmatrix} 1 & 0 \\ 0 & 1 \end{bmatrix}$$

$$BA = \begin{bmatrix} 5 & -3 \\ -3 & 2 \end{bmatrix}\begin{bmatrix} 2 & 3 \\ 3 & 5 \end{bmatrix} = \begin{bmatrix} 5(2)+(-3)(3) & 5(3)+(-3)(5) \\ -3(2)+2(3) & -3(3)+2(5) \end{bmatrix} = \begin{bmatrix} 1 & 0 \\ 0 & 1 \end{bmatrix}$$

Since both product matrices are the 2×2 identity matrix for multiplication, A and B are inverses of one another.

Inclusion Strategies

ENGLISH LANGUAGE DEVELOPMENT Have students write a brief report in which they relate the concept of the inverse of a matrix to other types of inverses that they have studied. Suggest that they focus on their perceptions of the ways in which inverses of matrices are similar to other inverses and the ways in which they are different. Remind them to include a description of the special characteristics a matrix must have in order to have an inverse and a description of the product of a matrix and its inverse.

CHECKPOINT ✔

$\begin{bmatrix} a & b \\ c & d \end{bmatrix}$

ADDITIONAL EXAMPLE 1

Let $C = \begin{bmatrix} 2 & 5 \\ 3 & 7 \end{bmatrix}$ and $D = \begin{bmatrix} -7 & 5 \\ 3 & -2 \end{bmatrix}$.

Show that C and D are inverses of one another.

$CD = \begin{bmatrix} 2(-7)+5(3) & 2(5)+5(-2) \\ 3(-7)+7(3) & 3(5)+7(-2) \end{bmatrix} = \begin{bmatrix} 1 & 0 \\ 0 & 1 \end{bmatrix}$

$DC = \begin{bmatrix} -7(2)+5(3) & -7(5)+5(7) \\ 3(2)+-2(3) & 3(5)+-2(7) \end{bmatrix} = \begin{bmatrix} 1 & 0 \\ 0 & 1 \end{bmatrix}$

Teaching Tip

TECHNOLOGY To access the **MATRX** menu on a TI-82 or TI-83, press MATRX, and to access it on a TI-83 Plus, press 2nd x^{-1}. To find the inverse of a matrix entered into **[A]**, access the **MATRX** menu, select **NAMES**, choose **1:[A]**, press ENTER, and then press x^{-1} ENTER. If the inverse does not exist, the error message **ERR:SINGULAR MAT** will be shown. Select **1:Quit** to return to the home screen. If students use ^ (–) 1 to find the inverse of a matrix, the error message **ERR:DOMAIN** will be shown.

ADDITIONAL EXAMPLE 2

Use a graphics calculator to find the inverse of each matrix.

a. $R = \begin{bmatrix} 2 & 4 \\ 3 & 5 \end{bmatrix}$

$R^{-1} = \begin{bmatrix} -2.5 & 2 \\ 1.5 & -1 \end{bmatrix}$

b. $S = \begin{bmatrix} 2 & 4 & 0 \\ 0 & 1 & 2 \\ -6 & -1 & -3 \end{bmatrix}$

$S^{-1} = \begin{bmatrix} 0.02 & -0.24 & -0.16 \\ 0.24 & 0.12 & 0.08 \\ -0.12 & 0.44 & -0.04 \end{bmatrix}$

c. $T = \begin{bmatrix} 2 & 3 \\ 4 & 6 \end{bmatrix}$

no inverse

You can use the equation $AB = I$ to find the inverse of a matrix. For example, let $A = \begin{bmatrix} 1 & 2 \\ 3 & 5 \end{bmatrix}$ and $B = \begin{bmatrix} a & b \\ c & d \end{bmatrix}$. Then write the equation $AB = I$, and proceed as follows:

$$\begin{bmatrix} 1 & 2 \\ 3 & 5 \end{bmatrix} \begin{bmatrix} a & b \\ c & d \end{bmatrix} = \begin{bmatrix} 1 & 0 \\ 0 & 1 \end{bmatrix}$$

$$\begin{bmatrix} a + 2c & b + 2d \\ 3a + 5c & 3b + 5d \end{bmatrix} = \begin{bmatrix} 1 & 0 \\ 0 & 1 \end{bmatrix}$$

Set the corresponding entries equal to each other, and solve the two resulting systems.

$$\begin{cases} a + 2c = 1 \\ 3a + 5c = 0 \end{cases} \qquad \begin{cases} b + 2d = 0 \\ 3b + 5d = 1 \end{cases}$$

$a = -5$ and $c = 3$ $\qquad$ $b = 2$ and $d = -1$

Thus, the inverse of matrix A does exist. $B = A^{-1} = \begin{bmatrix} -5 & 2 \\ 3 & -1 \end{bmatrix}$.

You can find the inverse of $A = \begin{bmatrix} 1 & 2 \\ 3 & 5 \end{bmatrix}$ on most graphics calculators. Enter the matrix, and use the x^{-1} key. The display at right shows matrices A and A^{-1}, the inverse of A.

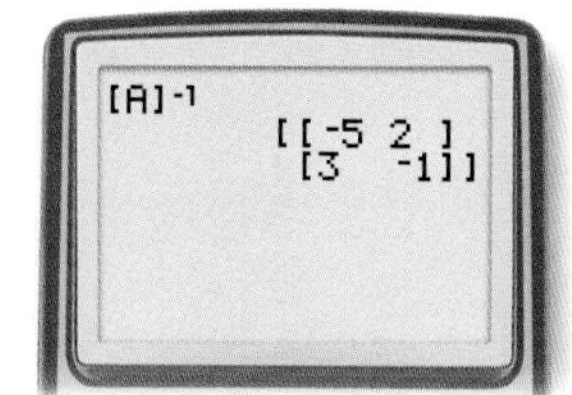

Finding the inverse of a 3 × 3 matrix or verifying that there is one is a very lengthy process. For matrices larger than 2 × 2, you will find that a graphics calculator especially useful.

EXAMPLE 2

Use a graphics calculator to find the inverse of each matrix.

a. $A = \begin{bmatrix} 6 & 8 \\ 5 & 7 \end{bmatrix}$ **b.** $B = \begin{bmatrix} 2 & -1 & 1 \\ -1 & 3 & 4 \\ -2 & 1 & 0 \end{bmatrix}$ **c.** $C = \begin{bmatrix} 8 & 4 \\ 6 & 3 \end{bmatrix}$

TECHNOLOGY GRAPHICS CALCULATOR Keystroke Guide, page 269

SOLUTION

To find each inverse, enter the matrix and use the x^{-1} key.

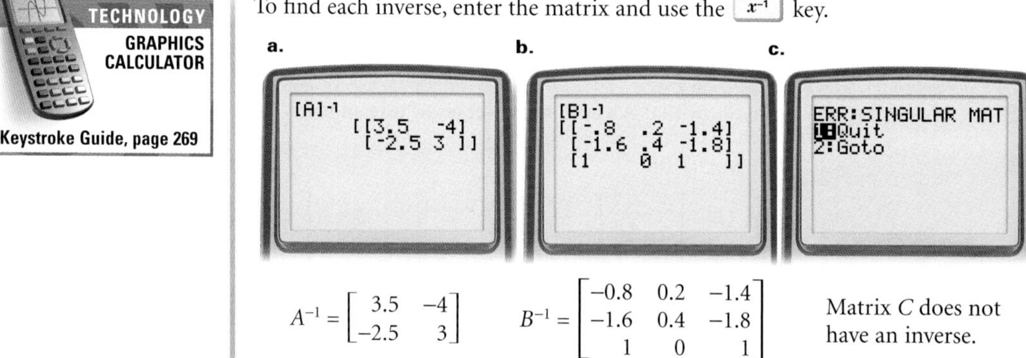

$A^{-1} = \begin{bmatrix} 3.5 & -4 \\ -2.5 & 3 \end{bmatrix}$ $\qquad B^{-1} = \begin{bmatrix} -0.8 & 0.2 & -1.4 \\ -1.6 & 0.4 & -1.8 \\ 1 & 0 & 1 \end{bmatrix}$ $\qquad$ Matrix C does not have an inverse.

Matrix C in Example 2 is called *non-invertible* because it *does not* have an inverse. An *invertible* matrix *does* have an inverse.

Enrichment

Give students matrices A and B below.

$$A = \begin{bmatrix} 2 & 3 & 5 \\ 4 & 6 & -1 \\ 0 & 2 & 4 \end{bmatrix} \quad B = \begin{bmatrix} a & d & g \\ b & e & h \\ c & f & i \end{bmatrix}$$

Have students write the systems of equations that must be solved in order to find the values of the variables that make B the inverse of A. Then have them solve the systems and use the solutions to write the inverse matrix, B.

$$\begin{cases} 2a + 3b + 5c = 1 \\ 4a + 6b - c = 0 \\ 2b + 4c = 0 \end{cases} \qquad \begin{cases} 2d + 3e + 5f = 0 \\ 4d + 6e - f = 1 \\ 2e + 4f = 0 \end{cases}$$

$$\begin{cases} 2g + 3h + 5i = 0 \\ 4g + 6h - i = 0 \\ 2h + 4i = 1 \end{cases} \qquad B = \begin{bmatrix} \frac{13}{22} & -\frac{1}{22} & -\frac{3}{4} \\ -\frac{4}{11} & \frac{2}{11} & \frac{1}{2} \\ \frac{2}{11} & -\frac{1}{11} & 0 \end{bmatrix}$$

CRITICAL THINKING If $A^{-1} = \begin{bmatrix} 2 & 3 \\ 5 & 7 \end{bmatrix}$, find A. Then find $(A^{-1})^{-1}$. Explain how the three matrices are related.

APPLICATION
CRYPTOGRAPHY

Using the table at the beginning of the lesson, assign a number to each letter and space in a message. If the message has an uneven number of characters, then add a zero to the end. Then choose an invertible matrix, A, that can multiply B to encode the message. A^{-1} can then be used to decode the message.

For example, the message GO BOB, represented by 7 | 15 | 0 | 2 | 15 | 2, would be translated into the 2 × 3 matrix B shown below:

$$B = \begin{bmatrix} 7 & 15 & 0 \\ 2 & 15 & 2 \end{bmatrix}$$

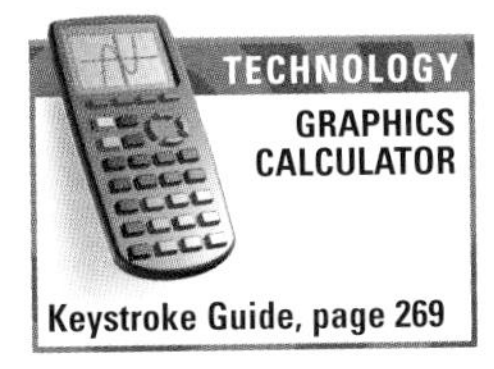

TECHNOLOGY
GRAPHICS CALCULATOR
Keystroke Guide, page 269

You can use an invertible matrix, such as $A = \begin{bmatrix} 6 & 5 \\ 7 & 6 \end{bmatrix}$, to encode the message by multiplying A and B.

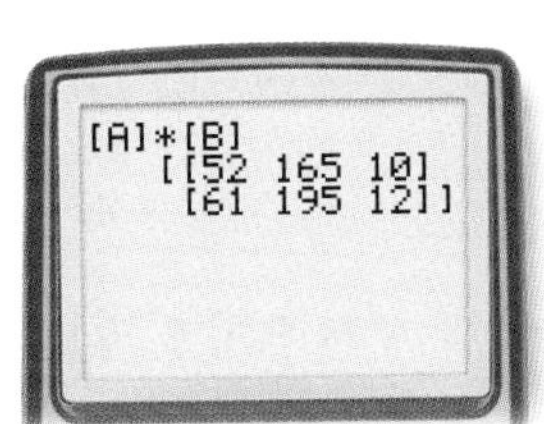

During World War II, the United States used a rotor machine called the ECM Mark II, also known as SIGABA, to encrypt messages. The SIGABA was so well designed that its codes were never broken.

Example 3 shows how you can then use A^{-1} to decode a message that was encoded by A.

EXAMPLE 3 **Use A^{-1} to decode the message 52 | 165 | 10 | 61 | 195 | 12 which was encoded by matrix $A = \begin{bmatrix} 6 & 5 \\ 7 & 6 \end{bmatrix}$.**

APPLICATION
CRYPTOGRAPHY

PROBLEM SOLVING

SOLUTION

Work backward. To decode the message, insert the code numbers into a 2 × 3 matrix C. Multiply this matrix by A^{-1}.

$$C = \begin{bmatrix} 52 & 165 & 10 \\ 61 & 195 & 12 \end{bmatrix}$$

TECHNOLOGY
GRAPHICS CALCULATOR
Keystroke Guide, page 269

Enter the coding matrix A. Then find $A^{-1}C$.

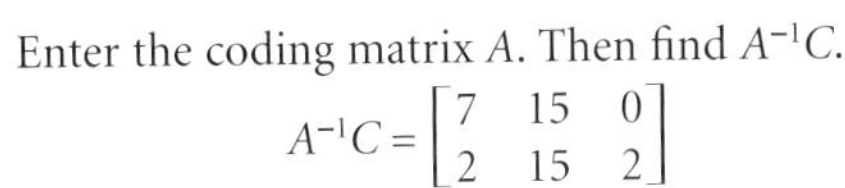

$$A^{-1}C = \begin{bmatrix} 7 & 15 & 0 \\ 2 & 15 & 2 \end{bmatrix}$$

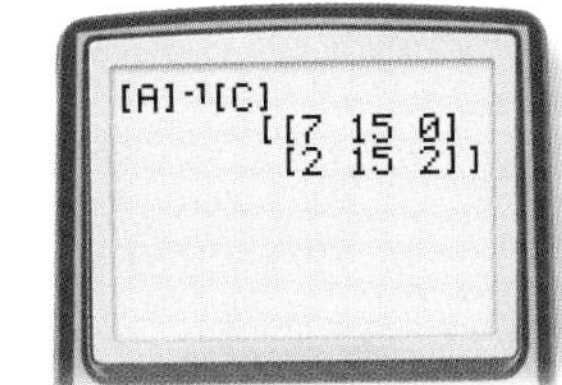

The matrix product gives the decoded message 7 | 15 | 0 | 2 | 15 | 2. These numbers translate to the original message, GO BOB.

CRITICAL THINKING

$A = \begin{bmatrix} -7 & 3 \\ 5 & -2 \end{bmatrix} = (A^{-1})^{-1}$;

A and $(A^{-1})^{-1}$ are both inverses of A^{-1}.

Teaching Tip

Remind students that matrix multiplication is not commutative. That is, $AB \neq BA$ for most matrices A and B. When multiplying matrix C by the inverse of matrix A in Example 3, students must compute $A^{-1}C$, not CA^{-1}.

ADDITIONAL EXAMPLE 3

Use Z^{-1} to decode the message 51 | 112 | 161 | 206 | 189 | 234 | 45 | 99 | 141 | 183 | 166 | 208, which was encoded by matrix Z below.

$$Z = \begin{bmatrix} 8 & 9 \\ 7 & 8 \end{bmatrix}$$

CESAR CHAVEZ

Teaching Tip

TECHNOLOGY When using a TI-82, TI-83, or TI-83 Plus to compute the determinant of a matrix, access the **MATRX** menu, select **MATH**, choose **1:det(**, access the **MATRX** menu again, choose the name of the matrix desired, and press **)** **ENTER**.

Reteaching the Lesson

COOPERATIVE LEARNING Have students work in groups of three. Tell each student to create four 2 × 2 matrices that have inverses, calling them A, B, C, and D. Each student should copy the four matrices onto a clean sheet of paper.

Now have each student pass the copy of the four matrices to the group member seated at his or her right. All students should find the inverses of the four matrices that they have received. They should write the inverses on a clean sheet of paper, identifying them as A^{-1}, B^{-1}, C^{-1}, and D^{-1}.

Now each student should pass the four inverses to the group member at his or her right. All students should find the inverses of the four inverses and write the results on a clean sheet of paper, identifying them as A, B, C, and D.

Each student should again pass the four matrices to the group member at his or her right. This time, all group members should receive the matrices that they originally created. If this has not happened, the group should work together to find the source of the error.

Determinants

Each square matrix can be assigned a real number called the *determinant of the matrix.* The determinant of a 2 × 2 matrix is defined below.

Determinant of a 2 × 2 Matrix

Let $A = \begin{bmatrix} a & b \\ c & d \end{bmatrix}$. The **determinant** of A, denoted by $\det(A)$ or $\begin{vmatrix} a & b \\ c & d \end{vmatrix}$, is defined as $\det(A) = \begin{vmatrix} a & b \\ c & d \end{vmatrix} = ad - bc$.

Matrix A has an inverse if and only if $\det(A) \neq 0$.

EXAMPLE 4 Find the determinant, and tell whether each matrix has an inverse.

a. $G = \begin{bmatrix} 7 & 8 \\ 6 & 7 \end{bmatrix}$ b. $H = \begin{bmatrix} 1 & 1 \\ 2 & 2 \end{bmatrix}$

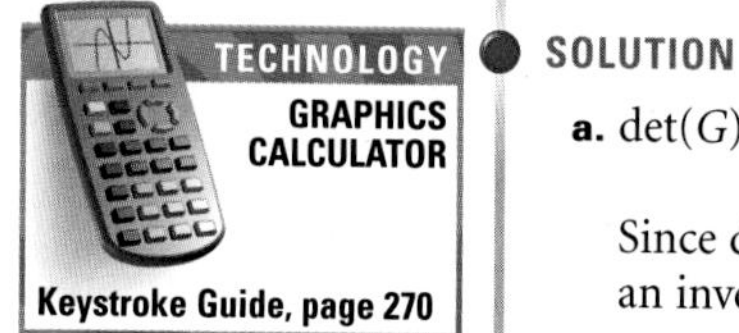

SOLUTION

a. $\det(G) = (7)(7) - (8)(6) = 1$

Since $\det(G) \neq 0$, matrix G has an inverse.

b. $\det(H) = (1)(2) - (1)(2) = 0$

Since $\det(H) = 0$, matrix H has no inverse.

TRY THIS Find the determinant, and tell whether each matrix has an inverse.

a. $S = \begin{bmatrix} -3 & 4 \\ -2 & 9 \end{bmatrix}$ b. $T = \begin{bmatrix} -6 & 12 \\ 2 & -4 \end{bmatrix}$

Activity Exploring Codes

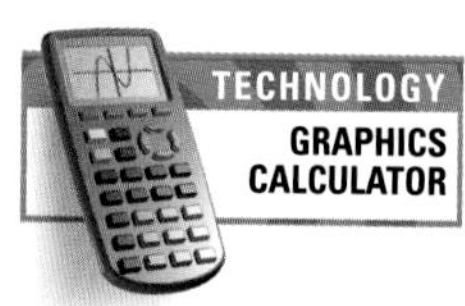

PROBLEM SOLVING

You will need: a graphics calculator

1. Create a 2 × 2 matrix that you would like to use to encode messages.
2. **Guess and check.** Use your calculator to verify that your matrix has an inverse. If it does not, modify your matrix so that it does. (Hint: Create your matrix so that its determinant does not equal 0.)
3. Write a brief message. Use the assignment table from page 234 to translate it into numbers.
4. Use your matrix to encode the message. Write the coded message.
5. Use your inverse to decode your message. (CHECKPOINT ✔)
6. Explain why a matrix must be square and invertible in order to be an encoding matrix.

ADDITIONAL EXAMPLE 4

Find the determinant, and tell whether each matrix has an inverse.

a. $J = \begin{bmatrix} 13 & 8 \\ 5 & 3 \end{bmatrix}$

$\det(J) = (13)(3) - (8)(5) = -1$

Since $\det(J) \neq 0$, matrix J has an inverse.

b. $K = \begin{bmatrix} 6 & 2 \\ 9 & 3 \end{bmatrix}$

$\det(K) = (6)(3) - (2)(9) = 0$

Since $\det(K) = 0$, matrix K has no inverse.

TRY THIS

a. $\det(S) = -19$; has an inverse

b. $\det(T) = 0$; has no inverse

Activity Notes

In this Activity, students create their own coding matrix and use it to encode a message of their choosing. They should come to the realization that in order for a given matrix to be a coding matrix, it must be square and invertible.

CHECKPOINT ✔

Answers may vary. Sample answers to the Activity:

Message matrix (GO TEAM):

$$\begin{bmatrix} 7 & 15 & 0 & 20 \\ 5 & 1 & 13 & 0 \end{bmatrix}$$

Coding matrix: $A = \begin{bmatrix} 1 & 2 \\ 3 & 7 \end{bmatrix}$

Coded message:

$$\begin{bmatrix} 1 & 2 \\ 3 & 7 \end{bmatrix}\begin{bmatrix} 7 & 15 & 0 & 20 \\ 5 & 1 & 13 & 0 \end{bmatrix} = \begin{bmatrix} 17 & 17 & 26 & 20 \\ 56 & 52 & 91 & 60 \end{bmatrix}$$

Decoded message:

$$\begin{bmatrix} 7 & -2 \\ -3 & 1 \end{bmatrix}\begin{bmatrix} 17 & 17 & 26 & 20 \\ 56 & 52 & 91 & 60 \end{bmatrix} = \begin{bmatrix} 7 & 15 & 0 & 20 \\ 5 & 1 & 13 & 0 \end{bmatrix}$$

Exercises

Communicate

1. Look up the words *encryption* and *decryption* in a dictionary, and explain how matrices are used to encrypt and decrypt messages.
2. Describe each product matrix without multiplying.
   a. $\begin{bmatrix} 1 & 3 \\ 2 & -1 \end{bmatrix}\begin{bmatrix} 1 & 0 \\ 0 & 1 \end{bmatrix}$ b. $\begin{bmatrix} 1 & 0 \\ 0 & 1 \end{bmatrix}\begin{bmatrix} -2 & 1 \\ 5 & 3 \end{bmatrix}$
3. How do you find the inverse of a matrix with a graphics calculator?
4. Explain one way in which determinants can be used.

Guided Skills Practice

5. Show that the matrices $\begin{bmatrix} 2 & 1 \\ 1 & 1 \end{bmatrix}$ and $\begin{bmatrix} 1 & -1 \\ -1 & 2 \end{bmatrix}$ are inverses. ***(EXAMPLE 1)***
6. Find the inverse of $\begin{bmatrix} 9 & 5 \\ 7 & 4 \end{bmatrix}$. ***(EXAMPLE 2)*** $\begin{bmatrix} 4 & -5 \\ -7 & 9 \end{bmatrix}$

APPLICATION

7. **CRYPTOGRAPHY** Use the inverse matrix you found in Exercise 6 to decode the message 282 | 9 | 260 | 75 | 180 | 221 | 7 | 203 | 60 | 140. ***(EXAMPLE 3)***
   WAY TO GO

Find the determinant, and tell whether each matrix has an inverse. ***(EXAMPLE 4)***

8. $\begin{bmatrix} 2 & 1 \\ 6 & 3 \end{bmatrix}$ 0; no
9. $\begin{bmatrix} -5 & 2 \\ 3 & -2 \end{bmatrix}$ 4; yes

Practice and Apply

Determine whether each pair of matrices are inverses of each other.

10. $\begin{bmatrix} 4 & 0 \\ 0 & 3 \end{bmatrix}, \begin{bmatrix} \frac{1}{4} & 0 \\ 0 & \frac{1}{3} \end{bmatrix}$ yes
11. $\begin{bmatrix} 1 & 2 \\ 3 & 4 \end{bmatrix}, \begin{bmatrix} -2 & 1 \\ \frac{3}{2} & -\frac{1}{2} \end{bmatrix}$ yes
12. $\begin{bmatrix} 4 & 3 \\ 4 & 2 \end{bmatrix}, \begin{bmatrix} 2 & -3 \\ -4 & 6 \end{bmatrix}$ no

Find the determinant, and tell whether each matrix has an inverse.

13. $\begin{bmatrix} 2 & -5 \\ -1 & 3 \end{bmatrix}$ 1; yes
14. $\begin{bmatrix} 1 & 2 \\ 1 & 3 \end{bmatrix}$ 1; yes
15. $\begin{bmatrix} 2 & 3 \\ 4 & 6 \end{bmatrix}$ 0; no
16. $\begin{bmatrix} -3 & 2 \\ 9 & -6 \end{bmatrix}$ 0; no
17. $\begin{bmatrix} 3 & 4 \\ 6 & 8 \end{bmatrix}$ 0; no
18. $\begin{bmatrix} 5 & 6 \\ 2 & 2 \end{bmatrix}$ −2; yes
19. $\begin{bmatrix} 2 & 2 \\ 3 & 4 \end{bmatrix}$ 2; yes
20. $\begin{bmatrix} 7 & 6 \\ 9 & 8 \end{bmatrix}$ 2; yes

internet connect
Homework Help Online
Go To: **go.hrw.com**
Keyword:
MB1 Homework Help for Exercises 13–20

Find the inverse matrix, if it exists. If the inverse matrix does not exist, write *no inverse*.

21. $\begin{bmatrix} -2 & 2 & -1 \\ 3 & -5 & 4 \\ 5 & -6 & 4 \end{bmatrix}$
22. $\begin{bmatrix} 1 & 6 & 2 \\ -2 & 3 & 5 \\ 7 & 12 & -4 \end{bmatrix}$ no inverse
23. $\begin{bmatrix} 1 & -2 & 1 \\ -2 & 4 & -2 \\ 3 & 5 & 3 \end{bmatrix}$ no inverse

21. $\begin{bmatrix} 4 & -2 & 3 \\ 8 & -3 & 5 \\ 7 & -2 & 4 \end{bmatrix}$

Assess

Selected Answers

Exercises 5–9, 11–67 odd

ASSIGNMENT GUIDE	
In Class	1–9
Core	11–19 odd, 21–23, 25–45 odd
Core Plus	10–20 even, 21–23, 24–34 even, 36–55
Review	56–68
Preview	69

Extra Practice can be found beginning on page 940.

Mid-Chapter Assessment for Lessons 4.1 through 4.3 can be found on page 46 of the *Assessment Resources.*

5. $\begin{bmatrix} 2 & 1 \\ 1 & 1 \end{bmatrix}\begin{bmatrix} 1 & -1 \\ -1 & 2 \end{bmatrix} = \begin{bmatrix} 1 & 0 \\ 0 & 1 \end{bmatrix}$
   $\begin{bmatrix} 1 & -1 \\ -1 & 2 \end{bmatrix}\begin{bmatrix} 2 & 1 \\ 1 & 1 \end{bmatrix} = \begin{bmatrix} 1 & 0 \\ 0 & 1 \end{bmatrix}$

Error Analysis

Some students may become confused and believe that a determinant of 0 indicates that a matrix *does* have an inverse. Encourage them to use the matrix editor and the inverse function of a graphics calculator to check the accuracy of their answers.

24. $\begin{bmatrix} -\frac{1}{2} & \frac{3}{2} \\ \frac{1}{2} & -\frac{1}{2} \end{bmatrix}$

25. $\begin{bmatrix} \frac{1}{2} & -\frac{1}{6} \\ 0 & \frac{1}{3} \end{bmatrix}$

26. no inverse

27. $\begin{bmatrix} \frac{10}{33} & -\frac{1}{11} \\ -\frac{3}{11} & \frac{2}{11} \end{bmatrix}$

28. $\begin{bmatrix} 2 & -3 \\ -1 & 2 \end{bmatrix}$

29. $\begin{bmatrix} 5 & -7 \\ -2 & 3 \end{bmatrix}$

30. $\begin{bmatrix} -1 & 3 \\ -\frac{1}{2} & 1 \end{bmatrix}$

Practice

NAME ______________ CLASS ________ DATE ________

Practice
4.3 *The Inverse of a Matrix*

Determine whether each pair of matrices are inverses of each other.

1. $\begin{bmatrix} 4 & -3 \\ -5 & 4 \end{bmatrix}, \begin{bmatrix} 4 & 3 \\ 5 & 4 \end{bmatrix}$ ___yes___
2. $\begin{bmatrix} 5 & -2 \\ -17 & 7 \end{bmatrix}, \begin{bmatrix} 7 & 2 \\ 17 & 5 \end{bmatrix}$ ___yes___
3. $\begin{bmatrix} 6 & 9 \\ 2 & 3 \end{bmatrix}, \begin{bmatrix} 3 & -9 \\ -2 & 6 \end{bmatrix}$ ___no___
4. $\begin{bmatrix} 3 & -3\frac{2}{3} \\ -4 & 5 \end{bmatrix}, \begin{bmatrix} 15 & 11 \\ 12 & 9 \end{bmatrix}$ ___yes___
5. $\begin{bmatrix} 12 & 5 \\ 14 & 6 \end{bmatrix}, \begin{bmatrix} 3 & -2.5 \\ -7 & 6 \end{bmatrix}$ ___yes___
6. $\begin{bmatrix} \frac{1}{2} & -\frac{1}{8} \\ -\frac{1}{4} & \frac{3}{16} \end{bmatrix}, \begin{bmatrix} 3 & 2 \\ 4 & 8 \end{bmatrix}$ ___yes___

Find the determinant and the inverse of each matrix, if it exists.

7. $\begin{bmatrix} 7 & 5 \\ 4 & 3 \end{bmatrix}$ ___$D = 1$; inverse: $\begin{bmatrix} 3 & -5 \\ -4 & 7 \end{bmatrix}$___
8. $\begin{bmatrix} 9 & 7 \\ 5 & 4 \end{bmatrix}$ ___$D = 1$; inverse: $\begin{bmatrix} 4 & -7 \\ -5 & 9 \end{bmatrix}$___
9. $\begin{bmatrix} 8 & 5 \\ 7 & 5 \end{bmatrix}$ ___$D = 5$; inverse: $\begin{bmatrix} 1 & -1 \\ -1.4 & 1.6 \end{bmatrix}$___
10. $\begin{bmatrix} 11 & 6 \\ 7 & 4 \end{bmatrix}$ ___$D = 2$; inverse: $\begin{bmatrix} 2 & -3 \\ -3.5 & 5.5 \end{bmatrix}$___
11. $\begin{bmatrix} 7\frac{1}{2} & 5 \\ 12 & 8 \end{bmatrix}$ ___$D = 0$; no inverse___
12. $\begin{bmatrix} 13 & 3 \\ 16 & 4 \end{bmatrix}$ ___$D = 4$; inverse: $\begin{bmatrix} 1 & -0.75 \\ -4 & 3.25 \end{bmatrix}$___

Find the inverse matrix, if it exists. If the inverse matrix does not exist, write *no inverse*.

13. $\begin{bmatrix} 2 & 1 \\ 3 & 1 \end{bmatrix}$ ___$\begin{bmatrix} -1 & 1 \\ 3 & -2 \end{bmatrix}$___
14. $\begin{bmatrix} 4 & 6 \\ 5 & 7 \end{bmatrix}$ ___$\begin{bmatrix} -3.5 & 3 \\ 2.5 & -2 \end{bmatrix}$___
15. $\begin{bmatrix} 6 & 4 \\ -3 & -2 \end{bmatrix}$ ___no inverse___
16. $\begin{bmatrix} 5 & 3 \\ 2 & 1 \end{bmatrix}$ ___$\begin{bmatrix} -1 & 3 \\ 2 & -5 \end{bmatrix}$___
17. $\begin{bmatrix} \frac{2}{3} & 2 \\ 4 & 12 \end{bmatrix}$ ___no inverse___
18. $\begin{bmatrix} 1.5 & -2.5 \\ -1 & 2 \end{bmatrix}$ ___$\begin{bmatrix} 4 & 5 \\ 2 & 3 \end{bmatrix}$___

internet connect
Activities Online
Go To: go.hrw.com
Keyword:
MB1 Magic Squares

If $A = \begin{bmatrix} a & b \\ c & d \end{bmatrix}$ and $\det(A) \neq 0$, then A^{-1} is given by the formula $A^{-1} = \frac{1}{\det(A)}\begin{bmatrix} d & -b \\ -c & a \end{bmatrix}$. Use this formula to find the inverse of each matrix, if it exists. If the inverse does not exist, write *no inverse*.

24. $\begin{bmatrix} 1 & 3 \\ 1 & 1 \end{bmatrix}$ **25.** $\begin{bmatrix} 2 & 1 \\ 0 & 3 \end{bmatrix}$ **26.** $\begin{bmatrix} -4 & 8 \\ 2 & -4 \end{bmatrix}$ **27.** $\begin{bmatrix} 6 & 3 \\ 9 & 10 \end{bmatrix}$

28. $\begin{bmatrix} 2 & 3 \\ 1 & 2 \end{bmatrix}$ **29.** $\begin{bmatrix} 3 & 7 \\ 2 & 5 \end{bmatrix}$ **30.** $\begin{bmatrix} 2 & -6 \\ 1 & -2 \end{bmatrix}$ **31.** $\begin{bmatrix} 2 & 1 \\ 1 & 1 \end{bmatrix}$

32. $\begin{bmatrix} 1 & 3 \\ 2 & 7 \end{bmatrix}$ **33.** $\begin{bmatrix} 5 & -7 \\ -2 & 3 \end{bmatrix}$ **34.** $\begin{bmatrix} \frac{1}{2} & \frac{3}{8} \\ 1 & \frac{1}{4} \end{bmatrix}$ **35.** $\begin{bmatrix} \frac{1}{3} & \frac{1}{6} \\ \frac{5}{6} & \frac{2}{3} \end{bmatrix}$

Find the inverse matrix, if it exists. Round entries to the nearest hundredth. If the inverse matrix does not exist, write *no inverse*.

36 $\begin{bmatrix} 2 & -4 \\ -3 & 6 \end{bmatrix}$ **37** $\begin{bmatrix} \frac{1}{2} & 0 \\ 1 & \frac{1}{4} \end{bmatrix}$ **38** $\begin{bmatrix} \frac{1}{2} & \frac{1}{10} \\ \frac{3}{2} & \frac{1}{5} \end{bmatrix}$

39 $\begin{bmatrix} \frac{1}{3} & \frac{1}{2} \\ \frac{1}{2} & \frac{1}{6} \end{bmatrix}$ **40** $\begin{bmatrix} 2 & -10 \\ -1 & 10 \end{bmatrix}$ **41** $\begin{bmatrix} -2 & -1 & 1 \\ 1 & -2 & 3 \\ 4 & 1 & 2 \end{bmatrix}$

42 $\begin{bmatrix} 2 & 0 & 5 \\ -3 & 1 & -5 \\ 0 & 2 & 4 \end{bmatrix}$ **43** $\begin{bmatrix} \pi & 2 & -1 \\ 1 & 5 & \pi \\ 2 & -3 & 4 \end{bmatrix}$ **44** $\begin{bmatrix} 2\pi & 3 & -1 \\ 0 & -2 & \pi \\ 3 & 0 & -5 \end{bmatrix}$

45. The determinant of a 3 × 3 matrix, $A = \begin{bmatrix} a & b & c \\ d & e & f \\ g & h & i \end{bmatrix}$, can be found by using the formula shown below.

$$\det(A) = \begin{vmatrix} a & b & c \\ d & e & f \\ g & h & i \end{vmatrix} = a\begin{vmatrix} e & f \\ h & i \end{vmatrix} - b\begin{vmatrix} d & f \\ g & i \end{vmatrix} + c\begin{vmatrix} d & e \\ g & h \end{vmatrix}$$

Use this formula to find the determinant of each 3 × 3 matrix.

a. $\begin{bmatrix} 2 & -1 & 3 \\ 4 & 0 & 1 \\ 2 & 0 & 3 \end{bmatrix}$ 10 **b.** $\begin{bmatrix} 3 & 1 & -6 \\ -5 & 2 & 10 \\ 4 & 2 & -8 \end{bmatrix}$ 0 **c.** $\begin{bmatrix} 1 & 2 & 0 \\ -3 & 7 & -2 \\ 2 & 1 & 5 \end{bmatrix}$ 59

CHALLENGE

46. a. Find x, y, u, and v in terms of a, b, c, and d such that

$$\begin{bmatrix} a & b \\ c & d \end{bmatrix}\begin{bmatrix} x & u \\ y & v \end{bmatrix} = I.$$

b. Substitute the values of x, y, u, and v from part **a** in $\begin{bmatrix} x & u \\ y & v \end{bmatrix}$. Divide out the common factor (a rational expression) from each term, and place it in front of the matrix as a scalar multiplier. How does this expression compare with the formula for A^{-1} given for Exercises 24–35?

31. $\begin{bmatrix} 1 & -1 \\ -1 & 2 \end{bmatrix}$

32. $\begin{bmatrix} 7 & -3 \\ -2 & 1 \end{bmatrix}$

33. $\begin{bmatrix} 3 & 7 \\ 2 & 5 \end{bmatrix}$

34. $\begin{bmatrix} -1 & \frac{3}{2} \\ 4 & -2 \end{bmatrix}$

35. $\begin{bmatrix} 8 & -2 \\ -10 & 4 \end{bmatrix}$

36. no inverse

37. $\begin{bmatrix} 2 & 0 \\ -8 & 4 \end{bmatrix}$

38. $\begin{bmatrix} -4 & 2 \\ 30 & -10 \end{bmatrix}$

39. $\begin{bmatrix} -0.86 & 2.57 \\ 2.57 & -1.71 \end{bmatrix}$

CONNECTION

47. COORDINATE GEOMETRY The matrix $A = \begin{bmatrix} 2 & 0 \\ 0 & 1 \end{bmatrix}$ horizontally stretches an object in the coordinate plane by a factor of 2. Find A^{-1}. Verify that A^{-1} horizontally compresses an object in the coordinate plane by a factor of $\frac{1}{2}$ by applying A^{-1} to the square with vertices at the points (2, 2), (2, −2), (−2, −2), and (−2, 2). Graph your results.

APPLICATION

CRYPTOGRAPHY Let $A = \begin{bmatrix} 5 & 3 \\ 3 & 2 \end{bmatrix}$. Use matrix A to encode each message.

48. MOVE OUT

49. HEAD NORTH

50. FALL IN

51. CEASE FIRE

Given $A = \begin{bmatrix} 5 & 3 \\ 3 & 2 \end{bmatrix}$, find A^{-1} and decode each message.

52. 97 | 70 | 68 | 80 | 24 | 62 | 45 | 45 | 52 | 16 — 52. HEAD SOUTH

53. 77 | 50 | 139 | 42 | 47 | 33 | 88 | 28 — 53. MAN DOWN

54. 8 | 160 | 100 | 17 | 18 | 124 | 42 | 5 | 100 | 60 | 11 | 11 | 79 | 28 — 54. ATTACK AT DAWN

55. 18 | 100 | 5 | 132 | 80 | 70 | 42 | 11 | 64 | 3 | 82 | 51 | 45 | 28 — 55. CHARGE AT NOON

Look Back

Write an equation in slope-intercept form for the line containing the indicated points. ***(LESSON 1.3)***

56. (2, −2) and (0, −1) $y = -\frac{1}{2}x - 1$

57. (0, 3) and (−2, −6) $y = \frac{9}{2}x + 3$

Solve each equation. ***(LESSON 1.6)***

58. $8.91 + x = 11.09$ 2.18

59. $\frac{1}{4} = \frac{3}{4} + x$ $-\frac{1}{2}$

60. $5\frac{1}{2}x = -62$ — 60. $-\frac{124}{11}$, or $-11\frac{3}{11}$

61. $\frac{1}{5}x = 0.3$ 1.5

62. $\frac{2}{3}x + 1 = x + 3$ −6

63. $\frac{1}{2}x + \frac{1}{4} = 2\frac{3}{4} - \frac{1}{3}x$ 3

64. Evaluate $f(x) = 2 - 5x + x^2$ for $x = 3$ and $x = -4$. −4; 38 ***(LESSON 2.3)***

65. Use any method to solve the system. $\begin{cases} 5x + 7y = 32 \\ 2x - 14y = 6 \end{cases}$ $\left(\frac{35}{6}, \frac{17}{42}\right)$ ***(LESSONS 3.1 AND 3.2)***

Graph the solution to each system of linear inequalities. ***(LESSON 3.4)***

66. $\begin{cases} 2x + y \ge 2 \\ y \ge 3x + 2 \end{cases}$

67. $\begin{cases} x + \frac{1}{2}y \le 2 \\ 2x + 3y < 2 \end{cases}$

68. $\begin{cases} 3x + 2y \ge 1 \\ 2x + 3y < 2 \\ x < 3 \end{cases}$

Look Beyond

69. a. Write the system of equations $\begin{cases} 3x + 7y = 4 \\ 2x + 5y = 1 \end{cases}$ in matrix form.
(Hint: $\begin{bmatrix} ? & ? \\ ? & ? \end{bmatrix}\begin{bmatrix} x \\ y \end{bmatrix} = \begin{bmatrix} 4 \\ 1 \end{bmatrix}$.) $\begin{bmatrix} 3 & 7 \\ 2 & 5 \end{bmatrix}\begin{bmatrix} x \\ y \end{bmatrix} = \begin{bmatrix} 4 \\ 1 \end{bmatrix}$

b. Let $A = \begin{bmatrix} 3 & 7 \\ 2 & 5 \end{bmatrix}$. Find A^{-1}. Find the product on each side of the equation $A^{-1}\begin{bmatrix} 3 & 7 \\ 2 & 5 \end{bmatrix}\begin{bmatrix} x \\ y \end{bmatrix} = A^{-1}\begin{bmatrix} 4 \\ 1 \end{bmatrix}$. What is the resulting equation?

c. Explain the connection between parts **a** and **b** of this exercise.

69a. $\begin{bmatrix} 3 & 7 \\ 2 & 5 \end{bmatrix}\begin{bmatrix} x \\ y \end{bmatrix} = \begin{bmatrix} 4 \\ 1 \end{bmatrix}$

b. $\begin{bmatrix} 5 & -7 \\ -2 & 3 \end{bmatrix}$; $\begin{bmatrix} x \\ y \end{bmatrix} = \begin{bmatrix} 13 \\ -5 \end{bmatrix}$

c. Part b gives the solution to the system in part a.

Look Beyond

In Exercise 69, students are guided step-by-step through an exploration in which they see how matrices can represent a system of linear equations and its solution. Students will learn how to use matrices to solve systems of linear equations in Lesson 4.4. You may wish to use this exercise as a transition to that lesson.

40. $\begin{bmatrix} 1 & 1 \\ 0.1 & 0.2 \end{bmatrix}$

41. $\begin{bmatrix} -0.54 & 0.23 & -0.08 \\ 0.77 & -0.62 & 0.54 \\ 0.69 & -0.15 & 0.38 \end{bmatrix}$

42. $\begin{bmatrix} -7 & -5 & 2.5 \\ -6 & -4 & 2.5 \\ 3 & 2 & -1 \end{bmatrix}$

43. $\begin{bmatrix} 0.27 & -0.05 & 0.10 \\ 0.02 & 0.13 & -0.10 \\ -0.12 & 0.12 & 0.12 \end{bmatrix}$

44. $\begin{bmatrix} 0.12 & 0.18 & 0.09 \\ 0.11 & -0.33 & -0.23 \\ 0.07 & 0.11 & -0.15 \end{bmatrix}$

The answers to Exercises 47 and 66–68 can be found in Additional Answers beginning on page 1002.

Student Technology Guide

NAME ______________ CLASS ________ DATE ________

Student Technology Guide
4.3 The Inverse of a Matrix

If a matrix has an inverse, that inverse can easily be found by using a graphics calculator. You can also use a graphics calculator to compute determinants.

Let $A = \begin{bmatrix} 6 & -4 \\ 3 & 1 \end{bmatrix}$. Suppose that you want to find A^{-1}.

- To enter A, press MATRX EDIT 1:[A] ENTER 2 ENTER 2 ENTER 6 ENTER (−) 4 ENTER 3 ENTER 1 ENTER.
- Press 2nd MODE to exit matrix edit mode.
- To find A^{-1}, press MATRX NAMES 1:[A] ENTER x^{-1} ENTER.
- To view A^{-1} in fractional form, press MATH MATH 1:▶Frac ENTER ENTER.

```
[A]-1
[[.0555555556  ...
 [-.1666666667 ...
Ans▶Frac
    [[1/18  2/9]
     [-1/6  1/3]]
```

From the display at the right above, $A^{-1} = \begin{bmatrix} \frac{1}{18} & \frac{2}{9} \\ -\frac{1}{6} & \frac{1}{3} \end{bmatrix}$.

Let $B = \begin{bmatrix} 3 & -5 \\ -9 & 15 \end{bmatrix}$. When you enter matrix B and try to find B^{-1}, you will get an error message. This happens because the determinant of matrix B equals 0. To find the determinant of matrix B, enter the dimensions and elements of B, and then press MATRX MATH 1:det(ENTER MATRX NAMES 2:[B] ENTER) ENTER. Since the determinant of matrix B equals 0, you do not need to attempt to find B^{-1}.

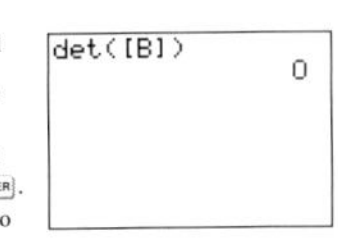

Use a graphics calculator to find the inverse matrix, if it exists. If the inverse matrix does not exist, write *no inverse*.

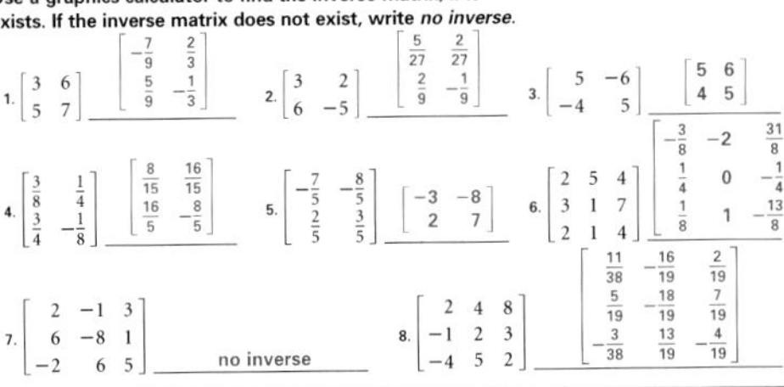

1. $\begin{bmatrix} 3 & 6 \\ 5 & 7 \end{bmatrix}$ $\begin{bmatrix} -\frac{7}{9} & \frac{2}{3} \\ \frac{5}{9} & -\frac{1}{3} \end{bmatrix}$

2. $\begin{bmatrix} 3 & 2 \\ 6 & -5 \end{bmatrix}$ $\begin{bmatrix} \frac{5}{27} & \frac{2}{27} \\ \frac{2}{9} & -\frac{1}{9} \end{bmatrix}$

3. $\begin{bmatrix} 5 & -6 \\ -4 & 5 \end{bmatrix}$ $\begin{bmatrix} 5 & 6 \\ 4 & 5 \end{bmatrix}$

4. $\begin{bmatrix} \frac{3}{8} & \frac{1}{4} \\ \frac{3}{4} & -\frac{1}{8} \end{bmatrix}$ $\begin{bmatrix} \frac{8}{15} & \frac{16}{15} \\ \frac{16}{5} & -\frac{8}{5} \end{bmatrix}$

5. $\begin{bmatrix} -\frac{7}{5} & -\frac{8}{5} \\ \frac{2}{5} & \frac{3}{5} \end{bmatrix}$ $\begin{bmatrix} -3 & -8 \\ 2 & 7 \end{bmatrix}$

6. $\begin{bmatrix} 2 & 5 & 4 \\ 3 & 1 & 7 \\ 2 & 1 & 4 \end{bmatrix}$ $\begin{bmatrix} -\frac{3}{8} & -2 & \frac{31}{8} \\ \frac{1}{4} & 0 & -\frac{1}{4} \\ \frac{1}{8} & 1 & -\frac{13}{8} \end{bmatrix}$

7. $\begin{bmatrix} 2 & -1 & 3 \\ 6 & -8 & 1 \\ -2 & 6 & 5 \end{bmatrix}$ no inverse

8. $\begin{bmatrix} 2 & 4 & 8 \\ -1 & 2 & 3 \\ -4 & 5 & 2 \end{bmatrix}$ $\begin{bmatrix} \frac{11}{38} & -\frac{16}{19} & \frac{2}{19} \\ \frac{5}{19} & -\frac{18}{19} & \frac{7}{19} \\ -\frac{3}{38} & \frac{13}{19} & -\frac{4}{19} \end{bmatrix}$

46a. $x = \frac{d}{ad - bc}$; $y = \frac{-c}{ad - bc}$; $u = \frac{-b}{ad - bc}$; $v = \frac{a}{ad - bc}$

b. $\begin{bmatrix} \frac{d}{ad - bc} & \frac{-b}{ad - bc} \\ \frac{-c}{ad - bc} & \frac{a}{ad - bc} \end{bmatrix} = \frac{1}{ad - bc}\begin{bmatrix} d & -b \\ -c & a \end{bmatrix}$; it is the same as above.

48. $\begin{bmatrix} 65 & 120 & 173 & 85 \\ 39 & 75 & 108 & 55 \end{bmatrix} \to$ 65 | 120 | 173 | 85 | 39 | 75 | 108 | 55

49. $\begin{bmatrix} 82 & 70 & 59 & 80 & 24 \\ 52 & 45 & 39 & 52 & 16 \end{bmatrix} \to$ 82 | 70 | 59 | 80 | 24 | 52 | 45 | 39 | 52 | 16

50. $\begin{bmatrix} 30 & 32 & 102 & 60 \\ 18 & 21 & 64 & 36 \end{bmatrix} \to$ 30 | 32 | 102 | 60 | 18 | 21 | 64 | 36

51. $\begin{bmatrix} 15 & 43 & 32 & 149 & 40 \\ 9 & 27 & 21 & 93 & 25 \end{bmatrix} \to$ 15 | 43 | 32 | 149 | 40 | 9 | 27 | 21 | 93 | 25

Focus

The factoring of a 155-digit number was a surprising mathematical feat that raised questions about the security of modern coding systems. To see how such systems work, students use small numbers to model the encoding and decoding of a secret message. This activity demonstrates the importance of number theory in security codes that are used to restrict access to important information.

Motivate

Before students begin the article, have them discuss codes. They might share their experiences with codes that they have seen. Ask them how they think codes are used in real life.

Have students read the article. Then point out some of the advantages, such as the two given below, of the coding system that is described.

- This type of coding system can be used by many people. Even if you know how the system works, you cannot decode a message unless you know the factors of p, which are numerous for large numbers.
- This type of system allows for a *public key*. A person can publish the key (a value of p) to be used to encode messages. Anyone can send this person a coded message. However, this person is the only one who knows the factors of p, so this person is the only one who can decode those messages.

EYEWITNESS MATH

HOW SECRET IS SECRET?

Biggest Division a Giant Leap in Math

BY GINA KOLATA

In a mathematical feat that seemed impossible a year ago, a group of several hundred researchers using about 1,000 computers has broken a 155-digit number down into three smaller numbers that cannot be further divided.

The latest finding could be the first serious threat to systems used by banks and other organizations to encode secret data before transmission, cryptography experts said yesterday.

These systems are based on huge numbers that cannot be easily factored, or divided into two numbers that cannot be divided further.

In 1977, a group of three mathematicians devised a way of making secret codes that involves scrambling messages according to a mathematical formula based on factoring. Now, such codes are used in banking, for secure telephone lines and by the Defense Department.

In making these codes, engineers have to strike a delicate balance when they select the numbers used to scramble messages. If they choose a number that is easy to factor, the code can be broken. If they make the number much larger, and much harder to factor, it takes much longer for the calculations used to scramble a message.

For most applications outside the realm of national security, cryptographers have settled on numbers that are about 150 digits long.

Dr. Mark Manasse of the Digital Equipment Corporation's Systems Research Center in Palo Alto, Calif., calculates that if a computer could perform a billion divisions a second, it would take 10 to the 60th years, or 1 with 60 zeros after it, to factor the number simply by trying out every smaller number that might divide into it easily. But with a newly discovered factoring method and with a world-wide collaborative effort, the number was cracked in a few months.

Connection Machine® is one of the most powerful high-performance computers in the world.

Factoring a 155-Digit Number:

13,407,807,929,942, 597,099,574,024,998, 205,846,127,479,365, 820,592,393,377,723, 561,443,721,764,030, 073,546,976,801,874, 298,166,903,427,690, 031,858,186,486,050, 853,753,882,811,946, 569,946,433,649,006, 084,097

equals

2,424,833

times

7,455,602,825,647, 884,208,337,395,736, 200,454,918,783,366, 342,657

times

741,640,062,627,530, 801,524,787,141,901, 937,474,059,940,781, 097,519,023,905,821, 306,144,415,759,504, 705,008,092,818,711, 693,940,737

[*Source:* New York Times, *June 20, 1990*]

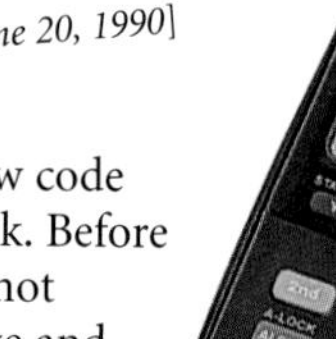

You can use small numbers to get an idea of how code systems like the one described in the article work. Before you begin, you need a key (a number, p, that is not prime). You will use p to encode a secret message and then use p and its factors to decode the message.

X secret message → **Sender uses p to encode the secret message.** → **Y** coded message → **Receiver uses p and its factors to decode the message.** → **X** secret message

Cooperative Learning

For this activity, use 55 for p. You will use a special algorithm to encode and decode the message. To keep things simple, let your secret message, X, be a two-digit number between 11 and 50.

1. To encode a secret message, X, follow the steps below.

a. Calculate X^3. — Let X be 42. $\rightarrow X^3 = 74{,}088$

b. Divide X^3 by 55. Multiply the whole-number part of the quotient by 55. Subtract that product from X^3. The difference is the remainder. Use the remainder as the coded message Y.

$74{,}088 \div 55 = 1347.054545\ldots$

$1347 \times 55 = 74{,}085$

$74{,}088 - 74{,}085 = 3 \rightarrow Y = 3$

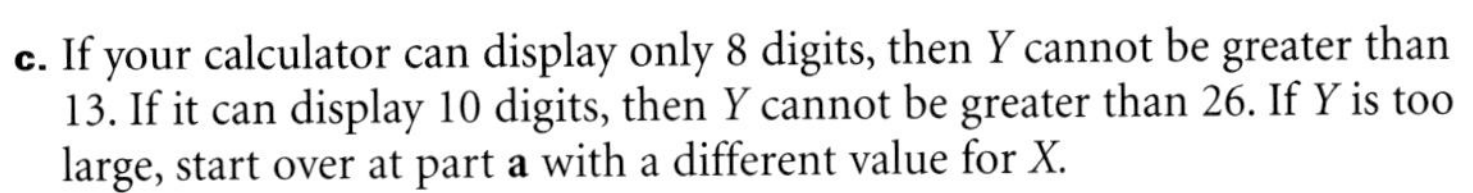

c. If your calculator can display only 8 digits, then Y cannot be greater than 13. If it can display 10 digits, then Y cannot be greater than 26. If Y is too large, start over at part **a** with a different value for X.

2. To decode your secret message, use the factors of 55. Follow these steps.

a. Find Y^3. Divide Y^3 by 5. Multiply the whole-number part of the quotient by 5, and subtract this product from Y^3. Call the remainder a.

Y is 3. $\rightarrow Y^3 = 27$

$27 \div 5 = 5.4$

$5 \times 5 = 25$

$27 - 25 = 2 \rightarrow a = 3$

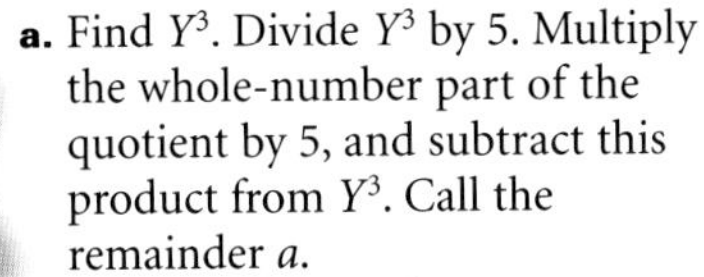

b. Find Y^7. Divide Y^7 by 11. Multiply the whole-number part of the quotient by 11, and subtract this product from Y^7. Call the remainder b.

Y is 3. $\rightarrow Y^7 = 2187$

$2187 \div 11 = 198.8181\ldots$

$198 \times 11 = 2178$

$2187 - 2178 = 9 \rightarrow b = 9$

c. Evaluate $11a + 45b$, and divide the result by 55. Multiply the whole-number part of the quotient by 55, and subtract this product from the value of $11a + 45b$. How does the remainder compare with X, the secret message?

$11a + 45b = 11(3) + 45(9) = 427$

$427 \div 55 = 7.736\ldots$

$7 \times 55 = 385$

$427 - 385 = 42$

3. Explain how you used the factors of p to encode or decode the message. Why would it be important for the factors of p to be secret?

4. How do you think coding systems may be affected if efficient methods for factoring very large numbers are found?

The enigma machine was used in World War II for breaking coded messages.

1. Check students' work.

2. Check students' work. The remainder after the last step should be the original "message."

3. Answers may vary. Sample answer: The factors are essential for decoding the message. Someone who could factor p would be able to decode a secret message not intended for him or her. This could have serious consequences if the message contained sensitive information. It could jeopardize a person's or a country's security or cause large amounts of money to be lost.

4. Answers may vary.

Cooperative Learning

You may wish to have students work in groups of two or three. Each group can work together on Step 1 and Step 2, studying the examples and discussing how the encoding and decoding processes work.

For Step 1, each group should code a short message and give it to another group. Remind them to use numbers whose values for Y do not exceed the capabilities of their calculators or spreadsheet software. In Step 2, each group should decode the message they receive. Have the groups complete Steps 3 and 4 and then discuss their findings with the entire class.

Discuss

Have groups report any problems that they encountered in encoding or decoding messages and describe how they resolved the problems. If the group was unable to resolve a problem, discuss it as class.

If you wish, you can use the activity on this page to introduce the terms and notation of modular arithmetic. For example, *mod 55* denotes a counting system in which you go back to 0 each time you reach 55. In this system, the number 56 is equivalent to the number 1, which is written as $56 \equiv 1 \pmod{55}$. Note that when you divide 56 by 55, the remainder is 1, so modular notation provides a way to describe classes of numbers that yield the same remainder.

You might want to suggest that students research the role of codes in literature, such as in *The Gold Bug* by Edgar Allen Poe or *The Adventure of the Dancing Men* by Sir Arthur Conan Doyle.

Prepare

NCTM PRINCIPLES & STANDARDS 1–4, 6–10

QUICK WARM-UP

Find each product.

1. $\begin{bmatrix} 4 & 1 \\ -6 & 9 \end{bmatrix}\begin{bmatrix} 0 & 12 \\ -6 & -5 \end{bmatrix}$

$\begin{bmatrix} -6 & 43 \\ -54 & -117 \end{bmatrix}$

2. $\begin{bmatrix} 0 & -4 \\ 2 & -2 \end{bmatrix}^{-1}\begin{bmatrix} 8 & 16 \\ 36 & -20 \end{bmatrix}$

$\begin{bmatrix} 16 & -14 \\ -2 & -4 \end{bmatrix}$

Also on Quiz Transparency 4.4

Teach

Why Review with students the techniques that they have learned for solving systems of equations: graphing, substitution, and elimination. Elicit their perceptions of the advantages and disadvantages of each method. Tell them that in this lesson, they will learn the matrix method, which also has advantages and disadvantages.

CHECKPOINT ✔

$\begin{cases} x + y = 100{,}000 \\ 0.05x + 0.14y = 10{,}000 \end{cases}$

$\begin{bmatrix} 1 & 1 \\ 0.05 & 0.14 \end{bmatrix}\begin{bmatrix} x \\ y \end{bmatrix} = \begin{bmatrix} 100{,}000 \\ 10{,}000 \end{bmatrix}$

Solving Systems With Matrix Equations

Objective

- Use matrices to solve systems of linear equations in mathematical and real-world situations.

Why *Many real-world situations, such as investment options, can be represented by a system of linear equations that can be solved quite efficiently by solving a matrix equation.*

APPLICATION INVESTMENTS

A financial manager wants to invest \$50,000 for a client by putting some of the money in a low-risk investment that earns 5% per year and some of the money in a high-risk investment that earns 14% per year. How much money should be invested at each interest rate to earn \$5000 in interest per year?
You will answer this question in Example 1.

PROBLEM SOLVING

Let x represent the amount invested at 5%, and let y represent the amount invested at 14%. **Write a system of linear equations** to represent the situation.

$$\begin{cases} x + y = 50{,}000 \\ 0.05x + 0.14y = 5000 \end{cases}$$

The system $\begin{cases} x + y = 50{,}000 \\ 0.05x + 0.14y = 5000 \end{cases}$ can be written as a **matrix equation**, $AX = B$.

Coefficient matrix, A	Variable matrix, X	Constant matrix, B

Let $A = \begin{bmatrix} 1 & 1 \\ 0.05 & 0.14 \end{bmatrix}$, $X = \begin{bmatrix} x \\ y \end{bmatrix}$, and $B = \begin{bmatrix} 50{,}000 \\ 5000 \end{bmatrix}$.

Then $AX = B$ is $\begin{bmatrix} 1 & 1 \\ 0.05 & 0.14 \end{bmatrix}\begin{bmatrix} x \\ y \end{bmatrix} = \begin{bmatrix} 50{,}000 \\ 5000 \end{bmatrix}$.

CHECKPOINT ✔ Use the interest rates mentioned above. If the client wants to invest \$100,000 and earn \$10,000 in interest, what system of linear equations and corresponding matrix equation represent this investment?

Alternative Teaching Strategy

WORKING BACKWARD Display the system at right on the board or overhead. Have students graph the equations and identify the point where the lines intersect. (1, −2) Remind them that the coordinates of the point of intersection represent the solution of the system. Tell them to write the matrix that represents the point.

$$\begin{cases} x - 3y = 7 \\ x - y = 3 \end{cases}$$

$\begin{bmatrix} 1 \\ -2 \end{bmatrix}$

Now ask students to perform the following matrix multiplication:

$\begin{bmatrix} 1 & -3 \\ 1 & -1 \end{bmatrix}\begin{bmatrix} 1 \\ -2 \end{bmatrix}$ $\begin{bmatrix} 7 \\ 3 \end{bmatrix}$

Display the multiplication and the product on the board or overhead near the system. Elicit the students' observations about the relationship between the system and the multiplication. Use their observations to initiate a presentation of the matrix method for solving systems of equations.

Solving a matrix equation of the form $AX = B$, where $X = \begin{bmatrix} x \\ y \end{bmatrix}$, is similar to solving a linear equation of the form $ax = b$, where a, b, and x are real numbers and $a \neq 0$.

Real Numbers	Matrices
$ax = b$	$AX = B$
$\frac{1}{a}(ax) = \frac{1}{a}(b)$	$A^{-1}(AX) = A^{-1}B$
$\left(\frac{1}{a} \cdot a\right)x = \frac{b}{a}$	$(A^{-1}A)\,X = A^{-1}B$
	$IX = A^{-1}B$
$x = \frac{b}{a}$	$X = A^{-1}B$

Just as $\frac{1}{a}$ must exist in order to solve $ax = b$ (where $a \neq 0$), A^{-1} must exist to solve $AX = B$.

CHECKPOINT ✔ When solving the matrix equation $AX = B$, does it matter whether you calculate $A^{-1}B$ or BA^{-1}? Explain.

EXAMPLE 1

APPLICATION
INVESTMENTS

Refer to the investment options described at the beginning of the lesson.

How much money should the manager invest at each interest rate to earn $5000 in interest per year?

SOLUTION

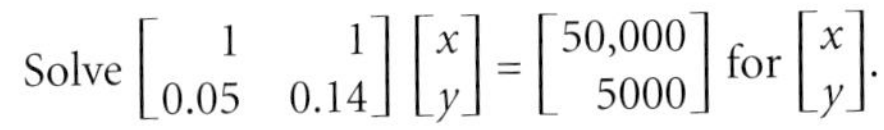

Solve $\begin{bmatrix} 1 & 1 \\ 0.05 & 0.14 \end{bmatrix} \begin{bmatrix} x \\ y \end{bmatrix} = \begin{bmatrix} 50{,}000 \\ 5000 \end{bmatrix}$ for $\begin{bmatrix} x \\ y \end{bmatrix}$.

TECHNOLOGY
GRAPHICS CALCULATOR
Keystroke Guide, page 270

Enter the coefficient matrix, $A = \begin{bmatrix} 1 & 1 \\ 0.05 & 0.14 \end{bmatrix}$, and the constant matrix, $B = \begin{bmatrix} 50{,}000 \\ 5000 \end{bmatrix}$, into your calculator. Solve for the variable matrix, X, by finding the product $A^{-1}B$.

$$X = A^{-1}B$$

$$\begin{bmatrix} x \\ y \end{bmatrix} = \begin{bmatrix} \$22{,}222.22 \\ \$27{,}777.78 \end{bmatrix}$$

The manager should invest $22,222.22 at 5% and $27,777.78 at 14% to achieve the earned interest goal of $5000 per year.

TRY THIS How much money should the manager invest at each interest rate to earn $4000 in interest per year?

CRITICAL THINKING Suppose that the earned interest goal is $10,000 per year for an investment of $50,000. What happens when you try to find how much to invest at 5% and how much to invest at 14%? Explain your response.

CHECKPOINT ✔
Yes, order matters because matrix multiplication is not commutative.

ADDITIONAL EXAMPLE 1

Amy has saved $5000 during the past three summers. Now she wants to put some of the money into an investment that earns 4% per year and some into an investment that earns 11% per year. **How much money should Amy invest at each interest rate to earn $500 in interest per year?**

$714.29 at 4% and $4285.71 at 11%

Teaching Tip

TECHNOLOGY If students get an error message when performing the matrix multiplication for Example 1, check that they have entered the multiplication in the correct order. Remind them that matrix multiplication is not commutative, so the order that they must use is $A^{-1}B$.

TRY THIS
$33,333.33 at 5% and $16,666.67 at 14%

CRITICAL THINKING
If the interest goal is $10,000, the answer contains a negative entry. It is not possible to earn that much interest on $50,000 in one year at those rates.

Interdisciplinary Connection

HEALTH Whole milk is about 3.25% butterfat, and skim milk is about 1% butterfat. How many ounces of whole milk and skim milk would you have to combine in order to make 1 gallon (128 ounces) of a "low fat" milk that is about 2% butterfat?
whole milk: about 51 oz; skim milk: about 77 oz

Inclusion Strategies

COOPERATIVE LEARNING Some students interpret any calculator error message as an indication that they have made a mistake. For these students, the error message that is received when working with the matrices for an inconsistent or dependent system may be disturbing. You may wish to present Example 3 by having students work in small groups, with each student entering the matrices and performing the multiplication on his or her own calculator. When students compare and discuss their results, it will be clear that the error message is appropriate.

ADDITIONAL EXAMPLE 2

Refer to the system of equations below.

$$\begin{cases} -3x + 6y - 4z = 8 \\ x - 4y + 2z = -3 \\ 8y - z = 0 \end{cases}$$

a. Write the system as a matrix equation.

$$\begin{bmatrix} -3 & 6 & -4 \\ 1 & -4 & 2 \\ 0 & 8 & -1 \end{bmatrix}\begin{bmatrix} x \\ y \\ z \end{bmatrix} = \begin{bmatrix} 8 \\ -3 \\ 0 \end{bmatrix}$$

b. Solve the matrix equation.

$$\begin{bmatrix} x \\ y \\ z \end{bmatrix} = \begin{bmatrix} -1.8 \\ -0.1 \\ -0.8 \end{bmatrix}$$

Thus, the solution of the system is $x = -1.8$, $y = -0.1$, and $z = -0.8$.

TRY THIS

a.
$$\begin{aligned} 4x - 2y + z &= 3 \\ 2x - y + 3z &= -6 \\ -2x + 3y - 2z &= 1 \end{aligned}$$

b.
$$\begin{bmatrix} 4 & -2 & 1 \\ 2 & -1 & 3 \\ -2 & 3 & -2 \end{bmatrix}\begin{bmatrix} x \\ y \\ z \end{bmatrix} = \begin{bmatrix} 3 \\ -6 \\ 1 \end{bmatrix}$$

$$\begin{bmatrix} x \\ y \\ z \end{bmatrix} = \begin{bmatrix} 1 \\ -1 \\ 3 \end{bmatrix}$$

Activity Notes

In this Activity, students explore the effects on the graph of a system when its constant matrix varies. They should discover that as the constant matrix changes, the y-intercepts of the lines change, but the slopes of the lines remain the same. This means that if the system has a unique solution for any one constant matrix, it will have a unique solution for all others.

CHECKPOINT ✔

6. Answers may vary. The system has a unique solution. The solution depends on the values of r and s.

Just as you can use a 2×2 matrix and its inverse to solve a system of two linear equations in two variables, you can use a 3×3 matrix and its inverse to solve a system of three equations in three variables, as shown in Example 2.

EXAMPLE 2 **Refer to the system of equations at right.**

a. Write the system as a matrix equation.
b. Solve the matrix equation.

$$\begin{cases} 2y - z = -7 - 5x \\ x - 2y + 2z = 0 \\ 3y = 17 - z \end{cases}$$

SOLUTION

a. First write the equations of the system in *standard form.*

In the third equation, use 0 as the coefficient of x.

$$\begin{cases} 5x + 2y - z = -7 \\ x - 2y + 2z = 0 \\ 3y + z = 17 \end{cases} \rightarrow \begin{bmatrix} 5 & 2 & -1 \\ 1 & -2 & 2 \\ 0 & 3 & 1 \end{bmatrix}\begin{bmatrix} x \\ y \\ z \end{bmatrix} = \begin{bmatrix} -7 \\ 0 \\ 17 \end{bmatrix}$$

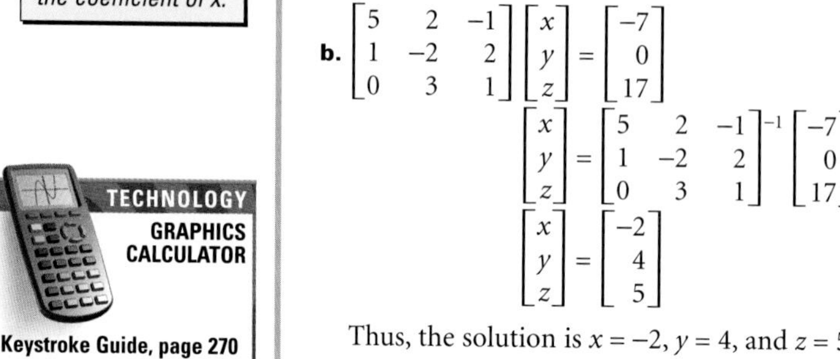

b.
$$\begin{bmatrix} 5 & 2 & -1 \\ 1 & -2 & 2 \\ 0 & 3 & 1 \end{bmatrix}\begin{bmatrix} x \\ y \\ z \end{bmatrix} = \begin{bmatrix} -7 \\ 0 \\ 17 \end{bmatrix}$$

$$\begin{bmatrix} x \\ y \\ z \end{bmatrix} = \begin{bmatrix} 5 & 2 & -1 \\ 1 & -2 & 2 \\ 0 & 3 & 1 \end{bmatrix}^{-1}\begin{bmatrix} -7 \\ 0 \\ 17 \end{bmatrix}$$

$$\begin{bmatrix} x \\ y \\ z \end{bmatrix} = \begin{bmatrix} -2 \\ 4 \\ 5 \end{bmatrix}$$

TECHNOLOGY GRAPHICS CALCULATOR
Keystroke Guide, page 270

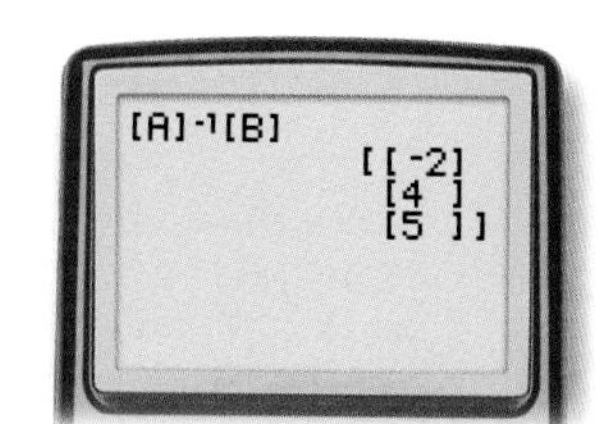

Thus, the solution is $x = -2$, $y = 4$, and $z = 5$.

TRY THIS Refer to the system at right.

a. Write the system as a matrix equation.
b. Solve the matrix equation.

$$\begin{cases} 2y - z = 4x - 3 \\ 2x + 3z = y - 6 \\ 3y - 1 = 2x + 2z \end{cases}$$

Activity
Exploring Slopes and Solutions

TECHNOLOGY GRAPHICS CALCULATOR
Keystroke Guide, page 270

You will need: a graphics calculator

Let $A = \begin{bmatrix} 4 & 9 \\ 2 & 5 \end{bmatrix}$ be the coefficient matrix for the system $\begin{cases} 4x + 9y = r \\ 2x + 5y = s \end{cases}$.

1. Find the determinant of matrix A. Does the matrix have an inverse? Justify your response.
2. Choose values for r and s. Write a matrix equation and use the inverse of matrix A to find x and y.
3. Find the slopes of the lines represented by the equations with your values for r and s. Do the slopes depend on the values of r and s?
4. Graph these lines. Based on the slopes and graphs, what can you conclude about any solution(s) of the system?

PROBLEM SOLVING

5. **Guess and check.** Choose other values for r and s, and graph the new equation. Do you think that this system will have a unique solution regardless of the values for r and s?

CHECKPOINT ✔

6. Summarize what you know about the solutions of this system.

Enrichment

Identify two systems of linear equations in two variables that can be solved by using the matrix equation above.

$$\begin{bmatrix} 1 & -2 \\ 2 & -1 \end{bmatrix}\begin{bmatrix} x & w \\ y & z \end{bmatrix} = \begin{bmatrix} 6 & 7 \\ 9 & 5 \end{bmatrix}$$

$x - 2y = 6$ $\quad w - 2z = 7$
$2x - y = 9$ $\quad 2w - z = 5$

Give the solution of the system.

$$\begin{bmatrix} x & w \\ y & z \end{bmatrix} = \begin{bmatrix} 1 & -2 \\ 2 & -1 \end{bmatrix}^{-1}\begin{bmatrix} 6 & 7 \\ 9 & 5 \end{bmatrix} = \begin{bmatrix} 4 & 1 \\ -1 & -3 \end{bmatrix}$$

So, $x = 4$, $y = -1$, $w = 1$, and $z = -3$.

Now write a matrix equation that can be used to solve *three* systems of two linear equations in two variables. Give the solution of each system.

sample: $\begin{bmatrix} 2 & -1 \\ 3 & -2 \end{bmatrix}\begin{bmatrix} x & w & u \\ y & z & v \end{bmatrix} = \begin{bmatrix} 0 & 2 & -5 \\ -1 & 4 & -9 \end{bmatrix}$

$$\begin{bmatrix} x & w & u \\ y & z & v \end{bmatrix} = \begin{bmatrix} 2 & -1 \\ 3 & -2 \end{bmatrix}^{-1}\begin{bmatrix} 0 & 2 & -5 \\ -1 & 4 & -9 \end{bmatrix} = \begin{bmatrix} 1 & 0 & -1 \\ 2 & -2 & 3 \end{bmatrix}$$

So, $x = 1$, $y = 2$, $w = 0$, $z = -2$, $u = -1$, and $v = 3$.

As you have learned, not all systems of linear equations have solutions. In a matrix equation of the form $AX = B$, if the coefficient matrix, A, does not have an inverse, then the system represented by the matrix equation does not have a unique solution. This is shown in Example 3.

EXAMPLE 3 **Solve $\begin{cases} -3x + 4y = 3 \\ -6x + 8y = 18 \end{cases}$, if possible, by using a matrix equation. If not possible, classify the system.**

SOLUTION

The given system is represented by $\underset{A}{\begin{bmatrix} -3 & 4 \\ -6 & 8 \end{bmatrix}} \underset{X}{\begin{bmatrix} x \\ y \end{bmatrix}} = \underset{B}{\begin{bmatrix} 3 \\ 18 \end{bmatrix}}$.

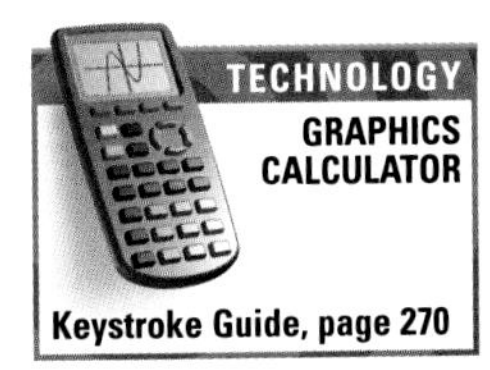

When you try to find $A^{-1}B$ on a graphics calculator, you will get an error. The inverse of A does not exist. Notice that $\det(A) = (-3)(8) - (4)(-6) = 0$, which also indicates that the inverse of A does not exist.

Therefore, there is no *unique* solution for the system, and the system is either dependent or inconsistent.

In this case, the system is inconsistent because the lines are parallel.

$$\begin{cases} -3x + 4y = 3 \\ -6x + 8y = 18 \end{cases} \rightarrow \begin{cases} y = \frac{3}{4}x + \frac{3}{4} \\ y = \frac{3}{4}x + 3 \end{cases}$$

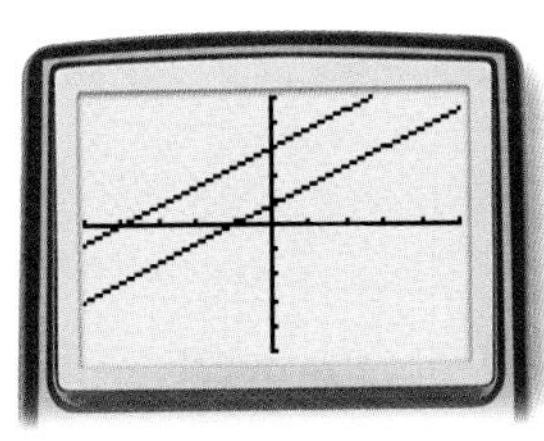

TRY THIS Solve $\begin{bmatrix} -3 & 4 \\ -6 & 8 \end{bmatrix} \begin{bmatrix} x \\ y \end{bmatrix} = \begin{bmatrix} 3 \\ 6 \end{bmatrix}$, if possible, by using a matrix equation. If not possible, classify the system.

CRITICAL THINKING Consider the system $\begin{cases} ax + by = e \\ (na)x + (nb)y = f \end{cases}$, where b and n are nonzero. Find the slope of each equation. Find the determinant of the coefficient matrix. How are the slopes and the determinant related?

Reteaching the Lesson

COOPERATIVE LEARNING

Have students work in groups of two or three. $\begin{bmatrix} 3 & n \\ -2 & 4 \end{bmatrix} \begin{bmatrix} x \\ y \end{bmatrix} = \begin{bmatrix} a \\ b \end{bmatrix}$ Display the matrix equation above on the board or overhead. Tell students that the matrix equation represents a system of two linear equations in two variables. Ask them to find values of n, a, and b so that the system satisfies each condition given at right. When appropriate, they should give the solution of the system.

Sample answers are given.

1. The system has a unique solution.
   $n = -1, a = 3, b = 8; x = 2, y = 3$
2. The system has no solution.
   $n = -6, a = 12, b = 12$
3. The system has infinitely many solutions.
   $n = -6, a = 12, b = -8$; all ordered pairs (x, y) that satisfy the equation $y = 0.5x - 2$

If students need additional practice, repeat the activity with a similar matrix equation.

ADDITIONAL EXAMPLE 3

Solve $\begin{cases} 9x - 3y = 27 \\ -6x + 2y = 18 \end{cases}$, if possible, by using a matrix equation. If not possible, classify the system.

The system is dependent. There is no unique solution.

Teaching Tip

TECHNOLOGY If you are using a TI-82 or TI-83 graphics calculator in Example 3, the error message gives the choice **1: Quit** or **2: Goto**. The first choice simply returns you to the home screen, whereas the second takes you to the exact place where the error occurred. In this case, if you choose **2: Goto**, the cursor will flash on the exponent, −1, to indicate that no inverse could be calculated for matrix A.

TRY THIS

No inverse; the system is dependent.

CRITICAL THINKING

$-\frac{a}{b}, -\frac{a}{b}$; $\det\begin{bmatrix} a & b \\ na & nb \end{bmatrix} = 0$

Parallel lines result in a coefficient matrix with a determinant of 0. Therefore, the system has no unique solution.

Assess

Selected Answers

Exercises 7–10, 11–41 odd

ASSIGNMENT GUIDE	
In Class	1–10
Core	11–29 odd, 33, 35
Core Plus	11–31, 32–36 even
Review	37–42
Preview	43–46

Extra Practice can be found beginning on page 940.

11. $\begin{bmatrix} 3 & -5 \\ 2 & 1 \end{bmatrix}\begin{bmatrix} x \\ y \end{bmatrix} = \begin{bmatrix} 1 \\ -2 \end{bmatrix}$

12. $\begin{bmatrix} -3 & 1 \\ 6 & -12 \end{bmatrix}\begin{bmatrix} x \\ y \end{bmatrix} = \begin{bmatrix} -3 \\ 6 \end{bmatrix}$

13. $\begin{bmatrix} 2 & 4 \\ 1 & -1 \end{bmatrix}\begin{bmatrix} a \\ b \end{bmatrix} = \begin{bmatrix} -3 \\ 9 \end{bmatrix}$

14. $\begin{bmatrix} 4 & 1 & -2 \\ 3 & 0 & 5 \\ 8 & 3 & -1 \end{bmatrix}\begin{bmatrix} x \\ y \\ z \end{bmatrix} = \begin{bmatrix} 10 \\ 14 \\ 23 \end{bmatrix}$

15. $\begin{bmatrix} 0 & 1 & 5 \\ -2 & 3 & -1 \\ 6 & 0 & -3 \end{bmatrix}\begin{bmatrix} x \\ y \\ z \end{bmatrix} = \begin{bmatrix} -14 \\ 2 \\ 21 \end{bmatrix}$

Practice

NAME ____________ CLASS ________ DATE ________

Practice

4.4 Solving Systems With Matrix Equations

Write the matrix equation that represents each system.

1. $\begin{cases} 3x + y - z = -19 \\ -x - y + 3z = 21 \\ 2x + 2y + z = -7 \end{cases}$ $\begin{bmatrix} 3 & 1 & -1 \\ -1 & -1 & 3 \\ 2 & 2 & 1 \end{bmatrix}\begin{bmatrix} x \\ y \\ z \end{bmatrix} = \begin{bmatrix} -19 \\ 21 \\ -7 \end{bmatrix}$

2. $\begin{cases} 5x - 2y + z = 13 \\ -x + 4y - z = -1 \\ 4x - 8y + 3z = 6 \end{cases}$ $\begin{bmatrix} 5 & -2 & 1 \\ -1 & 4 & -1 \\ 4 & -8 & 3 \end{bmatrix}\begin{bmatrix} x \\ y \\ z \end{bmatrix} = \begin{bmatrix} 13 \\ -1 \\ 6 \end{bmatrix}$

3. $\begin{cases} 9x - 5y + z = 6 \\ 3x + y - z = 2 \\ 4x - 3y - 2x = -1 \end{cases}$ $\begin{bmatrix} 9 & -5 & 1 \\ 3 & 1 & -1 \\ 4 & -3 & -2 \end{bmatrix}\begin{bmatrix} x \\ y \\ z \end{bmatrix} = \begin{bmatrix} 6 \\ 2 \\ -1 \end{bmatrix}$

Write the system of equations represented by each matrix equation.

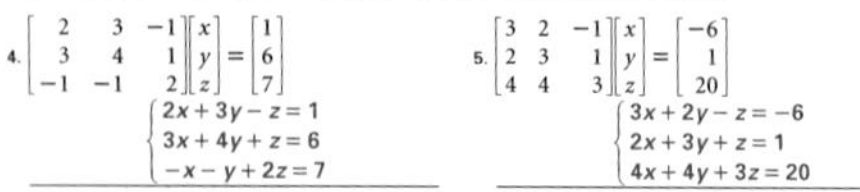

4. $\begin{bmatrix} 2 & 3 & -1 \\ 3 & 4 & 1 \\ -1 & -1 & 2 \end{bmatrix}\begin{bmatrix} x \\ y \\ z \end{bmatrix} = \begin{bmatrix} 1 \\ 6 \\ 7 \end{bmatrix}$ $\begin{cases} 2x + 3y - z = 1 \\ 3x + 4y + z = 6 \\ -x - y + 2z = 7 \end{cases}$

5. $\begin{bmatrix} 3 & 2 & -1 \\ 2 & 3 & 1 \\ 4 & 4 & 3 \end{bmatrix}\begin{bmatrix} x \\ y \\ z \end{bmatrix} = \begin{bmatrix} -6 \\ 1 \\ 20 \end{bmatrix}$ $\begin{cases} 3x + 2y - z = -6 \\ 2x + 3y + z = 1 \\ 4x + 4y + 3z = 20 \end{cases}$

Write the matrix equation that represents each system, and solve the system, if possible, by using a matrix equation.

6. $\begin{cases} 8x + 7y = 5 \\ 4x - 9y = 65 \end{cases}$ $\begin{bmatrix} 8 & 7 \\ 4 & -9 \end{bmatrix}\begin{bmatrix} x \\ y \end{bmatrix} = \begin{bmatrix} 5 \\ 65 \end{bmatrix}$; (5, −5)

7. $\begin{cases} 7x + 5y = 14 \\ 4x + 3y = 9 \end{cases}$ $\begin{bmatrix} 7 & 5 \\ 4 & 3 \end{bmatrix}\begin{bmatrix} x \\ y \end{bmatrix} = \begin{bmatrix} 14 \\ 9 \end{bmatrix}$; (−3, 7)

8. $\begin{cases} 3x - 7y = 25 \\ 5x - 8y = 27 \end{cases}$ $\begin{bmatrix} 3 & -7 \\ 5 & -8 \end{bmatrix}\begin{bmatrix} x \\ y \end{bmatrix} = \begin{bmatrix} 25 \\ 27 \end{bmatrix}$; (−1, −4)

9. $\begin{cases} 4x + y + z = 1 \\ 8x - 4y - 7z = 2 \\ 5y - 9z = 3 \end{cases}$ $\begin{bmatrix} 4 & 1 & 1 \\ 8 & -4 & -7 \\ 0 & 5 & 9 \end{bmatrix}\begin{bmatrix} x \\ y \\ z \end{bmatrix} = \begin{bmatrix} 1 \\ 2 \\ 3 \end{bmatrix}$; (0.5, −3, 2)

10. $\begin{cases} 3x - 3y + 5z = 13 \\ 5x + 6y - 2z = 10 \\ 7x + 5y = 18 \end{cases}$ $\begin{bmatrix} 3 & -3 & 5 \\ 5 & 6 & -2 \\ 7 & 5 & 0 \end{bmatrix}\begin{bmatrix} x \\ y \\ z \end{bmatrix} = \begin{bmatrix} 13 \\ 10 \\ 18 \end{bmatrix}$; (4, −2, −1)

11. $\begin{cases} x - 2y - 3z = 3 \\ 3x + y + z = 12 \\ 3x - 2y - 4z = 15 \end{cases}$ $\begin{bmatrix} 1 & -2 & -3 \\ 3 & 1 & 1 \\ 3 & -2 & -4 \end{bmatrix}\begin{bmatrix} x \\ y \\ z \end{bmatrix} = \begin{bmatrix} 3 \\ 12 \\ 15 \end{bmatrix}$; (3, 9, −6)

12. $\begin{cases} x - 2y + z = 15 \\ 3x + y - 2x = 8 \\ 5x - 10y + 5z = 21 \end{cases}$ $\begin{bmatrix} 1 & -2 & 1 \\ 3 & 1 & -2 \\ 5 & -10 & 5 \end{bmatrix}\begin{bmatrix} x \\ y \\ z \end{bmatrix} = \begin{bmatrix} 15 \\ 8 \\ 21 \end{bmatrix}$; no solution

13. $\begin{cases} 12x + 7y + z = -5 \\ 3x + 4y + 2z = 3 \\ 5x + 3y - 3z = 12 \end{cases}$ $\begin{bmatrix} 12 & 7 & 1 \\ 3 & 4 & 2 \\ 5 & 3 & -3 \end{bmatrix}\begin{bmatrix} x \\ y \\ z \end{bmatrix} = \begin{bmatrix} -5 \\ 3 \\ 12 \end{bmatrix}$; (−3, 5, −4)

14. $\begin{cases} 8x + y - z = 0 \\ 5x + 2y - 9z = -3 \\ 12x + y + 5z = 8 \end{cases}$ $\begin{bmatrix} 8 & 1 & -1 \\ 5 & 2 & -9 \\ 12 & 1 & 5 \end{bmatrix}\begin{bmatrix} x \\ y \\ z \end{bmatrix} = \begin{bmatrix} 0 \\ -3 \\ 8 \end{bmatrix}$; (−1, 10, 2)

Exercises

Communicate

1. Explain how to write a system of equations as a matrix equation.

2. Describe how to represent the system at right by using a matrix equation. $\begin{cases} x - y = 5 \\ -z + y = -6 \\ 2x - z = 2 \end{cases}$

3. Discuss how solving the matrix equation $AX = B$ is similar to solving the linear equation $ax = b$, where a, b, and x are real numbers and $a \neq 0$.

4. Describe the steps involved in using a matrix equation to solve a system of linear equations such as $\begin{cases} 2x - 5y = 0 \\ x + y = -2 \end{cases}$.

5. How can you verify that $\begin{bmatrix} -1 \\ -3 \\ 2 \end{bmatrix}$ is a solution of $\begin{bmatrix} 1 & 1 & 3 \\ 2 & -1 & -2 \\ 3 & 2 & -2 \end{bmatrix}\begin{bmatrix} x \\ y \\ z \end{bmatrix} = \begin{bmatrix} 2 \\ -3 \\ -13 \end{bmatrix}$?

6. If coefficient matrix A in the matrix equation $AX = B$ does not have an inverse, what do you know about the related system of equations?

Guided Skills Practice

APPLICATION

7 **INVESTMENTS** A total of $10,000 was invested in two certificates of deposit that earned 6% per year and 8% per year. If the investments earned $750 in interest each year, find the amount invested at each rate. ***(EXAMPLE 1)***
$2500 at 6%, $7500 at 8%

Write each system of equations as a matrix equation. Then solve the system, if possible, by using a matrix equation. If not possible, classify the system. ***(EXAMPLES 2 AND 3)***

8 $\begin{cases} x + y = 8 \\ 2x + y = 1 \end{cases}$ **(−7, 15)**

9 $\begin{cases} x + 3y = 7 \\ x + 3y = -2 \end{cases}$ **inconsistent**

10 $\begin{cases} 3x - 2y = 11 \\ 6x - 4y = 5 \end{cases}$ **inconsistent**

Practice and Apply

internet connect
Homework Help Online
Go To: **go.hrw.com**
Keyword: **MB1 Homework Help** for Exercises 11–16, 19–30

Write the matrix equation that represents each system.

11. $\begin{cases} 3x - 5y = 1 \\ 2x + y = -2 \end{cases}$

12. $\begin{cases} -3x + y = -3 \\ 6x - 12y = 6 \end{cases}$

13. $\begin{cases} 2a + 4b = -3 \\ a - b = 9 \end{cases}$

14. $\begin{cases} 4x + y - 2z = 10 \\ 3x + 5z = 14 \\ 8x + 3y - z = 23 \end{cases}$

15. $\begin{cases} y + 5z = -14 \\ -2x + 3y - z = 2 \\ 6x - 3z = 21 \end{cases}$

16. $\begin{cases} 12x + y - z = -7 \\ 11x + 2y = -2 \\ -x + 9y = -9 \end{cases}$

Write the system of equations represented by each matrix equation.

17. $\begin{bmatrix} 2 & -1 & 3 \\ -3 & 0 & -1 \\ 1 & -3 & 1 \end{bmatrix}\begin{bmatrix} x \\ y \\ z \end{bmatrix} = \begin{bmatrix} 4 \\ 1 \\ 5 \end{bmatrix}$

18. $\begin{bmatrix} -2 & 2 & 8 \\ 4 & 3 & 5 \\ 5 & 1 & 0 \end{bmatrix}\begin{bmatrix} x \\ y \\ z \end{bmatrix} = \begin{bmatrix} 3 \\ -2 \\ 1 \end{bmatrix}$

16. $\begin{bmatrix} 12 & 1 & -1 \\ 11 & 2 & 0 \\ -1 & 9 & 0 \end{bmatrix}\begin{bmatrix} x \\ y \\ z \end{bmatrix} = \begin{bmatrix} -7 \\ -2 \\ -9 \end{bmatrix}$

17. $\begin{cases} 2x - y + 3z = 4 \\ -3x - z = 1 \\ x - 3y + z = 5 \end{cases}$

18. $\begin{cases} -2x + 2y + 8z = 3 \\ 4x + 3y + 5z = -2 \\ 5x + y = 1 \end{cases}$

Write the matrix equation that represents each system, and solve the system, if possible, by using a matrix equation.

19 $\begin{cases} x + y - z = 14 \\ 4x - y + 5z = -22 \\ 2x + 2y - 3z = 35 \end{cases}$ **(4, 3, −7)**

20 $\begin{cases} -2x + y + 6z = 18 \\ 5x + 8z = -16 \\ 3x + 2y - 10z = -3 \end{cases}$ **(−4, 7, 0.5)**

21 $\begin{cases} 3x + 6y - 6z = 9 \\ 2x - 5y + 4z = 6 \\ -x + 16y + 14z = -3 \end{cases}$ **(3, 0, 0)**

22 $\begin{cases} x + 3y - 2z = 4 \\ 4x - y + z = -1 \\ 3x - 4y + 3z = -5 \end{cases}$ **no solution**

23 $\begin{cases} x - 2y + 3z = 11 \\ 4x - z = 4 \\ 2x - y + 3z = 10 \end{cases}$ **(1.4, −2.4, 1.6)**

24 $\begin{cases} x + 2y - z = 5 \\ 3x + 9y - z = 8 \\ 2x + 10y - 2z = -2 \end{cases}$ **(8.5, −2, −0.5)**

25 $\begin{cases} x + y - 2z = -2 \\ 2x - 3y + z = 1 \\ 2x + y - 3z = -2 \end{cases}$ **no solution**

26 $\begin{cases} x - \frac{1}{2}y - 3z = -9 \\ 8z = -16 - 5x \\ \frac{3}{5}x + \frac{2}{5}y - 2z = -\frac{3}{5} \end{cases}$ **(−4, 7, 0.5)**

27 $\begin{cases} 2.5x + y - z = -6 \\ -3.5y + 2.5z = 2.5 \\ 5x + 4y - 2z = -12 \end{cases}$ **(−2, 0, 1)**

28 $\begin{cases} x + 2y = -6 \\ y + 2z = 11 \\ 2x + z = 16 \end{cases}$ **(4, −5, 8)**

Write the matrix equation that represents each system, and solve the system, if possible, by using a matrix equation.

29 $\begin{cases} x + y + z + w = 10 \\ 2x - y + z - 3w = -9 \\ 3x + y - z - w = -2 \\ 2x - 3y + z - w = -5 \end{cases}$

30 $\begin{cases} x + 2y - 6z + w = 12 \\ -2x - 3y + 9z + w = -19 \\ x + 2y - 5z + 2w = 15 \\ 2x + 4y - 12z + 3w = 24 \end{cases}$

CHALLENGE

31. The matrix $\begin{bmatrix} -1 & 0 \\ 0 & 1 \end{bmatrix}$ is its own inverse. Find another matrix, other than the identity matrix, which is its own inverse.

CONNECTION

32 **GEOMETRY** The measure of the largest angle of a certain triangle is 3 times the measure of the smallest angle. The measure of the remaining angle of the triangle is the average of the measures of the largest and smallest angles. Write the system of equations that describes the measure of each angle of the triangle. Then solve the system by using a matrix equation.

APPLICATION

33 **ENTERTAINMENT** One hundred and twenty people attended a musical. The total amount of money collected for tickets was \$1515. Prices were \$15 for regular adult admission, \$12 for children, and \$10 for senior citizens. Twice as many children's tickets as regular adult tickets were sold. Write a system of equations to find the number of children, adults, and senior citizens that attended the musical. Then solve the system by using a matrix equation and the inverse matrix.

internet connect go.hrw.com

Activities Online
Go To: **go.hrw.com**
Keyword:
MB1 Cramer

29. $x = 1$, $y = 2$, $z = 3$, $w = 4$

30. $x = 2$, $y = 14$, $z = 3$, $w = 0$

32. 30°, 60°, 90°

33. 70 children, 35 adults, 15 seniors

Error Analysis

When creating the coefficient matrix for a system of equations, students sometimes forget that a subtraction sign means that the coefficient of the following term is a negative number. Suggest that they begin each exercise by first rewriting the system and changing each subtraction to the equivalent addition. This should focus their attention on the signs of the entries in the coefficient matrix. Remind students that the coefficient of any missing variable is 0.

31. Answers may vary. sample answers:

$\begin{bmatrix} 1 & 0 \\ 0 & -1 \end{bmatrix}, \begin{bmatrix} 0 & -1 \\ -1 & 0 \end{bmatrix},$ or $\begin{bmatrix} 0 & 1 \\ 1 & 0 \end{bmatrix}$

Student Technology Guide

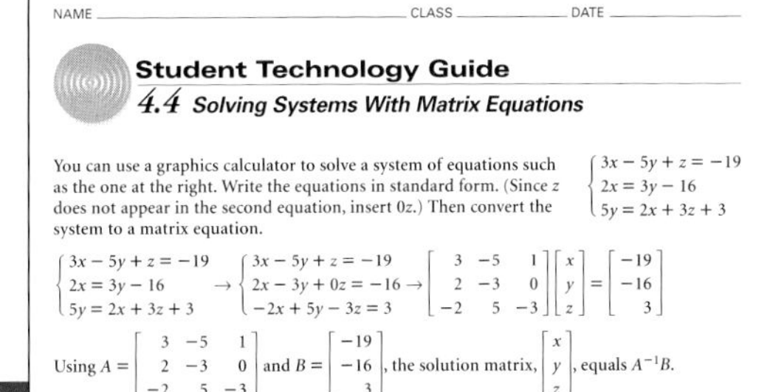

NAME ______ CLASS ______ DATE ______

Student Technology Guide
4.4 Solving Systems With Matrix Equations

You can use a graphics calculator to solve a system of equations such as the one at the right. Write the equations in standard form. (Since z does not appear in the second equation, insert $0z$.) Then convert the system to a matrix equation.

$\begin{cases} 3x - 5y + z = -19 \\ 2x = 3y - 16 \\ 5y = 2x + 3z + 3 \end{cases}$

$\begin{cases} 3x - 5y + z = -19 \\ 2x = 3y - 16 \\ 5y = 2x + 3z + 3 \end{cases} \rightarrow \begin{cases} 3x - 5y + z = -19 \\ 2x - 3y + 0z = -16 \\ -2x + 5y - 3z = 3 \end{cases} \rightarrow \begin{bmatrix} 3 & -5 & 1 \\ 2 & -3 & 0 \\ -2 & 5 & -3 \end{bmatrix}\begin{bmatrix} x \\ y \\ z \end{bmatrix} = \begin{bmatrix} -19 \\ -16 \\ 3 \end{bmatrix}$

Using $A = \begin{bmatrix} 3 & -5 & 1 \\ 2 & -3 & 0 \\ -2 & 5 & -3 \end{bmatrix}$ and $B = \begin{bmatrix} -19 \\ -16 \\ 3 \end{bmatrix}$, the solution matrix, $\begin{bmatrix} x \\ y \\ z \end{bmatrix}$, equals $A^{-1}B$.

- To enter A, press MATRX EDIT 1:[A] ENTER 3 ENTER 3 ENTER 3 ENTER (−) 5 … (−) 3 ENTER.
- Follow a similar procedure to enter B.
- Press 2nd MODE to exit matrix edit mode.
- To find $A^{-1}B$, press MATRX NAMES 1:[A] ENTER x^{-1} MATRX NAMES 2:[B] ENTER ENTER. From the display at right, the solution is $x = -2$, $y = 4$, and $z = 7$.

```
[A]-1[B]
        [[-2]
         [4 ]
         [7 ]]
```

If A^{-1} does not exist, the system has no unique solution, and the calculator gives an error message.

Write the matrix equation that represents each system. Then solve the system, if possible, by using a matrix inverse.

1. $\begin{cases} 2x - 5y = -19 \\ 3x + 4y = 6 \end{cases}$ $\begin{bmatrix} 2 & -5 \\ 3 & 4 \end{bmatrix}\begin{bmatrix} x \\ y \end{bmatrix} = \begin{bmatrix} -19 \\ 6 \end{bmatrix}$; $x = -2$, $y = 3$
2. $\begin{cases} 2x + 7 = 3y \\ 6x - 5y = -41 \end{cases}$ $\begin{bmatrix} 2 & -3 \\ 6 & -5 \end{bmatrix}\begin{bmatrix} x \\ y \end{bmatrix} = \begin{bmatrix} -7 \\ -41 \end{bmatrix}$; $x = -11$, $y = -5$
3. $\begin{cases} 4x - 5z = -64 \\ 2y = 18 \\ 3x + 7y - z = 26 \end{cases}$ $\begin{bmatrix} 4 & 0 & -5 \\ 0 & 2 & 0 \\ 3 & 7 & -1 \end{bmatrix}\begin{bmatrix} x \\ y \\ z \end{bmatrix} = \begin{bmatrix} -64 \\ 18 \\ 26 \end{bmatrix}$; $x = -11$, $y = 9$, $z = 4$
4. $\begin{cases} x - 5y + 2z = -31 \\ 3x + 4y - z = 35 \\ 6x - 7y + 3z = -30 \end{cases}$ $\begin{bmatrix} 1 & -5 & 2 \\ 3 & 4 & -1 \\ 6 & -7 & 3 \end{bmatrix}\begin{bmatrix} x \\ y \\ z \end{bmatrix} = \begin{bmatrix} -31 \\ 35 \\ -30 \end{bmatrix}$; $x = 3$, $y = 6$, $z = -2$
5. $\begin{cases} 9x + y - 3z = 46 \\ 2x - 4y = -6 - 6z \\ 7y - 3z = 2x + 33 \end{cases}$ $\begin{bmatrix} 9 & 1 & -3 \\ 2 & -4 & 6 \\ -2 & 7 & -3 \end{bmatrix}\begin{bmatrix} x \\ y \\ z \end{bmatrix} = \begin{bmatrix} 46 \\ -6 \\ 33 \end{bmatrix}$; $x = 5$, $y = 7$, $z = 2$

In Exercises 43–46, students solve systems of equations by using a technique that employs determinants. Students will learn more about this method, known as *Cramer's rule*, if they take a course in advanced mathematics.

APPLICATIONS

34. 1.2 liters of 3%, 0.8 liter of 8%

35. 1 liter of 4%, 2 liters of 7%

34 **CHEMISTRY** A nurse is mixing a 3% saline solution and an 8% saline solution to get 2 liters of a 5% saline solution. How many liters of each solution must be combined?

35 **CHEMISTRY** A solution of 4% acid and a solution of 7% acid are to be mixed to create 3 liters of a 6% acid solution. How many liters of the 3% solution and 7% solution must be combined?

36 **INVESTMENTS** A brokerage firm invested in three mutual-fund companies, A, B, and C. The firm invested \$16,110 in low-risk funds, \$9016.25 in medium-risk funds, and \$5698.75 in high-risk funds, distributed among the three companies as shown below.

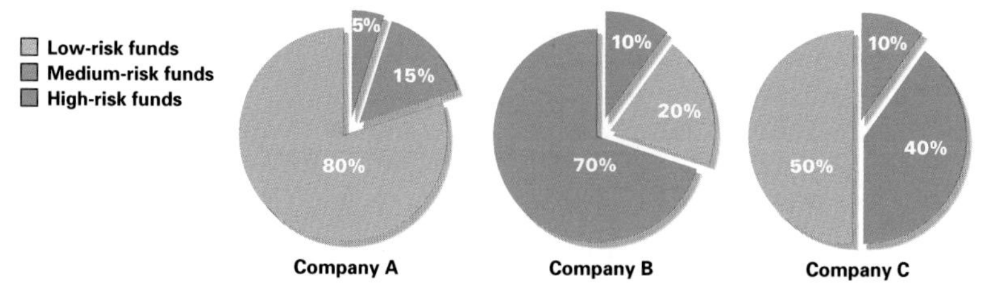

a. Write a system of equations to find the amount that the firm invested in each company.
b. Write the matrix equation that represents this system.
c. Solve the matrix equation by using the inverse matrix.

Look Back

37. State the domain and range of $f(x) = |x - 5| + 2$. ***(LESSON 2.3)***

Let $f(x) = 2x + 3$ and $g(x) = x^2 - 3x + 1$. ***(LESSONS 2.3 AND 2.4)***

38. Find $g(-3)$. **19** **39.** Find $g \circ f$. $\mathbf{4x^2 + 6x + 1}$ **40.** Find $f \circ g$. $\mathbf{2x^2 - 6x + 5}$

Graph each function. ***(LESSON 2.6)***

41. $f(x) = 3[x]$ **42.** $f(x) = 2\lceil x \rceil - 3$

Look Beyond

A method called *Cramer's rule* allows you to solve systems of equations by using determinants. Cramer's rule states that if $\begin{cases} ax + by = e \\ cx + dy = f \end{cases}$ is a system of equations and $\begin{vmatrix} a & b \\ c & d \end{vmatrix} \neq 0$, then the solution of the system can be found as follows:

$$x = \frac{\begin{vmatrix} e & b \\ f & d \end{vmatrix}}{\begin{vmatrix} a & b \\ c & d \end{vmatrix}} \quad \text{and} \quad y = \frac{\begin{vmatrix} a & e \\ c & f \end{vmatrix}}{\begin{vmatrix} a & b \\ c & d \end{vmatrix}}$$

Use Cramer's rule to solve each system.

43. $\begin{cases} 2x + y = 6 \\ x + 2y = 9 \end{cases}$ **44.** $\begin{cases} x + y = 3 \\ 3x + 2y = 4 \end{cases}$ **45.** $\begin{cases} x - 2y = 5 \\ 2x + 2y = 4 \end{cases}$ **46.** $\begin{cases} 2x + y = 10 \\ 3x + 3y = 21 \end{cases}$

43. (1, 4)
44. (−2, 5)
45. (3, −1)
46. (3, 4)

36a. $\begin{cases} 0.80x + 0.20y + 0.50z = 16{,}110.00 \\ 0.15x + 0.70y + 0.10z = 9016.25 \\ 0.05x + 0.10y + 0.40z = 5698.75 \end{cases}$

b. $\begin{bmatrix} 0.80 & 0.20 & 0.50 \\ 0.15 & 0.70 & 0.10 \\ 0.05 & 0.10 & 0.40 \end{bmatrix} \begin{bmatrix} x \\ y \\ z \end{bmatrix} = \begin{bmatrix} 16{,}110.00 \\ 9016.25 \\ 5698.75 \end{bmatrix}$

c. \$11,275 in company A, \$8950 in company B, and \$10,600 in company C.

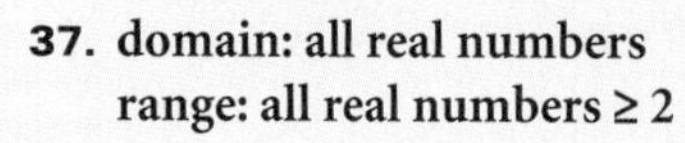

37. domain: all real numbers
range: all real numbers ≥ 2

41.

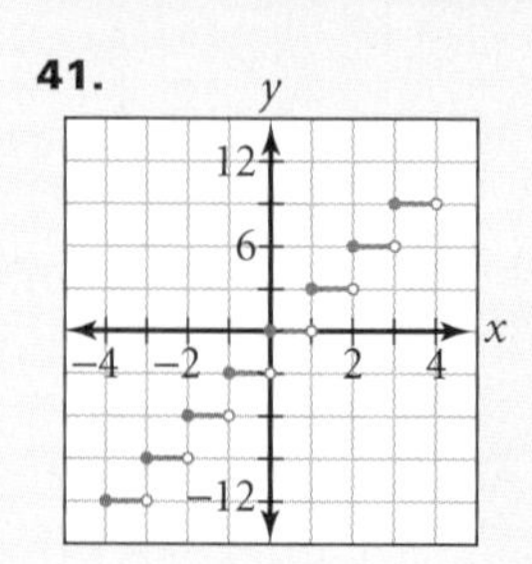

42.

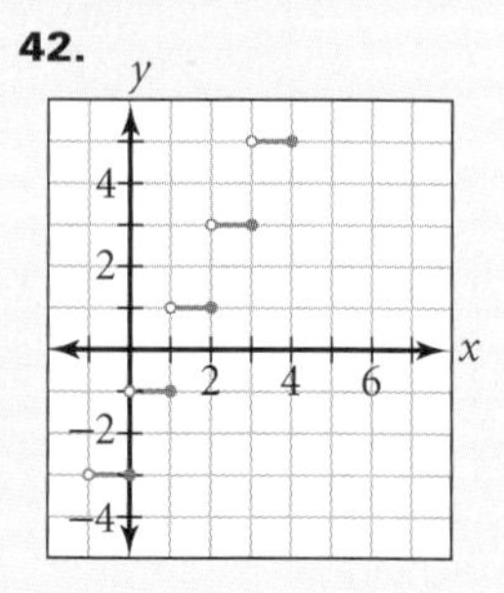

Using Matrix Row Operations

Objectives

- Represent a system of equations as an augmented matrix.
- Solve a system of linear equations by using elementary row operations.

Why *You can use matrix operations to learn whether a system of linear equations that has no unique solution is dependent or inconsistent.*

APPLICATION
SMALL BUSINESS

Maya and her friends Amit and Nina have a lawn care business offering three services:

- lawn mowing and edging
- fertilizing and weeding
- trimming shrubs and small trees

They charge a flat rate for each service. The three partners divide up the work for a particular customer as shown below.

Service	Workers and Hours	Cost
Mowing	Maya—1 hr Amit—1 hr Nina—1 hr	\$21
Fertilizing	Maya—2 hr Amit—1 hr	\$23
Trimming	Amit—1 hr Nina—3 hr	\$25

How much does each partner earn per hour and how much will each partner earn in total for his or her work for this customer? *You will solve this problem in Example 1.*

Recall from Lesson 4.4 that if the coefficient matrix of a matrix equation has an inverse, the system represented by the matrix equation is consistent and independent. If there is no inverse, the system is either dependent or inconsistent, but you cannot determine which one.

The **row-reduction method** of solving a system allows you to determine whether the system is independent, dependent, or inconsistent.

The row-reduction method is performed on an *augmented matrix*. An **augmented matrix** consists of the coefficients and constant terms in the system of equations.

Prepare

NCTM PRINCIPLES & STANDARDS 1–10

QUICK WARM-UP

Write each system as a matrix equation. Then solve the system, if possible, by using the matrix equation.

1. $\begin{cases} x + 7y = 5 \\ -3x + 2y = 8 \end{cases}$

$\begin{bmatrix} 1 & 7 \\ -3 & 2 \end{bmatrix} \begin{bmatrix} x \\ y \end{bmatrix} = \begin{bmatrix} 5 \\ 8 \end{bmatrix}$

$x = -2$ and $y = 1$

2. $\begin{cases} 3x - 6y = -3 \\ -4x + 8y = -4 \end{cases}$

$\begin{bmatrix} 3 & -6 \\ -4 & 8 \end{bmatrix} \begin{bmatrix} x \\ y \end{bmatrix} = \begin{bmatrix} -3 \\ -4 \end{bmatrix}$

no solution

Also on Quiz Transparency 4.5

Teach

Why Discuss with students the method of solving systems by using inverse matrices. Ask for their perceptions of its advantages and disadvantages. Point out that if the determinant of the coefficient matrix is 0, it is not immediately obvious whether the system is dependent or inconsistent. In this lesson, they will learn a technique that yields more information about the solution of a system.

Alternative Teaching Strategy

USING COGNITIVE STRATEGIES Define *augmented matrix*. Then give students the set of matrices at right, one at a time, and ask them to write a system of linear equations in x and y that the matrix represents. Each time they write a system, ask them to explain why it is equivalent to the system that preceded it. Have students verify that $x = 1$ and $y = 2$ are solutions of each system in the series. Use this activity as a basis for a presentation of matrix row operations.

(1) $\left[\begin{array}{cc:c} 3 & 1 & 5 \\ -1 & 2 & 3 \end{array}\right]$ $\begin{cases} 3x + y = 5 \\ -x + 2y = 3 \end{cases}$

(2) $\left[\begin{array}{cc:c} 3 & 1 & 5 \\ -3 & 6 & 9 \end{array}\right]$ $\begin{cases} 3x + y = 5 \\ -3x + 6y = 9 \end{cases}$

(3) $\left[\begin{array}{cc:c} 3 & 1 & 5 \\ 0 & 7 & 14 \end{array}\right]$ $\begin{cases} 3x + y = 5 \\ 7y = 14 \end{cases}$

(4) $\left[\begin{array}{cc:c} 3 & 1 & 5 \\ 0 & 1 & 2 \end{array}\right]$ $\begin{cases} 3x + y = 5 \\ y = 2 \end{cases}$

(5) $\left[\begin{array}{cc:c} 3 & 0 & 3 \\ 0 & 1 & 2 \end{array}\right]$ $\begin{cases} 3x = 3 \\ y = 2 \end{cases}$

(6) $\left[\begin{array}{cc:c} 1 & 0 & 1 \\ 0 & 1 & 2 \end{array}\right]$ $\begin{cases} x = 1 \\ y = 2 \end{cases}$

The system of equations and the corresponding augmented matrix that represent the lawn-care problem are shown below. Let m, a, and n represent the hourly wages for Maya, Amit, and Nina, respectively.

System

$$\begin{cases} m + a + n = 21 \\ 2m + a = 23 \\ a + 3n = 25 \end{cases}$$

Augmented Matrix

$$\left[\begin{array}{ccc|c} 1 & 1 & 1 & 21 \\ 2 & 1 & 0 & 23 \\ 0 & 1 & 3 & 25 \end{array}\right]$$

coefficients

constants

The goal of the row-reduction method is to transform, if possible, the coefficient columns into columns that form an identity matrix. This is called the **reduced row-echelon form** of an augmented matrix if the matrix represents an independent system. If the identity matrix can be formed, then the resulting constants will represent the unique solution to the system.

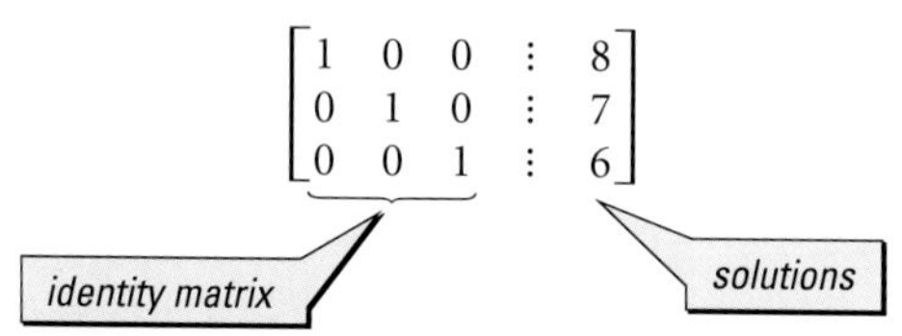

$$\left[\begin{array}{ccc|c} 1 & 0 & 0 & 8 \\ 0 & 1 & 0 & 7 \\ 0 & 0 & 1 & 6 \end{array}\right]$$

This final matrix is said to be in *reduced row-echelon* form.

To transform an augmented matrix into reduced row-echelon form, use the elementary row operations described below.

Elementary Row Operations

The following operations produce equivalent matrices, and may be used in any order and as many times as necessary to obtain reduced row-echelon form.

- Interchange two rows.
- Multiply all entries in one row by a nonzero number.
- Add a multiple of one row to another row.

You may use *row operation notation* to keep a record of the row operations that you perform.

ROW OPERATION	NOTATION
• Interchange rows 1 and 2.	$R_1 \leftrightarrow R_2$
• Multiply each entry in row 3 by −2.	$-2R_3 \rightarrow R_3$
• Replace row 1 with the sum of row 1 and 4 times each entry in row 2.	$4R_2 + R_1 \rightarrow R_1$

CRITICAL THINKING Explain why the row operation $R_2 - 4R_1 \rightarrow R_1$ results in an equivalent matrix.

CRITICAL THINKING
Performing the operation $R_2 - 4R_1 \rightarrow R_1$ is the same as performing the following two row operations in the given order: $-4R_1 \rightarrow R_1$ and $R_2 + R_1 \rightarrow R_1$.

Interdisciplinary Connection

CHEMISTRY Suppose that you need 6 ounces of an 8% acid solution for an experiment. In the supply cabinet, there is 1 quart (48 ounces) of a 10% acid solution and 1 quart of a 15% acid solution. Your lab partner says that it is not possible to mix only these solutions to obtain the solution that you need.

Write a system of equations to find the amount of each solution that is needed. Solve the system by using augmented matrices to either prove or disprove your partner's statement.

Let x represent the number of ounces of the 10% acid solution and y represent the number of ounces of the 15% acid solution.

$$\left[\begin{array}{cc|c} 1 & 1 & 6 \\ 0.1 & 0.15 & 0.48 \end{array}\right] \rightarrow \left[\begin{array}{cc|c} 1 & 0 & 8.4 \\ 0 & 1 & -2.4 \end{array}\right]$$

The lab partner's statement is correct. To make the required solution, the reduced row-echelon form of the matrix indicates that you would need 8.4 oz of the 10% solution and −2.4 oz of the 15% solution, which is not possible.

EXAMPLE 1 **Refer to the lawn-care problem described at the beginning of the lesson.**

a. Use the row-reduction method to solve the system.

b. Find the hourly wages for Maya, Amit, and Nina. Then find the total amount that each partner earns for this job.

APPLICATION
SMALL BUSINESS

SOLUTION

System	Augmented Matrix
$\begin{cases} m + a + n = 21 \\ 2m + a = 23 \\ a + 3n = 25 \end{cases}$	$\left[\begin{array}{ccc:c} 1 & 1 & 1 & 21 \\ 2 & 1 & 0 & 23 \\ 0 & 1 & 3 & 25 \end{array}\right]$

a. Perform row operations.

- Inspect column 1.

The first row begins with 1, but the 2 in the second row needs to become 0.

$-2R_1 + R_2 \rightarrow R_2$ — *Replace row 2 with the sum of row 2 and −2 times row 1.*

$$\left[\begin{array}{ccc:c} 1 & 1 & 1 & 21 \\ \textcircled{0} & -1 & -2 & -19 \\ 0 & 1 & 3 & 25 \end{array}\right]$$

- Inspect column 2.

Row 1: Change the entry to 0. $R_2 + R_1 \rightarrow R_1$

$$\left[\begin{array}{ccc:c} 1 & \textcircled{0} & -1 & 2 \\ 0 & -1 & -2 & -19 \\ 0 & 1 & 3 & 25 \end{array}\right]$$

Row 2: Change the entry to 1. $-1R_2 \rightarrow R_2$

$$\left[\begin{array}{ccc:c} 1 & 0 & -1 & 2 \\ 0 & \textcircled{1} & 2 & 19 \\ 0 & 1 & 3 & 25 \end{array}\right]$$

Row 3: Change the entry to 0. $-1R_2 + R_3 \rightarrow R_3$

$$\left[\begin{array}{ccc:c} 1 & 0 & -1 & 2 \\ 0 & 1 & 2 & 19 \\ 0 & \textcircled{0} & 1 & 6 \end{array}\right]$$

- Inspect column 3.

Row 1: Change the entry to 0. $R_3 + R_1 \rightarrow R_1$

$$\left[\begin{array}{ccc:c} 1 & 0 & \textcircled{0} & 8 \\ 0 & 1 & 2 & 19 \\ 0 & 0 & 1 & 6 \end{array}\right]$$

Row 2: Change the entry to 0. $-2R_3 + R_2 \rightarrow R_2$

$$\left[\begin{array}{ccc:c} 1 & 0 & 0 & 8 \\ 0 & 1 & \textcircled{0} & 7 \\ 0 & 0 & 1 & 6 \end{array}\right]$$

The matrix is now in reduced row-echelon form.

$$\left[\begin{array}{ccc:c} 1 & 0 & 0 & 8 \\ 0 & 1 & 0 & 7 \\ 0 & 0 & 1 & 6 \end{array}\right] \quad \begin{array}{l} m = 8 \\ a = 7 \\ n = 6 \end{array}$$

b. Maya receives \$8 an hour; since she works 3 hours, she will earn \$24. Amit receives \$7 an hour and works 3 hours, so he will earn \$21. Nina receives \$6 an hour and works 4 hours, so she will earn \$24.

CHECKPOINT ✔ Check the solution to Example 1 by using substitution.

ADDITIONAL EXAMPLE 1

Marble High School sponsored a debate contest between three teams. There were three types of questions—a, b, and c—each worth a certain number of points. The table summarizes each team's performance.

Team	Question type and no. correct	Total points
1	a—3 b—1 c—1	24
2	a—0 b—1 c—2	8
3	a—1 b—2 c—1	16

The system and augmented matrix below represent the performance of the three teams.

System

$$\begin{cases} 3a + b + c = 24 \\ b + 2c = 8 \\ a + 2b + c = 16 \end{cases}$$

Augmented matrix

$$\left[\begin{array}{ccc:c} 3 & 1 & 1 & 24 \\ 0 & 1 & 2 & 8 \\ 1 & 2 & 1 & 16 \end{array}\right]$$

a. Use the row-reduction method to solve the system.
$a = 6, b = 4, c = 2$

b. What is the point value for each type of question?
a: 6 points
b: 4 points
c: 2 points

CHECKPOINT ✔

$8 + 7 + 6 = 21$ True
$2(8) + 7 = 23$ True
$7 + 3(6) = 25$ True

Inclusion Strategies

LINGUISTIC LEARNERS Students who have a more verbal learning style may feel overwhelmed by the symbolism of row operation notation. Spend some time focusing on just the notation until they become comfortable with it.

For example, give students a sheet of paper with this matrix written at the top. Show several transformations of the matrix beneath it.

$$\left[\begin{array}{ccc:c} 2 & 4 & -10 & -2 \\ 3 & 9 & -21 & 0 \\ 1 & 5 & -12 & 1 \end{array}\right]$$

Beneath each transformation, have students write a verbal description of the operation that was performed. Then have them represent the operation symbolically. Here are two samples.

$$\left[\begin{array}{ccc:c} 1 & 2 & -5 & -1 \\ 3 & 9 & -21 & 0 \\ 1 & 5 & -12 & 1 \end{array}\right]$$

Each element of row 1 was multiplied by 0.5.
$R_1 \rightarrow 0.5R_1$

$$\left[\begin{array}{ccc:c} 1 & 5 & -12 & 1 \\ 3 & 9 & -21 & 0 \\ 2 & 4 & -10 & -2 \end{array}\right]$$

Row 1 and row 3 were interchanged.
$R_1 \leftrightarrow R_3$

ADDITIONAL
EXAMPLE 2

Use row reduction to solve the system below. Then classify the system as independent, dependent, or inconsistent.

$$\begin{cases} x + 5y + 2z = 0 \\ -2x - 3y = 0 \\ x - y + 3z = 0 \end{cases}$$

$x = 0, y = 0, z = 0$; independent

Teaching Tip

TECHNOLOGY When using a TI-83 or TI-83 Plus graphics calculator, you can find the reduced row-echelon form of a matrix by accessing the **MATRX** menu, selecting MATH, choosing B:rref(, accessing the **MATRX** menu again, selecting the matrix's name, and pressing) ENTER. The number of columns must be greater than or equal to the number of rows to use this feature.

TRY THIS

$(8z - 28, 18 - 5z, z)$ or $x = 8z - 28, y = 18 - 5z, z = z$; dependent

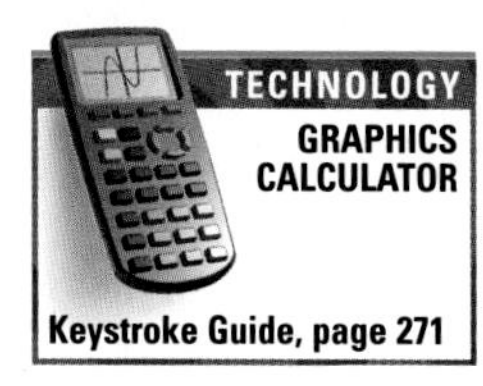

Keystroke Guide, page 271

On many graphics calculators, you can enter an augmented matrix, and the calculator will give you the reduced row-echelon form. The displays below show the augmented matrix and the reduced row-echelon form for the system in Example 1.

Sometimes row operations do not result in an identity matrix in the coefficient columns. Examples 2 and 3 illustrate two possible alternative results and how they show that the system is inconsistent or dependent.

EXAMPLE 2 **Use the row-reduction method to solve the system below. Then classify the system as independent, dependent, or inconsistent.**

$$\begin{cases} x - 2y - 2z = 6 \\ 3x - 4y + z = -1 \\ 5x - 8y - 3z = 11 \end{cases}$$

SOLUTION

For the given system, the augmented matrix is shown below.

$$\begin{bmatrix} 1 & -2 & -2 & \vdots & 6 \\ 3 & -4 & 1 & \vdots & -1 \\ 5 & -8 & -3 & \vdots & 11 \end{bmatrix}$$

The reduced row-echelon form of the matrix and the corresponding simplified system of equations are shown below.

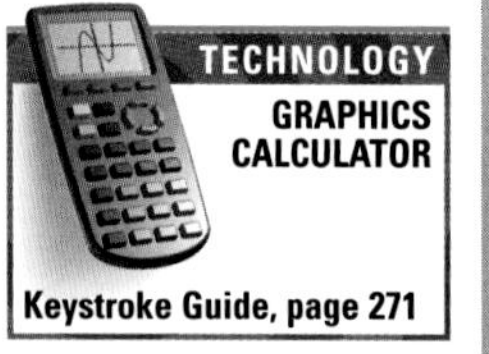

Keystroke Guide, page 271

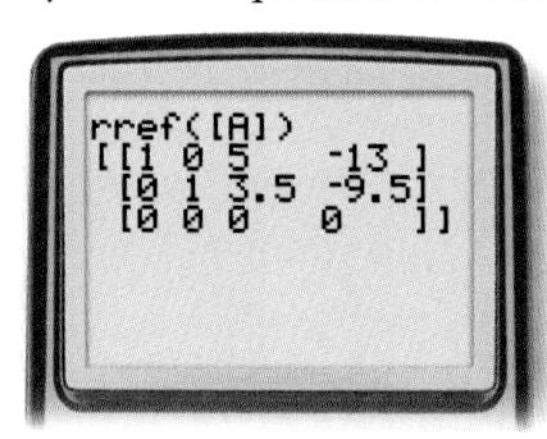

$$\rightarrow \begin{cases} x + 5z = -13 \\ y + 3.5z = -9.5 \\ 0z = 0 \end{cases} \rightarrow \begin{cases} x = -13 - 5z \\ y = -9.5 - 3.5z \\ 0z = 0 \end{cases}$$

This equation tells you that z can be any real number.

The system is dependent because it has infinitely many solutions. You can describe the solution as $(-13 - 5z, -9.5 - 3.5z, z)$, where z can be any real number.

TRY THIS Use the row-reduction method to solve the system below. Then classify the system as independent, dependent, or inconsistent.

$$\begin{cases} 2x + 3y - z = -2 \\ x + 2y + 2z = 8 \\ 5x + 9y + 5z = 22 \end{cases}$$

Enrichment

Have students work in pairs. Give them an augmented matrix in reduced row-echelon form. Tell them that it represents the solution of a system of linear equations. Each student should now work independently to "unreduce" the given matrix, creating the matrix for a system whose solution it might represent.

Now have the partners exchange the matrices that they created. Each should reduce the other's matrix. The partners can then compare results and resolve any difficulties. Encourage them to compare the processes of "unreducing" and reducing.

Repeat the activity three times, having students work with an augmented matrix for an independent system, an augmented matrix for a dependent system, and an augmented matrix for an inconsistent system.

EXAMPLE 3 **Use the row-reduction method to solve the system below. Then classify the system as independent, dependent, or inconsistent.**

$$\begin{cases} 2x + 6y - 4z = 1 \\ x + 3y - 2z = 4 \\ 2x + y - 3z = -7 \end{cases}$$

SOLUTION

Write the augmented matrix.

$$\left[\begin{array}{ccc|c} 2 & 6 & -4 & 1 \\ 1 & 3 & -2 & 4 \\ 2 & 1 & -3 & -7 \end{array}\right]$$

Find the reduced row-echelon form and the simplified system as shown below.

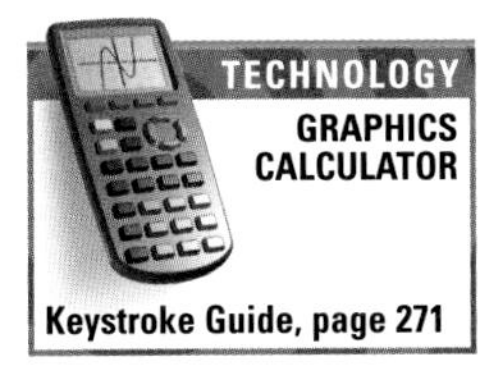


```
rref([A])
 [[1 0 -1.4 0]
  [0 1 -.2  0]
  [0 0 0    1]]
```

$$\rightarrow \begin{cases} x - 1.4z = 0 \\ y - 0.2z = 0 \\ 0 = 1 \end{cases}$$

This false statement tells you that there is no solution.

The system is inconsistent.

TRY THIS Use the row-reduction method to solve the system below. Then classify the system as independent, dependent, or inconsistent.

$$\begin{cases} 4x - 4y - 3z = 2 \\ 4x + 3z = 3 \\ 4y + 6z = 3 \end{cases}$$

Activity: Exploring Systems of Three Equations

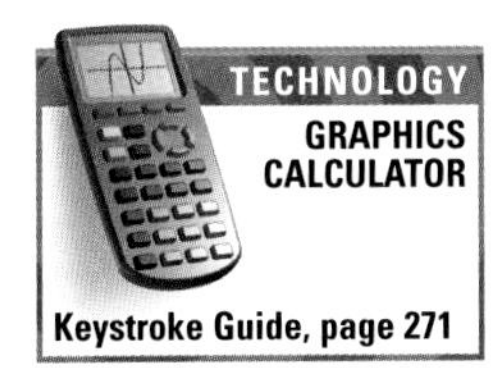

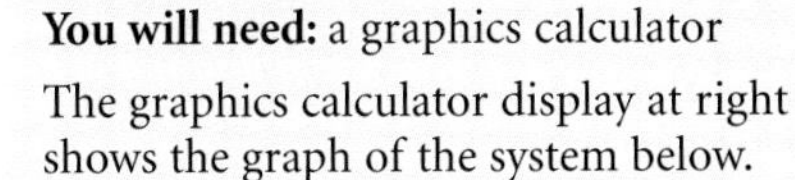

You will need: a graphics calculator

The graphics calculator display at right shows the graph of the system below.

$$\begin{cases} -5x + 2y = 6 \\ x + 2y = -8 \\ x + 2y = 8 \end{cases}$$

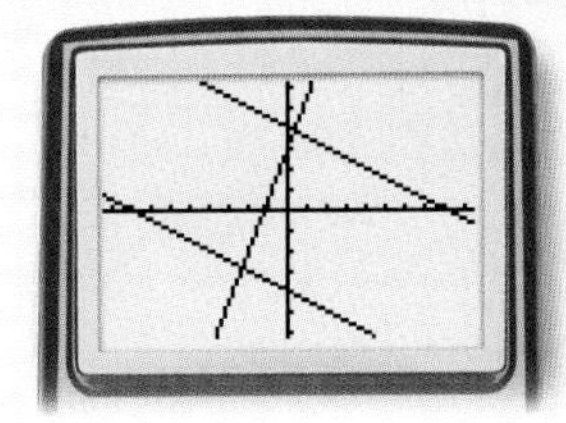

CONNECTION GEOMETRY

1. How can you tell from the system of equations that two of the lines are parallel?
2. Modify the equations in the system so that the intersections of the graphs form the vertices of a triangle. Graph the new system on a graphics calculator. What must be true of the system you wrote in order for the vertices of a triangle to be formed?

CHECKPOINT ✔

3. Modify the equations in the original system so that the intersections of the graphs form the vertices of a right triangle. Graph the new system on a graphics calculator. Explain your strategy for changing the equations.

Reteaching the Lesson

INVITING PARTICIPATION Have students work in groups of four. Assign each group a place at the board.

Display a system of linear equations on the board or overhead. Each group should write the augmented matrix for the system on the board and then use row reductions to solve the system.

When all the groups have finished, have one group present the steps that they used to solve the system. Then have other groups present alternative approaches. Lead the class in a discussion of similarities and differences among all the approaches presented.

Repeat the activity as many times as needed. Be sure to include at least one example each of independent, dependent, and inconsistent systems.

ADDITIONAL EXAMPLE 3

Use row reduction to solve the system below. Then classify the system as independent, dependent, or inconsistent.

$$\begin{cases} 4x + 12y - 8z = 2 \\ 2x + 6y - 4z = 8 \\ 4x + 2y - 6z = -14 \end{cases}$$

no solution; inconsistent

TRY THIS

no solution; inconsistent

Activity Notes

In this Activity, students are asked to explain how they know that two equations will produce parallel lines. They should note that the last two equations are identical except for the constant terms. In Step 2, they must modify the equation of one of the parallel lines in order to get a triangle from the points of intersection. In Step 3, suggest that they change the first equation to $-4x + 2y = 6$ so that it is perpendicular to the second equation and then determine the equation for the third line.

CHECKPOINT ✔

3. **Answers may vary. Sample answer: Leave the second equation as written. Create an equation perpendicular to the second equation by changing the first equation to $-4x + 2y = 6$. The third equation can be any line not parallel to either of the first two equations, such as $x = 4$.**

Assess

Selected Answers

Exercises 5–7, 9–53 odd

ASSIGNMENT GUIDE	
In Class	1–7
Core	8–12, 13–33 odd, 37–41 odd
Core Plus	8–12, 14–30 even, 31–41
Review	42–53
Preview	54

Extra Practice can be found beginning on page 940.

5a. $\begin{cases} x + y = 120 \\ 200x + 50y = 12{,}750 \end{cases}$

b. 45 leather, 75 imitation leather

Exercises

Communicate

1. Describe how the second and third elementary row operations listed on page 252 may correspond to the operations you perform on equations in a system when using the elimination method.

2. Explain how to write the augmented matrix for the system of equations at right.

$$\begin{cases} x - 4y + 7z = 17 \\ 2x + y - z = -5 \\ x + 4z = 13 \end{cases}$$

3. Write the system of equations represented by the augmented matrix at right.

$$\begin{bmatrix} -3 & -4 & 0 & \vdots & 2 \\ 4 & 2 & -3 & \vdots & 6 \\ -2 & 0 & 1 & \vdots & -6 \end{bmatrix}$$

4. State which row operations were applied to the first matrix to obtain the second matrix.

$$\begin{bmatrix} 1 & 2 & 3 & \vdots & 1 \\ 2 & 5 & 7 & \vdots & 3 \\ 3 & 2 & 1 & \vdots & 1 \end{bmatrix} \rightarrow \begin{bmatrix} 2 & 4 & 6 & \vdots & 2 \\ 2 & 5 & 7 & \vdots & 3 \\ 5 & 7 & 8 & \vdots & 4 \end{bmatrix}$$

Guided Skills Practice

APPLICATION

5. MANUFACTURING A company makes a total of 120 leather and imitation-leather jackets per week. The leather jackets cost the company \$200 each to produce and the imitation-leather cost the company \$50 each to produce. The company spends \$12,750 per week on costs for producing jackets. ***(EXAMPLE 1)***

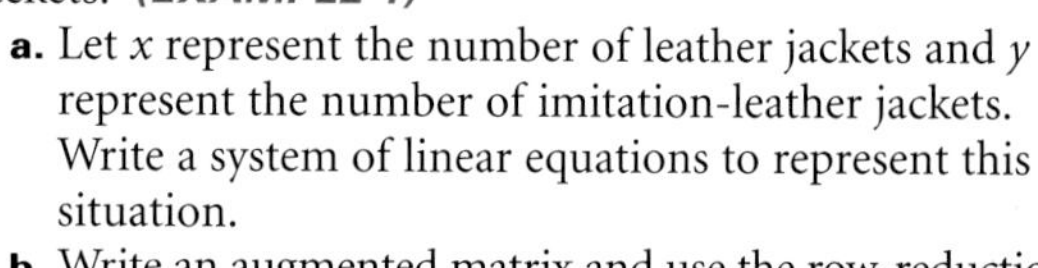

a. Let x represent the number of leather jackets and y represent the number of imitation-leather jackets. Write a system of linear equations to represent this situation.

b. Write an augmented matrix and use the row-reduction method to solve the system and find the number of each type of jacket made each week.

Use the row-reduction method to solve each system. Then classify each system as independent, dependent, or inconsistent. ***(EXAMPLES 2 AND 3)***

6. $\begin{cases} x + 2y + z = 3 \\ y + 2z = 3 \\ y + 2z = 5 \end{cases}$ no solution; incons.

7. $\begin{cases} x + y = 3 \\ 3x + y = 15 \\ 5x + y + z = 25 \end{cases}$ (6, −3, −2); indep.

internet connect

Homework Help Online
Go To: **go.hrw.com**
Keyword:
MB1 Homework Help for Exercises 8–10, 19–30

Practice and Apply

Write the augmented matrix for each system of equations.

8. $\begin{cases} x + 3y = 23 \\ 4x - 2y = -6 \end{cases}$

9. $\begin{cases} -x + 2y - 5z = 23 \\ 2x + 7z = 19 \\ 5x - 2y + z = -10 \end{cases}$

10. $\begin{cases} 2x - 5y - z = -32 \\ -x + 4y + 2z = 34 \\ 3x + 7y - 3z = -2 \end{cases}$

8. $\begin{bmatrix} 1 & 3 & \vdots & 23 \\ 4 & -2 & \vdots & -6 \end{bmatrix}$

9. $\begin{bmatrix} -1 & 2 & -5 & \vdots & 23 \\ 2 & 0 & 7 & \vdots & 19 \\ 5 & -2 & 1 & \vdots & -10 \end{bmatrix}$

10. $\begin{bmatrix} 2 & -5 & -1 & \vdots & -32 \\ -1 & 4 & 2 & \vdots & 34 \\ 3 & 7 & -3 & \vdots & -2 \end{bmatrix}$

Perform the indicated row operations on matrix A.

11. $R_2 = [0 \;\; 13 \;\; -17 \;\vdots\; 15]$
12. $R_3 = [0 \;\; -12 \;\; 41 \;\vdots\; -39]$

11. $3R_1 + R_2 \rightarrow R_2$

12. $-8R_1 + R_3 \rightarrow R_3$

$$A = \begin{bmatrix} -1 & 2 & -5 & \vdots & 4 \\ 3 & 7 & -2 & \vdots & 3 \\ -8 & 4 & 1 & \vdots & -7 \end{bmatrix}$$

Find the reduced row-echelon form of each matrix.

13. $\begin{bmatrix} 2 & -2 & \vdots & -2 \\ 0 & 1 & \vdots & 3 \end{bmatrix}$ **14.** $\begin{bmatrix} 3 & 3 & 6 & \vdots & 30 \\ 0 & 1 & 0 & \vdots & 3 \\ 0 & 0 & 2 & \vdots & 2 \end{bmatrix}$ **15.** $\begin{bmatrix} 4 & 4 & \vdots & 32 \\ 1 & 3 & \vdots & 16 \end{bmatrix}$

16. $\begin{bmatrix} 3 & 0 & 3 & \vdots & 24 \\ 1 & 2 & 3 & \vdots & 28 \\ 0 & 0 & 2 & \vdots & 12 \end{bmatrix}$ **17.** $\begin{bmatrix} 5 & 2 & 1 & \vdots & -4 \\ 7 & 4 & 1 & \vdots & -4 \\ 3 & 2 & 1 & \vdots & 0 \end{bmatrix}$ **18.** $\begin{bmatrix} 2 & 1 & 2 & \vdots & 19 \\ 3 & 3 & 3 & \vdots & 33 \\ 2 & 2 & 4 & \vdots & 30 \end{bmatrix}$

Solve each system of equations by using the row-reduction method. Show each step.

19. $\begin{cases} 4x + 3y = 1 \\ 3x - 2y = 5 \end{cases}$ (1, −1) **20.** $\begin{cases} x + 2y = 16 \\ 2x + y = 11 \end{cases}$ (2, 7) **21.** $\begin{cases} 3x - y = 4 \\ x + 4y = -3 \end{cases}$ (1, −1)

24. $\left(\frac{18}{7}, \frac{8}{7}, \frac{2}{7}\right)$

22. $\begin{cases} x + 4y - 3z = -13 \\ -2y + z = 1 \\ -6z = -30 \end{cases}$ (−6, 2, 5) **23.** $\begin{cases} 4x - 7y + 5z = -52 \\ 3y + 8z = 7 \\ -z = 1 \end{cases}$ (−3, 5, −1) **24.** $\begin{cases} 2x - 3y + z = 2 \\ x - y + 2z = 2 \\ x + 2y - 3z = 4 \end{cases}$

25. $\left(-\frac{5}{7}, \frac{27}{7}, -\frac{1}{7}\right)$

25. $\begin{cases} 2x + y + 3z = 2 \\ x + y + 8z = 2 \\ x + y + z = 3 \end{cases}$ **26.** $\begin{cases} 2x - y + z = 1 \\ 2x + 2z = 4 \\ x + y + z = 4 \end{cases}$ (1, 2, 1) **27.** $\begin{cases} 2x + 5z = 5 \\ x - 3y + 2z = 2 \\ 3x + y + 3z = 3 \end{cases}$ (0, 0, 1)

28. $\begin{cases} y + 2z = \frac{3}{2} \\ 2x + 2y + 2z = 4 \\ x + y = 2 \end{cases}$ $\left(\frac{1}{2}, \frac{3}{2}, 0\right)$ **29.** $\begin{cases} 3x + 3y = -2 \\ x + z = 4 \\ 2x + y = 0 \end{cases}$ $\left(\frac{2}{3}, -\frac{4}{3}, \frac{10}{3}\right)$ **30.** $\begin{cases} 2x + y + 4z = 4 \\ x - 3y + 2z = 2 \\ 3x + y + 6z = 6 \end{cases}$ $(2 - 2z, 0, z)$

Classify each system as inconsistent, dependent, or independent.

31. dependent
32. inconsistent
33. independent
34. independent
35. dependent
36. dependent

31 $\begin{cases} x + y = 3 \\ 2x + 2y = 6 \end{cases}$ **32** $\begin{cases} x + y = 0 \\ x + y = 1 \end{cases}$ **33** $\begin{cases} 2x + 3y = 8 \\ 3x + 2y = 7 \end{cases}$

34 $\begin{cases} 2x + y + 4z = 1 \\ 3x - y + z = 2 \\ x + 2y - z = -1 \end{cases}$ **35** $\begin{cases} x + y + z = 2 \\ 3x + 2y + z = 3 \\ 6x + 4y + 2z = 6 \end{cases}$ **36** $\begin{cases} 3x + 2y + z = 1 \\ x - y - z = -5 \\ 6x + 4y + 2z = 2 \end{cases}$

CHALLENGE

37a. $E = \begin{bmatrix} 2 & 0 \\ 0 & 1 \end{bmatrix}$; $EA = \begin{bmatrix} 2 & 4 \\ 3 & 4 \end{bmatrix}$

b. $E = \begin{bmatrix} 0 & 1 \\ 1 & 0 \end{bmatrix}$; $EA = \begin{bmatrix} 3 & 4 \\ 1 & 2 \end{bmatrix}$

c. $E = \begin{bmatrix} 1 & 0 \\ 2 & 1 \end{bmatrix}$; $EA = \begin{bmatrix} 1 & 2 \\ 5 & 8 \end{bmatrix}$

37. To perform a row operation on an $n \times n$ matrix A, you can perform the row operations on the identity matrix I_n to obtain a matrix E. The product EA is same as the product obtained by performing the row operations on A directly.

Let $A = \begin{bmatrix} 1 & 2 \\ 3 & 4 \end{bmatrix}$. Perform each operation below by using this technique.

First find matrix E. Then find EA. Verify that EA is equivalent to the matrix that results from performing the operations on A directly.

a. Multiply the entries in row 1 by 2.
b. Interchange rows 1 and 2.
c. Replace row 2 with the sum of 2 times the entries in row 1 and the entries in row 2.

Error Analysis

When creating an augmented matrix, some students forget to use a coefficient of 0 for terms that are "missing." You may want to suggest that they label each column of the matrix with the corresponding variable.

13. $\begin{bmatrix} 1 & 0 & \vdots & 2 \\ 0 & 1 & \vdots & 3 \end{bmatrix}$

14. $\begin{bmatrix} 1 & 0 & 0 & \vdots & 5 \\ 0 & 1 & 0 & \vdots & 3 \\ 0 & 0 & 1 & \vdots & 1 \end{bmatrix}$

15. $\begin{bmatrix} 1 & 0 & \vdots & 4 \\ 0 & 1 & \vdots & 4 \end{bmatrix}$

16. $\begin{bmatrix} 1 & 0 & 0 & \vdots & 2 \\ 0 & 1 & 0 & \vdots & 4 \\ 0 & 0 & 1 & \vdots & 6 \end{bmatrix}$

17. $\begin{bmatrix} 1 & 0 & 0 & \vdots & -2 \\ 0 & 1 & 0 & \vdots & 2 \\ 0 & 0 & 1 & \vdots & 2 \end{bmatrix}$

18. $\begin{bmatrix} 1 & 0 & 0 & \vdots & 4 \\ 0 & 1 & 0 & \vdots & 3 \\ 0 & 0 & 1 & \vdots & 4 \end{bmatrix}$

38a. Numbers in the units place that are 5 or more are represented by one horizontal rod, which is worth 5, and the number of vertical rods necessary to bring the sum to the required number.

b. Numbers in the tens place are in a separate column from numbers in the units place. Here, each horizontal rod is worth 10.

c. The first three columns represent the coefficient matrix; the remaining column represents the constant matrix.

38. CULTURAL CONNECTION: CHINA Over 2000 years ago, the Chinese used a counting board to solve systems of equations. The counting board later evolved into the abacus.

The sticks, or rod numerals, arranged on the counting board at right represent the system of equations below.

$$\begin{cases} 3x + 2y + z = 39 \\ 2x + 3y + z = 34 \\ x + 2y + 3z = 26 \end{cases}$$

a. How are numbers greater than 5 represented?

b. How are the numerals in the tens place represented?

c. How is the Chinese counting board similar to an augmented matrix?

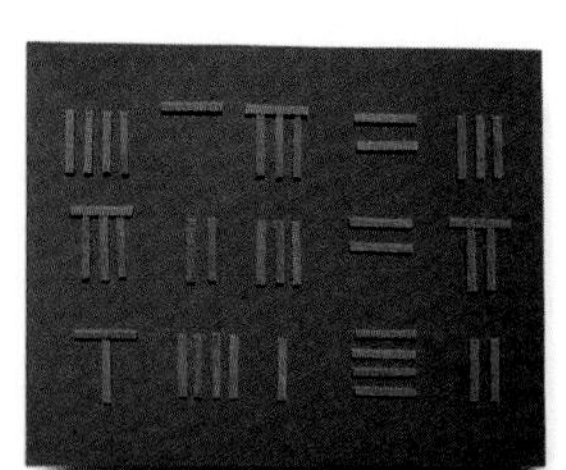

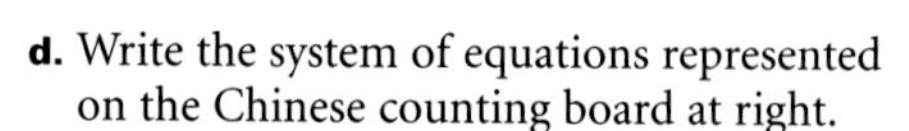

d. Write the system of equations represented on the Chinese counting board at right.

$$\begin{cases} 4x+5y+8z=23 \\ 8x+2y+3z=27 \\ 6x+4y+z=42 \end{cases}$$

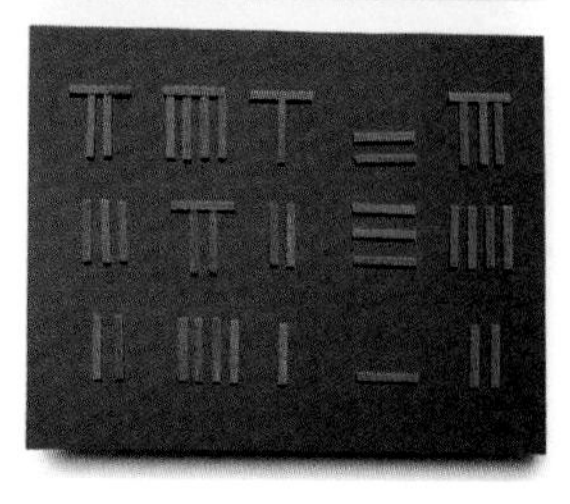

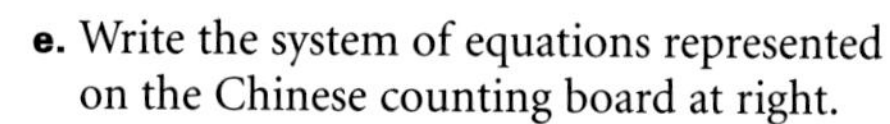

e. Write the system of equations represented on the Chinese counting board at right.

$$\begin{cases} 7x+9y+6z=28 \\ 3x+7y+2z=34 \\ 2x+4y+z=12 \end{cases}$$

CONNECTION

39. $p_1 = \frac{2}{3}$, $p_2 = \frac{1}{6}$, $p_3 = \frac{1}{6}$

39. PROBABILITY Suppose that a certain experiment has three possible outcomes with probabilities p_1, p_2, and p_3. The sum of p_1, p_2, and p_3 is 1. If $3p_1 + 18p_2 - 12p_3 = 3$ and $p_1 - 2p_2 - 2p_3 = 0$, find the probabilities of the three outcomes by solving a system of 3 equations.

APPLICATION

40. MANUFACTURING A tool company manufactures pliers and scissors. In one hour the company uses 140 units of steel and 290 units of aluminum. Each pair of scissors requires 1 unit of steel and 3 units of aluminum. Each pair of pliers contains 2 units of steel and 4 units of aluminum. How many scissors and how many pliers can the tool company make in one hour?
65 pliers, 10 scissors

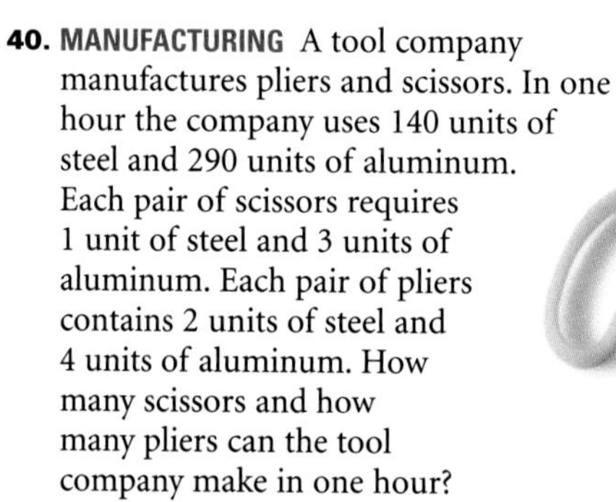

Practice

NAME ______ CLASS ______ DATE ______

Practice
4.5 Using Matrix Row Operations

Write the augmented matrix for each system of equations.

1. $\begin{cases} 4x + 5y + z = 2 \\ 7x + 9y + 2z = 7 \\ x + y + z = 2 \end{cases}$ $\left[\begin{array}{ccc|c} 4 & 5 & 1 & 2 \\ 7 & 9 & 2 & 7 \\ 1 & 1 & 1 & 2 \end{array}\right]$

2. $\begin{cases} 2x + y - 3z = 4 \\ 5x + 6y + z = 6 \\ 7x + 8y - 3z = 2 \end{cases}$ $\left[\begin{array}{ccc|c} 2 & 1 & -3 & 4 \\ 5 & 6 & 1 & 6 \\ 7 & 8 & -3 & 2 \end{array}\right]$

3. $\begin{cases} x + y + z = -1 \\ 2x + 3y + 3z = 4 \\ 6x + 7y + 3z = 8 \end{cases}$ $\left[\begin{array}{ccc|c} 1 & 1 & 1 & -1 \\ 2 & 3 & 3 & 4 \\ 6 & 7 & 3 & 8 \end{array}\right]$

Find the reduced row-echelon form of each matrix.

4. $\left[\begin{array}{ccc|c} 1 & 1 & 0 & -1 \\ 0 & 2 & -1 & -3 \\ 3 & 0 & -2 & 5 \end{array}\right]$ $\left[\begin{array}{ccc|c} 1 & 0 & 0 & 1 \\ 0 & 1 & 0 & -2 \\ 0 & 0 & 1 & -1 \end{array}\right]$

5. $\left[\begin{array}{ccc|c} 1 & 0 & -2 & 0 \\ 0 & 3 & 1 & 2 \\ -2 & 1 & 0 & 5 \end{array}\right]$ $\left[\begin{array}{ccc|c} 1 & 0 & 0 & -2 \\ 0 & 1 & 0 & 1 \\ 0 & 0 & 1 & -1 \end{array}\right]$

6. $\left[\begin{array}{ccc|c} 3 & 2 & -1 & 7 \\ 1 & -4 & 1 & -9 \\ 5 & 0 & -3 & -1 \end{array}\right]$ $\left[\begin{array}{ccc|c} 1 & 0 & 0 & 1 \\ 0 & 1 & 0 & 3 \\ 0 & 0 & 1 & 2 \end{array}\right]$

7. $\left[\begin{array}{ccc|c} 1 & 1 & 1 & 5 \\ 2 & 0 & 1 & 8 \\ 0 & 3 & 2 & 5 \end{array}\right]$ $\left[\begin{array}{ccc|c} 1 & 0 & 0 & 2 \\ 0 & 1 & 0 & -1 \\ 0 & 0 & 1 & 4 \end{array}\right]$

8. $\left[\begin{array}{ccc|c} -3 & -4 & 5 & 4 \\ 2 & 5 & 3 & -1 \\ -4 & -1 & 2 & -8 \end{array}\right]$ $\left[\begin{array}{ccc|c} 1 & 0 & 0 & 3 \\ 0 & 1 & 0 & -2 \\ 0 & 0 & 1 & 1 \end{array}\right]$

9. $\left[\begin{array}{ccc|c} 5 & 1 & -1 & 9 \\ 1 & -2 & -1 & -6 \\ 3 & 4 & 5 & 0 \end{array}\right]$ $\left[\begin{array}{ccc|c} 1 & 0 & 0 & 0 \\ 0 & 1 & 0 & 5 \\ 0 & 0 & 1 & -4 \end{array}\right]$

Solve each system of equations by using the row-reduction method.

10. $\begin{cases} 2x + y = 2 \\ 3x + 2y = 7 \end{cases}$ (−3, 8)

11. $\begin{cases} 3x + 11y = 10 \\ 2x - 5y = 19 \end{cases}$ (7, −1)

12. $\begin{cases} 3x + 4y = 1 \\ 8x + 11y = 4 \end{cases}$ (−5, 4)

13. $\begin{cases} x + y - 3z = -21 \\ 2x - y + z = 12 \\ 3x + 2y + 2z = 7 \end{cases}$ (1, −4, 6)

14. $\begin{cases} 2x + 5y - 3z = -11 \\ 3x - 2y + 4z = 7 \\ 2x + 3y - 2x = -10 \end{cases}$ (−3, 2, 5)

15. $\begin{cases} 3x + 6y - 4z = -42 \\ 2x + 2y + 3z = 14 \\ 4x + 3y - 5z = -34 \end{cases}$ (2, −4, 6)

16. $\begin{cases} x + y + z = 6 \\ 2x - 3y + 5z = -11 \\ x + 3y - 4z = 19 \end{cases}$ (3, 4, −1)

17. $\begin{cases} 3x - 2y + z = 16 \\ x + 3y + 4z = 9 \\ 2x - y + 3z = 15 \end{cases}$ (4, −1, 2)

18. $\begin{cases} 2x - 4y + 3z = -8 \\ x + 3y - 2z = 9 \\ 3x + 2y + z = 13 \end{cases}$ (1, 4, 2)

APPLICATION

41. TRAVEL A traveler is going south along the west coast of South America through the Andes mountain range. While in Peru, the traveler spent \$20 per day on housing and \$30 per day on food and travel. Passing through Bolivia, the traveler spent \$30 per day on housing and \$20 per day on food and travel. Finally, while following the long coast of Chile, the traveler spent \$20 per day both for housing and for food and travel. In each country, the traveler spent \$10 per day on miscellaneous items. The traveler spent a total of \$220 on housing, \$230 on food and travel, and \$100 on miscellaneous items. Write and solve a system of linear equations in three variables to find the number of days the traveler spent in each country.
3 days in Peru, 2 in Bolivia, 5 in Chile

Look Back

Tell whether each equation is linear. *(LESSON 1.1)*

42. $y + 1 = 5$ **yes** **43.** $y + x^2 = 6$ **no** **44.** $y + x = 5$ **yes**

State whether each relation represents a function. Explain. *(LESSON 2.3)*

45. $\{(-1, 6), (0, 6), (1, 6), (2, 6)\}$ **yes** **46.** $\{(0, 14), (1, 12), (2, 10), (3, 8)\}$ **yes**

47. $\{(-2, 0), (0, 0), (2, 0), (4, 0)\}$ **yes** **48.** $\{(-1, 0), (0, -1), (0, 1), (1, 0)\}$ **no**

49. Graph the piecewise function $f(x) = \begin{cases} x^2 - 1 & \text{if } 0 \le x < 5 \\ 3x + 9 & \text{if } 5 \le x < 10 \end{cases}$. *(LESSON 2.6)*

BUSINESS A movie theater charges \$5 for an adult ticket and \$3 for a child's ticket. The theater needs to sell at least \$2500 worth of tickets to cover its expenses. Graph the solution to each scenario. *(LESSON 3.3)*

50. The theater sells less than \$2500 worth of tickets.

51. The theater breaks even, selling exactly \$2500 worth of tickets.

52. The theater makes a profit, selling more than \$2500 worth of tickets.

53. Let $A = \begin{bmatrix} 3 & 2 & -2 \\ 1 & 1 & 4 \\ -1 & 2 & 3 \end{bmatrix}$ and $B = \begin{bmatrix} 2 & 3 & 1 \\ -1 & -2 & 2 \\ 4 & 1 & -4 \end{bmatrix}$. *(LESSON 4.2)*

a. Find AB. **b.** Find BA. **c.** Is AB equal to BA?

53a. $\begin{bmatrix} -4 & 3 & 15 \\ 17 & 5 & -13 \\ 8 & -4 & -9 \end{bmatrix}$

b. $\begin{bmatrix} 8 & 9 & 11 \\ -7 & 0 & 0 \\ 17 & 1 & -16 \end{bmatrix}$

c. no

Look Beyond

54. Let the points (0, 0), (1, 4) and (2, 12) be on the graph of the function $f(x) = ax^2 + bx + c$. Write a system of three equations in terms of a, b, and c. Solve the system and use the values to write the function f.

54. $f(x) = 2x^2 + 2x$

Look Beyond

In Exercise 54, students learn how to use a system of equations to find an equation for a parabola when they are given three points on it. You may wish to use this exercise as a transition to Chapter 5, where students will study quadratic functions and their graphs. Students will study this specific technique in greater detail in Lesson 5.7.

49.

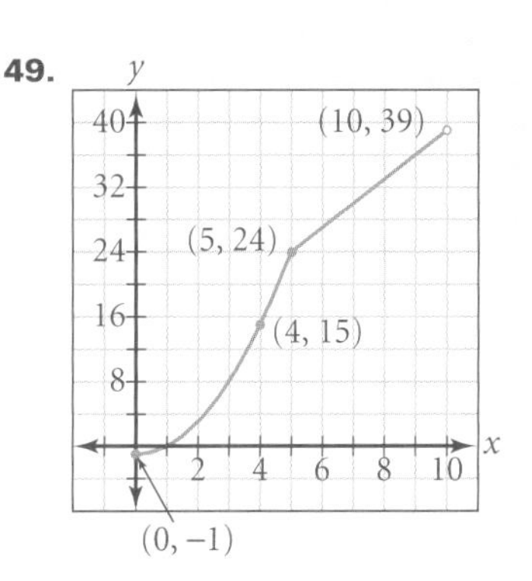

50.

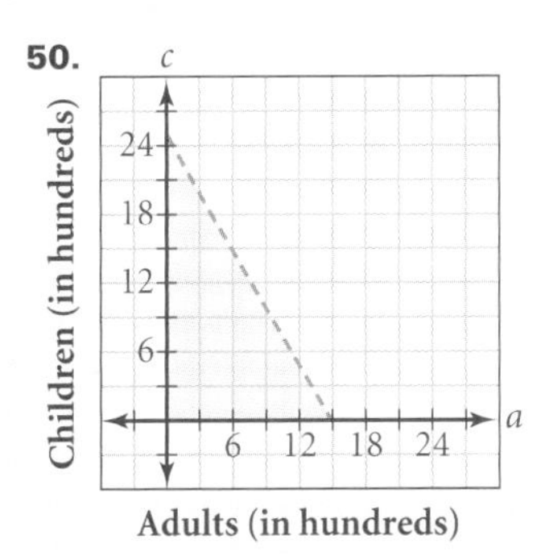

51.

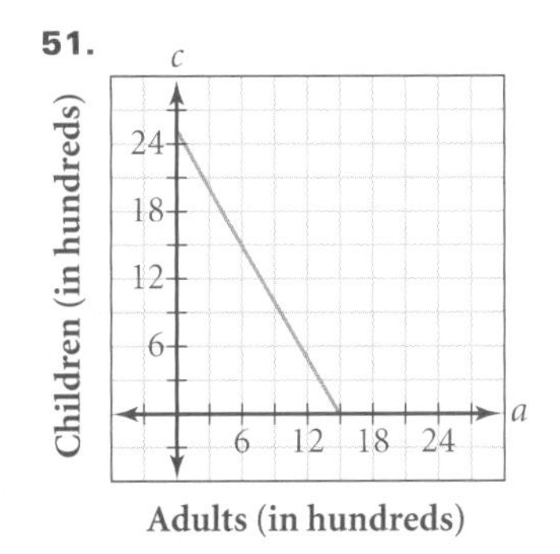

52.

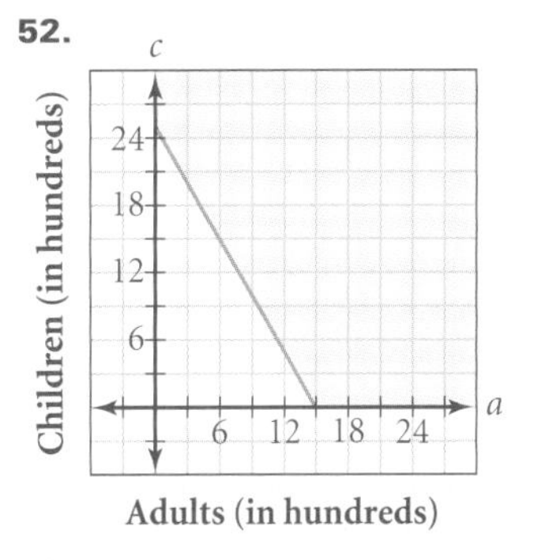

Student Technology Guide

NAME ______ CLASS ______ DATE ______

Student Technology Guide
4.5 Using Matrix Row Operations

Finding the reduced row-echelon form of an augmented matrix by hand can be a lengthy process filled with possible arithmetic errors. A graphics calculator can quickly give you the correct reduced row-echelon form. However, you must still be able to interpret the display the calculator gives you.

A system of equations in three variables and its augmented matrix, A, are shown at right.

$\begin{cases} x + y + z = 21 \\ 2x + y = 23 \\ y + 3z = 25 \end{cases} \longrightarrow A = \left[\begin{array}{ccc|c} 1 & 1 & 1 & 21 \\ 2 & 1 & 0 & 23 \\ 0 & 1 & 3 & 25 \end{array}\right]$

To find the reduced row-echelon form:
- Enter matrix A by pressing: MATRX EDIT 1:[A] ENTER 3 ENTER 4 ENTER 1 ENTER 1 ENTER ...3 ENTER 25 ENTER 2nd MODE
(The graphics calculator does not show a separator between columns.)
- Find the reduced row-echelon form for A by pressing MATRX MATH B:rref(ENTER MATRX NAMES 1:[A] ENTER) ENTER.

```
rref([A])
 [[1 0 0 8]
  [0 1 0 7]
  [0 0 1 6]]
```

The reduced row-echelon form of the matrix shows that $x = 8$, $y = 7$, and $z = 6$. This system of equations is *independent* (and *consistent*), because it has a unique solution.

You can determine whether a system is *dependent* (infinitely many solutions) or *inconsistent* (no solution) by looking at the reduced row-echelon form of its augmented matrix.

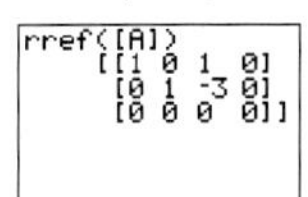

```
rref([A])
 [[1 0 1 0]
  [0 1 -3 0]
  [0 0 0 0]]
```

The last row of the matrix implies that $0z = 0$. This is true for any real number z, so the system has infinitely many solutions; it is *dependent*.

```
rref([A])
 [[1 0 -1.4 0]
  [0 1 -.2 0]
  [0 0 0 1]]
```

The last row of the matrix implies that that $0z = 1$. This is never true for any real number, so the system has no solution; it is *inconsistent*.

Solve each system. Then classify the system as independent, dependent, or inconsistent.

1. $\begin{cases} x + 2y + z = 5 \\ -y + 3z = 8 \\ -2x - 4y - 2z = -3 \end{cases}$ no solution; inconsistent

2. $\begin{cases} x - 2y + z = 7 \\ 3x + y = 5 \\ y - z = -4 \end{cases}$ $x = 2$, $y = -1$, and $z = 3$; consistent, independent

Project

Focus

Spell-checker software employs directed networks to connect words that have matching letters. Students will create a simple network of the type used in such software and use matrices to find the number of paths linking several words.

Motivate

Ask students if they have ever used the spell-checker feature of a word processor. If so, have them describe the procedure for using it and the type of information that it produces.

If time permits and the equipment is available, set up a classroom demonstration of the spell-checker feature of a word processor. Have students experiment with the various types of feedback that can be obtained. Be sure that they see the confirming message that appears when all words are spelled correctly as well as the list of suggestions provided for a misspelled word.

Activity 1

1 and 2.

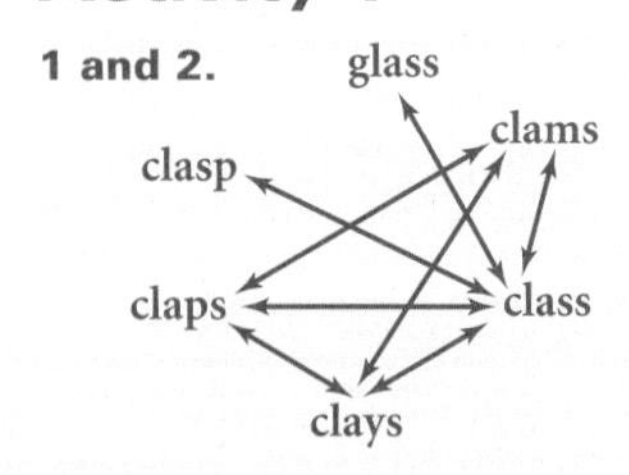

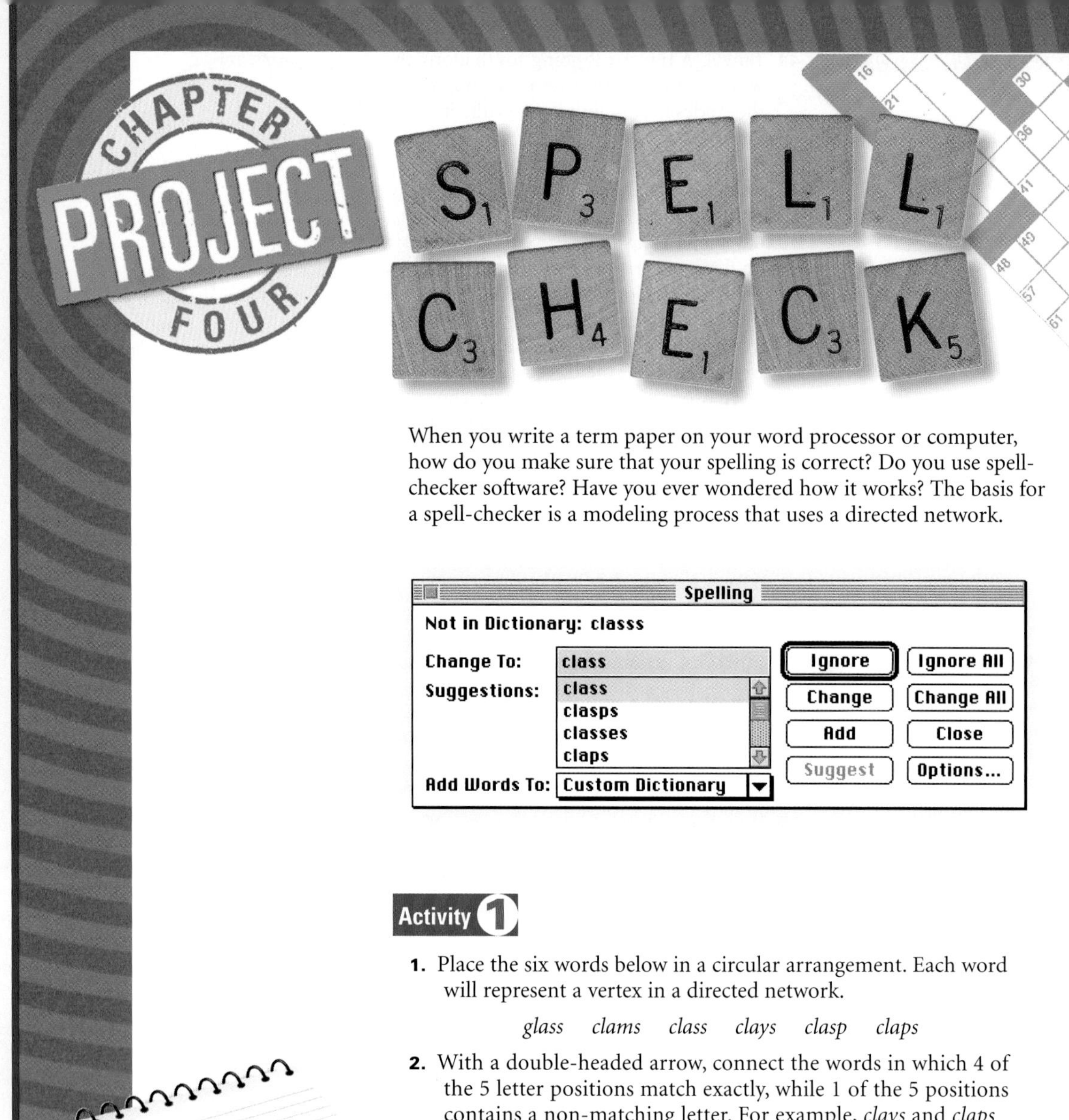

When you write a term paper on your word processor or computer, how do you make sure that your spelling is correct? Do you use spell-checker software? Have you ever wondered how it works? The basis for a spell-checker is a modeling process that uses a directed network.

Activity 1

1. Place the six words below in a circular arrangement. Each word will represent a vertex in a directed network.

   glass *clams* *class* *clays* *clasp* *claps*

2. With a double-headed arrow, connect the words in which 4 of the 5 letter positions match exactly, while 1 of the 5 positions contains a non-matching letter. For example, *clays* and *claps* can be joined with a double-headed arrow, but *clays* and *clasp* cannot be.

3. Represent the directed network in an adjacency matrix, W. Record a 1 if there is an arrow joining two words, or vertices. Otherwise record a 0.

4. Which row or column in W represents the paths leading to the word *clasp*?

5. Which row or column in W represents the paths going from the word *clays*?

3.

$$W = \begin{array}{c} \\ \text{Glass} \\ \text{Clams} \\ \text{Class} \\ \text{Clays} \\ \text{Claps} \\ \text{Clasp} \end{array} \begin{array}{c} \begin{array}{cccccc} \text{Glass} & \text{Clams} & \text{Class} & \text{Clays} & \text{Claps} & \text{Clasp} \end{array} \\ \begin{bmatrix} 0 & 0 & 1 & 0 & 0 & 0 \\ 0 & 0 & 1 & 1 & 1 & 0 \\ 1 & 1 & 0 & 1 & 1 & 1 \\ 0 & 1 & 1 & 0 & 1 & 0 \\ 0 & 1 & 1 & 1 & 0 & 0 \\ 0 & 0 & 1 & 0 & 0 & 0 \end{bmatrix} \end{array}$$

4. 6

5. 4

Activity 2

1. Using your results from Activity 1, find the number of two-stage paths from *class* to *clams*. List these paths.
2. Find the number of two-stage paths from *clams* to *clams*. List these paths.

Activity 3

The matrix $W + W^2$ gives the total number of one-stage and two-stage paths from one word to another.

1. Find $W + W^2$.
2. In how many ways can *clays* be connected to *claps* by using only a one-stage or a two-stage path? Which entry in the matrix gives this number? List the paths.

Cooperative Learning

You may wish to have students work in pairs. In **Activity 1**, partners can work together to create the adjacency matrix, with one student counting matching letters and the other student creating the matrix.

For Step 1 of **Activity 2**, one partner can find the number of paths by using the matrix while the other draws the network and counts the paths directly. For Step 2, the partners should switch roles.

In **Activity 3**, have students first do Steps 1 and 2 independently. Then partners should compare their results and resolve any discrepancies.

Discuss

Bring the class together to discuss their results. Ask students to display their diagrams. Each pair of students will probably have a slightly different visualization of the network.

As an extension, you might have students keep a record of the words that a spell checker suggests for a misspelled word. Have them share their results with the class. Ask students to make observations about the number of matched letters in the misspelled word and the suggestions.

Activity 2

1. $W^2 = \begin{bmatrix} 1 & 1 & 0 & 1 & 1 & 1 \\ 1 & 3 & 2 & 2 & 2 & 1 \\ 0 & 2 & 5 & 2 & 2 & 0 \\ 1 & 2 & 2 & 3 & 2 & 1 \\ 1 & 2 & 2 & 2 & 3 & 1 \\ 1 & 1 & 0 & 1 & 1 & 1 \end{bmatrix}$;

$n_{32} = 2$; class → clays → clams; class → claps → clams

2. $n_{22} = 3$; clams → class → clams; clams → clays → clams; clams → claps → clams

Activity 3

1. $W + W^2 = \begin{bmatrix} 1 & 1 & 1 & 1 & 1 & 1 \\ 1 & 3 & 3 & 3 & 3 & 1 \\ 1 & 3 & 5 & 3 & 3 & 1 \\ 1 & 3 & 3 & 3 & 3 & 1 \\ 1 & 3 & 3 & 3 & 3 & 1 \\ 1 & 1 & 1 & 1 & 1 & 1 \end{bmatrix}$

2. 3; row 4, column 5; clays → claps; clays → class → claps; clays → clams → claps

Chapter Review

1. $\begin{bmatrix} 5 & -7 & 5 \\ 1 & 5 & 10 \end{bmatrix}$

2. $\begin{bmatrix} 7 & -11 & 3 \\ 1 & -5 & 4 \end{bmatrix}$

3. $\begin{bmatrix} 0 & 0 & 0 \\ 0 & 0 & 0 \end{bmatrix}$

4. $\begin{bmatrix} 5 & -6 & 7 \\ -1 & 1 & 3 \end{bmatrix}$

4 Chapter Review and Assessment

VOCABULARY

Key Skills & Exercises

LESSON 4.1

Key Skills

Add and subtract matrices, and find the scalar product of a number and a matrix.

$$\begin{bmatrix} 0 & 7 \\ 1 & 2 \end{bmatrix} - \begin{bmatrix} -2 & 1 \\ 3 & 9 \end{bmatrix} + 2\begin{bmatrix} 2 & 8 \\ 0 & -1 \end{bmatrix}$$

$$= \begin{bmatrix} 0-(-2) & 7-1 \\ 1-3 & 2-9 \end{bmatrix} + \begin{bmatrix} 2(2) & 2(8) \\ 2(0) & 2(-1) \end{bmatrix}$$

$$= \begin{bmatrix} 2 & 6 \\ -2 & -7 \end{bmatrix} + \begin{bmatrix} 4 & 16 \\ 0 & -2 \end{bmatrix}$$

$$= \begin{bmatrix} 2+6 & 4+16 \\ -2+0 & -7-2 \end{bmatrix}$$

$$= \begin{bmatrix} 8 & 20 \\ -2 & -9 \end{bmatrix}$$

Exercises

Let $A = \begin{bmatrix} -1 & 2 & 1 \\ 0 & 5 & 3 \end{bmatrix}$, $B = \begin{bmatrix} 6 & -9 & 4 \\ 1 & 0 & 7 \end{bmatrix}$, and $C = \begin{bmatrix} 0 & -1 & -2 \\ 2 & 4 & 7 \end{bmatrix}$. Perform the indicated operations.

1. $A + B$
2. $B - A$
3. $C - C$
4. $B + A - C$
5. $11C$
6. $-3B$
7. $C - 3A$
8. $0.5A - 3B$
9. $-C - 2B$
10. $A + 2C - B$

LESSON 4.2

Key Skills

Multiply matrices.

In order to multiply matrices, the inner dimensions must be the same. The dimensions of the product matrix are the result of the outer dimensions. For example, the product of a 3×2 matrix and a 2×1 matrix is a 3×1 matrix.

$$\begin{bmatrix} 4 & 0 \\ -1 & 3 \\ 2 & -5 \end{bmatrix}\begin{bmatrix} 1 \\ -3 \end{bmatrix} = \begin{bmatrix} 4(1) + 0(-3) \\ -1(1) + 3(-3) \\ 2(1) + (-5)(-3) \end{bmatrix}$$

$$= \begin{bmatrix} 4 \\ -10 \\ 17 \end{bmatrix}$$

Exercises

Let $A = \begin{bmatrix} -9 & 2 \\ 4 & 1 \end{bmatrix}$, $B = \begin{bmatrix} 4 & 5 \\ -2 & 1 \\ 3 & 0 \end{bmatrix}$, and $C = \begin{bmatrix} 1 & -2 & 1 \\ 0 & 3 & -4 \end{bmatrix}$. Find each product, if possible.

11. AB
12. BC
13. BA
14. AC
15. CA
16. CB

---

5. $\begin{bmatrix} 0 & -11 & -22 \\ 22 & 44 & 77 \end{bmatrix}$

6. $\begin{bmatrix} -18 & 27 & -12 \\ -3 & 0 & -21 \end{bmatrix}$

7. $\begin{bmatrix} 3 & -7 & -5 \\ 2 & -11 & -2 \end{bmatrix}$

8. $\begin{bmatrix} -18.5 & 28 & -11.5 \\ -3 & 2.5 & -19.5 \end{bmatrix}$

9. $\begin{bmatrix} -12 & 19 & -6 \\ -4 & -4 & -21 \end{bmatrix}$

10. $\begin{bmatrix} -7 & 9 & -7 \\ 3 & 13 & 10 \end{bmatrix}$

11. not possible

12. $\begin{bmatrix} 4 & 7 & -16 \\ -2 & 7 & -6 \\ 3 & -6 & 3 \end{bmatrix}$

13. $\begin{bmatrix} -16 & 13 \\ 22 & -3 \\ -27 & 6 \end{bmatrix}$

14. $\begin{bmatrix} -9 & 24 & -17 \\ 4 & -5 & 0 \end{bmatrix}$

Chapter Test, Form A

NAME ________ CLASS ________ DATE ________

Chapter Assessment

Chapter 4, Form A, page 1

Write the letter that best answers the question or completes the statement.

__d__ 1. Let $M = \begin{bmatrix} 2 & 15 & 1 & 9 & 0 \\ 12 & 6 & 5 & 4 & 3 \\ 7 & 10 & 11 & 8 & 14 \end{bmatrix}$. What is the entry in location m_{31}?

a. 1 b. 3 c. 4 d. 7

__b__ 2. Which value of x makes $3\begin{bmatrix} 6 & 2 \\ 0 & x \end{bmatrix} - 5\begin{bmatrix} 1 & 0 \\ 2 & x \end{bmatrix} = \begin{bmatrix} 13 & 12 \\ -10 & 6 \end{bmatrix}$ true?

a. 3 b. -3 c. 5 d. -5

__c__ 3. If $AB = \begin{bmatrix} 3 & 2 & 9 & 7 \\ 5 & 1 & 0 & 6 \end{bmatrix}$ and A is a 2×3 matrix, what are the dimensions of B?

a. 2×4 b. 4×2 c. 3×4 d. 4×3

__c__ 4. Which value of x makes $\begin{bmatrix} 2 & 1 & 0 \\ 5 & x & 2 \end{bmatrix}\begin{bmatrix} 3 & 3 \\ 0 & 4 \\ 4 & x \end{bmatrix} = \begin{bmatrix} 6 & 10 \\ 23 & -9 \end{bmatrix}$ true?

a. 6 b. 1 c. -4 d. -9

__d__ 5. Let $A = \begin{bmatrix} 6 & 7 \\ 1 & 2 \end{bmatrix}$. Which of the following is A^{-1}?

a. $\begin{bmatrix} -6 & -7 \\ -1 & -2 \end{bmatrix}$ b. $\begin{bmatrix} 1 & 2 \\ 6 & 7 \end{bmatrix}$ c. $\begin{bmatrix} -1.2 & 0.2 \\ 1.4 & -0.4 \end{bmatrix}$ d. $\begin{bmatrix} 0.4 & -1.4 \\ -0.2 & 1.2 \end{bmatrix}$

__a__ 6. If the reduced row-echelon form of the augmented matrix for a system of equations is $\left[\begin{array}{ccc:c} 1 & 0 & 2 & 0 \\ 0 & 1 & 1 & 0 \\ 0 & 0 & 0 & 1 \end{array}\right]$, then the system is

a. inconsistent. b. consistent and independent.
c. consistent and dependent. d. none of these.

__b__ 7. Find the matrix form of the system $\begin{cases} 3x + 2y = 8 \\ 4x - y = -5 \end{cases}$.

a. $\begin{bmatrix} 3 & 2 \\ 4 & -1 \end{bmatrix}\begin{bmatrix} 8 \\ -5 \end{bmatrix} = \begin{bmatrix} x \\ y \end{bmatrix}$ b. $\begin{bmatrix} 3 & 2 \\ 4 & -1 \end{bmatrix}\begin{bmatrix} x \\ y \end{bmatrix} = \begin{bmatrix} 8 \\ -5 \end{bmatrix}$

c. $\begin{bmatrix} 3 & 4 \\ 2 & -1 \end{bmatrix}\begin{bmatrix} x \\ y \end{bmatrix} = \begin{bmatrix} 8 \\ -5 \end{bmatrix}$ d. $\begin{bmatrix} 3 & 4 \\ 2 & -1 \end{bmatrix}\begin{bmatrix} 8 \\ -5 \end{bmatrix} = \begin{bmatrix} x \\ y \end{bmatrix}$

NAME ________ CLASS ________ DATE ________

Chapter Assessment

Chapter 4, Form A, page 2

Write the letter that best answers the question or completes the statement.

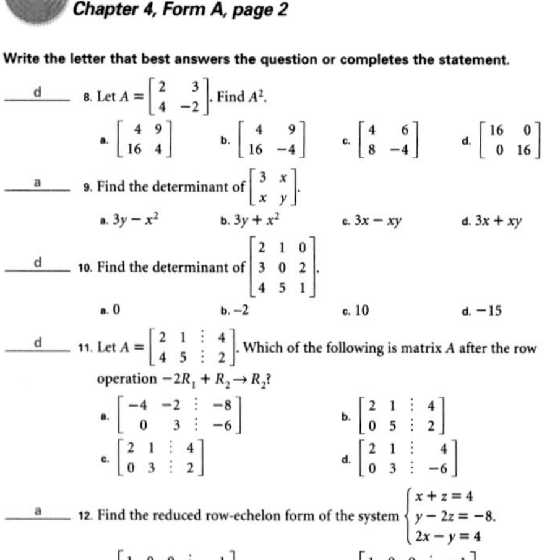

__d__ 8. Let $A = \begin{bmatrix} 2 & 3 \\ 4 & -2 \end{bmatrix}$. Find A^2.

a. $\begin{bmatrix} 4 & 9 \\ 16 & 4 \end{bmatrix}$ b. $\begin{bmatrix} 4 & 9 \\ 16 & -4 \end{bmatrix}$ c. $\begin{bmatrix} 4 & 6 \\ 8 & -4 \end{bmatrix}$ d. $\begin{bmatrix} 16 & 0 \\ 0 & 16 \end{bmatrix}$

__a__ 9. Find the determinant of $\begin{bmatrix} 3 & x \\ x & y \end{bmatrix}$.

a. $3y - x^2$ b. $3y + x^2$ c. $3x - xy$ d. $3x + xy$

__d__ 10. Find the determinant of $\begin{bmatrix} 2 & 1 & 0 \\ 3 & 0 & 2 \\ 4 & 5 & 1 \end{bmatrix}$.

a. 0 b. -2 c. 10 d. -15

__d__ 11. Let $A = \left[\begin{array}{cc:c} 2 & 1 & 4 \\ 4 & 5 & 2 \end{array}\right]$. Which of the following is matrix A after the row operation $-2R_1 + R_2 \rightarrow R_2$?

a. $\left[\begin{array}{cc:c} -4 & -2 & -8 \\ 0 & 3 & -6 \end{array}\right]$ b. $\left[\begin{array}{cc:c} 2 & 1 & 4 \\ 0 & 5 & 2 \end{array}\right]$

c. $\left[\begin{array}{cc:c} 2 & 1 & 4 \\ 0 & 3 & 2 \end{array}\right]$ d. $\left[\begin{array}{cc:c} 2 & 1 & 4 \\ 0 & 3 & -6 \end{array}\right]$

__a__ 12. Find the reduced row-echelon form of the system $\begin{cases} x + z = 4 \\ y - 2z = -8. \\ 2x - y = 4 \end{cases}$

a. $\left[\begin{array}{ccc:c} 1 & 0 & 0 & 1 \\ 0 & 1 & 0 & -2 \\ 0 & 0 & 1 & 3 \end{array}\right]$ b. $\left[\begin{array}{ccc:c} 1 & 0 & 0 & -1 \\ 0 & 1 & 0 & 2 \\ 0 & 0 & 1 & -3 \end{array}\right]$

c. $\left[\begin{array}{ccc:c} 1 & 0 & 1 & 1 \\ 0 & 1 & 0 & -2 \\ 1 & 0 & 1 & -3 \end{array}\right]$ d. $\left[\begin{array}{ccc:c} 0 & 0 & 1 & 1 \\ 0 & 1 & 0 & -2 \\ 1 & 0 & 0 & 3 \end{array}\right]$

__b__ 13. Solve the equation $\begin{bmatrix} 2 & 3 & 4 \\ 5 & -2 & 3 \\ -3 & 5 & -2 \end{bmatrix}\begin{bmatrix} x \\ y \\ z \end{bmatrix} = \begin{bmatrix} -6 \\ 1 \\ 8 \end{bmatrix}$.

a. $(2, -4, -5)$ b. $(4, 2, -5)$ c. $(-5, 2, 4)$ d. $(4, -2, 5)$

LESSON 4.3

Key Skills

Find the inverse and the determinant of a matrix.

Let $C = \begin{bmatrix} 4 & 2 \\ -1 & -3 \end{bmatrix}$. Find the determinant and the inverse of the matrix, if it exists.

$$\det(C) = (4)(-3) - (2)(-1) = -10$$

If $\det(C) \neq 0$, then C^{-1} exists.

Use a graphics calculator to find the inverse of the matrix.

$$C^{-1} = \begin{bmatrix} 0.3 & 0.2 \\ -0.1 & -0.4 \end{bmatrix}$$

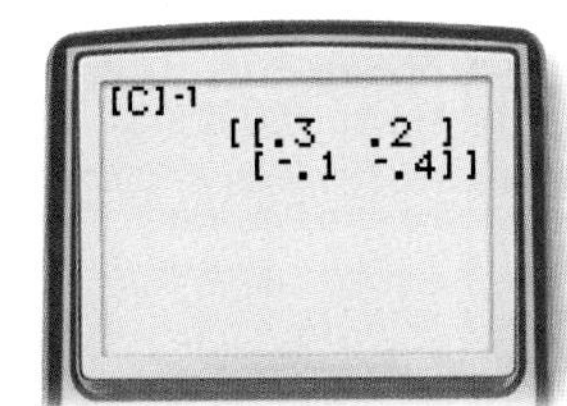

Exercises

Find the determinant and inverse of each matrix. If the inverse matrix does not exist, write *no inverse*.

17. $\begin{bmatrix} -2 & 3 \\ 5 & 0 \end{bmatrix}$ **18.** $\begin{bmatrix} 3 & 1 \\ 7 & 9 \end{bmatrix}$ **19.** $\begin{bmatrix} 2 & 2 \\ 2 & 2 \end{bmatrix}$

20. $\begin{bmatrix} 3 & 3 \\ 1 & 1 \end{bmatrix}$ **21.** $\begin{bmatrix} 5 & -1 \\ -1 & 5 \end{bmatrix}$ **22.** $\begin{bmatrix} 0 & 7 \\ -2 & 0 \end{bmatrix}$

23. $\begin{bmatrix} -3 & -4 \\ -1 & -3 \end{bmatrix}$ **24.** $\begin{bmatrix} 5 & 6 \\ 7 & 8 \end{bmatrix}$ **25.** $\begin{bmatrix} -9 & 12 \\ 3 & -4 \end{bmatrix}$

26. Let $A = \begin{bmatrix} 5 & 7 \\ 7 & 10 \end{bmatrix}$ and $B = \begin{bmatrix} 10 & -7 \\ -7 & 5 \end{bmatrix}$. Show that A and B are inverses of one another.

LESSON 4.4

Key Skills

Use matrices to solve systems of linear equations.

Write the system as a matrix equation, $AX = B$. Insert 0 for any missing variables in an equation.

$$\begin{cases} x + 2y - z = -3 \\ -2x + y - 3z = -9 \\ y + 2z = 3 \end{cases} \rightarrow \overset{A}{\begin{bmatrix} 1 & 2 & -1 \\ -2 & 1 & -3 \\ 0 & 1 & 2 \end{bmatrix}} \times \overset{X}{\begin{bmatrix} x \\ y \\ z \end{bmatrix}} = \overset{B}{\begin{bmatrix} -3 \\ -9 \\ 3 \end{bmatrix}}$$

Solve the matrix equation by using an inverse matrix.

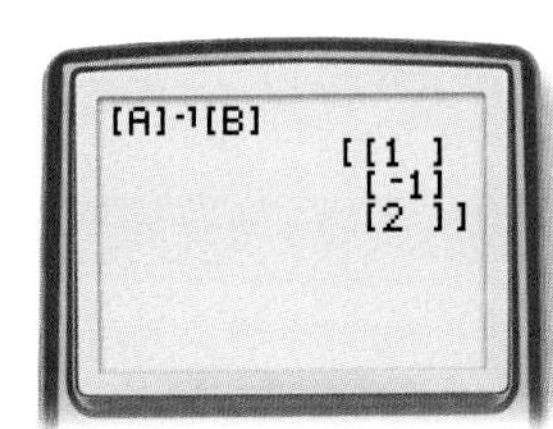

$$X = A^{-1}B \rightarrow \begin{bmatrix} x \\ y \\ z \end{bmatrix} = \begin{bmatrix} 1 \\ -1 \\ 2 \end{bmatrix}$$

$x = 1$, $y = -1$, and $z = 2$

Exercises

Use a matrix equation to solve each system of linear equations, if possible.

27. $\begin{cases} 6x + 4y = 12 \\ 3x + 2y = 6 \end{cases}$ no solution

28. $\begin{cases} 2x + y = 3 \\ 3x - 2y = 8 \end{cases}$ (2, −1)

29. $\begin{cases} 3x - y = -1 \\ 2x - y + z = -6 \\ x + 4y - z = 9 \end{cases}$ (0, 1, −5)

30. $\begin{cases} 2x - 3y + 5z = 5 \\ x - y - 2z = 2 \\ -3x + 3y + 6z = 5 \end{cases}$ no solution

31. $\begin{cases} -\frac{1}{6}y = -\frac{1}{6} - \frac{1}{3}x \\ x + \frac{1}{2}z = -3 + \frac{1}{2}y \\ \frac{1}{5}x + \frac{4}{5}y = \frac{9}{5} + \frac{1}{5}z \end{cases}$ (0, 1, −5)

32. JEWELRY A jeweler plans to combine two silver alloys to make 50 grams of a new alloy that is 75% silver and contains 37.5 grams of pure silver (75% of 50 grams). If one silver alloy is 80% silver and the other is 60% silver, how many grams of each are needed?

37.5 g of 80%, 12.5 g of 60%

Chapter Test, Form B

NAME ______ CLASS ______ DATE ______

Chapter Assessment

Chapter 4, Form B, page 1

Perform each operation for the given matrices, if possible.

$A = \begin{bmatrix} 5 & 2 \\ 9 & 4 \end{bmatrix}$ $B = \begin{bmatrix} -4 & -1 \\ 0 & -2 \end{bmatrix}$ $C = \begin{bmatrix} 3 & 1 \\ 2 & 4 \\ 0 & 1 \end{bmatrix}$ $D = \begin{bmatrix} 1 & 2 & 0 \\ 5 & 3 & 1 \\ 0 & 6 & 2 \end{bmatrix}$

1. $-2B + 3A$ $\begin{bmatrix} 23 & 8 \\ 27 & 16 \end{bmatrix}$
2. C^2 not possible
3. DC $\begin{bmatrix} 7 & 9 \\ 21 & 18 \\ 12 & 26 \end{bmatrix}$
4. A^{-1} $\begin{bmatrix} 2 & -1 \\ -4.5 & 2.5 \end{bmatrix}$

Find x in each equation.

5. $\begin{bmatrix} 4-2x & 7 \\ 9 & 3 \end{bmatrix} = \begin{bmatrix} 2x+16 & 7 \\ 9 & 3 \end{bmatrix}$ $x = -3$
6. $\begin{bmatrix} 1 & 3x & 2 \\ 3 & 0 & 5 \end{bmatrix}\begin{bmatrix} 4 & 1 \\ 2 & 0 \\ 0 & 1 \end{bmatrix} - \begin{bmatrix} 16 & 3 \\ 6 & 2 \end{bmatrix} = 3\begin{bmatrix} x-2 & 0 \\ 2 & 2 \end{bmatrix}$ $x = 2$

7. a. Find the determinant of the matrix $B = \begin{bmatrix} 7 & 8 & 9 \\ 5 & 10 & 15 \\ 8 & 10 & 12 \end{bmatrix}$. 0

   b. Find B^{-1}, if it exists. does not exist

8. a. Write the system as a matrix equation. $\begin{cases} 2x + 3y + z = 12 \\ 3x + y - 4z = -5 \\ -4x - y + 3z = 6 \end{cases}$

   $\begin{bmatrix} 2 & 3 & 1 \\ 3 & 1 & -4 \\ -4 & -1 & 3 \end{bmatrix}\begin{bmatrix} x \\ y \\ z \end{bmatrix} = \begin{bmatrix} 12 \\ -5 \\ 6 \end{bmatrix}$

   b. Solve the system by using a matrix inverse. (−2, 5, 1)

9. Find x and y in the equation $\begin{bmatrix} x & y \\ 0 & 4 \end{bmatrix}\begin{bmatrix} 3 & 5 \\ 2 & 3 \end{bmatrix} = \begin{bmatrix} 17 & 27 \\ 8 & 12 \end{bmatrix}$. (3, 4)

NAME ______ CLASS ______ DATE ______

Chapter Assessment

Chapter 4, Form B, page 2

10. Find matrix $A = \left[\begin{array}{ccc|c} 2 & 3 & 4 & 1 \\ 2 & 1 & 3 & -1 \\ 6 & 2 & 0 & 5 \end{array}\right]$ after the row operation $-3R_2 + R_3 \rightarrow R_3$. $\left[\begin{array}{ccc|c} 2 & 3 & 4 & 1 \\ 2 & 1 & 3 & -1 \\ 0 & -1 & -9 & 8 \end{array}\right]$

11. a. Write an augmented matrix for the system $\begin{cases} x - 2z = 2 \\ -2x + 2y + 6z = 2 \\ 3y + 3z = 9 \end{cases}$. $\left[\begin{array}{ccc|c} 1 & 0 & -2 & 2 \\ -2 & 2 & 6 & 2 \\ 0 & 3 & 3 & 9 \end{array}\right]$

    b. Write your matrix from part a in reduced row-echelon form. $\left[\begin{array}{ccc|c} 1 & 0 & -2 & 2 \\ 0 & 1 & 1 & 3 \\ 0 & 0 & 0 & 0 \end{array}\right]$

    c. Classify the system as inconsistent, consistent and independent, or consistent and dependent. Write all solutions to the system. consistent and dependent; $(2z + 2, 3 - z, z)$

12. Represent the network at right as an adjacency matrix A.

	A	B	C
A	0	2	1
B	2	0	1
C	1	1	1

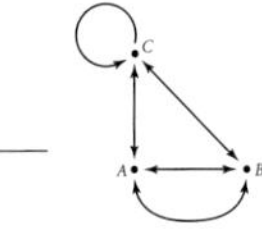

13. The measure of a triangle's largest angle is 6 times the measure of its smallest angle. The measure of the largest angle is also twice the sum of the measures of the other two angles.

    a. Write a system of three equations using the measures of the angles of the triangle. $\begin{cases} x + y + z = 180 \\ x = 6z \\ x = 2(y + z) \end{cases}$

    b. Find the measures of the angles. 120°, 40°, and 20°

14. A grocery store sells peanuts for \$3.20 per pound and cashews for \$8 per pound. The grocer wants to make 100 pounds of a mixture of peanuts and cashews that can be sold for \$4.40 per pound. How many pounds of each type of nut should the grocer use? 75 lb of peanuts and 25 lb of cashews

15. not possible

16. $\begin{bmatrix} 11 & 3 \\ -18 & 3 \end{bmatrix}$

17. -15; $\begin{bmatrix} 0 & 0.2 \\ 0.3 & 0.1\overline{3} \end{bmatrix}$

18. 20; $\begin{bmatrix} 0.45 & -0.05 \\ -0.35 & 0.15 \end{bmatrix}$

19. 0; no inverse

20. 0; no inverse

21. 24; $\begin{bmatrix} 0.208\overline{3} & 0.041\overline{6} \\ 0.041\overline{6} & 0.208\overline{3} \end{bmatrix}$

22. 14; $\begin{bmatrix} 0 & -0.5 \\ 0.142857 & 0 \end{bmatrix}$

23. 5; $\begin{bmatrix} -0.6 & 0.8 \\ 0.2 & -0.6 \end{bmatrix}$

24. -2; $\begin{bmatrix} -4 & 3 \\ 3.5 & -2.5 \end{bmatrix}$

25. 0; no inverse

26. $AB = BA = I$, so A and B are inverses.

LESSON 4.5

Key Skills

Solve a system of linear equations by using elementary row operations.

Write the system as an augmented matrix.

$$\begin{cases} 2x + 3y - z = -7 \\ x + y - z = -4 \\ 3x - 2y - 3z = -7 \end{cases} \rightarrow \begin{bmatrix} 2 & 3 & -1 & \vdots & -7 \\ 1 & 1 & -1 & \vdots & -4 \\ 3 & -2 & -3 & \vdots & -7 \end{bmatrix}$$

Use a calculator to obtain the reduced row-echelon form.

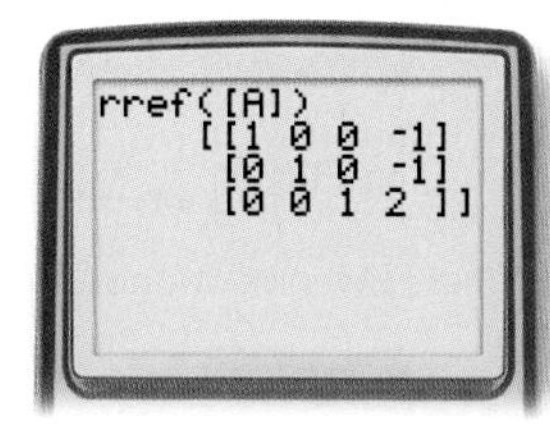

The solution for this independent system is $x = -1$, $y = -1$, and $z = 2$.

From the reduced row-echelon form of an augmented matrix, you can classify the system as dependent or inconsistent.

Dependent system

$$\begin{bmatrix} 1 & 0 & 5 & \vdots & -13 \\ 0 & 1 & 3.5 & \vdots & -9.5 \\ 0 & 0 & 0 & \vdots & 0 \end{bmatrix}$$

0 = 0 is always a true statement.

Inconsistent system

$$\begin{bmatrix} 1 & 0 & 1 & \vdots & 0 \\ 0 & 1 & 3 & \vdots & 0 \\ 0 & 0 & 0 & \vdots & 1 \end{bmatrix}$$

0 = 1 is always a false statement.

Exercises

Use the row-reduction method to solve each system, if possible. Then classify each system as independent, dependent, or inconsistent.

33. $\begin{cases} 3x - 2y + z = -5 \\ -2x + 3y - 3z = 12 \\ 3x - 2y - 2z = 4 \end{cases}$ (0, 1, –3); indep.

34. $\begin{cases} 3x - y + 2z = 9 \\ x - 2y - 3z = -1 \\ 2x - 3y + z = 10 \end{cases}$ (1, –2, 2); indep.

35. $\begin{cases} x - 2y + z = -2 \\ 2x + 6y = 12 \\ 3x - y + 2z = 4 \end{cases}$ no solution; inconsistent

36. $\begin{cases} 3x - y - 2z = 0 \\ x + 2y - 4z = 0 \\ 2x - 10y + 12z = 0 \end{cases}$ $\left(\frac{8}{7}z, \frac{10}{7}z, z\right)$; dependent

37. $\begin{cases} x - 3y - z = 0 \\ 2x - y - 4z = 0 \\ -2x + 6y - 4z = 0 \end{cases}$ (0, 0, 0); indep.

38. $\begin{cases} 4x + 6y - 2z = 10 \\ 4x - 5y + 5z = -3 \\ 3x - y + 2z = 1 \end{cases}$ $\left(\frac{8}{11} - \frac{5}{11}z, \frac{13}{11} + \frac{7}{11}z, z\right)$; dependent

Application

39. **FUND-RAISING** Two hundred and ten people attended a school carnival. The total amount of money collected for tickets was \$710. Prices were \$5 for regular admission, \$3 for students, and \$1 for children. The number of regular tickets sold was 10 more than twice the number of child tickets sold. Write a system of equations to find the number of regular tickets, student tickets, and child tickets sold. Solve the system by using a matrix equation. 70 regular, 110 student, 30 child

Chapter Test

For Items 1–4, let $A = \begin{bmatrix} 2 & 5 & -3 \\ 4 & -1 & 7 \end{bmatrix}$, $B = \begin{bmatrix} -4 & 1 & 0 \\ 2 & -5 & -1 \end{bmatrix}$, **and** $C = \begin{bmatrix} 0 & -2 & 1 \\ 1 & 0 & 3 \end{bmatrix}$. **Perform the indicated operations.**

1. $A + B$ **2.** $2C - B$

3. $A - B + 3C$ **4.** $-2C$

5. COORDINATE GEOMETRY Represent the quadrilateral $ABCD$ as a matrix P.

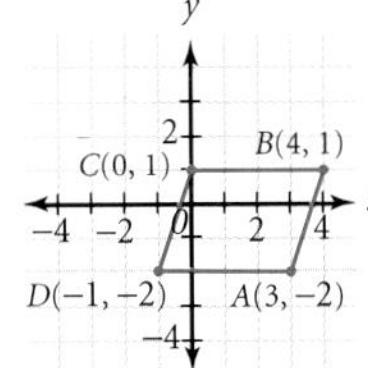

For Items 6–11, let $A = \begin{bmatrix} 2 & 3 & -1 \\ 0 & -5 & 4 \end{bmatrix}$, $B = \begin{bmatrix} 1 & 0 \\ 2 & -1 \end{bmatrix}$, **and** $C = \begin{bmatrix} 1 & -3 \\ 0 & 2 \\ -2 & -1 \end{bmatrix}$. **Find each product, if possible.**

6. AB not possible **7.** BA

8. AC **9.** CB

10. CA **11.** CBA

Find the determinant and inverse of each matrix. If the inverse matrix does not exist, write *no inverse*.

12. $\begin{bmatrix} 1 & 2 \\ 3 & 4 \end{bmatrix}$ **13.** $\begin{bmatrix} -5 & 4 \\ 10 & -8 \end{bmatrix}$ 0; no inverse

14. $\begin{bmatrix} 1 & 0 & -3 \\ 0 & 2 & -1 \\ 1 & 3 & 0 \end{bmatrix}$ **15.** $\begin{bmatrix} 2 & 0 & 3 \\ 4 & 0 & 1 \\ 2 & -1 & 3 \end{bmatrix}$

Use matrices to solve each system of linear equations, if possible.

16. $\begin{cases} 3x + 4y = -1 \\ 4x - 3y = 32 \end{cases}$ (5, −4)

17. $\begin{cases} 7x + 3y = 25 \\ 9x - 2y = -3 \end{cases}$ (1, 6)

18. $\begin{cases} x + y + z = 6 \\ x + y - z = 0 \\ 3x + 2y + z = 10 \end{cases}$ (1, 2, 3)

19. CHEMISTRY A radiator in a large truck has a capacity of 10 gallons. Currently the radiator fluid is 20% antifreeze. How much fluid should be drained out and replaced with a 75% antifreeze solution to bring the solution in the radiator to 50% antifreeze? Set up and solve a system of equations to answer this question. 5.45 gal.

Use the row-reduction method to solve each system, if possible. Then classify each system as independent, dependent, or inconsistent.

20. $\begin{cases} 5x - 7y + 4z = -3 \\ 3x - y + 2z = 1 \\ -2x - 3y + 5z = 2 \end{cases}$ $\begin{bmatrix} 1 & 0 & 0 & 0 \\ 0 & 1 & 0 & 1 \\ 0 & 0 & 1 & 1 \end{bmatrix}$; (0,1,1); independent

21. $\begin{cases} x + 6y + 2z = 4 \\ -2x + 3y + 5z = 9 \\ 7x + 12y - 4z = 0 \end{cases}$ $\begin{bmatrix} 1 & 0 & -1.6 & 0 \\ 0 & 1 & 0.6 & 0 \\ 0 & 0 & 0 & 1 \end{bmatrix}$; inconsistent

22. $\begin{cases} 2x - y + 3z = -5 \\ x + 3y - 4z = 9 \\ 3x + 2y - z = 4 \end{cases}$

23. ENTERTAINMENT Concert tickets are \$24 for adults, \$15 for children and \$12 for seniors. The revenue from the concert was \$5670. Five times as many adults attended as seniors. Twice as many children attended as seniors. Set up and solve a system of equations to determine how many of each ticket were sold. $a = 175$, $c = 70$, $s = 35$

23. $\begin{cases} 24a + 15c + 12s = 5{,}670 \\ a = 5s \\ c = 2s \end{cases}$

$a = 175$, $c = 70$, $s = 35$

Chapter Test

1. $\begin{bmatrix} -2 & 6 & -3 \\ 6 & -6 & 6 \end{bmatrix}$

2. $\begin{bmatrix} 4 & -5 & 2 \\ 0 & 5 & 7 \end{bmatrix}$

3. $\begin{bmatrix} 6 & -2 & 0 \\ 5 & 4 & 17 \end{bmatrix}$

4. $\begin{bmatrix} 0 & 4 & -2 \\ -2 & 0 & -6 \end{bmatrix}$

5. $P = \begin{bmatrix} 3 & 4 & 0 & -1 \\ -2 & 1 & 1 & -2 \end{bmatrix}$

7. $\begin{bmatrix} 2 & 3 & -1 \\ 4 & 11 & -6 \end{bmatrix}$

8. $\begin{bmatrix} 4 & 1 \\ -8 & -14 \end{bmatrix}$

9. $\begin{bmatrix} -5 & 3 \\ 4 & -2 \\ -4 & 1 \end{bmatrix}$

10. $\begin{bmatrix} 2 & 18 & -13 \\ 0 & -10 & 8 \\ -4 & -1 & -2 \end{bmatrix}$

11. $\begin{bmatrix} -10 & -30 & 17 \\ 8 & 22 & -12 \\ -8 & -17 & 8 \end{bmatrix}$

12. -2; $\begin{bmatrix} -2 & 1 \\ 1.5 & -0.5 \end{bmatrix}$

14. 9; $\begin{bmatrix} \frac{1}{3} & -1 & \frac{2}{3} \\ -\frac{1}{9} & \frac{1}{3} & \frac{1}{9} \\ -\frac{2}{9} & -\frac{1}{3} & \frac{2}{9} \end{bmatrix}$

15. -10; $\begin{bmatrix} -0.1 & 0.3 & 0 \\ 1 & 0 & -1 \\ 0.4 & -0.2 & 0 \end{bmatrix}$

19. $0.20x + 0.75y = 0.50(10)$
$x + y = 10$
$x = 4.55$, $y = 5.45$
5.45 gal.

22. $\begin{bmatrix} 1 & 0 & \frac{5}{7} & -\frac{6}{7} \\ 0 & 1 & -\frac{11}{7} & \frac{23}{7} \\ 0 & 0 & 0 & 0 \end{bmatrix}$

$\left(-\frac{5}{7}z - \frac{6}{7}, \frac{11}{7}z + \frac{23}{7}, z\right)$

dependent

College Entrance Exam Practice

Multiple-Choice and Quantitative-Comparison Samples

The first half of the Cumulative Assessment contains two types of items found on standardized tests—multiple-choice questions and quantitative-comparison questions. Quantitative-comparison items emphasize the concepts of equality, inequality, and estimation.

Free-Response Grid Samples

The second half of the Cumulative Assessment is a free-response section. This part of the Cumulative Assessment requires student-produced response items like those commonly found on college entrance exams. These questions require the use of machine-scored answer grids. You may wish to have students practice answering these items in preparation for standardized tests.

CHAPTERS 1–4 CUMULATIVE ASSESSMENT

College Entrance Exam Practice

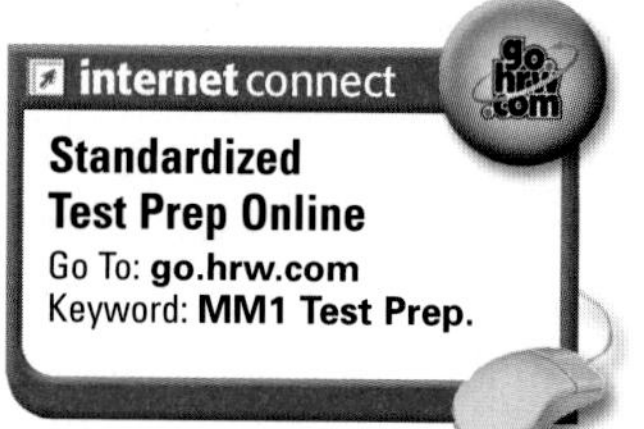

QUANTITATIVE COMPARISON For Items 1–4, write
A if the quantity in Column A is greater than the quantity in Column B;
B if the quantity in Column B is greater than the quantity in Column A;
C if the two quantities are equal; or
D if the relationship cannot be determined from the given information.

	Column A	Column B	Answers
1. A	$\begin{bmatrix} 2 & 3x+1 \\ 2y-1 & 5 \end{bmatrix} = \begin{bmatrix} 2x-2 & -3y+1 \\ -5 & 5 \end{bmatrix}$ x	y	Ⓐ Ⓑ Ⓒ Ⓓ [Lesson 4.1]
2. B	$\lfloor 5.001 \rfloor - \lceil 3.125 \rceil$	$\lceil 1.75 \rceil - \lfloor 0.99 \rfloor$	Ⓐ Ⓑ Ⓒ Ⓓ [Lesson 2.6]
3. A	The value of x $\frac{x}{2} - 1 = 3$	$1 - 2x = 5$	Ⓐ Ⓑ Ⓒ Ⓓ [Lesson 1.6]
4. A	$\begin{cases} y = 10 - 3x \\ 2x + 3y = -5 \end{cases}$ x	y	Ⓐ Ⓑ Ⓒ Ⓓ [Lesson 3.1]

5. Find the slope of the line containing the points (4, 5) and (6, 3). ***(LESSON 1.3)***
d
a. −2 **b.** 2 **c.** 1 **d.** −1

6. Simplify $\left(\frac{-2x^4y^{-1}}{5x^{-2}y^4}\right)^3$. ***(LESSON 2.2)***
b
a. $\frac{8}{5}x^{14}y^{-7}$ **b.** $-\frac{8}{125}x^{18}y^{-15}$
c. $\frac{8}{125}x^{-18}y^{15}$ **d.** $\frac{8}{5}x^{18}y^{-9}$

7. If $A = \begin{bmatrix} 3 & 0 & -2 \\ -1 & 2 & 1 \\ 1 & 5 & -1 \end{bmatrix}$ and $B = \begin{bmatrix} 0 & -3 & -1 \\ -1 & 4 & 3 \\ 2 & 2 & 2 \end{bmatrix}$, what is $A - B$? ***(LESSON 4.1)***
a
a. $\begin{bmatrix} 3 & 3 & -1 \\ 0 & -2 & -2 \\ -1 & 3 & -3 \end{bmatrix}$ **b.** $\begin{bmatrix} 3 & -3 & -3 \\ -2 & 6 & 4 \\ 3 & 7 & 1 \end{bmatrix}$

c. $\begin{bmatrix} 3 & -3 & -3 \\ 0 & -2 & 4 \\ -1 & 7 & 1 \end{bmatrix}$ **d.** $\begin{bmatrix} 3 & 3 & -1 \\ -2 & -2 & -2 \\ 3 & 3 & -3 \end{bmatrix}$

8. Which equation in x and y represents this system of parametric equations? ***(LESSON 3.6)***
b
$\begin{cases} x(t) = 2t + 1 \\ y(t) = t - 2 \end{cases}$
a. $y = x - 5$
b. $y = \frac{1}{2}x - \frac{5}{2}$
c. $y = x - t - 3$
d. $y = \frac{1}{2}x + 3$

9. Which set of ordered pairs represents a function? ***(LESSON 2.3)***
d
a. {(8, −9), (−9, 8), (8, 9)}
b. {(0.1, 5), (0.2, 5), (0.1, 0)}
c. {(−1, 4), (3, −2), (3, 2), (4, −1)}
d. {(0, 1), (2, −2), (−2, 2), (1, 0)}

10. What is the inverse of {(0, 1), (2, −1), (3, 2), (4, −1)}? Is the inverse a function? ***(LESSON 2.5)***
d
a. {(0, 1), (2, −1), (3, 2), (4, −1)}; yes
b. {(0, 1), (2, −1), (3, 2), (4, −1)}; no
c. {(1, 0), (−1, 2), (2, 3), (−1, 4)}; yes
d. {(1, 0), (−1, 2), (2, 3), (−1, 4)}; no

11. Let $A = \begin{bmatrix} 3 & -1 \\ -4 & 2 \end{bmatrix}$. Find A^{-1}. ***(LESSON 4.3)***

b

a. $\begin{bmatrix} 3 & 1 \\ 4 & 2 \end{bmatrix}$ **b.** $\begin{bmatrix} 1 & 0.5 \\ 2 & 1.5 \end{bmatrix}$

c. $\begin{bmatrix} 2 & 1 \\ 4 & 3 \end{bmatrix}$ **d.** $\begin{bmatrix} 1.5 & 0.5 \\ 2 & 1 \end{bmatrix}$

ANATOMY The table below lists the height and weight of 10 young adult males with medium frames. The heights are in inches, and the weights are in pounds. ***(LESSON 1.5)***

Height	65	66	70	69	67	68	67	66	68	71
Weight	146	145	157	158	145	149	155	141	154	159

12. Create a scatter plot, and identify the correlation as positive, negative, or none. pos.

13. Use the least-squares line to estimate the weight of a 74-inch-tall male with a medium frame. about 169 lb

14. Solve $3 - 2x \geq -x + 2$, and graph the solution on a number line. ***(LESSON 1.7)***

For Items 15 and 16, find $f \circ g$ and $g \circ f$.
(LESSON 2.4)

15. $f(x) = 3x$, $g(x) = 2x - 1$ $6x - 3$; $6x - 1$

16. $f(x) = 2 - x$, $g(x) = x + 3$ $-x - 1$; $-x + 5$

17. Find the inverse of $f(x) = -5x + 3$. Is the inverse a function? ***(LESSON 2.5)***

18. Graph the piecewise function below.

$$f(x) = \begin{cases} x - 6 & \text{if } 0 \leq x < 4 \\ -2x & \text{if } 4 \leq x < 10 \end{cases}$$ ***(LESSON 2.6)***

19. Identify the transformation from the parent function $f(x) = |x|$ to $g(x) = -2|x - 4|$. ***(LESSON 2.7)***

20. Graph the solution to $2(1 - 2x) \geq -y + 4$. ***(LESSON 3.3)***

21. Graph the solution to the system of linear inequalities below. ***(LESSON 3.4)***

$$\begin{cases} x + 2y \leq 4 \\ -2x + 3y \geq 6 \end{cases}$$

22. Solve the system below, if possible, by using a matrix equation. ***(LESSON 4.4)***

$$\begin{cases} x + y - 2z = -1 \\ 2x + 2y + z = 3 \\ -3x - 2y - 3z = -4 \end{cases}$$ (−1, 2, 1)

FREE-RESPONSE GRID The following questions may be answered by using a free-response grid such as that commonly used by standardized test services.

23. If the inverse function for $f(x) = \frac{2}{3}x - 7$ is written in the form $g(x) = ax + b$, what is the value of a? ***(LESSON 2.5)*** $\frac{3}{2}$

24. Find the maximum value of the objective function $P = 2x + 3y$ given the constraints below. ***(LESSON 3.5)***

$$\begin{cases} x + y \leq 4 \\ 2x + y \geq 2 \\ x \geq 0, y \geq 0 \end{cases}$$ 12

25. Find the constant of variation, k, if y varies directly as x and $y = 102$ when $x = 3$. ***(LESSON 1.4)*** 34

SMALL BUSINESS Joshua wants to mix two types of candy. Candy A costs \$2.50 per pound and candy B costs \$4.50 per pound. Ten pounds of the combined candy mixture cost \$37.00. ***(LESSON 3.1)***

26. How many pounds of candy A are in the mixture? 4

27. How many pounds of candy B are in the mixture? 6

SPORTS The cost of an adult ticket to a football game is \$4.00, and a student ticket is \$2.50. The total amount received from 600 tickets was \$1830. ***(LESSON 3.2)***

28. How many adult tickets were sold? 220

29. How many student tickets were sold? 380

14. $x \leq 1$

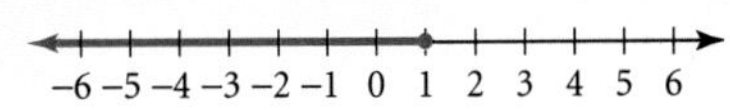

17. $f^{-1}(x) = -\frac{x - 3}{-5}$; yes, it is a function.

18.

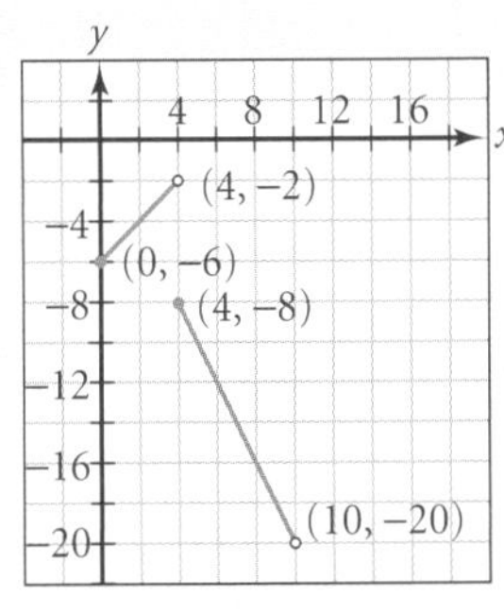

19. horizontal translation 4 units to the right, vertical stretch by a factor of 2, and reflection across the x-axis

20. $y \geq 4x + 2$

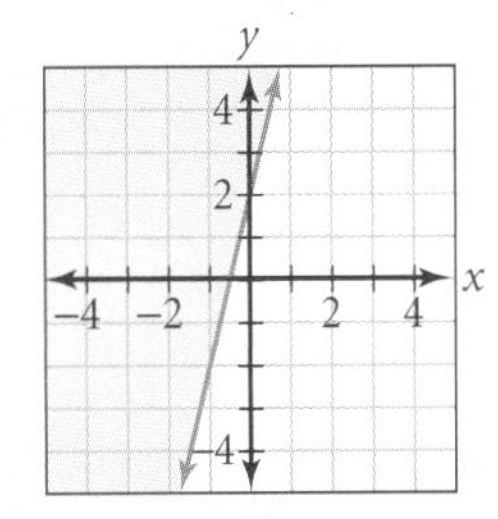

21.

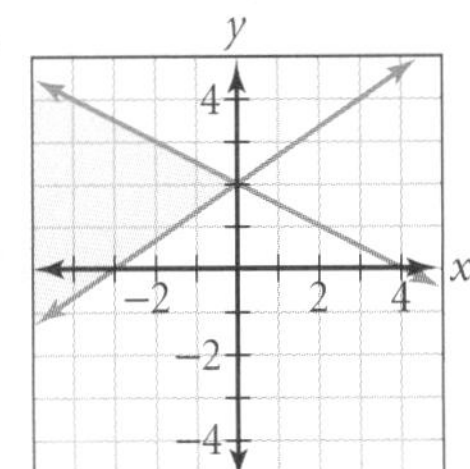

Keystroke Guide for Chapter 4

Essential keystroke sequences (using the model TI-82 or TI-83 graphics calculator) are presented below for all Activities and Examples found in this chapter that require or recommend the use of a graphics calculator.

internet connect

For Keystrokes of other graphing calculator models, visit the HRW web site at **go.hrw.com** and enter the keyword **MB1 CALC**.

LESSON 4.1

EXAMPLE 3 Page 217

Let $A = \begin{bmatrix} -2 & 0 & 1 \\ 5 & -7 & 8 \end{bmatrix}$ and $B = \begin{bmatrix} 5 & 7 & -1 \\ 0 & 2 & -8 \end{bmatrix}$. **For part a, find $A + B$.**

For TI-83 Plus, press 2nd x^{-1} (MATRX) to access the matrix menu.

Enter the matrices:

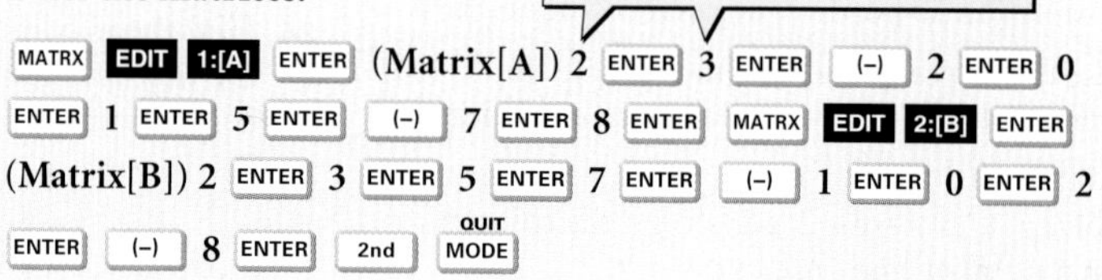

MATRX EDIT 1:[A] ENTER (Matrix[A]) 2 ENTER 3 ENTER (–) 2 ENTER 0 ENTER 1 ENTER 5 ENTER (–) 7 ENTER 8 ENTER MATRX EDIT 2:[B] ENTER (Matrix[B]) 2 ENTER 3 ENTER 5 ENTER 7 ENTER (–) 1 ENTER 0 ENTER 2 ENTER (–) 8 ENTER 2nd MODE (QUIT)

Add the matrices:

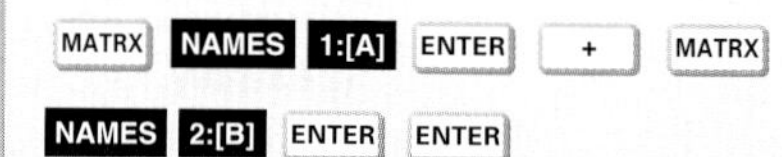

MATRX NAMES 1:[A] ENTER + MATRX NAMES 2:[B] ENTER ENTER

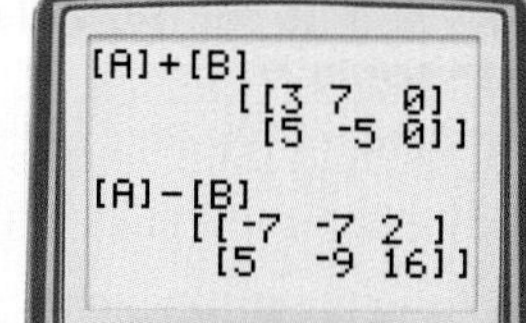

For part **b**, find $A - B$ by using a similar keystroke sequence to subtract the matrices.

LESSON 4.2

EXAMPLE 3 Page 228

Enter matrix A, and find A^2.

Enter the matrix:

Use a keystroke sequence similar to that in Example 3 of Lesson 4.1.

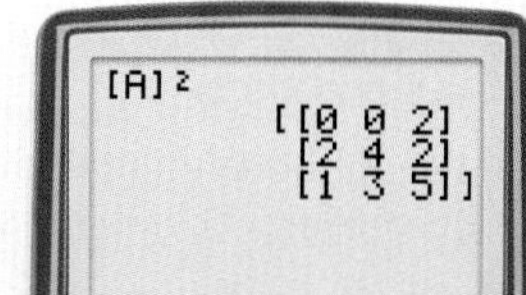

Square the matrix:

MATRX NAMES 1:[A] ENTER x^2 ENTER

TECHNOLOGY
Page 236

Enter matrix $A = \begin{bmatrix} 1 & 2 \\ 3 & 5 \end{bmatrix}$, and find its inverse.

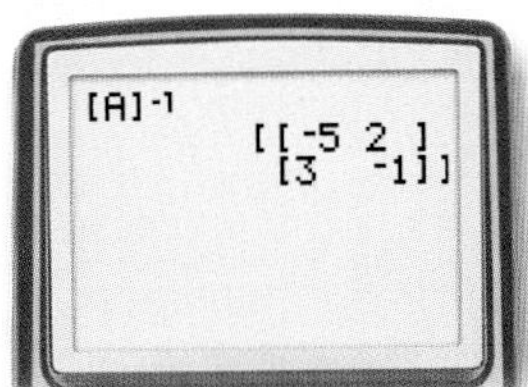

Enter the matrix:

Use a keystroke sequence similar to that in Example 3 of Lesson 4.1.

Find the inverse of matrix A:

EXAMPLE 2
Page 236

For part a, enter matrix $A = \begin{bmatrix} 6 & 8 \\ 5 & 7 \end{bmatrix}$, and find its inverse.

Enter the matrix:

Use a keystroke sequence similar to that in Example 3 of Lesson 4.1.

Find the inverse of matrix A:

For parts **b** and **c**, use a similar keystroke sequence.

TECHNOLOGY
Page 237

Let $A = \begin{bmatrix} 6 & 5 \\ 7 & 6 \end{bmatrix}$ and $B = \begin{bmatrix} 7 & 15 & 0 \\ 2 & 15 & 2 \end{bmatrix}$. Find the product AB.

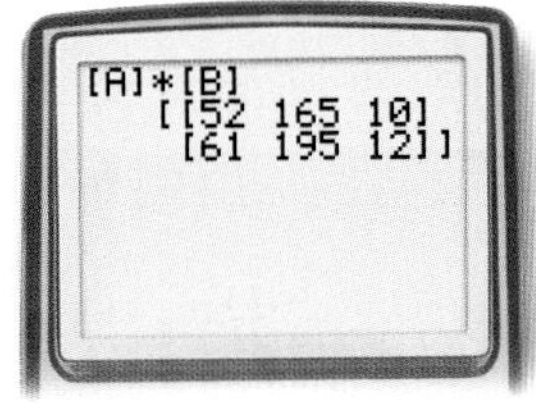

Enter the matrices:

Use a keystroke sequence similar to that in Example 3 of Lesson 4.1.

Find the product AB:

EXAMPLE 3
Page 237

Let $A = \begin{bmatrix} 6 & 5 \\ 7 & 6 \end{bmatrix}$ and $C = \begin{bmatrix} 52 & 165 & 10 \\ 61 & 195 & 12 \end{bmatrix}$. Find $A^{-1}C$.

Enter the matrices:

Use a keystroke sequence similar to that in Example 3 of Lesson 4.1.

Find the product $A^{-1}C$:

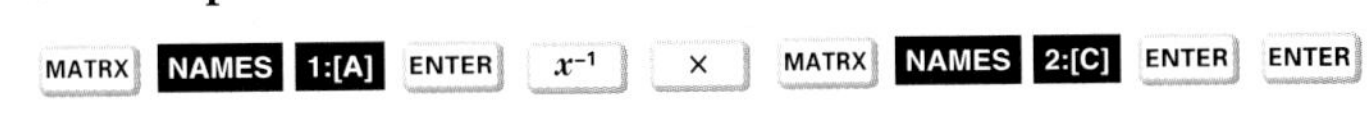

EXAMPLE 4 **For part a, find the determinant of $\begin{bmatrix} 7 & 8 \\ 6 & 7 \end{bmatrix}$.**

Page 238

Enter the matrix:

Use a keystroke sequence similar to that in Example 3 of Lesson 4.1.

Find the determinant:

MATRX MATH 1:det(ENTER MATRX NAMES 1:[A] ENTER ENTER

LESSON 4.4

EXAMPLE 1 **Solve $\begin{bmatrix} 1 & 1 \\ 0.05 & 0.14 \end{bmatrix}\begin{bmatrix} x \\ y \end{bmatrix} = \begin{bmatrix} 50{,}000 \\ 5000 \end{bmatrix}$ for $\begin{bmatrix} x \\ y \end{bmatrix}$.**

Page 245

Enter the matrices:

Enter the coefficient matrix, $A = \begin{bmatrix} 1 & 1 \\ 0.05 & 0.14 \end{bmatrix}$, and the constant matrix, $B = \begin{bmatrix} 50{,}000 \\ 5000 \end{bmatrix}$. Use a keystroke sequence similar to that in Example 3 of Lesson 4.1.

Do not enter the comma when you enter the number 50,000.

Find the product $A^{-1}B$:

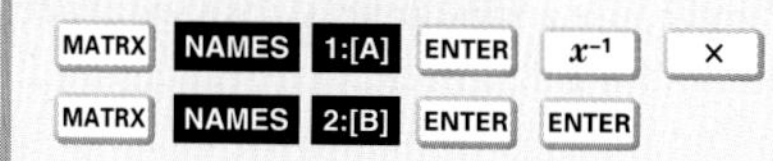

EXAMPLE 2 **Solve $\begin{bmatrix} 5 & 2 & -1 \\ 1 & -2 & 2 \\ 0 & 3 & 1 \end{bmatrix}\begin{bmatrix} x \\ y \\ z \end{bmatrix} = \begin{bmatrix} -7 \\ 0 \\ 17 \end{bmatrix}$ for $\begin{bmatrix} x \\ y \\ z \end{bmatrix}$.**

Page 246

Enter the matrices:

Enter matrix $A = \begin{bmatrix} 5 & 2 & -1 \\ 1 & -2 & 2 \\ 0 & 3 & 1 \end{bmatrix}$ and matrix $B = \begin{bmatrix} -7 \\ 0 \\ 17 \end{bmatrix}$.

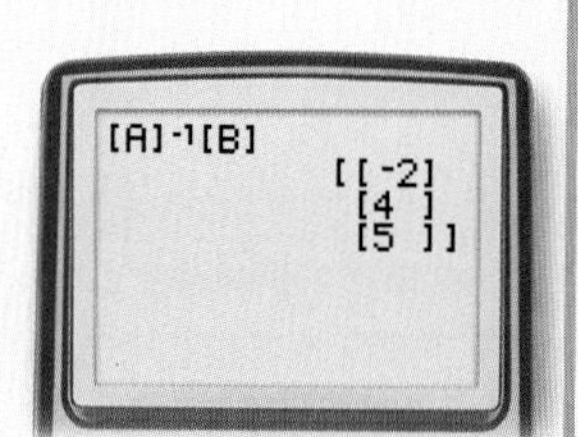

Use a keystroke sequence similar to that in Example 3 of Lesson 4.1.

Find the product $A^{-1}B$:

Use a keystroke sequence similar to that in Example 1 of Lesson 4.4.

Page 246

For Step 1, find the determinant of matrix $A = \begin{bmatrix} 4 & 9 \\ 2 & 5 \end{bmatrix}$.

Enter the matrix:

Use a keystroke sequence similar to that in Example 3 of Lesson 4.1.

Find the determinant:

MATRX MATH 1:det(ENTER MATRX NAMES 1:[A] ENTER ENTER

EXAMPLE 3 Solve $\begin{bmatrix} -3 & 4 \\ -6 & 8 \end{bmatrix}\begin{bmatrix} x \\ y \end{bmatrix} = \begin{bmatrix} 3 \\ 18 \end{bmatrix}$ for $\begin{bmatrix} x \\ y \end{bmatrix}$.

Page 247

Enter matrix $A = \begin{bmatrix} -3 & 4 \\ -6 & 8 \end{bmatrix}$ and matrix $B = \begin{bmatrix} 3 \\ 18 \end{bmatrix}$, and find $A^{-1}B$. Use a keystroke sequence similar to that in Example 1 of Lesson 4.4.

Solve $-3x + 4y = 3$ and $-6x + 8y = 18$ for y, and graph the lines:

Use viewing window [−5, 5] by [−5, 5].

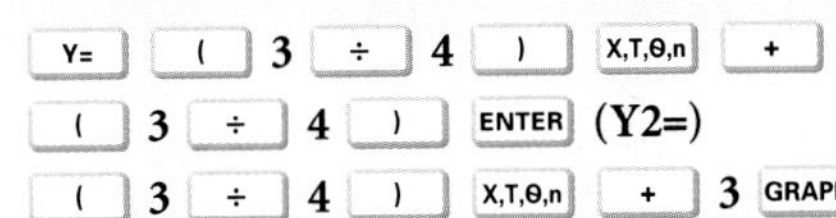

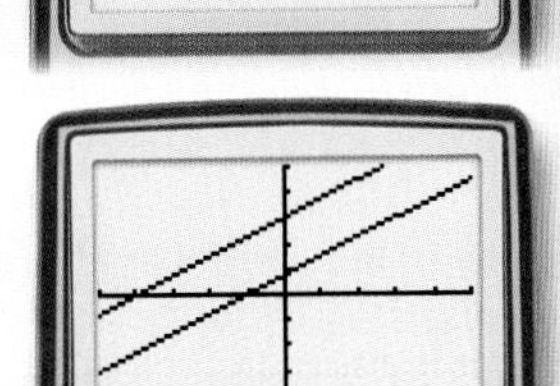

LESSON 4.5

TECHNOLOGY

Page 254

Find the reduced row-echelon form of the matrix $\begin{bmatrix} 1 & 1 & 1 & \vdots & 21 \\ 2 & 1 & 0 & \vdots & 23 \\ 0 & 1 & 3 & \vdots & 25 \end{bmatrix}$.

Enter the matrix:

Enter the augmented matrix as a 3×4 matrix without the column of dots. Use a keystroke sequence similar to that in Example 3 of Lesson 4.1.

The reduced row-echelon form can not be computed with one command on the TI-82.

Find the reduced row-echelon form:

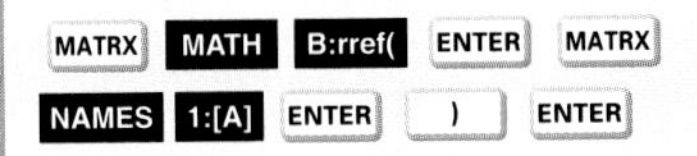

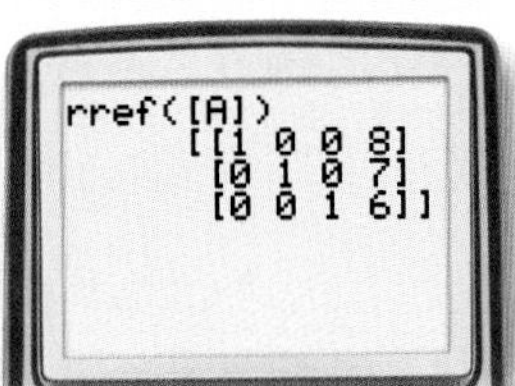

EXAMPLES 2 and 3 Find the reduced row-echelon form of the augmented matrix in each example.

Pages 254 and 255

Use a keystroke sequence similar to that in the Technology example above.

Page 255

Graph the system of equations $\begin{cases} -5x + 2y = 6 \\ x + 2y = -8 \\ x + 2y = 8 \end{cases}$.

Use square viewing window [−9.4, 9.4] by [−6.2, 6.2].

Solve each equation for y, and use a keystroke sequence similar to that in Example 3 of Lesson 4.4.

5 Quadratic Functions

Lesson Presentation CD-ROM
PowerPoint® presentations for each lesson 5.1–5.8

CHAPTER PLANNING GUIDE

Lesson	5.1	5.2	5.3	5.4	5.5	5.6	5.7	5.8	Project and Review
Pupil's Edition Pages	274–280	281–289	290–298	299–306	307–313	314–321	322–329	330–337	338–347
Practice and Assessment									
Extra Practice (Pupil's Edition)	955	955	956	956	957	957	958	958	
Practice Workbook	27	28	29	30	31	32	33	34	
Practice Masters Levels A, B, and C	79–81	82–84	85–87	88–90	91–93	94–96	97–99	100–102	
Standardized Test Practice Masters	31	32	33	34	35	36	37	38	39
Assessment Resources	55	56	57	58	60	61	62	63	59, 64–69
Visual Resources									
Lesson Presentation Transparencies Vol. 1	105–108	109–112	113–116	117–120	121–124	125–128	129–132	133–136	
Teaching Transparencies		17	18	19		20			
Answer Key Transparencies	167–169	170–172	173–176	177–179	180–183	184–190	191–192	193–199	200–207
Quiz Transparencies	5.1	5.2	5.3	5.4	5.5	5.6	5.7	5.8	
Teacher's Tools									
Reteaching Masters	53–54	55–56	57–58	59–60	61–62	63–64	65–66	67–68	
Make-Up Lesson Planner for Absent Students	27	28	29	30	31	32	33	34	
Student Study Guide	27	28	29	30	31	32	33	34	
Spanish Resources	27	28	29	30	31	32	33	34	
Block Scheduling Handbook									10–11
Activities and Extensions									
Lesson Activities	27	28	29	30	31	32	33	34	
Enrichment Masters	27	28	29	30	31	32	33	34	
Cooperative-Learning Activities	27	28	29	30	31	32	33	34	
Problem Solving/ Critical Thinking	27	28	29	30	31	32	33	34	
Student Technology Guide	27	28	29	30	31	32	33	34	
Long Term Projects									17–20
Writing Activities for Your Portfolio									13–15
Tech Prep Masters									21–24
Building Success in Mathematics									12–14

LESSON PACING GUIDE

Lesson	5.1	5.2	5.3	5.4	5.5	5.6	5.7	5.8	Project and Review
Traditional	1 day	1 day	2 days	2 days	2 days	2 days	2 days	2 days	2 days
Block	$\frac{1}{2}$ day	$\frac{1}{2}$ day	1 day	1 day	1 day	1 day	1 day	1 day	1 day
Two-Year	2 days	2 days	4 days	4 days	4 days	4 days	4 days	4 days	4 days

CONNECTIONS AND APPLICATIONS

Lesson	5.1	5.2	5.3	5.4	5.5	5.6	5.7	5.8	Review
Algebra	274–280	281–289	290–298	299–306	307–313	314–321	322–329	330–337	338–347
Geometry		284, 285, 286, 287, 288	296, 297	305		318, 321			345
Maximum/Minimum	276, 277, 278				312			335	
Patterns in Data							328		
Statistics							323, 325		
Transformations	279	289		302, 304, 306		321			345
Business and Economics	279	286, 288, 289	298		307, 309, 311, 312	321	328	330, 332, 336, 337	347
Life Skills	279			305		321	322, 325, 327		
Science and Technology	274	281, 283, 285, 286, 288	290, 295	299, 303, 305		314	328, 329	336, 337	344, 345, 347
Sports and Leisure		288, 289	297, 298	304, 306	312, 313		329	336, 337	344
Cultural Connection: Europe						315			
Cultural Connection: Africa				306					
Cultural Connection: Asia		284							

BLOCK SCHEDULING GUIDE

Day	Lesson	Teacher Directed: Lesson Examples, Teaching Transparencies	Student Guided Activity, Try This	Cooperative-Learning Activity, Lesson Activity, Student Technology Guide	Practice: Practice & Apply, Extra Practice, Practice Workbook	Assessment: Quiz, Mid-Chapter Assessment	Problem Solving, Reteaching
1	5.1	10 min	10 min	8 min	25 min	8 min	8 min
	5.2	10 min	10 min	7 min	25 min	7 min	7 min
2	5.3	20 min	20 min	15 min	50 min	15 min	15 min
3	5.4	10 min	15 min	15 min	65 min	15 min	15 min
4	5.5	10 min	15 min	15 min	65 min	15 min	15 min
5	5.6	20 min	20 min	15 min	50 min	15 min	15 min
6	5.7	10 min	15 min	15 min	65 min	15 min	15 min
7	5.8	10 min	15 min	15 min	65 min	15 min	15 min
8	Assess.	50 min **PE:** Chapter Review	90 min **PE:** Chapter Project, Writing Activities	90 min Tech Prep Masters	65 min **PE:** Chapter Assessment, Test Generator	30 min Chap. Assess. (A or B), Alt. Assess. (A or B), Test Generator	

PE: Pupil's Edition

Alternative Assessment

The following suggest alternative assessments for students who may benefit from a different type of assessment than the regular chapter quizzes and the mid-chapter/end-of-chapter test. Visit the HRW web site to get additional Alternative Assessment material.

internet connect

Alternative Assessment
Go To: **go.hrw.com**
Keyword: **MB1 Alt Assess**

Performance Assessment

1. a. Solve each quadratic equation by using any method.

$3x^2 - 15 = 20 \quad x^2 - 11x = 0 \quad (x + 1)^2 = -1$

$2x^2 - 7x + 5 = 0 \quad 3(x + 5)^2 + 1 = 7$

b. State the quadratic formula. Explain why it is not the best solution method for solving all of the equations above.

2. a. Outline a solution method for solving a quadratic inequality in x.

b. Apply the method you outlined in part **a** to solve and graph $x^2 - 7x + 6 > 0$.

c. Explain how to modify the graph of your solution if the $>$ sign is changed to a $<$ sign.

Portfolio Project

Suggest that students choose one of the following projects for inclusion in their portfolios.

1. a. Explain how to extend the methods for solving quadratic equations to solve each of the equations below.

$(ax)^4 - b^4 = 0 \qquad x^4 + bx^2 + c = 0$

b. Use examples to show how these types of equations can be solved.

2. Just as you can add, subtract, multiply, and divide complex numbers, you can find square roots of complex numbers.

a. Let $(x + yi)(x + yi) = 2i$. Use multiplication of complex numbers and the properties of equality to find x and y. What are the square roots of $2i$?

b. Experiment with the method suggested in part **a** to find the square roots of $4i$ and $6i$.

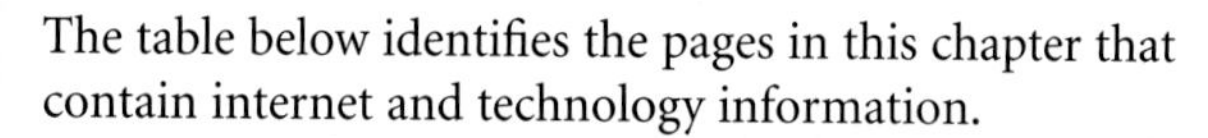

The table below identifies the pages in this chapter that contain internet and technology information.

Content Links

Activities Online	**pages 289, 311, 319, 329, 334**
Portfolio Extensions	**pages 298, 306, 337**
Homework Help Online	**pages 278, 286, 296, 304, 311, 320, 327, 335**
Graphic Calculator Support	**page 348**

Resource Links

Parents can go online and find concepts that students are learning—lesson by lesson—and questions that pertain to each lesson, which facilitate parent-student discussion.

Go To: **go.hrw.com**
Keyword: **MB1 Parent Guide**

Technical Support

The following may be used to obtain technical support for any HRW software product.

Online Help: **www.hrwtechsupport.com**
e-mail: **tschrw@hrwtechsupport.com**

HRW Technical Support Center: **(800)323-9239**
7 AM to 10 PM Monday through Friday Central Time

Visit the HRW math web site at: **www.hrw.com/math**

Technology

Lesson Suggestions and Calculator Examples

(Keystrokes are based on a TI-83 calculator.)

Lesson 5.1 Introduction to Quadratic Functions

To explore the features of a quadratic function, consider having students follow the steps listed below.

1. Enter a quadratic function.
2. Graph the function, and use the TRACE key to approximate the vertex.
3. Press 2nd WINDOW to set up a table, using x-values that surround the one found in Step 2.
4. Press 2nd GRAPH to view the table of values and get a better approximation of the coordinates of the vertex.

Lesson 5.2 Introduction to Solving Quadratic Equations

The solutions to $f(x) = g(x)$ are the same as the solutions to $f(x) - g(x) = 0$. For example, to solve the equation $4x^2 + 13 = 252$, students can graph $f(x) = 4x^2 + 13 - 252$ and approximate its zeros.

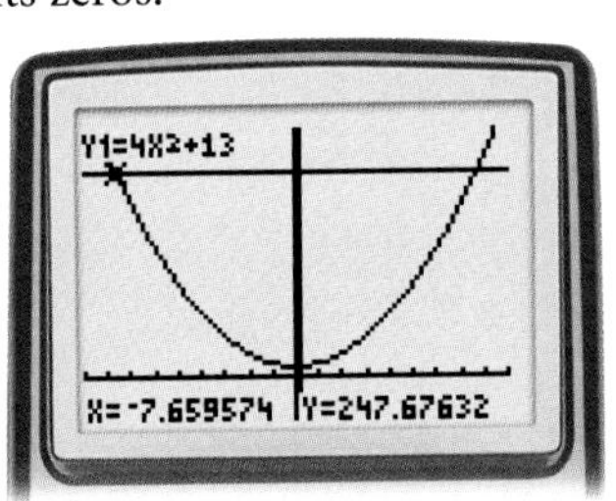

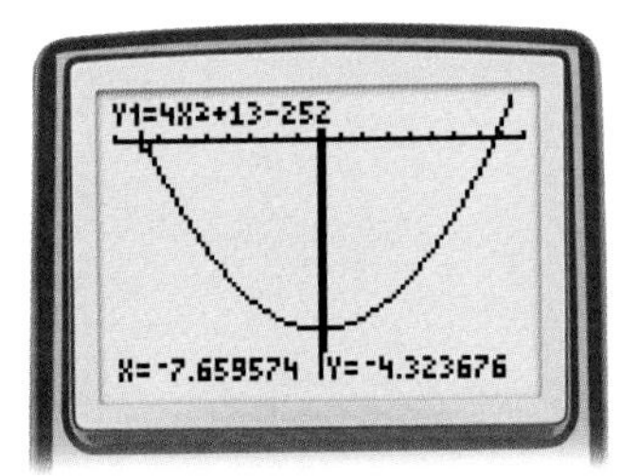

Lesson 5.3 Factoring Quadratic Expressions

The technology discussion on page 293 is based on the fact that if two functions form an identity, then the same set of ordered pairs defines both functions. Therefore, the graphs of the two functions are identical.

Ask students to use the reasoning outlined above to determine whether $2x^2 + 7x + 3$ can be factored as $(2x + 3)(x + 1)$ or as $(2x + 1)(x + 3)$. The second product is the correct one.

Lesson 5.4 Completing the Square

The display at right shows the graph of $y = x^2 - 2x + 3$. Tracing suggests that the coordinates of the vertex are (1, 2). Students should verify this by completing the square. This method should give the equation $y = (x - 1)^2 + 2$.

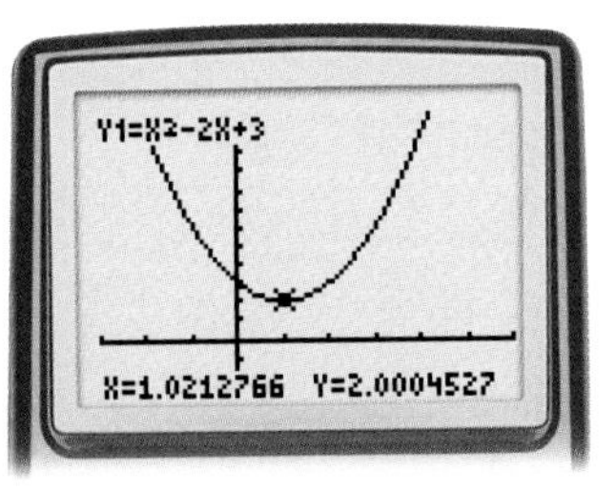

Lesson 5.5 The Quadratic Formula

Have students use the quadratic equation format, such as **(–4 + √(4²–4*2*(–6)))/(2*2).** After they obtain the first solution, they can press 2nd ENTER to recall the previous entry and use

◄ to change + to – in order to get the other solution.

Lesson 5.6 Quadratic Equations and Complex Numbers

To perform complex arithmetic on a graphics calculator, press MODE and select **a+bi** on the second line from the bottom of the screen. To access i, press 2nd . .

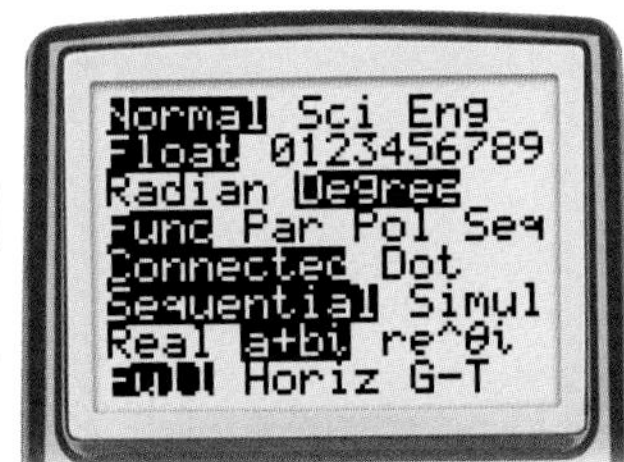

Lesson 5.7 Curve Fitting With Quadratic Models

To find a curve of best fit, students should use these steps.

1. Enter the data.
2. Use a scatter plot to verify that a quadratic model is reasonable.
3. Obtain the quadratic regression equation by pressing STAT and selecting **CALC** **5:QuadReg** .

Lesson 5.8 Solving Quadratic Inequalities

Discuss with students what inequality problems they can solve by using the graphics calculator displays below.

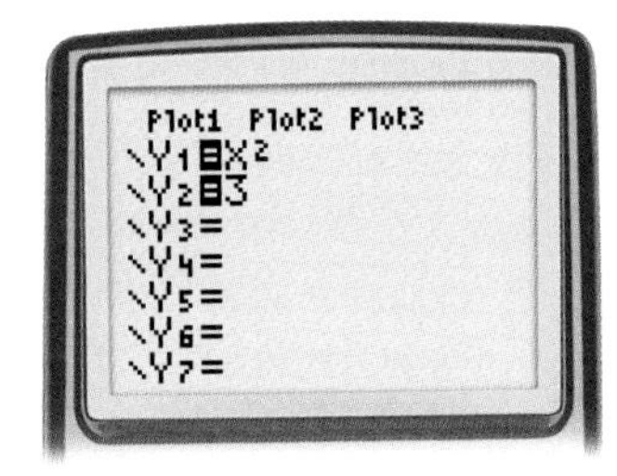

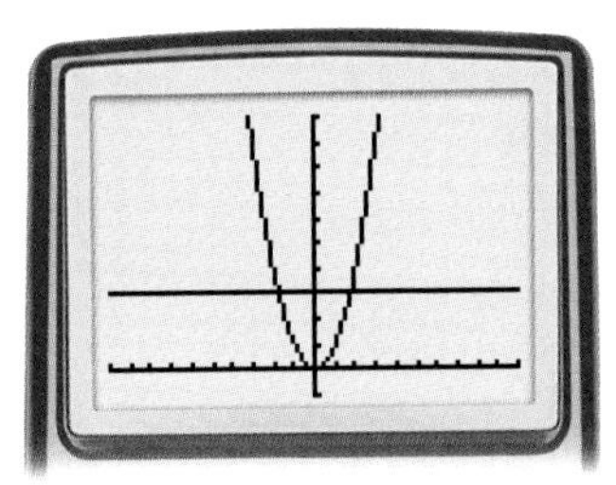

For further information, refer to the

- technology discussions in the lessons.
- lesson-related teacher's commentary in the side columns of this *Teacher's Edition.*
- lesson-related *Student Technology Guide* masters.
- *HRW Technology Handbook.*

internet connect

For keystrokes of other graphing calculators models, visit the HRW web site at **go.hrw.com** and enter the keyword **MB1 CALC**.

About Chapter 5

Background Information

Students will find solutions of quadratic equations by taking the square root of both sides of an equation, by factoring, by graphing, by completing the square, and by using the quadratic formula. Many connections between algebra and geometry are noted, and the discriminant is used to determine the number of real solutions. Complex numbers and their operations are introduced. Finally, curve fitting with quadratic models and quadratic inequalities is presented.

CHAPTER RESOURCES

- Block-Scheduling Handbook
- Writing Activities for Your Portfolio
- Tech Prep Masters
- Long-Term Project
- Assessment Resources:
  - Mid-Chapter Assessment
  - Chapter Assessments
  - Alternative Assessments
- Test Generator
- Technology Handbook

Chapter Objectives

- Define, identify, and graph quadratic functions. [5.1]
- Multiply linear binomials to produce a quadratic expression. [5.1]
- Solve quadratic equations by taking square roots. [5.2]
- Use the Pythagorean theorem to solve problems involving right triangles. [5.2]
- Factor a quadratic expression. [5.3]

Quadratic Functions

QUADRATIC FUNCTIONS HAVE IMPORTANT applications in science and engineering. For example, the parabolic path of a bouncing ball is described by a quadratic function. In fact, the motion of all falling objects can be described by quadratic functions. In this chapter, various techniques for solving quadratic equations are included, such as factoring and the quadratic formula.

Lessons

About the Photos

The path of a falling object can be modeled by a quadratic equation because gravity accelerates the object. That is, the object's rate of fall increases over time. When distance is measured in feet, gravity accelerates the object's rate of fall by 32 feet per second each second, and the equation $d = 16t^2$ gives the distance in feet that an object falls in t seconds. When distance is measured in meters, an object falls 9.8 meters per second faster every second.

The distance traveled by an object that is not simply dropped is given by $f(t) = -\frac{1}{2}gt^2 + v_0t + h_0$, where g is the acceleration due to gravity, v_0 is the initial velocity of the object, and h_0 is the initial height of the object. The acceleration due to gravity is negative because gravity is always directed downward. Note that $f(t)$ is the distance an object has fallen, and it usually does not represent the object's path. Parametric equations are useful for modeling the path of an object that is not thrown vertically or dropped.

A basketball game is about to begin. The referee tosses the ball vertically into the air. A video camera follows the motion of the ball as it rises to its maximum height and then begins to fall.

The following table and graph represent the height of the ball in feet at 0.1-second intervals. After 1.1 seconds, one of the players makes contact with the ball by tapping it out of its vertical path to a teammate.

Time, x	Height, y
0.0	6.00
0.1	7.84
0.2	9.36
0.3	10.61
0.4	11.47
0.5	12.00
0.6	12.30
0.7	12.19
0.8	11.83
0.9	11.12
1.0	9.98
1.1	8.64

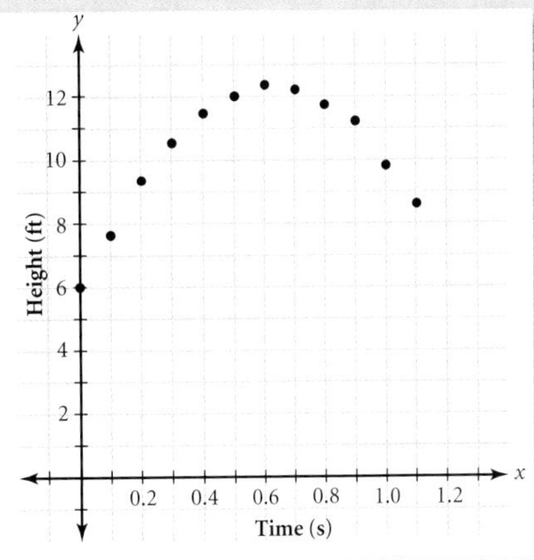

About the Chapter Project

The Chapter Project, *Out of This World*, extends the idea presented in the Portfolio Activities. By using different accelerations caused by the gravity on other planets, the height of the ball after a vertical toss on different planets can be compared with its height after a similar toss on Earth.

After completing the Chapter Project, you will be able to do the following:

- Use the function $h(t) = \frac{1}{2}gt^2 + v_0t + h_0$ to model the vertical motion of a basketball.
- Compare and contrast algebraic models of the form $h(t) = \frac{1}{2}gt^2 + v_0t + h_0$ for the vertical motion of the basketball on different planets.

About the Portfolio Activities

Throughout the chapter, you will be given opportunities to complete Portfolio Activities that are designed to support your work on the Chapter Project.

- Finding a reasonable model for the basketball data is included in the Portfolio Activity on page 280.
- Using various algebraic methods to answer questions about the height of the basketball along its path is in the Portfolio Activities on pages 298, 306, and 313.
- Comparing the algebraic model from physics with the quadratic regression model for the basketball data is included in the Portfolio Activity on page 329.
- Solving quadratic inequalities to answer questions about the height of the basketball along its path is in the Portfolio Activity on page 337.

internet connect

Chapter Internet Features and Online Activities

Lesson	Keyword	Page
5.1	MB1 Homework Help	278
5.2	MB1 Homework Help	286
	MB1 Tallest	289
5.3	MB1 Homework Help	296
	MB1 FactorQuad	298
5.4	MB1 Homework Help	304
	MB1 CompleteSq	306
5.5	MB1 Homework Help	311
	MB1 QuadFormula	311

Lesson	Keyword	Page
5.6	MB1 Homework Help	320
	MB1 Fractals	319
5.7	MB1 Homework Help	327
	MB1 Schools	329
5.8	MB1 Homework Help	335
	MB1 Videos	334
	MB1 QInequalities	337

- Use factoring to solve a quadratic equation and find the zeros of a quadratic function. [**5.3**]
- Use completing the square to solve a quadratic equation. [**5.4**]
- Use the vertex form of a quadratic function to locate the axis of symmetry of its graph. [**5.4**]
- Use the quadratic formula to find real roots of quadratic equations. [**5.5**]
- Use the roots of a quadratic equation to locate the axis of symmetry of a parabola. [**5.5**]
- Classify and find all roots of a quadratic equation. [**5.6**]
- Graph and perform operations on complex numbers. [**5.6**]
- Find a quadratic function that exactly fits three data points. [**5.7**]
- Find a quadratic model to represent a data set. [**5.7**]
- Write, solve, and graph a quadratic inequality in one variable. [**5.8**]
- Write, solve, and graph a quadratic inequality in two variables. [**5.8**]

In addition to the Portfolio Activities presented on this page, Portfolio Activities appear at the end of Lessons 5.1, 5.3, 5.4, 5.5, 5.7, and 5.8. Each serves as preparation for the Chapter Project. The Portfolio Activities as well as the Chapter Project Activities are appropriate for inclusion in the student's portfolio. Students should be encouraged to include in their portfolios any other work in which they feel a sense of pride or a sense of accomplishment.

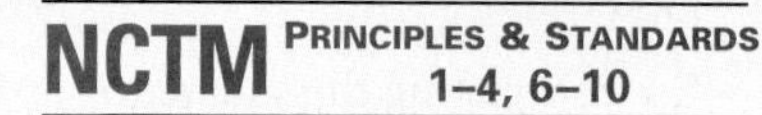

NCTM PRINCIPLES & STANDARDS 1–4, 6–10

QUICK WARM-UP

Use the distributive property to find each product.

1. $(x+1)(x-1)$
   $x^2 - 1$
2. $(x+2)(x+9)$
   $x^2 + 11x + 18$
3. $(x+5)(4x-7)$
   $4x^2 + 13x - 35$
4. $(3x-1)(5x+4)$
   $15x^2 + 7x - 4$
5. $(-x-3)(-2x-3)$
   $2x^2 + 9x + 9$

Also on Quiz Transparency 5.1

Teach

Why Relationships that occur in real-world situations often cannot be modeled linearly. Understanding nonlinear relationships can be very important. Ask students to visualize the emergency braking of the car pictured on page 274. Have them estimate the stopping distance of the car at different speeds.

5.1 Introduction to Quadratic Functions

Why
Many real-world situations, such as the total stopping distance for a car, can be modeled by quadratic functions.

Objectives

- Define, identify, and graph quadratic functions.
- Multiply linear binomials to produce a quadratic expression.

APPLICATION PHYSICS

Recall from Lesson 2.4 that the total stopping distance of a car on certain types of road surfaces is modeled by the function

$$d(x) = \frac{11}{10}x + \frac{1}{19}x^2,$$

where x is the speed of the car in miles per hour at the moment the hazard is observed and $d(x)$ is the distance in feet required to bring the car to a complete stop.

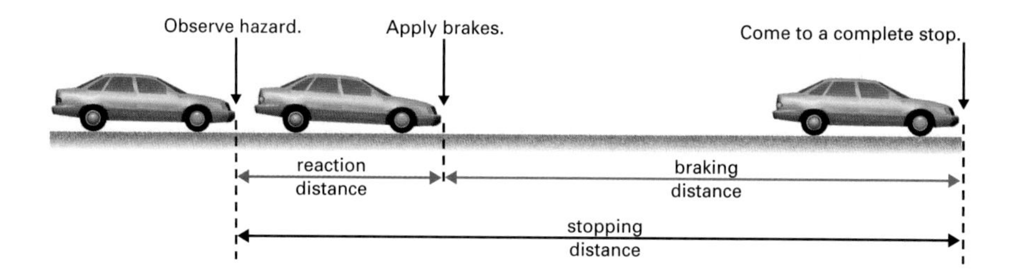

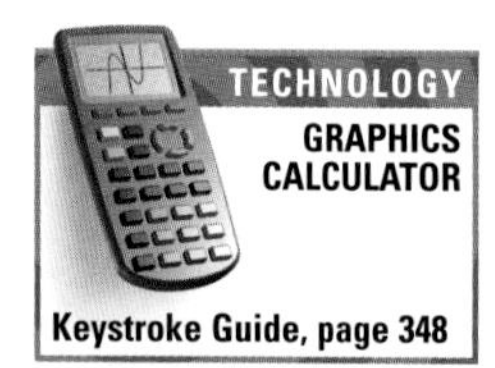

Observe hazard.
Apply brakes.

A table of values for the function d shows that a motorist driving at 20 miles per hour requires about 43 feet to come to a complete stop. However, a motorist traveling at 60 miles per hour requires over 255 feet to stop. Although the speed tripled, the total stopping distance increased by about 6 times. Clearly, the function for stopping distance is not a linear function.

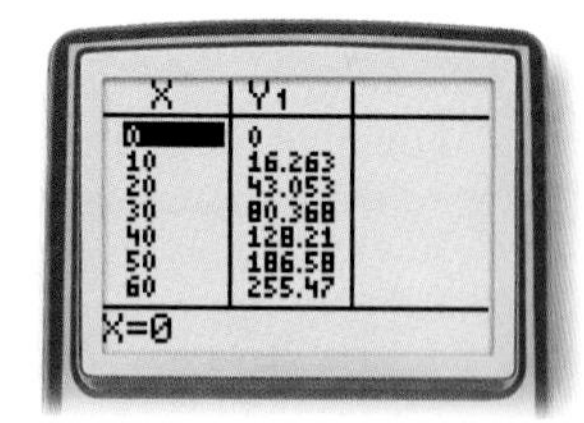

X	Y1
0	0
10	16.263
20	43.053
30	80.368
40	128.21
50	186.58
60	255.47

X=0

Alternative Teaching Strategy

USING TABLES Give students the quadratic function $y = x^2$ and have them list several points on the graph by picking a value for x and find the corresponding y-value. Have them plot the points and connect the points with a smooth curve. Discuss the terminology of this lesson as you label appropriate parts of the graph, and tell students that all quadratic functions have the same basic shape. Then have them graph $y = -x^2$, $y = x^2 - 1$, and $y = -x^2 + 2x + 3$ by using the same method and label the vertex and the x-intercepts. Discuss whether the graph opens up or down and whether it has a maximum or a minimum.

Now have them graph $y = x^2$, $y = 2x^2$, $y = -2x^2$, $y = \frac{1}{2}x^2$, and $y = -\frac{1}{2}x^2$ on overhead transparencies. Compare each graph with that of $y = x^2$ by overlaying the transparencies. Lead them to see that multiplying by a constant will stretch or shrink the graph and that the sign of the leading coefficient will determine whether the graph opens up or down. Tell them that the standard form of a quadratic function is $f(x) = ax^2 + bx + c$.

Investigating Quadratic Functions

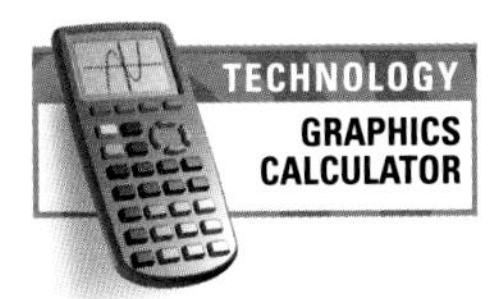

You will need: a graphics calculator

1. Graph functions f and g from the first row of the table below on the same screen. Describe f and g and their graphs.
2. Graph $f \cdot g$ with f and g from the first row of the table on the same screen. Describe $f \cdot g$ and its graph. Then clear all three functions.

f	g	$f \cdot g$
$f(x) = 2x - 2$	$g(x) = 2x + 1$	$(f \cdot g)(x) = (2x - 2)(2x + 1)$
$f(x) = x + 1$	$g(x) = x + 1$	$(f \cdot g)(x) = (x + 1)(x + 1)$
$f(x) = 2x$	$g(x) = -2x + 1$	$(f \cdot g)(x) = 2x(-2x + 1)$
$f(x) = -x + 2$	$g(x) = 0.5x + 1$	$(f \cdot g)(x) = (-x + 2)(0.5x + 1)$

3. Repeat Steps 1 and 2 for the functions in the other rows of the table.

CHECKPOINT ✔ 4. In what ways do the graphs of f and g differ from the graph of $f \cdot g$?

CHECKPOINT ✔ 5. How are the x-intercepts of the graphs of f and g related to the x-intercepts of the graph of $f \cdot g$? Explain.

As the Activity suggests, when you multiply two linear functions with nonzero slopes, the result is a *quadratic function.*

In general, a **quadratic function** is any function that can be written in the form $f(x) = ax^2 + bx + c$, where $a \neq 0$. It is defined by a **quadratic expression**, which is an expression of the form $ax^2 + bx + c$, where $a \neq 0$. The stopping-distance function $d(x) = \frac{11}{10}x + \frac{1}{19}x^2$, or $d(x) = \frac{1}{19}x^2 + \frac{11}{10}x$, is an example of a quadratic function.

CHECKPOINT ✔ Identify a, b, and c for the stopping-distance function, d, on page 274.

EXAMPLE 1 **Let $f(x) = (2x - 1)(3x + 5)$. Show that f represents a quadratic function. Identify a, b, and c when the function is written in the form $f(x) = ax^2 + bx + c$.**

SOLUTION

Method 1

$$\begin{aligned} f(x) &= (2x - 1)(3x + 5) \\ &= (2x - 1)3x + (2x - 1)5 \\ &= 6x^2 - 3x + 10x - 5 \\ &= 6x^2 + 7x - 5 \end{aligned}$$

Method 2

$$\begin{aligned} f(x) &= (2x - 1)(3x + 5) \\ &= 2x(3x + 5) + (-1)(3x + 5) \\ &= 6x^2 + 10x - 3x - 5 \\ &= 6x^2 + 7x - 5 \end{aligned}$$

Since $f(x) = 6x^2 + 7x - 5$ has the form $f(x) = ax^2 + bx + c$, f is a quadratic function with $a = 6$, $b = 7$, and $c = -5$.

TRY THIS Let $g(x) = (2x - 5)(x - 2)$. Show that g represents a quadratic function. Identify a, b, and c when the function is written in the form $g(x) = ax^2 + bx + c$.

Activity Notes

In this Activity, students will graph linear functions and the product of these linear functions. Students should discover that the x-intercepts of the linear functions are also the x-intercepts of the quadratic functions formed by the product of the linear functions.

CHECKPOINT ✔

4. **The graphs of f and g are straight lines. The graph of $f \cdot g$ is not a straight line but is a U-shaped curve.**

CHECKPOINT ✔

5. **The x-intercepts of the U-shaped graph are the same as the x-intercepts of the corresponding lines.**

CHECKPOINT ✔

$a = \frac{1}{19}$, $b = \frac{11}{10}$, and $c = 0$

ADDITIONAL EXAMPLE 1

Let $g(x) = (4x + 3)(x - 6)$. **Show that g represents a quadratic function. Identify a, b, and c when the function is written in the form $g(x) = ax^2 + bx + c$.**

$g(x) = 4x^2 - 21x - 18$

$a = 4, b = -21, c = -18$

TRY THIS

$g(x) = 2x^2 - 9x + 10$ has the form $g(x) = ax^2 + bx + c$, where $a = 2$, $b = -9$, and $c = 10$.

Enrichment

Describe the characteristics common to the graphs of all quadratic functions of the given form.

1. $f(x) = (x - a)(x - a)$ **The graph is a parabola that opens up and has its vertex on the x-axis at the point $(a, 0)$.**
2. $f(x) = -x^2 + b$ **The graph is a parabola that opens down and has its vertex on the y-axis at the point $(0, b)$.**

Inclusion Strategies

USING PATTERNS Have students find the first and second differences of the data found on page 274. Remind students that constant first differences denote a linear pattern. Have students round to the nearest tenth in order to get the constant, 10.5.

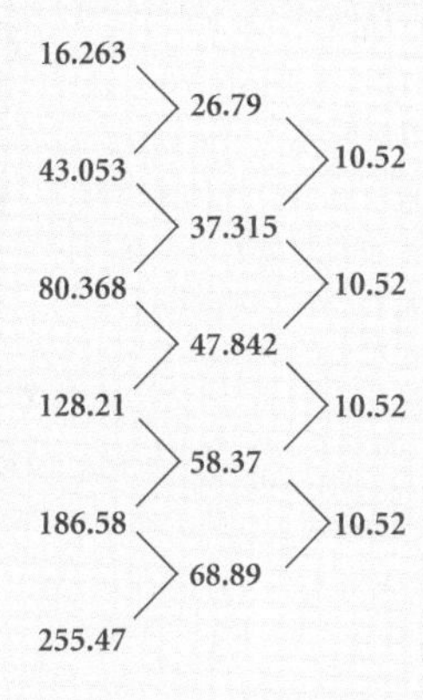

Teaching Tip

When discussing expressions of the form $ax^2 + bx + c$, refer to ax^2 as the *quadratic term*, bx as the *linear term*, and c as the *constant term*.

ADDITIONAL EXAMPLE 2

Identify whether the function $g(x) = -2x^2 + 4x + 1$ has a maximum value or a minimum value at the vertex. Then give the approximate coordinates of the vertex.

maximum; (1, 3)

Math CONNECTION

MAXIMUM/MINIMUM Encourage students to notice that if the parabola opens up, the function has a minimum value at the vertex, and if the parabola opens down, the function has a maximum value at the vertex.

Teaching Tip

TECHNOLOGY When using a TI-83 or a TI-82 to graph the function in Example 2, press Y=, enter **X^2–X+1** for **Y1**, and set the viewing window as follows:

Xmin=–4.7	Ymin=–2
Xmax=4.7	Ymax=6
Xscl=1	Yscl=1

Press TRACE, position the cursor on the vertex by using the arrow keys, and read the coordinates of the indicated point at the bottom of the screen. For other methods of finding the coordinates of the vertex on a TI-83 or a TI-82, see the Teaching Tip on page 348 in the Keystroke Guide.

TRY THIS

maximum; (–1, 3)

CRITICAL THINKING

Because $f(0) = f(1)$, the axis of symmetry is halfway between the lines $x = 0$ and $x = 1$; $x = \frac{1}{2}$.

The graph of a quadratic function is called a **parabola.** Two types of parabolas are graphed below. Notice that each parabola has an **axis of symmetry**, a line that divides the parabola into two parts that are mirror images of each other. The **vertex of a parabola** is either the lowest point on the graph or the highest point on the graph.

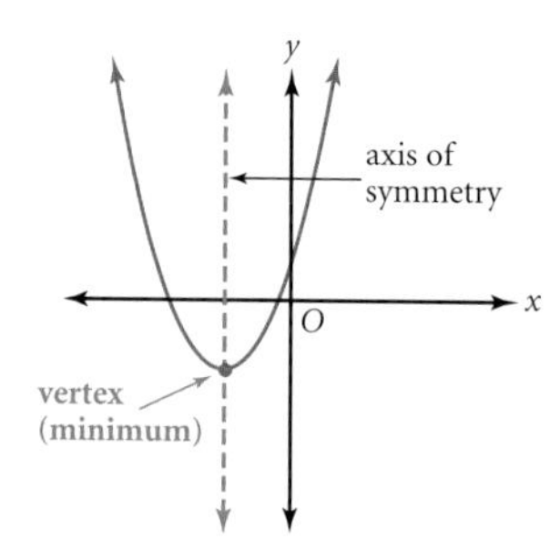

Notice that the axis of symmetry passes through the vertex of the parabola.

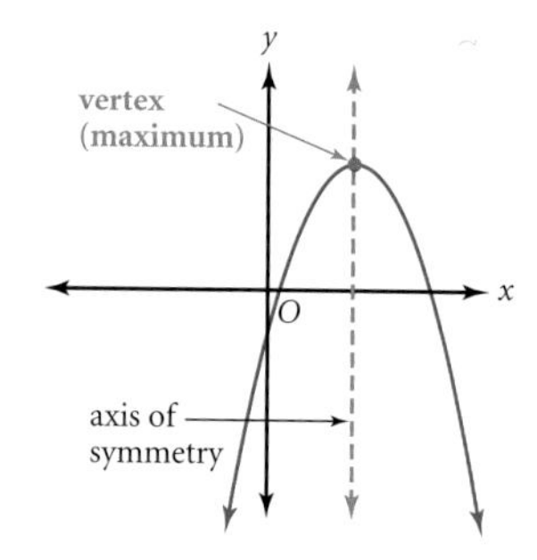

The domain of any quadratic function is the set of all real numbers. The range is either the set of all real numbers greater than or equal to the minimum value of the function (when the graph opens up) or the set of all real numbers less than or equal to the maximum value of the function (when the graph opens down).

EXAMPLE 2 **Identify whether $f(x) = x^2 - x + 1$ has a maximum value or a minimum value at the vertex. Then give the approximate coordinates of the vertex.**

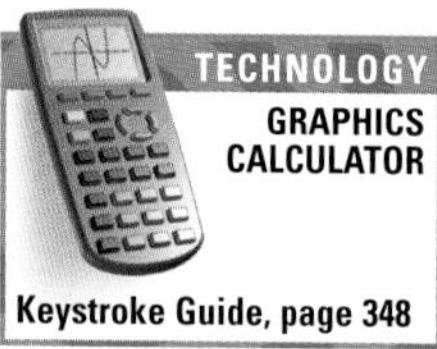

SOLUTION

Method 1 Use a graph.
From the graph, you can see that the function has a minimum value.

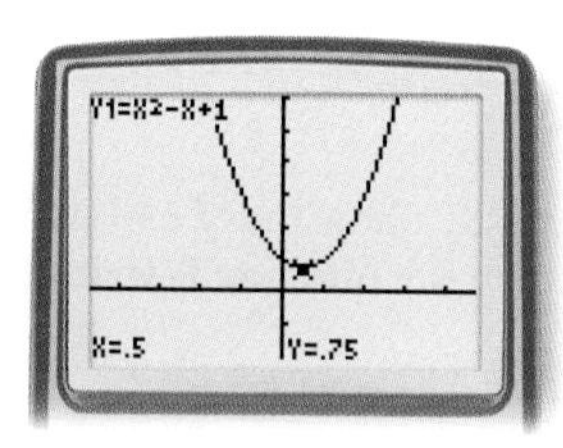

Tracing the graph, the coordinates of the vertex appear to be (0.5, 0.75).

Method 2 Use a table.
From a table of values you can see that an x-value between 0 and 1 gives the minimum value of the function.

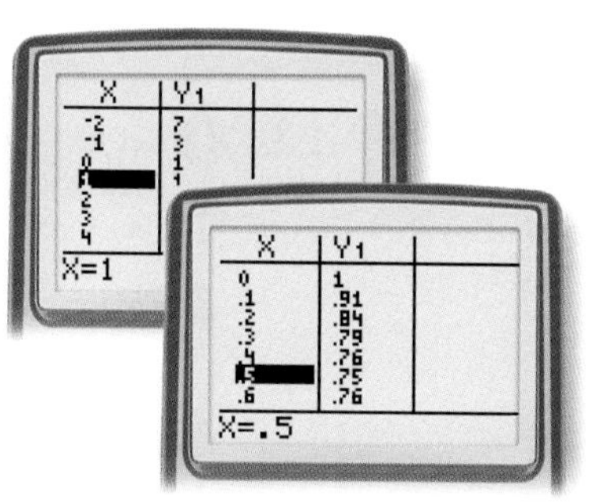

The coordinates of the vertex appear to be (0.5, 0.75).

TRY THIS Identify whether $f(x) = -2x^2 - 4x + 1$ has a maximum value or a minimum value at the vertex. Then give the approximate coordinates of the vertex.

CRITICAL THINKING Refer to solution Methods 1 and 2 in Example 2. If you know that $f(0) = f(1)$ for $f(x) = x^2 - x + 1$, describe how you can find the equation for the axis of symmetry.

Reteaching the Lesson

WORKING BACKWARDS Show students the graphs of the functions at right and have them identify the following characteristics of each function:

- axis of symmetry
- vertex
- x-intercept(s)
- opens up or down
- maximum or minimum
- range of the function

Then give them the function for each graph, and ask them to state the characteristics they see in the algebraic representation.

1. $y = x^2$
2. $y = x^2 + 1$
3. $y = (x + 2)(x - 4)$
4. $y = -(x - 3)^2 + 1$
5. $y = -x^2 + 2x + 3$

Axis	Vertex	x-intercept(s)	Opens	Max/min	Range
$x = 0$	(0, 0)	(0, 0)	up	min	$y > 0$
$x = 0$	(0, 1)	none	up	min	$y > 1$
$x = 1$	(1, –9)	(0, –2), (0, 4)	up	min	$y > -9$
$x = 3$	(3, 1)	(2, 0), (4, 0)	down	max	$y < 1$
$x = 1$	(1, 4)	(–1, 0), (3, 0)	down	max	$y < 4$

By examining a in $f(x) = ax^2 + bx + c$, you can identify whether the function has a maximum or a minimum value.

Minimum and Maximum Values

Let $f(x) = ax^2 + bx + c$, where $a \neq 0$. The graph of f is a parabola.

If $a > 0$, the parabola opens up and the vertex is the lowest point. The y-coordinate of the vertex is the **minimum value** of f.

If $a < 0$, the parabola opens down and the vertex is the highest point. The y-coordinate of the vertex is the **maximum value** of f.

EXAMPLE 3

CONNECTION
MAXIMUM/MINIMUM

State whether the parabola opens up or down and whether the y-coordinate of the vertex is the minimum value or the maximum value of the function. Then check by graphing.

a. $f(x) = x^2 + x - 6$

b. $g(x) = 5 + 4x - x^2$

SOLUTION

a. In $f(x) = x^2 + x - 6$, the coefficient of x^2 is 1. Because $a > 0$, the parabola opens up and the function has a minimum value at the vertex.

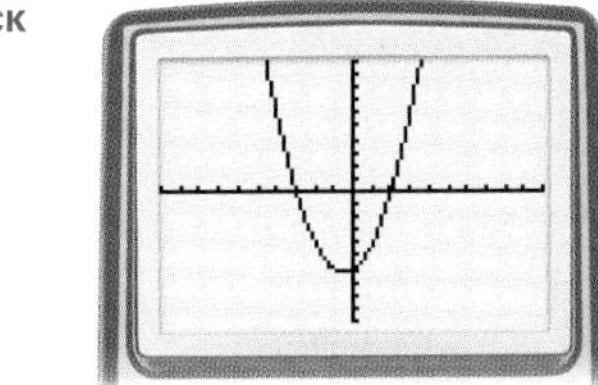

b. In $g(x) = 5 + 4x - x^2$, the coefficient of x^2 is -1. Because $a < 0$, the parabola opens down and the function has a maximum value at the vertex.

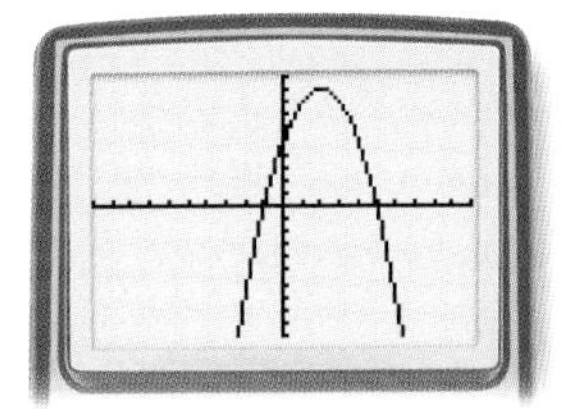

TECHNOLOGY
GRAPHICS CALCULATOR

Keystroke Guide, page 348

Exercises

Communicate

1. Describe differences between the graphs of linear and quadratic functions.
2. Explain the difference between the expressions that define linear and quadratic functions.
3. How can you determine whether a quadratic function has a minimum value or a maximum value?

ADDITIONAL EXAMPLE 3

State whether the parabola opens up or down and whether the y-coordinate of the vertex is the minimum value or the maximum value of the function. Then check by graphing.

a. $f(x) = -5x + 2x^2 + 2$

The parabola opens up, and the function has a minimum at the vertex.

b. $g(x) = 7 - 6x - 2x^2$

The parabola opens down, and the function has a maximum at the vertex.

Math CONNECTION

MAXIMUM/MINIMUM Remind students that the vertex of a parabola is the lowest point or the highest point on the graph.

Teaching Tip

TECHNOLOGY When using a TI-83 or TI-82 to check the minimum or maximum values in Example 3, set the viewing window as follows:

Xmin=–10	**Ymin=–10**
Xmax=10	**Ymax=10**
Xscl=1	**Yscl=1**

After entering each function into the **Y=** editor, you may find the vertex by pressing 2nd TRACE and selecting **3:minimun** or **4:maximum**. Key in –1 for the left bound, 0 for the right bound, and –0.5 for your guess. If more than one function's graph is shown, use the up and down arrow keys to switch between the graphs.

Teaching Tip

If students have trouble using the distributive property to multiply binomials, then they may use the FOIL method. Using this method, students will multiply the first terms of each factor, **F**, the outside terms of each factor, **O**, the inner terms of each factor, **I**, and the last terms of each factor, **L**. For example,

$$(x+2)(x-1)$$

F O I L

$$(x)(x)+(x)(-1)+(2)(x)+(2)(-1)$$
$$x^2-x+2x-2$$
$$x^2+x-2$$

5. $g(x)=x^2+7x+10$; $a=1$, $b=7$, $c=10$
6. $f(x)=6x^2+17x+5$; $a=6$, $b=17$, $c=5$

CONNECTIONS

13. $f(x)=x^2+5x-24$; $a=1$, $b=5$, $c=-24$
14. $k(x)=x^2-2x-15$; $a=1$, $b=-2$, $c=-15$
15. $g(x)=-x^2-3x+28$; $a=-1$, $b=-3$, $c=28$
16. $g(x)=-x^2+6x+40$; $a=-1$, $b=6$, $c=40$
17. $g(x)=-x^2-4x+12$; $a=-1$, $b=-4$, $c=12$
18. $f(x)=-x^2+6x+27$; $a=-1$, $b=6$, $c=27$
19. $f(x)=3x^2-3x-6$; $a=3$, $b=-3$, $c=-6$
20. $h(x)=6x^2-2x-8$; $a=6$, $b=-2$, $c=-8$
21. $h(x)=x^2-3x$; $a=1$, $b=-3$, $c=0$
22. $f(x)=2x^2+10x$; $a=2$, $b=10$, $c=0$
23. $g(x)=-2x^2+5x+12$; $a=-2$, $b=5$, $c=12$
24. $f(x)=-4x^2+15x+4$; $a=-4$, $b=15$, $c=4$
25. $h(x)=x^2-16$; $a=1$, $b=0$, $c=-16$
26. $f(x)=x^2-36$; $a=1$, $b=0$, $c=-36$

internet connect

Homework Help Online
Go To: **go.hrw.com**
Keyword: **MB1 Homework Help** for Exercises 13–26

Guided Skills Practice

Show that each function is a quadratic function by writing it in the form $f(x)=ax^2+bx+c$ and identifying a, b, and c. *(EXAMPLE 1)*

4. $f(x)=(x+1)(x-7)$ $f(x)=x^2-6x-7$; $a=1$, $b=-6$, $c=-7$
5. $g(x)=(x+2)(x+5)$
6. $f(x)=(2x+5)(3x+1)$

MAXIMUM/MINIMUM Identify whether each function has a maximum or minimum value. Then give the approximate coordinates of the vertex. *(EXAMPLE 2)*

7 $g(x)=x^2-3x+5$ min at (1.5, 2.75)
8 $f(x)=2-3x-x^2$ max at (−1.5, 4.25)
9 $g(x)=x^2+5x+3$ min at (−2.5, −3.25)

MAXIMUM/MINIMUM State whether the parabola opens up or down and whether the y-coordinate of the vertex is the maximum value or the minimum value of the function. Then check by graphing. *(EXAMPLE 3)*

10. $f(x)=x^2-2x+7$ up; min
11. $g(x)=-x^2+8x+14$ down; max
12. $g(x)=-2x^2-5x+1$ down; max

Practice and Apply

Show that each function is a quadratic function by writing it in the form $f(x)=ax^2+bx+c$ and identifying a, b, and c.

13. $f(x)=(x-3)(x+8)$
14. $k(x)=(x+3)(x-5)$
15. $g(x)=(4-x)(7+x)$
16. $g(x)=(10-x)(x+4)$
17. $g(x)=-(x-2)(x+6)$
18. $f(x)=-(x+3)(x-9)$
19. $f(x)=3(x-2)(x+1)$
20. $h(x)=2(x+1)(3x-4)$
21. $h(x)=x(x-3)$
22. $f(x)=2x(x+5)$
23. $g(x)=(2x+3)(4-x)$
24. $f(x)=(4x+1)(4-x)$
25. $h(x)=(x-4)(x+4)$
26. $f(x)=(x-6)(x+6)$

Identify whether each function is a quadratic function. Use a graph to check your answers.

27. $f(x)=3-x^2$ yes
28. $g(s)=3-s$ no
29. $f(t)=\frac{1}{4}t^2+\frac{1}{2}t-\frac{2}{3}$ yes
30. $h(x)=\frac{3x^2+4x+1}{x+1}$ no
31. $g(t)=t^2-t^2(t+7)$ no
32. $h(x)=|x^2+5x-2|$ no

State whether the parabola opens up or down and whether the y-coordinate of the vertex is the minimum value or the maximum value of the function.

33. $f(x)=-2x^2-2x$ down; max
34. $f(x)=8x^2-x$ up; min
35. $g(x)=-(3x^2-x+3)$ down; max
36. $f(x)=2+3x-5x^2$ down; max
37. $h(x)=1-9x-x^2$ down; max
38. $g(x)=-(x^2+x-12)$ down; max
39. $g(x)=3(x+8)(-x+9)$ down; max
40. $h(x)=-(4x+1)(x+4)$ down; max

Graph each function and give the approximate coordinates of the vertex.

41 $f(x)=x^2-x+9$ (0.5, 8.75)
42 $g(x)=9-2x-x^2$ (−1, 10)
43 $g(x)=4x^2-2x+2$ (0.25, 1.75)
44 $f(x)=-0.5(x+4)^2$ (−4, 0)
45 $f(x)=(x-2)^2-1$ (2, −1)
46 $f(x)=-(x-2)(x+6)$ (−2, 16)

Practice

NAME ____________ CLASS ______ DATE ______

Practice
5.1 Introduction to Quadratic Functions

Show that each function is a quadratic function by writing it in the form $f(x)=ax^2+bx+c$ and identifying a, b, and c.

1. $f(x)=(x-3)(x-5)$ $f(x)=x^2-8x+15$; $a=1$, $b=-8$, $c=15$
2. $g(x)=(7-x)(9-x)$ $g(x)=x^2-16x+63$; $a=1$, $b=-16$, $c=63$
3. $k(x)=-3(x-11)(x+1)$ $k(x)=-3x^2+30x+33$; $a=-3$, $b=30$, $c=33$
4. $h(x)=(2x+5)(3x-1)$ $h(x)=6x^2+13x-5$; $a=6$, $b=13$, $c=-5$
5. $d(x)=(x-3)^2-4$ $d(x)=x^2-6x+5$; $a=1$, $b=-6$, $c=5$

Identify whether each function is quadratic. Use a graph to check your answers.

6. $f(x)=-4x+x^2$ yes
7. $k(x)=\frac{1}{x}$ no
8. $h(x)=\frac{2x^3+x}{x^2-1}$ no
9. $g(x)=16-3x$ no
10. $b(x)=x^2-2x(x+1)$ yes
11. $m(x)=3x-x(x+9)$ yes

State whether the parabola opens up or down and whether the y-coordinate of the vertex is the minimum value or the maximum value of the function.

12. $f(x)=5x^2-3x$ up; minimum value
13. $g(x)=4x^2+7x-2$ up; minimum value
14. $h(x)=(5-x)(2-3x)$ up; minimum value
15. $q(x)=(4-x)(2+7x)$ down; maximum value

Graph each function and give the approximate coordinates of the vertex.

16. $k(x)=4x^2-3$ (0, −3)
17. $h(x)=-x^2-x+6$ (−0.5, 6.25)
18. $p(x)=-(x+4)(x-0.5)$ (−1.75, 5.0625)

CHALLENGE

47. Describe a way to find the exact coordinates of the vertex of a parabola given by $f(x) = (x + a)(x - a)$. **$(0, -a^2)$**

CONNECTION

48. TRANSFORMATIONS Graph each function.

$f(x) = (x + 2)(x - 4)$ $\quad g(x) = 2(x + 2)(x - 4)$ $\quad h(x) = \frac{1}{2}(x + 2)(x - 4)$

$i(x) = -(x + 2)(x - 4)$ $\quad j(x) = -2(x + 2)(x - 4)$ $\quad k(x) = -\frac{1}{2}(x + 2)(x - 4)$

a. What do all of the graphs have in common? **same x-intercepts**

b. Which of the functions have a maximum value? ***i, j, k***

c. Which of the functions have a minimum value? ***f, g, h***

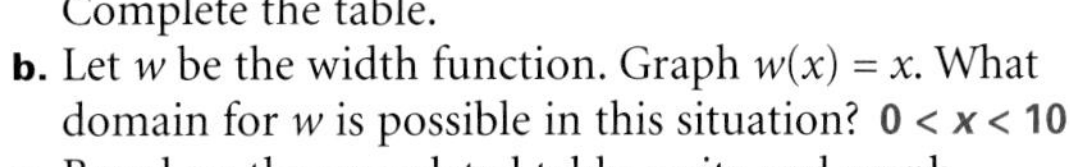

APPLICATIONS

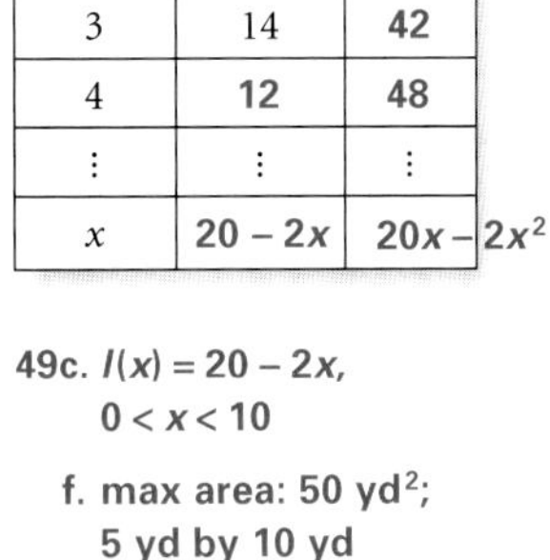

Width (yd)	Length (yd)	Area (yd²)
1	18	18
2	16	32
3	14	**42**
4	**12**	**48**
⋮	⋮	⋮
x	**$20 - 2x$**	**$20x - 2x^2$**

49c. **$l(x) = 20 - 2x$, $0 < x < 10$**

f. max area: 50 yd²; 5 yd by 10 yd

49. CONSTRUCTION Carly plans to build a rectangular pen against an existing fence for her dog. She will buy 20 yards of fence material.

a. The table at left shows some different widths, lengths, and resulting areas that are possible with 20 yards of fence material. Complete the table.

b. Let w be the width function. Graph $w(x) = x$. What domain for w is possible in this situation? **$0 < x < 10$**

c. Based on the completed table, write and graph a linear function, l, for the length. What domain for l is possible in this situation?

d. Let $A(x) = w(x) \cdot l(x)$ be the area function. Show that the area function is quadratic. **$A(x) = -2x^2 + 20x$**

e. What domain for A is possible in this situation? What range is possible? **$0 < x < 10$; $0 < A(x) < 50$**

f. What is the maximum area possible for the pen? What width and length will produce the maximum area?

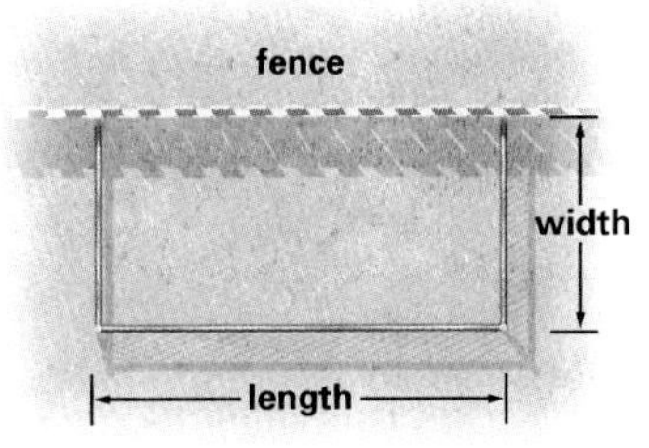

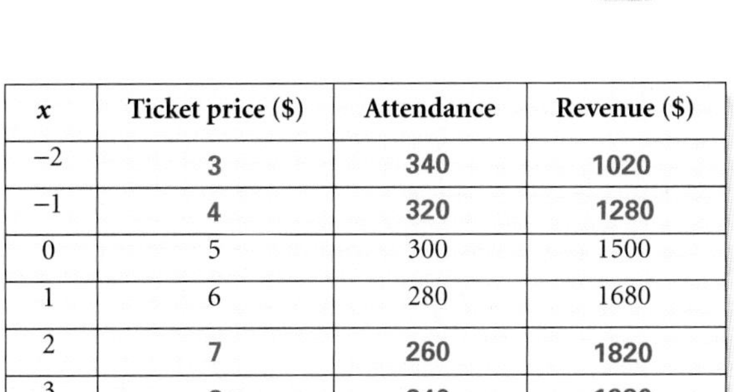

x	Ticket price ($)	Attendance	Revenue ($)
−2	**3**	**340**	**1020**
−1	**4**	**320**	**1280**
0	5	300	1500
1	6	280	1680
2	**7**	**260**	**1820**
3	**8**	**240**	**1920**
4	**9**	**220**	**1980**
⋮	⋮	⋮	⋮
x	**$5 + x$**	**$300 - 20x$**	**$(5 + x)(300 - 20x)$**

50. FUND-RAISING The student council plans to run a talent show to raise money. Last year tickets sold for $5 each and 300 people attended. This year, the student council wants to make an even bigger profit than last year. They estimate that for each $1 increase in the ticket price, attendance will drop by 20 people, and for each $1 decrease in the ticket price, attendance will increase by 20 people.

a. Let x be the change in the ticket price, in dollars. Copy and complete the table at left.

Assess

Selected Answers

Exercises 4–12, 13–53 odd

ASSIGNMENT GUIDE

In Class	1–12
Core	13–45 odd, 49
Core Plus	14–46 even, 47, 48
Review	51–54
Preview	55

Extra Practice can be found beginning on page 940.

47. The axis of symmetry of the parabola will lie midway between the values of a and $-a$. Therefore, the coordinates of the vertex are $x = 0$ and $y = f(0) = -a^2$, or $(0, -a^2)$.

Student Technology Guide

NAME ________ CLASS ________ DATE ________

Student Technology Guide

5.1 Introduction to Quadratic Functions

As you saw in Lesson 5.1 of your textbook, the graph of a quadratic function is called a *parabola*. The trace function on a graphics calculator can help you approximate the coordinates of key points on a parabola: the vertex, the x-intercepts (if any), and the y-intercept. Consider $f(x) = x^2 + 4x + 7$.

- Enter this function by pressing Y= (−) X,T,θ,n x² + 4 X,T,θ,n + 7.
- Press GRAPH.
- To explore the parabola in more detail, press TRACE. Press (−) 2 ENTER to move the cursor to (−2, −5). Then use ► to move along the graph.
- Notice the following:

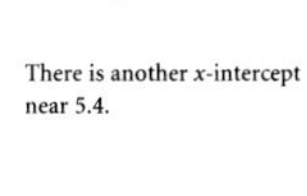

There is an x-intercept near −1.2. The y-intercept is 7.

The vertex is near (1.9, 11.0). The value of y increases until x is about 1.9. Then the value of y decreases.

There is another x-intercept near 5.4.

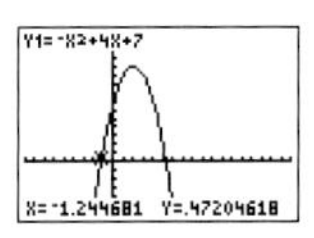

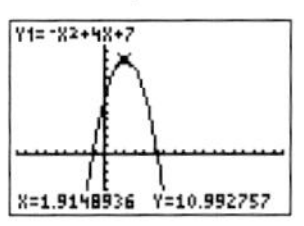

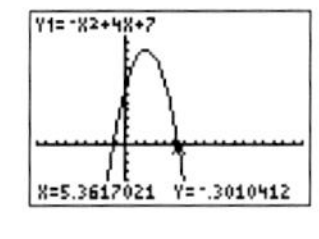

Graph each function in an appropriate window. Approximate the x-intercepts of the graph, the coordinates of the vertex, and the y-intercept. Round answers to the nearest tenth, if necessary.

1. $f(x) = x^2 - 4x + 3$ x-intercepts: 1 and 3; vertex: (2, −1); y-intercept: 3
2. $f(x) = -2x^2 + 10x - 5$ x-intercepts: 0.6 and 4.4; vertex: (2.5, 7.5); y-intercept: −5
3. $f(x) = -x^2 - 10x - 15$ x-intercepts: −8.2 and −1.8; vertex: (−5, 10); y-intercept: −15
4. $f(x) = 5x^2 - 14x + 12$ no x-intercepts; vertex: (1.4, 2.2); y-intercept: 12

Error Analysis

Students may check their factoring by multiplying the factors or by choosing a value for the variable and substituting it into both the unfactored and factored expressions. Both expressions should yield the same value. While this procedure will not ensure that the factoring was done correctly, it will identify an error if the values are not equal.

Look Beyond

In Exercise 55, students will graph three functions on the same screen—one with no x-intercept, another with one x-intercept, and a third with two x-intercepts. This concept will be studied further in the next lesson.

ALTERNATIVE

Assessment

Portfolio Activity

The Portfolio Activity can be used as preparation for the Chapter Project or as a separate activity. In the Portfolio Activity on this page, students use graphics calculators to simulate the height of a ball thrown into the air with different initial velocities.

Answers to Portfolio Activities can be found in Additional Answers of the Teacher's Edition.

54.

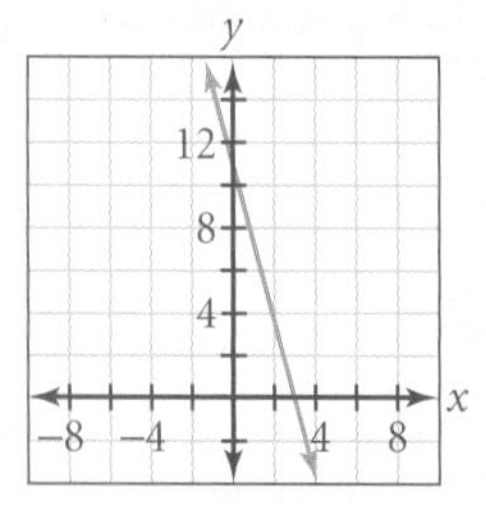

50b. $t(x) = 5 + x$; linear; $x > -5$

c. $a(x) = 300 - 20x$; linear; $-5 < x < 15$

b. Write the function for the ticket price, $t(x)$. What type of function is t? What domain for t is possible in this situation?

c. Write the function for the attendance, $a(x)$. What type of function is a? What domain for a is possible in this situation?

d. Let $R(x) = t(x) \cdot a(x)$ be the function for the revenue. Show that this function is quadratic. $R(x) = -20x^2 + 200x + 1500$

e. What domain for R is possible in this situation? $-5 < x < 15$

f. What is the maximum revenue possible for the talent show? What ticket price and attendance will produce the maximum revenue? **\$2000; \$10, 200 people**

Look Back

For Exercises 51–54, let $y = -4x + 11$. *(LESSON 1.2)*

51. Identify the slope m. -4

52. What is the x-intercept? $\frac{11}{4}$

53. What is the y-intercept? 11

54. Graph the line.

Look Beyond

55 Graph $y = x^2 - 3x + 5$, $y = x^2 + 7x + 6$, and $y = x^2 - 14x + 49$ on the same screen. How many x-intercepts are possible for the graph of a quadratic function? **0, 1, or 2**

SPORTS Refer to the basketball toss described on page 273.

1. Create a scatter plot of the data.

2. The height of a basketball thrown vertically into the air can be modeled by the function $h(t) = \frac{1}{2}gt^2 + v_0t + h_0$, where g is the acceleration due to gravity (−32 feet per second squared), v_0 is the initial velocity in feet per second, and h_0 is the initial height in feet of the ball. Thus, $h(t) = -16t^2 + v_0t + h_0$. Substitute 6 for the initial height, h_0, of the ball into $h(t) = -16t^2 + v_0t + h_0$. Then graph the function on the same screen as the scatter plot, using different values for the initial velocity, v_0, until you find a model that provides a reasonably good fit for the data.

3. Use your best model to answer the questions below.
   a. What is the value of v_0?
   b. What is the maximum height achieved by the basketball?
   c. At what time does the basketball reach its maximum height?

4. Solve for v_0 algebraically by substituting the coordinates of one of the data points into $h(t) = -16t^2 + v_0t + h_0$. Graph the resulting function on the same screen as the scatter plot. Use this model to answer the questions below.
   a. What is the value of v_0?
   b. What is the maximum height achieved by the basketball?
   c. At what time does the basketball reach its maximum height?

WORKING ON THE CHAPTER PROJECT

You should now be able to complete Activity 1 of the Chapter Project.

Introduction to Solving Quadratic Equations

Why *You can solve many real-world problems, such as those involving the force of gravity on a falling object, by solving a quadratic equation.*

Objectives

- Solve quadratic equations by taking square roots.
- Use the Pythagorean Theorem to solve problems involving right triangles.

APPLICATION
RESCUE

A rescue helicopter hovering 68 feet above a boat in distress drops a life raft. The height in feet of the raft above the water can be modeled by $h(t) = -16t^2 + 68$, where t is the time in seconds after it is dropped. How many seconds after the raft is dropped will it hit the water? Solving this problem involves finding square roots. *You will answer this question in Example 3.*

If $x^2 = a$ and $a \geq 0$, then x is called a square root of a. If $a > 0$, the number a has two square roots, $\sqrt{a}$ and $-\sqrt{a}$. The positive square root of a, $\sqrt{a}$, is called the **principal square root** of a. If $a = 0$, then $\sqrt{0} = 0$. When you solve a quadratic equation of the form $x^2 = a$, you can use the rule below.

Solving Equations of the Form $x^2 = a$

If $x^2 = a$ and $a \geq 0$, then $x = \sqrt{a}$ or $x = -\sqrt{a}$, or simply $x = \pm\sqrt{a}$.

The expression $\pm\sqrt{a}$ is read as "plus or minus the square root of a." To use the rule above, you may need to transform a given equation so that it is in the form $x^2 = a$. You can also use the *Properties of Square Roots* below to simplify the resulting square root.

Properties of Square Roots

Product Property of Square Roots If $a \geq 0$ and $b \geq 0$: $\sqrt{ab} = \sqrt{a} \cdot \sqrt{b}$

Quotient Property of Square Roots If $a \geq 0$ and $b > 0$: $\sqrt{\frac{a}{b}} = \frac{\sqrt{a}}{\sqrt{b}}$

Prepare

NCTM PRINCIPLES & STANDARDS 1–4, 6–10

QUICK WARM-UP

Simplify.

1. $\sqrt{81}$ — 9
2. $-\sqrt{144}$ — -12
3. $\pm\sqrt{1}$ — $1, -1$
4. $\sqrt{25 \cdot 9}$ — 15
5. $\sqrt{16} \cdot \sqrt{121}$ — 44
6. $\sqrt{16} \cdot \sqrt{11}$ — $4\sqrt{11}$

Also on Quiz Transparency 5.2

Teach

Why Have students imagine that they are standing next to a water well. Ask them if they can determine its depth by finding the time it takes a falling rock to hit the water and determine whether the rock travels at a constant rate or has an increasing velocity as it falls. Have them estimate the depth of the well if it takes about 4 seconds for the rock to hit the water. 256 ft

Use Teaching Transparency 17.

Alternative Teaching Strategy

USING TABLES Use Exercise 10 to show students the tabular method for approximating a solution to an equation. Tell them to begin by making a table of values for the function $h(t) = -16t^2 + 110$ by substituting integer values of t.

t	0	1	2	3
$h(t)$	110	94	46	−34

According to the table, $h(2) = 46$ and $h(3) = -34$, so h will equal 0 somewhere between $t = 2$ and $t = 3$. Have them make a new table in which t is shown in increments of one-tenth.

t	2.5	2.6	2.7
$h(t)$	10	1.84	−6.64

Tell students the new table shows that 2.6 is an approximate zero to the nearest tenth since 1.84 is closer to 0 than −6.64. Ask students to approximate the positive zero to the nearest hundredth. ≈2.62

EXAMPLE 1 Solve $4x^2 + 13 = 253$. Give exact solutions. Then approximate the solutions to the nearest hundredth.

SOLUTION

$4x^2 + 13 = 253$	
$4x^2 = 240$	*Subtract 13 from each side.*
$x^2 = 60$	*Divide each side by 4.*
$x = \pm\sqrt{60}$	*Take the square root of each side.*
$x = \sqrt{60}$ *or* $x = -\sqrt{60}$	*Exact solution*
$x \approx 7.75$ $\quad x \approx -7.75$	*Approximate solution*

CHECK
Graph $y = 4x^2 + 13$ and $y = 253$ on the same screen, and find any points of intersection.

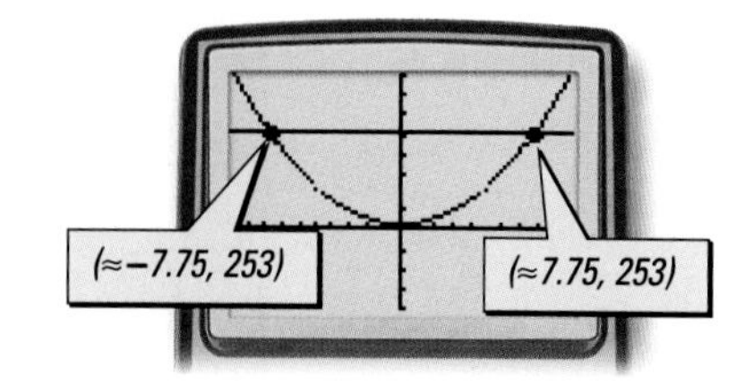

TRY THIS Solve $5x^2 - 19 = 231$. Give exact solutions. Then approximate the solutions to the nearest hundredth.

CHECKPOINT ✔ Use the Product Property of Square Roots to show that $\sqrt{60} = 2\sqrt{15}$. Then use a calculator to approximate $2\sqrt{15}$.

EXAMPLE 2 Solve $9(x - 2)^2 = 121$.

SOLUTION

$9(x-2)^2 = 121$	
$(x-2)^2 = \frac{121}{9}$	*Divide each side by 9.*
$x - 2 = \pm\sqrt{\frac{121}{9}}$	*Take the square root of each side.*
$x = 2 + \sqrt{\frac{121}{9}}$ *or* $x = 2 - \sqrt{\frac{121}{9}}$	
$x = 2 + \frac{\sqrt{121}}{\sqrt{9}}$ $\quad x = 2 - \frac{\sqrt{121}}{\sqrt{9}}$	*Use the Quotient Property of Square Roots.*
$x = 2 + \frac{11}{3}$ $\quad x = 2 - \frac{11}{3}$	
$x = \frac{17}{3}$, or $5\frac{2}{3}$ $\quad x = -\frac{5}{3}$, or $-1\frac{2}{3}$	

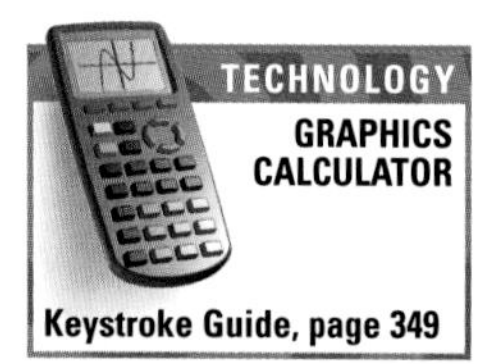

CHECK
Graph $y = 9(x - 2)^2$ and $y = 121$ on the same screen, and find any points of intersection.

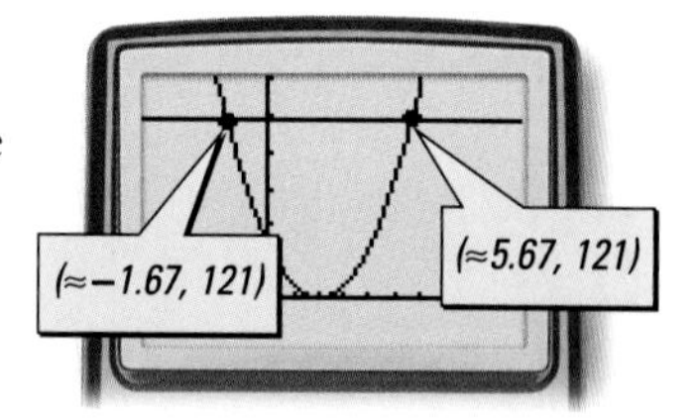

TRY THIS Solve $4(x + 2)^2 = 49$.

ADDITIONAL EXAMPLE 1

Solve $7x^2 - 23 = 103$. Give exact solutions in radical form. Then give approximate solutions to the nearest hundredth.

$3\sqrt{2}$ and $-3\sqrt{2}$;
4.24 and −4.24

Teaching Tip

TECHNOLOGY When using a TI-82 or TI-83 graphics calculator for the check in Example 1, set the viewing window as follows:

Xmin=−10	**Ymin=−200**
Xmax=10	**Ymax=400**
Xscl=1	**Yscl=50**

Enter **Y1=4X^2 + 13** and **Y2=253.** Then press GRAPH to display the graphs. To approximate the solutions, press 2nd TRACE, choose **5:intersect**, use ▲ or ▼ to place the cursor on the first graph, and then press ENTER. Use the same procedure to identify the second graph. Press ◄ or ► to move the cursor to an approximate point of intersection, and then press ENTER. The coordinates of the point of intersection will be shown at the bottom of the screen.

TRY THIS
$x = \pm 5\sqrt{2} \approx \pm 7.07$

CHECKPOINT ✔
$\sqrt{60} = \sqrt{4 \cdot 15} = \sqrt{4} \cdot \sqrt{15}$
$= 2\sqrt{15}$, or ≈ 7.75

ADDITIONAL EXAMPLE 2

Solve $64(x + 5)^2 = 289$.

$x = -\frac{23}{8}$ *or* $x = -\frac{57}{8}$

TRY THIS
$x = -\frac{11}{2}$ *or* $x = \frac{3}{2}$

Enrichment

Have students simulate motion along a vertical line by using parametric equations. Ask them to refer to Example 3. Tell them that graphing $x(t) = 2$ and $y(t) = -16t^2 + 68$ will simulate the motion along the vertical line $x = 2$ and that graphing $x(t) = t$ and $y(t) = -16t^2 + 68$ is the same as graphing h. If they are using a TI-82 or TI-83 graphics calculator, have them press MODE and choose **PAR**, **DOT**, and **SIMUL**. Then have students set the viewing window as shown at right above.

Tmin=0	**Xmin=0**	**Ymin=0**
Tmax=5	**Xmin=3**	**Ymax=70**
Tstep=.1	**Xscl=1**	**Yscl=10**

Tell them to enter the following equations:

X1T = 2	**Y1T= −16T^2 + 68**
X1T= T	**Y1T= −16T^2 + 68**

Press GRAPH to view the motion along the line. Ask students to explain why the spacing of the dots varies. **The rate of change is not constant.**

Exploring Solutions to Quadratic Equations

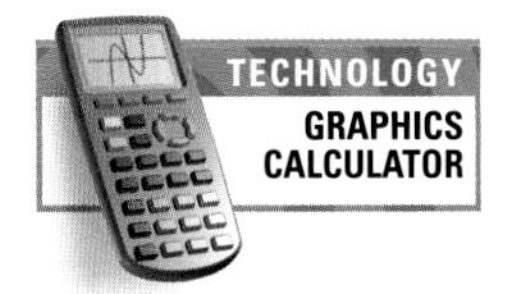

You will need: a graphics calculator or graph paper

1. Copy and complete the second and third columns of the table below.

Equation	Exact solution(s)	Number of solutions	Related function	Number of x-intercepts
$x^2 - 7 = 0$	$x = \pm\sqrt{7}$	2	$f(x) = x^2 - 7$	2
$x^2 - 2 = 0$			$f(x) = x^2 - 2$	
$x^2 = 0$			$f(x) = x^2$	
$-x^2 + 2 = 0$			$f(x) = -x^2 + 2$	
$-x^2 + 7 = 0$			$f(x) = -x^2 + 7$	
$-x^2 = 0$			$f(x) = -x^2$	

2. Graph the related quadratic function for each equation, and complete the last column of the table.

CHECKPOINT ✔ 3. What is the relationship between the number of solutions to a quadratic equation and the number of x-intercepts of the related function?

EXAMPLE 3

APPLICATION
RESCUE

Refer to the rescue-helicopter problem described at the beginning of the lesson.

After how many seconds will the raft dropped from the helicopter hit the water?

SOLUTION

The raft will hit the water when its height above the water is 0 feet, or when $h(t) = 0$.

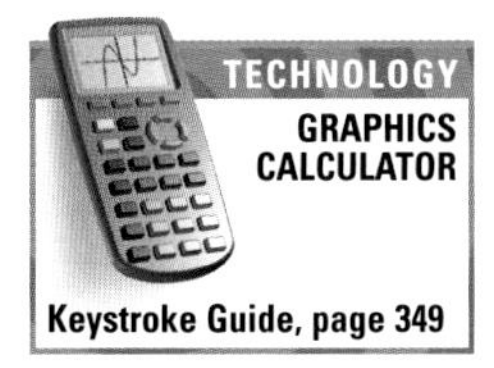

Method 1 Use algebra.
Let $h(t) = 0$.

$$-16t^2 + 68 = 0$$
$$-16t^2 = -68$$
$$t^2 = \frac{-68}{-16}$$
$$t^2 = \frac{17}{4}$$
$$t = \pm\sqrt{\frac{17}{4}}$$
$$t \approx \pm 2.1$$

Method 2 Use the graph.
Graph $h(t) = -16t^2 + 68$, and find the reasonable t-value for which $h(t) = 0$.

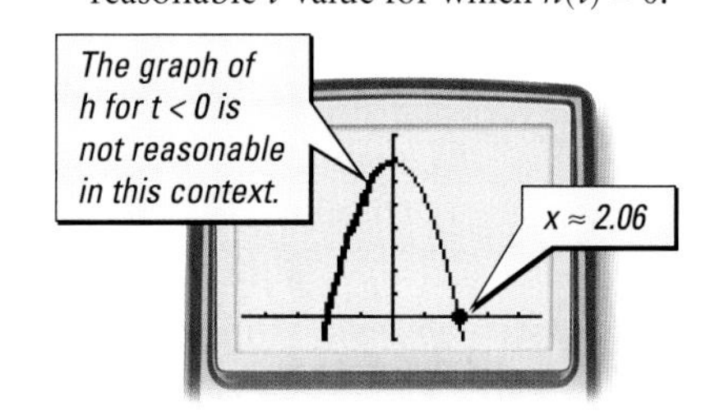

Since t must be greater than 0, the raft will hit the water about 2.1 seconds after it is dropped.

TRY THIS How many seconds will it take the raft to hit the water if the helicopter drops the raft from a height of 34 feet?

Activity Notes

In this Activity, students will use the positive and negative solutions of each quadratic equation to determine the number of x-intercepts for each related function. Students will discover that the number of real solutions of an equation is equal to the number of x-intercepts for the related function.

CHECKPOINT ✔

3. The number of solutions to a quadratic equation and the number of x-intercepts of the related function are equal.

ADDITIONAL EXAMPLE 

A humanitarian group provides assistance to people who need help in obtaining resources. The group has decided to send a shipment of food by helicopter to some families living in a remote area. The helicopter, which is 85 feet above the designated drop-off area, releases a box containing food. The height of the box above the ground can be modeled by $h(t) = -16t^2 + 85$, where t is the time in seconds after it has been released. **After how many seconds will the box land on the ground?**
≈ 2.3 s

TRY THIS
≈ 1.5 s

Inclusion Strategies

VISUAL LEARNERS Students who have trouble with abstractions may not be able to distinguish between solving an equation such as $3(x^2 + 1) = 12$ and solving an equation such as $3(x + 1)^2 = 12$. Have students copy the equation to be solved and highlight the variable or expression that is squared so that they can clearly see the term that should be isolated.

$3(x^2 + 1) = 12$	$3(x + 1)^2 = 12$
$3x^2 + 3 = 12$	$(x + 1)^2 = 4$
$3x^2 = 9$	$x + 1 = \pm 2$
$x^2 = 3$	$x = -1 \pm 2$
$x = \pm\sqrt{3}$	$x = -3$ *or* $x = 1$
$x \approx \pm 1.73$	

Math CONNECTION

GEOMETRY You can use the Pythagorean theorem to find the length of one side of a right triangle if you know the lengths of the other two sides.

ADDITIONAL EXAMPLE 4

Find the unknown length in each right triangle. Give your answers to the nearest tenth.

a.

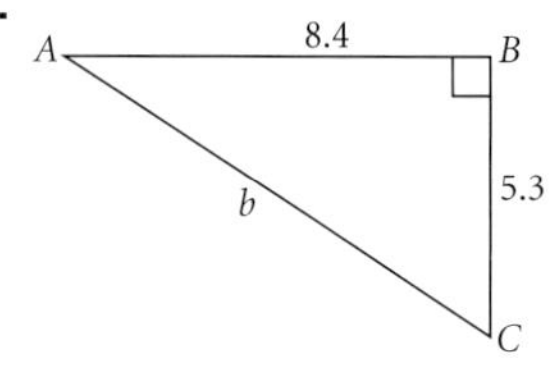

≈9.9 units

b.

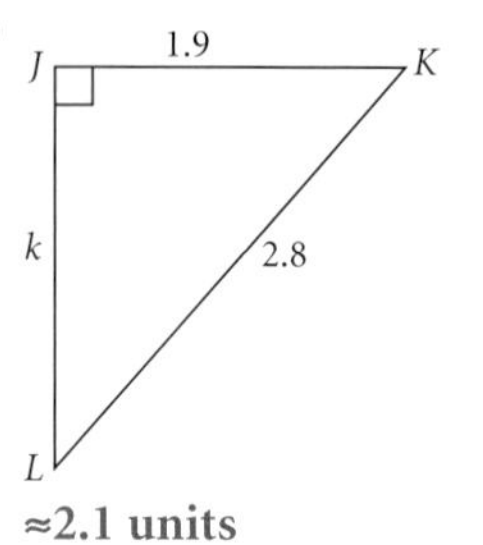

≈2.1 units

Using the Pythagorean Theorem

Greek mathematician Pythagoras (around 580–500 B.C.E.)

CULTURAL CONNECTION: ASIA
Sometime between 1900 B.C.E. and 1600 B.C.E. in ancient Babylonia (now Iraq), a table of numbers was inscribed on a clay tablet. When archeologists discovered the tablet, part of it was missing, so the meaning of the numbers on it remained a mystery. The sets of numbers on the tablet are believed to be triples, called *Pythagorean triples*, that form a special right-triangle relationship. This relationship, named after the Greek mathematician Pythagoras, is commonly called the *Pythagorean Theorem.*

Ancient tablet believed to contain Pythagorean triples

When you sketch a right triangle, use capital letters to name the angles and corresponding lowercase letters to name the lengths of the sides opposite the angles. For example, $\overline{BC}$ is labeled a because it is opposite angle A.

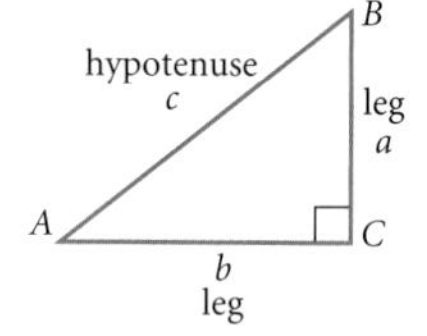

Pythagorean Theorem

If $\triangle ABC$ is a right triangle with the right angle at C, then $a^2 + b^2 = c^2$.

When you apply the Pythagorean Theorem, use the principal square root because distance and length cannot be negative.

EXAMPLE  **Find the unknown length in each right triangle. Give answers to the nearest tenth.**

a.

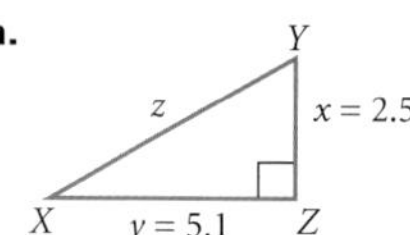

b.

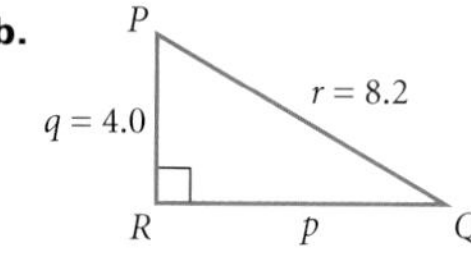

SOLUTION

PROBLEM SOLVING

Use a formula.

a.

$$x^2 + y^2 = z^2$$
$$2.5^2 + 5.1^2 = z^2$$
$$z^2 = 5.1^2 + 2.5^2$$
$$z = \sqrt{5.1^2 + 2.5^2}$$
$$z \approx 5.68$$

z is about 5.7 units.

b.

$$p^2 + q^2 = r^2$$
$$p^2 + 4.0^2 = 8.2^2$$
$$p^2 = 8.2^2 - 4.0^2$$
$$p = \sqrt{8.2^2 - 4.0^2}$$
$$p \approx 7.16$$

p is about 7.2 units.

TRY THIS Find the unknown length in each right triangle. Give answers to the nearest tenth.

a.

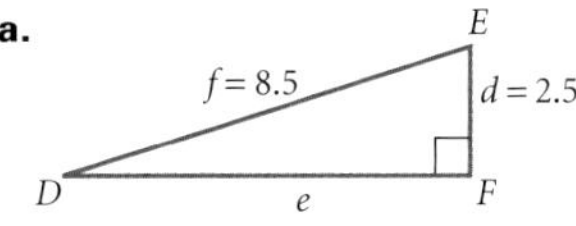

b.

R, s = 3.4, t, T, r = 8.7, S

CRITICAL THINKING Suppose that $\triangle ABC$ is a right triangle with the right angle at C. Write a formula for b in terms of a and c, assuming that a and c are known. Then write a formula for a in terms of b and c, assuming that b and c are known.

Sometimes you may need to apply the Pythagorean Theorem twice in order to find the solution to a problem. This is shown in Example 5.

EXAMPLE 5

APPLICATION: ENGINEERING

The diagram shows support wires $\overline{AD}$ and $\overline{BD}$ for a tower.

How far apart are the support wires where they contact the ground? Give your answer to the nearest whole foot.

D, 900 ft, 760 ft, 850 ft, C, B, A

[Not drawn to scale]

SOLUTION

CONNECTION: GEOMETRY

The distance between the support wires where they contact the ground is AB.

1. Apply the Pythagorean Theorem to $\triangle ADC$ to find AC.

$$(AC)^2 + (CD)^2 = (AD)^2$$
$$(AC)^2 + 760^2 = 900^2$$
$$(AC)^2 = 900^2 - 760^2$$

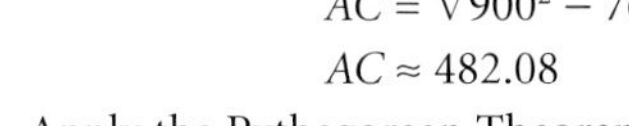

$$AC = \sqrt{900^2 - 760^2}$$
$$AC \approx 482.08$$

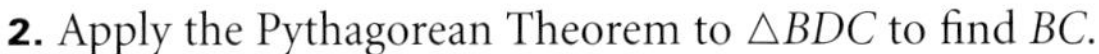

2. Apply the Pythagorean Theorem to $\triangle BDC$ to find BC.

TECHNOLOGY: SCIENTIFIC CALCULATOR

$$(BC)^2 + (CD)^2 = (BD)^2$$
$$(BC)^2 + 760^2 = 850^2$$
$$(BC)^2 = 850^2 - 760^2$$
$$BC = \sqrt{850^2 - 760^2}$$
$$BC \approx 380.66$$

√(900²-760²)
482.0788317
√(850²-760²)
380.6573262

You can edit the previous entry by replacing 900 with 850.

3. Find $AC - BC = AB$.

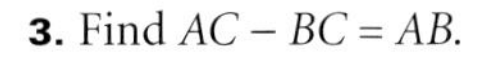

$$482.08 - 380.66 = 101.42$$

The wires are about 101 feet apart at ground level.

TRY THIS In the diagram below, find PQ. Give your answer to the nearest whole meter.

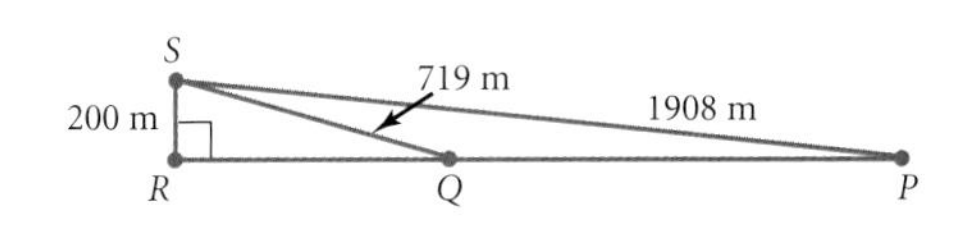

TRY THIS
a. ≈**8.1 units**
b. ≈**9.3 units**

CRITICAL THINKING
$b = \sqrt{c^2 - a^2}$; $a = \sqrt{c^2 - b^3}$

ADDITIONAL EXAMPLE 5

The diagram below represents a tower, $\overline{RS}$, and support wires, $\overline{PS}$ and $\overline{QS}$. **What must be the length of $\overline{QS}$ so that it contacts the ground halfway between the base of the tower and the point where $\overline{PS}$ contacts the ground?**

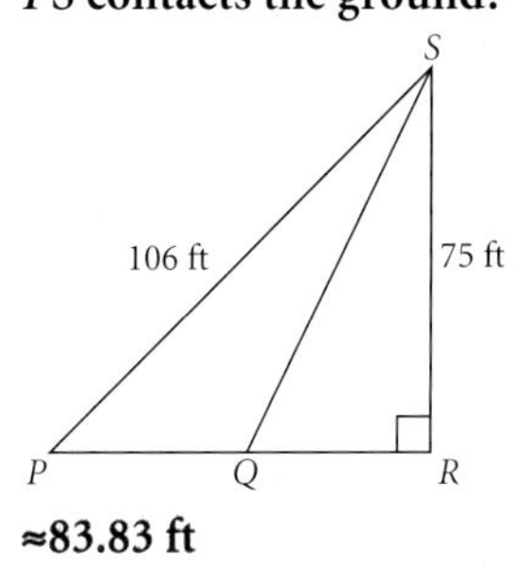

≈**83.83 ft**

Math CONNECTION

GEOMETRY In Example 5, students find indirect measurements from a diagram by applying the Pythagorean theorem.

TRY THIS
$PQ \approx$ **1207 m**

Reteaching the Lesson

USING COGNITIVE STRATEGIES Demonstrate the following method for solving $x^2 = 6$ by guessing:

Because $4 < 6 < 9$, or $2^2 < 6 < 3^2$, x must be between 2 and 3.

Guess that $x = 2.4$ and evaluate: $2.4^2 = 5.76 < 6$.
Guess that $x = 2.5$ and evaluate: $2.5^2 = 6.25 > 6$.
So, $2.4 \le x \le 2.5$.

Guess that $x = 2.44$ and evaluate: $2.44^2 = 5.9536 < 6$.
Guess that $x = 2.45$ and evaluate: $2.45^2 = 6.0025 > 6$.
So, $2.44 < x < 2.45$.

Guess that $x = 2.445$ and evaluate:
$2.445^2 = 5.978025 < 6$.
So, $2.445 < x < 2.45$.

Therefore, $x \approx 2.45$ to the nearest hundredth.

Give students the following quadratic equations to solve by guessing: $x^2 = 8$, $x^2 - 15 = 0$, and $2x^2 - 5 = 0$. Have them verify their guesses by solving the equations algebraically.

Assess

Selected Answers

Exercises 4–13, 15–69 odd

ASSIGNMENT GUIDE	
In Class	1–13
Core	15–43 odd, 49, 53, 55
Core Plus	14–60 even
Review	60–69
Preview	70–72

Extra Practice can be found beginning on page 940.

Exercises

Communicate

1. Describe the procedure you would use to solve $5(x+3)^2 = 12$.

2. Describe three situations in which it makes sense to consider only the principal root as a solution.

3. How can you find the length of the hypotenuse of a right triangle with legs that are 3 and 4 units long?

Guided Skills Practice

4. $\pm\sqrt{29} \approx \pm 5.39$

5. $\pm\sqrt{11} \approx \pm 3.32$

Solve each equation. Give exact solutions. Then approximate each solution to the nearest hundredth, if necessary. ***(EXAMPLES 1 AND 2)***

4. $x^2 = 29$ **5.** $2x^2 - 4 = 18$ **6.** $(x+1)^2 = 9$ **−4, 2**

7. $3(x-2)^2 = 21$ $2 \pm \sqrt{7} \approx -0.65, 4.65$ **8.** $2(x^2-4)+3 = 15$ $\pm\sqrt{10} \approx \pm 3.16$ **9.** $\frac{1}{2}(x^2+6) - 5 = 10$ $\pm\sqrt{24} = \pm 2\sqrt{6} \approx \pm 4.90$

APPLICATION

10. AVIATION A crate of blankets and clothing is dropped without a parachute from a helicopter hovering at an altitude of 110 feet. The crate's height in feet above the ground is modeled by $h(t) = -16t^2 + 110$, where t is the time in seconds after it is dropped. How long will it take for the crate to reach the ground? ***(EXAMPLE 3)*** **about 2.6 sec**

CONNECTION

GEOMETRY Find the unknown length in each right triangle. Give your answer to the nearest tenth. ***(EXAMPLE 4)***

11. 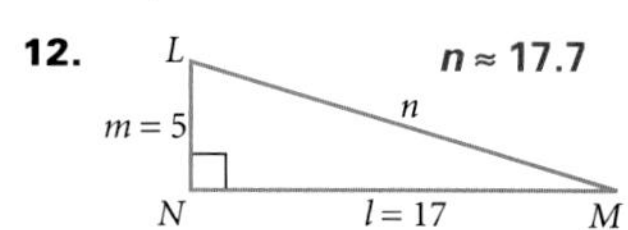

e ≈ 3.7

12. 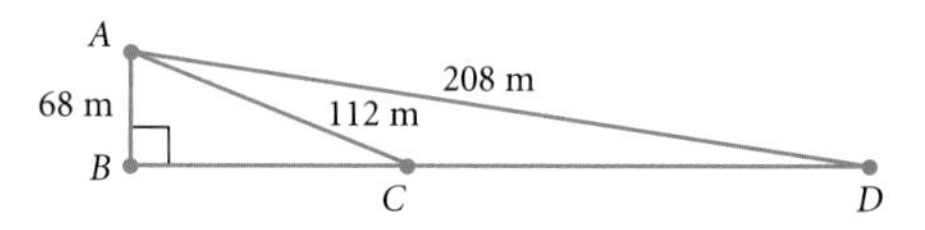

n ≈ 17.7

APPLICATION

13. CONSTRUCTION Two support wires and their lengths are shown in the diagram below. What is the distance in meters between the wires, represented by $\overline{CD}$, where they are attached to the ground? Give your answer to the nearest tenth of a meter. ***(EXAMPLE 5)*** **107.6 m**

A, 68 m, B, 112 m, 208 m, C, D

internet connect
Homework Help Online
Go To: **go.hrw.com**
Keyword:
MB1 Homework Help for Exercises 14–31

Practice and Apply

Solve each equation. Give exact solutions. Then approximate each solution to the nearest hundredth, if necessary.

14. $x^2 = 121$ **15.** $x^2 = 32$ **16.** $3x^2 = 49$

17. $4x^2 = 20$ **18.** $4t^2 = 1$ **19.** $\frac{1}{2}x^2 = 6$

14. ± 11

15. $\pm\sqrt{32} = \pm 4\sqrt{2} \approx \pm 5.66$

16. $\pm\sqrt{\frac{49}{3}} = \pm\frac{7\sqrt{3}}{3} = \pm 4.04$

17. $\pm\sqrt{5} \approx \pm 2.24$

18. $\pm\frac{1}{2}$

19. $\pm\sqrt{12} = \pm 2\sqrt{3} \approx \pm 3.46$

20. $\pm\sqrt{\frac{39}{2}} \approx \pm 4.42$
21. ± 6
22. $\pm\sqrt{37} \approx \pm 6.08$
23. $\pm\sqrt{\frac{11}{2}} \approx \pm 2.35$
24. $\pm\sqrt{\frac{15}{4}} = \pm\frac{\sqrt{15}}{2} \approx \pm 1.94$
25. $\pm\sqrt{2} \approx \pm 1.41$
26. $\pm\sqrt{\frac{11}{2}} \approx \pm 2.35$
27. 1, 9
28. $-2 \pm \sqrt{7} \approx -4.65, 0.65$
29. $\pm\sqrt{126} = \pm 3\sqrt{14} \approx \pm 11.22$
30. $\pm\sqrt{5} \approx \pm 2.24$
31. $-1 \pm \sqrt{5} \approx -3.24, 1.24$

20. $\frac{2}{3}a^2 = 13$ **21.** $x^2 + 5 = 41$ **22.** $x^2 - 37 = 0$

23. $2t^2 - 5 = 6$ **24.** $4x^2 + 5 = 20$ **25.** $10 - 3x^2 = 4$

26. $8 - 2x^2 = -3$ **27.** $(x - 5)^2 = 16$ **28.** $(t + 2)^2 = 7$

29. $\frac{1}{3}(t^2 - 15) = 37$ **30.** $4(s^2 + 7) - 9 = 39$ **31.** $7 = 2(r + 1)^2 - 3$

Find the unknown length in each right triangle. Give answers to the nearest tenth.

32. $b \approx 5.5$

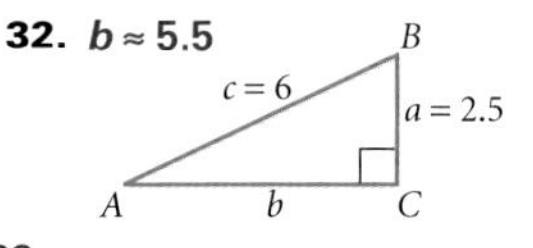

33. $t \approx 2.2$

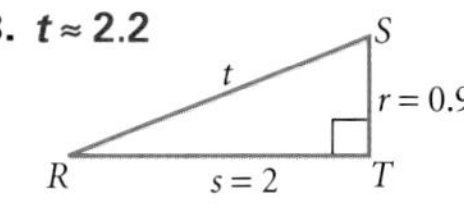

34. $j \approx 19.3$

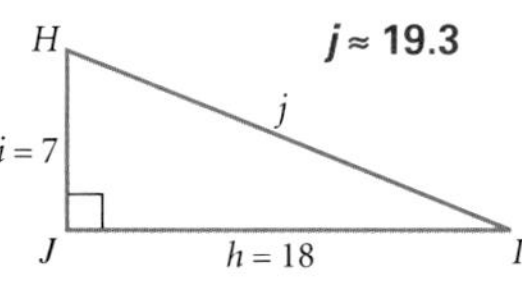

35. $d \approx 5.3$

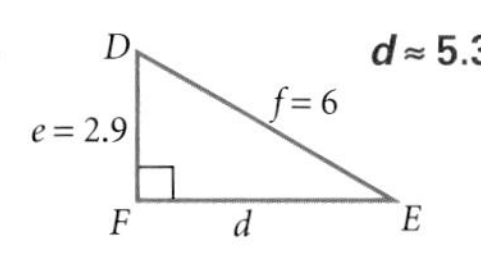

36. $e \approx 0.9$

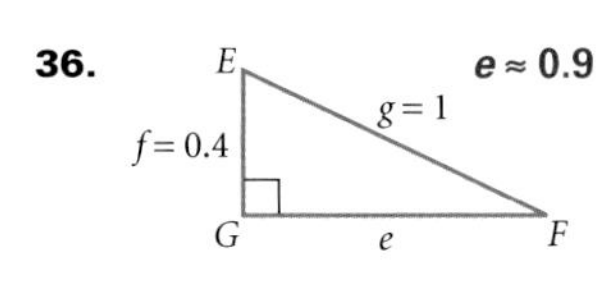

37. $v \approx 24.2$

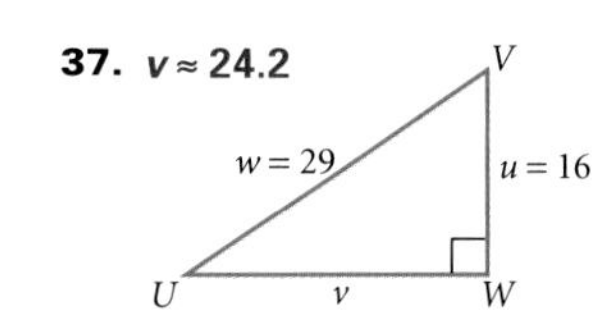

Find the missing side length in right triangle *ABC*. Give answers to the nearest tenth, if necessary.

38. a is 9 and b is 2. $c \approx 9.2$ **39.** a is 8 and b is 4. $c \approx 8.9$

40. c is 5 and b is 3. $a = 4$ **41.** c is $\sqrt{29}$ and b is 5. $a = 2$

42. a is 9 and c is $\sqrt{90}$. $b = 3$ **43.** a is 7 and c is $\sqrt{74}$. $b = 5$

CHALLENGE

Write a quadratic equation for each pair of solutions.

44. 7 and −7 **45.** $\sqrt{17}$ and $-\sqrt{17}$ **46.** $\sqrt{2001}$ and $-\sqrt{2001}$

44. $x^2 - 49 = 0$
45. $x^2 - 17 = 0$
46. $x^2 - 2001 = 0$

CONNECTIONS

47. GEOMETRY The area of a circle is 20π square inches. Find the radius of the circle. (Hint: The area of a circle is given by $A = \pi r^2$.) **about 4.47 in.**

48. GEOMETRY Copy and complete the table.

Area of square	4	5	6	7	8	...	A
Side of square	2	$\sqrt{5}$	$\sqrt{6}$	$\sqrt{7}$	$\sqrt{8}$	...	$\sqrt{A}$
Diagonal of square	$\sqrt{8}$	$\sqrt{10}$	$\sqrt{12}$	$\sqrt{14}$	4	...	$\sqrt{2A}$

49a. 4.24 ft
b. 5.20 ft

49. GEOMETRY A cube measures 3 feet on each edge.

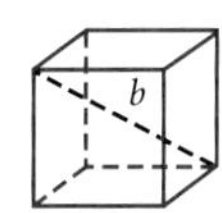

a. What is the length of a diagonal, such as a, along one of the faces of the cube?

b. What is the length of a diagonal, such as b, that passes through the interior of the cube?

Error Analysis

Some students only find the positive square root when solving an equation of the form $x^2 = k$. Remind students that solving the equation $x^2 = 9$ is different from evaluating $\sqrt{9}$. There are two solutions for $x^2 = 9$, 3 and −3, but only one solution for $\sqrt{9}$, 3.

50. GEOMETRY A right pyramid that is 12 feet tall has a square base whose side length is 5 feet.
a. What is its slant height? **12.26 ft**
b. What is the length of its lateral edges? **12.51 ft**

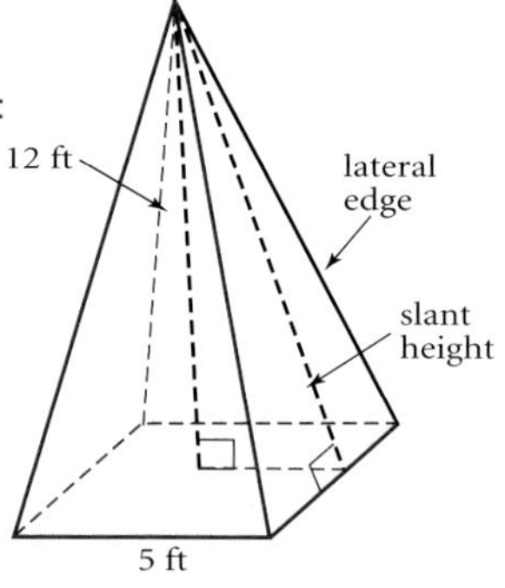

APPLICATIONS

51. PHYSICS A worker drops a hammer from a second-story roof that is 10 meters above the ground. If the hammer's height in meters above the ground is modeled by $h(t) = -4.9t^2 + 10$, where t represents time in seconds after the hammer is dropped, about how long will it take the hammer to reach the ground? **1.43 sec**

52. RECREATION A child at a swimming pool jumps off a 12-foot platform into the pool. The child's height in feet above the water is modeled by $h(t) = -16t^2 + 12$, where t is the time in seconds after the child jumps. How long will it take the child to reach the water? **0.87 sec**

53. SPORTS A baseball diamond is a square with sides of 90 feet. To the nearest foot, how long is a throw to first base from third base? **127.28 ft**

54. NAVIGATION A hiker leaves camp and walks 5 miles east. Then he walks 10 miles south. How far from camp is the hiker? **11.18 mi**

55. TELECOMMUNICATIONS The cable company buries a line diagonally across a rectangular lot. The lot measures 105 feet by 60 feet. How long is the cable line? **120.93 ft**

56. ENGINEERING The velocity, v, in centimeters per second, of a fluid flowing in a pipe varies with respect to the radius, x, of the pipe according to the equation $v(x) = 16 - x^2$.
a. Find x for $v = 7$.
b. Find x for $v = 12$.**3 cm**
2 cm

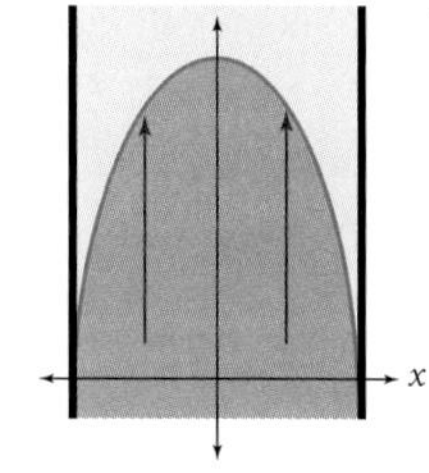

Practice

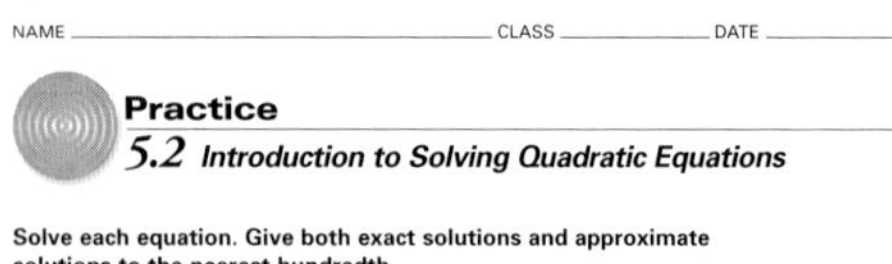

NAME ______ CLASS ______ DATE ______

Practice 5.2 Introduction to Solving Quadratic Equations

Solve each equation. Give both exact solutions and approximate solutions to the nearest hundredth.

1. $x^2 = 100$ — **−10 or 10**
2. $12x^2 = 36$ — $-\sqrt{3} \approx -1.73$ or $\sqrt{3} \approx 1.73$
3. $(x + 3)^2 = 81$ — **−12 or 6**
4. $5x^2 - 4 = 96$ — $-2\sqrt{5} \approx -4.47$ or $2\sqrt{5} \approx 4.47$
5. $x^2 - 12 = 4$ — **−4 or 4**
6. $6x^2 + 15 = 23$ — $\frac{2\sqrt{3}}{3} \approx 1.15$ or $-\frac{2\sqrt{3}}{3} \approx -1.15$
7. $4x^2 - 9 = 17$ — $-\sqrt{6.5} \approx -2.55$ or $\sqrt{6.5} \approx 2.55$
8. $12 = 4(x - 2)^2 - 8$ — $2 - \sqrt{5} \approx -0.24$ or $2 + \sqrt{5} \approx 4.24$
9. $14 = 0.5x^2 + 5$ — $-3\sqrt{2} \approx -4.24$ or $3\sqrt{2} \approx 4.24$
10. $7(x + 1)^2 = 161$ — $-1 - \sqrt{23} \approx -5.80$ or $-1 + \sqrt{23} \approx 3.80$

Find the unknown length in each right triangle. Round answers to the nearest tenth.

11. 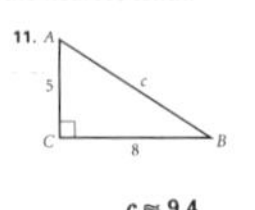

$c \approx 9.4$

12. 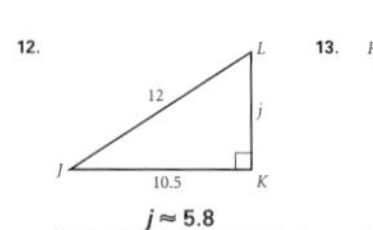

$j \approx 5.8$

13. 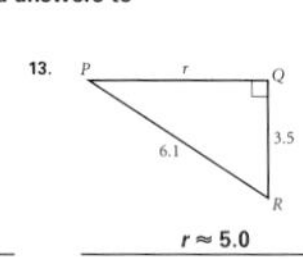

$r \approx 5.0$

Find the missing side length in right triangle *ABC*. Round answers to the nearest tenth.

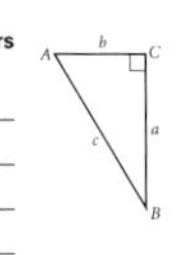

14. $a = 15$ and $b = 7$ — $c \approx 16.6$
15. $a = 2.4$ and $c = 7.3$ — $b \approx 6.9$
16. $b = 2$ and $c = \sqrt{10}$ — $a \approx 2.4$
17. $a = 9.1$ and $b = 7$ — $c \approx 11.5$

APPLICATIONS

57. CONSTRUCTION The bottom of a 20-foot ladder is placed $4\frac{1}{2}$ feet from the base of a house, as shown at right. At what height does the ladder touch the house? **19.49 ft**

58. SPORTS A cliff diver stands on a cliff overlooking water. To approximate his height above the water, he drops a pebble and times its fall. If the pebble takes about 3 seconds to strike the water, approximately how high is the diver above the water? Use the model $h(t) = -4.9t^2 + h_0$, where h is the pebble's height in meters above the ground, t is the time in seconds after the pebble is dropped, and h_0 is the height of the cliff in meters. **44.1 m**

59. CONSTRUCTION Find the length of the rafter that provides for a rise of 5 feet over a run of 24 feet. Allow for a 1-foot overhang in the length of the rafter. **14 ft**

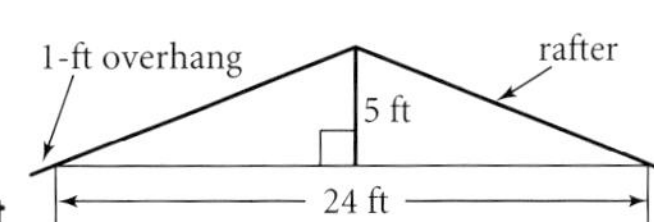

Look Back

Determine whether each table represents a linear relationship between *x* and *y*. If the relationship is linear, write the next ordered pair that would appear in the table. *(LESSON 1.1)*

60. **no**

x	−5	−3	−1	0	2	4
y	2	1	0	−1	−2	−3

61. **yes; (9, 15)**

x	−3	−1	1	3	5	7
y	−9	−5	−1	3	7	11

Find the slope of the line passing through the given points, and write the equation of the line in slope-intercept form. *(LESSON 1.2)*

62. $(-4, -2)$ and $(6, -7)$ **$m = -\frac{1}{2}$; $y = -\frac{1}{2}x - 4$**

63. $(3, 5)$ and $(-6, 1)$

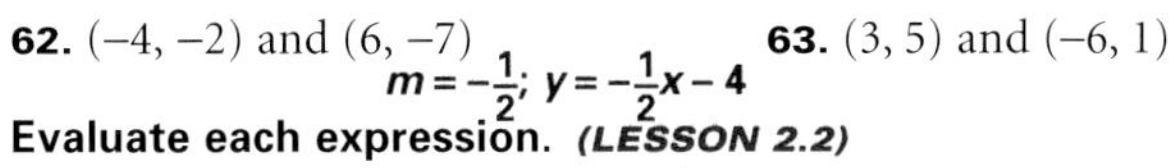

Evaluate each expression. *(LESSON 2.2)*

64. $4^5 \cdot 4^{-3}$ **16**

65. $(5^4)^{-\frac{1}{2}}$ **$\frac{1}{25}$**

66. $(6^4 \cdot 6)^0$ **1**

Give the domain and range of each function. *(LESSON 2.3)*

67. $f(x) = 3x^2 - 7$
domain: all real
range: all real ≥ −7

68. $f(x) = \frac{x - 6}{5}$
domain: all real
range: all real

69. $f(x) = 3\left(\frac{x}{2}\right)^2$
domain: all real
range: all real ≥ 0

Look Beyond

CONNECTION

TRANSFORMATIONS **Graph each pair of functions, find the vertices, and describe how the graphs are related.**

70 $f(x) = (x - 3)^2 - 5$ and $g(x) = (x - 3)^2 + 5$

71 $f(x) = (x + 5)^2 - 4$ and $g(x) = (x - 2)^2 - 4$

72 $f(x) = x^2$ and $g(x) = (x - 3)^2 - 4$

70. *f*: (3, −5), *g*: (3, 5); *g* is *f* shifted 10 units up.

71. *f*: (−5, −4), *g*: (2, −4); *g* is *f* shifted 7 units to the right.

72. *f*: (0, 0), *g*: (3, −4); *g* is *f* shifted 4 units down and 3 units to the right.

Look Beyond

These exercises require students to examine relationships between graphs of quadratic functions written in vertex form. Transformations of quadratic functions will be studied in depth in Lesson 5.4.

Student Technology Guide

NAME ______ CLASS ______ DATE ______

Student Technology Guide
5.2 Introduction to Solving Quadratic Equations

You can solve a quadratic equation by using the zero feature on a graphics calculator.

Example: Solve $2(x + 2)^2 = 35$.

- First, set one side of the equation equal to zero. To solve $2(x + 2)^2 = 35$, write it as $2(x + 2)^2 - 35 = 0$.
- Enter and graph the equation in a standard viewing window by pressing Y= 2 (X,T,θ,n + 2) x² − 35 ZOOM 6:ZStandard ENTER.
- Press TRACE. Use arrow keys to move just above a zero, as shown at left below.
- Press 2nd TRACE 2:zero ENTER.
- When asked for the left bound, press ENTER to enter the coordinates of the trace point. See the display at left below.
- When asked for the right bound, trace just below the zero. Press ENTER. See the display in the middle below.
- When asked for a guess, trace as close as you can to the zero. Press ENTER. See the display at right below.

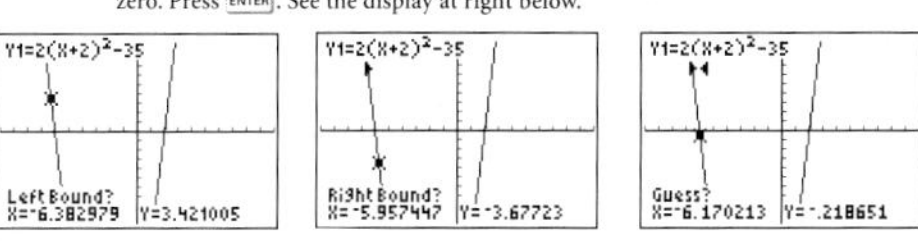

As you can see, one zero is near $x = -6.18$. Repeat the process to find that the second zero is near $x = 2.18$.

Solve each equation by graphing. Round solutions to the nearest hundredth.

1. $3x^2 = 37$ **−3.51, 3.51**
2. $2(x - 1)^2 = 17$ **−1.92, 3.92**
3. $5(x + 2.7)^2 = 11.4$ **−4.21, −1.19**

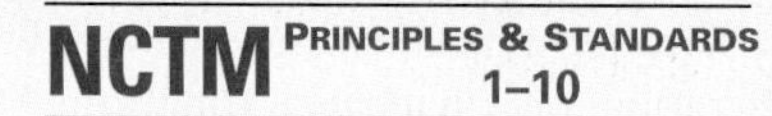

Prepare

NCTM PRINCIPLES & STANDARDS 1–10

QUICK WARM-UP

List all the factors of each number.

1. 10 — 1, 2, 5, 10

2. 48 — 1, 2, 3, 4, 6, 8, 12, 16, 24, 48

3. 7 — 1, 7

Find the greatest common factor (GCF) of each set of numbers.

4. 6, 14 — 2

5. 12, 18, 30 — 6

6. 4, 8, 15, 20 — 1

Also on Quiz Transparency 5.3

Teach

Why Ask students if they have ever watched the water from a fountain as it flows up into the air, forming an arc. Ask them to discuss how they would model the curve of water coming from the fountain. Have them consider such ideas as making a physical model, collecting data, and obtaining measurements.

5.3 Factoring Quadratic Expressions

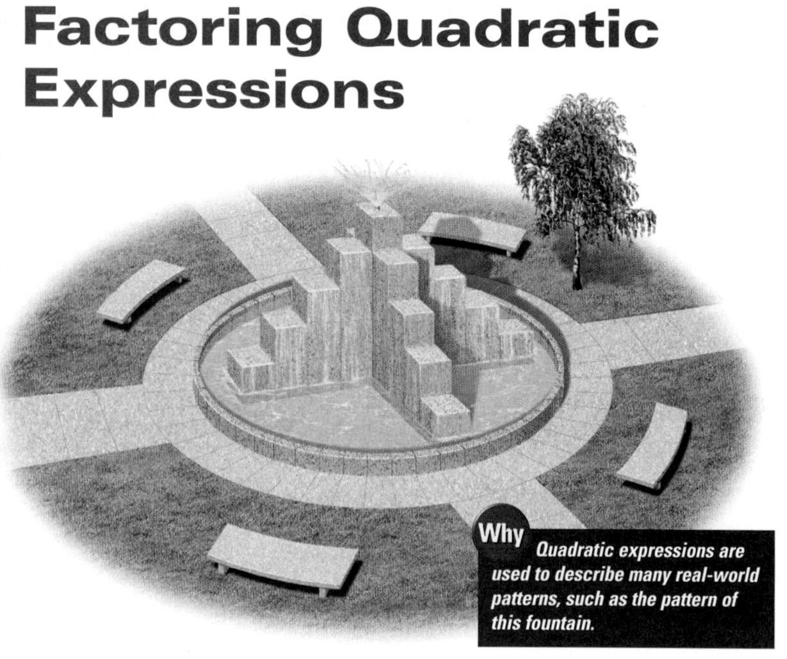

Why *Quadratic expressions are used to describe many real-world patterns, such as the pattern of this fountain.*

Objectives

- Factor a quadratic expression.
- Use factoring to solve a quadratic equation and find the zeros of a quadratic function.

APPLICATION
ARCHITECTURE

An architect created a proposal for the fountain shown above. Each level (except the top one) is an X formed by cubes. The number of cubes in each of the four parts of the X is one less than the number on the level below. A formula for the total number of cubes, c, in the fountain is given by $c = 2n^2 - n$, where n is the number of levels in the fountain. How many levels would a fountain consisting of 66 cubes have? *You will answer this question in Example 7.*

Factoring Quadratic Expressions

Multiplying →
$2x(x + 3) = 2x^2 + 6x$
← Factoring

When you learned to multiply two expressions like $2x$ and $x + 3$, you learned how to write a product as a sum. **Factoring** reverses the process, allowing you to write a sum as a product.

To factor an expression containing two or more terms, factor out the *greatest common factor* (GCF) of the two expressions, as shown in Example 1.

EXAMPLE 1 **Factor each quadratic expression.**

a. $3a^2 - 12a$

b. $3x(4x + 5) - 5(4x + 5)$

SOLUTION

Factor out the GCF for all of the terms.

a. $3a^2 = 3a \cdot a$ and $12a = 3a \cdot 4$
The GCF is $3a$.
$3a^2 - 12a = 3a(a) - 3a(4)$
$= (3a)(a - 4)$

b. The GCF is $4x + 5$.
$3x(4x + 5) - 5(4x + 5)$
$= (3x - 5)(4x + 5)$

TRY THIS Factor $5x^2 + 15x$ and $(2x - 1)4 + (2x - 1)x$.

Alternative Teaching Strategy

USING PATTERNS Remind students that $n \cdot 0 = 0$ regardless of the value of n and that if a product is 0, then at least one of the factors is 0. In this regard, 0 is a special number in the set of real numbers because it is the only number that acts this way when multiplied. State the zero-product property and ask students to solve $3x = 0$, $3(x - 2) = 0$, and $(x + 3)(x - 2) = 0$, pointing out that if the product is 0, then one or both of the factors must be 0. Now ask students to speculate about how to solve $x^2 + x - 6 = 0$.

Then remind students that any equation can be solved by graphing, and that when one side of the equation is 0, they will be looking for the x-intercepts. Once students know the x-intercepts, they can write the linear factors of the equation. That is, the x-intercepts, or zeros, of the function $y = f(x)$ are related to the linear factors of the expression $f(x)$ and to the solutions to the related equation, $f(x) = 0$. Use the graph of $y = 3x^2 - 5x + 2$ to factor $3x^2 - 5x + 2$ into $(x - 1)\left(x - \frac{2}{3}\right)$ by finding the x-intercepts, 1 and $\frac{2}{3}$.

An expression of the form $ax^2 + bx + c$, where $a \neq 0$, is often called a quadratic *trinomial.* In the Activity below, you will investigate how to factor this type of expression.

Activity
Factoring With Algebra Tiles

You will need: algebra tiles

You can model a quadratic expression that can be factored with algebra tiles, as shown below.

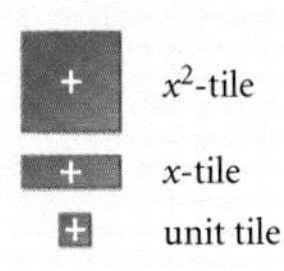

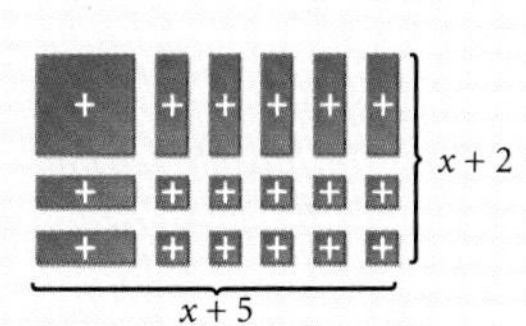

The rectangular region formed by algebra tiles above illustrates that the total area, $x^2 + 7x + 10$, can be represented as the product $(x + 5)(x + 2)$.

1. Use tiles to determine whether $x^2 + 4x + 4$ can be represented as the product of two linear factors. Justify and illustrate your response.
2. Use tiles to determine whether $x^2 + 6x + 8$ can be represented as the product of two linear factors. Justify and illustrate your response.
3. Use tiles to determine whether $x^2 + 7x + 12$ can be represented as the product of two linear factors. Justify and illustrate your response.

CHECKPOINT ✔ 4. Describe how algebra tiles can be used to factor a quadratic trinomial.

Many quadratic expressions can be factored algebraically. Examine the factored expressions below.

PROBLEM SOLVING **Look for a pattern.** Notice how the sums and products of the constants in the *binomial* factors are related to the last two terms in the unfactored expression.

$x^2 + 7x + 10 = (x + 5)(x + 2)$ — $5 + 2 = 7$; $5 \times 2 = 10$

$x^2 - 7x + 10 = (x - 5)(x - 2)$ — $-5 - 2 = -7$; $-5 \times (-2) = 10$

$x^2 + 3x - 10 = (x + 5)(x - 2)$ — $5 - 2 = 3$; $5 \times (-2) = -10$

$x^2 - 3x - 10 = (x - 5)(x + 2)$ — $-5 + 2 = -3$; $-5 \times 2 = -10$

The patterns shown above suggest a rule for factoring quadratic expressions of the form $x^2 + bx + c$.

Factoring $x^2 + bx + c$

To factor an expression of the form $ax^2 + bx + c$ where $a = 1$, look for integers r and s such that $r \cdot s = c$ and $r + s = b$. Then factor the expression.

$$x^2 + bx + c = (x + r)(x + s)$$

Interdisciplinary Connection

EARTH SCIENCE The output, in megawatts, of a geothermal power plant between midnight and noon is approximated by $P = 4h^2 - 69h + 750$, where h is the hour of the day on a 24-hour clock. At what time will the output be 670 megawatts?

$h = 16$ *or* $\frac{5}{4}$

$h = 16$, or 4:00 P.M., is not between midnight and noon, so $h = \frac{5}{4}$, or 1:15 A.M.

ADDITIONAL EXAMPLE 1

Factor each quadratic expression below.

a. $27c^2 - 18c$ $\quad 9c(3c - 2)$

b. $5z(2z + 1) - 2(2z + 1)$
$(2z + 1)(5z - 2)$

TRY THIS

a. $5x(x + 3)$

b. $(2x - 1)(4 + x)$

Activity Notes

In this Activity, students use algebra tiles to show the factorization of a quadratic trinomial by forming a rectangular region with the tiles. The sides of the rectangle formed represent the linear factors of the product, while the area of the region represents the expanded product.

CHECKPOINT ✔

4. After arranging the tiles in the pattern for the product polynomial, the factors should appear as the tiles in the right column and bottom row.

ADDITIONAL
EXAMPLE 2

Factor $x^2 + 12x + 27$.
$(x+9)(x+3)$

TRY THIS
$(x+4)(x+5)$

ADDITIONAL
EXAMPLE 3

Factor $x^2 - 15x - 54$.
$(x+3)(x-18)$

TRY THIS
$(x+1)(x-11)$

Teaching Tip

Remind students that many quadratic trinomials will not factor over the set of integers. Quadratic trinomials of the form $ax^2 + bx + c$ can be factored over the set of integers if and only if there are two integers whose product is ac and whose sum is b. The sign of ac determines whether the integers will have the same or different signs. To see if $2x^2 - x - 6$ will factor, list all integer factors of $2(-6)$, or -12, in pairs: 1 and 12, 2 and 6, 3 and 4. Look for a pair that will *subtract* to make -1 because the sign of ac is *negative*. Because $3 - 4 = -1 = b$ and $3(-4) = -12 = ac$, the expression will factor: $2x^2 - x - 6 = (2x+3)(x-2)$. Notice that in this case, 3 and -4 do *not* appear in the factorization.

ADDITIONAL
EXAMPLE 4

Factor $5x^2 + 14x + 8$. Check by graphing.
$5x^2 + 10x + 4x + 8$
$5x(x+2) + 4(x+2)$
$(x+2)(5x+4)$

When c is positive in $x^2 + bx + c$ $(c > 0)$, test factors with the same sign.

EXAMPLE 2 **Factor $x^2 + 5x + 6$.**

SOLUTION

PROBLEM SOLVING

Guess and check. Begin with $(x \quad)(x \quad)$. Find the factors of 6 that result in $5x$ for the bx-term.

$(x+1)(x+6)$	$(x+2)(x+3)$	$(x-1)(x-6)$	$(x-2)(x-3)$
$1x + 6x = 5x$	$2x + 3x = 5x$	$-1x + (-6x) = 5x$	$-2x + (-3x) = 5x$
False	**True**	**False**	**False**

Thus, $x^2 + 5x + 6 = (x+2)(x+3)$.

TRY THIS

Factor $x^2 + 9x + 20$.

When c is negative in $x^2 + bx + c$ $(c < 0)$, test factors with opposite signs.

EXAMPLE 3 **Factor $x^2 - 7x - 30$.**

SOLUTION

PROBLEM SOLVING

Guess and check. Begin with $(x \quad)(x \quad)$. Find the factors of -30 that result in $-7x$ for the bx-term.

$(x-1)(x+30)$	$(x+1)(x-30)$	$(x-2)(x+15)$
$-1x + 30x = -7x$	$1x + (-30x) = -7x$	$-2x + 15x = -7x$
False	**False**	**False**
$(x+2)(x-15)$	$(x-3)(x+10)$	$(x+3)(x-10)$
$2x + (-15x) = -7x$	$-2x + 10x = -7x$	$3x + (-10x) = -7x$
False	**False**	**True**

Thus, $x^2 - 7x - 30 = (x+3)(x-10)$.

TRY THIS

Factor $x^2 - 10x - 11$.

You can use guess-and-check to factor an expression of the form $ax^2 + bx + c$, where $a \neq 1$.

EXAMPLE 4 **Factor $6x^2 + 11x + 3$. Check by graphing.**

SOLUTION

PROBLEM SOLVING

Guess and check. The positive factors of a, are 1, 6, 3, and 2. Begin with $(6x \quad)(x \quad)$ or $(3x \quad)(2x \quad)$. Find the factors of 3 that produce $11x$ for the bx-term.

$(6x+3)(x+1)$	$(6x+1)(x+3)$	$(3x+3)(2x+1)$	$(3x+1)(2x+3)$
$3x + 6x = 11x$	$1x + 18x = 11x$	$6x + 3x = 11x$	$2x + 9x = 11x$
False	**False**	**False**	**True**

Thus, $6x^2 + 11x + 3 = (3x+1)(2x+3)$.

Enrichment

Give students the following trinomials:

1. $x^2 + kx + 81$
2. $x^2 + 14x + k$
3. $kx^2 - 10x - 1$

Have students find all values of k that will make each trinomial a perfect-square trinomial.

1. **$k = 18$ or $k = -18$**
2. **$k = 49$**
3. **no solution**

Have them find a value of k so that each trinomial will factor over the set of integers.

1. **$k \in \{\pm 82, \pm 30, \pm 18\}$**
2. **$k = n(14 - n)$, where n is an integer**
3. **$k \in \{-25, -24, -21, -16, -9, 11, 24, 39, 56, \ldots\}$**

CHECK

Graph $y = 6x^2 + 11x + 3$ and $y = (3x + 1)(2x + 3)$.

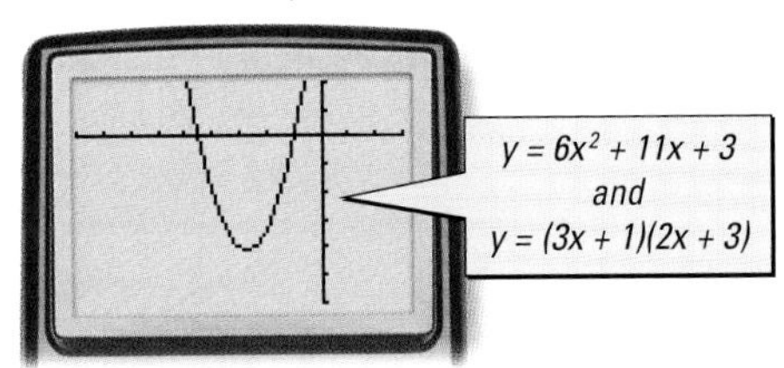

The graphs appear to coincide. Thus, $6x^2 + 11x + 3 = (3x + 1)(2x + 3)$.

TRY THIS Factor $3x^2 + 11x - 20$. Check by graphing.

Examine the product when $x + 3$ and $x - 3$ are multiplied.

$$(x + 3)(x - 3) = x^2 + 3x - 3x + 9$$
$$= x^2 - 9$$
$$= x^2 - 3^2 \quad \text{difference of two squares}$$

Factoring the Difference of Two Squares

$$a^2 - b^2 = (a + b)(a - b)$$

Examine the products when $x + 3$ and $x - 3$ are squared.

$$(x + 3)^2 = (x + 3)(x + 3) \qquad (x - 3)^2 = (x - 3)(x - 3)$$
$$= x^2 + 3x + 3x + 9 \qquad = x^2 - 3x - 3x + 9$$
$$= x^2 + 2(3x) + 9 \qquad = x^2 - 2(3x) + 9$$

perfect-square trinomials

Factoring Perfect-Square Trinomials

$$a^2 + 2ab + b^2 = (a + b)^2 \qquad a^2 - 2ab + b^2 = (a - b)^2$$

EXAMPLE 5 Factor each expression.

a. $x^4 - 16$

b. $4x^2 - 24x + 36$

SOLUTION

a. $x^4 - 16 = (x^2 + 4)(x^2 - 4)$
$= (x^2 + 4)(x + 2)(x - 2)$

b. $4x^2 - 24x + 36 = 4(x^2 - 6x + 9)$
$= 4[x^2 - 2(3)x + 3^2]$
$= 4(x - 3)^2$

TRY THIS Factor $9x^2 - 49$ and $3x^2 + 6x + 3$.

Teaching Tip

TECHNOLOGY If you are using a TI-82 or TI-83 graphics calculator as a check for Example 4, set the viewing window as follows:

Xmin=–3	Ymin=1
Xmax=–3	Ymax=1
Xscl=1	Yscl=1

TRY THIS

$(3x - 4)(x + 5)$

ADDITIONAL EXAMPLE 5

Factor each expression.

a. $16x^4 - 81$

$(4x^2 + 9)(2x - 3)(2x + 3)$

b. $2x^2 - 24x + 72$

$2(x - 6)^2$

TRY THIS

$(3x + 7)(3x - 7)$

$3(x + 1)^2$

Inclusion Strategies

ENGLISH LANGUAGE DEVELOPMENT Ask students to write a general description of how to solve quadratic equations by factoring and graphing and how the x-intercepts of the function relate to the linear factors of the expression in the function and the roots of the equation.

CHECKPOINT ✓
$x = -5$ *or* $x = -1$

Use Teaching Transparency 18.

ADDITIONAL EXAMPLE 6

Use the zero-product property to find the zeros of each quadratic function.

a. $p(x) = 5x^2 + 9x$
0 and $-\frac{9}{5}$

b. $q(x) = x^2 + 21x + 108$
-9 and -12

TRY THIS

a. $x = 0$ *or* $x = -4$

b. $x = -7$ *or* $x = 3$

CRITICAL THINKING

$$ax^2 + bx = 0$$
$$x(ax + b) = 0$$
$$x = 0 \quad or \quad ax + b = 0$$
$$ax = -b$$
$$x = -\frac{b}{a}$$

Using Factoring to Solve Quadratic Equations

You can sometimes use factoring to solve an equation and to find zeros of a function. A **zero of a function** f is any number r such that $f(r) = 0$.

Zero-Product Property

If $pq = 0$, then $p = 0$ *or* $q = 0$.

An equation of the form $ax^2 + bx + c = 0$ is called the **general form of a quadratic equation**. If a quadratic equation is in standard form and the expression $ax^2 + bx + c$ can be factored, then the Zero-Product Property can be applied to solve the equation. To apply the Zero-Product Property, write the equation as a factored expression equal to zero. For example, $x^2 + 6x = -5$ must first be rewritten in standard form as $x^2 + 6x + 5 = 0$ and then factored as $(x + 5)(x + 1) = 0$.

CHECKPOINT ✓ What is the solution to the equation $(x + 5)(x + 1) = 0$?

Example 6 shows you how to use the Zero-Product Property to find the zeros of a quadratic function.

EXAMPLE 6 **Use the Zero-Product Property to find the zeros of each quadratic function.**

a. $f(x) = 2x^2 - 11x$ **b.** $g(x) = x^2 - 14x + 45$

SOLUTION

Set each function equal to zero, and use the Zero-Product Property to solve the resulting equation.

a.
$$2x^2 - 11x = 0$$
$$x(2x - 11) = 0$$
$$x = 0 \quad or \quad 2x - 11 = 0$$
$$x = 0 \qquad x = \frac{11}{2}$$

b.
$$x^2 - 14x + 45 = 0$$
$$(x - 5)(x - 9) = 0$$
$$x - 5 = 0 \quad or \quad x - 9 = 0$$
$$x = 5 \qquad x = 9$$

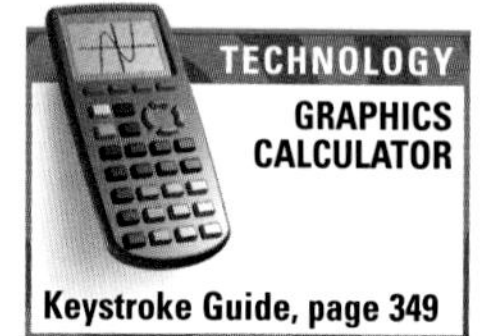

CHECK

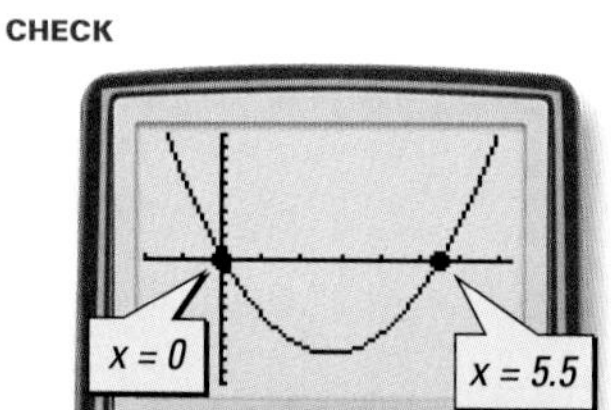

CHECK

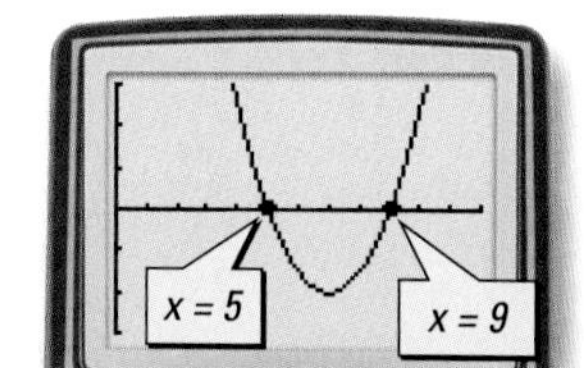

TRY THIS Use the Zero-Product Property to find the zeros of each function.

a. $h(x) = 3x^2 + 12x$ **b.** $j(x) = x^2 + 4x - 21$

CRITICAL THINKING Show that $f(x) = ax^2 + bx$, where $a \neq 0$, has two zeros, namely 0 and $-\frac{b}{a}$.

Reteaching the Lesson

USING ALGORITHMS Give students the general guidelines for factoring quadratic expressions.

Factoring Quadratic Expressions

1. Look for a common factor among all the terms and divide out that factor.
2. Look for the difference of two squares.
3. Look for a perfect-square trinomial.
4. Look for two integer factors of the constant whose sum is the coefficient of the linear term.
5. Look at each factor to see if it can be factored.
6. Always check the factorization by multiplying.

To factor $2x^2 - 17x - 9$, find two factors of $2(9)$, or 18, that subtract (based on the sign of constant term) to make 17, the coefficient of x. Rewrite $-17x$ using those two numbers: $2x^2 - 18x + x - 9$. Check that $-18x + x = -17x$ and that the sign of the linear term in the original expression is correct. Factor the first two terms and the last two terms: $2x(x - 9) + 1(x - 9)$. Use the distributive property to factor out the common factor by grouping: $(x - 9)(2x + 1)$. If the sign of the constant term is *positive*, find two numbers that *add* to make the linear term coefficient and follow the same procedure.

EXAMPLE 7 Refer to the fountain problem discussed at the beginning of the lesson.

APPLICATION
ARCHITECTURE

How many levels would a fountain consisting of 66 cubes have?

SOLUTION

Method 1 Use algebra.
Solve $2n^2 - n = 66$ by factoring.

$2n^2 - n - 66 = 0$ *Write in standard form.*
$(2n + 11)(n - 6) = 0$ *Factor $2n^2 - n - 66$.*
$2n + 11 = 0$ *or* $n - 6 = 0$
$n = -5.5$ $\quad n = 6$

Because the number of levels must be a positive integer, −5.5 cannot be a solution. The fountain would have 6 levels.

TECHNOLOGY
GRAPHICS CALCULATOR
Keystroke Guide, page 349

Method 2 Use a table.
Make a table of values for $y = 2x^2 - x - 66$. From the table at right you can see that the function has a zero at $x = 6$.

Therefore, the fountain would have 6 levels.

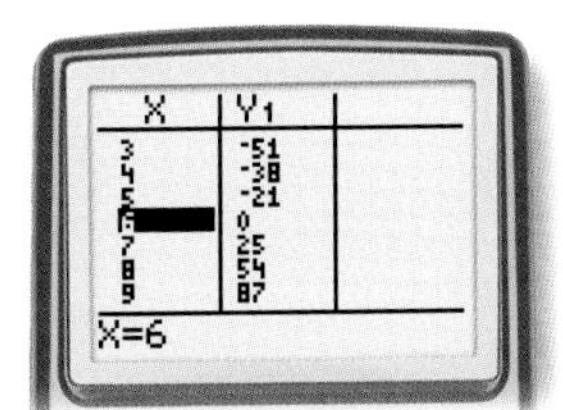

ADDITIONAL EXAMPLE 7

Refer to the fountain problem discussed at the beginning of the lesson. **How many levels would a fountain consisting of 91 cubes have?**
7 levels

Teaching Tip

TECHNOLOGY If you are using a TI-82 or TI-83 graphics calculator to create the table shown on this page, press 2nd WINDOW to access the **TABLE SETUP** feature, and choose the following settings:

TblStart=3 (or **TblMin=3**)
ΔTbl=1
Indpnt: Auto
Depend: Auto

Press Y=, enter **2x²–X–66** into **Y1**, and press 2nd GRAPH to view the table.

Exercises

Communicate

1. If $x^2 + 34x + 285$ is factored as $(x + q)(x + s)$, how do you find q and s?
2. What do you know about the factors of $x^2 + bx + c$ when c is positive? when c is negative? What information does the sign of b give you in each case?
3. State what must be true about the numbers p and q if $pq = 0$.

Guided Skills Practice

Factor each quadratic expression. ***(EXAMPLE 1)***

4. $2x^2 - 8x$ **$2x(x - 4)$**
5. $2y^2 - 6y$ **$2y(y - 3)$**
6. $5ax^2 - 15a^2x$ **$5ax(x - 3a)$**
7. $4x(x + 3) - 7(x + 3)$ **$(4x - 7)(x + 3)$**
8. $(4r + 7)3 - (4r + 7)2r$ **$(4r + 7)(3 - 2r)$**
9. $(9s - 5)8s + 3(9s - 5)$ **$(9s - 5)(8s + 3)$**

Factor each quadratic expression. ***(EXAMPLES 2, 3, AND 4)***

10. $x^2 + 5x + 6$
11. $x^2 + 8x + 7$
12. $y^2 - 5y + 4$
13. $x^2 - 4x - 12$
14. $y^2 - 9y - 36$
15. $x^2 + 10x - 24$
16. $2x^2 + 9x + 10$
17. $3x^2 + 5x + 2$
18. $5x^2 + 13x - 6$
19. $8x^2 + 24x - 14x - 42$
20. $12r^2 + 21r - 8r - 14$
21. $72s^2 - 56s - 36s + 28$

10. $(x + 3)(x + 2)$
11. $(x + 7)(x + 1)$
12. $(y - 4)(y - 1)$
13. $(x + 2)(x - 6)$
14. $(y + 3)(y - 12)$
15. $(x + 12)(x - 2)$
16. $(2x + 5)(x + 2)$
17. $(3x + 2)(x + 1)$
18. $(x + 3)(5x - 2)$
19. $2(4x - 7)(x + 3)$
20. $(3r - 2)(4r + 7)$
21. $4(2s - 1)(9s - 7)$

Assess

Selected Answers
Exercises 4–29, 31–113 odd

ASSIGNMENT GUIDE	
In Class	1–29
Core	31–93 odd, 97–101 odd
Core Plus	30–102 even
Review	103–114
Preview	115–117

Extra Practice can be found beginning on page 940.

Error Analysis

When solving quadratic equations, students may incorrectly apply the zero-product property to products other than zero. For example, some students may try to solve $3x^2 - 5x = 2$ with the following incorrect method:

$$3x^2 - 5x = 2$$
$$x(3x - 5) = 2$$
$$x = 2 \quad or \quad 3x - 5 = 2$$
$$3x = 7$$
$$x = \frac{7}{3}$$

Show them this is not correct by graphing $y = 3x^2 - 5x$ and $y = 2$. Point out that the points of intersection occur when $x = \frac{2}{3}$ and when $x = 1$.

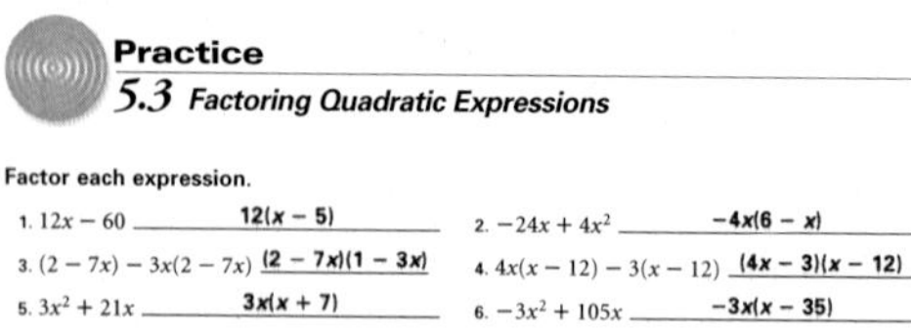

Practice

NAME ______ CLASS ______ DATE ______

Practice
5.3 Factoring Quadratic Expressions

Factor each expression.

1. $12x - 60$ **12(x − 5)**
2. $-24x + 4x^2$ **−4x(6 − x)**
3. $(2 - 7x) - 3x(2 - 7x)$ **(2 − 7x)(1 − 3x)**
4. $4x(x - 12) - 3(x - 12)$ **(4x − 3)(x − 12)**
5. $3x^2 + 21x$ **3x(x + 7)**
6. $-3x^2 + 105x$ **−3x(x − 35)**

Factor each quadratic expression.

7. $x^2 + 17x + 52$ **(x + 4)(x + 13)**
8. $x^2 - x - 20$ **(x − 5)(x + 4)**
9. $x^2 - 7x - 18$ **(x − 9)(x + 2)**
10. $x^2 + 11x + 28$ **(x + 4)(x + 7)**
11. $x^2 - x - 90$ **(x − 10)(x + 9)**
12. $x^2 + 11x - 26$ **(x + 13)(x − 2)**
13. $4x^2 + 4x + 1$ **$(2x + 1)^2$**
14. $3x^2 + 5x - 2$ **(3x − 1)(x + 2)**
15. $2x^2 - 3x - 2$ **(2x + 1)(x − 2)**

Solve each equation by factoring and applying the Zero-Product Property.

16. $x^2 - 256 = 0$ **−16 or 16**
17. $5x^2 - 125 = 0$ **−5 or 5**
18. $x^2 + 9x + 14 = 0$ **−7 or −2**
19. $3x^2 - 6x + 3 = 0$ **1**
20. $4x^2 - 12x + 9 = 0$ **1.5**
21. $6x^2 - x - 1 = 0$ **$-\frac{1}{3}$ or $\frac{1}{2}$**
22. $9x^2 - 4 = 0$ **$-\frac{2}{3}$ or $\frac{2}{3}$**
23. $5x^2 + 1 = 6x$ **1 or $\frac{1}{5}$**
24. $7x^2 - 2 = 5x$ **$-\frac{2}{7}$ or 1**

Use factoring and the Zero-Product Property to find the zeros of each quadratic function.

25. $f(x) = x^2 - x - 12$ **−3, 4**
26. $g(x) = 2x^2 + 3x - 5$ **−2.5, 1**
27. $h(x) = x^2 + 12x - 45$ **−15, 3**
28. $b(x) = x^2 - 13x + 42$ **6, 7**
29. $k(x) = 2x^2 + 3x - 2$ **−2, 0.5**
30. $q(x) = 4x^2 + 12x + 9$ **−1.5**

Factor each quadratic expression. ***(EXAMPLE 5)***

22. $x^4 - 81$ **$(x^2 + 9)(x + 3)(x - 3)$**
23. $2x^2 - 8$ **$2(x + 2)(x - 2)$**
24. $16x^2 - 25$ **$(4x + 5)(4x - 5)$**
25. $x^2 + 8x + 16$ **$(x + 4)^2$**

Use the Zero-Product Property to find the zeros of each quadratic function. ***(EXAMPLE 6)***

26. $f(x) = x^2 + 7x$ **0, −7**
27. $g(x) = x^2 + 6x + 9$ **−3**
28. $f(t) = t^2 + 3t - 10$ **−5, 2**

CONNECTION

29. GEOMETRY Line segments are drawn to connect n points with one another. The number of connecting segments is described by the function $h(n) = \frac{n(n-1)}{2}$. If 36 connecting segments are drawn, how many points are there? ***(EXAMPLE 7)*** **9**

Practice and Apply

Factor each expression.

30. $3x + 6$ **$3(x + 2)$**
31. $3x^2 + 18$ **$3(x^2 + 6)$**
32. $10n - n^2$ **$n(10 - n)$**
33. $x - 4x^2$ **$x(1 - 4x)$**
34. $6x - 2x^2$ **$2x(3 - x)$**
35. $-3y^2 - 15y$ **$-3y(y - 5)$**
36. $5x(x - 2) - 3(x - 2)$ **$(5x - 3)(x - 2)$**
37. $(x + 3)(2x) + (x + 3)(7)$ **$(x + 3)(2x + 7)$**
38. $a^2x + 5a^2x^2 - 2ax$ **$ax(x + 5ax - 2)$**
39. $4ab^2 - 6a^2b$ **$2ab(2b - 3a)$**

Factor each quadratic expression.

40. $x^2 - 16x + 15$ **$(x - 15)(x - 1)$**
41. $x^2 + 8x + 16$ **$(x + 4)^2$**
42. $x^2 - 26x + 48$ **$(x - 24)(x - 2)$**
43. $x^2 + 4x - 32$ **$(x + 8)(x - 4)$**
44. $x^2 + 7x - 30$ **$(x + 10)(x - 3)$**
45. $x^2 - 10x - 24$ **$(x - 12)(x + 2)$**
46. $-22x - 48 + x^2$ **$(x - 24)(x + 2)$**
47. $2x + x^2 - 24$ **$(x + 6)(x - 4)$**
48. $x^2 - 56 - 10x$ **$(x + 4)(x - 14)$**
49. $56 + 10x - x^2$ **$-(x + 4)(x - 14)$**
50. $30 + x - x^2$ **$-(x + 5)(x - 6)$**
51. $24 + 10x - x^2$ **$-(x + 2)(x - 12)$**
52. $3x^2 + 10x + 3$ **$(3x + 1)(x + 3)$**
53. $2x^2 + 5x + 2$ **$(2x + 1)(x + 2)$**
54. $2x^2 + 3x + 1$ **$(2x + 1)(x + 1)$**
55. $3x^2 + 7x + 2$ **$(3x + 1)(x + 2)$**
56. $12x^2 - 3x - 9$ **$3(4x + 3)(x - 1)$**
57. $3x^2 - 5x - 2$ **$(3x + 1)(x - 2)$**

internet connect
Homework Help Online
Go To: **go.hrw.com**
Keyword:
MB1 Homework Help
for Exercises 58–87

Solve each equation by factoring and applying the Zero-Product Property.

58. $15x^2 = 7x + 2$ **$-\frac{1}{5}, \frac{2}{3}$**
59. $3x^2 - 5x = 2$ **$-\frac{1}{3}, 2$**
60. $4x - 4 = -15x^2$ **$-\frac{2}{3}, \frac{2}{5}$**
61. $3x^2 + 3 = 10x$ **$\frac{1}{3}, 3$**
62. $2x^2 - 15 = -7x$ **$-5, \frac{3}{2}$**
63. $6x^2 - 17x = -12$ **$\frac{4}{3}, \frac{3}{2}$**
64. $x^2 - 36 = 0$ **± 6**
65. $t^2 - 9 = 0$ **± 3**
66. $x^4 - 81 = 0$ **± 3**
67. $x^4 - 1 = 0$ **± 1**
68. $4x^2 - 9 = 0$ **$\pm\frac{3}{2}$**
69. $25x^2 - 16 = 0$ **$\pm\frac{4}{5}$**
70. $x^2 - 2x + 1 = 0$ **1**
71. $x^2 + 4x + 4 = 0$ **−2**
72. $9x^2 = -6x - 1$ **$-\frac{1}{3}$**
73. $4x^2 + 1 = 4x$ **$\frac{1}{2}$**
74. $-4 + 20x - 25x^2 = 0$ **$\frac{2}{5}$**
75. $40x + 25 = -16x^2$ **$-\frac{5}{4}$**
76. $64 + 16x + x^2 = 0$ **−8**
77. $9 - 6x + x^2 = 0$ **3**

Use factoring and the Zero-Product Property to find the zeros of each quadratic function.

78. $f(x) = x^2 - 7x + 10$ **5, 2**

79. $g(t) = t^2 - 2t - 15$ **−3, 5**

80. $f(x) = 4x^2 + 4x - 24$ **−3, 2**

81. $g(x) = 6x^2 + 3x - 9$ $\mathbf{-\frac{3}{2}, 1}$

82. $f(t) = t^2 + 7t - 60$ **5, −12**

83. $h(x) = x^2 - 15x + 56$ **8, 7**

84. $f(x) = x^2 + 8x + 12$ **−6, −2**

85. $g(x) = x^2 - 3x - 40$ **−5, 8**

86. $g(x) = 6x^2 + 20x - 16$ $\mathbf{-4, \frac{2}{3}}$

87. $h(x) = 4x^2 - 8x + 3$ $\mathbf{\frac{3}{2}, \frac{1}{2}}$

Use graphing to find the zeros of each function.

88 $f(n) = n^2 - n - 30$ **−5, 6**

89 $g(t) = 24 + 8t - 2t^2$ **−2, 6**

90 $f(x) = 2x^2 + 13x + 15$ **−5, −1.5**

91 $f(x) = 5x^2 + 30x + 40$ **−4, −2**

92 $g(a) = 24a^2 + 36a - 24$ **−2, 0.5**

93 $h(x) = 6x^2 - 33x - 18$ **−0.5, 6**

CHALLENGE

Factor each expression. **94.** $\mathbf{8ab(a^2 + b^2)}$

94. $(a + b)^4 - (a - b)^4$

95. $x^{2n} - 1$ $\mathbf{(x^n + 1)(x^n - 1)}$

96. $x^{2n} - 2x^n + 1$ $\mathbf{(x^n - 1)^2}$

CONNECTIONS

GEOMETRY The area of a triangle is $A = \frac{1}{2}bh$, where b is the base and h is the altitude. Use this information for Exercises 97 and 98.

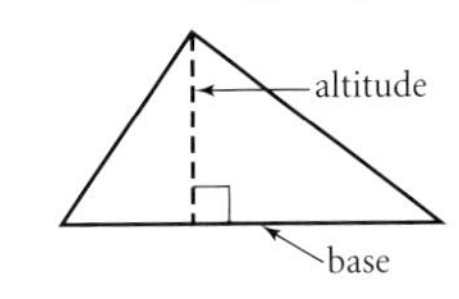

97. The base of a triangle is 5 centimeters longer than its altitude. If the area of the triangle is 42 square centimeters, what is its altitude? **7 cm**

98. The altitude of a triangle is 5 centimeters shorter than its base. If the area of the triangle is 12 square centimeters, what is its base? **8 cm**

99. GEOMETRY The area of a circle is given by $A = \pi r^2$, where r is the radius. If the radius of a circle is increased by 4 inches, the area of the resulting circle is 100π square inches. Find the radius of the original circle. **6 in.**

100. GEOMETRY The length of one leg of a right triangle is 7 centimeters longer than the other leg, and the hypotenuse is 13 centimeters. Find the lengths of the two legs. **5 cm, 12 cm**

APPLICATIONS

101. SPORTS A soccer ball is kicked from the ground, and its height in meters above ground is modeled by the function $h(t) = -4.9t^2 + 19.6t$, where t represents the time in seconds after the ball is kicked. How long is the ball in the air? **4 sec**

102. SPORTS A golf ball is hit from the ground, and its height in feet above the ground is modeled by the function $h(t) = -16t^2 + 180t$, where t represents the time in seconds after the ball is hit. How long is the ball in the air? **11.25 sec**

Student Technology Guide

NAME ______ CLASS ______ DATE ______

Student Technology Guide
5.3 Factoring Quadratic Expressions

You can find the zeros of a quadratic function by factoring the defining quadratic expression. With a graphics calculator, you can locate the zeros first, and then you can write the factored form.

Example: Find the zeros of $f(x) = x^2 - 2x - 8$. Then factor $x^2 - 2x - 8$.

- Press Y= X,T,θ,n x² − 2 X,T,θ,n − 8 GRAPH.
- Press TRACE and use ◄ and ► to move as close to a zero as you can. There appears to be a zero near $x = -2$. To check whether the zero is really at this value, press (−) 2 ENTER. This moves the trace point to $x = -2$. We see that y is indeed zero when $x = -2$.
- If there is another zero, repeat the step above. There is a second zero at $x = 4$.
- Write the factors. If −2 and 4 are zeros, then $[x - (-2)] = (x + 2)$ and $(x - 4)$ are factors. Thus, $x^2 - 2x - 8 = (x + 2)(x - 4)$.

Y1=X2-2X-8
X=-2 Y=0

If a, the coefficient of x^2, is not 1, begin by factoring out a. The factorization will be a times the binomial factors, as shown in the example below.

Example: Find the zeros of $f(x) = 3x^2 + x - 14$. Then factor $3x^2 + x - 14$.

- First factor out 3: $f(x) = 3\left(x^2 + \frac{1}{3}x - \frac{14}{3}\right)$
- Find the zeros of $g(x) = x^2 + \frac{1}{3}x - \frac{14}{3}$. Note that these will also be the zeros of f. Using the method shown in the example above, the zeros are 2 and $-\frac{7}{3}$. Therefore, $x^2 + \frac{1}{3}x - \frac{14}{3} = (x - 2)\left(x + \frac{7}{3}\right)$.
- Multiply by 3 to find the factorization of the original expression. $3x^2 + x - 14 = 3(x - 2)\left(x + \frac{7}{3}\right) = (x - 2)(3x + 7)$

Find the zeros of each function. Then factor the quadratic trinomial.

1. $f(x) = x^2 - 5x - 24$ **−3 and 8; (x + 3)(x − 8)**
2. $f(x) = x^2 + 2x - 35$ **−7 and 5; (x + 7)(x − 5)**
3. $g(x) = -2x^2 - 5x + 12$ **−4 and 1.5; −2(x + 4)(x − 1.5)**

Use the zeros of the function to factor each quadratic trinomial.

4. $f(x) = x^2 + 7x + 12$ **(x + 3)(x + 4)**
5. $f(x) = 3x^2 - 6x - 9$ **3(x − 3)(x + 1)**
6. $g(x) = -9x^2 - 30x + 24$ **−3(x + 4)(3x − 2)**

Look Beyond

Encourage students to recognize that Exercises 115 and 117 show the pattern for the difference of two squares. Exercise 116 represents the sum of two squares. This quadratic cannot be factored over the real numbers because its related quadratic function has no x-intercepts. Quadratic equations and complex numbers will be studied further in Lesson 5.6.

ALTERNATIVE Assessment

Portfolio Activity

The Portfolio Activity can be used as preparation for the Chapter Project or as a separate activity. In the Portfolio Activity on this page, students are asked to determine the time that it takes for a ball to return to a given height by solving the related quadratic equation. Then they are asked to find the axis of symmetry and use it to estimate the time that a ball takes to return to that height.

Answers to Portfolio Activities can be found in Additional Answers of the Teacher's Edition.

Look Back

Solve each inequality, and graph the solution on a number line. ***(LESSON 1.7)***

103. $2x - 4 > 12 + 5x$ $x < -5\frac{1}{3}$

104. $2x - \frac{3}{4} \geq 7$ $x \geq 3\frac{7}{8}$

105. $3(3x + 7) - 12 \leq 8 - \left(\frac{1}{2}x + 9\right)$ $x \leq -1\frac{1}{19}$

106. $-2\left(\frac{2}{3}x + 5\right) - 13 < -6$ $x > -12\frac{3}{4}$

APPLICATION

107. ADVERTISING An advertising agency can spend no more than $1,270,000 for television advertising. The two available time slots for commercials cost $40,000 and $25,000, and at least 40 commercials are desired. ***(LESSON 3.4)***

a. Write a system of linear inequalities to represent how many commercials at each price can be purchased.

b. Graph the region in which the solution to the system of inequalities can be found.

c. Find the coordinates of a point that is a solution to the system. What do the coordinates represent?

For Exercises 108–111, let $A = \begin{bmatrix} -3 & 1 & 0 \\ 2 & 6 & -4 \\ 5 & -2 & 1 \end{bmatrix}$ **and** $B = \begin{bmatrix} 0 & 2 & 8 \\ 1 & 4 & 3 \\ -2 & 0 & -1 \end{bmatrix}$.

Evaluate each expression. ***(LESSON 4.1)***

108. $3A$ **109.** $-\frac{1}{2}B$ **110.** $A - B$ **111.** $2B + A$

108. $\begin{bmatrix} -9 & 3 & 0 \\ 6 & 18 & -12 \\ 15 & -6 & 3 \end{bmatrix}$

109. $\begin{bmatrix} 0 & -1 & -4 \\ -\frac{1}{2} & -2 & -\frac{3}{2} \\ 1 & 0 & \frac{1}{2} \end{bmatrix}$

110. $\begin{bmatrix} -3 & -1 & -8 \\ 1 & 2 & -7 \\ 7 & -2 & 2 \end{bmatrix}$

111. $\begin{bmatrix} -3 & 5 & 16 \\ 4 & 14 & 2 \\ 1 & -2 & -1 \end{bmatrix}$

Find each product. ***(LESSON 5.1)***

112. $(3x + 4)(-x - 5)$ $-3x^2 - 19x - 20$

113. $(-2x + 9)(-4x + 7)$ $8x^2 - 50x + 63$

114. $\left(\frac{1}{3}x + \frac{1}{4}\right)(-5x - 2)$ $-\frac{5}{3}x^2 - \frac{23}{12}x - \frac{1}{2}$

Look Beyond

Factor each quadratic expression, if possible.

115. $(x + 2)^2 - 4$ $x(x + 4)$

116. $(x + 9)^2 + 36$ not possible

117. $(x - 1)^2 - 16$ $(x + 3)(x - 5)$

internet connect
Portfolio Extension
Go To: go.hrw.com
Keyword: MB1 FactorQuad

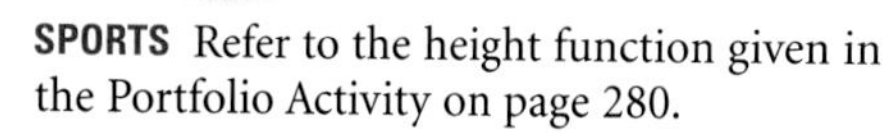

SPORTS Refer to the height function given in the Portfolio Activity on page 280.

1. Find the time when the basketball will return to a height of 6 feet by solving the related quadratic equation for $h(t) = 6$.
2. Use the data on page 273 to estimate the equation of the axis of symmetry of the function. Using symmetry, estimate when the basketball will return to a height of 6 feet.
3. Compare your answers to Steps 1 and 2 above. Explain any difference you find.

103. $x < -5\frac{1}{3}$

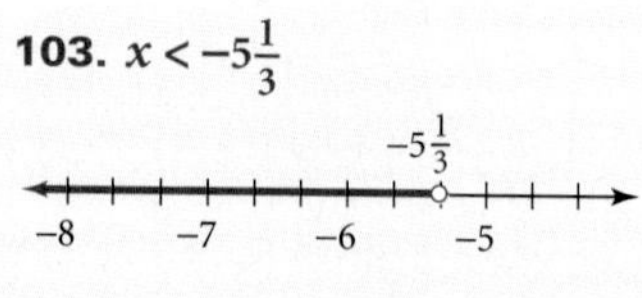

104. $x \geq 3\frac{7}{8}$

105. $x \leq -1\frac{1}{19}$

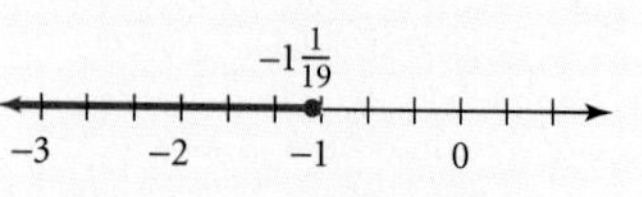

106. $x \geq -12\frac{3}{4}$

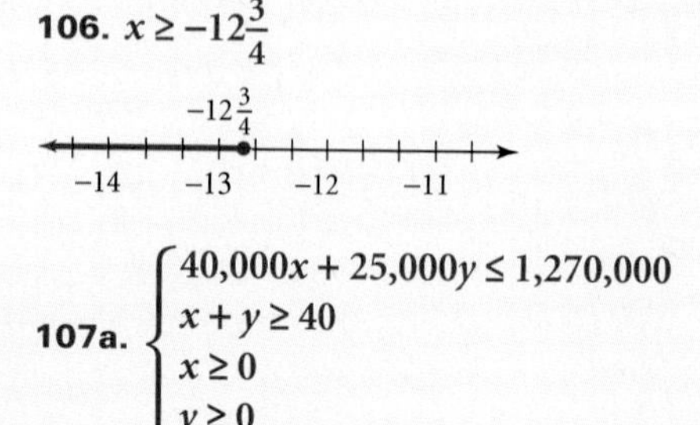

107a. $\begin{cases} 40{,}000x + 25{,}000y \leq 1{,}270{,}000 \\ x + y \geq 40 \\ x \geq 0 \\ y \geq 0 \end{cases}$

b.

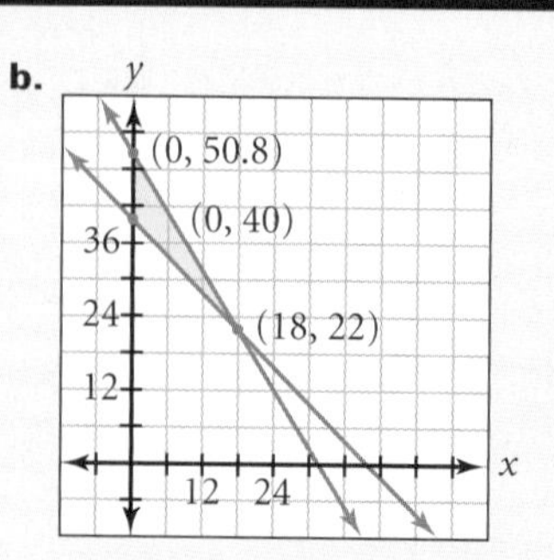

c. Answers may vary. Sample answer: The solution, (18, 22), represents buying 18 commercials in the $40,000 time slot and 22 in the $25,000 time slot.

Completing the Square

Why
You can solve many real-world problems, such as finding the lowest point on the cable of a suspension bridge, by using a method called "completing the square" to solve quadratic equations.

Objectives

- Use completing the square to solve a quadratic equation.
- Use the vertex form of a quadratic function to locate the axis of symmetry of its graph.

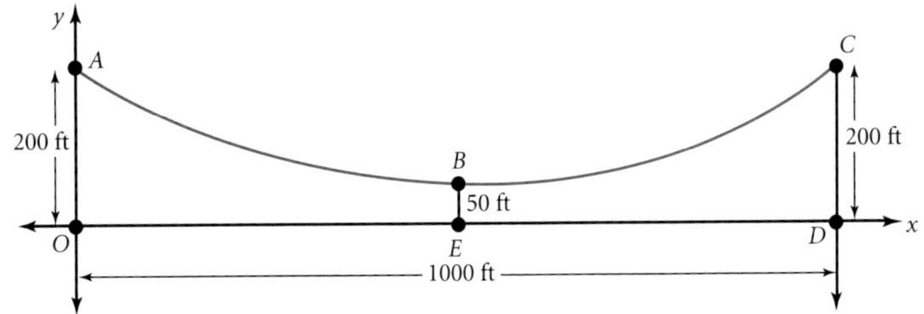

APPLICATION
ENGINEERING

Engineers are planning to build a cable suspension bridge like the one shown in the diagram above. The cable forms a *catenary* curve, which can be approximated by the quadratic function $f(x) = \frac{3}{5000}x^2 - \frac{3}{5}x + 200$, where $0 \leq x \leq 1000$. Write this quadratic function in a form that allows the coordinates of the lowest point on the cable to be easily identified. You will solve this problem in Example 5.

Completing the Square

Activity
Completing the Square With Tiles

You will need: algebra tiles

1. Can you arrange 1 x^2-tile and 4 x-tiles to form a square? Explain your response.
2. What is the smallest number of unit tiles you need to add to your tiles to form a square?
3. Write an expression in both standard and factored form for the complete set of tiles that form the square.

Alternative Teaching Strategy

USING ALGORITHMS Show students that the solutions, $x = 3 \pm \sqrt{14}$, to the quadratic equation $(x-3)^2 = 14$ can be found by taking the square root of both sides of the quadratic equation. Then ask the students to compare the process shown at left below with the process shown at right below.

↓	$(x-3)^2 = 14$	↑	$x^2 - 6x = 5$
↓	$x^2 - 6x + 9 = 14$	↑	$x^2 - 6x + 9 = 5 + 9$
↓	$x^2 - 6x + 9 - 9 = 14 - 9$	↑	$x^2 - 6x + 9 = 14$
↓	$x^2 - 6x = 5$	↑	$(x-3)^2 = 14$

Be sure they notice that the right column is the same as the left column done in reverse. The solution of $x^2 - 6x = 5$ must be the same as the solution of $(x-3)^2 = 14$. Now show the students the process of completing the square for Example 2. Tell students that the process of completing the square can *always* be used to solve a quadratic equation, while factoring works only on equations that can be factored.

Prepare

NCTM PRINCIPLES & STANDARDS 1–4, 6–10

QUICK WARM-UP

Factor each trinomial.

1. $x^2 + 14x + 49$ $(x+7)^2$
2. $x^2 - 22x + 121$ $(x-11)^2$
3. $x^2 - 12x - 64$ $(x+4)(x-16)$

Solve each equation.

4. $d^2 - 100 = 0$ $d = 10$ *or* $d = -10$
5. $z^2 - 2z + 1 = 0$ $z = 1$
6. $t^2 + 16 = -8t$ $t = -4$

Also on Quiz Transparency 5.4

Teach

Why The shape of the suspension cables on bridges such as the George Washington Bridge and the Golden Gate Bridge can be approximated by a quadratic function. Ask students to describe the curvature of the cables and discuss how they think this shape provides the necessary support for these bridges.

Activity Notes

In this Activity, students are given a quadratic term and a linear term and are asked to arrange algebra tiles to form a square by adding the correct number of unit tiles to complete the square. Encourage students to write the area of each new square in the forms $x^2 + bx + c$ and $(x+h)^2$, noting that half of b is h and that c is h squared.

Use Teaching Transparency 19.

Cooperative Learning

Have students work in pairs. One student should work with the tiles while the other records the number of unit tiles needed to complete the square and writes the standard and factored forms of the area of the completed square.

CHECKPOINT ✔

6. Divide the number of x-tiles by two and square the result.

$\left(\frac{8}{2}\right)^2 = 4^2 = 16$

ADDITIONAL EXAMPLE 1

Complete the square for each quadratic expression to form a perfect-square trinomial. Then write the expression as a squared binomial.

a. $x^2 - 12x$

$x^2 - 12x + 36 = (x - 6)^2$

b. $x^2 + 27x$

$x^2 + 27x + \left(\frac{27}{2}\right)^2 = \left(x + \frac{27}{2}\right)^2$

TRY THIS

a. $x^2 - 7x + \left(\frac{7}{2}\right)^2 = \left(x - \frac{7}{2}\right)^2$

b. $x^2 + 16x + 8^2 = (x + 8)^2$

4. Repeat Steps 1–3 with 1 x^2-tile and 6 x-tiles.
5. Repeat Steps 1–3 with 1 x^2-tile and 8 x-tiles.

CHECKPOINT ✔

6. Explain how to find the smallest number of unit tiles needed to complete a square with 1 x^2-tile and 8 x-tiles.

Recall from Lesson 5.2 that you can solve equations of the form $x^2 = k$ by taking the square root of each side.

$$x^2 = 9$$
$$\sqrt{x^2} = \pm\sqrt{9}$$
$$x = \pm 3$$

The same is true for equations of the form $(x + a)^2 = k$.

$$(x + 3)^2 = 16$$
$$\sqrt{(x + 3)^2} = \pm\sqrt{16}$$
$$x + 3 = \pm 4$$
$$x = 1 \quad or \quad x = -7$$

When a quadratic equation does not contain a perfect square, you can create a perfect square in the equation by *completing the square.* **Completing the square** is a process by which you can force a quadratic expression to factor.

Examine the relationship between terms in a perfect-square trinomial.

Specific case

$$x^2 + 8x + 16 = (x + 4)^2$$

$\frac{1}{2}(8) = 4 \rightarrow 4^2 = 16$

General case

$$x^2 + bx + \left(\frac{b}{2}\right)^2 = \left(x + \frac{b}{2}\right)^2$$

$\frac{1}{2}(b) = \frac{b}{2} \rightarrow \left(\frac{b}{2}\right)^2$

In general, the constant term in a perfect square trinomial is the square of one-half the coefficient of the second term.

EXAMPLE 1 **Complete the square for each quadratic expression to form a perfect-square trinomial. Then write the new expression as a binomial squared.**

a. $x^2 - 6x$ **b.** $x^2 + 15x$

SOLUTION

a. The coefficient of x is -6.

$\frac{1}{2}(-6) = -3 \rightarrow (-3)^2 = \mathbf{9}$

Thus, the perfect-square trinomial is $x^2 - 6x + \mathbf{9}$.

$$x^2 - 6x + 9 = (x - 3)^2$$

b. The coefficient of x is 15.

$\frac{1}{2}(15) = \frac{15}{2} \rightarrow \left(\frac{15}{2}\right)^2$

Thus, the perfect-square trinomial is $x^2 + 15x + \left(\frac{15}{2}\right)^2$.

$$x^2 + 15x + \left(\frac{15}{2}\right)^2 = \left(x + \frac{15}{2}\right)^2$$

TRY THIS Complete the square for each quadratic expression to form a perfect-square trinomial. Then write the new expression as a binomial squared.

a. $x^2 - 7x$ **b.** $x^2 + 16x$

Interdisciplinary Connection

PHYSICS The distance that a ball bearing rolls down an elevated track can be approximated by the quadratic function $s(t) = 0.5t^2 + 4t$, where t is the time in seconds and $s(t)$ is the distance in feet. Ask students to determine how long it takes the ball to travel 10 feet. **2 s** Students can model this situation by making an elevated track and recording the time it takes for a ball bearing to roll different distances. They can enter the data into a table, graph the data to see the parabolic shape, and find a quadratic regression equation that fits the data.

Enrichment

Extend Example 5 by having students work in pairs to approximate a catenary curve with a quadratic function. If you are using a TI-82 or TI-83, first press Y=. Then press 2nd 0 for the catalog list. Choose **cosh(.** To the right of the parenthesis, enter **X).** Then press GRAPH to display the catenary curve. Ask students to find a quadratic function that approximates this curve. $\mathbf{y = x^2 + 1}$ Repeat this activity, using a wide range of values of a for the catenary function $f(x) = a\cosh\left(\frac{x}{a}\right)$.

Solving Equations by Completing the Square

Examples 2 and 3 show you how to solve a quadratic equation by completing the square and by using square roots.

EXAMPLE 2 **Solve $x^2 + 6x - 16 = 0$ by completing the square.**

SOLUTION

$$x^2 + 6x - 16 = 0$$
$$x^2 + 6x = 16$$
$$x^2 + 6x + \left(\frac{6}{2}\right)^2 = 16 + \left(\frac{6}{2}\right)^2 \quad \text{Add } \left(\frac{6}{2}\right)^2 \text{ to each side of the equation.}$$
$$x^2 + 6x + 9 = 25$$
$$(x + 3)^2 = 25$$
$$x + 3 = \pm\sqrt{25} \quad \text{Take the square root of each side.}$$
$$x + 3 = \pm 5$$
$$x = 5 - 3 \quad \text{or} \quad x = -5 - 3$$
$$x = 2 \qquad\qquad x = -8$$

TRY THIS Solve $x^2 + 10x - 24 = 0$ by completing the square.

EXAMPLE 3 **Solve $2x^2 + 6x = 7$.**

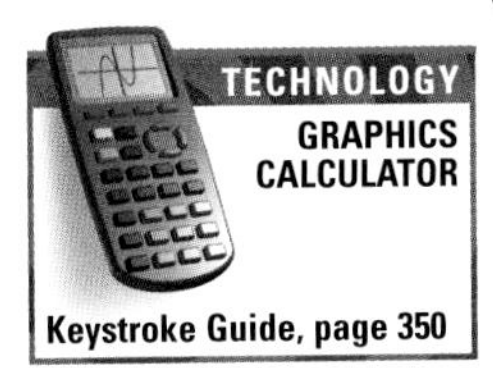

SOLUTION

Method 1 Use algebra.
Solve by completing the square.

$$2x^2 + 6x = 7$$
$$2(x^2 + 3x) = 7$$
$$x^2 + 3x = \frac{7}{2}$$
$$x^2 + 3x + \left(\frac{3}{2}\right)^2 = \frac{7}{2} + \left(\frac{3}{2}\right)^2$$
$$\left(x + \frac{3}{2}\right)^2 = \frac{7}{2} + \frac{9}{4}$$
$$x + \frac{3}{2} = \pm\sqrt{\frac{23}{4}}$$
$$x = -\frac{3}{2} + \sqrt{\frac{23}{4}} \quad \text{or} \quad x = -\frac{3}{2} - \sqrt{\frac{23}{4}}$$
$$x \approx 0.90 \qquad\qquad x \approx -3.90$$

Thus, the exact solution is $x = -\frac{3}{2} + \sqrt{\frac{23}{4}}$ or $x = -\frac{3}{2} - \sqrt{\frac{23}{4}}$.

Method 2 Use a graph.
Graph $y = 2x^2 + 6x$ and $y = 7$, and find the x-coordinates of any points of intersection.

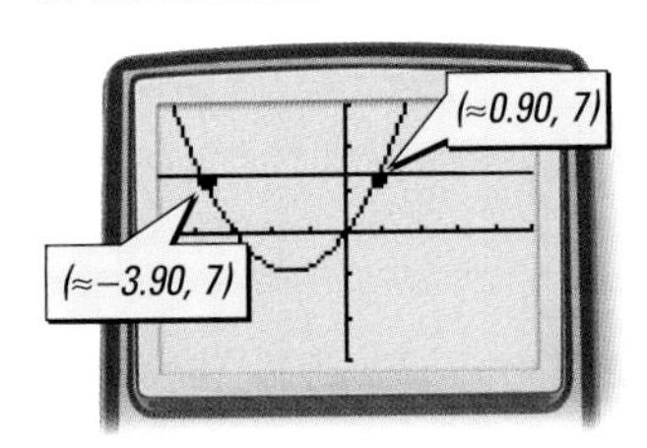

Or graph $y = 2x^2 + 6x - 7$, and find any zeros.

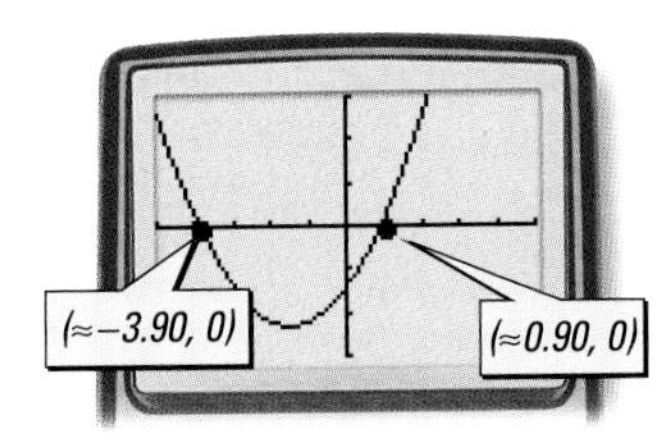

TRY THIS Solve $2x^2 + 10x = 6$.

ADDITIONAL EXAMPLE 2

Solve $x^2 + 18x - 40 = 0$ by completing the square.
$x = 2$ or $x = -20$

TRY THIS
$x = -12$ or $x = 2$

ADDITIONAL EXAMPLE 3

Solve $3x^2 - 6x = 5$.
$x \approx -0.63$ or $x \approx 2.63$

Teaching Tip

TECHNOLOGY If you are using a TI-82 or TI-83 graphics calculator in Example 3, set the viewing window as follows:

Xmin=–5	**Ymin=–12**
Xmax=5	**Ymax=12**
Xscl=1	**Yscl=4**

Press Y= and enter **Y1=2X^2+6x** and **Y2=7.** To find the x-coordinates of any points of intersection, press 2nd TRACE and choose **5:intersect**. Then use the arrow keys to move the cursor on the graphs and press ENTER to identify the two functions and your guess.

TRY THIS
$x = \frac{-5 \pm \sqrt{37}}{2}$; $x \approx 0.54$ or $x \approx -5.54$

Inclusion Strategies

LINGUISTIC LEARNERS Have students write a step-by-step procedure in their own words for solving a general quadratic equation by completing the square. Have them share their procedure with the class, and then outline the method on the board as follows:

1. If necessary, rewrite the equation so that the constant term is isolated on the right side of the equation.
2. If the coefficient of the x^2-term is a number other than 1, multiply both sides of the equation by the reciprocal of the coefficient.
3. Using the rewritten equation, calculate half of the coefficient of the x-term and square it. Add the result to both sides.
4. Factor the left side as the square of a binomial.
5. Solve by taking the square root of both sides.
6. Check all solutions in the original equation.

ADDITIONAL
EXAMPLE 4

Given $h(x) = 2x^2 + 16x + 23$, write the function in vertex form, and give the coordinates of the vertex and the equation of the axis of symmetry. Then describe the transformations from $f(x) = x^2$ to h.

$h(x) = 2[x - (-4)]^2 + (-9)$; the vertex is $(-4, -9)$; the axis of symmetry is $x = -4$; the graph of h is obtained by vertically stretching the graph of $f(x) = x^2$ by a factor of 2 and translating the resulting graph 4 units to the left and 9 units down.

Math CONNECTION

TRANSFORMATIONS As with other forms of equations, $f(x - k)$ translates the graph of the parent function h units to the right or left, depending on the sign of h, while $f(x) + k$ translates the graph of the parent function k units up or down depending of the sign of k. The transformation $af(x)$ stretches or shrinks the graph of the parent function by a factor of a.

Teaching Tip

TECHNOLOGY If you are using a TI-82 or TI-83 graphics calculator to graph Example 4, set the viewing window as follows:

Xmin=−10	Ymin=−7
Xmax=10	Ymax=13
Xscl=1	Yscl=1

Vertex Form

You know that the graph of $y = ax^2 + bx + c$, where $a \neq 0$, is a parabola. Using the method of completing the square, you can write the quadratic function in a form that contains the coordinates (h, k) of the vertex of the parabola.

Vertex Form

If the coordinates of the vertex of the graph of $y = ax^2 + bx + c$, where $a \neq 0$, are (h, k), then you can represent the parabola as $y = a(x - h)^2 + k$, which is the **vertex form** of a quadratic function.

Recall from Lesson 2.7 that if $y = f(x)$, then

- $y = af(x)$ gives a vertical stretch or compression of f,
- $y = f(bx)$ gives a horizontal stretch or compression of f,
- $y = f(x) + k$ gives a vertical translation of f, and
- $y = f(x - h)$ gives a horizontal translation of f.

EXAMPLE 4 **Given $g(x) = 2x^2 + 12x + 13$, write the function in vertex form, and give the coordinates of the vertex and the equation of the axis of symmetry. Then describe the transformations from $f(x) = x^2$ to g.**

SOLUTION

$g(x) = 2x^2 + 12x + 13$
$= 2(x^2 + 6x) + 13$ — *Factor 2 from the x^2- and x-terms.*
$= 2(x^2 + 6x + \mathbf{9}) + 13 - \mathbf{2(9)}$ — *Complete the square.*
$= 2(x + 3)^2 - 5$ — *Simplify.*
$= 2[x - (-3)]^2 + (-5)$ — *Write in vertex form.*

The coordinates (h, k) of the vertex are $(-3, -5)$, and the equation for the axis of symmetry is $x = -3$.

Notice from the vertex form of the function that $g(x) = 2f(x + 3) - 5$. There are three transformations from f to g:

- a vertical stretch by a factor of 2
- a horizontal translation of 3 units to the left
- a vertical translation of 5 units down

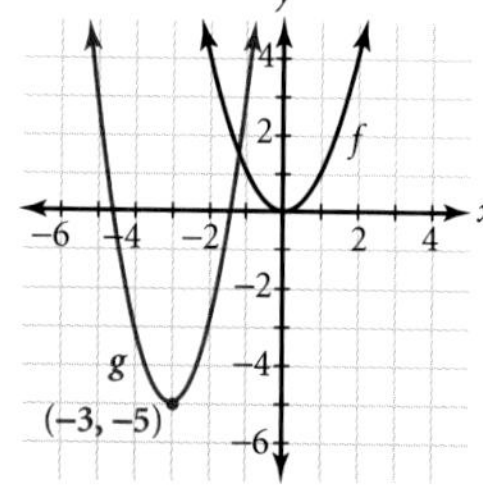

TRY THIS Given $g(x) = 3x^2 - 9x - 2$, write the function in vertex form, and give the coordinates of the vertex and the equation of the axis of symmetry. Then describe the transformations from $f(x) = x^2$ to g.

Reteaching the Lesson

USING PATTERNS Show students that the solutions of a quadratic equation of the form $ax^2 + bx + c = 0$ with real coefficients can be a single rational root, two distinct rational roots, two distinct irrational roots, or have no real solutions. Ask students to solve the following equations by using graphs. Then solve each equation algebraically by using an appropriate method.

1. $x^2 + 6x + 9 = 0$
   $x = -3$
2. $x^2 + 6x + 5 = 0$
   $x = -1$ *or* $x = 5$
3. $x^2 + 6x + 3 = 0$
   $x = -3 \pm \sqrt{6}$
4. $x^2 + 6x + 15 = 0$
   no real solution

EXAMPLE 5

APPLICATION
ENGINEERING

Refer to the suspension bridge described at the beginning of the lesson.

Complete the square, and write $f(x) = \frac{3}{5000}x^2 - \frac{3}{5}x + 200$ in vertex form. Then find the coordinates of the lowest point on the cable.

SOLUTION

Method 1 Use algebra.

$$\begin{aligned} f(x) &= \frac{3}{5000}x^2 - \frac{3}{5}x + 200 \\ &= \frac{3}{5000}(x^2 - 1000x) + 200 \\ &= \frac{3}{5000}\left[x^2 - 1000x + \left(\frac{1000}{2}\right)^2\right] + 200 - \frac{3}{5000}\left(\frac{1000}{2}\right)^2 \\ &= \frac{3}{5000}(x^2 - 1000x + 500^2) + 200 - \frac{3}{5000}(500^2) \\ &= \frac{3}{5000}(x - 500)^2 + 50 \end{aligned}$$

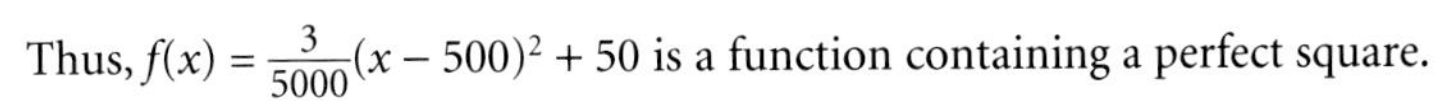

Thus, $f(x) = \frac{3}{5000}(x - 500)^2 + 50$ is a function containing a perfect square.

The lowest point on the cable is represented by the vertex (500, 50).

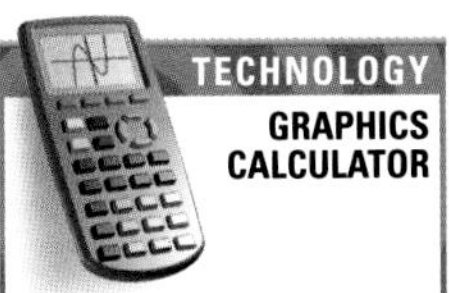

TECHNOLOGY
GRAPHICS CALCULATOR

Keystroke Guide, page 350

Method 2 Use a graph.

Graph $y = \frac{3}{5000}x^2 - \frac{3}{5}x + 200$, and find the coordinates of the lowest point.

From the graph of the function, the lowest point on the cable appears to be (500, 50).

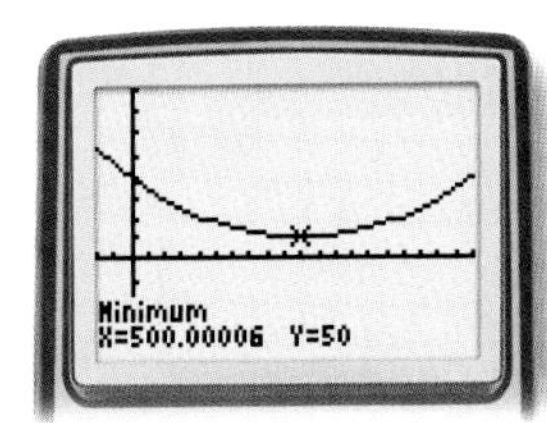

CRITICAL THINKING

Explain why you cannot write an equation for a parabola when given only the coordinates of the vertex.

Exercises

Communicate

1. Explain how to solve the equation $x^2 + 4x - 13 = 0$ by completing the square.
2. Explain how to solve the equation $2x^2 + 4x = 15$ by completing the square.
3. Refer to Example 3. Describe two ways of solving $2x^2 + 4x = 15$ by graphing.
4. Explain what h and k represent in the vertex form of a parabola.

ADDITIONAL EXAMPLE 5

A telephone cable is attached to two 250-foot poles. The cable forms a catenary curve, such as the one that is described at the beginning of the lesson. The equation of the curve can be approximated by the function $h(d) = \frac{3}{5000}d^2 - \frac{3}{5}d + 250$. **Complete the square, and write h as a function containing a perfect square. Then find the coordinates of the lowest point on the cable.**

$h(d) = \frac{3}{5000}(d - 500)^2 + 100$; (500, 100)

Teaching Tip

TECHNOLOGY If you are using a TI-82 or TI-83 graphics calculator, set the viewing window as follows:

Xmin=–100	**Ymin=–150**
Xmax=1000	**Ymax=400**
Xscl=50	**Yscl=50**

After entering the equation in the **Y=** window, press 2nd TRACE and choose **3:minimum** to locate the minimum. Enter 485 for a left bound and 508 for a right bound. Press ENTER to accept the guess. **Minimum X=500.00006 Y=50** is displayed at the bottom of the screen.

CRITICAL THINKING

Whether the parabola opens up or down cannot be determined from the coordinates of the vertex alone. Another point is required to determine the equation.

Assess

Selected Answers

Exercises 5–10, 11–65 odd

ASSIGNMENT GUIDE	
In Class	1–10
Core	11–51 odd
Core Plus	12–52 even
Review	53–66
Preview	67–69

Extra Practice can be found beginning on page 940.

Mid-Chapter Assessment for Lessons 5.1 through 5.4 can be found on page 59 of the *Assessment Resources.*

9. $g(x) = [x - (-6)]^2 + (-16)$; vertex: $(-6, -16)$; axis of symmetry: $x = -6$; the graph of g is the graph of f shifted vertically 16 units down and horizontally 6 units to the left.

32. $x = -\frac{15}{2} \pm \frac{\sqrt{217}}{4}$

33. $x = -\frac{7}{2} \pm \frac{\sqrt{41}}{4}$

34. $x = 16 \pm \sqrt{272}$

Practice

NAME ________ CLASS ________ DATE ________

Practice
5.4 Completing the Square

Complete the square for each quadratic expression in order to form a perfect-square trinomial. Then write the new expression as a binomial squared.

1. $x^2 + 24x$ — $x^2 + 24x + 144$; $(x + 12)^2$
2. $x^2 - 40x$ — $x^2 - 40x + 400$; $(x - 20)^2$
3. $x^2 - 20x$ — $x^2 - 20x + 100$; $(x - 10)^2$
4. $x^2 + 5x$ — $x^2 + 5x + 6.25$; $(x + 2.5)^2$
5. $x^2 + 9x$ — $x^2 + 9x + 20.25$; $(x + 4.5)^2$
6. $x^2 - 19x$ — $x^2 - 19x + 90.25$; $(x - 9.5)^2$

Solve by completing the square. Round your answers to the nearest tenth, if necessary.

7. $x^2 - 2x - 7 = 0$ — −1.8 or 3.8
8. $x^2 - 8x + 13 = 0$ — 2.3 or 5.7
9. $x^2 - 14x - 1 = 0$ — −0.1 or 14.1
10. $x^2 + 20x = 3$ — −20.1 or 0.1
11. $x^2 + 1 = 5x$ — 0.2 or 4.8
12. $x^2 - 4 = 6x$ — −0.6 or 6.6
13. $2x^2 - 13 = 2x$ — −2.1 or 3.1
14. $x^2 + 7x + 2 = 0$ — −6.7 or −0.3
15. $2x^2 + 16x = 3$ — −8.2 or 0.2

Write each quadratic function in vertex form. Find the coordinates of the vertex and the equation of the axis of symmetry.

16. $f(x) = -\frac{1}{2}x^2$ — $f(x) = -\frac{1}{2}x^2$; $(0, 0)$; $x = 0$
17. $f(x) = 7 - 3x^2$ — $f(x) = -3x^2 + 7$; $(0, 7)$; $x = 0$
18. $f(x) = x^2 - 12x - 3$ — $f(x) = (x - 6)^2 - 39$; $(6, -39)$; $x = 6$
19. $f(x) = x^2 - 2x - 10$ — $f(x) = (x - 1)^2 - 11$; $(1, -11)$; $x = 1$
20. $f(x) = x^2 - 10x - 10$ — $f(x) = (x - 5)^2 - 35$; $(5, -35)$; $x = 5$
21. $f(x) = 3x^2 + 15x - 2$ — $f(x) = 3(x + 2.5)^2 - 20.75$; $(-2.5, -20.75)$; $x = -2.5$

Guided Skills Practice

Complete the square for each quadratic expression to form a perfect-square trinomial. Then write the new expression as a binomial squared. ***(EXAMPLE 1)***

5. $x^2 - 12x$ **$x^2 - 12x + 36$; $(x - 6)^2$**

6. $x^2 + 5x$ **$x^2 + 5x + \frac{25}{4}$; $\left(x + \frac{5}{2}\right)^2$**

7. Solve $x^2 - 4x - 21 = 0$ by completing the square. ***(EXAMPLE 2)*** **−3, 7**

8. Solve $2x^2 + 5x = 3$. ***(EXAMPLE 3)*** **$-3, \frac{1}{2}$**

CONNECTION

9. TRANSFORMATIONS Given $g(x) = x^2 + 12x + 20$, write the function in vertex form, and give the coordinates of the vertex and the equation of the axis of symmetry. Then describe the transformations from $f(x) = x^2$ to g. ***(EXAMPLE 4)***

APPLICATION

10. SPORTS A softball is thrown upward with an initial velocity of 32 feet per second from 5 feet above ground. The ball's height in feet above the ground is modeled by $h(t) = -16t^2 + 32t + 5$, where t is the time in seconds after the ball is released. Complete the square and rewrite h in vertex form. Then find the the maximum height of the ball. ***(EXAMPLE 5)***
$h(t) = -16(t - 1)^2 + 21$; 21 ft

11. $x^2 + 10x + 25$; $(x + 5)^2$

12. $x^2 - 14x + 49$; $(x - 7)^2$

13. $x^2 - 8x + 16$; $(x - 4)^2$

14. $x^2 + 2x + 1$; $(x + 1)^2$

15. $x^2 + 13x + \frac{169}{4}$; $\left(x + \frac{13}{2}\right)^2$

internet connect
Homework Help Online
Go To: **go.hrw.com**
Keyword:
MB1 Homework Help for Exercises 17–37

Practice and Apply

Complete the square for each quadratic expression to form a perfect-square trinomial. Then write the new expression as a binomial squared.

11. $x^2 + 10x$ **12.** $x^2 - 14x$ **13.** $x^2 - 8x$
14. $x^2 + 2x$ **15.** $x^2 + 13x$ **16.** $x^2 + 7x$

Solve each equation by completing the square. Give exact solutions.

17. $x^2 - 8x = 3$ **$4 \pm \sqrt{19}$**
18. $x^2 + 2x = 13$ **$-1 \pm \sqrt{14}$**
19. $x^2 - 5x - 1 = 4 - 3x$ **$1 \pm \sqrt{6}$**
20. $0 = x^2 - 6x + 3$
21. $0 = x^2 + 7x - 26$
22. $0 = x^2 - 3x - 6$
23. $x^2 + 7x + 10 = 0$
24. $x^2 + 10x + 16 = 0$
25. $x^2 - x = 30$
26. $0 = 3x^2 - 2x - 12$
27. $-2x^2 + 14x + 60 = 0$
28. $0 = 3x^2 - 11x + 6$
29. $-10 = x^2 - 8x + 2$
30. $x^2 + 16x = 2$
31. $4 - x^2 = 10x$
32. $x^2 = 23 - 15x$
33. $8x - 2 = x^2 + 15x$
34. $-32x = 16 - x^2$
35. $2x^2 = 22x - 11$
36. $4x^2 - 8 = -13x$
37. $2x^2 - 12 = 3x$

16. $x^2 + 7x + \frac{49}{4}$; $\left(x + \frac{7}{2}\right)^2$
20. $3 \pm \sqrt{6}$
21. $-\frac{7}{2} \pm \sqrt{\frac{153}{4}}$
22. $\frac{3}{2} \pm \sqrt{\frac{33}{4}}$
23. −5, −2
24. −8, −2
25. −5, 6
26. $\frac{1}{3} \pm \sqrt{\frac{37}{9}}$
27. −3, 10
28. $\frac{2}{3}$, 3
29. 2, 6
30. $-8 \pm \sqrt{66}$
31. $-5 \pm \sqrt{29}$

Write each quadratic function in vertex form. Give the coordinates of the vertex and the equation of the axis of symmetry. Then describe the transformations from $f(x) = x^2$ to g.

38. $g(x) = 3x^2$ **39.** $g(x) = -x^2 + 2$ **40.** $g(x) = x^2 - 5x$
41. $g(x) = x^2 + 8x + 11$ **42.** $g(x) = x^2 - 6x - 2$ **43.** $g(x) = -x^2 + 4x + 2$
44. $g(x) = x^2 + 7x + 3$ **45.** $g(x) = -3x^2 + 6x - 9$ **46.** $g(x) = -2x^2 + 12x + 13$

47. Write three different quadratic functions that each have a vertex at (2, 5).

---

35. $x = \frac{11}{2} \pm \frac{\sqrt{99}}{4}$

36. $x = -\frac{13}{8} \pm \frac{\sqrt{397}}{64}$

37. $x = \frac{3}{4} \pm \frac{\sqrt{105}}{16}$

38. $g(x) = 3(x - 0)^2 + 0$; $(0, 0)$; $x = 0$

39. $g(x) = -(x - 0)^2 + 2$; $(0, 2)$; $x = 0$

40. $g(x) = \left(x - \frac{5}{2}\right)^2 + \left(-\frac{25}{4}\right)$; $\left(\frac{5}{2}, -\frac{25}{4}\right)$; $x = \frac{5}{2}$

41. $g(x) = [x - (-4)]^2 + (-5)$; $(-4, -5)$; $x = -4$

42. $g(x) = (x - 3)^2 + (-11)$; $(3, -11)$; $x = 3$

43. $g(x) = -(x - 2)^2 + 6$; $(2, 6)$; $x = 2$

44. $g(x) = \left[x - \left(-\frac{7}{2}\right)\right]^2 + \left(-\frac{37}{4}\right)$; $\left(-\frac{7}{2}, -\frac{37}{4}\right)$; $x = -\frac{7}{2}$

45. $g(x) = -3(x - 1)^2 + (-6)$; $(1, -6)$; $x = 1$

46. $g(x) = -2(x - 3)^2 + 31$; $(3, 31)$; $x = 3$

47. Answers may vary.

Descriptions of the transformations from f to g for Exercises 38–46 can be found in Additional Answers beginning on page 1002.

CHALLENGE

48. Write an equation for the quadratic function that has a vertex at (2, 5) and contains the point (1, 8). **$f(x) = 3(x-2)^2 + 5 = 3x^2 - 12x + 17$**

CONNECTIONS

For Exercises 49 and 50, give both an exact answer and an approximate answer rounded to the nearest tenth.

49. GEOMETRY Each side of a square is increased by 2 centimeters, producing a new square whose area is 30 square centimeters. Find the length of the sides of the original square. width: **$-2 + \sqrt{30} \approx 3.5$ cm**

50. GEOMETRY The length of a rectangle is 6 feet longer than its width. If the area is 50 square feet, find the length and the width of the rectangle. **width: $-3 + \sqrt{59} \approx 4.7$ ft; length: $3 + \sqrt{59} \approx 10.7$ ft**

APPLICATIONS

51. PHYSICS The power, in megawatts, produced between midnight and noon by a power plant is given by $P = h^2 - 12h + 210$, where h is the hour of the day.

a. At what time does the minimum power production occur? **6:00 A.M.**
b. What is the minimum power production? **174 megawatts**
c. During what hour(s) is the power production 187 megawatts? **3rd hour (2:00–3:00 A.M.) and 10th hour (9:00–10:00 A.M.)**

52. FUND-RAISING Each year a school's booster club holds a dance to raise funds. In the past, the profit the club made after paying for the band and other costs has been modeled by the function $P(t) = -16t^2 + 800t - 4000$, where t represents the ticket price in dollars.

a. What ticket price gives the maximum profit? **\$25**
b. What is the maximum profit? **\$6000**
c. What ticket price(s) would generate a profit of \$5424? **\$19, \$31**

Look Back

Solve each equation. ***(LESSON 1.6)***

53. $5x + 3 = 2x + 18$ **5** **54.** $\frac{2(x+3)}{5} = x - 3$ **7** **55.** $20 = 6x - 10$ **5**

For Exercises 56–59, determine whether each set of ordered pairs represents a function. ***(LESSON 2.3)***

56. $\{(11, 0), (12, -1), (21, -2)\}$ **yes** **57.** $\{(0, 0), (2, 5), (3, 3)\}$ **yes**

58. $\{(1, -1), (1, -2), (1, -3)\}$ **no** **59.** $\{(4, 1), (5, 2), (6, 3)\}$ **yes**

60. Evaluate $f(x) = \frac{1}{3}x - 2$ for $x = 2$ and $x = -3$. ***(LESSON 2.3)*** **$-\frac{4}{3}$; -3**

61. Evaluate $g(x) = 7 - 4x$ for $x = 2$ and $x = -3$. ***(LESSON 2.3)*** **-1; 19**

Error Analysis

Have students write the equation in standard form before completing the square, and remind them to always divide both sides by the leading coefficient if it is not 1. Many quadratic equations have no solutions, as indicated by the fact that the graph of the related function does not cross the x-axis. Point out that a quadratic equation written in the form $(x-h)^2 = \frac{k}{a}$ has no real solution when the isolated constant term, $\frac{k}{a}$, is negative because the square root of a negative number is not a real number.

47. Answers may vary. sample answer:

$f(x) = (x-2)^2 + 5$
$= x^2 - 4x + 9$

$g(x) = -2(x-2)^2 + 5$
$= -2x^2 + 8x - 3$

$h(x) = 5(x-2)^2 + 5$
$= 5x^2 - 20x + 25$

Student Technology Guide

NAME ______ CLASS ______ DATE ______

Student Technology Guide
5.4 Completing the Square

You can use a graphics calculator to find the vertex of a parabola. Then you can write the vertex form of the related quadratic function.

Example: Write $y = 3x^2 - 12x + 10$ in vertex form.

- Note the value of a, the coefficient of x^2. In this example, $a = 3$.
- Graph the equation. Press Y= 3 X,T,θ,n x^2 − 12 X,T,θ,n + 10 GRAPH.
- Press TRACE. Use ◄ and ► to move to a point just to the left of the vertex. Press 2nd TRACE 3:minimum ENTER. (Use 4:maximum if the vertex is a maximum point.)
- When asked for the left bound, press ENTER to enter the coordinates of the trace point.
- When asked for the right bound, trace just to the right of the vertex. Press ENTER.
- When asked for a guess, trace as close as you can to the vertex. Press ENTER.
- The calculator displays the approximate coordinates of the vertex. (Press 2 ENTER to be sure the coordinates are actually (2, −2).) Thus, $h = 2$, $k = -2$, and $a = 3$, so the vertex form of the equation is $y = 3(x-2)^2 - 2$.

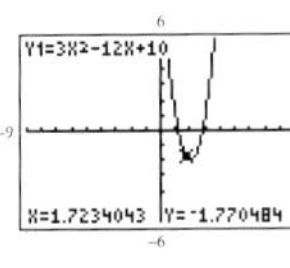

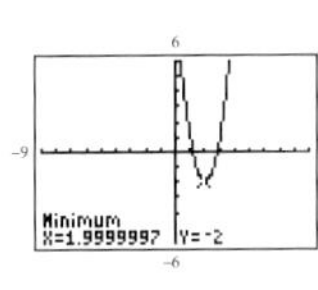

Write each quadratic function in vertex form.

1. $f(x) = x^2 - 6x + 3$ $y = (x-3)^2 - 6$
2. $f(x) = x^2 + 8x - 11$ $y = (x+4)^2 - 27$
3. $g(x) = -x^2 - 6x + 12$ $y = -(x+3)^2 + 21$
4. $m(x) = x^2 + 11x + 27$ $y = (x+5.5)^2 - 3.25$
5. $f(x) = 2x^2 - 6x - 15$ $y = 2(x-1.5)^2 - 19.5$
6. $g(x) = -3x^2 - 12x - 3$ $y = -3(x+2)^2 + 9$

In Exercises 67 through 69 students are asked to find solutions to specific quadratic equations by using the method developed by al-Khowarizmi over 1000 years ago. The method uses algebra tiles to complete the square by arranging the x-tiles along the sides of an x^2-tile and then filling in the corners of the incomplete square with unit tiles. The length of the rectangular tiles is a solution to the original equation.

You may want to have students do additional research about the history of al-Khowarizmi, the history of Arabic mathematics, and the adoption of the Hindu numerals we use today. Encourage students to make posters illustrating their findings and display them around the classroom.

ALTERNATIVE
Assessment

Portfolio Activity

The Portfolio Activity can be used as preparation for the Chapter Project or as a separate activity. In the Portfolio Activity on this page, students will complete the square for a function that relates to the basketball problem presented at the beginning of the chapter and will compare the new procedure with the procedures used in Activities 1 and 2 on pages 280 and 298.

Answers to Portfolio Activities can be found in Additional Answers of the Teacher's Edition.

CONNECTION

TRANSFORMATIONS For Exercises 62–65, write the equation for the graph described. ***(LESSON 2.7)***

62. the graph of $f(x) = |x|$ translated 7 units to the left **$g(x) = |x + 7|$**

63. the graph of $f(x) = x^2$ translated 6 units down **$g(x) = x^2 - 6$**

64. the graph of $f(x) = x^2$ stretched vertically by a factor of 8 **$g(x) = 8x^2$**

65. the graph of $f(x) = |x|$ reflected through the x-axis and stretched horizontally by a factor of 3 **$g(x) = -|3x|$**

66. Write a quadratic function whose zeros are −2 and 6. ***(LESSON 5.1)*** **$f(x) = x^2 - 4x - 12$**

Look Beyond

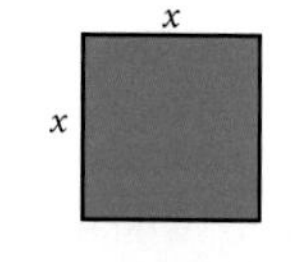

CULTURAL CONNECTION: AFRICA One of the first algebra books was written in Arabic by al-Khowarizmi around 800 C.E. This book used a method similar to algebra tiles to complete the square and solve for x.

To complete the square for the equation $x^2 + 12x = 45$, the al-Khowarizmi method begins with one square that is x units long on each side and 12 rectangles that are 1 unit wide and x units long.

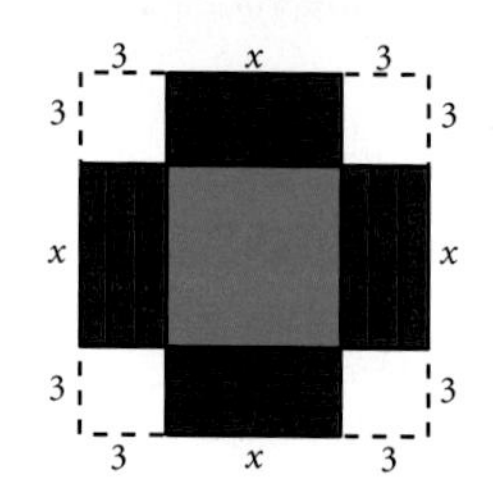

Step 1 Divide the **12 rectangles** into 4 groups, and arrange them on the sides of the square. From the equation $x^2 + 12x = 45$, you know that the area of this shape is still 45 square units.

Step 2 To complete the square, add 3×3, or 9 units to each of the 4 corners:

$$9 \text{ units} \times 4 = 36 \text{ units}$$

The new area is $45 + 36 = 81$. If the area is 81, then the side length is 9. To find the length of x, solve $3 + x + 3 = 9$. Thus, $x = 3$.

67. Solve $x^2 + 20x = 125$ by using the al-Khowarizmi method. **5**

68. Solve $x^2 + 32x = 33$ by using the al-Khowarizmi method. **1**

69. Solve $x^2 + 56x = 116$ by using the al-Khowarizmi method. **2**

SPORTS Refer to the height function in the Portfolio Activity on page 280.

Complete the square for the height function and answer the questions below.

1. What is the maximum height achieved by the basketball?
2. How long does it take the basketball to reach its maximum height?
3. What is the equation of the axis of symmetry for the height function?
4. How long would it take the basketball to drop to the ground if it were not tapped by a player?
5. Compare your answers to the questions above with those obtained in Portfolio Activities on page 280 and 298.

The Quadratic Formula

Objectives

- Use the quadratic formula to find real roots of quadratic equations.
- Use the roots of a quadratic equation to locate the axis of symmetry of a parabola.

APPLICATION
CONSTRUCTION

Why *You can solve many real-world problems, such as finding the dimensions of this patio, by using the quadratic formula to solve quadratic equations.*

After watching a home-improvement show, the Wilkersons decided to build a patio along two sides of their home, as shown here. The patio will have the same width along both sides.

Find the width of the patio if the Wilkersons have enough material to cover a surface area of 650 square feet. You can use the *quadratic formula* to solve this problem. You will solve this problem in Example 3.

Using the method of completing the square, you can derive a formula that can be used to solve any quadratic equation in standard form.

$ax^2 + bx + c = 0$	*Assume that $a \neq 0$.*
$x^2 + \frac{b}{a}x + \frac{c}{a} = 0$	*Divide each side by a.*
$x^2 + \frac{b}{a}x = -\frac{c}{a}$	*Subtract $\frac{c}{a}$ from each side.*
$x^2 + \frac{b}{a}x + \left(\frac{b}{2a}\right)^2 = -\frac{c}{a} + \left(\frac{b}{2a}\right)^2$	*Complete the square.*
$\left(x + \frac{b}{2a}\right)^2 = \frac{-4ac + b^2}{4a^2}$	*Simplify.*
$\sqrt{\left(x + \frac{b}{2a}\right)^2} = \pm\sqrt{\frac{b^2 - 4ac}{4a^2}}$	*Take the square root of each side.*
$x + \frac{b}{2a} = \pm\frac{\sqrt{b^2 - 4ac}}{2a}$	*Simplify.*
$x = -\frac{b}{2a} \pm \frac{\sqrt{b^2 - 4ac}}{2a}$	*Subtract $\frac{b}{2a}$ from each side.*
$x = \frac{-b \pm \sqrt{b^2 - 4ac}}{2a}$	*Simplify.*

Alternative Teaching Strategy

COOPERATIVE LEARNING Write each step of the derivation of the quadratic formula on a slip of paper. Clearly label the standard form of the quadratic equation as 1 to signify the first step. Then mix up the strips of paper.

Divide the class into groups of three or four. Give each group a complete set of mixed-up strips of paper. Have the students in each group work together to arrange the steps in the proper order. Bring the class together in a discussion to confirm that each group got the correct order and to discuss the significance of the equation arrived at in the last step.

Prepare

NCTM PRINCIPLES & STANDARDS 1–10

QUICK WARM-UP

Solve each equation by factoring and then applying the zero-product property.

1. $x^2 + 13x + 36 = 0$
   $x = -4$ *or* $x = -9$
2. $x^2 + 12x + 36 = 0$
   $x = -6$
3. $x^2 = 4x + 21$
   $x = -3$ *or* $x = 7$
4. $2x^2 + 5x = 12$
   $x = -4$ *or* $x = \frac{3}{2}$
5. Solve $x^2 - 4x - 7 = 0$ by completing the square. Round your answer to the nearest tenth.
   $x \approx 5.3$ *or* $x \approx -1.3$

Also on Quiz Transparency 5.5

Teach

Why Tell students that a local builder donated 325 feet of fencing to a neighborhood organization to help them set up a rectangular garden. Ask students to consider how they would determine the area that can be enclosed and discuss the limitations on the length, width, and area to be fenced.

Quadratic Formula

If $ax^2 + bx + c = 0$ and $a \neq 0$, then the solutions, or roots, are

$$x = \frac{-b \pm \sqrt{b^2 - 4ac}}{2a}.$$

Example 1 shows how to use the quadratic formula to solve a quadratic equation that is in standard form.

EXAMPLE 1 **Use the quadratic formula to find the roots of $x^2 + 5x - 14 = 0$.**

SOLUTION

In $x^2 + 5x - 14 = 0$, $a = 1$, $b = 5$, and $c = -14$.

$$x = \frac{-b \pm \sqrt{b^2 - 4ac}}{2a}$$

$$x = \frac{-5 \pm \sqrt{5^2 - 4(1)(-14)}}{2(1)} \quad \textit{Use the quadratic formula.}$$

$$x = \frac{-5 + \sqrt{81}}{2} \quad or \quad x = \frac{-5 - \sqrt{81}}{2}$$

$$x = \frac{-5 + 9}{2} \qquad x = \frac{-5 - 9}{2} \quad \textit{Substitute 9 for } \sqrt{81}.$$

$$x = 2 \qquad x = -7$$

TRY THIS Use the quadratic formula to solve $x^2 - 7x + 6 = 0$.

CHECKPOINT ✔ Solve $x^2 + 5x - 14 = 0$ by factoring to check the solution to Example 1.

The solutions to a quadratic equation can be irrational numbers. This is shown in Example 2.

EXAMPLE 2 **Use the quadratic formula to solve $4x^2 = 8 - 3x$. Give exact solutions and approximate solutions to the nearest tenth.**

SOLUTION

PROBLEM SOLVING **Use a formula.** Write the equation in standard form. Then use the quadratic formula.

$$4x^2 = 8 - 3x$$

$$4x^2 + 3x - 8 = 0$$

$$x = \frac{-3 \pm \sqrt{3^2 - 4(4)(-8)}}{2(4)}$$

$$x = \frac{-3 \pm \sqrt{137}}{8}$$

$$x = \frac{-3 + \sqrt{137}}{8} \quad or \quad x = \frac{-3 - \sqrt{137}}{8} \quad \textit{Exact solution}$$

$$x \approx 1.1 \qquad x \approx -1.8 \quad \textit{Approximate solution}$$

TRY THIS Use the quadratic formula to solve $2x^2 - 6x = -3$. Give exact solutions and approximate solutions to the nearest tenth.

Teaching Tip

As you work through the derivation of the quadratic formula, work through an example with specific coefficients next to the one with arbitrary numbers and explain each step of completing the square as you go in both the specific and arbitrary versions.

All the equations in this lesson have real roots. Complex roots will be presented in Lesson 5.6.

ADDITIONAL EXAMPLE 1

Use the quadratic formula to solve $x^2 - 16x - 36 = 0$.

$x = 18$ *or* $x = -2$

TRY THIS

$x = 1$ *or* $x = 6$

CHECKPOINT ✔

$(x + 7)(x - 2) = 0$

$x + 7 = 0$ *or* $x - 2 = 0$

$x = -7$ *or* $x = 2$

ADDITIONAL EXAMPLE 2

Use the quadratic formula to solve $5x^2 + 1 = 7x$. Give exact solutions and approximate solutions to the nearest tenth.

$x = \frac{7 \pm \sqrt{29}}{10}$

$x \approx 1.2$ *or* $x \approx 0.2$

TRY THIS

$x = \frac{3 \pm \sqrt{3}}{2}$

$x \approx 2.4$ *or* $x \approx 0.6$

Interdisciplinary Connection

CHEMISTRY In chemistry, quadratic equations are used to describe the equilibrium concentrations of ions in aqueous solutions. To avoid rounding errors, the formulas below are used to solve quadratic equations in standard form, where $b > 0$.

$$x_1 = \frac{-2c}{b + \sqrt{b^2 - 4ac}}, \; x_2 = \frac{-b - \sqrt{b^2 - 4ac}}{2a}$$

Ask students to use these two alternate forms of the quadratic formula to find the roots of the quadratic equation in Example 1. Then have them compare the results.

Enrichment

Have students find the solutions of quadratic equations graphically with the following method developed by Karl Georg Christian von Staudt (1798–1867): Given the quadratic equation $x^2 + bx + c = 0$, plot the points $M\left(-\frac{c}{b}, 0\right)$ and $N\left(-\frac{4}{b}, 2\right)$ on a coordinate system. Draw a circle centered at (0, 1) with a radius of 1. Draw $\overline{MN}$ and label the points of intersection with the circle as S and T. From (0, 2), draw line segments through S and T to the x-axis. The points where the line segments intersect the x-axis are the solutions to the equation.

EXAMPLE 3 Refer to the patio described at the beginning of the lesson.

APPLICATION
CONSTRUCTION

Find the width of the patio if there is enough material to cover a surface area of 650 square feet.

SOLUTION

PROBLEM SOLVING

Write an equation.

$$A(x) = 25x + 30x + x^2$$
$$= x^2 + 55x$$

Solve $x^2 + 55x = 650$ for x.

$$x^2 + 55x = 650$$
$$x^2 + 55x - 650 = 0$$
$$x = \frac{-55 \pm \sqrt{55^2 - 4(1)(-650)}}{2(1)}$$
$$x = \frac{-55 \pm 75}{2}$$
$$x = 10 \quad or \quad x = -65$$

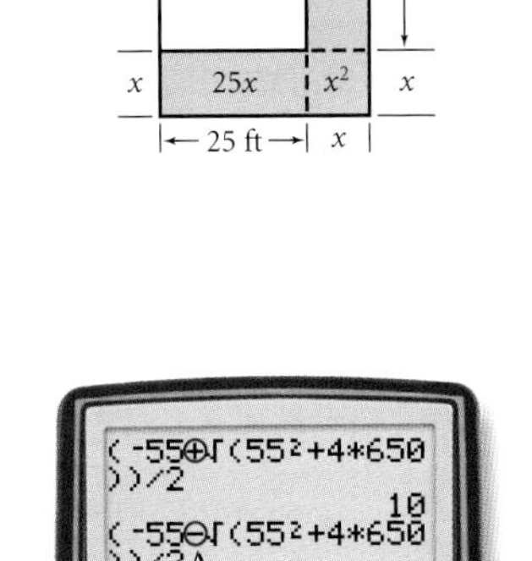

A shortcut on a graphics calculator is to edit the previous entry by changing the plus to a minus.

Since the width must be positive, -65 cannot be a solution. The patio should be 10 feet wide.

TRY THIS Find the width of the patio described at the beginning of the lesson if there is enough cement to cover a surface area of 500 square feet.

Recall from Lesson 5.3 that real-number solutions of a quadratic equation $ax^2 + bx + c = 0$ are also the x-intercepts of the related quadratic function $f(x) = ax^2 + bx + c$. You can examine this in the Activity below.

Activity: Exploring Roots of Equations

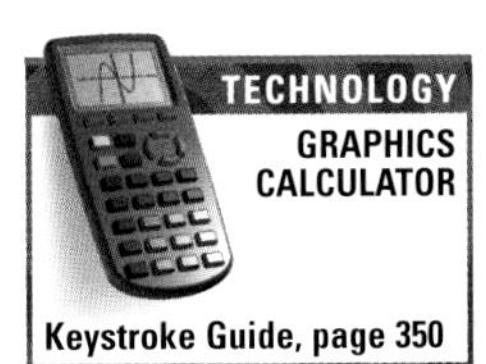

Keystroke Guide, page 350

You will need: a graphics calculator

1. Copy the table below. Complete the second and third columns of the table by using any method to find the roots of each equation.
2. Graph each related function and find the x-coordinate of the vertex. Then complete the last two columns of the table.

Equation	Roots	Average of roots	Related function	x-coordinate of vertex
$x^2 + 2x = 0$	$0, -2$	-1	$f(x) = x^2 + 2x$	-1
$-x^2 + 4 = 0$				
$x^2 + 4x + 4 = 0$				
$2x^2 + 5x - 3 = 0$				
$-x^2 - 3x + 4 = 0$				

CHECKPOINT ✔ 3. Write a brief explanation that tells how to find the x-coordinate of the vertex for the graph of a quadratic function.

ADDITIONAL EXAMPLE 3

After seeing the Wilkersons' patio, the Santiagos decided to build a similar patio, as shown below. The patio will have an equal width along both sides of the house. **Find the width of the patio if the Santiagos have enough bricks to cover 600 square feet.**

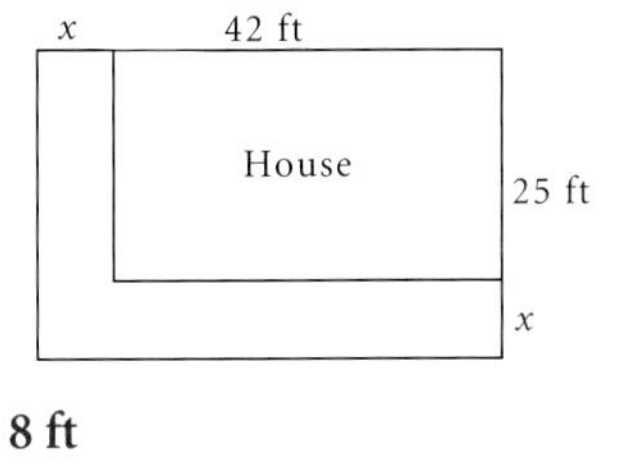

8 ft

TRY THIS
≈7.9 ft

Activity Notes

Tell students that this Activity gives them a general method for finding the x-coordinate of the vertex by averaging the roots of the equations. Once the x-coordinate is known, the y-coordinate can be found by evaluating the related function for that value of x.

CHECKPOINT ✔
Find the roots of the related equation and average them to get the x-coordinate of the vertex of the graph of a quadratic function.

Inclusion Strategies

INVITING PARTICIPATION Ask students to share any "tricks" they use to remember the quadratic formula—mnemonics, a chant, a riddle, etc.

Reteaching the Lesson

USING SYMBOLS To help students identify the values of a, b, and c in a quadratic equation in standard form, $ax^2 + bx + c = 0$, give them the quadratic formula in the following form:

$$x = \frac{-(b) \pm \sqrt{(b)^2 - (4)(a)(c)}}{(2)(a)}$$

Now give them the quadratic equation $x^2 - 2x - 4 = 0$. Have them rewrite the equation as $(1)x^2 + (-2)x + (-4) = 0$, identify a, b, and c from the equation, substitute these numbers into the formula, and simplify the result. $\mathbf{x = 1 \pm \sqrt{5}}$

Teaching Tip

Point out to students that if the graph of a quadratic function does not intersect the x-axis, the procedure shown in the Activity on this page will not yield the x-coordinate of the vertex of the graph, and that if the graph of a function intersects the x-axis at only one point, that point is the vertex of the graph.

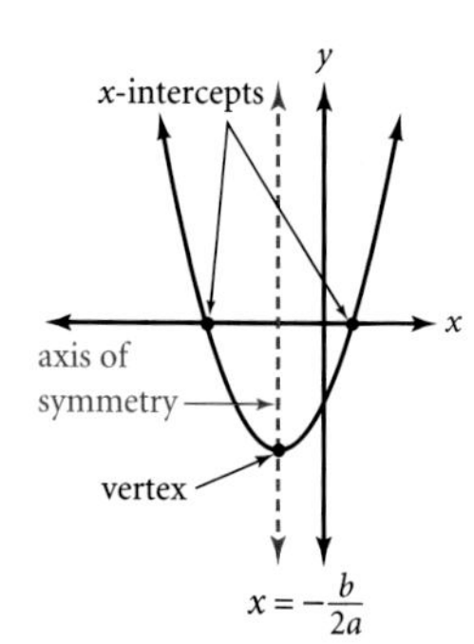

Recall from Lesson 5.1 that the equation of the axis of symmetry of a parabola is obtained from the x-coordinate of the vertex of the parabola. Using the symmetry of the parabola, an equation for the axis of symmetry can be found by taking the average of the two roots found with the quadratic formula.

$$\frac{\left(\frac{-b + \sqrt{b^2 - 4ac}}{2a}\right) + \left(\frac{-b - \sqrt{b^2 - 4ac}}{2a}\right)}{2} = -\frac{b}{2a}$$

Axis of Symmetry of a Parabola

If $y = ax^2 + bx + c$, where $a \neq 0$, then the equation for the axis of symmetry of the parabola is $x = -\frac{b}{2a}$.

EXAMPLE 4 **Let $f(x) = 19 + 8x + 2x^2$. Write the equation for the axis of symmetry of the graph, and find the coordinates of the vertex.**

SOLUTION

To find a and b, rewrite the function as $f(x) = 2x^2 + 8x + 19$. Then $a = 2$ and $b = 8$.

PROBLEM SOLVING

Use a formula. The axis of symmetry is $x = -\frac{b}{2a}$.

$x = -\frac{8}{2(2)} = -2$

$f(x) = 19 + 8x + 2x^2$

$f(-2) = 19 + 8(-2) + 2(-2)^2$ *Substitute −2 for x.*

$f(-2) = 11$

Thus, the equation for the axis of symmetry is $x = -2$, and the coordinates of the vertex are $(-2, 11)$.

TRY THIS Let $g(x) = x^2 - 4x + 1$. Write the equation for the axis of symmetry of the graph, and find the coordinates of the vertex.

CRITICAL THINKING What do you know about an equation that has integer solutions when it is solved by using the quadratic formula?

Exercises

Communicate

1. Describe at least three methods you can use to find the x-intercepts of the parabola described by $f(x) = x^2 + 2x - 3$.
2. Describe two ways to find the vertex of a parabola.
3. How is the axis of symmetry related to the vertex of a parabola?

ADDITIONAL EXAMPLE 4

Let $f(x) = -2x + 3 + 2x^2$. Write an equation for the axis of symmetry of the graph, and find the coordinates of the vertex.

$x = \frac{1}{2}$; $\left(\frac{1}{2}, \frac{5}{2}\right)$

TRY THIS

$x = 2$; $(2, -3)$

CRITICAL THINKING

When the solutions of a quadratic equation are integers,

- the equation can be factored over the set of integers,
- the discriminant, $b^2 - 4ac$, must be a perfect square, and
- the numerator of the formula must be a multiple of $2a$.

Practice

NAME ______ CLASS ______ DATE ______

Practice
5.5 The Quadratic Formula

Use the quadratic formula to solve each equation. Round your answer to the nearest tenth.

1. $x^2 - 10x + 3 = 0$ — 0.3 or 9.7
2. $x^2 - 4x + 1 = 0$ — 0.3 or 3.7
3. $x^2 - 11 = 0$ — −3.3 or 3.3
4. $x^2 + 3x - 15 = 0$ — −5.7 or 2.7
5. $x^2 = 9 - 4x$ — −5.6 or 1.6
6. $x^2 + 7x - 13 = 2x$ — −6.9 or 1.9
7. $14 = 2x^2 + x$ — −2.9 or 2.4
8. $(x - 2)(x - 5) = 2$ — 1.4 or 5.6
9. $2x^2 - 6x = 9$ — −1.1 or 4.1
10. $3x^2 - 6 = 4x$ — −0.9 or 2.2
11. $-2x^2 + 3x + 16 = 0$ — −2.2 or 3.7
12. $4x^2 + x - 1 = 0$ — −0.6 or 0.4

For each quadratic function, find the equation for the axis of symmetry and the coordinates of the vertex. Round your answers to the nearest tenth, if necessary.

13. $y = 2x^2 + 4x - 3$ — $x = -1$; $(-1, -5)$
14. $y = -3x^2 + 9x + 5$ — $x = 1.5$; $(1.5, 11.75)$
15. $y = 6x^2 - 12x + 5$ — $x = 1$; $(1, -1)$
16. $y = 3x^2 + 4x - 9$ — $x = -\frac{2}{3}$; $\left(-\frac{2}{3}, -10\frac{1}{3}\right)$
17. $y = 4x^2 - 8x + 1$ — $x = 1$; $(1, -3)$
18. $y = 5x^2 + 4x - 1$ — $x = -\frac{2}{5}$; $\left(-\frac{2}{5}, -1\frac{4}{5}\right)$

Guided Skills Practice

Use the quadratic formula to find the roots of each equation. ***(EXAMPLE 1)***

4. $x^2 - 5x + 4 = 0$ **1, 4**

5. $2x^2 - 5x = 3$ $-\frac{1}{2}, 3$

6. $\frac{3 \pm \sqrt{57}}{6}$; $-0.8, 1.8$

6. Use the quadratic formula to solve $3x^2 - 3x = 4$. Give exact solutions and approximate solutions to the nearest tenth. ***(EXAMPLE 2)***

APPLICATION

7. CONSTRUCTION If the Wilkersons have enough material to cover a surface area of 700 square feet, then the equation becomes $x^2 + 55x = 700$. Find the width of the patio, to the nearest tenth of a foot, from this equation. ***(EXAMPLE 3)*** **10.7 ft**

For each function, write the equation for the axis of symmetry of the graph, and find the coordinates of the vertex. ***(EXAMPLE 4)***

8. $f(x) = x^2 - x - 2$ $x = \frac{1}{2}; \left(\frac{1}{2}, -\frac{9}{4}\right)$

9. $f(x) = 2x^2 - 12x + 11$ $x = 3; (3, -7)$

Practice and Apply

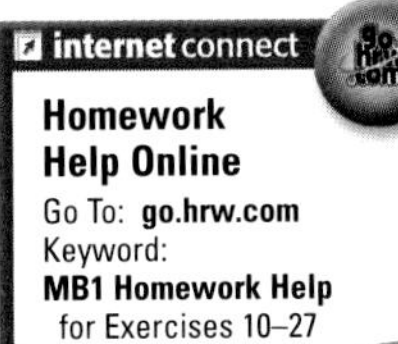

Use the quadratic formula to solve each equation. Give exact solutions.

10. $x^2 + 7x + 9 = 0$ $\frac{-7 \pm \sqrt{13}}{2}$

11. $x^2 + 6x = 0$ **−6, 0**

12. $(x + 1)(x - 2) = 5$ $\frac{1 \pm \sqrt{29}}{2}$

13. $(x - 4)(x + 5) = 7$ $\frac{-1 \pm \sqrt{109}}{2}$

14. $t^2 - 9t + 5 = 0$ $\frac{9 \pm \sqrt{61}}{2}$

15. $x^2 - 3x - 1 = 0$ $\frac{3 \pm \sqrt{13}}{2}$

16. $x^2 + 9x - 2 = -16$ **−7, −2**

17. $x^2 - 5x - 6 = 18$ **−3, 8**

18. $5x^2 + 16x - 6 = 3$ $-\frac{8}{5} \pm \frac{\sqrt{109}}{5}$

19. $4x^2 = -8x - 3$ $-\frac{3}{2}, -\frac{1}{2}$

20. $3x^2 - 3 = -5x - 1$ $-2, \frac{1}{3}$

21. $x^2 + 3x = 2 - 2x$ $\frac{-5 \pm \sqrt{33}}{2}$

22. $x^2 + 6x + 5 = 0$ **−5, −1**

23. $x^2 + 10x = 5$ $-5 \pm \sqrt{30}$

24. $-2x^2 + 4x = -2$ $1 \pm \sqrt{2}$

25. $5x^2 - 2x - 3 = 0$ $-\frac{3}{5}, 1$

26. $-6x^2 + 3x + 19 = 0$ $\frac{1}{4} \pm \frac{\sqrt{465}}{2}$

27. $-x^2 - 3x + 1 = 0$ $\frac{3 \pm \sqrt{13}}{-2}$

For each quadratic function, write the equation for the axis of symmetry, and find the coordinates of the vertex.

internet connect
Activities Online
Go To: go.hrw.com
Keyword:
MB1 QuadFormula

28. $y = 7x^2 + 6x - 5$ $x = -\frac{3}{7}; \left(-\frac{3}{7}, -\frac{44}{7}\right)$

29. $y = x^2 + 9x + 14$ $x = -\frac{9}{2}; \left(-\frac{9}{2}, -\frac{25}{4}\right)$

30. $y = 3 + 7x + 2x^2$ $x = -\frac{7}{4}; \left(-\frac{7}{4}, -\frac{25}{8}\right)$

31. $y = 10 - 5x^2 - 15x$ $x = -\frac{3}{2}; \left(-\frac{3}{2}, -\frac{85}{4}\right)$

32. $y = 3x^2 + 6x - 18$ $x = -1; (-1, -21)$

33. $y = 14 + 8x - 2x^2$ $x = 2; (2, 22)$

34. $y = 4 - 10x + 5x^2$ $x = 1; (1, -1)$

35. $y = -x^2 - 6x + 2$ $x = -3; (-3, 11)$

36. $y = 3x^2 + 21x - 4$ $x = -\frac{7}{2}; \left(-\frac{7}{2}, -\frac{163}{4}\right)$

37. $y = -2x^2 + 3x - 1$ $x = \frac{3}{4}; \left(\frac{3}{4}, \frac{1}{8}\right)$

38. $y = 3x^2 - 18x + 22$ $x = 3; (3, -5)$

39. $y = -2x^2 + 8x + 13$ $x = 2; (2, 21)$

40. $y = 3x - 2x^2 + 2$ $x = \frac{3}{4}; \left(\frac{3}{4}, \frac{25}{8}\right)$

41. $y = -1 - 8x + 12x^2$ $x = \frac{1}{3}; \left(\frac{1}{3}, -\frac{7}{3}\right)$

42. $x = \frac{6}{7}; \left(\frac{6}{7}, -\frac{22}{7}\right)$

46. $x = \frac{5}{2}; \left(\frac{5}{2}, \frac{25}{4}\right)$

42. $y = 7x^2 - 12x + 2$

43. $y = 2x - 2 + x^2$ $x = -1; (-1, -3)$

44. $y = 4x^2 - 3x - 8$ $x = \frac{3}{8}; \left(\frac{3}{8}, -\frac{137}{16}\right)$

45. $y = 9 - 3x^2$ $x = 0; (0, 9)$

46. $y = 5x - x^2$

47. $y = 5x^2 + 2x - 3$ $x = -\frac{1}{5}; \left(-\frac{1}{5}, -\frac{16}{5}\right)$

CHALLENGE

48. Prove that if the roots of $ax^2 + bx + c = 0$, where $a \neq 0$, are reciprocals, then $a = c$.

Assess

Selected Answers

Exercises 5–9, 11–73 odd

ASSIGNMENT GUIDE	
In Class	1–9
Core	11–57 odd
Core Plus	10–58 even, 59
Review	60–73
Preview	74

Extra Practice can be found beginning on page 940.

48. Proofs may vary. Sample: Suppose the two roots of $ax^2 + bx + c = 0$ are n and $\frac{1}{n}$, where $n \neq 0$.

$$ax^2 + bx + c$$
$$= (x - n)\left(x - \frac{1}{n}\right)$$
$$= x^2 - nx - \frac{1}{n}x + 1$$
$$= nx^2 - n^2x - x + n$$
$$= nx^2 - (n^2 + 1)x + n$$

Therefore, $a = n$ and $c = n$, so $a = c$.

Student Technology Guide

NAME ______________ CLASS ________ DATE ________

Student Technology Guide
5.5 The Quadratic Formula

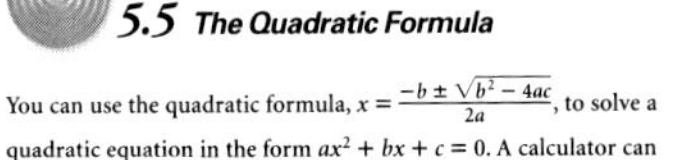

You can use the quadratic formula, $x = \frac{-b \pm \sqrt{b^2 - 4ac}}{2a}$, to solve a quadratic equation in the form $ax^2 + bx + c = 0$. A calculator can help you evaluate the formula.

Example: Use the quadratic formula to solve $10x + 4 = 4x^2$. Round answers to the nearest hundredth.

- Rewrite $10x + 4 = 4x^2$ in standard form by subtracting $4x^2$ from each side. This gives $-4x^2 + 10x + 4 = 0$.
- Identify a, b, and c: $a = -4$, $b = 10$, $c = 4$.
- Substituting a, b, and c into the quadratic formula, $x = \frac{-10 \pm \sqrt{(10)^2 - 4(-4)(4)}}{2(-4)}$.
- Simplify: $x = \frac{-10 \pm \sqrt{164}}{-8}$.
- Use a calculator to find $\frac{-10 + \sqrt{164}}{-8}$.

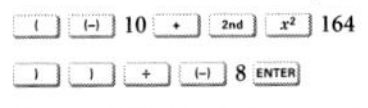

((−) 10 + 2nd x^2 164
)) ÷ (−) 8 ENTER

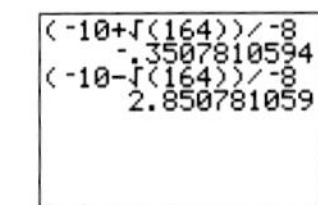

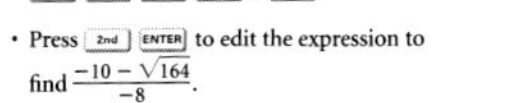

- Press 2nd ENTER to edit the expression to find $\frac{-10 - \sqrt{164}}{-8}$.

The solutions to $10x + 4 = 4x^2$ are approximately equal to -0.35 and 2.85.

Use the quadratic formula to solve each equation. Round answers to the nearest hundredth when necessary.

1. $x^2 - 5x + 4 = 0$ — 1, 4
2. $-x^2 + 5x + 9 = 0$ — −1.41, 6.41
3. $2x^2 = -3x + 5$ — −2.5, 1
4. $2x^2 - 11x + 3 = 0$ — 0.29, 5.21
5. $-x^2 + 1.25x = -9.75$ — −2.56, 3.81
6. $0.1x^2 - 0.17 = -0.2x$ — −2.64, 0.64
7. The first floor of a home has an area of 1500 square feet. The floor plan is rectangular, with a length 10 feet longer than the width. Find the length and width of the home. Round your answers to the nearest tenth of a foot.

w; $w + 10$

length ≈ 44.1 feet; width ≈ 34.1 feet

Error Analysis

Many students use the quadratic formula without putting the equation into standard form, which results in incorrect solutions. Remind them that before they can determine the values for a, b, and c, they must set the quadratic expression in the equation equal to 0. Additionally, some students will divide the first term in the numerator, $-b$, by $2a$ but fail to divide the value of the radical by $2a$. When reading the formula, be sure to stress that all terms in the numerator are to be divided by $2a$ by stating "negative b plus or minus the square root of the quantity b squared minus $4ac$ all over $2a$."

CONNECTION

MAXIMUM/MINIMUM A professional pyrotechnician shoots fireworks vertically into the air from the ground with an initial velocity of 192 feet per second. The height in feet of the fireworks is given by $h(t) = -16t^2 + 192t$.

49. How long does it take for the fireworks to reach the maximum height? **6 sec**

50. What is the maximum height reached by the fireworks? **576 ft**

51. Certain fireworks shoot sparks for 2.5 seconds, leaving a trail. After how many seconds might the pyrotechnician want the fireworks to begin firing in order for the sparks to be at the maximum height? Explain. **3.5 seconds**

APPLICATIONS

x

x

x

x

12 in.

MANUFACTURING To form a rectangular rain gutter from a flat sheet of aluminum that is 12 inches wide, an equal amount of aluminum, x, is bent up on each side, as shown at left. The area of the cross section is 18 square inches.

52. Write a quadratic equation to model the area of the cross section. $\mathbf{12x - 2x^2 = 18}$

53. Find the depth of the gutter. **3 in.**

BUSINESS The owner of a company that produces handcrafted music stands hires a consultant to help set the selling price for the product. The consultant analyzes the production costs and consumer demand for the stands and arrives at a function for the profit, $P(x) = -0.3x^2 + 75x - 2000$, where x represents the selling price of the stands.

54. $125 each

54. At what price should the stands be sold to earn the maximum profit?

55. According to the function given, what is the maximum profit that the company can make? **$2687.50**

56. What are the break-even points (the selling prices for which the profit is 0)? Give your answers to the nearest cent. **$30.35, $219.65**

57. For what values of x does the company make a profit? $\mathbf{30.35 < x < 219.65}$

58. For what values of x does the company suffer a loss? $\mathbf{0 \le x < 30.35}$ ***or*** $\mathbf{x > 219.65}$

CHALLENGE

59. ART In the art of many cultures, a ratio called the *golden ratio*, or *golden mean*, has been used as an organizing principle because it produces designs that are deemed naturally pleasing to the eye. The golden mean is based on the division of a line segment into two parts, a and b, such that the ratio of the longer segment, a, to the shorter segment, b, equals the ratio of the whole segment, $a + b$, to the longer segment, a.

a. Solve the golden ratio for a in terms of b.

b. Find the value of the golden ratio. Give the exact value and the approximate value to the nearest hundredth. (Hint: Divide each side of the equation from part **a** by b, and simplify.)

59a. $a = \frac{b \pm b\sqrt{5}}{2}$

b. ≈ 1.62

Look Back

Write an equation in slope-intercept form for the line that contains the given point and is perpendicular to the given line. ***(LESSON 1.3)***

60. $(-2, 3), y = x - 5$ **$y = -x + 1$**

61. $(4, -6), 2x - y = 1$ **$y = -\frac{1}{2}x - 4$**

Write an equation in slope-intercept form for the line that contains the given point and is parallel to the given line. ***(LESSON 1.3)***

62. $(8, -1), y = -3x + 12$ **$y = -3x + 23$**

63. $(-4, -2), 5x = 4 - y$ **$y = -5x - 22$**

Solve each absolute-value inequality. Graph the solution on a number line. ***(LESSON 1.8)***

64. $|x + 6| > 2$ ***$x < -8$ or $x > -4$***

65. $|x - 3| < 5$ ***$x > -2$ and $x < 8$***

66. $|-4x| \le 8$ ***$x \ge -2$ and $x \le 2$***

67. $|8 - 2x| \ge 6$ ***$x \le 1$ or $x \ge 7$***

Find the inverse of each matrix, if it exists. Round numbers to the nearest hundredth. Indicate if the matrix does not have an inverse. ***(LESSON 4.3)***

68 $\begin{bmatrix} -3 & 2 \\ 12 & 9 \end{bmatrix}$

69 $\begin{bmatrix} -1 & 8 \\ 4 & -7 \end{bmatrix}$

70 $\begin{bmatrix} 6 & -4 & 18 \\ 21 & -3 & 19 \\ 4 & 5 & -2 \end{bmatrix}$

68 $\begin{bmatrix} -0.18 & 0.04 \\ 0.24 & 0.06 \end{bmatrix}$

69 $\begin{bmatrix} 0.28 & 0.32 \\ 0.16 & 0.04 \end{bmatrix}$

70 $\begin{bmatrix} -0.08 & 0.07 & -0.02 \\ 0.11 & -0.08 & 0.24 \\ 0.11 & -0.04 & 0.06 \end{bmatrix}$

Solve each equation. ***(LESSON 5.2)***

71. $-2x^2 = -16$ **$\pm 2\sqrt{2}$**

72. $-3x^2 + 15 = -6$ **$\pm\sqrt{7}$**

73. $32 = 2x^2 - 4$ **$\pm 3\sqrt{2}$**

Look Beyond

74. Use the quadratic formula to solve $2x^2 + 5x + 6 = 0$. Can you find a real-number solution? Explain. **$\frac{-5 \pm \sqrt{-23}}{4}$; no, $\sqrt{-23}$ is not a real number.**

SPORTS Use the quadratic formula and the formula for the axis of symmetry of a parabola to answer the questions below based on the height function from the Portfolio Activity on page 280.

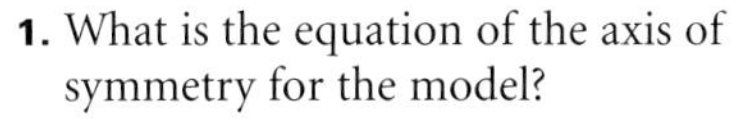

1. What is the equation of the axis of symmetry for the model?

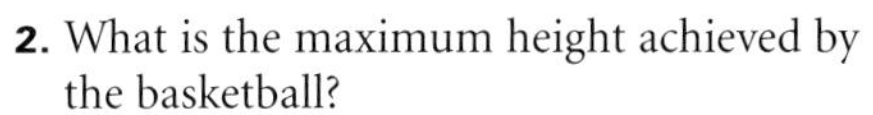

2. What is the maximum height achieved by the basketball?
3. How long does it take the basketball to reach its maximum height?
4. How much time would it take the basketball to drop to the ground if it were not tapped?
5. Compare your answers to the questions above with those obtained in the Portfolio Activities on pages 280, 298, and 306.

WORKING ON THE CHAPTER PROJECT

You should now be able to complete Activity 2 of the Chapter Project.

Look Beyond

In Exercise 74, the equation has no real solutions because the discriminant is negative: $b^2 - 4ac = -23$. Since there is no real number whose square is negative, there is no real number that equals $\sqrt{-23}$. In Lesson 5.6, students will learn about operations with complex numbers.

ALTERNATIVE
Assessment

Portfolio Activity

The Portfolio Activity can be used as preparation for the Chapter Project or as a separate activity. In the Portfolio Activity on this page, students use the equation of the axis of symmetry and the quadratic formula to answer questions about the height function and compare the procedures used to answer questions in this activity with the procedures used in Activities 1, 2, and 3 on pages 280, 298, and 306.

Answers to Portfolio Activities can be found in Additional Answers of the Teacher's Edition.

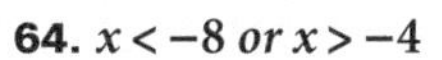

64. $x < -8$ *or* $x > -4$

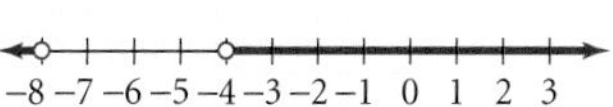

65. $x > -2$ *and* $x < 8$

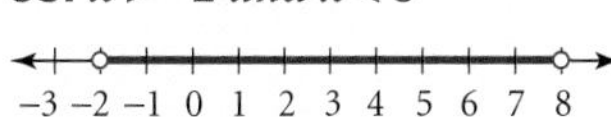

66. $x \ge -2$ *and* $x \le 2$

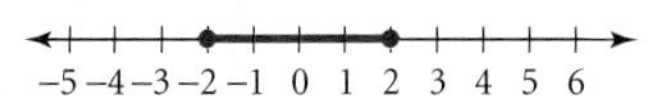

67. $x \le 1$ *or* $x \ge 7$

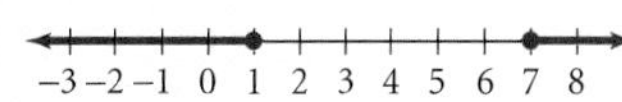

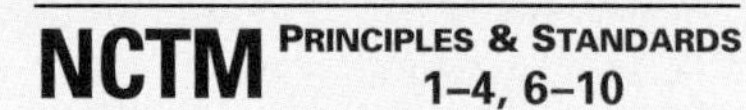

Prepare

NCTM PRINCIPLES & STANDARDS 1–4, 6–10

QUICK WARM-UP

Use the quadratic formula to solve each equation.

1. $x^2 + 12x + 35 = 0$
   $x = -7$ *or* $x = -5$
2. $x^2 + 81 = 18x$
   $x = 9$
3. $x^2 + 4x - 9 = 0$
   $x = -2 \pm \sqrt{13}$
4. $2x^2 = 5x + 9$
   $x = \frac{5 \pm \sqrt{97}}{4}$

Also on Quiz Transparency 5.6

Teach

Why With the the imaginary unit, *i*, you can find solutions to all quadratic equations. The concept of imaginary numbers will be new to most students. Ask them to discuss how they use the term *imaginary* and what they think it might mean as a mathematical concept.

Use Teaching Transparency 20.

Quadratic Equations and Complex Numbers

Objectives

- Classify and find all roots of a quadratic equation.
- Graph and perform operations on complex numbers.

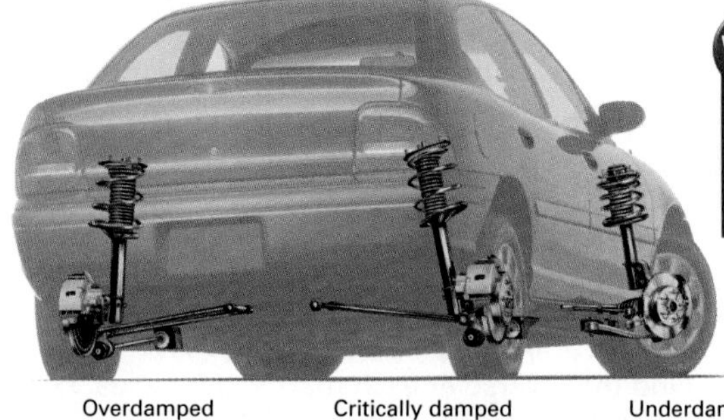

Why *Solutions to many real-world problems, such as classifying a shock absorber spring system, involve complex numbers.*

Overdamped	Critically damped	Underdamped
2 real roots	1 real root	0 real roots

Each diagram represents the motion of a shock absorber spring over a period of 2 seconds from left to right.

APPLICATION ENGINEERING

Carlos and Keiko are mechanical engineers who are analyzing car shock absorbers. They know that the motion of the spring is affected by a damping force. Three situations are possible depending on how the mass of the car, the spring, and the damping force are related. The roots of $x^2 + mx + n = 0$, where m and n depend on the car's mass, the spring, and the damping force, help them to classify a spring system. You will classify a spring system after Example 1.

The Discriminant

When you apply the quadratic formula to any quadratic equation, you will find that the value of $b^2 - 4ac$ is either positive, negative, or 0. The expression $b^2 - 4ac$ is called the **discriminant** of a quadratic equation.

You can see from the quadratic formula that if $b^2 - 4ac > 0$, the formula will give two different answers. If $b^2 - 4ac = 0$, there will be one answer, called a **double root**. If $b^2 - 4ac < 0$, the radical will be undefined for real numbers, so the formula gives no real solutions.

Solutions of a Quadratic Equation

Let $ax^2 + bx + c = 0$, where $a \neq 0$.

- If $b^2 - 4ac > 0$, then the quadratic equation has 2 distinct real solutions.
- If $b^2 - 4ac = 0$, then the equation has 1 real solution, a double root.
- If $b^2 - 4ac < 0$, then the equation has 0 real solutions.

Alternative Teaching Strategy

USING COGNITIVE STRATEGIES Tell students that the roots of quadratic equations include all types of numbers. Then have them complete a chart like the one below.

Equation	Number of roots	Type of root
$x^2 = 25$	2	rational
$x^2 = 21$	2	irrational
$x^2 = -21$	2	complex

Now ask students to solve $x^2 = -21$ by taking the square root of each side. Have students write the solutions as $x = \sqrt{-21}$ *or* $x = -\sqrt{-21}$. Tell them they need a new type of number to describe the solutions. Then define $i = \sqrt{-1}$ and show the solutions as $x = \sqrt{-21} = \sqrt{-1} \cdot \sqrt{21} = i\sqrt{21}$ *or* $x = -\sqrt{-21} = -(\sqrt{-1} \cdot \sqrt{21}) = -i\sqrt{21}$.

EXAMPLE 1 **Find the discriminant for each equation. Then determine the number of real solutions for each equation by using the discriminant.**

a. $2x^2 + 4x + 1 = 0$ **b.** $2x^2 + 4x + 2 = 0$ **c.** $2x^2 + 4x + 3 = 0$

SOLUTION

a. $2x^2 + 4x + 1 = 0$

$b^2 - 4ac$

$= 4^2 - 4(2)(1)$

$= 8$

Because $b^2 - 4ac > 0$, the equation has 2 real solutions.

b. $2x^2 + 4x + 2 = 0$

$b^2 - 4ac$

$= 4^2 - 4(2)(2)$

$= 0$

Because $b^2 - 4ac = 0$, the equation has 1 real solution.

c. $2x^2 + 4x + 3 = 0$

$b^2 - 4ac$

$= 4^2 - 4(2)(3)$

$= -8$

Because $b^2 - 4ac < 0$, the equation has no real solutions.

CHECK

Graph the related functions to check.

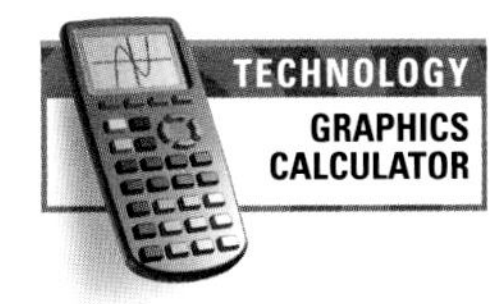

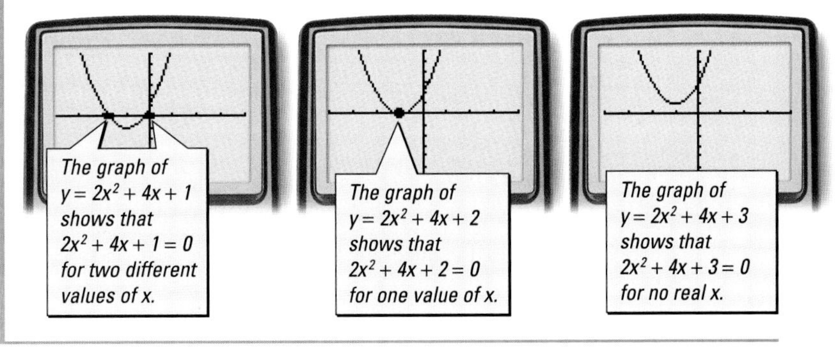

The graph of $y = 2x^2 + 4x + 1$ shows that $2x^2 + 4x + 1 = 0$ for two different values of x.
The graph of $y = 2x^2 + 4x + 2$ shows that $2x^2 + 4x + 2 = 0$ for one value of x.
The graph of $y = 2x^2 + 4x + 3$ shows that $2x^2 + 4x + 3 = 0$ for no real x.

TRY THIS Identify the number of real solutions to $-3x^2 - 6x + 15 = 0$.

CHECKPOINT ✔ Refer to the shock-absorber situation described on page 314. Classify a spring system in which $m = 8$ and $n = 24$.

Complex Numbers

Leonhard Euler

CULTURAL CONNECTION: EUROPE Whether it is possible to take the square root of a negative number is a question that puzzled mathematicians for a long time. In the sixteenth century, Italian mathematician Girolamo Cardano (1501–1576) was the first to use complex numbers to solve quadratic equations. Leonhard Euler (1707–1783) defined the **imaginary unit, *i***, such that $i = \sqrt{-1}$ and $i^2 = -1$. In the nineteenth century, the theoretical basis of complex numbers was rigorously developed.

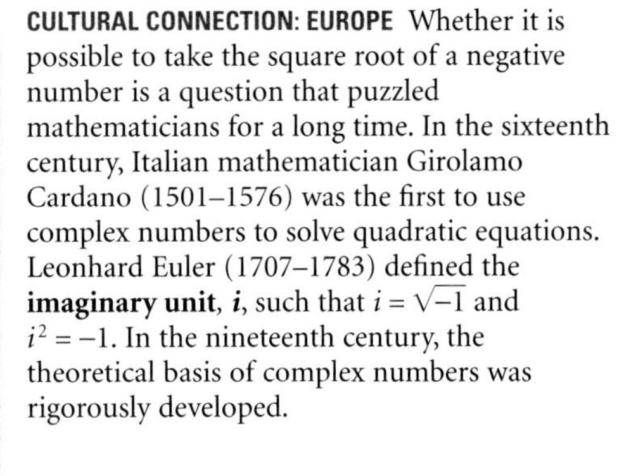

Girolamo Cardano

Interdisciplinary Connection

PHYSICS In a simplified electric circuit, the resistance to the flow of the current is called *impedance*, *Z*, and is often expressed as a complex number. In a parallel circuit, the total impedance, measured in ohms, is given by $Z_T = \frac{Z_1 Z_2}{Z_1 + Z_2}$, and in a series circuit, the total impedance is given by $Z_T = Z_1 + Z_2$, where Z_1 and Z_2 are the impedances of the individual circuits. Have students find the total impedance in a parallel circuit and a series circuit if $Z_1 = 5 + 2i$ ohms and $Z_2 = 5 - 2i$ ohms. **$\frac{29}{10}$ ohms in parallel and 10 ohms in series**

ADDITIONAL EXAMPLE 1

Find the discriminant for each equation. Then determine the number of real solutions for each equation by using the discriminant.

a. $3x^2 - 6x + 4 = 0$
$b^2 - 4ac = -12 < 0$; no real solutions

b. $3x^2 - 6x + 3 = 0$
$b^2 - 4ac = 0$; one real solution

c. $3x^2 - 6x + 2 = 0$
$b^2 - 4ac = 12 > 0$; two real solutions

Teaching Tip

TECHNOLOGY If you are using a TI-82 or TI-83 graphics calculator as a check for Example 1, set the viewing window as follows:

Xmin=–4	Ymin=–6
Xmax=4	Ymax=6
Xscl=1	Yscl=1

TRY THIS
$b^2 - 4ac = 216 > 0$; two real solutions

CHECKPOINT ✔
The system is underdamped because there are no real roots.

ADDITIONAL EXAMPLE 2

Use the quadratic formula to solve $6x^2 - 3x + 1 = 0$.
$x = \frac{3 \pm i\sqrt{15}}{12}$

TRY THIS
$x = \frac{5 \pm i\sqrt{23}}{8}$

ADDITIONAL EXAMPLE 3

Find x and y such that $-3x + 4iy = 21 - 16i$.
$x = -7$ and $y = -4$

TRY THIS
$x = -4$ and $y = \frac{10}{3}$

Imaginary Numbers

If $r > 0$, then the **imaginary number** $\sqrt{-r}$ is defined as follows:
$$\sqrt{-r} = \sqrt{-1} \cdot \sqrt{r} = i\sqrt{r}$$

For example, $\sqrt{-4} = \sqrt{-1} \cdot \sqrt{4} = 2i$ and $\sqrt{-6} = \sqrt{-1} \cdot \sqrt{6} = i\sqrt{6}$.

With the quadratic formula and the imaginary unit, i, you can find solutions to any quadratic equation. This is shown in Example 2.

EXAMPLE 2 **Use the quadratic formula to solve $3x^2 - 7x + 5 = 0$.**

SOLUTION

$$x = \frac{-(-7) \pm \sqrt{(-7)^2 - 4(3)(5)}}{2(3)}$$ *Substitute a = 3, b = −7, and c = 5.*

$$x = \frac{7 \pm \sqrt{-11}}{6}$$

$$x = \frac{7}{6} + \frac{\sqrt{-11}}{6} \quad \text{or} \quad x = \frac{7}{6} - \frac{\sqrt{-11}}{6}$$

$$x = \frac{7}{6} + \frac{i\sqrt{11}}{6} \qquad x = \frac{7}{6} - \frac{i\sqrt{11}}{6}$$ *Replace $\sqrt{-11}$ with $i\sqrt{11}$.*

The numbers $\frac{7}{6} + \frac{i\sqrt{11}}{6}$ and $\frac{7}{6} - \frac{i\sqrt{11}}{6}$ are complex numbers.

TRY THIS Use the quadratic formula to solve $-4x^2 + 5x - 3 = 0$.

Complex Numbers

A **complex number** is any number that can be written as $a + bi$, where a and b are real numbers and $i = \sqrt{-1}$; a is called the **real part** and b is called the **imaginary part**.

The form $a + bi$ is called the *standard form* of a complex number. Real numbers are complex numbers for which $b = 0$. A complex number is called a *pure* imaginary number if its real part, a, is 0.

EXAMPLE 3 **Find x and y such that $7x - 2iy = 14 + 6i$.**

SOLUTION

Two complex numbers are equal if their real parts are equal and their imaginary parts are equal.

Real parts	Imaginary parts
$7x = 14$	$-2y = 6$
$x = 2$	$y = -3$

Thus, $x = 2$ and $y = -3$.

TRY THIS Find x and y such that $2x + 3iy = -8 + 10i$.

Enrichment

Tell students the complex number $a + bi$ is expressed in matrix form as $\begin{bmatrix} a & b \\ -b & a \end{bmatrix}$. Operations can be performed on complex numbers by using matrix operations. Show them how to multiply the two complex numbers in Example 5 by using matrix multiplication.

$(2 + i)(-5 - 3i) = -7 - 11i$

$$\begin{bmatrix} 2 & 1 \\ -1 & 2 \end{bmatrix}\begin{bmatrix} -5 & -3 \\ 3 & -5 \end{bmatrix} = \begin{bmatrix} -7 & -11 \\ 11 & -7 \end{bmatrix}$$

Now have students divide the complex numbers in Example 6 by multiplying $\begin{bmatrix} 2 & 5 \\ -5 & 2 \end{bmatrix}\begin{bmatrix} 2 & -3 \\ 3 & 2 \end{bmatrix}^{-1}$.

Give students the following pairs of complex numbers to multiply and divide by using matrix operations:

1. $2 - 3i$ and $4 + i$ — **$11 - 10i, \frac{5}{17} - \frac{14}{17}i$**
2. $3i$ and $2 + i$ — **$-3 + 6i, \frac{3}{5} + \frac{6}{5}i$**
3. $3 + 2i$ and -3 — **$-9 - 6i, -1 - \frac{2}{3}i$**
4. $\frac{1}{2} - \frac{3}{4}i$ and $-\frac{1}{2} + \frac{1}{4}i$ — **$-\frac{1}{16} + \frac{1}{2}i, -\frac{7}{5} + \frac{4}{5}i$**

Operations With Complex Numbers

EXAMPLE 4 **Find each sum or difference.**

a. $(-3 + 5i) + (7 - 6i)$ **b.** $(-3 - 8i) - (-2 - 9i)$

SOLUTION

Add or subtract the corresponding real parts and imaginary parts.

a. $(-3 + 5i) + (7 - 6i)$
$= (-3 + 7) + (5i - 6i)$
$= 4 - 1i$
$= 4 - i$

b. $(-3 - 8i) - (-2 - 9i)$
$= (-3 + 2) - (8i - 9i)$
$= -1 + 1i$
$= -1 + i$

Two complex numbers whose real parts are opposites and whose imaginary parts are opposites are called *additive inverses.*

$$(4 + 3i) + (-4 - 3i) = 0 + 0i = 0$$

CHECKPOINT ✔ What is the additive inverse of $2i - 12$?

EXAMPLE 5 **Multiply $(2 + i)(-5 - 3i)$.**

SOLUTION

$(2 + i)(-5 - 3i) = 2(-5 - 3i) + i(-5 - 3i)$ *Apply the Distributive Property.*
$= -10 - 6i - 5i - 3i^2$
$= -10 - 11i - 3(-1)$ *Replace i^2 with -1.*
$= -7 - 11i$

TRY THIS Multiply $(6 - 4i)(5 - 4i)$.

Activity: Exploring Powers of i

You will need: no special materials

1. Copy and complete the table below. (Hint: Recall that $i^2 = -1$.)

i	$i^2 =$	$i^3 =$	$i^4 =$
$i^5 =$	$i^6 =$	$i^7 =$	$i^8 =$

2. Observe the patterns in the table above. Use your observations to complete the table below.

$i^9 =$	$i^{10} =$	$i^{11} =$	$i^{12} =$
$i^{13} =$	$i^{14} =$	$i^{15} =$	$i^{16} =$

CHECKPOINT ✔ 3. Describe the pattern that occurs in the powers of i. Explain how to use the pattern to evaluate i^{41}, i^{66}, i^{75}, and i^{100}.

Inclusion Strategies

VISUAL LEARNERS To help students see the relationship between complex, real, and imaginary numbers, have them draw a diagram similar to the one at right. Have students write an example of each type of number under each heading. Be sure they see that every real number is a complex number.

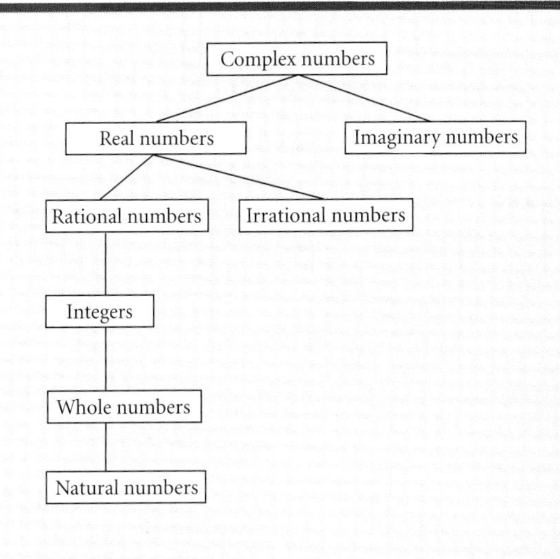

ADDITIONAL EXAMPLE 4

Find each sum or difference.

a. $(-10 - 6i) + (8 - i)$
$-2 - 7i$

b. $(-9 + 2i) - (3 - 4i)$
$-12 + 6i$

CHECKPOINT ✔
$-2i + 12$

ADDITIONAL EXAMPLE 5

Multiply: $(2 - i)(-3 - 4i)$.
$10 - 5i$

TRY THIS
$14 - 44i$

Activity Notes

In this Activity, students find several powers of i. They should see that all powers of i are equal to $1, i, -1,$ or $-i$, and that i^n is the sames as i^r, where r is the remainder when n is divided by 4. They are asked to use the pattern to evaluate four large powers of i.

CHECKPOINT ✔

3. The powers of i are cyclic and repeat in a pattern of four numbers: $i, -1, -i,$ and 1. To find i^n, divide n by 4 and match the remainder to one of the powers of i in the table below.

$i^1 = i$	$i^2 = -1$
$i^3 = -i$	$i^4 = i^0 = 1$

$i^{41} = i^{4(10)+1} = i^1 = i$
$i^{66} = i^{4(16)+2} = i^2 = -1$
$i^{75} = i^{4(18)+3} = i^3 = -i$
$i^{100} = i^{4(25)+0} = i^0 = 1$

ADDITIONAL
EXAMPLE 6

Simplify $\frac{3-2i}{-4+i}$. Write your answer in standard form.
$-\frac{14}{17}+\frac{5}{17}i$

Teaching Tip

TECHNOLOGY If you are using a TI-83 graphics calculator, you can enter i by pressing 2nd [.]. To divide the complex numbers in Example 6, enter **(2 + 5i)/(2 − 3i)** and press ENTER.

To show the result in fractional form, press MATH, select **1:>Frac**, and press ENTER.

TRY THIS
$\frac{2}{5}-\frac{11}{5}i$

CRITICAL THINKING
If $a + bi = a - bi$, then $a = a$ and $b = -b$. $b = -b$ is true only when $b = 0$. Thus, $a + bi = a - bi$ is true only when $a + bi = a + 0i = a$, a real number.

Math CONNECTION

COORDINATE GEOMETRY Students are shown the complex plane with the real and imaginary axes labeled. Complex numbers are plotted as points, where the first coordinate represents the real part and the second coordinate represents the imaginary part.

Remind students that the distance between two points $A(x_1, y_1)$ and $B(x_2, y_2)$ is given by the equation $AB = \sqrt{(x_2 - x_1)^2 + (y_2 - y_1)^2}$.

In order to simplify a fraction containing complex numbers, you often need to use the *conjugate of a complex number*. For example, the conjugate of $2 + 5i$ is $2 - 5i$ and the conjugate of $1 - 3i$ is $1 + 3i$.

Conjugate of a Complex Number

The **conjugate** of a complex number $a + bi$ is $a - bi$. The conjugate of $a + bi$ is denoted $\overline{a + bi}$.

To simplify a quotient with an imaginary number in the denominator, multiply by a fraction equal to 1, using the conjugate of the denominator, as shown in Example 6. This process is called **rationalizing the denominator.**

EXAMPLE 6 **Simplify $\frac{2+5i}{2-3i}$. Write your answer in standard form.**

SOLUTION

$$\frac{2+5i}{2-3i} = \frac{2+5i}{2-3i} \cdot \frac{2+3i}{2+3i}$$ *Multiply by 1, using the conjugate of the denominator.*

$$= \frac{(2+5i)(2+3i)}{(2-3i)(2+3i)}$$
$$= \frac{4+10i+6i+15i^2}{4-6i+6i-9i^2}$$
$$= \frac{-11+16i}{13}$$
$$= -\frac{11}{13}+\frac{16}{13}i$$

CHECK
In complex mode, enter the expression. Then express the answer with fractions.

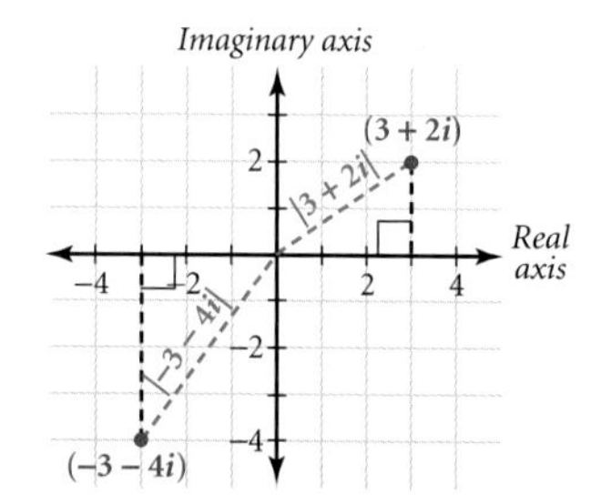

(2+5i)/(2-3i)
-.8461538462+1....
Ans▸Frac
-11/13+16/13i
Note that the last term is $\frac{16}{13}i$, not $\frac{16}{13i}$.

TRY THIS Simplify $\frac{3-4i}{2+i}$. Write your answer in standard form.

CRITICAL THINKING Show that if a complex number is equal to its conjugate, the number is real.

CONNECTION
COORDINATE GEOMETRY

Complex numbers are graphed in the *complex plane*. In the **complex plane**, the horizontal axis is called the **real axis** and the vertical axis is called the **imaginary axis.** To graph the complex number $a + bi$, plot the point (a, b). For example, the point (3, 2) represents the complex number $3 + 2i$ and the point (−3, −4) represents the complex number $-3 - 4i$.

Imaginary axis; Real axis; (3 + 2i); |3 + 2i|; |−3 − 4i|; (−3 − 4i)

The absolute value of a real number is its distance from zero on the number line. Likewise, the **absolute value of a complex number** $a + bi$, denoted by $|a + bi|$, is its distance from the origin in the complex plane. By the Pythagorean Theorem, $|a + bi| = \sqrt{a^2 + b^2}$. For example, $|3 + 2i| = \sqrt{3^2 + 2^2} = \sqrt{13}$ and $|-3 - 4i| = \sqrt{(-3)^2 + (-4)^2} = 5$.

Reteaching the Lesson

CONNECTING TO PRIOR KNOWLEDGE Ask students to find the sum $(3 + 2x) + (7 - 3x)$ and the sum $(3 + 2i) + (7 - 3i)$. **$10 - x$ and $10 - i$** Have them describe how the processes are similar. Be sure they notice they are collecting like terms when adding.

Now ask them to multiply $(3 + 2x)(7 - 3x)$ and $(3 + 2i)(7 - 3i)$. **$21 + 5x - 6x^2$ and $21 + 5i - 6i^2$** Have them describe how the processes are similar. Be sure they notice that the distributive property is used in both cases.

EXAMPLE 7 Evaluate $|-2-3i|$. Sketch a diagram that shows $-2-3i$ and $|-2-3i|$.

SOLUTION

$$|-2-3i| = \sqrt{(-2)^2+(-3)^2} = \sqrt{4+9} = \sqrt{13}$$

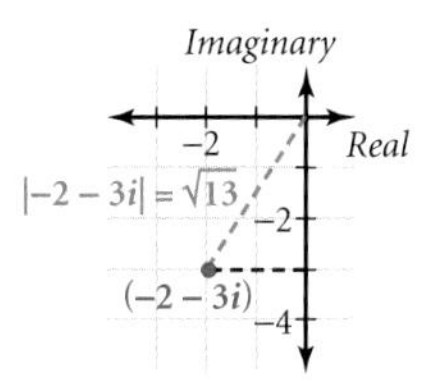

TRY THIS Evaluate $|-3+5i|$. Sketch a diagram that shows $-3+5i$ and $|-3+5i|$.

CRITICAL THINKING The real numbers are said to be "well-ordered," meaning that any real number is either larger or smaller than any other given real number. Are the complex numbers well-ordered? Explain.

ADDITIONAL EXAMPLE 7

Find $|3-5i|$. Sketch a right-triangle diagram showing $3-5i$ and $|3-5i|$.

$\sqrt{34}$

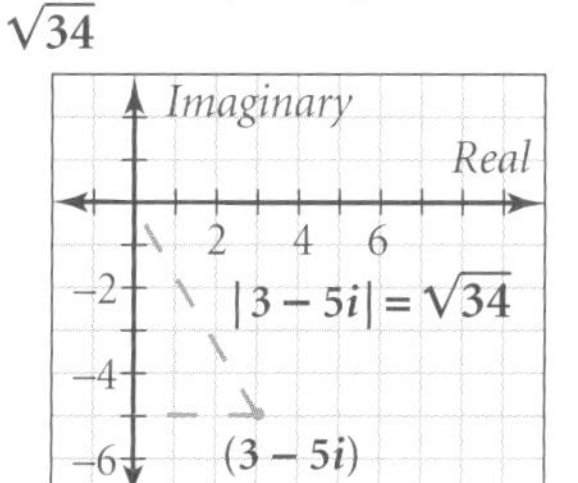

TRY THIS

$\sqrt{34}$

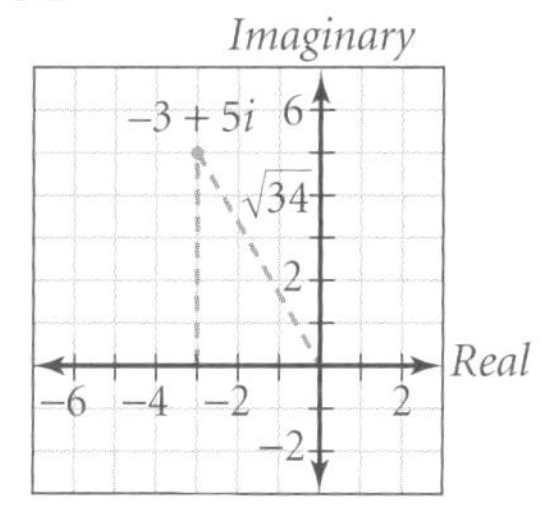

CRITICAL THINKING

Answers may vary. Sample answer: No, the complex numbers are not well-ordered because they cannot be represented on a number line.

Exercises

Communicate

1. What can the discriminant tell you about the solutions of a quadratic equation?
2. How do you simplify a rational expression that contains a complex number in the denominator?
3. How do you graph a real number in the complex plane? a pure imaginary number?

Guided Skills Practice

Determine the number of real solutions for each equation. *(EXAMPLE 1)*

4. $x^2+2x+1=0$ 1
5. $2x^2+4x+5=0$ 0
6. $x^2+3x+1=0$ 2
7. Solve the equation $2x^2+5x+4=0$. *(EXAMPLE 2)* $-\frac{5}{4} \pm \frac{i\sqrt{7}}{4}$
8. Find x and y such that $-2x+3yi=2+6i$. *(EXAMPLE 3)* $x=-1$, $y=2$

internet connect
Activities Online
Go To: **go.hrw.com**
Keyword: **MB1 Fractals**

Simplify each expression.

9. $(2+3i)+(4+7i)$ *(EXAMPLE 4)* $6+10i$
10. $(8+4i)-(3+2i)$ *(EXAMPLE 4)* $5+2i$
11. $(-1+2i)(3+4i)$ *(EXAMPLE 5)* $-11+2i$
12. $\frac{-1+4i}{2+3i}$ *(EXAMPLE 6)* $\frac{10}{13}+\frac{11}{13}i$
13. Evaluate $|3+4i|$. Sketch a diagram that shows $3+4i$ and $|3+4i|$. 5 *(EXAMPLE 7)*

13. 5

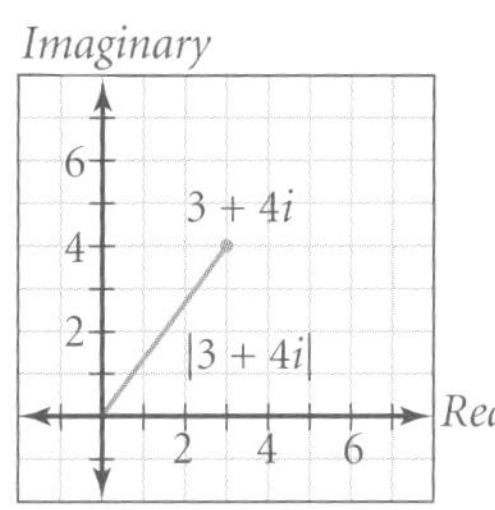

Assess

Selected Answers

Exercises 4–13, 15–107 odd

ASSIGNMENT GUIDE	
In Class	1–13
Core	15–87 odd, 97
Core Plus	4–100 even
Review	101–108
Preview	109–112

Extra Practice can be found beginning on page 940.

Error Analysis

Some students mistakenly think that the discriminant of $ax^2 + bx + c = 0$ is $\sqrt{b^2 - 4ac}$ instead of $b^2 - 4ac$. Remind students that the expression under the radical in the quadratic formula is the discriminant and that it lets them determine the type of solution they can expect from a quadratic equation without finding the actual roots.

76.

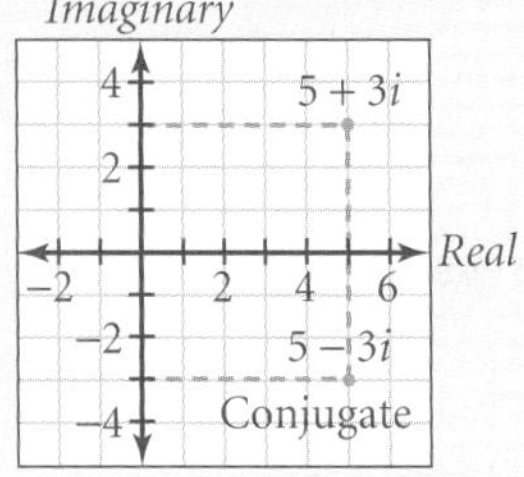

77.

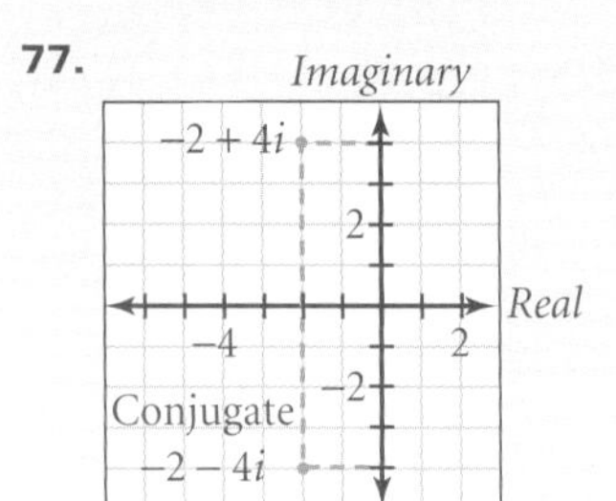

24. $-7; 0; -\frac{5}{2} \pm \frac{i\sqrt{7}}{2}$

25. $-23; 0; \frac{5}{6} \pm \frac{i\sqrt{23}}{6}$

27. $-15; 0; \frac{1}{2} \pm \frac{i\sqrt{15}}{10}$

28. $-12; 0; 4 \pm i\sqrt{3}$

29. $37; 2; \frac{3}{2} \pm \frac{\sqrt{37}}{2}$

internet connect

Homework Help Online
Go To: **go.hrw.com**
Keyword:
MB1 Homework Help
for Exercises 24–41

30. $-20; 0; \frac{5}{2} \pm i\frac{\sqrt{5}}{2}$

31. $-4; 0; \frac{3}{2} \pm \frac{i}{2}$

32. $33; 2; -\frac{1}{8} \pm \frac{\sqrt{33}}{8}$

33. $9; 2; 0, -\frac{3}{2}$

34. $-84; 0; -\frac{2}{5} \pm \frac{i\sqrt{21}}{5}$

35. $-12; 0; -\frac{1}{2} \pm \frac{i\sqrt{3}}{2}$

38. $-4; 0; 2 \pm i$

45. $-3 - i$

46. $-3i$

47. $1 + \frac{1}{5}i$

54. $-1 + 21i$

70. $\frac{6}{5} + \frac{12}{5}i$

Practice and Apply

Identify the real and imaginary parts of each complex number.

14. $-5 + 6i$ real: −5, imag.: 6 **15.** $2 + i$ real: 2, imag.: 1 **16.** 6 real: 6, imag.: 0 **17.** $4i$ real: 0, imag.: 4

Simplify.

18. $\sqrt{-36}$ $6i$ **19.** $\sqrt{-100}$ $10i$ **20.** $\sqrt{-13}$ $i\sqrt{13}$

21. $\sqrt{-17}$ $i\sqrt{17}$ **22.** $(-3i)^2$ -9 **23.** $(-7i)^2$ -49

Find the discriminant, and determine the number of real solutions. Then solve.

24. $x^2 + 5x + 8 = 0$ **25.** $3x^2 - 5x + 4 = 0$ **26.** $x^2 - 7x = -10$ $9; 2; 2, 5$

27. $5x^2 - 5x + 2 = 0$ **28.** $-x^2 + 8x - 19 = 0$ **29.** $x^2 - 3x = 7$

30. $-2x^2 + 10x = 15$ **31.** $2x^2 - 6x = -5$ **32.** $4x^2 + x - 2 = 0$

33. $2x^2 + 3x = 0$ **34.** $5x^2 + 4x = -5$ **35.** $2x^2 + 2x + 2 = 0$

36. $16 - 8x = -x^2$ $0; 1; 4$ **37.** $x^2 + 49 = 14x$ $0; 1; 7$ **38.** $-x^2 + 4x - 5 = 0$

39. $8x^2 + 5x + 2 = 0$ $-39; 0; -\frac{5}{16} \pm \frac{i\sqrt{39}}{16}$ **40.** $4x^2 + 9 = 12x$ $0; 1; \frac{3}{2}$ **41.** $1 + 9x^2 = 6x$ $0; 1; \frac{1}{3}$

Find *x* and *y*.

42. $6x + 7iy = 18 - 21i$ $x = 3, y = -3$ **43.** $3x - 4iy = 4 + 4i$ $x = \frac{4}{3}, y = -1$ **44.** $2x + 5i = 8 + 20yi$ $x = 4, y = \frac{1}{4}$

Perform the indicated addition or subtraction.

45. $(-2 + 3i) + (-1 - 4i)$ **46.** $(1 + 2i) - (1 + 5i)$ **47.** $\left(\frac{1}{2} + \frac{2}{5}i\right) + \left(\frac{1}{2} - \frac{1}{5}i\right)$

48. $\left(\frac{3}{8} + \frac{2}{3}i\right) - \left(\frac{1}{4} - \frac{1}{3}i\right)$ $\frac{1}{8} + i$ **49.** $(3 + i) + (6 - 2i)$ $9 - i$ **50.** $(8 - 6i) - (4 - 3i)$ $4 - 3i$

Multiply.

51. $i(2 + i)$ $-1 + 2i$ **52.** $(1 + i)(1 - i)$ 2 **53.** $(-5 - i)(-2 + 2i)$ $12 - 8i$

54. $(-5 + 3i)(2 - 3i)$ **55.** $(6 - 7i)^2$ $-13 - 84i$ **56.** $(2 - 4i)^2$ $-12 - 16i$

57. $(-1 + i\sqrt{5})^2$ $-4 - 2i\sqrt{5}$ **58.** $(2 + i\sqrt{3})^2$ $1 + 4i\sqrt{3}$ **59.** $(2 - 3i\sqrt{2})^2$ $-14 - 12i\sqrt{2}$

Write the conjugate of each complex number.

60. $2 + 3i$ $2 - 3i$ **61.** 8 8 **62.** $-4 - i$ $-4 + i$ **63.** $8 - 3i$ $8 + 3i$

Simplify.

64. $\frac{4 + 2i}{2 + i}$ 2 **65.** $\frac{3 + 2i}{5 + i}$ $\frac{17}{26} + \frac{7}{26}i$ **66.** $\left(\frac{1}{2} + i\right)\left(\frac{2}{3} + \frac{1}{4}i\right)$ $\frac{1}{12} + \frac{19}{24}i$

67. $\frac{4 + 2i}{\frac{2}{3} + \frac{1}{2}i}$ $\frac{132}{25} - \frac{24}{25}i$ **68.** $2(3 - 2i) + 5(1 + i)$ $11 + i$ **69.** $\frac{1}{4}(1 + i\sqrt{3})(1 - i\sqrt{3})$ 1

70. $(1 + i)^2 + \frac{2 + 2i}{2 + i}$ **71.** $\frac{3 + i}{4 + i} + 1 + i$ $\frac{30}{17} + \frac{18}{17}i$ **72.** $(-3i)(3i) - (2 + 2i)$ $7 - 2i$

73. $i(1 + i) - (3 + i)$ -4 **74.** $i^3(5 + i) + 7i$ $1 + 2i$ **75.** $\left(\frac{\sqrt{2}}{2} + \frac{\sqrt{2}}{2}i\right)^2$ i

Graph each number and its conjugate in the complex plane.

76. $5 + 3i$ **77.** $-2 + 4i$ **78.** $-3 - 5i$ **79.** $2 - 3i$

80. $-2i$ **81.** $6i$ **82.** $-2 + 7i$ **83.** $-7 - 2i$

84. -4 **85.** $5 + i$ **86.** $3 - 5i$ **87.** 3

78.

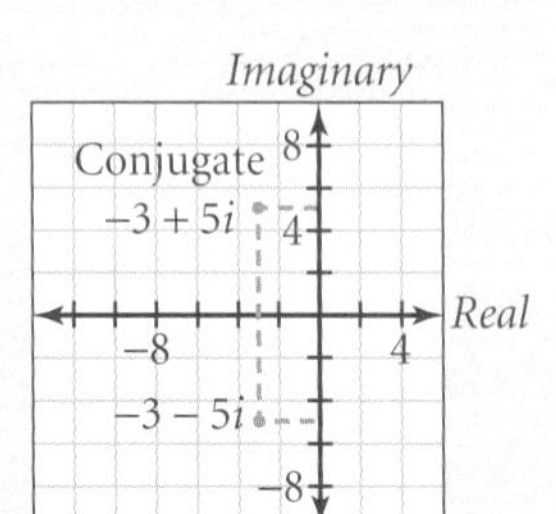

79.

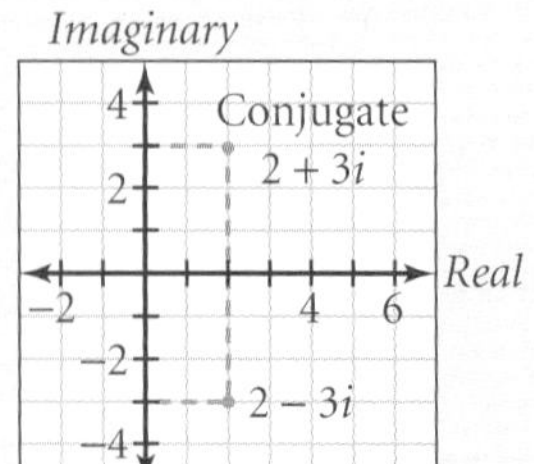

The answers to Exercises 80–87 can be found in Additional Answers beginning on page 1002.

Practice

NAME ________ CLASS ________ DATE ________

Practice 5.6 *Quadratic Equations and Complex Numbers*

Find the discriminant, and determine the number of real solutions. Then solve.

1. $-2x^2 + 5x - 3 = 0$ — 1; two; $x = 1.5$ or $x = 1$
2. $x^2 + 3x + 9$ — -27; none; $x = -\frac{3}{2} - \frac{3i\sqrt{3}}{2}$ or $x = -\frac{3}{2} + \frac{3i\sqrt{3}}{2}$
3. $3x^2 - 2x + 6 = 0$ — -68; none; $x = \frac{1}{3} - \frac{i\sqrt{17}}{3}$ or $x = \frac{1}{3} + \frac{i\sqrt{17}}{3}$
4. $4x^2 + 4 = x$ — -63; none; $x = \frac{1}{8} - \frac{3i\sqrt{7}}{8}$ or $x = \frac{1}{8} + \frac{3i\sqrt{7}}{8}$
5. $6x^2 - 3x + 4 = 0$ — -87; none; $x = \frac{1}{4} - \frac{i\sqrt{87}}{12}$ or $x = \frac{1}{4} + \frac{i\sqrt{87}}{12}$
6. $4x = 4x^2 + 7$ — -96; none; $x = \frac{1}{2} - \frac{i\sqrt{6}}{2}$ or $x = \frac{1}{2} + \frac{i\sqrt{6}}{2}$

Perform the indicated addition or subtraction.

7. $(-6 + 12i) + (4 - i)$ — $-2 + 11i$
8. $(3 - i) - (-4 + 9i)$ — $7 - 10i$
9. $(11 + i) + (2 + 8i)$ — $13 + 9i$
10. $(-8 + 4i) + (7 - i)$ — $-1 + 3i$
11. $(-8 + 4i) - (7 - i)$ — $-15 + 5i$
12. $(12 + 16i) - (12 + 11i)$ — $5i$
13. $(-7 - 2i) + (3 + 3i)$ — $-4 + i$
14. $(4 - 12i) + 7i$ — $4 - 5i$
15. $(-7 + 13i) - (1 + 6i)$ — $-8 + 7i$

Write the conjugate of each complex number.

16. $15i$ — $-15i$
17. $27 - 4i$ — $27 + 4i$
18. $-12 + 19i$ — $-12 - 19i$

Simplify.

19. $4i(-7 + i)$ — $-4 - 28i$
20. $\frac{5 - i}{5 + i}$ — $\frac{12}{13} - \frac{5i}{13}$
21. $(3 - i)(9 + 3i)$ — 30
22. $\frac{14 - 2i}{3 + i}$ — $4 - 2i$
23. $(2 + 5i)^2$ — $-21 + 20i$
24. $(6 - 3i)(2 + 2i)$ — $18 + 6i$
25. $\frac{5 - 4i}{i}$ — $-4 - 5i$
26. $\frac{-2 + 3i}{-3 - 2i}$ — $-i$

91. $\sqrt{1.0001}$

96. $3 - i, -2 + 4i, -3 - 3i$

97. $(3, 1), (-2, -4), (-3, 3)$

Evaluate. Then sketch a diagram that shows the absolute value.

88. $|1 + i|$ $\sqrt{2}$ **89.** $|i|$ 1 **90.** $\left|\frac{3}{5} + \frac{4}{5}i\right|$ 1 **91.** $|1 + 0.01i|$

92. $|2i|$ 2 **93.** $|2 + 3i|$ $\sqrt{13}$ **94.** $\left|\frac{1}{\sqrt{2}} + \frac{1}{\sqrt{2}}i\right|$ 1 **95.** $\left|\frac{1}{\sqrt{3}} + \frac{\sqrt{2}}{\sqrt{3}}i\right|$ 1

96. Identify the complex numbers graphed on the complex plane at right.

97. Identify the coordinates of the conjugate of each complex number graphed at right.

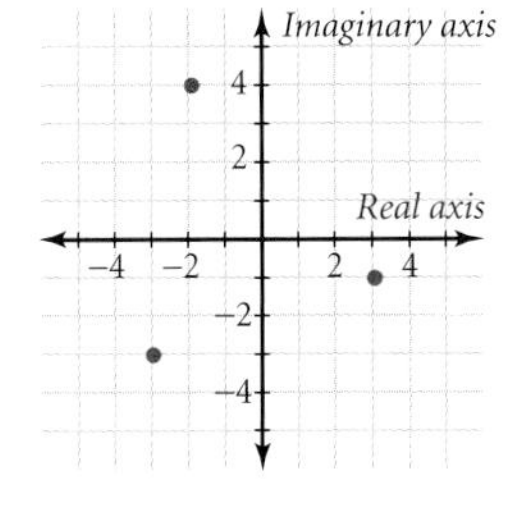

CHALLENGES

98. If $\overline{c + di} = -(c + di)$, what can you say about the complex number $c + di$? (Hint: Solve for c.) **$c = 0$; $c + di$ must be pure imaginary.**

99. Let $a > 0$ and $b > 0$. Plot $a + bi$, $a - bi$, $-(a + bi)$, and $-a + bi$ in the complex plane. Connect the points with vertical and horizontal lines. What is the result? **rectangle**

CONNECTION

100. TRANSFORMATIONS Describe the relationship between the graph of a complex number and the graph of its conjugate as a translation, rotation, or reflection. **The graph of a complex number and the graph of its conjugate are reflections of each other across the horizontal, or *real*, axis.**

Look Back

Graph each function and its inverse on the same coordinate plane. ***(LESSON 2.5)***

101. $f(x) = 2x - 3$ **102.** $f(x) = -3x + 5$ **103.** $f(x) = 2x$

Let $f(x) = 3x - 5$. Find the indicated function. ***(LESSONS 2.4 AND 2.5)***

104. $f^{-1}(x) = \frac{1}{3}x + \frac{5}{3}$ **105.** $(f \circ f^{-1})(x) = x$ **106.** $(f^{-1} \circ f)(x) = x$

APPLICATIONS

107. BUSINESS Gene ordered 8 prints for his new restaurant. Each unframed print cost \$50, and each framed print cost \$98. The total cost of the prints was \$640. How many framed prints did Gene buy? **5** ***(LESSONS 3.1 AND 3.2)***

108. RECYCLING A recycling center pays \$1.25 for 100 pounds of newspaper and \$0.40 for 1 pound of aluminum. Linda took a 460-pound load of aluminum and newspaper to the recycling center and was paid \$21.82. How many pounds of aluminum and how many pounds of newspaper did Linda have? ***(LESSONS 3.1 AND 3.2)***
418.5 pounds of newspaper, 41.5 pounds of aluminum

Look Beyond

109. 2, −3, −5
110. 0, 6, −1
111. 0, −2, 1
112. 0, −1

Predict the zeros and verify by graphing.

109 $f(x) = (x - 2)(x + 3)(x + 5)$ **110** $f(x) = x(x - 6)(x + 1)$

111 $f(x) = x^3 + x^2 - 2x$ **112** $f(x) = x^3 + 2x^2 + x$

90. *Imaginary*

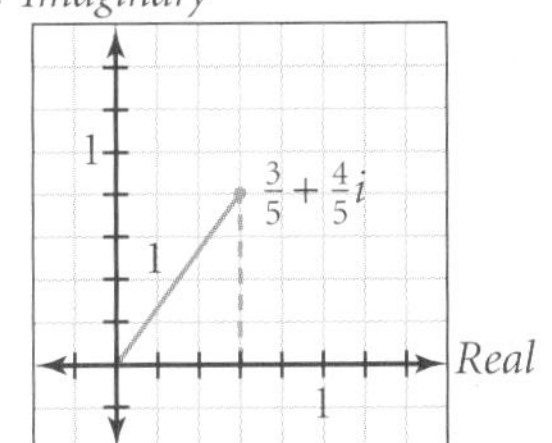

91. *Imaginary*

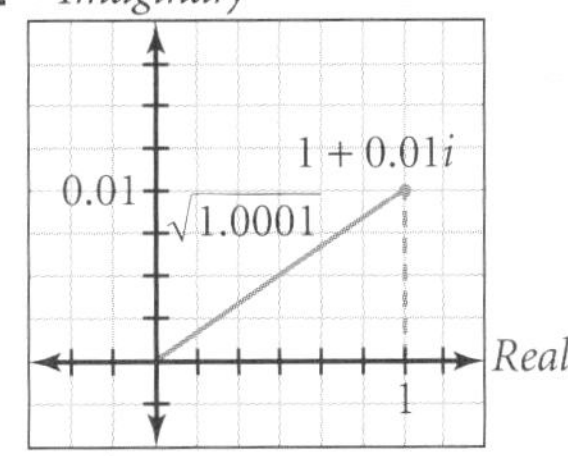

The answers to Exercises 92–95 and 101–103 can be found in Additional Answers beginning on page 1002.

Look Beyond

Exercises 109 and 110 introduce students to polynomial functions expressed as a product of linear factors. Exercises 111 and 112 represent a polynomial function written in expanded form. In Chapter 7, students will convert the expanded form of the polynomial to its factored form in order to obtain the roots of a polynomial.

88. *Imaginary*

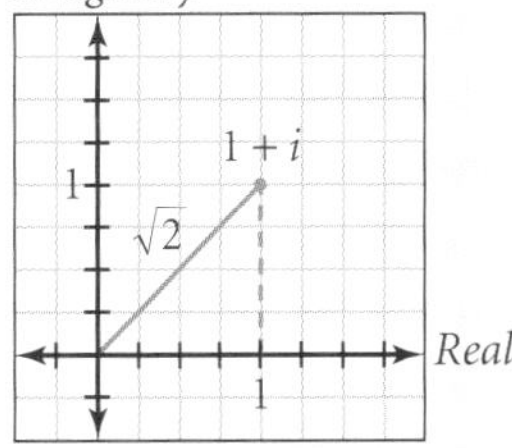

89. *Imaginary*

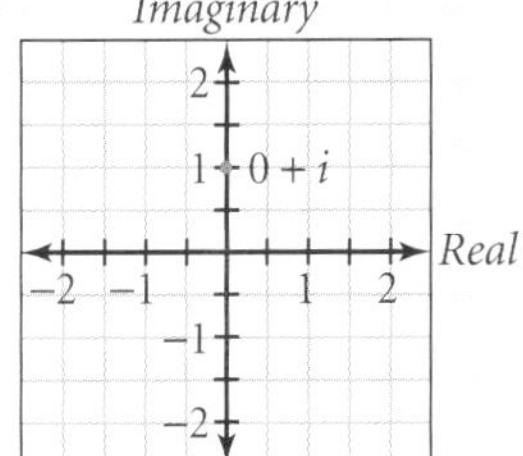

Student Technology Guide

NAME ______ CLASS ______ DATE ______

Student Technology Guide
5.6 Quadratic Equations and Complex Numbers

Many graphics calculators have a mode for complex numbers. This mode will allow you to simplify the square root of a negative number and perform complex-number arithmetic.

Example: Simplify $\sqrt{-27}$.

- Put the calculator into complex-number mode by pressing MODE a+bi ENTER. You can see this in the display at left below.
- Press 2nd MODE to exit the mode menu.
- Enter the square root you want to evaluate. Press 2nd x^2 (−) 27) ENTER.

√(-27)
5.196152423i

Thus, $\sqrt{-27} \approx 5.20i$.

Simplify. Round answers to the nearest hundredth when necessary.

1. $\sqrt{-16}$ $4i$ 2. $\sqrt{-47}$ $6.86i$ 3. $\sqrt{-135}$ $11.62i$ 4. $\sqrt{-2.89}$ $1.7i$

Example: Find the sum, difference, product, and quotient of $2 + 3i$ and $4 - 2i$.

- To find the sum, press (2 + 3 2nd .) + (4 − 2 2nd .) ENTER. Notice that the i is above the . key.
- To find the difference, product, and quotient, press 2nd ENTER and edit the expression as appropriate. Notice that when finding the quotient, pressing MATH 1:▶Frac ENTER ENTER after the expression displays the result in standard form.

```
(2+3i)+(4-2i)
            6+i
(2+3i)-(4-2i)
           -2+5i
(2+3i)*(4-2i)
           14+8i
(2+3i)/(4-2i)▶Fr
ac
       1/10+4/5i
```

Find the sum, difference, product, and quotient of each pair of complex numbers.

5. $8 + 2i$ and $-6 + 4i$
sum: $2 + 6i$; difference: $14 - 2i$;
product: $-56 + 20i$; quotient: $-\frac{10}{13} - \frac{11}{13}i$

6. $2 - 5i$ and $-6 - i$
sum: $-4 - 6i$; difference: $-8 - 4i$;
product: $-17 + 28i$; quotient: $-\frac{7}{37} - \frac{32}{37}i$

Prepare

NCTM PRINCIPLES & STANDARDS 1–10

QUICK WARM-UP

1. Write the system of equations represented by $\begin{bmatrix} 2 & 2 \\ 1 & -3 \end{bmatrix}\begin{bmatrix} x \\ y \end{bmatrix} = \begin{bmatrix} 16 \\ -4 \end{bmatrix}$.
$\begin{cases} 2x + 2y = 16 \\ x - 3y = -4 \end{cases}$
2. Write the matrix equation that represents $\begin{cases} x + y = 6 \\ 2x = 4 \end{cases}$.
$\begin{bmatrix} 1 & 1 \\ 2 & 0 \end{bmatrix}\begin{bmatrix} x \\ y \end{bmatrix} = \begin{bmatrix} 6 \\ 4 \end{bmatrix}$
3. Find the inverse of $\begin{bmatrix} -3 & 1 \\ 4 & -2 \end{bmatrix}$.
$\begin{bmatrix} -1 & -0.5 \\ -2 & -1.5 \end{bmatrix}$
4. Solve $\begin{cases} x = 4 \\ -x + y = 5 \end{cases}$ by using a matrix inverse.
$x = 4, y = 9$

Also on Quiz Transparency 5.7

Teach

Why Ask students to list objects or situations that could be modeled by a parabola. Examples may include the cables of a suspension bridge, the arc of water in a fountain, or the path of an arrow. Ask them how they would find the equation that represents the object or situation.

5.7 Curve Fitting With Quadratic Models

Objectives

- Find a quadratic function that exactly fits three data points.
- Find a quadratic model to represent a data set.

Why *You can model real-world situations, such as the total stopping distance of a car, by using a process called curve fitting to find an appropriate model.*

Traffic speeds are reduced in school zones for improved safety of students.

APPLICATION
HIGHWAY SAFETY

Fitting a set of data with a quadratic model is an example of *curve fitting*. In Lesson 5.1, you examined a quadratic model for the total stopping distance, d, in feet as a function of a car's speed, x, in miles per hour.

$$d(x) = \frac{11}{10}x + \frac{1}{19}x^2$$

This function models actual data that is provided in Lesson 2.4. Another set of data for speed and stopping distance is shown at right. How can you find a quadratic model for this data? *You will answer this question in Example 3.*

Speed (mph)	Stopping distance (ft)
10	12.5
20	36.0
30	69.5
40	114.0
50	169.5
60	249.0
70	325.5

Some graphics calculators can fit a parabola to three noncollinear points. Example 1 illustrates another method that you can use.

EXAMPLE 1 **Find a quadratic function whose graph contains the points (1, 3), (2, –3), and (6, 13).**

SOLUTION

1. To find a, b, and c in $f(x) = ax^2 + bx + c$, write and solve a system of three linear equations in three variables, a, b, and c.

Point	Substitution	Equation
(1, 3)	$a(1)^2 + b(1) + c = 3$	$a + b + c = 3$
(2, –3)	$a(2)^2 + b(2) + c = -3$	$4a + 2b + c = -3$
(6, 13)	$a(6)^2 + b(6) + c = 13$	$36a + 6b + c = 13$

$$\begin{cases} a + b + c = 3 \\ 4a + 2b + c = -3 \\ 36a + 6b + c = 13 \end{cases}$$

Alternative Teaching Strategy

HANDS-ON STRATEGIES Give students a set of several points that can be modeled by a quadratic function. The data should include points that are close to but do not lie on a unique parabola. Have them plot the points on a coordinate plane and draw a "parabola of best fit" through the points. Tell them that three noncollinear points determine a unique parabola. Have them estimate the equation that models the curve they drew by picking three points from the data and finding the coefficients for the equation by solving the related 3×3 system. Then have them use the statistics feature of a graphics calculator to find a quadratic regression equation that fits the set of points given. They should graph their equation along with the quadratic regression equation and compare their estimated values with the regression values. If their calculator permits, have them view the table of values for both equations.

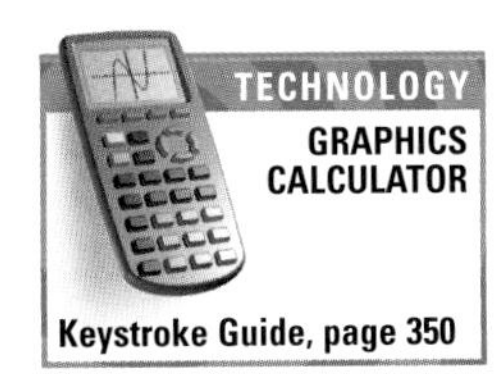

2. Use a matrix equation to solve the system.

$$\begin{bmatrix} 1 & 1 & 1 \\ 4 & 2 & 1 \\ 36 & 6 & 1 \end{bmatrix}\begin{bmatrix} a \\ b \\ c \end{bmatrix} = \begin{bmatrix} 3 \\ -3 \\ 13 \end{bmatrix} \rightarrow \begin{bmatrix} a \\ b \\ c \end{bmatrix} = \begin{bmatrix} 2 \\ -12 \\ 13 \end{bmatrix}$$

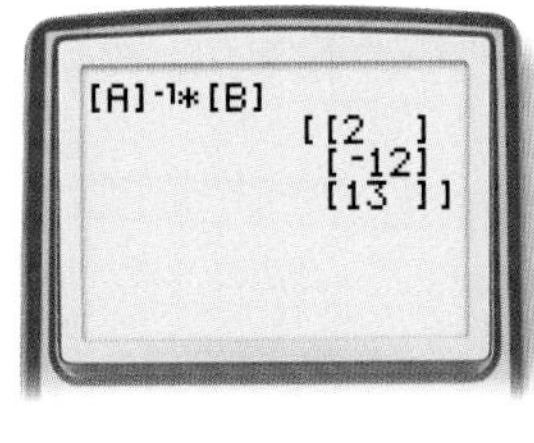

The solution is $a = 2$, $b = -12$, and $c = 13$.

3. Write the quadratic function.

$f(x) = ax^2 + bx + c$
$f(x) = 2x^2 - 12x + 13$

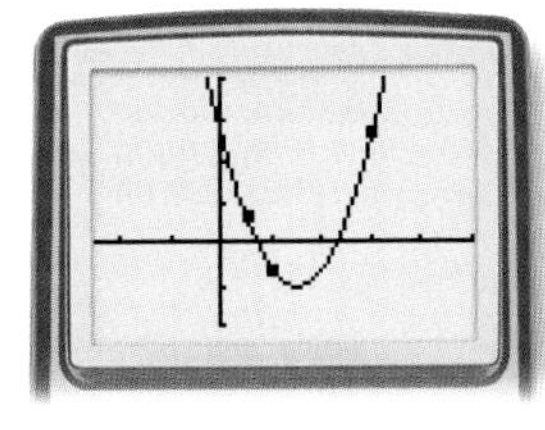

CHECK
Create a scatter plot of the three data points.
Graph $f(x) = 2x^2 - 12x + 13$ on the same screen.
The graph should contain the data points.

TRY THIS Find a quadratic function whose graph contains the points $(2, -3)$, $(4, 3)$, and $(6, 1)$.

When variables in a table represent a quadratic relationship, a constant difference in the consecutive x-values results in a constant *second difference* in the respective y-values. This fact is used in Example 2 to help determine whether a quadratic model is appropriate for a set of data.

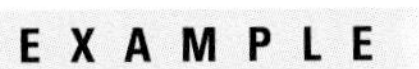

EXAMPLE 2 **Refer to the pattern of dots below, in which each set of dots except the first is formed by adding a bottom row containing 1 more dot than the previous bottom row.**

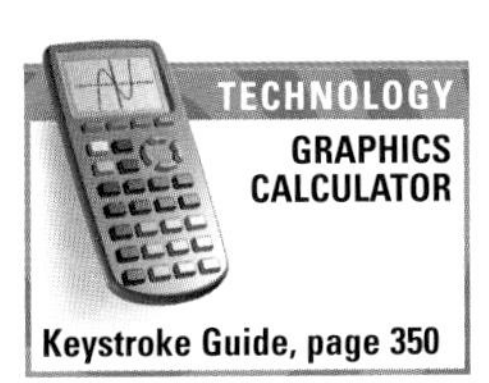

$d = 1$, $n = 1$; $d = 3$, $n = 2$; $d = 6$, $n = 3$; $d = 10$, $n = 4$; $d = 15$, $n = 5$; and so on...

a. Explain why a quadratic model is suitable for relating the number of dots, d, in the triangle to the number of dots, n, on each side of the triangle.
b. Find a quadratic function for d in terms of n.
c. Use the model to predict the number of dots in a triangle with 10 dots on each side.

SOLUTION

a. Make a scatter plot of points, (n, d). Use $(1, 1)$, $(2, 3)$, $(3, 6)$, $(4, 10)$, and $(5, 15)$. It appears that a curve rather than a straight line will fit the data.

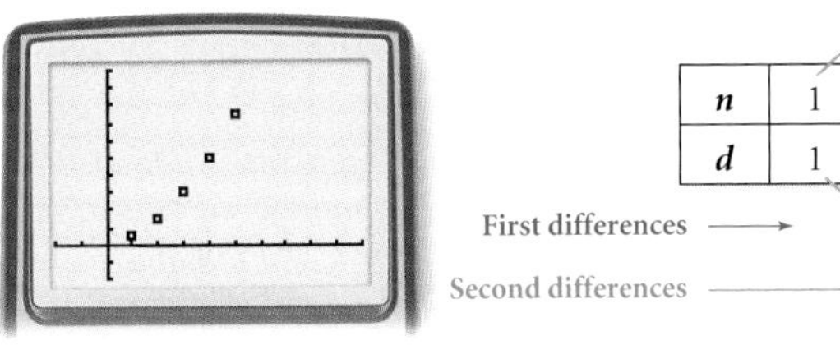

n	1	2	3	4	5
d	1	3	6	10	15

(differences in n: 1, 1, 1, 1)
First differences → 2, 3, 4, 5
Second differences → 1, 1, 1

PROBLEM SOLVING

Look for a pattern. Because the second differences are constant, there will be a quadratic equation that models the data exactly.

ADDITIONAL EXAMPLE 1

Find a quadratic function whose graph contains the points $(-3, 16)$, $(2, 6)$, and $(1, -4)$. $f(x) = 3x^2 + x - 8$

TRY THIS
$f(x) = -x^2 + 9x - 17$

ADDITIONAL EXAMPLE 2

Refer to the oblong numbers represented by the set of dot diagrams below.

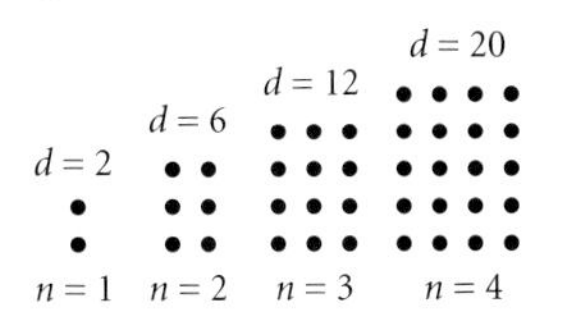

a. Explain why a quadratic model is suitable for relating the number of dots, d, in each rectangle to the number of dots, n, on the shorter side of the rectangle.
From a scatter plot of the points $(1, 2)$, $(2, 6)$, $(3, 12)$, and $(4, 20)$, it appears that a curve rather than a straight line will fit the data.

b. Find a quadratic function for d in terms of n.
$d(n) = n^2 + n$

c. Use the model to predict the number of dots in a rectangle with 15 dots on its shorter side. 240

Interdisciplinary Connection

BUSINESS The owner of a small business decided to find a relationship between the selling price, x, and the number of items sold, $n(x)$, and between the selling price, x, and the income, $I(x)$. Give students the table below. Have them make a scatter plot of the points $(x, n(x))$. Ask them to explain why a linear function is a suitable model for this data. Now have them make a scatter plot of the points $(x, I(x))$. Ask them to explain why a quadratic function is a suitable model for this data.

x	1	2	3	4	5
$n(x)$	90	80	70	60	50
$I(x)$	90	160	210	240	250

Enrichment

Ask students to find a quadratic function whose graph contains the points $(1, 3)$ and $(2, -3)$. Students should see that many quadratic functions include these points. Now have them find a quadratic function whose graph contain the two points above and $(0, -1)$. $y = -5x^2 + 9x - 1$ They should notice that there is only one parabola that contains all three points. Now require a fourth point, such as $(3, -5)$, to be on the graph. They should conclude that no parabola contains these four points.

Math CONNECTION

STATISTICS Scatter plots allow data to be plotted and viewed in order to determine which type of regression is appropriate to model the data.

Teaching Tip

Be sure students examine the tabular data and the shape of the plotted data. Point out that quadratic functions are characterized by constant second differences for equally spaced inputs.

Teaching Tip

TECHNOLOGY When using a TI-83 or TI-82 graphics calculator, press STAT, choose **1:Edit...**, and enter the x-values for **L1** and the corresponding y-values for **L2.** To find a quadratic regression model for the data, press STAT and select **CALC** **5:QuadReg**. **QuadReg** is then displayed on home screen. Press ENTER to display the values for a, b, and c.

When you have the data in **L1** and **L2,** you do not need to enter the names of the lists. But if the data are in other lists, enter the list names, separated by a comma, after **QuadReg.** For example, if the data are in **L1** and **L3,** after pressing STAT, selecting **CALC**, and choosing **5:QuadReg**, enter **L1, L3** by pressing 2nd 1 , 2nd 3. Then press ENTER.

CRITICAL THINKING

Three noncollinear points allow the equation to be solved for the three variables—a, b, and c.

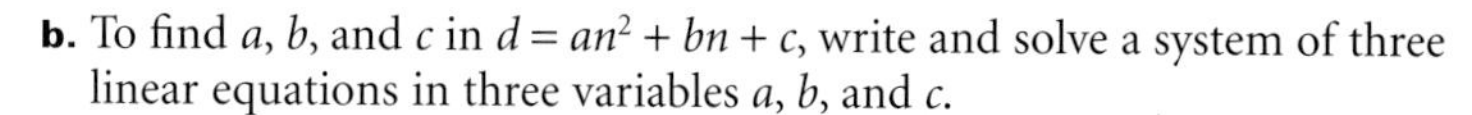

b. To find a, b, and c in $d = an^2 + bn + c$, write and solve a system of three linear equations in three variables a, b, and c.

Point	Evaluation	Equation
(1, 1)	$a(1)^2 + b(1) + c = 1$	$a + b + c = 1$
(2, 3)	$a(2)^2 + b(2) + c = 3$	$4a + 2b + c = 3$
(3, 6)	$a(3)^2 + b(3) + c = 6$	$9a + 3b + c = 6$

$$\begin{cases} a + b + c = 1 \\ 4a + 2b + c = 3 \\ 9a + 3b + c = 6 \end{cases}$$

TECHNOLOGY
GRAPHICS CALCULATOR
Keystroke Guide, page 351

Use a matrix equation to solve the system.

$$\begin{bmatrix} 1 & 1 & 1 \\ 4 & 2 & 1 \\ 9 & 3 & 1 \end{bmatrix}\begin{bmatrix} a \\ b \\ c \end{bmatrix} = \begin{bmatrix} 1 \\ 3 \\ 6 \end{bmatrix} \rightarrow \begin{bmatrix} a \\ b \\ c \end{bmatrix} = \begin{bmatrix} 0.5 \\ 0.5 \\ 0 \end{bmatrix}$$

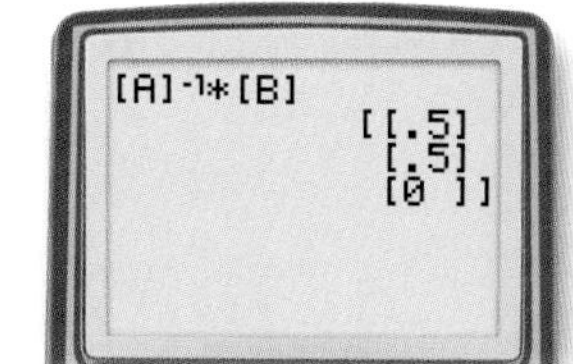

The solution is $a = 0.5$, $b = 0.5$, and $c = 0$.

Write the quadratic function.

$$d(n) = 0.5n^2 + 0.5n$$

CHECK

Check that the fourth and fifth data points satisfy $d(n) = 0.5n^2 + 0.5n$.

$$d(n) = 0.5n^2 + 0.5n \qquad d(n) = 0.5n^2 + 0.5n$$
$$10 \stackrel{?}{=} 0.5(4)^2 + 0.5(4) \qquad 15 \stackrel{?}{=} 0.5(5)^2 + 0.5(5)$$
$$10 = 8 + 2 \quad \textbf{True} \qquad 15 = 12.5 + 2.5 \quad \textbf{True}$$

Thus, $d(n) = 0.5n^2 + 0.5n$ models the pattern of dots.

c. Evaluate d for $n = 10$.

$$d(n) = 0.5n^2 + 0.5n$$
$$d(10) = 0.5(10)^2 + 0.5(10)$$
$$= 55$$

Thus, a triangle with 10 dots on each side would contain a total of 55 dots.

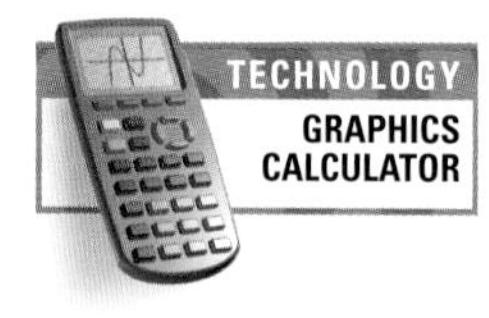

In Example 1, part **b**, you can also use a statistical model to find a quadratic function for d in terms of n. Enter the data, and use the keystrokes below to find a quadratic regression model for the dot pattern.

The quadratic regression equation is also $d(n) = 0.5n^2 + 0.5n$. Because the second differences for the number of dots in the triangle pattern are constant, this quadratic model fits the data exactly.

CRITICAL THINKING Explain why you can always find a quadratic function to fit any three noncollinear points in the coordinate plane.

Inclusion Strategies

ENGLISH LANGUAGE DEVELOPMENT Have students describe the vertical motion model $h(t) = -\frac{1}{2}gt^2 + v_0t + h_0$ for specific values of g, v_0, and h_0. Remind them that in the metric system, acceleration due to gravity is 9.8 meters per second squared, while the corresponding acceleration in the English system is 32 feet per second squared. Have them describe the different motions represented by positive and negative values of v_0 and describe situations in which h_0 is positive and h_0 is negative.

$v_0 < 0$ implies that the initial velocity of an object was directed downward; $v_0 > 0$ implies that the initial velocity of an object was directed upward; $h_0 < 0$ implies that the initial height of the object was below ground level; $h_0 > 0$ implies that the initial height of the object was above ground level.

Modeling Real-World Data

Real-world data typically do not conform perfectly to a particular mathematical model. Just as you can use a linear function to represent data that show a linear pattern, you can use a quadratic function to model data that follow a parabolic pattern. In Example 3, you find a quadratic regression model to represent real-world data.

EXAMPLE 3 **Make a scatter plot of the data below. Find a quadratic model to represent this data.**

APPLICATION
HIGHWAY SAFETY

Speed (mph)	Stopping distance (ft)
10	12.5
20	36.0
30	69.5
40	114.0
50	169.5
60	249.0
70	325.5

SOLUTION

The scatter plot is shown at right.

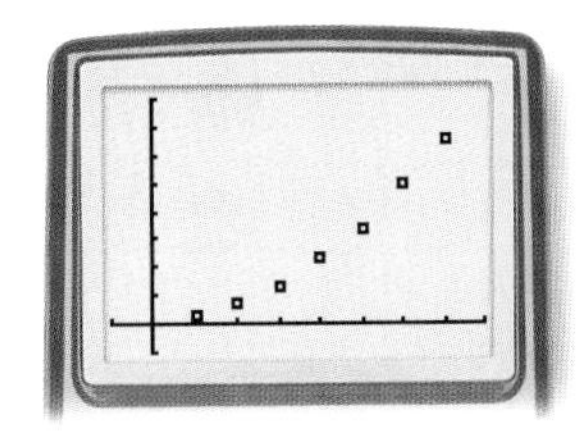

The data appear to follow a parabolic pattern. Use the quadratic regression feature of a graphics calculator to find a quadratic model.

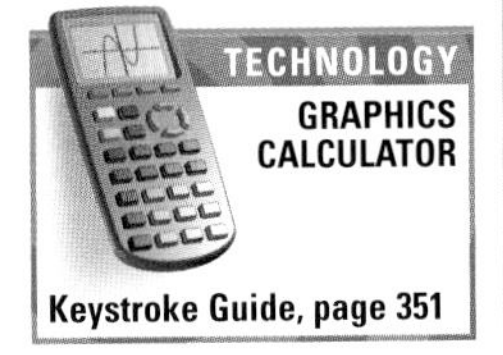

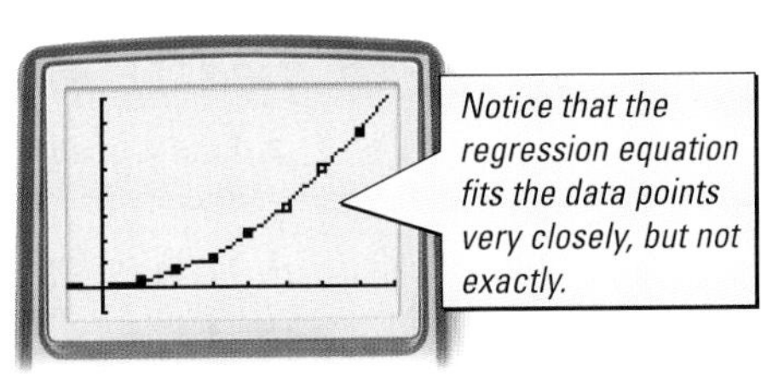

Notice that the regression equation fits the data points very closely, but not exactly.

According to this data, the quadratic model for the stopping distance, y, in terms of the car's speed, x, is $y \approx 0.06x^2 + 0.31x + 4$.

CHECKPOINT ✔ Explain why a quadratic model for stopping distance depends on the actual data from which it is obtained. Discuss the circumstances of the situation, such as the road conditions, the reaction time of the driver, and the condition of the brakes.

Math CONNECTION

STATISTICS Regression equations are usually used to predict values of unavailable data. If the data do not closely model a particular type of function, then that type of function is not appropriate to use in statistical regression. The more closely data fit a model, the more accurate predicted values will be.

ADDITIONAL EXAMPLE 3

Make a scatter plot of the data in the table below. Find a quadratic model to represent the data.

x	y
25	150
50	178
75	216
100	265
125	323
150	392
175	470.4

$y = 0.008x^2 + 0.524x + 131.7$

CHECKPOINT ✔

Different road conditions or faulty brakes may change the data points which, in turn, would result in different values of a, b, and c.

Reteaching the Lesson

WORKING BACKWARD Have students work in pairs. Each student should write a quadratic equation and, on a separate sheet of paper, write three ordered pairs that satisfy the equation. Have the partners trade the sets of ordered pairs. Each student should work backward from the points to identify the other student's equation.

Teaching Tip

TECHNOLOGY If you are using a TI-83 or TI-82 graphics calculator to graph the line of best fit, use the following keystrokes to generate the quadratic regression equation after entering the data into **L1** and **L2**:

After the calculator generates the regression equation, it stores the equation in a variable named **RegEQ.** To graph the regression equation, use the following keystrokes:

1:RegEq (or 7:RegEq) GRAPH

Note that the variable **RegEQ** holds the last regression equation generated by the calculator.

Activity Notes

In this Activity, students will use a motion-detecting unit to collect data on a free-falling object. Students will discover that the graph of the data has a parabolic shape.

CHECKPOINT ✓

3. Answers may vary.

Assess

Selected Answers

Exercises 3–5, 7–43 odd

ASSIGNMENT GUIDE

In Class	1–5
Core	7–21 odd
Core Plus	6–30 even
Review	32–44
Preview	45 and 46

✎ Extra Practice can be found beginning on page 940.

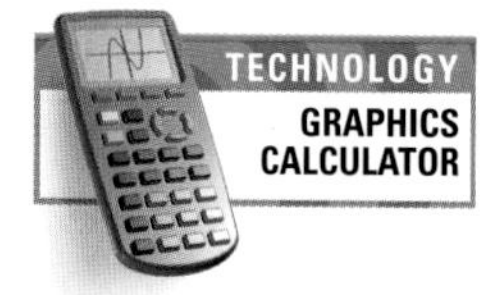

Activity
Collecting and Modeling Data Electronically

You will need: a graphics calculator and a motion detector

This Activity involves tossing a ball vertically into the air, using a motion detector to measure its height over the elapsed time, using a graph to represent the data, and finding a quadratic model for the data.

1. Using a graphics calculator with a motion detector, collect data (elapsed time and height) for the vertical toss of a ball, such as a beach ball.
2. Make a scatter plot for the data. Do the data indicate a linear or quadratic relationship? Explain your response.

CHECKPOINT ✓

3. Find a quadratic model for the data. Give values of a, b, and c to the nearest hundredth.
4. Acceleration due to gravity is about 32 feet per second squared. The value of a in your model represents half of the ball's acceleration due to gravity. Is the value of a in your model reasonable? Explain.

Exercises

Communicate

1. Let the points (1, 6), (−3, 8), and (5, −2) be on a parabola.
   a. Explain how to write a system of equations representing these points.
   b. Explain how to find a quadratic function that fits these points.
2. Describe how to find first and second differences of a data set. What type of function is indicated by a constant first difference? a constant second difference?

Guided Skills Practice

3. Find a quadratic function whose graph contains the points (2, −8), (5, 1), and (0, 6). ***(EXAMPLE 1)*** $f(x) = 2x^2 - 11x + 6$
4. Suppose that everyone in a room is required to shake hands with everyone else in the room. ***(EXAMPLE 2)***

Number of people, n	Number of handshakes, h
2	1
3	3
4	6
5	**10**
6	**15**

   a. Copy and complete the table at left. Explain why a quadratic model is suitable for relating the number of handshakes, h, to the number of people, n.
   b. Find a quadratic function for h in terms of n.
   c. Use the model to predict the number of handshakes when there are 10 people in the room.

4a. The first differences for h are 2, 3, 4, 5; all of the second differences are 1, indicating a quadratic relationship between n and h.
b. $h(n) = 0.5n^2 - 0.5n$
c. 45

APPLICATION

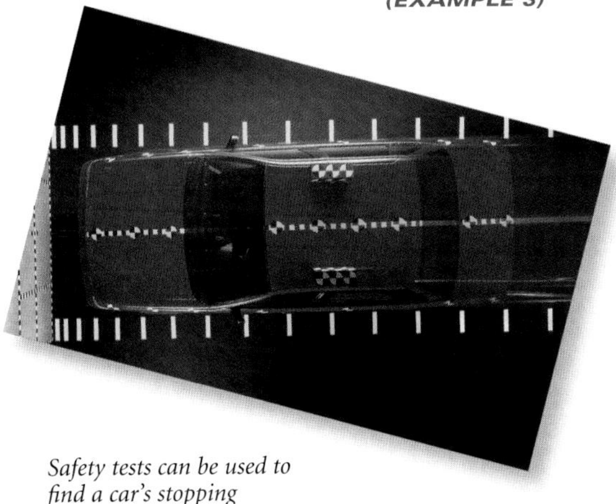

Safety tests can be used to find a car's stopping distance.

5. HIGHWAY SAFETY The table below appears in a driver's manual. *(EXAMPLE 3)*

Stopping Distance on Dry Concrete

Speed (mph)	Reaction distance before applying brakes (ft)*	Braking distance (ft)	Stopping distance (ft)
20	22	22	44
30	33	50	83
40	44	88	132
50	55	138	193
60	66	198	264
70	77	270	347

*Assume that perception time is zero seconds and reaction time

a. Make a scatter plot for stopping distance versus speed.
b. Does it appear that a quadratic function would fit the data? If so, find a quadratic function for the stopping distance, d, in terms of the speed, x.
yes; $d \approx 0.06x^2 + 1.10x + 0.06$

Practice and Apply

6. If the point (2, 3) is on a parabola whose equation is in the form $y = ax^2 + bx + c$, which linear equation in three variables represents the parabola?

$y = 2x^2 + 3x + c$ $\quad y = 4a + 2b + c$ $\quad 2 = 9a + 3b + c$
$3 = 4x^2 + 2x + c$ $\quad$ (circled) $3 = 4a + 2b + c$

Solve a system of equations in order to find a quadratic function that fits each set of data points exactly.

7. (1, −1), (2, 5), (3, 13)
8. (1, 13), (4, 7), (5, −3)
9. (2, 18), (6, 10), (8, −6)
10. (−2, −7), (1, 8), (2, 21)
11. (0, 4), (1, 5), (3, 25)
12. (−3, 7), (−1, −5), (6, 16)
13. (0, 15), (2, 5), (3, 6)
14. (1, 7), (−2, 4), (3, 19)
15. (5, 9), (2, 21), (4, 9)
16. (−1, 5), (4, 5), (8, −13)
17. (0, 7), (−2, −3), (2, 13)
18. (0, 4), (2, 1), (−2, 3)
19. (−2, 7), (4, 10), (1, 4)
20. (4, −4), (−2, 5), (0, 6)

21. Find a quadratic model for the data points (1, 4), (2, 5), and (3, 10) by using two methods: (1) writing a system of three equations in three variables and solving it and (2) using a graphics calculator to find a quadratic regression equation. **$y = 2x^2 - 5x + 7$**

7. $y = x^2 + 3x - 5$
8. $y = -2x^2 + 8x + 7$
9. $y = -x^2 + 6x + 10$
10. $y = 2x^2 + 7x - 1$
11. $y = 3x^2 - 2x + 4$
12. $y = x^2 - 2x - 8$
13. $y = 2x^2 - 9x + 15$
14. $y = x^2 + 2x + 4$
15. $y = 2x^2 - 18x + 49$
16. $y = -0.5x^2 + 1.5x + 7$
17. $y = -0.5x^2 + 4x + 7$
18. $y = -0.5x^2 - 0.5x + 4$
19. $y = 0.5x^2 - 0.5x + 4$
20. $y = -0.5x^2 - 0.5x + 6$

internet connect
Homework Help Online
Go To: **go.hrw.com**
Keyword:
MB1 Homework Help for Exercises 21–31

Teaching Tip

Many students may try to graph a regression equation on a graphics calculator before the calculator has computed such an equation. Remind them to follow the steps below.

1. Enter the data into lists by using the **STAT** feature of a graphics calculator.
2. On a TI-83 or TI-82, compute the regression equation for the data by pressing STAT, selecting **CALC**, and choosing the type of regression model desired.
3. If the data are not in **L1** and **L2**, the names of the lists containing the data must be entered following **QuadReg.** Separate the two list names with a comma by using 2nd 1 for **L1**, 2nd 2 for **L2**, and so on.
4. Enter the equation into **Y1** by pressing VARS, selecting **5:Statistics**, choosing **EQ** and **1:RegEq** (or **7:RegEq**), and pressing ENTER.

5a.

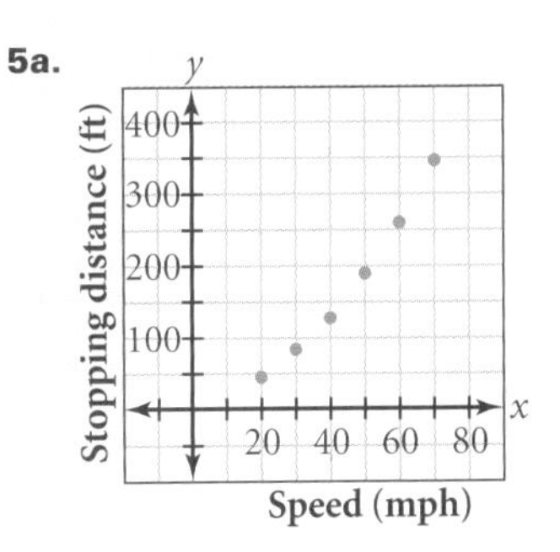

Error Analysis

Many students select an inappropriate regression to model a set of data. Remind them to create a scatter plot and determine which model best fits the data before finding the regression equation. They should graph the regression equation and the scatter plot on the same coordinate plane to see how well the regression equation fits the data. If it does not fit well, they should select another type of regression model.

Also, many students confuse the *quadratic* regression feature of a calculator with the *quartic* regression feature. Remind them to choose **QuadReg,** not **QuartReg,** to obtain a quadratic regression equation. The quartic regression equation is a fourth-degree polynomial function.

Practice

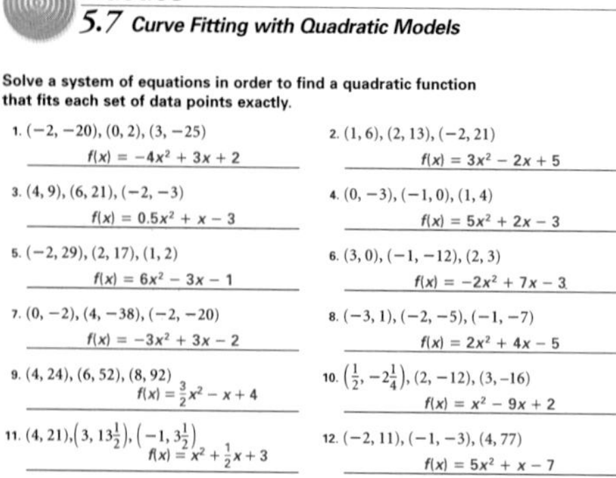

NAME ______ CLASS ______ DATE ______

Practice

5.7 *Curve Fitting with Quadratic Models*

Solve a system of equations in order to find a quadratic function that fits each set of data points exactly.

1. $(-2, -20), (0, 2), (3, -25)$ $f(x) = -4x^2 + 3x + 2$
2. $(1, 6), (2, 13), (-2, 21)$ $f(x) = 3x^2 - 2x + 5$
3. $(4, 9), (6, 21), (-2, -3)$ $f(x) = 0.5x^2 + x - 3$
4. $(0, -3), (-1, 0), (1, 4)$ $f(x) = 5x^2 + 2x - 3$
5. $(-2, 29), (2, 17), (1, 2)$ $f(x) = 6x^2 - 3x - 1$
6. $(3, 0), (-1, -12), (2, 3)$ $f(x) = -2x^2 + 7x - 3$
7. $(0, -2), (4, -38), (-2, -20)$ $f(x) = -3x^2 + 3x - 2$
8. $(-3, 1), (-2, -5), (-1, -7)$ $f(x) = 2x^2 + 4x - 5$
9. $(4, 24), (6, 52), (8, 92)$ $f(x) = \frac{3}{2}x^2 - x + 4$
10. $\left(\frac{1}{2}, -2\frac{1}{4}\right), (2, -12), (3, -16)$ $f(x) = x^2 - 9x + 2$
11. $(4, 21), \left(3, 13\frac{1}{2}\right), \left(-1, 3\frac{1}{2}\right)$ $f(x) = x^2 + \frac{1}{2}x + 3$
12. $(-2, 11), (-1, -3), (4, 77)$ $f(x) = 5x^2 + x - 7$

A baseball player throws a ball. The table shows the height, *y*, of the ball *x* seconds after it is thrown.

Time (seconds)	Height (feet)
0.25	7
0.5	9
1	7

13. Find a quadratic function to model the data. $f(x) = -16x^2 + 20x + 3$
14. What was the maximum height reached by the ball? 9.25 feet
15. How long did it take the ball to reach its maximum height? about 0.6 second
16. Use your model to predict the height of the ball 1.25 seconds after it was thrown. 3 feet
17. Use your model to determine how many seconds it took for the ball to hit the ground. about 1.4 seconds

CONNECTION

22. PATTERNS IN DATA Cubes of various sizes are to be built from unit cubes. The surfaces of the resulting figures are to be painted green.

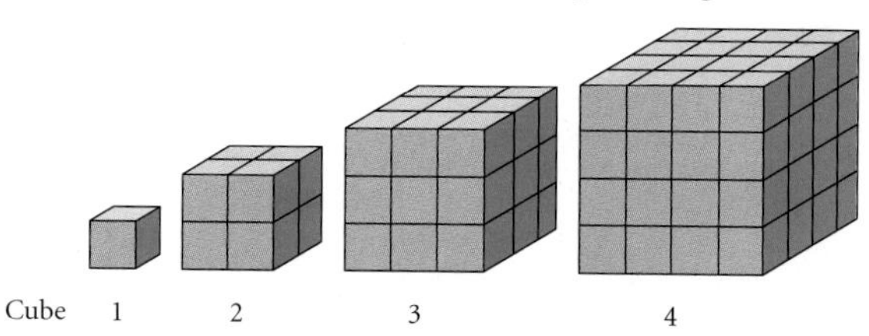

a. Copy and complete the table below.

Cube	Unit cubes with exactly 3 green faces	Unit cubes with exactly 2 green faces	Unit cubes with exactly 1 green face
2	8	0	0
3	8	12	6
4	**8**	**24**	**24**
5	**8**	**36**	**54**
6	**8**	**48**	**96**

b. Find an appropriate model for the data in each column.

c. Use your models to add rows to the table for cubes 7, 8, and 20.

b. n, **8, 12(**n **– 2), 6(**n **– 2)**^{2}

c. 7, 8, 60, 150;
8, 8, 72, 216;
20, 8, 216, 1944

APPLICATIONS

PHYSICS Imagine that an experiment is conducted on Mars in which an object is launched vertically from a height of h_0 above the surface of the planet and with an initial velocity of v_0 in feet per second. The height of the object is measured at three different points in time, and the data are recorded in the table shown at right.

Time (s)	Height (ft)
1	19
2	21
3	11

23. Find a quadratic function that fits the data by solving a system.

24. Find the initial height and the initial velocity of the object. **5 ft; 20 ft/s**

25. How long will it take for the object to reach its maximum height? $\mathbf{1\frac{2}{3}}$ **s**

26. What is the maximum height reached by the object? $\mathbf{21\frac{2}{3}}$ **ft**

27. How long after the object is launched will it return to the surface of Mars?

28. Use your function to predict the height of the object when t is 2.5 seconds.

29. Use your function to predict the time(s) when the object is 16 feet above the surface of Mars. **0.69 s, 2.64 s**

23. $h(t) = -6t^2 + 20t + 5$

27. about 3.57 s

28. 17.5 ft

30. BUSINESS A small computer company keeps track of its monthly production and profit over three months. The data are shown in the table at left. $P(n) = -n^2 + 160n - 400$

Number of computers produced	Profit
50	$5100
100	$5600
150	$1100

a. Use the data in the table to find a quadratic function that describes the profit as a function of the number of computers produced.

b. Use your function to predict the level of production that will maximize the profit. **80 computers**

c. Use your function to predict the maximum profit, assuming that all business conditions stay the same. **$6000**

APPLICATION

31. PHYSICS With the use of a regulation-size girls' basketball, a graphics calculator, a CBL, and a motion detector, the data below was collected and rounded to the nearest hundredth. Find a quadratic model to represent the data. $h(t) \approx -15.11t^2 + 15.19t + 1.65$

Time (s)	0.01	0.11	0.21	0.31	0.41	0.51	0.61	0.71	0.81	0.91
Height (ft)	1.88	3.05	4.13	4.91	5.37	5.50	5.31	4.82	4.00	2.97

Look Back

Determine whether the inverse of each function below is also a function. *(LESSON 2.5)*

32. $f(x) = 1 - 2x^2$ no

33. $g(x) = x - 2$ yes

34. $h(x) = -\frac{1}{2}x + 3$ yes

35. $f(x) = \frac{1}{3}x - 3$ yes

36. $g(x) = \frac{3 - x}{2}$ yes

37. $h(x) = \frac{2}{x}$ yes

internet connect

Activities Online

Go To: **go.hrw.com**
Keyword: **MB1 Schools**

Evaluate. *(LESSON 2.6)*

38. $[-0.99] + [1.99]$ 0

39. $[-2.1 - 1.1]$ −4

40. $[-0.3] - [3.7]$ −4

Find the determinant and tell whether each matrix has an inverse. *(LESSON 4.3)*

41. $\begin{bmatrix} 6 & -4 \\ -1 & 2 \end{bmatrix}$ 8; yes

42. $\begin{bmatrix} \frac{1}{3} & 0 \\ 1 & \frac{1}{2} \end{bmatrix}$ $\frac{1}{6}$; yes

43. $\begin{bmatrix} \frac{1}{2} & \frac{1}{5} \\ \frac{2}{3} & \frac{1}{10} \end{bmatrix}$ $-\frac{1}{12}$; yes

44. $\begin{bmatrix} 1 & 0 \\ 0 & 1 \end{bmatrix}$ 1; yes

Look Beyond

45. Which is the solution to the inequality $x^2 - 16 < 0$? c

a. $x < 4$
b. $x > -4$
c. $-4 < x < 4$
d. $x > 4$ *or* $x < -4$

46. Which is the solution to the inequality $x^2 - 2x - 15 < 0$? a

a. $-3 < x < 5$
b. $x < -3$ *or* $x > 5$
c. $-5 < x < 3$
d. $x < -5$ *or* $x > 3$

SPORTS Refer to the basketball data given on page 273 to answer the questions below.

1. Write and solve a system of equations to find a quadratic function whose graph contains any three points from the basketball data.
2. Use your calculator to find a quadratic regression model for the three data points.
3. Do your quadratic models from Steps 1 and 2 agree? How do these models compare with the models used in the Portfolio Activities on pages 280, 298, 306, and 313? Explain.

WORKING ON THE CHAPTER PROJECT

You should now be able to complete the Chapter Project.

Exercises 45 and 46 preview methods for solving quadratic inequalities. This topic will be studied in the next lesson.

ALTERNATIVE

Assessment

Portfolio Activity

The Portfolio Activity can be used as preparation for the Chapter Project or as a separate activity. In the Portfolio Activity on this page, students will use three data points to find a quadratic function that describes the vertical motion of a basketball in free fall. They will compare the algebraic model from physics with the quadratic regression model.

Answers to Portfolio Activities can be found in Additional Answers of the Teacher's Edition.

Student Technology Guide

NAME ____________ CLASS ________ DATE ________

Student Technology Guide
5.7 *Curve Fitting With Quadratic Models*

A best-fit line is not always the best mathematical model for a data set. A graphics calculator can use other functions, including quadratic functions, to model data.

Example: The table gives carbon monoxide (CO) emissions in the United States for several years from 1940 to 1990. Emissions are measured in millions of tons. Make a scatter plot of the data, using 1940 as year 0. Find a quadratic regression equation for the data. Use the equation to predict the amount of carbon monoxide emitted in 2000.

Years since 1940	0	10	20	30	40	50
CO emissions	90.9	98.8	103.8	128.1	115.6	103.8

- First enter the data. Press STAT 1:Edit... ENTER. Enter the years into the L1 column. Enter the CO emissions into the L2 column.
- To see the scatter plot, press 2nd Y= 1: Plot1...On ENTER On Type: ENTER GRAPH.
- To find the constants for the regression equation, press STAT CALC 5:QuadReg ENTER ENTER. Rounding to hundredths, $y \approx 0.003x^2 + 1.90x + 86.9$ is the quadratic regression equation that best models this data.

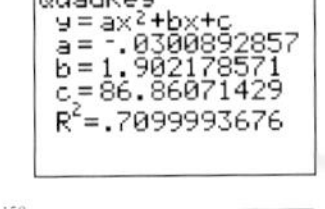

- Press Y= VARS 5:Statistics... ENTER EQ 1: RegEQ ENTER to enter the regression equation. Press GRAPH to see how well this equation fits the data.
- To predict CO emissions in 2000 (year 60), press TRACE, use ▲ or ▼ to toggle to the graph of the function, and then press 60 ENTER.

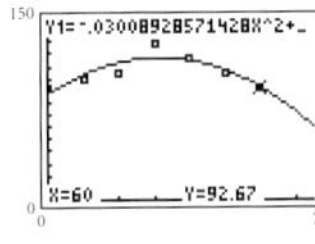

The model predicts about 92.7 million tons of CO emissions in the year 2000.

The following table gives the median sales price, in thousands of dollars, for a home in the United States in recent years. Make a scatter plot and find a quadratic regression equation for the data. Use the equation to predict the median sale price of a home in 2005. $y \approx 0.11x^2 + 2.61x + 23.8$; about $254,000

Years since 1970	0	5	10	15	20
Home sale price	23.4	39.3	64.6	84.3	122.9

Prepare

NCTM PRINCIPLES & STANDARDS 1–10

QUICK WARM-UP

Solve each compound inequality, and graph the solution on a number line.

1. $-3p \leq 6$ *and* $p - 3 \leq 1$
   $p \geq -2$ *and* $p \leq 4$

2. $5 + 2z < -1$ *or* $9z + 7 > 16$
   $z < -3$ *or* $z > 1$

3. $8 \leq 4t$ *or* $17 + 4t < 3 - 3t$
   $t \geq 2$ *or* $t < -2$

Also on Quiz Transparency 5.8

Teach

Why Tell students that in a business, a profit is made when the revenue from goods sold is greater than the cost of producing those goods. A loss occurs when cost is greater than the revenue. Many times the expression for the profit is a quadratic expression written in terms of the selling price. A procedure to solve quadratic inequalities is needed to determine when profit occurs.

Solving Quadratic Inequalities

Why *You can solve many real-world problems, such as those involving business profits related to revenue and cost, by solving quadratic inequalities.*

Objectives

- Write, solve, and graph a quadratic inequality in one variable.
- Write, solve, and graph a quadratic inequality in two variables.

APPLICATION
SMALL BUSINESS

Katie makes and sells T-shirts. A consultant found that her monthly costs, C, are related to the selling price, p, of the shirts by the function $C(p) = 75p + 2500$. The revenue, R, from the sale of the shirts is represented by $R(p) = -25p^2 + 700p$. Her profit, P, is the difference between the revenue and the costs each month.

$$\begin{aligned} P(p) &= R(p) - C(p) \\ &= -25p^2 + 700p - (75p + 2500) \\ &= -25p^2 + 625p - 2500 \end{aligned}$$

For what range of prices can Katie sell the shirts in order to make a profit? That is, for what values of p will $-25p^2 + 625p - 2500 > 0$? *You will answer this question in Example 2.*

One-Variable Quadratic Inequalities

Activity Exploring Quadratic Inequalities

TECHNOLOGY
GRAPHICS CALCULATOR
Keystroke Guide, page 351

You will need: a graphics calculator

The display at right shows the values of $f(x) = x^2 - 2x - 3$ for integer values of x between -2 and 4 inclusive.

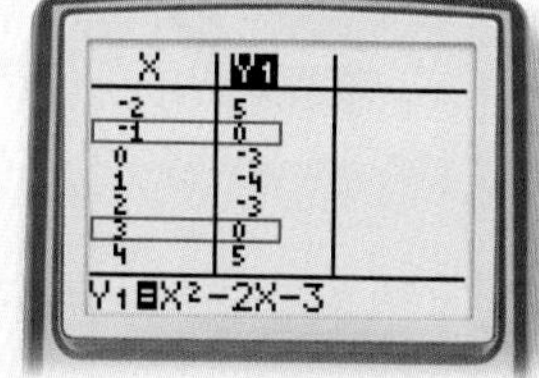

The table suggests the following three cases:

- When $x = -1$ or $x = 3$, $f(x) = 0$.
- When $x < -1$ or $x > 3$, $f(x) > 0$.
- When $-1 < x < 3$, $f(x) < 0$.

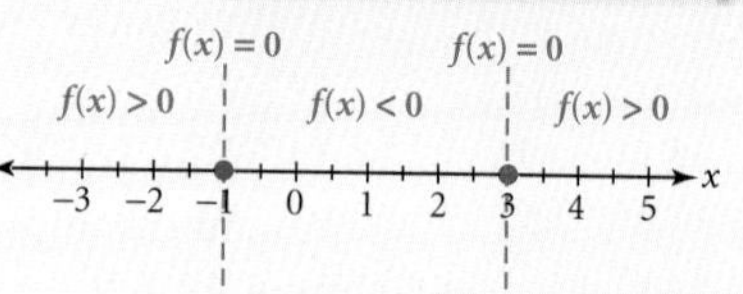

Alternative Teaching Strategy

USING VISUAL MODELS Given the graph of $y = x^2 - 4x - 5$ below, ask students to estimate the value(s) of x when $x^2 - 4x - 5 > 0$, when $x^2 - 4x - 5 < 0$, and when $x^2 - 4x - 5 = 0$. Relate the intervals on the x-axis where $y > 0$ to the intervals on the real-number line where $x^2 - 4x - 5 > 0$ by picking several values of x within the intervals $(-\infty, -1)$, $(-1, 5)$, and $(5, +\infty)$ and evaluating $x^2 - 4x - 5$ for those values. Evaluate the expression for $x = -1$ and $x = 5$ and lead students to determine that these values mark the boundaries of the intervals where $x^2 - 4x - 5 > 0$ and where $x^2 - 4x - 5 < 0$ because they occur when $x^2 - 4x - 5 = 0$. Remind them that the x-intercepts can be used to write the quadratic expression as a product of linear factors and that they can solve quadratic inequalities by graphing the related function and looking at the y-values within the intervals determined by the x-intercepts.

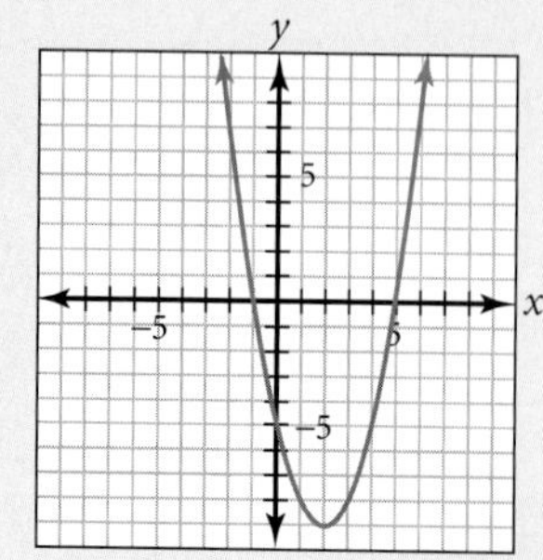

1. Copy and complete the table below. What values of x satisfy each equation or inequality?

Function	Number of x-intercepts	$f(x) = 0$	$f(x) > 0$	$f(x) < 0$
$f(x) = x^2 - 4$	2			
$f(x) = -x^2 + 2x + 3$				

2. Repeat Step 1 for the functions in the table below.

Function	Number of x-intercepts	$f(x) = 0$	$f(x) > 0$	$f(x) < 0$
$f(x) = x^2$	1			
$f(x) = -x^2$				

3. Repeat Step 1 for the functions in the table below.

Function	Number of x-intercepts	$f(x) = 0$	$f(x) > 0$	$f(x) < 0$
$f(x) = -x^2 + x - 1$	0			
$f(x) = x^2 + x + 3$				

CHECKPOINT ✔ 4. **a.** If the graph of a quadratic function crosses the x-axis at 2 distinct points, the graph separates the x-axis into _?_ distinct interval(s).
b. If the graph of a quadratic function crosses or touches the x-axis at 1 point, the graph separates the x-axis into _?_ distinct interval(s).
c. If the graph of a quadratic function does not cross the x-axis, the graph separates the x-axis into _?_ distinct interval(s).

You can determine the solution to a given inequality by finding the roots of the related quadratic equation or by using the graph of the related quadratic equation.

EXAMPLE 1 **Solve $x^2 - 2x - 15 \geq 0$. Graph the solution on a number line.**

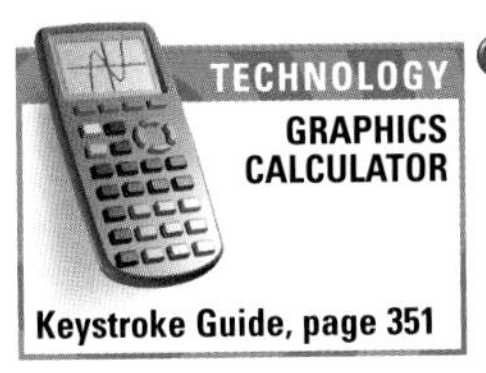

SOLUTION

The graph of $y = x^2 - 2x - 15$ indicates that the solution has two parts.

$x \leq$ smaller root *or* **$x \geq$ larger root**

$$x^2 - 2x - 15 = 0$$
$$(x + 3)(x - 5) = 0$$
$$x = -3 \quad or \quad x = 5$$

x is less than or equal to the lesser root.

x is greater than or equal to the greater root.

Therefore, the solution to the given inequality is $x \leq -3$ *or* $x \geq 5$.

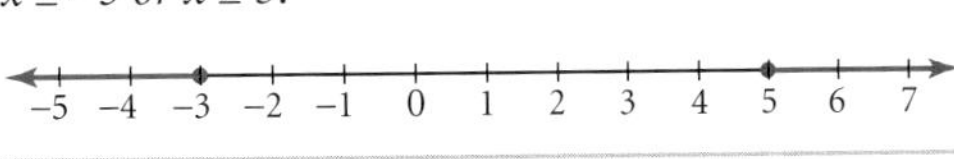

TRY THIS Solve $x^2 - 8x + 12 \leq 0$. Graph the solution on a number line.

Activity Notes

Given a table of values for $f(x) = x^2 - 2x - 3$, students are asked to find the values of x where $f(x) = 0$, $f(x) > 0$, and $f(x) < 0$. By picking several values of x within each interval, lead students to see that every value of x within a given interval either makes $x^2 - 2x - 3$ greater then 0 or makes $x^2 - 2x - 3$ less than 0. Note also when $x = -1$ or $x = 3$, $x^2 - 2x - 3$ is equal to 0.

CHECKPOINT ✔
4a. 3 **b.** 2 **c.** 1

Teaching Tip

TECHNOLOGY When using the TI-83 or TI-82 graphics calculator to create the graph in the solution of Example 1, set the viewing window as follows:

Xmin=–5	**Ymin=–20**
Xmax=7	**Ymax=15**
Xscl=1	**Yscl=1**

ADDITIONAL EXAMPLE 1

Solve $x^2 + 3x - 4 < 0$. Graph the solution on a number line.

$-4 < x < 1$

–4 –2 0 2 4

TRY THIS

$2 \leq x \leq 6$

Interdisciplinary Connection

ECONOMICS A *production possibilities curve* is a tool used by economists for expressing alternatives in making a choice. It can also be used by students to help them allocate time spent working and studying. Give students the table below, in which x represents the number of hours worked and y represents the quiz grade. Tell them the table shows the possible ways in which a student can allocate 4 hours of time for working and studying. The student must choose either more work and a higher income or more study time and a higher quiz grade.

Work: x	0	1	2	3	4
Grade: y	100	89	77	66	54

Have students make a scatter plot of the data and then find a quadratic model for y in terms of x. $\mathbf{y = -0.07x^2 - 11.21x + 100.06}$ Have them use their model to approximate the values of x for which $y > 70$. **$0 < x < 2.63$, or about 2 hours and 38 minutes**

EXAMPLE 2 Refer to Katie's T-shirt business from the beginning of the lesson.

At what price range can Katie sell her T-shirts in order to make a profit?

APPLICATION
SMALL BUSINESS

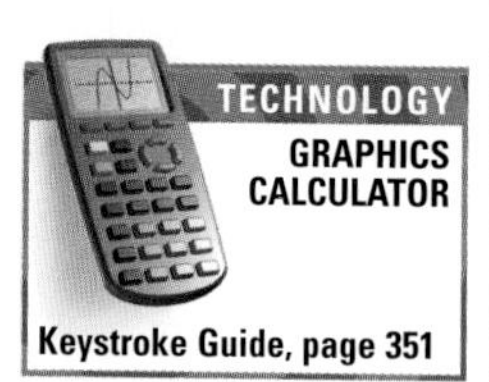

SOLUTION

Use the quadratic formula to find the roots of $-25x^2 + 625x - 2500 = 0$.

$$p = \frac{-b \pm \sqrt{b^2 - 4ac}}{2a}$$

$$p = \frac{-625 \pm \sqrt{625^2 - 4(-25)(-2500)}}{2(-25)}$$

$$p = 5 \quad or \quad p = 20$$

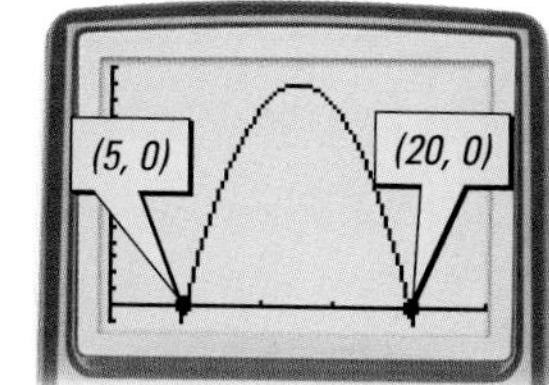

The graph of $y = -25x^2 + 625x - 2500$ indicates that the profit is positive *between* the roots of the related equation. If Katie sells her shirts at a price between \$5 and \$20, she will make a profit.

Special types of solutions to quadratic inequalities are shown in Example 3.

EXAMPLE 3 **Solve each inequality.**

a. $(x - 2)^2 \geq 0$ **b.** $(x + 3)^2 < 0$ **c.** $2(x + 1)^2 > 0$

SOLUTION

a. The square of every real number is greater than or equal to 0. Therefore, the solution is all real numbers.

$y = (x - 2)^2$

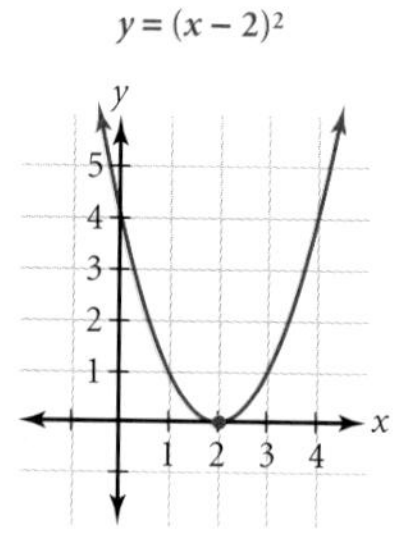

The graph of $y = (x - 2)^2$ indicates that all y-values of $y = (x - 2)^2$ are greater than or equal to zero for all real numbers x.

b. The square of a real number cannot be negative. Therefore, there is no solution.

$y = (x + 3)^2$

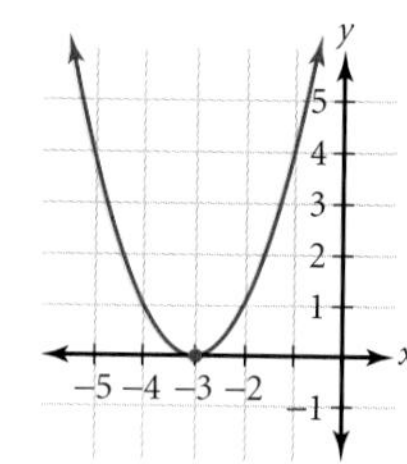

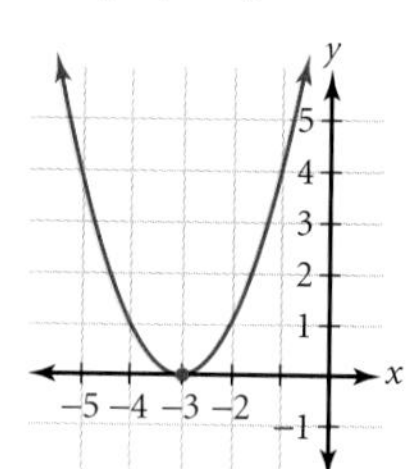

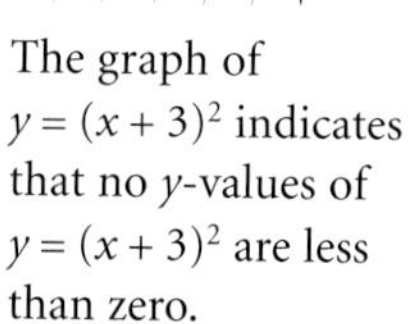

The graph of $y = (x + 3)^2$ indicates that no y-values of $y = (x + 3)^2$ are less than zero.

c. The square of any *nonzero* real number is greater than 0. Thus, $2(x + 1)^2 > 0$ is always true unless $x + 1 = 0$, or $x = -1$.

$y = 2(x + 1)^2$

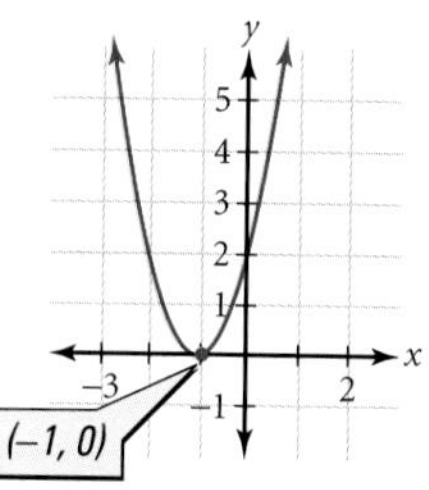

The graph of $y = 2(x + 1)^2$ indicates that the y-values of $y = 2(x + 1)^2$ are greater than zero for all real numbers x except $x = -1$.

TRY THIS Find all solutions of $-(x + 1)^2 > 0$, if any exist.

ADDITIONAL EXAMPLE 2

The owner of a small business in a resort area sells souvenir hats to tourists. The profit, P, for selling the hats as a function of the selling price, p, is given by the function $P(p) = -15p^2 + 357p - 1380$. **At what prices can the owner sell the hats in order to make a profit?**

between \$4.86 and \$18.94

Teaching Tip

TECHNOLOGY If you are using a TI-83 or TI-82 graphics calculator to create the graph for the profit in Example 2, set the viewing window as follows:

Xmin=0	Ymin=–100
Xmax=25	Ymax=1500
Xscl=5	Yscl=100

ADDITIONAL EXAMPLE 3

Solve each inequality.

a. $3(x + 1)^2 < 0$
no solution

b. $(x - 5)^2 > 0$
all real numbers except 5

c. $(x + 4)^2 \geq 0$
all real numbers

TRY THIS
no solution

Inclusion Strategies

TACTILE LEARNERS To help students see the relationship between the number-line solution to a quadratic inequality and the graph of the related quadratic function, have students graph $y = x^2 - 2x - 15$ and use the x-axis as a number line to graph the solution to $x^2 - 2x - 15 < 0$. $-3 < x < 5$ They should see that the solution to $x^2 - 2x - 15 < 0$ corresponds to the interval on the x-axis where the graph of $y = x^2 - 2x - 15$ is below the x-axis, that is, where $y < 0$. Repeat the same procedure for $x^2 - 2x - 15 > 0$. $x < -3$ or $x > 5$

Enrichment

Ask students to use the quadratic equations $y = x^2$ and $y = x^2 + 10$ to create a system of quadratic inequalities whose intersection contains

a. no points.
b. infinitely many points.

a. $y \leq x^2$ and $y \geq x^2 + 10$
b. $x^2 \leq y \leq x^2 + 10$

Two-Variable Quadratic Inequalities

A **quadratic inequality in two variables** is an inequality that can be written in one of the forms below, where a, b, and c are real numbers and $a \neq 0$.

$$y \geq ax^2 + bx + c \qquad y > ax^2 + bx + c$$
$$y \leq ax^2 + bx + c \qquad y < ax^2 + bx + c$$

Example 4 shows how to graph a quadratic inequality in two variables.

EXAMPLE 4 **Graph the solution to $y \geq (x-2)^2 + 1$.**

SOLUTION

1. Graph the related equation, $y = (x-2)^2 + 1$. Use a solid curve because the inequality symbol is $\geq$.
2. Test $(0, 0)$ to see if this point satisfies the given inequality.

   $y \geq (x-2)^2 + 1$
   $0 \overset{?}{\geq} (0-2)^2 + 1$
   $0 \geq 5$ **False**

   Test a point inside the parabola, such as $(2, 3)$.

   $y \geq (x-2)^2 + 1$
   $3 \overset{?}{\geq} (2-2)^2 + 1$
   $3 \geq 1$ **True**

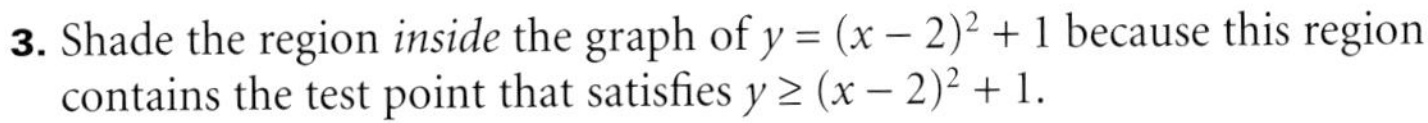

3. Shade the region *inside* the graph of $y = (x-2)^2 + 1$ because this region contains the test point that satisfies $y \geq (x-2)^2 + 1$.

TRY THIS Graph the solution to $y < (x+2)^2 - 3$.

CRITICAL THINKING Explain why you should not use $(0, 0)$ as a test point when graphing the solution to an inequality such as $y > x^2 + 2x$.

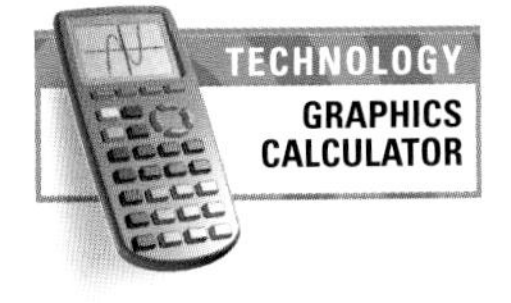

On most graphics calculators, you can graph a quadratic inequality in two variables. The graph of $y \geq (x-2)^2 + 1$ from Example 4 is shown below. The keystrokes for a TI-83 model are also given.

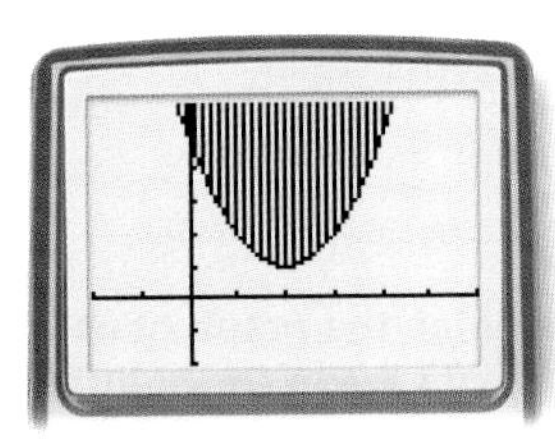

Use viewing window [–2, 6] by [–2, 6].

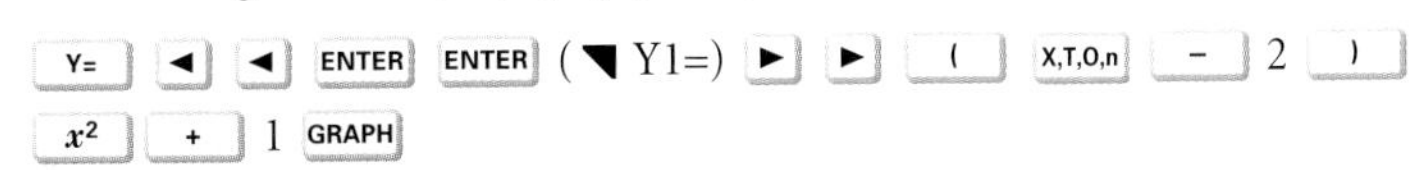

Reteaching the Lesson

CONNECTING TO PRIOR KNOWLEDGE Show students that the graphing techniques used to solve quadratic inequalities are similar to those used to solve linear inequalities. Ask them to graph the solution to the linear inequality $y < -2x + 3$ by shading the region below the line. Then ask them to graph the solution to the quadratic inequality $y < (-2x+3)^2$ by shading the region below the parabola. Now have them compare the steps. They should note that a dashed line or a dashed parabola is used to show that those values are not included in the solution. Have them choose a test point from the region they shaded to see whether the coordinates satisfy the given inequality.

ADDITIONAL EXAMPLE 4

Graph the solution to $y \leq (x-1)^2 - 5$.

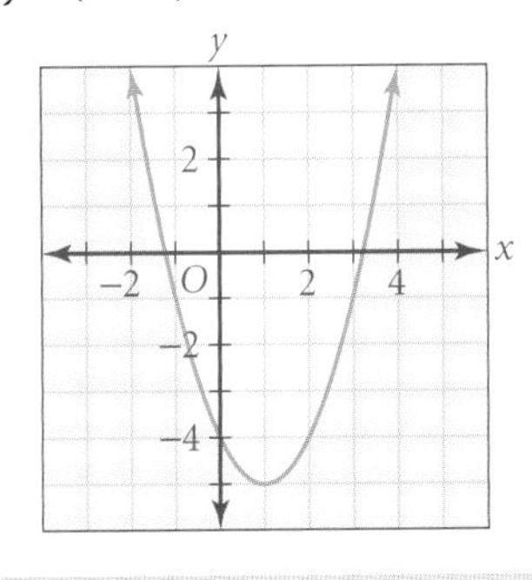

TRY THIS

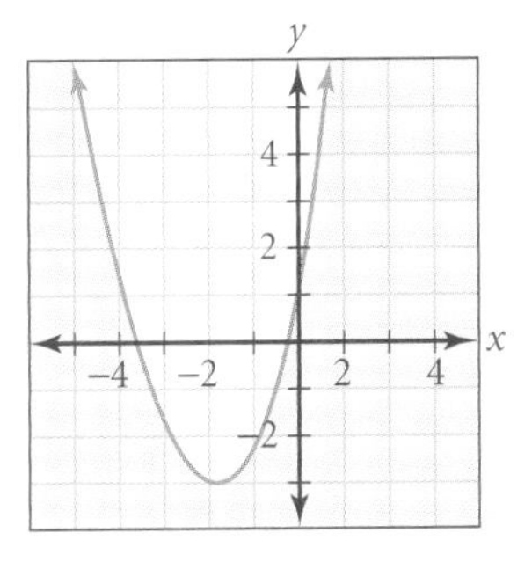

CRITICAL THINKING

The point $(0, 0)$ lies on the curve $y = x^2 + 2x$ and thus is neither in the region above the curve nor the region below the curve, so it cannot be used to test either region.

Teaching Tip

TECHNOLOGY When using the TI-83 or TI-82 graphics calculator to graph the solution to the quadratic inequality in Example 4, set the viewing window as follows:

Xmin=–4	Ymin=–4
Xmax=4	Ymax=4
Xscl=1	Yscl=1

If you are using a TI-83, press Y= and enter **(x – 2)^2 + 1**. Press ◄ to move the cursor to the graph style icon to the left of the equal sign. Press ENTER ENTER to select the "shade above graph" style. Then press GRAPH. If you are using a TI-82, press 2nd PRGM to access the **DRAW** menu. Choose **7:Shade(**. To the right of the parenthesis, enter **(x – 2)^2 + 1,6)**. Then press ENTER.

Assess

Selected Answers

Exercises 6–11, 13–81 odd

ASSIGNMENT GUIDE	
In Class	1–11
Core	13–57 odd, 61, 66, 67
Core Plus	12–70 even
Review	71–81
Preview	82

 Extra Practice can be found beginning on page 940.

11.

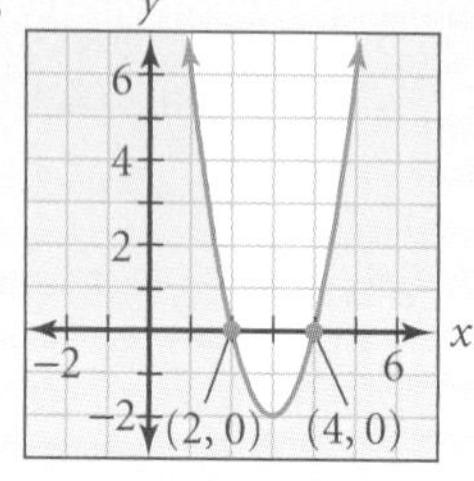

Exercises

internet connect
Activities Online
Go To: **go.hrw.com**
Keyword:
MB1 Videos

Communicate

1. Explain how to solve $x^2 - 2x - 8 > 0$.

2. Explain how a graph can assist you when solving $x^2 - 2x - 8 > 0$.

3. Explain how to graph an inequality such as $y < (x - 2)^2 + 2$.

4. Explain how to test whether the correct area has been shaded in the graph of an inequality.

5. Explain how to determine the possible solutions to $(x + 7)^2 < 0$ without solving the inequality.

Guided Skills Practice

6. Solve $x^2 - 7x + 12 \geq 0$. Graph the solution on a number line. ***(EXAMPLE 1)*** $x \leq 3$ *or* $x \geq 4$

7. For what integer values of x is $-2x^2 + 25x - 72 > 0$ true? ***(EXAMPLE 2)*** **5, 6, and 7**

Solve each inequality. *(EXAMPLE 3)*

8. $(x - 3)^2 < 0$ **no solution**

9. $(x - 5)^2 > 0$ **all real numbers ≠ 5**

10. $(x - 1)^2 < 0$ **no solution**

11. Graph the solution to $y \leq 2(x - 3)^2 - 2$. ***(EXAMPLE 4)***

Practice and Apply

Solve each inequality. Graph the solution on a number line.

12. $x^2 - 1 \geq 0$

13. $-x^2 + 5x - 6 > 0$

14. $x^2 - 8x + 12 \leq 0$

15. $x^2 - 4x - 5 < 0$

16. $x^2 - 7x + 10 \leq 0$

17. $50 - 15x > -x^2$

18. $x^2 \leq \frac{3}{4} + x$

19. $x^2 - x - 12 \leq 0$

20. $-x^2 + \frac{4}{3}x - \frac{5}{9} > 0$

21. $x^2 - 4x - 12 > 0$

22. $x^2 - 2x - 99 > 0$

23. $x^2 + x - 6 \leq 0$

24. $-x^2 - x + 20 < 0$

25. $x^2 \leq 7x - 6$

26. $x^2 + 35 > -12x$

27. $10 - x^2 \geq 9x$

28. $x^2 + 10x + 25 > 0$

29. $x^2 + 3x - 18 > 0$

30. $x^2 - 2 > x$

31. $x^2 + 6x \geq 7$

32. $15 - 8x \leq -x^2$

33. $-x^2 + 3x + 6 < 0$

34. $4x - 1 > 8 - x^2$

35. $x^2 + 5x - 7 < 4x$

Sketch the graph of each inequality. Then decide which of the given points are in the solution region.

36. $y \geq (x - 1)^2 + 5$; $A(4, 1)$, $B(4, 14)$, $C(4, 20)$ ***B* and *C***

37. $y > -(x - 3)^2 + 8$; $A(5, 1)$, $B(5, 4)$, $C(5, 6)$ ***C***

38. $y < (x - 2)^2 + 6$; $A(3, 1)$, $B(3, 7)$, $C(3, 10)$ ***A***

39. $y \leq -(x - 4)^2 + 7$; $A(6, 1)$, $B(6, 3)$, $C(6, 5)$ ***A* and *B***

12. $x \leq -1$ *or* $x \geq 1$
13. $2 < x < 3$
14. $2 \leq x \leq 6$
15. $-1 < x < 5$
16. $2 \leq x \leq 5$
17. $x < 5$ *or* $x > 10$
18. $-\frac{1}{2} \leq x \leq \frac{3}{2}$
19. $-3 \leq x \leq 4$
20. no solution
21. $x < -2$ *or* $x > 6$
22. $x < -9$ *or* $x > 11$
23. $-3 \leq x \leq 2$
24. $x < -5$ *or* $x > 4$
25. $1 \leq x \leq 6$
26. $x < -7$ *or* $x > -5$
27. $-10 \leq x \leq 1$
28. $x < -5$ *or* $x > -5$
29. $x < -6$ *or* $x > 3$
30. $x < -1$ *or* $x > 2$
31. $x \leq -7$ *or* $x \geq 1$
32. $3 \leq x \leq 5$
33. $x < \frac{3}{2} - \frac{\sqrt{33}}{2}$ *or* $x > \frac{3}{2} + \frac{\sqrt{33}}{2}$
34. $x < -2 - \sqrt{13}$ *or* $x > -2 + \sqrt{13}$
35. $-\frac{1}{2} - \frac{\sqrt{29}}{2} < x < -\frac{1}{2} + \frac{\sqrt{29}}{2}$

12. $x \leq -1$ *or* $x \geq 1$

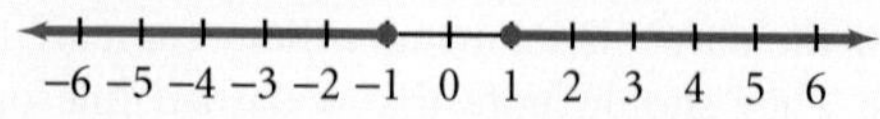

13. $2 < x < 3$

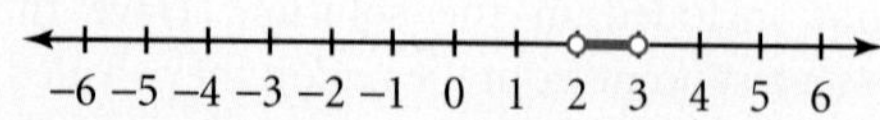

14. $2 \leq x \leq 6$

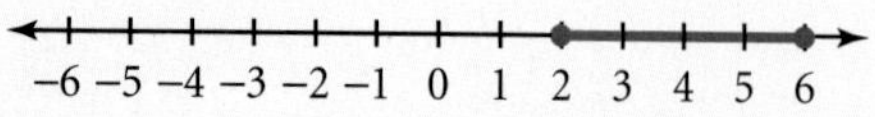

−6 −5 −4 −3 −2 −1 0 1 2 3 4 5 6

15. $-1 < x < 5$

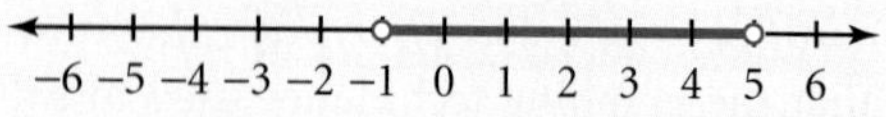

−6 −5 −4 −3 −2 −1 0 1 2 3 4 5 6

16. $2 \leq x \leq 5$

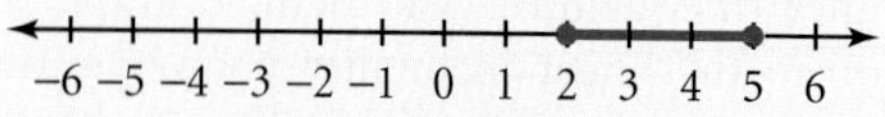

−6 −5 −4 −3 −2 −1 0 1 2 3 4 5 6

17. $x < 5$ *or* $x > 10$

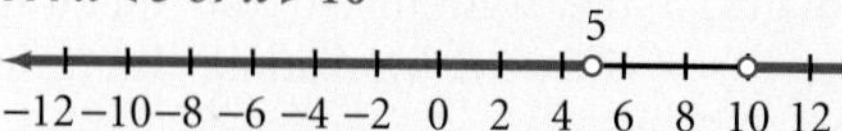

The answers to Exercises 6 and 18–39 can be found in Additional Answers beginning on page 1002.

Graph each inequality.

40. $y \le (x-2)^2 + 2$ **41.** $y \ge (x+2)^2$ **42.** $y < (x-5)^2 + 1$

43. $y > 2(x+3)^2 - 5$ **44.** $y \le \left(x - \frac{1}{2}\right)^2 + 1$ **45.** $y \ge (x+1)^2 + 2$

46. $y \le x^2 + 2x + 1$ **47.** $y < x^2 - 3x + 2$ **48.** $y \ge 2x^2 + 5x + 1$

49. $y > x^2 + 4x + 2$ **50.** $y - 3 \le x^2 - 6x$ **51.** $y - 1 < x^2 - 4x$

52. $y \le (x - \pi)^2 + 1$ **53.** $y \le -\left(x - \frac{5}{7}\right)^2 + 2$ **54.** $y + 3 < (x-1)^2$

55. $y > x^2 + 12x + 35$ **56.** $x + y > x^2 - 6$ **57.** $y - 2x \le x^2 - 8$

CHALLENGES

58. Create a quadratic function in which $f(x) \ge 0$ for values of x between 2 and 6 inclusive. $f(x) = -x^2 + 8x - 12$

59. $x^2 - 10x + 21 > 0$

59. Write a quadratic inequality whose solution is $x < 3$ *or* $x > 7$.

CONNECTION

60. MAXIMUM/MINIMUM Jon is a sales representative for a winter sports equipment wholesaler. The price per snowboard varies based on the number of snowboards purchased in each order. Beginning with a price of \$124 for one snowboard, the price for each snowboard is reduced by \$1 when additional snowboards are purchased.

a. Copy and complete the table.

b. What is the function for the revenue? $R(x) = -x^2 + 125x$

c. What is the maximum revenue per order? \$3906

d. How many snowboards must be sold per order to attain the maximum revenue? 62 or 63

e. Assume that it costs the wholesaler \$68 to produce each snowboard and that John spends an average of \$128 in fixed costs (travel expenses, phone calls, and so on) per order. Based on these two factors alone, what is the function for the costs? Is the function linear or quadratic? $C(x) = 68x + 128$; linear

Number of snowboards purchased	Price per board (\$)	Revenue per order (\$)
1	124	124
2	123	246
3	122	366
4	121	484
5	120	600
⋮	⋮	⋮
x	$125 - x$	$x(125 - x)$

f. Graph the revenue and cost functions on the same coordinate plane. In order for the revenue to be greater than the costs, how many snowboards does Jon need to sell?

g. What is the function for profit per order?

h. Graph the profit function on the same coordinate plane as the functions for revenue per order and cost per order. How many snowboards does Jon need to sell per order to make a profit?

i. What is the maximum profit per order? How many snowboards must be sold to earn the maximum profit per order?

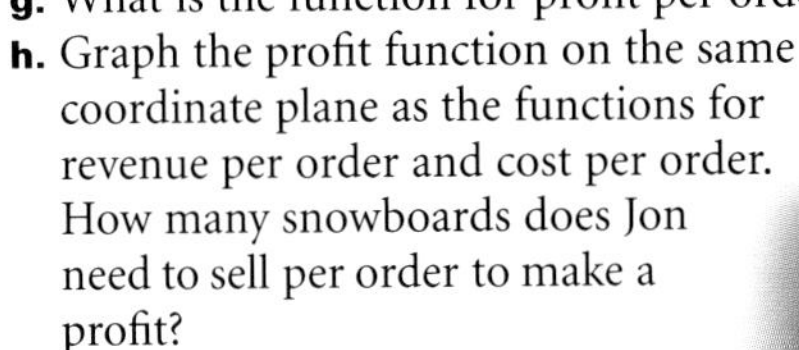

60f. between 3 and 54, inclusive

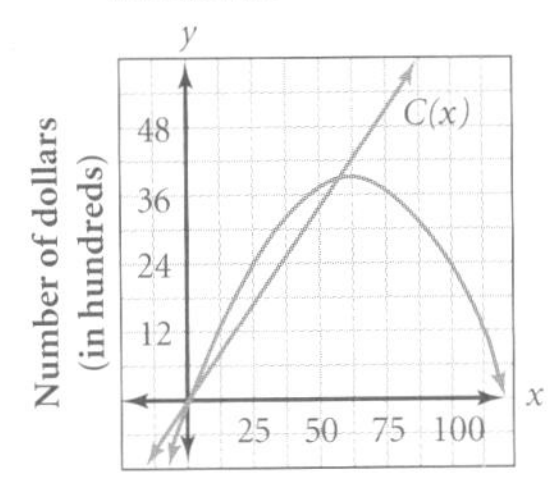

g. $P(x) = -x^2 + 57x - 128$

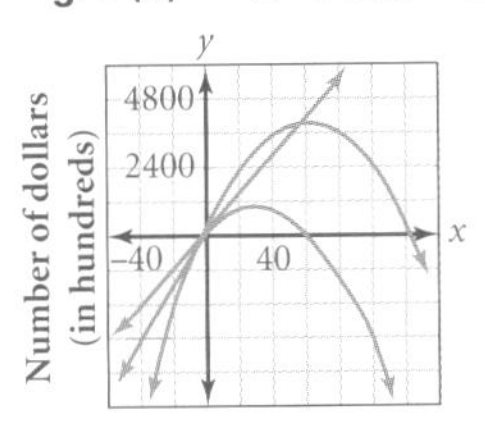

h. between 3 and 54, inclusive

i. \$684; 28 or 29 snowboards

Error Analysis

Some students graph the boundary line when graphing an inequality of the form $y > ax^2 + bx + c$ or $y < ax^2 + bx + c$. Remind students to use a dashed parabola for inequalities with < or > and a solid parabola for inequalities with ≤ or ≥.

If students are using a TI-83 or TI-82 graphics calculator, remind them to set the calculator to dot mode for inequalities with < or >. Also, tell them to check a test point from the region they shade to see whether the coordinates make the inequality true. A test point is easily found by using the arrow keys and noting the coordinates at the bottom of the screen.

40.

41.

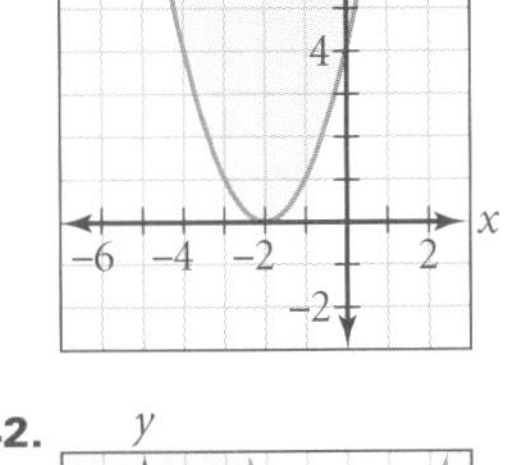

42.

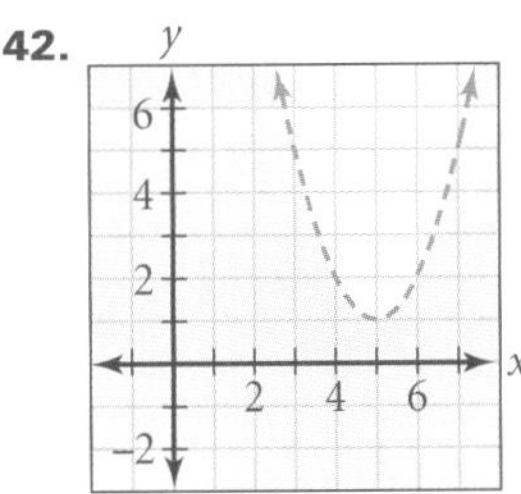

43.

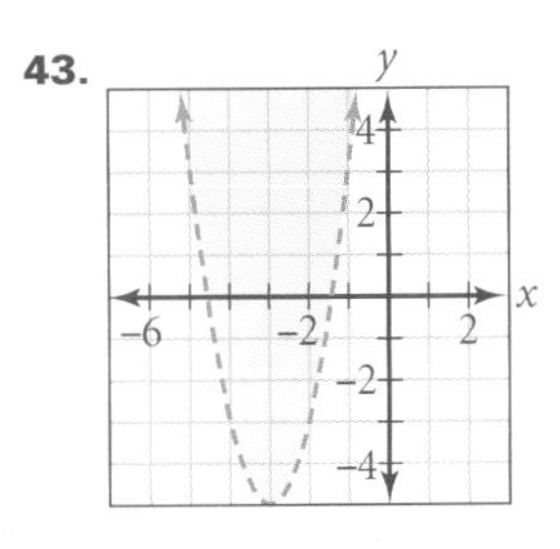

44.

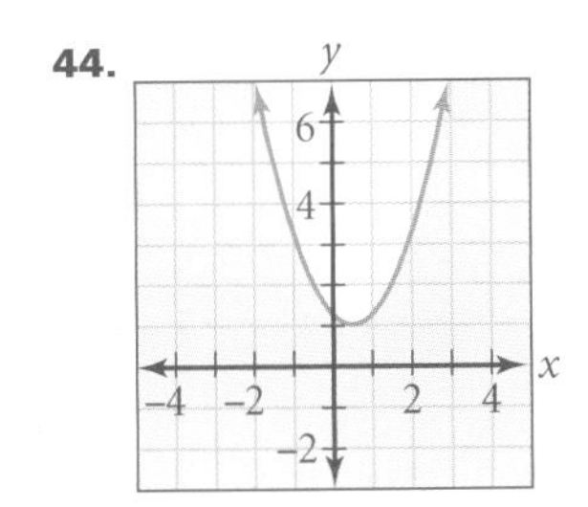

45.

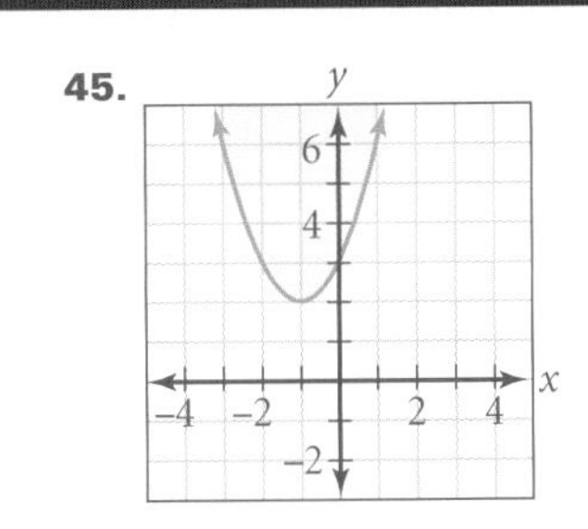

The answers to Exercises 46–57 can be found in Additional Answers beginning on page 1002.

In Exercise 82, a polynomial with an arbitrary constant term and a point are given. Students are asked to determine the value of the constant term when the given point is on the graph of the polynomial. They should substitute the coordinates of the point into the polynomial and solve for the constant term. Polynomial functions will be studied in Chapter 7.

63d.

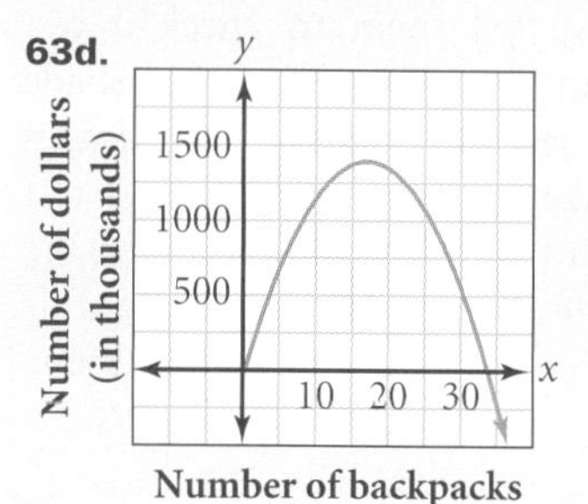

Practice

NAME ______ CLASS ______ DATE ______

Practice
5.8 Solving Quadratic Inequalities

Solve each inequality. Graph the solution on a number line. Round irrational numbers to the nearest hundredth.

1. $x^2 - 16 \geq 0$ $x \leq -4$ or $x \geq 4$
2. $x^2 + 2x - 8 \leq 0$ $-4 \leq x \leq 2$
3. $x^2 + 7x + 10 \leq 0$ $-5 \leq x \leq -2$
4. $x^2 - 2x + 4 < 0$ no solution
5. $x^2 - 3x - 1 > 0$ $x < -0.30$ or $x > 3.30$
6. $x^2 + 10x - 3 < 7x$ $-3.79 < x < 0.79$

Graph each inequality and shade the solution region.

7. $y > x^2 + \frac{1}{2}x$
8. $y \leq (x + 3)^2 - 2$
9. $y > -(x - 1)^2 + 1$
10. $y \geq x^2 - 4x - 5$
11. $y \leq -(x - 3)^2 + 3$
12. $y < 2x^2 - x - 1$

APPLICATIONS

61. SPORTS At the beginning of a basketball game, the referee tosses the ball vertically into the air. Its height, h, in feet after t seconds is given by $h(t) = -16t^2 + 24t + 5$. During what time interval (to the nearest tenth of a second) is the height of the ball greater than 9 feet? **$0.2 \leq t \leq 1.3$**

62a. 7
b. Yes; the profit is greater than \$100 for $30 < x < 50$.

62. SMALL BUSINESS Suppose that the profit, p, for selling x bumper stickers is given by $p(x) = -0.1x^2 + 8x - 50$.
a. What is the minimum number of bumper stickers that must be sold to make a profit?
b. Is it possible for the profit to be greater than \$100? Justify your answer algebraically and graphically.

63a. from 0 to 33,000
b. 20,000
c. No; cost increases as the number of backpacks increases.
e. from 1000 to 33,000; same as part **a**
f. 17,000; yes
g. after 33,000

63. BUSINESS A camping supplies company has determined cost and revenue information for the production of their backpacks.

The cost is given by $C(x) = 50 + 30x$, and the revenue is given by $R(x) = 5x(40 - x)$, where x is the number of backpacks sold in thousands. Both the cost and revenue are given in thousands of dollars.

The profit is given by $P(x) = R(x) - C(x)$.

Use the graph at right to give approximate answers to the questions below.

y; 2000, 1800, 1600, 1400, 1200, 1000, 800, 600, 400, 200; Revenue; Cost; 10, 20, 30, 40; x; Number of dollars (in thousands); Number of backpacks sold (in thousands)

a. To make a profit, revenue must be greater than cost. What is the range for the quantity of backpacks that the company must sell in order to make a profit?
b. At what number of backpacks sold is the revenue maximized?
c. Is there a greatest cost? Why or why not?
d. Graph the profit function.

e. What is the range for the quantity of backpacks that the company must sell to make a profit? Compare this answer to your answer from part **a.** Would you expect the answers to be the same? Why or why not?
f. For what number of backpacks sold is the profit maximized? Compare this answer to your answer from part **b.** Would you expect the answers to be the same? Why or why not?
g. At what point will the company start to lose money by producing too many backpacks?

PHYSICS The approximate length of a pendulum, l, in feet is related to the time, t, in seconds required for one complete swing as given by the formula $l \approx 0.81t^2$.

64. For what values of t is $l > 2$? **approx. $t > 1.57$**

65. For what values of t is $l < 5$? **approx. $0 \leq t \leq 2.48$**

APPLICATIONS

PHYSICS An object is dropped from a height of 1000 feet. Its height, h, in feet after t seconds is given by the function $h(t) = -16t^2 + 1000$.

66. For approximately how long, to the nearest hundredth of a second, will the height of the object be above 500 feet? **approx. 5.59 seconds**

67. About how long, to the nearest hundredth of a second, does it take the object to fall from 500 feet to the ground? **about 2.32 seconds**

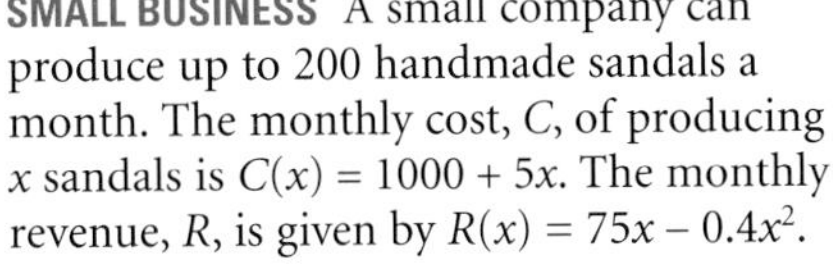

SMALL BUSINESS A small company can produce up to 200 handmade sandals a month. The monthly cost, C, of producing x sandals is $C(x) = 1000 + 5x$. The monthly revenue, R, is given by $R(x) = 75x - 0.4x^2$.

68. For what values of x is the revenue greater than the cost?

69. At what production levels will the company make a profit?

70. At what production levels will the company lose money?

between 0 and 15 or between 160 and 200 pairs, inclusive

68. between 16 and 159, inclusive

69. between 16 and 159, inclusive

Look Back

Graph each equation, and state whether y is a function of x. *(LESSON 2.3)*

71. $y = |x|$ **yes** **72.** $x = |y|$ **no** **73.** $y = -|x|$ **yes** **74.** $x = y^2$ **no**

Solve each equation. Give exact solutions. *(LESSON 5.2)*

75. $-2x^2 = -16$ $\pm\sqrt{8}$ **76.** $-3x^2 + 15 = -6$ $\pm\sqrt{7}$ **77.** $32 = 2x^2 - 4$ $\pm\sqrt{18}$

Simplify each expression, where $i = \sqrt{-1}$. *(LESSON 5.6)*

78. $(8 - 2i)(6 + 3i)$ **$54 + 12i$** **79.** $(1 + 5i) - (2 + i)$ **$-1 + 4i$**

80. $\frac{2 + i}{2 + 3i}$ **$\frac{7}{13} - \frac{4}{13}i$** **81.** $(3 + 2i) + (3 - i)$ **$6 + i$**

internet connect

Portfolio Extension

Go To: **go.hrw.com**
Keyword:
MB1 QInequalities

Look Beyond

82 Let $f(x) = x^3 - 2x^2 + 3x + d$. If the graph of f contains the point $(1, 9)$, find the value of d. Verify your answer by substituting this value for d in the function and graphing it. **7**

PORTFOLIO ACTIVITY

SPORTS Refer to the basketball data on page 273.

1. At what point(s) in time is the basketball at a height of 10 feet?
2. During what period(s) of time is the height of the basketball above 10 feet?
3. During what period(s) of time is the height of the basketball below 10 feet?
4. Use a graph to show that your answers to Steps 1–3 are correct.

ALTERNATIVE
Assessment

Portfolio Activity

The Portfolio Activity can be used as preparation for the Chapter Project or as a separate activity. In the Portfolio Activity on this page, students use quadratic inequalities to describe the height of a basketball as it rises vertically and begins to fall.

Answers to Portfolio Activities can be found in Additional Answers of the Teacher's Edition.

Student Technology Guide

NAME ______ CLASS ______ DATE ______

Student Technology Guide
5.8 *Solving Quadratic Inequalities*

You can use a graphics calculator to graph a quadratic inequality.

Example: Graph the quadratic inequality $y > x^2 - 5x + 3$. Tell whether the boundary line should be solid or dashed.

- Enter the equation of the boundary line as Y1. Here, this is $y = x^2 - 5x + 3$. Press [Y=] [X,T,θ,n] [x^2] [–] 5 [X,T,θ,n] [+] 3.
- Use [◄] to move the cursor to the left of the Y1 label. This inequality is satisfied by y-values *greater than* $x^2 - 5x + 3$, so press [ENTER] until you see ◥ to the left of Y1.
- Press [GRAPH]. The result for $y > x^2 - 5x + 3$ is shown at right.
- Decide whether points on the boundary line satisfy the inequality. Here, the inequality involves >, so the boundary should be dashed.

Graph each quadratic inequality and tell whether the boundary should be dashed or solid. Describe the solution region.

1. $y \geq x^2 + 3x - 3$ region above; solid
2. $y < -3x^2 - 5x + 2$ region below; dashed
3. $y \leq x^2 + 5x + 7$ region below; solid
4. $y \geq \frac{1}{4}x^2 + x - 5$ region above; solid
5. $y > -10x^2 - x + 4$ region above; dashed
6. $y \leq 0.02x^2$ region below; solid

You can also show a system of inequalities on a graphics calculator. To graph $y \geq 2x^2 + 4x + 1$ and $y < 2x - 3$, press [Y=] and enter $y = -2x^2 + 4x + 1$ as Y1 and $y < 2x - 3$ as Y2. Set the shading for Y1 to shade above the graph and the shading for Y2 to shade below the graph. Press [GRAPH]. The region that is shaded both ways, that is, with squares, represents the solution to the system.

7. Graph the system of inequalities $y < -x^2 + 2x + 4$ and $y > \frac{1}{2}x - 1$. Describe the solution region. the region below the parabola and above the line, having a dashed boundary all around

Focus

Gravity's effect on the motion of a basketball or on all planets in our solar system can be described by a quadratic model. Students will be asked to complete a table that provides the acceleration due to gravity near the surface of each planet in terms of the gravity on Earth. They will then plot the data points and compare the results.

Motivate

Ask students whether the referee's action of tossing the ball upward at the beginning of a game, often called a jump ball, is important. Could it be done incorrectly? If so, how could that give an advantage to one of the teams? How could it influence the outcome of a game? What are the rules concerning a jump ball? Tell students they will find a quadratic function to model the vertical motion of the basketball in a jump-ball situation.

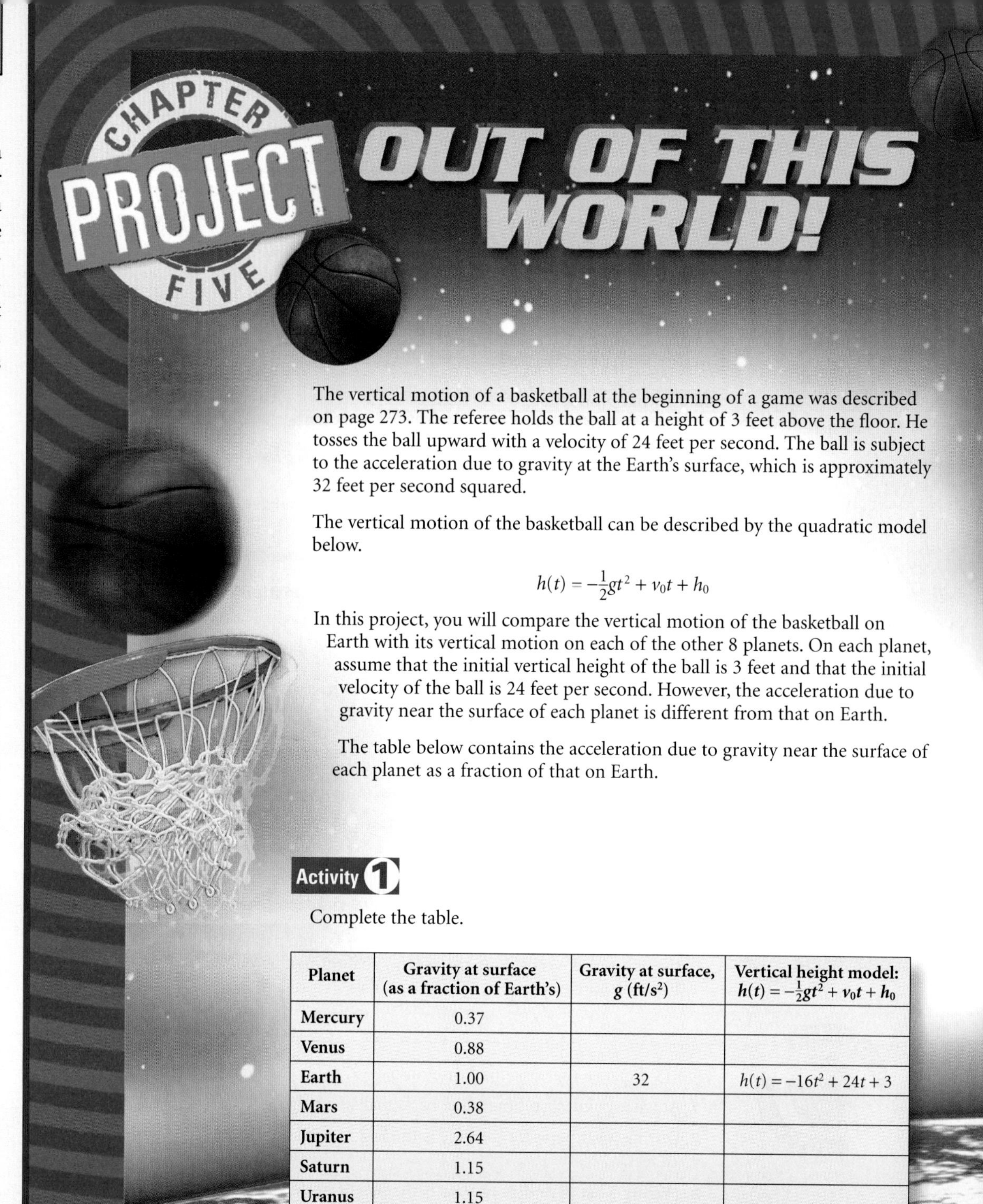

CHAPTER FIVE PROJECT

OUT OF THIS WORLD!

The vertical motion of a basketball at the beginning of a game was described on page 273. The referee holds the ball at a height of 3 feet above the floor. He tosses the ball upward with a velocity of 24 feet per second. The ball is subject to the acceleration due to gravity at the Earth's surface, which is approximately 32 feet per second squared.

The vertical motion of the basketball can be described by the quadratic model below.

$$h(t) = -\frac{1}{2}gt^2 + v_0t + h_0$$

In this project, you will compare the vertical motion of the basketball on Earth with its vertical motion on each of the other 8 planets. On each planet, assume that the initial vertical height of the ball is 3 feet and that the initial velocity of the ball is 24 feet per second. However, the acceleration due to gravity near the surface of each planet is different from that on Earth.

The table below contains the acceleration due to gravity near the surface of each planet as a fraction of that on Earth.

Activity 1

Complete the table.

Planet	Gravity at surface (as a fraction of Earth's)	Gravity at surface, g (ft/s²)	Vertical height model: $h(t) = -\frac{1}{2}gt^2 + v_0t + h_0$
Mercury	0.37		
Venus	0.88		
Earth	1.00	32	$h(t) = -16t^2 + 24t + 3$
Mars	0.38		
Jupiter	2.64		
Saturn	1.15		
Uranus	1.15		
Neptune	1.12		
Pluto	0.04		

Activity 1

Planet	Gravity at surface (as a fraction of Earth's)	Gravity at surface, g (ft/s²)	Vertical height model, V $h(t) = -\frac{1}{2}gt^2 + v_0t + h_0$
Mercury	0.37	11.84	$h(t) = -5.92t^2 + 24t + 3$
Venus	0.88	28.16	$h(t) = -14.08t^2 + 24t + 3$
Earth	1.00	32	$h(t) = -16t^2 + 24t + 3$
Mars	0.38	12.16	$h(t) = -6.08t^2 + 24t + 3$
Jupiter	2.64	84.48	$h(t) = -42.24t^2 + 24t + 3$
Saturn	1.15	36.80	$h(t) = -18.40t^2 + 24t + 3$
Uranus	1.15	36.80	$h(t) = -18.40t^2 + 24t + 3$
Neptune	1.12	35.84	$h(t) = -17.92t^2 + 24t + 3$
Pluto	0.04	1.28	$h(t) = -0.64t^2 + 24t + 3$

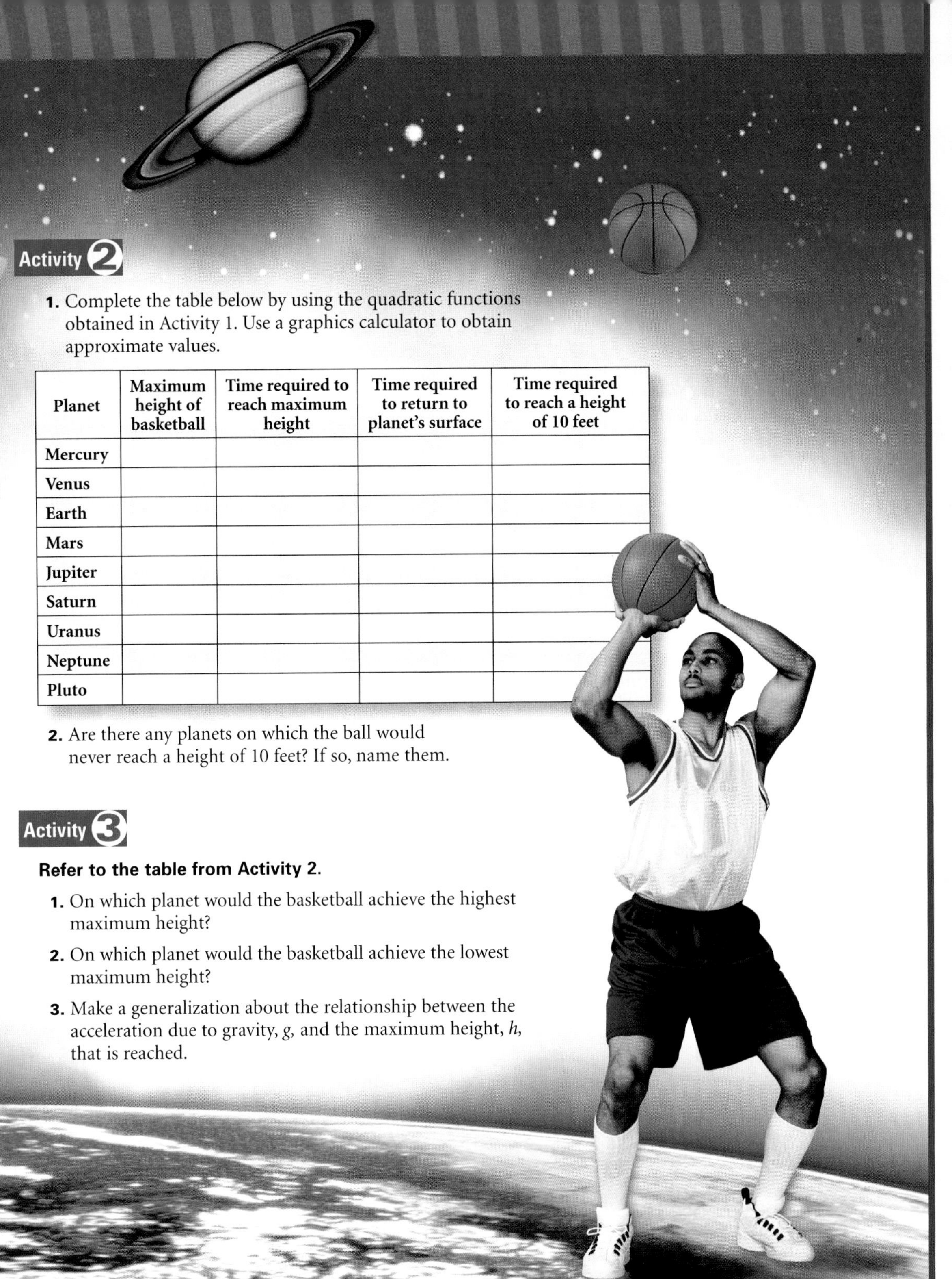

Activity 2

1. Complete the table below by using the quadratic functions obtained in Activity 1. Use a graphics calculator to obtain approximate values.

Planet	Maximum height of basketball	Time required to reach maximum height	Time required to return to planet's surface	Time required to reach a height of 10 feet
Mercury				
Venus				
Earth				
Mars				
Jupiter				
Saturn				
Uranus				
Neptune				
Pluto				

2. Are there any planets on which the ball would never reach a height of 10 feet? If so, name them.

Activity 3

Refer to the table from Activity 2.

1. On which planet would the basketball achieve the highest maximum height?
2. On which planet would the basketball achieve the lowest maximum height?
3. Make a generalization about the relationship between the acceleration due to gravity, g, and the maximum height, h, that is reached.

Cooperative Learning

Suppose that the formation of an Interplanetary Basketball League is being considered. Form the class into groups of four to do a feasibility study of this idea. Their first task will be to study the gravitational factor on each planet as compared with gravity on Earth. Within a group, each member will be assigned two planets and will provide the data for those planets in the table in **Activity 1** and calculate the values for the table in **Activity 2.** However, group members should assist each other so that the group can compare the graphs for all the planets. The group should work together to complete the remaining steps in the project. The group should write a summary that includes possible rule changes that might be required on each planet due to the different gravitational forces.

Discuss

Each group should present its summary to the class. Based on this information, the class can discuss possible rule changes that might be required on each planet. For instance, a lower gravitational force might mean that players could jump longer and higher and that the ball could go farther when thrown upward. Would that require bigger basketball courts and a higher basket? What changes would the group recommend?

Activity 2

1.

Planet	Maximum height of basketball	Time required to reach maximum height	Time required to return to planet's surface	Time required to reach a height of 10 ft
Mercury	27.3 ft	2.03 s	4.17 s	0.32 s, 3.74 s
Venus	13.2 ft	0.85 s	1.82 s	0.37 s, 1.33 s
Earth	12.0 ft	0.75 s	1.62 s	0.40 s, 1.10 s
Mars	26.7 ft	1.97 s	4.07 s	0.32 s, 3.63 s
Jupiter	6.4 ft	0.28 s	0.67 s	not possible
Saturn	10.8 ft	0.65 s	1.42 s	0.44 s, 0.86 s
Uranus	10.8 ft	0.65 s	1.42 s	0.44 s, 0.86 s
Neptune	11.0 ft	0.67 s	1.45 s	0.43 s, 0.91 s
Pluto	228 ft	18.75 s	37.62 s	0.29 s, 37.21 s

2. Yes; Jupiter

Activity 3

1. Pluto
2. Jupiter
3. As g increases, h decreases; as g decreases, h increases.

5 Chapter Review and Assessment

VOCABULARY

Key Skills & Exercises

LESSON 5.1

Key Skills

Multiply linear binomials, and identify and graph a quadratic function.

$$f(x) = (x + 4)(x - 1)$$
$$= x(x - 1) + 4(x - 1)$$
$$= x^2 + 3x - 4$$

The function f is a quadratic function because it can be written in the form $f(x) = ax^2 + bx + c$, where $a = 1$, $b = 3$, and $c = -4$.

Since $a > 0$ in $f(x) = x^2 + 3x - 4$, the parabola opens up and the vertex contains the minimum value of the function.

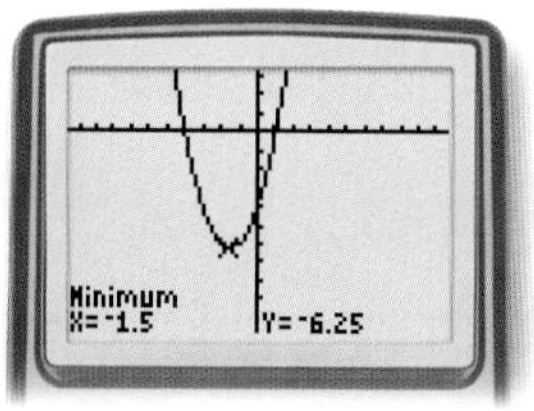

The coordinates of the vertex are (−1.5, −6.25).

The equation of the axis of symmetry is $x = -1.5$.

Exercises

Show that each function is a quadratic function by writing it in the form $f(x) = ax^2 + bx + c$ and identifying a, b, and c.

1. $f(x) = -(x + 1)(x - 4)$ **$f(x) = -x^2 + 3x + 4$; $a = -1$, $b = 3$, $c = 4$**
2. $f(x) = 5(2x - 1)(3x + 2)$ **$f(x) = 30x^2 + 5x - 10$; $a = 30$, $b = 5$, $c = -10$**

Graph each function and give the approximate coordinates of the vertex.

3. $f(x) = -x^2 + 3x - 1$ **(1.5, 1.25)**
4. $f(x) = 5x^2 - x - 12$ **(0.1, −12.05)**

State whether the parabola opens up or down and whether the y-coordinate of the vertex is the maximum or the minimum value of the function.

5. $f(x) = -x^2 - x - 1$ **down; max**
6. $f(x) = (x - 3)(x + 2)$ **up; min**

Chapter Review

Chapter Test, Form A

NAME ______________ CLASS ______________ DATE ______________

Chapter Assessment
Chapter 5, Form A, page 1

Write the letter that best answers the question or completes the statement.

c 1. The graph of which of the following quadratic functions opens down?
a. $f(x) = -6 + 5x^2 - 8x$ b. $f(x) = 3(x - 3)^2 - 9$
c. $f(x) = (5 + 2x)(3 - x)$ d. $f(x) = -3x(5 - x) + 7$

d 2. Find the solutions to the equation $2(x - 5)^2 + 7 = 79$.
a. $x = 6$ or $x = -6$ b. $x = 31$ or $x = 41$
c. $x = 11.56$ or $x = -1.56$ d. $x = 11$ or $x = -1$

d 3. Find the length of $\overline{BC}$.
a. $\sqrt{500}$ meters
b. $\sqrt{756}$ meters
c. 14 meters
d. 18 meters

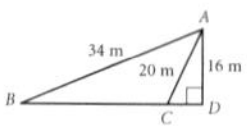

d 4. Factor $6x^2 - 17x + 12$.
a. $(x + 6)(6x + 2)$ b. $(3x - 4)(2x + 3)$
c. $(2x - 6)(3x - 2)$ d. $(2x - 3)(3x - 4)$

a 5. Factor $64x^2 - 49$.
a. $(8x + 7)(8x - 7)$ b. $(8x - 7)(8x - 7)$
c. $(-8x + 7)(-8x + 7)$ d. $(8x + 7)(8x + 7)$

d 6. Find the zeros of the quadratic function $f(x) = x^2 + 10x + 24$.
a. 8 and 3 b. 5 and 2
c. −12 and −2 d. −4 and −6

b 7. Which of the following completes the square for the expression $x^2 - 7x$?
a. $\frac{7}{2}$ b. $\frac{49}{4}$ c. 49 d. −49

a 8. Find the equation of the axis of symmetry for the graph of $f(x) = 3(x + 2)^2 - 5$.
a. $x = -2$ b. $x = -3$ c. $x = -4$ d. $x = -5$

c 9. Find the vertex of $f(x) = 2x^2 - 20x + 18$.
a. (2, −20) b. (−20, 18) c. (5, −32) d. (1, 9)

d 10. Find the solutions to $3x^2 + 15x + 17 = 0$.
a. $x = \frac{15 \pm \sqrt{429}}{6}$ b. $x = \frac{15 \pm \sqrt{765}}{6}$ c. $x = \frac{-15 \pm \sqrt{-429}}{6}$ d. $x = \frac{-15 \pm \sqrt{21}}{6}$

NAME ______________ CLASS ______________ DATE ______________

Chapter Assessment
Chapter 5, Form A, page 2

a 11. Find the discriminant for the equation $5x^2 = 3x - 7$.
a. −131 b. −149 c. 131 d. 149

b 12. How many real solutions does the equation $36 = 24x - 4x^2$ have?
a. 0 b. 1 c. 2 d. 3

a 13. Find the solutions to $2x^2 - 6x + 17 = 0$.
a. $x = \frac{3}{2} \pm \frac{5}{2}i$ b. $x = 4$ or $x = -1$
c. $x = \frac{6 \pm \sqrt{172}}{4}$ d. $x = 3 \pm 5i$

c 14. What is the complex conjugate of $-5 + 7i$?
a. $5 - 7i$ b. $25 + 49i$ c. $-5 - 7i$ d. $7 - 5i$

d 15. Find the product $(3 + 6i)(4 - 7i)$.
a. $12 - 42i$ b. $-30 + 45i$ c. $-44 - 45i$ d. $54 + 3i$

c 16. Find the quotient $\frac{4 + 6i}{2 - 3i}$.
a. $-\frac{5}{26} - \frac{6}{13}i$ b. $2 - 2i$
c. $-\frac{10}{13} + \frac{24}{13}i$ d. none of these

b 17. Which of the following is $|12 - 5i|$?
a. 7 b. 13 c. 17 d. 60

d 18. Find the quadratic function whose graph contains the points (1, −6), (−2, 15), and (3, 30).
a. $f(x) = 2x^2 - 9x + 5$ b. $f(x) = 9x^2 - 2x - 5$
c. $f(x) = 5x^2 - 9x - 2$ d. $f(x) = 5x^2 - 2x - 9$

b 19. Which of the following is the solution to the inequality $x^2 + 6x \geq 7$?
a. $x \geq 7$ *or* $x \leq -1$ b. $x \leq -7$ *or* $x \geq 1$
c. $x \leq 7$ *and* $x \geq -1$ d. $x \geq -7$ *and* $x \leq 1$

a 20. Which inequality is graphed at right?
a. $y > (x - 3)^2 + 1$
b. $y < (x + 3)^2 - 1$
c. $y \geq (x - 3)^2 + 1$
d. $y \leq (x + 3)^2 - 1$

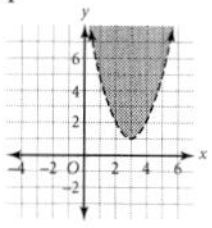

7. $x = \pm\sqrt{8}$; $x \approx \pm 2.83$

8. $x = \pm\sqrt{20}$; $x \approx \pm 4.47$

9. $x \approx \pm 7$

10. $x = \pm\sqrt{5}$; $x \approx \pm 2.24$

11. $x = -5$ *or* $x = 11$

12. $x = 5 \pm \sqrt{48}$; $x \approx -1.93$ *or* $x \approx 11.93$

LESSON 5.2

Key Skills

Solve quadratic equations by taking square roots.

$$16(x+3)^2 = 81$$
$$(x+3)^2 = \frac{81}{16}$$
$$x + 3 = \pm\sqrt{\frac{81}{16}}$$
$$x = -3 + \frac{9}{4} \quad or \quad x = -3 - \frac{9}{4}$$
$$x = -\frac{3}{4} \qquad\qquad x = -5\frac{1}{4}$$

Use the Pythagorean Theorem to solve problems involving right triangles.

Find the unknown length, b, in right triangle ABC.

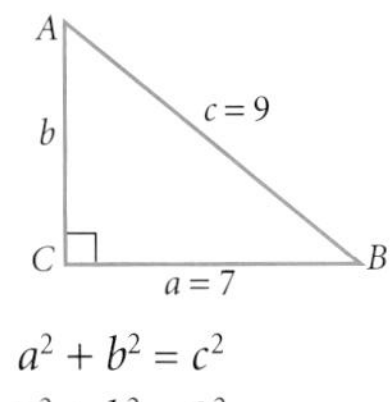

$$a^2 + b^2 = c^2$$
$$7^2 + b^2 = 9^2$$
$$b = \sqrt{9^2 - 7^2}$$
$$b = \sqrt{32}$$
$$b \approx 5.66$$

Exercises

Solve each equation, giving both exact solutions and approximate solutions to the nearest hundredth.

7. $x^2 = 8$
8. $3x^2 = 60$
9. $x^2 - 3 = 46$
10. $x^2 + 4 = 9$
11. $(x-3)^2 = 64$
12. $(x-5)^2 = 48$
13. $7(x+1)^2 = 54$
14. $6(x+2)^2 = 30$

Find the unknown length in right triangle *ABC*. Give your answers to the nearest tenth.

15. $a = 4, b = 5$ **$c \approx 6.4$**
16. $c = 4, a = 1$ **$b \approx 3.9$**
17. $b = 7, c = 12$ **$a \approx 9.7$**
18. $a = 12, c = 15$ **$b = 9$**
19. $c = 25, b = 5$ **$a \approx 24.5$**
20. $b = 6, a = 6$ **$c \approx 8.5$**
21. $a = 0.2, c = 0.75$ **$b \approx 0.7$**
22. $b = 3.2, c = 5.8$ **$a \approx 4.8$**

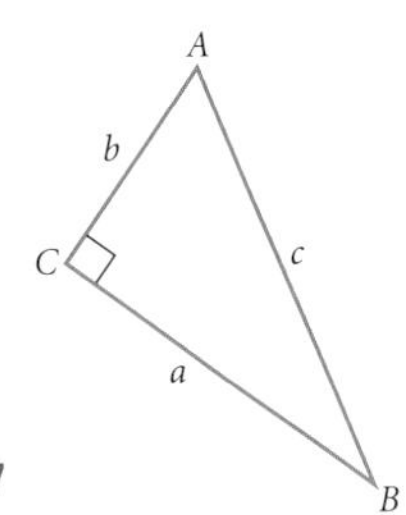

LESSON 5.3

Key Skills

Use factoring to solve a quadratic equation and to find the zeros of a quadratic function.

$$6x^2 + 9x = 6$$
$$6x^2 + 9x - 6 = 0$$
$$3(2x^2 + 3x - 2) = 0$$
$$3(2x - 1)(x + 2) = 0$$
$$x = \frac{1}{2} \quad or \quad x = -2$$

The zeros of the related quadratic function, $f(x) = 6x^2 + 9x - 6$, are $\frac{1}{2}$ and -2.

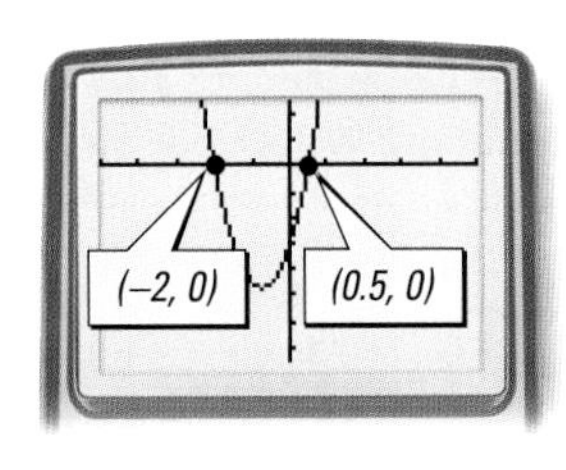

Exercises

Factor each expression.

23. $7x^2 - 21x$
24. $6n - 4n^2$
25. $x^2 + 7x + 10$
26. $x^2 + 11x + 28$
27. $t^2 - 5t - 24$
28. $x^2 - 7x + 12$
29. $x^2 - 8x - 20$
30. $y^2 - 6y - 27$
31. $x^2 + x - 20$
32. $x^2 + 4x - 21$
33. $3y^2 - y - 2$
34. $2x^2 - 5x - 25$
35. $16 - 9x^2$
36. $4x^2 - 49$
37. $x^2 - 16x + 64$
38. $4a^2 + 4a + 1$

Use the Zero-Product Property to find the zeros of each function.

39. $f(x) = x^2 - 10x + 24$ **4, 6**
40. $g(x) = 2x^2 - 3x - 2$ **$-\frac{1}{2}, 2$**
41. $h(t) = 6t^2 + 11t - 10$ **$-\frac{5}{2}, \frac{2}{3}$**

---

13. $x = -1 \pm \frac{\sqrt{54}}{7}$; $x \approx -3.78$ *or* $x \approx 1.78$

14. $x = -2 \pm \sqrt{5}$; $x \approx -4.24$ *or* $x \approx 0.24$

23. $7x(x-3)$

24. $2n(3-2n)$

25. $(x+5)(x+2)$

26. $(x+7)(x+4)$

27. $(t+3)(t-8)$

28. $(x-3)(x-4)$

29. $(x+2)(x-10)$

30. $(y+3)(y-9)$

31. $(x+5)(x-4)$

32. $(x+7)(x-3)$

33. $(3y+2)(y-1)$

34. $(2x+5)(x-5)$

35. $(4+3x)(4-3x)$

36. $(2x+7)(2x-7)$

37. $(x-8)^2$

38. $(2a+1)^2$

Chapter Test, Form B

NAME ______ CLASS ______ DATE ______

Chapter Assessment

Chapter 5, Form B, page 1

1. Solve $\frac{3}{5}(x^2 - 75) - 17 = 73$. $x = \pm 15$

2. In the right triangle shown, find the length of the third side. 72 centimeters

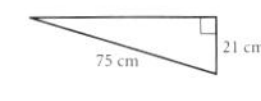

Find the coordinates of the vertex and the equation of the axis of symmetry for each parabola.

3. $y = 5(x-6)^2 + 12$ (6, 12); $x = 6$
4. $y = 2x^2 + 12x + 10$ (−3, −8); $x = -3$

Find the x-intercepts for each parabola.

5. $y = 3x^2 - 9x + 12$ no x-intercepts
6. $y = 4x^2 + 7x - 15$ (1.25, 0) and (−3, 0)

Simplify each expression.

7. $(5+3i)^2$ $16 + 30i$
8. $\frac{2+5i}{6-8i}$ $-\frac{7}{25} + \frac{23}{50}i$
9. $2(4+3i) - (5-2i)$ $3 + 8i$

Factor each expression.

10. $144x^2 - 25$ $(12x+5)(12x-5)$
11. $x^2 - 3x - 40$ $(x+5)(x-8)$
12. $3x^2 - 16x + 5$ $(3x-1)(x-5)$

13. Solve $3x^2 + 24x + 36 = 0$ by completing the square. $x = -2$ *or* $x = -6$

14. Write $f(x) = x^2 - 6x + 16$ in vertex form. $f(x) = (x-3)^2 + 7$

15. Find the discriminant and the number of real solutions for the equation $7x^2 = 10x - 9$. −152; no real solutions

NAME ______ CLASS ______ DATE ______

Chapter Assessment

Chapter 5, Form B, page 2

Solve each equation by using the quadratic formula. Give exact answers and approximate answers to the nearest hundredth.

16. $5x^2 + 6x + 9 = 0$ $x = -\frac{3}{5} \pm \frac{6}{5}i$
17. $x^2 + 11 = 7x$ $x = \frac{7 \pm \sqrt{5}}{2}$; $x \approx 4.62$ *or* $x \approx 2.38$

18. Find a quadratic model for y in terms of x that fits the data in the table below .

x	−1	1	2	4	7
y	−6	8	9	−1	−46

$y = -2x^2 + 7x + 3$

19. a. Graph $-2 - 5i$ on the complex plane.
    b. Find $|-2 - 5i|$. $\sqrt{29} \approx 5.4$
    c. Graph the conjugate of $-2 - 5i$ on the complex plane.

Imaginary
Real

20. Solve $x^2 + 5x < 36$. Graph your answer on the number line at right. $x > -9$ *and* $x < 4$

21. Graph $y \leq -2(x-4)^2 - 1$.

22. The manager of a factory has found that the output per worker at the factory can be modeled by the function $T(n) = -0.2n^2 + 8n - 24$, where n is the number of workers at the factory on a given day and $T(n)$ is the output per worker. Find the number of workers needed in order to maximize the output per worker. What is the maximum output per worker at this factory?

20 workers; maximum output per worker is 56

50. $y = 2(x-4)^2 + 1$; $(4, 1)$

51. $y = -3(x+1)^2 - 4$; $(-1, -4)$

52. $y = -\left(x + \frac{5}{2}\right)^2 + \frac{17}{4}$; $\left(-\frac{5}{2}, \frac{17}{4}\right)$

53. $y = 4\left(x - \frac{9}{8}\right)^2 - \frac{49}{16}$; $\left(\frac{9}{8}, -\frac{49}{16}\right)$

54. $x = \frac{1}{2} \pm \frac{\sqrt{12}}{4}$

55. $x = -\frac{3}{5} \pm \frac{\sqrt{76}}{10}$

56. $x = 2$ *or* $x = 5$

57. $x = -4$ *or* $x = -2$

58. $x = \frac{11}{10} \pm \frac{\sqrt{181}}{10}$

59. $x = \frac{1}{12} \pm \frac{\sqrt{73}}{12}$

60. $x = -\frac{5}{2} \pm \frac{\sqrt{37}}{2}$

61. $x = -\frac{1}{2} \pm \frac{\sqrt{5}}{2}$

62. $\left(-\frac{7}{2}, -\frac{25}{4}\right)$

63. $\left(\frac{1}{2}, -\frac{49}{4}\right)$

64. $(-1, -4)$

65. $(-6, -31)$

70. $x = 3 - 4i$ *or* $x = 3 + 4i$

71. $x = -5 - 3i$ *or* $x = -5 + 3i$

72. $x = -4 - 2i$ *or* $x = -4 + 2i$

73. $x = 3 \pm \frac{i\sqrt{8}}{2}$

74. $x = \frac{1}{4} \pm \frac{i\sqrt{12}}{8}$

75. $x = \frac{1}{3} \pm \frac{i\sqrt{20}}{6}$

LESSON 5.4

Key Skills

Use completing the square to solve a quadratic equation.

$$3x^2 - 8x = 48$$
$$x^2 - \frac{8}{3}x = 16$$
$$x^2 - \frac{8}{3}x + \left(\frac{8}{6}\right)^2 = 16 + \left(\frac{8}{6}\right)^2$$
$$\left(x - \frac{8}{6}\right)^2 = \frac{160}{9}$$
$$x = \frac{8}{6} \pm \frac{\sqrt{160}}{3}$$
$$x = \frac{4 + \sqrt{160}}{3} \quad or \quad x = \frac{4 - \sqrt{160}}{3}$$

The coordinates of the vertex of the graph of a quadratic function in vertex form, $y = a(x-h)^2 + k$, are (h, k).

Exercises

Solve each quadratic equation by completing the square.

42. $x^2 - 6x = 27$ $-3, 9$
43. $5x^2 = 2x + 1$ $\frac{1}{5} \pm \frac{\sqrt{6}}{5}$
44. $x^2 - 10x + 21 = 0$ $3, 7$
45. $x^2 + 5x = 84$ $-12, 7$
46. $x^2 - 7x - 8 = 0$ $-1, 8$
47. $2x^2 + 7x = 4$ $-4, \frac{1}{2}$
48. $4x^2 + 4 = 17x$ $\frac{1}{4}, 4$
49. $2x + 8 = 3x^2$ $-\frac{4}{3}, 2$

Write each function in vertex form, and identify the coordinates of the vertex.

50. $y = 2x^2 - 16x + 33$
51. $y = -3x^2 - 6x - 7$
52. $y = -x^2 - 5x - 2$
53. $y = 4x^2 - 9x + 2$

LESSON 5.5

Key Skills

Use the quadratic formula to find the real roots of quadratic equations.

Solve $2x^2 + x = 10$.

$2x^2 + x - 10 = 0 \rightarrow a = 2, b = 1, c = -10$

$$x = \frac{-b \pm \sqrt{b^2 - 4ac}}{2a}$$
$$x = \frac{-1 \pm \sqrt{1^2 - 4(2)(-10)}}{2(2)}$$
$$x = \frac{-1 \pm \sqrt{81}}{4}$$
$$x = 2 \quad or \quad x = -\frac{10}{4}$$

The coordinates of the vertex of the graph of $f(x) = ax^2 + bx + c$ are $\left(-\frac{b}{2a}, f\left(-\frac{b}{2a}\right)\right)$.

Exercises

Use the quadratic formula to solve each equation.

54. $2x + 1 = 2x^2$
55. $6x = 2 - 5x^2$
56. $x^2 - 7x = -10$
57. $x^2 + 6x = -8$
58. $11x = 5x^2 - 3$
59. $x = 6x^2 - 3$
60. $3 = x^2 + 5x$
61. $x^2 = 1 - x$

For each function, find the coordinates of the vertex of the graph.

62. $f(x) = x^2 + 7x + 6$
63. $f(x) = x^2 - x - 12$
64. $g(x) = x^2 + 2x - 3$
65. $f(n) = n^2 + 12n + 5$

LESSON 5.6

Key Skills

Find and classify all roots of a quadratic equation.

Solve $x^2 + 8 = 0$. *$a = 1$, $b = 0$, and $c = 8$*

$$b^2 - 4ac = 0^2 - 4(1)(8) = -32$$

Because the discriminant is less than zero, the solutions are imaginary.

$$x^2 + 8 = 0$$
$$x = \pm\sqrt{-8}$$
$$x = \pm i\sqrt{8}, \text{ or } \pm 2i\sqrt{2}$$

Exercises

Determine the number of real solutions for each equation by using the discriminant.

66. $4x^2 - 20x = -25$ 1
67. $9x^2 + 12x = -2$ 2
68. $x^2 = 21x - 110$ 2
69. $-x^2 + 6x = 10$ 0

Solve each equation. Write your answers in the form *a* + *bi*.

70. $x^2 - 6x + 25 = 0$
71. $x^2 + 10x + 34 = 0$
72. $x^2 + 8x + 20 = 0$
73. $x^2 - 6x + 11 = 0$
74. $4x^2 = 2x - 1$
75. $3x^2 + 2 = 2x$

76. $2 - 12i$ 77. $1 - 7i$ 80. $4 + 13i$ 81. $6 - 9i$

Graph and perform operations on complex numbers.

a. $(3 + 2i) + (-4 + i)$
$= -1 + 3i$

b. $(3 + 2i) - (-4 + i)$
$= 7 + i$

c. $(3 + 2i)(-4 + i)$
$= -12 - 8i + 3i + 2i^2$
$= -14 - 5i - 2$
$= -16 - 5i$

d. $\frac{(3 + 2i)}{-4 + i} = \frac{(3 + 2i)(-4 - i)}{(-4 + i)(-4 - i)}$
$= -\frac{10}{17} - \frac{11}{17}i$

e. $|3 - 3i| = \sqrt{3^2 + (-3)^2}$
$= \sqrt{18}$

The graph of $|3 - 3i|$ is shown at right.

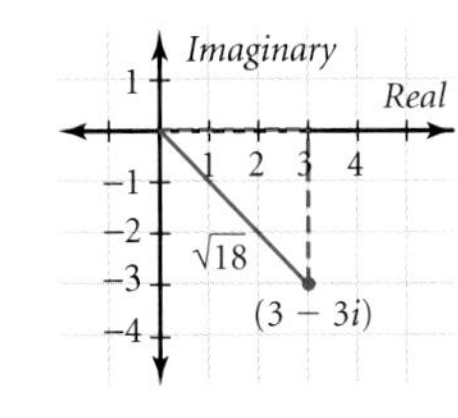

Perform the indicated operation.

76. $(3 - 4i) + (-1 - 8i)$

77. $(-2 - i)(1 + 3i)$

78. $\frac{5 - 2i}{-4 + i}$ $-\frac{22}{17} + \frac{3}{17}i$

79. $\frac{-3 + 3i}{3 - 3i}$ -1

80. $(2 + 7i) - (-2 - 6i)$

81. $(9 - 3i) + (-3 - 6i)$

82. $(2 + i)(4 - 8i)$ $16 - 12i$

83. $(8 - 2i) - (6 + 2i)$ $2 - 4i$

Find each value. Then sketch a diagram that shows the absolute value.

84. $|4 - 2i|$ $\sqrt{20}$

85. $|-3 - i|$ $\sqrt{10}$

86. $|5 - 2i|$ $\sqrt{29}$

87. $|1 + i|$ $\sqrt{2}$

84. $\sqrt{20}$

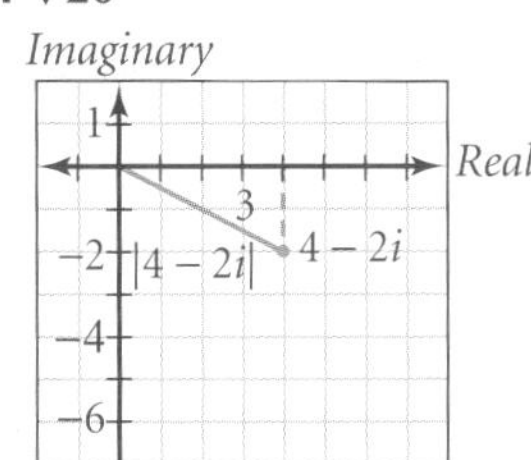

85. $\sqrt{10}$

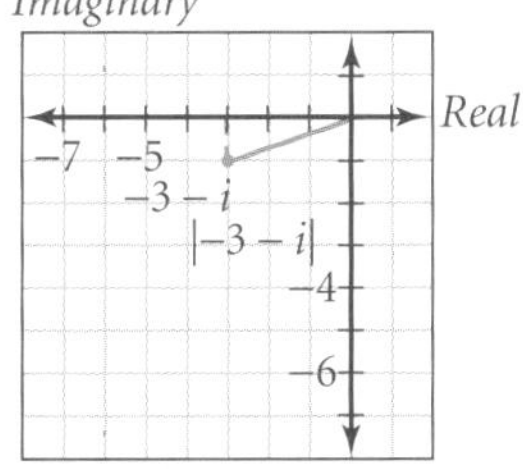

86. $\sqrt{29}$

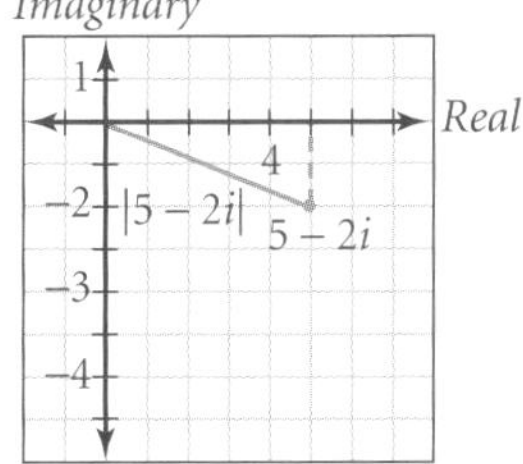

87. $\sqrt{2}$

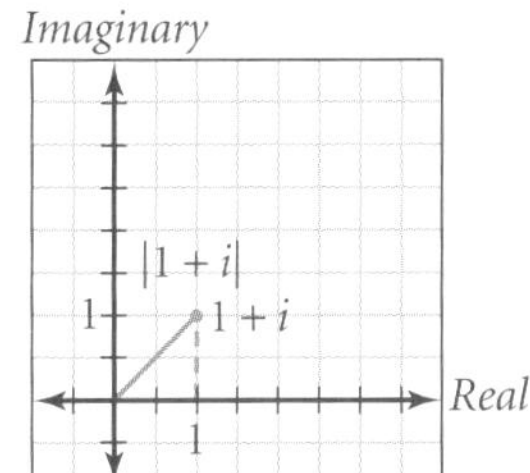

LESSON 5.7

Key Skills

Find a quadratic function that exactly fits three data points.

Find a quadratic function whose graph contains the points $(-1, 2)$, $(0, 3)$, and $(3, -6)$.

$$\begin{aligned} a(-1)^2 + b(-1) + c &= 2 \\ a(0)^2 + b(0) + c &= 3 \\ a(3)^2 + b(3) + c &= -6 \end{aligned} \quad \rightarrow \quad \begin{cases} a - b + c = 2 \\ c = 3 \\ 9a + 3b + c = -6 \end{cases}$$

$$\begin{bmatrix} 1 & -1 & 1 \\ 0 & 0 & 1 \\ 9 & 3 & 1 \end{bmatrix} \begin{bmatrix} a \\ b \\ c \end{bmatrix} = \begin{bmatrix} 2 \\ 3 \\ -6 \end{bmatrix} \quad \rightarrow \quad \begin{bmatrix} a \\ b \\ c \end{bmatrix} = \begin{bmatrix} -1 \\ 0 \\ 3 \end{bmatrix}$$

The quadratic function is $f(x) = -x^2 + 3$.

Find a quadratic model to represent a data set.

If the points of a given data set form an approximately parabolic shape, find the quadratic regression equation.

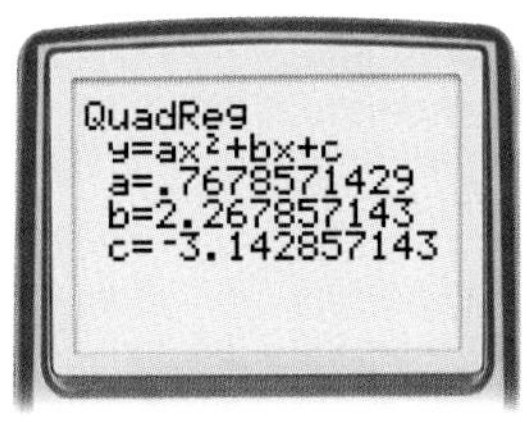

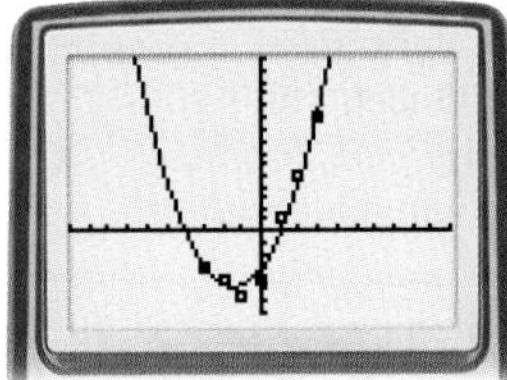

The quadratic regression equation shown above is $f(x) \approx 0.8x^2 + 2.3x - 3.1$.

Exercises

Find a quadratic function that fits each set of data points exactly.

88. $(-10, 185)$, $(-5, 70)$, $(1, -2)$ $f(x) = x^2 - 8x + 5$

89. $(0, -9)$, $(3, 42)$, $(-5, 106)$ $f(x) = 5x^2 + 2x - 9$

90. $(-2, -7)$, $(1, 2)$, $(-1, -2)$ $f(x) = -x^2 + 2x + 1$

91. $(0, 6)$, $(1, 0)$, $(-2, -12)$ $f(x) = -5x^2 - x + 6$

92. $(-1, 4)$, $(0, 4)$, $(4, 84)$ $f(x) = 4x^2 + 4x + 4$

93. $(-1, 3)$, $(1, -3)$, $(0, -3)$ $f(x) = 3x^2 - 3x - 3$

94. $(-2, -1)$, $(3, -6)$, $(0, 9)$ $f(x) = -2x^2 + x + 9$

95. $(-1, 6)$, $(1, 6)$, $(0, -1)$ $f(x) = 7x^2 - 1$

Find a quadratic model to represent each data set.

96 $f(x) \approx 6.8x^2 + 3.9x + 4.2$

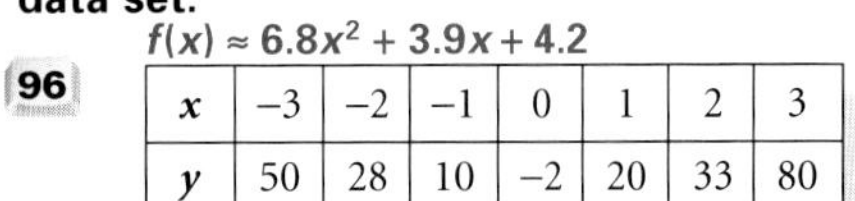

x	−3	−2	−1	0	1	2	3
y	50	28	10	−2	20	33	80

97

x	−3	−2	−1	0	1	2	3
y	−12	−10	−8	−2	−1	−1	−6

98

x	−25	−10	0	15	30
y	1500	275	49	800	2015

97. $f(x) \approx -0.7x^2 + 1.5x - 3.1$
98. $f(x) \approx 2.1x^2 + 0.8x + 134.4$

99. $x < 2$ *or* $x > 6$

100. $-2 < x < 5$

101. $x \leq -5$ *or* $x \geq -2$

102. $-3 < x < \frac{5}{2}$

103. $x < -\frac{3}{4}$ *or* $x > 3$

104. $x \leq 0$ *or* $x \geq \frac{9}{2}$

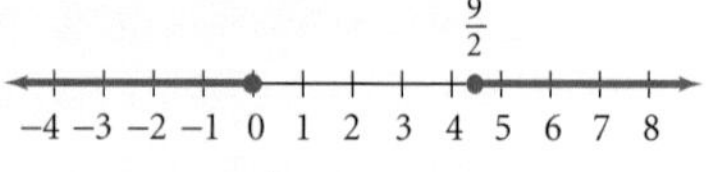

105. $0 \leq x \leq \frac{5}{2}$

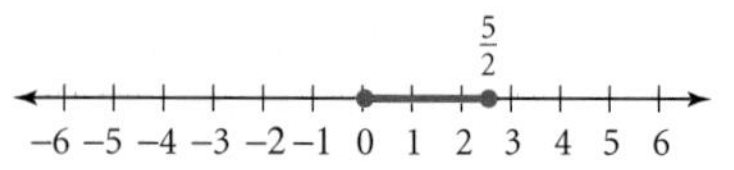

106. $2 \leq x \leq 4$

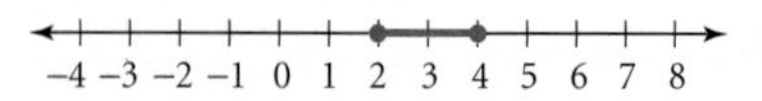

107.

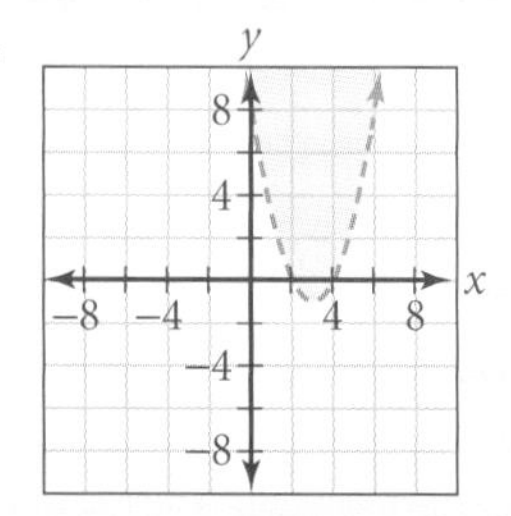

108.

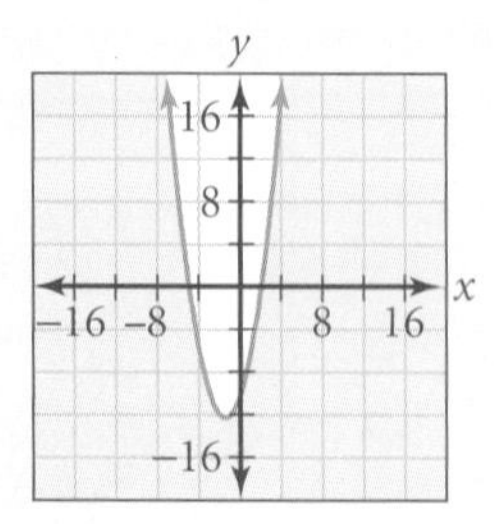

109.

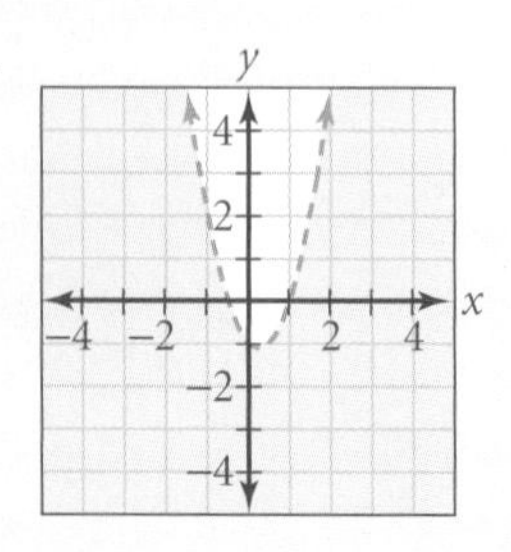

LESSON 5.8

Key Skills

Solve and graph quadratic inequalities in one variable.

$$x^2 - x < 12$$
$$x^2 - x - 12 < 0$$
$$(x + 3)(x - 4) < 0$$

The roots of the related equation, $x^2 - x - 12 = 0$, are -3 and 4.

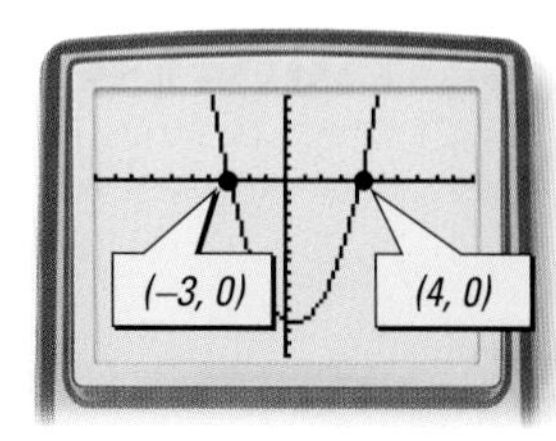

The graph of $y = x^2 - x - 12$ indicates that $x^2 - x - 12$ is negative when x is between the roots. The solution is $-3 < x < 4$.

Solve and graph quadratic inequalities in two variables.

To graph $y \leq x^2 - 2x - 3$, graph $y = x^2 - 2x - 3$, and test a point, such as (0, 0).

$y \leq x^2 - 2x - 3$

$0 \overset{?}{\leq} 0^2 - 2(0) - 3$

$0 \leq -3$ **False**

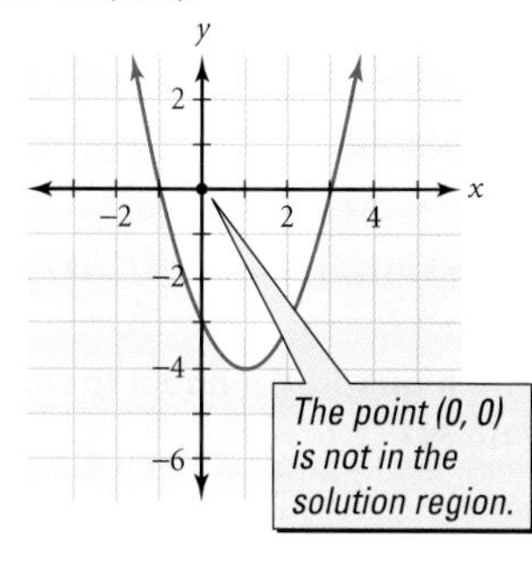

The point (0, 0) is not in the solution region.

Exercises

Solve each quadratic inequality, and graph the solution on a number line.

99. $x^2 - 8x + 12 > 0$ **$x < 2$ *or* $x > 6$**

100. $x^2 - 3x - 10 < 0$ **$-2 < x < 5$**

101. $x^2 + 7x + 10 \geq 0$ **$x \leq -5$ *or* $x \geq -2$**

102. $2x^2 + x < 15$ **$-3 < x < \frac{5}{2}$**

103. $4x^2 > 9x + 9$ **$x < -\frac{3}{4}$ *or* $x > 3$**

104. $2x^2 \geq 9x$ **$x \leq 0$ *or* $x \geq \frac{9}{2}$**

105. $4x^2 \leq 10x$ **$0 \leq x \leq \frac{5}{2}$**

106. $-x^2 + 6x \geq 8$ **$2 \leq x \leq 4$**

Graph each quadratic inequality on a coordinate plane.

107. $y > x^2 - 6x + 8$

108. $y \leq x^2 + 3x - 10$

109. $y - 2x^2 < -x - 1$

110. $y - 5x \leq 2x^2 - 3$

111. $y + 4x + 21 \geq x^2$

112. $y - 5x > 6x^2 + 1$

Applications

113. AVIATION A crate of blankets and clothing is dropped without a parachute from a helicopter hovering at a height of 125 feet. The altitude of the crate, a, in feet, is modeled by $a(t) = -16t^2 + 125$, where t is the time, in seconds, after it is released. How long will it take for the crate to reach the ground? **about 2.8 sec**

114. RECREATION Students are designing an archery target with one ring around the bull's-eye. The bull's-eye has a radius of 6 inches. The area of the outer ring should be 5 times that of the bull's-eye. What should be the radius of the outer circle? **14.7 in.**

110.

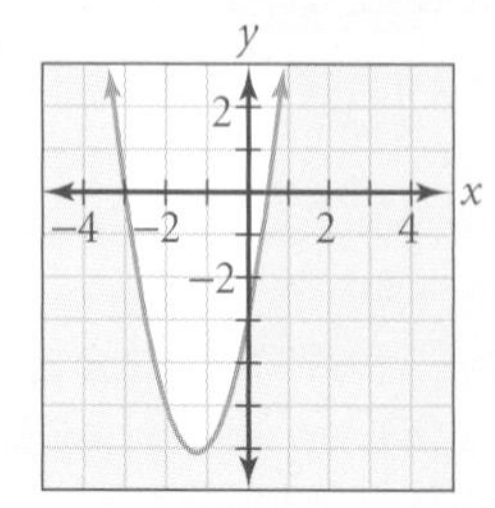

111.

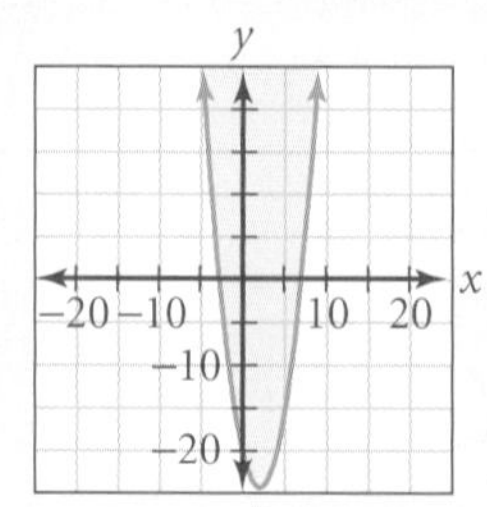

112.

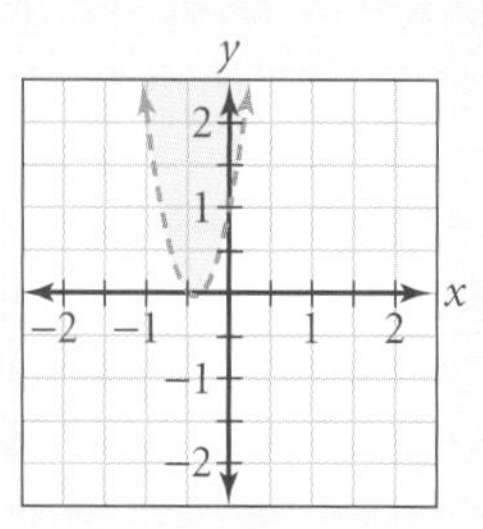

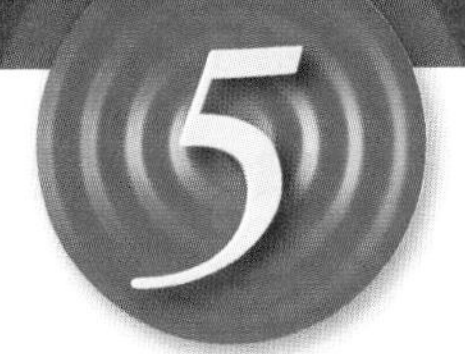

Chapter Test

Write each function in the form $f(x) = ax^2 + bx + c$ and identify *a*, *b*, and *c*. State whether the parabola opens up or down and whether the *y*-coordinate of the vertex is the maximum or the minimum.

1. $f(x) = (x + 3)(x - 4)$
2. $f(x) = -5(x + 1)(x - 7)$
3. $f(x) = -2(x + 3)(3x)$

Solve each equation giving both exact and approximate solutions to the nearest hundredth.

4. $3x^2 = 81$ $\pm 3\sqrt{3}$; ± 5.20
5. $(x - 7)^2 = 12$ $7 \pm 2\sqrt{3}$; 10.46; 3.54

Find the unknown length in the right triangle ABC to the nearest tenth.

6. $a = 7, b = 9$ $\sqrt{130} \approx 11.4$
7. $a = 2, c = 4$ $2\sqrt{3} \approx 3.5$
8. $b = 8.4, c = 9.2$ $\sqrt{14.08} \approx 3.8$

Use factoring and the Zero-Product Property to find the zeros of each quadratic function.

9. $f(x) = x^2 - 9x$ $x(x - 9)$; 0 and 9
10. $f(x) = 4x^2 - 64$ $4(x - 4)(x + 4)$; 4 and −4
11. $f(x) = 4x^2 - 4x + 1$ $(2x - 1)^2$; $\frac{1}{2}$
12. $f(x) = x^2 - 3x - 10$ $(x + 2)(x - 5)$; −2 and 5
13. **NUMBER THEORY** The product of two numbers is 90. One number is 3 more than twice the other number. Model these numbers with a quadratic equation. Solve the equation by factoring and using the Zero-Product Property. $x(2x + 3) = 90$; 6 and 15; −7.5 and −12

Solve each quadratic equation by completing the square.

14. $x^2 - 8x + 4 = 0$ $4 \pm 2\sqrt{3}$
15. $2x^2 - 11x + 5 = 0$ 5, $\frac{1}{2}$
16. **GEOMETRY** The area of a triangle is 24 square inches. The height is 4 inches shorter than the base. Find the height and the base of the triangle. h = 5.21 in., b = 9.21 in.

Use the quadratic formula to solve each equation.

17. $x^2 + 2x - 5 = 0$ $-1 \pm \sqrt{6}$
18. $-3x^2 + 15 = 12x$ 1 or −5

For each quadratic function, write the equation for the axis of symmetry, and find the coordinates of the vertex.

19. $y = x^2 - 7x + 10$ $x = \frac{7}{2}$; $\left(\frac{7}{2}, -\frac{9}{4}\right)$
20. $y = 3x^2 + 18x + 6$ $x = -3$; $(-3, -21)$

Use the discriminant to determine the number of real solutions.

21. $x^2 + 2x + 5 = 0$ 0
22. $-3x^2 = 5 + 3x$ 0
23. $4x^2 = 27$ 2

Perform the indicated operations.

24. $(3 + 2i) + (5 - 7i)$ $8 - 5i$
25. $(2 + i) - (6 - 3i)$ $-4 + 4i$
26. $3i(7 + 3i)$ $-9 + 21i$
27. $(-2 + i)(3 - 4i)$ $-2 + 11i$
28. $\frac{2 + 3i}{1 - i}$ $\frac{-1 + 5i}{2}$
29. $|5 + 12i|$ 13

Find a quadratic function that fits each set of data points exactly.

30. $(-1, -6), (2, 3), (1, -2)$ $y = x^2 + 2x - 5$
31. $(2, -11), (3, 9), (-1, -23)$ $y = 4x^2 - 27$

Solve each quadratic inequality. Graph the solution on a number line.

32. $x^2 - x - 12 > 0$
33. $15x^2 - 2x - 8 \geq 0$

Graph each quadratic inequality on a coordinate plane.

34. $y \leq x^2 + 4x - 5$
35. $y + 1 > x^2 - 2x - 7$

Chapter Test

1. $f(x) = x^2 - x - 12$; $a = 1, b = -1, c = -12$; up; minimum
2. $f(x) = -5x^2 - 30x + 35$; $a = -5, b = 30, c = 35$; down; maximum
3. $f(x) = -6x^2 - 18x$; $a = -6, b = -18, c = 0$; down; maximum

32. $x < -3$ or $x > 4$

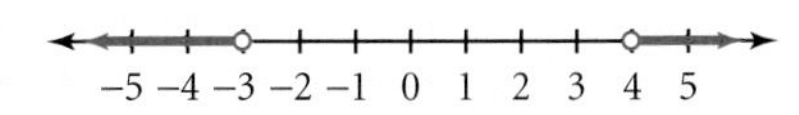

33. $x \leq -\frac{2}{3}$ or $x \geq \frac{4}{5}$

34.

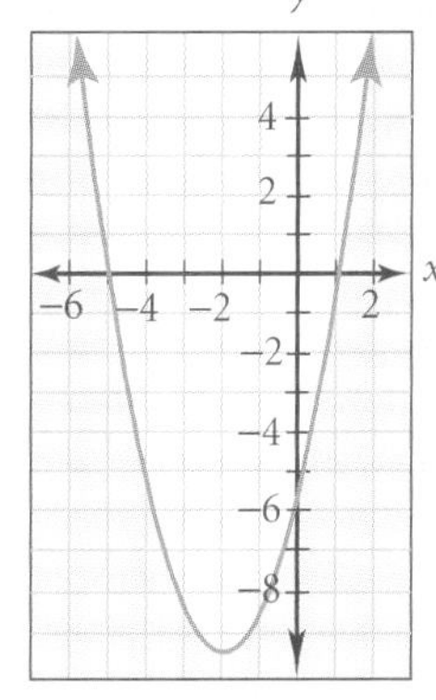

35.

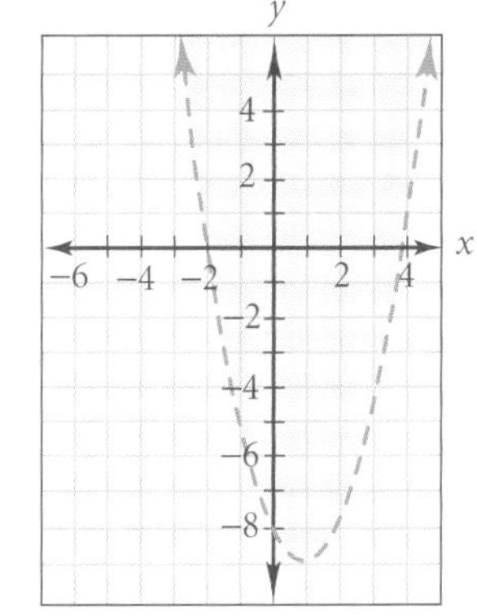

College Entrance Exam Practice

Multiple-Choice and Quantitative-Comparison Samples

The first half of the Cumulative Assessment contains two types of items found on standardized tests—multiple-choice questions and quantitative-comparison questions. Quantitative-comparison items emphasize the concepts of equality, inequality, and estimation.

Free-Response Grid Samples

The second half of the Cumulative Assessment is a free-response section. This part of the Cumulative Assessment requires student-produced response items like those commonly found on college entrance exams. These questions require the use of machine-scored answer grids. You may wish to have students practice answering these items in preparation for standardized tests.

College Entrance Exam Practice

QUANTITATIVE COMPARISON For Items 1–6, write
A if the quantity in Column A is greater than the quantity in Column B;
B if the quantity in Column B is greater than the quantity in Column A;
C if the quantities are equal; or
D if the relationship cannot be determined from the given information.

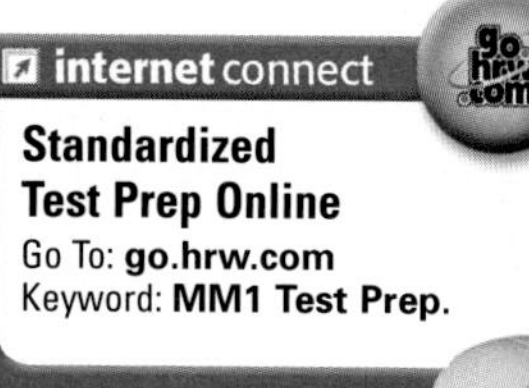

	Column A	Column B	Answers
1. B	$27^{\frac{1}{3}}$	4	(A) (B) (C) (D) [Lesson 2.2]
2. B	$\lvert 2i \rvert$	$\lvert -3i \rvert$	(A) (B) (C) (D) [Lesson 5.6]
3. C	$\left(\frac{1}{2}\right)^{-2}$	$(-2)^2$	(A) (B) (C) (D) [Lesson 2.2]
4. B	Entry m_{11} of the matrix product $M = \begin{bmatrix} 1 & 2 \\ 3 & 1 \end{bmatrix}\begin{bmatrix} 1 & 2 \\ 1 & 2 \end{bmatrix}$	Entry m_{11} of the matrix product $M = \begin{bmatrix} 2 & 3 \\ 3 & 2 \end{bmatrix}\begin{bmatrix} 1 & 2 \\ 1 & 2 \end{bmatrix}$	(A) (B) (C) (D) [Lesson 4.2]
5. B	The largest root of the equation $x^2 - 3x + 2 = 0$	The largest root of the equation $x^2 - 4x + 3 = 0$	(A) (B) (C) (D) [Lesson 5.3]
6. C	The value that completes the square for $x^2 + 16x$	8	(A) (B) (C) (D) [Lesson 5.4]

7. Find $2R - N$ given $R = \begin{bmatrix} 3 & 5 \\ -2 & 0 \end{bmatrix}$ and $N = \begin{bmatrix} 1 & -1 \\ 3 & 1 \end{bmatrix}$. ***(LESSON 4.1)***
d

a. $\begin{bmatrix} -2 & 6 \\ -5 & -1 \end{bmatrix}$ **b.** $\begin{bmatrix} 2 & 4 \\ 1 & -1 \end{bmatrix}$
c. $\begin{bmatrix} 2 & 6 \\ -5 & -1 \end{bmatrix}$ **d.** $\begin{bmatrix} 5 & 11 \\ -7 & -1 \end{bmatrix}$

8. How many roots does the equation $5x^2 + 2x + 1 = 0$ have? ***(LESSON 5.6)***
b

a. 2 real roots **b.** no real roots
c. 1 real root **d.** 2 complex roots

9. Which is the value of i^{13}? ***(LESSON 5.6)***
c

a. 1 **b.** -1
c. i **d.** $-i$

10. Simplify: $\frac{a^2b^{-1}}{a^{-3}b^2}$. ***(LESSON 2.2)***
b

a. $\frac{a}{b}$ **b.** $\frac{a^5}{b^3}$
c. $\frac{b^2}{a}$ **d.** $\frac{b^3}{a^5}$

11. Which is the solution of the system? $\begin{cases} 5x + y = 11 \\ 3x + 2y = 8 \end{cases}$ ***(LESSON 3.2)***
b

a. $(3, 2)$ **b.** $(2, 1)$
c. $(-1, 2)$ **d.** $(5, 6)$

15.

$-24 \quad -22 \quad -20 \quad -18 \quad -16 \quad -14 \quad -12$

12. Which is a correct factorization of $x^2 + 5x + 6$? **(LESSON 5.3)** b

a. $(x + 1)(x + 6)$ **b.** $(x + 2)(x + 3)$
c. $(x - 1)(x - 6)$ **d.** $(x - 2)(x - 3)$

13. Which term describes the system? b

$\begin{cases} 2x + 5y = 3 \\ 4x + 10y = 6 \end{cases}$ **(LESSON 3.1)**

a. inconsistent **b.** dependent
c. independent **d.** incompatible

14. Which is the inverse of the function $f(x) = 3x + 2$? **(LESSON 2.5)** d

a. $g(x) = 6x$ **b.** $g(x) = \frac{x-3}{2}$
c. $g(x) = 3 + \frac{x}{2}$ **d.** $g(x) = \frac{x-2}{3}$

15. Graph $-\frac{1}{3}x \leq 6$. **(LESSON 3.3)**

16. Find $(2 + i)(3 + 2i)$. **(LESSON 5.6)** $4 + 7i$

17. Solve $\begin{cases} 3x - 2y = 2 \\ x + y = 4 \end{cases}$. **(LESSON 3.1)** (2, 2)

18. Let $A = \begin{bmatrix} 3 & 2 \\ 1 & 4 \end{bmatrix}$ and $B = \begin{bmatrix} 1 & 4 \\ 2 & 3 \end{bmatrix}$. Find $A + B$. **(LESSON 4.1)** $\begin{bmatrix} 4 & 6 \\ 3 & 7 \end{bmatrix}$

19. Solve $x^2 + 3x + 1 = 0$. **(LESSON 5.5)** $-\frac{3}{2} \pm \frac{\sqrt{5}}{2}$

20. Let $A = \begin{bmatrix} 1 & 4 \\ 3 & 1 \end{bmatrix}$ and $B = \begin{bmatrix} 0 & 1 \\ 2 & 3 \end{bmatrix}$. Find $A + B$. **(LESSON 4.1)** $\begin{bmatrix} 1 & 5 \\ 5 & 4 \end{bmatrix}$

21. Write the pair of parametric equations $\begin{cases} x(t) = 1 - t \\ y(t) = 2 - t \end{cases}$ as a single equation in x and y. **(LESSON 3.6)** $y = 1 + x$

22. Write the function for the graph of $f(x) = x^2$ translated 3 units to the left. **(LESSON 2.7)** $g(x) = (x + 3)^2$

23. Let $f(x) = 3x + 1$ and $g(x) = x^2$. Find $(f \cdot g)(x)$. **(LESSON 2.4)** $(f \cdot g)(x) = 3x^3 + x^2$

24. Find the product $\begin{bmatrix} 2 & 3 \\ 2 & 1 \end{bmatrix}\begin{bmatrix} 3 & 2 \\ 1 & 0 \end{bmatrix}$. **(LESSON 4.2)** $\begin{bmatrix} 9 & 4 \\ 7 & 4 \end{bmatrix}$

25. Find the value of $i^2 + i^4$. **(LESSON 5.6)** 0

26. Evaluate $h(x) = 11 - \frac{1}{2}x$ for $x = -6$. **(LESSON 2.3)** 14

27. Given $5\begin{bmatrix} 1 & x \\ x - y & 5 \end{bmatrix} = \begin{bmatrix} 5 & 15 \\ 20 & 25 \end{bmatrix}$, find x and y. **(LESSON 4.1)** $x = 3$, $y = -1$

28. CHEMISTRY A scientist wants to create 60 milliliters of a 5% salt solution from a 2% salt solution and a 12% salt solution. How much of each should be used? **(LESSON 3.1)** 42 mL of 2%, 18 mL of 12%

FREE RESPONSE GRID The following questions may be answered by using a free-response grid such as that commonly used by standardized-test services.

	/	/	
.	.	.	.
	0	0	0
1	1	1	1
2	2	2	2
3	3	3	3
4	4	4	4
5	5	5	5
6	6	6	6
7	7	7	7
8	8	8	8
9	9	9	9

29. Evaluate $8^{\frac{2}{3}}$. **(LESSON 2.2)** 4

30. Simplify $\frac{(3^2 - 7)^2}{3^{(2^2 - 2)}}$. **(LESSON 2.1)** $\frac{4}{9}$

31. Find $\left|\frac{\sqrt{2}}{4} + \frac{\sqrt{2}}{4}i\right|$. **(LESSON 5.6)** $\frac{1}{2}$

32. Find the discriminant for $x^2 + 4x + 1 = 0$. **(LESSON 5.6)** 12

33. What is the maximum value of $f(x) = -x^2 + 2x + 1$? **(LESSON 5.1)** 2

34. Let $A = \begin{bmatrix} 3 & 2 \\ 4 & 1 \end{bmatrix}$ and $B = \begin{bmatrix} 2x & 2 \\ 4 & 1 \end{bmatrix}$. For what value of x does $A = B$? **(LESSON 4.1)** $\frac{3}{2}$

35. Find the maximum value of the objective function $P = 2x + 3y$ that satisfies the given constraints. **(LESSON 3.5)** 12

$$\begin{cases} x \geq 0, y \geq 0 \\ x + y \leq 4 \\ 2x + y \geq 2 \end{cases}$$

36. PHYSICS A ball is dropped from a height of 10 feet. If the ball's height is modeled by $h(t) = -16t^2 + 10$, where h represents the height of the ball in feet and t represents time in seconds, how many seconds, to the nearest tenth, will it take the ball to reach the ground? **(LESSON 5.2)** 0.87

37. BUSINESS A company's profit on sales of digital pagers is modeled by the function $P(x) = -x^2 + 90x + 497{,}975$, where x represents the price of a pager in dollars. To the nearest dollar, what price gives the maximum profit? **(LESSON 5.4)** 45

Keystroke Guide for Chapter 5

Essential keystroke sequences (using the model TI-82 or TI-83 graphics calculator) are presented below for Activities and Examples found in this chapter that require or recommend the use of a graphics calculator.

internet connect

For Keystrokes of other graphing calculator models, visit the HRW web site at **go.hrw.com** and enter the keyword **MB1 CALC**.

LESSON 5.1

TECHNOLOGY
Page 274

Create a table of values for $y = \frac{11}{10}x + \frac{1}{19}x^2$.

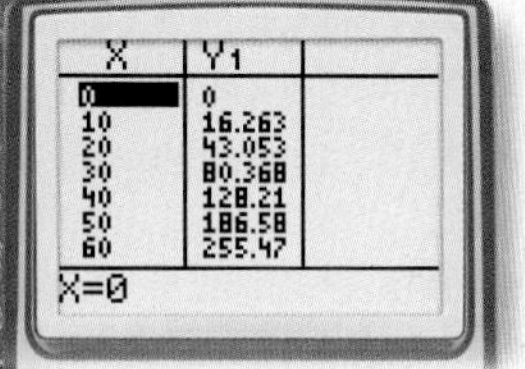

Enter the function:

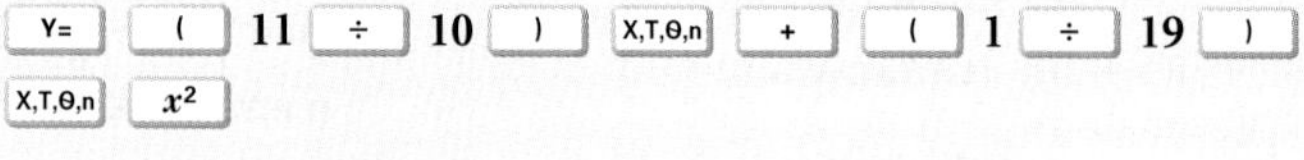

Create a table of values:

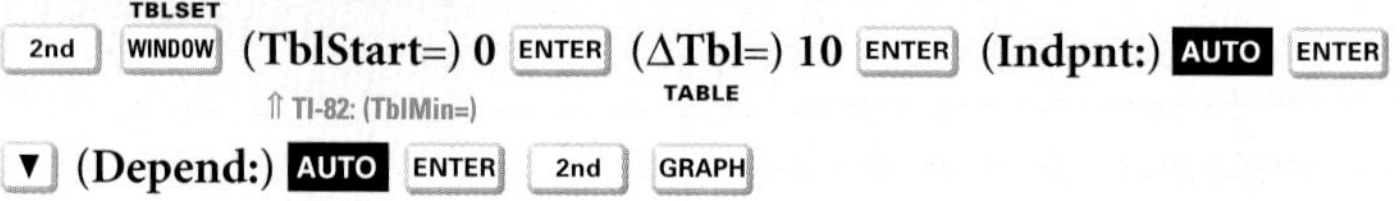

TBLSET
2nd WINDOW (TblStart=) 0 ENTER (ΔTbl=) 10 ENTER (Indpnt:) AUTO ENTER
⇑ TI-82: (TblMin=)
TABLE
▼ (Depend:) AUTO ENTER 2nd GRAPH

EXAMPLES 2 and 3
Pages 276 and 277

For Example 2, graph $y = x^2 - x + 1$, and find the maximum or minimum value at the vertex.

Use friendly viewing window [–4.7, 4.7] by [–2, 6].

Graph the function:

Find the minimum value:
Press TRACE, and use your cursor.

Create a table of values:
Use a keystroke sequence similar to that used in the Technology example above. First use TblStart = –2 and ΔTbl = 1. Then refine the table by using TblStart = 0 and ΔTbl = 0.1.

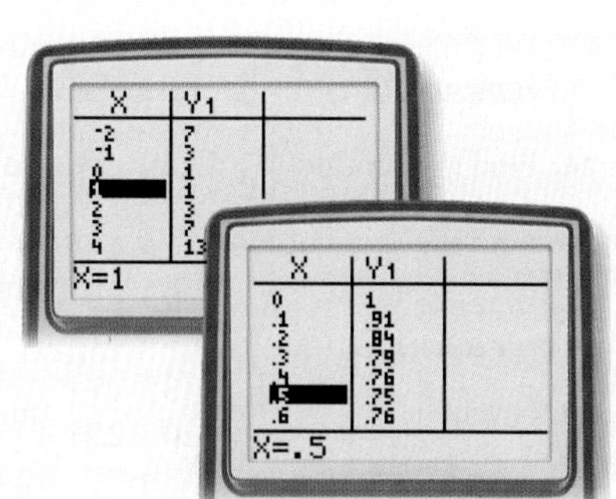

For Example 3, use a keystroke sequence similar to that above to graph each function. Use viewing window [–10, 10] by [–10, 10].

LESSON 5.2

EXAMPLES 1 and 2 For Example 1, graph $y = 4x^2 + 13$ and $y = 253$ on the same screen, and find any points of intersection.

Page 282

Use viewing window [–10, 10] by [–150, 400].

Graph the functions:
Use a keystroke sequence similar to that in Example 2 of Lesson 5.1.

Find any points of intersection:

Move your cursor as indicated.

2nd TRACE (CALC) 5:intersect (First curve?)
ENTER (Second curve?) ENTER
(Guess?) ENTER

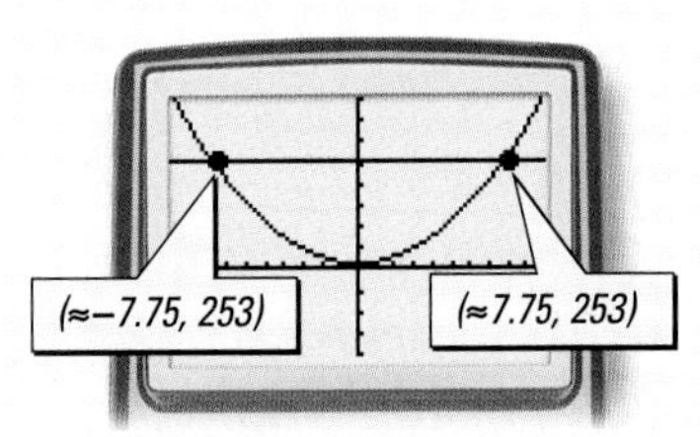

For Example 2, use viewing window [–5, 10] by [–25, 150]. Use a keystroke sequence similar to that above.

EXAMPLE 3 Graph $y = -16x^2 + 68$, and find the reasonable x-intercept.

Page 283

Use viewing window [–5, 5] by [–10, 80].

Graph the functions:
Use a keystroke sequence similar to that in Example 2 of Lesson 5.1.

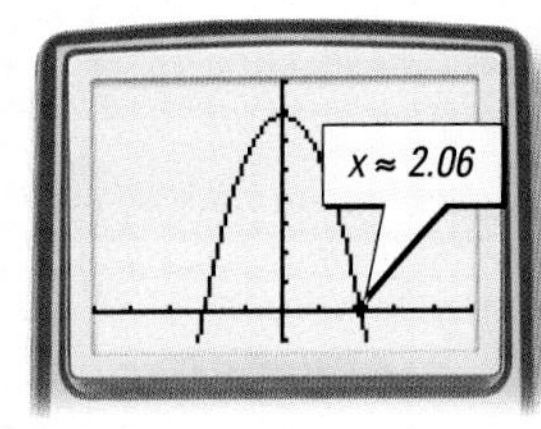

Find the x-intercepts:

Move your cursor as indicated.

2nd TRACE (CALC) 2:zero (Left Bound?) ENTER
(TI-82: 2:root)
(Right Bound?) ENTER (Guess?) ENTER

LESSON 5.3

EXAMPLES 4 and 6 For Example 4, use viewing window [–9, 3] by [–7, 2].
For part **a** of Example 6, use viewing window [–2, 7] by [–20, 8].
For part **b** of Example 6, use viewing window [0, 12] by [–6, 6].

Pages 293 and 294

To graph the functions, use a keystroke sequence similar to that in Example 2 of Lesson 5.1. To find the zeros for Example 6, use a keystroke sequence similar to finding the x-intercepts in Example 3 of Lesson 5.2. Repeat for each zero.

EXAMPLE 7 Make a table of values for $y = 2x^2 - x - 66$.

Page 295

Use a keystroke sequence similar to that used in the Technology example of Lesson 5.1. Use TblStart = 3 and ΔTbl = 1.

LESSON 5.4

EXAMPLE 3 **Solve $2x^2 + 6x = 7$ by graphing.**

Page 301

Use viewing window [–5, 5] by [–12, 12].

To graph $y = 2x^2 - 6x$ and $y = 7$ and find the x-coordinates of any points of intersection, use keystroke sequences similar to those in Example 2 of Lesson 5.1 and Example 1 of Lesson 5.2.

To graph $y = 2x^2 - 6x - 7$ and find any zeros, use keystroke sequences similar to those in Example 2 of Lesson 5.1 and Example 6 of Lesson 5.3.

EXAMPLE 5 **Graph $y = \frac{3}{5000}x^2 - \frac{3}{5}x + 200$, and find the coordinates of the lowest point.**

Page 303

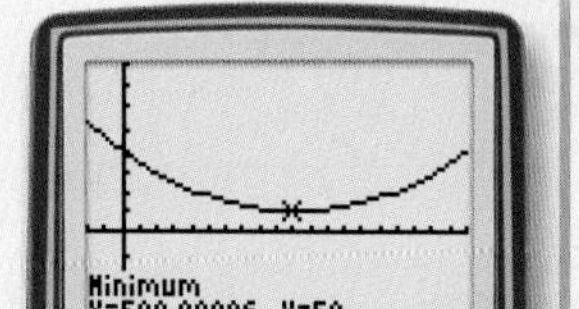

Minimum
X=500.00006 Y=50

Use viewing window [–100, 1000] by [–150, 40].

Find the minimum value:

2nd CALC TRACE 3:minimum (Left Bound?) ENTER (Right Bound?) ENTER

⇑ TI-82: (Lower Bound?) ⇑ TI-82: (Upper Bound?)

(Guess)? ENTER

LESSON 5.5

Activity

Page 309

For Step 2, use friendly viewing window [–9.4, 9.4] by [–7, 7]. Press TRACE, and use your cursor to find the coordinates of each vertex.

LESSON 5.6

EXAMPLE 6 **Evaluate the expression $\frac{2+5i}{2-3i}$, and express the answer with fractions.**

Page 318

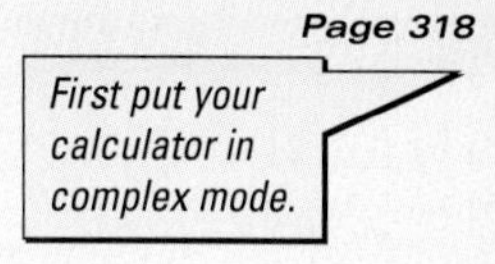

First put your
calculator in
complex mode.

MODE a+bi ENTER 2nd QUIT MODE (2 + 5 2nd i .) ÷ (2 – 3 2nd i .) ENTER MATH 1:▷Frac ENTER ENTER

The TI-82 does not have a complex mode.

LESSON 5.7

EXAMPLES 1 and 2 **For Step 2 of Example 1, solve $\begin{bmatrix} 1 & 1 & 1 \\ 4 & 2 & 1 \\ 36 & 6 & 1 \end{bmatrix}\begin{bmatrix} a \\ b \\ c \end{bmatrix} = \begin{bmatrix} 3 \\ -3 \\ 13 \end{bmatrix}$.**

Pages 323 and 324

Enter the coefficient matrix and the constant matrix:

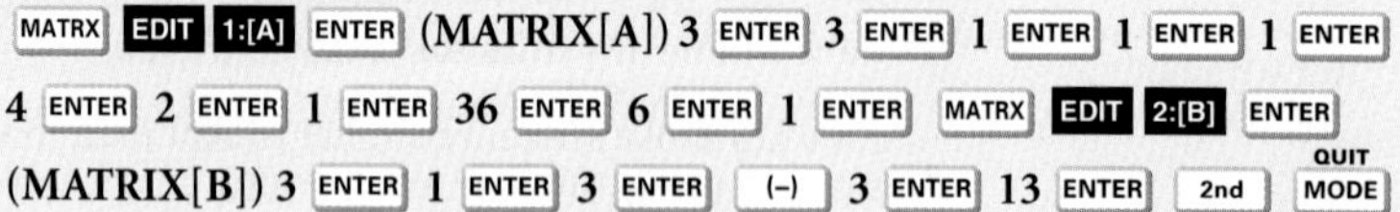

MATRX EDIT 1:[A] ENTER (MATRIX[A]) 3 ENTER 3 ENTER 1 ENTER 1 ENTER 1 ENTER 4 ENTER 2 ENTER 1 ENTER 36 ENTER 6 ENTER 1 ENTER MATRX EDIT 2:[B] ENTER (MATRIX[B]) 3 ENTER 1 ENTER 3 ENTER (–) 3 ENTER 13 ENTER 2nd QUIT MODE

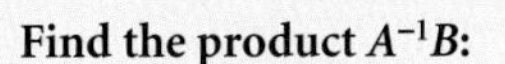

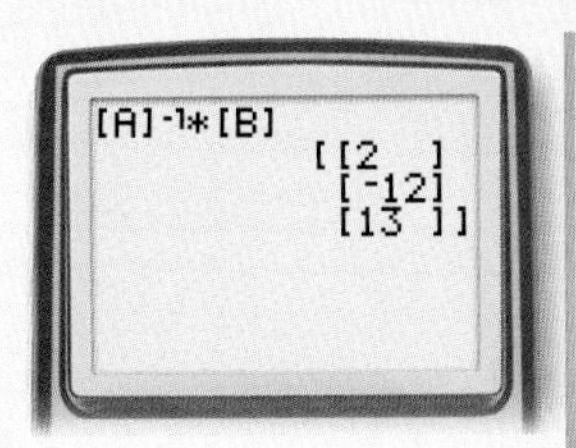

Find the product $A^{-1}B$:

MATRX NAMES 1:[A] ENTER x^{-1} × MATRX NAMES 2:[B] ENTER ENTER

Plot points (1, 3), (2, –3), and (6, 13), and make a scatter plot:

Use viewing window [–5, 10] by [–10, 20].

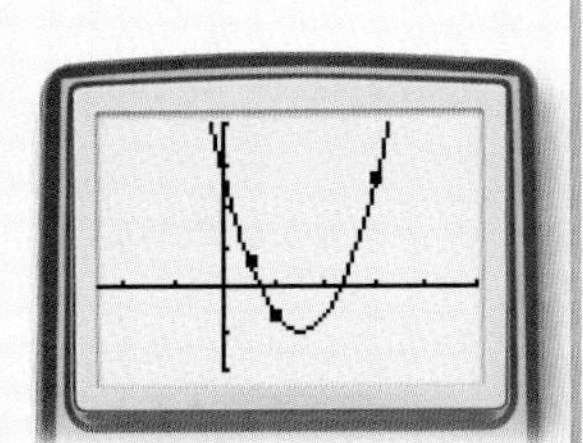

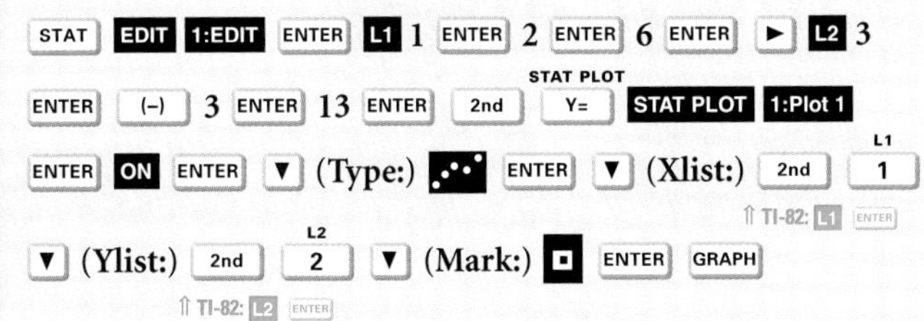

STAT EDIT 1:EDIT ENTER L1 1 ENTER 2 ENTER 6 ENTER ► L2 3

ENTER (–) 3 ENTER 13 ENTER 2nd Y= (STAT PLOT) STAT PLOT 1:Plot 1

ENTER ON ENTER ▼ (Type:) [scatter] ENTER ▼ (Xlist:) 2nd 1 (L1)
⇑ TI-82: L1 ENTER

▼ (Ylist:) 2nd 2 (L2) ▼ (Mark:) ▫ ENTER GRAPH
⇑ TI-82: L2 ENTER

To graph $f(x) = 2x^2 - 12x + 13$, use a keystroke sequence similar to that in Example 2 of Lesson 5.1.

For Example 2, use a similar keystroke sequence. Use viewing window [–2, 10] by [–5, 20].

EXAMPLE 3 Page 325

Create a scatter plot of the given data, and find a quadratic model to represent the data.

Create the scatter plot:
Use a keystroke sequence similar to that in Example 2 of this lesson.

Find a quadratic model:
Use a keystroke sequence similar to that given on page 324.

LESSON 5.8

Page 330

For Step 2, use a keystroke sequence similar to that used in the Technology example of Lesson 5.1. Use TblStart = –2 and ΔTbl = 1.

EXAMPLES 1 and 2 Pages 331 and 332

For Example 1, graph $y = x^2 - 2x - 15$ and find the zeros of the function.

Use viewing window [–5, 7] by [–20, 15].

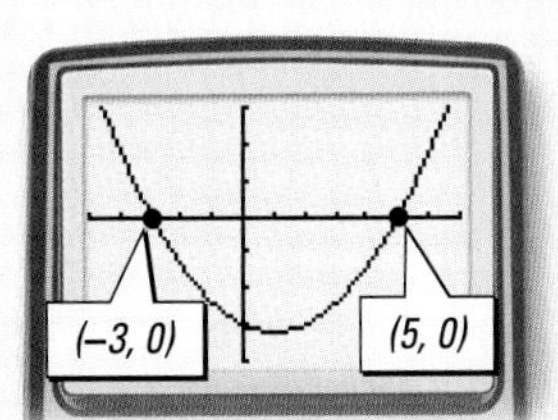

Graph the function:
Use a keystroke sequence similar to that in Example 2 of Lesson 5.1.

Find the zeros of the function:
Use a keystroke sequence similar to that used to find the x-intercepts in Example 3 of Lesson 5.2.

For Example 2, use a similar keystroke sequence. Use viewing window [0, 25] by [–100, 1500].

6 Exponential and Logarithmic Functions

Lesson Presentation CD-ROM
PowerPoint® presentations for each lesson 6.1–6.7

CHAPTER PLANNING GUIDE

Lesson	6.1	6.2	6.3	6.4	6.5	6.6	6.7	Project and Review
Pupil's Edition Pages	354–361	362–369	370–376	377–384	385–391	392–399	402–409	400–401, 410–417
Practice and Assessment								
Extra Practice (Pupil's Edition)	959	959	960	960	961	961	962	
Practice Workbook	35	36	37	38	39	40	41	
Practice Masters Levels A, B, and C	103–105	106–108	109–111	112–114	115–117	118–120	121–123	
Standardized Test Practice Masters	40	41	42	43	44	45	46	47
Assessment Resources	70	71	72	73	75	76	77	74, 78–83
Visual Resources								
Lesson Presentation Transparencies Vol. 1	137–140	141–144	145–148	149–152	153–156	157–160	161–164	
Teaching Transparencies		21, 22	23, 24					
Answer Key Transparencies	208–210	211–215	216–220	221–223	224–225	226–231	232–234	
Quiz Transparencies	6.1	6.2	6.3	6.4	6.5	6.6	6.7	
Teacher's Tools								
Reteaching Masters	69–70	71–72	73–74	75–76	77–78	79–80	81–82	
Make-Up Lesson Planner for Absent Students	35	36	37	38	39	40	41	
Student Study Guide	35	36	37	38	39	40	41	
Spanish Resources	35	36	37	38	39	40	41	
Block Scheduling Handbook								12–13
Activities and Extensions								
Lesson Activities	35	36	37	38	39	40	41	
Enrichment Masters	35	36	37	38	39	40	41	
Cooperative-Learning Activities	35	36	37	38	39	40	41	
Problem Solving/ Critical Thinking	35	36	37	38	39	40	41	
Student Technology Guide	35	36	37	38	39	40	41	
Long Term Projects								21–24
Writing Activities for Your Portfolio								16–18
Tech Prep Masters								25–28
Building Success in Mathematics								15–16

LESSON PACING GUIDE

Lesson	6.1	6.2	6.3	6.4	6.5	6.6	6.7	Project and Review
Traditional	1 day	1 day	1 day	1 day	1 day	1 day	2 days	2 days
Block	$\frac{1}{2}$ day	$\frac{1}{2}$ day	$\frac{1}{2}$ day	$\frac{1}{2}$ day	$\frac{1}{2}$ day	$\frac{1}{2}$ day	1 day	1 day
Two-Year	2 days	2 days	2 days	2 days	2 days	2 days	4 days	4 days

CONNECTIONS AND APPLICATIONS

Lesson	6.1	6.2	6.3	6.4	6.5	6.6	6.7	Review
Algebra	354–361	362–369	370–376	377–384	385–391	392–399	402–409	410–417
Geometry								415
Patterns in Data	355, 359							
Statistics		366, 368						
Transformations		363, 364, 368	373, 375		386	398		
Business and Economics	361	365, 367, 368		384		392, 394, 396, 397, 398, 399	409	412, 414, 415, 417
Science and Technology	354, 357, 358, 359, 360		373, 374, 375, 376	383	386, 387, 389, 390, 391	396, 397, 398, 399	402, 403, 405, 407, 408	413, 414, 415
Social Studies	356, 358, 359, 360						408	415

BLOCK SCHEDULING GUIDE

Day	Lesson	Teacher Directed: Lesson Examples, Teaching Transparencies	Student Guided Activity, Try This	Cooperative-Learning Activity, Lesson Activity, Student Technology Guide	Practice: Practice & Apply, Extra Practice, Practice Workbook	Assessment: Quiz, Mid-Chapter Assessment	Problem Solving, Reteaching
1	6.1	10 min	10 min	8 min	25 min	8 min	8 min
	6.2	10 min	10 min	7 min	25 min	7 min	7 min
2	6.3	10 min	10 min	8 min	25 min	8 min	8 min
	6.4	10 min	10 min	7 min	25 min	7 min	7 min
3	6.5	10 min	10 min	8 min	25 min	8 min	8 min
	6.6	10 min	10 min	7 min	25 min	7 min	7 min
4	6.7	10 min	15 min	15 min	65 min	15 min	15 min
5	Assess.	50 min **PE:** Chapter Review	90 min **PE:** Chapter Project, Writing Activities	90 min Tech Prep Masters	65 min **PE:** Chapter Assessment, Test Generator	30 min Chap. Assess. (A or B), Alt. Assess. (A or B), Test Generator	15 min

PE: Pupil's Edition

Alternative Assessment

The following suggest alternative assessments for students who may benefit from a different type of assessment than the regular chapter quizzes and the mid-chapter/end-of-chapter test. Visit the HRW web site to get additional Alternative Assessment material.

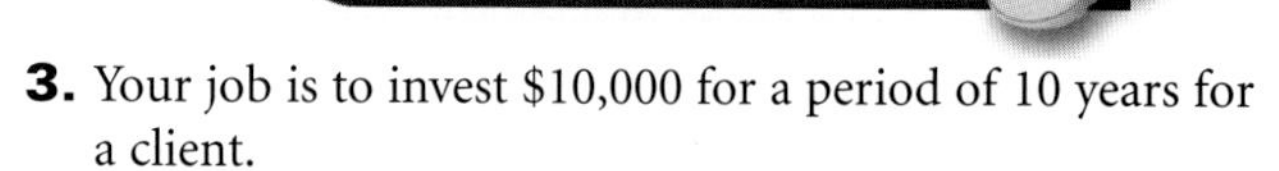

internet connect

Alternative Assessment
Go To: **go.hrw.com**
Keyword: **MB1 Alt Assess**

Performance Assessment

1. **a.** Describe an exponential function.
   b. Distinguish an exponential function from other types of functions.
   c. Identify and explain how an exponential function can be used to model at least two real-world situations. In your responses, show how the evaluation of exponential functions, equations, and inequalities helps solve real-world problems.
2. The *Richter scale* provides a way to measure the intensity of an earthquake.
   a. Describe how the Richter scale is logarithmic.
   b. Explain how the intensity of two earthquakes with Richter scale readings of 5 and 6 are related to one another.
   c. How would you determine the intensity of an earthquake whose Richter scale reading is 5.5?
3. Your job is to invest $10,000 for a period of 10 years for a client.
   a. How would you calculate the growth of the money in an account that pays 6% annual interest compounded quarterly?
   b. How would you calculate the growth of the money in an account that pays 6% annual interest compounded continuously?

Portfolio Project

Suggest that students include the following project in their portfolios.

1. **a.** Write a summary of the the major definitions you learned in this chapter. Include definitions related to functions and their properties.
   b. List various applications of exponential and logarithmic functions. Include illustrations if appropriate.

internet connect

The table below identifies the pages in this chapter that contain internet and technology information.

Content Links

Activities Online	**pages 366, 390, 406**
Portfolio Extensions	**page 409**
Homework Help Online	**pages 358, 367, 375, 382, 389, 397, 407**
Graphic Calculator Support	**page 418**

Resource Links

Parents can go online and find concepts that students are learning–lesson by lesson–and questions that pertain to each lesson, which facilitate parent-student discussion.

Go To: **go.hrw.com**
Keyword: **MB1 Parent Guide**

Technical Support

The following may be used to obtain technical support for any HRW software product.

Online Help: **www.hrwtechsupport.com**
e-mail: **tschrw@hrwtechsupport.com**
HRW Technical Support Center: **(800)323-9239**
7 AM to 10 PM Monday through Friday Central Time

Visit the HRW math web site at: **www.hrw.com/math**

Technology

Lesson Suggestions and Calculator Examples

(Keystrokes are based on a TI-83 calculator.)

Lesson 6.1 Exponential Growth and Decay

Below is a graphics calculator exploration of exponential growth and decay. Have students choose an initial number, such as 2.4, and then choose a multiplier greater than 1 to model growth. Then have them choose a number between 0 and 1 to model decay.

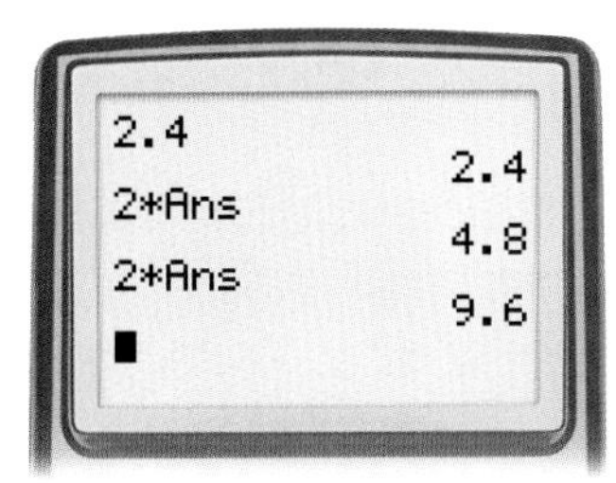

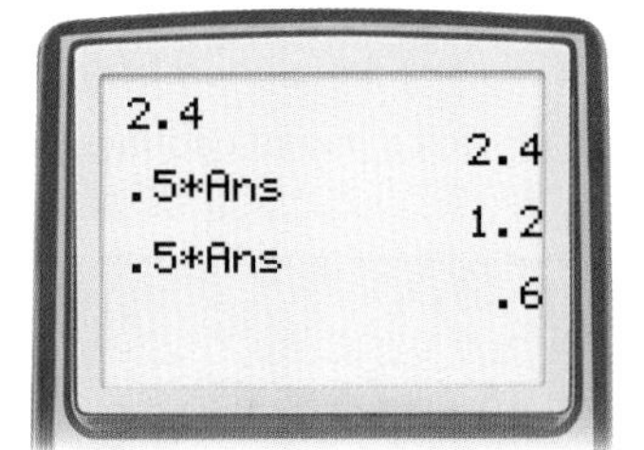

Lesson 6.2 Exponential Functions

Consider having students graph the three functions below.

$f(x) = 2x$ $\quad$ $g(x) = x^2$ $\quad$ $h(x) = 2^x$

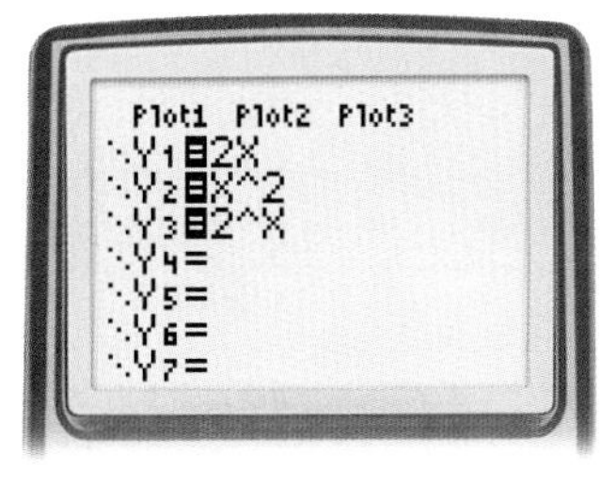

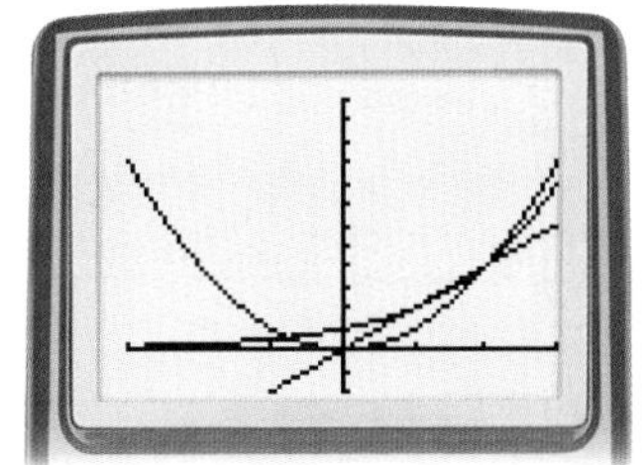

Ask students to compare and contrast the graph of $h(x) = 2^x$ with those of the other functions.

Have students graph both exponential growth and exponential decay on the same display, using $f(x) = 2^x$ and $g(x) = 2^{-x}$. Compare and contrast the graphs.

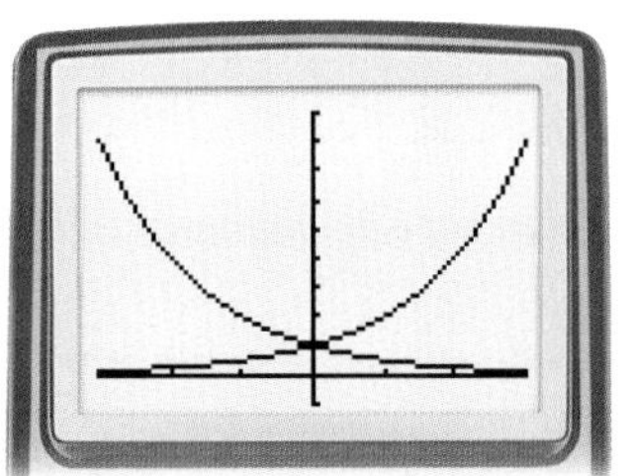

Lesson 6.3 Logarithmic Functions

A logarithmic function can have any positive number, except 1, as its base. The calculator, however, only offers two choices for a base: base 10 (Lesson 6.5) and base e (Lesson 6.6). Students can enter base-10 logarithms in the calculator by pressing LOG.

To help students understand that a logarithm is an exponent, have students evaluate expressions by using powers of 10 and logarithms in base 10. For example, the display at right shows that $\log(10^3) = 3$ and $10^{\log 1000} = 1000$.

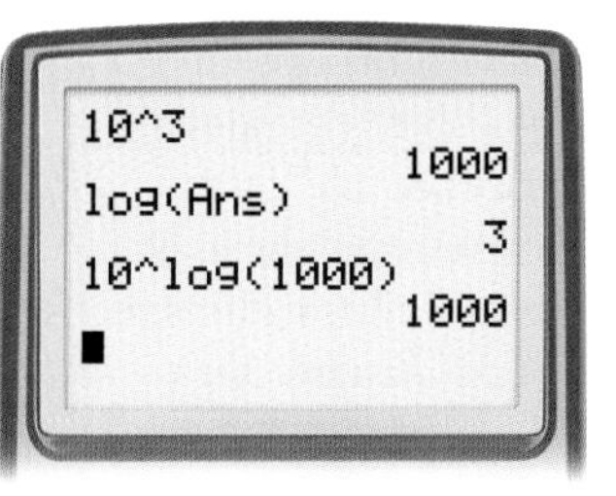

Lesson 6.4 Properties of Logarithmic Functions

You can extend the Activity on page 378 by using a calculator display like the one at right. Have students choose values for a and b in the equation $\log a + \log b = \log ab$. The calculator will complete the calculation.

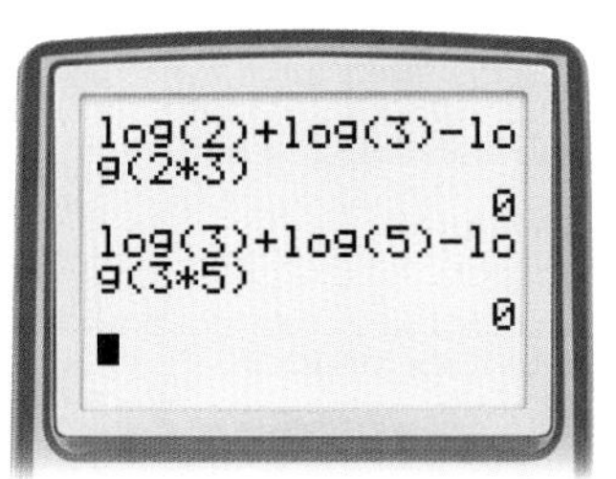

Lesson 6.5 Applications of Common Logarithms

In this lesson, students see the graphical approach used to check an algebraic solution. Ask students how they might solve the equation $\log x = 10^{-x}$. Ask what happens when they try to apply an algebraic approach.

Lesson 6.6 The Natural Base, *e*

Explain to students that they can access e^x by pressing 2nd LN. The notation ln x stands for a natural logarithmic function with base e. Have students evaluate the natural exponential function $f(x) = e^x$ and the natural logarithmic function $g(x) = \ln x$ for specific values of x.

Lesson 6.7 Solving Equations and Modeling

For the graphical approach, have students explore each of the following:

- TRACE
- 2nd TRACE 5:intersect
- 2nd GRAPH

Ask students to describe how each of these features provides information about the solution(s).

For further information, refer to the

- technology discussions in the lessons.
- lesson-related teacher's commentary in the side columns of this *Teacher's Edition.*
- lesson-related *Student Technology Guide* masters.
- *HRW Technology Handbook.*

internet connect

For keystrokes of other graphing calculators models, visit the HRW web site at **go.hrw.com** and enter the keyword **MB1 CALC**.

About Chapter 6

Background Information

In Chapter 6, students look at a variety of applications that are modeled by logarithmic and exponential functions, learn about the properties of these functions, and learn how to solve equations involving logarithms and exponents. Students investigate the uses of base 10 and base *e* logarithms, the common and natural logarithms.

CHAPTER RESOURCES

- Block-Scheduling Handbook
- Writing Activities for Your Portfolio
- Tech Prep Masters
- Long-Term Project
- Assessment Resources:
  Mid-Chapter Assessment
  Chapter Assessments
  Alternative Assessments
- Test and Practice Generator
- Technology Handbook

Chapter Objectives

- Determine the multiplier for exponential growth and decay. [**6.1**]
- Write and evaluate exponential expressions to model growth and decay situations. [**6.1**]
- Classify an exponential function as representing exponential growth or exponential decay. [**6.2**]
- Calculate the growth of investments under various conditions. [**6.2**]
- Write equivalent forms for exponential and logarithmic equations. [**6.3**]

Exponential and Logarithmic Functions

EXPONENTIAL AND LOGARITHMIC FUNCTIONS model many scientific phenomena. Some applications of exponential functions include population growth, compound interest, and radioactive decay. Radioactive decay is used dating ancient objects found at archeological sites. Applications of logarithmic functions include the pH scale in chemistry, sound intensity, and Newton's law of cooling.

Lessons

About the Photos

In living material, the ratio of radioactive carbon isotopes (C^{14}) to the number of nonradioactive carbon isotopes (C^{12}) is approximately 1 to 10^{12}. When organic material dies, its C^{12} content remains fixed, while its C^{14} content decays with a half-life of approximately 5730 years. Scientists use these facts to estimate the age of dead *organic* material and use the formula shown to calculate the amount, $A(t)$, of C^{14} remaining.

$$A(t) = \left(\frac{1}{2}\right)^{\frac{t}{5730}}$$

Other types of radioactive isotopes and their respective half-lives are shown below.

Isotope	Half-life
iodine-131	8 days
potassium-40	1.3×10^9 years
radon-222	3.8 days
radium-226	1.6×10^3 years
uranium-235	7.0×10^8 years

Background: Prehistoric rock art from the Canyon de Chelly National Monument, Arizona;

Right: Anasazi sandal, 700–900 years old, found at Navajo National Monument, Arizona

About the Chapter Project

The heating and cooling of objects can be modeled by functions. Throughout this chapter and in the Chapter Project, *Warm Ups*, you will model the heating and cooling of a temperature probe over several temperature ranges in order to find an appropriate general model for these phenomena.

After completing the Chapter Project, you will be able to do the following:

- Collect real-world data on the heating and cooling of an object, and determine an appropriate exponential function to model the heating and cooling of an object.
- Make predictions about the temperature of an object that is heating or cooling to a constant surrounding temperature.
- Verify Newton's law of cooling.

In the Portfolio Activities for Lessons 6.1 and 6.4 and in the Chapter Project, you will need to use a program like the one shown on the calculator screen at right to collect temperature data with a CBL.

About the Portfolio Activities

Throughout the chapter, you will be given opportunities to complete Portfolio Activities that are designed to support your work on the Chapter Project.

- Using a CBL to collect cooling temperature data in a laboratory setting is included in the Portfolio Activity on page 361.
- Comparing different models for the cooling temperature data is included in the Portfolio Activity on page 369.
- Using a CBL to collect warming temperature data and performing appropriate transformations on regression equations are included in the Portfolio Activity on page 384.
- Comparing Newton's law of cooling with regression models from empirical data is included in the Portfolio Activity on page 409.

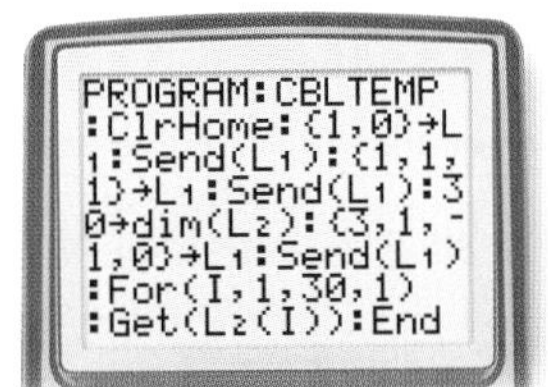

internet connect go.hrw.com

Chapter Internet Features and Online Activities

Lesson	Keyword	Page
6.1	MB1 Homework Help	358
6.2	MB1 Homework Help	367
	MB1 Medicine	366
6.3	MB1 Homework Help	375
6.4	MB1 Homework Help	382
6.5	MB1 Homework Help	389
	MB1 Stock Closings	390
6.6	MB1 Homework Help	397
6.7	MB1 Homework Help	407
	MB1 Spacecraft	406
	MB1 Newton	409

- Use the definitions of exponential and logarithmic functions to solve equations. [**6.3**]
- Simplify and evaluate expressions involving logarithms. [**6.4**]
- Solve equations involving logarithms. [**6.4**]
- Define and use the common logarithmic function to solve exponential and logarithmic equations. [**6.5**]
- Evaluate logarithmic expressions by using the change-of-base formula. [**6.5**]
- Evaluate natural exponential and natural logarithmic functions. [**6.6**]
- Model exponential growth and decay processes. [**6.6**]
- Solve logarithmic and exponential equations by using algebra and graphs. [**6.7**]
- Model and solve real-world problems involving exponential and logarithmic relationships. [**6.7**]

Portfolio Activities appear at the end of Lessons 6.1, 6.2, 6.4, and 6.7. Each serves as preparation for the Chapter Project. The Portfolio Activities as well as the Chapter Project Activities are appropriate for inclusion in the student's portfolio. Students should be encouraged to include in their portfolios any other work in which they feel a sense of pride or a sense of accomplishment.

Prepare

NCTM PRINCIPLES & STANDARDS 1, 2, 4, 6, 8–10

QUICK WARM-UP

Write as a decimal.

1. 8% 0.08 **2.** 2.4% 0.024 **3.** 0.01% 0.0001

Evaluate.

4. 3^6 729 **5.** $3^4 \cdot 4^3$ 5184 **6.** $24 \cdot 2^3$ 192

Also on Quiz Transparency 6.1

Teach

Why Have students discuss ways to model the bacterial growth described in the text. Have them cut a large circle out of paper to represent one bacterium. Then have students cut the circle in half to model division of the bacterium. In a table, have students record the number of bacteria after each division. Have students continue cutting each piece in half and recording the number of bacteria represented.

Exponential Growth and Decay

Objectives

- Determine the multiplier for exponential growth and decay.
- Write and evaluate exponential expressions to model growth and decay situations.

Why *Exponential growth and decay can be used to model a number of real-world situations, such as population growth of bacteria and the elimination of medicine from the bloodstream.*

APPLICATION BIOLOGY

Bacteria are very small single-celled organisms that live almost everywhere on Earth. Most bacteria are not harmful to humans, and some are helpful, such as the bacteria in yogurt.

Bacteria reproduce, or grow in number, by dividing. The total number of bacteria at a given time is referred to as the population of bacteria. When each bacterium in a population of bacteria divides, the population doubles.

Keystroke Guide, page 418

Activity: Modeling Bacterial Growth

You will need: a calculator

You can use a calculator to model the growth of 25 bacteria, assuming that the entire population doubles every hour.

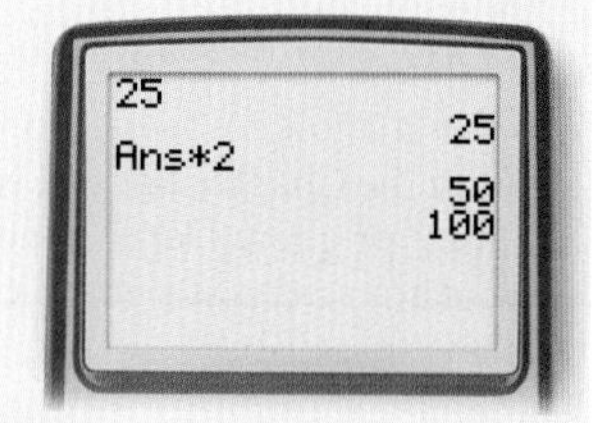

First enter 25. Then multiply this number by 2 to find the population of bacteria after 1 hour. Repeat this doubling procedure to find the population after 2 hours.

Alternative Teaching Strategy

USING VISUAL MODELS Give students the following scenario: The teacher knows a secret and tells two students; then each of those two students tell exactly two other students, and so on. Ask students to guess how many stages it would take for the entire class to know the secret. What about the entire school? Ask students to make a diagram showing the number of people who know the secret at each stage. Then have students calculate the number of new students that learn the secret in each stage.

Ask students to find an expression to model the number pattern. 2^n How many new students would know the secret after 10 stages? 1024 Ask students to add another column to the right of the numbers in the figure above. Calculate the total number of students who know the rumor in each stage, and write an expression describing this new pattern. $2^{n+1} - 1$

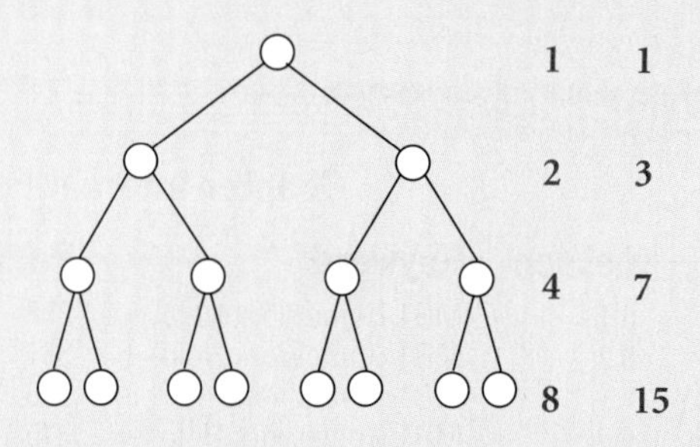

1. Copy and complete the table below.

Time (hr)	0	1	2	3	4	5	6
Population	25	50	100				

2. Write an algebraic expression that represents the population of bacteria after n hours. (Hint: Factor out 25 from each population figure.)
3. Use your algebraic expression to find the population of bacteria after 10 hours and after 20 hours.

CHECKPOINT ✔ 4. Suppose that the initial population of bacteria was 75 instead of 25. Find the population after 10 hours and after 20 hours.

You can represent the growth of an initial population of 100 bacteria that doubles every hour by creating a table.

+1 +1 +1 +1

Time (hr)	0	1	2	3	4	...	n
Population	100	200	400	800	1600	...	$100(2)^n$

×2 ×2 ×2 ×2

The bar chart at right illustrates how the doubling pattern of growth quickly leads to large numbers.

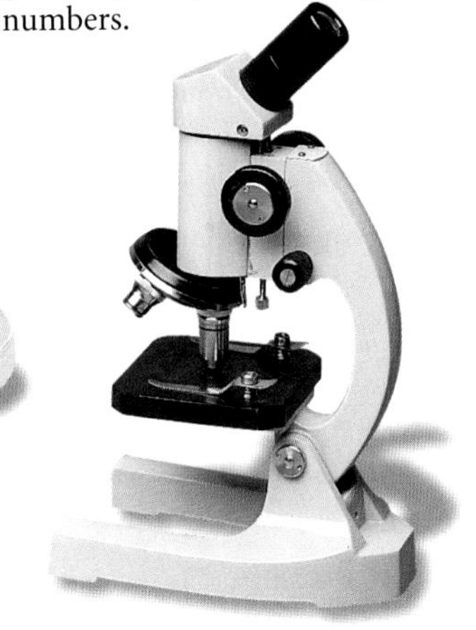

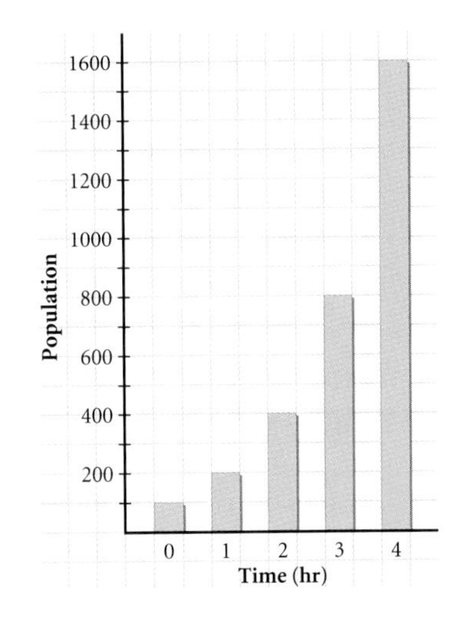

CHECKPOINT ✔ Assuming an initial population of 100 bacteria, predict the population of bacteria after 5 hours and after 6 hours.

The population after n hours can be represented by the following *exponential expression*:

$$100 \times \overbrace{2 \times 2 \times 2 \times \cdots \times 2}^{n \text{ times}} = 100 \times 2^n$$

This expression, $100 \cdot 2^n$, is called an **exponential expression** because the exponent, n, is a variable and the base, 2, is a fixed number. The base of an exponential expression is commonly referred to as the **multiplier**.

Activity Notes

Students use a calculator to model bacterial growth. If they have trouble finding an algebraic expression, suggest that they divide each population value by 25, the initial population. This will make the expression $25 \cdot 2^n$ more understandable. After 10 hours, the population will be 25,600, and after 20 hours, the population will be 26,214,400.

Cooperative Learning

You may wish to have students perform this Activity in pairs. One student can enter values into the calculator while the other records the results in the table.

CHECKPOINT ✔
76,000; 78,643,200

Teaching Tip

TECHNOLOGY When using a TI-83 or TI-82 to calculate the successive populations of bacteria in the Activity, press **25** ENTER and then press × 2 ENTER. Now every time the students press ENTER, it will double the previous result. Make sure that students count the number of years correctly. Year zero is represented by the initial population entry, 25. The first year is represented by the first time the initial population is multiplied by 2.

Interdisciplinary Connection

BIOLOGY The bacterial growth example is a theoretical model of population growth. However, populations cannot continue to grow exponentially. Restraints such as food supply, living space, and predators create a natural limit for a population. This type of growth can be modeled by a logistic curve. For example, the population of mice in a certain area is given by the function below.

$$P(t) = \frac{1000}{1 + e^{42 - 5t}}$$

Have students use a graphics calculator to analyze the graph of this function. Ask students to use the trace feature of the calculator to find the maximum population of mice in this area. **1000** Discuss possible reasons why the population levels off. **overpopulation, inadequate food supply, predators**

Math CONNECTION

PATTERNS IN DATA A bar graph is used to display the population growth data.

CHECKPOINT ✔
3200; 6400

ADDITIONAL
EXAMPLE 1

Suppose that you invested \$1000 in a company's stock at the end of 1999 and that the value of the stock increased at a rate of about 15% per year. **Predict the value of the stock, to the nearest cent, at the end of the years 2004 and 2009.** \$2011.36; \$4045.56

TRY THIS
188,700,000; 194,400,000

CRITICAL THINKING
≈10.5%

Modeling Human Population Growth

APPLICATION
DEMOGRAPHICS

Human populations grow much more slowly than bacterial populations. Bacterial populations that double each hour have a growth rate of 100% per hour. The population of the United States in 1990 was growing at a rate of about 8% per decade.

In Example 1, you will use this growth rate to make predictions.

EXAMPLE 1 The population of the United States was 248,718,301 in 1990 and was projected to grow at a rate of about 8% per decade. [*Source: U.S. Census Bureau*]

Predict the population, to the nearest hundred thousand, for the years 2010 and 2025.

TECHNOLOGY
SCIENTIFIC CALCULATOR

SOLUTION

1. To obtain the multiplier for exponential growth, add the growth rate to 100%.

$$100\% + 8\% = 108\%, \text{ or } 1.08$$

2. Write the expression for the population n decades after 1990.

$$248{,}718{,}301 \cdot (1.08)^n$$

3. Since the year 2010 is 2 decades after 1990, substitute 2 for n.

$$\begin{aligned} &248{,}718{,}301(1.08)^n \\ &= 248{,}718{,}301(1.08)^2 \\ &= 290{,}105{,}026.3 \end{aligned}$$

To the nearest hundred thousand, the predicted population for 2010 is 290,100,000.

Since the year 2025 is 3.5 decades after 1990, substitute 3.5 for n.

$$\begin{aligned} &248{,}718{,}301(1.08)^n \\ &= 248{,}718{,}301(1.08)^{3.5} \\ &= 325{,}604{,}866 \end{aligned}$$

To the nearest hundred thousand, the predicted population for 2025 is 325,600,000.

These predictions are based on the assumption that the growth rate remains a constant 8% per decade.

TRY THIS The population of Brazil was about 162,661,000 in 1996 and was projected to grow at a rate of about 7.7% per decade. Predict the population, to the nearest hundred thousand, of Brazil for 2016 and 2020. [*Source: U.S. Census Bureau*]

CRITICAL THINKING If a population's growth rate is 1% per *year*, what is the population's growth rate per *decade*?

Enrichment

Have students enter the data from various examples in the lesson into a graphics calculator and create a scatter plot of the data. Then have them enter the algebraic expression that corresponds to the data in the function editor of the calculator. The graph should fit the data in the scatter plot closely. Ask students to make predictions about growth of the function beyond the data given for each situation.

Inclusion Strategies

USING MANIPULATIVES Have students use small manipulative cubes to model the data throughout the lesson. For example, with population growth, have students begin with 1 cube that represents 25 bacteria. At the next stage, students should double the number of cubes by adding another cube, making 50 bacteria, and then double again for a total of 4 cubes and a population of 100 bacteria. Let students use cubes as much as they feel it is necessary to help them visualize the growth or decay represented in a given data set.

Modeling Biological Decay

APPLICATION
HEALTH

Caffeine is eliminated from the bloodstream of a child at a rate of about 25% per hour. This exponential decrease in caffeine in a child's bloodstream is shown in the bar chart.

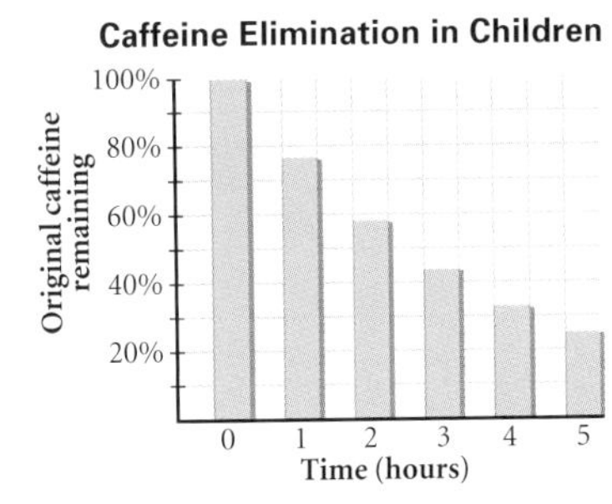

A *rate of decay* can be thought of as a negative growth rate. To obtain the multiplier for the decrease in caffeine in the bloodstream of a child, subtract the rate of decay from 100%. Thus, the multiplier is 0.75, as calculated below.

$$100\% - 25\% = 75\%, \text{ or } 0.75$$

EXAMPLE 2 The rate at which caffeine is eliminated from the bloodstream of an adult is about 15% per hour. An adult drinks a caffeinated soda, and the caffeine in his or her bloodstream reaches a peak level of 30 milligrams.

Predict the amount, to the nearest tenth of a milligram, of caffeine remaining 1 hour after the peak level and 4 hours after the peak level.

Caffeine is an ingredient in coffee, tea, chocolate, and some soft drinks.

SOLUTION

1. To obtain the multiplier for exponential decay, subtract the rate of decay from 100%. The multiplier is found as follows:

$$100\% - 15\% = 85\%, \text{ or } 0.85$$

2. Write the expression for the caffeine level x hours after the peak level.

$$30(0.85)^x$$

TECHNOLOGY
SCIENTIFIC CALCULATOR

3. Substitute 1 for x.

$$30(0.85)^x$$
$$= 30(0.85)^1$$
$$= 25.5$$

The amount of caffeine remaining 1 hour after the peak level is 25.5 milligrams.

Substitute 4 for x.

$$30(0.85)^x$$
$$= 30(0.85)^4$$
$$\approx 15.7$$

The amount of caffeine remaining 4 hours after the peak level is about 15.7 milligrams.

TRY THIS A vitamin is eliminated from the bloodstream at a rate of about 20% per hour. The vitamin reaches a peak level in the bloodstream of 300 milligrams. Predict the amount, to the nearest tenth of a milligram, of the vitamin remaining 2 hours after the peak level and 7 hours after the peak level.

ADDITIONAL EXAMPLE 2

Suppose that you buy a car for $15,000 and that its value decreases at a rate of about 8% per year. **Predict the value of the car, to the nearest cent, after 4 years and after 7 years.** $10,745.89; $8367.70

TRY THIS
192 mg; 62.9 mg

Reteaching the Lesson

TECHNOLOGY Have students use the **STAT** feature of a graphics calculator to generate exponential graphs. In **L1,** have students enter values from −10 to 10. In **L2,** have students go to the list header by using the up-arrow key and define **L2** as **2^L1.** Ask students to scroll through the numerical values and describe what is happening to the numbers. Then have students create a scatter plot to see the growth (or decay) graphically. Make sure that students set an appropriate window for their data. Students can also define other lists, perhaps including a beginning value, as shown in the Activity in the lesson.

Assess

Selected Answers

Exercises 5–14, 15–69 odd

ASSIGNMENT GUIDE

In Class	1–14
Core	15–45 odd, 53
Core Plus	16–52 even
Review	54–70
Preview	71

Extra Practice can be found beginning on page 940.

Practice

NAME ______ CLASS ______ DATE ______

Practice

6.1 ***Exponential Growth and Decay***

Find the multiplier for each rate of exponential growth or decay.

1. 1% growth 1.01
2. 1% decay 0.9
3. 7% decay 0.93
4. 12% growth 1.12
5. 10% growth 1.1
6. 3% decay 0.97
7. 5.2% decay 0.948
8. 7.5% growth 1.075
9. 0.4% growth 1.004
10. 5.9% decay 0.941

Evaluate each expression to the nearest thousandth for the given value of *x*.

11. 2^x for $x = 0.5$ 1.414
12. $10(2^x)$ for $x = \frac{2}{3}$ 15.874
13. $\left(\frac{1}{2}\right)^{3x}$ for $x = 2$ 0.016
14. $7(0.5)^x$ for $x = -2$ 28
15. $42 \cdot 2^{x-1}$ for $x = \frac{3}{4}$ 35.318
16. $20 \cdot 2^{2x}$ for $x = 1.75$ 226.274
17. $15\left(\frac{1}{2}\right)^{2x-1}$ for $x = 2$ 1.875
18. $66\left(\frac{1}{2}\right)^x$ for $x = 3$ 8.25
19. $512(2)^{3x}$ for $x = 0.1$ 630.346
20. $12(2)^{x-2}$ for $x = 6.5$ 271.529

Predict the result in each situation.

21. The population of a city in 1990 was 1,215,112. The population was growing at a rate of about 5% per decade. Predict the population of the city
    a. in the year 2000. 1,275,868
    b. in the year 2005. 1,307,375

22. The initial population of bacteria in a lab test is 400. The number of bacteria doubles every 30 minutes. Predict the bacteria population at the end of
    a. two hours. 6400
    b. three hours. 25,600

Exercises

Communicate

1. What type of values of n are possible in the bacterial growth expression $25 \cdot 2^n$ and in the United States population growth expression $248{,}718{,}301 \cdot (1.08)^n$?
2. Explain how the United States population growth expression $248{,}718{,}301 \cdot (1.08)^n$ incorporates the growth rate of 8% per decade.
3. What assumption(s) do you make about a population's growth when you make predictions by using an exponential expression?
4. Describe the difference between the procedures for finding the multiplier for a growth rate of 5% and for a decay rate of 5%.

Guided Skills Practice

Find the multiplier for each rate of exponential growth or decay. ***(EXAMPLES 1 AND 2)***

5. 5.5% growth **1.055**
6. 0.25% growth **1.0025**
7. 3% decay **0.97**
8. 0.5% decay **0.995**

Evaluate each expression for $x = 3$. ***(EXAMPLES 1 AND 2)***

9. 2^x **8**
10. $50(3)^x$ **1350**
11. 0.8^x **0.512**
12. $100(0.75)^x$ **42.1875**

APPLICATIONS

13. **DEMOGRAPHICS** The population of Tokyo-Yokohama, Japan, was about 28,447,000 in 1995 and was projected to grow at an annual rate of 1.1%. Predict the population, to the nearest hundred thousand, for the year 2004. [*Source: U.S. Census Bureau*] ***(EXAMPLE 1)*** **31,400,000**
14. **HEALTH** A certain medication is eliminated from the bloodstream at a rate of about 12% per hour. The medication reaches a peak level in the bloodstream of 40 milligrams. Predict the amount, to the nearest tenth of a milligram, of the medication remaining 2 hours after the peak level and 3 hours after the peak level. ***(EXAMPLE 2)*** **31.0 mg; 27.3 mg**

Practice and Apply

Find the multiplier for each rate of exponential growth or decay.

15. 7% growth **1.07**
16. 9% growth **1.09**
17. 6% decay **0.94**
18. 2% decay **0.98**
19. 6.5% growth **1.065**
20. 8.2% decay **0.918**
21. 0.05% decay **0.9995**
22. 0.08% growth **1.0008**
23. 0.075% growth **1.00075**

internet connect
Homework Help Online
Go To: **go.hrw.com**
Keyword: **MB1 Homework Help** for Exercises 24–35

Given $x = 5$, $y = \frac{3}{5}$, and $z = 3.3$, evaluate each expression.

24. 2^x **32**
25. 3^y **1.9**
26. 2^{2x} **1024**
27. $50(2)^{3x}$ **1,638,400**
28. $25(2)^z$ **246.2**
29. $25(2)^y$ **37.9**
30. $100(3)^{x-1}$ **8100**
31. $10(2)^{z+2}$ **394.0**
32. 2^{2y-1} **1.1**
33. $100(2)^{4z}$ **941,013.7**
34. $100(0.5)^{3z}$ **0.1**
35. $75(0.5)^{2y}$ **32.6**

Predict the population of bacteria for each situation and time period.

36. 55 bacteria that double every hour
a. after 3 hours **440** **b.** after 5 hours **1760**

37. 125 bacteria that double every hour
a. after 6 hours **8000** **b.** after 8 hours **32,000**

38. 33 *E. coli* bacteria that double every 30 minutes
a. after 1 hour **132** **b.** after 6 hours **135,168**

39. 75 *E. coli* bacteria that double every 30 minutes
a. after 2 hours **1200** **b.** after 3 hours **4800**

40. 225 bacteria that triple every hour
a. after 1 hour **675** **b.** after 3 hours **6075**

41. 775 bacteria that triple every hour
a. after 2 hours **6975** **b.** after 4 hours **62,775**

CHALLENGE

42. Suppose that you put $2500 into a retirement account that grows with an interest rate of 5.25% compounded once each year. After how many years will the balance of the account be at least $15,000? **36 years**

CONNECTION

PATTERNS IN DATA **Determine whether each table represents a linear, quadratic, or exponential relationship between *x* and *y*.**

43. **exp.**

x	y
0	2
1	4
2	8
3	16

44. **exp.**

x	y
1	1
2	3
3	9
4	27

45. **linear**

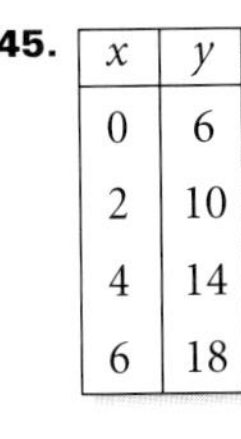

x	y
0	6
2	10
4	14
6	18

46. **quad.**

x	y
0	−2
3	7
6	34
9	79

APPLICATIONS

47. 278,700,000

47. DEMOGRAPHICS The population of Indonesia was 191,256,000 in 1990 and was growing at a rate of 1.9% per year. Predict the population, to the nearest hundred thousand, of Indonesia in 2010. [*Source: U.S. Census Bureau*]

Bali, Indonesia

48. 0.09 grams

48. HEALTH A dye is injected into the pancreas during a certain medical procedure. A physician injects 0.3 grams of the dye, and a healthy pancreas will secrete 4% of the dye each minute. Predict the amount of dye remaining, to the nearest hundredth of a gram, in a healthy pancreas 30 minutes after the injection.

49. 1,359,600,000; 1,399,800,000

49. DEMOGRAPHICS The population of China was 1,210,005,000 in 1996 and was growing at a rate of about 6% per decade. Predict the population, to the nearest hundred thousand, of China in 2016 and in 2021. [*Source: U.S. Census Bureau*]

Error Analysis

For exercises in which the exponent is a product or a sum, make sure students find the product or sum first or, if using a calculator, put parentheses around the exponential expression. For example, $2^{2+3} = 2^5 = 32$; on a calculator, this would be entered as **2^(2+3).** If students simply enter **2^2+3,** the calculator will calculate $2^2 + 3 = 7$.

Student Technology Guide

NAME ______ CLASS ______ DATE ______

Student Technology Guide
6.1 Exponential Growth and Decay

You can use a graphics calculator to explore exponential growth and decay.

Example: A population of 12 rabbits doubles every year. Find the population of the rabbits in years 0 through 5. Show your results in a table.

- Type 12 into the calculator. Press ENTER.
- Press 2 ×. Then press 2nd (−) ENTER.
- Press ENTER four more times. The table below shows the results.

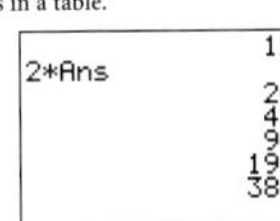

Year	0	1	2	3	4	5
Population	12	24	48	96	192	384

Example: Between 1990 and 1992, the population of one city decreased at a rate of 2.1% per year. Its population in 1990 was 1,585,577. Use this information to estimate the population of the city from 1990 to 1994.

If the population *decreased* by 2.1% per year, the multiplier is 0.979.

- Type 1,585,577 into the calculator. Press ENTER.
- Press .979 × 2nd (−) ENTER.
- Press ENTER three more times. The table below shows the results.

Year	1990	1991	1992	1993	1994
Population	1,585,577	1,552,280	1,519,682	1,487,769	1,456,526

1. There are 50 bacteria in a dish. The population triples every hour. Find the population of the bacteria for hours 0 through 5.

Year	0	1	2	3	4	5
Population	50	150	450	1350	4050	12,150

2. Between 1990 and 1992, the population of a city decreased at a rate of 3.9% per year. Its population in 1990 was 574,283. Use this information to estimate the city's population from 1990 to 1993.

Year	1990	1991	1992	1993
Population	574,283	551,886	530,362	509,678

APPLICATIONS

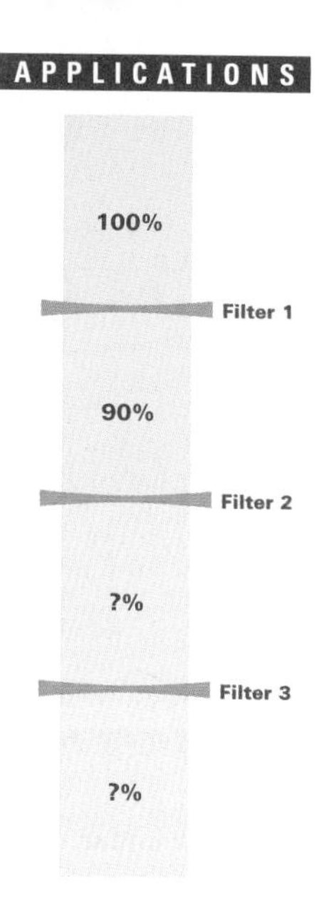

50. PHYSICAL SCIENCE Suppose that a camera filter transmits 90% of the light striking it, as illustrated at left.

a. If a second filter of the same type is added, what portion of light is transmitted through the combination of the two filters? **81%**

b. Write an expression to model the portion of light that is transmitted through n filters. $(0.9)^n$ **c. 65.61%; 59.049%; 53.14%**

c. Calculate the portion of light transmitted through 4, 5, and 6 filters.

51. DEMOGRAPHICS The population of India was 952,108,000 in 1996 and was growing at a rate of about 1.3% per year. [*Source: U.S. Census Bureau*]

a. Predict the population, to the nearest hundred thousand, of India in 2000 and in 2010. **1,002,600,000; 1,140,800,000**

b. Find the growth rate per decade that corresponds to the growth rate of 1.3% per year. **13.79%**

c. Suppose that the population growth rate of India slows to 1% per year after the year 2000. What is the predicted population, to the nearest hundred thousand, of India in 2010? **1,107,500,000**

52. CHEMISTRY A dilution is commonly used to obtain the desired concentration of a sample. For example, suppose that 1 milliliter of hydrochloric acid, or HCl, is combined with 9 milliliters of a buffer. The concentration of the resulting mixture is $\frac{1}{10}$ of the original concentration of HCl.

a. Suppose that this dilution is performed again with 1 millimeter of the already diluted mixture and 9 milliliters of buffer. What is the concentration of the resulting mixture (compared with the original concentration)? $\frac{1}{100}$

b. Write an expression to model the concentration of HCl in the resulting mixture after repeated dilutions as described in part **a.** $\left(\frac{1}{10}\right)^n$

c. What is the concentration of the resulting mixture (compared to the original concentration) after 5 repeated dilutions? $\frac{1}{100{,}000}$

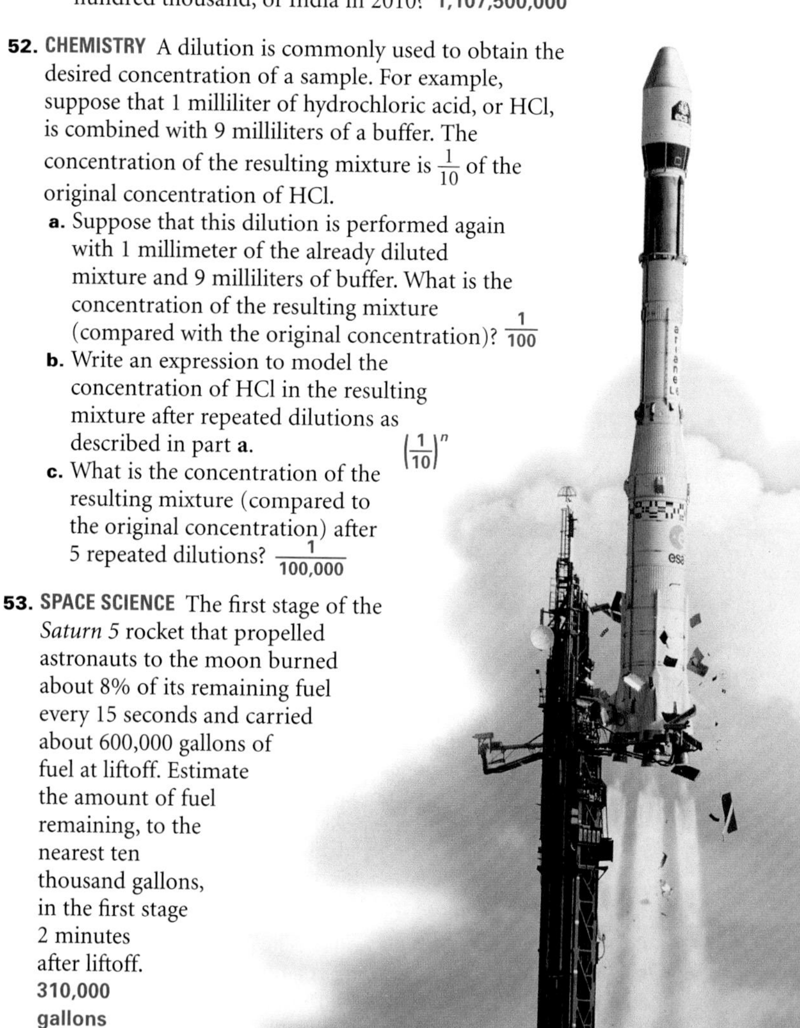

53. SPACE SCIENCE The first stage of the *Saturn 5* rocket that propelled astronauts to the moon burned about 8% of its remaining fuel every 15 seconds and carried about 600,000 gallons of fuel at liftoff. Estimate the amount of fuel remaining, to the nearest ten thousand gallons, in the first stage 2 minutes after liftoff. **310,000 gallons**

62. vertical stretch by a factor of 6
63. reflection across y-axis, horiz. compression by $\frac{1}{2}$
64. reflection across x-axis, ver. comp. by $\frac{1}{2}$, vertical trans. 1 unit up
65. reflection across x-axis, horiz. stretch by 2, vert. trans. 3 units up
66. horiz. trans. 3 units to the right, vert. trans. 2 units up
67. reflection across x-axis, vert. stretch by 5, horiz. trans. 2 units to the right, vert. trans. 2 units down

Evaluate each expression. *(LESSON 2.2)*

54. 4^{-2} $\frac{1}{16}$ **55.** $\left(\frac{1}{2}\right)^{-1}$ 2 **56.** $25^{\frac{3}{2}}$ 125 **57.** $49^{\frac{1}{2}}$ 7

Simplify each expression, assuming that no variable equals zero. Write your answer with positive exponents only. *(LESSON 2.2)*

58. $\left(\frac{2x^3}{x^{-2}}\right)^2$ $4x^{10}$ **59.** $\left(\frac{m^{-1}n^2}{n^{-3}}\right)^{-3}$ $\frac{m^3}{n^{15}}$ **60.** $\left(\frac{2a^3b^{-2}}{-a^2b^{-3}}\right)^{-1}$ $-\frac{1}{2ab}$ **61.** $\frac{(2y^2y)^{-2}}{3xy^{-4}}$ $\frac{1}{12xy^2}$

Identify each transformation from the graph of $f(x) = x^2$ to the graph of g. *(LESSON 2.7)*

62. $g(x) = 6x^2$ **63.** $g(x) = (-2x)^2$ **64.** $g(x) = -\frac{1}{2}x^2 + 1$

65. $g(x) = -(0.5x)^2 + 3$ **66.** $g(x) = (x - 3)^2 + 2$ **67.** $g(x) = -5(x - 2)^2 - 4$

State whether each parabola opens up or down and whether the y-coordinate of the vertex is the maximum or minimum value of the function. *(LESSON 5.1)*

68. $f(x) = \frac{1}{2}x^2$ up; min **69.** $f(x) = -2x^2 - x + 1$ down; max **70.** $f(x) = 3 - 5x - x^2$ down; max

APPLICATION

71. INVESTMENTS Suppose that you want to invest $100 in a bank account that earns 5% interest *compounded once* at the end of each year. Determine the balance after 10 years. **$162.89**

Refer to the discussions of the Portfolio Activities and Chapter Project on page 353 for background on this activity.

You will need a CBL with a temperature probe, a glass of ice water, and a graphics calculator.

1. First use the CBL to find the temperature of the air. Then place the probe in the ice water for 2 minutes. Record 30 CBL readings taken at 2-second intervals. Take a final reading at the end of the 2 minutes.
2. **a.** Use the linear regression feature on your calculator to find a linear function that models your first 30 readings. (Use the variable t for the time in seconds).
   b. Use your linear function to predict the temperature of the probe after 2 minutes, or 120 seconds. Compare this prediction with your actual 2-minute reading.
   c. Discuss the usefulness of your linear function for modeling the cooling process. (You may want to illustrate your answer with graphs.)

Save your data and results for use in the remaining Portfolio Activities.

WORKING ON THE CHAPTER PROJECT

You should now be able to complete Activity 1 of the Chapter Project.

In Exercise 71, students investigate compound interest. They will study this topic in greater detail in Lesson 6.2.

ALTERNATIVE
Assessment

Portfolio Activity

The Portfolio Activity can be used as preparation for the Chapter Project or as a separate activity. In the Portfolio Activity on this page, students use a CBL unit to find the rate at which a temperature probe cools from room temperature to the temperature of ice water. Students then find a linear regression equation to fit the data and discuss the usefulness of such a model.

Answers to Portfolio Activities can be found in Additional Answers of the Teacher's Edition.

Prepare

NCTM PRINCIPLES & STANDARDS 1–10

QUICK WARM-UP

Evaluate.

1. 3^4 81 **2.** 3^0 1

3. $(-4)^2$ 16 **4.** 4^{-2} $\frac{1}{16}$

Evaluate each expression for $x = 2$.

5. x^5 32 **6.** 5^x 25

7. 5^{-x} $\frac{1}{25}$ **8.** $\left(\frac{1}{5}\right)^x$ $\frac{1}{25}$

Also on Quiz Transparency 6.2

Teach

Why Discuss with students the difference between compound interest and simple interest. Ask them which type of interest would be more beneficial to them for a savings account and why. The formulas for simple and compound interest are $A = P(1 + rt)$ and $A = P\left(1 + \frac{r}{n}\right)^{nt}$, respectively.

Use Teaching Transparency 21.

Exponential Functions

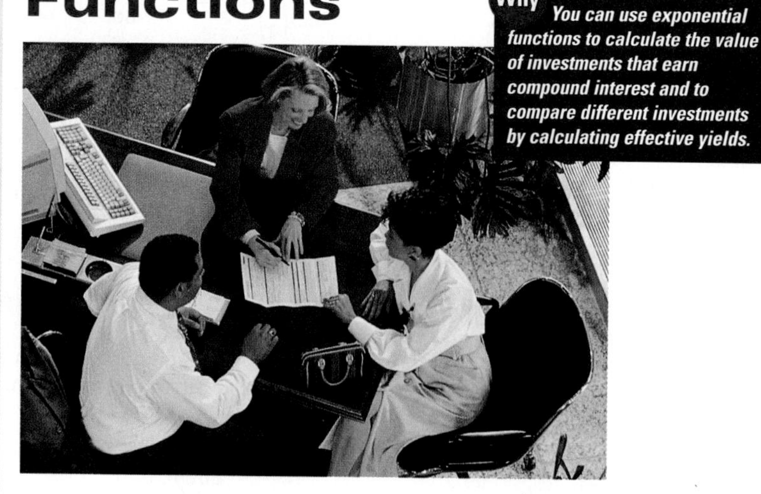

Why *You can use exponential functions to calculate the value of investments that earn compound interest and to compare different investments by calculating effective yields.*

Objectives

- Classify an exponential function as representing exponential growth or exponential decay.
- Calculate the growth of investments under various conditions.

$$f(x) = b^x$$

EXPONENT

BASE

Consider the function $y = x^2$ and $y = 2^x$. Both functions have a base and an exponent. However, $y = x^2$ is a quadratic function, and $y = 2^x$ is an *exponential function.* In an exponential function, the base is fixed and the exponent is variable.

Exponential Function

The function $f(x) = b^x$ is an **exponential function** with **base** b, where b is a positive real number other than 1 and x is any real number.

---

x	$y = 2^x$
-3	$2^{-3} = \frac{1}{8}$
-2	$2^{-2} = \frac{1}{4}$
-1	$2^{-1} = \frac{1}{2}$
0	$2^0 = 1$
1	$2^1 = 2$
$\sqrt{2}$	$2^{\sqrt{2}} \approx 2.67$
2	$2^2 = 4$
3	$2^3 = 8$

Examine the table at left and the graph at right of the exponential function $y = 2^x$.

Notice that as x-values decrease, the y-values for $y = 2^x$ get closer and closer to 0, approaching the x-axis as an *asymptote.* An **asymptote** is a line that a graph approaches (but does not reach) as its x- or y-values become very large or very small.

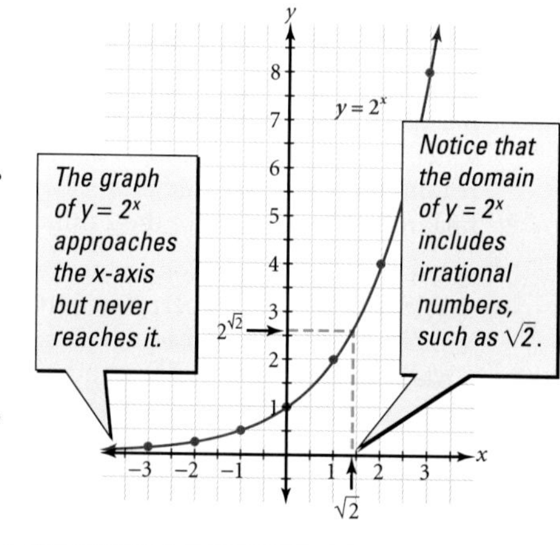

y
x
y = 2^x
2^√2
√2
The graph of y = 2^x approaches the x-axis but never reaches it.
Notice that the domain of y = 2^x includes irrational numbers, such as √2.

Alternative Teaching Strategy

USING PATTERNS Write the following functions on the board or overhead:

Bacteria model: $B(x) = 100 \cdot (2^x)$

Population model: $P(x) = 248{,}718{,}301 \cdot (1.08^x)$

Caffeine-elimination model: $C(x) = 30 \cdot (0.85^x)$

Ask students to discuss the similarities of the functions and then to use a graphics calculator to find the shape of the graphs. Be sure to discuss subtle differences in the function and graph of the caffeine-elimination model. Point out to students that this function still fits the general form of a constant raised to a variable exponent and that the graph is similar in shape to the others but is reflected across the y-axis. The difference in this example is that the base is between 0 and 1. Have students use these observations to write a definition of exponential function and describe the graphs of exponential functions.

Investigating Exponential Functions

You will need: a graphics calculator

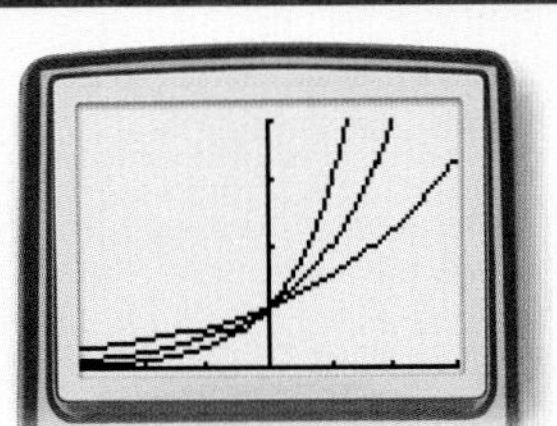

1. Graph $y_1 = 3^x$, $y_2 = 2^x$, and $y_3 = (1.5)^x$ on the same screen.
2. For what value of x is $y_1 = y_2 = y_3$ true? For what values of x is $y_1 > y_2 > y_3$ true? For what values of x is $y_1 < y_2 < y_3$ true?
3. Graph $y_4 = \left(\frac{1}{3}\right)^x$, $y_5 = \left(\frac{1}{2}\right)^x$, and $y_6 = \left(\frac{1}{1.5}\right)^x$ on the same screen as y_1, y_2, and y_3.

PROBLEM SOLVING

4. **Look for a pattern.** Examine each corresponding pair of functions.

$$y_1 = 3 \text{ and } y_4 = \left(\frac{1}{3}\right)^x \qquad y_2 = 2^x \text{ and } y_5 = \left(\frac{1}{2}\right)^x$$

$$y_3 = (1.5)^x \text{ and } y_6 = \left(\frac{1}{1.5}\right)^x$$

How are the graphs of each corresponding pair of functions related? How are the bases of each corresponding pair of functions related?

CHECKPOINT ✔

5. For what values of b does the graph of $y = b^x$ rise from left to right? For what values of b does the graph of $y = b^x$ fall from left to right?

The graphs of $f(x) = 2^x$ and $g(x) = \left(\frac{1}{2}\right)^x$ exhibit the two typical behaviors for exponential functions.

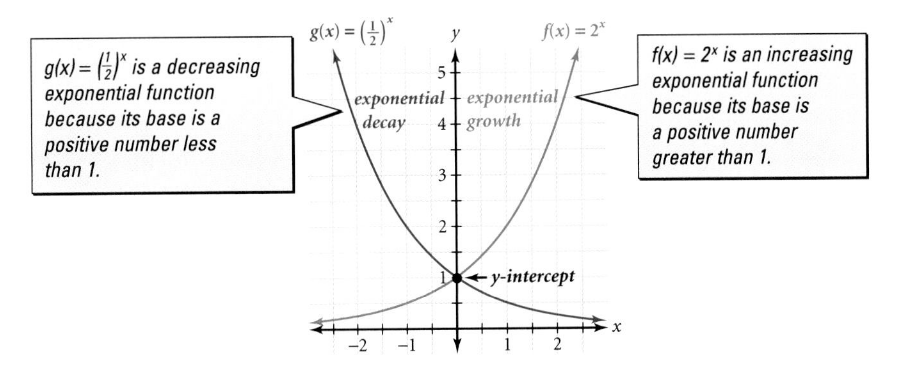

Recall from Lesson 2.7 that the graphs of f and g are reflections of one another across the y-axis because $g(x) = f(-x) = 2^{-x} = \left(\frac{1}{2}\right)^x$.

Exponential Growth and Decay

When $b > 1$, the function $f(x) = b^x$ represents **exponential growth**.
When $0 < b < 1$, the function $f(x) = b^x$ represents **exponential decay**.

Interdisciplinary Connection

CHEMISTRY Many radioactive substances have known half-lives. A half-life is the amount of time necessary for half of the substance to decay. Radioactive decay is an example of exponential decay, and it is generally modeled by a function of the form $A(t) = A_0e^{-kt}$, where A_0 is the initial amount of the substance, t is the amount of time the substance has been decaying, and k is a constant that varies depending on the substance. Have students analyze the decay of a substance defined by the function $A(t) = 500e^{-0.002t}$, where A is measured in grams and t is measured in years. What is the initial amount of the substance? 500 g Use the trace feature of the graphics calculator to estimate the half-life of this substance. ≈347 yr

Activity Notes

In this Activity, students use a graphics calculator to investigate the behavior of exponential functions. They should observe that all functions of the form $y = a^x$ pass through the point (0, 1). Students should also observe that for $x > 0$, the larger the base, the faster the function grows and that for $x < 0$, the opposite is true. In Steps 3 and 4, students investigate exponential decay models that are a reflection of the growth models across the y-axis.

Teaching Tip

TECHNOLOGY When using a TI-83 or TI-82 graphics calculator for the Activity, set the viewing window as follows:

Xmin=–3	Ymin=0
Xmax=3	Ymax=4
Xscl=1	Yscl=1

Using different line styles for equations graphed in the same window can help students identify which equation corresponds with which graph. To change the line style of **Y1**, use the left arrow key to move the cursor to the left of **Y1** and press ENTER until the desired line style appears.

CHECKPOINT ✔
5. $b > 1$; $0 < b < 1$

Use Teaching Transparency 22.

Math CONNECTION

TRANSFORMATIONS If the bases of two exponential functions are reciprocals of each other, their graphs are reflections of each other across the y-axis. Reflections preserve the shape of a graph but change its orientation.

Teaching Tip

TECHNOLOGY When using a TI-83 or TI-82 graphics calculator to produce the graphs for Example 1, set the viewing window as follows:

Xmin=–5	Ymin=–2
Xmax=5	Ymax=12
Xscl=1	Yscl=1

ADDITIONAL EXAMPLE 1

Graph $f(x) = 2^x$ along with each function below. Tell whether each function represents exponential growth or exponential decay. Then give the y-intercept.

a. $\frac{1}{2} \cdot f(-x)$
exponential decay; $\left(0, \frac{1}{2}\right)$

b. $4 \cdot f(x)$
exponential growth; $(0, 4)$

Math CONNECTION

TRANSFORMATIONS When an exponential function is multiplied by a constant, the graph is transformed by a dilation (stretch or compression).

TRY THIS

a. exponential growth; $\left(0, \frac{1}{3}\right)$

b. exponential decay; $\left(0, \frac{1}{4}\right)$

CHECKPOINT ✔

The graph is reflected across the x-axis.

CRITICAL THINKING

When $b > 1$ and b gets larger, the graph rises more quickly as x increases. When $0 < b < 1$ and b gets smaller, the graph falls more quickly as x increases.

Teaching Tip

TECHNOLOGY Exponential functions are not defined for bases that are not positive, that is, for $b \leq 0$.

Exponential growth functions and exponential decay functions of the form $y = b^x$ have the same domain, range, and y-intercept. For example:

Function	Domain	Range	y-intercept
$f(x) = 2^x$	all real numbers	all positive real numbers	1
$g(x) = \left(\frac{1}{2}\right)^x$	all real numbers	all positive real numbers	1

Recall from Lesson 2.7 that $y = a \cdot f(x)$ represents a vertical stretch or compression of the graph of $y = f(x)$. This transformation is applied to exponential functions in Example 1.

EXAMPLE 1

Graph $f(x) = 2^x$ along with each function below. Tell whether each function represents exponential growth or exponential decay. Then give the y-intercept.

a. $y = 3 \cdot f(x)$

b. $y = 5 \cdot f(-x)$

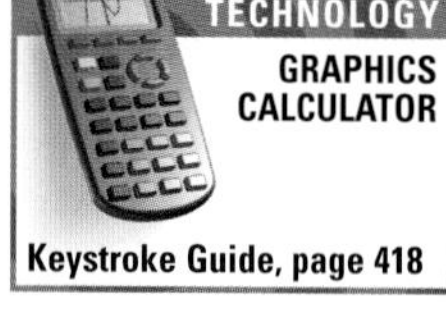

TECHNOLOGY
GRAPHICS CALCULATOR
Keystroke Guide, page 418

CONNECTION
TRANSFORMATIONS

SOLUTION

a. $y = 3 \cdot f(x) = 3 \cdot 2^x$

The function $y = 3 \cdot 2^x$ represents exponential growth because the base, 2, is greater than 1.

The y-intercept is 3 because the graph of $f(x) = 2^x$, which has a y-intercept of 1, is stretched by a factor of 3.

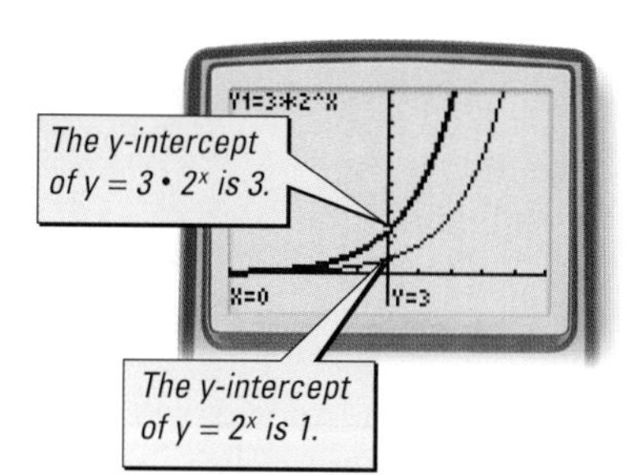

b. $y = 5 \cdot f(-x) = 5 \cdot 2^{-x} = 5 \cdot \left(\frac{1}{2}\right)^x$

The function $y = 5 \cdot \left(\frac{1}{2}\right)^x$ represents exponential decay because the base, $\frac{1}{2}$, is less than 1.

The y-intercept is 5 because the graph of $f(x) = 2^x$, which has a y-intercept of 1, is stretched by a factor of 5.

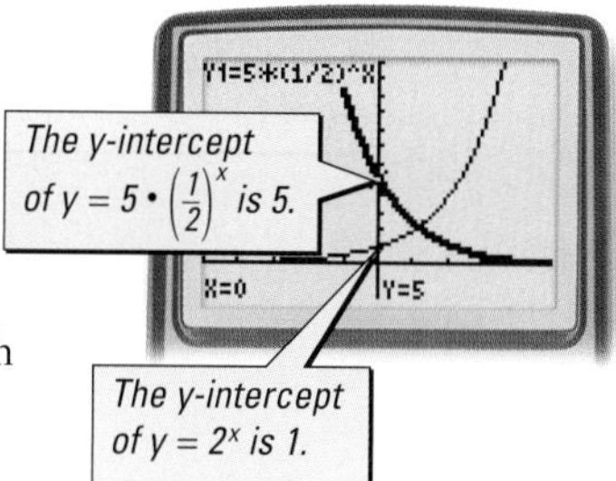

TRY THIS Graph $f(x) = 2^x$ along with each function below. Tell whether each function represents exponential growth or exponential decay. Then give the y-intercept.

a. $y = \frac{1}{3} \cdot f(x)$

b. $y = \frac{1}{4} \cdot f(-x)$

CHECKPOINT ✔ What transformation of f occurs when $a < 0$ in $y = a \cdot f(x)$?

CRITICAL THINKING Describe the effect on the graph of $f(x) = b^x$ when $b > 1$ and b increases. Describe the effect on the graph of $f(x) = b^x$ when $0 < b < 1$ and b decreases.

Enrichment

Have students find data in an almanac or on the Internet that seems to exhibit exponential growth or decay. (If students have trouble finding data, suggest that they look for population models.) Have students enter the data into their graphics calculator and create a scatter plot of the data. Then have students use the calculator to find an exponential regression equation that fits the data. Using this equation, ask students to predict future values for their data and to describe the validity of their predictions.

Inclusion Strategies

INVITING PARTICIPATION Many students have difficulty linking the graphical and algebraic representations of exponential functions to their previous knowledge of exponents. Ask students to write everything they know about exponents. Then combine the lists to make a cumulative class list. Using this list, make connections to exponential functions. For example, raising a nonzero number to a power of 0 always yields 1. This explains why the y-intercept of an exponential function of the form $f(x) = b^x$ is 1.

Compound Interest

APPLICATION
INVESTMENTS

The growth in the value of investments earning compound interest is modeled by an exponential function.

Compound Interest Formula

The total amount of an investment, A, earning compound interest is

$$A(t) = P\left(1 + \frac{r}{n}\right)^{nt},$$

where P is the principal, r is the annual interest rate, n is the number of times interest is compounded per year, and t is the time in years.

EXAMPLE 2 **Find the final amount of a $100 investment after 10 years at 5% interest compounded annually, quarterly, and daily.**

SOLUTION

In this situation, the principal is $100, the annual interest rate is 5%, and the time period is 10 years. Thus, $P = 100$, $r = 0.05$, and $t = 10$. The table shows calculations for $n = 1$, $n = 4$, and $n = 365$.

Compounding period	n	$A(10) = 100\left(1 + \frac{0.05}{n}\right)^{n \cdot 10}$	Final amount
annually	1	$A(10) = 100\left(1 + \frac{0.05}{1}\right)^{1 \cdot 10}$	$162.89
quarterly	4	$A(10) = 100\left(1 + \frac{0.05}{4}\right)^{4 \cdot 10}$	$164.36
daily	365	$A(10) = 100\left(1 + \frac{0.05}{365}\right)^{365 \cdot 10}$	$164.87

CHECKPOINT ✔ Describe what happens to the final amount as the number of compounding periods increases.

Effective Yield

APPLICATION
INVESTMENTS

Suppose that you buy an item for $100 and sell the item one year later for $105. In this case, the *effective yield* of your investment is 5%. The **effective yield** is the annually compounded interest rate that yields the final amount of an investment. You can determine the effective yield by fitting an exponential regression equation to two points.

Teaching Tip

TECHNOLOGY When using a TI-83 or TI-82 graphics calculator to find the total amount for different compounding periods, you do not need to reenter the entire expression each time. Begin by entering **100(1+.05/1)^(1*10)** for the total amount when compounded annually. To calculate the amount earned when compounded quarterly, press 2nd ENTER to display the last entry, use the arrow keys to edit the last entry by replacing 1 with 4 in the expression, and press ENTER. By using a similar procedure, you can replace 4 with 365, but press 2nd DEL to insert the last two digits of 365 so that you do not overwrite part of the expression.

ADDITIONAL EXAMPLE 2

Find the final amount of a $500 investment after 8 years at 7% interest compounded annually, quarterly, monthly, and daily.

annually: $859.09
quarterly: $871.11
monthly: $873.91
daily: $875.29

CHECKPOINT ✔

It approaches a maximum amount. In Example 2, it approaches $164.87.

Reteaching the Lesson

USING TABLES Ask students to make a table to find the final amount of a $300 investment at 4% interest compounded annually. The table should give the balance for each of 10 years. At the end of the first year, the amount in the account can be found by calculating $300 + 0.04 \cdot 300 = 312$. The balance after 2 years can be found by calculating $312 + 0.04 \cdot 312 = 324.48$. Once students have created their table, ask them to analyze the data and make observations about the rate at which the amounts grow. Have students enter the values from the table into a graphics calculator and make a scatter plot. Then have them find an exponential regression equation to fit the data. Finally, have students compare the compound interest formula, leaving t as a variable, with the exponential regression equation they found.

ADDITIONAL EXAMPLE 3

A collector buys an antique stove for $500 at the beginning of 1990 and sells it for $875 at the beginning of 1998. **Use an exponential regression equation to find the effective yield.** ≈7.25%

Teaching Tip

TECHNOLOGY Effective yield per compounding period is given by $\left(\frac{A(t)}{P}\right)^{\frac{1}{nt}} - 1$. When using a TI-83 or TI-82 graphics calculator to find the effective yield for Example 3, enter the following:

(150/100)^(1/8)–1

Math CONNECTION

STATISTICS Students use regression equations and correlation factors to find mathematical models for the given data. The correlation factor shows students the accuracy of the model. To show r^2, the coefficient of determination, on a TI-83 graphics calculator, view the **Catalog** menu by pressing 2nd 0, select **DiagnosticOn,** and then press ENTER twice.

Teaching Tip

TECHNOLOGY When using a TI-83 or TI-82 graphics calculator, have students clear any lists by pressing STAT, selecting 4:ClrList, and then pressing 2nd followed by the number of the list. Then press ENTER. The home screen will display **Done.** To clear several lists, select 4:ClrList and then enter the list names separated by commas.

TRY THIS
7.2%

EXAMPLE 3

A collector buys a painting for $100,000 at the beginning of 1995 and sells it for $150,000 at the beginning of 2000.
Use an exponential regression equation to find the effective yield.

SOLUTION

1. Find the exponential equation that represents this situation.

   To find effective yield, the interest is compounded annually, so $n = 1$.

   From 1995 to 2000 is 5 years, so $t = 5$.

$$A(t) = P\left(1 = \frac{r}{n}\right)^{nt}$$

$$150{,}000 = 100{,}000\left(1 + \frac{r}{1}\right)^{1 \cdot 5}$$

$$150{,}000 = 100{,}000(1 + r)^5$$

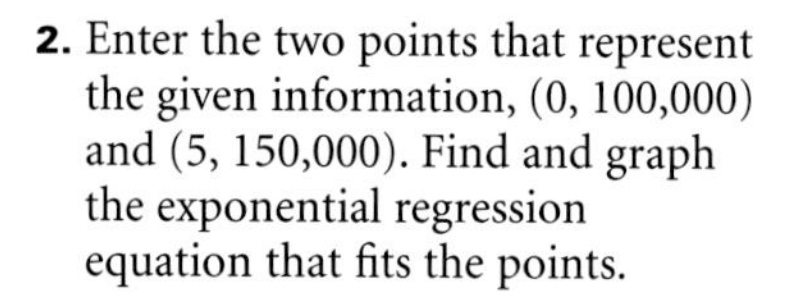

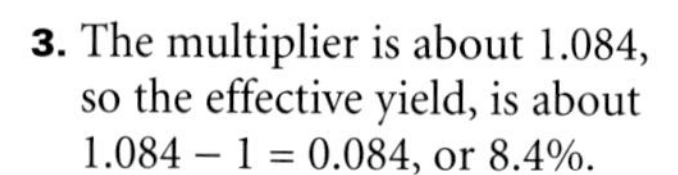

2. Enter the two points that represent the given information, (0, 100,000) and (5, 150,000). Find and graph the exponential regression equation that fits the points.

3. The multiplier is about 1.084, so the effective yield, is about $1.084 - 1 = 0.084$, or 8.4%.

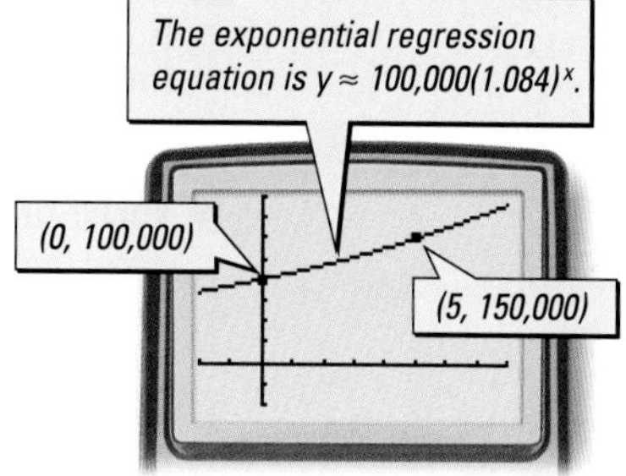

TRY THIS Find the effective yield for a painting bought for $100,000 at the end of 1994 and sold for $200,000 at the end of 2004.

Exercises

Communicate

1. If $b > 0$ and the graph of $y = b^x$ falls from left to right, describe the possible values of b.
2. Compare the domain and range of $y = 3^x$ with the domain and range of $y = \left(\frac{1}{3}\right)^x$.
3. Describe how the y-intercept of the graph of $f(x) = 2(5)^x$ is related to the value of a in $f(x) = ab^x$.
4. How are the functions $y = x^2$ and $y = 2^x$ similar, and how are they different?

Guided Skills Practice

Tell whether each function represents exponential growth or exponential decay, and give the *y*-intercept. ***(EXAMPLE 1)***

5. $f(x) = \left(\frac{1}{2}\right)^x$ **decay; 1** **6.** $g(x) = 3(2)^x$ **growth; 3** **7.** $k(x) = 5(0.5)^x$ **decay; 5**

APPLICATIONS

8. $334.56; $336.71; $337.46

9. 4.7%

8. INVESTMENTS Find the final amount of a $250 investment after 5 years at 6% interest compounded annually, quarterly, and daily. ***(EXAMPLE 2)***

9. INVESTMENTS Find the effective yield for a $2000 investment that is worth $4000 after 15 years. ***(EXAMPLE 3)***

Practice and Apply

Identify each function as linear, quadratic, or exponential.

10. $g(x) = 10x + 3$ **linear** **11.** $k(x) = (77 - x)x$ **quad.** **12.** $f(x) = 12(2.5)^x$ **exp.**

13. $k(x) = 0.5^x - 3.5$ **exp.** **14.** $g(x) = (2200)^{3.5x}$ **exp.** **15.** $h(x) = 0.5x^2 + 7.5$ **quad.**

Tell whether each function represents exponential growth or decay.

16. growth
17. growth
18–24. decay

16. $y(x) = 12(2.5)^x$ **17.** $k(x) = 500(1.5)^x$ **18.** $y(t) = 45\left(\frac{1}{4}\right)^t$

19. $d(x) = 0.125\left(\frac{1}{2}\right)^x$ **20.** $g(x) = 0.25(0.8)^x$ **21.** $s(k) = 0.5(0.5)^k$

22. $m(x) = 222(0.9)^x$ **23.** $f(k) = 722^{-k}$ **24.** $g(x) = 0.5(787)^{-x}$

Match each function with its graph.

25. $y = 2^x$ **d** **26.** $y = 2(3)^x$ **c**

27. $y = 2\left(\frac{1}{3}\right)^x$ **b** **28.** $y = \left(\frac{1}{2}\right)^x$ **a**

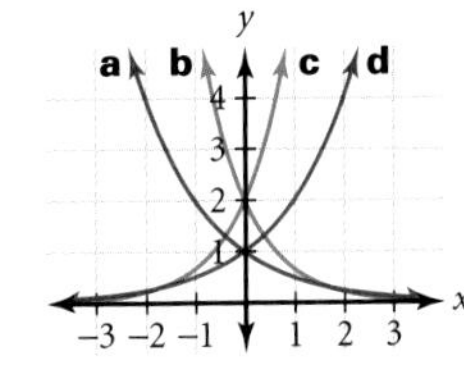

Find the final amount for each investment.

29. $1000 at 6% interest compounded annually for 20 years **$3207.14**

30. $1000 at 6% interest compounded semiannually for 20 years **$3262.04**

31. $750 at 10% interest compounded quarterly for 10 years **$2013.80**

32. $750 at 5% interest compounded quarterly for 10 years **$1232.71**

33. $1800 at 5.65% interest compounded daily for 3 years **$2132.45**

34. $1800 at 5.65% interest compounded daily for 6 years **$2526.31**

35. Graph $f(x) = 2^x$, $g(x) = 5^x$, and $h(x) = 8^x$.
a. Which function exhibits the fastest growth? the slowest growth?
b. What is the y-intercept of each function?
c. State the domain and range of each function.

36. Graph $a(x) = \left(\frac{1}{2}\right)^x$, $b(x) = \left(\frac{1}{5}\right)^x$, and $c(x) = \left(\frac{1}{8}\right)^x$.
a. Which function exhibits the fastest decay? the slowest decay?
b. What is the y-intercept of each function?
c. State the domain and range of each function.

CHALLENGE

37. Describe when the graph of $f(x) = ab^x$ is a horizontal line.

Teaching Tip

TECHNOLOGY When using a TI-83 or TI-82 graphics calculator to compute the exponential regression equation, press STAT, select CALC, choose 0:ExpReg (or A:ExpReg), and press ENTER. If you want to find the regression equation for data contained in **L1** or **L2,** press ENTER; otherwise, enter the names of the two lists separated by a comma and press ENTER. **L3** through **L6** are entered by pressing 2nd 1 through 2nd 6.

Press Y= and set **Y1=** to the most recently found regression equation by pressing or selecting the following:

(or 7:RegEQ) ENTER

To produce the graphs for Example 3, set the viewing window as follows:

Xmin=–2	**Ymin=–20000**
Xmax=7	**Ymax=200000**
Xscl=1	**Yscl=10000**

35a. $h(x) = 8^x$; $f(x) = 2^x$
b. 1
c. for all 3 functions: domain: all reals; range: all positive reals

36a. $c(x) = \left(\frac{1}{8}\right)^x$; $a(x) = \left(\frac{1}{2}\right)^x$
b. 1
c. for all 3 functions: domain: all reals; range: all positive reals

37. when $a = 0$ or $b = 1$

Assess

Selected Answers

Exercises 5–9, 11–69 odd

ASSIGNMENT GUIDE

In Class	1–9
Core	11–45 odd, 49, 50
Core Plus	10–52 even
Review	54–69
Preview	70

Extra Practice can be found beginning on page 940.

Error Analysis

The order of operations is important in transformations of the exponential parent functions and applications of the compound interest formula. If students are using a calculator to perform these calculations, make sure that they use parentheses appropriately.

38. g is f stretched vertically by a factor of 5.

39. g is f compressed vertically by a factor of $\frac{1}{2}$.

Practice

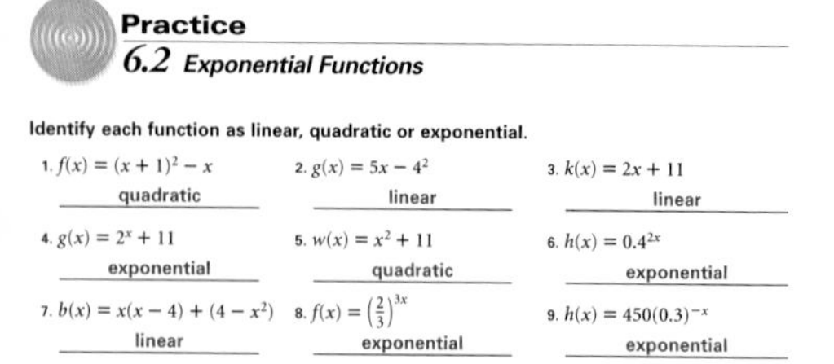

NAME ______________ CLASS ________ DATE ________

Practice
6.2 Exponential Functions

Identify each function as linear, quadratic or exponential.

1. $f(x) = (x + 1)^2 - x$ — quadratic
2. $g(x) = 5x - 4^2$ — linear
3. $k(x) = 2x + 11$ — linear
4. $g(x) = 2^x + 11$ — exponential
5. $w(x) = x^2 + 11$ — quadratic
6. $h(x) = 0.4^{2x}$ — exponential
7. $b(x) = x(x - 4) + (4 - x^2)$ — linear
8. $f(x) = \left(\frac{2}{3}\right)^{3x}$ — exponential
9. $h(x) = 450(0.3)^{-x}$ — exponential

Tell whether each function represents exponential growth or decay.

10. $f(x) = 5.9(2.6)^x$ — exponential growth
11. $b(x) = 13(0.7)^x$ — exponential decay
12. $k(x) = 22(0.15)^x$ — exponential decay
13. $m(x) = 51(4.3)^x$ — exponential growth
14. $w(x) = 0.72 \cdot 2^x$ — exponential growth
15. $z(x) = 47(0.55)^x$ — exponential decay
16. $h(x) = 2.5(0.8)^x$ — exponential decay
17. $g(x) = 0.8(3.2)^x$ — exponential growth
18. $a(x) = 150(1.1)^x$ — exponential growth

Find the final amount for each investment.

19. \$1300 earning 5% interest compounded annually for 10 years — \$2117.56
20. \$850 earning 4% interest compounded annually for 6 years — \$1075.52
21. \$720 earning 6.2% interest compounded semiannually for 5 years — \$977.06
22. \$1100 earning 5.5% interest compounded semiannually for 2 years — \$1226.08
23. \$300 earning 4.5% interest compounded quarterly for 3 years — \$343.10
24. \$1000 earning 6.5% interest compounded quarterly for 4 years — \$1294.22
25. \$5000 earning 6.3% interest compounded daily for 1 year — \$5325.11
26. \$2000 earning 5.5% interest compounded daily for 3 years — \$2358.76

CONNECTIONS

TRANSFORMATIONS Graph each pair of functions and describe the transformations from f to g.

38. $f(x) = \left(\frac{1}{2}\right)^x$ and $g(x) = 5\left(\frac{1}{2}\right)^x$

39. $f(x) = \left(\frac{1}{10}\right)^x$ and $g(x) = 0.5\left(\frac{1}{10}\right)^x$

40. $f(x) = 2^x$ and $g(x) = 3(2)^x + 1$

41. $f(x) = 10^x$ and $g(x) = 2(10)^x - 3$

42. $f(x) = 10^x$ and $g(x) = 3(10)^{x+2}$

43. $f(x) = 2^x$ and $g(x) = 5(2)^{x-1}$

44. $f(x) = 3\left(\frac{1}{2}\right)^x$ and $g(x) = 3(2^x)$

45. $f(x) = \left(\frac{1}{3}\right)^x$ and $g(x) = 2(3)^{-x}$

46. **TRANSFORMATIONS** Describe how each transformation of $f(x) = b^x$ affects the domain and range, the asymptotes, and the intercepts.

a. a vertical stretch
b. a vertical compression
c. a horizontal translation
d. a vertical translation
e. a reflection across the y-axis

46. no changes except as noted:
a. y-intercept increases.
b. y-intercept decreases.
c. y-intercept changes depending on direction of trans.
d. Range becomes all real numbers greater than the value of translation; the equation of horiz. asymp. is y = value of trans.; y-intercept becomes value of the trans. plus 1.
e. no changes

STATISTICS Use an exponential regression equation to find the effective yield for each investment. Assume that interest is compounded only once each year.

47 a \$1000 mutual fund investment made at the beginning of 1990 that is worth \$1450 at the beginning of 2000 **3.8%**

48 a house that is bought for \$75,000 at the end of 1995 and that is worth \$95,000 at the end of 2005 **2.4%**

STATISTICS Use an exponential regression equation to model the annual rate of inflation, or percent increase in price, for each item described.

49 a half-gallon of milk cost \$1.37 in 1989 and \$1.48 in 1995 [*Source: U.S. Bureau of Labor Statistics*] **1.3%**

50 a gallon of regular unleaded gasoline cost \$0.93 in 1986 and \$1.11 in 1993 [*Source: U.S. Bureau of Labor Statistics*] **2.6%**

APPLICATIONS

51. **INVESTMENTS** Find the final amount of a \$2000 certificate of deposit (CD) after 5 years at an annual interest rate of 5.51% compounded annually. **\$2615.16**

Certificate of Deposit
5.51%
Annual Percentage Yield*
\$2,000 Minimum

52. **INVESTMENTS** Consider a \$1000 investment that is compounded annually at three different interest rates: 5%, 5.5%, and 6%.

a. Write and graph a function for each interest rate over a time period from 0 to 60 years.
b. Compare the graphs of the three functions.
c. Compare the shapes of the graphs for the first 10 years with the shapes of the graphs between 50 and 60 years.

52a. $y = 1000\ (1.05)^x$; $y = 1000\ (1.055)^x$; $y = 1000\ (1.06)^x$

53. **INVESTMENTS** The final amount for \$5000 invested for 25 years at 10% annual interest compounded semiannually is \$57,337.

a. What is the effect of doubling the amount invested?
b. What is the effect of doubling the annual interest rate?
c. What is the effect of doubling the investment period?
d. Which of the above has the greatest effect on the final amount of the investment?

53a. Final amount is doubled: \$114,674
b. Final amount is more than 10 times larger: \$586,954
c. Final amount is more than 11 times larger: \$657,506
d. doubling the investment period

40. g is f stretched vertically by a factor of 3 and translated vertically 1 unit up.

41. g is f stretched vertically by a factor of 2 and translated vertically 3 units down.

42. g is f shifted 2 units to the left and stretched vertically by a factor of 3.

43. g is f shifted 1 unit to the right and stretched vertically by a factor of 5.

44. g is f reflected across the y-axis.

45. g is f stretched vertically by a factor of 2.

52b. They look fairly similar until about 30 years; then it is most apparent that the growth of the investment at 5% is much slower than that of the investment at 6%.

c. The shapes are almost identical for the first 10 years, but the shapes of the graphs between 50 and 60 years are different, namely the graph of the investment at 6% is steeper.

Look Back

Find the inverse of each function. State whether the inverse is a function. ***(LESSON 2.5)***

54. $\{(-2, 4), (-3, -1), (2, 2), (3, 4)\}$ **$\{(4, -2), (-1, -3), (2, 2), (4, 3)\}$; no**

55. $\{(7, 2), (3, -1), (2, 2), (0, 0)\}$ **$\{(2, 7), (-1, 3), (2, 2), (0, 0)\}$; no**

56. $y = 2(x + 3)$ **$y = \frac{1}{2}x - 3$; yes**

57. $y = 3x^2$ **$y = \pm\frac{\sqrt{3x}}{3}$; no**

58. $y = x^2 + 2$ **$y = \pm\sqrt{x - 2}$; no**

59. $y = -x^2$ **$y = \pm\sqrt{-x}$; no**

Graph each piecewise function. ***(LESSON 2.6)***

60. $f(x) = \begin{cases} 9 & \text{if } 0 \le x < 5 \\ 2x - 1 & \text{if } 5 \le x < 10 \end{cases}$

61. $g(x) = \begin{cases} x & \text{if } 0 \le x < 2 \\ -3x + 8 & \text{if } 2 \le x < 5 \\ -5 & \text{if } 5 \le x < 10 \end{cases}$

Let $A = \begin{bmatrix} 3 & 4 & -1 \\ -2 & -8 & 6 \\ 10 & 8 & 0 \end{bmatrix}$, $B = \begin{bmatrix} 0 & 2 \\ 7 & -2 \\ -5 & 1 \end{bmatrix}$, and $C = \begin{bmatrix} 2 & -6 & -2 \\ 3 & -1 & 4 \end{bmatrix}$. **Find each product matrix, if it exists.** ***(LESSON 4.2)***

62. AB **63.** BA **64.** AC **65.** CA **66.** BC **67.** CB

62. $\begin{bmatrix} 33 & -3 \\ -86 & 18 \\ 56 & 4 \end{bmatrix}$

63. does not exist

64. does not exist

65. $\begin{bmatrix} -2 & 40 & -38 \\ 51 & 52 & -9 \end{bmatrix}$

66. $\begin{bmatrix} 6 & -2 & 8 \\ 8 & -40 & -22 \\ -7 & 29 & 14 \end{bmatrix}$

67. $\begin{bmatrix} -32 & 14 \\ -27 & 12 \end{bmatrix}$

Find a quadratic function to fit each set of points exactly. ***(LESSON 5.7)***

68. $(1, -1), (2, -5), (3, 13)$ **$y = 11x^2 - 37x + 25$**

69. $(0, 4), (1, 5), (3, 25)$ **$y = 3x^2 - 2x + 4$**

Look Beyond

70. Use guess-and-check to find x such that $10^x = 50$. **$x \approx 1.699$**

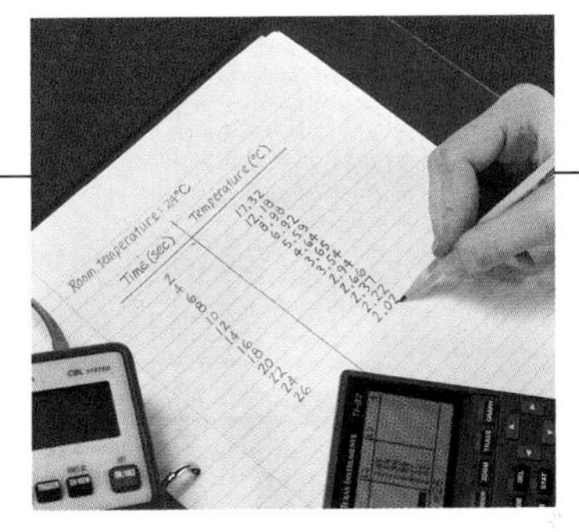

For this activity, use the data collected in the Portfolio Activity on page 361.

1. **a.** Use the quadratic regression feature on your calculator to find a quadratic function that models your first 30 readings.
   b. Use your linear function to predict the temperature of the probe after 2 minutes. Compare this prediction with your actual 2-minute reading.
   c. Discuss the usefulness of your quadratic function for modeling the cooling process.
2. Now use the exponential regression feature on your calculator to find an exponential function that models your first 30 readings, and repeat parts **b** and **c** of Step 1.

Save your data and results to use in the remaining Portfolio Activities.

WORKING ON THE CHAPTER PROJECT

You should now be able to complete Activity 2 of the Chapter Project.

Look Beyond

In Exercise 70, students use guess-and-check to solve an exponential equation. In Lesson 6.7, students will learn how to solve equations like this by using logarithms.

ALTERNATIVE

Assessment

Portfolio Activity

The Portfolio Activity can be used to as preparation for the Chapter Project or as a separate activity. In the Portfolio Activity in Lesson 6.1, students found a linear regression equation for data from an experiment. In the Portfolio Activity on this page, students will find a quadratic regression equation and an exponential regression equation for the same data used in Lesson 6.1.

Answers to Portfolio Activities can be found in Additional Answers of the Teacher's Edition.

Student Technology Guide

NAME ________ CLASS ________ DATE ________

Student Technology Guide
6.2 **Exponential Functions**

You can use a graphics calculator to find an exponential regression model for a data set. An exponential model is the exponential equation whose graph best fits the data points.

Consider the data for expenditures for health services and supplies in the United States shown in the table.

Year (year 0: 1970)	0	5	10	15	20
Health expenditures (billions of dollars)	69.1	124.7	238.9	407.2	652.4

- Press STAT 1:Edit... ENTER. As shown at left below, enter the years into the L1 column and the health expenditures into the L2 column. Press ENTER after each data value.
- To find a regression equation, press STAT CALC 0:ExpReg ENTER ENTER. Rounding to the nearest hundredth, $y \approx 71.58(1.12)^x$ is the exponential regression equation that best fits this data.

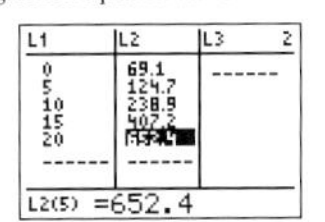

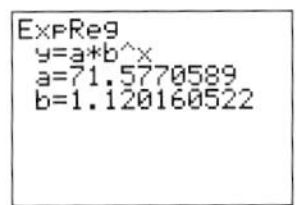

Use a graphics calculator to find the exponential regression equation whose graph best fits each data set. Use 1990 as year 0.

1. Applications for U.S. citizenship

Year	1990	1991	1992	1993	1994	1995
Number	233,843	206,668	342,269	522,298	543,353	1,021,969

$y \approx 191{,}282.83(1.36)^x$

2. Money spent for technology by public schools in billions of dollars

Year	1992	1993	1994	1995	1996
Amount	2.1	2.5	2.8	3.6	4.0

$y \approx 1.51(1.18)^x$

60.

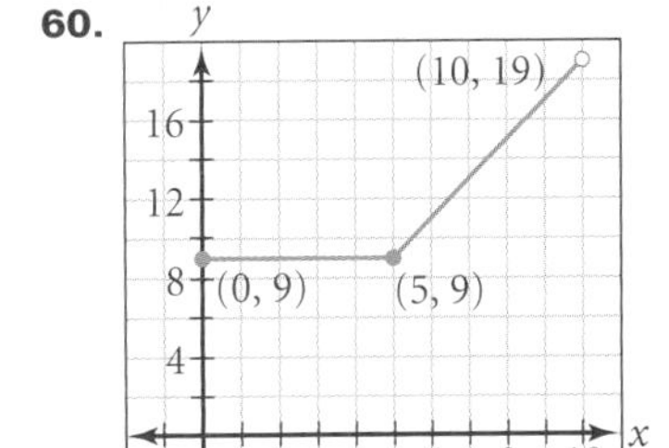

61.

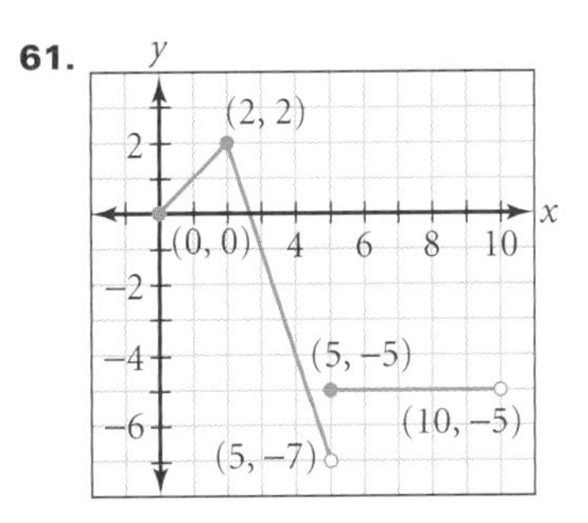

Prepare

NCTM PRINCIPLES & STANDARDS 1–4, 6–10

QUICK WARM-UP

Find the inverse of each function.

1. $f(x) = x + 10$
   $f^{-1}(x) = x - 10$
2. $g(x) = 3x$
   $g^{-1}(x) = \frac{x}{3}$
3. $h(x) = 5x + 3$
   $h^{-1}(x) = \frac{1}{5}x - \frac{3}{5}$
4. $j(x) = \frac{1}{4}x + 2$
   $j^{-1}(x) = 4x - 8$

Also on Quiz Transparency 6.3

Teach

Why Ask students to share what they know about inverse functions. In this lesson, students will use that foundation to explore the inverses of exponential functions.

6.3 Logarithmic Functions

Objectives

- Write equivalent forms for exponential and logarithmic equations.
- Use the definitions of exponential and logarithmic functions to solve equations.

Why *Logarithmic functions are widely used in measurement scales such as the pH scale, which ranges from 0 to 14.*

Substance	pH
gastric fluid	1.8
lemon juice	2.2–2.4
vinegar	2.4–3.4
banana	4.8
saliva	6.5–7.5
water	7
egg white	7.6–8.0
Rolaids, Tums	9.9
milk of magnesia	10.5

The pH of an acidic solution is less than 7, the pH of a basic solution is greater than 7, and the pH of a neutral solution is 7.

Logarithms are used to find unknown exponents in exponential models.

Logarithmic functions define many measurement scales in the sciences, including the pH, decibel, and Richter scales.

Activity: Approximating Exponents

TECHNOLOGY GRAPHICS CALCULATOR
Keystroke Guide, page 419

You will need: a graphics calculator

Use the table below to complete this Activity.

x	-3	-2	-1	0	1	2	3
$y = 10^x$	$\frac{1}{1000}$	$\frac{1}{100}$	$\frac{1}{10}$	1	10	100	1000

1. How are the x-values in the table related to the y-values?

CHECKPOINT ✔
2. Use the table above to find the value of x in each equation below.
   a. $10^x = 1000$ **b.** $10^x = \frac{1}{100}$ **c.** $10^x = \frac{1}{1000}$ **d.** $10^x = 1$

PROBLEM SOLVING
3. **Make a table** of values for $y = 10^x$. Use the table to approximate the solution to $10^x = 7$ to the nearest hundredth.

CHECKPOINT ✔
4. Use a table of values to approximate the solution to $10^x = 85$ to the nearest hundredth.

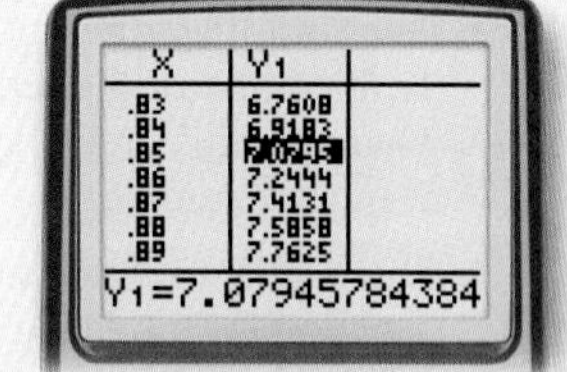

Alternative Teaching Strategy

USING MODELS Have students graph the function $f(x) = 10^x$ on graph paper, using an appropriate scale for the y-axis. Then have them draw the inverse of the function by using a geomirror to reflect f across the line $y = x$. Ask students to label four points on the graph of $f(x) = 10^x$ and find their corresponding coordinates on the graph of the inverse. For example, the point (0, 1) on the graph of f corresponds to the point (1, 0) on the graph of f^{-1}. Ask students to discuss the differences between the two graphs. Students should observe that f has a horizontal asymptote at $y = 0$ and f^{-1} has a vertical asymptote at $x = 0$; the values of $f(x)$ are powers of 10 and grow at an increasing rate, and the values of $f^{-1}(x)$ are the same as the value of the exponent and grow more slowly. Define $f^{-1}(x)$ as $\log_{10} x$. Have students use a graphics calculator to graph various exponential functions and their inverse logarithmic functions.

A table of values for $y = 10^x$ can be used to solve equations such as $10^x = 1000$ and $10^x = \frac{1}{100}$. However, to solve equations such as $10^x = 85$ or $10^x = 2.3$, a *logarithm* is needed. With logarithms, you can write an exponential equation in an equivalent logarithmic form.

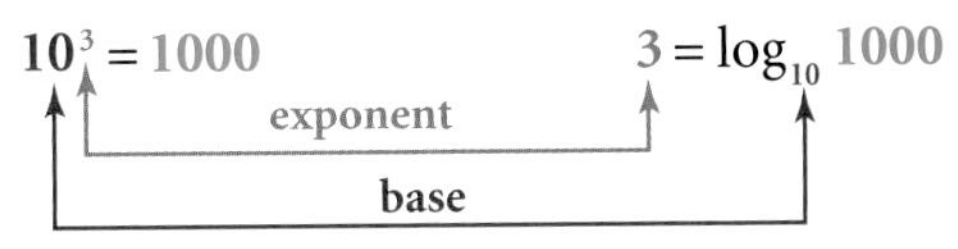

Equivalent Exponential and Logarithmic Forms

For any positive base b, where $b \neq 1$:
$b^x = y$ if and only if $x = \log_b y$

EXAMPLE 1

a. Write $5^3 = 125$ in logarithmic form.
b. Write $\log_3 81 = 4$ in exponential form.

SOLUTION

a. $5^3 = 125 \quad \rightarrow \quad 3 = \log_5 125$ *3 is the exponent and 5 is the base.*
b. $\log_3 81 = 4 \quad \rightarrow \quad 3^4 = 81$ *3 is the base and 4 is the exponent.*

TRY THIS Copy and complete each column in the table below.

Exponential form	$2^5 = 32$	?	$3^{-2} = \frac{1}{9}$	?
Logarithmic form	?	$\log_{10} 1000 = 3$	?	$\log_{16} 4 = \frac{1}{2}$

You can evaluate logarithms with a base of 10 by using the LOG key on a calculator.

EXAMPLE 2 **Solve for $10^x = 85$ for x. Round your answer to the nearest thousandth.**

SOLUTION

Write $10^x = 85$ in logarithmic form, and use the LOG key.

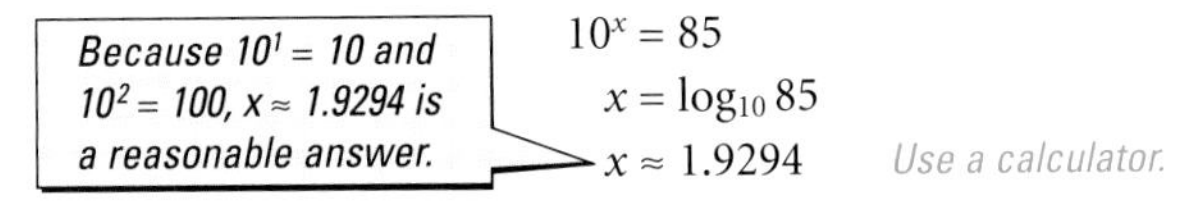

Because 10¹ = 10 and 10² = 100, x ≈ 1.9294 is a reasonable answer.

$$10^x = 85$$
$$x = \log_{10} 85$$
$$x \approx 1.9294 \quad \textit{Use a calculator.}$$

TRY THIS Solve $10^x = \frac{1}{109}$ for x. Round your answer to the nearest thousandth.

Activity Notes

In this Activity, students use tables of values to solve exponential equations. Students are given the y-value of the equation and asked to find the x-value. They should look through the y-values and read the corresponding x-values from the table.

Teaching Tip

TECHNOLOGY When using a TI-83 or TI-82 graphics calculator to produce the table for Step 3 of the Activity, students can enter **10^X** into **Y1** and then use the **Table Setup** feature to produce the table of values by pressing 2nd WINDOW and entering the following values:

TblStart=0 (or **TblMin=0**)
ΔTbl=1
Indpnt: Auto Ask
Depend: Auto Ask

To display the table, press 2nd GRAPH. Letting **ΔTbl=1** allows students to get a general understanding of the range of y-values. Then have them set **ΔTbl=.01,** view the table again, and approximate the value of x when y is 7 and when y is 85 to answer Steps 3 and 4, respectively.

CHECKPOINT ✔
2a. $x = 3$ **b.** $x = -2$
c. $x = -3$ **d.** $x = 0$

CHECKPOINT ✔
4. $x \approx 1.93$

Use Teaching Transparency 23.

☞ For Additional Examples 1 and 2 and answer to Try This, see page 372.

Interdisciplinary Connection

ASTRONOMY The *limiting magnitude* of a telescope is the magnitude of the faintest star that can be seen with the telescope. The limiting magnitude, L, of a telescope depends on the diameter of the lens, d, in centimeters and is given by the function $L(d) = 7.1 + 5.1 \log d$. Find the limiting value of a telescope with a lens diameter of 20 centimeters. **13.7** Find the lens diameter of a telescope whose limiting magnitude is 12.3. **10.5 cm** Have students research the lens diameter of the Hubble telescope and calculate its limiting magnitude.

Enrichment

Have students gain a sense for the values of logarithms by using estimation. For example, to estimate $\log_3 10$, find the closest power of 3 to 10. This would be $9 = 3^2$, so $\log_3 10 > \log_3 9 = \log_3 3^2 = 2$. Using this reasoning, we can estimate $\log_3 10$ to be slightly greater than 2. Students can use their calculators to check the accuracy of their estimate by calculating $3^{2.1}$. Have students estimate the following logarithms:

1. $\log_4 15$ **slightly less than 2, or ≈1.953**
2. $\log_7 52$ **slightly more than 2, or ≈2.031**
3. $\log_2 12$ **≈3.585**

Use Teaching Transparency 24.

ADDITIONAL EXAMPLE 1

a. Write $6^4 = 1296$ in logarithmic form.
$\log_6 1296 = 4$

b. Write $\log_8 512 = 3$ in exponential form.
$8^3 = 512$

TRY THIS
$10^3 = 1000$; $16^{\frac{1}{2}} = 4$

$\log_2 32 = 5$; $\log_3 \frac{1}{9} = -2$

ADDITIONAL EXAMPLE 

Solve $10^x = 14.5$ for x. Round your answer to the nearest thousandth.
$x \approx 1.161$

TRY THIS
−2.037

CRITICAL THINKING
The graph is the vertical line $x = 1$. No, $y = \log_1 x$ is not a function.

Definition of Logarithmic Function

The inverse of the exponential function $y = 10^x$ is $x = 10^y$. To rewrite $x = 10^y$ in terms of y, use the equivalent logarithmic form, $y = \log_{10} x$.

Examine the tables and graphs below to see the inverse relationship between $y = 10^x$ and $y = \log_{10} x$.

x	$y = 10^x$
−3	$\frac{1}{1000}$
−2	$\frac{1}{100}$
−1	$\frac{1}{10}$
0	1
1	10
2	100
3	1000

x	$y = \log_{10} x$
$\frac{1}{1000}$	−3
$\frac{1}{100}$	−2
$\frac{1}{10}$	−1
1	0
10	1
100	2
1000	3

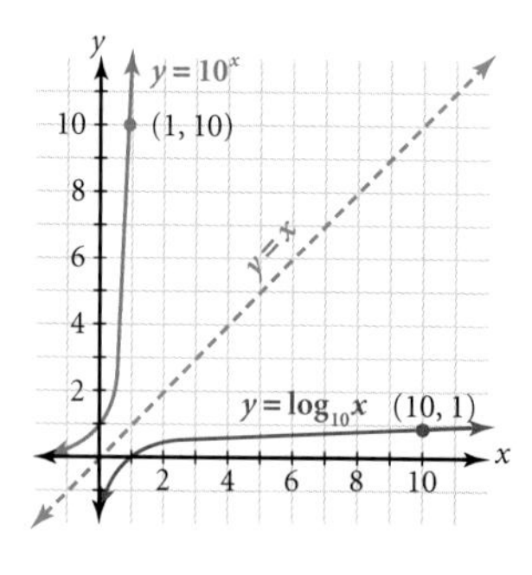

The table below summarizes the relationship between the domain and range of $y = 10^x$ and of $y = \log_{10} x$.

Function	Domain	Range
$y = 10^x$	all real numbers	all positive real numbers
$y = \log_{10} x$	all positive real numbers	all real numbers

Logarithmic Functions

The **logarithmic function** $y = \log_b x$ with base b, or $x = b^y$, is the inverse of the exponential function $y = b^x$, where $b \neq 1$ and $b > 0$.

CRITICAL THINKING Describe the graph that results if $b = 1$ in $y = \log_b x$. Is $y = \log_1 x$ a function?

Because $y = \log_b x$ is the inverse of the exponential function $y = b^x$ and $y = \log_b x$ is a function, the exponential function $y = b^x$ is a one-to-one function. This means that for each element in the domain of an exponential function, there is exactly one corresponding element in the range. For example, if $3^x = 3^2$, then $x = 2$. This is called the *One-to-One Property of Exponents.*

One-to-One Property of Exponents

If $b^x = b^y$, then $x = y$.

Inclusion Strategies

ENGLISH LANGUAGE DEVELOPMENT Many students have trouble adjusting to logarithmic notation and how it corresponds to exponential form. To help students see the two forms as equivalent, have them verbalize the outcome of taking a logarithm. Students should focus on the fact that the logarithm gives an exponent as its value. For example, when evaluating $\log_3 27$, students should ask themselves what power they must raise 3 to in order to get 27. To help students remember this relationship, write the log equation as "$\log_{base}$ value = exponent" stressing that the expression "$\log_{base}$ value" represents an exponent.

EXAMPLE 3 **Find the value of v in each equation.**

a. $v = \log_{125} 5$ **b.** $5 = \log_v 32$ **c.** $4 = \log_3 v$

SOLUTION

Write the equivalent exponential form, and solve for v.

a. $v = \log_{125} 5$

$125^v = 5$

$(5^3)^v = 5$

$5^{3v} = 5^1$

$3v = 1$ — *Apply the One-to-One Property.*

$v = \frac{1}{3}$

b. $5 = \log_v 32$

$v^5 = 32$

$v^5 = 2^5$

$v = 2$

c. $4 = \log_3 v$

$3^4 = v$

$81 = v$

TRY THIS Find the value of v in each equation.

a. $v = \log_4 64$ **b.** $2 = \log_v 25$ **c.** $6 = \log_3 v$

Recall from Lesson 2.7 that the graph of $y = -f(x)$ is the graph of $y = f(x)$ reflected across the x-axis. The graph of $y = \log_{10} x$ and of its reflection across the x-axis, $y = -\log_{10} x$, are shown at right.

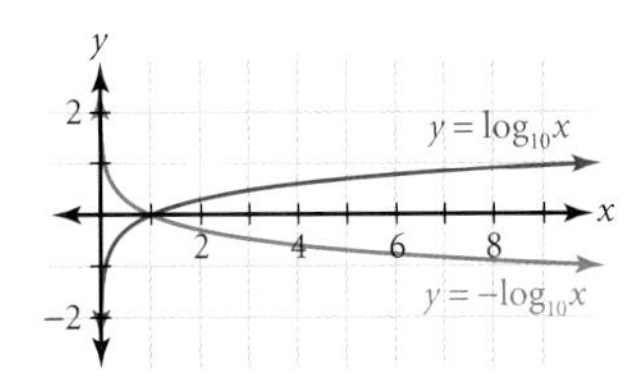

The function $y = -\log_{10} x$ is used in chemistry to measure pH levels. The pH of a solution describes its acidity. Substances that are more acidic have a lower pH, while substances that are less acidic, or basic, have a higher pH. The pH of a substance is defined as $\text{pH} = -\log_{10}[H^+]$, where $[H^+]$ is the hydrogen ion concentration of a solution in moles per liter.

EXAMPLE 4 The pH of a carbonated soda is 3.

What is $[H^+]$ for this soda?

SOLUTION

$\text{pH} = -\log_{10}[H^+]$

$3 = -\log_{10}[H^+]$ *Substitute 3 for pH.*

$-3 = \log_{10}[H^+]$

$10^{-3} = [H^+]$ *Write the equivalent exponential equation.*

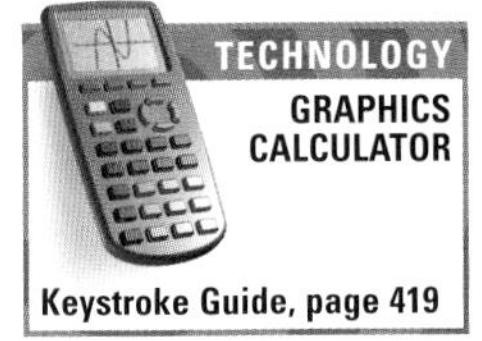

CHECK

Graph $y = -\log_{10} x$ and $y = 3$ on the same screen, and find the point of intersection. The window at right shows x-values between 0 and 0.01.

Thus, there is $\frac{1}{1000}$, or 0.001, moles of hydrogen ions in a liter of carbonated soda that has a pH of 3.

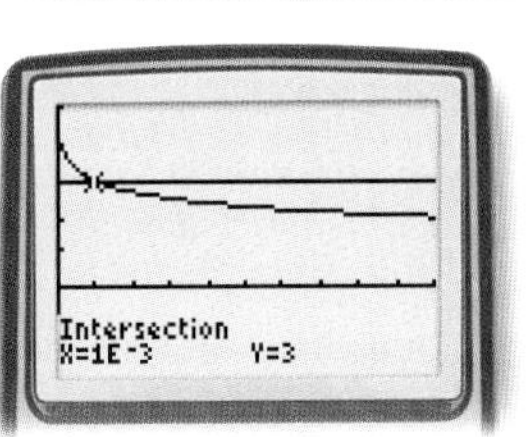

TRY THIS Find $[H^+]$ for orange juice that has a pH of 3.75.

Reteaching the Lesson

USING TECHNOLOGY On a TI-83 or TI-82 graphics calculator, have students enter values from 1 to 10 into list **L1**, use the up arrow key to select the heading **L2**, and then define **L2=LOG(L1).** Next have them define **L3** as **10^L2** by using a similar procedure. **L3** should give the exact values listed in **L1.** Have students discuss why this happens. Make sure students mention the inverse relationship between logarithms and exponents. Emphasize to students that **L3** came from using the values in **L2** as exponents. This reinforces the idea that the value of a logarithm is an exponent.

ADDITIONAL **EXAMPLE 3**

Find the value of the variable in each equation:

a. $r = \log_2 1$ $r = 0$

b. $\frac{1}{2} = \log_y 9$ $y = 81$

c. $3 = \log_7 D$ $D = 343$

TRY THIS

a. $v = 3$

b. $v = 5$

c. $v = 729$

Math CONNECTION

TRANSFORMATIONS Students reflect a graph across the x-axis by changing every y-value to its opposite. This is written algebraically as $(x, f(x)) \rightarrow (x, -f(x))$.

ADDITIONAL **EXAMPLE 4**

The pH of a sample of sea water is 8.3. **What is $[H^+]$ for this sea water?** $10^{-8.3}$

Teaching Tip

TECHNOLOGY When using a TI-83 or TI-82 graphics calculator to find the point of intersection in Example 4, press 2nd TRACE, choose 5:intersect, and press ENTER twice to identify the two graphs. Identify a guess by using ◄ or ► and then pressing ENTER. Set the viewing window as follows:

Xmin=0	Ymin=–2
Xmax=.01	Ymax=5
Xscl=.001	Yscl=1

TRY THIS

$[H^{375}] = 10^{-3.75}$, or ≈ 0.00018 moles/L

Teaching Tip

Throughout this lesson, remind students that a logarithm is the *exponent* to which the base must be raised to yield a given number. Relate this statement to $y = \log_b x$ and $x = b^y$ and stress that the expression $\log_b x$ represents the exponent to which b is raised in order to yield x.

Assess

Selected Answers

Exercises 5–11, 13–111 odd

ASSIGNMENT GUIDE	
In Class	1–11
Core	13–85 odd, 91
Core Plus	12–86 even, 88–100 even
Review	102–112
Preview	113

Extra Practice can be found beginning on page 940.

Practice

NAME ________________ CLASS ________ DATE ________

Practice
6.3 Logarithmic Functions

Write each equation in logarithmic form.

1. $19^2 = 361$ — $\log_{19} 361 = 2$
2. $20^3 = 8000$ — $\log_{20} 8000 = 3$
3. $3375^{\frac{1}{3}} = 15$ — $\log_{3375} 15 = \frac{1}{3}$
4. $\left(\frac{3}{4}\right)^{-3} = 64$ — $\log_{\frac{3}{4}} 64 = -3$
5. $\left(\frac{3}{7}\right)^3 = \frac{27}{343}$ — $\log_{\frac{3}{7}} \frac{27}{343} = 3$
6. $11^{-3} = \frac{1}{1331}$ — $\log_{11} \frac{1}{1331} = -3$

Write each equation in exponential form.

7. $\log_{12} 144 = 2$ — $12^2 = 144$
8. $\log_5 15{,}625 = 6$ — $5^6 = 15{,}625$
9. $\log_{21} 9261 = 3$ — $21^3 = 9261$
10. $\log_{3600} 60 = \frac{1}{2}$ — $3600^{\frac{1}{2}} = 60$
11. $\log_{11} \frac{1}{14{,}641} = -4$ — $11^{-4} = \frac{1}{14{,}641}$
12. $\log_{\frac{1}{5}} 625 = -4$ — $\left(\frac{1}{5}\right)^{-4} = 625$

Solve each equation for *x*. Round your answers to the nearest hundredth.

13. $10^x = 35$ — $x \approx 1.54$
14. $10^x = 91$ — $x \approx 1.96$
15. $10^x = 0.2$ — $x \approx -0.70$
16. $10^x = 1.8$ — $x \approx 0.26$
17. $10^x = 0.08$ — $x \approx -1.10$
18. $10^x = 1055$ — $x \approx 3.02$

Find the value of *v* in each equation.

19. $v = \log_{10} 1000$ — $v = 3$
20. $v = \log_{15} 225$ — $v = 2$
21. $v = \log_{12} 144$ — $v = 2$
22. $8 = \log_2 v$ — $v = 256$
23. $-4 = \log_5 v$ — $v = \frac{1}{625}$
24. $-3 = \log_7 v$ — $v = \frac{1}{343}$
25. $-2 = \log_v \frac{1}{100}$ — $v = 10$
26. $\log_v 729 = 6$ — $v = 3$
27. $\log_v \frac{1}{256} = -4$ — $v = 4$

Exercises

Communicate

1. Describe the relationship between logarithmic functions and exponential functions.
2. State the domain and range of logarithmic functions. How are they related to the domain and range of exponential functions?
3. Explain how to approximate the value of x in $2^x = 58$ by using the table feature of a graphics calculator.

Guided Skills Practice

4. Write $4^2 = 16$ in logarithmic form. ***(EXAMPLE 1)*** **$\log_4 16 = 2$**
5. Write $\log_5 25 = 2$ in exponential form. ***(EXAMPLE 1)*** **$5^2 = 25$**

Solve each equation for *x*. Round your answers to the nearest thousandth. *(EXAMPLE 2)*

6. $10^x = 568$ **2.754**
7. $10^x = \frac{1}{500}$ **−2.699**

Find the value of *v* in each equation. *(EXAMPLE 3)*

8. $v = \log_7 49$ **2**
9. $2 = \log_v 144$ **12**
10. $2 = \log_4 v$ **16**

APPLICATION

11. **CHEMISTRY** The pH of black coffee is 5. What is $[H^+]$ for this coffee? ***(EXAMPLE 4)*** **10^{-5}, or 0.00001, moles per liter**

Practice and Apply

Write each equation in logarithmic form.

12. $11^2 = 121$
13. $5^4 = 625$
14. $3^5 = 243$
15. $6^3 = 216$
16. $6^{-2} = \frac{1}{36}$
17. $7^{-2} = \frac{1}{49}$
18. $27^{\frac{1}{3}} = 3$
19. $16^{\frac{1}{4}} = 2$
20. $\left(\frac{1}{4}\right)^{-3} = 64$
21. $\left(\frac{1}{9}\right)^{-2} = 81$
22. $\left(\frac{1}{3}\right)^2 = \frac{1}{9}$
23. $\left(\frac{1}{2}\right)^3 = \frac{1}{8}$

Answers:
12. $\log_{11} 121 = 2$
13. $\log_5 625 = 4$
14. $\log_3 243 = 5$
15. $\log_6 216 = 3$
16. $\log_6 \frac{1}{36} = -2$
17. $\log_7 \frac{1}{49} = -2$
18. $\log_{27} 3 = \frac{1}{3}$
19. $\log_{16} 2 = \frac{1}{4}$
20. $\log_{\frac{1}{4}} 64 = -3$
21. $\log_{\frac{1}{9}} 81 = -2$
22. $\log_{\frac{1}{3}} \frac{1}{9} = 2$
23. $\log_{\frac{1}{2}} \frac{1}{8} = 3$

Write each equation in exponential form.

24. $\log_6 36 = 2$ **$6^2 = 36$**
25. $\log_{10} 1000 = 3$ **$10^3 = 1000$**
26. $\log_{10} 0.001 = -3$ **$10^{-3} = 0.001$**
27. $\log_{10} 0.1 = -1$ **$10^{-1} = 0.1$**
28. $3 = \log_9 729$ **$9^3 = 729$**
29. $3 = \log_7 343$ **$7^3 = 343$**
30. $\log_3 \frac{1}{81} = -4$ **$3^{-4} = \frac{1}{81}$**
31. $\log_2 \frac{1}{32} = -5$ **$2^{-5} = \frac{1}{32}$**
32. $-2 = \log_2 \frac{1}{4}$ **$2^{-2} = \frac{1}{4}$**
33. $-3 = \log_3 \frac{1}{27}$ **$3^{-3} = \frac{1}{27}$**
34. $\log_{121} 11 = \frac{1}{2}$ **$121^{\frac{1}{2}} = 11$**
35. $\log_{144} 12 = \frac{1}{2}$ **$144^{\frac{1}{2}} = 12$**

Find the approximate value of each logarithmic expression.

36. $\log_{10} 1026$ **3**
37. $\log_{10} 79$ **2**
38. $\log_{10} 8$ **1**
39. $\log_{10} 21{,}050$ **4**
40. $\log_{10} 0.08$ **−1**
41. $\log_{10} 0.9$ **0**
42. $\log_{10} 0.002$ **−3**
43. $\log_{10} 0.00013$ **−4**

Solve each equation for *x*. Round your answers to the nearest hundredth.

44. $10^x = 31$ **1.49**
45. $10^x = 12$ **1.08**
46. $10^x = 72$ **103.86**
47. $10^x = 3588$ **3.55**
48. $10^x = 1.498$ **0.18**
49. $10^x = 1.89$ **0.28**
50. $10^x = 0.0054$ **−2.27**
51. $10^x = 0.035$ **−1.46**
52. $10^x = \frac{3}{49}$ **−1.21**
53. $10^x = \frac{1}{1085}$ **−3.04**
54. $10^x = \sqrt{7.4}$ **0.43**
55. $10^x = \frac{1}{\sqrt{500}}$ **−1.35**

Find the value of *v* in each equation.

56. $v = \log_{10} 1000$ **3**
57. $v = \log_4 64$ **3**
58. $v = \log_7 343$ **3**
59. $v = \log_{17} 289$ **2**
60. $v = \log_3 3$ **1**
61. $v = \log_7 7$ **1**
62. $v = \log_{10} 0.001$ **−3**
63. $v = \log_{10} 0.01$ **−2**
64. $v = \log_2 \frac{1}{4}$ **−2**
65. $v = \log_{10} \frac{1}{100}$ **−2**
66. $v = \log_4 1$ **0**
67. $v = \log_9 1$ **0**
68. $3 = \log_6 v$ **216**
69. $2 = \log_7 v$ **49**
70. $1 = \log_5 v$ **5**
71. $1 = \log_3 v$ **3**
72. $\frac{1}{2} = \log_9 v$ **3**
73. $\frac{1}{3} = \log_8 v$ **2**
74. $-2 = \log_6 v$ $\frac{1}{36}$
75. $-3 = \log_4 v$ $\frac{1}{64}$
76. $0 = \log_{13} v$ **1**
77. $0 = \log_2 v$ **1**
78. $\log_v 16 = 2$ **4**
79. $\log_v 125 = 3$ **5**
80. $\log_v 9 = \frac{1}{2}$ **81**
81. $\log_v 4 = \frac{1}{3}$ **64**
82. $\log_v \frac{1}{16} = -4$ **2**
83. $\log_v \frac{1}{8} = -3$ **2**
84. $\log_v 216 = 3$ **6**
85. $\log_v 243 = 5$ **3**

86. Graph $f(x) = 3^x$ along with f^{-1}. Make a table of values that illustrates the relationship between f and f^{-1}.

87. Graph $f(x) = 3^{-x}$ along with f^{-1}. Make a table of values that illustrates the relationship between f and f^{-1}.

CHALLENGE

Find the value of each expression.

88. $\log_{27} \sqrt{3}$ $\frac{1}{6}$
89. $\log_2 16\sqrt{2}$ **4.5**
90. $\log_{\frac{1}{2}} 8$ **−3**

internet connect
Homework Help Online
Go To: **go.hrw.com**
Keyword:
MB1 Homework Help for Exercises 91–96

TRANSFORMATIONS **Let $f(x) = \log_{10} x$. For each function, identify the transformations from *f* to *g*.**

91. $g(x) = 3 \log_{10} x$
92. $g(x) = -5 \log_{10} x$
93. $g(x) = \frac{1}{2} \log_{10} x + 1$
94. $g(x) = 0.25 \log_{10} x - 2$
95. $g(x) = -\log_{10}(x - 2)$
96. $g(x) = \log_{10}(x + 5) - 3$

CHEMISTRY Calculate $[H^+]$ for each of the following:

97. household ammonia with a pH of about 10 **10^{-10} mole per liter**

98. distilled water with a pH of 7 **10^{-7} mole per liter**

99. human blood with a pH of about 7.4 **3.98×10^{-8} mole per liter**

100. CHEMISTRY How much greater is $[H^+]$ for lemon juice, which has a pH of 2.1, than $[H^+]$ for water, which has a pH of 7.0?

100. **$[H^+]$ for lemon juice is 79,433 times greater than $[H^+]$ for water.**

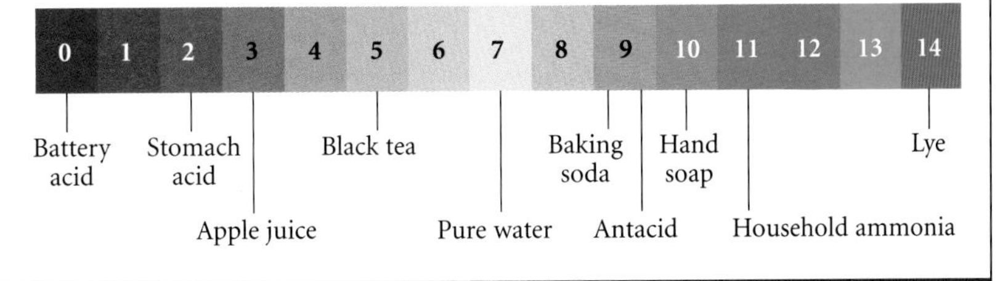

pH paper turns red in an acidic solution, $0 < pH < 7$; the paper turns green in a neutral solution, indicating a pH of 7; and the paper turns blue in a basic solution, $7 < pH < 14$.

Error Analysis

For Exercises 91–96, students may need to be reminded about the general transformations of functions. One particular transformation is $f(x - a)$, as in Exercises 95 and 96. Students often perform the horizontal shift in the wrong direction.

91. stretched vertically by a factor of 3

92. reflected across the *x*-axis and stretched vertically by a factor of 5

93. compressed vertically by a factor of $\frac{1}{2}$ and translated vertically 1 unit up

94. compressed vertically by a factor of $\frac{1}{4}$ and translated vertically 2 units down

95. reflected across the *x*-axis and translated horizontally 2 units to the right

96. translated horizontally 5 units to the left and vertically 3 units down

86.

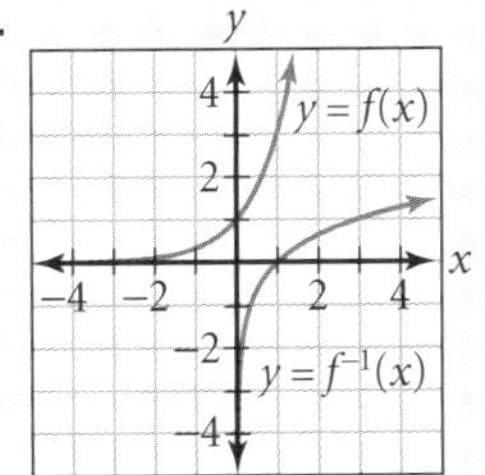

Tables may vary. Sample tables are provided in Additional Answers beginning on page 1002.

87.

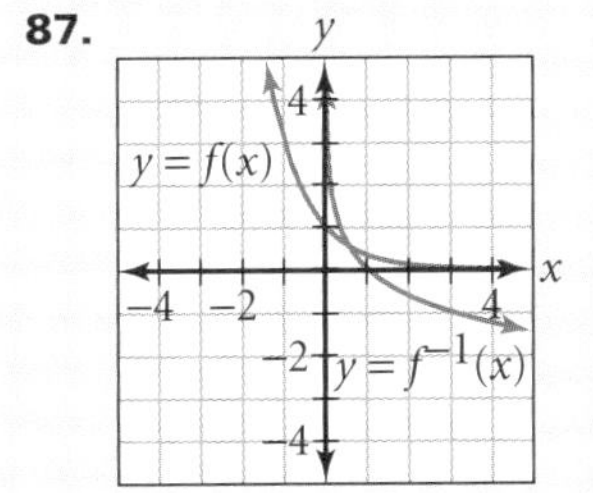

Tables may vary. Sample tables are provided in Additional Answers beginning on page 1002.

Student Technology Guide

NAME ____________ CLASS ________ DATE ________

Student Technology Guide
6.3 Logarithmic Functions

Logarithmic functions are inverse exponential functions. On most calculators, you will find both of these functions on the same key, with one being the 2nd function for the key.

Example: Solve $10^x = 125$. Round your answer to the nearest thousandth.

- If $10^x = 125$, then $x = \log 125$.
  Press LOG 125) ENTER.
  The solution to $10^x = 125$ is $x \approx 2.097$.

```
log(125)
        2.096910013
```

Alternatively, you can solve $10^x = 125$ by using a graphical approach.

- Enter 10^x as Y1 and 125 as Y2. Y= 10 ^ X,T,θ,n ENTER 125 ENTER
- Press WINDOW. Choose Xmin = 0, Xmax = 4, Ymin = 0, and Ymax = 150.
- Press GRAPH.
- Use TRACE and ◄ and ► to trace to the point where the graphs intersect. The intersection point appears to be near (2.1, 125), as shown at left below.
- You can use the intersect feature of the calculator to get a better estimate. With the trace point near the point of intersection as shown, press 2nd TRACE 5:intersect ENTER ENTER ENTER ENTER.

The solution is approximately 2.097.

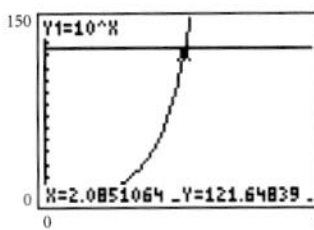

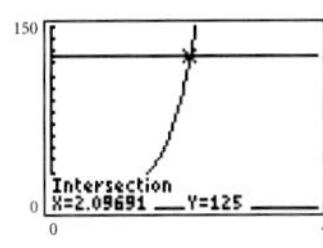

Solve each equation, if possible. Round your answers to the nearest thousandth.

1. $10^x = 42$ 1.623
2. $10^x = 15$ 1.176
3. $10^x = 180$ 2.255
4. $10^x = 100{,}000$ 5
5. $10^x = 7.26$ 0.861
6. $10^x = \frac{1}{451}$ −2.654
7. $10^x = \sqrt{14.5}$ 0.581
8. $10^x = -22.9$ no real solution

Look Beyond

In Exercise 113, students calculate the value of two logarithmic expressions, a sum and a difference, that simplify to $\log_2 16$ and then compare the values. In Lesson 6.4, students will learn properties of logarithms that will allow them to calculate sums and differences even when the logarithms cannot be simplified.

APPLICATION

101. PHYSICS Earth's atmosphere is like an "ocean" of air with the upper layers of air pressing down on the lower layers of air. The weight of the layers of air creates atmospheric air pressure. At sea level (altitude of zero), the average air pressure is about 14.7 pounds per square inch. The air pressure, P, decreases with altitude, a, in feet according to the function $P = 14.7(10)^{-0.000018a}$. Find the altitude that corresponds to the air pressure commonly found in commercial airplanes, 11.82 pounds per square inch. **5261.1 feet**

Look Back

102. Write a linear equation for a line with a slope of 4 and a y-intercept of 3. ***(LESSON 1.2)*** **$y = 4x + 3$**

State the property that is illustrated in each statement. All variables represent real numbers. ***(LESSON 2.1)***

103. $1 \cdot (5xy) = 5xy$

104. $(2 + z) + y = 2 + (z + y)$

105. $2(3x) = 3x(2)$

106. $-x + 0 = -x$

107. $\frac{a}{2} \cdot \frac{2}{a} = 1$, where $a \neq 0$

108. $-3 + x = x + (-3)$

109 Find the inverse of the matrix $\begin{bmatrix} 2 & 3 \\ 1 & -4 \end{bmatrix}$. ***(LESSON 4.3)***

110. Solve the quadratic equation $x^2 - 6x + 9 = 0$. ***(LESSONS 5.2 AND 5.4)***

111. State the two solutions of the equation $x^2 + 1 = 0$. ***(LESSON 5.6)***

112. If an interest rate is 7.3%, what is the multiplier? ***(LESSON 6.1)***

Look Beyond

113. Calculate $\log_2 2 + \log_2 8$ and $\log_2 32 - \log_2 2$. Then compare these values with the value of $\log_2 16$.

103. Identity Prop. of Mult.

104. Assoc. Prop. of Add.

105. Comm. Prop. of Mult.

106. Ident. Prop. of Add.

107. Inverse Prop. of Mult.

108. Comm. Prop. of Add.

109. $A^{-1} \approx \begin{bmatrix} 0.364 & 0.273 \\ 0.091 & -0.182 \end{bmatrix}$

110. $x = 3$

111. $\pm i$

112. 1.073

113. all answers are 4

Properties of Logarithmic Functions

Why *The properties of logarithms allow you to simplify logarithmic expressions, which makes evaluating the expressions easier.*

Objectives

- Simplify and evaluate expressions involving logarithms.
- Solve equations involving logarithms.

John Napier (1550–1617)

Title page and calculations from Napier's Mirifici Logarithmorum Canonis Descriptio

In the seventeenth century, a Scottish mathematician named John Napier developed methods for efficiently performing calculations with large numbers. He found a method for finding the *product* of two numbers by *adding* two corresponding numbers, which he called logarithms.

John Napier's contributions to mathematics are contained in two essays: *Mirifici Logarithmorum Canonis Descriptio* (Description of the Marvelous Canon of Logarithms), published in 1614, and *Mirifici Logarithmorum Canonis Constructio* (Construction of the Marvelous Canon of Logarithms), published in 1619, two years after his death.

Prepare

NCTM PRINCIPLES & STANDARDS 1, 2–9

QUICK WARM-UP

Simplify each expression. Assume that no variable has a value of zero.

1. $z^3 \cdot z^4$ z^7
2. $(t^3)^4$ t^{12}
3. $\frac{c^5}{c^2}$ c^3
4. $\left(\frac{m}{n}\right)^3$ $\frac{m^3}{n^3}$
5. $(ab)^6$ a^6b^6
6. $(r^2s)^4$ r^8s^4

Also on Quiz Transparency 6.4

Teach

Why From the text, students learn that Napier found a way to express a product as a sum. Ask students when they have expressed a product as a sum. when finding the product of exponential expressions with a common base

Ask students to discuss the relationship between logarithms and exponents.

Alternative Teaching Strategy

USING ALGORITHMS Begin by discussing the exponential-logarithmic inverse properties. Remind students that they observed the inverse relationship between exponential and logarithmic functions in the previous lesson. Based on their knowledge of exponents, ask students how $\log(m \cdot n)$ can be rewritten. $\log m + \log n$ As a class, justify the product property of logarithms, as shown at right.

Let $x = \log_b m$ and $y = \log_b n$. Then $m = b^x$, $n = b^y$, and $mn = b^x b^y = b^{x+y}$. By taking the base b logarithm of both sides of the last equation, $\log_b(mn) = \log_b b^{x+y} = x + y = \log_b m + \log_b n$. Therefore, $\log_b(mn) = \log_b m + \log_b n$.

Have students replace b, m and n with numbers and work through the justification. Then have students work in groups to verify the quotient and power properties of logarithms by using related exponential properties.

Activity Notes

In this Activity, students look at several examples of logarithms of a product. By finding values for each pair of expressions in the Activity, students should see that the first and second expressions are equivalent, thereby proving the product and quotient properties.

CHECKPOINT ✔
2. $\log_2(a \cdot b) = \log_2 a + \log_2 b$

CHECKPOINT ✔
4. $\log_2 \frac{a}{b} = \log_2 a - \log_2 b$

Product and Quotient Properties of Logarithms

The Product, Quotient, and Power Properties of Exponents are as follows:

$a^m \cdot a^n = a^{m+n}$ Product Property
$\frac{a^m}{a^n} = a^{m-n}$ Quotient Property
$(a^m)^n = a^{m \cdot n}$ Power Property

Each property of exponents has a corresponding property of logarithms.

Activity
Exploring Properties of Logarithms

You will need: no special tools

Use the following table to complete the activity:

x	2	4	8	16	32	64	128
$y = \log_2 x$	1	2	3	4	5	6	7

1. The expression $\log_2(2 \cdot 4)$ can be written as $\log_2 8$. Use this fact and the table above to evaluate each expression below.
a. $\log_2(2 \cdot 4) =$? and $\log_2 2 + \log_2 4 =$?
b. $\log_2(2 \cdot 8) =$? and $\log_2 2 + \log_2 8 =$?
c. $\log_2(2 \cdot 16) =$? and $\log_2 2 + \log_2 16 =$?
d. $\log_2(2 \cdot 32) =$? and $\log_2 2 + \log_2 32 =$?

CHECKPOINT ✔ **2.** In Step 1, how is the first expression in each pair related to the second expression? Use this pattern to make a conjecture about $\log_2(a \cdot b)$.

3. The expression $\log_2 \frac{16}{2}$ can be written as $\log_2 8$. Use this fact and the table above to evaluate each expression below.
a. $\log_2 \frac{16}{2} =$? and $\log_2 16 - \log_2 2 =$?
b. $\log_2 \frac{64}{32} =$? and $\log_2 64 - \log_2 32 =$?
c. $\log_2 \frac{32}{8} =$? and $\log_2 32 - \log_2 8 =$?
d. $\log_2 \frac{8}{4} =$? and $\log_2 8 - \log_2 4 =$?

CHECKPOINT ✔ **4.** In Step 3, how is the first expression in each pair related to the second expression? Use this pattern to make a conjecture about $\log_2 \frac{a}{b}$.

The patterns explored in the Activity illustrate the *Product and Quotient Properties of Logarithms* given below.

Product and Quotient Properties of Logarithms

For $m > 0$, $n > 0$, $b > 0$, and $b \neq 1$:

Product Property $\log_b(mn) = \log_b m + \log_b n$
Quotient Property $\log_b \frac{m}{n} = \log_b m - \log_b n$

Interdisciplinary Connection

ASTRONOMY The difference between the apparent brightness of a star, m, and its true brightness, M, is given by the formula $m - M = 5 \log_{10} \frac{r}{10}$, where r is the star's distance from Earth in kiloparsecs.

a. Given $m = 10$ and $r = 100$, find M. 5

b. Given $m = 5$ and $M = 10$, find r. 1

Enrichment

Have students graph $y = \log x$ on their calculators, using the following window:

Xmin=0 **Ymin=–1**
Xmax=10,000 **Ymax=5**
Xscl=1000 **Yscl=1**

Then have them graph $y = \log(10 \cdot x)$ and $y = \log 10 + \log x$. Describe what happens to the graphs in each case. Next ask students to compare $y = \log \frac{10}{x}$ with $y = \log 10 - \log x$. Students should relate their responses to the properties of logarithms studied in this lesson.

You can use the Product and Quotient Properties of Logarithms to evaluate logarithmic expressions. This is shown in Example 1.

EXAMPLE 1 **Given $\log_2 3 \approx 1.5850$, approximate the value of each expression below by using the Product and Quotient Properties of Logarithms.**

a. $\log_2 12$ **b.** $\log_2 1.5$

SOLUTION

a. $\log_2 12 = \log_2 (2 \cdot 2 \cdot 3)$
$= \log_2 2 + \log_2 2 + \log_2 3$
$\approx 1 + 1 + 1.5850$
≈ 3.5850

b. $\log_2 1.5 = \log_2 \frac{3}{2}$
$= \log_2 3 - \log_2 2$
$\approx 1.5850 - 1$
≈ 0.5850

TRY THIS Given that $\log_2 3 = 1.5850$, approximate each expression below by using the Product and Quotient Properties of Logarithms.

a. $\log_2 18$ **b.** $\log_2 \frac{3}{4}$

Example 2 demonstrates how to use the properties of logarithms to rewrite a logarithmic expression as a single logarithm.

EXAMPLE 2 **Write each expression as a single logarithm. Then simplify, if possible.**

a. $\log_3 10 - \log_3 5$ **b.** $\log_b u + \log_b v - \log_b uw$

SOLUTION

a. $\log_3 10 - \log_3 5 = \log_3 \frac{10}{5}$
$= \log_3 2$

b. $\log_b u + \log_b v - \log_b uw = \log_b uv - \log_b uw$
$= \log_b \frac{uv}{uw}$
$= \log_b \frac{v}{w}$

TRY THIS Write each expression as a single logarithm. Then simplify if possible.

a. $\log_4 18 - \log_4 6$ **b.** $\log_b 4x - \log_b 3y + \log_b y$

The Power Property of Logarithms

Examine the process of rewriting the expression $\log_b(a^4)$.

$$\log_b(a^4) = \log_b(a \cdot a \cdot a \cdot a)$$
$$= \log_b a + \log_b a + \log_b a + \log_b a$$
$$= 4 \cdot \log_b a$$

This illustrates the *Power Property of Logarithms* given below.

Power Property of Logarithms

For $m > 0$, $b > 0$, $b \neq 1$, and any real number p:
$$\log_b m^p = p \log_b m$$

ADDITIONAL EXAMPLE 1

Given $\log_2 5 \approx 2.3219$, approximate the value for each expression below by using the product and quotient properties of logarithms.

a. $\log_2 10$ ≈ 3.3219

b. $\log_2 \frac{5}{2}$ ≈ 1.3219

TRY THIS

a. 4.17

b. −0.4150

ADDITIONAL EXAMPLE 2

Write each expression as a single logarithm. Then simplify, if possible.

a. $\log_8 12 - \log_8 4$ $\log_8 3$

b. $\log_z 2a - \log_z b + \log_z bc$

$\log_z 2ac$

TRY THIS

a. $\log_4 3$

b. $\log_b \frac{4x}{3}$

Inclusion Strategies

VISUAL LEARNERS For some students, the algebraic derivations of the properties of logarithms are abstract. In order for students to see numerically that the properties of logarithms hold true, they can use the list feature of a graphics calculator. Have students enter values from 1 to 10 into **L1** and values from 5 to 14 into **L2.** Next, have students define **L3** as log (**L1** • **L2**) and **L4** as log **L1** + log **L2.** When students look at **L3** and **L4,** they can see that the values are exactly the same. Have students redefine **L3** and **L4** to illustrate the other properties of logarithms so that they can see the numerical equivalency for each property.

ADDITIONAL
EXAMPLE 3

Evaluate $\log_3 27^8$.
24

TRY THIS
300

ADDITIONAL
EXAMPLE 4

Evaluate each expression.
a. $\log_7 7^4 + 5^{\log_5 3}$ 7

b. $9^{\log_9 2} - \log_4 \frac{1}{4}$ 3

TRY THIS
a. 7
b. 13

CRITICAL THINKING
Letting $y = b^x$, $x = \log_b y$ by equivalent exponential-logarithm forms, and $x = \log_b b^x$ by substitution.

Letting $y = \log_b x$, $b^y = x$ by equivalent exponential-logarithm forms, and $b^{\log_b x} = x$ by substitution.

In Example 3, the Power Property of Logarithms is used to simplify powers.

EXAMPLE 3 Evaluate $\log_5 25^4$.

SOLUTION

$$\log_5 25^4 = 4 \log_5 25 \quad \textit{Use the Power Property of Logarithms.}$$
$$= 4 \cdot 2$$
$$= 8$$

TRY THIS Evaluate $\log_3 27^{100}$.

Exponential-Logarithmic Inverse Properties

Recall from Lesson 2.5 that functions f and g are inverse functions if and only if $(f \circ g)(x) = x$ and $(g \circ f)(x) = x$. The functions $f(x) = \log_b x$ and $g(x) = b^x$ are inverses, so $(f \circ g)(x) = \log_b b^x = x$ and $(g \circ f)(x) = b^{\log_b x} = x$.

Exponential-Logarithmic Inverse Properties

For $b > 0$ and $b \neq 1$:
$\log_b b^x = x$ and $b^{\log_b x} = x$ for $x > 0$

EXAMPLE 4 Evaluate each expression.
a. $3^{\log_3 4} + \log_5 25$
b. $\log_2 32 - 5^{\log_5 3}$

SOLUTION

a. $3^{\log_3 4} + \log_5 25$
$= 4 + \log_5 5^2$
$= 4 + 2$
$= 6$

b. $\log_2 32 - 5^{\log_5 3}$
$= \log_2 2^5 - 3$
$= 5 - 3$
$= 2$

TRY THIS Evaluate each expression.
a. $7^{\log_7 11} - \log_3 81$
b. $\log_8 8^5 + 3^{\log_3 8}$

CRITICAL THINKING Verify the Exponential-Logarithmic Inverse Properties by using only the equivalent exponential and logarithmic forms given on page 371.

Because exponential functions and logarithmic functions are one-to-one functions, for each element in the domain of $y = \log x$, there is exactly one corresponding element in the range of $y = \log x$.

One-to-One Property of Logarithms

If $\log_b x = \log_b y$, then $x = y$.

Reteaching the Lesson

USING PATTERNS Have students evaluate $\log_2(64 \cdot 256)$ by rewriting the expression as $\log_2(2^6 \cdot 2^8)$. Let $x = \log_2(2^6 \cdot 2^8)$. Then rewrite the equation in exponential form: $2^x = 2^6 \cdot 2^8$. Use the product property of exponents to simplify the equation: $2^x = 2^{6+8}$, or $2^x = 2^{14}$. Thus, $x = 14$, or $\log_2(64 \cdot 256) = 14$. Have students explore the quotient and power properties of logarithms by using the numbers 64 and 256 to form expressions such as $\log_2 \frac{64}{256}$, $\log_2 \frac{256}{64}$, $\log_2 64^3$, and $\log_2 256^4$. Then compare the differences and similarities between the properties.

EXAMPLE 5 **Solve $\log_3(x^2 + 7x - 5) = \log_3(6x + 1)$ for x. Check your answers.**

SOLUTION

$\log_3(x^2 + 7x - 5) = \log_3(6x + 1)$

$x^2 + 7x - 5 = 6x + 1$ *Use the One-to-One Property of Logarithms.*

$x^2 + x - 6 = 0$

$(x - 2)(x + 3) = 0$

$x = 2$ *or* $x = -3$ *Use the Zero Product Property.*

CHECK

Let $x = 2$.	Let $x = -3$.
$\log_3(x^2 + 7x - 5) \stackrel{?}{=} \log_3(6x + 1)$	$\log_3(x^2 + 7x - 5) \stackrel{?}{=} \log_3(6x + 1)$
$\log_3 13 = \log_3 13$	$\log_3(-17) = \log_3(-17)$
True	**Undefined**

Since the domain of a logarithmic function excludes negative numbers, the solution cannot be −3. Therefore, the solution is 2.

ADDITIONAL **EXAMPLE 5**

Solve $\log_2(2x^2 + 8x - 11) = \log_2(2x + 9)$ for x. Check your answers. $x = 2$

Exercises

Communicate

1. Given that $\log_{10} 5 \approx 0.6990$, explain how to approximate the values of $\log_{10} 0.005$ and $\log_{10} 500$.
2. Explain how to write an expression such as $\log_7 32 - \log_7 4$ as a single logarithm.
3. Explain how to evaluate $4^{\log_4 8}$ and $\log_2 2^7$. Include the names of the properties you would use.
4. Explain why you must check your answers when solving an equation such as $\log_2 3x = \log_2(x + 4)$ for x.

Guided Skills Practice

Given $\log_3 7 \approx 1.7712$, approximate the value for each logarithm by using the Product and Quotient Properties of Logarithms. *(EXAMPLE 1)*

5. $\log_3 49$ 3.5424
6. $\log_3 \frac{3}{7}$ −0.7712

Write each expression as a single logarithm. Then simplify, if possible. *(EXAMPLE 2)*

7. $\log_3 x - \log_3 y + \log_3 z$ $\log_3 \frac{xz}{y}$
8. $\log_2 3 + \log_2 6 - \log_2 10$ $\log_2 \frac{9}{5}$

Evaluate each expression. *(EXAMPLES 3 AND 4)*

9. $\log_4 16^8$ 16
10. $3^{\log_3 12}$ 12
11. $\log_7 7^3$ 3

$x = 4$ 12. Solve $\log_3 x = \log_3(2x - 4)$ for x, and check your answers. *(EXAMPLE 5)*

Assess

Selected Answers

Exercises 5–12, 13–81 odd

ASSIGNMENT GUIDE	
In Class	1–12
Core	13–67 odd
Core Plus	14–72 even, 75
Review	76–80
Preview	86

Extra Practice can be found beginning on page 940.

Mid-Chapter Assessment for Lessons 6.1 through 6.4 can be found on page 74 of the *Assessment Resources.*

Error Analysis

Students often use the product and power properties of exponents incorrectly because they think that $\log_b(m+n) = \log_b m \cdot \log_b n$ is true. Similarly, students may think that $\log_b(m-n) = \log_b \frac{m}{n}$ is true. Neither $\log_b(m+n)$ nor $\log_b(m-n)$ can be expanded as $(\log_b m)(\log_b n)$, and $\frac{\log_b m}{\log_b n}$ cannot be condensed.

55. $\log_2 7x = \log_2(x^2+12)$

$7x = x^2 + 12$

One-to-one prop. of logs

$x^2 - 7x + 12 = 0$

Inverse prop. of add.

$(x-3)(x-4) = 0$

Factor.

$x = 3$ *or* $x = 4$

56. $\log_5(3x^2 - 1) = \log_5 2x$

$3x^2 - 1 = 2x$

One-to-one prop. of logs

$3x^2 - 2x - 1 = 0$

Inverse prop. of add.

$(3x+1)(x-1) = 0$

Factor.

$x = -\frac{1}{3}$ *or* $x = 1$

$\left(-\frac{1}{3}\text{ is extraneous}\right)$

Practice

NAME ______ CLASS ______ DATE ______

Practice 6.4 Properties of Logarithmic Functions

Write each expression as a sum or a difference of logarithms. Then simplify, if possible.

1. $\log_{10}(4 \cdot 100)$ — $\log_{10} 4 + \log_{10} 100 = \log_{10} 4 + 2$
2. $\log_5 \frac{72}{25}$ — $\log_5 72 - \log_5 25 = \log_5 72 - 2$
3. $\log_7(5 \cdot 3 \cdot 4)$ — $\log_7 5 + \log_7 3 + \log_7 4$
4. $\log_3 15q$ — $\log_3 15 + \log_3 q$
5. $\log_8 \frac{64}{y}$ — $\log_8 64 - \log_8 4 = 2 - \log_8 4$
6. $\log_9 \frac{3a}{7}$ — $\log_9 3 + \log_9 a - \log_9 7 = 0.5 + \log_9 a - \log_9 7$

Write each expression as a single logarithm. Then simplify, if possible.

7. $\log_3 5 + \log_3 6$ — $\log_3 30$
8. $\log_5 x - \log_5 2$ — $\log_5 \frac{x}{2}$
9. $\log_8 2 + \log_8 32$ — $\log_8 64 = 2$
10. $\log_9 5 + \log_9 y - \log_9 4$ — $\log_9 \frac{5y}{4}$
11. $2\log_{12} 6 + \log_{12} 4$ — $\log_{12} 144 = 2$
12. $\frac{1}{2}\log_3 81 + \log_3 15$ — $\log_3 135$
13. $\log_b m + \log_b 2 - \log_b x$ — $\log_b \frac{2m}{x}$
14. $3\log_b x - (\log_b 4 + \log_b x)$ — $\log_b \frac{x^3}{4x} = \log_b \frac{x^2}{4}$
15. $3\log_b z + \log_b y - 4\log_b z$ — $\log_b \frac{z^3 y}{4z} = \log_b \frac{z^2 y}{4}$

Evaluate each expression.

16. $5^{\log_5 12}$ — 12
17. $12^{\log_{12} 73}$ — 73
18. $\log_3 3^{2.5}$ — 2.5
19. $\log_2 2^{4.7}$ — 4.7
20. $\log_4 4^3 - \log_3 81$ — −1
21. $9^{\log_9 15} - \log_3 3^5$ — 10

Solve for x and check your answers.

22. $\log_2(10x) = \log_2(3x+14)$ — 2
23. $2\log_3 x = \log_3 4$ — 2
24. $\log_5(4x-3) = \log_5(x+1)$ — $\frac{4}{3}$
25. $\log_7(x^2-1) = \log_7 8$ — −3, 3
26. $\log_8(x^2-3x) = \log_8(2x+6)$ — −1, 6
27. $2\log_2(x+2) = \log_2(3x+16)$ — 3
28. $\log_b 8 = \log_b x + \log_b(x-2)$ — 4
29. $2\log_b(x+1) = \log_b(-x+11)$ — 2

13. $\log_8 5 + 1$
14. $3 + \log_2 x + \log_2 y$
15. $\log_3 x - 2$
16. $\log_4 x - \frac{5}{2}$

internet connect
Homework Help Online
Go To: **go.hrw.com**
Keyword:
MB1 Homework Help for Exercises 29–42

37. $\log_2 \frac{m^5}{n^2}$
38. $\log_3 \frac{y^7}{x^4}$
39. $\log_b \frac{m^4 n^{\frac{1}{2}}}{8p^3}$
40. $\log_b \frac{(3c)^{\frac{1}{2}}(4d)^{\frac{1}{2}}}{(5e)^2}$
41. $\log_7 \frac{7}{x^2}$
42. $\log_3 9x^4$

55. $x = 3$ *or* $x = 4$
56. $x = 1$
57. $x = 8$
58. $x = 1$ *or* $x = 4$
59. $x = 3$
60. $x = \frac{3}{2}$
61. $x = 3$
62. $t = 1$

69. always
70. sometimes
71. sometimes

Practice and Apply

Write each expression as a sum or difference of logarithms. Then simplify, if possible.

13. $\log_8(5 \cdot 8)$ **14.** $\log_2 8xy$ **15.** $\log_3 \frac{x}{9}$ **16.** $\log_4 \frac{x}{32}$

Use the values given below to approximate the value of each logarithmic expression in Exercises 17–28.

$\log_2 7 \approx 2.8074$	$\log_2 5 \approx 2.3219$	$\log_4 5 \approx 1.1610$
$\log_4 3 \approx 0.7925$	$\log_2 3 \approx 1.5850$	$\log_{10} 8.3 \approx 0.9191$

17. $\log_4 15$ **1.9535** **18.** $\log_2 35$ **−0.5145** **19.** $\log_2 28$ **4.8074**

20. $\log_4 12$ **1.7925** **21.** $\log_4 60$ **2.9535** **22.** $\log_2 105$ **6.7143**

23. $\log_{10} 830$ **2.9191** **24.** $\log_{10} 0.0083$ **−2.0809** **25.** $\log_4 \frac{3}{5}$ **−0.3685**

26. $\log_2 \frac{7}{10}$ **−0.5145** **27.** $\log_4 \frac{5}{4}$ **0.1610** **28.** $\log_2 \frac{2}{7}$ **−1.8074**

Write each expression as a single logarithm. Then simplify, if possible.

29. $\log_2 5 + \log_2 7$ **$\log_2 35$** **30.** $\log_4 8 + \log_4 2$ **2**

31. $\log_3 45 - \log_3 9$ **$\log_3 5$** **32.** $\log_2 14 - \log_2 7$ **1**

33. $\log_2 5 + \log_2 x - \log_2 10$ **$\log_2 \frac{x}{2}$** **34.** $\log_3 x + \log_3 4 - \log_3 2$ **$\log_3 2x$**

35. $\log_7 3x - \log_7 9x + \log_7 6y$ **$\log_7 2y$** **36.** $\log_5 6s - \log_5 s + \log_5 4t$ **$\log_5 24t$**

37. $5\log_2 m - 2\log_2 n$ **38.** $7\log_3 y - 4\log_3 x$

39. $4\log_b m + \frac{1}{2}\log_b n - 3\log_b 2p$ **40.** $\frac{1}{2}\log_b 3c + \frac{1}{2}\log_b 4d - 2\log_b 5e$

41. $1 - 2\log_7 x$ **42.** $2 + 4\log_3 x$

Evaluate each expression.

43. $3^{\log_3 8}$ **8** **44.** $9^{\log_9 2}$ **2** **45.** $\log_4 4^5$ **5**

46. $\log_{10} 10^2$ **2** **47.** $7^{\log_7 9} + \log_2 8$ **12** **48.** $5^{\log_5 7} + \log_3 9$ **9**

49. $\log_9 9^{11} - \log_4 64$ **8** **50.** $\log_3 3^5 + \log_5 125$ **8** **51.** $6^{\log_6 3} - \log_5 \frac{1}{25}$ **5**

52. $2^{\log_2 3} + \log_6 \frac{1}{36}$ **1** **53.** $\log_3 \frac{1}{9} - 2^{\log_2 3}$ **−5** **54.** $\log_2 \frac{1}{8} - 4^{\log_4 7}$ **−10**

Solve for x, and check your answers. Justify each step in the solution process.

55. $\log_2 7x = \log_2(x^2+12)$ **56.** $\log_5(3x^2-1) = \log_5 2x$

57. $\log_b(x^2-15) = \log_b(6x+1)$ **58.** $\log_{10}(5x-3) - \log_{10}(x^2+1) = 0$

59. $2\log_a x + \log_a 2 = \log_a(5x+3)$ **60.** $\log_b(x^2-2) + 2\log_b 6 = \log_b 6x$

61. $2\log_3 x + \log_3 5 = \log_3(14x+3)$ **62.** $\log_5 2 + 2\log_5 t = \log_5(3-t)$

State whether each equation is always true, sometimes true, or never true. Assume that x is a positive real number.

63. $\log_3 9 = 2\log_3 3$ **always** **64.** $\log_2 8 - \log_2 2 = 2$ **always** **65.** $\log x^2 = 2\log x$ **always**

66. $\log x - \log 5 = \log \frac{x}{5}$ **always** **67.** $\frac{\log 3}{\log x} = \log 3 - \log x$ **never** **68.** $\log(x-2) = \frac{\log x}{\log 2}$ **never**

69. $\frac{1}{2}\log x = \log\sqrt{x}$ **70.** $\log 12x = 12\log x$ **71.** $\log_3 x + \log_3 x = \log_3 2x$

57. $\log_b(x^2-15) = \log_b(6x+1)$

$x^2 - 15 = 6x + 1$

One-to-one prop. of logs

$x^2 - 6x - 16 = 0$

Inverse prop. of add.

$(x+2)(x-8) = 0$

Factor.

$x = -2$ *or* $x = 8$

(-2 is extraneous)

58. $\log_5(5x-3) - \log_5(x^2+1) = 0$

$\log_5(5x-3) = \log_5(x^2+1)$

Inverse prop. of add.

$5x - 3 = x^2 + 1$

One-to-one prop. of logs

$x^2 - 5x + 4 = 0$

Inverse prop. of add.

$(x-1)(x-4) = 0$

Factor.

$x = 1$ *or* $x = 4$

CHALLENGE **Solve each equation.**

72. $\log_4(\log_3 x) = 0$ $x = 3$

73. $\log_6[\log_5(\log_3 x)] = 0$ $x = 243$

APPLICATIONS

74. HEALTH The surface area of a person is commonly used to calculate dosages of medicines. The surface area of a child is often calculated with the following formula, where S is the surface area in square centimeters, W is the child's weight in kilograms, and H is the child's height in centimeters.

$$\log_{10} S = 0.425 \log_{10} W + 0.725 \log_{10} H + \log_{10} 71.84$$

Use the properties of logarithms to write a formula for S without logarithms.

75. PHYSICS Atmospheric air pressure, P, in pounds per square inch and altitude, a, in feet are related by the logarithmic equation $a = -55{,}555.56 \log_{10} \frac{P}{14.7}$. Use properties of logarithms to find how much greater the air pressure at the top of Mount Whitney in the United States is compared with the air pressure at the top of Mount Everest on the border of Tibet and Nepal. The altitude of Mount Everest is 29,028 feet, and the altitude of Mount Whitney is 14,495 feet. (Hint: Find the ratio of the air pressures.)

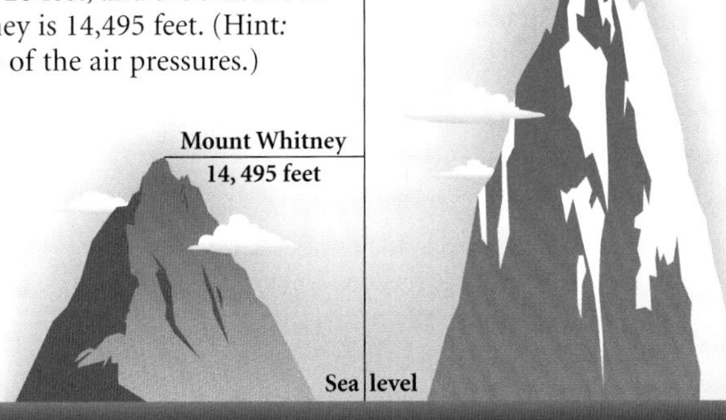

Look Back

Minimize each objective function under the given constraints. *(LESSON 3.5)*

76. Objective function: $C = 2x + 5y$

Constraints: $\begin{cases} x + 5y \geq 8 \\ y - 3x \leq 14 \\ x \geq 0 \\ y \geq 0 \end{cases}$

77. Objective function: $C = x + 4y$

Constraints: $\begin{cases} x - y \geq 12 \\ 7y - x \leq 12 \\ x \geq 0 \\ y \geq 0 \end{cases}$

Write the matrix equation that represents each system. *(LESSON 4.4)*

78. $\begin{cases} 2x - y = 5 \\ 3x + 4y = -3 \end{cases}$

79. $\begin{cases} 3x + 2y - z = 7 \\ 5x + 3y - 2z = -12 \\ 3y - z + 2x = -5 \end{cases}$

80. $\begin{cases} x + y + z = 18 \\ \frac{1}{4}x + \frac{1}{2}y + \frac{1}{3}z = 6 \\ 2y + 3z = 33 \end{cases}$

Answers

74. $S = 71.84W^{0.425}H^{0.725}$

75. 1.83 times greater

76. 8

77. 12

78. $\begin{bmatrix} 2 & -1 \\ 3 & 4 \end{bmatrix}\begin{bmatrix} x \\ y \end{bmatrix} = \begin{bmatrix} 5 \\ -3 \end{bmatrix}$

79. $\begin{bmatrix} 3 & 2 & -1 \\ 5 & 3 & -2 \\ 2 & 3 & -1 \end{bmatrix}\begin{bmatrix} x \\ y \\ z \end{bmatrix} = \begin{bmatrix} 7 \\ -12 \\ -5 \end{bmatrix}$

80. $\begin{bmatrix} 1 & 1 & 1 \\ \frac{1}{4} & \frac{1}{2} & \frac{1}{3} \\ 0 & 2 & 3 \end{bmatrix}\begin{bmatrix} x \\ y \\ z \end{bmatrix} = \begin{bmatrix} 18 \\ 6 \\ 33 \end{bmatrix}$

59. $2 \log_a x + \log_a 2 = \log_a(5x + 3)$

$\log_a x^2 + \log_a 2 = \log_a(5x + 3)$
Power prop. of logs

$\log_a 2x^2 = \log_a(5x + 3)$
Prod. prop. of logs

$2x^2 = 5x + 3$
One-to-one prop. of logs

$2x^2 - 5x - 3 = 0$
Inverse prop. of add.

$(2x + 1)(x - 3) = 0$
Factor.

$x = -\frac{1}{2}$ *or* $x = 3$

$\left(-\frac{1}{2}\text{ is extraneous}\right)$

60. $\log_b(x^2 - 2) + 2 \log_b 6 = \log_b 6x$

$\log_b(x^2 - 2) + \log_b 6^2 = \log_b 6x$
Power prop. of logs

$\log_b 36(x^2 - 2) = \log_b 6x$
Prod. prop. of logs

$36(x^2 - 2) = 6x$
One-to-one prop. of logs

$36x^2 - 6x - 72 = 0$
Inverse prop. of add.

$(6x + 8)(6x - 9) = 0$
Factor.

$x = -\frac{4}{3}$ *or* $x = \frac{3}{2}$

$\left(-\frac{4}{3}\text{ is extraneous}\right)$

61. $2 \log_3 x + \log_3 5 = \log_3(14x + 3)$

$\log_3 x^2 + \log_3 5 = \log_3(14x + 3)$
Power prop. of logs

$\log_3 5x^2 = \log_3(14x + 3)$
Prod. prop. of logs

$5x^2 = 14x + 3$
One-to-one prop. of logs

$5x^2 - 14x - 3 = 0$
Inverse prop. of add.

$(5x + 1)(x - 3) = 0.$
Factor.

$x = -\frac{1}{5}$ *or* $x = 3$

$(-5$ is extraneous$)$

62. $\log_5 2 + 2 \log_5 t = \log_5(3 - t)$

$\log_5 2 + \log_5 t^2 = \log_5(3 - t)$
Power prop. of logs

$\log_5 2t^2 = \log_5(3 - t)$
Prod. prop. of logs

$2t^2 = 3 - t$
One-to-one prop. of logs

$2t^2 + t - 3 = 0$
Inverse prop. of add.

$(2t + 3)(t - 1) = 0$
Factor.

$t = -\frac{3}{2}$ *or* $t = 1$

$\left(-\frac{3}{2}\text{ is extraneous}\right)$

Student Technology Guide

NAME ______________ CLASS __________ DATE __________

Student Technology Guide
6.4 *Properties of Logarithmic Functions*

You can use a calculator to evaluate logarithms. For example, to find the base 10 logarithm of a number such as log 2, press LOG 2) ENTER.

Complete this table. Round entries to the nearest thousandth.

x	2	3	6	12	18	36
$\log x$	0.301	1. 0.477	2. 0.778	3. 1.079	4. 1.255	5. 1.556

Use the numbers from the table above to fill in the blanks in the equations below. Look for patterns.

6. log 2 + log 3 = log 6
7. log 3 + log 6 = log 18
8. log 2 + log 6 = log 12
9. log 3 + log 12 = log 36
10. log 6 − log 3 = log 2
11. log 12 − log 2 = log 6
12. log 18 − log 6 = log 3
13. log 36 − log 2 = log 18
14. log 36 − log 3 = log 12
15. Use the pattern from Exercises 6–14 to write a general rule for $\log m + \log n$. $\log mn$
16. Use the pattern from Exercises 6–14 to write a general rule for $\log m - \log n$. $\log \frac{m}{n}$

Write each expression as a single logarithm.

17. log 4 + log 10 — log 40
18. log 5 + log 6 — log 30
19. log 32 − log 8 — log 4
20. log 208 − log 4 — log 52

Write each expression as a sum or a difference of logarithms.

21. $\log(3 \cdot 14)$ — $\log 3 + \log 14$
22. $\log\left(\frac{5}{3}\right)$ — $\log 5 - \log 3$
23. $\log_7(4 \cdot 8 \cdot 9)$ — $\log_7 4 + \log_7 8 + \log_7 9$
24. $\log_{15}\left(\frac{34}{19}\right)$ — $\log_{15} 34 - \log_{15} 19$

To see that logarithmic and exponential functions are inverses:

- Enter $y = 10^x$ as Y1 by pressing 2nd LOG X,T,θ,n). To turn this graph off, move the cursor to the equal sign beside Y1 and press ENTER.
- Move to Y2. Press LOG VARS Y-VARS 1:Function... ENTER 1:Y1 ENTER). This sets Y2 equal to the log of Y1.
- Press GRAPH. The result is the line $y = x$. This shows that $\log 10^x = x$.

25. Use a graphics calculator to show that $10^{\log x} = x$. This will complete the verification that $y = 10^x$ and $y = \log x$ are inverses of one another.

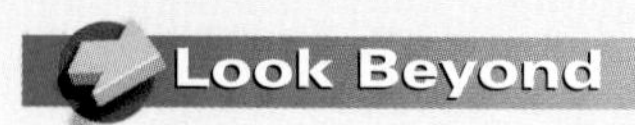

In Exercise 86, students use natural logarithms to solve an exponential equation. In Lesson 6.6, students will explore the value of e and will learn that the natural logarithm, ln (pronounced as the two letters "ell en"), is a logarithm with base e.

ALTERNATIVE

Assessment

Portfolio Activity

The Portfolio Activity can be used as preparation for the Chapter Project or as a separate activity. In the Portfolio Activity on this page, students use a CBL to collect temperature data and a graphics calculator to look at various mathematical models that might fit the data. Students then examine shifts in the function they found.

Answers to Portfolio Activities can be found in Additional Answers of the Teacher's Edition.

81. $\begin{bmatrix} -3 & 2 & \vdots & 11 \\ 4 & 1 & \vdots & 5 \end{bmatrix}$

82. $\begin{bmatrix} 3 & -6 & 3 & \vdots & 4 \\ 1 & -2 & 1 & \vdots & 1 \\ 2 & -4 & 2 & \vdots & 5 \end{bmatrix}$

83. $\begin{bmatrix} 0.5 & 0.3 & 0 & \vdots & 2.2 \\ 0 & -8.5 & 1.2 & \vdots & -24.4 \\ 1.3 & 0 & 3.3 & \vdots & 29 \end{bmatrix}$

Write the augmented matrix for each system of equations. ***(LESSON 4.5)***

81. $\begin{cases} -3x + 2y = 11 \\ 4x = 5 - y \end{cases}$

82. $\begin{cases} 3x - 6y + 3z = 4 \\ x - 2y = 1 - z \\ 2x - 4y + 2z = 5 \end{cases}$

83. $\begin{cases} 0.5x + 0.3y = 2.2 \\ -8.5y + 1.2z = -24.4 \\ 3.3z + 1.3x = 29 \end{cases}$

APPLICATIONS

84. INVESTMENTS An investment of $100 earns an annual interest rate of 5%. Find the amount after 10 years if the interest is compounded annually, quarterly, and daily. ***(LESSON 6.2)*** **$162.89; $164.36; $164.87**

85. INVESTMENTS Find the final amount after 8 years of a $500 investment that is compounded semiannually at 6%, 7%, and 8% annual interest. ***(LESSON 6.2)*** **$802.35; $866.99; $936.49**

Look Beyond

86 e is an irrational number between 2 and 3. The expression $\log_e x$ is commonly written as $\ln x$. Use the LN key to solve $2e^{3x} = 5$ for x to the nearest hundredth. **0.31**

In this activity, you will use the CBL to collect data as a warm probe cools in air.

1. First record the air temperature for reference. Then place the temperature probe in hot water until it reaches a reading of at least 60°C. Remove the probe from the water and record the readings as in Step 1 of the Portfolio Activity on page 361.
2. a. Using the regression feature on your calculator, find linear, quadratic, and exponential functions that model this data. (Use the variable t for time in seconds.)
   b. Use each function to predict the temperature of the probe after 2 minutes (120 seconds). Compare the predictions with the actual 2-minute reading.
   c. Discuss the usefulness of each function for modeling the cooling process.
3. Create a new function to model the cooling process by performing the steps below.
   a. Subtract the air temperature from each temperature recorded in your data list. Store the resulting data values in a new list.
   b. Use the exponential regression feature on your calculator to find an exponential function of the form $y = a \cdot b^t$ that models this new data set.
   c. Add the air temperature to the function you found in part **b**. Graph the resulting function, $y = a \cdot b^t + c$, which will be called the approximating function.
   d. Repeat parts **b** and **c** from Step 2 with the approximating function.

Save your data and results to use in the last Portfolio Activity.

WORKING ON THE CHAPTER PROJECT

You should now be able to complete Activity 3 of the Chapter Project.

Applications of Common Logarithms

Objectives

- Define and use the common logarithmic function to solve exponential and logarithmic equations.
- Evaluate logarithmic expressions by using the change-of-base formula.

Why *Common logarithmic functions are used to define many real-world measurement scales, such as the decibel scale for the relative intensities of sounds.*

The human ear is sensitive to a wide range of sound intensities.

Type of sound	Relative intensity, *R* (in dB)
threshold of hearing	0
whisper	≈20
soft music	≈30
conversation	≈65
rock band	≈100
threshold of pain	120

The base-10 logarithm is called the *common logarithm*. The **common logarithm**, $\log_{10} x$, is usually written as $\log x$.

A table of values and a graph for $y = \log x$ are given below. Notice that the values in the domain increase quickly (by a factor of 10), while values in the range increase slowly (by adding 1). In general, logarithmic functions are used to assign large values in the domain to small values in the range.

x	10	100	1000	10,000	100,000	...
$y = \log x$	1	2	3	4	5	...

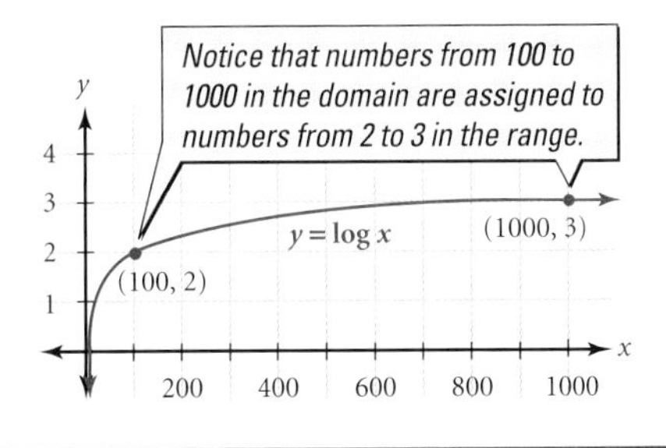

Prepare

NCTM PRINCIPLES & STANDARDS 1–4, 6–10

QUICK WARM-UP

Solve each equation for *x*. Round your answers to the nearest hundredth.

1. $10^x = 32$ $\quad x \approx 1.51$
2. $10^x = 1.76$ $\quad x \approx 0.25$
3. $10^x = \frac{1}{200}$ $\quad x \approx -2.30$

Find the value of *v* in each equation below.

4. $-3 = \log_{10} v$ $\quad v = 0.001$
5. $2 = \log_v 49$ $\quad v = 7$
6. $v = \log_{16} 4$ $\quad v = \frac{1}{2}$

Also on Quiz Transparency 6.5

Teach

Why Have students use a calculator-based laboratory system with a microphone probe to experiment with various sound intensities. Have students model whispering, normal conversations, and loud music.

Alternative Teaching Strategy

USING TECHNOLOGY Review with students the properties of logarithms studied in Lessons 6.3 and 6.4. Stress the concepts of writing equivalent exponential and logarithmic forms of an expression and the power property. Next, point out the LOG key on a calculator, and explain that it calculates the common logarithm, $\log_{10}$. Have students experiment with this key to see that $\log 100 = 2$, $\log 1000 = 3$, and so forth. Then give students a problem such as $3^x = 10$, and ask them to estimate a value for x. $x \approx 2.096$ Point out that the equation is difficult to solve because the unknown is the exponent. Students should know that the solution does not change if the same operation is applied to both sides of an equation. Suggest that they take the common logarithm, $\log_{10}$, of both sides of the equation and then use the power property and their calculator to solve for x. Some students might suggest rewriting the exponential equation in logarithmic form as $\log_3 10 = x$, but at this point, students do not have a way to evaluate $\log_3 10$. This shows the need for the change-of-base formula.

Math CONNECTION

TRANSFORMATIONS When the graph of $y = f(x)$ is multiplied by a constant, such as in $y = a \cdot f(x)$, the graph is stretched or compressed.

ADDITIONAL EXAMPLE 1

The intensity of a certain dog's bark is about 600 times as loud as the threshold of hearing, I_0. **Find the relative intensity, R, of this dog's bark in decibels.**

≈28 dB

CONNECTION
TRANSFORMATIONS

Recall from Lesson 2.7 that the graph of $y = a \cdot f(x)$ is the graph of $y = f(x)$ stretched by a factor of a. Therefore, the graph of $y = 10 \log x$ is the graph of $y = \log x$ stretched by a factor of 10.

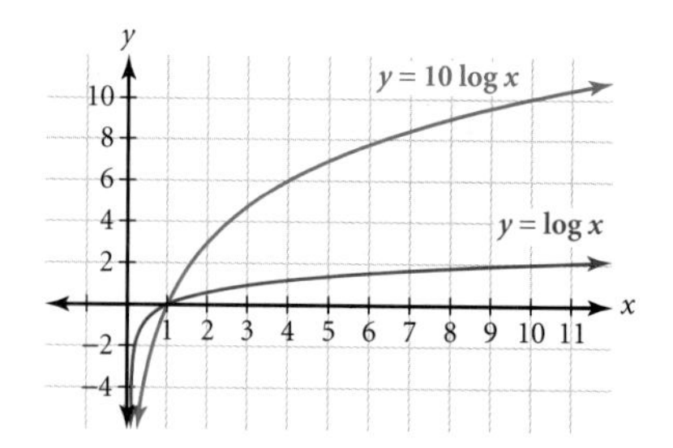

APPLICATION
PHYSICS

The function $y = 10 \log x$ models the relative intensities of sound. The intensity of the faintest sound audible to the human ear, called the *threshold of hearing*, is 10^{-12} watts per meters squared (W/m²). Sound intensities, I, which are between 0 and 1, are often compared with the threshold of hearing, I_0, yielding a ratio of sound intensities, $\frac{I}{I_0}$, between 1 and 10^{12}. On the decibel scale, the relative intensity, R, of a sound in decibels (dB) is given by the function $R = 10 \log \frac{I}{I_0}$.

The bel, the unit of measure for the intensity of sound, was named after Alexander Graham Bell, 1847–1922.

Domain of R — Range of R

I (in W/m²)	$\frac{I}{I_0}$	$R = 10 \log \frac{I}{I_0}$ (in dB)
10^{-12}	1	0
10^{-11}	10	10
10^{-10}	100	20
⋮	⋮	⋮
10^{-2}	10^{10}	100
10^{-1}	10^{11}	110
$10^0 = 1$	10^{12}	120

EXAMPLE 1

The intensity of a whisper is about 300 times as loud as the threshold of hearing, I_0.

Find the relative intensity, R, of this whisper in decibels.

SOLUTION

PROBLEM SOLVING

Identify the wanted, given, and unknown information. In this problem, the ratio of I to I_0 is given and you need to find R.

TECHNOLOGY
SCIENTIFIC CALCULATOR

$R = 10 \log \frac{I}{I_0}$

$R = 10 \log \frac{300I_0}{I_0}$ — *Substitute $300I_0$ for the intensity, I.*

$R = 10 \log 300$

$R \approx 25$ — *Use the LOG key on a calculator.*

Interdisciplinary Connection

SCIENCE Have students visit the library to find examples of the use of logarithmic equations in many different areas. For example, they might look in books about astronomy, optics, archeology, geology, or chemistry. Students may also explore periodicals and journals, such as *Scientific American*, *Science*, *Physics Teacher*, and so on. They should create a poster that summarizes an example they found and hang the poster in the classroom. When all of the posters are up, have students walk around and examine the various examples.

Enrichment

Have students mark the left most vertical line of a sheet of graph paper as the y-axis and the lowest horizontal line as the x-axis. From left to right, the marks along the x-axis will represent 1, 10, 100, 1000, and so on. From bottom to top, the marks along the y-axis will represent 0, 1, 2, 3, 4, and so on. Graph $y = \log_{10} x$ on these axes. Have students describe the graph and explain why the graph is shaped the way it is. **The graph is a straight line at a 45° angle to the x-axis.** The graph paper that students created is known as semilog paper.

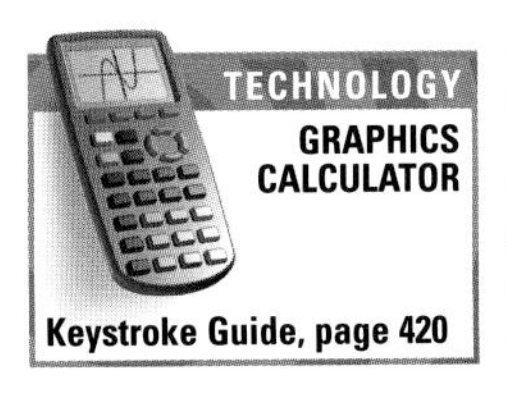

CHECK

Use a graphics calculator to graph $y = 10 \log x$. You find that when x is 300, y is about 25.

The relative intensity of this whisper is about 25 decibels.

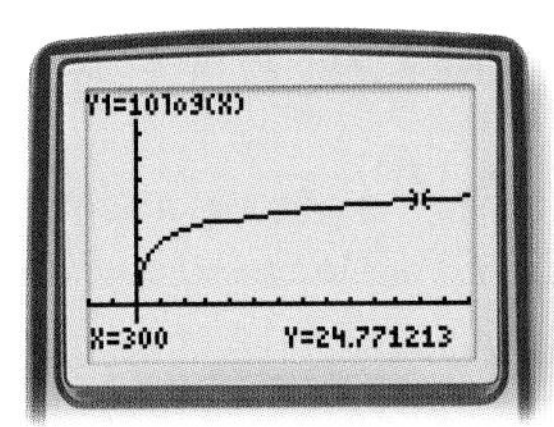

Teaching Tip

TECHNOLOGY For Example 1, use the following window:

Xmin=–50	Ymin=–10
Xmax=350	Ymax=50
Xscl=25	Yscl=5

EXAMPLE 2 The relative intensity, R, of a running vacuum cleaner is about 70 decibels.

APPLICATION

PHYSICS

Compare the intensity of this running vacuum cleaner with the threshold of hearing.

SOLUTION

PROBLEM SOLVING

Identify the wanted, given, and unknown information. In this problem, R is given and you need to find the ratio of I to I_0.

$$R = 10 \log \frac{I}{I_0}$$

$$70 = 10 \log \frac{I}{I_0} \quad \textit{Substitute 70 for the relative intensity, R.}$$

$$7 = \log \frac{I}{I_0}$$

$$10^7 = \frac{I}{I_0} \quad \textit{Write the equivalent exponential equation.}$$

$$10^7 \cdot I_0 = I \quad \textit{Write the intensity, I, in terms of } I_0.$$

This running vacuum cleaner is about 10^7, or 10,000,000, times as loud as the threshold of hearing.

ADDITIONAL EXAMPLE 2

The relative intensity, R, of a certain mosquito's buzzing is about 40 decibels. **Compare the relative intensity of this buzzing with a sound at the threshold of hearing.**
The buzzing is about 10^4, or 10,000, times as loud as a sound at the threshold of hearing.

If x and y are positive real numbers and $x = y$, then $\log x = \log y$ by substitution. This is used to solve an equation in Example 3.

EXAMPLE 3 **Solve $5^x = 62$ for x. Round your answer to the nearest hundredth.**

SOLUTION

$$5^x = 62$$

$$\log 5^x = \log 62 \quad \textit{Take the common logarithm of both sides.}$$

$$x \log 5 = \log 62 \quad \textit{Apply the Power Property of Logarithms.}$$

$$x = \frac{\log 62}{\log 5}$$

$$x \approx 2.56 \quad \textit{Use a calculator to evaluate.}$$

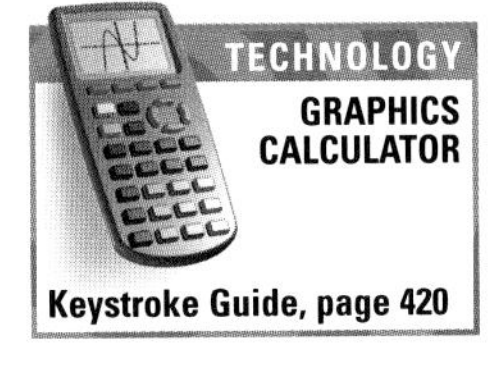

CHECK

Graph $y = 5^x$ and $y = 62$ in the same window, and find the point of intersection.

Thus, the solution to $5^x = 62$ is approximately 2.56.

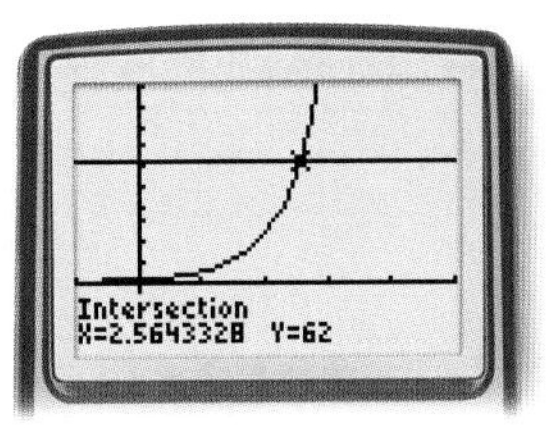

TRY THIS Solve $8^x = 792$ for x. Round your answer to the nearest hundredth.

ADDITIONAL EXAMPLE 3

Solve $6^x = 82$ for x. Round your answer to the nearest hundredth. $x \approx 2.46$

Teaching Tip

TECHNOLOGY For Example 3, use the following viewing window:

Xmin=–1	Ymin=–40
Xmax=5	Ymax=100
Xscl=1	Yscl=10

Students can use the intersect feature in the **CALC** menu by pressing 2nd TRACE, selecting **5:intersect**, and pressing ENTER three times to designate the two graphs and the guess.

TRY THIS

$x \approx 3.21$

Inclusion Strategies

USING TECHNOLOGY Many students need a concrete example of the change-of-base formula along with the algebraic derivation. Group students into pairs, and give each pair a set of simple logarithmic expressions to simplify, such as $\log_2 8$, $\log_4 64$, $\log_2 32$, and $\log_{10} 1000$. In each group, one student should apply the change-of-base formula to the expression and use a calculator to evaluate the expression. The other student should rewrite the expression in exponential form and find the answer. For example, $\log_2 8 = x$ is equivalent to $2^x = 8$, so $x = 3$. Students should alternate roles for each expression. Notice that for the last expression given above, $\log_{10} 1000$, students do not have to use the change-of-base formula to enter the expression into their calculator. Many students may apply the formula, forgetting that LOG on their calculator is the same as $\log_{10}$. In this case, the denominator of the change-of-base formula would be $\log_{10} 10$, or 1.

Activity Notes

In this Activity, students are led through a derivation of the change-of-base formula, using a numerical example. Students should find that $\log_3 81 = x = \frac{\log 81}{\log 3}$. Then they should find $x = \log_b y$ and $x = \frac{\log y}{\log b}$. Therefore, $\log_b y = \frac{\log y}{\log b}$.

CHECKPOINT ✔

4. $\log_b y = x;$
$b^x = y$
$\log b^x = \log y$
$x \log b = \log y$
$x = \frac{\log y}{\log b};$
$\log_b y = \frac{\log y}{\log b}$

CHECKPOINT ✔

$\log_9 27 = \frac{\log_3 27}{\log_3 9} = \frac{3}{2}$

ADDITIONAL EXAMPLE 4

Evaluate $\log_8 97$. Round your answer to the nearest hundredth. ≈ 2.20

TRY THIS

≈ 1.72

CRITICAL THINKING

a. $(\log_a b)(\log_b c)$
$= (\log_a b) \cdot \frac{\log_a c}{\log_a b}$
$= \log_a c$

b. $\log_a b = \frac{\log_b b}{\log_b a}$
$= \frac{1}{\log_b a}$

The Activity below leads to a method for evaluating logarithmic expressions with bases other than 10.

Activity: Exploring Change of Base

You will need: a scientific calculator

1. Write $3^x = 81$ as a logarithmic expression for x in base 3.
2. Write $3^x = 81$ as a logarithmic expression for x in base 10. (Hint: Refer to Example 3.)
3. Set your expressions for x from Steps 1 and 2 equal to each other.

CHECKPOINT ✔

4. Write $b^x = y$ in logarithmic form. Then solve $b^x = y$ for x, and give the result as a quotient of logarithms. Set your two resulting expressions equal to each other.

The answer to Step 4 in the Activity suggests a *change-of-base formula,* shown below, for writing equivalent logarithmic expressions with different bases.

Change-of-Base Formula

For any positive real numbers $a \neq 1$, $b \neq 1$, and $x > 0$:

$$\log_b x = \frac{\log_a x}{\log_a b}$$

CHECKPOINT ✔ Write $\log_9 27$ as a base 3 expression.

You can use the change-of-base formula to change a logarithmic expression of any base to base 10 so that you can use the LOG key on a calculator. This is shown in Example 4.

EXAMPLE 4 **Evaluate $\log_7 56$. Round your answer to the nearest hundredth.**

SOLUTION

Use the change-of-base formula to change from base 7 to base 10.

$$\log_7 56 = \frac{\log 56}{\log 7} \approx \frac{1.748}{0.845} \approx 2.07$$

Use a calculator to evaluate.

TRY THIS Evaluate $\log_8 36$. Round your answer to the nearest hundredth.

CRITICAL THINKING Use the change-of-base formula to justify each formula below.

a. $(\log_a b)(\log_b c) = \log_a c$ **b.** $\log_a b = \frac{1}{\log_b a}$

Reteaching the Lesson

USING TECHNOLOGY Have students explore the graphs of logarithmic functions with various bases. To do this, students need to use the change-of-base formula when entering the function into their graphics calculator. For example, to graph $y = \log_2 x$, students should enter **Y1=log(X)/log(2).** Have students graph $y = \log_2 x$, $y = \log_3 x$, $y = \log_4 x$, and $y = \log_5 x$. Use the following window:

Xmin=0	**Ymin=–2**
Xmax=4	**Ymax=2**
Xscl=1	**Yscl=1**

Have students discuss how the graphs are similar and how they relate to transformations of the graphs of functions. Because our number system is a base-10 system, it is useful to use $\log_{10}$ to write many exponential and logarithmic expressions.

Exercises

Communicate

1. Explain why a common logarithmic function is appropriate to use for the decibel scale of sound intensities.
2. Describe the steps you would take to solve $6^x = 39$ for x.
3. Explain how to evaluate $\log_4 29$ by using a calculator.

Guided Skills Practice

APPLICATIONS

4. **PHYSICS** Suppose that a soft whisper is about 75 times as loud as the threshold of hearing, I_0. Find the relative intensity, R, of this whisper in decibels. ***(EXAMPLE 1)*** **18.75**
5. **PHYSICS** The relative intensity, R, of a loud siren is about 130 decibels. Compare the intensity of this siren with the threshold of hearing, I_0. ***(EXAMPLE 2)*** **10^{13} times louder**

Solve each exponential equation for x. Round your answers to the nearest hundredth. ***(EXAMPLE 3)***

6. $8^x = 4$ **0.67**
7. $4^x = 72$ **3.08**

Evaluate each logarithmic expression. Round your answers to the nearest hundredth. ***(EXAMPLE 4)***

8. $\log_2 46$ **5.52**
9. $\log_5 2$ **0.43**

Practice and Apply

internet connect
Homework Help Online
Go To: **go.hrw.com**
Keyword:
MB1 Homework Help for Exercises 10–27

Solve each equation. Round your answers to the nearest hundredth.

10. $4^x = 17$ **2.04**
11. $2^x = 49$ **5.61**
12. $7^x = 908$ **3.50**
13. $8^x = 240$ **2.64**
14. $3.5^x = 28$ **2.66**
15. $7.6^x = 64$ **2.05**
16. $25^x = 0.04$ **−1**
17. $3^x = 0.26$ **−1.23**
18. $2^{-x} = 0.045$ **4.47**
19. $7^{-x} = 0.022$ **1.96**
20. $3^x = 0.45$ **−0.73**
21. $5^x = 1.29$ **0.16**
22. $2^{x+1} = 30$ **3.91**
23. $3^{x-6} = 81$ **10**
24. $11 - 6^x = 3$ **1.16**
25. $67 - 2^x = 39$ **4.81**
26. $8 + 3^x = 10$ **0.63**
27. $1 + 5^x = 360$ **3.66**

Assess

Selected Answers

Exercises 4–9, 11–73 odd

ASSIGNMENT GUIDE

In Class	1–9
Core	11–45 odd, 51
Core Plus	10–60 even
Review	61–73
Preview	74

Extra Practice can be found beginning on page 940.

Error Analysis

When solving exponential equations by using logarithms, students sometimes forget to take the logarithm of both sides of the equation and instead take only the logarithm of the exponential expression. Remind students that, as with other algebraic equations, they must keep the equation balanced by performing the same operation to both sides of the equation. For Exercises 24–27, remind students that they will need to isolate the exponential expression before taking the logarithm of both sides. Additionally, they may want to *incorrectly* simplify $\log(1 + 5^x)$ as $\log 1 + \log 5^x$, but these two expressions are not equivalent.

Practice

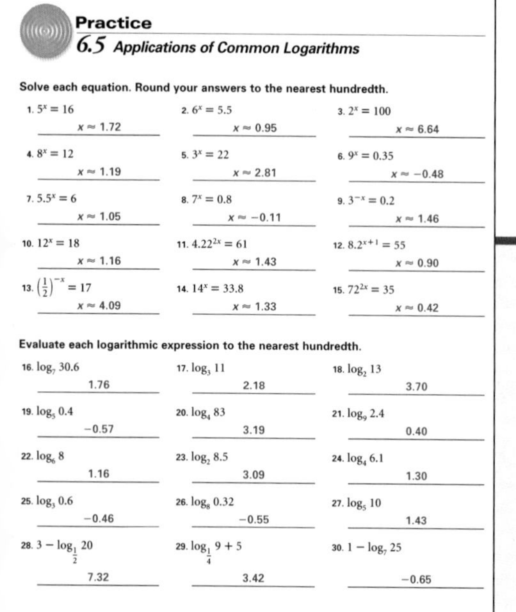

NAME ______ CLASS ______ DATE ______

Practice
6.5 Applications of Common Logarithms

Solve each equation. Round your answers to the nearest hundredth.

1. $5^x = 16$ — $x \approx 1.72$
2. $6^x = 5.5$ — $x \approx 0.95$
3. $2^x = 100$ — $x \approx 6.64$
4. $8^x = 12$ — $x \approx 1.19$
5. $3^x = 22$ — $x \approx 2.81$
6. $9^x = 0.35$ — $x \approx -0.48$
7. $5.5^x = 6$ — $x \approx 1.05$
8. $7^x = 0.8$ — $x \approx -0.11$
9. $3^{-x} = 0.2$ — $x \approx 1.46$
10. $12^x = 18$ — $x \approx 1.16$
11. $4.22^{2x} = 61$ — $x \approx 1.43$
12. $8.2^{x+1} = 55$ — $x \approx 0.90$
13. $\left(\frac{1}{2}\right)^{-x} = 17$ — $x \approx 4.09$
14. $14^x = 33.8$ — $x \approx 1.33$
15. $72^{2x} = 35$ — $x \approx 0.42$

Evaluate each logarithmic expression to the nearest hundredth.

16. $\log_7 30.6$ — 1.76
17. $\log_3 11$ — 2.18
18. $\log_2 13$ — 3.70
19. $\log_5 0.4$ — −0.57
20. $\log_4 83$ — 3.19
21. $\log_9 2.4$ — 0.40
22. $\log_6 8$ — 1.16
23. $\log_2 8.5$ — 3.09
24. $\log_4 6.1$ — 1.30
25. $\log_3 0.6$ — −0.46
26. $\log_8 0.32$ — −0.55
27. $\log_5 10$ — 1.43
28. $3 - \log_{\frac{1}{2}} 20$ — 7.32
29. $\log_{\frac{1}{4}} 9 + 5$ — 3.42
30. $1 - \log_7 25$ — −0.65

Evaluate each logarithmic expression to the nearest hundredth.

28. $\log_4 92$ **3.26**
29. $\log_6 87$ **2.49**
30. $\log_6 18$ **1.61**
31. $\log_3 15$ **2.46**
32. $\log_6 3$ **0.61**
33. $\log_5 2$ **0.43**
34. $\log_9 4$ **0.63**
35. $\log_8 3$ **0.53**
36. $\log_4 0.37$ **−0.72**
37. $\log_9 1.43$ **0.16**
38. $\log_{\frac{1}{3}} 9$ **−2**
39. $\log_{\frac{1}{2}} 8$ **−3**
40. $\log_8 \frac{1}{4}$ **−0.67**
41. $\log_7 \frac{1}{50}$ **−2.01**
42. $8 - \log_2 64$ **2**
43. $1 - \log_5 21$ **−0.89**
44. $9 + \log_3 27$ **12**
45. $4 + \log_5 125$ **7**

CHALLENGE

46. Prove that $\log_{(b^n)} x = \frac{1}{n} \log_b x$ is true.

46. $\log_{(b^n)} x = \frac{\log_b x}{\log_b b^n} = \frac{\log_b x}{n} = \frac{1}{n} \log_b x$

APPLICATIONS

47. PHYSICS The sound of a leaf blower is about $10^{10.5}$ times the intensity of the threshold of hearing, I_0. Find the relative intensity, R, of this leaf blower in decibels. **105 decibels**

48. PHYSICS The sound of a conversation is about 350,000 times the intensity of the threshold of hearing, I_0. Find the relative intensity, R, of this conversation in decibels. **55.4 decibels**

49. PHYSICS Suppose that the relative intensity, R, of a rock band is about 115 decibels. Compare the intensity of this band with that of the threshold of hearing, I_0. **$10^{11.5}$ times louder**

50. PHYSICS The relative intensity, R, of an automobile engine is about 55 decibels. Compare the intensity of this engine with that of the threshold of hearing, I_0. **$10^{5.5}$ times louder**

51. PHYSICS Suppose that background music is adjusted to an intensity that is 1000 times as loud as the threshold of hearing. What is the relative intensity of the music in decibels? **30 decibels**

52. PHYSICS Suppose that a burglar alarm has a rating of 120 decibels. Compare the intensity of this decibel rating with that of the threshold of hearing, I_0. **10^{12} times louder**

53. PHYSICS Simon Robinson set the world record for the loudest scream by producing a scream of 128 decibels at a distance of 8 feet and 2 inches. Compare the intensity of this decibel rating with that of the threshold of hearing, I_0. [*Source: The Guinness Book of World Records, 1997*]

53. **$10^{12.8}$ times louder**

54. PHYSICS A small jet engine produces a sound whose intensity is one billion times as loud as the threshold of hearing. What is the relative intensity of the engine's sound in decibels? **90 decibels**

internet connect
Activities Online
Go To: **go.hrw.com**
Keyword:
MB1 Stock Closings

APPLICATIONS

CHEMISTRY In chemistry, pH is defined as pH = –log [H⁺], where [H⁺] is the hydrogen ion concentration in moles per liter.

55. $10^{-14} < [H^+] < 10^{-7}$

56. $10^{-7} < [H^+] < 10^{0}$

57. ≈1.3

58. ≈7.8

59. ≈10.5

60a. ≈4.2
b. ≈1.5
c. ≈476 times

55. An alkaline solution has a pH in the range $7 < pH < 14$. Determine the corresponding range of $[H^+]$ for alkaline substances.

56. An acidic solution has a pH in the range $0 < pH < 7$. Determine the corresponding range of $[H^+]$ for acidic substances.

57. For hydrochloric acid, $[H^+]$ is about 5×10^{-2} moles per liter. Find the pH for this strong acid to the nearest tenth.

58. For chicken eggs, $[H^+]$ is about 1.6×10^{-8} moles per liter. Find the pH of chicken eggs to the nearest tenth.

59. For milk of magnesia, $[H^+]$ is about 3.2×10^{-11} moles per liter. Find the pH of milk of magnesia to the nearest tenth.

60. CHEMISTRY Find the pH that corresponds to each hydrogen ion concentration.
a. $[H^+]$ is about 6.3×10^{-5} moles per liter for tomato juice.
b. $[H^+]$ is about 0.03 moles per liter for gastric juice.
c. How much more concentrated is the gastric juice than the tomato juice? (Hint: Make a ratio of their hydrogen ion concentrations.)

 Look Back

Solve each system of equations. *(LESSONS 3.1 AND 3.2)*

61. (5, 2)
62. (2, 7)
63. (–3, –2)

61. $\begin{cases} x + y = 7 \\ 2x - 3y = 4 \end{cases}$ **62.** $\begin{cases} x + 3y = 23 \\ 4x - 2y = -6 \end{cases}$ **63.** $\begin{cases} -2x + 5y = -4 \\ 3x - y = -7 \end{cases}$

Show that each function is a quadratic function by writing it in the form $f(x) = ax^2 + bx + c$. *(LESSON 5.1)*

64. $h(x) = -11x^2 + 55x$
65. $g(x) = 2x^2 - 8x - 10$
66. $k(x) = 4x^2 - 100$
67. $f(x) = -3x^2 - 2x + 1$

64. $h(x) = 11x(5 - x)$ **65.** $g(x) = (2x - 10)(x + 1)$

66. $k(x) = 4(x + 5)(x - 5)$ **67.** $f(x) = -(x + 1)(3x - 1)$

Solve each equation for x. *(LESSON 5.6)*

68. $x = -3$ *or* $x = -2$
69. $x = -5$ *or* $x = 3$
70. $x \approx -3.85$ *or* $x \approx -2.85$

68. $x^2 + 5x = -6$ **69.** $x^2 + 2x - 15 = 0$ **70.** $(x - 2)(x + 3) = 5$

Identify each function as representing exponential growth or exponential decay. *(LESSON 6.2)*

71. $f(t) = 1000(2.5)^t$ growth **72.** $f(t) = 55(0.5)^t$ decay **73.** $f(t) = 0.005(8)^t$ growth

 Look Beyond

74 Graph $y = 2^x$, $y = e^x$, and $y = 3^x$. Describe how the graph of e^x compares to the others. Use your calculator to find a value for e to four decimal places.

 Look Beyond

In Exercise 74, students begin to explore the number e. Natural logarithms and e will be studied further in Lesson 6.6.

74. Answers may vary. Sample answer: The graph of $y = e^x$ looks similar to the graphs of $y = 2^x$ and $y = 3^x$. It rises more steeply than 2^x and less steeply than 3^x; 2.7183.

Student Technology Guide

NAME ______ CLASS ______ DATE ______

Student Technology Guide
6.5 Applications of Common Logarithms

Most calculators have only two types of logarithms: common logarithms, which you saw in Lesson 6.5 of the textbook, and natural logarithms, which you will learn about in Lesson 6.6. However, there is a formula that allows you to change logarithms with other bases to common logarithms.

The change-of-base formula states the following:

For positive real numbers $a \neq 1, b \neq 1$, and $x > 0$, $\log_b x = \frac{\log_a x}{\log_a b}$.

To find $\log_3 25$, use $a = 10$, $b = 3$, and $x = 25$, as shown at right.

$$\log_3 25 = \frac{\log 25}{\log 3}$$

Example: Evaluate $\log_5 37$. Round your answer to the nearest hundredth.

- Use the change-of-base formula.

$$\log_5 37 = \frac{\log 37}{\log 5}$$

- Use a calculator to evaluate the fraction. Press LOG 37) ÷ LOG 5) ENTER.

Thus, $\log_5 37 \approx 2.24$.

```
log(37)/log(5)
      2.243589445
```

Evaluate each logarithmic expression. Round your answers to the nearest hundredth.

1. $\log_6 14$ 1.47 2. $\log_3 9$ 2 3. $\log_8 2$ 0.33 4. $\log_4 8$ 1.5
5. $\log_{\frac{1}{2}} 9$ –3.17 6. $\log_7\left(\frac{1}{4}\right)$ –0.71 7. $\log_6\left(\frac{1}{216}\right)$ –3 8. $\log_{100} 15.75$ 0.60

The pH of a substance is defined as $-\log[H^+]$, where $[H^+]$ is the concentration of hydrogen ions in moles per liter. To find the pH of a substance, use common logarithms.

Example: For lemon juice, $[H^+]$ is about 5×10^{-3} moles per liter. Find the pH of lemon juice.

- To find the pH, press (–) LOG 5 2nd , (–) 3) ENTER. (The pair of keys 2nd , is the way to enter numbers in scientific notation.)

The pH of lemon juice is about 2.3 moles per liter.

```
-log(5E-3)
      2.301029996
```

9. For household ammonia, $[H^+]$ is about 3.2×10^{-12} moles per liter. Find the pH of ammonia. 11.5 moles/lite

Prepare

NCTM PRINCIPLES & STANDARDS 1–4, 6–10

QUICK WARM-UP

Evaluate log x for each value.

1. $x = 10$ 1
2. $x = \frac{1}{10}$ -1
3. $x = -10$ undefined
4. $x = 1$ 0

Find the final amount for each investment.

5. \$1000 earning 4% annual interest compounded annually for 10 years \$1480.24
6. \$1000 earning 4% annual interest compounded quarterly for 10 years \$1488.86

Also on Quiz Transparency 6.6

Teach

Why Have students locate e on their calculators and enter e^1. Ask students to share the values given by the calculators. Most calculators will show nine places after the decimal, so $e \approx 2.718281828$. Students may assume that the decimal will repeat. Discuss e as an irrational number and compare e with π.

6.6 The Natural Base, e

The model below shows the embryo inside an 18-inch dinosaur egg, the largest known.

Objectives

- Evaluate natural exponential and natural logarithmic functions.
- Model exponential growth and decay processes.

Why *The exponential function with base e and its inverse, the natural logarithmic function, have a wide variety of real-world applications. For example, these functions are used to estimate the ages of artifacts found at archaeological digs.*

The natural base, e, is used to estimate the ages of artifacts and to calculate interest that is compounded continuously. Recall from Lesson 6.2 the compound interest formula, $A(t) = P\left(1 + \frac{r}{n}\right)^{nt}$, where P is the principal, r is the annual interest rate, n is the number of compounding periods per year, and t is the time in years. This formula is used in the Activity below.

APPLICATION INVESTMENTS

Activity

Investigating the Growth of \$1

You will need: a scientific calculator

1. Copy and complete the table below to investigate the growth of a \$1 investment that earns 100% annual interest ($r = 1$) over 1 year ($t = 1$) as the number of compounding periods per year, n, increases. Use a calculator, and record the value of A to five places after the decimal point.

Compounding schedule	n	$1\left(1 + \frac{1}{n}\right)^n$	Value, A
annually	1	$1\left(1 + \frac{1}{1}\right)^1$	2.00000
semiannually	2	$1\left(1 + \frac{1}{2}\right)^2$	
quarterly	4		
monthly	12		
daily	365		
hourly			
every minute			
every second			

CHECKPOINT ✔ 2. Describe the behavior of the sequence of numbers in the *Value* column.

As n becomes very large, the value of $1\left(1 + \frac{1}{n}\right)^n$ approaches the number 2.71828 . . . , named e. Because e is an irrational number like π, its decimal expansion continues forever without repeating patterns.

Alternative Teaching Strategy

USING TECHNOLOGY Have students enter the function $y = \left(1 + \frac{1}{x}\right)^x$ into a graphics calculator and use the following window:

Xmin=0	**Ymin=–2**
Xmax=10	**Ymax=5**
Xscl=1	**Yscl=1**

Ask them to describe what happens to the y-values as the x-values get larger. To analyze this behavior further, have students change **Xmax** to **500** and use the trace feature and the table feature of their calculators to get a visual and numerical understanding of the function. Suggest that they set the table minimum to 200. Then ask students to estimate the limit of this function to four decimal places. ≈2.7183 Next have students clear the **Y=** menu and graph the function $y = e^x$ in a standard viewing window. Ask students to use logarithms to write the inverse of the exponential function $y = e^x$. $y = \log_e x$ Logarithms of base e are called natural logarithms and are written in the form ln x. Have students graph $y = \ln x$ in the same window as $y = e^x$.

The Natural Exponential Function

The exponential function with base e, $f(x) = e^x$, is called the **natural exponential function** and e is called the **natural base.** The function $f(x) = e^x$ is graphed at right. Notice that the domain is all real numbers and the range is all positive real numbers.

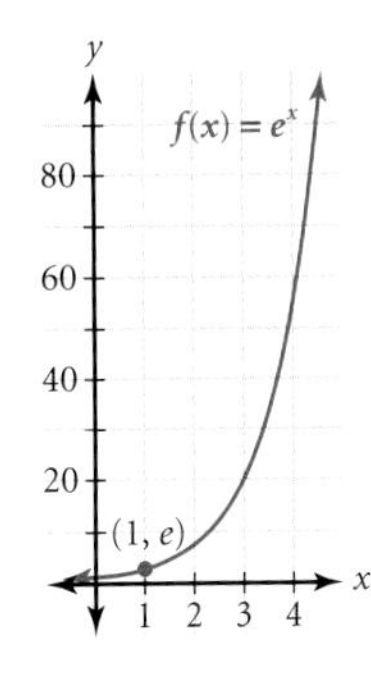

CHECKPOINT ✔ What is the y-intercept of the graph of $f(x) = e^x$?

Natural exponential functions model a variety of situations in which a quantity grows or decays continuously. Examples that you will solve in this lesson include continuous compounding interest and continuous radioactive decay.

EXAMPLE 1 **Evaluate $f(x) = e^x$ to the nearest thousandth for each value of x below.**

a. $x = 2$ **b.** $x = \frac{1}{2}$ **c.** $x = -1$

SOLUTION

a. $f(2) = e^2 \approx 7.389$

b. $f\left(\frac{1}{2}\right) = e^{\frac{1}{2}} \approx 1.649$

c. $f(-1) = e^{-1} \approx 0.368$

CHECK

Use a table of values for $y = e^x$ or a graph of $y = e^x$ to verify your answers.

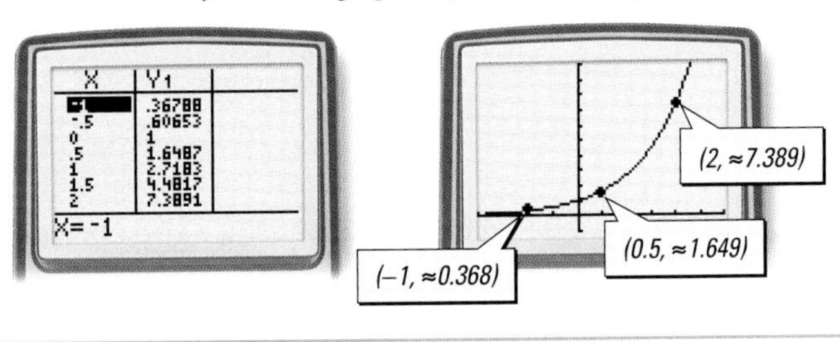

TRY THIS Evaluate $f(x) = e^x$ to the nearest thousandth for $x = 6$ and $x = -\frac{1}{3}$.

Many banks compound the interest on accounts daily or monthly. However, some banks compound interest continuously, or at every instant, by using the *continuous compounding formula*, which includes the number e.

Continuous Compounding Formula

If P dollars are invested at an interest rate, r, that is compounded continuously, then the amount, A, of the investment at time t is given by

$$A = Pe^{rt}.$$

Interdisciplinary Connection

BIOLOGY The spread of a disease can be modeled by an exponential function. The model assumes that the rate at which the number of infected people changes is proportional to the number of infected people and is given by $y = y_0e^{kt}$, where y_0 is the original number of infected people and y is the number of people infected after t hours. Suppose that 1 infected person visits a town and that 5 hours later, 3 other people are infected.

1. Find the value of k for this situation. **≈0.22**
2. How many people will be infected after 24 hours? **≈196**

Enrichment

Populations are often modeled by exponential functions. However, it is not realistic for populations to grow forever, so exponential models are accurate only for a short time. A logistic model shows the leveling off of a population. Suppose that the flu spread through a school according to the function $P(t) = \frac{100}{1 + e^{3-t}}$, where t represents the time since the infection began and $P(t)$ represents the number of people infected. Sketch a graph of this function. What is the maximum number of students who may be infected? **100**

Activity Notes

In this Activity, students use a real-world situation to discover the number e. The sequence of values will approach $e \approx 2.71828182846$.

Cooperative Learning

You may wish to have students perform this Activity in pairs so that they can check each other's results for computational errors.

Teaching Tip

TECHNOLOGY Students will need a scientific calculator because the evaluation of the function is complicated. However, students could use a graphics calculator. Enter **Y1 = (1 + 1/X)^X,** change the table setup to ask for the independent variable, and find the y-values in the table. Students can also graph the function to get a visual sense of its end behavior.

CHECKPOINT ✔

2. The A-values become larger as n becomes larger, but the rate of increase becomes smaller as the values approach 2.71828.

CHECKPOINT ✔

(0, 1)

☞ For Additional Example 1 and the answer to Try This, see page 394.

ADDITIONAL
EXAMPLE 1

Evaluate the function $f(x) = e^x$ to the nearest thousandth for each x-value.

a. $x = 3$ ≈ 20.086

b. $x = 0.25$ ≈ 1.284

c. $x = -2$ ≈ 0.135

TRY THIS
403.429; 0.717

ADDITIONAL
EXAMPLE 2

$2000 is invested at an annual interest rate of 9%. **Compare the final amounts after 12 years for interest compounded quarterly and continuously.**
quarterly: ≈$5819.28
continuously: ≈$5889.36
Compounding continuously results in a final amount that is about $70 more than that from compounding quarterly.

TRY THIS
$716.66

CHECKPOINT ✔
$y = e^x$: domain is all real numbers; range is all positive real numbers.

$y = \ln x$: domain is all positive real numbers; range is all real numbers.

ADDITIONAL
EXAMPLE 3

Evaluate $f(x) = \ln x$ to the nearest thousandth for each value of x below.

a. $x = 5$ ≈ 1.609

b. $x = 0.85$ ≈ -0.163

c. $x = 1$ 0

EXAMPLE 2 An investment of $1000 earns an annual interest rate of 7.6%.

APPLICATION
INVESTMENTS

Compare the final amounts after 8 years for interest compounded quarterly and for interest compounded continuously.

SOLUTION

Substitute 1000 for P, 0.076 for r, and 8 for t in the appropriate formulas.

Compounded quarterly	Compounded continuously
$A = P\left(1+\frac{r}{n}\right)^{nt}$	$A = Pe^{rt}$
$A = 1000\left(1+\frac{0.076}{4}\right)^{4 \cdot 8}$	$A = 1000e^{0.076 \cdot 8}$
$A \approx 1826.31$	$A \approx 1836.75$

Interest that is compounded continuously results in a final amount that is about $10 more than that for the interest that is compounded quarterly.

TRY THIS Find the value of $500 after 4 years invested at an annual interest rate of 9% compounded continuously.

The Natural Logarithmic Function

The **natural logarithmic function,** $y = \log_e x$, abbreviated $y = \ln x$, is the inverse of the natural exponential function, $y = e^x$. The function $y = \ln x$ is graphed along with $y = e^x$ at right.

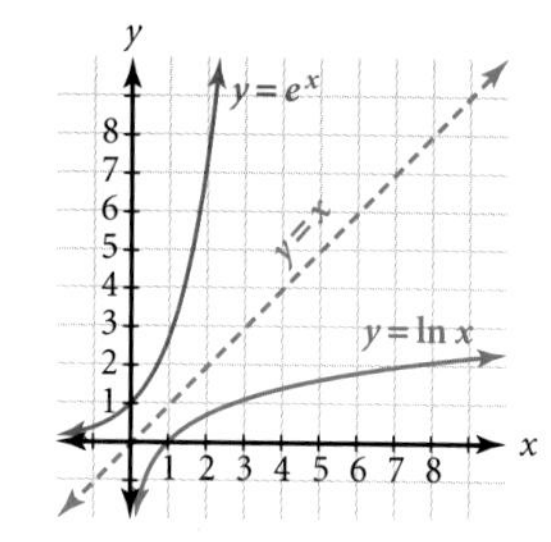

CHECKPOINT ✔ State the domain and range of $y = e^x$ and of $y = \ln x$.

EXAMPLE 3 **Evaluate $f(x) = \ln x$ to the nearest thousandth for each value of x below.**

a. $x = 2$ **b.** $x = \frac{1}{2}$ **c.** $x = -1$

SOLUTION

a. $f(2) = \ln 2 \approx 0.693$

b. $f\left(\frac{1}{2}\right) = \ln \frac{1}{2} \approx -0.693$

c. $f(-1) = \ln(-1)$ is undefined.

TECHNOLOGY
GRAPHICS CALCULATOR
Keystroke Guide, page 420

CHECK
Use a table of values for $y = \ln x$ or a graph of $y = \ln x$ to verify your answers.

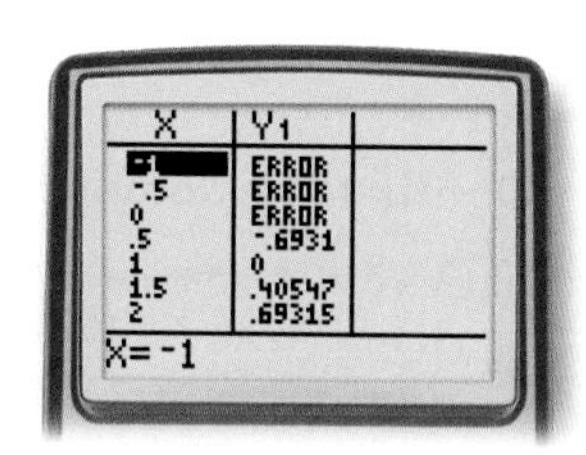

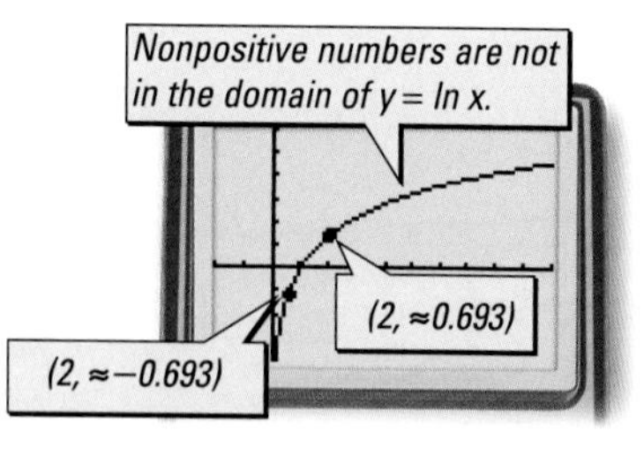

Inclusion Strategies

USING TABLES Many students get overwhelmed by the symbolic notation often associated with applications of exponential functions and natural logarithms. When solving the problems, encourage students to enter the equations into the **Y=** menu of their graphics calculator and to use the table of values to estimate their solution. This will give students confidence in the solutions they obtained by solving the problem algebraically.

The natural logarithmic function can be used to solve an equation of the form $A = Pe^{rt}$ for the exponent t in order to find the time it takes for an investment that is compounded continuously to reach a specific amount. This is shown in Example 4.

EXAMPLE 4 **How long does it take for an investment to double at an annual interest rate of 8.5% compounded continuously?**

SOLUTION

PROBLEM SOLVING

Use the formula $A = Pe^{rt}$ with $r = 0.085$.

$A = Pe^{0.085t}$

$2 \cdot P = Pe^{0.085t}$ — *When the investment doubles, $A = 2 \cdot P$.*

$2 = e^{0.085t}$

$\ln 2 = \ln e^{0.085t}$ — *Take the natural logarithm of both sides.*

$\ln 2 = 0.085t$ — *Use the Exponential-Logarithmic Inverse Property.*

$t = \frac{\ln 2}{0.085}$

$t \approx 8.15$

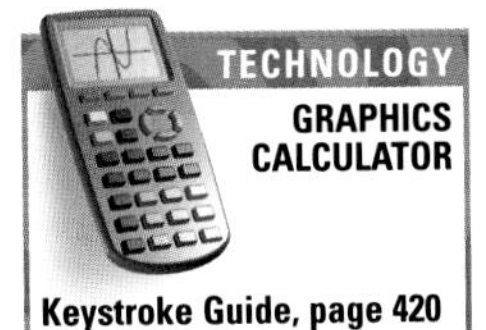

CHECK

Graph $y = e^{0.085x}$ and $y = 2$, and find the point of intersection.

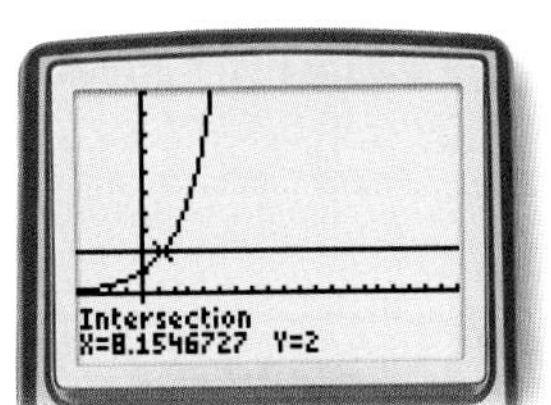

Notice that the graph of $y = e^{0.085x}$ is a horizontal stretch of the function $y = e^x$ by a factor of $\frac{1}{0.085}$, or almost 12.

Thus, it takes about 8 years and 2 months to double an investment at an annual interest rate of 8.5% compounded continuously.

TRY THIS How long does it take for an investment to triple at an annual interest rate of 7.2% compounded continuously?

CRITICAL THINKING Explain why the time required for the value of an investment to double or triple does not depend on the amount of principal.

Radioactive Decay

Most of the carbon found in the Earth's atmosphere is the isotope carbon-12, but a small amount is the radioactive isotope carbon-14. Plants absorb carbon dioxide from the atmosphere, and animals obtain carbon from the plants they consume. When a plant or animal dies, the amount of carbon-14 it contains decays in such a way that exactly half of its initial amount is present after 5730 years. The function below models the decay of carbon-14, where N_0 is the initial amount of carbon-14 and $N(t)$ is the amount present t years after the plant or animal dies.

$$N(t) = N_0 e^{-0.00012t}$$

ADDITIONAL EXAMPLE 4

How long does it take for an investment to triple at an annual interest rate of 7.5% compounded continuously?

≈14 years and 8 months

Teaching Tip

TECHNOLOGY Graph **Y1=e^0.075X** and **Y2=3** by using the following window:

Xmin=–5	**Ymin=–2**
Xmax=40	**Ymax=20**
Xscl=5	**Yscl=2**

TRY THIS

≈15.26 years, or ≈15 years and 3 months

CRITICAL THINKING

Because $3A = Ae^{rt}$ is equivalent to $3 = e^{rt}$, the amount of time required does not depend on the principal.

Reteaching the Lesson

USING DISCUSSION Have students use their calculators to find e by calculating e^1, and have them enter the following functions into a graphics calculator: **Y1=2^X, Y2=e^X,** and **Y3=3^X.** Use a window with **Xmin=0, Xmax=10, Ymin=–10,** and **Ymax=500,** and have them compare the graphs. Then discuss that just as $y = \log_2 x$ is the inverse of $y = 2^x$, $y = \log_e x$, or $\ln x$, is the inverse of $y = e^x$. We call this special logarithm the natural logarithm.

Solving equations involving e and natural logarithms is no different from solving equations involving other bases. Natural logarithms are helpful when solving exponential equations involving e because $\ln e = 1$. Have students discuss why this is true. For examaple, when solving $0.5 = e^{0.04t}$, you could take the common logarithm, $\log_{10}$, of each side, as in the previous lesson, but you would be left with $\log_{10} e$ to evaluate with a calculator. Using natural logarithms instead, you get $\ln 0.5 = 0.04t \ln e$, or $\ln 0.5 = 0.04t(1)$. Thus, $t = \frac{\ln 0.5}{0.04} \approx -17.33$.

ADDITIONAL
EXAMPLE 5

Suppose that a jar containing grain is found at an archaeological dig and that archaeologists claim that it is 6500 years old. Tests indicate that the grain contains 62% of its original carbon-14. **Could the grain be 6500 years old? Explain.**

No; using the given formula, $t \approx 3983.6$, so it appears that the grain is only about 4000 years old.

Assess

Selected Answers

Exercises 5–11, 13–91 odd

ASSIGNMENT GUIDE	
In Class	1–11
Core	13–59 odd, 67, 71, 73, 75
Core Plus	12–66 even, 70–80 even
Review	81–92
Preview	93

Extra Practice can be found beginning on page 940.

Example 5 shows how *radiocarbon dating* is used to estimate the age of an archaeological artifact.

EXAMPLE 5

APPLICATION
ARCHAEOLOGY

Suppose that archaeologists find scrolls and claim that they are 2000 years old. Tests indicate that the scrolls contain 78% of their original carbon-14.

Could the scrolls be 2000 years old?

SOLUTION

Since the scrolls contain 78% of their original carbon-14, substitute $0.78N_0$ for $N(t)$.

$$N(t) = N_0e^{-0.00012t}$$
$$0.78N_0 = N_0e^{-0.00012t} \quad \text{Substitute } 0.78N_0 \text{ for } N(t).$$
$$0.78 = e^{-0.00012t}$$
$$\ln 0.78 = -0.00012t \quad \text{Take the natural logarithm of each side.}$$
$$-0.00012t = \ln 0.78$$
$$t = \frac{\ln 0.78}{-0.00012}$$
$$t \approx 2070.5$$

Thus, it appears that the scrolls are about 2000 years old.

Exercises

Communicate

1. Compare the natural and exponential logarithmic functions with the base-10 exponential and logarithmic functions.
2. Give a real-world example of an exponential growth function and of an exponential decay function that each have the base e.
3. State the continuous compounding formula, and describe what each variable represents.
4. Describe how the continuous compounding formula can represent continuous growth as well as continuous decay.

Guided Skills Practice

APPLICATION

Evaluate $f(x) = e^x$ to the nearest thousandth for each value of x. ***(EXAMPLE 1)***

5. $x = 3$ **20.086**
6. $x = 3.5$ **33.115**

7. **INVESTMENTS** An investment of $1500 earns an annual interest rate of 8.2%. Compare the final amounts after 5 years for interest compounded quarterly and for interest compounded continuously. ***(EXAMPLE 2)***

7. **$2250.88; $2260.23**

Evaluate $f(x) = \ln x$ to the nearest thousandth for each value of x. *(EXAMPLE 3)*

8. $x = 5$ 1.609

9. $x = 2.5$ 0.916

APPLICATIONS

10. ≈9.24 years

11. ≈20,066 years old

10. INVESTMENTS How long does it take an investment to double at an annual interest rate of 7.5% compounded continuously? *(EXAMPLE 4)*

11. ARCHAEOLOGY A piece of charcoal from an ancient campsite is found in an archaeological dig. It contains 9% of its original amount of carbon-14. Estimate the age of the charcoal. *(EXAMPLE 5)*

Practice and Apply

internet connect

Homework Help Online
Go To: go.hrw.com
Keyword:
MB1 Homework Help
for Exercises 12–31

Evaluate each expression to the nearest thousandth. If the expression is undefined, write *undefined*.

12. e^6 403.429
13. e^9 8103.084
14. $e^{1.2}$ 3.320
15. $e^{3.4}$ 29.964
16. $2e^{0.3}$ 2.700
17. $3e^{0.05}$ 3.154
18. $2e^{-0.5}$ 1.213
19. $3e^{-0.257}$ 2.320
20. $e^{\sqrt{2}}$ 4.113
21. $e^{\frac{1}{4}}$ 1.284
22. $\ln 3$ 1.099
23. $\ln 7$ 1.946
24. $\ln 10{,}002$
25. $\ln 99{,}999$
26. $\ln 0.004$
27. $\ln 0.994$
28. $\ln \frac{1}{5}$ −1.609
29. $\ln \sqrt{5}$ 0.805
30. $\ln(-2)$ undefined
31. $\ln(-3)$ undefined

24. 9.211
25. 11.513
26. −5.521
27. −0.006

For Exercises 32–35, write the expressions in ascending order.

32. $e^2, e^5, \ln 2, \ln 5$
33. $e, e^0, \ln 1, \ln \frac{1}{2}$
34. $e^{2.5}, \ln 2.5, 10^{2.5}, \log 2.5$
35. $e^{1.3}, \ln 1.3, 10^{1.3}, \log 1.3$

32. $\ln 2, \ln 5, e^2, e^5$
33. $\ln \frac{1}{2}, \ln 1, e^0, e$
34. $\log 2.5, \ln 2.5, e^{2.5}, 10^{2.5}$
35. $\log 1.3, \ln 1.3, e^{1.3}, 10^{1.3}$

State whether each equation is always true, sometimes true, or never true.

36. $e^{5x} \cdot e^3 = e^{15x}$
37. $\left(e^{4x}\right)^3 = e^{12x}$
38. $e^{6x-4} = e^{6x} \cdot e^{-4}$
39. $\frac{e^{8x}}{e^4} = e^{2x}$

36. never true
37. always true
38. always true
39. sometimes true

Simplify each expression.

40. $e^{\ln 2}$ 2
41. $e^{\ln 5}$ 5
42. $e^{3\ln 2}$ 8
43. $e^{2\ln 5}$ 25
44. $\ln e^3$ 3
45. $\ln e^4$ 4
46. $3 \ln e^2$ 6
47. $2 \ln e^4$ 8

Write an equivalent exponential or logarithmic equation.

48. $e^x = 30$ $x = \ln 30$
49. $e^x = 1$ $\ln 1 = x$
50. $\ln 2 \approx 0.69$ $2 \approx e^{0.69}$
51. $\ln 5 \approx 1.61$ $5 \approx e^{1.61}$
52. $e^{\frac{1}{3}} \approx 1.40$ $\ln 1.40 \approx \frac{1}{3}$
53. $e^{0.69} \approx 1.99$ $\ln 1.99 \approx 0.69$

Solve each equation for x by using the natural logarithm function. Round your answers to the nearest hundredth.

54. $35^x = 30$ 0.96
55. $1.3^x = 8$ 7.93
56. $3^{-3x} = 17$ −0.86
57. $36^{2x} = 20$ 0.42
58. $0.42^{-x} = 7$ 2.24
59. $2^{-\frac{1}{3}x} = 10$ −9.97

CHALLENGE

60. Sketch $f(x) = e^x$ for $-1 \le x \le 2$. A line that intersects a curve at only one point is called a *tangent line* of the curve.
a. Sketch lines that are tangent to the graph of $f(x) = e^x$ at $x = 0.5$, $x = 0$, $x = 1$, and $x = 2$.
b. Find the approximate slope of each tangent line. Compare the slope of each tangent line with the corresponding y-coordinate of the point where the tangent line intersects the graph.
c. Make a conjecture about the slope of $f(x) = e^x$ as x increases.

Error Analysis

Remind students to use both the opening and closing parentheses when entering fractional exponents into the calculator. Even though the opening parenthesis is included in exponential and logarithmic features on some calculators, the closing parenthesis must be entered to insure correct results.

Also, remind students about the difference between the negative key, (-), and the minus key, -. When an exponent is negative, students must use the negative key, (-).

60. $f(x) = e^x$

a.

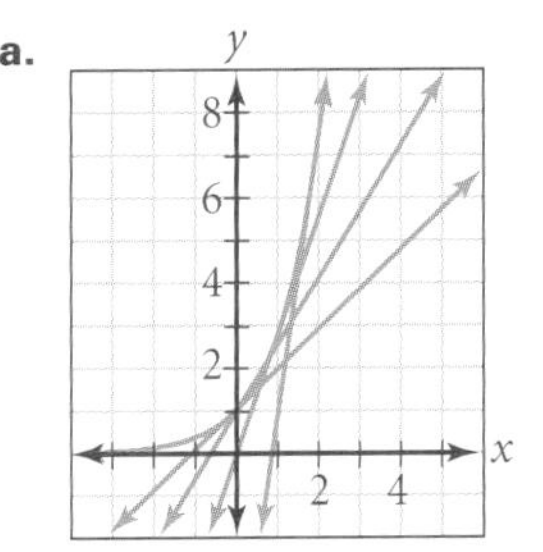

b. at $x = 0.5$, slope ≈ 1.6, $e^{0.5} \approx 1.65$

at $x = 0$, slope ≈ 1, $e^0 \approx 1$

at $x = 1$, slope ≈ 2.7, $e^1 \approx 2.72$

c. As x gets larger, the slope of $f(x) = e^x$ gets larger, and the slope equals e^x.

61. vertical stretch by a factor of 6 and vertical translation 1 unit up

62. vertical compression by a factor of $\frac{3}{4}$ and vertical translation 4 units down

63. horizontal translation 1 unit to the left, horizontal compression by a factor of $\frac{1}{4}$, and vertical compression by a factor of 0.25

64. horizontal translation 2 units to the right, horizontal compression by a factor of $\frac{1}{2}$, and vertical stretch by a factor of 3

65. vertical stretch by a factor of 3 and horizontal translation 1 unit to the left

66. reflection across the x-axis, vertical stretch by a factor of 2, and horizontal translation 1 unit to the right

67. horizontal compression by a factor of $\frac{1}{5}$, vertical compression by a factor of 0.5, and vertical translation 2 units down

68. horizontal stretch by a factor of 4, vertical stretch by a factor of 5, and vertical translation 1 unit down

Practice

NAME ______________________ CLASS ______________ DATE ______________

Practice 6.6 The Natural Base, e

Evaluate each expression to the nearest thousandth.

1. e^8 — 2980.958
2. $e^{2.5}$ — 12.182
3. $e^{5.2}$ — 181.272
4. $2e^4$ — 109.196
5. $\ln 35$ — 3.555
6. $\ln 12.6$ — 2.534
7. $\ln(-1.4)$ — not defined
8. $\ln \sqrt{12}$ — 1.242
9. $\ln 112$ — 4.718

Write an equivalent exponential or logarithmic equation.

10. $e^x \approx 55$ — $\ln 55 \approx x$
11. $\ln 44 \approx 3.78$ — $e^{3.78} \approx 44$
12. $e^{-3} \approx 0.05$ — $\ln 0.05 \approx -3$
13. $\ln 10 \approx 2.30$ — $e^{2.30} \approx 10$
14. $e^4 \approx 54.6$ — $\ln 54.6 \approx 4$
15. $\ln 125 \approx 4.83$ — $e^{4.83} \approx 125$
16. $e^5 \approx 148$ — $\ln 148 \approx 5$
17. $\ln 1 = 0$ — $e^0 = 1$
18. $e^{-0.8} \approx 0.45$ — $\ln 0.45 \approx -0.8$

Solve each equation for x by using the natural logarithm function. Round your answers to the nearest hundredth.

19. $33^x = 74$ — $x \approx 1.23$
20. $15^x = 19.5$ — $x \approx 1.10$
21. $4.8^x = 30$ — $x \approx 2.17$
22. $0.7^x = 22$ — $x \approx -8.67$
23. $1.5^x = 70$ — $x \approx 10.48$
24. $4^{\frac{2}{3}x} = 0.5$ — $x \approx -0.75$
25. $15^{-x} = 24$ — $x \approx -1.17$
26. $0.25^{2x} = 41$ — $x \approx -1.34$
27. $44^x = 19$ — $x \approx 0.78$

28. $1000 is deposited in an account with an interest rate of 6.5%. Interest is compounded continuously, and no deposits or withdrawals are made. Find the amount in the account at the end of three years. — $1215.31

CONNECTIONS

TRANSFORMATIONS Let $f(x) = e^x$. For each function, describe the transformations from f to g.

61. $g(x) = 6e^x + 1$
62. $g(x) = 0.75e^x - 4$
63. $g(x) = 0.25e^{(4x+4)}$
64. $g(x) = 3e^{(2x-4)}$

TRANSFORMATIONS Let $f(x) = \ln x$. For each function, describe the transformations from f to g.

65. $g(x) = 3\ln(x + 1)$
66. $g(x) = -2\ln(x - 1)$
67. $g(x) = 0.5\ln(5x) - 2$
68. $g(x) = 5\ln(0.25x) - 1$

$f(x) = e^{-2x}$, $g(x) = e^{-x}$, $i(x) = e^{2x}$, $h(x) = e^x$

69. **TRANSFORMATIONS** The graphs of $f(x) = e^{-2x}$, $g(x) = e^{-x}$, $h(x) = e^x$, and $i(x) = e^{2x}$ are shown on the same coordinate plane at left. What transformations relate each function, f, g, and i, to h?

70. **TRANSFORMATIONS** For $f(x) = e^x$, describe how each transformation affects the domain, range, asymptotes, and y-intercept.
 a. a vertical stretch
 b. a horizontal stretch
 c. a vertical translation
 d. a horizontal translation

71. **TRANSFORMATIONS** For $f(x) = \ln x$, describe how each transformation affects the domain, range, asymptotes, and x-intercept.
 a. a vertical stretch
 b. a horizontal stretch
 c. a vertical translation
 d. a horizontal translation

APPLICATIONS

72. 22.06 grams

72. **PHYSICS** The amount of radioactive strontium-90 remaining after t years decreases according to the function $N(t) = N_0e^{-0.0238t}$. How much of a 40-gram sample will remain after 25 years?

73a. $285,000,000
 b. during 1999

73. **ECONOMICS** The factory sales of pagers from 1990 through 1995 can be modeled by the function $S = 116e^{0.18t}$, where $t = 0$ in 1990 and S represents the sales in millions of dollars. [*Source: Electronic Market Data Book*]
 a. According to this function, find the factory sales of pagers in 1995 to the nearest million.
 b. If the sales of pagers continued to increase at the same rate, when would the sales be double the 1995 amount?

74. 3%: $3644.24
 5%: $5436.56

74. **INVESTMENTS** Compare the growth of an investment of $2000 in two different accounts. One account earns 3% annual interest, the other earns 5% annual interest, and both are compounded continuously over 20 years.

75. **ARCHAEOLOGY** A wooden chest is found and is said to be from the second century B.C.E. Tests on a sample of wood from the chest reveal that it contains 92% of its original carbon-14. Could the chest be from the second century B.C.E.? **No; according to the tests, the chest is only about 700 years old.**

69. f to h: reflection across the y-axis and horizontal compression by a factor of $\frac{1}{2}$
g to h: reflection across the y-axis
i to h: horizontal compression by a factor of $\frac{1}{2}$

70. The only changes are the following:
 a. y-intercept: the value of the factor of the vertical stretch
 b. no changes
 c. y-intercept: changes
 d. range: real numbers greater than the value of the translation; asymptotes: y is the value of the translation; y-intercept: value of the translation plus 1

The answer to Exercise 71 can be found in Additional Answers beginning on page 1002.

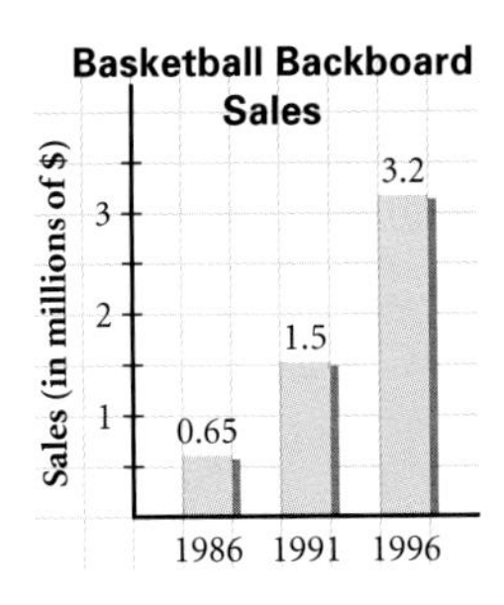

76. a. \$3,655,000
b. during 2000

77. about 11 years and 7 months

78. about 6 years and 11 months

76. BUSINESS Sales of home basketball backboards from 1986 to 1996 can be modeled by $S = 0.65e^{0.157t}$, where S is the sales in millions of dollars, t is time in years, and $t = 0$ in 1986. [*Source: Huffy Sports*]

a. Use this model to estimate the sales of backboards in 1997 to the nearest thousand.

b. If the sales of basketball backboards continued to increase at the same rate, when would the sales of basketball backboards be double the amount of 1996?

INVESTMENTS For Exercises 77–79, assume that all interest rates are compounded continuously.

77. How long will it take an investment of \$5000 to double if the annual interest rate is 6%?

78. How long will it take an investment to double at 10% annual interest?

79. If it takes a certain amount of money 3.7 years to double, at what annual interest rate was the money invested? 18.7%

80. AGRICULTURE The percentage of farmers in the United States workforce has declined since the turn of the century. The percent of farmers in the workforce, f, can be modeled by the function $f(t) = 29e^{-0.036t}$, where t is time in years and $t = 0$ in 1920. Find the percent of farmers in the workforce in 1995. [*Source: Bureau of Labor Statistics*] 1.95%

Look Back

80. $x \geq 5$ or $x \leq -5$

81. $x \leq -\frac{8}{3}$ or $x \geq \frac{16}{3}$

82. $x \geq 6$ or $x \leq -\frac{22}{3}$

Solve each inequality, and graph the solution on a number line. *(LESSON 1.8)*

81. $|-3x| \geq 15$ **82.** $|3x - 4| \geq 12$ **83.** $\left|\frac{3x+2}{-4}\right| \geq 5$

Graph each function. *(LESSON 2.7)*

84. $f(x) = \frac{1}{2}|x|$ **85.** $g(x) = -\lceil x \rceil$ **86.** $h(x) = [x - 3]$

Graph each system. *(LESSON 3.4)*

87. $\begin{cases} y > 2x - 1 \\ 4 - 3x \geq y \end{cases}$ **88.** $\begin{cases} 1 - 3x > y + 4 \\ 2x + 3y \leq 8 \end{cases}$ **89.** $\begin{cases} y + 2x \geq 0 \\ 4 - 2x > y \\ x \leq 3 \end{cases}$

90. $(x - 5)(x + 2)$

91. $3(x - 1)^2$

92. $(x - 7)(x + 7)$

Factor each expression. *(LESSON 5.3)*

90. $x^2 - 3x - 10$ **91.** $3x^2 - 6x + 3$ **92.** $x^2 - 49$

Look Beyond

93 Solve $\ln x + \ln(x + 2) = 5$ by graphing $y = \ln x + \ln(x + 2)$ and $y = 5$ and finding the x-coordinate of the point of intersection. $x \approx 11.22$

Look Beyond

In Exercise 89, students solve a logarithmic equation by finding the point of intersection of the graphs of two expressions, which were created by separating the two sides of the equation. In Lesson 6.7, students will learn more about solving logarithmic equations.

81. $x \geq 5$ or $x \leq -5$

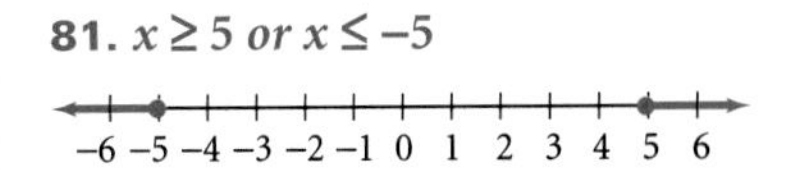

82. $x \leq -\frac{8}{3}$ or $x \geq \frac{16}{3}$

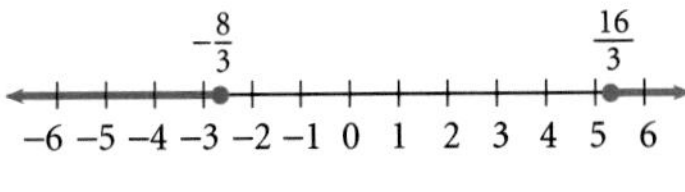

83. $x \geq 6$ or $x \leq -\frac{22}{3}$

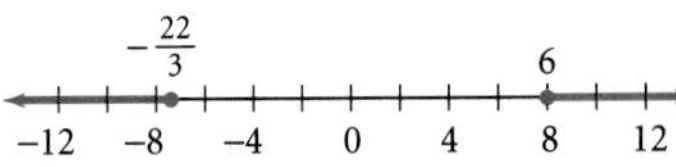

84.

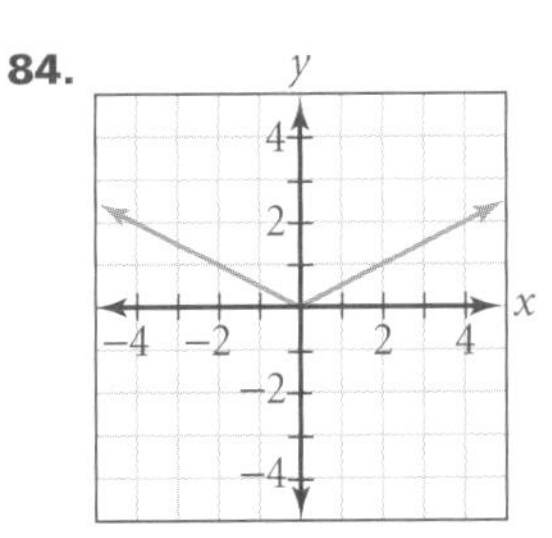

85.

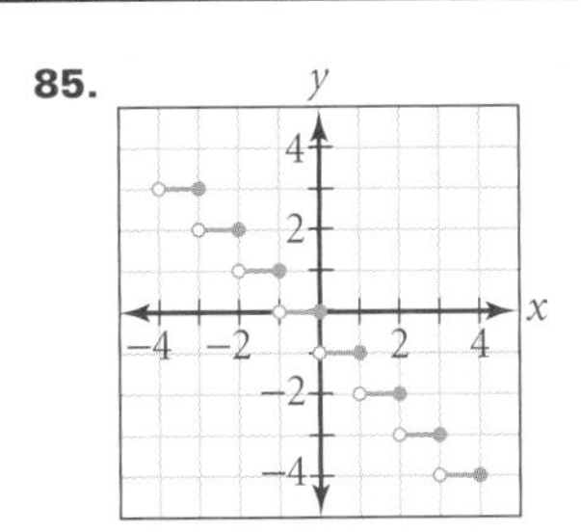

86.

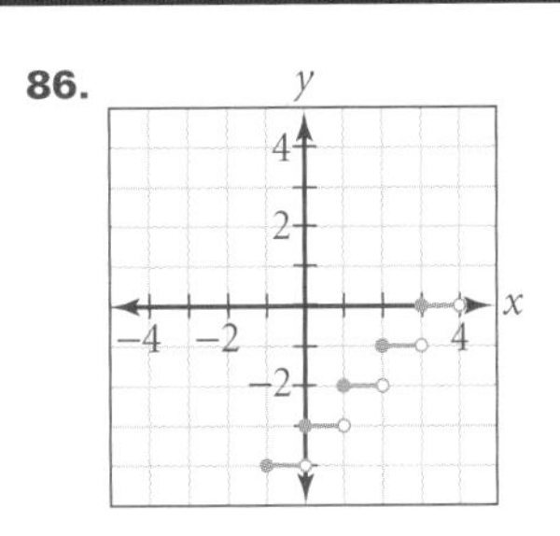

Student Technology Guide

NAME ______ CLASS ______ DATE ______

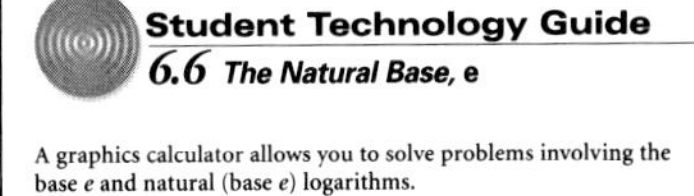

Student Technology Guide
6.6 The Natural Base, e

A graphics calculator allows you to solve problems involving the base e and natural (base e) logarithms.

Example: Evaluate each expression. Round answers to the nearest hundredth.

a. ln 15
Press LN 15) ENTER. Thus, ln 15 ≈ 2.71.

b. e^7
Press 2nd LN 7) ENTER or 2nd ÷ ^ 7 ENTER. Thus, e^7 ≈ 1096.63.

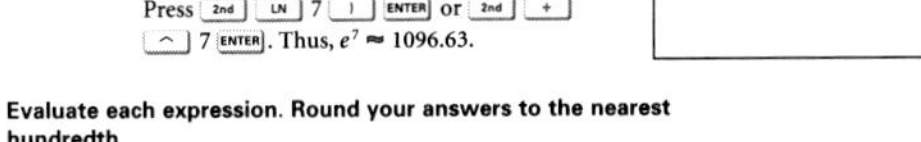

```
ln(15)
        2.708050201
e^(7)
        1096.633158
```

Evaluate each expression. Round your answers to the nearest hundredth.

1. ln 224 5.41 2. e^3 20.09 3. ln e 1 4. $e^{-1.7}$ 0.18

If you invest P dollars in an account that pays interest compounded continuously, then $A = Pe^{rt}$ gives the amount A in the account after t years. In the formula, r is the annual rate of interest expressed as a decimal.

Example: How long does it take for an investment to double if it has an annual interest rate of 6.4% compounded continuously?

- The investment has doubled when $A = 2P$. The formula above becomes $2P = Pe^{0.064t}$, or $2 = e^{0.064t}$.
- To enter $y = e^{0.064x}$ as Y1, press Y= 2nd LN .064 X,T,θ,n). Enter $y = 2$ as Y2. Press GRAPH.
- Press TRACE and trace to the intersection. For a more precise answer, press 2nd TRACE 5:intersect ENTER ENTER ENTER ENTER.

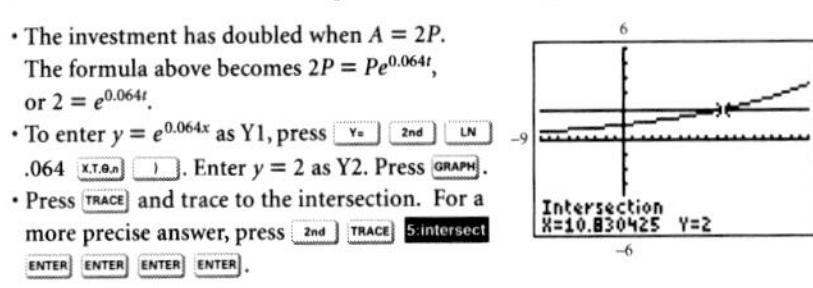

The investment will double in about 10.8 years.

If the given interest rate is compounded continuously, how long will it take an investment to double? Round your answers to the nearest tenth.

5. 2% interest 34.7 years 6. 4.4% interest 15.8 years 7. 11.5% interest 6.0 years 8. 14.2% interest 4.9 years

The answers to Exercises 87–89 can be found in Additional Answers beginning on page 1002.

Focus

An inverted *catenary curve* is a mathematical abstraction based on the sum of functions that have the form $y = e^{ax}$. This stable and sweeping curve was used in the design of the Gateway Arch in St. Louis. Students will have the opportunity to explore the function, graph the curve, examine its unusual properties, and experiment with some transformations. Because the curve is similar to a parabola, students will also compare a parabola with a catenary curve.

Motivate

Before studying the catenary curve, have students experiment with the functions $f(x) = e^x$ and $g(x) = e^x + e^{-x}$ by using a graphics calculator or graphics software. Have the students do research to learn about the Gateway Arch and its construction.

EYEWITNESS MATH

Meet "e" in St. Louis

How does a 630-foot-high arch stand up to the forces of nature? Will it last for its projected life of 1000 years? How does it withstand winds up to 150 miles per hour?

The secret is in the shape of the arch, which transfers forces downward through its legs into huge underground foundations. You can learn more about this remarkable shape, called a **catenary curve,** by looking at some of its simpler forms.

The general equation for a catenary curve is $y = \frac{a}{2}\left(e^{\frac{x}{a}} + e^{-\frac{x}{a}}\right)$, where a is a real nonzero constant.

ARCH COMPLETED

St. Louis Can Now Boast of the Nation's Highest Memorial

With the joining of the stainless steel legs of the Gateway Arch Thursday, St. Louis became the location of the tallest—630 feet—national memorial in the United States. More than 10,000 persons—including hundreds of school children who attended the "topping out" as a field trip—look upward as the leg-locking segment is hoisted off the ground and begins the long trip to the Arch.

[*Source:* St. Louis Globe-Democrat, *October 29, 1965*]

1a. $y = e^{\frac{x}{2}} + e^{\frac{-x}{2}}$

b.

x	0	1	2	3	4	5
y	2.00	2.26	3.09	4.70	7.52	12.26

c. $y = 2.26$ for $x = -1$; this is the same y-value as that for $x = 1$.

d. If you substitute $-x$ for x in the equation, you get $y = e^{\frac{-x}{2}} + e^{\frac{x}{2}}$, which is equivalent to the original equation.

2a. The curve looks like a parabola-type curve that opens up and has a vertex at (0, 2).

b. Multiply the function by −1; $y = -\left(e^{\frac{-x}{2}} + e^{\frac{x}{2}}\right)$.

The shape of the St. Louis arch is approximately that of an inverted, weighted catenary curve—inverted because it is upside down and weighted because the lower sections of the legs are wider than the upper sections.

Cooperative Learning

1. Begin exploring the catenary by letting $a = 2$.
   a. Write the equation for a catenary curve with $a = 2$.
   b. Copy and complete the table below.

x	0	1	2	3	4	5
y						

   c. What is y when $x = -1$? Compare this value for y with the value of y when $x = 1$.
   d. Why will the y-values for $-x$ and x always be equal for this equation?

2. Now explore the graph of the equation for the curve with $a = 2$.
   a. Graph the equation for a catenary curve with $a = 2$. Describe the graph.
   b. Your graph should look like an upside-down arch. What can you do to the equation to invert the graph? (Hint: Think about how to invert, or reflect, a parabola.) Check your new equation by graphing it.

3. Your catenary curve may look similar to a parabola. To see how it is different, follow the steps below.
   a. Write an equation and draw the graph for a parabola that resembles your graph from part **a** of Step 2. (Hint: How do the values of a, h, and k in a quadratic equation of the form $y = a(x - h)^2 + k$ affect the location and shape of the parabola?)
   b. Compare the graph of your parabola with that of the catenary curve.

3a. Answers may vary. Since the vertex is (0, 2) and the curve is very wide, a possible parabola is $y = 0.5x^2 + 2$.

b. Answers may vary. Using the answer given in C1, the catenary is wider than the parabola near the vertex, but as x gets larger, the catenary becomes narrower than the parabola.

Cooperative Learning

Have groups of students use graphics calculators to explore the catenary curve as described on this page. Have each student choose a different value of a. One student should be the recorder, keeping a table of the results for each value of a. When writing the equation of the parabola that is similar to the catenary curve, have students verify the equation with the graphics calculator.

Discuss

Discuss the results from the Cooperative Learning task above. Have students make conjectures and generalizations based on their observations. Then discuss how the function for the basic catenary curve must be changed to translate the graph of the function horizontally.

Additional discussion can focus on the function $f(x) = e^x$. The history, application, and calculation of this function can stimulate interest and motivate research projects. One approximation of the value of e is given by the following series:

$$e = 1 + \frac{1}{1!} + \frac{1}{2!} + \frac{1}{3!} + \cdots + \frac{1}{n!} \approx 2.718$$

Prepare

NCTM PRINCIPLES & STANDARDS 1–4, 6, 8–10

QUICK WARM-UP

Solve each equation for x. Round your answer to the nearest hundredth, if necessary.

1. $\log x = 3$ — $x = 1000$
2. $\log x = 0.477$ — $x \approx 3.00$
3. $\ln x = 0$ — $x = 1$
4. $\ln x = 1.61$ — $x \approx 5.00$
5. $10^x = 0.1$ — $x = -1$
6. $10^x = 8$ — $x \approx 0.90$
7. $e^x = 1$ — $x = 0$
8. $e^x = 8$ — $x \approx 2.08$

Also on Quiz Transparency 6.7

Teach

Why Discuss with students the value of using logarithms to scale down very large numbers. Give students examples involving the measurements of the energy released by an earthquake. For example, a recent earthquake in Mexico City released 8.91×10^{23} ergs of energy. Have students discuss why logarithms are helpful in graphing earthquake magnitudes.

Solving Equations and Modeling

Objectives

- Solve logarithmic and exponential equations by using algebra and graphs.
- Model and solve real-world problems involving exponential and logarithmic relationships.

Why *Physicists, chemists, and geologists use exponential and logarithmic equations to model various phenomena, such as the magnitude of earthquakes.*

RICHTER SCALE RATINGS		
Magnitude	Result near the epicenter	Approximate number of occurrences per year
8–9	near total damage	0.2
7.0–7.9	serious damage to buildings	14
6.0–6.9	moderate damage to buildings	185
5.0–5.9	slight damage to buildings	1000
4.0–4.9	felt by most people	2800
3.0–3.9	felt by some people	26,000
2.0–2.9	not felt but recorded	800,000

APPLICATION GEOLOGY

On the Richter scale, the magnitude, M, of an earthquake depends on the amount of energy, E, released by the earthquake as follows:

$$M = \frac{2}{3} \log \frac{E}{10^{11.8}}$$

The amount of energy, measured in ergs, is based on the amount of ground motion recorded by a seismograph at a known distance from the epicenter of the quake.

The logarithmic function for the Richter scale assigns very large numbers for the amount of energy, E, to numbers that range from 1 to 9. A rating of 2 on the Richter scale indicates the smallest tremor that can be detected. Destructive earthquakes are those rated greater than 6 on the Richter scale.

Alternative Teaching Strategy

INVITING PARTICIPATION Divide students into eight groups and assign each group one of the properties in the summary of exponential and logarithmic properties on page 403. Each group should write three equations that can be solved by using their assigned property. Have students put their equations on the board, on the overhead, or on a poster and then walk around room to examine each set of equations. Once everyone has viewed the equations, ask each group questions about the properties. Next place an equation such as $\log_3 \frac{3}{5} + \log_3(x + 4) = 2$ on the board and ask each group to decide whether their assigned property should be used in solving for the variable.

$\log_3\left[\frac{3}{5}(x + 4)\right] = 2$ — **Product prop.**

$9 = \frac{3}{5}x + \frac{12}{5}$ — **Exp.-log prop.**

$45 = 3x + 12$ — **Simplify.**

$33 = 3x$

$x = 11$

EXAMPLE 1

APPLICATION
GEOLOGY

One of the strongest earthquakes in recent history occurred in Mexico City in 1985 and measured 8.1 on the Richter scale.

Find the amount of energy, E, released by this earthquake.

A seismogram produced by a seismograph

SOLUTION

PROBLEM SOLVING **Use a formula.**

$$M = \frac{2}{3}\log\frac{E}{10^{11.8}}$$

$$8.1 = \frac{2}{3}\log\frac{E}{10^{11.8}} \quad \text{Substitute 8.1 for the magnitude, } M.$$

$$12.15 = \log\frac{E}{10^{11.8}}$$

$$10^{12.15} = \frac{E}{10^{11.8}} \quad \text{Use the definition of logarithm.}$$

$$10^{11.8} \cdot 10^{12.15} = E$$

$$8.91 \times 10^{23} \approx E \quad \text{Write the answer in scientific notation.}$$

The amount of energy, E, released by this earthquake was approximately 8.91×10^{23} ergs. *In physics, an erg is a unit of work or energy.*

To solve the logarithmic equation in Example 1, you must use the definition of a logarithm. However, solving exponential and logarithmic equations often requires a variety of the definitions and properties from this chapter. A summary of the definitions and properties that you have learned is given below.

Copy these properties and definitions into your notebook for reference.

SUMMARY
Exponential and Logarithmic Definitions and Properties

Definition/Property	Statement
Definition of logarithm	$y = \log_b x$ if and only if $b^y = x$
Product Property	$\log_b mn = \log_b m + \log_b n$
Quotient Property	$\log_b\left(\frac{m}{n}\right) = \log_b m - \log_b n$
Power Property	$\log_b m^p = p \log_b m$
Exponential-Logarithmic Inverse Properties	$b^{\log_b x} = x$ for $x > 0$; $\log_b b^x = x$ for all x
One-to-One Property of Exponents	If $b^x = b^y$, then $x = y$.
One-to-One Property of Logarithms	If $\log_b x = \log_b y$, then $x = y$.
Change-of-base formula	$\log_c a = \frac{\log_b a}{\log_b c}$

CHECKPOINT ✔ Show how to solve $M = \frac{2}{3}\log\frac{E}{10^{11.8}}$ for E.

CRITICAL THINKING Use the properties of exponents and logarithms to show that $\log_a\left(\frac{1}{x}\right) = \log_{\frac{1}{a}} x$.

ADDITIONAL EXAMPLE 1

Suppose that the magnitude of an earthquake measures 7.5 on the Richter scale. **Find the amount of energy, E, released by this earthquake.**

$\approx 1.12 \times 10^{23}$ ergs

CHECKPOINT ✔

Answers may vary.

$$\frac{3}{2}M = \log\left(\frac{E}{10^{11.8}}\right)$$

$$\frac{3}{2}M = \log E - \log 10^{11.8}$$

$$\frac{3}{2}M - 11.8 = \log E$$

$$10^{\left(\frac{3}{2}M - 11.8\right)} = E$$

CRITICAL THINKING

$$\log_{\frac{1}{a}} x = \frac{\log_a x}{\log_a \frac{1}{a}} = \frac{\log_a x}{\log_a a^{-1}} = \frac{\log_a x}{-1} = -\log_a x = \log_a x^{-1} = \log_a \frac{1}{x}$$

Interdisciplinary Connection

PHYSICS Electricity is drained from a capacitor according to the formula $V(t) = V_0 e^{-0.025t}$, where V is measured in volts, t is measured in seconds, and V_0 is the initial voltage capacity in volts. Use this formula to find how long it will take for the voltage to drop to 10% of its original value.

≈ 92.1 s

Enrichment

Have students find the ratio of energy released by earthquakes whose magnitudes measure 1 and 2 on the Richter scale. Repeat this process for magnitudes that measure 2 and 3, 3 and 4, and 7 and 8. Have students discuss the ratios that they found.

ADDITIONAL
EXAMPLE 2

Solve the logarithmic equation $\log x + \log(x + 3) = 1$ for x by using the following:

a. graphing

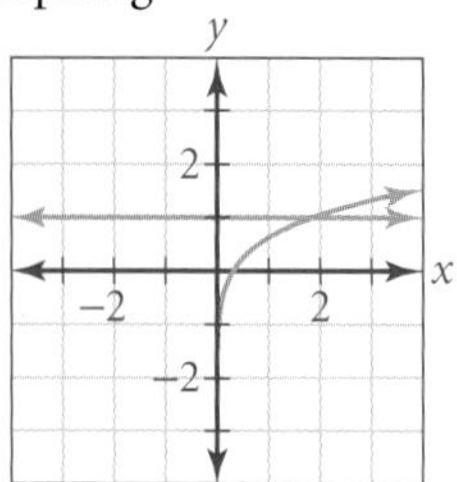

b. algebra $x = 2$

Teaching Tip

TECHNOLOGY Students should choose an appropriate window, such as the ones below, to see the solutions in Examples 2 and 3 when using a graphics calculator.

Example 2:

Xmin=–2	**Ymin=–1**
Xmax=8	**Ymax=5**
Xscl=1	**Yscl=1**

Example 3:

Xmin=–1	**Ymin=–30**
Xmax=4	**Ymax=100**
Xscl=1	**Yscl=10**

TRY THIS

$x = 2$; $x = -50$ is extraneous

ADDITIONAL
EXAMPLE 3

Solve $8e^{2x-5} = 56$ for x by using the following:

a. graphing

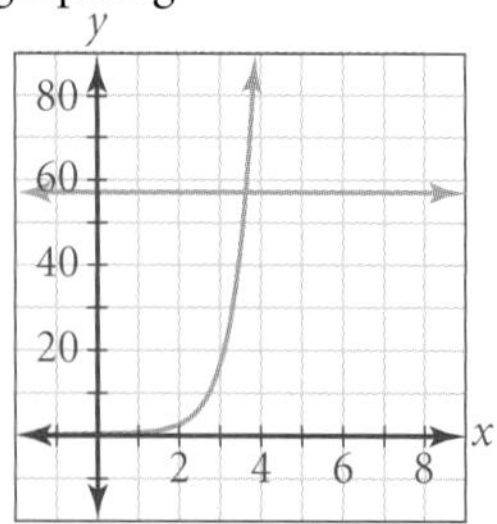

b. algebra $x \approx 3.47$

EXAMPLE 2 Solve $\log x + \log(x - 3) = 1$ for x.

SOLUTION

Method 1 Use algebra.

$$\log x + \log(x - 3) = 1$$
$$\log[x(x - 3)] = 1 \quad \textit{Apply the Product Property of Logarithms.}$$
$$x(x - 3) = 10^1 \quad \textit{Write the equivalent exponential equation.}$$
$$x^2 - 3x - 10 = 0$$
$$(x - 5)(x + 2) = 0$$
$$x = 5 \quad or \quad x = -2$$

CHECK

Let $x = 5$.	Let $x = -2$.
$\log x + \log(x - 3) = 1$	$\log x + \log(x - 3) = 1$
$\log 5 + \log 2 \stackrel{?}{=} 1$	$\log(-2) + \log(-5) = 1$ **Undefined**
$1 = 1$ **True**	

Since the domain of a logarithmic function excludes negative numbers, the only solution is 5.

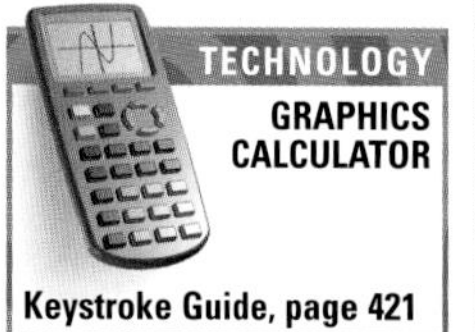

Method 2 Use a graph.
Graph $y = \log x + \log(x - 3)$ and $y = 1$, and find the point of intersection.

The coordinates of the point of intersection are (5, 1), so the solution is 5.

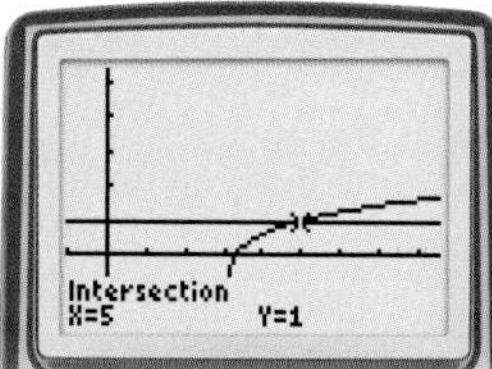

TRY THIS Solve $\log(x + 48) + \log x = 2$ by using algebra and a graph.

EXAMPLE 3 Solve $4e^{3x-5} = 72$ for x.

SOLUTION

Method 1 Use algebra.

$$4e^{3x-5} = 72$$
$$e^{3x-5} = 18$$
$$\ln e^{3x-5} = \ln 18 \quad \textit{Take the natural logarithm of each side.}$$
$$3x - 5 = \ln 18 \quad \textit{Use Exponential-Logarithmic Inverse Properties.}$$
$$x = \frac{\ln 18 + 5}{3} \quad \textit{Exact solution}$$
$$x \approx 2.63 \quad \textit{Approximate solution}$$

Method 2 Use a graph.
Graph $y = 4e^{3x-5}$ and $y = 72$, and find the point of intersection.

The coordinates of the point of intersection are approximately (2.63, 72), so the solution is approximately 2.63.

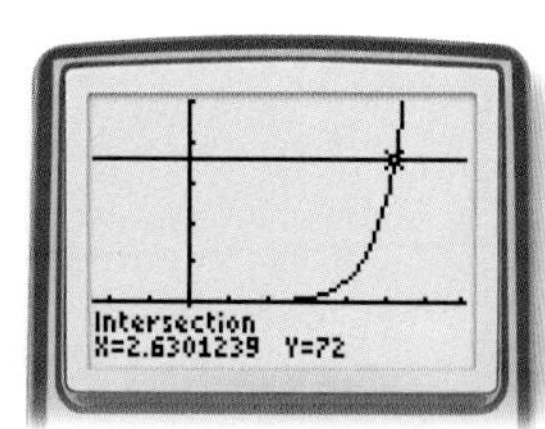

Inclusion Strategies

ENGLISH LANGUAGE DEVELOPMENT It may help students to think about the goal of solving an equation. In general, the goal is to find all values of x that make the equation true. The equation represents operations performed on x, and every step in the solution process must undo those operations. Just as subtraction undoes addition, exponents undo logarithms, and logarithms undo exponents. This way of viewing operations may help students decide which properties to use at various stages in the solution.

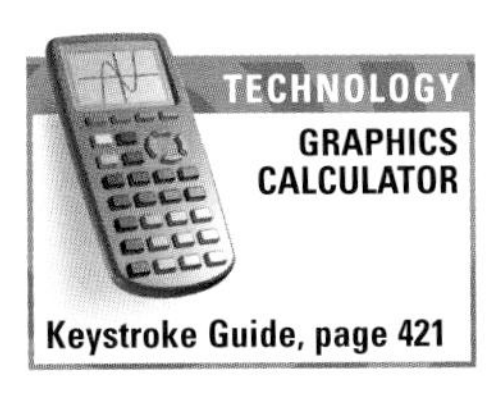

Activity: Solving Exponential Inequalities

You will need: a graphics calculator

1. Graph $y_1 = \log x + \log(x + 21)$ and $y_2 = 2$ on the same screen.
2. For what value(s) of x is $y_1 = y_2$? $y_1 < y_2$? $y_1 > y_2$?

CHECKPOINT ✔ 3. Explain how you can use a graph to solve $\log x + \log(x + 21) > 2$.

4. Graph $y_1 = 2e^{4x-1}$ and $y_2 = 38$ on the same screen.
5. For what approximate value(s) of x is $y_1 = y_2$? $y_1 < y_2$? $y_1 > y_2$?

CHECKPOINT ✔ 6. Explain how you can use a graph to solve $2e^{4x-1} < 38$.

Newton's Law of Cooling

An object that is hotter than its surroundings will cool off, and an object that is cooler than its surroundings will warm up. **Newton's law of cooling** states that the temperature difference between an object and its surroundings decreases exponentially as a function of time according to the following:

$$T(t) = T_s + (T_0 - T_s)e^{-kt}$$

T_0 is the initial temperature of the object, T_s is the temperature of the object's surroundings (assumed to be constant), t is the time, and $-k$ represents the constant rate of decrease in the temperature difference $(T_0 - T_s)$.

EXAMPLE 4

APPLICATION
PHYSICS

When a container of milk is taken out of the refrigerator, its temperature is 40°F. An hour later, its temperature is 50°F. Assume that the temperature of the air is a constant 70°F.

a. Write the function for the temperature of this container of milk as a function of time, t.

b. What is the temperature of the milk after 2 hours?

c. After how many hours is the temperature of the milk 65°F?

SOLUTION

a. First substitute 40 for T_0 and 70 for T_s, and simplify.

$$T(t) = T_s + (T_0 - T_s)e^{-kt}$$
$$T(t) = 70 + (40 - 70)e^{-kt}$$
$$T(t) = 70 + (-30)e^{-kt}$$

Since $T(1) = 50$, substitute 1 for t and 50 for $T(t)$, and solve for $-k$.

$$50 = 70 - 30e^{-k}$$
$$30e^{-k} = 20$$
$$e^{-k} = \frac{2}{3}$$
$$\ln e^{-k} = \ln \frac{2}{3}$$
$$-k = \ln \frac{2}{3}$$

Reteaching the Lesson

USING ALGORITHMS Ask students to generate a list of all the properties of logarithms they have studied in the chapter. Write this list on the board. Ask students to apply these properties to solve for x in the equation $3^x = 4$. To help students get started, remind them that they can apply an algebraic operation to both sides of an equation. Also point out that in this equation, having x as an exponent is troublesome. Students should see that the power property will help to remove the exponent. Solve the equation with the class by taking a logarithm of both sides. Next, have students apply the properties to solve the equation $\log_3 x + \log_3(x - 2) = 2$. Remind students that they cannot apply the definition of a logarithm to a sum of logarithms. They must first write the left side of the equation as a single logarithm by using the product property. Solve the equation as a class.

Activity Notes

In this Activity, students solve logarithmic and exponential inequalities by using a graphics calculator.

CHECKPOINT ✔

3. Find the point of intersection of the following:
Y1=log(X) + log(X+21)
Y2=2
The solution is all values of x such that Y1 is above Y2.

CHECKPOINT ✔

6. Find the point of intersection of the following:
Y1=2e^(4X–1)
Y2=38
The solution is all values of x such that Y1 is below Y2.

Teaching Tip

TECHNOLOGY Students should choose an appropriate window, such as the one below, to see the solutions.

Xmin=–1	Ymin=–1
Xmax=6	Ymax=3

ADDITIONAL EXAMPLE 4

When a container of water is taken out of the freezer, its temperature is 32°F. An hour later, its temperature is 44°F. Assume that the temperature of the air is a constant 72°F.

a. Write the function for the temperature of this container of water as a function of time, t.
$T(t) = 72 - 40e^{-0.3567t}$

b. What is the temperature of the water after 2 hours?
≈52.4°F

c. After how many hours is the temperature of the water 60°F?
≈3.38 hr, or 3 hr and 23 min

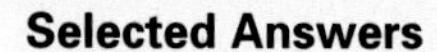

Assess

Selected Answers

Exercises 4–7, 9–47 odd

ASSIGNMENT GUIDE	
In Class	1–7
Core	9–25 odd, 29
Core Plus	8–34 even
Review	36–47
Preview	48

Extra Practice can be found beginning on page 940.

Substitute $\ln \frac{2}{3}$ for $-k$ and simplify to get the function for the temperature of this container of milk.

$$T(t) = 70 - 30e^{-kt}$$
$$T(t) = 70 - 30e^{\left(\ln \frac{2}{3}\right)t}$$
$$T(t) = 70 - 30\left(\frac{2}{3}\right)^t \quad \textit{Apply the Exponential-Logarithmic Inverse Property.}$$

The function for the temperature of this container of milk is $T(t) = 70 - 30\left(\frac{2}{3}\right)^t$.

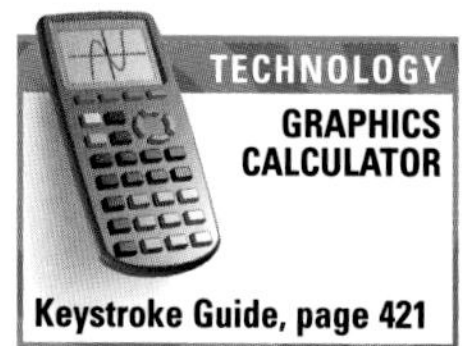

Keystroke Guide, page 421

b. Find $T(2)$.

$$T(t) = 70 - 30\left(\frac{2}{3}\right)^t$$
$$T(2) = 70 - 30\left(\frac{2}{3}\right)^2 \approx 56.7$$

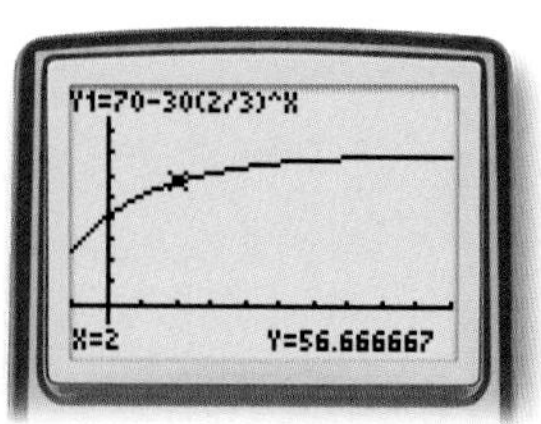

The temperature of the milk after 2 hours is approximately 56.7°F.

c. Substitute 65 for $T(t)$, and solve for t.

$$T(t) = 70 - 30\left(\frac{2}{3}\right)^t$$
$$65 = 70 - 30\left(\frac{2}{3}\right)^t$$
$$30\left(\frac{2}{3}\right)^t = 5$$
$$\left(\frac{2}{3}\right)^t = \frac{1}{6}$$
$$t \ln \frac{2}{3} = \ln \frac{1}{6}$$
$$t = \frac{\ln \frac{1}{6}}{\ln \frac{2}{3}} \approx 4.42$$

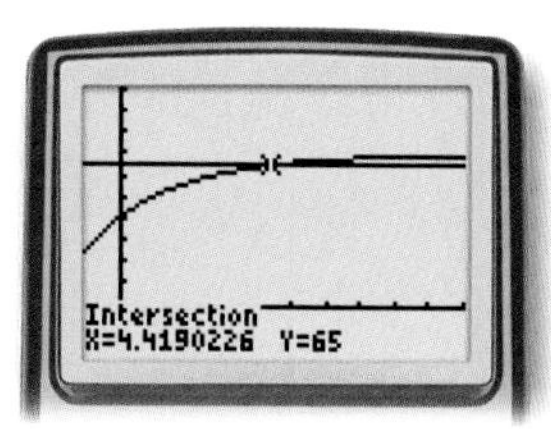

It will take approximately 4.42 hours, or about 4 hours and 25 minutes, for the milk to warm up to 65°F.

Exercises

Communicate

1. Explain how to solve the exponential equation $e^{x+7} = 98$ by algebraic methods.

2. How can you solve the logarithmic equation $\log_2 x + \log_2 (x + 3) = 2$ by algebraic methods?

3. Explain how to solve exponential and logarithmic equations by graphing.

Guided Skills Practice

APPLICATIONS

4. 2.82×10^{22} ergs
5. $x = 100$
6. $x \approx 54.78$
7a. $T(t) = 70 + 100e^{-0.7133t}$, or $T(t) = 70 + 100(0.7)^{2t}$
b. 119°F
c. about 2.26 hours, or about 2 hours and 16 minutes

4. GEOLOGY In 1989, an earthquake that measured 7.1 on the Richter scale occurred in San Francisco, California. Find the amount of energy, E, released by this earthquake. *(EXAMPLE 1)*

5. Solve $\log(x - 90) + \log x = 3$ for x. *(EXAMPLE 2)*

6. Solve $0.5e^{0.08t} = 40$ for x. *(EXAMPLE 3)*

7. PHYSICS When the air temperature is a constant 70°F, an object cools from 170°F to 140°F in one-half hour. *(EXAMPLE 4)*

a. Write the function for the temperature of this object, T, as a function of time, t.

b. What is the temperature of this object after 1 hour?

c. After how many hours is the temperature of this object 90°F?

Practice and Apply

internet connect
Homework Help Online
Go To: go.hrw.com
Keyword: MB1 Homework Help for Exercises 8–25

Solve each equation for x. Write the exact solution and the approximate solution to the nearest hundredth, when appropriate.

8. $3^x = 3^4$ 4	**9.** $3^{2x} = 81$ 2	**10.** $5^{x-2} = 25$ 4
11. $x = \log_3 \frac{1}{27}$ −3	**12.** $x = \log_4 \frac{1}{64}$ −3	**13.** $\log_x \frac{1}{16} = -2$ 4
14. $4 = \log_x \frac{1}{16}$ $\frac{1}{2}$	**15.** $e^{2x} = 20$ $\frac{\ln 20}{2} \approx 1.50$	**16.** $e^{-2(x+1)} = 2$
17. $\ln(2x - 3) = \ln 21$	**18.** $\ln(x + 3) = 2\ln 4$	**19.** $10^{2x} + 75 = 150$
20. $e^{-4x} - 22 = 56$	**21.** $3\ln x = \ln 4 + \ln 2$	**22.** $\ln x + \ln(x + 1) = \ln 2$
23. $2\ln x + 2 = 1$	**24.** $3\ln x + 3 = 1$	**25.** $3\log x + 7 = 5$

CHALLENGE

26. no real solution
27. 1; $10^{-\sqrt{3}} \approx 0.02$; $10^{\sqrt{3}} \approx 53.96$

Solve each equation for x. Write the exact solution and the approximate solution to the nearest hundredth, when appropriate.

26. $\ln(3\sqrt{x}) = \sqrt{\ln x}$

27. $\log x^3 = (\log x)^3$

APPLICATIONS

28. 7.08×10^{18} ergs
29. 5.4
30a. $T(t) = 160e^{-0.0327t}$
b. about 3.16°C
c. about 2.6 minutes, or 2 minutes and 36 seconds

28. GEOLOGY On May 10, 1997, a light earthquake with a magnitude of 4.7 struck the Calaveras Fault 10 miles east of San Jose, California. Find the amount of energy, E, released by this earthquake.

29. GEOLOGY In 1976, an earthquake that released about 8×10^{19} ergs of energy occurred in San Salvador, El Salvador. Find the magnitude, M, of this earthquake to the nearest tenth.

30. PHYSICS A hot coal (at a temperature of 160°C) is immersed in ice water (at a temperature of 0°C). After 30 seconds, the temperature of the coal is 60°C. Assume that the ice water is kept at a constant temperature of 0°C.

a. Write the function for the temperature, T, of this object as a function of time, t, in seconds.

b. What will be the temperature of this coal after 2 minutes (120 seconds)?

c. After how many minutes will the temperature of the coal be 1°C?

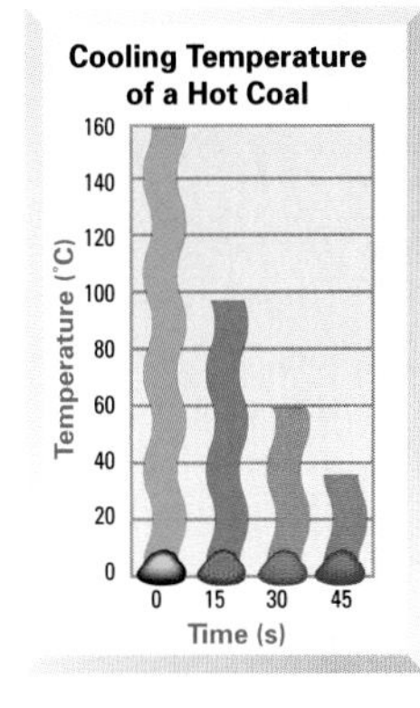

Error Analysis

When solving logarithmic and exponential equations, make sure that students use the properties of exponents and logarithms to get a single logarithmic or exponential expression on one side of the equation. For example, in Exercise 19, make sure that students subtract 75 from both sides before taking the logarithm of both sides.

16. $\frac{\ln 2 + 2}{-2} \approx -1.35$
17. 12
18. 13
19. $\frac{\log 75}{2} \approx 0.94$
20. $\frac{\ln 78}{-4} \approx -1.09$
21. 2
22. 1
24. $e^{-\frac{1}{2}} \approx 0.61$
23. $e^{-\frac{2}{3}} \approx 0.51$
25. $10^{-\frac{2}{3}} \approx 0.22$

APPLICATIONS

31. GEOLOGY Compare the amounts of energy released by earthquakes that differ by 1 in magnitude. In other words, how much more energy is released by an earthquake of magnitude 6.8 than an earthquake of magnitude 5.8?
about 31.6 times as much energy

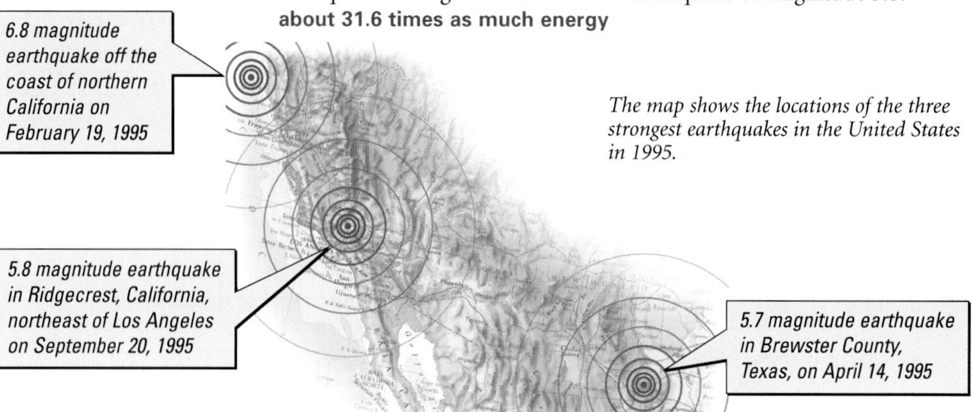

6.8 magnitude earthquake off the coast of northern California on February 19, 1995
5.8 magnitude earthquake in Ridgecrest, California, northeast of Los Angeles on September 20, 1995
5.7 magnitude earthquake in Brewster County, Texas, on April 14, 1995

The map shows the locations of the three strongest earthquakes in the United States in 1995.

32a. 82
b. 72
c. about 67 months

32. PSYCHOLOGY Educational psychologists sometimes use mathematical models of memory. Suppose that some students take a chemistry test. After a time, t (in months), without a review of the material, they take an equivalent form of the same test. The mathematical model $a(t) = 82 - 12\log(t + 1)$, where a is the average score at time t, is a function that describes the students' retention of the material.

a. What is the average score when the students first took the test ($t = 0$)?
b. What is the average score after 6 months?
c. After how many months is the average score 60?

33a. $P(t) = 10{,}000e^{0.46t}$
b. 2,500,000
c. 623,000,000

33. BIOLOGY A population of bacteria grows exponentially. A population that initially consists of 10,000 bacteria grows to 25,000 bacteria after 2 hours.

a. Use the exponential growth function, $P(t) = P_0e^{kt}$, to find the value of k. Then write a function for this population of bacteria in terms of time, t. Round the value of k to the nearest hundredth.
b. How many bacteria will the population consist of after 12 hours, rounded to the nearest hundred thousand?
c. How many bacteria will the population consist of after 24 hours?

34a. $P(t) = 574{,}220{,}000e^{0.026t}$
b. 1,252,600,000
c. during 2011

34. DEMOGRAPHICS The population of India was estimated to be 574,220,000 in 1974 and 746,388,000 in 1984. Assume that this population growth is exponential. Let $t = 0$ represent 1974 and $t = 10$ represent 1984.

a. Use the exponential growth function, $P(t) = P_0e^{kt}$, to find the value of k. Then write the function for this population as a function of time, t.
b. Estimate the population in 2004, rounded to the nearest hundred thousand.
c. Use the function you wrote in part **a** to estimate the year in which the population will reach 1.5 billion.

35a. $0.40N_0 = N_0e^{-0.00012t}$
b. ≈7636 years old

35. ARCHEOLOGY Refer to the discussion of radioactive decay on page 395. Suppose that an animal bone is unearthed and it is determined that the amount of carbon-14 it contains is 40% of the original amount.

a. Use the decay function for carbon-14, $N(t) = N_0e^{-0.00012t}$, to write an equation using the percentage of carbon-14 given above.
b. Use the equation you wrote in part **a** to find the approximate age, t, of the bone.

Practice

NAME ______ CLASS ______ DATE ______

Practice
6.7 Solving Equations and Modeling

Solve each equation for x. Write the exact solution and the approximate solution to the nearest hundredth, when appropriate.

1. $7^x = 7^4$ — $x = 4$
2. $5^{3x} = 25$ — $x = \frac{2}{3} \approx 0.67$
3. $\log_4 x = \frac{1}{2}$ — $x = 2$
4. $\log x = 3$ — $x = 1000$
5. $5 = \log_x \frac{1}{32}$ — $x = \frac{1}{2}$
6. $9^x = 6$ — $x = \frac{\ln 6}{\ln 9} \approx 0.82$
7. $e^{3x} = 15$ — $x = \frac{1}{3}\ln 15 \approx 0.90$
8. $\ln(2x - 7) = \ln 3$ — $x = 5$
9. $10^x + 4 = 32$ — $x = \log 28 \approx 1.45$
10. $\log_3(2x + 1) = 2$ — $x = 4$
11. $3\ln x = \ln 16 + \ln 4$ — $x = 4$
12. $\ln 2x - \ln(x - 2) = \ln 3$ — $x = 6$
13. $\log_x \frac{1}{9} = -2$ — $x = 3$
14. $\log_{\frac{1}{8}} \frac{1}{16} = x$ — $x = \frac{4}{3}$
15. $\ln 2x = 3$ — $x = \frac{e^3}{2} \approx 10.0$
16. $5\left(1 + e^{\frac{x}{3}}\right) = 8.2$ — $x = 3\ln 0.64 \approx -1.34$

In Exercises 17 and 18, use the equation $M = \frac{2}{3}\log\frac{E}{10^{11.8}}$.

17. On January 17, 1994, an earthquake with a magnitude of 6.6 injured more than 8000 people and caused an estimated $13–20 billion of damage to the San Fernando Valley in California. Find the amount of energy released by the earthquake. — 5.0×10^{21} ergs
18. On January 17, 1995, an earthquake struck Osaka, Kyoto, and Kobe, Japan, injuring more than 36,000 people and causing an estimated $100 billion of damage. The quake released about 3.98×10^{22} ergs of energy. Find the earthquake's magnitude on the Richter scale. Round your answer to the nearest tenth. — 7.2

Look Back

41. $x = 5$

42. $x = 9$

43. $x = \frac{1}{2}$ or $x = 2$

44. $x = 3$

Graph each system of linear inequalities. *(LESSON 4.8)*

36. $\begin{cases} 2x - 5y < 4 \\ -3x \geq 2y \end{cases}$ **37.** $\begin{cases} -x \leq -3 \\ y - 5 > -3 \end{cases}$ **38.** $\begin{cases} -y + 3 \leq 12 \\ -x < y + 8 \end{cases}$

Solve each equation for x. *(LESSON 5.2)*

39. $x^2 - 3 = 46$ $x = 7$ *or* $x = -7$

40. $7 - x^2 = 4$ $x = \sqrt{3}$ *or* $x = -\sqrt{3}$

Solve for x, and check your answers. *(LESSON 6.4)*

41. $\log_b(x^2 - 11) = \log_b(2x + 4)$

42. $\log_{10}(8x + 1) = \log_{10}(x^2 - 8)$

43. $\log_a(x^2 + 1) + 2 \log_a 4 = \log_a 40x$

44. $2 \log_b x - \log_b 3 = \log_b(2x - 3)$

APPLICATION

45. about 13.86 years

46. about 8.66 years

47. about 21.7%

INVESTMENTS Assume that all interest rates are compounded continuously in Exercises 45–47. *(LESSON 6.6)*

45. How long will it take an investment of $5000 to double if the annual interest rate is 5%?

46. How long will it take an investment to double at 8% annual interest?

47. If it takes a certain amount of money 3.2 years to double, at what annual interest rate was the money invested?

Portfolio Extension
Go To: go.hrw.com
Keyword: MB1 Newton

Look Beyond

48. Graph each function and compare the shapes of the graphs.
a. $y = x^2$ **b.** $y = x^3 - 2x$ **c.** $y = x^4 - 2x^2$

This activity requires the data collected for the Portfolio Activities on pages 361, 369, and 384.

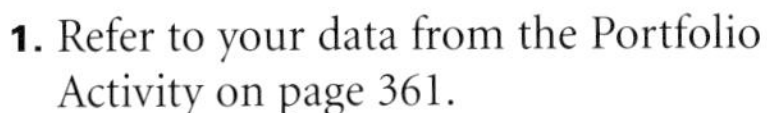

1. Refer to your data from the Portfolio Activity on page 361.
a. Use Newton's law of cooling, found in Example 4 on page 405, to write a function that models the temperature of the probe as it cooled to the temperature of ice water.
b. Compare the graph of this function with the graph of the exponential function that you generated for the same data in the Portfolio Activity on page 369. (Hint: You can use the table function on your graphics calculator to compare the y-values of these functions with the original values.)

2. a. Repeat part **a** of Step 1, using the data collected in the Portfolio Activity on page 384.
b. Repeat part **b** from Step 1, comparing your new graph with the graphs of both the exponential function and the approximating function from the Portfolio Activity on page 384.

WORKING ON THE PROJECT

You should now be able to complete the Chapter Project.

Look Beyond

In Exercise 48, students describe the graphs of various polynomial functions. In Lesson 7.2, students will study the graphs of polynomial functions in more detail.

ALTERNATIVE
Assessment

Portfolio Activity

The Portfolio Activity can be used as preparation for the Chapter Project or as a separate activity. In the Portfolio Activity on this page, students use data and Newton's law of cooling to create a mathematical model for the rate at which a probe cools from room temperature to the temperature of ice water.

48a. turns once, U-shaped
b. turns twice, S-shaped
c. turns 3 times, W-shaped

Student Technology Guide

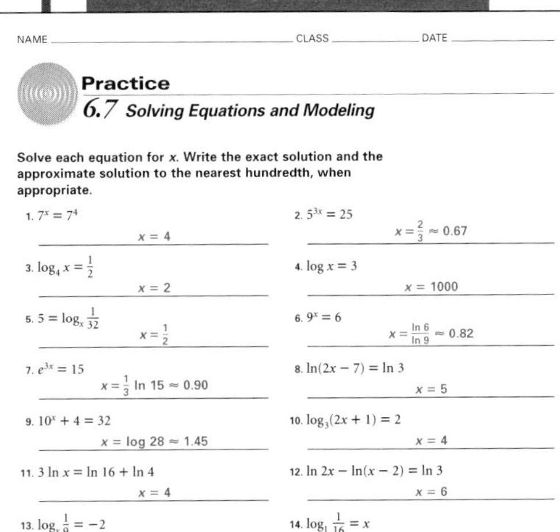

NAME ______ CLASS ______ DATE ______

Practice
6.7 Solving Equations and Modeling

Solve each equation for x. Write the exact solution and the approximate solution to the nearest hundredth, when appropriate.

1. $7^x = 7^4$ — $x = 4$
2. $5^{3x} = 25$ — $x = \frac{2}{3} \approx 0.67$
3. $\log_4 x = \frac{1}{2}$ — $x = 2$
4. $\log x = 3$ — $x = 1000$
5. $5 = \log_x \frac{1}{32}$ — $x = \frac{1}{2}$
6. $9^x = 6$ — $x = \frac{\ln 6}{\ln 9} \approx 0.82$
7. $e^{3x} = 15$ — $x = \frac{1}{3} \ln 15 \approx 0.90$
8. $\ln(2x - 7) = \ln 3$ — $x = 5$
9. $10^x + 4 = 32$ — $x = \log 28 \approx 1.45$
10. $\log_3(2x + 1) = 2$ — $x = 4$
11. $3 \ln x = \ln 16 + \ln 4$ — $x = 4$
12. $\ln 2x - \ln(x - 2) = \ln 3$ — $x = 6$
13. $\log_x \frac{1}{9} = -2$ — $x = 3$
14. $\log_{\frac{1}{8}} \frac{1}{16} = x$ — $x = \frac{4}{3}$
15. $\ln 2x = 3$ — $x = \frac{e^3}{2} \approx 10.0$
16. $5\left(1 + e^{\frac{x}{3}}\right) = 8.2$ — $x = 3 \ln 0.64 \approx -1.34$

In Exercises 17 and 18, use the equation $M = \frac{2}{3} \log \frac{E}{10^{11.8}}$.

17. On January 17, 1994, an earthquake with a magnitude of 6.6 injured more than 8000 people and caused an estimated $13–20 billion of damage to the San Fernando Valley in California. Find the amount of energy released by the earthquake. — 5.0×10^{21} ergs

18. On January 17, 1995, an earthquake struck Osaka, Kyoto, and Kobe, Japan, injuring more than 36,000 people and causing an estimated $100 billion of damage. The quake released about 3.98×10^{22} ergs of energy. Find the earthquake's magnitude on the Richter scale. Round your answer to the nearest tenth. — 7.2

36.

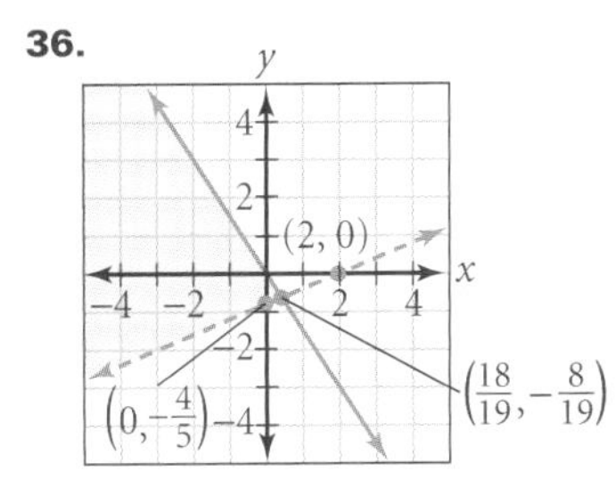

37.

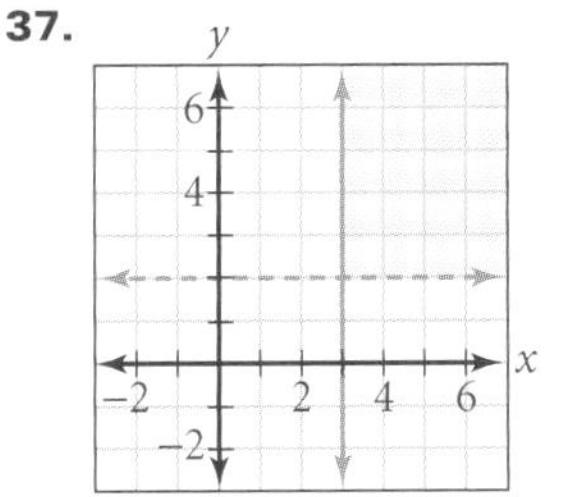

38.

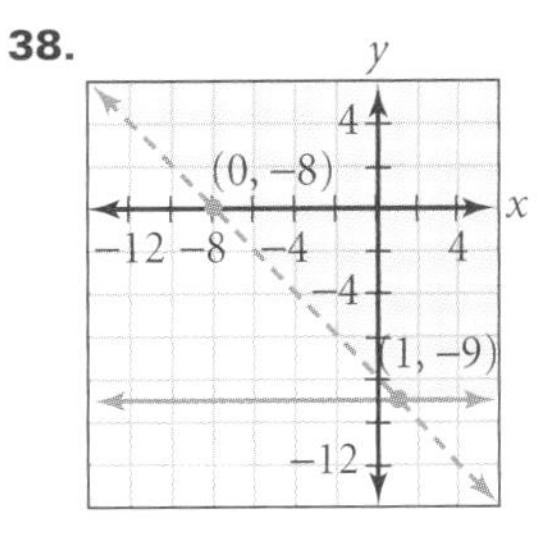

Focus

Students collect data from a temperature probe warming from freezing to room temperature and use various regression equations to fit the data collected. Students then use the regression equations to make predictions. Finally, students use Newton's law of cooling to write a function.

Motivate

Once the experiment has been described to the class, ask students to think about what the graph of the temperature might look like. For example, will it be linear or nonlinear? increasing or decreasing? curving upward or downward? Will the curve have limits? Ask students to explain why they think that the curve will look the way they imagine.

You will need a CBL with a temperature probe, a glass of ice water, and a graphics calculator. Refer to the discussion of the Chapter Project on page 358.

Step 1: First use the CBL to measure the air temperature. Record this temperature.

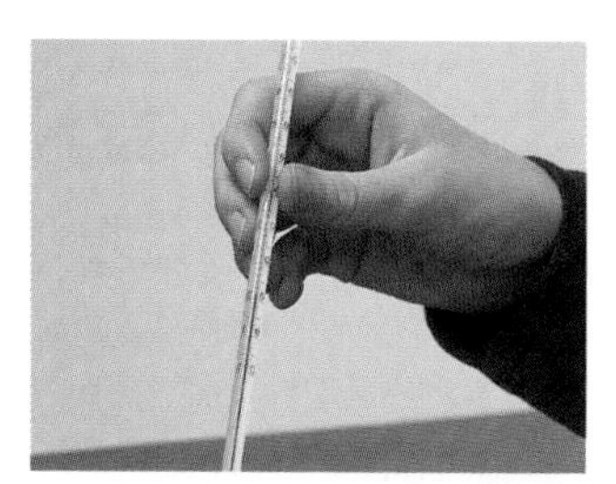

Step 2: Cool the temperature probe in the ice water to near 0°C.

Step 3: Remove the probe from the water. Record 30 CBL readings taken at 2-second intervals. These readings will be your data set.

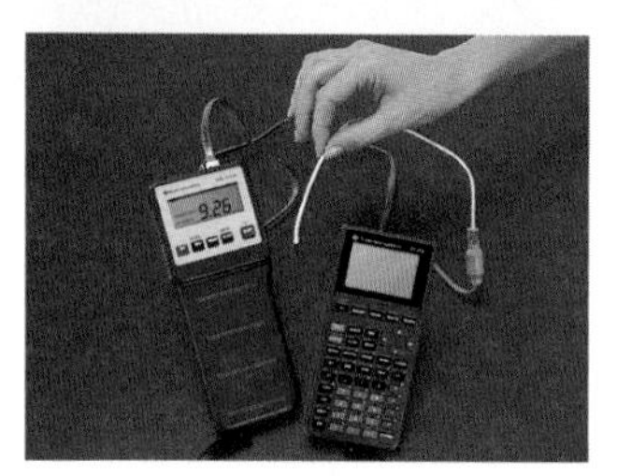

Step 4: Take and record a final reading at 2 minutes.

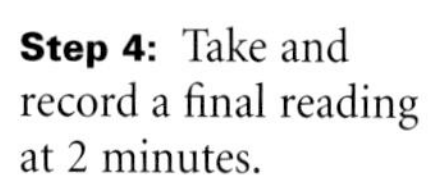

Activity 1

1. Answers will vary. Sample answer: The linear regression function is $y = 0.1641733871t + 2.939314516$.
2. Answers will vary. Sample: Using the linear regression model, the predicted temperature after 120 seconds would be $y = 0.1641733871(120) + 2.939314516 \approx 22.6$, which is significantly higher than the actual reading.

Activity 2

1. Answers will vary. Sample: A function that models quadratic regression for the data is $y = -0.0015536518t^2 + 0.2573924942t + 2.038196481$. An exponential regression equation for the data is $y = 3.945191182 \times (1.021425848)^t$.

Activity 1

1. Find the function that models the linear regression of your data set. Use the variable t for time in seconds.
2. Use the linear model to predict the temperature of the probe after 2 minutes. Compare the prediction to your actual 2-minute reading. Discuss the usefulness of a linear function for modeling the warming process.

Activity 2

1. Use the regression feature on your calculator to find a quadratic function and an exponential function that model your data set.
2. Predict the temperature of the probe after 2 minutes by using the quadratic and exponential models. Compare the predictions with your actual 2-minute reading. Discuss the usefulness of each function for modeling the warming process.

Activity 3

1. Transform your data by taking the opposite of each temperature in the data set and add the air temperature to it. Store this data in a new list.
2. Use the exponential regression feature on your calculator to find an exponential function to model this transformed data. Discuss the usefulness of this function for modeling the warming process.
3. Add the air temperature to the exponential function in Step 2. This new function will be called the "approximating function."
4. Use the approximating function to predict the temperature of the probe after 2 minutes. Discuss the usefulness of this function for modeling the warming process.

Activity 4

1. Use your data and Newton's law of cooling (from page 405) to write a function for the temperature of the probe as it warms to room temperature. Use this function to predict the temperature of the probe after 2 minutes.
2. Compare the accuracy of the prediction from Step 1 with the accuracy of the prediction obtained with the approximating function in Step 4 of Activity 3.

Cooperative Learning
Put students in groups of three to work on this project. The experiment with the CBL unit will require each student to take an active role. Let one student be the temperature reader, who is responsible for recording the room temperature and making sure that the probe is initially set at 0°C. Let another student be the technician, who is responsible for the CBL and graphics calculator setup. Let the third student be the coordinator, who is responsible for coordinating the efforts of all the team members and making sure that the technician begins recording the data at the appropriate time.

Discuss

After each activity, bring students together to discuss their results. Have students discuss how well the model found in the activity fits the data. For example, for the linear model, students should point out that the data seems to have a limit but that a linear function does not. As students progress through the activities, compare the results of the predictions obtained by using the various models.

2. Answers will vary. Sample: Using the quadratic regression model, the predicted temperature after 120 seconds would be $y = -0.0015536518(120)^2 + 0.2573924942(120) + 2.038196481 \approx 10.6$, which is significantly lower than the actual reading. Using the exponential regression model, the predicted temperature after 120 seconds would be $y = 3.945191182 \times (1.021425848)120 \approx 50.1$, which is significantly higher than the actual reading.

3. Answers will vary. Sample: The quadratic model produces increasing values for the temperature until around 80 seconds, and then the temperature values drop. The actual temperature levels off at air temperature. The exponential model represents the warming behavior fairly well except that it produces temperature values that are higher than the air temperature.

The Chapter Project table and answers to Activities 3–4 can be found in Additional Answers beginning on page 1002.

6 Chapter Review and Assessment

VOCABULARY

Key Skills & Exercises

LESSON 6.1

Key Skills

Write and evaluate exponential expressions.

The world population rose to about 5,734,000,000 in 1995. The world population was increasing at an annual rate of 1.6%. Write and evaluate an expression to predict the world population in 2020. [*Source: Worldbook Encyclopedia*]

$$5{,}734{,}000{,}000(1.016)^x$$
$$5{,}734{,}000{,}000(1.016)^{25} \approx 8{,}527{,}000{,}000$$

The projected world population for 2020 is about 8.5 billion people.

Exercises

1. **INVESTMENTS** The value of a painting is $12,000 in 1990 and increases by 8% of its value each year. Write and evaluate an expression to estimate the painting's value in 2005. **$12{,}000(1.08)^t$; ≈$38,066**

2. **DEPRECIATION** The value of a new car is $23,000 in 1998; it loses 15% of its value each year. Write and evaluate an expression to estimate the car's value in 2005. **$23{,}000(0.85)^t$; ≈$7373**

LESSON 6.2

Key Skills

Classify an exponential function as exponential growth or exponential decay.

When $b > 1$, the function $f(x) = b^x$ represents exponential growth.

When $0 < b < 1$, the function $f(x) = b^x$ represents exponential decay.

Calculate the growth of investments.

The total amount of an investment, A, earning compound interest is $A(t) = P\left(1 + \frac{r}{n}\right)^{nt}$, where P is the principal, r is the annual interest rate, n is the number of times interest is compounded per year, and t is the time in years.

Exercises

Identify each function as representing exponential growth or decay.

3. $f(x) = 4(0.89)^x$ **decay**
4. $g(x) = \frac{1}{3}(1.06)^x$ **growth**
5. $h(x) = 5(1.06)^x$ **growth**
6. $j(x) = 25\left(\frac{2}{5}\right)^x$ **decay**

INVESTMENTS For each compounding period below, find the final amount of a $2400 investment after 12 years at an annual interest rate of 4.5%.

7. annually **$4070.12**
8. quarterly **$4106.02**
9. daily **$4118.28**

Chapter Test, Form A

NAME ______ CLASS ______ DATE ______

Chapter Assessment

Chapter 6, Form A, page 1

Write the letter that best answers the question or completes the statement.

__c__ 1. Which of the following is not an exponential function?
a. $f(x) = 5(2.7)^x$ b. $f(n) = 125e^{0.12n}$
c. $f(x) = 7x^2$ d. $f(x) = 6.5\left(\frac{3}{4}\right)^{-x}$

__d__ 2. Which of the following is a model for exponential decay?
a. $f(x) = 0.8(3)^x$ b. $f(x) = 10\left(\frac{1}{3}\right)^{-x}$
c. $f(x) = 5 + 3x^{-3}$ d. $f(x) = 15(0.7)^x$

__b__ 3. The population of Starke County is 72,000 and is growing at a rate of 4.8% per decade. Which of the following is an expression for the population of Starke county after n decades?
a. $72{,}000(1.48)^n$ b. $72{,}000(1.048)^n$
c. $72{,}000(4.8)^n$ d. $72{,}000 + (4.8)^n$

__c__ 4. Kim invests $3000 at 7.8% annual interest compounded monthly. Find the final amount of Kim's investment after 15 years.
a. $9557.74 b. $9593.62 c. $9629.45 d. $9665.98

__b__ 5. Jamie invests $800 at 8.2% annual interest compounded continuously. Find the final amount of Jamie's investment after 20 years.
a. $3869.32 b. $4124.14 c. $4356.63 d. $4409.87

__d__ 6. Which of the following is the logarithmic form of $3^5 = 243$?
a. $\log_5 243 = 3$ b. $\log_{243} 5 = 3$
c. $\log 243 = 5$ d. $\log_3 243 = 5$

__d__ 7. Find the solution to the equation $\log_2 x = -4$.
a. $x = -8$ b. $x = -16$ c. $x = \frac{8}{3}$ d. $x = \frac{1}{16}$

__a__ 8. Find the solution to the equation $\log_x 8 = \frac{1}{3}$.
a. $x = 512$ b. $x = 24$ c. $x = \frac{1}{8}$ d. $x = 2$

__c__ 9. Find the value of the expression $\log_3 81 - \log_3\left(\frac{1}{9}\right)$.
a. 2 b. 3 c. 6 d. 9

__b__ 10. Find the approximate solution to the equation $3(1.5^x) + 10 = 280$.
a. $x = 0.1$ b. $x = 11.1$ c. $x = 34.8$ d. $x = 60$

NAME ______ CLASS ______ DATE ______

Chapter Assessment

Chapter 6, Form A, page 2

__a__ 11. Find the value of the expression $\log_5 125^{-3}$.
a. -9 b. -6 c. -1 d. $-\frac{3}{5}$

__c__ 12. Find the value of the expression $e^{\ln 2.5}$.
a. 30 b. 12.2 c. 2.5 d. 0.9

__b__ 13. Find the approximate value of the expression $\log_{16} 160$.
a. 0.54 b. 1.83 c. 2.20 d. 10

__d__ 14. What is another way to write $4 \log_3 a + 3 \log_3(2b) - \log_3 m$?
a. $\log_3\left(\frac{4a + 6b}{m}\right)$ b. $\log_3\left(\frac{24ab}{m}\right)$
c. $\log_3\left(\frac{2a^4b^3}{m}\right)$ d. $\log_3\left(\frac{8a^4b^3}{m}\right)$

__c__ 15. Find the solution to the equation $2 \log_a 3 + \log_a(x - 4) = \log_a(x + 8)$.
a. $x = 7.2$ b. $x = 6.4$ c. $x = 5.5$ d. $x = 4$

__a__ 16. Find the approximate solution to the equation $8.4 = \frac{3}{4} \log\left(\frac{x}{10^{2.4}}\right)$.
a. 3.98×10^{13} b. 5.01×10^8
c. $1.12 \times 10^{3.4}$ d. $6.3 \times 10^{2.4}$

__b__ 17. Find the solution to the equation $\ln(x^2 + 3x) - \ln 10 = 0$.
a. $x = 5$ or $x = -2$ b. $x = -5$ or $x = 2$
c. $x = 5$ d. $x = 2$

__d__ 18. The amount of a pollutant, measured in parts per million, in Pine Lake can be modeled by the function $A(t) = 14e^{-0.16t}$, where t is the number of years since a program to clean up the lake began. Approximately how long will it take for the amount of the pollutant in Pine Lake to reach 7 parts per million?
a. 83.51 years b. 12.52 years c. 7.85 years d. 4.33 years

__c__ 19. The number of industrial jobs in Fulton County is decreasing by 9% per year. If there are 12,600 industrial jobs in Fulton County this year, estimate the number of industrial jobs, to the nearest hundred, in 10 years.
a. 29,800 b. 11,300 c. 4900 d. 1300

LESSON 6.3

Key Skills

Write equivalent forms of exponential and logarithmic equations.

$3^4 = 81$ is $\log_3 81 = 4$ in logarithmic form.

$\log_2 64 = 6$ is $2^6 = 64$ in exponential form.

Use the definitions of exponential and logarithmic functions to solve equations.

$v = \log_6 36$	$3 = \log_4 v$	$4 = \log_v 81$
$6^v = 36$	$4^3 = v$	$v^4 = 81$
$6^v = 6^2$	$64 = v$	$v^4 = 3^4$
$v = 2$		$v = 3$

Exercises

10. Write $5^2 = 25$ in logarithmic form. **$\log_5 25 = 2$**

11. Write $\log_3 27 = 3$ in exponential form. **$3^3 = 27$**

12. Write $\log_3 \frac{1}{9} = -2$ in exponential form. **$3^{-2} = \frac{1}{9}$**

Find the value of v in each equation.

13. $v = \log_8 64$ **2**

14. $\log_v 4 = 2$ **2**

15. $2 = \log_{12} v$ **144**

16. $3 = \log_v 1000$ **10**

17. $\log_2 v = -3$ **$\frac{1}{8}$**

18. $\log_{27} 3 = v$ **$\frac{1}{3}$**

19. $\log_v 49 = 2$ **7**

20. $\log_4 \frac{1}{16} = v$ **−2**

LESSON 6.4

Key Skills

Use the Product, Quotient, and Power Properties of Logarithms to simplify and evaluate expressions involving logarithms.

Given $\log_3 7 \approx 1.7712$, $\log_3 66$ can be approximated as shown below.

$$\log_3 63 = \log_3 9 + \log_3 7$$
$$= 2 + 2.1827 \approx 4.1827$$

$$\log_5 25^7 = 7 \log_5 25 = 7 \cdot 2 = 14$$

Exercises

Given $\log_7 5 \approx 0.8271$ and $\log_7 9 \approx 1.1292$, approximate the value of each logarithm.

21. $\log_7 45$ **1.9563**

22. $\log_7 \frac{5}{9}$ **−0.3021**

23. $\log_7 35$ **1.8271**

Write each expression as a single logarithm. Then simplify, if possible.

24. $\log_5 3 + \log_5 6 + \log_5 9$ **$\log_5 162$**

25. $\log 6 - \log 3 + 2 \log 7$ **$\log 98$**

Evaluate each expression.

26. $2^{\log_2 12}$ **12**

27. $\log_7 7^3$ **3**

28. $\log_6 36^7$ **14**

LESSON 6.5

Key Skills

Use the common logarithmic function to solve exponential and logarithmic equations.

$4 = \log x$	$2^x = 34$
$10^4 = x$	$\log 2^x = \log 34$
$10{,}000 = x$	$x \log 2 = \log 34$
	$x = \frac{\log 34}{\log 2}$
	$x \approx 5.09$

Apply the change-of-base formula to evaluate logarithmic expressions.

The change-of-base formula is $\log_a x = \frac{\log_b x}{\log_b a}$, where $a \neq 1$, $b \neq 1$, and $x > 0$.

$$\log_2 5 = \frac{\log 5}{\log 2} \approx 2.32$$

Exercises

Solve each equation. Give your answers to the nearest hundredth.

29. $\log x = 8$

30. $\log 0.01 = x - 5$

31. $5^x + 100 = 98$

32. $5 - 2^x = 40$

33. $7 + 3^{2x-1} = 154$

34. $3 + 7^{3x+1} = 346$

Evaluate each logarithmic expression to the nearest hundredth.

35. $\log_3 14$ **2.40**

36. $\log_{16} 3$ **0.40**

37. $\log_{0.5} 6$ **−2.58**

38. $\log_{1.5} 10$ **5.68**

39. CHEMISTRY What is $[H^+]$ of a carbonated soda if its pH is 2.5? **$10^{-2.5}$ moles per liter**

29. 100,000,000

30. 3

31. no solution

32. no solution

33. 2.77

34. $\frac{2}{3}$

Chapter Test, Form B

NAME ______ CLASS ______ DATE ______

Chapter Assessment

Chapter 6, Form B, page 1

1. Write $4^5 = 1024$ in logarithmic form. $\log_4 1024 = 5$

Evaluate each expression. If necessary, round your answer to the nearest tenth.

2. $\log_9 125$ 2.2
3. $5^{\log_5 7.2}$ 7.2
4. $\log_4 64^5$ 15

Write each expression as a single logarithm.

5. $3 \log_7 x + 2 \log_7(3y) + \log_7 w$ $\log_7(9x^3y^2w)$
6. $3(\log_a n - 2 \log_a m)$ $\log_a\left(\frac{n^3}{m^6}\right)$
7. Find the final amount of a $1500 investment after 25 years at 6.6% annual interest compounded quarterly and monthly. compounded quarterly: $7706.01; compounded monthly: $7775.24
8. a. Find the final amount of a $200 investment after 18 years at 7.3% annual interest compounded continuously. $774.21

   b. How long does it take for an investment to triple in value at an annual interest rate of 7.3% compounded continuously? Round your answer to the nearest hundredth. about 15.05 years

Solve each equation for x. If necessary, round your answers to the nearest hundredth.

9. $10^x = 8520$ $x \approx 3.93$
10. $14^x = 960$ $x \approx 2.60$
11. $\log_3 x = 6$ $x = 729$
12. $\log_x 5 = \frac{1}{4}$ $x = 625$
13. $\ln(5x - 3) + \ln 2 = \ln(24 - 2x)$ $x = 2.5$
14. $\log_5(3x + 10) - 3 \log_5 4 = 2$ $x = 530$
15. $8 = 5 + 2 \log\left(\frac{x}{4}\right)$ $x \approx 126.49$
16. $\log_6(x + 4) + \log_6(x - 1) = 2$ $x = 5$

Find the multiplier for each situation.

17. 24% growth 1.24
18. 0.8% decay 0.992

NAME ______ CLASS ______ DATE ______

Chapter Assessment

Chapter 6, Form B, page 2

19. Find the y-intercept of the graph of $f(x) = -3(0.65)^{-x}$. (0, −3)
20. It cost Kurt $430 to maintain his car this year. Kurt's mechanic says that Kurt can expect his maintenance cost to increase by 15% per year for the next 8 years.

   a. Write a function for the maintenance cost per year for Kurt's car after n years. $C(n) = 430(1.15)^n$

   b. When will Kurt need to spend $1000 to maintain his car for the year? Round your answer to the nearest year. 6 years

   c. Use your function to predict how much it will cost Kurt to maintain his car in 8 years. about $1315
21. The managers of a lake resort decide to stock the lake with fish. The managers buy 3000 fish and put them into the lake. The fish suppliers tell the managers that the population of fish will decline by 9% per month for the next year.

   a. Write a function for the number of fish in the lake after n months. $F(n) = 3000(0.91)^n$

   b. Predict the number of fish in the lake after 3 months. about 2261 fish

   c. Predict how long it will be until half of the 3000 fish are left in the lake. Round your answer to the nearest tenth. about 7.3 months
22. The remaining amount of a 100-milligram dose of a particular painkiller in a person's bloodstream after t hours can be modeled by the function $A(t) = 100e^{-0.21t}$.

   a. Find the remaining amount of the painkiller, to the nearest tenth of a milligram, after 4 hours. $A(4) \approx 43.2$ milligrams

   b. How long will it be until only 10 milligrams of the painkiller remain in the bloodstream? Round your answer to the nearest hundredth. about 10.96 hours
23. A vintage car was bought for $30,000 and was sold 10 years later for $50,000. Find the effective yield. 5.24%

LESSON 6.6

Key Skills

Evaluate exponential functions of base *e* and natural logarithms.

Using a calculator and rounding to the nearest thousandth, $e^{2.5} \approx 12.182$ and $\ln 3.5 \approx 1.253$.

Model exponential growth and decay processes by using base *e*.

The continuous compounding formula is $A = Pe^{rt}$, where A is the final amount when the principal P is invested at an annual interest rate of r for t years.

Exercises

Evaluate each expression to the nearest thousandth.

40. $e^{0.5}$ **1.649**

41. e^{-5} **0.007**

42. $\ln 5$ **1.609**

43. $\ln 0.05$ **−2.996**

44. **INVESTMENTS** Sharon invests $2500 at an annual interest rate of 9%. How much is the investment worth after 10 years if the interest is compounded continuously? **$6149.01**

LESSON 6.7

Key Skills

Solve logarithmic and exponential equations.

$$\log_x \frac{1}{32} = -5$$
$$x^{-5} = \frac{1}{32}$$
$$x^{-5} = 2^{-5}$$
$$x = 2$$

$$\ln x^3 + 5 = 1$$
$$3 \ln x = -4$$
$$\ln x = -\frac{4}{3}$$
$$x = e^{-\frac{4}{3}}$$
$$x \approx 0.264$$

$$\ln(x + 6) = 2 \ln 3$$
$$\ln(x + 6) = \ln 9$$
$$x + 6 = 9$$
$$x = 3$$

Exercises

Solve each equation for *x*. Write the exact solution and the approximate solution to the nearest hundredth, when appropriate.

45. $\log_x \frac{1}{128} = -7$ **2**

46. $\ln(2x) = 4 \ln 2$ **8**

47. $x \log \frac{1}{6} = \log 6$ **−1**

48. $\ln \sqrt{x} - 3 = 1$ $\mathbf{e^8 \approx 2980.96}$

49. **HEALTH** The normal healing of a wound can be modeled by $A = A_0e^{-0.35n}$, where A is the area of the wound in square centimeters after n days. After how many days is the area of the wound half of its original size, A_0? **2 days**

Applications

50. **BIOLOGY** Given favorable living conditions, fruit fly populations can grow at the astounding rate of 28% per day. If a laboratory selects a population of 25 fruit flies to reproduce, about how big will the population be after 3 days? after 5 days? after 1 week? **52; 86; 141**

51. **PHYSICS** Suppose that the sound of busy traffic on a four-lane street is about $10^{8.5}$ times the intensity of the threshold of hearing, I_0. Find the relative intensity, R, in decibels of the traffic on this street. **85 decibels**

52. **PHYSICS** Radon is a radioactive gas that has a half-life of about 3.8 days. This means that only half of the original amount of radon gas will be present after about 3.8 days. Using the exponential decay function $A = Pe^{-kt}$, find the value of k to the nearest hundredth, and write the function for the amount of radon remaining after t days. **0.18;** $\mathbf{A = Pe^{-0.18t}}$

Fruit fly

6 Chapter Test

1. **DEMOGRAPHICS** The population of Petoskey, Michigan, was 6076 in 1990 and was growing at the rate of 3.7% per year. The city planners want to know what the population will be in the year 2025. Write and evaluate an expression to estimate this population. **$P(t) = 6076(1.037)^t$; 21,671**

2. **INCOME TAX** The government allows for linear depreciation of capital expenditures for income tax purposes at the rate of 10% per year. What will be the value of a $150,000 tool and die machine after 7 years of use? **$71,744.54**

Tell whether each function represents exponential growth or decay.

3. $f(x) = 3.6(1.01)^x$ **growth**

4. $g(t) = 0.015(1.23)^t$ **growth**

5. $h(t) = \left(7\frac{3}{4}\right)\left(\frac{5}{8}\right)^t$ **decay**

6. $j(x) = 2500(0.25)^x$ **decay**

INVESTMENTS For each compounding period below, find the final amount of a $5000 investment after 10 years at a 5.6% annual interest rate.

7. daily **$8752.99**
8. monthly **$8741.97**
9. quarterly **$8719.43**
10. annually **$8622.02**

Write each logarithmic equation in exponential form and each exponential equation in logarithmic form.

11. $\log_3 81 = 4$ **$3^4 = 81$**
12. $2^8 = 256$ **$\log_2 256 = 8$**
13. $\left(\frac{1}{4}\right)^{-5} = 1024$ **$\log_{\frac{1}{4}} 1024 = -5$**
14. $\log_5 \frac{1}{625} = -4$ **$5^{-4} = \frac{1}{625}$**

Find the value of v in each equation.

15. $3 = \log_v 343$ **7**

16. $\log_9 729 = v$ **3**

17. $\log_6 v = 5$ **7776**

Write each expression as a single logarithm. Then simplify, if possible.

18. $\log_2 5 - 3\log_2 3 + \log_2 6$ **$\log_2 \frac{10}{9}$**

19. $\log_7 \frac{1}{4} + 2\log_7 4 - \frac{1}{2}\log_7 16$ **0**

Evaluate each expression.

20. $5^{\log_5 32}$ **32**
21. $\log_6 36$ **2**
22. $\log_9 9^{\frac{2}{3}}$ **$\frac{2}{3}$**
23. $\log_b b^{(x-2)}$ **$x - 2$**

Solve each equation. Give your answers to the nearest hundredth.

24. $\log_4 x = 6.2$ **5404.70**
25. $3^{x+2} = 238$ **2.98**
26. $274 - 5^x = 198$ **2.69**
27. $\log_6 468 = x$ **3.43**

28. **SEISMOLOGY** The amount of energy E, in ergs, released by an earthquake of magnitude M is given by the formula $E = 10^{(1.5M + 11.8)}$. What is the difference in the amount of energy released by an earthquake of magnitude 6.5 and one of magnitude 8.7?

Evaluate each expression to the nearest thousandth.

29. $e^{3.4}$ **29.964**
30. $\ln\pi$ **1.145**
31. $e^{-3.25}$ **0.039**
32. $\ln(e^{1.618})$ **1.618**

33. **ARCHAEOLOGY** The age of an artifact can be determined using carbon-14 dating with the equation $N(t) = N_0 e^{-0.00012t}$. What is the approximate age of an artifact if a sample reveals that it contains 34% of its original carbon-14? **8990 years**

Solve each equation for x. Write the exact solution and the approximate solution to the nearest hundredth, when appropriate.

34. $3^x = 5^{2.3}$ **3.37**
35. $\ln(x + 1) = 2\ln 4$ **15**
36. $\log x + \log(x + 3) = 1$ **2**

Chapter Test

28. E for 6.5 magnitude = 3.55×10^{21} ergs; E for 8.7 magnitude = 7.08×10^{24} ergs; difference ≈ 7.076×10^{24} ergs

College Entrance Exam Practice

Multiple-Choice and Quantitative-Comparison Samples

The first half of the Cumulative Assessment contains two types of items found on standardized tests—multiple-choice questions and quantitative-comparison questions. Quantitative-comparison items emphasize the concepts of equality, inequality, and estimation.

Free-Response Grid Samples

The second half of the Cumulative Assessment is a free-response section. This part of the Cumulative Assessment requires student-produced response items like those commonly found on college entrance exams. These questions require the use of machine-scored answer grids. You may wish to have students practice answering these items in preparation for standardized tests.

College Entrance Exam Practice

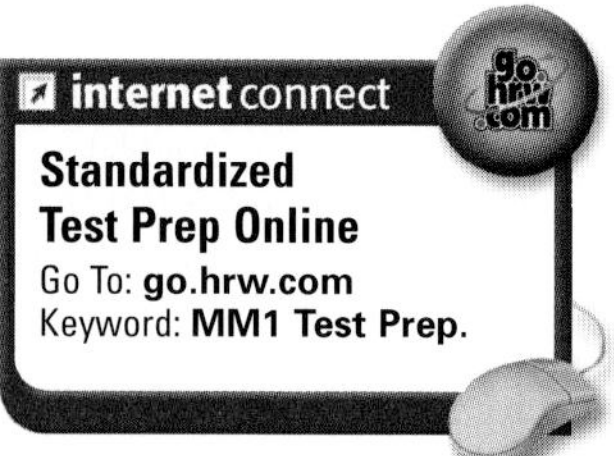

QUANTITATIVE COMPARISON For Items 1–5, write
A if the quantity in Column A is greater that the quantity in Column B;
B if the quantity in Column B is greater that the quantity in Column A;
C if the quantities are equal; or
D if the relationship cannot be determined from the given information.

	Column A	Column B	Answers
1. A	The minimum value in the range of the function $f(x) = \|x - 2\|$	The minimum value in the range of the function $f(x) = \|x\| - 2$	Ⓐ Ⓑ Ⓒ Ⓓ [Lesson 2.6]
2. B	$(5 - 2i)(5 + 2i)$ where $i = \sqrt{-1}$	$6i^2 \cdot 7i^2$ where $i = \sqrt{-1}$	Ⓐ Ⓑ Ⓒ Ⓓ [Lesson 5.6]
3. A	$\frac{1}{5}(x - 2)^2 = 1$ The absolute value of the difference of the solutions	The absolute value of the sum of the solutions	Ⓐ Ⓑ Ⓒ Ⓓ [Lesson 5.2]
4. B	Let $f(x) = 3x$ and $g(x) = x^2 + 2$ $(f \circ g)(-2)$	$(g \circ f)(-2)$	Ⓐ Ⓑ Ⓒ Ⓓ [Lesson 2.4]
5. A	The number of real solutions $2x^2 - 3x = 3$	$4x^2 - 4x = -1$	Ⓐ Ⓑ Ⓒ Ⓓ [Lesson 5.6]

6. What is the slope of the line that contains the points (6, –8) and (–2, –4)? ***(LESSON 1.2)***
c
a. 2 **b.** –2
c. $-\frac{1}{2}$ **d.** $\frac{1}{2}$

7. Which term describes the system of equations below? ***(LESSON 3.1)***
b

$$\begin{cases} 2x - y = 7 \\ 2y - 4x = -14 \end{cases}$$

a. inconsistent **b.** dependent
c. independent **d.** incompatible

8. What are the coordinates of the vertex for the graph of $y = (x - 2)(x + 1)$? ***(LESSON 5.5)***
c
a. (–0.5, –2.25) **b.** (–1, 0)
c. (0.5, –2.25) **d.** (–0.5, 3.75)

9. In triangle *ABC* below, what is the value of *a*? ***(LESSON 5.2)***
b
a. 28
b. $2\sqrt{7}$
c. 2
d. 4

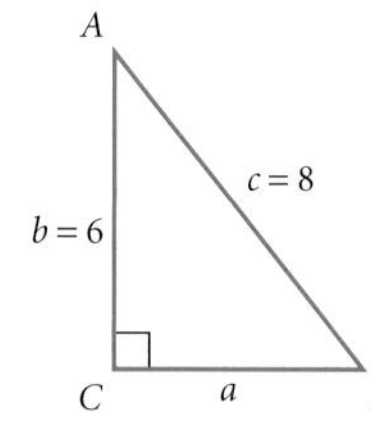

10. Which is a solution for the system below?
d *(LESSON 3.1)*

$$\begin{cases} y = x + 3 \\ y - x = 4 \end{cases}$$

a. (3, 7) **b.** (2, 5)
c. both **a** and **b** **d.** neither **a** nor **b**

11. Simplify $\left(\frac{-x^{-3}y^{5}}{2xy^{-2}}\right)^{4}$. *(LESSON 2.2)*
c

a. $\frac{y^{28}}{8x^{16}}$ **b.** $-\frac{y^{28}}{8x^{16}}$

c. $\frac{y^{28}}{16x^{16}}$ **d.** $\frac{1}{16}x^{11}y^{22}$

12. Determine which set of ordered pairs
d represents a function. *(LESSON 2.3)*

a. $\{(1, -1), (-1, 1), (1, 1)\}$
b. $\{(0, 1), (1, 0), (0, 0)\}$
c. $\{(-1, 1), (1, -1), (1, 0), (0, -1)\}$
d. $\{(0, 1), (1, 2), (2, 1), (-1, 0)\}$

13. Solve $\begin{cases} 2x + 8y = -10 \\ -3x + 12y = 3 \end{cases}$. $\left(-3, -\frac{1}{2}\right)$

(LESSONS 3.1 AND 3.2)

14. Solve $2x^2 - 7 = 121$ for x. *(LESSON 5.2)* $x = 8$ *or* $x = -8$

15. Write the equation in vertex form for the parabola described by $f(x) = 2x^2 - 3x + 7$. *(LESSON 5.4)* $f(x) = 2\left(x - \frac{3}{4}\right)^2 + \frac{47}{8}$

16. If $f(x) = x^2 - 2x$, find the inverse of f. *(LESSON 2.5)* $f^{-1}(x) = 1 \pm \sqrt{1 + x}$

17. Find the matrix product below. *(LESSON 4.2)*

$$\begin{bmatrix} 3 & -6 \\ 5 & 4 \end{bmatrix}\begin{bmatrix} 1 & 0 \\ 0 & 1 \end{bmatrix} \quad \begin{bmatrix} 3 & -6 \\ 5 & 4 \end{bmatrix}$$

18. Factor to find the zeros of $f(x) = x^2 - x - 6$. *(LESSON 5.3)* $x = 3$ *or* $x = -2$

19. Let $f(x) = 2x - 3$ and $g(x) = -3x$. Find $(f \cdot g)(x)$. *(LESSON 2.4)* $-6x^2 + 9x$

20. Let $A = \begin{bmatrix} 2 & 3 & 8 \\ -4 & 5 & -8 \\ 0 & 6 & -5 \end{bmatrix}$ and let

$B = \begin{bmatrix} 0 & 3 & -1 \\ -1 & 4 & 3 \\ 2 & -7 & 2 \end{bmatrix}$. Find $A - B$. $\begin{bmatrix} 2 & 0 & 9 \\ -3 & 1 & -11 \\ -2 & 13 & -7 \end{bmatrix}$

(LESSON 4.1)

FREE-RESPONSE GRID The following questions may be answered by using a free-response grid such as that commonly used by standardized-test services.

21. In the formula $F = \frac{9}{5}C + 32$, find the value of C if $F = 95$. *(LESSON 1.6)* 35

22. What is the value of $\log_2 \frac{1}{8}$? *(LESSON 6.3)* -3

23. What is the minimum value of $f(x) = x^2 - 5x + 8$? *(LESSON 5.1)* $\frac{7}{4}$

24. Solve the equation $5e^{3-x} = 2$ for x. *(LESSON 6.6)* about 3.92

25. **INVESTMENTS** If $2500 is invested at 6.9% compounded daily, determine the value of the investment after 12 years. *(LESSON 6.2)* $5721.39

26. Simplify $\frac{(9-3)^2}{2(3-1)}$. *(LESSON 2.1)* 9

27. Evaluate $[3.2] - [4.99]$. *(LESSON 2.6)* -1

28. Find the determinant of $\begin{bmatrix} 3 & 6 \\ -1 & 4 \end{bmatrix}$. 18 *(LESSON 4.3)*

BUSINESS A skate store makes two kinds of in-line skates: regular skates and those with custom boots. The store can make 90 pairs of skates per month and can spend, at most, no more than $11,400 per month to produce them. Each pair of regular skates costs $80 to make and brings a profit of $60. Each pair of skates with custom boots costs $150 to make and brings a profit of $70. *(LESSON 3.5)*

29. Find the number of regular in-line skates that the skate store needs to sell in order to maximize its profit. 30

30. Find the number of in-line skates with custom boots that the skate store needs to sell in order to maximize its profit. 60

Keystroke Guide for Chapter 6

Essential keystroke sequences (using the model TI-82 or TI-83 graphics calculator) are presented below for all Activities and Examples found in this chapter that require or recommend the use of a graphics calculator.

internet connect

For Keystrokes of other graphing calculator models, visit the HRW web site at **go.hrw.com** and enter the keyword **MB1 CALC**.

LESSON 6.1

Page 354

Use a calculator to model the population growth of 25 bacteria that double every hour for 6 hours.

25 [ENTER] [X] 2 [ENTER] [ENTER] [ENTER] [ENTER] [ENTER] [ENTER]

LESSON 6.2

Page 363

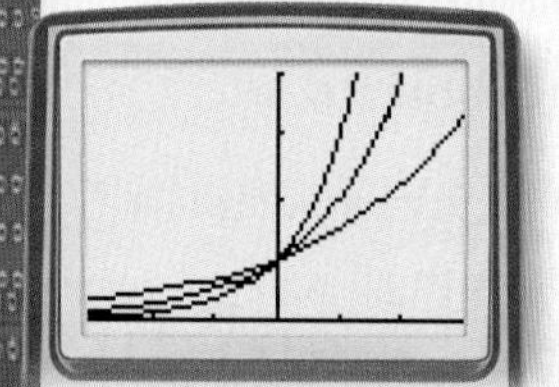

For Step 1, graph $y_1 = 3^x$, $y_2 = 2^x$, and $y_3 = (1.5)^x$ on the same screen.

Use viewing window [−3, 3] by [0, 4].

[Y=] 3 [^] [X,T,θ,n] [ENTER] (Y2 =) 2 [^] [X,T,θ,n] [ENTER]
(Y3 =) 1.5 [^] [X,T,θ,n] [ENTER]

For Step 3, graph $y_4 = \left(\frac{1}{3}\right)^x$, $y_5 = \left(\frac{1}{2}\right)^x$, and $y_6 = \left(\frac{1}{1.5}\right)^x$ on the same screen as the graphs above.

[Y=] (Y4=) [(] 1 [D] 3 [)] [^] [X,T,θ,n] [ENTER]
(Y5=) [(] 1 [D] 2 [)] [^] [X,T,θ,n] [ENTER]
(Y6=) [(] 1 [D] 1.5 [)] [^] [X,T,θ,n] [ENTER]

EXAMPLE 1

Page 364

For part a, graph $y_1 = 2^x$ and $y_2 = 3 \cdot 2^x$ together, and find the y-intercepts.

Use viewing window [−5, 5] by [−2, 12].

To graph the functions, use a keystroke sequence similar to that in the Activity for this lesson.

Find the y-intercepts:

[2nd] [TRACE] (CALC) [1: value] [ENTER] [ENTER] (X=) 0 [ENTER] [▼]

For part **b**, use a similar keystroke sequence.

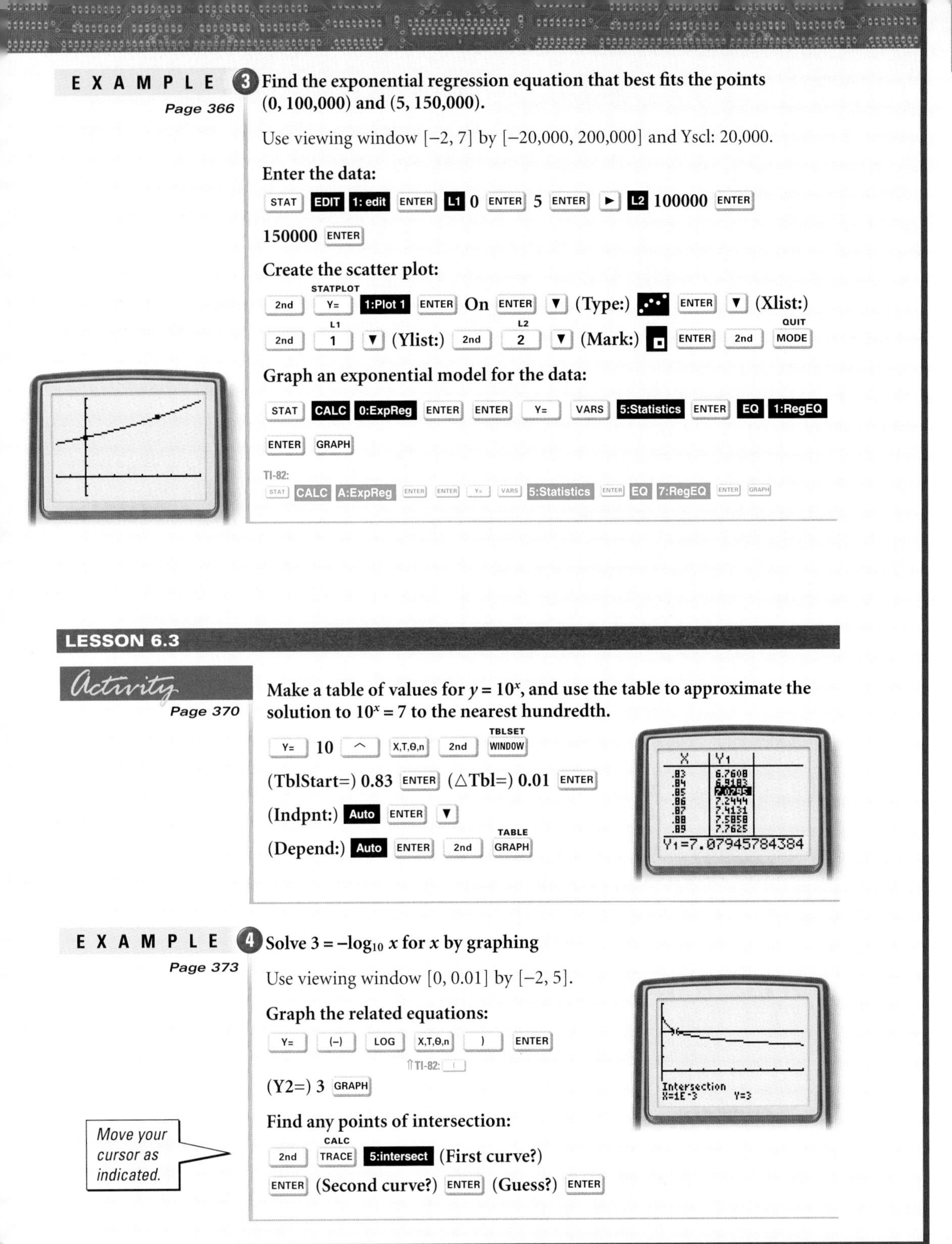

EXAMPLE 3 Find the exponential regression equation that best fits the points (0, 100,000) and (5, 150,000).

Page 366

Use viewing window [−2, 7] by [−20,000, 200,000] and Yscl: 20,000.

Enter the data:

STAT EDIT 1: edit ENTER L1 0 ENTER 5 ENTER ► L2 100000 ENTER

150000 ENTER

Create the scatter plot:

2nd Y= (STATPLOT) 1:Plot 1 ENTER On ENTER ▼ (Type:) ⁘ ENTER ▼ (Xlist:)

2nd 1 (L1) ▼ (Ylist:) 2nd 2 (L2) ▼ (Mark:) ▫ ENTER 2nd MODE (QUIT)

Graph an exponential model for the data:

STAT CALC 0:ExpReg ENTER ENTER Y= VARS 5:Statistics ENTER EQ 1:RegEQ

ENTER GRAPH

TI-82:
STAT CALC A:ExpReg ENTER ENTER Y= VARS 5:Statistics ENTER EQ 7:RegEQ ENTER GRAPH

LESSON 6.3

Activity

Page 370

Make a table of values for $y = 10^x$, and use the table to approximate the solution to $10^x = 7$ to the nearest hundredth.

Y= 10 ^ X,T,θ,n 2nd WINDOW (TBLSET)

(TblStart=) 0.83 ENTER (ΔTbl=) 0.01 ENTER

(Indpnt:) Auto ENTER ▼

(Depend:) Auto ENTER 2nd GRAPH (TABLE)

X Y1
.83 6.7608
.84 6.9183
.85 7.0795
.86 7.2444
.87 7.4131
.88 7.5858
.89 7.7625
Y1=7.07945784384

EXAMPLE 4 Solve $3 = -\log_{10} x$ for x by graphing

Page 373

Use viewing window [0, 0.01] by [−2, 5].

Graph the related equations:

Y= (−) LOG X,T,θ,n) ENTER

⇑TI-82: (

(Y2=) 3 GRAPH

Find any points of intersection:

Move your cursor as indicated.

2nd TRACE (CALC) 5:intersect (First curve?)

ENTER (Second curve?) ENTER (Guess?) ENTER

LESSON 6.5

EXAMPLE 1 Page 387

Graph $y = 10 \log x$, and evaluate y for $x = 300$.

Use viewing window [−50, 350] by [−10, 50].

Graph the function:

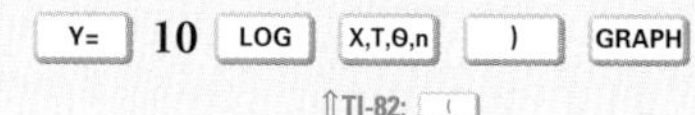

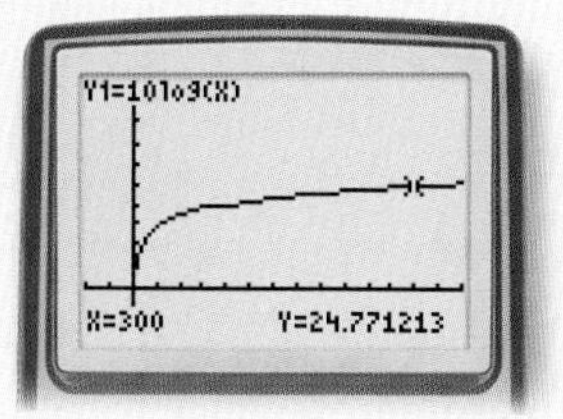

Find the point on the graph where $x = 300$:

2nd CALC TRACE 1: value ENTER (X=) 300 ENTER

EXAMPLE 3 Page 387

Solve $5^x = 62$ by graphing.

Use viewing window [−1, 5] by [−40, 100].

Graph the related equations, and use a keystroke sequence similar to that in Example 4 of Lesson 6.3 to find any points of intersection.

LESSON 6.6

EXAMPLES 1 and 3 Pages 393 and 394

For part a of Example 1, evaluate $y = e^x$ to the nearest thousandth for $x = 2$.

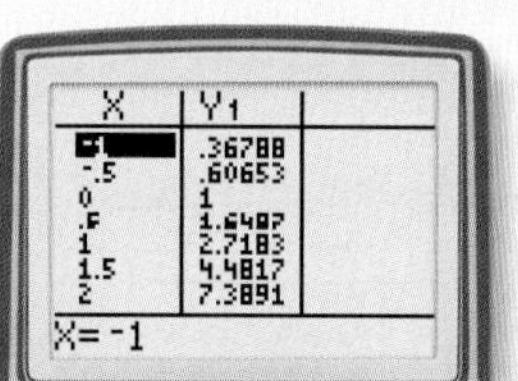

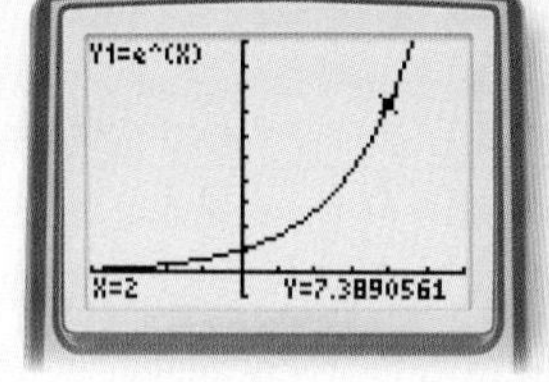

Use viewing window [−2, 3] by [−1, 10].

Enter the equation: Y= 2nd e^x LN X,T,θ,n) ENTER

⇑TI-82:

Create a table of values:
Use a keystroke sequence similar to that in the Activity in Lesson 6.3. Use TblStart = −1 and △Tbl = 0.5.

Find the point on the graph where $x = 2$:
Use a keystroke sequence similar to that in Example 1 of Lesson 6.5.

For parts **b** and **c** of Example 1, use a similar keystroke sequence.

For Example 3, use viewing window [−2, 10] by [−2, 4] and a similar keystroke sequence.

EXAMPLE 4 Page 395

Solve $2 = e^{0.085x}$ by graphing.

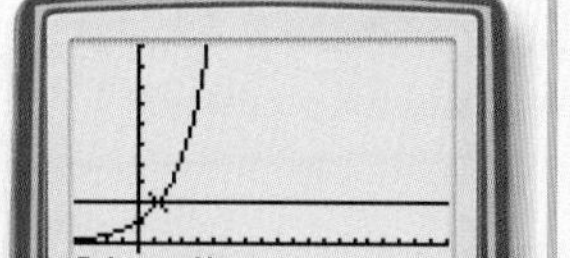

Use viewing window [−20, 100] by [−3, 10].

Graph the related equations:

Y= 2 ENTER (Y2=) 2nd e^x LN .085 X,T,θ,n) ENTER

⇑TI-82:

Find the point of intersection:
Use a keystroke sequence similar to that in Example 4 of Lesson 6.3.

LESSON 6.7

EXAMPLE 2 Page 404

Solve $\log x + \log(x - 3) = 1$ by graphing.

Use viewing window [−2, 8] by [−1, 5].

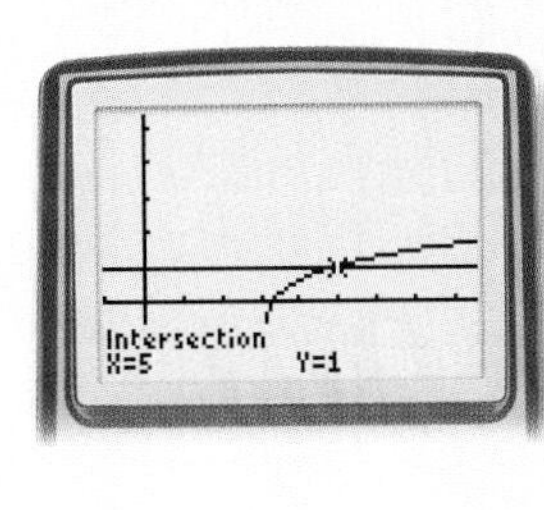

Graph the related equations:

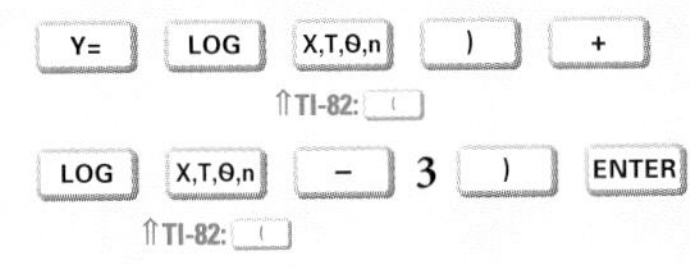

Y= LOG X,T,θ,n) +
⇑TI-82: (

LOG X,T,θ,n − 3) ENTER
⇑TI-82: (

(Y2=) 1 GRAPH

Find the point of intersection:
Use a keystroke sequence similar to that in Example 4 of Lesson 6.3.

EXAMPLE 3 Page 404

Solve $4e^{3x-5} = 72$ by graphing.

Use viewing window [−1, 4] by [−30, 100].

Graph the related equations:

Y= 4 2nd LN (e^x) 3 X,T,θ,n − 5)
ENTER (Y2=) 72 GRAPH

Intersection
X=2.6301239 Y=72

Find the point of intersection:
Use a keystroke sequence similar to that in Example 4 of Lesson 6.3.

Page 405

For Step 1, graph $y_1 = \log x + \log(x + 21)$ and $y_2 = 2$ together, and find any points of intersection.

Use viewing window [−1, 6] by [−1, 3].

Use a keystroke sequence similar to that in Example 2 of Lesson 6.7.

For Step 4, use viewing window [−1, 4] by [−30, 100] and a keystroke sequence similar to that in Example 3 of Lesson 6.7.

EXAMPLE 4 Page 406

For part b, evaluate $y = 70 - 30\left(\frac{2}{3}\right)^x$ for $x = 2$.

Use viewing window [−1, 10] by [−20, 100].

Graph the function:

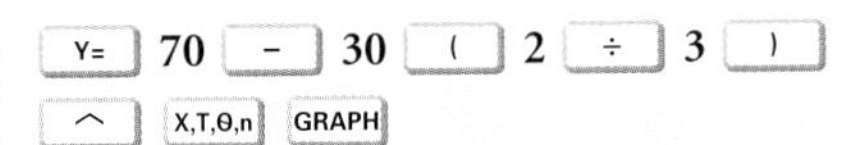

Y= 70 − 30 (2 ÷ 3)
^ X,T,θ,n GRAPH

Y1=70-30(2/3)^X
X=2 Y=56.666667

Find the y-value for $x = 2$:
Use a keystroke sequence similar to that in Example 1 of Lesson 6.2.

For part **c**, use a keystroke sequence similar to that in Example 4 of Lesson 6.3.

7 Polynomial Functions

Lesson Presentation CD-ROM
PowerPoint® presentations for each lesson 7.1–7.5

CHAPTER PLANNING GUIDE

Lesson	7.1	7.2	7.3	7.4	7.5	Project and Review
Pupil's Edition Pages	424–431	432–439	440–447	448–455	458–465	456–457, 466–473
Practice and Assessment						
Extra Practice (Pupil's Edition)	962	963	963	964	964	
Practice Workbook	42	43	44	45	46	
Practice Masters Levels A, B, and C	124–126	127–129	130–132	133–135	136–138	
Standardized Test Practice Masters	48	49	50	51	52	53
Assessment Resources	84	85	86	88	89	87, 90–95
Visual Resources						
Lesson Presentation Transparencies Vol. 1	165–168	169–172	173–176	177–180	181–184	
Teaching Transparencies		25	26	27	28	
Answer Key Transparencies	238–241	242–245	246–248	249–251	252–254	255–258
Quiz Transparencies	7.1	7.2	7.3	7.4	7.5	3.6
Teacher's Tools						
Reteaching Masters	83–84	85–86	87–88	89–90	91–92	
Make-Up Lesson Planner for Absent Students	42	43	44	45	46	
Student Study Guide	42	43	44	45	46	
Spanish Resources	42	43	44	45	46	
Block Scheduling Handbook						14–15
Activities and Extensions						
Lesson Activities	42	43	44	45	46	
Enrichment Masters	42	43	44	45	46	
Cooperative-Learning Activities	42	43	44	45	46	
Problem Solving/ Critical Thinking	42	43	44	45	46	
Student Technology Guide	42	43	44	45	46	
Long Term Projects						25–28
Writing Activities for Your Portfolio						19–21
Tech Prep Masters						31–34
Building Success in Mathematics						17–19

LESSON PACING GUIDE

Lesson	7.1	7.2	7.3	7.4	7.5	Project and Review
Traditional	1 day	1 day	2 days	1 day	1 day	2 days
Block	$\frac{1}{2}$ day	$\frac{1}{2}$ day	1 day	$\frac{1}{2}$ day	$\frac{1}{2}$ day	1 day
Two-Year	2 days	2 days	4 days	2 days	2 days	4 days

CONNECTIONS AND APPLICATIONS

Lesson	7.1	7.2	7.3	7.4	7.5	Review
Algebra	424–431	432–439	440–447	448–455	458–465	466–473
Geometry	430			448		
Statistics		436, 438				
Transformations					462	
Business and Economics	424, 426, 430	438	477	454	458, 461	470, 471, 473
Science and Technology				453, 454	465	473
Social Studies		432			464	
Sports and Leisure		439				
Cultural Connection: Asia				454		

BLOCK SCHEDULING GUIDE

Day	Lesson	Teacher Directed: Lesson Examples, Teaching Transparencies	Student Guided Activity, Try This	Cooperative-Learning Activity, Lesson Activity, Student Technology Guide	Practice: Practice & Apply, Extra Practice, Practice Workbook	Assessment: Quiz, Mid-Chapter Assessment	Problem Solving, Reteaching
1	7.1	10 min	10 min	8 min	25 min	8 min	8 min
	7.2	10 min	10 min	7 min	25 min	7 min	7 min
2	7.3	20 min	20 min	15 min	50 min	15 min	15 min
3	7.4	10 min	10 min	8 min	25 min	8 min	8 min
	7.5	10 min	10 min	7 min	25 min	7 min	7 min
4	Assess.	50 min **PE:** Chapter Review	90 min **PE:** Chapter Project, Writing Activities	90 min Tech Prep Masters	65 min **PE:** Chapter Assessment, Test Generator	30 min Chap. Assess. (A or B), Alt. Assess. (A or B), Test Generator	

PE: Pupil's Edition

Alternative Assessment

The following suggest alternative assessments for students who may benefit from a different type of assessment than the regular chapter quizzes and the mid-chapter/end-of-chapter test. Visit the HRW web site to get additional Alternative Assessment material.

internet connect

Alternative Assessment
Go To: **go.hrw.com**
Keyword: **MB1 Alt Assess**

Performance Assessment

1. Let $P(x) = x^4 - 6x^3 - 3x^2 + 52x - 60$.
- **a.** Using synthetic division, write $P(x)$ in factored form.
- **b.** Describe the characteristics of P, including intervals over which it increases and decreases, end behavior, intercepts, local maxima, and local minima.
- **c.** Graph the function by using the information you found in part **b.**

2. Let $P(x) = 2x^4 - 3x^3 + 5x^2 + 11x - 15$.
- **a.** Outline an algebraic strategy for finding all the real zeros of P.
- **b.** Outline a graphical strategy for finding *all* the zeros of P.
- **c.** Use either strategy to find all zeros of P.

Portfolio Project

Suggest that students choose one of the following projects for inclusion in their portfolios.

1. Let $P(x) = (x - a)(x - b)(x - c)(x - d)$, where $a \le b \le c \le d$.
- **a.** Describe the graph of P if a, b, c, and d are all different.
- **b.** Choose values for a, b, c, and d so that the graph has a different number of zeros from the one you described in part **a.** Look through the chapter to find polynomial functions in factored form to help you.

2. The graph of $P(x) = ax^3 + bx^2 + cx + d$ contains points $K(1, 18)$, $L(2, 0)$, $M(4, 0)$, and $N(5, 126)$.
- **a.** Write and solve a system of equations involving a, b, c, and d. Use a calculator with matrix capability to solve for a, b, c, and d.
- **b.** Test the following conjecture: Given the coordinates of four points in a coordinate plane, you can determine an equation for a unique cubic function.

internet connect

The table below identifies the pages in this chapter that contain internet and technology information.

Content Links

Activities Online	**pages 437, 445, 452**
Portfolio Extensions	**pages 431, 455**
Homework Help Online	**pages 430, 438, 446, 453, 463**
Graphic Calculator Support	**474**

Resource Links

Parents can go online and find concepts that students are learning—lesson by lesson—and questions that pertain to each lesson, which facilitate parent-student discussion.

Go To: **go.hrw.com**
Keyword: **MB1 Parent Guide**

Technical Support

The following may be used to obtain technical support for any HRW software product.

Online Help: **www.hrwtechsupport.com**
e-mail: **tschrw@hrwtechsupport.com**

HRW Technical Support Center: **(800)323-9239**
7 AM to 10 PM Monday through Friday Central Time

Visit the HRW math web site at: **www.hrw.com/math**

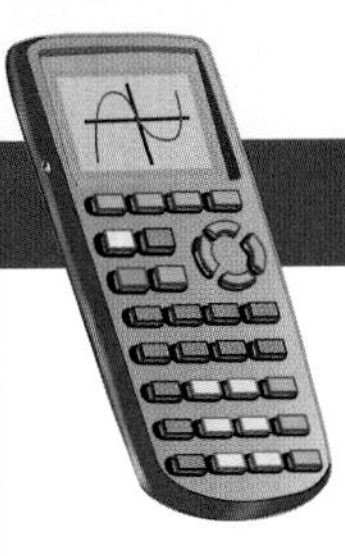

Technology

Lesson Suggestions and Calculator Examples

(Keystrokes are based on a TI-83 calculator.)

Lesson 7.1 An Introduction to Polynomials

In this lesson, students will explore polynomial functions, such as the cubic and quartic polynomial functions whose graphs are shown below.

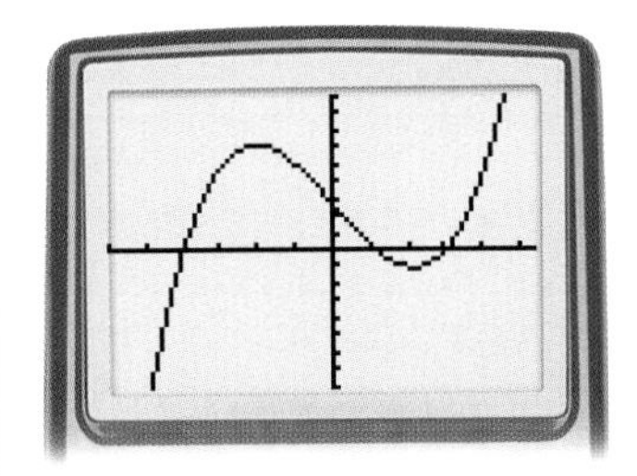

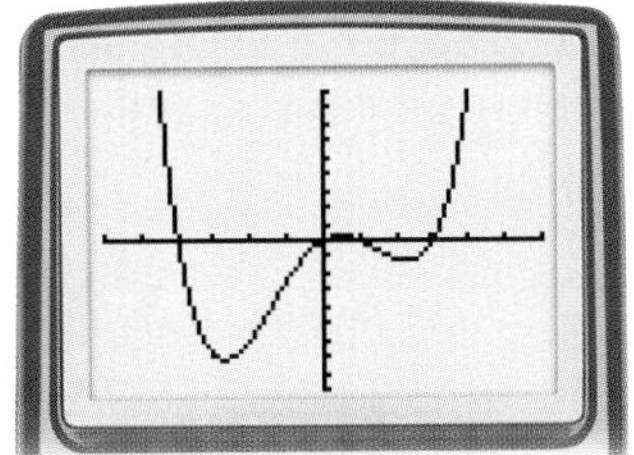

Give students quartic equations of the form $f(x) = (x - a)(x - b)(x - c)(x - d)$ and ask them what happens when two or more of the variables a, b, c, and d are equal.

Lesson 7.2 Polynomial Functions and Their Graphs

The graph of $P(x) = x^3 + 3x^2 - x - 3$, which is given in Example 1 on page 434, is shown at right.

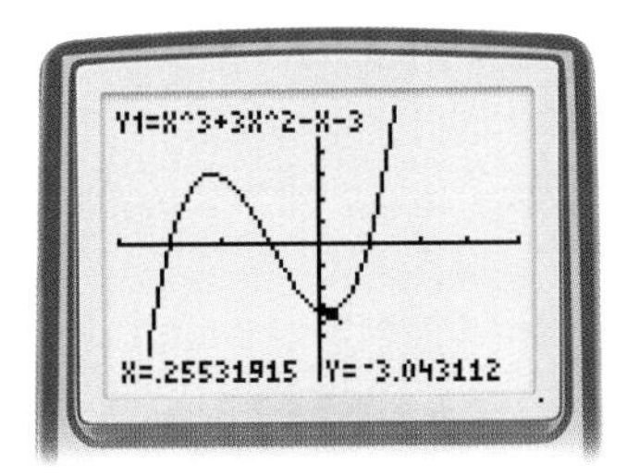

Have students make a table whose headings are *local minimum, local maximum, y-intercept, x-intercept(s), interval(s) of increase, and interval(s) of decrease.* Have students use TRACE to gather information to fill in the table.

Alternatively, consider using the graphics calculator to find an equation whose graph best fits a set of points. Using (0, 0), (1, 2), (2, 0), (3, −2) as an example, have students follow the steps below.

1. Enter the data into **L1** and **L2**.
2. Press STAT and select **CALC** **6:CubicReg** .

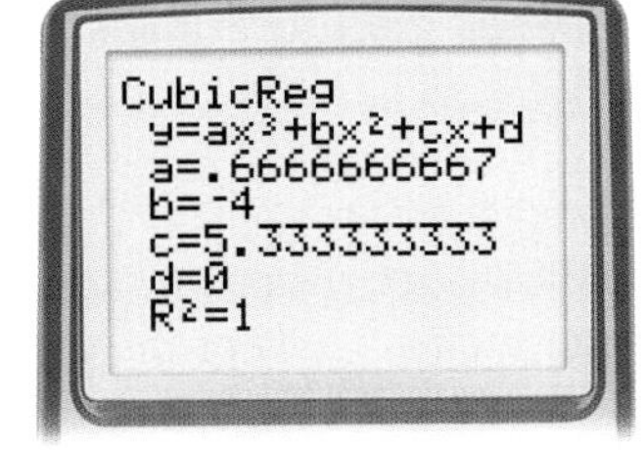

From this display, an equation for the line of best fit is $f(x) \approx 0.67x^3 - 4x^2 + 5.33x$.

Lesson 7.3 Products and Factors of Polynomials

This lesson introduces important connections between algebraic and geometric representations of functions. When using the graphical approach to find x-intercepts, remind students that the intercepts may be integers even though TRACE reports a decimal close to an integer. The graph of the function $P(x) = x^3 - 2x^2 - 5x + 6$, given in Example 4 on page 442, is shown at right above.

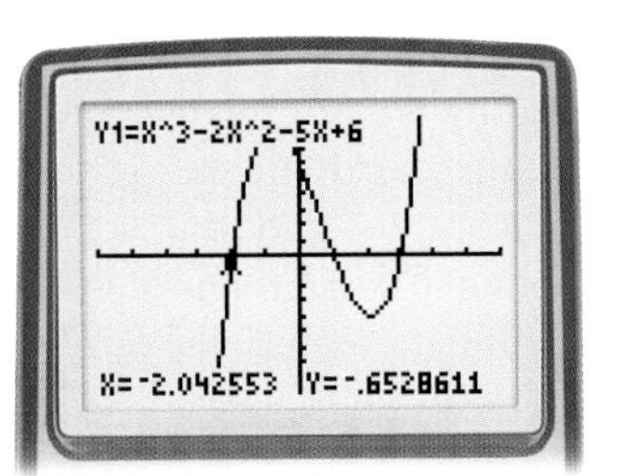

Notice that the display shows an x-intercept at approximately −2. Students should ask if the x-intercept might be −2 and should verify their conjecture.

The graphics calculator can also be used to perform synthetic division or synthetic substitution thereby providing students with exposure to an algorithm, or a finite iterative process. The display at right shows the repetitive process of multiplying and adding that constitutes the synthetic division of $x^3 + 3x^2 - 4x - 12$ by $x - 2$. From the display, students can see that the coefficients of the quotient are 1, 5, and 6 and the remainder is 0 (synthetic division) or just that the remainder is 0 (synthetic substitution).

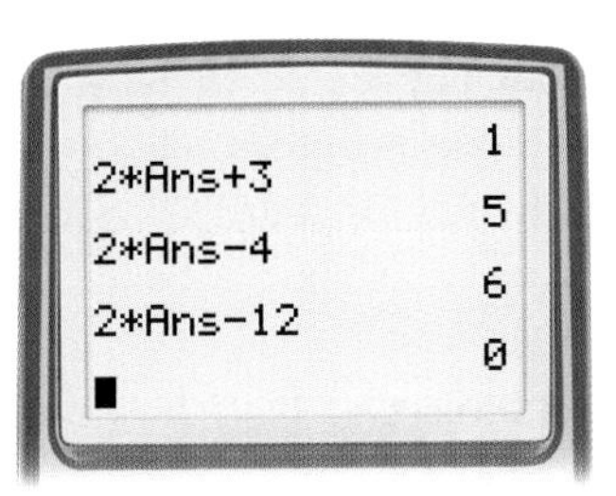

Lesson 7.4 Solving Polynomial Equations

This lesson is a review about how to use a graphics calculator to explore and solve polynomial functions.

Lesson 7.5 Zeros of Polynomial Functions

The graph of $P(x) = 3x^3 - 10x^2 + 10x - 4$ is shown at right.

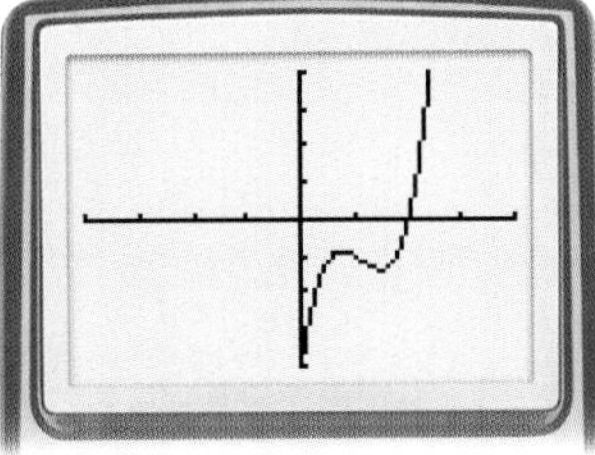

Ask students the following:

1. How many real zeros are showing?
2. What is the total number of zeros?
3. How can the nonreal zeros be found?

For further information, refer to the

- technology discussions in the lessons.
- lesson-related teacher's commentary in the side columns of this *Teacher's Edition.*
- lesson-related *Student Technology Guide* masters.
- *HRW Technology Handbook.*

internet connect

For keystrokes of other graphing calculators models, visit the HRW web site at **go.hrw.com** and enter the keyword **MB1 CALC**.

About Chapter 7

Background Information

Students add and subtract polynomials, factor polynomials of degree greater than 2, divide polynomials by using long division and synthetic division, and learn the characteristic shapes of the graphs of polynomial functions. They use the rational root theorem, the complex conjugate theorem, and the fundamental theorem of algebra to find the zeros of polynomial functions and model real-world data with polynomials.

CHAPTER RESOURCES

- Block-Scheduling Handbook
- Writing Activities for Your Portfolio
- Tech Prep Masters
- Long-Term Project
- Assessment Resources:
  - Mid-Chapter Assessment
  - Chapter Assessments
  - Alternative Assessments
- Test and Practice Generator
- Technology Handbook

Chapter Objectives

- Identify, evaluate, add, and subtract polynomials. [7.1]
- Classify polynomials and describe the shapes of their graphs. [7.1]
- Identify and describe the important features of the graph of a polynomial function. [7.2]

7 Polynomial Functions

POLYNOMIAL FUNCTIONS CAN BE USED TO model many real-world situations, including annuities and volumes. For example, the volumes of the irregularly shaped buildings shown here can be modeled by polynomial functions.

Polynomial functions are important in algebra. In this chapter, you will learn how to combine polynomial functions, how to graph them, and how to find their roots.

Lessons

Atomium in Brussels, Belgium

Montreal Expo Habitat in Montreal, Canada

About the Photos

The Atomium, a metal structure built in the shape of an elementary iron crystal with its nine atoms magnified 150 billion times, was created for the Expo of 1958 in order to celebrate Belgium's metal and iron industry and the belief in atomic power. Each sphere has a diameter of 18 m (≈59 ft), and the entire structure is 102 m high (≈335 ft). The upper sphere has an observation area and a restaurant, and other spheres contain exhibits. The monument is scheduled for restoration and cleaning in the near future, and the maintenance company will need to know the volume and surface area of the structure. The volume of each sphere can be found by using the formula $V = \frac{4}{3}\pi r^3$, where r is the radius of the sphere, and the surface area of each sphere can be found by using $SA = 4\pi r^2$. The total volume of all nine spheres is $V = 9\left(\frac{4}{3}\pi 9^3\right)$, or about 27,483 m^3 (≈970,552 ft^3), and the total surface area of the nine spheres is $SA = 9(4\pi 9^2)$, or about 9161 m^2 (≈98,608 ft^2).

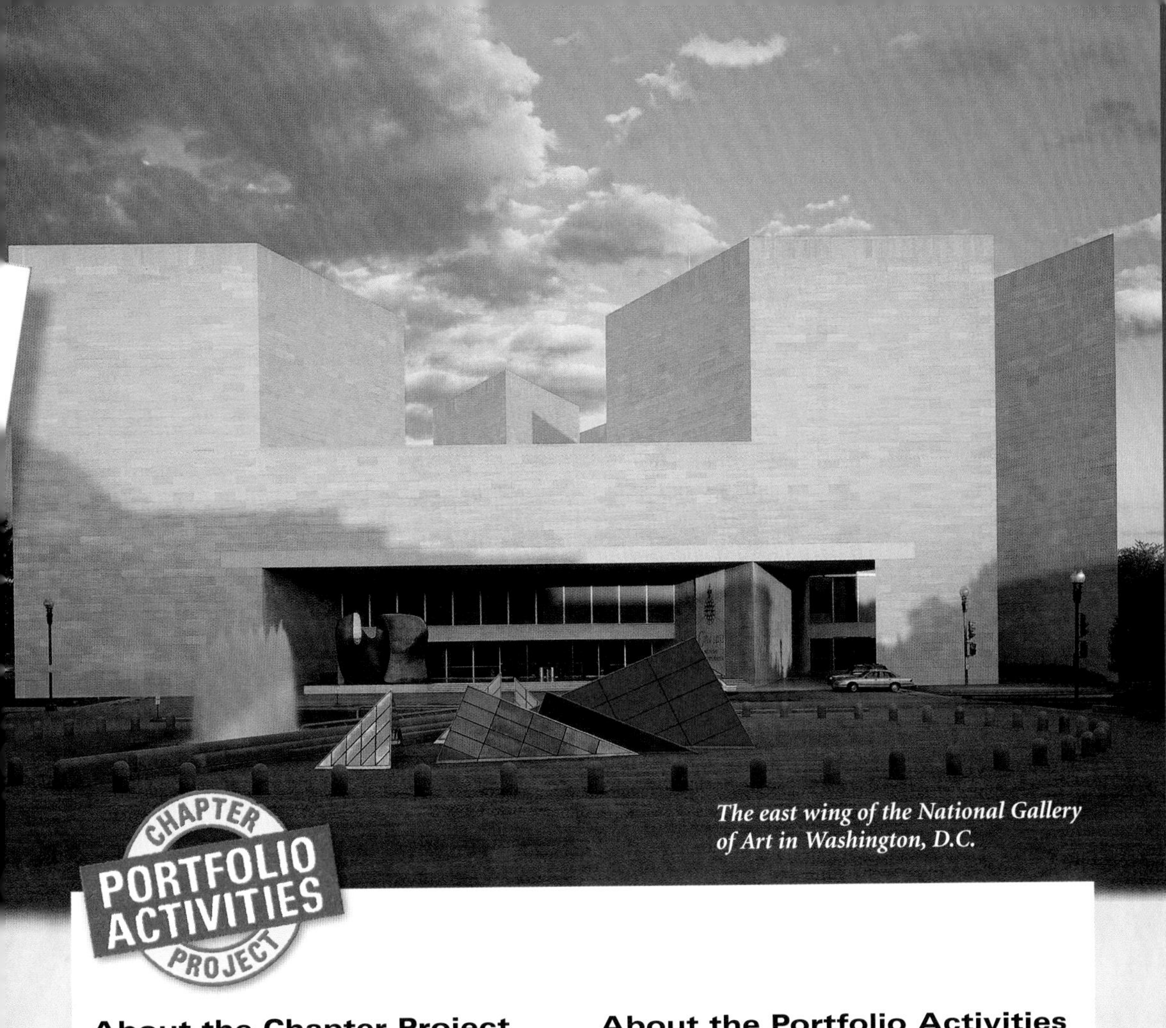

The east wing of the National Gallery of Art in Washington, D.C.

About the Chapter Project

In this chapter, you will use polynomial functions to model real-world data. A good model must consistently provide answers to the question or problem it was created to solve. In the Chapter Project, *Fill It Up!*, you will predict the shape of containers by using polynomial models that are created from the relationship between the volume of water contained and the height of water in the container.

After completing the Chapter Project, you will be able to do the following:

- Collect and organize data.
- Determine a polynomial model that best fits a data set.
- Test your polynomial model.

About the Portfolio Activities

Throughout the chapter, you will be given opportunities to complete Portfolio Activities that are designed to support your work on the Chapter Project.

- Creating a polynomial model for the volume of an irregularly shaped container is included in the Portfolio Activity on page 431.
- Analyzing the behavior of the polynomial model is included in the Portfolio Activity on page 439.
- Creating a polynomial model and using it to solve problems is included in the Portfolio Activity on page 455.

- Use a polynomial function to model real-world data. [7.2]
- Multiply polynomials, and divide one polynomial by another by using long division and synthetic division. [7.3]
- Use the remainder and factor theorems to solve problems. [7.3]
- Solve polynomial equations. [7.4]
- Find the real zeros of polynomial functions and state the multiplicity of each. [7.4]
- Use the rational root theorem and the complex conjugate root theorem to find the zeros of a polynomial function. [7.5]
- Use the fundamental theorem of algebra to write a polynomial function given sufficient information about its zeros. [7.5]

Portfolio Activities appear at the end of Lessons 7.1, 7.2, and 7.4. Each serves as preparation for the Chapter Project. The Portfolio Activities as well as the Chapter Project Activities are appropriate for inclusion in the student's portfolio. Students should be encouraged to include in their portfolios any other work in which they feel a sense of pride or a sense of accomplishment.

internet connect go.hrw.com

Chapter Internet Features and Online Activities

Lesson	Keyword	Page
7.1	MB1 Homework Help	430
	MB1 Polynomials	431
7.2	MB1 Homework Help	438
	MB1 Traffic	437
7.3	MB1 Homework Help	446
	MB1 Ceramics	445
7.4	MB1 Homework Help	453
	MB1 Biosphere	452
	MB1 PolySolve	455
7.5	MB1 Homework Help	463

Prepare

NCTM PRINCIPLES & STANDARDS 1–4, 6–10

QUICK WARM-UP

Evaluate each expression for $x = -2$.

1. $-x + 1$ 3 **2.** $x^2 - 5$ -1

3. $-(x - 6)$ 8 **4.** $7 - x^3$ 15

Find the opposite of each expression.

5. $x - 3$ $-x + 3$

6. $-x^2 + 5x + 1$ $x^2 - 5x - 1$

Simplify each expression.

7. $8x - x$ $7x$

8. $(x + 5) + (2x + 3)$ $3x + 8$

9. $(x + 9) - (4x + 6)$ $-3x + 3$

10. $(-x^2 - 2) - (x^2 - 2)$ $-2x^2$

Also on Quiz Transparency 7.1

Teach

Why Polynomial functions can be used to model the value of a type of investment known as an annuity. Ask students what they know about various investment alternatives. Have them discuss savings accounts, retirement plans, and insurance policies. What factors do they think would determine the value of these investment alternatives?

An Introduction to Polynomials

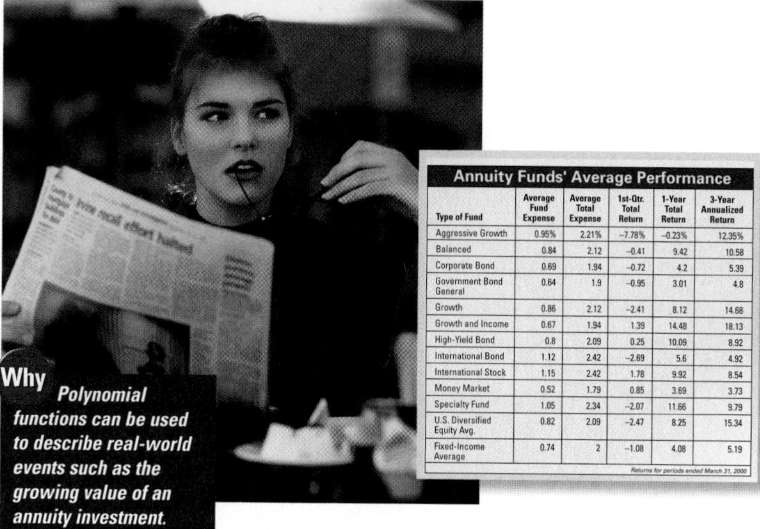

Annuity Funds' Average Performance

Type of Fund	Average Fund Expense	Average Total Expense	1st-Qtr. Total Return	1-Year Total Return	3-Year Annualized Return
Aggressive Growth	0.95%	2.21%	–7.78%	–0.23%	12.35%
Balanced	0.84	2.12	–0.41	9.42	10.58
Corporate Bond	0.69	1.94	–0.72	4.2	5.39
Government Bond General	0.64	1.9	–0.95	3.01	4.8
Growth	0.86	2.12	–2.41	8.12	14.68
Growth and Income	0.67	1.94	1.39	14.48	18.13
High-Yield Bond	0.8	2.09	0.25	10.09	8.92
International Bond	1.12	2.42	–2.69	5.6	4.92
International Stock	1.15	2.42	1.78	9.92	8.54
Money Market	0.52	1.79	0.85	3.69	3.73
Specialty Fund	1.05	2.34	–2.07	11.66	9.79
U.S. Diversified Equity Avg.	0.82	2.09	–2.47	8.25	15.34
Fixed-Income Average	0.74	2	–1.08	4.08	5.19

Returns for periods ended March 31, 2000

Why *Polynomial functions can be used to describe real-world events such as the growing value of an annuity investment.*

Objectives

- Identify, evaluate, add, and subtract polynomials.
- Classify polynomials, and describe the shapes of their graphs.

APPLICATION
INVESTMENTS

Periodically adding a fixed amount of money to an account that pays compound interest is an investment that people often use to prepare for the future. Such an investment option is called an *annuity*.

Consider an annuity of \$1000 invested at the beginning of each year in an account that pays 5% interest compounded annually. The diagram below illustrates the growth of this investment over 4 years.

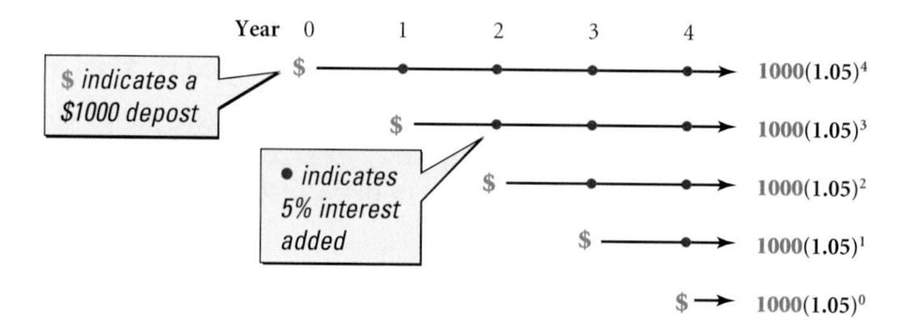

The total value of this investment after 4 years can be written as the sum below.

$$1000(1.05)^4 + 1000(1.05)^3 + 1000(1.05)^2 + 1000(1.05)^1 + 1000(1.05)^0$$

To consider different interest rates, replace 1.05 with 1.06 for 6%, 1.07 for 7%, and so on. In general, the value of this investment after 4 years can be represented by the *polynomial* expression $1000x^4 + 1000x^3 + 1000x^2 + 1000x + 1000$, where x is the multiplier determined by the annual interest rate.

How much will the investment be worth at the time the last payment is made if the interest rate is 6%? *You will answer this question in Example 2.*

Alternative Teaching Strategy

USING PATTERNS You can show students how polynomial addition is a generalization of addition of whole numbers. Write the number 1111, the number 4321, and the sum 5432 in expanded notation. Then use a vertical format to add the polynomials $x^3 + x^2 + x + 1$ and $4x^3 + 3x^2 + 2x + 1$.

$$\begin{aligned} 1111 &= 1 \cdot 10^3 + 1 \cdot 10^2 + 1 \cdot 10 + 1 \\ +\ 4321 &= 4 \cdot 10^3 + 3 \cdot 10^2 + 2 \cdot 10 + 1 \\ \hline 5432 &= 5 \cdot 10^3 + 4 \cdot 10^2 + 3 \cdot 10 + 2 \end{aligned}$$

$$\begin{array}{r} x^3 + x^2 + x + 1 \\ +\ 4x^3 + 3x^2 + 2x + 1 \\ \hline 5x^3 + 4x^2 + 3x + 2 \end{array}$$

Point out that in addition of whole numbers, the place value of each digit is indicated by a power of 10. In the polynomial expression, the power of the variable x indicates the degree of the term. Have students write an example that shows the relationship between the subtraction of whole numbers and the subtraction of polynomials. Then tell students they can add or subtract polynomials by using a vertical or horizontal format.

A **monomial** is a numeral, a variable, or the product of a numeral and one or more variables. A monomial with no variables, such as -1 or $\frac{2}{3}$, is called a **constant**.

A **coefficient** is the numerical factor in a monomial. For example:

- x has a coefficient of 1.
- $-2t$ has a coefficient of -2.
- $\frac{-x^3y^2}{3}$ has a coefficient of $-\frac{1}{3}$ because it can be written as $-\frac{1}{3}x^3y^2$.
- $-ab$ has a coefficient of -1.

The **degree of a monomial** is the sum of the exponents of its variables. For example, x^2yz is of degree 4 because $x^2yz = x^2y^1z^1$ and $2 + 1 + 1 = 4$. A nonzero constant such as 3 is of degree 0 because it can be written as $3x^0$.

A **polynomial** is a monomial or a sum of terms that are monomials. In this chapter, you will study polynomials in one variable. Polynomials can be classified by the number of terms they contain. A polynomial with two terms is a **binomial**. A polynomial with three terms is a **trinomial**.

The **degree of a polynomial** is the same as that of its term with the greatest degree. Polynomials can also be classified by degree, as shown below.

CLASSIFICATION OF A POLYNOMIAL BY DEGREE

Degree	Name	Example
$n = 0$	constant	3
$n = 1$	linear	$5x + 4$
$n = 2$	quadratic	$-x^2 + 11x - 5$
$n = 3$	cubic	$4x^3 - x^2 + 2x - 3$
$n = 4$	quartic	$9x^4 + 3x^3 + 4x^2 - x + 1$
$n = 5$	quintic	$-2x^5 + 3x^4 - x^3 + 3x^2 - 2x + 6$

EXAMPLE 1 Classify each polynomial by degree and by number of terms.

a. $2x^3 - 3x + 4x^5$

b. $-2x^3 + 3x^4 + 2x^3 + 5$

SOLUTION

a. The greatest exponent of x is 5, so the degree is 5.

The polynomial has three terms, so it is a trinomial.

The polynomial is a quintic trinomial.

b. When simplified to $3x^4 + 5$, the greatest exponent of x is 4, so the degree is 4.

The simplified polynomial has two terms, so it is a binomial.

The polynomial is a quartic binomial.

TRY THIS Classify each polynomial by degree and by number of terms.

a. $x^2 + 4 - 8x - 2x^3$

b. $3x^3 + 2 - x^3 - 6x^5$

Teaching Tip

Point out that a polynomial is made of several monomials, also called terms, which are separated by plus or minus signs. Note that the sign preceding a term is included in that term, and a leading plus sign is understood when no sign is written. For example, $2x - 3$ is a linear binomial which contains the terms $+2x$ and -3.

ADDITIONAL EXAMPLE 1

Classify each polynomial by degree and by number of terms.

a. $5x + 2x^3 - 2x^2$
cubic trinomial

b. $x^5 - 4x^3 - x^5 + 3x^2 + 4x^3$
quadratic monomial

TRY THIS

a. cubic polynomial

b. quintic trinomial

Interdisciplinary Connection

ECONOMICS An annuity requires that the periodic payments always be the same amount and that the interval between the payments always be the same. It also assumes that the interest is compounded once each time period. Payments can be made at the beginning or end of a time period. An annuity is classified as an *ordinary annuity* if the payments occur at the end of the period and as an *annuity due* if the payments occur at the beginning of the period.

Example 2 illustrates an annuity due. To have students see how the value of an annuity can be affected by the alternative chosen, have them represent the value of an ordinary annuity with a polynomial function and evaluate the function for $x = 1.06$. Compare this amount with the result of the annuity due obtained in Example 2. **The value of the ordinary annuity is $4374.62.**

ADDITIONAL EXAMPLE 2

Refer to the annuity described at the beginning of the lesson. **How much will the investment be worth at the time the last payment is made if the annual interest rate is 7%?** $5750.74

Teaching Tip

TECHNOLOGY If you are using a TI-83 or TI-82 graphics calculator, you can create the table shown on this page by letting **Y1=1000*(x⁴+x³+ x²+x+1),** pressing 2nd WINDOW to access the **TABLE SETUP** menu, and entering the following settings:

TblStart=1.03
(or **TblMin=1.03**)
ΔTbl=.01
Indpnt: Auto Ask
Depend: Auto Ask

Then use the following window to view the graph:

Xmin=0	**Ymin=–1000**
Xmax=1.5	**Ymax=7000**
Xscl=.1	**Yscl=500**

TRY THIS
17.6875

ADDITIONAL EXAMPLE 3

Find the sum: $(6x^3 + 3x^2 - 4) + (10 - 3x - 5x^2 + 2x^3)$.
$8x^3 - 2x^2 - 3x + 6$

TRY THIS
$4x^3 - 7x^2 + 13x - 12$

CRITICAL THINKING
$P = -2x^2 + 3x - 5$

Evaluating Polynomials

Example 2 shows you how polynomials can be used for calculations in real-world situations.

EXAMPLE 2 Refer to the annuity described at the beginning of the lesson.

APPLICATION — INVESTMENTS

How much will the investment be worth at the time the last payment is made if the annual interest rate is 6%?

SOLUTION

Method 1 Use substitution.

$1000x^4 + 1000x^3 + 1000x^2 + 1000x + 1000$
$= 1000(1.06)^4 + 1000(1.06)^3 + 1000(1.06)^2 + 1000(1.06) + 1000$
$= 5637.09$

Method 2 Use a table or a graph.

PROBLEM SOLVING

Enter $y = 1000x^4 + 1000x^3 + 1000x^2 + 1000x + 1000$ into a graphics calculator. **Use a table of values or a graph** to find the y-value for $x = 1.06$.

TECHNOLOGY — GRAPHICS CALCULATOR
Keystroke Guide, page 474

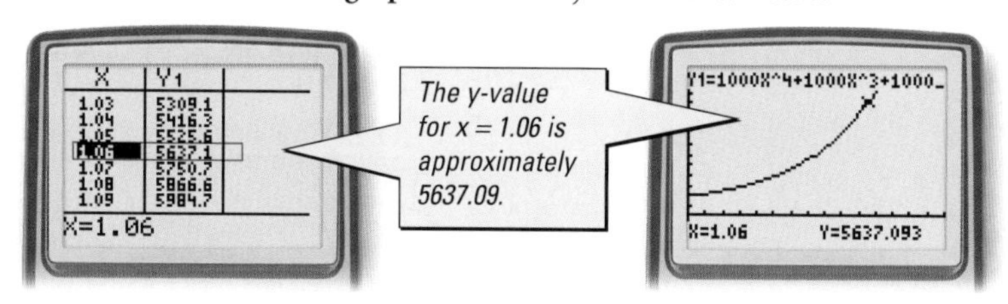

The investment will be worth $5637.09.

TRY THIS Evaluate the polynomial $3x^4 + 2x^2 + 2x - 5$ for $x = 1.5$.

Adding and Subtracting Polynomials

To add and subtract polynomials, combine like terms. Recall that *like terms* have the same degree, or exponent, of the variable. When you write your answer, use *standard form*. The **standard form** of a polynomial expression is written with the exponents in *descending order* of degree.

EXAMPLE 3 **Find the sum.** $(-2x^2 - 3x^3 + 5x + 4) + (-2x^3 + 7x - 6)$

SOLUTION

$(-2x^2 - 3x^3 + 5x + 4) + (-2x^3 + 7x - 6)$
$= (-3x^3 - 2x^3) + (-2x^2) + (5x + 7x) + (4 - 6)$ *Combine like terms.*
$= -5x^3 - 2x^2 + 12x - 2$ *Write in standard form.*

TRY THIS Find the sum. $(2x^4 + 4x^3 + 5x - 2) + (-2x^4 - 7x^2 + 8x - 10)$

CRITICAL THINKING Find a polynomial expression P such that $(2x^2 - 3x + 5) + P = 0$.

Enrichment

Have students evaluate $6x^3 - 2x^2 + 4x - 5$ when $x = 10$ by using substitution. 5835 To show students a more efficient way to evaluate the polynomial when $x = 10$, have them rewrite the polynomial as $6x^3 - 2x^2 + 4x - 5 = (6x^2 - 2x + 4)x - 5 = [(6x - 2)x + 4]x - 5$. Solve the expression inside the parentheses first and then work outward, using the following steps:

1. Multiply 10 by 6.	60
2. Subtract 2.	58
3. Multiply by 10.	580
4. Add 4.	584
5. Multiply by 10.	5840
6. Subtract 5.	5835

Point out that the rewritten form of the polynomial is obtained by working from right to left, factoring out x at each step by using the distributive property. This procedure is called *synthetic substitution* or *synthetic division* and will be studied in Lesson 7.3. Give students other polynomials to evaluate for specific values by using this procedure.

EXAMPLE 4 Find the difference. $(-6x^3 - 6x^2 + 7x - 1) - (3x^3 - 5x^2 - 2x + 8)$

SOLUTION

$(-6x^3 - 6x^2 + 7x - 1) - (3x^3 - 5x^2 - 2x + 8)$

$= (-6x^3 - 3x^3) + (-6x^2 + 5x^2) + (7x + 2x) + (-1 - 8)$ *Combine like terms.*

$= -9x^3 - x^2 + 9x - 9$ *Write in standard form.*

TRY THIS Find the difference. $(3x^3 - 12x^2 - 5x + 1) - (-x^2 + 5x + 8)$

Graphing Polynomial Functions

A **polynomial function** is a function that is defined by a polynomial expression. In the Activity below, you will explore the characteristic shapes of the graphs of some polynomial functions.

Activity: Exploring Graphs of Polynomial Functions

TECHNOLOGY GRAPHICS CALCULATOR

Keystroke Guide, page 474

You will need: a graphics calculator

Graph each function below in a viewing window that shows all of the "U-turns" in the graph. Copy the table and record the degree and number of U-turns.

	Function	Degree	Number of U-turns in the graph
1.	$y = x^2 + x - 2$	2	1
2.	$y = 3x^3 - 12x + 4$		
3.	$y = -2x^3 + 4x^2 + x - 2$		
4.	$y = x^4 + 5x^3 + 5x^2 - x - 6$		
5.	$y = x^4 + 2x^3 - 5x^2 - 6x$		

PROBLEM SOLVING

6. **Look for a pattern.** Make a conjecture about the degree of a function and the number of U-turns in its graph.

Graph each function below in a viewing window that shows all of the U-turns in the graph. Copy the table and record the degree and number of U-turns.

	Function	Degree	Number of U-turns in the graph
7.	$y = x^3$		
8.	$y = x^3 - 3x^2 + 3x - 1$		
9.	$y = x^4$		

CHECKPOINT ✔

10. Now make another conjecture about the degree of a function and the number of U-turns in its graph. Is this conjecture different from the conjecture you made in Step 6? Explain.

Inclusion Strategies

ENGLISH LANGUAGE DEVELOPMENT To make sure that students recognize the terminology introduced in this lesson, instruct them to add and subtract polynomials that are described to them verbally. For example, ask them to write a quartic and quadratic binomial whose sum is a quartic binomial and to write two cubic polynomials whose difference is a linear polynomial.

sample answers:
$(2x^4 - 3x^2) + (3x^2 + 4) = 2x^4 + 4$;
$(x^3 + 2x^2 - x + 3) - (x^3 + 2x^2 + 3x - 5) = -2x + 8$

ADDITIONAL EXAMPLE 4

Find the difference below.
$(5x^2 - 6x - 11) - (-8x^3 + x^2 + 2)$
$8x^3 + 4x^2 - 6x - 13$

TRY THIS
$3x^3 - 11x^2 - 10x - 7$

Activity Notes

In this Activity, students estimate the number of "U-turns" in a graph by looking at eight polynomial functions. After graphing the first five functions, they may guess that the number of U-turns is one less than the degree of the polynomial. After graphing the last three polynomials, they should note that the number of U-turns is *less than or equal to* one less than the degree of the polynomial.

Point out to students that a U-turn in a graph happens when the function values change from increasing to decreasing or vice versa and indicates either a local maximum or a local minimum.

Teaching Tip

TECHNOLOGY If you are using a TI-82 or TI-83 graphics calculator for this Activity, the following window values will show all of the U-turns in the graphs:

Xmin=–5	Ymin=–15
Xmax=3	Ymax=15
Xscl=1	Yscl=1

CHECKPOINT ✔

10. The number of U-turns in a graph is no greater than one less than the degree of the function.

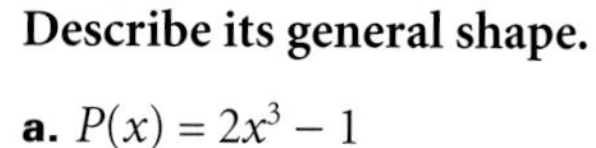

ADDITIONAL EXAMPLE 5

Graph each function. Describe its general shape.

a. $P(x) = 2x^3 - 1$

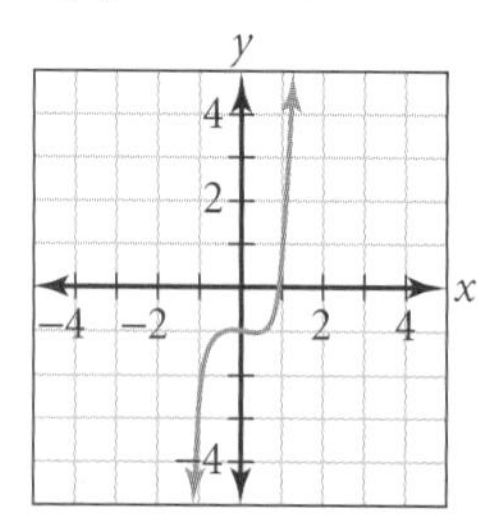

The graph of this cubic function is a curve that always rises to the right.

b. $Q(x) = -3x^4 + 2$

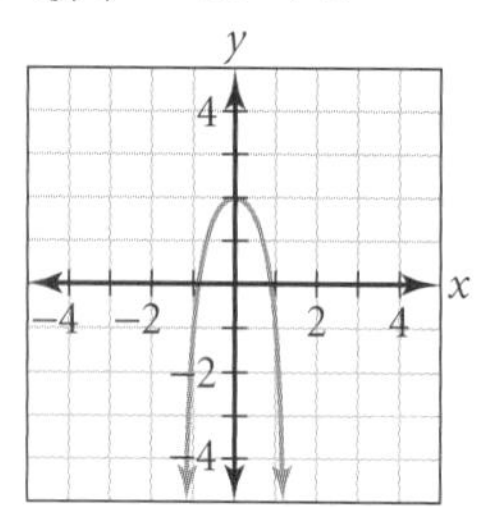

The graph of this quartic function is a U-shape that has one turn.

TRY THIS

a.

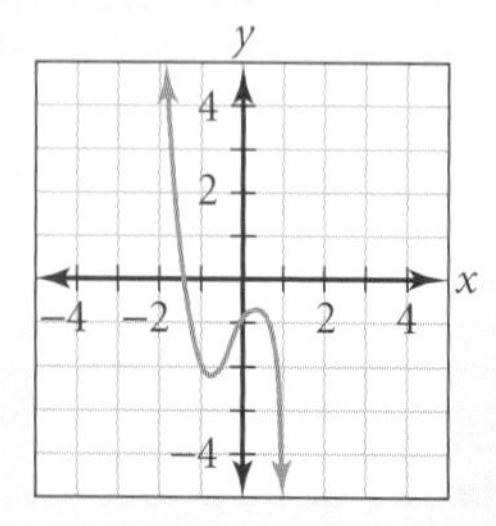

The graph of this function is an S-shape that has 2 turns.

b.

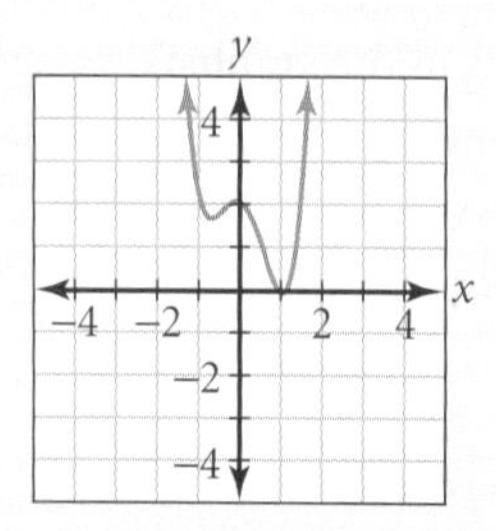

The graph of this function is a W-shape that has 3 turns.

CHECKPOINT ✔

odd degree range: $-\infty < f(x) > \infty$; even degree range: $f(x) > c$, where c is the minimum function value

EXAMPLE 5 **Graph each function. Describe its general shape.**

a. $P(x) = 3x^3 - 5x^2 - 2x + 1$ **b.** $Q(x) = x^4 - 8x^2$

SOLUTION

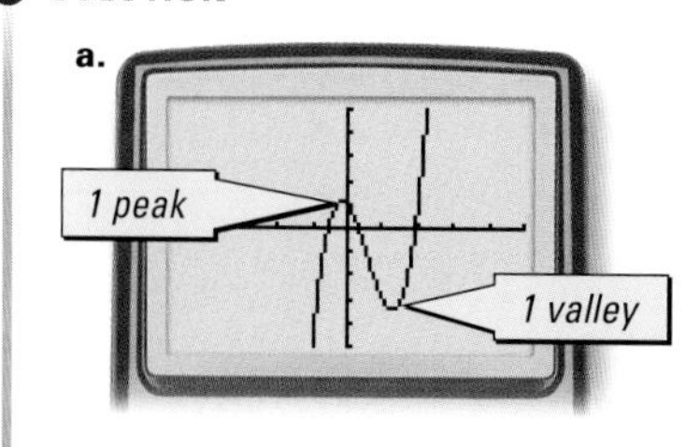

a. The graph of this cubic function has an S-shape with 2 U-turns.

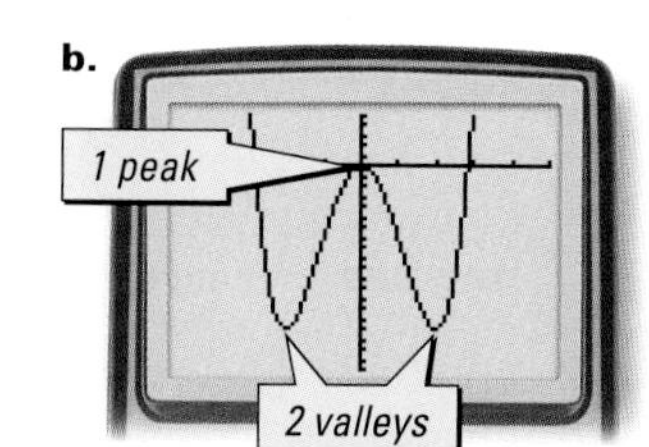

b. The graph of this quartic function has a W-shape with 3 U-turns.

TRY THIS Graph each function. Describe its general shape.

a. $P(x) = -3x^3 - 2x^2 + 2x - 1$ **b.** $Q(x) = 2x^4 - 3x^2 - x + 2$

Examine the shapes of the linear, quadratic, cubic, and quartic functions shown below.

Linear

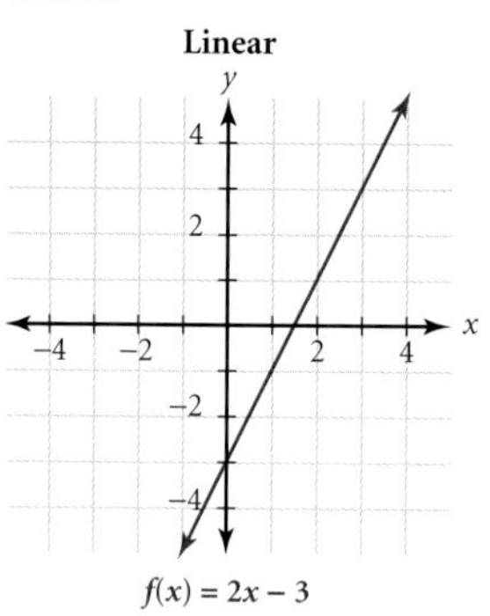

$f(x) = 2x - 3$

Quadratic

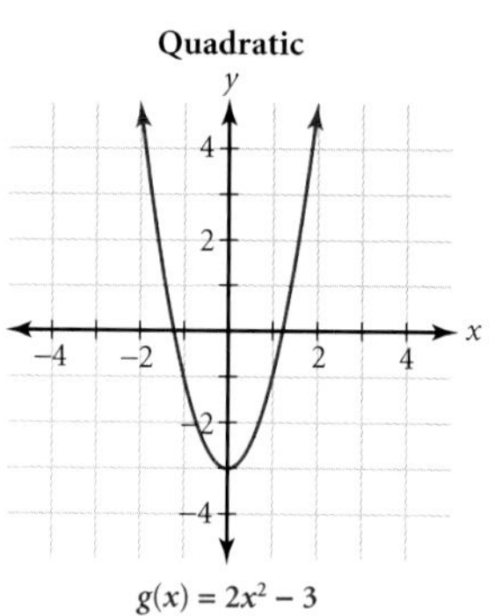

$g(x) = 2x^2 - 3$

Cubic

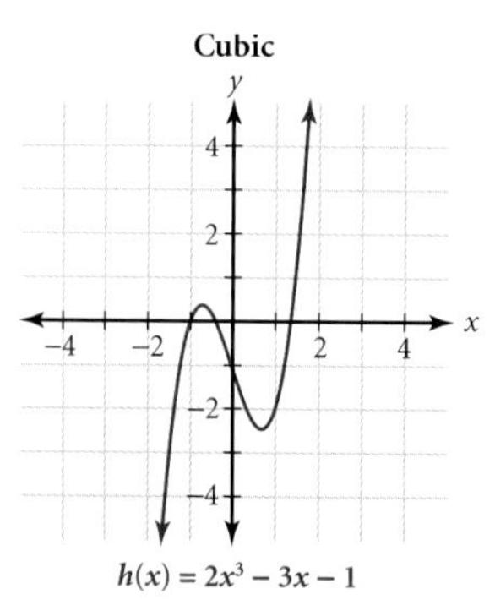

$h(x) = 2x^3 - 3x - 1$

Quartic

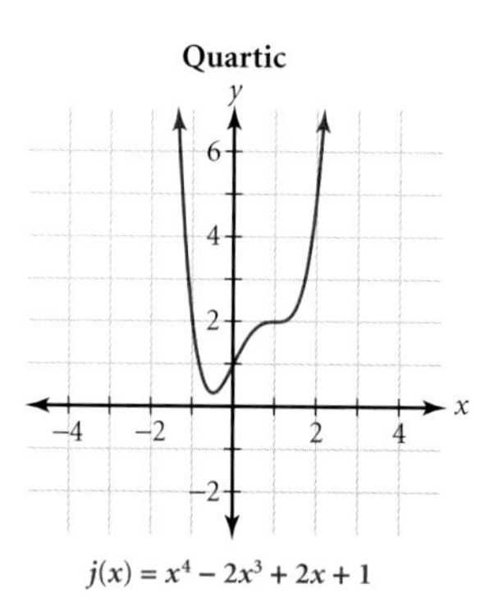

$j(x) = x^4 - 2x^3 + 2x + 1$

CHECKPOINT ✔ What is the range of the polynomial functions above that are of an odd degree? What can you say about the ranges of the polynomial functions above that are of an even degree?

In Lesson 7.2, you will learn more about the graphs of polynomial functions.

Reteaching the Lesson

USING PATTERNS Give students the polynomials $4x^3 + 5x^2 - 2x + 3$ and $x^3 - 4x - 7$. Show them the use of the commutative and distributive properties in combining like terms to add the polynomials.

$(4x^3 + 5x^2 - 2x + 3) + (x^3 - 4x + 7)$
$= (4x^3 + 1x^3) + (5x^2 - 4x^2) + (-2x + 0x) + (3 + 7)$
$= (4 + 1)x^3 + (5 - 4)x^2 + (-2 + 0)x + (3 + 7)$
$= 5x^3 + 1x^2 + (-2x) + 10$
$= 5x^3 + x^2 - 2x + 10$

Show them how to use the same procedure to subtract the polynomial $x^2 - 7x + 9$ from $5x^3 + 2x^2 - 3x - 8$.

$(5x^3 + 2x^2 - 3x - 8) - (x^2 - 7x + 9)$
$= (5x^3 + 2x^2 - 3x - 8) + (-x^2 + 7x - 9)$
$= (5 + 0)x^3 + (2 - 1)x^2 + (-3 + 7)x + (-8 - 9)$
$= 5x^3 + x + 4x - 17$

Give students other polynomials to add and subtract by using this procedure.

Exercises

Communicate

1. In your own words, define a *polynomial.*
2. Describe how to determine the degree of a polynomial function.
3. Use the definition of a polynomial function to explain how you know that a quadratic function is a polynomial function.

Guided Skills Practice

Classify each polynomial by degree and by number of terms. ***(EXAMPLE 1)***

4. $x^4 + 3x^3 - x^2$
5. $x^5 + 3x^4 - 4x^2 + x - 1$
6. Evaluate the polynomial $x^3 + 2x^2 - x + 1$ for $x = 2$. ***(EXAMPLE 2)***
7. Find the sum. $(2x^3 + 3x^2 - x + 2) + (-3x^2 + 4x + 5)$ ***(EXAMPLE 3)***
8. Find the difference. $(6x^3 - 5x^2 + 14x + 3) - (3x^3 - 2x^2 + 7x - 2)$ ***(EXAMPLE 4)***

Graph each function and describe the general shape. ***(EXAMPLE 5)***

9. $f(x) = x^3 - x$
10. $f(x) = x^4 - x^2 + 1$

4. quartic trinomial
5. quintic polynomial
6. 15
7. $2x^3 + 3x + 7$
8. $3x^3 - 3x^2 + 7x + 5$
9. S-shape, 2 turns
10. W-shape, 3 turns

Practice and Apply

Write each polynomial in standard form.

11. $5x^3 + 4x + 2x^2 + 1$
12. $4x^4 + x^2 + x^3 + x + 1$
13. $2.7x^3 + 3.3x^8 + 4.1x^2$
14. $9.1x^2 + 5.4x^5 + 3.3x^2 + 2.1$
15. $\frac{x^7}{13} + \frac{x^9}{7} - \frac{2}{3}$
16. $\frac{13}{15}x^4 + \frac{5}{7}x^3 + \frac{3}{5}x^5 + \frac{1}{2}$

Determine whether each expression is a polynomial. If so, classify the polynomial by degree and by number of terms.

17. $7x^5 + 3x^3 - 2x + 4$
18. $-4x^2 + 3x^3 - 5x^6 + 4$
19. $3^x + 2^x - x - 7$
20. $4^{2x} + 5^x - x + 1$
21. $0.35x^4 + 2x^2 + 3.8x$
22. $7.81x^4 + 8.9x^3 + 2.5x^2$
23. $\frac{3}{x^2} + \frac{5}{x} + 6$
24. $\frac{8}{x^3} - \frac{7}{x^2} + x$
25. $\frac{5}{7}x^6 + \frac{2}{3}x^4 + 5$
26. $\frac{x^5}{5} - \frac{x^3}{3}$
27. $\sqrt{x} - 1$
28. $7\sqrt{x} + 4$

Evaluate each polynomial expression for the indicated value of *x*.

29. $x^3 + x^2 + 1$ for $x = -3$ **−17**
30. $x^4 + 2x^3 + 2$ for $x = -2$ **2**
31. $-2x^3 - 3x + 2$ for $x = 4$ **−138**
32. $-4x^3 + 1 + x$ for $x = 3$ **−104**
33. $3x^3 + x^2 + 2x + 4$ for $x = 5$ **414**
34. $5x^3 + 2x^2 - 5x + 2$ for $x = 6$ **1124**
35. $\frac{1}{4}x^4 + \frac{1}{8}x^3 + \frac{3}{8}x^2 + \frac{5}{8}x + \frac{7}{8}$ for $x = 2$ **$8\frac{5}{8}$**
36. $\frac{3}{10}x^3 + \frac{7}{10}x^2 + \frac{1}{10}x + \frac{9}{10}$ for $x = 10$ **$371\frac{9}{10}$**
37. $1 + x^2 - 3x^3$ for $x = 2.5$ **−39.625**
38. $5x^3 + 4x + 2x^2 + 1$ for $x = 3.8$ **319.44**

11. $5x^3 + 2x^2 + 4x + 1$
12. $4x^4 + x^3 + x^2 + x + 1$
13. $3.3x^8 + 2.7x^3 + 4.1x^2$
14. $5.4x^5 + 12.4x^2 + 2.1$
15. $\frac{x^9}{7} + \frac{x^7}{13} - \frac{2}{3}$
16. $\frac{3}{5}x^5 + \frac{13}{15}x^4 + \frac{5}{7}x^3 + \frac{1}{2}$
17. yes; quintic polynomial
18. yes; polynomial of degree 6
19. no
20. no
21. yes; quartic trinomial
22. yes; quartic trinomial
23. no
24. no
25. yes; trinomial of degree 6
26. yes; quintic binomial
27. no
28. no

Teaching Tip

TECHNOLOGY If you are using a TI-82 or TI-83 graphics calculator to create the graph in part **a** of Example 5, set the viewing window as follows:

Xmin=–5	Ymin=–5
Xmax=5	Ymax=5
Xscl=1	Yscl=1

To create the graph in part **b** of Example 5, set the viewing window as follows:

Xmin=–5	Ymin=–20
Xmax=5	Ymax=5
Xscl=1	Yscl=1

Assess

Selected Answers

Exercises 4–9, 11–79 odd

ASSIGNMENT GUIDE

In Class	1–10
Core	11–57 odd, 63
Core Plus	10–62 even
Review	64–80
Preview	81

Extra Practice can be found beginning on page 940.

Error Analysis

When subtracting polynomials in horizontal form, some students will change the sign of only the leading coefficient of the second polynomial instead of changing the sign of every term of the second polynomial by distributing –1 throughout.

Remind students that when subtracting integers, the sign of the second number is changed and the operation is changed to addition. This procedure applies to polynomial subtraction when the sign of every term in the second polynomial is changed.

51. S-shape with 2 turns
52. S-shape with 2 turns
53. S-shape with 2 turns
54. W-shape with 3 turns
55. S-shape with 2 turns
56. S-shape with 2 turns
57. W-shape with 3 turns
58. W-shape with 3 turns

Practice

NAME ______________________ CLASS __________ DATE __________

Practice
7.1 ***An Introduction to Polynomials***

Determine whether each expression is a polynomial. If so, classify the polynomial by degree and by number of terms.

1. $5x^2 - 22x^5 + 17x$ — yes; quintic trinomial
2. $\frac{x}{2} - \frac{x^2}{2} + 13$ — yes; quadratic trinomial
3. $\frac{7}{x^2} + \frac{13}{x^3}$ — no
4. $-2x^3 - 4x^2 - 15x + 7$ — yes; cubic polynomial
5. $\sqrt[3]{x} + 12\sqrt{x}$ — no
6. $43x^{-6} + 9x^{-7} + 12x^{-1}$ — no

Evaluate each polynomial expression for the indicated value of *x*.

7. $2x^3 - 3x^2 + 4x, x = -2$ — −36
8. $-x^4 + 3x^3 - 2x^2 + 4, x = -1$ — −2
9. $x^5 + x^4 + x^3 + x^2 + 1, x = 2$ — 61
10. $0.5x^3 - 0.6x^2 - 3x, x = 10$ — 410
11. $7x^2 - 19x, x = 5$ — 80
12. $0.75x^3 - 15x^2 + 10x, x = 4$ — −152

Write each sum or difference as a polynomial in standard form.

13. $(3x^4 + 12x^3 - 2x^2) + (5x^4 - x^3 + 7x^2)$ — $8x^4 + 11x^3 + 5x^2$
14. $(-7x^4 + 24x^5 - 3x^2 + 9) - (2x^5 + 6x^4 + x +$ — $22x^5 - 13x^4 - 3x^2 - x + 8$
15. $(8.8x + 2 + 3x^2 - x^4) - (5x^3 + 10x - 7x^2)$ — $-x^4 - 5x^3 + 10x^2 - 1.2x + 2$
16. $(7.1x^3 + 3.2x^2 - 7x + 8) + (9x^2 - 2x^3 + 18)$ — $5.1x^3 + 12.2x^2 - 7x + 26$

Sketch the graph of each function. Describe the general shape of the graph.

17. $a(x) = -2x^4 + 5x^3 - 2$
18. $k(x) = 4x^4 + 4x^3 - 6x^2$
19. $f(x) = \frac{3}{4}x^3 + 2x^2 + 1$

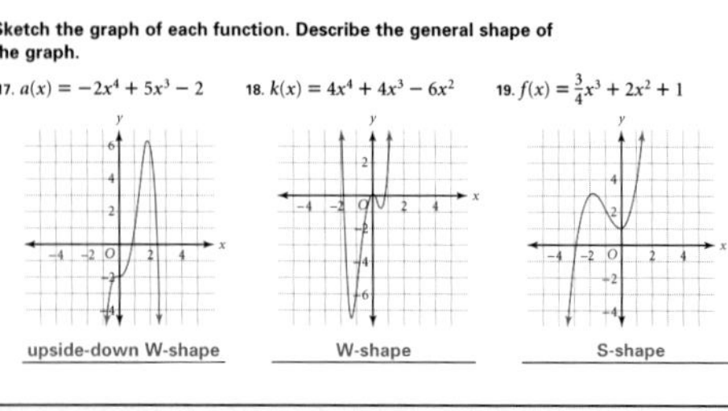

39. cubic polynomial
40. quintic polynomial
41. quartic polynomial
42. cubic polynomial
43. cubic polynomial
44. cubic polynomial
45. cubic trinomial
46. quartic polynomial
47. cubic polynomial
48. cubic polynomial
49. quartic polynomial
50. quartic polynomial

Write each sum or difference as a polynomial in standard form. Then classify the polynomial by degree and by number of terms.

39. $(x^3 + x^2 + x + 1) + (2x^3 + 3x^2 + x + 3)$ $\mathbf{3x^3 + 4x^2 + 2x + 4}$

40. $(x^5 + x^3 + x) + (x^4 + x^2 + 1)$ $\mathbf{x^5 + x^4 + x^3 + x^2 + x + 1}$

41. $(1 - 5x + x^3) - (2x^4 + 5x^3 - 10x^2)$ $\mathbf{-2x^4 - 4x^3 + 10x^2 - 5x + 1}$

42. $(5x^3 + 3x^2 + 8x + 2) - (2x^2 + 4x + 7)$ $\mathbf{5x^3 + x^2 + 4x - 5}$

43. $(2x^2 - 5x + 3) + (4x^3 + 6x^2 - 2x + 5)$ $\mathbf{4x^3 + 8x^2 - 7x + 8}$

44. $(x^2 - 5x^3 + 7) + (6x + x^3 + 3x^2)$ $\mathbf{-4x^3 + 4x^2 + 6x + 7}$

45. $(x^4 + 5x^2 + x) - (x^4 + 2x^3 + x - 4)$ $\mathbf{-2x^3 + 5x^2 + 4}$

46. $(8x^2 + x^3 + 1 - 3x) + (2x^3 + 11x^4)$ $\mathbf{11x^4 + 3x^3 + 8x^2 - 3x + 1}$

47. $\left(\frac{2}{3}x + \frac{2}{3}x^3 + 1\right) + \left(\frac{2}{3} + \frac{1}{3}x^2 + \frac{1}{3}x\right)$ $\mathbf{\frac{2}{3}x^3 + \frac{1}{3}x^2 + x + \frac{5}{3}}$

48. $\left(\frac{2}{7}x^2 + \frac{1}{7}x + \frac{3}{7}\right) - \left(\frac{4}{7}x^3 + \frac{6}{7}x^2 + \frac{2}{7}\right)$ $\mathbf{-\frac{4}{7}x^3 - \frac{4}{7}x^2 + \frac{1}{7}x + \frac{1}{7}}$

49. $(-3.2x^2 + 2.7x^3 + 7.8x) + (4.9x^3 + 2.5x^4)$ $\mathbf{2.5x^4 + 7.6x^3 - 3.2x^2 + 7.8x}$

50. $(4.1x^2 + 5.6x + 7.8) - (x^4 + 7.6x^2 + 9.8x)$ $\mathbf{-x^4 - 3.5x^2 - 4.2x + 7.8}$

Graph each function. Describe its general shape.

51 $f(x) = x^3 - 3x^2 - 3x + 9$

52 $b(x) = x^3 - 4x^2 - 2x + 8$

53 $k(x) = x^4 - x^3 - x^2$

54 $m(x) = x^4 - 10x^2 + 9$

55 $r(x) = -4x^3 + 4x^2 + 19x - 10$

56 $s(x) = -2x^3 + x^2 + 10x - 5$

57 $j(x) = -x^4 + 7x^2 - 6$

58 $k(x) = -x^4 + 2x^3 + 13x^2 - 14x - 24$

CHALLENGES

59. When $4x^3 - 3ax + 5$ is subtracted from $11x^3 + ax^2 - x + b$, the result is $cx^3 - 2x^2 + dx - 1$. Find *a*, *b*, *c*, and *d*. **$a = -2$, $b = 4$, $c = 7$, $d = -7$**

60. The expression $ax^3 + 2x^2 + cx + 1$ is $5x^3 - 3$ greater than $3x^3 + bx^2 + d - 7x$. Find *a*, *b*, *c*, and *d*. **$a = 8$, $b = 2$, $c = -7$, $d = 4$**

CONNECTION

61. GEOMETRY Find the total area of the faces of the rectangular prism at right. **$28x^2 + 54x$**

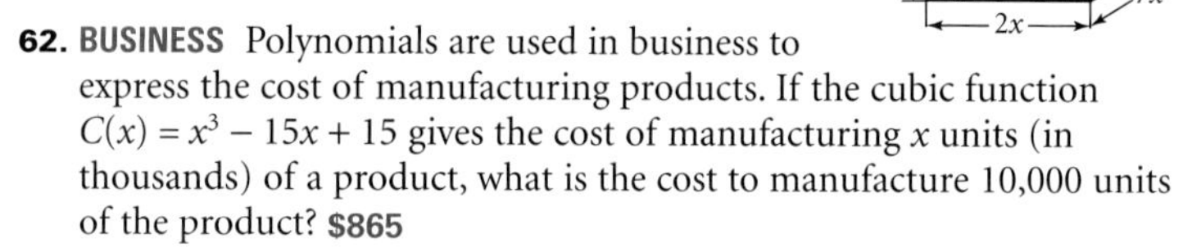

APPLICATIONS

CDs on a conveyor belt

62. BUSINESS Polynomials are used in business to express the cost of manufacturing products. If the cubic function $C(x) = x^3 - 15x + 15$ gives the cost of manufacturing *x* units (in thousands) of a product, what is the cost to manufacture 10,000 units of the product? **$865**

63. BUSINESS The cost of manufacturing a certain product can be represented by $C(x) = 3x^3 - 18x + 45$, where *x* is the number of units of the product in hundreds. What is the cost to manufacture 20,000 units of the product? **$23,996,445**

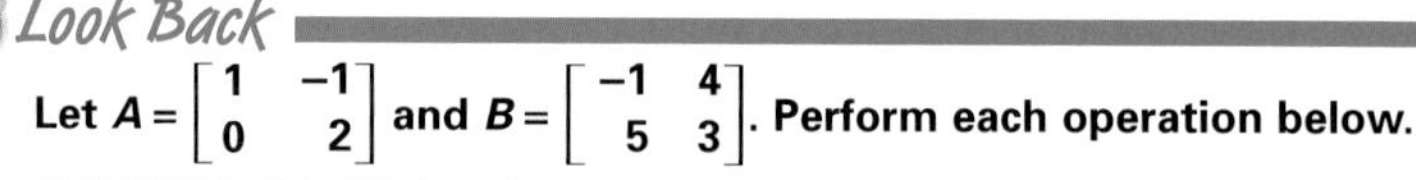

Look Back

Let $A = \begin{bmatrix} 1 & -1 \\ 0 & 2 \end{bmatrix}$ and $B = \begin{bmatrix} -1 & 4 \\ 5 & 3 \end{bmatrix}$. Perform each operation below.

(LESSONS 4.1 AND 4.2)

64. $A - B$ **65.** $B - A$ **66.** $2A$ **67.** $2B$

68. AB **69.** BA **70.** $A - 4B$ **71.** $B - 3A$

64. $\begin{bmatrix} 2 & -5 \\ -5 & -1 \end{bmatrix}$

65. $\begin{bmatrix} -2 & 5 \\ 5 & 1 \end{bmatrix}$

66. $\begin{bmatrix} 2 & -2 \\ 0 & 4 \end{bmatrix}$

67. $\begin{bmatrix} -2 & 8 \\ 10 & 6 \end{bmatrix}$

68. $\begin{bmatrix} -6 & 1 \\ 10 & 6 \end{bmatrix}$

69. $\begin{bmatrix} -1 & 9 \\ 5 & 1 \end{bmatrix}$

70. $\begin{bmatrix} 5 & -17 \\ -20 & -10 \end{bmatrix}$

71. $\begin{bmatrix} -4 & 7 \\ 5 & -3 \end{bmatrix}$

Solve each equation by factoring. ***(LESSON 5.2)***

72. $x^2 + 12x + 11 = 0$ **−11, −1**

73. $x^2 - 2x - 15 = 0$ **−3, 5**

74. $x^2 + 14x + 48 = 0$ **−8, −6**

75. $x^2 + 17x + 72 = 0$ **−9, −8**

76. Give an example of a perfect-square trinomial and then write it as a binomial squared. ***(LESSON 5.3)*** **Answers may vary but should be of the form $x^2 + 2ax + a^2 = (x + a)^2$ or $x^2 - 2ax + a^2 = (x - a)^2$ where $a \neq 0$.**

Use the quadratic formula to solve each equation. ***(LESSONS 5.5 AND 5.6)***

77. $3x^2 + 7x + 2 = 0$ **$-2, -\frac{1}{3}$**

78. $-2x^2 + 4x + 5 = 0$ **$\frac{2 \pm \sqrt{14}}{2}$**

79. $5x^2 + 2x + 4 = 0$ **$\frac{-1 \pm i\sqrt{19}}{5}$**

80. $-6x^2 + 5x - 4 = 0$ **$\frac{5 \pm i\sqrt{71}}{12}$**

Look Beyond

81. Sketch the graph of a cubic polynomial that intersects the x-axis at exactly the number of points indicated. Write *impossible* if appropriate.

a. 3 points **b.** 2 points **c.** 1 point **d.** 0 points

The bottle below has a circular base, a flat bottom, and curved sides. There is no geometric formula for the volume of a solid with this shape. In this activity, you will find a polynomial function that models the volume of a solid with this shape.

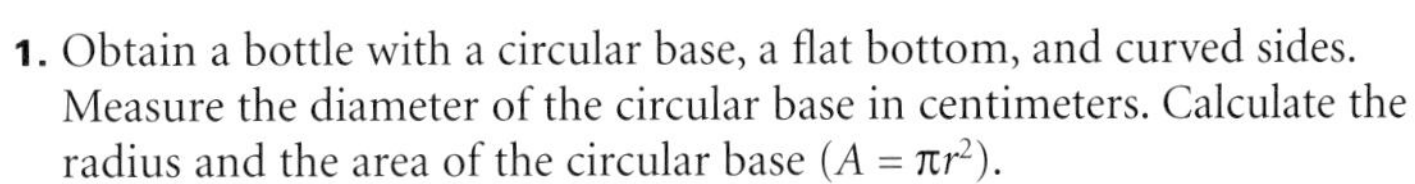

1. Obtain a bottle with a circular base, a flat bottom, and curved sides. Measure the diameter of the circular base in centimeters. Calculate the radius and the area of the circular base ($A = \pi r^2$).

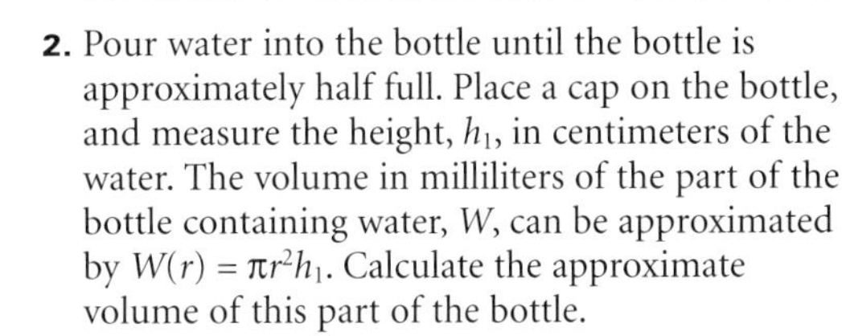

2. Pour water into the bottle until the bottle is approximately half full. Place a cap on the bottle, and measure the height, h_1, in centimeters of the water. The volume in milliliters of the part of the bottle containing water, W, can be approximated by $W(r) = \pi r^2 h_1$. Calculate the approximate volume of this part of the bottle.

3. Turn the bottle upside-down, and measure the height, h_2, in centimeters of the air space above the water. The volume of the air space, A, in milliliters can be approximated by $A(r) = \pi r^2 h_2$. Calculate the approximate volume of the air space.

4. The total volume in milliliters of the bottle, V, can be modeled by the function below. Find the total volume of the bottle.

$$V(r) = W(r) + A(r) = \pi r^2 h_1 + \pi r^2 h_2$$

WORKING ON THE CHAPTER PROJECT

You should now be able to complete Activity 1 of the Chapter Project.

81. Answers may vary. Sample answers:

a. $x^3 - 9x$

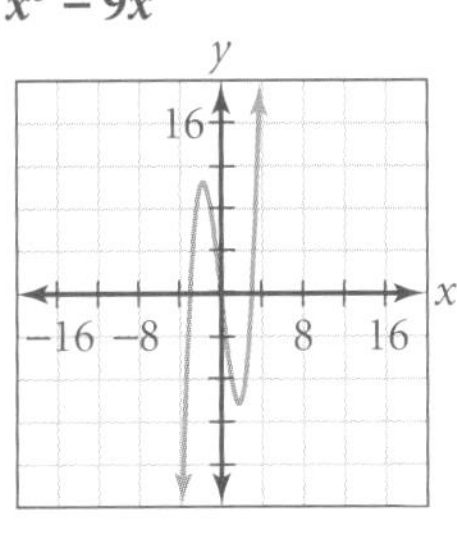

b. $x^3 - x^2 - x + 1$

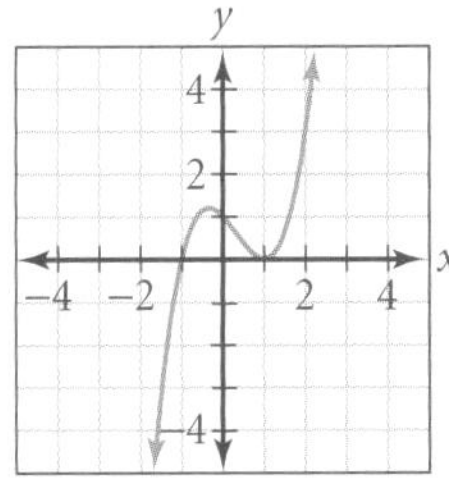

c. $x^3 - 1$

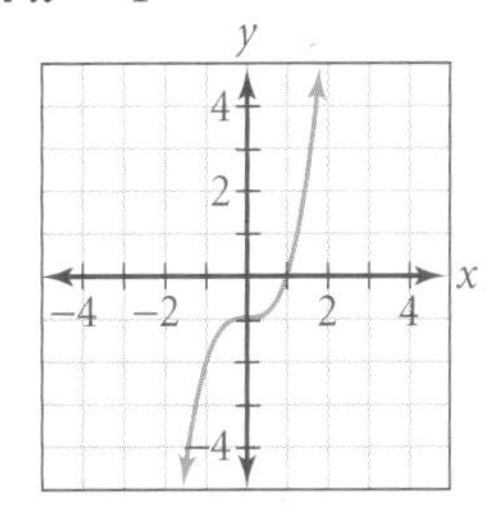

d. impossible

Look Beyond

In Exercise 81, students are asked to produce graphs of cubic polynomial functions that have three, two, one, or zero x-intercepts. The fundamental theorem of algebra, which will be studied in Lesson 7.5, can be used to answer the exercise.

ALTERNATIVE Assessment

Portfolio Activity

The Portfolio Activity can be used as preparation for the Chapter Project or as a separate activity. In the Portfolio Activity on this page, students use an empirical method to create a polynomial model for the total volume of an irregularly shaped bottle.

Student Technology Guide

NAME ________ CLASS ________ DATE ________

Student Technology Guide

7.1 An Introduction to Polynomials

In Lesson 7.1 of your textbook, you saw many important features of polynomial functions and their graphs. The table and graph features of a graphics calculator are excellent tools for investigating polynomial functions.

Example: Evaluate the polynomial expression $x^4 - x^3 - 6x^2 + 4x + 5$ for $x = 2.3$. Round your answer to the nearest hundredth.

- Enter the function by pressing Y= X,T,θ,n ^ 4 − X,T,θ,n ^ 3 − 6 X,T,θ,n x² + 4 X,T,θ,n + 5.
- To evaluate the expression by using a table, first press 2nd WINDOW and type 2.0 for TblStart and .1 for ΔTbl. These settings will ensure that 2.3 is in the first table you see. Then press 2nd GRAPH. As seen at left below, the table shows that the value of the expression is about −1.72 for $x = 2.3$.
- To evaluate the expression by using a graph, press GRAPH. Then press TRACE and use ◄ and ► to move as close to $x = 2.3$ as you can. To get a more precise answer, press 2.3 ENTER. The graph confirms the answer seen in the table.

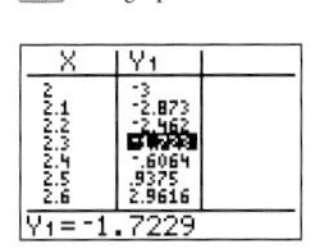

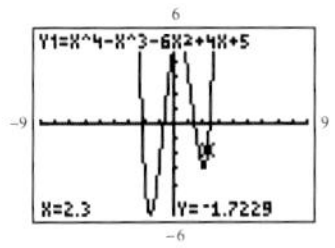

Note: When using a graphics calculator to evaluate an expression, you may need to press WINDOW and change the settings so that the point you are interested in appears in the viewing window.

Evaluate each polynomial expression for the indicated value of x. Round your answers to the nearest hundredth.

1. $x^3 - 2x^2 - 5x + 5$; $x = -2$ **−1**

2. $x^4 - x^2 - 4$; $x = 1.2$ **−3.37**

3. $0.1x^4 + 0.2x^3 - 0.3x^2 + 0.4x - 6$; $x = 3$ **−7.2**

4. $1 - \frac{1}{4}x^2 - \frac{1}{8}x^3 + \frac{1}{16}x^4$; $x = 50$ **374,376**

Prepare

NCTM PRINCIPLES & STANDARDS 1–10

QUICK WARM-UP

Graph the function below. Identify whether it has a maximum value or a minimum value, and approximate the coordinates of the vertex.

$g(x) = -3x^2 + 6x + 4$

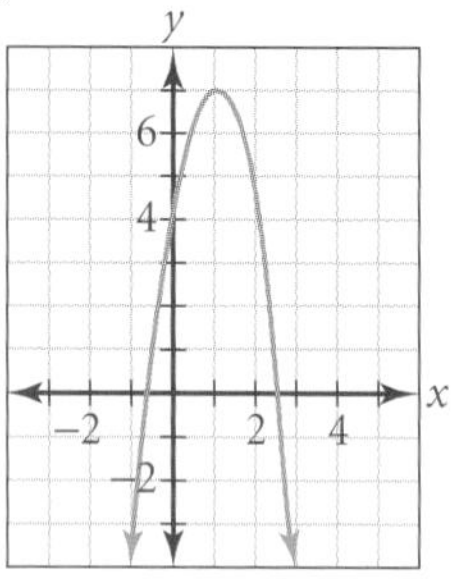

maximum; (1, 7)

Also on Quiz Transparency 7.2

Teach

Why Polynomial functions can be used to model many different types of real-world situations in which data changes either from increasing to decreasing or from decreasing to increasing more than once. Students are familiar with such patterns from everyday life. Ask them to discuss situations in which their performance seems to be improving, then goes into a slump, and then with extra effort improves again.

CHECKPOINT ✔
The number of graduates begins to increase again in about 1991, so a parabola would not model the data.

7.2 Polynomial Functions and Their Graphs

Why *You can use a polynomial function to model real-world data such as the number of high school graduates in the United States.*

Objectives

- Identify and describe the important features of the graph of a polynomial function.
- Use a polynomial function to model real-world data.

The table below gives the number of students who graduated from high school in the United States from 1960 to 1994. The scatter plot of the data indicates an increase and then a decrease in the number of graduates since 1960. Find a quartic regression model for the data in this table. *You will solve this problem in Example 3.*

APPLICATION
EDUCATION

Year	Graduates	Year	Graduates
1960	1679	1978	3161
1961	1763	1979	3160
1962	1838	1980	3089
1963	1741	1981	3053
1964	2145	1982	3100
1965	2659	1983	2964
1966	2612	1984	3012
1967	2525	1985	2666
1968	2606	1986	2786
1969	2842	1987	2647
1970	2757	1988	2673
1971	2872	1989	2454
1972	2961	1990	2355
1973	3059	1991	2276
1974	3101	1992	2398
1975	3186	1993	2338
1976	2987	1994	2517
1977	3140		

[*Source: Statistical Abstract of the United States, 1996*]

Number of High School Graduates in the United States

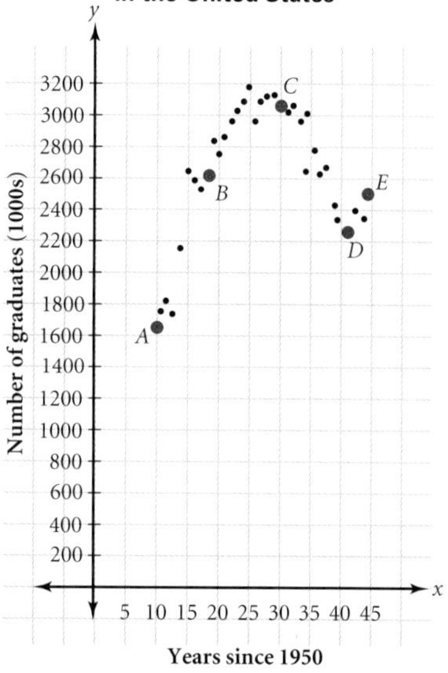

CHECKPOINT ✔ Explain why a quadratic model with a vertex near point C in the scatter plot would not be suitable for predicting the number of graduates after 1994.

Alternative Teaching Strategy

TECHNOLOGY Have students work in small groups, and give each group the following list of polynomial functions:

1. $g(x) = -3x^2 - 2x + 5$
2. $h(x) = 2x^3 - 14x - 12$
3. $k(x) = 2x^4 - 4x^3 - 11x^2 + 16x - 4$

One student should graph a function on a graphics calculator without the others seeing it and then verbally describe the graph. The other students should sketch the described graph on paper. The students in each group will then exchange roles so that each student has the opportunity to both describe and sketch the graph. After sketching all the functions, the group should list the characteristics used to describe the graphs.

After all groups have graphed the functions, lead the class in a discussion about the vocabulary of the lesson. Then have students repeat the activity with the quartic function $f(x) = 0.15x^4 + 0.25x^3 - x^2 - 0.5x$.

Graphs of Polynomial Functions

When a function rises and then falls over an interval from left to right, the function has a *local maximum.* If the function falls and then rises over an interval from left to right, it has a *local minimum.*

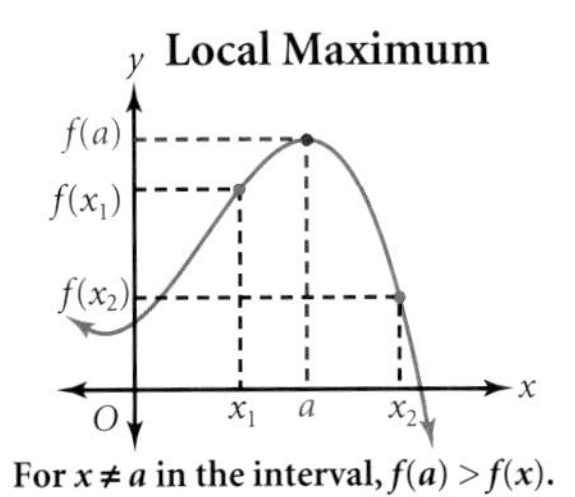

For $x \neq a$ in the interval, $f(a) > f(x)$.

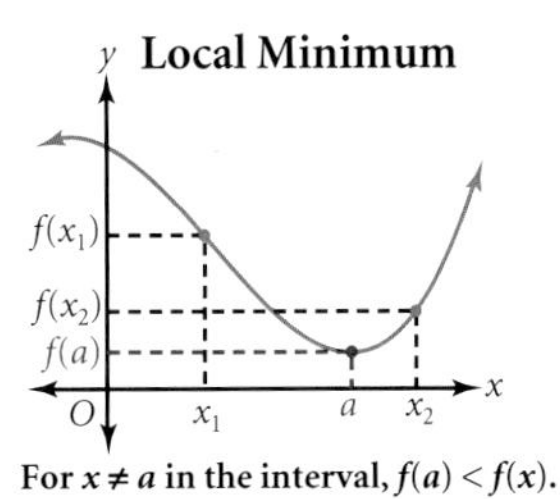

For $x \neq a$ in the interval, $f(a) < f(x)$.

Use Teaching Transparency 25.

Local Maximum and Minimum

$f(a)$ is a **local maximum** (plural, *local maxima*) if there is an interval around a such that $f(a) > f(x)$ for all values of x in the interval, where $x \neq a$.

$f(a)$ is a **local minimum** (plural, *local minima*) if there is an interval around a such that $f(a) < f(x)$ for all values of x in the interval, where $x \neq a$.

The points on the graph of a polynomial function that correspond to local maxima and local minima are called **turning points.** Functions change from *increasing* to *decreasing* or from *decreasing* to *increasing* at turning points. A cubic function has at most 2 turning points, and a quartic function has at most 3 turning points. In general, a polynomial function of degree n has at most $n - 1$ turning points.

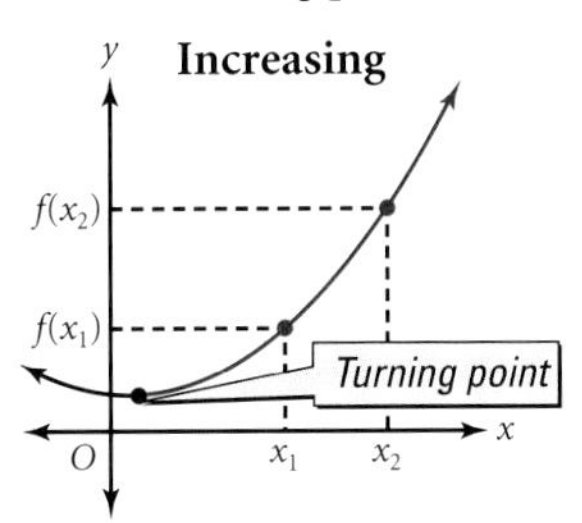

For every $x_1 < x_2$ in the interval, $f(x_1) < f(x_2)$.

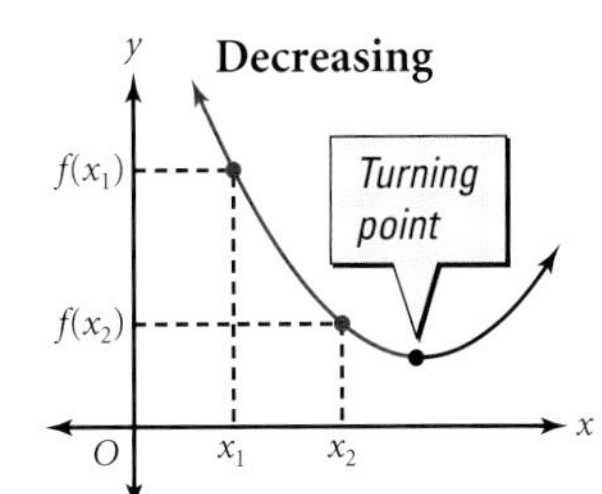

For every $x_1 < x_2$ in the interval, $f(x_1) > f(x_2)$.

Increasing and Decreasing Functions

Let x_1 and x_2 be numbers in the domain of a function, f.

The function f is **increasing** over an open interval if for every $x_1 < x_2$ in the interval, $f(x_1) < f(x_2)$.

The function f is **decreasing** over an open interval if for every $x_1 < x_2$ in the interval, $f(x_1) > f(x_2)$.

Interdisciplinary Connection

HOME ECONOMICS The table below gives the average number of kilowatt-hours of electrical use per day for each month of the year for a private residence.

Jan.	Feb.	Mar.	April	May	June
34	29	20	32	47	60

July	Aug.	Sept.	Oct.	Nov.	Dec.
74	68	47	33	26	30

Let x represent the months, and let y represent the average number of kilowatt-hours. Have students make a scatter plot of the data and find a quartic regression model.
$\mathbf{y = 0.11x^4 - 2.99x^3 + 26.39x^2 - 79.45x + 94.48}$
($x = 1$ represents January.)

Ask them to explain whether this is a good model to use for calculating the average number of kilowatt-hours of electrical use per day for each month in the future. $\mathbf{r \approx 0.96}$

EXAMPLE 1

Graph $P(x) = -2x^3 - x^2 + 5x + 6$.

a. Approximate any local maxima or minima to the nearest tenth.

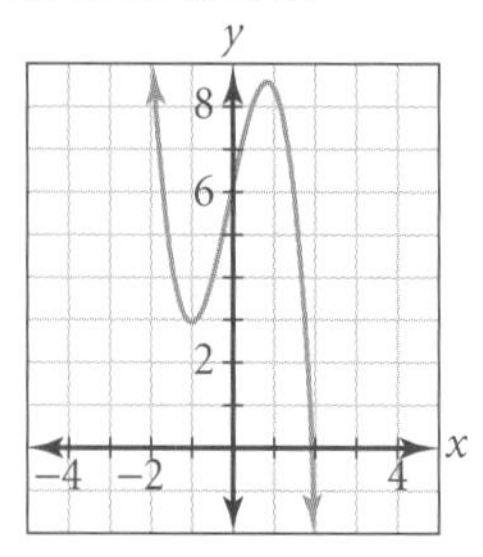

local maximum: (0.8, 8.3);
local minimum: (−1.1, 2.0)

b. Find the intervals over which the function is increasing and decreasing.

increasing: $-1.1 < x < 0.8$;
decreasing: $x < -1.1$ and $x > 0.8$

Teaching Tip

TECHNOLOGY If you are using a TI-83 or TI-82 graphics calculator in Example 1, enter the function as **Y1** and use the following viewing window:

Xmin=−4	**Ymin=−8**
Xmax=2	**Ymax=6**
Xscl=1	**Yscl=1**

To produce the table on the left, press 2nd WINDOW and choose the following settings in the **TABLE SETUP** screen:

TblStart=−3 (or **TblMin=0**)
ΔTbl=1
Indpnt: Auto Ask
Depend: Auto Ask

To produce the table on the right, choose the following settings:

TblStart=−2.5 (or **TblMin=0**)
ΔTbl=.1

TRY THIS

a. local maximum (1.5, 6.6)
local minimum (−0.2, 3.9)

b. decreases for all values of x except over the interval of approximately $-0.2 \le x \le 1.5$, where it increases

EXAMPLE 1 Graph $P(x) = x^3 + 3x^2 - x - 3$.

a. Approximate any local maxima or minima to the nearest tenth.
b. Find the intervals over which the function is increasing and decreasing.

SOLUTION

a. The graph of P has 2 turning points, a local maximum of about 3.1, and a local minimum of about −3.1.
b. The function increases for all values of x except over the interval of approximately $-2.2 < x < 0.2$, where it decreases.

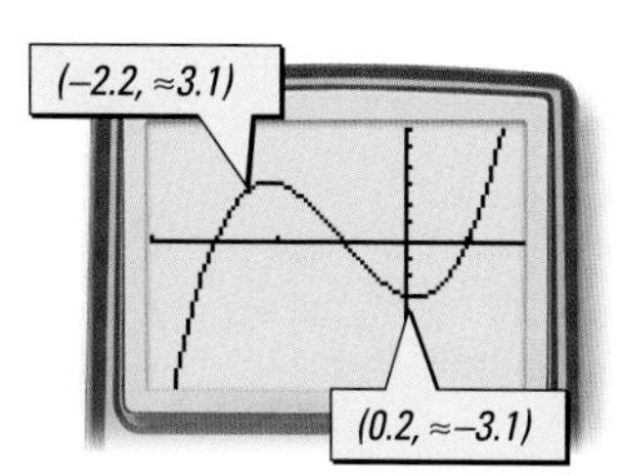

CHECK

You can use the table feature to verify the approximate coordinates of the turning points.

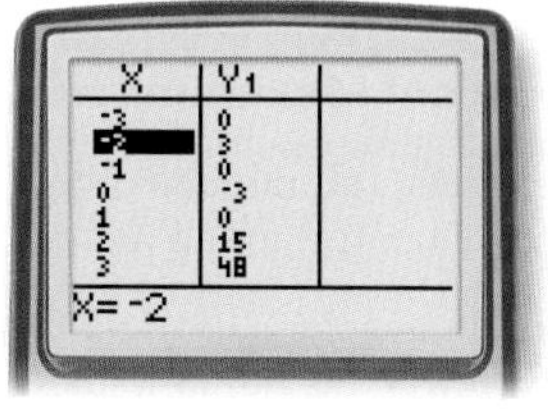

X	Y1
-2.5	2.625
-2.4	2.856
-2.3	3.003
-2.2	3.072
-2.1	3.069
-2	3
-1.9	2.871

X=-2.2

In the table above, when $x = -2$, the y-value, 3, is greater than the neighboring y-values. This indicates that the maximum point has an x-value between −3 and −1.

In the table above, the local maximum can be approximated more closely as 3.1, to the nearest tenth.

A similar procedure can be used to check the local minimum.

TRY THIS Graph $Q(x) = -x^3 + 2x^2 + x + 4$.

a. Approximate any local maxima or minima to the nearest tenth.
b. Find the intervals over which the function is increasing and decreasing.

Recall from Lesson 5.3 that a zero of a function $P(x)$ is any number r such that $P(r) = 0$. The real zeros of a polynomial function correspond to the x-intercepts of the graph of the function. For example, the graph of $P(x) = x^3 + 3x^2 - x - 3$ in Example 1 above shows zeros of P at x-values of −3, −1, and 1.

Polynomial functions are one type of *continuous functions.* The graph of a **continuous function** is unbroken. The graph of a **discontinuous function** has breaks or holes in it.

Continuous Functions

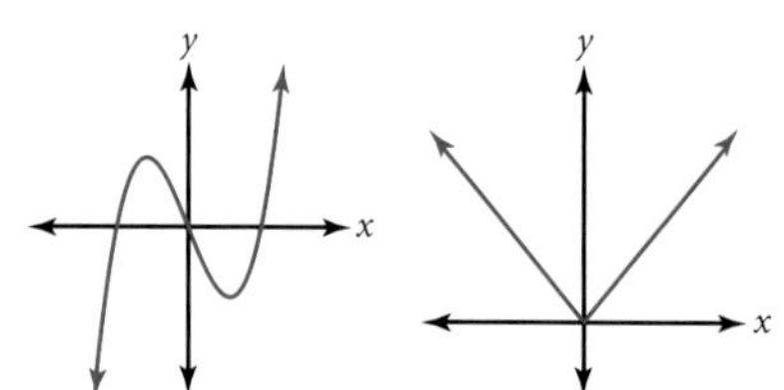

Discontinuous Functions

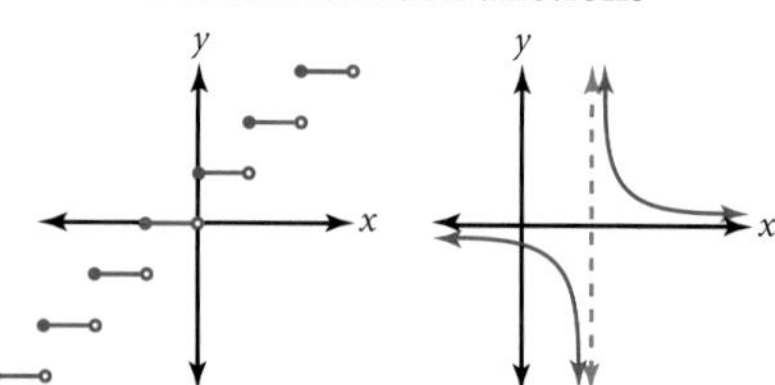

Enrichment

Have students graph a cubic polynomial with clearly visible turning points, such as $f(x) = 4x^3 + 6x^2 - 5$, on a graphics calculator. Remind them to make sure that the viewing window includes the x- and y-intercepts. Ask them to replace the constant term with other integers, one at a time, while keeping the other coefficients the same. Students should notice that changing the constant term produces a vertical shift of the graph of f. You may also want them to vary the values of the leading coefficient and describe the stretch, and then replace x with $x - h$ for different values of h and have them describe the horizontal transformations.

Continuity of a Polynomial Function

Every polynomial function $y = P(x)$ is continuous for all values of x.

The domain of every polynomial function is the set of all real numbers. As a result, the graph of a polynomial function extends infinitely. What happens to a polynomial function as its domain values get very small and very large is called the **end behavior** of a polynomial function.

Activity

Exploring the End Behavior of $f(x) = ax^n$

TECHNOLOGY
GRAPHICS CALCULATOR
Keystroke Guide, page 475

You will need: a graphics calculator

Graph each function separately. For each function, answer parts a–c.

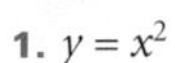

1. $y = x^2$	**2.** $y = x^4$	**3.** $y = 2x^2$	**4.** $y = 2x^4$
5. $y = x^3$	**6.** $y = x^5$	**7.** $y = 2x^3$	**8.** $y = 2x^5$
9. $y = -x^2$	**10.** $y = -x^4$	**11.** $y = -2x^2$	**12.** $y = -2x^4$
13. $y = -x^3$	**14.** $y = -x^5$	**15.** $y = -2x^3$	**16.** $y = -2x^5$

a. Is the degree of the function even or odd?
b. Is the leading coefficient positive or negative?
c. Does the graph rise or fall on the left? on the right?

CHECKPOINT ✔ **17.** Write a conjecture about the end behavior of a function of the form $f(x) = ax^n$ for each pair of conditions below. Then test each conjecture.
a. when $a > 0$ and n is even   **b.** when $a < 0$ and n is even
c. when $a > 0$ and n is odd   **d.** when $a < 0$ and n is odd

If a polynomial function is written in standard form,

$$f(x) = a_n x^n + a_{n-1}x^{n-1} + \cdots + a_1 x + a_0,$$

the **leading coefficient** is a_n. That is, the leading coefficient is the coefficient of the term of greatest degree in the polynomial.

The end behavior of a polynomial function depends on the sign of its leading coefficient and whether the degree of the polynomial is odd or even. Let P be a polynomial function of degree n and with a leading coefficient of a. There are four possible end behaviors for P. The end behavior of P can rise on the left and rise on the right (↖ ↗), fall on the left and fall on the right (↙ ↘), fall on the left and rise on the right (↙ ↗), or rise on the left and fall on the right (↖ ↘). These four possible end behaviors are summarized below.

END BEHAVIOR OF A POLYNOMIAL FUNCTION, $f(x) = ax^n + \cdots$

	$a > 0$		$a < 0$	
	left	right	left	right
n is even.	↖ rise	rise ↗	↙ fall	fall ↘
n is odd.	↙ fall	rise ↗	↖ rise	fall ↘

Inclusion Strategies

HANDS-ON STRATEGIES Have students graph $y = x^2$ on a transparency sheet and overlay it with another sheet that shows a coordinate grid with the axes marked. Ask students to move the graph vertically or horizontally to show zero, one, and two x-intercepts. Have students graph $y = x^3$ and $y = x^3 + x^2 - 2x$ on additional transparency sheets and show possible placements of the graphs for one, two, and three x-intercepts by moving the graphs vertically or horizontally. Ask if they can position either of these cubic functions so that there is no x-intercept. Relate the terms *x-intercepts* and *zeros of the function* with solutions of the related equation.

The graph of $y = x^3$ will have exactly one x-intercept no matter where it is positioned. The graph of $y = x^3 + x^2 - 2x$ may have 1, 2, or 3 x-intercepts depending on where it is positioned.

Activity **Notes**

Point out to students that the values of a_n and n determine the end behavior of the graph of any polynomial function. Students should find the following:

- When the degree of the function is odd and the leading coefficient is positive, the graph falls on the left and rises on the right.
- When the degree of the function is odd and the leading coefficient is negative, the graph rises on the left and falls on the right.
- When the degree of the function is even and the leading coefficient is positive, the graph rises on the left and the right.
- When the degree of the function is even and the leading coefficient is negative, the graph falls on the left and right.

Cooperative Learning

Have students do the Activity in pairs. After each pair has completed the Activity, have them share their conjecture and reasoning with the class.

Teaching Tip

TECHNOLOGY If students are using a TI-82 or TI-83 graphics calculator for the Activity, encourage them to graph each function on successively larger viewing windows.

CHECKPOINT ✔

17a. rises on left, rises on right
b. falls on left, falls on right
c. falls on left, rises on right
d. rises on left, falls on right

ADDITIONAL
EXAMPLE 2

Describe the end behavior of each function.

a. $V(x) = x^3 - 2x^2 - 5x + 3$
falls on the left and rises on the right

b. $R(x) = 1 + x - x^2 - x^3 + 2x^4$
rises on the left and the right

TRY THIS

a. falls on the left and rises on the right

b. rises on the left and the right

Teaching Tip

TECHNOLOGY When using a TI-83 or TI-82 for Example 3, enter the x-values into **L1** and the y-values into **L2.** Then press 2nd Y= to access the **STAT PLOT** menu and choose **1:Plot1...On**. Select the options highlighted below.

Set the viewing window as follows:

Xmin=0	Ymin=0
Xmax=60	Ymax=5000
Xscl=5	Yscl=500

Press GRAPH to view the scatter plot. To calculate the quartic regression equation, press STAT and select **CALC** **7:QuartReg** (or **8:QuartReg**) ENTER. To graph the regression equation, enter the following into **Y1**:

Math CONNECTION

STATISTICS Cubic and quartic regression models fit polynomials to data whose scatter plots resemble the basic shapes of third- and fourth-degree polynomials, respectively.

EXAMPLE 2 **Describe the end behavior of each function.**

a. $P(x) = -x^3 + x^2 + 3x - 1$ **b.** $Q(x) = -x + 3 - x^4 + 3x^2 + x^3$

SOLUTION

a. $P(x)$ is written in standard form.

$$P(x) = -x^3 + x^2 + 3x - 1$$

The degree of P is 3, which is odd, so the graph will rise at one end and fall at the other end. The leading coefficient is negative, so the graph rises on the left and falls on the right.

b. Write $Q(x)$ in standard form.

$$Q(x) = -x^4 + x^3 + 3x^2 - x + 3$$

The degree of Q is 4, which is even, so its graph will either rise at both ends or fall at both ends. The leading coefficient is negative, so the graph falls on the left and the right.

CHECK
Graph $y = -x^3 + x^2 + 3x - 1$.

TECHNOLOGY
GRAPHICS CALCULATOR
Keystroke Guide, page 475

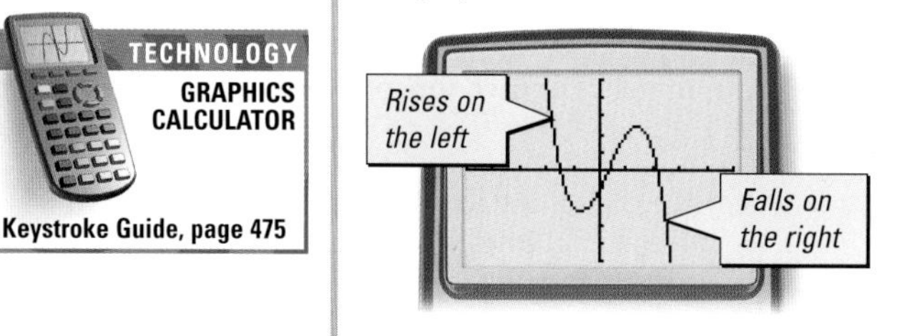

CHECK
Graph $y = -x^4 + x^3 + 3x^2 - x + 3$.

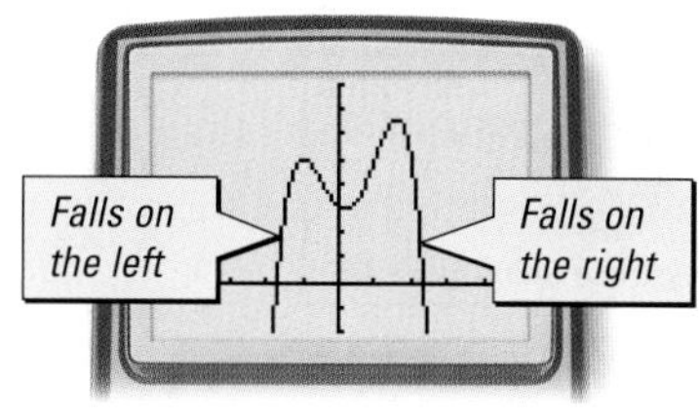

TRY THIS Describe the end behavior of each function.

a. $P(x) = -6x^2 + x^3 + 32$ **b.** $Q(x) = x^4 + x^3 - 2x^2 + 1$

Modeling With Polynomial Functions

You can use a polynomial of degree 3 or higher to fit a data set that has more than one local maximum and minimum.

EXAMPLE 3 **Selected data from the table at the beginning of the lesson are given at right.**

	A	B	C	D	E
x	10	18	30	41	44
y	1679	2606	3089	2276	2517

a. Find a quartic regression model for the number of high school graduates in the United States from 1960 to 1994.

b. Use your quartic regression model to estimate the number of high school graduates in 1970. Compare the value given by the regression model with the actual data value given for 1970.

TECHNOLOGY
GRAPHICS CALCULATOR
Keystroke Guide, page 475

SOLUTION

a. Make a scatter plot of the selected data points, and find the quartic regression model. The calculator gives the following quartic regression model:

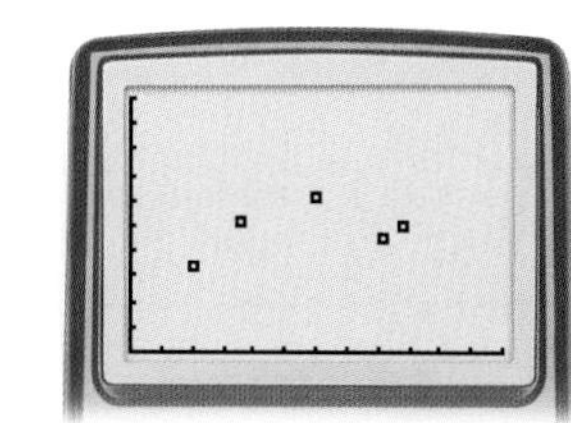

$$P(x) \approx 0.02x^4 - 1.94x^3 + 63.67x^2 - 723.74x + 4296.30$$

Reteaching the Lesson

TECHNOLOGY Tell students to graph the functions below on a graphics calculator and to list the following characteristics of each:

a. the number of turning points
b. the coordinates of each turning point
c. the end behavior
d. the x-intercepts of each function

1. $y = x^3 + 4x^2 + x - 6$
2. $y = -x^3 - 4x^2 - x + 6$
3. $y = x^4 + 2x^3 - 7x^2 - 8x + 12$
4. $y = -x^4 - 2x^3 + 7x^2 + 8x - 12$

1a. 2 **b.** min: (−0.13, −6.06), max: (−2.5, 0.88) **c.** falls on left, rises on right **d.** −3, −2, 1
2a. 2 **b.** min: (−2.5, −0.88), max:(−0.13, 6.06) **c.** rises on left, falls on right **d.** −3, −2, 1
3a. 3 **b.** min: (−2.6, −4) and (1.6, −4), max: (−0.5, 14.1) **c.** rises on left and right **d.** −3, −2, 1, 2
4a. 3 **b.** min: (−0.5, −14.1), max: (−2.6, 4) and (1.6, 4) **c.** falls on left and right **d.** −3, −2, 1, 2

b. Because x represents the number of years since 1950, $x = 20$ represents the year 1970. Find $P(20)$.

The quartic model gives the number of high school graduates (in thousandths) in 1970 as 2850. The actual value given in the table is 2757, which differs by only 93 students.

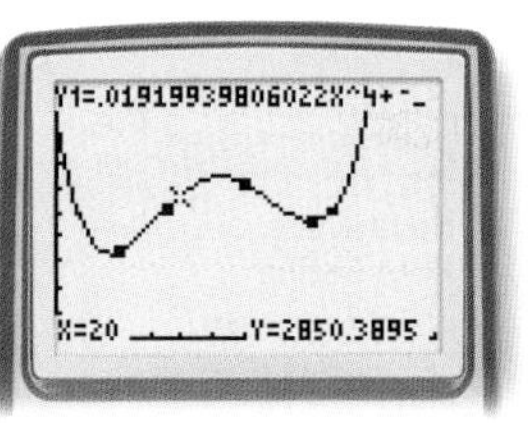

TRY THIS Find a quartic regression model for the data in the table below.

x	2	5	6	9	11
y	4	16	11	14	9

CRITICAL THINKING The function in Example 3 models the number of high school graduates for the period from 1960 to 1994. Explain why this function may not be a good model for estimating the number of high school graduates after 1994.

Exercises

Communicate

internet connect

Activities Online
Go To: **go.hrw.com**
Keyword:
MB1 Traffic

1. Describe the graph of $f(x) = 2x^2 + x^3 + 3x + 1$. Include any turning points, its continuity, and its end behavior.

2. Using your own words, define a local maximum and a local minimum.

3. Describe the four possibilities for the end behavior of the graph of a polynomial function.

4. Using your own words, define increasing and decreasing functions.

Guided Skills Practice

5 Graph $P(x) = x^3 + x^2 - 2x$. Approximate any local maxima or minima to the nearest tenth. Find the intervals over which the function is increasing and decreasing. ***(EXAMPLE 1)***

Describe the end behavior of each function. ***(EXAMPLE 2)***

6. $P(x) = x^6 + x^4 + x + 1$

7. $P(x) = x^4 + 1 + x^3 - x^5$

8 Find a quartic regression model for the data in the table below. ***(EXAMPLE 3)***

x	1	2	3	4	5
y	2	3	2	1	5

5. max ≈ 2.1, min ≈ −0.6; decreases over $-1.2 < x < 0.5$, increases elsewhere

6. rises on left and right

7. rises on left, falls on right

8. $f(x) \approx 0.125x^4 - 0.917x^3 + 1.375x^2 + 1.417x$

TRY THIS
$y = -0.10x^4 + 2.74x^3 - 26.16x^2 + 100.55x - 112.79$

CRITICAL THINKING
Sample answer: It is not likely that the number of high school graduates will always increase.

ADDITIONAL EXAMPLE 3

The table below gives the number of students who participated in the American College Testing (ACT) program during selected years from 1970 to 1995. The variable x represents the number of years since 1960, and y represents the number of participants in thousands.

x	y
10	714
15	822
20	836
25	739
30	817
35	945

[*Source: Statistical Abstract of the United States, 1997*]

a. Find a quartic regression model for the number of students who participated in the ACT program during the given years.
$A(x) = -0.004x^4 + 0.44x^3 - 17.96x^2 + 293.83x - 836$

b. Use the quartic regression model to estimate the number of students who participated in the ACT program in 1985. Compare the value given by the regression model with the actual value given in the table.
The model gives the number of participants in 1985 as 767,000. The actual value from the table is 739,000.

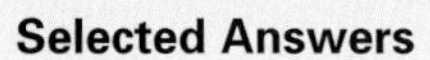

Assess

Selected Answers

Exercises 5–8, 9–55 odd

ASSIGNMENT GUIDE

In Class	1–8
Core	9–43 odd
Core Plus	10–36 even, 37, 38, 40, 42–44
Review	45–55
Preview	56, 57

Extra Practice can be found beginning on page 940.

Error Analysis

A graphics calculator is helpful when graphing functions of high degree. Be sure that students use the appropriate viewing window on their calculators to show all of the turning points. Give students sufficient sample graphs of cubic and quartic polynomial functions so that they can visualize the shape of such curves before graphing them.

Practice

NAME ______ CLASS ______ DATE ______

Practice

7.2 ***Polynomial Functions and Their Graphs***

Graph each function and approximate any local maxima or minima to the nearest tenth.

1. $P(x) = x^2 + 3x + 4$ — min: (−1.5, 1.75)
2. $P(x) = 6 + x - 3x^2$ — max: (0.2, 6)
3. $P(x) = 2x^3 - 2x^2 + 1$ — max: (0, 1), min: (0.7, 0.7)
4. $P(x) = x^4 + x^3 - 4x^2 - 2x + 2$ — max: (−0.2, 2.2), min: (−1.7, 2.7) and (1.3, −2.3)

Graph each function. Find any local maxima or minima to the nearest tenth. Find the intervals over which the function is increasing and decreasing.

5. $P(x) = 4x^3 - 3x^2 + 2, -6 \le x \le 6$ — max: (0, 2), min: (0.5, 1.8); incr.: $-6 \le x \le 0$ and $0.5 < x \le 6$, decr.: $0 < x \le 0.5$
6. $P(x) = 0.3x^4 + x^3 - x, -4 \le x \le 4$ — max: (−0.7, 0.4), min: (−2.3, −1.5) and (0.5, −0.4); incr.: $-2.3 \le x \le -0.7$ and $0.5 \le x \le 4$, decr.: $-4 \le x \le -2.3$ and $-0.7 \le x \le 0.5$
7. $P(x) = x^3 + 1.2x^2 - 2, -5 \le x \le 5$ — max: (−0.8, −1.7), min: (0, −2); incr.: for $-5 \le x \le -0.8$ and $0 \le x \le 4$, decr.: $-0.8 \le x \le 0$
8. $P(x) = -x^4 + 2.5x^3 - x^2 + 1, -4 \le x \le 4$ — max: (0, 1) and (1.6, 2.1), min: (0.3, 1.0); incr.: $-4 \le x \le 0$ and $0.3 \le x \le 1.6$, decr.: $0 \le x \le 0.3$ and $1.6 \le x \le 4$

Describe the end behavior of each function.

9. $12 - 4.2x^3 + x^2$ — rises on the left, falls on the right
10. $3.3x^3 - 2x^2 - 5x + 1$ — falls on the left, rises on the right
11. $5x^3 - 6x^4 + x^2 + 1$ — falls on the left and the right
12. $1.1x^4 - 2.2x^3 + 3.3x^2 - 4$ — rises on the left and the right

13. Factory sales of passenger cars, in thousands, in the United States are shown in the table below. Find a quartic regression model for the data by using $x = 0$ for 1990. (*Source: Bureau of the Census*)

1990	1991	1992	1993	1994	1995
6050	5407	5685	5969	6549	6310

$f(x) = -4.17x^4 - 30.83x^3 + 473.42x^2 - 1019.27x + 6039.64$

internet connect

Homework Help Online
Go To: **go.hrw.com**
Keyword: **MB1 Homework Help** for Exercises 9–28

Practice and Apply

Graph each function and approximate any local maxima or minima to the nearest tenth.

9. $P(x) = 2x^3 - 5x$
10. $P(x) = x^3 - 3x + 1$
11. $P(x) = 2x^3 - 4x + 1$
12. $P(x) = 3x - 3 - 3x^3$
13. $P(x) = -2x + 3 + x^2$
14. $P(x) = -x^2 + 6x - 11$
15. $P(x) = x^4 - 5x^2 + 2$
16. $P(x) = -x^4 + x^3 + 4x^2 - 3$
17. $P(x) = -3x^3 + 3x + x^4 + 3$
18. $P(x) = 3x^3 - x^4 - 3x - 3$

Graph each function. Find any local maxima or minima to the nearest tenth. Find the intervals over which the function is increasing and decreasing.

19. $P(x) = x^3 - 4x; -8 \le x \le 8$
20. $P(x) = -2x^3 + 3x; -5 \le x \le 5$
21. $P(x) = x^4 - 2x^2 + 2; -5 \le x \le 5$
22. $P(x) = -x^4 + 3x^2 + 3; -5 \le x \le 5$
23. $P(x) = -x^2 + 4x - 1; -5 \le x \le 10$
24. $P(x) = x^2 - 6x + 7; -5 \le x \le 10$
25. $P(x) = x^4 - 3x^3 + 3x + 3; -2 \le x \le 5$
26. $P(x) = -x^4 + 3x^3 - 3x - 3; -2 \le x \le 5$
27. $P(x) = x^3 - 3x + 3; -3 \le x \le 4$
28. $P(x) = -x^3 + 4x - 2; -3 \le x \le 3$

Describe the end behavior of each function.

29. $P(x) = 2x^3 + x^2 + 3x + 2$
30. $P(x) = -3x^3 + 5x^2 + x + 2$
31. $P(x) = 6x + 1 - x^2$
32. $P(x) = x^2 - 8x + 3$
33. $P(x) = 4x^4 + x^5 + 1 + 3x^3$
34. $P(x) = x^6 + x^4 + 3x^2 + 2$
35. $P(x) = 7x^3 + 2 - 8x^5$
36. $P(x) = 5x^4 - 6x^6 + 3x + 2$

37. The function $y = 10x^3 - 25x^2 + x^4 - 10x + 24$ is graphed at left. Explain how you can tell that the viewing window chosen here does not show all of the important characteristics of the graph. Find a viewing window that does show all of the important characteristics of the graph.

38. Solve a system of equations by using a matrix equation to find the coefficients a_3, a_2, a_1, and a_0 such that the polynomial function $P(x) = a_3x^3 + a_2x^2 + a_1x + a_0$ passes through the points (0, 0), (1, 1), (2, 0), and (3, 2).

STATISTICS **Find a quartic regression model for each data set.**

39.

x	1	2	3	4	5
y	1	−2	3	0	1

40.

x	1	2	3	4	5
y	0	4	3	1	4

41.

x	2	4	6	8	10
y	3	−1	4	−3	4

42.

x	−1	0	2	4	5
y	2	1	1	6	3

43. **REAL ESTATE** The median monthly rent (in dollars) in the United States for 1988 through 1996 is given in the table below. Find a quartic regression model for the data by using $x = 0$ for 1980. [*Source: U.S. Bureau of Census*]

1988	1989	1990	1991	1992	1993	1994	1995	1996
343	346	371	398	411	430	429	438	444

Answers (side column)

9. max: ≈3.0
   min: ≈−3.0
10. max: ≈1.1
    min: ≈−1.25
11. max: ≈3.2
    min: ≈−1.2
12. max: ≈−1.8
    min: ≈−4.2
13. min: ≈2.0
    no max
14. max: ≈−2.0
    no min

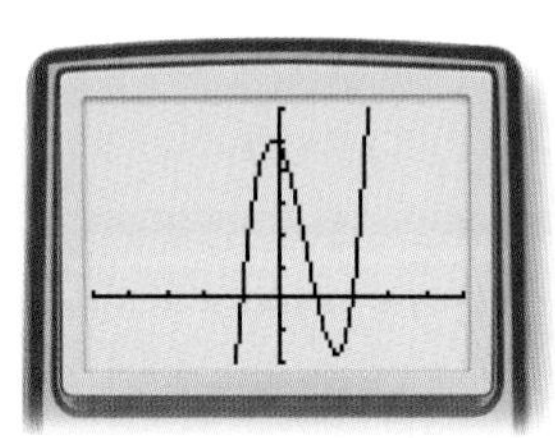

CHALLENGE

15. max: ≈2.0
    min: ≈−4.3, ≈−4.3
16. max: ≈−1.0, ≈5.3
    min: ≈−3.0
17. max: ≈4.3
    mins: ≈1.9, ≈1.0
18. max: ≈−1.9, ≈−1.0
    min: ≈−4.3

19. maximum of 3.1; minimum of −3.1; increases for all values of x except over the interval of approximately $-1.2 < x < 1.2$, where it decreases

20. maximum of 1.4; minimum of −1.4; decreases for all values of x except over the interval of approximately $-0.7 < x < 0.7$, where it increases

21. maximum of 2.0; minima of 1.0 and 1.0; decreases for $-\infty < x < -1.0$ and $0.0 < x < 1.0$; increases for $-1.0 < x < 0.0$ and $1.0 < x < \infty$

22. maxima of 5.3 and 5.3; minimum of 3.0; increases for $-\infty < x < -1.2$ and $0.0 < x < 1.2$; decreases for $-1.2 < x < 0.0$ and $1.2 < x < \infty$

23. maximum of 3.0; no minimum; increases for $-\infty < x < 2.0$; decreases for $2.0 < x < \infty$

24. minimum of −2.0; no maximum; decreases for $-\infty < x < 3.0$; increases for $3.0 < x < \infty$

25. maximum of 4.3; minima of 1.9 and 1.0; decreases for $-\infty < x < -0.5$ and $0.7 < x < 2.1$; increases for $-0.5 < x < 0.7$ and $2.1 < x < \infty$

APPLICATION

44 **TRAVEL** The number of business and pleasure travelers (in thousands) to the United States from Canada is given in the table below for certain years. Find a quartic regression model for the data by using $x = 0$ for 1980. [*Source: U.S. Travel and Tourism Administration*]

1985	1989	1990	1991	1992	1993	1994	1995
10,721	15,325	17,263	19,113	18,596	17,293	14,970	13,668

Look Back

Find the inverse of each matrix, if it exists. If the inverse does not exist, write *no inverse*. *(LESSON 4.3)*

45 $\begin{bmatrix} 1 & 1 \\ 1 & 1 \end{bmatrix}$ **46** $\begin{bmatrix} 1 & 3 \\ 1 & 2 \end{bmatrix}$ **47** $\begin{bmatrix} 2 & 5 \\ 4 & 11 \end{bmatrix}$ **48** $\begin{bmatrix} 2 & 2 \\ 4 & 4 \end{bmatrix}$

45. no inverse

46. $\begin{bmatrix} -2 & 3 \\ 1 & -1 \end{bmatrix}$

47. $\begin{bmatrix} 5.5 & -2.5 \\ -2 & 1 \end{bmatrix}$

48. no inverse

Solve each quadratic equation. Give exact solutions. *(LESSON 5.2)*

49. $x^2 - 50 = 0$ **$\pm 5\sqrt{2}$** **50.** $(x + 3)^2 - 36 = 0$ **$-9, 3$** **51.** $3(x - 1)^2 = 12$ **$-1, 3$**

Solve each logarithmic equation. *(LESSON 6.3)*

52. $\log 1000 = x$ **3** **53.** $\log_x 8 = 3$ **2** **54.** $\log_3 x = 2$ **9** **55.** $\log_x 1 = 0$ **any positive number**

Look Beyond

56. Find $(x + y)(x^2 - xy + y^2)$. **$x^3 + y^3$**

57. Find $(x - y)(x^2 + xy + y^2)$. **$x^3 - y^3$**

Examining the behavior of a polynomial function can often provide insight into the real-world situation that the function models.

1. Graph each function in a viewing window that includes all of the x- and y-intercepts. For each function, answer questions **a** and **b** below.

$f(x) = x^2 - 2x + 1$ $\quad g(x) = x^3 - 3x + 2$

$h(x) = x^4 - 4x + 3$ $\quad j(x) = x^5 - 5x + 4$

a. Describe the shape of the graph, and state the number of local maxima and minima.

b. Does the function have any zeros that are local maxima or minima?

2. Compare the graphs of the functions from Step 1. How are they alike? How are they different?

3. Refer to the bottle from the Portfolio Activity on page 431. If height h_1 is 1 unit less than the radius, r, and height h_2 is 2 units less than the radius, r, write a function for the total volume, V, in terms of r.

4. Graph the function you wrote in Step 3, and describe the graph. Explain how the graph relates to the real-world situation that it models.

WORKING ON THE CHAPTER PROJECT

You should now be able to complete Activity 2 of the Chapter Project.

These exercises require students to calculate products of linear and quadratic factors that result in the sum or difference of two cubes. These types of products are discussed in Lesson 7.3.

ALTERNATIVE
Assessment

Portfolio Activity

The Portfolio Activity can be used as preparation for the Chapter Project or as a separate activity. The Portfolio Activity on this page shows how a polynomial model can be used to analyze the relationship among the radius, height, and volume of a bottle.

Answers to Portfolio Activities can be found in Additional Answers of the Teacher's Edition.

44. $f(x) \approx 17.41x^4 - 761.72x^3 + 11{,}751.83x^2 - 74{,}233.04x + 172{,}429.44$

Student Technology Guide

NAME ______ CLASS ______ DATE ______

Student Technology Guide
7.2 *Polynomial Functions and Their Graphs*

A graphics calculator can help you describe the important features of the graph of a polynomial function. These include the local maxima and minima and the end behavior of the function.

Example: Graph $f(x) = x^3 - 2x^2 - 4x + 12$. Describe the end behavior of the function. Give approximate locations for all local maxima and minima, rounding coordinates to the nearest tenth.

- Press Y= X,T,θ,n ^ 3 − 2 X,T,θ,n x^2 − 4 X,T,θ,n + 12. Press GRAPH.
- Press TRACE. Use ◄ and ► to move to a point just to the left of the local maximum. Press 2nd TRACE 4:maximum ENTER.

- When asked for the left bound, press ENTER to enter the coordinates of the trace point.
- When asked for the right bound, trace just to the right of the maximum. Press ENTER.
- When asked for a guess, trace as close as you can to the maximum. Press ENTER.

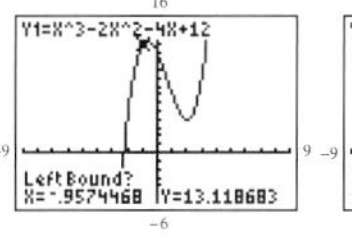

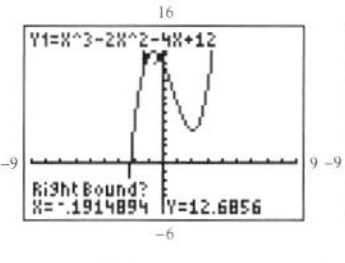

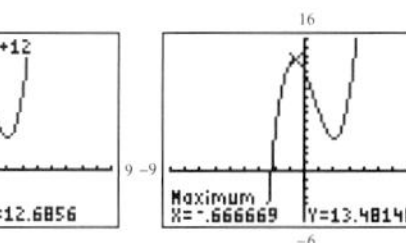

- Follow a similar procedure to find the minimum. (Select 3:minimum instead of 4:maximum.)

The function has a local maximum near $(-0.7, 13.5)$ and a local minimum at $(2, 4)$. The graph rises on the left and on the right.

Graph each function and describe its end behavior. Approximate any maxima or minima to the nearest tenth.

1. $f(x) = x^4 - 2x^3 - 2x^2 + 2x + 6$ **rises on the right and the left; local min. near (−0.7, 4.5) and (1.9, 1.9); local max. near (0.4, 6.4)**

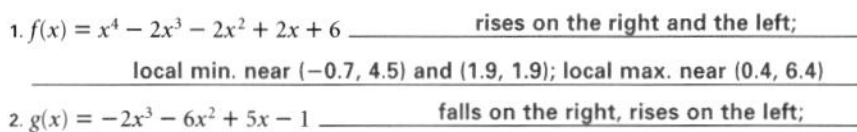

2. $g(x) = -2x^3 - 6x^2 + 5x - 1$ **falls on the right, rises on the left; local min. near (−2.4, −19.9); local max. near (0.4, −0.1)**

26. maxima of -1.9 and -1.0; minimum of -4.3 decreases for $-0.5 < x < 0.7$ and $2.1 < x < \infty$; increases for $-\infty < x < -0.5$ and $0.7 < x < 2.1$

27. maximum of 5.0; minimum of 1.0; increases for all values of x except over the interval of approximately $-1.0 < x < 1.0$, where it decreases

28. maximum of 1.1; minimum of -5.1; decreases for all values of x except over the interval of approximately $-1.2 < x < 1.2$, where it increases

29. falls on the left and rises on the right

30. rises on the left and falls on the right

31. falls on the left and the right

32. rises on the left and the right

33. falls on the left and rises on the right

34. rises on the left and the right

35. rises on the left and falls on the right

36. falls on the left and the right

The answers to Exercises 37–43 can be found in Additional Answers beginning on page 1002.

Prepare

NCTM PRINCIPLES & STANDARDS 1–4, 6–10

QUICK WARM-UP

Factor each expression.

1. $6x^2 - 15x$ $\quad 3x(2x - 5)$
2. $x^2 + 3x - 28$ $\quad (x + 7)(x - 4)$
3. $x^2 - 64$ $\quad (x + 8)(x - 8)$
4. $4x^2 - 8x + 4$ $\quad 4(x - 1)^2$
5. $6x^2 - 5x - 6$ $\quad (3x + 2)(2x - 3)$
6. $16x^4 - 1$ $\quad (4x^2 + 1)(2x + 1)(2x - 1)$

Also on Quiz Transparency 7.3

Teach

Why Polynomial functions can model the relationship between the dimensions of a rectangular box and its volume. To show how valuable this can be, ask students to consider the room that they are sitting in. They should see that it is actually a box. Ask them what would happen if the height of the room were doubled. Would the room's volume be doubled? If the length of all the walls were reduced by $\frac{1}{3}$, would the volume be reduced by $\frac{1}{3}$? Now ask the students to compare their estimated volumes with the actual volumes.

Products and Factors of Polynomials

Why *You can use the factored form of a polynomial function to create a model for the volume of an open-top box.*

Objectives

- Multiply polynomials, and divide one polynomial by another by using long division and synthetic division.
- Use the Remainder and Factor Theorems to solve problems.

Making an open-top box out of a single rectangular sheet involves cutting and folding square flaps at each of the corners. These flaps are then pasted to the adjacent side to provide reinforcement for the corners.

The dimensions of the rectangular sheet and the square flaps determine the volume of the resulting box. For the 12-inch-by-16-inch sheet shown, the volume function is $V(x) = x(16 - 2x)(12 - 2x)$, where x is the side length in inches of the square flap.

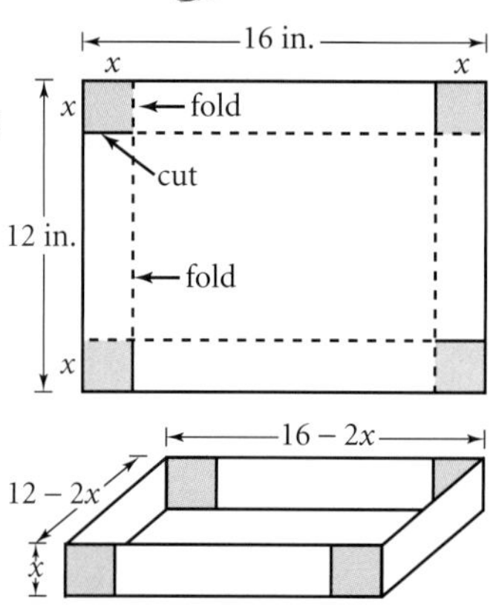

Multiplying Polynomials

EXAMPLE 1 **Write the volume function for the open-top box, $V(x) = x(16 - 2x)(12 - 2x)$, as a polynomial function in standard form.**

SOLUTION

$$
\begin{aligned}
x(16 - 2x)(12 - 2x) &= x[(16 - 2x)(12 - 2x)] \\
&= x[16(12 - 2x) - 2x(12 - 2x)] \\
&= x(192 - 32x - 24x + 4x^2) \\
&= x(192 - 56x + 4x^2) \\
&= 192x - 56x^2 + 4x^3 \\
&= 4x^3 - 56x^2 + 192x
\end{aligned}
$$

Alternative Teaching Strategy

USING VISUAL MODELS Have students graph $y = x^3 - 2x^2 - 5x + 6$ on a graphics calculator and identify the x-intercepts. For each x-intercept, k, form a linear factor of the form $(x - k)$, and multiply all the factors. Compare the product with the original function. Ask students how to find the value of $x^3 - 2x^2 - 5x + 6$, or $(x - 1)(x - 3)(x + 2)$, when x equals an x-intercept without actually substituting a value into the original function. Discuss the x-intercepts and zeros of the function $f(x) = x^3 - 2x^2 - 5x + 6$, solutions of the equation $x^3 - 2x^2 - 5x + 6 = 0$, and the linear factors of the expression $x^3 - 2x^2 - 5x + 6$. Compare and contrast the terms *x-intercept, zero, solution, root,* and *linear factor* in the different contexts, stressing the differences among the terms *function, equation,* and *expression.*

Ask students to identify the x-intercepts of $f(x) = x^3 - 2.5x^2 - 12x + 13.5$, factor the expression $x^3 - 2.5x^2 - 12x + 13.5$, and solve the equation $x^3 - 2.5x^2 - 12x + 13.5 = 0$.

$(-3, 0)$, $(1, 0)$, $(4.5, 0)$; $(x + 3)(x - 1)(x - 4.5)$; $x = -3$, $x = 1$, *or* $x = 4.5$

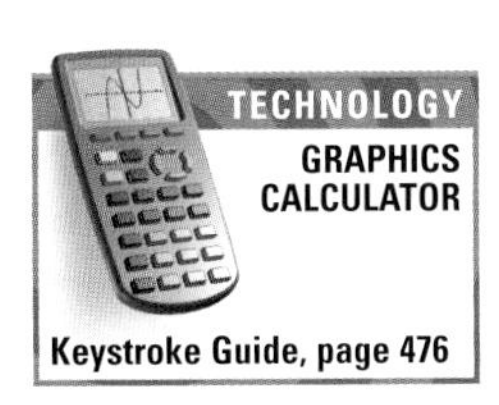

CHECK

You can check your multiplication by graphing the polynomial function in factored form and in standard form. If the graphs coincide, then the functions are equivalent.

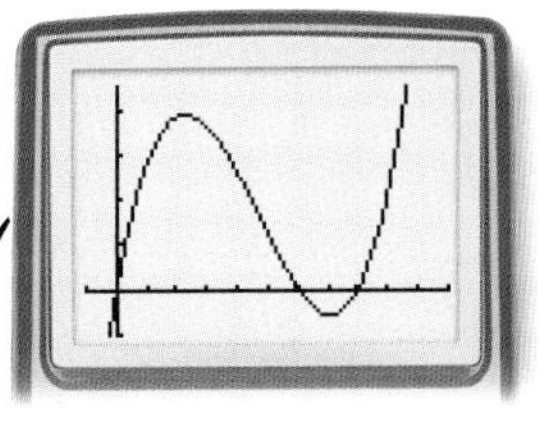

The graphs of $Y1 = x(16 - 2x)(12 - 2x)$ and $Y2 = 4x^3 - 56x^2 + 192x$ appear to coincide.

TRY THIS Write $f(x) = 2x^2(x^2 + 2)(x - 3)$ as a polynomial in standard form.

Factoring Polynomials

Just as a quadratic expression is factored by writing it as a product of two factors, a polynomial expression of a degree greater than 2 is factored by writing it as a product of more than two factors.

EXAMPLE 2 **Factor each polynomial.**

a. $x^3 - 5x^2 - 6x$ **b.** $x^3 + 4x^2 + 2x + 8$

SOLUTION

a. $x^3 - 5x^2 - 6x = x(x^2 - 5x - 6)$ *Factor out the GCF, x.*
$= x(x - 6)(x + 1)$ *Factor the trinomial into binomials.*

b. The polynomial $x^3 + 4x^2 + 2x + 8$ can be factored in pairs.

$x^3 + 4x^2 + 2x + 8 = (x^3 + 4x^2) + (2x + 8)$ *Group the terms in pairs.*
$= x^2(x + 4) + 2(x + 4)$ *Factor each pair of terms.*
$= (x^2 + 2)(x + 4)$ *Factor (x + 4) from each term.*

TRY THIS Factor the polynomials $x^3 - 9x$ and $x^3 - x^2 + 2x - 2$.

Factoring the Sum and Difference of Two Cubes

$a^3 + b^3 = (a + b)(a^2 - ab + b^2)$ $a^3 - b^3 = (a - b)(a^2 + ab + b^2)$

EXAMPLE 3 **Factor each polynomial.**

a. $x^3 + 27$ **b.** $x^3 - 1$

SOLUTION

a. $x^3 + 27 = x^3 + 3^3$
$= (x + 3)(x^3 - 3x + 9)$

b. $x^3 - 1 = x^3 - 1^3$
$= (x - 1)(x^2 + x + 1)$

TRY THIS Factor the polynomials $x^3 + 1000$ and $x^3 - 125$.

Interdisciplinary Connection

BUSINESS In a trade discount, each discount is applied successively to the declining balance in order to arrive at the final invoice price.

Suppose that the list price of an item is \$100. If three successive discounts are given, then the final price of an item, P, can be modeled by the polynomial function $P(x) = 100(1 - x)^3$, where x is the percent of discount expressed as a decimal. Have students write this function in standard form. Then have them find the cost of the item after a trade discount of 10%.

$P(x) = -100x^3 + 300x^2 - 300x + 100$; \$72.90

Use Teaching Transparency 26.

ADDITIONAL EXAMPLE 1

Write the function $f(x) = (x - 1)(x + 4)(x - 3)$ as a polynomial function in standard form.

$f(x) = x^3 - 13x + 12$

Teaching Tip

TECHNOLOGY If you are using a TI-82 or TI-83 graphics calculator to show that the graphs in Example 1 coincide, set the viewing window as follows:

Xmin=–1	Ymin=–30
Xmin=10	Ymax=200
Xscl=1	Yscl=50

TRY THIS

$2x^5 - 6x^4 + 4x^3 - 12x^2$

ADDITIONAL EXAMPLE 2

Factor each polynomial.

a. $x^3 - 16x^2 + 64x$
$x(x - 8)^2$

b. $x^3 + 6x^2 - 5x - 30$
$(x^2 - 5)(x + 6)$

TRY THIS

$x(x + 3)(x - 3)$; $(x^2 + 2)(x - 1)$

ADDITIONAL EXAMPLE 3

Factor each polynomial.

a. $x^3 + 125$
$(x + 5)(x^2 - 5x + 25)$

b. $x^3 - 27$
$(x - 3)(x^2 + 3x + 9)$

TRY THIS

$(x + 10)(x^2 - 10x + 100)$;
$(x - 5)(x^2 + 5x + 25)$

ADDITIONAL EXAMPLE 4

Use substitution to determine whether $x - 1$ is a factor of $x^3 - x^2 - 5x - 3$.

If $f(x) = x^3 - x^2 - 5x - 3$, $f(1) = 1^3 - 1^2 - 5(1) - 3 = -8 \neq 0$. Because $f(1) \neq 0$, by the factor theorem, $x - 1$ is not a factor of $x^3 - x^2 - 5x - 3$.

Teaching Tip

TECHNOLOGY If you are using a TI-82 or TI-83 graphics calculator to confirm that −2 is a zero of $f(x)$ in Example 4, set the viewing window as follows:

Xmin=–5	Ymin=–10
Xmax=5	Ymax=10
Xscl=1	Yscl=1

TRY THIS

$(-3)^3 - 3(-3)^2 - 6(-3) + 8$
$= -27 - 27 + 18 + 8$
$= -28 \neq 0,$
$x + 3$ is not a factor.

The *Factor Theorem*, given below, states the relationship between the linear factors of a polynomial expression and the zeros of the related polynomial function.

Factor Theorem

$x - r$ is a factor of the polynomial expression that defines the function P if and only if r is a solution of $P(x) = 0$, that is, if and only if $P(r) = 0$.

With the Factor Theorem, you can test for linear factors involving integers by using substitution.

EXAMPLE 4 **Use substitution to determine whether $x + 2$ is a factor of $x^3 - 2x^2 - 5x + 6$.**

TECHNOLOGY GRAPHICS CALCULATOR

Keystroke Guide, page 476

SOLUTION

Write the related function, $f(x) = x^3 - 2x^2 - 5x + 6$.

Write $x + 2$ as $x - (-2)$.

Find $f(-2)$.

$f(-2) = (-2)^3 - 2(-2)^2 - 5(-2) + 6$
$= 0$

Because $f(-2) = 0$, the Factor Theorem states that $x + 2$ is a factor of $x^3 - 2x^2 - 5x + 6$.

CHECK

Graph $y = x^3 - 2x^2 - 5x + 6$.

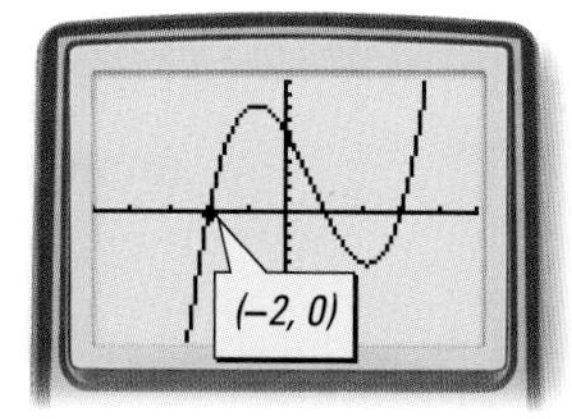

The graph confirms that −2 is a zero of the related function.

TRY THIS Use substitution to determine whether $x + 3$ is a factor of $x^3 - 3x^2 - 6x + 8$.

Dividing Polynomials

A multiplication equation can be rewritten as two or more division equations.

$$x^3 + 3x^2 - 4x - 12 = (x^2 + 5x + 6)(x - 2)$$

$$\frac{x^3 + 3x^2 - 4x - 12}{x^2 + 5x + 6} = x - 2 \qquad \frac{x^3 + 3x^2 - 4x - 12}{x - 2} = x^2 + 5x + 6$$

A polynomial can be divided by a divisor of the form $x - r$ by using **long division** or a shortened form of long division called **synthetic division**.

Long division of polynomials is similar to long division of real numbers. Examine the division of $\frac{x^3 + 3x^2 - 4x - 12}{x - 2}$ shown at right.

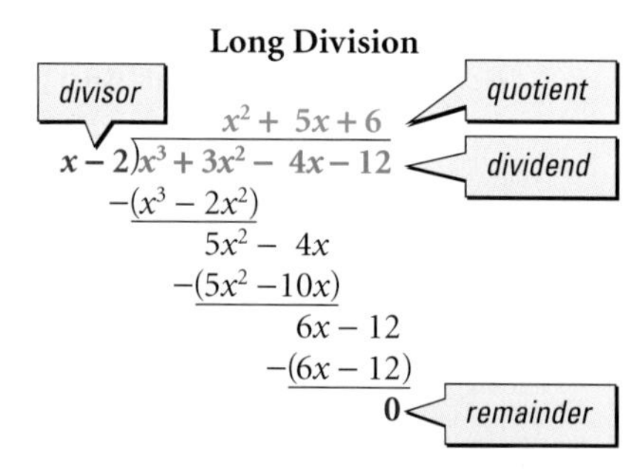

Enrichment

A mountain has approximately the same shape as the graph of $f(x) = -x^4 + 3x^2 - 3x + 6$ between $x = -3$ and $x = 2.0$. Have students use a graphics calculator to graph f and answer the following questions:

a. The zeros of the function represent sea level. What are the approximate zeros of f? **−2.32, 1.79**

b. Use the trace feature to find the approximate maximum value of f. **≈12.24**

c. If the maximum height of the mountain is 3200 feet, find a vertical scale factor that transforms the function values to the actual heights. **1 unit ≈261 ft**

d. If the vertical and horizontal scales are the same, how long is the mountain as represented by the function? **≈1073 ft**

In synthetic division, you do not write the variables.

Synthetic Division

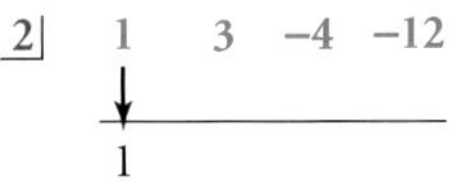

Step 1: Write the coefficients of the polynomial $x^3 + 3x^2 - 4x - 12$, and then write the r-value, 2, of the divisor, $x - 2$, on the left. Write the first coefficient, 1, below the line.

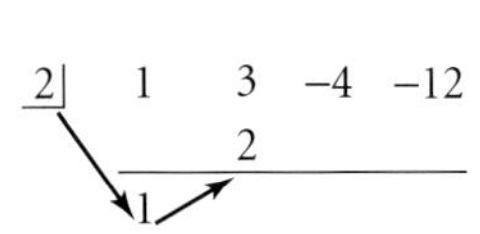

Step 2: Multiply the r-value, 2, by the number below the line, and write the product below the next coefficient.

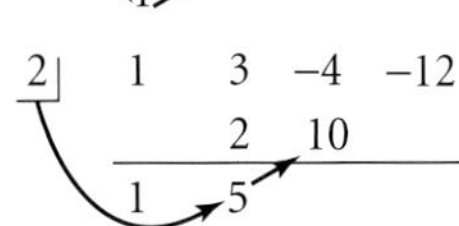

Step 3: Write the *sum* (not the difference) of 3 and 2 below the line. Multiply 2 by the number below the line, and write the product below the next coefficient.

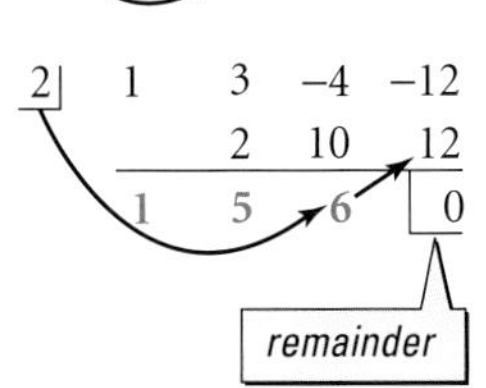

Step 4: Write the sum of −4 and 10 below the line. Multiply 2 by the number below the line, and write the product below the next coefficient.

The remainder is 0, and the resulting numbers, 1, 5, and 6, are the coefficients of the quotient, $x^2 + 5x + 6$.

Synthetic division can be used to divide a polynomial only by a linear binomial of the form $x - r$. When dividing by nonlinear divisors, long division must be used. This is shown in Example 5.

EXAMPLE 5 **Find the quotient.** $(x^3 + 3x^2 + 3x + 2) \div (x^2 + x + 1)$

SOLUTION

Step 1: Divide the first term of the **dividend** by the first term of the **divisor**: $x^3 \div x^2 = x$.

Step 2: Write x in the **quotient** and use it to multiply the **divisor**: $x(x^2 + x + 1)$.

Step 3: Subtract the product, $x^3 + x^2 + x$, from the **dividend**.

Step 4: Repeat Steps 1–3, using the difference from Step 3 as the new dividend.

$$\begin{array}{r} x + 2 \\ x^2 + x + 1\overline{)x^3 + 3x^2 + 3x + 2} \\ -(x^3 + x^2 + x) \\ 2x^2 + 2x + 2 \\ -(2x^2 + 2x + 2) \\ 0 \end{array}$$

(quotient: $x + 2$; dividend: $x^3 + 3x^2 + 3x + 2$; divisor: $x^2 + x + 1$; remainder: 0)

$(x^3 + 3x^2 + 3x + 2) \div (x^2 + x + 1) = x + 2$

TRY THIS Find the quotient. $(x^3 + 3x^2 - 13x - 15) \div (x^2 - 2x - 3)$

If there is a nonzero remainder after dividing with either method, it is usually written as the numerator of a fraction, with the divisor as the denominator.

$$\begin{array}{r|rrrr} -3 & 1 & 0 & 0 & 48 \\ & & -3 & 9 & -27 \\ \hline & 1 & -3 & 9 & \boxed{21} \end{array}$$

$(x^3 + 48) \div (x + 3) = x^2 - 3x + 9 + \frac{21}{x+3}$

ADDITIONAL EXAMPLE 5

Find the quotient:
$(x^3 - x^2 - 4) \div (x^2 + x + 2)$

$x - 2$

TRY THIS

$x + 5$

Teaching Tip

It is assumed throughout this chapter that factoring is done over the set of integers. Discuss with students factoring over the set of real numbers, and show them an example of synthetic division with a noninteger divisor such as the following:

$(x^3 + 0.5x^2 + 0.5x - 0.5) \div (x - 0.5)$

$$\begin{array}{r|rrrr} 0.5 & 1 & 0.5 & 0.5 & -0.5 \\ & & 0.5 & 0.5 & 0.5 \\ \hline & 1 & 1.0 & 1.0 & \boxed{0} \end{array}$$

$\frac{x^3 + 0.5x^2 + 0.5x - 0.5}{x - 0.5}$

$\frac{(x - 0.5)(x^2 + x + 1)}{x - 0.5} = x^2 + x + 1$

Be sure students know that synthetic division works only with linear divisors. If the divisor is not linear, long division must be used to find the quotient.

Inclusion Strategies

USING ALGORITHMS To help students see the connection between long division and synthetic division, give the following pattern:

$$\begin{array}{r|rrrr} -1 & 1 & -2 & 0 & -6 \\ & & -1 & 3 & -3 \\ \hline & 1 & -3 & 3 & \boxed{-9} \end{array}$$

Have students write the divisor, the dividend, the quotient, and the remainder represented by the synthetic division as polynomial expressions. Now have them check their result by multiplying the quotient by the divisor and adding the remainder to see whether they get the dividend.

$x + 1; x^3 - 2x - 6; x^2 - 3x + 3; -9;$
$(x + 1)(x^2 - 3x + 3) + (-9)$
$= x^3 - 2x^2 + 3 - 9$
$= x^3 - 2x^2 - 6$

ADDITIONAL EXAMPLE 6

Given that 2 is a zero of $P(x) = x^3 - 3x^2 + 4$, use division to factor $x^3 - 3x^2 + 4$.
$(x + 1)(x - 2)^2$

TRY THIS
$(x + 3)(x - 4)(x + 1)$

CRITICAL THINKING
Answers may vary. Sample answer: In long division, the products contain multiples of $-r$, while in synthetic division, the products are multiples of r. Subtracting $-r$ or adding r produces the same result.

ADDITIONAL EXAMPLE 7

Given the function $P(x) = 3x^3 - 4x^2 + 9x + 5$, find $P(6)$ by using both synthetic division and substitution.

a. Using synthetic division:

6	3	−4	9	5
		18	84	558
	3	14	93	563

The remainder is 563, so $P(6) = 563$.

b. Using substitution:

$P(6)$
$= 3(6)^3 - 4(6)^2 + 9(6) + 5$
$= 648 - 144 + 54 + 5$
$= 563$

TRY THIS
$P(3) = 94$

EXAMPLE 6 **Given that 2 is a zero of $P(x) = x^3 + x - 10$, use division to factor $x^3 + x - 10$.**

SOLUTION

Method 1 Use long division.

$$\begin{array}{r} x^2 + 2x + 5 \\ x - 2\overline{)x^3 + 0x^2 + x - 10} \\ -(x^3 - 2x^2) \\ 2x^2 + x \\ -(2x^2 - 4x) \\ 5x - 10 \\ -(5x - 10) \\ 0 \end{array}$$

Method 2 Use synthetic division.

2	1	0	1	−10
		2	4	10
	1	2	5	0

Notice that zeros are used for the coefficients of terms that do not exist.

Quotient: $x^2 + 2x + 5$

Thus, $x^3 + x - 10 = (x - 2)(x^2 + 2x + 5)$. *$x^2 + 2x + 5$ cannot be factored.*

TRY THIS Given that −3 is a zero of $P(x) = x^3 - 13x - 12$, use division to factor $x^3 - 13x - 12$.

CRITICAL THINKING Explain why you add the products in synthetic division instead of subtracting them as in long division.

Given $P(x) = 2x^3 + 7x^2 + 2x + 1$, the *Remainder Theorem* states that $P(-3)$ is the value of the remainder when $2x^3 + 7x^2 + 2x + 1$ is divided by $x + 3$.

synthetic division:

−3	2	7	2	1
		−6	−3	3
	2	1	−1	4

The remainder is 4.

substitution:
$P(x) = 2x^3 + 7x^2 + 2x + 1$
$P(-3) = 2(-3)^3 + 7(-3)^2 + 2(-3) + 1 = 4$

Remainder Theorem

If the polynomial expression that defines the function of P is divided by $x - a$, then the remainder is the number $P(a)$.

EXAMPLE 7 Given $P(x) = 2x^3 + 7x^2 + 2x + 1$, find $P(5)$.

SOLUTION

Method 1 Use synthetic division.

5	2	7	2	1
		10	85	435
	2	17	87	436

Method 2 Use substitution.

$P(x) = 2x^3 + 7x^2 + 2x + 1$
$P(5) = 2(5)^3 + 7(5)^2 + 2(5) + 1$
$= 250 + 175 + 10 + 1$
$= 436$

Thus, $P(5) = 436$.

TRY THIS Given $P(x) = 3x^3 + 2x^2 - 3x + 4$, find $P(3)$.

Reteaching the Lesson

HANDS-ON STRATEGIES Have students bring to class an open rectangular box, such as a cereal box with the top flaps removed or the bottom of a shoe box. Have them measure the length, width, and height of the box in inches and then calculate the volume in cubic inches. Now have them cut the seams of the box to make it a single sheet. Ask students to trace the sheet on a piece of paper and draw in the square flaps that were cut from each corner. Then have them measure the outline of the box, including the square flaps, to find the length and width of the rectangular sheet. Now have them write a polynomial function in standard form for the volume, where x is the side length in inches of the square flaps. Have them graph the volume function, and ask them to estimate the value of x that gives the largest volume.

Exercises

Communicate

1. Describe how the Factor Theorem can be used to determine whether $x + 1$ is a factor of $x^3 - 2x^2 - 8x - 5$.
2. Describe the condition necessary to use synthetic division to divide polynomials.
3. Explain how to use the Remainder Theorem to evaluate $P(5)$ if P is a polynomial function.

Guided Skills Practice

4. Write $P(x) = x(10 - x)(2 + x)$ as a polynomial function in standard form. ***(EXAMPLE 1)*** $P(x) = -x^3 + 8x^2 + 20x$

Factor each polynomial. ***(EXAMPLES 2 AND 3)***

5. $x^3 - 5x^2 + 6x$
6. $x^3 + 5x^2 + 3x + 15$
7. $x^3 - 216$

8. Use substitution to determine whether $x + 2$ is a factor of $x^3 + 4x^2 + 5x + 2$. ***(EXAMPLE 4)*** $f(-2) = (-2)^3 + 4(-2)^2 + 5(-2) + 2 = 0$, yes
9. Find the quotient. $(x^3 + 4x^2 + 4x + 3) \div (x^2 + x + 1)$ ***(EXAMPLE 5)*** $x + 3$

Given that −3 is a zero of $P(x) = x^3 - 14x - 15$, use each method below to factor $x^3 - 14x - 15$. ***(EXAMPLE 6)***

10. long division $(x+3)(x^2 - 3x - 5)$
11. synthetic division $(x+3)(x^2 - 3x - 5)$

Given $P(x) = 2x^3 + 3x^2 + 4x + 1$, find $P(2)$ by using each method below. ***(EXAMPLE 7)***

12. synthetic division 37
13. substitution 37

Practice and Apply

Write each product as a polynomial in standard form.

14. $3x^2(4x^3 - 2x^2 + 5x + 2)$
15. $2x^3(4x^3 - 2x^2 + x + 3)$
16. $(2x - 3)(x + 4)$
17. $(x + 7)(5x - 3)$
18. $(x + 2)(x^2 + 4x + 1)$
19. $(x + 3)(2x^3 + 3x^2 + 1)$
20. $(2x + 3)(x^3 - 5x^2 + 4)$
21. $(2x + 1)(x^2 - 4x - 3)$
22. $(x - 4)(2x^3 - 3x^2 + 2)$
23. $(x - 5)(-3x^3 - 4x - 1)$
24. $(x - 3)(2 - x)(x - 1)$
25. $(x - 2)(2x + 3)(3 - x)$
26. $(x + 1)^2(x - 2)$
27. $(2x - 4)(x + 1)^2$
28. $(2x + 1)^3$
29. $(3x + 2)^3$
30. $(x - 1)^2(x^2 - 3x + 2)$
31. $(-3x^2 - x + 2)(x + 1)^2$
32. $\left(x - \frac{5}{7}\right)\left(\frac{2}{5}x^2 - \frac{1}{5}x + \frac{3}{5}\right)$
33. $\left(x - \frac{1}{4}\right)\left(\frac{2}{3}x^2 + \frac{1}{3}x + \frac{2}{3}\right)$

Answers (margin)

5. $x(x - 2)(x - 3)$
6. $(x^2 + 3)(x + 5)$
7. $(x - 6)(x^2 + 6x + 36)$
14. $12x^5 - 6x^4 + 15x^3 + 6x^2$
15. $8x^6 - 4x^5 + 2x^4 + 6x^3$
16. $2x^2 + 5x - 12$
17. $5x^2 + 32x - 21$
18. $x^3 + 6x^2 + 9x + 2$
19. $2x^4 + 9x^3 + 9x^2 + x + 3$
20. $2x^4 - 7x^3 - 15x^2 + 8x + 12$
21. $2x^3 - 7x^2 - 10x - 3$
22. $2x^4 - 11x^3 + 12x^2 + 2x - 8$
23. $-3x^4 + 15x^3 - 4x^2 + 19x + 5$
24. $-x^3 + 6x^2 - 11x + 6$
25. $-2x^3 + 7x^2 + 3x - 18$
26. $x^3 - 3x - 2$
27. $2x^3 - 6x - 4$
28. $8x^3 + 12x^2 + 6x + 1$
29. $27x^3 + 54x^2 + 36x + 8$
30. $x^4 - 5x^3 + 9x^2 - 7x + 2$
31. $-3x^4 - 7x^3 - 3x^2 + 3x + 2$
32. $\frac{2}{5}x^3 - \frac{17}{35}x^2 + \frac{26}{35}x - \frac{3}{7}$
33. $\frac{2}{3}x^3 + \frac{1}{6}x^2 + \frac{7}{12}x - \frac{1}{6}$

Assess

Selected Answers

Exercises 4–13, 15–101 odd

ASSIGNMENT GUIDE

In Class	1–13
Core	15–97 odd, 101
Core Plus	14–102 even
Review	103–114
Preview	115

Extra Practice can be found beginning on page 940.

Mid-Chapter Assessment for Lessons 7.1 through 7.3 can be found on page 87 of the *Assessment Resources.*

Error Analysis

Some students may incorrectly factor polynomial expressions of the form $a^3 + b^3$ or the form $a^3 - b^3$ as $(a+b)(a+b)(a+b)$ or as $(a-b)(a-b)(a-b)$, respectively. Encourage students to check their factoring by graphing the related polynomial function in factored form and in standard form or by multiplying out the factored expression. Use polynomial long division to show students that the correct factorizations of the sum of two cubes and difference of two cubes are $(a+b)(a^2 - ab + b^2)$ and $(a-b)(a^2 + ab + b^2)$, respectively.

50. $(x-4)(x^2+4x+16)$

51. $(x+10)(x^2-10x+100)$

52. $(x^2+5)(x^4-5x^2+25)$

53. $(x^2+3)(x^4-3x^2+9)$

54. $(x-2)(x^2+2x+4)$

55. $(x-6)(x^2+6x+36)$

56. $(2x-1)(4x^2+2x+1)$

57. $(3x-5)(9x^2+15x+25)$

58. $(x+1)(x^2-x+1)(x-1)(x^2+x+1)$

59. $(4-x)(16+4x+x^2)$

60. $(3+2x)(9-6x+4x^2)$

Practice

NAME ______ CLASS ______ DATE ______

Practice
7.3 ***Products and Factors of Polynomials***

Write each product as a polynomial in standard form.

1. $0.5x(16x^4 - 10x^3 + 6x)$ — $8x^5 - 5x^4 + 3x^2$
2. $(x-10)(2x+3)$ — $2x^2 - 17x - 30$
3. $(x-4)(5x^2+3x+7)$ — $5x^3 - 17x^2 - 5x - 28$
4. $(x+2)(x-8)(x-1)$ — $x^3 - 7x^2 - 10x + 16$
5. $(2x-5)(x+1)^2$ — $2x^3 - x^2 - 8x - 5$
6. $(3x-1)^3$ — $27x^3 - 27x^2 + 9x - 1$

Use substitution to determine whether the given linear expression is a factor of the given polynomial.

7. $x^2 + 2x - 12; x + 4$ — no
8. $3x^2 - x - 4; x + 1$ — yes
9. $x^3 - 9x + 1; x - 3$ — no
10. $2x^3 - 11x^2 + 8x - 15; x - 5$ — yes
11. $2x^3 + 10x^2 - 28x; x + 7$ — yes
12. $3x^3 - 2x^2 - 6x - 2; x - 2$ — no

Divide by using long division.

13. $(2x^2 + 7x - 30) \div (x + 6)$ — $2x - 5$
14. $(6x^2 + 2x - 5) \div (3x + 5)$ — $2x - 1 - \frac{5x}{3x+5}$
15. $(8x^3 + 12x^2 + 6x + 5) \div (2x + 1)$ — $4x^2 + 4x + 1 + \frac{4}{2x+1}$
16. $(5x^3 + x^2 - x + 3) \div (x + 1)$ — $5x^2 - 4x + 3$

Divide by using synthetic division.

17. $(x^3 - x^2 + x - 21) \div (x^2 + 2x + 7)$ — $x - 3$
18. $(x^3 - 5x^2 - 20x - 32) \div (x - 8)$ — $x^2 + 3x + 4$
19. $(x^3 + 4x^2 + 4x + 3) \div (x^2 + x + 1)$ — $x + 3$
20. $(x^5 - x^3 - 3) \div (x^2 - 3)$ — $x^3 + 2x + \frac{6x-3}{x^2-3}$

For each function below, use synthetic division and substitution to find the indicated value.

21. $P(x) = x^2 + 3x + 1; P(2)$ — 11
22. $P(x) = x^3 - 2x + 4; P(3)$ — 25
23. $P(x) = 2x^4 - 3x^3 + 2x^2 - 6; P(2)$ — 10
24. $P(x) = 3x^4 - 4x^3 + x^2 - 1; P(-2)$ — 83

internet connect
Homework Help Online
Go To: go.hrw.com
Keyword: **MB1 Homework Help** for Exercises 34–60

34. $x(x+3)(x+5)$
35. $x(x+2)(x+4)$
36. $x(x-6)(x-4)$
37. $x(x+3)(x-1)$
38. $x(x-6)(x+5)$
39. $x(x-3)(x+1)$
40. $3x(x+10)(x-10)$
41. $2x(3x-5)^2$
42. $(x^2-2)(x+3)$
43. $(x^2+4)(x-3)$
44. $(x-1)(x+1)(x+3)$
45. $(x^2-5)(x-2)$
46. $(x^2+1)(x+1)$
47. $(1+x^2)(1-x)$
48. $x(x+7)(x+2)$
49. $(x^2+2)(x+1)$

73. $x^2 - x - 6$
74. $x^2 + 6x + 9$
75. $x - 1$
76. $x - 2$
77. $x - 6$
78. $x - 4$

Factor each polynomial.

34. $x^3 + 8x^2 + 15x$ **35.** $x^3 + 6x^2 + 8x$ **36.** $x^3 - 10x^2 + 24x$
37. $x^3 + 2x^2 - 3x$ **38.** $x^3 - x^2 - 30x$ **39.** $x^3 - 2x^2 - 3x$
40. $3x^3 - 300x$ **41.** $18x^3 - 60x^2 + 50x$ **42.** $x^3 + 3x^2 - 2x - 6$
43. $x^3 - 3x^2 + 4x - 12$ **44.** $x^3 + 3x^2 - x - 3$ **45.** $x^3 - 2x^2 - 5x + 10$
46. $x^3 + x^2 + x + 1$ **47.** $1 - x + x^2 - x^3$ **48.** $x^3 + 2x^2 + 14x + 7x^2$
49. $x^3 + x^2 + 2 + 2x$ **50.** $x^3 - 64$ **51.** $x^3 + 1000$
52. $x^6 + 125$ **53.** $x^6 + 27$ **54.** $x^3 - 8$
55. $x^3 - 216$ **56.** $8x^3 - 1$ **57.** $27x^3 - 125$
58. $x^6 - 1$ **59.** $64 - x^3$ **60.** $27 + 8x^3$

Use substitution to determine whether the given linear expression is a factor of the polynomial.

61. $x^2 + x + 1; x - 1$ **no**
62. $x^2 + 2x + 1; x + 2$ **no**
63. $x^3 + 3x^2 - 33x - 35; x + 1$ **yes**
64. $x^3 + 5x^2 - 18x - 48; x + 6$ **no**
65. $x^3 + 3x^2 - 18x - 40; x - 4$ **yes**
66. $x^3 - 8x^2 + 9x + 18; x - 6$ **yes**
67. $x^3 + 6x^2 - x - 30; x - 2$ **yes**
68. $x^3 - x^2 - 17x - 15; x + 3$ **yes**
69. $2x^3 + 9x^2 + 6x + 8; x + 4$ **yes**
70. $2x^3 - x^2 - 12x - 9; x - 3$ **yes**

Divide by using long division.

71. $(x^2 + 4x + 4) \div (x + 2)$ $\boldsymbol{x + 2}$
72. $(x^2 - 3x + 2) \div (x - 1)$ $\boldsymbol{x - 2}$
73. $(x^3 - 7x - 6) \div (x + 1)$
74. $(x^3 + 11x^2 + 39x + 45) \div (x + 5)$
75. $(3x^2 - x + x^3 - 3) \div (x^2 + 4x + 3)$
76. $(x^3 + 6x^2 - x - 30) \div (x^2 + 8x + 15)$
77. $(x^3 - 43x + 42) \div (x^2 + 6x - 7)$
78. $(10x - 5x^2 + x^3 - 24) \div (x^2 - x + 6)$
79. $\left(x^2 - \frac{1}{6}x - \frac{1}{6}\right) \div \left(x - \frac{1}{2}\right)$ $\boldsymbol{x + \frac{1}{3}}$
80. $\left(x^2 + \frac{1}{2}x - \frac{3}{16}\right) \div \left(x + \frac{3}{4}\right)$ $\boldsymbol{x - \frac{1}{4}}$

Divide by using synthetic division.

81. $(x^2 - 4x - 12) \div (x - 4)$
82. $(x^2 - 3x + 2) \div (x - 1)$
83. $(x^3 + x^2 - 9x - 9) \div (x + 1)$
84. $(x^3 - 2x^2 - 22x + 40) \div (x - 4)$
85. $(x^3 + 5x^2 - 18) \div (x + 3)$
86. $(x^3 - 27) \div (x - 3)$
87. $(x^3 + 3) \div (x - 1)$
88. $(x^2 - 6) \div (x + 4)$
89. $(x^4 - 3x + 2x^3 - 6) \div (x - 2)$
90. $(x^5 + 6x^3 - 5x^4 + 5x - 15) \div (x - 3)$

For each function below, use synthetic division and substitution to find the indicated value.

91. $P(x) = x^2 + 1; P(1)$ **2**
92. $P(x) = x^2 + 1; P(2)$ **5**
93. $P(x) = x^2 + x; P(2)$ **6**
94. $P(x) = x^2 + x; P(1)$ **2**
95. $P(x) = 4x^2 - 2x + 3; P(3)$ **33**
96. $P(x) = 3x^3 + 2x^2 + 3x + 1; P(-2)$ **−21**
97. $P(x) = 2x^4 + x^3 - 3x^2 + 2x; P(-4)$ **392**
98. $P(x) = 2x^3 - 3x^2 + 2x - 2; P(3)$ **31**

CHALLENGE

Find the value of *k* that makes the linear expression a factor of the cubic expression.

99. $x^3 + 3x^2 - x + k; x - 2$ **−18**
100. $kx^3 - 2x^2 + x - 6; x + 3$ **−1**

81. $x^2 - \frac{12}{x-4}$
82. $x - 2$
83. $x^2 - 9$
84. $x^2 + 2x - 14 - \frac{16}{x-4}$
85. $x^2 + 2x - 6$
86. $x^2 + 3x + 9$
87. $x^2 + x + 1 + \frac{4}{x-1}$
88. $x - 4 + \frac{10}{x+4}$
89. $x^3 + 4x^2 + 8x + 13 + \frac{20}{x-2}$
90. $x^4 - 2x^3 + 5$

APPLICATIONS

101a. $4x^3 - 92x^2 + 448x$
b. 560 cubic inches

102a. $V(x) = 5x^3 - \frac{175}{2}x^2 + 380x$
b. 450 cubic inches
c. 481.25 cubic inches

101. MANUFACTURING An open-top box is made from a 14-inch-by-32-inch piece of cardboard, as shown at right. The volume of the box is represented by $V(x) = x(14 - 2x)(32 - 2x)$, where x is the height of the box.

a. Write the volume of the box as a polynomial function in standard form.

b. Find the volume of the box if the height is 2 inches.

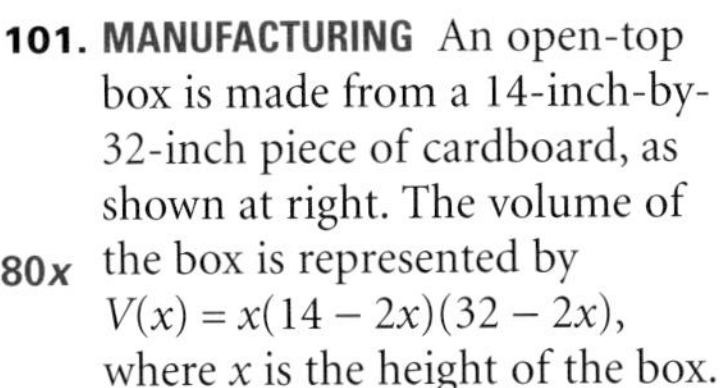
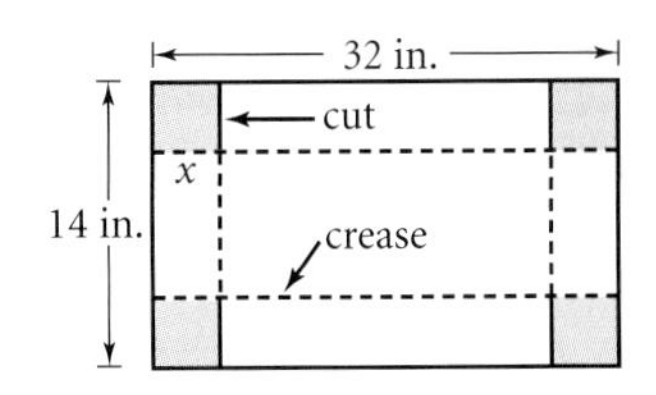

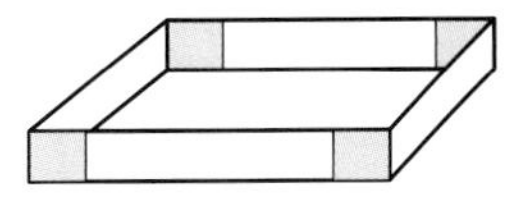

102. PACKAGING A pizza box with a lid is made from a 19-inch-by-40-inch piece of cardboard, as shown below. The volume of the pizza box is represented by $V(x) = \frac{1}{2}x(19 - 2x)(40 - 5x)$, where x is the height of the box.

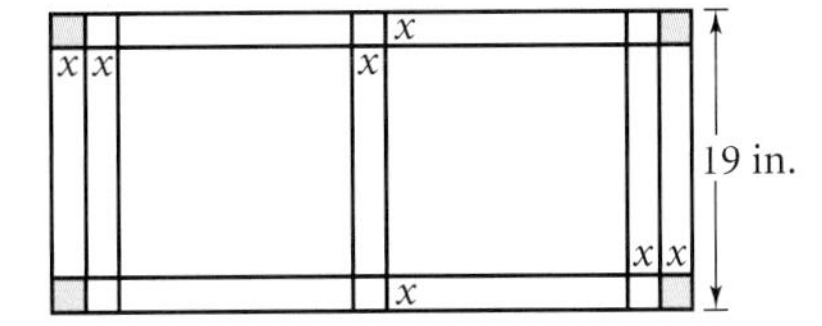

a. Write the volume of the box as a polynomial function in standard form.

b. Find the volume of the box if its height is 2 inches.

c. Find the volume of the box if its height is 2.5 inches.

Look Back

103. Solve $x + 3 \leq 3(x - 1)$ for x. Graph the solution on a number line. ***(LESSON 1.7)*** **$3 \leq x$**

104. Does the set of ordered pairs $\{(2, 3), (3, 5), (4, 3)\}$ represent a function? Explain. ***(LESSON 2.3)*** **Yes; each domain element is paired with exactly one range element.**

Factor each expression. ***(LESSON 5.3)***

105. $5a^2 - 5b^2$	**106.** $2x^2 - 32y^2$	**107.** $n^2 + n - 12$
108. $5 - 6s + s^2$	**109.** $4x^2 + 4x + 1$	**110.** $2x^2 + 11x + 15$

105. $5(a + b)(a - b)$

106. $2(x + 4y)(x - 4y)$

107. $(n + 4)(n - 3)$

108. $(5 - s)(1 - s)$

109. $(2x + 1)^2$

110. $(x + 3)(2x + 5)$

Solve for x. Round your answers to the nearest hundredth. ***(LESSONS 6.3 AND 6.6)***

111. $10^x = 32$ **≈1.51**	**112.** $3^x = 7$ **≈1.77**	**113.** $e^x = 5$ **≈1.61**	**114.** $e^{2x} = 7$ **≈0.97**

Look Beyond

115. If $f(x) = x^3 + 4x^2 - 3x - 18$ and $f(2) = 0$, how many other values of x are zeros of f? How many times does the graph of f cross the x-axis?

Look Beyond

In this exercise, students examine a cubic function to determine the number of real zeros. In Lesson 7.5, students will use the rational root theorem, the complex conjugate root theorem, and the fundamental theorem of algebra to find the zeros of a polynomial function.

103. $3 \leq x$

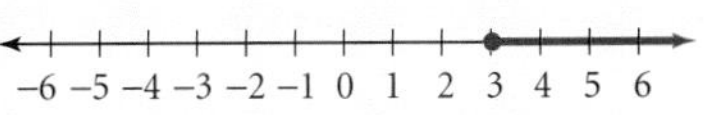

115. There are at most two other values of x that are zeros of f. The graph of f crosses the x-axis once (at $x = 2$) and touches the x-axis at $x = -3$.

Student Technology Guide

NAME ______ CLASS ______ DATE ______

Student Technology Guide
7.3 **Products and Factors of Polynomials**

In Lesson 7.3 of your textbook, you factored polynomial expressions algebraically. You can also factor graphically by using a graphics calculator to find the zeros of the related function, $P(x)$.

Example: Factor $x^3 - x^2 - 16x + 16$.

- Let $P(x) = x^3 - x^2 - 16x + 16$. Press Y= X,T,θ,n ^ 3 − X,T,θ,n x² − 16 X,T,θ,n + 16. Press GRAPH.
- Three zeros appear in this viewing window. A cubic function can only change directions twice, so these three points are all the zeros.
- Press TRACE. Use ◄ and ► to move as close as you can to the zero that is farthest to the left. This zero appears to be near $x = -4$.
- To check, press (−) 4 ENTER. As shown, −4 is a zero of $P(x)$. This means $[x - (-4)] = (x + 4)$ is a factor of $x^3 - x^2 - 16x + 16$.
- Repeat these steps in order to find the other zeros of the function. In this example, the zeros are $x = 1$ and $x = 4$, so $(x - 1)$ and $(x - 4)$ are the other factors of $x^3 - x^2 - 16x + 16$. Thus, $x^3 - x^2 - 16x + 16 = (x + 4)(x - 1)(x - 4)$.

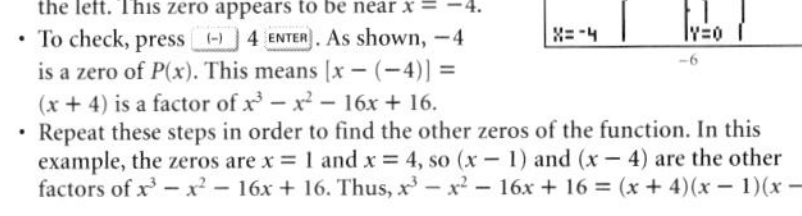

Factor each polynomial.

1. $x^3 - 8x^2 + 12x$ — $x(x - 6)(x - 2)$
2. $x^3 - 7x - 6$ — $(x - 3)(x + 1)(x + 2)$
3. $x^3 + 6x^2 - x - 30$ — $(x - 2)(x + 3)(x + 5)$

Some calculators, such as the TI-92, can factor expressions involving variables. To factor $x^3 - 11x^2 + 20x + 32$, press F2 Algebra 2:factor(ENTER X ^ 3 − 11 X ^ 2 + 20 X + 32) ENTER.

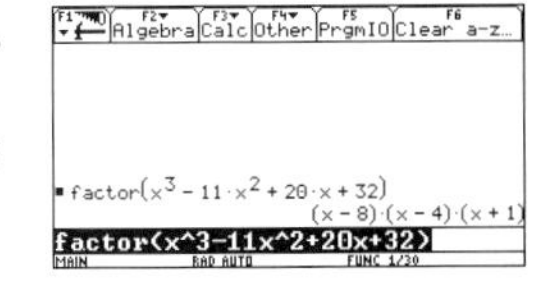

Factor each polynomial.

4. $2x^3 - 10x^2 - 16x + 24$ — $2(x - 6)(x - 1)(x + 2)$
5. $x^4 + 17x^3 + 99x^2 + 223x + 140$ — $(x + 1)(x + 4)(x + 5)(x + 7)$
6. $x^4 + 4x^3 - 2x^2 - 12x + 9$ — $(x - 1)^2(x + 3)^2$

Prepare

NCTM PRINCIPLES & STANDARDS 1–4, 6–10

QUICK WARM-UP

In Exercises 1–4, use factoring and the zero-product property to solve each equation.

1. $x^2 - 5x - 14 = 0$
   $x = -2$ *or* $x = 7$
2. $x^2 + 9x = 0$
   $x = 0$ *or* $x = -9$
3. $x^2 - 121 = 0$
   $x = -11$ *or* $x = 11$
4. $x^2 - 16 = 6x$
   $x = -2$ *or* $x = 8$
5. Factor $2x^3 - 5x^2 - 7x$.
   $x(2x - 7)(x + 1)$
6. Find the quotient.
   $(x^3 - 2x^2 - 5x + 6) \div (x + 2)$
   $x^2 - 4x + 3$

Also on Quiz Transparency 7.4

Teach

Why Polynomial functions can be applied to many real-world situations. For example, an architect or construction engineer can model the relationship between the dimensions and volume of a proposed warehouse, office building, or sports arena. Ask students to discuss what advantages and cost savings could be gained by using this information.

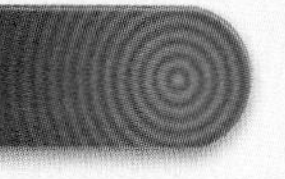

7.4 Solving Polynomial Equations

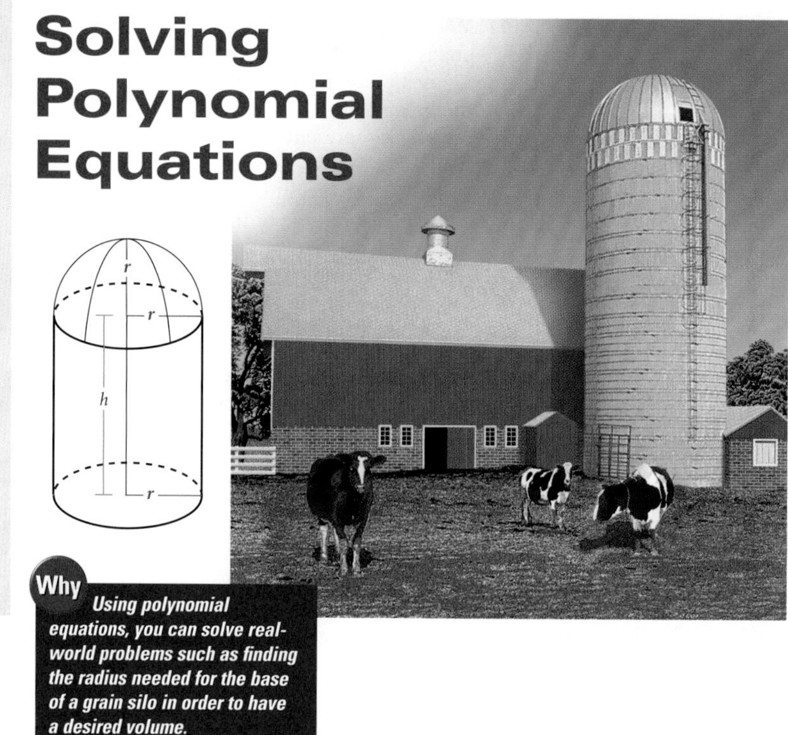

Objectives

- Solve polynomial equations.
- Find the real zeros of polynomial functions and state the multiplicity of each.

Why ***Using polynomial equations, you can solve real-world problems such as finding the radius needed for the base of a grain silo in order to have a desired volume.***

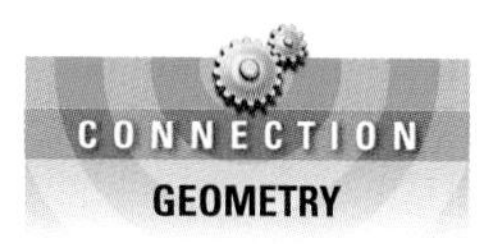

CONNECTION **GEOMETRY**

A silo that stores grain is often shaped as a cylinder with a hemispherical top (dome), as shown above. From geometry, you know that the volume of the cylinder, C, is represented by $C(r) = \pi r^2 h$ and that the volume of the hemispherical top, H, is represented by $H(r) = \left(\frac{1}{2}\right)\left(\frac{4}{3}\pi r^3\right)$, or $\frac{2}{3}\pi r^3$. Therefore, the total volume of the silo is given by the function T below.

$$\begin{aligned} T(r) &= H(r) + C(r) \\ &= \tfrac{2}{3}\pi r^3 + \pi r^2 h \end{aligned}$$

A farmer wants to design a silo whose cylindrical part has a height of 20 feet. Approximately what radius of the cylinder and hemispherical top will give a total volume of 1830 cubic feet? To answer this question, you will solve a polynomial equation. *You will answer this question in Example 4.*

EXAMPLE 1 **Use factoring to solve $2x^3 - 7x^2 + 3x = 0$.**

SOLUTION

$$2x^3 - 7x^2 + 3x = 0$$
$$x(2x^2 - 7x + 3) = 0 \qquad \textit{Factor out the GCF.}$$
$$x(2x - 1)(x - 3) = 0 \qquad \textit{Factor the remaining trinomial.}$$
$$x = 0 \quad or \quad 2x - 1 = 0 \quad or \quad x - 3 = 0 \qquad \textit{Use the Zero-Product Property.}$$
$$x = \tfrac{1}{2} \qquad x = 3$$

Alternative Teaching Strategy

USING COGNITIVE STRATEGIES To help students see the relationship between the number of times a factor occurs in the factorization of a polynomial expression and the x-intercepts on the graph of the related polynomial function, have students graph $f(x) = (x - 1)(x + 1)(x - 3)$. Point out that the graph crosses the x-axis at $(-1, 0)$, $(1, 0)$ and $(3, 0)$. Now have them graph the following functions on the same screen as the graph of $f(x)$:

$g(x) = (x - 1)(x + 1)(x - 2)$
$h(x) = (x - 1)(x + 1)(x - 1.5)$
$j(x) = (x - 1)(x + 1)(x - 1)$

Have students describe what they notice. Be sure they see that the graph of $f(x)$ crosses the x-axis three times and that the graph of $j(x)$ touches the x-axis at $(1, 0)$ but does not cross the x-axis at that x-intercept. Now have them write a third-degree polynomial function in factored form with the following characteristics: the left end turns upward, the right end turns downward, one x-intercept crosses the x-axis, and one x-intercept touches but does not cross the x-axis.
sample answer: $f(x) = -(x + 2)(x - 1)^2$

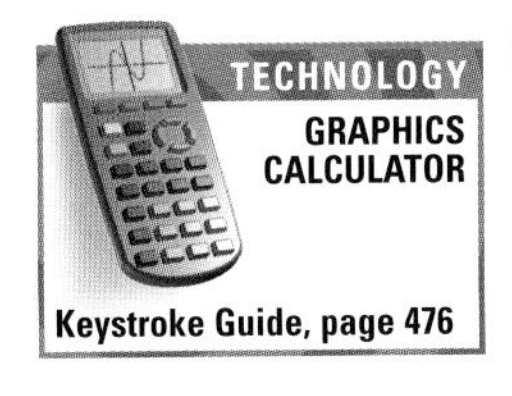

CHECK

Graph $y = 2x^3 - 7x^2 + 3x$, and look for any zeros of the function.

The graph of $y = 2x^3 - 7x^2 + 3x$ confirms that the solutions of $2x^3 - 7x^2 + 3x = 0$ are $x = 0$, $x = \frac{1}{2}$, and $x = 3$.

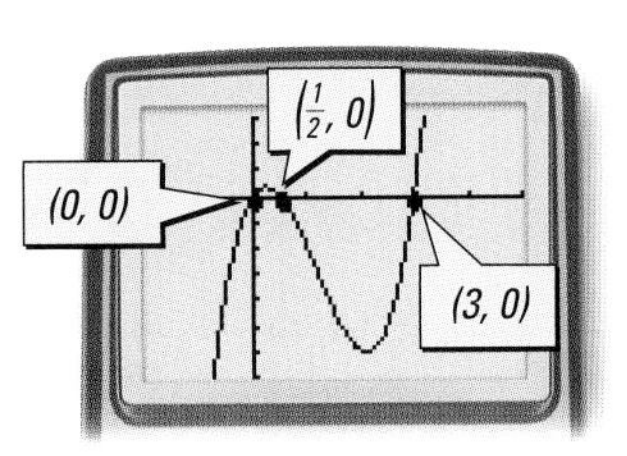

TRY THIS Use factoring to solve $2x^3 + x^2 - 6x = 0$.

The polynomial equation in Example 1 has three linear factors and three distinct solutions, or **roots.** Some polynomial equations have factors (and roots) that occur more than once, as shown in Example 2.

EXAMPLE 2 **Use a graph, synthetic division, and factoring to find all of the roots of $x^3 - 7x^2 + 15x - 9 = 0$.**

SOLUTION

PROBLEM SOLVING

Use a graph of the related function to approximate the roots. Then use synthetic division to test your choices.

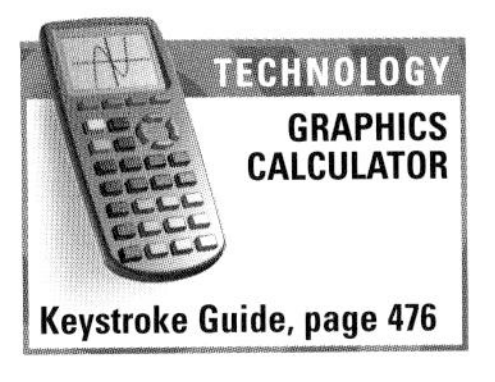

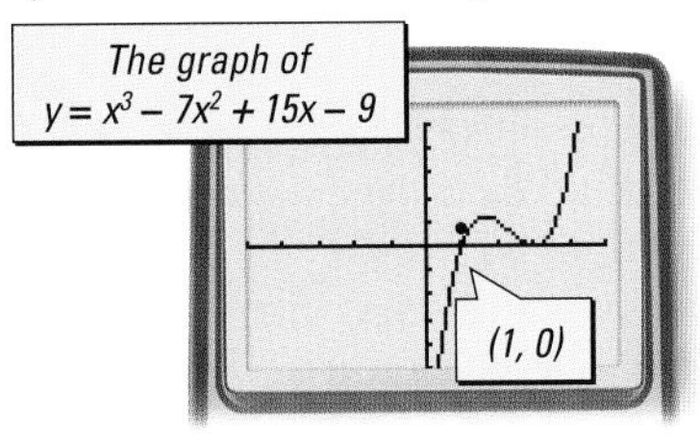

1	1	−7	15	−9
		1	−6	9
	1	−6	9	0

The remainder is 0.

The quotient is $x^2 - 6x + 9$.

Since the remainder is 0, $x - 1$ is a factor of $x^3 - 7x^2 + 15x - 9$.

$$x^3 - 7x^2 + 15x - 9 = 0$$
$$(x - 1)(x^2 - 6x + 9) = 0 \qquad \textit{Factor out } x - 1.$$
$$(x - 1)(x - 3)^2 = 0 \qquad \textit{Factor the remaining trinomial.}$$
$$x - 1 = 0 \quad \text{or} \quad x - 3 = 0 \quad \text{or} \quad x - 3 = 0 \qquad \textit{Use the Zero-Product Property.}$$
$$x = 1 \qquad x = 3 \qquad x = 3$$

The roots of $x^3 - 7x^2 + 15x - 9 = 0$ are 1 and 3, with the root 3 occurring twice.

TRY THIS Use a graph, synthetic division, and factoring to find all of the roots of $x^3 + 2x^2 - 4x - 8 = 0$.

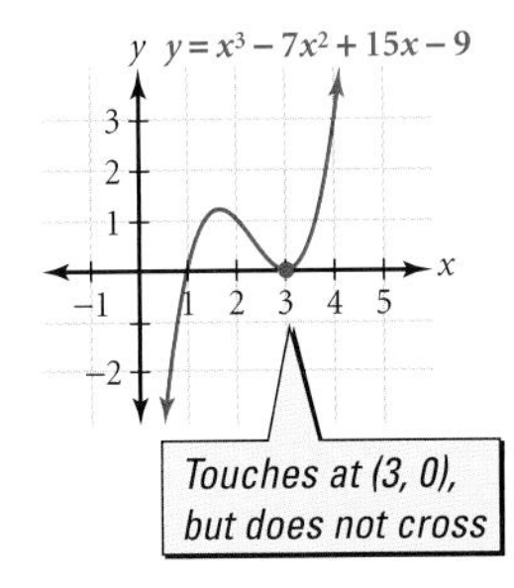

If $x - r$ is a factor that occurs m times in the factorization of a polynomial expression, P, then r is a root with **multiplicity** m of the related polynomial equation, $P = 0$. In Example 2 above, 3 is a root with multiplicity 2 of the equation $x^3 - 7x^2 + 15x - 9 = 0$.

When r is a root with *even multiplicity*, then the graph of the related function *will touch but not cross* the x-axis at $(r, 0)$. This is shown at left for the related function from Example 2 above.

In general, you cannot tell the multiplicity of a root by a graph of the related function alone. The graph may appear to touch but actually cross the x-axis.

Math CONNECTION

GEOMETRY Remind students that the formula for the volume of a cylinder equals the area of a base times the height of the cylinder, $V = \pi r^2 h$, and the volume of a sphere is given by $V = \frac{4}{3}\pi r^3$.

ADDITIONAL EXAMPLE 1

Use factoring to solve $5x^3 - 12x^2 + 4x = 0$.

$x = 0$ *or* $x = \frac{2}{5}$ *or* $x = 2$

TRY THIS

$x = 0$ *or* $x = -2$ *or* $x = \frac{3}{2}$

ADDITIONAL EXAMPLE 2

Use a graph, synthetic division, and factoring to find all roots of $x^3 + 3x^2 - 4 = 0$.

1 and −2, with the root −2 occurring twice

Teaching Tip

TECHNOLOGY If you are using a TI-82 or TI-83 graphics calculator to create the graph for Example 2, set the viewing window as follows:

Xmin=−5	Ymin=−5
Xmax=5	Ymax=5
Xscl=1	Yscl=1

To locate the y-value that corresponds to an x-value of 1, press TRACE 1 ENTER. At the bottom of the screen you will see that the y-value is 0.

TRY THIS

$x = -2$ *or* $x = 2$

Interdisciplinary Connection

GRAPHIC ARTS The real roots of quartic and cubic polynomial equations are useful in computer animation. To generate objects with curved surfaces, these equations must be solved exactly, rather than approximately. The algorithms used by programmers for solving cubic and quartic polynomial equations stem from the sixteenth century. The algorithm used to solve cubic equations is called Cardano's method (Girolamo Cardano, 1501–1576), and the algorithm used to solve quartic polynomial equations is called Ferrari's method (Lodovico Ferrari, 1522–1565). Encourage students to research these methods. Be sure they understand that solving a quartic equation depends on solving a cubic equation and solving a cubic equation depends on solving a quadratic equation.

ADDITIONAL

EXAMPLE 3

Use variable substitution and factoring to find all roots of $x^4 - 5x^2 - 6 = 0$.
$x = \sqrt{6}$ *or* $x = -\sqrt{6}$

Teaching Tip

TECHNOLOGY When using a TI-82 or TI-83 graphics calculator to create the graph for Example 3, set the viewing window as follows:

Xmin=–5	Ymin=–5
Xmax=5	Ymax=5
Xscl=1	Yscl=1

To locate the y-value that corresponds to an x-value of $-\sqrt{3}$, press TRACE (–) 2nd x^2 3) ENTER. At the bottom of the screen, you will see that the corresponding y-value is 0.

Remind students that the negative key, which is represented by (–) and is located to the left of ENTER, is different from the subtraction key, which is represented by – and is located above +.

TRY THIS
$x = \sqrt{7}$ *or* $x = -\sqrt{7}$ *or* $x = \sqrt{2}$ *or* $x = -\sqrt{2}$

CRITICAL THINKING
The roots are 0 and $\frac{-b \pm \sqrt{b^2 - 4ac}}{2a}$; the roots are complex if $b^2 - 4ac < 0$, and are real otherwise.

Sometimes polynomials can be factored by using *variable substitution*. This is shown in Example 3.

EXAMPLE 3 **Use variable substitution and factoring to find all of the roots of $x^4 - 4x^2 + 3 = 0$.**

SOLUTION

PROBLEM SOLVING

1. **Solve a simpler problem.** The expression $x^4 - 4x^2 + 3$ can be put in the form of a factorable quadratic expression by substituting u for x^2. Solve the resulting equation for u.

$$x^4 - 4x^2 + 3 = 0$$
$$(x^2)^2 - 4(x^2) + 3 = 0$$
$$u^2 - 4u + 3 = 0$$
$$(u - 1)(u - 3) = 0$$
$$u = 1 \text{ or } u = 3$$

2. Replace u with x^2, and solve for x by using the Zero-Product Property.

$$x^2 = 1 \quad \text{or} \quad x^2 = 3$$
$$x = \pm\sqrt{1} \qquad x = \pm\sqrt{3}$$

The roots of $x^4 - 4x^2 + 3 = 0$ are $\sqrt{1}, -\sqrt{1}, \sqrt{3}$, and $-\sqrt{3}$.

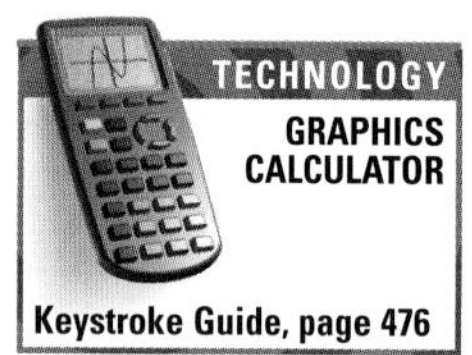

CHECK

Graph $y = x^4 - 4x^2 + 3$, and look for any zeros of the function.

Because $\sqrt{3} \approx 1.73$ and $\sqrt{1} = 1$, the graph of $y = x^4 - 4x^2 + 3$ confirms that the roots of the related equation are $-\sqrt{3}, -\sqrt{1}, \sqrt{1}$, and $\sqrt{3}$.

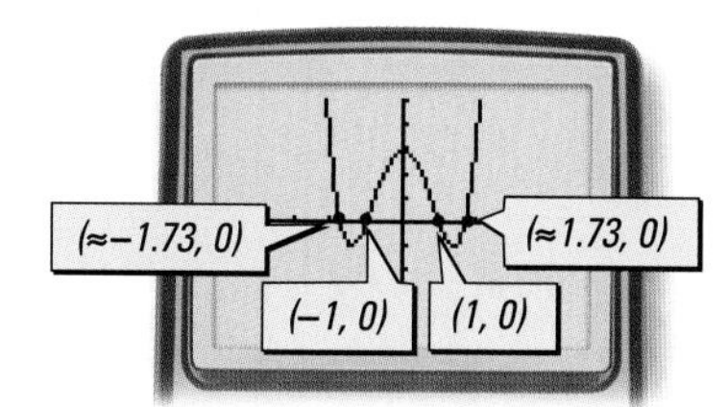

TRY THIS Use variable substitution and factoring to find all of the roots of $x^4 - 9x^2 + 14 = 0$.

CRITICAL THINKING Let a, b, and c be real numbers. Find all roots of $ax^3 + bx^2 + cx = 0$, where $a \neq 0$. Classify the roots as real or complex.

Finding Real Zeros

The *Location Principle*, given below, can be used to find real zeros.

Location Principle

If P is a polynomial function and $P(x_1)$ and $P(x_2)$ have opposite signs, then there is a real number r between x_1 and x_2 that is a zero of P, that is, $P(r) = 0$.

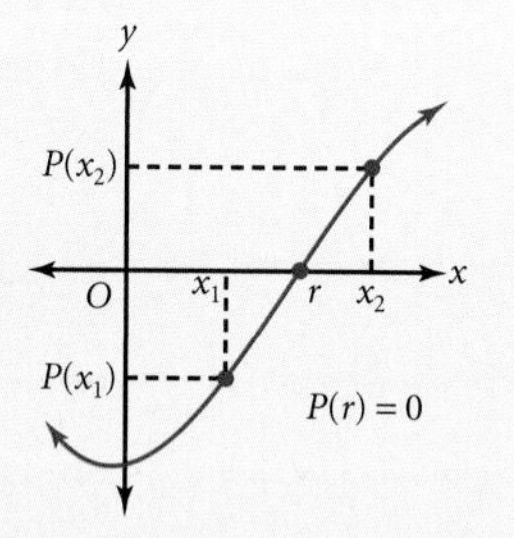

Inclusion Strategies

KINESTHETIC Have students make three or four rectangular prisms. They should measure the sides and determine the volume. Then they should label one edge length x, write expressions for the other edge lengths in terms of x, and write an algebraic expression for the volume. Have students exchange labeled prisms and then find the volume by using an algebraic method.

Enrichment

Some polynomial expressions and equations can be written in quadratic form. For example, $2x^4 - 5x^2 + 3$ can be written as $(2x^2)^2 - 5(x)^2 + 3$ and factored as $(2x^2 - 3)(x^2 - 1) = (2x^2 - 3)(x + 1)(x - 1)$; the equation $x - 6\sqrt{x} + 5 = 0$ can be written as $(\sqrt{x})^2 - 6\sqrt{x} + 5 = (\sqrt{x} - 5)(\sqrt{x} - 1) = 0$. Have students find the real solutions to the following polynomial equations by quadratic methods:

1. $x^4 + 10x^2 - 11 = 0$ — **$x = 1, x = -1$**
2. $x^6 - 8x^3 - 9 = 0$ — **$x = -1, x = \sqrt[3]{9}$**
3. $x + \sqrt{x} - 6 = 0$ — **$x = 4$**
4. $5x + 11\sqrt{x} + 2 = 0$ — **no solution**

The graph of $P(x) = x^3 - 4x + 2$ at right shows that when x is 1, $P(x) < 0$, and when x is 2, $P(x) > 0$.

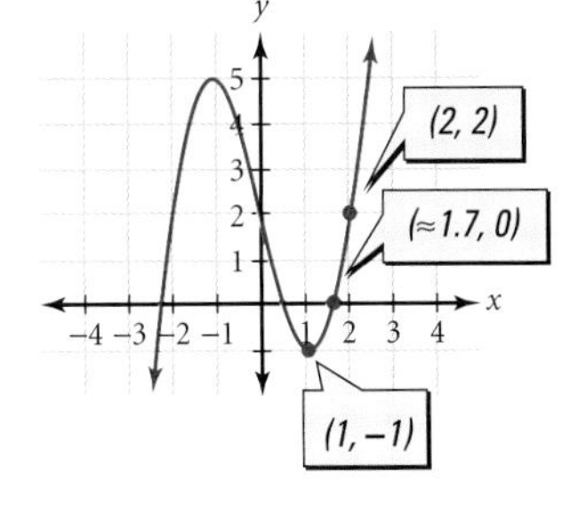

According to the Location Principle, there is a zero of $P(x) = x^3 - 4x + 2$ somewhere between 1 and 2. It is about 1.7.

When you use a graph to find a zero of a continuous function, you are actually using the Location Principle.

CHECKPOINT ✔ Graph $y = x^2 + 2$. Use the graph and the Location Principle to explain why $y = x^2 + 2$ has no real zeros.

EXAMPLE 4 Refer to the silo described at the beginning of the lesson.

What radius of the cylinder and hemispherical top gives a total volume of 1830 cubic feet if the cylinder's height is 20 feet?

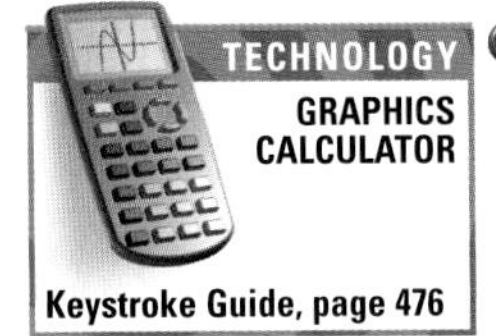

Keystroke Guide, page 476

SOLUTION

The total volume is represented by $T(r) = \frac{2}{3}\pi r^3 + \pi r^2 h$. Substituting 20 for the height and 1830 for the total volume gives $1830 = \frac{2}{3}\pi r^3 + 20\pi r^2$.

Solve for r by graphing the related function, $y = \frac{2}{3}\pi x^3 + 20\pi x^2 - 1830$, and approximating the real zeros.

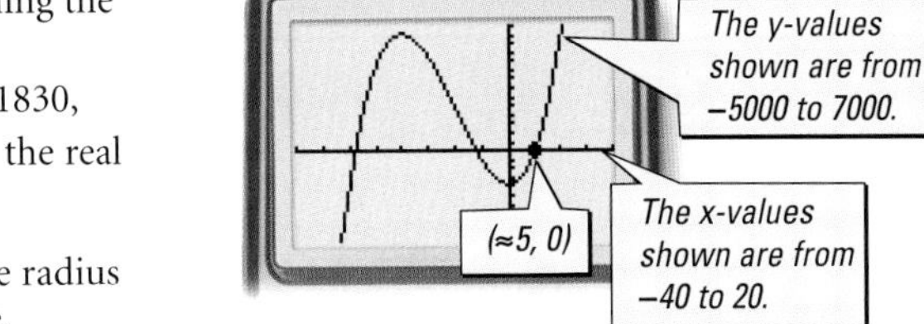

The only reasonable radius is a positive x-value.

Thus, a radius of about 5 feet gives a volume of approximately 1830 cubic feet.

The Activity below involves using a table of values and the Location Principle to find the zeros of a function.

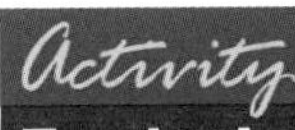

Exploring With Tables

TECHNOLOGY GRAPHICS CALCULATOR
Keystroke Guide, page 477

You will need: a graphics calculator

Let $P(x) = x^4 - 3x^3 - 5x^2 + 13x + 6$.

1. Use the table feature of a calculator to find $P(x)$ for each x-value listed in the table at right.
2. From the table, what integer zeros of P can you find?
3. Use the table feature of a graphics calculator and the Location Principle to locate another zero of P to the nearest hundredth.

x	$P(x)$
−3	
−2	
−1	
0	
1	
2	
3	

Teaching Tip

TECHNOLOGY When using a TI-82 or TI-83 graphics calculator to create the graph of $P(x)$, set the viewing window as follows:

Xmin=−5	Ymin=−2
Xmax=5	Ymax=6

To locate the zero, press 2nd TRACE and select 2:zero. Enter 1 for the left bound, 2 for the right bound, and 1.5 for the guess. At the bottom of the screen you will see that the zero is $x \approx 1.68$.

CHECKPOINT ✔

The graph never crosses or touches the x-axis.

ADDITIONAL

EXAMPLE 4

A water trough is shaped like a half-cylinder with a half-hemisphere attached at each end, as shown below. The volume, V, of the trough is represented by $V(x) = \frac{2}{3}\pi r^3 + 5\pi r^2$, where r is the radius of the cylinder and the hemispherical ends. **What radius gives a total volume of 385 cubic feet?**

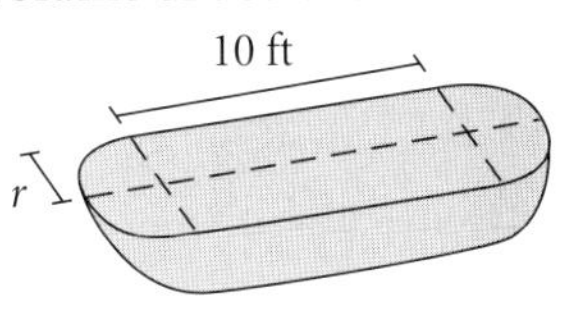

≈4 ft

Teaching Tip

TECHNOLOGY When using a TI-82 or TI-83 graphics calculator for Example 4, set the viewing window as follows:

Xmin=−40	Ymin=−5000
Xmax=20	Ymax=7000
Xscl=5	Yscl=500

☞ For Activity Notes, see page 452.

Reteaching the Lesson

USING VISUAL MODELS To encourage students to summarize what they have learned about real roots and zeros, give them a sentence such as "0 is the remainder when $x^3 - 4x^2 - x + 4$ is divided by $x - 4$." Have students refer to the summary of roots and zeros on page 452 and write three true statements that follow from the given statement. One statement should be that 4 is a solution, or root, of $x^3 - 4x^2 - x + 4 = 0$. Now instruct students to find the other two solutions and, using the summary box as a guide, to write three statements arising from each of these solutions.

$x - 4$ is a factor of $x^3 - 4x^2 - x + 4$; 4 is an x-intercept of the graph of $x^3 - 4x^2 - x + 4$; 1 and −1 are solutions of $x^3 - 4x^2 - x + 4$; $(x - 1)$ and $(x + 1)$ are factors of $x^3 - 4x^2 - x + 4$; when $x^3 - 4x^2 - x + 4$ is divided by $x - 1$ or divided by $x + 1$, the remainder is 0; 1 and −1 are x-intercepts of the graph of $x^3 - 4x^2 - x + 4$.

Activity Notes

Students should notice that a sign change in consecutive y-values indicates that the graph of the polynomial function will cross the x-axis within the interval determined by the corresponding x-values. This verifies the existence of a real zero of the polynomial function.

CHECKPOINT ✔

5. **A table can be used to locate whole-number zeros and sign changes. After the sign changes are identified, a graph can approximate the zeros more closely. The zeros of the function are the real roots of the related equation.**

Assess

Selected Answers

Exercises 4–10, 11–79 odd

ASSIGNMENT GUIDE

In Class	1–10
Core	11–59 odd, 63
Core Plus	12–64 even
Review	65–79
Preview	80

Extra Practice can be found beginning on page 940.

4. Now graph $P(x) = x^4 - 3x^3 - 5x^2 + 13x + 6$ in a viewing window that shows x-values from -5 to 5 and y-values from -15 to 15. Use the graph to locate a fourth zero.

CHECKPOINT ✔ **5.** Explain how to use a graph and a table to find the real zeros of a polynomial function. Explain how to use a graph and a table to find the real roots of a polynomial equation.

Below is a summary of what you have learned about real roots and zeros.

SUMMARY

Roots and Zeros

The real number r is a zero of $f(x)$ if and only if all of the following are true:

- r is a solution, or root, of $f(x) = 0$.
- $x - r$ is a factor of the expression that defines f (that is, $f(r) = 0$).
- When the expression that defines f is divided by $x - r$, the remainder is 0.
- r is an x-intercept of the graph of f.

Exercises

Communicate

1. If a polynomial has an even number of repeated factors, what do you know about its graph?

2. Describe how the Location Principle can help you to find the zeros of a polynomial function.

3. Explain how these terms are related: zeros, solutions, roots, factors, and x-intercepts.

Guided Skills Practice

Use factoring to solve each equation. ***(EXAMPLE 1)***

4. $x^3 - x^2 - 12x = 0$ **0, 4, −3**

5. $y^3 + 15y^2 + 54y = 0$ **0, −6, −9**

Use a graph, synthetic division, and factoring to find all of the roots of each equation. ***(EXAMPLE 2)***

6 $x^3 - 5x^2 + 3x = 0$ **$0, \frac{5 \pm \sqrt{13}}{2}$**

7 $x^3 - 3x - 2 = 0$ **2, −1, −1**

Use variable substitution and factoring to find all of the roots of each equation. ***(EXAMPLE 3)***

8. $x^4 - 8x^2 + 16 = 0$ **±2**

9. $x^4 - 2x^2 + 1 = 0$ **±1**

APPLICATION

10 **AGRICULTURE** The volume of a cylindrical silo with a cone-shaped top is represented by $V(x) = \frac{1}{3}\pi r^3 + 25\pi r^2$, where r is the radius of the silo in feet. Find the radius to the nearest tenth of a foot that gives a volume of 2042 cubic feet. ***(EXAMPLE 4)*** **4.9 feet**

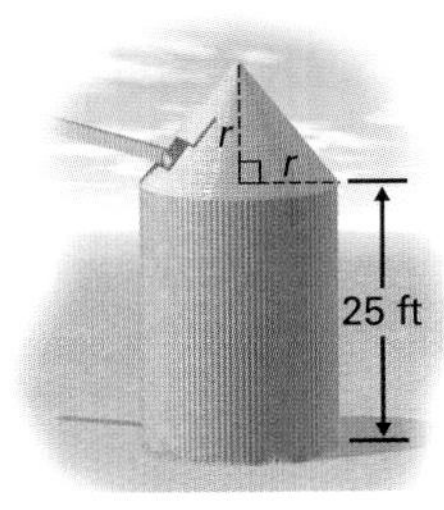

Practice and Apply

internet connect
Homework Help Online
Go To: **go.hrw.com**
Keyword:
MB1 Homework Help for Exercises 11–26

Use factoring to solve each equation.

11. $x^3 + 2x^2 - 35x = 0$ **0, 5, −7**
12. $x^3 + 2x^2 - 48x = 0$ **0, 6, −8**
13. $y^3 - 6y^2 - 27y = 0$ **0, −3, 9**
14. $a^3 - 8a^2 - 48a = 0$ **0, −4, 12**
15. $x^3 - 13x^2 + 40x = 0$ **0, 5, 8**
16. $x^3 - 7x^2 + 10x = 0$ **0, 5, 2**
17. $x^3 = 25x$ **0, 5, −5**
18. $y^3 = 49y$ **0, 7, −7**
19. $2x^3 - 10x^2 - 100x = 0$ **0, −5, 10**
20. $16x - 6x^2 - x^3 = 0$ **0, −8, 2**
21. $3y^3 + 9y^2 - 162y = 0$ **0, −9, 6**
22. $20a^2 + 5a^3 - 60a = 0$ **0, −6, 2**
23. $110x - 2x^3 = 12x^2$ **0, −11, 5**
24. $3y^3 + 36y^2 = 3y^4$ **0, 4, −3**
25. $28a - 5a^2 - 3a^3 = 0$ **0, $\frac{7}{3}$, −4**
26. $15y - 4y^2 - 3y^3 = 0$ **0, $\frac{5}{3}$, −3**

Use a graph, synthetic division, and factoring to find all of the roots of each equation.

27 $a^3 - a^2 - 5a - 3 = 0$ **3, −1**
28 $x^3 + 5x^2 + 7x + 3 = 0$ **−1, −3**
29 $x^3 + 5x^2 + 3x - 9 = 0$ **1, −3**
30 $x^3 - 4x^2 - 3x + 18 = 0$ **3, −2**
31 $2b^3 + 16b^2 + 32b = 0$ **0, −4**
32 $5h^3 - 60h^2 + 180h = 0$ **6, 0**
33 $x^3 - 3x - 2 = 0$ **2, −1**
34 $x^3 - 3x + 2 = 0$ **−2, 1**
35 $x^3 - 2x^2 - 9x + 18 = 0$ **−3, 2, 3**
36 $x^3 + 3x^2 - 4x - 12 = 0$ **2, −3, −2**
37 $n^3 + 8 = 2n^2 + 4n$ **2, −2**
38 $x^3 + 3x^2 = 27 + 9x$ **3, −3**

Use variable substitution and factoring to find all of the roots of each equation.

39. $x^4 - 4x^2 + 4 = 0$ **$\pm\sqrt{2}$**
40. $x^4 - 6x^2 + 9 = 0$ **$\pm\sqrt{3}$**
41. $y^4 - 18y^2 + 81 = 0$ **±3**
42. $a^4 - 24a^2 + 144 = 0$ **$\pm 2\sqrt{3}$**
43. $x^4 - 13x^2 + 36 = 0$ **±3, ±2**
44. $x^4 - 9x^2 + 18 = 0$ **$\pm\sqrt{6}, \pm\sqrt{3}$**
45. $x^5 - 9x^3 + 8x = 0$ **$\pm 2\sqrt{2}$, ±1, 0**
46. $z^5 - 28z^3 + 27z = 0$ **$\pm 3\sqrt{3}$, ±1, 0**
47. $x^4 - 12x^2 = -36$ **$\pm\sqrt{6}$**
48. $x^4 - 14x^2 = -49$ **$\pm\sqrt{7}$**
49. $h^4 + 12 = 7h^2$ **$\pm\sqrt{3}$, ±2**
50. $t^4 + 14 = 9t^2$ **$\pm\sqrt{7}, \pm\sqrt{2}$**

Use a graph and the Location Principle to find the real zeros of each function. Give approximate values to the nearest hundredth, if necessary.

51 $f(x) = 9x^3 - x^4 - 23$ **≈8.97, ≈1.45**
52 $g(x) = x^3 - 3x - 2$ **−1, 2**
53 $f(a) = a^3 - a^2 - 8a + 12$ **−3, 2**
54 $h(x) = x^3 - 36x^2 + 18x - 27$ **≈35.51**
55 $m(n) = n^3 - 6n^2 + 12n - 8$ **2**
56 $f(t) = 64t^2 - 80t + 25$ **0.625**
57 $g(x) = 2x^4 - 2x^2 + 3x - 1$
58 $h(x) = 13x^4 - 21x + 7$
59 $f(x) = x^2 - 5x^3 + 3x$
60 $a(x) = x^2 + 12x^3 - 3x$

57. ≈−1.49, ≈0.44
58. ≈0.34, ≈1.03
59. 0, ≈−0.68, ≈0.88
60. 0, ≈−0.54, ≈0.46

Error Analysis

Some students will solve the equation $x^3 = 25x$ by dividing both sides of the equation by x. If this is done, one of the roots, 0, is omitted. To avoid this error, remind students to write the equation in standard form and use factoring and the zero-product property.

CHALLENGE

61. Find the range of values of c such that the function $f(x) = 2x^3 - x^2 - 6x + c$ has a zero between $x = 0$ and $x = 1$. **$0 < c < 5$**

62 **CULTURAL CONNECTION: ASIA** Omar Khayyam, 1050–1122 C.E., of Persia (modern-day Iran) is usually remembered as a poet and the author of *The Rubaiyat,* although he was also a scientist and mathematician. Khayyam developed a method for finding the zeros of a cubic polynomial function of the form $f(x) = x^3 - bx - a$, where $a > 0$ and $b > 0$, by finding the x-coordinates of the intersection points of the graphs of familiar curves.

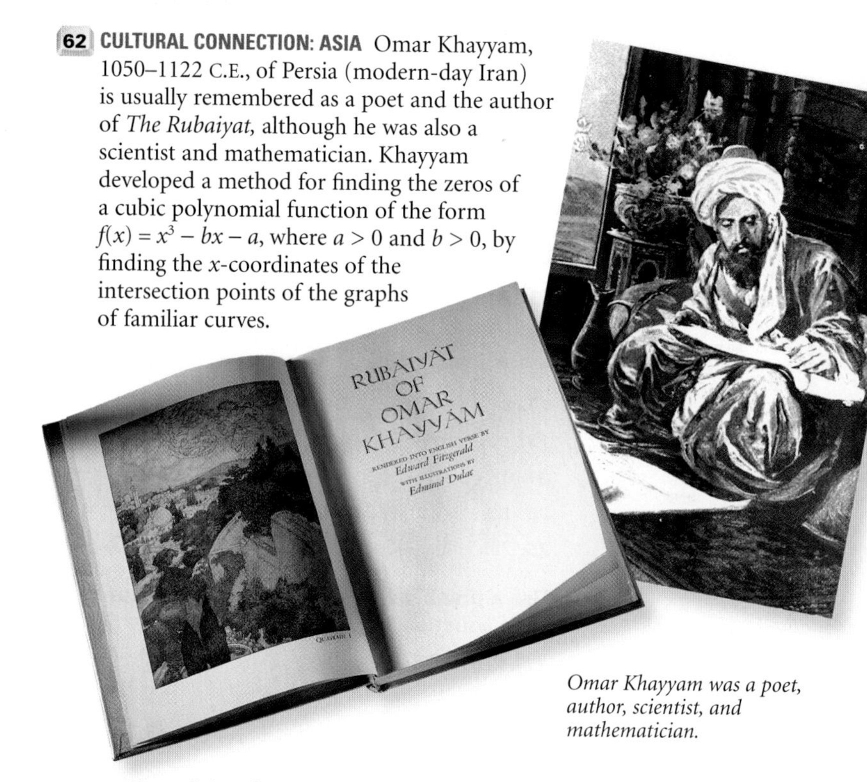

Omar Khayyam was a poet, author, scientist, and mathematician.

a. Given $f(x) = x^3 - 7x - 6$, identify the values of a and b.

b. Graph $y = -\frac{1}{\sqrt{b}}x^2$, $y = \sqrt{x^2 + \frac{a}{b}x}$, and $y = -\sqrt{x^2 + \frac{a}{b}x}$, using the values of a and b from part **a**.

c. Find all nonzero x-coordinates of the intersection points of these graphs. The x-coordinates represent the zeros of the original function, f.

d. Graph $f(x) = x^3 - 7x - 6$ and find all of its zeros to verify that the x-coordinates from part **c** are the zeros.

62a. $a = 6$, $b = 7$

b. $y = -\frac{1}{\sqrt{7}}x^2$,

$y = \pm\sqrt{x^2 + \frac{6}{7}x}$

c–d. $-2, -1, 3$

63. 2 feet × 6 feet × 3 feet

64. 2.005 millimeters

APPLICATIONS

63. MANUFACTURING The volume of a cedar chest whose length is 3 times its width and whose height is 1 foot greater than its width is represented by $V(x) = 3x^3 + 3x^2$, where V is the volume in cubic feet and x is the width in feet. If the volume of the chest is 36 cubic feet, what are its dimensions?

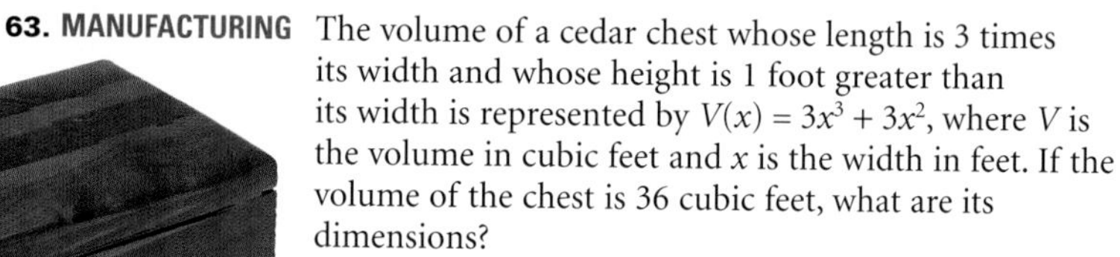

64. MEDICINE The volume of a cylindrical vitamin with a hemispherical top and bottom can be represented by the function $V(x) = 10\pi r^2 + \frac{4}{3}\pi r^3$, where V is the volume in cubic millimeters and r is the radius in millimeters. What should be the radius of the cylinder and hemispherical ends (to the nearest thousandth) so that the total volume is 160 cubic millimeters?

Practice

NAME ______ CLASS ______ DATE ______

Practice
7.4 ***Solving Polynomial Equations***

Use factoring to solve each equation.

1. $x^3 - 81x = 0$ — -9, 0, and 9
2. $x^3 - 11x^2 + 10x = 0$ — 0, 1, and 10
3. $2x^3 - x^2 - x = 0$ — $-\frac{1}{2}$, 0, and 1
4. $x^3 + 2x^2 = 15x$ — -5, 0, and 3
5. $2x^3 - 2x^2 - 24x = 0$ — -3, 0, and 4
6. $3x^3 + x = 4x^2$ — 0, $\frac{1}{3}$, and 1

Use graphing, synthetic division, and factoring to find all of the roots of each equation.

7. $x^3 - 3x^2 - 4x + 12 = 0$ — -2, 2, and 3
8. $x^3 + 4x^2 + x = 6$ — -3, -2, and 1
9. $3x^2 + 2x^2 - 37x + 12$ — 3, -4, and $\frac{1}{3}$
10. $x^3 + 29x + 42 = 12x^2$ — -1, 6, and 7
11. $x^3 - 11x^2 + 24x + 36 = 0$ — -1 and 6 (multiplicity 2)
12. $x^3 + 64 = 4x^2 + 16x$ — -4 and 4 (multiplicity 2)

Use variable substitution and factoring to find all of the roots of each equation. If necessary, leave your answers in radical form.

13. $x^4 - 10x^2 + 24 = 0$ — $-2, 2, -\sqrt{6}$, and $\sqrt{6}$
14. $x^4 - 10x^2 + 21 = 0$ — $-\sqrt{3}, \sqrt{3}, -\sqrt{7}$, and $\sqrt{7}$
15. $x^4 + 54 = 15x^2$ — $-\sqrt{6}, \sqrt{6}, -3$, and 3
16. $x^4 - 7x^2 = -10$ — $-\sqrt{2}, \sqrt{2}, -\sqrt{5}$, and $\sqrt{5}$
17. $x^4 - 17x^2 + 16 = 0$ — $-4, 4, -1, 1$
18. $x^4 + 20 = 12x^2$ — $-\sqrt{2}, \sqrt{2}, -\sqrt{10}$, and $\sqrt{10}$

Use a graph and the Location Principle to find the real zeros of each function. Give approximate values to the nearest tenth, if necessary.

19. $P(x) = 2x^3 - 4x + 1$ — 1.3, 0.3, and -1.5
20. $P(x) = 1.5x^3 + 2x^2 - 0.25$ — -1.2, -0.4, and 0.3
21. $P(x) = 2.5x^4 - 2x^2$ — -0.9, 0 (multiplicity 2), and 0.9
22. $P(x) = 12x^3 - 15x^2 + x + 1$ — -0.2, 0.4, and 1.1
23. $P(x) = 8x^3 - 6x^2 - 2x + 1$ — -0.4, 0.3, and 0.9
24. $P(x) = 0.5x^4 + 2x^3 - 5x - 1$ — -2.9, -2.4, -0.2, and 1.4

Solve each equation by factoring and applying the Zero-Product Property. *(LESSON 5.3)*

65. $x^2 - 2x = 63$ **9, −7** **66.** $2x^2 - 5 = -9x$ **$\frac{1}{2}$, −5** **67.** $3x^2 + 60 = 27x$ **4, 5**

Simplify. *(LESSON 5.6)*

68. $\sqrt{-49}$ **$7i$** **69.** $\sqrt{-81}$ **$9i$** **70.** $\sqrt{-11}$ **$i\sqrt{11}$**

Write the conjugate of each complex number. *(LESSON 5.6)*

71. $6 - 2i$ **$6 + 2i$** **72.** $4i - 2$ **$-2 - 4i$** **73.** 5 **5**

Plot each number and its conjugate in the complex plane. *(LESSON 5.6)*

74. $-4 - 3i$ **75.** $-1 - 3i$ **76.** $2i - 4$

Use the quadratic formula to find the solutions to each equation. Write your answers in the form $a + bi$. *(LESSON 5.6)*

77. $x^2 + 2x + 7 = 0$ **$-1 \pm i\sqrt{6}$** **78.** $4x^2 - 3x + 5 = 0$ **$\frac{3}{8} \pm \frac{\sqrt{71}}{8}i$** **79.** $x^2 - 3x = -9$ **$\frac{3}{2} \pm \frac{3\sqrt{3}}{2}i$**

Portfolio Extension
Go To: **go.hrw.com**
Keyword:
MB1 PolySolve

Look Beyond

80. Let $P(x) = x^3 - 2x^2 + 5x + 26$.
  a. Verify that $x - (2 + 3i)$ and $x - (2 - 3i)$ are factors of P.
  b. Find the third factor and write P in factored form.
  $(x + 2)$; $P(x) = (x + 2)[x - (2 + 3i)][x - (2 - 3i)]$

In the Portfolio Activity on page 431, you took measurements and made calculations to find a polynomial model for a cylindrical container with a flat circular base. In this activity, you will find a model for the volume of a bottle with a different shape.

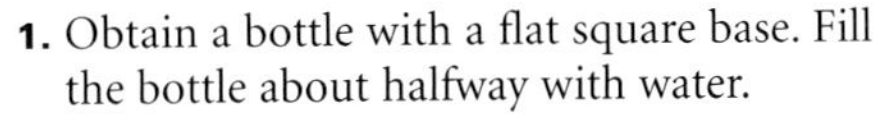

1. Obtain a bottle with a flat square base. Fill the bottle about halfway with water.
2. Write a function that approximates the volume, in milliliters, of the part of the bottle containing water, W.
3. Turn the bottle upside-down, and write a function that approximates the volume, in milliliters, of the air space in the bottle, A.
4. Write a function that models the total volume, in milliliters, of the bottle, V.
5. Suppose that the height of the water is equal to the length of each side of the square base. If the height of the air space above the water is 2 centimeters, what side length of the base will give a total volume of 96 milliliters?

WORKING ON THE CHAPTER PROJECT

You should now be able to complete Activity 3 of the Chapter Project.

The polynomial $P(x)$ is factored over the set of complex numbers. In Lesson 7.5, students will use the rational root theorem and the quadratic formula to write a cubic polynomial as a product of exactly three linear factors.

ALTERNATIVE

Assessment

Portfolio Activity

The Portfolio Activity can be used as preparation for the Chapter Project or as a separate activity. In the Portfolio Activity on this page, students will create a polynomial model for a bottle with a flat square base and will use the model to determine what length is needed for a given total volume.

74.

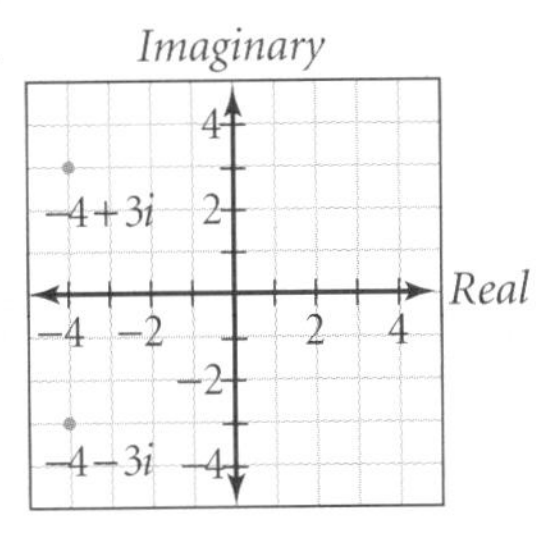

75.

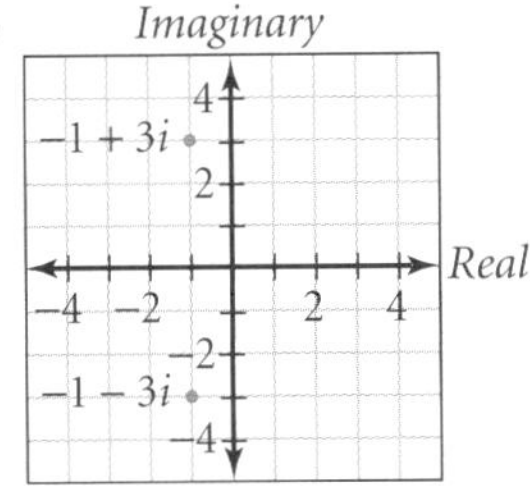

76.

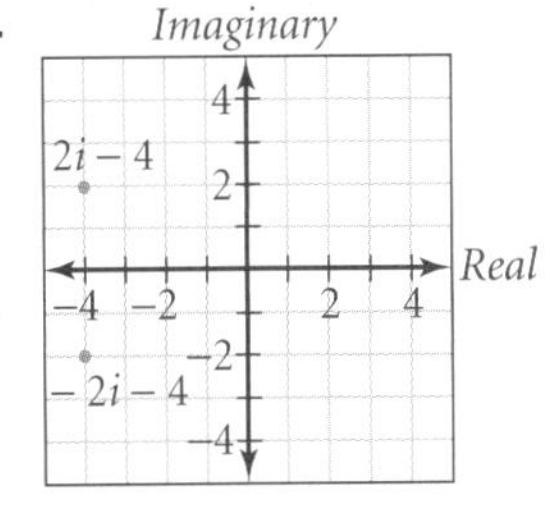

The answer to Exercise 80a can be found in Additional Answers beginning on page 1002.

Student Technology Guide

NAME ____________ CLASS ________ DATE ________

Student Technology Guide
7.4 Solving Polynomial Equations

You can approximate the roots of a polynomial equation by looking at the graph of the related function. The zero feature of a graphics calculator provides a quick way to locate the zeros.

Example: Find all roots of $x^3 - 7x^2 + 15x - 9 = 0$.

- Graph the related function. In this example, let $P(x) = x^3 - 7x^2 + 15x - 9$. Press Y= X,T,θ,n ^ 3 − 7 X,T,θ,n x² + 15 X,T,θ,n − 9. Then press GRAPH.
- Press TRACE. Use ◄ and ► to move to a point just to the left of a zero, as shown at left below. Press 2nd TRACE 2:zero ENTER.

- When asked for the left bound, press ENTER to enter the coordinates of the trace point.
- When asked for the right bound, trace just to the right of the zero. Press ENTER.
- When asked for a guess, trace as close as you can to the zero. Press ENTER.

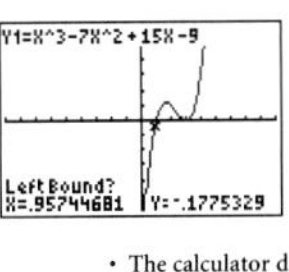

Y1=X^3-7X^2+15X-9
Right Bound?
X=1.1489362 Y=.51032045

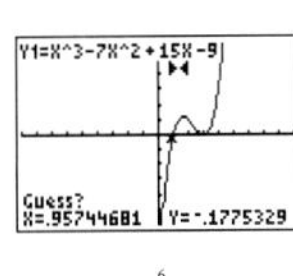

- The calculator displays the approximate coordinates of the zero. In this example, 1 is a zero of the related function, so it is a root of the polynomial.
- Repeat the process for all other zeros. In this example, 3 is the only other zero of the function.

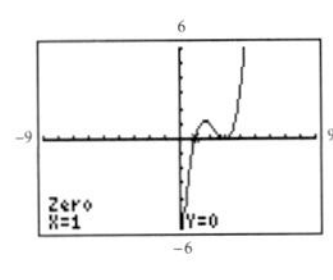

The roots of $x^3 - 7x^2 + 15x - 9 = 0$ are 1 and 3.

Find the roots of each equation. Round answers to the nearest hundredth.

1. $x^3 - 3x^2 - 13x + 15 = 0$ **−3, 1, and 5**
2. $\frac{1}{10}x^3 - 2x^2 + 10x = 0$ **0 and 10**
3. $x^3 + 5x^2 - 7x - 10 = 0$ **−5.90, −0.93, and 1.83**

Focus

This article describes the highest, fastest, and steepest roller coaster in the world. Students will sketch a graph of a polynomial function with at least three turning points. They will identify where the graph is increasing, identify where it is decreasing, and write the function that models their roller coaster track. Their design will include considerations for height, shape, friction, and wind resistance.

Motivate

Have students discuss roller coasters based on either their personal experience or from what they have seen on television. Ask them to estimate the average height, steepness, and speed of a typical roller coaster. Then ask them how high a roller coaster might ultimately be constructed, how steep it would be, and how fast it would be. Also ask them what safety considerations they think would limit the size and speed of roller coasters as they are currently constructed.

Suggest that students do research in the library or use other sources to find out more about the construction of roller coasters, including any restrictions that actually exist. Have them share the results of their research with the class.

EYEWITNESS MATH
SCREAM MACHINE!
'Highest, fastest, steepest' in the world
[Source: Sandusky Register, January 31, 1989]
Coasting to Records
On June 9, 1988, Cedar Point's Magnum XL-200 roller coaster was certified as the fastest roller coaster in the world, with the longest vertical drop. As a result, it was listed in the 1990 edition of the Guinness Book of World Records. To enter the record books, the coaster needed two witnesses: one was Lt. Gov. Paul Leonard, who observed the ride's 72 miles per hour top speed and vertical drop of 194 feet, 8 inches.
[Source: Sandusky Register, July 27, 1989]

1–2. One example sets the height of the hill 1 as 225 ft, hill 2 at 100 ft, and hill 3 at 49 ft.

Bottom of hill 1	
Theoretical speed	120 ft/s
Loss of speed (5%)	6 ft/s
Actual speed:	114 ft/s (≈ 78 mph)

Top of hill 2	
Theoretical speed	34 ft/s
Loss of speed (5%)	5.7 ≈ 6 ft/s
Actual speed:	28 ft/s (≈ 19 mph)

Bottom of hill 2	
Theoretical speed	108 ft/s
Loss of speed (5%)	5.4 ≈ 5 ft/s
Actual speed:	103 ft/s (≈ 70 mph)

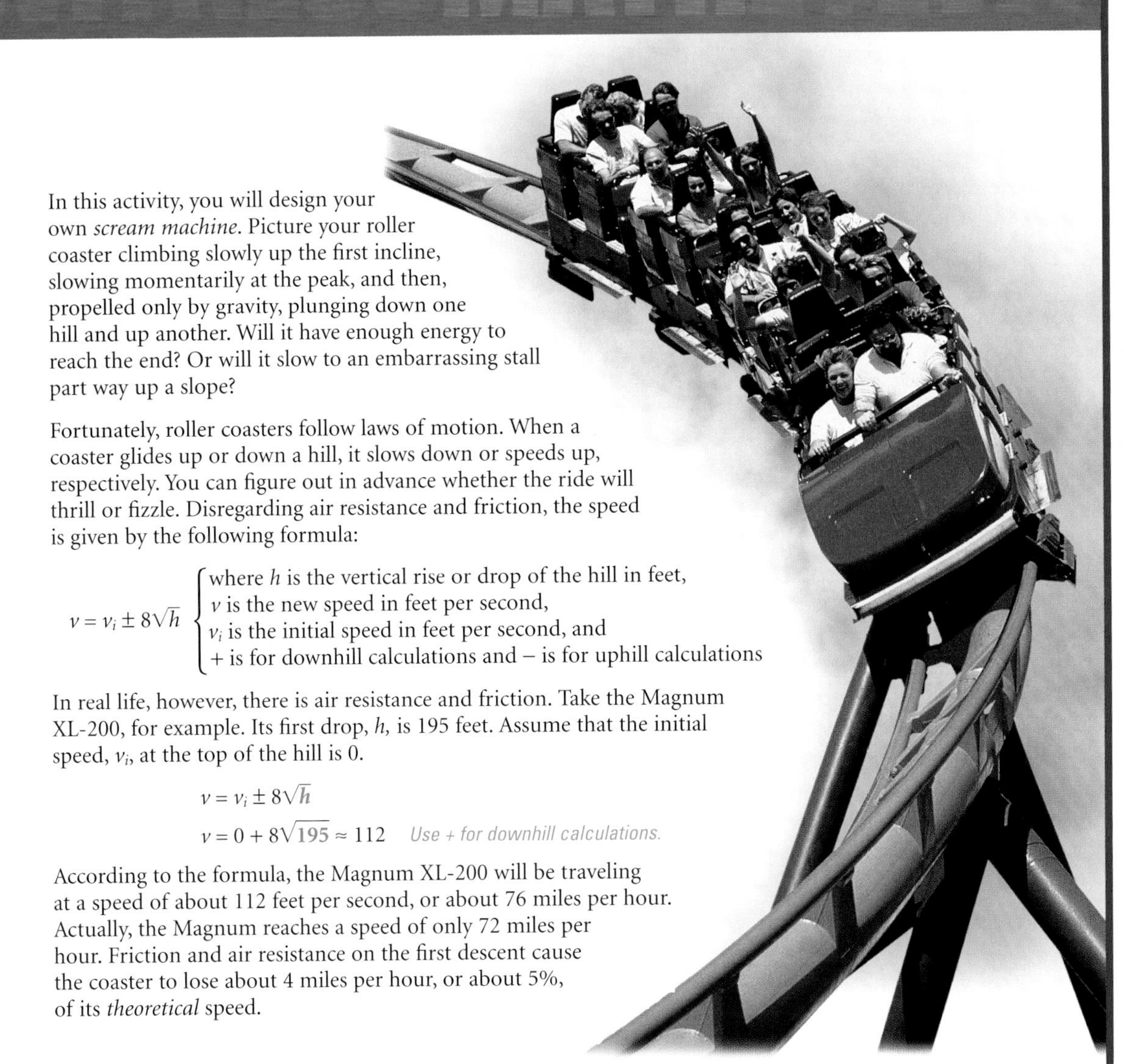

In this activity, you will design your own *scream machine*. Picture your roller coaster climbing slowly up the first incline, slowing momentarily at the peak, and then, propelled only by gravity, plunging down one hill and up another. Will it have enough energy to reach the end? Or will it slow to an embarrassing stall part way up a slope?

Fortunately, roller coasters follow laws of motion. When a coaster glides up or down a hill, it slows down or speeds up, respectively. You can figure out in advance whether the ride will thrill or fizzle. Disregarding air resistance and friction, the speed is given by the following formula:

$$v = v_i \pm 8\sqrt{h} \quad \begin{cases} \text{where } h \text{ is the vertical rise or drop of the hill in feet,} \\ v \text{ is the new speed in feet per second,} \\ v_i \text{ is the initial speed in feet per second, and} \\ + \text{ is for downhill calculations and } - \text{ is for uphill calculations} \end{cases}$$

In real life, however, there is air resistance and friction. Take the Magnum XL-200, for example. Its first drop, h, is 195 feet. Assume that the initial speed, v_i, at the top of the hill is 0.

$$v = v_i \pm 8\sqrt{h}$$

$$v = 0 + 8\sqrt{195} \approx 112 \quad \textit{Use + for downhill calculations.}$$

According to the formula, the Magnum XL-200 will be traveling at a speed of about 112 feet per second, or about 76 miles per hour. Actually, the Magnum reaches a speed of only 72 miles per hour. Friction and air resistance on the first descent cause the coaster to lose about 4 miles per hour, or about 5%, of its *theoretical* speed.

Cooperative Learning

Design your own roller-coaster track. Include at least three hills. Assume that the coaster loses 5% of its theoretical speed each time it goes up or down a hill. Make sure your roller coaster doesn't stall going up a hill.

1. Determine the vertical rise of each hill.
2. Assuming that the speed at the top of the first hill is 0 feet per second, determine the actual speeds at the top and the bottom of each hill.
3. What would it mean if the speed at the top of a hill was negative? How would you change the design to fix this situation?

Top of hill 3	
Theoretical speed	47 ft/s
Loss of speed (5%)	5.15 ≈ 5 ft/s
Actual speed:	42 ft/s (≈ 29 mph)

Bottom of hill 3	
Theoretical speed	98 ft/s
Loss of speed (5%)	4.9 ≈ 5 ft/s
Actual speed:	93 ft/s (≈ 63 mph)

3. A negative speed would mean the coaster would not make it up the hill. You could fix that by making that hill lower or by making the first hill higher.

Cooperative Learning

Divide the class into design teams of four members each. Each team member will create at least two track designs. The team will consider all of the designs created by its individual members and select the one that is best. The team will then write a short summary of its work, stating why it chose that design. The summary should also include a brief review of the designs that were not chosen.

Discuss

Each team will present its summary and the design it selected to the class. The presentation should explain how the team arrived at its final decision and highlight the advantages of the design that caused it to be selected. The report should also include the answers to the questions in Step 3.

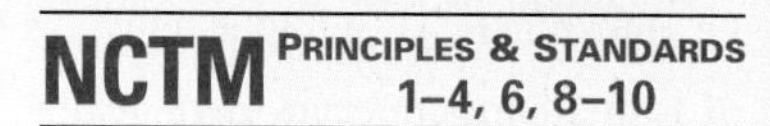

NCTM PRINCIPLES & STANDARDS 1–4, 6, 8–10

QUICK WARM-UP

List all the factors of each number.

1. 21 1, 3, 7, 21
2. 60 1, 2, 3, 4, 5, 6, 10, 12, 15, 20, 30, 60

Solve each equation.

3. $x^2 - 4x + 2 = 0$ $x = 2 \pm \sqrt{2}$
4. $2x^2 + 3x - 5 = 0$ $x = 1$ *or* $x = -\frac{5}{2}$

State whether the given linear expression is a factor of the given polynomial?

5. $x^3 - 2x^2 - 5x + 6; x - 3$ yes
6. $3x^3 - 4x^2 - x + 5; x + 1$ no

Also on Quiz Transparency 7.5

Teach

Why Polynomial functions provide a versatile tool for modeling problem situations in many areas. Ask students to imagine that they just inherited \$20,000. The terms of the inheritance will allow them to take the \$20,000 immediately or place it in an account for 10 years, at which time they would receive \$40,000. Ask them how they would determine which alternative would be more beneficial to them.

Zeros of Polynomial Functions

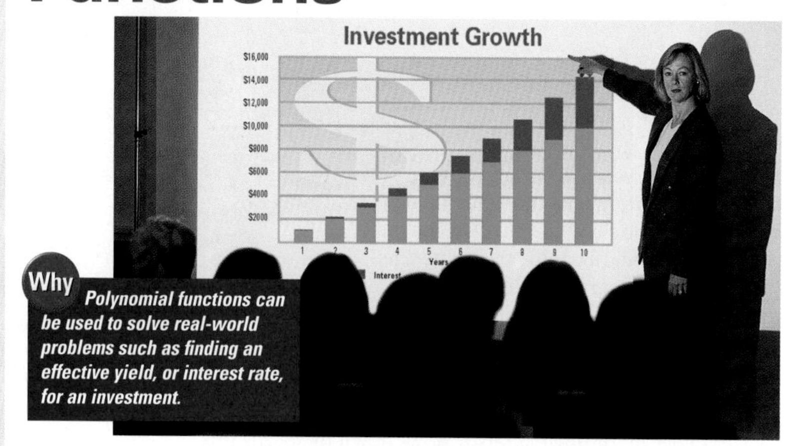

Why ***Polynomial functions can be used to solve real-world problems such as finding an effective yield, or interest rate, for an investment.***

Objectives

- Use the Rational Root Theorem and the Complex Conjugate Root Theorem to find the zeros of a polynomial function.
- Use the Fundamental Theorem to write a polynomial function given sufficient information about its zeros.

APPLICATION INVESTMENTS

Polynomial functions can be used to solve problems in a variety of fields. Consider the following example from the field of finance:

> \$1000 is invested at the beginning of each year in a fund that earns a variable interest rate. At the end of 10 years, the account is worth \$14,371.56. To find the *effective interest rate* of this investment over these 10 years, you can solve the polynomial equation
>
> $$1000x^{10} + 1000x^9 + \cdots + 1000x^2 + 1000x = 14{,}371.56,$$
>
> or find the roots of
>
> $$1000x^{10} + 1000x^9 + \cdots + 1000x^2 + 1000x - 14{,}371.56 = 0.$$

The *Rational Root Theorem* can be used to identify possible roots of polynomial equations with integer coefficients. To see how this theorem can be used, examine the solution to the equation $3x^2 + 10x - 8 = 0$ below.

$$3x^2 + 10x - 8 = 0$$
$$(3x - 2)(x + 4) = 0$$
$$3x - 2 = 0 \quad \text{or} \quad x + 4 = 0$$
$$x = \frac{2}{3} \qquad x = -4, \text{ or } \frac{-4}{1}$$

Notice that the numerators, 2 and −4, are factors of the constant term, −8, in the polynomial. Also notice that the denominators, 3 and 1, are the factors of the leading coefficient, 3, in the polynomial.

Rational Root Theorem

Let P be a polynomial function with integer coefficients in standard form. If $\frac{p}{q}$ (in lowest terms) is a root of $P(x) = 0$, then

- p is a factor of the constant term of P and
- q is a factor of the leading coefficient of P.

Alternative Teaching Strategy

USING VISUAL MODELS Ask students to graph the following: $f(x) = x^3 - 3x^2 + 3$, $g(x) = x^3 - 3x^2$, and $h(x) = x^3 - 3x^2 + 6$. Have them determine the number of real zeros each function has by using graphs, and then write each function in factored form. Give them the fundamental theorem of algebra and ask them what information it provides. Have them find any complex zeros for each of the given functions. Be sure they see that complex zeros occur in conjugate pairs, and state the complex conjugate root theorem. Now tell them that 1 and $2 + 3i$ are roots of a polynomial equation. Ask them to find the smallest degree of a polynomial that can have the two given roots, and have them write that polynomial in factored and standard form with the given roots. $f(x) = x^3 - 5x^2 + 17x - 13$ Then discuss the rational root theorem, and ask them to describe the number and nature of the zeros of $f(x) = 2x^3 - x^2 - x - 3$, list the possible rational zeros, and write the function in factored form. Finally summarize the fundamental theorem of algebra, noting that a polynomial function of degree n has exactly n complex zeros, counting multiplicities.

EXAMPLE 1 Find all of the rational roots of $10x^3 + 9x^2 - 19x + 6 = 0$.

SOLUTION

Write the related function, $P(x) = 10x^3 + 9x^2 - 19x + 6$. According to the Rational Root Theorem, $\frac{p}{q}$ is a root of $10x^3 + 9x^2 - 19x + 6 = 0$ if p is a factor of the constant term, 6, and q is a factor of the leading coefficient, 10.

PROBLEM SOLVING

Make an organized list. Form all quotients that have factors of 6 in the numerator and factors of 10 in the denominator.

factors of 6: ±1, ±2, ±3, ±6
factors of 10: ±1, ±2, ±5, ±10

$\pm\frac{1}{1}, \pm\frac{1}{2}, \pm\frac{1}{5}, \pm\frac{1}{10}$	$\pm\frac{2}{1}, \pm\frac{2}{2}, \pm\frac{2}{5}, \pm\frac{2}{10}$	$\pm\frac{3}{1}, \pm\frac{3}{2}, \pm\frac{3}{5}, \pm\frac{3}{10}$	$\pm\frac{6}{1}, \pm\frac{6}{2}, \pm\frac{6}{5}, \pm\frac{6}{10}$

Notice that some rational numbers are repeated, such as $\frac{3}{1}$ and $\frac{6}{2}$.

PROBLEM SOLVING

Use a graph. Rather than testing each quotient to see which ones satisfy $10x^3 + 9x^2 - 19x + 6 = 0$, you can examine the graph of the related function, $P(x) = 10x^3 + 9x^2 - 19x + 6$, and use the process of elimination.

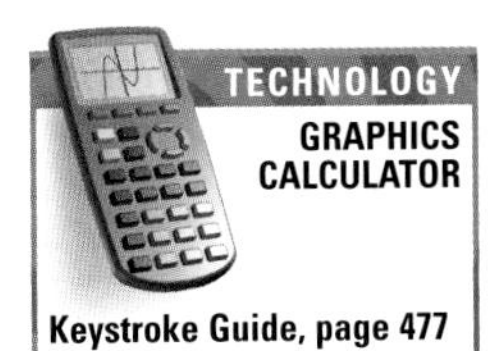

One root of $P(x) = 0$ appears to be -2. Test whether $P(-2) = 0$ is true.

```
-2| 10    9  -19    6
        -20   22   -6
    10  -11    3  |0   ← remainder
```

Since the remainder is 0, $P(-2) = 0$ is true and -2 is a root of $P(x) = 0$.

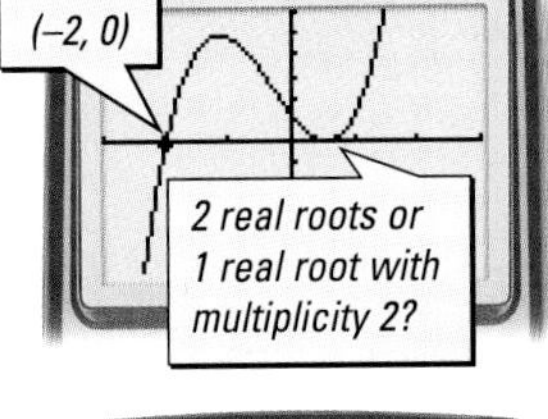

From the graph of the related function, there appears to be two real zeros or one real zero with a multiplicity of 2 between 0 and 1.

A closer look between 0 and 1 on the graph of the related function shows two zeros at $\frac{1}{2}$ and $\frac{3}{5}$, which are possible roots of $P(x) = 0$ according to the Rational Root Theorem.

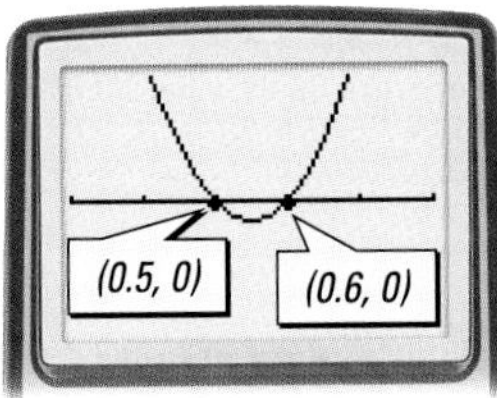

Test whether $P\left(\frac{1}{2}\right) = 0$ and $P\left(\frac{3}{5}\right) = 0$ are true.

```
1/2| 10    9  -19    6        3/5| 10    9  -19    6
           5    7   -6                   6    9   -6
     10   14  -12  |0  ← remainder   10   15  -10  |0  ← remainder
```

Thus, there are three rational roots: -2, $\frac{1}{2}$, and $\frac{3}{5}$.

TRY THIS Find all of the rational roots of $3x^3 - 17x^2 + 59x - 65 = 0$.

ADDITIONAL EXAMPLE 1

Find all rational roots of $8x^3 + 10x^2 - 11x + 2 = 0$.

$-2, \frac{1}{4}, \frac{1}{2}$

Teaching Tip

TECHNOLOGY When using a TI-82 or TI-83 graphics calculator to examine the graph in Example 1, set the viewing window as follows:

Xmin=−3	Ymin=−30
Xmax=3	Ymax=30
Xscl=1	Yscl=5

To examine the graph between $x = 0$ and $x = 1$, set the viewing window as follows:

Xmin=.3	Ymin=−.5
Xmax=.8	Ymax=.5
Xscl=.1	Yscl=.1

TRY THIS
$x = \frac{5}{3}$

Interdisciplinary Connection

ECONOMICS Many credit card companies compound interest monthly on outstanding balances. If \$1000 is borrowed and repaid by equal monthly payments of \$80, the balance, B, at the end of twelve months can be represented by the polynomial function $B(x) = 1000x^{12} - 80(x^{11} + x^{10} + x^9 + x^8 + x^7 + x^6 + x^5 + x^4 + x^3 + x^2 + x + 1)$, where x is 1 plus the monthly interest rate. Ask students to replace x with 1.0099, for a monthly interest rate of 0.99%, and determine the remaining balance. **\$111.45** You can suggest that students choose a specific monthly interest rate, vary the amount borrowed, and find B to determine the balance at the end of 12 months.

ADDITIONAL
EXAMPLE 2

Find all of the zeros of $Q(x) = x^3 + 4x^2 - 6x - 12$.
$2, -3 + \sqrt{3}, -3 - \sqrt{3}$

Teaching Tip

TECHNOLOGY When using a TI-82 or TI-83 graphics calculator to create the graph in Example 2, set the viewing window as follows:

Xmin=–5	Ymin=–5
Xmax=5	Ymax=5
Xscl=1	Yscl=1

To test whether $Q(-1) = 0$, press TRACE (-) 1 ENTER. At the bottom of the screen you should see the following:

X=–1	Y=0

TRY THIS
$2, 2 + \sqrt{5}$, and $2 - \sqrt{5}$

ADDITIONAL
EXAMPLE 3

Find all of the zeros of $P(x) = -4x^3 + 2x^2 - x + 3$.
$1, -\frac{1}{4} + \frac{i\sqrt{11}}{4}, -\frac{1}{4} - \frac{i\sqrt{11}}{4}$

Teaching Tip

TECHNOLOGY When using a TI-82 or TI-83 graphics calculator to create the graph in Example 3, set the viewing window as follows:

Xmin=–3	Ymin=–6
Xmax=5	Ymax=4
Xscl=1	Yscl=1

To test whether $P(2)=0$, press TRACE 2 ENTER.

TRY THIS
$5, 2 + 5i$, and $2 - 5i$

You can often use the quadratic formula to solve a polynomial equation.

EXAMPLE 2 Find all of the zeros of $Q(x) = x^3 - 2x^2 - 2x + 1$.

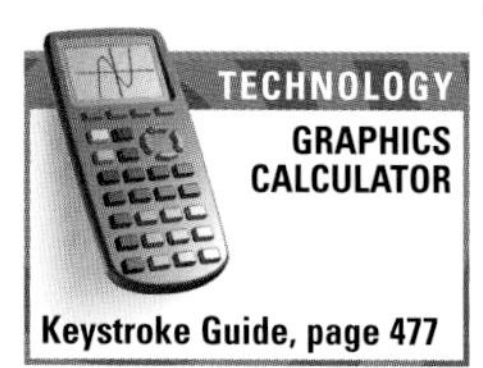

SOLUTION

From the Rational Root Theorem and the graph of $Q(x) = x^3 - 2x^2 - 2x + 1$, you know that a zero may occur at -1. Test whether $Q(-1) = 0$ is true.

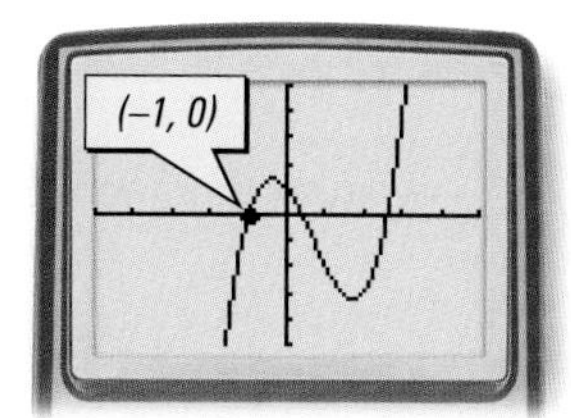

$$\begin{array}{r|rrrr} -1 & 1 & -2 & -2 & 1 \\ & & -1 & 3 & -1 \\ \hline & 1 & -3 & 1 & \boxed{0} \end{array}$$

Because $Q(-1) = 0$, $x + 1$ is a factor of $x^3 - 2x^2 - 2x + 1$.

$$x^3 - 2x^2 - 2x + 1 = 0$$
$$(x + 1)(x^2 - 3x + 1) = 0 \qquad \textit{Factor out } x + 1.$$
$$x + 1 = 0 \quad \text{or} \quad x^2 - 3x + 1 = 0 \qquad \textit{Apply the Zero-Product Property.}$$
$$x = -1 \qquad x = \frac{3 \pm \sqrt{9 - 4}}{2} \qquad \textit{Use the quadratic formula.}$$
$$x = \frac{3 \pm \sqrt{5}}{2}$$

Notice that $\frac{3+\sqrt{5}}{2}$ and $\frac{3-\sqrt{5}}{2}$ are conjugates.

The zeros of $Q(x) = x^3 - 2x^2 - 2x + 1$ are -1, $\frac{3 + \sqrt{5}}{2}$, and $\frac{3 - \sqrt{5}}{2}$.

TRY THIS Find all of the zeros of $P(x) = x^3 - 6x^2 + 7x + 2$.

EXAMPLE 3 Find all of the zeros of $P(x) = 3x^3 - 10x^2 + 10x - 4$.

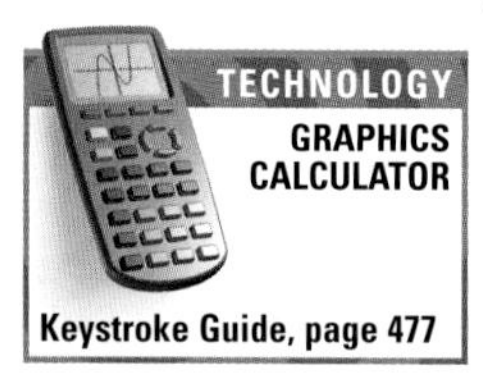

SOLUTION

From the Rational Root Theorem and the graph of $P(x) = 3x^3 - 10x^2 + 10x - 4$, you know that a zero may occur at 2. Test whether $P(2) = 0$ is true.

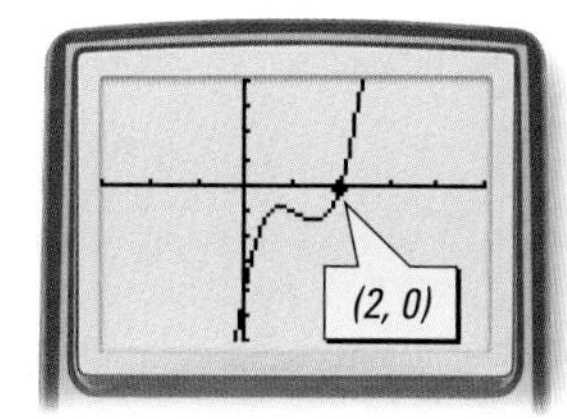

$$\begin{array}{r|rrrr} 2 & 3 & -10 & 10 & -4 \\ & & 6 & -8 & 4 \\ \hline & 3 & -4 & 2 & \boxed{0} \end{array}$$

Because $P(2) = 0$, $x - 2$ is a factor of $3x^3 - 10x^2 + 10x - 4$.

$$3x^3 - 10x^2 + 10x - 4 = 0$$
$$(x - 2)(3x^2 - 4x + 2) = 0 \qquad \textit{Factor out } x - 2.$$
$$x - 2 = 0 \quad \text{or} \quad 3x^2 - 4x + 2 = 0 \qquad \textit{Apply the Zero-Product Property.}$$
$$x = 2 \qquad x = \frac{4 \pm \sqrt{16 - 24}}{6} \qquad \textit{Use the quadratic formula.}$$
$$x = \frac{2 \pm i\sqrt{2}}{3}$$

Notice that $\frac{2+i\sqrt{2}}{3}$ and $\frac{2-i\sqrt{2}}{3}$ are conjugates.

The zeros of $P(x) = 3x^3 - 10x^2 + 10x - 4$ are 2, $\frac{2 + i\sqrt{2}}{3}$, and $\frac{2 - i\sqrt{2}}{3}$.

TRY THIS Find all of the zeros of $P(x) = x^3 - 9x^2 + 49x - 145$.

Enrichment

Give students the following puzzle:

> I am a polynomial function of third degree.
> My name is Q.
> One of my roots is $4 + i$.
> $Q(4) = 6$
> What is my function rule?

$$Q(x) = (x - a)(x - 4 + i)(x - 4 - i)$$
$$= x^3 - (a + 8)x^2 + (8a + 17)x - 17a$$

$Q(4) = 6$ implies that
$$64 - 16(a + 8) + 4(8a + 17) - 17a = 6,$$
or $a = -2$.

Thus, $Q(x) = x^3 - 6x^2 + x + 34$ satisfies the conditions.

After students have solved the puzzle, have them work in pairs to create several other "What is my function rule?" puzzles. Be sure they understand that the goal is to describe the function with the fewest number of statements.

Complex Conjugate Root Theorem

If P is a polynomial function with real-number coefficients and $a + bi$ (where $b \neq 0$) is a root of $P(x) = 0$, then $a - bi$ is also a root of $P(x) = 0$.

Exploring Zeros of Cubic Functions

You will need: a graphics calculator

1. How many real zeros does the function in Example 2 have? How many real zeros does the function in Example 3 have?
2. Graph $R(x) = 2x^3 - x^2 - 4x$. How many real zeros does R have?
3. Graph $S(x) = 2x^3 - x^2 - 4x + 3$. How many real zeros does S have?
4. Graph $T(x) = 2x^3 - x^2 - 4x + 6$. How many real zeros does T have?

CHECKPOINT ✔

5. Write a conjecture about the number of real zeros that a cubic function can have. How does the Complex Conjugate Root Theorem help you to determine whether your conjecture is true?

EXAMPLE 4

APPLICATION
MANUFACTURING

Jasmine is a manufacturing engineer who is trying to find out if she can use a 12-inch-by-20-inch sheet of cardboard and a 15-inch-by-16-inch sheet of cardboard to make boxes with the same height and volume.

Is such a pair of boxes possible?

SOLUTION

Box 1: 12 inches by 20 inches
$V_1(x) = x(12 - 2x)(20 - 2x)$

Box 2: 15 inches by 16 inches
$V_2(x) = x(15 - 2x)(16 - 2x)$

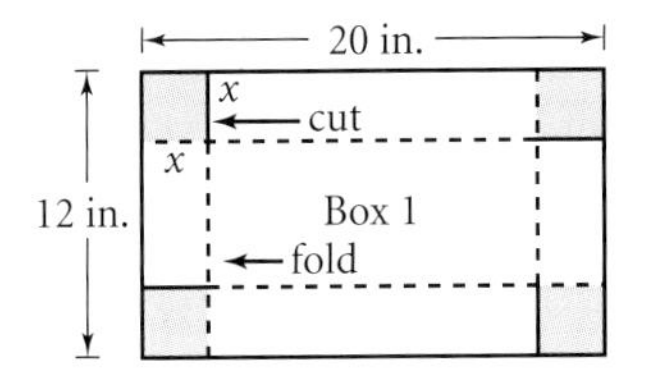

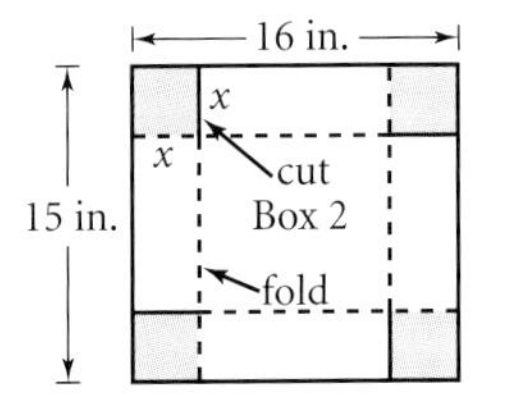

Find x such that $V_1(x) = V_2(x)$. Because x represents height, find $x > 0$ such that $x(20 - 2x)(12 - 2x) = x(16 - 2x)(15 - 2x)$.

Graph $y = x(20 - 2x)(12 - 2x)$ and $y = x(16 - 2x)(15 - 2x)$ on the same screen.

TECHNOLOGY
GRAPHICS CALCULATOR
Keystroke Guide, page 477

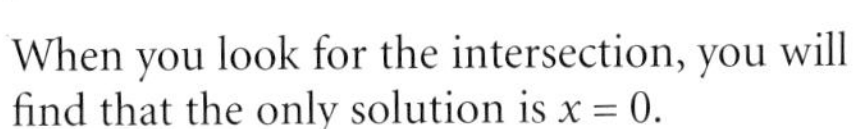

When you look for the intersection, you will find that the only solution is $x = 0$.

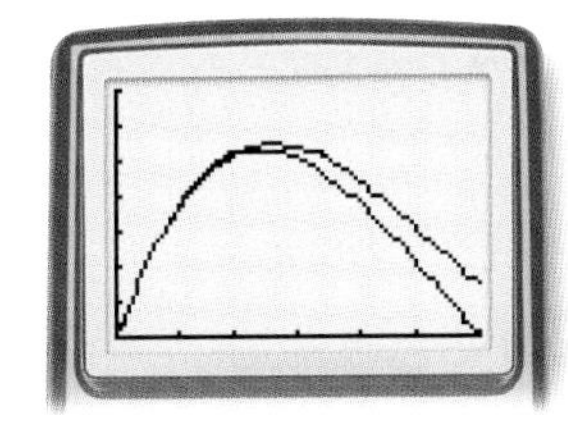

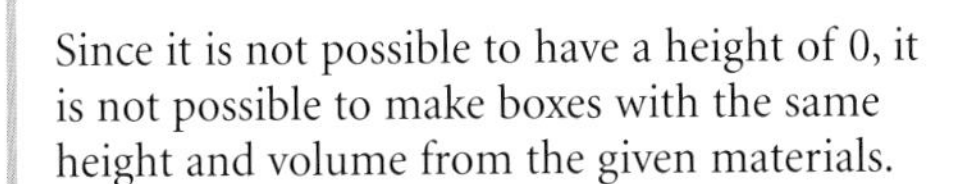

Since it is not possible to have a height of 0, it is not possible to make boxes with the same height and volume from the given materials.

Inclusion Strategies

USING VISUAL MODELS To show students the relationship between the degree of a polynomial function and the number of x-intercepts of its graph, provide students with various graphs of polynomial functions, such as the one at right. Ask students to determine the lowest possible degree of the function graphed, and have them write a possible function represented by the graph.
$f(x) = x(x + 2)(x - 1)(x - 3)$, or
$f(x) = x^4 - x^3 - 4x^2 + 4x$

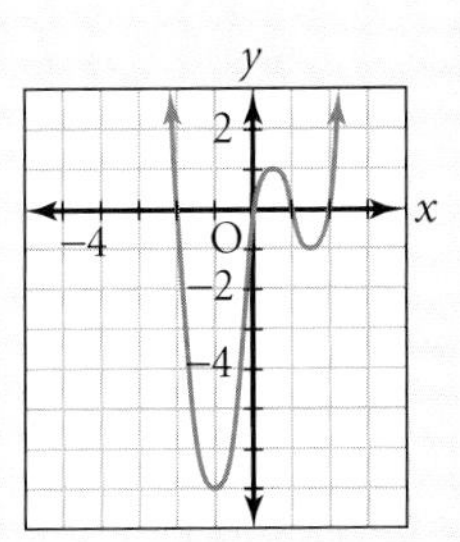

Be sure students notice that the graph of a polynomial function of odd degree must cross the x-axis at least once. The graph of a function of even degree may not cross the x-axis.

Activity Notes

Remind students that the graph of a cubic function has either no turning points or two turning points. Students should find that a cubic equation has at least one real zero and may have two or three real zeros.

Teaching Tip

TECHNOLOGY When using a TI-82 or TI-83 graphics calculator to create the graph for R, S, or T in the Activity, set the viewing window as follows:

Xmin=–5	Ymin=–3
Xmax=5	Ymax=8
Xscl=1	Yscl=1

CHECKPOINT ✔

5. A cubic function must have at least one real zero, can have two real zeros when one zero has a multiplicity of 2, and can have at most three real zeros. By the complex conjugate theorem, if a cubic function has one complex zero, it must have another, the conjugate of the complex zero.

Use Teaching Transparency 28.

Teaching Tip

TECHNOLOGY When using a TI-82 or TI-83 graphics calculator to create the graph of V_1 and V_2 in Example 4, set the viewing window as follows:

Xmin=0	Ymin=0
Xmax=6	Ymax=350
Xscl=1	Yscl=50

☞ For Additional Example 4, see page 462.

ADDITIONAL EXAMPLE 4

Vernon has a 9-inch-by-15-inch sheet of cardboard and an 8-inch-by-24-inch sheet of cardboard. He wants to know if he can make a box with the same height and volume from each sheet. **Is such a pair of boxes possible?**

Yes, when the height is 3.5625 inches, each sheet of cardboard can be made into a box with a volume of approx. 52.6 cubic inches.

ADDITIONAL EXAMPLE 5

The zeros of a fifth-degree polynomial function, Q, include -5 (with a multiplicity of 3) and $3 + i$. In addition, $Q(0) = 2500$. **Write the function in factored form and in standard form.**

$Q(x) = 2(x + 5)^3(x^2 - 6x + 10)$;
$Q(x) = 2x^5 + 18x^4 - 10x^3 - 350x^2 + 2500$

Math CONNECTION

TRANSFORMATIONS Be sure students understand that multiplying P by a number greater than 1 stretches the graph vertically but the x-intercepts remain the same.

Teaching Tip

Throughout this chapter, the phrase "factor the polynomial function" indicates that the factorization should be done over the set of integers. Point out to students that polynomials can be factored over the set of real numbers and over the set of complex numbers as well. For example, $x^4 - 25$ is $(x^2 + 5)(x^2 - 5)$ when factored over the integers, is $(x^2 + 5)(x + \sqrt{5})(x - \sqrt{5})$ when factored over the set of real numbers, and is $(x + 5i)(x - 5i)(x + \sqrt{5})(x - \sqrt{5})$ when factored over the set of complex numbers.

In this chapter, you have learned several different methods of finding the zeros of polynomial functions. The following theorem, along with its important corollary, tells you how many zeros a function has.

Fundamental Theorem of Algebra

Every polynomial function of degree $n \geq 1$ has at least one complex zero.

Corollary: Every polynomial function of degree $n \geq 1$ has exactly n complex zeros, counting multiplicities.

CULTURAL CONNECTION: EUROPE At the age of 20, German mathematician Karl Frederick Gauss, 1777–1855, earned his Ph.D. by proving that every polynomial equation has at least one complex root. The proof is beyond the scope of this book, but it remains one of the outstanding accomplishments of modern mathematics.

Karl Frederick Gauss
1777–1855

When given enough information about a polynomial function, you can find and write the function. One possibility is shown in Example 5.

EXAMPLE 5 The zeros of a fourth-degree polynomial function, P, include 3, with a multiplicity of 2, and $5 + 3i$. Also, $P(0) = 1224$.

Write the function in factored form and in standard form.

SOLUTION

The four factors of the polynomial are as follows:

$$(x - 3) \quad (x - 3) \quad [x - (5 + 3i)] \quad [x - (5 - 3i)]$$

The simplest possible polynomial is $(x - 3)^2[x - (5 + 3i)][x - (5 - 3i)]$.

Because the graph of P can be stretched vertically by any nonzero constant factor and retain the same zeros, let a represent the stretch factor for this polynomial.

$$P(x) = a(x - 3)^2[x - (5 + 3i)][x - (5 - 3i)], \text{ where } a \neq 0$$
$$P(x) = a(x - 3)^2(x^2 - 10x + 34)$$

CONNECTION
TRANSFORMATIONS

Since $P(0) = 1224$, substitute 0 for x and 1224 for $P(x)$. Solve for a.

$$\begin{aligned} P(x) &= a(x - 3)^2(x^2 - 10x + 34) \\ 1224 &= a(0 - 3)^2[0^2 - 10(0) + 34] \\ 1224 &= a(-3)^2(34) \\ 1224 &= 306a \\ 4 &= a \end{aligned}$$

The function in factored form is $P(x) = 4(x - 3)^2(x^2 - 10x + 34)$.

Multiplying $4(x - 3)^2(x^2 - 10x + 34)$ gives the standard form, $P(x) = 4x^4 - 64x^3 + 412x^2 - 1176x + 1224$.

Reteaching the Lesson

USING COGNITIVE STRATEGIES Give students polynomial functions with incorrect zeros and ask them to explain why each of the given zeros cannot be a zero of the function. Then have them list all possible rational zeros of the function and determine all real and complex zeros. Finally, have students factor the function over the set of integers and over the set of complex numbers. For example, give students $f(x) = 2x^3 - 3x^2 - 3x - 5$. Have students explain why -3 is not a zero of the function, list all possible rational zeros of the function, and test each possible rational zero. Have them write the function factored over the set of integers, identify the complex zeros and, optionally, factor the function over the set of complex numbers.

$f(-3) \neq 0$

Possible rational zeros: $\pm 1, \pm 5, \pm\frac{1}{2}, \pm\frac{5}{2}$

Rational zero: $x = \frac{5}{2}$

$f(x) = (2x - 5)(x^2 + x + 1)$

Complex zeros: $x = \frac{-1 + i\sqrt{3}}{2}, x = \frac{-1 - i\sqrt{3}}{2}$

$f(x) = (2x - 5)\left(x - \frac{-1 + i\sqrt{3}}{2}\right)\left(x - \frac{-1 - i\sqrt{3}}{2}\right)$

TRY THIS The zeros of a third-degree polynomial function, P, include 1 and $3 + 4i$. Also, $P(0) = 50$. Write the function in factored form and standard form.

TRY THIS
$P(x) = -2(x - 1)(x^2 - 6x + 25)$
$= -2x^3 + 14x^2 - 62x + 50$

Exercises

Communicate

1. What are the possible rational roots for the polynomial equation $2x^3 + a_2x^2 + a_1x + 3 = 0$, where a_2 and a_1 are integers?
2. If 0 and $3 - 2i$ are roots of a polynomial equation with rational coefficients, what other numbers must also be roots of the equation?
3. How many zeros does $P(x) = x^5 + x - 3$ have? Are all of the zeros distinct? Why or why not? How many are real zeros?
4. Let $P(x) = x(x - 1)(x - 2i)$ and $Q(x) = x(x - 1)(x - 2i)(x + 2i)$. Does P have real-number coefficients? Does Q have real-number coefficients? Multiply to verify your responses. How do your answers support the Complex Conjugate Root Theorem?

Guided Skills Practice

5. Find all of the rational roots of $12x^3 - 32x^2 - 145x + 25 = 0$. ***(EXAMPLE 1)***
6. Find all of the zeros of $Q(x) = x^3 + 3x^2 - 8x - 4$. ***(EXAMPLE 2)***
7. Find all of the zeros of $Q(x) = 2x^3 - 4x^2 - 5x - 3$. ***(EXAMPLE 3)***
8. Let $P(x) = (x - 1)^3$ and $Q(x) = (x - 3)^2$. Find all real values of x such that $P(x) = Q(x)$. ***(EXAMPLE 4)***
9. The zeros of a fourth-degree polynomial function P include 3, with a multiplicity of 2, and $3 + 4i$. Also, $P(0) = -6$. Write the function in factored form and in standard form. ***(EXAMPLE 5)***

5. $5, -\frac{5}{2}, \frac{1}{6}$
6. $2, -\frac{5}{2} \pm \frac{\sqrt{17}}{2}$
7. $3, -\frac{1}{2} \pm \frac{1}{2}i$
8. 2
9. $P(x) = -\frac{2}{75}(x - 3)^2(x^2 - 6x + 25) = -\frac{2}{75}x^4 + \frac{8}{25}x^3 - \frac{28}{15}x^2 + \frac{136}{25}x - 6$

Practice and Apply

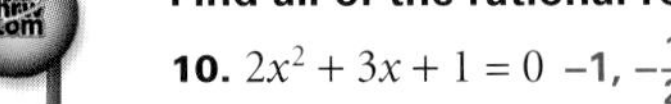

internet connect
Homework Help Online
Go To: **go.hrw.com**
Keyword: **MB1 Homework Help** for Exercises 10–21

Find all of the rational roots of each polynomial equation.

10. $2x^2 + 3x + 1 = 0$ **$-1, -\frac{1}{2}$**
11. $6x^3 - 29x^2 - 45x + 18 = 0$ **$6, -\frac{3}{2}, \frac{1}{3}$**
12. $4x^3 - 13x^2 + 11x - 2 = 0$ **$\frac{1}{4}, 1, 2$**
13. $3x^3 - 2x^2 - 12x + 8 = 0$ **$-2, 2, \frac{2}{3}$**
14. $3x^3 + 3x^2 - 4x + 4 = 0$ **-2**
15. $4x^3 + 3x^2 + x + 2 = 0$ **-1**
16. $15a^3 + 38a^2 + 17a + 2 = 0$
17. $10x^3 + 69x^2 - 9x - 14 = 0$
18. $18c^3 - 23c + 6 + 9c^2 = 0$
19. $50x^3 - 7x + 35x^2 - 6 = 0$
20. $18x^4 + 15x + 15x^3 - 34x^2 - 2 = 0$ **$-2, \frac{1}{2}, \frac{1}{3}$**
21. $10x^4 - 103x^3 - 207x + 294x^2 - 54 = 0$ **$3, 6, -\frac{1}{5}, \frac{3}{2}$**

16. $-2, -\frac{1}{3}, -\frac{1}{5}$
17. $-7, -\frac{2}{5}, \frac{1}{2}$
18. $-\frac{3}{2}, \frac{1}{3}, \frac{2}{3}$
19. $-\frac{3}{5}, -\frac{1}{2}, \frac{2}{5}$

Teaching Tip

Remind students that an nth-degree polynomial function must have n complex zeros counting multiplicities, and the related polynomial will have n linear factors of the form $x - c$, where c is a complex zero of the related function.

Assess

Selected Answers
Exercises 5–9, 11–65 odd

ASSIGNMENT GUIDE	
In Class	1–9
Core	11–49 odd, 53
Core Plus	10–50 even, 51, 52
Review	54–66
Preview	67, 68

Extra Practice can be found beginning on page 940.

Error Analysis

Some students are overwhelmed by the total number of possible rational solutions that must be tested. Tell them that the task of finding the rational roots can be simplified by examining the graph of the related function. From the appearance of the graph, students can locate the possible rational crossing points and test any zeros that appear to be correct. If a function has an x-intercept and none of the possible rational zeros are correct, the zero must be irrational.

Also, some students think that complex zeros appear on the graph of a polynomial function. Remind them that the real zeros are the x-intercepts of a polynomial function and that complex zeros cannot be determined from the graph.

22. $4, \pm\sqrt{3}$
23. $\frac{1}{3}, \pm 2\sqrt{2}$
24. $5, \pm\frac{\sqrt{3}}{2}$
25. $-8, \pm\frac{\sqrt{5}}{3}$
26. $-6, 4 \pm \sqrt{3}$
27. $5, \pm 12i$
28. $3, \pm 2i$
29. $5, \pm 3i$
30. $-5, \pm i\sqrt{2}$
31. $3, \pm 2i\sqrt{3}$
32. $-1, \frac{1}{2}, \pm\sqrt{5}$
33. $-5, 1, \pm 2i$

37. $0, -1.65$
38. $\pm 2.07, \pm 0.83$
39. $\pm 2.14, \pm 0.66$
40. ± 1.21

Find all zeros of each polynomial function.

22. $B(x) = x^3 - 4x^2 - 3x + 12$
23. $P(x) = 3x^3 - x^2 - 24x + 8$
24. $f(x) = 4x^3 - 20x^2 - 3x + 15$
25. $f(x) = 9x^3 + 72x^2 - 5x - 40$
26. $t(x) = x^3 - 2x^2 + 78 - 35x$
27. $g(x) = x^3 - 5x^2 + 144x - 720$
28. $N(x) = x^3 - 3x^2 + 4x - 12$
29. $f(b) = b^3 - 5b^2 + 9b - 45$
30. $G(x) = x^3 + 5x^2 + 10 + 2x$
31. $a(x) = x^3 - 3x^2 + 12x - 36$
32. $W(x) = x^3 + 2x^4 - 5x + 5 - 11x^2$
33. $H(x) = x^4 + 16x - x^2 + 4x^3 - 20$

Find all real values of x for which the functions are equal. Give your answers to the nearest hundredth.

34. $P(x) = x^2$, $Q(x) = x^3 + 3x^2 + 3x + 1$ **−0.43**
35. $P(x) = x^2 - 2x + 1$, $Q(x) = -x^3 + 3x^2 - 3x + 3$ **2**
36. $P(x) = x^4 - 6x + 3$, $Q(x) = -0.2x^4$ **0.51, 1.49**
37. $P(x) = x^3 - 5x$, $Q(x) = 0.5x^4$
38. $P(x) = x^4 - 4x^2 + 3$, $Q(x) = x^2$
39. $P(x) = x^4 - 5x^2 + 4$, $Q(x) = 2$
40. $P(x) = 0.25x^4$, $Q(x) = -x^2 + 2$

Write a polynomial function, P, in factored form and in standard form by using the given information.

41. P is of degree 2; $P(0) = 12$; zeros: 2, 3
42. P is of degree 2; $P(0) = 4$; zeros: −1, 4
43. P is of degree 3; $P(0) = 20$; zeros: −2, 1, 2
44. P is of degree 3; $P(0) = 24$; zeros: −1, 2, 4
45. P is of degree 4; $P(0) = 1$; zeros: 1 (multiplicity 2), 2 (multiplicity 2)
46. P is of degree 5; $P(0) = 2$; zeros: 1 (multiplicity 3), 2 (multiplicity 2)
47. P is of degree 3; $P(0) = -1$; zeros: 1, i
48. P is of degree 3; $P(0) = 4$; zeros: −1, $2i$
49. P is of degree 4; $P(0) = 3$; zeros: 1, 2, $5i$
50. P is of degree 4; $P(0) = -3$; zeros: 2, 5, $3i$

CHALLENGE

51. Use the Rational Root Theorem and the polynomial $P(x) = x^2 - 3$ to show that $\sqrt{3}$ is irrational.

APPLICATION

52. **GEOGRAPHY** A mountain ridge is drawn to scale on the wall of a visitors' center in West Virginia. The ridge has approximately the same shape as the graph of the function $f(x) = -x^4 + 3x^2 - 3x + 6$ between x-values of −2.5 and 2.0.
   a. The zeros of the function represent sea level. What are the approximate zeros of f?
   b. Find the approximate maximum value of f.
   c. Find the vertical scale factor that would place the maximum height of the ridge at 3200 feet above sea level.

y

x

52a. −2.32, 1.79
b. 12.24
c. 261.4

41. $P(x) = 2(x - 2)(x - 3)$
$= 2x^2 - 10x + 12$

42. $P(x) = -(x + 1)(x - 4)$
$= -x^2 + 3x + 4$

43. $P(x) = 5(x + 2)(x - 1)(x - 2)$
$= 5x^3 - 5x^2 - 20x + 20$

44. $P(x) = 3(x + 1)(x - 2)(x - 4)$
$= 3x^3 - 15x^2 + 6x + 24$

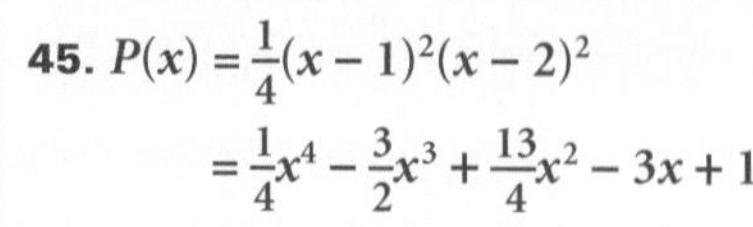

45. $P(x) = \frac{1}{4}(x - 1)^2(x - 2)^2$
$= \frac{1}{4}x^4 - \frac{3}{2}x^3 + \frac{13}{4}x^2 - 3x + 1$

46. $P(x) = -\frac{1}{2}(x - 1)^3(x - 2)^2$
$= -\frac{1}{2}x^5 + \frac{7}{2}x^4 - \frac{19}{2}x^3 + \frac{25}{2}x^2 - 8x + 2$

47. $P(x) = (x - 1)(x - i)(x + i)$
$= x^3 - x^2 + x - 1$

48. $P(x) = (x + 1)(x - 2i)(x + 2i)$
$= x^3 + x^2 + 4x + 4$

49. $P(x) = \frac{3}{50}(x - 1)(x - 2)(x - 5i)(x + 5i)$
$= \frac{3}{50}x^4 - \frac{9}{50}x^3 + \frac{81}{50}x^2 - \frac{9}{2}x + 3$

50. $P(x) = -\frac{1}{30}(x - 2)(x - 5)(x - 3i)(x + 3i)$
$= -\frac{1}{30}x^4 + \frac{7}{30}x^3 - \frac{19}{30}x^2 + \frac{21}{10}x - 3$

51. Since $P(\sqrt{3}) = (\sqrt{3})^2 - 3 = 3 - 3 = 0$, $\sqrt{3}$ is a root of P. By the Rational Root Theorem, if X3 $\sqrt{3} = \frac{p}{q}$ for some integers p and q, then p is a factor of −3 and q is a factor of 1. However, the only possible values of $\frac{p}{q}$ are ±3 and ±1. Therefore, $\sqrt{3}$ is irrational.

Practice

NAME ______________ CLASS ______ DATE ______

Practice
7.5 Zeros of Polynomial Functions

Find all the rational roots of each polynomial equation.

1. $P(x) = 5x^2 - 6x + 1$ — $\frac{1}{5}$ and 1
2. $P(x) = 3x^3 - x^2 - 12x + 4$ — $-2, \frac{1}{3}$, and 2
3. $P(x) = 6x^3 + x^2 - 4x + 1$ — $-1, \frac{1}{3}$, and $\frac{1}{2}$
4. $P(x) = 4x^3 - 11x^2 + 5x + 2$ — $-\frac{1}{4}$, 1, and 2
5. $P(x) = 2x^4 - 5x^3 - 12x^2 - x + 4$ — −1 (multiplicity 2), $\frac{1}{2}$, and 4
6. $P(x) = 18x^4 - 9x^3 - 17x^2 + 4x + 4$ — $-\frac{2}{3}, -\frac{1}{2}, \frac{2}{3}$, and 1

Find all the zeros of each polynomial function.

7. $P(x) = x^3 + 5x^2 + 2x + 10$ — $-5, i\sqrt{2}$, and $-i\sqrt{2}$
8. $P(x) = x^3 + 3x^2 + 4x + 12$ — $-3, 2i$, and $-2i$
9. $P(x) = x^3 - 3x^2 + 12x - 36$ — $3, 2i\sqrt{3}$, and $-2i\sqrt{3}$
10. $P(x) = x^3 - 5x^2 + 9x - 45$ — $5, 3i$ and $-3i$
11. $P(x) = x^3 + 2x^2 - 16$ — $2, -2 + 2i$, and $-2 - 2i$
12. $P(x) = x^4 - 5x^3 + 5x^2 - 5x + 4$ — $1, 4, i$, and $-i$

Find all real values of x for which the functions are equal. Round your answers to the nearest hundredth.

13. $P(x) = x^4 - 5$, $Q(x) = x + 1$ — −1.46 and 1.66
14. $P(x) = x^3 - 5x + 3$, $Q(x) = -x^3 + 5x + 3$ — −2.24, 0, and 2.24
15. $P(x) = x^3 + 2x + 1$, $Q(x) = 4x + 1$ — −1.41, 0, and 1.41
16. $P(x) = x^4 + x^3 + 2x - 1$, $Q(x) = x^2 + 2x + 1$ — −1.85 and 1.16

Write a polynomial function, P, in standard form by using the given information.

17. The zeros of $P(x)$ are −3, 2, and 4, and $P(0) = 120$. $P(x) = 5x^3 - 15x^2 - 50x + 120$
18. $P(0) = -96$, and two of the three zeros are −3 and $4i$. $P(x) = -2x^3 - 6x^2 - 32x - 96$

APPLICATION

53. THERMODYNAMICS A hot-air balloon rises at an increasing rate as its altitude increases. Its altitude, in feet, can be modeled by the function $y = 0.025t^2 + 2t$, where t is time in seconds. How long will it take for the balloon to reach an altitude of 800 feet? **about 143 seconds**

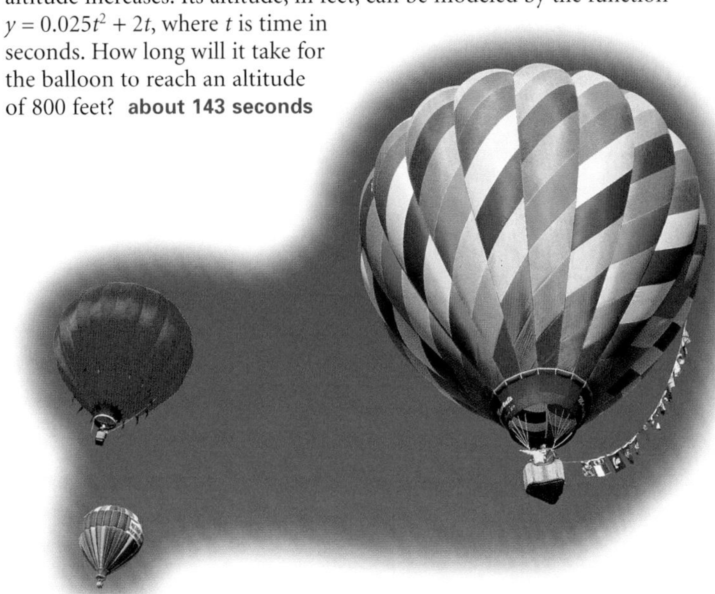

Look Back

Let $f(x) = 5x + 4$ and $g(x) = 2 - x$. Find or evaluate each composite function. ***(LESSON 2.4)***

54. $f \circ g$ **$14 - 5x$** **55.** $g \circ f$ **$-2 - 5x$** **56.** $f \circ f$ **$25x + 24$**

57. $(f \circ g)(-2)$ **24** **58.** $(g \circ f)(5)$ **–27** **59.** $(f \circ f)(0.2)$ **29**

Write the equation for the axis of symmetry and find the coordinates of the vertex. ***(LESSON 5.5)***

60. $y = 6x^2 - x - 12$ **61.** $y = 2x^2 + 5x + 2$ **62.** $y = x^2 + 3x - 2$

63 Graph $f(x) = x^3 - 3x^2 + 4x - 5$, and describe its end behavior. ***(LESSON 7.2)***

64. Describe the end behavior of the function $f(x) = -3(x + 1)^2(x - 1)^3(x - 2)^4$. ***(LESSON 7.2)***

In Exercises 65 and 66, divide by using synthetic division. ***(LESSON 7.3)***

65. $\dfrac{3x^4 - 4x^2 + 2x - 1}{x - 1}$ **$3x^3 + 3x^2 - x + 1$** **66.** $\dfrac{x^4 + 4x^3 + 5x^2 - 5x - 14}{x + 2}$ **$x^3 + 2x^2 + x - 7$**

Look Beyond

67. Use factoring to simplify the expression $\dfrac{x^2 + 5x + 6}{x^2 + 7x} \cdot \dfrac{x^2 - 2x}{x^2 - 4}$. **$\dfrac{x+3}{x+7}$**

68. Solve the equation $\sqrt{x + 1} = \sqrt{2x}$. Begin by squaring both sides. Check by substituting your solution(s) into the original equation. **$x = 1$**

60. $x = \frac{1}{12}$; $\left(\frac{1}{12}, -\frac{289}{24}\right)$

61. $x = -\frac{5}{4}$; $\left(-\frac{5}{4}, -\frac{9}{8}\right)$

62. $x = -\frac{3}{2}$; $\left(-\frac{3}{2}, -\frac{17}{4}\right)$

63. falls on left, rises on right

64. rises on the left, falls on the right

Look Beyond

Exercise 67 is an example of multiplying two rational expressions, and Exercise 68 is an example of solving a radical equation. Multiplying and dividing rational expressions will be studied in depth in Lesson 8.3, and solving radical equations will be presented in Lesson 8.8.

Student Technology Guide

NAME ______ CLASS ______ DATE ______

Student Technology Guide
7.5 Zeros of Polynomial Functions

A graphics calculator, combined with your knowledge of the Rational Root Theorem, allows you to quickly identify the rational roots of a polynomial function.

The Rational Root Theorem states that if $\frac{p}{q}$ is a root of $P(x) = 0$, then p is a factor of the constant term of P and q is a factor of the leading coefficient of P.

Example: Find all the rational zeros of $y = 8x^3 - 34x^2 + 31x + 10$.

- Identify the possible rational zeros of the function. Eliminating "repeats," the possible rational zeros are $\pm\frac{1}{1}, \pm\frac{2}{1}, \pm\frac{5}{1}, \pm\frac{10}{1}, \pm\frac{1}{2}, \pm\frac{10}{2}, \pm\frac{5}{2}, \pm\frac{1}{4}, \pm\frac{5}{4}, \pm\frac{1}{8}$, and $\pm\frac{5}{8}$.
- Graph the function. Press Y= 8 X,T,θ,n ^ 3 − 34 X,T,θ,n x² + 31 X,T,θ,n + 10. Press GRAPH.
- Examine the graph and use a process of elimination to find the rational zeros. This graph indicates three zeros—one between −1 and 0, one near 2, and one between 2 and 3. The zero between −1 and 0 could be $-\frac{1}{2}, -\frac{1}{4}, -\frac{1}{8}$, or $-\frac{5}{8}$.
- Press TRACE. Use ◄ and ► to move as close to the zero as you can. The zero is near $x = -0.255$, which may mean that $-\frac{1}{4}$ is a rational zero. To be sure, press (−) .25 ENTER.
- Repeat this process to look for other rational zeros.

The rational zeros are $-\frac{1}{4}$, 2, and $\frac{5}{2}$.

Find all the rational zeros of the following polynomial functions. In some cases, you may need to zoom in by pressing ZOOM 2:Zoom In ENTER ENTER.

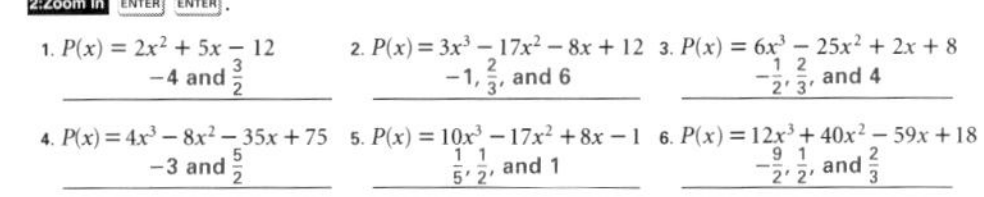

1. $P(x) = 2x^2 + 5x - 12$ **-4 and $\frac{3}{2}$**
2. $P(x) = 3x^3 - 17x^2 - 8x + 12$ **$-1, \frac{2}{3}$, and 6**
3. $P(x) = 6x^3 - 25x^2 + 2x + 8$ **$-\frac{1}{2}, \frac{2}{3}$, and 4**
4. $P(x) = 4x^3 - 8x^2 - 35x + 75$ **-3 and $\frac{5}{2}$**
5. $P(x) = 10x^3 - 17x^2 + 8x - 1$ **$\frac{1}{5}, \frac{1}{2}$, and 1**
6. $P(x) = 12x^3 + 40x^2 - 59x + 18$ **$-\frac{9}{2}, \frac{1}{2}$, and $\frac{2}{3}$**

Focus

For businesses, a critical aspect of product development is packaging. The design of a package, including its shape and volume, can have a great impact on ease of handling, consumer acceptance, and overall cost of the product. Students will use polynomial models to represent the shape of different containers. They will collect and organize data about the heights and volumes of the containers, draw a scatter plot of the data, and determine the polynomial model that best fits the data. They will then test the reliability of their model.

Motivate

Ask students to consider the role that packaging plays in the manufactured products they use in their daily lives. Have them discuss the advantages of shapes of the containers for their favorite products and what problems could arise if those containers were drastically changed. Suppose that rectangular containers were changed to cylindrical containers, or were made much taller or much wider. How might this affect the amount of product the container holds? Ask students if they have ever thought about the planning and experimentation required to develop product packaging. Suggest that they research the subject.

CHAPTER SEVEN PROJECT

FILL IT UP!

In this project you will find polynomial models to represent the shapes of various containers. You will perform experiments that involve adding water to a container in equal increments until the container is full. After each addition of water, the height of the water in the container is measured. The volume of water in the container and the height of the water are both recorded. Ordered pairs are formed from the data and displayed in a scatter plot. A polynomial regression model is selected to represent the data.

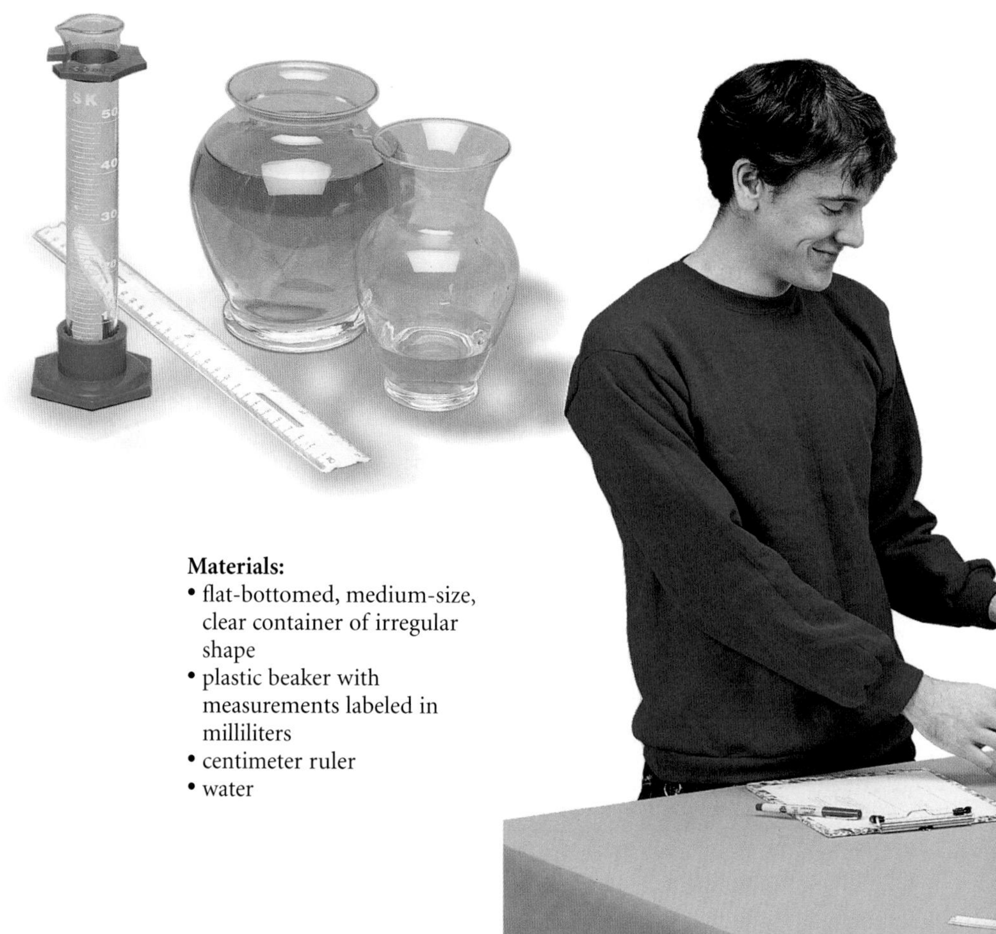

Materials:
- flat-bottomed, medium-size, clear container of irregular shape
- plastic beaker with measurements labeled in milliliters
- centimeter ruler
- water

Activity 1

1. **Answers will vary. Sample: The container used was a vase with volume of 1200 mL. The volume divided by 10 is $1200 \div 10 = 120$ mL.**

2. **Answers will vary. Sample for the vase of 1200 mL:**

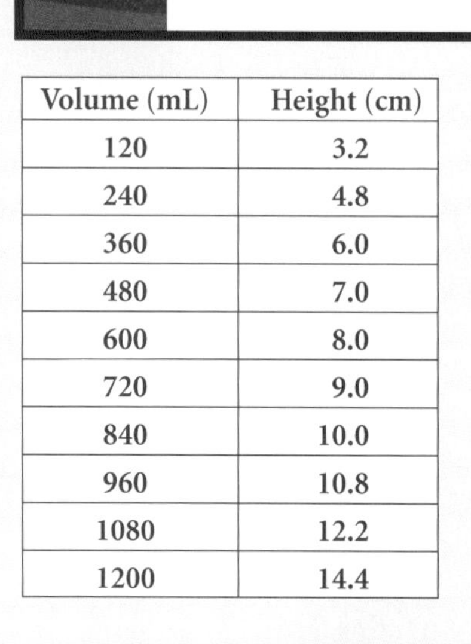

Volume (mL)	Height (cm)
120	3.2
240	4.8
360	6.0
480	7.0
600	8.0
720	9.0
840	10.0
960	10.8
1080	12.2
1200	14.4

Activity 2

1. **Answers will vary. Students' plots should show an increasing trend in the data. Sample graph for the data from Activity 1:**

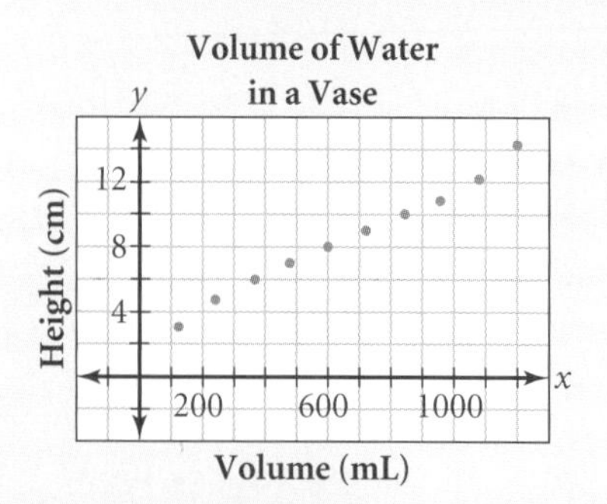

Activity 1

1. Each student should have a flat-bottomed, clear container of irregular shape. First determine the total volume of the container. Divide the total volume by 10, and round to the nearest whole number. (For example, if the total volume is 347 milliliters, divide by 10 and round 34.7 to 35.)
2. Add water to the container in 10 equal increments based on your answer from Step 1. After each increment is added, measure the height of the water in the container. Record your data in a table like the one below. Continue this procedure until the container is full.

[Note: 1 cm^3 = 1 mL]

	Volume (mL)	Height (cm)
1.		
2.		
3.		
4.		
5.		
6.		
7.		
8.		
9.		
10.		

Activity 2

1. Using the data in your table from Activity 1, let x represent the volume in milliliters, and let y represent the height in centimeters. Make a scatter plot of the data in the table.
2. Use the regression feature on your graphics calculator to find a cubic polynomial model and a quadratic polynomial model for the data you collected.
3. Compare the two models. Choose the model that appears to best fit the data.
4. Can you use this model to predict data points not plotted? Explain.

Activity 3

In this activity, you will test the validity of the polynomial model that you chose as the best fit in Activity 2.

1. In small groups, mix up the containers and graphs so that the graphs are not matched with the containers.
2. Trade containers and graphs with another group. Each group must match the containers they receive with the corresponding graphs.
3. Discuss how each group's matching choices were made.

Cooperative Learning

Divide the class into groups of four. Each member should have a medium-sized, flat-bottomed, clear container, a plastic beaker with measurements labeled in milliliters, and a centimeter ruler. Be sure to have different shapes of containers in each group. Each group member will provide the information required in **Activities 1 and 2.** The group should do the rematching exercise in **Activity 3** together.

Discuss

As a class, discuss each group's rematching choices. The project can be extended further by asking students additional questions such as the following: Why are all the data points in the first quadrant? Would it make sense to extend the polynomial model to include the other quadrants? Did you see any patterns in the polynomial models? How are they the same? How are they different? Would the model be the same if the water were added in less than 10 equal increments? In which shape of container did the height increase most rapidly? What may have caused this rapid increase?

2. Answers will vary. Students should give both cubic and quadratic models. Sample: A cubic regression model is $y = 0.00000001x^3 - 0.0000199x^2 + 0.01966x + 1.0733$; a quadratic regression model is $y = 0.00000099x^2 + 0.0081x + 2.637$.

3. Answers will vary, depending on the shape of the container. Sample: The graphs of the two functions compared with the scatter plot indicate that the cubic model is the best fit.

4. Answers will vary. Sample: You can use the model to predict points that lie between the data points. You cannot predict points for greater volumes because the vase has been filled.

Activity 3

1–3. Answers will vary. When the container widens, the graph increases more slowly; when the container narrows, the graph increases more rapidly.

7 Chapter Review and Assessment

VOCABULARY

Key Skills & Exercises

LESSON 7.1

Key Skills

Classify polynomials.

$$5x - 4x^4 + -5x^3 - 9x + 4x + 1$$
$$= 4x^4 + 5x^3 + 1 \quad \textit{Combine like terms.}$$

The polynomial is a quartic trinomial.

Evaluate, add, and subtract polynomial functions.

Evaluate $g(x) = x^3 - 2x^2 + 4x$ for $x = -5$.

$$g(-5) = (-5)^3 - 2(-5)^2 + 4(-5)$$
$$= -125 - 50 - 20 = -195$$

Simplify $(-2x^3 + 4x^2 - 9x + 5) - (5x^3 - x + 7)$.

$$(-2x^3 + 4x^2 - 9x + 5) - (5x^3 - x + 7)$$
$$= -2x^3 - 5x^3 + 4x^2 - 9x + x + 5 - 7$$
$$= -7x^3 + 4x^2 - 8x - 2$$

Exercises

Classify each polynomial by degree and by number of terms.

1. $3a^3 + 11a^2 - 2a + 1$
2. $8x^5 - 6x^2 + 10x^3$
3. $-b^2 + 8b - 5b^4 - 3$
4. $-2x^2 - x^3 + 7x^4$

Evaluate each polynomial for $x = 2$ and $x = -1$.

5. $-x^3 + 4x^2 - 2$ **6; 3**
6. $x^3 + 2x^2 - 1$ **15; 0**
7. $x^4 - 22$ **-6; -21**
8. $19 - x^2 - x^3$ **7; 19**

Write each sum or difference as a polynomial in standard form. Then classify the polynomial by degree and by number of terms.

9. $(3x^3 - 5x^2 + 8x + 1) + (11x^3 - x^2 + 2x - 3)$
10. $(7x^3 - 8x^2 + 2x - 3) - (2x^3 + x^2 - 6)$

LESSON 7.2

Key Skills

Identify and describe the important features of the graph of a polynomial function.

Let $f(x) = -x^3 - 3x^2 + 10$. Because the leading coefficient of f is negative and the degree of f is odd, its graph rises on the left and falls on the right.

Exercises

Describe the end behavior of each function.

11. $c(a) = -a^2 - 2a + 22$ **falls on both ends**
12. $d(x) = -2x^3 + 3x^2 - 7$ **rises on left, falls on right**
13. $f(x) = x^4 + 7x + 1$ **rises on both ends**
14. $f(x) = 4x + x^3 - 6$ **falls on left, rises on right**

---

1. cubic polynomial
2. quintic trinomial
3. quartic polynomial
4. quartic trinomial

9. $14x^3 - 6x^2 + 10x - 2$; cubic polynomial
10. $5x^3 - 9x^2 + 2x + 3$; cubic polynomial

Chapter Test, Form A

NAME ______________ CLASS ________ DATE ________

Chapter Assessment

Chapter 7, Form A, page 1

Write the letter that best answers the question or completes the statement.

__c__ 1. Which of the following is a quartic trinomial?
a. $4x^2 + 4x + 4$ b. $6x^4 - 5x^2$ c. $12x + 3x^4 - 25$ d. $3x^5 + 7x^4 + 6x^2$

__d__ 2. Write $(2x^2 - 3x^4 - 6) - (5x^4 - 2x^2 + 4)$ in standard form.
a. $-3x^2 - x^4 - 10$ b. $4x^2 - 8x^4 - 10$ c. $-8x^4 - 4x^2 - 2$ d. $-8x^4 + 4x^2 - 10$

__d__ 3. Write$(x - 3)(x + 2)^2$ in standard form.
a. $x^3 - x^2 + 8x - 12$ b. $x^3 - x^2 - 8x - 12$ c. $x^3 + x^2 - 8x + 12$ d. $x^3 + x^2 - 8x - 12$

__a__ 4. Find the value of $\frac{2}{3}x^3 + \frac{3}{4}x^2 - 2x + 1$ when $x = -6$.
a. -104 b. -128 c. -158 d. 160

__a__ 5. Which of the following is the end behavior of the graph of $f(x) = 3x^2 + 6 - 2x^5$?
a. The graph rises on the left and falls on the right.
b. The graph rises on the left and the right.
c. The graph falls on the left and the right.
d. The graph falls on the left and rises on the right.

__c__ 6. Find the local minimum for $f(x) = x^3 - 5x^2 + 3x + 2$.
a. $(0.3, 2.5)$ b. $(0.3, -2.5)$ c. $(3, -7)$ d. $(-3, 7)$

__b__ 7. Factor $8x^3 - 125$.
a. $(2x + 5)(4x^2 - 10x + 25)$ b. $(2x - 5)(4x^2 + 10x + 25)$ c. $(2x + 5)(4x^2 + 10x - 25)$ d. $(2x - 5)(4x^2 - 10x + 25)$

__c__ 8. Find the remainder when $3x^3 - 9x^2 + 7x + 4$ is divided by $x - 3$.
a. -179 b. -17 c. 25 d. 142

__d__ 9. Find the quotient when $4x^3 + 12x^2 - 14x + 8$ is divided by $x + 4$.
a. $4x^3 - 4x^2 + 2x$ b. $4x^2 + 4x - 2$ c. $4x^3 + 4x^2 - 2x$ d. $4x^2 - 4x + 2$

NAME ______________ CLASS ________ DATE ________

Chapter Assessment

Chapter 7, Form A, page 2

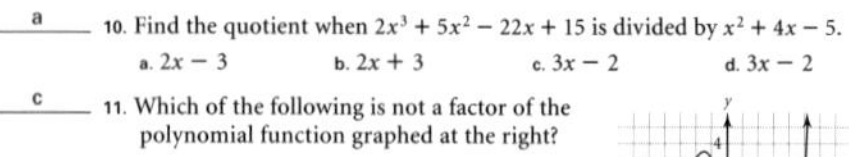

__a__ 10. Find the quotient when $2x^3 + 5x^2 - 22x + 15$ is divided by $x^2 + 4x - 5$.
a. $2x - 3$ b. $2x + 3$ c. $3x - 2$ d. $3x - 2$

__c__ 11. Which of the following is not a factor of the polynomial function graphed at the right?
a. x b. $x + 3$ c. $x + 4$ d. $x - 5$

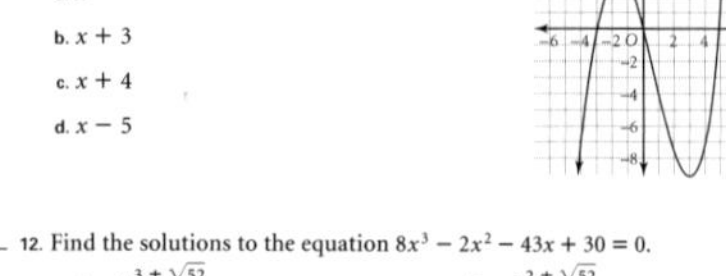

__c__ 12. Find the solutions to the equation $8x^3 - 2x^2 - 43x + 30 = 0$.
a. 2 and $\frac{3 \pm \sqrt{52}}{8}$ b. 2 and $\frac{2 \pm \sqrt{52}}{16}$ c. $2, -\frac{5}{2}$, and $\frac{3}{4}$ d. 1, 2, and -3

__b__ 13. Which of the following is a zero of the polynomial function graphed at the right?
a. -3 b. 3 c. -6 d. 6

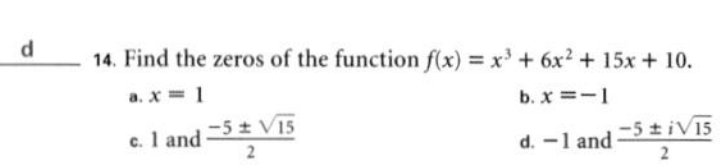

__d__ 14. Find the zeros of the function $f(x) = x^3 + 6x^2 + 15x + 10$.
a. $x = 1$ b. $x = -1$ c. 1 and $\frac{-5 \pm \sqrt{15}}{2}$ d. -1 and $\frac{-5 \pm i\sqrt{15}}{2}$

__a__ 15. If $P(x) = x^3 - 5x^2 + 24x - 20$ and one root of $P(x) = 0$ is $2 + 4i$, which of the following is another root?
a. $2 - 4i$ b. $-2 - 4i$ c. $4 + 2i$ d. $4 - 2i$

__c__ 16. Which of the following could not be a rational root of $12x^4 - 28x^3 + 13x^2 + 7x - 4 = 0$?
a. 1 b. $\frac{4}{3}$ c. $\frac{3}{5}$ d. $\frac{1}{2}$

The graph of $f(x) = -x^3 - 3x^2 + 10$ below shows that f increases for $-2 < x < 0$ and f decreases for $x < -2$ and for $x > 0$.

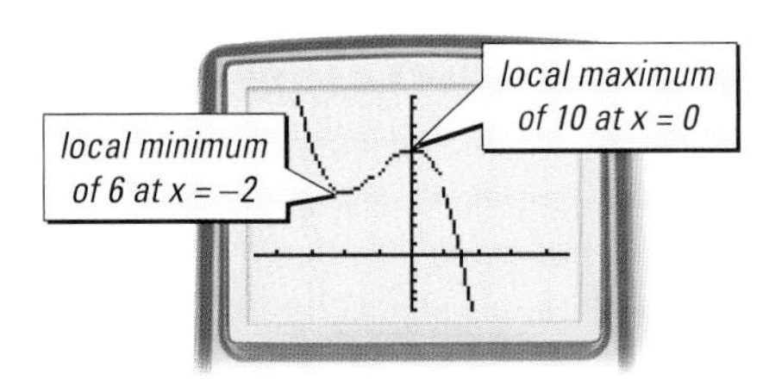

Graph each function. Find any local maxima and minima to the nearest tenth and the intervals over which the function is increasing and decreasing.

15 $f(x) = x^2 - 2x + 9$

16 $g(x) = -x^2 + 5x - 4$

17 $f(a) = 2a^3 + 5a^2 - 1$

18 $m(x) = -2x^3 + x^2 - 9$

LESSON 7.3

Key Skills

Use long division and synthetic division to find factors and remainders of polynomials.

Use long division to find whether $x - 5$ is a factor of $x^3 - 2x^2 - 13x - 10$.

$$
\begin{array}{r}
x^2 + 3x + 2 \\
x - 5\overline{)x^3 - 2x^2 - 13x - 10} \\
-(x^3 - 5x^2) \\
\hline
3x^2 - 13x \\
-(3x^2 - 15x) \\
\hline
2x - 10 \\
-(2x - 10) \\
\hline
0
\end{array}
$$

The remainder is 0, so $x - 5$ is a factor and $x^3 - 2x^2 - 13x - 10 = (x - 5)(x^2 + 3x + 2)$.

Because $P(5) = 0$, $x - 5$ is a factor.

Use synthetic division to find whether $x - 2$ is a factor of $x^3 + 2x - 3$.

$$
\begin{array}{r|rrrr}
2 & 1 & 0 & 2 & -3 \\
 & & 2 & 4 & 12 \\
\hline
 & 1 & 2 & 6 & \boxed{9}
\end{array}
$$

The remainder is 9, so $x - 2$ is not a factor of $x^3 + 2x - 3$.

The Remainder Theorem states that if the expression that defines $P(x) = x^3 + 2x - 3$ is divided by $x - 2$, the remainder is the value of $P(2)$.

$P(2) = 2^3 + 2(2) - 3 = 9$

Exercises

Write each product as a polynomial in standard form.

19. $-2x^3(5x^4 - 3x + x^2 - 6 - x^3)$

20. $(x + 4)(x^3 - 7)(x + 1)$

Factor each polynomial.

21. $x^3 + 4x^2 - 5x$

22. $x^3 - 3x^2 - 10x$

23. $x^3 - 125$

24. $27x^3 + 1$

Use substitution to tell whether the given binomial is a factor of the polynomial.

25. $(x^3 - 7x^2 + 4x + 12)$; $(x - 2)$ **yes**

26. $(x^3 - 5x^2 - 2x + 24)$; $(x + 2)$ **yes**

Divide by using long division.

27. $(x^3 + 6x^2 - x - 30) \div (x - 2)$ $\mathbf{x^2 + 8x + 15}$

28. $(x^3 - 2x^2 - 11x + 12) \div (x^2 - x - 12)$ $\mathbf{x - 1}$

Divide by using synthetic division.

29. $(x^3 + 3x^2 - 34x + 48) \div (x - 3)$ $\mathbf{x^2 + 6x - 16}$

30. $(x^3 + x^2 - 22x - 40) \div (x + 4)$ $\mathbf{x^2 - 3x - 10}$

LESSON 7.4

Key Skills

Find all of the roots of a polynomial equation.

Use variable substitution to find all of the roots of $x^4 - 7x^2 + 12 = 0$.

Substitute u for x^2: $(u)^2 - 7(u) + 12 = 0$
Factor and solve for u: $u = 3$ *or* $u = 4$
Substitute x^2 for u: $x^2 = 3$ *or* $x^2 = 4$

Thus, the roots are $\sqrt{3}$, $-\sqrt{3}$, 2, and -2.

Exercises

Find all of the real roots of each polynomial equation.

31. $x^4 - 8x^2 + 16 = 0$ $\mathbf{\pm 2}$

32. $x^4 - 10x^2 + 24 = 0$ $\mathbf{\pm 2, \pm\sqrt{6}}$

33. $x^4 - 10x^2 + 9 = 0$ $\mathbf{\pm 1, \pm 3}$

34. $x^4 - 13x^2 + 12 = 0$ $\mathbf{\pm 1, \pm 2\sqrt{3}}$

35. $x^3 - x^2 - 6x + 6 = 0$ $\mathbf{\pm\sqrt{6}, 1}$

15. The graph of f has a local minimum of 8 at $x = 1$, is decreasing when $x < 1$, and is increasing when $x > 1$.

16. The graph of g has a local maximum of 2.25 at $x = 2.5$, is increasing when $x < 2.5$, and is decreasing when $x > 2.5$.

17. The graph of f has a local maximum of about 3.6 at $a = -1\frac{2}{3}$ and a local minimum of -1 at $a = 0$. The graph is increasing when $a = -1\frac{2}{3}$ and when $a > 0$. The graph is decreasing when $-1\frac{2}{3} < a < 0$.

18. The graph of m has a local minimum of -9 at $x = 0$ and a local maximum of about -8.96 at $x = \frac{1}{3}$. The graph is decreasing when $x < 0$ and when $x > \frac{1}{3}$. The graph is increasing when $0 < x < \frac{1}{3}$.

19. $-10x^7 + 2x^6 - 2x^5 + 6x^4 + 12x^3$

20. $x^5 + 5x^4 + 4x^3 - 7x^2 - 35x - 28$

21. $x(x + 5)(x - 1)$

22. $x(x - 5)(x + 2)$

23. $(x - 5)(x^2 + 5x + 25)$

24. $(3x + 1)(9x^2 - 3x + 1)$

Chapter Test, Form B

NAME ______ CLASS ______ DATE ______

Chapter Assessment
Chapter 7, Form B, page 1

1. Classify $f(x) = 2x^5 + 6x^2 + 3$ by degree and number of terms. quintic trinomial
2. Evaluate $f(x) = -\frac{3}{4}x^4 + \frac{5}{2}x^3 - \frac{5}{8}x^2 + 2$ when $x = -4$. $f(-4) = -360$

Perform each operation. Write your answer in standard form.

3. $(-4x^3 + 7x^2 - 3x - 2) - (6x^4 + 7x^3 - 5x^2 + 4)$ $-6x^4 - 11x^3 + 12x^2 - 3x - 6$
4. $(x + 2)^2(x - 4)$ $x^3 - 12x - 16$

Write in factored form.

5. $3x^3 - 6x^2 - 24x$ $3x(x + 2)(x - 4)$
6. $x^3 + y^3$ $(x + y)(x^2 - xy + y^2)$
7. $x^3 + 2x^2 - 19x - 20$ $(x + 5)(x + 1)(x - 4)$
8. $125x^3 - 27$ $(5x - 3)(25x^2 + 15x + 9)$
9. Use a graphics calculator to graph $f(x) = x^4 - 2x^3 - 4x^2 + 6x + 5$. Sketch the graph on the grid at right. Approximate the coordinates of any local maxima and minima to the nearest tenth.
local minima at $(-1.2, -2.4)$ and $(2.1, 0.9)$;
local maximum at $(0.6, 6.9)$

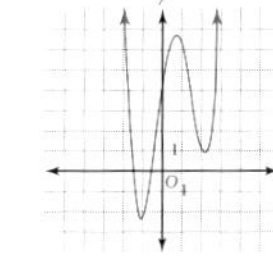

Describe the end behavior of the graph of each function.

10. $f(x) = 3x^2 + 4x^3 - 5x^4$ falls on the left and the right
11. $f(x) = -4x^7 + 6x^5 + 8x^3$ rises on the left and falls on the right
12. Find the remainder when $P(x) = 2x^4 - 8x^3 - 3x + 7$ is divided by $x - 4$. $P(4) = -5$
13. Find a cubic model for the data in the table below.

x	-2	0	2	4	6
y	-11	8	3	70	305

$f(x) = 2x^3 - 3x^2 - 4.5x + 8$

NAME ______ CLASS ______ DATE ______

Chapter Assessment
Chapter 7, Form B, page 2

14. Let $P(x) = x^3 - 3x^2 + 2$ and $Q(x) = x^2 - 6x + 11$. Find all values of x such that $P(x) = Q(x)$. $x = 3$

Use synthetic division to find the quotient.

15. $(x^3 - 28x - 48) \div (x - 6)$ $x^2 + 6x + 8$
16. $(x^4 + 6x^3 + 3x^2 - 23x - 12) \div (x + 4)$ $x^3 + 2x^2 - 5x - 3$

Use long division to find the quotient.

17. $(5x^3 - 17x^2 + 10x - 12) \div (x - 3)$ $5x^2 - 2x + 4$
18. $(3x^3 - 24) \div (x^2 + 2x + 4)$ $3x - 6$

Find all roots to each equation.

19. $x^3 - 13x^2 + 56x - 78 = 0$ 3, $5 + i$, and $5 - i$
20. $x^4 - 11x^2 + 18 = 0$ ± 3 and $\pm\sqrt{2}$
21. List all possible rational zeros of the function $f(x) = 5x^7 - 7x^3 + 12x - 4$. $\pm 1, \pm 2, \pm 4, \pm\frac{1}{5}, \pm\frac{2}{5}, \pm\frac{4}{5}$
22. Find all the zeros of the function $f(x) = 5x^3 + 14x^2 - 7x - 12$. $-\frac{4}{5}, -3$, and 1
23. A quadratic polynomial, $P(x)$, has zeros of -2 and 4. In addition, $P(0) = 40$. Write $P(x)$ in factored form and in standard form. $P(x) = -5(x + 2)(x - 4) = -5x^2 + 10x + 40$
24. The zeros of a fourth-degree polynomial, $P(x)$, with real coefficients include 3, with a multiplicity of 2, and $3 + 4i$. In addition, $P(0) = 675$. Write $P(x)$ in factored form with real-number factors. $P(x) = 3(x - 3)^2(x^2 - 6x + 25)$

47. $P(x) = (x-2)(x+1)^2(x-3)$
$= x^4 - 3x^3 - 3x^2 + 7x + 6$

48. $P(x) = 2(x+2)^3(x-1)$
$= 2x^4 + 10x^3 + 12x^2 - 8x - 16$

49. $P(x) = -(x-3)^2(x-(2+i))(x-(2-i))$
$= -x^4 + 10x^3 - 38x^2 + 66x - 45$

50. $P(x) = 9(x-5)(x+7)\left(x+\frac{2}{3}i\right)\left(x-\frac{2}{3}i\right)$
$= 9x^4 + 18x^3 - 311x^2 + 8x - 140$

Use factoring to find all of the roots of $x^3 - 4x^2 - 3x + 12 = 0$.

Factor by grouping:

$$(x^3 - 12x^2) - (3x - 12) = 0$$
$$x^2(x-4) - 3(x-4) = 0$$
$$(x^2 - 3)(x-4) = 0$$

Thus, the roots are $\sqrt{3}, -\sqrt{3}$, and 4.

Given that one root of $x^3 - 3x - 2 = 0$ is 2, use synthetic division to find all of the other roots.

$$\begin{array}{r|rrrr} 2 & 1 & 0 & -3 & -2 \\ & & 2 & 4 & 2 \\ \hline & 1 & 2 & 1 & 0 \end{array} \rightarrow \begin{array}{r} x^3 - 3x - 2 = 0 \\ (x-2)(x^2+2x+1) = 0 \\ (x-2)(x+1)^2 = 0 \end{array}$$

Thus, the remaining root is -1 (multiplicity 2).

36. $x^5 - x^3 - 8x^2 + 8 = 0$ **±1, 2**

37. $x^4 + 2x^3 - x - 2 = 0$ **1, −2**

38. $x^5 - 9x^3 - x^2 + 9 = 0$ **1, ±3**

Given the indicated root, find all of the other roots of the polynomial equation.

39. -3; $x^3 + 7x^2 + 16x + 12 = 0$ **−2**

40. -3; $x^3 + 3x^2 - 16x - 48 = 0$ **−4, 4**

41. 4; $x^3 - 11x^2 + 38x - 40 = 0$ **2, 5**

42. 6; $x^3 - 6x^2 - x + 6 = 0$ **−1, 1**

LESSON 7.5

Key Skills

Find all rational roots of a polynomial equation.

Find all rational roots of $9x^2 - 9x + 2 = 0$. The possible values are $\pm\frac{1}{1}, \pm\frac{1}{3}, \pm\frac{1}{9}, \pm\frac{2}{1}, \pm\frac{2}{3}$, and $\pm\frac{2}{9}$. Checking each possibility reveals that the rational roots are $\frac{1}{3}$ and $\frac{2}{3}$.

Write a polynomial function given sufficient information.

The zeros of a fourth degree polynomial include 1 (multiplicity 2) and i. Also, $P(0) = 4$. Write the function in factored form and in standard form.

$$P(x) = a(x-1)^2(x-i)(x+i)$$
$$4 = a(0-1)^2(0-i)(0+i)$$
$$4 = a$$

$$P(x) = 4(x-1)^2(x-i)(x+i)$$
$$= 4x^4 - 8x^3 + 8x^2 - 8x + 4$$

Exercises

Find all of the rational roots of each polynomial equation.

43. $15x^2 + x - 2 = 0$ $-\frac{2}{5}, \frac{1}{3}$

44. $9x^3 - 27x^2 - 34x - 8 = 0$ $-\frac{2}{3}, -\frac{1}{3}$, **4**

45. $9x^3 - 18x^2 + 2x - 4 = 0$ **2**

46. $x^3 - 3x^2 + 4x - 12 = 0$ **3**

Write a polynomial function of degree 4 in factored form and in standard form by using the given information.

47. $P(0) = 6$; zeros: 2, -1(multiplicity 2), and 3

48. $P(0) = -16$; zeros: -2 (multiplicity 3) and 1

49. $P(0) = -45$; zeros: 3 (multiplicity 2) and $2 + i$

50. $P(0) = -140$; zeros: 5, -7, and $-\frac{2}{3}i$

Applications

51. **INVESTMENTS** Suppose that \$500 is invested at the end of every year for 5 years. One year after the last payment, the investment is worth \$3200. Use the polynomial equation $500x^5 + 500x^4 + 500x^3 + 500x^2 + 500x = 3200$ to find the effective interest rate of this investment. **8.3%**

52. **MANUFACTURING** Using standard form, write the polynomial function that represents the volume of a crate with a length of x feet, a width of $8 - x$ feet, and a height of $x - 3$ feet. Then find the maximum volume of the crate and the dimensions for this volume. $V = -x^3 + 11x^2 - 24x$; **36 ft³; 6 ft × 2 ft × 3 ft**

Chapter Test

Evaluate each polynomial for $x = 3$ and $x = -2$.

1. $x^3 - 2x^2 + 5$ **14; −11**
2. $x^4 - x^2 + 3x - 4$ **77; 2**
3. $5x^2 - 3x + 1$ **37; 27**
4. $7x^3 + 4x^2 - 3$ **222; −43**

Write each sum or difference as a polynomial in standard form. Then classify the polynomial by degree and number of terms.

5. $(5x^3 - 3x^2 + x - 7) + (3x^2 - x + 6)$ **$5x^3 - 1$; cubic binomial**
6. $(2x^5 + 9x^3 - 7x + 4) - (9x^3 + 3x^2 + 4)$ **$2x^5 - 3x^2 - 7x$; quintic trinomial**
7. **FINANCE** An annuity of \$500 invested at the beginning of each year earns 7% interest, compounded annually. How much will the investment be worth at the time of the fifth payment? **\$2875.37**

Describe the end behavior of each function.

8. $P(x) = x^4 - 3x^2 + x$ **rises at both ends**
9. $P(x) = 12x^3 - 7x^5$ **rises on left, falls on right**
10. $P(x) = 3x^3 - x^2 - 1$ **falls on left, rises on right**
11. $P(x) = 3 - 2x$ **rises on left, falls on right**

Use a graphing calculator to graph each function. Find any local maxima or minima to the nearest tenth and find the intervals over which the function is increasing and decreasing.

12. $P(x) = 3 - 2x - x^2$
13. $P(x) = x^3 + 3x^2 + 4$
14. $P(x) = x^4 - 3x^2 - 4$
15. $P(x) = 5 - 3x^2 - x^3$

Factor each polynomial.

16. $5x^4 - 180x^2$ **$5x^2(x - 6)(x + 6)$**
17. $4x^3 - 5x^2 - 8x + 10$ **$(4x - 5)(x + \sqrt{2})(x - \sqrt{2})$**
18. $2x^3 + 128$ **$2(x + 4)(x^2 - 4x + 16)$**
19. $x^4 - 7x^3 + 12x^2$ **$x^2(x - 3)(x - 4)$**

Divide by using long division.

20. $(2x^4 - 7x^3 - 15x^2 + 8x + 12) \div (2x + 3)$ **$x^3 - 5x^2 + 4$**
21. $(x^3 + 3x^2 - 2x - 6) \div (x^2 - 2)$ **$x + 3$**

Divide by using synthetic division.

22. $(-x^3 + 6x^2 - 11x + 6) \div (x - 3)$ **$-x^2 + 3x - 2$**
23. $(x^3 + 6x^2 - 20) \div (x + 3)$ **$x^2 + 3x - 9 + \frac{7}{x + 3}$**
24. **MANUFACTURING** A box with an open top has a volume given by the formula $V(x) = x(14 - 2x)(32 - 2x)$. Write the volume of the box as a polynomial function in standard form. Then find the volume when $x = 3$. **$V(x) = 4x^3 - 92x^2 + 448x$; 624u³**

Find all of the real roots of each polynomial.

25. $-2x^3 + 7x^2 + 3x - 18 = 0$ **$3, 2, -\frac{3}{2}$**
26. $x^4 + 2x^3 - 7x^2 - 14x = 0$ **$-2, 0, \pm\sqrt{7}$**
27. $x^4 - 6x^2 + 8 = 0$ **$\pm 2, \pm\sqrt{2}$**
28. $8x^3 + 1 = 0$ **$-\frac{1}{2}$**
29. **ENERGY** A liquid propane tank has the shape of a cylinder with hemispherical ends. The volume of the tank is represented by the function $V(x) = 15\pi x^2 + \frac{4}{3}\pi x^3$, where x is the radius. If the volume is approximately 540 cubic feet, what is the radius? **≈ 3 ft.**

Find all of the rational roots of each polynomial equation.

30. $6x^2 + 11x - 10 = 0$ **$\frac{2}{3}, -\frac{5}{2}$**
31. $6x^3 - 13x^2 - 41x - 12 = 0$ **$4, -\frac{1}{3}, -\frac{3}{2}$**

Write a polynomial function *P* in factored form and in standard form by using the given information.

32. degree = 2; $P(0) = 3$; zeros: $1, \frac{3}{7}$
33. degree = 3; $P(0) = -18$; zeros: $-3, -1, 3$
34. degree = 3; $P(0) = 30$; zeros: $-3, -1, 2$
35. degree = 4; $P(0) = 15$; zeros: $-1 + i, -3, 5$

Chapter Test

12. maxima at $(-1, 4)$; increases for $x < -1$; decreases for $x > -1$
13. maxima at $(-2, 8)$; minima at $(0, 4)$; increases for $x < -2$ and $x > 0$; decreases for $-2 < x < 0$
14. maxima at $(0, -4)$; minima at $(\pm 1.2, -6.3)$; increases for $-1.2 < x < 0$ and $x > 1.2$; decreases for $x < -1.2$ and $0 < x < 1.2$
15. minima at $(-2, 1)$; maxima at $(0, 5)$; increases for $-2 < x < 0$; decreases for $x < -2$ and $x > 0$

32. $P(x) = (x - 1)(7x - 3) = 7x^2 - 10x + 3$
33. $P(x) = 2(x + 1)(x + 3)(x - 3) = 2x^3 + 2x^2 - 18x - 18$
34. $P(x) = -5(x + 3)(x + 1)(x - 2) = -5x^3 - 10x^2 + 25x + 30$
35. $P(x) = -\frac{1}{2}(x + 3)(x - 5)(x^2 + 2x + 2) = -\frac{1}{2}x^4 + \frac{17}{2}x^2 + 17x + 15$

College Entrance Exam Practice

Multiple-Choice and Quantitative-Comparison Samples

The first half of the Cumulative Assessment contains two types of items found on standardized tests—multiple-choice questions and quantitative-comparison questions. Quantitative-comparison items emphasize the concepts of equality, inequality, and estimation.

Free-Response Grid Samples

The second half of the Cumulative Assessment is a free-response section. This part of the Cumulative Assessment requires student-produced response items like those commonly found on college entrance exams. These questions require the use of machine-scored answer grids. You may wish to have students practice answering these items in preparation for standardized tests.

1–7 College Entrance Exam Practice

QUANTITATIVE COMPARISON For Items 1–5, write
A if the quantity in Column A is greater than the quantity in Column B;
B if the quantity in Column B is greater than the quantity in Column A;
C if the two quantities are equal; or
D if the relationship cannot be determined from the given information.

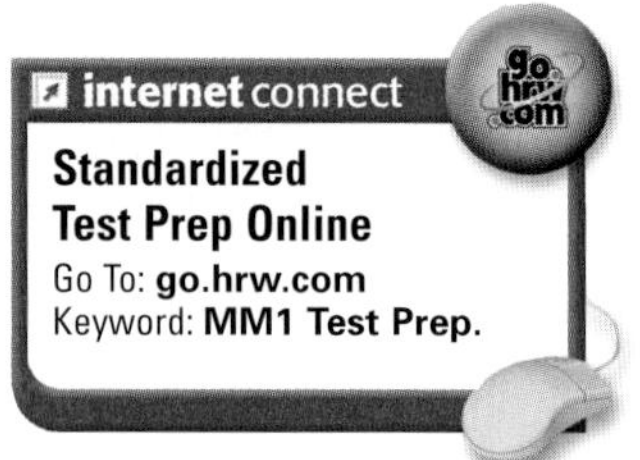

	Column A	Column B	Answers
1. A	The degree of $x^4 + 2x^3 + x + 1$	The degree of $3x^2 + 2x + 1$	A B C D [Lesson 7.1]
2. B	$\ln \frac{1}{2}$	$\ln 2$	A B C D [Lesson 6.6]
3. C	$e^{\ln 2}$	2	A B C D [Lessons 6.4 and 6.6]
4. C	The number of factors of $x^3 - 4x$	The number of factors of $x^3 - 5x^2 + 6x$	A B C D [Lesson 7.5]
5. C	The minimum value of $f(x) = x^2 + 1$	The minimum value of $f(x) = (x - 1)^2 + 1$	A B C D [Lesson 5.1]

6. Which is a solution of the system below?
b $\begin{cases} y \geq -x \\ y \geq 3x + 2 \end{cases}$ ***(LESSON 3.4)***
a. (1, −5) **b.** (0, 5)
c. both **a** and **b** **d.** neither **a** nor **b**

7. Find the slope of the line $3x + 4y = 2$.
c ***(LESSON 1.2)***
a. 3 **b.** $\frac{2}{3}$
c. $-\frac{3}{4}$ **d.** 4

8. Which equation contains the point (1, −3)
b and is perpendicular to $y = 2x - 2$?
(LESSON 1.3)
a. $2y = -x + 5$ **b.** $2y = -x - 5$
c. $y = -\frac{1}{2}x + 6$ **d.** $y = -\frac{1}{2}x + \frac{3}{2}$

9. Which is the solution of the system below?
b $\begin{cases} x + 2y = 4 \\ 2x + y = 5 \end{cases}$ ***(LESSON 3.1)***
a. (2, 3) **b.** (2, 1)
c. (−3, 2) **d.** (0, 1)

10. Which set of ordered pairs represents a
b function? ***(LESSON 2.3)***
a. {(0, 3), (1, 4), (1, 3)}
b. {(2, 1), (4, −1), (6, 2)}
c. {(3, 5), (2, 10), (3, 15)}
d. {(10, 4), (10, 5), (20, 6)}

11. For which function does the vertex give a
b maximum value? ***(LESSON 5.1)***
a. $a(x) = 3x^2 + 5x$ **b.** $b(x) = 7x + 5x - 3x^2$
c. $c(x) = 3 + 5x + \frac{1}{3}x^2$ **d.** $d(x) = \frac{1}{3}x^2$

12. How many solutions does a consistent system of linear equations have? ***(LESSON 3.1)*** c

a. 0 **b.** 1
c. at least 1 **d.** infinitely many

13. What is the y-intercept for the graph of the equation $x - 5y = 15$? ***(LESSON 1.2)*** d

a. 15 **b.** −1
c. 3 **d.** −3

14. Which is the solution of the inequality $4x + 2 < 2x + 1$? ***(LESSON 1.7)*** d

a. $x \geq 1$ **b.** $x > 2$
c. $x < \frac{1}{3}$ **d.** $x < -\frac{1}{2}$

15. Which is the solution to the inequality $|x| \leq 5$? ***(LESSON 1.8)*** a

a. $-5 \leq x \leq 5$ **b.** $-2 \leq x \leq 2$
c. $-5 \geq x \geq 5$ **d.** $-3 \leq x \leq 3$

16. Which is the factorization of $x^2 - 5x + 6$? ***(LESSON 5.3)*** a

a. $(x - 2)(x - 3)$ **b.** $(x + 2)(x - 3)$
c. $(x + 1)(x + 6)$ **d.** $(x - 1)(x - 6)$

17. Write $(x + 1)(x + 2)(x - 4)$ in standard form. ***(LESSON 7.3)*** $x^3 - x^2 - 10x - 8$

18. Find the product. $\begin{bmatrix} 2 & 3 \\ 4 & 3 \end{bmatrix}\begin{bmatrix} 1 & 1 \\ 2 & 3 \end{bmatrix}$ $\begin{bmatrix} 8 & 11 \\ 10 & 13 \end{bmatrix}$ ***(LESSON 4.2)***

19. Let $f(x) = 3x + 5$ and $g(x) = x^2 + x + 5$. Find $f + g$. ***(LESSON 2.4)*** $x^2 + 4x + 10$

20. Solve $\ln x = 5$ for x. ***(LESSONS 6.3 AND 6.6)*** $x = e^5 \approx 148.41$

21. Write the equation in vertex form for the parabola described by $f(x) = 2x^2 - 8x + 9$. ***(LESSON 5.4)*** $y = 2(x - 2)^2 + 1$

22. Find the sum. $(2x^3 + 3x^2 + 1) + (5x^2 - 2x + 2)$ ***(LESSON 7.1)*** $2x^3 + 8x^2 - 2x + 3$

23. Find the difference below. ***(LESSON 7.1)*** $(5x^3 + 4x^2 - x) - (x^3 + 2x^2 - 1)$ $4x^3 + 2x^2 - x + 1$

24. Let $f(x) = 3x + 2$. Find the inverse of f. ***(LESSON 2.5)*** $f^{-1}(x) = \frac{1}{3}x - \frac{2}{3}$

25. Find all rational roots of $3x^2 + 5x - 2 = 0$. ***(LESSON 7.5)*** $-2, \frac{1}{3}$

26. Multiply. $(4 - 2i)(4 + 2i)$ ***(LESSON 5.6)*** 20

27. $-y(3y + 5)$ 28. $(x + 4)(x - 9)$
29. $(8x - 9)(3x + 4)$ 30. $2(9x + 2)(2x - 3)$

Factor each quadratic expression, if possible. *(LESSON 5.3)*

27. $-3y^2 - 5y$ **28.** $x^2 - 5x - 36$
29. $24x^2 + 5x - 36$ **30.** $36x^2 - 46x - 12$

FREE-RESPONSE GRID The following questions may be answered by using a free-response grid such as that commonly used by standardized-test services.

31. Evaluate $\lceil 3.1 \rceil + \lceil 2.5 \rceil$. 6 ***(LESSON 2.6)***

32. Find the value of $\log_5 25$. 2 ***(LESSON 6.3)***

33. Solve the equation $\ln x + \ln(2 - x) = 0$ for x. 1 ***(LESSON 6.7)***

34. Simplify $5[3 - (3 - 2)]^2$. ***(LESSON 2.1)*** 20

35. Find the rational zeros of the polynomial function $f(x) = 2x^3 - x^2 - 4x + 2$. $\frac{1}{2}$ ***(LESSON 7.5)***

36. Solve the proportion $\frac{x + 2}{2} = \frac{2x}{3}$ for x. 6 ***(LESSON 1.4)***

37. Evaluate $|-2.5| - |3.2|$. ***(LESSON 2.6)*** −0.7

38. Solve $-3x + 4 = 5 - x$ for x. ***(LESSON 1.6)*** $-\frac{1}{2}$

39. **INVESTMENTS** Find the final amount in dollars of a $1000 investment after 5 years at an interest rate of 8% compounded annually. ***(LESSON 6.2)*** 1469.33

40. **ARCHAEOLOGY** A sample of wood found at an archaeological site contains 60% of its original carbon-14. Use the function $N(t) = N_0e^{-0.00012t}$, which gives the amount of remaining carbon-14, to estimate the age of the sample of wood. ***(LESSON 6.6)***

41. **INVESTMENTS** If $1000 is invested at 8% annual interest, compounded continuously, find the dollar amount of the investment after 5 years. ***(LESSON 6.6)*** 1491.82

40. 4257

Keystroke Guide for Chapter 7

Essential keystroke sequences (using the model TI-82 or TI-83 graphics calculator) are presented below for all Activities and Examples found in this chapter that require or recommend the use of a graphics calculator.

internet connect

For Keystrokes of other graphing calculator models, visit the HRW web site at **go.hrw.com** and enter the keyword **MB1 CALC**.

LESSON 7.1

EXAMPLE 2 Page 426

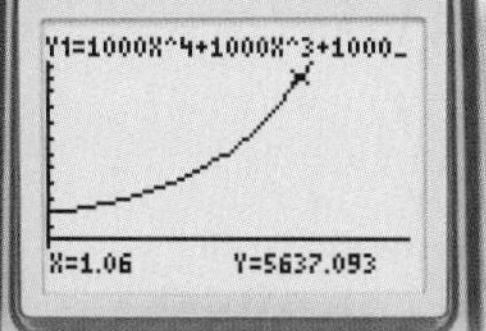

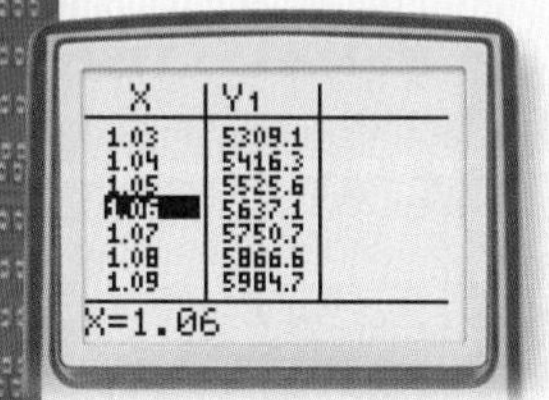

Graph $y = 1000x^4 + 1000x^3 + 1000x^2 + 1000x + 1000$, and evaluate for $x = 1.06$.

Use viewing window [0, 1.5] by [−1000, 7000].

Graph the function:

Y= 1000 X,T,θ,n ^ 4 + 1000 X,T,θ,n ^ 3 + 1000 X,T,θ,n x^2 + 1000 X,T,θ,n + 1000 GRAPH

Use the graph:

2nd TRACE (CALC) 1:valu ENTER (X=) 1.06 ENTER

Use a table:

2nd WINDOW (TBLSET) (TblStart=) 1.03 ▼ (△Tbl=) .01 ▼ (Indpt:) Aut ENTER ▼

⇑ TI-82: (TblMin=)

(Depend:) Aut ENTER ▼ 2nd GRAPH (TABLE)

Activity Page 427

For Step 1, graph $y = x^2 + x - 2$.

Use viewing window [−15, 15] by [−15, 15].

Y= X,T,θ,n x^2 + X,T,θ,n − 2 GRAPH

Use a similar keystroke sequence for Steps 2–9. For Steps 7–9, use viewing window [−4, 4] by [−4, 4].

EXAMPLE 5 Page 428

For part a, graph $y = 3x^3 - 5x^2 - 2x + 1$.

Use viewing window [−5, 5] by [−5, 5].

Y= 3 X,T,θ,n ^ 3 − 5 X,T,θ,n x^2 − 2 X,T,θ,n + 1 GRAPH

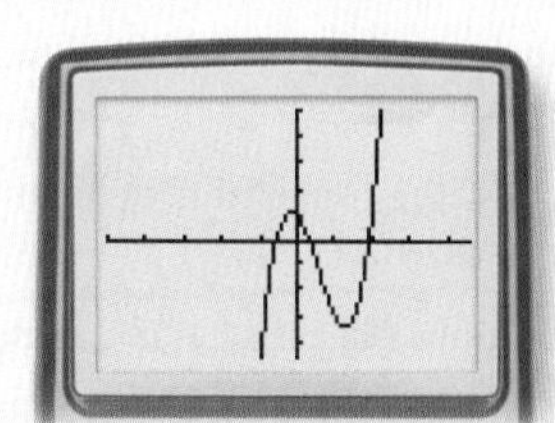

For part **b**, use a similar keystroke sequence. Use viewing window [−5, 5] by [−20, 5].

LESSON 7.2

EXAMPLE 1 Page 434

Graph $y = x^3 + 3x^2 - x - 3$, and approximate the coordinates of any local maxima or minima.

Use friendly viewing window [−4.7, 4.7] by [−10, 10].

Graph the function:

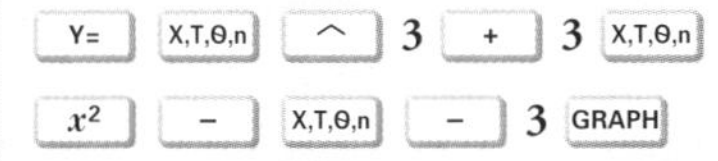

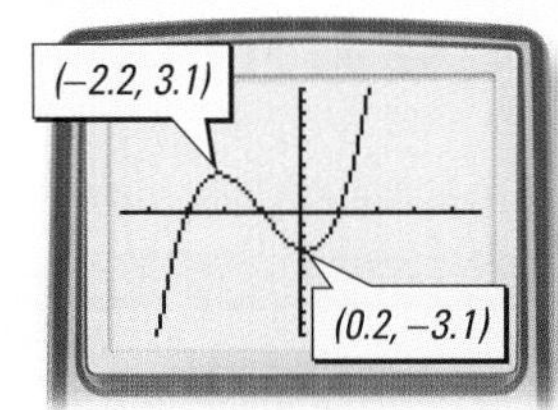

Find the coordinates of any local maxima or minima:

Press TRACE, and move the cursor.

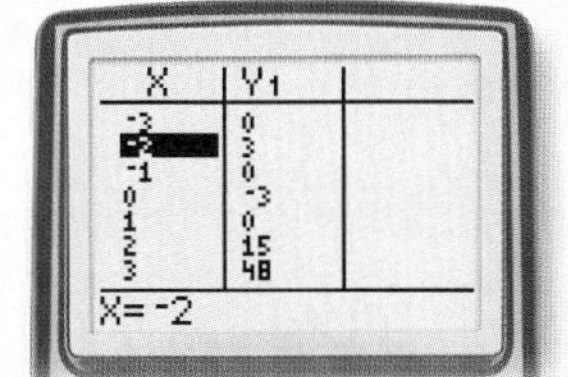

Make a table of values:

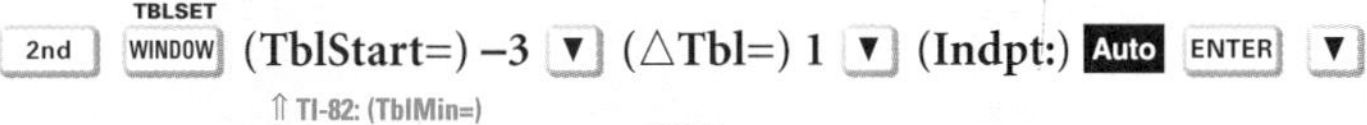

(Depend:) Auto ENTER ▼ 2nd GRAPH (TABLE)

To approximate more closely, change to TblStart= −2.5 and △Tbl= 0.1.

Page 435

For Steps 1–16, graph each function by using a keystroke sequence similar to that in Example 2 of Lesson 7.1.

Use viewing window [−5, 5] by [−5, 5].

EXAMPLE 2 Page 436

For part a, graph $y = -x^3 + x^2 + 3x - 1$.

Use viewing window [−5, 5] by [−5, 5].

For part **b**, use a similar keystroke sequence to graph $y = -x^4 + x^3 + 3x^2 - x + 3$. Use viewing window [−5, 5] by [−2, 8].

EXAMPLE 3 Page 436

Enter the given data, and find a quartic regression model for the data. Then estimate y when x is 20.

Use viewing window [0, 60] by [0, 5000].

Enter the data:

STAT EDIT 1:Edit ENTER L1 10 ENTER 18 ENTER 30 ENTER 41 ENTER 44

ENTER ► L2 1679 ENTER 2606 ENTER 3089 ENTER 2276 ENTER 2517 ENTER

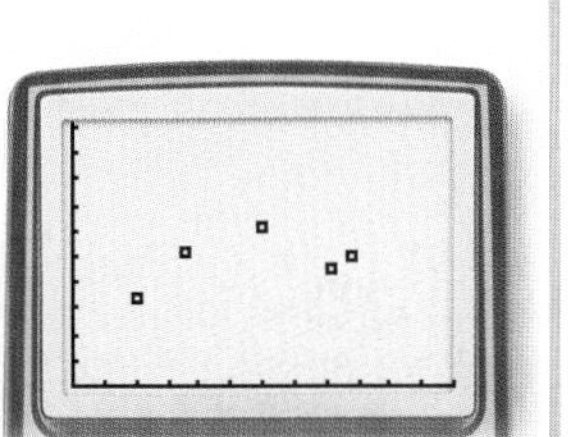

Graph the scatter plot:

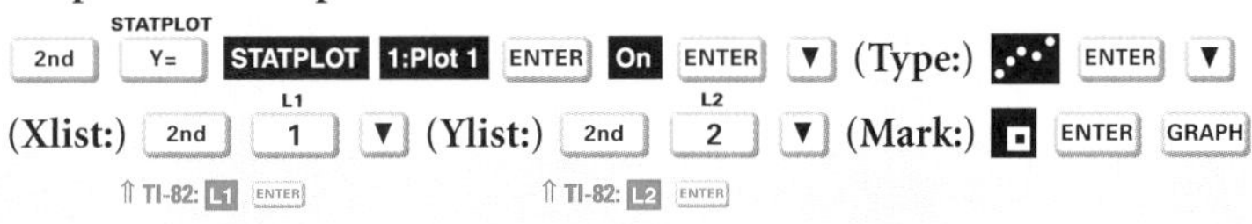

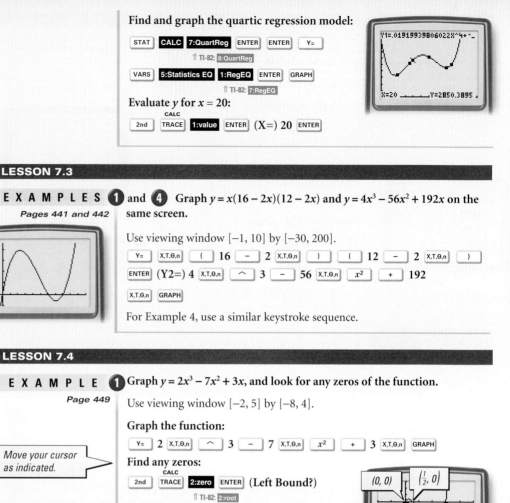

Find and graph the quartic regression model:

STAT **CALC** **7:QuartReg** ENTER ENTER Y=

⇑ TI-82: 8:QuartReg

VARS **5:Statistics EQ** **1:RegEQ** ENTER GRAPH

⇑ TI-82: 7:RegEQ

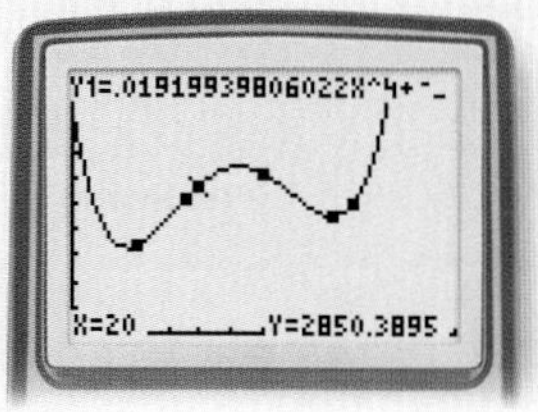

Evaluate y for $x = 20$:

2nd TRACE (CALC) **1:value** ENTER (X=) 20 ENTER

LESSON 7.3

EXAMPLES 1 and 4 Graph $y = x(16 - 2x)(12 - 2x)$ and $y = 4x^3 - 56x^2 + 192x$ on the same screen.

Pages 441 and 442

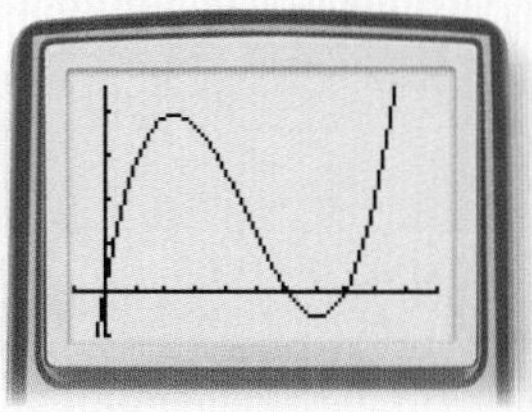

Use viewing window [−1, 10] by [−30, 200].

Y= X,T,θ,n (16 − 2 X,T,θ,n) (12 − 2 X,T,θ,n) ENTER (Y2=) 4 X,T,θ,n ^ 3 − 56 X,T,θ,n x^2 + 192 X,T,θ,n GRAPH

For Example 4, use a similar keystroke sequence.

LESSON 7.4

EXAMPLE 1 Graph $y = 2x^3 - 7x^2 + 3x$, and look for any zeros of the function.

Page 449

Use viewing window [−2, 5] by [−8, 4].

Graph the function:

Y= 2 X,T,θ,n ^ 3 − 7 X,T,θ,n x^2 + 3 X,T,θ,n GRAPH

Move your cursor as indicated.

Find any zeros:

2nd TRACE (CALC) **2:zero** ENTER (Left Bound?)

⇑ TI-82: 2:root

ENTER (Right Bound?) ENTER (Guess?) ENTER

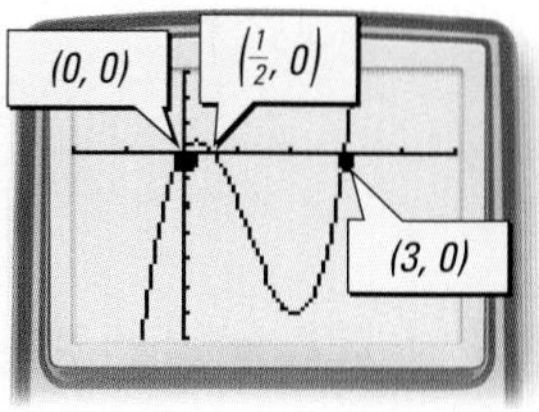

EXAMPLES 2, 3, and 4 Graph each function, and find any zeros.

Pages 449–451

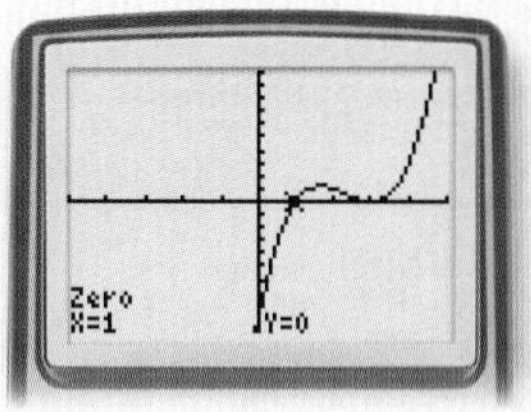

For Examples 2 and 3, use viewing window [−5, 5] by [−5, 5].
For Example 4, use viewing window [−40, 20] by [−5000, 7000].

Graph the function:

Use a keystroke sequence similar to that in Example 2 of Lesson 7.1.

Find any zeros:

Use a keystroke sequence similar to that in Example 1 of Lesson 7.4.

Activity

Page 451

For Step 1, enter $y = x^4 - 3x^3 - 5x^2 + 13x + 6$, and use a table of values to find the corresponding y-value for each x-value in the table.

Use viewing window [−5, 5] by [−15, 15].

Enter and graph the function:
Use a keystroke sequence similar to that in Example 2 of Lesson 7.1.

Make a table of values:
Use a keystroke sequence similar to that in Example 1 of Lesson 7.2.

LESSON 7.5

EXAMPLES 1, 2, and 3 **For Example 1, graph $y = 10x^3 + 9x^2 - 19x + 6$, and verify that -2, $\frac{1}{2}$, and $\frac{3}{5}$ are zeros.**

Pages 459 and 460

(−2, 0)

Use viewing window [−3, 3] by [−30, 30].

Graph the function:
Use a keystroke sequence similar to that in Example 1 of Lesson 7.1.

Evaluate y for x-values of -2, $\frac{1}{2}$, and $\frac{3}{5}$:
Use a keystroke similar to that used in Example 2 of Lesson 7.2. For a closer look around $x = \frac{1}{2}$ and $x = \frac{3}{5}$, use viewing window [0.3, 0.8] by [−1, 1].

For Example 2, graph $y = x^3 - 2x^2 - 2x + 1$, and evaluate for $x = -1$. Use viewing window [−5, 5] by [−5, 5].

For Example 3, graph $y = 3x^3 - 10x^2 + 10x - 4$, and evaluate for $x = 2$. Use viewing window [−3, 5] by [−6, 4].

Activity

Page 461

For Step 2, graph $y = 2x^3 - x^2 - 4x$.

Use viewing window [−5, 5] by [−3, 8].

[Y=] 2 [X,T,θ,n] [^] 3 [−] [X,T,θ,n] [x^2] [−] 4 [X,T,θ,n] [GRAPH]

For Steps 3 and 4, use a similar keystroke sequence.

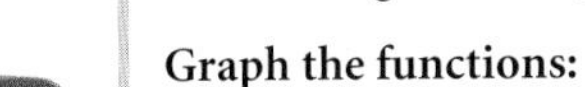

EXAMPLE 4 **Graph $y = x(20 - 2x)(12 - 2x)$ and $y = x(16 - 2x)(15 - 2x)$ on the same screen, and look for any points of intersection.**

Page 461

Use viewing window [0, 6] by [0, 350].

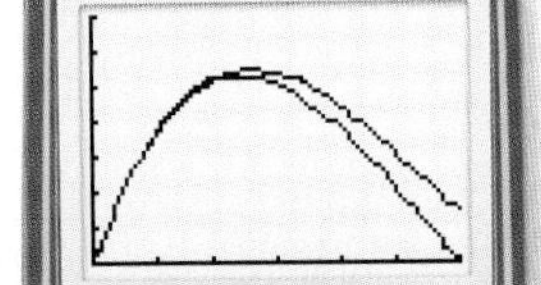

Graph the functions:
Use a keystroke sequence similar to that in Example 1 of Lesson 7.3.

Look for any points of intersection:

[2nd] CALC [TRACE] [5:intersect] (First curve?) [ENTER] (Second curve?) [ENTER] (Guess?) [ENTER]

Rational Functions and Radical Functions

Lesson Presentation CD-ROM
PowerPoint® presentations for each lesson 8.1–1.8

CHAPTER PLANNING GUIDE

Lesson	8.1	8.2	8.3	8.4	8.5	8.6	8.7	8.8	Project and Review
Pupil's Edition Pages	480–488	489–497	498–504	505–511	512–519	520–527	528–535	536–543	544–553
Practice and Assessment									
Extra Practice (Pupil's Edition)	965	965	966	966	967	967	968	968	
Practice Workbook	47	48	49	50	51	52	53	54	
Practice Masters Levels A, B, and C	139–141	142–144	145–147	148–150	151–153	154–156	157–159	160–162	
Standardized Test Practice Masters	54	55	56	57	58	59	60	61	62
Assessment Resources	96	97	98	99	101	102	103	104	100, 105–110
Visual Resources									
Lesson Presentation Transparencies Vol. 1	1–4	5–8	9–12	13–16	17–20	21–24	25–28	29–32	
Teaching Transparencies		29, 30, 31							
Answer Key Transparencies	259–262	263–270	271–273	274–276	277–279	280–286	287–289	290–292	293–298
Quiz Transparencies	8.1	8.2	8.3	8.4	8.5	8.6	8.7	8.8	
Teacher's Tools									
Reteaching Masters	93–94	95–96	97–98	99–100	101–102	103–104	105–106	107–108	
Make-Up Lesson Planner for Absent Students	47	48	49	50	51	52	53	54	
Student Study Guide	47	48	49	50	51	52	53	54	
Spanish Resources	47	48	49	50	51	52	53	54	
Block Scheduling Handbook									16–17
Activities and Extensions									
Lesson Activities	47	48	49	50	51	52	53	54	
Enrichment Masters	47	48	49	50	51	52	53	54	
Cooperative-Learning Activities	47	48	49	50	51	52	53	54	
Problem Solving/ Critical Thinking	47	48	49	50	51	52	53	54	
Student Technology Guide	47	48	49	50	51	52	53	54	
Long Term Projects									29–32
Writing Activities for Your Portfolio									22–24
Tech Prep Masters									35–38
Building Success in Mathematics									20–21

LESSON PACING GUIDE

Lesson	1.1	1.2	1.3	1.4	1.5	1.6	1.7	1.8	Project and Review
Traditional	2 days	2 days	2 days	2 days	2 days	2 days	2 days	2 days	2 days
Block	1 day	1 day	1 day	1 day	1 day	1 day	1 day	1 day	1 day
Two-Year	4 days	4 days	4 days	4 days	4 days	4 days	4 days	4 days	4 days

CONNECTIONS AND APPLICATIONS

Lesson	8.1	8.2	8.3	8.4	8.5	8.6	8.7	8.8	Review
Algebra	480–488	489–497	498–504	505–511	512–519	520–527	528–535	536–543	544–553
Geometry	482, 483, 485	496	503	510	518	526, 527	534		
Maximum/Minimum									553
Statistics	487								
Transformations	481	493				521, 526			
Business and Economics	480	496	501, 503				533		
Life Skills			498				528, 530, 535		
Science and Technology	484, 487	489, 495, 496, 497	504	510, 511	518	520, 523 525, 527		536, 538 541, 542	550
Sports and Leisure	485, 487			505, 508, 509, 510	512, 517, 518			543	553
Cultural Connection: Asia					518				

BLOCK SCHEDULING GUIDE

Day	Lesson	Teacher Directed: Lesson Examples, Teaching Transparencies	Student Guided Activity, Try This	Cooperative-Learning Activity, Lesson Activity, Student Technology Guide	Practice: Practice & Apply, Extra Practice, Practice Workbook	Assessment: Quiz, Mid-Chapter Assessment	Problem Solving, Reteaching
1	8.1	10 min	15 min	15 min	65 min	15 min	15 min
2	8.2	10 min	15 min	15 min	65 min	15 min	15 min
3	8.3	10 min	15 min	15 min	65 min	15 min	15 min
4	8.4	10 min	15 min	15 min	65 min	15 min	15 min
5	8.5	10 min	15 min	15 min	65 min	15 min	15 min
6	8.6	10 min	15 min	15 min	65 min	15 min	15 min
7	8.7	10 min	15 min	15 min	65 min	15 min	15 min
8	8.8	10 min	15 min	15 min	65 min	15 min	15 min
9	Assess.	50 min **PE:** Chapter Review	90 min **PE:** Chapter Project, Writing Activities	90 min Tech Prep Masters	65 min **PE:** Chapter Assessment, Test Generator	30 min Chap. Assess. (A or B), Alt. Assess. (A or B), Test Generator	

PE: Pupil's Edition

Alternative Assessment

The following suggest alternative assessments for students who may benefit from a different type of assessment than the regular chapter quizzes and the mid-chapter/end-of-chapter test. Visit the HRW web site to get additional Alternative Assessment material.

internet connect

Alternative Assessment
Go To: **go.hrw.com**
Keyword: **MB1 Alt Assess**

Performance Assessment

1. Let $f(x) = \frac{2x+6}{5x-2}$.

a. For what value(s) of x is f undefined? What are the equations for the horizontal and vertical asymptotes?

b. Graph the function, showing the asymptotes.

c. Graph $g(x) = \frac{(5x-2)^2}{5x-2}$. How is the graph of g different from that of f?

2. a. Find the product $\frac{x^2-6x+9}{x^2-1} \cdot \frac{2x-2}{x-3}$. Simplify your answer.

State all necessary restrictions on x.

b. Find the difference $\frac{2a+1}{2a-1} - \frac{a-1}{a+1}$. Simplify your answer.
State all necessary restrictions on a.

c. Explain how addition of rational expressions is similar to addition of rational numbers.

Portfolio Project

Suggest that students choose one of the following projects for inclusion in their portfolios.

1. *Boyle's law, Charles's law* and the *ideal-gas equation* relate pressure, temperature, volume, and quantity of contained gas. Research these laws and equations. What types of variation exist among the quantities? Present your research in an illustrated report.

2. The quotient $\frac{f(x+h)-f(h)}{h}$, where $h > 0$, is called the *difference quotient* for function f. If $P(x, f(x))$ and $Q(x+h, f(x+h))$ are on the graph of f, then the difference quotient gives the slope of $\overleftrightarrow{PQ}$.

a. Write and simplify the difference quotient if $f(x) = 2x - 3$.

b. Explain how the difference quotient tells you that a given line has exactly one slope.

internet connect

The table below identifies the pages in this chapter that contain internet and technology information.

Content Links

Activities Online	**page 541**
Portfolio Extensions	**pages 488, 497, 519**
Homework Help Online	**pages 486, 495, 502, 509, 518, 525, 534, 542**
Graphic Calculator Support	**page 554**

Resource Links

Parents can go online and find concepts that students are learning—lesson by lesson—and questions that pertain to each lesson, which facilitate parent-student discussion.

Go To: **go.hrw.com**
Keyword: **MB1 Parent Guide**

Technical Support

The following may be used to obtain technical support for any HRW software product.

Online Help: **www.hrwtechsupport.com**
e-mail: **tschrw@hrwtechsupport.com**
HRW Technical Support Center: **(800)323-9239**
7 AM to 10 PM Monday through Friday Central Time

Visit the HRW math web site at: **www.hrw.com/math**

Technology

Lesson Suggestions and Calculator Examples

(Keystrokes are based on a TI-83 calculator.)

Lesson 8.1 Inverse, Joint, and Combined Variation

Point out to students that any ordered pair that belongs to an inverse-variation relationship of the form $y = \frac{k}{x}$ can be used to calculate the constant of variation, k. That is, if (x_1, y_1) is such an ordered pair, then $y_1 = \frac{k}{x_1}$ and $k = x_1y_1$. Students can generate a table of y-values for a set of x-values on a calculator. Make sure that students know how to

- enter an expression using Y= ,
- set up a table by using 2nd WINDOW , and
- display the table by using 2nd GRAPH .

Lesson 8.2 Rational Functions and Their Graphs

In this lesson, students will learn about rational functions that are based on polynomial functions. Have students graph the function $f(x) = x$ and its reciprocal $g(x) = \frac{1}{x}$. The display at right shows the difference between the graphs of rational functions and the graphs of polynomial functions.

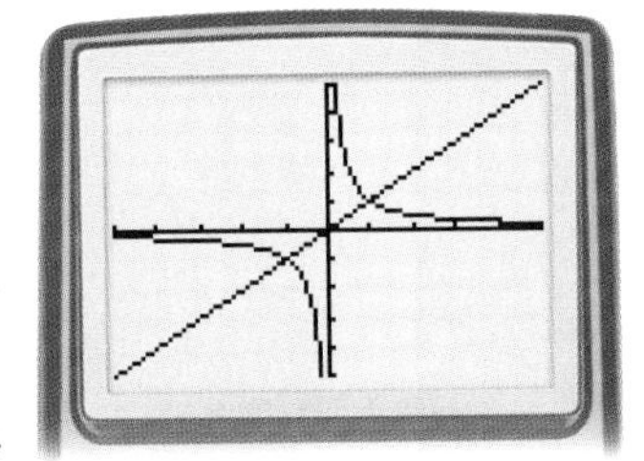

Lesson 8.3 Multiplying and Dividing Rational Expressions

Although most of Lesson 8.3 requires students to do calculations by hand, graphics calculators can be used also. In the problem-solving discussion on page 500, students are reminded how to use a graph to determine whether a polynomial expression can be factored.

Lesson 8.4 Adding and Subtracting Rational Expressions

In Example 3 on page 507, students are shown how to check their work by determining whether the graphs of two expressions coincide. This is easiest to see if the thicknesses of the two lines are different. To select the style and thickness of each line, move the cursor to the left of each function in the **Y=** editor and press ENTER until the desired style is shown.

Lesson 8.5 Solving Rational Equations and Inequalities

If $f(x) = g(x)$, then $f(x) - g(x) = 0$. Graphing $f - g$, instead of graphing both $f(x)$ and $g(x)$, makes scanning the graph easier because students need to look for the zeros of one function.

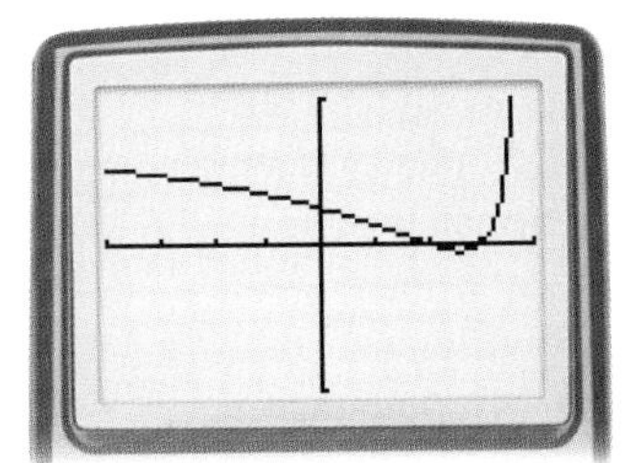

For example, the display at right shows the graph of $y = \frac{x}{x-6} - \frac{1}{x-4}$. The zeros are located at $x = 2$ and $x = 3$. Therefore, $\frac{x}{x-6} = \frac{1}{x-4}$ at these values.

Lesson 8.6 Radical Expressions and Radical Functions

Suggest that students look at each transformation separately. For example, consider $y = -2\sqrt{x+1} + 4$.

Graph each of the following steps:

$$\sqrt{x} \rightarrow \sqrt{x+1} \rightarrow 2\sqrt{x+1} \rightarrow -2\sqrt{x+1} \rightarrow -2\sqrt{x+1} + 4$$

Lesson 8.7 Simplifying Radical Expressions

Ask students to explore the following statement on a calculator:

$$\sqrt[n]{ab} = \sqrt[n]{a} \cdot \sqrt[n]{b}$$

The display at right shows the evaluation of the statement for $n = 2$ when a and b are positive numbers. Ask students to explore cases for $n = 2$ when a and b are negative numbers. Then have them use the same process to explore cases for $n = 3$ when a and b are negative and positive numbers.

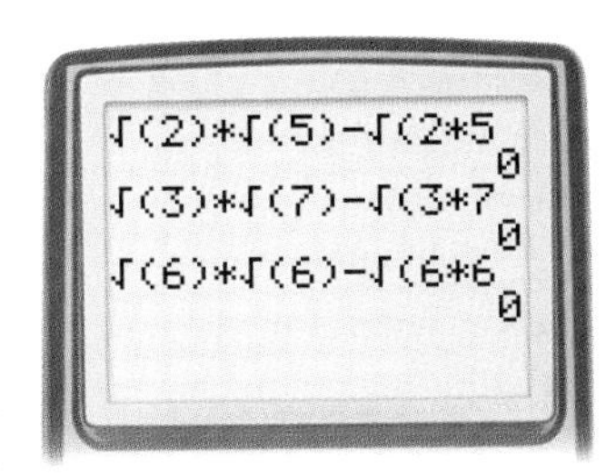

Lesson 8.8 Solving Radical Equations and Inequalities

Ask students to solve the equations below and to discuss whether a graphical approach is appropriate.

$\sqrt{x^2 - 2} = 4 \qquad \sqrt{x^2 - 2} = \sqrt[3]{x^3 + 3}$

For further information, refer to the
- technology discussions in the lessons.
- lesson-related teacher's commentary in the side columns of this *Teacher's Edition*.
- lesson-related *Student Technology Guide* masters.
- *HRW Technology Handbook*.

internet connect

For keystrokes of other graphing calculators models, visit the HRW web site at **go.hrw.com** and enter the keyword **MB1 CALC**.

About Chapter 8

Background Information

In Chapter 8, students work with rational and radical expressions. Students begin by looking at rational functions in the context of joint, inverse, and combined variation. Then students take a functional approach to rational equations and explore the graphs of such functions. Students learn to manipulate rational functions and to solve equations and inequalities involving rational expressions. Students then learn to simplify and manipulate radical expressions so that they can solve radical equations and inequalities.

CHAPTER RESOURCES

- Block-Scheduling Handbook
- Writing Activities for Your Portfolio
- Tech Prep Masters
- Long-Term Project
- Assessment Resources:
  - Mid-Chapter Assessment
  - Chapter Assessments
  - Alternative Assessments
- Test and Practice Generator
- Technology Handbook

Chapter Objectives

- Identify inverse, joint, and combined variations, find the constant of variation, and write an equation for the variation. [**8.1**]
- Solve real-world problems involving inverse, joint, or combined variation. [**8.1**]

8 Rational Functions and Radical Functions

IN THIS CHAPTER, YOU WILL STUDY RATIONAL functions and radical functions. Rational functions are ratios of polynomials. Radical functions are formed from the *n*th roots of numbers. The square-root function studied in Chapter 5 is an example of a radical function. Rational functions and radical functions have applications in physics, chemistry, engineering, business, economics, and many other fields.

Lessons

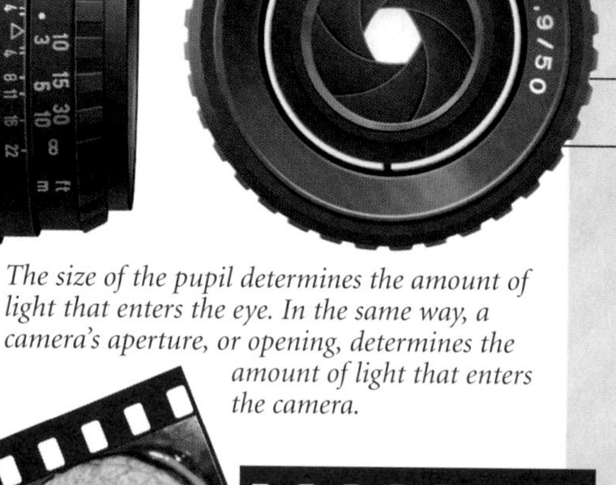

The size of the pupil determines the amount of light that enters the eye. In the same way, a camera's aperture, or opening, determines the amount of light that enters the camera.

About the Photos

A pendulum is a bob attached to an arm that is suspended from a fixed point so that the pendulum can swing freely under the action of gravity. A pendulum's frequency is the number of times it returns to the same position in a given time period. This frequency does not depend on the mass of the object but instead depends on the length of the pendulum and on the acceleration due to gravity. The formula for the frequency of a pendulum is given by $f = \frac{1}{2\pi}\sqrt{\frac{g}{L}}$, where g is acceleration due to gravity and L is the length of the pendulum. In the English system, $g = 32$ ft/s^2 and L is measured in feet. So, the frequency of a pendulum that is 8 in., or $\frac{2}{3}$ ft, long is $\frac{1}{2\pi}\sqrt{\frac{32}{\frac{2}{3}}}$, or about 1.1 cycles per second. Frictional forces can cause the pendulum to lose energy, so energy must be added continually if the pendulum is to swing for an extended time. The same idea can be applied to a child's swing—you must keep pushing the swing to keep it moving.

Galileo is credited as the first person to notice that the motion of a pendulum depends only upon its length.

About the Chapter Project

Finding an average is something that most people can do almost instinctively. Currency exchange, hourly wage, car mileage, and average speed are common daily topics of discussion. You can measure an average in various ways. Two common averages are the arithmetic mean and the harmonic mean. In the Chapter Project, *Means to an End,* you will use the data provided to determine the most appropriate average.

After completing the Chapter Project, you will be able to do the following:

- Find the arithmetic mean and the harmonic mean of a set of data.
- Determine the relationship between the arithmetic and harmonic mean.
- Determine which of the averages—arithmetic mean, harmonic mean, or weighted harmonic mean—best represents a data set.

About the Portfolio Activities

Throughout the chapter, you will be given opportunities to complete Portfolio Activities that are designed to support your work on the Chapter Project.

- Exploring a historical representation of harmonic means is included in the Portfolio Activity on page 488.
- Exploring a geometric representation of harmonic means is included in the Portfolio Activity on page 497.
- Extending the definition of harmonic mean to n numbers is included in the Portfolio Activity on page 511.
- Using harmonic means to find average speeds is included in the Portfolio Activity on page 519.

internet connect go.hrw.com

Chapter Internet Features and Online Activities

Lesson	Keyword	Page
8.1	MB1 Homework Help	486
	MB1 MeanMean	488
8.2	MB1 Homework Help	495
	MB1 Rational Graph	497
8.3	MB1 Homework Help	502
8.4	MB1 Homework Help	509
8.5	MB1 Homework Help	518
	MB1 QInequal	519
8.6	MB1 Homework Help	525
8.7	MB1 Homework Help	534
8.8	MB1 Homework Help	542
	MB1 Relativity	541

- Identify and evaluate rational functions. **[8.2]**
- Graph a rational function, find its domain, write equations for its asymptotes, and identify any holes in its graph. **[8.2]**
- Multiply and divide rational expressions. **[8.3]**
- Simplify rational expressions, including complex fractions. **[8.3]**
- Add and subtract rational expressions. **[8.4]**
- Solve a rational equation or inequality by using algebra, a table, or a graph. **[8.5]**
- Solve problems by using a rational equation or inequality. **[8.5]**
- Analyze the graphs of radical functions, and evaluate radical expressions. **[8.6]**
- Find the inverse of a quadratic function. **[8.6]**
- Add, subtract, multiply, divide, and simplify radical expressions. **[8.7]**
- Rationalize a denominator. **[8.7]**
- Solve radical equations. **[8.8]**
- Solve radical inequalities. **[8.8]**

Portfolio Activities appear at the end of Lessons 8.1, 8.2, 8.4, and 8.5. Each serves as preparation for the Chapter Project. The Portfolio Activities as well as the Chapter Project Activities are appropriate for inclusion in the student's portfolio. Students should be encouraged to include in their portfolios any other work in which they feel a sense of pride or a sense of accomplishment.

Prepare

NCTM PRINCIPLES & STANDARDS 1–4, 6–10

QUICK WARM-UP

The variable y varies directly as x. Find the constant of variation, k, and write an equation of direct variation that relates the two variables.

1. $y = -6$ when $x = 3$
   $k = -2$; $y = -2x$
2. $y = 3$ when $x = -6$
   $k = -0.5$; $y = -0.5x$
3. $y = 3.75$ when $x = 0.3$
   $k = 12.5$; $y = 12.5x$

The variable a varies directly as b.

4. If a is 36 when b is –9, find a when b is 12.
   $a = -48$
5. If a is 36 when b is –9, find b when a is 12.
   $b = -3$

Also on Quiz Transparency 8.1

Teach

Why Ask students to discuss situations in which an increase in one quantity corresponds to a decrease in another quantity. For example, as supply increases, price decreases, and as speed increases, the time to travel a given distance decreases.

Inverse, Joint, and Combined Variation

Objectives

- Identify inverse, joint, and combined variations, find the constant of variation, and write an equation for the variation.
- Solve real-world problems involving inverse, joint, or combined variation.

Why *Inverse, joint, and combined variation relationships occur frequently in the real world. For example, the per-person cost of renting property decreases as the number of people paying increases.*

APPLICATION
BUSINESS

A ranch can be rented for rodeos for a flat fee of \$6000 per day. Janet Flores, a promoter, wants to rent the ranch for a 1-day rodeo. She knows that an admission price of \$20 or less will be acceptable to the general public.

The table at right indicates what the per-person charge, c, would be if n people attend the rodeo. From the table, any number of attendees greater than 300 people will enable the promoter to cover the ranch's rental fee.

Notice that as the number of attendees increases, the per-person charge decreases. That is, as n increases, c decreases, which is characteristic of an *inverse-variation relationship.*

n	$c = \frac{6000}{n}$
300	20.00
320	18.75
340	≈17.65
360	≈16.67
380	≈15.79
400	15.00
420	≈14.29
440	≈13.64
460	≈13.04
480	12.50
500	12.00
520	≈11.54
540	≈11.11

Inverse Variation

Two variables, x and y, have an **inverse-variation** relationship if there is a nonzero number k such that $xy = k$, or $y = \frac{k}{x}$. The **constant of variation** is k.

Alternative Teaching Strategy

USING TABLES Ask students to recall the definition of a multiplicative inverse. **two numbers whose product is 1** Have students work in pairs to make a table, labeling one column x and the other y, where x and y are multiplicative inverses of each other. Then have students plot each ordered pair on a piece of graph paper and sketch a curve through the points. Have students write a general expression for y in terms of x. $y = \frac{1}{x}$ Have students graph this function on a graphics calculator and compare it with the one they drew on graph paper.

Next have each pair of students make a table in which the product of x and y is a nonzero constant other than 1. Make sure that students choose a variety of constants, including fractions and negative numbers. Have students graph these ordered pairs on the same graph as above, using a pencil or pen of a different color. Ask each pair of students to describe their graph, compare it with the graph of the *parent function*, $y = \frac{1}{x}$, and write the function represented by each set of data.

Activity: Exploring Inverse Variation

You will need: no special materials

1. Copy and complete the table below for $y = \frac{1}{x}$.

x	$\frac{1}{10}$	$\frac{1}{4}$	$\frac{1}{2}$	1	2	3	4	5	6
y									
xy									

PROBLEM SOLVING

2. **Look for a pattern.** What can you say about the values of y as the values of x increase? What can you say about the values of y as the values of x decrease?
3. Repeat Step 1 for $y = \frac{2}{x}$ and $y = \frac{4}{x}$. Do you think that the pattern you identified in Step 2 will also be true for $y = \frac{3}{x}$? Repeat Step 1 for $y = \frac{3}{x}$.
4. CHECKPOINT ✔ Describe the behavior of $y = \frac{k}{x}$, where $k > 0$, as x increases and as x decreases.
5. Let $y = \frac{k}{x}$, where $k > 0$. What happens when $x = 0$?

The functions $y = \frac{1}{x}$, $y = \frac{2}{x}$, and $y = \frac{4}{x}$ all represent inverse-variation relationships. The typical graphical behavior of such a relationship can be observed at right.

Notice that the graphs of $y = \frac{2}{x}$ and $y = \frac{4}{x}$ are vertical stretches of the graph of the parent function $y = \frac{1}{x}$. That is, if $f(x) = \frac{1}{x}$, $g(x) = \frac{2}{x}$, and $h(x) = \frac{4}{x}$, then $g(x) = 2 \cdot f(x)$ and $h(x) = 4 \cdot f(x)$.

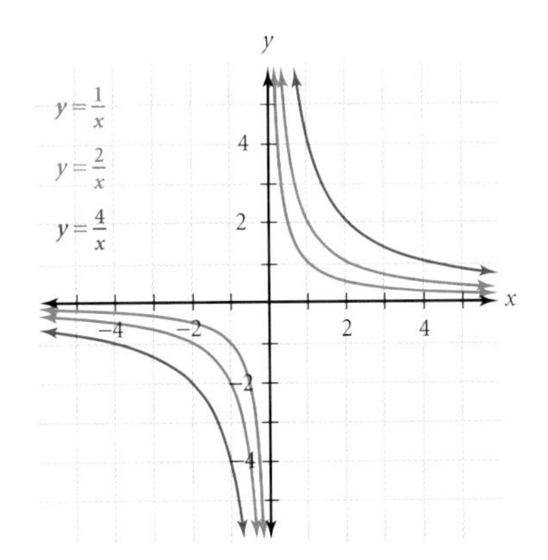

CHECKPOINT ✔ Describe where the graph of $y = \frac{3}{x}$ would lie in relation to the graphs shown above.

If you know one ordered pair (x, y) in a given inverse-variation relationship, you can find the constant of variation, k, and write the inverse-variation equation, $y = \frac{k}{x}$. In real-world situations, the x- and y-values are usually positive, and as the x-values increase, the y-values decrease. This is shown in Example 1 on the next page.

Interdisciplinary Connection

PHYSICS If a seesaw is balanced, each person's distance from the fulcrum varies inversely as his or her weight. If a 120-pound person sits 5 feet from a seesaw fulcrum, where must a 100-pound person sit to balance the seesaw?

$W_1d_1 = W_2d_2$

$120(5) = 100d_2$

$d_2 = 6$ **ft from the fulcrum**

What is the constant of variation? $k = 600$

Activity Notes

In the Activity, students use a graphics calculator to explore the behavior of functions in the form of $y = \frac{k}{x}$. Students discover that when $k > 0$, y decreases as x increases.

Teaching Tip

TECHNOLOGY To display the table on a TI-83 or TI-82 graphics calculator, let **Y1=1/X**, press 2nd WINDOW to access the **TABLE SETUP** screen, and set **Indpnt** to **Ask** and **Depend** to **Auto.** The values of **TblStart** (or **TblMin**) and **ΔTbl** are not used when **Indpnt** is set to **Ask.** Then press 2nd GRAPH and enter the x-values in the **X** column. The y-values will appear in the **Y1** column.

CHECKPOINT ✔

4. For $k > 0$, as x increases, $\frac{k}{x}$ decreases; as x decreases, $\frac{k}{x}$ increases.

Math CONNECTION

TRANSFORMATIONS When a function is multiplied by a positive number, the graph is stretched or compressed vertically.

CHECKPOINT ✔

The graph of $y = \frac{3}{x}$ would lie between the graphs of $y = \frac{2}{x}$ and $y = \frac{4}{x}$.

ADDITIONAL
EXAMPLE 1

The variable y varies inversely as x, and $y = 22.5$ when $x = 6.2$.

a. Find the constant of variation, and write an equation for the relationship.
$k = 139.5$; $y = \frac{139.5}{x}$

b. Find y when x is 0.1, 0.2, 0.3, 0.4, and 0.5.
1395, 697.5, 465, 348.75, 279

Teaching Tip

TECHNOLOGY When using a TI-83 or TI-82 graphics calculator to produce the table, set **Y1=60.75/X**, press 2nd WINDOW, and use the following settings:

TblStart=0
ΔTbl=.5
Indpnt: Auto Ask
Depend: Auto Ask

Press 2nd GRAPH to view the table.

TRY THIS
$k = 780; y = \frac{780}{x}; 520; 173.3\overline{3}; 97.5; 62.4; \approx 55.7$

ADDITIONAL
EXAMPLE 2

a. Write an equation for the volume, V, of the rectangular prism shown below. Identify the type of variation, the variable(s) involved, and the constant of variation.

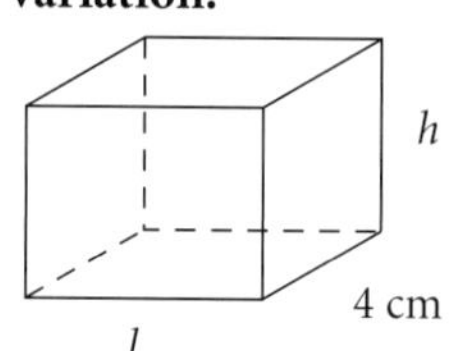

$V = 4lh$; the volume varies jointly as the length, l, and the height, h, and the constant of variation is 4.

b. Find the volume if the length is 5.5 cm and the height is 3 cm. 66 cm^3

EXAMPLE 1 **The variable y varies inversely as x, and $y = 13.5$ when $x = 4.5$.**
a. Find the constant of variation, and write an equation for the relationship.
b. Find y when x is 0.5, 1, 1.5, 2, and 2.5.

SOLUTION

a.
$$xy = k$$
$$(4.5)(13.5) = k$$
$$k = 60.75$$

Thus, an equation for the relationship is $y = \frac{60.75}{x}$.

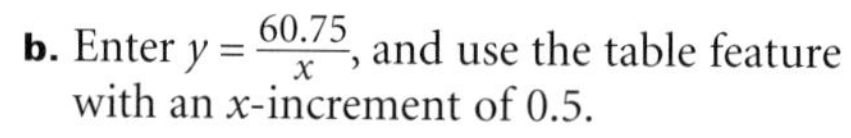

TECHNOLOGY
GRAPHICS CALCULATOR
Keystroke Guide, page 554

b. Enter $y = \frac{60.75}{x}$, and use the table feature with an x-increment of 0.5.

From the table, you can see that the x-values of 0.5, 1, 1.5, 2, and 2.5 have corresponding y-values of 121.5, 60.75, 40.5, 30.375, and 24.3, respectively.

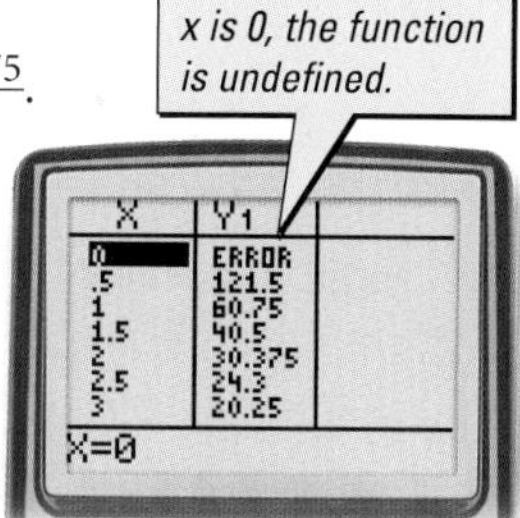

TRY THIS The variable y varies inversely as x, and $y = 120$ when $x = 6.5$. Find the constant of variation, and write an equation for the relationship. Then find y when x is 1.5, 4.5, 8, 12.5, and 14.

Recall from Lesson 1.4 that in a direct-variation relationship, y varies directly as x for any nonzero value of k such that $y = kx$. In *joint variation,* one quantity varies directly as two quantities.

Joint Variation

If $y = kxz$, then y **varies jointly** as x and z, and the constant of variation is k.

EXAMPLE 2 **Refer to the rectangular prism shown at right.**
a. Write an equation for the volume of the prism, and identify the type of variation and the constant of variation.
b. Find the volume of the prism if the length of the base is 4 inches and the width of the base is 2 inches.

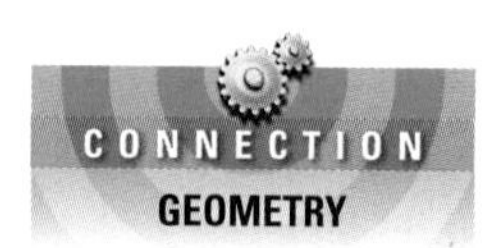

CONNECTION
GEOMETRY

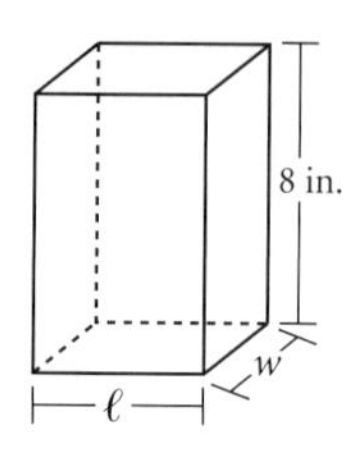

SOLUTION

a. The volume of a rectangular prism is $V = \ell wh$.
$$V = \ell wh$$
$$V = \ell w(8), \text{ or } 8\ell w$$
Volume varies jointly as the length, ℓ, and width, w. The constant of variation is 8.

b. $V = 8\ell w = 8(4)(2) = 64$ The volume is 64 cubic inches.

Enrichment

Ask students to draw the reflection of the graph of $y = \frac{1}{x}$ across the x-axis on a sheet of graph paper and to describe the graph. **The branches of the graph are in the second and fourth quadrants.** Have students write an equation for the reflected graph. $y = -\frac{1}{x}$ Ask students to discuss the effects on the graph of $y = \frac{k}{x}$ when $k < 0$. **The graph is reflected across the x-axis.**

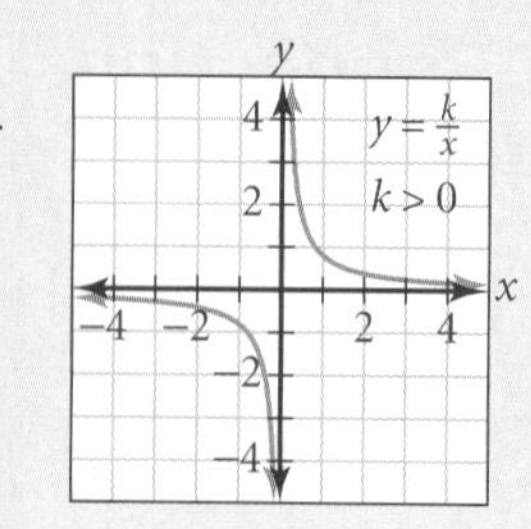

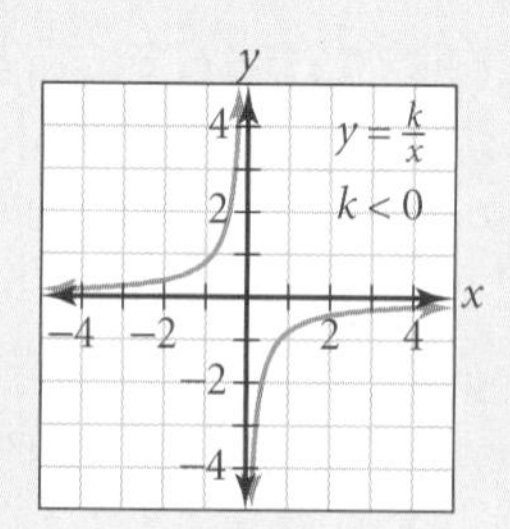

TRY THIS

a. Write an equation for the volume of a rectangular prism whose base has a length of 12 inches. Identify the type of variation and the constant of variation.
b. Find the volume of the prism if the width of the base is 2 inches and the height of the prism is 4 inches.

If $y = kx^2$, where k is a nonzero constant, then y varies directly as the square of x. Many geometric relationships involve this type of joint variation, as shown in Example 3.

EXAMPLE 3

a. Write an equation to represent the area, A, of the isosceles right triangle shown at right. Identify the type of variation and the constant of variation.
b. Find the area of the triangle when x is 1.5, 2.5, 3.5, and 4.5.

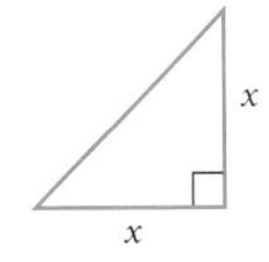

SOLUTION

a. The equation for the area is $A = \frac{1}{2}bh$, where b is the base and h is the height. Since the base is x and the height is x, A varies directly as the square of x. The constant of variation is $\frac{1}{2}$.

$$A = \frac{1}{2}bh$$
$$A = \frac{1}{2}(x)(x)$$
$$A = \frac{1}{2}x^2$$

TECHNOLOGY
GRAPHICS CALCULATOR
Keystroke Guide, page 554

b. Enter $y = \frac{1}{2}x^2$ into a graphics calculator, and use the table feature. Begin with $x = 1.5$, and use an x-increment of 1.

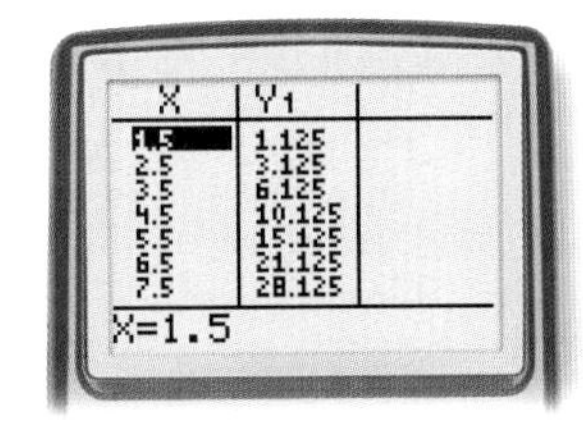

From the table, you can read the areas.

x	Area
1.5	1.125 square units
2.5	3.125 square units
3.5	6.125 square units
4.5	10.125 square units

TRY THIS

a. Write the formula for the area, A, of a circle whose radius is r. Identify the type of variation and the constant of variation.
b. Find the area of the circle when r is 1.5, 2.5, 3.5, and 4.5.

CRITICAL THINKING

Let y vary as the square of x. How does y change when x is doubled, tripled, and quadrupled? Justify your response.

Math CONNECTION

GEOMETRY In Example 2, students see that if the height of a rectangular prism is held constant at 8 inches, then the volume of the prism varies jointly as the length and the width of the base.

TRY THIS

a. $V = 12wh$; volume varies jointly as the width, w, and the height, h; $k = 12$.

b. 96 in.^{3}

ADDITIONAL EXAMPLE 

a. Write an equation to represent the surface area, *SA*, of the sphere shown below. Identify the type of variation, the variable(s) involved, and the constant of variation.

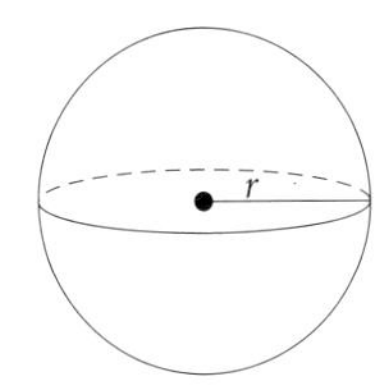

$SA = 4\pi r^2$; SA varies directly as r^2; $k = 4\pi$

b. Find the surface area of the sphere when r is 0.5, 1.5, 2.5, 3.5, and 4.5.
≈ 3.14; ≈ 28.27; ≈ 78.54; ≈ 153.94

Math CONNECTION

GEOMETRY Students write the area of an isosceles triangle as a function of x, which represents both the base and height of the triangle.

TRY THIS

a. $A = \pi r^2$; area varies directly as the square of r.

b. ≈ 7.07; ≈ 19.63; ≈ 38.48; ≈ 63.62

CRITICAL THINKING

y is multiplied by 4;
y is multiplied by 9;
y is multiplied by 16.

Inclusion Strategies

COOPERATIVE LEARNING In order to give students a better understanding of inverse variation, have them work in groups of six to perform the task described below. Each group will need a pack of index cards. Separate the cards into three stacks, and have three students count the cards while one of the other students records the time in seconds needed to count the cards. Then separate the cards into two stacks and have two students count the cards while another student times them and another records the time. Repeat this counting and timing procedure three more times with 1, 4, and 5 students counting the cards. Have students make a graph of their data, letting the horizontal axis represent the number of students counting and the vertical axis represent the time spent counting the cards. Students should see that it takes less time to count the cards as more students participate in the counting. Have them determine if the equation $y = \frac{6}{x}$ represents a good approximation of the relationship.

ADDITIONAL
EXAMPLE 4

A bicycle's pedal gear has 58 teeth and rotates at 70 revolutions per minute. A chain links the pedal gear to a rear-wheel gear that has 32 teeth and is attached to a 14-inch wheel. **At what speed, in miles per hour, is the bicycle traveling?**
≈5.3 mph

TRY THIS
≈8.5 mph

CHECKPOINT ✔
The rear-wheel gear will turn more slowly than the pedal gear.
60 teeth: ≈4.4 mph
80 teeth: ≈3.3 mph
100 teeth: ≈2.6 mph

CRITICAL THINKING
The speed of the rear-wheel gear is $\frac{n}{m}$ times the speed of the pedal gear.

When more than one type of variation occurs in the same equation, the equation represents a **combined variation.**

APPLICATION
MACHINES

The rotational speeds, s_A and s_B, of gear A with t_A teeth and gear B with t_B teeth are related as indicated below.

$$t_A s_A = t_B s_B$$

Gear B Gear A

The combined-variation equation for the rotational speed of gear B in terms of t_A, s_A, and t_B is as follows:

$$s_B = \frac{t_A s_A}{t_B}$$

EXAMPLE 4

A bicycle's pedal gear has 52 teeth and is rotating at 65 revolutions per minute. A chain links the pedal gear to a rear-wheel gear that has 18 teeth and is attached to a 26-inch wheel.

At what speed, in miles per hour, is the bicycle traveling?

SOLUTION

Let the pedal gear be gear A and the rear-wheel gear be gear B.

1. Find the rotational speed in revolutions per minute for gear B.

$$s_B = \frac{t_A s_A}{t_B}$$
$$s_B = \frac{(52)(65)}{18}$$
$$s_B \approx 188$$

The rear-wheel gear is rotating at about 188 revolutions per minute.

2. Convert revolutions per minute to miles per hour to find the speed of the bicycle.

A 26-inch wheel (including tire) has a circumference of 26π inches. Therefore, the bicycle is traveling at about $26\pi \times 188$ inches per minute. To convert to miles per hour, multiply by fractions equal to 1.

$$\frac{(26\pi)(188)\ \cancel{\text{in.}}}{1\ \cancel{\text{min}}} \times \frac{1\ \cancel{\text{ft}}}{12\ \cancel{\text{in.}}} \times \frac{1\text{ mi}}{5280\ \cancel{\text{ft}}} \times \frac{60\ \cancel{\text{min}}}{1\text{ hr}} \approx 14.5 \text{ miles per hour}$$

TRY THIS A bicycle's pedal gear has 46 teeth and is rotating at 55 revolutions per minute. If the pedal gear is linked to a rear-wheel gear that has 24 teeth and is attached to a 27-inch wheel, at what speed, in miles per hour, is the bicycle traveling?

CHECKPOINT ✔ Predict the effect of a rear-wheel gear with more teeth than the pedal gear. Verify your prediction by reworking Example 4 for a rear-wheel gear that has 60 teeth, 80 teeth, and 100 teeth.

CRITICAL THINKING Suppose that the pedal gear in Example 4 has n teeth, the rear-wheel gear has m teeth, and $n > m$. What is the rotational speed of the rear-wheel gear relative to the speed of the pedal gear?

Reteaching the Lesson

HANDS-ON STRATEGIES Have students work in groups to design and conduct a fulcrum experiment (as described in the Interdisciplinary Connection at the bottom of page 481). Materials could include two objects with different weights, a meter stick as a lever, a fulcrum made from a piece of strong cardboard that is bent in the middle or a small wooden pyramid, and a scale. Have students weigh one of the objects. After balancing two objects on the lever and determining the distance of the objects from the fulcrum, have students calculate the weight of the second object. Use the scale to confirm the weight. This experiment can be conducted several times, using different objects each time. Have students write the inverse-variation equation for each trial in the experiment. How are these equations similar?

Exercises

Communicate

1. Explain how direct variation and joint variation are related.

2. Let $xy = k$, where k is a constant greater than zero. Explain how you know that y must decrease when x increases.

CONNECTION

GEOMETRY Two formulas from geometry are shown below. Explain how to identify the type of variation found in each formula.

3. area, A, of a rhombus with diagonals of lengths p and q

$$A = \frac{1}{2}pq$$

4. area, A, of an equilateral triangle with sides of length s

$$A = \frac{\sqrt{3}}{4}s^2$$

5. Identify and explain how z varies as x, y, and d in the equation $z = \frac{2xy}{d^2}$.

6. Given the proportion $\frac{a}{b} = \frac{c}{d}$, write a related combined-variation equation. Explain how to form a different proportion from the combined-variation equation you wrote.

Guided Skills Practice

7. $k = 1980$; $y = \frac{1980}{x}$; 1320, 990, 792, 660, ≈565.71

7. The variable y varies inversely as x, and $y = 132$ when $x = 15$. Find the constant of variation, and write an equation for the relationship. Then find y when x is 1.5, 2.0, 2.5, 3.0, and 3.5. ***(EXAMPLE 1)***

CONNECTIONS

GEOMETRY Refer to the rectangular prism at right. ***(EXAMPLE 2)***

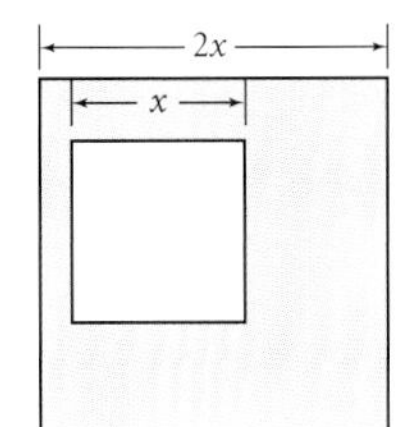

8. $V = 2.5\ell h$; volume varies jointly as length, ℓ, and height, h; $k = 2.5$

8. Write an equation for the volume of the prism, and identify the type of variation and the constant of variation.

9. 40 cm^3

9. Find the volume of the prism if the length of the base is 2 centimeters and the height is 8 centimeters.

GEOMETRY Refer to the figure at right, which consists of one square inside a larger square. ***(EXAMPLE 3)***

10. $A = 3x^2$; area varies directly as the square of the side, x; $k = 3$

10. Write an equation to represent the area, A, of the shaded region. Identify the type of variation and the constant of variation.

11. Find the area of the shaded region for x-values of 1.5, 2.5, 3.5, 4.5, 5.5, and 6.5.

APPLICATION

12. RECREATION A bicycle's pedal gear has 52 teeth and is rotating at 145 revolutions per minute. At what speed, in miles per hour, is the bicycle traveling if the rear-wheel gear has 28 teeth and the wheels are 26 inches in diameter? ***(EXAMPLE 4)*** **≈20.8 miles per hour**

Assess

Selected Answers

Exercises 7–12, 13–61 odd

ASSIGNMENT GUIDE	
In Class	1–12
Core	13–39 odd, 43, 45
Core Plus	14–40 even, 41, 42–46 even
Review	47–62
Preview	63

Extra Practice can be found beginning on page 940.

11. 6.75, 18.75, 36.75, 60.75, 90.75, and 126.75 sq. units

Error Analysis

In Exercises 13–28, check the students' variation equations to make sure that they are using the correct equation to find the unknowns. Remind them that joint variation involves a product and that inverse variation involves a quotient. Closely monitor Exercises 29–34 because writing combined-variation equations correctly is often difficult for students.

Practice and Apply

internet connect go.hrw.com

Homework Help Online
Go To: go.hrw.com
Keyword: **MB1 Homework Help** for Exercises 13–48

For Exercises 13–20, y varies inversely as x. Write the appropriate inverse-variation equation, and find y for the given values of x.

13. $y = 36$ when $x = 9$; x-values: 3, 4, 5, 6, and 7

14. $y = 10$ when $x = 5$; x-values: 2.5, 3, and 3.5 **$y = \frac{50}{x}$; 20, ≈16.7, ≈14.3**

15. $y = 0.5$ when $x = 8$; x-values: 5, 4, 3, 2, and 1 **$y = \frac{4}{x}$; 0.8, 1, ≈1.3, 2, 4**

16. $y = 0.25$ when $x = 0.3$; x-values: 0.1, 0.2, 0.3, and 0.4

17. $y = 14$ when $x = 8$; x-values: 10, 15, and 20 **$y = \frac{112}{x}$; 11.2, ≈7.47, 5.6**

18. $y = 2.25$ when $x = 1.5$; x-values: 0.1, 0.2, 0.3, and 0.4

19. $y = 1000$ when $x = 0.2$; x-values: 10^2, 10^3, and 10^4 **$y = \frac{200}{x}$; 2, 0.2, 0.02**

20. $y = 10^{-2}$ when $x = 50$; x-values: 10^2, 10^3, and 10^4 **$y = \frac{0.5}{x}$; 0.005, 0.0005, 0.00005**

13. $y = \frac{324}{x}$; 108, 81, 64.8, 54, ≈46.3
16. $y = \frac{0.075}{x}$; 0.75, 0.375, 0.25, 0.1875
18. $y = \frac{3.375}{x}$; 33.75, 16.875, 11.25, 8.4375

For Exercises 21–28, y varies jointly as x and z. Write the appropriate joint-variation equation, and find y for the given values of x and z.

21. $y = -108$ when $x = -4$ and $z = 3$; $x = 6$ and $z = -2$ **$y = 9xz$; −108**

22. $y = -315$ when $x = 5$ and $z = 9$; $x = -7$ and $z = 8$ **$y = -7xz$; 392**

23. $y = 15$ when $x = 9$ and $z = 1.5$; $x = 18$ and $z = 3$ **$y = \frac{10}{9}xz$; 60**

24. $y = 0.5$ when $x = 10$ and $z = 3$; $x = 1.8$ and $z = 6$ **$y = \frac{1}{60}xz$; 0.18**

25. $y = 120$ when $x = 8$ and $z = 20$; $x = 54$ and $z = 7$ **$y = \frac{3}{4}xz$; 283.5**

26. $y = 0.1$ when $x = 0.1$ and $z = 5$; $x = 0.2$ and $z = 0.4$ **$y = 0.2xz$; 0.016**

27. $y = 10^4$ when $x = 10^2$ and $z = 10^1$; $x = 10^5$ and $z = 10^2$ **$y = 10xz$; 10^8**

28. $y = 2 \times 10^5$ when $x = 10^1$ and $z = 10^3$; $x = 3 \times 10^4$ and $z = 1.5 \times 10^3$ **$y = 20xz$; 9×10^8**

For Exercises 29–34, z varies jointly as x and y and inversely as w. Write the appropriate combined-variation equation, and find z for the given values of x, y, and w.

29. $z = 3$ when $x = 3$, $y = -2$, and $w = -4$; $x = 6$, $y = 7$, and $w = -4$

30. $z = 10$ when $x = 5$, $y = -2$, and $w = 3$; $x = 8$, $y = 6$, and $w = -12$

31. $z = 36$ when $x = 9$, $y = 10$, and $w = 15$; $x = 10$, $y = 18$, and $w = 5$

32. $z = 15$ when $x = 3$, $y = 4$, and $w = 9$; $x = 1.5$, $y = 20.5$, and $w = 5.4$

33. $z = 100$ when $x = 100$, $y = 7$, and $w = 2$; $x = 3.5$, $y = 24$, and $w = 27$

34. $z = 54$ when $x = 8$, $y = 10$, and $w = 1.5$; $x = 1.5$, $y = 2.4$, and $w = 3$

29. $z = \frac{2xy}{w}$; −21
30. $z = \frac{-3xy}{w}$; 12
31. $z = \frac{6xy}{w}$; 216
32. $z = \frac{11.25xy}{w}$; 64.0625
33. $z = \frac{2xy}{7w}$; $\frac{8}{9}$
34. $z = \frac{81xy}{80w}$; 1.215

If (x_1, y_1) and (x_2, y_2) satisfy $xy = k$, then $x_1y_1 = x_2y_2$. Find x or y as indicated.

35. $(x, 2.5)$ and $(6, 4)$ **$x = 9.6$**

36. $(4.5, y)$ and $(12, 6)$ **$y = 16$**

37. $(3.6, 5)$ and $(7.2, y)$ **$y = 2.5$**

38. $(3, y)$ and $(18, 6)$ **$y = 36$**

39. $(18, 2)$ and (y, y) **$y = \pm 6$**

40. (x, x) and $(5, 125)$ **$x = \pm 25$**

CHALLENGE

41. Show that if (x_1, y_1) and (x_2, y_2) satisfy the inverse-variation equation $xy = k$ and if x_1, y_1, x_2, and y_2 are nonzero, then $\frac{x_1}{x_2} = \frac{y_2}{y_1}$ and $y_2 = y_1\left(\frac{x_1}{x_2}\right)$.

41. If $x_1y_1 = k$ and $x_2y_2 = k$, then by substitution, $x_1y_1 = x_2y_2$. Dividing both sides by x_2y_1 gives $\frac{x_1}{x_2} = \frac{y_2}{y_1}$. Multiplying both sides by y_1 gives $y_2 = y_1\left(\frac{x_1}{x_2}\right)$.

Practice

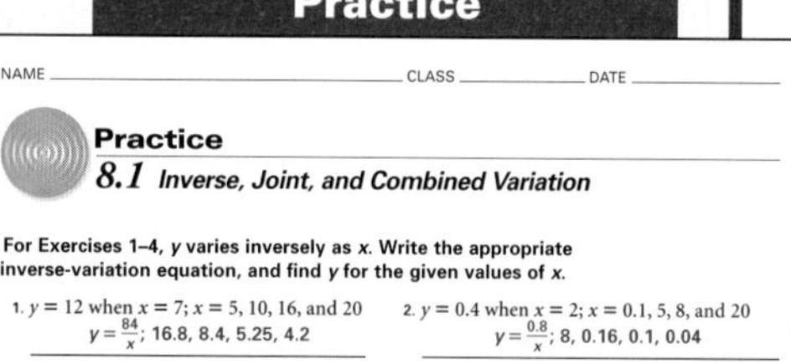

NAME ______ CLASS ______ DATE ______

Practice
8.1 ***Inverse, Joint, and Combined Variation***

For Exercises 1–4, y varies inversely as x. Write the appropriate inverse-variation equation, and find y for the given values of x.

1. $y = 12$ when $x = 7$; $x = 5, 10, 16$, and 20 — $y = \frac{84}{x}$; 16.8, 8.4, 5.25, 4.2
2. $y = 0.4$ when $x = 2$; $x = 0.1, 5, 8$, and 20 — $y = \frac{0.8}{x}$; 8, 0.16, 0.1, 0.04
3. $y = 3\frac{1}{3}$ when $x = 45$; $x = 6, 15, 20$, and 60 — $y = \frac{150}{x}$; 25, 10, 7.5, 2.5
4. $y = 12$ when $x = 0.4$; $x = 0.5, 6, 10$, and 16 — $y = \frac{4.8}{x}$; 9.6, 0.8, 0.48, 0.3

For Exercises 5–8, y varies jointly as x and z. Write the appropriate joint-variation equation, and find y for the given values of x and z.

5. $y = 16$ when $x = 4$ and $z = 0.5$; $x = 2$ and $z = 0.25$ — $y = 8xz$; 4
6. $y = 120$ when $x = 2.5$ and $z = 2$; $x = 3$ and $z = 2$ — $y = 24xz$; 144
7. $y = 12$ when $x = 4$ and $z = 5$; $x = 6$ and $z = 3$ — $y = 0.6xz$; 10.8
8. $y = 192$ when $x = 2$ and $z = 3$ $x = 0.6$ and $z = 5$ — $y = 32xz$; 96

For Exercises 9–12, z varies jointly as x and y and inversely as w. Write the appropriate combined-variation equation, and find z for the given values of x, y, and w.

9. $z = 320$ when $x = 4$, $y = 10$, and $w = 2.5$; $x = 5$, $y = 6$, and $w = 8$ — $z = \frac{20xy}{w}$; 75
10. $z = 3.2$ when $x = 0.2$, $y = 8$, and $w = 4$; $x = 3$, $y = 6$, and $w = 16$ — $z = \frac{8xy}{w}$; 9
11. $z = 3.75$ when $x = 6$, $y = 12$, and $w = 48$; $x = 0.05$, $y = 40$, and $w = 0.5$ — $z = \frac{5xy}{2w}$; 10
12. $z = 4.8$ when $x = 0.2$, $y = 10$, and $w = 5$; $x = 2\frac{1}{3}$, $y = 5$, and $w = 8$ — $z = \frac{12xy}{w}$; 17.5

13. The *apothem* of a regular polygon is the perpendicular distance from the center of the polygon to a side. The area, A, of a regular polygon varies jointly as the apothem, a, and the perimeter, p. A regular triangle with an apothem of 3 inches and a perimeter of 31.2 inches has an area of 46.8 square inches. Find the constant of variation and write a joint-variation equation. Then find the area of a regular triangle with an apothem of 2.3 inches and a perimeter of 12 inches. — 0.5; $A = 0.5ap$; 13.8 in.^{2}

CONNECTION

42a. $y = 0.25x^2$
b. 144, 156.25, 169, 182.25, 196, 210.25, 225

43. 6 feet

42 **STATISTICS** Refer to the table and scatter plot below.

x	0	1	2	3	4	5	6
y	0.00	0.25	1.00	2.25	4.00	6.25	9.00

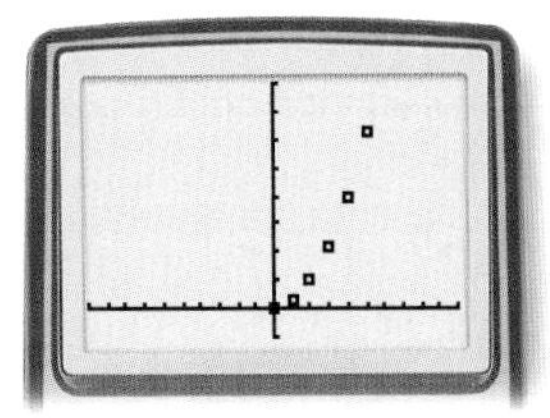

a. Use a graphics calculator to find an equation of the form $y = kx^2$ to represent the data.

b. Find y for x-values of 24, 25, 26, 27, 28, 29, and 30.

APPLICATIONS

43. PHOTOGRAPHY Under certain conditions of artificial light, the exposure time required to photograph an object varies directly as the candlepower of the light and varies inversely as the square of the object's distance from the light source. If the exposure time is 0.01 second when the light source is 6.00 feet from the object, how far from the object should the light source be when both the candlepower and the exposure time are doubled?

44. AVIATION The lifting force exerted by the atmosphere on the wings of an airplane in flight varies directly as the surface area of the wings and the square of the plane's airspeed. A small private plane has a cruising airspeed of 250 miles per hour. In order to obtain 3 times the lifting force, a new plane is designed with a wing surface area twice that of the older model. What cruising speed, to the nearest mile per hour, is planned for the new model? **306 miles per hour**

45. PHYSICS Heat loss, h, in calories per hour through a glass window varies jointly as the difference, d, between the inside and outside temperatures and as the area, A, of the window and inversely as the thickness, t, of the pane of glass. If the temperature difference is 30°F, there is a heat loss of 9000 calories per hour through a window with an area of 1500 cubic centimeters and thickness of 0.25 centimeter. Find the heat loss through a window with the same area and a thickness of 0.2 centimeter when the temperature difference is 15°F. **5625 calories**

Double-crested cormorant

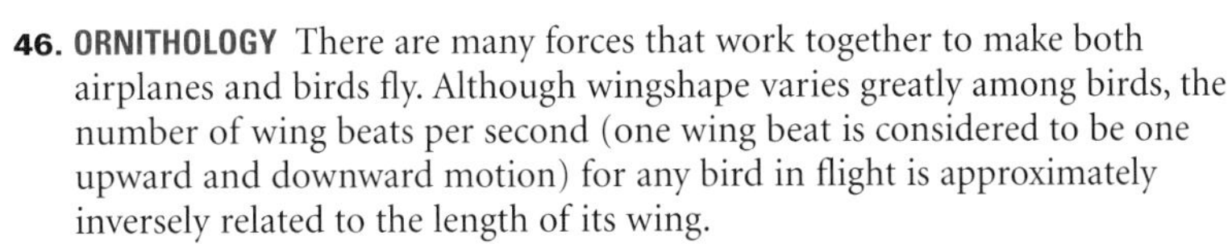

46. ORNITHOLOGY There are many forces that work together to make both airplanes and birds fly. Although wingshape varies greatly among birds, the number of wing beats per second (one wing beat is considered to be one upward and downward motion) for any bird in flight is approximately inversely related to the length of its wing.

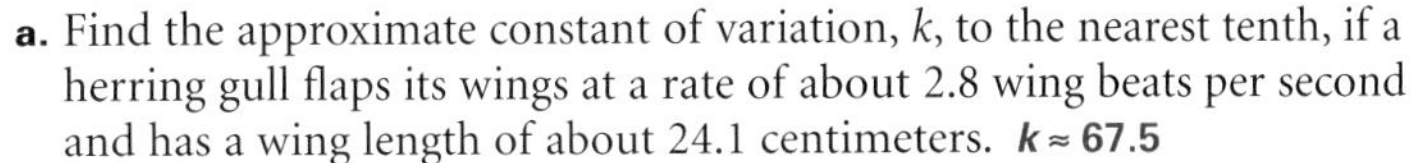

a. Find the approximate constant of variation, k, to the nearest tenth, if a herring gull flaps its wings at a rate of about 2.8 wing beats per second and has a wing length of about 24.1 centimeters. **$k \approx 67.5$**

b. Use the approximate value of k from part **a** to find the approximate number of wing beats per second for a starling that has a wing length of about 13.2 centimeters. **5.1 beats/s**

c. Use the approximate value of k from part **a** to find the approximate wing length of a cormorant who flaps its wings at a rate of about 1.9 wing beats per second. **35.5 cm**

Starling

Student Technology Guide

NAME ______ CLASS ______ DATE ______

Student Technology Guide
***8.1** Inverse, Joint, and Combined Variation*

You can use the table feature of a graphics calculator to test for relationships between variables.

Example: Identify the type of relationship represented by the data. Write an equation for the relationship between x and y.

x	1	2	3	4	6	12
y	5	20	45	80	180	720

- First enter the data. Press STAT 1:Edit... ENTER.
- Enter the x-values into the L1 column.
- Use ► to move to L2. Enter the y-values into the L2 column.
- Because y increases as x increases, you may suspect a direct variation, where $y = kx$, or $\frac{y}{x} = k$. To see whether $\frac{y}{x}$ is constant, make each number in L3 the quotient of the corresponding numbers in L1 and L2. Use ► to move to L3 and ▲ to move to the L3 label. Press 2nd 2 ÷ 2nd 1 ENTER. As seen at left below, the numbers in L3 are not constant.
- A second look at the data shows that y increases more quickly than x, suggesting that $y = kx^2$ may be a better guess. To test this, move to the L3 label and press 2nd 2 ÷ 2nd 1 x^2 ENTER.

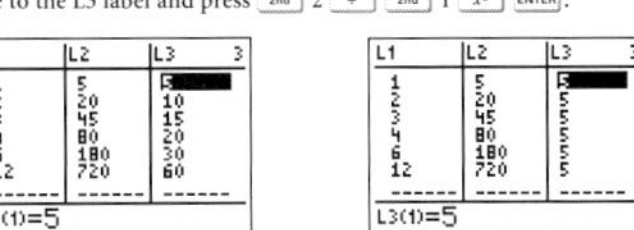

As shown above at right, y varies directly as the square of x. An equation for the relationship is $y = 5x^2$.

Identify the type of relationship represented by each data set. Write an equation for the relationship between the variables.

1.

x	2	3	6	9
y	54	36	18	12

inverse variation; $xy = 108$

2.

x	2	3	6	8
y	4	13.5	108	256

y varies directly as the cube of x; $y = \frac{x^3}{2}$

Look Beyond

In Exercise 63, students explore the graph of a shifted rational function that has a vertical asymptote at $x = -2$. Students will learn more about the graphs of rational functions and restrictions on their domains in Lesson 8.2.

ALTERNATIVE

Assessment

Portfolio Activity

The Portfolio Activity can be used as preparation for the Chapter Project or as a separate activity. In the Portfolio Activity on this page, students compare harmonic and arithmetic means. The harmonic mean involves rational expressions. In the activity, students look at these two means geometrically.

Answers to Portfolio Activities can be found in Additional Answers of the Teacher's Edition.

50. $\frac{b^3d}{a^2c^5}$
51. $\frac{1}{x^{18}}$
52. (0, 5); $x = 0$
53. (−1, −4); $x = -1$
54. $\left(-\frac{5}{2}, \frac{49}{4}\right)$; $t = -\frac{5}{2}$
55. (0, 2); $x = 0$
56. $\left(-\frac{1}{2}, \frac{3}{4}\right)$; $t = -\frac{1}{2}$
57. $\left(\frac{3}{4}, \frac{7}{8}\right)$; $t = \frac{3}{4}$

internet connect

Portfolio Extension
Go To: **go.hrw.com**
Keyword:
MB1 MeanMean

Look Back

Rewrite each expression with positive exponents only. *(LESSON 2.2)*

47. x^{-1} $\frac{1}{x}$ **48.** ab^{-3} $\frac{a}{b^3}$ **49.** $\left(\frac{x}{y}\right)^{-2}$ $\frac{y^2}{x^2}$ **50.** $a^{-2}b^3c^{-5}d$ **51.** $[(x^{-3})^{-2}]^{-3}$

Identify the vertex and the axis of symmetry in the graph of each function. *(LESSON 5.4)*

52. $f(x) = -3x^2 + 5$ **53.** $h(x) = x^2 + 2x - 3$ **54.** $b(t) = -t^2 - 5t + 6$

55. $g(x) = x^2 + 2$ **56.** $d(t) = t^2 + t + 1$ **57.** $r(t) = 2t^2 - 3t + 2$

Give the degree of each polynomial. *(LESSON 7.1)*

58. $3x^5 - 2x^4 + x^2 - 1$ 5 **59.** $2 - 5x + 7x^2 - x^3$ 5 **60.** $-5x^3 - x^4 + 1$ 4

Describe the end behavior of each function. *(LESSON 7.2)*

61. $f(x) = x^4 + 2x^3 - x^2 + 2$
rises on left and right

62. $g(x) = -3x^3 - 2x^2 + x - 4$
rises on left, falls on right

Look Beyond

63. Make a table of values for $f(x) = \frac{1}{x + 2}$ for values of x near −2. Use this table to describe the values of f near $x = -2$.

As early as 300 C.E., Pappus of Alexandria represented the arithmetic mean and the harmonic mean geometrically by constructing the figure shown below. In the figure, arc ADC is a semicircle of circle O.

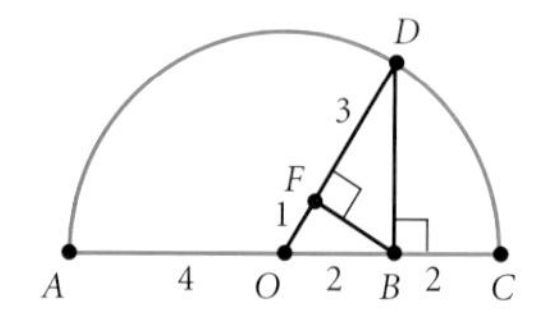

For any two numbers a and b, the arithmetic mean is $\frac{a+b}{2}$.

For any two nonzero numbers a and b, the harmonic mean is $\frac{2}{\frac{1}{a} + \frac{1}{b}}$.

1. Find the arithmetic mean and the harmonic mean of AB and BC.
2. Identify the side of a triangle in the figure whose length equals the arithmetic mean that you found in Step 1.
3. Identify the side of a triangle in the figure whose length equals the harmonic mean that you found in Step 1.

WORKING ON THE CHAPTER PROJECT

You should now be able to complete Activity 1 of the Chapter Project.

63. As x approaches −2 from the left, the values of f decrease rapidly.

x	−3	−2.9	−2.8	−2.7	−2.6	−2.5	−2.4	−2.3	−2.2	−2.1
$f(x)$	−1	−1.111	−1.25	−1.429	−1.667	−2	−2.5	−3.333	−5	−10

As x approaches −2 from the right, the values of f increase rapidly.

x	−1	−1.1	−1.2	−1.3	−1.4	−1.5	−1.6	−1.7	−1.8	−1.9
$f(x)$	1	1.1111	1.25	1.4286	1.6667	2	2.5	3.3333	5	10

Rational Functions and Their Graphs

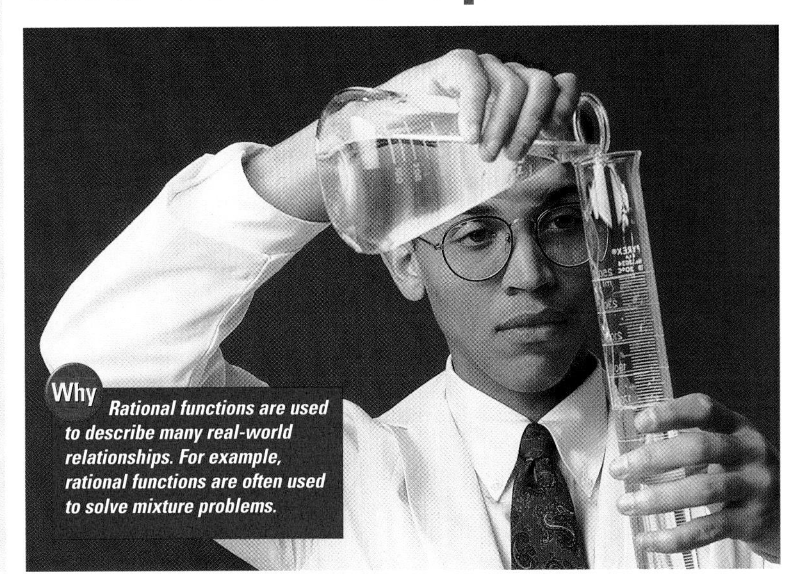

Why *Rational functions are used to describe many real-world relationships. For example, rational functions are often used to solve mixture problems.*

Objectives

- Identify and evaluate rational functions.
- Graph a rational function, find its domain, write equations for its asymptotes, and identify any holes in its graph.

Dane is a chemist who is varying the salt concentration of a solution. He can use a *rational function* to represent the salt concentration of the solution.

EXAMPLE 1 **Dane begins with 65 milliliters of a 10% saline solution. He adds x milliliters of distilled water to the container holding the saline solution.**

APPLICATION
CHEMISTRY

a. Write a function, C, that represents the salt concentration of the solution.
b. What is the salt concentration of the solution if 100 millimeters of distilled water is added?

SOLUTION

a. Original concentration of salt solution:

$$10\% \text{ of } 65 = 6.5 \quad \frac{6.5}{65} \begin{array}{l} \leftarrow \text{salt} \\ \leftarrow \text{solution} \end{array}$$

New concentration of salt solution:

$$\text{Add } x \text{ millimeters of distilled water.} \quad \frac{6.5}{65+x} \begin{array}{l} \leftarrow \text{salt} \\ \leftarrow \text{solution} \end{array}$$

Function for the salt concentration in this solution:

$$C(x) = \frac{6.5}{65+x}$$

b. Salt concentration when 100 milliliters of distilled water is added:

$$C(100) = \frac{6.5}{65+100} \approx 0.039, \text{ or } 3.9\%$$

A **rational expression** is the quotient of two polynomials. A **rational function** is a function defined by a rational expression. The function $C(x) = \frac{6.5}{65+x}$ from Example 1 is a rational function.

CHECKPOINT ✔ Explain why $y = \frac{e^x}{x-1}$ and $y = \frac{x^2+2}{|x|}$ are not rational functions.

Alternative Teaching Strategy

USING PATTERNS Ask students to graph each set of functions below on the same set of axes in a standard window by using a graphics calculator. For each set, have students discuss similarities and differences among the graphs. Ask students to pay particular attention to the domain of each function.

- $y_1 = \frac{1}{x}, y_2 = \frac{1}{x-1}, y_3 = \frac{1}{x+2}, y_4 = \frac{1}{x-3}$
- $y_1 = \frac{1}{(x-2)(x+3)}, y_2 = \frac{1}{x(x+3)}, y_3 = \frac{1}{(x+1)^2}$

Have students make conjectures about vertical asymptotes based on the sets of functions above and test these conjectures on the set of functions below.

- $y_1 = \frac{x-1}{x^2+3x+2}, y_2 = \frac{x-1}{x^2+x-2}$

Next have students look at horizontal asymptotes that occur in the following sets of functions:

- $y_1 = \frac{1}{x}, y_2 = \frac{x}{x^2-1}, y_3 = \frac{1}{x^2}, y_4 = \frac{x^2}{x^3-2x^2+x}$
- $y_1 = \frac{x}{x-1}, y_2 = \frac{3x}{x-1}, y_3 = \frac{2x^2+1}{3x^2}, y_4 = \frac{x^4}{x^3+3x-1}$

Have students discuss how to determine the horizontal asymptotes of a rational function.

Prepare

NCTM PRINCIPLES & STANDARDS 1–4, 6–10

QUICK WARM-UP

Solve each equation.

1. $x + 5 = 0$ — $x = -5$
2. $5x = 0$ — $x = 0$
3. $5x + 2 = 0$ — $x = -\frac{2}{5}$
4. $x^2 - 5x = 0$ — $x = 0$ *or* $x = 5$
5. $x^2 - 5x - 14 = 0$ — $x = 7$ *or* $x = -2$
6. $x^3 + 3x^2 - 54x = 0$ — $x = 0$ *or* $x = -9$ *or* $x = 6$
7. $-1 + 2x - x^2 = 0$ — $x = 1$

Also on Quiz Transparency 8.2

Teach

Why Ask students to read the setup for Example 1 and to think about what a graph of the amount of water added versus the salt concentration would look like. Students should do this without actually determining the algebraic function and without using a graphics calculator, but instead should give a general description of the graph.

☞ For Additional Example 1 and answer to Checkpoint, see page 490.

The rational function $f(x) = \frac{1}{x}$ is undefined when $x = 0$. In general, the domain of a rational function is the set of all real numbers except those real numbers that make the denominator equal to zero.

EXAMPLE 2 **Find the domain of $g(x) = \frac{x^2 - 7x + 12}{x^2 + 9x + 20}$.**

SOLUTION

Find the values of x for which the denominator, $x^2 + 9x + 20$, equals 0.

$$x^2 + 9x + 20 = 0$$
$$(x + 4)(x + 5) = 0$$
$$x = -4 \quad or \quad x = -5$$

The domain is all real numbers except −4 and −5.

TECHNOLOGY GRAPHICS CALCULATOR

Keystroke Guide, page 554

CHECK

Enter $y = \frac{x^2 - 7x + 12}{x^2 + 9x + 20}$ into a graphics calculator. Use the table feature to verify that −4 and −5 are not included in the domain.

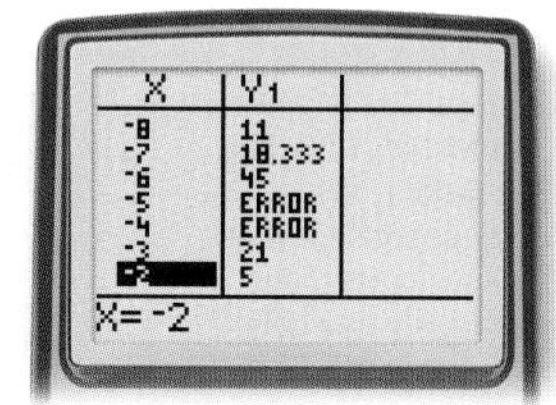

TRY THIS Find the domain of $j(x) = \frac{3x^2 + x - 2}{x^2 + 2x - 3}$.

Vertical Asymptotes

Recall from Lesson 6.2 that exponential functions, such as $y = 2^x$, have a horizontal asymptote. Rational functions can have horizontal *and* vertical asymptotes. In the Activity below, you will explore the vertical asymptotes of some rational functions.

Activity: Exploring Vertical Asymptotes

TECHNOLOGY GRAPHICS CALCULATOR

Keystroke Guide, page 555

You will need: a graphics calculator

1. Enter the function $y = \frac{1}{x - 2}$ into a graphics calculator.

   a. Copy the table below. Use the table feature to complete it.

x	1.0	1.1	1.2	1.3	1.4	1.5	1.6	1.7	1.8	1.9
y										

   b. Copy the table below. Use the table feature to complete it.

x	3.0	2.9	2.8	2.7	2.6	2.5	2.4	2.3	2.2	2.1
y										

Use Teaching Transparency 29.

ADDITIONAL EXAMPLE 1

William begins with 75 milliliters of a 15% acid solution. He adds x milliliters of distilled water to the container holding the acid solution.

a. Write a function, C, that represents the acid concentration of the solution in terms of x.

$C(x) = \frac{11.25}{75 + x}$

b. What is the acid concentration of the solution if 35 milliliters of distilled water is added?

≈10.2%

CHECKPOINT ✔

e^x and $|x|$ are not polynomials.

ADDITIONAL EXAMPLE 2

Find the domain of $h(x) = \frac{x^2 - 4x - 21}{x^2 - 9x - 36}$.

$x \neq -3$ and $x \neq 12$

TRY THIS

all real numbers except −3 and 1

Activity Notes

In this Activity, students investigate the behavior of a rational function around a point of discontinuity. They should find that when x approaches 2 from the left, the value of y approaches larger and larger negative numbers. As x approaches 2 from the right, the value of y approaches larger and larger positive numbers. There is a vertical asymptote at $x = 2$.

Interdisciplinary Connection

CHEMISTRY The concentration of acid in an acid-base mixture is given by the formula $C(A) = \frac{A}{A + B}$, where A is the amount of acid and B is the amount of base. Let B be any constant value. Have students describe the graph of C and discuss the meaning of the graph in terms of the amount of acid, A, and the concentration of the mixture, $C(A)$. **As A increases, $C(A)$ also increases and approaches 1.** Now let A be any constant value. Have students describe the new graph of $C(B)$ and discuss its meaning. **As B increases, $C(B)$ decreases and approaches 0.** Have students discuss the behavior of each graph. **The graphs increase or decrease rapidly at first and then level off to approach their respective asymptotes.**

CHECKPOINT ✔ 2. What can you say about y as x approaches 2 from values of x less than 2? What can you say about y as x approaches 2 from values of x greater than 2? What do you think is the value of y when $x = 2$?

CHECKPOINT ✔ 3. Let $y = \frac{1}{x+3}$. Use the table feature of a graphics calculator to evaluate y for values of x a little less than −3 and for values of x a little greater than −3. What can you say about y as x approaches −3 from values of x less than −3? What can you say about y as x approaches −3 from values of x greater than −3? What do you think is the value of y when $x = -3$?

Real numbers for which a rational function is not defined are called **excluded values**.

CHECKPOINT ✔ Find the excluded values for the function $y = \frac{x+3}{x^2 - x - 6}$.

At an excluded value, a rational function *may* have a vertical asymptote. The necessary conditions for a vertical asymptote are given below.

Vertical Asymptote

If $x - a$ is a factor of the denominator of a rational function but not a factor of its numerator, then $x = a$ is a vertical asymptote of the graph of the function.

EXAMPLE 3 **Identify all vertical asymptotes of the graph of $r(x) = \frac{2x}{x^2 - 1}$.**

SOLUTION

Factor the denominator: $r(x) = \frac{2x}{x^2 - 1} = \frac{2x}{(x-1)(x+1)}$

Since neither factor of the denominator is a factor of the numerator, equations for the vertical asymptotes are $x = -1$ and $x = 1$.

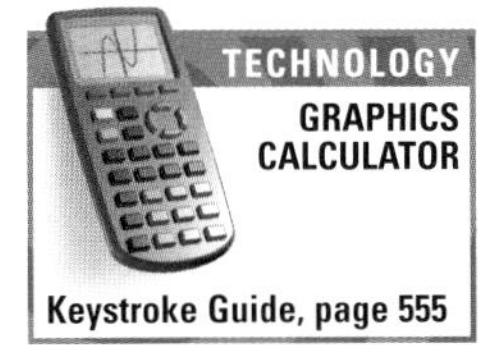

CHECK

Graph $y = \frac{2x}{x^2 - 1}$, and look for the vertical asymptotes, $x = -1$ and $x = 1$.

Note: Depending on the viewing window used, the calculator may display lines that look like vertical asymptotes but are actually lines that connect consecutive points on each side of the asymptotes.

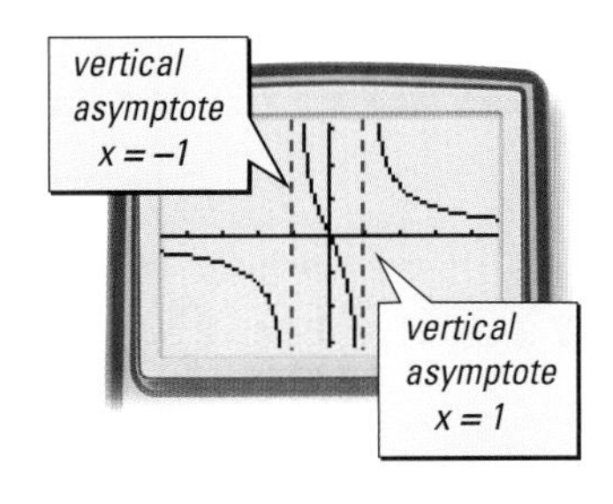

TRY THIS Identify all vertical asymptotes of the graph of $r(x) = \frac{x}{x^2 + 5x + 6}$.

CRITICAL THINKING Let P be a polynomial expression. Write a rational function of the form $R(x) = \frac{1}{P}$ that has vertical asymptotes at $x = -2$, $x = 0$, and $x = 2$.

Enrichment

Remind students how to use long division to divide polynomials. When describing the end behavior of a rational function, it is useful to write it in quotient-remainder form. For example, $f(x) = \frac{3x^2 - x - 1}{x - 1}$ can be rewritten as $f(x) = 3x + 2 + \frac{1}{x-1}$. As x gets very large, the graph of $f(x)$ approaches the graph of the line $y = 3x + 2$. This line is called a *slant asymptote*. Have students rewrite the functions at right in quotient-remainder form and give the equation of the slant asymptote of the function.

1. $g(x) = \frac{x}{x-1}$ — $g(x) = 1 + \frac{1}{x-1}$; $y = 1$
2. $h(x) = \frac{2x^3 + 3x}{x^2 + 1}$ — $h(x) = 2x + \frac{x}{x^2+1}$; $y = 2x$
3. $k(x) = \frac{x^3 + 1}{x^2}$ — $k(x) = x - \frac{1}{x^2}$

(Note: The idea of a slant asymptote as an end behavior is an extension of the topics presented in Lesson 7.2, where end behavior was introduced.)

Teaching Tip

TECHNOLOGY When using a TI-83 or TI-82 graphics calculator to produce the tables in the Activity, set **Y1=1/(X−2).** Then press 2nd WINDOW and set both **Indpnt** and **Depend** to **Auto, TblStart=1** (or **TblMin=1**), and **ΔTbl=.1** for part **a,** and **TblStart=3** (or **TblMin=3**) and **ΔTbl=−.1** for part **b.**

CHECKPOINT ✔
2. y decreases; y increases; y is undefined.

CHECKPOINT ✔
3. y decreases; y increases; y is undefined.

CHECKPOINT ✔
−2 and 3

ADDITIONAL EXAMPLE 3

Identify all vertical asymptotes of the graph of $r(x) = \frac{3x}{x^2 - 3x + 2}$.

$x = 1$ and $x = 2$

Teaching Tip

TECHNOLOGY When using a TI-83 or TI-82 graphics calculator for Example 3, set the viewing window as follows:

Xmin=−4.7	Ymin=−3.1
Xmax=4.7	Ymax=3.1
Xscl=1	Yscl=1

TRY THIS
$x = -2$ and $x = -3$

CRITICAL THINKING
$R(x) = \frac{1}{x(x+2)(x-2)} = \frac{1}{x^3 - 4x}$

Teaching Tip

TECHNOLOGY When using a TI-83 or TI-82 graphics calculator to produce the tables on this page, enter the function into **Y1**, press 2nd WINDOW, set **ΔTbl=10,** and choose **Auto** for both **Indpnt** and **Depend.** For the table on the left, set **TblStart=10** (or **TblMin=10**). For the table on the right, set **TblStart=–10** (or **TblMin=–10**). Press 2nd GRAPH to view each table.

Horizontal Asymptotes

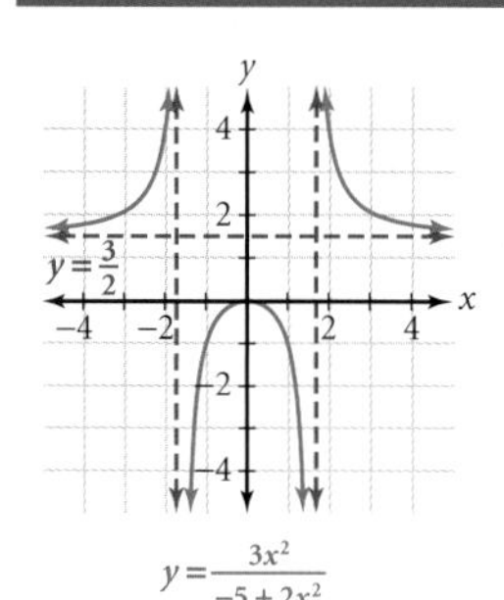

$y=\frac{3x^2}{-5+2x^2}$

To examine the *horizontal asymptotes* of rational functions, consider the graph of $y=\frac{3x^2}{-5+2x^2}$ shown at left. The graph shows a horizontal asymptote at $y=\frac{3}{2}$, or $y = 1.5$. The tables below confirm that as the x-values get farther away from 0, the y-values approach 1.5.

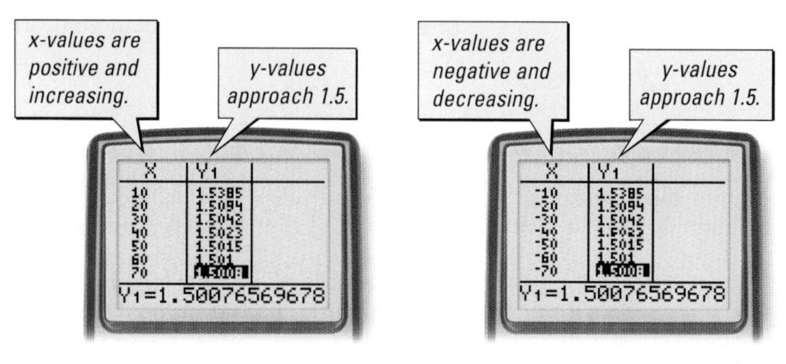

Because both the numerator and the denominator of the function $y=\frac{3x^2}{-5+2x^2}$ have the same degree, you can use the leading coefficients, 3 and 2, of the numerator and denominator to write the equation for the horizontal asymptote of its graph, $y=\frac{3}{2}$.

Horizontal Asymptote

Let $R(x)=\frac{P}{Q}$ be a rational function, where P and Q are polynomials.

- If the degree of P is less than the degree of Q, then $y = 0$ is the equation of the horizontal asymptote of the graph of R.
- If the degree of P equals the degree of Q and a and b are the leading coefficients of P and Q, respectively, then $y=\frac{a}{b}$ is the equation of the horizontal asymptote of the graph of R.
- If the degree of P is greater than the degree of Q, then the graph of R has no horizontal asymptote.

EXAMPLE 4 **Let $R(x)=\frac{x}{x^2-2x-3}$. Identify all vertical asymptotes and all horizontal asymptotes of the graph of R.**

SOLUTION

1. To find the equations of all vertical asymptotes, factor the denominator, and look for excluded values of the domain.

$$R(x)=\frac{x}{(x-3)(x+1)}$$

Since neither factor of the denominator is a factor of the numerator, the equations for the vertical asymptotes are $x = 3$ and $x = -1$.

ADDITIONAL EXAMPLE 4

Let $R(x)=\frac{x^3}{x^2+x-20}$. Identify all vertical asymptotes and all horizontal asymptotes.

vertical asymptotes: $x = 4$ and $x = -5$; horizontal asymptotes: none

Teaching Tip

TECHNOLOGY When using a TI-83 or TI-82 graphics calculator for Example 4, use the following viewing window:

Xmin=–4.7	**Ymin=–3.1**
Xmax=4.7	**Ymax=3.1**
Xscl=1	**Yscl=1**

When the calculator is in **CONNECTED** mode, lines appear on the graph where the vertical asymptotes should be. The calculator *is not* drawing the asymptotes; it is just connecting points on the graph. Press MODE and select **DOT** for a more accurate graph.

Inclusion Strategies

USING VISUAL MODELS Have students work in pairs. Each pair should place a transparency sheet over a piece of graph paper that has the axes labeled. Have them graph the rational function $f(x)=\frac{1}{x-a}+b$ on the transparency for $a = 0$ and $b = 0$. Then have them move the transparency to represent changes in a and b, and write the function that represents the new placement of the graph. Students may check these functions with a graphics calculator.

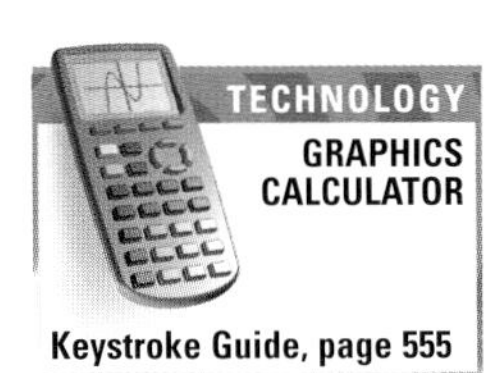

2. Because the degree of the numerator is less than the degree of the denominator, the equation for the horizontal asymptote is $y = 0$.

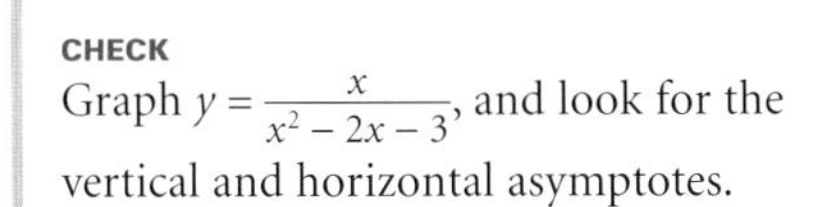

CHECK

Graph $y = \frac{x}{x^2 - 2x - 3}$, and look for the vertical and horizontal asymptotes.

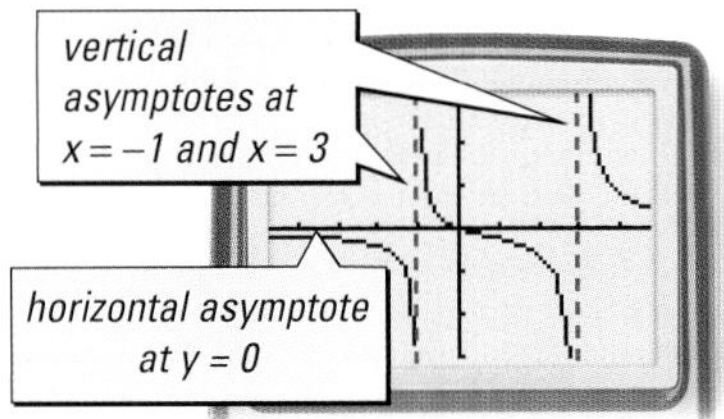

TRY THIS Let $R(x) = \frac{2x^2 - 2x + 1}{x^2 - x - 12}$. Find the equations of all vertical asymptotes and all horizontal asymptotes of the graph of R.

Using Asymptotes to Graph

You can graph a rational function by using the asymptotes, as shown in Example 5.

EXAMPLE 5 **Graph $y = \frac{x+2}{x-2}$, showing all asymptotes.**

SOLUTION

Write equations for the asymptotes, and graph them as dashed lines.

vertical asymptote: $x = 2$
horizontal asymptote: $y = 1$

PROBLEM SOLVING

Use a table. To help obtain an accurate graph, plot some points on each branch of the graph, and sketch the curves through these points.

x	−1	0	1	3	4	5
y	$-\frac{1}{3}$	−1	−3	5	3	$2\frac{1}{3}$

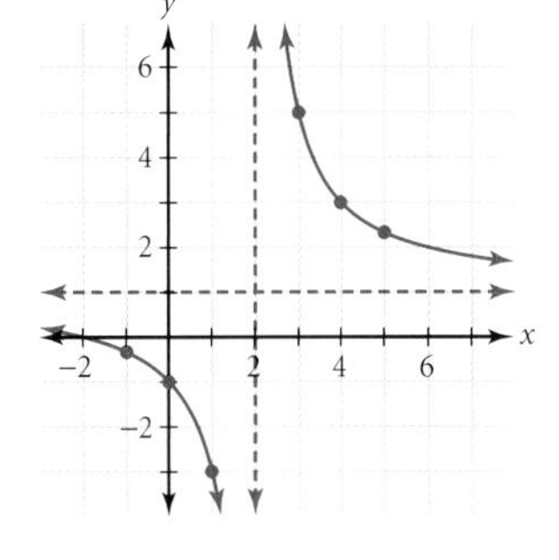

By dividing, you can rewrite $y = \frac{x+2}{x-2}$ from Example 5 as $y = 1 + \frac{4}{x-2}$.

From this form, you can see the transformations of the graph of $y = \frac{1}{x}$.

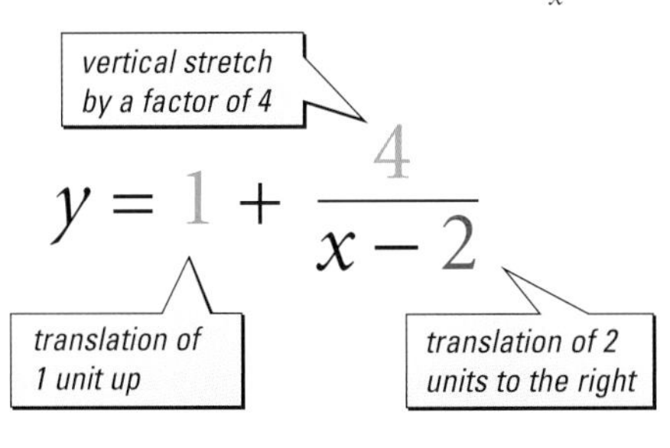

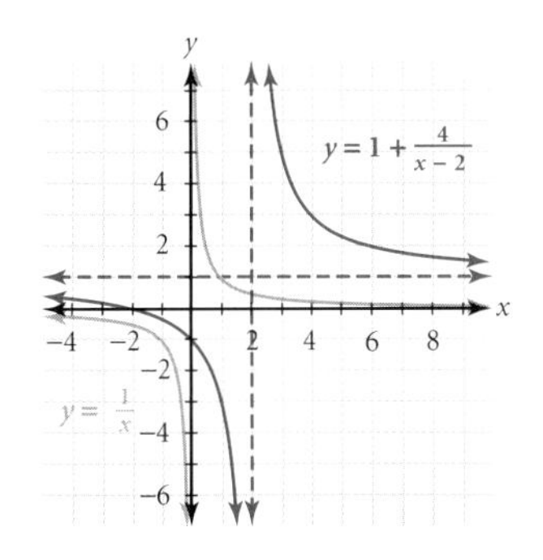

TRY THIS

$x = -3$ and $x = 4$; $y = 2$

ADDITIONAL EXAMPLE 5

Graph $y = \frac{2x-1}{x+4}$, showing all asymptotes.

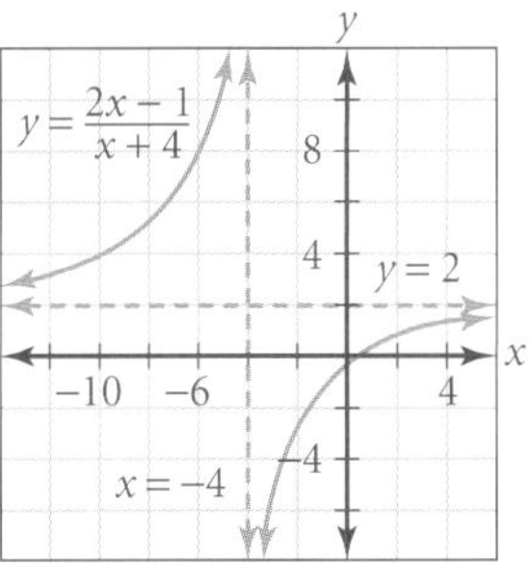

Math CONNECTION

TRANSFORMATIONS In this example, students apply transformations to the graph of a function.

Use Teaching Transparency 30.

Use Teaching Transparency 31.

Reteaching the Lesson

COOPERATIVE LEARNING Have students work in pairs, with one graphics calculator per pair. One student in each pair should graph a rational function and describe its graph to his or her partner. The description should include any vertical and horizontal asymptotes, discontinuities, and symmetries. The partner then writes a function that fits the given description. The pair should work together to rewrite the function that was entered into the calculator based only on descriptions of the graph. Each pair should do several examples, alternating roles.

ADDITIONAL EXAMPLE 6

Let $y = \frac{3x - 28 + x^2}{x^2 + 12x + 35}$. Identify all asymptotes and holes in the graph.

vertical asymptote: $x = -5$, horizontal asymptote: $y = 1$, hole when $x = -7$

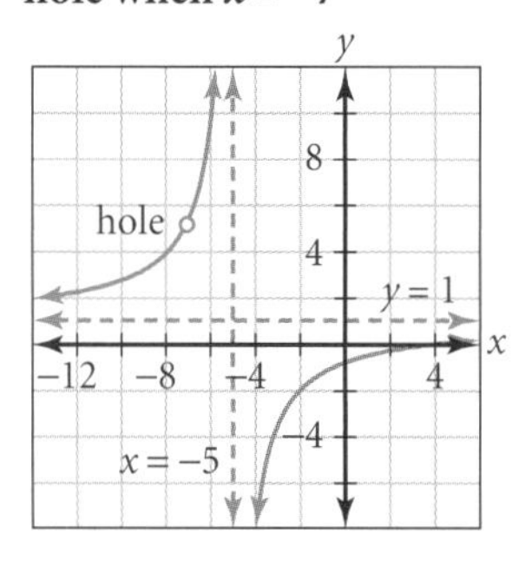

Teaching Tip

TECHNOLOGY When using a TI-83 or TI-82 calculator to produce a table of values for a rational function, a **Y** column in the table will show **ERROR** for values of x where there is a hole in the graph as well as for values of x that are horizontal asymptotes.

TRY THIS

hole at $x = -3$; vertical asymptotes: $y = 1$; no horizontal asymptote

Holes in Graphs

The graph of a rational function may have a *hole* in it. For example, $f(x) = \frac{x^2 - 9}{x - 3}$ can be written as $f(x) = \frac{(x-3)(x+3)}{x-3}$. Because $x - 3$ is a factor of *both* the numerator *and* the denominator, the graph of f has a hole when $x = 3$.

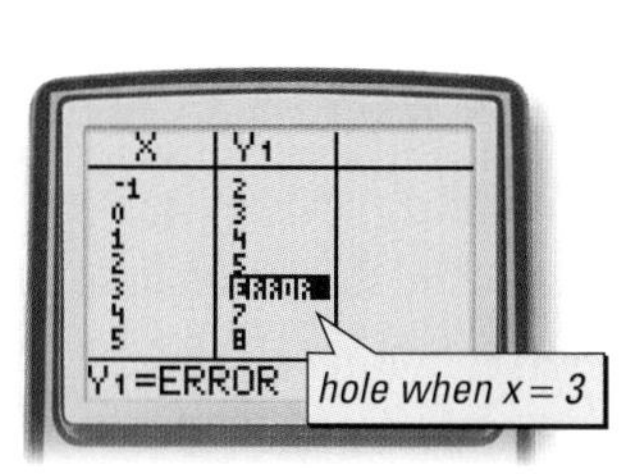

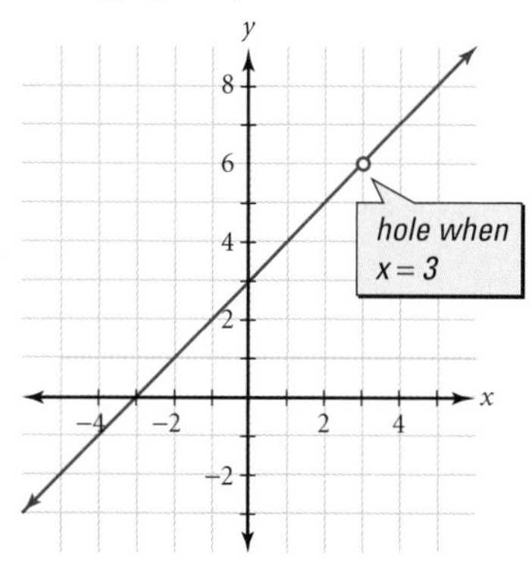

Hole in the Graph

If $x - b$ is a factor of the numerator and the denominator of a rational function, then there is a **hole in the graph** of the function when $x = b$ unless $x = b$ is a vertical asymptote.

EXAMPLE 6 **Let $y = \frac{3 - 2x - x^2}{x^2 + x - 2}$. Identify all asymptotes and holes in the graph.**

SOLUTION

Factor the numerator and denominator.

$$y = \frac{3 - 2x - x^2}{x^2 + x - 2} = \frac{-(x^2 + 2x - 3)}{(x-1)(x+2)} = \frac{-(x+3)\cancel{(x-1)}}{\cancel{(x-1)}(x+2)}$$

Because $x - 1$ is a factor of *both* the numerator and the denominator, the graph has a hole when $x = 1$.

Because $x + 2$ is a factor of *only* the denominator, there is a vertical asymptote at $x = -2$.

Because the degree of the numerator equals the degree of the denominator, there is a horizontal asymptote at $y = \frac{-1}{1} = -1$.

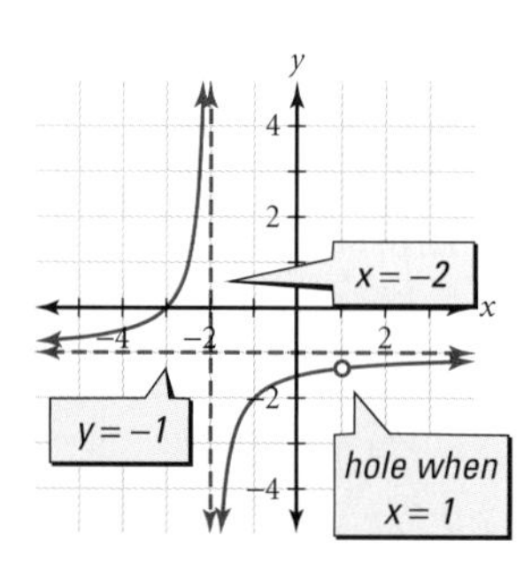

TRY THIS Let $y = \frac{3x^2 + x^3}{x^2 + 2x - 3}$. Identify all asymptotes and holes in the graph.

Exercises

Communicate

internet connect
Activities Online
Go To: go.hrw.com
Keyword:
MB1 Electricity

1. Why is the quotient of two polynomials described as "rational"?
2. Describe how to find the excluded values of a rational function.
3. Explain how to decide whether a linear factor of the denominator of a rational function is related to a vertical asymptote of the graph of the function or to a hole in the graph of the function.
4. Explain how to use the asymptotes of the graph of $g(x) = \frac{x-5}{x-3}$ to sketch the graph of the function.

Guided Skills Practice

APPLICATION

5. $C(x) = \frac{13.5}{90 + x}$; 5.6%

5. **CHEMISTRY** Refer to the salt-concentration problem at the beginning of the lesson. If Dane begins with 90 milliliters of a 15% saline solution and adds x milliliters of distilled water, what function represents the salt concentration of the new solution? What is the concentration of the solution if 150 milliliters of distilled water is added? ***(EXAMPLE 1)***

6. $x \neq 3, 4$

6. Find the domain of $h(x) = \frac{2x^2 - 5}{x^2 - 7x + 12}$. ***(EXAMPLE 2)***

7. $x = -\frac{3}{2}$, $x = \frac{3}{2}$, $y = 0$; no holes
8. $x = -3$, $x = 3$, $y = 2$; no holes
9. $x = 2$, $y = 1$; hole ($x = 3$)

Identify all asymptotes and holes in the graph of each rational function. ***(EXAMPLES 3, 4, AND 6)***

7. $r(x) = \frac{3x - 1}{4x^2 - 9}$
8. $R(x) = \frac{2x^2 - 1}{x^2 - 9}$
9. $f(x) = \frac{(x-3)^2}{x^2 - 5x + 6}$

10. Graph $f(x) = \frac{2x+1}{x-3}$, showing all asymptotes. ***(EXAMPLE 5)***

Practice and Apply

14. rat.; $x \neq \frac{7}{2}, -3$
15. Not rat.; 5^x is not a polynomial.
16. not rat.; no polynomials
17. $x = 2$, $y = 3$; no holes
18. $x = 0$, $y = 0$; no holes

internet connect
Homework Help Online
Go To: go.hrw.com
Keyword:
MB1 Homework Help for Exercises 17–40

Determine whether each function is a rational function. If so, find the domain. If the function is not rational, state why not.

11. $f(x) = \frac{x}{2x - 7}$ rat.; $x \neq \frac{7}{2}$
12. $g(x) = \frac{x+2}{2x}$ rat.; $x \neq 0$
13. $h(x) = \frac{x^{-\frac{1}{2}}}{x^2}$ not rat.; $x^{-\frac{1}{2}}$ not poly.
14. $f(x) = \frac{x}{(2x-7)(x+3)}$
15. $g(x) = \frac{5^x}{x^5}$
16. $h(x) = \frac{|x^2 - 4|}{|x + 2|}$

Identify all asymptotes and holes in the graph of each rational function.

17. $f(x) = \frac{3x+5}{x-2}$
18. $g(x) = \frac{x+2}{2x^2}$
19. $d(x) = \frac{x^2 - 4}{x^2 - 4x + 4}$
20. $g(x) = \frac{(x+2)^2}{x^2 + 5x + 6}$
21. $r(x) = \frac{x^2 - 16}{4 - 5x + x^2}$
22. $b(x) = \frac{x^2 - 2x + 1}{x^2 + x - 2}$

Assess

Selected Answers

Exercises 5–10, 11–61 odd

ASSIGNMENT GUIDE

In Class	1–10
Core	11–39 odd, 43, 45
Core Plus	12–40 even, 41, 42–46 even
Review	47–62
Preview	63–65

Extra Practice can be found beginning on page 940.

10.

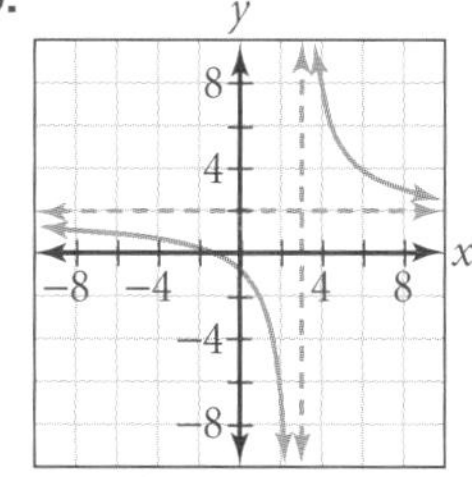

19. $x = 2, y = 1$; no holes
20. $x = -3, y = 1$; hole ($x = -2$)
21. $x = 1, y = 1$; hole ($x = 4$)
22. $x = -2, y = 1$; hole ($x = 1$)

Error Analysis

Students must factor polynomials correctly in order to find the correct asymptotes and holes in the graph. Make sure that students check their factors by multiplying if they are having trouble.

Also, many students fail to use parentheses when entering rational functions into a calculator. Remind them to enclose both the numerator and denominator within parentheses to ensure correct results.

23. domain of all real numbers except $x = -1$; vertical asymptote: $x = -1$; horizontal asymptote: $y = 1$ no holes

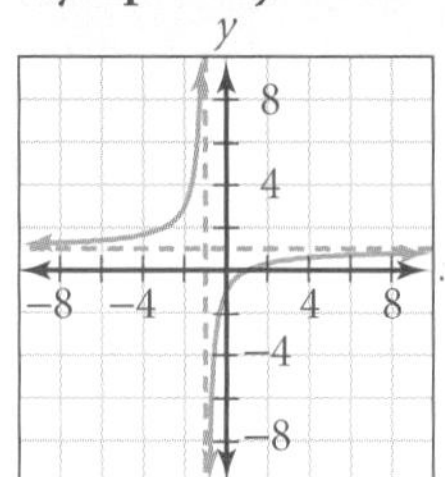

Find the domain of each rational function. Identify all asymptotes and holes in the graph of each rational function. Then graph.

23. $h(x) = \dfrac{2x-2}{2x+2}$ **24.** $r(x) = \dfrac{2x}{2x(x-5)}$ **25.** $w(x) = \dfrac{(3x-1)(x+2)}{x+2}$

26. $a(x) = \dfrac{x+1}{x^2+4x-21}$ **27.** $d(x) = \dfrac{3x-1}{9x^2-36}$ **28.** $d(x) = \dfrac{7x+8}{x^2-10x+25}$

29. $f(x) = \dfrac{7x+1}{5x^2+3}$ **30.** $r(x) = \dfrac{x^2-4}{x^2+4}$ **31.** $m(x) = \dfrac{x(x^2-4)}{x^2-7x+6}$

32. $t(x) = \dfrac{x^2-4x}{x^3-x^2-20x}$ **33.** $g(x) = \dfrac{5x^2-3x}{x^3-8x^2+16x}$ **34.** $t(x) = \dfrac{2x+1}{x^3-27}$

Write a rational function with the given asymptotes and holes.

35. $x = 2$ and $y = 3$ $f(x) = \dfrac{3x}{x-2}$ **36.** $x = -2$ and $y = 0$ $f(x) = \dfrac{1}{x+2}$

37. $x = \pm 1$, $y = 1.5$, hole when $x = 0$ **38.** $x = 1$, $y = 3$, hole when $x = 2$

39. holes when $x = 0$ and $x = 2$ **40.** holes when $x = 0$, $x = 2$, and $x = 3$

37. $f(x) = \dfrac{3x^3}{2x^3-2x}$

38. $f(x) = \dfrac{3x^2-6x}{x^2-3x+2}$

39. $f(x) = \dfrac{x^3-2x^2}{x^2-2x}$

40. $f(x) = \dfrac{x^4-5x^3+6x^2}{x^3-5x^2+6x}$

CHALLENGE

41. Let $f(x) = \dfrac{1}{x^2-3x+c}$. Find c such that the graph of f has the given number of vertical asymptotes. Justify your responses.

a. none $c > \frac{9}{4}$ **b.** one $c = \frac{9}{4}$ **c.** two $c < \frac{9}{4}$

CONNECTIONS

42. GEOMETRY Refer to the rectangle at right.

a. Write a rational function, R, to represent the ratio of the perimeter, P, to the area, A.

b. For what values of x is P defined? is A defined? is R defined?

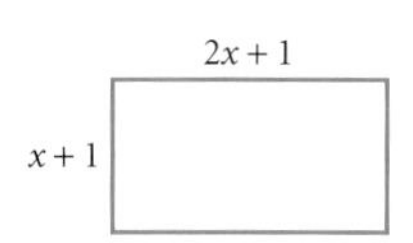

42a. $R(x) = \dfrac{6x+4}{2x^2+3x+1}$

b. $x > -0.5$; $x > -0.5$; $x > -0.5$

43. GEOMETRY Refer to the circle at right.

a. Write a rational function, R, to represent the ratio of the circumference, C, to the area, A.

b. For what values of x is C defined? is A defined? is R defined?

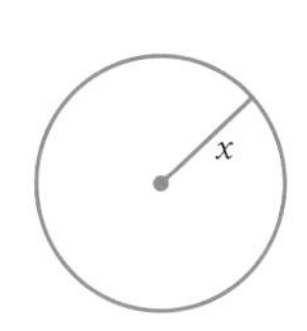

43a. $R(x) = \dfrac{2}{x}$

b. $x > 0$; $x > 0$; $x > 0$

APPLICATIONS

44. CHEMISTRY Leon begins with 72 milliliters of an 8% saline solution. He adds x milliliters of distilled water to the saline solution.

a. What function represents salt concentration, C, of the new solution?

b. What is the approximate concentration if 720 milliliters of distilled water is added?

44a. $C(x) = \dfrac{5.76}{72+x}$

b. 0.7%

45. ECONOMICS Max owns a small florist shop. His fixed costs are \$250 per week, and his average variable costs are \$11.45 per arrangement.

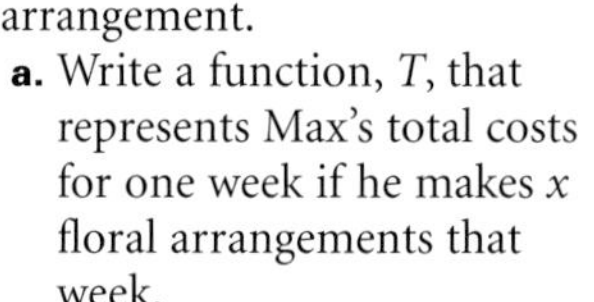

a. Write a function, T, that represents Max's total costs for one week if he makes x floral arrangements that week.

b. Write a function in terms of T and x for the cost, C, per floral arrangement during that week.

45a. $T(x) = 11.45x + 250$

b. $C(x) = \dfrac{T(x)}{x} = \dfrac{11.45x+250}{x}$

24. domain of all real numbers except $x = 0$ and $x = 5$; vertical asymptote: $x = 5$; horizontal asymptote: $y = 0$; hole when $x = 0$

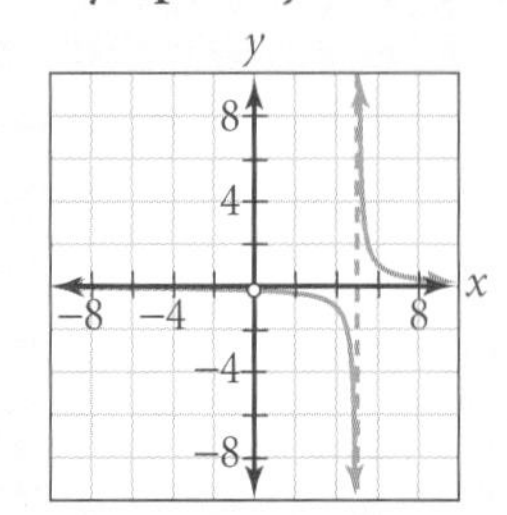

25. domain of all real numbers except $x = -2$; no vertical asymptote; no horizontal asymptote; hole when $x = -2$

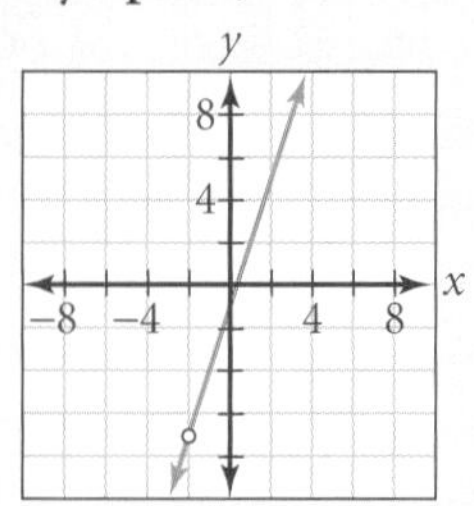

Practice

NAME ______ CLASS ______ DATE ______

Practice
8.2 *Rational Functions and Their Graphs*

Determine whether each function below is a rational function. If so, find the domain. If the function is not rational, state why not.

1. $f(x) = \dfrac{x^3-5x+7}{x^2-3}$ yes; all real numbers except $-\sqrt{3}$ and $\sqrt{3}$
2. $h(x) = \dfrac{x+2}{|x|-2}$ No; $|x| - 2$ is not a polynomial.
3. $w(x) = \dfrac{12-2x}{x^2-1}$ yes; all real numbers except -1 and 1

Identify all vertical and horizontal asymptotes of the graph of each rational function.

4. $k(x) = \dfrac{2x+1}{x-9}$ vertical: $x = 9$; horizontal: $y = 2$
5. $p(x) = \dfrac{2x^2+3}{(x-1.5)^2}$ vertical: $x = 1.5$; horizontal: $y = 0$
6. $m(x) = \dfrac{3x-8}{x^2-7}$ vertical: $x = -\sqrt{7}$ and $x = \sqrt{7}$; horizontal: $y = 0$

Find the domain of each rational function. Identify all asymptotes and holes in the graph of each rational function.

7. $h(x) = \dfrac{4x-3}{x^2-6x}$ all real numbers except 0 and 6; vertical: $x = 0$, $x = 6$; horizontal: $y = 0$
8. $g(x) = \dfrac{x-1}{x^2+4x-5}$ all real numbers except 1 and -5; vertical: $x = -5$; horizontal: $y = 0$; hole at 1
9. $n(x) = \dfrac{3x^2+12x}{x^2+7x+12}$ all real numbers except -3 and -4; vertical: $x = -3$, horizontal: $y = 3$; hole at -4

Sketch the graph of each rational function. Identify all asymptotes and holes in the graph of the function.

10. $a(x) = \dfrac{3x}{x-4}$ 11. $f(x) = \dfrac{x+2}{2x^2+3x-2}$ 12. $b(x) = \dfrac{x+x^2}{x^2-1}$

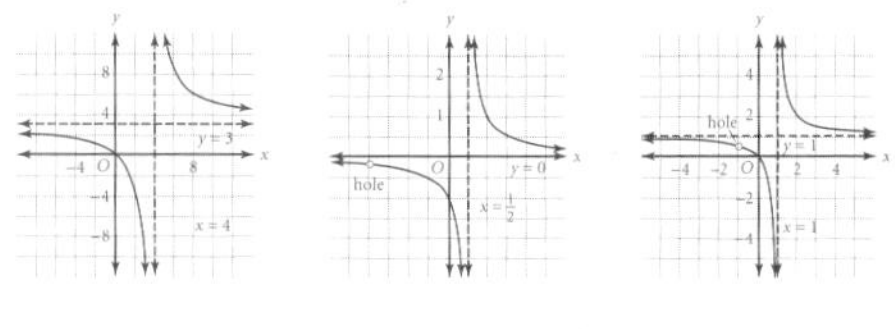

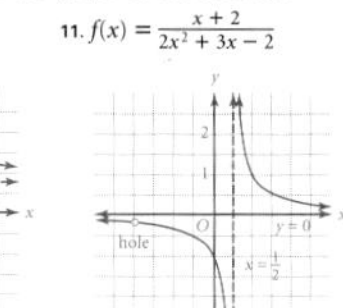

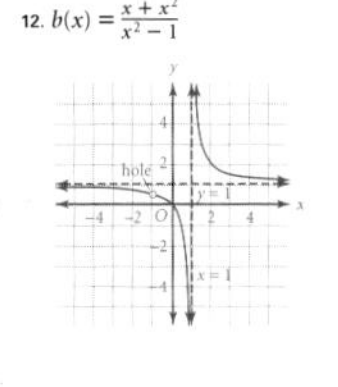

APPLICATION

46a. It can be written as quotient of two polynomials.
b. $0.997w_0$; $0.994w_0$; $0.969w_0$
c. ≈ 2651 km

46. PHYSICS An object weighing w_0 kilograms on Earth is h kilometers above the surface of the Earth. Then the function for the object's weight at that altitude is $w(h) = w_0\left(\frac{6400}{6400 + h}\right)^2$.

a. Explain why w is a rational function of h.
b. Use a table to find w for h-values of 10, 20, and 100.
c. At about what altitude will an object weigh half of what it weighs on Earth?

Look Back

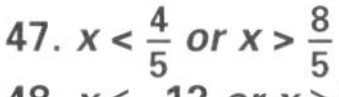

47. $x < \frac{4}{5}$ or $x > \frac{8}{5}$
48. $x \le -12$ or $x \ge 2$
49. $-\frac{4}{5} \le x \le 2$
50. no solution
51. $-36x^2 + 24x$
52. $18x^2 - 27x + 7$
53. $-5x^2 + 49x - 36$
54. $2x^2 - 7x - 15$
55. $9x^2 - 16$
56. $-4x^2 + 24x - 36$

Solve. *(LESSON 1.8)*

47. $|5x - 6| > 2$ **48.** $|x + 5| \ge 7$ **49.** $\left|\frac{3}{2} - \frac{5}{2}x\right| \le \frac{7}{2}$ **50.** $\left|\frac{3}{2} - \frac{5}{2}x\right| \le -\frac{7}{2}$

Write each expression in standard form, $ax^2 + bx + c$. *(LESSON 5.1)*

51. $-12x(3x - 2)$ **52.** $(3x - 1)(6x - 7)$ **53.** $(4 - 5x)(x - 9)$
54. $(x - 5)(2x + 3)$ **55.** $(3x - 4)(3x + 4)$ **56.** $-4(x - 3)^2$

Factor each expression. *(LESSON 5.3)*

57. $3x^2 - 6x$ $3x(x - 2)$ **58.** $1 - 25y^2$ $(1 + 5y)(1 - 5y)$ **59.** $9x^2 - 49$ $(3x + 7)(3x - 7)$
60. $t^2 - 5t - 24$ $(t + 3)(t - 8)$ **61.** $x^2 + 12x + 36$ $(x + 6)^2$ **62.** $x^2 - 16x + 64$ $(x - 8)^2$

Look Beyond

Simplify.

63. $\frac{9}{3}$ 3 **64.** $\frac{x^2}{x}$ x **65.** $\frac{x^2 + 4x + 4}{x + 2}$ $x + 2$

In this activity you will use rational functions to show that if a rectangle and a square have the same ratio of area to perimeter, then the side length of the square is the harmonic mean of the width and length of the rectangle.

Refer to the rectangle and square shown at left.

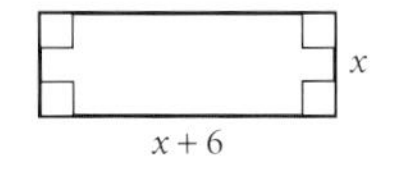

x + 2
x + 2

1. Write a rational function, R, to represent the ratio of the area, A, of the rectangle to its perimeter, P.
2. Find a value of x such that the ratio of the area to the perimeter is 2.
3. Use this value to find the length and width of the rectangle.
4. Write a rational function, S, to represent the ratio of the area, A, of the square to its perimeter, P.
5. Find a value of x such that the ratio of the area to the perimeter is 2.
6. Use this value to find the side length of the square.
7. Show that the side length of the square is the harmonic mean of the length and width of the rectangle.

26. domain of all real numbers except $x = -7$ and $x = 3$; vertical asymptotes: $x = -7$ and $x = 3$; horizontal asymptote: $y = 0$; no holes

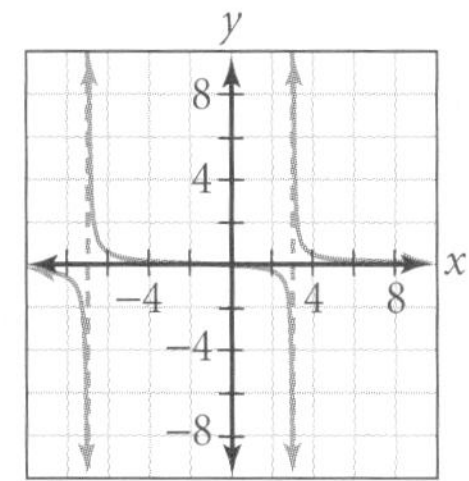

27. domain of all real numbers except $x = -2$ and $x = 2$; vertical asymptotes: $x = -2$ and $x = 2$; horizontal asymptote: $y = 0$; no holes

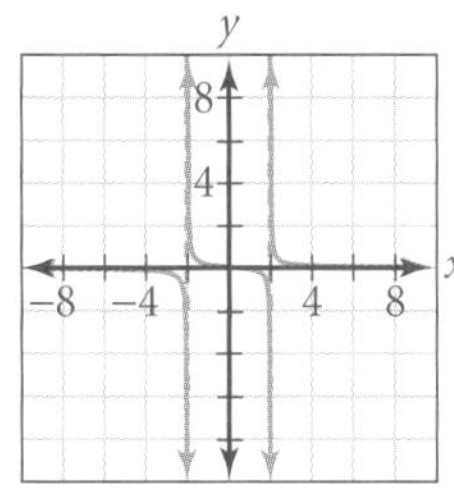

The answers to Exercises 28–34, 41, and the Portfolio Activity can be found in Additional Answers beginning on page 1002.

Look Beyond

In Exercises 63–65, students are asked to simplify numerical fractions as well as rational expressions. These skills will be used in Lesson 8.3 as students learn to divide rational expressions.

ALTERNATIVE Assessment

Portfolio Activity

The Portfolio Activity can be used as preparation for the Chapter Project or as a separate activity. In the Portfolio Activity on this page, students use rational expressions that relate the area and perimeter of a rectangle and square in order to represent the harmonic mean of the sides of the rectangle.

Student Technology Guide

NAME ______ CLASS ______ DATE ______

Student Technology Guide
8.2 Rational Functions and Their Graphs

As seen in Lesson 8.2 of your textbook, graphs of many rational functions have *asymptotes*—lines that the graph approaches but never reaches. The graphing and table features of a graphics calculator can help you find asymptotes and holes of a rational function.

Example: Find equations for all vertical and horizontal asymptotes of $f(x) = \frac{3x + 1}{x - 2}$.

- Press Y= (3 X,T,θ,n + 1) ÷ (X,T,θ,n − 2) GRAPH. (Note: You must use parentheses to group the numerator and denominator.)
- There appears to be a vertical asymptote near $x = 2$ and a horizontal asymptote near $y = 3$. You can use tables to help verify these asymptotes.
- Press 2nd GRAPH. Use the arrow keys to scroll through the table to find any error messages in the Y1 column. This table shows "ERROR" at $x = 2$, indicating a hole or a vertical asymptote at that x-value. Based on the graph, $x = 2$ is a vertical asymptote.

X	Y1
0	-.5
1	-4
2	ERROR
3	10
4	6.5
5	5.3333
6	4.75

Y1=ERROR

- To look for horizontal asymptotes, press 2nd WINDOW and use ▼ to move to ΔTbl. Entering a value of −50 for ΔTbl can help you investigate negative x-values, as shown at left below, and entering a value of 50 can help you look at positive x-values, as shown at right below. Press 2nd GRAPH. The line $y = 3$ is the horizontal asymptote.

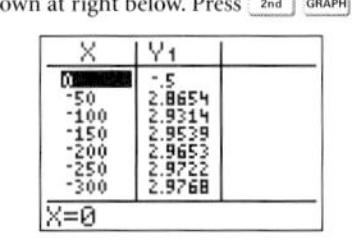

X	Y1
0	-.5
-50	2.8654
-100	2.9314
-150	2.9539
-200	2.9653
-250	2.9722
-300	2.9768

X=0

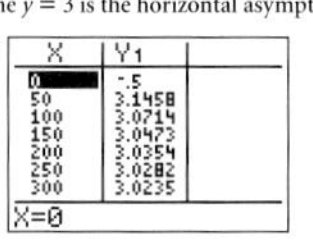

X	Y1
0	-.5
50	3.1458
100	3.0714
150	3.0473
200	3.0354
250	3.0282
300	3.0235

X=0

Write equations for the vertical and horizontal asymptotes of each rational function.

1. $f(x) = \frac{1}{x + 3}$ vertical: $x = -3$, horizontal: $y = 0$
2. $g(x) = \frac{3x + 1}{5x + 10}$ vertical: $x = -2$, horizontal: $y = 0.6$
3. $k(x) = \frac{5x}{4x + 6}$ vertical: $x = -1.5$, horizontal: $y = 1.25$
4. $d(x) = \frac{x^2 - 9}{x^2 + 4x + 4}$ vertical: $x = -2$, horizontal: $y = 1$

Prepare

NCTM PRINCIPLES & STANDARDS 1–3, 6–10

QUICK WARM-UP

Evaluate.

1. $\frac{2}{3} \cdot \frac{3}{7}$ $\quad \frac{2}{7}$
2. $\frac{3}{4} \cdot \frac{8}{15}$ $\quad \frac{2}{5}$
3. $\frac{4}{9} \div \frac{5}{12}$ $\quad \frac{16}{15}$, or $1\frac{1}{15}$
4. $\frac{9}{10} \div \frac{3}{10}$ $\quad 3$

Factor each expression.

5. $x^2 - 81$
   $(x + 9)(x - 9)$
6. $x^2 - 16x - 36$
   $(x - 18)(x + 2)$

Also on Quiz Transparency 8.3

Teach

Why Have students discuss the economic analysis given, and ask them to describe what the revenue-to-cost ratio tells them. What does it mean if the ratio is greater than 1? less than 1?

ADDITIONAL EXAMPLE 1

Simplify $\frac{x^2 - 7x - 18}{x^2 - 8x - 9}$. $\quad \frac{x+2}{x+1}$

TRY THIS

$\frac{b+7}{b-1}$

8.3 Multiplying and Dividing Rational Expressions

Why *Multiplying and dividing rational expressions are sometimes used to solve real-world problems, such as analyzing the revenue and cost for a fund-raising activity.*

Objectives

- Multiply and divide rational expressions.
- Simplify rational expressions, including complex fractions.

APPLICATION FUND-RAISING

To analyze the revenue and costs from the sale of school-spirit ribbons, members of the Jamesville High School Home Economics Club used the revenue-to-cost ratio below.

$$\frac{\text{revenue from the sale of each ribbon}}{\text{cost of making each ribbon}}$$

They represented the number of ribbons produced and sold by x and the total production cost in dollars by $0.8x + 25$. If the revenue for each ribbon was $3, for how many ribbons was the revenue-to-cost ratio 1.5 or greater? Finding the answer to this question involves writing and simplifying a rational expression. *You will answer this question in Example 6.*

Simplifying Rational Expressions

To *simplify* a rational expression, divide the numerator and the denominator by a common factor. The expression is simplified when you can no longer divide the numerator and denominator by a common factor other than 1.

EXAMPLE 1 Simplify $\frac{x^2 + 5x - 6}{x^2 - 36}$.

SOLUTION

$$\frac{x^2 + 5x - 6}{x^2 - 36} = \frac{(x+6)(x-1)}{(x-6)(x+6)} \quad \textit{Factor the numerator and denominator.}$$

$$= \frac{\cancel{(x+6)}(x-1)}{(x-6)\cancel{(x+6)}} \quad \textit{Divide out the common factor.}$$

$$= \frac{x-1}{x-6}$$

Note that 6 and −6 are excluded values of x in the original expression.

TRY THIS Simplify $\frac{b^2 - 49}{b^2 - 8b + 7}$.

Alternative Teaching Strategy

INVITING PARTICIPATION Have students work in pairs. One student should write a rational expression containing a product of factors, making sure that there is at least one common factor in the numerator and denominator. Then each student should give the unsimplified rational expression to his or her partner to simplify. Have students alternate roles. Students can check their work by entering the related functions for the simplified and unsimplified expressions into a graphics calculator. The graphs of both functions should be the same except that holes in the unsimplified function will be filled in on the simplified function. Point out to students that undefined points or holes do not always show on graphics calculators. Students can use the **TRACE** feature to note the absence of a y-value at the hole in the graph. Also, pressing **ZOOM** and selecting **4:ZDecimal** when the calculator is in **DOT** mode often produces graphs with visible holes.

Multiplying Rational Expressions

Multiplying rational expressions is similar to multiplying rational numbers.

Rational Numbers	Rational Expressions
$\frac{15}{4} \cdot \frac{14}{9} = \frac{3 \cdot 5}{4} \cdot \frac{2 \cdot 7}{9} = \frac{35}{6}$	$\frac{15}{x^2} \cdot \frac{4x^4}{21} = \frac{3 \cdot 5}{x^2} \cdot \frac{4 \cdot x^4}{3 \cdot 7} = \frac{20x^2}{7}$

EXAMPLE 2 Simplify $\frac{3}{4x^2} \cdot \frac{4x^3}{21} \cdot \frac{14}{4x^5}$.

SOLUTION

$$\frac{3}{4x^2} \cdot \frac{4x^3}{21} \cdot \frac{14}{4x^5} = \frac{3 \cdot 4 \cdot 2 \cdot 7}{4 \cdot 3 \cdot 7 \cdot 2 \cdot 2} \cdot \frac{x^3}{x^7} = \frac{1}{2x^4}$$

TRY THIS Simplify $\frac{28}{4a^3} \cdot \frac{4a^5}{21} \cdot \frac{3}{49a^4}$.

To multiply one rational expression by another, multiply as with fractions.

$$\frac{a}{b} \cdot \frac{c}{d} = \frac{ac}{bd}, \text{ where } b \neq 0 \text{ and } d \neq 0$$

You can simplify the product by dividing out the common factors in the numerator and denominator before or after multiplying.

EXAMPLE 3 Simplify $\frac{x+1}{x^2+2x-3} \cdot \frac{x^2+x-6}{x^2-2x-3}$.

SOLUTION

$$\frac{x+1}{x^2+2x-3} \cdot \frac{x^2+x-6}{x^2-2x-3} = \frac{x+1}{(x+3)(x-1)} \cdot \frac{(x+3)(x-2)}{(x-3)(x+1)}$$

$$= \frac{x-2}{(x-1)(x-3)}, \text{ or } \frac{x-2}{x^2-4x+3}$$

TRY THIS Simplify $\frac{x^2-25}{x^2-5x+6} \cdot \frac{x^2-4}{x^2+2x-15}$.

Dividing Rational Expressions

Dividing one rational expression by another is similar to dividing one rational number by another.

Rational Numbers	Rational Expressions
$\frac{6}{8} \div \frac{12}{32} = \frac{6}{8} \cdot \frac{32}{12}$ (Multiply by the reciprocal of $\frac{12}{32}$.)	$\frac{6}{x^3} \div \frac{12}{x^5} = \frac{6}{x^3} \cdot \frac{x^5}{12}$ (Multiply by the reciprocal of $\frac{12}{x^5}$.)
$= \frac{6}{8} \cdot \frac{32}{12}$	$= \frac{6}{x^3} \cdot \frac{x^5}{12}$
$= \frac{4}{2}$, or 2	$= \frac{x^2}{2}$, or $\frac{1}{2}x^2$

Enrichment

Consider the rational function below.

$$f(x) = \frac{x^3 - 2x^2 - 5x + 6}{x - A}$$

Find all values of A for which the graph of f has no vertical asymptotes. **–2, 1, and 3**

ADDITIONAL EXAMPLE 2

Simplify $\frac{9x^5}{5} \cdot \frac{6}{x^7} \cdot \frac{5x}{18}$. $\frac{3}{x}$

TRY THIS

$\frac{4}{49a^2}$

Teaching Tip

A quick way to check whether the result of an operation involving rational expressions is correct is to evaluate the original problem and the result for the same value of the variable. If the two values differ, the result is incorrect. This method is not an absolute check, but it will usually identify incorrect results.

ADDITIONAL EXAMPLE 3

Simplify

$\frac{x^2+4x-12}{x^2+11x+30} \cdot \frac{x^2-2x-35}{x+4}$.

$\frac{(x-2)(x-7)}{x+4}$, or $\frac{x^2-9x+14}{x+4}$

TRY THIS

$\frac{x^2-3x-10}{x^2-6x+9}$

ADDITIONAL EXAMPLE 4

Simplify

$\frac{x^2-25}{x^2+2x-3} \div \frac{x+5}{x^2-3x-18}$.

$\frac{(x-5)(x-6)}{x-1} = \frac{x^2-11x+30}{x-1}$

TRY THIS

$x+3$

ADDITIONAL EXAMPLE 5

Simplify the complex fraction

$\dfrac{\frac{b^2-4}{b^2-2b+1}}{\frac{b+2}{b-1}}$. $\frac{b-2}{b-1}$

TRY THIS

$\frac{x^2-x-6}{x-2}$

CRITICAL THINKING

1

ADDITIONAL EXAMPLE 6

Pearson High School's art club is selling caricatures to raise money for their annual trip. The revenue from the sale of each caricature is $10. The variable x represents the number of caricatures produced and sold, and $C(x) = 4.5x + 60$ represents the total production cost in dollars. **For how many caricatures was the revenue-to-cost ratio 1.5 or greater?**

28 or more caricatures

To divide one rational expression by another, multiply by the reciprocal of the divisor.

$$\frac{a}{b} \div \frac{c}{d} = \frac{a}{b} \cdot \frac{d}{c} = \frac{ad}{bc}, \text{ where } b \neq 0, c \neq 0, \text{ and } d \neq 0$$

Simplify by dividing out common factors in the numerator and denominator.

EXAMPLE 4 Simplify $\frac{x-4}{(x-2)^2} \div \frac{x^2-3x-4}{x^2-4}$.

SOLUTION

$$\frac{x-4}{(x-2)^2} \div \frac{x^2-3x-4}{x^2-4} = \frac{x-4}{(x-2)^2} \cdot \frac{x^2-4}{x^2-3x-4}$$ *Multiply by the reciprocal.*

$$= \frac{x-4}{(x-2)(x-2)} \cdot \frac{(x-2)(x+2)}{(x-4)(x+1)}$$ *Divide out common factors.*

$$= \frac{x+2}{(x-2)(x+1)}, \text{ or } \frac{x+2}{x^2-x-2}$$

TRY THIS Simplify $\frac{(x+3)^2}{x-5} \div \frac{x^2-9}{x^2-8x+15}$.

PROBLEM SOLVING You can **use a graph** to identify polynomials in a rational expression that cannot be factored with real numbers. For example, to determine whether $x^2 - x + 1$ in the rational expression $\frac{x-4}{x-1} \cdot \frac{x^2-x+1}{x^2}$ can be factored with real numbers, look for x-intercepts in the graph of $y = x^2 - x + 1$.

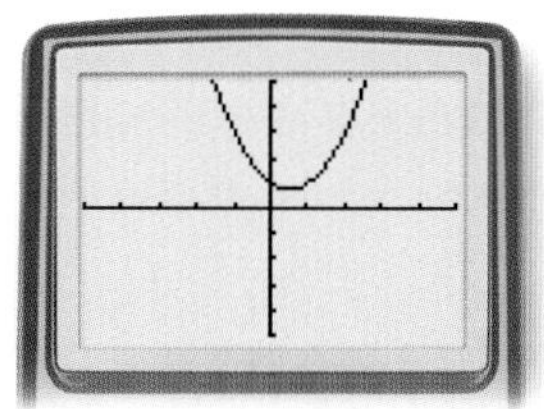

The graph of $y = x^2 - x + 1$ has no x-intercepts, so $y = x^2 - x + 1$ has no real zeros and $x^2 - x + 1$ cannot be factored with real numbers. Thus, the rational expression $\frac{x-4}{x-1} \cdot \frac{x^2-x+1}{x^2}$ cannot be simplified further.

Complex Fractions

A **complex fraction** is a quotient that contains one or more fractions in the numerator, the denominator, or both.

EXAMPLE 5 Simplify the complex fraction $\dfrac{\frac{4a^2-1}{a^2-4}}{\frac{2a-1}{a+2}}$.

SOLUTION

$$\dfrac{\frac{4a^2-1}{a^2-4}}{\frac{2a-1}{a+2}} = \frac{4a^2-1}{a^2-4} \cdot \frac{a+2}{2a-1}$$ *Multiply by the reciprocal.*

$$= \frac{(2a-1)(2a+1)}{(a-2)(a+2)} \cdot \frac{a+2}{2a-1}$$ *Factor.*

$$= \frac{(2a-1)(2a+1)(a+2)}{(a-2)(a+2)(2a-1)}$$ *Divide out common factors.*

$$= \frac{2a+1}{a-2}$$

Inclusion Strategies

ENGLISH LANGUAGE DEVELOPMENT Before students can work with algebraic expressions, they need to have a firm understanding of similar operations with numbers. Have students write an explanation of how fractions are multiplied and divided. Then have them explain how complex fractions are simplified. Have students share their explanations with the class in open discussion to make sure that all students understand operations on fractions before discussing operations on rational expressions.

TRY THIS Simplify $\dfrac{\frac{(x+2)^2}{x-3}}{\frac{x^2-4}{(x-3)^2}}$.

CRITICAL THINKING Use mental math to simplify $\dfrac{\frac{x+y}{x-y}}{\frac{y+x}{y-x}}$.

EXAMPLE 6 Refer to the revenue-to-cost ratio given at the beginning of the lesson.

APPLICATION
ECONOMICS

For how many ribbons was the revenue-to-cost ratio 1.5 or greater?

SOLUTION

$$\frac{\text{revenue from the sale of each ribbon}}{\text{cost of making each ribbon}} = \frac{3}{\frac{0.8x+25}{x}}$$

Simplify the complex fraction.

$$\frac{3}{\frac{0.8x+25}{x}} = 3 \cdot \frac{x}{0.8x+25} = \frac{3x}{0.8x+25}$$

TECHNOLOGY GRAPHICS CALCULATOR
Keystroke Guide, page 555

Enter $y = \frac{3x}{0.8x+25}$ into a graphics calculator, and examine a table of values.

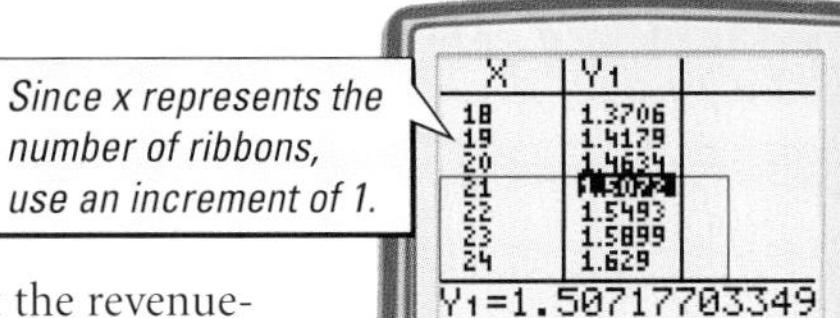

Since x represents the number of ribbons, use an increment of 1.
X | Y1
18 | 1.3706
19 | 1.4179
20 | 1.4634
21 | 1.5072
22 | 1.5493
23 | 1.5899
24 | 1.629
Y1=1.50717703349

From the table, you can see that the revenue-to-cost ratio, y, is greater than 1.5 when 21 or more ribbons, x, were produced and sold.

Activity: Exploring Excluded Values in Quotients

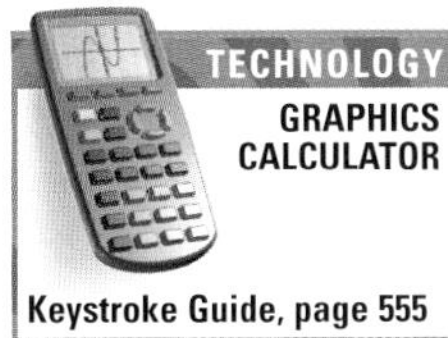
TECHNOLOGY GRAPHICS CALCULATOR
Keystroke Guide, page 555

You will need: a graphics calculator

1. Let $f(x) = \dfrac{\frac{x-3}{x+2}}{\frac{x-2}{x+2}}$. When the complex fraction that defines f is simplified, it becomes $\frac{x-3}{x-2}$. Let $g(x) = \frac{x-3}{x-2}$. Graph f and g on the same screen. What observations can you make about these graphs?

PROBLEM SOLVING

2. **Make a table** to evaluate f and g for x-values of -3, -2, -1, 0, 1, 2, and 3. How do the entries in the table compare?
3. Repeat Steps 1 and 2 for the functions $f(x) = \dfrac{\frac{x^2-4}{x-1}}{\frac{x^2-9}{x-1}}$ and $g(x) = \dfrac{x^2-4}{x^2-9}$.

CHECKPOINT ✔

4. Let $f(x) = \dfrac{\frac{P}{Q}}{\frac{R}{Q}}$ and $g(x) = \frac{P}{R}$, where P, Q, and R are polynomials. Explain how to find the excluded values of f and why you should not try to find those values by examining g.

Teaching Tip

TECHNOLOGY In Example 6, students might want to view the graph to get a general idea of where the graph reaches a y-value of 1.5. Use the following window:

Xmin=0	Ymin=0
Xmax=50	Ymax=2
Xscl=5	Yscl=1

Activity Notes

In this Activity, students learn that a simplified rational expression alone is not always equivalent to the original rational expression. The original expression must be used to find all excluded values in the expression. As students work through this Activity, they will notice that the graphics calculator gives error messages in the table for the excluded values of the original rational expression but not for those of the simplified expression.

Teaching Tip

TECHNOLOGY When using a TI-83 or TI-82 graphics calculator to analyze the graphs in the Activity, use the following viewing window:

Xmin=–4.7	Ymin=–3.1
Xmax=4.7	Ymax=3.1
Xscl=1	Yscl=1

CHECKPOINT ✔

4. Find the values of x that make $Q = 0$ and the values that make $R = 0$. If the fraction is reduced by cancellation, some excluded factors will be lost.

Reteaching the Lesson

INVITING PARTICIPATION Give each student a modified bingo card whose squares contain various unsimplified rational expressions. Call out the simplified form of each expression on the cards in random order. Students should place a chip over any equivalent expression on their card. Write the simplified expressions on the board as they are called out so that the bingo winners can be verified. Once a student has a row, column, or diagonal covered, he or she must verify the bingo for the class.

Assess

Selected Answers

Exercises 4–8, 9–63 odd

ASSIGNMENT GUIDE	
In Class	1–8
Core	9–45 odd
Core Plus	10–46 even
Review	48–63
Preview	64–67

Extra Practice can be found beginning on page 940.

Exercises

Communicate

1. In what ways is the multiplication of two rational expressions similar to the multiplication of two rational numbers?
2. In what ways is the division of two rational expressions similar to the division of two rational numbers?
3. Explain how to simplify a complex fraction such as $\dfrac{\frac{x^2}{x^2-1}}{\frac{x}{x^2+2x-3}}$. Compare the excluded values of x in the complex fraction and in its simplified form.

Guided Skills Practice

Simplify each rational expression. ***(EXAMPLES 1 AND 2)***

4. $\dfrac{x^2-25}{x^2-10x+25}$ $\mathbf{\dfrac{x+5}{x-5}}$
5. $\dfrac{4x^2}{5} \cdot \dfrac{30}{x^4} \cdot \dfrac{20x^3}{60}$ $\mathbf{8x}$

Simplify each rational expression. ***(EXAMPLES 3 AND 4)***

6. $\dfrac{x^2+8x+12}{x^2+2x-15} \cdot \dfrac{x^2+8x+15}{x^2+9x+18}$ $\mathbf{\dfrac{x+2}{x-3}}$
7. $\dfrac{x^2-2x+1}{x^2+6x+8} \div \dfrac{x^2-1}{x^2+3x+2}$ $\mathbf{\dfrac{x-1}{x+4}}$
8. Simplify the complex fraction $\dfrac{\frac{2x-6}{x^2+9x+20}}{\frac{x^2-9}{x^2+5x+4}}$. ***(EXAMPLES 5 AND 6)*** $\mathbf{\dfrac{2(x+1)}{(x+5)(x+3)}}$, **or** $\mathbf{\dfrac{2x+2}{x^2+8x+15}}$

Practice and Apply

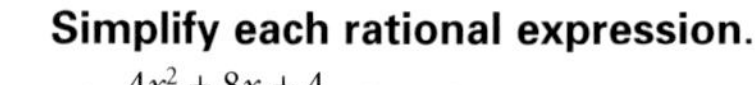

internet connect
Homework Help Online
Go To: **go.hrw.com**
Keyword:
MB1 Homework Help for Exercises 9–43

Simplify each rational expression.

9. $\dfrac{4x^2+8x+4}{x+1}$ $\mathbf{4x+4}$
10. $\dfrac{x^2-6x+9}{x^2-9}$ $\mathbf{\dfrac{x-3}{x+3}}$
11. $\dfrac{15}{x^2} \cdot \dfrac{x^5}{12} \cdot \dfrac{4}{x}$ $\mathbf{5x^2}$
12. $\dfrac{36x}{9x^2} \cdot \dfrac{12x^7}{2x} \cdot \dfrac{5}{x^2}$ $\mathbf{120x^3}$
13. $\dfrac{x^2-10x+9}{x^2+2x-3}$ $\mathbf{\dfrac{x-9}{x+3}}$
14. $\dfrac{-x^2-x+6}{x^2-5x+6}$ $\mathbf{\dfrac{-x-3}{x-3}}$
15. $\dfrac{x}{9x^8} \cdot \dfrac{x^7}{2x} \cdot \dfrac{45}{x^4}$ $\mathbf{\dfrac{5}{2x^5}}$
16. $\dfrac{-5}{x^3} \cdot \dfrac{-x^5}{3} \cdot \dfrac{-4}{x} \cdot \dfrac{20}{x^3}$ $\mathbf{\dfrac{-400}{3x^2}}$
17. $\dfrac{x^2-4x-5}{x^2-3x+2} \cdot \dfrac{x^2-4}{x^2-3x-10}$ $\mathbf{\dfrac{x+1}{x-1}}$
18. $\dfrac{x^2-9}{x^2-4x+4} \cdot \dfrac{x^2-4}{x^2-x-6}$ $\mathbf{\dfrac{x+3}{x-2}}$
19. $\dfrac{2x^2-2x}{x^2-9} \div \dfrac{x^2+x-2}{x^2+2x-3}$ $\mathbf{\dfrac{2x^2-2x}{x^2-x-6}}$
20. $\dfrac{4x^2+20x}{9+6x+x^2} \div \dfrac{x+5}{x^2-9}$ $\mathbf{\dfrac{4x^2-12x}{x+3}}$
21. $\dfrac{x^4+2x^3+x^2}{x^2+x-6} \cdot \dfrac{x^2-x-2}{x^4-x^2}$ $\mathbf{\dfrac{x^2+2x+1}{x^2+2x-3}}$
22. $\dfrac{x^5-4x^3}{x^2-x-2} \cdot \dfrac{x^2-1}{x^5-x^4-2x^3}$ $\mathbf{\dfrac{x^2+x-2}{x^2-x-2}}$
23. $\dfrac{4x^3-9x}{2x-7} \div \dfrac{3x^3+2x^2}{4x^2-14x}$ $\mathbf{\dfrac{8x^2-18}{3x+2}}$
24. $\dfrac{x^4-4x^2}{x^2-9} \div \dfrac{4x^2-4x^3+x^4}{x^2-6x+9}$ $\mathbf{\dfrac{x^2-x-6}{x^2+x-6}}$
25. $\dfrac{ax-bx+ay-by}{ax+bx+ay+by}$ $\mathbf{\dfrac{a-b}{a+b}}$
26. $\dfrac{x^2-y^2-4x+4y}{x^2-y^2+4x-4y}$ $\mathbf{\dfrac{x+y-4}{x+y+4}}$
27. $\dfrac{x^2}{4} \cdot \left(\dfrac{xy}{6}\right)^{-1} \cdot \dfrac{2y^2}{x}$ $\mathbf{3y}$
28. $2rs \div \dfrac{2r^2}{s} \div \dfrac{2s^2}{r}$ $\mathbf{\dfrac{1}{2}}$

Practice

NAME ______________ CLASS ________ DATE ________

Practice
8.3 Multiplying and Dividing Rational Expressions

Simplify each rational expression.

1. $\dfrac{2x^4}{x^5} \cdot \dfrac{6x}{x^3} \cdot \dfrac{x}{4}$ $\dfrac{3}{x^2}$
2. $\dfrac{x^2-8x+7}{x^2+6x-7}$ $\dfrac{x-7}{x+7}$
3. $\dfrac{9x^2+12x+4}{9x^2-4}$ $\dfrac{3x+2}{3x-2}$
4. $\dfrac{3x}{x^{10}} \cdot \dfrac{x^3}{27} \cdot \dfrac{9x^4}{2}$ $\dfrac{1}{2x^2}$
5. $\dfrac{x^2+7x+12}{x^2+x-6}$ $\dfrac{x+4}{x-2}$
6. $\dfrac{4x^4}{9x} \cdot \dfrac{9x^3}{10x} \cdot \dfrac{15x^2}{2x}$ $3x^6$
7. $\dfrac{x^2-5x+6}{x+4} \cdot \dfrac{3x+12}{x-2}$ $3x-9$
8. $\dfrac{2x-3}{5x+1} \div \dfrac{6x^2-13x+6}{15x^2-7x-2}$ 1
9. $\dfrac{4x-8}{x^2-x-6} \div \dfrac{x^3+x^2-6x}{x^2-9}$ $\dfrac{4}{x+2}$
10. $\dfrac{x^3-9x}{x^2+11x+24} \cdot \dfrac{x^2+7x-8}{x^2-4x+3}$ x
11. $\dfrac{\frac{x^2-16}{x-3}}{\frac{x+4}{x^2-9}}$ x^2-x-12
12. $\dfrac{\frac{x-5}{x^2-100}}{\frac{x^2-25}{x+10}}$ $\dfrac{1}{x^2-5x-50}$
13. $\dfrac{\frac{x^2+10x-11}{x^2+6x+5}}{\frac{x^2+9x-22}{x^2+3x-10}}$ $\dfrac{x-1}{x+1}$
14. $\dfrac{\frac{x^4-81}{3x^2+27}}{\frac{x^2-x-12}{x}}$ $\dfrac{x^2-3x}{3x-12}$
15. $\dfrac{x-6}{x+2} \cdot \dfrac{\frac{2x-1}{x-6}}{\frac{x-2}{x+2}}$ $\dfrac{2x-1}{x-2}$
16. $\dfrac{\frac{x^2-5x+6}{x^2-8x+15}}{\frac{x-2}{x-5}} \div \dfrac{x^2-9}{x^2+3x}$ $\dfrac{x}{x-3}$
17. $\dfrac{x-4}{x-7} \cdot \dfrac{\frac{x^2-49}{3x-12}}{\frac{x^2+14x+49}{x+5}}$ $\dfrac{x+5}{3x+21}$
18. $\dfrac{\frac{x^2-y^2}{5x^3y^2}}{\frac{4x+4y}{15x^2y^5}}$ $\dfrac{3y^3(x-y)}{4x}$

Simplify each expression.

29. $\dfrac{\frac{(x+2)^2}{(x+3)^2}}{\frac{x+3}{x+2}}$ **$\dfrac{(x+2)^3}{(x+3)^3}$**

30. $\dfrac{\frac{x^2-4}{x^2-9}}{\frac{(x-2)^2}{(x-3)^2}}$

31. $\dfrac{\frac{x^2-9x+14}{x^2-6x+5}}{\frac{x^2-8x+7}{x^2-7x+10}}$

32. $\dfrac{\frac{x^2+4x+3}{x^2+6x+8}}{\frac{x^2+9x+18}{x^2+7x+10}}$

33. $\dfrac{x+2}{x+5} \cdot \dfrac{\frac{x^2}{x+2}}{\frac{x+1}{x+5}}$ **$\dfrac{x^2}{x+1}$**

34. $\dfrac{\frac{1}{x+3}}{\frac{x^2}{x-7}} \cdot \dfrac{x}{x-7}$ **$\dfrac{1}{x^2+3x}$**

35. $\dfrac{2x+3}{x-1} \div \dfrac{\frac{x}{x-1}}{\frac{3x}{2x+3}}$ **3**

36. $\dfrac{x}{x^2-1} \div \dfrac{\frac{x+1}{x-1}}{\frac{x}{x+1}}$ **$\dfrac{x^2}{(x+1)^3}$**

37. $\dfrac{\frac{x+3}{x-1}}{x(x-1)^{-1}}$ **$\dfrac{x+3}{x}$**

38. $\dfrac{\frac{2y+6}{y-7}}{(y+2)(y+3)^{-1}}$

39. $\dfrac{\frac{(x+y)^2}{(x+y)^3}}{\frac{x+y}{x^2+2xy+y^2}}$ **1**

40. $\dfrac{\frac{x+2y}{2x^2+3xy+y^2}}{\frac{2x^2+5xy+2y^2}{x+y}}$

41. $\dfrac{1-7x^{-1}-18x^{-2}}{1-4x^{-2}}$ **$\dfrac{x-9}{x-2}$**

42. $\dfrac{1+12x^{-1}+27x^{-2}}{x^{-1}+9x^{-2}}$ **$x+3$**

43. $\dfrac{(x+y)y^{-1}-2x(x+y)^{-1}}{(x-y)y^{-1}+2x(x-y)^{-1}}$

CHALLENGE

44. Find the rational expression R whose numerator and denominator have degree 2 and leading coefficients of 1 such that $\dfrac{x^2+3x-10}{x^2-8x+15} \cdot R = \dfrac{x-2}{x-3}$.

CONNECTION

45. GEOMETRY An open-top box is to be made from a sheet of cardboard that is 20 inches by 16 inches. Squares with sides of x inches are to be cut on one side and creased on another to form tabs. When the sides are folded up, these tabs are glued to the adjacent sides to provide reinforcement.

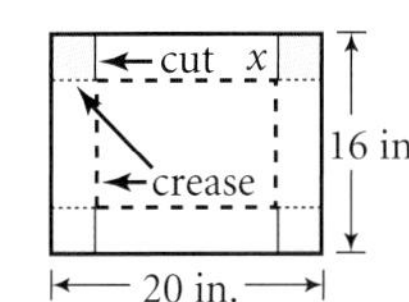
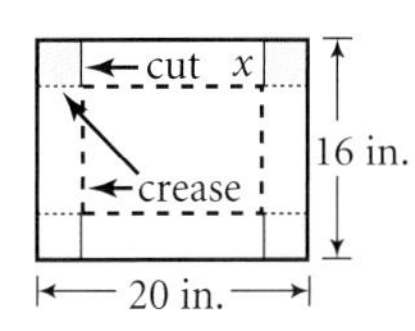

a. Show that $x(20-2x)(16-2x)$ represents the volume of the box.

b. Show that $320-4x^2$ represents the surface area of the bottom and sides of the inside of the box.

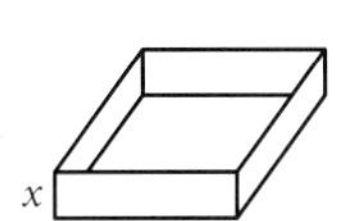

c. Write and simplify an expression for the ratio of the volume of the box to the inside surface area of the box.

d. How does the ratio from part **c** change as x increases?

APPLICATION

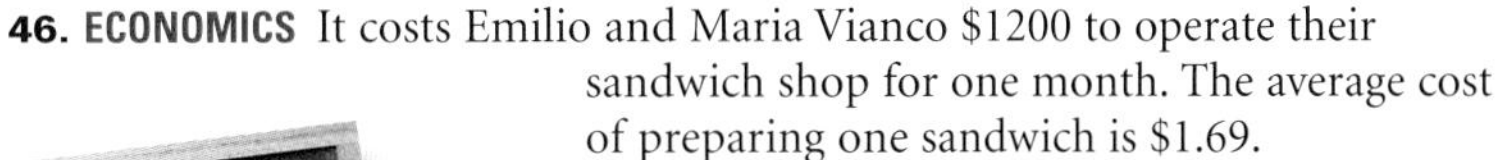

46. **ECONOMICS** It costs Emilio and Maria Vianco \$1200 to operate their sandwich shop for one month. The average cost of preparing one sandwich is \$1.69.

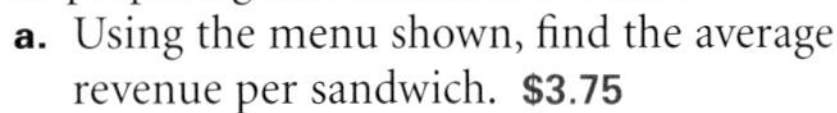
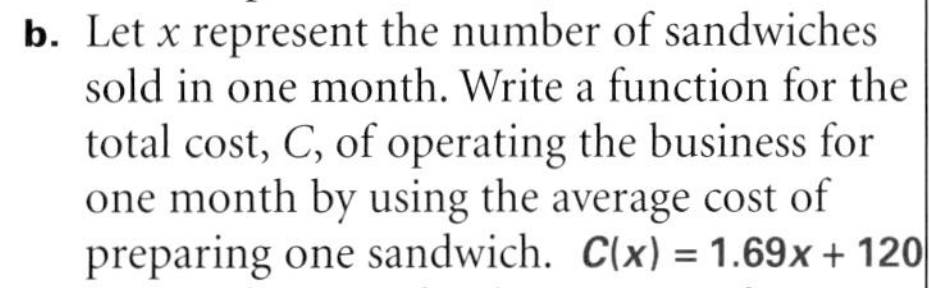

a. Using the menu shown, find the average revenue per sandwich. **\$3.75**

b. Let x represent the number of sandwiches sold in one month. Write a function for the total cost, C, of operating the business for one month by using the average cost of preparing one sandwich. **$C(x) = 1.69x + 1200$**

c. Write a function for the ratio, R, of average revenue per sandwich to the average cost per sandwich. **$R(x) = \dfrac{3.75x}{1.69x + 1200}$**

Answers

30. $\dfrac{x^2-x-6}{x^2+x-6}$

31. $\dfrac{x^2-4x+4}{x^2-2x+1}$

32. $\dfrac{x^2+6x+5}{x^2+10x+24}$

38. $\dfrac{2(y+3)^2}{(y-7)(y+2)}$

40. $\dfrac{1}{4x^2+4xy+y^2}$

43. $\dfrac{x-y}{x+y}$

44. $R = \dfrac{x^2-5x}{x^2+5x}$

45a. $V = \ell wh$
$= x(20-2x)(16-2x)$

c. $R(x) = \dfrac{x^3-18x^2+80x}{80-x^2}$

d. The ratio increases.

45b. total area:
$A = (20-2x)(16-2x) +$
$2x(20-2x) + 2x(16-2x)$
$= 320 - 40x - 32x + 4x^2 +$
$40x - 4x^2 + 32x - 4x^2$
$= 320 - 4x^2$

Error Analysis

When dividing fractions, remind students to multiply by the reciprocal of the divisor. Also remind them that they should not divide out common factors until the division problem has been rewritten as a product and that they should always remove common factors before multiplying to ensure the result is written in simplified form.

Student Technology Guide

NAME ______ CLASS ______ DATE ______

Student Technology Guide
8.3 ***Multiplying and Dividing Rational Expressions***

Simplifying rational expressions, their products, or their quotients involves factoring polynomials. You can use a graphics calculator to help decide whether a polynomial can be factored.

If a polynomial can be factored, the graph of its related function must have one or more x-intercepts.

Example: Simplify $\dfrac{x^2-x-6}{x^2+4x+6}$.

Graph the related function for each polynomial in the rational expression.

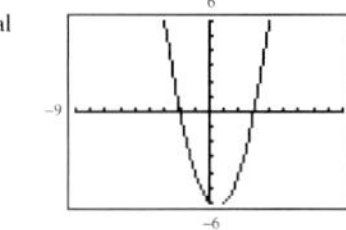

- Press Y= X,T,θ,n x² − X,T,θ,n − 6 GRAPH. The graph, shown at right, has x-intercepts at $x = -2$ and $x = 3$, so $x^2 - x - 6 = (x+2)(x-3)$.
- Press Y= X,T,θ,n x² + 4 X,T,θ,n + 6 GRAPH. The graph has no x-intercepts, so $x^2 + 4x + 6$ cannot be factored.

Thus, $\dfrac{x^2-x-6}{x^2+4x+6}$ cannot be simplified because the numerator and denominator have no common factors.

Simplify.

1. $\dfrac{x^2+3x-18}{x^2-36}$ **$\dfrac{x-3}{x-6}$**

2. $\dfrac{x+2}{x^2+2x+5} \cdot \dfrac{x+1}{x^2+7x+10}$ **See above.**

2. $\dfrac{x+1}{(x^2+2x+5)(x+5)} = \dfrac{x+1}{x^3+7x^2+15x+25}$

3. $\dfrac{3x^2+1}{x^2-3x-4} \div \dfrac{1}{x+1}$ **$\dfrac{3x^2+1}{x-4}$**

Some calculators, such as the TI-92, can simplify products and quotients of rational expressions. To simplify $\dfrac{x^2-9}{x^2+6x+8} \div \dfrac{x^2+6x+9}{x+4}$, press ((X ^ 2 − 9) ÷ (X ^ 2 + 6 X + 8)) ÷ ((X ^ 2 + 6 X + 9) ÷ (X + 4)) ENTER.

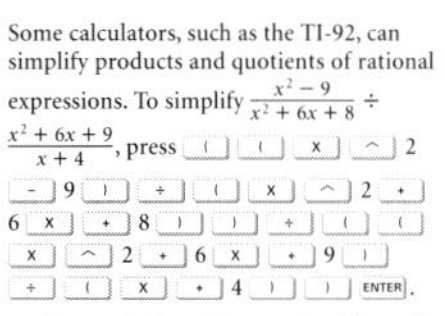
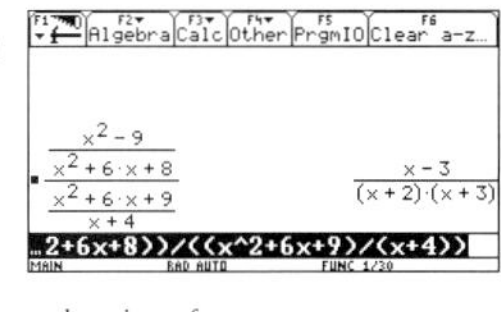

4. Use a calculator that can simplify products and quotients of rational expressions to check your answers to Exercises 1–3.

In Exercises 64–67, students use their skills in adding fractions to find the sum of simple rational expressions. This concept will be developed more fully in Lesson 8.4.

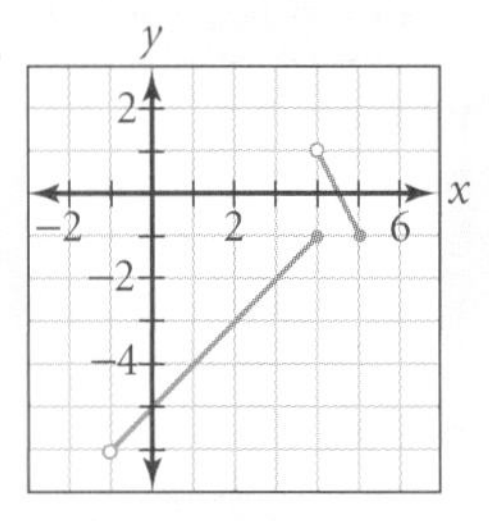

51.

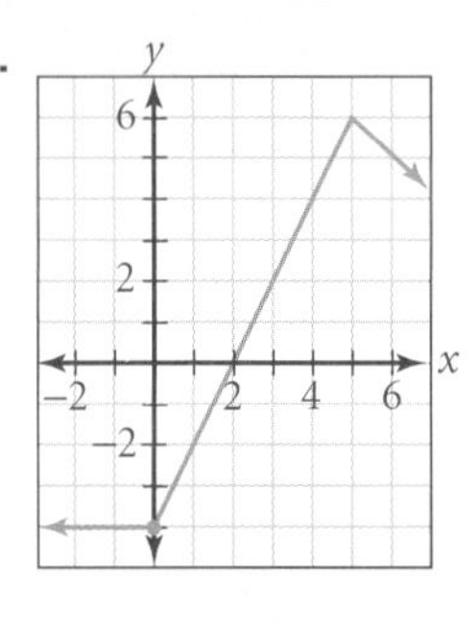

APPLICATION

47. **PHYSICS** The diagram below illustrates an ambulance traveling a definite distance in a specific period of time. The average acceleration, a, is defined as the change in velocity over the corresponding change in time.

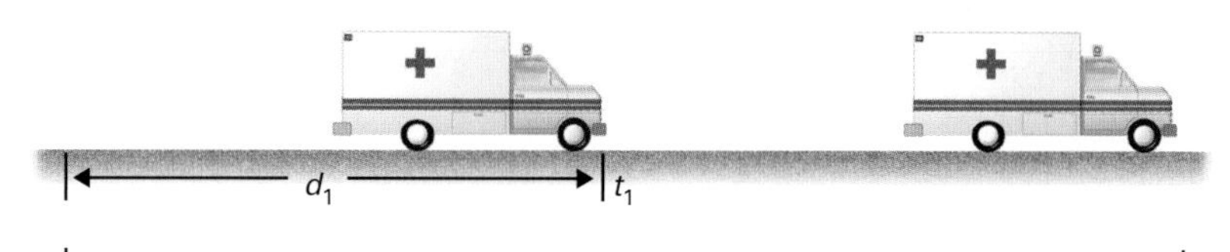

47a. $a = \frac{d_2t_1 - d_1t_2}{t_1t_2(t_2 - t_1)}$

b. feet per second per second

a. Simplify the expression at right that defines a.
b. If the distance, d, is measured in feet and the time, t, is measured in seconds, in what units is acceleration measured?

$$a = \frac{\frac{d_2}{t_2} - \frac{d_1}{t_1}}{t_2 - t_1}$$

Look Back

Write an equation in slope-intercept form for the line that contains the given point and is perpendicular to the given line. *(LESSON 1.3)*

48. $(8, -4)$, $y = -6x - 1$ **$y = \frac{1}{6}x - 5\frac{1}{3}$**

49. $(3, 5)$, $y = \frac{1}{5}x - 11$ **$y = -5x + 20$**

Graph each piecewise function. *(LESSON 2.6)*

50. $f(x) = \begin{cases} x - 5 & \text{if } -1 < x \leq 4 \\ 9 - 2x & \text{if } 4 < x \leq 5 \end{cases}$

51. $g(x) = \begin{cases} -4 & \text{if } x < 0 \\ 2x - 4 & \text{if } 0 \leq x \leq 5 \\ -\frac{2}{5}x + 8 & \text{if } x > 5 \end{cases}$

Factor each expression. *(LESSON 5.3)*

52. $8x^2 - 4x$ **$4x(2x - 1)$**

53. $12x^2 - 3x + 6$ **$3(4x^2 - x + 2)$**

54. $12 - 4a - 22a^2$ **$2(6 - 2a - 11a^2)$**

Simplify each expression. Write your answer in the standard form for a complex number. *(LESSON 5.6)*

55. $\frac{2+i}{3+2i}$ **$\frac{8}{13} - \frac{1}{13}i$**

56. $\frac{4-i}{6-3i}$ **$\frac{3}{5} + \frac{2}{15}i$**

57. $\frac{3-2i}{5+i}$ **$\frac{1}{2} - \frac{1}{2}i$**

Write each product as a polynomial expression in standard form. *(LESSON 7.3)*

58. $x^2(x^3 - x^2 - 6x + 2)$ **$x^5 - x^4 - 6x^3 + 2x^2$**

59. $(x - 2)(3x^3 - 6x - x^2)$ **$3x^4 - 7x^3 - 4x^2 + 12x$**

60. $(x^2 + 1)(2x^3 - 9)$ **$2x^5 + 2x^3 - 9x^2 - 9$**

Factor each polynomial expression. *(LESSON 7.3)*

61. $x^3 - 1$ **$(x - 1)(x^2 + x + 1)$**

62. $125x^3 + 27$ **$(5x + 3)(25x^2 - 15x + 9)$**

63. $x^3 - 6x^2 - 8x$ **$x(x^2 - 6x - 8)$**

Look Beyond

Simplify.

64. $\frac{5}{8} + \frac{1}{8}$ **$\frac{3}{4}$**

65. $\frac{3}{x} + \frac{1}{x}$ **$\frac{4}{x}$**

66. $\frac{3}{2x} + \frac{1}{x}$ **$\frac{5}{2x}$**

67. $\frac{3}{2x} + \frac{1}{3x}$ **$\frac{11}{6x}$**

Adding and Subtracting Rational Expressions

Why
Adding and subtracting rational expression can be used to solve real-world problems, such as calculating the average rate of speed for an entire trip.

Objective

- Add and subtract rational expressions.

APPLICATION
TRAVEL

A cab driver drove from the airport to a passenger's home at an average speed of 55 miles per hour. He returned to the airport along the same highway at an average speed of 45 miles per hour. What was the cab driver's average speed over the entire trip? The answer is not the average of 45 and 55. To answer this question, you need to add two rational expressions. *You will answer this question in Example 5.*

Adding two rational expressions with the same denominator is similar to adding two rational numbers with the same denominator.

Rational Numbers	Rational Expressions
$\frac{1}{7}+\frac{3}{7}=\frac{1+3}{7}=\frac{4}{7}$ (Common denominator)	$\frac{3}{x^2}+\frac{5}{x^2}=\frac{3+5}{x^2}=\frac{8}{x^2}$ (Common denominator)

EXAMPLE 1 Simplify.

a. $\frac{2x}{x+3}+\frac{5}{x+3}$

b. $\frac{x^2}{x-3}-\frac{9}{x-3}$

SOLUTION

a. $\frac{2x}{x+3}+\frac{5}{x+3}=\frac{2x+5}{x+3}$

b. $\frac{x^2}{x-3}-\frac{9}{x-3}=\frac{x^2-9}{x-3}$

$=\frac{(x+3)(x-3)}{x-3}$

$=x+3$

Note that 3 is an excluded value of x in the original expression.

TRY THIS Simplify.

a. $\frac{3x-1}{2x-1}+\frac{5+2x}{2x-1}$

b. $\frac{2x}{x-5}-\frac{10}{x-5}$

Prepare

NCTM PRINCIPLES & STANDARDS 1–4, 6–10

QUICK WARM-UP

Simplify.

1. $-\frac{5}{8}-\left(-\frac{1}{8}\right)$ $\quad -\frac{1}{2}$
2. $-\frac{2}{3}+\frac{1}{9}$ $\quad -\frac{5}{9}$
3. $\frac{1}{x}+\frac{1}{x}$ $\quad \frac{2}{x}$
4. $\frac{1}{2n}+\frac{1}{2n}$ $\quad \frac{1}{n}$
5. $\frac{1}{a}+\frac{1}{b}$ $\quad \frac{a+b}{ab}$
6. $\frac{x}{y}+\frac{x}{y}$ $\quad \frac{2x}{y}$

Also on Quiz Transparency 8.4

Teach

Why Have students read the travel scenario given on this page. Ask students discuss why the average speed for the trip cannot be the average of 45 and 55. **Jack travels at 45 mph longer than at 55 mph.** Without doing any algebra, have students estimate what they think the average speed would be. **The average rate is 49.5 mph.**

ADDITIONAL EXAMPLE 1

Simplify.

a. $\frac{9x}{x+7}+\frac{2x}{x+7}$ $\quad \frac{11x}{x+7}$

b. $\frac{7x}{x-4}-\frac{4x+12}{x-4}$ $\quad 3x-12,\ x\neq 4$

TRY THIS

a. $\frac{5x+4}{2x-1}$

b. $2, x\neq 5$

Alternative Teaching Strategy

USING TECHNOLOGY Have students use graphing technology to graph the following sums of rational functions:

1. $y=\frac{1}{x}+\frac{4}{x}$
2. $y=\frac{1}{x-3}+\frac{1}{x+2}$
3. $y=\frac{1}{x}+\frac{1}{x(x+2)}$

Then have students graph the functions below.

a. $y=\frac{2x-1}{x^2-x-6}$ **b.** $y=\frac{5}{x}$ **c.** $y=\frac{x+3}{x^2+2x}$

Have students match the equivalent functions and then compare them. **1 matches b, 2 matches a, and 3 matches c.** Suggest that they factor the denominator of functions **a** and **c** to see where the common denominator came from. Have students use what they know about adding numerical fractions to help them understand how each pair of rational functions is equivalent.

ADDITIONAL EXAMPLE 2

Simplify $\frac{x}{x+6}+\frac{72}{x^2-36}$.

$\frac{x-12}{x-6}, x \neq 6$

Teaching Tip

TECHNOLOGY Point out that, in general, the domain of the original expression is not the domain of the simplified expression. That is, the original and simplified expressions are equivalent except for those domain values that are excluded from the original expression but not excluded from the simplified form. For example, $\frac{x^2-x-2}{x+1}=x-2$ when $x \neq -1$; therefore, the graph of $y=\frac{x^2-x-2}{x+1}$ is the same as the line $y=x-2$ with a hole at $(-1, -3)$.

TRY THIS

$\frac{x-10}{x-5}, x \neq -5$

CHECKPOINT ✔

Factoring the polynomials may make the LCD more apparent by revealing common factors; $\frac{x^2-3x+5}{x^2-6x+9}$.

To add two rational expressions with unlike denominators, you first need to find common denominators. The **least common denominator** (**LCD**) of two rational expressions is the *least common multiple* of the denominators. The **least common multiple** (**LCM**) of two polynomials is the polynomial of lowest degree that is divisible by each polynomial.

Finding the LCM for two rational expressions is similar to finding the LCM for two rational numbers. Compare the procedures for rational numbers and for rational expressions shown below.

Rational Numbers

$$\frac{7}{300}+\frac{1}{90}=\frac{7}{300}\left(\frac{3}{3}\right)+\frac{1}{90}\left(\frac{10}{10}\right)$$
$$=\frac{21+10}{900}$$
$$=\frac{31}{900}$$

Least common denominator

Rational Expressions

$$\frac{7}{3x^2}+\frac{1}{9x}=\frac{7}{3x^2}\left(\frac{3}{3}\right)+\frac{1}{9x}\left(\frac{x}{x}\right)$$
$$=\frac{21+x}{9x^2}$$

Least common denominator

Adding and Subtracting Rational Expressions

To add or subtract two rational expressions, find a common denominator, rewrite each expression by using the common denominator, and then add or subtract. Simplify the resulting rational expression.

EXAMPLE 2 Simplify $\frac{x}{x-2}+\frac{-8}{x^2-4}$.

SOLUTION

$$\frac{x}{x-2}+\frac{-8}{x^2-4}=\frac{x}{x-2}+\frac{-8}{(x-2)(x+2)}$$

$$=\frac{x}{x-2}\left(\frac{x+2}{x+2}\right)+\frac{-8}{(x-2)(x+2)}$$ *The LCD is (x – 2)(x + 2).*

$$=\frac{x(x+2)-8}{(x-2)(x+2)}$$ *Add the fractions.*

$$=\frac{x^2+2x-8}{(x-2)(x+2)}$$ *Write the numerator in standard form.*

$$=\frac{(x+4)(x-2)}{(x-2)(x+2)}$$ *Factor the numerator.*

$$=\frac{(x+4)\cancel{(x-2)}^{1}}{\cancel{(x-2)}_{1}(x+2)}$$ *Divide out the common factors.*

$$=\frac{x+4}{x+2}$$

Note that 2 and –2 are excluded values of x in the original expression.

TRY THIS Simplify $\frac{x}{x+5}+\frac{-50}{x^2-25}$.

CHECKPOINT ✔ Explain how factoring a polynomial can help you to add two rational expressions. Illustrate your response by simplifying $\frac{x}{x-3}+\frac{5}{x^2-6x+9}$.

Interdisciplinary Connection

PHYSICAL EDUCATION A triathlon requires athletes to run 5 miles, bike 20 miles, and swim 2 miles. Alice estimates her average speeds, in miles per minute, for each event as follows:

running: $r_r=\frac{1}{7}$ biking: $r_b=\frac{1}{4}$

swimming: $r_s=\frac{1}{30}$.

Write an expression for the total time it will take Alice to finish the race, and find her finish time.

$t_{\text{total}}=\frac{d_r}{r_r}+\frac{d_b}{r_b}+\frac{d_s}{r_s}$; **175 min, or 2 hr and 55 min**

Enrichment

Have students investigate *partial fractions*. Suppose that $\frac{4x+1}{x^2+x}$ is the sum of two rational expressions. Factor the denominator and create two partial fractions having those factors as denominators: $\frac{4x+1}{x(x+1)}=\frac{A}{x}+\frac{B}{x+1}=\frac{A(x+1)+Bx}{x(x+1)}$. Set the numerators equal to each other, and combine like terms: $4x+1=(A+B)x+A$. Solving $\begin{cases}A+B=4\\A=1\end{cases}$ yields $A=1$ and $B=3$. This process is valid when the denominator consists of *nonrepeating linear* factors.

EXAMPLE 3 Simplify $\frac{6x}{3x-1} - \frac{4x}{2x+5}$.

SOLUTION

$$\frac{6x}{3x-1} - \frac{4x}{2x+5} = \frac{6x}{3x-1}\left(\frac{2x+5}{2x+5}\right) - \frac{4x}{2x+5}\left(\frac{3x-1}{3x-1}\right)$$ *The LCD is (3x – 1)(2x + 5).*

$$= \frac{6x(2x+5) - 4x(3x-1)}{(3x-1)(2x+5)}$$ *Subtract.*

$$= \frac{12x^2 + 30x - 12x^2 + 4x}{(3x-1)(2x+5)}$$

$$= \frac{34x}{(3x-1)(2x+5)}, \text{ or } \frac{34x}{6x^2+13x-5}$$

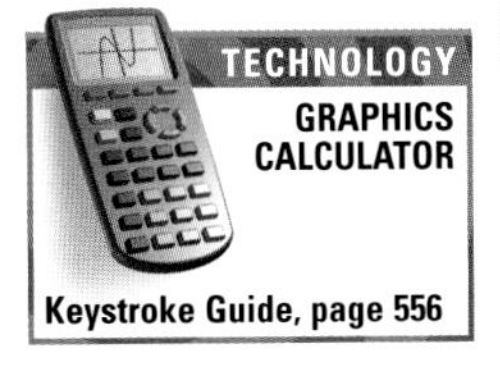

CHECK

Graph $y = \frac{6x}{3x-1} - \frac{4x}{2x+5}$ and $y = \frac{34x}{6x^2+13x-5}$ together on the same screen to see if the graphs are the same.

You can also use a table of values to verify that the corresponding y-values are the same.

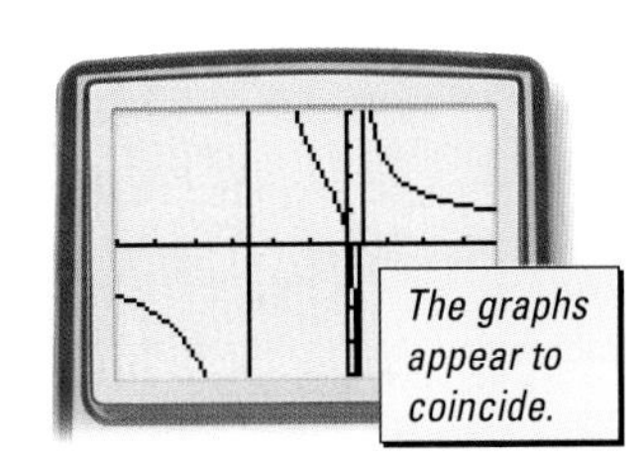

The graphs appear to coincide.

CHECKPOINT ✔ Identify the excluded values of x for the original expression and for the simplified expression in Example 3. Are they the same or different? Explain.

TRY THIS Simplify $\frac{6}{x^2-2x} - \frac{1}{x^2-4}$.

Sometimes you need to rewrite complex fractions as rational expressions in order to add or subtract them, as shown in Example 4.

EXAMPLE 4 Simplify $\frac{1}{1+\frac{1}{a}} + \frac{1}{1-\frac{1}{a}}$.

SOLUTION

$$\frac{1}{1+\frac{1}{a}} + \frac{1}{1-\frac{1}{a}} = \frac{1}{\frac{a+1}{a}} + \frac{1}{\frac{a-1}{a}}$$ *Add or subtract within the denominators.*

$$= 1 \cdot \frac{a}{a+1} + 1 \cdot \frac{a}{a-1}$$ *Multiply by the reciprocals.*

$$= \frac{a}{a+1} + \frac{a}{a-1}$$

$$= \frac{a}{a+1}\left(\frac{a-1}{a-1}\right) + \frac{a}{a-1}\left(\frac{a+1}{a+1}\right)$$ *The LCD is (a + 1)(a – 1).*

$$= \frac{a^2 - a + a^2 + a}{(a+1)(a-1)}$$ *Multiply, and then add.*

$$= \frac{2a^2}{a^2-1}$$

TRY THIS Simplify $\frac{a}{a-\frac{1}{a}} - \frac{a}{a+\frac{1}{a}}$.

ADDITIONAL EXAMPLE 3

Simplify $\frac{x}{3x-3} - \frac{x}{x^2-1}$.

$\frac{x^2-2x}{3(x+1)(x-1)}$, or $\frac{x^2-2x}{3x^2-3}$

Teaching Tip

TECHNOLOGY When using a TI-83 or TI-82 graphics calculator for Example 3, set the viewing window to the following:

Xmin=–6	Ymin=–4
Xmax=4	Ymax=4
Xscl=1	Yscl=1

CHECKPOINT ✔

The excluded values in the original expression and in the simplified expression are $\frac{1}{3}$ and $-\frac{5}{2}$. The excluded values are the same for both because none of the factors in the denominator can be cancelled.

TRY THIS

$\frac{5x+12}{x^3-4x}$

ADDITIONAL EXAMPLE 4

Simplify $\frac{2}{3+\frac{5}{a}} - \frac{2}{3-\frac{5}{a}}$.

$\frac{-20a}{9a^2-25}$

TRY THIS

$\frac{2a^2}{a^4-1}$

Inclusion Strategies

TACTILE LEARNERS Before students work with rational expressions, they need to have a firm understanding of similar operations with rational numbers, and many students at this level still have difficulty adding numerical fractions. Use fraction bars to strengthen students' understanding of adding fractions and have them make general statements about their procedures. For example, have them describe how to use fraction bars to find a common denominator and to find an equivalent fraction with a new denominator. Once students are able to describe these procedures in general, give them simple rational expressions to add and subtract.

ADDITIONAL

EXAMPLE 5

Angela drove from home to work at an average speed of 65 miles per hour. When she left work at 5:00 P.M., she drove home at an average speed of 50 miles per hour. **What was her average speed over the entire trip?**

≈56.5 mph

CRITICAL THINKING

Let d represent the distance to the rider's home. Then Jack's average speed is given below.

$$\frac{d+d}{\frac{d}{a}+\frac{d}{b}} = \frac{2d}{\frac{d(a+b)}{ab}} = \frac{2ab}{a+b}$$

$\frac{2ab}{a+b}$ is not equal to $\frac{a+b}{2}$, the average of a and b, unless $a = b$.

Assess

Selected Answers

Exercises 4–9, 11–61 odd

ASSIGNMENT GUIDE	
In Class	1–9
Core	11–41 odd, 47
Core Plus	10–40 even, 42–45, 46–50 even
Review	51–62
Preview	63

Extra Practice can be found beginning on page 940.

Mid-Chapter Assessment for Lessons 8.1 through 8.4 can be found on page 100 of the *Assessment Resources.*

EXAMPLE 5 Refer to the cab driver's round trip described at the beginning of the lesson.

What is the cab driver's average speed for the entire trip?

APPLICATION
TRAVEL

SOLUTION

Let d represent the length of the trip one way, let t_1 represent the cab driver's travel time to the passenger's home, and let t_2 represent his travel time back to the airport.

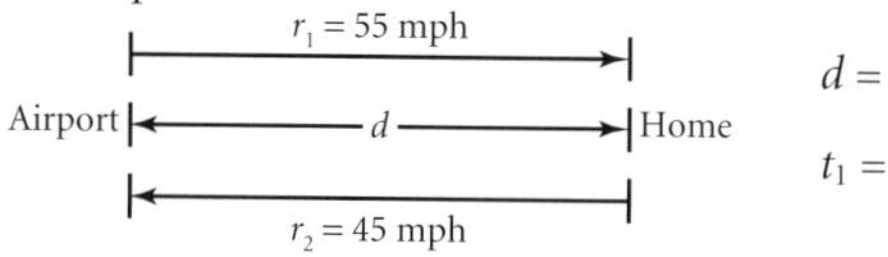

$d = r_1t_1$ and $d = r_2t_2$

$t_1 = \frac{d}{r_1} = \frac{d}{55}$ $\qquad t_2 = \frac{d}{r_2} = \frac{d}{45}$

$$\text{average speed} = \frac{\text{total distance}}{\text{total time}}$$

$$= \frac{d+d}{t_1+t_2}$$

$$= \frac{d+d}{\frac{d}{55}+\frac{d}{45}} \qquad \textit{Substitute } \tfrac{d}{55} \textit{ for } t_1 \textit{ and } \tfrac{d}{45} \textit{ for } t_2.$$

$$= \frac{2d}{\frac{9d+11d}{495}} \qquad \textit{The LCD for } \tfrac{d}{55} \textit{ and } \tfrac{d}{45} \textit{ is 495.}$$

$$= \frac{2d}{\frac{20d}{495}}$$

$$= 2d \times \frac{495}{20d} \qquad \textit{Multiply by the reciprocal.}$$

$$= 49.5$$

Thus, the cab driver's average speed was 49.5 miles per hour. The average speed was less than the average of 45 and 55 because he spent more time driving at 45 miles per hour than at 55 miles per hour.

CRITICAL THINKING Suppose that the cab driver travels to the passenger's home at a miles per hour and returns along the same route at b miles per hour. Show that his average speed for the entire trip is not simply $\frac{a+b}{2}$.

Exercises

Communicate

1. Explain how to find the least common denominator in order to add $\frac{x+5}{x^2-7x+6} + \frac{x-1}{x^2-36}$.
2. Explain how to use a graph to check your answer when you add two rational expressions.
3. Choose the two expressions below that are equivalent and explain why they are equivalent.

a. $\frac{3+7}{x^2+4}$ **b.** $\frac{10}{x^2} + \frac{10}{4}$ **c.** $\frac{3}{x^2} + \frac{7}{4}$ **d.** $\frac{3}{x^2+4} + \frac{7}{x^2+4}$

Reteaching the Lesson

COOPERATIVE LEARNING Have students work in groups of four, giving each group a set of rational expressions to add. Have two students simplify the sum algebraically, while the other two students graph the original and simplified related functions on a graphics calculator. The group should compare the graphs, and if the two graphs do not coincide, the group should work together to find the mistake(s).

Guided Skills Practice

Simplify. *(EXAMPLES 1 AND 2)*

4. $\frac{3x}{x-1} + \frac{2}{x-1}$ $\quad \frac{3x+2}{x-1}$

5. $\frac{3x+5}{x+2} - \frac{x+1}{x+2}$ $\quad 2$

6. $\frac{12}{x^2-1} + \frac{4}{x+1}$ $\quad \frac{4x+8}{x^2-1}$

Simplify. *(EXAMPLES 3 AND 4)*

7. $\frac{x+1}{2x-1} - \frac{2x+1}{x-1}$ $\quad \frac{-3x^2}{2x^2-3x+1}$

8. $\frac{1}{1-\frac{1}{t}}$ $\quad \frac{t}{t-1}$

APPLICATION

9. TRAVEL Refer to the cab driver's round trip described at the beginning of the lesson. What is the average speed for the entire trip if he drives to the passenger's home at 52 miles per hour and returns to the airport along the same route at 38 miles per hour? *(EXAMPLE 5)* 43.9 mph

Practice and Apply

internet connect

Homework Help Online
Go To: go.hrw.com
Keyword: **MB1 Homework Help** for Exercises 13–15, 17-18, 21, 24, 34–41

Simplify.

10. $\frac{2x-3}{x+1} + \frac{6x+5}{x+1}$ $\quad \frac{8x+2}{x+1}$

11. $\frac{7x-13}{2x-1} + \frac{x+9}{2x-1}$ $\quad 4$

12. $\frac{r+9}{4} + \frac{r-3}{2}$ $\quad \frac{3r+3}{4}$

13. $\frac{x+7}{3} - \frac{4x+1}{9}$ $\quad \frac{-x+20}{9}$

14. $\frac{x}{x^2-4} - \frac{2}{x-2}$ $\quad \frac{-x-4}{x^2-4}$

15. $\frac{2x}{x+3} - \frac{x-3}{x^2+6x+9}$

16. $\frac{-4}{x-5} + \frac{5}{x+3}$

17. $\frac{2}{x+2} - \frac{6}{x-2}$

18. $\frac{3}{x-1} - \frac{2}{x+1}$

19. $\frac{8}{3x-5} + \frac{7}{2x+3}$

20. $\frac{2x+3}{x+3} + \frac{x}{x-2}$

21. $\frac{x+2}{2x-1} - \frac{2x}{x-1}$

22. $x^2 + \frac{2x}{3x-5}$

23. $\frac{x+1}{(x-1)^2} + \frac{x-2}{x-1}$

24. $2x^2 - 1 - \frac{x-1}{x+2}$

25. $\frac{3}{\frac{2x-1}{x}}$ $\quad \frac{3x}{2x-1}$

26. $\frac{\frac{1}{3x+1}}{2}$ $\quad \frac{1}{6x+2}$

27. $\frac{\frac{4}{x-1}}{\frac{2}{x-1}} + \frac{3}{x-1}$ $\quad \frac{2x+1}{x-1}$

28. $\frac{\frac{4}{x+2}}{\frac{x+2}{3}} - \frac{3}{x+2}$

29. $\frac{\frac{x+2}{x+5}}{\frac{x-1}{x+5}} + \frac{1}{x+1}$

30. $\frac{\frac{2x+10}{x-1}}{\frac{x+5}{x^2-1}} - \frac{4}{x+1}$

31. $\frac{1-xy^{-1}}{x^{-1}-y^{-1}}$ $\quad x$

32. $\frac{x-y}{x^{-1}-y^{-1}}$ $\quad -xy$

33. $\frac{\frac{1}{a^2}-\frac{1}{b^2}}{a^{-2}+2(ab)^{-1}+b^{-2}}$ $\quad \frac{b-a}{b+a}$

Write each expression as a single rational expression in simplest form.

34. $\frac{3x}{x-1} + \frac{5x+2}{x-1} - \frac{10}{x-1}$ $\quad 8$

35. $\frac{7x}{x^2-1} - \frac{x}{x^2-1} + \frac{6}{x^2-1}$ $\quad \frac{6}{x-1}$

36. $\frac{7}{x+7} + \frac{-x}{x-7} + \frac{2x}{x^2-49}$ $\quad \frac{-x^2+2x-49}{x^2-49}$

37. $\frac{x}{x-3} - \frac{3}{x+4} + \frac{7}{x^2+x-12}$ $\quad \frac{x^2+x+16}{x^2+x-12}$

38. $(a-b)^{-1} - (a+b)^{-1}$ $\quad \frac{2b}{a^2-b^2}$

39. $(a-b)^{-2} - (a+b)^{-2}$ $\quad \frac{4ab}{a^4-2a^2b^2+b^4}$

40. $\frac{x}{x-y} - \frac{x^2+y^2}{x^2-y^2} + \frac{y}{x+y}$ $\quad \frac{2y}{x+y}$

41. $\frac{3r}{2r-s} - \frac{2r}{2r+s} + \frac{2s^2}{4r^2-s^2}$ $\quad \frac{2r^2+5rs+2s}{4r^2-s^2}$

CHALLENGE

Find numbers *A*, *B*, *C*, and *D* such that the given rational expression equals the sum of the two simpler rational expressions, as indicated.

42. $\frac{3x+2}{x-5} = \frac{Ax}{x-5} + \frac{D}{x-5}$ $\quad A=3, D=2$

43. $\frac{-x+1}{(x-2)(x-3)} = \frac{B}{x-2} + \frac{D}{x-3}$ $\quad B=1, D=-2$

44. $\frac{2x^2+5}{x^2+11x+30} = \frac{Ax}{x+5} + \frac{Cx+D}{x+6}$

45. $\frac{x^2-7}{x^2+2x-3} = \frac{Ax}{x+3} + \frac{Cx+D}{x-1}$

15. $\frac{2x^2+5x+3}{x^2+6x+9}$

16. $\frac{x-37}{x^2-2x-15}$

17. $\frac{-4x-16}{x^2-4}$

18. $\frac{x+5}{x^2-1}$

19. $\frac{37x-11}{6x^2-x-15}$

20. $\frac{3x^2+2x-6}{x^2+x-6}$

21. $\frac{-3x^2+3x-2}{2x^2-3x+1}$

22. $\frac{3x^3-5x^2+2x}{3x-5}$

44. $A=-11, C=13, D=1$

45. $A=\frac{1}{6}, C=\frac{5}{6}, D=-\frac{7}{3}$

Error Analysis

When subtracting rational expressions, students often forget to distribute the negative sign to each term in the numerator of the subtracted expression. For example, in Exercise 13, students may incorrectly simplify the numerator as follows:

$$3(x+7) - (4x+1) = 3x + 21 - 4x + 1 = -x + 22$$

The correct simplification is shown below.

$$3(x+7) - (4x+1) = 3x + 21 - 4x - 1 = -x + 20$$

Have students use numerical examples to see that the two methods give very different solutions.

23. $\frac{x^2-2x+3}{x^2-2x+1}$

24. $\frac{2x^3+4x^2-2x-1}{x+2}$

28. $\frac{-3x+6}{x^2+4x+4}$

29. $\frac{x^2+4x+1}{x^2-1}$

30. $\frac{2x^2+4x-2}{x+1}$

CONNECTION

46. GEOMETRY In the diagram at right, square A is 1 unit on a side, square B is $\frac{1}{2}$ of a unit on a side, square C is $\frac{1}{4}$ of a unit on a side, and so on.

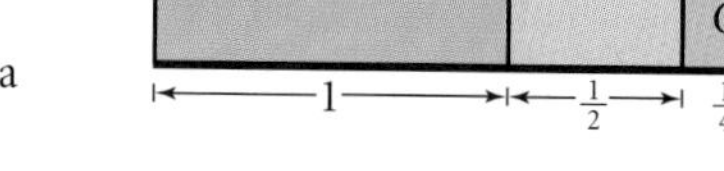

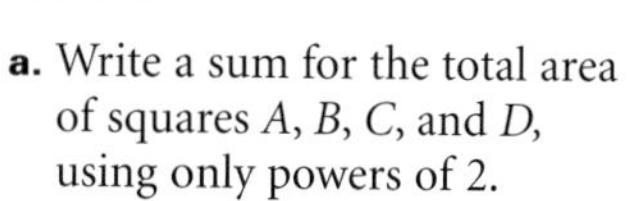

a. Write a sum for the total area of squares A, B, C, and D, using only powers of 2.

b. Rewrite the sum you wrote in part **a** as a single rational number.

c. Suppose that two more squares, E and F, are added to the set of squares, continuing the pattern. Write a single rational number for the total area of squares A through F.

d. Convert your answers from parts **b** and **c** to decimals rounded to the nearest ten-thousandth. What common fraction do the answers appear to be getting closer and closer to?

46a. $\frac{1}{2^0} + \frac{1}{2^2} + \frac{1}{2^4} + \frac{1}{2^6}$

b. $\frac{85}{64}$

c. $\frac{1365}{1024}$

d. 1.3281, 1.3330; $\frac{4}{3}$

APPLICATIONS

47. ELECTRICITY The effective resistance, R_T, of parallel resistors in an electric circuit equals the reciprocal of the sum of the reciprocals of the individual resistances.

$$R_T = \frac{1}{\frac{1}{R_A} + \frac{1}{R_B} + \frac{1}{R_C}}$$

Resistance in an electric circuit is measured in ohms.

a. A circuit has three parallel resistors, R_A, R_B, and R_C. Find R_T to the nearest hundredth, given $R_A = 5$ ohms, $R_B = 8$ ohms, and $R_C = 12$ ohms.

b. Write R_T as a rational expression with no fractions in the denominator.

47a. ≈2.45 ohms

b. $R_T = \frac{R_A R_B R_C}{R_A R_B + R_A R_C + R_B R_C}$

TRAVEL The diagram below shows the parts of a trip that Justine recently took. The distances between A and B, B and C, and C and D are all equal. The speed in each direction is shown in the diagram. Find each average speed listed below to the nearest tenth of a mile per hour.

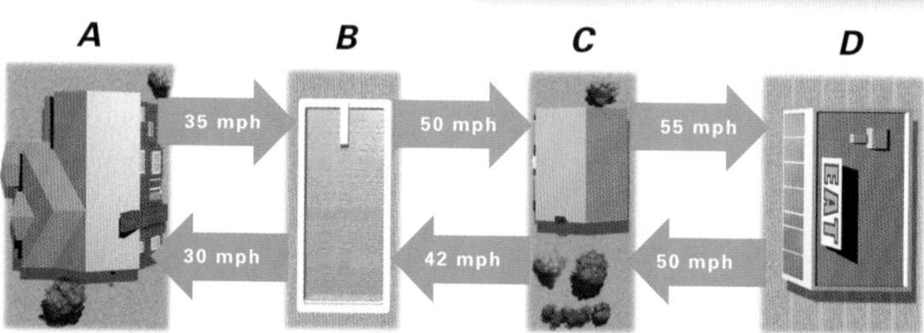

≈37.8 mph **48.** Justine's average speed for a trip from A to C and back to A

≈48.8 mph **49.** Justine's average speed for a trip from B to D and back to B

≈41.7 mph **50.** Justine's average speed for a trip from A to D and back to A

Practice

NAME ______ CLASS ______ DATE ______

Practice
8.4 *Adding and Subtracting Rational Expressions*

Simplify.

1. $\frac{x-7}{3} + \frac{x+2}{4}$ $\frac{7x-22}{12}$

2. $\frac{5x-4}{x^3+1} - \frac{2x+3}{x^3+1}$ $\frac{3x-7}{x^3+1}$

3. $\frac{3x+4}{3x} - \frac{2x+1}{2x}$ $\frac{5}{6x}$

4. $\frac{x-2}{x^2-4} + \frac{2}{3x-6}$ $\frac{5x-2}{3x^2-12}$

5. $\frac{x-2}{x+3} + \frac{x+3}{x-2}$ $\frac{2x^2+2x+13}{x^2+x-6}$

6. $\frac{4x}{x^2-16} - \frac{4}{x+4}$ $\frac{16}{x^2-16}$

7. $\frac{x+4}{2x^2-2x} - \frac{5}{2x-2}$ $-\frac{2}{x}$

8. $\frac{x-2}{x+8} - \frac{x-2}{x^2+6x-16}$ $\frac{x-3}{x+8}$

9. $\frac{x+5}{x^2+10x+25} - \frac{2x}{x^2-25}$ $-\frac{1}{x-5}$

10. $\frac{\frac{x+14}{x-3}}{2}$ $\frac{2x+28}{x-3}$

11. $\frac{\frac{12}{x-2}}{\frac{3}{x-2}} + \frac{7}{x-2}$ $\frac{4x-1}{x-2}$

12. $\frac{\frac{x}{x-6}}{\frac{2x-1}{x-6}} - \frac{3}{2x-1}$ $\frac{x-3}{2x-1}$

13. $\frac{\frac{10}{3x+1}}{\frac{5x}{3x+1}} + \frac{x^2+1}{x}$ $\frac{x^2+3}{x}$

14. $\frac{\frac{x}{x-2}}{\frac{2x^2}{2-x}} + \frac{1}{x+1}$ $\frac{x-1}{2x(x+1)}$

15. $\frac{\frac{x-3}{2x+1}}{\frac{2x-1}{x+3}} - \frac{x^2-9}{4x^2-1}$ 0

Write each expression as a single rational expression in simplest form.

16. $\frac{5x-2}{x^2-49} + \frac{x-15}{x^2-49} - \frac{3x+4}{x^2-49}$ $\frac{3}{x+7}$

17. $\frac{2x+3}{x^2-9} - \frac{2x-3}{x^2-9} + \frac{1}{x+9}$ $\frac{1}{x+3}$

18. $\frac{x}{x-5} - \frac{x^2+25}{25-x^2} + \frac{5}{x+5}$ $\frac{2x}{x-5}$

19. $\frac{x+1}{x-2} + \frac{x+2}{x-4} + \frac{16-5x}{x^2-6x+8}$ $\frac{2(x-2)}{x-4}$

20. $\frac{5x}{x^2-9} - \frac{4}{x+3} + \frac{2}{3-x}$ $\frac{-x+6}{(x+3)(x-3)}$

21. $\frac{-2x^2-5x}{x^2+7x} + \frac{x-2}{x+7} + \frac{2x-3}{x}$ $\frac{x-3}{x}$

22. $\frac{2x-3}{3x^2-13x-10} + \frac{2x+1}{5-x} + \frac{1}{3x+2}$ $\frac{-2(3x^2+2x+5)}{(3x+2)(x-5)}$

23. $\frac{5}{xy+3y-2x-6} + \frac{4}{x+3} - \frac{2}{2-y}$ $\frac{3+4y+2x}{(x+3)(y-2)}$

Look Back

State the property that is illustrated in each statement. All variables represent real numbers. ***(LESSON 2.1)***

51. $-8x(5x + 2) = -40x^2 - 16x$ **Distributive Prop.**

52. $(3 - x)12x = 12x(3 - x)$ **Commutative Prop.**

Find the discriminant, and determine the number of real solutions. Then solve. ***(LESSON 5.6)***

53. $0 = x^2 - 3x + 4$ **-7; 0 sol.; $x = \frac{3}{2} \pm \frac{i\sqrt{7}}{2}$**

54. $x^2 - 2x + 1 = 0$ **0; 1 sol.; $x = 1$**

55. $-2x^2 - 5x + 12 = 0$ **121; 2 sol.; $x = -4, \frac{3}{2}$**

APPLICATION

56. PHYSICAL SCIENCE When sunlight strikes the surface of the ocean, the intensity of the light beneath the surface decreases exponentially with the depth of the water. If the intensity of the light is *reduced* by 75% for each meter of depth, what expression represents the intensity of light beneath the surface? ***(LESSON 6.2)*** **$I_0(0.25)^d$, where d is meters below surface**

Write each expression as a single logarithm. Then evaluate. ***(LESSON 6.4)***

57. $\log_2 32 - \log_2 8$ **$\log_2 4$; 2**

58. $\log_2 4^3 + \log_2 16$ **$\log_2 1024$; 10**

Use a graph and the Location Principle to find the real zeros of each function. ***(LESSON 7.4)***

59. $d(x) = x^3 - 6x^2 + 5x + 12$

60. $f(x) = x^3 - 2x^2 - 11x + 12$

61. $f(x) = x^3 + 2x^2 - 5x - 6$ **$x = -3, -1, 2$**

62. $g(x) = x^3 + 8x^2 + 4x - 48$ **$x = -6, -4, 2$**

59. $x = -1, 3, 4$

60. $x = -3, 1, 4$

Look Beyond

63 Find all real solutions of the rational equation $1.4 = \frac{(x + 3)(x - 1)}{x^2 - 1}$. Be sure to check for excluded values. **$x = 4$**

The definition of a harmonic mean may be extended for 3, 4, or n numbers. For any three numbers a, b, and c, the harmonic mean is $\frac{3}{\frac{1}{a} + \frac{1}{b} + \frac{1}{c}}$.

1. Simplify this complex fraction.
2. Find the harmonic mean of the numbers 3, 4, and 5.

For any four numbers a, b, c, and d, the harmonic mean is $\frac{4}{\frac{1}{a} + \frac{1}{b} + \frac{1}{c} + \frac{1}{d}}$.

3. Simplify this complex fraction.
4. Find the harmonic mean of the numbers 2, 4, 6, and 8.

WORKING ON THE CHAPTER PROJECT

You should now be able to complete Activity 2 of the Chapter Project.

Look Beyond

In Exercise 63, students solve a rational equation. Students will learn more techniques for solving rational equations in Lesson 8.5.

ALTERNATIVE

Assessment

Portfolio Activity

The Portfolio Activity can be used as preparation for the Chapter Project or as a separate activity. In the Portfolio Activity on this page, students find the harmonic mean of three numbers and four numbers. This involves sums of rational expressions and complex rational expressions.

Answers to Portfolio Activities can be found in Additional Answers of the Teacher's Edition.

Student Technology Guide

NAME ________ CLASS ________ DATE ________

Student Technology Guide

8.4 Adding and Subtracting Rational Expressions

You can use a graphics calculator to check a sum or difference of rational expressions.

Example: Vertify that $\frac{2}{x-4} + \frac{5}{x+3}$ is equal to $\frac{7x - 14}{x^2 - x - 12}$.

- To enter $\frac{2}{x-4} + \frac{5}{x+3}$ as Y1, press [Y=] 2 [÷] [(] [X,T,θ,n] [-] 4 [)] [+] 5 [÷] [(] [X,T,θ,n] [+] 3 [)].
- Move to Y2 and press [(] 7 [X,T,θ,n] [-] 14 [)] [÷] [(] [X,T,θ,n] [x^2] [-] [X,T,θ,n] [-] 12 [)].
- Press [GRAPH]. The graphs of Y1 and Y2 look as if they overlap exactly, suggesting that $\frac{2}{x-4} + \frac{5}{x+3} = \frac{7x - 14}{x^2 - x - 12}$.
- For more evidence, press [TRACE]. Use [►] and [◄] to move to different x-values and use [▲] and [▼] to switch between Y1 and Y2. Notice that the y-values do not change when you switch between the functions.

Simplify. Check your answers by graphing.

1. $\frac{2}{x+1} + \frac{3}{x-2}$ $\frac{5x - 1}{x^2 - x - 2}$
2. $\frac{x}{x^2 - 4} + \frac{2}{x+2}$ $\frac{3x - 4}{x^2 - 4}$
3. $\frac{x}{5} - \frac{5}{x}$ $\frac{x^2 - 25}{5x}$
4. $\frac{7}{x-2} - \frac{2}{x+4}$ $\frac{5x + 32}{x^2 + 2x - 8}$

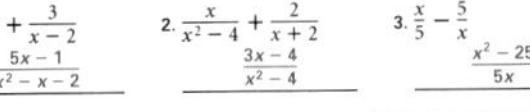

Some calculators, such as the TI-92, can simplify sums and differences of rational expressions.

To simplify $\frac{x + 2}{x^2 + 4x + 3} + \frac{4}{x+3}$, press

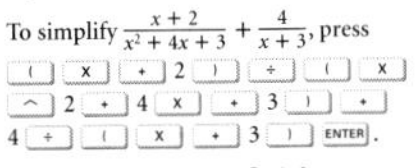

[(] [X] [+] 2 [)] [÷] [(] [X] [^] 2 [+] 4 [X] [+] 3 [)] [+] 4 [÷] [(] [X] [+] 3 [)] [ENTER].

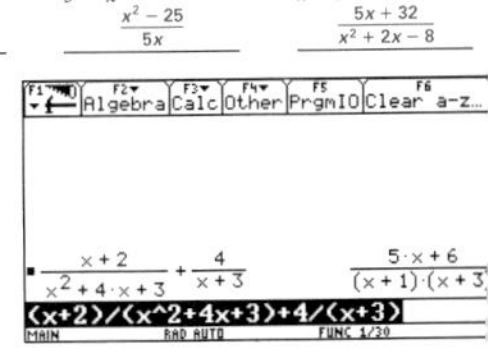

Notice that the sum, $\frac{5x + 6}{(x+1)(x+3)}$, is shown with the denominator in factored form. If the denominators have no common factors, press [F2] [6:comDenom(] before entering the expression.

Simplify.

5. $\frac{x}{x+5} + \frac{4}{x^2 - 25}$ $\frac{x^2 - 5x + 4}{(x-5)(x+5)} = \frac{x^2 - 5x + 4}{x^2 - 25}$

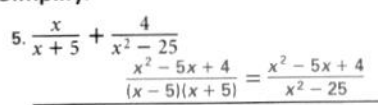

6. $\frac{x}{x^2 - 9} - \frac{4}{x^2 + 6x + 9} + \frac{2}{x+3}$ $\frac{3x^2 - x - 6}{(x-3)(x+3)^2} = \frac{3x^2 - x - 6}{x^3 + 3x^2 - 9x - 27}$

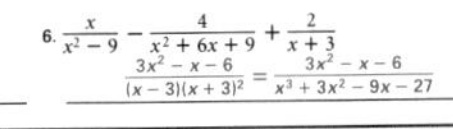

Prepare

NCTM PRINCIPLES & STANDARDS 1–4, 6–10

QUICK WARM-UP

Solve.

1. $-6 < 8c$ $\quad c > -0.75$
2. $-2r \geq 15$ $\quad r \leq -7.5$
3. $x^2 + 5x - 36 = 0$ $\quad x = -9$ or $x = 4$
4. $3x^2 - 4 = x$ $\quad x = -1$ or $x = \frac{4}{3}$
5. $\frac{16}{5} = \frac{m}{3}$ $\quad m = 9.6$
6. $\frac{15}{g} = \frac{2.5}{4}$ $\quad g = 24$
7. $\frac{5}{4} = \frac{3}{2a}$ $\quad a = 1.2$
8. $\frac{1}{x} + \frac{1}{2x} = \frac{1}{3}$ $\quad x = 4.5$

Also on Quiz Transparency 8.5

Teach

Why Using the application in the text as a start, have students discuss other situations in which a rational equation would be helpful. Students should think back to applications in previous lessons of the chapter, such as concentration of mixtures, resistors, and travel.

8.5 Solving Rational Equations and Inequalities

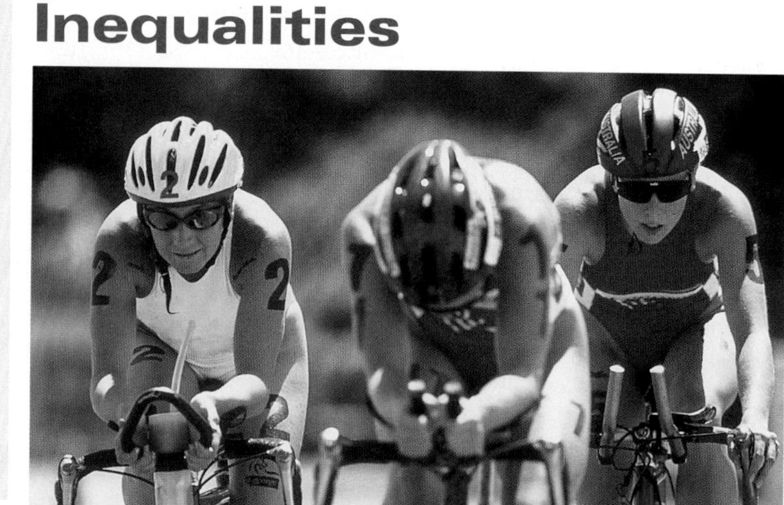

Objectives

- Solve a rational equation or inequality by using algebra, a table, or a graph.
- Solve problems by using a rational equation or inequality.

Why *There are many events in the real world that can be represented by a rational equation or inequality. For example, you can write a rational equation to represent speed and distance information for a triathlon.*

APPLICATION SPORTS

Rachel finished a triathlon involving swimming, bicycling, and running in 2.5 hours. Rachel's bicycling speed was about 6 times her swimming speed, and her running speed was about 5 miles per hour greater than her swimming speed. To find the speeds at which Rachel competed, you can solve a *rational equation.* A **rational equation** is an equation that contains at least one rational expression.

	Distance (mi)	Speed (mph)
Swimming	$d_s = 0.5$	s
Bicycling	$d_b = 25$	$6s$
Running	$d_r = 6$	$s + 5$

EXAMPLE 1 **Find the speeds at which Rachel competed if she finished the triathlon in 2.5 hours.**

SOLUTION

1. Find the time for each part of the triathlon.

Swimming time	Bicycling time	Running time
$d_s = rt_s$	$d_b = rt_b$	$d_r = rt_r$
$\mathbf{0.5} = \mathbf{s}t_s$	$\mathbf{25} = (\mathbf{6s})t_b$	$\mathbf{6} = (\mathbf{s+5})t_r$
$\frac{0.5}{s} = t_s$	$\frac{25}{6s} = t_b$	$\frac{6}{s+5} = t_r$

Alternative Teaching Strategy

USING COGNITIVE STRATEGIES Ask students to solve a simple proportion such as $\frac{7}{4} = \frac{x}{2}$. Tell students to write every step they use to solve the equation, no matter how trivial. Now give students a rational equation, such as $\frac{x-3}{x+1} = \frac{x-2}{x-1}$, and have them use the same steps to solve for x. $x = \frac{5}{3}$ Now give students another rational equation, such as $\frac{6}{x-1} - \frac{3}{x+1} = \frac{3}{x}$. In this case, make sure students realize that they first have to simplify the sum on the left side of the equation to use the same steps as for the simple equation. $x = -\frac{1}{3}$ Use a similar set of examples for solving rational inequalities. Make sure students remember that when multiplying an inequality by a negative value, the inequality sign is reversed.

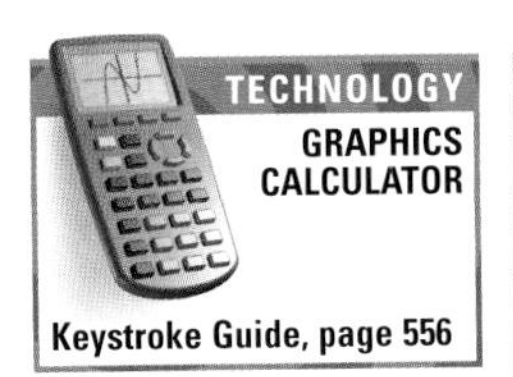

2. Write a rational function to represent the total time, T, in hours for the triathlon in terms of the swimming speed, s, in miles per hour.

$$T(s) = t_s + t_b + t_r$$

$$T(s) = \frac{0.5}{s} + \frac{25}{6s} + \frac{6}{s+5}$$

$$T(s) = \frac{0.5}{s}\left[\frac{6(s+5)}{6(s+5)}\right] + \frac{25}{6s}\left(\frac{s+5}{s+5}\right) + \frac{6}{s+5}\left(\frac{6s}{6s}\right) \quad \textit{The LCD is } 6s(s+5).$$

$$T(s) = \frac{3s + 15 + 25s + 125 + 36s}{6s(s+5)}$$

$$T(s) = \frac{64s + 140}{6s(s+5)}$$

3. Solve the rational equation $2.5 = \frac{64s + 140}{6s(s+5)}$.

Graph $y = \frac{64x + 140}{6x(x+5)}$ and $y = 2.5$, and find the x-coordinate of the point of intersection.

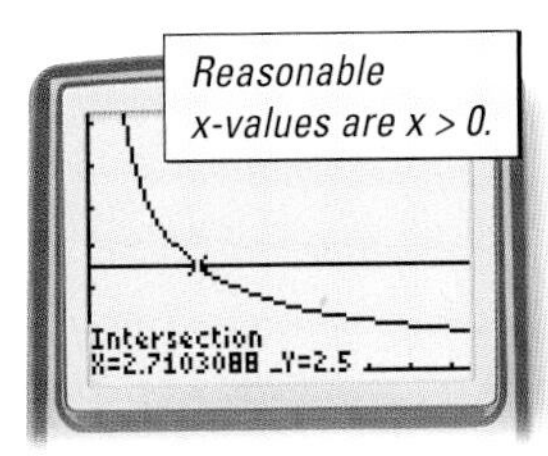

Thus, Rachel swam at about 2.7 miles per hour, bicycled at about 6 • 2.7, or 16.2, miles per hour, and ran at about 5 + 2.7, or 7.7, miles per hour.

CHECKPOINT ✔ Examine the table of values at right for $y = \frac{64x + 140}{6x(x+5)}$. Describe Rachel's triathlon times if her swimming speed (x-value) were less than 2.7 miles per hour and if her swimming speed were greater than 2.7 miles per hour.

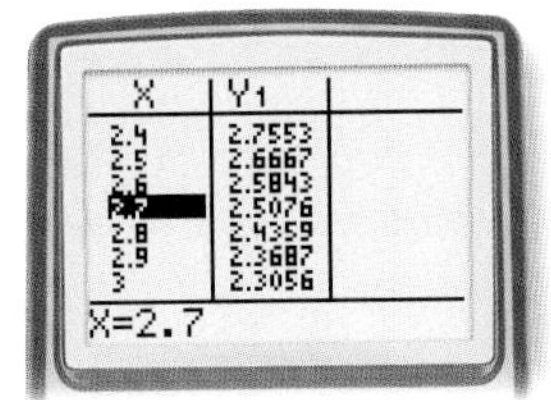

X
Y1
2.4
2.5
2.6
2.7
2.8
2.9
3
2.7553
2.6667
2.5843
2.5076
2.4359
2.3687
2.3056
X=2.7

CRITICAL THINKING How can you solve $2.5 = \frac{64x + 140}{6x(x+5)}$ by using the quadratic formula?

EXAMPLE 2 Solve $\frac{x}{x-6} = \frac{1}{x-4}$.

SOLUTION

Method 1 Use algebra.

$$\frac{x}{x-6} = \frac{1}{x-4} \qquad x \neq 6, x \neq 4$$

$$x(x-4) = 1(x-6)$$

$$x^2 - 4x = x - 6$$

$$x^2 - 5x + 6 = 0$$

$$(x-2)(x-3) = 0$$

$$x = 2 \quad or \quad x = 3$$

CHECK

Let $x = 2$.

$$\frac{x}{x-6} = \frac{1}{x-4}$$

$$\frac{2}{2-6} \stackrel{?}{=} \frac{1}{2-4}$$

$$-\frac{1}{2} = -\frac{1}{2} \quad \textbf{True}$$

Let $x = 3$.

$$\frac{x}{x-6} = \frac{1}{x-4}$$

$$\frac{3}{3-6} \stackrel{?}{=} \frac{1}{3-4}$$

$$-1 = -1 \quad \textbf{True}$$

The solutions are 2 and 3.

Method 2 Use a graph.

Because it is not easy to see the intersection of $y = \frac{x}{x-6}$ and $y = \frac{1}{x-4}$, use another graphing method.

Write $\frac{x}{x-6} = \frac{1}{x-4}$ as $\frac{x}{x-6} - \frac{1}{x-4} = 0$.

Then graph $y = \frac{x}{x-6} - \frac{1}{x-4}$, and find the zeros of the function.

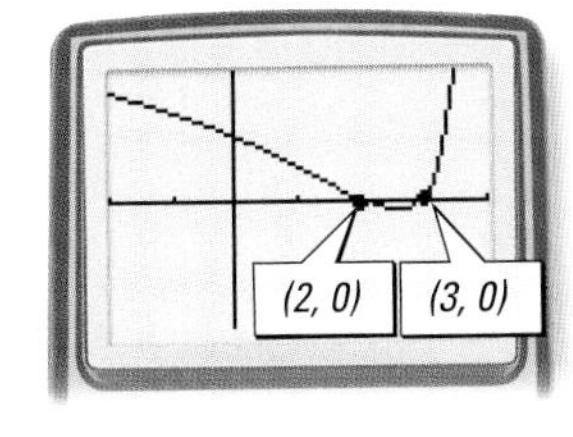

TRY THIS Solve $\frac{x}{3} = \frac{1}{x-2}$.

Interdisciplinary Connection

PHYSICAL EDUCATION A jogger runs 2 miles per hour faster than a hiker walks. If the jogger ran 15 miles in the same time that it took the hiker to walk 10 miles, what was the speed of each?

$r_{jogger} = 6$ **mph**, $r_{hiker} = 4$ **mph**

Enrichment

Have students solve the rational inequality $\frac{-3}{x^2} > \frac{2}{x^2+1}$. $x^2 < \frac{-3}{5}$; **no solution** Have them discuss why there is no solution to this inequality. Note that when solving the inequality, there is no need to have two cases because both denominators are nonnegative for all values of x.

Have students work in pairs to create one or more rational inequalities that have no solution. Then have pairs of students exchange inequalities. Each student should then determine whether the inequality received has no solution.

ADDITIONAL EXAMPLE 1

In the men's triathlon, Delmar's bicycling speed was 3 times his swimming speed, and his running speed was 7 miles per hour greater than his swimming speed. The distances involved were 1.5 miles for swimming, 50 miles for bicycling, and 13 miles for running. Delmar finished the race in 4.15 hours. **Find Delmar's speed for each activity.**

swimming: ≈5.8 mph
bicycling: ≈17.4 mph
running: ≈12.8 mph

Teaching Tip

TECHNOLOGY When using a TI-83 or TI-82 graphics calculator for Example 1, enter the function as **Y1** and set the viewing window as follows:

Xmin=0	**Ymin=0**
Xmax=5	**Ymax=7**
Xscl=1	**Yscl=1**

To produce the table, press 2nd WINDOW and use the following:

TblStart=2.4 (or **TblMin=2.4**)
ΔTbl=.1
Indpnt: Auto Ask
Depend: Auto Ask

Press 2nd GRAPH to view the table.

CHECKPOINT ✔
For $x < 2.7$ mph, $y > 2.5$ hr.
For $x > 2.7$ mph, $y < 2.5$ hr.

CRITICAL THINKING
Multiply both sides of the equation by $6x(x+5)$, subtract $64x + 140$ from both sides, and use the quadratic formula to solve the resulting equation.

ADDITIONAL EXAMPLE 2

Solve $\frac{x}{x+3} = \frac{6}{x-1}$.

$x = 9$ *or* $x = -2$

☞ For Example 2 Teaching Tip and answer to Try This, see page 514.

Teaching Tip

TECHNOLOGY So that students can clearly see the roots of the rational equation in Example 2, set the viewing window as follows:

Xmin=–2	Ymin=–.5
Xmax=4	Ymax=.5
Xscl=1	Yscl=1

ADDITIONAL EXAMPLE 3

Solve $\frac{x}{x+2} - \frac{8}{x^2-4} = \frac{2}{x-2}$.

$x = 6$

TRY THIS

$x = -1$ *or* $x = 3$

Teaching Tip

TECHNOLOGY For Example 3, use the following viewing window:

Xmin=–12.4	Ymin=–8
Xmax=6.4	Ymax=8
Xscl=1	Yscl=1

TRY THIS

$x = -\frac{1}{2}$

CRITICAL THINKING

Multiplying each side of the equation by $(x-2)(x-3)$, a quantity that equals 0 when x is 2 or 3, can introduce extraneous solutions because multiplying each side by 0 does not produce an equivalent equation.

Sometimes solving a rational equation introduces *extraneous solutions.* An **extraneous solution** is a solution to a resulting equation that is not a solution to the original equation. Therefore, it is important to check your answers, as shown in Example 3.

EXAMPLE 3 Solve $\frac{x}{x-3} + \frac{2x}{x+3} = \frac{18}{x^2-9}$.

SOLUTION

Method 1 Use algebra.

Multiply each side of the equation by the LCD, $(x-3)(x+3)$, or $x^2 - 9$.

$$\frac{x}{x-3} + \frac{2x}{x+3} = \frac{18}{x^2-9}, \text{ where } x \neq 3 \text{ and } x \neq -3$$

$$\frac{x}{x-3}(x+3)(x-3) + \frac{2x}{x+3}(x+3)(x-3) = \frac{18}{x^2-9}(x+3)(x-3)$$

$$x(x+3) + 2x(x-3) = 18$$

$$x^2 + 3x + 2x^2 - 6x = 18$$

$$3x^2 - 3x - 18 = 0$$

$$3(x^2 - x - 6) = 0$$

$$3(x-3)(x+2) = 0$$

$$x = 3 \quad or \quad x = -2$$

CHECK

Since $x = 3$ is an excluded value of x in the original equation, it is an extraneous solution. Check $x = -2$.

$$\frac{x}{x-3} + \frac{2x}{x+3} = \frac{18}{x^2-9}$$

$$\frac{-2}{-2-3} + \frac{2(-2)}{-2} + 3 \stackrel{?}{=} \frac{18}{(-2)^2-9}$$

$$-3\tfrac{3}{5} = \frac{-18}{5} \quad \textbf{True}$$

Thus, the only solution is $x = -2$.

Method 2 Use a graph.

Because it is not easy to see the intersection of $y = \frac{x}{x-3} + \frac{2x}{x+3}$ and $y = \frac{18}{x^2-9}$, use another graphing method.

Write $\frac{x}{x-3} + \frac{2x}{x+3} = \frac{18}{x^2-9}$ as $\frac{x}{x-3} + \frac{2x}{x+3} - \frac{18}{x^2-9} = 0$. Then graph $y = \frac{x}{x-3} + \frac{2x}{x+3} - \frac{18}{x^2-9}$, and find any zeros of the function.

TECHNOLOGY GRAPHICS CALCULATOR Keystroke Guide, page 556

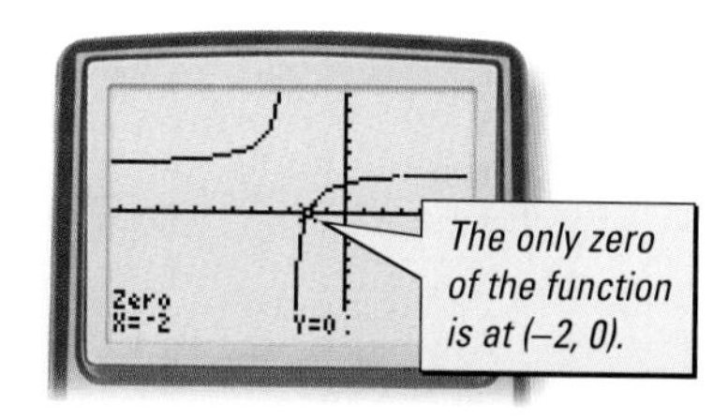

Zero
X=-2
Y=0
The only zero of the function is at (–2, 0).

The solution is –2.

TRY THIS Solve $\frac{x}{x-2} + \frac{x}{x-3} = \frac{3}{x^2-5x+6}$.

CRITICAL THINKING Explain why an extraneous solution is obtained in Example 3 above.

Inclusion Strategies

USING DISCUSSION Have students explain the goals and procedures involved in solving rational equations and inequalities. You might also have them explain how to identify a rational equation. Point out to students that one goal when solving a rational equation is to reduce the problem to a type of equation they already know how to solve. Have them explain the importance of checking all solutions in the original equation or inequality.

A **rational inequality** is an inequality that contains at least one rational expression.

Solving Rational Inequalities

You will need: a graphics calculator

1. Graph $y_1 = \frac{x+2}{x-4}$.
2. Use the table feature and the graph to identify the values of x for which y_1 is 0, y_1 is undefined, y_1 is positive, and y_1 is negative.
3. On the same screen, graph $y_2 = 2x - 11$.
4. For what values of x is $y_1 = y_2$? $y_1 < y_2$? $y_1 > y_2$?

CHECKPOINT ✔

5. Explain how to use a graph and a table of values to solve $\frac{x+2}{x-4} < 2x - 11$ and $\frac{x+2}{x-4} > 2x - 11$.

EXAMPLE 4 Solve $\frac{x}{2x-1} \le 1$.

SOLUTION

Method 1 Use algebra.

To clear the inequality of fractions, multiply each side by $2x - 1$. You must consider both cases: $2x - 1$ is positive *or* $2x - 1$ is negative.

$\frac{x}{2x-1} \le 1$, where $2x - 1 > 0$

$\frac{x}{2x-1} \le 1$

$x \le 2x - 1$

$-x \le -1$

$x \ge 1$ *Change ≤ to ≥.*

For this case, $x > \frac{1}{2}$ because $2x - 1 > 0$.
Therefore, the solution must satisfy $x \ge 1$ *and* $x > \frac{1}{2}$.

For this case, $x \ge 1$.

or

$\frac{x}{2x-1} \le 1$, where $2x - 1 < 0$

$\frac{x}{2x-1} \le 1$

$x \ge 2x - 1$ *Change ≤ to ≥.*

$-x \ge -1$

$x \le 1$ *Change ≥ to ≤.*

For this case, $x < \frac{1}{2}$ because $2x - 1 < 0$.
Therefore, the solution must satisfy $x \le 1$ *and* $x < \frac{1}{2}$.

For this case, $x < \frac{1}{2}$.

Thus, the solution is $x \ge 1$ *or* $x < \frac{1}{2}$.

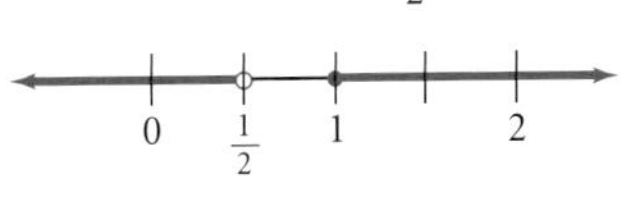

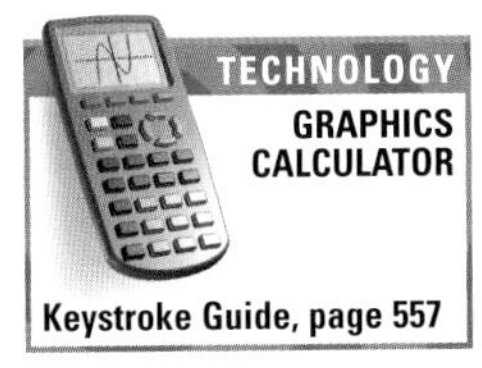

Method 2 Use a graph.

Graph $y = \frac{x}{2x-1}$ and $y = 1$, and find the values of x for which the graph of $y = \frac{x}{2x-1}$ is below the graph of $y = 1$.

Thus, the solution is $x \ge 1$ *or* $x < \frac{1}{2}$.

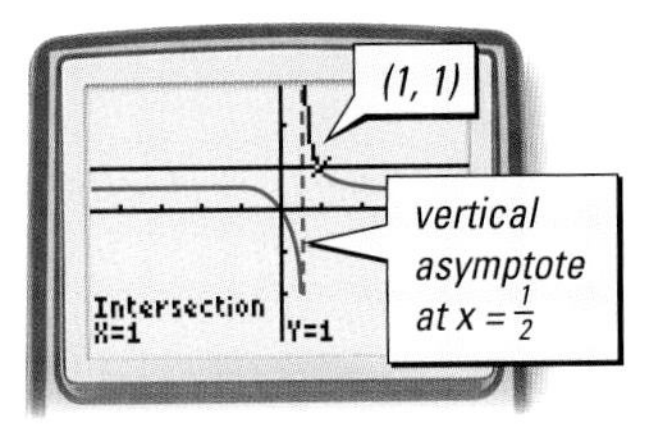

TRY THIS Solve $\frac{x-1}{x+2} < 3$.

Reteaching the Lesson

INVITING PARTICIPATION Make a list of several rational equations and inequalities and their solutions. Place a problem and an unrelated answer on an index card, and display the group of cards randomly around the classroom. Have pairs of students begin at any card. They should determine the solution to the problem shown and find the card with that answer. They will then find the solution of the new card's problem and find the card with this new solution. If they solve all of the problems correctly, they will have been to each displayed card, ending with the card where they started. When developing the cards, make sure to note the links between problems and solutions to ensure that students will visit each card. These cards can be laminated and used from year to year.

Activity Notes

In this Activity, students graph each side of a rational inequality on a graphics calculator. Using the graph and the table of values, students can see when the inequality is true.

Teaching Tip

TECHNOLOGY When using a TI-83 or TI-82 graphics calculator to produce a table for the Activity, enter the functions into **Y1** and **Y2**, press 2nd WINDOW to access the **Table Setup** screen, and set the following:

TblStart=–5 (or **TblMin=–5**)
ΔTbl=1
Indpnt: Auto Ask
Depend: Auto Ask

Set the viewing window to the following:

Xmin=–6.8	**Ymin=–10**
Xmax=12	**Ymax=10**
Xscl=1	**Yscl=1**

CHECKPOINT ✔

5. Graph each function (Steps 1 and 3), and then compare the graphs or the y-values in the table for various values of x.

ADDITIONAL EXAMPLE 4

Solve $\frac{4x}{x-3} \le 6$.

$x < 3$ *or* $x \ge 9$

Teaching Tip

TECHNOLOGY For Example 4, use the following viewing window:

Xmin=–4.7	**Ymin=–3**
Xmax=4.7	**Ymax=3**
Xscl=1	**Yscl=1**

TRY THIS

$x < -\frac{7}{2}$ *or* $x > -2$

ADDITIONAL
EXAMPLE 5

Solve $\frac{2}{x-1} > \frac{3}{x+4}$.

$x < -4$ *or* $x > 1$

TRY THIS

$x < 1$

Teaching Tip

TECHNOLOGY When using a TI-83 or TI-82 graphics calculator, use the following viewing window:

Xmin=–9.4	**Ymin=–3.1**
Xmax=9.4	**Ymax=3.1**
Xscl=1	**Yscl=1**

EXAMPLE 5 Solve $\frac{x-2}{2(x-3)} > \frac{x}{x+3}$.

SOLUTION

In order to clear the inequality of fractions, you can multiply each side by the LCD, $2(x-3)(x+3)$. You must consider four possible cases:

Case 1: $x - 3$ is positive *and* $x + 3$ is positive,
or
Case 2: $x - 3$ is positive *and* $x + 3$ is negative,
or
Case 3: $x - 3$ is negative *and* $x + 3$ is positive,
or
Case 4: $x - 3$ is negative *and* $x + 3$ is negative.

The algebraic method of solution is beyond the scope of this textbook, but with a graphics calculator, the solution is much easier to find.

Rewrite $\frac{x-2}{2(x-3)} > \frac{x}{x+3}$ as $\frac{x-2}{2(x-3)} - \frac{x}{x+3} > 0$.

Graph $y = \frac{x-2}{2(x-3)} - \frac{x}{x+3}$, and find the values of x for which $y > 0$.

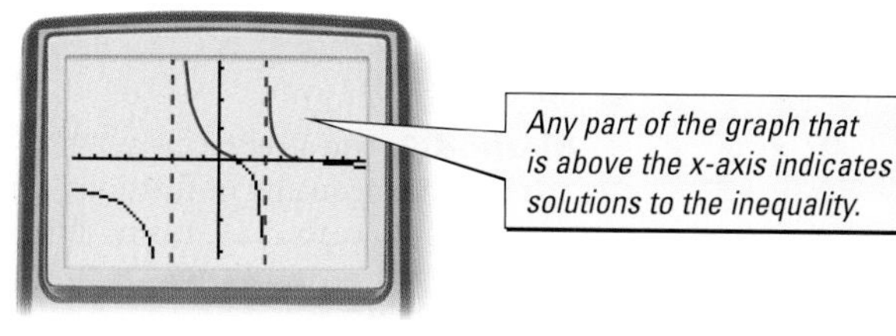

Any part of the graph that is above the x-axis indicates solutions to the inequality.

The graph shows that there are two intervals of x for which $y > 0$. These two intervals can be found by using a table of values.

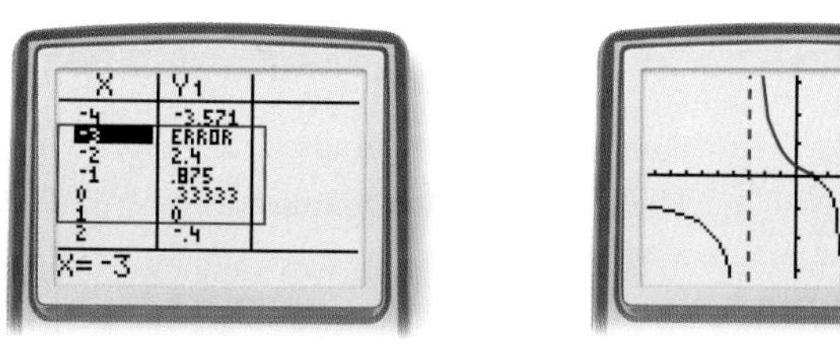

One interval for which $y > 0$ is $-3 < x < 1$, as shown above. The other interval for which $y > 0$ is $3 < x < 6$, as shown below.

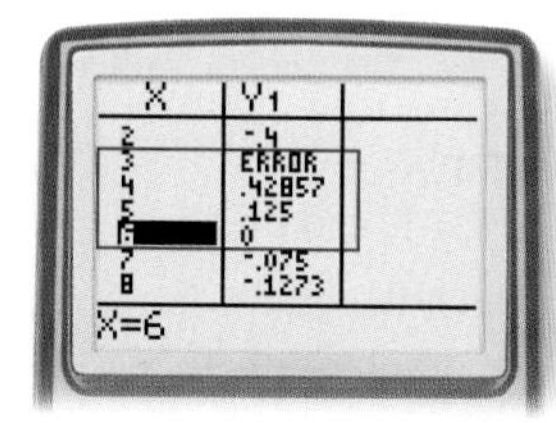

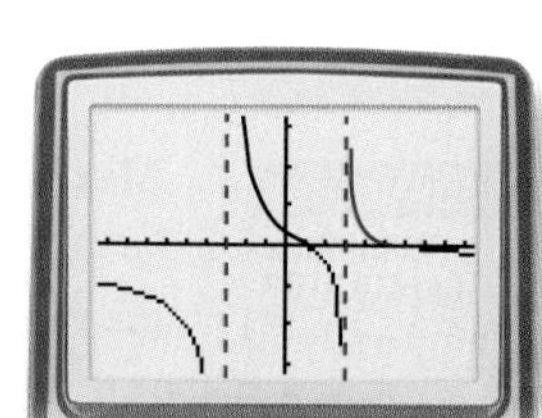

Thus, the solution is $-3 < x < 1$ *or* $3 < x < 6$.

TRY THIS Solve $\frac{x+1}{x-1} < \frac{x}{x-1}$.

Exercises

Communicate

1. Explain what an extraneous solution is and how you can tell whether a solution to a rational equation is extraneous.
2. Explain how to use a graph to check the solutions to a rational equation that are obtained by using algebra.
3. Explain how to use the graphs of $y = \frac{x-1}{x+2}$ and $y = 3$, shown at right, to solve $\frac{x-1}{x+2} < 3$ and $\frac{x-1}{x+2} > 3$.

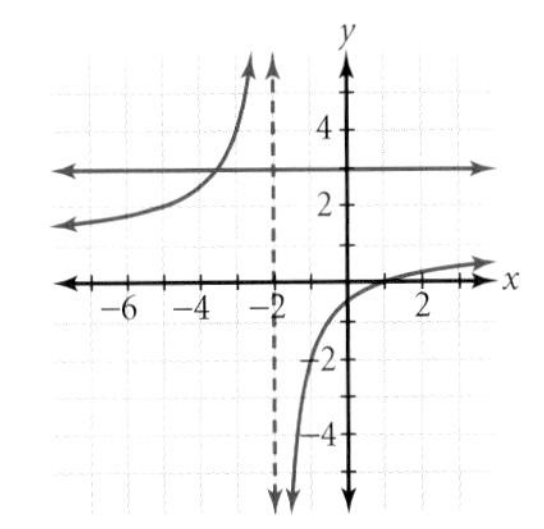

Guided Skills Practice

APPLICATION

4. **SPORTS** Refer to Rachel's triathlon information given at the beginning of the lesson. At what swimming, bicycling, and running speeds must Rachel compete in order to finish the triathlon in 2 hours? ***(EXAMPLE 1)***

4. swim at ≈3.6 mph, bike at ≈21.6 mph, and run at ≈8.6 mph

Solve each equation.

5. $\frac{2x-1}{x} = \frac{3}{x+2}$ ***(EXAMPLE 2)*** $x = \pm 1$

6. $\frac{2}{x-1} + \frac{2}{x+1} = \frac{-4}{x^2-1}$ ***(EXAMPLE 3)*** no solution

Solve each inequality.

7. $\frac{2x-3}{x} \geq 2$ ***(EXAMPLE 4)*** $x < 0$

8. $\frac{1}{x+2} < \frac{1}{x+3}$ ***(EXAMPLE 5)*** $-3 < x < -2$

Practice and Apply

Solve each equation. Check your solution.

9. $\frac{x+3}{2x} = \frac{5}{8}$ $x = 12$

10. $\frac{2y-1}{4y} = \frac{4}{6}$ $y = -\frac{3}{2}$

11. $\frac{4}{n+4} = 1$ $n = 0$

12. $\frac{-6}{m-3} = 1$ $m = -3$

13. $\frac{1}{3z} + \frac{1}{8} = \frac{4}{3z}$ $z = 8$

14. $\frac{1}{t} + \frac{1}{3} = \frac{8}{3t}$ $t = 5$

15. $\frac{y+3}{y-1} = \frac{y+2}{y-3}$ $y = -7$

16. $\frac{2n+1}{3n+4} = \frac{2n-8}{3n+8}$ $n = -\frac{8}{7}$

17. $\frac{x+3}{x} + 1 = \frac{x+5}{x}$ $x = 2$

18. $\frac{2x}{x+3} - 1 = \frac{x}{x+3}$

19. $\frac{x+1}{x-1} + \frac{2}{x} = \frac{x}{x+1}$

20. $\frac{3}{x+2} - \frac{x}{1} = \frac{4}{3}$

18. no solution

19. $x = -\frac{1}{10} \pm \frac{\sqrt{41}}{10}$

20. $x = -\frac{5}{3} \pm \frac{2\sqrt{7}}{3}$

21. $\frac{1}{6} - \frac{1}{x} = \frac{4}{3x^2}$ $x = 3 \pm \sqrt{17}$

22. $\frac{1}{1+c} - \frac{1}{2+c} = \frac{1}{4}$ $c = -\frac{3}{2} \pm \frac{\sqrt{17}}{2}$

23. $\frac{2x+3}{x-1} - \frac{2x-3}{x+1} = \frac{10}{x^2-1}$ no solution

24. $\frac{x-4}{x+2} + \frac{2}{x-2} = \frac{17}{x^2-4}$ $x = -1, 5$

25. $\frac{b}{b+3} - \frac{b}{b-2} = \frac{10}{b^2+b-6}$ $b = -2$

26. $\frac{3z}{z-1} + \frac{2z}{z-6} = \frac{5z^2-15z+20}{z^2-7z+6}$ $z = -4$

27. $\frac{3}{x+2} + \frac{12}{x^2-4} = \frac{-1}{x-2}$ no solution

28. $\frac{x+2}{2x-3} - \frac{x-2}{2x+3} = \frac{21}{4x^2-9}$ no solution

Assess

Selected Answers

Exercises 4–8, 9–71 odd

ASSIGNMENT GUIDE

In Class	1–8
Core	9–51 odd, 55
Core Plus	10–52 even, 53, 54–56 even
Review	58–71
Preview	72

Extra Practice can be found beginning on page 940.

Error Analysis

Remind students that when multiplying an inequality by a negative number, the inequality sign is reversed. When mulitplying both sides of an inequality by an expression containing a variable, students must consider the cases in which the expression is positive and cases in which the expression is negative. When the expression is negative, the inequality sign must be reversed when multiplying both sides by the expression. Stress that students can avoid using cases to solve an inequality by solving the related equation, finding the critical values, and testing points in each interval determined by the critical values to find the interval(s) where the inequality is true. Also, remind students that the domain of the original equation or inequality applies to the solution.

internet connect

Homework Help Online
Go To: **go.hrw.com**
Keyword:
MB1 Homework Help for Exercises 29–46

Solve each inequality. Check your solution.

29. $\frac{x+3}{3x} > 2$ **30.** $\frac{x+5}{4x} > 3$ **31.** $\frac{x-5}{3x} < -3$

32. $\frac{x-5}{3x} < 3$ **33.** $\frac{2x+1}{x-2} > 4$ **34.** $3 < \frac{3x+4}{2+1}$

35. $\frac{x+1}{x} \le \frac{1}{2}$ **36.** $-\frac{1}{2} \ge \frac{1}{x-4}$ **37.** $\frac{7x}{3x+2} < 2$

Use a graphics calculator to solve each rational inequality. Round answers to the nearest tenth.

38 $\frac{1}{2} > x^2$ **39** $\frac{1}{x} \le x^2 - 1$ **40** $\frac{x-2}{x-1} \ge 2x$

41 $x^2 - 4 \le \frac{1}{x^2}$ **42** $2x + 1 \ge \frac{1}{2x+1}$ **43** $\frac{t}{t-1} - \frac{2}{t+1} \le \frac{5}{t^2-1}$

44 $\frac{x+1}{x-1} + \frac{2}{x} \ge 1$ **45** $\frac{a-3}{3a} \ge \frac{1}{3a^2+9a} + \frac{1}{a+3}$ **46** $\frac{x^2+1}{(x-1)^2} > \frac{1}{x}$

38. $-0.7 < x < 0.7$
39. $x < 0$ *or* $x \ge 1.3$
40. $x < 1$
41. $-2.1 \le x < 0$ *or* $0 < x \le 2.1$
42. $-1 \le x < -\frac{1}{2}$ *or* $x \ge 0$
43. $-1.3 \le t < -1$ *or* $1 < t \le 2.3$
44. $0 < x \le \frac{1}{2}$ *or* $x > 1$
45. $a < -3$ *or* $-2 \le a < 0$ *or* $a \ge 5$
46. $x < 0$ *or* $x > 1$ *or* $0.4 < x < 1$

State whether each equation is always true, sometimes true, or never true.

47. $\frac{2x+8}{x^2-16} = \frac{2}{x-4}, x \ne \pm 4$ **always**

48. $\frac{1-5x^{-1}+4x^{-2}}{1-16x^{-2}} = \frac{x-1}{x+4}, x \ne 0 \text{ or } \pm 4$ **always**

49. $\frac{3}{x+2} + \frac{12}{x^2-4} = \frac{-1}{x-2}$ **never**

50. $\frac{2x+3}{x-1} - \frac{2x-3}{x+1} = \frac{10}{x^2-1}$ **never**

51. $\frac{x}{x+4} > 2x + 6$ **sometimes**

52. $\frac{x-6}{x^2-2x-8} + \frac{3}{x-4} \le \frac{2}{x+2}$ **sometimes**

CHALLENGE

53. Solve $\frac{3}{(x-1)^2} > 0$ by using mental math. **all real numbers except $x = 1$**

54. CULTURAL CONNECTION: ASIA A ninth-century Indian mathematician, Mahavira, posed the following problem:

There are four pipes leading into a well. Individually, the four pipes can fill the well in $\frac{1}{2}$, $\frac{1}{3}$, $\frac{1}{4}$, and $\frac{1}{5}$ of a day. How long would it take for the pipes to fill the well if they were all working simultaneously, and what fraction of the well would be filled by each pipe? **$\frac{1}{14}$ of a day, or ≈1.7 hours; $\frac{1}{7}, \frac{3}{14}, \frac{2}{7}, \frac{5}{14}$**

56a. $T(s) = \frac{0.6}{s} + \frac{15}{9s} + \frac{8}{s+6}$
b. swims at 3.5 mph, bikes at 31.5 mph, runs at 9.5 mph

CONNECTION

57. ≈4530 km

55. GEOMETRY The length of a rectangle is 5 more than its width. Find the length and the width of the rectangle if the ratio of the length to the width is at least 1.5 and no more than 3. **$2.5 \le w \le 10$, $7.5 \le \ell \le 15$**

APPLICATIONS

Satellite in orbit over Earth.

56. SPORTS Michael is training for a triathlon. He swims 0.6 miles, bicycles 15 miles, and runs 8 miles. Michael bicycles about 9 times as fast as he swims, and he runs about 6 miles per hour faster than he swims.

a. Write a rational function, in terms of swimming speed, for the total time it takes Michael to complete his workout.

b. Find the speeds at which Michael must swim, run, and bike to complete his workout in 1.5 hours.

57. PHYSICS An object weighing w_0 kilograms on Earth is h kilometers above Earth. The function that represents the object's weight at that altitude is $w(h) = w_0\left(\frac{6400}{6400+h}\right)^2$. Find the approximate altitude of a satellite that weighs 3500 kilograms on Earth and 1200 kilograms in space.

Practice

NAME ______ CLASS ______ DATE ______

Practice

8.5 Solving Rational Equations and Inequalities

Solve each equation. Check your solution.

1. $\frac{2x+1}{4x-4} = \frac{4}{5}$ — 3.5
2. $\frac{x-5}{x-8} = \frac{x+1}{x-5}$ — 11
3. $\frac{x-15}{x+5} = \frac{x-12}{x}$ — 7.5
4. $\frac{x-8}{x+5} = \frac{x-1}{2x+10}$ — 15
5. $\frac{x^2+1}{x+2} = 3x - 1$ — $\frac{1}{2}, -3$
6. $\frac{x-10}{2x+1} = \frac{4x}{3x+4}$ — $-2, -4$
7. $\frac{x-2}{x} - 1 = \frac{2x+3}{x}$ — -2.5
8. $\frac{3}{4} - \frac{1}{x} = \frac{1}{2x}$ — 2
9. $\frac{x}{x-2} - \frac{x-5}{5} = \frac{x-2}{5}$ — 7, 1
10. $\frac{3}{x-1} + 4 = \frac{1}{1-x^2}$ — $0, -\frac{3}{4}$
11. $\frac{x+3}{x-2} - \frac{14}{x+2} = \frac{3x-2}{x^2-4}$ — 6
12. $\frac{3}{x-2} + \frac{5}{x+2} = \frac{4x^2}{x^2-4}$ — 1

Solve each inequality. Check your solution.

13. $\frac{x}{x-2} < 2$ — $x < 2$ *or* $x > 4$
14. $\frac{x}{x-6} < 2$ — $x < 6$ *or* $x > 12$
15. $\frac{x}{x+1} < \frac{x}{x-1}$ — $-1 < x < 0$ *or* $x > 1$
16. $\frac{x}{x-3} > \frac{4}{x-2}$ — $x < 2$ *or* $x > 3$
17. $\frac{x+1}{x-1} > 2$ — $1 < x < 3$
18. $\frac{x+1}{x+2} - \frac{x}{x+3} \le \frac{7}{x^2+5x+6}$ — $x < -3$ *or* $-2 < x < 2$
19. $\frac{x}{x+1} - \frac{2}{x-1} > 1$ — $x < -1$ *or* $-\frac{1}{3} < x < 1$
20. $\frac{x}{x+3} + \frac{1}{x-4} < 1$ — $-3 < x < 4$ *or* $x > \frac{15}{2}$
21. $\frac{x+1}{x+1} - \frac{x}{x+1} > \frac{2}{x^2-1}$ — $x < -1$ *or* $x > 1$

Use a graphics calculator to solve each rational inequality. Round answers to the nearest tenth.

22. $\frac{2}{x-3} \le x + 3$ — $-2.2 \le x < 3$ *or* $x > 3$
23. $\frac{x+2}{x+4} < 1 - x$ — $-4 < x < 0.5$ *or* $x < -4.5$
24. $\frac{x-3}{x-4} > x$ — $x < 0.7$ *or* $4 < x < 4.3$
25. $\frac{x+1}{x-2} < \frac{1}{x-3}$ — $0 < x < 2$
26. $\frac{x-4}{x} - \frac{x}{x-4} < 1$ — $x < -6.5$ *or* $0 < x < 2.5$ *or* $x > 4$
27. $\frac{2x-3}{x} - \frac{3}{x-2} > 5$ — $-1.4 < x < 0$ *or* $1.4 < x < 2$

29. $0 < x < \frac{3}{5}$

30. $0 < x < \frac{5}{11}$

31. $0 < x < \frac{1}{2}$

32. $x < -\frac{1}{2}$ *or* $x > 0$

33. $2 < x < \frac{9}{2}$

34. $-\frac{1}{2} < x < \frac{1}{3}$

35. $-2 \le < x < 0$

36. $2 \le < x < 4$

37. $-\frac{2}{3} < x < 4$

Look Back

Evaluate each expression. ***(LESSON 2.2)***

58. $81^{\frac{1}{2}}$ 9 **59.** 13^0 1 **60.** $9^{\frac{3}{2}}$ 27 **61.** $27^{\frac{1}{3}}$ 3

Find the inverse of each function. State whether the inverse is a function. ***(LESSON 2.5)***

62. $\{(3, 5), (2, 8), (1, 5), (0, 3)\}$

63. $\{(-1, -4), (-2, -3), (-3, -2), (0, -1)\}$

64. $g(x) = \frac{1}{4}x - 5$ $g^{-1}(x) = 4x + 20$; function

65. $h(x) = \frac{5 - x}{2}$ $h^{-1}(x) = 5 - 2x$; function

Identify each transformation from the parent function $f(x) = x^2$ to g. ***(LESSON 2.7)***

66. $g(x) = -2x^2$

67. $g(x) = (x - 2)^2$

68. $g(x) = \frac{1}{2}(x + 3)^2$

69. $g(x) = 3x^2 - 5$

70. $g(x) = (-2x)^2 + 1$

71. $g(x) = 2(4 - x)^2 - 6$

Look Beyond

72 Graph the functions $f(x) = \sqrt{x}$, $g(x) = \sqrt[3]{x}$, $h(x) = \sqrt[4]{x}$, and $k(x) = \sqrt[5]{x}$. How are they alike? How are they different? (Hint: Use the fact that $\sqrt[n]{x} = x^{\frac{1}{n}}$.)

PORTFOLIO ACTIVITY

Refer to Example 5 on page 508. Notice that the cab driver's average speed is the total distance divided by the total time. The harmonic mean of the two speeds, 45 miles per hour and 55 miles per hour, gives the average speed for the entire trip.

1. Find the harmonic mean of 45 and 55.
2. How does your answer to Step 1 compare with the average speed found in Example 5?

Justin cycles for $4\frac{1}{2}$ hours. He cycles along a level road for 24 miles, and then he cycles up an incline for 24 miles more. Justin immediately turns around and cycles back to his starting point along the same route. Justin cycles on level ground at a rate of 24 miles per hour, uphill at a rate of 12 miles per hour, and downhill at a rate of 48 miles per hour.

3. Explain why Justin's average speed over the entire trip is the harmonic mean of 24, 12, 48, and 24.
4. Find Justin's average speed over the entire trip.

WORKING ON THE CHAPTER PROJECT

You should now be able to complete the Chapter Project.

62. $\{(5, 3), (8, 2), (5, 1), (3, 0)\}$; not a function

63. $\{(-4, -1), (-3, -2), (-2, -3), (-1, 0)\}$; function

66. vertical stretch by a factor of 2 and then reflection across the x-axis

67. horizontal translation 2 units to the right

68. horizontal translation 3 units to the left and then vertical compression by a factor of $\frac{1}{2}$

69. vertical stretch by a factor of 3 and then vertical translation 5 units down

70. vertical stretch by a factor of 4 and then vertical translation 1 unit up

71. horizontal translation 4 units to the right, reflection across the y-axis, vertical stretch by a factor of 2, and vertical translation 6 units down

72. They all have the same shape for $x \geq 0$, a half parabola turned sideways. If $x < 0$ and the root is even, there is no graph. If $x < 0$ and the root is odd, the graphs look like the graphs for $x \geq 0$ rotated 180° about (0, 0).

Look Beyond

In Exercise 72, students use a graphics calculator to explore radical functions by looking at the similarities and differences in the graphs of various radical functions. Radical functions and their graphs will be studied in Lesson 8.6.

ALTERNATIVE Assessment

Portfolio Activity

The Portfolio Activity can be used as preparation for the Chapter Project or as a separate activity. In the Portfolio Activity on this page, students compare the average speed for a trip with the harmonic mean of the speeds.

Student Technology Guide

NAME ______ CLASS ______ DATE ______

Student Technology Guide
8.5 Solving Rational Equations and Inequalities

You can use a graphics calculator to solve a rational equation or inequality. One way to do this is to rewrite the expression so that one side is equal to zero before you graph the related function.

Example: Solve $\frac{x}{x-2} = \frac{5}{x^2-4}$.

- First, rewrite the equation so that one side is equal to 0. By subtracting $\frac{5}{x^2-4}$ from each side, $\frac{x}{x-2} - \frac{5}{x^2-4} = 0$.
- Graph the related function $y = \frac{x}{x-2} - \frac{5}{x^2-4}$.
  Press Y= X,T,θ,n ÷ (X,T,θ,n - 2) - 5 ÷ (X,T,θ,n x^2 - 4) GRAPH.
- There appear to be two zeros, one between −4 and −3 and the other between 1 and 2. You can use the zero feature to get a good approximation of the zeros.
- Press TRACE. Use ◄ to move the cursor just to the left of the zero between −4 and −3.
- Press 2nd TRACE 2:zero ENTER.
- When asked for the left bound, press ENTER to enter the coordinates of the trace point.
- When asked for the right bound, trace just to the right of the zero. Press ENTER.
- When asked for a guess, trace as close as you can to the zero. Press ENTER. The zero is $x \approx -3.45$.
- Repeat this process for the other zero. You will find that the other zero is $x \approx 1.45$.

Using the solution to $\frac{x}{x-2} = \frac{5}{x^2-4}$, you can solve $\frac{x}{x-2} \geq \frac{5}{x^2-4}$. The solution to $\frac{x}{x-2} \geq \frac{5}{x^2-4}$ is the set of all real numbers where $\frac{x}{x-2} - \frac{5}{x^2-4}$ is positive, that is, $x \leq -3.45$ *or* $-2 < x \leq 1.45$ *or* $x > 2$.

Solve each equation and inequality. Round answers to the nearest hundredth.

1. $\frac{x+3}{x} = \frac{7}{4}$; $\frac{x+3}{x} \geq \frac{7}{4}$ $x = 4$; $0 \leq x \leq 4$

2. $\frac{2x-3}{x} = \frac{6}{x+5}$; $\frac{2x-3}{x} \geq \frac{6}{x+5}$ −3 and 2.5; $x < -5$ *or* $-3 \leq x < 0$ *or* $x \geq 2.5$

3. $\frac{4}{x-1} - 2 = \frac{x}{2}$; $\frac{4}{x-1} - 2 \geq \frac{x}{2}$ −5.27 and 2.27; $x \leq -5.27$ *or* $1 < x \leq 2.27$

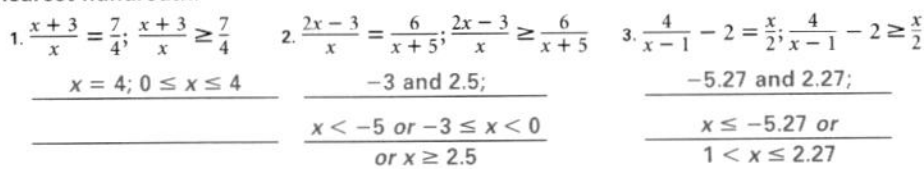

Prepare

NCTM PRINCIPLES & STANDARDS 1–4, 6–10

QUICK WARM-UP

Identify each transformation of the parent function $f(x) = x^2$.

1. $f(x) = x^2 + 5$ 5 units up
2. $f(x) = (x + 5)^2$ 5 units to the left
3. $f(x) = 5x^2$ vertical stretch by a factor of 5
4. $f(x) = -5x^2$ vertical stretch by a factor of 5, reflection across the x-axis
5. $f(x) = (5x)^2$ horizontal compression by $\frac{1}{5}$
6. $f(x) = \left(\frac{1}{5}x\right)^2$ horizontal stretch by 5

Also on Quiz Transparency 8.6

Teach

Why Have students consider the pendulum problem posed in the text. Ask them to think in general about the relationship between the length of the pendulum and the period. As the pendulum gets longer, would the time it takes to complete one cycle get shorter or longer? **longer** What does this indicate about the graph of the function of this model? **The function is increasing from left to right.**

8.6 Radical Expressions and Radical Functions

Objectives

- Analyze the graphs of radical functions, and evaluate radical expressions.
- Find the inverse of a quadratic function.

Why *Radical functions are used to model many real-world relationships . For example, the relationship between the length of a pendulum and the time it takes to complete one full swing is described by a radical function.*

APPLICATION PHYSICS

Nancy noticed that a long pendulum swings more slowly than a short pendulum. The time it takes for a pendulum to complete one full swing, or cycle, is called the *period.* The relationship between the period, T, (in seconds) of the pendulum and its length, x, (in meters) is given below.

$$T(x) = 2\pi\sqrt{\frac{x}{9.8}}$$

Find the period for pendulums whose lengths are 0.1 meter, 0.2 meter, and 0.3 meter. *You will do this in Example 4.*

Square-Root Functions

The **square root** of a number, $\sqrt{x}$, is a number that when multiplied by itself produces the given number, x. Recall from Lesson 5.6 that the expression $\sqrt{-4}$ is not a real number.

Since the domain of a function is the set of all real-number values of x for which a function, f, is defined, the domain of the square-root function, $f(x) = \sqrt{x}$, does not include negative numbers.

Alternative Teaching Strategy

HANDS-ON STRATEGIES Put students in groups of four to simulate the pendulum experiment. Each group will need a piece of string, a washer, tape, a ruler, and a stopwatch. Group members should take the following roles: pendulum operator, timer, pendulum watcher, and recorder. Students should tie the washer to one end of the string and tape the other end of the string to the edge of their desk so that the string and washer hang straight down. The pendulum operator should lift the washer to the side to start the pendulum, let the pendulum swing a few times so that the pendulum watcher can get the rhythm, and then signal the start and stop of a cycle. The pendulum operator then measures the length of the string and gives this data to the recorder who also gets the time from the timer. The group should repeat the experiment several times, varying the length of the string each time, and then make a scatter plot of the data with the length of the string as the independent variable and the period of the pendulum as the dependent variable. They should calculate a power regression equation for the data and compare it with the formula given in the text.

The domain of $f(x) = \sqrt{x}$ is all nonnegative real numbers, and the range of $f(x) = \sqrt{x}$ is all nonnegative real numbers.

The graph of $f(x) = \sqrt{x}$ is shown at right.

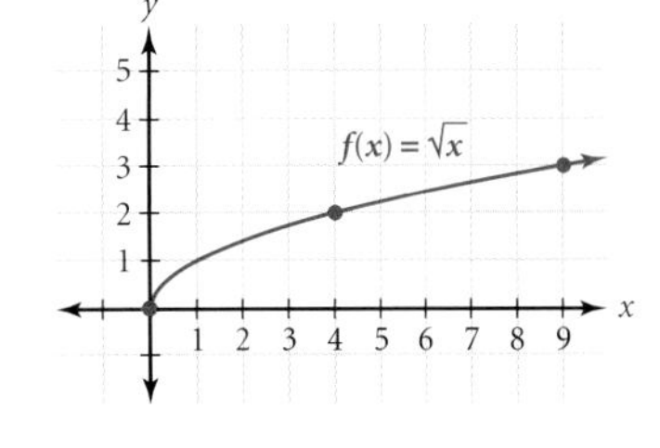

Example 1 shows you how to determine the domain of a square-root function.

EXAMPLE 1 **Find the domain of $f(x) = \sqrt{2x - 5}$.**

SOLUTION

The domain is all real numbers x that do not make $2x - 5$ negative. Solve $2x - 5 \geq 0$ to find the domain of $f(x) = \sqrt{2x - 5}$.

$$2x - 5 \geq 0$$
$$x \geq \frac{5}{2}, \text{ or } 2.5$$

TECHNOLOGY GRAPHICS CALCULATOR
Keystroke Guide, page 557

CHECK

Graph $y = \sqrt{2x - 5}$.

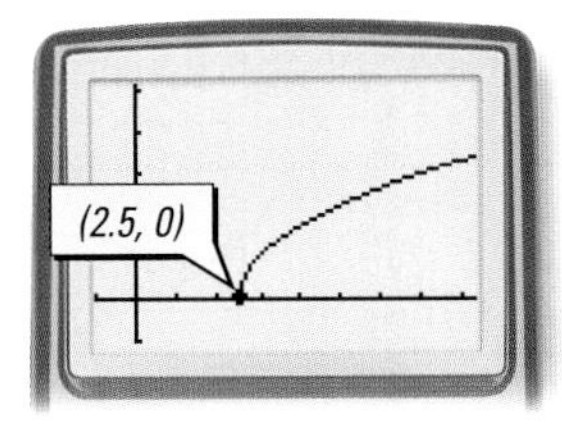

The domain of $f(x) = \sqrt{2x - 5}$ is $x \geq \frac{5}{2}$.

TRY THIS Find the domain of $g(x) = \sqrt{5x + 18}$.

The transformations given in Lesson 2.7 are summarized below for the square-root parent function, $y = \sqrt{x}$.

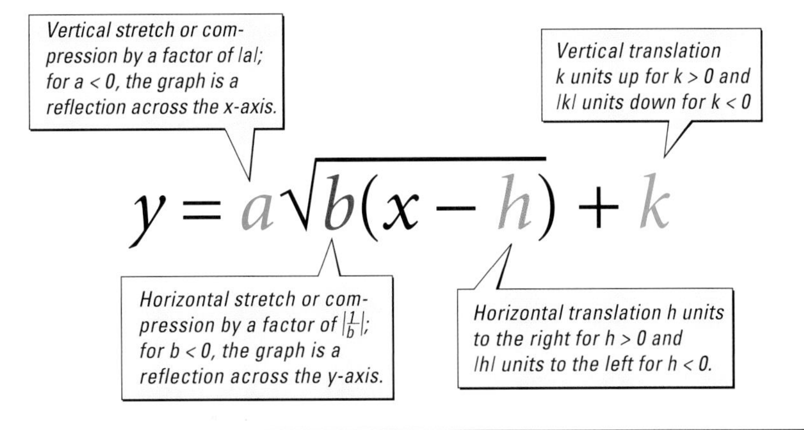

Vertical stretch or compression by a factor of |a|; for a < 0, the graph is a reflection across the x-axis.
Vertical translation k units up for k > 0 and |k| units down for k < 0
y = a√b(x − h) + k
Horizontal stretch or compression by a factor of |1/b|; for b < 0, the graph is a reflection across the y-axis.
Horizontal translation h units to the right for h > 0 and |h| units to the left for h < 0.

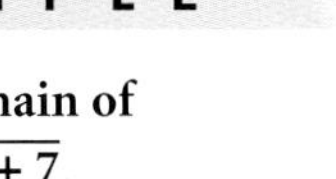

ADDITIONAL EXAMPLE 1

Find the domain of $h(x) = \sqrt{-4x + 7}$.

$x \leq \frac{7}{4}$

Teaching Tip

For Example 1, set the viewing window as follows:

Xmin=−1	**Ymin=−1**
Xmax=8	**Ymax=5**
Xscl=1	**Yscl=1**

TRY THIS

$x \geq -3.6$, or $-\frac{18}{5}$

Math CONNECTION

TRANSFORMATIONS Students apply geometric transformations to the square-root function.

Interdisciplinary Connection

PHYSICAL SCIENCE In meteorology, the windchill can be calculated by using the equation $C = 0.0817(3.71\sqrt{V} + 5.81 - 0.25V)(t - 91.4) + 91.4$, where V is the wind speed in miles per hour and t is the air temperature in degrees Fahrenheit. Find the windchill when the wind speed is 10 miles per hour and the temperature is 30°F. $\approx 16°$

ADDITIONAL

EXAMPLE 2

For each function, describe the transformations applied to $y = \sqrt{x}$.

a. $y = 3\sqrt{x-2} - 5$
vertical stretch by a factor of 3, horizontal translation 2 units to the right, vertical translation 5 units down

b. $y = -\sqrt{-2x+1} + 8$
reflection across the x-axis, horizontal compression by a factor of $\frac{1}{2}$, reflection across the y-axis, horizontal translation $\frac{1}{2}$ unit to the right, vertical translation 8 units up

Teaching Tip

TECHNOLOGY In Examples 2 and 3, use the following viewing window:

Xmin=–2.7	Ymin=–3
Xmax=6.7	Ymax=6

TRY THIS

a. vertical stretch by a factor of 3, vertical translation 2 units down, horizontal translation 1 unit to the right

b. horizontal compression by a factor of $\frac{1}{2}$, vertical translation 3 units up, and horizontal translation $\frac{1}{2}$ unit to the left

ADDITIONAL

EXAMPLE 3

Find the inverse of $y = x^2 + 1$. Then graph the function and its inverse together.
$y = \pm\sqrt{x-1}$

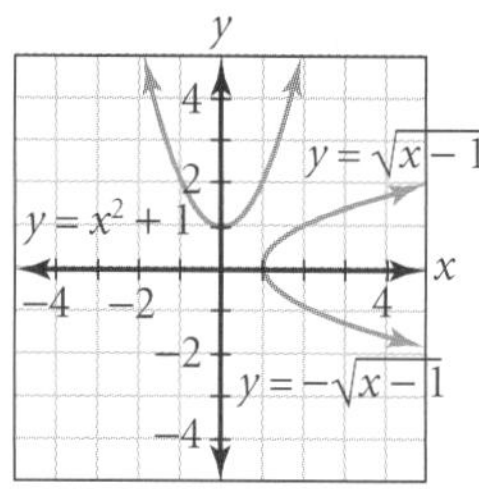

EXAMPLE 2 **For each function, describe the transformations applied to $y = \sqrt{x}$.**

a. $y = -2\sqrt{x+1} + 4$ **b.** $y = \sqrt{4x-3} - 1$

SOLUTION

a. Use the form $y = a\sqrt{b(x-h)} + k$. **b.** Use the form $y = a\sqrt{b(x-h)} + k$.

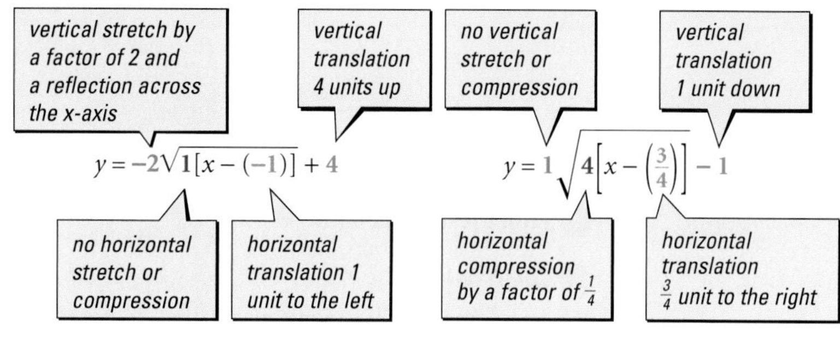

CHECK

Graph $y = -2\sqrt{x+1} + 4$ and $y = \sqrt{x}$ on the same screen, and compare the graphs.

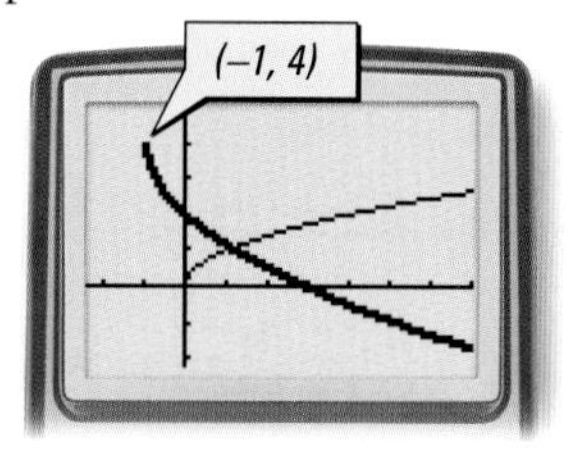

CHECK

Graph $y = \sqrt{4x-3} - 1$ and $y = \sqrt{x}$ on the same screen, and compare the graphs.

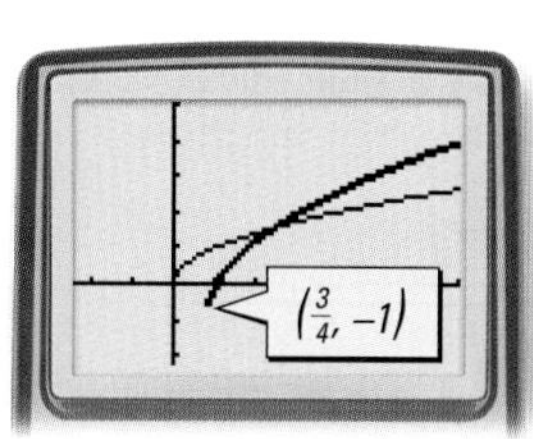

TRY THIS For each function, describe the transformations applied to $y = \sqrt{x}$.

a. $y = 3\sqrt{x-1} - 2$ **b.** $y = \sqrt{2x+1} + 3$

Recall from Lesson 2.5 that you can find the inverse of a function by interchanging x and y and then solving for y. This is shown for a quadratic function in Example 3.

EXAMPLE 3 **Find the inverse of $y = x^2 - 2x$. Then graph the function and its inverse together.**

SOLUTION

1. Interchange x and y.

$$y = x^2 - 2x \quad \rightarrow \quad x = y^2 - 2y$$

2. Solve $x = y^2 - 2y$ for y.

$$x = y^2 - 2y$$

$y^2 - 2y - x = 0$ *Write as a quadratic equation in terms of y.*

$y = \frac{-(-2) \pm \sqrt{(-2)^2 - 4(1)(-x)}}{2(1)}$ *Apply the quadratic formula for a = 1, b = −2, and c = −x.*

$y = \frac{2 \pm \sqrt{4+4x}}{2}$

$y = 1 \pm \sqrt{1+x}$

Enrichment

TACTILE LEARNERS Have students find a power regression equation **(PwrReg)** and its inverse based on the table below, which gives the speed, s, in miles per hour and the air resistance, R, in pounds for a bicyclist in a race.

R (lb)	s(mph)
1	11.66
2	16.49
5	26.08
10	36.88
15	45.17

$s = 11.66\sqrt{R}$; $R = \frac{s^2}{135.96}$

Have students answer the following:

1. What is the speed of a bicyclist who is experiencing 30 pounds of air resistance? $s \approx 63.9$

2. What happens to air resistance when a bicyclist's speed is doubled? **The air resistance is multiplied by 4.**

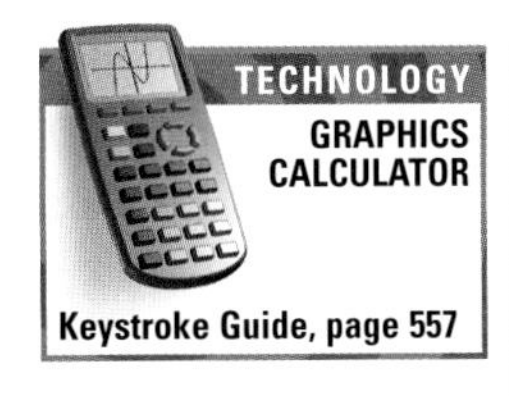

Keystroke Guide, page 557

3. Graph $y = x^2 - 2x$ and its inverse, $y = 1 + \sqrt{1 + x}$ and $y = 1 - \sqrt{1 + x}$, along with $y = x$.

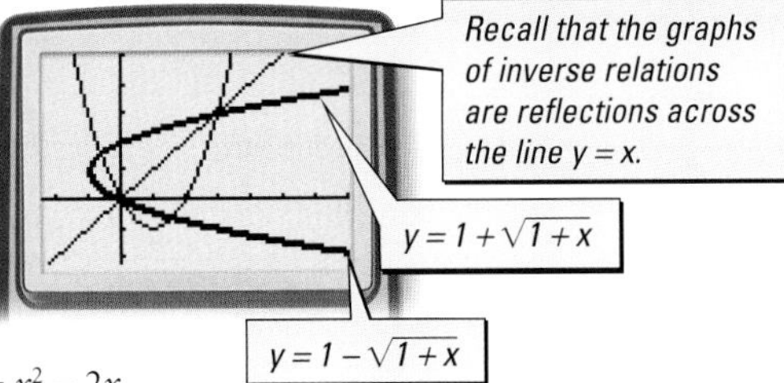

Thus, the inverse of $y = x^2 - 2x$ is $y = 1 \pm \sqrt{1 + x}$.

TRY THIS Find the inverse of $y = x^2 + 3x - 4$. Then graph the function and its inverse together.

CRITICAL THINKING Show that if $y = ax^2 + bx + c$, then $x = \dfrac{-b \pm \sqrt{b^2 - 4a(c - y)}}{2a}$.

EXAMPLE 4

APPLICATION
PHYSICS

Recall the pendulum problem described at the beginning of the lesson. The relationship between the period, T, (in seconds) of a pendulum and its length, x, (in meters) is $T(x) = 2\pi\sqrt{\dfrac{x}{9.8}}$.

Find the period for pendulums whose lengths are 0.1 meter, 0.2 meter, and 0.3 meter.

SOLUTION

Enter the function $y = 2\pi\sqrt{\dfrac{x}{9.8}}$ into a graphics calculator, and use the table feature with an x-increment of 0.1.

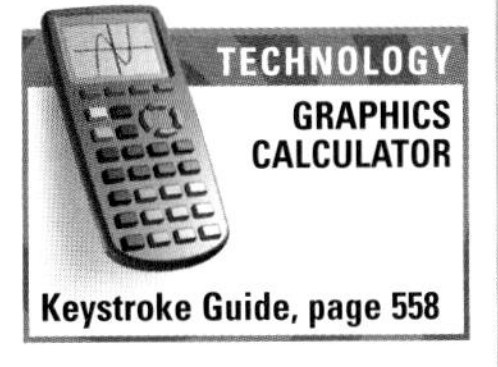

Keystroke Guide, page 558

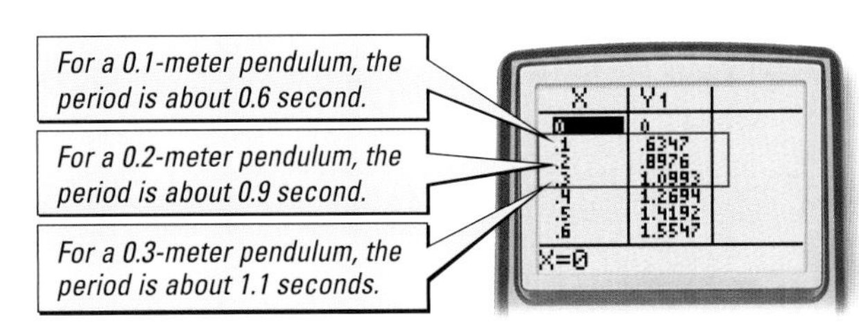

Cube-Root Functions

The **cube root** of a number, $\sqrt[3]{x}$, is a number that when multiplied by itself 3 times produces the given number, x. Recall from Lesson 2.2 that the expressions $\sqrt[3]{-8}$ and $\sqrt[3]{8}$ are both defined.

$$(-2)(-2)(-2) = -8 \text{ and } \sqrt[3]{-8} = -2$$

$$(2)(2)(2) = 8 \text{ and } \sqrt[3]{8} = 2$$

The domain of $f(x) = \sqrt[3]{x}$ is all real numbers, and the range of f is all real numbers, as shown in the graph at right.

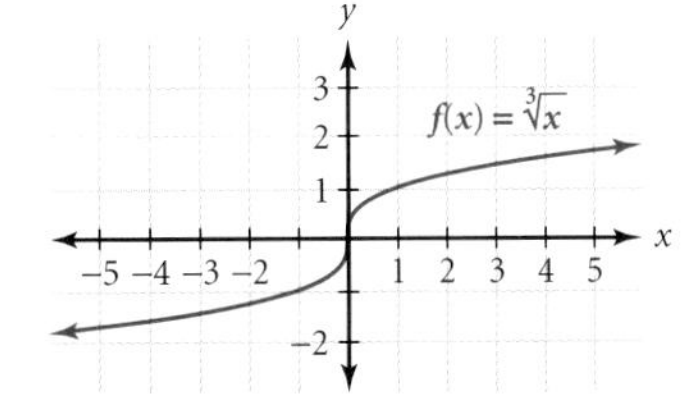

Inclusion Strategies

USING TABLES Some students find it difficult to understand why root functions with even indices have a restricted domain and root functions with odd indices do not. They will often memorize this fact without understanding why. To help build this understanding, have students make a table with column headings of x, $y = x^2$, $y = x^3$, $y = x^4$, $y = x^5$, $y = x^6$, and $y = x^7$. Have students list integers from –5 to 5 in the column for x. Then have them use a scientific calculator to fill in the rest of the table. Ask students to make comparisons and observations about the values in the table. Students should notice that the even powers are always positive, while the odd powers are negative when x is negative and positive when x is positive. Have them make a similar table for the inverses of the given functions. Discuss with students that radicals with even indices "undo" the even powers, so the domain of $y = \sqrt{x}$, for example, comes from the column for $y = x^2$ in the original table.

Teaching Tip

TECHNOLOGY Notice that the inverse of the quadratic function given in Example 3 is not a function. Graphics calculators are capable of graphing only functions, so each branch of the inverse must be graphed separately.

TRY THIS

$y = \dfrac{-3 \pm \sqrt{4x + 25}}{2}$

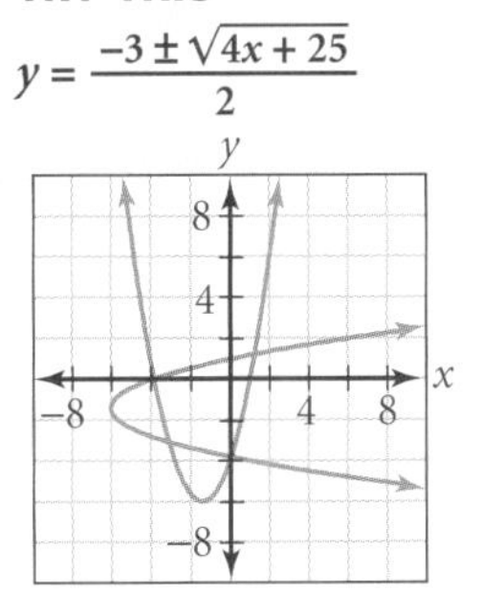

CRITICAL THINKING

$ax^2 + bx + c = y$

$ax^2 + bx + (c - y) = 0$

$x = \dfrac{-b \pm \sqrt{b^2 - 4a(c - y)}}{2a}$

ADDITIONAL EXAMPLE 4

You can use the formula $D(x) = \sqrt{1.5x}$ to approximate the maximum distance, D, in miles that you can see from a height, x, in feet. **Find the maximum distance you can see from heights of 10 feet, 20 feet, and 30 feet.**

≈3.9 mi, ≈5.5 mi, ≈6.7 mi

Teaching Tip

TECHNOLOGY When using a TI-83 or TI-82 graphics calculator to solve Example 4, enter the function into **Y1,** press 2nd WINDOW, and use the following settings:

TblStart=0 (or **TblMin=0**)
ΔTbl=.1
Indpnt: Auto Ask
Depend: Auto Ask

Then press 2nd GRAPH to view the table.

ADDITIONAL EXAMPLE 5

Evaluate each expression.

a. $-6\sqrt[3]{-8}-2$ 10

b. $4\left(\sqrt[3]{216}\right)^2$ 144

TRY THIS

0, 26

Activity Notes

In this Activity, students look at the graphs of basic radical functions. Students should notice that when n is even, the domain of the function is restricted to $x \geq 0$ and that for the functions given, the range is $y \geq 0$. However, when n is odd, the domain and range are always all real numbers.

Teaching Tip

TECHNOLOGY Make sure that students use the viewing window shown below so that they will see all four quadrants. Otherwise, they may miss a part of a graph.

Xmin=–5	Ymin=–5
Xmax=5	Ymax=5
Xscl=1	Yscl=1

CHECKPOINT ✔

5. **If n is a positive even integer, the domain of $y=\sqrt[n]{x}$ is $x \geq 0$ and the range is $y \geq 0$. If n is a positive odd integer, the domain of $y=\sqrt[n]{x}$ is all real numbers and the range is all real numbers.**

EXAMPLE 5 Evaluate each expression.

a. $3\sqrt[3]{27}-5$

b. $2\left(\sqrt[3]{-64}\right)^2+7$

SOLUTION

a. $3\sqrt[3]{27}-5$
$= 3(3)-5$ *$3^3 = 27$*
$= 4$

b. $2\left(\sqrt[3]{-64}\right)^2+7$
$= 2(-4)^2+7$ *$(-4)^3 = -64$*
$= 39$

TRY THIS Evaluate $-2\sqrt[3]{-125}-10$ and $6\left(\sqrt[3]{8}\right)^2+2$.

A **radical function** is a function that is defined by a *radical expression.* A **radical expression** is an expression that contains at least one **radical symbol,** such as $2\pi\sqrt{\frac{x}{9.80}}$, $\sqrt[4]{x}$, $\sqrt[7]{x+1}$, and $\sqrt[3]{2x+5}$. In the radical expression $\sqrt[3]{2x+5}$, $2x+5$ is the **radicand** and the number 3 is the **index.** Read the expression $\sqrt[3]{2x+5}$ as "the cube root of $2x+5$."

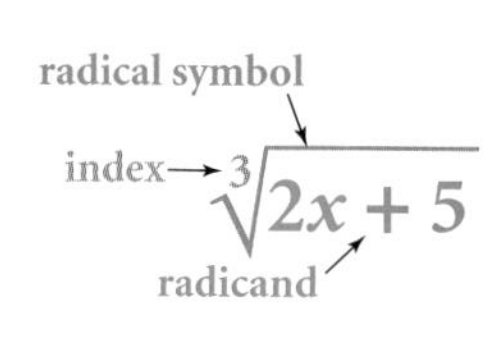

Index	Root	Symbol
2	square	$\sqrt{}$
3	cube	$\sqrt[3]{}$
4	fourth	$\sqrt[4]{}$
n	nth	$\sqrt[n]{}$

The index of the square root is usually omitted.

Recall from Lesson 2.2 that the definition of a rational exponent $\frac{1}{n}$, where n is a positive integer, is as follows: $a^{\frac{1}{n}} = \sqrt[n]{a}$. This definition is used when entering a radical expression into a calculator. For example, to evaluate $3\sqrt[3]{27}-5$ with a calculator, enter 3 [(] 27 [^] [(] 1 [÷] 3 [)] [)] [–] 5.

The Activity below introduces two *families* of radical functions.

Activity: Comparing Radical Functions

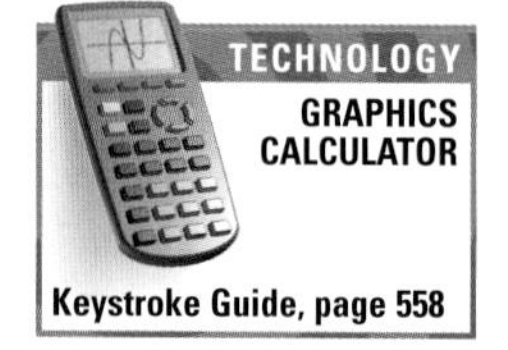

You will need: a graphics calculator

1. Graph $y=\sqrt{x}$, $y=\sqrt[3]{x}$, $y=\sqrt[4]{x}$, $y=\sqrt[5]{x}$, $y=\sqrt[6]{x}$, and $y=\sqrt[7]{x}$ one at a time. Sketch the shape of each graph.
2. Which functions appear similar?
3. Which functions have a domain of $x \geq 0$?
4. Which functions have a domain of all real numbers?

CHECKPOINT ✔ 5. Make a conjecture about the domain and range of $y=\sqrt[n]{x}$ when n is a positive even integer and when n is a positive odd integer.

Reteaching the Lesson

USING DISCUSSION Have students discuss transformations of a function in general, concentrating on vertical and horizontal shifts and stretches. Students may want to refer back to Lesson 2.7 to help their discussion. Then have students discuss the domain and range of a function. From these two discussions, have students describe any problems associated with the domain of radical functions.

Give students several examples of radical functions, and have a student lead the class in identifying the domain and range of the function, naming the parent function, and specifying the transformations of the parent function. Have students graph the parent function and the function on a graphics calculator and compare the graphs.

Exercises

Communicate

internet connect
Activities Online
Go To: **go.hrw.com**
Keyword:
MB1 RadicalGraphs

1. Describe the transformations applied to the graph of $f(x) = \sqrt{x}$ in order to obtain the graph of $y = 3\sqrt{x - 4}$.

2. How do you determine the domain of a radical function?

3. Describe the procedure for finding the inverse of $y = 4x^3 + 2$.

Guided Skills Practice

4. Find the domain of $f(x) = \sqrt{-2x + 3}$. ***(EXAMPLE 1)***

4. $x \leq \frac{3}{2}$

For each function, describe the transformations applied to $y = \sqrt{x}$. ***(EXAMPLE 2)***

5. $y = 2\sqrt{x - 1} - 2$ **6.** $y = \sqrt{2x + 1} + 2$

7. Find the inverse of $y = 4x^2 - x$. Then graph the function and its inverse together. ***(EXAMPLE 3)***

7. $y = \frac{1 \pm \sqrt{1 + 16x}}{8}$

APPLICATION

8. PHYSICS The vibration period of a mass on a spring is the time required for the mass to make a complete cycle in its motion. The relationship between the vibration period, T, (in seconds) of a certain spring and its mass, m, (in kilograms) is represented by $T(m) = 2\pi\sqrt{\frac{m}{200}}$. Find the period for a spring with masses of 1.0, 1.5, and 2.0 kilograms. Give answers to the nearest tenth. ***(EXAMPLE 4)*** **0.4, 0.5, and 0.6 second**

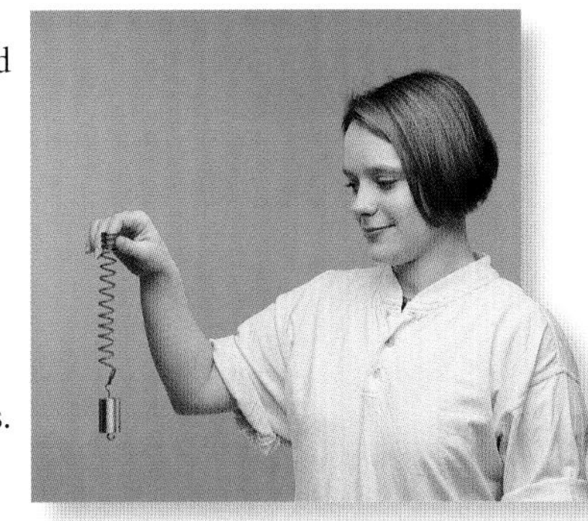

Evaluate each expression. ***(EXAMPLE 5)***

9. $4\sqrt[3]{-8} + 3$ **−5** **10.** $-2\left(\sqrt[3]{64}\right)^2 - 3$ **−35**

Practice and Apply

internet connect
Homework Help Online
Go To: **go.hrw.com**
Keyword:
MB1 Homework Help for Exercises 11–24

Find the domain of each radical function.

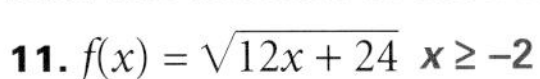

11. $f(x) = \sqrt{12x + 24}$ $x \geq -2$ **12.** $f(x) = \sqrt{3x - 2}$ $x \geq \frac{2}{3}$

13. $f(x) = \sqrt{3(x - 2)}$ $x \geq 2$ **14.** $f(x) = \sqrt{3(x + 2) - 1}$ $x \geq -\frac{5}{3}$

15. $f(x) = \sqrt{2 - 3(x + 1)}$ $x \leq -\frac{1}{3}$ **16.** $f(x) = \sqrt{3 - 3(x - 4)}$ $x \leq \frac{11}{2}$

17. $f(x) = \sqrt{x^2 - 25}$ $x \leq -5$ *or* $x \geq 5$ **18.** $f(x) = \sqrt{9x^2 - 16}$ $x \leq -\frac{4}{3}$ *or* $x \geq \frac{4}{3}$

19. $f(x) = \sqrt{x^2 + 5x + 6}$ **20.** $f(x) = \sqrt{x^2 + 10x - 25}$

21. $f(x) = \sqrt{2x^2 + 5x - 12}$ **22.** $f(x) = \sqrt{3x^2 + 7x + 2}$

23. $f(x) = \sqrt{6x^2 - 13x + 5}$ **24.** $f(x) = \sqrt{8x^2 - 10x - 3}$

19. $x \leq -3$ *or* $x \geq -2$

20. $x \leq -5 - 5\sqrt{2}$ *or* $x \geq -5 + 5\sqrt{2}$

21. $x \leq -4$ *or* $x \geq \frac{3}{2}$

22. $x \leq -2$ *or* $x \geq -\frac{1}{3}$

23. $x \leq \frac{1}{2}$ *or* $x \geq \frac{5}{3}$

24. $x \leq -\frac{1}{4}$ *or* $x \geq \frac{3}{2}$

Assess

Selected Answers

Exercises 4–10, 11–79 odd

ASSIGNMENT GUIDE	
In Class	1–10
Core	11–51 odd, 55–63 odd
Core Plus	12–66 even
Review	67–79
Preview	80

Extra Practice can be found beginning on page 940.

5. a horizontal translation 1 unit to the right, a vertical stretch by a factor of 2, and a vertical translation 2 units down

6. a horizontal translation $\frac{1}{2}$ unit to the left, a horizontal compression by a factor of $\frac{1}{2}$, and a vertical translation 2 units up

7. $y = \frac{1 \pm \sqrt{1 + 16x}}{8}$

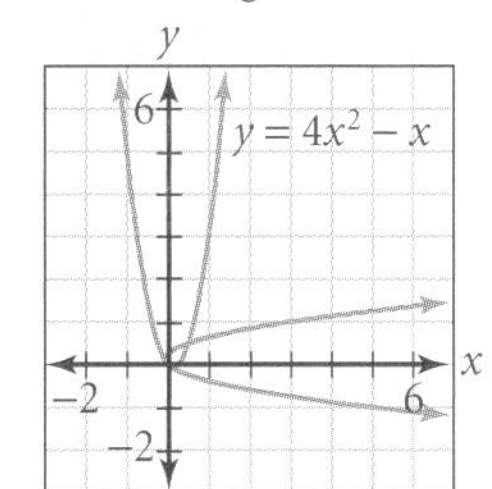

Error Analysis

When finding the domain of a radical function with a quadratic radicand, students need to factor the radicand and examine the cases in which the product is positive. Remind students that the product of two negatives is a positive and that this case must be used when finding the domain of a radical function with a factored quadratic radicand.

The graphs to Exercises 25–36 can be found in Additional Answers beginning on page 1002.

54. a vertical translation 4 units down

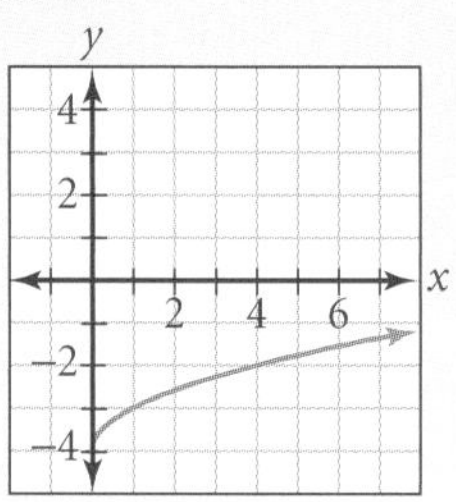

Find the inverse of each quadratic function. Then graph the function and its inverse in the same coordinate plane.

25. $y = x^2 - 1$ $y = \pm\sqrt{x+1}$

26. $y = x^2 + 3$ $y = \pm\sqrt{x-3}$

27. $y = 3x^2 + x$ $y = \frac{-1 \pm \sqrt{1+12x}}{6}$

28. $y = x^2 + 2x$ $y = -1 \pm \sqrt{1+x}$

29. $y = x^2 + 4x + 4$ $y = -2 \pm \sqrt{x}$

30. $y = x^2 + 6x + 9$ $y = -3 \pm \sqrt{x}$

31. $y = x^2 - 2x + 1$ $y = 1 \pm \sqrt{x}$

32. $y = x^2 - 7x + 12$ $y = \frac{7 \pm \sqrt{1+4x}}{2}$

33. $y = 4 - x^2 + 3x$ $y = \frac{3 \pm \sqrt{25-4x}}{2}$

34. $y = 2x + x^2 - 8$ $y = -1 \pm \sqrt{9+x}$

35. $y = x + 2x^2 - 1$ $y = \frac{-1 \pm \sqrt{9+8x}}{4}$

36. $y = 3 + x - 2x^2$ $y = \frac{1 \pm \sqrt{25-8x}}{4}$

Evaluate each expression.

37. $\frac{1}{2}\left(\sqrt[3]{8}\right)^6$ 32

38. $2\left(\sqrt[4]{625}\right)^2$ 50

39. $\frac{1}{2}\sqrt[3]{-8} + 3$ 2

40. $-\frac{2}{5}\sqrt[3]{-125} - 3$ −1

41. $\frac{1}{8}\left(\sqrt[3]{-8^3}\right)^2$ 8

42. $2\left(\sqrt[3]{8}\right)^2 - 3$ 5

43. $-2\sqrt[3]{216} - 3$ −15

44. $3\sqrt[3]{343} + 1$ 22

45. $10\left(\sqrt[3]{1000}\right)^3 - 1$ 9999

46. $\frac{1}{6}\left(\sqrt[3]{\frac{54}{4}}\right)^3 + 5$ $\frac{29}{4}$

47. $\frac{1}{2}\left(\sqrt[3]{-216}\right)^4$ 648

48. $2\left(\sqrt[3]{-343}\right)^2$ 98

49. $\frac{1}{2}\sqrt[3]{216}$ 3

50. $\left(\sqrt[4]{81} - 1\right)^2$ 4

51. $-\frac{1}{3}\left(\sqrt[4]{1296}\right)^2$ −12

CHALLENGES

52. State the domain of $y = \sqrt{\frac{1}{x^2+1}}$. Graph the function to check. **all reals**

53. Let $f(x) = ax^2 + bx + c$, where $a \neq 0$. Find an equation for the axis of symmetry for the graph of the inverse of f. $y = -\frac{b}{2a}$

CONNECTIONS

TRANSFORMATIONS For each function, describe the transformations applied to $f(x) = \sqrt{x}$. Then graph each transformed function.

54. $g(x) = \sqrt{x} - 4$

55. $h(x) = \sqrt{x-4}$

56. $p(x) = 3\sqrt{x-4}$

57. $r(x) = -\frac{1}{5}\sqrt{x-4}$

58. $s(x) = -2\sqrt{x-4}$

59. $g(x) = \sqrt{x+2} + 5$

60. $h(x) = 2\sqrt{x+3} - 1$

61. $p(x) = 2(\sqrt{x+3} - 1)$

62. $k(x) = \frac{1}{2}\sqrt{x-1} + 4$

63. $m(x) = \frac{1}{3}\sqrt{x+2} + 3$

64. GEOMETRY The volume, V, of a sphere with a radius of r is given by $V = \frac{4}{3}\pi r^3$.

a. Solve this equation for r in terms of V.

b. Show that if V increases, then r must increase.

c. A spherical hot-air balloon has a volume of 12,000 cubic feet. Find the approximate radius of the balloon by using the equation you wrote in part **a** and a table. Give your answer to the nearest tenth of a foot.

64a. $r = \sqrt[3]{\frac{3V}{4\pi}}$

b. As V increases, $\sqrt[3]{V}$ increases. Because 3, 4, and π are constants, $\sqrt[3]{\frac{3V}{4\pi}}$ increases as V increases.

c. 14.2 feet

Practice

NAME ______ CLASS ______ DATE ______

Practice

8.6 Radical Expressions and Radical Functions

Find the domain of each radical function.

1. $f(x) = \sqrt{12x - 30}$ $x \geq 2.5$
2. $f(x) = \sqrt{7(x-4)}$ $x \geq 4$
3. $f(x) = \sqrt{x^2 - 36}$ $x \leq -6$ or $x \geq 6$
4. $f(x) = \sqrt{4x^2 - 25}$ $x \leq -2.5$ or $x \geq 2.5$
5. $f(x) = \sqrt{x^2 - 10x + 25}$ $x \geq 5$
6. $f(x) = \sqrt{x^2 + 4x + 3}$ $x \leq -3$ or $x \geq -1$

Find the inverse of each quadratic function. Then graph the function and its inverse in the same coordinate plane.

7. $y = x^2 - 6x + 8$

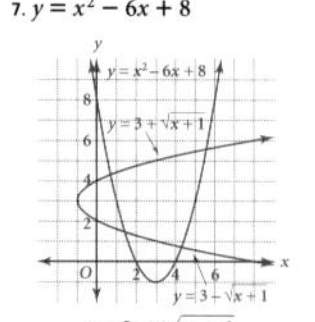

$y = 3 \pm \sqrt{x+1}$

8. $y = 2 - x^2$

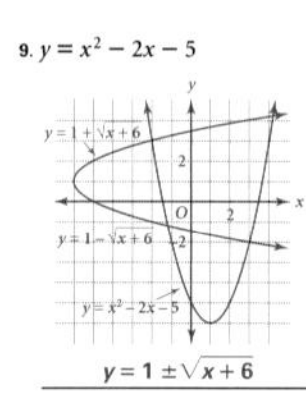

$y = \pm\sqrt{2-x}$

9. $y = x^2 - 2x - 5$

$y = 1 \pm \sqrt{x+6}$

Evaluate each expression. Give exact answers.

10. $\sqrt[3]{\frac{192}{3}}$ 4
11. $\frac{3}{4}\sqrt[4]{10{,}000}$ 7.5
12. $15\sqrt[3]{-\frac{8}{125}}$ −6
13. $4\sqrt[3]{-216}$ −24
14. $-8\sqrt[3]{-\frac{1}{8}}$ 4
15. $\frac{2}{3}\sqrt[3]{-27}$ −2

16. The volume of a sphere with diameter d is given by the equation $V = \frac{1}{6}\pi d^3$. Solve this equation for d in terms of V. Then use your equation to find the diameter, to the nearest foot, of a sphere with a volume of 1000 cubic feet.

$d = \sqrt[3]{\frac{6V}{\pi}}$; 12.4 feet

55. a horizontal translation 4 units to the right

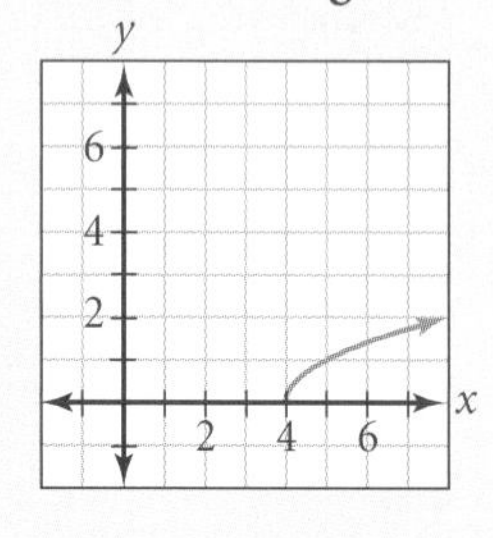

56. a horizontal translation 4 units to the right and a vertical stretch by a factor of 3

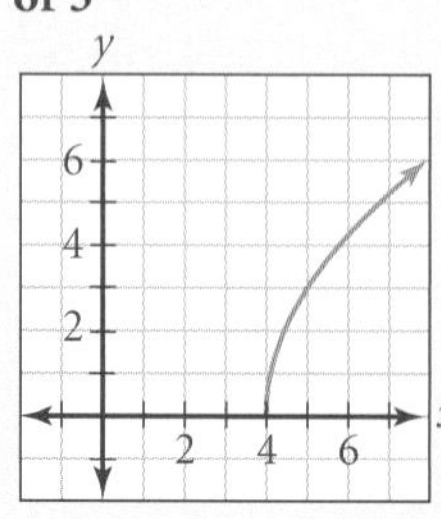

57. a horizontal translation 4 units to the right, a vertical compression by a factor of $\frac{1}{5}$, and a reflection across the x-axis

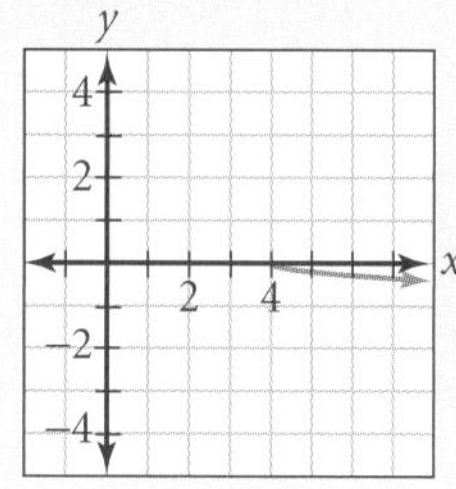

APPLICATIONS

65a. $T \approx \frac{29{,}250 \pm \sqrt{5(12{,}500 + 29E)}}{290}$

b. 99°C

66a. $r = \frac{\sqrt[4]{8LNf}}{\pi p}$; r varies directly as the 4th root of f.

b. f varies directly as p. If p is doubled, f is doubled.

c. f varies directly as the 4th power of r. If r is doubled, f is multiplied by 2^4, or 16.

65. PHYSICS If E represents the elevation in meters above sea level and T represents the boiling point of water in degrees Celsius at that elevation, then E and T are related by the equation below.

$$E \approx 1000(100 - T) + 580(100 - T)^2$$

a. Solve the equation above for T. (Hint: Begin by substituting x for $100 - T$. Then solve for x.)

b. What is the approximate boiling point of water at an elevation of 1600 meters?

66. PHYSICS In the 1840s, a French physiologist, Jean Marie Poiseuville, found a relationship between the flow rate, f, of a liquid flowing through a cylinder whose length is L and whose radius is r. The flow rate also depends on the pressure, p, exerted on the liquid and the coefficient of viscosity, N, of the liquid. (*Viscosity* is a measure of the resistance of a fluid to flow.)

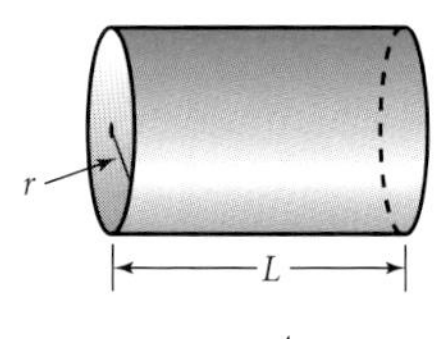

$$f = \frac{\pi p r^4}{8LN}$$

a. Solve the equation at right above for r in terms of the other variables. How does r vary as f, given that p, L, and N are constant?

b. Assume that r, L, and N are constant. How does f vary as p? How does f change if p is doubled?

c. Assume that p, L, and N are constant. How does f vary as r? How does f change if r is doubled?

Alaskan oil pipeline

Look Back

Simplify each expression. Assume that no variable equals zero. *(LESSON 2.2)*

67. $(-2y^3y^5)^2$ $\mathbf{4y^{16}}$

68. $2a^4(-3ab^2)^3$ $\mathbf{-54a^7b^6}$

69. $(5x^{-2}y^4)^{-1}$ $\mathbf{\frac{x^2}{5y^4}}$

70. $\left(\frac{-3xy^3}{x^{-4}y^5}\right)^3$ $\mathbf{\frac{-27x^{15}}{y^6}}$

71. $\left(\frac{-4m^4n^3}{m^2n^3}\right)^{-1}$ $\mathbf{-\frac{1}{4m^2}}$

72. $\left(\frac{3m^4n^{-2}}{2m^0n^3}\right)^{-2}$ $\mathbf{\frac{4n^{10}}{9m^8}}$

Perform the indicated addition or subtraction. *(LESSON 5.6)*

73. $2i - (4 - 5i)$ $\mathbf{-4 + 7i}$

74. $(-2 + 4i) + (-1 - i)$ $\mathbf{-3 + 3i}$

75. $(-4 + i) - (3 - 2i)$ $\mathbf{-7 + 3i}$

Write each expression as a sum or difference of logarithms. *(LESSON 6.4)*

79. $\log x + \log z - \log y$

76. $\log(6 \cdot 3)$ **log 6 + log 3**

77. $\log 2x$ **log 2 + log x**

78. $\log\left(\frac{17}{8}\right)$ **log 17 − log 8**

79. $\log\left(\frac{xz}{y}\right)$

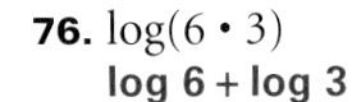

Look Beyond

CONNECTION

80 COORDINATE GEOMETRY Let $x^2 + y^2 = 1$. Solve for y. Graph the two resulting equations together. Describe the shape that the pair of graphs creates.

Look Beyond

In Exercise 80, students look at the equation of a circle. Students graph the circle by solving for y and entering the two resulting functions into a graphics calculator. Circles and other conic sections will be studied in Chapter 9.

58. a horizontal translation 4 units to the right, a vertical stretch by a factor of 2, and a reflection across the x-axis

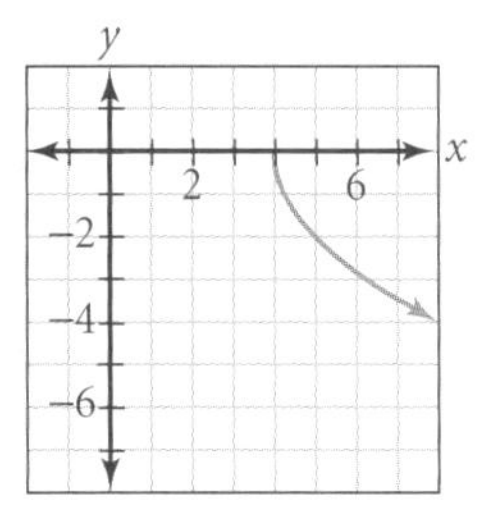

59. a horizontal translation 2 units to the left and a vertical translation 5 units up

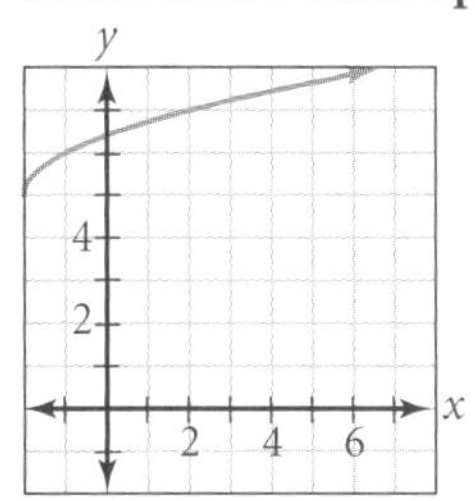

60. a horizontal translation 3 units to the left, a vertical stretch by a factor of 2, and a vertical translation 1 unit down

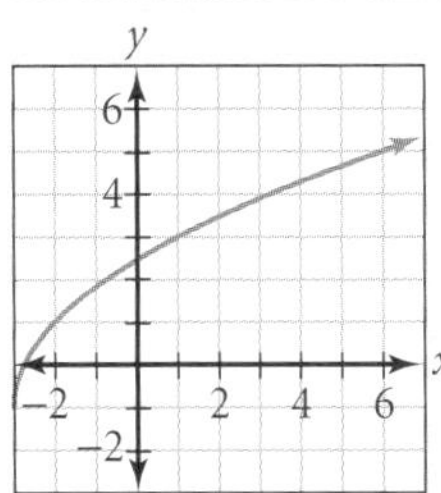

61. a horizontal translation 3 units to the left, a vertical translation 2 units down, and a vertical stretch by a factor of 2

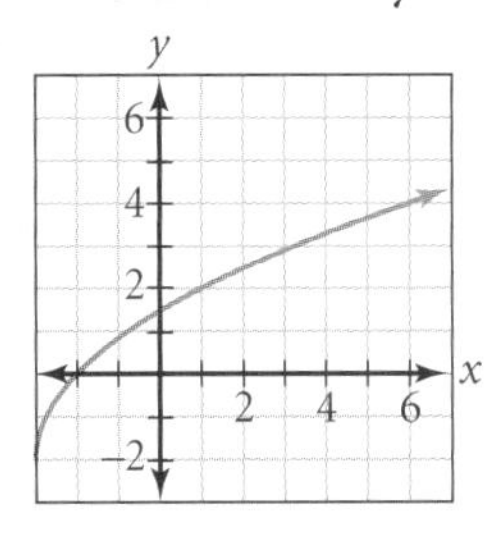

The answers to Exercises 62–63 and 80 can be found in Additional Answers beginning on page 1002.

Student Technology Guide

NAME ______ CLASS ______ DATE ______

Student Technology Guide

8.6 Radical Expressions and Radical Functions

The domain of a square-root function includes all x-values that make the radicand positive or zero. You can use a graphics calculator to identify the domain of a radical function.

Example: Find the domain of $f(x) = \sqrt{2x - 5}$.

- To graph the function, press Y= 2nd x² 2 X,T,θ,n − 5) GRAPH. (Note: The calculator may provide the first parenthesis for you, but you must close the parentheses.)
- Press TRACE. Press ▶ until the trace cursor appears. Notice that the cursor first appears near the end of the graph where $x \approx 2.68$. So, a first estimate for the domain is $x \geq 2.68$.
- To get a better estimate, press ZOOM 2:Zoom In ENTER ENTER. Press TRACE again and use the arrow keys to move as close as you can to the endpoint. Repeat this process until you have the accuracy you need.

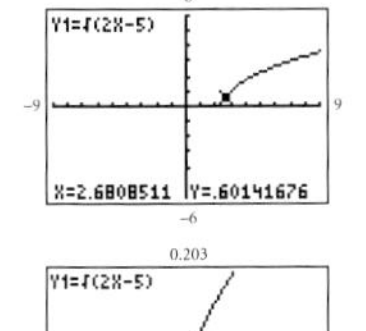

The left endpoint of the graph occurs when $x = 2.5$. From the reasoning above and the graph, the domain is $x \geq 2.5$. The endpoint of the domain of a radical function is often (though not always) an exact decimal value, so a decimal window can be a good place to start when finding the domain of radical functions. To get a decimal window, press ZOOM 4:ZDecimal ENTER. You can use TRACE to confirm that the x-coordinate of the endpoint of the function in the example is exactly 2.5.

Find the domain of each function. If necessary, round answers to the nearest hundredth.

1. $f(x) = \sqrt{5x + 15}$ $x \geq -3$
2. $f(x) = \sqrt{4x - 3}$ $x \geq 0.75$
3. $f(x) = \sqrt{x^2 - 16}$ $x \leq -4$ or $x \geq 4$
4. $f(x) = -3\sqrt{12 - 5x}$ $x \leq 2.4$
5. $f(x) = -\sqrt{(7x + 2)} + 1$ $x \geq -0.43$
6. $f(x) = \sqrt{x^2 + 3x - 8}$ $x \leq -4.70$ or $x \geq 1.70$

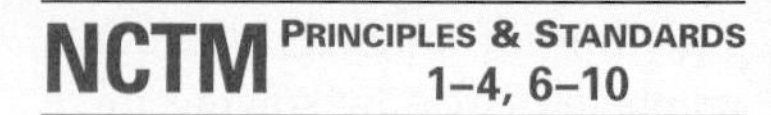

Prepare

NCTM PRINCIPLES & STANDARDS 1–4, 6–10

QUICK WARM-UP

Evaluate each expression.

1. $36^{\frac{1}{2}}$ ± 6
2. $-125^{\frac{2}{3}}$ -25

Simplify each expression.

3. $(ab^4)(a^2b)$ a^3b^5
4. $(m^2n)^3$ m^6n^3
5. $\frac{r^2s^3}{r^5s}$ $\frac{s^2}{r^3}$
6. $\left(\frac{cd^2}{c^4d^3}\right)^2$ $\frac{1}{c^6d^2}$
7. $(2x-1)(3x+2)$ $6x^2+x-2$

Also on Quiz Transparency 8.7

Teach

Why Have students recall the formula of the volume of a cube. $V = s^3$ Ask students what the volume of a cube with sides of 2 ft would be. **8 ft³** Using trial and error and cubing numbers, have students give a range of lengths for the sides of each box discussed on this page. **54-ft³ box: 3–4 ft; 128-ft³ box: 5–6 ft; 250-ft³ box: 6–7 ft**

Simplifying Radical Expressions

Objectives

- Add, subtract, multiply, divide, and simplify radical expressions.
- Rationalize a denominator.

Why *Solving real-world problems sometimes involves operations with radical expressions. For example, if you know the volumes of cube-shaped boxes, you can determine the height of a stack of them by using a radical expression.*

APPLICATION
PACKAGING

Danielle is loading moving boxes onto a truck. She has three large cube-shaped boxes whose volumes are 54 cubic feet, 128 cubic feet, and 250 cubic feet. If she stacks them on top of one another, how tall will the stack be? Answering this question involves working with radical expressions. *You will answer this question in Example 3.*

In the Activity below, you will investigate some properties of radicals.

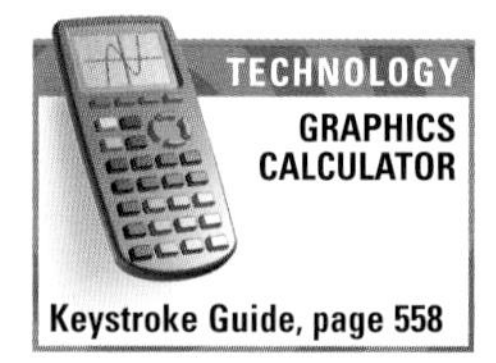

TECHNOLOGY
GRAPHICS CALCULATOR
Keystroke Guide, page 558

Activity Exploring Properties of Radicals

You will need: a graphics calculator

Complete the following steps, using the fact that $\sqrt[n]{x} = x^{\frac{1}{n}}$:

1. Graph $y = \sqrt{x^2}$. What function does the graph represent?
2. **Guess and check.** Use your work from Step 1 to predict the appearance of the graph of $y = \sqrt[4]{x^4}$. Verify your response. (PROBLEM SOLVING)
3. Predict the appearance of the graph of $y = \sqrt[n]{x^n}$, where n is a positive even integer. Then illustrate your prediction. (CHECKPOINT ✔)
4. Graph $y = \sqrt[3]{x^3}$. What function does the graph represent?
5. Use your work from Step 4 to predict the appearance of the graph of $y = \sqrt[5]{x^5}$. Verify your response.
6. Predict the appearance of the graph of $y = \sqrt[n]{x^n}$, where n is a positive odd integer. Then illustrate your prediction, and explain your reasoning. (CHECKPOINT ✔)

Alternative Teaching Strategy

MAKING CONNECTIONS Discuss with students the relationship between the radical form of an expression and the exponential form, and have students recall the properties of exponents from Lesson 2.2. Radical expressions are an extension of exponential expressions, but some exponents may be fractional. Ask students to simplify the expression $(x^4y^6)^2$, describing each step. x^8y^{12} Now have students simplify $\sqrt{x^4y^6}$ by changing the expression from radical form to exponential form and following the same steps as for the first expression. $(x^4y^6)^{\frac{1}{2}} = x^2y^3$ Again, have students verbalize the process. Next have students simplify the expression $3x^2 + 2y^3 + 6x^2 - 3y^2$. $9x^2 + 2y^3 - 3y^2$ Have students discuss their steps, pointing out that sums or differences can be combined only if the monomials have the same base and power. Now have students simplify another expression such as $3\sqrt{x} + 2\sqrt[3]{y} + 6\sqrt{x} - 3\sqrt{y}$. $9\sqrt{x} + 2\sqrt[3]{y} - 3\sqrt{y}$ Continue to give students radical expressions to simplify and have students make analogies to expressions that contain nonfractional exponents.

Just as there are Properties of Exponents that you can use to simplify exponential expressions, there are *Properties of Radicals* that are used to simplify radical expressions.

Properties of Radicals

For any real number a,

$\sqrt[n]{a^n} = |a|$ if n is a positive even integer, and

$\sqrt[n]{a^n} = a$ if n is a positive odd integer.

For example, $\sqrt{(-3)^2} = |-3| = 3$ and $\sqrt[3]{(-3)^3} = -3$.

EXAMPLE 1 **Simplify each expression by using the Properties of Radicals.**

a. $\sqrt{49x^2y^5z^6}$ **b.** $\sqrt[3]{-27x^7y^3z^2}$

SOLUTION

a. $\sqrt{49x^2y^5z^6} = \sqrt{7^2x^2y^4yz^6}$
$= 7|x|y^2|z^3|\sqrt{y}$

b. $\sqrt[3]{-27x^7y^3z^2} = \sqrt[3]{(-3)^3x^6xy^3z^2}$
$= -3x^2y\sqrt[3]{xz^2}$

TRY THIS Simplify each expression by using the Properties of Radicals.

a. $\sqrt{64a^4bc^3}$ **b.** $\sqrt[5]{-32f^6g^5h^2}$

Recall from Lesson 2.2 that $a^{\frac{m}{n}} = \left(a^{\frac{1}{n}}\right)^m = \left(\sqrt[n]{a}\right)^m$ and that $a^{\frac{m}{n}} = (a^m)^{\frac{1}{n}} = \sqrt[n]{a^m}$.

You can write an expression with a rational exponent in *radical form*.

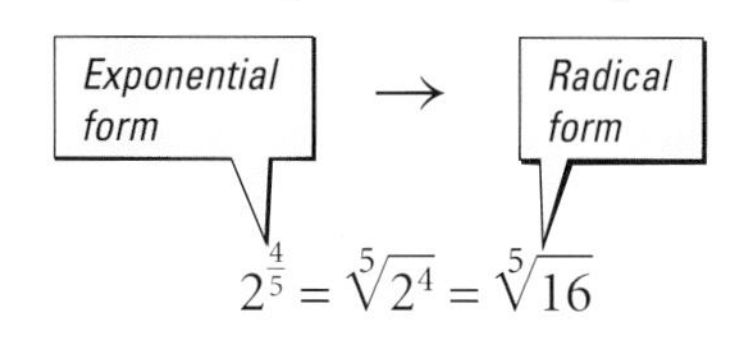

$2^{\frac{4}{5}} = \sqrt[5]{2^4} = \sqrt[5]{16}$

You can write an expression with a radical in *exponential form*.

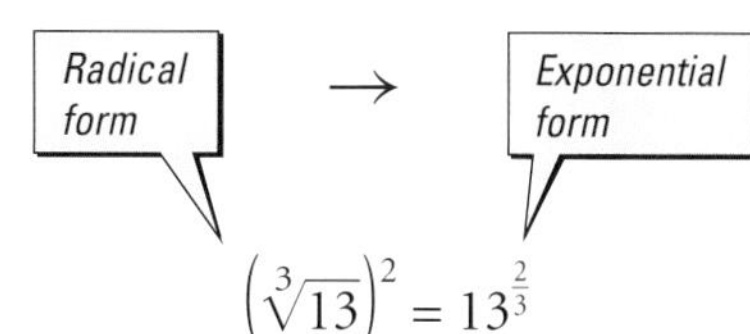

$\left(\sqrt[3]{13}\right)^2 = 13^{\frac{2}{3}}$

CHECKPOINT ✔ Write $2^{\frac{3}{5}}$ in radical form.

You can use the *Product and Quotient Properties of Radicals* to multiply, divide, and simplify radical expressions.

Product and Quotient Properties of Radicals

For $a \geq 0$, $b \geq 0$, and a positive integer n:

$\sqrt[n]{ab} = \sqrt[n]{a} \cdot \sqrt[n]{b}$ and $\sqrt[n]{\frac{a}{b}} = \frac{\sqrt[n]{a}}{\sqrt[n]{b}}$ $(b \neq 0)$

Interdisciplinary Connection

PHYSICS Use the Pythagorean theorem to solve the following problem. Draw and label a picture that represents the situation.

An airplane is flying due west at 400 miles per hour while the wind is blowing due south at 30 miles per hour. What would be the speed of the plane with no wind? **≈401.12 mph**

Enrichment

Recall the rule for factoring a difference of two squares: $(a^2 - b^2) = (a - b)(a + b)$. Use this pattern to factor the following expressions over the set of real numbers:

1. $x^2 - 9$ $(x-3)(x+3)$
2. $x^4 - 16$ $(x-2)(x+2)(x^2+4)$
3. $x^2 - 5$ $(x-\sqrt{5})(x+\sqrt{5})$
4. $x^2 - 8$ $(x-2\sqrt{2})(x+2\sqrt{2})$
5. $x^2 - a$ $(x-\sqrt{a})(x+\sqrt{a})$

Have students check each problem by multiplying the factors.

Activity Notes

In this Activity, students discover the properties of $y = \sqrt[n]{x^n}$ by graphing related functions for cases when n is even and when n is odd. Students may have trouble verbalizing the absolute-value function when n is even, and some students may have written a piecewise function to describe the related graph. If so, this is a wonderful way to reinforce the piecewise definition of absolute value: $|a| = a$ for $a \geq 0$ and $|a| = -a$ for $a < 0$.

Cooperative Learning

Have students work in groups of two or three. Each group should choose a reporter to summarize the members' description of the graphs. The other members will enter the data and discuss methods of predicting the graph's appearance. When all steps of the Activity are completed, each group should report its findings to the class.

Teaching Tip

TECHNOLOGY In the Activity, make sure that students use a standard viewing window so that they can see all four quadrants. Encourage students to check their answer by graphing any function(s) they think have the same graph.

CHECKPOINT ✔

3. **the graph of $y = |x|$**

CHECKPOINT ✔

6. **the graph of $y = x$**

☞ For Additional Example 1 and the Checkpoint, see page 530.

ADDITIONAL EXAMPLE 1

Simplify each expression.

a. $\sqrt{81x^3y^4z^6}$ $9y^2|xz^3|\sqrt{x}$

b. $\sqrt[3]{125x^6yz^5}$ $5x^2z\sqrt[3]{yz^2}$

TRY THIS

a. $8a^2|c|\sqrt{bc}$ b. $-2fg\sqrt[5]{fh^2}$

CHECKPOINT ✔

$2^{\frac{3}{5}} = \sqrt[5]{2^3}$, or $\sqrt[5]{8}$

ADDITIONAL EXAMPLE 2

Simplify each expression. Assume that the value of each variable is positive.

a. $(64ab^5)^{\frac{1}{2}} \cdot \sqrt{8ab^6}$
$16ab^5\sqrt{2b}$

b. $\dfrac{9\sqrt[3]{48x^8}}{(2x^3)^{\frac{1}{3}}}$ $18x\sqrt[3]{3x^2}$

TRY THIS

a. $3rs^2\sqrt[3]{r^2s}$ b. $3|x|\sqrt{2y}$

ADDITIONAL EXAMPLE 3

Darius is organizing his garage. He bought three storage boxes, each in the shape of a cube. He wants to stack the boxes on top of each other. The volumes of the boxes are 24 cubic feet, 81 cubic feet, and 192 cubic feet. **How tall will the stack of boxes be? Give both an exact answer and an answer rounded to the nearest hundredth of a foot.**

$9\sqrt[3]{3}$, or ≈12.98, ft

EXAMPLE 2 **Simplify each expression. Assume that the value of each variable is positive.**

a. $(27ab^3)^{\frac{1}{3}} \cdot \sqrt[3]{5a^4b}$ **b.** $\dfrac{8(54x^5)^{\frac{1}{2}}}{4\sqrt{3x^3}}$

SOLUTION

a. $(27ab^3)^{\frac{1}{3}} \cdot \sqrt[3]{5a^4b}$

$= \sqrt[3]{27ab^3} \cdot \sqrt[3]{5a^4b}$

$= \sqrt[3]{(27ab^3)(5a^4b)}$

$= \sqrt[3]{(3^3ab^3)(5a^4b)}$

$= \sqrt[3]{3^3a^3b^3 \cdot 5^1a^2b^1}$ *Associative Property*

$= \sqrt[3]{3^3} \cdot \sqrt[3]{a^3} \cdot \sqrt[3]{b^3} \cdot \sqrt[3]{5a^2b}$

$= 3ab\sqrt[3]{5a^2b}$

b. $\dfrac{8(54x^5)^{\frac{1}{2}}}{4\sqrt{3x^3}}$

$= \dfrac{8\sqrt{54x^5}}{4\sqrt{3x^3}}$

$= 2\sqrt{\dfrac{54x^5}{3x^3}}$

$= 2\sqrt{18x^2}$

$= 2\sqrt{3^2 \cdot x^2 \cdot 2}$ *Associative Property*

$= 2 \cdot 3 \cdot |x|\sqrt{2}$

$= 6x\sqrt{2}$ *Assume x is positive.*

TRY THIS Simplify each expression. Assume that the value of each variable is positive.

a. $\sqrt[3]{3r^2s^3} \cdot (9r^3s^4)^{\frac{1}{3}}$ **b.** $\dfrac{\sqrt{54x^3y^3}}{(3xy^2)^{\frac{1}{2}}}$

Operations With Radical Expressions

When you add or subtract radical expressions, it is helpful to write your answers in *simplest radical form*. The expression $\sqrt{a}$ is in **simplest radical form** if no factor of a is a perfect square.

EXAMPLE 3 Refer to the cube-shaped boxes described at the beginning of the lesson. The volumes are 54 cubic feet, 128 cubic feet, and 250 cubic feet.

APPLICATION PACKAGING

How tall will the stack of boxes be? Give both an exact answer and an answer rounded to the nearest hundredth of a foot.

SOLUTION

The formula for the volume, V, of a cube is $V = e^3$. The length of one edge is given by the radical expression $e = \sqrt[3]{V}$. The height of the stack is the sum of all three edge lengths.

$$\sqrt[3]{54} + \sqrt[3]{128} + \sqrt[3]{250} = \sqrt[3]{3^3 \cdot 2} + \sqrt[3]{4^3 \cdot 2} + \sqrt[3]{5^3 \cdot 2}$$
$$= 3\sqrt[3]{2} + 4\sqrt[3]{2} + 5\sqrt[3]{2}$$
$$= 12\sqrt[3]{2}$$

The stack of boxes will be $12\sqrt[3]{2}$, or about 15.12, feet tall.

Inclusion Strategies

USING SYMBOLS Some students find it difficult to simplify a radical expression with an index of n when the radicand is not a perfect nth root, such as $\sqrt[3]{x^5}$. To help, use numbers to illustrate the fact that a factor under a radical with an index of n must be repeated n times in order to remove the factor. For example, $\sqrt[5]{64} = \sqrt[5]{(2 \cdot 2 \cdot 2 \cdot 2 \cdot 2) \cdot 2} = 2\sqrt[5]{2}$. Have students circle the five factors of 2 that can be replaced by a single 2 outside the radical in order to get $2\sqrt[5]{2}$. Remind students to write the radicand as the product of prime factors to ensure that they can see all repeated factors. Students can also use this process with variables. To simplify $\sqrt[3]{x^5}$, rewrite the expression as $\sqrt[3]{(x \cdot x \cdot x) \cdot x \cdot x} = x\sqrt[3]{x^2}$. Give students more complicated examples for practice, and encourage them to use this technique throughout their homework.

EXAMPLE 4 Simplify each sum or difference.

a. $(6+\sqrt{12})+(-7+\sqrt{75})$ **b.** $(-5-\sqrt{18})-(6+\sqrt{50})$

SOLUTION

a. $(6+\sqrt{12})+(-7+\sqrt{75})$

$=(6+\sqrt{2^2 \cdot 3})+(-7+\sqrt{5^2 \cdot 3})$

$=(6+2\sqrt{3})+(-7+5\sqrt{3})$

$=6-7+2\sqrt{3}+5\sqrt{3}$ *Simplify.*

$=-1+7\sqrt{3}$ *Combine like terms.*

b. $(-5-\sqrt{18})-(6+\sqrt{50})$

$=(-5-\sqrt{3^2 \cdot 2})-(6+\sqrt{5^2 \cdot 2})$

$=(-5-3\sqrt{2})-(6+5\sqrt{2})$

$=-5-6-3\sqrt{2}-5\sqrt{2}$ *Simplify.*

$=-11-8\sqrt{2}$ *Combine like terms.*

TRY THIS Simplify each sum or difference.

a. $(-3+\sqrt{32})+(6+\sqrt{98})$ **b.** $(8-\sqrt{45})-(-2+\sqrt{20})$

CHECKPOINT ✔ Since $\sqrt{3}$ is between $\sqrt{1}=1$ and $\sqrt{4}=2$, you can estimate that the value of $-1+7\sqrt{3}$ is between $-1+7\sqrt{1}=6$ and $-1+7\sqrt{4}=13$. Use this method of estimation to show that $-1+7\sqrt{3}$ is a reasonable answer for the sum $(6+\sqrt{12})+(-7+\sqrt{75})$ in part **a** of Example 4.

EXAMPLE 5 Simplify each product.

a. $(-3+5\sqrt{2})(4+2\sqrt{2})$ **b.** $(4-\sqrt{3})(2\sqrt{3}+5)$

SOLUTION

a. $(-3+5\sqrt{2})(4+2\sqrt{2})$

$=(-3)(4)+(-3)(2\sqrt{2})+(5\sqrt{2})(4)+(5\sqrt{2})(2\sqrt{2})$ *Distributive Property*

$=-12-6\sqrt{2}+20\sqrt{2}+20$ *Simplify.*

$=8+14\sqrt{2}$

b. $(4-\sqrt{3})(2\sqrt{3}+5)$

$=(4)(2\sqrt{3})+(4)(5)+(-\sqrt{3})(2\sqrt{3})+(-\sqrt{3})(5)$ *Distributive Property*

$=8\sqrt{3}+20-6-5\sqrt{3}$ *Simplify.*

$=14+3\sqrt{3}$

TRY THIS Simplify each product.

a. $(3-5\sqrt{5})(-4+6\sqrt{5})$ **b.** $(-4\sqrt{6}+1)(5-3\sqrt{6})$

CRITICAL THINKING Show that $(a+b\sqrt{2})(a-b\sqrt{2})=a^2-2b^2$ is true.

Reteaching the Lesson

COOPERATIVE LEARNING Have each student create five radical expressions that need to be simplified. Then have students trade papers and simplify each others' expressions. The writer of the expressions should check the answers, and the simplifier may dispute any differing answers. Partners should justify their answers to each other. Any simplifications that cannot be resolved between the partners should be discussed with the class.

ADDITIONAL EXAMPLE 4

Simplify each sum or difference.

a. $(3-\sqrt{24})+(8-\sqrt{96})$
$11-6\sqrt{6}$

b. $(4+\sqrt{27})-(-15+\sqrt{48})$
$19-\sqrt{3}$

TRY THIS

a. $3+11\sqrt{2}$ **b.** $10-5\sqrt{5}$

CHECKPOINT ✔

Because $\sqrt{12}$ is between $\sqrt{9}=3$ and $\sqrt{16}=4$, we know that $6+\sqrt{12}$ is between 9 and 10. Because $\sqrt{75}$ is between $\sqrt{64}=8$ and $\sqrt{81}=9$, we know that $-7+\sqrt{75}$ is between 1 and 2. So, $(6+\sqrt{12})+(-7+\sqrt{75})$ is between 10 and 12. Because we know that $-1+7\sqrt{3}$ is between 6 and 13, it is a reasonable answer.

ADDITIONAL EXAMPLE 5

Simplify.

a. $(2+3\sqrt{3})(9-7\sqrt{3})$
$-45+13\sqrt{3}$

b. $(3-\sqrt{2})(3\sqrt{2}+1)$
$-3+8\sqrt{2}$

TRY THIS

a. $-162+38\sqrt{5}$ **b.** $77-23\sqrt{6}$

CRITICAL THINKING

$(a+b\sqrt{2})(a-b\sqrt{2})$
$=a^2+ab\sqrt{2}-ab\sqrt{2}-(b\sqrt{2})^2$
$=a^2-2b^2$

ADDITIONAL
EXAMPLE 6

Write each expression with a rational denominator.

a. $\frac{2}{\sqrt{6}}$ $\frac{\sqrt{6}}{3}$

b. $\frac{4}{3-\sqrt{7}}$ $6+2\sqrt{7}$

TRY THIS

a. $\frac{3\sqrt{5}}{5}$ **b.** $-6-2\sqrt{2}$

CRITICAL THINKING

$\frac{a-b\sqrt{2}}{a^2-2b^2}$

Assess

Selected Answers

Exercises 4–13, 15–109 odd

ASSIGNMENT GUIDE	
In Class	1–13
Core	15–89 odd, 93
Core Plus	14–94 even
Review	96–110
Preview	111–113

Extra Practice can be found beginning on page 940.

Rationalizing a denominator is a procedure for transforming a quotient with a radical in the denominator into an expression with no radical in the denominator. Use this procedure so that your answers are in the same form as the answers in this textbook.

EXAMPLE 6 **Write each expression with a rational denominator.**

a. $\frac{1}{\sqrt{3}}$ **b.** $\frac{2}{1+\sqrt{3}}$

SOLUTION

a. $\frac{1}{\sqrt{3}} = \frac{1}{\sqrt{3}}\left(\frac{\sqrt{3}}{\sqrt{3}}\right)$ *Multiply by 1.*

$= \frac{\sqrt{3}}{\sqrt{3}\cdot\sqrt{3}}$

$= \frac{\sqrt{3}}{3}$

b. $\frac{2}{1+\sqrt{3}} = \frac{2}{1+\sqrt{3}}\left(\frac{1-\sqrt{3}}{1-\sqrt{3}}\right)$ *Use the conjugate of $1+\sqrt{3}$ to multiply by 1.*

$= \frac{2-2\sqrt{3}}{1-\sqrt{3}+\sqrt{3}-3}$

$= \frac{2-2\sqrt{3}}{-2}$

$= -1+\sqrt{3}$

TRY THIS Write each expression with a rational denominator.

a. $\frac{3}{\sqrt{5}}$ **b.** $\frac{-14}{3-\sqrt{2}}$

CRITICAL THINKING Let a and b represent nonzero integers. Write $\frac{1}{a+b\sqrt{2}}$ as an expression with a rational denominator.

Exercises

Communicate

1. Explain how to simplify $(1+2\sqrt{2})(2-3\sqrt{2})$. Include a description about how the Distributive Property is applied.
2. Explain how to simplify the radical expression $\sqrt{4x^3}$.
3. Explain how to rationalize the denominator of $\frac{5+3\sqrt{2}}{4+7\sqrt{2}}$.

Guided Skills Practice

Simplify each expression by using the Properties of Radicals. *(EXAMPLE 1)*

4. $\sqrt{128ab^2c^5}$ **$8|b|c^2\sqrt{2ac}$**
5. $\sqrt[3]{-54x^5y^9}$ **$-3xy^3\sqrt[3]{2x^2}$**

Simplify each expression. Assume that the value of each variable is positive. ***(EXAMPLE 2)***

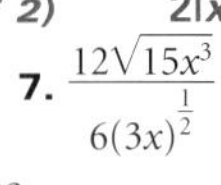

6. $\sqrt[3]{27a^4b^3}(81a^2b)^{\frac{1}{3}}$ $\mathbf{9a^2b\sqrt[3]{3b}}$

7. $\dfrac{12\sqrt{15x^3}}{6(3x)^{\frac{1}{2}}}$ $\mathbf{2|x|\sqrt{5}}$

APPLICATION

8. $10\sqrt[3]{3}$, or ≈14.42 inches

8. BUSINESS How tall is a stack of three cube-shaped boxes, one on top of the other, given that their volumes are 24 cubic inches, 81 cubic inches, and 375 cubic inches? Give both an exact answer and an answer rounded to the nearest hundredth of an inch. ***(EXAMPLE 3)***

Simplify each sum or difference. ***(EXAMPLE 4)***

9. $(-3+\sqrt{32})+(3+2\sqrt{98})$ $\mathbf{18\sqrt{2}}$

10. $(16+\sqrt{75})-(-4+10\sqrt{3})$ $\mathbf{20-5\sqrt{3}}$

11. Simplify the product $(-2+3\sqrt{5})(3-6\sqrt{5})$. ***(EXAMPLE 5)*** $\mathbf{-96+21\sqrt{5}}$

Write each expression with a rational denominator. ***(EXAMPLE 6)***

12. $\dfrac{2}{\sqrt{7}}$ $\mathbf{\dfrac{2\sqrt{7}}{7}}$

13. $\dfrac{-3}{5-\sqrt{3}}$ $\mathbf{\dfrac{-15-3\sqrt{3}}{22}}$

Practice and Apply

Simplify each radical expression by using the Properties of *n*th Roots.

14. $\sqrt{50}$ $\mathbf{5\sqrt{2}}$

15. $\sqrt{128}$ $\mathbf{8\sqrt{2}}$

16. $\sqrt[3]{-54}$ $\mathbf{-3\sqrt[3]{2}}$

17. $-32\sqrt[3]{-48}$ $\mathbf{-2\sqrt[3]{6}}$

18. $\sqrt{32x^3}$ $\mathbf{4|x|\sqrt{2x}}$

19. $\sqrt{18x^3}$ $\mathbf{3|x|\sqrt{2x}}$

20. $\sqrt[3]{-27x^5}$ $\mathbf{-3x\sqrt[3]{x^2}}$

21. $\sqrt[3]{-81x^7}$ $\mathbf{-3x^2\sqrt[3]{3x}}$

22. $\sqrt{27b^3c^4}$ $\mathbf{3|b|c^2\sqrt{3b}}$

23. $\sqrt{50a^3b^4}$ $\mathbf{5|a|b^2\sqrt{2a}}$

24. $\sqrt[3]{24x^5y^3z^9}$ $\mathbf{2xyz^3\sqrt[3]{3x^2}}$

25. $\sqrt[3]{250r^7s^2t^3}$ $\mathbf{5r^2t\sqrt[3]{2rs^2}}$

26. $\sqrt{98x^8y^3z}$ $\mathbf{7x^4|y|\sqrt{2yz}}$

27. $(16x^6)^{\frac{1}{4}}$ $\mathbf{2|x|\sqrt[4]{x^2}}$

28. $(40a^7)^{\frac{1}{3}}$ $\mathbf{2a^2\sqrt[3]{5a}}$

Simplify each product or quotient. Assume that the value of each variable is positive.

29. $\sqrt{2x^3}\cdot\sqrt{4x^3}$ $\mathbf{2x^3\sqrt{2}}$

30. $\sqrt[3]{3y^3}\cdot\sqrt[3]{9y^2}$ $\mathbf{3y\sqrt[3]{y^2}}$

31. $\sqrt[4]{25x^2}\cdot(25x^2)^{\frac{1}{4}}$ $\mathbf{5x}$

32. $(16x^2)^{\frac{1}{3}}\cdot\sqrt[3]{4x}$ $\mathbf{4x}$

33. $(24rs)^{\frac{1}{2}}\cdot\sqrt{6r^3s^4}\cdot\sqrt{rs^2}$ $\mathbf{12r^2s^3\sqrt{rs}}$

34. $\sqrt[3]{3a^2b^4}\cdot(a^3b^5)^{\frac{1}{3}}\cdot\sqrt[3]{ab}$ $\mathbf{a^2b^3\sqrt[3]{3b}}$

35. $\dfrac{(64y^7)^{\frac{1}{3}}}{\sqrt[3]{y^3}}$ $\mathbf{4y\sqrt[3]{y}}$

36. $\dfrac{(42a^4)^{\frac{1}{2}}}{\sqrt{8a^5}}$ $\mathbf{\dfrac{\sqrt{21a}}{2a}}$

37. $\dfrac{\sqrt{24x^5}}{(6x^3)^{\frac{1}{2}}}$ $\mathbf{2x}$

38. $\dfrac{\sqrt[3]{32x^7}}{(4x^2)^{\frac{1}{3}}}$ $\mathbf{2x\sqrt[3]{x^2}}$

39. $\dfrac{(64x^7)^{\frac{1}{4}}}{\sqrt[4]{x}}$ $\mathbf{2x\sqrt[4]{4x^2}}$

40. $\dfrac{(24x^5z^2)^{\frac{1}{2}}}{\sqrt{12x^2z^2}}$ $\mathbf{x\sqrt{2x}}$

41. $(21x^2y^3)^{\frac{1}{2}}\sqrt{3x^4y^8}$ $\mathbf{3x^3y^5\sqrt{7y}}$

42. $(81d^4f^4)^{\frac{1}{2}}\sqrt{5d^2f^8}\sqrt{d^2f^2}$ $\mathbf{9d^4f^7\sqrt{5}}$

43. $\sqrt[3]{64y^7c^3t^5}(yct)^{-\frac{2}{3}}$ $\mathbf{4yt\sqrt[3]{y^2c}}$

44. $\sqrt[3]{162a^7b^3c^5}(54abc)^{-\frac{1}{3}}$ $\mathbf{a^2c\sqrt[3]{3b^2c}}$

Error Analysis

In Exercises 84–90, students will often incorrectly multiply the numerator and the denominator by only the radical part of the denominator. Show students by example that this still leaves a radical in the denominator. They must use a binomial that will create a difference of squares in the denominator to eliminate the radical.

Find each sum, difference, or product. Give your answer in simplest radical form.

45. $(3+\sqrt{3})+(3+\sqrt{3})$ $\mathbf{6+2\sqrt{3}}$

46. $(4+\sqrt{7})-(-3+2\sqrt{7})$ $\mathbf{7-\sqrt{7}}$

47. $(3+\sqrt{18})+(-1-4\sqrt{2})$ $\mathbf{2-\sqrt{2}}$

48. $(-6-\sqrt{6})-(-1-3\sqrt{24})$ $\mathbf{-5+5\sqrt{6}}$

49. $(3+\sqrt{32})-(4+2\sqrt{98})$ $\mathbf{-1-10\sqrt{2}}$

50. $(5+\sqrt{125})+(-10+10\sqrt{5})$

51. $(\sqrt{12}-4)-(8+\sqrt{27})$ $\mathbf{-12-\sqrt{3}}$

52. $(-3\sqrt{5}+2)-(3+2\sqrt{20})$ $\mathbf{-1-7\sqrt{5}}$

53. $(2+\sqrt{3})(-1+\sqrt{3})$ $\mathbf{1+\sqrt{3}}$

54. $(\sqrt{5}-8)(-1+3\sqrt{5})$ $\mathbf{23-25\sqrt{5}}$

55. $(8+\sqrt{12})(3-\sqrt{12})$ $\mathbf{12-10\sqrt{3}}$

56. $(-6+\sqrt{5})(-4+2\sqrt{5})$ $\mathbf{34-16\sqrt{5}}$

57. $(3+\sqrt{2})(3+\sqrt{2})$ $\mathbf{11+6\sqrt{2}}$

58. $(2+3\sqrt{7})(2-3\sqrt{7})$ $\mathbf{-59}$

59. $(\sqrt{100}-6)(-4+\sqrt{12})$ $\mathbf{-16+8\sqrt{3}}$

60. $(5+\sqrt{18})(-\sqrt{16}-3)$ $\mathbf{-35-21\sqrt{2}}$

61. $(4-2\sqrt{27})(1+\sqrt{75})$ $\mathbf{-86+14\sqrt{3}}$

62. $(-3+2\sqrt{8})(\sqrt{20}-5)$

63. $(13+\sqrt{2})-(-3+2\sqrt{2})+3\sqrt{2}$

64. $(1+\sqrt{75})+(-2+\sqrt{125})-7\sqrt{3}$

65. $(3-5\sqrt{2})-(-4+\sqrt{2})$ $\mathbf{7-6\sqrt{2}}$

66. $(4+3\sqrt{7})-(5+3\sqrt{7})$ $\mathbf{9-6\sqrt{7}}$

67. $2\sqrt{6}(\sqrt{24}-7)$ $\mathbf{24-14\sqrt{6}}$

68. $3\sqrt{5}(-\sqrt{20}+2)$ $\mathbf{-30+6\sqrt{5}}$

69. $(\sqrt{75}-8)\sqrt{12}$ $\mathbf{30-16\sqrt{3}}$

70. $(3\sqrt{12}+4)\sqrt{27}$ $\mathbf{54+12\sqrt{3}}$

71. $4\sqrt{2}(\sqrt{12}-3\sqrt{2}+4\sqrt{8})$ $\mathbf{40-8\sqrt{6}}$

72. $2\sqrt{3}(7\sqrt{3}-\sqrt{8}+2\sqrt{5})$

50. $\mathbf{-5+15\sqrt{5}}$

62. $\mathbf{15-20\sqrt{2}-6\sqrt{5}+8\sqrt{10}}$

63. $\mathbf{16+2\sqrt{2}}$

64. $\mathbf{-1-2\sqrt{3}+5\sqrt{5}}$

72. $\mathbf{42-4\sqrt{6}+4\sqrt{15}}$

Write each expression with a rational denominator and in simplest form.

73. $\frac{2}{\sqrt{5}}$ $\mathbf{\frac{2\sqrt{5}}{5}}$

74. $\frac{1}{\sqrt{2}}$ $\mathbf{\frac{\sqrt{2}}{2}}$

75. $\frac{4}{\sqrt{6}}$ $\mathbf{\frac{2\sqrt{6}}{3}}$

76. $\frac{5}{\sqrt{15}}$ $\mathbf{\frac{\sqrt{15}}{3}}$

77. $\frac{6}{\sqrt{12}}$ $\mathbf{\sqrt{3}}$

78. $\frac{15}{\sqrt{18}}$ $\mathbf{\frac{5\sqrt{2}}{2}}$

79. $\frac{\sqrt{5}}{\sqrt{75}}$ $\mathbf{\frac{\sqrt{15}}{15}}$

80. $\frac{\sqrt{3}}{\sqrt{27}}$ $\mathbf{\frac{1}{3}}$

81. $\frac{\sqrt{24}}{\sqrt{6}}$ $\mathbf{2}$

82. $\frac{\sqrt{60}}{\sqrt{5}}$ $\mathbf{2\sqrt{3}}$

83. $\frac{\sqrt{18}}{\sqrt{12}}$ $\mathbf{\frac{\sqrt{6}}{2}}$

84. $\frac{1}{\sqrt{3}+3}$ $\mathbf{\frac{3-\sqrt{3}}{6}}$

85. $\frac{8}{\sqrt{2}+4}$ $\mathbf{\frac{16-4\sqrt{2}}{7}}$

86. $\frac{12}{\sqrt{5}-1}$ $\mathbf{3\sqrt{5}+3}$

87. $\frac{3}{2+\sqrt{3}}$ $\mathbf{6-3\sqrt{3}}$

88. $\frac{14}{\sqrt{5}+\sqrt{3}}$ $\mathbf{7\sqrt{5}-7\sqrt{3}}$

89. $\frac{10}{\sqrt{7}+\sqrt{2}}$ $\mathbf{2\sqrt{7}-2\sqrt{2}}$

90. $\frac{14}{\sqrt{3}-\sqrt{2}}$ $\mathbf{14\sqrt{3}+14\sqrt{2}}$

CHALLENGE

Simplify.

91. $\frac{x}{3-5\sqrt{2}}-(2+3\sqrt{2})$

92. $\frac{x}{4+\sqrt{2}}-(-1+3\sqrt{2})$ $\mathbf{\frac{4x-x\sqrt{2}-42\sqrt{2}+14}{14}}$

91. $\mathbf{-\frac{3x+5x\sqrt{2}+82+123\sqrt{2}}{41}}$

CONNECTIONS

93. GEOMETRY Refer to the figure at right. Find the length of the red path from point A to point B to point C. Give an exact answer and an answer rounded to the nearest hundredth. $\mathbf{6\sqrt{2}}$**, or ≈8.49 miles**

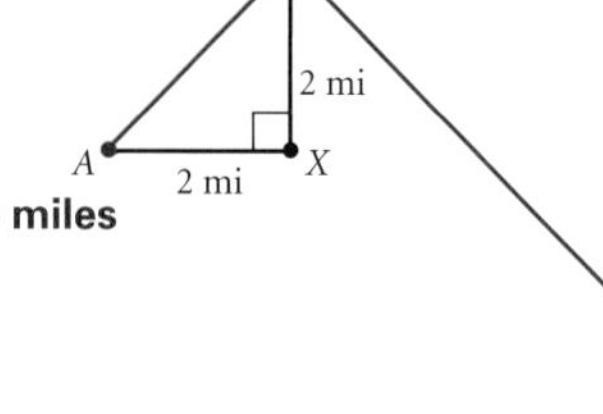

94. GEOMETRY Refer to the figure at right. Find the length of the red path from point A to point B to point C if $AX = BX = a$ and $BY = CY = 2a$, where $a > 0$. Give your answer in simplest radical form. $\mathbf{3a\sqrt{2}}$

Practice

NAME ______ CLASS ______ DATE ______

Practice
8.7 Simplifying Radical Expressions

Simplify each radical expression by using the Properties of *n*th Roots.

1. $\sqrt[5]{32}$ **2**
2. $\sqrt[4]{81}$ **3**
3. $\sqrt{288x^2y^4}$ $\mathbf{12|x|y^2\sqrt{2}}$
4. $\sqrt[3]{343x^5y^9z}$ $\mathbf{7xy^3\sqrt[3]{x^2z}}$
5. $(80x^5)^{\frac{1}{2}}$ $\mathbf{4x^2(5x)^{\frac{1}{2}}}$
6. $(-16x^3y^4)^{\frac{1}{3}}$ $\mathbf{-2xy(2y)^{\frac{1}{3}}}$

Simplify each product or quotient. Assume that the value of each variable is positive.

7. $\sqrt[5]{16x^6}\cdot\sqrt[5]{2x^4}$ $\mathbf{2x^2}$
8. $\sqrt[3]{16x^2y^5}\cdot\sqrt[3]{4x^2y}$ $\mathbf{4xy^2\sqrt[3]{x}}$
9. $\sqrt{2x^3y}\cdot\sqrt{5x^3y^3}\cdot\sqrt{10x^2y}$ $\mathbf{10x^4y^2\sqrt{y}}$
10. $\frac{\sqrt[5]{64x^3y^7z^3}}{\sqrt[5]{2xy}}$ $\mathbf{2y\sqrt[5]{x^2yz^3}}$
11. $\frac{\sqrt[3]{96x^2y^5z^4}}{\sqrt[3]{4yx}}$ $\mathbf{2yz\sqrt[3]{3xyz}}$
12. $\frac{\sqrt[4]{64x^{10}y^{10}}}{\sqrt[4]{2x}}$ $\mathbf{2x^2y^2\sqrt[4]{2xy^2}}$

Find each sum, difference, or product. Give your answer in simplest radical form.

13. $(16+3\sqrt{2})+(9-\sqrt{2})$ $\mathbf{25+2\sqrt{2}}$
14. $(5+7\sqrt{3})-(2-3\sqrt{12})$ $\mathbf{3+13\sqrt{3}}$
15. $(11-3\sqrt{18})-(6-4\sqrt{8})$ $\mathbf{5-\sqrt{2}}$
16. $(4+3\sqrt{2})(3-6\sqrt{2})$ $\mathbf{-24-15\sqrt{2}}$
17. $(4-7\sqrt{5})(3+2\sqrt{5})$ $\mathbf{-58-13\sqrt{5}}$
18. $8\sqrt{2}(\sqrt{8}-3\sqrt{2}+7\sqrt{32})$ $\mathbf{432}$

Write each expression with a rational denominator and in simplest form.

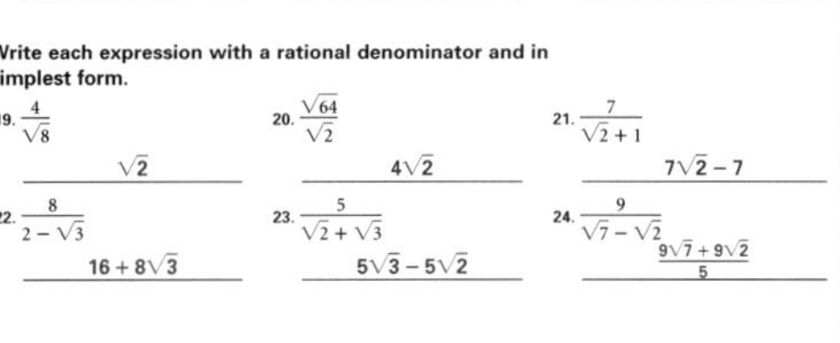

19. $\frac{4}{\sqrt{8}}$ $\mathbf{\sqrt{2}}$
20. $\frac{\sqrt{64}}{\sqrt{2}}$ $\mathbf{4\sqrt{2}}$
21. $\frac{7}{\sqrt{2}+1}$ $\mathbf{7\sqrt{2}-7}$
22. $\frac{8}{2-\sqrt{3}}$ $\mathbf{16+8\sqrt{3}}$
23. $\frac{5}{\sqrt{2}+\sqrt{3}}$ $\mathbf{5\sqrt{3}-5\sqrt{2}}$
24. $\frac{9}{\sqrt{7}-\sqrt{2}}$ $\mathbf{\frac{9\sqrt{7}+9\sqrt{2}}{5}}$

APPLICATION

Coefficients of Friction

	concrete	tar
wet	$f = 0.4$	$f = 0.5$
dry	$f = 0.8$	$f = 1.0$

95. **TRANSPORTATION** Accident investigators can usually estimate a motorist's speed, s, in miles per hour by examining the length, d, in feet of the skid marks on the highway. The estimate of the speed also depends on the road surface and weather conditions. If f represents the coefficient of friction between rubber and concrete or tar, then $s = \sqrt{30fd}$ gives an estimate of the motorist's speed in miles per hour.

a. Write a function for s in terms of d under wet conditions on a concrete road. Give an answer in simplest radical form, and give an approximation to the nearest tenth.

95a. $s = 2\sqrt{3d}$, or $\approx 3.5\sqrt{d}$, mph

b. Estimate a motorist's speed under wet conditions on a concrete road if the skid marks are estimated to be 200 feet long. Give your estimate to the nearest whole number of miles per hour.

b. 49 mph

c. Compare the speed of a motorist whose skid marks are 200 feet long with that of a motorist whose skid marks are 400 feet long, both under wet conditions on a concrete road.

c. Speed is not double; $\approx$69 mph.

Look Back

Solve each equation, giving both exact and approximate solutions. ***(LESSON 5.2)***

96. $x^2 = 34$ $\pm\sqrt{34}$; ± 5.83

97. $2x^2 - 6 = 26$ ± 4

98. $(x - 3)^2 + 4 = 14$ $3 \pm \sqrt{10}$; ≈ -0.16, ≈ 6.16

Factor each expression. ***(LESSON 5.3)***

99. $x^2 + 6x + 5$ $(x + 5)(x + 1)$

100. $x^2 - 2x - 15$ $(x + 3)(x - 5)$

101. $x^2 - 3x - 40$ $(x + 5)(x - 8)$

Solve each equation by factoring and applying the Zero Product Property. ***(LESSON 5.3)***

102. $x^2 + 3x = 28$ 4, –7

103. $x^2 - 11x = -30$ 5, 6

104. $-x^2 = 9x + 20$ –4, –5

Find the zeros of each quadratic function. ***(LESSON 5.3)***

105. $f(x) = x^2 + 9x + 18$ –6, –3

106. $f(x) = x^2 - 2x - 8$ –2, 4

107. $f(x) = x^2 - 3x - 18$ –3, 6

Find the domain of each radical function. ***(LESSON 8.6)***

108. $f(x) = \sqrt{3x - 1}$ $x \geq \frac{1}{3}$

109. $g(x) = 2\sqrt{5 - x}$ $x \leq 5$

110. $h(x) = -\sqrt{2(3 - 5x)}$ $x \leq \frac{3}{5}$

Look Beyond

Rationalize each denominator. (Hint: $(1 + a)(1 - a + a^2) = 1 + a^3$)

111. $\frac{1}{\sqrt[3]{5}}$ $\frac{\sqrt[3]{25}}{5}$

112. $\frac{\sqrt[3]{6}}{\sqrt[3]{3}}$ $\sqrt[3]{2}$

113. $\frac{2}{1 + \sqrt[3]{x}}$ $\frac{2 - 2\sqrt[3]{x} + 2\sqrt[3]{x^2}}{1 + x}$

In Exercises 111–113, students rationalize radical expressions with a cube root of a variable in the denominator. Students must use the rule for the difference of cubes and an understanding of cube roots in order to rationalize these expressions.

Student Technology Guide

NAME ______________ CLASS ______________ DATE ______________

Student Technology Guide

8.7 Simplifying Radical Expressions

Most calculators do not have a special key for evaluating radicals that are not square roots. However, there are other ways to use calculators to evaluate such expressions.

Example: Evaluate $\sqrt[4]{16^5}$.

There are different ways to rewrite this expression, including $16^{\frac{5}{4}}$ and $16^{1.25}$.

- To evaluate $\sqrt[4]{16^5}$, press 4 MATH 5:x√ ENTER (16 ^ 5) ENTER.
- To evaluate $16^{\frac{5}{4}}$, press 16 ^ (5 ÷ 4) ENTER.
- To evaluate $16^{1.25}$, press 16 ^ 1.25 ENTER.

```
4 ×√(16^5)
                32
16^(5/4)
                32
16^1.25
                32
```

Note that the answer is the same for each expression.

Evaluate each radical expression. Round answers to the nearest tenth.

1. $\sqrt[5]{243^3}$ 27
2. $114^{\frac{7}{4}}$ 3977.3
3. $12^{3.18}$ 2702.7
4. $\sqrt[3]{177^8}$ 987,632.5

Some calculators, such as the TI-92, can simplify radical expressions involving variables.

To simplify $\frac{\sqrt[3]{x^8}}{\sqrt[3]{x^4}}$, press (x ^ 8) ^ (1 ÷ 3) ÷ (x ^ 4) ^ (1 ÷ 3) ENTER.

The calculator gives the answer as an exponential expression. In simplest radical form, the answer is $x^3\sqrt{x}$.

```
F1  F2▾     F3▾  F4▾   F5     F6
    Algebra Calc Other PrgmIO Clear a-z...

■ (x^8)^(1/3) / (x^4)^(1/3)        x^(4/3)
(x^8)^(1/3)/(x^4)^(1/3)
MAIN        RAD AUTO        FUNC 1/30
```

Simplify. Assume that the value of each variable is positive.

5. $\sqrt{x^3} \cdot \sqrt{x^8}$ $\sqrt{x^{11}} = x^5\sqrt{x}$
6. $\sqrt[3]{x^8} \cdot \sqrt[3]{x^5}$ $\sqrt[3]{x^{13}} = x^4\sqrt[3]{x}$
7. $\frac{\sqrt{x^7}}{\sqrt{x^3}}$ x^2

Prepare

NCTM PRINCIPLES & STANDARDS 1–4, 6–10

QUICK WARM-UP

Solve.

1. $2(a + 9) = 10$ $a = -4$
2. $6 - t = t - 3$ $t = 4.5$
3. $y^2 + 4y = 0$ $y = 0$ or $y = -4$
4. $m^2 - 2m - 15 = 0$
   $m = -3$ or $m = 5$
5. $x^2 + 2x = 4x + 7$
   $x = 1 \pm 2\sqrt{2}$
6. $\frac{2}{x-1} + \frac{x}{x+1} = \frac{4}{x^2-1}$
   $x = -2$

Also on Quiz Transparency 8.8

Teach

Why Referring to the diagram in the text, ask students to describe what length is given by $\sqrt{L^2 - h^2}$. Then have students explain why the expression $\frac{d}{\sqrt{L^2 - h^2}}$ gives one less than the number of vertical supports. **$\sqrt{L^2 - h^2}$ represents the base of a right triangle whose height is h; $2\sqrt{L^2 - h^2}$ represents the length of the base of the corresponding isosceles triangle; $\frac{d}{2\sqrt{L^2 - h^2}}$ represents the number of isosceles triangles contained within the distance, d. The number of isosceles triangles is one less than the number of vertical supports.**

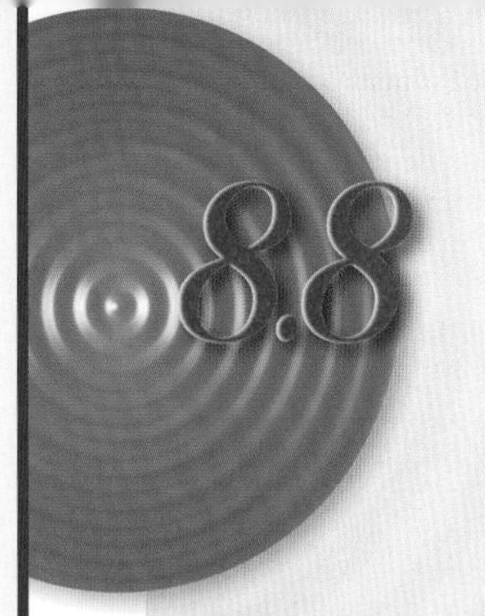

8.8 Solving Radical Equations and Inequalities

Why *Sometimes solving a real-world problem involves solving a radical equation or inequality. For example, designing a bridge may involve solving a radical equation.*

Objectives

- Solve radical equations.
- Solve radical inequalities.

APPLICATION
ENGINEERING

Engineers and town planners are proposing a new bridge to span a river. The diagram below shows a side view of the proposed bridge. The distance, d, across the river is 286 feet. Each vertical support, h, above the roadbed is 15 feet high, and there are 6 equally spaced piers below the roadbed.

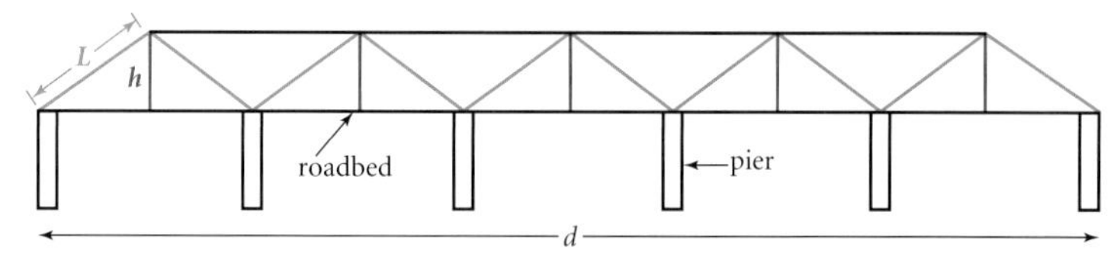

The function $n = \frac{d}{2\sqrt{L^2 - h^2}} + 1$ relates the number of piers, n, below the roadbed to d, L, and h. What is the length, L, of each slanted support? To answer this question you need to solve a *radical equation.* You will answer this question in Example 4.

A **radical equation** is an equation that contains at least one radical expression with a variable under the radical symbol. To solve a radical equation, you can raise each side of the equation to the same power.

Principle of Powers

If $a = b$ and n is a positive integer, then $a^n = b^n$.

Alternative Teaching Strategy

USING TECHNOLOGY Have students enter both sides of a radical equation or inequality as **Y1** and **Y2,** respectively, in a graphics calculator. Demonstrate each step in the solution of the equation or inequality, and have students graph each side of the resulting equation or inequality. At each step, students should note where the two graphs intersect. They should see that the points of intersection of the graphs lie on a vertical line. The x-value of the points on this line represents the solution. Students will see that squaring both sides of an equation changes only the y-values, not the solution. Using this procedure, students have a visual model that they can use to see when extraneous roots appear.

To solve a radical equation with a square root on one side, square each side of the equation. Recall from Lesson 8.5 that an extraneous solution is a solution to a resulting equation that is not a solution to the original equation. Raising the expression on each side of an equation to an even power may introduce extraneous solutions, so it is important to check your solutions.

EXAMPLE 1 **Solve $2\sqrt{x+5} = 8$. Check your solution.**

SOLUTION

$$2\sqrt{x+5} = 8$$
$$\sqrt{x+5} = 4$$
$$\left(\sqrt{x+5}\right)^2 = 4^2 \quad \textit{Square each side of the equation.}$$
$$x + 5 = 16$$
$$x = 11$$

CHECK

$$2\sqrt{x+5} = 8$$
$$2\sqrt{(11)+5} \stackrel{?}{=} 8$$
$$2 \cdot 4 = 8 \quad \textbf{True}$$

TECHNOLOGY GRAPHICS CALCULATOR

Keystroke Guide, page 558

You can also check by graphing. Graph $y = 2\sqrt{x+5}$ and $y = 8$ on the same screen. The x-coordinate of the point of intersection is 11.

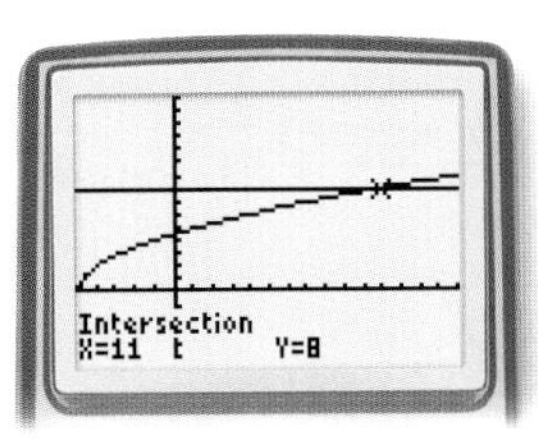

The solution to $2\sqrt{x+5} = 8$ is 11.

TRY THIS Solve $3\sqrt{2x-1} = 6$. Check your solution.

A radical equation may contain the variable on each side of the equation, as shown in Example 2 below and Example 3 on the next page.

EXAMPLE 2 **Solve $\sqrt[3]{x-5} = \sqrt[3]{7-x}$. Check your solution.**

SOLUTION

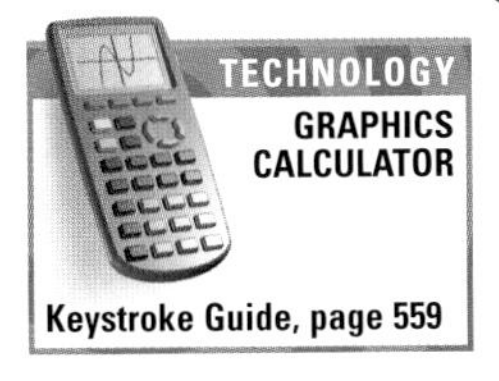

Keystroke Guide, page 559

Method 1 Use algebra.

$$\sqrt[3]{x-5} = \sqrt[3]{7-x}$$
$$\left(\sqrt[3]{x-5}\right)^3 = \left(\sqrt[3]{7-x}\right)^3$$
$$x - 5 = 7 - x$$
$$x = 6$$

CHECK

$$\sqrt[3]{x-5} = \sqrt[3]{7-x}$$
$$\sqrt[3]{(6)-5} \stackrel{?}{=} \sqrt[3]{7-(6)}$$
$$\sqrt[3]{1} = \sqrt[3]{1} \quad \textbf{True}$$

Method 2 Use a graph.

Graph $y = \sqrt[3]{x-5}$ and $y = \sqrt[3]{7-x}$ on the same screen, and find the x-coordinates of any points of intersection.

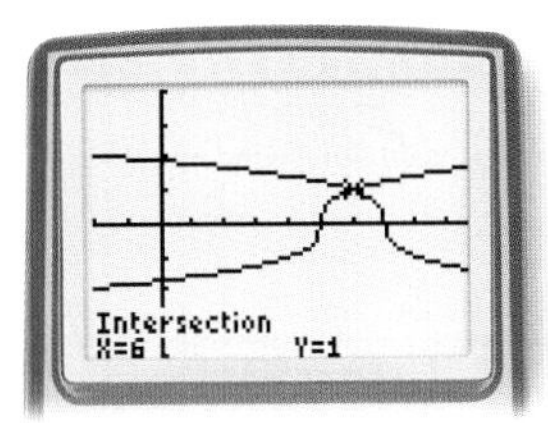

TRY THIS Solve $\sqrt{x+2} = \sqrt{5-2x}$. Check your solution.

ADDITIONAL EXAMPLE 1

Solve $12 = 4\sqrt{2x+1}$. Check your solution.

$x = 4$

Teaching Tip

TECHNOLOGY Students can use the following viewing window:

Xmin=–5	Ymin=–6
Xmax=15	Ymax=15
Xscl=1	Yscl=1

TRY THIS

$x = \frac{5}{2}$

ADDITIONAL EXAMPLE 2

Solve $\sqrt[3]{9x+1} = \sqrt[3]{11+8x}$. Check your solution.

$x = 10$

Teaching Tip

The viewing window for Example 2 is as follows:

Xmin=–2	Ymin=–4
Xmax=10	Ymax=4

To enter a cubic root, students can either change the expression to exponential form and raise the expression to the $\frac{1}{3}$ power, or press MATH and select 4:∛(.

TRY THIS

$x = 1$

Interdisciplinary Connection

PHYSICS Sound travels through solid objects at various speeds, depending on characteristics of the solid. The speed of a sound wave traveling through a thin rod is given by the formula $v = \sqrt{\frac{Y}{\rho}}$, where v is the speed of the wave in meters per second, Y is Young's modulus in pascals, and ρ is the density of the solid in kilograms per cubic meter. Suppose that a sound wave travels through a solid at a speed of 3220 meters per second and Young's modulus for the solid is 8.0×10^{10} pascals. Find the density of the solid. (Note: 1 pascal (Pa) = 1 kg/m/s^2) $\rho \approx 7715.75$ **kg/m^3**

ADDITIONAL EXAMPLE 3

Solve $\sqrt{4x+2} - 3x = 1$. Check your solution.

$x \approx 0.24$

Teaching Tip

TECHNOLOGY Use the following viewing window:

Xmin=–2	Ymin=–1
Xmax=10	Ymax=10
Xscl=1	Yscl=1

TRY THIS

$x = \frac{5-\sqrt{5}}{2}$, or ≈ 1.38

ADDITIONAL EXAMPLE 4

Refer to the bridge described at the beginning of the lesson. **Use a graph and a table to find the length, L, of each slanted support given that the distance across the river is 300 feet, the height of each vertical support above the roadbed is 20 feet, and there are 8 equally spaced vertical supports below the roadbed.**

$L \approx 29.31$ ft

EXAMPLE 3 **Solve $\sqrt{x+1} + 3 = 2x$. Check your solution.**

SOLUTION

$$\sqrt{x+1} + 3 = 2x$$

$$\sqrt{x+1} = 2x - 3 \quad \textit{Isolate the radical before squaring.}$$

$$x + 1 = (2x-3)^2 \quad \textit{Square each side of the equation.}$$

$$x + 1 = 4x^2 - 12x + 9$$

$$4x^2 - 13x + 8 = 0 \quad \textit{Write in standard form with } a = 4,\ b = -13,\ c = 8.$$

$$x = \frac{-(-13) \pm \sqrt{13^2 - 4(4)(8)}}{2(4)} \quad \textit{Apply the quadratic formula.}$$

$$x = \frac{13+\sqrt{41}}{8} \quad \text{or} \quad x = \frac{13-\sqrt{41}}{8}$$

$$x \approx 2.43 \qquad\qquad x \approx 0.82$$

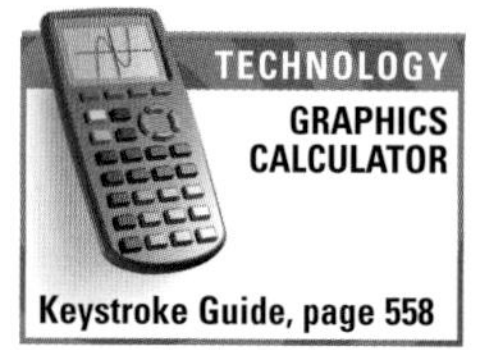

CHECK

Graph $y = \sqrt{x+1} + 3$ and $y = 2x$ on the same screen, and find the x-coordinates of any points of intersection. The graphs intersect at only one point, where $x \approx 2.43$.

Thus, $\frac{13-\sqrt{41}}{8}$ is an extraneous solution; the only solution is $\frac{13+\sqrt{41}}{8}$, or approximately 2.43.

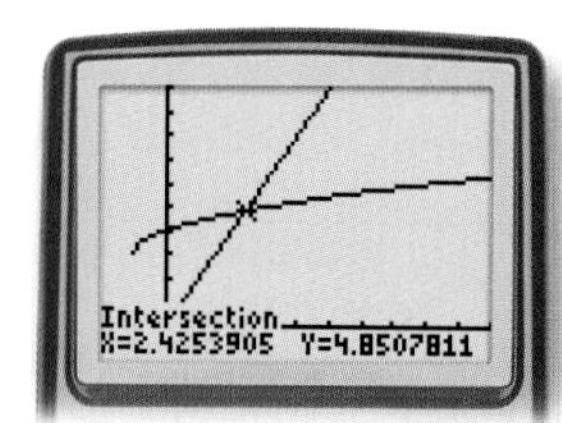

TRY THIS Solve $\sqrt{x-1} = -x + 2$. Check your solution.

EXAMPLE 4 Refer to the bridge described at the beginning of the lesson. **Find the length, L, of each slanted support.**

APPLICATION ENGINEERING

SOLUTION

Substitute 6 for n, 286 for d, and 15 for h in the function $n = \frac{d}{2\sqrt{L^2 - h^2}} + 1$ to get $6 = \frac{286}{2\sqrt{L^2 - 15^2}} + 1$. Solve for L.

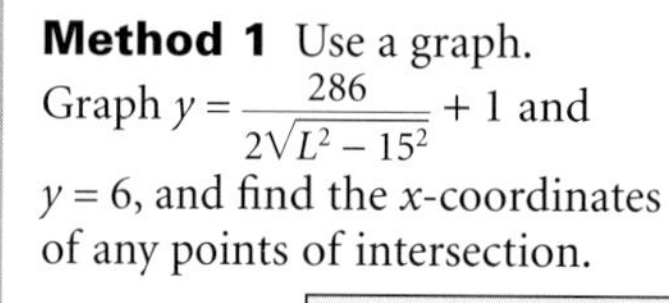

Method 1 Use a graph.

Graph $y = \frac{286}{2\sqrt{L^2 - 15^2}} + 1$ and $y = 6$, and find the x-coordinates of any points of intersection.

Method 2 Use a table.

Examine the table of values for the function $y = \frac{286}{2\sqrt{L^2 - 15^2}} + 1$.

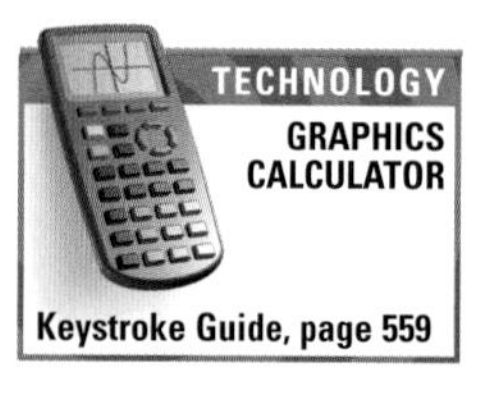

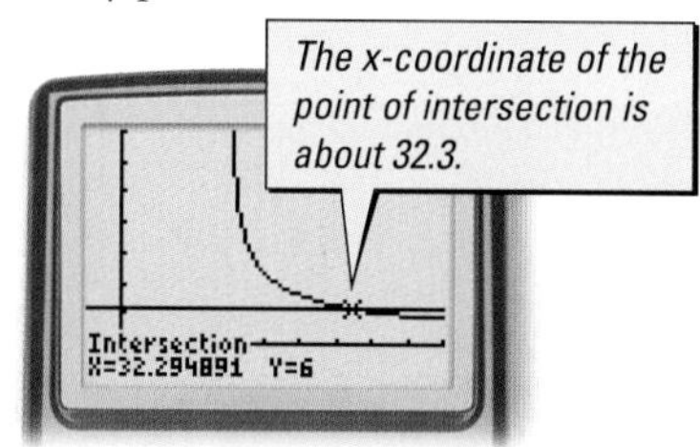

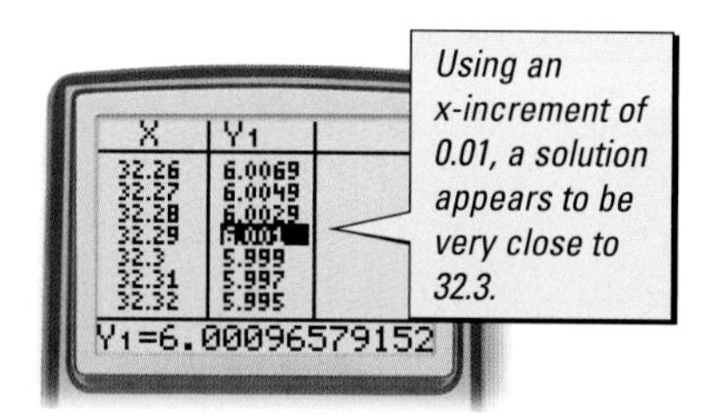

X
Y1
32.26
32.27
32.28
32.29
32.3
32.31
32.32
6.0069
6.0049
6.0029
5.999
5.997
5.995
Y1=6.00096579152
Using an x-increment of 0.01, a solution appears to be very close to 32.3.

Thus, the length of each slanted support is about 32.3 feet.

Enrichment

Some continued radicals can represent integers. For example, $\sqrt{12 + \sqrt{12 + \sqrt{12 + \sqrt{12 + \cdots}}}} = 4$. To determine what number is represented by a continued radical, let x equal the continued radical, write an equation in terms of x, and solve the equation. Letting $x = \sqrt{12 + \sqrt{12 + \sqrt{12 + \sqrt{12 + \cdots}}}}$ and noting that x is contained under the radical because the radical repeats forever, write the equation as $x = \sqrt{12 + x}$. Solving this equation yields $x = 4$. Have students evaluate the continued radicals at right above by solving a related equation.

1. $\sqrt{6 + \sqrt{6 + \cdots}}$ 3
2. $\sqrt{2 + \sqrt{2 + \sqrt{2 + \cdots}}}$ 2

For what values of k does $\sqrt{k + \sqrt{k + \sqrt{k + \cdots}}}$ represent an integer? **k is an integer which has two integer factors that differ by 1. The integer represented by the continued radical is the larger of the two factors. Sample answers: For $k = 12, 20, 30$, and 42, $x = 4, 5, 6$, and 7.** Have students verify the value of the continued radicals by calculating an approximation obtained by entering several iterations of the radicand into a calculator.

Some radical equations have no solutions, as shown in Example 5.

EXAMPLE 5 **Solve $\sqrt{x} - 1 = \sqrt{2x + 1}$. Check your solution.**

SOLUTION

$$\sqrt{x} - 1 = \sqrt{2x+1}$$

$$\left(\sqrt{x} - 1\right)^2 = \left(\sqrt{2x+1}\right)^2 \quad \textit{Square each side of the equation.}$$

$$x - 2\sqrt{x} + 1 = 2x + 1$$

$$-2\sqrt{x} = x \quad \textit{Simplify.}$$

$$\left(-2\sqrt{x}\right)^2 = x^2 \quad \textit{Square each side of the equation again.}$$

$$4x = x^2$$

$$x^2 - 4x = 0$$

$$x(x - 4) = 0$$

$$x = 0 \quad or \quad x = 4$$

CHECK

$\sqrt{0} - 1 \stackrel{?}{=} \sqrt{2(0) + 1}$ $\sqrt{4} - 1 \stackrel{?}{=} \sqrt{2(4) + 1}$

$-1 = 1$ **False** $1 = 3$ **False**

The equation $\sqrt{x} - 1 = \sqrt{2x+1}$ has no real solutions.

The graphs of $y = \sqrt{x} - 1$ and $y = \sqrt{2x+1}$ are shown on the same screen at right.

Because the graphs do not intersect, the equation $\sqrt{x} - 1 = \sqrt{2x+1}$ has no real solutions.

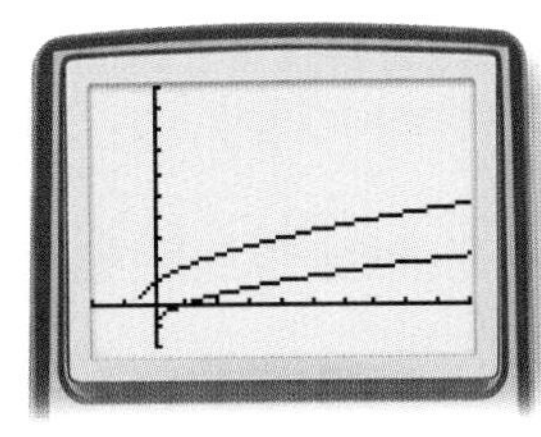

TRY THIS Solve $3\sqrt{x} + 2 = \sqrt{3x}$. Check your solution.

SUMMARY

Solving Radical Equations

To solve a radical equation, follow these steps:

1. If possible, write the equation with the radical expression isolated on one side.
2. Raise the expression on each side of the equation to the appropriate power.
3. Repeat Steps 1 and 2 as needed to obtain an equation with no radical expressions. Solve this equation.
4. Check your solutions and discard extraneous solutions.

CRITICAL THINKING Describe the algebraic strategy that you would use to solve the equation $\sqrt{x-1} = \sqrt{x+1} - 1$.

ADDITIONAL EXAMPLE 5

Solve $\sqrt{3x+4} = \sqrt{x} - 2$.
Check your solution.
no real solution

Teaching Tip

TECHNOLOGY Use the following viewing window:

Xmin=–2	Ymin=–2
Xmax=10	Ymax=10

The keystrokes used in Example 5 are not shown in the Keystroke Guide at the end of this chapter, but the procedure to graph the equations is similar to that for the examples on pages 558 and 559 of the Keystroke Guide.

TRY THIS
no real solution

CRITICAL THINKING
Square both sides of the equation, isolate the radical expression, and then square both sides again. Check answers for extraneous solutions; $x = \frac{5}{4}$

Inclusion Strategies

COOPERATIVE LEARNING Have students work in pairs on examples from the lesson or from the Guided Skills Practice section of the exercises. While one student solves the problem algebraically, the other should solve the same problem graphically. Each student should write down the reason for each step taken to arrive at the solution. Have students alternate roles with each new problem. This process will help students see alternative approaches to solving problems while analyzing the reasons for each approach.

Notes

In this Activity, students use a graphics calculator to explore the solutions to radical inequalities. Students graph both sides of an inequality and identify the intervals on the x-axis where one graph is greater than or less than the other graph.

Teaching Tip

TECHNOLOGY Have students use the following viewing window:

Xmin=–4.7	Ymin=–3.2
Xmax=4.7	Ymax=3.2
Xscl=1	Yscl=1

Cooperative Learning

Have students work in groups of two or three and have each group discuss how the model they see on the calculator relates to the algebraic solution of radical equations. Then have them discuss how they can use their knowledge of solving radical equations to help them solve radical inequalities. Then lead a class discussion in which the groups summarize their findings.

CHECKPOINT ✔

3. Yes; for $a > 0$, $\sqrt{x} \geq a$ has solutions where $x \geq a^2$, and $\sqrt{x} \leq a$ has solutions where $0 \leq x \leq a^2$.

CHECKPOINT ✔

4. Graph $y = \sqrt{x}$ and $y = x - 2$ and determine the points of intersection. The values of x for which $\sqrt{x}$ lies on or above $x - 2$ are solutions to $\sqrt{x} \geq x - 2$. The values of x for which $\sqrt{x}$ lies on or below $x - 2$ are solutions to $\sqrt{x} \geq x - 2$.

Solving Radical Inequalities

A **radical inequality** is an inequality that contains at least one radical expression. To explore solutions to radical inequalities, complete the Activity below.

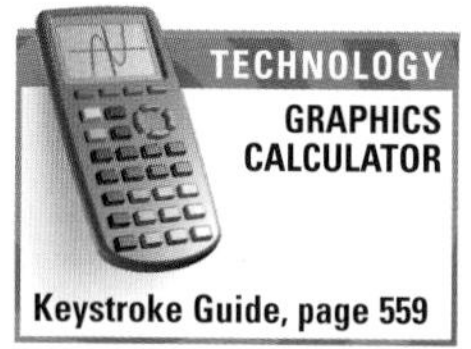

Activity: Exploring Radical Inequalities

You will need: a graphics calculator

1. **a.** Let $f(x) = \sqrt{x}$ and $g(x) = 1$. Graph f and g on the same screen.
   b. For what values of x is $f(x) \geq g(x)$?
   c. For what values of x is $f(x) \leq g(x)$?
2. Explain why the answer to part **b** of Step 1 is not $x \leq 1$.

CHECKPOINT ✔ 3. Do $\sqrt{x} \geq a$ and $\sqrt{x} \leq a$, where $a > 0$, always have solutions? Explain.

CHECKPOINT ✔ 4. Explain how to use a graph to solve $\sqrt{x} \geq x - 2$ and $\sqrt{x} \leq x - 2$. Then state the solutions.

The fact below is helpful for solving radical inequalities.

$$\text{If } a \geq 0,\ b \geq 0, \text{ and } a \geq b, \text{ then } a^n \geq b^n.$$

This fact allows you to square the expression on each side of an inequality, as shown in Example 6.

EXAMPLE 6 **Solve $\sqrt{x+1} < 2$. Check your solution.**

SOLUTION

Because the radicand of a radical expression cannot be negative, first solve $x + 1 \geq 0$.

$$x + 1 \geq 0$$
$$x \geq -1$$

Then solve the original inequality.

$$\sqrt{x+1} < 2$$
$$\left(\sqrt{x+1}\right)^2 < 2^2$$
$$x + 1 < 4$$
$$x < 3$$

It appears that the solution is $x \geq -1$ *and* $x < 3$, or $-1 \leq x < 3$.

CHECK

Graph $y = \sqrt{x+1}$ and $y = 2$ on the same screen.

The graph verifies that the solution is $-1 \leq x < 3$.

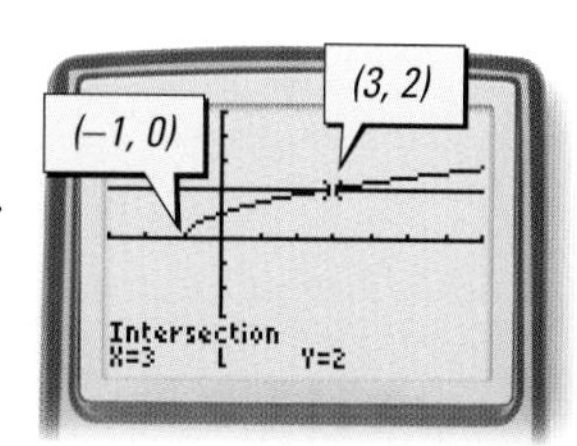

TRY THIS Solve $\sqrt{2x-3} < 5$. Check your solution.

Reteaching the Lesson

INVITING PARTICIPATION Have the class play a modified tic-tac-toe game. Choose nine students, and place their names in a three-by-three square grid on the board. Split the remainder of the class into two teams, identified as O's and X's. Have each team choose a student's name from the square grid. Both the team and the student chosen should solve a radical equation or inequality given by the teacher. The student chosen should state the solution; the team may agree or disagree. If they disagree, they must state why. If they agree, the student in the square should share his or her solution process with the class. If the team correctly determined the solution, the square is marked with the symbol for the team. If the team did not correctly determine the solution, no symbol is marked in the square. The object of the game is for a team to get three of its symbols in a row across, down, or diagonally in the square grid.

EXAMPLE 7 **Solve $x < 3\sqrt[3]{x+1}$ by graphing. Give the solution to the nearest tenth.**

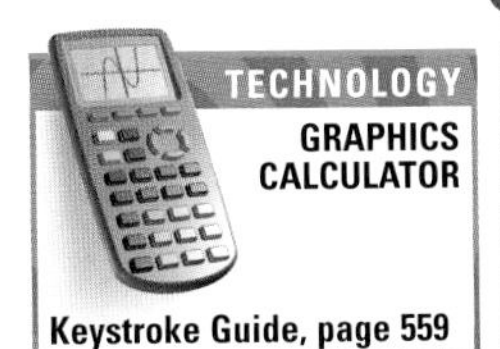

SOLUTION

Graph $y = x$ and $y = 3\sqrt[3]{x+1}$ on the same screen.

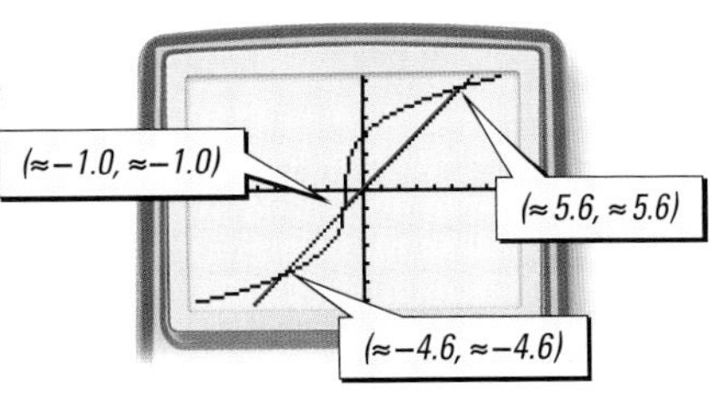

The graph of $y = x$ is below that of $y = 3\sqrt[3]{x+1}$ for approximately $x < -4.6$ and for approximately $-1.0 < x < 5.6$.

Thus, the solution to $x < 3\sqrt[3]{x+1}$ is approximately $x < -4.6$ *or* $-1.0 < x < 5.6$.

TRY THIS Solve $x - 1 < 3\sqrt[3]{2x+1}$ by graphing. Give the solution to the nearest tenth.

Exercises

Communicate

1. Briefly describe two methods discussed in the lesson for solving a radical equation or inequality such as $\sqrt{x} = 3\sqrt{x-4}$ or $\sqrt{x} \leq 3\sqrt{x-4}$.
2. Explain why it is necessary to check any solutions to a radical equation that you obtain algebraically.
3. Explain how to determine that $\sqrt{x} = \sqrt{x+1}$ has no solution. Use algebra and a graph in your explanation.

internet connect
Activities Online
Go To: **go.hrw.com**
Keyword:
MB1 Relativity

Guided Skills Practice

4. Solve $3\sqrt{2x-5} = 20$. Check your solution. ***(EXAMPLE 1)*** $\frac{445}{18}$, or $24\frac{13}{18}$
5. Solve $\sqrt[3]{3x+1} = \sqrt[3]{2x+3}$. Check your solution. ***(EXAMPLE 2)*** 2
6. Solve $\sqrt{5x+7} - 2 = x$. Check your solution. ***(EXAMPLE 3)*** $\frac{1 \pm \sqrt{13}}{2}$
7. **ENGINEERING** Refer to the bridge described at the beginning of the lesson. Find the length of each slanted support given that there are 6 piers below the roadbed, the distance across the river is 560 feet, and the vertical supports above the roadbed are 17 feet tall. Give your answer to the nearest tenth of a foot. ***(EXAMPLE 4)*** 58.5 feet
8. Solve $2\sqrt{x+1} = \sqrt{x} - 3$. Check your solution. ***(EXAMPLE 5)*** no solution
9. Solve $\sqrt{3x-2} \leq 8$. Check your solution. ***(EXAMPLE 6)*** $\frac{2}{3} \leq x \leq 22$
10. Solve $-0.5x + 1 \leq 4\sqrt[3]{3x-2}$ by graphing. Give the solution to the nearest tenth. ***(EXAMPLE 7)*** $x \geq 0.7$

APPLICATION

ADDITIONAL EXAMPLE 6

Solve $\sqrt{x+7} \geq 4$. Check your solution.
$x \geq 9$

Teaching Tip

TECHNOLOGY In Example 6, use the following viewing window:

Xmin=−3	Ymin=−5
Xmax=7	Ymax=5
Xscl=1	Yscl=1

TRY THIS
$\frac{3}{2} \leq x < 14$

ADDITIONAL EXAMPLE 7

Solve $4\sqrt[3]{x-1} < x$ by graphing. Give the solution to the nearest tenth.
$-8.5 < x < 1.0$ *or* $x > 7.4$

Teaching Tip

TECHNOLOGY When using a TI-83 or TI-82 graphics calculator for Example 7, set the viewing window as follows:

Xmin=−9.4	Ymin=−6.2
Xmax=9.4	Ymax=6.2
Xscl=1	Yscl=1

TRY THIS
$x < -5.4$ *or* $-0.6 < x < 9$

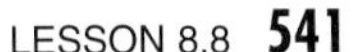

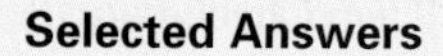

Assess

Selected Answers

Exercises 4–10, 11–77 odd

ASSIGNMENT GUIDE	
In Class	1–10
Core	11–59 odd, 63
Core Plus	12–62 even
Review	64–77
Preview	78

Extra Practice can be found beginning on page 940.

62b.

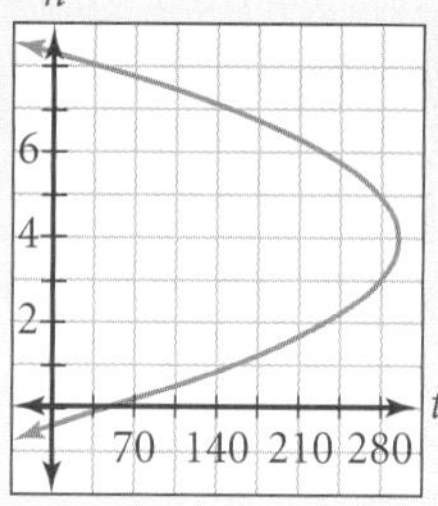

Practice

NAME ______ CLASS ______ DATE ______

Practice

8.8 ***Solving Radical Equations and Inequalities***

Solve each radical equation by using algebra. If the equation has no solution, write *no solution*. Check your solution.

1. $\sqrt{x+5}=10$ 95
2. $\sqrt{x-6}=2$ 10
3. $\sqrt{x^2-4}=2\sqrt{3}$ −4, 4
4. $\sqrt{x+2}=\sqrt{x}$ no solution
5. $\sqrt{2x+3}=x+1$ $-\sqrt{2}, \sqrt{2}$
6. $2\sqrt{x-2}=x-2$ 2, 6
7. $3\sqrt{2x+3}=\sqrt{x-7}$ no solution
8. $\sqrt{3x-5}=x-1$ 2, 3
9. $\sqrt[3]{x+4}=\sqrt[3]{3x-6}$ 5

Solve each radical inequality by using algebra. If the inequality has no solution, write *no solution*. Check your solution.

10. $\sqrt{x-3}\le 3$ $3\le x\le 12$
11. $\sqrt{x+2}>4$ $x>14$
12. $\sqrt{3x}<5$ $0\le x<8\frac{1}{3}$
13. $\sqrt{x^2-2x+1}>1.5$ $x<-0.5$ *or* $x>2.5$
14. $\sqrt[4]{x+3}<\sqrt{x+1}$ $x>1$
15. $\sqrt{x-4}>x-10$ $8<x<13$
16. $\sqrt{x+5}<\sqrt{x-3}$ no solution
17. $\sqrt{3-2x}>4$ $x<-6.5$
18. $\sqrt{4x-3}\le 7$ $\frac{3}{4}\le x\le 13$

Solve each radical equation or inequality by using a graph. Round solutions to the nearest tenth. Check your solution by any method.

19. $\sqrt{3x-4}<x-2$ $x>5.6$
20. $\sqrt{3-x}\ge x^2-1$ $-1.8\le x\le 1.5$
21. $\sqrt{x+4}>x^3$ $-4\le x\le 1.3$
22. $\sqrt{2x+3}<x-2$ $x\ge 5.8$
23. $\sqrt[3]{x+7}<\sqrt{2x+3}$ $x>0.4$
24. $\sqrt{x^2+2}\ge\sqrt{7x-3}$ $0.4\le x\le 0.8$ *or* $x\ge 6.2$
25. $\sqrt{x+9}\le x^2-3x$ $-9\le x\le -0.8$ *or* $x\ge 3.9$
26. $\sqrt{x^3+1}>x+3$ $x>3.4$
27. $\sqrt{7x-1}\le x-2$ $x\ge 10.5$

Practice and Apply

internet connect
Homework Help Online
Go To: **go.hrw.com**
Keyword:
MB1 Homework Help
for Exercises 11–38

Solve each radical equation by using algebra. If the equation has no real solution, write *no solution*. Check your solution.

11. $\sqrt{x}=4$ **16**

12. $\sqrt{x-3}=2$ **7**

13. $5=\sqrt{x^2+16}$ **−3, 3**

14. $3=\sqrt{x^2-16}$ **−5, 5**

15. $\sqrt{x+2}=4\sqrt{x}$ $\mathbf{\frac{2}{15}}$

16. $2\sqrt{x}=3\sqrt{x-2}$ $\mathbf{\frac{18}{5}}$

17. $\sqrt{3x-2}=x-2$ **6**

18. $x+1=\sqrt{6x-7}-1$ **no solution**

19. $\sqrt[3]{x-2}=\sqrt[3]{2x+1}$ **−3**

20. $\sqrt[3]{2x-3}=\sqrt[3]{2-x}$ $\mathbf{\frac{5}{3}}$

21. $\sqrt[3]{x+2}=\sqrt[3]{x+3}$ **no solution**

22. $3\sqrt{2x+1}=2\sqrt{2x}-1$ **no solution**

23. $\sqrt{2x+1}=x+1$ **0**

24. $\sqrt{3x+2}=x-2$ $\mathbf{\frac{7+\sqrt{41}}{2}}$

Solve each radical inequality by using algebra. If the inequality has no real solution, write *no solution*. Check your solution.

25. $\sqrt{2x-1}\ge 1$ $\mathbf{x\ge 1}$

26. $\sqrt{3x+4}\ge 2$ $\mathbf{x\ge 0}$

27. $\sqrt{2x-1}\le 1$ $\mathbf{\frac{1}{2}\le x\le 1}$

28. $\sqrt{3x+4}\le 2$ $\mathbf{-\frac{4}{3}\le x\le 0}$

29. $-\sqrt{x}\le 2$ $\mathbf{x\ge 0}$

30. $-\sqrt{x}\ge 2$ **no solution**

31. $1>3\sqrt{3x-1}$ $\mathbf{\frac{1}{3}\le x<\frac{10}{27}}$

32. $\sqrt{2x+1}-3<0$ $\mathbf{-\frac{1}{2}\le x<4}$

33. $3\sqrt{3x-1}<-1$ **no solution**

34. $\sqrt{2x+1}+3<0$ **no solution**

35. $\sqrt{2x+2}>\sqrt{3x}$ $\mathbf{0\le x<2}$

36. $\sqrt{9x+7}<\sqrt{14x}$ $\mathbf{x>\frac{7}{5}}$

37. $x\le\sqrt{x}$ $\mathbf{0\le x\le 1}$

38. $\frac{1}{8}\le x\le\sqrt{x}$ $\mathbf{\frac{1}{8}\le x\le 1}$

Solve each radical equation or inequality by using a graph. Round solutions to the nearest tenth. Check your solution by any method.

39 $\sqrt{2x-1}=x^2$ **0.5, 1**

40 $(x+1)^2=\sqrt{2x+1}$ **−0.5, 0**

41 $\sqrt[3]{2x+1}=x^2+2x$ **−1, 0.5**

42 $x^2-x=\sqrt[3]{x-2}$ **no solution**

43 $\sqrt{x}=\sqrt[3]{x}$ **0, 1**

44 $\sqrt[3]{x}=\sqrt[4]{x}$ **0, 1**

45 $\sqrt{x}-2x\ge 0$ $\mathbf{0\le x\le 0.3}$

46 $2x>\sqrt{3-x}$ $\mathbf{0.8<x\le 3}$

47 $\sqrt[3]{x+2}>\sqrt[3]{x-2}+2$ **−1.9, 1.9**

48 $\sqrt[3]{x}<\sqrt{x}+3$ $\mathbf{x\ge 0}$

49 $\sqrt{x}\le x^2+\sqrt{x+1}-2$ $\mathbf{x\ge 1.3}$

50 $0.5\sqrt[4]{x+2}>\sqrt[3]{x}$ $\mathbf{-2\le x<0.2}$

51 $2\sqrt{x^2-1}\le\sqrt{x+5}$
$\mathbf{-1.4\le x\le -1}$ ***or*** $\mathbf{1\le x\le 1.6}$

52 $2\sqrt{x^2-2}-\sqrt{3x+7}<0$
$\mathbf{-1.6\le x\le -1.4}$ ***or*** $\mathbf{1.4\le x<2.3}$

Identify whether each statement is always true, sometimes true, or never true.

53. $\sqrt{x+3}=-12$ **never**

54. $\sqrt{x-6}=3+\sqrt{x}$ **never**

55. $\sqrt{3x-6}<0$ **never**

56. $2\sqrt{2x-3}+4=1$ **never**

57. $2\sqrt[4]{3x}=\sqrt[4]{3x+15}$ **sometimes**

58. $\sqrt{x+4}+\sqrt{x-4}>4$ **sometimes**

59. $\sqrt{2x+5}-7\le x-2$ **always**

60. $\sqrt{x+5}-\sqrt{5x-21}\le 4$ **always**

CHALLENGE

61. Find a value of a that makes each statement below true. Use a graph to confirm your response.

a. The equation $\sqrt{x+a}=\sqrt{2x-a}$ has no solution. $\mathbf{a<0}$

b. The equation $\sqrt{x+a}=\sqrt{2x-a}$ has one solution. $\mathbf{a\ge 0}$

APPLICATIONS

62. **PHYSICS** The function $h = -16t^2 + 128t + 50$ models the height, h, in feet above the ground of a projectile after t seconds of flight.

a. Solve for t in terms of h. **$t = 4 \pm \frac{\sqrt{306 - h}}{4}$**

b. Graph the equation you wrote in part **a**.

c. After how many seconds will the height of the projectile above the ground be 270 feet? **2.5 seconds and 5.5 seconds**

63. **SIGHTSEEING** On a clear day, the approximate distance, d, in feet that a sightseer standing at the top of a building h feet tall can see is given by $d = 6397.2\sqrt{h}$.

a. On a clear day, what distance in feet can a sightseer see from the top of a 900-foot building? Give your answer to the nearest foot. **191,916 feet**

b. Convert your answer from part **a** to miles. (Note: 1 mile equals 5280 feet.) **36.3 miles**

c. How tall is a building from which a sightseer can see 32 miles on a clear day? Give your answer to the nearest foot. **698 feet**

Look Back

Write each pair of parametric equations as a single equation in x and y. ***(LESSON 3.6)***

64. $\begin{cases} x(t) = 3t \\ y(t) = t - 2 \end{cases}$ **$y = \frac{x}{3} - 2$**

65. $\begin{cases} x(t) = t + 4 \\ y(t) = 2t + 1 \end{cases}$ **$y = 2x - 7$**

66. $\begin{cases} x(t) = 3t + 1 \\ y(t) = -t \end{cases}$ **$y = \frac{-x + 1}{3}$**

Use the quadratic formula to solve each equation. Give your answers to the nearest tenth if necessary. ***(LESSON 5.5)***

67. $2x^2 + 10x^2 + 12 = 0$ **$-3, -2$**

68. $2x^2 - 11x = 6$ **$-\frac{1}{2}, 6$**

69. $3x^2 + 4x = 5$ **$-2.1, 0.8$**

Divide by using synthetic division. ***(LESSON 7.3)***

70. $(x^3 + 7x^2 - 10x - 16) \div (x - 2)$ **$x^2 + 9x + 8$**

71. $(x^3 - 63x - 162) \div (x + 6)$ **$x^2 - 6x - 27$**

Simplify each radical expression. Assume that the value of each variable is positive. ***(LESSON 8.7)***

72. $\sqrt{28a^{16}}$ **$2a^8\sqrt{7}$**

73. $4x^3y\sqrt{72x^7y^{10}}$ **$24x^6y^6\sqrt{2x}$**

74. $\sqrt[3]{12s^2t} \cdot \sqrt[3]{36st^7}$ **$6st^2\sqrt[3]{2t^2}$**

75. $\frac{\sqrt{30x^3}}{\sqrt{6x}}$ **$x\sqrt{5}$**

76. $\frac{\sqrt{4a^2}}{\sqrt{125b^3}}$ **$\frac{2a\sqrt{5b}}{25b^2}$**

77. $\frac{\sqrt{80x^3}}{\sqrt{2x}}$ **$2x\sqrt{10}$**

Look Beyond

78. Use the Pythagorean Theorem to find the distance between the two points graphed at right. **$\sqrt{41}$**

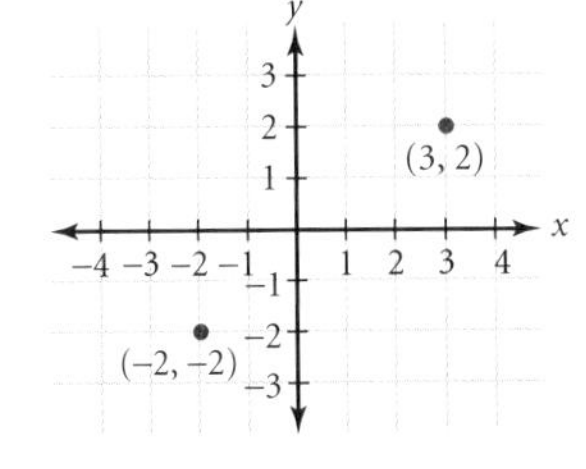

Error Analysis

When students raise both sides of an equation to a power, make sure that they raise everything on both sides to that power. If there is a sum on one side of the equation, students must expand that sum raised to the power by using the distributive property. If there is a number multiplied by a radical expression, that number is also raised to the power. Also, many students do not check all solutions and fail to exclude extraneous roots.

Look Beyond

In Exercise 78, to find the distance between two given points, students should draw a right triangle on a coordinate plane and use their knowledge of right triangles to find the length of the hypotenuse. The development of the distance formula is given in Lesson 9.1.

Student Technology Guide

NAME ______ CLASS ______ DATE ______

Student Technology Guide

8.8 Solving Radical Equations and Inequalities

You can use a graphics calculator to find an approximate solution to a radical equation.

Example: Solve $5 = \sqrt{3x + 2} + 1$. Round your answer to the nearest tenth.

Method 1: Use a table

- Press 2nd WINDOW. Set TblStart to 0 and ΔTbl to 1. Let the others be Auto.
- To enter the function $y = \sqrt{3x + 2} + 1$ as Y1, press Y= 2nd x^2 3 X,T,θ,n + 2) + 1.
- Press 2nd GRAPH. As shown at left below, the required x-value is between 4 and 5.
- Press 2nd WINDOW. Change TblStart to 4 and ΔTbl to 0.1 to narrow the search. Press 2nd GRAPH again.
- Use ▼ to scroll through the table. The required x-value is between 4.6 and 4.7.
- Press 2nd WINDOW. Change TblStart to 4.6 and ΔTbl to 0.01 to narrow the search. Press 2nd GRAPH again, and scroll through the table.

The solution is between 4.66 and 4.67. So, to the nearest tenth, the solution is $x \approx 4.7$.

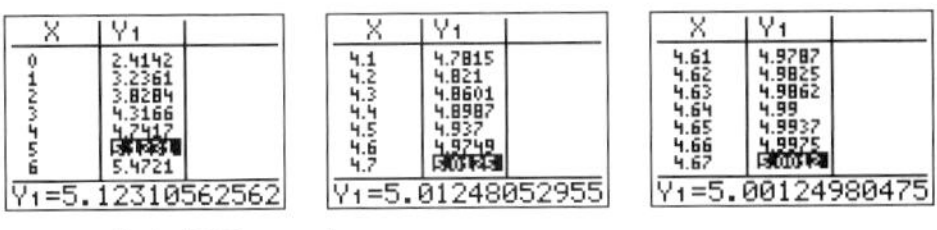

X	Y1
0	2.4142
1	3.2361
2	3.8284
3	4.3166
4	4.7417
5	[illegible]
6	5.4721

Y1=5.12310562562

X	Y1
4.1	4.7815
4.2	4.821
4.3	4.8601
4.4	4.8987
4.5	4.937
4.6	4.9749
4.7	[illegible]

Y1=5.01248052955

X	Y1
4.61	4.9787
4.62	4.9825
4.63	4.9862
4.64	4.99
4.65	4.9937
4.66	4.9975
4.67	[illegible]

Y1=5.00124980475

Method 2: Use a graph

- You have already entered $y = \sqrt{3x + 2} + 1$ as Y1. Press Y= and enter 5 as Y2. Press GRAPH.
- Press TRACE. Use ◄ and ► to move as close as you can to the intersection point.
- Press 2nd TRACE 5:intersect ENTER ENTER ENTER ENTER.

The intersection point occurs at $x \approx 4.7$.

Using the solution to $5 = \sqrt{3x + 2} + 1$, you can solve $5 > \sqrt{3x + 2} + 1$. Because $\sqrt{3x + 2} + 1$ is defined only when $x \geq \frac{2}{3}$, the solution is $\frac{2}{3} \leq x < 4.7$.

Solve each radical equation and inequality. Round solutions to the nearest tenth.

1. $\sqrt{7x - 3} - 4 = 6$; $\sqrt{7x - 3} - 4 \geq 6$ **$x \approx 14.7$; $x \geq 14.7$**

2. $\sqrt{9 - 2x} + 2 = 4$; $\sqrt{9 - 2x} + 2 \leq 4$ **$x = 2.5$; $2.5 \leq x \leq 4.5$**

Focus

Students investigate two types of averages commonly used in the real world. These are the arithmetic mean and the harmonic mean. Students consider various situations in which the arithmetic mean is misleading and the harmonic mean is a better measure for describing the data.

Motivate

Have students describe what the average reveals about a set of numbers, and remind them that the arithmetic mean refers to dividing a sum by the number of items used to compute the sum. Ask them if there is more than one way to find a average and if there are times when a weighted average is needed. Before beginning the project, have students read a few of the situations given and discuss those as a class to reach consensus that there are useful types of averages other than the arithmetic mean.

Chapter Project Eight: Means to an End

In problem-solving situations that are based on the calculation of an average, the choice of which average to use—arithmetic or harmonic—is critical. The results obtained from each average can be significantly different.

For any two numbers a and b, the *arithmetic mean* is $\frac{a+b}{2}$ and the *harmonic mean* is $\frac{2}{\frac{1}{a}+\frac{1}{b}}$.

To provide consumers with a standard to compare fuel economy for new cars, the Environmental Protection Agency (EPA) requires that estimates of fuel consumption be attached to the window of each new car.

For example, the EPA estimates that a certain car with a 1.9-liter engine and a 4-speed transmission can travel 24 miles per gallon (mpg) in the city and 34 miles per gallon on the highway.

The *Fuel Economy Guide,* published by the U.S. Department of Energy, is an aid to consumers who are considering the purchase of a new vehicle. The guide estimates the fuel consumption, in miles per gallon, for each vehicle available for the new model year. In the *Fuel Economy Guide,* the consumer is advised as follows:

> *Please be cautioned that simply averaging the mpg for city and highway driving and then looking up a single value in estimating may result in inaccurate estimates of the annual fuel cost.*

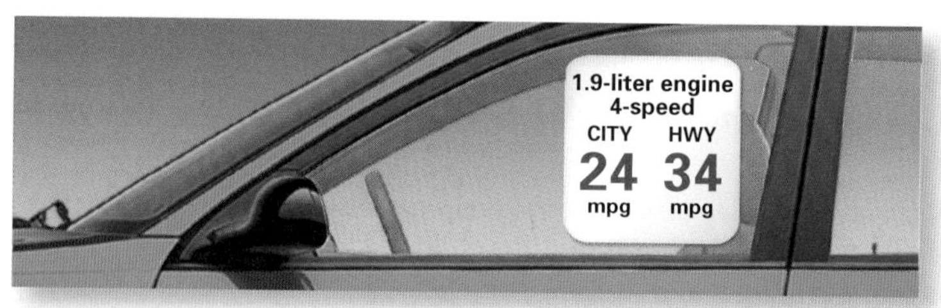

Activity 1

1. For the car described above, find the arithmetic mean of the fuel consumption in miles per gallon for city driving and for highway driving.
2. Find the harmonic mean for the city driving rate and highway driving rate.
3. Examine each mean. Explain why there is a warning about "simply averaging" the city rate and the highway rate when finding the annual fuel cost.

Activity 1

1. **29 mpg**
2. **about 28 mpg**
3. **"Simply averaging" the numbers, or finding the arithmetic mean, will give a different amount than finding the harmonic mean.**

Cooperative Learning

Have students work in groups of three or four. For each of the Activities, a different group member should be responsible for summarizing the findings of the group.

Discuss

After all groups have completed the Activities, bring the class together to compare and discuss the results. Encourage students to share any insights they have gained about finding averages. Point out to students that there are other ways to determine the average of a set of numbers. Show them an example of a geometric mean, which is the nth root of the product of n numbers. As an extension to this project, have each group design another scenario in which the harmonic mean would be preferred over the arithmetic mean.

Activity 2

The number of miles that a person drives in the city may not be the same as the number of miles that the person drives on the highway. Therefore, to find the average annual fuel consumption for a car, you need to use a *weighted harmonic mean.*

Let d_1 represent the number of miles driven in the city in one year.

Let d_2 represent the number of miles driven on the highway in one year.

Then the number of gallons of fuel used per year for each type of driving is as follows:

City: $(d_1 \text{ miles})\left(\frac{\text{gallon}}{24 \text{ miles}}\right) = \frac{d_1}{24}$ gallons

Highway: $(d_2 \text{ miles})\left(\frac{\text{gallon}}{34 \text{ miles}}\right) = \frac{d_2}{34}$ gallons

A rational function, a, for the average annual fuel consumption for the car described on the previous page can be expressed in terms of the total miles driven.

$$a(d) = \frac{\text{total distance}}{\text{total number of gallons}} = \frac{d_1 + d_2}{\frac{d_1}{24} + \frac{d_2}{34}}$$

1. Suppose that you purchased this car and drove it 12,000 miles in one year—8000 miles in the city and 4000 miles on the highway. Find the average annual fuel consumption for this car.
2. Determine the total fuel cost for the year if gasoline costs $1.099 per gallon.

Activity 3

Jennifer is considering a strategy for an upcoming 2-mile bicycle race. During practice she maintains a speed of 20 miles per hour for the first mile, but fatigue reduces her speed to 10 miles per hour for the second mile.

1. Explain why Jennifer's average speed over these 2 miles is not the same as the arithmetic mean of 20 miles per hour and 10 miles per hour. Find Jennifer's average speed for these 2 miles.
2. Determine the speed at which Jennifer must travel during the second mile if she rides 20 miles per hour during the first mile and she wants her average speed for the entire 2-mile trip to be 15 miles per hour.

Activity 2

1. about 26.6 mpg
2. about $495.65

Activity 3

1. Jennifer's average speed is the harmonic mean, which is not necessarily equal to the arithmetic mean. Her average speed is 13.3 mph.
2. 12 mph

8 Chapter Review and Assessment

VOCABULARY

Key Skills & Exercises

LESSON 8.1

Key Skills

Solve problems involving inverse, joint, or combined variation.

inverse variation: $y = \frac{k}{x}$ or $xy = k$

joint variation: $y = kxz$

combined variation: $y = \frac{kz}{x}$

A variable n varies jointly as x and y and varies inversely as the cube of z.

$$n = k\frac{xy}{z^3}$$

If $n = -14$ when $x = 3$, $y = 7$, and $z = -3$, what is n when $x = 4$, $y = 5$, and $z = 3$?

First find the constant, k, by solving $n = k\frac{xy}{z^3}$ for k and using substitution.

$$n = k\frac{xy}{z^3} \quad \rightarrow \quad k = \frac{nz^3}{xy}$$

$$k = \frac{nz^3}{xy} = \frac{(-14)(-3)^3}{(3)(7)} = 18$$

Now find n when $x = 4$, $y = 5$, and $z = 3$.

$$n = k\frac{xy}{z^3} = (18)\frac{(4)(5)}{(3)^3} = \frac{40}{3}$$

Exercises

1. A variable y varies jointly as x and z. If $y = 2$ when $x = 4$ and $z = 6$, what is y when $x = 3$ and $z = 8$? **$y = 2$**
2. A variable m varies directly as a and inversely as b. If $m = 6$ when $a = 7$ and $b = 4$, what is m when $a = 9$ and $b = 12$? **$m = \frac{18}{7}$**
3. A variable a varies directly as b and inversely as the square of c. If $a = 3$ when $b = 18$ and $c = 2$, what is a when $b = 20$ and $c = 6$? **$a = \frac{10}{27}$**
4. A variable f varies jointly as g and the square of h and inversely as j. If $f = -14$ when $g = 5$, $h = 8$, and $j = 20$, what is f when $g = 4$, $h = 6$, and $j = 9$? **$f = -14$**

Chapter Test, Form A

NAME ______ CLASS ______ DATE ______

Chapter Assessment
Chapter 8, Form A, page 1

Write the letter that best answers the question or completes the statement.

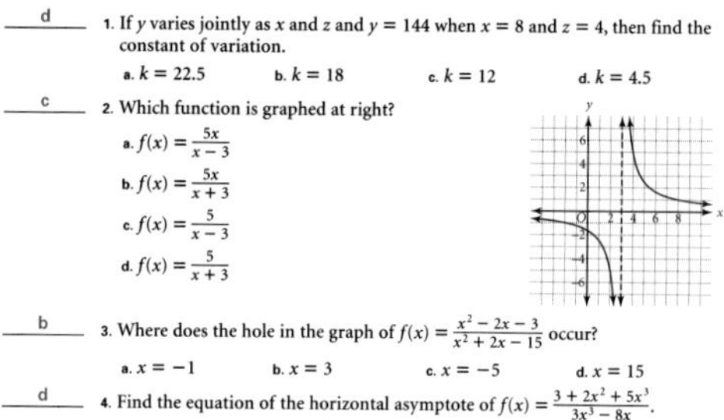

d 1. If y varies jointly as x and z and $y = 144$ when $x = 8$ and $z = 4$, then find the constant of variation.
a. $k = 22.5$ b. $k = 18$ c. $k = 12$ d. $k = 4.5$

c 2. Which function is graphed at right?
a. $f(x) = \frac{5x}{x-3}$
b. $f(x) = \frac{5x}{x+3}$
c. $f(x) = \frac{5}{x-3}$
d. $f(x) = \frac{5}{x+3}$

b 3. Where does the hole in the graph of $f(x) = \frac{x^2-2x-3}{x^2+2x-15}$ occur?
a. $x = -1$ b. $x = 3$ c. $x = -5$ d. $x = 15$

d 4. Find the equation of the horizontal asymptote of $f(x) = \frac{3+2x^2+5x^3}{3x^3-8x}$.
a. $y = 0$ b. $y = \frac{5}{8}$ c. $y = 1$ d. $y = \frac{5}{3}$

d 5. Which of the following is the simplest form of $\frac{5x}{x^4} \cdot \frac{4x^5}{8x^2} \cdot \frac{x^6}{2}$?
a. $4x^2$ b. $\frac{5x^2}{4}$ c. $\frac{5x^3}{4}$ d. $\frac{5x^6}{4}$

a 6. Which of the following is the simplest form of $\frac{x^2+2x}{x^2-9} \div \frac{x^2}{x^2+5x+6}$?
a. $\frac{(x+2)^2}{x(x-3)}$ b. $\frac{x^3}{(x-3)(x+3)^2}$
c. $\frac{10x+6}{-9}$ d. $-\frac{4}{3}$

b 7. Write $\frac{5}{\sqrt{7}+2}$ with a rational denominator.
a. $\frac{5\sqrt{7}}{9}$ b. $\frac{5\sqrt{7}-10}{3}$ c. $\frac{5\sqrt{7}+10}{3}$ d. $\frac{5\sqrt{7}-10}{9}$

b 8. If y varies inversely as x and $y = 12$ when $x = 3$, find x when $y = 18$.
a. $x = 1.5$ b. $x = 2$ c. $x = 36$ d. $x = 45$

c 9. Which of the following is the simplest form of $\sqrt[3]{-64x^5y^{12}}$?
a. $-4x^2y^4$ b. $-4|x|y^4\sqrt[3]{x}$ c. $-4xy^4\sqrt[3]{x^2}$ d. $-4xy^3\sqrt[3]{2xy}$

NAME ______ CLASS ______ DATE ______

Chapter Assessment
Chapter 8, Form A, page 2

b 10. Which of the following is the simplest form of $\frac{5}{x+2} - \frac{8}{x+4}$?
a. $\frac{-3x+36}{x^2+6x+8}$ b. $\frac{-3x+4}{x^2+6x+8}$ c. $\frac{-5}{x+4}$ d. $\frac{-3x+4}{x^2+8}$

d 11. If the value of each variable is positive, then which of the following is the simplest form of $\frac{(36x^7y^9)^{\frac{1}{2}}}{\sqrt{2x^2y^3}}$?
a. $18x^5y^6$ b. $3x^2y\sqrt{xy}$ c. $3x^4y^2\sqrt{2xy}$ d. $3x^2y^3\sqrt{2x}$

a 12. Multiply $(3+\sqrt{6})(4-2\sqrt{6})$.
a. $-2\sqrt{6}$ b. $12-2\sqrt{12}$ c. $24-2\sqrt{6}$ d. 0

c 13. Which of the following is the domain of $f(x) = \sqrt{x^2-4}$?
a. $x \geq 4$ b. $-2 \leq x \leq 2$ c. $x \leq -2$ or $x \geq 2$ d. $x \leq 2$

b 14. Which transformation was not applied to $f(x) = \sqrt{x}$ to obtain $f(x) = -3\sqrt{2(x+3)}$?
a. reflection across the x-axis b. vertical translation of 3 units up
c. vertical stretch by a factor of 3 d. horizontal compression by a factor of $\frac{1}{2}$

a 15. Which of the following is the solution to the equation $\frac{x+3}{x} - \frac{7}{x+2} = \frac{14}{x^2+2x}$?
a. $x = 4$ b. $x = -10$ c. $x = 4$ or $x = -2$ d. $x = -10$ or $x = 2$

a 16. Which of the following is the solution to the equation $\sqrt{x+17} + 3 = x$?
a. $x = 8$ b. $x = -17$ c. $x = 3$ or $x = -17$ d. $x = 8$ or $x = -1$

a 17. Which of the following is the solution to the inequality $\frac{15}{x+4} > 3$?
a. $-4 < x < 1$ b. $x < -4$ c. $x > 1$ or $x < -4$ d. $x > 1$

d 18. Which of the following is the solution to the inequality $4 < \sqrt{2x+6}$?
a. $-3 < x < 5$ b. $x < -3$ c. $x < -3$ or $x > 5$ d. $x > 5$

b 19. If a motorcycle driver travels 40 miles at m miles per hour and then 50 miles at $m + 10$ miles per hour, find the average speed, $s(m)$.
a. $s(m) = \frac{40}{m} + \frac{50}{m+10}$ b. $s(m) = \frac{9m^2+90m}{9m+40}$
c. $s(m) = \frac{8100m+3600}{m^2+10m}$ d. $s(m) = \frac{90}{m^2+10m}$

LESSON 8.2

Key Skills

Identify all excluded values, asymptotes, and holes in the graph of a rational function.

$$y = \frac{4x^2 + 12x}{x^2 + x - 6} = \frac{4x(x+3)}{(x-2)(x+3)}$$

The excluded values are $x = 2$ and $x = -3$.

$x + 3$ is a factor of the numerator and the denominator, so the graph has a hole when $x = -3$.

$x - 2$ is a factor of only the denominator, so the vertical asymptote is $x = 2$.

The degree of the numerator is equal to the degree of the denominator, so the horizontal asymptote is $y = \frac{4}{1} = 4$.

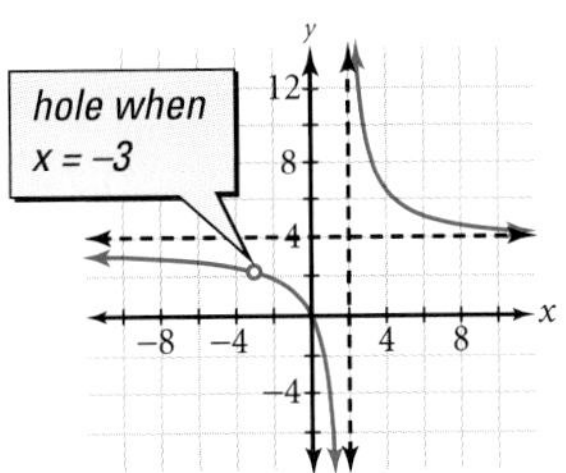

Exercises

Identify all excluded values, asymptotes, and holes in the graph of each rational function.

5. $R(x) = \frac{2x - 3}{x^2 - 8x + 12}$
6. $g(x) = \frac{3x - 5}{x^2 - 25}$
7. $f(x) = \frac{x^2 - x - 42}{x^2 + 5x - 14}$
8. $r(a) = \frac{a^2 + 4a - 12}{3a^2 - 12}$
9. $s(x) = \frac{x^2 - 9}{3x + 5}$
10. $M(x) = \frac{x^4 - 10x + 9}{3x^2 - 27}$
11. $h(y) = \frac{2y}{6y^4 - 18y^3}$
12. $r(t) = \frac{t^3 - t^2 - 4t + 4}{t^2 + t - 2}$

LESSON 8.3

Key Skills

Multiply, divide, and simplify rational expressions, including complex fractions.

Simplify $\frac{x^2 + 2x - 3}{x^2 + 5x + 6} \div \frac{4x^2 - 4x}{x^2 + 3x + 2}$.

$$\frac{x^2 + 2x - 3}{x^2 + 5x + 6} \div \frac{4x^2 - 4x}{x^2 + 3x + 2}$$
$$= \frac{x^2 + 2x - 3}{x^2 + 5x + 6} \cdot \frac{x^2 + 3x + 2}{4x^2 - 4x}$$
$$= \frac{(x+3)(x-1)}{(x+2)(x+3)} \cdot \frac{(x+1)(x+2)}{4x(x-1)}$$
$$= \frac{x+1}{4x}$$

Simplify complex fractions.

$$\frac{\frac{x^2 - 1}{2x^2 - x - 15}}{\frac{4x + 4}{x^2 - 3x}} = \frac{x^2 - 1}{2x^2 - x - 15} \div \frac{4x + 4}{x^2 - 3x}$$
$$= \frac{(x+1)(x-1)}{(x-3)(2x+5)} \cdot \frac{x(x-3)}{4(x+1)}$$
$$= \frac{x(x-1)}{4(2x+5)}, \text{ or } \frac{x^2 - x}{8x + 20}$$

Exercises

Simplify each expression.

13. $\frac{x^2 + 6x}{10} \cdot \frac{4}{x^2 - 36}$ $\quad \frac{2x}{5x - 30}$
14. $\frac{3x^2 + 10x - 8}{3x^2 - 17x + 10} \cdot \frac{2x^2 + 9x - 5}{x^2 + 3x - 4}$ $\quad \frac{2x^2 + 9x - 5}{x^2 - 6x + 5}$
15. $\frac{4a + 8}{5a - 20} \div \frac{a^2 + 3a - 10}{a^2 - 4a}$ $\quad \frac{4a^2 + 8a}{5a^2 + 15a - 50}$
16. $\frac{x^2 - 9}{6} \div \frac{4x - 12}{x}$ $\quad \frac{x^2 + 3x}{24}$
17. $\frac{\frac{z}{z+1}}{\frac{z+2}{z}}$ $\quad \frac{z^2}{z^2 + 3z + 2}$
18. $\frac{\frac{a+1}{a^2}}{\frac{(a-1)^2}{a}}$ $\quad \frac{a+1}{a^3 - 2a^2 + a}$
19. $\frac{\frac{x+1}{x}}{\frac{(x+1)^2}{x+2}}$ $\quad \frac{x+2}{x^2 + x}$
20. $\frac{\frac{4x^2}{6x - 3}}{\frac{15x}{2x - 1}}$ $\quad \frac{4x}{45}$

5. excluded values: $x = 2$ and $x = 6$; no holes; vertical asymptotes: $x = 2$ and $x = 6$; horizontal asymptote: $y = 0$
6. excluded values: $x = -5$ and $x = 5$; no holes; vertical asymptotes: $x = -5$ and $x = 5$; horizontal asymptote: $y = 0$
7. excluded values: $x = -7$ and $x = 2$; no holes; vertical asymptotes: $x = -7$ and $x = 2$; horizontal asymptote: $y = 1$
8. excluded values: $a = -2$ and $a = 2$; hole when $a = 2$; vertical asymptote: $a = -2$; horizontal asymptote: $r = \frac{1}{3}$
9. excluded value: $x = -\frac{5}{3}$; no holes; vertical asymptote: $x = -\frac{5}{3}$; no horizontal asymptote
10. excluded values: $x = -3$ and $x = 3$; no holes; vertical asymptotes: $x = -3$ and $x = 3$; no horizontal asymptote
11. excluded values: $y = 0$ and $y = 3$; hole when $y = 0$; vertical asymptote: $y = 3$; horizontal asymptote: $h = 0$
12. excluded values: $t = -2$ and $t = 1$; holes when $t = -2$ and $t = 1$; no vertical asymptote; no horizontal asymptote

Chapter Test, Form B

NAME ______ CLASS ______ DATE ______

Chapter Assessment

Chapter 8, Form B, page 1

1. If y varies inversely as x and $y = 15$ when $x = 24$, find the constant of variation. Then find y when $x = 9$.
   $k = 360$; $y = 40$
2. Write the equations of all asymptotes of the graph of the function $f(x) = \frac{x^2 + 7x + 10}{4x^2 - 9}$.
   $y = \frac{1}{4}$; $x = \pm\frac{3}{2}$
3. Graph $f(x) = \frac{x^2 + 5x + 6}{x + 3}$ on the grid at right.

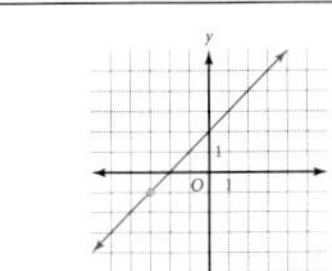

Simplify each expression.

4. $(3 + 2\sqrt{5})(2 - 3\sqrt{5}) + (15 + \sqrt{20})$
   $-9 - 3\sqrt{5}$
5. $\sqrt[3]{27x^8y^{15}}$
   $3x^2y^5\sqrt[3]{x^2}$
6. $\frac{x^2 + 7x + 10}{x^2 - 6x} \div \frac{x^3 - 4x}{x^2 - 8x + 12}$ $\quad \frac{x + 5}{x^2}$
7. $\frac{3x}{2x + 1} - \frac{5}{3x + 2}$ $\quad \frac{9x^2 - 4x - 5}{6x^2 + 7x + 2}$
8. $\frac{\frac{x+2}{x^2 + 9x + 20}}{\frac{x+2}{x+5}}$
   $\frac{1}{x + 4}$
9. $\frac{(24x^8y^6)^{\frac{1}{2}}}{\sqrt{3x^2y^4}}$
   $2|x^3y|\sqrt{2}$
10. Describe the transformations applied to $f(x) = \sqrt{x}$ in order to obtain $g(x) = 5\sqrt{-3(x + 2)} - 4$.
    vertical stretch by a factor of 5, vertical translation 4 units down, reflection across the y-axis, horizontal compression by a factor of $\frac{1}{3}$, horizontal translation of 2 units to the left.

NAME ______ CLASS ______ DATE ______

Chapter Assessment

Chapter 8, Form B, page 2

Find the domain of each function.

11. $f(x) = \frac{3x^2 + 8x + 12}{x^2 - 3x - 28}$
    all real numbers except 7 and −4
12. $g(x) = \sqrt{4x - 2}$ all real numbers greater than or equal to $\frac{1}{2}$
13. Evaluate $-5(\sqrt[4]{81})^3 + 25$. Give an exact answer. $\quad -110$
14. Write $\frac{5}{4 - \sqrt{10}}$ with a rational denominator. $\quad \frac{20 + 5\sqrt{10}}{6}$

Solve each equation.

15. $\frac{5x + 4}{5} = \frac{12 - 2x}{6}$
    $x = 0.9$
16. $\sqrt[3]{5x - 8} = \sqrt[3]{12 - 3x}$
    $x = 2.5$
17. $1 + \frac{4}{x} = \frac{12}{x^2}$
    $x = 2$ or $x = -6$
18. $\sqrt{2x - 3} = x - 3$
    $x = 6$
19. $\frac{2}{x + 7} + \frac{5}{x - 7} = \frac{x^2 + 3}{x^2 - 49}$
    $x = 9$ or $x = -2$
20. $\sqrt{2x} + 1 = \sqrt{x + 7}$
    $x = 2$

Solve each inequality.

21. $\frac{4x}{3x - 6} < 2$
    $x < 2$ or $x > 6$
22. $6 > \sqrt{9 - 3x}$
    $-9 < x < 3$
23. Find the inverse of the function $f(x) = x^2 + 3x$. $\quad y = \frac{-3 \pm \sqrt{9 + 4x}}{2}$
24. The amount of interest earned by a bank account varies jointly as the amount left in the account and the time for which it is left in years. If \$400 is left in the account for 3 years and earns \$90 in interest, find the amount of interest earned if \$500 is left in the account for 8 years.
    \$300
25. Construct a rational function with horizontal asymptote $y = 2$ and vertical asymptote $x = -2$.
    Sample answer: $y = \frac{2x + 3}{x + 2}$

21. $\dfrac{23y-29}{10y-30}$

22. $\dfrac{y^2+10y-3}{y^2-11y+18}$

23. $\dfrac{-3x^2+x-3}{x^2-3x}$

24. $\dfrac{-24}{b^2-7b+10}$

LESSON 8.4

Key Skills

Add and subtract rational expressions.

Simplify $\dfrac{2a}{a-5} - \dfrac{5a}{3a+2}$.

$$\frac{2a}{a-5} - \frac{5a}{3a+2} = \frac{2a}{a-5}\left(\frac{3a+2}{3a+2}\right) - \frac{5a}{3a+2}\left(\frac{a-5}{a-5}\right)$$
$$= \frac{2a(3a+2)-5a(a-5)}{(a-5)(3a+2)}$$
$$= \frac{a^2+29a}{3a^2-13a-10}$$

Exercises

Simplify each expression.

21. $\dfrac{3y-5}{2y-6} + \dfrac{4y-2}{5y-15}$

22. $\dfrac{9y+3}{y^2-11y+18} + \dfrac{y+3}{y-9}$

23. $\dfrac{2x-3}{x^2-3x} - \dfrac{3x+1}{x-3}$

24. $\dfrac{3b-39}{b^2-7b+10} - \dfrac{3}{b-2}$

25. $\dfrac{\frac{2}{x}}{4} + \dfrac{\frac{5}{x}}{3}$ $\dfrac{23}{x}$

26. $\dfrac{x}{3+\frac{5}{x}} - \dfrac{4}{1+\frac{2}{x}}$ $\dfrac{x^3-10x^2-20x}{3x^2+11x+10}$

LESSON 8.5

Key Skills

Solve rational equations.

Solve $\dfrac{1}{x^2} = x$.

$$\frac{1}{x^2} = x$$
$$1 = x^3$$
$$x^3 - 1 = 0$$
$$(x-1)(x^2+x+1) = 0$$
$$x = 1 \text{ or } x = -\frac{1}{2} + \frac{\sqrt{3}}{2}i \text{ or } x = -\frac{1}{2} - \frac{\sqrt{3}}{2}i$$

The only real solution is $x = 1$. Therefore, the only point of intersection for the graphs of $y = \frac{1}{x^2}$ and $y = x$ occurs when $x = 1$.

Solve rational inequalities.

Solve $\dfrac{x}{1+x} \le 2$.

$\dfrac{x}{1+x} \le 2,\ 1+x>0$ *or* $\dfrac{x}{1+x} \le 2,\ 1+x<0$

$\dfrac{x}{1+x} \le 2$	$\dfrac{x}{1+x} \le 2$
$x \le 2(1+x)$	$x \ge 2(1+x)$
$x \le 2+2x$	$x \ge 2+2x$
$-x \le 2$	$-x \ge 2$
$x \ge -2$	$x \le -2$
If $1+x>0$, then $x>-1$. Thus, $x>-1$ *and* $x \ge -2$, or simply $x>-1$.	If $1+x<0$, then $x<-1$. Thus, $x<-1$ *and* $x \le -2$, or simply $x \le -2$.

The solution is $x>-1$ *or* $x \le -2$.

The graphs of $y = \dfrac{x}{1+x}$ and $y = 2$ verify this solution.

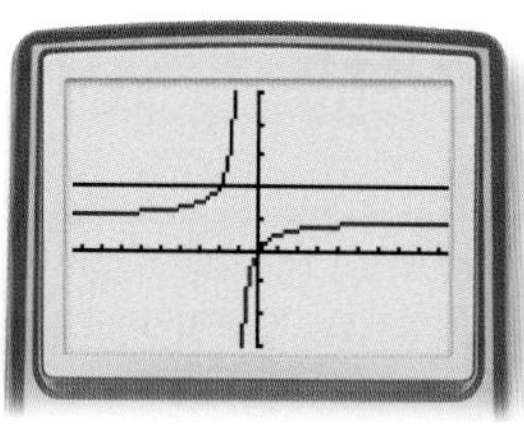

Exercises

Solve each equation.

27. $\dfrac{1}{x^2+1} = \dfrac{1}{2}$ $x = \pm 1$

28. $\dfrac{4}{x^2+1} = 1$ $x = \pm\sqrt{3}$

29. $\dfrac{3x-1}{x^2+2x} = -1$ $x = \dfrac{-5 \pm \sqrt{29}}{2}$

30. $\dfrac{2}{1-x^2} = \dfrac{x^2}{x^2+1}$ no sol.

31. $\dfrac{1}{1-x^2} = -1$ $x = \pm\sqrt{2}$

32. $\dfrac{1}{x} = \dfrac{x+2}{x+1}$ $x = -\dfrac{1}{2} \pm \dfrac{\sqrt{5}}{2}$

Solve each inequality by using algebra.

33. $\dfrac{1}{x} < 1$ $x > 1$ *or* $x < 0$

34. $\dfrac{1}{x} \ge 2$ $0 < x \le \dfrac{1}{2}$

35. $\dfrac{1}{x^2+1} < \dfrac{1}{2}$ $x < -1$ *or* $x > 1$

36. $\dfrac{1}{x^2+1} \ge \dfrac{1}{3}$ $-\sqrt{2} \le x \le \sqrt{2}$

37. $\dfrac{1+x}{2x+3} < 1$ $x < -2$ *or* $x > \dfrac{-3}{2}$

38. $\dfrac{1+2x}{2x-1} < 2$ $x < \dfrac{1}{2}$ *or* $x > \dfrac{3}{2}$

Solve each inequality by graphing.

39. $\dfrac{1}{x} \ge x$ $x \le -1$ *or* $0 < x \le 1$

40. $\dfrac{1}{x} < 2x$ $x > \dfrac{\sqrt{2}}{2}$ *or* $-\dfrac{\sqrt{2}}{2} < x < 0$

41. $\dfrac{x^2+x+1}{x^2+3x+2} \ge x$

42. $\dfrac{x^3+2}{x^2+2x+1} \le 3x$

43. $\dfrac{1}{x^2+2x+1} > 2$

44. $\dfrac{1}{x^2-x+2} < x$

41. $x < -2$ *or* $-1 < x \le 0.5$
42. $-2 \le x \le 1.4$ *or* $x \ge 0.4$
43. $-1.7 < x < -1$ *or* $-1 < x < -0.3$
44. $x > 0.6$

LESSON 8.6

Key Skills

Find the inverse of a quadratic function.

Find the inverse of $y = x^2 - 7x + 10$. Interchange x and y, and solve for y by applying the quadratic formula.

$$x = y^2 - 7y + 10$$
$$y = \frac{-(-7) \pm \sqrt{(-7)^2 - 4(1)(10 - x)}}{2(1)}$$
$$y = \frac{7 \pm \sqrt{9 + 4x}}{2}$$

Describe the transformations applied to the square-root parent function, $f(x) = \sqrt{x}$.

Describe the transformations applied to $f(x) = \sqrt{x}$ to obtain $y = 2\sqrt{3x - 3} + 4$.

$$g(x) = 2\sqrt{3x - 3} + 4 = 2\sqrt{3(x - 1)} + 4$$

The parent function is stretched vertically by a factor of 2, compressed horizontally by a factor of $\frac{1}{3}$, translated horizontally 1 unit to the right, and translated vertically 4 units up.

Exercises

Find the inverse of each quadratic function.

45. $y = 3x + x^2$ 46. $y = 8x + 12 + x^2$

47. $y = 3x^2 - 16x + 5$ 48. $y = 2x^2 + 7x + 6$

For each function, describe the transformations applied to $f(x) = \sqrt{x}$.

49. $g(x) = \frac{1}{3}\sqrt{x}$ 50. $h(x) = 3\sqrt{x} - 5$

51. $k(x) = \sqrt{2x - 3}$ 52. $g(x) = 4\sqrt{2x + 1} + 2$

53. $h(x) = -2\sqrt{3x} - 6$ 54. $r(x) = 5\sqrt{3(x - 1)} + 1$

Evaluate each expression.

55. $5\left(\sqrt[3]{-27}\right)^2$ 45 56. $\frac{1}{2}\sqrt[3]{8} + 1$ 2

45. $y = \frac{-3 \pm \sqrt{9 + 4x}}{2}$

46. $y = -4 \pm \sqrt{4 + x}$

47. $y = \frac{8 \pm \sqrt{49 + 3x}}{3}$

48. $y = \frac{-7 \pm \sqrt{1 + 8x}}{4}$

LESSON 8.7

Key Skills

Simplify expressions involving radicals.

Simplify $\frac{(24a^8b^5)^{\frac{1}{4}} \cdot \sqrt[4]{4a^3b^2}}{\sqrt[4]{3ab^2}}$.

$$\frac{(24a^8b^5)^{\frac{1}{4}} \cdot \sqrt[4]{4a^3b^2}}{\sqrt[4]{3ab^2}} = \frac{\sqrt[4]{24a^8b^5} \cdot \sqrt[4]{4a^3b^2}}{\sqrt[4]{3ab^2}}$$
$$= \frac{\sqrt[4]{96a^{11}b^7}}{\sqrt[4]{3ab^2}}$$
$$= \sqrt[4]{32a^{10}b^5}$$
$$= \sqrt[4]{2^4a^8b^4 \cdot 2a^2b}$$
$$= 2a^2|b|\sqrt[4]{2a^2b}$$

Rationalize the denominators of expressions.

Write $\frac{1}{2 + \sqrt{2}}$ with a rational denominator.

$$\frac{1}{2 + \sqrt{2}} = \frac{1}{2 + \sqrt{2}}\left(\frac{2 - \sqrt{2}}{2 - \sqrt{2}}\right) = \frac{2 - \sqrt{2}}{2}$$

Exercises

Simplify each radical expression. Assume that the value of each variable is positive.

57. $\sqrt{6x^2y^4} \cdot (3x^5y)^{\frac{1}{2}}$ $3x^3y^2\sqrt{2xy}$ 58. $(5a^3b^5)^{\frac{1}{3}} \cdot \sqrt[3]{4a^4b}$ $a^2b^2\sqrt[3]{20a}$

59. $\frac{\sqrt[3]{42c^4d^{17}}}{(6cd^{11})^{\frac{1}{3}}}$ $cd^2\sqrt[3]{7}$ 60. $\frac{(45s^3t^6)^{\frac{1}{2}}}{\sqrt{3t^2}}$ $st^2\sqrt{15s}$

61. $\frac{(6x^5y^7)^{\frac{1}{2}} \cdot \sqrt{3x^2y^4}}{\sqrt{2x}}$ $3x^3y^5\sqrt{y}$ 62. $\frac{(24m^9n)^{\frac{1}{3}} \cdot \sqrt[3]{9m^3n^7}}{\sqrt[3]{3mn^2}}$ $2m^3n^2\sqrt[3]{9m^2}$

Write each expression with a rational denominator and in simplest form.

63. $\frac{1}{\sqrt{5}}$ $\frac{\sqrt{5}}{5}$ 64. $\frac{1}{\sqrt{7}}$ $\frac{\sqrt{7}}{7}$

65. $\frac{3}{2 - \sqrt{3}}$ $6 + 3\sqrt{3}$ 66. $\frac{4}{-2 + \sqrt{5}}$ $8 + 4\sqrt{5}$

67. $\frac{1 + \sqrt{2}}{3 - \sqrt{3}}$ $\frac{3 + \sqrt{6} + \sqrt{3} + 3\sqrt{2}}{6}$ 68. $\frac{2 - \sqrt{3}}{3 + \sqrt{2}}$ $\frac{6 + \sqrt{6} - 3\sqrt{3} - 2\sqrt{2}}{7}$

49. The parent function is vertically compressed by a factor of $\frac{1}{3}$.

50. The parent function is vertically stretched by a factor of 3 and vertically translated 5 units down.

51. The parent function is horizontally translated $\frac{3}{2}$ units to the right and horizontally compressed by a factor of $\frac{1}{2}$.

52. The parent function is translated $\frac{1}{2}$ unit to the left, horizontally compressed by a factor $\frac{1}{2}$, vertically stretched by the factor 4, and vertically translated 2 units up.

53. The parent function is horizontally compressed by a factor of $\frac{1}{3}$, vertically stretched by a factor 2, reflected across the x-axis, and vertically translated 6 units down.

54. The parent function is horizontally translated one unit to the right, horizontally compressed by a factor of $\frac{1}{3}$, vertically stretched by a factor of 5, and vertically translated 1 unit up.

LESSON 8.8

Key Skills

Solve radical equations.

Solve $2x = \sqrt{3 - x}$.

$$2x = \sqrt{3 - x}$$
$$4x^2 = 3 - x$$
$$4x^2 + x - 3 = 0$$
$$(4x - 3)(x + 1) = 0$$
$$x = \frac{3}{4} \quad or \quad x = -1$$

Check for extraneous solutions.

$$2x = \sqrt{3 - x} \qquad 2x = \sqrt{3 - x}$$
$$2\left(\frac{3}{4}\right) \stackrel{?}{=} \sqrt{3 - \left(\frac{3}{4}\right)} \qquad 2(-1) \stackrel{?}{=} \sqrt{3 - (-1)}$$
$$\frac{3}{2} = \frac{3}{2} \text{ True} \qquad -2 = 2 \text{ False}$$

Solve radical inequalities.

To solve $\sqrt{2x - 1} \le 1$, first solve $2x - 1 \ge 0$.

$$2x - 1 \ge 0$$
$$x \ge \frac{1}{2}$$

Then solve the original inequality.

$$\sqrt{2x - 1} \le 1$$
$$\left(\sqrt{2x - 1}\right)^2 \le 1^2$$
$$2x - 1 \le 1$$
$$2x \le 2$$
$$x \le 1$$

Thus, $x \ge \frac{1}{2}$ *and* $x \le 1$, or $\frac{1}{2} \le x \le 1$. The solution can be verified by graphing.

Exercises

Solve each radical equation by using algebra. If the inequality has no real solution, write *no solution*. Check your solution.

69. $\sqrt{x + 2} = -2$ **no sol.**

70. $3\sqrt{x + 7} + 8 = 6$ **no sol.**

71. $\sqrt[3]{x + 2} = -2$ **−10**

72. $3\sqrt[3]{x + 7} + 8 = 6$ $-\frac{197}{27}$

73. $\sqrt{x} = 2x$ **0,** $\frac{1}{4}$

74. $\sqrt{x + 2} = 3$ **7**

75. $\sqrt{x} = \sqrt{-x + 3}$ $\frac{3}{2}$

76. $\sqrt{2x + 1} = \sqrt{4x - 4}$ $\frac{5}{2}$

77. $\sqrt[3]{4 - x} = \sqrt[3]{3x}$ **1**

78. $\sqrt[5]{2x} = \sqrt[5]{x + 3}$ **3**

79. $\sqrt{x} - 2 = \sqrt{x - 2}$ **no solution**

80. $\sqrt{3x - 1} = \sqrt{x} + 2$ $2 + \frac{\sqrt{15}}{2}$

Solve each radical inequality by using algebra. Check your solution.

81. $\sqrt{x} \le 5$ $0 \le x \le 25$

82. $\sqrt{x - 1} < 2$ $1 \le x < 5$

83. $\sqrt{x} \ge 5$ $x \ge 25$

84. $\sqrt{x - 1} > 2$ $x > 5$

85. $\sqrt[4]{x - 2} \ge 1$ $x \ge 3$

86. $\sqrt[3]{x - 1} < 1$ $x < 2$

87. $\sqrt{2x} + 2 > 4$ $x > 2$

88. $-2\sqrt{x - 2} < -1$ $x > \frac{9}{4}$

89. $\sqrt{6x} < 0$ **no solution**

90. $4\sqrt{5x - 1} < 0$ **no solution**

Solve each radical inequality by graphing.

91 $\sqrt[3]{x - 2} \le \sqrt{x}$ $x \ge 0$

92 $\sqrt[5]{2x + 1} \ge 2$ $x \ge \frac{31}{2}$

Applications

PHYSICS The weight of an object varies inversely as the square of the distance from the object to the center of Earth, whose radius is approximately 4000 miles.

93. If an astronaut weighs 175 pounds on Earth, what will the astronaut weigh at a point 60 miles above Earth's surface? **≈169.9 pounds**

94. If an astronaut weighs 145 pounds at a point 80 miles above the Earth's surface, how much does the astronaut weigh on Earth? **≈150.9 pounds**

Chapter Test

1. y varies jointly as x and z. If $y = -63$ when $x = 7$ and $z = -9$, find y when $x = -15$ and $z = -\frac{1}{2}$. **7.5**

2. **CONSTRUCTION** The strength of a beam varies directly with the width of a beam and inversely as the cube of the depth. If a beam 10 mm wide by 20 mm deep will support 1200 kg, how much will a beam 8 mm by 25 mm support? **491.52 kg**

Identify all excluded values, asymptotes, and holes in the graph of each rational function.

3. $f(x) = \frac{x-4}{x^2-16}$

4. $h(x) = \frac{x^2+2x-15}{2x^2-18}$

5. $g(x) = \frac{2x^3-16}{x^3-2x^2-9x+18}$

Simplify each expression.

6. $\frac{x^2-9}{2x^2-8x+6} \cdot \frac{4x^2-12x+36}{x^3+27}$ $\frac{2}{x-1}$

7. $\dfrac{\frac{x^3}{3x^2-12}}{\frac{x^3+5x^2}{3x^2+9x-30}}$ $\frac{x}{x+2}$

8. $\frac{3x}{x-2} \div \frac{6x^2}{2x^2-8} \cdot \frac{5x+1}{2x+4}$ $\frac{5x+1}{2x}$

Simplify each expression.

9. $\frac{4}{x^2-4} + \frac{x+3}{x+2}$

10. $\frac{x-37}{x^2-2x-15} - \frac{5}{x+3}$

11. **GEOMETRY** Find the area of the shaded region of the figure at right if the largest triangle has an area $A = x$. $\frac{85x}{256}$ u²

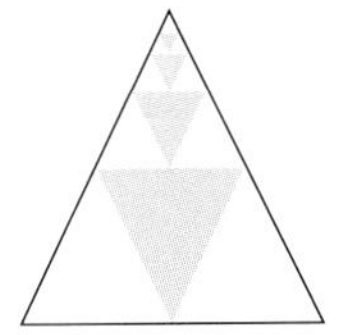

Solve each equation or inequality.

12. $\frac{x+3}{x-1} = 2$ $x = 5$

13. $\frac{z-4}{z+2} + \frac{z-5}{z-4} = 1$ $z = 7$ or $z = 2$

14. $\frac{3}{x+4} \leq 5$

15. $\frac{3}{x+4} < \frac{5}{x+7}$ $-7 < x < -4$ or $x > \frac{1}{2}$

For each function, describe the transformations applied to $f(x) = \sqrt{x}$.

16. $g(x) = \sqrt{x-4}$

17. $h(x) = -2\sqrt{x} + 3$

Find the inverse of each quadratic function.

18. $y = x^2 + x$ $y = \frac{-1 \pm \sqrt{4x+1}}{2}$

19. $y = 5x^2 - 3x - 4$ $y = \frac{3 \pm \sqrt{20x+89}}{10}$

Evaluate each expression.

20. $(3\sqrt[4]{81})^2 - 31$ **50**

21. $\frac{1}{5}\left((\sqrt{9})^3 + (\sqrt[3]{64})^2 + 2\right)$ **9**

Simplify each expression. Assume that the value of each variable is positive.

22. $5\sqrt{8x^3y^6} \cdot (2x^5y)^{\frac{1}{2}}$ $20x^4y^3\sqrt{y}$

23. $\frac{8\sqrt{5r^7s^9}}{\sqrt{25r^3s^5t}}$ $\frac{8r^2s^2\sqrt{5t}}{5t}$

24. $(5 - \sqrt{12}) - (2\sqrt{27} + 8)$ $-3 - 8\sqrt{3}$

25. $(2 + \sqrt{5})(3 - 2\sqrt{5})$ $-4 - \sqrt{5}$

Write each expression with a rational denominator and in simplest form.

26. $\frac{4}{\sqrt{11}}$ $\frac{4\sqrt{11}}{11}$

27. $\frac{1}{2+\sqrt{5}}$ $\sqrt{5} - 2$

28. $\frac{2-\sqrt{3}}{5+\sqrt{7}}$

Solve each radical equation or inequality. If no solution, write no solution.

29. $\sqrt{2x+7} = -3$ no solution

30. $\sqrt[4]{3x} = \sqrt[4]{4x-7}$ $x = 7$

31. $\sqrt{x-7} < 5$ $7 \leq x < 32$

32. $\sqrt[3]{2x+1} \geq 3$ $x \geq 13$

Chapter Test

3. excluded values: $x = 4, -4$; asymptotes: $x = -4$ and $y = 0$; holes: $x = 4$
4. excluded values: $x = 3, -3$; asymptotes: $x = -3$ and $y = \frac{1}{2}$; holes: $x = 3$
5. excluded values: $x = 2, 3, -3$; asymptotes: $x = 3$, $x = -3$ and $y = 2$; holes: $x = 2$

9. $\frac{x-1}{x-2}$

10. $\frac{-4}{x-5}$

14. $x \geq -\frac{17}{5}$ or $x < -4$

16. a horizontal translation to the right

17. a vertical translation up 3, a vertical stretch by a factor of 2, and a reflection across the x-axis

28. $\frac{\sqrt{21} - 2\sqrt{7} - 5\sqrt{3} + 10}{18}$

College Entrance Exam Practice

Multiple-Choice and Quantitative-Comparison Samples

The first half of the Cumulative Assessment contains two types of items found on standardized tests—multiple-choice questions and quantitative-comparison questions. Quantitative-comparison items emphasize the concepts of equality, inequality, and estimation.

Free-Response Grid Samples

The second half of the Cumulative Assessment is a free-response section. This part of the Cumulative Assessment requires student-produced response items like those commonly found on college entrance exams. These questions require the use of machine-scored answer grids. You may wish to have students practice answering these items in preparation for standardized tests.

College Entrance Exam Practice

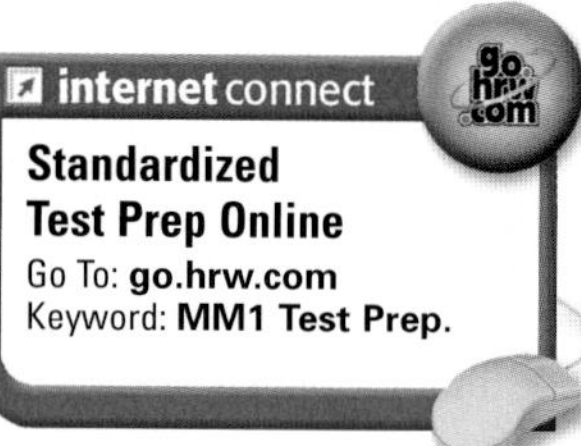

QUANTITATIVE COMPARISON. For Items 1–6, write
A if the quantity in Column A is greater than the quantity in Column B;
B if the quantity in Column B is greater than the quantity in Column A;
C if the two quantities are equal; or
D if the relationship cannot be determined from the given information.

	Column A	Column B	Answers
1. D	Let $f(x) = 3x - 5$ and $g(x) = x^2 + 1$. $f \circ g$	$g \circ f$	Ⓐ Ⓑ Ⓒ Ⓓ [Lesson 2.6]
2. D	$\frac{1}{x}$	x	Ⓐ Ⓑ Ⓒ Ⓓ [Lesson 2.1]
3. A	4	$\log_3 50$	Ⓐ Ⓑ Ⓒ Ⓓ [Lesson 6.3]
4. D	$-5x < 10$ x	0	Ⓐ Ⓑ Ⓒ Ⓓ [Lesson 1.7]
5. A	$x \neq 0$ $\frac{1}{x^2}$	$\frac{1}{x^2 + 1}$	Ⓐ Ⓑ Ⓒ Ⓓ [Lesson 8.2]
6. B	$\sqrt[3]{-\frac{1}{3}}$	$\sqrt[4]{\frac{1}{3}}$	Ⓐ Ⓑ Ⓒ Ⓓ [Lesson 5.3]

7. If $\frac{a}{b} = \frac{c}{d}$, which of the following is *not* always true? ***(LESSON 1.4)***
c
a. $ad = bc$ **b.** $ad = cb$
c. $\frac{a}{d} = \frac{c}{b}$ **d.** $\frac{a-b}{b} = \frac{c-d}{d}$

8. Solve $2^x = \frac{1}{64}$ for x. ***(LESSON 6.3)***
d
a. 32 **b.** 6
c. −32 **d.** −6

9. Simplify $(6\sqrt{8} - 6\sqrt{2})(2\sqrt{2} + 1)$. ***(LESSON 8.7)***
b
a. $6\sqrt{2} - 24$ **b.** $6\sqrt{2} + 24$
c. $12\sqrt{2} - 23$ **d.** $6\sqrt{2} - 23$

10. How many solutions does the system $\begin{cases} y = 3x + 2 \\ y = 3x - 2 \end{cases}$ have? ***(LESSON 3.1)***
a
a. 0 **b.** 1
c. 2 **d.** infinite

11. Find the remainder when $2x^2 - 5x + 8$ is divided by $x + 4$. ***(LESSON 7.3)***
a
a. 60 **b.** −44
c. 0 **d.** 20

12. Simplify $(a^3b^{-2})^{-2}$. Assume that no variable equals zero. ***(LESSON 2.2)***
a
a. $\frac{b^4}{a^6}$ **b.** $\frac{1}{a^2b}$
c. $\frac{a^6}{b^4}$ **d.** a^2b

13. Which of the following is *not* a root of $x^3 + x^2 - 9x - 9 = 0$? ***(LESSON 7.4)*** d

a. -3 **b.** 3 **c.** -1 **d.** 1

14. Which would you add to $x^2 - 10x$ to complete the square? ***(LESSON 5.4)*** c

a. 5 **b.** -5 **c.** 25 **d.** -25

15. Find the product $\begin{bmatrix} 3 & 6 & 1 \end{bmatrix}\begin{bmatrix} 0 & 5 \\ 1 & -2 \\ -3 & 4 \end{bmatrix}$. ***(LESSON 4.2)*** $[3\ 7]$

16. Write an equation of the line that contains the points $(-3, 8)$ and $(9, -4)$. ***(LESSON 1.3)*** $y = -x + 5$

17. Write the function for the graph of $f(x) = |x|$ translated 2 units to the left and 1 unit down. ***(LESSON 2.7)*** $f(x) = |x + 2| - 1$

18. Factor $25x^2 - 60x + 36$ completely. ***(LESSON 5.3)*** $(5x - 6)^2$

19. Find the inverse of $f(x) = \frac{2}{3}x + 6$. ***(LESSON 2.5)*** $f^{-1}(x) = \frac{3x - 18}{2}$

20. Simplify $\frac{1 + x}{x^2 + 3x - 4} \div \frac{x^3}{x^2 + 4x}$. ***(LESSON 8.3)*** $\frac{1 + x}{x^3 - x^2}$

21. Find x if $\log_{10} x + \log_{10} 8 = \log_{10} 16$. ***(LESSON 6.4)*** $x = 2$

22. Write a polynomial function in standard form by using the information given below. ***(LESSON 7.5)*** $P(x) = x^4 + 3x^3 - 9x^2 - 23x - 12$

$P(0) = -12$; zeros: $-4, 3, -1$ (multiplicity 2)

23. Write the parametric equations $\begin{cases} x(t) = 3 - t^2 \\ y(t) = \frac{5}{2}t \end{cases}$ as a single equation in x and y. ***(LESSON 3.6)*** $x = 3 - \frac{4}{25}y^2$

24. Find the inverse, if one exists, of the matrix $\begin{bmatrix} 2 & 4 \\ 3 & 5 \end{bmatrix}$. If the inverse does not exist, write *no inverse*. ***(LESSON 4.3)*** $\begin{bmatrix} -2.5 & 2 \\ 1.5 & -1 \end{bmatrix}$

25. Write with a rational denominator $\frac{8}{5 - 3\sqrt{2}}$. ***(LESSON 8.6)*** $\frac{40 + 24\sqrt{2}}{7}$

26. **TRAVEL** An airplane travels from Hawaii to Los Angeles at 580 miles per hour with a tailwind and returns to Hawaii at 460 miles per hour against a headwind. What is the average speed of the airplane over the entire trip? ***(LESSON 8.4)*** ≈513.1 mph

27. **MAXIMUM/MINIMUM** At the Springfield City Fourth of July celebration, a fireworks shell is shot upward with an initial velocity of 250 feet per second. Find its maximum height and the time required to reach it. Use the formula $h = v_0t - 16t^2$, where v_0 is the initial velocity, t is time in seconds, and h is the height in feet. ***(LESSON 5.2)*** ≈976.6 feet; ≈7.8 seconds

FREE-RESPONSE GRID

The following questions may be answered by using a free-response grid such as that commonly used by standardized-test services.

28. Find x if $\log_2 x = 4$. ***(LESSON 6.3)*** 16

29. What is the greatest integer x for which $-6x - 1 > 10$? ***(LESSON 1.7)*** -2

30. Solve the system for z. ***(LESSON 4.4)*** 3

$$\begin{cases} x - 2y - z = 2 \\ x - y + 2z = 9 \\ 2x + y + z = 3 \end{cases}$$

31. Find $|7 - 24i|$. ***(LESSON 5.6)*** 25

32. Solve $2\sqrt{x - 4} = \sqrt{x + 2}$ for x. ***(LESSON 8.8)*** 6

Solve each equation. ***(LESSONS 1.6 AND 1.8)***

33. $1 - 2x = 5$ -2

34. $-5x + 3 = \frac{1}{2}x - 1$ $\frac{8}{11}$

35. Solve $\log_x 27 = -3$ for x. Give an approximate solution to the nearest hundredth. ***(LESSON 6.7)*** $\frac{1}{3}$

36. Approximate to the nearest tenth the real zero of $f(x) = x^3 - 2x^2 + 3x - 1$. ***(LESSON 7.5)*** 0.4

37. Simplify $\sqrt[3]{64^{\frac{1}{2}}}$. ***(LESSON 8.7)*** 2

38. **ENTERTAINMENT** A rectangular stage is 20 meters wide and 38 meters long. To make room for additional seating, two adjacent sides of the stage are shortened by the same amount. If this reduces the stage by 265 square meters, by how many meters are each of the two adjacent sides of the stage shortened? ***(LESSON 5.2)*** 5

Keystroke Guide for Chapter 8

Essential keystroke sequences (using the model TI-82 or TI-83 graphics calculator) are presented below for all Activities and Examples found in this chapter that require or recommend the use of a graphics calculator.

internet connect

For Keystrokes of other graphing calculator models, visit the HRW web site at **go.hrw.com** and enter the keyword **MB1 CALC**.

LESSON 8.1

EXAMPLES 1 and 3 For Example 1, enter $y = \frac{60.75}{x}$, and use a table to find the y-values for the given x-values.

Pages 482 and 483

Enter the function:

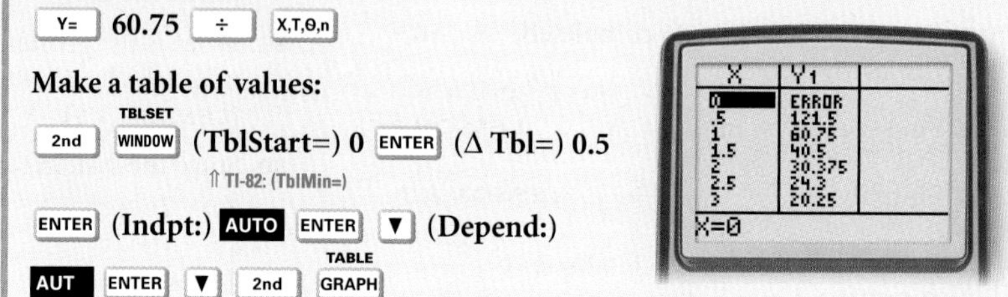

Y= 60.75 ÷ X,T,θ,n
Make a table of values:
TBLSET
2nd WINDOW (TblStart=) 0 ENTER (Δ Tbl=) 0.5
⇑ TI-82: (TblMin=)
ENTER (Indpt:) AUTO ENTER ▼ (Depend:)
TABLE
AUT ENTER ▼ 2nd GRAPH
X Y1
0 ERROR
.5 121.5
1 60.75
1.5 40.5
2 30.375
2.5 24.3
3 20.25
X=0

For Example 3, use a similar keystroke sequence. Use (TblStart=) 1.5 and (Δ Tbl=) 1.

LESSON 8.2

EXAMPLE 2 Enter $y = \frac{x^2 - 7x + 12}{x^2 + 9x + 20}$, and use a table of values to verify that –4 and –5 are not in the domain.

Page 490

Enter the function:

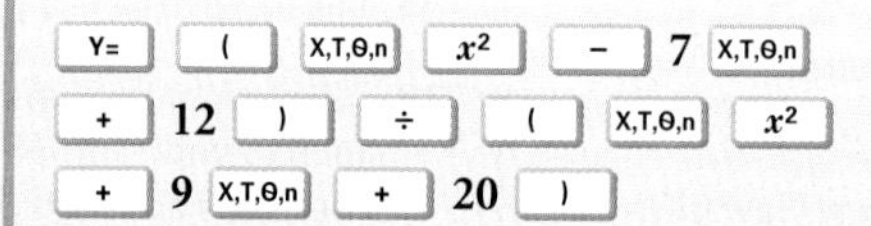

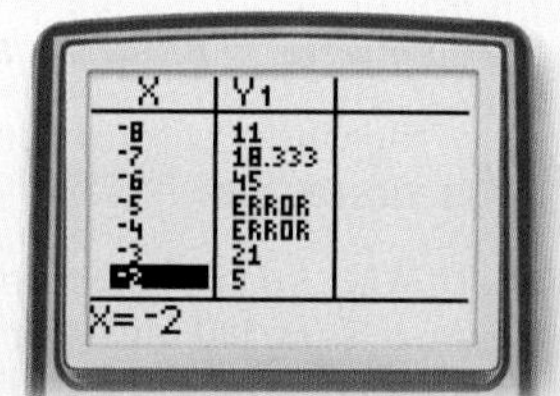

Make a table of values:

Use a keystroke sequence similar to that used in Example 1 of Lesson 8.1. Use (TblStart=) –8 and (Δ Tbl=)1.

Activity Page 490

For Step 1, enter $y = \frac{1}{x-2}$, and use a table to find the y-values for the given x-values.

Use a keystroke sequence similar to that used in Example 1 of Lesson 8.1. For part **a**, use (TblStart=) 1 and (△ Tbl=) 0.1. For part **b**, use (TblStart=) 3 and (△ Tbl=) −0.1.
For Step 3, use a similar keystroke sequence.

EXAMPLES 3 and 4 Pages 491–493

Graph the function, and look for asymptotes.

Use friendly viewing window [−4.7, 4.7] by [−3.1, 3.1].

Enter the function by using a keystroke sequence similar to that in Example 2 of Lesson 8.2. Then graph.

EXAMPLE 6 Page 501

Enter $y = \frac{3x}{0.8x + 25}$, and make a table of values.

Enter the function:

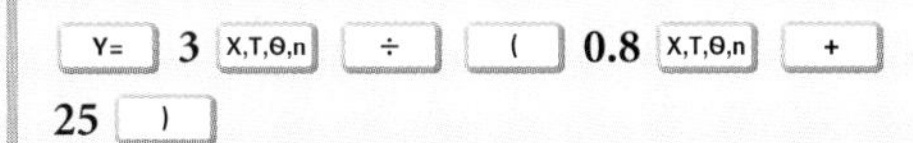

25)

Make a table of values:

Use a keystroke sequence similar to that used in Example 1 of Lesson 8.1. Use (TblStart=) 18 and (△ Tbl=) 1.

Activity Page 501

For Steps 1 and 2, graph $y = \frac{\frac{x-3}{x+2}}{\frac{x-2}{x+2}}$ and $y = \frac{x-3}{x-2}$ on the same screen. Then make a table of values with the given x-values.

Use friendly viewing window [−4.7, 4.7] by [−3.1, 3.1].

Graph the functions:

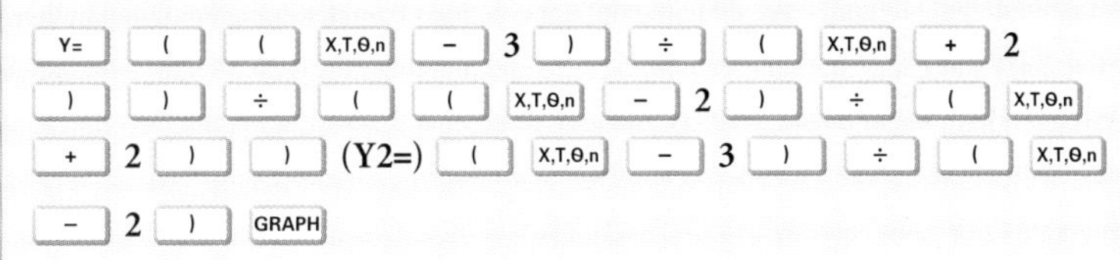

Make a table of values:

Use a keystroke sequence similar to that used in Example 1 of Lesson 8.1. Use (TblStart=) −3 and (△ Tbl=) 1.

LESSON 8.4

EXAMPLE 3 Page 507

Graph $y = \frac{6x}{3x-1} - \frac{4x}{2x+5}$ and $y = \frac{34x}{6x^2+13x-5}$ on the same screen.

Use friendly viewing window [–6, 4] by [–4, 4].

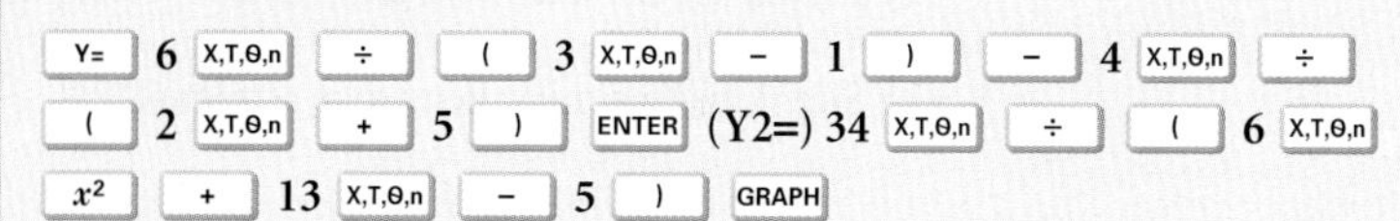

Y= 6 X,T,θ,n ÷ (3 X,T,θ,n – 1) – 4 X,T,θ,n ÷
(2 X,T,θ,n + 5) ENTER (Y2=) 34 X,T,θ,n ÷ (6 X,T,θ,n
x2 + 13 X,T,θ,n – 5) GRAPH

LESSON 8.5

EXAMPLE 1 Pages 512 and 513

Graph $y = \frac{64x+140}{6x(x+5)}$ and $y = 2.5$ on the same screen, and find any points of intersection.

Use viewing window [0, 5] by [0, 7].

Intersection X=2.7103088 Y=2.5

Graph the functions:

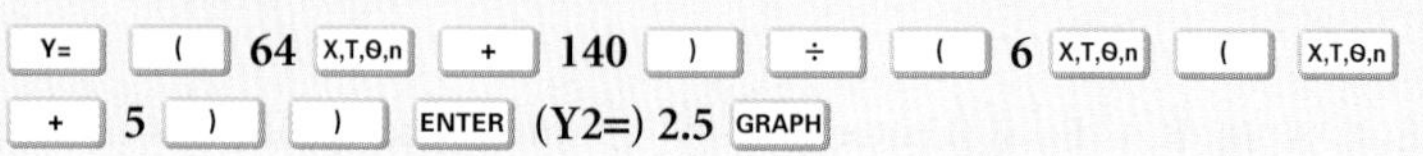

Find any points of intersection:

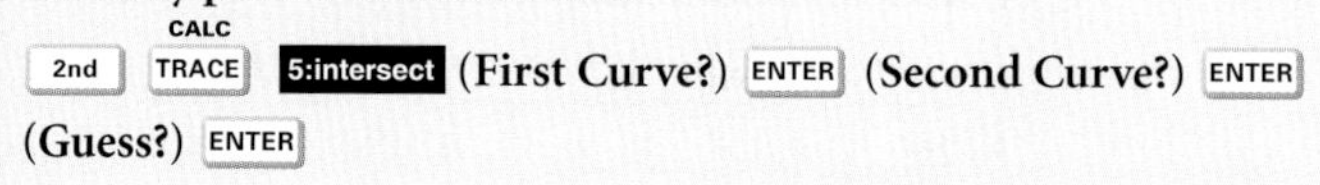

(Guess?) ENTER

EXAMPLES 2 and 3 Pages 513 and 514

For Example 2, graph $y = \frac{x}{x-6} - \frac{1}{x-4}$ and find any zeros.

Use viewing window [–2, 4] by [–0.5, 0.5].

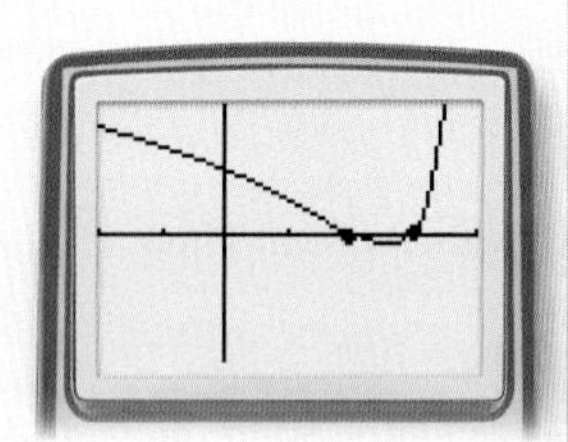

Graph the function:

Use a keystroke sequence similar to that in Example 3 of Lesson 8.4.

Find any zeros: 2nd TRACE (CALC) 2:intersect (Left Bound?) ENTER (Right Bound?) ENTER (Guess?) ENTER

⇑ TI-82: 2:root

For Example 3, use a similar keystroke sequence. Use friendly viewing window [–12.4, 6.4] by [–8, 8].

Activity Page 515

For Step 1, graph $y = \frac{x+2}{x-4}$, and make a table of values.

Use friendly viewing window [–6.8, 12] by [–10, 10].

Graph the function:

Y= (X,T,θ,n + 2) ÷ (X,T,θ,n – 4) GRAPH

Make a table of values: Use a keystroke sequence similar to that used in Example 1 of Lesson 8.1. Use (TblStart=) –5 and (△ Tbl=) 1.

EXAMPLE 4 Graph $y = \frac{x}{2x - 1}$ and $y = 1$, and find any points of intersection.

Page 515

Use friendly viewing window [−4.7, 4.7] by [−3, 3].

Graph the functions:

Y= X,T,Θ,n ÷ (2 X,T,Θ,n − 1) ENTER (Y2=) 1 GRAPH

Find any points of intersection:

Use a keystroke sequence similar to that used in Example 1 of Lesson 8.5.

EXAMPLE 5 Graph $y = \frac{x-2}{2(x-3)} - \frac{x}{x+3}$, and find any zeros of the function. Then make a table of values.

Page 516

Use friendly viewing window [−9.4, 9.4] by [−3.1, 3.1].

Graph the function:

Y= (X,T,Θ,n − 2) ÷ (2 (X,T,Θ,n − 3)

⇑ TI-82: ( )

) − X,T,Θ,n ÷ (X,T,Θ,n + 3) GRAPH

⇑ TI-82: ( )

Find any zeros:

Use a keystroke sequence similar to that in Example 2 of Lesson 8.5.

Make a table of values:

Use a keystroke sequence similar to that in Example 1 of Lesson 8.1. Use (TblStart=) −4 and (△ Tbl=) 1.

LESSON 8.6

EXAMPLE 1 Graph $y = \sqrt{2x - 5}$.

Page 521

Use viewing window [−1, 8] by [−1, 5].

Y= 2nd x^2 ($\sqrt{\ }$ ⇓ TI-82: ( ) 2 X,T,Θ,n − 5) GRAPH

EXAMPLES 2 and 3 For Example 2, part a, graph $y = \sqrt{x}$ and $y = -2\sqrt{x+1} + 4$ on the same screen, and compare the graphs.

Pages 522 and 523

Use friendly viewing window [−2.7, 6.7] by [−3, 6].

Graph the functions:

Y= 2nd x^2 ($\sqrt{\ }$ ⇓ TI-82: ( ) X,T,Θ,n) ENTER (Y2=) ◄ ◄ ENTER (⧹Y2=) ► ► (TI-82: omit)

(−) 2 2nd x^2 ($\sqrt{\ }$ ⇓ TI-82: ( ) X,T,Θ,n + 1) + 4 GRAPH

For part **b** of Example 2, use a similar keystroke sequence and the same friendly viewing window.

For Example 3, use friendly viewing window [−2.7, 6.7] by [−1.6, 4.6].

EXAMPLE 4 Enter the function $y = 2\pi\sqrt{\frac{x}{9.8}}$, and make a table of values.

Page 523

Enter the function:

[Y=] 2 [2nd] [^] (π) [2nd] [x^2] ($\sqrt{\ }$) [X,T,θ,n] [÷] 9.8 [)]

Make a table of values:

Use a keystroke sequence similar to that used in Example 1 of Lesson 8.1. Use (TblStart=) 0 and (△ Tbl=) 0.1.

Activity

Page 524

For Step 1, graph each radical function.

Use viewing window [–5, 5] by [–5, 5]. Use the fact that $\sqrt[n]{x} = x^{\frac{1}{n}}$ to graph a radical function for $n > 2$. For example, graph $y = \sqrt[3]{x}$:

[Y=] [X,T,θ,n] [^] [(] 1 [÷] 3 [)] [GRAPH]

LESSON 8.7

Page 528

For Step 1, graph $y = \sqrt{x^2}$.

Use viewing window [–10, 10] by [–10, 10].

[Y=] [2nd] [x^2] ($\sqrt{\ }$) [X,T,θ,n] [x^2] [)] [GRAPH]

⇑ TI-82: omit

For Step 2, graph $y = \sqrt[4]{x^4}$.

[Y=] [(] [X,T,θ,n] [^] 4 [)] [^] [(] 1 [÷] 4 [)] [GRAPH]

For Steps 3–6, use a keystroke sequence similar to that used in Step 2.

LESSON 8.8

EXAMPLES 1 and 3 For Example 1, graph $y = 2\sqrt{x+5}$ and $y = 8$ on the same screen, and find any points of intersection.

Pages 537 and 538

Use viewing window [–5, 20] by [–2, 15].

Graph the functions:

[Y=] 2 [2nd] [x^2] ($\sqrt{\ }$) [X,T,θ,n] [+] 5 [)] [ENTER] (Y2=) 8 [GRAPH]

Find any points of intersection:

Use a keystroke sequence similar to that used in Example 1 of Lesson 8.5.

For Example 3, use a similar keystroke sequence and viewing window [–2, 10] by [–1, 10].

EXAMPLE 2 Page 537

Graph $y = \sqrt[3]{x-5}$ and $y = \sqrt[3]{7-x}$ on the same screen, and find any points of intersection.

Use viewing window [−2, 10] by [−4, 4].

Enter the functions:

Y= (X,T,θ,n − 5) ^ (1 ÷ 3) ENTER (Y2=)

(7 − X,T,θ,n) ^ (1 ÷ 3) GRAPH

Find any points of intersection:

Use a keystroke sequence similar to that used in Example 1 of Lesson 8.5.

EXAMPLE 4 Page 538

Graph $y = \frac{286}{2\sqrt{x^2 - 15^2}} + 1$ and $y = 6$ on the same screen, and find any points of intersection. Then make a table of values for $y = \frac{286}{2\sqrt{x^2 - 15^2}} + 1$.

Use viewing window [−2, 50] by [−1, 20].

Graph the functions:

Y= 286 ÷ (2 2nd x^2 ($\sqrt{\ }$) X,T,θ,n x^2 − 15 x^2)

↑ TI-82: (

) + 1 ENTER (Y2=) 6 GRAPH

Find any points of intersection:

Use a keystroke sequence similar to that used in Example 1 of Lesson 8.5.

Make a table of values:

Use a keystroke sequence similar to that used in Example 1 of Lesson 8.1. Use (TblStart=) 32.25 and (△ Tbl=) 0.01.

Activity Page 540

Graph $y = \sqrt{x}$ and $y = 1$ on the same screen, and find any points of intersection.

Use viewing window [−5, 5] by [−5, 5].

Use a keystroke sequence similar to that used in Example 1 of Lesson 8.8.

EXAMPLES 6 and 7 Pages 540 and 541

For Example 6, graph $y = \sqrt{x+1}$ and $y = 2$ on the same screen, and find any points of intersection.

Use viewing window [−3, 7] by [−5, 5].

Use a keystroke sequence similar to that used in Example 1 of Lesson 8.8.

For Example 7, use a keystroke sequence similar to that in Example 2 of this lesson. Use viewing window [−9.4, 9.4] by [−6.2, 6.2].

9 Conic Sections

Lesson Presentation CD-ROM
PowerPoint® presentations for each lesson 9.1–9.6

CHAPTER PLANNING GUIDE

Lesson	9.1	9.2	9.3	9.4	9.5	9.6	Project and Review
Pupil's Edition Pages	562–569	570–578	579–585	586–594	595–603	606–613	604–605, 614–621
Practice and Assessment							
Extra Practice (Pupil's Edition)	569	569	570	570	571	571	
Practice Workbook	55	56	57	58	59	60	
Practice Masters Levels A, B, and C	163–165	166–168	169–171	172–174	175–177	178–180	
Standardized Test Practice Masters	63	64	65	66	67	68	69
Assessment Resources	111	112	113	115	116	117	114, 118–123
Visual Resources							
Lesson Presentation Transparencies Vol. 2	33–36	37–40	41–44	45–48	49–52	53–56	
Teaching Transparencies		32	33, 34	35	36		
Answer Key Transparencies	299–303	304–311	312–318	319–327	328–340	341–348	349–359
Quiz Transparencies	9.1	9.2	9.3	9.4	9.5	9.6	
Teacher's Tools							
Reteaching Masters	109–110	111–112	113–114	115–116	117–118	119–120	
Make-Up Lesson Planner for Absent Students	55	56	57	58	59	60	
Student Study Guide	55	56	57	58	59	60	
Spanish Resources	55	56	57	58	59	60	
Block Scheduling Handbook							18–19
Activities and Extensions							
Lesson Activities	55	56	57	58	59	60	
Enrichment Masters	55	56	57	58	59	60	
Cooperative-Learning Activities	55	56	57	58	59	60	
Problem Solving/ Critical Thinking	55	56	57	58	59	60	
Student Technology Guide	55	56	57	58	59	60	
Long Term Projects							33–36
Writing Activities for Your Portfolio							25–27
Tech Prep Masters							41–44
Building Success in Mathematics							22–24

LESSON PACING GUIDE

Lesson	9.1	9.2	9.3	9.4	9.5	9.6	Project and Review
Traditional	2 days	2 days	2 days	2 days	2 days	2 days	2 days
Block	1 day	1 day	1 day	1 day	1 day	1 day	1 day
Two-Year	4 days	4 days	4 days	4 days	4 days	4 days	4 days

CONNECTIONS AND APPLICATIONS

Lesson	9.1	9.2	9.3	9.4	9.5	9.6	Review
Algebra	562–569	570–578	579–586	587–594	595–603	606–613	616–621
Geometry	563, 564, 566, 568, 569	577	584			612	
Transformations	577	584					
Business and Economics					608, 612, 613		
Science and Technology		570, 573, 577	579, 581, 582, 585	588, 591, 593	595	610	618
Sports and Leisure		576, 577					621
Other	564, 567				602		

BLOCK SCHEDULING GUIDE

Day	Lesson	Teacher Directed: Lesson Examples, Teaching Transparencies	Student Guided Activity, Try This	Cooperative-Learning Activity, Lesson Activity, Student Technology Guide	Practice: Practice & Apply, Extra Practice, Practice Workbook	Assessment: Quiz, Mid-Chapter Assessment	Problem Solving, Reteaching
1	9.1	10 min	15 min	15 min	65 min	15 min	15 min
2	9.2	10 min	15 min	15 min	65 min	15 min	15 min
3	9.3	10 min	15 min	15 min	65 min	15 min	15 min
4	9.4	10 min	15 min	15 min	65 min	15 min	15 min
5	9.5	10 min	15 min	15 min	65 min	15 min	15 min
6	9.6	10 min	15 min	15 min	65 min	15 min	15 min
7	Assess.	50 min **PE:** Chapter Review	90 min **PE:** Chapter Project, Writing Activities	90 min Tech Prep Masters	65 min **PE:** Chapter Assessment, Test Generator	30 min Chap. Assess. (A or B), Alt. Assess. (A or B), Test Generator	

PE: Pupil's Edition

Alternative Assessment

The following suggest alternative assessments for students who may benefit from a different type of assessment than the regular chapter quizzes and the mid-chapter/end-of-chapter test. Visit the HRW web site to get additional Alternative Assessment material.

internet connect

Alternative Assessment
Go To: **go.hrw.com**
Keyword: **MB1 Alt Assess**

Performance Assessment

1. Write a paragraph to explain each statement below. If you use a diagram or an equation, be sure to clearly label any elements to which you refer.
   a. An ellipse is very flat when the eccentricity is close to 1.
   b. An ellipse resembles a circle when the eccentricity is close to 0.
2. Use the system $\begin{cases} y = x^2 \\ x^2 + y^2 = r^2 \end{cases}$, where $r > 0$, to answer the following:
   a. Classify the conic sections involved in the system.
   b. Solve the system for $r = 1$, 2, and 3.
   c. Determine whether the following statement is true: The system always has two solutions. Justify your response.

Portfolio Project

Suggest that students choose one of the following projects for inclusion in their portfolios.

1. Make models of conic sections by constructing many cones of the same size and then cutting the cones to make models of a circle, an ellipse, a hyperbola, and a parabola. Tools and materials might include a compass, construction paper, tape, scissors, markers, and the like. Be exact and inventive.
2. Consider $x^n + y^n = 1$ and $x^n - y^n = 1$, where n is a natural number.
   a. Describe the graph of each equation for $n = 1$ and for $n = 2$.
   b. Investigate the graphs of these equations for other values of n. Present your conclusions by using illustrations of any conjectures that you make.

internet connect

The table below identifies the pages in this chapter that contain internet and technology information.

Content Links

Activities Online	**pages 566, 582, 591**
Portfolio Extensions	**pages 578, 594**
Homework Help Online	**pages 567, 577, 583, 592, 601, 611**
Graphic Calculator Support	**page 622**

Resource Links

Parents can go online and find concepts that students are learning–lesson by lesson–and questions that pertain to each lesson, which facilitate parent-student discussion.

Go To: **go.hrw.com**
Keyword: **MB1 Parent Guide**

Technical Support

The following may be used to obtain technical support for any HRW software product.

Online Help: **www.hrwtechsupport.com**
e-mail: **tschrw@hrwtechsupport.com**

HRW Technical Support Center: **(800)323-9239**
7 AM to 10 PM Monday through Friday Central Time

Visit the HRW math web site at: **www.hrw.com/math**

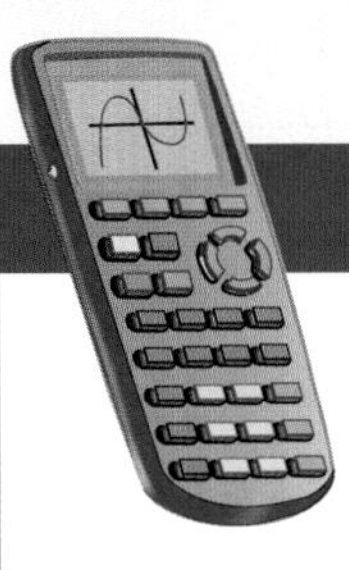

Technology

Lesson Suggestions and Calculator Examples

(Keystrokes are based on a TI-83 calculator.)

Lesson 9.1 Introduction to Conic Sections

A graphics calculator cannot graph an equation that does not represent a function. Discuss with students the strategy of breaking an equation into two equations, as shown below.

$$x^2 + y^2 = 25 \rightarrow \begin{cases} y = \sqrt{25 - x^2} \\ y = -\sqrt{25 - x^2} \end{cases}$$

This skill will be needed throughout the chapter. Also point out that students will need a square viewing window in order to get an accurate picture of a conic section, but rounding errors may produce gaps in some graphs.

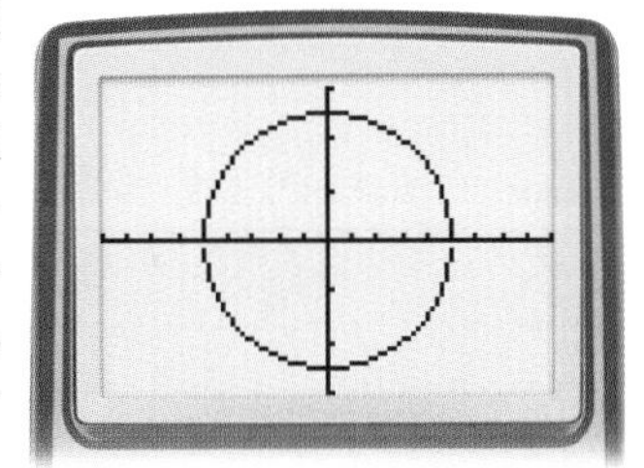

Lesson 9.2 Parabolas

In this lesson, students will explore parabolas that do not represent quadratic functions; that is, parabolas that open to the left or right. The discussion on page 575 describes the graph of the equation $x - 1 = -\frac{1}{8}(y - 4)^2$. Point out to students that the equation must be separated into two related equations because a graphics calculator accepts only functions. Again, rounding errors may produce a gap.

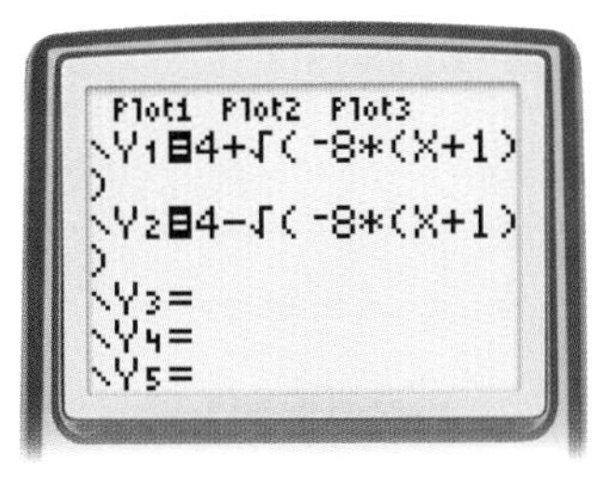

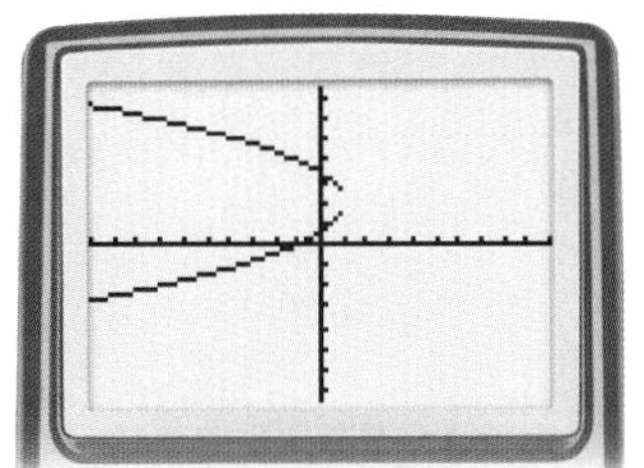

Lesson 9.3 Circles

Refer to Example 3 on page 581. To test whether a given point $P(a, b)$ is outside, inside, or on the circle that has its center at $(25, -30)$ and a radius of 50, students can use the two expressions below.

$$(x - 25)^2 + (y + 30)^2 \text{ and } 50^2$$

Substitute the specific values of a and b for x and y, respectively. If the expression on the left is *greater than* 50^2, P is outside the circle. If the expression on the left is *less than* 50^2, P is inside the circle. If the expression on the left is *equal to* 50^2, P is on the circle.

Lesson 9.4 Ellipses

Students can write an equation for an ellipse by using analysis of known facts and then using a graphics calculator to verify that the equation is correct. The graph of the ellipse with a horizontal major axis 10 units long, a minor axis 6 units long, and a center at (2, 2) is shown at right. Students can use TRACE to verify that the vertices are $(2 \pm 5, 2)$ and that the co-vertices are $(2, \pm 3)$.

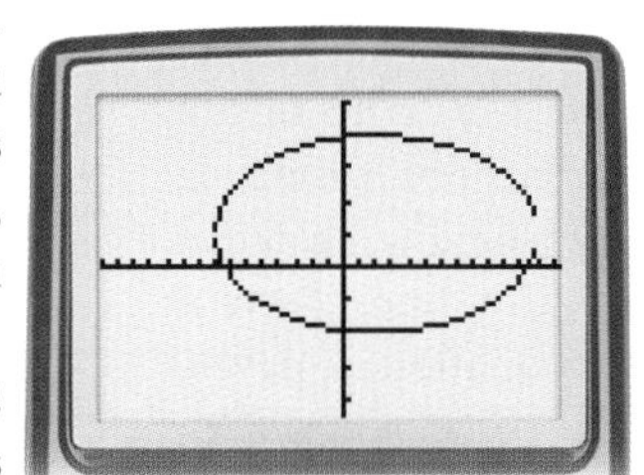

Lesson 9.5 Hyperbolas

To help students reinforce the notion that a hyperbola has asymptotes, ask students to graph $x^2 - y^2 = 1$ and its asymptotes, $y = \pm x$, as shown at right. Alternatively, have students graph the equation of a hyperbola whose transverse and conjugate axes are reversed.

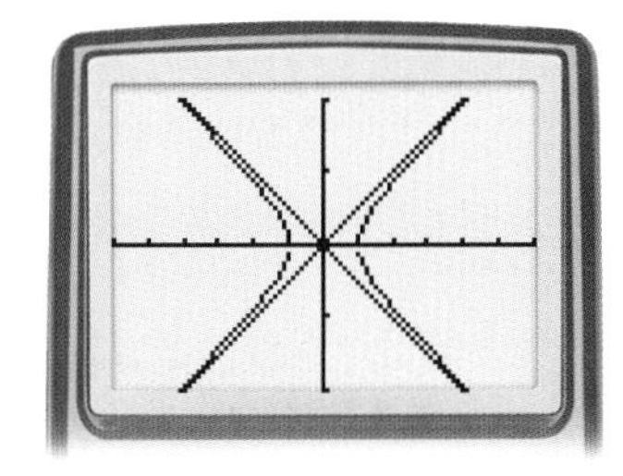

Lesson 9.6 Solving Nonlinear Systems

After the students have completed the Activity on page 606, ask them to design their own graphics calculator activity in order to explore nonlinear systems.

Have students graph the circle $(x + 2)^2 + y^2 = 4$. Then ask them to write and graph an equation of a circle for each condition below.

1. intersects the given circle in two points
2. intersects the given circle in one point
3. does not intersect the given circle

Ask students to apply what they know about translations in order to find the required equation.

For further information, refer to the
- technology discussions in the lessons.
- lesson-related teacher's commentary in the side columns of this *Teacher's Edition.*
- lesson-related *Student Technology Guide* masters.
- *HRW Technology Handbook.*

internet connect

For keystrokes of other graphing calculators models, visit the HRW web site at **go.hrw.com** and enter the keyword **MB1 CALC**.

About Chapter 9

Background Information

In this chapter, students investigate the types of curves formed by the intersection of a plane and a double-napped cone. These curves, called *conic sections*, are the parabola, circle, ellipse, and hyperbola. Each conic section is studied as a figure in a coordinate plane, and it is described in terms of the sum or difference of distances between points or the distance from a point to a line. Then a second-degree equation is developed for each conic section by using the definition of the conic section and the distance formula. Previous knowledge of solving systems of equations is extended to solving systems of nonlinear equations both graphically and algebraically.

CHAPTER RESOURCES

- Block-Scheduling Handbook
- Writing Activities for Your Portfolio
- Tech Prep Masters
- Long-Term Project
- Assessment Resources:
  Mid-Chapter Assessment
  Chapter Assessments
  Alternative Assessments
- Test and Practice Generator
- Technology Handbook

Chapter Objectives

- Classify a conic section as the intersection of a plane and a double cone. [**9.1**]
- Use the distance and midpoint formulas. [**9.1**]
- Write and graph the standard equation of a parabola given sufficient information. [**9.2**]

9

Conic Sections

CONIC SECTIONS ARE CURVES THAT INCLUDE circles, parabolas, ellipses, and hyperbolas. Parabolas are used to describe the shape of satellite dishes. Ellipses describe the paths of planets around the Sun and of comets that return to the solar system on a regular basis, such as Halley's comet. The lens of a telescope can be either parabolic or hyperbolic. The primary mirror in the Hubble Space Telescope is parabolic.

Lessons

About the Photos

One of the earliest examples of the study of conic sections is an eight-volume treatise by Apollonius of Perga (262–190 B.C.E.), who lived in ancient Greece and studied in Alexandria, Egypt. Before the time of Apollonius, the ellipse, parabola, and hyperbola were derived as sections of three distinctly different types of cones. Apollonius systematically showed that all three varieties of conic sections can be formed simply by varying the inclination of a plane when cutting a single cone. Apollonius introduced the double-napped cone used today and associated the names ellipse and hyperbola with the two curves.

Apollonius's mathematical development of an ellipse provided the foundation for the principles of modern astronomy. Johannes Kepler (1571–1630) discovered that the planets move in elliptical orbits, and Sir Isaac Newton (1642–1727) later verified Kepler's theory and created many of the principles of modern astronomy.

About the Chapter Project

In this chapter, you will study the conic sections, their equations, and some of their special properties. In the Chapter Project, *Focus on This!*, you will use an alternative method of graphing to create parabolas, ellipses, and hyperbolas based on their definitions.

After completing the Chapter Project, you will be able to do the following:

- Describe the properties of ellipses, parabolas, and hyperbolas.
- Create ellipses, parabolas, and hyperbolas by using an alternative method of graphing.

About the Portfolio Activities

Throughout the chapter, you will be given opportunities to complete Portfolio Activities that are designed to support your work on the Chapter Project.

- Using wax paper to create a parabolic conic section is included in the Portfolio Activity on page 578.
- Using wax paper to create an elliptical conic section is included in the Portfolio Activity on page 594.
- Using wax paper to create a hyperbolic conic section is included in the Portfolio Activity on page 603.

internet connect

Chapter Internet Features and Online Activities

Lesson	Keyword	Page
9.1	MB1 Homework Help	567
	MB1 Hospital	566
9.2	MB1 Homework Help	577
	MB1 Parabola	578
9.3	MB1 Homework Help	583
	MB1 TV	582
9.4	MB1 Homework Help	592
	MB1 Comet	591
	MB1 Ellipse	594
9.5	MB1 Homework Help	601
9.6	MB1 Homework Help	611

- Given an equation of a parabola, graph it and label the vertex, focus, and directrix. **[9.2]**
- Write an equation for a circle given sufficient information. **[9.3]**
- Given an equation of a circle, graph it and label the radius and the center. **[9.3]**
- Write the standard equation for an ellipse given sufficient information. **[9.4]**
- Given an equation of an ellipse, graph it and label the center, vertices, co-vertices, and foci. **[9.4]**
- Write the standard equation for a hyperbola given sufficient information. **[9.5]**
- Graph the equation of a hyperbola, and identify the center, foci, vertices, and co-vertices. **[9.5]**
- Solve a system of equations containing first- or second-degree equations in two variables. **[9.6]**
- Identify a conic section from its equation. **[9.6]**

Portfolio Activities appear at the end of Lessons 9.2, 9.4, and 9.5. Each serves as preparation for the Chapter Project. The Portfolio Activities as well as the Chapter Project Activities are appropriate for inclusion in the student's portfolio. Students should be encouraged to include in their portfolios any other work in which they feel a sense of pride or a sense of accomplishment.

Prepare

NCTM PRINCIPLES & STANDARDS 1–4, 6–10

QUICK WARM-UP

Solve each equation for y.

1. $3x + y = 7$ $y = -3x + 7$
2. $6x + 3y = -15$ $y = -2x - 5$
3. $2x - 2y = 8$ $y = x - 4$

Solve each equation for x.

4. $x^2 = 16$ $x = \pm 4$
5. $x^2 = 7$ $x = \pm\sqrt{7}$
6. $x^2 - 4 = 11$ $x = \pm\sqrt{15}$

Find the unknown lengths. Round your answers to the nearest hundredth.

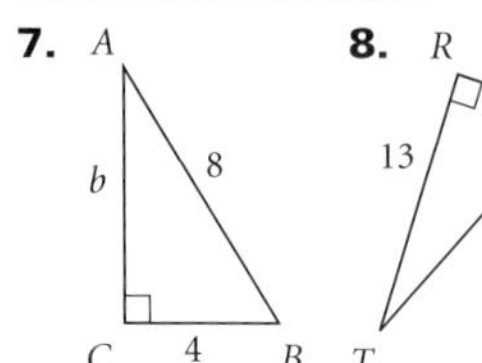

$b \approx 6.93$ $r \approx 13.93$

Also on Quiz Transparency 9.1

Teach

Why Situations that require a calculation of the distance between two points can be modeled with the distance formula. Ask students to discuss how they plan their route to a destination when they are running late. Tell students that the distance formula is used to define all conic sections.

9.1 Introduction to Conic Sections

Objectives

- Classify a conic section as the intersection of a plane and a double cone.
- Use the distance and midpoint formulas.

Why *The lines or curves that can be created by the intersection of a plane and a double cone, called conic sections, are shapes that can be found in the world around us.*

In the diagram shown at right, a slanted line is revolved all the way around a vertical line, the axis, in three-dimensional space. Because the two lines intersect, the result is a pair of cones that have one point in common. Although the diagram cannot show it, the two cones extend indefinitely both upward and downward, forming a double-napped cone.

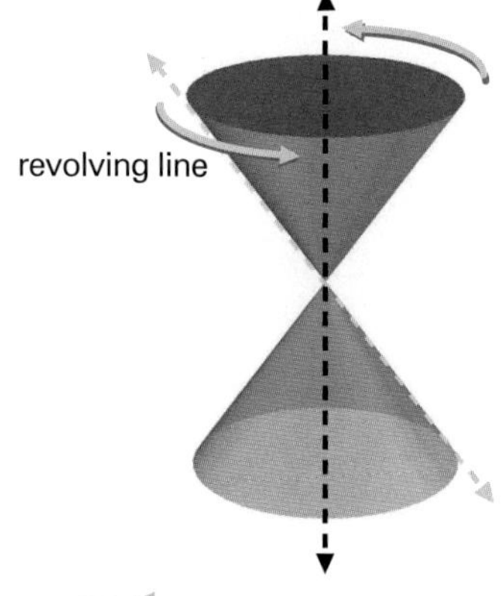

The intersection of a double cone and a plane is called a **conic section**. Three conic sections are illustrated below.

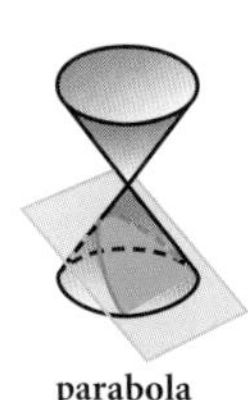
parabola

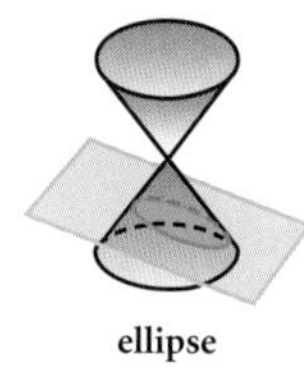
ellipse

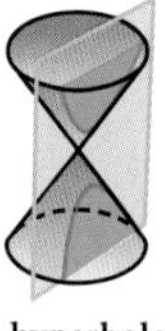
hyperbola

A circle, a point, a line, and a pair of intersecting lines are special cases of the three conic sections shown above.

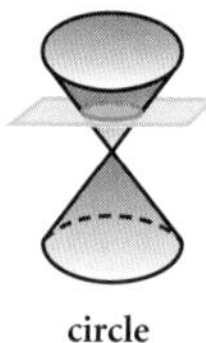
circle (ellipse)

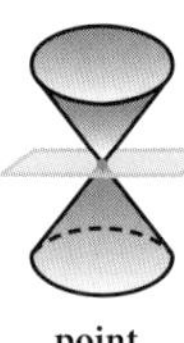
point (ellipse)

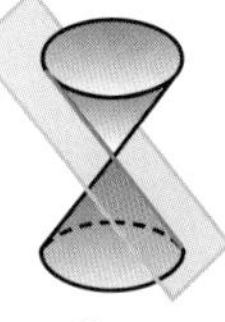
line (hyperbola)

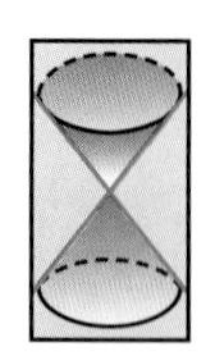
pair of intersecting lines (hyperbola)

Since all conic sections are plane figures, you can represent them in a coordinate plane.

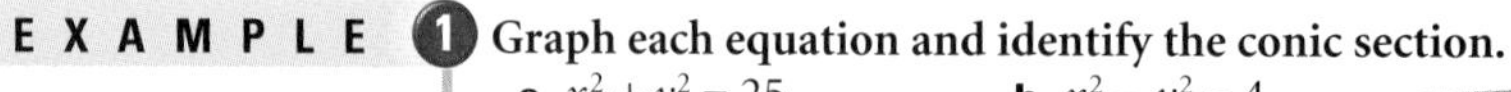
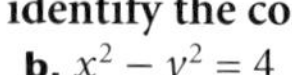
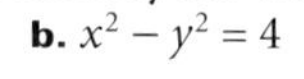

EXAMPLE 1 **Graph each equation and identify the conic section.**

a. $x^2 + y^2 = 25$ **b.** $x^2 - y^2 = 4$

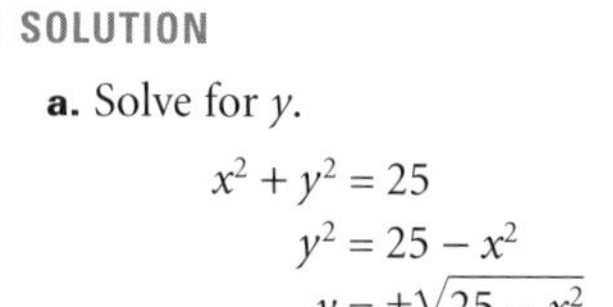

SOLUTION

a. Solve for y.

$$x^2 + y^2 = 25$$
$$y^2 = 25 - x^2$$
$$y = \pm\sqrt{25 - x^2}$$

Graph $y = \sqrt{25 - x^2}$ and $y = -\sqrt{25 - x^2}$ together on the same screen.

Be sure to use a square viewing window.

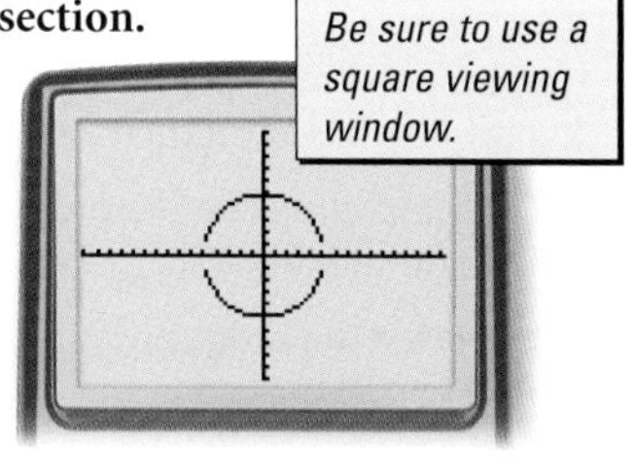

The equation $x^2 + y^2 = 25$ represents a circle.

Alternative Teaching Strategy

USING VISUAL MODELS Tell students that they will study polynomial functions and relations that can be written as second-degree equations of the form $Ax^2 + By^2 + Cx + Dy + E = 0$, where A, B, C, D and E are constants. Equations of this form can be illustrated by the conic sections formed when a double-napped cone is cut by a plane. Discuss the shapes of a parabola, circle, ellipse, and hyperbola. Then have students write an equation from the general form of a second-degree relation by using the values at right. For each equation given, have them solve for y, graph the resulting function(s), and identify the figure as a circle, ellipse, parabola, or hyperbola. Then have students repeat this process for other values. Discuss the results with the class.

1. $A = 1, B = 1, C = 0, D = 0$, and $E = -4$
   $y = \pm\sqrt{4 - x^2}$; circle
2. $A = 1, B = -1, C = 0, D = 0$, and $E = -4$
   $y = \pm\sqrt{x^2 - 4}$; hyperbola
3. $A = 4, B = 25, C = 0, D = 0$, and $E = -100$
   $y = \pm\frac{1}{5}\sqrt{100 - 4x^2}$; ellipse
4. $A = 2, B = 0, C = -3, D = -1$, and $E = -35$
   $y = 2x^2 - 3x - 35$; parabola

b. Solve for y.

$$x^2 - y^2 = 4$$
$$-y^2 = 4 - x^2$$
$$y^2 = x^2 - 4$$
$$y = \pm\sqrt{x^2 - 4}$$

Graph $y = \sqrt{x^2 - 4}$ and $y = -\sqrt{x^2 - 4}$ together on the same screen.

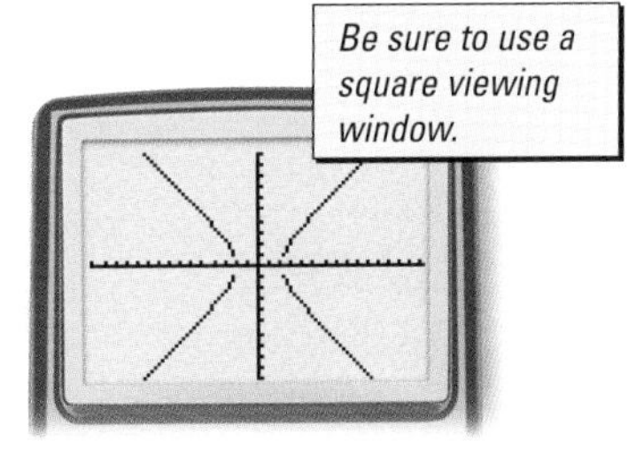

The equation $x^2 - y^2 = 4$ represents a hyperbola.

TRY THIS Graph each equation and identify the conic section.

a. $4x^2 + 9y^2 = 36$ **b.** $6x - y^2 = 0$

Using the Distance Formula

When conic sections are studied as figures in a coordinate plane, they can be described in terms of distances between points or between points and a line. The *distance formula* will play a role in the definition of each conic section studied in this chapter.

COORDINATE GEOMETRY

Given points $P(-1, 4)$ and $Q(3, -2)$ in the coordinate plane, you can use the Pythagorean Theorem to find the distance, d, between them.

$$(PQ)^2 = (PR)^2 + (RQ)^2$$
$$d^2 = 6^2 + 4^2$$
$$d = \sqrt{52}$$
$$\approx 7.2$$

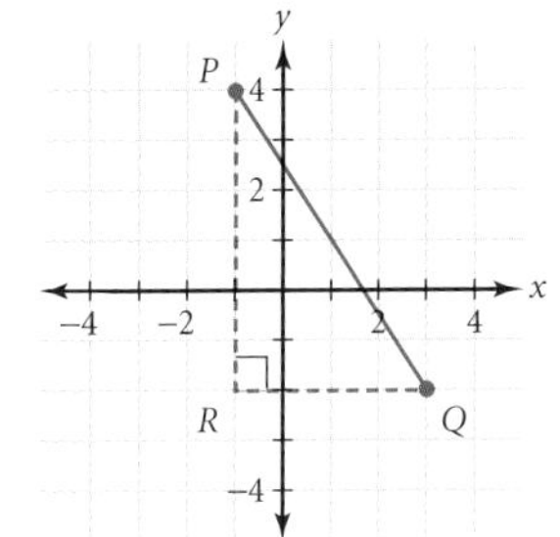

In general, given points $P(x_1, y_1)$ and $Q(x_2, y_2)$ in the coordinate plane, you can use the Pythagorean Theorem to find a formula for the distance, d, between them.

$$(PQ)^2 = (PR)^2 + (RQ)^2$$
$$d^2 = (x_2 - x_1)^2 + (y_2 - y_1)^2$$
$$d = \sqrt{(x_2 - x_1)^2 + (y_2 - y_1)^2}$$

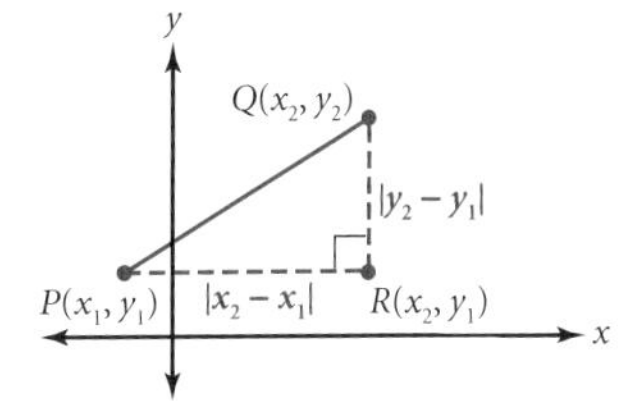

Distance Formula

The distance, d, between $P(x_1, y_1)$ and $Q(x_2, y_2)$ is

$$d = \sqrt{(x_2 - x_1)^2 + (y_2 - y_1)^2}.$$

ADDITIONAL EXAMPLE 1

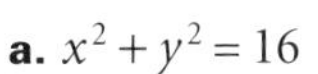

Graph each equation and identify the conic section.

a. $x^2 + y^2 = 16$

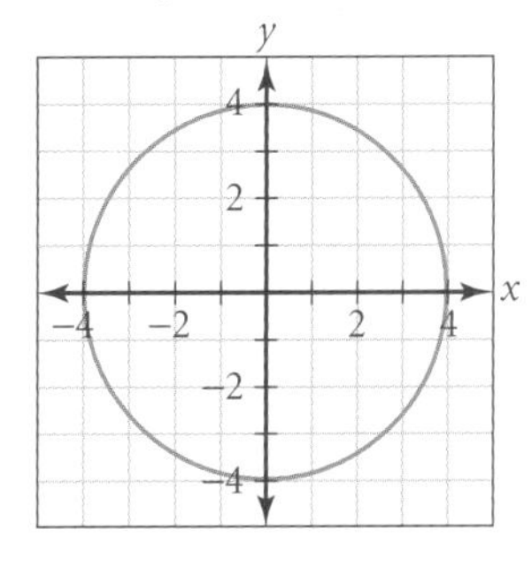

circle

b. $9x^2 + 16y^2 = 144$

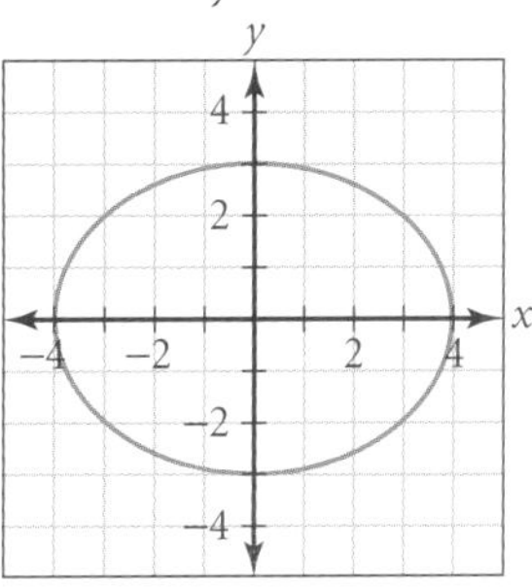

ellipse

Teaching Tip

TECHNOLOGY When using a TI-83 or TI-82 graphics calculator, graph the circle and hyperbola of Example 1 in a square viewing window by pressing 2nd TRACE Zoom and selecting 5:ZSquare.

TRY THIS

a. $y = \pm\frac{\sqrt{36 - 4x^2}}{3}$**; ellipse**

b. $y = \pm\sqrt{6x}$**; parabola**

Interdisciplinary Connection

GEOGRAPHY Tell students that a square grid has been superimposed on a map. One city is near the grid location (3, 19), and a second city is near the grid location (25, 46). The scale on the map indicates that each unit of the grid represents 40 miles. Ask students to estimate the distance between the cities. **≈1393 mi**

Suggest that students obtain a map of their state, superimpose a suitable square grid on it, and use the method described to estimate distances between cities.

Enrichment

In Example 2, the distance between $P(-1, 2)$ and $Q(6, 4)$ is $\sqrt{53}$. Show students that the distance between $P(-1, 2)$ and $Q(-8, 0)$ is also $\sqrt{53}$. Have them name two other possible locations for point Q such that the distance between P and Q is $\sqrt{53}$. **sample: (−8, 4), (6, 0)**

Math CONNECTION

COORDINATE GEOMETRY Remind students that the Pythagorean theorem states that in a right triangle, the length of the hypotenuse, c, is related to the lengths of the legs, a and b, by the equation $c^2 = a^2 + b^2$.

Activity
Exploring the Distance Formula

CONNECTION
COORDINATE GEOMETRY

You will need: no special materials

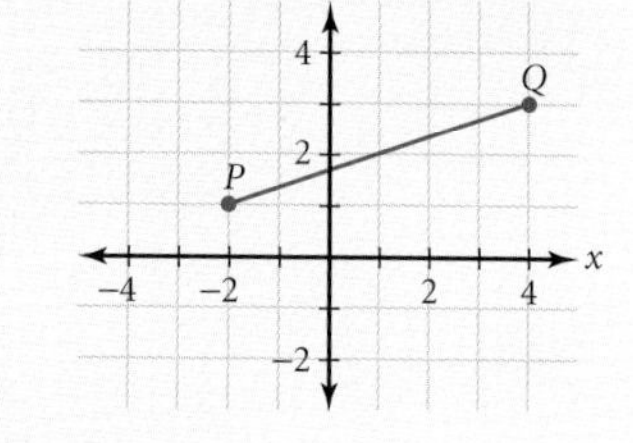

1. Let $P(x_1, y_1)$ be the point $(-2, 1)$ and $Q(x_2, y_2)$ be the point $(4, 3)$ in the graph at right. What is x_1? x_2? $(x_2 - x_1)$? $(x_2 - x_1)^2$? Use the distance formula to find PQ.
2. Let $P(x_1, y_1)$ be the point $(4, 3)$ and $Q(x_2, y_2)$ be the point $(-2, 1)$. What is x_1? x_2? $(x_2 - x_1)$? $(x_2 - x_1)^2$? Use the distance formula to find PQ.
3. Is $(x_2 - x_1)$ the same in Steps 1 and 2? Is $(x_2 - x_1)^2$ the same in Steps 1 and 2? Is PQ the same in Steps 1 and 2?

CHECKPOINT ✓

4. What seems to happen to the distance between P and Q when you relabel points P and Q, as in Steps 1 and 2? Explain why this happens.

EXAMPLE 2

Find the distance between $P(-1, 2)$ and $Q(6, 4)$. Give an exact answer and an approximate answer rounded to the nearest hundredth.

SOLUTION

Let (x_1, y_1) be $(-1, 2)$ and let (x_2, y_2) be $(6, 4)$. Use the distance formula.

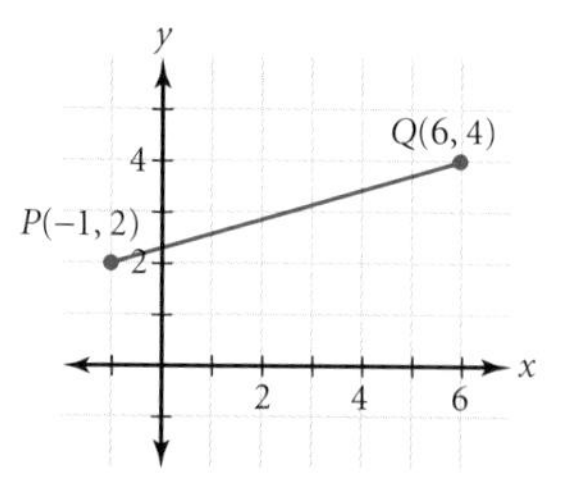

$$d = \sqrt{(x_2 - x_1)^2 + (y_2 - y_1)^2}$$
$$d = \sqrt{[(6 - (-1)]^2 + (4 - 2)^2}$$
$$d = \sqrt{53} \qquad \textit{Exact answer}$$
$$d \approx 7.28 \qquad \textit{Approximate answer}$$

Thus, the distance between P and Q is $\sqrt{53}$, or about 7.28.

TRY THIS Find the distance between $P(2, 5)$ and $Q(-3, -8)$. Give an exact answer and an approximate answer rounded to the nearest hundredth.

EXAMPLE 3

APPLICATION
EMERGENCY SERVICES

An EMS helicopter is stationed at Hospital 1, which is 4 miles west and 2 miles north of an automobile accident. Another EMS helicopter is stationed at Hospital 2, which is 3 miles east and 3 miles north of the accident.

Which helicopter is closer to the accident?

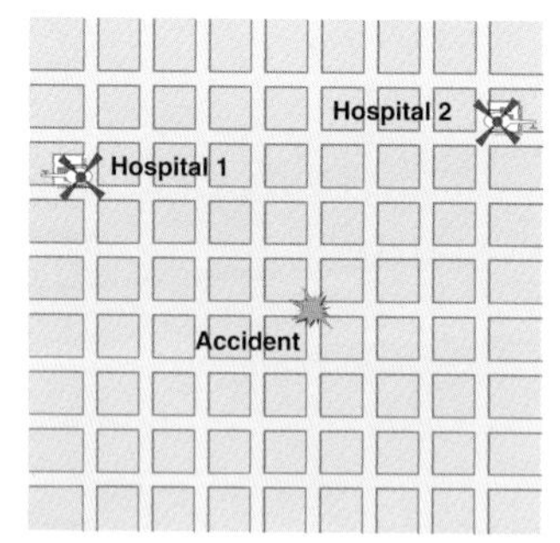

Activity Notes

In this Activity, students are given a pair of points. They are asked to identify x_1, x_2, $x_2 - x_1$, $(x_2 - x_1)^2$, and the distance between the points when the points are given in one order and then in the reverse order. They are asked to explain what happens to the distance between the points when the order of the points is reversed. Students should find that the distance formula gives the same results regardless of the order in which the points are taken. Point out that $(x_1 - x_2)^2$ is the square of the horizontal distance between the points and that $(y_1 - y_2)^2$ is the square of the vertical distance between the points.

Math CONNECTION

GEOMETRY Be sure students notice that they can locate point $A(4, 1)$ in such a way that $\triangle PQA$ is a right triangle. They can then find the length of the two legs by using the Pythagorean theorem.

CHECKPOINT ✓

4. Answers may vary. Sample answer: The distance stays the same. This happens because $(x_2 - x_1)^2 = (x_1 - x_2)^2$ and $(y_2 - y_1)^2 = (y_2 - y_1)^2$.

ADDITIONAL EXAMPLE 2

Find the distance between $S(3, 5)$ and $T(-4, -2)$. Give an exact answer and an approximate answer rounded to the nearest hundredth.
$\sqrt{98} = 7\sqrt{2} \approx 9.90$

TRY THIS
$\sqrt{194}$, or ≈ 13.93

Inclusion Strategies

TACTILE LEARNERS The shape of the light beam from a flashlight is a cone. When that cone of light hits a flat surface, the outline is shaped like one of the conic sections. Have students experiment with generating these conic sections by reflecting the light from a flashlight on a wall of a darkened room.

Have them first hold the flashlight perpendicular to the wall. Ask what conic section they see. **circle** Then have them hold the flashlight at an angle of about 75° to the wall and ask again what conic section they see. **ellipse**

Now have students investigate how to position the flashlight in order to create a parabola on the wall. **Position the flashlight so that a slant height of the cone of light is parallel to the wall and perpendicular to the ceiling or floor.**

Finally, have students hold the flashlight so that it is parallel to the wall. Note that one branch of a hyperbola is formed. Tell them to describe a "flashlight" that could generate the entire hyperbola. **sample: one light between two conical reflectors that are facing in opposite directions**

SOLUTION

1. Represent the locations as points in the coordinate plane.

   Since the locations of the hospitals are given with reference to the accident, place the accident at the origin. Then $A(-4, 2)$ represents the location of Hospital 1 and $B(3, 3)$ represents the location of Hospital 2.

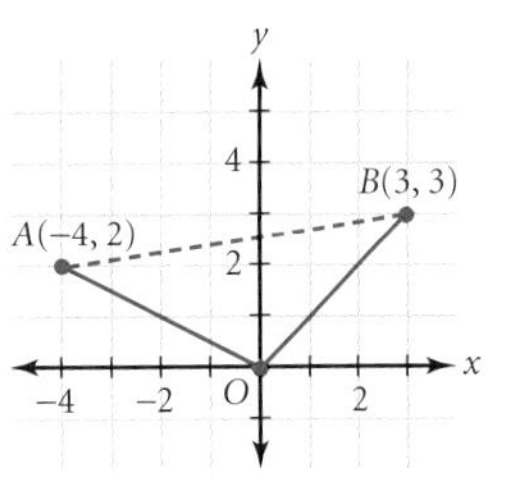

2. Find the distances OA and OB.

$$d = \sqrt{(x_2 - x_1)^2 + (y_2 - y_1)^2}$$
$$OA = \sqrt{[(-4) - 0]^2 + (2 - 0)^2}$$
$$OA = \sqrt{20} \text{ miles, or } \approx 4.47 \text{ miles}$$

$$d = \sqrt{(x_2 - x_1)^2 + (y_2 - y_1)^2}$$
$$OB = \sqrt{(3 - 0)^2 + (3 - 0)^2}$$
$$OB = \sqrt{18} \text{ miles, or } \approx 4.24 \text{ miles}$$

Since $OB < OA$, the helicopter at Hospital 2 is closer to the accident.

Using the Midpoint Formula

The coordinates of the midpoint between two points can be found by using the coordinates of the points.

Midpoint Formula

The coordinates of the midpoint, M, between two points, $P(x_1, y_1)$ and $Q(x_2, y_2)$, are $M\left(\frac{x_1 + x_2}{2}, \frac{y_1 + y_2}{2}\right)$.

Notice that the x-coordinate of M is the average of the x-coordinates of P and Q and that the y-coordinate of M is the average of the y-coordinates of P and Q. The midpoint formula is used in Example 4.

EXAMPLE 4 **Find the coordinates of the midpoint, M, of the line segment whose endpoints are $P(-4, 6)$ and $Q(5, 10)$.**

SOLUTION

Let (x_1, y_1) be $(-4, 6)$ and (x_2, y_2) be $(5, 10)$.

$$\frac{x_1 + x_2}{2} = \frac{(-4) + 5}{2} = \frac{1}{2} \qquad \frac{y_1 + y_2}{2} = \frac{6 + 10}{2} = 8$$

Thus, $M\left(\frac{1}{2}, 8\right)$ is the midpoint.

The graph of points P, M, and Q indicates that the answer is reasonable.

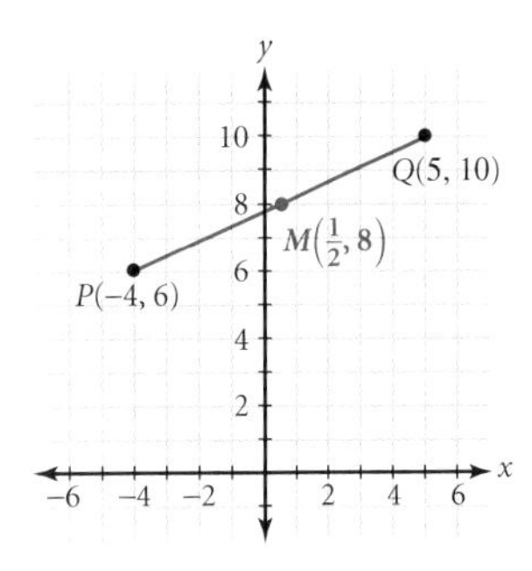

TRY THIS Find the coordinates of the midpoint, M, of the line segment whose endpoints are $A(2.5, 5.5)$ and $B(8.5, -4.5)$.

ADDITIONAL EXAMPLE 3

Jan lives 5 blocks west and 3 blocks south of Pison Junior High School. Amy lives 2 blocks east and 6 blocks north of the same school. Both Jan and Amy can reach the school by a straight route. **Which person, Jan or Amy, lives closest to the school?** Jan

ADDITIONAL EXAMPLE 4

Find the coordinates of the midpoint, M, of the line segment whose endpoints are $S(-2, 1)$ and $T(8, 3)$.
$M(3, 2)$

TRY THIS
(5.5, 0.5)

Reteaching the Lesson

USING PATTERNS Give students the coordinates of the vertices of a triangle drawn in the coordinate plane, such as $A(-3, 4)$, $B(5, 6)$, and $C(3, 2)$. Have them find the midpoint of each side of the triangle and connect the midpoints to form a new inscribed triangle. Have students find the vertices and lengths of the sides of the original triangle and the inscribed triangle, and ask them to describe how the inscribed triangle is related to the original triangle. Have students repeat the process for a rectangle.

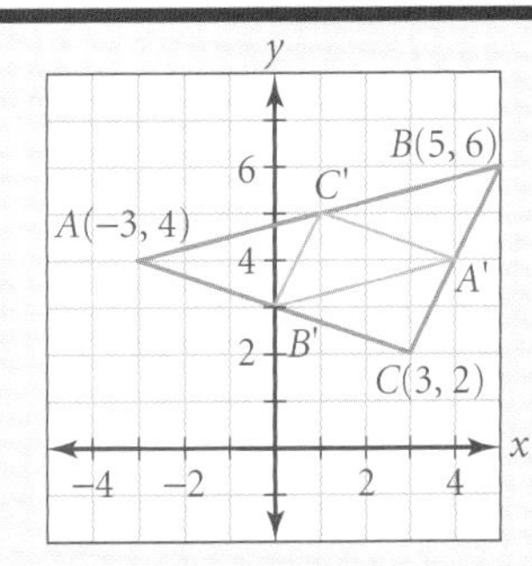

Vertices: $A'(4, 4)$, $B'(0, 3)$, $C'(1, 5)$
Length: $AB = 2\sqrt{17}$, $AC = 2\sqrt{10}$, $BC = 2\sqrt{5}$
$A'B' = \sqrt{17}$, $A'C' = \sqrt{10}$, $C'B' = \sqrt{5}$
The inscribed triangle is similar to the original triangle, but the lengths of its sides are half the length of the sides of the original triangle.

CRITICAL THINKING A line segment has the endpoints $A(-5, -6)$ and $B(7, 4)$. Find the coordinates of points R, S, and T on segment AB such that $AR = RS = ST = TB$.

Recall from geometry that a chord of a circle is a segment with endpoints on the circle. A **diameter** of a circle is a chord that contains the center. A **radius** of a circle is a segment with one endpoint at the center of the circle and the other endpoint on the circle. The length of the radius is one-half the length of the diameter.

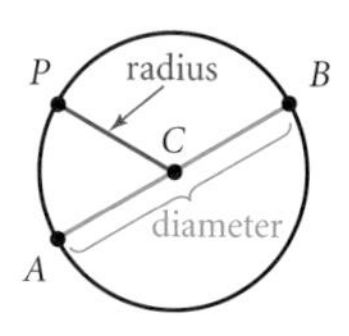

EXAMPLE 5 The endpoints of a diameter of the circle at right are $A(-2, 3)$ and $B(6, 9)$.

Find the center, circumference, and area of the circle.

SOLUTION

1. Use the midpoint formula to find the coordinates of the center, C.
   $C\left(\frac{x_1 + x_2}{2}, \frac{y_1 + y_2}{2}\right) \rightarrow C\left(\frac{-2 + 6}{2}, \frac{3 + 9}{2}\right) \rightarrow C(2, 6)$
2. Use the distance formula to find the radius.
   $CA = \sqrt{(-2 - 2)^2 + (3 - 6)^2} = \sqrt{25} = 5$ — *You can find CA or CB to find the radius.*
3. Find the circumference and area.
   Circumference $= 2\pi r = 2r(5) = 10\pi$
   Area $= \pi r^2 = \pi(5)^2 = 25\pi$

Thus, the center is $(2, 6)$, the circumference is 10π units, and the area is 25π square units.

TRY THIS The endpoints of a diameter of a circle are $A(-3, 2)$ and $B(5, 4)$. Find the center, circumference, and area of the circle.

Exercises

Communicate

1. Illustrate and explain how a plane can intersect a double cone to produce a circle, a parabola, an ellipse, and a hyperbola.
2. Explain why it does not matter which set of coordinates you subtract from the other set when using the distance formula.
3. Explain how to find the circumference and the area of a circle if you are given the coordinates of the center and of a point on the circle.

CRITICAL THINKING
$R(-2, -3.5)$, $S(1, -1)$, $T(4, 1.5)$

Math CONNECTION

GEOMETRY Remind students that a *circle* is defined as the set of all points in a plane that are a given distance from a point in the plane. The point is called the *center* of the circle.

ADDITIONAL EXAMPLE 5

The endpoints of a diameter of a circle have the coordinates $F(4, -3)$ and $G(-1, 9)$. **Find the center, the circumference, and the area of the circle.**
center: (1.5, 3)
circumference: 13π units
area: 42.25π units^2

TRY THIS
center: (1, 3);
$C = 2\sqrt{17}\pi$ units;
$A = 17\pi$ units^2

Assess

Selected Answers
Exercises 4–9, 11–71 odd

ASSIGNMENT GUIDE

In Class	1–9
Core	11–57 odd
Core Plus	10–58 even, 59
Review	60–72
Preview	73

Extra Practice can be found beginning on page 940.

Guided Skills Practice

Graph each equation and identify the conic section. ***(EXAMPLE 1)***

4 $9x^2 - 4y^2 = 100$ **hyperbola**

5 $x^2 + y^2 = 100$ **circle**

6. Find the distance between $P(-7, 12)$ and $Q(6, 5)$. Give an exact answer and an approximate answer rounded to the nearest hundredth. ***(EXAMPLE 2)***

7. EMERGENCY SERVICES If a fire station is 3 miles east and 2 miles south of a fire and the trucks can travel along a straight route to the fire from the station, how long is their route? ***(EXAMPLE 3)***

8. Find the coordinates of the midpoint, M, of the line segment whose endpoints are $P(-3, 10)$ and $Q(6, -12)$. ***(EXAMPLE 4)***

9. The endpoints of a diameter of a circle are $A(-5, 2)$ and $B(3, 4)$. Find the center, circumference, and area of the circle. ***(EXAMPLE 5)***

6. $\sqrt{218} \approx 14.76$

APPLICATION

7. $\sqrt{13} \approx 3.61$ miles

8. $M\left(\frac{3}{2}, -1\right)$

9. $(-1, 3)$; $C = 2\pi\sqrt{17}$; $A = 17\pi$

Practice and Apply

Graph each equation and identify the conic section.

10 $x^2 + y^2 = 36$ **circle**

11 $x^2 + y^2 = 121$ **circle**

12 $x^2 - y^2 = 36$ **hyperbola**

13 $4x^2 - 9y^2 = 36$ **hyperbola**

14 $5x^2 + 9y^2 = 45$ **ellipse**

15 $36x^2 + 5y^2 = 180$ **ellipse**

16 $x^2 - 4y = 0$ **parabola**

17 $y^2 + 4x = 0$ **parabola**

18 $y^2 - 4x = 0$ **parabola**

19 $x^2 + y^2 = 1$ **circle**

20 $x^2 + 9y^2 = 9$ **ellipse**

21 $25x^2 - 9y^2 = 225$ **hyperbola**

Find the distance between P and Q and the coordinates of M, the midpoint of $\overline{PQ}$. Give exact answers and approximate answers to the nearest hundredth when appropriate.

22. $P(-3, -2)$ and $Q(5, -2)$

23. $P(-5, -2)$ and $Q(-5, 5)$

24. $P(7, 0)$ and $Q(8, 0)$

25. $P(-3, -4)$ and $Q(5, -5)$

26. $P(7, -2)$ and $Q(-5, -1)$

27. $P(\frac{5}{2}, 3)$ and $Q(-\frac{3}{2}, 2)$

28. $P(10, 7)$ and $Q(1, 8)$

29. $P(10, 2)$ and $Q(8, 0)$

30. $P(7.6, 10.1)$ and $Q(4.6, 3.1)$

31. $P(2\sqrt{2}, 5)$ and $Q(\sqrt{2}, 0)$

32. $P\left(\frac{1}{2}, \frac{7}{8}\right)$ and $Q\left(3, -\frac{3}{8}\right)$

33. $P\left(\frac{3}{2}, \frac{1}{4}\right)$ and $Q\left(-6, \frac{3}{4}\right)$

34. $P(5, 5\sqrt{2})$ and $Q(6, \sqrt{2})$

35. $P(2\sqrt{2}, \sqrt{7})$ and $Q(\sqrt{2}, 5\sqrt{7})$

36. $P(0, 0)$ and $Q(a, 2a + 1)$

37. $P(2a, a)$ and $Q(a, -3a)$

22. 8; $M(1, -2)$

23. 7; $M\left(-5, \frac{3}{2}\right)$

24. 1; $M\left(\frac{15}{2}, 0\right)$

25. $\sqrt{65} \approx 8.06$; $M\left(1, -\frac{9}{2}\right)$

26. $\sqrt{145} \approx 12.04$; $M\left(1, -\frac{3}{2}\right)$

internet connect
go hrw .com
Homework Help Online
Go To: go.hrw.com
Keyword:
MB1 Homework Help
for Exercises 22–37

27. $\sqrt{17} \approx 4.12$; $M\left(\frac{1}{2}, \frac{5}{2}\right)$

28. $\sqrt{82} \approx 9.06$; $M\left(\frac{11}{2}, \frac{15}{2}\right)$

29. $2\sqrt{2} \approx 2.83$; $M(9, 1)$

Error Analysis

Some students mistakenly find the difference between the coordinates rather than the average when finding the coordinates of the midpoint. Remind students that the midpoint coordinates are found by averaging the x-coordinates and averaging the y-coordinates. Therefore, they should find the sum of the x-coordinates and then divide by 2.

Reinforce the proper use of addition and subtraction when using the distance formula by reminding students that $|x_2 - x_1|$ represents the horizontal distance between the two points, while $|y_2 - y_1|$ represents the vertical distance between the two points. The vertical and horizontal distances are squared and added together before the square root is taken. This is because the distance formula is based on the Pythagorean theorem, $a^2 + b^2 = c^2$, where a represents the horizontal distance and b represents the vertical distance between the two points.

30. $\sqrt{58} \approx 7.62$; $M(6.1, 6.6)$

31. $3\sqrt{3} \approx 5.20$; $M\left(\frac{3\sqrt{2}}{2}, \frac{5}{2}\right)$, or $\approx M(2.12, 2.5)$

32. $\frac{5\sqrt{5}}{4} \approx 2.80$; $M\left(\frac{7}{4}, \frac{1}{4}\right)$

33. $\sqrt{\frac{113}{2}} \approx 7.52$; $M\left(-\frac{9}{4}, \frac{1}{2}\right)$

34. $\sqrt{33} \approx 5.74$; $M\left(\frac{11}{2}, 3\sqrt{2}\right)$, or $\approx M(5.5, 4.24)$

35. $\sqrt{114} \approx 10.68$; $M\left(\frac{3\sqrt{2}}{2}, 3\sqrt{7}\right)$, or $\approx M(2.12, 7.94)$

36. $\sqrt{5a^2 + 4a + 1}$; $M\left(\frac{a}{2}, a + \frac{1}{2}\right)$

37. $|a|\sqrt{17} \approx 4.12|a|$; $M\left(\frac{3a}{2}, -a\right)$

Find the center, circumference, and area of each circle described below.

38. diameter with endpoints $P(-4, -3)$ and $Q(-10, 5)$ **$(-7, 1)$; $C = 10\pi$; $A = 25\pi$**

39. diameter with endpoints $P(2, 4)$ and $Q(-3, 16)$ **$\left(-\frac{1}{2}, 10\right)$; $C = 13\pi$; $A = \frac{169\pi}{4}$**

40. diameter with endpoints $P(0, 0)$ and $Q(50, 50)$

41. diameter with endpoints $P(-12, -8)$ and $Q(0, 0)$

42. diameter with endpoints $P(-2, 2)$ and $Q(4, 6)$ **$(1, 4)$; $C = 2\pi\sqrt{13}$; $A = 13\pi$**

43. diameter with endpoints $P(8, -3)$ and $Q(-2, 1)$ **$(3, -1)$; $C = 2\pi\sqrt{29}$; $A = 29\pi$**

40. $(25, 25)$; $C = 50\pi\sqrt{2}$; $A = 1250\pi$

41. $(-6, -4)$; $C = 4\pi\sqrt{13}$; $A = 52\pi$

For $\overline{PQ}$, the coordinates of P and M, the midpoint of $\overline{PQ}$, are given. Find the coordinates of Q.

44. $P(-2, 3)$ and $M(5, 1)$ **$(12, -1)$**

45. $P(2, -3)$ and $M(-5, 1)$ **$(-12, 5)$**

46. $P(3, 11)$ and $M(0, 0)$ **$(-3, -11)$**

47. $P(0, 0)$ and $M(-7, 7)$ **$(-14, 14)$**

Three points, A, B, and C, are collinear if they lie on the same line. If A, B, and C are collinear and B is between A and C, then $AB + BC = AC$. For each set of points A, B, and C given below, find AB, BC, and AC, and determine whether the three points are collinear.

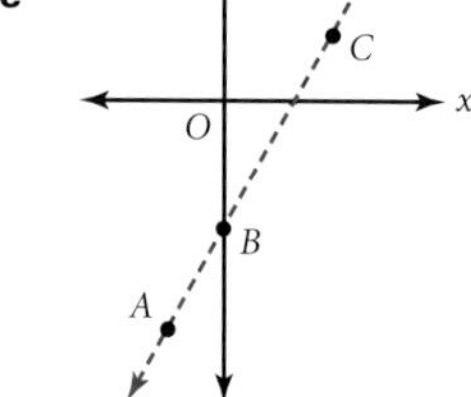

48. $A(0, 0)$, $B(2, 4)$, and $C(3, 6)$

49. $A(0, 0)$, $B(2, 4)$, and $C(3, 7)$

50. $A(-4, 5)$, $B(-2, 2)$, and $C(0, -1)$

51. $A(-3, 1)$, $B(2, 9)$, and $C(7, 17)$

48. $AB = 2\sqrt{5}$; $BC = \sqrt{5}$; $AC = 3\sqrt{5}$; collinear

49. $AB = 2\sqrt{5}$; $BC = \sqrt{10}$; $AC = \sqrt{58}$; not collinear

50. $AB = \sqrt{13}$; $BC = \sqrt{13}$; $AC = 2\sqrt{13}$; collinear

51. $AB = \sqrt{89}$; $BC = \sqrt{89}$; $AC = 2\sqrt{89}$; collinear

CHALLENGE

52. You are given $P(-3, 2)$, $Q(3, 1)$, and $R(0, 6)$. Find the coordinates of point A such that $PA = QA = RA$, that is, $(PA)^2 = (QA)^2 = (RA)^2$.

52. $\left(\frac{11}{54}, \frac{49}{18}\right)$

CONNECTIONS

COORDINATE GEOMETRY **For Exercises 53–55, refer to the coordinate plane at right.**

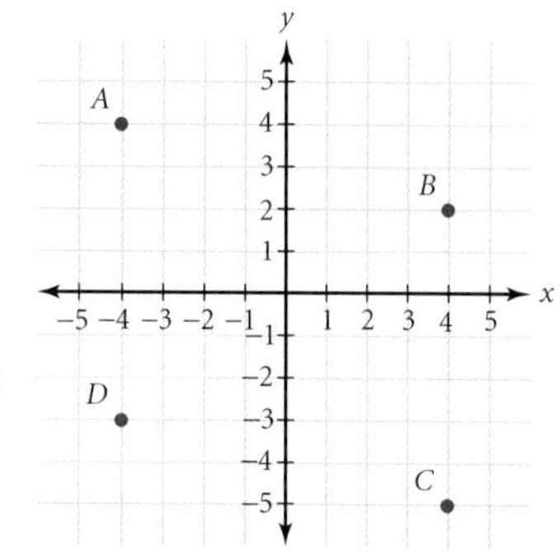

53. a. Find AB, BC, CD, and DA.

b. Based on your answers to part **a**, are the opposite sides of quadrilateral $ABCD$ equal in length?

54. a. Find the coordinates of the midpoints of $\overline{AC}$ and of $\overline{BD}$.

b. Based on your answers to part **a**, what conclusion can you draw about the diagonals of quadrilateral $ABCD$?

55. a. Find the slopes of $\overline{AB}$ and $\overline{CD}$ and of $\overline{AD}$ and $\overline{BC}$.

b. Based on your answers to part **a**, are the opposite sides of quadrilateral $ABCD$ parallel? Explain.

56. COORDINATE GEOMETRY Determine whether the triangle with vertices at $C(-2, 3)$, $D(1, -1)$, and $E(3, 3)$ is isosceles, equilateral, or both.

57. COORDINATE GEOMETRY Determine whether the triangle with vertices at $F(7, 3)$, $H(6, 9)$, and $L(2, 3)$ is isosceles or scalene. (Hint: No two sides of a scalene triangle are equal in length.)

53a. $AB = 2\sqrt{17}$; $BC = 7$; $CD = 2\sqrt{17}$; $DA = 7$

b. yes

54a. AC: $M\left(0, -\frac{1}{2}\right)$; BD: $M\left(0, -\frac{1}{2}\right)$

b. The diagonals intersect at their common midpoint.

55a. AB: $m = -\frac{1}{4}$

CD: $m = -\frac{1}{4}$

AD: m is undefined

BC: m is undefined

b. Yes, because AB and CD have the same slope and AD and BC are both vertical.

56. isosceles ($CD = CE \neq DE$)

57. scalene

Practice

NAME ______ CLASS ______ DATE ______

Practice
***9.1* Introduction to Conic Sections**

Solve each equation for y, graph the resulting equation, and identify the conic section.

1. $x^2 - 3y = 0$ — $y = \frac{1}{3}x^2$

2. $x^2 + y^2 = 400$ — $y = \pm\sqrt{400 - x^2}$

3. $9x^2 - y^2 = 9$ — $y = \pm 3\sqrt{x^2 - 1}$

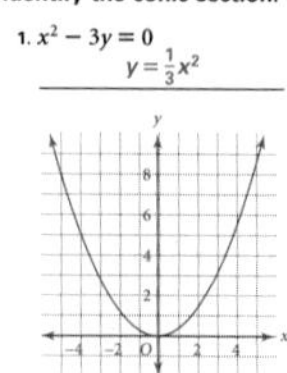

parabola

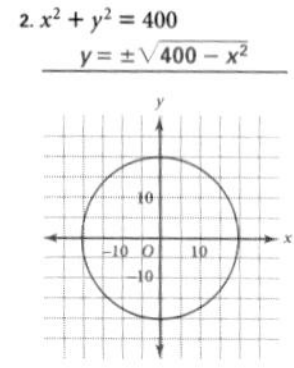

circle

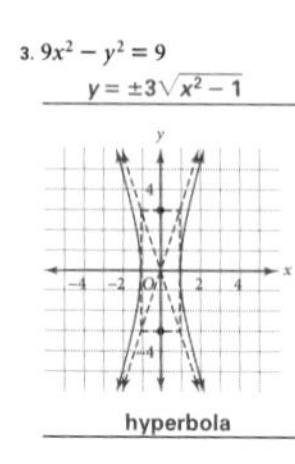

hyperbola

Find the distance between P and Q, and find the coordinates of M, the midpoint of $\overline{PQ}$. Give exact answers and approximate answers to the nearest hundredth when appropriate.

4. $P(0, 0)$ and $Q(5, 12)$ — 13; $(2.5, 6)$

5. $P(4, 1)$ and $Q(12, -5)$ — 10; $(8, -2)$

6. $P(12, 4)$ and $Q(-8, 2)$ — $2\sqrt{101} \approx 20.10$; $(2, 3)$

7. $P(7.5, 3)$ and $Q(-1.5, 5)$ — $\sqrt{85} \approx 9.22$; $(3, 4)$

8. $P(-8, -8)$ and $Q(4, 4)$ — $12\sqrt{2} \approx 16.97$; $(-2, -2)$

9. $P(-1, -1)$ and $Q(1, 2)$ — $\sqrt{13} \approx 3.61$; $(0, 0.5)$

Find the center, circumference, and area of the circle whose diameter has the given endpoints.

10. $P(6, 20)$ and $Q(12, 8)$ — $(9, 14)$; 10π; 25π

11. $P(0, 0)$ and $Q(9, 40)$ — $(4.5, 20)$; 41π; $\frac{1681\pi}{4}$

12. $P(4, 16)$ and $Q(-4, 1)$ — $(0, 8.5)$; 17π; $\frac{289\pi}{4}$

13. $P(3, 7)$ and $Q(4, -5)$ — $(3.5, 1)$; $\pi\sqrt{145}$; $\frac{145\pi}{4}$

14. $P(10, 5)$ and $Q(20, 6)$ — $(15, 5.5)$; $\pi\sqrt{101}$; $\frac{101\pi}{4}$

15. $P(-8, 8)$ and $Q(13, -3)$ — $(2.5, 2.5)$; $\pi\sqrt{562}$; $\frac{562\pi}{4}$

CONNECTION

58. $y = \frac{3}{10}x + \frac{41}{10}$

CHALLENGE

59. $y = \frac{7}{5}x - \frac{4}{5}$

58. GEOMETRY A *midsegment* of a triangle is the line segment whose endpoints are the midpoints of two sides of the triangle. A triangle has vertices whose coordinates are $A(0, 0)$, $B(6, 10)$, and $C(10, 3)$. Find an equation for the line containing the midsegment formed by the midpoints of $\overline{AB}$ and $\overline{BC}$.

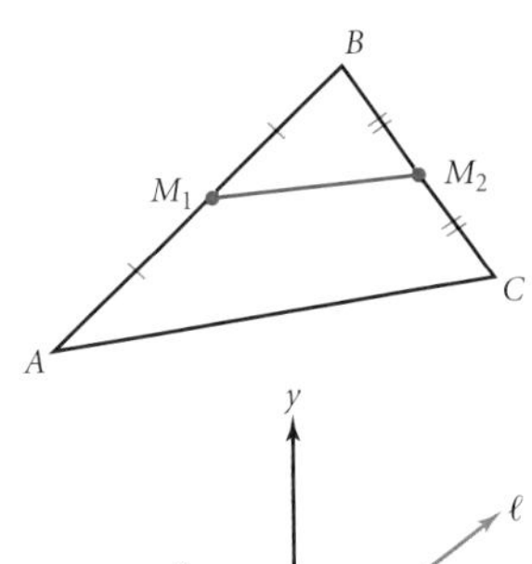

59. COORDINATE GEOMETRY The perpendicular bisector of $\overline{PQ}$ in a plane is the line, ℓ, that contains the midpoint of $\overline{PQ}$ and is perpendicular to $\overline{PQ}$. If B is on ℓ, then $BP = BQ$ and $(BP)^2 = (BQ)^2$. Write the equation in standard form for the perpendicular bisector of $\overline{PQ}$ for $P(-5, 7)$ and $Q(9, -3)$.

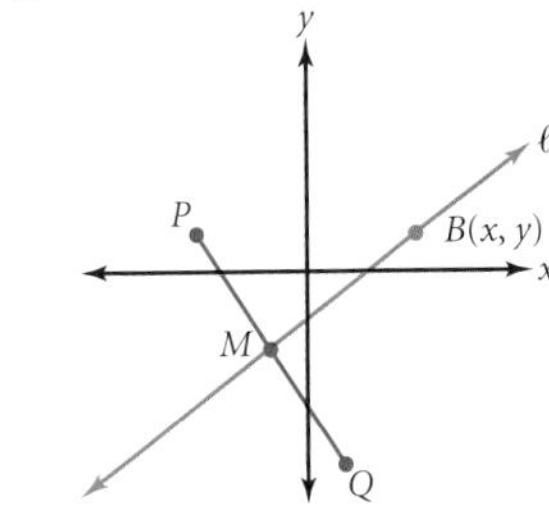

Look Back

Evaluate. *(LESSON 2.6)*

60. $\lceil -5.1 \rceil - \lceil -3.5 \rceil$ **−2** **61.** $[-0.33] + [2.99]$ **1** **62.** $\lceil 4.4 \rceil + \lfloor -0.5 \rfloor$ **4**

Solve by completing the square. Give exact solutions. *(LESSON 5.4)*

63. $x^2 = 8x - 15$ **3, 5** **64.** $x^2 - 8x = 48$ **−4, 12** **65.** $x^2 - 6x - 20 = 0$ **$3 \pm \sqrt{29}$**

Solve by using the quadratic formula. *(LESSON 5.5)*

66. $x^2 - 5x = 50$ **−5, 10** **67.** $6x^2 - 7x = -1$ **$\frac{1}{6}$, 1** **68.** $2a^2 - 7a + 6 = 0$ **$\frac{3}{2}$, 2**

Divide by using synthetic division. *(LESSON 7.3)*

69. $(4x^4 - 11x^3 + 8x^2 - 3x - 2) \div (x - 2)$ **$4x^3 - 3x^2 + 2x + 1$**
70. $(5x^4 + 5x^3 + x^2 + 2x + 1) \div (x + 1)$ **$5x^3 + x + 1$**

Let z vary jointly as x and y. Use the given information to find the value of z for the given values of x and y. *(LESSON 8.1)*

71. $z = 3$ when $x = 2$ and $y = 3$; given $x = 5$ and $y = 10$ **25**
72. $z = 8$ when $x = 1$ and $y = 4$; given $x = 6$ and $y = 8$ **96**

73. Let $f(x) = x^2$. The point $(1, 1)$ is on the graph of f.
a. Find the distance between the point $(1, 1)$ and the point $\left(0, \frac{1}{4}\right)$.
b. Find the shortest distance between the point $(1, 1)$ and the line $y = -\frac{1}{4}$.
c. Compare the two distances found in parts **a** and **b**.
d. Pick another point on the graph of f. Repeat parts **a**, **b**, and **c** for your point.
e. Make a conjecture about the relationship between points on the graph of f, the point $\left(0, \frac{1}{4}\right)$, and the line $y = -\frac{1}{4}$.

In Exercise 73, students should discover that every point of the parabola is equidistant to the point $\left(0, \frac{1}{4}\right)$ and the line $y = -\frac{1}{4}$. In Lesson 9.2, they will learn that the point is the *focus* of the parabola, the line is the *directrix*, and both the focus and directrix are required to define a parabola in analytic geometry. A parabola is the set of all points in the plane equidistant to a fixed point (the focus) and a fixed line (the directrix).

73a. $\frac{5}{4}$

b. $\frac{5}{4}$

c. They are the same.

d. Answers may vary. Sample answer: Using $(2, 4)$, both distances are $\frac{17}{4}$.

e. For any point (x, x^2) on the graph of f, the distance between the point and $\left(0, \frac{1}{4}\right)$ is the same as the distance between the point and the line $y = -\frac{1}{4}$.

Student Technology Guide

NAME ______ CLASS ______ DATE ______

Student Technology Guide
9.1 Introduction to Conic Sections

Many conic sections are not functions. For this reason, you need to use a special method to graph such conic sections on a graphics calculator.

To make an undistorted graph of a conic section, use a square viewing window. You can do this by pressing ZOOM 5:ZSquare ENTER or by pressing WINDOW and changing Xmin, Xmax, Ymin, and Ymax so that the range for the x-values is about 1.5 times the range for the y-values.

Example: Graph $x^2 + y^2 = 36$. Identify the conic section it represents.

- First solve for y: $y = \sqrt{36 - x^2}$ or $y = -\sqrt{36 - x^2}$.
- Enter the first equation as Y1. Press Y= 2nd x^2 36 − X,T,θ,n x^2).
- Set Y2 equal to the opposite of Y1 by moving to Y2 and pressing (−) VARS Y-VARS 1:Function... ENTER 1:Y1 ENTER.
- Press GRAPH. The conic section is a circle.

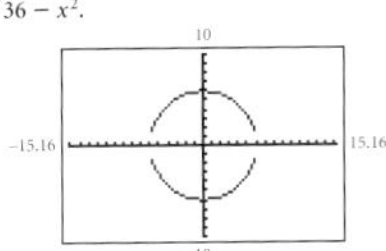

Solve each equation for y, graph the resulting equation, and identify the conic section.

1. $x^2 + y^2 = 81$ **$y = \pm\sqrt{81 - x^2}$; circle**
2. $x^2 + 4y^2 = 36$ **$y = \pm\sqrt{9 - \frac{x^2}{4}}$; ellipse**
3. $2x^2 - y^2 = 20$ **$y = \pm\sqrt{2x^2 - 20}$; hyperbola**
4. $16x^2 + 9y^2 = 144$ **$y = \pm\sqrt{16 - \frac{16x^2}{9}}$; ellipse**

You can also use a graphics calculator to check the midpoint of a line segment.

Example: Given points $P(-6, 7)$ and $Q(2, 4)$, find the coordinates of M, the midpoint of $\overline{PQ}$.

- Use the midpoint formula.
  $M\left(\frac{-6+2}{2}, \frac{7+4}{2}\right) = M(-2, 5.5)$
- As a check, use a drawing command to plot points. First, press Y= and clear any equations.
- Press 2nd PRGM POINTS 1:Pt-On(ENTER (−) 6 , 7) ALPHA : . Repeat two more times, entering the coordinates of Q and M. Press ENTER. $(-2, 5.5)$ appears to be the midpoint of $\overline{PQ}$.

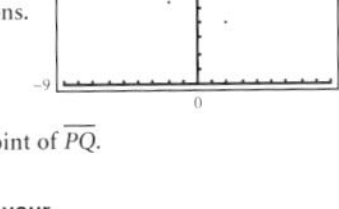

Find the coordinates of M, the midpoint of $\overline{PQ}$. Check your answers by graphing.

5. $P(-2, 3)$ and $Q(6, 9)$ **(2, 6)**
6. $P(-6, -3)$ and $Q(-6, 11)$ **(−6, 4)**

Prepare

NCTM PRINCIPLES & STANDARDS 1–4, 6–10

QUICK WARM-UP

The graph of each function given below is a parabola. **For each parabola, find an equation for the axis of symmetry and the coordinates of the vertex, state whether the parabola opens up or down, and whether the y-coordinate of the vertex is the minimum or maximum value of the function.**

1. $f(x) = -x^2$
   $x = 0$; $(0, 0)$;
   down; maximum
2. $f(x) = x^2 - 4x$
   $x = 2$; $(2, -4)$;
   up; minimum
3. $f(x) = 3 - 8x - 4x^2$
   $x = -1$; $(-1, 7)$;
   down; maximum

Also on Quiz Transparency 9.2

Teach

Why The equation of a parabola is used to model the cross section of objects such as a telescope mirror, headlight reflector, or satellite-dish receiver. Ask students how a parabolic shape is used in these objects.

Parabolas

Why *Parabolas are used in the design of many real-world objects, such as parabolic reflectors that are used to pick up sounds on the playing field of a sports event.*

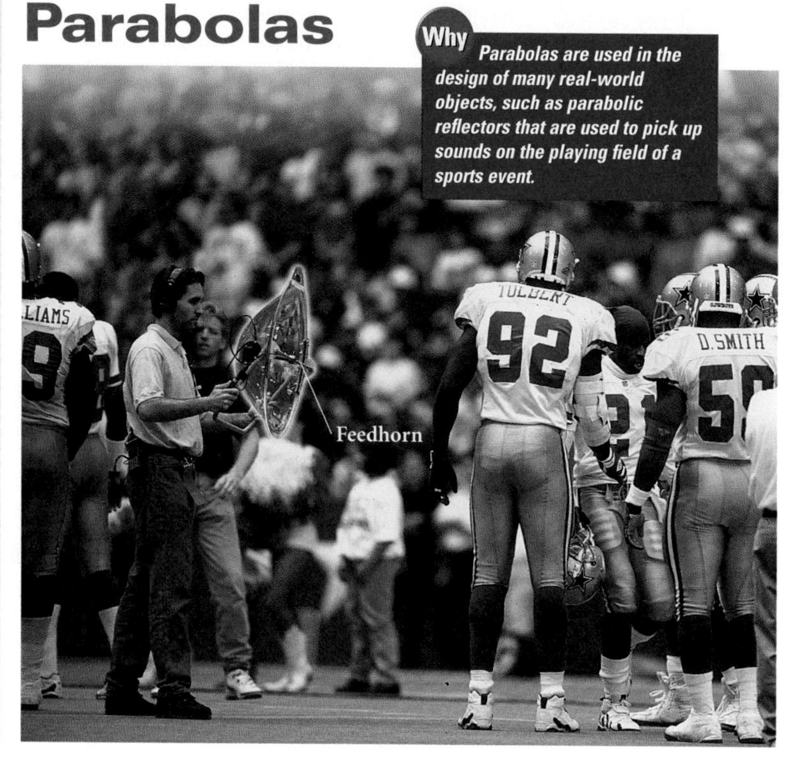

Objectives

- Write and graph the standard equation of a parabola given sufficient information.
- Given an equation of a parabola, graph it and label the vertex, focus, and directrix.

APPLICATION
COMMUNICATIONS

Sports broadcast technicians often use a parabolic reflector to pick up sounds on the playing field during a sports event. The picture above shows a parabolic reflector whose feedhorn is 10 inches long. The reflector focuses the incoming sounds at the end of the feedhorn. Write the standard equation of the parabola that is a cross section of the reflector. *You will solve this problem in Example 3.*

A parabola is defined in terms of a fixed point, called the **focus**, and a fixed line, called the **directrix**.

In a parabola, the distance from any point, P, on the parabola to the focus, F, is equal to the shortest distance from P to the directrix. That is, $PF = PD$ for any point, P, on the parabola.

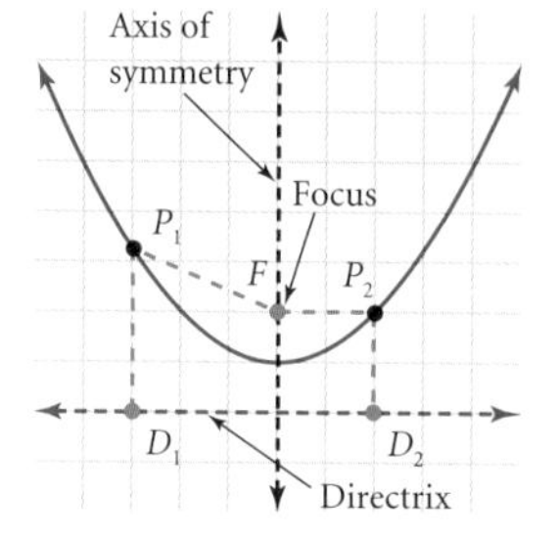

$P_1F = P_1D_1$ and $P_2F = P_2D_2$

Definition of Parabola

A **parabola** is the set of all points $P(x, y)$ in the plane whose distance to a fixed point, called the focus, equals its distance to a fixed line, called the directrix.

Alternative Teaching Strategy

USING COGNITIVE STRATEGIES On the board or on overhead, copy the figure on page 571 that illustrates the general case of the horizontal directrix. Develop the general form of the equation for this type of parabola by working through the proof shown at right. Have students provide a justification for each step.

$$\begin{aligned} PF &= PD \\ \sqrt{(x-0)^2 + (y-p)^2} &= \sqrt{(x-x)^2 + (y+p)^2} \\ x^2 + (y-p)^2 &= (y+p)^2 \\ x^2 + y^2 - 2py + p^2 &= y^2 + 2py + p^2 \\ x^2 &= 4py \\ y &= \frac{1}{4p}x^2 \end{aligned}$$

Now refer students to the figure on page 571 that illustrates a vertical directrix. Have them write a similar proof for the general form of the equation for this type of parabola.

If the point $F(0, 3)$ is the focus of a parabola with its vertex at the origin, then the equation for the directrix is $y = -3$. This is shown in the figure at right.

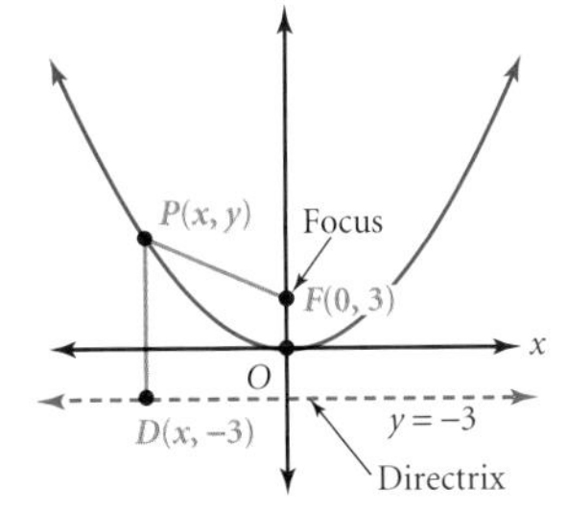

You can find an equation for this parabola by using the definition of a parabola and the distance formula, as shown below.

$$PF = PD$$
$$\sqrt{(x-0)^2 + (y-3)^2} = \sqrt{(x-x)^2 + (y+3)^2}$$
$$x^2 + (y-3)^2 = (y+3)^2 \quad \text{Square each side.}$$
$$x^2 = (y+3)^2 - (y-3)^2$$
$$x^2 = y^2 + 6y + 9 - (y^2 - 6y + 9) \quad \text{Expand each binomial.}$$
$$x^2 = 12y$$
$$y = \frac{1}{12}x^2$$

In general, if F is the point $(0, p)$ and the directrix is $y = -p$, then the equation of the parabola is $y = \frac{1}{4p}x^2$.

Recall from Lesson 5.1 that the **axis of symmetry** of a parabola goes through the vertex and divides the parabola into two equal parts. The axis of symmetry also contains the focus and is perpendicular to the directrix. The **vertex** of a parabola is the midpoint between the focus and the directrix.

Standard Equation of a Parabola

The standard equation of a parabola with its vertex at the origin is given below.

Horizontal directrix	**Vertical directrix**
$y = \frac{1}{4p}x^2$	$x = \frac{1}{4p}y^2$
$p > 0$: opens upward	$p > 0$: opens right
$p < 0$: opens downward	$p < 0$: opens left
focus: $(0, p)$	focus: $(p, 0)$
directrix: $y = -p$	directrix: $x = -p$
axis of symmetry: y-axis	axis of symmetry: x-axis

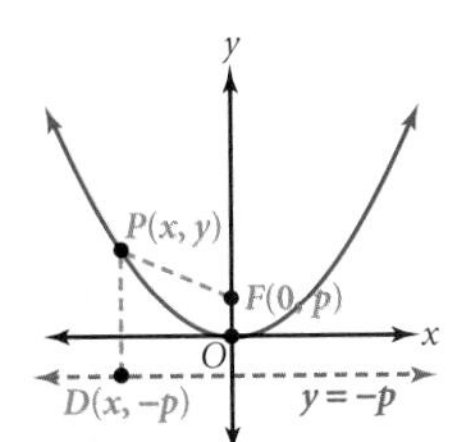

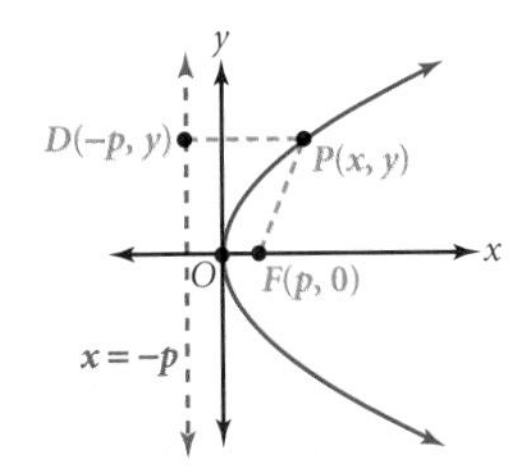

Teaching Tip

Remind students of the characteristics of parabolas that were studied in Chapter 5, where representations of parabolas were functions that opened up or down. Students were shown how to locate the vertex, axis of symmetry, and x-intercepts of quadratic functions of the form $y = ax^2 + bx + c$ by writing the functions in vertex form and in x-intercept form. The characteristics of each form are as follows:

- vertex form: $y = a(x - h)^2 + k$, where (h, k) is the vertex and $x = h$ is the axis of symmetry
- x-intercept form: $y = a(x - s)(x - t)$, where $(s, 0)$ and $(t, 0)$ are the x-intercepts

Relate the form $y = ax^2$ to the new form $y = \frac{1}{4p}x^2$. Point out that $a = \frac{1}{4p}$ or $p = \frac{1}{4a}$. To help students visualize the graph, tell them that a parabola always bends away from the directrix.

Interdisciplinary Connection

PHYSICS A suspension bridge supports a roadway with huge cables that extend from one end of the bridge to the other. These cables rest on top of high towers and are secured at each end. Because of the manner in which a suspension bridge supports its load, the main cable connecting the towers can be approximated by a parabolic curve. Have students write the standard equation of the parabola that models the bridge given that its focus is (100, 70) and its directrix is $y = -30$. Then have them illustrate the bridge by graphing the parabola.

$\mathbf{y - 20 = \frac{1}{200}(x - 100)^2}$

ADDITIONAL
EXAMPLE 1

Graph $x = \frac{1}{4}y^2$. Label the vertex, focus, and directrix.

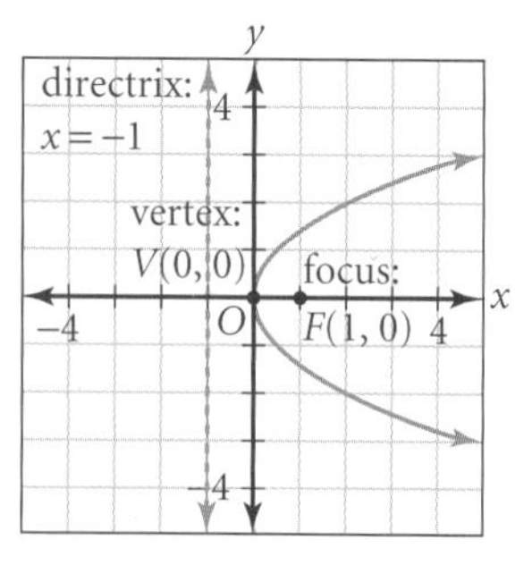

TRY THIS

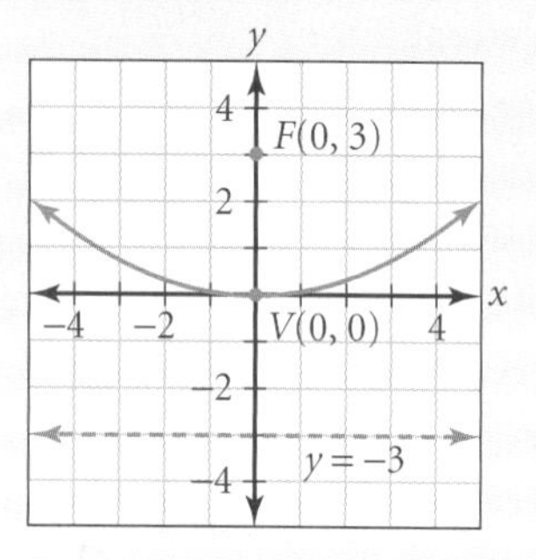

ADDITIONAL
EXAMPLE 2

Write the standard equation of the parabola with its vertex at the origin and the directrix $y = -6$.

$y = \frac{1}{24}x^2$

TRY THIS

$x = -\frac{1}{16}y^2$

CRITICAL THINKING

Apply a vertical compression by a factor of $\frac{1}{2}$ to $x = y^2$:

$x = (2y)^2$, or $x = 4y^2$

Apply a horizontal stretch by a factor of 4 to $x = y^2$:

$\left(\frac{1}{4}x\right) = y^2$, or $x = 4y^2$

Both transformations result in the same equation.

Example 1 shows how to graph a parabola from its equation.

EXAMPLE 1 **Graph $x = -\frac{1}{8}y^2$. Label the vertex, focus, and directrix.**

SOLUTION

1. Identify p. Rewrite $x = -\frac{1}{8}y^2$ as $x = \frac{1}{4(-2)}y^2$. Thus, $p = -2$.

2. Identify the vertex, focus, and directrix.
Since $p < 0$, the parabola opens to the left.
Vertex: (0, 0)
Focus: $(p, 0)$, or $(-2, 0)$
Directrix: $x = -p = -(-2)$, or $x = 2$

PROBLEM SOLVING

3. **Use a table** of values to sketch the graph.

y	−4	−2	0	2	4
x	−2	$\frac{1}{2}$	0	$\frac{1}{2}$	−2

The graph is shown at right.

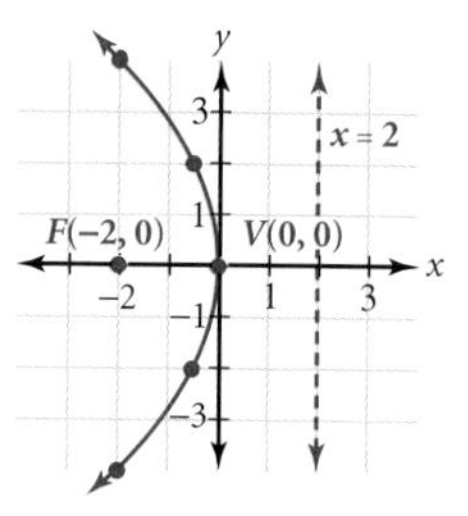

TRY THIS Graph $y = \frac{1}{12}x^2$. Label the vertex, focus, and directrix.

When you know the locations of any two of the three main characteristics of a parabola (focus, vertex, and directrix), you can write an equation for the parabola. This is shown in Example 2.

EXAMPLE 2 **Write the standard equation of the parabola with its vertex at the origin and with the directrix $y = 4$.**

SOLUTION

PROBLEM SOLVING

Draw a diagram.

Because the vertex is below the directrix, the parabola opens downward. Since $y = -p$ and $y = 4$, $p = -4$.

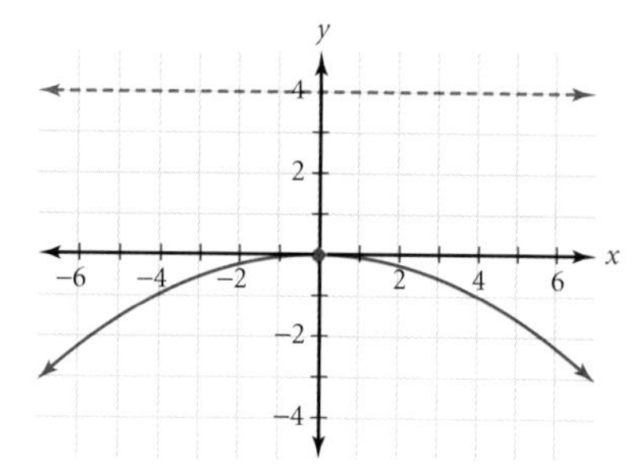

Thus, $y = \frac{1}{4(-4)}x^2$, or $y = -\frac{1}{16}x^2$, is the equation of the parabola.

TRY THIS Write the standard equation of the parabola with its vertex at the origin and with the directrix $x = 4$.

CRITICAL THINKING Explain how applying a vertical compression by a factor of $\frac{1}{2}$ to $x = y^2$ produces the same graph as applying a horizontal stretch by a factor of 4 to the graph of $x = y^2$.

Enrichment

Have students determine the equation of the line that is tangent to the parabola $y = 4x^2$ at the point (1, 4). Letting m be the slope of the tangent line, the equation of the tangent line has the form $y = m(x - 1) + 4$. Because the tangent line intersects the parabola at only one point, the following system has only one solution:

$$\begin{cases} y = m(x-1) + 4 \\ y = 4x^2 \end{cases}$$

Solving the system yields the following:

$$\begin{aligned} m(x-1) + 4 &= 4x^2 \\ m(x-1) &= 4(x^2 - 1) \\ m(x-1) - 4(x-1)(x+1) &= 0 \\ (x-1)[m - 4(x+1)] &= 0 \\ x - 1 = 0 \quad &or \quad m - 4(x+1) = 0 \\ x = 1 \quad &\quad\quad m = 4(x+1) \end{aligned}$$

Because $x = 1$ at the point of intersection and there can be only one point of intersection, $m = 4(1 + 1)$ or $m = 8$. Therefore, the equation of the line tangent to the parabola at the point (1, 4) is $y = 8x - 4$.

EXAMPLE 3 Refer to the parabolic reflector described at the beginning of the lesson.

Write the standard equation of the parabola that is a cross section of the reflector.

APPLICATION
COMMUNICATIONS

SOLUTION

For convenience, place the vertex of the parabola at the origin. You can let the parabola open upward, downward, to the right, or to the left. If it opens to the right, then the equation is of the form $x = \frac{1}{4p}y^2$.

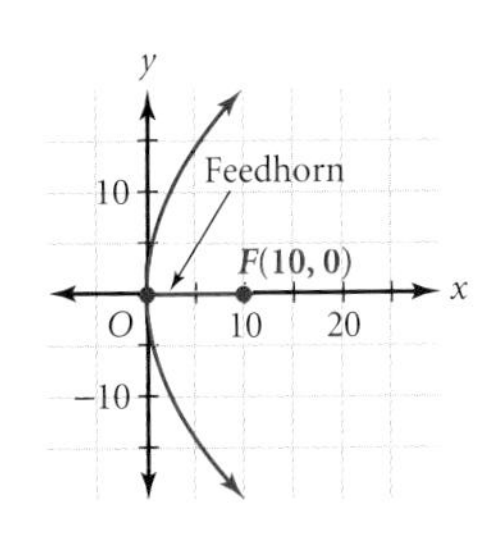

The feedhorn is 10 inches long, so the focus at the end of the feedhorn has coordinates (10, 0) and $p = 10$.

$$x = \frac{1}{4p}y^2$$

$$x = \frac{1}{4(10)}y^2 \quad \text{Substitute 10 for } p.$$

$$x = \frac{1}{40}y^2$$

Thus, the equation that represents a cross section of the parabolic reflector is $x = \frac{1}{40}y^2$.

The vertex of a parabola can be anywhere in the coordinate plane. In the Activity below, you can explore some of these translations.

Activity: Exploring Translations of Parabolas

TECHNOLOGY
GRAPHICS CALCULATOR

Keystroke Guide, page 622

You will need: a graphics calculator

1. Graph the parabola described by the equation $y = \frac{1}{4}x^2$.
2. Solve $y - 3 = \frac{1}{4}(x - 2)^2$ for y, and graph the parabola. Find the coordinates of the vertex.

CHECKPOINT ✔

3. Describe the translation from the graph of $y = \frac{1}{4}x^2$ in Step 1 to the graph of $y - 3 = \frac{1}{4}(x - 2)^2$ in Step 2.
4. Compare the equation $y - 5 = \frac{1}{4}(x - 4)^2$ and the coordinates of the vertex of this parabola. Describe the relationship that you observe.

CHECKPOINT ✔

5. Predict the coordinates of the vertex of the parabola given by the equation $y + 2 = \frac{1}{4}(x - 1)^2$. Graph the parabola to verify your prediction.

Inclusion Strategies

USING VISUAL MODELS Show students how to construct a parabola by using a straightedge, a piece of cardboard, string, a thumbtack, a right-angle drawing triangle, and a pencil. Tape the straightedge on the cardboard to represent the directrix, and place the thumbtack above the directrix to represent the focus. Place one leg of the drawing triangle along the straightedge, tie a piece of string that is the same length as the vertical leg of the drawing triangle to the thumbtack, and tape the other end of the string to the vertex at the top of the triangle. Press the string against the edge of the triangle with the tip of the pencil, and slide the triangle along the straightedge, keeping the string taut. Students may want to mark points on the parabola rather than draw a smooth curve. They can then connect the points to obtain the curve. This will create half of the parabola. To create the other half, flip the drawing triangle over the parabola's line of symmetry and repeat the procedure. Have students discuss why this procedure creates a parabola.

ADDITIONAL
EXAMPLE 3

Refer to the parabolic reflector described at the beginning of the lesson. **Write the standard equation of the parabola that is a cross section of a similar reflector whose feedhorn is 16 inches long.**

If the vertex of the parabola is at the origin and the parabola opens to the right, the standard equation is $x = \frac{1}{64}y^2$.

Activity Notes

In the Activity, students graph several parabolas, find the vertices, and describe the transformations from the related parent function. They should see that if a function is written in the form $y - k = \frac{1}{4}p(x - h)^2$ or $x - h = \frac{1}{4}p(y - k)^2$, the vertex is (h, k) and the graph is a translation of $y = \frac{1}{4}px^2$ or $x = \frac{1}{4}py^2$, respectively. Have students state whether the parabola opens vertically or horizontally and give the coordinates of the vertex before graphing the parabola. When using a graphics calculator, use a standard viewing window.

Cooperative Learning

Students can work in small groups for this Activity. Have each member graph the parabola and the translation. Have members compare their graphs. Then have each group summarize their findings.

CHECKPOINT ✔

3. 2 units right, 3 units up

CHECKPOINT ✔

5. vertex: (1, −2)

ADDITIONAL EXAMPLE 4

Write the standard equation of the parabola graphed below, with a focus at $F(-3, 2)$ and directrix $y = 4$.

$y - 3 = -\frac{1}{16}(x + 3)^2$

Standard Equation of a Translated Parabola

The standard equation of a parabola with its vertex at (h, k) is given below.

Horizontal Directrix

$$y - k = \frac{1}{4p}(x - h)^2$$

$p > 0$: opens upward
$p < 0$: opens downward

focus: $(h, k + p)$
directrix: $y = k - p$
axis of symmetry: $x = h$

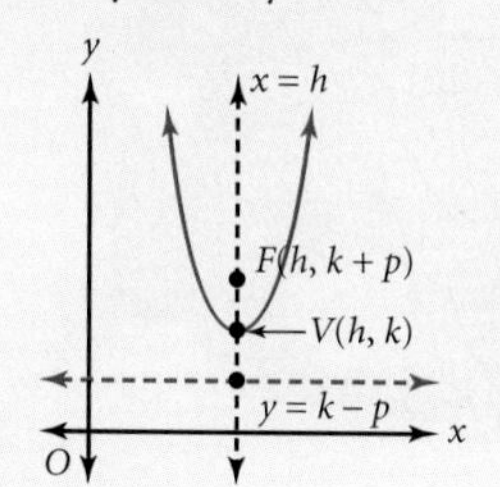

Vertical Directrix

$$x - h = \frac{1}{4p}(y - k)^2$$

$p > 0$: opens right
$p < 0$: opens left

focus: $(h + p, k)$
directrix: $x = h - p$
axis of symmetry: $y = k$

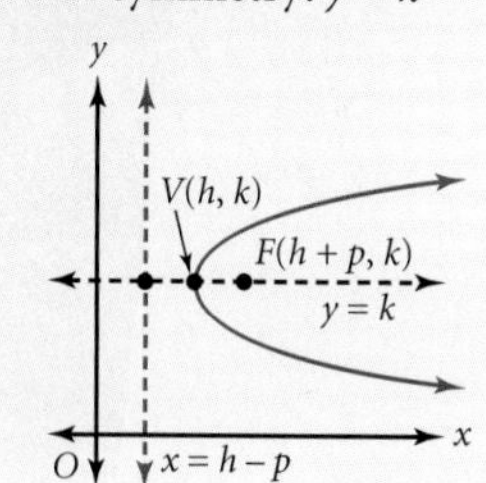

EXAMPLE 4 **Write the standard equation of the parabola graphed at right.**

SOLUTION

The parabola opens upward, so the equation is of the form $y - k = \frac{1}{4p}(x - h)^2$ and $p > 0$.

The vertex is halfway between the focus and the directrix. Use the midpoint formula.

Vertex: $\left(2, \frac{\frac{10}{3} + \frac{8}{3}}{2}\right)$, or $(2, 3)$ $\quad h = 2$ and $k = 3$

The focus is $(h, k + p)$, or $\left(2, \frac{10}{3}\right)$. Solve for p.

$$k + p = \frac{10}{3}$$
$$3 + p = \frac{10}{3}$$
$$p = \frac{1}{3}$$

Write the standard equation.

$$y - k = \frac{1}{4p}(x - h)^2$$
$$y - 3 = \frac{1}{4\left(\frac{1}{3}\right)}(x - 2)^2$$
$$y - 3 = \frac{3}{4}(x - 2)^2$$

Thus, the standard equation of the parabola is $y - 3 = \frac{3}{4}(x - 2)^2$.

Reteaching the Lesson

COOPERATIVE LEARNING Provide students with several copies of the "Parabola Profile" shown at right. Have them work in pairs. Each student should write the standard equation of a parabola on the profile sheet. Instruct the partners to exchange profile sheets. Each partner should then complete the profile of the other's parabola.

Parabola Profile

standard equation:	________	
vertex:	(☐, ☐)	
focus:	(☐, ☐)	
directrix:	x = ☐	or y = ☐
axis of symmetry:	x = ☐	or y = ☐
opening:	up down	left right

TRY THIS Write the standard equation of the parabola with its focus is at $(-6, 4)$ and with the directrix $x = 2$.

EXAMPLE 5 **Graph the parabola $y^2 - 8y + 8x + 8 = 0$. Label the vertex, focus, and directrix.**

SOLUTION

1. Complete the square to find the standard equation.

$$y^2 - 8y + 8x + 8 = 0$$

$$y^2 - 8y = -8x - 8 \qquad \textit{Isolate the y-terms.}$$

$$y^2 - 8y + (-4)^2 = -8x - 8 + (-4)^2 \qquad \textit{Complete the square.}$$

$$(y-4)^2 = -8x + 8$$

$$(y-4)^2 = -8(x-1)$$

$$-\frac{1}{8}(y-4)^2 = x - 1 \qquad \textit{Divide each side by 8.}$$

$$x - 1 = -\frac{1}{8}(y-4)^2 \qquad \textit{Write the standard equation.}$$

2. From the standard equation, the vertex is $(1, 4)$. Find p.

$$\frac{1}{4p} = -\frac{1}{8}$$

$$-4p = 8$$

$$p = -2$$

Focus: $(h + p, k)$
$= (1 + (-2), 4)$
$= (-1, 4)$

Directrix: $x = h - p$
$x = 1 - (-2)$
$x = 3$

PROBLEM SOLVING

3. Use a table of values. Choose y-values from 0 to 8.

$$x = -\frac{1}{8}(y-4)^2 + 1$$

y	0	2	4	6	8
x	-1	$\frac{1}{2}$	1	$\frac{1}{2}$	-1

The graph is shown at right.

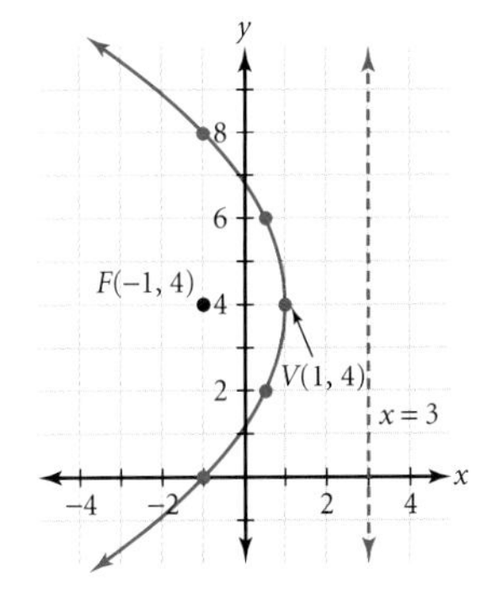

TRY THIS Graph the parabola $x^2 - 6x + 6y + 18 = 0$. Label the vertex, focus, and directrix.

TECHNOLOGY GRAPHICS CALCULATOR
Keystroke Guide, page 623

To graph the equation $x - 1 = -\frac{1}{8}(y-4)^2$ from Example 5 on a graphics calculator, first solve for y.

$$x - 1 = -\frac{1}{8}(y-4)^2$$

$$-8(x-1) = (y-4)^2$$

$$\pm\sqrt{-8(x-1)} = y - 4$$

$$4 \pm \sqrt{-8(x-1)} = y$$

Graph $y = 4 + \sqrt{-8(x-1)}$ and $y = 4 - \sqrt{-8(x-1)}$ together on the same screen.

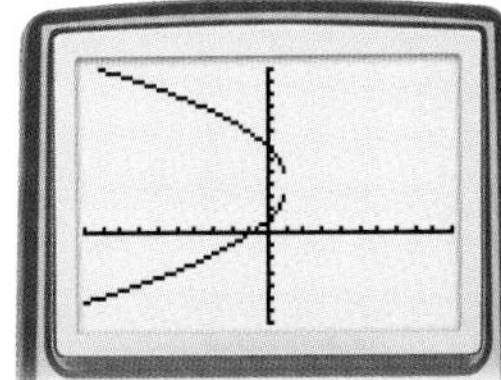

TRY THIS

$x + 2 = -\frac{1}{8}(y-4)^2$

ADDITIONAL EXAMPLE 5

Graph the parabola $-\frac{1}{4}y^2 - y + x - 2 = 0$. Label the vertex, focus, and directrix.

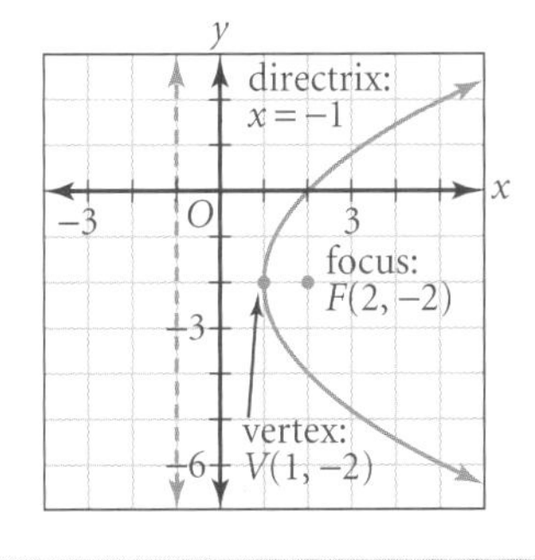

TRY THIS

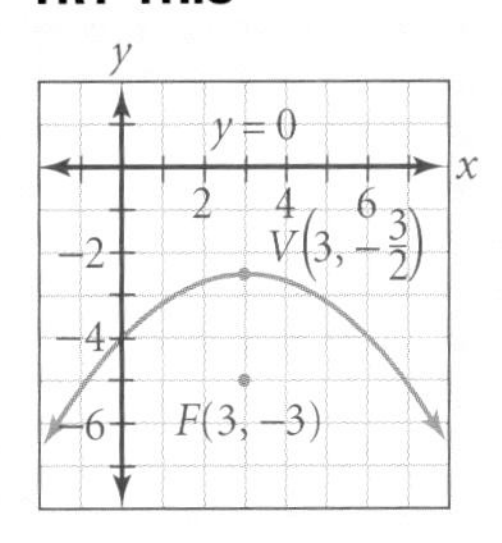

Teaching Tip

TECHNOLOGY If you are using a TI-82 or TI-83 graphics calculator to create the graph for Example 5, set the viewing window as follows:

Xmin=−10	Ymin=−7
Xmax=10	Ymax=13
Xscl=1	Yscl=1

Tell students that the gap in the graph is due to calculator rounding errors.

Assess

Selected Answers

Exercises 4–8, 9–67 odd

ASSIGNMENT GUIDE	
In Class	1–8
Core	9–49 odd, 53
Core Plus	10–52 even
Review	54–67
Preview	68

Extra Practice can be found beginning on page 940.

Error Analysis

Some students will have a difficult time visualizing a parabola from its equation. Remind them that a parabola opens away from the directrix and toward the focus. This may help them in determining whether a parabola opens vertically or horizontally. Additionally, urge students to find the vertex before graphing a parabola so that they can set the viewing window to display all of the main characteristics of the graph.

Practice

NAME ______ CLASS ______ DATE ______

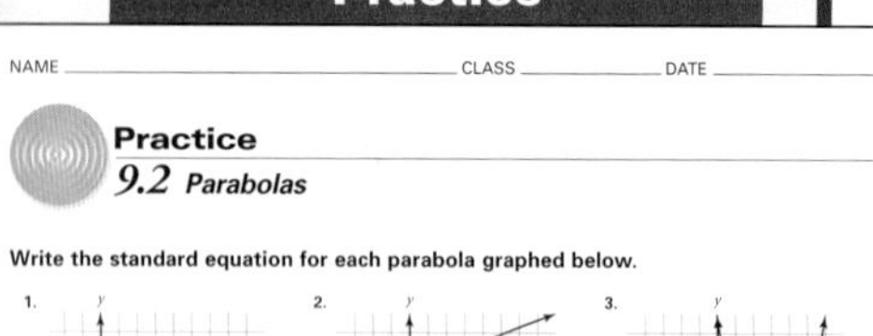

Practice 9.2 Parabolas

Write the standard equation for each parabola graphed below.

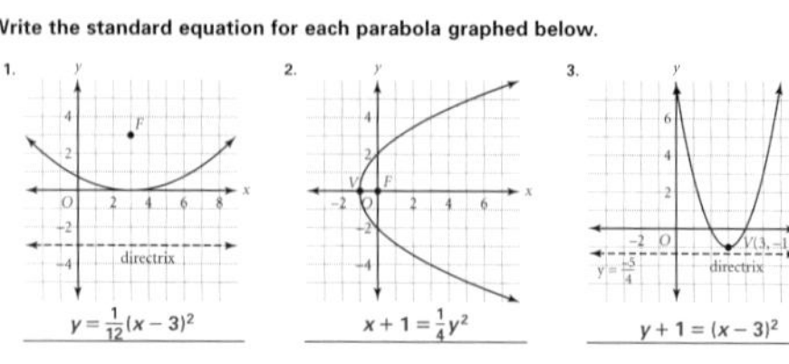

1. $y = \frac{1}{12}(x-3)^2$
2. $x + 1 = \frac{1}{4}y^2$
3. $y + 1 = (x-3)^2$

Graph each equation. Label the vertex, focus, and directrix.

4. $x = \frac{1}{4}y^2$
5. $y - 4 = \frac{1}{4}(x-1)^2$
6. $x - 1 = \frac{1}{8}(y-2)^2$

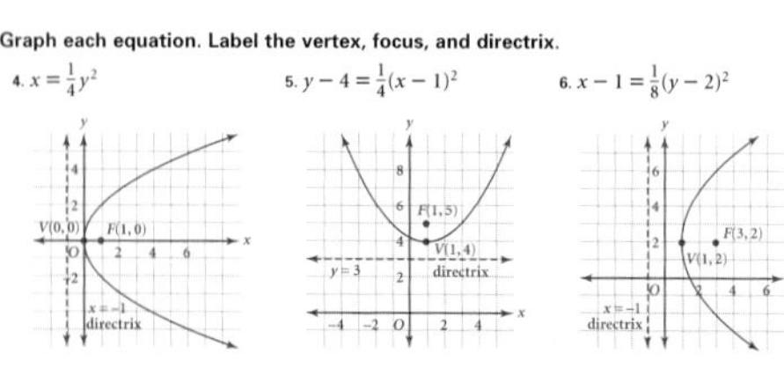

Write the standard equation for the parabola with the given characteristics.

7. vertex: (0, 0); focus: (0, 6) $y = \frac{1}{24}x^2$
8. vertex: (10, 0); directrix: $x = 8$ $x - 10 = \frac{1}{8}y^2$
9. focus: (3, 0); directrix: $x = -3$ $x = \frac{1}{12}y^2$
10. vertex: (5, 2); directrix: $y = 1$ $y - 2 = \frac{1}{4}(x-5)^2$
11. vertex: (6, −7); focus: (4, −7) $x - 6 = -\frac{1}{8}(y+7)^2$
12. focus: (9, 5); directrix: $y = -5$ $y = \frac{1}{20}(x-9)^2$

Exercises

Communicate

1. Explain how to tell from the standard equation of a parabola whether the graph opens upward, downward, to the left, or to the right.
2. How can you determine from the standard equation of a parabola the locations of the focus, vertex, and directrix?
3. Explain how to graph $x = \frac{3}{4}y^2$ on a graphics calculator.

Guided Skills Practice

4. Graph $x = y^2$. Label the vertex, focus, and directrix. *(EXAMPLE 1)*
5. Write the standard equation of the parabola with its vertex at the origin and with the directrix $x = 3$. *(EXAMPLE 2)*

5. $x = -\frac{1}{12}y^2$

APPLICATION

6. **SPORTS** Refer to the parabolic reflector described at the beginning of the lesson. Write the standard equation of the parabola that is a cross section of a reflector whose feedhorn is 12 inches long. *(EXAMPLE 3)*
7. Write the standard equation for the parabola with its focus at (−2, 3) and with the directrix $x = 3$. *(EXAMPLE 4)*
8. Graph the parabola $x^2 + 10x + 16y - 7 = 0$. Label the vertex, focus, and directrix. *(EXAMPLE 5)*

6. $x = \frac{1}{48}y^2$
7. $x - \frac{1}{2} = -\frac{1}{10}(y-3)^2$

Practice and Apply

Write the standard equation for each parabola graphed below.

9. 10.

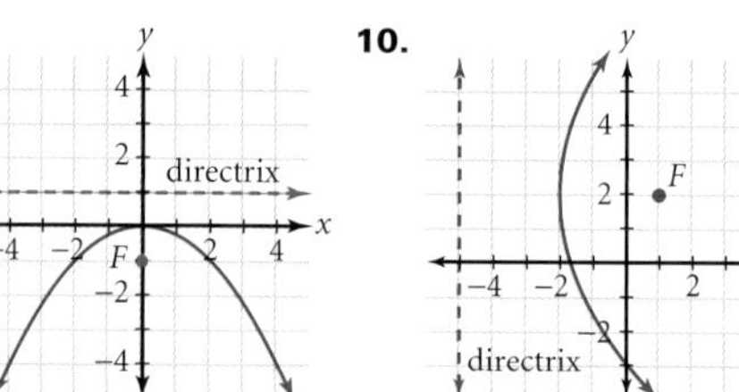

11.

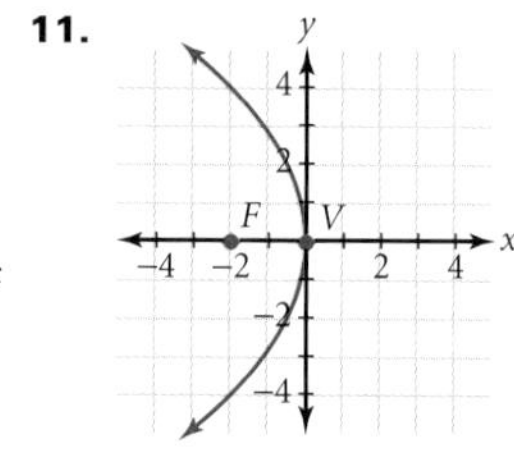

12.

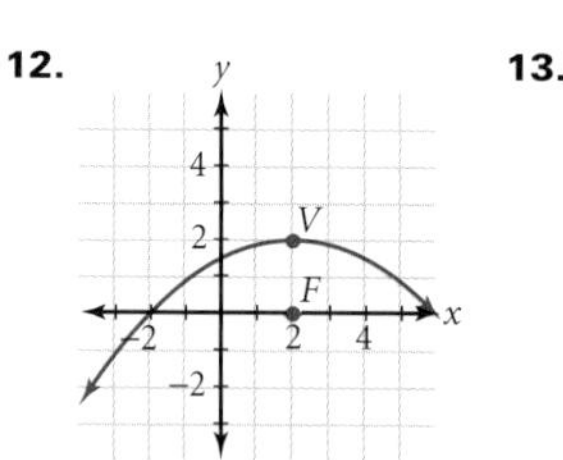

13.

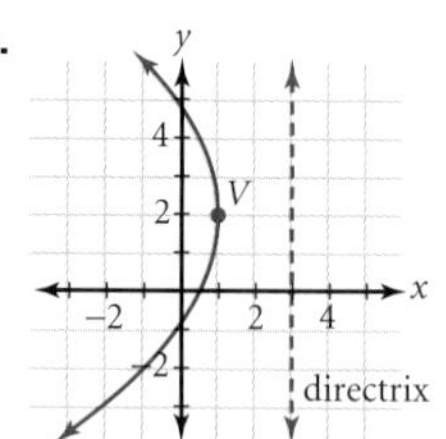

14.

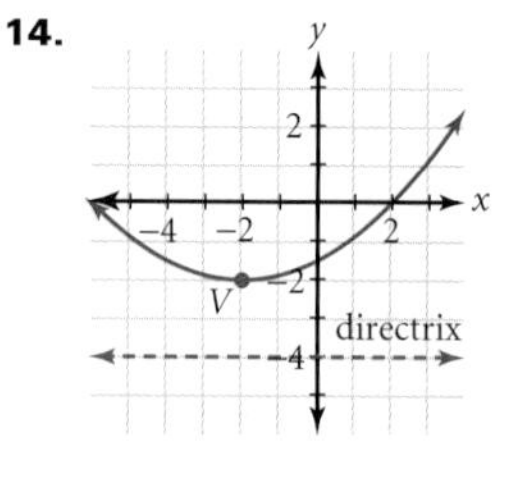

9. $y = -\frac{1}{4}x^2$
10. $x + 2 = \frac{1}{12}(y-2)^2$
11. $x = -\frac{1}{8}y^2$
12. $y - 2 = -\frac{1}{8}(x-2)^2$
13. $x - 1 = -\frac{1}{8}(y-2)^2$
14. $y + 2 = \frac{1}{8}(x+2)^2$

4.

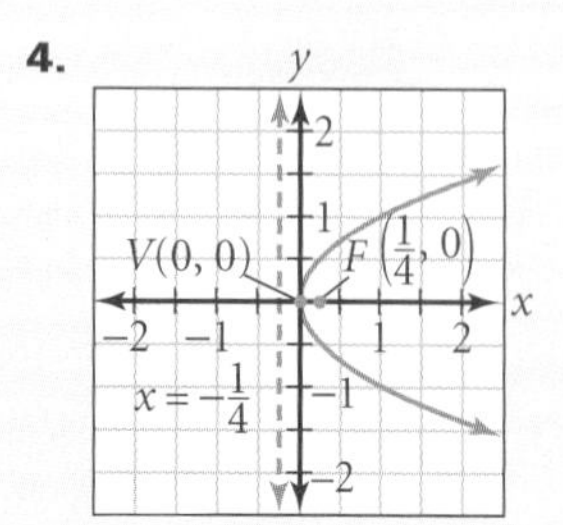

8.

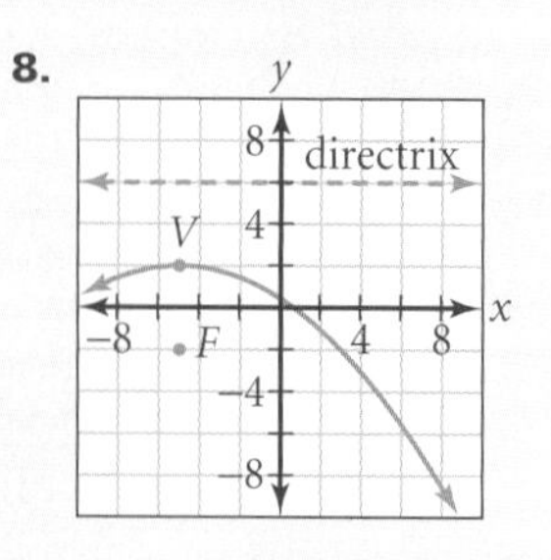

vertex: (−5, 2); focus: (−5, −2); directrix: $y = 6$

Graph each equation. Label the vertex, focus, and directrix.

15. $y = \frac{1}{4}x^2$ **16.** $y = \frac{1}{8}x^2$ **17.** $x = \frac{1}{20}y^2$

18. $x = \frac{1}{40}y^2$ **19.** $y = x^2$ **20.** $y = 2x^2$

21. $y + 3 = \frac{1}{8}(x + 2)^2$ **22.** $x - 1 = \frac{1}{12}(y + 2)^2$ **23.** $y - 4 = -(x - 1)^2$

24. $y - 1 = \frac{1}{4}(x - 1)^2$ **25.** $y = \frac{1}{8}(x - 1)^2$ **26.** $x - 3 = -\frac{1}{8}(y + 1)^2$

27. $y + 3 = \frac{1}{12}x^2$ **28.** $-12y = (x + 2)^2$ **29.** $x - 1 = \frac{1}{2}(y + 2)^2$

30. $x^2 + 4x - 6y = -10$ **31.** $x^2 - 6x + 10y = 1$ **32.** $x^2 - 8x - y + 20 = 0$

33. $4x + y^2 + 3y = -5$ **34.** $4x + y^2 - 6y = 9$ **35.** $-14x + 2y^2 - 8y = 20$

Write the standard equation for the parabola with the given characteristics.

36. vertex: (0, 0)
focus: (−4, 0)

37. vertex: (0, 0)
focus: (0, −5)

38. vertex: (0, 0)
directrix: $y = -1$

39. vertex: (0, 0)
directrix: $x = 4$

40. vertex: (0, 0)
focus: (0, 3)

41. vertex: (0, 0)
focus: (2, 0)

42. vertex: (0, 0)
directrix: $x = -3$

43. vertex: (0, 0)
directrix: $y = 12$

44. directrix: $y = -4$
focus: (0, 4)

45. focus: (3, 0)
directrix: $x = -3$

46. focus: (0, −5)
directrix: $y = 5$

47. directrix: $x = 8$
focus: (−8, 0)

36. $x = -\frac{1}{16}y^2$
37. $y = -\frac{1}{20}x^2$
38. $y = \frac{1}{4}x^2$
39. $x = -\frac{1}{16}y^2$
40. $y = \frac{1}{12}x^2$
41. $x = \frac{1}{8}y^2$
42. $x = \frac{1}{12}y^2$
43. $y = -\frac{1}{48}x^2$
44. $y = \frac{1}{16}x^2$
45. $x = \frac{1}{12}y^2$
46. $y = -\frac{1}{20}x^2$
47. $x = -\frac{1}{32}y^2$

48. TRANSFORMATIONS A parabola defined by the equation $x + 3 = \frac{1}{8}(y + 2)^2$ is translated 4 units down and 3 units to the right. Write the standard equation of the resulting parabola. $\mathbf{x = \frac{1}{8}(y + 6)^2}$

49. TRANSFORMATIONS A parabola defined by the equation $4x + y^2 - 6y = 9$ is translated 2 units up and 4 units to the left. Write the standard equation of the resulting parabola. $\mathbf{x - \frac{1}{2} = -\frac{1}{4}(y - 5)^2}$

CHALLENGE

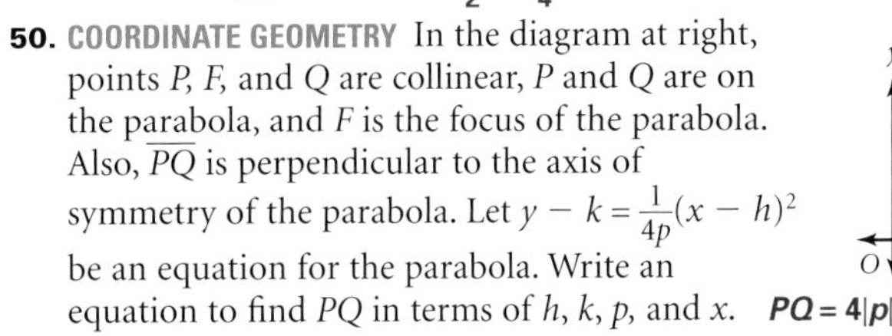

50. COORDINATE GEOMETRY In the diagram at right, points P, F, and Q are collinear, P and Q are on the parabola, and F is the focus of the parabola. Also, $\overline{PQ}$ is perpendicular to the axis of symmetry of the parabola. Let $y - k = \frac{1}{4p}(x - h)^2$ be an equation for the parabola. Write an equation to find PQ in terms of h, k, p, and x. $\mathbf{PQ = 4|p|}$

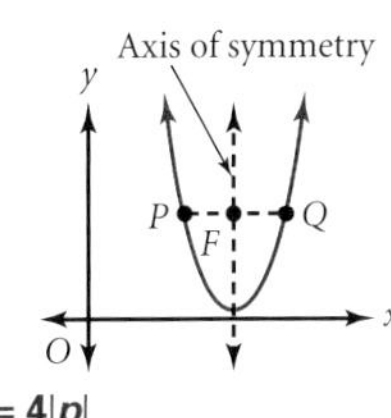

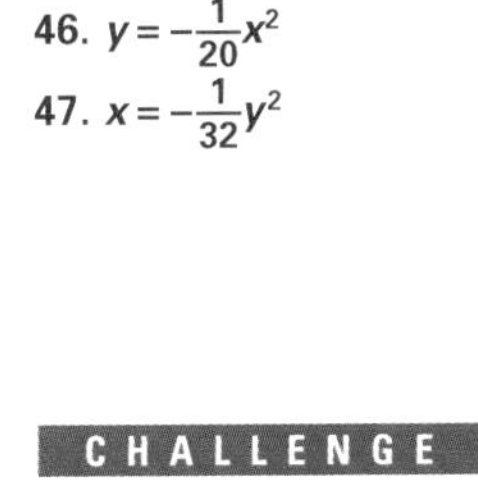

Cross section of a parabolic satellite dish

51. COMMUNICATIONS Write an equation for the cross section of a parabolic satellite dish whose focus is 1 foot from its vertex.

52. SPORTS Suppose that a golf ball travels a distance of 600 feet as measured along the ground and reaches an altitude of 200 feet. If the origin represents the tee and the ball travels along a parabolic path that opens downward, find an equation for the path of the golf ball.

53. LIGHTING The lightsource of a flashlight is $\frac{1}{2}$ inch from the vertex of the parabolic reflector and is located at the focus. Assuming that the parabolic reflector is directed upward and the vertex is at the origin, write an equation for a cross section of the reflector. $\mathbf{y = \frac{1}{2}x^2}$

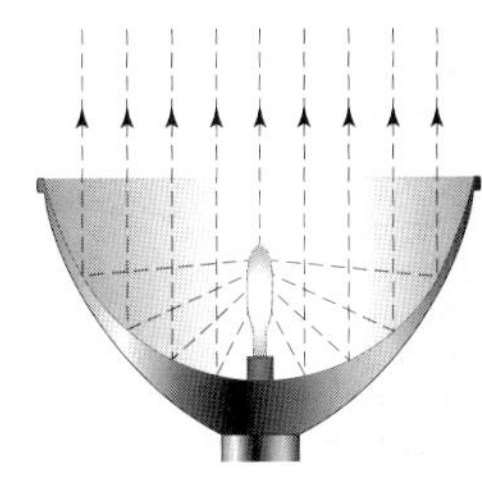

15.

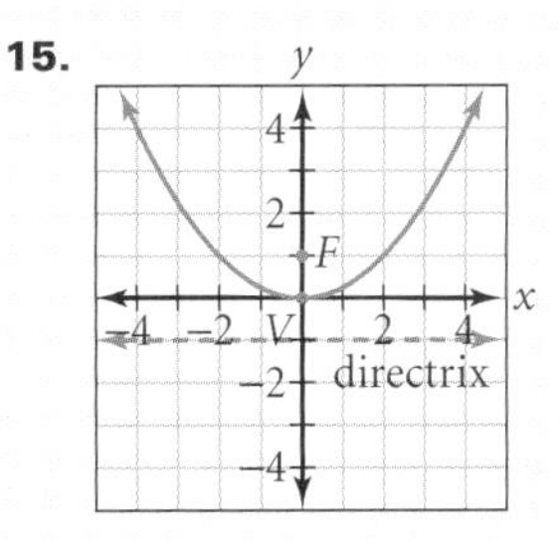

vertex: (0, 0); focus: (0, 1); directrix: $y = -1$

16.

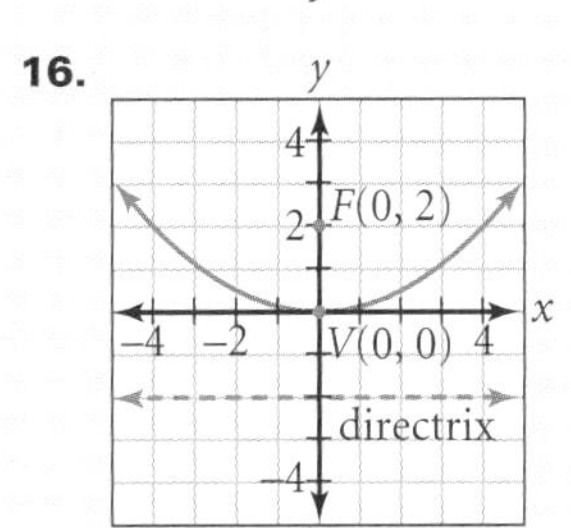

vertex: (0, 0); focus: (0, 2); directrix: $y = -2$

17.

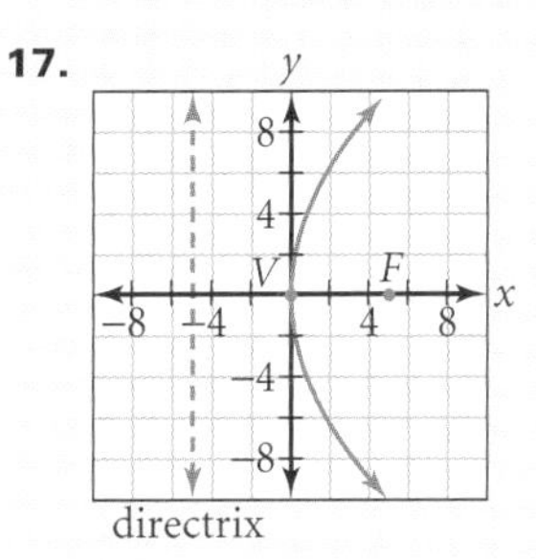

vertex: (0, 0); focus: (5, 0); directrix: $x = -5$

NAME ______ CLASS ______ DATE ______

Student Technology Guide
9.2 **Parabolas**

You are familiar with the graphs of quadratic functions—parabolas that open up or down. However, parabolas can also open in other directions. You can use a graphics calculator to graph parabolas that open to the left and the right as well as those that open up or down.

To see an undistorted graph of a conic section, make sure that you are using a square viewing window. You can do this by pressing ZOOM 5:ZSquare ENTER or by pressing WINDOW and changing Xmin, Xmax, Ymin, and Ymax so that the range for the x-values is about 1.5 times the range for the y-values.

Example: Graph $y = -7x^2 - 2x + 4$.

- Press Y= (-) 7 X,T,θ,n x² − 2 X,T,θ,n + 4 ENTER.

The graph opens downward, is narrow, and has a y-intercept of 4. This makes sense because $a = -7$ and $c = 4$.

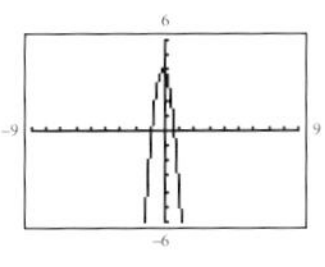

Example: Graph $y^2 - 4y + x + 3 = 0$.

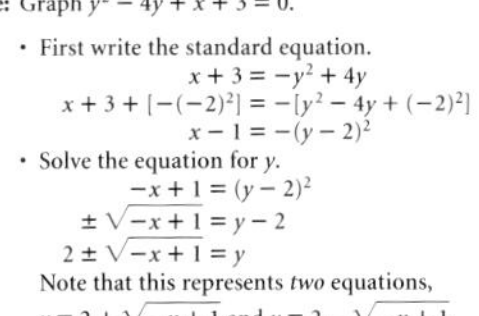

- First write the standard equation.
$x + 3 = -y^2 + 4y$
$x + 3 + [-(-2)^2] = -[y^2 - 4y + (-2)^2]$
$x - 1 = -(y - 2)^2$
- Solve the equation for y.
$-x + 1 = (y - 2)^2$
$\pm\sqrt{-x + 1} = y - 2$
$2 \pm \sqrt{-x + 1} = y$

Note that this represents *two* equations,
$y = 2 + \sqrt{-x + 1}$ and $y = 2 - \sqrt{-x + 1}$.

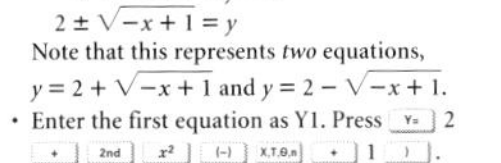

- Enter the first equation as Y1. Press Y= 2 + 2nd x² (-) X,T,θ,n + 1) .
- Move to Y2 and press 2 − 2nd x² (-) X,T,θ,n + 1) .
- Press GRAPH.

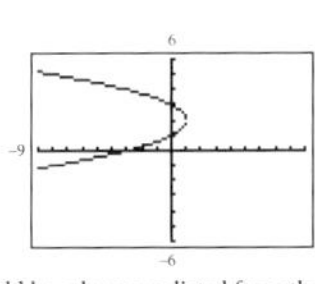

Notice that the parabola opens to the left. This could have been predicted from the a-value of −1 in the standard form of the equation.

Graph each equation. In which direction does the graph open?

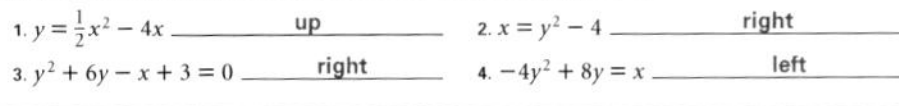

1. $y = \frac{1}{2}x^2 - 4x$ ____up____ 2. $x = y^2 - 4$ ____right____
3. $y^2 + 6y - x + 3 = 0$ ____right____ 4. $-4y^2 + 8y = x$ ____left____

18.

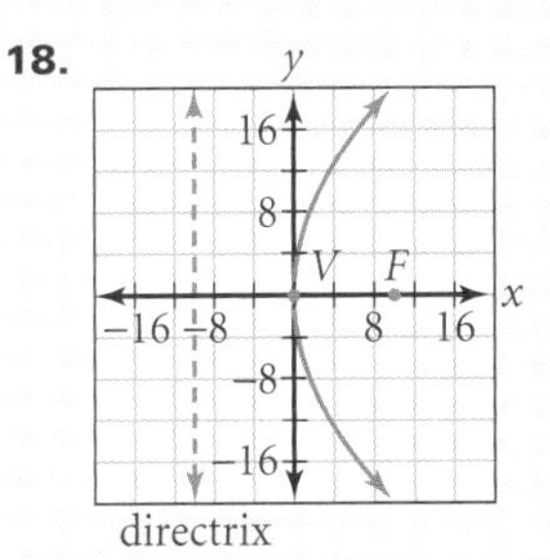

vertex: (0, 0); focus: (10, 0); directrix: $x = -10$

19.

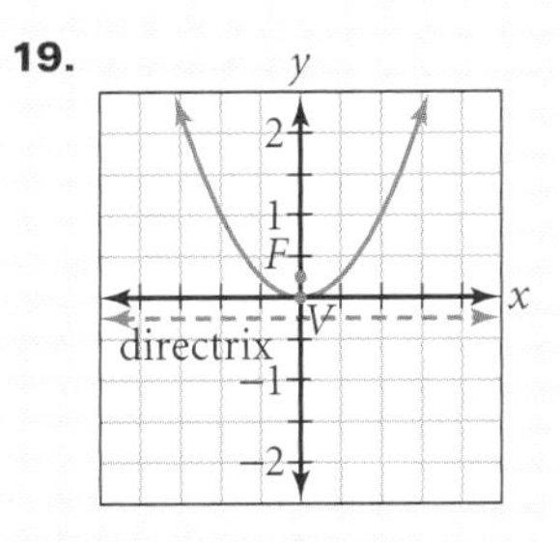

vertex: (0, 0); focus: $\left(0, \frac{1}{4}\right)$; directrix: $y = -\frac{1}{4}$

The answers to Exercises 20–35 and 51–52 can be found in Additional Answers beginning on page 1002.

This exercise introduces students to a nonlinear system of equations. In Lesson 9.6, students will learn how to solve a system of equations in which at least one equation is not linear.

ALTERNATIVE
Assessment

Portfolio Activity

The Portfolio Activity can be used as preparation for the Chapter Project or as a separate activity. In the Portfolio Activity on this page, students create a parabola by folding patty paper marked with a focus and a directrix. They are asked to make a conjecture about how to make a narrower or wider parabola.

Answers to Portfolio Activities can be found in Additional Answers fo the Teacher's Edition.

Factor each expression. *(LESSON 5.3)*

54. $16x^2 - 1$ $(4x+1)(4x-1)$

55. $3a^2 + 3a$ $3a(a+1)$

56. $6x^2y - 18xy^2$ $6xy(x-3y)$

57. $m^2n + 7m^2n^2 - 3mn$ $mn(m+7mn-3)$

Factor each quadratic expression. *(LESSON 5.3)*

58. $(x-4)^2$
59. $(y-2)(y-10)$
60. $(7x-2)(x-2)$

58. $x^2 - 8x + 16$

59. $y^2 - 12y + 20$

60. $7x^2 - 16x + 4$

61. $5x^2 - 15x + 10$ $5(x-2)(x-1)$

62. $12w^2 - w - 6$ $(3w+2)(4w-3)$

63. $6y^2 - 5y - 25$ $(3y+5)(2y-5)$

Simplify each rational expression. *(LESSON 8.3)*

64. $\frac{x^2-x-2}{x^2-2x-8} \cdot \frac{x+2}{x^2+5x+4}$ $\frac{x-2}{x^2-16}$

65. $\frac{x^2-x-6}{x-1} \cdot \frac{x^3-1}{x^2-2x-3}$ $\frac{x^3+3x^2+3x+2}{x+1}$

66. $\frac{x^2-4}{x^2+x-6} \div \frac{x+2}{x^2+4x+3}$ $x+1$

67. $\frac{x^2+7x+10}{x^2+8x+15} \div \frac{x^2+4x+4}{x+2}$ $\frac{1}{x+3}$

Look Beyond

internet connect
Portfolio Extension
Go To: **go.hrw.com**
Keyword:
MB1 Parabola

68 Graph $y = 2(x-3)^2 + 4$ and $x = \frac{1}{4}(y-3)^2 + 1$ together on the same screen, and find any points of intersection. $(\approx 2.22, \approx 5.21)$ and $(\approx 4.13, \approx 6.54)$

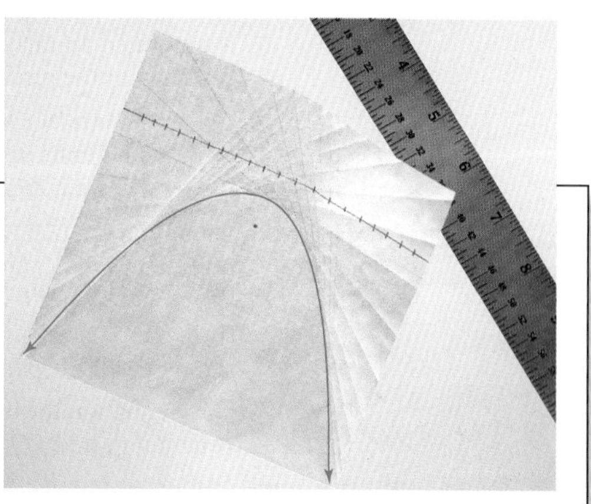

1. Draw a straight line from one side of the patty paper, or wax paper, to the other. This will be the directrix.
2. Place the focus anywhere except on the directrix.
3. Fold the paper so that the focus lies on one end of the directrix, and crease the paper. Then unfold it.
4. Move the focus along the directrix, making folds as you go, until you come to the other end of the directrix. You should make between 15 and 25 folds.
5. Compare the parabola formed by your creases with your classmates' parabolas. Make a conjecture about how to make a narrower or wider parabola. Verify your conjecture by folding a second parabola.
6. Fold your parabola in half along its axis of symmetry.
7. Explain how the definition of a parabola is related to your folded parabola.

WORKING ON THE CHAPTER PROJECT

You should now be able to complete Activity 1 of the Chapter Project.

Circles

Why Circles are used to describe the receiving area of radio signals.

Objectives

- Write an equation for a circle given sufficient information.
- Given an equation of a circle, graph it and label the radius and the center.

APPLICATION
COMMUNICATIONS

A radio tower is located 25 miles east and 30 miles south of Lorne's home. The radio signal is strong enough to reach homes within a 50 mile radius. Write an equation that represents all ground locations 50 miles from the radio tower. Can someone living 10 miles east and 5 miles north of Lorne receive the radio signal? This is an example of a *locus problem.* You will solve this problem in Example 3.

A **locus problem** involves an equation that represents a set of points that satisfy certain conditions. For example, an equation for the set of all points $P(x, y)$ in a plane that are a fixed distance, r, from $O(0, 0)$ can be obtained by using the distance formula.

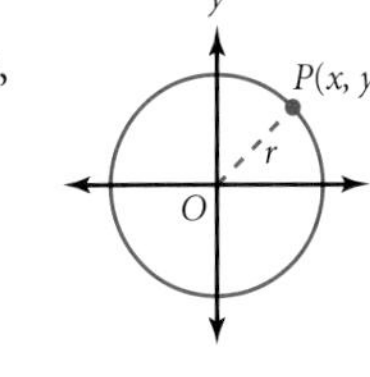

$$OP = r$$
$$\sqrt{(x-0)^2 + (y-0)^2} = r$$
$$\sqrt{x^2 + y^2} = r$$
$$x^2 + y^2 = r^2$$

A **circle** is the set of all points in a plane that are a constant distance, called the radius, from a fixed point, called the **center.**

Standard Equation of a Circle

An equation for the circle with its center at (0, 0) and a radius of r is

$$x^2 + y^2 = r^2.$$

Alternative Teaching Strategy

USING TABLES Tell students to draw a set of coordinate axes on graph paper then draw a circle with its center at the origin and a radius of 8 units. Give students the table at right. Have them complete the table for points on or near the circle. Ask them to explain why all the entries in the right column are at or near 8. **These are distances from the origin, and the circle was drawn as the set of all points 8 units from the origin.** Lead students to see that an equation of the circle must be $\sqrt{x^2 + y^2} = 8$, or $x^2 + y^2 = 64$.

x	y	x^2+y^2	$\sqrt{x^2+y^2}$
8	0	64	8
−8	0	64	8
0	8	64	8
0	−8	64	8
4	7	65	≈8.1
4	−7	65	≈8.1
−7	4	65	≈8.1
−7	−4	65	≈8.1
7.5	3	65.25	≈8.1
7.5	−3	65.25	≈8.1

Prepare

NCTM PRINCIPLES & STANDARDS 1–4, 6–10

QUICK WARM-UP

Find the distance between P and Q.

1. $P(-4, 10)$ and $Q(5, -2)$ 15
2. $P(1, 2)$ and $Q(3, 5)$ $\sqrt{13}$

Solve by completing the square.

3. $x^2 - 4x = 12$
   $x = 6$ *or* $x = -2$
4. $x^2 + 6x + 1 = 0$
   $x = -3 \pm 2\sqrt{2}$

Graph each equation and identify the conic section.

5. $x^2 + 9y^2 = 9$ ellipse
6. $x^2 + y^2 = 49$ circle
   Check students' graphs.

Also on Quiz Transparency 9.3

Teach

Why The circle has been a key consideration in religious and philosophical thinking, as well as the sciences and mathematics. Have students discuss ways in which the circle plays a part in the world around them.

ADDITIONAL EXAMPLE 1

Write the standard equation of the circle whose center is at the origin and whose radius is 4. Sketch the graph.

$x^2 + y^2 = 16$

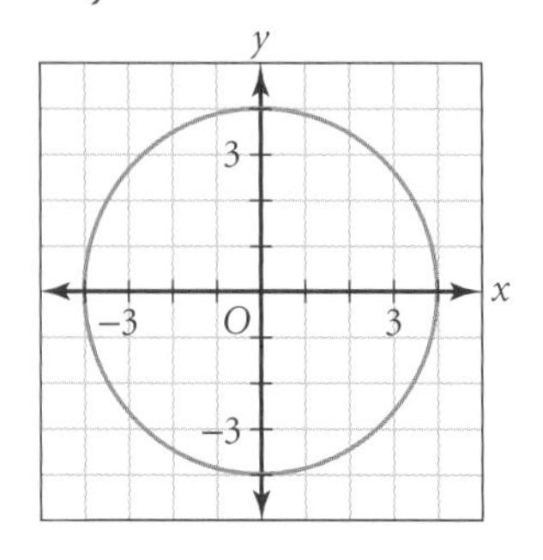

TRY THIS

$x^2 + y^2 = 4$

Teaching Tip

TECHNOLOGY If you are using a TI-82 or TI-83 graphics calculator to create the circle in Example 1, use the following square viewing window:

Xmin=–4.7	Ymin=–3.1
Xmax=4.7	Ymax=3.1
Xscl=1	Yscl=1

Activity Notes

Remind students that to graph a circle on a graphics calculator, the equation must be solved for y and graphed in two parts, and the circle should be viewed in a square viewing window to appear circular. Students should find that vertical and horizontal transformation techniques studied previously apply to circles.

CHECKPOINT ✔

3. It is translated 3 units to the left and 2 units up.

CHECKPOINT ✔

5. $r = 3$; $C(2, -1)$

EXAMPLE 1 **Write the standard equation of the circle whose center is at the origin and whose radius is 3. Sketch the graph.**

SOLUTION

$$x^2 + y^2 = r^2$$
$$x^2 + y^2 = 3^2$$

Thus, an equation for the circle is $x^2 + y^2 = 9$.

Plot the points (3, 0), (0, 3), (–3, 0), and (0, –3), and sketch a circle through these points.

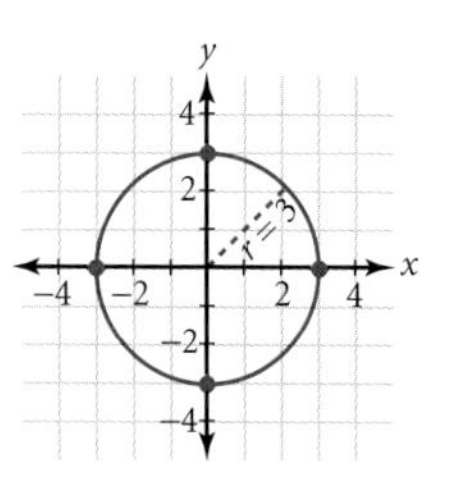

TRY THIS Write the standard equation of the circle whose center is at the origin and whose radius is 2. Sketch the graph.

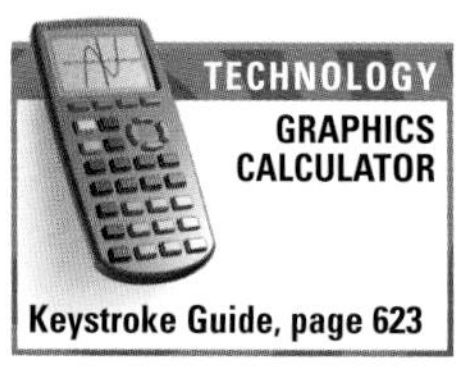

Recall from Lesson 9.1 that you can use a graphics calculator to graph a circle. To graph $x^2 + y^2 = 4$, solve for y and graph the two resulting equations together.

$$x^2 + y^2 = 4 \rightarrow \begin{cases} y = \sqrt{4 - x^2} \\ y = -\sqrt{4 - x^2} \end{cases}$$

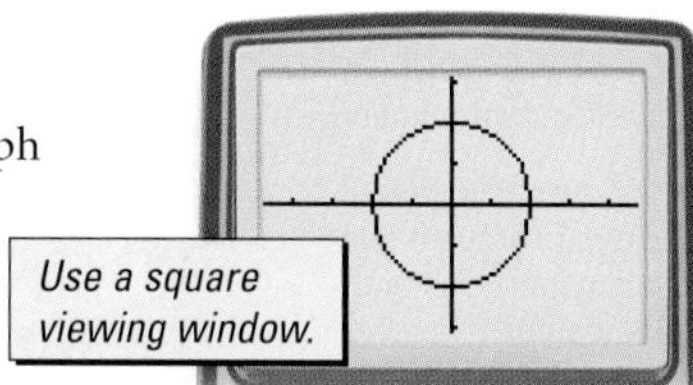

Use a square viewing window.

Activity: Exploring Translations of Circles

You will need: a graphics calculator

1. Graph the circle defined by $x^2 + y^2 = 16$. Find its radius.
2. Graph the circle defined by $(x + 3)^2 + (y - 2)^2 = 16$. Find its radius, and approximate the coordinates of the center.
3. Describe the transformation of the graph of $x^2 + y^2 = 16$ in Step 1 to the graph of $(x + 3)^2 + (y - 2)^2 = 16$ in Step 2. **CHECKPOINT ✔**
4. Compare the equation of the circle $(x + 3)^2 + (y - 2)^2 = 16$ with the coordinates of its center. Describe the relationship you observe.
5. Predict the radius and center of the circle defined by the equation $(x - 2)^2 + (y + 1)^2 = 9$. Graph the circle to support your prediction. **CHECKPOINT ✔**

The standard equation for a translated circle is given below.

Standard Equation of a Translated Circle

The standard equation for a circle with its center at (h, k) and a radius of r is

$$(x - h)^2 + (y - k)^2 = r^2.$$

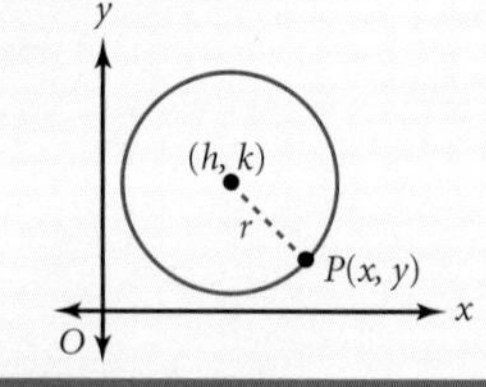

Interdisciplinary Connection

EARTH SCIENCE Geostationary satellites stay above the same point on Earth at all times. This means the satellite must complete one revolution in exactly the same time that Earth takes for one revolution. To do this, a satellite is placed in a circular orbit at an altitude of approximately 22,300 miles above Earth's surface.

Assume that the center of Earth is at the origin of a coordinate plane and that Earth's radius is about 3960 miles. Find an equation of a communication satellite's orbit. $x^2 + y^2 = 26{,}260^2$

Enrichment

Tell students that tangent circles touch each other at one point. Ask them to write the equation of a circle centered at the origin and find the equation of a tangent circle. They should graph both the original circle and the tangent circle. Then they should find the point of tangency both graphically and algebraically. Some advanced students may be able to make a couple of generalizations about all circles tangent to the given circle.

EXAMPLE 2 **Write the standard equation for the translated circle graphed at right.**

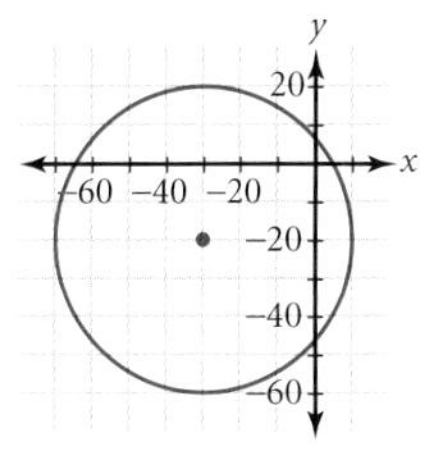

SOLUTION

The center of the circle is (−30, −20).
Substitute −30 for h and −20 for k.
Since the radius is 40, substitute 40 for r.

$$(x-h)^2+(y-k)^2=r^2$$
$$[x-(-30)]^2+[y-(-20)]^2=40^2$$
$$(x+30)^2+(y+20)^2=1600 \quad \text{Write the standard equation.}$$

EXAMPLE 3 **Refer to the radio-signal problem described at the beginning of the lesson.**

APPLICATION
COMMUNICATIONS

a. Write an equation that represents all ground locations 50 miles from the radio tower, given that Lorne's home is located at (0, 0).

b. Can someone who lives 10 miles east and 5 miles north of Lorne's home receive the radio signal?

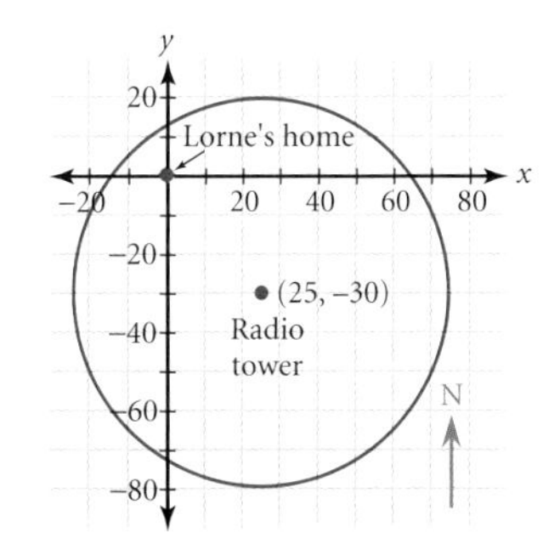

SOLUTION

PROBLEM SOLVING

a. Select appropriate notation. Since the origin represents Lorne's home, the radio tower is represented by the point (25, −30). If the radio signal is strong enough to reach homes within a 50-mile radius, then $r = 50$.

Write the standard equation of a circle with its center at (25, −30) and a radius of 50.

$$(x-h)^2+(y-k)^2=r^2$$
$$(x-25)^2+[y-(-30)]^2=50^2$$
$$(x-25)^2+(y+30)^2=2500$$

In order to listen to a radio station, you must be able to receive the radio signal.

b. The coordinates of a point that lie within the circle will satisfy the inequality $(x-25)^2+(y+30)^2<2500$. The point (10, 5) represents a location 10 miles east and 5 miles north of Lorne's home. Test (10, 5) to see if it satisfies the inequality $(x-25)^2+(y+30)^2<2500$.

$$(x-25)^2+(y+30)^2<2500$$
$$(10-25)^2+(5+30)^2 \overset{?}{<} 2500$$
$$1450<2500 \quad \textbf{True}$$

Thus, someone who lives 10 miles east and 5 miles north of Lorne's home can receive the radio signal.

TRY THIS Tell whether $A(2, 1)$ is inside, outside, or on the circle whose center is at (−2, 3) and whose radius is 4.

CRITICAL THINKING Explain how to tell whether $A(s, t)$ is inside, outside, or on the circle centered at (h, k) with a radius of r.

Inclusion Strategies

VISUAL LEARNERS Have students create a "color template" of the standard equation of a translated circle as shown at the bottom of page 580. They should copy the equation on a sheet of paper, writing h in one color, k in another, and r in another. Below this they should copy the graph, using the same colors to label h, k, and r. Students can then utilize this color scheme each time they write the standard equation of a circle, using the template colors for the specific values of h, k, and r. They can then use the template graph as a guide when graphing the given equation.

Teaching Tip

TECHNOLOGY When using a TI-83 or TI-82 graphics calculator to graph the circle and its transformation in Steps 1–4 of the Activity, enter the following:

Y1=√(16–X²)
Y2=–√(16–X²)
Y3=√(16–(X+3)²)+2
Y4=–√(16–(X+3)²)+2

Use the following square viewing window:

Xmin=–9.4	Ymin=–6.2
Xmax=9.4	Ymax=6.2
Xscl=1	Yscl=1

ADDITIONAL EXAMPLE 2

Write the standard equation for the translated circle graphed below.

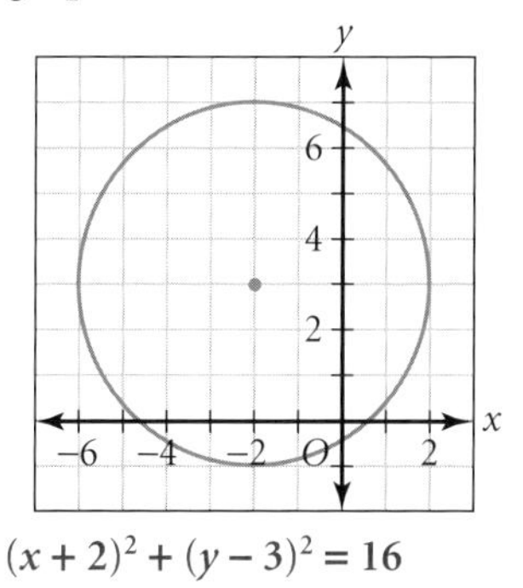

$(x+2)^2+(y-3)^2=16$

For Additional Example 3, see page 582.

TRY THIS
outside

CRITICAL THINKING
Evaluate the quantity $(s-h)^2+(t-k)^2$. If the result is less than r^2, the point is inside the circle. If the result is greater than r^2, the point is outside the circle. If the result is equal to r^2, the point is on the circle.

ADDITIONAL
EXAMPLE 3

A radio tower for a local station is located 20 miles north and 35 miles west of the high school. The radio signal is strong enough to reach locations within a 40-mile radius of the tower. **With the high school at the origin of a coordinate system, write an equation that represents all ground locations 40 miles from the radio tower. Can someone who lives 2 miles west and 4 miles south of the high school receive the signal? Justify your response.**

$(x + 35)^2 + (y - 20)^2 = 1600$; no; $(-2, -4)$ is farther than 40 mi from the station.

ADDITIONAL
EXAMPLE 4

Write the standard equation for the circle given by $x^2 + y^2 - 8x + 7 = 0$. State the coordinates of its center and give its radius. Then sketch the graph.

$(x - 4)^2 + y^2 = 9$; center: $(4, 0)$; radius: 3

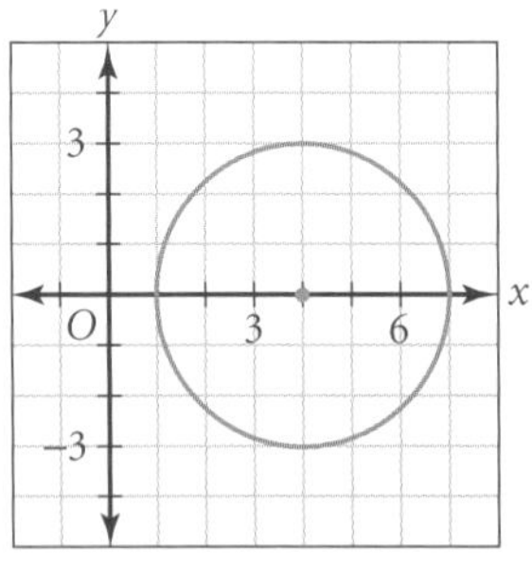
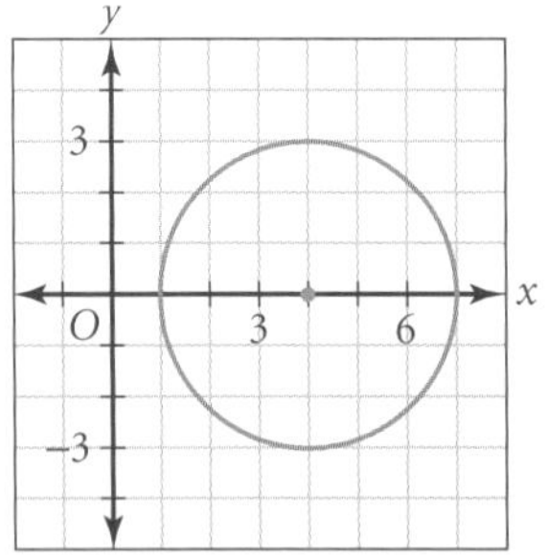

TRY THIS
$(x - 1)^2 + (y + 1)^2 = 9$;
center: $(1, -1)$; radius: 3

To write the standard equation of a translated circle, you may need to complete the square. This is shown in Example 4.

EXAMPLE 4 **Write the standard equation for the circle $x^2 + y^2 + 4x - 6y - 3 = 0$. State the coordinates of its center and give its radius. Then sketch the graph.**

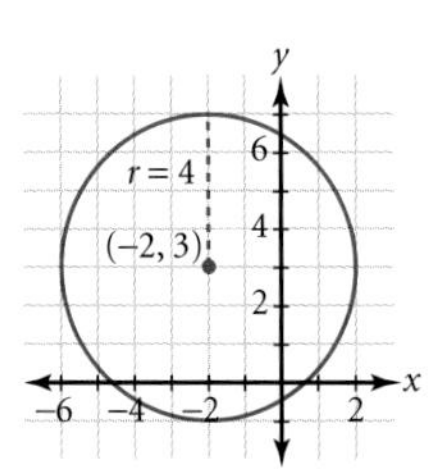

SOLUTION

$$x^2 + y^2 + 4x - 6y - 3 = 0$$
$$x^2 + 4x + y^2 - 6y = 3$$
$$(x^2 + 4x + 4) + (y^2 - 6y + 9) = 3 + 4 + 9 \qquad \textit{Complete the squares.}$$
$$(x + 2)^2 + (y - 3)^2 = 16 \qquad \textit{Write the standard equation.}$$

Thus, the standard equation is $(x + 2)^2 + (y - 3)^2 = 16$. The center is at $(-2, 3)$ and the radius is 4.

TRY THIS Write the standard equation for the circle $x^2 + y^2 - 2x + 2y - 7 = 0$. State the coordinates of its center and give its radius. Then sketch the graph.

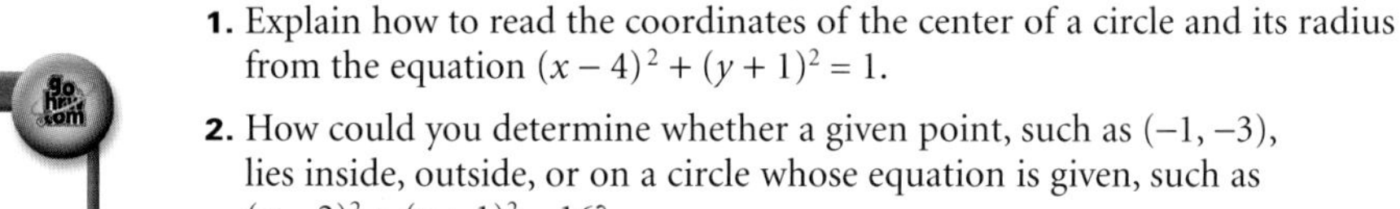

Communicate

internet connect
Activities Online
Go To: **go.hrw.com**
Keyword:
MB1 TV

1. Explain how to read the coordinates of the center of a circle and its radius from the equation $(x - 4)^2 + (y + 1)^2 = 1$.
2. How could you determine whether a given point, such as $(-1, -3)$, lies inside, outside, or on a circle whose equation is given, such as $(x - 2)^2 + (y + 1)^2 = 16$?
3. How would you determine whether the equation $x^2 + 8x + y^2 + 16 = 0$ represents a circle?

Guided Skills Practice

4. $x^2 + y^2 = 36$

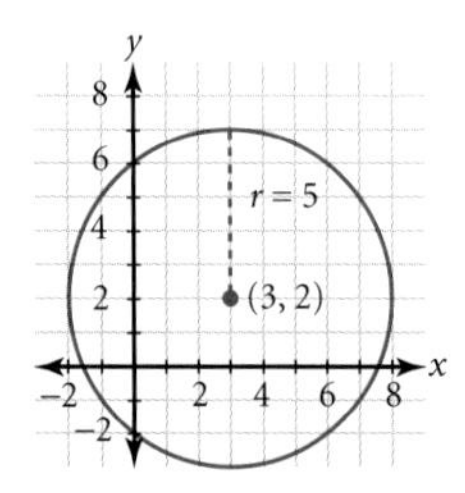

4. Write an equation for the circle whose center is at the origin and whose radius is 6. Sketch the graph. ***(EXAMPLE 1)***
5. Write the standard equation for the translated circle graphed at left. ***(EXAMPLE 2)*** $(x - 3)^2 + (y - 2)^2 = 25$
6. **COMMUNICATIONS** Refer to the radio-signal problem posed at the beginning of the lesson. Can someone who lives 9 miles west and 5 miles south of Lorne's home receive the radio signal? ***(EXAMPLE 3)*** yes

Reteaching the Lesson

USING DISCUSSION Give students the equations listed below.

a. $x^2 + y^2 = 36$ **b.** $x^2 - 4x + y^2 - 32 = 0$
c. $x^2 + (y + 1)^2 = 25$ **d.** $(x - 2)^2 + (y + 1)^2 = 25$
e. $(x - 2)^2 + y^2 = 25$ **f.** $x^2 + y^2 + 2y - 15 = 0$
g. $x^2 - y^2 = 25$ **h.** $(x + 2)^2 + (y - 1)^2 = 36$
i. $x^2 + y^2 = 49$ **j.** $(x + 2)^2 - (y + 1)^2 = 49$

Ask students to classify the equations as follows:

- circles with the same center a, i; b, e; c, f
- circles with the same radius a, b, h; c, d, e
- not circles g, j

Encourage them to justify their classifications.

7. Write the standard equation for the circle $x^2 + 6x + y^2 - 4y - 3 = 0$. State the coordinates of its center and give its radius. Then sketch the graph. ***(EXAMPLE 4)*** $(x + 3)^2 + (y - 2)^2 = 16$; $C(-3, 2)$; $r = 4$

Practice and Apply

Write the standard equation for each circle graphed below.

8. $x^2 + y^2 = 16$
9. $x^2 + y^2 = 1$
10. $(x + 5)^2 + (y + 3)^2 = 4$

8.

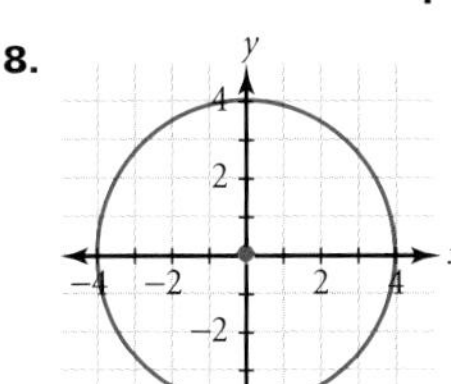

9.

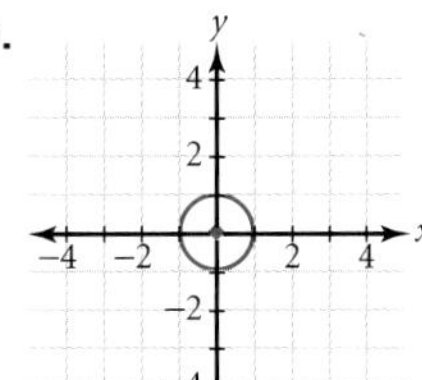

10.

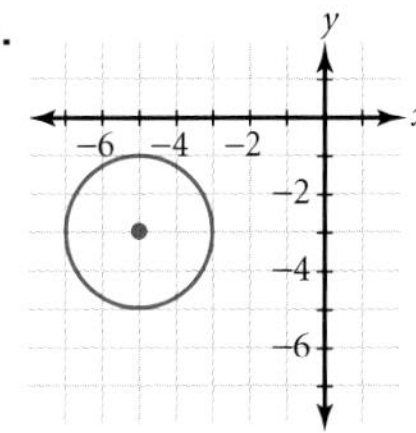

11.

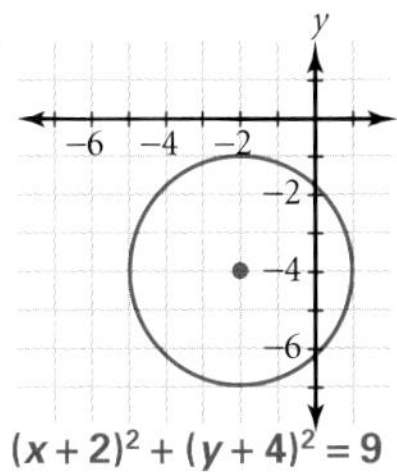

$(x + 2)^2 + (y + 4)^2 = 9$

12.

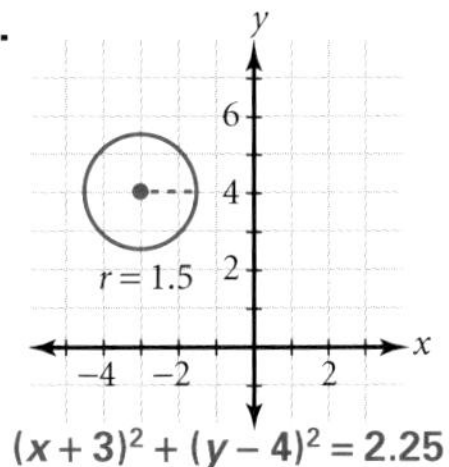

$(x + 3)^2 + (y - 4)^2 = 2.25$

13.

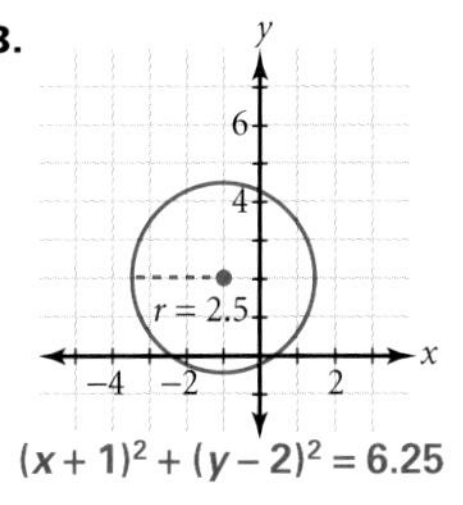

$(x + 1)^2 + (y - 2)^2 = 6.25$

Write the standard equation of a circle with each given radius and center.

14. $r = 4$; $C(0, 0)$ $x^2 + y^2 = 16$

15. $r = 5$; $C(0, 0)$ $x^2 + y^2 = 25$

16. $r = 11$; $C(0, 0)$ $x^2 + y^2 = 121$

17. $r = 7$; $C(0, 0)$ $x^2 + y^2 = 49$

18. $r = 1$; $C(2, 3)$ $(x - 2)^2 + (y - 3)^2 = 1$

19. $r = 12$; $C(3, 5)$

20. $r = 10$; $C(-2, -7)$

21. $r = 5$; $C(-5, -1)$

22. $r = 4$; $C(-2, 8)$

23. $r = 15$; $C(-6, 9)$

24. $r = 2$; $C(0, 12)$ $x^2 + (y - 12)^2 = 4$

25. $r = 3$; $C(0, 4)$ $x^2 + (y - 4)^2 = 9$

26. $r = \frac{1}{3}$; $C(-2, -2)$ $(x + 2)^2 + (y + 2)^2 = \frac{1}{9}$

27. $r = 2$; $C(3, 3)$ $(x - 3)^2 + (y - 3)^2 = 4$

28. $r = \frac{1}{2}$; $C(2, 0)$ $(x - 2)^2 + y^2 = \frac{1}{4}$

29. $r = \frac{1}{4}$; $C(1, 0)$ $(x - 1)^2 + y^2 = \frac{1}{16}$

30. $r = 1$; $C(a, a)$, where $a > 0$
$(x - a)^2 + (y - a)^2 = 1$

31. $r = 2$; $C(a, -2a)$, where $a > 0$
$(x - a)^2 + (y + 2a)^2 = 4$

19. $(x - 3)^2 + (y - 5)^2 = 144$
20. $(x + 2)^2 + (y + 7)^2 = 100$
21. $(x + 5)^2 + (y + 1)^2 = 25$
22. $(x + 2)^2 + (y - 8)^2 = 16$
23. $(x + 6)^2 + (y - 9)^2 = 225$

Graph each equation. Label the center and the radius.

32. $x^2 + y^2 = 9$

33. $x^2 + y^2 = 49$

34. $(x - 2)^2 + y^2 = 4$

35. $(x + 5)^2 + y^2 = 36$

36. $x^2 + (y + 3)^2 = 16$

37. $x^2 + (y - 2)^2 = 81$

38. $(x + 1)^2 + (y + 5)^2 = 100$

39. $(x + 6)^2 + (y + 1)^2 = 4$

40. $(x - 2)^2 + (y + 2)^2 = 64$

41. $(x - 3)^2 + (y + 3)^2 = 25$

42. $(x + 4)^2 + (y - 3)^2 = 49$

43. $(x + 2)^2 + (y - 4)^2 = 16$

Assess

Selected Answers

Exercises 4–7, 9–93 odd

ASSIGNMENT GUIDE	
In Class	1–7
Core	9–75 odd, 79
Core Plus	8–82 even
Review	83–94
Preview	95

Extra Practice can be found beginning on page 940.

Mid-Chapter Assessment for Lessons 9.1 through 9.3 can be found on page 114 of the *Assessment Resources.*

Error Analysis

For a circle with the equation $(x - 3)^2 + (y + 1)^2 = 4$, students often give the center of the circle as $(-3, 1)$ and the radius as 4, instead of $(3, -1)$ and 2, respectively. To avoid this error, suggest that they write the equation as $[x - (3)]^2 + [y - (-1)]^2 = 2^2$ in order to determine the correct values for h, k, and r.

7. $(x + 3)^2 + (y - 2)^2 = 16$; $C(-3, 2)$; $r = 4$

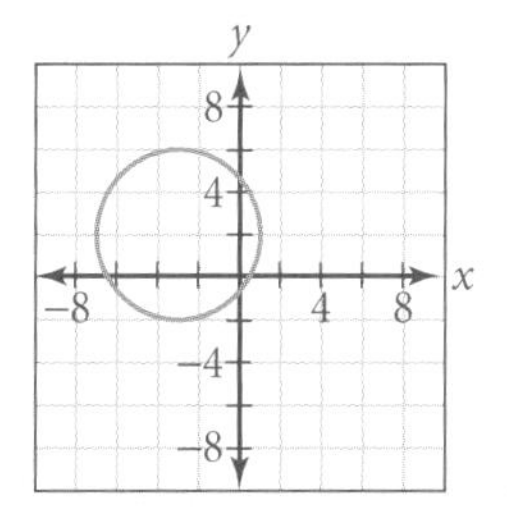

The answers to Exercises 35–43 can be found in Additional Answers beginning on page 1002.

32.

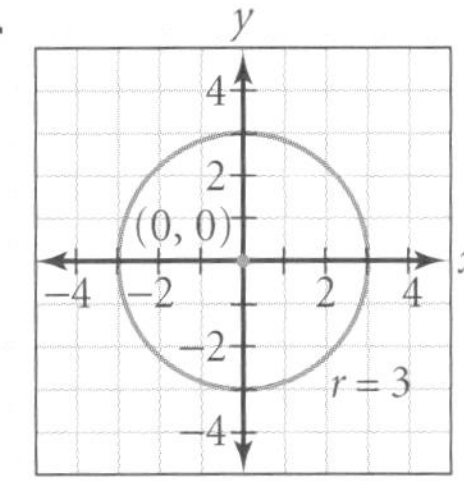

33.

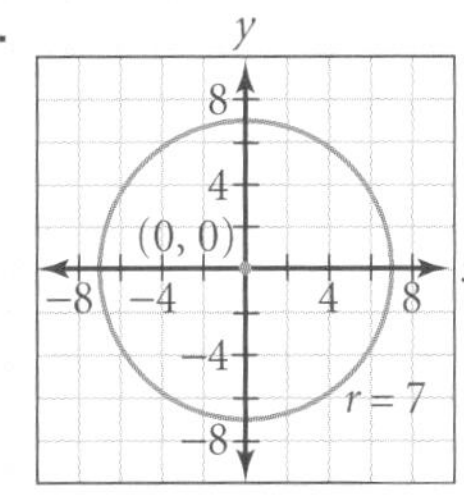

34.

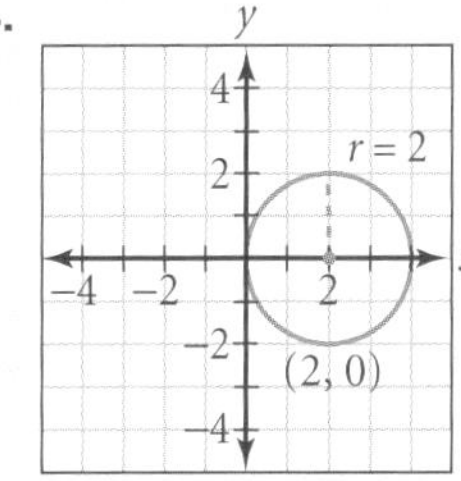

Exercise 95 demonstrates how small changes to the equation of a circle can produce a completely different curve, called an *ellipse.* Students will study ellipses in greater detail in Lesson 9.4, where they will learn how to determine the coordinates of the center, vertices, co-vertices, and foci from an equation of an ellipse.

52. $\left(x+\frac{1}{2}\right)^2+\left(y+\frac{1}{2}\right)^2=\frac{1}{2}$; $C\left(-\frac{1}{2}, -\frac{1}{2}\right)$; $r=\frac{\sqrt{2}}{2}\approx 0.71$

53. $\left(x-\frac{1}{2}\right)^2+\left(y+\frac{1}{2}\right)^2=\frac{1}{2}$; $C\left(\frac{1}{2}, -\frac{1}{2}\right)$; $r=\frac{\sqrt{2}}{2}\approx 0.71$

54. $(x-0.5)^2+(y+1.5)^2=10$; $C(0.5, -1.5)$; $r=\sqrt{10}\approx 3.16$

55. $(x-0.5)^2+(y+3.5)^2=25$; $C(0.5, -3.5)$; $r=5$

56. $(x-6)^2+(y-1)^2=45$; $C(6, 1)$; $r=3\sqrt{5}\approx 6.71$

57. $(x-3)^2+(y-5)^2=36$; $C(3, 5)$; $r=6$

58. $(x+3)^2+(y-7)^2=100$; $C(-3, 7)$; $r=10$

59. $(x-5)^2+(y+3)^2=34$; $C(5, -3)$; $r=\sqrt{34}\approx 5.83$

Practice

NAME ______ CLASS ______ DATE ______

Practice 9.3 Circles

Write the standard equation for each circle graphed below.

1. 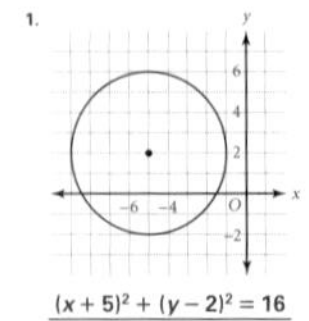

$(x+5)^2+(y-2)^2=16$

2. $(x-6)^2+(y-3)^2=\frac{25}{4}$

3. 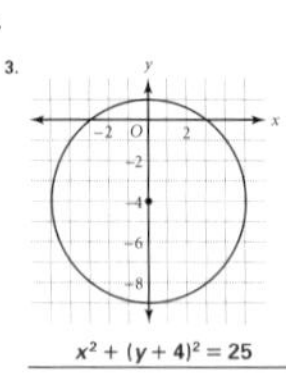

$x^2+(y+4)^2=25$

Write the standard equation of a circle with the given radius and center.

4. $r=\frac{3}{4}$; $C(0, 0)$ $x^2+y^2=\frac{9}{16}$
5. $r=2.5$; $C(-2, 1)$ $(x+2)^2+(y-1)^2=6.25$
6. $r=24$; $C(-3, -3)$ $(x+3)^2+(y+3)^2=576$

Graph each equation. Label the center and the radius.

7. $x^2+y^2=256$

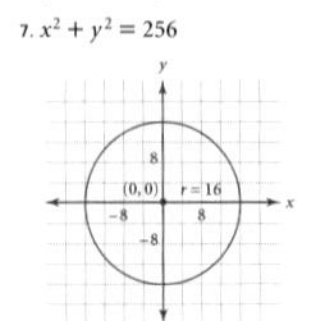

8. $x^2+(y-5)^2=16$

9. $(x-3)^2+(y+3)^2=9$

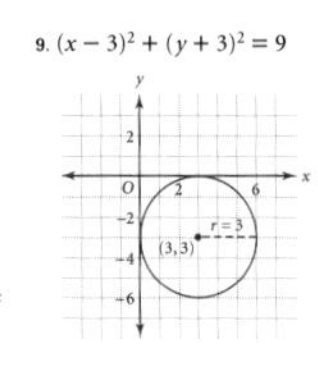

Write the standard equation for each circle. Then state the coordinates of its center, and give its radius.

10. $x^2+y^2-10x-16y+88=0$ $(x-5)^2+(y-8)^2=1$; $C(5, 8)$; $r=1$
11. $x^2+y^2+22x-2y=-120$ $(x+11)^2+(y-1)^2=2$; $C(-11, 1)$; $r=\sqrt{2}$

44. $x^2+(y+2)^2=16$; $C(0, -2)$; $r=4$

45. $(x-1)^2+y^2=9$; $C(1, 0)$; $r=3$

46. $(x+1)^2+(y+1)^2=4$; $C(-1, -1)$; $r=2$

47. $(x+1)^2+(y+3)^2=16$; $C(-1, -3)$; $r=4$

48. $(x-5)^2+(y-1)^2=49$; $C(5, 1)$; $r=7$

49. $(x-6)^2+(y+3)^2=64$; $C(6, -3)$; $r=8$

50. $(x+3)^2+y^2=26$; $C(-3, 0)$; $r=\sqrt{26}\approx 5.10$

51. $x^2+(y-10)^2=81$; $C(0, 10)$; $r=9$

Write the standard equation for each circle. Then state the coordinates of its center and give its radius.

44. $x^2+y^2+4y=12$ **45.** $x^2-2x+y^2=8$

46. $x^2+2x+y^2+2y=2$ **47.** $x^2+2x+y^2+6y=6$

48. $x^2+y^2-10x-2y=23$ **49.** $x^2+y^2-12x+6y=19$

50. $x^2+y^2+6x-17=0$ **51.** $x^2+y^2-20y+19=0$

52. $x^2+y^2+x+y=0$ **53.** $x^2+y^2-x+y=0$

54. $x^2+y^2-x+3y=7.5$ **55.** $x^2+y^2-x+7y=12.5$

56. $x^2+y^2-12x-2y-8=0$ **57.** $x^2+y^2-6x-10y-2=0$

58. $x^2+y^2+6x-14y-42=0$ **59.** $x^2+y^2-10x+6y=0$

State whether the graph of each equation is a parabola or a circle. Justify your response.

60. $y=x^2$ **parabola** **61.** $x^2=12-y^2$ **circle**

62. $y^2=12-x^2$ **circle** **63.** $x=y^2$ **parabola**

64. $x^2=4-(y-2)^2$ **circle** **65.** $10y^2=5(x-2)$ **parabola**

66. $4x=8(y+3)^2$ **parabola** **67.** $(y+2)^2=15-(x-2)^2$ **circle**

State whether the given point is inside, outside, or on the circle whose equation is given. Justify your response.

70. inside
71. inside
72. inside
73. outside

68. $P(2, 2)$; $x^2+y^2=9$ **inside** **69.** $P(1, 6)$; $x^2+y^2=49$ **inside**

70. $P(5, 1)$; $x^2-6x+y^2+8y=24$ **71.** $P(2, -3)$; $x^2-4x+y^2+6y=12$

72. $P(0, 0)$; $x^2+10x+y^2+2y=10$ **73.** $P(12, 3)$; $x^2-12x+y^2+2y=12$

74. $P(0.5, 0.5)$; $x^2+y^2=1$ **inside** **75.** $P(1.5, 3.5)$; $x^2+y^2=6$ **outside**

CHALLENGE

76. Tell whether $P(a, a)$ is inside, outside, or on the circle defined by the equation $x^2-2ax+y^2+4ay=4a^2$. Justify your response. **on the circle**

CONNECTIONS

77. COORDINATE GEOMETRY The figure at right shows two circles with centers at C_1 and C_2. An equation for the circle with its center at C_1 is $(x-3)^2+(y-3)^2=(r_1)^2$, and an equation with the circle for its center at C_2 is $(x-3)^2+(y-1)^2=(r_2)^2$.

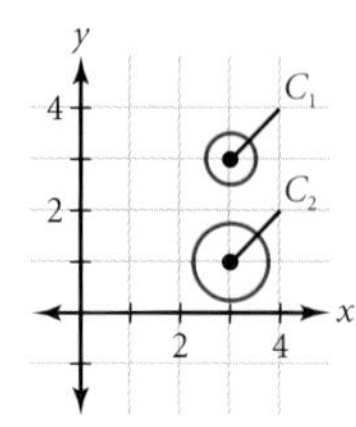

a. Find one value of r_1 and one value of r_2 such that the circle with its center at C_1 is completely enclosed by the circle with its center at C_2. Justify your response.

b. Find one value of r_1 and one value of r_2 such that the circles intersect at two points. Justify your response.

c. Find one value of r_1 and one value of r_2 such that the circles intersect at exactly one point. Justify your response.

78. TRANSFORMATIONS The circle defined by $(x-1)^2+(y+4)^2=16$ is translated 3 units to the left and 2 units down. Write the standard equation for the resulting circle. $(x+2)^2+(y+6)^2=16$

79. TRANSFORMATIONS The circle defined by $x^2+y^2-8x-4y+9=0$ is translated 4 units to the right and 6 units up. Write the standard equation for the resulting circle. $(x-8)^2+(y-8)^2=11$

60. Parabola; the equation has the form $y=\frac{1}{4p}x^2$, where $p=\frac{1}{4}$.

61. Circle; the equation can be written as $x^2+y^2=12$, which has the form $x^2+y^2=r^2$, where $r=\sqrt{12}=2\sqrt{3}$.

62. Circle; the equation can be written as $x^2+y^2=12$, which has the form $x^2+y^2=r^2$, where $r=\sqrt{12}=2\sqrt{3}$.

63. Parabola; the equation has the form $x=\frac{1}{4p}y^2$, where $p=\frac{1}{4}$.

64. Circle; the equation can be written as $x^2+(y-2)^2=4$, which has the form $(x-h)^2+(y-k)^2=r^2$, where $h=0$, $k=2$, and $r=2$.

65. Parabola; the equation can be written as $x-2=2y^2$, which has the form $(x-h)=\frac{1}{4p}(y-k)^2$, where $h=2$, $k=0$, and $p=\frac{1}{8}$.

66. Parabola; the equation can be written as $x=2(y+3)^2$, which has the form $(x-h)=\frac{1}{4p}(y-k)^2$, where $h=0$, $k=-3$, and $p=\frac{1}{8}$.

APPLICATIONS

80. $x^2 + y^2 = (26{,}260)^2$

80. SPACE SCIENCE A satellite in a stationary orbit rotates once each day about the Earth. Assume that the satellite is 22,300 miles above the surface of Earth and that Earth's radius is 3960 miles. Write an equation that represents the orbit of this satellite on a coordinate plane with the origin representing the center of the Earth.

81. yes

81. GEOLOGY Vibrations from a certain earthquake were noticeable up to 120 miles away from the earthquake's epicenter. The center of the closest city to the earthquake's epicenter is 85 miles east and 90 miles south of the epicenter. If Alexandra was 15 miles west of this city, could she have noticed the vibrations at the time of the earthquake?

82. no

82. COMMUNICATIONS A radio program is broadcast from a van that is located 40 miles east and 30 miles north of a radio tower. The van sends the radio signal to the tower, which then transmits the signal a maximum distance of 70 miles. If Jessica is 30 miles east of the van, is she able to receive the radio program?

Many radio stations have the equipment needed to broadcast from a van.

92. Translate the graph 2 units to the left and then stretch it vertically by a factor of 4.

93. Translate the graph 4 units to the left, stretch it horizontally by a factor of 2, and then translate it 3 units down.

94. Translate the graph 1 unit to the right, compress it horizontally by a factor of 0.5, compress it vertically by a factor of 0.5, and then translate it 6 units up.

Look Back

83. $x = \frac{11}{2}$; $\left(\frac{11}{2}, -\frac{9}{4}\right)$

84. $x = -\frac{3}{4}$; $\left(-\frac{3}{4}, -\frac{9}{8}\right)$

85. $x = -\frac{3}{2}$; $\left(-\frac{3}{2}, \frac{49}{4}\right)$

For each quadratic function, find the equation for the axis of symmetry and give the coordinates of the vertex. ***(LESSON 5.5)***

83. $y = -11x + 28 + x^2$ **84.** $y = 2x^2 + 3x$ **85.** $y = -3x + 10 - x^2$

86. $x = \frac{1}{3}\left(\frac{\ln 748}{\ln 7} + 2\right) \approx 1.80$

87. $x = -\frac{\ln 12}{3} \approx -0.83$

88. $x = \frac{\ln 7}{\ln 2} \approx 2.81$

Solve each equation. Write the exact solution and the approximate solution to the nearest hundredth, when appropriate. ***(LESSON 6.7)***

86. $7^{3x-2} + 2 = 750$ **87.** $e^{-3x} = 12$ **88.** $\log_2 7 = x$

89. $\log_a 8 = \frac{3}{2}$ $a = 4$ **90.** $\log_3 x = 4$ $x = 81$ **91.** $\ln(x + 1) = 2 \ln 5$ $x = 24$

For each function, describe the transformations applied to $f(x) = \sqrt{x}$. ***(LESSON 8.6)***

92. $g(x) = 4\sqrt{x - 2}$ **93.** $a(x) = \sqrt{0.5x + 4} - 3$ **94.** $b(x) = 0.5\sqrt{2x - 1} + 6$

Look Beyond

95. Sample answer: The graphs resemble circles that have been stretched vertically. (They are ellipses.)

95 In this lesson you learned that the graph of $x^2 + y^2 = r^2$, where r is a positive constant, is a circle centered at the origin. Using a square viewing window, graph $4x^2 + 2y^2 = 12$, and describe the shape of the graph. Graph $8x^2 + y^2 = 24$, and describe the shape of the graph.

67. Circle; the equation can be written as $(x - 2)^2 + (y + 2) = 15$, which has the form $(x - h)^2 + (y - k)^2 = r^2$, where $h = 2$, $k = -2$, and $r = \sqrt{15}$.

68. inside because $2^2 + 2^2 = 8 < 9$

69. inside because $1^2 + 6^2 = 37 < 49$

70. inside because $(5 - 3)^2 + (1 + 4)^2 = 29 < 49$

71. inside because $P(2, -3)$ is the center

72. inside because $(0 + 5)^2 + (0 + 1)^2 = 26 < 36$

73. outside because $(12 - 6)^2 + (3 + 1)^2 = 52 > 49$

74. inside because $0.5^2 + 0.5^2 = 0.5 < 1$

75. outside because $1.5^2 + 3.5^2 = 14.5 > 6$

76. on the circle because $(a - a)^2 + (a + 2a)^2\ (3a)^2 = 9a^2$

The answer to Exercise 77 can be found in Additional Answers beginning on page 1002.

Student Technology Guide

NAME ______ CLASS ______ DATE ______

Student Technology Guide
9.3 Circles

There are several ways to graph a circle on a graphics calculator. One method relies on solving the equation of the circle for y. Another method uses a special drawing feature.

Example: Graph $x^2 + 8x + y^2 - 6y = 0$.

- First complete the squares.
  $(x^2 + 8y + 16) + (y^2 - 6y + 9) = 16 + 9$
  $(x + 4)^2 + (y - 3)^2 = 25$
  Notice that the center of this circle is at $(-4, 3)$, and the radius of the circle is 5.
- Solve for y.
  $(y - 3)^2 = 25 - (x + 4)^2$
  $y = 3 \pm \sqrt{25 - (x + 4)^2}$
- The ± sign means you need to graph two equations. Enter the first equation as Y1. Press Y= 3 + 2nd x^2 25 − (X,T,θ,n + 4) x^2).
- Enter the second equation as Y2. Move to Y2 and press 3 − 2nd x^2 25 − (X,T,θ,n + 4) x^2).
- Press GRAPH.

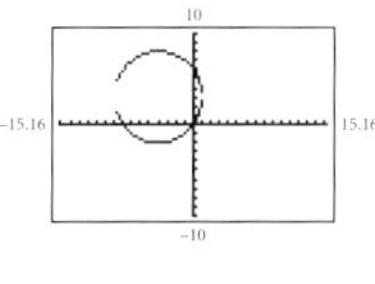

Remember to use a square viewing window to see an undistorted graph.

Graph each equation. For each circle, state the coordinates of its center and give the radius.

1. $x^2 + y^2 + 4y = 0$ center (0, −2); radius 2
2. $x^2 - 8x + y^2 - 6y + 9 = 0$ center (4, 3); radius 4
3. $x^2 + y^2 - 2x + 8y - 13 = 0$ center (1, −4); radius $\sqrt{30}$

You can use a drawing feature to draw a circle with a given center and radius. (Note: Press Y= and clear any equations before using a drawing command.)

To draw a circle with a center at $(2, -3)$ and a radius of 6, press 2nd PRGM 9:Circle ENTER 2 , (−) 3 , 6) ENTER.

To erase a drawing, press 2nd PRGM 1:ClrDraw ENTER.

There are some disadvantages to using the circle drawing feature instead of graphing the equations. You cannot use the trace, intersect, or zero feature to explore the circle.

Prepare

NCTM PRINCIPLES & STANDARDS 1–4, 6–10

QUICK WARM-UP

Write the standard equation of the circle with the given radius and center.

1. 9; (0, 0)
   $x^2 + y^2 = 81$
2. 1; (0, 5)
   $x^2 + (y - 5)^2 = 1$
3. 4; (−8, −1)
   $(x + 8)^2 + (y + 1)^2 = 16$
4. 5; (−4, 2)
   $(x + 4)^2 + (y - 2)^2 = 25$
5. Write the standard equation of the circle defined by the given equation. Then state the coordinates of its center and give its radius.

   $x^2 + y^2 + 2x - 6y = 6$
   $(x + 1)^2 + (y - 3)^2 = 16$
   center: (−1, 3); radius: 4

Also on Quiz Transparency 9.4

Teach

Why The orbital movements of planets in our solar system can be modeled by elliptical equations. Ask students to discuss the shapes of the path of the Moon around the Earth and the Earth around the Sun.

Ellipses

Objectives

- Write the standard equation for an ellipse given sufficient information.
- Given an equation of an ellipse, graph it and label the center, vertices, co-vertices, and foci.

Why *Ellipses are used to describe the paths of planets around the Sun.*

In the Activity below, you will explore the definition of an ellipse by modeling ellipses.

Activity: Modeling Ellipses

You will need: 3 pieces of corrugated cardboard at least 8 inches by 8 inches, 2 tacks, 36 inches of string, and a pencil

Draw each ellipse on a separate piece of cardboard, and write on the cardboard the length of string that was used to create the ellipse.

1. Place two tacks 4 inches apart into a piece of corrugated cardboard. Tie the ends of a 10-inch piece of string together and loop the string around the tacks. Hold the tacks down with one hand, and with the other hand, use the pencil to pull the string taut as shown above. Move your pencil along the path that keeps the string taut at all times. When you return to your starting point, the path of the pencil will have formed an ellipse.

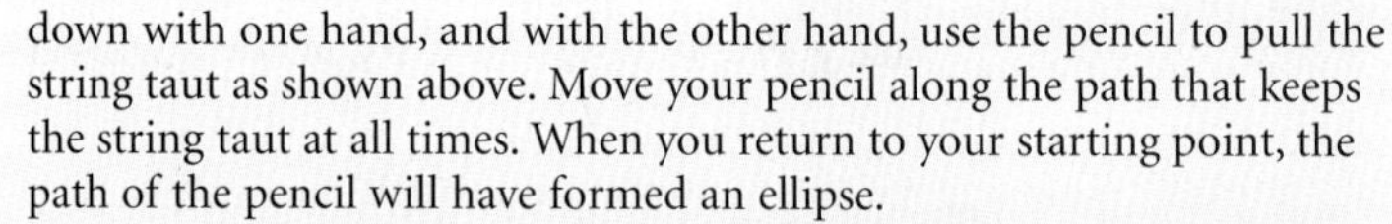

2. Repeat Step 1 with the same locations for the tacks but a 12-inch string.
3. Repeat Step 1 with the same locations for the tacks but a 14-inch string.

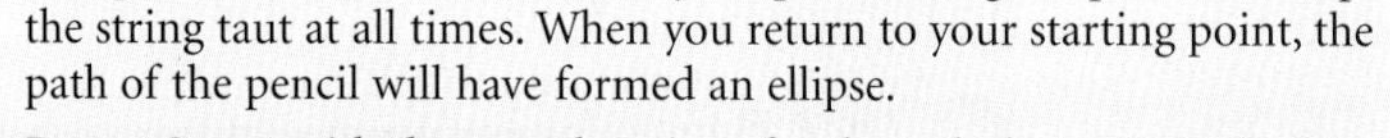

CHECKPOINT ✔ 4. How does the length of the string affect the shape of the ellipse that is drawn?

Alternative Teaching Strategy

USING PATTERNS Have students plot the points (5, 0), (−5, 0), (3, 3.2), (−4, 2.4), (0, 4), (−3, −3.2), (0, −4), and (4, −2.4) and connect them with a curve. Tell students that the curve they graphed is called an ellipse. Now give them the fixed points $F_1(-3, 0)$ and $F_2(3, 0)$. Ask them to use the distance formula to find the sum of the distances from the two fixed points, F_1 and F_2, to (3, 3.2). Repeat for the points (0, −4), (−4, 2.4), (−3, −3.2), and (−5, 0). Be sure they notice that all of the sums equal 10. Now define an ellipse and have them identify the center, foci, vertices, and co-vertices of the ellipse whose equation is $\frac{x^2}{25} + \frac{y^2}{16} = 1$.

center: (0, 0); foci: $F_1(-3, 0)$ and $F_2(3, 0)$; vertices: (−5, 0), (5, 0); co-vertices: (0, −4), (0, 4)

Definition of Ellipse

An **ellipse** is the set of all points P in a plane such that the sum of the distances from P to two fixed points, F_1 and F_2, called the **foci**, is a constant.

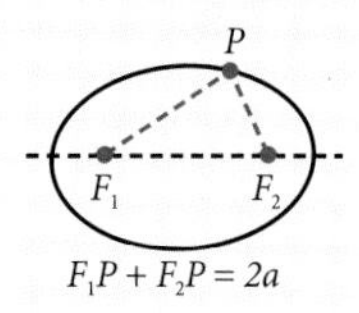

You can use the definition of an ellipse and the distance formula to write an equation for an ellipse whose foci are $F_1(-4, 0)$ and $F_2(4, 0)$ and whose constant sum is 10. Let $P(x, y)$ be any point on the ellipse.

$$F_1P + F_2P = 10$$

$$\sqrt{[x-(-4)]^2 + (y-0)^2} + \sqrt{(x-4)^2 + (y-0)^2} = 10$$

$$\sqrt{(x+4)^2 + y^2} = 10 - \sqrt{(x-4)^2 + y^2}$$

Square each side. $(x+4)^2 + y^2 = 100 - 20\sqrt{(x-4)^2 + y^2} + [(x-4)^2 + y^2]$

Simplify. $16x - 100 = -20\sqrt{(x-4)^2 + y^2}$

Divide each side by 4. $4x - 25 = -5\sqrt{(x-4)^2 + y^2}$

Square each side again. $16x^2 - 200x + 625 = 25[(x-4)^2 + y^2]$

$$16x^2 - 200x + 625 = 25x^2 - 200x + 400 + 25y^2$$

$$225 = 9x^2 + 25y^2$$

Divide each side by 225. $1 = \frac{x^2}{25} + \frac{y^2}{9}$

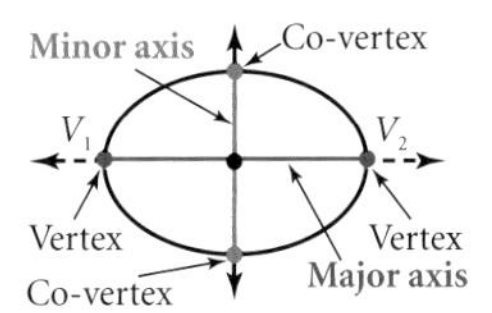

An ellipse has two axes of symmetry. The **major axis** is the longer axis of the ellipse and the **minor axis** is the shorter axis of the ellipse. The endpoints of the major axis are called the **vertices** of the ellipse. The endpoints of the minor axis are called the **co-vertices** of the ellipse. The foci are always on the major axis. The point of intersection of the major and minor axes is called the **center**.

Standard Equation of an Ellipse

The standard equation of an ellipse centered at the origin is given below.

Horizontal major axis

$$\frac{x^2}{a^2} + \frac{y^2}{b^2} = 1$$

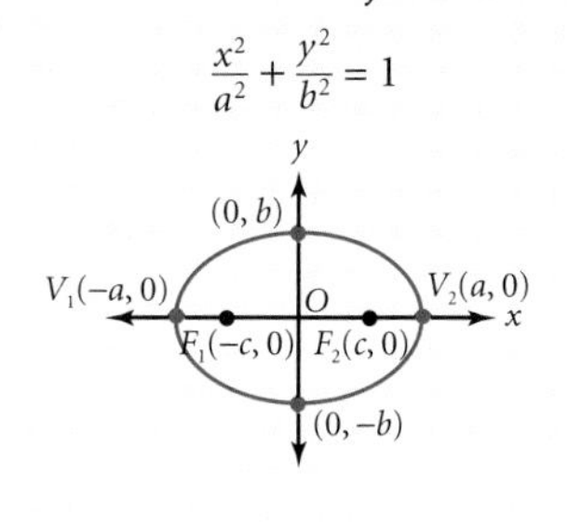

Vertical major axis

$$\frac{x^2}{b^2} + \frac{y^2}{a^2} = 1$$

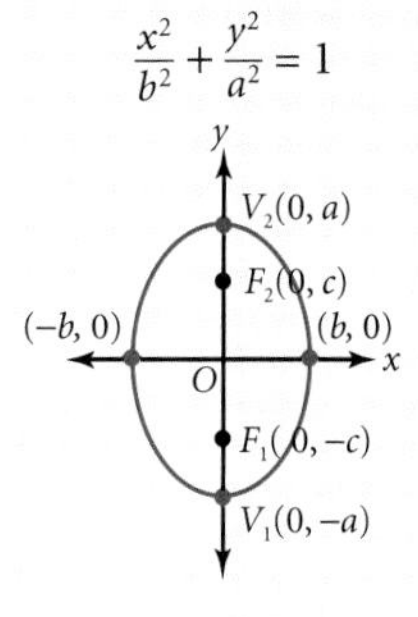

In each case:
- $a^2 > b^2$, and $a^2 - b^2 = c^2$,
- the length of the major axis is $2a$, and
- the length of the minor axis is $2b$.

Activity Notes

Be sure students notice that the length of the string attached to the two tacks is constant. Therefore, the sum of the distances from any point on the ellipse (the pencil point) to each of the two fixed points (the tacks) is constant.

Cooperative Learning

Students can work together in groups of four for this Activity. Be sure that each member writes an explanation of why the path of the pencil is an ellipse.

CHECKPOINT ✔

4. As the length of the string increases, the ellipse becomes larger and more closely resembles a circle.

Teaching Tip

The area of an ellipse is πab, where $2a$ is the length of the major axis and $2b$ is the length of the minor axis. Have students relate this to the area of a circle, πr^2.

Interdisciplinary Connection

ARCHITECTURE NASA uses a large elliptical tunnel in Langley, Virginia, to study the dynamics of air flow. The Transonic Tunnel is 82 feet wide and 58 feet high. Have students write an equation that models the tunnel, with the center of the ellipse located at the origin. Then have them graph the equation and label the vertices, the co-vertices, and both foci. Then have them state the eccentricity of the ellipse.

equation: $\frac{x^2}{1681} + \frac{y^2}{841} = 1$

vertices: $(-41, 0)$ and $(41, 0)$

co-vertices: $(0, -29)$ and $(0, 29)$

foci: $(-2\sqrt{210}, 0)$ and $(2\sqrt{210}, 0)$

$e \approx 0.71$

ADDITIONAL EXAMPLE 1

Write the standard equation for an ellipse with foci at (−8, 0) and (8, 0) and with a major axis of 20. Sketch the graph.

$\frac{x^2}{100} + \frac{y^2}{36} = 1$

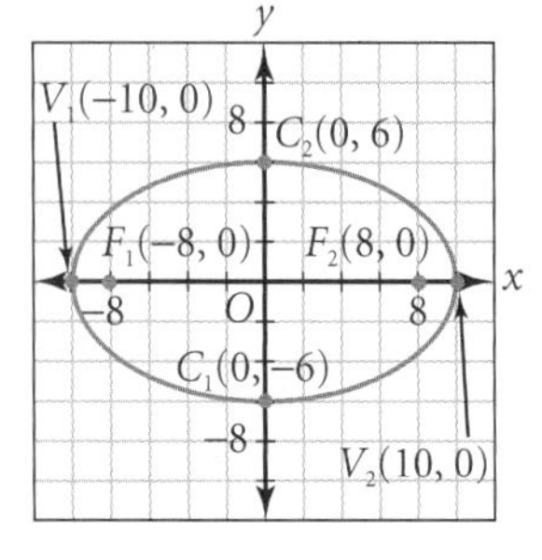

TRY THIS

$\frac{x^2}{169} + \frac{y^2}{25} = 1$

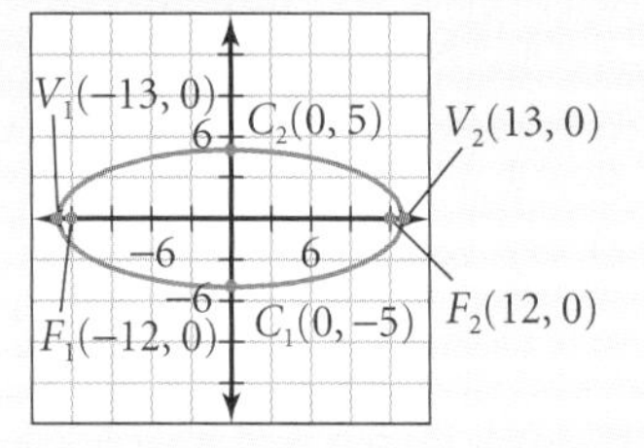

ADDITIONAL EXAMPLE 2

The planet Mercury orbits the Sun in an elliptical path whose minimum distance from the Sun is approximately 29 million miles and whose maximum distance from the Sun is approximately 43 million miles. **Write the standard equation of this path, where the center of the ellipse is at the origin and the foci are points on the x-axis.**

$\frac{x^2}{1296} + \frac{y^2}{1247} = 1$

EXAMPLE 1 **Write the standard equation for an ellipse with foci at (0, −4) and (0, 4) and with a minor axis of 6. Sketch the graph.**

SOLUTION

The coordinates of the foci are (0, −4) and (0, 4), so $c = 4$. The length of the minor axis is $2b$ and $2b = 6$, so $b = 3$.

Substitute 4 for c and 3 for b to find a.

$$a^2 - b^2 = c^2$$
$$a^2 - 3^2 = 4^2$$
$$a = 5$$

Substitute 5 for a and 3 for b.

$$\frac{x^2}{b^2} + \frac{y^2}{a^2} = 1 \rightarrow \frac{x^2}{9} + \frac{y^2}{25} = 1$$

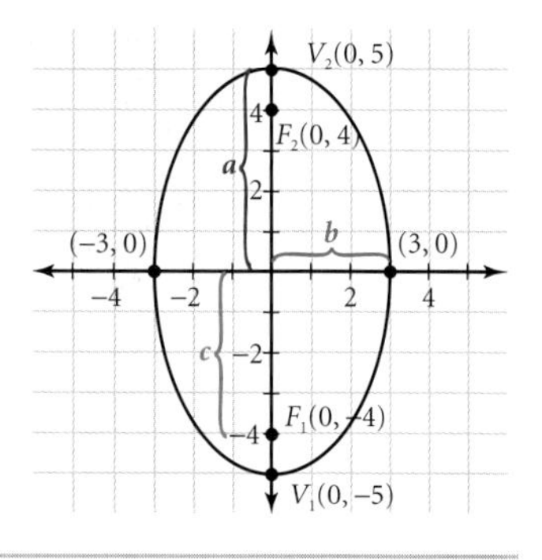

To sketch the graph, plot the vertices, (0, −5) and (0, 5), and the co-vertices, (−3, 0) and (3, 0). Connect them to form an ellipse.

TRY THIS Write the standard equation for an ellipse with foci at (−12, 0) and (12, 0) and with a major axis of 26. Sketch the graph.

EXAMPLE 2 Mars orbits the Sun in an elliptical path whose minimum distance from the Sun is 129.5 million miles and whose maximum distance from the Sun is 154.4 million miles. The Sun represents one focus of the ellipse.

APPLICATION
ASTRONOMY

Write the standard equation for the elliptical orbit of Mars around the Sun, where the center of the ellipse is at the origin.

SOLUTION

PROBLEM SOLVING

Draw a diagram.

An image of Mars, composed from 102 Viking Orbiter images

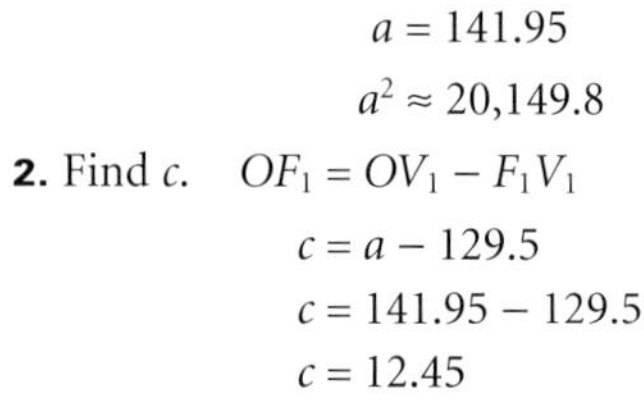

1. Find a^2. $V_1V_2 = 154.4 + 129.5$
$2a = 283.9$
$a = 141.95$
$a^2 \approx 20{,}149.8$

2. Find c. $OF_1 = OV_1 - F_1V_1$
$c = a - 129.5$
$c = 141.95 - 129.5$
$c = 12.45$

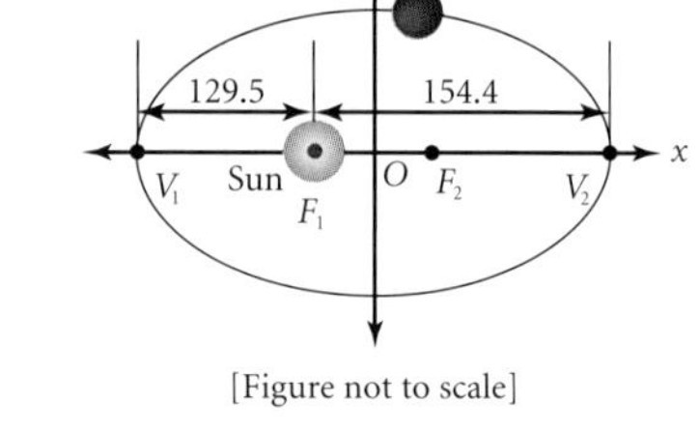

[Figure not to scale]

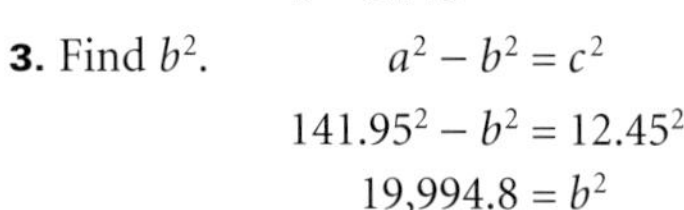

3. Find b^2. $a^2 - b^2 = c^2$
$141.95^2 - b^2 = 12.45^2$
$19{,}994.8 = b^2$

The standard equation for the orbit of Mars around the Sun in millions of miles is approximately $\frac{x^2}{20{,}149.8} + \frac{y^2}{19{,}994.8} = 1$.

Enrichment

A chord of an ellipse which passes through a focus and is perpendicular to the major axis of the ellipse is called a *focal chord.* (See the figure at right.) Have students find the length of a focal chord of the ellipse $\frac{x^2}{a^2} + \frac{y^2}{b^2} = 1$. **Let $x^2 = c^2$, or $a^2 - b^2$, and find the corresponding y to determine the focal chord's length above the x-axis: $\frac{a^2 - b^2}{a^2} + \frac{y^2}{b^2} = 1$; $y = \frac{b^2}{a}$. So, the length of the focal chord is $\frac{2b^2}{a}$.** Then have students find the length of a focal chord of the ellipse $9x^2 + 16y^2 = 144$. $\frac{9}{2}$

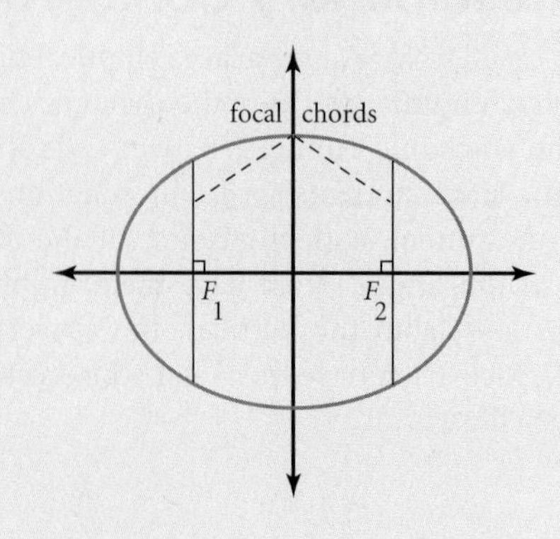

TRY THIS Venus orbits the Sun in an elliptical path whose minimum distance from the Sun is 66.7 million miles and whose maximum distance from the Sun is 67.6 million miles. The Sun represents one focus of the ellipse. Write the standard equation for the elliptical orbit of Venus around the Sun, where the center of the ellipse is at the origin.

TRY THIS

$$\frac{x^2}{4509.1} + \frac{y^2}{4508.9} = 1$$

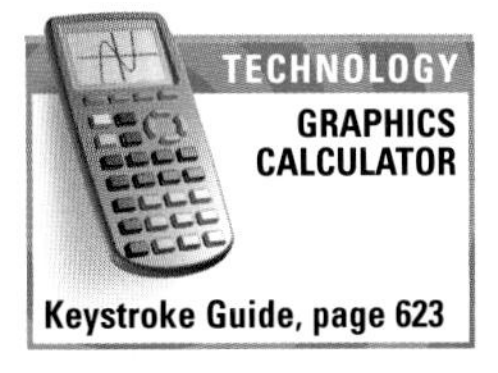

You can graph ellipses with a graphics calculator. To graph $\frac{x^2}{25} + \frac{y^2}{9} = 1$, solve for y, and graph the two resulting equations together.

You may wish to simplify the equations before entering them into the graphics calculator, but this is not necessary.

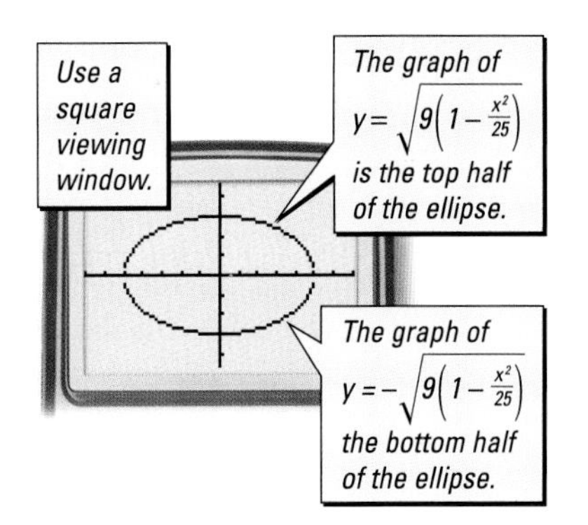

Teaching Tip

TECHNOLOGY If you are using a TI-82 or TI-83 to create the graph of the ellipse on this page, you can set a square viewing window as follows:

Xmin=–7	**Ymin=–4.6**
Xmax=7	**Ymax=4.6**
Xscl=1	**Yscl=1**

The center of an ellipse can be anywhere in a coordinate plane.

Standard Equation of a Translated Ellipse

The standard equation of an ellipse centered at (h, k) is given below.

Horizontal major axis

$$\frac{(x-h)^2}{a^2} + \frac{(y-k)^2}{b^2} = 1$$

Vertical major axis

$$\frac{(x-h)^2}{b^2} + \frac{(y-k)^2}{a^2} = 1$$

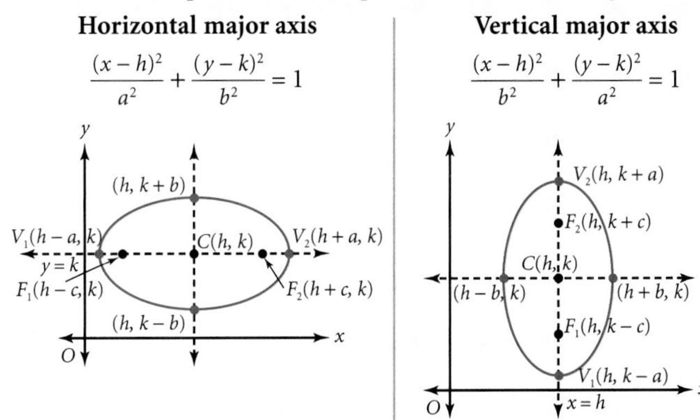

In each case:
- $a^2 > b^2$, and $a^2 - b^2 = c^2$,
- the length of the major axis is $2a$, and
- the length of the minor axis is $2b$.

The **eccentricity** of an ellipse is a measure of how round or flat it is. The eccentricity, E, is the ratio of the distance, c, between the center and a focus to the distance, a, between the center and a vertex.

$$E = \frac{c}{a}, \text{ where } c = \sqrt{a^2 - b^2}$$

If $c = 0$, then $E = 0$ and the ellipse is a circle. As the value of c approaches the value of a, the value of E approaches 1, and the ellipse becomes flatter.

Inclusion Strategies

KINESTHETIC LEARNERS If there is an accessible field or empty lot near the school, students can make a large model of an ellipse. You will need to provide several sturdy wooden stakes (or dowels), a tool such as a rubber mallet to drive the stakes into the ground, and a ball of string.

Tell students that the goal is to create an ellipse that a landscaper could use to dig a flower bed. Have them decide on suitable lengths for the major and minor axes of the ellipse.

Now show them the materials and have them work together to devise a plan for creating the ellipse. Be sure that all students feel comfortable with the plan before going outdoors to make the ellipse.

Begin by identifying a, b, and c. Drive two stakes into the ground $2c$ feet apart. Tie one end of a piece of string that is $2a$ feet long to each stake. Hold the string taut and drive stakes into the ground at several locations around the ellipse. Then connect the stakes at the marked locations with string.

ADDITIONAL
EXAMPLE 3

Write the standard equation for the ellipse with its center at (4, 3) and with a vertical major axis of 10 and minor axis of 8. Sketch the graph of the ellipse.

$\frac{(x-4)^2}{16}+\frac{(y-3)^2}{25}=1$

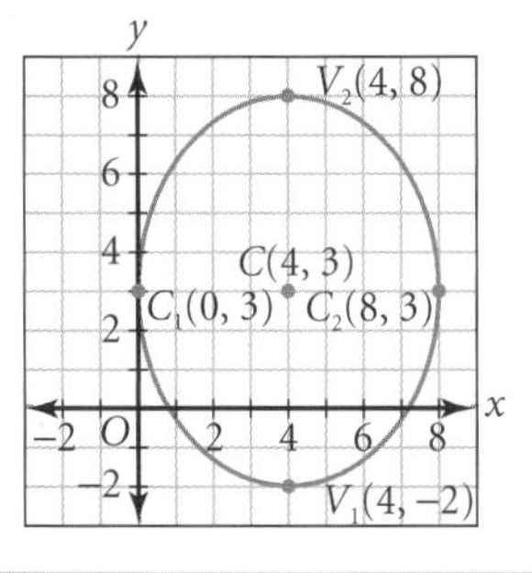

TRY THIS

$\frac{(x+1)^2}{4}+\frac{(y+2)^2}{16}=1$

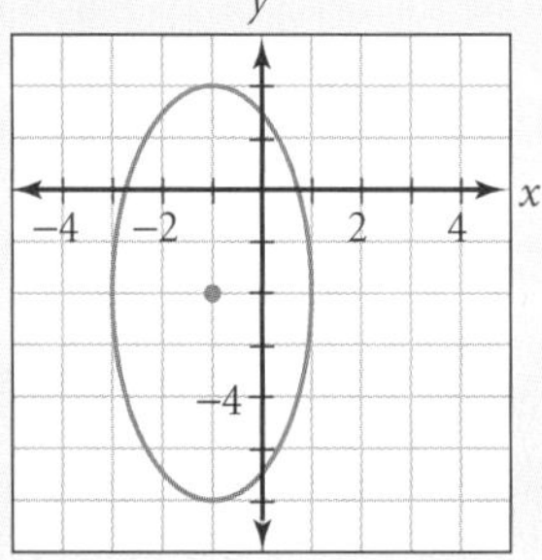

EXAMPLE 3 **Write the standard equation for an ellipse with its center at (2, −4) and with a horizontal major axis of 10 and minor axis of 6. Sketch the graph.**

SOLUTION

The center, (h, k), is at $(2, -4)$.
The length of the major axis is $2a = 10$, so $a = 5$.
The length of the minor axis is $2b = 6$, so $b = 3$.

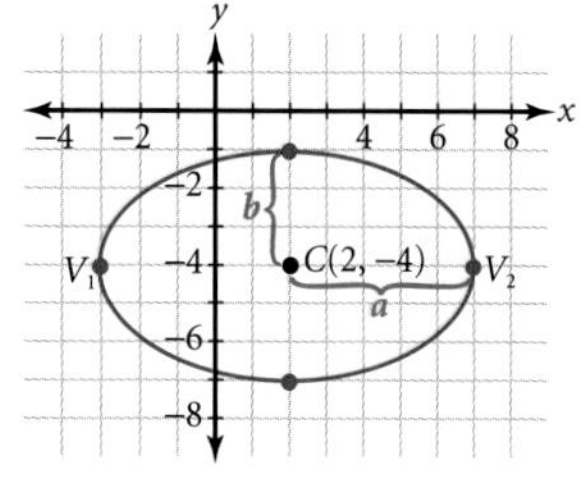

The major axis is horizontal.

$$\frac{(x-h)^2}{a^2}+\frac{(y-k)^2}{b^2}=1$$
$$\frac{(x-2)^2}{25}+\frac{(y+4)^2}{9}=1$$

Use the value of a to find the vertices, $(-3, -4)$ and $(7, -4)$.

Use the value of b to find the co-vertices, $(2, -7)$ and $(2, -1)$.

TRY THIS Write the standard equation for an ellipse with its center at (−1, −2) and with a vertical major axis of 8 and minor axis of 4. Sketch the graph.

EXAMPLE 4 **An ellipse is defined by $4x^2 + y^2 + 24x - 4y + 36 = 0$. Write the standard equation, and identify the coordinates of its center, vertices, co-vertices, and foci. Sketch the graph.**

SOLUTION

$$4x^2 + y^2 + 24x - 4y + 36 = 0$$
$$4x^2 + 24x + y^2 - 4y = -36$$
$$4(x^2 + 6x) + (y^2 - 4y) = -36$$
$$4(x^2 + 6x + 9) + (y^2 - 4y + 4) = -36 + 4(9) + 4 \quad \textit{Complete the squares.}$$
$$4(x + 3)^2 + (y - 2)^2 = 4$$
$$\frac{(x+3)^2}{1}+\frac{(y-2)^2}{4}=1 \quad \textit{Divide each side by 4.}$$

From the equation, the center is at $(-3, 2)$, $a^2 = 4$, and $b^2 = 1$. Find c.

$$a^2 - b^2 = c^2$$
$$4 - 1 = c^2$$
$$\sqrt{3} = c$$

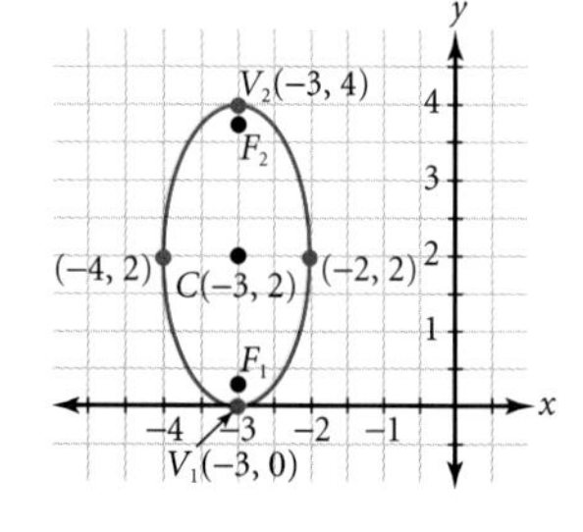

Use the value of c to find the foci.

$(-3, 2 - \sqrt{3})$ and $(-3, 2 + \sqrt{3})$

Use the value of a to find the vertices.

$(-3, 0)$ and $(-3, 4)$

Use the value of b to find the co-vertices.

$(-4, 2)$ and $(-2, 2)$

CRITICAL THINKING Identify the graph of $Ax^2 + Cy^2 + E = 0$ given that A and C are positive numbers.

Reteaching the Lesson

COOPERATIVE LEARNING To help students see how a, b, h, and k are related in the standard equation of a translated ellipse, divide the class into groups of four students. Have each group select a set of four values for each of the constants a, b, h, and k so that they can graph ellipses with four different shapes. Each group should choose values so that the graphs show an ellipse with a horizontal major axis, an ellipse with vertical major axis, an elongated ellipse, and a nearly circular ellipse.

Exercises

Communicate

1. Describe the information you can obtain from the standard equation of a translated ellipse, such as $\frac{x^2}{16} + \frac{(y-2)^2}{25} = 1$.

2. Describe the procedure used to write the standard equation of the ellipse defined by $9x^2 + 4y^2 + 18x - 40y + 73 = 0$.

3. Explain how to graph $\frac{(x-4)^2}{3^2} + \frac{(y-5)^2}{2^2} = 1$.

4. Describe the graph of $\frac{x^2}{25} + \frac{y^2}{25} = 1$.

internet connect
Activities Online
Go To: **go.hrw.com**
Keyword: **MB1 Comet**

Guided Skills Practice

5. Write the standard equation for an ellipse centered at the origin with foci at (5, 0) and (−5, 0) and with a minor axis of 8. Sketch the graph. ***(EXAMPLE 1)***

6. ASTRONOMY The diagram at left gives the minimum and maximum distances (in millions of miles) from Jupiter to the Sun. Write the standard equation for Jupiter's elliptical orbit around the Sun. ***(EXAMPLE 2)***

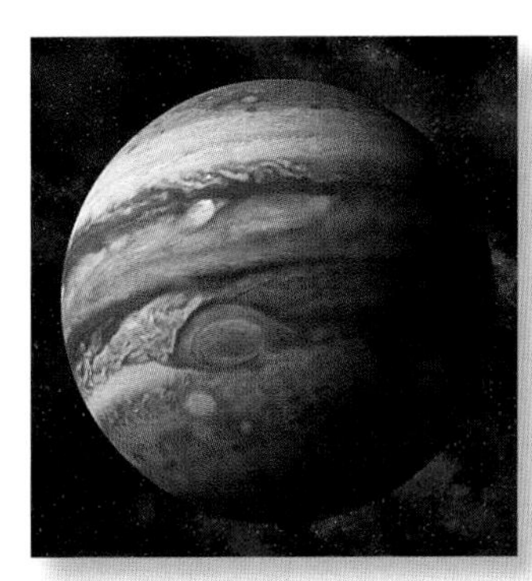

Jupiter as seen by Voyager 1

7. Write the standard equation for an ellipse centered at (1, −2) with a vertical major axis of 6 and a minor axis of 4. Sketch the graph. ***(EXAMPLE 3)***

8. An ellipse is defined by the equation $25x^2 + 4y^2 + 50x - 8y - 71 = 0$. Write the standard equation, and identify the coordinates of the center, vertices, co-vertices, and foci. Sketch the graph. ***(EXAMPLE 4)*** **$\frac{(x+1)^2}{4} + \frac{(y-1)^2}{25} = 1$; center: (−1, 1); vertices: (−1, −4), (−1, 6); co-vertices: (−3, 1), (1, 1); foci: $(-1, 1-\sqrt{21})$, $(-1, 1+\sqrt{21})$**

Practice and Apply

Find the vertices and co-vertices of each ellipse.

9. $\frac{x^2}{25} + \frac{y^2}{9} = 1$ **10.** $\frac{x^2}{16} + \frac{y^2}{49} = 1$ **11.** $\frac{x^2}{81} + \frac{y^2}{4} = 1$

12. $\frac{x^2}{9} + \frac{y^2}{36} = 1$ **13.** $\frac{x^2}{1} + \frac{y^2}{64} = 1$ **14.** $\frac{x^2}{1} + \frac{y^2}{4} = 1$

Write the standard equation of each ellipse. Find the coordinates of the center, vertices, co-vertices, and foci.

15. $3x^2 + 12y^2 = 12$ **16.** $50x^2 + 2y^2 = 50$ **17.** $3x^2 + 7y^2 = 28$

18. $5x^2 + 20y^2 = 80$ **19.** $\frac{x^2}{8} + \frac{y^2}{18} = 2$ **20.** $\frac{x^2}{3} + \frac{y^2}{12} = 3$

5. $\frac{x^2}{41} + \frac{y^2}{16} = 1$

6. $\frac{x^2}{233{,}578.89} + \frac{y^2}{233{,}026.64} = 1$

APPLICATION

Jupiter; y; 459.8; 506.8; x; V_1 Sun; F_1; F_2; V_2

[Not to scale]

7. $\frac{(x-1)^2}{4} + \frac{(y+2)^2}{9} = 1$

9. vertices: (−5, 0), (5, 0) co-vertices: (0, −3), (0, 3)

10. vertices: (0, −7), (0, 7) co-vertices: (−4, 0), (4, 0)

11. vertices: (−9, 0), (9, 0) co-vertices: (0, −2), (0, 2)

12. vertices: (0, −6), (0, 6) co-vertices: (−3, 0), (3, 0)

13. vertices: (0, −8), (0, 8) co-vertices: (−1, 0), (1, 0)

14. vertices: (0, −2), (0, 2) co-vertices: (−1, 0), (1, 0)

ADDITIONAL EXAMPLE 4

An ellipse is defined by the equation $4x^2 + 25y^2 - 24x + 200y + 336 = 0$. **Write the standard equation and identify the coordinates of the center, vertices, co-vertices, and foci. Sketch the graph of the ellipse.**

$\frac{(x-3)^2}{25} + \frac{(y+4)^2}{4} = 1$

center: $C(3, -4)$; vertices: $V_1(-2, -4)$ and $V_2(8, -4)$; co-vertices: $C_1(3, -6)$ and $C_2(3, -2)$; foci: $F_1(3-\sqrt{21}, -4)$ and $F_2(3+\sqrt{21}, -4)$

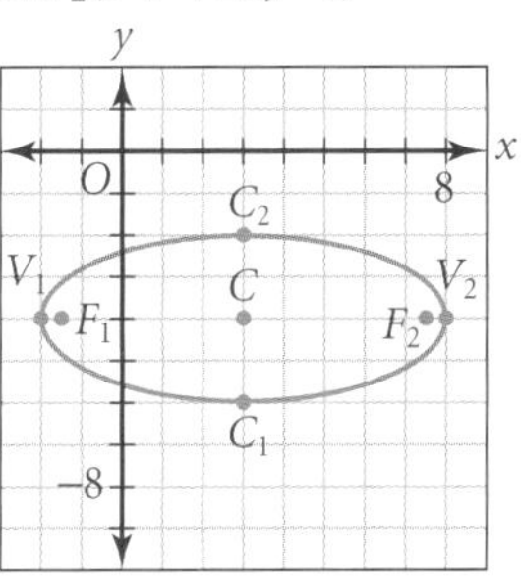

CRITICAL THINKING

If $E > 0$, there is no graph. If $E = 0$, the graph consists of the point (0, 0). If $E < 0$, the graph is an ellipse (or a circle, if $A = C$).

Assess

Selected Answers

Exercises 5–8, 9–73 odd

ASSIGNMENT GUIDE	
In Class	1–8
Core	9–57 odd, 63
Core Plus	10–64 even
Review	65–74
Preview	75

Extra Practice can be found beginning on page 940.

5. $\frac{x^2}{41} + \frac{y^2}{16} = 1$

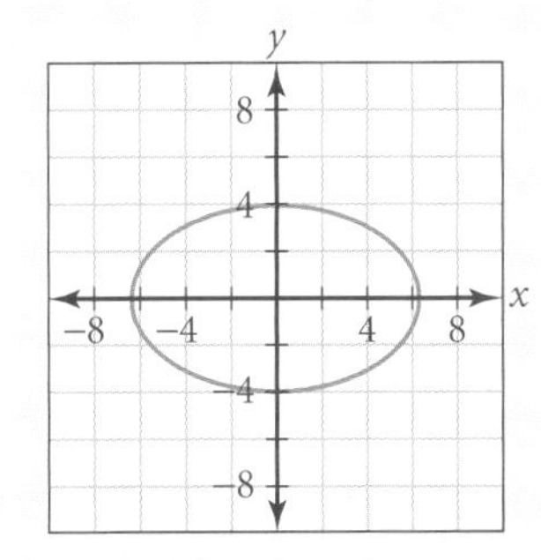

7. $\frac{(x-1)^2}{4} + \frac{(y+2)^2}{9} = 1$

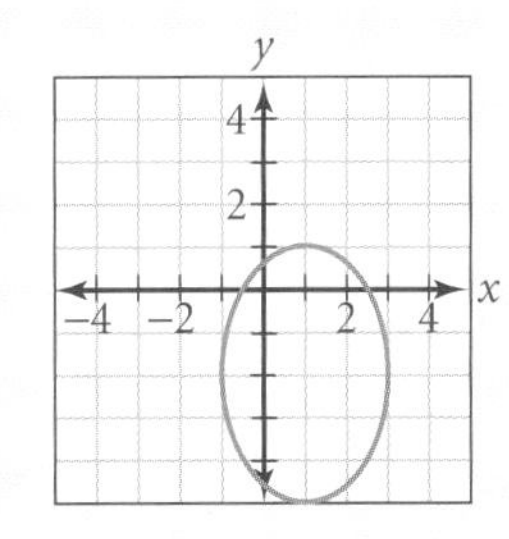

The answers to Exercises 15–20 can be found in Additional Answers beginning on page 1002.

8. $\frac{(x+1)^2}{4} + \frac{(y-1)^2}{25} = 1$; center: (−1, 1); vertices: (−1, −4) and (−1, 6); co-vertices: (−3, 1) and (1, 1); foci: $(-1, 1-\sqrt{21})$ and $(-1, 1+\sqrt{21})$

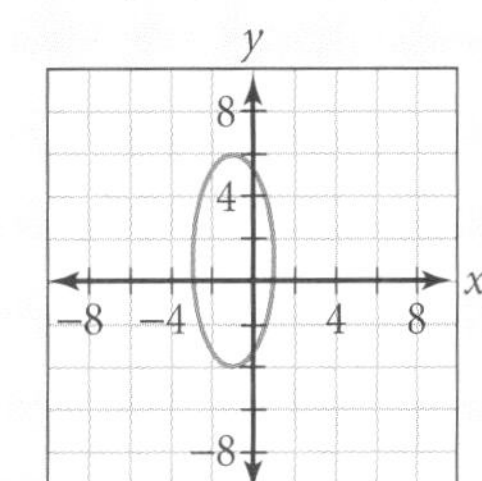

Error Analysis

Students sometimes mistakenly apply the relationship $c^2 = a^2 + b^2$ to ellipses. Remind them that the correct relationship among a, b, and c in an ellipse is $c^2 = a^2 - b^2$. Additionally, some students mistakenly identify the demoninator associated with the x-term with a^2 because it is listed first. Remind them that the value of a is the square root of the larger of the two denominators when an ellipse is written in standard form.

21. $\frac{x^2}{36} + \frac{y^2}{25} = 1$
22. $\frac{x^2}{9} + \frac{y^2}{16} = 1$
23. $\frac{(x-3)^2}{16} + \frac{(y-3)^2}{9} = 1$
24. $\frac{(x-1)^2}{25} + \frac{(y+3)^2}{9} = 1$
25. $\frac{(x+2)^2}{4} + \frac{(y-2)^2}{9} = 1$
26. $\frac{(x-3)^2}{9} + \frac{y^2}{36} = 1$

Write the standard equation for each ellipse.

21.

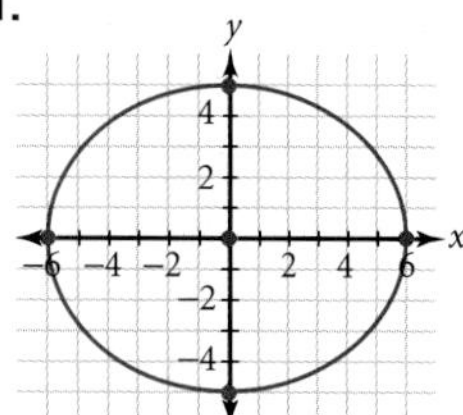

22.

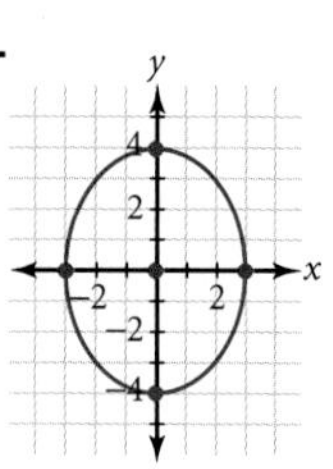

23.

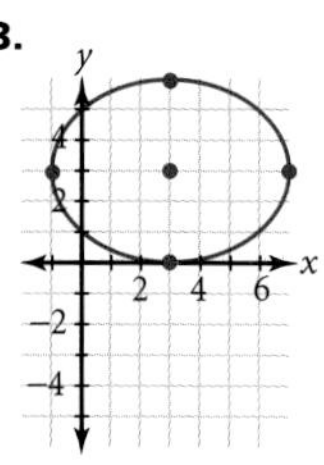

24.

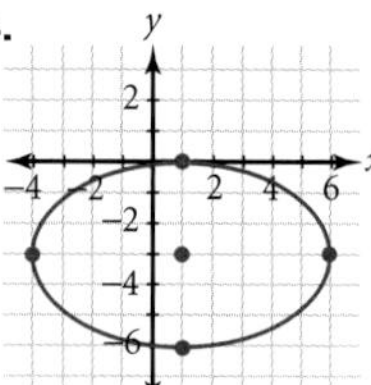

25.

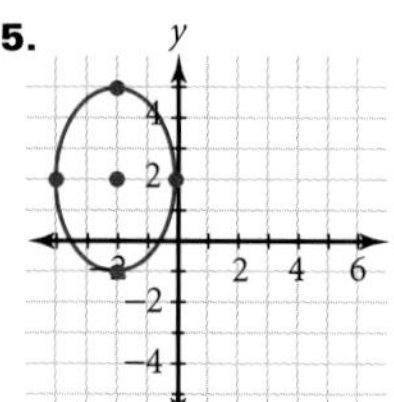

26.

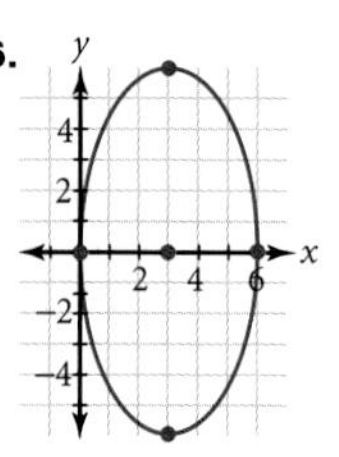

27. center: (0, 0); vertices: (−5, 0) and (5, 0); co-vertices: (0, −2) and (0, 2); foci: $(-\sqrt{21}, 0)$ and $(\sqrt{21}, 0)$

28. center: (0, 0); vertices: (0, −3) and (0, 3); co-vertices: (−1, 0) and (1, 0); foci: $(0, -2\sqrt{2})$ and $(0, 2\sqrt{2})$

29. center: (0, 0); vertices: (0, −3) and (0, 3); co-vertices: (−2, 0) and (2, 0); foci: $(0, -\sqrt{5})$ and $(0, \sqrt{5})$

30. center: (0, 0); vertices: (−4, 0) and (4, 0); co-vertices: (0, −1) and (0, 1); foci: $(-\sqrt{15}, 0)$ and $(\sqrt{15}, 0)$

Sketch the graph of each ellipse. Label the center, foci, vertices, and co-vertices.

27. $\frac{x^2}{25} + \frac{y^2}{4} = 1$ **28.** $\frac{x^2}{1} + \frac{y^2}{9} = 1$ **29.** $\frac{x^2}{4} + \frac{y^2}{9} = 1$

30. $\frac{x^2}{16} + \frac{y^2}{1} = 1$ **31.** $\frac{(x+2)^2}{4} + \frac{(y+1)^2}{9} = 1$ **32.** $\frac{(x-2)^2}{9} + \frac{(y-2)^2}{4} = 1$

33. $\frac{x^2}{1} + \frac{(y+2)^2}{9} = 1$ **34.** $\frac{(x+1)^2}{4} + \frac{y^2}{1} = 1$ **35.** $\frac{(x-1)^2}{4} + \frac{(y-1)^2}{4} = 1$

36. $16(x+1)^2 + 9(y-1)^2 = 144$ **37.** $9(x-1)^2 + 25(y+2)^2 = 225$

38. $4x^2 + 25y^2 = 100$ **39.** $25x^2 + 9y^2 = 225$

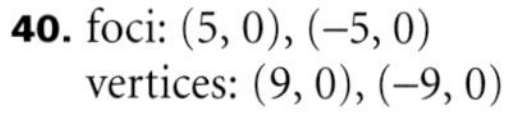

Homework Help Online
Go To: **go.hrw.com**
Keyword:
MB1 Homework Help for Exercises 40–45

Write the standard equation for the ellipse with the given characteristics.

40. foci: (5, 0), (−5, 0)
vertices: (9, 0), (−9, 0)

41. foci: (0, 4), (0, −4)
vertices: (0, 8), (0, −8)

42. foci: (7, 0), (−7, 0)
co-vertices: (0, 3), (0, −3)

43. foci: (0, 3), (0, −3)
co-vertices: (1, 0), (−1, 0)

44. co-vertices: (0, 2), (0, −2)
vertices: (3, 0), (−3, 0)

45. vertices: (5, 0), (−5, 0)
co-vertices: (0, 4), (0, −4)

40. $\frac{x^2}{81} + \frac{y^2}{56} = 1$
41. $\frac{x^2}{48} + \frac{y^2}{64} = 1$
42. $\frac{x^2}{58} + \frac{y^2}{9} = 1$
43. $\frac{x^2}{1} + \frac{y^2}{10} = 1$
44. $\frac{x^2}{9} + \frac{y^2}{4} = 1$
45. $\frac{x^2}{25} + \frac{y^2}{16} = 1$

State whether each equation represents a parabola, a circle, or an ellipse.

46. $\frac{x}{2} = \frac{(y-3)^2}{4}$ **parabola** **47.** $\frac{y}{4} = \frac{(x+2)^2}{2}$ **parabola**

48. $\frac{(x-1)^2}{12} = 6 - \frac{(y+5)^2}{9}$ **ellipse** **49.** $\frac{(y+4)^2}{6} = 8 - \frac{(x-1)^2}{4}$ **ellipse**

Write the standard equation for each ellipse. Identify the coordinates of the center, vertices, co-vertices, and foci.

50. $x^2 + 4y^2 + 6x - 8y = 3$ **51.** $16x^2 + 4y^2 + 32x - 8y = 44$

52. $x^2 + 16y^2 - 64y = 0$ **53.** $25x^2 + y^2 - 50x = 0$

54. $4x^2 + 9y^2 - 16x + 18y = 11$ **55.** $25x^2 + 9y^2 + 100x + 18y = 116$

56. $9x^2 + 16y^2 - 36x - 64y - 44 = 0$ **57.** $36x^2 + 25y^2 - 72x + 100y = 764$

Practice

NAME ______ CLASS ______ DATE ______

Practice
9.4 Ellipses

Write the standard equation for each ellipse.

1.

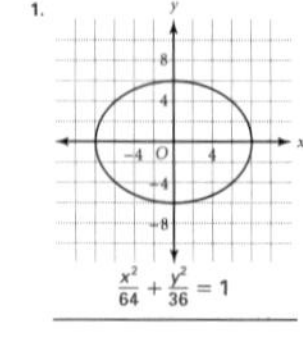

$\frac{x^2}{64} + \frac{y^2}{36} = 1$

2.

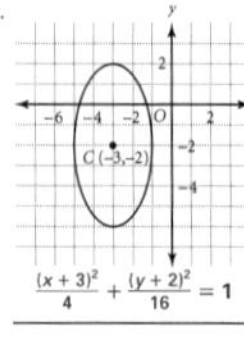

$\frac{(x+3)^2}{4} + \frac{(y+2)^2}{16} = 1$

3.

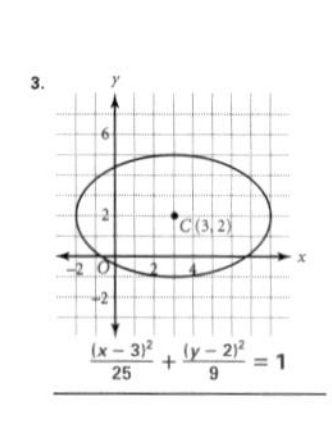

$\frac{(x-3)^2}{25} + \frac{(y-2)^2}{9} = 1$

Sketch the graph of each ellipse. Label the center, foci, vertices, and co-vertices.

4. $\frac{x^2}{4} + \frac{y^2}{81} = 1$

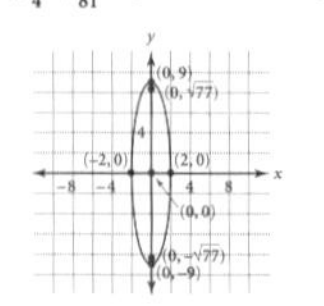

5. $\frac{x^2}{49} + \frac{(y-1)^2}{36} = 1$

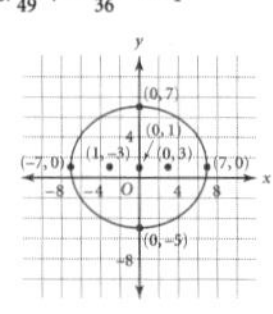

6. $\frac{(x-4)^2}{9} + \frac{(y+3)^2}{25} = 1$

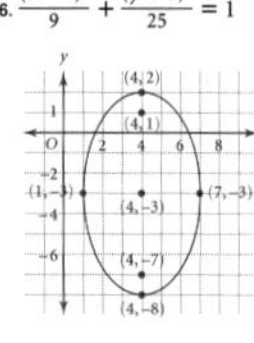

Write the standard equation for the ellipse with the given characteristics.

7. vertices: (−25, 0) and (25, 0); co-vertices: (0, −15) and (0, 15) $\frac{x^2}{625} + \frac{y^2}{225} = 1$

8. foci: (−10, 0) and (10, 0); co-vertices: (0, −3), (0, 3) $\frac{x^2}{109} + \frac{y^2}{9} = 1$

9. co-vertices: (−20, 0) and (20, 0); foci: (0, −8) and (0, 8) $\frac{x^2}{400} + \frac{y^2}{464} = 1$

10. An ellipse is defined by $x^2 + 4y^2 + 6x - 27 = 0$. Write the standard equation, and identify the coordinates of the center, vertices, co-vertices, and foci. $\frac{(x+3)^2}{36} + \frac{y^2}{9} = 1$;
center: (−3, 0); foci: $(-3 - 3\sqrt{3}, 0)$ and $(-3 + 3\sqrt{3}, 0)$; vertices: (−9, 0) and (3, 0); co-vertices (−3, −3), and (−3, 3)

31. center: (−2, −1); vertices: (−2, −4) and (−2, 2); co-vertices: (−4, −1) and (0, −1); foci: $(-2, -1 - \sqrt{5})$ and $(-2, -1 + \sqrt{5})$

32. center: (2, 2); vertices: (−1, 2) and (5, 2); co-vertices: (2, 0) and (2, 4); foci: $(2 - \sqrt{5}, 2)$ and $(2 + \sqrt{5}, 2)$

33. center: (0, −2); vertices: (0, −5) and (0, 1); co-vertices: (−1, −2) and (1, −2); foci: $(0, -2 - 2\sqrt{2})$ and $(0, -2 + 2\sqrt{2})$

34. center: (−1, 0); vertices: (−3, 0) and (1, 0); co-vertices: (−1, −1) and (−1, 1); foci: $(-1 - \sqrt{3}, 0)$ and $(-1 + \sqrt{3}, 0)$

35. The graph is a circle with its center at (1, 1) and a radius of 2.

36. center: (−1, 1); vertices: (−1, −3) and (−1, 5); co-vertices: (−4, 1) and (2, 1); foci: $(-1, 1 - \sqrt{7})$ and $(-1, 1 + \sqrt{7})$

37. center: (1, −2); vertices: (−4, −2) and (6, −2); co-vertices: (1, −5) and (1, 1); foci: (−3, −2) and (5, −2)

CONNECTIONS

58. $\frac{x^2}{36}+\frac{(y-2)^2}{64}=1$

59. $\frac{(x+10)^2}{4}+\frac{(y+5)^2}{16}=1$

58. TRANSFORMATIONS If the ellipse defined by the equation $\frac{(x+5)^2}{36}+\frac{(y-1)^2}{64}=1$ is translated 1 unit up and 5 units to the right, write the standard equation of the resulting ellipse.

59. TRANSFORMATIONS If the ellipse defined by the equation $16x^2+4y^2+96x+8y+84=0$ is translated 4 units down and 7 units to the left, write the standard equation of the resulting ellipse.

CHALLENGES

61. The point $(-2, 1)$

60. Use equations to explain why the eccentricity of an ellipse cannot equal 1.

61. Describe the graph of the equation $\frac{(x+2)^2}{3}+\frac{(y-1)^2}{6}=0$.

APPLICATIONS

62 a. $\frac{x^2}{(387{,}000)^2}+\frac{y^2}{(386{,}500)^2}=1$

b. 367,334 km

c. 406,666 km

d. 0.051

62. ASTRONOMY The Moon orbits Earth in an elliptical path with the center of the Earth at one focus. The major axis of the orbit is 774,000 kilometers, and the minor axis is 773,000 kilometers.

a. Using $(0, 0)$ as the center of the ellipse, write the standard equation for the orbit of the Moon around Earth.

b. How far from the center of Earth is the Moon at its closest point?

c. How far from the center of Earth is the Moon at its farthest point?

d. Find the eccentricity of the Moon's orbit around Earth.

James Irwin salutes the flag during the Apollo 15 *mission to the Moon.*

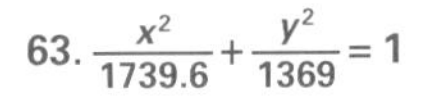

63. $\frac{x^2}{1739.6}+\frac{y^2}{1369}=1$

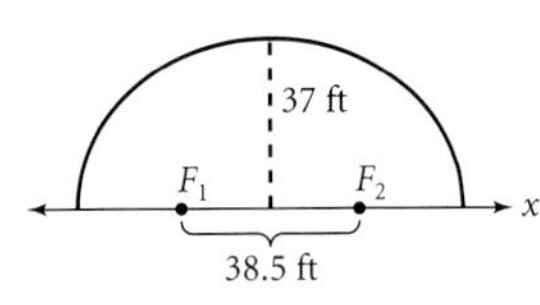

63. ARCHITECTURE The ceiling of the "whispering gallery" of the Statuary Hall in the United States Capitol Building can be approximated by a *semi-ellipse.* Because of the properties of reflection, the whispering of someone standing at one focus can be clearly heard by a person standing at the other focus. It is said that John Quincy Adams used this attribute of the Statuary Hall to eavesdrop on his adversaries. Suppose that the distance between the foci is 38.5 feet and the maximum height of the ceiling above ear level is 37 feet. Find the equation of an elliptical cross section of this gallery, assuming that the center is placed at the origin.

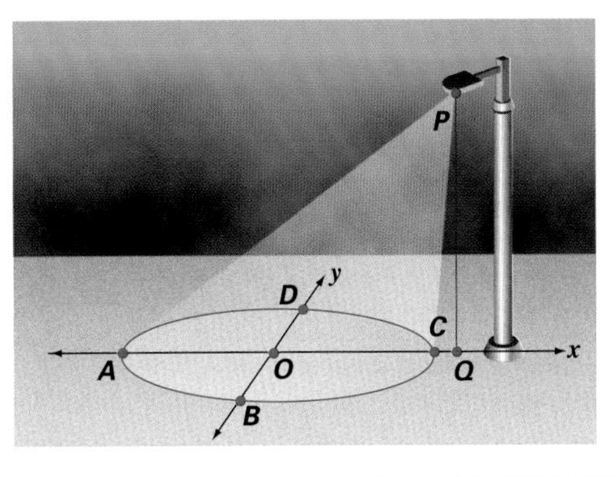

64. LIGHTING A light atop the pole represented by $\overline{PQ}$ illuminates an elliptical region at the base of the pole as shown in the illustration at left, where $PQ = 18$ feet, $CQ = 2$ feet, $AQ = 26$ feet, and $BD = 18$ feet.

a. Using the x- and y-axes shown, write an equation for the boundary of the elliptical region illuminated by the light.

b. Write an inequality in terms of x and y that represents the points in the illuminated area.

c. Describe the region that would be illuminated if the pole stood straight up at point O and the light were directed straight down.

53. $\frac{(x-1)^2}{1}+\frac{y^2}{25}=1$; center: $(1, 0)$; vertices: $(1, -5)$ and $(1, 5)$; co-vertices: $(0, 0)$ and $(2, 0)$; foci: $(1, -2\sqrt{6})$ and $(1, 2\sqrt{6})$

54. $\frac{(x-2)^2}{9}+\frac{(y+1)^2}{4}=1$; center: $(2, -1)$; vertices: $(-1, -1)$ and $(5, -1)$; co-vertices: $(2, -3)$ and $(2, 1)$; foci: $(2-\sqrt{5}, -1)$ and $(2+\sqrt{5}, -1)$

55. $\frac{(x+2)^2}{9}+\frac{(y+1)^2}{25}=1$; center: $(-2, -1)$; vertices: $(-2, -6)$ and $(-2, 4)$; co-vertices: $(-5, -1)$ and $(1, -1)$; foci: $(-2, -5)$ and $(-2, 3)$

56. $\frac{(x-2)^2}{16}+\frac{(y-2)^2}{9}=1$; center: $(2, 2)$; vertices: $(-2, 2)$ and $(6, 2)$; co-vertices: $(2, -1)$ and $(2, 5)$; foci: $(2-\sqrt{7}, 2)$ and $(2+\sqrt{7}, 2)$

57. $\frac{(x-1)^2}{25}+\frac{(y+2)^2}{36}=1$; center: $(1, -2)$; vertices: $(1, -8)$ and $(1, 4)$; co-vertices: $(-4, -2)$ and $(6, -2)$; foci: $(1, -2-\sqrt{11})$ and $(1, -2+\sqrt{11})$

38. center: $(0, 0)$; vertices: $(-5, 0)$ and $(5, 0)$; co-vertices: $(0, -2)$ and $(0, 2)$; foci: $(-\sqrt{21}, 0)$ and $(\sqrt{21}, 0)$

39. center: $(0, 0)$; vertices: $(0, -5)$ and $(0, 5)$; co-vertices: $(-3, 0)$ and $(3, 0)$; foci: $(0, -4)$ and $(0, 4)$

50. $\frac{(x+3)^2}{16}+\frac{(y-1)^2}{4}=1$; center: $(-3, 1)$; vertices: $(-7, 1)$ and $(1, 1)$; co-vertices: $(-3, -1)$ and $(-3, 3)$; foci: $(-3-2\sqrt{3}, 1)$ and $(-3+2\sqrt{3}, 1)$

51. $\frac{(x+1)^2}{4}+\frac{(y-1)^2}{16}=1$; center: $(-1, 1)$; vertices: $(-1, -3)$ and $(-1, 5)$; co-vertices: $(-3, 1)$ and $(1, 1)$; foci: $(-1, 1-2\sqrt{3})$ and $(-1, 1+2\sqrt{3})$

52. $\frac{x^2}{64}+\frac{(y-2)^2}{4}=1$; center: $(0, 2)$; vertices: $(-8, 2)$ and $(8, 2)$; co-vertices: $(0, 0)$ and $(0, 4)$; foci: $(-2\sqrt{15}, 2)$ and $(2\sqrt{15}, 2)$

The graphs for Exercises 27–39 and answers to Exercises 60 and 64 can be found in Additional Answers beginning on page 1002.

Student Technology Guide

NAME ______ CLASS ______ DATE ______

Student Technology Guide
9.4 *Ellipses*

It is easy to see that an ellipse fails the vertical line test, so an equation for an ellipse is not a function. To graph an ellipse on a graphing calculator, you can graph two functions, each representing half the ellipse.

Example: Graph $\frac{x^2}{9}+\frac{y^2}{16}=1$.

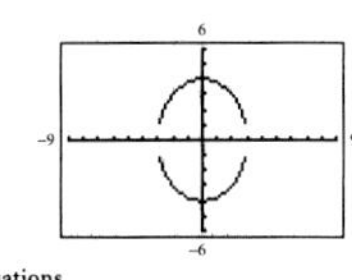

- Solve for y.
  $\frac{y^2}{16}=1-\frac{x^2}{9}$
  $y^2=16\left(1-\frac{x^2}{9}\right)$
  $y=\pm\sqrt{16\left(1-\frac{x^2}{9}\right)}$
- The ± sign means you need to graph two equations. Enter the first as Y1. Press [Y=] [2nd] [x^2] 16 [(] 1 [−] [X,T,θ,n] [x^2] [÷] 9 [)] [)].
- Make Y2 the opposite of Y1 by moving to Y2 and pressing [(−)] [VARS] Y-VARS 1:Function... [ENTER] 1:Y1 [ENTER].
- Press [GRAPH].

Remember to use a square viewing window to see an undistorted graph.

Graph each equation. Does the ellipse spread out horizontally or vertically?

1. $\frac{x^2}{25}+\frac{y^2}{4}=1$ horizontally
2. $\frac{x^2}{9}+\frac{y^2}{36}=1$ vertically
3. $x^2+\frac{y^2}{16}=1$ vertically
4. $x^2+4y^2-64=0$ horizontally
5. $6x^2+2y^2-12=0$ vertically
6. $\frac{(x-3)^2}{36}+\frac{y^2}{4}=1$ horizontally
7. Give the vertices of the ellipse in the example. Explain how you could have predicted these vertices from the equation $\frac{x^2}{9}+\frac{y^2}{16}=1$.
   $(-3, 0)$, $(3, 0)$, $(0, -4)$, $(0, 4)$; Explanations may vary. For an ellipse $\frac{x^2}{b^2}+\frac{y^2}{a^2}=1$ with center at $(0, 0)$ and major axis along the y-axis, the vertices are $(-b, 0)$, $(b, 0)$, $(0, -a)$, and $(0, a)$. For the ellipse in the example, $b = 3$ and $a = 4$.

In Exercise 75, students discover how replacing the plus sign with a minus sign in the equation of an ellipse produces a completely different curve called a *hyperbola*. Hyperbolas will be studied in depth in Lesson 9.5.

ALTERNATIVE
Assessment

Portfolio Activity

The Portfolio Activity can be used as preparation for the Chapter Project or as a separate activity. In the Portfolio Activity on this page, students will create a model of an ellipse by folding patty paper or waxed paper.

Answers to Portfolio Activities can be found in Additional Answers of the Teacher's Edition.

Simplify each expression, assuming that no variable equals zero. Write your answers with positive exponents only. *(LESSON 2.2)*

65. $(4x^3y^4)^2(2x^4y)^{-1}$ $\mathbf{8x^2y^7}$ **66.** $\left(\frac{a^5b^{-3}}{ab^4}\right)^{-2}$ $\mathbf{\frac{b^{14}}{a^8}}$ **67.** $\left(\frac{6xy^2z^{-5}}{5x^2y^{-2}z}\right)^2$ $\mathbf{\frac{36y^8}{25x^2z^{12}}}$

Solve each equation for x by using natural logarithms. Round answers to the nearest hundredth. *(LESSON 6.6)*

68. $10^x = 56$ $\mathbf{x \approx 1.75}$ **69.** $25^x = 123$ $\mathbf{x \approx 1.50}$ **70.** $0.3^{-x} = 0.81$ $\mathbf{x \approx -0.18}$

Write each sum or difference as a polynomial expression in standard form. *(LESSON 7.1)*

71. $(18a^3 - 5a^2 - 6a + 2) + (7a^3 - 8a + 9)$ $\mathbf{25a^3 - 5a^2 - 14a + 11}$

72. $2x^4 - 3x^2 + 2 - 5x) + (4x^2 + 2x - 7x^4 + 6)$ $\mathbf{-5x^4 + x^2 - 3x + 8}$

73. $(9x^2y^2 - 5xy + 25y^2) - (5x^2y^2 + 10xy - 9y^2)$ $\mathbf{4x^2y^2 - 15xy + 34y^2}$

74. $(-x^2 - y^2) - (-2x^2 + 3xy - 2y^2)$ $\mathbf{x^2 - 3xy + y^2}$

Look Beyond

75 Graph $\frac{x^2}{9} - \frac{y^2}{16} = 1$, and describe its shape. Explain why there are no real-number values of y for $-3 < x < 3$. **Sample answer: The graph consists of two curves, one for $x \geq 3$ and one for $x \leq -3$. There are no real-number values of y for $-3 < x < 3$ because $\frac{y^2}{16}$ cannot be negative if y is a real number.**

PORTFOLIO ACTIVITY

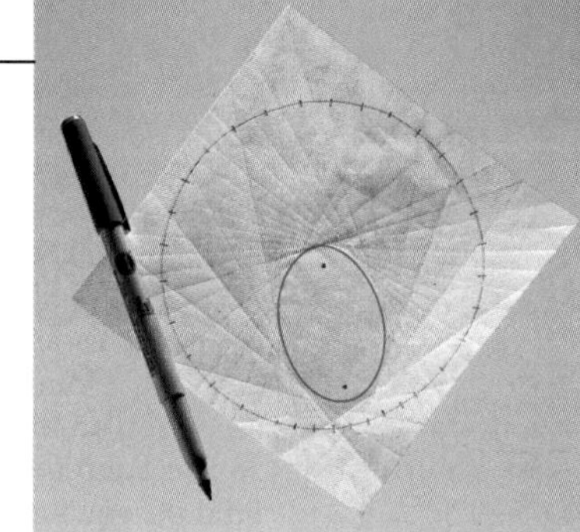

1. Draw a circle with a diameter of at least 4.5 inches in the middle of a piece of patty paper or wax paper.
2. Place at least 20 tick marks on the circle, approximately evenly spaced.
3. Place a point inside the circle anywhere except at the center. Fold the paper so that this inside point lies on one of the tick marks on the circle. Crease the paper, and then unfold it.
4. Repeat this procedure for each tick mark on the circle.
5. Compare the figure formed by the creases with the figures formed by your classmates. How do the figures vary? Could they all be classified as one type of conic section? If so, which one?
6. Make a conjecture about how to make a flatter or rounder figure. Verify your conjecture by repeating the process with the appropriate modification(s).
7. How does the point that you chose in Step 3 appear to be mathematically significant? Explain your reasoning.

WORKING ON THE CHAPTER PROJECT

You should now be able to complete Activity 2 of the Chapter Project.

Hyperbolas

Why *The principle of the LORAN navigation system used by ships at sea is based on the definition of a hyperbola.*

Objectives

- Write the standard equation for a hyperbola given sufficient information.
- Graph the equation of a hyperbola, and identify the center, foci, vertices, and co-vertices.

APPLICATION
RADIO NAVIGATION

Ships can navigate by using the LORAN radio navigation system, which is based on the time differences between receiving radio signals from pairs of transmitting stations that send signals at the same time.

For example, the ship at point P in the diagram at right can use the time difference $t_2 - t_1$ and the distance F_1F_2 to locate itself somewhere on the branches of the blue hyperbola. Then, using the time difference $t_3 - t_1$ and the distance F_1F_3, the ship can locate itself somewhere on the branches of the red hyperbola.

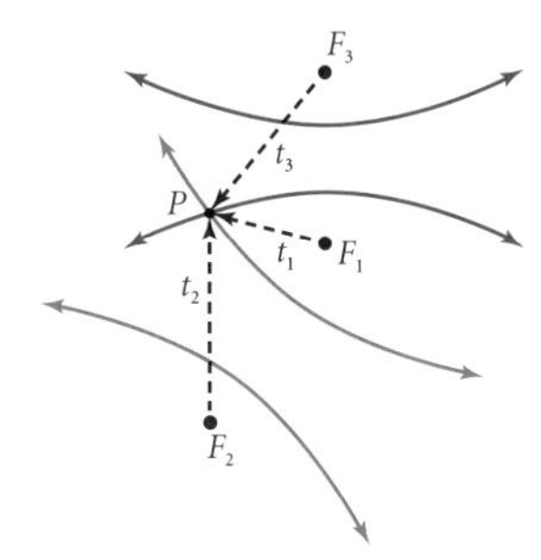

By finding the reasonable intersection of these two hyperbolas, the ship can determine its exact location.

The radio navigation system described utilizes the *definition of a hyperbola* to locate objects.

Definition of a Hyperbola

A **hyperbola** is the set of points $P(x, y)$ in a plane such that the absolute value of the difference between the distances from P to two fixed points in the plane, F_1 and F_2, called the **foci**, is a constant.

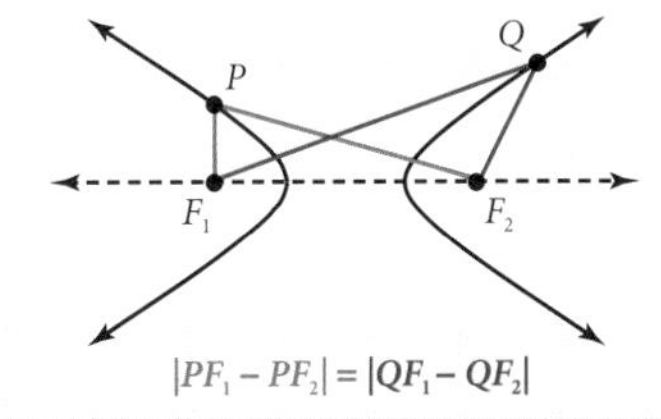

Prepare

NCTM PRINCIPLES & STANDARDS 1–4, 6–10

QUICK WARM-UP

Graph the equation and identify the conic section.

1. $x^2 + y^2 = 4$ circle
2. $x^2 + y = 4$ parabola
3. $x^2 + 4y^2 = 4$ ellipse
4. Write the ellipse defined by $4x^2 + 16x + 9y^2 - 54y = -61$ in standard form. Identify the coordinates of the center, vertices, co-vertices, and foci.
   $\frac{(x+2)^2}{9} + \frac{(y-3)^2}{4} = 1$; center: $(-2, 3)$; vertices: $(-5, 3)$ and $(1, 3)$; co-vertices: $(-2, 1)$ and $(-2, 5)$; foci: $(-2 - \sqrt{5}, 3)$ and $(-2 + \sqrt{5}, 3)$

Also on Quiz Transparency 9.5

Teach

Why Two hyperbolas serve as a coordinate system in the LORAN navigation system. A ship's position is determined from time differences among radio signals received from three stations located at the foci of two different hyperbolas that share one foci. Discuss how such signals can be used to locate a ship's position.

Alternative Teaching Strategy

HANDS-ON STRATEGIES Ask students to solve the equation $\frac{x^2}{4} - y^2 = 1$ for y. Have them find the corresponding y-coordinates when $x = \pm 2, \pm 2.5$, and ± 5.2 and then use these points to graph the equation. Now give them the fixed points $F_1(-\sqrt{5}, 0)$ and $F_2(\sqrt{5}, 0)$. Letting P represent the point $(2.5, 0.75)$, ask them to use the distance formula to find $PF_1 - PF_2$. Repeat the calculation, using the points $(5.2, 2.4)$ and $(-2, 0)$ for P. Be sure students notice that $PF_1 - PF_2$ is 4 in all three cases.

Now define a hyperbola. Using the hyperbola students have drawn, identify the center, vertices, co-vertices, and foci. **center: $(0, 0)$; vertices: $V_1(-2, 0)$, $V_2(2, 0)$; co-vertices: $C_1(0, 1)$, $C_2(0, -1)$; foci: $F_1(-\sqrt{5}, 0)$, $F_2(\sqrt{5}, 0)$** Point out that an alternative way of writing $\frac{x^2}{4} - y^2 = 1$ is $\frac{(x-0)^2}{4} + \frac{(y-0)^2}{1} = 1$. Show them how to use this form of the equation to identify the center, vertices, co-vertices, and foci. Relate the defining features of a hyperbola to the defining features of an ellipse and a parabola, noting any similarities and differences.

Teaching Tip

Remind students that the form of an equation for an ellipse is $\frac{x^2}{a^2} + \frac{y^2}{b^2} = 1$, that the transverse axis of an ellipse is determined by which value is larger, a or b, and that the values of a, b, and c are related by $a^2 - b^2 = c^2$. Geometrically, the distance from a vertex to the center of an ellipse, a, is the same as the length of the hypotenuse of the right triangle formed by the center, a focus, and a co-vertex.

Then point out that the transverse axis of a hyperbola and the value of a are determined by which term in the standard form of the equation, x^2 or y^2, is negative, and that the values of a, b, and c are related by $a^2 + b^2 = c^2$. Geometrically, the distance from a focus to the center of a hyperbola, c, is the length of the hypotenuse of the right triangle formed by the center, a vertex, and a co-vertex.

You can use the definition of a hyperbola and the distance formula to find an equation for the hyperbola that contains $P(x, y)$, has foci at $F_1(-5, 0)$ and $F_2(5, 0)$, and has a constant difference of 6.

$$|F_1P - F_2P| = 6$$
$$F_1P - F_2P = \pm 6$$
$$\sqrt{[x-(-5)]^2 + (y-0)^2} - \sqrt{(x-5)^2 + (y-0)^2} = \pm 6$$
$$\sqrt{(x-5)^2 + y^2} = \pm 6 + \sqrt{(x-5)^2 + y^2}$$

Square each side.
$$(x+5)^2 + y^2 = 36 \pm 12\sqrt{(x-5)^2 + y^2} + (x-5)^2 + y^2$$
$$x^2 + 10x + 25 + y^2 = 36 \pm 12\sqrt{(x-5)^2 + y^2} + x^2 - 10x + 25 + y^2$$

Simplify.
$$20x - 36 = \pm 12\sqrt{(x-5)^2 + y^2}$$

Divide each side by 4.
$$5x - 9 = \pm 3\sqrt{(x-5)^2 + y^2}$$

Square each side again.
$$25x^2 - 90x + 81 = 9x^2 - 90x + 225 + 9y^2$$
$$16x^2 - 9y^2 = 144$$

Divide each side by 144.
$$\frac{x^2}{9} - \frac{y^2}{16} = 1$$

Notice that the transverse axis can be shorter than the conjugate axis.

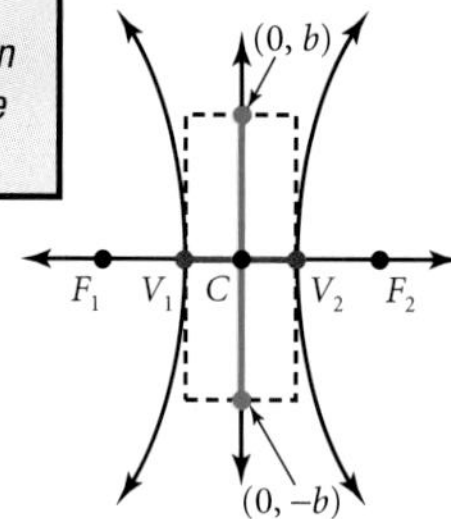

A hyperbola has two axes of symmetry. One axis contains the **transverse axis**, $\overline{V_1V_2}$, of the hyperbola, and the other axis contains the **conjugate axis**, from $(0, -b)$ to $(0, b)$, of the hyperbola. The endpoints of the transverse axis are called the **vertices** of the hyperbola. The endpoints of the conjugate axis are called the **co-vertices** of the hyperbola. The point of intersection of the transverse axis and the conjugate axis is called the **center** of the hyperbola.

Standard Equation of a Hyperbola

The standard equation of a hyperbola centered at the origin is given below.

Horizontal transverse axis

$$\frac{x^2}{a^2} - \frac{y^2}{b^2} = 1$$

Vertical transverse axis

$$\frac{y^2}{a^2} - \frac{x^2}{b^2} = 1$$

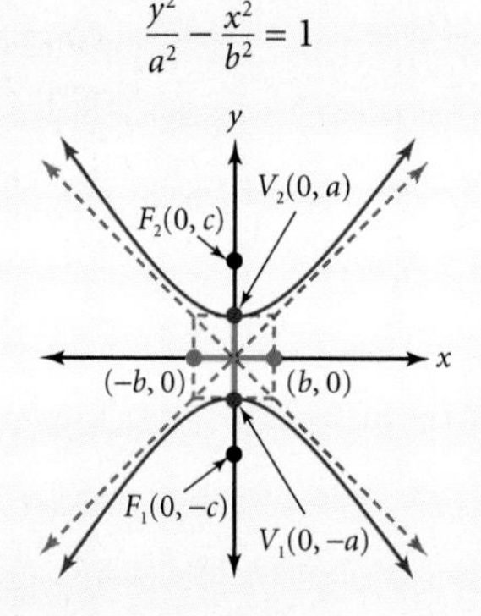

In each case:
- $a^2 + b^2 = c^2$,
- the length of the transverse axis is $2a$, and
- the length of the conjugate axis is $2b$.

Interdisciplinary Connection

ASTRONOMY A telescope may have a hyperbolic mirror with the property that a ray of light directed at one focus is reflected to the other focus. If the center is located at the origin, a focus has coordinates (5, 0), a vertex has coordinates (3, 0), and one end of the mirror is attached to the telescope at the point $\left(5, \frac{16}{3}\right)$, find the equation that defines the mirror. See the diagram.

$16x^2 - 9y^2 = 144$

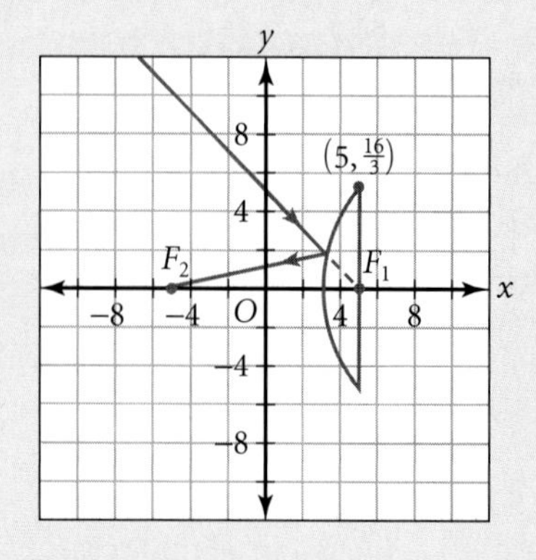

EXAMPLE 1 **Write the standard equation for the hyperbola with vertices at (0, −4) and (0, 4) and co-vertices at (−3, 0) and (3, 0). Then sketch the graph.**

SOLUTION

Because the vertices lie along the y-axis and the center is at the origin, the equation is of the form $\frac{y^2}{a^2} - \frac{x^2}{b^2} = 1$.

From the coordinates of the vertices, $a = 4$. From the coordinates of the co-vertices, $b = 3$. The equation is $\frac{y^2}{4^2} - \frac{x^2}{3^2} = 1$, or $\frac{y^2}{16} - \frac{x^2}{9} = 1$.

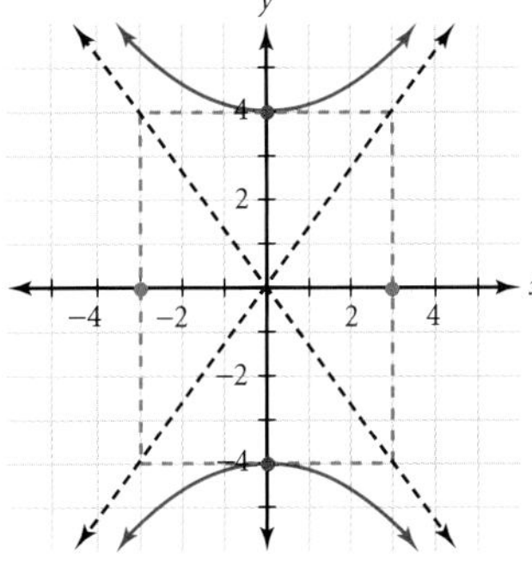

Plot the vertices and co-vertices, and draw the rectangle determined by these points. Sketch the lines containing the diagonals of this rectangle. Then sketch the branches of the hyperbola between the lines that contain the diagonals, as shown at right above.

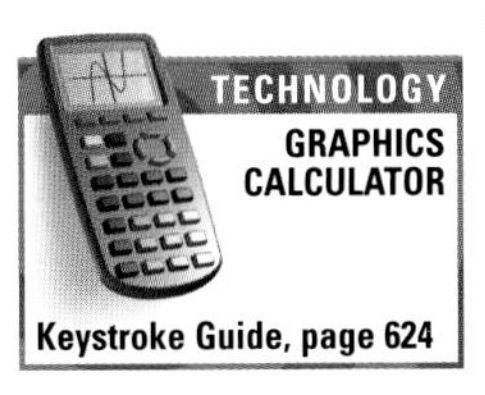

CHECK

To graph the hyperbola defined by $\frac{y^2}{16} - \frac{x^2}{9} = 1$, solve for y and graph the two resulting equations together.

$\frac{y^2}{16} - \frac{x^2}{9} = 1 \rightarrow y = \pm\sqrt{16\left(1 + \frac{x^2}{9}\right)}$

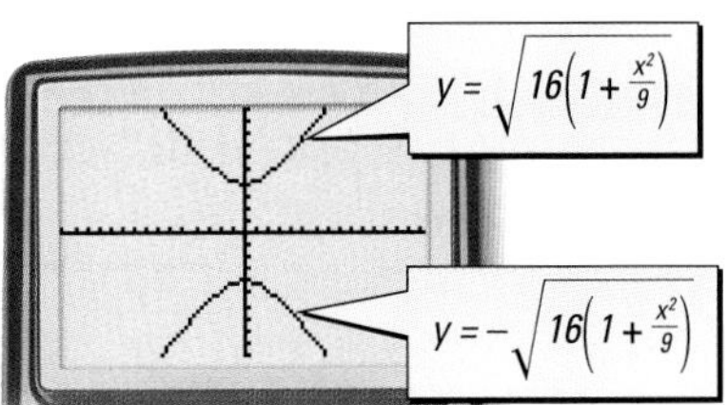

TRY THIS Write the standard equation for the hyperbola with vertices at (−7, 0) and (7, 0) and co-vertices at (0, −4) and (0, 4). Then sketch the graph.

CRITICAL THINKING Write the standard equation for the hyperbola that has a horizontal transverse axis, is centered at the origin, and passes through the points (1, 2) and (5, 12).

Notice that the hyperbola in Example 1 above is graphed with a pair of dashed lines that contain the diagonals of the rectangle that is determined by the vertices and co-vertices. These lines are **asymptotes of the hyperbola**. In the Activity below, you will explore the equations for the asymptotes of hyperbolas.

Exploring Asymptotes of Hyperbolas

You will need: a graphics calculator

1. Solve $\frac{x^2}{2^2} - \frac{y^2}{3^2} = 1$ for y, and graph the resulting equations with the lines $y = \frac{3}{2}x$ and $y = -\frac{3}{2}x$ on the same screen. Do the branches of the hyperbola appear to approach these lines?

Enrichment

Two hyperbolas in which the transverse axis of one is the conjugate axis of the other are called *conjugate hyperbolas.* For example, $\frac{y^2}{16} - \frac{x}{36} = 1$ and $\frac{x^2}{36} - \frac{y^2}{16} = 1$ are equations of conjugate hyperbolas. Ask students to find the equations for the asymptotes of each hyperbola and graph the conjugate hyperbolas on the same coordinate plane. Then ask them to make a conjecture about the asymptotes of conjugate hyperbolas.

$y = \pm\frac{2}{3}x$; the asymptotes of conjugate hyperbolas are the same.

ADDITIONAL EXAMPLE 1

Write the standard equation for the hyperbola whose vertices are at $V_1(-5, 0)$ and $V_2(5, 0)$ and whose co-vertices are at $C_1(0, -6)$ and $C_2(0, 6)$. Graph the hyperbola.

$\frac{x^2}{25} - \frac{y^2}{36} = 1$

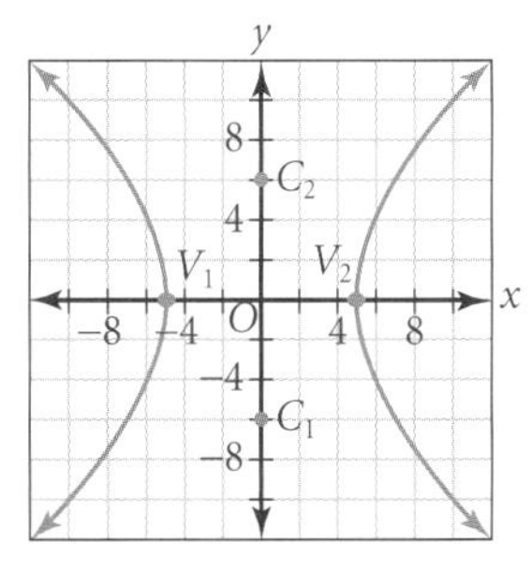

Teaching Tip

TECHNOLOGY When using a TI-82 or TI-83 graphics calculator to create the hyperbola in Example 1, use the following viewing window:

Xmin=−15	Ymin=−10
Xmax=15	Ymax=10
Xscl=1	Yscl=1

TRY THIS

$\frac{x^2}{49} - \frac{y^2}{16} = 1$

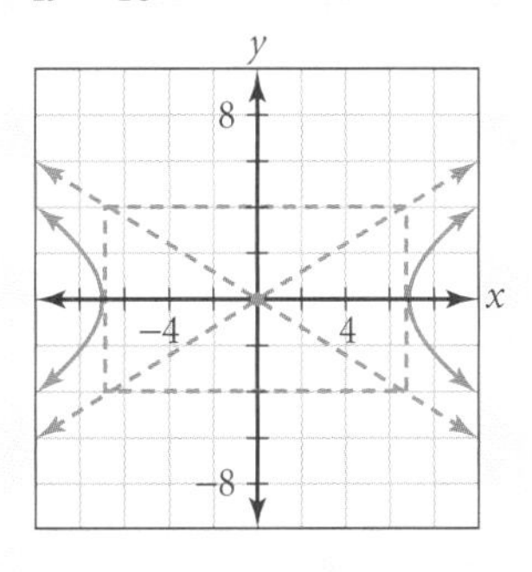

CRITICAL THINKING

$\frac{x^2}{\frac{11}{35}} - \frac{y^2}{\frac{11}{6}} = 1$

☞ For Activity Notes, see page 598.

Activity Notes

Some students may have difficulty trying to graph points on a hyperbola without first graphing the asymptotes, which are the extended diagonals of the rectangle formed by the vertices and the co-vertices. Suggest that they graph the asymptotes before graphing the hyperbola. Students should notice that as the points on a branch get farther from the foci, they approach but do not touch one of the asymptotes.

CHECKPOINT ✔

4. $y = \pm\frac{4}{5}x$

ADDITIONAL EXAMPLE 2

Find the equations of the asymptotes and the coordinates of the vertices for the graph of $\frac{y^2}{9} - \frac{x^2}{4} = 1$. Then graph the hyperbola.

asymptotes: $y = \pm 1.5x$;
vertices: $V_1(0, -3)$ and $V_2(0, 3)$

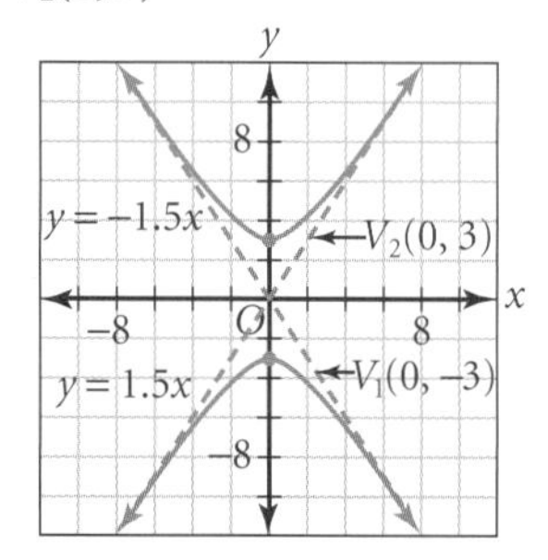

TRY THIS

$y = \pm\frac{4}{5}x$;

vertices: (−4, 0) and (4, 0)

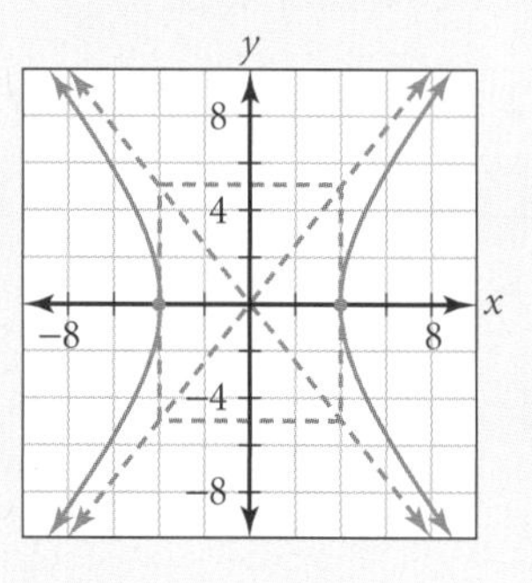

2. Copy and complete the table of values to the nearest hundredth.

x	$y = \sqrt{9\left(\frac{x^2}{4} - 1\right)}$	$y = -\sqrt{9\left(\frac{x^2}{4} - 1\right)}$	$y = \frac{3}{2}x$	$y = -\frac{3}{2}x$
10				
20				
30				
40				
50				
60				

3. Do the values in the table above suggest that the branches of this hyperbola approach the lines $y = \frac{3}{2}x$ and $y = -\frac{3}{2}x$? Explain your response.

CHECKPOINT ✔ 4. Consider $\frac{x^2}{5^2} - \frac{y^2}{4^2} = 1$. Predict the equations for the asymptotes of this hyperbola. Use a graph and a table to check your prediction.

Given the standard equation of any hyperbola with a horizontal or vertical transverse axis and with its center at the origin, you can write equations for the asymptotes.

Asymptotes of a Hyperbola

standard equation: $\frac{x^2}{a^2} - \frac{y^2}{b^2} = 1 \rightarrow$ asymptotes: $y = \pm\frac{b}{a}x$

standard equation: $\frac{y^2}{a^2} - \frac{x^2}{b^2} = 1 \rightarrow$ asymptotes: $y = \pm\frac{a}{b}x$

EXAMPLE 2 **Find the equations of the asymptotes and the coordinates of the vertices for the graph of $\frac{y^2}{16} - \frac{x^2}{36} = 1$. Then sketch the graph.**

SOLUTION

The equation is of the form $\frac{y^2}{a^2} - \frac{x^2}{b^2} = 1$, where $a = 4$ and $b = 6$.

The asymptotes, $y = \pm\frac{a}{b}x$, are $y = \pm\frac{4}{6}x$, or $y = \frac{2}{3}x$ and $y = -\frac{2}{3}x$.

The vertices are (0, −4) and (0, 4).

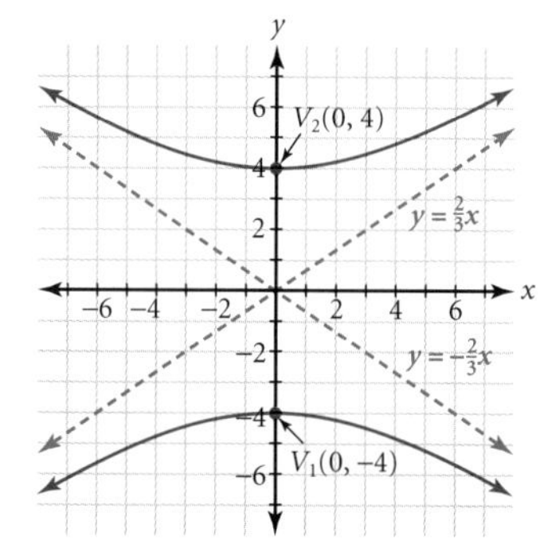

TRY THIS Find the equations of the asymptotes and the coordinates of the vertices for the graph of $\frac{x^2}{16} - \frac{y^2}{25} = 1$. Then sketch the graph.

Inclusion Strategies

USING VISUAL MODELS Have students work in groups of four to design clown masks that include the graphs (or parts of the graphs) of parabolas, circles, ellipses, and hyperbolas. Each group should design and produce three masks, showing happiness, sadness, and surprise. Each group can then exchange their mask equations with another group, which will draw the mask defined by the equations. Students may want to construct the masks and either wear them to class or display them around the classroom.

The center of a hyperbola can be anywhere in a coordinate plane.

Standard Equation of a Translated Hyperbola

The standard equation of a hyperbola centered at (h, k) is given below.

Horizontal transverse axis	Vertical transverse axis
$\frac{(x-h)^2}{a^2} - \frac{(y-k)^2}{b^2} = 1$	$\frac{(y-k)^2}{a^2} - \frac{(x-h)^2}{b^2} = 1$

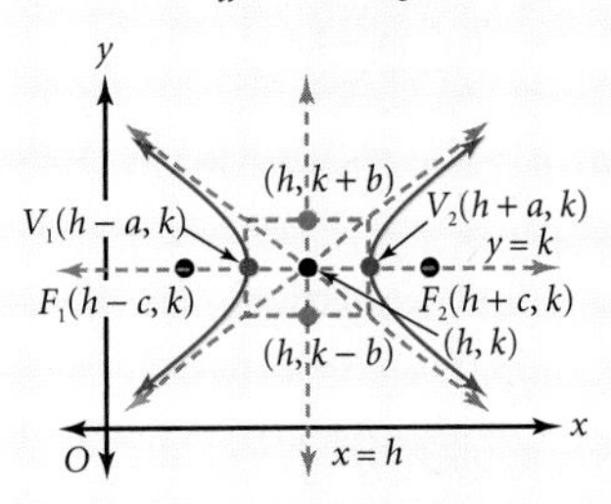

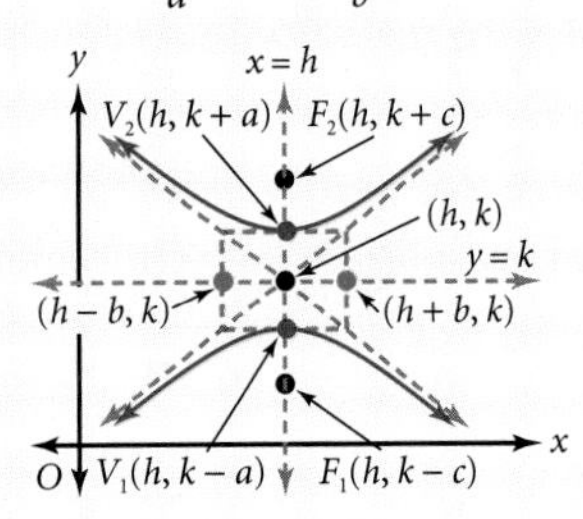

In each case:
- $a^2 + b^2 = c^2$,
- the length of the transverse axis is $2a$, and
- the length of the conjugate axis is $2b$.

Given sufficient information, you can write the standard equation of a translated hyperbola.

EXAMPLE 3 **Write the standard equation of the hyperbola with foci at $(-1, 1)$ and $(5, 1)$ and vertices at $(0, 1)$ and $(4, 1)$.**

SOLUTION

1. To find the center, (h, k), find the midpoint of the transverse axis.

$$\frac{x_1 + x_2}{2} = \frac{0 + 4}{2} = 2 \qquad \frac{y_1 + y_2}{2} = \frac{1 + 1}{2} = 1$$

The center is at $(2, 1)$, so $h = 2$ and $k = 1$.

2. Find a, b, and c. The transverse axis is horizontal and the standard equation is of the form $\frac{(x-h)^2}{a^2} - \frac{(y-k)^2}{b^2} = 1$.

The vertex $(0, 1)$ is $(h - a, k)$, so $h - a = 0$.

$$h - a = 0$$
$$2 - a = 0$$
$$a = 2$$

The focus $(-1, 1)$ is $(h - c, k)$, so $h - c = -1$.

$$h - c = -1$$
$$2 - c = -1$$
$$c = 3$$

$$a^2 + b^2 = c^2 \quad \rightarrow \quad 2^2 + b^2 = 3^2 \quad \rightarrow \quad b^2 = 5$$

Thus, the standard equation of this hyperbola is $\frac{(x-2)^2}{4} - \frac{(y-1)^2}{5} = 1$.

TRY THIS Write the standard equation of the hyperbola with foci at $(3, -3)$ and $(3, 7)$ and vertices at $(3, -1)$ and $(3, 5)$.

ADDITIONAL EXAMPLE 3

Write the standard equation of the hyperbola with foci at $(1, 5)$ and $(7, 5)$ and with vertices at $(2, 5)$ and $(6, 5)$.

$\frac{(x-4)^2}{4} - \frac{(y-5)^2}{5} = 1$

TRY THIS

$\frac{(y-2)^2}{9} - \frac{(x-3)^2}{16} = 1$

Reteaching the Lesson

COOPERATIVE LEARNING To help students see how a, b, h, and k are related in the standard equation of a translated hyperbola, divide the class into groups of four students. Give each group a set of four values for each constant, a, b, h, and k. From the values given, members should select values for the constants for each of four hyperbolas so that different hyperbolas will be created. Members should be able to explain how their graphs differ from that of the hyperbola whose equation is $x^2 - y^2 = 1$, describe the asymptotes of each hyperbola, write the standard equation of each hyperbola, and graph each hyperbola.

ADDITIONAL
EXAMPLE 4

The equation $x^2 - y^2 - 6x - 10y - 20 = 0$ represents a hyperbola. **Write the standard equation of the hyperbola. Give the coordinates of the center, vertices, co-vertices, and foci. Then graph the hyperbola.**

$\frac{(x-3)^2}{4} - \frac{(y+5)^2}{4} = 1$;
center: $C(3, -5)$;
vertices: $V_1(1, -5)$ and $V_2(5, -5)$;
co-vertices: $C_1(3, -3)$ and $C_2(3, -7)$;
foci: $F_1(3 + 2\sqrt{2}, -5)$ and $F_2(3 - 2\sqrt{2}, -5)$

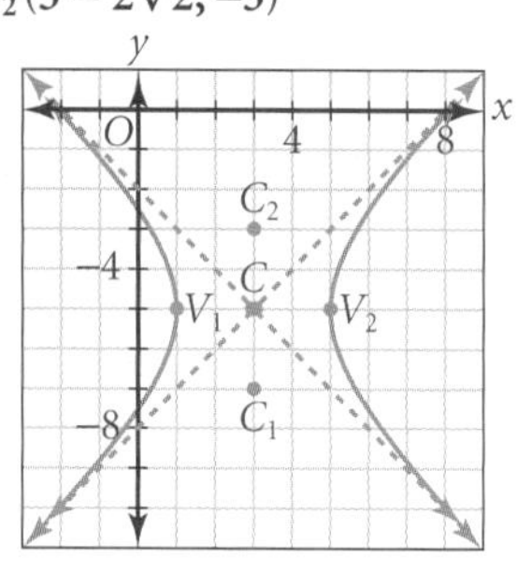

TRY THIS
$\frac{(x-1)^2}{25} - \frac{(y-2)^2}{4} = 1$;
center: $(1, 2)$;
vertices: $(-4, 2)$ and $(6, 2)$;
co-vertices: $(1, 0)$ and $(1, 4)$;
foci: $(1 + \sqrt{29}, 2)$ and $(1 - \sqrt{29}, 2)$

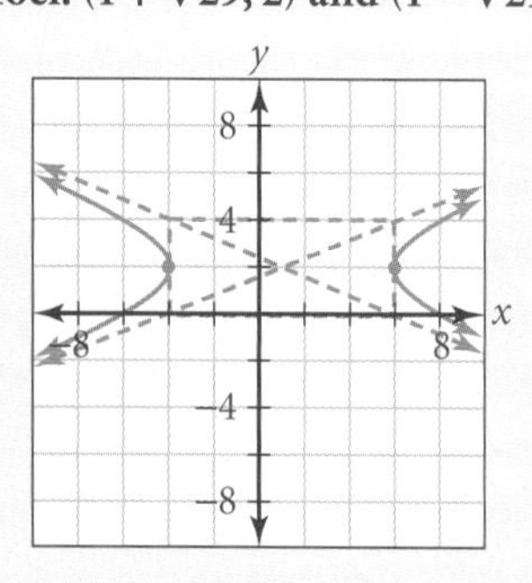

EXAMPLE 4 **The equation $-2x^2 + y^2 + 4x + 6y + 3 = 0$ represents a hyperbola. Write the standard equation of this hyperbola. Give the coordinates of the center, vertices, co-vertices, and foci. Then sketch the graph.**

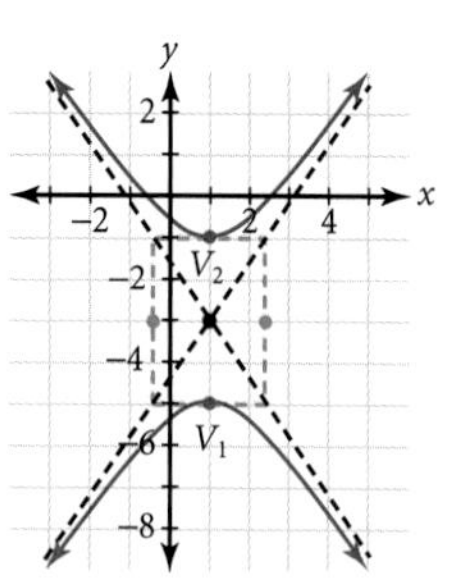

SOLUTION

$$-2x^2 + y^2 + 4x + 6y + 3 = 0$$
$$y^2 + 6y - 2x^2 + 4x = -3$$
$$(y^2 + 6y + 9) - 2(x^2 - 2x + 1) = -3 + 9 - 2 \quad \textit{Complete the squares.}$$
$$(y+3)^2 - 2(x-1)^2 = 4$$
$$\frac{(y+3)^2}{4} - \frac{(x-1)^2}{2} = 1$$

From the equation, $h = 1$, $k = -3$, $a = 2$, and $b = \sqrt{2}$. Then $c = \sqrt{4+2} = \sqrt{6}$.

The center, (h, k), is at $(1, -3)$.
Use the value of a to find the vertices, $(1, -5)$ and $(1, -1)$.
Use the value of b to find the co-vertices, $(1 - \sqrt{2}, -3)$ and $(1 + \sqrt{2}, -3)$.
Use the value of c to find the foci, $(1, -3 - \sqrt{6})$ and $(1, -3 + \sqrt{6})$.

TRY THIS The equation $4x^2 - 25y^2 - 8x + 100y - 196 = 0$ represents a hyperbola. Write the standard equation of this hyperbola. Give the coordinates of the center, vertices, co-vertices, and foci. Then sketch the graph.

Exercises

Communicate

1. Explain how to find the coordinates of the center, vertices, co-vertices, and foci when given the standard equation of a translated hyperbola.
2. Explain how to use the asymptotes and the coordinates of the vertices of a hyperbola to sketch it.

Guided Skills Practice

3. Write the standard equation for the hyperbola with vertices at $(-4, 0)$ and $(4, 0)$ and co-vertices at $(0, -2)$ and $(0, 2)$. Then sketch the graph. *(EXAMPLE 1)*
4. Find the equations of the asymptotes and the coordinates of the vertices of the graph of $\frac{x^2}{9} - \frac{y^2}{25} = 1$. Then sketch the graph. *(EXAMPLE 2)*
5. Write the standard equation of the hyperbola with foci at $(0, 4)$ and $(6, 4)$ and vertices at $(1, 4)$ and $(5, 4)$. *(EXAMPLE 3)*
6. The equation $x^2 - y^2 + 2x + 4y - 2 = 12$ represents a hyperbola. Write the standard equation of this hyperbola. Give the coordinates of the center, vertices, co-vertices, and foci. Then sketch the graph. *(EXAMPLE 4)*

3. $\frac{x^2}{16} - \frac{y^2}{4} = 1$

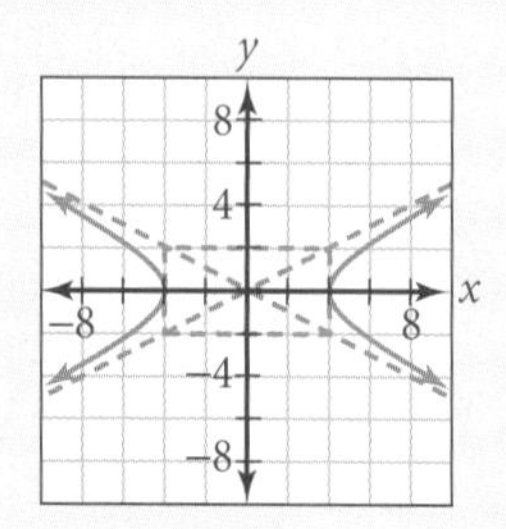

4. asymptotes: $y = \frac{5}{3}x$ and $y = -\frac{5}{3}x$; vertices: $(-3, 0)$ and $(3, 0)$

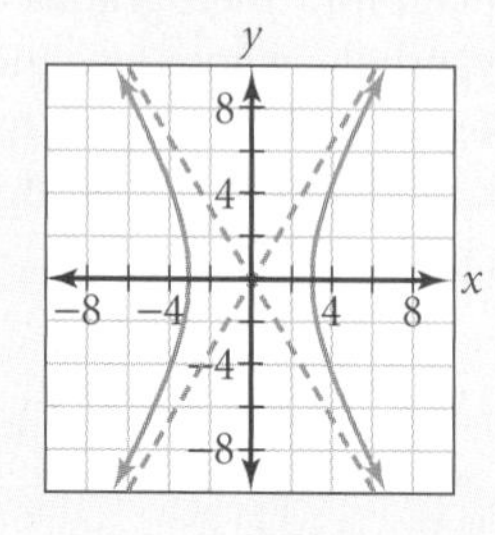

5. $\frac{(x-3)^2}{4} - \frac{(y-4)^2}{5} = 1$

6. $\frac{(x+1)^2}{11} - \frac{(y-2)^2}{11} = 1$; center: $(-1, 2)$; vertices: $(-1 - \sqrt{11}, 2)$ and $(-1 + \sqrt{11}, 2)$; co-vertices: $(-1, 2 - \sqrt{11})$ and $(-1, 2 + \sqrt{11})$; foci: $(-1 - \sqrt{22}, 2)$ and $(-1 + \sqrt{22}, 2)$

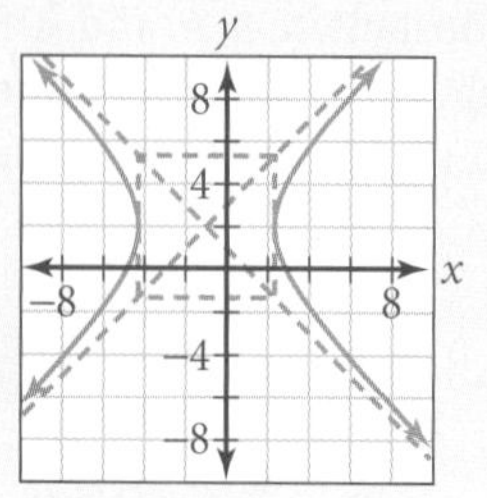

Practice and Apply

Write the standard equation for each hyperbola.

7.

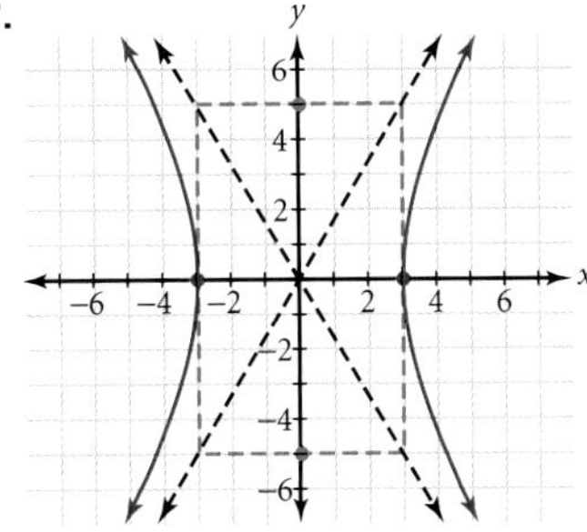

8.

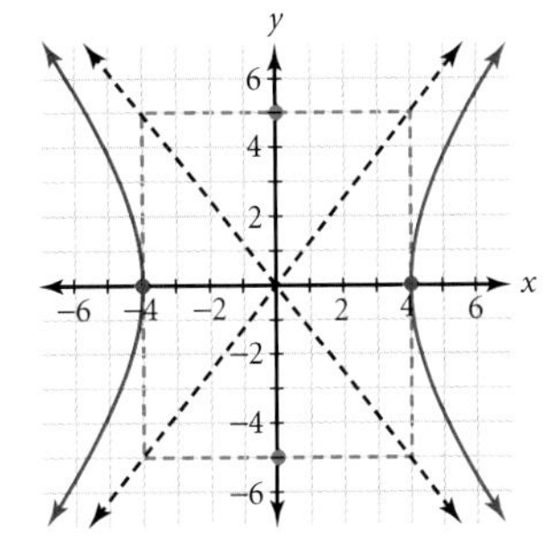

9.

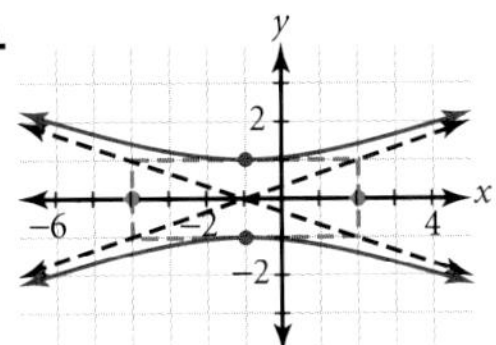

10.

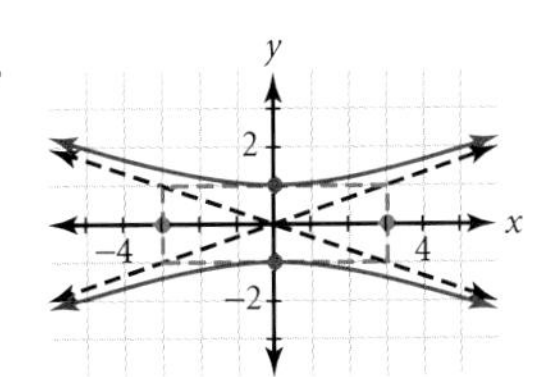

11.

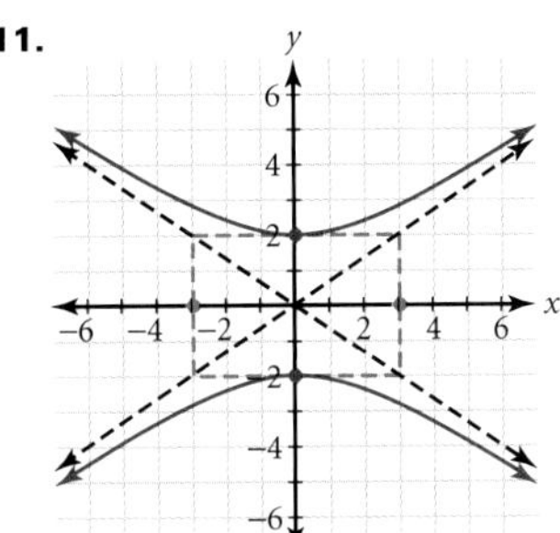

12.

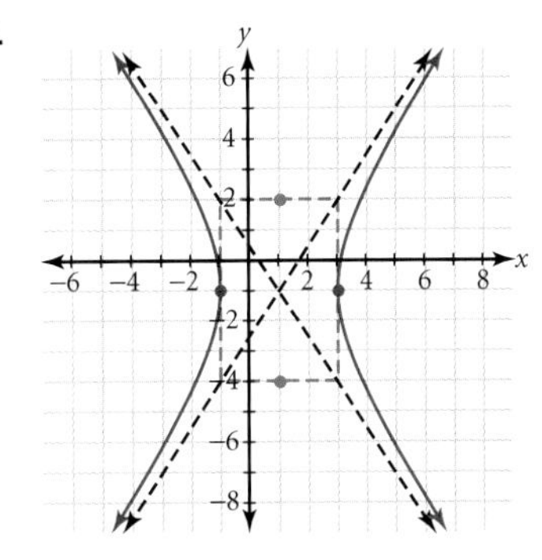

7. $\frac{x^2}{9} - \frac{y^2}{25} = 1$
8. $\frac{x^2}{16} - \frac{y^2}{25} = 1$
9. $\frac{y^2}{1} - \frac{(x+1)^2}{9} = 1$
10. $\frac{y^2}{1} - \frac{x^2}{9} = 1$
11. $\frac{y^2}{4} - \frac{x^2}{9} = 1$
12. $\frac{(x-1)^2}{4} - \frac{(y+1)^2}{9} = 1$

internet connect
Homework Help Online
Go To: **go.hrw.com**
Keyword:
MB1 Homework Help for Exercises 13–24

Graph each hyperbola. Label the center, vertices, co-vertices, foci, and asymptotes.

13. $x^2 - y^2 = 1$ **14.** $y^2 - x^2 = 1$ **15.** $\frac{y^2}{3^2} - \frac{x^2}{5^2} = 1$

16. $\frac{x^2}{2^2} - \frac{y^2}{3^2} = 1$ **17.** $x^2 - \frac{y^2}{2^2} = 1$ **18.** $y^2 - \frac{x^2}{3^2} = 1$

19. $\frac{y^2}{100} - \frac{x^2}{64} = 1$ **20.** $\frac{x^2}{25} - \frac{y^2}{36} = 1$ **21.** $4x^2 - 25y^2 = 100$

22. $36y^2 - 4x^2 = 144$ **23.** $\frac{(x-1)^2}{2^2} - \frac{(y+2)^2}{3^2} = 1$ **24.** $\frac{(x+2)^2}{3^2} - \frac{(y-2)^2}{4^2} = 1$

For Exercises 25–32, write the standard equation for the hyperbola with the given characteristics.

25. vertices: $(-3, 0)$ and $(3, 0)$; co-vertices: $(0, -5)$ and $(0, 5)$

26. vertices: $(0, -2)$ and $(0, 2)$; co-vertices: $(-4, 0)$ and $(4, 0)$

27. vertices: $(0, -4)$ and $(0, 4)$; foci: $(0, -5)$ and $(0, 5)$

28. vertices: $(-5, 0)$ and $(5, 0)$; foci: $(-7, 0)$ and $(7, 0)$

25. $\frac{x^2}{9} - \frac{y^2}{25} = 1$
26. $\frac{y^2}{4} - \frac{x^2}{16} = 1$
27. $\frac{y^2}{16} - \frac{x^2}{9} = 1$
28. $\frac{x^2}{25} - \frac{y^2}{24} = 1$

Assess

Selected Answers

Exercises 3–6, 7–63 odd

ASSIGNMENT GUIDE	
In Class	1–6
Core	7–45 odd, 49
Core Plus	8–50 even
Review	51–64
Preview	65

Extra Practice can be found beginning on page 940.

Error Analysis

Students are often confused about how the horizontal and vertical orientations of a hyperbola relate to its equation. Remind them that if the y^2-term is negative, the transverse axis is horizontal, and if the x^2-term is negative, the transverse axis is vertical.

13. center: $(0, 0)$; vertices: $(-1, 0)$ and $(1, 0)$; co-vertices: $(0, -1)$ and $(0, 1)$; foci: $(-\sqrt{2}, 0)$ and $(\sqrt{2}, 0)$; asymptotes: $y = -x$ and $y = x$

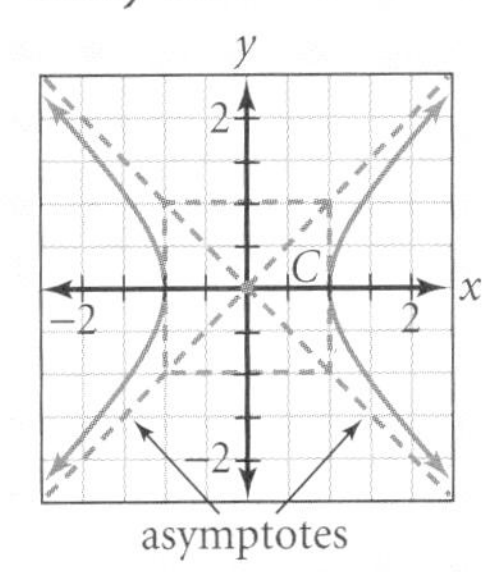

14. center: $(0, 0)$; vertices: $(0, -1)$ and $(0, 1)$; co-vertices: $(-1, 0)$ and $(1, 0)$; foci: $(0, -\sqrt{2})$ and $(0, \sqrt{2})$; asymptotes: $y = -x$ and $y = x$

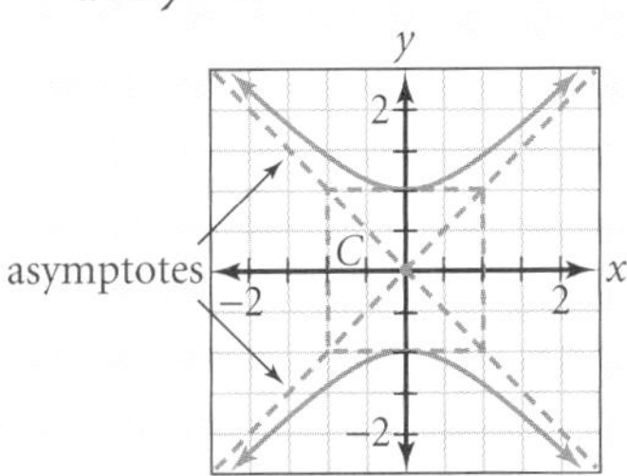

15. center: $(0, 0)$; vertices: $(0, -3)$ and $(0, 3)$; co-vertices: $(-5, 0)$ and $(5, 0)$; foci: $(0, -\sqrt{34})$ and $(0, \sqrt{34})$; asymptotes: $y = -\frac{3x}{5}$ and $y = \frac{3x}{5}$

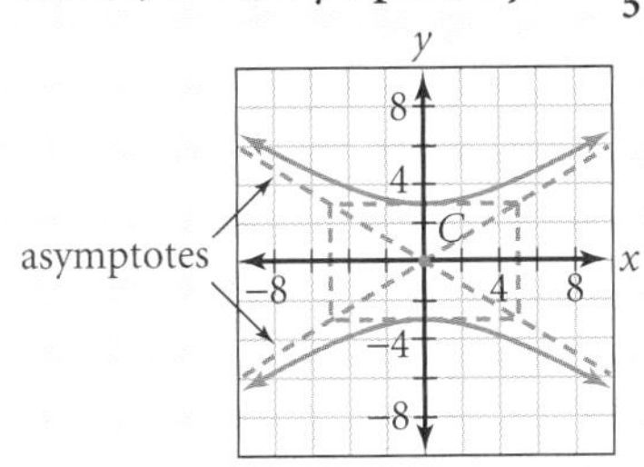

16. center: $(0, 0)$; vertices: $(-2, 0)$ and $(2, 0)$; co-vertices: $(0, -3)$ and $(0, 3)$; foci: $(-\sqrt{13}, 0)$ and $(\sqrt{13}, 0)$; asymptotes: $y = -\frac{3x}{2}$ and $y = \frac{3x}{2}$

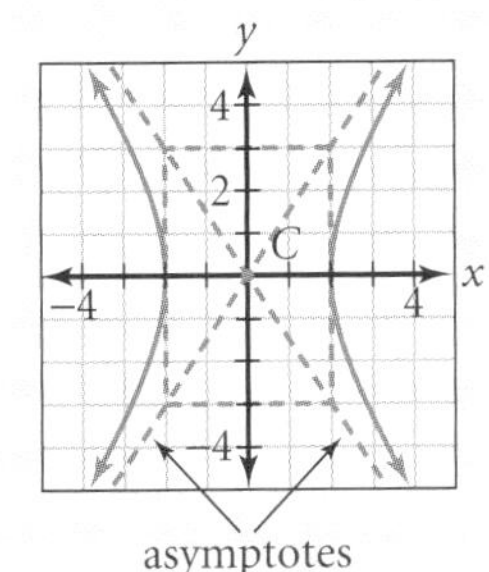

The answers to Exercises 17–24 can be found in Additional Answers beginning on page 1002.

33. $\frac{(x-1)^2}{9} - \frac{(y-3)^2}{4} = 1$; center: $(1, 3)$; vertices: $(-2, 3)$ and $(4, 3)$; co-vertices: $(1, 1)$ and $(1, 5)$; foci: $(1 - \sqrt{13}, 3)$ and $(1 + \sqrt{13}, 3)$

34. $\frac{(x-1)^2}{25} - \frac{(y-2)^2}{16} = 1$; center: $(1, 2)$; vertices: $(-4, 2)$ and $(6, 2)$; co-vertices: $(1, -2)$ and $(1, 6)$; foci: $(1 - \sqrt{41}, 2)$ and $(1 + \sqrt{41}, 2)$

35. $\frac{(y+1)^2}{36} - \frac{(x+1)^2}{4} = 1$; center: $(-1, -1)$; vertices: $(-1, -7)$ and $(-1, 5)$; co-vertices: $(-3, -1)$ and $(1, -1)$; foci: $(-1, -1 - 2\sqrt{10})$ and $(-1, -1 + 2\sqrt{10})$

36. $\frac{(y-1)^2}{16} - \frac{(x-2)^2}{25} = 1$; center: $(2, 1)$; vertices: $(2, -3)$ and $(2, 5)$; co-vertices: $(-3, 1)$ and $(7, 1)$; foci: $(2, 1 - \sqrt{41})$ and $(2, 1 + \sqrt{41})$

37. $\frac{(y-3)^2}{9} - \frac{(x+2)^2}{1} = 1$; center: $(-2, 3)$; vertices: $(-2, 0)$ and $(-2, 6)$; co-vertices: $(-3, 3)$ and $(-1, 3)$; foci: $(-2, 3 - \sqrt{10})$ and $(-2, 3 + \sqrt{10})$

29. $\frac{x^2}{5} - \frac{y^2}{4} = 1$
30. $\frac{y^2}{3} - \frac{x^2}{1} = 1$
31. $\frac{(x-2)^2}{9} - \frac{(y-3)^2}{25} = 1$
32. $\frac{(x+1)^2}{25} - \frac{(y+3)^2}{9} = 1$

29. co-vertices: $(0, -2)$ and $(0, 2)$; foci: $(-3, 0)$ and $(3, 0)$

30. co-vertices: $(-1, 0)$ and $(1, 0)$; foci: $(0, -2)$ and $(0, 2)$

31. center: $(2, 3)$; vertices: $(-1, 3)$ and $(5, 3)$; co-vertices: $(2, -2)$ and $(2, 8)$

32. center: $(-1, -3)$; vertices: $(-6, -3)$ and $(4, -3)$; co-vertices: $(-1, -6)$ and $(-1, 0)$

Write the standard equation for each hyperbola. Give the coordinates of the center, vertices, co-vertices, and foci.

33. $4x^2 - 9y^2 - 8x + 54y = 113$
34. $16x^2 - 25y^2 - 32x + 100y = 484$
35. $4y^2 - 36x^2 - 72x + 8y = 176$
36. $25y^2 - 16x^2 + 64x - 50y = 439$
37. $y^2 - 9x^2 - 6y = 36 + 36x$
38. $16x^2 - 9y^2 + 64x = 89 - 18y$
39. $16x^2 + 64y - 256 = 16y^2 - 64x$
40. $25y^2 + 100x - 100y - 625 = 25x^2$
41. $3y^2 + 20x = 23 + 5x^2 + 12y$
42. $7x^2 - 5y^2 = 48 - 20y - 14x$

Write the standard equations for both hyperbolas whose asymptotes contain the diagonals of rectangle *ABCD* and whose vertices lie on the sides of the given rectangle.

43. $A(-5, 4)$, $B(5, 4)$, $C(5, -4)$, and $D(-5, -4)$
44. $A(-2, 6)$, $B(2, 6)$, $C(2, -6)$, and $D(-2, -6)$
45. $A(1, 12)$, $B(11, 12)$, $C(11, 1)$, and $D(1, 1)$
46. $A(0, 7)$, $B(6, 7)$, $C(6, 4)$, and $D(0, 4)$

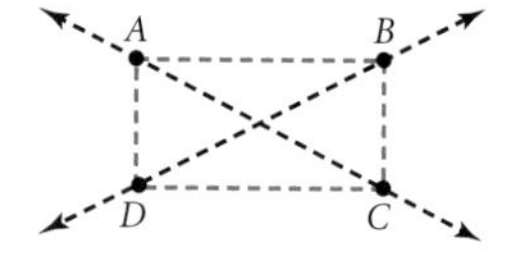

CHALLENGE

47. Let $\frac{x^2}{a^2} - \frac{y^2}{a^2} = 1$.
  a. Find the coordinates of the vertices and co-vertices.
  b. If the four points in part **a** are joined to form a quadrilateral, classify the quadrilateral.
  c. Justify your conclusion from part **b**.

CONNECTIONS

48. $\frac{(x-8)^2}{1} - \frac{(y+4)^2}{25} = 1$
49. $\frac{(x+9)^2}{4} - \frac{(y-3)^2}{9} = 1$

48. TRANSFORMATIONS If the hyperbola defined by the equation $\frac{(x-5)^2}{1} - \frac{(y-2)^2}{25} = 1$ is translated 6 units down and 3 units to the right, write the standard equation of the resulting hyperbola.

49. TRANSFORMATIONS Translate the hyperbola defined by the equation $9x^2 - 4y^2 + 54x + 8y + 41 = 0$ up 2 units and to the left 6 units. Write the standard equation of the resulting hyperbola.

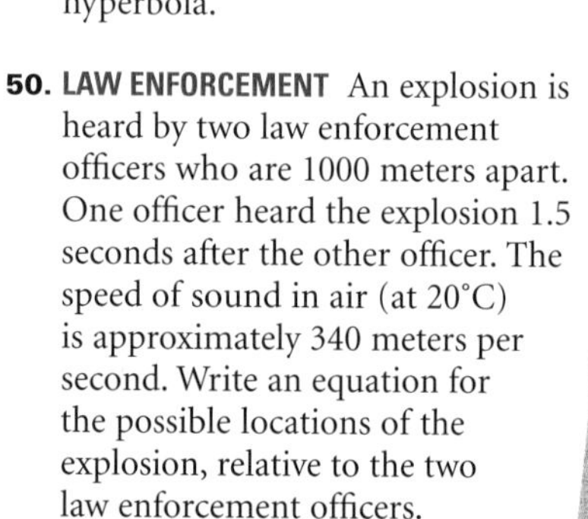

APPLICATION

50. Assuming the officers are located at (0, −500) and (0, 500), the equation is $\frac{x^2}{65{,}025} - \frac{y^2}{184{,}975} = 1$.

50. LAW ENFORCEMENT An explosion is heard by two law enforcement officers who are 1000 meters apart. One officer heard the explosion 1.5 seconds after the other officer. The speed of sound in air (at 20°C) is approximately 340 meters per second. Write an equation for the possible locations of the explosion, relative to the two law enforcement officers.

Practice

NAME ______________ CLASS ________ DATE ________

Practice
9.5 ***Hyperbolas***

Write the standard equation for each hyperbola.

1.

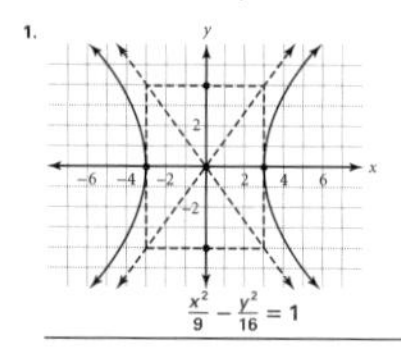

$\frac{x^2}{9} - \frac{y^2}{16} = 1$

2.

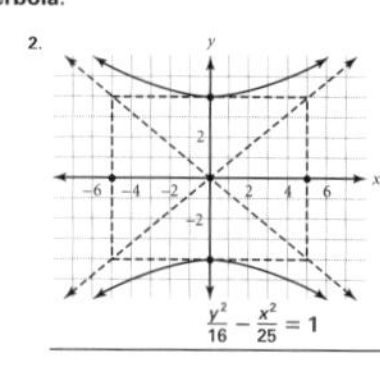

$\frac{y^2}{16} - \frac{x^2}{25} = 1$

Graph each hyperbola. Label the center, vertices, co-vertices, foci, and asymptotes.

3. $\frac{y^2}{9} - \frac{x^2}{25} = 1$

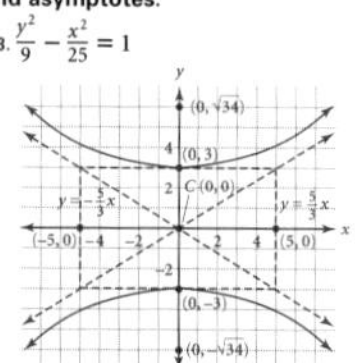

4. $\frac{(x-1)^2}{16} - \frac{(y-1)^2}{9} = 1$

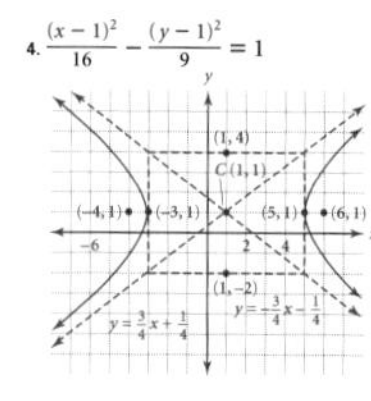

For Exercises 5–7, write the standard equation for the hyperbola with the given characteristics.

5. vertices: $(-\sqrt{10}, 0)$ and $(\sqrt{10}, 0)$; co-vertices: $(0, -\sqrt{15})$ and $(0, \sqrt{15})$ $\frac{x^2}{10} - \frac{y^2}{15} = 1$

6. foci: $(-5, -2)$ and $(5, -2)$; vertices: $(-3, 0)$ and $(3, 0)$ $\frac{x^2}{9} - \frac{(y+2)^2}{16} = 1$

7. center: $(1, 1)$; vertices: $(1, -4)$ and $(1, 6)$; co-vertices: $(13, 1)$ and $(-11, 1)$ $\frac{(y-1)^2}{25} - \frac{(x-1)^2}{144} = 1$

8. A hyperbola is defined by $x^2 - 4y^2 - 28x - 24y + 156 = 0$. Write the standard equation, and identify the coordinates of the center, vertices, co-vertices, and foci.
$\frac{(x-14)^2}{4} - (y+3)^2 = 1$; center: $(14, -3)$; vertices: $(12, -3)$ and $(16, -3)$; co-vertices: $(14, -4)$ and $(14, -2)$; foci: $(14 - \sqrt{5}, -3)$ and $(14 + \sqrt{5}, -3)$

38. $\frac{(x+2)^2}{9} - \frac{(y-1)^2}{16} = 1$; center: $(-2, 1)$; vertices: $(-5, 1)$ and $(1, 1)$; co-vertices: $(-2, -3)$ and $(-2, 5)$; foci: $(-7, 1)$ and $(3, 1)$

39. $\frac{(x+2)^2}{16} - \frac{(y-2)^2}{16} = 1$; center: $(-2, 2)$; vertices: $(-6, 2)$ and $(2, 2)$; co-vertices: $(-2, -2)$ and $(-2, 6)$; foci: $(-2 - 4\sqrt{2}, 2)$ and $(-2 + 4\sqrt{2}, 2)$

40. $\frac{(y-2)^2}{25} - \frac{(x-2)^2}{25} = 1$; center: $(2, 2)$; vertices: $(2, -3)$ and $(2, 7)$; co-vertices: $(-3, 2)$ and $(7, 2)$; foci: $(2, 2 - 5\sqrt{2})$ and $(2, 2 + 5\sqrt{2})$

41. $\frac{(y-2)^2}{5} - \frac{(x-2)^2}{3} = 1$; center: $(2, 2)$; vertices: $(2, 2 - \sqrt{5})$ and $(2, 2 + \sqrt{5})$; co-vertices: $(2 - \sqrt{3}, 2)$ and $(2 + \sqrt{3}, 2)$; foci: $(2, 2 - 2\sqrt{2})$ and $(2, 2 + 2\sqrt{2})$

42. $\frac{(x+1)^2}{5} - \frac{(y-2)^2}{7} = 1$; center: $(-1, 2)$; vertices: $(-1 - \sqrt{5}, 2)$ and $(-1 + \sqrt{5}, 2)$; co-vertices: $(-1, 2 - \sqrt{7})$ and $(-1, 2 + \sqrt{7})$; foci: $(-1 - 2\sqrt{3}, 2)$ and $(-1 + 2\sqrt{3}, 2)$

43. $\frac{x^2}{25} - \frac{y^2}{16} = 1$ or $\frac{y^2}{16} - \frac{x^2}{25} = 1$

Look Back

Use substitution to solve each system of equations. *(LESSON 3.1)*

51. $\begin{cases} x = y + 1 \\ x + 4y = 11 \end{cases}$ (3, 2) **52.** $\begin{cases} 9m + 8n = 21 \\ 2m = 7 - n \end{cases}$ (5, –3) **53.** $\begin{cases} 15x + 4y = 23 \\ 10x - y = -3 \end{cases}$ $\left(\frac{1}{5}, 5\right)$

Use elimination to solve each system of equations. *(LESSON 3.2)*

54. $\begin{cases} 5x - 2y = 30 \\ x + 2y = 6 \end{cases}$ (6, 0) **55.** $\begin{cases} 3x + 2y = 5 \\ 4x = 22 + 5y \end{cases}$ (3, –2) **56.** $\begin{cases} 5x - 2y = 3 \\ 2x + 7y = 9 \end{cases}$ (1, 1)

Let $A = \begin{bmatrix} -2 \\ 1 \\ 5 \end{bmatrix}$, $B = [4 \quad 0 \quad -5]$, **and** $C = \begin{bmatrix} 2 & 4 & 1 \\ -6 & 0 & -1 \\ 3 & 2 & 9 \end{bmatrix}$. **Find each product, if it exists.** *(LESSON 4.2)*

57. AB **58.** BA **59.** BC **60.** CA
61. CB **62.** $(AB)C$ **63.** $C(AB)$ **64.** $A(BC)$

57. $\begin{bmatrix} -8 & 0 & 10 \\ 4 & 0 & -5 \\ 20 & 0 & -25 \end{bmatrix}$
58. [–33]
59. [–7 6 –41]
60. $\begin{bmatrix} 5 \\ 7 \\ 41 \end{bmatrix}$

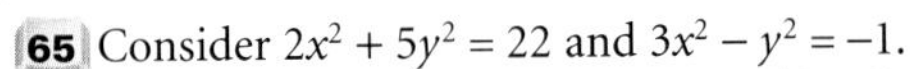

Look Beyond

65 Consider $2x^2 + 5y^2 = 22$ and $3x^2 - y^2 = -1$.

a. What two figures are represented by these equations?
b. Solve both equations for y.
c. Graph the equations from part **b** on your graphics calculator. Use a viewing window that allows you to see all points of intersection for the four graphs. Find all points of intersection.

65a. ellipse; hyperbola
b. $y = \pm\sqrt{\frac{22 - 2x^2}{5}}$; $y = \pm\sqrt{3x^2 + 1}$
c. (1, 2), (–1, 2), (1, –2), (–1, –2)

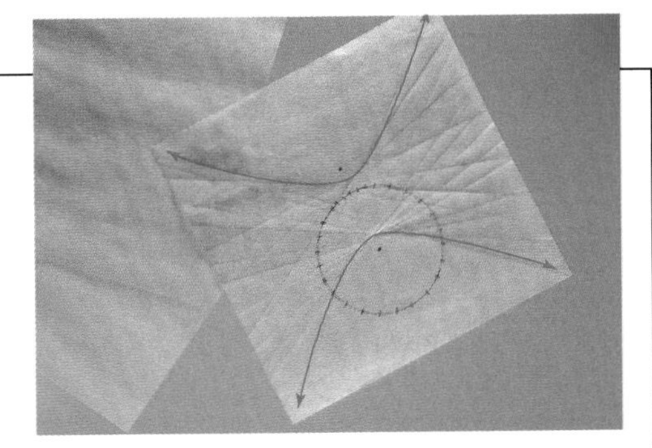

1. Draw a circle with a diameter between 2 and 4 inches on the left side of a piece of wax paper.
2. Place at least 20 tick marks on the circle, approximately evenly spaced.
3. Place a point outside and to the right of the circle. Fold the paper so that this outside point lies on one of the tick marks on the circle. Crease the paper, and then unfold it.
4. Repeat Step 3 for each tick mark on the circle.
5. Compare the figure formed by the creases with those of your classmates. How do the figures vary? Could they all be classified as one type of conic section? If so, which one?
6. Make a conjecture about how to make a flatter or more open figure. Verify your conjecture by repeating the process with the appropriate modification(s).
7. How does the point that you placed in Step 3 appear to be mathematically significant? Explain your reasoning.

WORKING ON THE CHAPTER PROJECT
You should now be able to complete Activity 3 of the Chapter Project.

Look Beyond

Exercise 65 represents a nonlinear system of equations. Students are asked to graph both equation in the system to determine the solutions. In Lesson 9.6, students will solve a system of nonlinear equations by applying the substitution method and the elimination method to the system.

ALTERNATIVE
Assessment

Portfolio Activity

The Portfolio Activity can be used as preparation for the Chapter Project or as a separate activity. In the Portfolio Activity on this page, students create a hyperbola by folding patty paper or wax paper.

Answers to Portfolio Activities can be found in Additional Answers of the Teachers Edition.

Student Technology Guide

NAME ________ CLASS ________ DATE ________

Student Technology Guide
9.5 Hyperbolas

You can use a graphics calculator to graph a hyperbola and to show its asymptotes.

Example: Graph $-\frac{x^2}{25} + \frac{y^2}{16} = 1$.

- Solve for y.
$\frac{y^2}{16} = 1 + \frac{x^2}{25}$
$y^2 = 16\left(1 + \frac{x^2}{25}\right) \longrightarrow y = \pm\sqrt{16\left(1 + \frac{x^2}{25}\right)}$
- The ± sign means you need to graph two equations. Enter the first as Y1. Press Y= 2nd x² 16 (1 + X,T,θ,n x² ÷ 25)).
- Make Y2 the opposite of Y1 by moving to Y2 and pressing (–) VARS Y-VARS 1:Function... ENTER 1:Y1 ENTER.
- Press GRAPH.

Remember to use a square viewing window to see an undistorted graph.

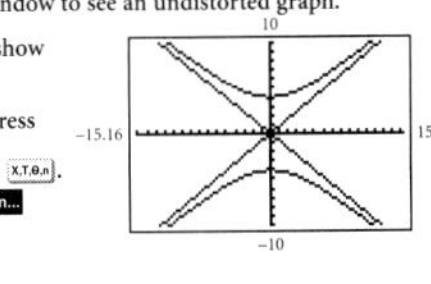

When you graph a hyperbola, you may want to show its asymptotes. In the example, these are $y = \pm\frac{\sqrt{16}}{\sqrt{25}}x = \pm\frac{4}{5}x$. To graph the asymptotes, press Y=. Move to Y3 and press (4 ÷ 5) X,T,θ,n. Move to Y4 and press (–) VARS Y-VARS 1:Function... ENTER 3:Y3 ENTER. Then press GRAPH.

Graph each equation. Show the asymptotes. Is the transverse axis horizontal or vertical?

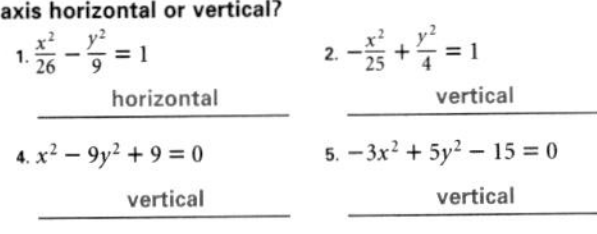

1. $\frac{x^2}{26} - \frac{y^2}{9} = 1$ horizontal
2. $-\frac{x^2}{25} + \frac{y^2}{4} = 1$ vertical
3. $\frac{x^2}{4} - \frac{y^2}{9} = 1$ horizontal
4. $x^2 - 9y^2 + 9 = 0$ vertical
5. $-3x^2 + 5y^2 - 15 = 0$ vertical
6. $\frac{x^2}{36} - \frac{y^2}{25} = 1$ horizontal

44. $\frac{x^2}{4} - \frac{y^2}{36} = 1$ or $\frac{y^2}{36} - \frac{x^2}{4} = 1$

45. $\frac{(x-6)^2}{25} - \frac{\left(y - \frac{13}{2}\right)^2}{\frac{121}{4}} = 1$ or $\frac{\left(y - \frac{13}{2}\right)^2}{\frac{121}{4}} - \frac{(x-6)^2}{25} = 1$

46. $\frac{(x-3)^2}{9} - \frac{\left(y - \frac{11}{2}\right)^2}{\frac{9}{4}} = 1$ or $\frac{\left(y - \frac{11}{2}\right)^2}{\frac{9}{4}} - \frac{(x-3)^2}{9} = 1$

61. does not exist

62. $\begin{bmatrix} 14 & -12 & 82 \\ -7 & 6 & -41 \\ -35 & 30 & -205 \end{bmatrix}$

63. $\begin{bmatrix} 20 & 0 & -25 \\ 28 & 0 & -35 \\ 164 & 0 & -205 \end{bmatrix}$

64. $\begin{bmatrix} 14 & -12 & 82 \\ -7 & 6 & -41 \\ -35 & 30 & -205 \end{bmatrix}$

The answer to Exercise 47 can be found in Additional Answers beginning on page 1002.

Focus

Traditionally, data analysis uses sets that are divided into distinct categories. This limits the effectiveness of sets because many events in the real world do not fall into such neatly defined groups. Students will be introduced to a more sophisticated approach, derived from computer science, called "fuzzy" logic. They will compare this new logic with the traditional approach by categorizing data first into traditional sets and then into "fuzzy" sets. Students will then evaluate the ease or difficulty of defining the boundaries for each type of set.

Motivate

Ask students to consider situations in their life in which they try to categorize things in order to see relationships or make decisions. Ask them how they would go about defining categories such as hot and cold, fast and slow, and big and small. Have them write down how they would rank animals, objects, or the like within these categories. Have them discuss any difficulties they had in doing this.

EYEWITNESS MATH

WHAT'S SO FUZZY?

Time for Some Fuzzy Thinking

By Philip Elmer-Dewitt

In the pages of *Books in Print,* listed among works like *Fuzzy Bear* and *Fuzzy Wuzzy Puppy,* are some strange sounding titles: *Fuzzy Systems, Fuzzy Set Theory* and *Fuzzy Reasoning & Its Applications.* The bedtime reading of scientists gone soft in the head? No, these academic tomes are the collected output of 25 years of mostly American research in fuzzy logic, a branch of mathematics designed to help computers simulate the various kinds of vagueness and uncertainty found in everyday life. Despite a distinguished corps of devoted followers, however, fuzzy logic has been largely relegated to the back shelves of computer science—at least in the U.S.

But not, it turns out, in Japan. Suddenly the term fuzzy and products based on principles of fuzzy logic seem to be everywhere in Japan: in television documentaries, in corporate magazine ads and in novel electronic gadgets ranging from computer-controlled air conditioners to golf-swing analyzers.

What is fuzzy logic? The original concept, developed in the mid-'60s by Lofti Zadeh, a Russian-born professor of computer science at the University of California, Berkeley, is that things in the real world do not fall into neat, crisp categories defined by traditional set theory, like the set of even numbers or the set of left-handed baseball players.

But this on-or-off, black-or-white, 0-or-1 approach falls apart when applied to many everyday classifications, like the set of beautiful women, the set of tall men or the set of very cold days.

This mathematics turns out to be surprisingly useful for controlling robots, machine tools and various electronic systems. A conventional air conditioner, for example, recognizes only two basic states: too hot or too cold. When geared for thermostat control, the cooling system either operates at full blast or shuts off completely. A fuzzy air conditioner, by contrast, would recognize that some room temperatures are closer to the human comfort zone than others. Its cooling system would begin to slow down gradually as the room temperature approached the desired setting. Result: a more comfortable room and a smaller electric bill.

Fuzzy logic began to find applications in industry in the early '70s, when it was teamed with another form of advanced computer science called the expert system. Expert systems solve complex problems somewhat like humans experts do—by applying rules of thumb. (Example: when the oven gets very hot, turn the gas down a bit.)

[*Source:* Time, *September 25, 1989*]

Fuzzy-logic controllers create a smoother ride for passengers and use less energy than human conductors or automated systems that are based on traditional logic.

Answers may vary for **1a** and **1c,** as well as for **2a.**

1b. The middle heights, 5′6″, 5′7″, and 5′8″, are hardest to place because students can easily fit in either category.

1d. There is no precise height above which someone is considered tall and below which someone is considered not tall. Naturally, people's opinions will differ more dramatically when the clarity of the boundaries between sets becomes more obscure.

2a.

Student	Height	Value
A	5′11″	1.0
B	5′3″	0.05
C	6′1″	1.0
D	4′11″	0.0
E	5′4″	0.1
F	5′10″	0.95
G	5′1″	0.0
H	5′8″	0.8
I	5′5″	0.3
J	5′7″	0.65
K	5′6″	0.5
L	5′9″	0.9

What is fuzzy logic? How can it make things work more efficiently? Does fuzzy logic resemble the way we think more than traditional logic does?

To answer questions like these, you need to explore the basis of fuzzy logic—something called *fuzzy sets.*

According to traditional logic, if someone belongs to the set of *people who are tall,* then they cannot belong to the set of *people who are not tall.* With fuzzy sets, the distinction is not so sharp, as you will see.

Traditional Logic

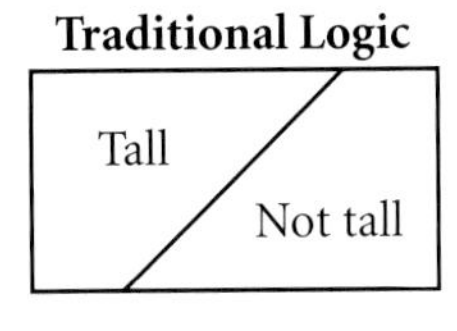

Cooperative Learning

For Steps 1 and 2, use the data in the table at right or use the heights of 12 students from your class.

Student	Height
A	5' 11"
B	5' 3"
C	6' 1"
D	4' 11"
E	5' 4"
F	5' 10"
G	5' 1"
H	5' 8"
I	5' 5"
J	5' 7"
K	5' 6"
L	5' 9"

1. Before looking at fuzzy sets, examine how you might use sets in traditional logic to categorize a group of students by height.
   a. Create a 2-column table. In the left column, list the heights of the students you think are tall. List the rest in the right column.
   b. Which heights were hardest to place? Why?
   c. At what height did you draw the line between *tall* and *not tall*? Compare your response with other groups. Describe the disagreements.
   d. Why would you expect there to be disagreements?
2. Now, try using fuzzy sets.
   a. For each student letter in the table, assign a value from 0 to 1 to indicate the extent to which that person belongs to the set *tall.* For example:
      0 for someone who definitely does not belong
      0.5 for someone who *may be* tall
      0.8 for someone who is *fairly* tall
      1 for someone who definitely belongs on the *tall* list
   b. On a graph, plot the 12 values you assigned in part **a.** Draw a smooth curve through the points.
   c. How does the graph show what you think the word *tall* means?
   d. Compare your graph with the graphs of other groups. How are they a like and how are they different?
3. What sort of difficulties can you run into when you apply traditional logic to real-world situations? How can fuzzy sets help resolve those difficulties?

2b.

y
1.0
0.8
0.6
0.4
0.2
Value
8 48 56 64 72 80 x
Height (inches)

2c. The graph shows that students with heights between about 5′1″ and 5′10″ can be considered tall or not tall to different extents. The taller you are in that range, the more you belong to the set *tall.*

2d and 3. Answers may vary.

Cooperative Learning

Divide the class into small groups of four or five students. Have each group work as a team to complete the table by following the steps outlined in Steps 1 and 2. One team member should be designated as the team leader. The team leader will make the final decision, if necessary, on any disagreements encountered in the categorizing process. Another team member should be designated as recorder to record the details of the group's analysis of the data, including any disagreements.

Discuss

Each team will present a report to the class. This will include the team's categorization of students' heights in the table and a discussion of the question posed in Step 3. The report will also review any disagreements recorded by the team recorder during the categorization process. Finally, each team should comment on the dynamics of the team process itself, such as how team members felt about working together and what positive and negative reactions they had to it.

Prepare

NCTM PRINCIPLES & STANDARDS 1–4, 6–10

QUICK WARM-UP

Solve each system.

1. $\begin{cases} x+y=6 \\ x-y=-4 \end{cases}$ (1, 5)

2. $\begin{cases} x+y=-3 \\ 3x-3y=9 \end{cases}$ (0, −3)

3. $\begin{cases} 2x-y=-3 \\ 4x-2y=6 \end{cases}$ no solution

4. $\begin{cases} 2x+y=-7 \\ 3x+2y=9 \end{cases}$ (−23, 39)

5. $\begin{cases} -2x+3y=1 \\ 3x-4y=-2 \end{cases}$ (−2, −1)

Also on Quiz Transparency 9.6

Teach

Why A system of nonlinear equations can be used to described the cross section of a three-dimensional figure. In medicine, cross sections of the body are analyzed to detect disease. Have students discuss their experience with the use of computed axial tomography (CAT scans), magnetic resonance imaging (MRIs), or positron-emission tomography (PET scans).

9.6 Solving Nonlinear Systems

Why *Solving nonlinear systems of equations is required to solve many real-world problems, such as finding the dimensions of a beam that is to be cut from a log.*

Objectives

- Solve a system of equations containing first- or second-degree equations in two variables.
- Identify a conic section from its equation.

A mill operator wants to cut a rectangular beam from a cylindrical log whose circular cross section has a diameter of 10 inches. The rectangular cross section of the beam is to have a length that is twice its width.

To find the dimensions of the beam, you need to solve a *system of nonlinear equations.* *You will solve this problem in Example 2.*

A **system of nonlinear equations** is a collection of equations in which at least one equation is not linear.

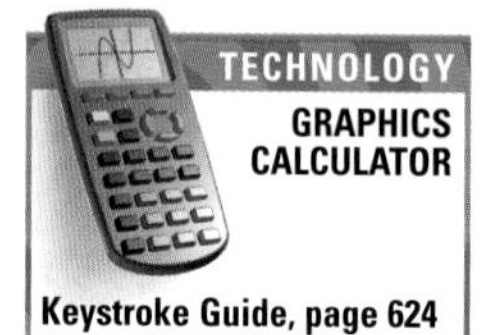

TECHNOLOGY GRAPHICS CALCULATOR

Keystroke Guide, page 624

Activity

Exploring Nonlinear Systems

You will need: a graphics calculator

1. Graph the system of equations in part **a.** Record the number of points of intersection. Repeat this for the systems of equations in parts **b, c,** and **d.**

 a. $\begin{cases} x^2+y^2=25 \\ y=2x \end{cases}$ **b.** $\begin{cases} x^2+y^2=25 \\ x^2-y^2=9 \end{cases}$ **c.** $\begin{cases} y=x^2 \\ y=2x+1 \end{cases}$ **d.** $\begin{cases} x^2+y^2=1 \\ y=x^2 \end{cases}$

CHECKPOINT ✔ 2. The equations in part **a** of Step 1 represent a circle and a line. Consider all of the ways in which a circle and a line can intersect. Make a conjecture about the number of possible intersection points for a circle and a line, and sketch an example of each.

CHECKPOINT ✔ 3. The equations in part **b** of Step 1 represent a circle and a hyperbola. Consider all of the ways in which a circle and a hyperbola can intersect. Make a conjecture about the number of possible intersection points for a circle and a hyperbola, and sketch an example of each.

CHECKPOINT ✔ 4. The equations in part **c** of Step 1 represent a parabola and a line. Consider all of the ways in which a parabola and a line can intersect. Make a conjecture about the number of possible intersection points for a parabola and a line, and sketch an example of each.

CHECKPOINT ✔ 5. The equations in part **d** of Step 1 represent a circle and a parabola. Consider all of the ways in which a circle and a parabola can intersect. Make a conjecture about the number of possible intersection points for a circle and a parabola, and sketch an example of each.

Alternative Teaching Strategy

USING VISUAL MODELS To show students a nonlinear system and its solutions, ask them to graph the equation $y = x^2 + 4$, locate the points (1, 5) and (−3, 13), write the equation of the line that contains the two points, and graph the line. Tell them the two equations, $y = x^2 + 4$ and $y = -2x + 7$, form a linear-quadratic system. The points of intersection, (1, 5) and (−3, 13), are the solutions of the system. Now ask them to locate the ordered pair (−2, 8) on the graph of the equation $y = x^2 + 4$. Have them write and graph the equation of a line that contains the point (−2, 8) and has a slope of −4. Tell students that the two equations, $y = x^2 + 4$ and $y = -4x$, form a linear-quadratic system and that the point of intersection, (−2, 8), is the solution of the system. Next have them write and graph an equation of a line that does not intersect the graph of $y = x^2 + 4$, forming a linear-quadratic system with no solution. Be sure students understand that there can be no more than two solutions for any linear-quadratic system. Now show students how they can also solve these systems of equations algebraically by using substitution and elimination.

A system of two second-degree equations in x and y can have no more than four real solutions, unless the system is dependent. This is demonstrated below for an ellipse and hyperbola.

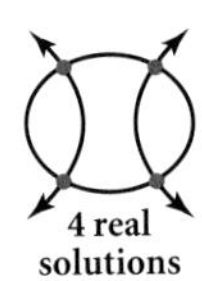

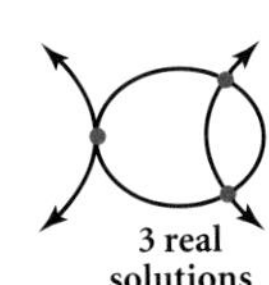

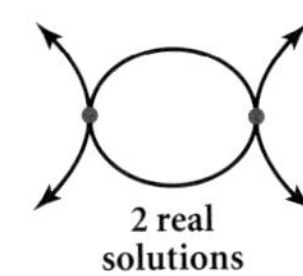

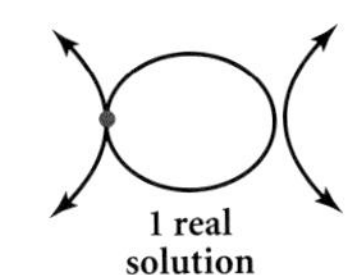

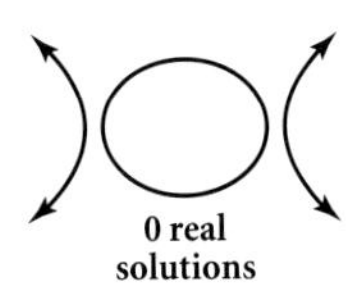

CRITICAL THINKING Give an example of a dependent system of nonlinear equations. How many solutions does it have?

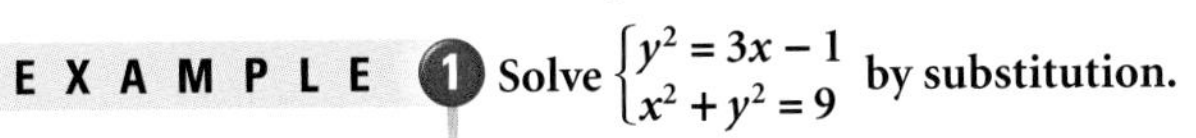

Solving Systems of Nonlinear Equations

Just as you can use substitution to solve some systems of linear equations, you can use substitution to solve some systems of nonlinear equations, as shown in Example 1 below.

EXAMPLE 1 **Solve $\begin{cases} y^2 = 3x - 1 \\ x^2 + y^2 = 9 \end{cases}$ by substitution.**

SOLUTION

Substitute $3x - 1$ for y^2 in $x^2 + y^2 = 9$.

$$x^2 + y^2 = 9$$
$$x^2 + (3x - 1) = 9$$
$$x^2 + 3x = 10$$
$$x^2 + 3x - 10 = 0$$
$$(x - 2)(x + 5) = 0$$
$$x = 2 \quad or \quad x = -5$$

Then find the corresponding y-values.

Substitute 2 for x.

$$y^2 = 3x - 1$$
$$y^2 = 3(2) - 1$$
$$y = \pm\sqrt{5}$$

Substitute -5 for x.

$$y^2 = 3x - 1$$
$$y^2 = 3(-5) - 1$$
$$y = \pm\sqrt{-16}$$
$$y = \pm 4i$$

The real solutions are $(2, \sqrt{5})$ and $(2, -\sqrt{5})$. The nonreal solutions, $(-5, 4i)$ and $(-5, -4i)$, are meaningless in this coordinate plane.

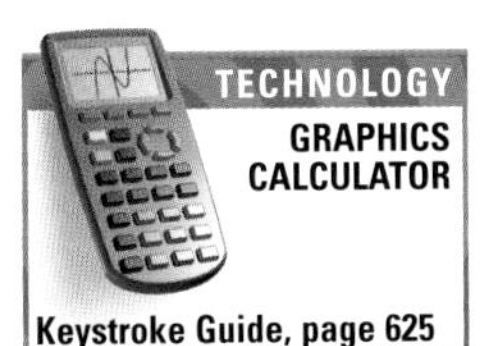

CHECK

Solve $y^2 = 3x - 1$ and $x^2 + y^2 = 9$ for y. Enter each of the four resulting functions separately, and graph them together in a square viewing window. From the graph, the only real solutions are $(2, \sqrt{5})$ and $(2, -\sqrt{5})$.

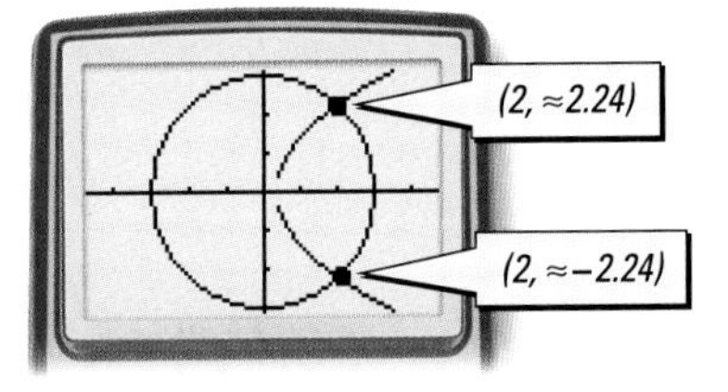

Interdisciplinary Connection

ASTRONOMY Two asteroids have elliptical orbits around the Sun represented by $\frac{x^2}{90000} + \frac{y^2}{62500} = 1$ and $\frac{x^2}{62500} + \frac{y^2}{90000} = 1$. Ask students whether it is possible for the asteroids to collide.
The asteroids could collide at four different positions in their orbits: (192, 192), (−192, 192), (−192, −192), (192, −192).

Enrichment

Consider the system $\begin{cases} y = ax^2 + bx + c \\ y = k \end{cases}$, where a, b, c, and k are constants. Have students find the value of k, in terms of a, b and c, that produces exactly one solution for the system. $k = \frac{4ac - b^2}{4a}$

Activity Notes

Remind students that each point of intersection of the graphs of the equations in a system represents a real solution of the system. Be sure students notice that a linear-quadratic system may have zero, one, or two real solutions and a quadratic-quadratic system may have zero, one, two, three, or four real solutions.

Teaching Tip

TECHNOLOGY When using a TI-82 or TI-83 graphics calculator for the Activity, you can use the following square viewing window:

Xmin=−9.6	**Ymin=−6.2**
Xmax=9.6	**Ymax=6.2**
Xscl=1	**Yscl=1**

CHECKPOINT ✔
2. A circle and a line can intersect at 0, 1, or 2 points.

CHECKPOINT ✔
3. A circle and a hyperbola can intersect at 0, 1, 2, 3, or 4 points.

CHECKPOINT ✔
4. A parabola and a line can intersect at 0, 1, or 2 points.

CHECKPOINT ✔
5. A circle and a parabola can intersect at 0, 1, 2, 3, or 4 points.

CRITICAL THINKING
Answers may vary. Sample answer: $\begin{cases} y^2 = x^2 + 4 \\ -2x^2 + 2y^2 = 8 \end{cases}$
This system has infinitely many solutions because both equations in the system define the same hyperbola.

☞ For Additional Example 1, see page 608.

ADDITIONAL
EXAMPLE 1

Solve $\begin{cases} y^2 = 4x + 13 \\ x^2 + y^2 = 25 \end{cases}$ by substitution.

real solutions: $(2, \sqrt{21})$, $(2, -\sqrt{21})$
complex solutions: $(-6, i\sqrt{11})$, $(-6, -i\sqrt{11})$

Teaching Tip

TECHNOLOGY Use the following viewing window for Example 1:

Xmin=-4.7	Ymin=-3.2
Xmax=4.7	Ymax=4.7
Xscl=1	Yscl=1

TRY THIS
real solutions: (4, 3), (4, −3)
complex solutions: $(-6, i\sqrt{11})$, $(-6, -i\sqrt{11})$

Teaching Tip

Because complex solutions are not shown on a graph, you may want to modify the directions for solving systems. Tell students to find the real solutions only.

The following is a system with three real solutions:

$\begin{cases} y = 3x^2 - 2 \\ x^2 + y^2 = 4 \end{cases}$

$(0, -2), \left(\frac{\sqrt{11}}{3}, \frac{5}{3}\right), \left(-\frac{\sqrt{11}}{3}, \frac{5}{3}\right)$

ADDITIONAL
EXAMPLE 2

A rectangular beam is to be cut from a cylindrical log whose circular cross section has a diameter of 20 inches. The rectangular cross section of the beam is to have a length that is three times its width. **Find the dimensions of the beam's cross sction to the nearest hundredth of an inch.**

The width of the beam's cross section is about 6.32 in., and its length is about 18.97 in.

TRY THIS Solve $\begin{cases} x^2 + y^2 = 25 \\ y^2 = 2x + 1 \end{cases}$ by substitution.

EXAMPLE 2 Refer to the mill problem described at the beginning of the lesson.

APPLICATION
BUSINESS

Find the dimensions of the rectangular cross section of the beam to the nearest hundredth of an inch.

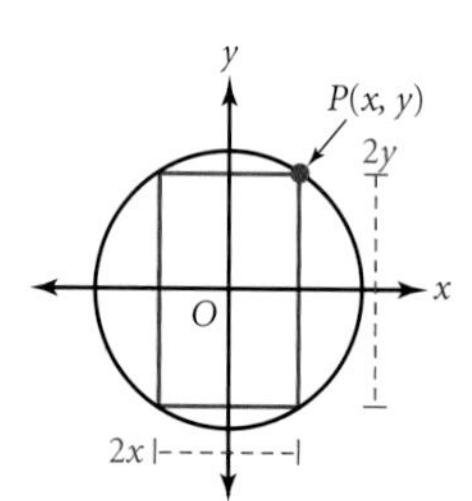

SOLUTION

The diagram at left shows the beam inscribed in the circular cross section of the log, with the center of the circle at the origin of the coordinate plane. Since the diameter is 10, $P(x, y)$ is on a circle with a radius of 5. The equation of the circle is $x^2 + y^2 = 25$. The length of the beam is twice the width, so $y = 2x$. Solve $\begin{cases} x^2 + y^2 = 25 \\ y = 2x \end{cases}$.

$$\begin{aligned} x^2 + y^2 &= 25 \\ x^2 + (2x)^2 &= 25 \quad \textit{Substitute } 2x \textit{ for } y. \\ 5x^2 &= 25 \\ x^2 &= 5 \\ x &= \pm\sqrt{5} \approx \pm 2.24 \end{aligned}$$

Since x represents a measurement, $\sqrt{5}$ is the only reasonable solution. The width is $2x$, or about 4.48 inches; the length is $2(2x)$, about 8.94 inches.

You can also use the elimination method to solve nonlinear systems, as shown in Example 3.

EXAMPLE 3 **Use the elimination method to solve $\begin{cases} 4x^2 + 25y^2 = 100 \\ x^2 + y^2 = 9 \end{cases}$. Give your answers to the nearest hundredth.**

SOLUTION

1. Multiply each side of the equation $x^2 + y^2 = 9$ by −4. Solve the resulting system by using elimination.

$$\begin{cases} 4x^2 + 25y^2 = 100 \\ x^2 + y^2 = 9 \end{cases} \rightarrow \begin{cases} 4x^2 + 25y^2 = 100 \\ -4x^2 - 4y^2 = -36 \end{cases}$$

$$\begin{aligned} 4x^2 + 25y^2 &= 100 \\ -4x^2 - 4y^2 &= -36 \\ \hline 21y^2 &= 64 \\ y &= \pm\sqrt{\frac{64}{21}} \approx \pm 1.75 \end{aligned}$$

2. Substitute $\frac{64}{21}$ for y^2 in $x^2 + y^2 = 9$.

$$\begin{aligned} x^2 + y^2 &= 9 \\ x^2 + \frac{64}{21} &= 9 \\ x &= \pm\sqrt{9 - \frac{64}{21}} \approx \pm 2.44 \end{aligned}$$

The approximate solutions are (2.44, 1.75), (−2.44, 1.75), (−2.44, −1.75), and (2.44, −1.75).

Inclusion Strategies

TACTILE LEARNERS In order to show students the relationship between the different ways in which two second-degree equations can intersect and the number of real solutions, have students draw a circle on graph paper and a parabola on tracing paper. Tell them such a system can have zero, one, two, three, or four real solutions. Instruct them to hold the circle stationary and move the parabola to illustrate each number of possible solutions. Have them repeat the procedure to illustrate the number of possible solutions to nonlinear systems consisting of two parabolas, two ellipses, an ellipse and a hyperbola, and two hyperbolas.

TECHNOLOGY GRAPHICS CALCULATOR

Keystroke Guide, page 625

CHECK

Solve each equation in the system for y, and enter each of the resulting functions separately, and graph them together.

$$\begin{cases} y = \pm\sqrt{\dfrac{100 - 4x^2}{25}} \\ y = \pm\sqrt{9 - x^2} \end{cases}$$

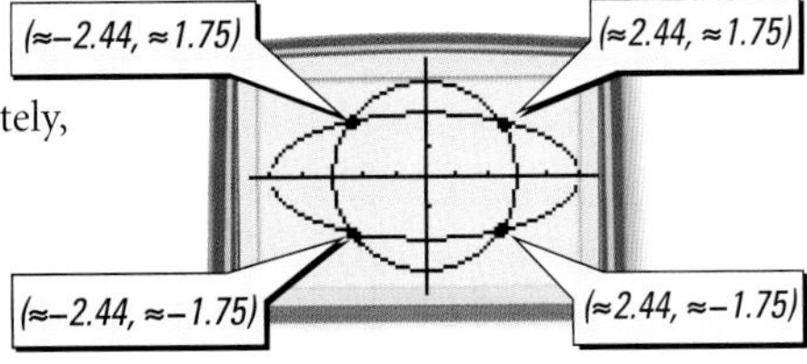

TRY THIS Use the elimination method to solve $\begin{cases} 4x^2 + 9y^2 = 36 \\ 9x^2 - y^2 = 9 \end{cases}$. Give your answers to the nearest hundredth.

Classifying a Conic Section

The equation $Ax^2 + Bxy + Cy^2 + Dx + Ey + F = 0$, where A, B, and C are not all equal to zero, represents a conic section. When $B = 0$, the axes of symmetry of the conic section are parallel to the x-axis or y-axis. When $B \neq 0$, the conic sections are rotated. This book discusses only the first case, in which $B = 0$. When $B = 0$, the equation reduces to $Ax^2 + Cy^2 + Dx + Ey + F = 0$.

CLASSIFYING A CONIC SECTION
$Ax^2 + Cy^2 + Dx + Ey + F = 0$

Type of conic	Coefficients
ellipse (or circle)	$AC > 0$
circle	$A = C, A \neq 0, C \neq 0$
parabola	$AC = 0$
hyperbola	$AC < 0$

Using the information in the table above, you can classify an equation of the form $Ax^2 + Cy^2 + Dx + Ey + F = 0$ without completing the square.

EXAMPLE 4 **Let $4x^2 + 8x = y^2 + 6y + 13$ be the equation of a conic section.**

a. Classify the conic section defined by this equation.

b. Write the standard equation for this conic section.

c. Sketch the graph.

SOLUTION

a. Rewrite $4x^2 + 8x = y^2 + 6y + 13$ in the form $Ax^2 + Cy^2 + Dx + Ey + F = 0$.

$4x^2 - y^2 + 8x - 6y - 13 = 0$, where $A = 4$ and $C = -1$

Because $4(-1) < 0$, the equation represents a hyperbola.

b.

$$4x^2 + 8x = y^2 + 6y + 13$$
$$4x^2 + 8x - y^2 - 6y = 13$$
$$4(x^2 + 2x + 4) - (y^2 + 6y + 9) = 13 + 4(4) - 9 \quad \text{Complete the squares.}$$
$$4(x + 2)^2 - (y + 3)^2 = 20$$
$$\frac{(x + 2)^2}{5} - \frac{(y + 3)^2}{20} = 1$$

Reteaching the Lesson

INVITING PARTICIPATION Give students the equation $x^2 + y^2 = 25$ and ask them to determine an equation of a line that intersects the circle at two points. **sample:** $y = \frac{4}{3}x$ Then give them the equation $\frac{x^2}{16} + \frac{y^2}{4} = 1$ and ask them to determine an equation of a circle that intersects the ellipse at two points. **sample:** $x^2 + y^2 = 4$ Now give them the equation $x = y^2 - 4$ and ask them to determine an equation of a hyperbola that does not intersect the parabola. **sample:** $(x + 7)^2 - y^2 = 1$

ADDITIONAL **EXAMPLE 3**

Use the elimination method to solve $\begin{cases} x^2 + 16y^2 = 144 \\ x^2 + y^2 = 25 \end{cases}$. Give your answers to the nearest hundredth.

The approximate solutions are (4.13, 2.82), (4.13, –2.82), (–4.13, –2.82), and (–4.13, 2.82).

Teaching Tip

TECHNOLOGY If you are using a graphics calculator to view the solutions in Example 3, set the square viewing window as follows:

Xmin=–5.875	Ymin=–3.875
Xmax=5.875	Ymax=3.875
Xscl=1	Yscl=1

TRY THIS

(≈1.17, ≈1.84), (≈1.17, ≈−1.84) (≈−1.17, ≈1.84), (≈−1.17, ≈−1.84)

ADDITIONAL **EXAMPLE 4**

Let $6x^2 + 12x = -3y - 9$ be the equation of a conic section.

a. Classify the conic section defined by this equation. **parabola**

b. Write the standard equation of this conic section. $y + 1 = -2(x + 1)^2$

c. Sketch the graph.

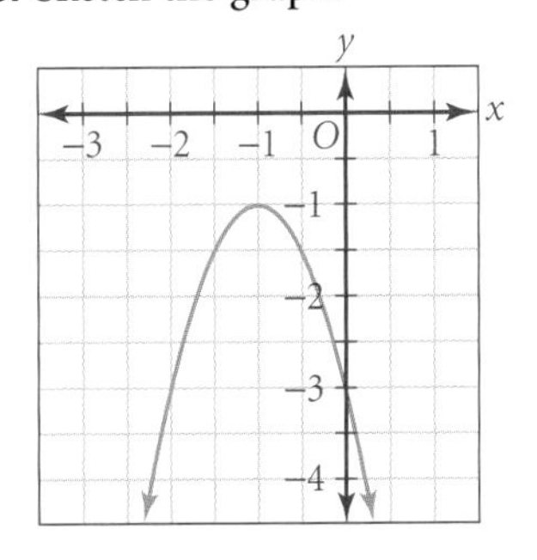

TRY THIS

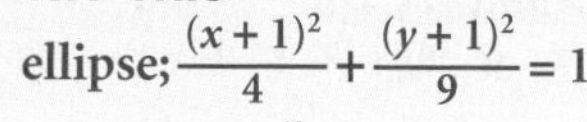

ellipse; $\frac{(x+1)^2}{4} + \frac{(y+1)^2}{9} = 1$

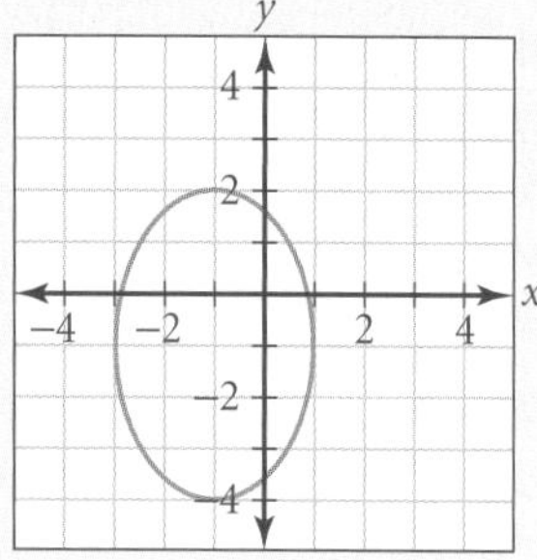

c. center: $(-2, -3)$

vertices: $(-2 - \sqrt{5}, -3)$ and $(-2 + \sqrt{5}, -3)$

co-vertices: $(-2, -3 + \sqrt{20})$ and $(-2, -3 - \sqrt{20})$

The graph of $4x^2 + 8x = y^2 + 6y + 13$, or $\frac{(x+2)^2}{5} - \frac{(y+3)^2}{20} = 1$, is shown at right.

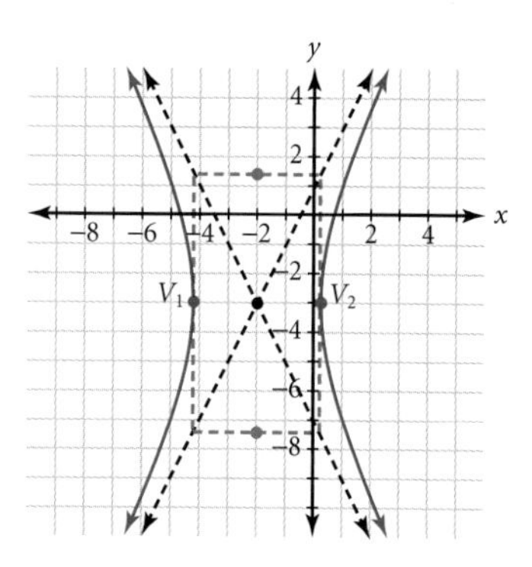

TRY THIS Let $9x^2 + 18x + 4y^2 + 8y = 23$ be the equation of a conic section. Classify the conic section defined by this equation, write the standard equation for this conic section, and sketch the graph.

Assess

Selected Answers

Exercises 5–8, 9–93 odd

ASSIGNMENT GUIDE	
In Class	1–8
Core	9–65 odd, 71–73
Core Plus	10–76 even
Review	78–93
Preview	94–99

Extra Practice can be found beginning on page 940.

Exercises

Communicate

1. Explain how to apply the substitution method to solve $\begin{cases} y = 2x \\ x^2 + y^2 = 10 \end{cases}$.

2. Explain how to apply the elimination method to solve $\begin{cases} 2x^2 - y^2 = 1 \\ x^2 + y^2 = 5 \end{cases}$.

3. Use an illustration to help describe all of the ways in which two ellipses can intersect.

4. Given the equation $Ax^2 + Cy^2 + Dx + Ey + F = 0$, what can be said about A and C if $AC > 0$? if $AC < 0$? if $AC = 0$?

Guided Skills Practice

5. Solve $\begin{cases} y^2 = 2x + 1 \\ x^2 + y^2 = 16 \end{cases}$ by substitution. ***(EXAMPLE 1)***

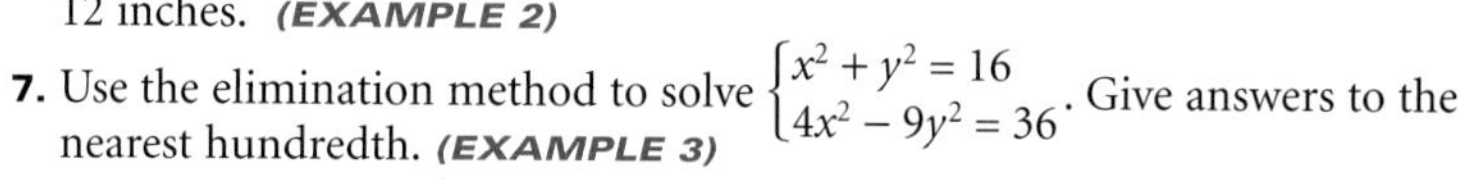

$(3, \sqrt{7}), (3, -\sqrt{7}), (-5, 3i), (-5, -3i)$

APPLICATION

6. **FORESTRY** Refer to the mill problem described at the beginning of the lesson. Find the dimensions of the rectangular cross section of the beam that can be cut from a log whose circular cross section has a diameter of 12 inches. ***(EXAMPLE 2)***

7. Use the elimination method to solve $\begin{cases} x^2 + y^2 = 16 \\ 4x^2 - 9y^2 = 36 \end{cases}$. Give answers to the nearest hundredth. ***(EXAMPLE 3)***

8. Let $4x^2 + 9y^2 + 8x + 18y - 23 = 0$ be the equation of a conic section. Classify the conic section defined by this equation, write the standard equation of this conic section, and sketch the graph. ***(EXAMPLE 4)***

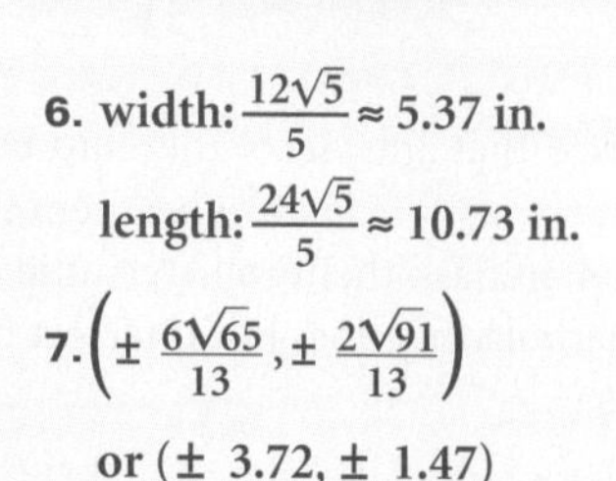

6. width: $\frac{12\sqrt{5}}{5} \approx 5.37$ in.

length: $\frac{24\sqrt{5}}{5} \approx 10.73$ in.

7. $\left(\pm \frac{6\sqrt{65}}{13}, \pm \frac{2\sqrt{91}}{13}\right)$ or $(\pm 3.72, \pm 1.47)$

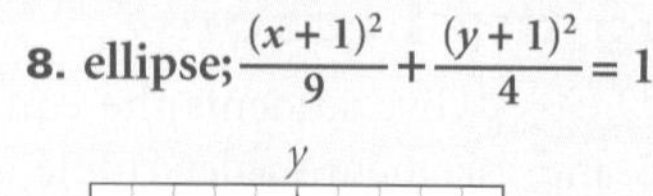

8. ellipse; $\frac{(x+1)^2}{9} + \frac{(y+1)^2}{4} = 1$

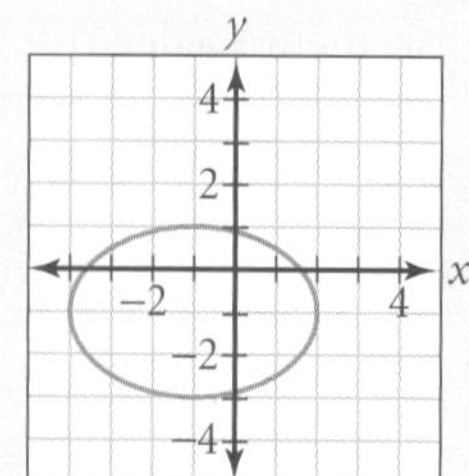

Practice and Apply

Use the substitution method to solve each system. If there are no real-number solutions, write *none*.

9. $\begin{cases} y = 5 \\ y = x^2 \end{cases}$ $(\pm\sqrt{5}, 5)$ **10.** $\begin{cases} y = 2 \\ y = x^2 \end{cases}$ $(\pm\sqrt{2}, 2)$ **11.** $\begin{cases} y = 2x \\ y = x^2 \end{cases}$ (0, 0), (2, 4) **12.** $\begin{cases} y = x \\ x = y^2 \end{cases}$ (0, 0), (1, 1)

13. $\begin{cases} y = x \\ x^2 + y^2 = 4 \end{cases}$ **14.** $\begin{cases} y = 3x \\ x^2 + y^2 = 9 \end{cases}$ **15.** $\begin{cases} y = x \\ x^2 - y^2 = 4 \end{cases}$ **16.** $\begin{cases} y = x^2 \\ x^2 - y^2 = 4 \end{cases}$

17. $\begin{cases} y = x^2 \\ x^2 + y^2 = 1 \end{cases}$ **18.** $\begin{cases} x^2 = 4 - y^2 \\ y = x^2 \end{cases}$ **19.** $\begin{cases} x^2 = y \\ x^2 = 4 + y^2 \end{cases}$ **20.** $\begin{cases} x = y \\ 9x^2 - 4y^2 = 36 \end{cases}$

Use the elimination method to solve each system. If there are no real-number solutions, write *none*.

21. $\begin{cases} x^2 + y^2 = 1 \\ 4x^2 + y^2 = 1 \end{cases}$ $(0, \pm1)$ **22.** $\begin{cases} x^2 + y^2 = 9 \\ 9x^2 + y^2 = 9 \end{cases}$ $(0, \pm3)$ **23.** $\begin{cases} x^2 + y^2 = 1 \\ x^2 + y^2 = 4 \end{cases}$ none

24. $\begin{cases} x^2 - y^2 = 1 \\ x^2 - y^2 = 4 \end{cases}$ none **25.** $\begin{cases} x^2 + y^2 = 25 \\ 4x^2 + 25y^2 = 100 \end{cases}$ $(\pm5, 0)$ **26.** $\begin{cases} x^2 + y^2 = 25 \\ 25x^2 + 4y^2 = 100 \end{cases}$ $(0, \pm5)$

27. $\begin{cases} x^2 - y^2 = 36 \\ 4y^2 - 9x^2 = 36 \end{cases}$ none **28.** $\begin{cases} 9x^2 + 4y^2 = 36 \\ 4x^2 + 9y^2 = 36 \end{cases}$ **29.** $\begin{cases} x^2 + y^2 = 36 \\ 4x^2 - 9y^2 = 36 \end{cases}$

30. $\begin{cases} 9x^2 + 4y^2 = 36 \\ 4x^2 - 9y^2 = 36 \end{cases}$ none **31.** $\begin{cases} x^2 + y^2 = 9 \\ 4x^2 + 9y^2 = 36 \end{cases}$ $(\pm3, 0)$ **32.** $\begin{cases} x^2 + y^2 = 4 \\ 4x^2 - 9y^2 = 36 \end{cases}$ none

internet connect
Homework Help Online
Go To: go.hrw.com
Keyword: **MB1 Homework Help** for Exercises 33–41

Use any method to solve each system. If there are no real-number solutions, write *none*.

33. $\begin{cases} y = x^2 \\ x = y^2 \end{cases}$ (0, 0), (1, 1) **34.** $\begin{cases} x^2 + y^2 = 16 \\ x + y = 1 \end{cases}$ **35.** $\begin{cases} y^2 - x^2 = 1 \\ 9x^2 - y^2 = 9 \end{cases}$ four solutions: $\left(\pm\frac{\sqrt{5}}{2}, \pm\frac{3}{2}\right)$

36. $\begin{cases} y = x^2 + 4 \\ x = y^2 + 4 \end{cases}$ none **37.** $\begin{cases} x^2 - 4y^2 = 20 \\ y^2 = 4 \end{cases}$ **38.** $\begin{cases} y - x^2 = 0 \\ x^2 = 20 - y^2 \end{cases}$ $(\pm2, 4)$

39. $\begin{cases} 4x^2 - 25y^2 = 100 \\ 4x^2 - 9y^2 = 36 \end{cases}$ none **40.** $\begin{cases} x^2 = y \\ x^2 = 4 - y^2 \end{cases}$ **41.** $\begin{cases} x = y^2 + 4 \\ y = x \end{cases}$ none

Solve each system by graphing. If there are no real-number solutions, write *none*.

42 $\begin{cases} x^2 + y^2 = 9 \\ 4x^2 - 4y^2 = 16 \end{cases}$ four approximate solutions: $(\pm2.55, \pm1.58)$ **43** $\begin{cases} 4y^2 + 25x^2 = 100 \\ 5x^2 - 2y^2 = 10 \end{cases}$ four approximate solutions: $(\pm1.85, \pm1.89)$

44 $\begin{cases} 4x^2 + 9y^2 = 36 \\ 16x^2 + 36y^2 = 144 \end{cases}$ infinitely many solutions **45** $\begin{cases} 25x^2 - 4y^2 = 9 \\ 100x^2 - 16y^2 = 36 \end{cases}$ infinitely many solutions

Classify the conic section defined by each equation. Write the standard equation of the conic section, and sketch the graph.

46. $x^2 + 2x + y^2 + 6y = 15$ **47.** $x^2 - 2x + y^2 - 2y - 6 = 0$

48. $4x^2 + 9y^2 - 8x - 18y - 23 = 0$ **49.** $4x^2 + 9y^2 + 16x - 36y = -16$

50. $4y^2 - 8y - x^2 - 4x - 4 = 0$ **51.** $4x^2 + 8x - 9y^2 + 36y - 68 = 0$

52. $4x^2 - 75 = 50y - 25y^2$ **53.** $9y^2 + 18y + 9 = 24x - 4x^2$

54. $9x^2 - 3 = 18x + 4y$ **55.** $y^2 - 2x = 6y - 5$

56. $x^2 - 8y - 16 = 4y^2 + 4x$ **57.** $-18x - 4y - 109 = y^2 - x^2$

13. $(-\sqrt{2}, -\sqrt{2})$ and $(\sqrt{2}, \sqrt{2})$
14. $\left(-\frac{3\sqrt{10}}{10}, -\frac{9\sqrt{10}}{10}\right)$ and $\left(\frac{3\sqrt{10}}{10}, \frac{9\sqrt{10}}{10}\right)$
15. none
16. none

28. four solutions: $\left(\pm\frac{6\sqrt{13}}{13}, \pm\frac{6\sqrt{13}}{13}\right)$
29. four solutions: $\left(\pm\frac{6\sqrt{130}}{13}, \pm\frac{6\sqrt{39}}{13}\right)$

34. $\left(\frac{1}{2} - \frac{\sqrt{31}}{2}, \frac{1}{2} + \frac{\sqrt{31}}{2}\right)$ and $\left(\frac{1}{2} + \frac{\sqrt{31}}{2}, \frac{1}{2} - \frac{\sqrt{31}}{2}\right)$
37. four solutions: $(\pm6, \pm2)$
40. $\left(\pm\sqrt{-\frac{1}{2} + \frac{\sqrt{17}}{2}}, -\frac{1}{2} + \frac{\sqrt{17}}{2}\right)$

46. circle; $(x + 1)^2 + (y + 3)^2 = 25$
47. circle; $(x - 1)^2 + (y - 1)^2 = 8$
48. ellipse; $\frac{(x-1)^2}{9} + \frac{(y-1)^2}{4} = 1$
49. ellipse; $\frac{(x+2)^2}{9} + \frac{(y-2)^2}{4} = 1$

Error Analysis

Students sometimes fail to find all of the possible solutions when solving systems of nonlinear equations. Encourage students to sketch the graphs, to check the number of real solutions.

17. $\left(\pm\sqrt{-\frac{1}{2} + \frac{\sqrt{5}}{2}}, -\frac{1}{2} + \frac{\sqrt{5}}{2}\right)$

18. $\left(\pm\sqrt{-\frac{1}{2} + \frac{\sqrt{17}}{2}}, -\frac{1}{2} + \frac{\sqrt{17}}{2}\right)$

19. none

20. $\left(-\frac{6\sqrt{5}}{5}, -\frac{6\sqrt{5}}{5}\right)$ and $\left(\frac{6\sqrt{5}}{5}, \frac{6\sqrt{5}}{5}\right)$

46. circle; $(x + 1)^2 + (y + 3)^2 = 25$

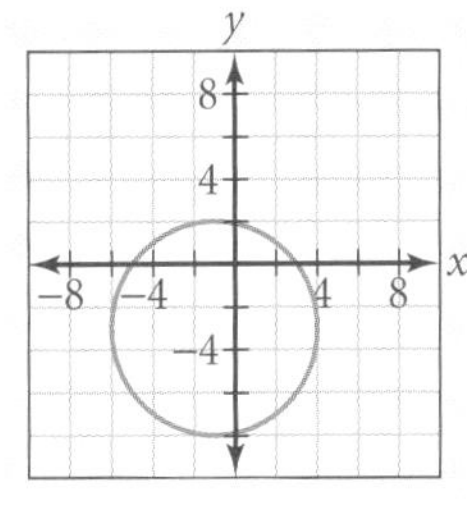

47. circle; $(x - 1)^2 + (y - 1)^2 = 8$

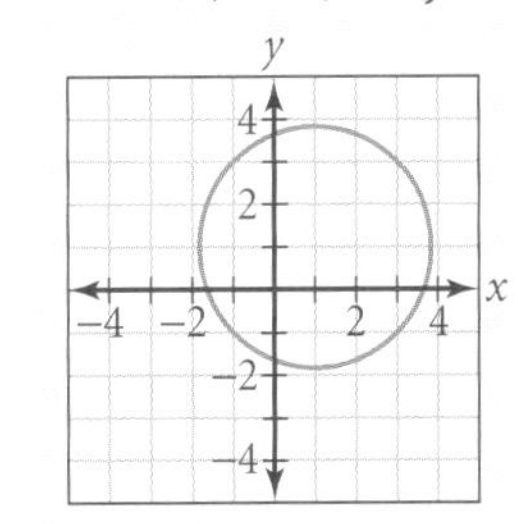

48. ellipse; $\frac{(x-1)^2}{9} + \frac{(y-1)^2}{4} = 1$

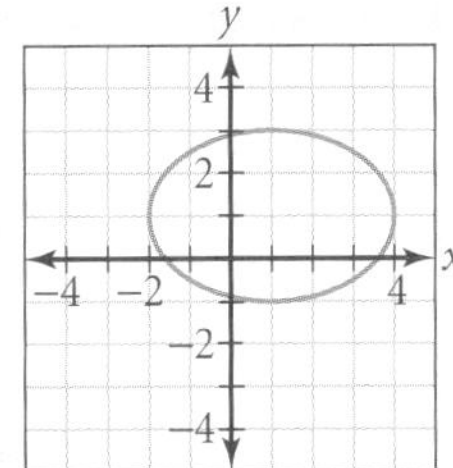

49. ellipse; $\frac{(x+2)^2}{9} + \frac{(y-2)^2}{4} = 1$

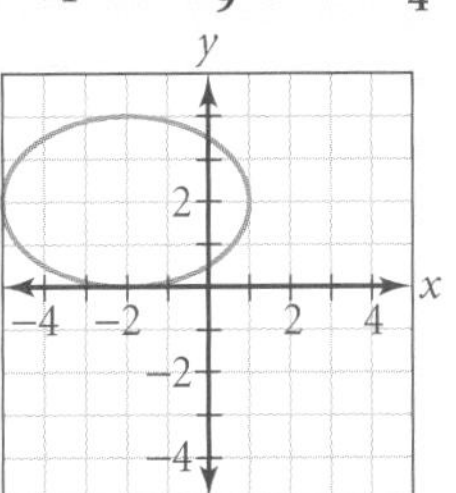

50. hyperbola; $\frac{(y-1)^2}{1} - \frac{(x+2)^2}{4} = 1$

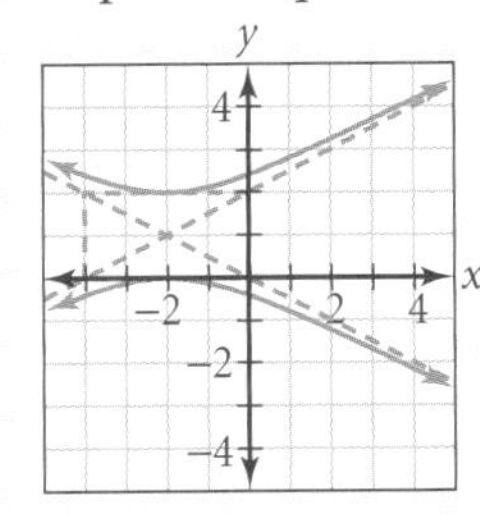

51. hyperbola; $\frac{(x+1)^2}{9} - \frac{(y-2)^2}{4} = 1$

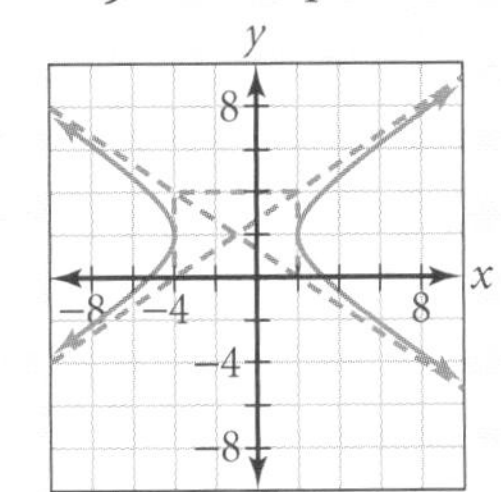

52. ellipse; $\frac{x^2}{25} + \frac{(y-1)^2}{4} = 1$

☞ The answers to Exercises 53–57 can be found on page 612.

53. ellipse; $\frac{(x-3)^2}{9}+\frac{(y+1)^2}{4}=1$

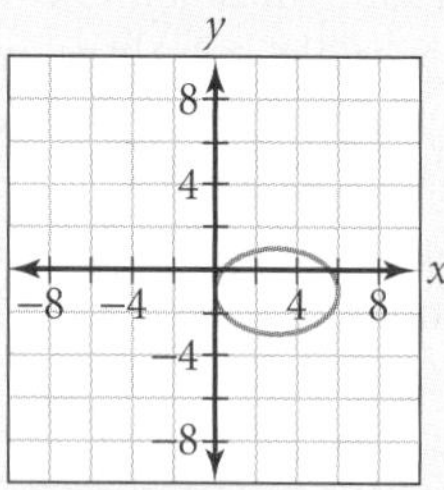

54. parabola; $y+3=\frac{9}{4}(x-1)^2$

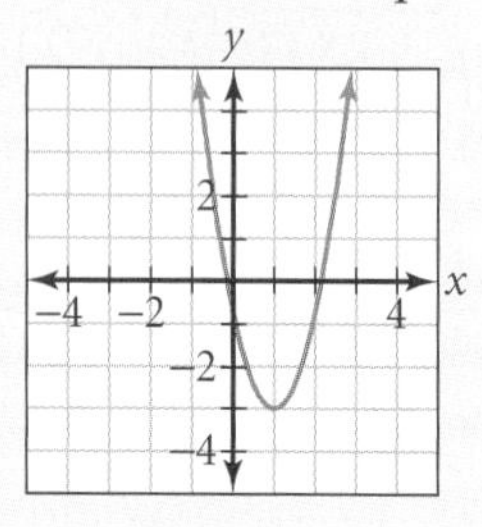

55. parabola; $x+2=\frac{1}{2}(y-3)^2$

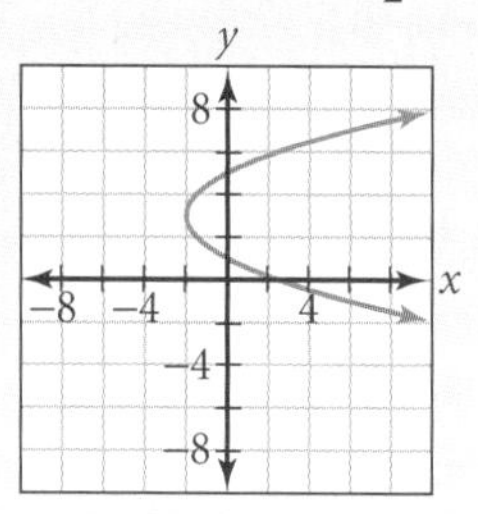

58. $\left(1 \pm \frac{3\sqrt{7}}{4}, \frac{5}{4}\right)$

59. $\left(-\frac{5}{3}, 2 \pm \frac{4\sqrt{5}}{3}\right)$

64. $\left(\frac{2}{5}+\frac{9\sqrt{6}}{10}, \frac{6}{5}-\frac{3\sqrt{6}}{10}\right)$ and $\left(\frac{2}{5}-\frac{9\sqrt{6}}{10}, \frac{6}{5}+\frac{3\sqrt{6}}{10}\right)$

65. $\left(\frac{5}{2}+\frac{5\sqrt{247}}{26}, \frac{1}{2}-\frac{\sqrt{247}}{26}\right)$ and $\left(\frac{5}{2}-\frac{5\sqrt{247}}{26}, \frac{1}{2}+\frac{\sqrt{247}}{26}\right)$

66a. circle and ellipse

b. no real solutions

c. $\begin{cases} x^2+y^2=16 \\ \frac{x^2}{9}+y^2=1 \end{cases}$

Solve each system of equations. If there are no real-number solutions, write *none*.

58. $\begin{cases}(x-1)^2+(y-1)^2=4\\(x-1)^2+(y+1)^2=9\end{cases}$

59. $\begin{cases}(x+2)^2+(y-2)^2=9\\(x-1)^2+(y-2)^2=16\end{cases}$

60. $\begin{cases}(x-1)^2+y^2=4\\(x-1)^2+y^2=1\end{cases}$ **none**

61. $\begin{cases}(x+2)^2+y^2=9\\(x+2)^2+y^2=25\end{cases}$ **none**

62. $\begin{cases}\frac{(x-1)^2}{9}+\frac{y^2}{4}=1\\\frac{(x-1)^2}{9}+y^2=1\end{cases}$ **(−2, 0) and (4, 0)**

63. $\begin{cases}\frac{(y-1)^2}{9}-x^2=1\\\frac{(y-1)^2}{9}-\frac{x^2}{9}=1\end{cases}$ **(0, −2) and (0, 4)**

64. $\begin{cases}x^2+y^2+2x+6y=15\\x^2+y^2-2x-6y=-1\end{cases}$

65. $\begin{cases}x^2+y^2-4x+4y=8\\x^2+y^2-6x-6y=-2\end{cases}$

66. The graphs of two nonlinear equations in x and y are shown at right.

a. What conic sections are graphed?

b. Describe any solutions to the system of equations represented.

c. Write the system of equations represented by the graphs.

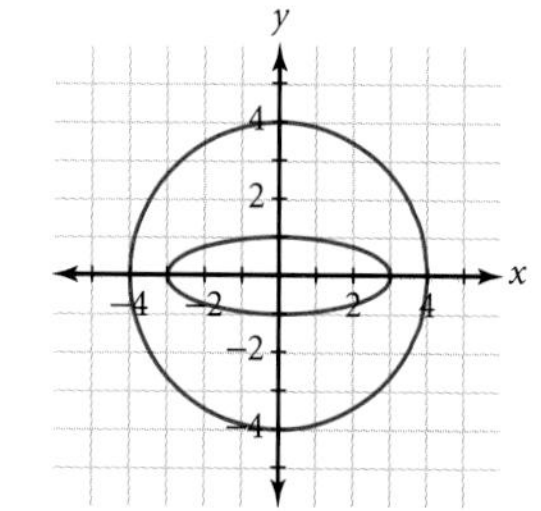

67. 51

67. A positive two-digit number is represented by $10t+u$. The sum of the squares of the digits is 26. If the number is decreased by the number with its digits reversed, the result is 36. Find the two-digit number.

68. −11 and −7

68. Find two negative numbers such that the sum of their squares is 170 and twice the square of the first minus 3 times the square of the second is 95.

CHALLENGES

69. $x^2=9$, $y^2=0$; $x=\pm 3$, $y=0$

69 You can write $\begin{cases}4x^2+9y^2=36\\x^2+y^2=9\end{cases}$ as $AX=B$, where $A=\begin{bmatrix}4 & 9\\1 & 1\end{bmatrix}$, $X=\begin{bmatrix}x^2\\y^2\end{bmatrix}$, and $B=\begin{bmatrix}36\\9\end{bmatrix}$. Use a matrix equation to find x^2 and y^2. Then find x and y.

70. Solve $\begin{cases}x^2+y^2<16\\x^2-6y\le 3\end{cases}$ by graphing.

CONNECTIONS

71. 3 feet by 4 feet

72. 28 meters

73. $6-2\sqrt{6}$ inches by $6+2\sqrt{6}$ inches, or about 1.10 by 10.90

71. GEOMETRY Find the dimensions of a rectangle with an area of 12 square feet and a diagonal of 5 feet.

72. GEOMETRY The area of a rectangle is 48 square meters. The length of a diagonal is 10 meters. Find the perimeter of the rectangle.

73. GEOMETRY The area of a rectangle is 12 square inches. The perimeter is 24 inches. Find the dimensions of the rectangle.

APPLICATION

74. 21,822 units

74. BUSINESS A company determines that its total monthly production cost, P, in thousands of dollars is defined by the equation $P=8x+4$. The company's monthly revenue, R, in thousands of dollars is defined by $8R-3x^2=0$. In both equations, x is the number of units, in thousands, of its product manufactured and sold per month. When the cost of manufacturing the product equals the revenue obtained by selling it, the company breaks even. How many units must the company produce in order to break even?

Practice

NAME ________ CLASS ________ DATE ________

Practice
9.6 Solving Nonlinear Systems

Use the substitution method to solve each system. If there are no real solutions, write *none*.

1. $\begin{cases}y=x^2+5\\y=5x+1\end{cases}$ (1, 6) and (4, 21)
2. $\begin{cases}y=x-2\\y=x^2\end{cases}$ none
3. $\begin{cases}y=x\\x^2+y^2=16\end{cases}$ (−2.83, −2.83) and (2.83, 2.83)

Use the elimination method to solve each system. If there are no real solutions, write *none*.

4. $\begin{cases}x^2+y^2=9\\4x^2-9y^2=36\end{cases}$ (−3, 0) and (3, 0)
5. $\begin{cases}x^2+y^2=8\\2x^2-3y^2=1\end{cases}$ (−2.24, −1.73), (−2.24, 1.73), (2.24, −1.73), and (2.24, 1.73)
6. $\begin{cases}x^2+2y^2=30\\3x^2-5y^2=24\end{cases}$ (−4.24, −2.45), (−4.24, 2.45), (4.24, −2.45), and (4.24, 2.45)

Solve each system by graphing. If there are no real solutions, write *none*.

7. $\begin{cases}x^2+2y^2=16\\4x^2+y^2=4\end{cases}$ none
8. $\begin{cases}25x^2-4y^2=100\\4x^2+9y^2=36\end{cases}$ (−2.08, −1.44), (−2.08, 1.44), (2.08, −1.44), and (2.08, 1.44)
9. $\begin{cases}9x^2-16y^2=144\\16x^2+9y^2=144\end{cases}$ none

Classify the conic section defined by each equation. Write the standard equation of the conic section, and sketch the graph.

10. $x^2-14x-4y+61=0$ parabola; $y-3=\frac{1}{4}(x-7)^2$

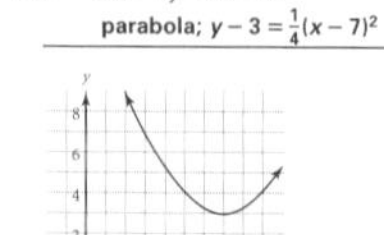

11. $4x^2-9y^2-40x+72y-80=0$ hyperbola; $\frac{(x-5)^2}{9}-\frac{(y-4)^2}{4}=1$

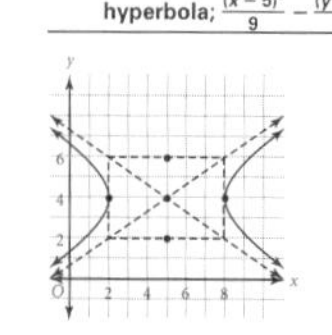

56. hyperbola; $\frac{(x-2)^2}{16}-\frac{(y+1)^2}{4}=1$

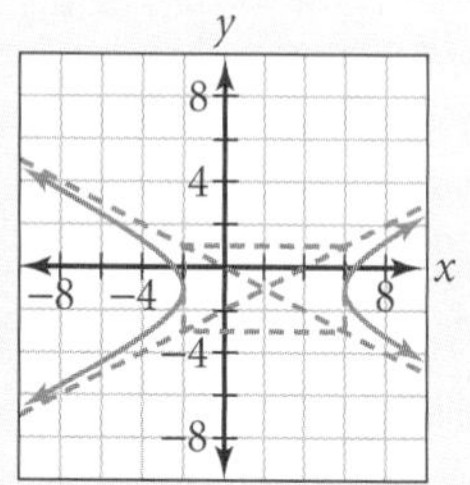

57. hyperbola; $\frac{x^2}{123}-\frac{(y+2)^2}{123}=1$

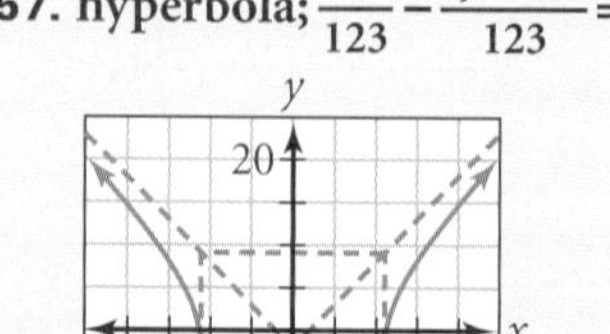

70.

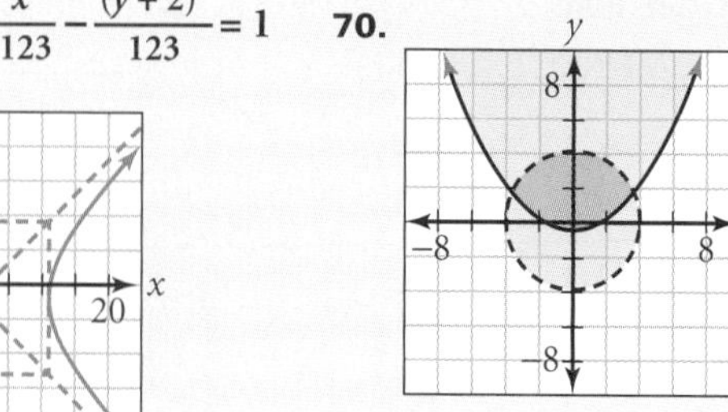

APPLICATION

BUSINESS The cross section of a log is circular. A mill operator wants to cut a beam from the log as shown below. Find the dimensions of each rectangular cross section described below to the nearest tenth of an inch.

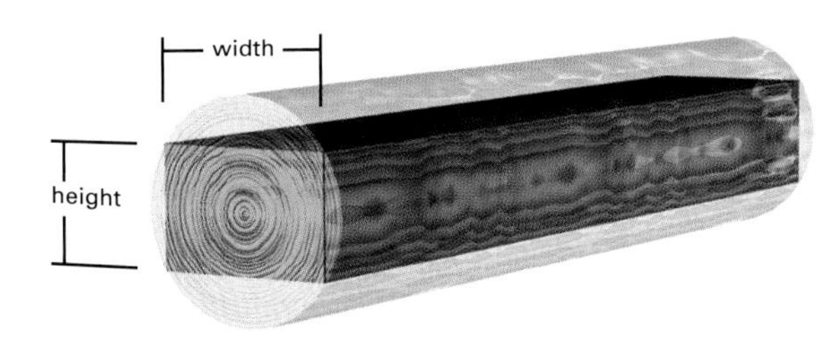

Answers to Exercises 75–77 are in inches.

75. $\frac{32\sqrt{5}}{5}$ by $\frac{16\sqrt{5}}{5}$, or about 14.31 by 7.16
76. $\frac{36\sqrt{5}}{5}$ by $\frac{18\sqrt{5}}{5}$, or about 16.10 by 8.05
77. $\frac{27\sqrt{10}}{5}$ by $\frac{9\sqrt{10}}{5}$, or about 17.08 by 5.69

75. Find the dimensions of the cross section if the circular cross section has a 16-inch diameter and the beam's height is $\frac{1}{2}$ its width.

76. Find the dimensions of the cross section if the circular cross section has an 18-inch diameter and the beam's height is $\frac{1}{2}$ its width.

77. Find the dimensions of the cross section if the circular cross section has an 18-inch diameter and the beam's width is 3 times its height.

Look Back

Find all rational zeros of each polynomial function. *(LESSON 7.5)*

78 $P(x) = x^3 - 8x^2 + 5x + 14$ **−1, 2, 7**

79 $P(x) = x^3 + 2x^2 - x - 2$ **−2, −1, 1**

80 $P(x) = x^4 + x^2 - 2$ **−1, 1**

81 $P(x) = x^4 + 7x^2 + 12$ **no rational zeros**

Find all zeros of each polynomial function. *(LESSON 7.5)*

82 $P(x) = x^3 + 4x^2 - 17x - 60$

83 $P(x) = 2x^3 - 5x^2 + 1$

84 $P(a) = 3a^4 - 5a^3 - 5a^2 - 19a - 6$

85 $P(a) = 3a^4 - 2a^3 + 8a^2 - 6a - 3$

82. −5, −3, and 4
83. $\frac{1}{2}$, $1 - \sqrt{2}$, and $1 + \sqrt{2}$
84. 3, $-\frac{1}{3}$, $-\frac{1}{2} - \frac{\sqrt{7}}{2}i$, and $-\frac{1}{2} + \frac{\sqrt{7}}{2}i$
85. $-\frac{1}{3}$, 1, $i\sqrt{3}$, $-i\sqrt{3}$

Simplify. *(LESSON 8.4)*

86. $\frac{x}{x+1} + \frac{1}{x^2 - 1}$ $\frac{x^2 - x + 1}{x^2 - 1}$

87. $\frac{2}{x(x-2)} - \frac{x+1}{x^2 - 4}$ $\frac{x^2 + 3x + 4}{x^3 - 4x}$

Solve each radical equation. *(LESSON 8.8)*

88. $\sqrt{x+4} = 2$ **0**

89. $\sqrt{x-1} = 3\sqrt{x-2}$ $\frac{17}{8}$

90. $\sqrt{-x} = 4\sqrt{-x-1}$ $-\frac{16}{15}$

Solve each radical inequality. *(LESSON 8.8)*

91. $\sqrt{3x+2} > 5$ $x > \frac{23}{3}$

92. $0 \geq 3\sqrt{x-2} - 2$ $2 \leq x \leq \frac{22}{9}$

93. $-6 > 2\sqrt{x-2} - 4$ no solution

Look Beyond

Classify the conic section defined by each equation. Then graph the conic section and describe its graph.

94 $4x^2 + y^2 = 0$

95 $y^2 = 4$

96 $x^2 = 4$

97 $y^2 - x^2 = 0$

98 $(x-1)^2 = (y+1)^2$

99 $(x-2)^2 + (y+2)^2 = 0$

94. point
95. horz. parallel lines
96. vert. parallel lines
97. intersecting lines
98. intersecting lines
99. point

Look Beyond

In Exercises 94–99, students are asked to classify the conic section defined by an equation, graph the conic section, and describe its graph.

94. The graph is the point (0, 0).

95. The graph consists of two parallel lines, $y = 2$ and $y = -2$.

96. The graph consists of two parallel lines, $x = 2$ and $x = -2$.

97. The graph consist of two intersecting lines, $y = x$ and $y = -x$.

98. The graph consists of two intersecting lines, $y = x - 2$ and $y = -x$.

99. The graph is the point $(-2, 2)$.

Student Technology Guide

NAME ____________ CLASS ________ DATE ________

Student Technology Guide
9.6 Solving Nonlinear Systems

You can use a graphics calculator to approximate the solution(s) to a nonlinear system of equations.

Example: Solve $\begin{cases} (x-3)^2 + y^2 = 25 \\ \frac{x^2}{16} + \frac{y^2}{49} = 1 \end{cases}$.

Notice that the first equation represents a circle, and the second represents an ellipse.

- Solve each equation for y.
  For the circle, $y = \pm\sqrt{25 - (x-3)^2}$. For the ellipse, $y = \pm\sqrt{49\left(1 - \frac{x^2}{16}\right)}$.
- To enter the halves of the circle as Y1 and Y2, press Y= 2nd x^2 25 - (X,T,θ,n - 3) x^2). Then make Y2 the opposite of Y1 by moving to Y2 and pressing (−) VARS Y-VARS 1:Function... ENTER 1:Y1 ENTER
- To enter the ellipse, move to Y3 and press 2nd x^2 49 (1 - X,T,θ,n x^2 ÷ 16)). Then move to Y4 and press (−) VARS Y-VARS 1:Function... ENTER 3:Y3 ENTER.
- Press GRAPH. Notice that there are two intersection points.
- To find the intersection of Y1 and Y3, press 2nd TRACE 5:intersect ENTER. When asked for the first curve, be sure that you are on Y1, then press ENTER.
- When asked for the second curve, use the down arrow key to move to Y3 and press ENTER. When asked for a guess, trace to the intersection and press ENTER. This solution is near (2.8, 5.0).
- Repeat this procedure to find the other solution, (2.8, −5.0), at the intersection of Y2 and Y4.

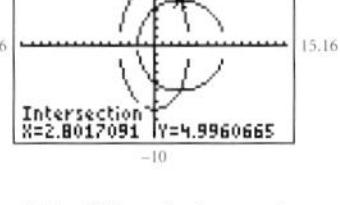

Solve each system by graphing. Round your answers to the nearest tenth.

1. $\begin{cases} x^2 + y^2 = 36 \\ \frac{x^2}{16} - \frac{y^2}{49} = 1 \end{cases}$ (−4.6, −3.9), (−4.6, 3.9), (4.6, −3.9), (4.6, 3.9)

2. $\begin{cases} 9x^2 + 25y^2 = 225 \\ -\frac{x^2}{2} + \frac{y^2}{3} = 1 \end{cases}$ (−1.8, −2.8), (−1.8, 2.8), (1.8, −2.8), (1.8, 2.8)

3. $\begin{cases} x^2 + y^2 = 36 \\ 4x^2 + 9y^2 = 144 \end{cases}$ (−6, 0) and (6, 0)

Focus

Special graph paper can help in the construction of the graphs of parabolas, ellipses, and hyperbolas. Students will construct focus-directrix and focus-focus graph paper and use it to graph a parabola, an ellipse, and a hyperbola. Students are asked to label the defining characteristics of each conic section and write its equation.

Motivate

To make students conscious of the significance of conic sections in the natural world, ask them to discuss what they know about the orbits of the planets around the Sun. Ask whether they know why a parabolic cross section is essential to the proper functioning of an automobile's head lamp reflectors or a satellite dish. Then have them share with the class any application of parabolic, elliptical, or hyperbolic curves that they have learned through books or movies.

The graph paper used in this project can be made with graphing software.

Parabolic radio telescope in New Mexico

Activity 1 Parabolas

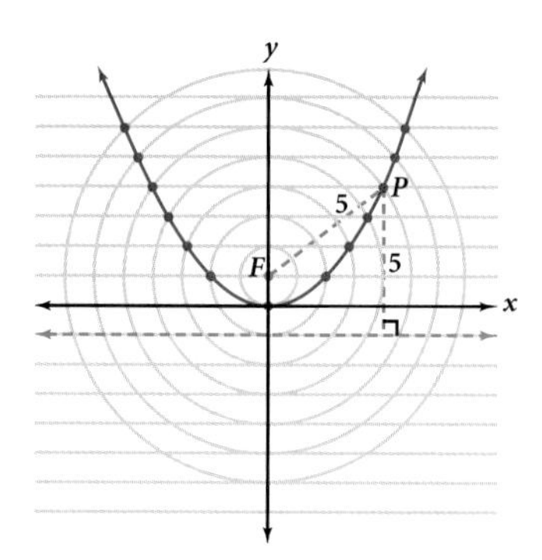

The figure at right shows a parabola with its focus at (0, 1) and a directrix of $y = -1$. Equally spaced concentric circles with their center at the parabola's focus enable you to measure distances from the focus. Equally spaced horizontal lines parallel to the parabola's directrix enable you to measure vertical distances from the directrix. This type of graph paper is called focus-directrix graph paper.

Notice that P is the point of intersection of the circle centered at (0, 1) with a radius of 5 and the horizontal line 5 units above the directrix. Thus, P is equidistant from the focus and directrix. Examine the figure above and note that all points on the parabola are equidistant from the focus and the directrix.

1. Create focus-directrix graph paper with the focus at (0, 3) and a directrix of $y = 3$. Label the focus, directrix, x-axis, and y-axis.
   - **a.** Plot points that are equidistant from the focus and directrix.
   - **b.** Identify the vertex of the resulting parabola.
   - **c.** Identify p as defined for the standard equation of a parabola.
   - **d.** Write the equation of the parabola.
2. Repeat Step 1 with the focus at (0, −3) and a directrix of $y = 3$.
3. Repeat Step 1 with the focus at (3, 0) and a directrix of $x = -3$. This time the focus-directrix graph paper should have concentric circles centered at (3, 0) and vertical lines parallel to the directrix, $x = -3$.
4. Repeat Step 1 with the focus at (−3, 0) and a directrix of $x = 3$.

Activity 1

1b. (0, 0)

c. $p = 3$

d. $y = \frac{1}{12}x^2$

2b. (0, 0)

c. $p = -3$

d. $y = -\frac{1}{12}x^2$

3b. (0, 0)

c. $p = 3$

d. $x = \frac{1}{12}y^2$

4b. (0, 0)

c. $p = -3$

d. $x = -\frac{1}{12}y^2$

The answers to Exercises 1a, 2a, 3a, and 4a can be found in Additional Answers beginning on page 1002.

Activity 2 Ellipses

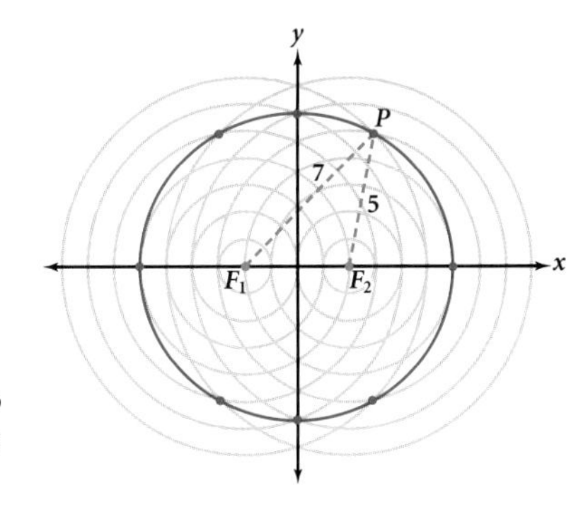

The figure at right shows an ellipse with foci at $F_1(-2, 0)$ and $F_2(2, 0)$. Equally spaced concentric circles centered at the two foci enable you to measure distances from the foci. This is called focus-focus graph paper.

Notice that P is the point of intersection of the two circles: one circle is centered at F_1 with a radius of 7, and the other circle is centered at F_2 with a radius of 5. Thus, $F_1P + F_2P = 7 + 5 = 12$. Notice that all points on the ellipse satisfy the equation $F_1P + F_2P = 12$.

1. Create focus-focus graph paper with foci at $F_1(-4, 0)$ and $F_2(4, 0)$. Label the foci, x-axis, and y-axis.
   a. Plot points that satisfy the equation $F_1P + F_2P = 10$.
   b. Identify the vertices and co-vertices of the resulting ellipse.
   c. Identify a, b, and c as defined for the standard equation of an ellipse, and find the lengths of the major and minor axes.
   d. Write the standard equation of the ellipse.
2. Repeat Step 1 with the foci at $F_1(0, -4)$and $F_2(0, 4)$.

Activity 3 Hyperbolas

The figure below shows a hyperbola with foci at $F_1(-6, 0)$ and $F_2(6, 0)$. Equally spaced concentric circles are centered at the foci. This is also called focus-focus graph paper.

Notice that P is the point of intersection of two circles: one circle is centered at $F_1(-6, 0)$ with a radius of 9, and the other circle is centered at $F_2(6, 0)$ with a radius of 5. Thus, $|F_1P - F_2P| = |9 - 5| = 4$. Notice that all points on the hyperbola satisfy the equation $|F_1P - F_2P| = 4$.

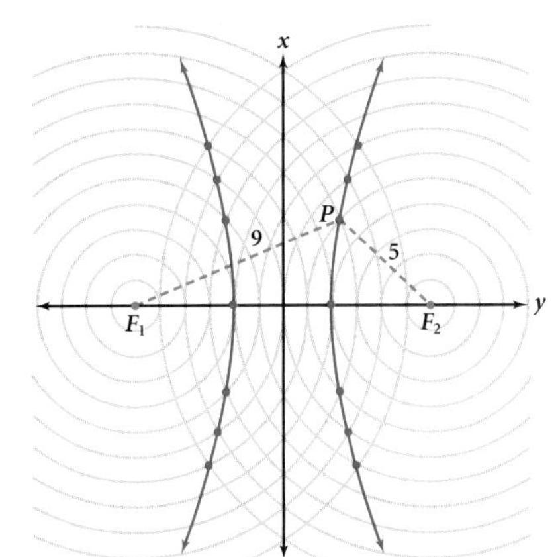

1. Create focus-focus graph paper with foci at $F_1(-5, 0)$ and $F_2(5, 0)$. Label the foci, x-axis, and y-axis.
   a. Plot points that satisfy the equation $|F_1P - F_2P| = 6$.
   b. Identify the vertices and co-vertices of the resulting hyperbola.
   c. Identify a, b, and c as defined for the standard equation of a hyperbola, and find the lengths of the transverse and conjugate axes.
   d. Write equations for the asymptotes of the hyperbola, and sketch them on your graph.
   e. Write the standard equation of the hyperbola.

Cooperative Learning

Divide the class into groups of four, with each pair of students completing half of **Activity 1 and 2.** For the hyperbola in **Activity 3,** each group member should complete all steps in the Activity. Members should be encouraged to assist each other with any problems so that each group can complete all of the steps. Each group should write a summary of the entire project, describing how the special graph paper helped in graphing each conic section.

Discuss

Each group should discuss the definition of each conic section, noting differences and similarities and the characteristics of the special graph paper that helped to graph each conic section. As an extension of the project, a family of conic sections can be drawn for each type. To draw a family of parabolas, have students use several different directrices (plural of directrix) with the same focus, such as (0, 3). For a family of ellipses, have students use different values for $F_1P + F_2P$ with the foci $F_1(-2, 0)$ and $F_2(2, 0)$. For a family of hyperbolas, have students use different values for $|F_1P - F_2P|$ with foci $F_1(-6, 0)$ and $F_2(6, 0)$. This should give students a better understanding of what affects the shape of the curves.

Activity 2

1b. vertices: (−5, 0) and (5, 0); co-vertices: (0, −3) and (0, 3)

c. $a = 5, b = 3, c = 4$

d. $\frac{x^2}{25} + \frac{y^2}{9} = 1$

2b. vertices: (0, −5) and (0, 5); co-vertices: (−3, 0) and (3, 0)

c. $a = 3, b = 5, c = 4$

d. $\frac{x^2}{9} + \frac{y^2}{25} = 1$

Activity 3

1a. The graph is shown in part **d**.

b. vertices: (−3, 0) and (3, 0); co-vertices: (0, −4) and (0, 4)

c. $a = 3,\ b = 4,\ c = 5$; transverse axis length: 6; conjugate axis length: 8

e. $\frac{x^2}{9} - \frac{y^2}{16} = 1$

The answers to Steps 1a and 2a of Activity 2 and Step 1d of Activity 3 can be found in Additional Answers beginning on page 1002.

Chapter Review

6. $y + 4 = (x - 2)^2$; vertex: $(2, -4)$; focus: $\left(2, -\frac{15}{4}\right)$; directrix: $y = -\frac{17}{4}$

7. $x + 23 = (y - 5)^2$; vertex: $(-23, 5)$; focus: $\left(-\frac{91}{4}, 5\right)$; directrix: $x = -\frac{93}{4}$

Chapter Test, Form A

NAME ______ CLASS ______ DATE ______

Chapter Assessment
Chapter 9, Form A, page 1

Write the letter that best answers the question or completes the statement.

c 1. Find the length of the segment with endpoints $P(6, -2)$ and $Q(3, 4)$.
a. $\sqrt{117}$ b. $\sqrt{65}$ c. $\sqrt{45}$ d. $\sqrt{13}$

d 2. One endpoint of $\overline{PQ}$ is $P(9, 3)$ and the midpoint, M, is $(1, 5)$. Find the coordinates of Q.
a. $Q(5, 4)$ b. $Q(4, -1)$ c. $Q(-4, 1)$ d. $Q(-7, 7)$

a 3. If a diameter of a circle has endpoints $(2, 5)$ and $(-8, 29)$, then find the center and radius of the circle.
a. center: $(-3, 17)$; radius: 13 b. center: $(-5, 17)$; radius: 26
c. center: $(5, 12)$; radius: 13 d. center: $(-3, 17)$; radius: $\sqrt{612}$

c 4. Find the equation of a circle with its center at $(5, -2)$ and a radius of 16.
a. $(x + 5)^2 + (y - 2)^2 = 256$ b. $(x + 5)^2 + (y - 2)^2 = 4$
c. $(x - 5)^2 + (y + 2)^2 = 256$ d. $(x - 5)^2 + (y + 2)^2 = 4$

a 5. Find the equation of the parabola shown at right.
a. $x + 2 = -\frac{1}{8}y^2$
b. $x - 2 = \frac{1}{8}y^2$
c. $y = \frac{1}{8}(x + 2)^2$
d. $y = -\frac{1}{8}(x - 2)^2$

b 6. Find the equation of the parabola with its focus at $(-4, 7)$ and directrix $y = 1$.
a. $y - 7 = \frac{1}{24}(x + 4)^2$ b. $y - 4 = \frac{1}{12}(x + 4)^2$
c. $y + 4 = -\frac{1}{12}(x - 4)^2$ d. $y - 3 = \frac{1}{24}(x + 4)^2$

d 7. Find the equation of the ellipse shown at right.
a. $\frac{(x-1)^2}{4} - \frac{(y+4)^2}{25} = 1$
b. $\frac{(x+1)^2}{4} + \frac{(y-4)^2}{25} = 1$
c. $\frac{(x-1)^2}{25} - \frac{(y+4)^2}{4} = 1$
d. $\frac{(x-1)^2}{25} + \frac{(y+4)^2}{4} = 1$

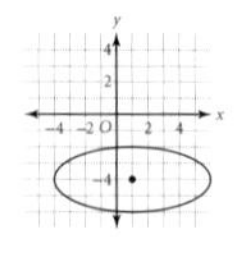

NAME ______ CLASS ______ DATE ______

Chapter Assessment
Chapter 9, Form A, page 2

b 8. Find the equation of the ellipse centered at the origin with foci at $(0, -3)$ and $(0, 3)$, and a major axis of 10.
a. $\frac{x^2}{25} + \frac{y^2}{16} = 1$ b. $\frac{x^2}{16} + \frac{y^2}{25} = 1$
c. $\frac{x^2}{25} + \frac{y^2}{9} = 1$ d. $\frac{x^2}{9} + \frac{y^2}{25} = 1$

c 9. Find the equation of the hyperbola shown at right.
a. $\frac{x^2}{9} - \frac{y^2}{4} = 1$
b. $\frac{x^2}{4} + \frac{y^2}{9} = 1$
c. $\frac{y^2}{4} - \frac{x^2}{9} = 1$
d. $\frac{y^2}{9} - \frac{x^2}{4} = 1$

a 10. Find the equation of the hyperbola with vertices at $(5, 4)$ and $(-3, 4)$ and co-vertices at $(1, 1)$ and $(1, 7)$.
a. $\frac{(x-1)^2}{16} - \frac{(y-4)^2}{9} = 1$ b. $\frac{(x+1)^2}{9} + \frac{(y+4)^2}{16} = 1$
c. $\frac{(x+1)^2}{9} - \frac{(y+4)^2}{16} = 1$ d. $\frac{(x-1)^2}{16} + \frac{(y+4)^2}{9} = 1$

d 11. Find the asymptotes of the hyperbola $\frac{y^2}{9} - \frac{x^2}{81} = 1$.
a. $y = \pm 9x$ b. $y = \pm\frac{1}{9}x$ c. $y = \pm 3x$ d. $y = \pm\frac{1}{3}x$

d 12. Which conic section is defined by the equation $4x^2 + 7x = 24 + 4y^2$?
a. parabola b. circle c. ellipse d. hyperbola

c 13. How many solutions does the system $\begin{cases} 4x^2 + 4y^2 = 100 \\ y - 2 = 2(x - 3)^2 \end{cases}$ have?
a. 0 b. 1 c. 2 d. 3

d 14. Write the standard equation of the circle defined by $x^2 + y^2 + 8x - 12y + 43 = 0$.
a. $(x - 6)^2 + (y + 4)^2 = 9$ b. $(x + 4)^2 + (y - 6)^2 = 95$
c. $(x + 4)^2 + (y - 6)^2 = 7$ d. $(x + 4)^2 + (y - 6)^2 = 9$

9 Chapter Review and Assessment

VOCABULARY

Key Skills & Exercises

LESSON 9.1

Key Skills

Use the distance and midpoint formulas.

Find the distance between points $P(1, 2)$ and $Q(-3, 5)$.

$$d = \sqrt{(x_2 - x_1)^2 + (y_2 - y_1)^2}$$

$$d = \sqrt{(-3 - 1)^2 + (5 - 2)^2} = \sqrt{25} = 5$$

Find the coordinates of M, the midpoint of $P(1, 2)$ and $Q(-3, 5)$.

$$M\left(\frac{x_1 + x_2}{2}, \frac{y_1 + y_2}{2}\right) = \left(\frac{1 + (-3)}{2}, \frac{2 + 5}{2}\right) = \left(-1, \frac{7}{2}\right)$$

Exercises

Find PQ and the coordinates of M, the midpoint of PQ.

1. $P(0, 4)$ and $Q(3, 0)$ **5; $M\left(\frac{3}{2}, 2\right)$**
2. $P(2, 7)$ and $Q(10, 13)$ **10; $M(6, 10)$**
3. $P(-3, 5)$ and $Q(3, 1)$ **$2\sqrt{13} \approx 7.21$; $M(0, 3)$**
4. $P(-1, 2)$ and $Q(-6, -8)$ **$5\sqrt{5} \approx 11.18$; $M\left(-\frac{7}{2}, -3\right)$**
5. $P(11, -9)$ and $Q(-2, -5)$ **$\sqrt{185} \approx 13.60$; $M\left(\frac{9}{2}, -7\right)$**

LESSON 9.2

Key Skills

Find the vertex, focus, and directrix of a parabola. Graph the parabola.

Given $y^2 - 6y - 12x + 57 = 0$, complete the square and write the standard equation of the parabola.

$$x - 4 = \frac{1}{12}(y - 3)^2$$

From the equation, $h = 4$, $k = 3$, and $p = 3$.

The vertex is at $(4, 3)$, the focus is at $(7, 3)$, and the directrix is $x = 1$.

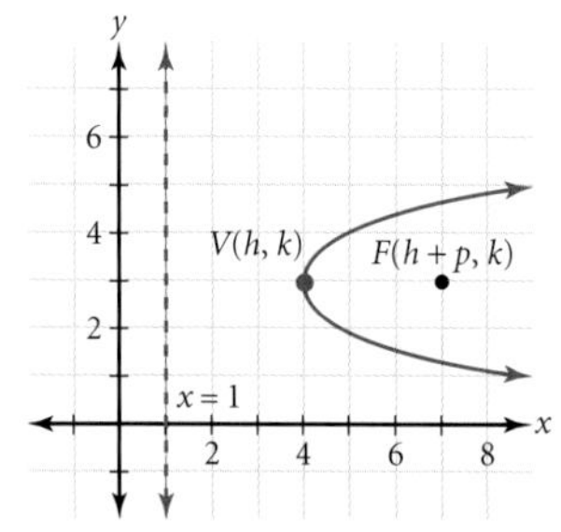

Exercises

Write the standard equation for each parabola. Find the vertex, focus, and directrix. Then sketch the graph.

6. $y = x^2 - 4x$ **$y + 4 = (x - 2)^2$**
7. $x - 2 = y^2 - 10y$ **$x + 23 = (y - 5)^2$**
8. $y = -2x^2 - 4x + 6$ **$y - 8 = -2(x + 1)^2$**
9. $x^2 + 8x - y + 20 = 0$ **$y - 4 = (x + 4)^2$**
10. $4y^2 - 8y - x + 1 = 0$ **$x + 3 = 4(y - 1)^2$**

8. $y - 8 = -2(x + 1)^2$; vertex: $(-1, 8)$; focus: $\left(-1, \frac{63}{8}\right)$; directrix: $y = \frac{65}{8}$

9. $y - 4 = (x + 4)^2$; vertex: $(-4, 4)$; focus: $\left(-4, \frac{17}{4}\right)$; directrix: $y = \frac{15}{4}$

10. $x + 3 = 4(y - 1)^2$; vertex: $(-3, 1)$; focus: $\left(-\frac{47}{16}, 1\right)$; directrix: $y = -\frac{49}{16}$

The graphs to Exercises 6–10 can be found in Additional Answers beginning on page 1002.

11. $x^2 + y^2 = 100$; $C(0, 0)$; $r = 10$

12. $x^2 + y^2 = 12$; $C(0, 0)$; $r = 2\sqrt{3}$

13. $(x - 1)^2 + (y - 49)^2 = 81$; $C(1, 49)$; $r = 9$

14. $(x - 5)^2 + (y + 4)^2 = 36$; $C(5, -4)$; $r = 6$

LESSON 9.3

Key Skills

Find the radius and center of a circle. Sketch the graph.

Given $x^2 + y^2 - 10x + 8y + 5 = 0$, complete the square and write the standard equation of the circle.

$(x - 5)^2 + (y + 4)^2 = 6^2$

The center is at $(5, -4)$, and the radius is 6.

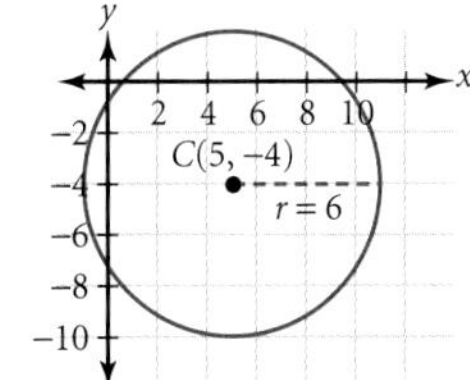

Exercises

Write the standard equation for each circle. Find the center and radius. Then sketch the graph.

11. $x^2 + y^2 = 100$
12. $3x^2 + 3y^2 = 36$
13. $(x - 1)^2 + (y - 49)^2 = 81$
14. $x^2 + y^2 - 10x + 8y + 5 = 0$
15. $x^2 + y^2 + 8y + 4x - 5 = 0$

LESSON 9.4

Key Skills

Find the coordinates of the center, vertices, co-vertices, and foci of an ellipse. Sketch the graph.

Given $16x^2 + 4y^2 - 96x + 8y + 84 = 0$, complete the square and write the standard equation.

$$\frac{(x-3)^2}{4} + \frac{(y+1)^2}{16} = 1$$

From the equation, $h = 3$, $k = -1$, $a = 4$, $b = 2$, and $c = 2\sqrt{3}$.

center: $(3, -1)$
vertices: $(3, -5)$, $(3, 3)$
co-vertices: $(1, -1)$, $(5, -1)$
foci: $(3, -1 + \sqrt{12})$, $(3, -1 - \sqrt{12})$

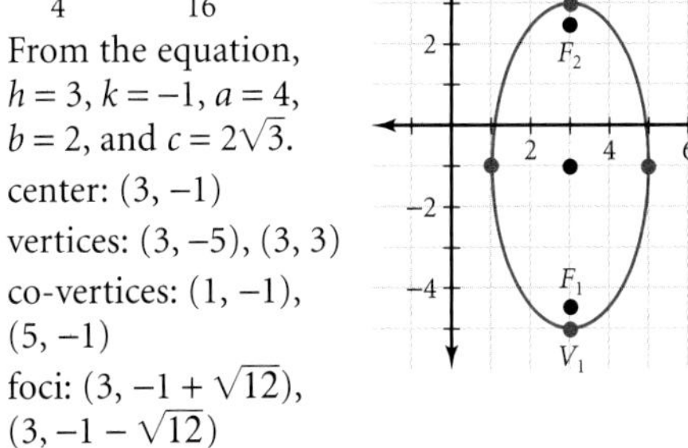

Exercises

Write the standard equation for each ellipse. Find the center, vertices, co-vertices, and foci. Then sketch the graph.

16. $\dfrac{(x+1)^2}{16} + \dfrac{(y-3)^2}{4} = 1$
17. $\dfrac{(x-4)^2}{9} + \dfrac{(y+1)^2}{25} = 1$
18. $25x^2 + 4y^2 = 100$
19. $x^2 + 4y^2 + 10x + 24y + 45 = 0$
20. $16x^2 + 4y^2 - 96x + 8y + 84 = 0$

18. $\dfrac{x^2}{4} + \dfrac{y^2}{25} = 1$
19. $\dfrac{(x+5)^2}{16} + \dfrac{(y+3)^2}{4} = 1$
20. $\dfrac{(x-3)^2}{4} + \dfrac{(y+1)^2}{16} = 1$

LESSON 9.5

Key Skills

Find the center, vertices, co-vertices, and foci of a hyperbola. Sketch the graph.

Given $4x^2 - y^2 + 24x + 4y + 28 = 0$, complete the square and write the standard equation.

$$\frac{(x+3)^2}{1} - \frac{(y-2)^2}{4} = 1$$

center: $(-3, 2)$
vertices: $(-4, 2)$, $(-2, 2)$
co-vertices: $(-3, 4)$, $(-3, 0)$
foci: $(-3 + \sqrt{5}, 2)$, $(-3 - \sqrt{5}, 2)$

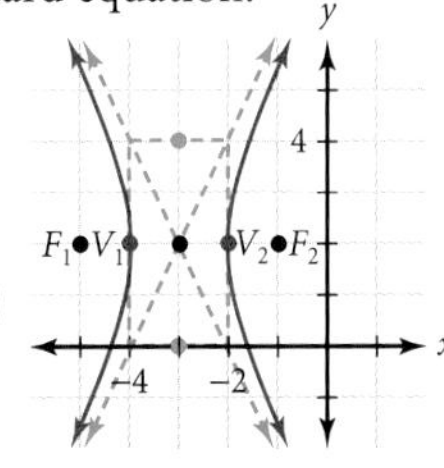

Exercises

Write the standard equation for each hyperbola. Find the center, vertices, co-vertices, and foci. Then sketch the graph.

21. $\dfrac{(x+5)^2}{36} - \dfrac{(y-1)^2}{64} = 1$
22. $\dfrac{(y+5)^2}{4} - \dfrac{(x-4)^2}{9} = 1$
23. $4y^2 - 25x^2 = 100$
24. $9x^2 - 16y^2 - 90x + 32y + 65 = 0$
25. $36y^2 - 4x^2 + 216y - 40x + 80 = 0$

15. $(x + 2)^2 + (y + 4)^2 = 25$; $C(-2, -4)$; $r = 5$

16. $\dfrac{(x+1)^2}{16} + \dfrac{(y-3)^2}{4} = 1$; center: $(-1, 3)$; vertices: $(-5, 3)$ and $(3, 3)$; co-vertices: $(-1, 1)$ and $(-1, 5)$; foci: $(-1 - 2\sqrt{3}, 3)$ and $(-1 + 2\sqrt{3}, 3)$

17. $\dfrac{(x-4)^2}{9} + \dfrac{(y+1)^2}{25} = 1$; center: $(4, -1)$; vertices: $(4, -6)$ and $(4, 4)$; co-vertices: $(1, -1)$ and $(7, -1)$; foci: $(4, -5)$ and $(4, 3)$

18. $\dfrac{x^2}{4} + \dfrac{y^2}{25} = 1$; center: $(0, 0)$; vertices: $(0, -5)$ and $(0, 5)$; co-vertices: $(-2, 0)$ and $(2, 0)$; foci: $(0, -\sqrt{21})$ and $(0, \sqrt{21})$

19. $\dfrac{(x+5)^2}{16} + \dfrac{(y+3)^2}{4} = 1$; center: $(-5, -3)$; vertices: $(-9, -3)$ and $(-1, -3)$; co-vertices: $(-5, -5)$ and $(-5, -1)$; foci: $(-5 - 2\sqrt{3}, -3)$ and $(-5 + 2\sqrt{3}, -3)$

The graphs to Exercises 11–20 and answers to Exercises 21–25 can be found in Additional Answers beginning on page 1002.

Chapter Test, Form B

NAME ______________ CLASS ________ DATE ________

Chapter Assessment
Chapter 9, Form B, page 1

1. For $\overline{PQ}$ with endpoints $P(1, 7)$ and $Q(9, -8)$, find PQ and the midpoint, M, of the segment.
   $PQ = 17$; $M(5, -0.5)$

2. One endpoint of $\overline{PQ}$ is $P(-6, 3)$ and the midpoint, M, is $(2, -12)$. Find the coordinates of Q.
   $Q(10, -27)$

3. Graph the circle defined by the equation $(x + 2)^2 + y^2 = 16$. Find the radius and coordinates of the center.
   radius: 4; center: $(-2, 0)$

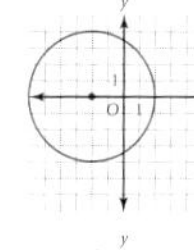

4. Graph $x - 3 = -\frac{1}{8}(y + 2)^2$. Write the coordinates of the vertex and the focus and the equation of the directrix.
   vertex: $(3, -2)$; focus: $(1, -2)$; directrix: $x = 5$

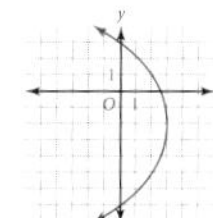

5. Graph $\frac{(x+4)^2}{4} + \frac{(y+1)^2}{25} = 1$. Write the coordinates of the center, vertices, and co-vertices.
   center: $(-4, -1)$; vertices: $(-4, 4)$ and $(-4, -6)$;
   co-vertices: $(-6, -1)$ and $(-2, -1)$

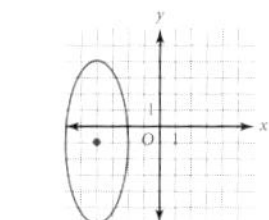

Find all solutions to each system of equations algebraically.

6. $\begin{cases} x^2 + y^2 = 25 \\ 3x^2 + 2y^2 = 59 \end{cases}$
   $(3, 4)$, $(3, -4)$, $(-3, 4)$, $(-3, -4)$

7. $\begin{cases} x^2 + y^2 = 10 \\ 3y = x^2 + 8 \end{cases}$
   $(1, 3)$, $(-1, 3)$

NAME ______________ CLASS ________ DATE ________

Chapter Assessment
Chapter 9, Form B, page 2

8. Graph $\frac{(y+1)^2}{16} - \frac{(x-2)^2}{9} = 1$. Write the coordinates of the center and the vertices.
   center: $(2, -1)$; vertices: $(2, 3)$ and $(2, -5)$

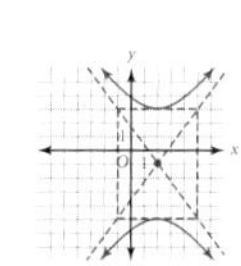

Write the standard equation of the conic section with the given characteristics.

9. circle with diameter endpoints at $(-8, 5)$ and $(4, 21)$
   $(x + 2)^2 + (y - 13)^2 = 100$

10. parabola with focus at $(6, 8)$ and directrix $y = -2$
   $y - 3 = \frac{1}{20}(x - 6)^2$

11. ellipse with center at $(5, -7)$, a major axis of 16, and foci at $(-1, -7)$ and $(11, -7)$
   $\frac{(x-5)^2}{64} + \frac{(y+7)^2}{28} = 1$

12. ellipse with vertices at $(2, 11)$ and $(2, -3)$ and co-vertices at $(7, 4)$ and $(-3, 4)$
   $\frac{(x-2)^2}{25} + \frac{(y-4)^2}{49} = 1$

13. hyperbola with foci at $(2, 0)$ and $(14, 0)$ and vertices at $(12, 0)$ and $(4, 0)$
   $\frac{(x-8)^2}{16} - \frac{y^2}{20} = 1$

14. hyperbola with vertices at $(3, -1)$ and $(3, -9)$ and co-vertices at $(-6, -5)$ and $(12, -5)$
   $\frac{(y+5)^2}{16} - \frac{(x-3)^2}{81} = 1$

15. Find the equations of the asymptotes of the hyperbola defined by the equation $\frac{x^2}{49} - \frac{y^2}{100} = 1$.
   $y = \pm\frac{10}{7}x$

Write the equation of each conic section in standard form.

16. $x^2 + y^2 - 10x + 6y + 18 = 0$
   $(x - 5)^2 + (y + 3)^2 = 16$

17. $x^2 + 9y^2 + 8x + 36y + 43 = 0$
   $\frac{(x+4)^2}{9} + \frac{(y+2)^2}{1} = 1$

18. Find the point(s) of intersection between the circle $x^2 + y^2 = 25$ and the line $x - y = -1$.
   $(3, 4)$ and $(-4, -3)$

19. If the circle $(x - 1)^2 + (y + 3)^2 = 16$ is translated 4 units to the left and 2 units upward, write the equation of the resulting circle.
   $(x + 3)^2 + (y + 1)^2 = 16$

26. 4 solutions: $(\pm3, \pm2)$

27. 4 solutions: $(\pm2, \pm\sqrt{2})$, or about $(\pm2, \pm1.41)$

28. none

29. $\left(-\frac{16}{5}, \frac{9}{5}\right)$

30. circle; $(x-1)^2 + (y+4)^2 = 25$

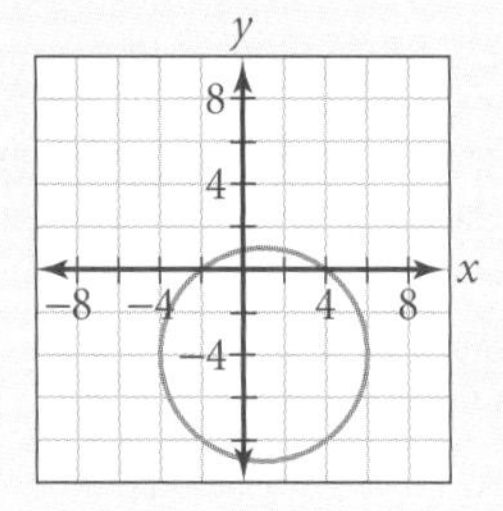

31. hyperbola; $\frac{(x-3)^2}{9} - \frac{(y+1)^2}{4} = 1$

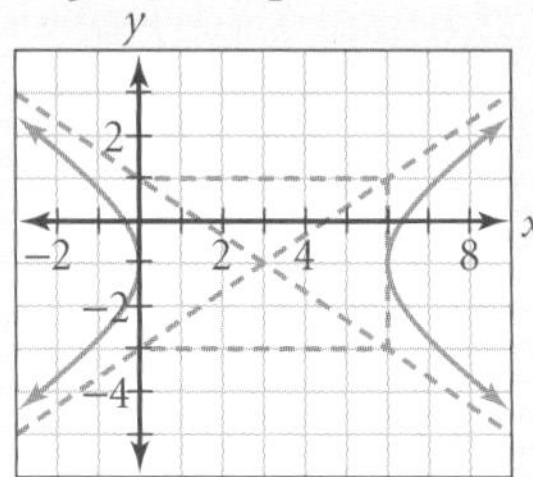

32. parabola; $x - 6 = -\frac{1}{4}(y-4)^2$

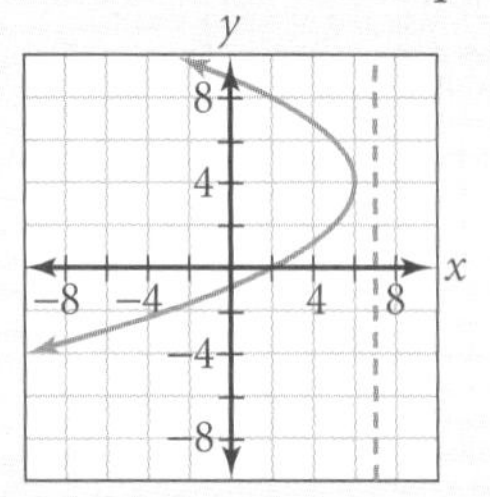

33. $\frac{2420}{3}$ watts, or about 806.7 watts

34. Answers may vary. Sample answer: Assuming that the major axis is horizontal and that x and y are measured in millions of miles, $\frac{x^2}{8649} + \frac{y^2}{8646.44} = 1$.

LESSON 9.6

Key Skills

Solve a system of nonlinear equations.

Use elimination or substitution to solve a nonlinear system algebraically. Independent systems of two conic sections can have 0, 1, 2, 3, or 4 solutions.

Identify a conic section from its equation.

$Ax^2 + Cy^2 + Dx + Ey + F = 0$ represents the following:

- a parabola if $AC = 0$
- a circle if $A = C$ ($A \neq 0$ and $C \neq 0$)
- an ellipse if $AC > 0$
- a hyperbola if $AC < 0$

Exercises

Solve each system. If there are no real-number solutions, write *none*.

26. $\begin{cases} 2x^2 + y^2 = 22 \\ x^2 + 3y^2 = 21 \end{cases}$

27. $\begin{cases} 4x^2 + 2y^2 = 20 \\ 3x^2 - 4y^2 = 4 \end{cases}$

28. $\begin{cases} x + y = 5 \\ x^2 + y^2 = 1 \end{cases}$

29. $\begin{cases} y - x = 5 \\ 9x^2 + 16y^2 = 144 \end{cases}$

Classify the conic section defined by each equation. Write the standard equation of the conic section, and sketch the graph.

30. $x^2 + y^2 - 2x + 8y - 8 = 0$

31. $4x^2 - 9y^2 - 24x - 18y - 9 = 0$

32. $y^2 - 8y + 4x - 8 = 0$

Applications

33. PHYSICS In a 220-volt electric circuit, the available power in watts, W, is given by the formula $W = 220I - 15I^2$, where I is the amount of current in amperes. Find the maximum amount of power available in the circuit.

34. ASTRONOMY Earth orbits the Sun in an elliptical path with the Sun at one focus of the ellipse. The closest that Earth gets to the Sun is 91.4 million miles and the farthest it gets is 94.6 million miles. Write an equation for Earth's orbit around the Sun with the center of the ellipse at the origin.

35. BIOLOGY To locate a whale, two microphones are placed 6000 feet apart in the ocean. One microphone picks up a whale's sound 0.5 second after the other microphone picks up the same sound. The speed of sound in water is about 5000 feet per second.

a. Find the equation of the hyperbola that describes the possible locations of the whale.

b. What is the shortest distance that the whale could be to either microphone?

36. GEOLOGY An earthquake transmits its energy in seismic waves that radiate from its underground focus in all directions. A seismograph station determines that the earthquake's epicenter (the point on the Earth's surface directly above the focus) is 100 miles from the station.

a. Write the standard equation for the possible locations of the epicenter, using (0, 0) as the location of the seismograph station.

b. A second station is 120 miles east and 160 miles south of the first station and 100 miles from the earthquake's epicenter. Write a second standard equation for the possible locations of the epicenter.

c. Find the coordinates of the epicenter by solving the system formed by the equations you wrote in parts **a** and **b**.

Tail of a humpback whale near Alaska

35a. Answers may vary. Sample answer: Assuming that distances are measured in feet and that the microphones are located on the x-axis, $\frac{x^2}{1{,}562{,}500} - \frac{y^2}{7{,}437{,}500} = 1$.

b. 1750 ft

36a. $x^2 + y^2 = 10{,}000$

b. $(x - 120)^2 + (y + 160)^2 = 10{,}000$

c. The epicenter is located at (60, −80).

Chapter Test

Find the distance *PQ* and the coordinates of *M*, the midpoint of *PQ*.

1. $P(2, 5)$ and $Q(4, 9)$ **$2\sqrt{5}$; (3,7)**
2. $P(-3, 7)$ and $Q(2, -4)$ **$\sqrt{146}$; $\left(-\frac{1}{2}, \frac{3}{2}\right)$**
3. $P(8, -12)$ and $Q(-13, -6)$ **$\sqrt{477}$; $\left(-\frac{5}{2}, -9\right)$**
4. **CARTOGRAPHY** Three cities found on a map are located at the coordinates of A(2°N, 5°W), B(8°N, 21°W), and C(17°N, 45°W). Are these three cities collinear? **yes**

Write the standard equation for the parabola with the given characteristics.

5. vertex: (0, 0), focus: (0, −2) **$y = -\frac{1}{8}x^2$**
6. vertex: (0, 0), directrix: $y = 3$ **$y = -\frac{1}{12}x^2$**
7. focus: (7, 0), directrix: $x = -7$ **$x = \frac{1}{28}y^2$**
8. **ASTRONOMY** Inside a telescope the focus of a parabolic mirror 10 centimeters across in a telescope lies 2 centimeters from the vertex. Find the equation of the mirror if the vertex is at the origin. **$y = \frac{1}{8}x^2$**

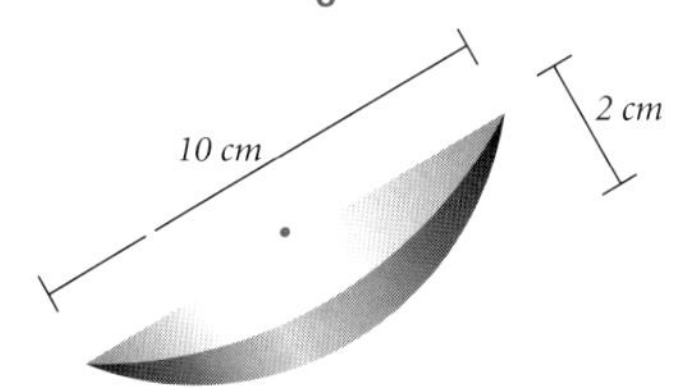

Write the standard equation for each circle. Find the radius and give the coordinates for the center.

9. $x^2 + y^2 - 144 = 0$
10. $5x^2 + 5y^2 - 125 = 0$
11. $x^2 + y^2 - 14x + 8y + 1 = 0$
12. **CIVIL DEFENSE** A civil defense system broadcasts an emergency signal within a circular area with a radius of 250 miles. If you live a distance of 150 miles east and 195 miles north of the broadcasting tower, will you be able to receive the signal? **yes**

Find the vertices and co-vertices of each ellipse. Sketch the graph for Exercise 14.

13. $\frac{x^2}{36} + \frac{y^2}{100} = 1$ **vertices: (0, 10), (0, −10); co-vertices: (6, 0), (−6, 0)**
14. $\frac{(x-3)^2}{25} + \frac{(y+4)^2}{4} = 1$

Write the standard equation for the ellipse; identify the coordinates of the center, vertices, co-vertices, and foci.

15. $4x^2 + y^2 + 16x - 6y = -9$

For each hyperbola, find the equations of the asymptotes and the coordinates of the vertices.

16. $\frac{x^2}{25} - y^2 = 1$
17. $\frac{y^2}{9} - \frac{x^2}{49} = 1$
18. Find the coordinates of the center, the vertices, and the co-vertices for the graph of $\frac{(x-3)^2}{4} - \frac{(y+5)^2}{9} = 1$.
19. Sketch the graph of $\frac{(y+1)^2}{1} - \frac{(x-2)^2}{4} = 1$

 Label the center, vertices, co-vertices, foci, and asymptotes.

Use any method to solve each system. If there are no real number solutions, write *none*.

20. $\begin{cases} y = \frac{4}{9}x^2 \\ x^2 + y^2 = 25 \end{cases}$ **(3, 4), (−3, 4)**
21. $\begin{cases} 4x^2 + 4y^2 = 4 \\ 4x^2 + y^2 = 1 \end{cases}$ **(0, 1), (0, −1)**
22. $\begin{cases} 64x = 10y^2 \\ x^2 - y^2 = 36 \end{cases}$ **(10, 8), (10, −8)**
23. $\begin{cases} 9x^2 + 4y^2 = 36 \\ 4x^2 - 9y^2 = 36 \end{cases}$ **no solution**

Classify the conic section defined by each equation.

24. $4x^2 + 4y^2 + 24x - 16y + 12 = 0$ **circle**
25. $3x + 8y^2 - 24 = 0$ **parabola**
26. $16x^2 - 9y^2 + 64x + 18y - 89 = 0$ **hyperbola**
27. $9x^2 + 16y^2 - 36x - 64y - 44 = 0$ **ellipse**

Chapter Test

9. $x^2 + y^2 = 144$; $C(0, 0)$; $r = 12$
10. $x^2 + y^2 = 25$; $C(0, 0)$; $r = 5$
11. $(x - 7)^2 + (y + 4)^2 = 64$; $C(7, -4)$; $r = 8$
14. vertices: (2, −4), (8, −4); co-vertices: (3, −6), (3, −2)

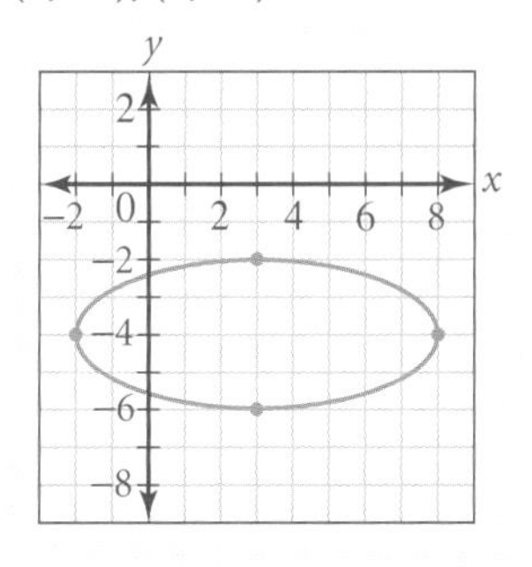

15. $\frac{(x+2)^2}{4} + \frac{(y-3)^2}{16} = 1$

 center: (−2, 3);
 vertices: (−2, −1), (−2, 7);
 co-vertices: (0, 3), (−4, 3);
 foci: $(-2, 3 - 2\sqrt{3})$, $(-2, 3 + 2\sqrt{3})$
16. asymptotes: $y = \frac{1}{5}x$ and $y = -\frac{1}{5}x$;

 vertices: (−5, 0), (5, 0)
17. asymptotes: $y = \frac{3}{7}x$ and $y = -\frac{3}{7}x$;

 vertices: (−0, −3), (0, 3)
18. center: (3, −5);
 vertices: (1, −5), (5, −5)
 co-vertices: (3, −2), (3, −8);
19.

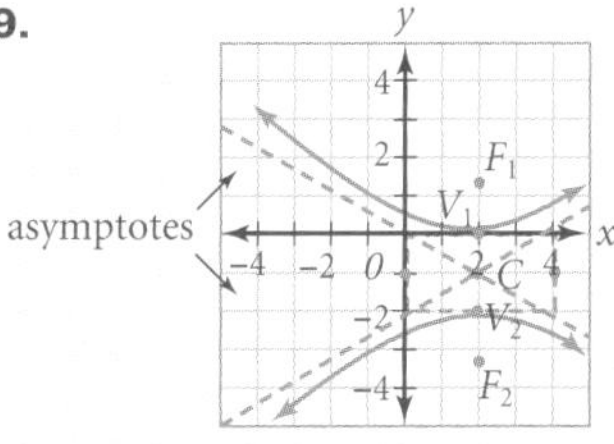

 center: (2, −1);
 vertices: (2, 0), (2, −2);
 co-vertices: (0, −1);
 foci: $(2, -1 + \sqrt{5})$, $(2, -1 - \sqrt{5})$;
 asymptotes: $y = \frac{1}{2}x$ and $y = -\frac{1}{2}x - 2$

College Entrance Exam Practice

Multiple-Choice and Quantitative-Comparison Samples

The first half of the Cumulative Assessment contains two types of items found on standardized tests—multiple-choice questions and quantitative-comparison questions. Quantitative-comparison items emphasize the concepts of equality, inequality, and estimation.

Free-Response Grid Samples

The second half of the Cumulative Assessment is a free-response section. This part of the Cumulative Assessment requires student-produced response items like those commonly found on college entrance exams. These questions require the use of machine-scored answer grids. You may wish to have students practice answering these items in preparation for standardized tests.

College Entrance Exam Practice

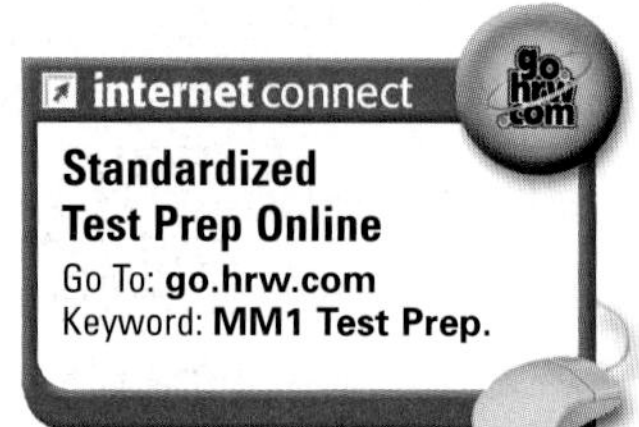

QUANTITATIVE COMPARISON For Items 1–6, write:
A if the quantity in Column A is greater than the quantity in Column B;
B if the quantity in Column B is greater than the quantity in Column A;
C if the quantities are equal; or
D if the relationship cannot be determined from the given information.

	Column A	Column B	Answers
1. A	The value of x: $\frac{1}{5}=\frac{7}{x}$	$\frac{-4x}{3}=x-1$	Ⓐ Ⓑ Ⓒ Ⓓ [Lesson 1.4]
2. B	$(-2)^3$	$\left(\frac{1}{2}\right)^{-3}$	Ⓐ Ⓑ Ⓒ Ⓓ [Lesson 2.2]
3. A	log 200	2	Ⓐ Ⓑ Ⓒ Ⓓ [Lesson 6.5]
4. A	$x > 0$: $\frac{x^2-25}{x+5}$	$x-1$	Ⓐ Ⓑ Ⓒ Ⓓ [Lesson 8.2]
5. B	The distance between the points: $(-2, 5)$ and $(-3, 2)$	$(6, -2)$ and $(-1, -4)$	Ⓐ Ⓑ Ⓒ Ⓓ [Lesson 9.1]
6. C	The degree of the polynomial function: $f(x) = 5x^2 - x^3 + 6x - 1$	$f(x) = 8x(x + 1)^2$	Ⓐ Ⓑ Ⓒ Ⓓ [Lesson 7.1]

7. How many solutions does an independent system of linear equations have? ***(LESSON 3.1)***
b
a. 0 **b.** 1
c. at least 1 **d.** infinitely many

8. Which statement is true for $f(x) = 3x - 6$ and $g(x) = 12 - 6x$? ***(LESSONS 2.4 AND 2.7)***
d
a. $f \circ g = g \circ f$ **b.** $2(f \circ g) = g \circ f$
c. $-2(f \circ g) = g \circ f$ **d.** none of these

9. Evaluate $f(x) = 2x^2 - 9x - 12$ for $x = -2$. ***(LESSON 2.3)***
b
a. 8 **b.** 14 **c.** 2 **d.** −22

10. Which describes the slope of a vertical line? ***(LESSON 1.2)***
b
a. 0 **b.** undefined
c. negative **d.** positive

11. For matrices A and B, each with dimensions $m \times n$, which statement is always true? ***(LESSONS 4.1 AND 4.2)***
b
a. $A = B$ **b.** $A + B = B + A$
c. $AB = BA$ **d.** $A - B = B - A$

12. Solve $7 - \sqrt{4a - 3} = 4$. ***(LESSON 8.8)***
c **a.** 0 **b.** −3 **c.** 3 **d.** $\sqrt{3}$

13. Which equation represents $5^x = 625$ in
b logarithmic form? ***(LESSON 6.3)***
a. $\log_x 5 = 625$ **b.** $\log_5 625 = x$
c. $\log_x 625 = 5$ **d.** $\log_{625} x = 5$

14. Which binomial is a factor of
c $x^3 - 2x^2 - 6x - 3$? ***(LESSON 7.3)***
a. $x - 3$ **b.** $x + 3$
c. $x + 1$ **d.** $x - 1$

15. Simplify $\frac{3+i}{1-3i}$. ***(LESSON 5.6)***
b **a.** $\frac{9}{10} + i$ **b.** i
c. $\frac{3}{10} + \frac{7}{10}i$ **d.** $\frac{3}{5} + i$

16. Which equation defines a circle with its center
c at (0, −2) and a radius of 5? ***(LESSON 9.3)***
a. $x^2 + (y - 2)^2 = 25$
b. $x^2 + (y - 2)^2 = 5$
c. $x^2 + (y + 2)^2 = 25$
d. $x^2 + (y + 2)^2 = 5$

17. Write an equation in slope-intercept form for the line that contains (−3, 4) and is parallel to $y = -\frac{1}{5}x + 2$. ***(LESSON 1.3)*** $y = -\frac{1}{5}x + \frac{17}{5}$

18. Solve the literal equation $A = p + prt$ for p. ***(LESSON 1.6)*** $p = \frac{A}{1 + rt}$

19. Use the quadratic formula to solve $x^2 + 4x - 1 = 0$. ***(LESSON 5.5)*** $x = -2 \pm \sqrt{5}$

20. Solve $\begin{cases} 3x - y = -4 \\ 2x - 3y = 2 \end{cases}$ by graphing. (−2, −2) ***(LESSON 3.1)***

21. Graph $2y > -1$. ***(LESSON 3.3)***

22. Find the determinant of $\begin{bmatrix} 4 & -2 \\ 3 & -1 \end{bmatrix}$. 2 ***(LESSON 4.3)***

23. Factor $9x^2 - 18x + 8$, if possible. ***(LESSON 5.3)*** $(3x - 4)(3x - 2)$

24. Write the function for the graph of $f(x) = x^2$ stretched vertically by a factor of 5 and translated 2 units up. ***(LESSON 2.7)*** $y = 5x^2 + 2$

25. Factor $6x^3 - 15x^2 - 4x + 10$, if possible. ***(LESSON 7.3)*** $(2x - 5)(3x^2 - 2)$

26. Write $(x + 2)(x^2 - 3x - 5)$ as a polynomial expression in standard form. ***(LESSON 7.3)*** $x^3 - x^2 - 11x - 10$

27. Write the standard equation for a circle with its center at (−5, 4) and a radius of 6. ***(LESSON 9.3)*** $(x + 5)^2 + (y - 4)^2 = 36$

28. Find the domain of $h(x) = \frac{3x^2}{-5 + 2x}$. $x \neq \frac{5}{2}$ ***(LESSON 8.2)***

29. Simplify $\frac{3x - 10}{x^2 - 8x + 12} - \frac{2}{x - 6}$. ***(LESSON 8.4)***

30. Find the domain of $h(x) = \sqrt{5x - 7}$. $x \geq \frac{7}{5}$ ***(LESSON 8.6)***

FREE RESPONSE GRID The following questions may be answered by using a free-response grid such as that commonly used by standardized-test services.

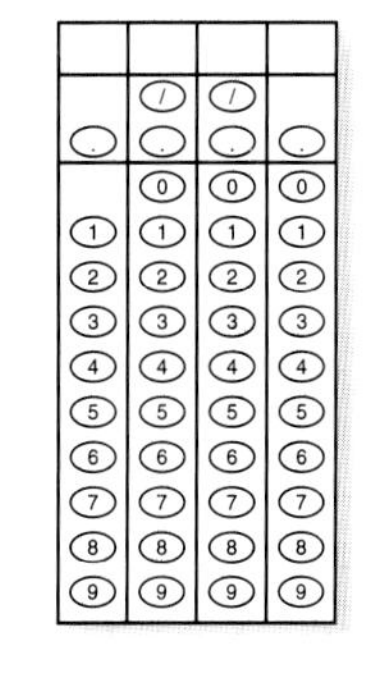

31. A truck travels 245 kilometers on 35 liters of gasoline. How many kilometers should the truck be able to travel on 100 liters? ***(LESSON 1.4)*** 700

32. Evaluate $2^{\log_2 9}$. ***(LESSON 6.4)*** 9

33. Solve $10^x = 222$ for x. Round your answer to the nearest hundredth. ***(LESSON 6.3)*** 2.35

34. Find the maximum value of the function $y = -2x^2 + 8x + 4$. ***(LESSON 5.1)*** 12

SPORTS A swimmer jumps from a platform that is 10 feet above the water. The vertical height of the swimmer at three different times is given in the table. ***(LESSON 5.7)***

Time (s)	Vertical height (ft)
0.0	10
0.5	8
0.75	4

35. Use a quadratic function to predict the time in seconds when the swimmer enters the water. 0.93

36. Find the maximum height of the swimmer. 10.25

21.

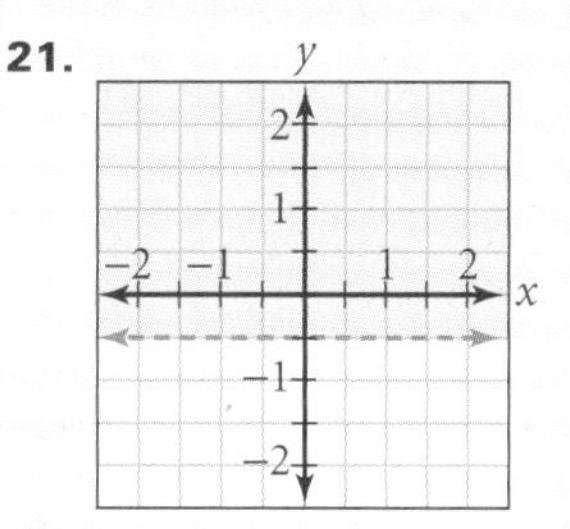

29. $\frac{1}{x - 2}$

Keystroke Guide for Chapter 9

Essential keystroke sequences (using the model TI-82 or TI-83 graphics calculator) are presented below for all Activities and Examples found in this chapter that require or recommend the use of a graphics calculator.

internet connect

For Keystrokes of other graphing calculator models, visit the HRW web site at **go.hrw.com** and enter the keyword **MB1 CALC**.

LESSON 9.1

EXAMPLE 1 Page 562

For part a, graph $y = \sqrt{25 - x^2}$ and $y = -\sqrt{25 - x^2}$ together.

Begin with viewing window [−10, 10] by [−10, 10].

Entering {−1, 1} implements the ± sign.

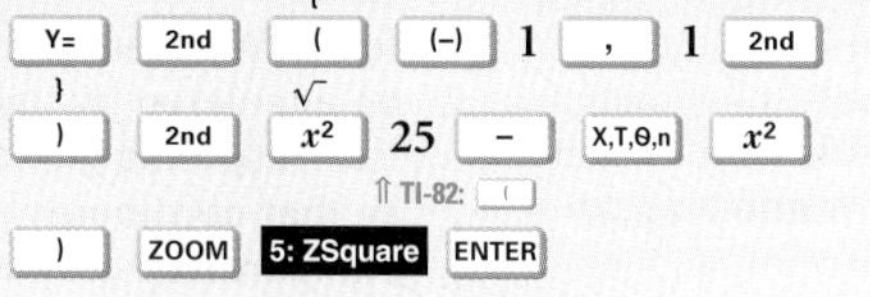

For part b, graph $y = \sqrt{x^2 - 4}$ and $y = -\sqrt{x^2 - 4}$ together.

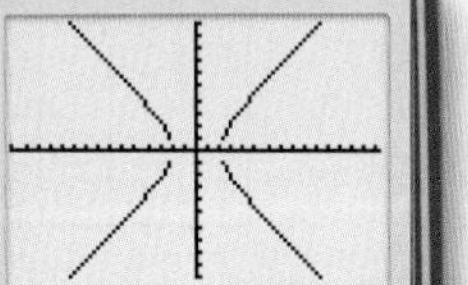

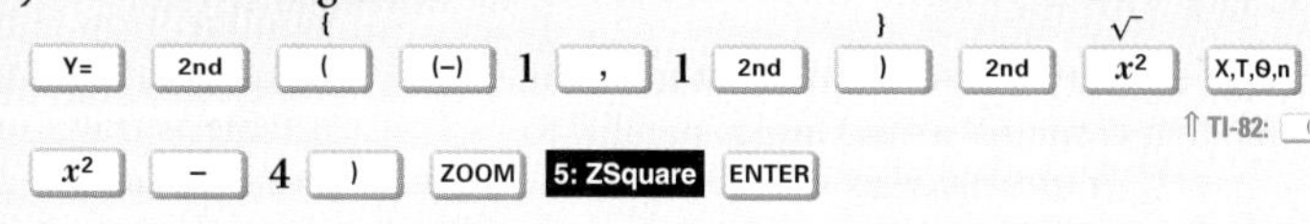

LESSON 9.2

Page 573

For Steps 1 and 2, graph $y = \frac{1}{4}x^2$ and $y = \frac{1}{4}(x - 2)^2 + 3$ on the same screen, and find the coordinates of the vertices.

Use viewing window [−10, 10] by [−10, 10].

Graph the functions:

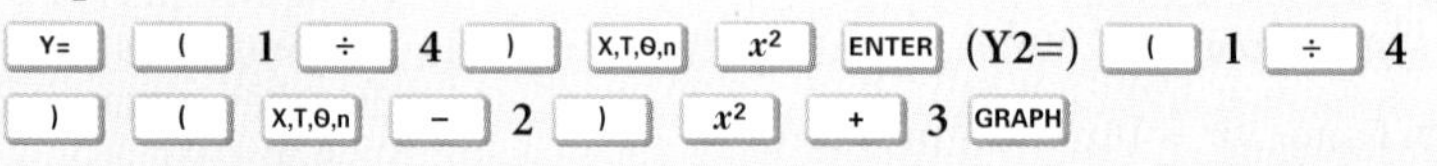

Find the vertex:

(RightBound?) ENTER **(Guess?)** ENTER

Repeat this keystroke sequence to find the other vertex.

TECHNOLOGY

Page 575

Graph $y = 4 + \sqrt{-8(x-1)}$ and $y = 4 - \sqrt{-8(x-1)}$ together.

Begin with viewing window [−10, 10] by [−7, 13].

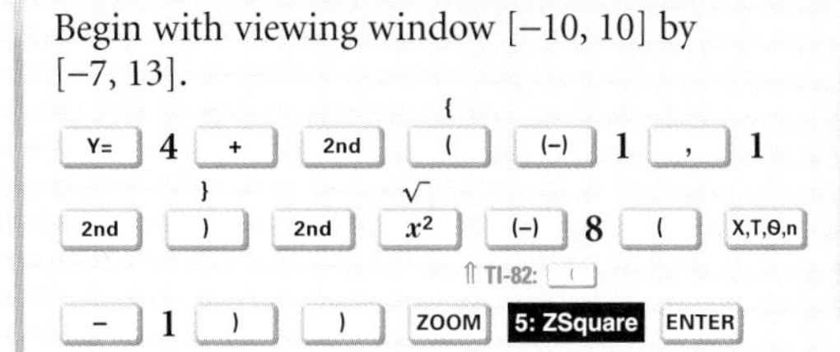

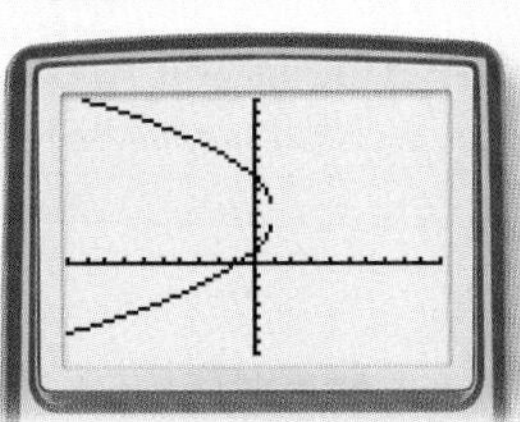

LESSON 9.3

TECHNOLOGY

Page 580

Graph $y = \sqrt{4 - x^2}$ and $y = -\sqrt{4 - x^2}$ together.

Begin with viewing window [−5, 5] by [−5, 5].

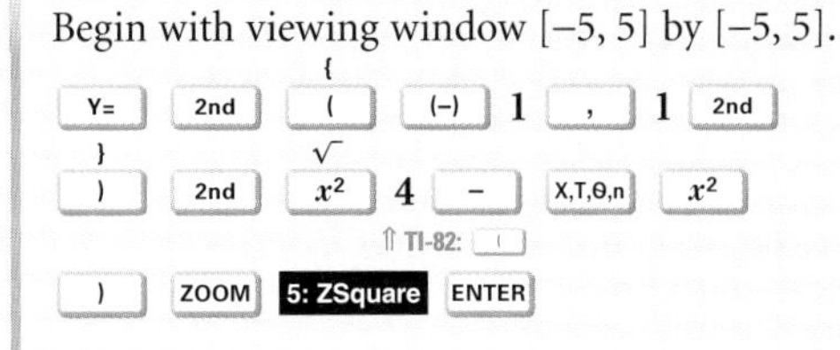

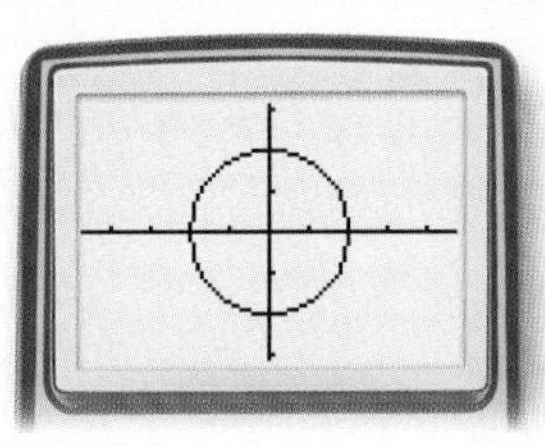

Activity

Page 580

Graph $x^2 + y^2 = 16$ and $(x + 3)^2 + (y - 2)^2 = 16$ on the same screen, and find the center of each.

Use friendly viewing window [−9.4, 9.4] by [−6.2, 6.2].

Graph the functions:

Solve for y, and use a keystroke sequence similar to that given in Example 1 of Lesson 9.1.

Find the centers:

Use the cursor keys ▲ ▼ ◄ ► to find the approximate coordinates of the center.

LESSON 9.4

TECHNOLOGY

Page 589

Graph $y = \sqrt{9\left(1 - \frac{x^2}{25}\right)}$ and $y = -\sqrt{9\left(1 - \frac{x^2}{25}\right)}$ together.

Begin with viewing window [−7, 7] by [−5, 5].

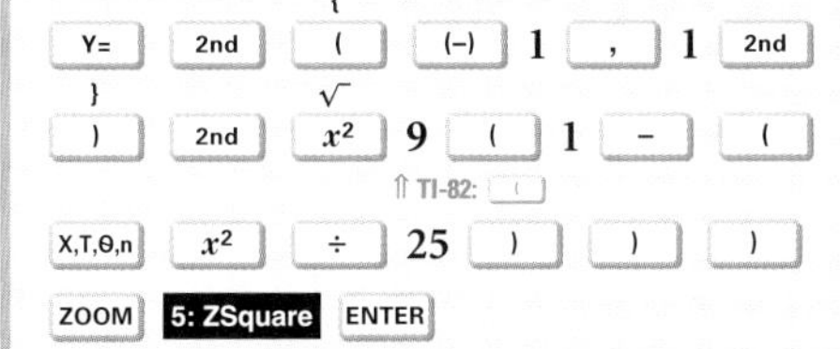

ZOOM 5: ZSquare ENTER

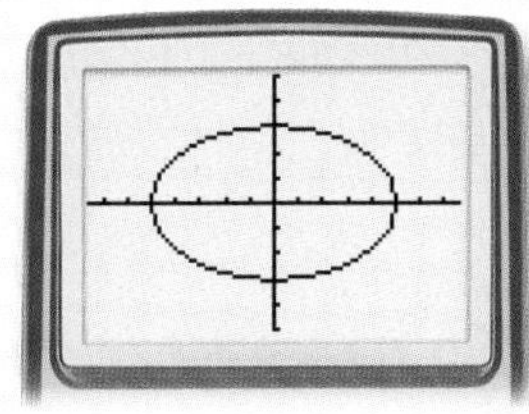

LESSON 9.5

EXAMPLE 1 Page 597

Graph $y = \sqrt{16\left(1 + \frac{x^2}{9}\right)}$ and $y = -\sqrt{16\left(1 + \frac{x^2}{9}\right)}$ together.

Begin with viewing window [−15, 15] by [−10, 10].

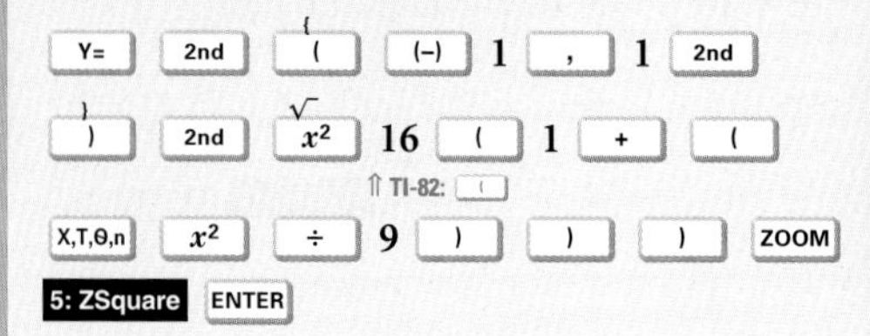

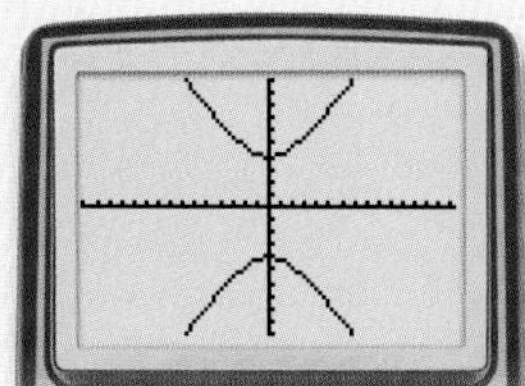

Activity Page 597

Graph $\frac{x^2}{2^2} - \frac{y^2}{3^2} = 1$, $y = \frac{3}{2}x$, and $y = -\frac{3}{2}x$ on the same screen.

Use viewing window [−10, 10] by [−10, 10].

Graph the functions:

Solve for y, and use a keystroke sequence similar to that given in Example 1 of this lesson.

LESSON 9.6

Activity Page 606

For part a of Step 1, graph $x^2 + y^2 = 25$ and $y = 2x$ on the same screen.

Begin with viewing window [−10, 10] by [−10, 10].

Solve for y, and use a keystroke sequence similar to that given in the Technology example of Lesson 9.3.

For part b of Step 1, graph $x^2 + y^2 = 25$ and $x^2 - y^2 = 9$ on the same screen.

Use viewing window [−10, 10] by [−10, 10].

Solve for y, and use keystroke sequences similar to those given in the Technology example of Lesson 9.3 and in Example 1 of Lesson 9.5.

For parts **c** and **d** of Step 1, use the same viewing window and similar keystroke sequences.

EXAMPLE 1 Page 607

Graph $y=\sqrt{3x-1}$, $y=-\sqrt{3x-1}$, $y=\sqrt{9-x^2}$, and $y=-\sqrt{9-x^2}$ on the same screen. Find any points of intersection.

If equations are not entered separately, it may be difficult to find the intersection points.

Begin with viewing window [−5, 5] by [−5, 5].

Graph the functions:

Y= 2nd x^2 (√) 3 X,T,Θ,n − 1) ENTER (Y2=) (−) 2nd x^2 (√) 3 X,T,Θ,n − 1) ENTER ZOOM 5: ZSquare ENTER (Y3=) 2nd x^2 (√) 9 − X,T,Θ,n x^2) ENTER (Y4=) (−) 2nd x^2 (√) 9 − X,T,Θ,n x^2)

⇑ TI-82: (

Find any points of intersection:

2nd TRACE 5: intersect (First curve?)
ENTER (Second curve?)
ENTER (Guess?) ENTER

(2, ≈2.24)
(2, ≈−2.24)

EXAMPLE 3 Page 609

Graph $y=\sqrt{\frac{100-4x^2}{25}}$, $y=-\sqrt{\frac{100-4x^2}{25}}$, $y=\sqrt{9-x^2}$, and $y=-\sqrt{9-x^2}$ on the same screen. Find any points of intersection.

Use viewing window [−6, 6] by [−4, 4].

Graph the functions:

Y= 2nd x^2 (√) (100 − 4 X,T,Θ,n x^2) ÷ 25) ENTER (Y2=) (−) 2nd x^2 (√) (100 − 4 X,T,Θ,n x^2) ÷ 25) ENTER (Y3=) 2nd x^2 (√) 9 − X,T,Θ,n x^2) ENTER (Y4=) (−) 2nd x^2 (√) 9 − X,T,Θ,n x^2)

⇑ TI-82: (

ZOOM 5:ZSquare ENTER

Find any points of intersection:

2nd TRACE 5:intersect (First curve?)
ENTER (Second curve?)
ENTER (Guess?) ENTER

(≈−2.44, ≈1.75)
(≈2.44, ≈1.75)
(≈−2.44, ≈−1.75)
(≈2.44, ≈−1.75)

10 Counting Principles and Probability

Lesson Presentation CD-ROM
PowerPoint® presentations for each lesson 10.1–10.7

CHAPTER PLANNING GUIDE

Lesson	10.1	10.2	10.3	10.4	10.5	10.6	10.7	Project and Review
Pupil's Edition Pages	628–635	636–642	643–649	652–658	659–663	664–670	671–677	650–651, 678–685
Practice and Assessment								
Extra Practice (Pupil's Edition)	972	972	973	973	974	974	975	
Practice Workbook	61	62	63	64	65	66	67	
Practice Masters Levels A, B, and C	181–183	184–186	187–189	190–192	193–195	196–198	199–200	
Standardized Test Practice Masters	70	71	72	73	74	75	76	77
Assessment Resources	124	125	126	127	129	130	131	128, 132–137
Visual Resources								
Lesson Presentation Transparencies Vol. 2	57–60	61–64	65–68	69–72	73–76	77–80	81–84	
Teaching Transparencies	37	38		39	40		41	
Answer Key Transparencies	360–363	364–366	367–369	370–372	373–374	375–377	378–379	380–381
Quiz Transparencies	10.1	10.2	10.3	10.4	10.5	10.6	10.7	
Teacher's Tools								
Reteaching Masters	121–122	123–124	125–126	127–128	129–130	131–132	133–134	
Make-Up Lesson Planner for Absent Students	61	62	63	64	65	66	67	
Student Study Guide	61	62	63	64	65	66	67	
Spanish Resources	61	62	63	64	65	66	67	
Block Scheduling Handbook								20–21
Activities and Extensions								
Lesson Activities	61	62	63	64	65	66	67	
Enrichment Masters	61	62	63	64	65	66	67	
Cooperative-Learning Activities	61	62	63	64	65	66	67	
Problem Solving/ Critical Thinking	61	62	63	64	65	66	67	
Student Technology Guide	61	62	63	64	65	66	67	
Long Term Projects								37–40
Writing Activities for Your Portfolio								28–30
Tech Prep Masters								45–48
Building Success in Mathematics								25–27

LESSON PACING GUIDE

Lesson	10.1	10.2	10.3	10.4	10.5	10.6	10.7	Project and Review
Traditional	2 days	1 day	1 day	2 days	1 day	1 day	2 days	2 days
Block	1 day	$\frac{1}{2}$ day	$\frac{1}{2}$ day	1 day	$\frac{1}{2}$ day	$\frac{1}{2}$ day	1 day	1 day
Two-Year	4 days	2 days	2 days	4 days	2 days	2 days	4 days	4 days

CONNECTIONS AND APPLICATIONS

Lesson	10.1	10.2	10.3	10.4	10.5	10.6	10.7	Review
Algebra	628–635	636–642	643–649	652–658	659–663	664–670	671–677	678–685
Geometry	630, 634	641		657		669	674, 675, 677	
Mimimum/Maximum		642						
Business and Economics	634	639, 641, 642	649					
Life Skills	632, 634		644, 645	656, 658		670	672, 673, 675	
Science and Technology	630, 632, 635	642	648		663	664, 667, 668, 670		682
Social Studies	634		645, 646	654, 657		669, 670		
Sports and Leisure		637, 641, 642	649	652, 655	660, 661, 663	665	671	685
Cultural Connection: Asia			648					

BLOCK SCHEDULING GUIDE

Day	Lesson	Teacher Directed: Lesson Examples, Teaching Transparencies	Student Guided Activity, Try This	Cooperative-Learning Activity, Lesson Activity, Student Technology Guide	Practice: Practice & Apply, Extra Practice, Practice Workbook	Assessment: Quiz, Mid-Chapter Assessment	Problem Solving, Reteaching
1	10.1	10 min	15 min	15 min	65 min	15 min	15 min
2	10.2	10 min	15 min	15 min	65 min	15 min	15 min
	10.3	10 min	15 min	15 min	65 min	15 min	15 min
3	10.4	10 min	15 min	15 min	65 min	15 min	15 min
4	10.5	10 min	15 min	15 min	65 min	15 min	15 min
	10.6	10 min	15 min	15 min	65 min	15 min	15 min
5	10.7	10 min	15 min	15 min	65 min	15 min	15 min
6	Assess.	50 min **PE:** Chapter Review	90 min **PE:** Chapter Project, Writing Activities	90 min Tech Prep Masters	65 min **PE:** Chapter Assessment, Test Generator	30 min Chap. Assess. (A or B), Alt. Assess. (A or B), Test Generator	

PE: Pupil's Edition

Alternative Assessment

The following suggest alternative assessments for students who may benefit from a different type of assessment than the regular chapter quizzes and the mid-chapter/end-of-chapter test. Visit the HRW web site to get additional Alternative Assessment material.

internet connect

Alternative Assessment
Go To: **go.hrw.com**
Keyword: **MB1 Alt Assess**

Performance Assessment

1. A committee consisting of 7 members is to be convened. Three members will be juniors and the rest will be seniors. There are 30 juniors and 40 seniors under consideration for selection. Decide whether permutations or combinations should be used to solve the problem. Find the total number of possible committees.
2. The following are probabilities related to events A and B, where one event is followed by the other.
   Situation 1: $P(A) = 0.50$ $P(A|A) = 0.90$
   Situation 2: $P(A) = 0.50$ $P(B|A) = 0.10$
   Situation 3: $P(B) = 0.50$ $P(A|B) = 0.80$
   Situation 4: $P(B) = 0.50$ $P(B|B) = 0.20$
   Could the events be independent? (For example, in situation 2, is there are possible value for $P(B)$ such that $P(A) \times P(B) = P(B|A)$?

Portfolio Project

1. **a.** In how many ways can you place 5 different-colored beach balls into 3 bins labeled A, B, and C? (Hint: Draw diagrams.)
   b. In how many ways can you place n beach balls into these 3 bins?
2. The diagram below shows the distribution of the elements in sets X, Y, and Z. Using these sets, find $P(X)$, $P(Y)$, $P(Z)$, $P(X \text{ and } Y)$, $P(X \text{ and } (Y \text{ or } Z))$, $P(\text{not } X)$, and $P(X|Y)$.

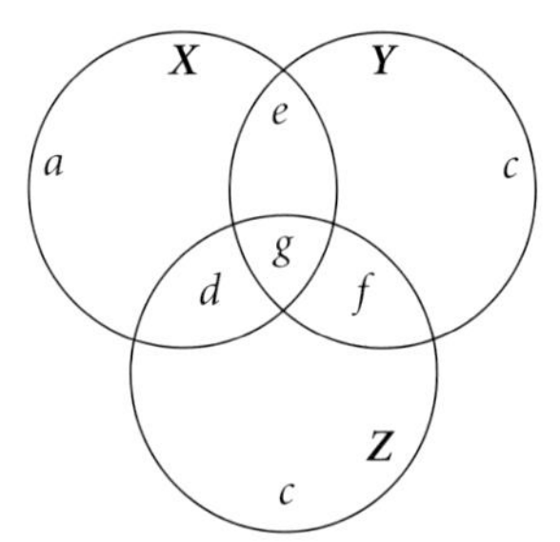

internet connect

The table below identifies the pages in this chapter that contain internet and technology information.

Content Links

Activities Online	**pages 640, 668, 675**
Portfolio Extensions	**pages 635, 649**
Homework Help Online	**pages 634, 641, 648, 656, 662, 668, 676**
Graphic Calculator Support	**page 686**

Resource Links

Parents can go online and find concepts that students are learning—lesson by lesson—and questions that pertain to each lesson, which facilitate parent-student discussion.

Go To: **go.hrw.com**
Keyword: **MB1 Parent Guide**

Technical Support

The following may be used to obtain technical support for any HRW software product.

Online Help: **www.hrwtechsupport.com**
e-mail: **tschrw@hrwtechsupport.com**
HRW Technical Support Center: **(800)323-9239**
7 AM to 10 PM Monday through Friday Central Time

Visit the HRW math web site at: **www.hrw.com/math**

Technology

Lesson Suggestions and Calculator Examples

(Keystrokes are based on a TI-83 calculator.)

Lesson 10.1 Introduction to Probability

In this lesson, students will learn the fundamental counting principle. Consider the following problem:

A password consists of three digits, the first of which is nonzero, followed by four letters of the alphabet, the first of which is not the letter *A*. How many passwords are possible?

Application of the fundamental counting principle gives the product shown on the display at right. Point out to students that the calculator can be helpful when a product has many factors and powers of large numbers.

Lesson 10.2 Permutations

To access the factorial and permutation functions, students press MATH, select PRB, and then select either 4:! or 2:nPr. The display below shows three different applications of these functions.

- permutations of 5 objects taken 5 at a time
- permutations of 5 objects taken 3 at a time
- permutations of 5 objects, with 2 alike and 3 alike

After students gain practice in using the factorial and permutation functions on the calculator, you might have students try the Activity on page 638.

Lesson 10.3 Combinations

Students can access the combination function by pressing MATH and selecting PRB 3:nCr.

After some practice, students can work Example 3. The calculator display at right shows the work required to solve the example.

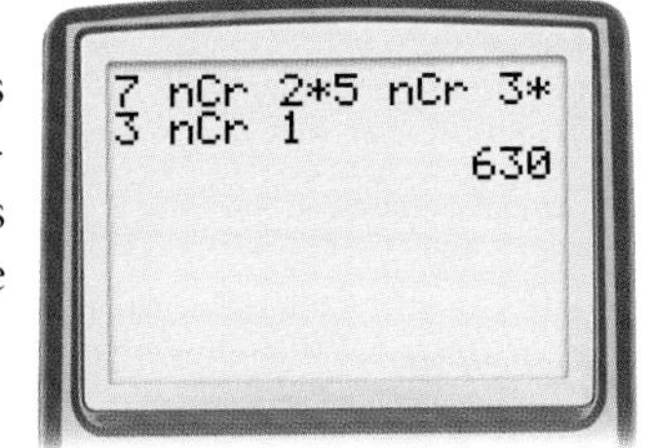

Lesson 10.4 Using Addition With Probability

In this lesson, students will enter a probability calculation, find the probability, and then edit the expression to find a different but related probability. After you work through Example 2, ask students how they would modify the calculator expression shown at right to find the probability *P*(man *and* favors).

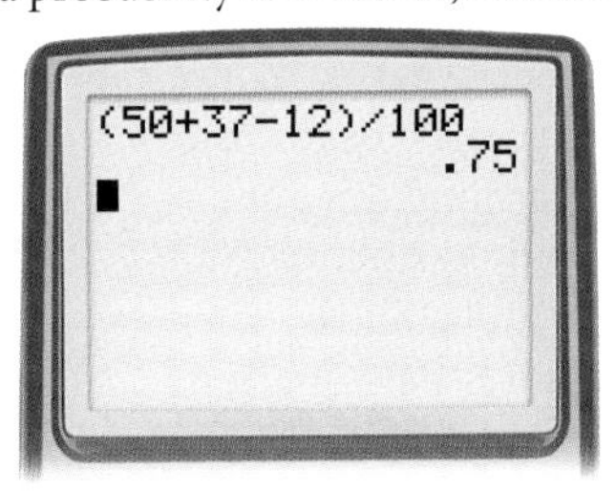

Lesson 10.5 Independent Events

A jar contains 5 red and 7 green marbles.

1. A marble is chosen at random and then replaced. A second marble is drawn. What is the probability that the 2 marbles chosen are red?
2. A marble is chosen at random but is not replaced. A second marble is drawn. What is the probability that the 2 marbles drawn are red?

The solutions shown at right indicate that the probability of choosing two red marbles is less when the first marble drawn is not replaced. Discuss why this happens.

Lesson 10.6 Dependent Events and Conditional Probability

Have students note the following: If *A* and *B* are independent events, then $P(A \text{ and } B) = P(A) \cdot P(B)$ and $P(A|B) = P(A)$. If *A* and *B* are dependent events, then $P(A \text{ and } B) \neq P(A) \cdot P(B)$. If *A* and *B* are mutually exclusive events, then $P(A \text{ and } B) = 0$, and *A* and *B* must be dependent events unless $P(A) = 0$ or $P(B) = 0$.

Lesson 10.7 Experimental Probability and Simulation Methods

Students can access a random-number generator by pressing MATH, selecting PRB 5:randInt(, and then typing the first and last integers of the desired range separated by a comma. This feature produces a random integer within the specified range each time you press ENTER. The display at right shows a range of 1 to 200, inclusive.

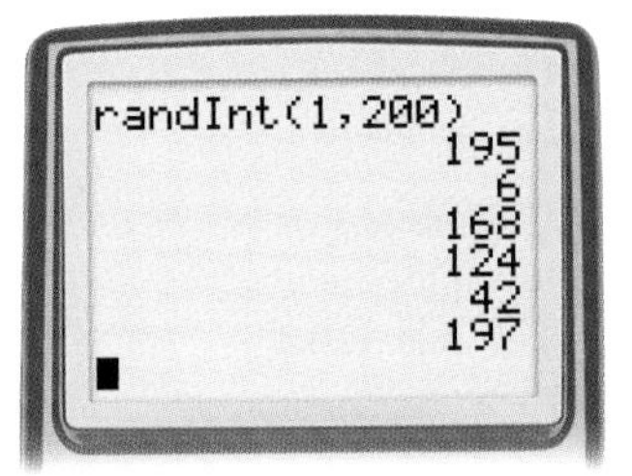

internet connect

For keystrokes of other graphing calculators models, visit the HRW web site at **go.hrw.com** and enter the keyword **MB1 CALC**.

About Chapter 10

Background Information

The probability of an outcome occurring is the ratio of the number of successful outcomes to the total number of possible outcomes. To solve problems involving probability, students need to develop efficient ways to count outcomes. Students begin this chapter by learning the fundamental counting principle and applying it to probabilities. Then students explore permutations and combinations of outcomes. Students then learn to calculate probabilities involving mutually exclusive events, dependent and independent events, and conditional probabilities. Finally students use random-number generators to perform simulations.

CHAPTER RESOURCES

- Block-Scheduling Handbook
- Writing Activities for Your Portfolio
- Tech Prep Masters
- Long-Term Project
- Assessment Resources:
  - Mid-Chapter Assessment
  - Chapter Assessments
  - Alternative Assessments
- Test and Practice Generator
- Technology Handbook

Chapter Objectives

- Find the theoretical probability of an event. [**10.1**]
- Apply the fundamental counting principle. [**10.1**]

DISCRETE MATHEMATICS

Counting Principles and Probability

PROBABILITY IS THE RELATIVE LIKELIHOOD, or chance, that an event will occur. Some events, such as those shown here, are unlikely, or have a low probability of occurence. In this chapter, you will use the Fundamental Counting Principle to find the number of ways in which an event can occur.

Probability is used in many real-world fields, such as insurance, medical research, law enforcement, and political science.

Lessons

About the Photos

Twins occur in about 1 out of every 38 births, and multiple births occur in about 1 out of every 36 births. That is, the probability of having twins is about 2.6%, and the probability of having multiple births is about 2.8%. The birth rate of identical twins has remained constant at 4 out of every 1000 births throughout history, while the birth rate of fraternal twins has increased because women are giving birth later in life and some are taking fertility drugs. Any change in the overall probability of having twins is due to fraternal twins only. Other statistics about twins are shown at right.

- Fraternal twins occur in 2.2% of all births.
- There are $\frac{2}{3}$ more fraternal twins than identical twins.
- $\frac{1}{3}$ of all twins are same-sex fraternal twins.
- $\frac{1}{3}$ of all twins are different-sex twins.
- Slightly more than half of all twins are male.
- About 25% of identical twins are mirror-image twins, where the right side of one twin will match the left side of the other twin.

[*Source: Report of Final Natality Statistics, 1996*]

- Solve problems involving linear permutations of distinct or indistinguishable objects. [10.2]
- Solve problems involving circular permutations. [10.2]
- Solve problems involving combinations. [10.3]
- Solve problems by distinguishing between permutations and combinations. [10.3]
- Find the probabilities of mutually exclusive events. [10.4]
- Find the probabilities of inclusive events. [10.4]
- Find the probability of two or more independent events. [10.5]
- Find conditional probabilities. [10.6]
- Use simulation methods to estimate or approximate the experimental probability of an event. [10.7]

Rubber ducks float down the Singapore River in the Great Duck Race fund-raiser.

About the Chapter Project

Making reasonable predictions about random events plays an important role in decision-making strategies from the simplest problems to the most complex problems. In many situations, the probability of a random event occurring can be determined empirically by observing the number of times the event occurs. When the random event is complex, the actual observation of the event may be impossible. In these cases, simulations are used. In the Chapter Project, *Next, Please . . .*, you will simulate random events and estimate probabilities.

After completing the Chapter Project, you will be able to do the following:

- Set up models that simulate random events.
- Use data from simulations to estimate probabilities.

About the Portfolio Activities

Throughout the chapter, you will be given opportunities to complete Portfolio Activities that are designed to support your work on the Chapter Project.

- You will use a spinner to model the possible outcomes of a geometric probability problem in the Portfolio Activity on page 635.
- You will use a random-number generator to model the possible outcomes of an event in the Portfolio Activity on page 649.
- You will use a random-number generator to model two-event probabilities in the Portfolio Activity on page 658.

Chapter Project

Portfolio Activities appear at the end of Lessons 10.1, 10.3, and 10.4. Each serves as preparation for the Chapter Project. The Portfolio Activities as well as the Chapter Project Activities are appropriate for inclusion in the student's portfolio. Students should be encouraged to include in their portfolios any other work in which they feel a sense of pride or a sense of accomplishment.

internet connect go.hrw.com

Chapter Internet Features and Online Activities

Lesson	Keyword	Page
10.1	MB1 Homework Help	634
	MB1 Probability	635
10.2	MB1 Homework Help	641
	MB1 Baseball	640
10.3	MB1 Homework Help	648
	MB1 Combinations	649
10.4	MB1 Homework Help	656
10.5	MB1 Homework Help	662
10.6	MB1 Homework Help	668
	MB1 Conditional	668
10.7	MB1 Homework Help	676
	MB1 Simulation	675

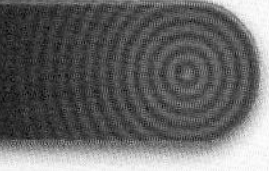

Prepare

NCTM PRINCIPLES & STANDARDS 1–6, 8–10

QUICK WARM-UP

Write as a percent. If necessary, round to the nearest tenth of a percent.

1. 0.157 — 15.7%
2. 10.122 — 1012.2%
3. $\frac{2}{5}$ — 40%
4. $\frac{17}{50}$ — 34%
5. $\frac{133}{200}$ — 66.5%
6. $\frac{3}{7}$ — 42.9%

Find the area of a circle with the given measure.

7. radius: 5 centimeters
   25π cm^2, or ≈ 78.5 cm^2
8. diameter: 6 inches
   9π in.^{2}, or ≈ 28.3 in.^{2}

Also on Quiz Transparency 10.1

Teach

Why Ask students approximately how many times a coin that is flipped 20 times shows heads. Now ask each student to flip a coin 20 times, keeping track of the number of times it shows heads. Discuss with students why there is a variation in the predicted number and the experimental results.

10.1 Introduction to Probability

Why *Probability is often studied using everyday objects, such as number cubes, coins, and darts.*

Objectives

- Find the theoretical probability of an event.
- Apply the Fundamental Counting Principle.

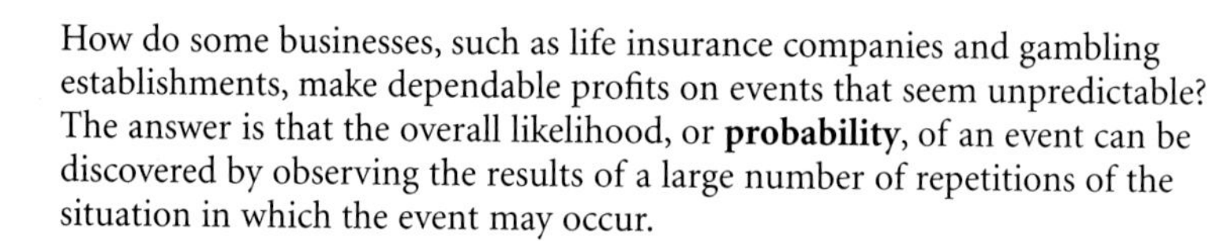

How do some businesses, such as life insurance companies and gambling establishments, make dependable profits on events that seem unpredictable? The answer is that the overall likelihood, or **probability**, of an event can be discovered by observing the results of a large number of repetitions of the situation in which the event may occur.

The terminology used to discuss probabilities is given below. An example related to rolling a number cube is given for each term.

DEFINITION	EXAMPLE
Trial: a systematic opportunity for an event to occur	rolling a number cube
Experiment: one or more trials	rolling a number cube 10 times
Sample space: the set of all possible outcomes of an event	1, 2, 3, 4, 5, 6
Event: an individual outcome or any specified combination of outcomes	rolling a 3 rolling a 3 *or* rolling a 5

Outcomes are **random** if all possible outcomes are equally likely. Although it is impossible to prove that the result of a real-world event is completely random, some outcomes are often assumed to be random. For example, the results of tossing a coin, the roll of a number cube, the outcome of a spinner, the selection of lottery numbers, and the gender of a baby are often assumed to be random.

Alternative Teaching Strategy

HANDS-ON STRATEGIES Discuss with students the difference between experimental and theoretical probabilities. Students should work in pairs, with one being the experimenter and the other being the recorder. Have students flip a coin 10 times, keeping track of the number of heads. Have students repeat the experiment four times, increasing the total number of flips to 20, 30, 40, and 50. Students should see that the more times the coin is flipped, the closer the experimental results come to the theoretical probability of 0.5. Students may wish to use a computer spreadsheet or a graphics calculator to simulate the experiment. Let 1 represent heads and 0 represent tails. On a TI graphics calculator, the command sequence MATH NUM 5:int(2 × MATH PRB 1:rand) will generate the integer 0 or 1. By entering this command once and then repeatedly pressing ENTER, the experiment may be simulated. Use results from the simulation and students' experiments to discuss the importance of sample size in experimental probability.

Probability is expressed as a number from 0 to 1, inclusive. It is often written as a fraction, decimal, or percent.

- An impossible event has a probability of 0.
- An event that must occur has a probability of 1.
- The sum of the probabilities of all outcomes in a sample space is 1.

In mathematics, the probability of an event can be assigned in two ways: *experimentally* (inductively) or *theoretically* (deductively).

Experimental probability is approximated by performing trials and recording the ratio of the number of occurrences of the event to the number of trials. As the number of trials in an experiment increases, the approximation of the experimental probability improves.

Theoretical probability is based on the assumption that all outcomes in the sample space occur randomly.

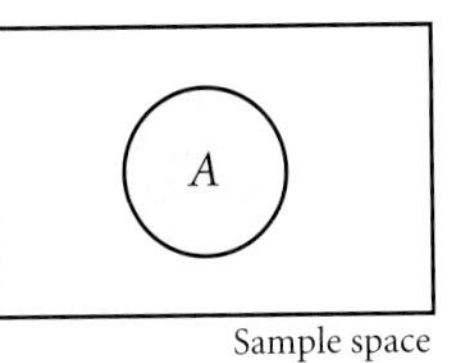

Sample space

Theoretical Probability

If all outcomes in a sample space are equally likely, then the theoretical probability of event A, denoted $P(A)$, is defined by

$$P(A) = \frac{\text{number of outcomes in event } A}{\text{number of outcomes in the sample space}}.$$

Use Teaching Transparency 37.

Teaching Tip

An event and an outcome may or may not be equivalent depending on the number of outcomes specified by the event. Emphasize to students that an event may include more than one outcome. For example, when rolling a number cube, an event might include two outcomes, such as rolling a 1 or rolling a 4.

ADDITIONAL EXAMPLE 1

Find the probability of randomly selecting an orange marble in one draw from a jar containing 8 blue marbles, 5 red marbles, and 2 orange marbles.

$\frac{2}{15}$, or ≈13.3%

TRY THIS

$\frac{4}{9}$, or ≈44.4%

EXAMPLE 1 **Find the probability of randomly selecting a red disk in one draw from a container that contains 2 red disks, 4 blue disks, and 3 yellow disks.**

SOLUTION

The event is selecting *1 red disk*.

Because there are 2 red disks, the number of outcomes in the event is 2.

The sample space is the set of all disks, so the total number of outcomes in the sample space is 2 + 4 + 3, or 9.

$$P(1 \text{ red disk}) = \frac{\text{number of outcomes in the event}}{\text{number of outcomes in the sample space}}$$

$$= \frac{2}{9}, \text{ or about } 22\%$$

Thus, the probability of randomly selecting a red disk in one draw is about 22%.

TRY THIS Find the probability of randomly selecting a blue disk in one draw from a container that contains 2 red disks, 4 blue disks, and 3 yellow disks.

Interdisciplinary Connection

HEALTH Suppose that a school cafeteria has decided to implement a healthy-eating campaign. Students will choose at least 1 entree, 1 vegetable, and 1 fruit. The cafeteria offers the following choices:

entree:	cheese lasagna, chicken
vegetable:	green beans, carrots, salad
fruit:	banana, apple, orange

How many different choices are possible for a lunch tray? 18

Inclusion Strategies

VISUAL LEARNERS When finding a probability, many students begin counting the possible outcomes of an event before organizing their work. This may cause them to miss an outcome or count an outcome twice. Tell students that a systematic way of counting possible outcomes provides a concrete way to solve many probability problems. Encourage them to check their tree diagrams by filling in the blanks with all possible outcomes. This should help them make a connection between a tree diagram and the fundamental counting principle.

In the next two examples, area models are used to find probabilities for situations in which the number of outcomes in the sample space is infinite, such as with area or time.

EXAMPLE 2

CONNECTION
GEOMETRY

Assume that a dart will land on the dartboard and that each point on the dartboard is equally likely to be hit.

Find the probability of a dart landing in region *A*, the outer ring.

SOLUTION

The event is landing in region *A*.

The sample space consists of all points on the dartboard.

$$P(A) = \frac{\text{area of region } A}{\text{total area}} = \frac{\pi(3)^2 - \pi(2)^2}{\pi(3)^2} = \frac{5\pi}{9\pi} = \frac{5}{9} \approx 0.556$$

The probability of a dart landing in region *A* is about 55.6%.

TRY THIS Find the probability of a dart landing in region *B* of the dartboard.

ADDITIONAL EXAMPLE 2

Refer to the dartboard in Example 2. **Find the probability of a dart landing in region C.**

$\frac{1}{9}$, or ≈11.1%

Math CONNECTION

GEOMETRY When calculating the probability of favorable outcomes for a dartboard, students will use the formula for the area of a circle, $A = \pi r^2$, to find the area of the desired region.

TRY THIS
$\frac{1}{3}$, or ≈33%

EXAMPLE 3

APPLICATION
COMPUTERS

Eduardo logs onto his electronic mail once during the time interval from 1:00 P.M. to 2:00 P.M.

Assuming that all times are equally likely, find the probability that he will log on during each time interval.

a. from 1:30 P.M. to 1:40 P.M.

b. from 1:30 P.M. to 1:35 P.M.

SOLUTION

a. The event is the interval from 1:30 P.M. to 1:40 P.M.

The sample space is the interval from 1:00 P.M. to 2:00 P.M.

Divide the sample space into 10-minute intervals to represent the equally likely events.

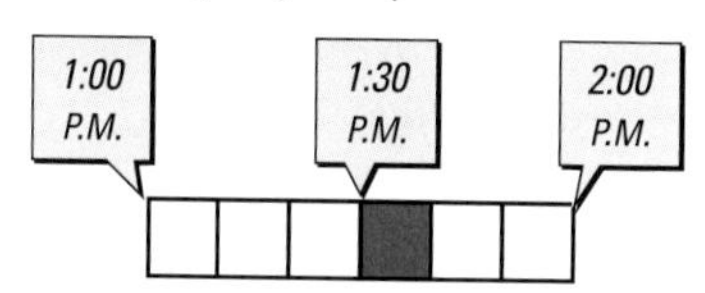

The event is $\frac{1}{6}$ of the sample space. Thus, the probability is $\frac{1}{6}$, or about 16.7%.

b. The event is the interval from 1:30 P.M. to 1:35 P.M.

The sample space is the interval from 1:00 P.M. to 2:00 P.M.

Divide the sample space into 5-minute intervals to represent the equally likely events.

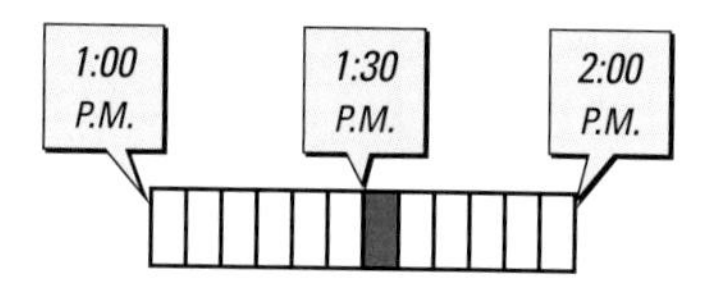

The event is $\frac{1}{12}$ of the sample space. Thus, the probability is $\frac{1}{12}$, or about 8.3%.

TRY THIS Janice leaves home for work sometime between 7:30 A.M. and 8:00 A.M. Assuming that all times are equally likely, find the probability that she will leave home during each time interval.

a. from 7:30 A.M. to 7:40 A.M.

b. from 7:30 A.M. to 7:32 A.M.

ADDITIONAL EXAMPLE 3

Jana sits down to dinner at some time between 6:00 P.M. and 7:00 P.M. **Assuming that all times are equally likely, find the probability that she will sit down to dinner during each given time interval.**

a. from 6:10 P.M. to 6:30 P.M.
$\frac{1}{3}$, or ≈33.3%

b. from 6:45 P.M. to 6:55 P.M.
$\frac{1}{6}$, or ≈16.7%

TRY THIS

a. $\frac{1}{3}$, or ≈33%

b. $\frac{1}{15}$, or ≈6.7%

Enrichment

Discuss the meaning of the following ideas:

1. an event with a probability of 0
2. an event with a probability of 1
3. an event with a probability between 0 and 1, inclusive
4. an event with a probability greater than 1
5. an event with a probability less than 0

In each case, distinguish between theoretical and experimental probability. Have students describe examples of each.

In the Activity below, a tree diagram is used to count all possible choices of pizza toppings.

Activity
Investigating Tree Diagrams

You will need: no special tools

A pizza shop offers a special price on a 2-topping pizza. You can choose 1 topping from each of the following groups:

- provolone cheese or extra mozzarella cheese
- pepperoni, sausage, or hamburger

1. Begin a tree diagram with the two cheese choices, as shown at right.

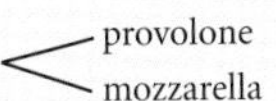

2. From each cheese choice, extend a line for each meat choice.
3. How many possible different combinations of two toppings are possible?

CHECKPOINT ✓ 4. If the special included a third topping of either onions or green peppers, how would you extend your diagram to show the additional possibilities? How many total 3-topping combinations are there?

There are several ways to determine the size of a sample space for an event that is a combination of two or more outcomes. One way is a tree diagram.

For example, a cafe's lunch special is a hamburger meal. It comes with a choice of beverage (soda or tea) and a choice of salad (garden, potato, or bean). The tree diagram below shows that there are 2×3, or 6, choices.

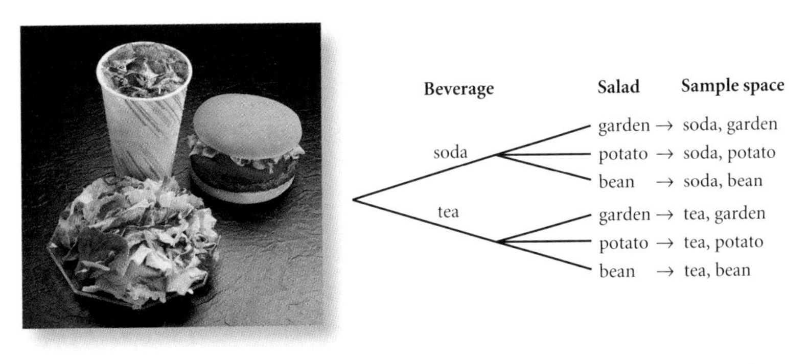

CHECKPOINT ✓ Make a tree diagram with the salad as the first choice and the beverage as the second choice. Does the order in which the choices are made affect the number of possible choices for the lunch special?

Tree diagrams illustrate the *Fundamental Counting Principle.*

Fundamental Counting Principle

If there are m ways that one event can occur and n ways that another event can occur, then there are $m \times n$ ways that both events can occur.

Activity Notes

In this Activity, students construct a tree diagram to illustrate the choices available when ordering a pizza with choices of 2 types of cheese and 3 meats. The diagram is then extended to include an additional choice of 2 vegetables. Students will construct a second tree diagram to illustrate the number of choices for a lunch special at a cafe.

CHECKPOINT ✓

4. Add a branch for onions and a branch for green peppers after each meat. 12 combinations are possible.

CHECKPOINT ✓

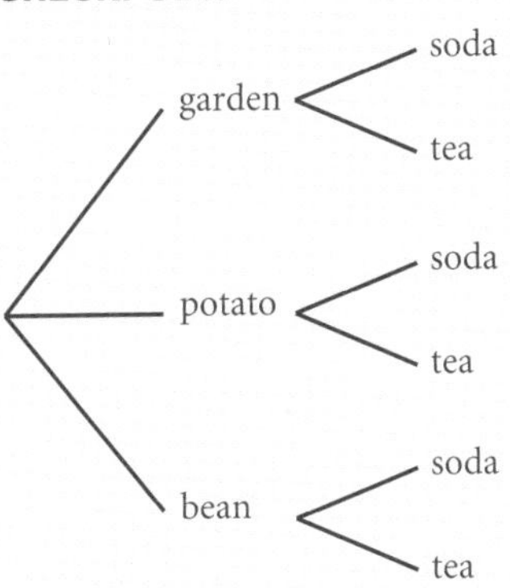

No, the order does not affect the number of choices.

Reteaching the Lesson

USING CHARTS Have students consider the possible sums when rolling two number cubes. For example, a roll could generate a 3 on one cube and a 4 on the other cube, resulting in a sum of 7. Have students make a chart, listing the outcomes of the first cube down the left side of the chart and the outcomes of the second cube across the top of the chart. Then students can fill in the chart with the sums of every possible roll. Have students determine the probability that the sum of two number cubes is 10 by finding the ratio of the number of ways that a sum of 10 can occur to the total number of sums in the chart. $\frac{3}{36} = \frac{1}{12}$

Have students use the chart to find other probabilities. Ask students to also use the fundamental counting principle to find the probabilities, and ask them to check their results.

ADDITIONAL EXAMPLE 4

Amy must choose a password for her voice-mail system. She cannot use the letters *A* and *Z* or the digits 0 and 9. Each letter or number may be used more than once. **How many possible passwords are there if the password is 3 letters followed by 3 digits?**
7,077,888 passwords

ADDITIONAL EXAMPLE 5

Three numbers from 0 to 9 are randomly selected. **Find the probability that the first number selected is 5 or 7, the second number is 2, and the third number is 3 or 9.**
0.4%

Assess

Selected Answers

Exercises 4–8, 9–51 odd

ASSIGNMENT GUIDE

In Class	1–8
Core	9–35 odd, 39–43 odd
Core Plus	10–36 even, 37, 38–44 even
Review	46–51
Preview	52

Extra Practice can be found beginning on page 940.

EXAMPLE 4 Ann is choosing a password for her access to the Internet. She decides not to use the digit 0 or the letter *O*. Each letter or number may be used more than once.

APPLICATION COMPUTERS

How many passwords of 2 letters followed by 4 digits are possible?

SOLUTION

Use the Fundamental Counting Principle. There are 25 possible letters and 9 possible digits.

1st letter		2nd letter		1st digit		2nd digit		3rd digit		4th digit
25	×	25	×	9	×	9	×	9	×	9

The number of possible passwords for Ann is $25^2 \times 9^4$, or 4,100,625.

EXAMPLE 5 A license plate consists of 2 letters followed by 3 digits. The letters, *A–Z*, and the numbers, 0–9, can be repeated.

APPLICATION TRANSPORTATION

Find the probability that your new license plate contains the initials of your first and last names in their proper order.

SOLUTION

1. Find the number of outcomes in the event.

	1st letter		*2nd letter*		*1st number*		*2nd number*		*3rd number*
Number of ways possible →	1	×	1	×	10	×	10	×	10

2. Find the number of outcomes in the sample space.

	1st letter		*2nd letter*		*1st number*		*2nd number*		*3rd number*
Number of ways possible →	26	×	26	×	10	×	10	×	10

3. Find the probability of this event.

$$P(A) = \frac{1 \times 1 \times 10 \times 10 \times 10}{26 \times 26 \times 10 \times 10 \times 10} = \frac{1000}{676{,}000} = \frac{1}{676} \approx 0.0015$$

Thus, the probability that your new license plate contains the initials of your first and last names is about 0.15%, which is more than 1 in 1000.

Exercises

Communicate

1. Give three examples of an event with more than one outcome.
2. How are theoretical and experimental probabilities similar? different?
3. Explain how an area model can be used to find probabilities.

Guided Skills Practice

4. Find the probability of randomly selecting a red marble in one draw from a bag of 5 blue marbles, 3 red marbles, and 1 white marble. ***(EXAMPLE 1)*** $\frac{3}{9}$, or $\frac{1}{3}$

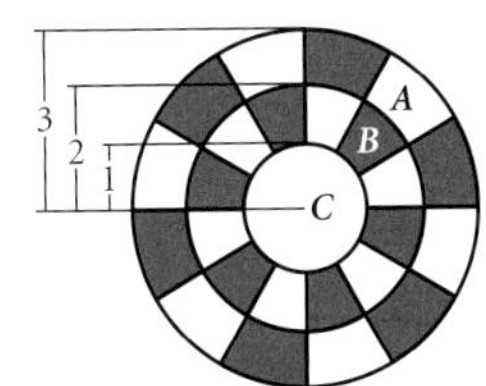

5. Assume that a dart will land on the dartboard and that each point on the dartboard is equally likely to be hit. Find the probability of a dart landing in region C, the inner ring. ***(EXAMPLE 2)*** $\frac{1}{9}$

6. Sue logs onto her electronic mail once during the time interval from 7:00 A.M. to 8:00 A.M. Assuming that all times are equally likely, find the probability that she will log on during the interval from 7:30 P.M. to 7:45 P.M. ***(EXAMPLE 3)*** $\frac{1}{4}$

7. John is deciding on a password for his access to the Internet. He decides not to use the digit 0 or the letter Q. How many possible passwords are there if he uses 2 letters followed by 3 digits? ***(EXAMPLE 4)*** 455,625

8. In a lottery, a 4-digit number from 0000 to 9999 is randomly selected. Find the probability that the number selected begins with 3 and ends in 2 or 0. ***(EXAMPLE 5)*** $\frac{1}{50}$

Practice and Apply

A bag contains 3 white cards, 2 black cards, and 5 red cards. Find the probability of each event for one draw.

9. a white card $\frac{3}{10}$ **10.** a black card $\frac{1}{5}$ **11.** a red card $\frac{1}{2}$

Calculate the probability of each event for one roll of a number cube.

12. 1 $\frac{1}{6}$

13. 4 $\frac{1}{6}$

14. an even number $\frac{1}{2}$

15. an odd number $\frac{1}{2}$

16. a number less than 3 $\frac{1}{3}$

17. a number greater than 3 $\frac{1}{2}$

18. a number greater than 6 **0**

19. a number less than 6 $\frac{5}{6}$

A bus arrives at Jason's house anytime from 8:00 A.M. to 8:05 A.M. If all times are equally likely, find the probability that Jason will catch the bus if he begins waiting at the given times.

20. 8:04 A.M. $\frac{1}{5}$ **21.** 8:02 A.M. $\frac{3}{5}$ **22.** 8:01 A.M. $\frac{4}{5}$ **23.** 8:03 A.M. $\frac{2}{5}$

For Exercises 24 and 25, create a tree diagram that shows the sample space for each event.

24. Involvement in one of each type of extracurricular activity

Sports: football, soccer, tennis
Arts: music, painting
Clubs: science, French

25. Involvement in one of each type of leisure activity

Outdoor: biking, gardening, rappeling
Indoor: reading, watching television, playing board games

Error Analysis

In Exercises 30–33, students learn how to find the odds of an event occurring. Students often confuse odds and probabilities because they probably hear about odds outside of the classroom more often than they hear about true probabilities. When students find odds, have them calculate the total number of possible outcomes by adding the numerator and denominator in the ratio of the odds. Point out that odds are ratios that compare the number of successful outcomes to the number of unsuccessful outcomes, while probabilities are ratios of the number of successful outcomes to the total number of possible outcomes. Then have them use their definition of probability to rewrite an odds expression as a probability of success.

Answers to Exercises 24–25 can be found in Additional Answers beginning on page 1002.

Find the number of possible passwords (with no letters or digits excluded) for each of the following conditions:

26. 2 digits followed by 3 letters followed by 1 digit **17,576,000**

27. 3 digits followed by 2 letters followed by 1 digit **6,760,000**

28. 3 letters followed by 3 digits **17,576,000**

29. 2 letters followed by 4 digits **6,760,000**

internet connect
Homework Help Online
Go To: **go.hrw.com**
Keyword:
MB1 Homework Help for Exercises 30–33

The odds in favor of an event are defined as the number of ways the event can happen divided by the number of ways it can fail to happen. If the odds in favor of an event are $\frac{a}{b}$, or a to b, then the probability of the event is $\frac{a}{a+b}$. Find the probability of each event, given the odds in favor of the event.

30. 3 to 8 $\frac{3}{11}$ **31.** 4 to 5 $\frac{4}{9}$ **32.** 3 to 7 $\frac{3}{10}$ **33.** 1 to 20 $\frac{1}{21}$

GEOMETRY To the nearest tenth of a percent, find or approximate the probability that a dart thrown at the square dartboard at right will land in the regions indicated below.

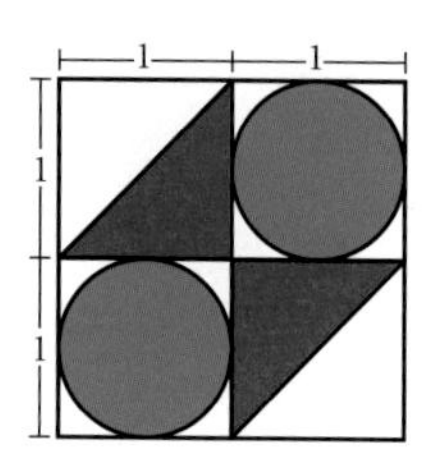

34. one of the circular regions **39.3%**

35. one of the red triangular regions **25%**

36. one of the white triangular regions **25%**

CHALLENGE

37. one of the white regions **35.7%**

APPLICATIONS

38. PUBLISHING When a book is published, it is assigned a number called an International Standard Book Number (ISBN). The number consists of 10 digits that provide information about the language in which the book is printed, the publisher of the book, the book itself, and a check digit. How many ISBN numbers are possible? **10 billion**

39. ACADEMICS A multiple-choice exam consists of 14 questions, each of which has 4 possible answers. How many different ways can all 14 questions on the exam be answered? **268,435,456**

DEMOGRAPHICS The table at right shows the 1993 college enrollment statistics, in thousands, for the United States. Based on this data, find the probability that a randomly selected person enrolled in college in the United States in 1993 is in the given age group. [*Source: Statistical Abstract of the United States, 1996*]

Age	Male	Female
14–17	83	93
18–19	1224	1416
20–21	1294	1414
22–24	1260	1263
25–29	950	1058
30–34	661	811
35 and over	955	1824

40. 18–24 **≈55%** **41.** 25–29 **≈14%**

42. 30–34 **≈10%** **43.** 30 or over **≈30%**

44. 962,391,456

44. TRANSPORTATION How many different license plates can be made if each plate consists of 2 letters followed by 2 digits (1 through 9) followed by 3 letters?

Practice

NAME ________ CLASS ________ DATE ________

Practice
***10.1* Introduction to Probability**

Find the probability of each event.

1. A blue card is drawn at random from a bag containing 2 white cards, 1 red card, and 7 blue cards. $\frac{7}{10} = 70\%$
2. Frederique, who arrives home at 6:42 P.M., is home to receive a call that can come at any time between 6:40 and 6:50. $\frac{4}{5} = 80\%$
3. A letter chosen at random from the letters of the word *permutation* is a vowel. $\frac{5}{11} \approx 45\%$
4. A card chosen at random from a standard 52-card deck is a heart or a diamond. $\frac{1}{2} = 50\%$
5. A card chosen at random from a standard deck is not an 8 or an ace. $\frac{11}{13} \approx 85\%$
6. A number cube is rolled, and a number greater than 3 and less than 6 results. $\frac{1}{3} \approx 33\%$
7. A letter chosen at random from the alphabet is not one of the 5 standard vowels. $\frac{21}{26} \approx 81\%$
8. A point on a 12-inch ruler is chosen at random and is located within an inch of an end of the ruler. $\frac{1}{6} \approx 17\%$

A spinner is divided into three colored regions. You spin the spinner a total of 150 times. The results are recorded in the table. Find the experimental probability of each event.

green	42
yellow	65
pink	43

9. green $\frac{42}{150} = 28\%$
10. yellow $\frac{65}{150} \approx 43\%$
11. pink $\frac{43}{150} \approx 29\%$
12. not pink $\frac{107}{150} \approx 71\%$
13. not yellow $\frac{85}{150} \approx 57\%$

Find the number of possible license plate numbers (with no letters or digits excluded) for each of the following conditions:

14. 6 digits 1,000,000
15. 2 letters followed by 3 digits 676,000
16. 4 letters followed by 3 digits 456,976,000
17. 5 digits followed by 2 letters 67,600,000
18. 2 digits followed by 2 letters followed by 2 digits 6,760,000

CHALLENGE

45. SECURITY A security specialist is designing a code for a security system. The code will use only the letters *A*, *B*, and *C*. If the specialist wants the probability of guessing the code at random to be less than 0.001, how long must the code be? **at least 7 letters long**

Look Back

46. $f^{-1}(x) = \frac{x-10}{3}$

47. $g^{-1}(x) = \frac{6x+2}{3}$

48. $h^{-1}(x) = -\frac{x}{5} + \frac{1}{10}$

For each function, find an equation of the inverse. Then use composition to verify that the equation you wrote is the inverse. *(LESSON 2.5)*

46. $f(x) = 3x + 10$ **47.** $g(x) = \frac{3x-2}{6}$ **48.** $h(x) = \frac{1}{2} - 5x$

For each function, describe the transformations from $f(x) = \sqrt{x}$ to g. Graph each transformation. *(LESSON 8.6)*

49. $g(x) = \frac{1}{2}\sqrt{x} - 5$ **50.** $g(x) = 3\sqrt{x-2}$ **51.** $g(x) = \sqrt{2x} + 1$

Look Beyond

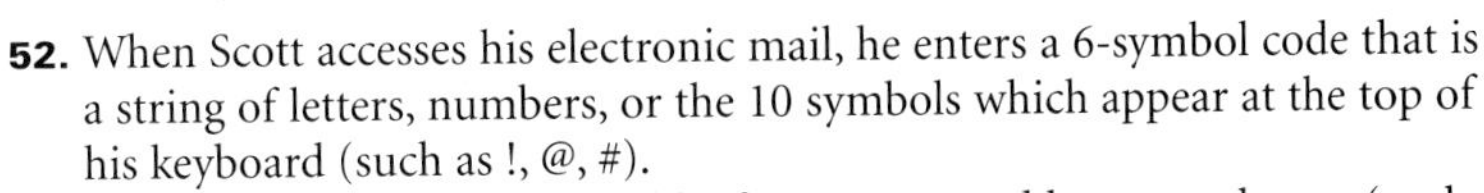

internet connect

Portfolio Extension

Go To: **go.hrw.com**
Keyword:
MB1 Probability

52. When Scott accesses his electronic mail, he enters a 6-symbol code that is a string of letters, numbers, or the 10 symbols which appear at the top of his keyboard (such as !, @, #).

a. How many codes are possible if uppercase and lowercase letters (such as *B* and *b*) are considered the same? **46^6**

b. How many codes are possible if uppercase and lowercase letters are considered different? **72^6**

c. How long would it take to try all of the codes in part **b** if you could enter one code per second? Give your answer in years. **4417.6 years**

Refer to Example 3 on page 630. Assuming that all the times are equally likely, you can perform an experiment with a spinner to model the probability that Eduardo logs onto his mail during the time interval from 1:30 P.M. to 1:40 P.M.

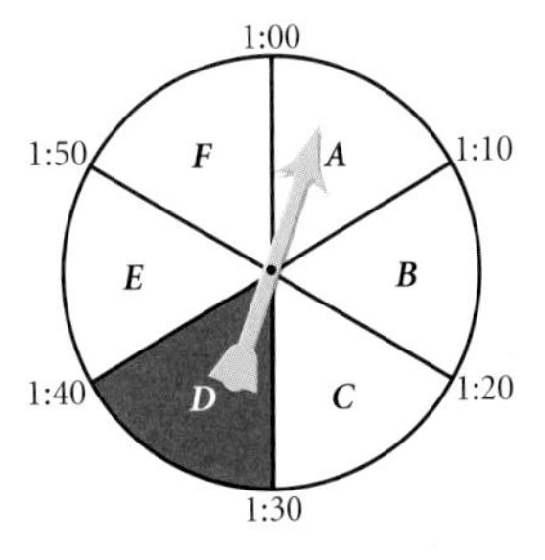

1. Divide a spinner into 6 equal sections, each representing a 10-minute interval. Let region *D* represent the 10-minute interval from 1:30 P.M. to 1:40 P.M.
2. Spin the spinner 40 times, and record the number of times that the spinner lands in region *D*.
3. Use your results to estimate the probability that Eduardo logs onto his mail in the interval from 1:30 P.M. to 1:40 P.M.
4. Compare your estimated probability with the theoretical probability obtained in the lesson. How well do you think the spinner models this problem? Explain.

WORKING ON THE CHAPTER PROJECT

You should now be able to complete Activity 1 of the Chapter Project.

Look Beyond

The questions in Exercise 52 lead students to think about ordered strings of characters. These are permutations because the order of the characters in the string is important. Permutations will be studied in Lesson 10.2.

ALTERNATIVE

Assessment

Portfolio Activity

The Portfolio Activity can be used as preparation for the Chapter Project or as a separate activity. In the Portfolio Activity on this page, students create a model to simulate when Eduardo checks his electronic mail (from Example 3 on page 630). Using the model, students conduct an experiment to find the probabilities asked for in the questions.

Student Technology Guide

NAME ______ CLASS ______ DATE ______

Student Technology Guide
10.1 Introduction to Probability

You can use the random number feature on a spreadsheet to perform a probability experiment.

Example: Use a spreadsheet to simulate 40 tosses of a fair coin. Give the experimental probabilities of heads and tails.

This experiment uses the RAND() feature. RAND() gives a random number such that 0 ≤ RAND() < 1.

- Let 1 = heads and 0 = tails. To generate a random 0 or 1, enter "=INT(2*RAND())" into cell A1. Multiplying RAND() by 2 gives a number greater than or equal to 0 and less than 2. Taking the integer, or "INT", part leaves a 0 or a 1.
- Click in cell A1 and drag to highlight the cells from A1 to A40. Use the "Fill Down" command in the EDIT menu to fill all these cells with the formula in A1.
- Click in cell A41, then click the sum (Σ) button. You will see "=SUM(A1:A40)". Click at the end of this expression, then type "/40" and press RETURN to find the experimental probability of heads. Click in cell A2 and enter "=1–A41" to find the experimental probability of tails.

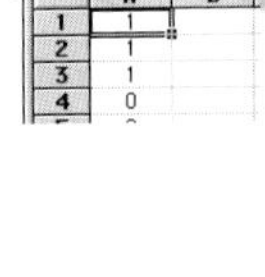

From the results shown at right above, the experimental probability of heads is 0.575, or 57.5%. The experimental probability of tails is 0.425, or 42.5%.

1. Use a spreadsheet to simulate 60 tosses of a fair coin. Give the probabilities of heads and tails for your experiment.
   Answers will vary.
2. Use a spreadsheet to simulate 150 tosses of a fair coin. Then simulate 500 tosses. Give the experimental probabilities of heads and tails for each experiment. Compare your results with those of your classmates. What happens to the experimental probabilities as the number of tosses increases?
   Simulations will vary. As the number of tosses increases, the experimental probabilities approach the theoretical probabilities of $\frac{1}{2}$, or 50%.

49. a vertical compression by a factor of $\frac{1}{2}$ and a vertical translation 5 units down

50. a horizontal translation 2 units to the right and a vertical stretch by a factor of 3

51. a horizontal compression by a factor of $\frac{1}{2}$ and a vertical translation 1 unit up

Prepare

NCTM PRINCIPLES & STANDARDS 1, 2, 4–10

QUICK WARM-UP

Find the number of passwords possible if the following patterns are used:

1. 1 letter followed by 3 digits **26,000**
2. 2 digits, excluding 1, followed by 2 letters, excluding B **50,625**
3. 1 digit followed by 3 letters followed by 2 digits **17,576,000**
4. 2 digits, excluding 0, followed by 1 letter, excluding Z, followed by 3 digits **2,025,000**

Also on Quiz Transparency 10.2

Teach

Why Have students consider the word *ant*. Ask them how many distinct arrangements of the letters are possible. **6** Have them consider the word *mom* and again find how many distinct arrangements of the letters are possible. **3** Discuss why the two answers are not the same.

Permutations

Objectives

- Solve problems involving linear permutations of distinct or indistinguishable objects.
- Solve problems involving circular permutations.

Why *There are many situations that involve an ordered arrangement, or permutation, of objects. For example, 12-tone music, developed by Arnold Schoenberg, consists of permutations of all 12 tones in an octave.*

Arnold Schoenberg, 1874–1951

In 12-tone music, pioneered by Arnold Schoenberg, each note of the chromatic scale must be used exactly once before any are repeated. A set of 12 tones is called a *tone row*. How many different tone rows are possible? *You will answer this question in Example 1.*

A **permutation** is an arrangement of objects in a specific order. When objects are arranged in a row, the permutation is called a **linear permutation**. Unless otherwise noted, the term *permutation* will be used to mean *linear permutations.*

PROBLEM SOLVING

Make an organized list. Each of the possible permutations of the letters A, B, C, and D are listed in the table at left.

ABCD	BACD	CABD	DABC
ABDC	BADC	CADB	DACB
ACBD	BCAD	CBAD	DBAC
ACDB	BCDA	CBDA	DBCA
ADBC	BDAC	CDAB	DCAB
ADCB	BDCA	CDBA	DCBA

The number of different permutations, 24, can be obtained by using the Fundamental Counting Principle, as shown below.

	1st choice		2nd choice		3rd choice		4th choice		
Number of possible choices →	4	×	3	×	2	×	1	=	24

Thus, there are $4 \times 3 \times 2 \times 1$, or 24, possible arrangements. You can use *factorial notation* to abbreviate this product: $4! = 4 \times 3 \times 2 \times 1 = 24$.

If n is a positive integer, then **n factorial**, written $n!$, is defined as follows:

$$n! = n \times (n-1) \times (n-2) \times \cdots \times 2 \times 1$$

Note that the value of $0!$ is defined to be 1.

You can find the number of permutations of any number of objects by using the Fundamental Counting Principle and factorials as follows:

Permutations of n Objects

The number of permutations of n objects is given by $n!$.

Alternative Teaching Strategy

USING TABLES Put students in groups of four, making sure that there is a mix of boys and girls in each group, and ask them to arrange themselves in a row of four chairs. **24 ways** The group should find all possible arrangements of their team members. Encourage students to keep a record of how they determine all of the possible permutations. One way that students can do this is to make a table with a column for each chair, put the number of possible choices in the column for each chair, and use the fundamental counting principle. Next, assign each team member a partner and stipulate that the partners must always sit next to each other. Have students find all possible seating arrangements. **8 ways** Again, a table is helpful. Now, have students repeat the two conditions above, but instead of arranging the chairs in a row, have students arrange them in a circle. **6 ways; 4 ways** Discuss the differences between linear and circular permutations. Finally, have students arrange themselves by gender, both linearly and circularly.

EXAMPLE 1

APPLICATION
MUSIC

In 12-tone music, each of the 12 notes in an octave must be used exactly once before any are repeated. A set of 12 tones is called a tone row.

How many different tone rows are possible?

SOLUTION

Find the number of permutations of 12 notes.

$$12! = 12 \times 11 \times 10 \times 9 \times 8 \times 7 \times 6 \times 5 \times 4 \times 3 \times 2 \times 1$$
$$= 479{,}001{,}600$$

There are 479,001,600 different tone rows for 12 tones.

TRY THIS How many different ways can the letters in the word *objects* be arranged?

The number of ways that you can listen to 3 different CDs from a selection of 10 CDs is a *permutation of 10 objects taken 3 at a time.*

	1st CD		2nd CD		3rd CD
Number of CDs available →	10	×	9	×	8

By the Fundamental Counting Principle, there are $10 \times 9 \times 8$, or 720, ways that you can listen to 3 different CDs from a selection of 10 CDs.

Permutations of *n* Objects Taken *r* at a Time

The number of permutations of n objects taken r at a time, denoted by $P(n, r)$ or ${}_nP_r$, is given by $P(n, r) = {}_nP_r = \frac{n!}{(n-r)!}$, where $r \leq n$.

Note the use of 0! in calculating the number of permutations of n objects taken n at a time.

$$P(n, n) = {}_nP_n = \frac{n!}{(n-n)!} = \frac{n!}{0!} = n!$$

EXAMPLE 2

APPLICATION
MUSIC

Find the number of ways to listen to 5 different CDs from a selection of 15 CDs.

SOLUTION

Find the number of permutations of 15 objects taken 5 at a time.

$${}_{15}P_5 = \frac{15!}{(15-5)!}$$
$$= \frac{15 \times 14 \times 13 \times 12 \times 11 \times \cancel{10!}}{\cancel{10!}}$$
$$= 15 \times 14 \times 13 \times 12 \times 11$$
$$= 360{,}360$$

TRY THIS Find the number of ways to listen to 4 CDs from a selection of 8 CDs.

Interdisciplinary Connection

BIOLOGY A biologist needs to classify 16,000 species of plants by assigning a sequence of 3 letters to each species. If repetition of letters is not acceptable, can the biologist complete the classification? Explain why or why not. **No, there are only 15,600 permutations of 26 letters taken 3 at a time.** If repetition of a letter is acceptable, can the biologist complete the classification? Explain why or why not. **Yes, with repetition there are 26^3, or 17,576, possibilities.**

Teaching Tip

If students wonder why 0! is defined to be 1, you can remind them that x^0 is defined to be 1 so that calculations produce consistent values. The formula for the number of permutations of n objects taken n at a time, ${}_nP_n = \frac{n!}{(n-n)!} = \frac{n!}{0!} = n!$, is a good example.

ADDITIONAL EXAMPLE 1

On a baseball team, 9 players are designated as the starting lineup. Before a game, the manager announces the order in which the 9 players will bat. **How many different orders are possible?**
362,880

TRY THIS
${}_7P_7 = 5040$

Teaching Tip

TECHNOLOGY When using a TI-83 or TI-82 graphics calculator, enter a number and use the following sequence to compute its factorial:

Use the following sequence to find the permutation ${}_nP_r$, where n and r represent numbers:

n   r

ADDITIONAL EXAMPLE 2

Find the number of ways to listen to 3 different CDs from a selection of 5 CDs.
60

TRY THIS
${}_8P_4 = 1680$

Activity Notes

In this Activity, students should find that the number of permutations of n things taken r at a time can be found by using the formula ${}_nP_r = \frac{n!}{(n-r)!}$ or by using the fundamental counting principal, where ${}_nP_r = n(n-1)(\ldots)(n-r+1)$.

Teaching Tip

TECHNOLOGY Many calculators have probability features. Permutations are usually denoted in a menu or on a keypad as **nPr**. On graphics calculators, it is often found in the **MATH** menu under the **PRB** (probability) submenu. To calculate ${}_6P_3$, students enter **6**, select the **nPr** feature, and then enter **3**. The screen will show **6 nPr 3**, and after pressing ENTER, the result of 120 will appear.

Most scientific calculators have a factorial feature. On the TI-83 graphics calculator, the factorial feature, *!*, can be found in the **MATH** menu under the **PRB** submenu. Remind students that when the denominator is a product of factorials, they must use parentheses to group the entire denominator when entering it into a calculator.

CHECKPOINT ✔

3. ${}_8P_4 = 8 \times 7 \times 6 \times 5 = 1680$
${}_8P_5 = 8 \times 7 \times 6 \times 5 \times 4 = 6720$
${}_{10}P_8 = 10 \times 9 \times 8 \times 7 \times 6 \times 5 \times 4 \times 3 = 1{,}814{,}400$

ADDITIONAL EXAMPLE 3

Ronald is organizing sports equipment for the physical education room. He has 15 balls that he must place in a line. **In how many ways can he line up 6 footballs, 2 soccer balls, 4 kickballs, and 3 basketballs?**
6,306,300

Activity Exploring Formulas for Permutations

You will need: a calculator

1. **a.** Evaluate ${}_7P_3$ and $7 \times 6 \times 5$. **b.** Evaluate ${}_7P_4$ and $7 \times 6 \times 5 \times 4$.
   c. Evaluate ${}_7P_5$ and $7 \times 6 \times 5 \times 4 \times 3$.
2. **a.** Write ${}_7P_r$ as a product. **b.** Write ${}_nP_r$ as a product.

CHECKPOINT ✔
3. Using your answer from part **b** of Step 2, write products to represent ${}_8P_4$, ${}_8P_5$, and ${}_{10}P_8$. Verify your results.

As you have seen, there are 4! permutations of the 4 letters *A*, *B*, *C*, and *D*. However, if 2 of the 4 letters are identical, as in OHIO, there will be less than 4! permutations because arrangements such as O_1HIO_2 and O_2HIO_1 are indistinguishable. To account for the 2 identical letters, divide by 2! because there are 2! ways of arranging the 2 Os.

$$\frac{4!}{2!} \begin{array}{l} \leftarrow \textit{permutation of 4 objects} \\ \leftarrow \textit{2 identical objects} \end{array}$$

Permutations With Identical Objects

The number of distinct permutations of n objects with r identical objects is given by $\frac{n!}{r!}$, where $1 \le r \le n$.
The number of distinct permutations of n objects with r_1 identical objects, r_2 identical objects of another kind, r_3 identical objects of another kind, . . . , and r_k identical objects of another kind is given by $\frac{n!}{r_1!r_2!r_3! \cdots r_k!}$.

EXAMPLE 3 T'anna is planting 11 colored flowers in a line.

In how many ways can she plant 4 red flowers, 5 yellow flowers, and 2 purple flowers?

SOLUTION

PROBLEM SOLVING

Use the formula for permutations with repeated objects.

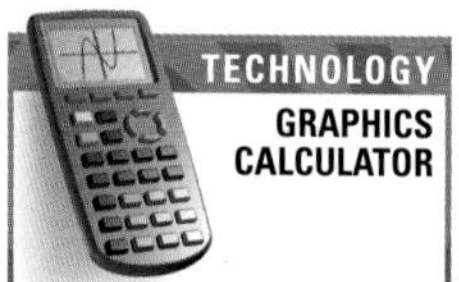

Keystroke Guide, page 686

$$\begin{aligned}
\frac{11!}{\underset{\text{red}}{4!} \times \underset{\text{yellow}}{5!} \times \underset{\text{purple}}{2!}} &= \frac{11 \times 10 \times 9 \times 8 \times 7 \times 6 \times \cancel{5!}}{4! \times \cancel{5!} \times 2!} \\
&= \frac{11 \times 10 \times 9 \times \cancel{8} \times 7 \times \cancel{6}}{(\cancel{4} \times \cancel{3} \times \cancel{2} \times 1) \times (\cancel{2} \times 1)} \\
&= \frac{11 \times 10 \times 9 \times 7}{1} \\
&= 6930
\end{aligned}$$

There are 6930 ways that T'anna can plant the flowers in a line.

Inclusion Strategies

TACTILE LEARNERS It is very difficult for students to simply memorize the formulas and know when to use them. Most students who have trouble with permutations do not think about the situation presented, but simply try to apply a formula. Encourage students to use manipulatives to model the situations given. For problems in which the numbers are too large to model physically, encourage students to simplify the problem by making a model based on fewer choices.

Enrichment

Have students use the formula to find the following permutations:

1. ${}_nP_0$ 1
2. ${}_0P_0$ 1

Ask students to create a situation that is modeled by each of these permutations.

TRY THIS In how many ways can T'anna plant 11 colored flowers if 5 are white and the remaining ones are red?

CHECKPOINT ✔ How does the formula for the number of permutations with repeated objects give the number of permutations of n objects when all of the objects are distinct?

CRITICAL THINKING A row of daisies consists of r yellow and s white flowers. In terms of r and s, write a formula for the number of permutations of all of the flowers.

In how many ways can you arrange 4 objects around the edge of a circular tray? The letters A, B, C, and D are arranged in a circle, as shown below. This type of a permutation is called a **circular permutation**.

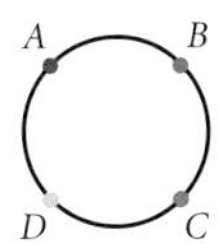

Circular permutations

A	B	C	D
B	C	D	A
C	D	A	B
D	A	B	C

Linear permutations

Notice that all 4 distinct linear permutations of the letters A, B, C, and D, give a single distinct circular permutation. Therefore, to find the number of *circular permutations* of 4 objects, divide the total number of linear permutations by 4. The result, 3!, is also $(4-1)!$, as shown below.

$$\frac{4!}{4} = \frac{\cancel{4} \times 3 \times 2 \times 1}{\cancel{4}} = 3!$$

Circular Permutations

If n distinct objects are arranged around a circle, then there are $(n-1)!$ circular permutations of the n objects.

EXAMPLE 4 **In how many ways can 7 different appetizers be arranged on a circular tray as shown at right?**

APPLICATION CATERING

SOLUTION

PROBLEM SOLVING **Use the formula** for circular permutations with $n = 7$.

$(n-1)! = (7-1)! = 6! = 720$

There are 720 distinct ways to arrange the appetizers on the tray.

TRY THIS In how many ways can seats be chosen for 12 couples on a Ferris wheel that has 12 double seats?

TRY THIS
462

CHECKPOINT ✔
$\frac{n!}{1!\,1!\,...\,1!} = n!$

CRITICAL THINKING
$\frac{(r+s)!}{r!s!}$

ADDITIONAL EXAMPLE 4

In how many different ways can 12 students attending a seminar be arranged in a circular seating pattern?
39,916,800

TRY THIS
39,916,800

Use Teaching Transparency 38.

Reteaching the Lesson

USING MODELS Most permutations can be solved with a counting model. For example, suppose that you want to know how many arrangements are possible for 5 people standing in a straight row. Draw 5 blanks, each representing one place in the row. How many choices are there to fill the first blank? 5 Write _5_ ___ ___ ___ ___. How many choices are left for the second blank? 4 Write _5_ _4_ ___ ___ ___. Continuing for the rest of the blanks, write _5_ _4_ _3_ _2_ _1_. Because there are no repeated students, just multiply the numbers in the blanks to find the solution. 120

You can apply a similar model to the following problem: How many ways are there to choose 3 balls from a bag of 8 different colored balls? There are 3 "spaces" to fill, so draw 3 blanks. How many choices are there for the first blank? 8 How many choices are there for the second blank? 7 Similarly, there are 6 choices for the third blank, so the result is _8_ × _7_ × _6_ = 336. This gives the same result as ${}_8P_2 = \frac{8!}{(8-3)!} = \frac{8 \times 7 \times 6 \times 5 \times 4 \times 3 \times 2 \times 1}{5 \times 4 \times 3 \times 2 \times 1}$
$= 8 \times 7 \times 6 = 336$.

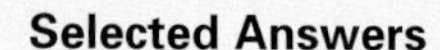

Assess

Selected Answers

Exercises 1–8, 9–73 odd

ASSIGNMENT GUIDE	
In Class	1–8
Core	9–53 odd, 57–63 odd
Core Plus	10–62 even
Review	64–74
Preview	75, 76

Extra Practice can be found beginning on page 940.

Practice

NAME ______ CLASS ______ DATE ______

Practice
***10.2* Permutations**

Find the number of permutations of the first 7 letters of the alphabet for each situation.

1. taking all 7 letters at a time 5040
2. taking 5 letters at a time 2520
3. taking 4 letters at a time 840
4. taking 3 letters at a time 840

In how many ways can 12 books be displayed on a shelf if the given number of books are available?

5. 12 books 479,001,600
6. 14 books $\approx 4.36 \times 10^{10}$
7. 15 books $\approx 2.18 \times 10^{11}$
8. 20 books $\approx 6.03 \times 10^{13}$

Find the number of permutations of the letters in each word.

9. *geometry* 20,160
10. *algebra* 2520
11. *addition* 10,080
12. *calculus* 5040
13. *mathematics* 4,989,600
14. *arithmetic* 907,200

15. Lizette decorates windows for a department store. She plans to design a baby's room with a row of stuffed elephants and monkeys along one wall. If she has 8 identical elephants and 10 identical monkeys, in how many different ways can the stuffed animals be displayed? 43,758
16. The 6 candidates for a student government office are invited to speak at an election forum. In how many different orders can they speak? 720
17. Representatives from 8 schools are represented at a school newspaper workshop. In how many different ways can the 8 representatives be seated around a circular table? 5040
18. Ten colleges are participating in a college fair. Booths will be positioned along one wall of a high school gymnasium. In how many different orders can the booths be arranged? 3,628,800

Exercises

Communicate

internet connect
Activities Online
Go To: **go.hrw.com**
Keyword:
MB1 Baseball

1. Explain how the Fundamental Counting Principle is used to find the number of permutation of 4 objects.

2. Which number of permutations of 4 letters is greater, 2 at a time or 3 at a time? Explain.

3. Explain why ${}_4P_4 = {}_4P_3$ is true. Is ${}_nP_n = {}_nP_{n-1}$ always true? Explain.

4. Does a circular permutation of 5 distinct objects always have fewer arrangements than a linear permutation of 5 distinct objects? Explain.

Guided Skills Practice

720 **5.** How many different ways can the letters in the word *orange* be arranged? ***(EXAMPLE 1)***

210 **6.** Find the number of ways to watch 3 videos from a selection of 7 videos. ***(EXAMPLE 2)***

2520 **7.** In how many ways can 5 pennies, 2 quarters, and 3 dimes be arranged in a straight line? ***(EXAMPLE 3)***

39,916,800 **8.** In how many ways can 12 different spices be arranged around a circular spice rack? ***(EXAMPLE 4)***

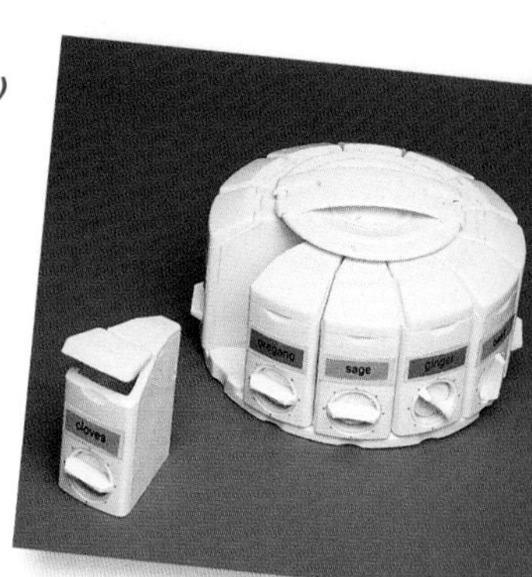

Practice and Apply

Evaluate each expression.

9. $7! - 5!$ **4920**
10. $6! - 4!$ **696**
11. $(7 - 5)!$ **2**
12. $(6 - 4)!$ **2**
13. $\frac{8!}{3! \times 5!}$ **56**
14. $\frac{10!}{4! \times 6!}$ **210**
15. $\frac{5! \times 0!}{(4-1)!}$ **20**
16. $\frac{(7-2)! \times 3!}{0!}$ **720**
17. ${}_{10}P_7$ **604,800**
18. ${}_7P_3$ **210**
19. ${}_{50}P_1$ **50**
20. ${}_{1000}P_1$ **1000**
21. $\frac{{}_{12}P_5}{{}_6P_5}$ **132**
22. $\frac{{}_4P_3}{{}_8P_3}$ **≈0.07**
23. ${}_4P_2 \times {}_7P_7$ **60,480**
24. ${}_{15}P_5 \times {}_5P_5$ **43,243,200**

Find the number of permutations of the first 8 letters of the alphabet for each situation.

25. taking 5 letters at a time **6720**
26. taking 3 letters at a time **336**
27. taking 4 letters at a time **1680**
28. taking 6 letters at a time **20,160**
29. taking 1 letter at a time **8**
30. taking all 8 letters at a time **40,320**

In how many ways can 8 new employees be assigned to the following number of vacant offices?

31. 8 **40,320**
32. 9 **362,880**
33. 10 **1,814,400**
34. 15

34. 259,459,200

35. 13,366,080
36. 96,909,120
37. 271,252,800
38. 427,518,000

internet connect — **Homework Help Online**
Go To: **go.hrw.com**
Keyword:
MB1 Homework Help for Exercises 39–47

47. 908,107,200

In how many ways can a teacher arrange 6 students in the front row of a classroom with the following number of students?

35. 18 **36.** 24 **37.** 28 **38.** 30

Find the number of permutations of the letters in each word.

39. *barley* **720** **40.** *pencil* **720** **41.** *trout* **60**

42. *circus* **360** **43.** *football* **10,080** **44.** *vignette* **10,080**

45. *bookkeeper* **151,200** **46.** *Mississippi* **34,650** **47.** *correspondence*

48. Five different stuffed animals are to be placed on a circular display rack in a department store. In how many ways can this be done? **24**

49. Seven different types of sunglasses are to be displayed on one level of a circular rack. In how many ways can they be arranged? **720**

50. Franklin High School has 4 valedictorians. In how many different ways can they give their graduation speeches? **24**

51. Suppose that a personal identification number (PIN) consists of 4 digits from 0 through 9.
a. How many PIN numbers are possible? **10,000**
b. How many PIN numbers are possible if no digit can be repeated? **5040**

52. In how many different ways can the expression $a^4b^2c^5d$ be written without exponents? **83,160**

53. A spinner is divided into 8 equal regions to represent the digits 1–8. In how many ways can the digits be arranged? **5040**

CHALLENGES

54. In how many different ways can 4 blue counters and 4 red counters be placed in a circle if red and blue counters are alternated? **1**

55. In how many ways can 3 males (Joe, Jerry, and John) and 3 females (Jamie, Jenny, and Jasmine) be seated in a row if the genders alternate down the row? **72**

56. In how many ways can 5 seniors, 3 juniors, 4 sophomores, and 3 freshmen be seated in a row if a senior must be seated at each end? Assume that the members of each class are distinct. **20 × 13! = 124,540,416,000**

CONNECTION

GEOMETRY **Geometric figures are often represented by the letters that identify each vertex. Determine the number of ways that each figure below can be named with the letters *A, B, C, D, E,* and *F*. Assume that each figure is irregular.**

57. triangle **120** **58.** quadrilateral **360**

59. hexagon **720** **60.** pentagon **720**

APPLICATIONS

61. SPORTS In a track meet, 7 runners compete for first, second, and third place. How many different ways can the runners place if there are no ties? **210**

62. SMALL BUSINESS A caterer is arranging a row of desserts. The row will contain 8 platters of cookies, 5 trays of fruit, and 3 pies. In how many distinct ways can the cookies, fruit, and pies, be arranged in a row, if each type of dessert is of the same kind? **720,720**

Error Analysis

Students often forget to divide out the repeated objects when finding permutations. To help students remember to do this, have them find the different permutations of words like *mom* or *bob* or *dad*. They can easily see that the repeated letters decrease the number of arrangements possible.

Student Technology Guide

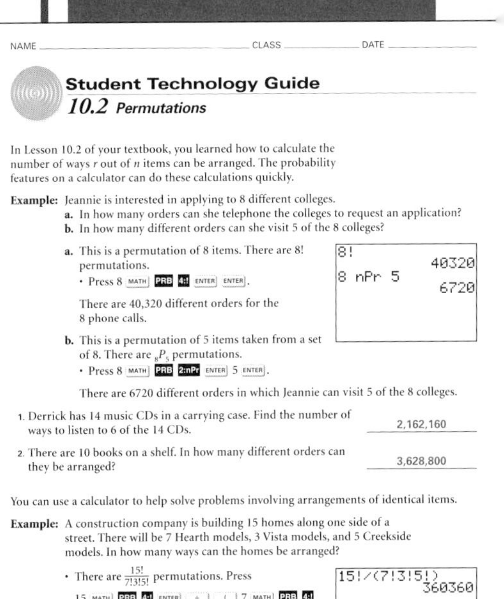

NAME ______ CLASS ______ DATE ______

Student Technology Guide
10.2 **Permutations**

In Lesson 10.2 of your textbook, you learned how to calculate the number of ways r out of n items can be arranged. The probability features on a calculator can do these calculations quickly.

Example: Jeannie is interested in applying to 8 different colleges.
a. In how many orders can she telephone the colleges to request an application?
b. In how many different orders can she visit 5 of the 8 colleges?

a. This is a permutation of 8 items. There are 8! permutations.
• Press 8 MATH PRB 4:! ENTER ENTER.

There are 40,320 different orders for the 8 phone calls.

b. This is a permutation of 5 items taken from a set of 8. There are $_8P_5$ permutations.
• Press 8 MATH PRB 2:nPr ENTER 5 ENTER.

There are 6720 different orders in which Jeannie can visit 5 of the 8 colleges.

1. Derrick has 14 music CDs in a carrying case. Find the number of ways to listen to 6 of the 14 CDs. 2,162,160

2. There are 10 books on a shelf. In how many different orders can they be arranged? 3,628,800

You can use a calculator to help solve problems involving arrangements of identical items.

Example: A construction company is building 15 homes along one side of a street. There will be 7 Hearth models, 3 Vista models, and 5 Creekside models. In how many ways can the homes be arranged?

• There are $\frac{15!}{7!3!5!}$ permutations. Press 15 MATH PRB 4:! ENTER ÷ (7 MATH PRB 4:! ENTER 3 MATH PRB 4:! ENTER 5 MATH PRB 4:! ENTER) ENTER.

There are 360,360 different ways to arrange the homes.

3. A gardener is planting a row of 18 trees. In how many ways can 9 elm, 3 eucalyptus, and 6 oak trees be arranged? 4,084,080

Look Beyond

In Exercises 75 and 76, students look at arrangements in which order does not matter. These are called *combinations* and are studied in Lesson 10.3.

APPLICATION

63. SPORTS A basketball team of 5 players is huddled in a circle along with their coach. In how many ways can the players and their coach be arranged in the huddle? **120**

Look Back

APPLICATION

64. NUTRITION You are given the diet information below. ***(LESSON 3.5)***

	Food X	Food Y	Required
Carbohydrates	2 units/ounce	3 units/ounce	at least 12 units
Fat	1 unit/ounce	2 units/ounce	at most 16 units
Protein	1 unit/ounce	3 units/ounce	at least 18 units
Cost	$0.20/ounce	$0.25/ounce	

Let x represent the required number of ounces of food X and let y represent the required number of ounces of food Y.

a. Write the set of constraints and the objective function.

b. Sketch the feasible region.

c. Find the minimum cost of a meal with foods X and Y.

64a. objective function: $C = 0.20x + 0.25y$

Constraints:

$$\begin{cases} 2x + 3y \geq 12 \\ x + 2y \leq 16 \\ x + 3y \geq 18 \\ x \geq 0 \\ y \geq 0 \end{cases}$$

64b.

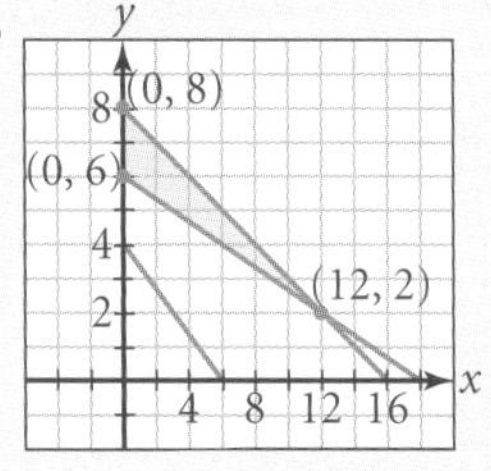

c. $1.50

Simplify each expression. *(LESSON 5.6)*

65. $(3i - 2) + (5 - 7i)$ **65. $3 - 4i$**

66. $(-1 + 4i)(2 - i)$ **$2 + 9i$**

67. $\frac{-1 - 8i}{2 + 5i}$ **$-\frac{42}{29} - \frac{11}{29}i$**

APPLICATION

68. SMALL BUSINESS The revenue earned from selling x items is given by the revenue function $R(x) = 6x$. The cost of producing x items is given by the cost function $C(x) = -0.1x^2 + 5x + 40$. Find the number of items that must be produced and sold to generate a profit (when revenue exceeds cost). **16** ***(LESSON 5.8)***

Write each expression as a single logarithm. Then simplify, if possible. *(LESSON 6.4)*

69. $\log_b 6 + \log_b 2 - \log_b 3$ **$\log_b 4$**

70. $3 \log 5^2 - 2 \log 5^3$ **0**

71. $\frac{1}{2}\log_2 36 - \log_2 12$ **-1**

72. $8 \log_5 x + \log_5 4$ **$\log_5 4x^8$**

CONNECTION

73 MAXIMUM/MINIMUM A rectangular piece of cardboard measures 12 inches by 16 inches. Find the maximum volume of an open-top box created by cutting squares of the same size from each corner and folding up the sides. ***(LESSON 7.3)*** **≈194.1 cubic inches**

74. Solve the nonlinear system at right. If there is no solution, write *none*. $\begin{cases} y = x^2 + 1 \\ 3x - y = -11 \end{cases}$ ***(LESSON 9.6)***

(−2, 5) and (5, 26)

Look Beyond

75. List all of the ways in which 2 of the letters A, B, C, D, and E can be chosen if the order in which letters are chosen is not important, that is, if AB is considered to be the same choice as BA.

76. Find the number of ways in which 2 of 26 letters can be chosen if the order in which letters are chosen is not important.

75. ***AB, AC, AD, AE, BC, BD, BE, CD, CE, DE***

76. 325

Combinations

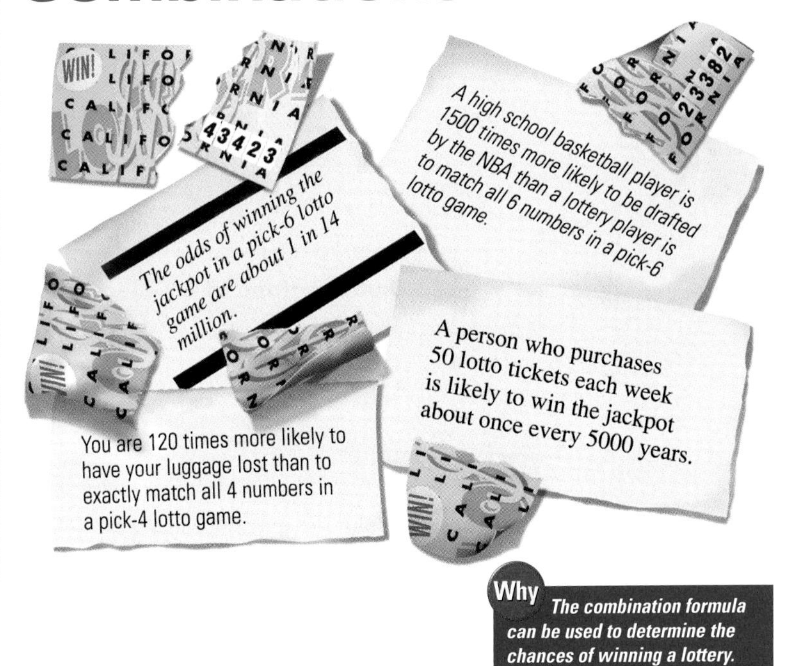

The odds of winning the jackpot in a pick-6 lotto game are about 1 in 14 million.
A high school basketball player is 1500 times more likely to be drafted by the NBA than a lottery player is to match all 6 numbers in a pick-6 lotto game.
A person who purchases 50 lotto tickets each week is likely to win the jackpot about once every 5000 years.
You are 120 times more likely to have your luggage lost than to exactly match all 4 numbers in a pick-4 lotto game.

Objectives

- Solve problems involving combinations.
- Solve problems by distinguishing between permutations and combinations.

Why *The combination formula can be used to determine the chances of winning a lottery. The results may surprise you.*

Prepare

NCTM PRINCIPLES & STANDARDS 1, 2, 4–10

QUICK WARM-UP

Evaluate each expression.

1. $6!$ 720
2. $4! \times 5!$ 2880
3. $\frac{8!}{3!}$ 6720
4. $\frac{9!}{4!5!}$ 126
5. ${}_7P_3$ 210
6. ${}_5P_2 \times {}_6P_4$ 7200

In how many different ways can 8 students be assigned to 8 seats that are arranged

7. in a line? 40,320
8. in a circle? 5040

Also on Quiz Transparency 10.3

Recall from Lesson 10.2 that a permutation is an arrangement of objects in a specific order. An arrangement of objects in which order is *not* important is called a **combination**.

In the Activity below, you can see the distinction between a permutation and a combination and learn how to count combinations.

Activity

Comparing Combinations and Permutations

APPLICATION
LOTTERY

You will need: no special tools

Consider a state lottery in which 3 numbers from 0 to 9 are selected. The numbers are not repeated. A lottery player can choose whether to play *exact match* or *any-order match.*

1. A person selects the numbers 8-4-1 and plays *exact match.* Write all of the ways that winning numbers can be drawn.
2. A person selects the numbers 8-4-1 and plays *any-order match.* Write all of the ways that winning numbers can be drawn.
3. Which has more ways to win: exact match or any-order match?
4. Explain why the prize is greater for winning with an exact match. CHECKPOINT ✔

Teach

Why In this lesson, students will study situations in which the order of arrangements does not matter. Have students discuss when such situations might occur. samples: numbers drawn in a bingo, members chosen for a committee

☞ For Activity Notes and answer to Checkpoint, see page 644.

Alternative Teaching Strategy

HANDS-ON STRATEGIES Give students 5 cubes of different colors and have them make as many combinations of 2 colors as possible. 10 Ask students to think about how this question is different from the questions about permutations studied in the last lesson.
Order does not matter here.

Have students draw two blanks to represent the cubes to be chosen. How many choices do they have for the first cube? 5 How many for the second cube? 4 Write _5_ × _4_ = 20. However, since a red-blue combination is the same as a blue-red combination, you must divide 20 by 2. Have students use similar reasoning to find all of the combinations of 3 colors chosen from 5 different colors. To account for the fact that a red-blue-green combination is the same as a blue-red-green combination, you must find the permutations of 3 colors and divide $5 \times 4 \times 3$ by that number. The number of combinations of 5 objects taken 3 at a time is $\frac{5 \times 4 \times 3}{3 \times 2 \times 1} = 10$. Have students model these combinations with the cubes.

Activity Notes

In this Activity, students investigate the difference between a permutation and a combination in a lottery situation. They should find that there are fewer ways to win an any-order match. Because there are more ways to win, the prize is greater for winning an exact match.

Cooperative Learning

Have students do the Activity in groups of three. Working in teams will help students be sure that they have written out all possible winning tickets for each scenario and will allow a discussion about the prize for each game.

CHECKPOINT ✔

4. **There are fewer ways to win with an exact match, so it is harder to do.**

ADDITIONAL EXAMPLE 1

Find the number of ways to rent 5 comedies from a collection of 30 comedies at a video store.

142,506

TRY THIS

36

CHECKPOINT ✔

$_{10}P_7$ is larger because rearranging the order of objects produces a different permutation.

The number of ways to listen to 2 of 5 CDs is $_5P_2 = 5 \times 4 = 20$. If you want to find the number of ways to purchase 2 of 5 CDs, their order does not matter. For example, choosing to purchase CD #1 and then CD #2 is no different than choosing CD #2 and then CD #1 because they are *combined* in the total purchase. To find the number of *combinations* of 2 CDs taken from 5 CDs, divide $_5P_2$ by 2 to compensate for duplicate combinations.

Notice that the formula for $_nC_r$ below is like the formula for $_nP_r$, except that it contains the factor $r!$ to compensate for duplicate combinations.

Combinations of *n* Objects Taken *r* at a Time

The number of combinations of n objects taken r at a time, is given by
$C(n, r) = {_nC_r} = \binom{n}{r} = \frac{n!}{r!(n-r)!}$, where $0 \le r \le n$.

The notations $C(n, r)$, $_nC_r$, and $\binom{n}{r}$ have the same meaning. All are read as "n choose r." In this chapter, we will use the notation $_nC_r$.

EXAMPLE 1 **Find the number of ways to purchase 3 different kinds of juice from a selection of 10 different juices.**

APPLICATION
SHOPPING

SOLUTION

The order in which the 3 juices are chosen is not important. Find the number of combinations of 10 objects taken 3 at a time.

$$\begin{aligned} _{10}C_3 &= \frac{10!}{3!(10-3)!} \\ &= \frac{10 \times 9 \times 8 \times 7!}{3! \times 7!} \\ &= \frac{10 \times 9 \times 8}{3 \times 2 \times 1} \\ &= 10 \times 3 \times 4 \\ &= 120 \end{aligned}$$

There are 120 ways to purchase 3 different kinds of juice from a selection of 10 different kinds.

TRY THIS Find the number of combinations of 9 objects taken 7 at a time.

CHECKPOINT ✔ Which is larger, $_{10}C_7$ or $_{10}P_7$? Does a given number of objects have more combinations or permutations? Explain.

Interdisciplinary Connection

GEOMETRY Select 6 distinct points on the circumference of a circle. Have students find the following:

a. How many chords can be drawn by joining pairs of points? $_6C_2 = 15$

b. How many triangles can be drawn by using 3 of these 6 points as vertices? $_6C_3 = 20$

c. How many quadrilaterals can be drawn by using 4 of these 6 points as vertices? $_6C_4 = 15$

d. How many pentagons can be drawn by using 5 of these 6 points as vertices? $_6C_5 = 6$

Inclusion Strategies

VISUAL LEARNERS Encourage students to draw diagrams or use manipulatives to model the problems given. Although this is useful when the number of combinations is small, it is ineffective when the number of combinations is large. However, if students have trouble deciding how to solve the problem, have them create a similar problem with the same structure but fewer items. For the simpler problem, they can use the same operations as those used to solve the complex problem.

When reading a problem, you need to determine whether the problem involves permutations or combinations.

EXAMPLE 2 **Use a permutation or combination to answer each question.**

APPLICATION
VOTING

a. How many ways are there to choose a committee of 3 people from a group of 5 people?

b. How many ways are there to choose 3 separate officeholders (chairperson, secretary, and treasurer) from a group of 5 people?

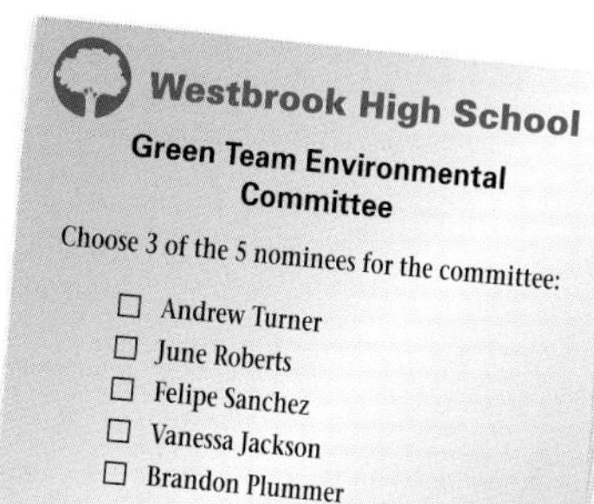

Westbrook High School
Green Team Environmental Committee
Choose 3 of the 5 nominees for the committee:
Andrew Turner
June Roberts
Felipe Sanchez
Vanessa Jackson
Brandon Plummer

SOLUTION

a. The members chosen for the committee will be members regardless of the order in which they are chosen. Find ${}_5C_3$.

$${}_5C_3 = \frac{5!}{3!2!} = \frac{5 \times 4}{2} = 10$$

There are 10 ways to choose a committee of 3 people from a group of 5 people.

b. The officeholders are chosen to fulfill particular positions. Therefore, order is important. Find ${}_5P_3$.

$${}_5P_3 = \frac{5!}{2!} = 5 \times 4 \times 3 = 60$$

There are 60 ways to choose a chairperson, a secretary, and a treasurer from a group of 5 people.

TRY THIS How many ways are there to choose a committee of 2 people from a group of 7 people? How many ways are there to choose a chairperson and a co-chairperson from a group of 7 people?

EXAMPLE 3 **How many different ways are there to purchase 2 CDs, 3 cassettes, and 1 videotape if there are 7 CD titles, 5 cassette titles, and 3 videotape titles from which to choose?**

APPLICATION
SHOPPING

SOLUTION

Consider the CDs, cassettes, and videotapes separately, and apply the Fundamental Counting Principle.

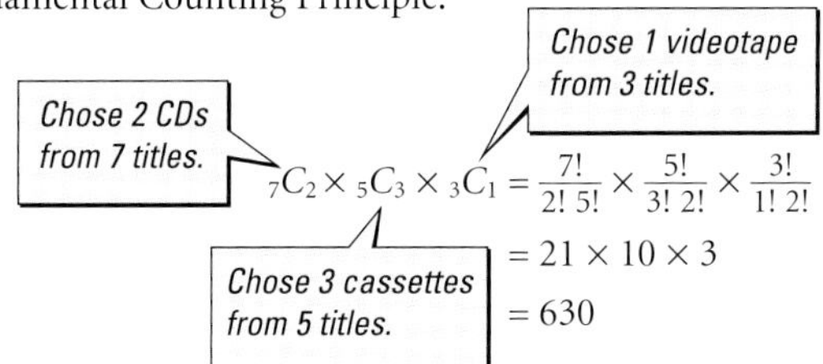

Chose 2 CDs from 7 titles.
Chose 1 videotape from 3 titles.
Chose 3 cassettes from 5 titles.

$${}_7C_2 \times {}_5C_3 \times {}_3C_1 = \frac{7!}{2!\,5!} \times \frac{5!}{3!\,2!} \times \frac{3!}{1!\,2!}$$
$$= 21 \times 10 \times 3$$
$$= 630$$

There are 630 different ways to make the purchase.

TRY THIS How many different ways are there to purchase 3 CDs, 4 cassettes, and 2 videotapes if there are 3 CD titles, 6 cassette titles, and 4 videotape titles from which to choose?

Enrichment

Consider the following combinations:

1. ${}_{10}C_{10}$ 1
2. ${}_{10}C_0$ 1
3. ${}_kC_0$, where k is a positive integer 1

Have students calculate each of these and discuss the results. Ask students to create an example for each of the situations given.

ADDITIONAL **EXAMPLE 2**

Use a permutation or a combination to answer each question.

a. How many ways are there to give 3 honorable mention awards to a group of 8 entrants in a contest? 56

b. How many ways are there to award first prize, second prize, and third prize to a group of 8 entrants in a contest? 336

Teaching Tip

TECHNOLOGY When using a TI-83 or TI-82 graphics calculator to find combinations and permutations, use the following, where n and r represent numbers:

For ${}_nP_r$, select
n MATH PRB 2:${}_nP_r$ r ENTER.

For ${}_nC_r$, select
n MATH PRB 3:${}_nC_r$ r ENTER.

${}_nP_r$ is often written as $P_{n,r}$, and ${}_nC_r$ is often written as $C_{n,r}$ or $\binom{n}{r}$.

TRY THIS
${}_7C_2 = 21$; ${}_7P_2 = 42$

ADDITIONAL **EXAMPLE 3**

Terry is buying paperback books to read while on vacation. **How many different ways are there for Terry to purchase 3 novels and 2 nonfiction books if there are 10 novels and 6 nonfiction books to choose from?**
1800

TRY THIS
${}_3C_3 \times {}_6C_4 \times {}_4C_2 = 1(15)(6) = 90$

ADDITIONAL

EXAMPLE 4

In a recent survey of 30 students, 25 students favored an earlier opening time for the school cafeteria and 5 opposed it. **Find the probability that in a random sample of 8 respondents from this survey, exactly 6 favored the earlier opening time and exactly 2 opposed it.**

≈0.30, or ≈30%

Teaching Tip

TECHNOLOGY When using a calculator to perform operations on combinations, it is important that students are precise in using parentheses to group combinations in the denominator.

TRY THIS

≈0.6%

CRITICAL THINKING

$\frac{{}_aC_r \times {}_bC_s}{{}_{a+b}C_{r+s}}$; $r \le a$ and $s \le b$

Using Combinations and Probability

Recall from Lesson 10.1 that you can find the probability of event A by using the following ratio:

$$P(A) = \frac{\text{number of outcomes in event } A}{\text{number of outcomes in the sample space}}$$

In many situations, you can find and evaluate the numerator and the denominator by applying the formula for combinations.

EXAMPLE 4

APPLICATION
SURVEYS

In a recent survey of 25 voters, 17 favor a new city regulation and 8 oppose it.

Find the probability that in a random sample of 6 respondents from this survey, exactly 2 favor the proposed regulation and 4 oppose it.

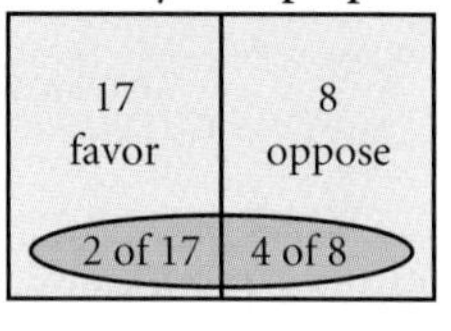

SOLUTION

1. Find the number of outcomes in the event. Use the Fundamental Counting Principle.

$${}_{17}C_2 \times {}_8C_4$$

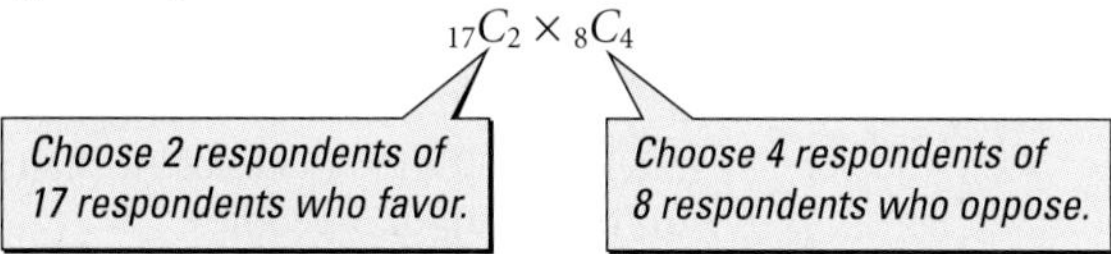

2. Find the number of outcomes in the sample space.

${}_{25}C_6$ ← Choose 6 people from the 25 respondents.

3. Find the probability.

$$\frac{\text{number of outcomes in event A}}{\text{number of outcomes in the sample space}} = \frac{{}_{17}C_2 \times {}_8C_4}{{}_{25}C_6} \approx 0.05$$

TECHNOLOGY
GRAPHICS CALCULATOR
Keystroke Guide, page 686

Thus, the probability of selecting exactly 2 respondents who favor the proposed regulation and 4 who oppose it in a randomly selected group of 6 respondents is about 0.05, or 5%.

TRY THIS

Find the probability that in a random sample of 10 respondents from the above survey, all 10 favor the proposed regulation.

CRITICAL THINKING

Sets A and B are two nonoverlapping sets. Set A contains a distinct objects and set B contains b distinct objects. You wish to choose x objects randomly from both set A and set B. Find the probability of choosing r objects from set A and s objects from set B. Are there any restrictions on r and s?

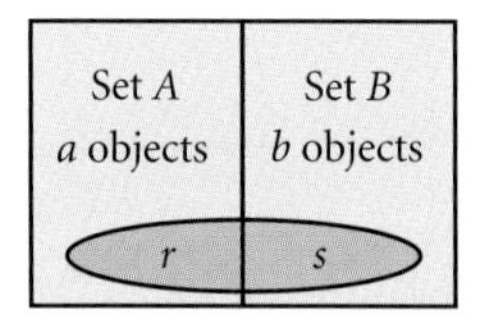

Reteaching the Lesson

INVITING PARTICIPATION Have groups of students design a game where they pick n numbers from a range of numbers, and have them test their design by using a random-number generator to obtain various sets of numbers. Each group should decide how many numbers are to be drawn for each set and the range of numbers from which the sets are to be drawn. To begin with, students should decide on the probability of winning a game in which the player must match 2 numbers from the given range of numbers. Have students use a random-number generator to simulate several rounds of the game. Then have students increase how many numbers are to be drawn from the range of numbers and compute the new probability of winning for each set. Have each group share their game with the class, and let the class decide whether the probability of winning that is given by the group is accurate.

A calculator with a random feature can be used as a random-number generator by taking the first one or two decimal places of each number generated. Any number not in the desired range of numbers should be ignored.

Exercises

Communicate

1. Describe the difference between a combination and a permutation. Give examples to illustrate your descriptions.
2. Describe the relationship between ${}_5P_3$ and ${}_5C_3$. Then explain how the formula for combinations is related to the formula for permutations.
3. Using the definition of a combination and the values for ${}_8C_8$ and ${}_8C_1$, find and describe ${}_nC_n$ and ${}_nC_1$ for any whole number n.

Guided Skills Practice

4. Find the number of ways to rent 4 comedy videos from a collection of 9 comedy videos. ***(EXAMPLE 1)*** **126**
5. **Use a permutation or combination to answer each question.** ***(EXAMPLE 2)***
   a. How many ways are there to choose a committee of 3 from a group of 12 people? **220**
   b. How many ways are there to choose 3 separate officeholders (chairperson, secretary, and treasurer) from a group of 12 people? **1320**
6. How many different ways are there to purchase 3 CDs, 4 cassettes, and 2 videotapes if there are 8 CD titles, 5 cassette titles, and 5 videotape titles from which to choose? ***(EXAMPLE 3)*** **2800**
7. A survey of 30 people showed that 19 people favor a new city regulation and that 11 people oppose it. Find the probability that in a random sample of 8 respondents from the survey, exactly 3 favor the proposed regulation and 5 oppose it. ***(EXAMPLE 4)*** **≈7.6%**

Practice and Apply

Find the value of each expression.

8. ${}_7C_4$ **35** **9.** ${}_8C_4$ **70** **10.** ${}_{10}C_7$ **120** **11.** ${}_9C_5$ **126**

12. ${}_9C_1$ **9** **13.** ${}_{11}C_1$ **11** **14.** ${}_{15}C_{15}$ **1** **15.** ${}_{12}C_{12}$ **1**

16. $\frac{6!}{2!4!} \times \frac{5!}{4!1!}$ **75** **17.** $\frac{4!}{3!1!} \times \frac{9!}{5!4!}$ **504** **18.** $\frac{{}_6C_5 \times {}_{15}C_2}{{}_{21}C_7}$ **19.** $\frac{{}_{14}C_5 \times {}_9C_7}{{}_{23}C_{12}}$

18. ≈0.005
19. ≈0.053

Find the number of ways in which each committee can be selected.

20. 3 people from a group of 5 **21.** 7 people from a group of 8

22. 8 people from a group of 12 **23.** 6 people from a group of 10

20. 10
21. 8
22. 495
23. 210

24. How many different 12-member juries can be chosen from a pool of 32 people? **225,792,840**

25. A test consists of 20 questions, and students are told to answer 15 of them. In how many different ways can they choose the 15 questions? **15,504**

Assess

Selected Answers

Exercises 4–7, 9–49 odd

ASSIGNMENT GUIDE	
In Class	1–7
Core	9–39 odd
Core Plus	8–42 even
Review	44–49
Preview	50

Extra Practice can be found beginning on page 940.

Error Analysis

Students often have difficulty deciding whether a situation involves a permutation or a combination, as required in Exercises 34–39. Encourage them to assign names to the items in the situations to help them decide whether the order matters. Role playing may help them make this decision.

40a. $_nC_n = \dfrac{n!}{n!(n-n)!} = \dfrac{n!}{n!0!}$
$= \dfrac{n!}{n!(1)} = 1$

b. $_nC_1 = \dfrac{n!}{1!(n-1)!} = \dfrac{n \times (n-1)!}{(n-1)!}$
$= n$

c. $_nC_0 = \dfrac{n!}{0!(n-0)!}$
$= \dfrac{n!}{n!} = 1$

d. $_nC_r = \dfrac{n!}{r!(n-r)!} = \dfrac{n!}{(n-r)!r!}$
$= \dfrac{n!}{(n-r)![n-(n-r)]!}$
$= {}_nC_{n-r}$

A pizza parlor offers a selection of 3 different cheeses and 9 different toppings. In how many ways can a pizza be made with the following ingredients?

26. 1 cheese and 3 toppings **252**
27. 1 cheese and 4 toppings **378**
28. 2 cheeses and 4 toppings **378**
29. 2 cheeses and 3 toppings **252**

A bag contains 5 white marbles and 3 green marbles. Find the probability of selecting each combination.

30. 1 green and 1 white **≈53.6%**
31. 2 green and 1 white **≈26.8%**
32. 2 green and 2 white **≈42.9%**
33. 3 green and 2 white **≈17.9%**

34. combination
35. combination
36. permutation
37. permutation
38. permutation

internet connect go.hrw.com
Homework Help Online
Go To: **go.hrw.com**
Keyword:
MB1 Homework Help
for Exercises 34–39

For Exercises 34–38, determine whether each situation involves a permutation or a combination.

34. Four recipes were selected for publication and 302 recipes were submitted.
35. Nine players are selected from a team of 15 to start the softball game.
36. Four out of 200 contestants were awarded prizes of $100, $75, $50, and $25.
37. A president and vice-president are elected for a class of 210 students.
38. The batting order for the 9 starting players is announced.

39. a. Find the number of different 5-card hands that can be dealt from a standard deck of 52 playing cards. **2,598,960**
b. Find the probability that 4 kings and a queen are randomly drawn from a standard deck of 52 playing cards. $\frac{1}{649{,}740} \approx 0.000154\%$

CHALLENGE

40. Use the formula for $_nC_r$ to verify each statement below.
a. $_nC_n = 1$ for all positive integer values of n
b. $_nC_1 = n$ for all positive integer values of n
c. $_nC_0 = 1$ for all positive integer values of n
d. $_nC_r = {}_nC_{n-r}$ for all positive integer values of n and r, where $0 \le r \le n$
e. Give an intuitive explanation for your answers to parts **a–d**.

41. $_6C_1 = 6$; $_6C_2 = 15$; $_6C_3 = 20$; $_6C_4 = 15$; $_6C_5 = 6$; $_6C_6 = 1$

41. CULTURAL CONNECTION: ASIA Around 600 B.C.E. in the Vedic period of Indian history, a writer named Sushruta found all combinations of the 6 different tastes: bitter, sour, salty, astringent, sweet, and hot. How many total combinations of the tastes are possible when taken 1 at a time, 2 at a time, etc?

APPLICATION

42a. 53,130
b. 12,600
c. ≈23.7%

42. HEALTH From a group of 10 joggers and 15 nonjoggers, a university researcher will choose 5 people to participate in a study of heart disease.
a. In how many ways can this be done if it does not matter how many of those chosen are joggers and how many are nonjoggers?
b. In how many ways can this be done if exactly 3 joggers must be chosen?
c. Find the probability that exactly 3 of the 5 people randomly selected from the group are joggers.

Practice

NAME ______ CLASS ______ DATE ______

Practice
***10.3* Combinations**

Find the number of ways in which each committee can be selected.

1. a committee of 5 people from a group of 8 people ______ 56
2. a committee of 2 people from a group of 16 people ______ 120
3. a committee of 4 people from a group of 7 people ______ 35
4. a committee of 8 people from a group of 15 people ______ 6435
5. a committee of 3 people from a group of 9 people ______ 84

At a luncheon, guests are offered a selection of 4 different grilled vegetables and 5 different relishes. In how many ways can the following items be chosen?

6. 2 vegetables and 3 relishes ______ 60
7. 3 vegetables and 2 relishes ______ 40
8. 4 vegetables and 4 relishes ______ 5
9. 3 vegetables and 3 relishes ______ 40

A bag contains 8 white marbles and 7 blue marbles. Find the probability of selecting each combination.

10. 2 white and 3 blue ______ about 33%
11. 3 white and 2 blue ______ about 39%
12. 4 white and 1 blue ______ about 16%

Determine whether each situation involves a permutation or a combination.

13. A high school offers 5 foreign language programs. In how many ways can a student choose 2 programs? ______ combination
14. In how many ways can 20 members be chosen from a 60-member chorus to sing the national anthem at a graduation ceremony? ______ combination
15. In how many ways can a captain, co-captain, and team manager be chosen from among the 18 members of a volleyball team? ______ permutation
16. First- through fourth-place prizes are to be awarded in an essay contest. In how many ways can the winners be selected from among 125 entries? ______ permutation

40e. Answers may vary. Sample answers:

Part a— There is only one way to choose all n objects.

Part b— When choosing one object, there are n possible choices.

Part c— There is only one way to choose none of the n objects.

Part d— Choosing r of the n objects is the same as excluding $n - r$ of the objects. Thus, for each set of n objects, there is a set of $n - r$ objects.

APPLICATION

43. LOTTERY In many lotteries, contestants choose 6 out of 50 numbers (without replacement).

Event (applies to U.S. population)	Experimental probability
Undergo an audit by the IRS this year	$\frac{1}{100}$
Being hit by lightning	$\frac{1}{9100}$
Being hit by a baseball in a major league game	$\frac{1}{300,000}$

[*Source: Les Krantz,* What the Odds Are, *Harper Perennial, 1992*]

a. Find the probability of winning this kind of lottery with one set of numbers. $\frac{1}{15,890,700}$

b. Determine how many times more likely than winning the lottery it is for each of the events in the table at left to occur.
undergo audit: 158,907 times
hit by lightening: 1746 times
hit by baseball: 53 times

Look Back

APPLICATION

BUSINESS A movie theater charges \$5 for adults and \$3 for children. The theater needs to sell at least \$2500 worth of tickets to cover expenses. Graph the solution for each situation below. ***(LESSON 3.3)***

44. The theater fails to sell at least \$2500 worth of tickets.

45. The theater breaks even, selling exactly \$2500 worth of tickets.

46. The theater makes a profit, selling more that \$2500 worth of tickets.

Graph the solution to each inequality. ***(LESSON 5.8)***

47. $y \leq x^2 - 5$ **48.** $y \geq (x + 1)^2$ **49.** $y < (x - 1)^2 + 3$

Look Beyond

50. A governor appointed 5 department heads from a group of people that consisted of 8 acquaintances and 22 others. Assuming that all choices are equally likely, find the probability of each event below.

a. All 5 were acquaintances. **≈0.039%**

b. Exactly 4 of the 5 were acquaintances. **≈1.1%**

c. At least 3 of the 5 were acquaintances. **≈10.239%**

PORTFOLIO ACTIVITY

Refer to Example 4 on page 646. Assign numbers from 1 to 25 to the survey respondents. Let the numbers from 1 to 17 represent the people who favor the new city regulation, and let the numbers from 18 to 25 represent the people who oppose it.

1. Generate 6 random integers from 1 to 25 inclusive. (Refer to page 687 of the Keystroke Guide.) This represents a random sample of 6 respondents. Repeat this generation of 6 random integers for a total of 30 trials. Record the outcomes.
2. For the 30 trials, count the total number of times that exactly 2 of the 6 random integers were between 1 and 17 inclusive. Use your outcomes to approximate the probability that exactly 2 of the 6 respondents chosen favor the proposed regulation.
3. Compare your approximate probability with the theoretical probability obtained in the lesson. How well do you think the random-number generator models this problem? Explain.

WORKING ON THE CHAPTER PROJECT

You should now be able to complete Activity 2 of the Chapter Project.

44.

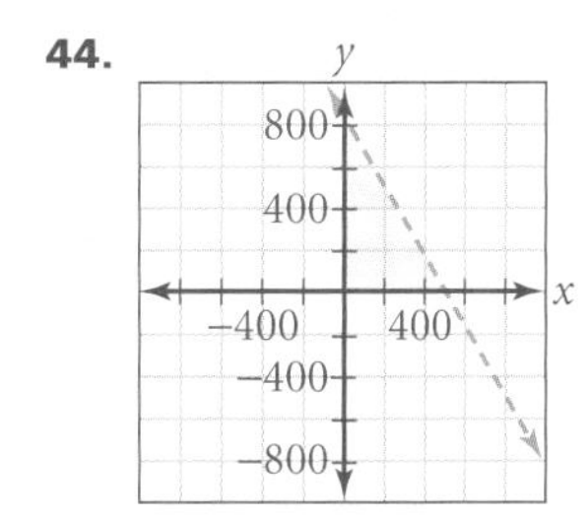

45.

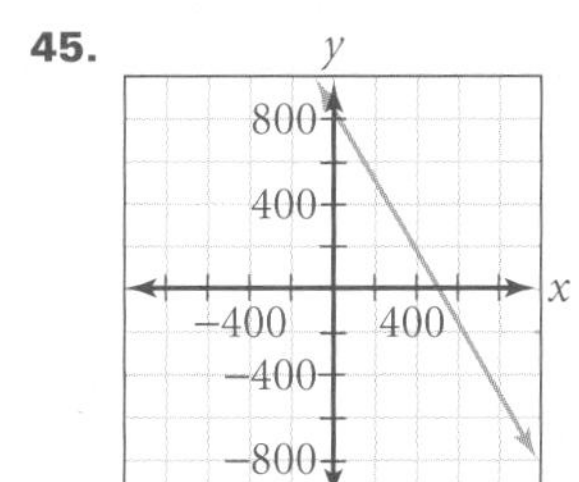

46.

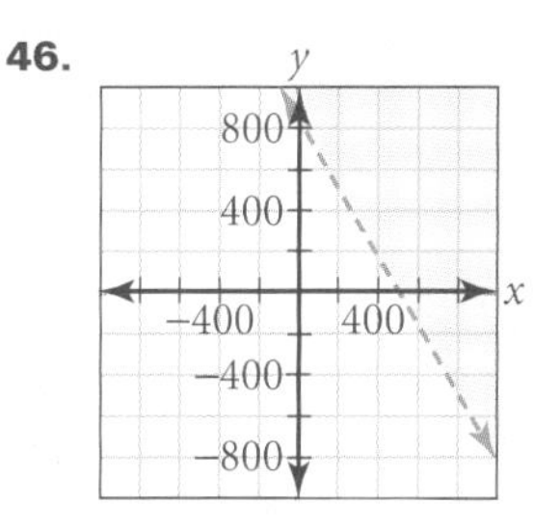

The answers to Exercises 47–49 can be found in Additional Answers beginning on page 1002.

In Exercise 50, students begin to explore situations in which probabilities must be combined due to conditions placed on the sample. Students will study these concepts further in Lesson 10.6.

ALTERNATIVE
Assessment

Portfolio Activity

The Portfolio Activity can be used as preparation for the Chapter Project or as a separate activity. In the Portfolio Activity on this page, students use a random-number generator to determine experimental probabilities. Students increase the number of trials to increase the accuracy. Then students compare their results with theoretical probabilities.

Student Technology Guide

NAME ______ CLASS ______ DATE ______

Student Technology Guide
10.3 ***Combinations***

In probability, a *combination* is an arrangement of objects in which order does not matter. Many calculators have a built-in feature for doing calculations involving combinations.

Example: The introductory offer for a book club allows a new member to choose 8 free books from a list of 45. Find the number of ways that the 8 free books can be chosen.

- This is a combination of 8 books taken from a set of 45. There are ${}_{45}C_8$ combinations. Press 45 MATH PRB 3:nCr ENTER 8 ENTER.

```
45 nCr 8
      215553195
```

There are 215,553,195 ways the free books can be chosen.

1. In how many ways can a basketball coach choose 5 starting players from a team of 11? 462
2. A bridge hand consists of 13 cards out of a standard deck of 52. How many different bridge hands are possible? about 6.35×10^{11}

In some probability problems, you can use combinations to find the number of favorable outcomes and the total number of outcomes.

Example: An animal shelter has 17 dogs and 25 cats available for adoption. Every Sunday, 8 of the pets are featured in the newspaper. If the featured pets are chosen at random, find the probability that exactly 5 of them are dogs and 3 are cats.

- Find the number of favorable outcomes. This is the number of ways that 5 of 17 dogs and 3 of 25 cats can be chosen, or ${}_{17}C_5 \times {}_{25}C_3$.
- Find the total number of outcomes. This is the number of ways 8 of the 42 pets can be chosen, or ${}_{42}C_8$.
- Use a calculator to find the probability. To divide the number of favorable outcomes by the total number of outcomes, press (17 MATH PRB 3:nCr ENTER 5) × (25 MATH PRB 3:nCr ENTER 3) ÷ (42 MATH PRB 3:nCr ENTER 8) ENTER.

```
(17 nCr 5)*(25 n
Cr 3)/(42 nCr 8)
      .1205827137
```

The probability that 5 of the pets are dogs and 3 are cats is about 0.12, or 12%.

3. A survey of 28 juniors showed that 20 of them take chemistry. Find the probability that, if 10 of the surveyed students are chosen at random, 6 of them take chemistry and 4 do not. about 0.21, or 21%

Focus

An actual controversy over a columnist's answer to a strategy question provides the basis for using probability theory to settle a dispute. Students use a simulation and examine a table of outcomes to determine who is right and why conclusions based on logical analysis sometimes go against our intuitive judgments, especially in probability.

Motivate

Before students read the excerpt from "Ask Marilyn," make sure that they understand the game-show situation. You may want to have them play a few rounds to demonstrate. Ask students whether the question in the column accurately describes the game-show situation. They may notice that the question does not state that the host must always show a losing door. You can discuss this now or leave it until the Activities have been completed.

EYEWITNESS MATH

You have reached the final round of a TV game show called *Let's Make a Deal*. Behind one of the three numbered doors is a new car. Behind each of the other two doors is a goat. You have chosen door number 1. The host, Monty Hall, knows what's behind each door. He opens door number 3 to show you a goat.

Behind one of these doors is a new car.

Should you stick with number 1 or switch to number 2?

Then he pops the question: "Do you want to change your mind?"

What would you do? Would you stick with door number 1 or switch to door number 2?

Many contestants faced such a dilemma on the show, which ran for over 25 years. In 1990, a question based on this situation was submitted to a columnist, Marilyn vos Savant, who is reported to have the highest IQ in the world.

BY MARILYN VOS SAVANT

Ask Marilyn®

Suppose you're on a game show, and you're given the choice of three doors: Behind one door is a car; behind the others, goats. You pick a door, say No. 1, and the host, who knows what's behind the doors, opens another door, say No. 3, which has a goat. He then says to you, "Do you want to pick door No. 2?" Is it to your advantage to switch your choice?

—Craig F. Whitaker, Columbus, Md.

Yes; you should switch. The first door has a one-third chance of winning, but the second door has a two-thirds chance. Here's a good way to visualize what happened. Suppose there are a million doors, and you pick door No. 1. Then the host, who knows what's behind the doors and will always avoid the one with the prize, opens them all except door #777,777. You'd switch to that door pretty fast, wouldn't you?

1. Answers may vary.

2. Answers may vary.

3.

Door with car	Door you choose	Door you are shown	Result if you switch	Result if you stick
1	1	2 or 3	lose	win
1	2	3	win	lose
1	3	2	win	lose
2	1	3	win	lose
2	2	1 or 3	lose	win
2	3	1	win	lose
3	1	2	win	lose
3	2	1	win	lose
3	3	1 or 2	lose	win

Nearly a year later, an article in the *New York Times* reported that Marilyn vos Savant had received about 10,000 letters in response to her answer. Most of the letters disagreed with her. Many came from mathematicians and scientists with arguments like this:

> ...You blew it! Let me explain: If one door is shown to be a loser, that information changes the probability of either remaining choice—neither of which has any reason to be more likely—to $1/2$. As a professional mathematician, I'm very concerned with the general public's lack of mathematical skills. Please help by confessing your error and, in the future, being more careful...

Cooperative Learning

1. Before you analyze the problem in detail, explain which strategy you think is best and why.
2. Make three cards (two goats, one car) to model the situation. Try 10 games in which you stick with your choice. Then try 10 games in which you switch. Compare the results.
3. Make a table with the column headings shown, and complete it using all possible options. According to your table, should you switch? Explain.
4. Suppose that you wrote the response for Marilyn vos Savant's column. How would you answer the question about the game show?

Door with car	Door you choose	Door you are shown	Result if you switch	Result if you stick
1	1	2 or 3	Lose	Win
1	2	3	Win	Lose
1				
2				
⋮				

4. Answers may vary. Sample answer: In the 9 possible cases, if you always switch your choice, you win 6 times and lose 3 times. If you always stick with your first choice, you win only 3 times and lose 6 times. So, if you always switch, you should win about $\frac{2}{3}$ of the time. If you always stick, you should win about $\frac{1}{3}$ of the time.

Cooperative Learning

Divide the class into groups of four, five, or six. Each group should work together to develop a strategy for the game based on probabilities. Have each group select two players. The other members of the team will be the hosts. Then have each group play several rounds of the game, using cards to simulate the doors. Encourage the hosts to try to manipulate the players away from the strategies discussed as a group. Note how the power of psychological manipulation can influence the players' choices.

Discuss

The situation is simple but ambiguous. Because the problem can have different interpretations, it is easy for some solutions to have subtle flaws. The original problem is not well-defined because it does not require the host to reveal a door in *every* case. In fact, in the actual game show, the host would use psychology to try to mislead contestants. For theoretical probability to be effective, it must be defined precisely. Students should discuss the effects of the host's psychological efforts on the players.

Prepare

NCTM PRINCIPLES & STANDARDS 1–10

QUICK WARM-UP

Swiss cheese was on sale at a local market. The table below shows the cheese purchases of 150 customers at this market.

Swiss	35
Other cheese	55
No cheese	60

Find the experimental probability of each event.

1. Swiss $\frac{35}{150}$, or ≈23.3%
2. no cheese $\frac{60}{150}$, or ≈40%

Of 600 customers at this market, how many can be expected to buy

3. Swiss cheese? 140
4. other cheese? 220

Also on Quiz Transparency 10.4

Teach

Why Have students discuss situations in which they would be interested in the probability of event A or event B occurring. For example, suppose that you want to know the chances of drawing a 3 or a 2 from a deck of cards.

10.4 Using Addition With Probability

Why *You can often use addition to determine the probability of two or more events occurring.*

Objectives

- Find the probabilities of mutually exclusive events.
- Find the probabilities of inclusive events.

APPLICATION
EXTRACURRICULAR ACTIVITIES

The drama, mathematics, and jazz clubs at Gloverdale High School have 32 members, 33 members, and 39 members, respectively. Some club members belong to more than one club, as indicated in the diagram at right. What is the probability that a randomly selected club member belongs to at least two clubs? *You will answer this question in Example 3.*

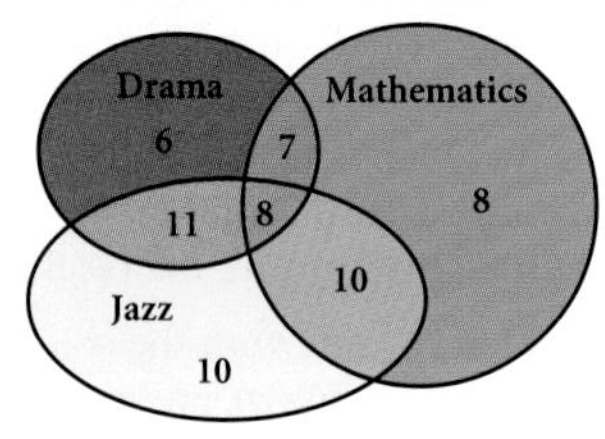

Events that can occur at the same time are called **inclusive events.** For example, a student can belong to more than one club at the same time.

Events that cannot occur at the same time are called **mutually exclusive events**. For example, if you flip a coin, you cannot get *both* heads and tails.

Activity: Exploring Two-Event Probabilities

You will need: a pair of number cubes, preferably of two different colors

1. Copy and complete the following table by rolling two number cubes 10 times:

Toss	1st cube	2nd cube	Sum	Product
1				
2				
3				
⋮				

Alternative Teaching Strategy

USING COGNITIVE STRATEGIES Have students consider a situation that has two mutually exclusive outcomes. Tossing a coin has two mutually exclusive outcomes because the coin will show either heads or tails but not both. Have students find the following probabilities: $P(H)$, $P(T)$, and $P(H \text{ or } T)$. **0.5; 0.5; 1** Have students compare $P(H \text{ or } T)$ to the individual probabilities $P(H)$ and $P(T)$. Now suppose that you are choosing a committee member from a pool of 3 people—John, Mary, and Alice. Have students find each of the following probabilities:

$P(\text{John})$ $\frac{1}{3}$

$P(\text{Mary})$ $\frac{1}{3}$

$P(\text{Alice})$ $\frac{1}{3}$

$P(\text{John } or \text{ Mary})$ $\frac{2}{3} = P(J) + P(M)$

$P(\text{John } or \text{ Mary } or \text{ Alice})$

$P(J) + P(M) + P(A) = \frac{1}{3} + \frac{1}{3} + \frac{1}{3} = 1$

Have students discuss each of the probabilities and compare them.

2. Using your 10 trials, copy and complete the table of experimental probabilities for each pair of events listed below.

   a. Let A be the event that the first cube is a 6 and let B be the event the first cube is a 3.

   b. Let A be the event that the first cube is a 6 and let B be the event that the sum is 7.

   c. Let A be the event that the sum is less than 5 and let B be the event that the product is greater than 5.

	a.	b.	c.
$P(A)$			
$P(B)$			
$P(A \text{ or } B)$			
$P(A) + P(B)$			

CHECKPOINT ✔ 3. Based on your results, does $P(A \text{ or } B) = P(A) + P(B)$? When can you be sure that this statement is true?

Mutually exclusive events and inclusive events are illustrated below with a number cube.

Mutually exclusive events

Let A represent an even number.

$P(A) = \frac{3}{6}$

Let B represent 3.

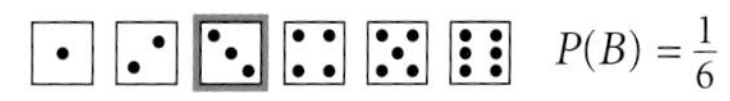

$P(B) = \frac{1}{6}$

P(an even number *or* 3)

$P(A \text{ or } B)$

Because A and B are mutually exclusive events, you can add $P(A)$ and $P(B)$ to find $P(A \text{ or } B)$.

$$P(A \text{ or } B) = \frac{3}{6} + \frac{1}{6} = \frac{4}{6}, \text{ or } \frac{2}{3}$$

Inclusive events

Let A represent an even number.

$P(A) = \frac{3}{6}$

Let C represent 4.

$P(C) = \frac{1}{6}$

P(an even number *or* 4)

$P(A \text{ or } C)$

Because A and C are inclusive events, you must *subtract* $P(A \text{ and } C)$ from the sum of $P(A)$ and $P(C)$ to find $P(A \text{ or } C)$.

$$P(A \text{ or } C) = \frac{3}{6} + \frac{1}{6} - \frac{1}{6} = \frac{3}{6}, \text{ or } \frac{1}{2}$$

A or B, exclusive

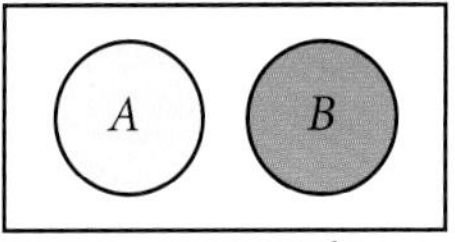

Sample space

A or B, inclusive

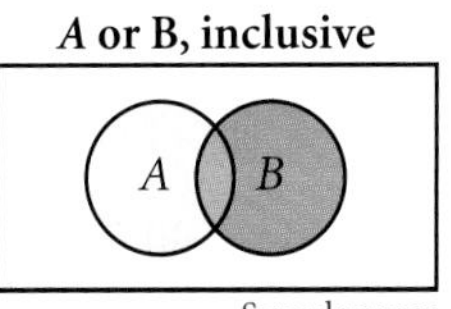

Sample space

Probability of *A or B*

Let A and B represent events in the same sample space.

If A and B are mutually exclusive events, then
$P(A \text{ or } B) = P(A) + P(B)$.

If A and B are inclusive events, then
$P(A \text{ or } B) = P(A) + P(B) - P(A \text{ and } B)$.

CHECKPOINT ✔ Which formula for $P(A \text{ or } B)$ can be used with any events A and B in the same sample space? Why?

Activity Notes

In this Activity, students use experimental data to show that $P(A \text{ or } B) = P(A) + P(B)$ when A and B are mutually exclusive events. Students use a pair of number cubes to find probabilities associated with different outcomes and should find that $P(A \text{ or } B) = P(A) + P(B)$ when A and B represent mutually exclusive events.

Cooperative Learning

You may wish to have students do the Activity in pairs, with one partner being the experimenter and the other being the recorder. Have students alternate roles during the Activity.

CHECKPOINT ✔

3. **$P(A \text{ or } B) = P(A) + P(B)$ is true when A and B are mutually exclusive.**

CHECKPOINT ✔

$P(A \text{ or } B) = P(A) + P(B) - P(A \text{ and } B)$ works for any two events in the same sample space.

$P(A \text{ and } B)$ is incorporated into $P(A)$ and into $P(B)$ and was added twice, so it must be subtracted out once.

Interdisciplinary Connection

BUSINESS A car dealer considers increasing the standard option package on a car. Looking at past sales records, the dealer finds that 20% of the customers added leather seats and 30% of the customers added a sunroof. In fact, 10% of all customers added leather seats and a sunroof. Based on these numbers, find the probability that a customer would add either leather seats or a sunroof.
0.40, or 40%

Enrichment

Have students consider tossing one coin and one number cube simultaneously. Find the probability of getting heads or a 3, $P(H \text{ or } 3)$. Have students toss the coin and number cube 20 times and record the number of tosses that result in heads or a 3. Combine the results from the class to get a larger sample space. Have students determine the probability of each outcome occurring independently. Have the class decide how these two probabilities are related to the single experimental probability. Then have students find the theoretical probability $P(H \text{ or } 3)$.
$\frac{1}{2} + \frac{1}{6} - \frac{1}{12} = \frac{7}{12}$, **or ≈58%**

Teaching Tip

Unless otherwise noted, all coins, number cubes, and spinners discussed in this chapter are assumed to be fair, and all outcomes from using these devices are assumed to be random.

Use Teaching Transparency 39.

ADDITIONAL EXAMPLE 1

A market research firm conducted a survey of 250 customers of long-distance carrier A and 250 customers of long-distance carrier B. The customers were asked whether they were satisfied with, dissatisfied with, or had no opinion about their long-distance service. The responses are given in the table below.

Group	A	B	Total
satisfied	65	180	245
dissatisfied	157	54	211
no opinion	28	16	44
Total	250	250	500

Find the probability that a randomly selected respondent to the survey is satisfied with or has no opinion about their long-distance service.

$\frac{289}{500}$, or 57.8%

TRY THIS

63%

ADDITIONAL EXAMPLE 2

Refer to the survey described above in Additional Example 1. **Find the probability that a randomly selected respondent to the survey is a customer of long-distance carrier B or is satisfied with the long-distance service.**

$\frac{315}{500}$, or 63%

TRY THIS

70%

CHECKPOINT ✔

$P(A) = 0.27$;

$P(A^C) = 0.37 + 0.36 = 0.73$;

$P(A) + P(A^C) = 0.27 + 0.73 = 1$

EXAMPLE 1 APPLICATION SURVEYS

In a survey about a change in public policy, 100 people were asked if they favor the change, oppose the change, or have no opinion about the change. The responses are indicated at right.

	Men	Women	Total
Favor	18	9	27
Oppose	12	25	37
No opinion	20	16	36
Total	50	50	100

Find the probability that a randomly selected respondent to the survey opposes *or* has no opinion about the change in policy.

SOLUTION

The events "oppose" and "no opinion" are mutually exclusive events.

$$P(\text{oppose } or \text{ no opinion}) = P(\text{oppose}) + P(\text{no opinion})$$
$$= \frac{37}{100} + \frac{36}{100}$$
$$= \frac{73}{100}, \text{ or } 73\%$$

The probability that a respondent opposes the change *or* has no opinion is 73%.

TRY THIS Find the probability that a randomly selected respondent to this survey favors *or* has no opinion about the change in policy.

EXAMPLE 2 APPLICATION SURVEYS

Refer to the survey results given in Example 1.

Find the probability that a randomly selected respondent to the survey is a man *or* opposes the change in policy.

SOLUTION

The events "man" and "opposes" are inclusive events.

$$P(\text{man } or \text{ opposes}) = P(\text{man}) + (\text{opposes}) - P(\text{man } and \text{ opposes})$$
$$= \frac{50}{100} + \frac{37}{100} - \frac{12}{100}$$
$$= \frac{75}{100}, \text{ or } 75\%$$

The probability that a respondent is a man *or* opposes the change is 75%.

TRY THIS Find the probability that a randomly selected respondent to this survey is a woman *or* has no opinion about the change in policy.

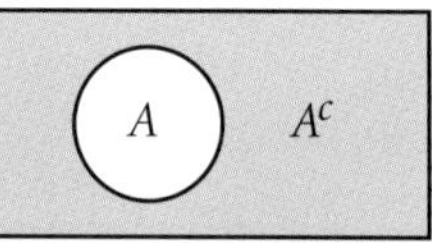

Sample space

The **complement** of event A consists of all outcomes in the sample space that are not in A and is denoted by A^c. For example, let A be the event "favor." Then the complement A^c is the event "oppose" *or* "no opinion." Conversely, if A is the event "oppose" *or* "no opinion," then A^c is the event "favor." The sum of the probabilities of all of the outcomes in a sample space is 1. Thus, $P(A) + P(A^c) = 1$.

CHECKPOINT ✔ Refer to the survey results given in Example 1. Let A be the event "favor." Verify that $P(A) + P(A^c) = 1$ is true.

Inclusion Strategies

VISUAL LEARNERS Some students may have difficulty deciding whether events are mutually exclusive or inclusive. Encourage students to draw models of the situation. In some cases, charts can be used. To show that a sum less than 5 and a product greater than 5 are mutually exclusive events when rolling two number cubes, have students create tables like the two shown at right where the desired sums and products are written in blue.

+	1	2	3	4	5	6
1	2	3	4	5	6	7
2	3	4	5	6	7	8
3	4	5	6	7	8	9
4	5	6	7	8	9	10
5	6	7	8	9	10	11
6	7	8	9	10	11	12

sums less than 5

×	1	2	3	4	5	6
1	1	2	3	4	5	6
2	2	4	6	8	10	12
3	3	6	9	12	15	18
4	4	8	12	16	20	24
5	5	10	15	20	25	30
6	6	12	18	24	30	36

products greater than 5

Other times, a Venn diagram that uses shading to show overlapping groups is useful.

Probability of the Complement of A

Let A represent an event in the sample space.

$$P(A) + P(A^c) = 1 \qquad P(A) = 1 - P(A^c) \qquad P(A^c) = 1 - P(A)$$

CRITICAL THINKING Let A be any event in a sample space. Why is $P(A \text{ or } A^c) = P(A) + P(A^c)$ true? Explain how this equation leads to the equations in the box above.

EXAMPLE 3 The drama, mathematics, and jazz clubs have 32 members, 33 members, and 39 members, respectively.

APPLICATION
EXTRACURRICULAR ACTIVITIES

Find the probability that a randomly selected club member belongs to at least two clubs.

SOLUTION

Let A represent membership in exactly one club. Then A^c represents membership in two or three clubs.

$$P(A^c) = 1 - P(A)$$
$$P(A^c) = 1 - \frac{6 + 8 + 10}{60}$$
$$= \frac{36}{60}$$
$$= 0.6$$

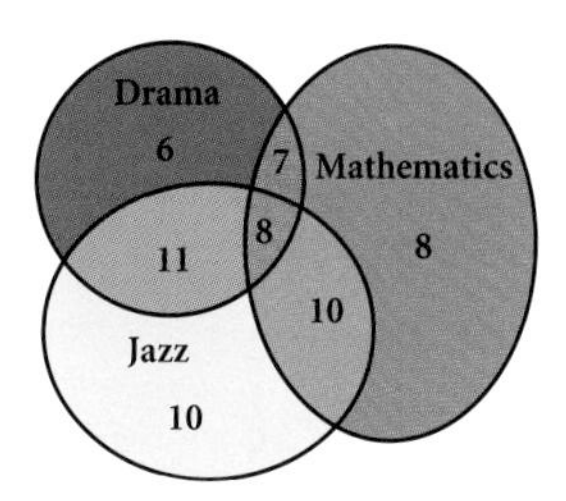

The probability that a randomly selected club member belongs to at least two clubs is 60%.

TRY THIS The science, art, and computer clubs have 45 members, 38 members, and 54 members, respectively. Some club members belong to more than one club, as indicated in the diagram at left. Find the probability that one of these club members selected at random will belong to more than one club.

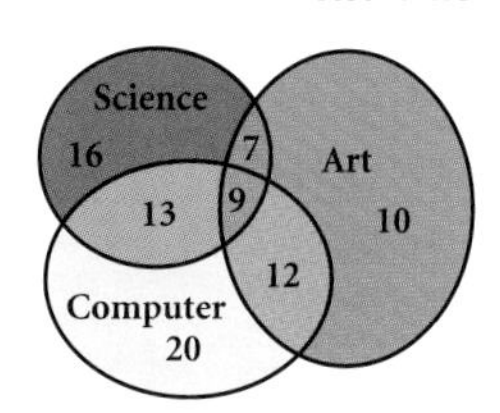

Exercises

Communicate

1. Explain the meaning of mutually exclusive events and of inclusive events. Give examples of each.
2. Describe how to find the complement of the event "rolling 1" *or* "rolling 2" on a number cube.
3. Explain how to find the probability of "rolling an odd number" *or* "rolling 3" on a number cube.

CRITICAL THINKING

Because of the way in which *complement* is defined, A and A^C are mutually exclusive events that include the entire sample space. So, $P(A) + P(A^C) = 1$. The equations $P(A) = 1 - P(A^C)$ and $P(A^C) = 1 - P(A)$ just involve subtracting either $P(A^C)$ or $P(A)$ from both sides of the equation.

ADDITIONAL EXAMPLE 3

The football, basketball, and track teams at Ridge High School have 36 members, 11 members, and 11 members, respectively. Some members belong to more than one team, as indicated in the diagram below.

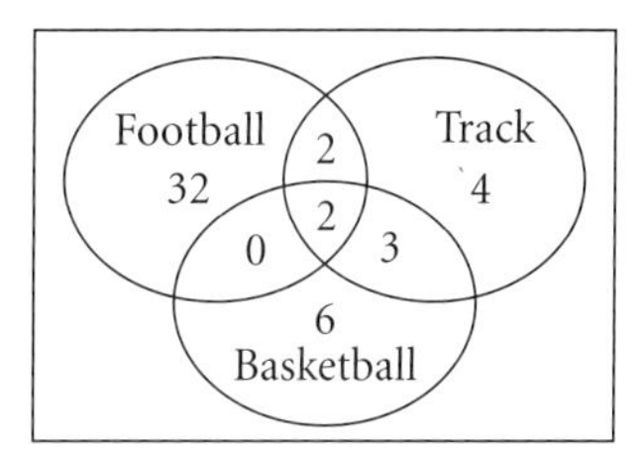

Find the probability that a randomly selected member belongs to at least two teams.

$\frac{7}{49}$, or ≈14.3%

TRY THIS

29.9%

Reteaching the Lesson

USING VISUAL MODELS Tell students to decide which television program to watch by randomly choosing one from a list of 20 programs. If 6 of the programs are educational, 11 are interesting, and 5 are both educational and interesting, have students illustrate the situation with a Venn diagram and find the probabilities listed at right.

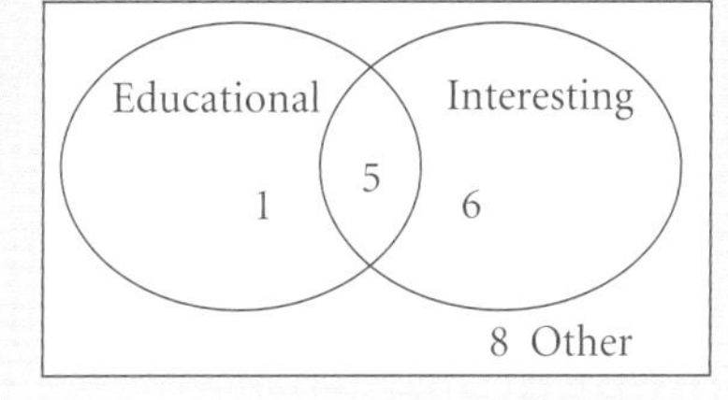

1. The chosen program will be educational.
   $P(E) = \frac{6}{20} = \frac{3}{10}$
2. The chosen program will be interesting.
   $P(I) = \frac{11}{20}$
3. The chosen program will be educational or interesting.
   $P(E \text{ or } I) = \frac{12}{20} = \frac{3}{5}$
4. The chosen program will not be educational.
   $P(E^c) = 1 - P(E) = 1 - \frac{3}{10} = \frac{7}{10}$
5. The chosen program will not be interesting.
   $P(I^c) = 1 - P(I) = 1 - \frac{11}{20} = \frac{9}{20}$

Assess

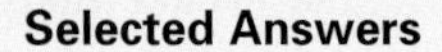

Selected Answers
Exercises 4–6, 7–57 odd

ASSIGNMENT GUIDE	
In Class	1–6
Core	7–39 odd, 43–49 odd
Core Plus	8–48 even
Review	50–58
Preview	59

Extra Practice can be found beginning on page 940.

Mid-Chapter Assessment for Lessons 10.1 through 10.4 can be found on page 128 of the *Assessment Resources*.

Guided Skills Practice

For Exercises 4 and 5, refer to the results of the survey about a change in public policy given on page 654.

4. Find the probability that a randomly selected respondent to the survey favors *or* opposes the change in policy. ***(EXAMPLE 1)*** **64%**

5. Find the probability that a randomly selected respondent to the survey is a man *or* favors the change in policy. ***(EXAMPLE 2)*** **59%**

APPLICATION

6. EDUCATION The economics, computer, and anatomy classes have 28 students, 29 students, and 24 students, respectively. Some students are taking more than one of these classes, as indicated in the diagram at right. Find the probability that a randomly selected student from these classes is taking at least two of the classes. ***(EXAMPLE 3)*** **≈60.9%**

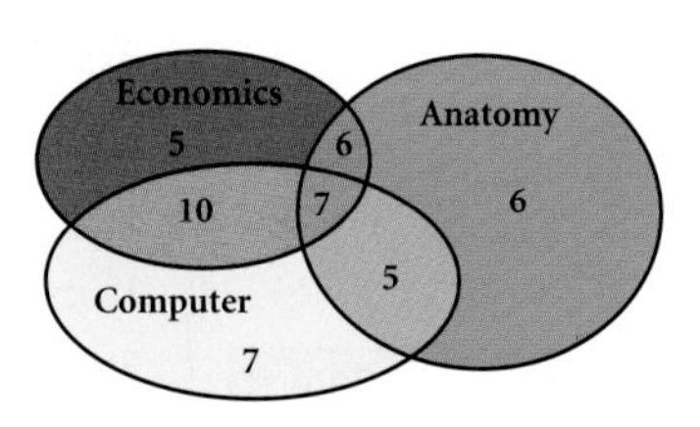

Practice and Apply

A number cube is rolled once, and the number on the top face is recorded. Find the probability of each event.

7. 5 *or* 6 $\frac{1}{3}$ **8.** 1 *or* 4 $\frac{1}{3}$ **9.** even *or* 3 $\frac{2}{3}$

10. odd *or* 2 $\frac{2}{3}$ **11.** less than 4 *or* 1 $\frac{1}{2}$ **12.** greater than 2 *or* 6 $\frac{2}{3}$

13. not 1 $\frac{5}{6}$ **14.** not even $\frac{1}{2}$ **15.** even *or* odd **1**

The table at right shows all of the possible outcomes of rolling two number cubes. Using the table, state whether the events in each pair below are inclusive or mutually exclusive. Then find the probability of each pair of events.

(1, 1)	(1, 2)	(1, 3)	(1, 4)	(1, 5)	(1, 6)
(2, 1)	(2, 2)	(2, 3)	(2, 4)	(2, 5)	(2, 6)
(3, 1)	(3, 2)	(3, 3)	(3, 4)	(3, 5)	(3, 6)
(4, 1)	(4, 2)	(4, 3)	(4, 4)	(4, 5)	(4, 6)
(5, 1)	(5, 2)	(5, 3)	(5, 4)	(5, 5)	(5, 6)
(6, 1)	(6, 2)	(6, 3)	(6, 4)	(6, 5)	(6, 6)

16. a sum of 2 *or* a sum of 4

17. a sum of 8 *or* a sum of 12

18. a sum of less than 3 *or* a sum of greater than 5

19. a sum of less than 7 *or* a sum of greater than 8

20. a sum of greater than 2 *or* a sum of greater than 6

21. a sum of less than 3 *or* a sum of less than 10

22. a sum of greater than 4 *or* a sum of less than 7

23. a sum of greater than 5 *or* a sum of less than 8

24. a product of greater than 8 *or* a product of less than 6

25. a product of greater than 16 *or* a product of less than 9

26. a product of greater than 6 *or* a product of less than 8

27. a product of greater than 9 *or* a product of less than 16

16. exclusive; $\frac{1}{9}$
17. exclusive; $\frac{1}{6}$
18. exclusive; $\frac{3}{4}$
19. exclusive; $\frac{25}{36}$
20. inclusive; $\frac{35}{36}$
21. inclusive; $\frac{5}{6}$
22. inclusive; 1
23. inclusive; 1
24. exclusive; $\frac{5}{6}$
25. exclusive; $\frac{13}{18}$
26. exclusive; 1
27. inclusive; 1

Practice

NAME ______ CLASS ______ DATE ______

Practice
10.4 ***Using Addition with Probability***

A card is drawn at random from a standard 52-card deck. Tell whether events *A* and *B* are inclusive or mutually exclusive. Then find *P*(*A or B*).

1. *A*: The card is a heart.
   B: The card is an 8.
   inclusive; $\frac{4}{13} \approx 31\%$

2. *A*: The card is a number less than 5.
   B: The card is a jack, a king, or a queen.
   mutually exclusive; $\frac{6}{13} \approx 46\%$

3. *A*: The card is black.
   B: The card is a number greater than 4.
   inclusive; $\frac{19}{26} \approx 73\%$

4. *A*: The card is not a diamond.
   B: The card is a spade.
   inclusive; $\frac{39}{52} = 75\%$

5. *A*: The card is red.
   B: The card is the ace of spades.
   mutually exclusive; $\frac{27}{52} \approx 52\%$

6. *A*: The card is a 2 or a 3.
   B: The card is not a heart.
   inclusive; $\frac{41}{52} \approx 79\%$

A spinner is divided into 8 congruent regions numbered 1 through 8. The spinner is spun once. Find the probability of each event.

7. The number is even or divisible by 3.
   $\frac{5}{8} = 62.5\%$

8. The number is odd or greater than 7.
   $\frac{5}{8} = 62.5\%$

9. The number is less than 2 or greater than 6.
   $\frac{3}{8} = 37.5\%$

10. The number is odd or divisible by 4.
   $\frac{3}{4} = 75\%$

Two number cubes are rolled, and the numbers on the top faces are added. The table at right shows the possible outcomes. Find each probability.

+	1	2	3	4	5	6
1	2	3	4	5	6	7
2	3	4	5	6	7	8
3	4	5	6	7	8	9
4	5	6	7	8	9	10
5	6	7	8	9	10	11
6	7	8	9	10	11	12

11. The sum is odd *or* greater than 11.
   $\frac{19}{36} \approx 53\%$

12. The sum is less than 6 *or* greater than 10.
   $\frac{13}{36} \approx 36\%$

13. The sum is even *or* less than 5.
   $\frac{5}{9} \approx 56\%$

14. The sum is less than 8 *or* a multiple of 6.
   $\frac{22}{36} \approx 61\%$

15. The sum is less than 4 *or* a multiple of 5.
   $\frac{5}{18} \approx 28\%$

28. **a.** How many integers from 1 to 600 are divisible by 2 or by 3? **400**
b. Find the probability that a random integer from 1 to 600 is divisible by neither 2 nor 3. $\frac{1}{3}$

29. **a.** How many integers from 1 to 3500 are divisible by 5 or by 7? **1100**
b. Find the probability that a random integer from 1 to 3500 is divisible by neither 5 nor 7. $\frac{24}{35}$

For Exercises 30–35, use the given probability to find $P(E^c)$.

30. $P(E) = \frac{1}{3}$ $\frac{2}{3}$ **31.** $P(E) = \frac{4}{11}$ $\frac{7}{11}$ **32.** $P(E) = 0.782$ **0.218**

33. $P(E) = 0.324$ **0.676** **34.** $P(E) = 0$ **1** **35.** $P(E) = 1$ **0**

Find the probability of each event.

36. 2 heads *or* 2 tails appearing in 2 tosses of a coin $\frac{1}{2}$

37. 3 heads *or* 2 tails appearing in 3 tosses of a coin $\frac{1}{2}$

38. at least 2 heads appearing in 3 tosses of a coin $\frac{1}{2}$

39. at least 3 heads appearing in 4 tosses of a coin $\frac{5}{16}$

CHALLENGE

40. A number cube is rolled 3 times. Find the probability of getting at least one 4. **≈42.1%**

41. ≈3.3%

41. In a group of 300 people surveyed, 46 like only cola A, 23 like only cola B, 18 like only cola C, 80 like both colas A and B, 66 like both colas A and C, 45 like both colas B and C, and 12 like all three colas. Find the probability that a randomly selected person from this survey likes none of these colas.

CONNECTION

42. GEOMETRY A circle with a radius of 3 and a 6×6 square are positioned inside a 10×10 square so that the top edge of the 6×6 square forms a diameter of the circle, as shown at right. A point inside the 10×10 square is selected at random. Find the

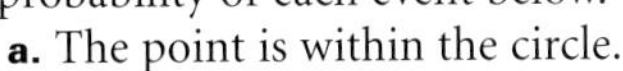
probability of each event below.
a. The point is within the circle.
b. The point is within the 6×6 square.
c. The point is within the 6×6 square *and* within the circle.
d. The point is within the 6×6 square *or* within the circle.

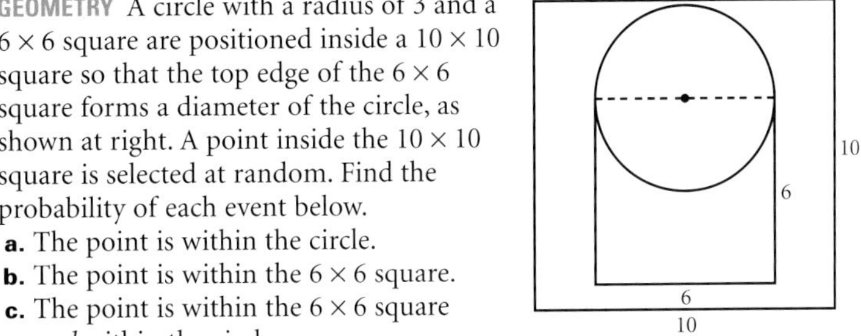

42a. ≈28.3%
b. 36%
c. ≈14.1%
d. ≈50.1%

APPLICATION

POLITICS The table shows the composition of the 106th Congress of the United States (1999–2001) according to political party.

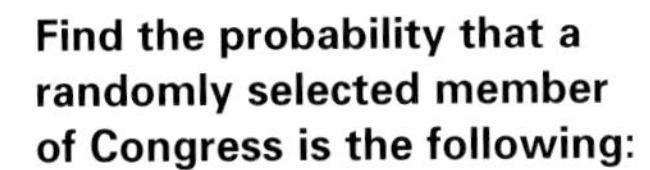
Find the probability that a randomly selected member of Congress is the following:

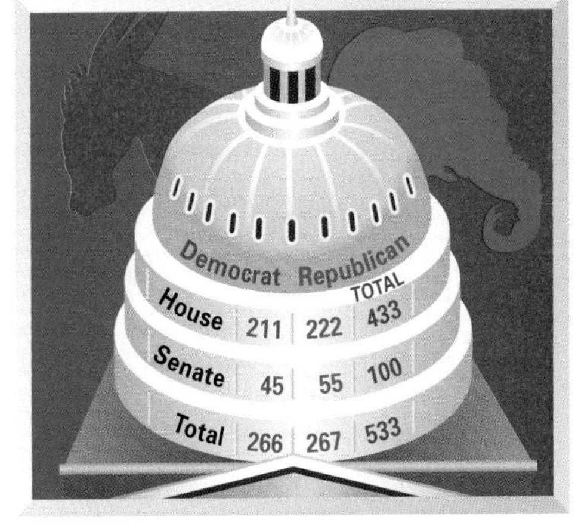

	Democrat	Republican	TOTAL
House	211	222	433
Senate	45	55	100
Total	266	267	533

43. ≈49.9%

43. a House Democrat *or* a Senate Republican

44. ≈50.1%

44. a House Republican *or* a Senate Democrat

45. ≈60.2%

45. a Democrat *or* a Senator

46. ≈58.5%

46. a Republican *or* a Senator

Error Analysis

In problems such as those presented in Exercises 16–27, students may have difficulty deciding whether the two outcomes are inclusive or mutually exclusive. Encourage them to make a sum or product chart and circle the outcomes to see whether there is an overlap.

Student Technology Guide

NAME ______ CLASS ______ DATE ______

Student Technology Guide
10.4 *Using Addition With Probability*

Finding the probability that one event *or* another happens can involve addition and subtraction of fractions. You can use a calculator to help evaluate these sums and differences.

Example: The table at right shows 1994 population data for the United States. For this data, find the probability that a randomly selected person is
a. under 20 or over 64.
b. female or over 64.

	0–19	20–64	over 64	Total
Male	38.4	75.2	13.5	127.1
Female	36.6	77.0	19.7	133.3
Total	75.0	152.2	33.2	260.3

(Rounded totals may not equal sums.)

a. The events "under 20" and "over 64" are mutually exclusive.
$P(\text{under 20 or over 64}) = \frac{75}{260.3} + \frac{33.2}{260.3}$.
Press (75 + 33.2) ÷ 260.3 ENTER
The probability that a person is under 20 or over 64 is about 0.42, or 42%.

b. The events "female" and "over 64" are not mutually exclusive.
$P(\text{female or over 64}) = \frac{133.3}{260.3} + \frac{33.2}{260.3} - \frac{19.7}{260.3}$.
Press (133.3 + 33.2 − 19.7) ÷ 260.3 ENTER.
The probability that a person is female or over 64 is about 0.56, or 56%.

1. Find the probability that a person in the survey is male or under 20. about 0.63, or 63%

Example: A survey of 107 people showed that 52 read science fiction books, 55 read mystery books, and 50 read romance books. As shown in the diagram at right, some people read more than one type of book. Find the probability that a person in the survey read at least two of these types of books.

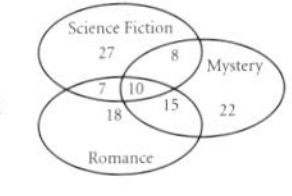

• Let A represent reading just one type of book. Then A^C represents reading two or three types.
$P(A^C) = 1 - P(A) = 1 - \frac{27 + 22 + 18}{107}$
Press 1 − (27 + 22 + 18) ÷ 107 ENTER.

The probability that a person in the survey read at least two of these types of books is about 0.37, or 37%.

2. Find the probability that a person from the survey has *not* read exactly two types of books. about 0.72, or 72%

In Exercise 59, students find a probability for a situation with given conditions. Students will study conditional probability in Lesson 10.6.

ALTERNATIVE

Assessment

Portfolio Activity

The Portfolio Activity can be used as preparation for the Chapter Project or as a separate activity. In the Portfolio Activity on this page, students conduct an experiment to find the probability of two mutually exclusive outcomes. Students increase the number of trials to see that the experimental probability approaches the theoretical probability.

Answers to Portfolio Activities can be found in Additional Answers of the Teacher's Edition.

50.

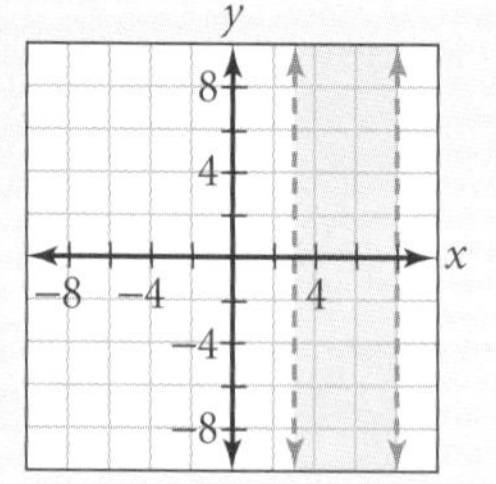

51.

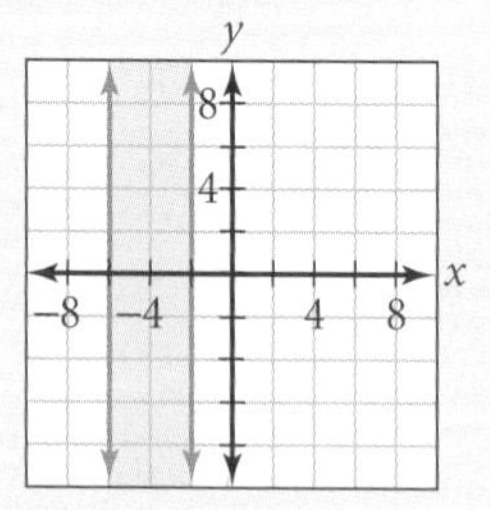

52.

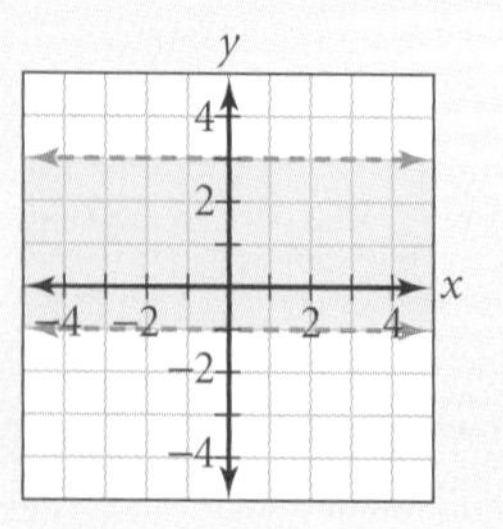

APPLICATIONS

47. ≈75.1%

48. ≈91.7%

49. ≈59.4%

47. QUALITY CONTROL A box contains 35 machine parts, 8 of which are defective. A quality control inspector randomly selects 5 of the 35 parts for testing. What is the probability that at least one part is defective?

PRODUCTION A shipment of 20 sets of skateboard wheels contains 7 sets that have a new type of wheel surface.

48. What is the probability that if 5 sets are chosen, at least 1 set will have the new surface?

49. What is the probability that if 5 sets are chosen, at least 2 sets will have the new surface?

Look Back

Graph each inequality in a coordinate plane. *(LESSON 3.4)*

50. $8 > x > 3$

51. $-6 \le x \le -2$

52. $3 > y > -1$

Factor each quadratic trinomial. *(LESSON 5.3)*

53. $x^2 - x - 42$ **$(x + 6)(x - 7)$**

54. $3x^2 - 16x - 12$ **$(3x + 2)(x - 6)$**

55. $81x^2 + 18x + 1$ **$(9x + 1)^2$**

Solve each equation. Check your solutions. *(LESSON 8.5)*

56. $\frac{x+2}{x} + 3 = \frac{x+6}{x}$ **$\frac{4}{3}$**

57. $\frac{3}{x-2} + x = \frac{17}{2}$ **$\frac{5}{2}, 8$**

58. $\frac{2y+5}{3y-2} = \frac{4y+3}{6y-1}$ **$-\frac{1}{27}$**

Look Beyond

59. If a card is drawn at random from a standard 52-card deck, what is the probability that the card is an ace? If a card is drawn at random from a standard deck and you are told that the card is a spade, does the probability that it is an ace change? Explain. **$\frac{1}{13}$; no; there is 1 ace of spades out of 13 possible spades in a 52-card deck.**

Let the sum of two random integers from 1 to 6 inclusive represent the sum of the roll of two number cubes.

1. Generate two random integers from 1 to 6 for a total of 40 trials. (Refer to the Keystroke Guide for Example 3 on page 687.) Record the outcomes.
2. Use your outcomes to estimate the probability of rolling "a sum of 7" *or* "a sum of 11." Describe how your results compare with the theoretical probability of rolling "a sum of 7" *or* "a sum of 11."
3. Perform another 40 trials for this experiment. Record the outcomes.
4. Use the outcomes of all 80 trials to estimate the probability of rolling "a sum of 7" *or* "a sum of 11." Compare your results with the theoretical probability of rolling "a sum of 7" *or* "a sum of 11." Describe what happens to your estimated probability as the number of trials increases.

WORKING ON THE CHAPTER PROJECT

You should now be able to complete the Chapter Project.

Independent Events

Objective

- Find the probability of two or more independent events.

Why *You can use the probability of independent events to find many interesting probabilities, such as the probability that two people at a party have the same birthday.*

In a group of 35 individuals, what is the probability that 2 or more individuals have the same birth month and day? *You will answer this question in Example 3.*

To answer this question, you need to know how to identify *independent* and *dependent events* and how to find probabilities of independent events. In the Activity below, you will investigate independent events.

Activity: Investigating Independent Events

You will need: no special tools

1. Let event A be the outcome of tossing heads on a coin. Find $P(A)$.
2. Let event B be the outcome of rolling a 3 on a number cube. Find $P(B)$.
3. Does event A have an effect on event B? Does event B have an effect on event A? Explain.
4. List all of the possible combinations for tossing a coin and rolling a number cube. Find the probability that both event A *and* event B occur, or $P(A \text{ and } B)$.
5. Find $P(A) \times P(B)$. Does $P(A \text{ and } B)$ equal $P(A) \times P(B)$?
6. Let even C be the outcome of rolling an odd number on a number cube. Find $P(C)$.
7. Find $P(A \text{ and } C)$ by using your list from Step 4. Does $P(A \text{ and } C)$ equal $P(A) \times P(C)$?

CHECKPOINT ✔ 8. What can you say about the probability that two events will both occur if they have no effect on one another?

Prepare

NCTM PRINCIPLES & STANDARDS 1, 2, 4–10

QUICK WARM-UP

A bag contains 2 black marbles, 7 white marbles, 6 green marbles, and 5 blue marbles. **Find the probability of drawing the following:**

1. 1 green marble **$\frac{6}{20}$, or 30%**
2. 1 yellow marble **$\frac{0}{20}$, or 0%**
3. 1 marble that is not blue **$\frac{3}{4}$, or 75%**
4. 1 marble that is not red **$\frac{20}{20}$, or 100%**
5. 1 black marble *or* 1 white marble **$\frac{9}{20}$, or 45%**

Also on Quiz Transparency 10.5

Teach

Why Sometimes we are interested in the probability that two things will occur independently of each other, such as rolling two number cubes. Ask students whether they think that the probability of two events occurring together would be greater than or less than the probability of each event occurring alone. **less than**

☞ For Activity Notes and answers to Checkpoints, see page 660.

Alternative Teaching Strategy

GUIDED ANALYSIS Consider rolling two number cubes. How many possible outcomes are there for each cube? **6** Use the fundamental counting principle to determine how many pairs of rolls are possible. **6 × 6, or 36** Make a chart showing all 36 possible pairs. The probability of rolling a specific pair is $\frac{1}{36}$. Tell students that as long as one event does not depend on the other, $P(A \text{ and } B) = P(A) \cdot P(B)$.

Suppose that you are given the following information: Event A can happen in 4 ways and event B can happen in 3 ways. How many ways can event A and event B both happen if the events are independent? **4 × 3, or 12** What is the probability that both event A and event B occur?

$$P(A \text{ and } B) = P(A) \times P(B) = \frac{1}{4} \times \frac{1}{3} = \frac{1}{12}$$

Two events are **independent** if the occurrence or non-occurrence of one event has no effect on the likelihood of the occurrence of the other event. For example, tossing two coins is an example of a pair of independent events. If one event does affect the occurrence of the other event, the events are **dependent**.

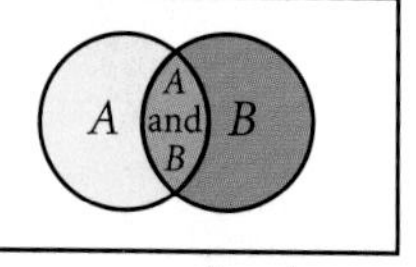

Sample space

Probability of Independent Events

Events A and B are independent events if and only if $P(A \text{ and } B) = P(A) \times P(B)$. Otherwise, A and B are dependent events.

EXAMPLE  Bag A contains 9 red marbles and 3 green marbles. Bag B contains 9 black marbles and 6 orange marbles.

Find the probability of selecting one green marble from bag A and one black marble from bag B in one draw from each bag.

SOLUTION

Bag A

$P(\text{green marble}) = \frac{3}{9+3} = \frac{3}{12} = \frac{1}{4}$

Bag B

$P(\text{black marble}) = \frac{9}{9+6} = \frac{9}{15} = \frac{3}{5}$

The events are independent.

$$P(\text{green marble } and \text{ black marble}) = \frac{1}{4} \times \frac{3}{5} = \frac{3}{20}$$

The probability of selecting a green marble from bag A and a black marble from bag B in one draw from each bag is 0.15, or 15%.

EXAMPLE 2 Two seniors, one from each government class, are to be randomly selected to travel to Washington, D.C. John is in a class of 18 students, and Peter is in another class of 20 students.

APPLICATION
EXTRACURRICULAR ACTIVITIES

Find the probability that both John and Peter will be selected.

SOLUTION

Event A
The probability that John will be chosen is $\frac{1}{18}$.

Event B
The probability that Peter will be chosen is $\frac{1}{20}$.

Events A and B are independent events.

$$P(A \text{ and } B) = P(A) \times P(B)$$
$$= \frac{1}{18} \times \frac{1}{20} = \frac{1}{360},$$
or about 0.003

The probability that both John and Peter will be selected is about 0.3%.

WASHINGTON, D.C.

Interdisciplinary Connection

BASKETBALL One-and-one foul shooting in basketball works as follows: If a player makes the first shot, he takes a second shot. If he misses the first shot, he does not take a second shot. Have students calculate the probabilities for a player scoring 0, 1, and 2 points in a one-and-one foul situation if the player's foul-shot average is 55%. Assume the foul shots are independent events.

$P(0 \text{ points}) = 1 - 0.55 = 0.45$, or 45%
$P(1 \text{ point}) = (0.55)(0.45) \approx 0.247$, or $\approx$25%
$P(2 \text{ points}) = (0.55)(0.55) = 0.3025$, or $\approx$30%

Inclusion Strategies

KINESTHETIC LEARNERS Have a student stand at the origin of a large number line drawn on the floor and toss a coin 10 times, moving 1 unit up for heads (H) and 1 unit down for tails (T). Have students find the probability that the student will end up in the positions given below. (Ask how many heads would be required to land on each position.)

1. 4 units up — **7H and 3T: ${}_{10}C_7\left(\frac{1}{2}\right)^{10} \approx 12\%$**
2. 3 units down — **cannot happen: 0%**
3. 2 units down — **4H and 6T: ${}_{10}C_4\left(\frac{1}{2}\right)^{10} \approx 21\%$**
4. at the origin — **5H and 5T: ${}_{10}C_5\left(\frac{1}{2}\right)^{10} \approx 25\%$**

Activity Notes

In this Activity, students compute probabilities of tossing heads on a coin and rolling a 3 on a number cube. They should realize that the result of tossing a coin has no effect on the result of rolling a number cube. They are asked to list all possible combinations of tossing a coin and rolling a number cube and to find the probability that two events occur. They should conclude that when two events have no effect on one another, the probability that both will occur is the product of their individual probabilities.

CHECKPOINT ✔

8. **If two events have no effect on one another, the probability that they both occur is the product of their individual probabilities.**

ADDITIONAL EXAMPLE 1

Michelle has 6 quarters and 4 dimes in her pocket. Darryl has 2 nickels, 1 quarter, and 5 pennies in his pocket. **Find the probability that Michelle selects 1 dime from her pocket and Darryl selects 1 penny from his pocket.**

$\frac{1}{4} = 0.25$, or 25%

ADDITIONAL EXAMPLE 2

Contestants for a local game show are randomly selected from 2 groups. Tina is a member of the first group of 8 people, and Janet is a member of the second group of 14 people. **Find the probability that both Tina and Janet will be selected.**

$\frac{1}{112} \approx 0.0089$, or 0.9%

Use Teaching Transparency 40.

The formula for the probability of independent events can be extended to 3 or more events. For example, the probability of obtaining 3 heads in 3 tosses of a coin is $\frac{1}{2} \times \frac{1}{2} \times \frac{1}{2} = \frac{1}{8} = 0.125$, or 12.5%.

CRITICAL THINKING Find the probability of obtaining 4 heads in 4 tosses of a coin. Write a formula for the probability of obtaining n heads in n tosses of a coin.

EXAMPLE 3 Refer to the birthday problem described at the beginning of the lesson.

APPLICATION ENTERTAINMENT

What is the probability that in a group of 35 people, 2 or more people have the same birth month and day?

SOLUTION

PROBLEM SOLVING

Change your point of view. Use the complementary event, which is that all birthdays are different. Use 365 days for a year (ignoring leap years).

The first person's birthday can be any day. $\frac{365}{365}$

The second person's birthday cannot be the same day. $\frac{364}{365}$

The third person's birthday cannot be the same day as either the first or the second person's. $\frac{363}{365}$

Continue this pattern for all 35 individuals. The probability that all individuals have different birthdays is as follows:

$$P(\text{different birthday}) = \frac{365}{365} \times \frac{364}{365} \times \frac{363}{365} \times \cdots \times \frac{331}{365} \approx 0.19$$

TECHNOLOGY GRAPHICS CALCULATOR

Keystroke Guide, page 687

Find the probability that 2 or more individuals have the same birthday.

$$P(\text{same birthday}) = 1 - P(\text{different birthday}) \approx 1 - 0.19 \approx 0.81$$

The probability that 2 or more individuals have the same birth month and day is about 81%.

TRY THIS What is the probability that in a group of 45 people, 2 or more people have the same birth month and day?

Exercises

Communicate

1. Give an example of independent events and of dependent events.
2. Explain how to find the probability of two independent events occurring.
3. Explain the difference between mutually exclusive events and independent events. Give an example.

CRITICAL THINKING

$P(4 \text{ heads in } 4 \text{ tosses}) = 6.25\%$

$P(n \text{ heads in } n \text{ tosses}) = \left(\frac{1}{2}\right)^n$

Teaching Tip

TECHNOLOGY When you calculate factorials contained in ratios, you must enter the individual ratios to prevent an overflow error on most calculators when the numbers are large. For Example 3, enter **365/365** and press ENTER. Then press *, enter **364/365,** and press ENTER again. Continue this process until the last ratio is entered. On a TI-83 or TI-82 graphics calculator, press 2nd (–) × 2nd ENTER, edit the last fraction by using the left arrow key, and press ENTER. The last numerator is one more than the difference between 365 and the number of group members. In Additional Example 3, the last numerator should be 365 – 45 + 1, or 321.

ADDITIONAL EXAMPLE 3

In a class of 20 students, what is the probability that 2 or more of the students have the same birthday?

about 41%

TRY THIS

≈94%

Assess

Selected Answers

Exercises 4–6, 7–35 odd

ASSIGNMENT GUIDE

In Class	1–6
Core	7–25 odd
Core Plus	8–26 even
Review	27–36
Preview	37

Extra Practice can be found beginning on page 940.

Enrichment

Suppose that A and B are independent events and consider situations in which $P(A \text{ and } B) = 0$ and $P(A \text{ and } B) = 1$. What would have to be true about the individual probabilities $P(A)$ and (B) in each situation? **At least one of the probabilities must be 0 in the first, and both of the probabilities must be 1 in the other.** Create an example of such a situation.

Reteaching the Lesson

HANDS-ON STRATEGIES Give students a bag containing 1 red and 3 blue crayons. Then have students find the probability of drawing 1 red crayon, replacing it, and then drawing 1 blue crayon. $P(R, B) = \frac{1}{4} \cdot \frac{3}{4} = \frac{3}{16}$ Have students find the probability of drawing 1 red crayon, without replacement, and then drawing 1 blue crayon. $P(R, B) = \frac{1}{4} \cdot \frac{3}{3} = \frac{1}{4}$ Have students discuss which drawing is independent. **Drawings with replacement are independent events.**

Error Analysis

In Exercises 17–20, make sure students realize that they are combining *and* and *or* statements. For example, finding the probability that 3 results from a spinner, with sections labeled 1 through 8, is 3 *or* greater than 5 and can be restated as finding the probability of spinning a 3 *or* 6 *or* 7 *or* 8 and a 3 *or* 6 *or* 7 *or* 8 and a 3 *or* 6 *or* 7 *or* 8.

Guided Skills Practice

4. Bag A contains 5 black marbles and 5 white marbles. Bag B contains 1 green marble and 2 red marbles. Find the probability of selecting a black marble from bag A and a green marble from bag B in two draws. ***(EXAMPLE 1)*** **$\frac{1}{6}$, or ≈16.7%**

5. Two students, one from each science class, are to be randomly selected to attend a national science conference. Melinda is in a class of 22 students, and Seth is in another class of 19 students. Find the probability that both Melinda and Seth will be selected. ***(EXAMPLE 2)*** **$\frac{1}{418}$, or ≈0.24%**

6. What is the probability that in a group of 40 people, 2 or more people have the same birth month and day? ***(EXAMPLE 3)*** **≈89%**

Practice and Apply

Events *A*, *B*, *C* and *D* are independent, and *P*(*A*) = 0.5, *P*(*B*) = 0.25, *P*(*C*) = 0.75, and *P*(*D*) = 0.1. Find each probability.

7. *P*(*A and B*) **0.125** **8.** *P*(*A and C*) **0.375** **9.** *P*(*C and B*) **0.1875**

10. *P*(*C and D*) **0.075** **11.** *P*(*A and D*) **0.05** **12.** *P*(*B and D*) **0.025**

Use the definition of independent events to determine whether the events below are independent or dependent.

13. the event *even* and the event *2 or 4* on one roll of a number cube

14. the event *even* and the event *1 or 4* on one roll of a number cube

15. the event *less than 5* and the event *6* on one roll of a number cube

16. the event *greater than 3* and the event *4* on one roll of a number cube

13. dependent
14. independent
15. dependent
16. dependent

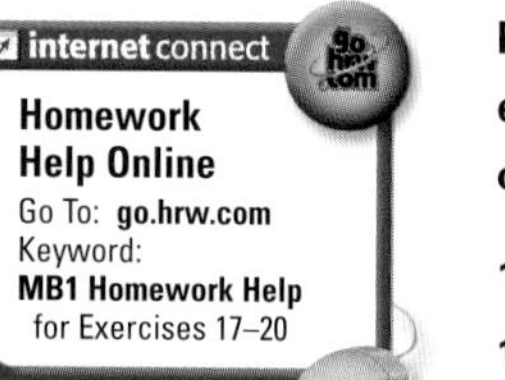

Refer to the spinner shown below in which each numbered section is exactly $\frac{1}{8}$ of the circle. Find the probability of each event in three spins of the spinner.

17. All three numbers are 3 *or* greater than 5.

18. All three numbers are 4 *or* less than 6.

19. Exactly one number is 5 *or* less than 7.

20. Exactly one number is 8 *or* greater than 3.

21. Suppose that the probability of Kevin coming to a party is 80% and the probability of Judy coming to a party is 95%. Assuming that these events are independent, what is the probability that they both will come to a party?

22. The integers 1 through 15 are written on slips of paper and placed into a box. One slip is selected at random and put back into the box, and then another slip is chosen at random.

a. What is the probability that the number 8 is selected both times?

b. What is the probability that the number 8 is selected exactly once? (Hint: Find the probability that an 8 is selected on the first or second draw, but not on *both* draws.)

17. $\frac{1}{8}$**, or 12.5%**
18. $\frac{125}{512}$**, or ≈24.4%**
19. $\frac{9}{64}$**, or ≈14.1%**
20. $\frac{135}{512}$**, or ≈26.4%**
21. 76%
22a. $\frac{1}{125}$**, or ≈0.44%**
b. $\frac{28}{125}$**, or ≈12.4%**

Practice

NAME ______ CLASS ______ DATE ______

Practice
***10.5* Independent Events**

Events *D*, *E*, *F* and *G* are independent, and *P*(*D*) = 0.2, *P*(*E*) = 0.1, *P*(*F*) = 0.4, and *P*(*G*) = 0.25. Find the probability of each combination of events.

1. *P*(*D and E*) 0.02 = 2%
2. *P*(*D and F*) 0.08 = 8%
3. *P*(*E and F*) 0.04 = 4%
4. *P*(*D and G*) 0.05 = 5%
5. *P*(*D and E and F*) 0.008 = 0.8%
6. *P*(*E and F and G*) 0.01 = 1%

A bag contains 3 white marbles, 2 red marbles, and 7 blue marbles. A marble is picked at random and is replaced. Then a second marble is picked at random. Find each probability.

7. Both marbles are blue. $\frac{49}{144}$ ≈ 34%
8. The first marble is white and the second marble is red. $\frac{1}{24}$ ≈ 4%
9. The first marble is white and the second marble is not white. $\frac{3}{16}$ = 18.75%
10. Neither marble is red. $\frac{25}{36}$ ≈ 69%
11. The first marble is blue and the second marble is red. $\frac{7}{72}$ ≈ 10%

A number cube is rolled twice. On each roll, the number on the top face is recorded. Find the probability of each event.

12. The first number is greater than 5 and the second is less than 3. $\frac{1}{18}$ ≈ 6%
13. Both numbers are greater than 4. $\frac{1}{9}$ ≈ 11%
14. The first number is even and the second number is odd. $\frac{1}{4}$ = 25%
15. Both numbers are less than 2. $\frac{1}{36}$ ≈ 3%
16. Neither number is greater than 4. $\frac{4}{9}$ ≈ 44%

A number cube is rolled, and two coins are tossed. Find the probability of each event.

17. The number on the cube is 2 and both coins are heads. $\frac{1}{24}$ ≈ 4%
18. The number on the cube is even, one coin shows heads, and one shows tails. $\frac{1}{8}$ = 12.5%
19. The number on the cube is greater than 4 and both coins are tails. $\frac{1}{12}$ ≈ 8%
20. The number on the cube is greater than 2 and the coins show different sides. $\frac{1}{3}$ ≈ 33%

APPLICATIONS

23. 89.24%

23. TRAVEL An airline's records show that its flights from Los Angeles to Dallas arrive on schedule 92% of the time. They also show that its flights from Dallas to Miami leave on schedule 97% of the time. If you fly from Los Angeles to Miami with a connection through Dallas, what is the probability that you will arrive at Dallas and leave from Dallas at your scheduled times?

A passenger checking flight information

SECURITY Suppose that a security system consists of four components: a motion detector, a glass-break detector, magnetic door and window contacts, and a video camera. The probabilities of escaping detection by each of the four devices are 0.2, 0.3, 0.4, and 0.6, respectively. Assume that all of the components act independently.

24. What is the probability that a thief can get past the magnetic contacts and the video camera? **0.24**

25. What is the probability that a thief can get past the glass-break detector and the motion detector? **0.06**

26. What is the probability that a thief can get past all four components? **0.0144**

Look Back

Simplify each expression. Assume that no variable equals zero. ***(LESSON 2.2)***

27. $(x^{-2}y^3)^2(3xy^0)^3$ $\mathbf{\frac{27y^6}{x}}$ **28.** $(2x^2y^{-2})^{-3}(-x^2y)^3$ $\mathbf{-\frac{y^9}{8}}$ **29.** $\left(\frac{3x^2y^{-2}}{5x^2y}\right)^2$ $\mathbf{\frac{9}{25y^6}}$

30. Find the maximum and minimum values of the objective function $C = 2x + 3y$ given the constraints $\begin{cases} y \ge x \\ y \le 5 \\ x \ge 0 \end{cases}$. ***(LESSON 3.5)*** **max = 25, min = 0**

Solve each equation for *x*. Round your answers to the nearest hundredth. ***(LESSONS 6.5 AND 6.6)***

31. $e^x = 3$ **1.10** **32.** $5^x = 11$ **1.49** **33.** $e^{x+1} = 9$ **1.20**

Write the standard equation of the parabola with the given characteristics. ***(LESSON 9.2)***

34. vertex: $(3, 2)$
focus: $(4, 2)$
$\mathbf{x = \frac{1}{4}(y-2)^2 + 3}$

35. vertex: $(1, 0)$
directrix: $y = 4$
$\mathbf{y = -\frac{1}{16}(x-1)^2}$

36. directrix: $x = -4$
focus: $(3, 0)$
$\mathbf{x = \frac{1}{14}y^2 - \frac{1}{2}}$

Look Beyond

37. Three coins are tossed. What is the probability of 3 heads appearing given the conditions below?

a. All three coins are regular, fair coins. $\mathbf{\frac{1}{8}}$

b. One of the three coins is a two-headed coin. $\mathbf{\frac{1}{4}}$

c. Two of the three coins are two-headed coins. $\mathbf{\frac{1}{2}}$

Look Beyond

In Exercise 37, students explore probabilities of tossing three coins under given conditions. Conditional probability will be studied in Lesson 10.6.

Student Technology Guide

NAME ______ CLASS ______ DATE ______

Student Technology Guide
10.5 ***Independent Events***

You can use a calculator to find the probability of independent events.

Example: Elias is playing a game with a number cube whose sides are numbered 1 through 6, and a number dodecahedron, a 12-sided solid whose sides are numbered 1 through 12. Find the probability that he rolls a 5 on the number cube and a number greater than 7 on the dodecahedron.

- Let event *A* be rolling a 5 on the number cube and event *B* be rolling a number greater than 7 on the dodecahedron. Thus, $P(A) = \frac{1}{6}$ and because the dodecahedron has 5 numbers greater than 7 (8, 9, 10, 11, and 12), $P(B) = \frac{5}{12}$.
- The rolls are independent events.
  $P(A \text{ and } B) = P(A) \times P(B) = \frac{1}{6} \times \frac{5}{12}$
  Press (1 ÷ 6) × (5 ÷ 12) ENTER.

```
(1/6)*(5/12)
          .0694444444
```

The probability that Elias rolls a 5 on the number cube and a number greater than 7 on the dodecahedron is about 0.07, or 7%.

1. Find the probability that Elias rolls any number except 1 on the number cube and a 2, 3, 4, 5, 6, 7, or 8 on the dodecahedron. **about 0.49, or 49%**

In some cases, it is easiest to find the probability of an independent event by subtracting the probability of the complementary event from 1.

Example: Mr. and Mrs. Chang plan to buy a house and live in it for 20 years. The neighborhood where the house is located has a 3% chance of a flood in any given year. Find the probability that the Changs will experience at least one flood. (Assume floods are independent events.)

- The probability that the house does *not* flood in any given year is 97%. So, the probability that it does not flood in a 20-year period is 0.97^{20}.
- The probability that the house *does* flood at least once is $1 - 0.97^{20}$. Press 1 − .97 ^ 20 ENTER.

```
1-.97^20
          .456205657
```

The probability that the Changs will experience at least one flood in a 20-year period is about 0.46, or 46%.

2. A voice recognition software program interprets 99.3% of spoken words correctly. Find the probability that this software misinterprets at least one word in 400 words of instructions. **about 0.94, or 94%**

Prepare

NCTM PRINCIPLES & STANDARDS 1–10

QUICK WARM-UP

Two number cubes are rolled. Find the probability of rolling each sum below.

1. a sum of 7 $\frac{6}{36}$, or ≈17%
2. a sum of 8 $\frac{5}{36}$, or ≈14%
3. a sum of 7 *or* a sum of 8 $\frac{11}{36}$, or ≈31%
4. a sum less than 7 $\frac{15}{36}$, or ≈42%
5. a sum less than 7 or a sum greater than 8 $\frac{25}{36}$, or ≈69%
6. a sum less than 12 or a sum greater than 7 $\frac{36}{36}$, or 100%

Also on Quiz Transparency 10.6

Teach

Why If a person tests positive for a certain disease, there is a chance that this person does not actually have the disease. This is known as a *false positive.* Have students discuss the implications a false positive result.

Dependent Events and Conditional Probability

Why *Conditional probability applies to many real-world situations in which the probability of one event is affected by the occurrence of another event.*

Objective

- Find conditional probabilities.

APPLICATION HEALTH

The ELISA test is used to screen donated blood for the presence of HIV antibodies. When HIV antibodies are present in the blood tested, ELISA gives a positive result 98% of the time. When HIV antibodies are not present in the blood tested, ELISA gives a positive result 7% of the time, which is called a *false positive.*

Suppose that 1 out of every 1000 units of donated blood actually contains HIV antibodies. What is the probability that a positive ELISA result is accurate? *You will answer this question in Example 4.*

To solve the problem above, you need to calculate a *conditional probability.* The Activity below will help you understand conditional probabilities.

Activity Exploring Conditional Probability

You will need: no special materials

Suppose that you select a card at random from a standard deck of playing cards. (Note: In a standard deck of 52 playing cards, 12 cards are face cards, and 4 of the face cards are kings.)

1. Find the probability that the card drawn is a king.
2. Find the probability that the card drawn is a king if you know that the card is a face card.
3. Find the probability that the card drawn is a king if you know that 2 queens were previously removed from the deck.
4. Find the probability that the card drawn is a king if you know that 2 kings were previously removed from the deck.

CHECKPOINT ✔

5. How does what you know in Steps 2 and 3 affect the resulting probabilities? How does what you know in Step 4 affect the resulting probability? Describe how probabilities can be affected by knowledge of a previous event.

Alternative Teaching Strategy

USING TABLES Give students the table below.

Garbage Collected in the United States (millions of tons)

Material	Recycled	Not recycled	Total
Paper	26.5	44.5	71
Other	18.5	110.5	129
Total	45	155	200

Tell students that a conditional probability occurs when some *condition* is known that may affect the probability of an event.

By using only the recycled column, find $P(\text{paper} \mid \text{recycled})$. $\frac{26.5}{45}$, or ≈59% Find $P(\text{recycled} \mid \text{paper})$ by using only the paper row. $\frac{26.5}{71}$, or ≈37% Then show students that

$$\frac{P(\text{recycled } \textit{and} \text{ paper})}{P(\text{recycled})} = \frac{\frac{26.5}{200}}{\frac{45}{200}} = \frac{26.5}{45} \text{ and}$$

$$\frac{P(\text{recycled } \textit{and} \text{ paper})}{P(\text{paper})} = \frac{\frac{26.5}{200}}{\frac{71}{200}} = \frac{26.5}{71}.$$

Lead students to state the general formula for a conditional probability.

Knowing the outcome of one event can affect the probability of another event.

EXAMPLE 1

APPLICATION
EXTRACURRICULAR ACTIVITIES

The band at Villesdale High School has 50 members, and the student council has 20 members. Five student council members are also in the band. Suppose that a student is randomly selected from these two groups.

Find the probability that the student is a member of the band if you know that he or she is on the student council.

SOLUTION

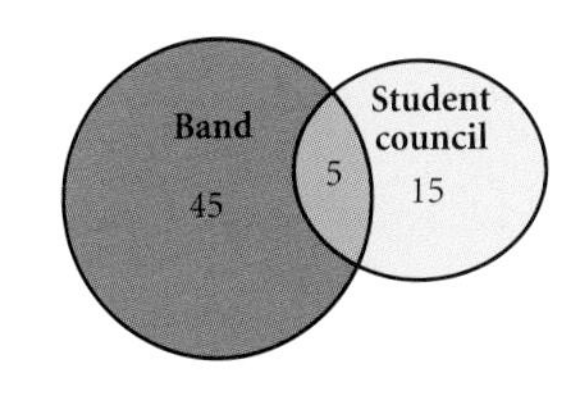

Because the student is on the student council, the sample space is 20 members.

$P(\text{band, given council}) = \frac{5}{20} = \frac{1}{4}$

Thus, the probability that the student is also a member of the band is 25%.

If a marble is selected from a bag of marbles and replaced and then a second marble is selected, the two selections are independent events. On the other hand, if a marble is selected from a bag of marbles and a second marble is selected *without* replacing the first marble, the second selection is a dependent event. That is, the probability of the second event changes, depending on the outcome of the first event.

EXAMPLE 2 **A bag contains 9 red marbles and 3 green marbles. For each case below, find the probability of randomly selecting a red marble on the first draw and a green marble on the second draw.**

a. The first marble is replaced.

b. The first marble is not replaced.

SOLUTION

a. If the first marble is replaced before the second marble is selected, then the events are independent.

$P(\text{red}) = \frac{9}{9+3} = \frac{9}{12} = \frac{3}{4}$

$P(\text{green}) = \frac{3}{9+3} = \frac{3}{12} = \frac{1}{4}$

$P(\text{red } \textit{and} \text{ green}) = \frac{3}{4} \times \frac{1}{4} = \frac{3}{16}$

The probability is $\frac{3}{16}$, or 18.75%.

b. If the first marble is *not* replaced before the second marble is selected, then the size of the sample space for the second event changes from 12 to 11.

$P(\text{red}) = \frac{9}{9+3} = \frac{9}{12} = \frac{3}{4}$

$P(\text{green}) = \frac{3}{8+3} = \frac{3}{11}$

$P(\text{red } \textit{and} \text{ green}) = \frac{3}{4} \times \frac{3}{11} = \frac{9}{44}$

The probability is $\frac{9}{44}$, or about 20.5%.

The probability of event *B*, given that event *A* has happened (or will happen), is called *conditional probability.*

Conditional Probability

The **conditional probability** of event *B*, given event *A*, denoted by $P(B|A)$, is given by $P(B|A) = \frac{P(A \text{ and } B)}{P(A)}$, where $P(A) \neq 0$.

Interdisciplinary Connection

ECONOMICS Analysts use conditional probability when making decisions about the stock market. Suppose the probability that the stock market will go up on Wednesday is 0.25. If it goes up on Wednesday, the probability that it will also go up on Thursday is 0.6. Find the probability that the stock market will go up on Wednesday and on Thursday. 15%

Inclusion Strategies

USING MANIPULATIVES Prepare some cards with two black sides, some with two white sides, and some with one white and one black side. Let students work in small groups, and give each group several cards. Have one student select a card at random and place it on the table. The other members of the group compute the probability that the hidden side is the same color as the visible side. They should guess the color of the other side and record whether their guess was correct. The group should record the number of trials they made and the number of correct guesses.

Activity Notes

In this Activity, students use a standard deck of cards to investigate conditional probabilities and should find that knowing additional information changes a probability.

CHECKPOINT ✔

5. The sample space is decreased so that the corresponding probabilities are increased; the number of outcomes in the event is decreased so that the probability is decreased; the number of outcomes in the event and/or the sample space may change.

ADDITIONAL EXAMPLE 1

The Wilmont High School honor society has 20 members, and the school's debate team has 18 members. Twelve members of the debate team are also in the honor society. Suppose that a student is randomly selected from these two groups. **Find the probability that the student is a member of the debate team if it is known that he or she is a member of the honor society.** 60%

ADDITIONAL EXAMPLE 2

A bag contains 5 blue and 8 yellow marbles. **For each case below, find the probability of randomly selecting a blue marble on the first draw and a yellow marble on the second draw.**

a. The first marble is replaced.
$\frac{40}{169}$, or ≈23.7%

b. The first marble is not replaced.
$\frac{40}{156}$, or ≈25.6%

CHECKPOINT ✔
$P(\text{band}|\text{council})$

$$= \frac{P(\text{band and council})}{P(\text{council})} = \frac{\frac{5}{65}}{\frac{20}{65}} = \frac{5}{20} = \frac{1}{4}$$

CHECKPOINT ✔
If A and B are independent events, then the following is true:

$$P(B|A) = \frac{P(A \text{ and } B)}{P(A)} = \frac{P(A) \times P(B)}{P(A)} = P(B)$$

ADDITIONAL EXAMPLE 3

A researcher is studying the development of babies and infants. The researcher selects a child from a local preschool by first selecting an age group and then randomly selecting a child from that age group. Each age group is equally likely to be chosen. That is, $P(1\text{ year}) = \frac{1}{4}$, $P(2\text{ years}) = \frac{1}{4}$, $P(3\text{ years}) = \frac{1}{4}$, and $P(4\text{ years}) = \frac{1}{4}$.

Find the probability that a boy is selected, given the data below about the enrollment at the preschool. Find the sum of probabilities of selecting a 1-year-old boy, a 2-year-old boy, a 3-year-old boy, and a 4-year-old boy.

Age	Girls	Boys	Total
1 year	13	17	30
2 years	11	15	26
3 years	8	10	18
4 years	7	9	16

$P(\text{boy}) \approx 57\%$;
$P(\text{1-yr boy}) + P(\text{2-yr boy}) + P(\text{3-yr boy}) + P(\text{4-yr boy}) = \frac{17}{90} + \frac{15}{90} + \frac{10}{90} + \frac{9}{90} = \frac{51}{90} \approx 57\%$

TRY THIS
$\approx 39.4\%$

CHECKPOINT ✔ Use the conditional probability formula to verify the solution in Example 1.

Using the Multiplication Property of Equality, the conditional probability formula $P(B|A) = \frac{P(A \text{ and } B)}{P(A)}$ can be rewritten as $P(A) \times P(B|A) = P(A \text{ and } B)$. You can use this form to verify the solutions in Example 2, as shown below.

Independent Events

The first marble is replaced.

Event A: red first
Event B: green second

$P(A) \times P(B|A) = P(A \text{ and } B)$
$\frac{3}{4} \times \frac{1}{4} = \frac{3}{16}$

Dependent Events

The first marble is not replaced.

Event A: red first
Event B: green second

$P(A) \times P(B|A) = P(A \text{ and } B)$
$\frac{3}{4} \times \frac{3}{11} = \frac{9}{44}$

CHECKPOINT ✔ Explain why $P(B|A) = P(B)$ if A and B are independent events.

EXAMPLE 3

APPLICATION
CONTEST

In a school contest, a class (sophomore, junior, or senior) will be selected according to the probabilities listed at right. Then a student from that class will be randomly selected. The distribution of the possible contest winners is shown in the table.

$P(\text{sophomore}) = \frac{1}{4}$
$P(\text{junior}) = \frac{1}{4}$
$P(\text{senior}) = \frac{1}{2}$

Find the probability that a girl is selected.

	Girls	Boys	Total
Sophomores	10	13	23
Juniors	7	4	11
Seniors	9	5	14

SOLUTION

To find the probability that a girl from any of these classes is selected, find the sum of probabilities of selecting a sophomore girl, a junior girl, and a senior girl.

Use the formula $P(A \text{ and } B) = P(A) \times P(B|A)$.

$$P(\text{sophomore } and \text{ girl}) = P(\text{sophomore}) \times P(\text{girl}|\text{sophomore}) = \frac{1}{4} \times \frac{10}{23} = \frac{5}{46}$$

$$P(\text{junior } and \text{ girl}) = P(\text{junior}) \times P(\text{girl}|\text{junior}) = \frac{1}{4} \times \frac{7}{11} = \frac{7}{44}$$

$$P(\text{senior } and \text{ girl}) = P(\text{senior}) \times P(\text{girl}|\text{senior}) = \frac{1}{2} \times \frac{9}{14} = \frac{9}{28}$$

Then add.

$$P(\text{girl}) = \frac{5}{46} + \frac{7}{44} + \frac{9}{28} = 0.589$$

The probability that a girl will be selected is about 58.9%.

TRY THIS Let $P(\text{sophomore}) = \frac{1}{6}$, $P(\text{junior}) = \frac{1}{3}$, and $P(\text{senior}) = \frac{1}{2}$ and let the distribution of the possible winners remain as shown in Example 3. Find the probability that a boy is selected.

Enrichment

Have student discuss when $P(A \text{ and } B) = P(A|B)$ and when $P(B|A) = P(A|B)$ based on the formula for conditional probability. **when $P(A \text{ and } B) = 0$ or $P(B) = 1$ and when $P(A) = P(B)$, respectively** Have students set up an experiment to model each relationship.

EXAMPLE 4 Refer to the donated blood problem from the beginning of the lesson.

APPLICATION
HEALTH

What is the probability that a positive ELISA result for a unit of donated blood is accurate?

SOLUTION

PROBLEM SOLVING

Draw a tree diagram to show each possibility.

$P(A) \times P(B|A) = P(B \text{ and } A)$

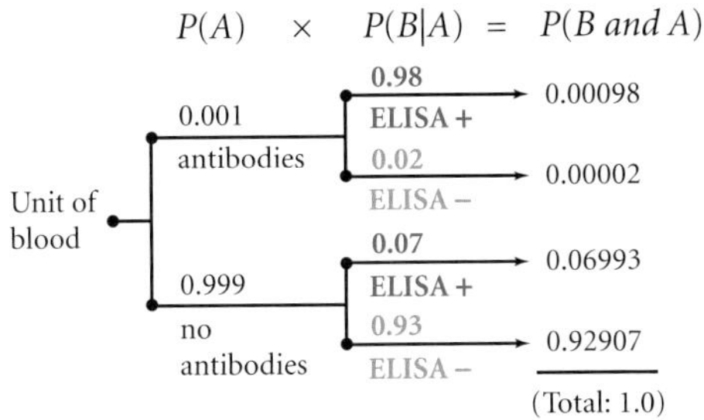

Use the formula for conditional probability.

$$P(\text{antibodies present}|\text{ELISA}+) = \frac{P(\text{antibodies present } and \text{ ELISA}+)}{P(\text{ELISA}+)}$$
$$= \frac{0.00098}{0.00098 + 0.06993}$$
$$= 0.014$$

Thus, the probability that a positive ELISA result is accurate is only about 1.4%. This surprising result is because most positive ELISA results are *false positives.*

TRY THIS Refer to Example 4. Suppose that 75 out of every 1000 units of donated blood are contaminated, instead of 1 out of every 1000 units. What is the probability that a positive ELISA result is accurate?

CRITICAL THINKING If 1 out of every 1000 units of donated blood is contaminated, what is the probability that a negative ELISA result is accurate? Which do you feel is more important, an accurate negative result or an accurate positive result? Why?

ADDITIONAL EXAMPLE 4

Refer to the problem at the beginning of the lesson. Suppose that 25 out of every 1000 people have the disease. **What is the probability that a randomly selected person who tests positive has the disease?**
about 26%

TRY THIS
≈53%

CRITICAL THINKING
≈99.998%; Sample answer: It is more important that the negative results are accurate so that people who receive negative test results are not misinformed or unaware that they have HIV antibodies.

Exercises

Communicate

1. Explain what the notation $P(B|A)$ represents.
2. Describe the difference between $P(A \text{ and } B)$ and $P(B|A)$.
3. Explain why $P(B|A) = 0$ if A and B are mutually exclusive events.

Reteaching the Lesson

HANDS-ON STRATEGIES Encourage students to use manipulatives to model situations when possible. For example, give students a bag with 4 red cubes and 4 blue cubes. Have students perform experiments to find the probability of drawing a red cube on the second draw, given the color of the first draw. For example, have students find the probability of drawing a red cube given that a blue cube has already been drawn. Students would take a blue cube out of the bag and draw several times to find the color of the second cube drawn. After looking at the experimental probability, students can place all 8 cubes in front of them, take away a blue cube, and find the probability of choosing a red cube from the remaining cubes. $\frac{4}{7}$

Assess

Selected Answers

Exercises 4–8, 9–55 odd

ASSIGNMENT GUIDE

In Class	1–8
Core	9–31 odd, 35, 39
Core Plus	10–40 even
Review	42–55
Preview	56–58

Extra Practice can be found beginning on page 940.

internet connect

Activities Online
Go To: **go.hrw.com**
Keyword:
MB1 Conditional

Guided Skills Practice

$\frac{5}{18}$ **4.** The band at Washington High School has 25 members, and the pep club has 18 members. Five students belong to both groups. Find the probability that a student randomly selected from these two groups is a member of the band if you know that he or she is in the pep club. ***(EXAMPLE 1)***

A bag contains 12 blue disks and 5 green disks. For each case below, find the probability of selecting a green disk on the first draw *and* a green disk on the second draw. *(EXAMPLE 2)*

5. The first disk is replaced. $\frac{25}{289}$

6. The first disk is *not* replaced. $\frac{5}{68}$

7. Refer to the table on page 666 about the student body and the respective probabilities for each class. Use the method shown in Example 3 to find the probability that a boy is selected. ***(EXAMPLE 3)*** **≈41.1%**

APPLICATION

8. HEALTH Refer to Example 4. Suppose that 5 out of every 1000 units of donated blood are contaminated with HIV antibodies. Find the probability that a positive ELISA result for a unit of donated blood is *not* accurate. ***(EXAMPLE 4)*** **≈93.4%**

Practice and Apply

internet connect

Homework Help Online
Go To: **go.hrw.com**
Keyword:
MB1 Homework Help
for Exercises 9–23

A bag contains 8 red disks, 9 yellow disks, and 5 blue disks. Two consecutive draws are made from the bag *without* replacement of the first draw. Find the probability of each event.

9. red first, red second $\frac{4}{33}$

10. yellow first, yellow second $\frac{12}{77}$

11. red first, blue second $\frac{20}{231}$

12. blue first, red second $\frac{20}{231}$

13. red first, yellow second $\frac{12}{77}$

14. yellow first, red second $\frac{12}{77}$

15. yellow first, blue second $\frac{15}{154}$

16. red first, blue second $\frac{20}{231}$

Two number cubes are rolled, and the first cube shows a 5. Find the probability of each event below for the two cubes.

17. a sum of 9 $\frac{1}{6}$

18. two odd numbers $\frac{1}{2}$

19. a sum of 7 or 9 $\frac{1}{3}$

For one roll of a number cube, let *A* be the event "even" and let *B* be the event "2." Find each probability.

20. a. $P(A)$ $\frac{1}{2}$ **b.** $P(A \text{ and } B)$ $\frac{1}{6}$ **c.** $P(B|A)$ $\frac{1}{3}$

21. a. $P(B)$ $\frac{1}{6}$ **b.** $P(B \text{ and } A)$ $\frac{1}{6}$ **c.** $P(A|B)$ **1**

For one roll of a number cube, let *A* be the event "odd" and let *B* be the event "1 *or* 3." Find each probability.

22. a. $P(A)$ $\frac{1}{2}$ **b.** $P(A \text{ and } B)$ $\frac{1}{3}$ **c.** $(B|A)$ $\frac{2}{3}$

23. a. $P(B)$ $\frac{1}{3}$ **b.** $P(B \text{ and } A)$ $\frac{1}{3}$ **c.** $P(A|B)$ **1**

24. Given $P(A \text{ and } B) = \frac{1}{4}$ and $P(A) = \frac{1}{2}$, find $P(B|A)$. $\frac{1}{2}$

25. Given $P(A \text{ and } B) = 0.38$ and $P(A) = 0.57$, find $P(B|A)$. $\frac{2}{3}$

26. Given $P(B|A) = \frac{1}{3}$ and $P(A) = \frac{1}{2}$, find $P(A \text{ and } B)$. $\frac{1}{6}$

27. Given $P(B|A) = 0.27$ and $P(A) = 0.76$, find $P(A \text{ and } B)$. **0.205**

28. Given $P(B|A) = 0.87$ and $P(A \text{ and } B) = 0.75$, find $P(A)$. **≈0.862**

29. Given $P(B|A) = \frac{3}{5}$ and $P(A \text{ and } B) = \frac{1}{2}$, find $P(A)$. $\frac{5}{6}$

Practice

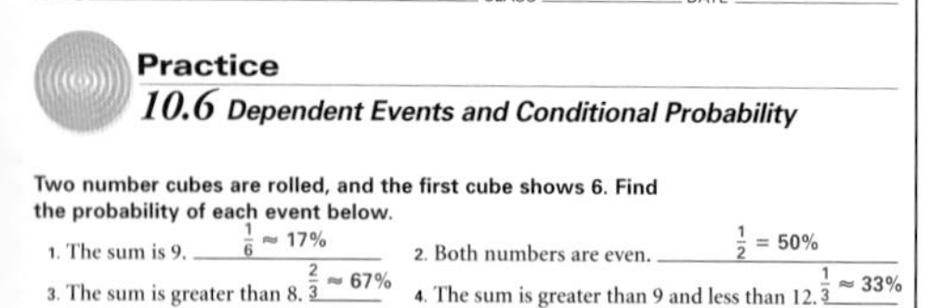

NAME ______ CLASS ______ DATE ______

Practice
***10.6* Dependent Events and Conditional Probability**

Two number cubes are rolled, and the first cube shows 6. Find the probability of each event below.

1. The sum is 9. $\frac{1}{6} \approx 17\%$
2. Both numbers are even. $\frac{1}{2} = 50\%$
3. The sum is greater than 8. $\frac{2}{3} \approx 67\%$
4. The sum is greater than 9 and less than 12. $\frac{1}{3} \approx 33\%$

A spinner that is divided into 8 congruent regions, numbered 1 through 8, is spun once. Let *A* be the event "even" and let *B* be the event "6." Find each of the following probabilities.

5. $P(A)$ $\frac{1}{2} = 50\%$
6. $P(B)$ $\frac{1}{8} = 12.5\%$
7. $P(A \text{ and } B)$ $\frac{1}{8} = 12.5\%$
8. $P(A \text{ or } B)$ $\frac{1}{2} = 50\%$
9. $P(A|B)$ $1 = 100\%$
10. $P(B|A)$ $\frac{1}{4} = 25\%$

A spinner that is divided into 5 congruent regions, numbered 1 through 5, is spun once. Let *A* be the event "odd" and let *B* be the event "less than 3." Find each of the following probabilities.

11. $P(A)$ $\frac{3}{5} = 60\%$
12. $P(B)$ $\frac{2}{5} = 40\%$
13. $P(A \text{ and } B)$ $\frac{1}{5} = 20\%$
14. $P(A \text{ or } B)$ $\frac{4}{5} = 80\%$
15. $P(A|B)$ $\frac{1}{2} = 50\%$
16. $P(B|A)$ $\frac{1}{3} \approx 33\%$

Let *A* and *B* represent events.

17. Given $P(A \text{ and } B) = 0.25$ and $P(A) = 0.4$, find $P(B|A)$. $0.625 = 62.5\%$
18. Given $P(A \text{ and } B) = \frac{3}{5}$ and $P(A) = \frac{2}{3}$, find $P(B|A)$. $\frac{9}{10} = 90\%$
19. Given $P(B|A) = \frac{4}{5}$ and $P(A) = \frac{5}{8}$, find $P(A \text{ and } B)$. $\frac{1}{2} = 50\%$
20. Given $P(B|A) = 0.4$ and $P(A) = 0.16$, find $P(A \text{ and } B)$. $0.064 = 6.4\%$
21. Given $P(B|A) = 0.5$ and $P(A \text{ and } B) = 0.2$, find $P(A)$. $0.4 = 40\%$
22. Given $P(B|A) = 0.8$ and $P(A \text{ and } B) = 0.45$, find $P(A)$. $0.5625 = 56.25\%$

CONNECTION

30a. $\frac{\sqrt{3}}{100}$, or ≈1.7%
b. $\frac{\sqrt{3}}{9}$, or ≈19.2%

31a. $\frac{1}{100}$, or 1%
b. $\frac{1}{4\pi}$, or ≈8.0%

GEOMETRY A point is randomly selected in the 10 × 10 square at right. Find the probability of each event.

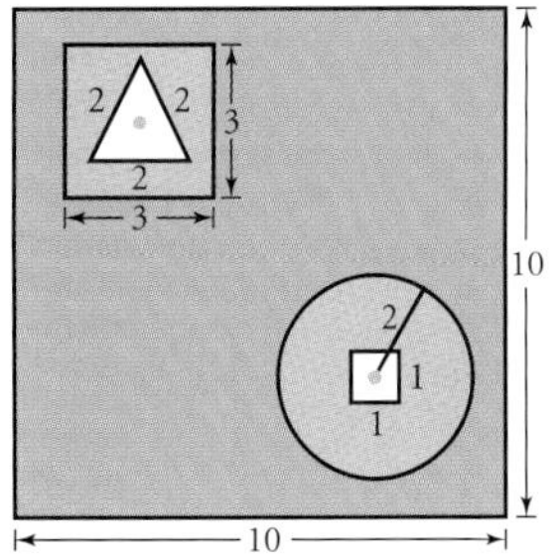
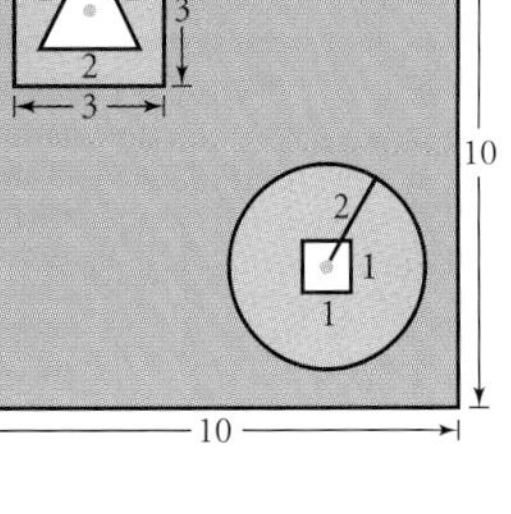

30. **a.** The point is inside the triangle.
b. The point is inside the triangle, given that it is within the 3 × 3 square.

31. **a.** The point is inside the 1 × 1 square.
b. The point is inside the 1 × 1 square, given that it is inside the circle.

APPLICATION

DEMOGRAPHICS The table below gives data from a survey on marital status in the United States for 1995.

Age	Number of persons (in thousands)				
	Total in age group	Never married	Married	Widowed	Divorced
18 to 19	7016	6643	357	2	13
20 to 24	18,142	13,372	4407	17	347
25 to 29	19,401	8373	9913	23	1090
30 to 34	21,988	5186	14,645	80	2077
35 to 39	22,241	3649	15,664	155	2773
40 to 44	20,094	2271	14,779	205	2838
45 to 54	30,694	2173	23,465	808	4248
55 to 64	20,756	961	15,640	1680	2474
65 to 74	18,214	750	12,120	4045	1299
75 and older	13,053	561	5670	6346	473
Total	191,599	43,939	116,660	13,361	17,632

[*Source: Statistical Abstract of the United States, 1996*]

Suppose that a person were chosen at random from the population in 1995. Find the probability of each event.

32. $\frac{4407}{18,142}$, or ≈24.3%

33. $\frac{14,320}{37,543}$, or ≈38.1%

34. $\frac{1437}{37,543}$, or ≈3.8%

35. $\frac{4850}{44,229}$, or ≈11.0%

36. $\frac{21,745}{37,543}$, or ≈57.9%

37. $\frac{11,106}{64,323}$, or ≈17.3%

32. The person is married, given that the person is 20 to 24 years old.

33. The person is married, given that the person is 20 to 29 years old.

34. The person is divorced, given that the person is 20 to 29 years old.

35. The person is divorced, given that the person is 30 to 39 years years old.

36. The person has never been married, given that the person is 20 to 29 years old.

37. The person has never been married, given that the person is 30 to 44 years old.

Error Analysis

Students often confuse independent events with mutually exclusive events. Remind them that when two events, A and B, are mutually exclusive, the intersection of these events is empty. Therefore, $P(A \text{ and } B) = 0$. If two events, C and D, are independent, then $P(C \text{ and } D) = P(C) \cdot P(D)$. So, if two events are mutually exclusive, they must be dependent unless the probability of one or both is 0.

Student Technology Guide

NAME ____________ CLASS ________ DATE ________

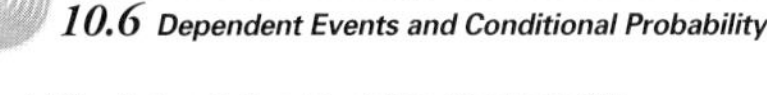

Student Technology Guide
10.6 Dependent Events and Conditional Probability

The probability of a dependent event is called *conditional probability*. You can use a calculator to help find conditional probabilities.

Example: Riverton has two high schools. North High has 2480 students, including 612 seniors. South High has 1998 students, including 537 seniors. Every year, the mayor flips a coin to choose a high school, then randomly chooses a student from that school to be the school representative to the city council. Find the probability that a senior is chosen.

- If a senior is chosen, he or she attends either North or South High.
  $P(\text{senior } and \text{ North}) = P(\text{North}) \cdot P(\text{senior}|\text{North}) = \frac{1}{2} \cdot \frac{612}{2480}$
  $P(\text{senior } and \text{ South}) = P(\text{South}) \cdot P(\text{senior}|\text{South}) = \frac{1}{2} \cdot \frac{537}{1998}$
- Add these probabilities to find the probability that a senior is chosen.
  Press (1 ÷ 2) × (612 ÷ 2480) + (1 ÷ 2) × (537 ÷ 1998) ENTER.

The probability a senior is chosen is about 0.26, or 26%.

1. North High has 580 sophomores and South High has 648 sophomores. Find the probability that a sophomore is chosen as school representative. about 0.28, or 28%

Example: A quality control test rejects defective machine parts 99% of the time. The test also rejects 3% of the acceptable parts. Seven out of every 1000 parts (0.7% of the parts) are defective. If a part fails the test, what is the probability that it is actually acceptable?

- Use the formula for conditional probability:
  $P(\text{acceptable}|\text{rejected}) = \frac{P(\text{acceptable } and \text{ rejected})}{P(\text{rejected})}$
  $P(\text{acceptable } and \text{ rejected}) = 0.993 \times 0.03$
  $P(\text{rejected}) = 0.007 \times 0.99 + 0.993 \times 0.03$
  Press (.993 × .03) ÷ (.007 × .99 + .993 × .03) ENTER.

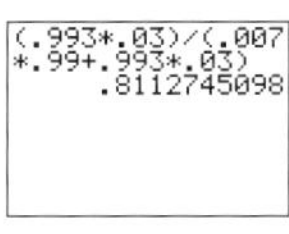

If a part fails the test, the probability that it is actually acceptable is about 0.81, or 81%.

2. In a test for a certain disease, 98% of people who have the disease test positive and 95% of those who do *not* have the disease test negative. If 23 out of every 10,000 people have the disease, what is the probability that a person who tests positive has the disease? about 0.96, or 96%

In Exercises 56–58, students find sums of combinations and see that they result in a powers of 2. This is a *geometric series*. Students will study geometric series in depth in Lesson 11.5.

54. ellipse; $\frac{(x-3)^2}{4}+\frac{(y+1)^2}{16}=1$

56. ${}_5C_0 + {}_5C_1 + {}_5C_2 + {}_5C_3 + {}_5C_4 + {}_5C_5$
$= 1 + 5 + 10 + 10 + 5 + 1$
$= 32$
$= 2^5$

57. ${}_6C_6 + {}_6C_5 + {}_6C_4 + {}_6C_3 + {}_6C_2 + {}_6C_1$
$= 1 + 6 + 15 + 20 + 15 + 6$
$= 63$
$= 64 - 1$
$= 2^6 - 1$

APPLICATIONS

38. ADVERTISING Suppose that 20% of a newspaper's readers see an ad for a new product. If 9% of the readers who see the ad purchase the product, what is the probability that a newspaper reader sees the ad and purchases the product? **1.8%**

39. Market research indicates that 77% of all computer owners buy their computers in an electronics store as opposed to a department or general merchandise store. Furthermore, 41% of all computer owners buy their computers and software from electronics stores. If Frances purchased a computer in an electronics store and is interested in buying software, what is the probability that he will buy it in an electronics store? **53.2%**

40. HEALTH Suppose that for a particular test, 97% of people who have the illness test positive and 98% of people who do not have the illness test negative. If the probability of having the illness is 0.004, find each probability.

a. that a positive test result is inaccurate **≈83.7%**

b. that a positive test result is accurate **≈16.3%**

CHALLENGE

41. On 3 tosses of a coin, 2 were heads. What is the probability that the first toss was heads? $\frac{2}{3}$

Look Back

Solve each equation. ***(LESSON 1.8)***

42. $|x-4| = 9$ **−5, 13** **43.** $|3x| - 6 = -2$ $-\frac{4}{3}, \frac{4}{3}$ **44.** $|3-4x| = 21$ **−4.5, 6**

45. Factor $72x - 24x^2 + 2x^3$, if possible. ***(LESSON 5.3)*** $2x(x-6)^2$

46. Factor $x^4 - 81$ completely. ***(LESSON 5.3)*** $(x-3)(x+3)(x^2+9)$

47. Use the quadratic formula to solve $a^2 - 5a - 2 = 0$. ***(LESSON 5.5)***

47. $a = \frac{5 \pm \sqrt{33}}{2}$

48. Find the zeros of the polynomial function $f(x) = x^3 - 7x^2 + 7x + 15$. ***(LESSON 7.5)*** **−1, 3, 5**

Find the domain of each radical function. ***(LESSON 8.6)***

49. $f(x) = \sqrt{3x-3}$ $x \ge 1$ **50.** $f(x) = \sqrt{3(x-3)}$ $x \ge 3$ **51.** $f(x) = \sqrt{4-3(x+1)}$ $x \le \frac{1}{3}$

Find *PQ* and the coordinates of *M*, the midpoint of $\overline{PQ}$. Give exact answers and approximate answers to the nearest hundredth when appropriate. ***(LESSON 9.1)***

52. $P(3, -4)$ and $Q(-2, -5)$ $M(0.5, -4.5)$ **53.** $P(-2, 7)$ and $Q(-8, 2)$ $M(-5, 4.5)$

54. Classify the conic section defined by $16x^2 + 4y^2 - 96x + 8y + 84 = 0$. Write the standard equation for this conic section and sketch the graph. ***(LESSON 9.6)*** ellipse; $\frac{(x-3)^2}{4}+\frac{(y+1)^2}{16}=1$

APPLICATION

55. VOTING A 3-person committee is to be elected from a group of 6 boys and 4 girls. If each person has an equal chance of being elected, find the probability that the elected committee consists of 2 boys and 1 girl. ***(LESSON 10.3)*** $\frac{1}{2}$

Look Beyond

56. Show that ${}_5C_0 + {}_5C_1 + {}_5C_2 + {}_5C_3 + {}_5C_4 + {}_5C_5 = 2^5$.

57. Show that ${}_6C_6 + {}_6C_5 + {}_6C_4 + {}_6C_3 + {}_6C_2 + {}_6C_1 = 2^6 - 1$.

CHALLENGE

58. Prove that ${}_nC_r + {}_nC_{r+1} = {}_{n+1}C_{r+1}$ is true for all integers r and n, where $0 \le r \le n$.

58.
$$
\begin{aligned}
{}_nC_r + {}_nC_{r+1} &= \frac{n!}{r!(n-r)!} + \frac{n!}{(r+1)![n-(r+1)]!} \\
&= \frac{(r+1)n!}{(r+1)r!(n-r)!} + \frac{n!(n-r)}{(r+1)!(n-r)[n-(r+1)]!} \\
&= \frac{n!(r+1)}{(r+1)!(n-r)!} + \frac{n!(n-r)}{(r+1)!(n-r)!} \\
&= \frac{n![(r+1)+(n-r)]}{(r+1)![(n+1)-(r+1)]!} \\
&= \frac{(n+1)!}{(r+1)![(n+1)-(r+1)]!} \\
&= {}_{n+1}C_{r+1}
\end{aligned}
$$

Experimental Probability and Simulation

Why *Simulations can be used to estimate probabilities. For example, you can use a simulation to estimate the probability that a family with 3 children has all boys or all girls.*

Objective

- Use simulation methods to estimate or approximate the experimental probability of an event.

Recall from Lesson 10.1 that the experimental probability of an event is approximated by performing trials and recording the ratio of the number of occurrences of the event to the number of trials.

Activity Exploring Experimental Probability

You will need: a coin

You can model, or *simulate*, two random events with coin tosses. Let H, heads, represent a female and let T, tails, represent a male. Each trial will represent a family with three children. Assume that genders are random.

Follow the steps below to investigate P(all males *or* all females).

1. Toss a coin three times, and record the outcomes, such as HTH or HHH, as a single trial. Perform 10 trials for the experiment. Find the number of occurrences of the event "TTT *or* HHH." What is the ratio of this number to the total number of trials?
2. Repeat Step 1 four more times.
3. Find the total number of occurrences of the event "TTT *or* HHH" in all five experiments. What is the ratio of this number to the total number of trials in all five experiments? (Let this be called the *average ratio.*)
4. Find the theoretical probability of the event "TTT *or* HHH."
5. Compare the experimental probability ratios that you obtained in each experiment with the theoretical probability. Compare your *average ratio* from Step 3 with the theoretical probability.

CHECKPOINT ✔

6. Which of your ratios is closest to the theoretical probability? Compare your results with those of your classmates. Make a generalization about the value of experimental probability ratios as the number of trials in an experiment increases.

Prepare

NCTM PRINCIPLES & STANDARDS 1–10

QUICK WARM-UP

A coin is tossed. What is the theoretical probability of

1. heads? $\frac{1}{2}$ 2. tails? $\frac{1}{2}$

The coin is tossed 20 times with the results below.

H H H T T T T H T T
H H T T H T H H H H

Using these results, what is the experimental probability of

3. heads? $\frac{11}{20}$ 4. tails? $\frac{9}{20}$

In 1000 tosses of this coin, how many heads can you expect

5. theoretically? 500
6. based on the results of the experiment? 550

Also on Quiz Transparency 10.7

Teach

Why With fast-speed technology, businesses can run simulations with very large numbers of trials to test their products. Have students brainstorm about possible simulations that a business might use.

Activity Notes

In the Activity, students should find that the *average ratio* better approximates the theoretical probability of having three children of the same gender. The theoretical probability is $\left(\frac{1}{2}\right)^3 = \frac{1}{8}$.

☞ For answer to Checkpoint, see page 673.

Alternative Teaching Strategy

USING TECHNOLOGY Have students simulate the Activity on page 671 by using a calculator with a random-number generator. Put students in groups of three, with one student operating the calculator, another tallying the results of the calculator, and the third student keeping track of the size of the trial. Have each group randomly generate the numbers 0 or 1 by entering **int(2*rand)** with 1 representing heads and 0 representing tails. (The commands **int** and **rand** may be found under the submenus **NUM** and **PRB** after pressing MATH.) Once the random-number generator command is entered, repeatedly pressing ENTER on most calculators will simulate coin tosses. The group should record each number generated and keep a tally of the size of the trial. Have students experiment with how large the trial must be for the experimental data to match the theoretical probability of obtaining 3 heads or 3 tails. Have each group share their results with the class and discuss the importance of large sample sizes when conducting a simulation.

ADDITIONAL
EXAMPLE 1

Traffic analysts counted the number of motorists per 100 that went in each direction at the intersection described in Example 1. The results are shown in the table below.

Straight	30
Left	45
Right	25

Use a simulation to estimate the probability that 2 or more of 5 motorists will go straight. Answer may vary.
sample simulation:
1–30 represents S,
31–75 represents L, and
76–100 represents R.

Random numbers	Result
11 69 98 63 31	SLRLL
8 99 19 39 11	SRSLS √
24 85 37 59 60	SRLLL
7 1 1 69 70	SSSLL √
88 37 11 45 98	RLSLR
69 54 63 92 67	LLLRL
79 55 33 21 62	RLLSL
88 12 45 46 28	RSLLS √
81 98 49 40 22	RRLLS
62 61 80 77 46	LLRRL

From the simulation, the probability is 30%.

Recall from Lesson 10.1 that as the number of trials in an experiment increases, the results will more closely approximate the actual probability.

One way to perform large numbers of trials is to use *simulation.* A **simulation** is a reproduction or representation of events that are likely to occur in the real world. Simulations are especially useful when actual trials would be difficult or impossible.

This lesson will concentrate on the use of *random-number generators* to perform simulations.

EXAMPLE 1

APPLICATION
TRANSPORTATION

Traffic analysts counted the number of motorists out of 200 that went in each direction at an intersection. The results are shown in the table below.

Straight	63
Left	89
Right	48

Use a simulation to estimate the probability that 3 or more out of 5 consecutive motorists will turn right.

SOLUTION

1. Copy the table, and add a third column. In the third column, write numbers from 1 to 200 according to the number of motorists recorded for each direction.

Straight	63	1–63	← The first 63 numbers
Left	89	64–152	← The next 89 numbers
Right	48	153–200	← The last 48 numbers

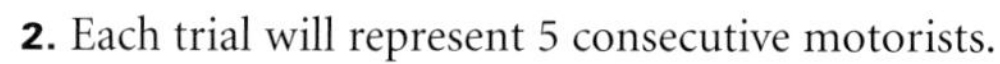

2. Each trial will represent 5 consecutive motorists.

   Generate 5 random integers from 1 to 200 inclusive. Categorize each resulting number as a motorist who goes straight, turns left, or turns right.

TECHNOLOGY
GRAPHICS CALCULATOR
Keystroke Guide, page 687

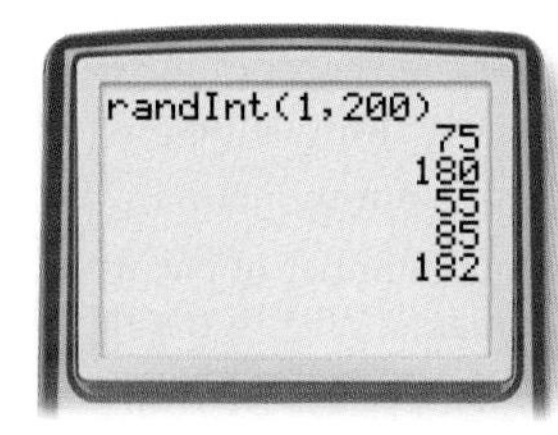

← Turns left (L)
← Turns right (R)
← Goes straight (S)
← Turns left (L)
← Turns right (R)

3. Perform 10 trials, and record your results in a table such as the one shown at right.

Trial	Result
1	LRSLR
2	LSLLS
3	LLSRR
4	SRRLR
5	LSLRL
6	RSRRR
7	LSLLS
8	RLSLL
9	LRRRS
10	SLSSR

4. Estimate the probability.

   In this experiment, there are 3 trials in which 3 or more motorists turned right.

   The estimated probability is $\frac{3}{10}$.

Interdisciplinary Connection

ENVIRONMENTAL SCIENCE The probabilities that there are 0, 1, 2, 3, or 4 significant leaks of toxins into a large bay on any one day are given by $P(x)$, where x represents the number of leaks. Suppose that $P(0) = 0.2418$, $P(1) = 0.3356$, $P(2) = 0.2814$, $P(3) = 0.1181$, and $P(4) = 0.0231$. Use a random-number generator to produce numbers from 1 to 10,000 in order to simulate the number of spills into the bay for 30 consecutive days. Compare the results of the simulation with the probabilities given.

Inclusion Strategies

TACTILE LEARNERS Students can find or make random-number generators. When 2 random numbers are needed, a coin toss serves as a random-number generator. For 6 outcomes, students can use a number cube. Have students look at board games to find number cubes with different shapes that can be used to generate other quantities of random numbers. A spinner can be used to generate different quantities of random numbers. Students can easily make a spinner from cardboard, a brad, and a paper clip. Have students discuss other ways to generate random numbers.

EXAMPLE 2

APPLICATION
MANAGEMENT

An airline statistician made the table at right for customer arrivals per minute at a ticket counter during the time period from 11:00 A.M. to 11:10 A.M. In the table, X represents the number of customers that may arrive at the counter during a one-minute interval, and $P(X)$ represents the probability of X customers arriving during that one-minute interval. For example, the probability that 5 customers arrive during a given minute is 0.189, or 18.9%.

X	$P(X)$
0	0.006
1	0.034
2	0.101
3	0.152
4	0.193
5	0.189
6	0.153
7	0.137
8	0.035

Find a reasonable estimate of the probability that 50 or more customers arrive at the counter between 11:00 A.M. and 11:10 A.M. inclusive.

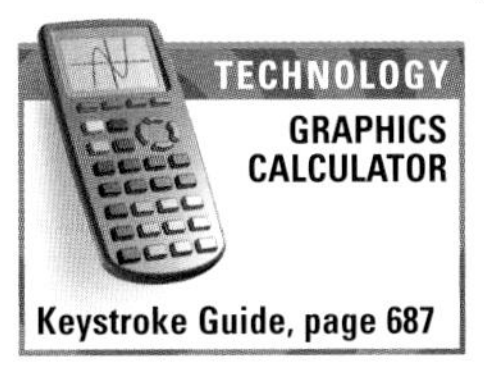

Keystroke Guide, page 687

SOLUTION

1. Copy the probability table and add a new column labeled *Random numbers*. In this column, write numbers from 1 to 1000 according to the given values of $P(X)$.

   For example:
   $P(X) = 0.006$: first 6 numbers, 1–6
   $P(X) = 0.034$: next 34 numbers, 7–40
   $P(X) = 0.101$: next 101 numbers, 41–141

X	$P(X)$	Random numbers
0	0.006	1–6
1	0.034	7–40
2	0.101	41–141
3	0.152	142–293
4	0.193	294–486
5	0.189	487–675
6	0.153	676–828
7	0.137	829–965
8	0.035	966–1000

2. Generate 10 random integers from 1 to 1000 inclusive. The results of the first trial are 684, 461, 145, 947, 832, 521, 302, 250, 947, and 926.

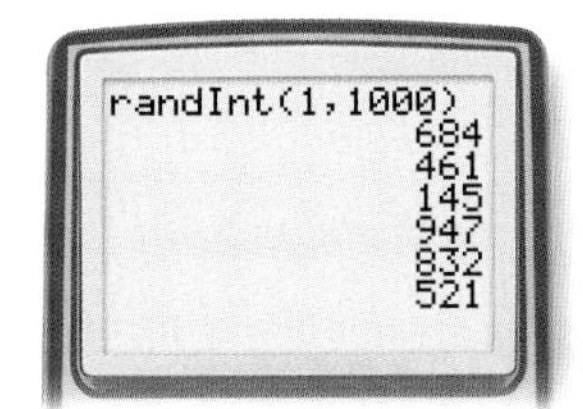

   Use the table to find the corresponding number of customers that arrived for this trial.

   6, 4, 3, 7, 7, 5, 4, 3, 7, 7

   Add these numbers together to find the total number of customers for this 10-minute interval.

   $$6 + 4 + 3 + 7 + 7 + 5 + 4 + 3 + 7 + 7 = 53$$

3. Perform 10 trials, and record your results as shown at left.

Trial	Total number of customers
1	53
2	40
3	55
4	42
5	62
6	32
7	45
8	58
9	64
10	29

4. For this simulation, there were 50 or more customers in 5 out of 10 trials, so an estimate of the probability is $\frac{5}{10}$, or $\frac{1}{2}$.

Enrichment

Have students use a spreadsheet to simulate the frequency of tossing from 0 to 10 heads in trials of 10 coin tosses each. They can use a separate spreadsheet column to represent each trial. Have students enter **=INT(2*RAND())**, which randomly produces 0 or 1, into a cell and then copy the formula by dragging the fill handle downward through 9 more cells to create 10 random outcomes for the first trial. Selecting all 10 cells in the first column and dragging the column's fill handle horizontally will create formulas in columns for several trials.

Let 1 represent heads and 0 represent tails. Students can have the spreadsheet count the number of heads in each trial by entering the following formula in the cell beneath each column: **=sum(*start cell, end cell*)**. Have students find the frequency of 0 heads, 1 heads, 2 heads, and so forth in each trial. Then have them create a frequency chart with the number of heads per trial shown on the horizontal axis and the frequency on the vertical axis. As students use technology to increase the number of trials, the graph should begin to approximate a normal distribution.

CHECKPOINT ✔
(From the Activity on page 671)

6. Answers may vary. In general, the average ratio should be closest to the theoretical probability. As the number of trials increases, the estimated probability should approach the theoretical probability.

ADDITIONAL EXAMPLE 2

Refer to the airline data from the simulation in Example 2. **Find a reasonable estimate of the probability that 20 or fewer customers arrive at the counter between 11:00 A.M. and 11:05 A.M., inclusive.**
Answers may vary.

Teaching Tip

TECHNOLOGY When using a TI-83 or TI-82 graphics calculator to generate random integers less than n, press MATH, select **NUM** and **5:int(**, enter n, press X and MATH, select **PRB** and **1:rand**, enter closing parenthesis, and press ENTER. Repeatedly pressing ENTER will randomly generate numbers less than n.

To generate random numbers between a and b on the TI-83, press MATH, select **5:randInt(**, enter a, press , , and enter b. On the TI-82, use the following sequence: MATH **4:int** MATH **PRB** **1:rand** $(a + 1) + b$. Repeatedly pressing ENTER on either calculator will randomly generate integers between a and b.

ADDITIONAL EXAMPLE 3

Refer to the dartboard shown in Example 3. **Use a simulation to estimate the probability that exactly 1 of the 3 darts lands in region *C*.**

Answers may vary. A sample simulation, where 1–5 represent *A*, 6–8 represent *B*, and 9 represents *C*, is given below.

Random integers	Result
4 6 6	0 in *C*
1 1 1	0 in *C*
7 7 8	0 in *C*
4 2 5	0 in *C*
9 7 5	1 in *C*
6 9 7	1 in *C*
8 5 4	0 in *C*
2 6 8	0 in *C*
2 5 5	0 in *C*
3 8 9	1 in *C*

Based on this simulation, an estimate of the probability is $\frac{3}{10}$, or 30%. The theoretical probability is ${}_3C_1\left(\frac{1}{9}\right)^1\left(\frac{8}{9}\right)^2$, or ≈26%. The binomial distribution is introduced in Lesson 11.8.

Using a Geometric Simulation

You can also use simulations to analyze geometric probability problems.

EXAMPLE 3

CONNECTION GEOMETRY

Suppose that 3 darts are thrown at the dartboard at right and that any point on the board is equally likely to be hit.

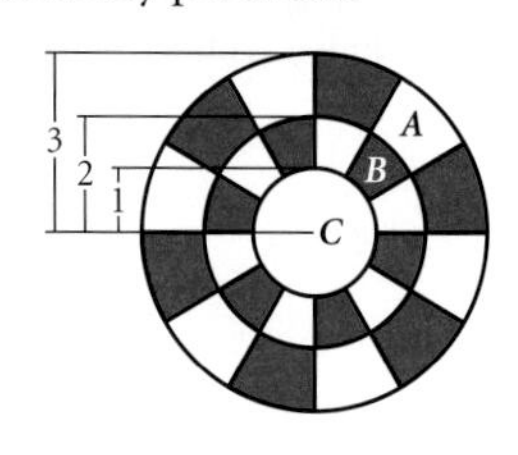

Use a simulation to estimate the probability that exactly 2 of the 3 darts land in ring *B*.

SOLUTION

1. Find the probability of a dart landing in region *A* (the outer ring), *B* (the inner ring), and *C* (the circle).

$$P(A) = \frac{\text{area of region } A}{\text{area of dartboard}} = \frac{9\pi - 4\pi}{9\pi} = \frac{5}{9}$$

$$P(B) = \frac{\text{area of region } B}{\text{area of dartboard}} = \frac{4\pi - \pi}{9\pi} = \frac{3}{9} = \frac{1}{3}$$

$$P(C) = \frac{\text{area of region } C}{\text{area of dartboard}} = \frac{\pi}{9\pi} = \frac{1}{9}$$

The area of a circle is πr^2.

2. Make a table of the probabilities and their respective random integers from 1 to 9 inclusive.

Probability	Random numbers
$P(A) = \frac{5}{9}$	1, 2, 3, 4, 5
$P(B) = \frac{3}{9}$	6, 7, 8
$P(C) = \frac{1}{9}$	9

Keystroke Guide, page 687

3. Generate a trial of 3 random integers. The results of the first trial are 1, 8, and 3. Therefore, there were 2 hits in region *A* (1 and 3), 1 hit in region *B* (8), and 0 hits in region *C*.

Trial	Result
1	1 in *B*
2	2 in *B*
3	0 in *B*
4	3 in *B*
5	2 in *B*
6	1 in *B*
7	2 in *B*
8	1 in *B*
9	0 in *B*
10	2 in *B*

4. Perform 10 trials, and record your results as shown in the table at right.

5. In 4 of the 10 trials, exactly 2 darts landed in region *B*, so an estimate of the probability is $\frac{4}{10}$, or $\frac{2}{5}$.

TRY THIS Refer to the dartboard in Example 3. Use a simulation to estimate the probability that exactly 2 of 3 darts land in ring *A*.

CRITICAL THINKING Calculate ${}_3C_2\left(\frac{1}{3}\right)^2\left(\frac{2}{3}\right)$, the theoretical probability of the event described in Example 3 above. Compare the experimental probability from Example 3 with the theoretical probability. How can you account for the discrepancy?

Math CONNECTION

GEOMETRY In Example 3, students use the formula for the area of a circle, $A = \pi r^2$, to help find the probabilities required.

TRY THIS

Answers may vary. sample: ≈0.41

CRITICAL THINKING

$\frac{2}{9}$; $\frac{2}{5} - \frac{2}{9} = \frac{8}{45}$, or ≈0.18; answers may vary.

Reteaching the Lesson

COOPERATIVE LEARNING Divide the class into four groups. Assign each group one of the examples from the lesson. Have each group go through the simulation in their example and create a new simulation table. Each group should discuss the procedure for creating the simulation, making sure that all students can produce a similar simulation. Each group should put their simulation results on a large piece of butcher paper to be displayed as they explain their example to the rest of the class. All group members should participate in the presentation to the class and answer any questions the class may have.

You can also use a geometric simulation to approximate π. Refer to the figure at right.

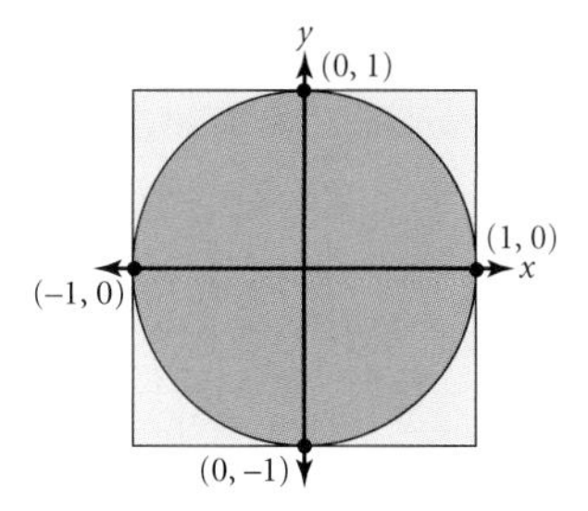

Let event R be a point in the square that is also in the circle.

$$P(R) = \frac{\text{area of circle}}{\text{area of square}} = \frac{\pi(1)^2}{2^2} = \frac{\pi}{4}$$

Make a spreadsheet that contains 20 random decimals between −1 and 1 inclusive in both columns A and B. For column C, use the following rule:

If $x^2 + y^2 \leq 1$, (where x and y are the values in columns A and B), print 1, otherwise print 0.

Then compute the experimental probability:

$$P(R) = \frac{\text{number of 1s}}{20}$$

The data from one simulation gives $P(R) = 0.8$. Thus, $\frac{\pi}{4} \approx 0.8$, or $\pi \approx 3.2$. The actual value of π is 3.14159 . . .

Workbook

	A	B	C
1	X	Y	
2	-0.79355678	0.34678998	1
3	0.82410872	-0.92525235	0
4	0.68198496	0.38564413	1
5			
17			
18			
19	0.31884303	0.45645644	0
20	0.56456678	-0.52534556	1
21	0.25647562	0.2947445	1
22			3.2

Sheet 1 / Sheet 2 / Sheet 3

Math CONNECTION

GEOMETRY In this example, students use area formulas and probability to estimate the number π. Students use formulas for the area of a circle, $A = \pi r^2$, and for the area of a square, $A = s^2$. The circular region is represented by $x^2 + y^2 \leq 1$.

Use Teaching Transparency 41.

Teaching Tip

TECHNOLOGY Most spreadsheets have a formula or function menu that can be used to find the random-number function. In most spreadsheets, entering **=RAND()** into a cell produces a random number between 0 and 1. To produce random integers within the interval $[0, n)$, enter **=INT(N*RAND())**. To produce random integers within the interval $[m, n)$ enter **=INT((N–M)*RAND()+M)**. Note that the lower bound may be produced, but the upper bound will not. The correct number and placement of parentheses in the formulas is critical.

Exercises

Communicate

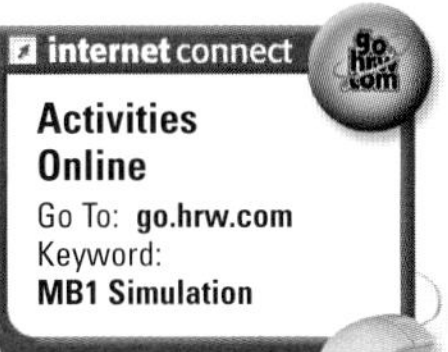

1. How can you use random numbers to simulate rolling a number cube?

2. For what type of event is flipping a coin an appropriate model? Explain.

3. Which experiment would generally produce better results, an experiment with 1 trial or an experiment with 100 trials? Explain.

Guided Skills Practice

APPLICATIONS

4. **TRANSPORTATION** Refer to the traffic data from page 672. Use a simulation with 10 trials to estimate the probability that 2 or more out of 6 consecutive motorists will go straight. ***(EXAMPLE 1)*** **≈60%**

5. **MANAGEMENT** Refer to the probability table for customer arrival on page 673. Use a simulation with 10 trials to estimate the probability that 30 or more customers will arrive during the given 10-minute period. **≈99%** ***(EXAMPLE 2)***

CONNECTION

6. **GEOMETRY** Refer to the dartboard on page 674. Use a simulation with 10 trials to estimate the probability that exactly 1 of 4 darts land in region C. ***(EXAMPLE 3)*** **≈30%**

Assess

Selected Answers

Exercises 4–6, 7–37 odd

ASSIGNMENT GUIDE	
In Class	1–6
Core	7–19 odd, 25, 27
Core Plus	8–26 even
Review	28–37
Preview	38

Extra Practice can be found beginning on page 940.

Error Analysis

Some students may make mistakes when assigning random numbers to categories in certain exercises based on the given probabilities. Encourage them to look over the ratio of each interval they assign to the total range of the intervals in order to make sure that their choices follow the given probabilities.

Practice

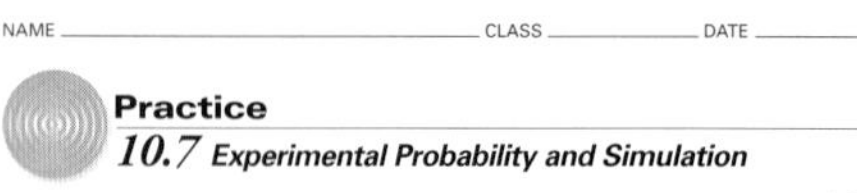

NAME ________________ CLASS ________ DATE ________

Practice
10.7 ***Experimental Probability and Simulation***

Use a simulation with 10 trials to find an estimate for each probability. 1–6. Simulation results will vary.

1. In 4 tosses of a coin, heads appears exactly 3 times.
2. In 5 tosses of a coin, tails appears more than 2 times.
3. In 4 rolls of a number cube, 3 appears twice.

Trial	Result
1	
2	
3	
4	
5	
6	
7	
8	
9	
10	

estimate: ________

Trial	Result
1	
2	
3	
4	
5	
6	
7	
8	
9	
10	

estimate: ________

Trial	Result
1	
2	
3	
4	
5	
6	
7	
8	
9	
10	

estimate: ________

Of 100 motorists observed at an intersection, 26 turned left, 47 went straight, and 27 turned right. Use a simulation with 10 trials to find an estimate for each probability.

4. Exactly 2 of every 4 motorists turn right.
5. At least 3 of every 4 motorists go straight.
6. Fewer than 2 of every 4 motorists turn left.

Trial	Result
1	
2	
3	
4	
5	
6	
7	
8	
9	
10	

estimate: ________

Trial	Result
1	
2	
3	
4	
5	
6	
7	
8	
9	
10	

estimate: ________

Trial	Result
1	
2	
3	
4	
5	
6	
7	
8	
9	
10	

estimate: ________

Practice and Apply

7. Use a simulation with 20 trials to estimate the number of coin tosses needed to obtain 2 consecutive heads. (Hint: Outcomes such as HH, THH, and THTHTHH meet the condition.) **about 6 tosses**

8. Use a simulation with 20 trials to estimate the number of coin tosses needed to obtain heads followed by tails. (Hint: Outcomes such as THT, HHT, TTHT, and THHHT meet the condition.) **about 4 tosses**

9. Use a simulation with 10 trials to estimate the number of times a 6-sided number cube must be rolled to obtain 2 consecutive numbers less than 5.

9. about 4 rolls

10. Use a simulation with 10 trials to estimate the number of times a 6-sided number cube must be rolled to obtain a number less than 5 followed by a 5 or 6.

10. about 4 or 5 rolls

Use the table of traffic data below and a simulation with 10 trials to estimate the probability of each event.

Straight	73
Left	53
Right	69

11. Exactly 2 out of 3 consecutive motorists turn right. **≈24%**

12. Exactly 3 out of 4 consecutive motorists turn left. **≈6%**

13. No more than 2 out of 5 consecutive motorists go straight. **≈73%**

14. At least 3 out of 5 consecutive motorists go straight. **≈27%**

A restaurant chain is giving away 1 of 4 different prizes with each purchase of a dinner. Assume that each prize is equally likely to be awarded.

15. Use a simulation with 10 trials to estimate the probability that you will have all 4 prizes after 5 dinner purchases. **≈25%**

16. Use a simulation with 10 trials to estimate the probability that you will have all 4 prizes after 8 dinner purchases. **≈60%**

17. Use a simulation with 10 trials to estimate how many dinners you must purchase to collect all 4 prizes. **about 8**

A multiple-choice test consists of 10 questions. Use the number of possible answers given below, and assume that each answer is a guess.

18. Suppose that each question has 4 possible answers. Use a simulation to find the probability of answering at least 6 questions correctly. **≈2%**

19. Suppose that each question has 3 possible answers. Use a simulation to find the probability of answering at least 6 questions correctly. **≈8%**

20. Suppose that each question has only 2 possible answers, true or false. Use a simulation to find the probability of answering at least 6 questions correctly. **≈38%**

internet connect
Homework Help Online
Go To: **go.hrw.com**
Keyword:
MB1 Homework Help for Exercises 21–23

Use a simulation to estimate the indicated area.

21. the area of the ellipse $\frac{x^2}{4} + \frac{y^2}{9} = 1$ **≈19 square units**

22. the area under the curve $y = e^x$ from $x = 0$ to $x = 5$ **≈147 square units**

23. the area under the curve $y = x^2 - 2x + 3$ from $x = 0$ to $x = 5$

23. ≈32 square units

In Exercise 38, students look at patterns to find the next number in a sequence. Sequences and series will be studied in detail in Chapter 11.

CONNECTION

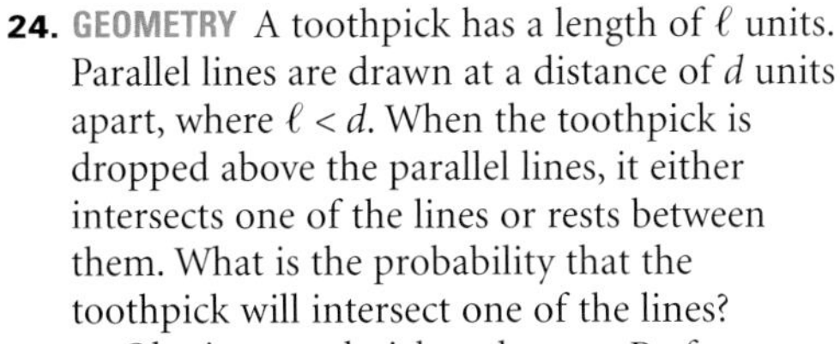

24. GEOMETRY A toothpick has a length of ℓ units. Parallel lines are drawn at a distance of d units apart, where $\ell < d$. When the toothpick is dropped above the parallel lines, it either intersects one of the lines or rests between them. What is the probability that the toothpick will intersect one of the lines?

a. Obtain a toothpick and paper. Perform a simulation with 20 trials to estimate the probability.

b. Compare your answer with French naturalist Compte de Buffon's theoretical value of $\frac{2\ell}{\pi d}$.

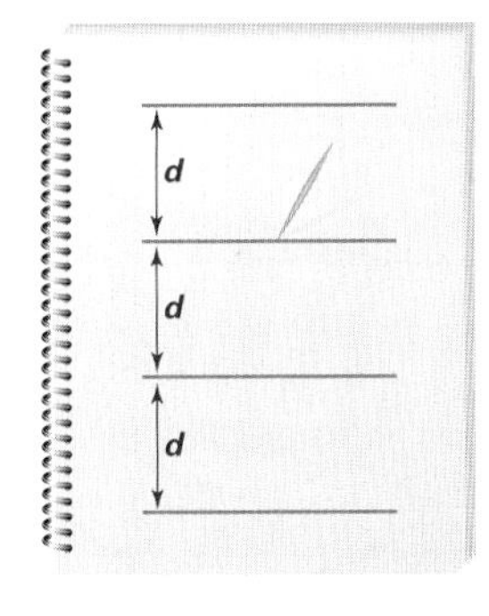

24a. about $\frac{2\ell}{\pi d}$
b. about $\frac{2\ell}{\pi d}$

APPLICATION

SPORTS A baseball player's batting statistics are given in the table at right. Use a simulation with 10 trials to find an approximate answer for each question below.

Batting Statistics

Outcome	Probability
Single	0.198
Double	0.061
Triple	0.020
Home run	0.032
Walk	0.118
Ground out	0.264
Fly out	0.187
Strike out	0.120
Total	**1.000**

25. How many times will the player bat before he gets a hit (a single, double, triple, or home run)? **≈3 times**

26. How many times will the player bat before he makes a ground out, fly out, or strike out? **≈2 times**

27. How many times will the player bat before he makes a ground out or fly out? **≈2 times**

Look Back

Solve each system of linear equations. *(LESSONS 3.1 AND 3.2)*

28. $\begin{cases} 21x + 2y = 3 \\ 3x + 4y = 5 \end{cases}$ $\left(\frac{1}{39}, \frac{16}{13}\right)$ **29.** $\begin{cases} -3x + y = -2 \\ 2x - y = 3 \end{cases}$ $(-1, -5)$ **30.** $\begin{cases} 7x - 3y = 1 \\ 2x + 5y = 2 \end{cases}$ $\left(\frac{11}{41}, \frac{12}{41}\right)$

Solve each nonlinear system of equations. *(LESSON 9.6)*

31. $\begin{cases} x^2 + y^2 = 4 \\ 3x - y = 0 \end{cases}$ **32.** $\begin{cases} 2x + y = 0 \\ x^2 + y^2 = 9 \end{cases}$ **33.** $\begin{cases} 2x^2 + \frac{y^2}{8} = 1 \\ x^2 + y^2 = 4 \end{cases}$

31. $\left(\frac{\sqrt{10}}{5}, \frac{3\sqrt{10}}{5}\right)$, $\left(-\frac{\sqrt{10}}{5}, -\frac{3\sqrt{10}}{5}\right)$
32. $\left(\frac{3\sqrt{5}}{5}, -\frac{6\sqrt{5}}{5}\right)$, $\left(-\frac{3\sqrt{5}}{5}, \frac{6\sqrt{5}}{5}\right)$
33. four solutions: $\left(\pm\frac{2\sqrt{15}}{15}, \pm\frac{2\sqrt{210}}{15}\right)$

34. How many 5-digit zip codes are possible? Assume that 00000 is not a valid zip code. *(LESSON 10.1)* **99,999**

A bag contains 4 red marbles, 7 white marbles, and 14 black marbles. Find the probability of each event for one random selection. *(LESSON 10.4)*

35. red *or* white $\frac{11}{25}$ **36.** red *or* black $\frac{18}{25}$ **37.** black *or* white $\frac{21}{25}$

Look Beyond

38. Find the next two terms in each sequence of numbers.
a. 1, 5, 9, 13, 17, . . . **21, 25** **b.** 20, 10, 5, 2.5, . . . **1.25, 0.625**

Student Technology Guide

NAME ________________ CLASS ________ DATE ________

Student Technology Guide
10.7 Experimental Probability and Simulation

Graphics calculators and spreadsheets can generate random numbers. You can use these random-number generators to conduct simulations.

Example: A telemarketing firm keeps statistics on its calls. Over a one-week period, one sales representative made 400 calls. Of these, 124 were not answered, 217 people decided not to buy the product, and 59 decided to buy the product. Find a reasonable estimate of the probability that at least 2 of 6 phone calls result in a sale.

- Make a table showing the outcomes and their frequencies as shown at right. The third column shows which numbers from 1 to 400 should represent each outcome.

Outcome	Frequency	Numbers (1–400)
Not answered	124	1–124
No sale	217	125–341
Sale	59	342–400

- Use a graphics calculator to generate one trial. Press MATH 5:randInt(ENTER 1 , 400) ENTER to generate an integer from 1 to 400. Press ENTER five more times to complete the trial.

```
randInt(1,400)
                310
                393
                167
                 41
                199
                121
```

- Use the table to classify the results. If N represents "not answered," X represents "no sale," and S represents "sale," the results of this trial are: XSXNXN.
- Perform additional trials as shown in the table at right.
- From the table, there are 2 trials where at least 2 of 6 calls resulted in a sale.

Trial	Result
1	XSXNXN
2	SNXNXS
3	XNNNXX
4	XNXNSX
5	XXNNXN
6	SXNNXN
7	NXNXXN
8	NXXNNS
9	XXNXXN
10	XNXXSS

The estimate of the probability that at least 2 of 6 calls result in a sale is $\frac{2}{10} = 20\%$.

1. Use a simulation to estimate the probability that at least 3 of 6 phone calls are not answered. Answers will vary.

2. Of 678 applicants to a small college, 349 were admitted, 284 were rejected, and 45 were placed on the waiting list. At George's high school, 5 students applied to this college. Use a simulation to estimate the probability that at least 2 of the 5 students were admitted. Describe how you performed this simulation.
Answers will vary.

Focus

An airline company is interested in analyzing the wait time for customers in the ticket line. In this project, students use information about the time it takes to process customers to create a simulation experiment. The simulation gives information about the probability of a customer waiting over 50 minutes.

Motivate

Ask students to describe times when they have traveled (by plane, bus, or train) during busy seasons. Have them describe the way the transportation company dealt with large crowds at the ticket counter, including any incentives the company provided for overbooked or delayed travelers. Ask whether any student has ever missed a travel connection because of a delay beyond their control. If so, how did the transportation company compensate them?

Next, Please . . .

A customer arrives at the ticket counter of an airline. A delay at the counter caused by a long line might result in the customer missing a flight or might cause the airline to delay the departure time so that the customer does not miss the flight. Both events are undesireable for the airline. Therefore, the airline is greatly concerned about its customer's arrival time and processing time in order to minimize delays.

Suppose that an airline is going to start operations in a new terminal. The airline must develop a strategy for providing the optimum number of check-in counters and service staff. Trial-and-error methods in this situation could be very expensive. Instead the airline decides to use a probability experiment.

Activity 1

Airline statisticians want to analyze the time that it takes to process customers who arrive at the ticket counter between 1:00 P.M. and 1:10 P.M. To do this, they begin by researching the time it takes to process each of 50 customers. As a result of this research, the following table of probabilities is generated:

Process Times (in seconds)

Processing time, t	10	20	30	40	50	60
Probability, $P(t)$	0.052	0.132	0.158	0.135	0.123	0.104

Processing time, t	70	80	90	100	110	120
Probability, $P(t)$	0.058	0.034	0.116	0.050	0.026	0.012

Another term for a waiting line is a queue. In real-world situations, queues can vary considerably. They grow for a while, then disappear, and then occur again.

The Answers to Activity 1 can be found in Additional Answers beginning on page 1002.

In the table, t represents the processing time to the nearest 10 seconds, and $P(t)$ represents the probability that processing a customer will take t seconds. The probabilities in the table are given to the nearest thousandth.

In order to set up a random-number simulation, first group numbers, N, from 1 to 1000 according to the given probabilities. For example, the first 52 numbers (1–52) correspond to $t = 10$. The next 132 numbers (53–184) correspond to $t = 20$, the next 158 numbers (185–342) correspond to $t = 30$, and so on. Complete the table in this manner.

Table for Processing Time Simulations

t	$P(t)$	N
0	0	000
10	0.052	001–052
20	0.132	053–184
30	0.158	185–342
⋮	⋮	⋮

Activity 2

Perform a simulation for 50 customers who arrive at the counter between 1:00 P.M. and 1:10 P.M.

1. Using a random-number generator, generate 50 random numbers between 1 and 1000 inclusive.
2. Create a table like the one shown at right, and record your random numbers in the *Number* column. Each random number represents a customer.
3. Refer to the table you created in Activity 1. For each random number that you generated, write the corresponding value of t in the *Time* column. The corresponding value of t represents the time it takes to process that customer. For example, if the first random number generated is 179, then the correspoı ding value of t is 20. This simulates a processing time of 20 seconds for the first customer.

Simulated Processing Time for 50 Consecutive Customers

Customer	Number, N	Time, t
1		
2		
3		
⋮		
49		
50		

Activity 3

1. Perform 10 trials of the simulation for 50 customers who arrive at the counter between 1:00 P.M. and 1:10 P.M. Estimate the probability that processing 50 customers takes longer than 50 minutes.
2. From the activities that you have completed in this project, you can see that efficient handling of customers, particularly in high-volume situations, is complicated and is not an exact science. From your own experiences, what are some situations in which these simulation techniques might be used to estimate probabilities? Describe how simulations might be designed for these situations.

Cooperative Learning

You may wish to have students do the project in groups of three or four. Have them alternate in the roles of random-number generator, recorder, and supervisor for each activity.

Discuss

Bring the class together after Activity 1 to discuss the setup of the problem situation and the data charts given. After Activity 2, have groups discuss the results of their simulation and how they set up the simulation. After Activity 3, have students discuss their results and share their ideas on other situations in which simulations might be used.

Activity 3

1. **Answers may vary but should be close to 3%.**
2. **Answers may vary. Sample answer: waiting times for telephone information services**

The Answers to Activity 2 can be found in Additional Answers beginning on page 1002.

Chapter Review

10 Chapter Review and Assessment

VOCABULARY

Key Skills & Exercises

LESSON 10.1

Key Skills

Find the theoretical probability of an event.

A bag contains 2 green marbles and 11 red marbles. The theoretical probability of drawing a green marble is $P(\text{green}) = \frac{2}{13}$, or about 15%.

A rehearsal is to begin at some time between 3:00 P.M. and 3:15 P.M. Assuming that all times are equally likely, find the probability that the rehearsal begins in the interval from 3:00 P.M. to 3:03 P.M.

Divide the 15-minute interval into 5 equal parts. The 3-minute interval from 3:00 P.M. to 3:03 P.M. is $\frac{1}{5}$ of the total interval. Thus, the probability is $\frac{1}{5}$.

Apply the Fundamental Counting Principle.

At a cafe, there are 5 choices of entrees and 4 choices of side dishes. By the Fundamental Counting Principle, there are 5×4, or 20, ways to choose an entree and a side dish.

Exercises

Find the probability of each event.

1. drawing a red marble from a bag that contains 3 red marbles and 5 purple marbles **$\frac{3}{8}$**
2. drawing a red marble from a bag that contains 4 red marbles and 10 black marbles **$\frac{2}{7}$**

A party is to begin at some time between 8:00 P.M. and 8:30 P.M. Assuming that all times are equally likely, find the probability that the first guest arrives during each given time interval.

3. from 8:00 P.M. to 8:05 P.M. **$\frac{1}{6}$**
4. from 8:12 P.M. to 8:18 P.M. **$\frac{1}{5}$**
5. from 8:21 P.M. to 8:24 P.M. **$\frac{1}{10}$**
6. If repetition is *not* allowed, how many 4-letter codes can be formed from only 5 letters of the alphabet? **120**
7. If repetition is allowed, how many 4-letter codes can be formed from 5 letters of the alphabet? **625**

LESSON 10.2

Key Skills

Find the number of linear permutations.

Five books on a shelf can be arranged $_5P_5 = 5! = 5 \times 4 \times 3 \times 2 \times 1 = 120$ different ways.

Two of the five books can be chosen and arranged in $_5P_2 = \frac{5!}{(5-2)!} = 20$ different ways.

Exercises

8. In how many ways can the letters in the word *pencil* be arranged? **720**

Find the number of ways that a coach can assign each number of basketball players to 5 distinct positions.

9. 8 **6720** 10. 10 **30,240** 11. 12 **95,040**

Chapter Test, Form A

NAME ____________ CLASS ________ DATE ________

Chapter Assessment
Chapter 10, Form A, page 1

Write the letter that best answers the question or completes the statement.

__c__ 1. A bag contains 8 red marbles, 12 blue marbles, and 17 green marbles. If one marble is randomly selected from the bag, what is the probability that the marble is red or green?
a. $\frac{8}{17}$ b. $\frac{25}{12}$ c. $\frac{25}{37}$ d. $\frac{20}{37}$

__c__ 2. If a point is randomly selected from the points inside the square shown below, find the probability that the point is inside the circle.
a. 0.314 b. 0.628 c. 0.785 d. 3.142

10 cm
10 cm

__a__ 3. If $P(B) = \frac{3}{8}$, find $P(B^c)$.
a. $\frac{5}{8}$ b. $\frac{11}{8}$ c. $-\frac{3}{8}$ d. $\frac{8}{3}$

__c__ 4. If A and B are independent events and $P(A) = 0.5$ and $P(B) = 0.4$, find $P(B|A)$.
a. 0.6 b. 0.5 c. 0.4 d. 0.2

__b__ 5. If A and B are independent events and $P(A) = 0.4$ and $P(B) = 0.25$, find $P(A \text{ or } B)$.
a. 0.15 b. 0.55 c. 0.65 d. 0.75

__b__ 6. The menu at a restaurant has 11 appetizers, 15 entrees, 10 desserts, and 8 beverages. The restaurant offers a dinner special that includes 1 appetizer, 1 entree, 1 dessert, and 1 beverage. In how many ways can a person order a dinner special at this restaurant?
a. 45,644 b. 13,200 c. 870 d. 44

__d__ 7. A CD has 10 songs on it. If a CD player randomly selects the order in which the songs are played, in how many different orders can the 10 songs be played?
a. 10 b. 34,670 c. 100,000 d. 3,628,800

__b__ 8. How many different permutations of the letters in *September* are possible?
a. 362,880 b. 60,480 c. 720 d. 6

__d__ 9. A baseball league that has 8 teams decides to add 3 new teams. The league then decides to divide into 2 divisions, one with 6 teams and the other with 5 teams. In how many ways can the 6 teams be chosen for the larger division?
a. 39,916,800 b. 332,640 c. 55,440 d. 462

NAME ____________ CLASS ________ DATE ________

Chapter Assessment
Chapter 10, Form A, page 2

__c__ 10. A pizzeria has 6 meat toppings and 9 vegetable toppings. In how many different ways can a pizza with 2 meat toppings and 3 vegetable toppings be ordered from this pizzeria?
a. 324 b. 924 c. 1260 d. 15,120

__d__ 11. If there are 16 students in a classroom that has 25 chairs, how many different seating arrangements are possible?
a. 25! b. $\frac{25!}{16!}$ c. $_{25}C_{16}$ d. $_{25}P_{16}$

__c__ 12. In how many different ways can 8 magazines be arranged on a circular carousel?
a. 362,880 b. 40,320 c. 5040 d. 720

__d__ 13. There are 15 dogs and 10 cats at an animal shelter. If a family decides to randomly select 5 animals to adopt, find the probability that the family selects 3 dogs and 2 cats.
a. 0.04 b. 0.167 c. 0.2 d. 0.385

__a__ 14. If two number cubes are rolled and the numbers that result are added, find the probability that the sum is an even number or a number less than 5.
a. $\frac{5}{9}$ b. $\frac{3}{8}$ c. $\frac{2}{3}$ d. $\frac{1}{2}$

__d__ 15. All of the sections on the circular spinner shown at right are equal in size. If the spinner is spun 3 times, what is the probability that an even number is spun all 3 times?
a. $\frac{6}{5}$ b. $\frac{3}{10}$ c. $\frac{6}{25}$ d. $\frac{8}{125}$

5 1 4 2 3

__a__ 16. There are 38 members on the basketball teams and 54 members on the track teams at Central High. There are 13 people who are on both a basketball team and a track team. If one member of these teams is randomly selected, find the probability that the person is on a track team given that the person is on a basketball team.
a. $\frac{13}{38}$ b. $\frac{13}{92}$ c. $\frac{54}{38}$ d. $\frac{38}{54}$

__c__ 17. If a number cube is rolled 4 times, what is the probability that the same number is rolled at least twice?
a. 0.09 b. 0.28 c. 0.72 d. 0.91

The letters in the word *hollow* can be arranged in $\frac{6!}{2!2!} = \frac{6!}{4} = 180$ different ways.

Find the number of circular permutations.

Five objects can be arranged around a circle in $(5 - 1)! = 4!$, or 24, different ways.

12. How many arrangements of the letters in the word *tomorrow* are possible? **3360**

13. In how many ways can 5 children be positioned around a merry-go-round? **24**

14. In how many ways can 8 employees be seated at a circular conference table? **5040**

LESSON 10.3

Key Skills

Find the number of combinations.

Find the number of ways to purchase 2 games from a display of 7 games.

$${}_7C_2 = \frac{7!}{2!(7-2)!} = \frac{7 \times 6 \times 5!}{2!5!} = 21$$

Exercises

15. Find the number of ways to choose 2 books from a set of 10 books. **45**

16. In how many ways can 2 of 27 ice cream flavors be chosen? **351**

17. In how many ways can 3 student representatives be chosen from 100 students? **161,700**

LESSON 10.4

Key Skills

Find the probability of event *A or B*.

For one roll of a number cube:

Mutually exclusive events: Find $P(2 \text{ or } 3)$.

$$P(A \text{ or } B) = P(A) + P(B)$$

$$P(2 \text{ or } 3) = \frac{1}{6} + \frac{1}{6} = \frac{2}{6} = \frac{1}{3}$$

Inclusive events: Find $P(\text{even or multiple of } 3)$.

$$P(A \text{ or } B) = P(A) + P(B) - P(A \text{ and } B)$$

$$P(\text{even or multiple of } 3) = \frac{1}{2} + \frac{1}{3} - \frac{1}{6} = \frac{4}{6} = \frac{2}{3}$$

Use the complement of an event to find a probability.

$$P(A) = 1 - P(A^c)$$

$$P(\text{less than } 6) = 1 - P(6)$$

$$= 1 - \frac{1}{6} = \frac{5}{6}$$

Exercises

Find the probability of each event for one roll of a number cube.

18. 4 *or* 7 $\frac{1}{3}$

19. 1 *or* 6 $\frac{1}{3}$

20. an odd number *or* a number greater than 4 $\frac{2}{3}$

21. an even number *or* a number less than 4 $\frac{5}{6}$

22. greater than 1 $\frac{5}{6}$

23. greater than 2 $\frac{2}{3}$

LESSON 10.5

Key Skills

Find the probability of independent events.

For 2 rolls of a number cube:

$$P(A \text{ and } B) = P(A) \times P(B)$$

$$P(6 \text{ first and even second}) = \frac{1}{6} \times \frac{3}{6} = \frac{3}{36} = \frac{1}{12}$$

Exercises

Find the probability of each event.

24. 3 heads on 3 tosses of a fair coin $\frac{1}{8}$

25. 2 even numbers on 2 rolls of a number cube $\frac{1}{4}$

Chapter Test, Form B

NAME ______ CLASS ______ DATE ______

Chapter Assessment
Chapter 10, Form B, page 1

For Exercises 1–3, evaluate each expression.

1. ${}_{14}P_0$ 1   2. ${}_{25}C_{10}$ 3,268,760   3. $\frac{15!}{5! \times 3!}$ 1,816,214,400

4. A car dealer has 8 exterior colors, 12 interior colors, and 7 options packages available for new cars. In how many different ways can a person buying a new car from this dealer select an exterior color, an interior color, and an options package? 672

5. A restaurant has a dinner special in which a party of four can order any 12 of the 40 items on the restaurant's menu for a fixed price. In how many different ways can a party of four order the dinner special in this restaurant? 5,586,853,480

6. How many permutations of the letters in *interstate* are possible? 302,400

7. A newspaper editor needs to cover stories in Paris, New York, and Lubbock. If the editor has 8 reporters available, in how many different ways can the editor assign reporters to the stories? 336

8. A bakery has 6 different kinds of cake and 12 different kinds of pie. If John and Mary come to this bakery, in how many different ways can they select 3 different cakes and 4 different pies for a party? 9900

9. If one point is randomly selected from the points inside the rectangle shown at right, find the probability, to the nearest percent, that the point is in a shaded area. ≈0.2146, or 21%

12 cm
6 cm

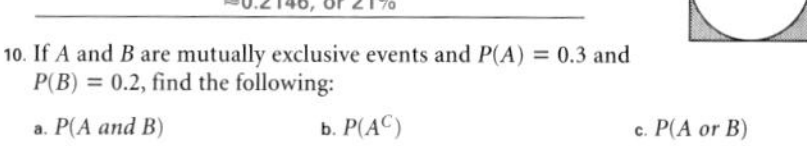

10. If A and B are mutually exclusive events and $P(A) = 0.3$ and $P(B) = 0.2$, find the following:

a. $P(A \text{ and } B)$ 0   b. $P(A^C)$ 0.7   c. $P(A \text{ or } B)$ 0.5

11. If A and B are independent events and $P(A) = 0.6$ and $P(B) = 0.5$, find the following:

a. $P(A \text{ and } B)$ 0.3   b. $P(B|A)$ 0.5   c. $P(A \text{ or } B)$ 0.8

12. A math teacher decides to grade 5 randomly selected problems from among the 20 assigned. If you completed only 8 of the problems, find the probability that the teacher will select 3 of the problems you completed and 2 of the problems you did not complete. ≈0.2384, or about 24%

NAME ______ CLASS ______ DATE ______

Chapter Assessment
Chapter 10, Form B, page 2

13. a. An anthropologist has trained a monkey to place sticks that are the same length end to end. The anthropologist then gives the monkey 5 sticks of different lengths. In how many ways can the monkey place the sticks of different lengths end to end? 120

b. If the monkey cannot distinguish between the different lengths of the sticks, what is the probability that the monkey will place the sticks in order from shortest to longest? $\frac{1}{120}$

14. For medical purposes, the managers of a company decide to record the blood type of all the employees. The results are shown in the table below.

	O	A	B	AB
Women	8	5	4	2
Men	12	6	2	1

Suppose that a single person is randomly selected from the people in this company.

a. Find the probability that the person has type O blood or type A blood. $\frac{31}{40}$, or 77.5%

b. Find the probability that the person is a woman or has type A blood. $\frac{5}{8}$, or 62.5%

c. Find the probability that the person is a man, given that the person has type O blood. $\frac{3}{5}$, or 60%

15. A wheel used on a game show has 7 spaces with dollar amounts written on them and 3 spaces with "Lose a Turn" written on them. All of the spaces on the wheel are equal in size. If a contestant on this show spins the wheel 3 times, find the probability that the contestant spins "Lose a Turn" all 3 times. $\frac{27}{1000}$, or 2.7%

16. The area of the rectangle shown at right is 1. The areas of the other regions are given in each region. Suppose that a point is randomly selected from the points in the rectangle.

A, B: 0.2, 0.1, 0.35

a. Find $P(B|A)$. $\frac{1}{3}$, or about 33.3%

b. Are picking a point in A and picking a point in B independent events? Explain. Answers may vary. Because $P(B) \neq P(B|A)$, A and B are not independent events.

LESSON 10.6

Key Skills

Find conditional probabilities.

For 1 roll of a number cube, find the probability that the number 2 is rolled if you know that the number is even.

$$P(2|\text{even}) = \frac{P(\text{even } and \text{ } 2)}{P(\text{even})} = \frac{\frac{1}{6}}{\frac{1}{2}} = \frac{1}{3}$$

Exercises

Find the probability of each event for one roll of a number cube.

26. 5, given that it is an odd number $\frac{1}{3}$

27. 1, given that it is *not* an odd number **0**

28. 2, given that it is less than or equal to 5 $\frac{1}{5}$

LESSON 10.7

Key Skills

Use simulations to estimate probabilities.

The probability table gives the probabilities of the number of traffic tickets written by one officer between 3:00 A.M. and 4:00 A.M.

X	$P(X)$
0	0.20
1	0.30
2	0.35
3	0.15

Use a simulation with 10 trials to estimate the probability that the officer writes less than 2 tickets during this time.

Add a column, N, as shown below. Generate 10 random integers.

X	$P(X)$	N	Random integers
0	0.20	1–20	14
1	0.30	21–50	38, 34, 23, 28
2	0.35	51–85	72, 57, 73
3	0.15	86–100	95, 97

In this experiment, less than 2 traffic tickets occurred 5 out of 10 times, so the probability is estimated to be $\frac{1}{2}$, or 50%.

Exercises

29. The table at right gives the probabilities, $P(X)$, of the number, X, of drinks ordered at a fast-food restaurant during a 1-minute interval between 3:00 P.M. and 3:20 P.M.

X	$P(X)$
0	0.18
1	0.37
2	0.22
3	0.10
4	0.07
5	0.06

Use a simulation with 10 trials to estimate the probability that more than 6 drinks are ordered between 3:00 P.M. and 3:20 P.M. inclusive. **Answers may vary; sample answer: ≈0.5.**

30. The table below gives the number of customers out of 100 that went up, down, or stayed on the same level after entering a mall.

Up	36
Down	20
Same	44

Use a simulation with 10 trials to estimate the probability that 3 or more out of 5 consecutive customers go up.

Answers may vary; sample answer: ≈0.25.

Applications

31. HEALTH A test of 100 adults showed that 40 of them consumed on average more than the recommended maximum of 2400 milligrams of sodium per day. Of those with the high sodium consumption, 50% had higher-than-normal blood pressure. Of the 60 adults whose sodium intake was at or below 2400 milligrams per day, only 15% had higher-than-normal blood pressure. Using this data as a sample of the general population, find the probability that a person with higher-than-normal blood pressure has a daily intake of more than 2400 milligrams of sodium. $\frac{20}{29}$, **or ≈69%**

Chapter Test

Find the probability of each event.

1. drawing a jack from a playing card deck of 52 cards **$\frac{1}{13}$**
2. drawing a green marble from a bag that contains 8 green marbles and 6 red marbles **$\frac{4}{7}$**
3. rolling an odd number on one roll of a number cube **$\frac{1}{2}$**
4. **LICENSING** A fishing license identification number consists of 2 letters and 8 numbers. How many possible license numbers are available? **67,600,000,000**

Evaluate the following expressions.

5. $12! - 7!$ **478,996,560**
6. $(12 - 7)!$ **120**
7. ${}_8P_3$ **336**
8. ${}_5P_3 \times {}_8P_5$ **403,200**
9. How many different ways can a president, vice president, and secretary be chosen from a group of 24 individuals? **12,144**
10. Find the number of permutations of the letters in the word *probability*. **9,979,200**
11. **SPORTS** The Indianapolis 500 race has a field of 32 cars. In how many different ways can the first three finishers be placed? **29,760**

Evaluate each expression.

12. ${}_8C_3$ **56**
13. ${}_8C_8$ **1**
14. $\frac{{}_8C_5}{{}_5C_2 \times {}_5C_3}$ **0.56**
15. $\frac{8!}{4!6!} \times \frac{5!}{3!4!}$ **1.944**
16. **SMALL BUSINESS** At a deli, a sub sandwich can be ordered with any of 7 different condiments. In how many different ways can a sandwich with exactly 3 condiments be ordered? **35**
17. **POLITICS** How many ways can a Senate committee of 12 members choose a subcommittee of 5 senators? **792**

Find the probability of each event for one roll of a 12-sided number cube labeled with the numbers 1–12.

18. a 7 *or* an even number **$\frac{7}{12}$**
19. a prime number *or* a multiple of 4 **$\frac{2}{3}$**
20. an odd number *or* a multiple of 3 **$\frac{2}{3}$**
21. a number greater than 8 *or* a multiple of 5 **$\frac{5}{12}$**
22. an even number *or* a number less than 6 **$\frac{3}{4}$**

Find the probability of each event.

23. heads on a coin and 5 on a 6-sided number cube, on one toss of a coin and one roll of a number cube **$\frac{1}{12}$**
24. a 7 and a face card with the draw of two cards at random from two different decks **$\frac{3}{169}$**
25. A time is chosen at random during the week to send off the next space shuttle. Find the probability of the liftoff time being on Tuesday between 5 A.M. and 6 A.M. **$\frac{1}{168}$**

Find the probability of each event for one draw of a card from a playing card deck of 52 cards.

26. a queen, given that it is a face card **$\frac{1}{3}$**
27. a diamond, given that it is a red card **$\frac{1}{2}$**
28. a jack, given that it is a red card **$\frac{1}{13}$**
29. a ten of clubs, given that it is a black card **$\frac{1}{26}$**
30. Given $P(A) = \frac{1}{3}$ and $P(B|A) = \frac{2}{3}$, what is $P(A \text{ and } B)$? **$\frac{2}{9}$**
31. Given $P(A \text{ and } B) = \frac{1}{6}$ and $P(A) = \frac{2}{5}$, what is $P(B|A)$? **$\frac{5}{12}$**
32. An ordinary thumbtack was dropped 100 times with the results shown in the table below. Use a simulation with 20 trials to estimate the probability the thumbtack lands with the point up.

point up	point down
68	32

≈ 0.50

College Entrance Exam Practice

Multiple-Choice and Quantitative-Comparison Samples

The first half of the Cumulative Assessment contains two types of items found on standardized tests—multiple-choice questions and quantitative-comparison questions. Quantitative-comparison items emphasize the concepts of equality, inequality, and estimation.

Free-Response Grid Samples

The second half of the Cumulative Assessment is a free-response section. This part of the Cumulative Assessment requires student-produced response items like those commonly found on college entrance exams. These questions require the use of machine-scored answer grids. You may wish to have students practice answering these items in preparation for standardized tests.

College Entrance Exam Practice

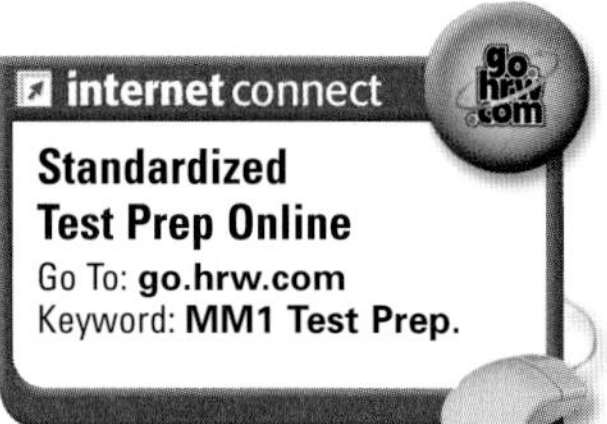

QUANTITATIVE COMPARISON For Items 1–5, write
A if the quantity in Column A is greater than the quantity in Column B;
B if the quantity in Column B is greater than the quantity in Column A;
C if the two quantities are equal; or
D if the relationship cannot be determined from the given information.

	Column A	Column B	Answers
1.	The value of x $-\frac{1}{3}x = 2x + 1$	$5 + \frac{1}{2}x = 6(x + 1)$	Ⓐ Ⓑ Ⓒ Ⓓ [Lesson 1.6]
2.	$\log 12$	10^{12}	Ⓐ Ⓑ Ⓒ Ⓓ [Lesson 6.3]
3.	$\begin{cases} y = -3x + 4 \\ y + 3x = -3 \end{cases}$ x	y	Ⓐ Ⓑ Ⓒ Ⓓ [Lesson 3.1]
4.	The number of real solutions $3x^2 - 2x + 1 = 0$	$2x^2 - 5x - 2 = 0$	Ⓐ Ⓑ Ⓒ Ⓓ [Lesson 5.6]
5.	$f(x) = 2x - 1$ $f(2)$	$f^{-1}(2)$	Ⓐ Ⓑ Ⓒ Ⓓ [Lesson 2.5]

6. Which of the following is a solution of the system $\begin{cases} 2y + x \le 6 \\ y - 3x \ge 4 \end{cases}$? ***(LESSON 3.4)***
b
a. $(0, 5)$ **b.** $(-1, 2)$
c. $(1, -1)$ **d.** $(0, 0)$

7. Solve $2(x + 2) - 7 < 8x + 15$. ***(LESSON 1.7)***
a
a. $x > -3$ **b.** $x < -3$
c. $x > 2$ **d.** $x < 2$

8. Which value would you add to $x^2 - 10x$ to complete the square? ***(LESSON 5.4)***
c
a. 5 **b.** -5 **c.** 25 **d.** -25

9. Simplify $\left(-\frac{1}{125}\right)^{-\frac{2}{3}}$. ***(LESSON 2.2)***
c
a. $\frac{1}{25}$ **b.** $-\frac{1}{25}$ **c.** 25 **d.** -25

10. If $f(x) = 2x - 6$ and $g(x) = 12 - 6x$, which statement is true? ***(LESSON 2.4)***
d
a. $f \circ g = g \circ f$ **b.** $2(f \circ g) = g \circ f$
c. $-2(f \circ g) = g \circ f$ **d.** none of these

11. Which of the following describes the relationship between the lines $y = \frac{1}{2}x$ and $y = -2x - 3$? ***(LESSON 1.3)***
c
a. horizontal **b.** vertical
c. perpendicular **d.** parallel

12. Solve $4y^2 + 7 = 0$. ***(LESSON 5.6)***
d
a. $y = \frac{7i}{2}$ **b.** $y = \frac{-\sqrt{7}}{4}$
c. $y = \frac{\sqrt{7}}{4}$ **d.** $y = \pm\frac{i\sqrt{7}}{2}$

13. Find the coordinates of the midpoint of the segment with endpoints at $(-4, -1)$ and $(2, -7)$. ***(LESSON 9.1)*** c

a. $(-1, -3)$ **b.** $(-3, 3)$

c. $(-1, -4)$ **d.** $(-3, -3)$

14. Which expression is not equivalent to the others? ***(LESSON 6.4)*** a

a. $2 \log_b \frac{5}{3}$ **b.** $\log_b \sqrt[3]{5^2}$

c. $\frac{1}{3} \log_b 25$ **d.** $\frac{2}{3} \log_b 5$

15. Write $(5x^3 - 2x^2 + x - 10) + (2x^3 - 3x - 1)$ in standard form. ***(LESSON 7.1)*** d

a. $3x^3 - 2x^2 - 4x - 9$
b. $3x^3 + 2x^2 + 4x - 9$
c. $7x^3 - 2x^2 - 2x - 9$
d. $7x^3 - 2x^2 - 2x - 11$

16. Simplify $\frac{a^2 + 3a - 4}{a^2} \bullet \frac{a^2 - 2a}{2a + 8}$. ***(LESSON 8.3)*** b

a. $-(a - 4)$ **b.** $\frac{(a-1)(a-2)}{2a}$

c. $\frac{(a+4)(a-2)}{2a}$ **d.** $\frac{(a+4)(a-1)}{a}$

17. Find the coordinates of the center of the circle defined by $x^2 + y^2 - 2x - 8y = 8$. ***(LESSON 9.6)*** c

a. $(-1, -4)$ **b.** $(1, 2)$

c. $(1, 4)$ **d.** $(-1, -8)$

18. Write an equation in slope-intercept form for the line containing the points $(3, -4)$ and $(2, 7)$. ***(LESSON 1.3)*** $y = -11x + 29$

19. Find the zeros of $f(x) = x^2 - 8x + 12$. 2, 6 ***(LESSON 5.3)***

20. Find the product $-2\begin{bmatrix} 7 & -6 & 0 \\ -3 & 5 & 1 \end{bmatrix}$, if it exists. ***(LESSON 4.1)*** $\begin{bmatrix} -14 & 12 & 0 \\ 6 & -10 & -2 \end{bmatrix}$

21. Factor $5x^2 + 10x - 40$, if possible. ***(LESSON 5.3)*** $5(x - 2)(x + 4)$

22. Find the inverse of $\{(2, -1), (4, 3), (6, 0), (3, 1)\}$. ***(LESSON 2.5)***

23. Simplify $(1 - 2i) - (3 - 4i)$. ***(LESSON 5.6)*** $-2 + 2i$

24. Factor $8x^3 + 64$. ***(LESSON 7.3)*** $8(x + 2)(x^2 - 2x + 4)$

25. Describe the end behavior of $P(x) = -2x^3 + x^2 - 11x + 4$. ***(LESSON 7.2)*** rises on the left, falls on the right

26. Simplify $\left(\sqrt{-36x^4}\right)^2$. ***(LESSON 8.7)*** $-36x^4$

27. Write equations for all vertical and horizontal asymptotes in the graph of $f(x) = \frac{(x+2)^2}{3x}$. ***(LESSON 8.2)*** vert. asymptote at $x = 0$

28. Write the standard equation for $x^2 + y^2 + 6x - 4y = 12$. ***(LESSON 9.6)***

29. Factor $6x^2 + 8x - 15x - 20$, if possible. ***(LESSON 7.3)*** $(2x - 5)(3x + 4)$

30. Simplify $\frac{\frac{x+4}{9x^3}}{\frac{x-6}{3x^4}}$. ***(LESSON 8.3)*** $\frac{x^2 + 4x}{3x - 18}$

FREE-RESPONSE GRID

The following questions may be answered by using a free-response grid such as that commonly used by standardized-test services.

31. Find the slope, m, of the line $y = 8$. ***(LESSON 1.2)*** 0

32. Find the determinant of $\begin{bmatrix} -3 & 5 \\ -1 & 6 \end{bmatrix}$. ***(LESSON 4.3)*** -13

33. Solve $\frac{6x + 2}{3x} = 6$. ***(LESSON 8.5)*** $\frac{1}{6}$

34. Solve $3^x = 9^5$. ***(LESSON 6.7)*** 10

35. What value would you add to $x^2 + 8x$ to complete the square? ***(LESSON 5.4)*** 16

36. Find the value of v in $v = \log_{10} \frac{1}{1000}$. -3 ***(LESSON 6.3)***

EXTRACURRICULAR ACTIVITIES The Outdoors Club has 12 students—5 girls and 7 boys. ***(LESSON 10.5)***

37. How many different committees of 6 students can be formed if at least 3 are girls? 462

38. How many different committees of 6 students can be formed if at least 3 are boys? 812

39. How many different committees of 6 students can be formed if no more than 3 are boys? 462

22. $\{(-1, 2), (3, 4), (0, 6), (1, 3)\}$

28. $(x + 3)^2 + (y - 2)^2 = 25$

Keystroke Guide for Chapter 10

Essential keystroke sequences (using the model TI-82 or TI-83 graphics calculator) are presented below for all Activities and Examples found in this chapter that require or recommend the use of a graphics calculator.

For Keystrokes of other graphing calculator models, visit the HRW web site at **go.hrw.com** and enter the keyword **MB1 CALC**.

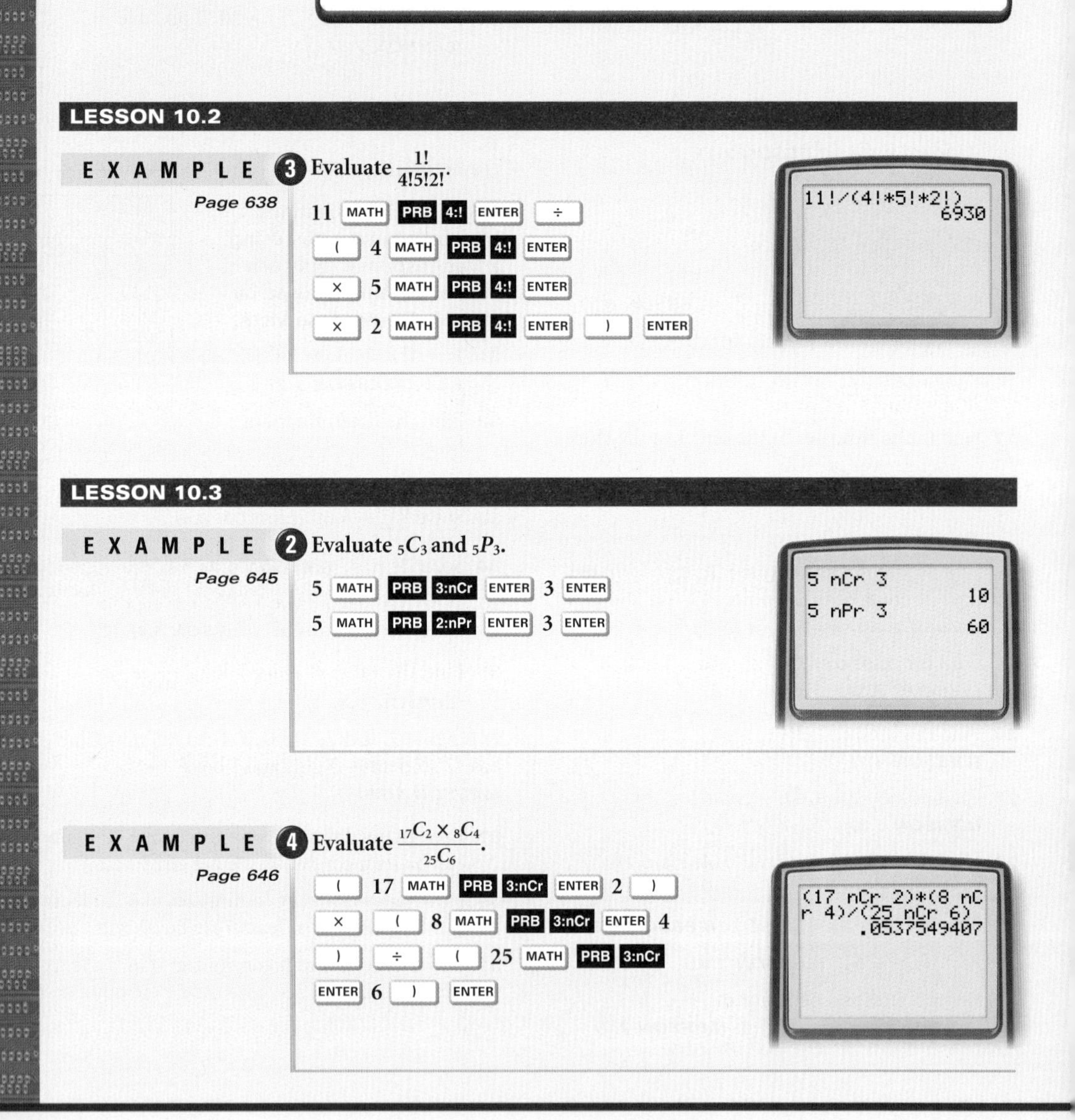

LESSON 10.2

EXAMPLE 3 Evaluate $\frac{1!}{4!5!2!}$.

Page 638

11 MATH PRB 4:! ENTER ÷

(4 MATH PRB 4:! ENTER

× 5 MATH PRB 4:! ENTER

× 2 MATH PRB 4:! ENTER) ENTER

LESSON 10.3

EXAMPLE 2 Evaluate ${}_5C_3$ and ${}_5P_3$.

Page 645

5 MATH PRB 3:nCr ENTER 3 ENTER

5 MATH PRB 2:nPr ENTER 3 ENTER

EXAMPLE 4 Evaluate $\frac{{}_{17}C_2 \times {}_8C_4}{{}_{25}C_6}$.

Page 646

(17 MATH PRB 3:nCr ENTER 2)

× (8 MATH PRB 3:nCr ENTER 4

) ÷ (25 MATH PRB 3:nCr

ENTER 6) ENTER

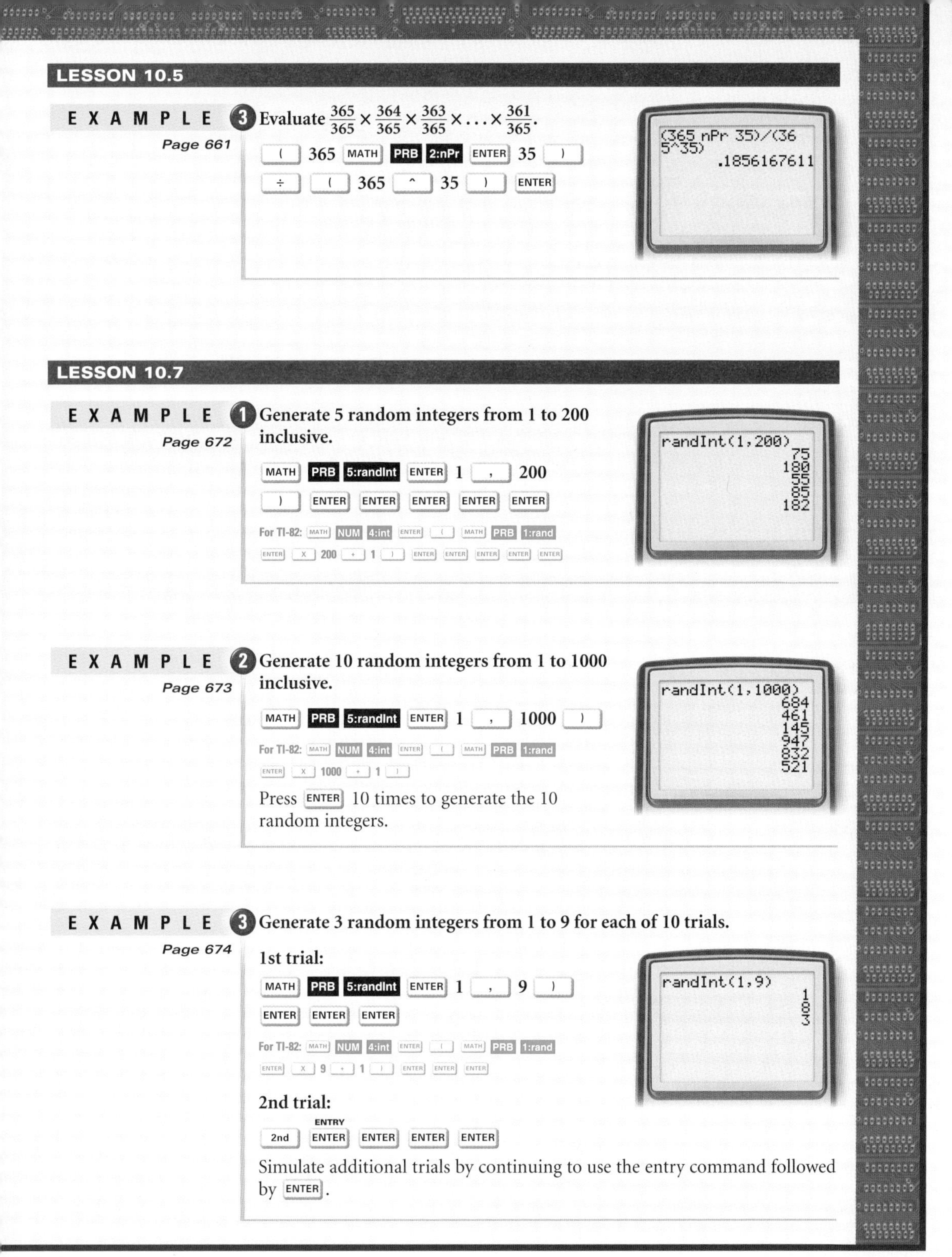

LESSON 10.5

EXAMPLE 3 Page 661

Evaluate $\frac{365}{365} \times \frac{364}{365} \times \frac{363}{365} \times \ldots \times \frac{361}{365}$.

(365 MATH PRB 2:nPr ENTER 35)
÷ (365 ^ 35) ENTER

(365 nPr 35)/(365^35)
.1856167611

LESSON 10.7

EXAMPLE 1 Page 672

Generate 5 random integers from 1 to 200 inclusive.

MATH PRB 5:randInt ENTER 1 , 200
) ENTER ENTER ENTER ENTER ENTER

For TI-82: MATH NUM 4:int ENTER (MATH PRB 1:rand ENTER X 200 + 1) ENTER ENTER ENTER ENTER ENTER

randInt(1,200)
75
180
55
85
182

EXAMPLE 2 Page 673

Generate 10 random integers from 1 to 1000 inclusive.

MATH PRB 5:randInt ENTER 1 , 1000)

For TI-82: MATH NUM 4:int ENTER (MATH PRB 1:rand ENTER X 1000 + 1)

Press ENTER 10 times to generate the 10 random integers.

randInt(1,1000)
684
461
145
947
832
521

EXAMPLE 3 Page 674

Generate 3 random integers from 1 to 9 for each of 10 trials.

1st trial:

MATH PRB 5:randInt ENTER 1 , 9)
ENTER ENTER ENTER

For TI-82: MATH NUM 4:int ENTER (MATH PRB 1:rand ENTER X 9 + 1) ENTER ENTER ENTER

randInt(1,9)
1
8
3

2nd trial:

2nd ENTER (ENTRY) ENTER ENTER ENTER

Simulate additional trials by continuing to use the entry command followed by ENTER.

Discrete Mathematics: Series and Patterns

Lesson Presentation CD-ROM
PowerPoint® presentations for each lesson 11.1–11.8

CHAPTER PLANNING GUIDE

Lesson	11.1	11.2	11.3	11.4	11.5	11.6	11.7	11.8	Project and Review
Pupil's Edition Pages	690–698	699–706	707–712	713–719	720–727	728–734	735–740	741–747	748–749, 750–757
Practice and Assessment									
Extra Practice (Pupil's Edition)	975	976	976	977	977	978	978	979	
Practice Workbook	68	69	70	71	72	73	74	75	
Practice Masters Levels A, B, and C	202–204	205–207	208–210	211–213	214–216	217–219	220–222	223–225	
Standardized Test Practice Masters	78	79	80	81	82	83	84	85	86
Assessment Resources	138	139	140	141	143	144	145	146	142, 147–152
Visual Resources									
Lesson Presentation Transparencies Vol. 2	85–88	89–92	93–96	97–100	101–104	105–108	109–111	112–117	
Teaching Transparencies	42	43		44		45, 46	47		
Answer Key Transparencies	382–384	385–388	389–390	391–394	395–403	404–407	408–410	411–415	416–421
Quiz Transparencies	11.1	11.2	11.3	11.4	11.5	11.6	11.7	11.8	
Teacher's Tools									
Reteaching Masters	135–136	137–138	139–140	141–142	143–144	145–146	147–148	149–150	
Make-Up Lesson Planner for Absent Students	68	69	70	71	72	73	74	75	
Student Study Guide	68	69	70	71	72	73	74	75	
Spanish Resources	68	69	70	71	72	73	74	75	
Block Scheduling Handbook									22–23
Activities and Extensions									
Lesson Activities	68	69	70	71	72	73	74	75	
Enrichment Masters	68	69	70	71	72	73	74	75	
Cooperative-Learning Activities	68	69	70	71	72	73	74	75	
Problem Solving/ Critical Thinking	68	69	70	71	72	73	74	75	
Student Technology Guide	68	69	70	71	72	73	74	75	
Long Term Projects									41–44
Writing Activities for Your Portfolio									31–33
Tech Prep Masters									51–54
Building Success in Mathematics									28–29

LESSON PACING GUIDE

Lesson	11.1	11.2	11.3	11.4	11.5	11.6	11.7	11.8	Project and Review
Traditional	2 days	1 day	1 day	2 days	2 days	1 day	1 day	2 days	2 days
Block	1 day	$\frac{1}{2}$ day	$\frac{1}{2}$ day	1 day	1 day	$\frac{1}{2}$ day	$\frac{1}{2}$ day	1 day	1 day
Two-Year	4 days	2 days	2 days	4 days	4 days	2 days	2 days	4 days	4 days

CONNECTIONS AND APPLICATIONS

Lesson	11.1	11.2	11.3	11.4	11.5	11.6	11.7	11.8	Review
Algebra	690–698	699–706	707–712	713–719	720–727	728–734	735–740	741–747	750–757
Geometry	697	705			725, 726	733		744, 745, 746	
Maximum/Minimum								747	
Patterns in Data			707, 708						
Probability							735, 738, 739		
Business and Economics		699, 701, 703, 705	710, 711, 712	713, 714, 717, 718	721, 724, 726	733, 734			754
Science	690, 692, 697	705			726				
Social Studies					726				
Sports and Leisure	698		712	719	727	728		741, 743, 746, 747	754
Cultural Connection: Europe	690								
Cultural Connection: Asia							740		

BLOCK SCHEDULING GUIDE

Day	Lesson	Teacher Directed: Lesson Examples, Teaching Transparencies	Student Guided Activity, Try This	Cooperative-Learning Activity, Lesson Activity, Student Technology Guide	Practice: Practice & Apply, Extra Practice, Practice Workbook	Assessment: Quiz, Mid-Chapter Assessment	Problem Solving, Reteaching
1	11.1	10 min	15 min	15 min	65 min	15 min	15 min
2	11.2	10 min	10 min	7 min	25 min	7 min	7 min
	11.3	10 min	10 min	8 min	25 min	8 min	8 min
3	11.4	10 min	15 min	15 min	65 min	15 min	15 min
4	11.5	10 min	15 min	15 min	65 min	15 min	15 min
5	11.6	10 min	10 min	7 min	25 min	7 min	7 min
6	11.7	10 min	10 min	8 min	25 min	8 min	8 min
7	11.8	10 min	15 min	15 min	65 min	15 min	15 min
8	Assess.	50 min **PE:** Chapter Review	90 min **PE:** Chapter Project, Writing Activities	90 min Tech Prep Masters	65 min **PE:** Chapter Assessment, Test Generator	30 min Chap. Assess. (A or B), Alt. Assess. (A or B), Test Generator	

PE: Pupil's Edition

Alternative Assessment

The following suggest alternative assessments for students who may benefit from a different type of assessment than the regular chapter quizzes and the mid-chapter/end-of-chapter test. Visit the HRW web site to get additional Alternative Assessment material.

internet connect

Alternative Assessment
Go To: **go.hrw.com**
Keyword: **MB1 Alt Assess**

Performance Assessment

1. The first term, t_1, of a sequence is 3.5, and $x = 2$.

a. What is the 10th term of the sequence if x is the constant difference? if x is the constant ratio?

b. What is the sum of the first 10 terms of the sequence if x is the constant difference? if x is the constant ratio?

c. Explain why there is no sum if x is the constant ratio and the series is infinite.

2. Expand $(2x - 3y)^4$. What is the sign of the term containing xy^3?

Portfolio Project

Suggest that students choose one of the following projects for inclusion in their portfolios.

1. Consider the sequence 1, 2, 3, 4, 5, . . .

a. Form a new sequence by selecting every kth term, starting with the first term. Classify the new sequence.

b. Consider the array of natural numbers below.

Row 1:			1		
Row 2:		2	3		
Row 3:		4	5	6	
Row 4:	7	8	9	10	
Row 5:	11	12	13	14	15

. . .

Explore patterns formed by rows and diagonals.

2. Consider the sequence $\frac{1}{n^0}, \frac{1}{n^1}, \frac{1}{n^2}, \frac{1}{n^3}, \frac{1}{n^4}, \frac{1}{n^5}, \ldots$, where n is a fixed natural number.

a. Create and classify the related series.

b. Suppose that m is another fixed natural number and that $m < n$. How do the two sequences compare term by term? How do the two series compare?

c. What can you say about the sequence and series formed by multiplying consecutive terms in the given sequence?

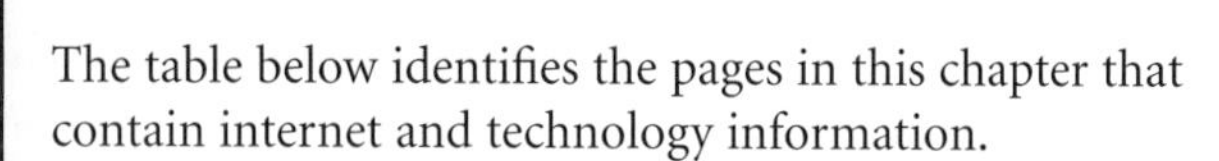

internet connect

The table below identifies the pages in this chapter that contain internet and technology information.

Content Links

Activities Online	**pages 695, 717, 745**
Portfolio Extensions	**pages 706, 734**
Homework Help Online	**pages 696, 703, 710, 717, 725, 732, 739, 746**
Graphic Calculator Support	**page 758**

Resource Links

Parents can go online and find concepts that students are learning—lesson by lesson—and questions that pertain to each lesson, which facilitate parent-student discussion.

Go To: **go.hrw.com**
Keyword: **MB1 Parent Guide**

Technical Support

The following may be used to obtain technical support for any HRW software product.

Online Help: **www.hrwtechsupport.com**
e-mail: **tschrw@hrwtechsupport.com**
HRW Technical Support Center: **(800)323-9239**
7 AM to 10 PM Monday through Friday Central Time

Visit the HRW math web site at: **www.hrw.com/math**

Technology

Lesson Suggestions and Calculator Examples

(Keystrokes are based on a TI-83 calculator.)

Lesson 11.1 Sequences and Series

Consider having students practice using the calculator to generate sequences. Ask students what the displays shown below represent.

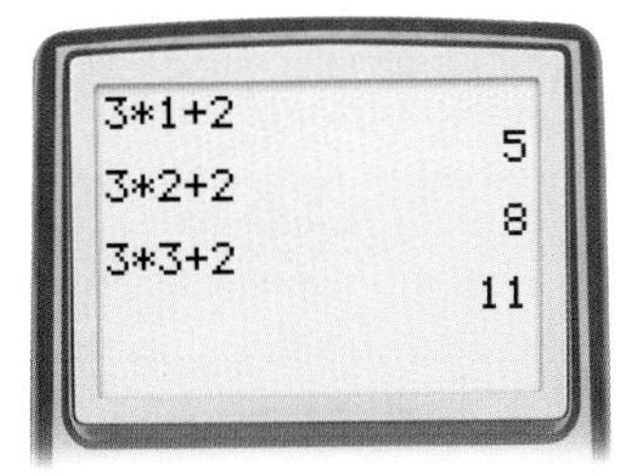

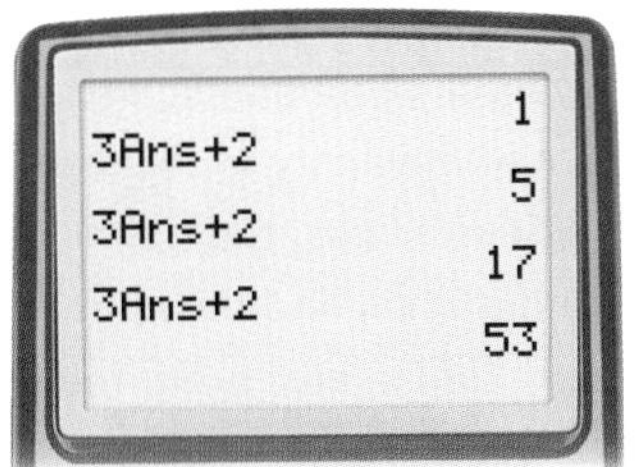

Lesson 11.2 Arithmetic Sequences

Consider having students find a particular term of an arithmetic sequence, once by using the recursive formula for the sequence and once by using the explicit formula. The displays below show finding the seventh term of the sequence 2, 5, 8, . . . by using each method. Discuss the approaches.

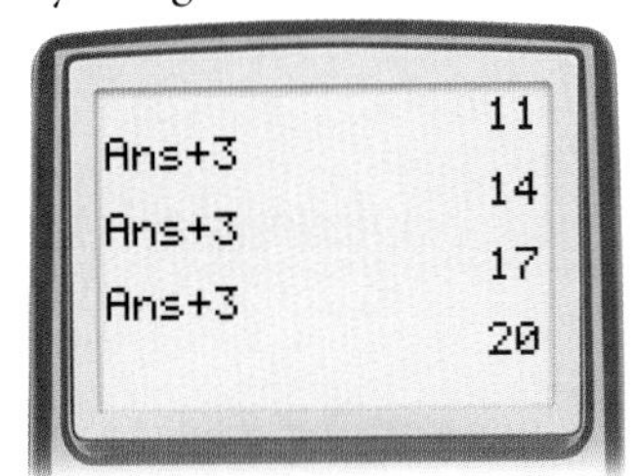

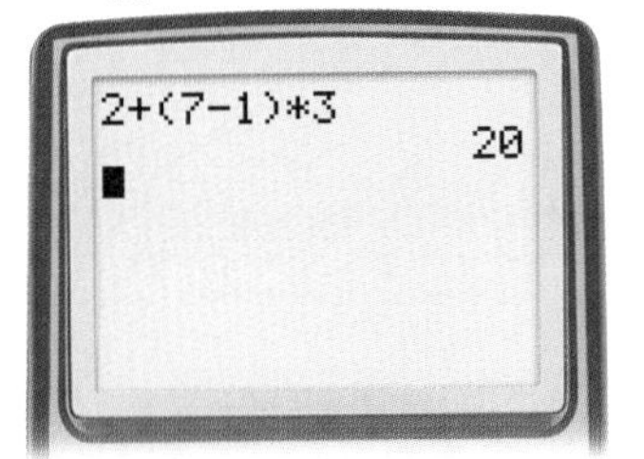

Lesson 11.3 Arithmetic Series

Students can generate a sequence by pressing **2nd** **STAT** and selecting **OPS** **5: seq(** . Students need to enter parameters in the following form and then press **ENTER**:

seq(*expression in* **X, X,** *start, end, increment)*

To find the sum of a sequence, students can press **2nd** **STAT**, select **MATH** **5: sum(**, press **2nd** **STAT**, select **OPS** **5: seq(** , and enter the parameters as shown above. The display at right shows that the sum of the previous sequence can be calculated by pressing **2nd** **(–)** to recall the previous answer without reentering the sequence commands.

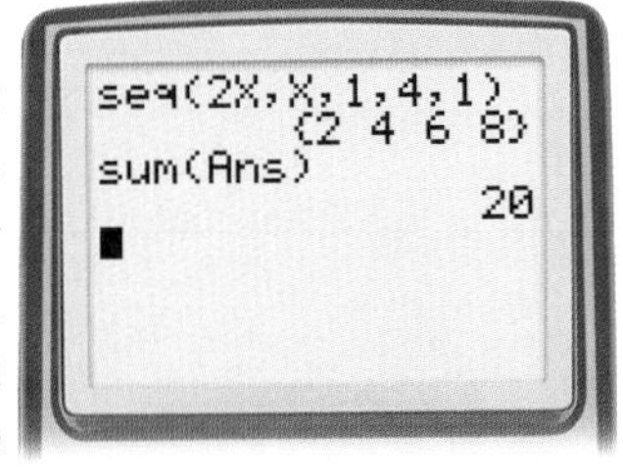

Lesson 11.4 Geometric Sequences

Students can generate the terms of a geometric sequence by changing the defining expression and following the procedure described above in Lesson 11.3.

Lesson 11.5 Geometric Series and Mathematical Induction

Students can adapt what they learned about using the calculator to find the sum of an arithmetic series to the new concept of finding the sum of a geometric series.

Consider the following exploration of conjectures with mathematical induction:

If the sum of the first n natural numbers is $\frac{n(n+1)}{2}$, then

$$\sum_{k=1}^{4} k - \frac{4(4+1)}{2} = 0 \text{ and } \sum_{k=1}^{5} k - \frac{5(5+1)}{2} = 0.$$

The display at right confirms this statement. Students should then use the steps for mathematical induction to prove the conjecture.

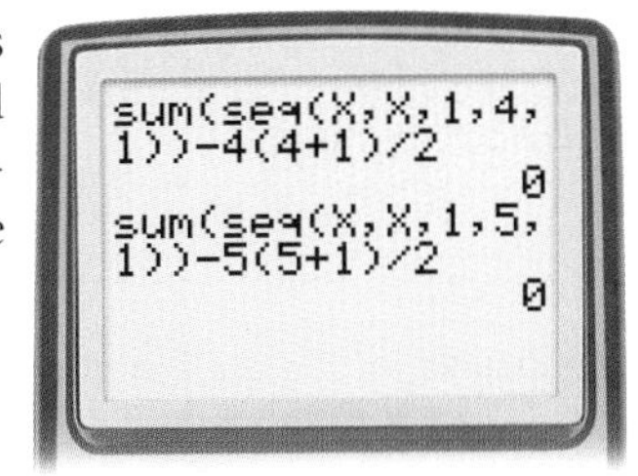

Lesson 11.6 Infinite Geometric Series

In this lesson, you will have the opportunity to show students how to test $\sum^{5} t_1 r^k$ for fixed values of t_1 and r and large values of n. Use the **sum(seq(** procedure from Lesson 11.3 to show the difference between $r > 1$ and $0 < r < 1$.

Lesson 11.7 Pascal's Triangle

In this lesson, you will have the opportunity to show students how the combination function on the calculator can be used in some binomial probability problems. Example 3 on page 738 provides a calculator example. You may also consider having students generate various entries in Pascal's triangle. The display above shows how the sequence function can help.

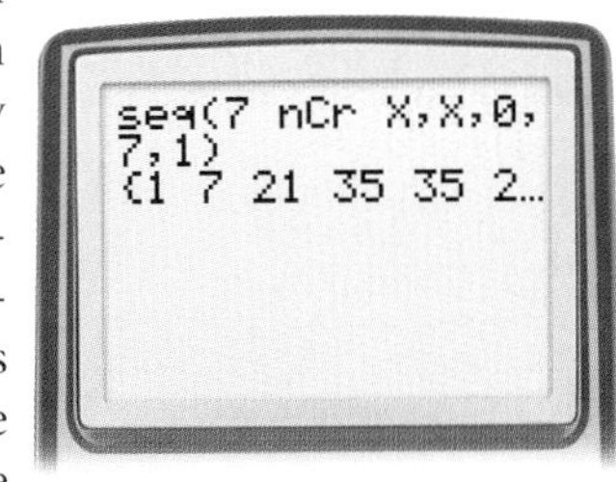

Lesson 11.8 The Binomial Theorem

Consider asking students how generating of a row of Pascal's triangle helps them write out the expansion of a binomial. After such a discussion, ask students what complications arise when $(x + y)^n$ is replaced by $(x + 2y)^n$ or by $(x - 2y)^n$.

internet connect

For keystrokes of other graphing calculators models, visit the HRW web site at **go.hrw.com** and enter the keyword **MB1 CALC**.

About Chapter 11

Background Information

Mathematics is the discovery and representation of patterns. Patterns can be finite or infinite and geometric or arithmetic. These patterns can be found in nature or can be invented by mathematicians. In this chapter, students will explore numerical patterns called *sequences* and *series*. Mathematical induction is introduced with geometric series, Pascal's triangle is introduced as a sequence whose terms can be found by using combination notation, and the binomial theorem is introduced in the expansion of binomial powers.

CHAPTER RESOURCES

- Block-Scheduling Handbook
- Writing Activities for Your Portfolio
- Tech Prep Masters
- Long-Term Project
- Assessment Resources:
  Mid-Chapter Assessment
  Chapter Assessments
  Alternative Assessments
- Test and Practice Generator
- Technology Handbook

Chapter Objectives

- Find the terms of a sequence. **[11.1]**
- Evaluate the sum of a series expressed in sigma notation. **[11.1]**
- Recognize arithmetic sequences, and find the indicated term of an arithmetic sequence. **[11.2]**
- Find arithmetic means between two numbers. **[11.2]**

11

DISCRETE MATHEMATICS

Series and Patterns

Lessons

DISCOVERING PATTERNS AND REPRESENTING them in sequences of numbers is an important mathematical skill. In this chapter, you will investigate different types of sequences, including the Fibonacci sequence. The Fibonacci sequence is used to describe a wide variety of structures in nature. For example, the spiral formed by a nautilus shell and the spiral pattern of sunflower seeds are both related to the Fibonacci sequence.

A Tohono O'odham basket from Arizona echoes the spiral pattern seen in nature.

About the Photos

In 1202, Leonardo of Pisa (better known as Fibonacci) published a book in which he discussed a very interesting set of numbers that are now known as the *Fibonacci numbers*. The number of petals on many flowers, the descendants of a pair of rabbits after one year, the spiral of a seashell, and the arrangement of seeds on flowerheads can all be modeled by Fibonacci numbers. The first two Fibonacci numbers are 1 and 1. Each subsequent Fibonacci number is the sum of the two previous numbers. So, the first few Fibonacci numbers are 1, 1, 2, 3, 5, 8, and 13, but the list of Fibonacci numbers is infinite. The entire infinite list is called the *Fibonacci sequence*. The equation below gives a way to find the nth Fibonacci number.

$$F_n = \frac{\left(\frac{1+\sqrt{5}}{2}\right)^n - \left(\frac{1-\sqrt{5}}{2}\right)^n}{\sqrt{5}}$$

This equation was published by Leonhard Euler in 1765, but Jacques Binet discovered it in 1843 and received all the credit.

It can be shown that the ratio of very large consecutive Fibonacci numbers approaches 1.618, which is commonly called the *golden ratio*.

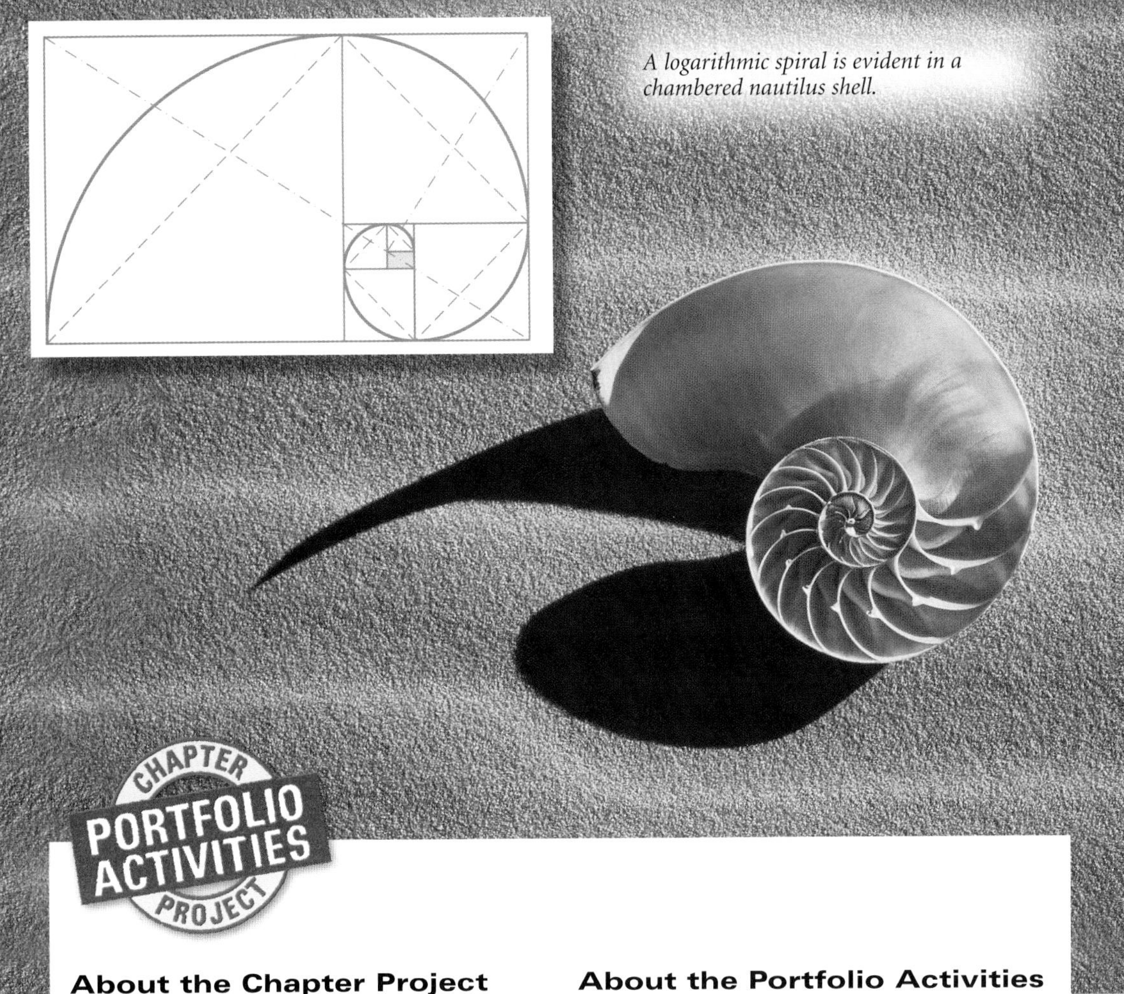

A logarithmic spiral is evident in a chambered nautilus shell.

About the Chapter Project

In this chapter, you will use series and sequences to model real-world situations.

In the Chapter Project, *Over the Edge,* you will investigate centers of gravity and perform experiments to develop a model involving sequences and series that you will use to determine whether a stack of objects will remain balanced.

After completing the Chapter Project, you will be able to do the following:

- Experimentally determine the center of gravity of an object.
- Model data from your experiments with a sequence or series.
- Determine whether your model gives predictions that are consistent with observations.

About the Portfolio Activities

Throughout the chapter, you will be given opportunities to complete Portfolio Activities that are designed to support your work on the Chapter Project.

- Finding the center of gravity of an object is included in the Portfolio Activity on page 706.
- Experimenting to determine some relationships that govern center of gravity is included inthe Portfolio Activity on page 727.
- Experimenting to determine the limits involved in keeping an object balanced is included in the Portfolio Activity on page 734.

internet connect go.hrw.com

Chapter Internet Features and Online Activities

Lesson	Keyword	Page	Lesson	Keyword	Page
11.1	MB1 Homework Help	696	11.5	MB1 Homework Help	725
	MB1 Fibonacci	695	11.6	MB1 Homework Help	732
11.2	MB1 Homework Help	703		MB1 System COG	734
	MB1 COG	706	11.7	MB1 Homework Help	739
11.3	MB1 Homework Help	710	11.8	MB1 Homework Help	746
11.4	MB1 Homework Help	717		MB1 Softball	745
	MB1 College	717			

- Find the sum of the first n terms of an arithmetic series. [11.3]
- Recognize geometric sequences, and find the indicated term of a geometric sequence. [11.4]
- Find geometric means between two numbers. [11.4]
- Find the sum of the first n terms of a geometric series. [11.5]
- Use mathematical induction to prove general statements about natural numbers. [11.5]
- Find the sum of an infinite geometric series, if one exists. [11.6]
- Write repeating decimals as fractions. [11.6]
- Find entries in Pascal's triangle. [11.7]
- Use Pascal's triangle to find combinations and probabilities. [11.7]
- Use the binomial theorem to expand $(x + y)^n$. [11.8]
- Use the binomial theorem to compute a probability. [11.8]

Portfolio Activities appear at the end of Lessons 11.2, 11.5, and 11.6. Each serves as preparation for the Chapter Project. The Portfolio Activities as well as the Chapter Project Activities are appropriate for inclusion in the student's portfolio. Students should be encouraged to include in their portfolios any other work in which they feel a sense of pride or a sense of accomplishment.

Prepare

NCTM PRINCIPLES & STANDARDS 1–4, 6–10

QUICK WARM-UP

Write the next three numbers in each pattern.

1. 15, 11, 7, 3, −1, . . .
   −5, −9, −13
2. −12, −7, −2, 3, 8, . . .
   13, 18, 23
3. 10, 11, 13, 16, 20, . . .
   25, 31, 38
4. 2, 6, 18, 54, 162, . . .
   486, 1458, 4374
5. 96, 48, 24, 12, 6, . . .
   3, 1.5, 0.75
6. 1, 4, 9, 16, 25, . . .
   36, 49, 81

Also on Quiz Transparency 11.1

Teach

Why Ordered lists of numbers are routinely used to describe real-world situations. Ask students to think of instances in which they use an ordered list of numbers to organize or plan their activities. They should think of a daily activity plan or the periods of a class schedule.

Sequences and Series

Objectives

- Find the terms of a sequence.
- Evaluate the sum of a series expressed in sigma notation.

Why *Sequences and series can be used to describe many patterns observed in the real world, such as the distances between the Sun and the planets.*

APPLICATION
ASTRONOMY

CULTURAL CONNECTION: EUROPE In the eighteenth century, German astronomers Johann Daniel Titius and Johann Elert Bode wrote a sequence of numbers to model the distances between the Sun and the planets.

They began with the number sequence below, in which each number (except 3) is double the preceding number.

$$0, 3, 6, 12, 24, 48, \ldots$$

From this sequence of numbers, Titius and Bode wrote a new sequence, called *Bode's sequence,* defined by adding 4 to each number in the original sequence and then dividing the result by 10.

Planet	Original number	Bode number
1st: Mercury	0	$\frac{0+4}{10} = 0.4$
2nd: Venus	3	$\frac{3+4}{10} = 0.7$
3rd: Earth	6	$\frac{6+4}{10} = 1.0$
4th: Mars	12	$\frac{12+4}{10} = 1.6$

Bode's law gives the distance from Earth to the Sun as 1 unit; each number in Bode's sequence represents the distance from a planet to the Sun in relation to that unit. Based on Bode's sequence, astronomers looked for a planet at the distance corresponding to the fifth Bode number, between Mars and Jupiter. The astronomers discovered Ceres, which turned out to be the largest member of the asteroid belt between Mars and Jupiter.

Alternative Teaching Strategy

USING PATTERNS Ask students to find the 10th term of the sequence 2, 6, 10, 14, 18, . . . Then ask them what method they used to get their answer. If they added 4 repeatedly, they used *recursive* thinking. Show them the recursive formula $t_1 = 2$ and $t_n = t_{n-1} + 4$. If they multiplied the term position number by 4 and subtracted 2, they used *explicit* thinking. Show them the explicit formula $t(n) = 4n - 2$. Now ask students to find the 10th term of the sequence −1, 2, 5, 8, 11, . . . by using the formula for the method they did not use with the first sequence. $t_{10} = 26$

recursive formula: $t_1 = -1$ and $t_n = t_n + 3$
explicit formula: $t(n) = 3n - 4$

Sequences

In mathematics, a **sequence** is an ordered list of numbers, called **terms**. The terms of a sequence are often arranged in a pattern. Some examples are shown below.

$2, 4, 6, 8, 10, \ldots \qquad 2, 4, 8, 16, 32, \ldots$

$5, 2, -1, -4, -7, \ldots \qquad 3, -9, 27, -81, 243, \ldots$

The three dots, called an ellipsis, indicate that a sequence is an **infinite sequence,** which continues without end. A **finite sequence** has a last term.

An infinite sequence can be defined by a function whose domain is the set of all natural numbers $\{1, 2, 3, \ldots, n, \ldots\}$ and a finite sequence can be defined by a function whose domain is the first n natural numbers $\{1, 2, 3, \ldots, n\}$. The range for both functions is the set of all real numbers. For example, the sequence 2, 4, 6, 8, 10, 12, . . . can be defined by the function t, shown below.

$t(n) = 2n$, where $n = 1, 2, 3, \ldots$

n	1	2	3	4	5	6	← *domain*
$t(n)$	2	4	6	8	10	12	← *range*

Each member of the range of a sequence is a term of the sequence. The terms of a sequence are often represented by the letter t with a subscript, as shown below.

$$t_1, t_2, t_3, \ldots, t_{n-1}, t_n$$

A formula that defines the nth term, or general term, of a sequence is called an **explicit formula.** With an explicit formula, each term of the sequence can be found by substituting the number of the term for n. This is shown in Example 1 below.

EXAMPLE 1 **Write the first six terms of the sequence defined by the explicit formula $t_n = -2n + 3$.**

SOLUTION

Evaluate $t_n = -2n + 3$ for n-values of 1, 2, 3, 4, 5, and 6.

n	1	2	3	4	5	6
t_n	1	−1	−3	−5	−7	−9

The first six terms of this sequence are 1, −1, −3, −5, −7, and −9.

TRY THIS Write the first six terms of the sequence defined by the explicit formula $t_n = 2n^2 - 1$.

A sequence can also be defined by a *recursive formula*. With a **recursive formula,** one or more previous terms are used to generate the next term. For example, the recursive formula for Example 1 is $t_1 = 1$ and $t_n = t_{n-1} - 2$, where $n \geq 2$.

$$t_n = t_{n-1} - 2 \text{ and } t_1 = 1$$
$$t_2 = t_1 - 2 = 1 - 2 = -1$$
$$t_3 = t_2 - 2 = -1 - 2 = -3$$
$$t_4 = t_3 - 2 = -3 - 2 = -5$$

Interdisciplinary Connection

GEOGRAPHY The population of a certain country grows at an annual rate of about 1%. This growth rate equals the birth rate minus the death rate. In addition, 15,000 people immigrate into the country each year. If the population in the country now is about 150,000,000, have students use a recursive method to write a sequence that gives the country's population for the first 3 years.
initial value (year 0): 150,000,000;
year 1: 151,515,000;
year 2: 153,045,150;
year 3: 154,590,602

Inclusion Strategies

USING PATTERNS The last two planets in our solar system, Neptune and Pluto, are located about 4,485,000,000 and 5,830,500,000 kilometers from the Sun, respectively. Have students determine whether these planets fit the general pattern of the Bode sequence. **Pluto fits the pattern as the ninth term, but Neptune does not fit the pattern.**

ADDITIONAL EXAMPLE 1

Write the first six terms of the sequence defined by $t_n = 4n + 5$. 9, 13, 17, 21, 25, 29

TRY THIS
1, 7, 17, 31, 49, 71

Teaching Tip

TECHNOLOGY When using a TI-83 or TI-82 graphics calculator, the **seq** feature may be entered by pressing 2nd STAT and selecting OPS and 5:seq(. The syntax for finding the terms of a sequence is **seq(*expression, variable, begin, end*).**

To produce the terms of the sequence for Example 1, use the following:

−2X+3,X,1,6) ENTER

You may use the left and right arrow keys to view a long sequence.

ADDITIONAL
EXAMPLE 2

Write the first six terms of the sequence defined by the following recursive formula: $t_1 = -2$ and $t_n = 2t_{n-1} - 1$, where $n \geq 2$.

$-2, -5, -11, -23, -47, -95$

TRY THIS

1, 2, 5, 14, 41, 122

ADDITIONAL
EXAMPLE 3

Refer to Bode's law, described at the beginning of the lesson.

a. Find the ninth term of the Bode's sequence given the ninth number of the original sequence.

9th: Pluto 384

38.8

b. Using Bode's law and 1.495×10^8 kilometers as the distance between Earth and the Sun, write the estimated distance in kilometers between the Sun and Pluto.

5.8006×10^9 km

CRITICAL THINKING

$t_n = \frac{t_{n-1} + 4}{10}$, for $n \geq 2$

EXAMPLE 2 **Write the first six terms of the sequence defined by the recursive formula $t_1 = 4$ and $t_n = 3t_{n-1} + 5$, where $n \geq 2$.**

SOLUTION

$$t_n = 3t_{n-1} + 5 \text{ and } t_1 = 4$$
$$t_2 = 3t_1 + 5 = 3(4) + 5 = 17$$
$$t_3 = 3t_2 + 5 = 3(17) + 5 = 56$$
$$t_4 = 3t_3 + 5 = 3(56) + 5 = 173$$
$$t_5 = 3t_4 + 5 = 3(173) + 5 = 524$$
$$t_6 = 3t_5 + 5 = 3(524) + 5 = 1577$$

The first six terms of this sequence are 4, 17, 56, 173, 524, and 1577.

TRY THIS Write the first six terms of the sequence defined by the recursive formula $t_1 = 1$ and $t_n = 3t_{n-1} - 1$, where $n \geq 2$.

EXAMPLE 3 **Refer to Bode's law, described at the beginning of the lesson.**

APPLICATION
ASTRONOMY

a. Find the next four terms of Bode's sequence, given the next four terms of the original sequence.

5th: Ceres	24
6th: Jupiter	48
7th: Saturn	96
8th: Uranus	192

b. Using Bode's law and 1.495×10^8 kilometers as the distance between Earth and the Sun, write the approximate distances in kilometers between the Sun and Ceres, Jupiter, Saturn, and Uranus.

SOLUTION

a.

Planet	Original number	Bode number
5th: Ceres	24	$\frac{24 + 4}{10} = 2.8$
6th: Jupiter	48	$\frac{48 + 4}{10} = 5.2$
7th: Saturn	96	$\frac{96 + 4}{10} = 10.0$
8th: Uranus	192	$\frac{192 + 4}{10} = 19.6$

b.

Planet	Bode number	Distance from planet to Sun
5th: Ceres	2.8	$2.8(1.495 \times 10^8) = 4.186 \times 10^8$
6th: Jupiter	5.2	$5.2(1.495 \times 10^8) = 7.774 \times 10^8$
7th: Saturn	10.0	$10.0(1.495 \times 10^8) = 1.495 \times 10^9$
8th: Uranus	19.6	$19.6(1.495 \times 10^8) = 2.9302 \times 10^9$

CRITICAL THINKING Write a recursive formula for the sequence of Bode numbers, beginning with the second term.

Enrichment

Show students the proof outlined at right of the second summation property shown on page 694. Have students justify each step by using the definition of summation notation and the commutative and the associative properties of real numbers.

$$\sum_{k=1}^{n}(a_k + b_k)$$
$$= (a_1 + b_1) + (a_2 + b_2) + \cdots + (a_n + b_n)$$
$$= a_1 + (b_1 + a_2) + b_2 + \cdots + (b_{n-1} + a_n) + b_n$$
$$= a_1 + (a_2 + b_1) + b_2 + \cdots + (a_n + b_{n-1}) + b_n$$
$$= (a_1 + a_2) + (b_1 + b_2) + \cdots + a_n + (b_{n-1} + b_n)$$
$$\vdots$$
$$= (a_1 + a_2 + \cdots + a_n) + (b_1 + b_2 + \cdots + b_n)$$
$$= \sum_{k=1}^{n} a_k + \sum_{k=1}^{n} b_k$$

Another famous sequence is the *Fibonacci sequence.* It is defined recursively as $a_1 = 1$, $a_2 = 1$, and $a_n = a_{n-2} + a_{n-1}$, where $n \geq 3$. The sequence is

$1+1=2 \quad 1+2=3 \quad 2+3=5 \quad 3+5=8 \quad 5+8=13 \quad 8+13=21$

$1, \; 1, \; 2, \; 3, \; 5, \; 8, \; 13, \; 21, \; \ldots$

CHECKPOINT ✔ Write the next four terms of the Fibonacci sequence.

The Fibonacci sequence, also called the *golden spiral,* has been used to study animal populations, relationships between elements in works of art, and various patterns in plants, such as the sunflower.

The seeds of a sunflower are arranged in spiral curves. The number of clockwise spirals and the number of counterclockwise spirals are always successive terms in the Fibonacci sequence.

Series

A **series** is an expression that indicates the sum of terms of a sequence. For example, if you add the terms of the sequence 2, 4, 6, 8, and 10, the resulting expression is the series 2 + 4 + 6 + 8 + 10.

Summation notation, which uses the Greek letter **sigma, Σ,** is a way to express a series in abbreviated form. For example, the series 2 + 4 + 6 + 8 + 10 can be represented as $\sum_{n=1}^{5} 2n$, which is read "the sum of $2n$ for values of n from 1 to 5."

Values of n from 1 to 5 are called the index. → $\sum_{n=1}^{5} 2n$ ← **Explicit formula for the general term of the related sequence**

EXAMPLE 4 **Write the terms of each series. Then evaluate.**

a. $\sum_{k=1}^{4} 5k$ **b.** $5\sum_{k=1}^{4} k$

SOLUTION

a. $\sum_{k=1}^{4} 5k = 5(1) + 5(2) + 5(3) + 5(4)$
$= 5 + 10 + 15 + 20$
$= 50$

b. $5\sum_{k=1}^{4} k = 5(1 + 2 + 3 + 4)$
$= 5(10)$
$= 50$

TRY THIS Write the terms of each series. Then evaluate.

a. $\sum_{k=1}^{5} 4k$ **b.** $\frac{1}{2}\sum_{k=1}^{4} k$

Reteaching the Lesson

USING PATTERNS To help students see the relationship between the recursive formula and the explicit formula for an arithmetic sequence, show them the process of finding the explicit formula of the sequence defined by using the recursive formula $t_1 = 3$ and $t_n = t_{n-1} + 5$.

$t_5 = t_4 + 5$
$= (t_3 + 5) + 5$
$= [(t_2 + 5) + 5] + 5$
$= \{[(t_1 + 5) + 5] + 5\} + 5$
$= t_1 + 4 \cdot 5$
$= t_1 + (n - 1) \cdot 5$
$= t_1 + 5n - 5$

Because $t_1 = 3$, $t_n = 3 + 5n - 5 = 5n - 2$.

Now ask students to repeat the process of finding the explicit formula for the sequence defined by the recursive formula $t_1 = 2$ and $t_n = t_{n-1} + 4$.
$t_n = 4n - 2$

CHECKPOINT ✔
34, 55, 89, 144

ADDITIONAL EXAMPLE 4

Write the terms of each series. Then evaluate.

a. $\sum_{k=1}^{6} 2k$

$2 + 4 + 6 + 8 + 10 + 12 = 42$

b. $2\sum_{k=1}^{6} k$

$2(1 + 2 + 3 + 4 + 5 + 6) = 42$

Teaching Tip

You may want to emphasize to students that summation notation represents a series, not the value of the sum of the series.

Use Teaching Transparency 42.

TRY THIS

a. $\sum_{k=1}^{5} 4k = 4 + 8 + 12 + 16 + 20$
$= 60$

b. $\frac{1}{2}\sum_{k=1}^{4} k = \frac{1}{2}(1 + 2 + 3 + 4)$
$= \frac{1}{2}(10)$
$= 5$

Teaching Tip

TECHNOLOGY When using a TI-83 or TI-82 graphics calculator, you may combine the **sum** and **seq** features. For part **a** of Example 4, use the following keystrokes:

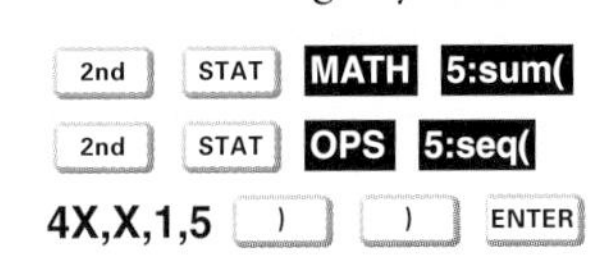

Exploring Summation Properties

You will need: no special materials

1. Evaluate $\sum_{k=1}^{4}(2k + k^2)$ and $\sum_{k=1}^{4} 2k + \sum_{k=1}^{4} k^2$.

CHECKPOINT ✔ 2. Are the two results from Step 1 the same? Explain.

3. Evaluate $\sum_{k=1}^{5} 2$, $\sum_{k=1}^{5} 3$, and $\sum_{k=1}^{5} 4$. (Hint: $\sum_{k=1}^{5} 1 = 1 + 1 + 1 + 1 + 1 = 5$)

CHECKPOINT ✔ 4. Find the pattern in the results from Step 3, and write a formula for $\sum_{k=1}^{n} c$.

Two summation properties are illustrated below.

$$\sum_{n=1}^{3} 4n^2 = 4 \cdot 1^2 + 4 \cdot 2^2 + 4 \cdot 3^2 = 4(1^2 + 2^2 + 3^2) = 4\sum_{n=1}^{3} n^2$$

$$\sum_{n=1}^{3}(n + n^2) = (1 + 1^2) + (2 + 2^2) + (3 + 3^2) = (1 + 2 + 3) + (1^2 + 2^2 + 3^2) = \sum_{n=1}^{3} + \sum_{n=1}^{3} n^2$$

These summation properties do not have names. Notice, however, the properties of real-number operations that are used in the illustrations. In the first case, the Distributive Property is used. In the second case, both the Associative and Commutative Properties are used.

Summation Properties

For sequences a_k and b_k and positive integer n:

1. $\sum_{k=1}^{n} ca_k = c\sum_{k=1}^{n} a_k$
2. $\sum_{k=1}^{n}(a_k + b_k) = \sum_{k=1}^{n} a_k + \sum_{k=1}^{n} b_k$

Series such as $\sum_{k=1}^{5} 2$ and $\sum_{k=1}^{5} 3$ are called *constant series*. The general term of a series may be defined by a constant, linear, or quadratic expression. The formulas below are used to find the sums of these series.

Summation Formulas

For all positive integers n:

Constant Series	Linear Series	Quadratic Series
$\sum_{k=1}^{n} c = nc$	$\sum_{k=1}^{n} k = \frac{n(n+1)}{2}$	$\sum_{k=1}^{n} k^2 = \frac{n(n+1)(2n+1)}{6}$

CHECKPOINT ✔ Verify the formulas for the sums of constant, linear, and quadratic series by using the series $2 + 2 + 2$, $1 + 2 + 3$, and $1 + 4 + 9$, respectively.

Activity Notes

Be sure that students understand the first directive. "Evaluate $\sum_{k=1}^{4}(2k + k^2)$" means to add all the terms obtained by replacing k in $2k + k^2$ with consecutive integers from 1 to 4 inclusive. They should see that $\sum_{k=1}^{4}(2k + k^2) = \sum_{k=1}^{4} 2k + \sum_{k=1}^{4} k^2$.

Cooperative Learning

Students can work in pairs. Have them complete Steps 1 and 3 independently. They should combine and summarize their results in Steps 2 and 4.

CHECKPOINT ✔

2. Yes; they both equal 50.

CHECKPOINT ✔

4. $\sum_{k=1}^{5} c = 5c$

CHECKPOINT ✔

$2 + 2 + 2 \stackrel{?}{=} \sum_{k=1}^{3} 2$

$6 \stackrel{?}{=} 3(2)$

$6 = 6$ True

$1 + 2 + 3 \stackrel{?}{=} \sum_{k=1}^{3} k$

$6 \stackrel{?}{=} \frac{3(3+1)}{2}$

$6 \stackrel{?}{=} \frac{12}{2}$

$6 = 6$ True

$1 + 4 + 9 \stackrel{?}{=} \sum_{k=1}^{3} k^2$

$14 \stackrel{?}{=} \frac{3(3+1)[2(3)+1]}{6}$

$14 = 14$ True

EXAMPLE 5 Evaluate $\sum_{m=1}^{5}(2m^2 + 3m + 2)$.

SOLUTION

$$\begin{aligned}\sum_{m=1}^{5}(2m^2 + 3m + 2) &= \sum_{m=1}^{5}2m^2 + \sum_{m=1}^{5}3m + \sum_{m=1}^{5}2 \\ &= 2\sum_{m=1}^{5}m^2 + 3\sum_{m=1}^{5}m + \sum_{m=1}^{5}2 \\ &= 2\left[\frac{5(6)(11)}{6}\right] + 3\left[\frac{5(6)}{2}\right] + 5 \cdot 2 \\ &= 110 + 45 + 10 \\ &= 165\end{aligned}$$

TRY THIS Evaluate $\sum_{j=1}^{5}(-j^2 + 2j + 5)$.

Exercises

Communicate

1. Explain the difference between a sequence and a series. Include examples in your response.
2. Explain the differences between the explicit formula $t_n = 2n + 1$ and the recursive formula $t_n = 2t_{n-1} + 1$.
3. Explain the difference between $\sum_{i=1}^{3}(i + 10)$ and $\sum_{i=1}^{3}i + 10$.

Guided Skills Practice

4. Write the first six terms of the sequence defined by the explicit formula $t_n = 3n - 2$. ***(EXAMPLE 1)*** **1, 4, 7, 10, 13, 16**
5. Write the first six terms of the sequence defined by the recursive formula below. ***(EXAMPLE 2)*** **1, 4, 13, 40, 121, 364**

   $t_1 = 1$ and $t_n = 3t_{n-1} + 1$, where $n \geq 2$
6. The first five terms of a sequence are 3, 5, 7, 9, and 11. ***(EXAMPLE 3)***
   a. Write the next five terms. **13, 15, 17, 19, 21**
   b. Write a recursive formula for the sequence. $t_1 = 3$, $t_n = t_{n-1} + 2$

Write the terms of each series. Then evaluate. ***(EXAMPLE 4)***

7. $\sum_{k=1}^{3}4k$ **24**
8. $5\sum_{k=1}^{4}k$ **50**
9. Evaluate $\sum_{k=1}^{4}(3k^2 + 2k + 4)$. ***(EXAMPLE 5)*** **126**

ADDITIONAL EXAMPLE 5

Evaluate $\sum_{n=1}^{5}(4n^2 - 2n + 7)$.

225

TRY THIS

$\sum_{j=1}^{5}(-j^2 + 2j + 5) = 0$

Assess

Selected Answers

Exercises 4–9, 11–95 odd

ASSIGNMENT GUIDE

In Class	1–9
Core	11–73 odd, 77–83 odd
Core Plus	10–82 even
Review	84–95
Preview	96–98

Extra Practice can be found beginning on page 940.

Error Analysis

Some students may draw conclusions from just a few terms in a sequence. For example, the first three terms of a sequence could represent the sequence 2, 4, 6, 8, 10, . . . , $2n$ or they could represent the sequence 2, 4, 6, 14, 34, . . . , $2n + (n - 1)(n - 2)(n - 3)$. Point out that given any finite number of terms, many different sequences can be determined.

Practice and Apply

Write the first four terms of each sequence defined by the given explicit formula.

10. $t_n = 2n + 3$ **5, 7, 9, 11**

11. $t_n = 4n + 1$ **5, 9, 13, 17**

12. $t_n = -2n + 1$ **−1, −3, −5, −7**

13. $t_n = -4n - 1$ **−5, −9, −13, −17**

14. $t_n = 6n + 2$ **8, 14, 20, 26**

15. $t_n = 5n - 1$ **4, 9, 14, 19**

16. $t_n = -7n + 3$ **−4, −11, −18, −25**

17. $t_n = -4n + 8$ **4, 0, −4, −8**

18. $t_n = 4n + 2$ **6, 10, 14, 18**

19. $t_n = \frac{1}{2}n + 1$ $\frac{3}{2}, 2, \frac{5}{2}, 3$

20. $t_n = \frac{1}{4}n + 2$ $\frac{9}{4}, \frac{5}{2}, \frac{11}{4}, 3$

21. $t_n = 8.75n + 3.67$

22. $t_n = 3.76n + 2.5$

23. $t_n = n^3$ **1, 8, 27, 64**

24. $t_n = (-1)^n$ **−1, 1, −1, 1**

25. $t_n = -2n^2$ **−2, −8, −18, −32**

21. 12.42, 21.17, 29.92, 38.67

22. 6.26, 10.02, 13.78, 17.54

Write the first six terms of each sequence defined by the given recursive formula.

26. $t_1 = 1$, $t_n = t_{n-1} + 3$ **1, 4, 7, 10, 13, 16**

27. $t_1 = 2$, $t_n = t_{n-1} + 2$ **2, 4, 6, 8, 10, 12**

28. $t_1 = 0$, $t_n = t_{n-1} - 4$ **0, −4, −8, −12, −16, −20**

29. $t_1 = -6$, $t_n = -2t_{n-1} + 3$ **−6, 15, −27, 57, −111, 225**

30. $t_1 = 7$, $t_n = 4t_{n-1} + 1$ **7, 29, 117, 469, 1877, 7509**

31. $t_1 = 10$, $t_n = 5t_{n-1} + 1$ **10, 51, 256, 1281, 6406, 32,031**

32. $t_1 = 10$, $t_n = 3t_{n-1} + 1$ **10, 31, 94, 283, 850, 2551**

33. $t_1 = 8$, $t_n = 3t_{n-1} - 2$ **8, 22, 64, 190, 568, 1702**

34. $t_1 = -2.24$, $t_n = 1.2t_{n-1} + 2.2$

35. $t_1 = 3.34$, $t_n = 2.2t_{n-1} - 1$

36. $t_1 = \frac{1}{3}$, $t_n = \frac{1}{2}t_{n-1} + 2$ $\frac{1}{3}, \frac{13}{6}, \frac{37}{12}, \frac{85}{24}, \frac{181}{48}, \frac{373}{96}$

37. $t_1 = \frac{5}{7}$, $t_n = \frac{1}{5}t_{n-1} + \frac{1}{3}$ $\frac{5}{7}, \frac{10}{21}, \frac{3}{7}, \frac{44}{105}, \frac{73}{175}, \frac{1094}{2625}$

34. −2.24, −0.488, 1.6144, 4.13728, 7.164736, 10.7976832

35. 3.34, 6.348, 12.9656, 27.52432, 59.553504, 130.0177088

internet connect

Homework Help Online

Go To: **go.hrw.com**

Keyword: **MB1 Homework Help** for Exercises 38–41

For each sequence below, write a recursive formula and find the next three terms.

38. 1, 5, 9, 13, ... **$t_1 = 1$, $t_n = t_{n-1} + 4$; 17, 21, 24**

39. 3, 9, 15, 21, ... **$t_1 = 3$, $t_n = t_{n-1} + 6$; 27, 33, 39**

40. 5, 9, 17, 33, ... **$t_1 = 5$, $t_n = 2t_{n-1} - 1$; 65, 129, 257**

41. 3, 7, 15, 31, ... **$t_1 = 3$, $t_n = 2t_{n-1} + 1$; 63, 127, 255**

Write the terms of each series. Then evaluate.

42. $\sum_{k=1}^{3} 4$

43. $\sum_{k=1}^{4} 10$

44. $\sum_{j=1}^{4} 3j$

45. $\sum_{k=1}^{3} 4k$

46. $\sum_{k=1}^{5} -2k$

47. $\sum_{k=1}^{4} -5k$

48. $\sum_{k=1}^{3} \frac{1}{2}k^2$

49. $\sum_{k=1}^{5} \frac{1}{3}k^2$

50. $\sum_{m=1}^{5} -\frac{1}{3}m^2 - m$

51. $\sum_{n=1}^{3} -\frac{1}{4}n^2 + n$

52. $\sum_{j=1}^{4} (2j + 3)$

53. $\sum_{k=1}^{3} (-3k + 1)$

54. $\sum_{m=1}^{4} (5m^2 + 1)$

55. $\sum_{k=1}^{3} (k^2 + k + 1)$

56. $\sum_{k=1}^{3} (2k^2 + 3k + 2)$

42. $4 + 4 + 4 = 12$

43. $10 + 10 + 10 + 10 = 40$

44. $3 + 6 + 9 + 12 = 30$

45. $4 + 8 + 12 = 24$

46. $-2 - 4 - 6 - 8 - 10 = -30$

47. $-5 - 10 - 15 - 20 = -50$

48. $\frac{1}{2} + 2 + \frac{9}{2} = 7$

49. $\frac{1}{3} + \frac{4}{3} + \frac{9}{3} + \frac{16}{3} + \frac{25}{3} = \frac{55}{3}$

50. $-\frac{4}{3} - \frac{10}{3} - 6 - \frac{28}{3} - \frac{40}{3} = -\frac{100}{3}$

51. $\frac{3}{4} + 1 + \frac{3}{4} = \frac{5}{2}$

52. $5 + 7 + 9 + 11 = 32$

53. $-2 - 5 - 8 = -15$

54. $6 + 21 + 46 + 81 = 154$

55. $3 + 7 + 13 = 23$

56. $7 + 16 + 29 = 52$

Practice

NAME ______________ CLASS ______ DATE ______

Practice 11.1 *Sequences and Series*

Write the first six terms of each sequence.

1. $b_n = 2.5n$ — 2.5, 5, 7.5, 10, 12.5, 15
2. $f_n = \frac{1}{2}n - \frac{1}{2}$ — $0, \frac{1}{2}, 1, 1\frac{1}{2}, 2, 2\frac{1}{2}$
3. $t_n = n^2 + 12$ — 13, 16, 21, 28, 37, 48
4. $a_1 = 20; a_n = 3a_{n-1} + 10$ — 20, 70, 220, 670, 2020, 6070
5. $a_1 = 1; a_n = a_{n-1} + 100$ — 1, 101, 201, 301, 401, 501
6. $a_1 = -5; a_n = 3a_{n-1}$ — −5, −15, −45, −135, −405, −1215

For each sequence below, write a recursive formula, and find the next three terms.

7. 1, 11, 121, 1331, ... — $a_1 = 1; a_n = 11a_{n-1}$; 14,641; 161,051; 1,771,561
8. 81, 78, 75, 72, ... — $a_1 = 81; a_n = a_{n-1} - 3$; 69, 66, 63
9. 2, −6, 18, −54, ... — $a_1 = 2; a_n = -3a_{n-1}$; 162, −486, 1458
10. $\frac{1}{4}, -1, 4, -16, \ldots$ — $a_1 = \frac{1}{4}; a_n = -4a_{n-1}$; 64, −256, 1024
11. 2, 11, 38, 119, ... — $a_1 = 2; a_n = 3a_{n-1} + 5$; 362, 1091, 3278
12. −2, −14, −74, −374, ... — $a_1 = -2; a_n = 5a_{n-1} - 4$; −1874; −9374; −46,874

Write the terms of each series. Then evaluate.

13. $\sum_{n=1}^{7} 4.5n$ — 4.5 + 9 + 13.5 + 18 + 22.5 + 27 + 31.5; 126
14. $\sum_{j=1}^{5} (j^2 + 8j + 2)$ — 11 + 22 + 35 + 50 + 67; 185

Evaluate.

15. $\sum_{m=1}^{6} 10m$ — 210
16. $\sum_{n=1}^{8} (2n - 12)$ — −24
17. $\sum_{j=1}^{5} (j - 3)^2$ — 10
18. $\sum_{k=1}^{10} (2k - 0.5)^2$ — 1432.5
19. $\sum_{a=1}^{6} (4a^2 + 3a - 5)$ — 397
20. $\sum_{n=1}^{12} (3.5n^2 - 5n + 2.2)$ — 1911.4

Evaluate the sum.

57. $\sum_{k=1}^{4} 3$ **12**

58. $\sum_{k=1}^{4} 2$ **8**

59. $\sum_{j=1}^{6} 4j$ **84**

60. $\sum_{k=1}^{7} 3k$ **84**

61. $\sum_{m=1}^{3} (5m+4)$ **42**

62. $\sum_{n=1}^{4} (2n+3)$ **32**

63. $\sum_{k=1}^{3} (4k+5)$ **39**

64. $\sum_{m=1}^{5} (2m^2+3m+2)$ **165**

65. $\sum_{j=1}^{4} (4j^2+4j+1)$ **164**

66. $\sum_{m=1}^{4} (-5m^2+2m+4)$ **−114**

67. $\sum_{n=1}^{4} (3-n)^2$ **6**

68. $\sum_{k=1}^{6} (k+2)^2$ **199**

69. $\sum_{n=1}^{3} \left(-\frac{2}{3}n^2+\frac{1}{2}n+\frac{5}{7}\right)$ **$-\frac{88}{21}$**

70. $\sum_{j=1}^{6} \left(\frac{1}{3}j^2+\frac{1}{5}j+2\right)$ **$\frac{698}{15}$**

71. $\sum_{m=1}^{4} (\pi m^2+2\pi m+2)$ **$50\pi+8$**

72. $\sum_{n=1}^{6} (\pi n^2+\pi n+4)$ **$112\pi+24$**

73. $\sum_{k=1}^{5} (0.7k^2+1.3k+2)$ **68**

74. $\sum_{j=1}^{5} (8.7j^2+8.6j+7.5)$ **645**

CHALLENGES

75. Find an explicit formula for the nth term of Bode's sequence.

75. $t_1 = 0.4$, t_n or $(n \leq 2 = 0.3(2^{n-2}) + 0.4$ for $n \geq 2$

76. A certain sequence is defined recursively by $t_1 = 1$, $t_2 = 2$, $t_{2n} = 2t_{2n-2}$, and $t_{2n+1} = 3t_{2n-1}$. Find the first eight terms of the sequence. **1, 2, 3, 4, 9, 8, 27, 16**

CONNECTION

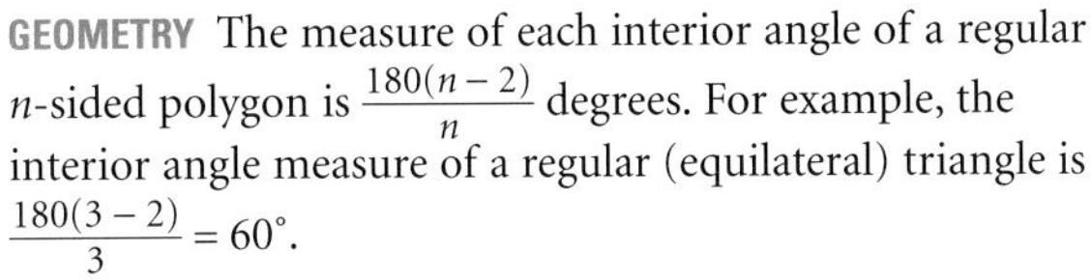

GEOMETRY The measure of each interior angle of a regular n-sided polygon is $\frac{180(n-2)}{n}$ degrees. For example, the interior angle measure of a regular (equilateral) triangle is $\frac{180(3-2)}{3} = 60°$.

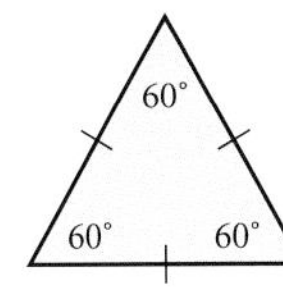

77. Find the interior angle measure of a square. **90°**

78. Find the interior angle measure of a regular pentagon. **108°**

79. Find the interior angle measure of a regular hexagon. **120°**

80. Does the interior angle measure of a regular n-sided polygon increase or decrease as n increases? **increases**

APPLICATIONS

81. 7.00, 7.30, 7.60, 7.90, 8.20, 8.50, 8.80, 9.10; $t_n = 7.00 + 0.3(n-1)$

81. INCOME Alan, a first-semester freshman, tutors in an afterschool program and earns \$7.00 an hour. Each semester, Alan gets a \$0.30 per hour raise. If he continues tutoring through his senior year, how much will he earn per hour in each semester of high school? Write an explicit formula to solve this problem.

82. BIOLOGY A single bacterium divides into 2 bacteria every 10 minutes. Assume that the same rate of division continues for 3 hours. Write a sequence that gives the number of bacteria after each 10-minute period.
$t_n = 2^n$, $1 \leq n \leq 18$

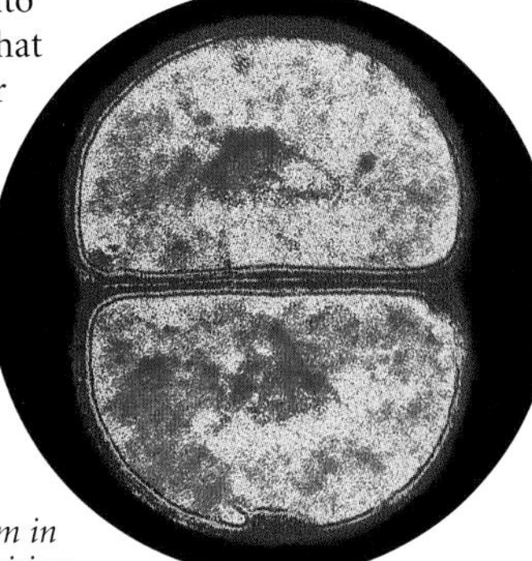

A spherical bacterium in the process of cell division

Look Beyond

Exercise 96 introduces students to a sequence of numbers in which the difference between any two consecutive terms is constant. This type of sequence is called an *arithmetic sequence.* In Lesson 11.2., Exercises 97 and 98 relate arithmetic sequences to linear functions.

Student Technology Guide

NAME ______ CLASS ______ DATE ______

Student Technology Guide
11.1 Sequences and Series

Although a graphics calculator can work with many sequences in Function mode, it also has a Sequence mode. You can use this mode to explore sequences defined by explicit and recursive formulas.

To change to sequence mode, press MODE. Use ▼ and ► to move to Seq. Press ENTER. In this mode, the X,T,θ,n key gives the variable n. The sequence names are u, v, and w (accessed by pressing 2nd and 7, 8, or 9).

Example: Write the first four terms of the sequence defined by the explicit formula $t_n = 3n - 4$.

- Enter the formula by pressing Y= 3 X,T,θ,n − 4.
- If u(n Min)= has a value, use the down arrow to move to that value and press CLEAR. Then press 2nd MODE.
- Press 2nd 7 (2nd (1 , 2 , 3 , 4 2nd)) ENTER. Terms 1 through 4 are shown at right below.

```
Plot1 Plot2 Plot3
nMin=1
\u(n)=3n-4
 u(nMin)=
\v(n)=
 v(nMin)=
\w(n)=
 w(nMin)=
```

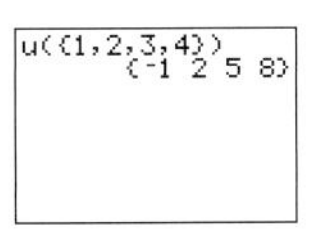

Example: Write the first four terms of the sequence defined by the recursive formula $a_1 = -3$ and $a_n = 2a_{n-1} + 4$, where $n \geq 2$.

- Enter the formula by pressing Y= 2 2nd 7 (X,T,θ,n − 1) + 4.
- Use the down arrow to move to u(n Min)= and press (−) 3. Then press 2nd MODE.
- Press 2nd 7 (2nd (1 , 2 , 3 , 4 2nd)) ENTER.

The first four terms of the sequence are −3, −2, 0, and 4.

Write the first four terms of each sequence.

1. $t_n = 4n + 1$ 5, 9, 13, 17
2. $f_n = -3n + 8$ 5, 2, −1, −4
3. $h_n = 3.14n + 2.72$ 5.86, 9, 12.14, 15.28
4. $t_n = (-1)^{(n+1)}$ 1, −1, 1, −1
5. $a_1 = 1, a_n = 2a_{n-1}$ 1, 2, 4, 8
6. $a_1 = 1, a_n = -3a_{n-1} + 5$ 1, 2, −1, 8
7. $a_1 = 3.4, a_n = 1.4a_{n-1} + 6.3$ 3.4, 11.06, 21.784, 36.7976

97.

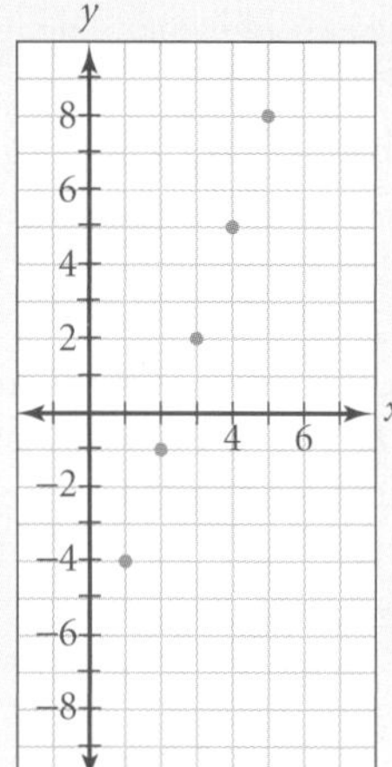

The points are collinear.

98.

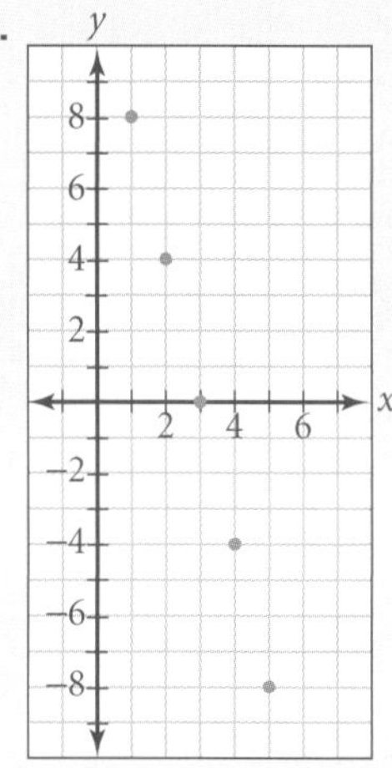

The points are collinear.

APPLICATION

83. RECREATION Anya is making paper flowers for the school prom. As she works, her speed increases. The first hour, she makes 12 flowers. During each additional hour that she works, she is able to make 4 more flowers than the hour before. If she works for 7 hours, how many flowers will she have made in total? Use sigma notation to model and solve this problem. (Hint: Start the index at 0.)

$\sum_{n=0}^{6}(12 + 4n) = 168$ **flowers**

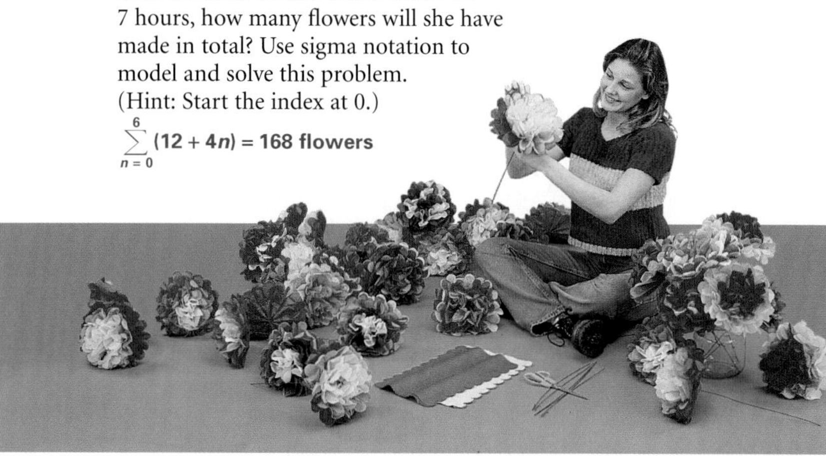

Look Back

Write an equation in slope-intercept form for the line that contains the given point and is parallel to the given line. ***(LESSON 1.3)***

84. $(-2, 5)$, $y = \frac{3}{4}x - 1$ $\;y = \frac{3}{4}x + \frac{13}{2}$

85. $(-4, 2)$, $2x - 3y = 1$ $\;y = \frac{2}{3}x + \frac{14}{3}$

Find the discriminant, and determine the number of real solutions. Then solve. ***(LESSON 5.6)***

86. $x^2 - 6x + 12 = 0$ **−12; 0; no real sol.**

87. $4x^2 - 4x + 1 = 0$ **0; 1 sol.;** $\frac{1}{2}$

88. $x^2 - 6x + 8 = 0$ **4; 2 sol.; 2, 4**

Solve each nonlinear system of equations and check your answers. ***(LESSON 9.6)***

89. $\begin{cases} 2x + y = 1 \\ 4x^2 + 9y^2 = 1 \end{cases}$ $\left(\frac{2}{5}, \frac{1}{5}\right)$, $\left(\frac{1}{2}, 0\right)$

90. $\begin{cases} 2y^2 = x \\ x - 2y = 12 \end{cases}$ **(8, −2), (18, 3)**

91. $\begin{cases} y^2 + 2x = 17 \\ x + 4y = -8 \end{cases}$ **(4, −3), (−52, 11)**

Calculate each permutation and combination below. ***(LESSONS 10.2 AND 10.3)***

92. ${}_8P_3$ **336**

93. ${}_9P_2$ **72**

94. ${}_{10}C_3$ **120**

95. ${}_8C_4$ **70**

Look Beyond

96. Find the next three terms of the sequence 5, 8, 11, 14, 17, . . . Then find an explicit formula for t_n, the nth term of the sequence. **20, 23, 26;** $t_n = 3(n - 1) + 5$

Graph each sequence below. What pattern do you see in the graphs?

97. **linear**

n	1	2	3	4	5
t_n	−4	−1	2	5	8

98. **linear**

n	1	2	3	4	5
t_n	8	4	0	−4	−8

Arithmetic Sequences

Objectives

- Recognize arithmetic sequences, and find the indicated term of an arithmetic sequence.
- Find arithmetic means between two numbers.

Why *Arithmetic sequences can be used to model real-world events such as the depreciation of a watering system.*

APPLICATION
DEPRECIATION

A new garden-watering system costs $389.95. As time goes on, the value of the watering system depreciates. Its value decreases by $42.50 per year. What is the value of the system after 9 years? To answer this question, you can use an arithmetic sequence. *You will answer this question in Example 2.*

Two examples of arithmetic sequences are shown below.

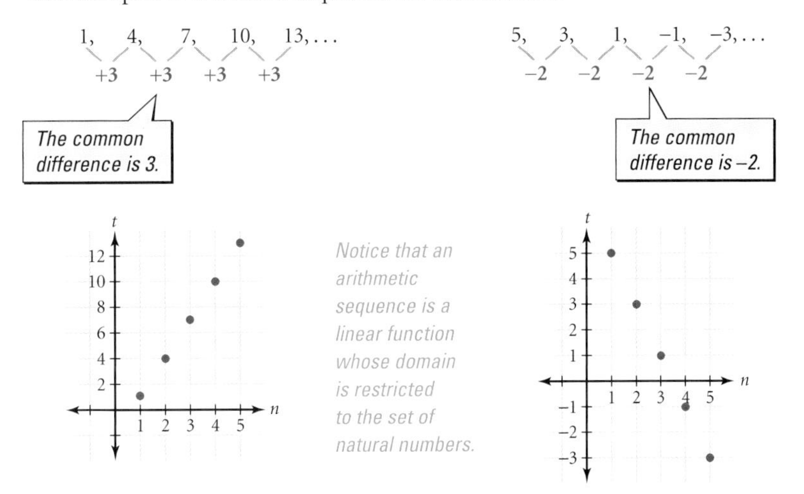

CHECKPOINT ✔ How does the common difference of an arithmetic sequence compare with the slope of the corresponding linear function?

Alternative Teaching Strategy

USING VISUAL MODELS You can help students recognize arithmetic sequences by showing them that they are related to linear equations. Have students refer to the arithmetic sequence at the top of page 700. Ask them to graph the explicit formula $t_n = 3 + (n - 1)4$ for values of n from 1 to 4. Be sure they notice that the graph consists of discrete points on the line $t(n) = 4n - 1$ and that the slope of the line, 4, is the constant difference of the sequence. Now give them the following problem: The fourth term of an arithmetic sequence is 10 and the common difference is 2. Have students write the recursive formula and use the graph of the related linear equation to find the explicit formula for the general term, t_n.
$t_1 = 4$ and $t_n = t_{n-1} + 2$; $t_n = 2n + 2$

Prepare

NCTM PRINCIPLES & STANDARDS 1–3, 6, 8–10

QUICK WARM-UP

Write the first five terms of each sequence.

1. $t_n = 3n$
   3, 6, 9, 12, 15
2. $t_n = 6n - 7$
   −1, 5, 11, 17, 23
3. $t_n = -9n + 1$
   −8, −17, −26, −35, −44
4. $a_1 = 2$; $a_n = -3a_{n-1}$
   2, −6, 18, −54, 162
5. $a_1 = 9$; $a_n = a_{n-1} - 4$
   9, 5, 1, −3, −7
6. $a_1 = -1$; $a_n = 3a_{n-1} + 5$
   −1, 2, 11, 38, 119

Also on Quiz Transparency 11.2

Teach

Why Patterns formed by sequences that have a common difference appear frequently in the real world. Have students consider situations in their own lives in which such a pattern occured. An exercise program in which the daily number of sit-ups or push-ups is increased by a constant number each week is an example.

CHECKPOINT ✔
The common difference of an arithmetic sequence is the slope of the corresponding linear function.

ADDITIONAL
EXAMPLE 1

Find the 10th term of the sequence defined by $t_1 = 7$ and $t_n = t_{n-1} + 6$. 61

Teaching Tip

TECHNOLOGY When using a TI-82 or TI-83 graphics calculator to find the fifth term of the sequence of Example 1, start by entering **–4** on the home screen and pressing ENTER. To add 3 to –4, press + **3** ENTER. To successively add 3 to the previous calculator value, press ENTER repeatedly.

TRY THIS
–15.5

Teaching Tip

Remind students that the value of the constant difference, d, in an arithmetic sequence may not be an integer. The sequence 5, 4.5, 4, 3.5, . . . is an example of an arithmetic sequence with a difference of –0.5.

An **arithmetic sequence** is a sequence whose successive terms differ by the same number, d, called the **common difference.** That is, if t_n, where $n \geq 2$, is any term in an arithmetic sequence, then the statement below is true.

$$t_n - t_{n-1} = d \quad \text{or} \quad t_n = t_{n-1} + d$$

The formula $t_n = t_{n-1} + d$, where $n \geq 2$, is a recursive formula because it gives the nth term of a sequence in relation to the previous term and d.

PROBLEM SOLVING

Look for a pattern. A pattern can be seen in the recursive formula for the arithmetic sequence defined by $t_1 = 3$ and $t_n = t_{n-1} + 4$, where $n \geq 2$.

Recursive formula	Pattern
$t_1 = 3$	$t_1 = 3 + (0)4$
$t_2 = 3 + 4$	$t_2 = 3 + (1)4$
$t_3 = (3 + 4) + 4$	$t_3 = 3 + (2)4$
$t_4 = [(3 + 4) + 4] + 4$	$t_4 = 3 + (3)4$

The pattern formed by the recursive formula gives the information needed to write an explicit formula for the same sequence.

$$t_n = 3 + (n - 1)4, \text{ where } n \geq 1$$

*n*th Term of an Arithmetic Sequence

The general term, t_n, of an arithmetic sequence whose first term is t_1 and whose common difference is d is given by the explicit formula $t_n = t_1 + (n - 1)d$.

If you know the first term of an arithmetic sequence and its common difference, you can use the explicit formula to find any term of the sequence. This is shown in Example 1.

EXAMPLE 1 **Find the fifth term of the sequence defined by the recursive formula $t_1 = -4$ and $t_n = t_{n-1} + 3$.**

SOLUTION

The sequence is arithmetic, in which $t_1 = -4$ and $d = 3$. Use the explicit formula for the general term of an arithmetic sequence.

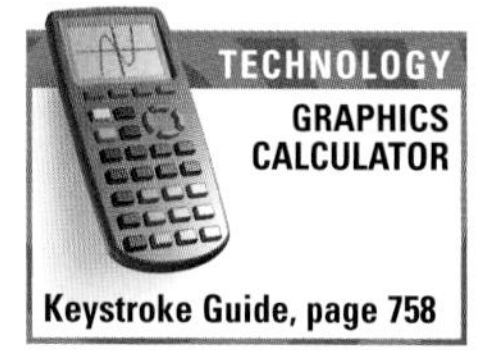

$$t_n = t_1 + (n - 1)d$$
$$t_5 = -4 + (5 - 1)3$$
$$t_5 = 8$$

Thus, the fifth term of the sequence is 8.

CHECK

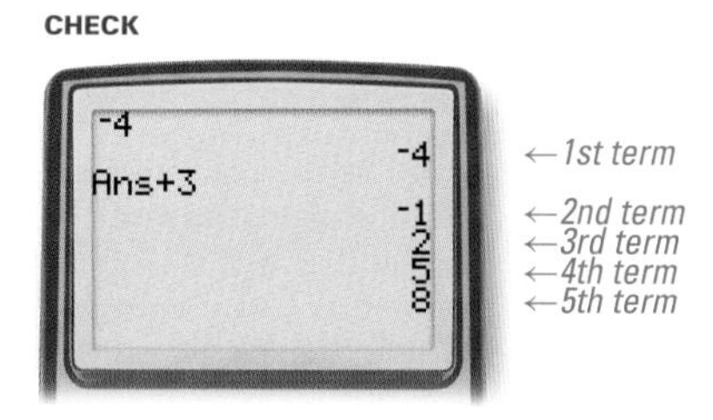

TRY THIS Find the seventh term of the sequence defined by the recursive formula $t_1 = 2.5$ and $t_n = t_{n-1} - 3$.

Interdisciplinary Connection

BUSINESS A school's drama club is producing a show. To accommodate a larger audience, they rented a theater with 21 rows of seats. Row 1, which is closest to the stage, has 18 seats. Each successive row, going toward the back of the theater, has one seat more than the previous row. The cost of a seat in rows 1–7 is 1.5 times the cost of a seat in rows 8–15 and 2 times the cost of a seat in rows 16–21. The total revenue when all the seats are sold is \$1824.75. Have students determine the cost of each type of seat. **\$4.50, \$3.00, and \$2.25**

Inclusion Strategies

INVITING PARTICIPATION To introduce arithmetic sequences, give students the table below.

n	1	2	3	4	5	6
t_n	3	8	13	18	23	28

Have them find the average of the first and second terms and the average of the second and third terms. Continue this procedure for all five pairs of terms. Ask students whether the averages they found form an arithmetic sequence. Have them explain their reasoning.

EXAMPLE 2 **Refer to the watering system described at the beginning of the lesson.**

APPLICATION
DEPRECIATION

a. Use a recursive formula to find the value of the system after 4 years.
b. Use the explicit formula for the *n*th term of an arithmetic sequence to find the value of the watering system after 9 years.

SOLUTION

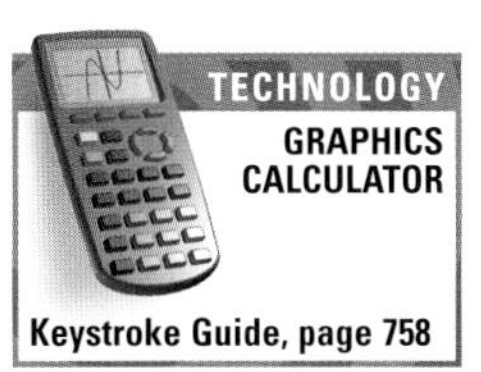

TECHNOLOGY
GRAPHICS CALCULATOR
Keystroke Guide, page 758

a. Initial value (year 0): 389.95
year 1: $389.95 - 42.5 = 347.45$
year 2: $347.45 - 42.5 = 304.95$
year 3: $304.95 - 42.5 = 262.45$
year 4: $262.45 - 42.5 = 219.95$

The value of the system after 4 years of use is \$219.95.

CHECK

389.95 → 389.95
Ans-42.5
347.45 ←1 year
304.95 ←2 years
262.45 ←3 years
219.95 ←4 years

b. At year 9, $n = 10$. Find t_{10}.

$t_n = t_1 + (n-1)d$
$t_{10} = 389.95 + (10-1)(-42.5)$
$t_{10} = 7.45$

The value of the system after 9 years of use is \$7.45.

CHECKPOINT ✔ Explain why you need to find the 10th term of the arithmetic sequence in Example 2 in order to find the value after 9 years.

TRY THIS A new washing machine costs \$352.65. It depreciates \$48.60 each year. Find the value of the washing machine after 7 years.

EXAMPLE 3 **Find the 10th term of the arithmetic sequence in which $t_3 = -5$ and $t_6 = 16$.**

SOLUTION

1. Find the common difference, *d*.

n	3	4	5	6
t_n	−5	?	?	16

d *d* *d* → $-5 + 3d = 16$
$d = 7$

2. Find t_1. Use $d = 7$ and either $t_3 = -5$ or $t_6 = 16$.

$t_n = t_1 + (n-1)d$
$-5 = t_1 + (3-1)7$ *Use $t_3 = -5$.*
$-19 = t_1$

$t_n = t_1 + (n-1)d$
$16 = t_1 + (6-1)7$ *Use $t_6 = 16$.*
$-19 = t_1$

TECHNOLOGY
GRAPHICS CALCULATOR
Keystroke Guide, page 758

3. Find the 10th term.

$t_n = t_1 + (n-1)d$
$t_{10} = -19 + (10-1)(7)$
$t_{10} = 44$

Thus, the 10th term of this arithmetic sequence is 44.

CHECK

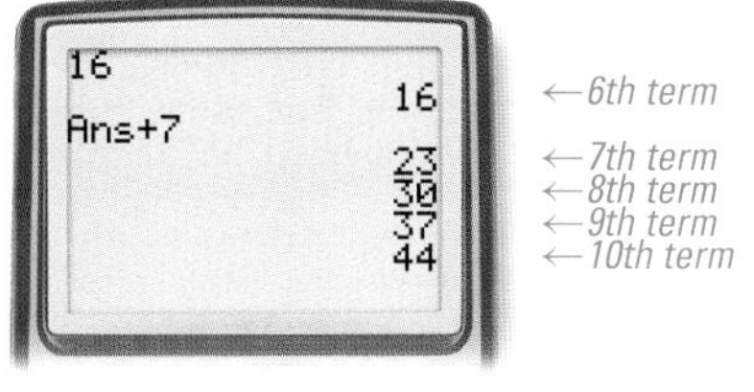

TRY THIS Find the 12th term of the arithmetic sequence in which $t_3 = 8$ and $t_7 = 20$.

Enrichment

To encourage students to look for a constant difference between successive terms in a given sequence, give them the sequence $x - 1$, $x + 3$, $3x - 1$. Ask them to find the value of x that would make the sequence arithmetic. $x = 4$ Then ask them to find the value of x that would make $3x$, $3x^2 - 2$, $4x^2 - 2$ an arithmetic sequence.

$x = 2$ *or* $x = -\frac{1}{2}$

Reteaching the Lesson

USING VISUAL MODELS To find the *n*th term, t_n, of an arithmetic sequence with $t_1 = 17$ and $d = -8$, have students graph the sequence by plotting the points (1, 17), (2, 9), (3, 1), and (4, −7). Now ask the students to find the slope of the line that contains the graph of the sequence and have them write an equation for the line. Point out that the equation $t_n = -8n + 25$ gives an explicit formula for the *n*th term of the sequence. Now ask students to find the 10th term of the sequence. −55

Use Teaching Transparency 43.

ADDITIONAL EXAMPLE 2

A new car costs \$16,000. The value of the car depreciates \$1500 per year.

a. **Use a recursive formula to find the value of the car after 5 years.**
\$8500

b. **Use the explicit formula for the *n*th term of an arithmetic sequence to find the value of the car after 7 years.**
\$5500

Teaching Tip

TECHNOLOGY When using a TI-82 or TI-83 graphics calculator to find the value of the system from Example 2 after 4 years of use, enter **389.95** on the home screen and press ENTER. To subtract 42.5, press **−42.5** ENTER. To successively subtract 42.5 from the previous calculator value, press ENTER repeatedly.

CHECKPOINT ✔
To produce the ninth difference that is added to t_1 in the formula $t_n = t_1 + (n-1)d$, n must be 10.

TRY THIS
\$12.45

ADDITIONAL EXAMPLE 3

Find the 15th term of the arithmetic sequence in which $t_5 = 7$ and $t_{10} = 22$.
37

TRY THIS
35

CRITICAL THINKING
$d = t_j - t_i$, for $j > i$;
$t_n = t_1 + (n - 1)d$

Activity Notes

Remind students that sequences can be represented as ordered pairs. The first coordinate in an ordered pair is the position number of the term, and the second coordinate is the term itself. Be sure students notice that the graph of any arithmetic sequence consists of discrete points on a line whose slope is equal to the common difference of the sequence.

Teaching Tip

TECHNOLOGY When using a TI-82 or TI-83 to graph the sequences in the Activity, you may use the disconnected sequence mode. Press MODE, select **Seq** by using the arrow keys, and press ENTER to select sequence mode. Then select **Dot** under the **MODE** menu and press ENTER. Press Y= to view the sequence entry screen, enter the value for the minimum value of the variable, ***n*Min=1**, and enter the sequence to be generated, such as **u(*n*)=5+2*n***. Then set the viewing window as follows:

***n*Min=1**	**Xmin=0**
***n*Max=6**	**Xmax=7**
PlotStart=1	**Xscl=1**
PlotStep=1	**Ymin=0**
	Ymax=18
	Yscl=1

Press GRAPH to view the graph of the sequence. Note that ***n*Min** in the **Y=** editor is the same as ***n*Min** in the viewing window editor when in **Seq** mode. This feature should only be used with recursive sequences. Reset the calculator to **Func** mode.

CHECKPOINT ✔

3. *a* and *b* are linear.
   c and *e* are exponential.
   d is reciprocal.
   f is quadratic.

CRITICAL THINKING In an arithmetic sequence, you are given consecutive terms t_i and t_j. Explain how to find d, and write an explicit formula for t_n, where n is any positive integer. Assume that $i < j$.

Activity Investigating Sequences

TECHNOLOGY GRAPHICS CALCULATOR
Keystroke Guide, page 758

You will need: a graphics calculator or graph paper

1. Graph each sequence for n-values from 1 to 6.
   a. $t_n = 5 + 2n$ **b.** $t_n = 8 - 0.5n$ **c.** $t_n = 2^n$
   d. $t_n = \frac{1}{n}$ **e.** $t_n = 3(0.1)^n$ **f.** $t_n = n^2$
2. Which of the sequences in Step 1 are arithmetic?

CHECKPOINT ✔

3. Classify the graph of each sequence in Step 1 as linear, quadratic, exponential, or reciprocal (restricted to the domain of natural numbers).

The terms between any two nonconsecutive terms of an arithmetic sequence are called the **arithmetic means** between the two nonconsecutive terms. For example, in the sequence 5, 11, 17, 23, 29, 35, . . . , the three arithmetic means between 5 and 29 are 11, 17, and 23. In the past, you have learned that an arithmetic mean is the average of two numbers. That definition still holds because each number in the sequence is the average of the numbers on each side.

EXAMPLE 4 **Find the four arithmetic means between 10 and –30.**

SOLUTION

PROBLEM SOLVING

Draw a diagram. Let $t_1 = 10$ and $t_6 = -30$.

$$\overset{t_1}{10}, \underline{?}, \underline{?}, \underline{?}, \underline{?}, \overset{t_6}{-30}$$

Find the common difference, d.

$$t_n = t_1 + (n - 1)d$$
$$-30 = 10 + (6 - 1)d \quad \textit{Because } t_6 = -30,\ n = 6 \textit{ and } t_n = -30.$$
$$-40 = 5d$$
$$\frac{-40}{5} = d$$
$$-8 = d$$

Use $d = -8$ to find the arithmetic means.

$$10 - 8 = 2$$
$$2 - 8 = -6$$
$$-6 - 8 = -14$$
$$-14 - 8 = -22$$

The four arithmetic means are 2, –6, –14, and –22.

TRY THIS Find the four arithmetic means between 24 and 39.

Exercises

Communicate

1. Explain how to write an explicit formula for the general term of the sequence −4, 2, 8, 14, . . .
2. Explain why $(n-1)d$ instead of nd is used to find the nth term of an arithmetic sequence.
3. Describe how the arithmetic mean of 4 and 20 and the three arithmetic means between 4 and 20 are related.

Guided Skills Practice

4. $t_4 = 2$

4. Find the fourth term of the sequence defined by $t_1 = -4$ and $t_n = t_{n-1} + 2$. *(EXAMPLE 1)*

APPLICATION

5. **DEPRECIATION** Sheryl purchased a sewing machine for her tailoring service. If the machine cost $1425.65 and depreciates at the rate of $85 per year, what will its value be after 10 years? *(EXAMPLE 2)* $575.65

6. Find the eighth term of the arithmetic sequence in which $t_4 = 2$ and $t_7 = 6$. *(EXAMPLE 3)* $\frac{22}{3}$
7. Find the four arithmetic means between 6 and 26. *(EXAMPLE 4)* 10, 14, 18, 22

Practice and Apply

internet connect

Homework Help Online
Go To: go.hrw.com
Keyword: **MB1 Homework Help** for Exercises 8–29

Based on the terms given, state whether or not each sequence is arithmetic. If it is, identify the common difference, *d*.

8. 6, 10, 14, 18, 22, . . . yes; $d = 4$
9. 2, 4, 6, 8, 10, . . . yes; $d = 2$
10. 8, 5, 2, −1, −4, . . . yes; $d = -3$
11. −3, 0, 3, 6, 9, . . . yes; $d = 3$
12. 5, −5, 5, −5, 5, . . . no
13. 1, 2, 4, 8, . . . no
14. 9, 7, 5, 3, 1, . . . yes; $d = -2$
15. 0, −6, −12, −18, −24, . . . yes; $d = -6$
16. 3, 6, 12, 24, . . . no
17. 3, 7, 12, 18, 25, . . . no
18. −1, 1, −1, 1, . . . no
19. 1, −3, 5, −7, . . . no
20. $0, \frac{1}{2}, 1, \frac{3}{2}, 2, \ldots$ yes; $d = \frac{1}{2}$
21. $\frac{1}{3}, \frac{2}{3}, 1, \frac{4}{3}, \frac{5}{3}, \ldots$ yes; $d = \frac{1}{3}$
22. $\frac{2}{7}, \frac{4}{7}, 1, \frac{11}{7}, \frac{16}{7}, \ldots$ no
23. $\frac{2}{5}, \frac{5}{6}, \frac{6}{7}, \frac{7}{8}, \frac{8}{9}, \ldots$ no
24. −2.8, 3.9, 5.0, 6.1, 12.2, . . . no
25. 4.23, 5.67, 6.01, . . . no
26. $\frac{\sqrt{2}}{\sqrt{3}}, \frac{2}{3}, \frac{4}{9}, \frac{16}{81}, \ldots$ no
27. 0.1, 0.01, 0.001, 0.0001, . . . no
28. $\pi, 2\pi, 3\pi, 4\pi, 5\pi, \ldots$ yes; $d = \pi$
29. $\pi, \pi^2, \pi^3, \pi^4, \pi^5, \ldots$ no

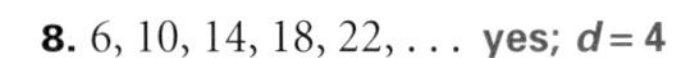

ADDITIONAL EXAMPLE 4

Find the five arithmetic means between 6 and 60.
15, 24, 33, 42, 51

TRY THIS
27, 30, 33, 36

Assess

Selected Answers

Exercises 4–7, 9–97 odd

ASSIGNMENT GUIDE	
In Class	1–7
Core	9–77 odd, 81
Core Plus	8–82 even
Review	83–97
Preview	98

Extra Practice can be found beginning on page 940.

Error Analysis

Some students may not recognize that a sequence is arithmetic when it is defined recursively. Suggest that they write out the first few terms of a sequence and decide whether the sequence is arithmetic based on those terms.

Use the recursive formula given to find the first four terms of each arithmetic sequence.

30. $t_1 = 5$, $t_n = t_{n-1} + 2$ **5, 7, 9, 11**

31. $t_1 = 18$, $t_n = t_{n-1} - 3$ **18, 15, 12, 9**

32. $t_1 = 0$, $t_n = t_{n-1} + 0.1$ **0, 0.1, 0.2, 0.3**

33. $t_1 = 1$, $t_n = t_{n-1} + 2$ **1, 3, 5, 7**

34. $t_1 = -5$, $t_n = t_{n-1} + 4$ **−5, −1, 3, 7**

35. $t_1 = -4$, $t_n = t_{n-1} + 3$ **−4, −1, 2, 5**

36. $t_1 = 3$, $t_n = t_{n-1} + 3$ **3, 6, 9, 12**

37. $t_1 = 7$, $t_n = t_{n-1} + 1$ **7, 8, 9, 10**

38. $t_1 = 6$, $t_n = t_{n-1} + 4$ **6, 10, 14, 18**

List the first four terms of each arithmetic sequence.

39. $t_n = 4 + (n-1)(3)$ **4, 7, 10, 13**

40. $t_n = -2 + (n-1)(4)$ **−2, 2, 6, 10**

41. $t_n = 3n - 4$ **−1, 2, 5, 8**

42. $t_n = -2n + 5$ **3, 1, −1, −3**

43. $t_n = -5n + 2$ **−3, −8, −13, −18**

44. $t_n = 4n - 2$ **2, 6, 10, 14**

45. $t_n = -3 + (n-1)(5)$ **−3, 2, 7, 12**

46. $t_n = 4 + (n-1)(-2)$ **4, 2, 0, −2**

47. $t_n = \frac{1}{3} + \frac{1}{3}n$ $\frac{2}{3}, 1, \frac{4}{3}, \frac{5}{3}$

48. $t_n = \frac{1}{5}n + \frac{4}{5}$ $1, \frac{6}{5}, \frac{7}{5}, \frac{8}{5}$

49. $t_n = \pi n + 4$ $\pi + 4, 2\pi + 4, 3\pi + 4, 4\pi + 4$

50. $t_n = \pi n + 5$ $\pi + 5, 2\pi + 5, 3\pi + 5, 4\pi + 5$

Find the indicated term given two other terms.

51. 5th term; $t_3 = 10$ and $t_7 = 26$ $t_5 = 18$

52. 5th term; $t_2 = -5$ and $t_6 = 7$ $t_5 = 4$

53. 10th term; $t_1 = 2.1$ and $t_4 = 1.83$ $t_{10} = 1.29$

54. 10th term; $t_1 = 2.1$ and $t_6 = -2.85$ $t_{10} = -6.81$

55. 1st term; $t_6 = -\frac{5}{6}$ and $t_8 = -\frac{3}{2}$ $t_1 = \frac{5}{6}$

56. 1st term; $t_2 = -\frac{13}{12}$ and $t_6 = -\frac{7}{4}$ $t_1 = -\frac{11}{12}$

Write an explicit formula for the *n*th term of each arithmetic sequence.

57. 6, 8, 10, 12, 14, . . .

58. 11, 15, 19, 23, 27, . . .

59. 1, −6, −13, −20, −27, . . .

60. 14, 9, 4, −1, −6, . . .

61. 23, 31, 39, 47, 55, . . .

62. 17, 22, 27, 32, 37, . . .

63. 20, 15, 10, 5, 0, . . .

64. 33, 24, 15, 6, −3, . . .

65. 100, 105, 110, 115, 120, . . .

66. 500, 520, 540, 560, 580, . . .

67. −50, −45, −40, −35, −30, . . .

68. −80, −76, −72, −68, −64, . . .

57. $t_n = 6 + (n-1)(2)$
58. $t_n = 11 + (n-1)(4)$
59. $t_n = 1 + (n-1)(-7)$
60. $t_n = 14 + (n-1)(-5)$
61. $t_n = 23 + (n-1)(8)$
62. $t_n = 17 + (n-1)(5)$
63. $t_n = 20 + (n-1)(-5)$
64. $t_n = 33 + (n-1)(-9)$
65. $t_n = 100 + (n-1)(5)$
66. $t_n = 500 + (n-1)(20)$
67. $t_n = -50 + (n-1)(5)$
68. $t_n = -80 + (n-1)(4)$

Find the indicated arithmetic means.

69. Find the three arithmetic means between 5 and 17. **8, 11, 14**

70. Find the four arithmetic means between 40 and 10. **34, 28, 22, 16**

71. Find the three arithmetic means between 18 and −10. **11, 4, −3**

72. Find the five arithmetic means between −40 and −10. **−35, −30, −25, −15**

73. Find the two arithmetic means between 5.26 and 6.34. **5.62, 5.98**

74. Find the three arithmetic means between 8.24 and 2.8. **6.88, 5.52, 4.16**

75. Find the five arithmetic means between 12 and −6. **9, 6, 3, 0, −3**

76. Find the six arithmetic means between −23 and 5. **−19, −15, −11, −7, −3, 1**

Practice

NAME ______ CLASS ______ DATE ______

Practice 11.2 Arithmetic Sequences

Based on the terms given, state whether or not each sequence is arithmetic. If so, identify the common difference, *d*.

1. 15,18, 21, 24, . . . yes; $d = 3$
2. 2, 5, 10, 17, . . . no
3. 7.2, 9.7, 12.2, 14.7, . . . yes; $d = 2.5$
4. 4, 6, 9, 13.5, . . . no
5. $1, 1\frac{3}{5}, 2\frac{1}{5}, 2\frac{4}{5}, \ldots$ yes; $d = \frac{3}{5}$
6. 8, 5.7, 3.4, 1.1, . . . yes; $d = -2.3$

Write an explicit formula for the *n*th term of each arithmetic sequence.

7. 16, 7, −2, −11, . . . $t_n = 25 - 9n$
8. −15, −7, 1, 9, . . . $t_n = 8n - 23$
9. 13, 16, 19, 22, . . . $t_n = 3n + 10$
10. −25, −13, −1, 11, . . . $t_n = 12n - 37$
11. 9, 20, 31, 42, . . . $t_n = 11n - 2$
12. 8.6, 7.3, 6, 4.7, . . . $t_n = 9.9 - 1.3n$

List the first four terms of each arithmetic sequence.

13. $t_1 = 50; t_n = t_{n-1} + 100$ 50, 150, 250, 350
14. $t_1 = 7.5; t_n = t_{n-1} + 2.5$ 7.5, 10, 12.5, 15
15. $t_1 = -20; t_n = t_{n-1} + 8$ −20, −12, −4, 4
16. $t_n = 40n - 15$ 25, 65, 105, 145
17. $t_n = 0.5n + 8$ 8.5, 9, 9.5, 10
18. $t_n = -12n - 3$ −15, −27, −39, −51

Find the indicated arithmetic means.

19. 3 arithmetic means between −12 and 16 −5, 2, and 9
20. 4 arithmetic means between 40 and 100 52, 64, 76, and 88
21. 2 arithmetic means between 50 and 86 62 and 74
22. 3 arithmetic means between 7 and 21 10.5, 14, and 17.5
23. 3 arithmetic means between 40 and 16 34, 28, and 22
24. 4 arithmetic means between −8 and 22 −2, 4, 10, and 16

77. Examine the pattern of dots below. How many dots will there be in the 14th figure? **27 dots**

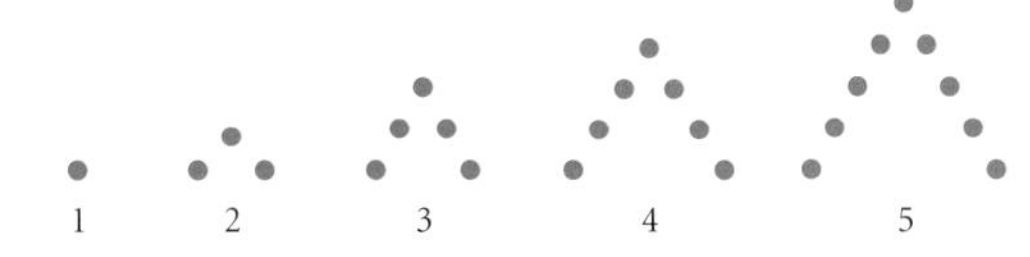

CHALLENGE

78. In a parking lot, each row has 3 more parking spaces than the previous row. If the 1st row has 20 spaces, how many spaces will the 15th row have? **62 spaces**

CONNECTION

79. **GEOMETRY** The lengths of the sides of a certain right triangle form an arithmetic sequence. Show that the triangle is similar to a right triangle with side lengths of 3, 4 and 5.

APPLICATIONS

79. Let a, $a + d$, and $a + 2d$ represent the side lengths ($d \neq 0$ in a right triangle).

80. **DEPRECIATION** A machine that puts labels on bottles is bought by a small company for $10,000. The machine depreciates at the beginning of each year at a rate of $1429 per year.

a. Write a formula for an arithmetic sequence that gives the value of the machine after n years. **$t_n = 10{,}000 - 1429n$**

b. Find the value of the machine at the beginning of the second, third, and sixth year. **$7142, $5713, $1426**

81. **INCOME** The starting salary for a teacher in one school district is $30,000. The salary increases by $800 each year.

a. Write a formula for an arithmetic sequence that gives the salary in the nth year. **$t_n = 30{,}000 + (n - 1)(800)$**

b. Find the salary for the 10th year. **$37,200**

82. **HEALTH** Amanda is beginning a fitness program. During the first week, she will do 25 abdominal exercises each day. Each week she will increase the number of daily exercises by 3.

a. Write a formula for an arithmetic sequence that gives the number of daily exercises done in the nth week. **$t_n = 25 + (n - 1)(3)$**

b. How many daily exercises will Amanda do after 20 weeks? **85 daily exercises**

Look Back

Complete the square for each quadratic expression to form a perfect square trinomial. Then write the new expression as a binomial squared. ***(LESSON 5.4)***

83. $x^2 + 6x$ **$(x + 3)^2$**

84. $x^2 + 20x$ **$(x + 10)^2$**

85. $x^2 + 14x$ **$(x + 7)^2$**

Solve each inequality. Check your solution. ***(LESSON 8.5)***

86. $\frac{-4}{2x + 5} > 0$ **$x < -\frac{5}{2}$**

87. $\frac{x - 1}{x - 4} \geq 2$ **$4 < x \leq 7$**

88 $\frac{(x-4)}{(x+2)} \leq 4$ **$x < -4$ or $x > -2$**

89 $\frac{x^2 - 3}{x^2 + 3} < -\frac{1}{2}$ **$-1 < x < 1$**

79. con't

Case 1: $d > 0$,

so $a < a + d < a + 2d$.
$a^2 + (a + d)^2 = (a + 2d)^2$
$a^2 + a^2 + 2ad + d^2 = a^2 + 4ad + 4d^2$
$a^2 - 2ad - 3d^2 = 0$
$(a - 3d)(a + 1d) = 0$
$(a - 3d) = 0$ or $(a + 1d) = 0$
$a = 3d$ or $a = -d$
Since $d > 0$, $a = -d$ is not a solution.
If $a = 3d$, then $a + d = 4d$ and $a + 2d = 5d$.
The proportion $3/3d = 4/4d = 5/5d$ is true, so the triangles are similar.

Case 2: $d < 0$, so
$a + 2d < a + d < a$.
$(a + 2d)^2 + (a + d)^2 = a^2$
$a^2 + 4ad + 4d^2 + a^2 + 2ad + d^2 = a^2$
$a^2 + 6ad + 5d^2 = 0$
$(a + 5d)(a + 1d) = 0$
$(a + 5d) = 0$ or $(a + 1d) = 0$
$a = -5d$ or $a = -d$
If $a = -d$, then $a + 2d = d$ is not a solution because $d < 0$.
If $a = -5d$, then $a + d = -4d$ and $a + 2d = -3d$. The proportion $3/-3d = 4/-4d = 5/-5d$ is true, so the triangles are similar.

Student Technology Guide

NAME ____________ CLASS ________ DATE ________

Student Technology Guide
11.2 ***Arithmetic Sequences***

Arithmetic sequences are generated by repeated addition (or subtraction). You can use a calculator to generate the terms of an arithmetic sequence.

Example: Find the first four terms of the sequence defined by the recursive formula $t_1 = 7$ and $t_n = t_{n-1} + 5$.

- Enter the first term. Press 7 ENTER.
- To get the second term, press 2nd (-) + 5 ENTER. Then press ENTER two more times in order to see the next two terms.

The first four terms of the sequence are 7, 12, 17, and 22, as shown at right.

```
7
                7
Ans+5
               12
               17
               22
```

Use the recursive formula given to find the first four terms of each sequence.

1. $t_1 = 4, t_n = t_{n-1} + 3$ **4, 7, 10, 13**
2. $t_1 = 0, t_n = t_{n-1} - 0.07$ **0, −0.07, −0.14, −0.21**
3. $t_1 = -3.1, t_n = t_{n-1} + 19.2$ **−3.1, 16.1, 35.3, 54.5**

Many real-world problems can be modeled by arithmetic sequences.

Example: A company buys a photocopier for $6100. If the value of the photocopier depreciates at a rate of $508.33 per year, find its value after 5 years.

- Enter the first term. Press 6100 ENTER.
- Press 2nd (-) - 508.33 ENTER ENTER ENTER ENTER ENTER.

```
             6100
Ans-508.33
          5591.67
          5083.34
          4575.01
          4066.68
          3558.35
```

The photocopier is worth $3558.35 after 5 years.

4. A company buys a pickup truck for $22,800. The truck depreciates at a rate of $3257.14 per year. Find the value of the truck after 4 years. **$9771.44**
5. A city plans to increase the size of its fire department. There are 184 firefighters now. Each month, 3 new firefighters will be added to the department. Find the number of firefighters after 16 months. **232 firefighters**
6. Each row of seats in a theater has 13 more seats than the previous row. Row 1 has 78 seats. Find the number of seats in rows 2, 3, 4, 5, 6, and 7. **91, 104, 117, 130, 143, 156**

Look Beyond

In Exercise 98, students are asked to find the sum of a finite number of terms of an arithmetic sequence. In Lesson 11.3, students will discover an algebraic method for finding the sum of a finite number of terms in an arithmetic sequence.

ALTERNATIVE
Assessment

Portfolio Activity

The Portfolio Activity can be used as preparation for the Chapter Project or as a separate activity. In the Portfolio Activity on this page, students will perform an experiment to find the center of gravity of an object.

Answers to Portfolio Activities can be found in Additional Answers of the Teacher's Edition.

90. $(x-1)^2 + y^2 = 1$; $C(1, 0)$; $r = 1$

91. $(x+2)^2 + (y-5)^2 = 16$; $C(-2, 5)$; $r = 4$

92. $\frac{x^2}{28} + \frac{y^2}{64} = 1$

93. $\frac{x^2}{25} + \frac{(y-2)^2}{9} = 1$

94. $\{F, F\}$; $\frac{1}{4}$

95. $\{MF, FM, FF\}$; $\frac{3}{4}$

Write the standard equation for each circle. Then state the coordinates for its center and give its radius. ***(LESSON 9.3)***

90. $x^2 - 2x + y^2 = 0$

91. $x^2 + 4x + y^2 - 10y + 13 = 0$

Write the standard equation for each ellipse. ***(LESSON 9.4)***

92. foci: $(0, 6), (0, -6)$
vertices: $(0, 8), (0, -8)$

93. foci: $(4, 2), (-4, 2)$
co-vertices: $(0, 5), (0, 1)$

Representing a male child with *M* and a female child with *F*, a sample space for families with 2 children can be written as {*MM, MF, FM, FF*}. Assume that each family below has 2 children. List each event below with the same notation as above. Then find the theoretical probability of each event. ***(LESSON 10.1)***

94. A family has exactly two girls.

95. A family has at least one girl.

96. A family has at most one girl.
$\{MM, MF, FM\}$; $\frac{3}{4}$

97. A family has exactly one girl.
$\{MF, FM\}$; $\frac{1}{2}$

Look Beyond

98. Find the sum of the first eight terms of the given arithmetic sequence.

$$\begin{cases} t_1 = 5 \\ t_n = t_{n-1} + 2 \end{cases}$$

96

Place a wooden 12-inch ruler on a desk so that one end of the ruler is aligned with the edge of the desk. Slowly slide the ruler over the edge of the desk as far as possible without the ruler falling off. The location on the ruler that is at the edge of the desk at this time is the ruler's center of gravity. Record the ruler's center of gravity.

All of the weight of an object can be considered to be concentrated at a single point called the center of gravity. Balancing an object is one way to locate an object's center of gravity.

WORKING ON THE CHAPTER PROJECT

You should now be able to complete Activity 1 of the Chapter Project.

Arithmetic Series

Why *Arithmetic series can be used to solve real-world problems such as finding the total number of cans needed to set up a certain triangular display.*

Objective

- Find the sum of the first n terms of an arithmetic series.

A pattern for stacking cans is shown above. To find the number of cans in a triangular display of 15 rows stacked in this pattern, you can use the sum of an *arithmetic series.* You will solve this problem in Example 1.

CONNECTION
PATTERNS IN DATA

Activity

Exploring Arithmetic Series

You will need: graph paper

The sum $1 + 2 + 3$ can be represented on graph paper as shown in Figure 1. Twice the sum can be represented as shown in Figure 2.

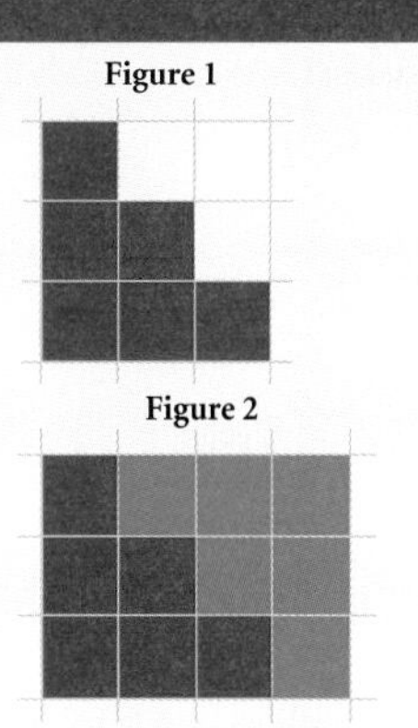

Figure 1

Figure 2

CHECKPOINT ✔

1. Explain how Figure 1 suggests the following:
$$1 + 2 + 3 = \frac{3(4)}{2} = \frac{3(1+3)}{2}$$
2. Draw a similar figure to represent twice the sum $1 + 2 + 3 + 4 + 5 + 6$, and write a similar equation for it.
3. Repeat Step 2 for twice the sum $1 + 2 + 3 + 4 + 5 + \cdots + 10$.

An **arithmetic series** is the indicated sum of the terms of an arithmetic sequence. Consider the sequence 1, 2, 3, . . . The sum of the first six terms of this sequence is denoted by S_6.

$$S_6 = 1 + 2 + 3 + 4 + 5 + 6 = 21$$

You can derive a formula for the sum of the first n terms of an arithmetic series by using the Addition Property of Equality, as shown below.

$$S_n = t_1 + (t_1 + d) + (t_1 + 2d) + \cdots + (t_n - d) + t_n$$

Because $t_n - t_{n-1} = d$, $t_{n-1} = t_n - d$.

$$S_n = t_n + (t_n - d) + (t_n - 2d) + \cdots + (t_1 + d) + t_1$$

$$2S_n = (t_1 + t_n) + (t_1 + t_n) + (t_1 + t_n) + \cdots + (t_1 + t_n) + (t_1 + t_n)$$

$(t_1 + t_n)$ is added n times

$$2S_n = n(t_1 + t_n)$$

$$S_n = \frac{n(t_1 + t_n)}{2}$$

Alternative Teaching Strategy

USING ALGORITHMS Have students refer to the pattern for stacking cans described at the beginning of the lesson. Write the series $1 + 2 + 3 + 4 + 5 + 6$ on the board or overhead. Have students find the sum of the first and last terms. $1 + 6 = 7$ Now have them sum the second and next-to-last-terms. $2 + 5 = 7$ Finally, have them sum the third and fourth terms. $3 + 4 = 7$ Be sure that students notice they obtained 3 sums of 7. Show them that the sum, S_6, is $3 \cdot 7 = 21$. Have the students find the sum of the first 15 counting numbers in the same manner. Now show students that they can simplify the process of finding the sum of an arithmetic sequence by multiplying the number of terms in the sequence by the average of the first and last terms of the sequence. Then generalize this pattern for the sum of the first n terms of an arithmetic sequence that represents the arithmetic series, S_n.

Prepare

NCTM PRINCIPLES & STANDARDS 1–3, 5, 8–10

QUICK WARM-UP

Find the sum of each series.

1. $\sum_{n=1}^{4} 6$ 24
2. $\sum_{k=1}^{6} (k + 2)$ 33
3. $\sum_{i=1}^{5} 4i$ 60
4. $\sum_{m=1}^{9} (2m - 5)$ 45

Find the indicated term of each arithmetic sequence.

5. 12th term, given $t_2 = 6$ and $t_8 = 24$ 36
6. 7th term, given $t_5 = 22$ and $t_{11} = 46$ 30

Also on Quiz Transparency 11.3

Teach

Why In many situations, it is the sum of the terms of an arithmetic sequence that best defines a situation rather than then the individual terms. Ask students to consider which would be better for them, getting a 10-week summer job that includes an automatic $5.00 raise each week, or receiving a $45.00 bonus.

☞ For Activity Notes and the answer to the Checkpoint, see page 708.

Math CONNECTION

PATTERNS IN DATA In the picture shown on page 707, students are shown cans stacked in a triangular pattern.

Activity Notes

You may want to point out that the sum of each finite series equals the product of the number of terms and the average of the first and last terms.

CHECKPOINT ✔

1. **Duplicate figure 1, turn the copy upside-down, and put the pieces together. They form a rectangle with side lengths of 3 and 4 units and an area of 12 square units. So, the area of Figure 1 is half that of the rectangle, or $\frac{3(3+1)}{2}$.**

ADDITIONAL EXAMPLE 1

Refer to the pattern for stacking cans described at the beginning of the lesson. **How many cans are in an 11-row display of cans stacked in this pattern?** 176

Teaching Tip

TECHNOLOGY When using a TI-83 or TI-82 graphics calculator to check the sum in Example 1, use the following keystrokes:

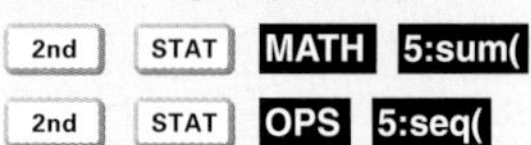

2nd STAT MATH 5:sum(

2nd STAT OPS 5:seq(

1+(X–1)3,X,1,15,1))

The last parameter in the entry, 1, represents the increment of the series. The increment is an optional parameter and, when omitted, defaults to 1.

Sum of the First n Terms of an Arithmetic Series

The sum, S_n, of the first n terms of an arithmetic series with first term t_1 and nth term t_n is given by $S_n = n\left(\frac{t_1 + t_n}{2}\right)$.

CHECKPOINT ✔ Using the sequence 1, 2, 3, . . . , find S_6 by following the procedure used in the derivation of S_n.

EXAMPLE 1 Refer to the pattern for stacking cans described at the beginning of the lesson. **How many cans are in a 15-row display of cans stacked in this pattern?**

CONNECTION
PATTERNS IN DATA

SOLUTION

1. Begin with the formula for the sum of the first n terms of an arithmetic series.

$$S_n = n\left(\frac{t_1 + t_n}{2}\right)$$

$$S_{15} = 15\left(\frac{1 + t_{15}}{2}\right) \quad \textit{Substitute 1 for } t_1 \textit{ and 15 for } n.$$

2. Find t_{15}, using the recursive formula.

$$t_n = t_1 + (n - 1)d$$

$$t_{15} = 1 + (15 - 1)3 \quad \textit{Notice from the pattern that } d = 3.$$

$$t_{15} = 43$$

3. Substitute 43 for t_{15} in the formula.

$$S_{15} = 15\left(\frac{1 + t_{15}}{2}\right)$$

$$S_{15} = 15\left(\frac{1 + 43}{2}\right)$$

$$S_{15} = 330$$

TECHNOLOGY
GRAPHICS CALCULATOR
Keystroke Guide, page 759

CHECK

Use the summation and the sequence commands of a graphics calculator. On this calculator, the summation command tells the calculator to add the items in parentheses, and the sequence command inside the parentheses describes the related sequence.

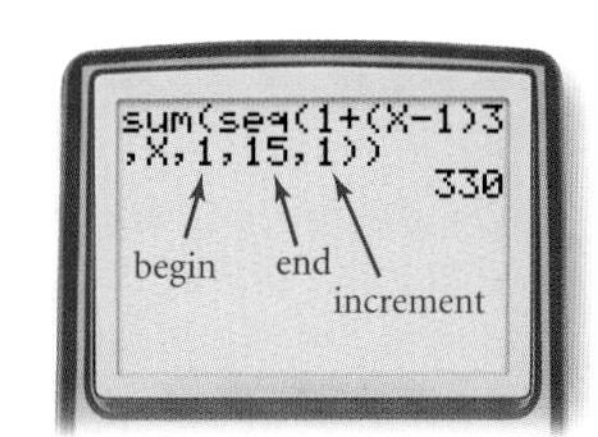

Thus, the 15-row display contains 330 cans.

The formula $S_n = n\left(\frac{t_1 + t_n}{2}\right)$ gives S_n in terms of n, t_1, and t_n. Notice that the sum is the number of terms, n, times the average term, $\frac{t_1 + t_n}{2}$. By substituting $t_1 + (n - 1)d$ for t_n in the formula for S_n, you can write a formula that can be used when the last term of the series is not known.

$$S_n = n\left(\frac{t_1 + t_n}{2}\right)$$

$$S_n = n\left[\frac{t_1 + t_1 + (n - 1)d}{2}\right]$$

$$S_n = n\left[\frac{2t_1 + (n - 1)d}{2}\right]$$

Interdisciplinary Connection

BUSINESS Tell students that a firm will start a new employee at $24,000 a year and will guarantee a $1500 raise each year, while a second firm will start a new employee at $26,000 and will guarantee an $1100 raise each year. Have students find how much each firm will pay a new employee over a 10-year period.
first firm: $307,500; second firm: $309,500

Inclusion Strategies

USING MANIPULATIVES Have students construct a trapezoidal display of cans in which each row has a constant number of fewer cans than the row below. Then have students restack the cans in an equivalent rectangular display. Have them discuss how the dimensions of the rectangular display relate to the dimensions of the trapezoidal display.

EXAMPLE 2 Given $3 + 12 + 21 + 30 + \cdots$, find S_{25}.

SOLUTION

Substitute 3 for t_1, 9 for d, and 25 for n.

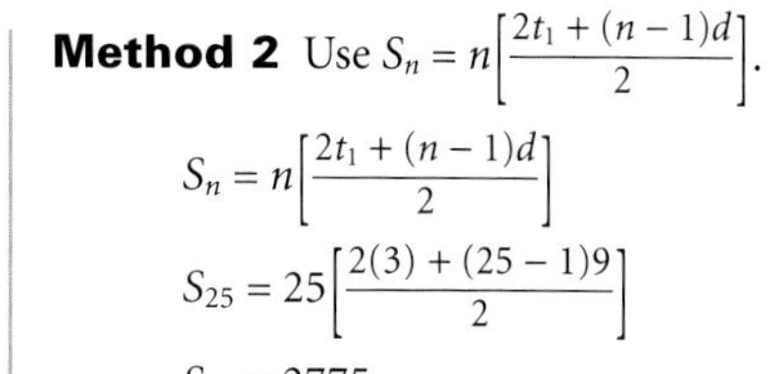

Method 1 Use $S_n = n\left(\frac{t_1 + t_n}{2}\right)$.

First find t_{25}.

$$t_n = t_1 + (n-1)d$$
$$t_{25} = 3 + (25-1)9$$
$$t_{25} = 219$$

Then find S_{25}.

$$S_{25} = 25\left(\frac{3+219}{2}\right) = 2775$$

Method 2 Use $S_n = n\left[\frac{2t_1 + (n-1)d}{2}\right]$.

$$S_n = n\left[\frac{2t_1 + (n-1)d}{2}\right]$$
$$S_{25} = 25\left[\frac{2(3) + (25-1)9}{2}\right]$$
$$S_{25} = 2775$$

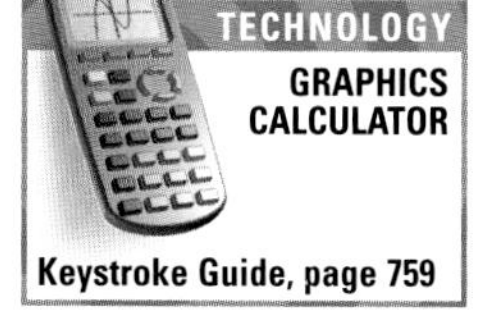

Keystroke Guide, page 759

CHECK

Use the summation and the sequence commands on a graphics calculator.

Thus, $S_{25} = 2775$.

TRY THIS Given $(-16) + (-12) + (-8) + (-4) + \cdots$, find S_{20}.

EXAMPLE 3 Evaluate $\sum_{k=1}^{12}(6 - 2k)$.

SOLUTION

Method 1

This summation notation describes the summation of the first 12 terms of the arithmetic series that begins $4 + 2 + 0 + (-2) + \cdots$, in which $t_1 = 4$ and $d = -2$.

First find t_{12}.

$$t_{12} = 4 + (12-1)(-2) = -18$$

Then find S_{12}.

$$S_{12} = 12\left[\frac{4 + (-18)}{2}\right] = -84$$

Method 2

You can use the properties of series and the formulas for constant and linear series to find the sum.

$$\sum_{k=1}^{12}(6-2k) = \sum_{k=1}^{12}6 + \sum_{k=1}^{12}-2k$$
$$= 12 \cdot 6 - 2\sum_{k=1}^{12}k$$
$$= 72 - 2\left[\frac{12(1+12)}{2}\right]$$
$$= 72 - 156$$
$$= -84$$

Keystroke Guide, page 759

CHECK

Use the sum and sequence commands on a graphics calculator. The sequence is given by $t_n = 6 - 2n$.

Thus, $\sum_{k=1}^{12}(6 - 2k) = -84$.

Enrichment

Tell students that an object dropped from rest falls a distance of $\frac{1}{2}g(2n - 1)$ feet during the nth second, where g is acceleration due to gravity. Let $g = 32$ feet per second per second, and have students find the total distance that the object falls in the first 4 seconds after it is dropped. 256 ft

Reteaching the Lesson

USING MANIPULATIVES Have students fold a strip of paper in half, unfold it, and count the number of rectangles. There are 1 + 2, or 3, rectangles. Now have them refold the strip in half, then fold in half again, and unfold the paper. The total number of rectangles after the second fold is 1 + 2 + 3 + 4, or 10. Have them fold the strip three successive times, guess the number of rectangles created, and then count them. 36 Point out that they can use the formula for the first n terms of an arithmetic series to find the number of rectangles after n successive folds.

ADDITIONAL EXAMPLE 2

Given 16, 12, 8, 4, . . . , find S_{11}.
−44

Teaching Tip

TECHNOLOGY When using a TI-82 or TI-83 graphics calculator to check the sum in Example 2, see the Teaching Tip after Additional Example 1 on the opposite page for the procedure to use when entering **sum(seq(9X–6,X,1,25,1)).** Press ENTER to display the sum, 2775.

TRY THIS
440

ADDITIONAL EXAMPLE 3

Evaluate $\sum_{k=1}^{21}(5 + 4k)$ 1029

Teaching Tip

TECHNOLOGY When using a TI-82 or TI-83 graphics calculator to check the sum in Example 3, see the Teaching Tip after Additional Example 1 on the previous page for the procedure to use when entering **sum(seq(6-2X,X,1,12,1)).** Press ENTER to display the sum, −84.

Teaching Tip

Students may wonder when to use summation notation and when to use S_n notation. Summation notation is used to represent an expression that is a series. S_n notation is used to represent a value that is the sum of a series.

TRY THIS Evaluate $\sum_{k=1}^{15}(22 - 7k)$.

CRITICAL THINKING If an arithmetic sequence has a nonzero constant difference, can any sum of terms in the corresponding arithmetic series be 0? Justify your response with examples.

Exercises

Communicate

1. What information about an arithmetic series is needed to find its sum?

2. Explain how to use summation notation to express the following series:

$$2 + 4 + 6 + 8 + 10 + 12$$

3. In how many different ways can you represent the series $2 + 4 + 6 + 8$? Explain.

Guided Skills Practice

APPLICATION

4. MERCHANDISING Refer to the pattern for stacking cans described at the beginning of the lesson. How many cans are in an 18-row triangular display of cans stacked in this pattern? ***(EXAMPLE 1)*** **477 cans**

5. Given $5 + 12 + 19 + 26 + \cdots$, find S_{26}. ***(EXAMPLE 2)*** **2405**

6. Evaluate $\sum_{k=1}^{10}(20 - 3k)$. ***(EXAMPLE 3)*** **35**

Practice and Apply

internet connect
Homework Help Online
Go To: **go.hrw.com**
Keyword:
MB1 Homework Help
for Exercises 7–35

Use the formula for an arithmetic series to find each sum.

7. $2 + 4 + 6 + 8 + 10$ **30**

8. $5 + 10 + 15 + 20$ **50**

9. $5 + 13 + 21 + 29$ **68**

10. $13 + 17 + 21 + 25$ **76**

11. $-100 + (-96) + (-92) + (-88)$ **−376**

12. $-50 + (-47) + (-44) + (-41)$ **−182**

13. $1 + 2 + 3 + 4 + \cdots + 11$ **66**

14. $15 + 21 + 27 + 33 + \cdots + 63$ **351**

15. $-4 + (-13) + (-22) + \cdots + (-76)$ **−360**

16. $10 + 8 + 6 + 4 + 2 + \cdots + (-4)$ **24**

17. Find the sum of the first 300 natural numbers. **45,150**

18. Find the sum of all even numbers from 2 to 200 inclusive. **10,100**

19. Find the sum of the multiples of 3 from 3 to 99 inclusive. **1683**

20. Find the sum of the multiples of 9 from 9 to 657 inclusive. **24,309**

TRY THIS
−510

CRITICAL THINKING
yes; samples: $\sum_{k=1}^{3}(-4 + 2k) = 0$ and $\sum_{n=1}^{5} -10 + (n - 1)5 = 0$

Assess

Selected Answers
Exercises 4–6, 7–65 odd

ASSIGNMENT GUIDE

In Class	1–6
Core	7–49 odd, 53
Core Plus	8–54 even
Review	56–66
Preview	67–69

Extra Practice can be found beginning on page 940.

Practice

NAME ________ CLASS ________ DATE ________

Practice
11.3 ***Arithmetic Series***

Use the formula for an arithmetic series to find each sum.

1. $62 + 66 + 70 + 74 + 78$ 350
2. $\frac{1}{2} + 2\frac{1}{2} + 3\frac{1}{2} + 4\frac{1}{2} + 5\frac{1}{2}$ $16\frac{1}{2}$
3. $-30 - 27 - 24 - 21 - \cdots - 3$ −165
4. $110 + 125 + 140 + \cdots + 305$ 2905
5. $14 + 17 + 20 + \cdots + 65$ 711
6. $33 + 38 + 43 + \cdots + 123$ 1482

Find each sum.

7. the sum of the first 225 natural numbers 25,425
8. the sum of the first 15 multiples of 3 360
9. the sum of the first 25 multiples of 4 1300
10. the sum of the multiples of 5 from 75 to 315, inclusive 9555
11. the sum of the multiples of 7 from 84 to 371, inclusive 9555

For each arithmetic series, find S_{22}.

12. $-6 + (-4) + (-2) + 0 + \cdots$ 330
13. $3 + 7 + 11 + 15 + \cdots$ 990
14. $-24 + (-21) + (-18) + (-15) + \cdots$ 165
15. $3 + 3\frac{3}{4} + 4\frac{1}{2} + 5\frac{1}{4} + \cdots$ 239.25
16. $18 + 8 + (-2) + (-12) + \cdots$ −1914
17. $3\sqrt{5} + 5\sqrt{5} + 7\sqrt{5} + 9\sqrt{5} + \cdots$ $528\sqrt{5} \approx 1180.64$

Evaluate.

18. $\sum_{n=1}^{6}(2n + 7)$ 84
19. $\sum_{j=1}^{8}(-3j - 3)$ −132
20. $\sum_{k=1}^{10}(10k + 4)$ 590
21. $\sum_{m=1}^{12}(-7 + 4m)$ 228
22. $\sum_{b=1}^{15}(13 + 5b)$ 795
23. $\sum_{i=1}^{9}(-8i + 1)$ −351

For each arithmetic series, find S_{25}.

21. $3 + 7 + 11 + 15 + \cdots$ **1275**

22. $25 + 24 + 23 + 22 + \cdots$ **325**

23. $4 + 14 + 24 + 34 + \cdots$ **3100**

24. $6 + 2 + (-2) + (-6) + \cdots$ **–1050**

25. $5 + 10 + 15 + 20 + \cdots$ **1625**

26. $3 + 6 + 9 + 12 + \cdots$ **975**

27. $-12 + (-6) + 0 + 6 + \cdots$ **1500**

28. $-17 + (-12) + (-7) + (-2) + 3 + \cdots$ **1075**

29. $10 + 20 + 30 + 40 + 50 + \cdots$ **3250**

30. $100 + 200 + 300 + 400 + 500 + \cdots$ **32,500**

31. $-10 + (-15) + (-20) + (-25) + (-30) + \cdots$ **–1750**

32. $-20 + (-22) + (-24) + (-26) + (-28) + \cdots$ **–1100**

33. $\sqrt{2} + 2\sqrt{2} + 3\sqrt{2} + 4\sqrt{2} + 5\sqrt{2} + \cdots$ **$325\sqrt{2}$**

34. $5\sqrt{3} + 10\sqrt{3} + 15\sqrt{3} + 20\sqrt{3} + \cdots$ **$1625\sqrt{3}$**

35. $2\pi + 3\pi + 4\pi + 5\pi + 6\pi + \cdots$ **350π**

Evaluate.

36. $\sum_{m=1}^{5}(15 - 2m)$ **45**

37. $\sum_{j=1}^{8}(30 - 2j)$ **168**

38. $\sum_{k=1}^{6}(10 + k)$ **81**

39. $\sum_{m=1}^{4}(5 + m)$ **30**

40. $\sum_{n=1}^{5}(100 - 5n)$ **425**

41. $\sum_{n=1}^{5}(60 - 4n)$ **240**

42. $\sum_{j=1}^{6}(1000 + 25j)$ **6525**

43. $\sum_{j=1}^{5}(500 + 2j)$ **2530**

44. $\sum_{k=1}^{10}(40 - 2k)$ **290**

45. $\sum_{m=1}^{10}(600 - 10m)$ **5450**

46. $\sum_{n=1}^{50}(500 + 20n)$ **50,500**

47. $\sum_{k=1}^{10}\left(\frac{2}{5} + \frac{1}{5}k\right)$ **15**

48. $\sum_{j=1}^{12}\left(\frac{1}{3} + \frac{1}{6}j\right)$ **17**

49. $\sum_{k=1}^{20}(2.2 + 3.1k)$ **695**

50. $\sum_{k=1}^{10}(1.2 - 4.1k)$ **–213.5**

CHALLENGES

51. For a certain arithmetic series, $S_4 = 50$ and $S_5 = 75$. Find the first five terms. **5, 10, 15, 20, 25**

52. Find the number of and values of the arithmetic means that should be inserted between 1 and 50 in order to make the sum of the resulting series equal to 459.

52. 16 means; $\frac{66}{17}, \frac{115}{17}, \frac{164}{17}, \ldots, \frac{801}{17}$

APPLICATION

53. INVENTORY Pipes are stacked as shown at right.

a. Find the number of pipes in a stack of 6 rows if the bottom row contains 9 pipes. **39 pipes**

b. Find the number of pipes in a stack of 7 rows if the bottom row contains 10 pipes. **49 pipes**

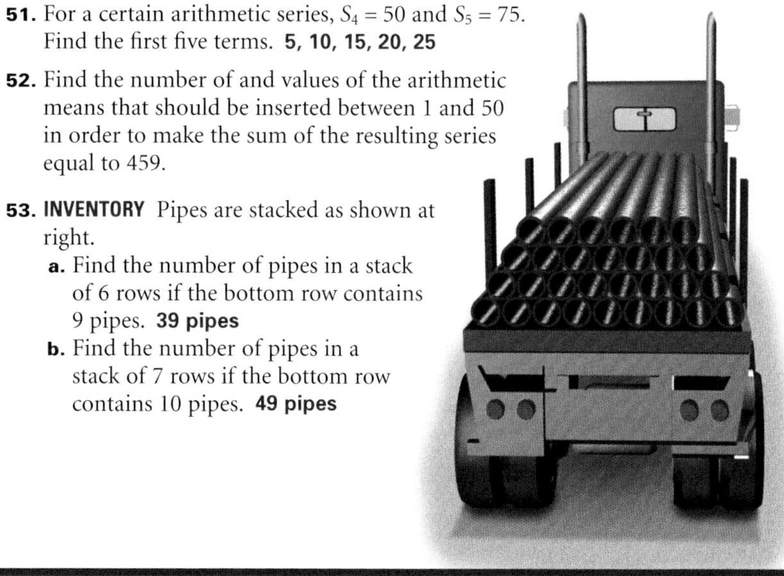

Error Analysis

Many students incorrectly use the terms *sequence*, *series*, and *sum*. To help students with the meaning of these terms, give them a recursive formula for an arithmetic sequence, such as $t_1 = 4$ and $t_n = t_{n-1} + 3$. Have them list some terms of the sequence and find the explicit formula. **4, 7, 11, 15, . . . ; $t_n = 3n + 1$** Then have them find the following:

- S_1 **4**
- S_2 **4 + 7 = 11**
- S_3 **4 + 7 + 10 = 21**
- S_4 **4 + 7 + 10 + 13 = 34**

The expressions 4, 4 + 7, 4 + 7 + 10, and 4 + 7 + 10 + 13 are *series*, while 4, 11, 21, and 34 are *sums*.

NAME ________ CLASS ________ DATE ________

Student Technology Guide
11.3 *Arithmetic Series*

An arithmetic series is a sum of terms from an arithmetic sequence. You can use the sequence and sum features of a graphics calculator to evaluate an arithmetic series.

Example: Find S_{16} for the sequence 1, 4, 7, 10, . . .

- Find a formula for the arithmetic sequence. For this sequence, the first term is 1 and the common difference is 3.
  $t_n = t_1 + (n - 1)d$
  $= 1 + (n - 1)3$
  $= 3n - 2$
- Press 2nd STAT MATH 5:sum(ENTER 2nd STAT OPS 5:seq(ENTER to access the sum and sequence features.
- Press 3 X,T,θ,n – 2 , X,T,θ,n , 1 , 16 , 1)) to enter the formula for the sequence, the variable name, the number of the first term in the series, the number of the last term in the series, and the increment by which the variable is increasing. Press ENTER.

```
sum(seq(3X-2,X,1
,16,1))
                376
```

$S_{16} = 376$.

For each arithmetic sequence, find S_{25}.

1. –10, –5, 0, 5, . . . **1250**
2. 19, 12, 5, –2, . . . **–1625**
3. –8, –10, –12, –14, . . . **–800**

Example: Evaluate $\sum_{k=1}^{20}(2k + 1)$.

- Press 2nd STAT MATH 5:sum(ENTER 2nd STAT OPS 5:seq(ENTER 2 X,T,θ,n + 1 , X,T,θ,n , 1 , 20 , 1)) ENTER.

```
sum(seq(2X+1,X,1
,20,1))
                440
```

$\sum_{k=1}^{20}(2k + 1) = 440$.

Find each sum.

4. $\sum_{k=1}^{14}(k + 4)$ **161**
5. $\sum_{k=1}^{25}(2k - 3)$ **575**
6. $\sum_{k=1}^{15}(-3k + 8)$ **–240**

Exercise 67 introduces students to geometric sequences. In Exercises 68 and 69, students graph a geometric sequence and describe the graph. Geometric sequences will be covered in Lesson 11.4.

68.

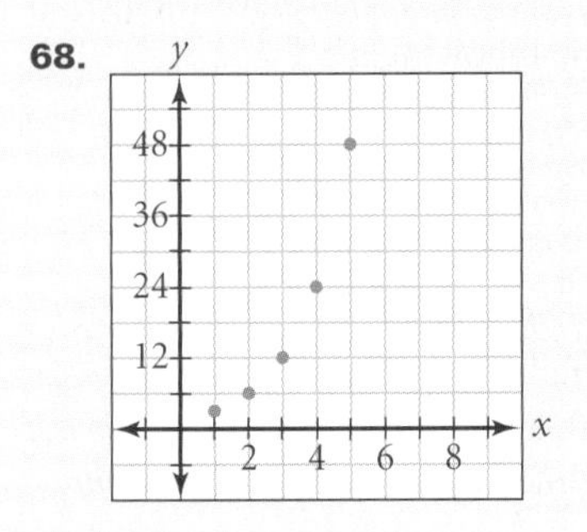

exponential curve

69.

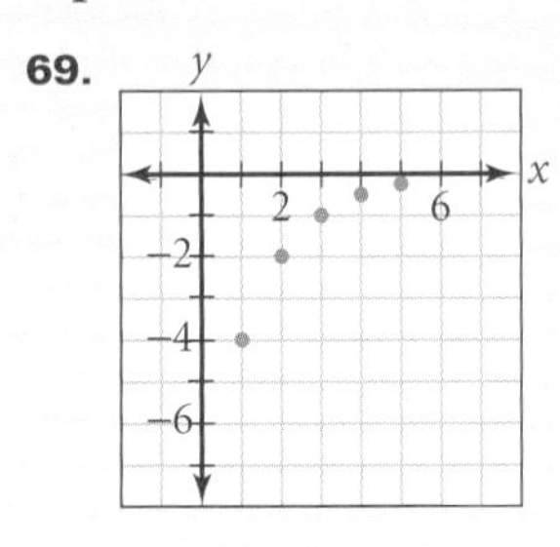

exponential curve

APPLICATIONS

54. MERCHANDISING A pattern for stacking cans is shown at right.

a. How many cans are in a 7-row display? **28 cans**

b. If 66 cans are to be stacked in this pattern, how many rows will the display have? **11 rows**

55. ENTERTAINMENT A marching band formation consists of 8 rows. The first row has 5 musicians, the second has 7, the third has 9, and so on. How many musicians are in the last row? How many musicians are there in all? **19 musicians; 96 musicians**

Look Back

Write each pair of parametric equations as a single equation in x and y. *(LESSON 3.6)*

56. $\begin{cases} x(t) = -3t \\ y(t) = t - 6 \end{cases}$ $\mathbf{y = -\frac{1}{3}x - 6}$

57. $\begin{cases} x(t) = t + 2 \\ y(t) = |t - 2| \end{cases}$ $\mathbf{y = |x - 4|}$

58. $\begin{cases} x(t) = 2t - 1 \\ y(t) = 3t + 1 \end{cases}$ $\mathbf{y = \frac{3}{2}x + \frac{5}{2}}$

Write each product as a polynomial expression in standard form. *(LESSON 7.3)*

59. $2x(3x^2 - 5x^3 + 2x - 6)$
$\mathbf{-10x^4 + 6x^3 + 4x^2 - 12x}$

60. $(x - 3)^2(x^2 - 2x + 5)$
$\mathbf{x^4 - 8x^3 + 26x^2 - 48x + 45}$

Write an equation to represent each relationship. Use k as the constant of variation. *(LESSON 8.1)*

61. y varies jointly as x and z and inversely as the square of m. $\mathbf{y = k\frac{xz}{m^2}}$

62. y varies directly as x^2 and inversely as z^3. $\mathbf{y = k\frac{x^2}{z^3}}$

Find the domain for each radical function. *(LESSON 8.6)*

63. $f(x) = 2\sqrt{x^2 + 36}$ **all real numbers**

64. $f(x) = \sqrt{2(x - 3)} + 1$ $\mathbf{x \geq 3}$

65. In how many ways can you choose 6 objects from a collection of 30 distinct objects, if order does not matter? ***(LESSON 10.3)*** **593,775**

66. Two number cubes are rolled. The sum of the numbers appearing on the top faces is recorded. What is the probability that the number rolled on one cube is 4 given that the sum of the numbers is 6? ***(LESSON 10.6)*** $\mathbf{\frac{2}{5}}$

Look Beyond

67. The first term of a certain sequence is 2. Each successive term is formed by doubling the previous term. Write the first eight terms of the sequence.
2, 4, 8, 16, 32, 64, 128, 256

Graph each sequence, and describe the graph.

68.

x	1	2	3	4	5
t_n	3	6	12	24	48

69.

x	1	2	3	4	5
t_n	−4	−2	−1	$-\frac{1}{2}$	$-\frac{1}{4}$

68. exponential
69. exponential

Geometric Sequences

Why *Geometric sequences can be used to model real-world events such as the depreciation of an automobile.*

Objectives

- Recognize geometric sequences, and find the indicated term of a geometric sequence.
- Find the geometric means between two numbers.

APPLICATION
DEPRECIATION

An automobile that cost $12,500 depreciates, and its value at the end of a given year is 80% of its value at the end of the preceding year. What is it worth after 10 years? You can answer this question by using a *geometric sequence*.
You will answer this question in Example 2.

Two examples of geometric sequences and their graphs are shown below.

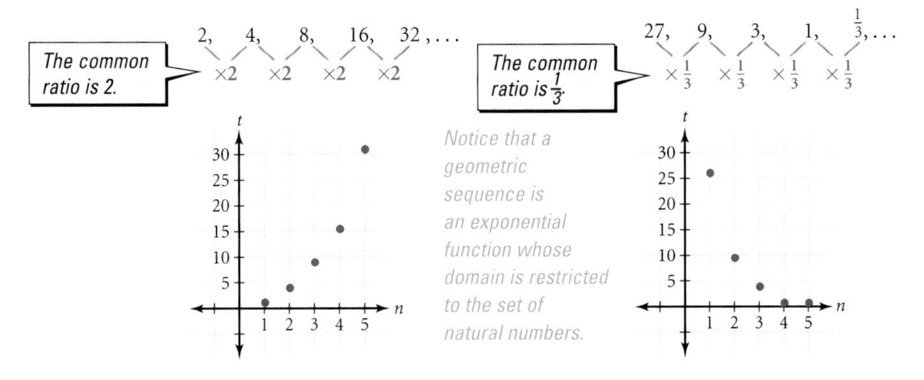

A **geometric sequence** is a sequence in which the ratio of successive terms is the same number, r, called the **common ratio**. That is, if t_n is any term in a geometric sequence, then the following is true:

$$\frac{t_n}{t_{n-1}} = r, \text{ or } t_n = rt_{n-1}, \text{ where } n \geq 2$$

The formula for the general term, $t_n = rt_{n-1}$, where $n \geq 2$, is a recursive formula because it gives the nth term of a sequence in relation to the previous term and r. You can also write an explicit formula for the nth term of a sequence by using the first term and r.

Alternative Teaching Strategy

USING TABLES Give students the table below.

n	1	2	3	4	5
t_n	6	3			

Show students that the ratio between consecutive terms is 0.5, and explain how to rewrite the first two values in the table in terms of this ratio. $6(0.5)^0$, $6(0.5)^1$ Then ask them to use the pattern to extend the table at right, giving the general formula for t_n in terms of t_1 and r, the common ratio.

1	2	3	4	...	n
t_1	t_2	t_3	t_4	...	t_n
6	3	1.5	0.75	...	
$6(0.5)^0$	$6(0.5)^1$	$6(0.5)^2$	$6(0.5)^3$	...	$t_1r^{n-1} = 6(0.5)^{n-1}$

Have them use the formula for t_n to find t_{10} and check their answer by extending the sequence. $t_{10} = 0.01171875$

Prepare

NCTM PRINCIPLES & STANDARDS 1–4, 6–10

QUICK WARM-UP

Find the fifth term of each arithmetic sequence.

1. $t_1 = 4$ and $t_n = t_{n-1} + 6$ 28
2. $t_1 = -7$ and $t_n = t_{n-1} - 3$ −19

Find the indicated term of each arithmetic sequence.

3. 11th term, given $t_2 = 10$ and $t_7 = 35$ 55
4. 15th term, given $t_6 = 5$ and $t_{12} = -19$ −31
5. Find three arithmetic means between −15 and 37. −2, 11, 24
6. Find three arithmetic means between 1 and −27. −6, −13, −20

Also on Quiz Transparency 11.4

Teach

Why Often in real-world situations you can find sequences whose consecutive terms have a common ratio. These are geometric sequences. Tell students that the decrease in a used car's value from year to year, which is called depreciation, is often determined by such a sequence. Discuss how used car prices are calculated.

Use Teaching Transparency 44.

ADDITIONAL EXAMPLE 1

Find the seventh term of the sequence defined by $t_1 = 5$ and $t_n = 2t_{n-1}$. 320

TRY THIS
−40,960

ADDITIONAL EXAMPLE 2

A car that costs $26,500 depreciates, and its value at the end of a given year is 90% of its value at the end of the preceding year.

a. **Use a recursive method to find the car's value after 5 years.**
$15,647.99

b. **Use the formula for the *n*th term of a geometric sequence to find the value of the car after 8 years.**
$11,407.38

Teaching Tip

There are several applications of series in which the initial value is the first term of the sequence, and the value after 1 year (or time period) is the second term. Students should be aware of this type of situation.

PROBLEM SOLVING **Look for a pattern** in the recursive formula for the geometric sequence defined by $t_1 = 3$ and $t_n = 4t_{n-1}$, where $n \geq 2$.

Recursive formula	Pattern
$t_1 = 3$	$t_1 = 4^0 \cdot 3$
$t_2 = 4 \cdot 3$	$t_2 = 4^1 \cdot 3$
$t_3 = 4(4 \cdot 3)$	$t_3 = 4^2 \cdot 3$
$t_4 = 4(4 \cdot 4 \cdot 3)$	$t_4 = 4^3 \cdot 3$

From the pattern, the explicit formula is $t_n = 4^{n-1} \cdot 3$, where $n \geq 1$.

*n*th Term of a Geometric Sequence

The *n*th term, t_n, of a geometric sequence whose first term is t_1 and whose common ratio is r is given by the explicit formula $t_n = t_1r^{n-1}$, where $n \geq 1$.

EXAMPLE 1 **Find the fifth term of the sequence defined by the recursive formula $t_1 = 8$ and $t_n = 3t_{n-1}$.**

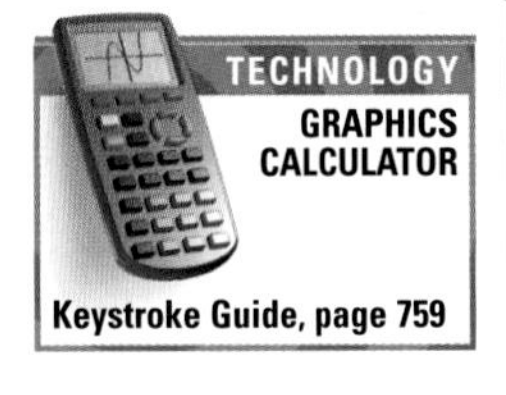

TECHNOLOGY GRAPHICS CALCULATOR
Keystroke Guide, page 759

SOLUTION

This is a geometric sequence in which $t_1 = 8$ and $r = 3$. Use the explicit formula for the *n*th term of a geometric sequence to find the fifth term.

$$t_n = t_1r^{n-1}$$
$$t_5 = 8(3)^{5-1}$$
$$t_5 = 648$$

The fifth term of the sequence is 648.

CHECK

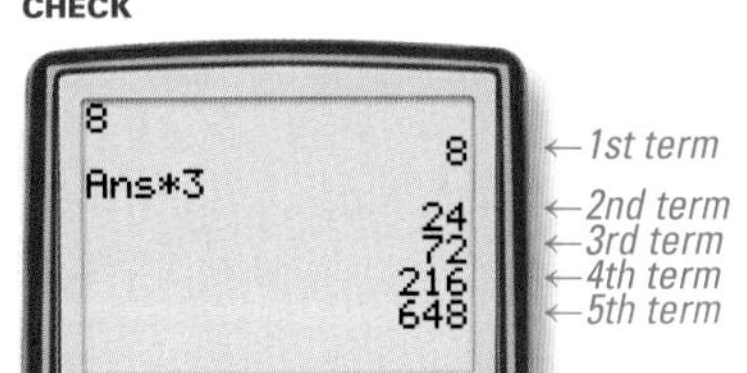

TRY THIS Find the eighth term of the sequence defined by $t_1 = 2.5$ and $t_n = -4t_{n-1}$.

EXAMPLE 2 **Refer to the automobile described at the beginning of the lesson.**

a. Use a recursive formula to find the value of the automobile after 4 years.
b. Use the explicit formula for the *n*th term of a geometric sequence to find the value of the automobile after 10 years.

APPLICATION
DEPRECIATION

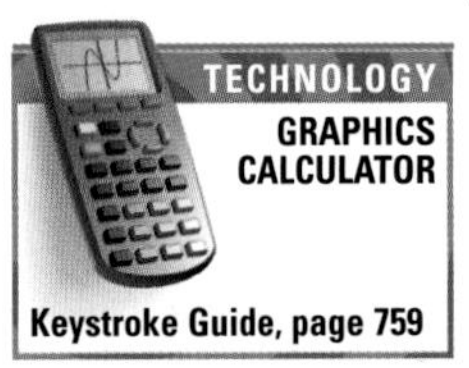

TECHNOLOGY GRAPHICS CALCULATOR
Keystroke Guide, page 759

SOLUTION

a. Initial value (year 0): 12,500

year 1: 12,500(0.80) = 10,000
year 2: 10,000(0.80) = 8000
year 3: 8000(0.80) = 6400
year 4: 6400(0.80) = 5120

The automobile's value after 4 years is $5120.

CHECK

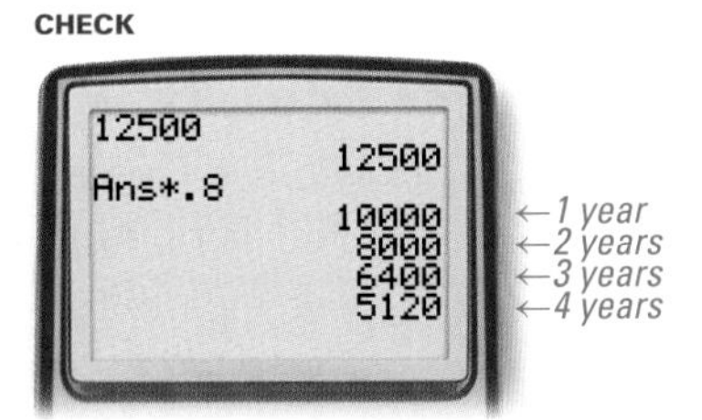

Interdisciplinary Connection

ECONOMICS An investment analyst studied a company's growth over the first 5 years of its existence. If the company's revenue for its first year of operation was $1,500,000 and its revenue increased at an annual rate of 1.5% per year, have students use a recursive method to find the company's annual revenue, to the nearest dollar, after 5 years. ≈$1,592,045

Inclusion Strategies

USING VISUAL MODELS Have students plot the ordered pairs (n, t_n) for the two sequences shown in Example 3 and compare the scatter plots with the graph of $f(x) = 4(3)^{x-1}$. Be sure they notice that if $r > 0$ and $r \neq 1$, the scatter plot of the geometric sequence lies on the graph of the related exponential function $f(x) = t_1r^{x-1}$. If $r < 0$, the scatter plot of the points (n, t_n) of the geometric sequence oscillate from below the *x*-axis to above the *x*-axis because even powers of r are positive and odd powers of r are negative. (Note: The function $f(x) = t_1r^{x-1}$ is undefined for $r < 0$.)

b. At year 10, $n = 11$. Find t_{11}.

$$t_n = t_1 r^{n-1}$$
$$t_{11} = 12{,}500(0.8^{11-1})$$
$$t_{11} \approx 1342.18$$

The automobile's value after 10 years is $1342.18.

TRY THIS An automobile that costs $12,500 depreciates such that its value at the end of a given year is 76% of its value at the end of the preceding year. Use the explicit formula for the *n*th term of a geometric sequence to find the automobile's value after 10 years.

TRY THIS
$803.61

ADDITIONAL EXAMPLE 3

Find the 10th term of the geometric sequence in which $t_2 = 48$ and $t_5 = 384$.
12,288

TRY THIS
$\frac{15}{64}$

If you know any two terms of a geometric sequence, you can often write all of the terms. However, Example 3 shows that it is possible for two geometric sequences to share terms.

EXAMPLE 3 **Find the eighth term of the geometric sequence in which $t_3 = 36$ and $t_5 = 324$.**

SOLUTION

1. Find the common ratio, *r*.

n	3	4	5
t_n	36	?	324

(ratio r between consecutive terms)

$$\rightarrow 36r^2 = 324$$
$$r = \pm\sqrt{\frac{324}{36}}$$
$$r = \pm 3$$

2. Find t_1 for both *r*-values. Because $t_5 = 324$, $n = 5$ and $t_n = 324$.

For $r = 3$:

$$t_n = t_1 r^{n-1}$$
$$324 = t_1(3)^{5-1}$$
$$\frac{324}{3^4} = t_1$$
$$4 = t_1$$

For $r = -3$:

$$t_n = t_1 r^{n-1}$$
$$324 = t_1(-3)^{5-1}$$
$$\frac{324}{(-3)^4} = t_1$$
$$4 = t_1$$

3. Find the eighth term.

For $r = 3$:

$$t_n = t_1 r^{n-1}$$
$$t_8 = 4(3)^{8-1}$$
$$t_8 = 8748$$

For $r = -3$:

$$t_n = t_1 r^{n-1}$$
$$t_8 = 4(-3)^{8-1}$$
$$t_8 = -8748$$

TECHNOLOGY GRAPHICS CALCULATOR
Keystroke Guide, page 759

CHECK

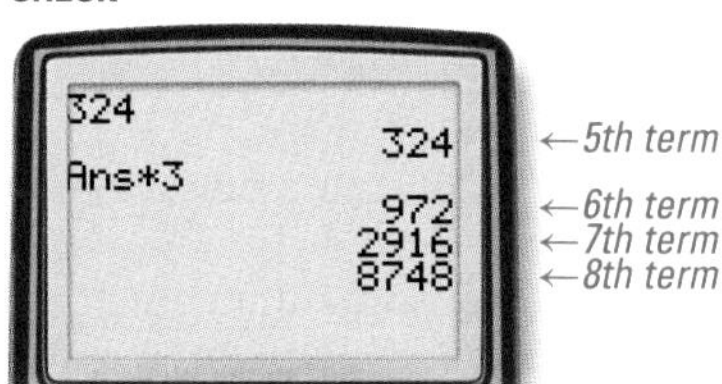

CHECK

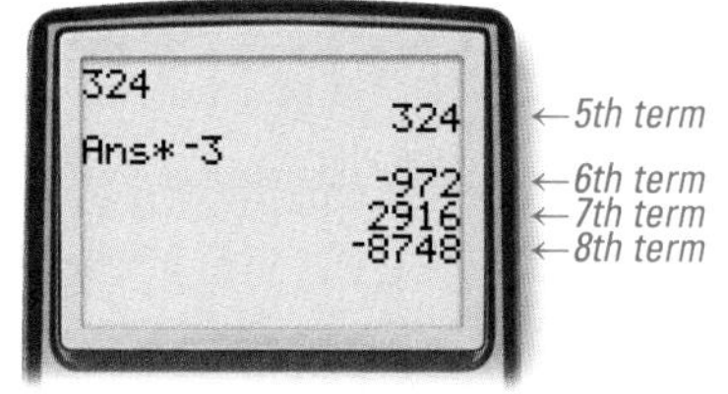

Thus, the eighth term of this geometric sequence is 8748 or −8748.

TRY THIS Find the 12th term of the geometric sequence in which $t_2 = 240$ and $t_5 = 30$.

Enrichment

To help students compare and contrast arithmetic and geometric sequences, give them the following problem: The sum of the first three terms in an arithmetic sequence is 24. If the first term is decreased by 1 and the second term is decreased by 2, the three terms are then the first terms of a geometric sequence. Ask them to find the first three terms of the arithmetic sequence(s).
4, 8, 12 or 13, 8, 3

Reteaching the Lesson

USING ALGORITHMS Have students refer to the geometric sequence defined by the recursive formula $t_1 = 3$ and $t_n = 4t_{n-1}$ and by the explicit formula $t_n = 3(4)^{n-1}$. Ask them to use both formulas to find t_8. **49,152** If $t_6 = 486$ and $t_9 = 13{,}122$, have students find both the recursive and explicit formulas. **$t_1 = 2$ and $t_n = 3t_{n-1}$, $t_n = 2(3)^{n-1}$** Then have them find the first three terms of the sequence and t_{13} by using both formulas. **2, 6, 18; $t_{13} = 1{,}062{,}882$** Ask them to explain which formula is easier to use when finding different types of terms.

The terms between any two nonconsecutive terms of a geometric sequence are called the **geometric means** between the two nonconsecutive terms. For example, in the sequence 5, **−10**, **20**, **−40**, 80, −160, . . . , the three geometric means between 5 and 80 are **−10**, **20**, and **−40**.

EXAMPLE 4 **Find three geometric means between 6 and 96.**

SOLUTION

PROBLEM SOLVING

Draw a diagram. Let $t_1 = 6$ and $t_5 = 96$.

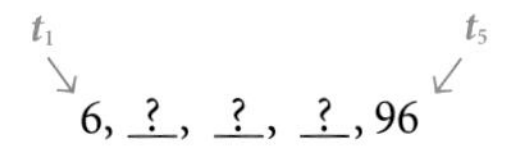

Find the common ratio, r.

$$t_n = t_1 r^{n-1}$$
$$96 = 6r^{5-1} \quad \textit{Because } t_5 = 96,\ n = 5 \textit{ and } t_n = 96.$$
$$16 = r^4$$
$$\pm 2 = r$$

Use $r = 2$ and $r = -2$ to find the two possible sets of geometric means.

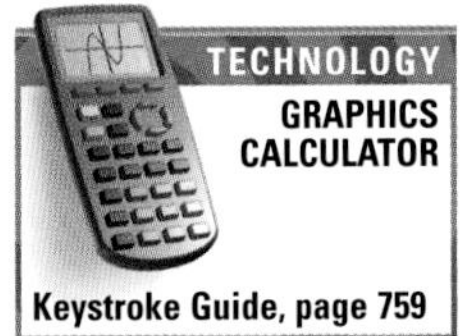

$r = 2$:

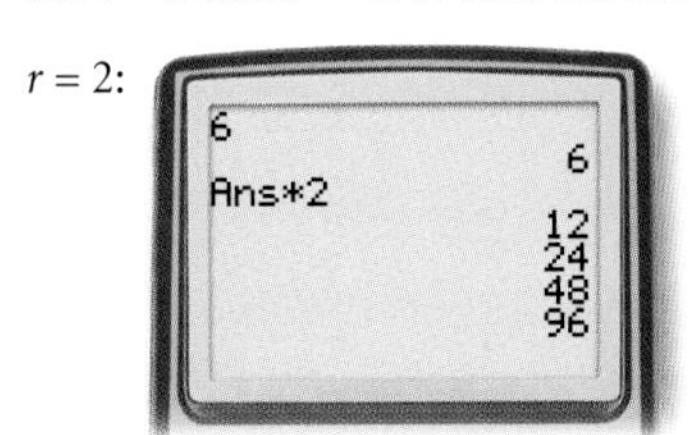

$r = -2$:

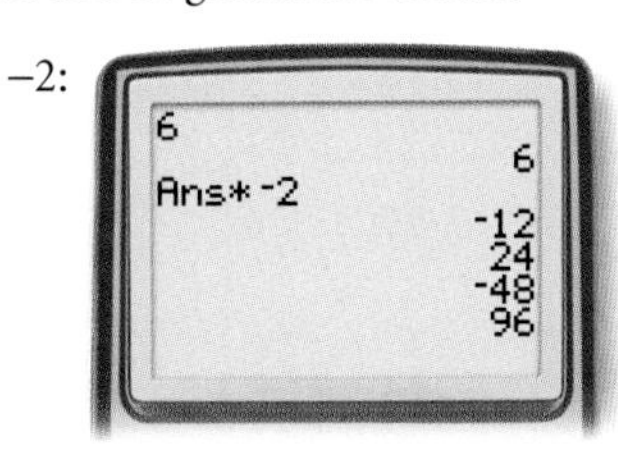

The three geometric means are 12, 24, and 48 or −12, 24, and −48.

TRY THIS Find three geometric means between 64 and 4.

CHECKPOINT ✔ How many sets of two geometric means are possible between $t_3 = -40$ and $t_6 = 5$? Explain.

CRITICAL THINKING In Example 4, there are two possible common ratios. If a geometric sequence were defined to include complex numbers, what two other common ratios would be possible?

Activity: Exploring Geometric Sequences

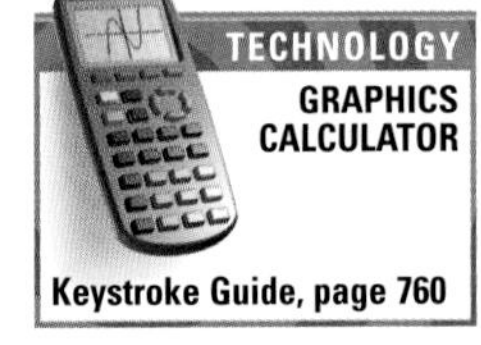

You will need: a graphics calculator

1. Write the first 5 terms of each geometric sequence below.
   a. $t_1 = 1, r = 10$ **b.** $t_1 = 1, r = 2$ **c.** $t_1 = 0.1, r = 1.1$
   d. $t_1 = 1, r = 1.01$ **e.** $t_1 = 1, r = 0.1$ **f.** $t_1 = 1, r = 0.9$
2. Which sequence from Step 1 has the largest 5th term? the smallest 5th term?

CHECKPOINT ✔

3. Describe how the value of r affects the terms of the sequence when $0 < r < 1$. Describe how the value of r affects the terms of the series when $r > 1$.

ADDITIONAL EXAMPLE 4

Find three geometric means between 324 and 4.

108, 36, 12 or −108, 36, −12

TRY THIS

32, 16, and 8 or −32, 16, and −8

CHECKPOINT

One set is possible because the cube root of a negative number is negative.

CRITICAL THINKING

$-2i$ or $2i$

Activity Notes

Encourage students to relate the value of r to the value of b in the exponential function $f(x) = ab^x$. Be sure students notice that the scatter plots of geometric sequences with positive ratios lie along an exponential curve whose base, b, is the same as the common ratio of the sequence, r. Remind students that an exponential function of the form $f(x) = ab^x$ is undefined for $b \leq 0$.

Teaching Tip

TECHNOLOGY When using a TI-82 or TI-83 graphics calculator, you may find the terms of a sequence by using the sequence command. To generate the first 20 terms of the geometric sequence for part **a** of Step 1 of the Activity, press 2nd STAT and select **OPS** **5:seq(**. To the right of the parenthesis, enter **10^X,X,1,20,1)**. Press ENTER to view the sequence, and press ► to view later terms of the sequence.

CHECKPOINT

3. If $0 < r < 1$, successive terms get smaller. If $r > 1$, successive terms get larger.

Exercises

Communicate

internet connect — Activities Online. Go To: go.hrw.com Keyword: MB1 College

1. Describe how to find the nth term of 1, 3, 9, 27, . . .
2. Explain what happens to the terms of a geometric sequence when the common ratio, r, is doubled. Justify your answer.
3. Explain what happens to the terms of a geometric sequence when the first term, t_1, is doubled.

Guided Skills Practice

4. Find the fifth term of the sequence defined by $t_1 = 2$ and $t_n = 2t_{n-1}$. ***(EXAMPLE 1)*** **32**

APPLICATION

5. **DEPRECIATION** A new automobile costs $14,000 and retains 85% of its value each year. Find the value of the automobile after 10 years. ***(EXAMPLE 2)*** **$2756.24**
6. Find the fifth term of the geometric sequence in which $t_4 = 768$ and $t_6 = 192$. ***(EXAMPLE 3)*** **384; –384**
7. Find three geometric means between 160 and 10. ***(EXAMPLE 4)*** **80, 40, and 20 or –80, 40, and –20**

8. yes; $r = 2$; 48, 96, 192
9. yes; $r = 2$; 320, 640, 1280
10. no
11. no
12. yes; $r = \frac{1}{5}$; $\frac{2}{125}, \frac{2}{625}, \frac{2}{3125}$
13. yes; $r = 5$; 1250, 6250, 31,250
14. yes; $r = 2$; 32, 64, 128
15. yes; $r = 3$; 162, 486, 1458
16. yes; $r = -4$; 256, –1024, 4096
17. no
18. yes; $r = 7$; 2058, 14,406, 100,842
19. yes; $r = 9$; 1458, 13,122, 118,098
20. yes; $r = \frac{1}{2}$; $\frac{5}{16}, \frac{5}{32}, \frac{5}{64}$
21. yes; $r = \frac{1}{3}$; $\frac{1}{9}, \frac{1}{27}, \frac{1}{81}$

Practice and Apply

Determine whether each sequence is a geometric sequence. If so, identify the common ratio, *r*, and give the next three terms.

8. 3, 6, 12, 24, . . .
9. 20, 40, 80, 160, . . .
10. 2, 4, 6, 8, . . .
11. 1, 3, 5, 7, . . .
12. 10, 2, $\frac{2}{5}, \frac{2}{25}$, . . .
13. 2, 10, 50, 250, . . .
14. 2, 4, 8, 16, . . .
15. 2, 6, 18, 54, . . .
16. 1, –4, 16, –64, . . .
17. 2, –5, 25, . . .
18. 6, 42, 294, . . .
19. 2, 18, 162, . . .
20. 5, $\frac{5}{2}, \frac{5}{4}, \frac{5}{8}$, . . .
21. 9, 3, 1, $\frac{1}{3}$, . . .
22. 9, 0.9, 0.09, 0.009, . . .
23. 12, 3, $\frac{3}{4}, \frac{3}{16}$, . . .
24. 2, 3, 4.5, 6.75, . . .
25. 16, 20, 25, 31.25, . . .

List the first four terms of each geometric sequence.

26. $t_1 = 3$, $t_n = 2t_{n-1}$ **3, 6, 12, 24**
27. $t_1 = -2$, $t_n = 4t_{n-1}$ **–2, –8, –32, –128**
28. $t_1 = 5$, $t_n = -2t_{n-1}$ **5, –10, 20, –40**
29. $t_1 = 4$, $t_n = -3t_{n-1}$ **4, –12, 36, –108**
30. $t_1 = 1$, $t_n = 4t_{n-1}$ **1, 4, 16, 64**
31. $t_1 = -1$, $t_n = -0.2t_{n-1}$ **–1, 0.2, –0.04, 0.008**
32. $t_1 = -3$, $t_n = -2.2t_{n-1}$
33. $t_1 = 3$, $t_n = 3.37t_{n-1}$
34. $t_1 = 3$, $t_n = -4.88t_{n-1}$

internet connect — Homework Help Online. Go To: go.hrw.com Keyword: MB1 Homework Help for Exercises 35–56

35. In a geometric sequence, $t_1 = 6$ and $r = 4$. Find t_7. **24,576**
36. In a geometric sequence, $t_1 = 5$ and $r = -2$. Find t_7. **320**
37. In a geometric sequence, $t_1 = 3$ and $r = \frac{1}{10}$. Find t_{20}. **3.0 × 10^{-19}**
38. In a geometric sequence, $t_1 = 3$ and $r = -\frac{1}{10}$. Find t_{20}. **–3.0 × 10^{-19}**

22. yes; $r = 0.1$; 0.0009, 0.00009, 0.000009

23. yes; $r = \frac{1}{4}$; $\frac{3}{64}, \frac{3}{256}, \frac{3}{1024}$

24. yes; $r = 1.5$; 10.125, 15.1875, 22.78125

22. yes; $r = 1.25$; 39.0625, 48.828125, 61.03515625

Assess

Selected Answers

Exercises 4–7, 9–91 odd

ASSIGNMENT GUIDE

In Class	1–7
Core	9–75 odd
Core Plus	8–78 even
Review	79–91
Preview	92

Extra Practice can be found beginning on page 940.

Mid-Chapter Assessment for Lessons 11.1 through 11.4 can be found on page 142 of the *Assessment Resources.*

Error Analysis

Some students have difficulty understanding why the formula for the nth term of a geometric sequence is given in terms of r raised to the $(n - 1)$ power. Have students list the first four terms of a geometric sequence. Then have them look for a pattern. Students may need to rewrite the explicit formula, $t_n = t_1r^{n-1}$, in recursive form, $t_n = rt_{n-1}$.

Tell students that the recursive form is the easiest to use for finding consecutive terms while the explicit form is the easiest to use for finding nonconsecutive terms.

32. –3, 6.6, –14.52, 31.944

33. 3, 10.11, 34.0707, 114.818259

34. 3, –14.64, 71.4432, –348.642816

60. 3, 9, and 27 or −3, 9, and −27

61. −1, $-\frac{1}{2}$, and $-\frac{1}{4}$ or 1, $-\frac{1}{2}$, and $\frac{1}{4}$

62. −3, $-\frac{3}{2}$, and $-\frac{3}{4}$ or 3, $-\frac{3}{2}$, and $\frac{3}{4}$

63. 162, 54, and 18 or −162, 54, and −18

64. −27, −18, and −12 or 27, −18, and 12

65. 125 and 375

66. 8, 16, 32, 64; geometric

67. 12, 36, 108, 324; geometric

68. 10, 5, $\frac{5}{2}$, $\frac{5}{4}$; geometric

69. $\frac{18}{5}$, $\frac{36}{25}$, $\frac{72}{125}$, $\frac{144}{625}$; geometric

70. 13, 103, 10,003; neither

71. 7, 27, 127, 627; neither

72. −30, −90, −270, −810; geometric

73. −400, −1600, −6400, −25,600; geometric

74. −150, 750, −3750, 18,750; geometric

Find the sixth term in the geometric sequence that includes each pair of terms.

39. $t_3 = 150$; $t_5 = 3750$ **±18,750**

40. $t_4 = 189$; $t_9 = 45{,}927$ **1701**

41. $t_4 = 36$; $t_8 = 2916$ **324**

42. $t_3 = 444$; $t_7 = 7104$ **±3552**

43. $t_5 = 24$; $t_8 = 3$ **12**

44. $t_7 = 10{,}935$; $t_{11} = 135$ **32,805**

45. $t_3 = -24$; $t_5 = -54$ **−81**

46. $t_7 = 4$; $t_{12} = 972$ $\mathbf{\frac{4}{3}}$

47. $t_3 = 12\frac{1}{2}$; $t_9 = \frac{25}{128}$ $\mathbf{\pm\frac{25}{16}}$

48. $t_2 = 25$; $t_4 = 2\frac{7}{9}$ $\mathbf{\pm\frac{25}{81}}$

Write an explicit formula for the *n*th term of each geometric sequence.

49. 2, 4, 8, 16, . . . $t_n = 2(2)^{n-1}$

50. 1, 3, 9, 27, . . . $t_n = 3^{n-1}$

51. $1, \frac{1}{2}, \frac{1}{4}, \frac{1}{8}, \ldots$ $t_n = \left(\frac{1}{2}\right)^{n-1}$

52. $1, \frac{1}{3}, \frac{1}{9}, \frac{1}{27}, \ldots$ $t_n = \left(\frac{1}{3}\right)^{n-1}$

53. $30, 10, 3\frac{1}{3}, 1\frac{1}{9}, \ldots$ $t_n = 30\left(\frac{1}{3}\right)^{n-1}$

54. $40, 10, 2\frac{1}{2}, \frac{5}{8}, \ldots$ $t_n = 40\left(\frac{1}{4}\right)^{n-1}$

55. $\sqrt{2}, 2, 2\sqrt{2}, 4, \ldots$ $t_n = \sqrt{2}(\sqrt{2})^{n-1}$

56. $\sqrt{3}, 3, 3\sqrt{3}, 9, \ldots$ $t_n = \sqrt{3}(\sqrt{3})^{n-1}$

Find the indicated geometric means.

57. Find two geometric means between 5 and 135. **15 and 45**

58. Find two geometric means between 4 and 13.5. **6 and 9**

59. Find two geometric means between 5 and 16.875. **7.5 and 11.25**

60. Find three geometric means between 1 and 81.

61. Find three geometric means between −2 and $-\frac{1}{8}$.

62. Find three geometric means between −6 and $-\frac{3}{8}$.

63. Find three geometric means between 486 and 6.

64. Find three geometric means between −40.5 and −8.

65. Find two geometric means between $41\frac{2}{3}$ and 1125.

List the first four terms of each sequence. Tell whether it is arithmetic, geometric, or neither.

66. $t_n = 4(2)^n$

67. $t_n = 4(3)^n$

68. $t_n = 20\left(\frac{1}{2}\right)^n$

69. $t_n = 9\left(\frac{2}{5}\right)^n$

70. $t_n = 3 + 10^n$

71. $t_n = 2 + 5^n$

72. $t_n = -10(3)^n$

73. $t_n = -100(4)^n$

74. $t_n = 30(-5)^n$

APPLICATION

REAL ESTATE An office building purchased for \$1,200,000 is appreciating because of rising property values in the city. At the end of each year its value is 105% of its value at the end of the previous year.

75. Use a recursive formula to determine what the value of the building will be 7 years after it is purchased. **\$1,688,520.51**

76. Use an explicit formula to find the value of the building 4 years after it is purchased. **\$1,458,607.50**

CHALLENGE

77. During the eighth year, the building begins to decrease in value at a rate of 8% per year. What would its value be after the 15th year? **\$866,580.59**

Practice

NAME ______________ CLASS ______ DATE ______

Practice 11.4 *Geometric Sequences*

Determine whether each sequence is geometric. If so, identify the common ratio, *r*, and give the next three terms.

1. 9, 25, 49, 81, . . . **no**
2. 200, 80, 32, 12.8, . . . **yes; $r = 0.4$**
3. $66\frac{2}{3}, -40, 24, -14\frac{2}{5}, \ldots$ **yes; $r = -\frac{3}{5}$**
4. 12, 18, 27, 40.5, . . . **yes; $r = 1.5$**
5. 54, 36, 24, 16, . . . **yes; $r = \frac{2}{3}$**
6. $1, \sqrt{2}, \sqrt{3}, 2, \ldots$ **no**

List the indicated terms of each geometric sequence.

7. $t_1 = 18$; $t_n = -2t_{n-1}$; first 4 terms **18, −36, 72, −144**
8. $t_1 = -4$; $t_n = 2.5t_{n-1}$; first 5 terms **−4, −10, −25, −62.5, −156.25**
9. $t_1 = 20$; $t_n = 0.5t_{n-1}$; first 4 terms **10, 5, 2.5, 1.25**
10. $t_2 = 40$; $t_4 = 2.5$; t_5 **0.625 or −0.625**
11. $t_2 = 16$; $t_6 = 64$; t_5 **$32\sqrt{2}$ or $-32\sqrt{2}$**
12. $t_3 = 20\frac{1}{4}$; $t_5 = 182\frac{1}{4}$; t_7 **$1640\frac{1}{4}$ or $-1640\frac{1}{4}$**

Write an explicit formula for the *n*th term of each geometric sequence.

13. 250, 100, 40, 16, . . . **$t_n = 250\left(\frac{2}{5}\right)^{n-1}$**
14. −30, 6, −1.2, 0.24, . . . **$t_n = -30(-0.2)^{n-1}$**
15. 40, 32, 25.6, 20.48, . . . **$t_n = 40(0.8)^{n-1}$**
16. 2, 5, 12.5, 31.25, . . . **$t_n = 2\left(\frac{5}{2}\right)^{n-1}$**
17. $20, 5, 1\frac{1}{4}, \frac{5}{16}, \ldots$ **$t_n = 80\left(\frac{1}{4}\right)^n$**
18. 1.5, −9, 54, −324, . . . **$t_n = -\frac{1}{4}(-6)^n$**

19. Find 2 geometric means between 7 and 875. **35 and 175**
20. Find 2 geometric means between −28 and −3.5. **−14 and −7**
21. Find 3 geometric means between 12 and 3072. **48, 192, and 768 or −48, 192, and −768**
22. Find 3 geometric means between 12.5 and 25.92. **15, 18, and 21.6 or −15, 18, and −21.6**
23. Find 3 geometric means between 12 and 7500. **60, 300, and 1500 or −60, 300, and −1500**
24. Find 4 geometric means between 4 and 972. **12, 36, 108, and 324**

APPLICATION

78. MUSIC A piano keyboard has 88 equally spaced musical notes. The first, and lowest, note is assigned the letter A, and it has a frequency of 27.5 hertz (Hz). The 13th note is also assigned the letter A, and it has a frequency twice that of the preceding A note, as does each of the subsequent A notes.

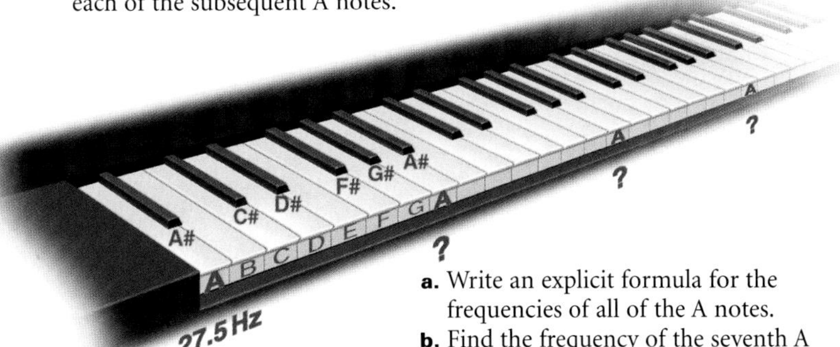

a. Write an explicit formula for the frequencies of all of the A notes.
b. Find the frequency of the seventh A note in the sequence defined in part **a**.
c. Find the frequencies of the 11 equally spaced notes between the first A note and the second A note. Round your answers to the nearest hundredth.

78a. $t_n = 27.5(2)^{n-1}$
b. 1760 Hz
c. frequencies in Hz: 29.13, 30.87, 32.70, 34.65, 36.71, 38.89, 41.20, 43.65, 46.25, 49.00, 51.91

Look Back

Write an equation in slope-intercept form for the line containing the indicated points. ***(LESSON 1.3)***

79. $(-2, 5)$ and $(0, -1)$ $y = -3x - 1$ **80.** $(9, -4)$ and $(-5, 3)$ $-\frac{1}{2}x + \frac{1}{2}$

Let $A = \begin{bmatrix} 1 & 0.5 \\ 2.4 & 3.8 \end{bmatrix}$, $B = \begin{bmatrix} 1.7 \\ 3.2 \end{bmatrix}$, **and** $C = [3.2 \quad 4.8]$. **Find each product, if it exists.** ***(LESSON 4.2)***

81. AB $\begin{bmatrix} 3.3 \\ 16.24 \end{bmatrix}$ **82.** CA $[14.72 \quad 19.84]$ **83.** BC $\begin{bmatrix} 5.44 & 8.16 \\ 10.24 & 15.36 \end{bmatrix}$ **84.** CB $[20.8]$

APPLICATION

85. SPORTS Liam plans to spend a certain amount of time every day training for the school track team. He wants to run 3 miles and ride his bicycle for 6 miles. If he rides his bicycle an average of 12 miles per hour faster than he runs, at what rate must he run and bike in order to complete his training workout in one hour? ***(LESSON 8.5)*** run: 4.68 mph, bike: 16.68 mph

Simplify. Write each expression with a rational denominator. ***(LESSON 8.7)***

86. $\sqrt{8} + \sqrt{98}$ $9\sqrt{2}$ **87.** $\frac{\sqrt{30} + \sqrt{14}}{\sqrt{2}}$ $\sqrt{15} + \sqrt{7}$ **88.** $\sqrt{6x} + \frac{\sqrt{2x}}{\sqrt{3}} - \sqrt{\frac{3x}{2}}$ $\frac{5}{6}\sqrt{6x}$

Find the number of permutations of the letters in each word below. ***(LESSON 10.2)***

89. *roommate* 10,080 **90.** *apple* 60 **91.** *apogee* 360

Look Beyond

92. Consider the geometric sequence 0.3, 0.03, 0.003, 0.0003, . . .
a. Find the sum of the first 4 terms and the sum of the first 10 terms.
b. What value are the sums approaching?

92a. 0.3333, 0.3333333333
b. $\frac{1}{3}$ or $0.\overline{3}$

Look Beyond

In Exercise 92, students explore the sum of the first n terms of a geometric sequence. Geometric series will be covered in Lesson 11.5.

Student Technology Guide

NAME ______ CLASS ______ DATE ______

Student Technology Guide
11.4 ***Geometric Sequences***

You can use a calculator to generate the terms of a geometric sequence.

Example: List the first four terms of the sequence defined by the formula $t_1 = 4$ and $t_n = -2t_{n-1}$.

- Enter the first term by pressing 4 ENTER.
- To find the second term, press 2nd (-) × (-) 2 ENTER. Press ENTER two more times.

The first four terms of the sequence are 4, −8, 16, and −32.

List the first four terms of each geometric sequence.

1. $t_1 = 2$ and $t_n = 4t_{n-1}$ 2, 8, 32, 128
2. $t_1 = -3$ and $t_n = -3t_{n-1}$ −3, 9, −27, 81
3. $t_1 = 100$ and $t_n = 1.5t_{n-1}$ 100, 150, 225, 337.5

Example: Find three geometric means between 4 and 2500.

- You need to fill in the blanks in the geometric sequence 4, ?, ?, ?, 2500.
- Let $t_1 = 4$ and $t_5 = 2500$. Find the common ratio, r.

$t_n = t_1r^{n-1}$
$2500 = 4r^{5-1} = 4r^4$
$\sqrt[4]{\frac{2500}{4}} = r$

- Press 4 MATH 5:$\sqrt[x]{}$ ENTER (2500 ÷ 4) ENTER. Notice that $r = 5$ is only the principal root. You must also consider the possibility that $r = -5$.
- Assume $r = 5$. Press 4 ENTER 2nd (-) × 5 ENTER ENTER ENTER ENTER.
- Assume $r = -5$. Press 4 ENTER 2nd (-) × (-) 5 ENTER ENTER ENTER ENTER.

```
4
            4
Ans*5
           20
          100
          500
         2500
```

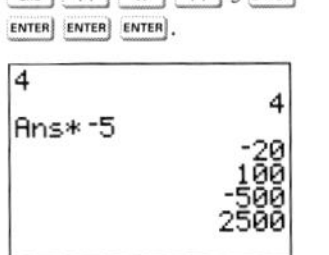

The three geometric means are 20, 100, and 500 or −20, 100, and −500.

Find three geometric means between each pair of numbers.

4. −2 and −162 −6, −18, −54 or 6, −18, 54
5. 1000 and 0.1 100, 10, 1 or −100, 10, −1
6. 1 and 5.0625 1.5, 2.25, 3.375 or −1.5, 2.25, −3.375

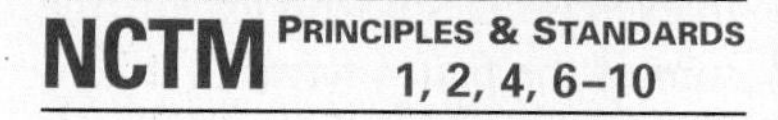

NCTM PRINCIPLES & STANDARDS 1, 2, 4, 6–10

QUICK WARM-UP

Find each sum.

1. $\sum_{m=1}^{8} 6$ 48

2. $\sum_{j=1}^{5} 4j$ 60

3. $\sum_{k=1}^{7} (10 - 2k)$ 14

4. $\sum_{n=1}^{6} (n^2 - n + 1)$ 76

Find the indicated sum for each arithmetic series.

5. $3 + 6 + 9 + 12 + 15 + \cdots$; S_{12} 234

6. $9 + 2 - 5 - 12 - 19 - \cdots$; S_{20} −1150

Also on Quiz Transparency 11.5

Teach

Why Often the sum of the terms of a geometric sequence best defines a real-world situation. Ask students how they would determine the number of their parental ancestors through 8 generations. $\sum_{k=1}^{8} 2^k = 5$

CHECKPOINT ✔
62

Geometric Series and Mathematical Induction

Objectives

- Find the sum of the first n terms of a geometric series.
- Use mathematical induction to prove statements about natural numbers.

Why *Geometric series can be used to solve real-world problems such as finding the right investment to save money for college.*

Mr. and Mrs. Sanchez want to invest money for their child's college education. They have decided to invest \$2000 at the beginning of every year for the next 10 years. If the investment is in an account that earns 8% annual interest, compounded once per year, how much will their investment be worth at the end of the 10th year? *You will solve this in Example 2.*

Geometric Series

A **geometric series** is the indicated sum of the terms of a geometric sequence. Consider the sequence 2, 4, 8, 16, 32, . . . The sum of the first five terms, denoted by S_5, is $S_5 = 2 + 4 + 8 + 16 + 32 = 62$.

You can derive a formula for the sum of a geometric series by using the Subtraction Property of Equality, as shown below.

$$S_n = t_1 + t_1 r + t_1 r^2 + \cdots + t_n r^{n-2} + t_n r^{n-1}$$

Each side is multiplied by the common ratio, r.

$$rS_n = t_1 r + t_1 r^2 + \cdots + t_n r^{n-2} + t_n r^{n-1} + t_n r^n$$

$$S_n - rS_n = t_1 + 0 + 0 + \cdots + 0 + 0 - t_n r^n$$

$$S_n(1 - r) = t_1 - t_1 r^n$$

To divide both sides of the equation by 1 − r, r cannot equal 1.

$$S_n = \frac{t_1(1 - r^n)}{1 - r}, \text{ or } t_1\left(\frac{1 - r^n}{1 - r}\right)$$

Notice that S_n is undefined when $r = 1$.

CHECKPOINT ✔ Using the sequence 2, 4, 8, . . . , find S_5 by following the procedure used in the derivation of S_n.

Sum of the First *n* Terms of a Geometric Series

The sum, S_n, of the first n terms of a geometric series is given by

$S_n = t_1\left(\frac{1 - r^n}{1 - r}\right)$, where t_1 is the first term, r is the common ratio, and $r \neq 1$.

Alternative Teaching Strategy

USING VISUAL MODELS You can introduce the topic of geometric series by asking students to draw an equilateral triangle with sides of 12 inches. Tell them to inscribe a second equilateral triangle inside the first by joining the midpoints of the sides of the first triangle. Then have them draw a third and fourth inscribed equilateral triangle by using the same procedure. Ask them to find the perimeter of each triangle and the sum of the perimeters of the four equilateral triangles.
$S_4 = 36 + 18 + 9 + 4.5 = 67.5$ in.

Have them rewrite S_4 in terms of the perimeter of the original triangle, 36, and the common ratio between the perimeters of consecutive triangles, 0.5. $S_4 = 36 + 36(0.5) + 36(0.5)^2 + 36(0.5)^3$ Show them the development of the formula for S_4 shown below.

$$S_4 = 36 + 36(0.5) + 36(0.5)^2 + 36(0.5)^3$$

$$+\,-(0.5)S_4 = \quad -36(0.5) - 36(0.5)^2 - 36(0.5)^3 - 36(0.5)^4$$

$$(1 - 0.5)S_4 = 36 \quad -36(0.5)^4$$

$$S_4 = \frac{36[1 - (0.5)^4]}{(1 - 0.5)}$$

$$= 67.5$$

Exploring Geometric Series

You will need: no special materials

Consider the geometric series $S_n = 1 + 2 + 4 + 8 + \cdots + 2^{n-1}$.

1. Copy and complete the table.

n	1	2	3	4	5	6	7	8
S_n	1	3	7	15	?	?	?	?

2. Compare each sum with 2^n.
3. Predict the sum of the first 10 terms. Check your prediction.

CHECKPOINT ✓ 4. Compare your answer from Step 3 with the formula for S_n when $t_1 = 1$ and $r = 2$.

EXAMPLE 1 **Given the series $3 + 4.5 + 6.75 + 10.125 + \cdots$, find S_{10} to the nearest tenth.**

SOLUTION

1. Find the common ratio, r. $r = \frac{4.5}{3} = \frac{6.75}{4.5} = 1.5$
2. Substitute 3 for t_1, 1.5 for r, and 10 for n in $S_n = t_1\left(\frac{1 - r^n}{1 - r}\right)$.

$$S_{10} = 3\left[\frac{1 - (1.5)^{10}}{1 - 1.5}\right] \approx 340.0$$

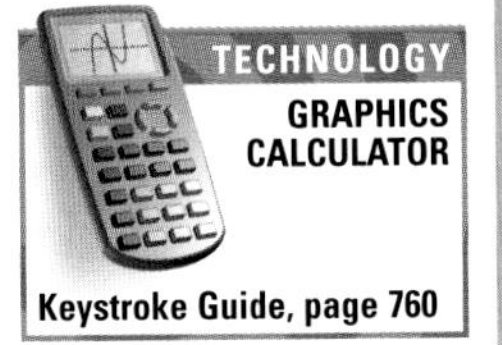

CHECK

Define the geometric sequence involved.

$t_n = t_1 r^{n-1}$

$t_n = 3 \times 1.5^{n-1}$

Then use the summation and sequence commands. Thus, $S_{10} \approx 340.0$.

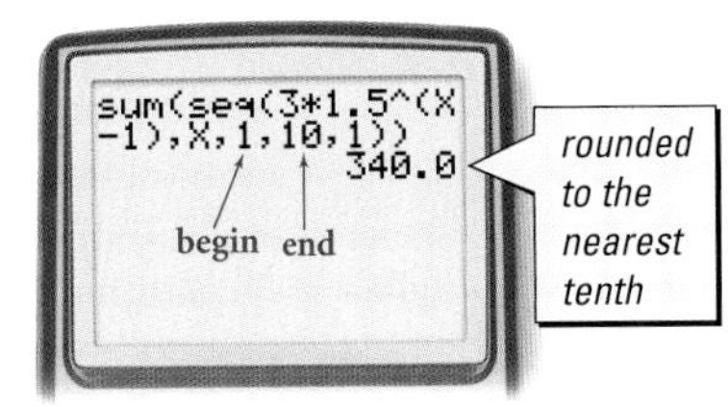

TRY THIS Given the series $400 + 300 + 225 + 168.75 + \ldots$, find S_{16} to the nearest tenth.

EXAMPLE 2 Refer to the investment situation described at the beginning of the lesson. **How much money will be in the account at the end of the 10th year?**

APPLICATION INVESTMENTS

SOLUTION

After year 1 → $S_1 = 2000(1.08)$

After year 2 → $S_2 = 2000(1.08) + 2000(1.08)^2$

After year 3 → $S_3 = 2000(1.08) + 2000(1.08)^2 + 2000(1.08)^3$

⋮

After year 10 → $S_{10} = 2000(1.08) + 2000(1.08)^2 + \cdots + 2000(1.08)^{10}$

This is a geometric series in which $t_1 = 2000(1.08) = 2160$, $r = 1.08$, and $n = 10$.

$$S_{10} = 2160\left[\frac{1 - (1.08)^{10}}{1 - 1.08}\right] \approx 31{,}290.97$$

At the end of the 10th year, the account will contain $31,290.97.

Interdisciplinary Connection

SPORTS Sixteen teams enter a single elimination tournament. Two teams play at a time without the possibility of a tie. Tell students that they are to make arrangements to provide referees for each game, and have them determine how many games will be required for the entire tournament. 15 Suggest that students use a tree diagram to list the possible outcomes of the tournament. Be sure they notice that in the first round, 8 games are played, in the second round, 4 games are played, and in the third round, 2 games are played. The fourth round is for the championship, in which the two remaining teams play to determine the tournament winner. Students should recognize that this is a geometric series.

Activity Notes

Be sure students notice that the sum of the geometric series in Step 1 is $2^n - 1$. Then encourage students to check their prediction by showing that if $t_1 = 1$, $r = 2$, and $n = 10$, then $S_{10} = \frac{1 - 2^{10}}{1 - 2} = 2^{10} - 1$.

Cooperative Learning

Have students work in pairs to show that $1 + 2 + 3 + 4 + 8 + \cdots + 2^{n-1} = 2^n - 1$. They should each complete the table individually but should decide on the answer and write down the result as a team. You can also have students repeat this Activity by comparing the sum of the geometric series $9 + 9(10) + 9(10)^2 + \cdots + 9(10)^{n-1}$ with $10^n - 1$.

CHECKPOINT ✓

4. They are the same: $2^{10} - 1$.

ADDITIONAL EXAMPLE 1

Given the series $6 - 14.4 + 34.56 - 82.944 + \cdots$, approximate S_8 to the nearest tenth.

≈ -1940.7

TRY THIS

≈ 1584.0

ADDITIONAL EXAMPLE 2

To save for a newborn child's college education, a family plans to invest $2500 at the beginning of every year for the next 18 years. Suppose that the investment earns 6% annual interest, compounded annually. **How much will the investment be worth at the end of the 18th year?**

$81,899.98

ADDITIONAL EXAMPLE 3

Evaluate $\sum_{k=1}^{7} 4(-5)^{k-1}$. 52,084

TRY THIS
16.5

CRITICAL THINKING
364; the result in Example 3 is two times 364; the sum of the series is multiplied by *c*.

Teaching Tip

You may want to explain to your students that the *basis step* in mathematical induction shows the truth of the statement for the smallest number in the set, the basis. The *induction step* shows that you can apply the pattern to each of the subsequent numbers by using the *induction hypothesis.*

EXAMPLE 3 Evaluate $\sum_{k=1}^{6} 2(3^{k-1})$.

SOLUTION

This summation notation indicates the sum of the first six terms of the geometric series in which $t_1 = 2$ and $r = 3$.

$$S_n = t_1\left(\frac{1-r^n}{1-r}\right)$$

$$S_6 = 2\left(\frac{1-3^6}{1-3}\right)$$

$$S_6 = 728$$

CHECK

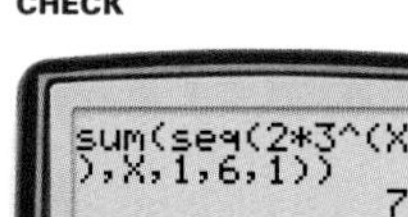

TRY THIS Evaluate $\sum_{k=1}^{5} 1.5[(-2)^{k-1}]$.

CRITICAL THINKING Evaluate $\sum_{k=1}^{6} (3^{k-1})$. Compare the result with the result in Example 3. Make a hypothesis about how a series is affected if each term of the corresponding sequence is multiplied by a constant number, *c*.

Mathematical Induction

How can you determine whether a general statement, such as $1 + 2 + 3 + \cdots + n = \frac{n(n+1)}{2}$, is true for every natural number? Because it is impossible to test every natural number, n, you can use a type of proof called *mathematical induction.*

The principle of mathematical induction is like a line of dominoes that fall over one by one after the first domino is pushed. If you can show that the statement is true for one natural number, then induction will prove it to be true for the next natural number and the next natural number and so on.

Mathematical Induction

To prove that a statement is true for all natural numbers n:

Basis Step: Show that the statement is true for $n = 1$.

Induction Step: Assume the statement is true for a natural number, k. Then prove that it is true for the natural number $k + 1$.

Inclusion Strategies

ENGLISH LANGUAGE DEVELOPMENT It is important that students are able to read problems involving the sum of a finite geometric series written in summation notation, which uses the Greek letter *sigma*, Σ. Have them verbally translate Exercises 37–54 of Practice and Apply in class. Start by reading Example 3 to the class. It should be read as "the summation of 2 times 3 raised to the $(k - 1)$ power, for k from 1 to 6."

Enrichment

Draw an 8 × 8 square checkerboard on the board or overhead. Tell students that you want to place 1 checker on the first square, 2 checkers on the second square, 4 checkers on the third square, and so on, continuing to double the number of checkers on each subsequent square until all 64 squares are covered. Have students find the number of checkers on the 64th square and how many checkers in total will be on the board.
$t_{64} = 2^{63}$, $S_{64} = 2^{64} - 1$

EXAMPLE 4 **Prove the following statement:**

For every natural number n, $1 + 2 + 3 + \cdots + n = \frac{n(n+1)}{2}$.

SOLUTION

1. Basis Step

Show that $1 + 2 + 3 + \cdots + n = \frac{n(n+1)}{2}$ is true for $n = 1$.

$$S_n = \frac{n(n+1)}{2}$$
$$1 \stackrel{?}{=} \frac{1(1+1)}{2}$$
$$1 = 1 \quad \textbf{True}$$

2. Induction Step

Assume the statement is true for a natural number, k.

$$S_k\text{: } 1 + 2 + 3 + \cdots + k = \frac{k(k+1)}{2}$$

Then prove that it is true for the next natural number, $k + 1$.

- Determine the statement to be proved, S_{k+1}:

  Add $k + 1$ to the left side, and substitute $k + 1$ for k on the right.

$$S_k\text{: } 1 + 2 + 3 + \cdots + k = \frac{k(k+1)}{2}$$
$$S_{k+1}\text{: } 1 + 2 + 3 + \cdots + k + (k+1) = \frac{(k+1)[(k+1)+1]}{2}$$
$$1 + 2 + 3 + \cdots + k + (k+1) = \frac{(k+1)(k+2)}{2} \quad \textit{Simplify.}$$

This is the statement to be proved.

- Begin with the statement assumed to be true, S_k, and use properties of equality to prove the statement that you want to prove, S_{k+1}:

$$S_k\text{: } 1 + 2 + 3 + \cdots + k = \frac{k(k+1)}{2}$$
$$1 + 2 + 3 + \cdots + k + (k+1) = \frac{k(k+1)}{2} + (k+1) \quad \textit{Addition Property of Equality}$$
$$= \frac{k(k+1)}{2} + \frac{2(k+1)}{2} \quad \textit{Common denominators}$$
$$= \frac{k^2 + k + 2k + 2}{2} \quad \textit{Add fractions.}$$
$$= \frac{k^2 + 3k + 2}{2} \quad \textit{Combine like terms.}$$
$$= \frac{(k+1)(k+2)}{2} \quad \textit{Factor.}$$

CHECK

Substitute any two consecutive natural numbers for k and $k + 1$.

$$\frac{k(k+1)}{2} + (k+1) = \frac{k(k+1)}{2} + \frac{2(k+1)}{2}$$
$$\frac{3(4)}{2} + 4 \stackrel{?}{=} \frac{3(4)}{2} + \frac{2(4)}{2}$$
$$10 = 10 \quad \textbf{True}$$

TRY THIS Prove the following statement:

For every natural number n, $4 + 8 + 12 + \cdots + 4n = 2n(n+1)$.

Reteaching the Lesson

USING PATTERNS Show students how to develop a formula for the sum, S_n, of the first n terms of a geometric series in terms of t_1, t_n, and r. Remind them that $rt_n = t_{n+1}$ and $t_n = r^{n-1}t_1$, so $t_{n+1} = r^n t_1$. Begin with the formula for S_n, given in terms of t_1, r, and n.

$$S_n = t_1\frac{1 - r^n}{1 - r}$$
$$= \frac{t_1 - r^n t_1}{1 - r}$$
$$= \frac{t_1 - t_{n+1}}{1 - r}$$
$$= \frac{t_1 - rt_n}{1 - r}$$

Now ask students to find S_{10} in Example 1 by using t_1, r, and t_{10}. $t_1 = 3$, $r = 1.5$, and $t_{10} = 3(1.5^9)$;

$$S_{10} = \frac{3 - 1.5(3 \cdot 1.5^9)}{1 - 1.5} \approx 340.0$$

ADDITIONAL EXAMPLE 4

Prove the following statement: For every natural number n, $5 + 11 + 17 + \cdots + (6n - 1) = n(3n + 2)$.

1. Basis Step

   Show that the statement is true for $n = 1$.

   $6(1) - 1 = 1[3(1) + 2]$

   $6 - 1 = 3 + 2$

   $5 = 5$ True

2. Induction Step

   Assume the statement is true for a natural number, k.

   $5 + \cdots + (6k - 1) = k(3k + 2)$

   Then prove that it is true for $k + 1$.

   $5 + \cdots + (6k - 1) + [6(k + 1) - 1]$

   $= k(3k + 2) + [6(k + 1) - 1]$

   $= 3k^2 + 2k + 6k + 6 - 1$

   $= 3k^2 + 8k + 5$

   $= (k + 1)(3k + 5)$

   $= (k + 1)[3(k + 1) + 2]$

   True

TRY THIS

1. Basis Step

   For $n = 1$, show that $4 + 8 + \cdots + 4n = 2n(n + 1)$ is true.

   $4 = 2(1)(2)$ True

2. Induction Step

   Assume the statement is true for a natural number, k.

   $4 + 8 + \cdots + 4k = 2k(k + 1)$

   Determine the statement to be proved by adding the next term, $4(k + 1)$, to the left side and substituting $k + 1$ for k on the right.

   $4 + 8 + 12 + \cdots + 4k + 4(k + 1)$

   $= 2(k + 1)[(k + 1) + 1]$

   $= 2(k + 1)(k + 2)$

   Add $4(k + 1)$ to both sides of the equation that is assumed to be true and rewrite the right side to produce the desired form.

   $4 + 8 + \cdots + 4k + 4(k + 1)$

   $= 2k(k + 1) + 4(k + 1)$

   $= (k + 1)(2k + 4)$

   $= 2(k + 1)(k + 2)$ True

Teaching Tip

Remind students that when you push over one domino in the middle of a line of dominoes, the ones in front of it remain standing because the line of dominoes falls in only one direction. The induction process also works in only one direction. That is why the smallest possible number, such as $n = 1$, should be chosen for the basis step. Choosing this number is similar to pushing over the first domino, and Communicate Exercise 4 on this page addresses this issue. The induction step shows that if any domino falls, then the next domino will also fall. That is, the basis step starts the dominoes falling, and the induction step shows that each domino that falls makes the next domino fall.

Exercises

Communicate

1. How is a geometric series different from an arithmetic series?
2. How would you use summation notation to express the series $2 + 4 + 8 + 16 + 32$? Show three different ways to do this.
3. Explain how to check the results of an induction proof.
4. If you use $n = 3$ for the basis step in an induction proof, is your general proof necessarily true for $n = 1$ and $n = 2$? Explain.

Guided Skills Practice

5. Given the series $2 + 5 + 12.5 + 31.25 + \cdots$, find S_{10} to the nearest tenth. ***(EXAMPLE 1)*** 12,714.3

APPLICATION

6. **INVESTMENTS** If a family deposits $2500 at the beginning of each year into an account earning 12% interest, compounded annually, how much would be in the account at the end of the 8th year? ***(EXAMPLE 2)*** $34,439.07
7. Evaluate $\sum_{k=1}^{6} 3(2^{k-1})$. ***(EXAMPLE 3)*** 189
8. Prove the following statement: ***(EXAMPLE 4)***
   For every natural number n, $\frac{1}{1(2)} + \frac{1}{2(3)} + \cdots + \frac{1}{n(n+1)} = \frac{n}{n+1}$.

Practice and Apply

Find each indicated sum of the geometric series $1 + 2 + 4 + 8 + \cdots$

9. S_3 7
10. S_5 31
11. S_8 255
12. S_{11} 2047

Find each indicated sum of the geometric series $2 + (-6) + 18 + (-54) + \cdots$

13. S_4 −40
14. S_6 −364
15. S_{15} 7,174,454
16. S_{20} −1,743,392,200

Find each indicated sum of the geometric series $4 + 6 + 9 + 13.5 + \cdots$ Give answers to the nearest tenth, if necessary.

17. S_2 10
18. S_3 19
19. S_6 83.1
20. S_7 128.7
21. S_{10} 453.3
22. S_{11} 684.0
23. S_{20} 26,594.1
24. S_{21} 39,895.1

Use the formula for the sum of the first *n* terms of a geometric series to find each sum. Give answers to the nearest tenth, if necessary.

25. $2 + 4 + 8 + 16 + 32$ 62
26. $3 + 9 + 27 + 81 + 243$ 363
27. $-1 + 2 + (-4) + 8$ 5
28. $-5 + 15 + (-45) + 135$ 100
29. $1 + 1.2 + 1.44 + 1.728$ 5.4
30. $2 + 4.6 + 10.58 + 24.334$ 41.5
31. $1 + \frac{2}{5} + \frac{4}{25} + \frac{8}{125}$ $\frac{203}{125}$
32. $\frac{2}{3} + \frac{1}{3} + \frac{1}{6} + \frac{1}{12} + \frac{1}{24}$ $\frac{31}{24}$

Assess

Selected Answers

Exercises 5–8, 9–83 odd

ASSIGNMENT GUIDE	
In Class	1–8
Core	9–61 odd, 65–73 odd
Core Plus	10–72 even
Review	74–83
Preview	84

Extra Practice can be found beginning on page 940.

8. 1. Basis Step

Show that $\frac{1}{1 \cdot 2} + \frac{1}{2 \cdot 3} + \cdots \frac{1}{n(n+1)} = \frac{n}{n+1}$ is true for $n = 1$.

$\frac{1}{1 \cdot 2} = \frac{1}{1+1}$ True

2. Induction Step

Assume that the statement is true for a natural number, k.

$\frac{1}{1 \cdot 2} + \frac{1}{2 \cdot 3} + \cdots + \frac{1}{k(k+1)} = \frac{k}{k+1}$

Determine the statement to be proved by adding $\frac{1}{(k+1)(k+2)}$ to the left side, and substituting $k + 1$ for k on the right.

$\frac{1}{1 \cdot 2} + \frac{1}{2 \cdot 3} + \cdots + \frac{1}{k(k+1)} + \frac{1}{(k+1)(k+2)} = \frac{k+1}{[(k+1)+1]}$

$\frac{1}{1 \cdot 2} + \frac{1}{2 \cdot 3} + \cdots + \frac{1}{k(k+1)} + \frac{1}{(k+1)(k+2)} = \frac{k+1}{k+2}$

Rewrite the left side by using the statement assumed to be true in order to obtain the right side.

$\frac{1}{1 \cdot 2} + \frac{1}{2 \cdot 3} + \cdots + \frac{1}{k(k+1)} + \frac{1}{(k+1)(k+2)}$

$= \frac{k}{k+1} + \frac{1}{(k+1)(k+2)}$

$= \frac{k}{k+1}\left(\frac{k+2}{k+2}\right) + \frac{1}{(k+1)(k+2)}$

$= \frac{k(k+2)+1}{(k+1)(k+2)} = \frac{k^2+2k+1}{(k+1)(k+2)}$

$= \frac{(k+1)^2}{(k+1)(k+2)} = \frac{k+1}{k+2}$ True

33. $\frac{3}{65{,}536} \approx 4.58(10)^{-5}$

34. $\frac{3}{68{,}719{,}476{,}736} \approx 4.37(10)^{-11}$

internet connect

Homework Help Online
Go To: **go.hrw.com**
Keyword:
MB1 Homework Help for Exercises 37–54

35. $\frac{1{,}048{,}575}{65{,}536} \approx 16.0$

36. $\frac{1{,}099{,}511{,}627{,}775}{68{,}719{,}476{,}736} \approx 16.0$

37. $t_1 = 5$; $r = 5$; $t_{10} = 9{,}765{,}625$; $S_{10} = 12{,}207{,}030$

38. $t_1 = 6$; $r = 2$; $t_{12} = 6144$; $S_{12} = 24{,}570$

39. $t_1 = 1$; $r = \frac{1}{11}$; $t_{12} = \frac{1}{285{,}311{,}670{,}611}$; $S_{12} = \frac{11}{10}$

40. $t_1 = 1$; $r = \frac{1}{4}$; $t_6 = \frac{1}{1024}$; $S_6 = \frac{1365}{1024}$

41. $t_1 = 2.76$; $r = 2.76$; $t_8 \approx 3367.23$; $S_8 \approx 5278.86$

42. $t_1 = 1$; $r = 7.65$; $t_{10} = 89{,}733{,}150.59$;

65a. $8, 4\sqrt{2}, 4, 2\sqrt{2}, 2, \sqrt{2}, 1$

b. $t_n = 8\left(\frac{\sqrt{2}}{2}\right)^{n-1}$

c. $S_7 = 15 + 7\sqrt{2} \approx 24.9$

d. $S_n = \sum_{k=1}^{n} 8\left(\frac{\sqrt{2}}{2}\right)^{k-1}$ or $S_n = 8\left(\frac{1 - \left(\frac{\sqrt{2}}{2}\right)^n}{1 - \frac{\sqrt{2}}{2}}\right)$

For Exercises 33–36, refer to the series $12 + 3 + \frac{3}{4} + \frac{3}{16} + \cdots$

33. Find t_{10}. **34.** Find t_{20}. **35.** Find S_{10}. **36.** Find S_{20}.

Identify t_1, r, and t_n, and evaluate the sum of each series.

37. $\sum_{k=1}^{10} 5^k$ **38.** $\sum_{k=1}^{12} (3 \cdot 2^k)$ **39.** $\sum_{j=1}^{12} \left(\frac{1}{11}\right)^{j-1}$

40. $\sum_{m=1}^{6} \left(\frac{1}{4}\right)^{m-1}$ **41.** $\sum_{n=1}^{8} 2.76^n$ **42.** $\sum_{t=1}^{10} 7.65^{t-1}$

Evaluate. Round answers to the nearest tenth, if necessary.

43. $\sum_{k=1}^{5} 4(2^{k-1})$ **124** **44.** $\sum_{k=1}^{10} 4(5^{k-1})$ **9,765,624** **45.** $\sum_{k=1}^{6} \left(\frac{1}{2}\right)^{k-1}$ $\frac{63}{32}$

46. $\sum_{d=1}^{6} \left(\frac{1}{3}\right)^{d-1}$ $\frac{364}{243}$ **47.** $\sum_{m=1}^{7} 3(0.2^{m-1})$ **3.8** **48.** $\sum_{k=1}^{6} 2(0.3)^{k-1}$ **2.9**

49. $\sum_{k=1}^{7} 5(4^{k-1})$ **27,305** **50.** $\sum_{k=1}^{8} 10(3^{k-1})$ **32,800** **51.** $\sum_{k=1}^{12} 6.92^{k-1}$ **2,036,814,259**

52. $\sum_{p=1}^{8} 2.87^{p-1}$ **2461.0** **53.** $\sum_{n=1}^{10} \left(\frac{1}{\pi}\right)^n$ **0.5** **54.** $\sum_{k=1}^{10} \left(\frac{1}{\pi}\right)^{k-1}$ **1.5**

Refer to the geometric series in which $t_1 = 5$ and $r = -5$.

55. Find S_4. **−520** **56.** Find S_6. **−13,020**

57. Given $t_n = 3125$, find n. $n = 5$ **58.** Given $t_n = 125$, find n. $n = 3$

Use mathematical induction to prove that each statement is true for all natural numbers, n.

59. $n < n + 1$ **60.** $2 \le n + 1$

61. $1 + 3 + 5 + \cdots + (2n - 1) = n^2$ **62.** $2 + 4 + 6 + \cdots + 2n = n(n + 1)$

Find S_{10} for each series. Round answers to the nearest tenth.

63. $2 + \sqrt{2} + 1 + \frac{\sqrt{2}}{2} + \cdots$ **6.6** **64.** $2 + 4\pi + 8\pi^2 + 16\pi^3 + \cdots$ **36,302,190.5**

65. GEOMETRY A side of a square is 8 centimeters long. A second square is inscribed in it by joining the midpoints of the sides of the first square. This process is continued as shown in the diagram at right.

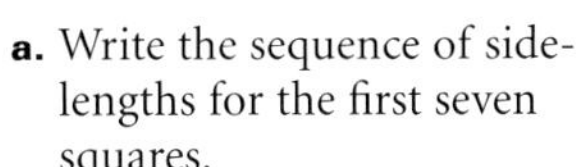

a. Write the sequence of side-lengths for the first seven squares.

b. Write an explicit formula for the sequence from part **a.**

c. Find the sum of the side lengths of the first seven squares.

d. Develop a formula in terms of t_1 and n for S_n, the sum of the side lengths of the first n squares.

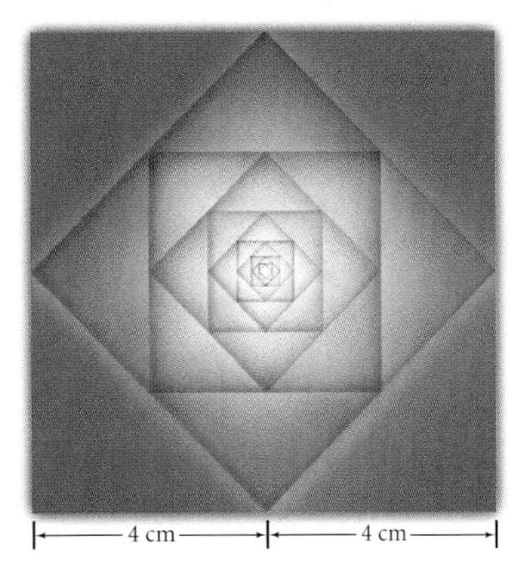

Error Analysis

Some students have difficulty evaluating the sum of a given geometric series when it is written in sigma notation. Have students identify the first term, the constant ratio, and the last term before evaluating the sum.

59. $n < n + 1$

1. Basis Step
Show that $n < n + 1$ is true for $n = 1$.
$1 < 1 + 1$ True

2. Induction Step
Assume that the statement is true for a natural number, k.
$k < k + 1$
Determine the statement to be proved.
$k + 1 < k + 2$
Rewrite the inequality and use the statement assumed to be true.
$k + 1 < k + 2$
$(k) + 1 < (k + 1) + 1$ True

60. $2 \le n + 1$

1. Basis Step
Show that $2 \le n + 1$ is true for $n = 1$.
$2 \le 1 + 1$ True

2. Induction Step
Assume that the statement is true for a natural number, k.
$2 \le k + 1$
Determine the statement to be proved.
$2 \le (k + 1) + 1$
$2 \le k + 2$
But if $2 \le k + 1$, by the addition property of inequality:
$2 \le 2 + 1 \le (k + 1) + 1$
$2 \le k + 2$ True

61. $1 + 3 + 5 + \cdots + (2n - 1) = n^2$

1. Basis Step
Show that $1 + 3 + 5 + \cdots + (2n - 1) = n^2$ is true for $n = 1$.
$1 = 1^2$ True

2. Induction Step
Assume the statement is true for a natural number, k.
$1 + 3 + 5 + \cdots + (2k - 1) = k^2$
Determine the statement to be proved by adding $2(k + 1) - 1$ to the left side and substituting $k + 1$ for k on the right side.
$1 + 3 + 5 + \cdots + (2k - 1) + [2(k + 1) - 1] = (k + 1)^2$
$1 + 3 + 5 + \cdots + (2k - 1) + (2k + 1) = (k + 1)^2$
Rewrite the left side by using the statement assumed to be true in order to obtain the right side.
$1 + 3 + 5 + \cdots + (2k - 1) + (2k + 1)$
$= k^2 + (2k + 1)$
$= k^2 + 2k + 1$
$= (k + 1)^2$ True

The answer to Exercise 62 can be found in Additional Answers beginning on page 1002.

74.

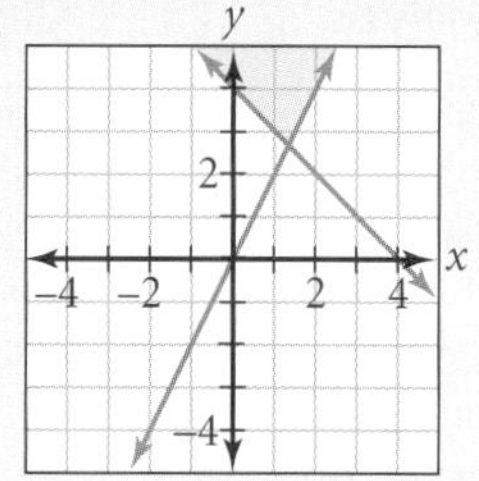

75.

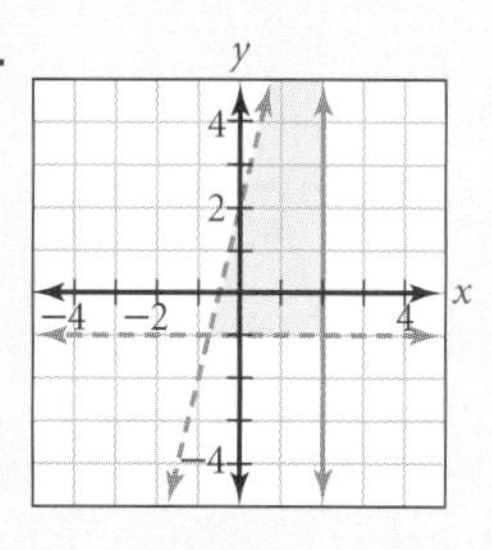

76.

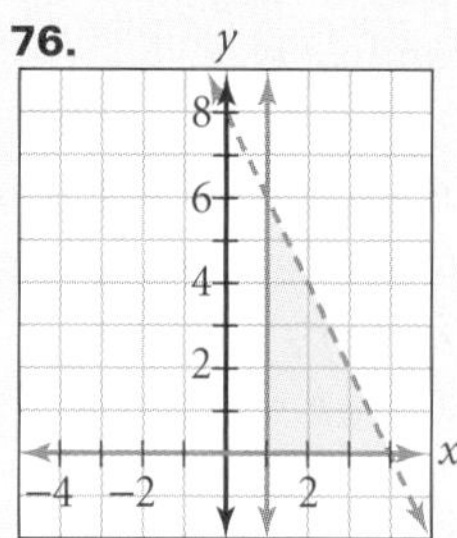

Practice

NAME ______________ CLASS ______________ DATE ______________

Practice

***11.5* Geometric Series and Mathematical Induction**

Find each sum. Round answers to the nearest tenth, if necessary.

1. S_{20} for the geometric series $4 + 12 + 36 + 108 + \cdots$ **6,973,568,800**
2. S_{15} for the geometric series $72 + 12 + 2 + \frac{1}{3} + \cdots$ **86.4**
3. S_6 for the series related to the geometric sequence $7, -14, 28, -56, \ldots$ **−147**
4. $\frac{7}{5} + \frac{7}{25} + \frac{7}{125} + \frac{7}{625} + \frac{7}{3125}$ **$\frac{5467}{3125} \approx 1.7$**
5. $1.3 - 5.2 + 20.8 - 83.2 + 332.8 - 1331.2$ **−1064.7**

For Exercises 6–9, refer to the geometric sequence 3, 6, 12, 24, . . .

6. Find t_{12}. **6144**
7. Find t_{20}. **1,572,864**
8. Find S_{12}. **12,285**
9. Find S_{20}. **3,145,725**

Evaluate. Round answers to the nearest hundredth, if necessary.

10. $\sum_{k=1}^{6} 6(2^{k-1})$ **378**
11. $\sum_{n=1}^{10} 4.8^{n-1}$ **1,708,554.00**
12. $\sum_{j=1}^{12} 5(0.25^k)$ **1.67**
13. $\sum_{m=1}^{15} \frac{2}{3}(3^m)$ **14,348,906**
14. $\sum_{p=1}^{6} 2(3)^{p-1}$ **728**
15. $\sum_{t=1}^{10} 3(-1)^{t+2}$ **0**
16. $\sum_{k=1}^{10} 5^k$ **12,207,030**
17. $\sum_{k=1}^{12} 3(2)^k$ **24,570**

Use mathematical induction to prove that the statement is true for every natural number, *n*.

18. $1^3 + 2^3 + 3^3 + \cdots + n^3 = \frac{n^2(n+1)^2}{4}$ Show the statement is true for $n = 1$: $1^3 = 1$ and $\frac{1^2(1+1)^2}{4} = 1$. Assume that the statement is true for a natural number k. Then $1^3 + 2^3 + 3^3 + \cdots + k^3 = \frac{k^2(k+1)^2}{4}$ and $1^3 + 2^3 + 3^3 + \cdots + k^3 + (k+1)^3 = \frac{k^2(k+1)^2}{4} + (k+1)^3 = \frac{k^2(k+1)^2 + 4(k+1)^3}{4} = \frac{(k+1)^2[k^2 + 4(k+1)]}{4} = \frac{(k+1)^2(k^2+4k+4)}{4} = \frac{(k+1)^2(k+2)^2}{4} = \frac{(k+1)^2((k+1)+1)^2}{4}$. Thus, the statement is true for $k + 1$.

CONNECTIONS

66. **GEOMETRY** Smaller and smaller squares are formed consecutively, as shown at right. Find the sum of the perimeters of the first nine squares if the first square is 40 inches wide. **319.375 in.**

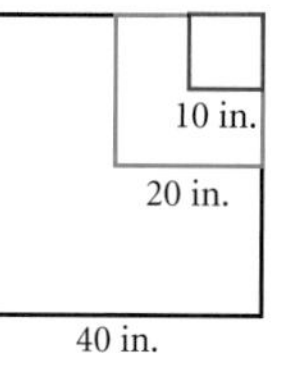

67. **GEOMETRY** Find the sum of the areas of the first six squares at right if the first square is 40 inches wide. **2132.8125 in.²**

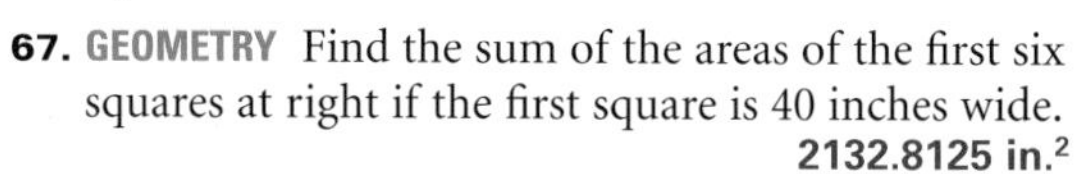

68. **GEOMETRY** A piece of wrapping paper is 0.0025 centimeter thick. Assuming that the result after folding is twice as thick, how thick will the paper be if it is folded on top of itself 5 times? **0.08 cm**

APPLICATIONS

INVESTMENTS Find the value of each investment at the end of the year of the last deposit.

69. \$3000 deposited at the beginning of every year for 10 years at 9% interest, compounded annually **\$49,680.88**
70. \$2000 deposited at the beginning of every year for 20 years at 8% interest, compounded annually **\$98,845.84**
71. \$1500 deposited at the beginning of every year for 50 years at 5% interest, compounded annually **\$329,723.09**

72. **DEMOGRAPHICS** The population of a city of 100,000 people increases 10% each year for 10 years. What will the population be after 10 years? **about 259,374**
73. **PHYSICS** A ball is dropped from a height of 8 feet. It rebounds to one-half of its original height and falls again. If the ball keeps rebounding in this manner, what is the total distance, to the nearest tenth, that the ball travels after 10 rebounds? (Hint: Draw a diagram, and notice the pattern.) **24.0 ft**

Look Back

Graph the solution to each system of linear inequalities. ***(LESSON 3.4)***

74. $\begin{cases} x + y \ge 4 \\ 2x \le y \end{cases}$

75. $\begin{cases} y < 4x + 2 \\ x \le 2 \\ y > -1 \end{cases}$

76. $\begin{cases} y < 8 - 2x \\ x \ge 1 \\ y \ge 0 \end{cases}$

APPLICATIONS

77. **DEPRECIATION** Personal computers often rapidly depreciate in value due to advances in technology. Suppose that a computer which originally cost \$3800 loses 10% of its value every 6 months. ***(LESSON 6.2)***
    a. What is the multiplier for this exponential decay function? **0.9**
    b. Write a formula for the value of the computer, $V(t)$, after t 6-month periods. **$V(t) = 3800(0.9)^t$**
    c. What is the value of the computer after 1 year? **\$3078**
    d. What is the value of the computer after 18 months? **\$2770.20**

Identify all asymptotes and holes in the graph of each rational function. ***(LESSON 8.2)***

78. $f(x) = \frac{2x+1}{x-3}$ **$x = 3$; $y = 2$; no holes**

79. $g(x) = \frac{x-5}{3x^2}$ **$x = 0$; $y = 0$; no holes**

80. $h(x) = \frac{x+4}{x^2+3x-4}$ **$x = 1$; $y = 0$; hole $(x = -4)$**

Simplify each rational function. ***(LESSON 8.3)***

81. $f(x) = \frac{x-3}{2x^2-5x-3}$ **$\frac{1}{2x+1}$**

82. $f(x) = \frac{x^2-5x+4}{x^2+2x-3}$ **$\frac{x-4}{x+3}$**

83. MUSIC Maria, a violinist, wants to form a string octet with friends from the school orchestra. She will need 3 more violinists, 2 violists, and 2 cellists. In the orchestra, there are 24 violinists (not including Maria), 7 violists, and 12 cellists. In how many different ways can she pick members of the octet? ***(LESSON 10.3)***
2,805,264 ways

Look Beyond

84. Consider the geometric sequence $\frac{1}{2}, \frac{1}{4}, \frac{1}{8}, \frac{1}{16}, \ldots$ Calculate S_1, S_2, S_3, S_4, and S_5. Plot these points on a number line. Describe the behavior of S_n as n increases. $\boldsymbol{\frac{1}{2}, \frac{3}{4}, \frac{7}{8}, \frac{15}{16}, \frac{31}{32}}$**; S_n approaches 1.**

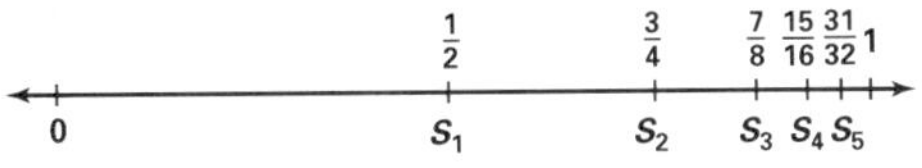

PORTFOLIO ACTIVITY

As tiles are moved over the edge of a desk, as shown at left, the center of gravity for the entire stack of tiles moves closer and closer to the edge of the desk. If the tiles are moved in such a way that the center of gravity for the entire stack is shifted beyond the edge of the desk, then the stack of tiles will fall over the edge.

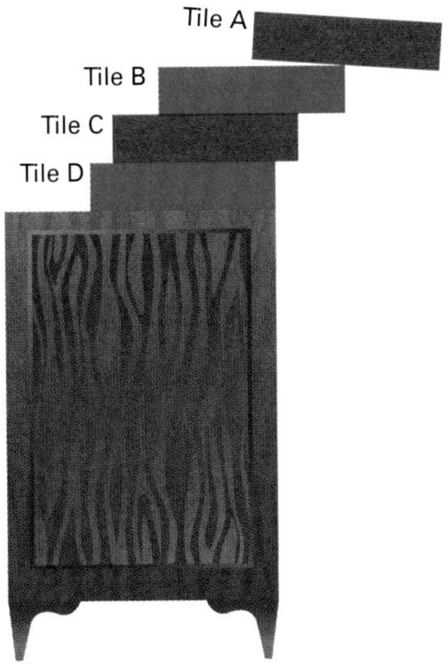

Place four congruent tiles in a neat stack on your desk so that the front edge of each tile is aligned with the edge of the desk. The original center of gravity for these tiles is located half of the length of a tile from the edge of the desk.

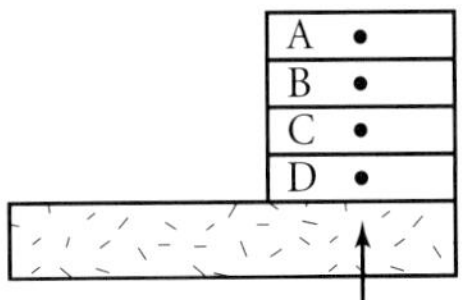

Extend tile A beyond tile B as far as possible while keeping the stack balanced. Keeping the position of tile A on tile B, extend tile B beyond tile C as far as possible while keeping the stack balanced. Continue this process with tiles C and D. Explain what happens to the distance that each successive tile can be extended.

WORKING ON THE CHAPTER PROJECT

You should now be able to complete Activity 2 of the Chapter Project.

Look Beyond

Exercise 84 allows students to explore the partial sums of an infinite geometric series. In Lesson 11.6, students will explore the limit of a geometric series.

ALTERNATIVE Assessment

Portfolio Activity

The Portfolio Activity can be used as preparation for the Chapter Project or as a separate activity. In the Portfolio Activity on this page, students will experimentally determine relationships between the center of gravity of stacked tiles.

Answers to Portfolio Activities can be found in Additional Answers of the Teacher's Edition.

Student Technology Guide

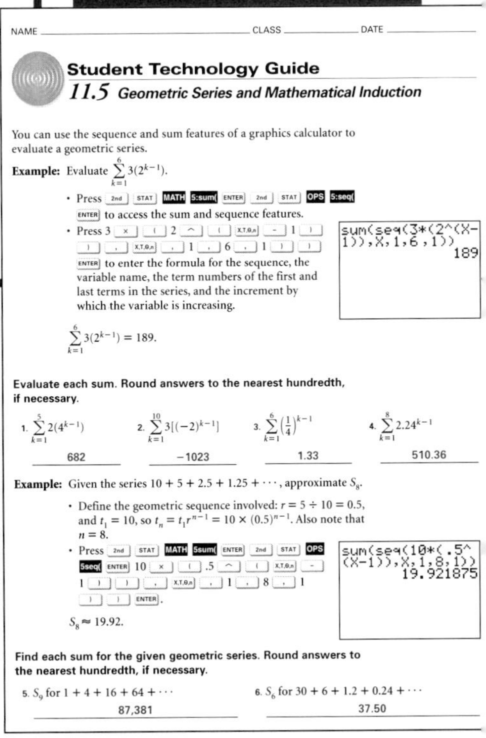

NAME ______ CLASS ______ DATE ______

Student Technology Guide
11.5 ***Geometric Series and Mathematical Induction***

You can use the sequence and sum features of a graphics calculator to evaluate a geometric series.

Example: Evaluate $\sum_{k=1}^{6} 3(2^{k-1})$.

- Press 2nd STAT MATH 5:sum(ENTER 2nd STAT OPS 5:seq(ENTER to access the sum and sequence features.
- Press 3 × (2 ^ (X,T,θ,n − 1)) , X,T,θ,n , 1 , 6 , 1)) ENTER to enter the formula for the sequence, the variable name, the term numbers of the first and last terms in the series, and the increment by which the variable is increasing.

```
sum(seq(3*(2^(X-
1)),X,1,6,1))
                189
```

$\sum_{k=1}^{6} 3(2^{k-1}) = 189$.

Evaluate each sum. Round answers to the nearest hundredth, if necessary.

1. $\sum_{k=1}^{5} 2(4^{k-1})$ 682
2. $\sum_{k=1}^{10} 3[(-2)^{k-1}]$ −1023
3. $\sum_{k=1}^{6} \left(\frac{1}{4}\right)^{k-1}$ 1.33
4. $\sum_{k=1}^{8} 2.24^{k-1}$ 510.36

Example: Given the series $10 + 5 + 2.5 + 1.25 + \cdots$, approximate S_8.

- Define the geometric sequence involved: $r = 5 \div 10 = 0.5$, and $t_1 = 10$, so $t_n = t_1 r^{n-1} = 10 \times (0.5)^{n-1}$. Also note that $n = 8$.
- Press 2nd STAT MATH 5sum(ENTER 2nd STAT OPS 5seq(ENTER 10 × (.5 ^ (X,T,θ,n − 1)) , X,T,θ,n , 1 , 8 , 1)) ENTER.

$S_8 \approx 19.92$.

```
sum(seq(10*(.5^
(X-1)),X,1,8,1))
          19.921875
```

Find each sum for the given geometric series. Round answers to the nearest hundredth, if necessary.

5. S_9 for $1 + 4 + 16 + 64 + \cdots$ 87,381
6. S_6 for $30 + 6 + 1.2 + 0.24 + \cdots$ 37.50

Prepare

NCTM PRINCIPLES & STANDARDS 1–4, 6–10

QUICK WARM-UP

Evaluate.

1. $\sum_{k=1}^{5}(7-k)$ 20
2. $\sum_{k=1}^{6}\left(\frac{1}{2}\right)^{k-1}$ $\frac{63}{32}$
3. $\sum_{k=1}^{4}5(-2)^{k-1}$ -25

Find the indicated sum.

4. $4 - 8 + 16 - 32 + 64 - \cdots$; S_{15} 43,692
5. $\frac{1}{5} + \frac{1}{10} + \frac{1}{20} + \frac{1}{40} + \cdots$; S_{11} $\frac{2047}{5120}$, or ≈ 0.4

Also on Quiz Transparency 11.6

Teach

Why Have students consider a football game in which the offense is on the defensive team's 2-yard line. On every subsequent play, the defensive team commits a violation and the ball is moved one-half of the remaining distance to the goal line. However, the rules will not allow the ball to cross the goal line for a touchdown as a result of a penalty. Ask students how close the ball can get to the goal line if these penalties continue indefinitely.

11.6 Infinite Geometric Series

Objective

- Find the sum of an infinite geometric series, if one exists.
- Write repeating decimals as fractions.

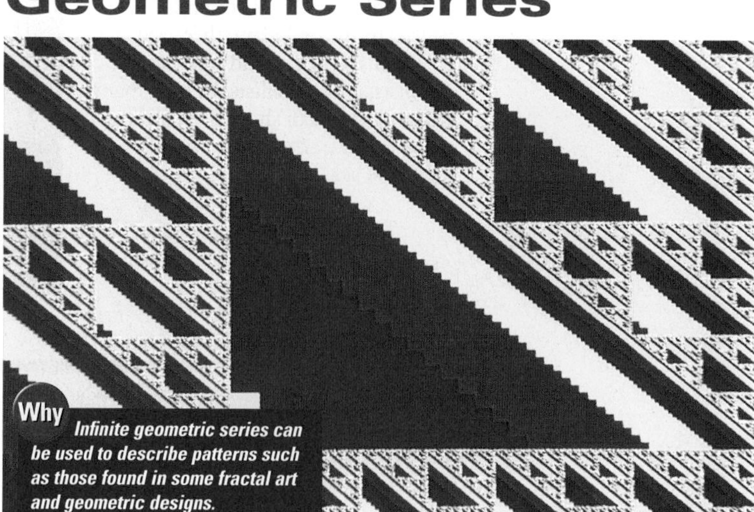

Why *Infinite geometric series can be used to describe patterns such as those found in some fractal art and geometric designs.*

From *Fractals Everywhere*, ©1988 Academic Press

APPLICATION ART

Denise is making a design using only equilateral triangles. The outer triangle is 20 inches long on each side. A second equilateral triangle is formed by joining the midpoints of the sides of the outer triangle. The process is continued. The length of the path shown in black in the figure at right can be modeled by the geometric series below.

$$20 + 10 + 5 + 2.5 + 1.25 + 0.625$$

If the path were to continue indefinitely, it could be modeled by the *infinite geometric series* below.

$$20 + 10 + 5 + 2.5 + \cdots + t_1r^{n-1} + \cdots$$

20 in.

An **infinite geometric series** is a geometric series with infinitely many terms. A **partial sum** of an infinite series is the sum of a given number of terms and not the sum of the entire series. Examine the table and the graph of partial sums for this infinite geometric series.

n	S_n
1	20
2	30
3	35
4	37.5
5	38.75
10	39.9609375
20	39.99996185
30	39.99999996
⋮	⋮

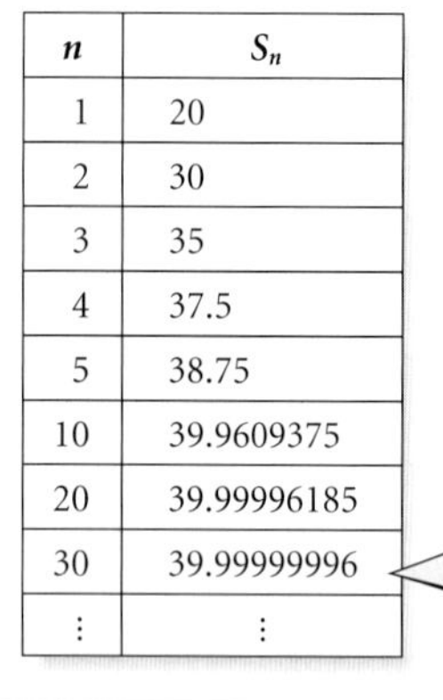

Notice that as n gets larger and larger, the sums get closer and closer to the number 40.

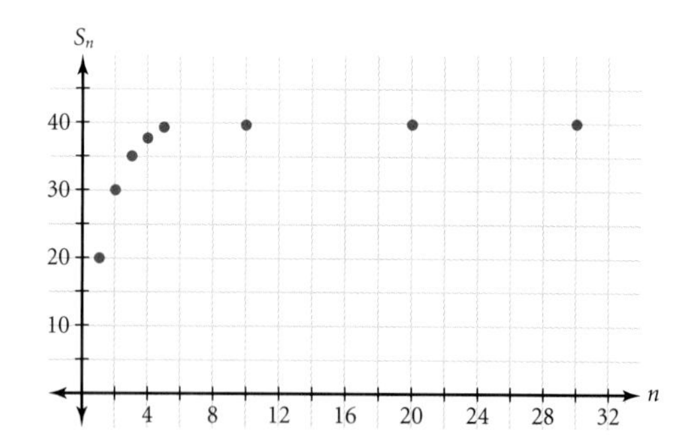

Alternative Teaching Strategy

USING MANIPULATIVES Have students cut a square, which is 1 foot on each side, from a sheet of paper. Have them cut the square in half to form two rectangles, each with an area of 0.5 square foot. Have them put one of the rectangles aside and record its area in a table similar to the one at right. Repeat the procedure of cutting the remaining piece in half, placing one of the new pieces aside, and recording in the table the total area set aside until the remaining piece is too small to cut.

Number of cuts	Combined area
1	0.5
2	0.5 + 0.25 = 0.75

Ask students to guess the combined area of the pieces if they could halve the pieces forever. They should see that the combined area of the pieces would eventually be 1 square foot. Now introduce the form of a infinite geometric series, and discuss when an infinite geometric series has a finite sum.

Examine the formula for the sum of a geometric series to see why this sum approaches 40.

$$S_n = t_1\left(\frac{1-r^n}{1-r}\right)$$

$$S_n = 20\left(\frac{1-0.5^n}{1-0.5}\right)$$

$$S_n = 20\left(\frac{1}{1-0.5} - \frac{0.5^n}{1-0.5}\right) \quad \text{Rewrite } \frac{1-0.5^n}{1-0.5}.$$

$$S_n = \frac{20}{1-0.5} - \frac{20(0.5^n)}{1-0.5}$$

What happens to 0.5^n as n gets larger?

As n gets larger, the rational expression $\frac{20(0.5^n)}{1-0.5}$ gets closer and closer to 0.

Therefore, the partial sums of this geometric series get closer and closer to $\frac{20}{1-0.5}$, or 40.

When the partial sums of an infinite series approach a fixed number as n increases, the infinite geometric series is said to **converge**.

When the partial sums of an infinite series do not approach a fixed number as n increases, the infinite geometric series is said to **diverge.**

Exploring Convergence

TECHNOLOGY GRAPHICS CALCULATOR
Keystroke Guide, page 760

You will need: a graphics calculator or paper and pencil

1. For each infinite geometric series indicated below, complete a table of values for n and S_n, using n-values of 1, 2, 3, 5, 10, and 100.
   a. $t_1 = -5$; $r = -0.25$ **b.** $t_1 = 5$; $r = -0.25$
   c. $t_1 = 2$; $r = \frac{1}{3}$ **d.** $t_1 = 2$; $r = -\frac{1}{3}$
2. For each infinite geometric series indicated below, complete a table of values for n and S_n, using n-values of 1, 2, 3, 5, 10, and 100.
   a. $t_1 = -5$; $r = -8$ **b.** $t_1 = 5$; $r = -8$
   c. $t_1 = 2$; $r = 3$ **d.** $t_1 = 2$; $r = -3$
3. Which of the infinite geometric series from Steps 1 and 2 converge? How are the common ratios of these series alike?
4. Which of the infinite geometric series from Steps 1 and 2 diverge? How are the common ratios of these series alike?
5. Make a conjecture about how the common ratio of an infinite geometric series determines whether or not the series converges. Test your conjecture. **CHECKPOINT ✔**

Sum of an Infinite Geometric Series

If a geometric sequence has common ratio r and $|r| < 1$, then the sum, S, of the related infinite geometric series is as follows:

$$S = \frac{t_1}{1-r}$$

Activity Notes

Before beginning each step of the Activity, ask students to write out the first few terms of the series and guess the sum. Then have them use calculators to compare their guesses with the calculated values for each series. Be sure they notice that if the partial sums of a geometric series approach a fixed number, then the sum of the infinite series is that number.

Teaching Tip

TECHNOLOGY When using a TI-82 or TI-83 graphics calculator, you can create a table of values for n and S_n by entering the formula for S_n in the **Y=** editor and then using the **TABLE SETUP** screen. To create a table of values for the Activity, let **Y1=T(1-R^X)/(1-R)** by pressing ALPHA × to enter **R**, and ALPHA 4 to enter **T**. Press 2nd WINDOW and make the following choices:

TblStart=1 (or **TblMin=1**)
△Tbl=1
Indpnt: Auto Ask
Depend: Auto Ask

Store values into **R** and **T** by entering a specific value and by pressing STO▶ ALPHA × ENTER for **R** values and by pressing STO▶ ALPHA 4 ENTER for **T** values. Then press 2nd GRAPH to view the table.

Cooperative Learning

Divide the class into groups of three. One member should develop the formula for each partial sum, one member should find the partial sums, and the third member should record each partial sum in a table.

CHECKPOINT ✔

5. If $|r| < 1$, the series converges. If $|r| \geq 1$, the series diverges.

Interdisciplinary Connection

GEOMETRY Begin with a number line that is 27 inches long. Beneath the number line, draw an identical line and erase the middle third, leaving the intervals [0, 9] and [18, 27]. Duplicate these intervals beneath the last intervals drawn, and remove the middle third from each duplicated interval. Repeat the process of removing the middle third of each remaining interval twice more. Have students determine the total length of all intervals drawn after the second, third, and fourth steps. $\left(\frac{2}{3}\right)^2$, $\left(\frac{2}{3}\right)^3$, **and** $\left(\frac{2}{3}\right)^4$ Have them generalize the total length of all intervals after n steps. $\left(\frac{2}{3}\right)^n$ Have students find the sum of lengths of all intervals if the process is repeated indefinitely. $\sum_{k=0}^{\infty}\left(\frac{2}{3}\right)^k = 3$

Use Teaching Transparency 45.

CHECKPOINT ✔
The signs of the partial sums alternate between positive and negative, but the absolute values of the partial sums increase as n increases.

CRITICAL THINKING
Answers may vary. S_n appears to be exponential because the table shows that the first and second differences are the same, but neither is constant. The shape of the graph reinforces the idea that the terms of the series are exponential.

ADDITIONAL EXAMPLE 1

Find the sum of the series $8 + 9.6 + 11.52 + 13.824 + \cdots$, if one exists.

The series diverges because $r = 1.2$ and $|1.2| > 1$.

Teaching Tip

TECHNOLOGY When using a TI-82 or TI-83 graphics calculator to check the sum of the infinite series in Example 1, access the **Y=** editor and enter **3(1–.4^X)/(1–.4)** for **Y1**. Then set the viewing window as follows:

Xmin=0	Ymin=0
Xmax=30	Ymax=10
Xscl=1	Yscl=1

After graphing the function, press TRACE and use the arrow keys to show that the terms approach 5 as x increases.

TRY THIS
$-\frac{8}{3}$

If a geometric sequence has common ratio r and $|r| > 1$, then the related infinite geometric series diverges and therefore does not have a sum. Examine the table of values and the graph of the infinite geometric series below in which $t_1 = 5$ and $r = 2$:

$$5 + 10 + 20 + 40 + \cdots + t_1r^{n-1} + \cdots$$

n	S_n
1	5
2	15
3	35
4	75
5	155
6	315
7	635
8	1275
⋮	⋮

Notice that as n gets larger and larger, the sums also get larger and larger.

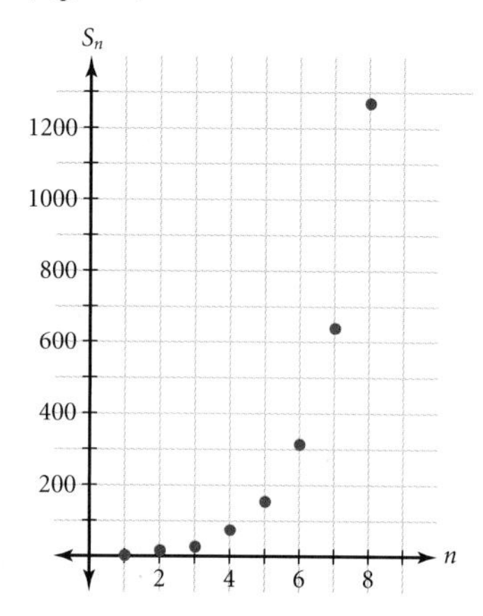

CHECKPOINT ✔ What happens to the partial sums of an infinite geometric series in which $t_1 = 5$ and $r = -2$ as n increases?

CRITICAL THINKING Examine the table and graph for S_n above. Does S_n increase exponentially? Explain.

EXAMPLE 1 **Find the sum of the infinite series $3 + 1.2 + 0.48 + 0.192 + \cdots$, if it exists.**

SOLUTION

1. Determine whether the infinite series is geometric.
$$\frac{t_2}{t_1} = \frac{1.2}{3} = 0.4 \qquad \frac{t_3}{t_2} = \frac{0.48}{1.2} = 0.4 \qquad \frac{t_4}{t_3} = \frac{0.192}{0.48} = 0.4$$
This is an infinite geometric series in which $t_1 = 3$ and $r = 0.4$.

2. Determine whether the series diverges or converges.
Since $|0.4| < 1$, the series converges. The sum is $S = \frac{t_1}{1-r}$.
$$S = \frac{t_1}{1-r} = \frac{3}{1-0.4} = 5$$

The sum of the series is 5. This seems reasonable because $S_4 = 4.872$.

CHECK
Graph the formula for the sum of the nth term of the geometric series in which $t_1 = 3$ and $r = 0.4$.

$$S_n = t_1\left(\frac{1-r^n}{1-r}\right), \text{ or } y = 3\left(\frac{1-0.4^x}{1-0.4}\right)$$

From the graph, you can see that the series converges to a sum of 5.

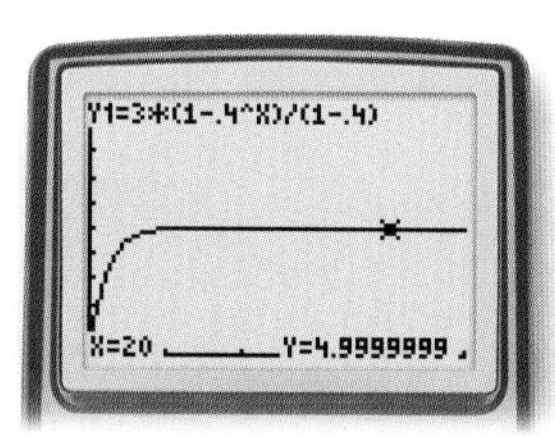

TRY THIS Find the sum of the infinite series $-4 + 2 + (-1) + 0.5 + \cdots$, if it exists.

Enrichment

Tell students that the formula for the sum of an infinite geometric series is true for complex numbers as well as real numbers. Then ask them to use this formula to find the sum of the series $\sum_{k=1}^{\infty}\left(\frac{i}{2}\right)^{k-1}$. $\frac{4}{5} + \frac{2}{5}i$

Inclusion Strategies

USING SYMBOLS To be sure that students understand summation notation, read Example 2 to them as "$\sum_{k=1}^{\infty}\frac{1}{3^{k+1}}$ represents the sum of $\frac{1}{3^{k+1}}$ for values of k from 1 to infinity." Have them use summation notation to express the sum of $(-2)^{1-k}$ for values of k from 1 to infinity.

The mathematical symbol for infinity is ∞. The notation $\sum_{k=1}^{\infty}$ indicates an infinite series.

EXAMPLE 2 **Find the sum of the infinite series $\sum_{k=1}^{\infty} \frac{1}{3^{k+1}}$, if it exists.**

SOLUTION

1. Determine whether the infinite series is geometric.

$$\frac{t_2}{t_1} = \frac{\frac{1}{27}}{\frac{1}{9}} = \frac{1}{3} \qquad \frac{t_3}{t_2} = \frac{\frac{1}{81}}{\frac{1}{27}} = \frac{1}{3}$$

This is a geometric series in which $t_1 = \frac{1}{9}$ and $r = \frac{1}{3}$.

2. Determine whether the series diverges or converges. Because $\left|\frac{1}{3}\right| < 1$, the series converges. The sum is $S = \frac{t_1}{1-r}$.

$$S = \frac{t_1}{1-r} = \frac{\frac{1}{9}}{1-\frac{1}{3}} = \frac{1}{6}$$

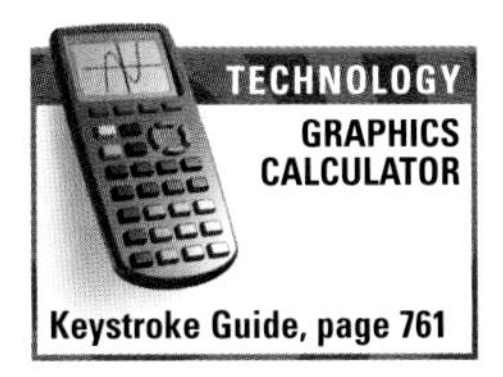

CHECK

Graph the formula for the sum of the nth term of the geometric series in which $t_1 = \frac{1}{9}$ and $r = \frac{1}{3}$.

From the graph, you can see that the series converges to a sum of $\frac{1}{6}$.

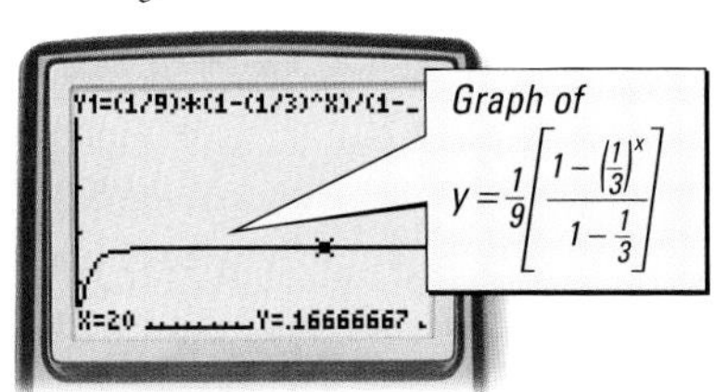

Graph of $y = \frac{1}{9}\left[\frac{1-\left(\frac{1}{3}\right)^x}{1-\frac{1}{3}}\right]$

TRY THIS Find the sum of the infinite series $\sum_{k=1}^{\infty} \frac{1}{2^{k+1}}$, if it exists.

Every repeating decimal is a rational number and can therefore be written in the form $\frac{p}{q}$, where p and q are integers with no common factors. Infinite series can be used to convert a repeating decimal into a fraction.

EXAMPLE 3 **Write $0.\overline{2}$ as a fraction in simplest form.**

SOLUTION

1. Write the repeating decimal as an infinite geometric series.

$$\begin{aligned} 0.\overline{2} &= 0.222\ldots \\ &= 0.2 + 0.02 + 0.002 + \cdots \\ &= \frac{2}{10} + \frac{2}{100} + \frac{2}{1000} + \cdots \end{aligned}$$

Notice that $0.\overline{2}$ can be written as an infinite geometric series of decimals or fractions. Thus, $t_1 = 0.2$, or $\frac{2}{10}$, and $r = 0.1$, or $\frac{1}{10}$.

2. Because $|0.1| < 1$, the series converges. The sum is $S = \frac{t_1}{1-r}$.

$$S = \frac{0.2}{1-0.1} = \frac{0.2}{0.9} = \frac{2}{9}$$

CHECK

Use a calculator: $2 \div 9 = 0.222\ldots$ Thus, $0.\overline{2}$ can be written as $\frac{2}{9}$.

Reteaching the Lesson

USING PATTERNS Have students refer to Example 1 and find S_1, S_2, S_3, and S_n. Be sure that they use the formula for the sum of the nth term of the geometric series to find S_n. **3, 4.2, 4.68, and $5(1 - 0.4^n)$** Now ask them to describe what happens to 0.4^n and to $5(1 - 0.4^n)$ when they substitute larger and larger values for n. **As n increases, 0.4^n approaches 0 and $5(1 - 0.4^n)$ approaches 5.** Tell them that since the sequence of sums converges to 5, they can say that the infinite series has a sum of 5.

ADDITIONAL EXAMPLE 2

Find the sum of the infinite series $\sum_{k=1}^{\infty} \frac{1}{5^{k+1}}$, if it exists. $\frac{1}{20}$

Teaching Tip

TECHNOLOGY When using a TI-82 or TI-83 graphics calculator to check the sum of the infinite series in Example 2, access the **Y=** editor and enter **1/9(1–(1/3)^X)/(1–(1/3))** for **Y1**. Then set the viewing window as follows:

Xmin=0	**Ymin=0**
Xmax=30	**Ymax=.5**
Xscl=1	**Yscl=1**

Graph the function, press TRACE, and use the arrow keys to show that the terms approach $\frac{1}{6}$, or ≈ 0.16667, as x increases.

TRY THIS

$\frac{1}{2}$

ADDITIONAL EXAMPLE 3

Write $0.0\overline{4}$ as a fraction in lowest terms. $\frac{2}{45}$

Teaching Tip

When using a TI-82 or TI-83 graphics calculator, you can change a repeating decimal to a fraction by pressing MATH, selecting 1:▶Frac, and pressing ENTER. Be sure to enter several decimal places to ensure the correct conversion.

TRY THIS Write $0.\overline{5}$ as a fraction in simplest form.

Exercises

Communicate

1. Explain how to determine whether an infinite geometric series has a sum.
2. How can you tell if the series $\sum_{k=1}^{\infty}\left(\frac{1}{2}\right)^k$ and the series $\sum_{k=1}^{\infty}2^k$ converge?
3. Explain how to write a repeating decimal as a fraction.

Guided Skills Practice

4. Find the sum of the infinite series $3 + 2 + \frac{4}{3} + \frac{8}{9} + \cdots$, if it exists. **9** ***(EXAMPLE 1)***
5. Find the sum of the infinite series $\sum_{k=1}^{\infty}\frac{1}{4^{k+1}}$, if it exists. ***(EXAMPLE 2)*** $\frac{1}{12}$
6. Write $0.\overline{3}$ as a fraction in simplest form. ***(EXAMPLE 3)*** $\frac{1}{3}$

Practice and Apply

Find the sum of each infinite geometric series, if it exists.

7. $\frac{1}{3} + \frac{1}{9} + \frac{1}{27} + \frac{1}{81} + \cdots$ $\frac{1}{2}$
8. $1 + \frac{1}{5} + \frac{1}{25} + \frac{1}{125} + \cdots$ $\frac{5}{4}$
9. $2 + 1.5 + 1.125 + 0.84375 + \cdots$ **8**
10. $3 + 1.2 + 0.48 + 0.192 + \cdots$ **5**
11. $1 + 2 + 4 + 8 + \cdots$ **none**
12. $2 + 6 + 18 + 54 + \cdots$ **none**
13. $\frac{11}{15} + \frac{1}{15} + \frac{1}{165} + \frac{1}{1815} + \cdots$ $\frac{121}{150}$
14. $\frac{9}{17} + \frac{3}{17} + \frac{1}{17} + \frac{1}{51} + \cdots$ $\frac{27}{34}$
15. $3 + 2.1 + 1.47 + 1.029 + \cdots$ **10**
16. $4 + 3.2 + 2.56 + 2.048 + \cdots$ **20**
17. $\frac{1}{3} + \frac{4}{9} + \frac{16}{27} + \frac{64}{81} + \cdots$ **none**
18. $\frac{2}{5} + \frac{12}{25} + \frac{72}{125} + \frac{432}{625} + \cdots$ **none**

internet connect
Homework Help Online
Go To: **go.hrw.com**
Keyword:
MB1 Homework Help
for Exercises 19–38

Find the sum of each infinite geometric series, if it exists.

19. $\sum_{k=0}^{\infty}\left(\frac{1}{10}\right)^k$ $\frac{10}{9}$
20. $\sum_{k=0}^{\infty}\left(\frac{5}{3}\right)^k$ **none**
21. $\sum_{k=1}^{\infty}\left(\frac{1}{8}\right)^k$ $\frac{1}{7}$
22. $\sum_{k=1}^{\infty}(-0.45)^k$ $-\frac{9}{29}$
23. $\sum_{k=0}^{\infty}\frac{7}{10^k}$ $\frac{70}{9}$
24. $\sum_{k=1}^{\infty}\frac{3}{5^{k+1}}$ $\frac{3}{20}$
25. $\sum_{k=0}^{\infty}\left(-\frac{3}{7}\right)^k$ $\frac{7}{10}$
26. $\sum_{j=0}^{\infty}\left(-\frac{4}{11}\right)^j$ $\frac{11}{15}$
27. $\sum_{k=1}^{\infty}2.9^k$ **none**
28. $\sum_{k=1}^{\infty}4.6^k$ **none**
29. $\sum_{k=1}^{\infty}0.7^k$ $\frac{7}{3}$
30. $\sum_{k=1}^{\infty}(-0.73)^k$ $-\frac{73}{173}$
31. $\sum_{n=0}^{\infty}3^n$ **none**
32. $\sum_{m=0}^{\infty}5^m$ **none**
33. $\sum_{k=1}^{\infty}\frac{3^{k-1}}{4^k}$ **1**
34. $\sum_{k=1}^{\infty}\frac{4^{k+1}}{3^k}$ **none**
35. $\sum_{k=0}^{\infty}\left(\frac{1}{\sqrt{2}}\right)^k$
36. $\sum_{j=0}^{\infty}\left(\frac{2}{\sqrt{3}}\right)^j$
37. $\sum_{j=0}^{\infty}\left(\frac{1}{\pi}\right)^j$
38. $\sum_{k=0}^{\infty}\left(\frac{\sqrt{2}}{\sqrt{3}}\right)^k$

35. $2 + \sqrt{2}$, or ≈ 3.4142

36. none

37. $\frac{\pi}{\pi - 1}$, or ≈ 1.4669

38. $3 + \sqrt{6}$, or ≈ 5.4495

TRY THIS
$\frac{5}{9}$

Assess

Selected Answers
Exercises 4–6, 7–69 odd

ASSIGNMENT GUIDE	
In Class	1–6
Core	7–63 odd
Core Plus	8–64 even
Review	65–70
Preview	71

Extra Practice can be found beginning on page 940.

Practice

NAME ______ CLASS ______ DATE ______

Practice
***11.6* Infinite Geometric Series**

Find the sum of each infinite geometric series, if it exists.

1. $60 + 84 + 117.6 + 164.64 + \cdots$ — does not exist
2. $\frac{4}{5} + \frac{4}{15} + \frac{4}{45} + \frac{4}{135} + \cdots$ — 1.2
3. $\frac{7}{8} + \frac{7}{12} + \frac{7}{18} + \frac{7}{27} + \cdots$ — 2.625
4. $5 + 4 + 3.2 + 2.56 + \cdots$ — 25

Find the sum of each infinite geometric series, if it exists.

5. $\sum_{n=1}^{\infty}0.8^n$ — 4
6. $\sum_{m=1}^{\infty}\left(\frac{11}{9}\right)^{m-1}$ — does not exist
7. $\sum_{k=1}^{\infty}11\cdot\left(\frac{1}{9}\right)^{k-1}$ — 12.375
8. $\sum_{j=1}^{\infty}0.75^j$ — 3
9. $\sum_{t=1}^{\infty}0.45^{t-1}$ — $1.\overline{81}$
10. $\sum_{x=1}^{\infty}0.92^x$ — 11.5
11. $\sum_{k=1}^{\infty}7.3^{k-1}$ — does not exist
12. $\sum_{b=1}^{\infty}49(0.02)^{b-1}$ — 50
13. $\sum_{n=1}^{\infty}20(0.1)^{n-1}$ — $22\frac{2}{9}$

Write each decimal as a fraction in simplest form.

14. $0.\overline{1}$ — $\frac{1}{9}$
15. $0.\overline{37}$ — $\frac{37}{99}$
16. $0.\overline{49}$ — $\frac{49}{99}$
17. $0.\overline{753}$ — $\frac{251}{333}$
18. $0.\overline{225}$ — $\frac{25}{111}$
19. $0.\overline{370}$ — $\frac{10}{27}$

Write an infinite geometric series that converges to the given number.

20. $0.0707070707\ldots$ — $\sum_{k=1}^{\infty}7\cdot\left(\frac{1}{100}\right)^k$
21. $0.9393939393\ldots$ — $\sum_{k=1}^{\infty}93\cdot\left(\frac{1}{100}\right)^k$
22. $0.1515151515\ldots$ — $\sum_{k=1}^{\infty}15\cdot\left(\frac{1}{100}\right)^k$
23. $0.358358358\ldots$ — $\sum_{k=1}^{\infty}358\cdot\left(\frac{1}{1000}\right)^k$
24. $0.011011011\ldots$ — $\sum_{k=1}^{\infty}11\cdot\left(\frac{1}{1000}\right)^k$
25. $0.445445445\ldots$ — $\sum_{k=1}^{\infty}445\cdot\left(\frac{1}{1000}\right)^k$

39. $\sum_{k=1}^{\infty} 19\left(\frac{1}{100}\right)^k$

40. $\sum_{k=1}^{\infty} 57\left(\frac{1}{100}\right)^k$

41. $\sum_{k=1}^{\infty} \left(\frac{1}{1000}\right)^k$

42. $\sum_{k=1}^{\infty} 219\left(\frac{1}{1000}\right)^k$

43. $\sum_{k=1}^{\infty} 35\left(\frac{1}{100}\right)^k$

44. $\sum_{k=1}^{\infty} 89\left(\frac{1}{100}\right)^k$

45. $\sum_{k=1}^{\infty} 819\left(\frac{1}{1000}\right)^k$

46. $\sum_{k=1}^{\infty} 733\left(\frac{1}{1000}\right)^k$

47. $\sum_{k=1}^{\infty} 121\left(\frac{1}{1000}\right)^k$

Write an infinite geometric series that converges to the given number.

39. 0.191919... **40.** 57.57575757... **41.** 0.001001001001...

42. 0.219219219... **43.** 0.353535... **44.** 0.898989...

45. 0.819819819... **46.** 0.733733733... **47.** 0.121121121...

Write each decimal as a fraction in simplest form.

48. $0.\overline{5}$ $\frac{5}{9}$ **49.** $0.\overline{4}$ $\frac{4}{9}$ **50.** $0.\overline{72}$ $\frac{8}{11}$ **51.** $0.\overline{36}$ $\frac{4}{11}$

52. $0.\overline{43}$ $\frac{43}{99}$ **53.** $0.\overline{54}$ $\frac{6}{11}$ **54.** $0.\overline{372}$ $\frac{124}{333}$ **55.** $0.\overline{586}$ $\frac{586}{999}$

56. $0.\overline{831}$ $\frac{277}{333}$ **57.** $0.\overline{474}$ $\frac{158}{333}$ **58.** $0.\overline{626}$ $\frac{626}{999}$ **59.** $0.\overline{031}$ $\frac{31}{999}$

60. Derive the formula for the sum of an infinite geometric series by using the examination of the formula shown on the top of page 729, in which n becomes increasingly large.

CONNECTIONS

61. GEOMETRY The midpoints of the sides of a square are joined to create a new square. This process is repeated for each new square. Find the sum of an infinite series of the areas of such squares if the side length of the original square is 10 centimeters. **200 square centimeters**

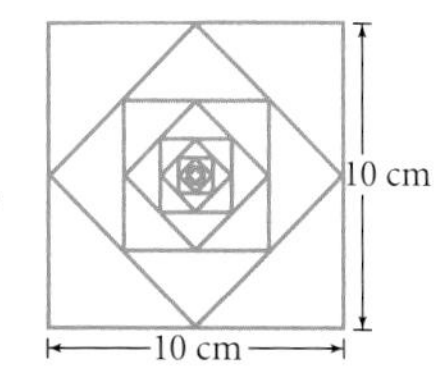

CHALLENGE

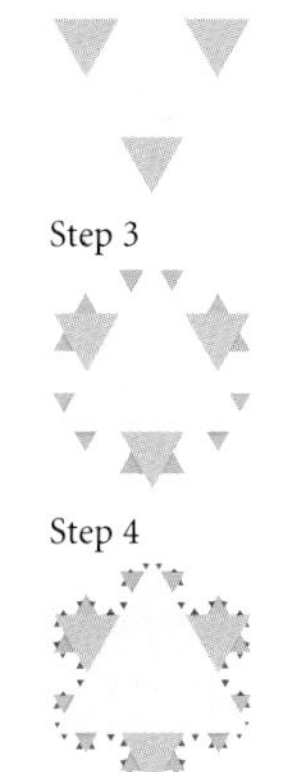

62. GEOMETRY A Koch curve can be constructed by taking an equilateral triangle and adding a smaller equilateral triangle along the middle third of each side (see Step 2 at left). This process is continued, or *iterated*, infinitely to form a *fractal*.

a. Assume that the original triangle has side lengths of 9. Copy and complete the table below. Step n may contain either recursive or explicit formulas.

	Step 1	Step 2	Step 3	Step 4	...	Step n
Number of sides	3	12	48	**192**	...	$3(4)^{n-1}$
Length of each side	9	3	1	$\frac{1}{3}$	...	$9\left(\frac{1}{3}\right)^{n-1}$
Perimeter	27	36	48	**64**	...	$27\left(\frac{4}{3}\right)^{n-1}$
Number of new triangles	1	3	12	**48**	...	$3(4)^{n-2}$
New area added	$\frac{81\sqrt{3}}{4}$	$\frac{9\sqrt{3}}{4}$	$\frac{\sqrt{3}}{4}$	$\frac{\sqrt{3}}{36}$	...	$\frac{81\sqrt{3}}{4}\left(\frac{1}{9}\right)^{n-1}$

b. Which of the formulas in Step n describe sequences? **all formulas**

c. If the formulas in Step n describe series, which are convergent and which are divergent? **conv.: length and area; div.: the rest**

d. What is the unusual relationship between the perimeter and area of this figure? **perimeter becomes infinite while area is finite**

APPLICATION

63. INVESTMENTS A *perpetuity* is an investment that pays a fixed amount of money at the end of every period forever. Perpetuities have a finite present value. For example, the present value of a perpetuity that earns 7% interest and that pays $500 at the end of every year is as follows:

$$P = 500\left(\frac{1}{1.07}\right) + 500\left(\frac{1}{1.07}\right)^2 + 500\left(\frac{1}{1.07}\right)^3 + 500\left(\frac{1}{1.07}\right)^4 + \cdots$$

a. Find P by finding the sum of the infinite geometric series. **$7142.86**

b. Find the present value of a perpetuity that earns 8% interest and that pays $1000 at the end of every year forever. **$12,500**

Use Teaching Transparency 46.

Error Analysis

Students should be reminded that the use of parentheses is especially important when using a calculator to evaluate fractions with complicated denominators. For example, when entering $\frac{3}{\left(1-\frac{1}{3}\right)}$ into a graphics calculator, enter 3/(1–(1/3)). Also, when raising a fraction to a power on a calculator, remind students to surround the fraction with parentheses so that both the numerator and denominator are raised to the power.

60. The sum of the first n terms of a geometric series is $S_n = t_1\left(\frac{1-r^n}{1-r}\right)$. This expression can be rewritten as $S_n = t_1\left(\frac{1}{1-r} - \frac{r^n}{1-r}\right) = \frac{t_1}{1-r} - \frac{t_1 r^n}{1-r}$. As n gets larger and larger, the value of r^n gets closer and closer to 0. Thus, the second term, $\frac{t_1 r^n}{1-r}$, gets closer and closer to 0. Therefore, the sum of an infinite geometric series is given by $S = \frac{t_1}{1-r}$.

Student Technology Guide

NAME ______ CLASS ______ DATE ______

Student Technology Guide
11.6 Infinite Geometric Series

An infinite geometric series *converges* if its sum approaches a particular number and *diverges* if it does not approach a particular number. You can use a graphics calculator to test whether a series converges, and, if it does, to find the number its sum approaches.

Example: Find the sum of the series 10 + 8 + 6.4 + 5.12 + ⋯, if one exists. Check your answer by graphing.

- For this series, $t_1 = 10$ and $r = 0.8$. Because $|0.8| < 1$, the series converges. Its sum is $S = \frac{t_1}{1-r} = \frac{10}{1-0.8} = 50$.
- Before checking by graphing, set up an appropriate viewing window. Press [WINDOW] and change Xmin to 0, Xmax to 94, Xscl to 10, Ymin to 0, Ymax to 60, and Yscl to 10. (This choice for Xmin and Xmax gives only whole-number values for x.)
- To check the answer by graphing, enter the formula for the sum of the series, $10\left(\frac{1-0.8^x}{1-0.8}\right)$.
- Press [Y=] 10 [×] [(] 1 [-] .8 [^] [X,T,θ,n] [)] [÷] [(] 1 [-] .8 [)] [GRAPH].
- Press [TRACE] and use [►] to move along the graph.

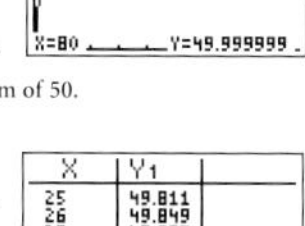

The graph confirms that the series converges to a sum of 50.

You can also use a table to explore infinite geometric series. With the formula for the series entered as Y1, press [2nd] [WINDOW]. Change TblStart to 1 and ΔTbl to 1. Be sure that the other two settings are set to Auto. Press [2nd] [GRAPH]. Use [▼] to move through the table.

X	Y1
25	49.811
26	49.849
27	49.879
28	49.903
29	49.923
30	49.938
31	49.95

X=31

The table confirms that the series converges and has a sum of 50.

Find the sum of each infinite geometric series, if one exists. Check your answers by graphing or by using a table.

1. 4 + 2.8 + 1.96 + 1.372 + ⋯ $13\frac{1}{3} \approx 13.33$
2. 1 + 1.5 + 2.25 + 3.375 + ⋯ Series diverges; sum does not exist.
3. $\frac{16}{15} + \frac{4}{15} + \frac{1}{15} + \frac{1}{60} + \cdots$ $1\frac{19}{45} \approx 1.42$

In Exercise 71, students are asked to write a repeating decimal from an infinite geometric series.

ALTERNATIVE

Assessment

Portfolio Activity

The Portfolio Activity can be used as preparation for the Chapter Project or as a separate activity. In the Portfolio Activity on this page, students will use experiments to determine the limits involved in keeping stacked objects balanced.

Answers to Portfolio Activities can be found in Additional Answers of the Teacher's Edition.

APPLICATION

64. PHYSICS A golf ball is dropped from a height of 81 inches. It rebounds to $\frac{2}{3}$ of its original height and continues rebounding in this manner. How far does it travel before coming to rest? **405 inches**

65 Use a matrix equation to solve the system at right. ***(LESSON 4.4)***

$$\begin{cases} -2x + y + 6z = 18 \\ 5x + 8z = -16 \\ 3x + 2y - 10z = -3 \end{cases}$$

(−4, 7, 0.5)

66. $x = 2 + \frac{\ln 23}{\ln 2} \approx 6.52$

67. $x = -\frac{\ln 19}{\ln 3} \approx -2.68$

68. $x = -\frac{\ln 0.5}{\ln 8} \approx -0.33$

69. $x = \frac{\ln 7.23}{\ln 2} \approx 2.85$

Solve each equation for *x*. Write the exact solution and the approximate solution to the nearest hundredth, when appropriate. *(LESSONS 6.6 AND 6.7)*

66. $2^{x-2} = 23$ **67.** $3^{-x} = 19$ **68.** $8^x = 0.5$ **69.** $2^x = 7.23$

APPLICATION

70. CONSTRUCTION A manager of a construction company is managing three projects. The probabilities that the three projects will be completed on schedule are 0.85, 0.72, and 0.94. If the events are independent, what is the probability, to the nearest hundredth, that all three projects will be completed on schedule? ***(LESSON 10.5)*** **0.58**

internet connect
Portfolio Extension
Go To: go.hrw.com
Keyword:
MB1 System COG

Look Beyond

71. a. Write out the first four terms of the infinite geometric series in which $t_1 = 0.14$ and $r_1 = 0.02$. **0.14, 0.1428, 0.142856, 0.14285712**

b. Find the sum of the infinite series from part **a.** Simplify your answer. $\frac{1}{7}$

c. Write the answer as a repeating decimal with bar notation. $0.\overline{142857}$

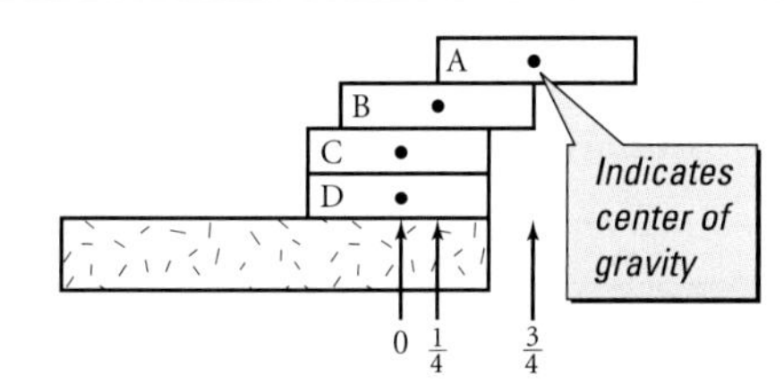

Refer to the diagram of tiles above. The center of gravity for tile A is $\frac{3}{4}$ of the length of a tile to the right of 0. The center of gravity for tile B is $\frac{1}{4}$ of the length of a tile to the right of 0. The centers of gravity for tiles C and D are both 0.

1. Position four tiles as shown in the diagram.
2. Keeping the position of tile A on tile B and tile B on tile C, find the maximum distance, in tile lengths, to the right of 0 that tile C can be moved without the stack of tiles falling. Record your results.
3. While keeping the position of tile A on tile B, tile B on tile C, and tile C on tile D (from Step 2), find the maximum distance to the right of 0 that tile D can be moved without the stack of tiles falling. Record your results.

WORKING ON THE CHAPTER PROJECT

You should now be able to complete Activity 3 of the Chapter Project.

Pascal's Triangle

In this painting by Stanley Meltzoff, do you notice anything unusual about the way that the pennies have landed?

Objectives

- Find entries in Pascal's triangle.
- Use Pascal's triangle to find combinations and probabilities.

Why *The patterns in Pascal's triangle can be used to solve real-world problems involving probability.*

Find the probability that exactly 3 heads or exactly 4 heads appear when a coin is tossed 7 times. *You will answer this question in Example 3.*

In this chapter you have studied the patterns in some sequences and series. A famous pattern in the form of an arithmetic triangle, called **Pascal's triangle,** is shown below.

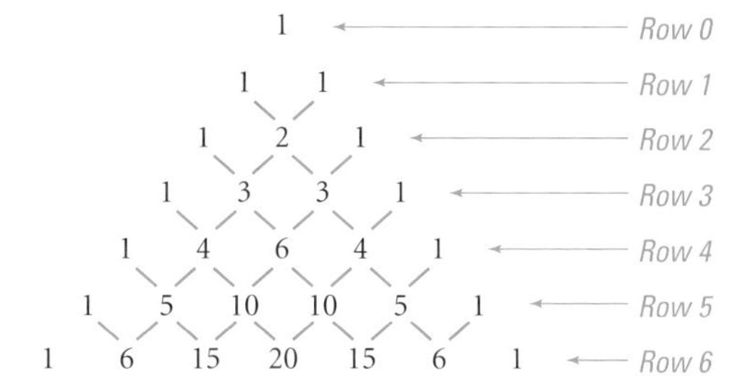

Blaise Pascal (1623–1662)

Notice that each row begins with 1 and ends with 1. Each number between the 1s is the sum of the pair of numbers above it in the previous row.

The arithmetic triangle above is called Pascal's triangle because the French mathematician and theologian Blaise Pascal analyzed the pattern extensively in the seventeenth century.

Prepare

NCTM PRINCIPLES & STANDARDS 1, 2, 4–6, 8–10

QUICK WARM-UP

Evaluate each expression.

1. ${}_{11}C_4$ 330 **2.** ${}_{11}C_1$ 11

3. ${}_{8}C_8$ 1 **4.** ${}_{8}C_0$ 1

5. ${}_{7}C_2$ 21 **6.** ${}_{7}C_5$ 21

A fair coin is tossed 3 times.

7. What is the total number of possible outcomes? 8

8. What is the probability of tossing

a. exactly 3 heads? $\frac{1}{8}$

b. exactly 2 heads? $\frac{3}{8}$

c. exactly 1 heads? $\frac{3}{8}$

d. 0 heads? $\frac{1}{8}$

Also on Quiz Transparency 11.7

Teach

Why Pascal's triangle is a pattern of numbers that is used to find the number of ways that an event can occur. Have students think about a football team with 40 players. Ask students how they would determine the number of ways in which a team of 11 players could be selected.
${}_{40}C_{11} = 2,311,801,440$

Alternative Teaching Strategy

USING TABLES To help students understand the relationship between Pascal's triangle and the ways in which n and k are related in successive values of ${}_nC_k$, give students the alternative form of Pascal's triangle in a table, as shown below.

	Taken 0 at a time	Taken 1 at a time	Taken 2 at a time	Taken 3 at a time
0 items	${}_0C_0 = 1$			
1 item	${}_1C_0 = 1$	${}_1C_1 = 1$		
2 items	${}_2C_0 = 1$	${}_2C_1 = 2$	${}_2C_2 = 1$	

Tell students that the first row is designated as row 0, the second row is designated as row 1, and so on. Have them read the first entry in row 0 as "the number of combinations of 0 items taken 0 at a time." At the same time have them use combination notation to write ${}_0C_0$. Now have them read the second entry in row 2 as "the number of combinations of 2 items taken 1 at a time" and write ${}_2C_1$. Have students complete the table through row 15 by using this pattern. When they have completed all the rows you can ask them to give the value of ${}_6C_2$ and other combinations. **${}_6C_2 = 15$**

Use Teaching Transparency 47.

Math CONNECTION

PROBABILITY You may want to show students how to list all of the possible outcomes of tossing a coin 7 times by using a tree diagram. By following each branch of the tree, you can show students the total possible outcomes. Then relate the frequency of each event to Pascal's triangle and to the probability of each outcome.

***Activity* Notes**

In Steps 5 and 6 of the Activity, students should see that the shaded regions are shaped like inverted equilateral triangles. The area to the left and right of each inverted triangle is the same shape as the triangle that appears immediately above it. If students have trouble seeing the shapes, recommend that they extend Pascal's triangle through row 24 and continue to shade the even numbers.

CHECKPOINT ✔

3. The sum of row n is 2^n.

CHECKPOINT ✔

4. Rows 2, 3, 7, 11, and 13 have this property too. Row n, where n is a prime number, will have this property.

CHECKPOINT ✔

6. The shaded numbers form inverted triangles.

In the Activity below, you can investigate some of the many interesting patterns in Pascal's triangle.

Activity Exploring Patterns in Pascal's Triangle

You will need: no special tools

1. Make several copies of Pascal's triangle through row 15.
2. Find the sum of the entries in each row of Pascal's triangle.
3. **CHECKPOINT ✔** What patterns do you see in the sums from Step 2?
4. **CHECKPOINT ✔** Notice that 5 divides all of the entries in row 5, except the first and last. Find three more rows in which the row number divides all the entries in that row (except the first and last). Find a rule to predict which rows have this property.
5. Shade all of the even numbers in the triangle.
6. **CHECKPOINT ✔** What shape(s) do you see in the shaded numbers?

The table below shows all of the possible outcomes when a coin is tossed 4 times. It also indicates how many ways each event can occur.

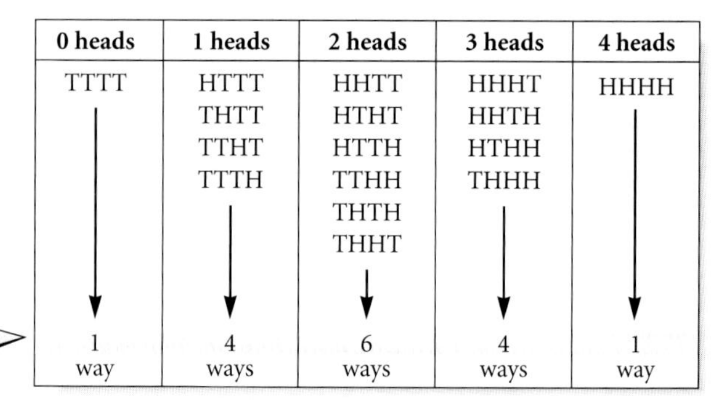

0 heads	1 heads	2 heads	3 heads	4 heads
TTTT	HTTT THTT TTHT TTTH	HHTT HTHT HTTH TTHH THTH THHT	HHHT HHTH HTHH THHH	HHHH
1 way	4 ways	6 ways	4 ways	1 way

Notice that these numbers are the numbers in row 4 of Pascal's Triangle.

The number of ways that each event above can occur can be found by using combinations. For example, the number of ways that 2 heads can occur when a coin is tossed 4 times is a combination of 4 objects taken 2 at a time, or ${}_4C_2$.

Pascal's triangle can also be expressed in combination notation, as shown below.

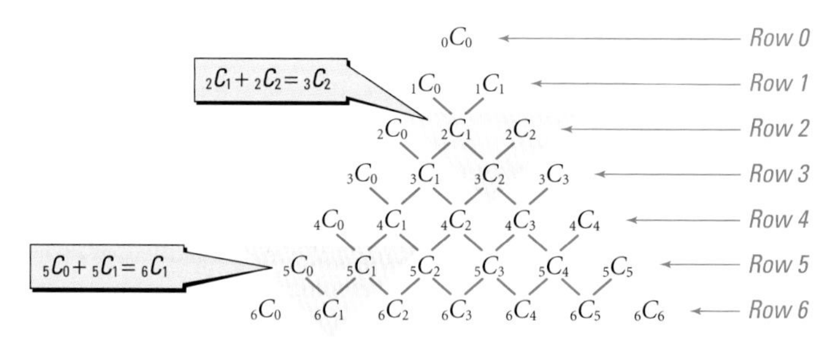

Interdisciplinary Connection

SPORTS In many sports, the regular season is followed by a playoff series. A "best-of-five series" is completed when one team wins 3 games. Have students record and count of all the ways a team can win the series. Be sure students realize that the series can be completed after 3 games, 4 games, or 5 games and that the final game must be a win for the winning team. **${}_2C_2 + {}_3C_2 + {}_4C_2 = 10$ ways** Have them do the same for a "best-of-seven series," in which 5 wins are needed to win the series. **${}_4C_4 + {}_5C_4 + {}_6C_4 = 21$ ways**

Inclusion Strategies

HANDS-ON STRATEGIES Suggest that students perform an experiment to simulate at least 60 tosses of 4 coins by using the random-number generator of a calculator. Have them record in a table the number of heads in each trial. Then have them write the theoretical frequency of each event in combination notation. They should see that their experimental results approximate the theoretical frequencies given by Pascal's triangle. Remind them that their experiment would more closely approximate the theoretical outcomes if the number of trials were increased.

The patterns listed below can be found in Pascal's triangle.

Patterns in Pascal's Triangle

- Row n of Pascal's triangle contains $n + 1$ entries.
- The kth entry in row n of Pascal's triangle is ${}_nC_{k-1}$.
- The sum of all the entries in row n of Pascal's triangle equals 2^n.
- ${}_nC_{k-1} + {}_nC_k = {}_{n+1}C_k$, where $0 < k \leq n$

CHECKPOINT ✔ Verify each equation below by evaluating combinations.

a. ${}_7C_4 + {}_7C_5 = {}_8C_5$ **b.** ${}_8C_1 + {}_8C_2 = {}_9C_2$ **c.** ${}_4C_2 + {}_4C_3 = {}_5C_3$

EXAMPLE 1 **Find the 4th and 10th entries in row 12 of Pascal's triangle.**

SOLUTION

TECHNOLOGY GRAPHICS CALCULATOR

Keystroke Guide, page 761

The kth entry in row n of Pascal's triangle is ${}_nC_{k-1}$.

4th entry in row 12:

$${}_nC_{k-1} = {}_{12}C_3 = \frac{12!}{9!3!} = 220$$

10th entry in row 12:

$${}_nC_{k-1} = {}_{12}C_9 = \frac{12!}{3!9!} = 220$$

The 4th and 10th entries in row 12 of Pascal's triangle are both 220.

CHECK

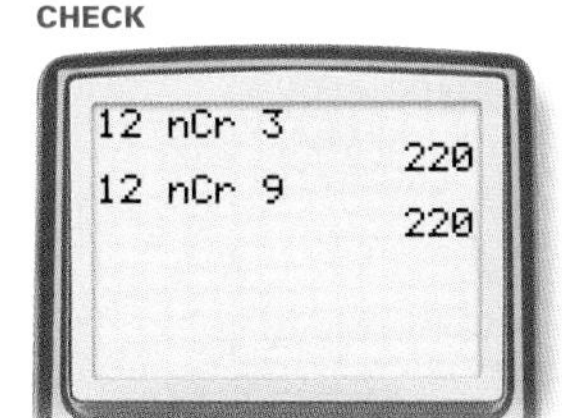

TRY THIS Find the 2nd and 12th entries in row 11 of Pascal's triangle.

You can solve probability problems by using Pascal's triangle.

EXAMPLE 2 **Suppose that a fair coin is tossed 7 times. In how many ways can exactly 0, 1, 2, 3, 4, 5, 6, and 7 heads appear?**

SOLUTION

Because the coin is tossed 7 times, use row 7 of Pascal's triangle.

Heads	0	1	2	3	4	5	6	7
Number of ways	${}_7C_0$ 1	${}_7C_1$ 7	${}_7C_2$ 21	${}_7C_3$ 35	${}_7C_4$ 35	${}_7C_5$ 21	${}_7C_6$ 7	${}_7C_7$ 1

TRY THIS Suppose that a fair coin is tossed 8 times. In how many ways can exactly 0, 1, 2, 3, 4, 5, 6, 7, and 8 heads appear?

Enrichment

Students will use a copy of Pascal's triangle for this activity. Be sure they see that all entries in Pascal's triangle, except those positioned at the edges of the sides, are surrounded by 6 other entries. For instance, in row 5, 10 is surrounded by the entries 4, 6, 10, 20, 15, and 5. Have students find the product of all the entries surrounding 20 in row 6. **15 • 10 • 10 • 15 • 35 • 35 = 27,562,500** Now ask them to find the product of alternating numbers around 20 (for example, 10, 15, and 35). **5250** Have students square this product. They should see that the two products are equal.

CHECKPOINT ✔

a. 35 + 21 = 56
b. 8 + 28 = 36
c. 6 + 4 = 10

ADDITIONAL EXAMPLE 1

Find the 6th and 11th entries in row 15 of Pascal's triangle. 3003 and 3003

TRY THIS

11 and 1

Teaching Tip

TECHNOLOGY When using a TI-82 or TI-83 graphics calculator to check the 4th and 10th entries in row 12 of Pascal's triangle, first enter **12** on the home screen, then press MATH, select **PRB** **3:nCr**, and enter **3**. Press ENTER. The result, 220, is displayed. Now press 2nd ENTER, move the cursor onto **3**, and enter **9**. Press ENTER. The result, 220, is displayed.

ADDITIONAL EXAMPLE 2

Suppose that a fair coin is tossed 5 times. **In how many ways can exactly 0, 1, 2, 3, 4, and 5 heads appear?**

Heads	0	1	2	3	4	5
Number of ways	1	5	10	10	5	1

TRY THIS

1, 8, 28, 56, 70, 56, 28, 8, 1

Teaching Tip

The formula below, known as Stirling's approximation, gives a good estimate of $n!$ for large values of n.

$$n! = \sqrt{2n\pi}\left(\frac{n}{e}\right)^n$$

Row n of Pascal's triangle indicates the number of ways that each possible outcome can occur when a coin is tossed n times. The sum of all the entries in row n of Pascal's triangle, 2^n, gives the total number of possible outcomes that can occur when a coin is tossed n times.

Pascal's Triangle and Two-Outcome Experiments

If a probability experiment with 2 equally likely outcomes is repeated in n independent trials, the probability $P(A)$ of event A occurring exactly k times is given by $P(A) = \frac{{}_nC_k}{2^n}$.

EXAMPLE 3 **Find the probability that exactly 3 heads or exactly 4 heads appear when a coin is tossed 7 times. Give your answer to the nearest hundredth.**

CONNECTION PROBABILITY

TECHNOLOGY GRAPHICS CALCULATOR

Keystroke Guide, page 761

SOLUTION

The events "3 heads" and "4 heads" are mutually exclusive.

$$P(3 \text{ heads } or \text{ } 4 \text{ heads}) = P(3 \text{ heads}) + P(4 \text{ heads})$$
$$= \frac{{}_7C_3}{2^7} + \frac{{}_7C_4}{2^7}$$
$$= \frac{35 + 35}{128}$$
$$\approx 0.55$$

The probability that exactly 3 heads or exactly 4 heads appear when a coin is tossed 7 times is about 0.55.

CHECK

TRY THIS Find the probability that exactly 4 heads or exactly 6 heads appear when a coin is tossed 8 times. Give your answer to the nearest hundredth.

Exercises

Communicate

1. Describe the patterns that you see in Pascal's triangle when it is written in ${}_nC_r$ notation.
2. Describe two patterns that you see in Pascal's triangle when it is written with integer entries.
3. **PROBABILITY** Explain how Pascal's triangle can be used to find the probability that exactly 4 tails will appear when a coin is tossed 6 times. CONNECTION

ADDITIONAL EXAMPLE 3

Find the probability of getting exactly 2 heads or exactly 3 heads when a fair coin is tossed 5 times. Give your answer to the nearest hundredth. 0.625

TRY THIS

≈0.38

Math CONNECTION

PROBABILITY Remind students that when tossing a coin, the event "heads" and the event "tails" are mutually exclusive because they cannot both occur in one toss of a coin. When tossing a coin 7 times, the events "3 heads" and "4 heads" are mutually exclusive for the same reason.

Practice

NAME ______ CLASS ______ DATE ______

Practice 11.7 Pascal's Triangle

State the location of each entry in Pascal's triangle. Then give the value of each expression.

1. ${}_7C_5$ sixth entry, row 7; 21
2. ${}_6C_3$ fourth entry, row 6; 20
3. ${}_8C_6$ seventh entry, row 8; 28
4. ${}_{10}C_5$ sixth entry, row 10; 252
5. ${}_{13}C_{10}$ eleventh entry, row 13; 286
6. ${}_{12}C_5$ sixth entry, row 12; 792

Find the indicated entries in Pascal's triangle.

7. fourth entry, row 10 120
8. seventh entry, row 13 1716
9. ninth entry, row 15 6435
10. third entry, row 18 153

A fair coin is tossed the indicated number of times. Find the probability of each event.

11. 5 tosses; exactly 3 heads 0.3125
12. 6 tosses; no more than 3 heads ≈0.66
13. 10 tosses; exactly 1 head ≈0.01
14. 8 tosses; fewer than 5 heads ≈0.64
15. 5 tosses; no fewer than 3 heads 0.5
16. 7 tosses; 2 *or* 3 heads 0.4375

A student guesses the answers to 6 questions on a true-false quiz. Find the probability that the indicated number of guesses are correct.

17. exactly 4 ≈0.23
18. fewer than 5 ≈0.89
19. no more than 2 ≈0.34
20. exactly 5 ≈0.09
21. at least 3 ≈0.66
22. at least 4 ≈0.34

Reteaching the Lesson

USING PATTERNS The town council wants to create a 3-person committee chosen from the mayor and 6 other council members. Have students determine how many committees are possible if the mayor must be a member and if the mayor cannot be a member. ${}_6C_2 = 15$, ${}_6C_3 = 20$ Now ask them to find how many 3-person committees are possible if all of the council (including the mayor) are eligible to be chosen. ${}_7C_3 = 35$ Point out that ${}_7C_3 = {}_6C_2 + {}_6C_3$. Now have them replace 6 with n and 3 with k in order to get the general form ${}_{n+1}C_k = {}_nC_{k-1} + {}_nC_k$, where $0 \leq k \leq n$. Then have them use this form to construct Pascal's triangle.

Guided Skills Practice

4. Find the third and fifth entries in row 10 of Pascal's triangle. ***(EXAMPLE 1)*** **45; 210**

CONNECTIONS

5. 1; 6; 15; 20; 15; 6; 1

5. PROBABILITY Suppose that a fair coin is tossed 6 times. In how many ways can exactly 0, 1, 2, 3, 4, 5, and 6 heads appear? ***(EXAMPLE 2)***

6. PROBABILITY Find the probability that exactly 2 heads or exactly 3 heads appear when a coin is tossed 6 times. Give your answer to the nearest hundredth. ***(EXAMPLE 3)*** **≈0.55**

Practice and Apply

State the location of each entry in Pascal's triangle. Then give the value of each expression.

7. ${}_5C_2$ **10** **8.** ${}_7C_3$ **35** **9.** ${}_8C_5$ **56** **10.** ${}_7C_4$ **35**

11. ${}_{11}C_2$ **55** **12.** ${}_{12}C_4$ **495** **13.** ${}_{10}C_3$ **120** **14.** ${}_9C_2$ **36**

15. ${}_8C_4$ **70** **16.** ${}_6C_4$ **15** **17.** ${}_7C_2$ **21** **18.** ${}_9C_2$ **36**

19. ${}_9C_6$ **84** **20.** ${}_{11}C_4$ **330** **21.** ${}_{13}C_7$ **1716** **22.** ${}_{12}C_8$ **495**

Find the 4th and 7th entries in the indicated row of Pascal's triangle.

23. row 7 **35; 7** **24.** row 9 **84; 84** **25.** row 11 **165; 462** **26.** row 13 **286; 1716**

CHALLENGES

internet connect
Homework Help Online
Go To: **go.hrw.com**
Keyword:
MB1 Homework Help for Exercise 27

27. In the figure below, the sum of each diagonal is at the end of the diagonal.
a. Find the missing sums. **5, 8, 13, 21, 34**
b. What is the name of the sequence formed by the sums? **Fibonacci**

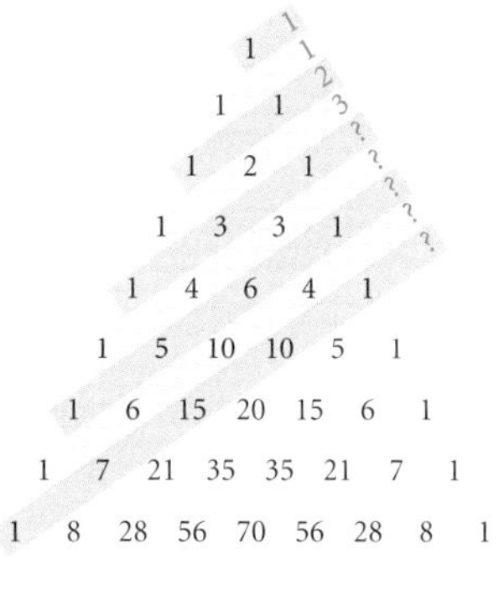

28. 121, 1331, 14641
a. The digits are the entries for rows 2, 3, and 4.

28. Evaluate 11^2, 11^3, and 11^4.
a. How are these numbers related to Pascal's triangle?
b. Formulate a rule for multiplying by 11 based on your answer to part **a.** Test your response by multiplying other numbers by 11.

CONNECTION

b. $11^n = {}_nC_0 10^n + {}_nC_1 10^{n-1} + {}_nC_2 10^{n-2} + \cdots + {}_nC_n 10^0$

PROBABILITY **Find the probability of each event.**

29. At least 4 heads appear in 6 tosses of a fair coin. **≈0.34**
30. At least 3 heads appear in 5 tosses of a fair coin. **0.5**
31. No more than 2 heads appear in 5 tosses of a fair coin. **0.5**
32. No more than 3 heads appear in 6 tosses of a fair coin. **≈0.66**
33. Either 4 *or* 5 heads appear in 8 tosses of a fair coin. **≈0.49**
34. Either 5 *or* 6 heads appear in 7 tosses of a fair coin. **≈0.22**

Teaching Tip

TECHNOLOGY When using a TI-82 or TI-83 graphics calculator to check the probability in Example 3, press MODE, move to the **Float** row, and select **2** to round the answer to the nearest hundredth. Then return to the home screen by pressing 2nd MODE. To calculate the solution for Example 3, enter **7** on the home screen. Press MATH and select PRB 3:nCr. Enter **3 + 7.** Press MATH and select PRB 3:nCr. Enter **4** and press ENTER. Press ÷ **2^7** and ENTER to display **0.55.** Reset the mode to **Float.**

7. ${}_5C_2$ **is the third entry in the fifth row,** ${}_5C_2 = 10$.

8. ${}_7C_3$ **is the fourth entry in the seventh row,** ${}_7C_3 = 35$.

9. ${}_8C_5$ **is the sixth entry in the eighth row,** ${}_8C_5 = 56$.

10. ${}_7C_4$ **is the fifth entry in the seventh row,** ${}_7C_4 = 35$.

Student Technology Guide

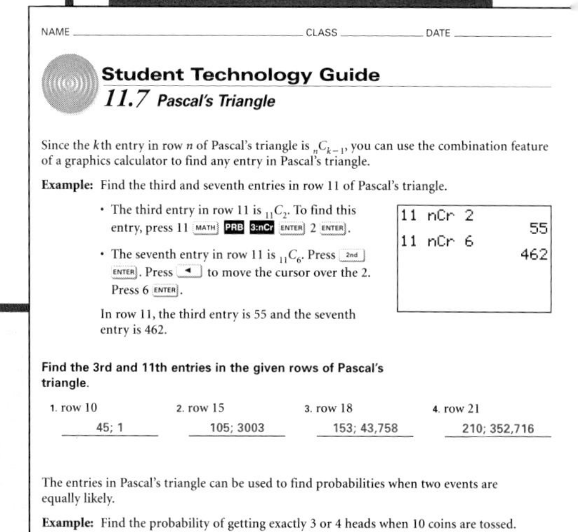
NAME ________ CLASS ________ DATE ________

Student Technology Guide
11.7 ***Pascal's Triangle***

Since the kth entry in row n of Pascal's triangle is ${}_nC_{k-1}$, you can use the combination feature of a graphics calculator to find any entry in Pascal's triangle.

Example: Find the third and seventh entries in row 11 of Pascal's triangle.

- The third entry in row 11 is ${}_{11}C_2$. To find this entry, press 11 MATH PRB 3:nCr ENTER 2 ENTER.
- The seventh entry in row 11 is ${}_{11}C_6$. Press 2nd ENTER. Press ◄ to move the cursor over the 2. Press 6 ENTER.

```
11 nCr 2
              55
11 nCr 6
             462
```

In row 11, the third entry is 55 and the seventh entry is 462.

Find the 3rd and 11th entries in the given rows of Pascal's triangle.

1. row 10 **45; 1**
2. row 15 **105; 3003**
3. row 18 **153; 43,758**
4. row 21 **210; 352,716**

The entries in Pascal's triangle can be used to find probabilities when two events are equally likely.

Example: Find the probability of getting exactly 3 or 4 heads when 10 coins are tossed.

- The probability of k heads out of n tosses is given by $\frac{{}_nC_k}{2^n}$. The events "3 heads" and "4 heads" are mutually exclusive, so $P(3 \text{ heads } or\ 4 \text{ heads}) = \frac{{}_{10}C_3 + {}_{10}C_4}{2^{10}}$.
- Press ((10 MATH PRB 3:nCr ENTER 3) + (10 MATH PRB 3:nCr ENTER 4)) ÷ 2 ^ 10 ENTER.

```
((10 nCr 3)+(10
nCr 4))/2^10
       .322265625
```

The probability of getting exactly 3 or 4 heads in 10 coin tosses is about 0.32, or 32%.

5. Find the probability of getting exactly 4 or 5 heads when 14 coins are tossed. **≈ 0.18, or 18%**

11. ${}_{11}C_2$ **is the third entry in the eleventh row,** ${}_{11}C_2 = 55$.

12. ${}_{12}C_4$ **is the fifth entry in the twelfth row,** ${}_{12}C_4 = 495$.

13. ${}_{10}C_3$ **is the fourth entry in the tenth row,** ${}_{10}C_3 = 120$.

14. ${}_9C_2$ **is the third entry in the ninth row,** ${}_9C_2 = 36$.

15. ${}_8C_4$ **is the fifth entry in the eighth row,** ${}_8C_4 = 70$.

16. ${}_6C_4$ **is the fifth entry in the sixth row,** ${}_6C_4 = 15$.

17. ${}_7C_2$ **is the third entry in the seventh row,** ${}_7C_2 = 21$.

18. ${}_9C_2$ **is the third entry in the ninth row,** ${}_9C_2 = 36$.

19. ${}_9C_6$ **is the seventh entry in the ninth row,** ${}_9C_6 = 84$.

The answers to Exercises 20–22 can be found in Additional Answers beginning on page 1002.

Assess

Selected Answers

Exercises 4–6, 7–53 odd

ASSIGNMENT GUIDE	
In Class	1–6
Core	7–25 odd, 29–39 odd
Core Plus	8–46 even
Review	48–54
Preview	55

✎ Extra Practice can be found beginning on page 940.

Error Analysis

Some students will give the wrong entries for the rows of Pascal's triangle because they will count the first row as row 1. Point out that the first row of Pascal's triangle is called row 0. Encourage students to label the rows when they use Pascal's triangle. Also, remind them that the kth entry in row n of Pascal's triangle represents ${}_nC_{k-1}$, not ${}_nC_k$.

Look Beyond

In Exercise 55, students expand a binomial and examine the coefficients of each term. In Lesson 11.8, students will be shown how the entries in Pascal's triangle match the coefficients in the binomial expansion.

50. $x < -3$ *or* $x > 2$

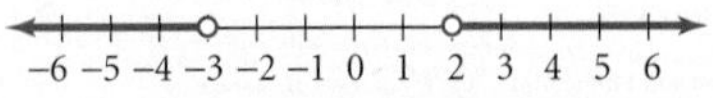

51. $-9 < x < 2$

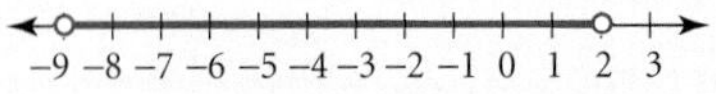

52. $x \leq -6$ *or* $x \geq 1$

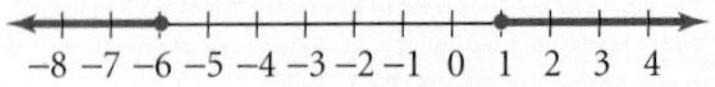

APPLICATIONS

GENETICS Assume that the genders of children are equally likely. A family has 5 children. Find the probability that the children are the following:

35. exactly 3 girls **≈0.31** **36.** exactly 5 girls **≈0.03** **37.** at least 4 boys **≈0.19**

38. at least 3 boys **0.5** **39.** at most 3 girls **≈0.81** **40.** at most 2 boys **0.5**

ACADEMICS A student guesses the answers for 5 items on a true-false quiz. Find the probability that the indicated number of answers are correct.

41. exactly 3 **≈0.31** **42.** exactly 4 **≈0.16** **43.** at least 4 **≈0.19**

44. at least 3 **0.5** **45.** at most 3 **≈0.81** **46.** at most 4 **≈0.97**

47a. yes;
−1, =2, ≡3, ≣4, ≣5, ⊢6, ⊨7, ⊨8, ≗10, ||||15, ≗20, ⊥21, ≟28, ||||35, ⊤56, ⫯70
b. sample answer: sum of the rows are powers of 2

47. CULTURAL CONNECTION: ASIA In 1303 the Chinese mathematician Chu Shih-Chieh published the book whose cover is shown at right. The caption reads "The Old Method Chart of the Seven Multiplying Squares."

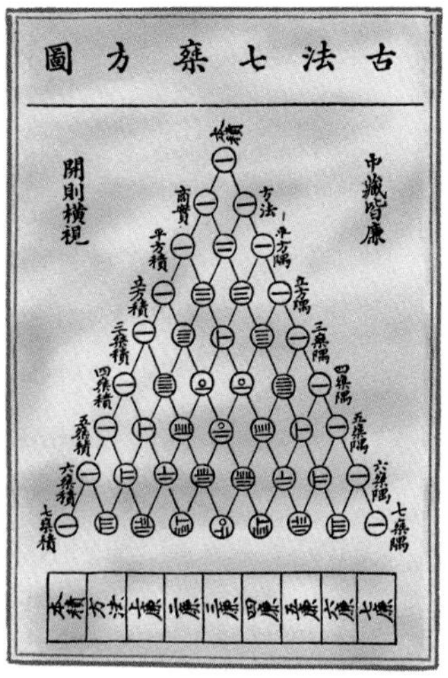

a. Do the numerical symbols appear to correspond to Pascal's triangle? If so, write the symbols that are equivalent to one another.

b. How can the name of the triangle be explained? (Hint: The third row from top to bottom, 1 2 1, is labeled as the first "multiplying square.")

Look Back

Find a quadratic function that fits each list of data points. *(LESSON 5.7)*

48. (0, 5), (1, 6), (3, 20) $f(x) = 2x^2 - x + 5$

49. (4, 4), (7, −2), (−3, −7)

49. $f(x) = -\frac{5}{14}x^2 + \frac{27}{14}x + 2$

Solve each inequality. Graph the solution on a number line. *(LESSON 5.8)*

50. $x^2 + x - 6 > 0$ $x < -3$ *or* $x > 2$

51. $x^2 + 7x - 18 < 0$ $-9 < x < 2$

52. $6 \leq 5x + x^2$ $x \leq -6$ *or* $x \geq 1$

State whether each situation involves a permutation or a combination. Then solve. *(LESSONS 10.2 AND 10.3)*

53. the number of ways to award 1st, 2nd, and 3rd prizes to a group of 10 floats entered in a homecoming parade **permutation; 720**

54. the number of ways to select a committee of 5 senators from a group of 100 senators **combination; 75,287,520**

Look Beyond

55. Use multiplication and the Distributive Property to expand $(x + y)^4$ as a polynomial in x and y with decreasing powers of x. Compare the coefficients with Pascal's triangle. Do the entries in one of the rows agree with the coefficients in the expansion? If so, which row?

55. $x^4 + 4x^3y + 6x^2y^2 + 4xy^3 + y^4$; **row 4**

The Binomial Theorem

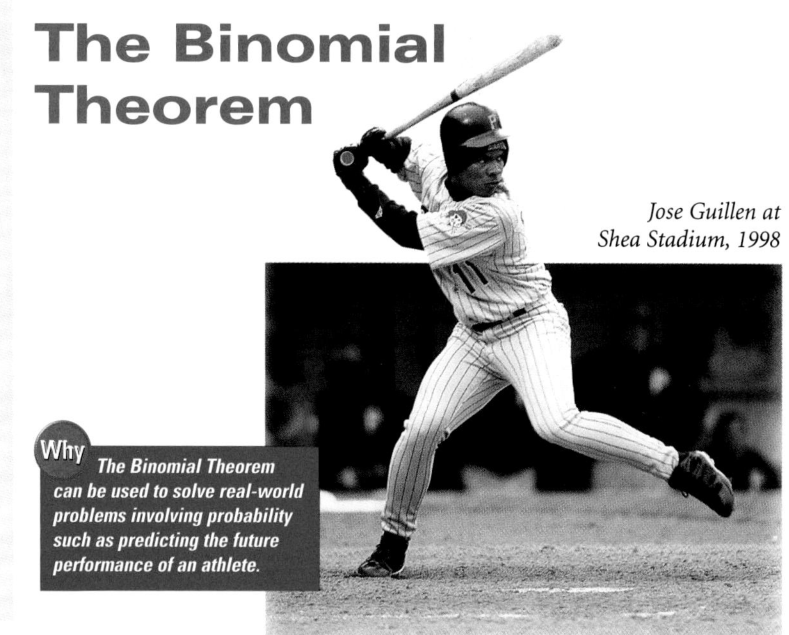

Jose Guillen at Shea Stadium, 1998

Objectives

- Use the Binomial Theorem to expand $(x + y)^n$.
- Use the Binomial Theorem to calculate a probability.

Why *The Binomial Theorem can be used to solve real-world problems involving probability such as predicting the future performance of an athlete.*

APPLICATION
SPORTS

At one time during the 1998 season, baseball player Jose Guillen had a season batting average of 0.337. If Guillen were to maintain his 0.337 average, what is the probability that he would get exactly 4 hits in the next 5 at bats? To answer this question, you can use the *Binomial Theorem.* You will solve this problem in Example 2.

In Chapters 5 and 7, you worked with linear binomials in x, such as $x + 3$ and $2x - 5$. In this lesson, you will work with linear binomials in x and y, such as $x + y$ and $3x - 2y$. You can use multiplication and the Distributive Property to *expand* linear binomials in x and y that are raised to a power.

$$\begin{aligned}(x+y)^4 &= (x+y)^2(x+y)^2\\ &= (x^2+2xy+y^2)(x^2+2xy+y^2)\\ &= x^2(x^2+2xy+y^2)+2xy(x^2+2xy+y^2)+y^2(x^2+2xy+y^2)\\ &= x^4+2x^3y+x^2y^2+2x^3y+4x^2y^2+2xy^3+x^2y^2+2xy^3+y^4\\ &= x^4+4x^3y+6x^2y^2+4xy^3+y^4\end{aligned}$$

Activity
Exploring the Binomial Theorem

You will need: no special tools

The table below shows the expansions of the first three nonnegative integral powers of $x + y$.

1. Copy the table at right.
2. Expand $(x + y)^3$ by multiplying $(x + y)^2$ by $(x + y)$. Add a row to the bottom of the table, and write the full expansion.

Product	Expansion
$(x+y)^0 =$	1
$(x+y)^1 =$	$1x + 1y$
$(x+y)^2 =$	$1x^2 + 2xy + 1y^2$

Prepare

NCTM PRINCIPLES & STANDARDS 1–4, 6–10

QUICK WARM-UP

Simplify.

1. $(x - 5)^2$ $x^2 - 10x + 25$
2. $(x + y)^2$ $x^2 + 2xy + y^2$
3. $(4m + 3n)^2$ $16m^2 + 24mn + 9n^2$
4. $(a + 3)(a^2 - 2a + 4)$ $a^3 + a^2 - 2a + 12$
5. $(r + 2)^3$ $r^3 + 6r^2 + 12r + 8$

Evaluate.

6. $\binom{8}{6}$ 28
7. $\binom{10}{0}$ 1
8. $\binom{4}{1}$ 4

Also on Quiz Transparency 11.8

Teach

Why Patterns imply predictability. Probability in mathematics deals with predictability, and the binomial theorem models real-world events in which only two outcomes are possible. Have students think of similar events, such as tossing a coin or guessing the answers on a true-false exam.

Activity Notes

Tell students that the objective of the Activity is to learn how to expand $(x + y)^n$, especially for large values of n, without having to multiply $x + y$ by itself n times. Students should see that each row of Pascal's triangle gives them the coefficients for the expansion of $(x + y)^n$.

Alternative Teaching Strategy

USING ALGORITHMS Have students try to find $(a + b)^7$ by multiplying. Then show them how to expand a binomial by filling in powers of the variable, putting in the binomial coefficients, and then evaluating the coefficients. For instance, to expand Example 1, have students write the following, leaving a blank for the coefficient of each term:

$$\begin{aligned}&(a+b)^7\\ &= \_a^7 + \_\,a^6b + \_a^5b^2 + \_a^4b^3\\ &\quad + \_a^3b^4 + \_a^2b^5 + \_ab^6 + \_b^7\end{aligned}$$

Now enter the binomial coefficients.

$$\begin{aligned}&(a+b)^7\\ &= \binom{7}{0}a^7 + \binom{7}{1}a^6b + \binom{7}{2}a^5b^2 + \binom{7}{3}a^4b^3\\ &\quad + \binom{7}{4}a^3b^4 + \binom{7}{5}a^2b^5 + \binom{7}{6}ab^6 + \binom{7}{7}b^7\end{aligned}$$

Finally, evaluate the coefficients by referring to the 7th row of Pascal's triangle, by using the formula for combinations, or by using a calculator to compute ${}_7C_k$ for $k = 0, 1, \ldots, 7$.

$$\begin{aligned}(a+b)^7 &= a^7 + 7a^6b + 21a^5b^2 + 35a^4b^3\\ &\quad + 35a^3b^4 + 21a^2b^5 + 7ab^6 + b^7\end{aligned}$$

3. Expand $(x+y)^4$ by multiplying $(x+y)^3$ by $(x+y)$. Write the full expansion below the row from Step 2.
4. Expand $(x+y)^5$ by multiplying $(x+y)^4$ by $(x+y)$. Write the full expansion below the row from Step 3.

CHECKPOINT ✔

5. Write conjectures about the number of terms and about symmetry in the terms of the expansion in any row of the table. Verify your conjecture by filling in the row that would follow Step 4.

Notice that each row of Pascal's triangle also gives you the coefficients for the expansion of $(x+y)^n$ for positive integers n.

The Binomial Theorem stated below enables you to expand a power of a binomial. Notice that the Binomial Theorem makes use of the number of combinations of n objects taken r at a time. Recall from Lesson 10.3 that the notation $\binom{n}{k}$ indicates the combination ${}_nC_k$.

Binomial Theorem

Let n be a positive integer.

$$(x+y)^n = \sum_{k=0}^{n}\binom{n}{k}x^{n-k}y^k$$

$$= \binom{n}{0}x^ny^0 + \binom{n}{1}x^{n-1}y^1 + \cdots + \binom{n}{n-1}x^1y^{n-1} + \binom{n}{n}x^0y^n$$

EXAMPLE 1 Expand $(x+y)^7$.

SOLUTION

Write the expansion. Then evaluate ${}_nC_k$ for each value of k.

$(x+y)^7$

$$= \binom{7}{0}x^7y^0 + \binom{7}{1}x^6y^1 + \binom{7}{2}x^5y^2 + \binom{7}{3}x^4y^3 + \binom{7}{4}x^3y^4 + \binom{7}{5}x^2y^5 + \binom{7}{6}x^1y^6 + \binom{7}{7}x^0y^7$$

$$= 1x^7y^0 + 7x^6y^1 + 21x^5y^2 + 35x^4y^3 + 35x^3y^4 + 21x^2y^5 + 7x^1y^6 + 1x^0y^7$$

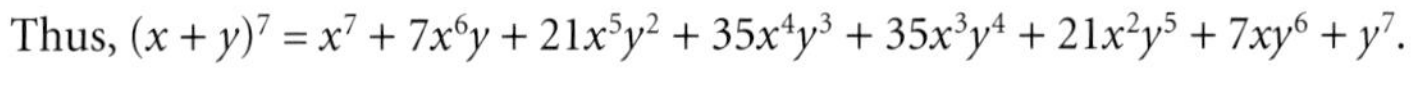

Thus, $(x+y)^7 = x^7 + 7x^6y + 21x^5y^2 + 35x^4y^3 + 35x^3y^4 + 21x^2y^5 + 7xy^6 + y^7$.

TECHNOLOGY GRAPHICS CALCULATOR

Keystroke Guide, page 761

CHECK

Use the sequence command of a graphics calculator to check the coefficients of your answer. Use the right arrow button to view all of the coefficients. Note that ${}_nC_r$ notation is used.

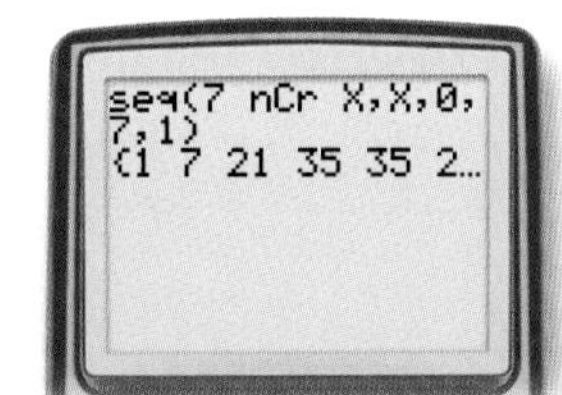

TRY THIS Expand $(m+n)^6$.

Cooperative Learning

Have students work in groups of two or three to complete the Activity. Each member should individually expand the binomials in Steps 2, 3, and 4. To ensure accuracy, the group should review each member's expansion. Then the group should work together to complete the diagram and answer the question in Step 5.

CHECKPOINT ✔

5. There are $n+1$ terms in the expansion of $(x+y)^n$, and the coefficients of the terms are symmetric about the center term(s); $(x+y)^6 = x^6 + 6x^5y + 15x^4y^2 + 20x^3y^3 + 15x^2y^4 + 6xy^5 + y^6$.

Teaching Tip

Remind students that $\binom{n}{r}$ and ${}_nC_r$ both represent the combination of n things taken r at a time.

ADDITIONAL EXAMPLE 1

Expand $(x+y)^9$.

$x^9 + 9x^8y + 36x^7y^2 + 84x^6y^3 + 126x^5y^4 + 126x^4y^5 + 84x^3y^6 + 36x^2y^7 + 9xy^8 + y^9$

TRY THIS

$(m+n)^6 = m^6 + 6m^5n + 15m^4n^2 + 20m^3n^3 + 15m^2n^4 + 6mn^5 + n^6$

Inclusion Strategies

USING VISUAL MODELS Show students how the visual pattern in a tree diagram can represent the expansion of a binomial expression. For instance, the tree diagram at right shows the expansion of $(a+b)^3$, where the beginning roots, a and b, are each multiplied by a to produce one branch and multiplied by b to produce a second branch. Each subsequent branch is produced the same way, as shown. Point out that the term b^2a is equivalent to ab^2 and ba^2 is equivalent to a^2b because of the commutative property of multiplication. The expanded products at each level can be read by adding the terms in that level. For example, the first level yields $(a+b)^1 = a + b$, the second level yields $(a+b)^2 = a^2 + 2ab + b^2$, and the third level yields $(a+b)^3 = a^3 + 3a^2b + 3ab^2 + b^3$.

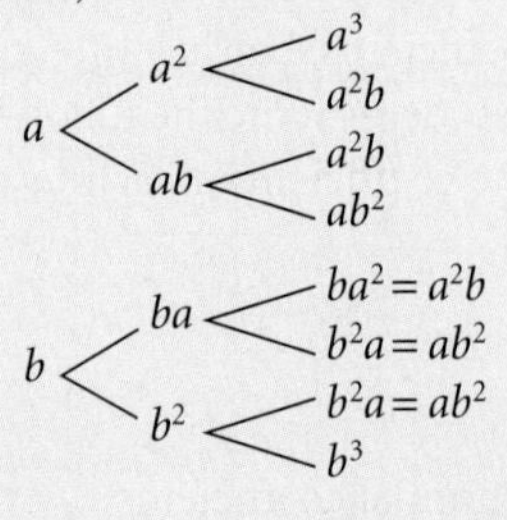

CHECKPOINT ✔ Write the power of the binomial $m + n$ given by the expansion $m^8 + 8m^7n + 28m^6n^2 + 56m^5n^3 + 70m^4n^4 + 56m^3n^5 + 28m^2n^6 + 8mn^7 + n^8$.

CHECKPOINT ✔
8th power

EXAMPLE 2

APPLICATION
SPORTS

Refer to the batting situation described at the beginning of the lesson.

If Jose Guillen were to maintain his 0.337 average, what is the probability that Jose Guillen would get exactly 4 hits in the next 5 at bats?

SOLUTION

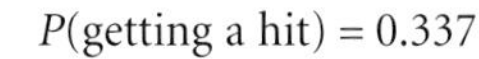

$P(\text{getting a hit}) = 0.337$ $\qquad$ $P(\text{not getting a hit}) = 0.663$

Use the Binomial Theorem because there are only two possible outcomes.

Substitute 0.337 for x and 0.663 for y in $(x + y)^n$. The number of times at bat, 5, is the exponent.

$$(0.337 + 0.663)^5 = \binom{5}{0}(0.337)^5 + \binom{5}{1}(0.337)^4(0.663) + \binom{5}{2}(0.337)^3(0.663)^2 + \binom{5}{3}(0.337)^2(0.663)^3 + \binom{5}{4}(0.337)(0.663)^4 + \binom{5}{5}(0.663)^5$$

Because you are looking for the probability of 4 *hits*, evaluate the term in which $(0.337)^4$ appears.

$$\binom{5}{1}(0.337)^4(0.663) \approx 0.04$$

Thus, if Jose Guillen maintains his 0.337 batting average in the current season, he has approximately a 4% chance of getting exactly 4 hits in the next 5 at bats.

TRY THIS What is the probability that Jose Guillen will get exactly 2 hits in the next 4 at bats?

ADDITIONAL EXAMPLE 2

In his rookie season, Ken Griffey, Jr. had a season batting average of 0.264. Suppose that this is also his batting average in the current season. **What is the probability that he will get exactly 3 hits in his next 5 at bats?**

$\binom{5}{3}(0.264)^3(0.736)^2 \approx 10\%$

TRY THIS
$\binom{4}{2}(0.337)^2(0.663)^2 \approx 0.30$

EXAMPLE 3

Find the fifth term in the expansion of $(x + y)^{10}$.

SOLUTION

1. Use $(x + y)^n = \sum_{k=0}^{n} \binom{n}{k} x^{n-k}y^k$ to identify n and k.
   For the expansion of $(x + y)^{10}$, $n = 10$.
   For the fifth term, $k = 4$ because k begins at 0.
2. Evaluate $\binom{n}{k} x^{n-k}y^k$ for $n = 10$ and $k = 4$.

$$\binom{n}{k} x^{n-k}y^k = \binom{10}{4} x^{10-4}y^4$$
$$= 210x^6y^4$$

Thus, the fifth term in the expansion of $(x + y)^{10}$ is $210x^6y^4$.

TRY THIS Find the sixth term in the expansion of $(r + s)^8$.

CHECKPOINT ✔ Explain why the expansion of $(x + y)^{12}$ cannot have a term containing x^6y^7. Explain why $24a^4b^5$ cannot be a term in the expansion of $(a + b)^9$.

ADDITIONAL EXAMPLE 3

Find the 4th term of $(a + b)^{11}$.

$165a^8b^3$

TRY THIS
${}_8C_5\, r^{8-5}s^5 = 56r^3s^5$

CHECKPOINT ✔
$6 + 7 \neq 12$;
$\binom{9}{5} = 126$, not 24

Enrichment

Have students pick a number, p, between 0 and 1 and pick a whole number, n. Let $q = 1 - p$. Have students write the expanded form of $\sum_{k=0}^{n} \binom{n}{k} p^{n-k}q^k$ and find the value of the sum. Have them pick another value of p between 0 and 1, choose another whole number for n, and repeat the process. They should see that the value of $\sum_{k=0}^{n} \binom{n}{k} p^{n-k}q^k$ is always 1.

Solutions may vary, but the sum is always 1. Sample answer for $p = 0.75$, $q = 0.25$, and $n = 2$ is given below.

$$\sum_{k=0}^{2} \binom{2}{k}(0.75)^{2-k}(0.25)^k$$
$$= (0.75)^2 + 2(0.75)(0.25) + (0.25)$$
$$= 0.5625 + 0.375 + 0.0625$$
$$= 1$$

ADDITIONAL
EXAMPLE 4

Expand $(x - 2)^6$.
$x^6 - 12x^5 + 60x^4 - 160x^3 + 240x^2 - 192x + 64$

TRY THIS
$(x + 2y)^5 = x^5 - 10x^4y + 40x^3y^2 - 80x^2y^3 + 80xy^4 - 32y^5$

CRITICAL THINKING
yes; $\frac{1}{a^4} + \frac{4}{a^3b} + \frac{6}{a^2b^2} + \frac{4}{ab^3} + \frac{1}{b^4}$

ADDITIONAL
EXAMPLE 5

Refer to the problem posed in Example 5. **Describe all of the pieces that must be added in order to increase the cube's length, width, and height by 0.5 unit each.**
1 piece with dimensions $s \times s \times s$ (the original cube), 3 pieces with dimensions $s \times s \times 0.5$, 3 pieces with dimensions $s \times 0.5 \times 0.5$, 1 piece with dimensions $0.5 \times 0.5 \times 0.5$

You can apply the Binomial Theorem to the expansion of a sum or difference of two monomials, as shown in Example 4.

EXAMPLE 4
Expand $(2x - 3y)^4$.

SOLUTION

Write $(2x - 3y)^4$ as $[2x + (-3y)]^4$.

$$[2x + (-3y)]^4$$
$$= \binom{4}{0}(2x)^4(-3y)^0 + \binom{4}{1}(2x)^3(-3y)^1 + \binom{4}{2}(2x)^2(-3y)^2 + \binom{4}{3}(2x)^1(-3y)^3 + \binom{4}{4}(2x)^0(-3y)^4$$
$$= 1(2x)^4(-3y)^0 + 4(2x)^3(-3y)^1 + 6(2x)^2(-3y)^2 + 4(2x)^1(-3y)^3 + 1(2x)^0(-3y)^4$$
$$= 1(16x^4)(1) + 4(8x^3)(-3y) + 6(4x^2)(9y^2) + 4(2x)(-27y^3) + 1(1)(81y^4)$$
$$= 16x^4 - 96x^3y + 216x^2y^2 - 216xy^3 + 81y^4$$

TRY THIS Expand $(x - 2y)^5$.

CRITICAL THINKING Can you apply the Binomial Theorem to $\left(\frac{1}{a} + \frac{1}{b}\right)^4$? What is the result?

EXAMPLE 5
A cube has dimensions of $s \cdot s \cdot s$.

Describe all of the pieces that need to be added in order to increase the cube's length, width, and height by 0.1 unit each.

SOLUTION

The expansion of $(s + 0.1)^3$ is shown below.

$$(s + 0.1)^3 = 1s^3(0.1)^0 + 3s^2(0.1)^1 + 3s^1(0.1)^2 + 1s^0(0.1)^3$$

The pieces indicated by this expansion are listed below:

$1s^3(0.1)^0$: 1 piece with dimensions of $s \cdot s \cdot s$ (the original cube)

$3s^2(0.1)^1$: 3 pieces with dimensions of $s \cdot s \cdot (0.1)$

$3s^1(0.1)^2$: 3 pieces with dimensions of $s \cdot (0.1) \cdot (0.1)$

$1s^0(0.1)^3$: 1 piece with dimensions of $(0.1) \cdot (0.1) \cdot (0.1)$

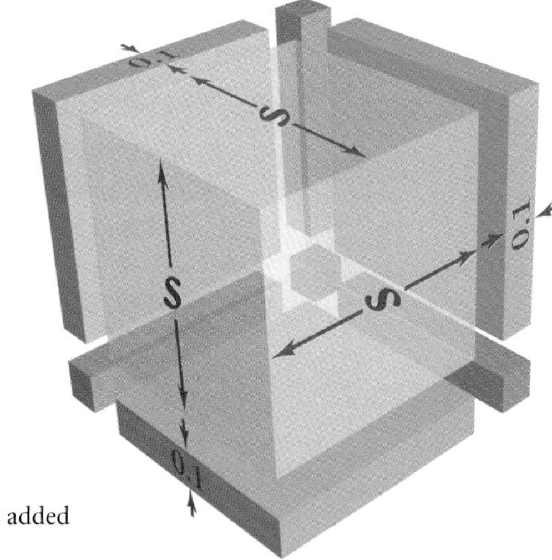

Thus, 7 pieces need to be added to the original cube.

Reteaching the Lesson

USING PATTERNS As you show students how to expand a binomial of the form $(x \pm y)^n$, point out the following:

- The number of terms in the expansion is $n + 1$.
- The first term is x^n, and the last term is y^n.
- For all but the first term, the coefficient of a term is the product of the coefficient of the preceding term and the exponent of x in the preceding term divided by the position number of the preceding term when the polynomial is written in standard form.
- The exponent of x in any term is one less than the exponent of x in the preceding term.
- The exponent of y in any term is one greater than the exponent of y in the preceding term.
- The sum of the exponents of x and y in each term is n.
- If the binomial is a sum, all of the terms are added. If the binomial is a difference, the signs of the terms alternate, with the even-numbered terms being subtracted.
- If the sum of the values of x and y is 1, then the sum of the expansion of the binomial is also 1.

Exercises

Communicate

1. Describe the connection between Pascal's triangle and the Binomial Theorem.
2. Explain how to use the Binomial Theorem to find the fourth term in the expansion of $(x + y)^6$.
3. Explain how to use a binomial expansion to find the probability of exactly 7 heads appearing in a total of 10 coin tosses.

Guided Skills Practice

4. Expand $(a + b)^5$. *(EXAMPLE 1)* $a^5 + 5a^4b + 10a^3b^2 + 10a^2b^3 + 5ab^4 + b^5$

APPLICATION

5. METEOROLOGY If there is a 0.30 probability of rain for each of the next 5 days, what is the probability of it raining exactly 3 out of the 5 days? *(EXAMPLE 2)* ≈ 0.13
6. Find the fourth term in the expansion of $(a + b)^{10}$. *(EXAMPLE 3)* $120a^7b^3$
7. Expand $(3x - 2y)^5$. *(EXAMPLE 4)* $243x^5 - 810x^4y + 1080x^3y^2 - 720x^2y^3 + 240xy^4 - 32y^5$

CONNECTION

8. GEOMETRY Refer to Example 5. Describe all of the pieces that need to be added to the cube in order to increase the length, width, and height by 0.2 unit each. *(EXAMPLE 5)* 1 piece with dimensions $s \cdot s \cdot s$
   3 pieces with dimensions $s \cdot s \cdot (0.2)$
   3 pieces with dimensions $s \cdot (0.2) \cdot (0.2)$
   1 piece with dimensions $(0.2) \cdot (0.2) \cdot (0.2)$

Practice and Apply

Expand each binomial raised to a power.

9. $(a + b)^5$
10. $(p + q)^6$
11. $(a + b)^8$
12. $(p + q)^7$
13. $(x + y)^4$
14. $(x + y)^5$
15. $(2 + x)^5$
16. $(y + 3)^4$
17. $(y + 4)^9$
18. $(6 + x)^6$
19. $(x - y)^4$
20. $(x - y)^5$

Write each summation as a binomial raised to a power. Then write it in expanded form.

21. $\sum_{k=0}^{4}\binom{4}{k}a^{4-k}b^k$ $(a + b)^4$
22. $\sum_{k=0}^{6}\binom{6}{k}a^{6-k}b^k$ $(a + b)^6$
23. $\sum_{k=0}^{5}\binom{5}{k}a^{5-k}b^k$ $(a + b)^5$
24. $\sum_{k=0}^{8}\binom{8}{k}a^{8-k}b^k$ $(a + b)^8$
25. $\sum_{k=0}^{9}\binom{9}{k}x^{9-k}y^k$ $(x + y)^9$
26. $\sum_{k=0}^{7}\binom{7}{k}x^{7-k}y^k$ $(x + y)^7$

19. $x^4 - 4x^3y + 6x^2y^2 - 4xy^3 + y^4$
20. $x^5 - 5x^4y + 10x^3y^2 - 10x^2y^3 + 5xy^4 - y^5$
21. $(a + b)^4 = a^4 + 4a^3b + 6a^2b^2 + 4ab^3 + b^4$
22. $(a + b)^6 = a^6 + 6a^5b + 15a^4b^2 + 20a^3b^3 + 15a^2b^4 + 6ab^5 + b^6$
23. $(a + b)^5 = a^5 + 5a^4b + 10a^3b^2 + 10a^2b^3 + 5ab^4 + b^5$
24. $(a + b)^8 = a^8 + 8a^7b + 28a^6b^2 + 56a^5b^3 + 70a^4b^4 + 56a^3b^5 + 28a^2b^6 + 8ab^7 + b^8$
25. $(x + y)^9 = x^9 + 9x^8y + 36x^7y^2 + 84x^6y^3 + 126x^5y^4 + 126x^4y^5 + 84x^3y^6 + 36x^2y^7 + 9xy^8 + y^9$
26. $(x + y)^7 = x^7 + 7x^6y + 21x^5y^2 + 35x^4y^3 + 35x^3y^4 + 21x^2y^5 + 7xy^6 + y^7$

Assess

Selected Answers

Exercises 4–8, 9–73 odd

ASSIGNMENT GUIDE	
In Class	1–8
Core	9–55 odd
Core Plus	10–64 even
Review	66–74
Preview	75

Extra Practice can be found beginning on page 940.

9. $a^5 + 5a^4b + 10a^3b^2 + 10a^2b^3 + 5ab^4 + b^5$
10. $p^6 + 6p^5q + 15p^4q^2 + 20p^3q^3 + 15p^2q^4 + 6pq^5 + q^6$
11. $a^8 + 8a^7b + 28a^6b^2 + 56a^5b^3 + 70a^4b^4 + 56a^3b^5 + 28a^2b^6 + 8ab^7 + b^8$
12. $p^7 + 7p^6q + 21p^5q^2 + 35p^4q^3 + 35p^3q^4 + 21p^2q^5 + 7pq^6 + q^7$
13. $x^4 + 4x^3y + 6x^2y^2 + 4xy^3 + y^4$
14. $x^5 + 5x^4y + 10x^3y^2 + 10x^2y^3 + 5xy^4 + y^5$
15. $32 + 80x + 80x^2 + 40x^3 + 10x^4 + x^5$
16. $y^4 + 12y^3 + 54y^2 + 108y + 81$
17. $y^9 + 36y^8 + 576y^7 + 5376y^6 + 32{,}256y^5 + 129{,}024y^4 + 344{,}064y^3 + 589{,}824y^2 + 589{,}824y + 262{,}144$
18. $46{,}656 + 45{,}656x + 19{,}440x^2 + 4320x^3 + 540x^4 + 36x^5 + x^6$

Error Analysis

Some students have difficulty expanding a binomial when the terms of the binomial have coefficients other than 1. To expand a binomial such as $(3x - 2)^3$, encourage students to write the binomial as $[(3x) + (-2)]^3$ and then expand it as a binomial of the form $(a + b)^3$. Then they may substitute $3x$ and -2 for a and b, respectively, before raising each factor to the correct power.

39. $1024a^5 + 3840a^4b + 5760a^3b^2 + 4320a^2b^3 + 1620ab^4 + 243b^5$

40. $a^4 + 8a^3b + 24a^2b^2 + 32ab^3 + 16b^4$

41. $x^4 - 8x^3y + 24x^2y^2 - 32xy^3 + 16y^4$

42. $32x^5 - 80x^4y + 80x^3y^2 - 40x^2y^3 + 10xy^4 - y^5$

43. $16x^4 + 96x^3 + 216x^2 + 216x + 81$

44. $x^4 + 12x^3y + 54x^2y^2 + 108xy^3 + 81y^4$

45. $243x^5 + 810x^4y + 1080x^3y^2 + 720x^2y^3 + 240xy^4 + 32y^5$

46. $x^4 - 8x^3y + 24x^2y^2 - 32xy^3 + 16y^4$

47. $\frac{1}{8}x^3 + \frac{1}{4}x^2y + \frac{1}{6}xy^2 + \frac{1}{27}y^3$

Practice

NAME ______________ CLASS ________ DATE ________

Practice 11.8 The Binomial Theorem

Expand each binomial raised to a power.

1. $(s + t)^5$ — $s^5 + 5s^4t + 10s^3t^2 + 10s^2t^3 + 5st^4 + t^5$
2. $(b + w)^6$ — $b^6 + 6b^5w + 15b^4w^2 + 20b^3w^3 + 15b^2w^4 + 6bw^5 + w^6$

For Exercises 3–5, refer to the expansion of $(r + s)^{15}$.

3. How many terms are in the expansion? — 16 terms
4. What is the exponent of r in the term that contains s^{12}? — 3
5. Write the term that contains r^5. — $3003r^5s^{10}$

Expand each binomial.

6. $(5x + y)^5$ — $3125x^5 + 3125x^4y + 1250x^3y^2 + 250x^2y^3 + 25xy^4 + y^5$
7. $\left(\frac{1}{2}w + 2z\right)^4$ — $\frac{1}{16}w^4 + w^3z + 6w^2z^2 + 16wz^3 + 16z^4$
8. $\left(\frac{3}{4}a - d\right)^5$ — $\frac{243}{1024}a^5 - \frac{405}{256}a^4d + \frac{135}{32}a^3d^2 - \frac{45}{8}a^2d^3 + \frac{15}{4}ad^4 - d^5$
9. $(2m - 3q)^6$ — $64m^6 - 576m^5q + 2160m^4q^2 - 4320m^3q^3 + 4860m^2q^4 - 2916mq^5 + 729q^6$

Ty Cobb was the American League batting champion for 12 of his 24 years in baseball. His lifetime batting average was 0.367. Use the Binomial Theorem to find each theoretical probability for Ty Cobb.

10. exactly 3 hits in 5 at bats — ≈ 0.20
11. at least 3 hits in 5 at bats — ≈ 0.26
12. no more than 2 hits in 5 at bats — ≈ 0.74
13. exactly 4 hits in 6 at bats — ≈ 0.11
14. fewer than 3 hits in 6 at bats — ≈ 0.61

For the expansion of $(r + s)^9$, find the indicated term.

27. third term $36r^7s^2$
28. fifth term $126r^5s^4$
29. sixth term $126r^4s^5$
30. second term $9r^8s$

For the expansion of $(x + 4)^8$, find the indicated term.

31. fourth term $3584x^5$
32. sixth term $57{,}344x^3$
33. seventh term $114{,}688x^2$
34. fifth term $17{,}920x^4$

For the expansion of $(x - y)^7$, find the indicated term.

35. fifth term $35x^3y^4$
36. fourth term $-35x^4y^3$
37. sixth term $-21x^2y^5$
38. third term $21x^5y^2$

internet connect
Homework Help Online
Go To: **go.hrw.com**
Keyword:
MB1 Homework Help
for Exercises 39–50

Expand each binomial.

39. $(4a + 3b)^5$
40. $(a + 2b)^4$
41. $(x - 2y)^4$
42. $(2x - y)^5$
43. $(2x + 3)^4$
44. $(x + 3y)^4$
45. $(3x + 2y)^5$
46. $(-x + 2y)^4$
47. $\left(\frac{1}{2}x + \frac{1}{3}y\right)^3$
48. $\left(\frac{2}{3}x + \frac{1}{2}y\right)^4$
49. $(0.7 + x)^4$
50. $(y + 1.2)^5$

51. Find the eighth term in the expansion of $(p + q)^{14}$. $3432p^7q^7$

52. How many terms are in the expansion of $(x + y)^{18}$? **19 terms**

53. In the expansion of $(a + b)^{10}$, a certain term contains b^3. What is the exponent of a in this term? What is the term? **7**; $120a^7b^3$

54. In the expansion of $(x + y)^{15}$, a certain term contains x^{10}. What is the exponent of y in this term? What is the term? **5**; $3003x^{10}y^5$

55. The term $36a^7b^2$ appears in the expansion of $(a + b)^n$. What is the value of n? $n = 9$

56. The value $\binom{10}{7}$ appears as a coefficient of two different terms in the expansion of $(a + b)^{10}$. What are the two terms? $120a^3b^7$; $120a^7b^3$

CHALLENGE

57. Let $a + bi = (1 + i)^{11}$, where $i = \sqrt{-1}$. Find a and b by using the Binomial Theorem. $a = -32$, $b = 32$

CONNECTION

58. **GEOMETRY** The side length of a cube is a. Describe the pieces that need to be added in order to increase its length, width, and height by 0.3 unit each. **1 piece: $a \cdot a \cdot a$; 3 pieces: $a \cdot a \cdot (0.3)$; 3 pieces: $a \cdot (0.3) \cdot (0.3)$; 1 piece: $(0.3) \cdot (0.3) \cdot (0.3)$**

APPLICATION

59. **SPORTS** Refer to the batting problem in Example 2.

a. The probability that Guillen will get at most 3 hits in the next 5 at bats is given by the sum of the last four terms of the expansion of $(0.337 + 0.663)^5$. Find this probability to the nearest hundredth. **0.95**

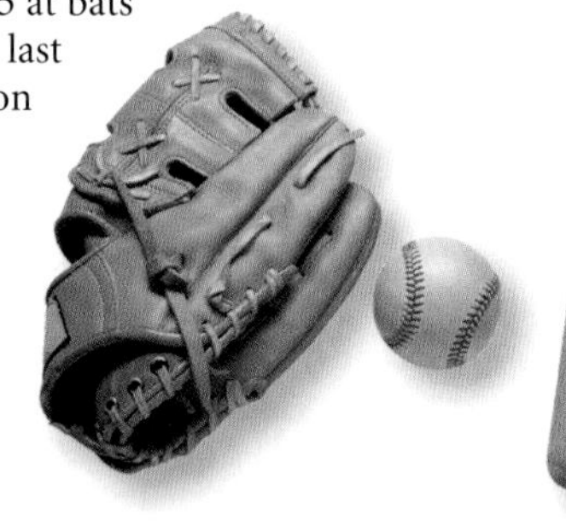

b. Find the probability that Guillen will get at least 3 hits in the next 5 at bats. **0.22**

48. $\frac{16}{81}x^4 + \frac{16}{27}x^3y + \frac{2}{3}x^2y^2 + \frac{1}{3}xy^3 + \frac{1}{16}y^4$

49. $0.2401 + 1.372x + 2.94x^2 + 2.8x^3 + x^4$

50. $y^5 + 6y^4 + 14.4y^3 + 17.28y^2 + 10.368y + 2.48832$

66b.

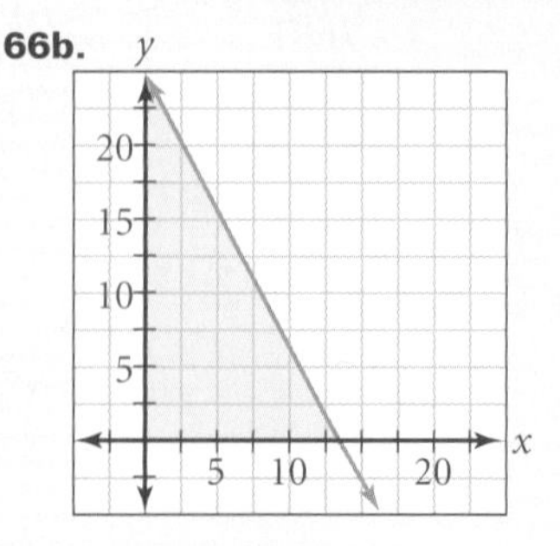

69.

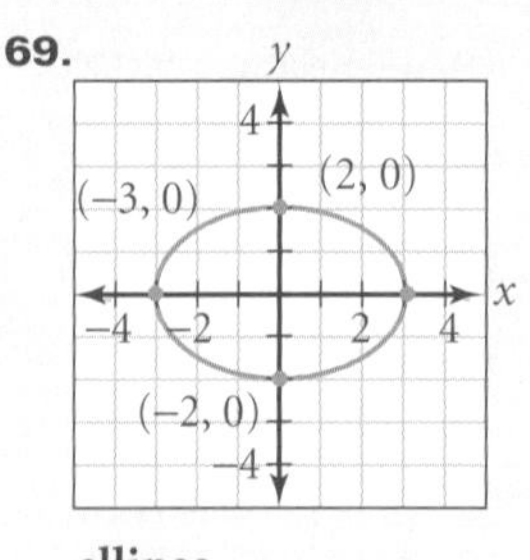

ellipse

APPLICATIONS

SPORTS If Guillen goes into a slump and his batting average drops to 0.285, what would be the probability of each event below?

60. exactly 3 hits in the next 6 at bats **0.17**

CHALLENGE

61. at least 3 hits in the next 6 at bats **0.23**

METEOROLOGY If there is a 20% probability of rain for each of the next 7 days, what is the probability of each event below?

0.37 **62.** It will rain exactly 1 of the next 7 days.

0.79 **63.** It will rain at least 1 of the next 7 days.

≈0 **64.** It will rain each of the next 7 days.

0.21 **65.** It will not rain each of the next 7 days.

Look Back

APPLICATION

66. ENTERTAINMENT Ricardo won a $100 gift certificate to a movie theater. The gift certificate can be used for movie tickets. Tickets for evening shows cost $7.50 and tickets for matinees cost $4. *(LESSON 3.3)*

a. Write an inequality that describes the total value of x tickets at $7.50 and y tickets at $4 that Ricardo could purchase with his gift certificate.

b. Graph the inequality.

c. Describe the values of the variables that are meaningful for this situation.

d. What is the maximum number of $7.50 tickets he can buy?

e. What is the maximum number of $4 tickets he can buy?

f. Identify three reasonable solutions of the inequality.

66a. $7.5x + 4y \le 100$
c. integer values of x and y such that $0 \le x \le 13$ and $0 \le y \le 25$
d. 13 tickets
e. 25 tickets
f. (5, 10), (10, 6), (8, 10)

CONNECTION

MAXIMUM/MINIMUM Maximize and minimize each objective function within the given constraints. *(LESSON 3.5)*

67. $S = 10x + 3y$

$$\begin{cases} 0 \le x \le 10 \\ y \ge 5 \\ y \le -0.3x + 10 \end{cases}$$

121; 15

68. $I = 100m + 160n$

$$\begin{cases} m + n \le 100 \\ 4m + 6n \le 500 \\ 20m + 10n \le 1600 \\ m \ge 0 \\ n \ge 0 \end{cases}$$

13,333.3; 0

Classify the conic section defined by each equation, and sketch its graph. *(LESSONS 9.2, 9.3, 9.4, AND 9.5)*

69. $4x^2 + 9y^2 = 36$ **ellipse**

70. $\frac{x^2}{100} + \frac{y^2}{36} = 1$ **ellipse**

71. $x = y^2 - 6y + 5$ **parabola**

72. $y = -x^2 + 5x + 6$ **parabola**

73. $(x + 2)^2 + (y - 5)^2 = 10$ **circle**

74. $3x^2 - 2y^2 = 30$ **hyperbola**

Look Beyond

75. Expand the expression $(x + y + z)^3$.

Look Beyond

Exercise 75 extends the concept of expansion of polynomials to include trinomials. Encourage students to examine the coefficients in the trinomial expansion and look for patterns.

70.

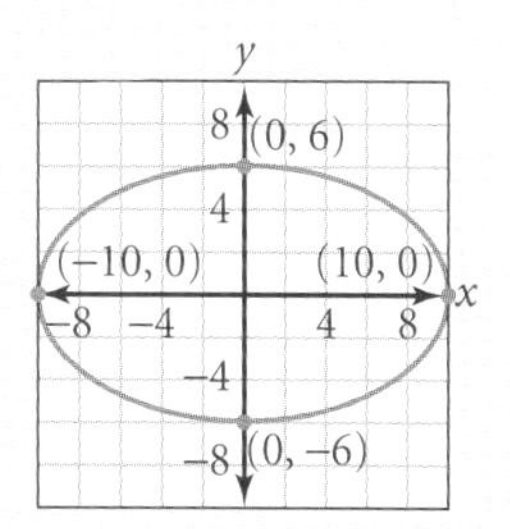

71.

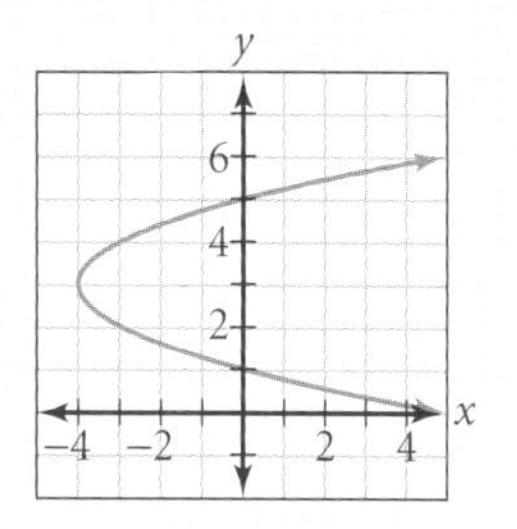

Student Technology Guide

NAME ____________ CLASS ________ DATE ________

Student Technology Guide
***11.8* The Binomial Theorem**

When you expand a binomial of the form $(x + y)^n$, you can use a graphics calculator to find the coefficients of the terms in the expansion.

Example: Expand $(a + b)^6$.

- Use the Binomial Theorem to write the expansion. Remember that $\binom{n}{k} = {}_nC_k$.

$(a + b)^6 = \binom{6}{0}a^6b^0 + \binom{6}{1}a^5b^1 + \binom{6}{2}a^4b^2 + \binom{6}{3}a^3b^3 + \binom{6}{4}a^2b^4 + \binom{6}{5}a^1b^5 + \binom{6}{6}a^0b^6$.

- Use the sequence and combination features to find the coefficients. Press 2nd STAT OPS 5:seq(ENTER 6 MATH PRB 3:nCr ENTER X,T,θ,n , X,T,θ,n , 0 , 6 , 1) ENTER.
- Use the coefficients 1, 6, 15, 20, 15, 6, and 1 to write the answer. (Notice that you need to use ► to see all of the numbers.)

seq(6 nCr X,X,0,6,1)
{1 6 15 20 15 6...

$(a + b)^6 = a^6 + 6a^5b + 15a^4b^2 + 20a^3b^3 + 15a^2b^4 + 5ab^5 + b^6$.

Expand each binomial.

1. $(x + y)^4$ $x^4 + 4x^3y + 6x^2y^2 + 4xy^3 + y^4$
2. $(m + n)^7$ $m^7 + 7m^6n + 21m^5n^2 + 35m^4n^3 + 35m^3n^4 + 21m^2n^5 + 7mn^6 + n^7$

Some calculators, such as the TI-92, can expand powers of binomials. This is especially useful when the terms in the binomial have coefficients other than 1.

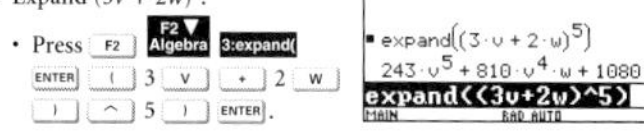

Example: Expand $(3v + 2w)^5$.

- Press F2 Algebra 3:expand(ENTER (3 V + 2 W) ^ 5) ENTER.

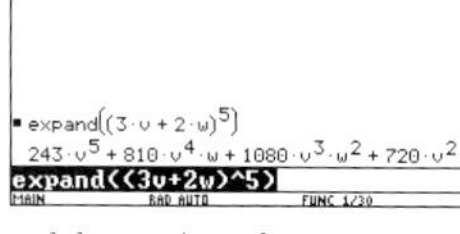

$(3v + 2w)^5 = 243v^5 + 810v^4w + 1080v^3w^2 + 720v^2w^3 + 240vw^4 + 32w^5$. (Note: you need to use the right arrow key to see all of the terms.)

3. Expand $(3x + 5y)^6$.
$729x^6 + 7290x^5y + 30,375x^4y^2 + 67,500x^3y^3 + 84,375x^2y^4 + 56,250xy^5 + 15,625y^6$

72.

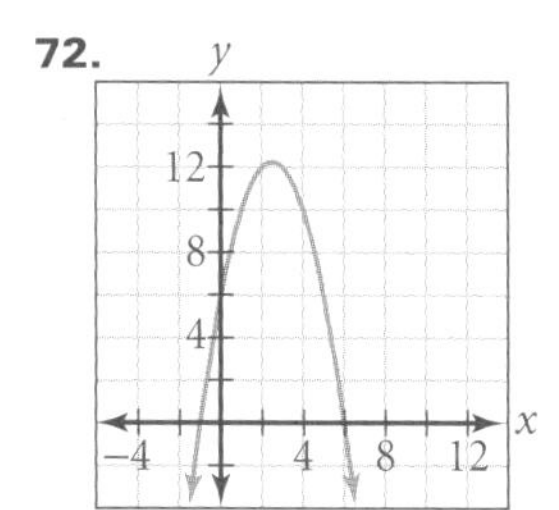

73.

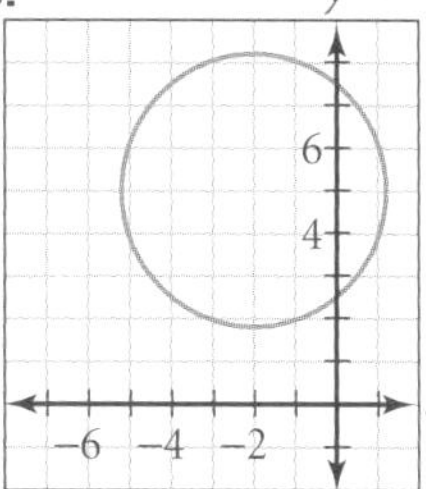

74.

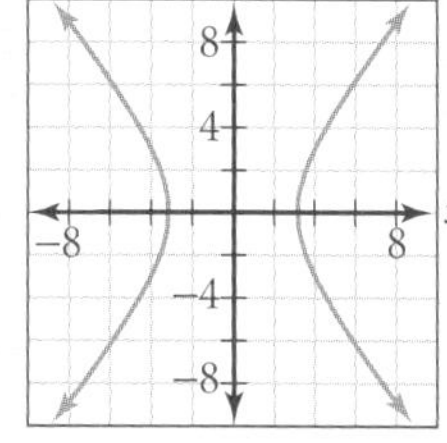

75. $x^3 + 3x^2y + 3x^2z + 3xy^2 + 6xyz + 3xz^2 + y^3 + 3y^2z + 3yz^2 + z^3$

Focus

The center of gravity of a stack of tiles can be determined by a mathematical model that involves sequences and series. Students will develop such a model and determine whether the model is predictable.

Motivate

Ask students to observe as you place a textbook, long side out, on the edge of a desk or table. Gradually push the book past the edge of the table until it falls. Now ask students what caused the book to fall and how they might determine how far the book could be pushed past the edge before it falls. Have them discuss how people maintain their balance when skiing or skating, and how this may be similar to balancing the book.

Notice the stack of 4 congruent tiles on the desk shown at right. What do you think is the maximum distance that each tile can be extended beyond the tile beneath it without falling over?

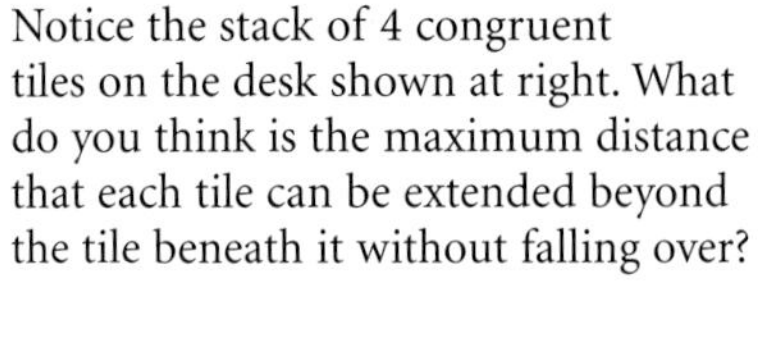

Activity 1

All of the weight of an object can be considered to be concentrated at a single point called the *center of gravity*. The Portfolio Activity on page 706 involved slowly sliding a 12-inch ruler over the edge of a desk to find its center of gravity, or balance point, which is located at the 6-inch mark. Use a 12-inch wooden ruler and a quarter for this activity.

1. Tape a quarter to the ruler at the locations described in the table below. Determine the center of gravity for the ruler with the quarter taped at each location. Record your results.

Location of the quarter	none	1 in.	2 in.	3 in.	4 in.	5 in.	6 in.
Location of the center of gravity	6 in.						

2. Let n represent the location of the quarter, and let c represent the location of the center of gravity for the ruler with the quarter. Write a recursive formula and an explicit formula for the sequence that models the relationship between the location of the quarter and the location of the center of gravity for the ruler with the quarter.

Activity 2

1. Place all of the tiles in a stack that is aligned with the edge of the desk so that the center of gravity is above 0. Extend tile A beyond the edge of tile B so that $\frac{1}{2}$ of its length is hanging over the edge of tile B. Now the center of gravity for tile A is located $\frac{1}{2}$ of a tile length to the right of 0.

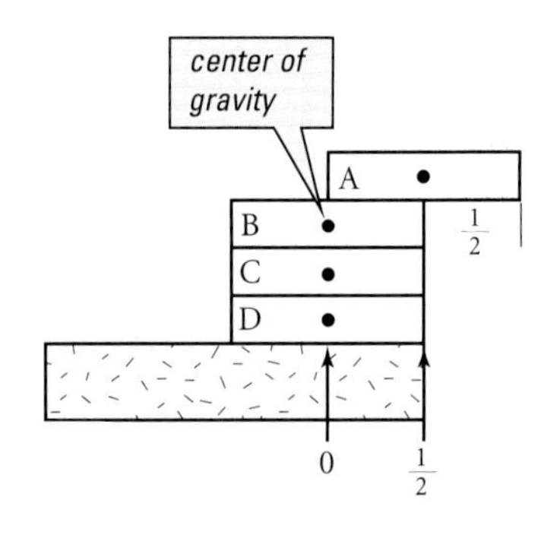

Activity 1

2. Recursive formula: $\begin{cases} c_1 = 5\frac{3}{8} \\ c_n = c_{n-1} + \frac{1}{8} \end{cases}$

Explicit formula: $c_n = 5\frac{2}{8} + \frac{1}{8}n$ or

$c_n = 5\frac{3}{8} + (n-1)\frac{1}{8}$

Activity 2

3. A: $\frac{7}{8}$ of a tile length to the right of 0

B: $\frac{3}{8}$ of a tile length to the right of 0

C: $\frac{1}{8}$ of a tile length to the right of 0

4. $\sum_{k=1}^{3} \left(\frac{1}{2}\right)^k$

The answer to Step 1 of Activity 1 can be found in Additional Answers beginning on page 1002.

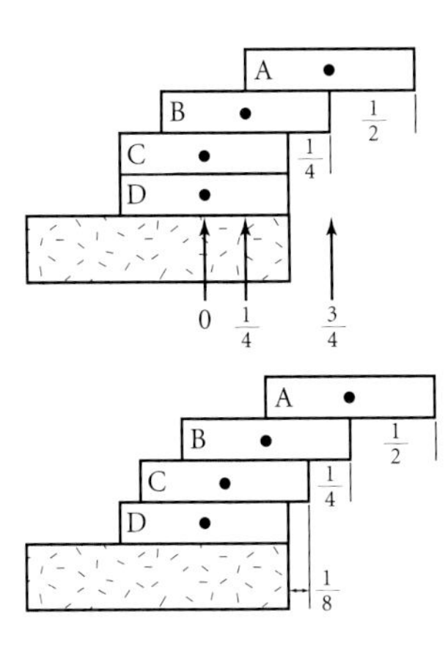

2. While keeping the same position of tile A on tile B, extend tile B so that $\frac{1}{4}$ of its length is hanging over the edge of tile C. The center of gravity for tile A is now $\frac{1}{4} + \frac{1}{2}$, or $\frac{3}{4}$, of a tile length to the right of 0, and the center of gravity for tile B is $\frac{1}{4}$ of a tile length to the right of 0.

3. While keeping the same position of tile A on tile B and tile B on tile C, extend tile C so that $\frac{1}{8}$ of its length is hanging over the edge of tile D. What is the center of gravity for tile A in this position? for tile B? for tile C?

4. Using summation notation, write a geometric series that models the shift in the center of gravity for tile A as each of these extensions is made.

Activity 3

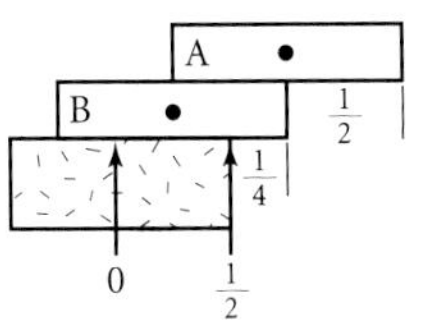

The center of gravity for an entire stack of tiles is the average of the centers of gravity for each tile in the stack. For example, the center of gravity for the 2-tile stack shown at left is to the right of 0 by the following fraction of a tile length:

$$\frac{\left(\frac{1}{2}+\frac{1}{4}\right)+\frac{1}{4}}{2}=\frac{1}{2}$$

The edge of the desk is $\frac{1}{2}$ of a tile length to the right of 0. If the center of gravity for a stack of tiles is less than or equal to $\frac{1}{2}$ of a tile length to the right of 0, then the stack will remain balanced. Since $\frac{1}{2} \le \frac{1}{2}$, the 2-tile stack remains balanced.

Let n represent the tile number, where $n = 1$ represents the top tile (A), $n = 2$ represents the next tile (B), and so on. Let x represent the fraction of a tile length that the tile extends beyond the tile directly below it. Then the equation $x = \frac{1}{2n}$ models the pattern of tile extensions.

1. Find x for n-values of 1, 2, and 3.
2. Show that the 3-tile stack described by your results in Step 1 remains balanced. Is each stack extended the maximum distance while maintaining the stack's balance?
3. Find x for n-values of 4, 5, and 6.
4. Show that the 6-tile stack described by your results from Steps 1 and 3 remains balanced. Is each tile extended the maximum distance while maintaining the stack's balance?
5. Using summation notation, write a series that models the shift in the center of gravity for the top tile as each of the extensions described in Steps 1 and 3 are made.
6. Place 6 tiles in the position described by your results above. Does the 6-tile stack remain balanced? Describe the relationship between the mathematical results and your experimental results.

Cooperative Learning

Organize the class into groups of three. The members of each group will take turns balancing the objects in **Activities 1, 2, and 3.** When not balancing the objects, the other two group members should act as observers, assisting to ensure that the experiments are performed correctly. All group members will work together to write the results of the experiments. They will create a series to model the experimental situations and will evaluate the model by using it in another experimental situation.

Discuss

Each group should give a report, describing the balancing experiments and whether the group's model kept the tiles balanced in Activity 3. Be sure that they include a description of the relationship between the mathematical results and experimental results. As an extension of this project, have students use the library or other sources to find additional experiments that deal with a center of gravity. You may want to point out that in geometry, the point of intersection of the medians of a triangle is called the *centroid* of the triangle because it is the center of gravity of the triangle.

Activity 3

1. $\frac{1}{2}, \frac{1}{4}, \frac{1}{6}$

2. $\frac{\left(\frac{1}{2}+\frac{1}{4}+\frac{1}{6}\right)+\left(\frac{1}{4}+\frac{1}{6}\right)+\frac{1}{6}}{3}=\frac{1}{2}$; yes

3. $\frac{1}{8}, \frac{1}{10}, \frac{1}{12}$

4. $\frac{\left(\frac{1}{2}+\frac{1}{4}+\frac{1}{6}+\frac{1}{8}+\frac{1}{10}+\frac{1}{12}\right)+\left(\frac{1}{4}+\frac{1}{6}+\frac{1}{8}+\frac{1}{10}+\frac{1}{12}\right)+\left(\frac{1}{6}+\frac{1}{8}+\frac{1}{10}+\frac{1}{12}\right)+\left(\frac{1}{8}+\frac{1}{10}+\frac{1}{12}\right)+\left(\frac{1}{10}+\frac{1}{12}\right)+\frac{1}{12}}{6}=\frac{1}{2}$

The tiles are extended the maximum distance because $\frac{1}{2} \le \frac{1}{2}$.

5. $\sum_{k=1}^{3} \frac{1}{2k}$

6. **The stack remains balanced. The experimental results confirm the mathematical results.**

11 Chapter Review and Assessment

VOCABULARY

Key Skills & Exercises

LESSON 11.1

Key Skills

Find the terms of a sequence.

You can represent a sequence as follows:

- ordered pairs in a table

n	1	2	3	⋯
t_n	2	5	8	⋯

- a list $\{2, 5, 8, \dots\}$
- an explicit formula $t_n = 3n - 1$
- a recursive formula $t_1 = 2$, $t_n = t_{n-1} + 3$, where $n > 2$

Write the first three terms of the sequence defined by the explicit formula $t_n = 2n + 1$.

$$t_1 = 2(1) + 1 = 3$$
$$t_2 = 2(2) + 1 = 5$$
$$t_3 = 2(3) + 1 = 7$$

Write the first three terms of the sequence defined by the recursive formula $t_1 = 4$ and $t_n = 3t_{n-1} - 1$.

$$t_1 = 4$$
$$t_2 = 3(4) - 1 = 11$$
$$t_3 = 3(11) - 1 = 32$$

Evaluate the sum of a series expressed in sigma notation.

Evaluate $\sum_{k=1}^{8}(k^2 + 3k + 1)$.

$$\sum_{k=1}^{8}(k^2 + 3k + 1) = \sum_{k=1}^{8} k^2 + 3\sum_{k=1}^{8} k + \sum_{k=1}^{8} 1$$
$$= \frac{8 \cdot 9 \cdot 17}{6} + 3\left(\frac{8 \cdot 9}{2}\right) + 8$$
$$= 204 + 108 + 8 = 320$$

Exercises

Write the first five terms of each sequence.

1. $t_n = 2n + 3$ **5, 7, 9, 11, 13**
2. $t_n = 3n - 4$ **–1, 2, 5, 8, 11**
3. $t_n = -0.2n + 0.1$ **–0.1, –0.3, –0.5, –0.7, –0.9**
4. $t_n = -2.22n + 1.3$ **–0.92, –3.14, –5.36, –7.58, –9.8**
5. $t_1 = 1$, $t_n = 2t_{n-1} + 2$ **1, 4, 10, 22, 46**
6. $t_1 = 1$, $t_n = 5t_{n-1} - 2$ **1, 3, 13, 63, 313**
7. $t_1 = 3$, $t_n = -3t_{n-1} + 1$ **3, –8, 25, –74, 223**
8. $t_1 = 2$, $t_n = -4t_{n-1} - 3$ **2, –11, 41, –167, 665**

Evaluate.

9. $\sum_{k=1}^{5} 6$ **30**
10. $\sum_{m=1}^{10} -5$ **–50**
11. $\sum_{k=1}^{8} -3k$ **–108**
12. $\sum_{k=1}^{6} 2k$ **42**
13. $\sum_{k=1}^{6}(2k^2 + k + 2)$ **215**
14. $\sum_{k=1}^{7}(3k^2 + 4k + 2)$ **546**

Chapter Test, Form A

NAME ______ CLASS ______ DATE ______

Chapter Assessment
Chapter 11, Form A, page 1

Write the letter that best answers the question or completes the statement.

__b__ 1. Find the value of $\sum_{k=1}^{8}(2k^2 + 3k - 5)$.
a. 329 b. 476 c. 556 d. 660

__d__ 2. Find a recursive formula for the sequence 5, 9, 17, 33, . . .
a. $t_n = 2^{n+1} + 1$ b. $t_n = 5 + 2(n - 1)$
c. $\begin{cases} t_1 = 5 \\ t_n = t_{n-1} + 4 \end{cases}$, where $n \geq 2$ d. $\begin{cases} t_1 = 5 \\ t_n = 2t_{n-1} - 1 \end{cases}$, where $n \geq 2$

__b__ 3. Find the 60th term of the arithmetic sequence in which $t_5 = -8$ and $t_{11} = -38$.
a. −251 b. −283 c. −288 d. −342

__c__ 4. Find three geometric means between 2 and 2592.
a. 3, 120, 1500 b. 16, 92, 624 c. 12, 72, 432 d. 649.5, 1297, 1944.5

__a__ 5. Which of the following is t_{120} in the sequence −60, −46, −32, −18, . . . ?
a. 1606 b. 1620 c. 1634 d. 1648

__c__ 6. Which of the following is a formula for the sequence 8, −20, 50, −125, . . . ?
a. $t_n = 8 + (n - 1)(-2.5)$ b. $t_n = 8 + (-2.5)^{n+1}$
c. $t_n = 8(-2.5)^{n-1}$ d. $t_n = 8(-2.5^{n-1})$

__a__ 7. Which of the following is S_{80} in the series 250 + 217 + 184 + 151 + ⋯ ?
a. −84,280 b. −72,600 c. −54,300 d. −42,100

__b__ 8. Suppose that the population of fish in a lake is decreasing by 40% per year because of the increase of pollution in the lake. If there are about 75,000 fish in the lake now, predict the number of fish in the lake after 7 years.
a. 3500 b. 2100 c. 1200 d. 300

__c__ 9. Sally decides to start doing sit-ups as part of her daily workout. She decides to do 15 sit-ups the first day and then increase the number by 4 each day after that. Find the total number of sit-ups Sally will do in the next 3 weeks if she sticks to her plan.
a. 95 b. 625 c. 1155 d. 1435

__d__ 10. It cost Kip $12 to maintain his car during the first month he owned it. Since then the cost has increased by $2.25 a month. If the cost continues to increase by the same amount each month, how much will it cost Kip to maintain his car during the 40th month he owns it?
a. $120 b. $104.40 c. $102 d. $99.75

NAME ______ CLASS ______ DATE ______

Chapter Assessment
Chapter 11, Form A, page 2

__d__ 11. Find the value of $\sum_{k=1}^{30} -4(1.08^{k-1})$ to the nearest tenth.
a. 349.8 b. 123.4 c. −415.9 d. −453.1

__c__ 12. Which of the following infinite series has a sum?
a. 12 + 10.25 + 8.5 + 6.75 + ⋯ b. −8 + (−12) + (−18) + (−27) + ⋯
c. 20 + (−16) + 12.8 + (−10.24) + ⋯ d. $\frac{2}{3} + \frac{1}{2} + \frac{3}{4} + \frac{5}{8} + \cdots$

__a__ 13. Find the sum of the infinite series 1300 + 455 + 159.25 + ⋯
a. 2000 b. 2952.6 c. 3714.3 d. 3000

__b__ 14. Which of the following is the eighth entry in row 17 of Pascal's triangle?
a. 12,870 b. 19,448 c. 24,310 d. 98,017,920

__d__ 15. Find the sum of the terms in row 12 of Pascal's triangle.
a. 144 b. 924 c. 2048 d. 4096

__c__ 16. In the expansion of $(a + b)^{13}$, which term contains b^9?
a. $715a^9b^9$ b. $1287a^9b^9$ c. $715a^4b^9$ d. $1287a^4b^9$

__b__ 17. In the expansion of $(3x - y)^7$, which term contains y^5?
a. $189x^2y^5$ b. $-189x^2y^5$ c. $63x^2y^5$ d. $-63x^5y^5$

__b__ 18. If 12 fair coins are tossed, in how many ways can 8 heads appear?
a. 96 b. 495 c. 792 d. 1264

__d__ 19. If a family deposits $2000 at the beginning of each year into a bank account that pays 6% annual interest, how much will the family have in the account after 18 years?
a. $5708.68 b. $38,160 c. $61,811.31 d. $65,519.98

__a__ 20. If a number cube is rolled 7 times, what is the probability that an even number occurs exactly 5 times?
a. 0.164 b. 0.225 c. 0.417 d. 0.523

__d__ 21. When Amy bowls, the probability that she will bowl one strike in a frame is 0.35. What is the probability that Amy will bowl exactly 4 strikes in the next 7 frames?
a. 1.40 b. 0.57 c. 0.27 d. 0.14

__d__ 22. The probability of being stopped at a particular traffic light is 0.21. If you go through this traffic light 10 times, what is the probability that you will be stopped 2 or fewer times?
a. 0.21 b. 0.30 c. 0.55 d. 0.65

LESSON 11.2

Key Skills

Find the indicated term of an arithmetic sequence.

Find the fifth term of the arithmetic sequence in which $t_1 = 2$ and $t_n = t_{n-1} + 3$.

$$t_n = t_1 + (n - 1)d$$
$$t_5 = 2 + (5 - 1)3 \quad n = 5,\ t_1 = 2, \text{ and } d = 3$$
$$t_5 = 14$$

Find arithmetic means between two nonconsecutive terms.

Find three arithmetic means between 10 and 18.

10, ?, ?, ?, 18

Substitute 10 for t_1, 18 for t_n, and 5 for n.

$$t_n = t_1 + (n - 1)d$$
$$18 = 10 + (5 - 1)d$$
$$d = 2$$

The three arithmetic means are 12, 14, and 16.

Exercises

Find the indicated term.

15. the sixth term of the arithmetic sequence in which $t_1 = 2$ and $t_n = t_{n-1} + 4$ **22**

16. the seventh term of the arithmetic sequence in which $t_1 = 1$ and $t_n = t_{n-1} - 2$ **-11**

17. the fifth term of the arithmetic sequence in which $t_1 = -3$ and $t_n = t_{n-1} - 4$ **-19**

18. the 10th term of the arithmetic sequence in which $t_1 = 3$ and $t_4 = -9$ **-33**

Find the indicated arithmetic means.

19. two arithmetic means between 20 and 50 **30 and 40**

20. three arithmetic means between 100 and 180 **120, 140, and 160**

21. four arithmetic means between −10 and −35 **−15, −20, −25, and −30**

22. three arithmetic means between −8 and 12 **−3, 2, and 7**

LESSON 11.3

Key Skills

Find the sum of the first n terms of an arithmetic series.

For the series $6 + 11 + 16 + 21 + \cdots$, find S_8.

Method 1:

First find t_8.

$$t_n = t_1 + (n - 1)d$$
$$t_8 = 6 + (8 - 1)5$$
$$t_8 = 41$$

Then find S_8.

$$S_n = n\left(\frac{t_1 + t_n}{2}\right)$$
$$S_8 = 8\left(\frac{6 + 41}{2}\right)$$
$$S_8 = 188$$

Method 2:

$$S_n = n\left[\frac{2t_1 + (n - 1)d}{2}\right]$$
$$S_8 = 8\left[\frac{12 + (8 - 1)5}{2}\right]$$
$$S_8 = 188$$

Exercises

For the series $2 + 5 + 8 + 11 + \cdots$, find the indicated sum.

23. S_5 **40**

24. S_{10} **155**

25. S_{12} **222**

26. S_{15} **345**

For each arithmetic series, find S_{20}.

27. $8 + 16 + 24 + 32 + \cdots$ **1680**

28. $14 + 16 + 18 + 20 + 22 + \cdots$ **660**

29. $-6 + (-12) + (-18) + (-24) + (-30) + \cdots$ **−1260**

30. $-7 + (-5) + (-3) + (-1) + \cdots$ **240**

Evaluate.

31. $\sum_{k=1}^{5} 6k$ **90**

32. $\sum_{j=3}^{7} (2j - 5)$ **25**

33. $\sum_{k=0}^{4} (5 - 2k)$ **5**

34. $\sum_{n=1}^{30} (7 - n)$ **-255**

Chapter Test, Form B

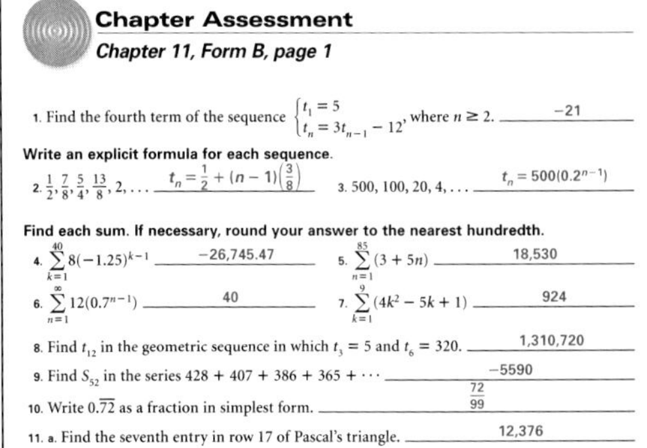

NAME ______ CLASS ______ DATE ______

Chapter Assessment

Chapter 11, Form B, page 1

1. Find the fourth term of the sequence $\begin{cases} t_1 = 5 \\ t_n = 3t_{n-1} - 12 \end{cases}$, where $n \geq 2$. **−21**

Write an explicit formula for each sequence.

2. $\frac{1}{2}, \frac{7}{8}, \frac{5}{4}, \frac{13}{8}, 2, \ldots$ **$t_n = \frac{1}{2} + (n - 1)\left(\frac{3}{8}\right)$**

3. 500, 100, 20, 4, ... **$t_n = 500(0.2^{n-1})$**

Find each sum. If necessary, round your answer to the nearest hundredth.

4. $\sum_{k=1}^{40} 8(-1.25)^{k-1}$ **−26,745.47**

5. $\sum_{n=1}^{85} (3 + 5n)$ **18,530**

6. $\sum_{n=1}^{\infty} 12(0.7^{n-1})$ **40**

7. $\sum_{k=1}^{9} (4k^2 - 5k + 1)$ **924**

8. Find t_{12} in the geometric sequence in which $t_3 = 5$ and $t_6 = 320$. **1,310,720**

9. Find S_{52} in the series $428 + 407 + 386 + 365 + \cdots$ **−5590**

10. Write $0.\overline{72}$ as a fraction in simplest form. **$\frac{72}{99}$**

11. a. Find the seventh entry in row 17 of Pascal's triangle. **12,376**
    b. Which other entry in row 17 has the same value as the seventh entry? **12th**
    c. Find the sum of all the entries in row 17 of Pascal's triangle. **131,072**

Expand each binomial.

12. $(r - t)^7$
    $r^7 - 7r^6t + 21r^5t^2 - 35r^4t^3 + 35r^3t^4 - 21r^2t^5 + 7rt^6 - t^7$

13. $(3x + 2y)^4$
    $81x^4 + 216x^3y + 216x^2y^2 + 96xy^3 + 16y^4$

14. In the expansion of $(x + 5)^{10}$, find the term containing x^8. **$1125x^8$**

15. a. If 12 fair coins are tossed, in how many ways can 9 heads appear? **220**
    b. If 12 fair coins are tossed, find the probability that 9 heads appear. **≈0.54**

16. Marie has just bought a new car for $14,000. She will make monthly payments on the car for 5 years.
    a. If the car retains 80% of its value each year, how much will Marie's car be worth after 5 years? **$4587.52**
    b. If the value of the car decreases by $1600 per year, how much will Marie's car be worth after 5 years? **$6000**

NAME ______ CLASS ______ DATE ______

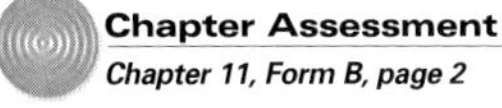

Chapter Assessment

Chapter 11, Form B, page 2

17. If, at the beginning of each year, a family puts $1500 into a bank account that pays 9% annual interest, how much money will the family have in the account after 12 years? **$32,930.08**

18. Bill has decided to start taking a daily walk in order to improve his health. He decides to walk 0.5 mile the first day and then increase the distance by 0.25 mile per day.
    a. If Bill sticks to his plan, how far will he walk on the 15th day? **4 miles**
    b. If Bill sticks to his plan for 4 weeks, find the total distance he will walk. **108.5 miles**

19. The owner of a new restaurant is disappointed because only 150 people came to the restaurant during the first week it was open. However, the owner believes that the number of people who come to the restaurant will increase by 10% per week.
    a. If the owner is correct, how many people will come to the restaurant during the 20th week it is open? **About 917 people**
    b. If the owner is correct, find the total number of people that will come to the restaurant during the first year it is open. **About 211,564 people**

20. If a person reserves a seat on a flight, the probability that the person shows up is 0.86. Suppose that 20 people have reserved seats on this flight. Find the following probabilities. Round your answers to the nearest hundredth.
    a. Find the probability that exactly 16 of the 20 people show up. **0.17**
    b. Find the probability that 18 or more of the 20 people show up. **0.46**

21. Use mathematical induction to prove that $3 + 7 + 11 + \cdots + (4n - 1) = n(2n + 1)$ is true for all natural numbers, n.
    1. Show that the statement is true for $n = 1$: $[4(1) - 1] = 1[2(1) + 1]$
    2. Assume $3 + 7 + 11 + \cdots + (4k - 1) = k(2k + 1)$ and show that $3 + 7 + 11 + \cdots + (4k - 1) + [4(k + 1) - 1] = (k + 1)[2(k + 1) + 1]$. $3 + 7 + 11 + \cdots + (4k - 1) + [4(k + 1) - 1] = k(2k + 1) + [4(k + 1) - 1] = 2k^2 + 5k + 3$ $= (k + 1)(2k + 3) = (k + 1)[2(k + 1) + 1]$

50. 1. Basis Step
Show that $n - 1 < n$ is true for $n = 1$.
$1 - 1 = 0 < 1$ True

2. Induction Step
Assume that the statement is true for a natural number, k.
$k - 1 < k$

Then prove that it is true for the next natural number, $k + 1$. Determine the statement to be proved.
$(k + 1) - 1 < k + 1$
$k < k + 1$

Add 1 to each side of the statement assumed true and simplify.
$(k - 1) + 1 < k + 1$
$k < k + 1$ True

LESSON 11.4

Key Skills

Find the indicated term of a geometric sequence.

Find the sixth term of the geometric sequence in which $t_1 = 4$ and $t_n = 2t_{n-1}$.

$$t_n = t_1 r^{n-1}$$
$$t_6 = 4 \cdot 2^{6-1}$$
$$t_6 = 128$$

Find geometric means between two nonconsecutive terms.

Find three geometric means between 10 and 2560.

10, ?, ?, ?, 2560

Substitute 10 for t_1 and 2560 for t_5.

$$t_n = t_1 r^{n-1}$$
$$2560 = 10 \cdot r^{5-1}$$
$$r = \pm 4$$

The three geometric means are 40, 160, and 640 or −40, 160, and −640.

Exercises

Find the indicated term.

35. the sixth term of the sequence in which $t_1 = -4$ and $t_n = 5t_{n-1}$ **−12,500**

36. the fifth term of the sequence in which $t_1 = 8$ and $t_n = -3t_{n-1}$ **648**

37. the 10th term of the sequence in which $t_1 = \frac{1}{3}$ and $t_n = \frac{1}{2}t_{n-1}$ $\frac{1}{1536}$

Find the indicated geometric means.

38. two geometric means between 10 and 1250

39. three geometric means between 5 and 100

40. three geometric means between 20 and 30

38. 50 and 250

39. $5(20)^{\frac{1}{4}}$, $5(20)^{\frac{1}{2}}$, and $5(20)^{\frac{3}{4}}$ or $-5(20)^{\frac{1}{4}}$, $5(20)^{\frac{1}{2}}$, and $-5(20)^{\frac{3}{4}}$

40. $20\left(\frac{3}{2}\right)^{\frac{1}{4}}$, $20\left(\frac{3}{2}\right)^{\frac{1}{2}}$, and $20\left(\frac{3}{2}\right)^{\frac{3}{4}}$ or $-20\left(\frac{3}{2}\right)^{\frac{1}{4}}$, $20\left(\frac{3}{2}\right)^{\frac{1}{2}}$, and $-20\left(\frac{3}{2}\right)^{\frac{3}{4}}$

LESSON 11.5

Key Skills

Find the sum of the first n terms of a geometric series.

Find the sum of the first eight terms in the series $1 + 2 + 4 + 8 + \cdots$

Substitute 2 for r, 1 for t_1, and 8 for n.

$$S_n = t_1\left(\frac{1 - r^n}{1 - r}\right)$$
$$S_8 = 1\left(\frac{1 - 2^8}{1 - 2}\right) = 255$$

Use mathematical induction to prove statements about the natural numbers.

To prove that a statement about natural numbers is true for all natural numbers, use the principle of mathematical induction stated below.

1. Show that the statement is true for $n = 1$.
2. Assume that the statement is true for a natural number k. Then prove that it is true for the natural number $k + 1$.

Exercises

For the series $1 + 1.5 + 2.25 + 3.375 + \cdots$, find each indicated sum to the nearest tenth.

41. S_4 8.1 **42.** S_5 13.2 **43.** S_{12} 257.5 **44.** S_{15} 873.8

Use the formula for the first n terms of a geometric series to find the sum of each geometric series. Give answers to the nearest tenth, if necessary.

45. $-1 + 10 + (-100) + 1000 + (-10{,}000)$ **−9091**

46. $0.4 + 4 + 40 + 400 + 4000$ **4444.4**

47. $3 + (-6) + 12 + (-24) + 48 + (-96)$ **−63**

Evaluate. Give answers to the nearest hundredth, if necessary.

48. $\sum_{p=1}^{4} \frac{1}{6}(3)^p$ 20 **49.** $\sum_{k=5}^{8} (0.2)^{k-4}$ 0.25

Use mathematical induction to prove each statement for all natural numbers, n.

50. $n - 1 < n$

51. $7 + 9 + 11 + \cdots + (2n + 5) = n(n + 6)$

51. 1. Basis Step
Show that $7 + 9 + 11 + \cdots + (2n + 5) = n(n + 6)$ is true for $n = 1$.
$7 = 1(1 + 6)$
$7 = 7$ True

2. Induction Step
Assume that the statement is true for a natural number, k.
$7 + 9 + 11 + \cdots + (2k + 5) = k(k + 6)$

Then prove that it is true for the next natural number, $k + 1$. Determine the statement to be proved.

$7 + 9 + 11 + \cdots + [2(k + 1) + 5] = (k + 1)[(k + 1) + 6]$ $7 + 9 + 11 + \cdots + (2k + 7) = (k + 1)(k + 7)$

Begin with the statement assumed to be true and use properties of equality.

$7 + 9 + 11 + \cdots + (2k + 5) + (2k + 7) = (k + 1)(k + 7)$ $k(k + 6) + (2k + 7) = (k + 1)(k + 7)$ $k^2 + 8k + 7 = (k + 1)(k + 7)$ $(k + 1)(k + 7) = (k + 1)(k + 7)$ True

LESSON 11.6

Key Skills

Find the sum of an infinite geometric series, if one exists.

Find the sum of the series $\sum_{k=0}^{\infty} \frac{2}{5^{k+2}}$, if it exists.

$$r = \frac{\frac{2}{125}}{\frac{2}{25}} = \frac{1}{5}$$

If a geometric sequence has common ratio r and $|r| < 1$, then the related infinite geometric series converges and has a sum.

The series converges because $|r| = \left|\frac{1}{5}\right| < 1$.

$$\sum_{k=0}^{\infty} \frac{2}{5^{k+2}} = \frac{t_1}{1-r}$$
$$= \frac{\frac{2}{25}}{1-\frac{1}{5}}$$
$$= \frac{1}{10}$$

Exercises

Find the sum of each infinite geometric series, if it exists.

52. $5 + \frac{5}{3} + \frac{5}{9} + \frac{5}{27} + \cdots$ 7.5

53. $8 + 4 + 2 + 1 + \cdots$ 16

54. $\frac{1}{2} + \left(-\frac{1}{4}\right) + \frac{1}{8} + \left(-\frac{1}{16}\right) + \cdots$ $\frac{1}{3}$

55. $0.5 + 0.05 + 0.005 + 0.0005 + \cdots$ $\frac{5}{9}$

56. $\sum_{k=1}^{\infty}\left(\frac{1}{3}\right)^k$ $\frac{1}{2}$ **57.** $\sum_{k=1}^{\infty}\left(\frac{1}{6}\right)^k$ $\frac{1}{5}$ **58.** $\sum_{k=1}^{\infty} 4^k$ none

59. $\sum_{k=0}^{\infty}\left(\frac{3}{7}\right)^k$ $\frac{7}{4}$ **60.** $\sum_{k=0}^{\infty}\left(\frac{5}{6}\right)^k$ 6 **61.** $\sum_{k=0}^{\infty}\left(\frac{5}{4}\right)^k$ none

Write each decimal as a fraction in simplest form.

62. $0.\overline{620}$ $\frac{620}{999}$ **63.** $0.\overline{032}$ $\frac{32}{999}$

LESSON 11.7

Key Skills

Find entries in Pascal's triangle.

Find the seventh entry in row 11 of Pascal's triangle.

The kth entry in the nth row is ${}_nC_{k-1}$.

$${}_{11}C_{7-1} = {}_{11}C_6$$
$$= \frac{11!}{5!6!}$$
$$= 462$$

Use Pascal's triangle to find probabilities.

Find the probability that exactly 1 heads *or* exactly 2 heads appears when a coin is tossed 5 times.

Number of possible outcomes = 2^5

Probability of exactly 1 heads = $\frac{{}_5C_1}{2^5}$

Probability of exactly 2 heads = $\frac{{}_5C_2}{2^5}$

$P(1 \text{ heads } or \text{ 2 heads}) = \frac{{}_5C_1}{2^5} + \frac{{}_5C_2}{2^5}$
≈ 0.47

Exercises

Find each entry of Pascal's triangle.

64. sixth entry of row 9 126

65. eighth entry of row 10 120

66. fifth entry of row 8 70

67. fourth entry of row 6 20

Find the probability of each event.

68. exactly 4 heads when a coin is tossed 4 times

69. exactly 3 tails when a coin is tossed 8 times

70. more than 6 tails when a coin is tossed 8 times

71. exactly 2 *or* exactly 4 heads when a coin is tossed 6 times

68. 0.06

69. 0.22

70. 0.035

71. 0.47

72. $625 + 500y + 150y^2 + 20y^3 + y^4$

73. $a^5 + 15a^4 + 90a^3 + 270a^2 + 405a + 243$

74. $32x^5 - 160x^4y + 320x^3y^2 - 320x^2y^3 + 160xy^4 - 32y^5$

75. $729x^6 - 1458x^5y + 1215x^4y^2 - 540x^3y^3 + 135x^2y^4 - 18xy^5 + y^6$

LESSON 11.8

Key Skills

Use the Binomial Theorem to expand a binomial raised to a power.

Expand $(a + b)^4$.

$$(a+b)^n = \sum_{k=0}^{n} \binom{n}{k} a^{n-k} b^k$$

$$(a+b)^4 = \binom{4}{0}a^4b^0 + \binom{4}{1}a^3b^1 + \binom{4}{2}a^2b^2 + \binom{4}{3}a^1b^3 + \binom{4}{4}a^0b^4$$

$$= a^4b^0 + 4a^3b^1 + 6a^2b^2 + 4a^1b^3 + a^0b^4$$

Find a term in a binomial expansion.

Find the seventh term in the expansion of $(a + b)^9$.

For the expansion of $(a + b)^9$, $n = 9$.

For the seventh term, $k = 6$.

$$\binom{n}{k}x^{n-k}y^k = \binom{9}{6}a^3b^6 = 84a^3b^6$$

Exercises

Expand each binomial raised to a power.

72. $(5 + y)^4$

73. $(a + 3)^5$

74. $(2x - 2y)^5$

75. $(3x - y)^6$

76. Find the sixth term in the expansion of $(x + y)^7$. **$21x^2y^5$**

77. Find the eighth term in the expansion of $(a + b)^{10}$. **$120a^3b^7$**

Write each summation as a binomial raised to a power. Then write it in expanded form.

78. $\sum_{k=0}^{5} \binom{5}{k} a^{5-k} b^k$

79. $\sum_{k=0}^{6} \binom{6}{k} a^{6-k} b^k$

78. $(a + b)^5 = a^5 + 5a^4b + 10a^3b^2 + 10a^2b^3 + 5ab^4 + b^5$

79. $(a + b)^6 = a^6 + 6a^5b + 15a^4b^2 + 20a^3b^3 + 15a^2b^4 + 6ab^5 + b^6$

Applications

80. **INVESTMENTS** Yolanda deposits \$1000 in a bank at the beginning of every year for 8 years. Her account earns 8% interest compounded annually. How much is in the account after the last year? **\$11,487.56**

81. **PHYSICS** A ball is dropped from a height of 12 meters. Each time it bounces, it rebounds $\frac{2}{3}$ of the distance it has fallen.
   - **a.** How far will it rebound after the third bounce? **≈3.6 meters**
   - **b.** Theoretically, what is the vertical distance that the ball will travel before coming to rest? **60 meters**

82. **MERCHANDISING** Boxes are stacked in a display of 21 rows with 2 boxes in the top row, 5 boxes in the second row, 8 boxes in the third row, and so on. How many boxes are there? **672**

83. **ENTERTAINMENT** A theater has 18 seats in the front row. Each succeeding row has 4 more seats than the row ahead of it. How many seats are there in the first 12 rows? 20 rows? **480 seats; 1120 seats**

84. **REAL ESTATE** A 5-year lease for an apartment calls for a rent of \$500 per month the first year with an increase of 6% for each remaining year of the lease. What will be the monthly rent during the last year of the lease? **\$631.24**

Chapter Test

Write the first five terms of each sequence.

1. $t_n = 5n + 7$ 12, 17, 22, 27, 32

2. $t_n = -3.14n + 1.6$ −1.54, −4.68, −7.82, −10.96, −14.1

3. $t_1 = 7$

$t_n = 3t_{n-1} + 2$ 7, 23, 71, 215, 647

Evaluate the sum.

4. $\sum_{k=1}^{7}(k-3)^2$ 35

5. $\sum_{k=1}^{5}(2k^2 - 3k + 4)$ 85

6. COMMUNICATION A telemarketing service uses a master computer to set up a chain of activating computers. The master computer links to 2 computers and activates them. Then the activated computers each link with 2 additional computers to activate them. At each step the activation process takes 10 minutes. After an hour how many computers have been activated? 126

Find the indicated term.

7. 8th term; given $t_1 = 7$ and $t_n = t_{n-1} + 7$ 56

8. 5th term; given $t_1 = -27$ and $t_n = t_{n-1} + 12$ 21

Find the indicated arithmetic means.

9. Find the three arithmetic means between 144 and 188. 155, 166, 177

10. Find the four arithmetic means between −12 and 18. −6, 0, 6, 12

For each arithmetic series, find S_{18}.

11. $3 + 10 + 17 + 24 + 31 + \cdots$ 1125

12. $5 + (-1) + (-7) + (-13) + (-19) + \cdots$ −828

Evaluate.

13. $\sum_{k=1}^{15} -4k$ −480

14. $\sum_{k=1}^{5}(2k - 9)$ −15

Find the indicated geometric means.

15. Find the three geometric means between 2 and 1250. 10, 50, 250, or −10, 50, −250

16. Write an explicit formula for the nth term of the geometric sequence.

3, 6, 12, 24... $t_n = 3(2)^{n-1}$

17. TECHNOLOGY A new laser printer costs $1800. Each year it loses 12% of its value from the previous year. Find its value after 8 years. $647.34

Evaluate. Round to the nearest tenth.

18. $\sum_{k=1}^{5} 3(2^{k-1})$ 93

19. $\sum_{k=1}^{6} \frac{1}{2}(4)^{k-1}$ 682.5

20. Refer to the geometric series in which $t_1 = 3$ and $r = 4$. Find S_6. 4095

21. BIOLOGY A bacterium reproduces by dividing into two bacteria. Find the number of bacteria after 10 such divisions. 1024

Find the sum of each infinite geometric series, if it exists.

22. $0.9 + 0.09 + 0.009 + 0.0009 + \cdots$ 1

23. $\frac{1}{3} + \frac{1}{9} + \frac{1}{27} + \frac{1}{81} + \cdots$ 0.5

24. $\sum_{k=0}^{\infty}(1.001)^k$ sum does not exist

25. $\sum_{k=0}^{\infty}\left(\frac{5}{9}\right)^k$ 2.25

Find each entry in Pascal's triangle.

26. fifth entry in the fifth row 5

27. ninth entry in the twelfth row 495

Find the probability of each event.

28. 4 girls in a family of 6 children 0.23

29. 8 right when randomly guessing on a true/false test with 12 questions 0.12

Expand each binomial.

30. $(x + y)^4$

31. $(2a - 3b)^6$

32. Find the fifth term in the expansion of $(r - 3t)^8$. $5670r^4t^4$

30. $x^4 + 4x^3y + 6x^2y^2 + 4xy^3 + y^4$

31. $64a^6 - 576a^5b + 2160a^4b^2 - 4320a^3b^3 + 4860a^2b^4 - 2916ab^5 + 729b^6$

College Entrance Exam Practice

Multiple-Choice and Quantitative-Comparison Samples

The first half of the Cumulative Assessment contains two types of items found on standardized tests—multiple-choice questions and quantitative-comparison questions. Quantitative-comparison items emphasize the concepts of equality, inequality, and estimation.

Free-Response Grid Samples

The second half of the Cumulative Assessment is a free-response section. This part of the Cumulative Assessment requires student-produced response items like those commonly found on college entrance exams. These questions require the use of machine-scored answer grids. You may wish to have students practice answering these items in preparation for standardized tests.

CHAPTERS CUMULATIVE ASSESSMENT

1–11

College Entrance Exam Practice

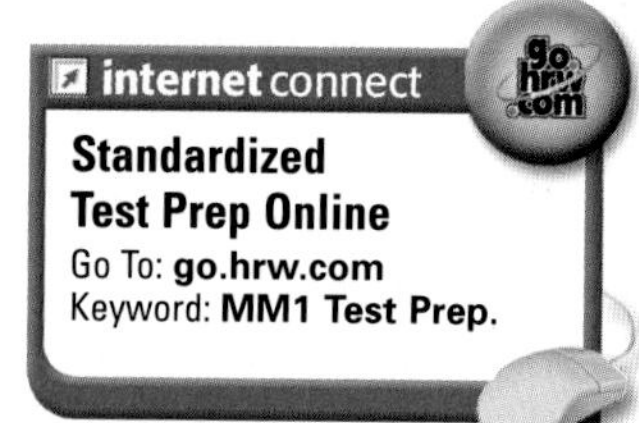

QUANTITATIVE COMPARISON For Items 1–5, write
A if the quantity in Column A is greater than the quantity in Column B;
B if the quantity in Column B is greater than the quantity in Column A;
C if the quantities are equal; or
D if the relationship cannot be determined from the given information.

	Column A	Column B	Answers
1. A	$f(x) = x - 5$, $g(x) = 3x$		
	$(f \circ g)(3)$	$(g \circ f)(3)$	Ⓐ Ⓑ Ⓒ Ⓓ [Lesson 2.4]
2. C	${}_{10}C_2$	${}_{10}C_8$	Ⓐ Ⓑ Ⓒ Ⓓ [Lesson 10.3]
3. B	$(5 - 2i)(5 + 2i)$	$6i^2 \cdot 7i^2$	Ⓐ Ⓑ Ⓒ Ⓓ [Lesson 5.6]
4. C	The probability of the event		
	rolling an odd number on one roll of a six-sided number cube	rolling a 1, 2, or 3 on one roll of a six-sided number cube	Ⓐ Ⓑ Ⓒ Ⓓ [Lesson 10.1]
5. A	The value of x		
	$\begin{cases} 2x + 3y = 21 \\ x - 2y = -7 \end{cases}$	$\begin{cases} -2x + y = 8 \\ 5x - 3y = -21 \end{cases}$	Ⓐ Ⓑ Ⓒ Ⓓ [Lesson 3.1]

6. Which function represents the graph of $f(x) = |x|$ translated 3 units to the right? ***(LESSON 2.7)*** a
   a. $g(x) = |x - 3|$ **b.** $g(x) = |x + 3|$
   c. $g(x) = |x| - 3$ **d.** $g(x) = |x| + 3$

7. Solve $|4t - 7| = 5t + 2$. ***(LESSON 1.8)*** c
   a. 9 **b.** −9 **c.** $\frac{5}{9}$ **d.** $9, -\frac{5}{9}$

8. Evaluate $-5^3 - 2 \cdot 3^2 \div 6$. ***(LESSON 2.1)*** b
   a. 122 **b.** −128 **c.** −122 **d.** −23

9. What are the coordinates of the vertex of the graph of $y = -x^2 + 2$? ***(LESSON 5.1)*** d
   a. (−1, 2) **b.** (0, −2)
   c. (−1, −2) **d.** (0, 2)

10. Find the slope of the line that contains (−2, 5) and (−5, −4). ***(LESSON 1.2)*** c
    a. $\frac{1}{3}$ **b.** −3 **c.** 3 **d.** $-\frac{1}{3}$

11. Which single equation represents the pair of parametric equations $\begin{cases} x(t) = 2 + t \\ y(t) = 3 + t \end{cases}$? ***(LESSON 3.6)*** c
    a. $y = x - 1$ **b.** $y = -x - 1$
    c. $y = x + 1$ **d.** $y = -x + 1$

12. What is the domain of $g(x) = -\frac{1}{2}x^2 - 4$? ***(LESSON 2.3)*** b
    a. $x > 0$ **b.** all real numbers
    c. $x > -4$ **d.** $x < -4$

13. Which is the greatest monomial factor of $2(6x^2 - 14x)$? ***(LESSON 5.3)*** d
    a. 2 **b.** 4 **c.** $2x$ **d.** $4x$

14. The graph of which ellipse has x-intercepts of 2 and -2 and y-intercepts of 5 and -5? **(LESSON 9.4)**
c

a. $\frac{x^2}{4} + \frac{y^2}{4} = 1$ **b.** $\frac{x^2}{4} - \frac{y^2}{25} = 1$
c. $\frac{x^2}{4} + \frac{y^2}{25} = 1$ **d.** $\frac{y^2}{25} - \frac{x^2}{4} = 1$

15. Which binomial is a factor of $x^3 - 3x^2 - 2x + 6$? **(LESSON 7.3)**
a

a. $x^2 - 2$ **b.** $x^2 + 2$
c. $x + 3$ **d.** $x + 1$

16. Which equation represents $\log_2 \frac{1}{32} = -5$ in exponential form? **(LESSON 6.3)**
a

a. $2^{-5} = \frac{1}{32}$ **b.** $2^{-5} = -32$
c. $2^{-5} = 32$ **d.** $2^{-5} = -\frac{1}{32}$

17. If y varies jointly as the cube of m and the square of n and inversely as p, which equation represents this relationship? **(LESSON 8.1)**
b

a. $y = \frac{p}{m^3n^2}$ **b.** $y = \frac{km^3n^2}{p}$
c. $y = m^3n^2p$ **d.** $y = km^3n^2p$

18. Which polynomial equation has -2 and $5i$ as roots? **(LESSON 7.4)**
c

a. $x^3 - 2x^2 + 25x - 50 = 0$
b. $x^3 + 2x^2 - 5x - 10 = 0$
c. $x^3 + 2x^2 + 25x + 50 = 0$
d. $x^3 - 2x^2 + 5x - 10 = 0$

19. Find the slope, m, of $x = -5$. **(LESSON 1.2)** undefined

20. Write the equation in slope-intercept form for the line that has a slope of $-\frac{3}{5}$ and contains the point $(-2, -5)$. **(LESSON 1.3)** $y = -\frac{3}{5}x - \frac{31}{5}$

21. Find the inverse of $\{(-1, -1), (-2, 2), (-3, 1), (0, 0)\}$. **(LESSON 2.5)** $\{(-1, -1), (2, -2), (1, -3), (0, 0)\}$

22. Find the zeros of $f(x) = 12x^2 - 8x - 15$. **(LESSON 5.3)** $x = \frac{3}{2}$ or $x = -\frac{5}{6}$

23. Graph $2y + x \le 6$. **(LESSON 3.3)**

24. Find the product $\begin{bmatrix} -1 & 2 & 3 \\ 4 & -2 & 3 \end{bmatrix} \begin{bmatrix} 2 & 1 & 3 \\ -1 & -2 & -3 \\ 4 & 1 & -2 \end{bmatrix}$, if it exists. **(LESSON 4.2)**

25. Solve $10^{-3x} = 125$ to the nearest hundredth. **(LESSON 6.7)** $x \approx -0.70$

26. Factor $6x^2 - 28x - 10$, if possible. **(LESSON 5.3)** $2(x - 5)(3x + 1)$

27. Find the 100th term of the sequence $-11, -5, 1, \ldots$ **(LESSON 11.2)** 583

28. Simplify $\frac{1 - 2i}{2 + i} + 3i$. **(LESSON 5.6)** $2i$

29. Graph $y \ge x^2 - 3x + 5$. **(LESSON 5.8)**

30. Solve $\begin{cases} 2x - 5y = 15 \\ 3x - 7y = 22 \end{cases}$ by using a matrix equation. **(LESSON 4.4)** $(5, -1)$

31. Write $(x - 1)^2(x^2 + 2x + 5)$ as a polynomial expression in standard form. **(LESSON 7.3)**

32. Simplify $\frac{x^2 + 3x - 18}{2x^2 - 5x - 3} \cdot \frac{1 - 4x^2}{x^2 - 36}$. **(LESSON 8.3)**

33. Use substitution to solve $\begin{cases} x + y = 18 \\ 4x - 4y = 5 \end{cases}$. $\left(\frac{77}{8}, \frac{67}{8}\right)$ **(LESSON 3.1)**

FREE-RESPONSE GRID **The following questions may be answered by using a free-response grid such as that commonly used by standardized-test services.**

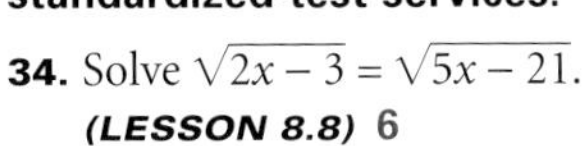

34. Solve $\sqrt{2x - 3} = \sqrt{5x - 21}$. **(LESSON 8.8)** 6

35. Evaluate $\log_{\frac{1}{3}} 27 + 3^{\log_3 5}$ **(LESSON 6.4)** 2

36. Find the value of v in $-1 = \log_{10} v$. **(LESSON 6.3)** 0.1

37. Find the perimeter, to the nearest hundredth, of a triangle whose vertices have the coordinates $(2, 0)$, $(0, 0)$, and $(0, -4)$. **(LESSON 9.1)** 10.47

38. How many 4 letter arrangements can be made with the letters in *meet*? **(LESSON 10.2)** 12

PHYSICS A ball is dropped from a height of 12 meters. Each time it bounces on the ground, it rebounds to $\frac{2}{3}$ the distance it has fallen.

39. How many meters will it rebound after the third bounce? **(LESSON 11.4)** 3.55

40. Theoretically, how many meters will the ball travel before coming to rest? **(LESSON 11.6)**
60

23.

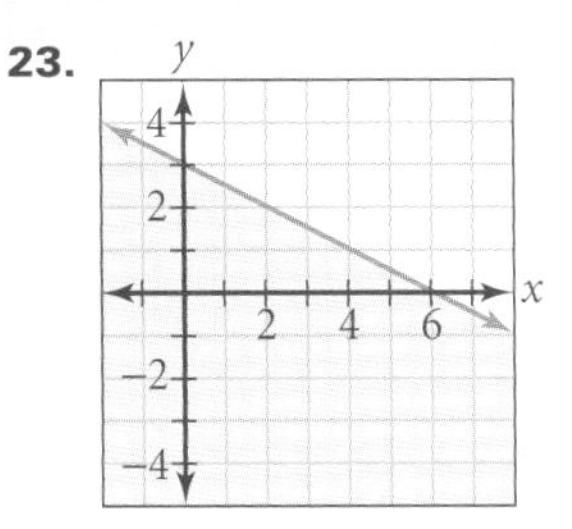

24. $\begin{bmatrix} 8 & -2 & -15 \\ 22 & 11 & 12 \end{bmatrix}$

29.

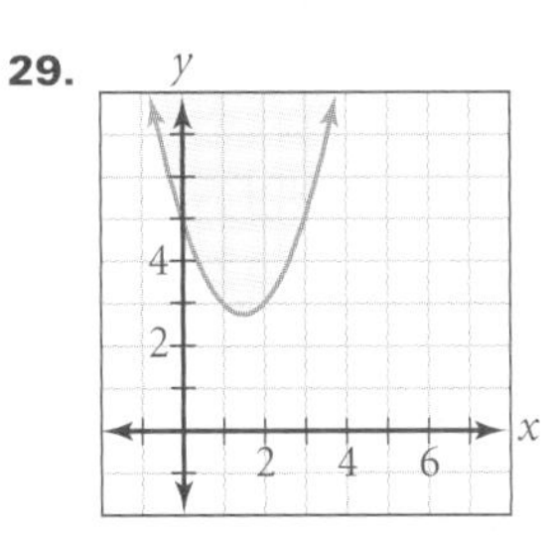

31. $x^4 + 2x^2 - 8x + 5$

32. $\frac{1 - 2x}{x - 6}$

Keystroke Guide for Chapter 11

Essential keystroke sequences (using the model TI-82 or TI-83 graphics calculator) are presented below for all Activities and Examples found in this chapter that require or recommend the use of a graphics calculator.

internet connect

For Keystrokes of other graphing calculator models, visit the HRW web site at **go.hrw.com** and enter the keyword **MB1 CALC**.

LESSON 11.2

EXAMPLE 1 *Page 700*

Find the fifth term of the sequence defined by $t_1 = -4$ and $t_n = t_{n-1} + 3$.

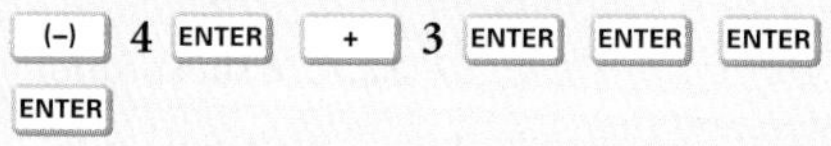

ENTER

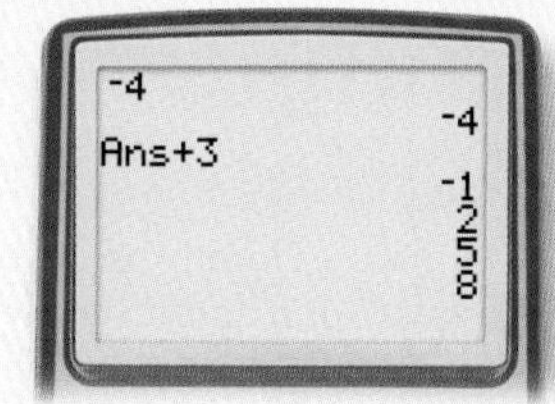

EXAMPLES 2 and 3 *Page 701*

For Example 2, find the fifth term of the sequence defined by $t_1 = 389.95$ and $t_n = t_{n-1} - 42.50$.

389.95 ENTER – 42.5 ENTER ENTER ENTER ENTER

Use a similar keystroke sequence for Example 3.

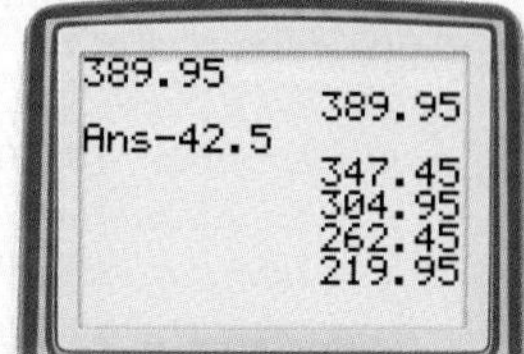

Activity *Page 702*

For Step 1, part a, use sequence mode to graph the sequence $t_n = 5 + 2n$ for n-values from 1 to 6.

Put the calculator in sequence and dot modes:

MODE Seq ENTER Dot ENTER 2nd MODE (QUIT)

Set the viewing window:

WINDOW (nMin=) 1 ENTER (nMax=) 6 ENTER (PlotStart=) 1 ENTER (PlotStep=) 1 ENTER (Xmin=) 0 ENTER (Xmax=) 7 ENTER (Xscl=) 1 ENTER (Ymin=) 0 ENTER (Ymax=) 20 ENTER (Yscl=) 4

TI-82:

WINDOW ENTER (U_nStart=) 7 ENTER ENTER (nStart=) 1 ENTER (nMin=) 1 ENTER (nMax=) 6 ENTER (Xmin=) 0 ENTER (Xmax=) 7 ENTER (Xsc1=) 1 ENTER (Ymin=) 0 ENTER (Ymax=) 20 ENTER (Ysc1=) 4

Graph the sequence:

Y= (u(*n*)=) 5 + 2 X,T,θ,n ENTER (u(*n*Min)=) 7 GRAPH

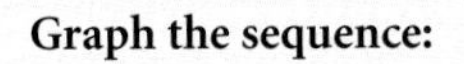

For part **b**, use the same viewing window. Use *n*Min = 1 and u(*n*Min) = 7.5.

For part **c**, change to Ymax = 40 in the viewing window. Graph the sequence by using *n*Min = 1 and u(*n*Min) = 2.

For part **d**, use Ymax = 2, Yscl = 0.5, *n*Min = 1, and u(*n*Min) = 1.

For part **e**, use Ymax = 0.4, Yscl = 0.05, *n*Min = 1, and u(*n*Min) = 0.3.

For part **f**, use Ymax = 40, Yscl = 4, *n*Min = 1, and u(*n*Min) = 1.

LESSON 11.3

EXAMPLES 1 and 2 Find S_{15} for the series in which $t_1 = 1$ and $d = 3$.

Pages 708 and 709

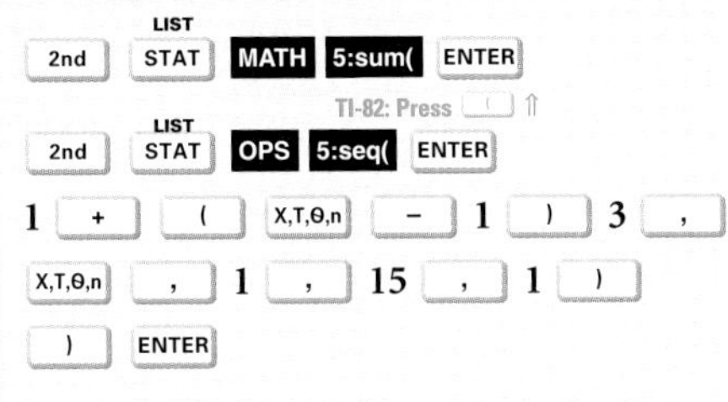

2nd STAT (LIST) MATH 5:sum(ENTER

TI-82: Press

2nd STAT (LIST) OPS 5:seq(ENTER

1 + (X,T,θ,n − 1) 3 ,

X,T,θ,n , 1 , 15 , 1)

) ENTER

Use a similar keystroke sequence for Example 2.

EXAMPLE 3 Evaluate $\sum_{k=1}^{12}(6-2k)$.

Page 709

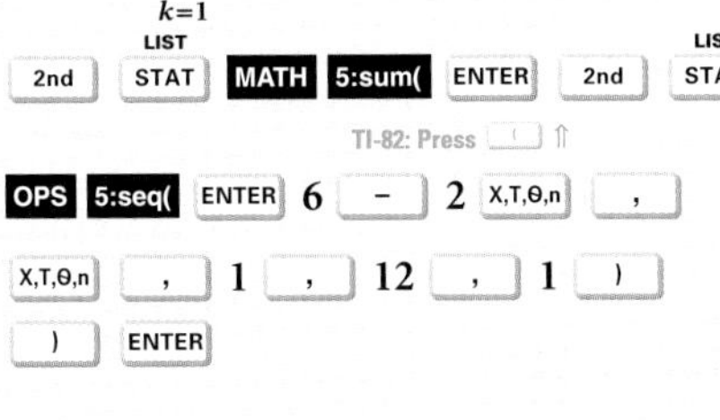

2nd STAT (LIST) MATH 5:sum(ENTER 2nd STAT (LIST)

TI-82: Press

OPS 5:seq(ENTER 6 − 2 X,T,θ,n ,

X,T,θ,n , 1 , 12 , 1)

) ENTER

LESSON 11.4

EXAMPLES 1, 2, 3, and 4 For Example 1, find the fifth term of the sequence defined by $t_1 = 8$ and $r = 3$.

Pages 714–716

8 ENTER × 3 ENTER ENTER ENTER ENTER

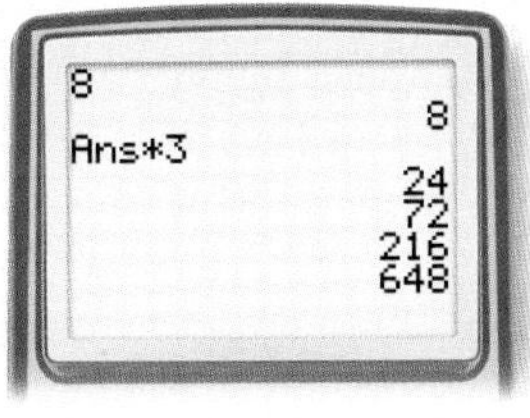

Use a similar keystroke sequence for Examples 2, 3, and 4.

Activity
Page 716

For Step 1, part a, find the first 5 terms of the geometric sequence defined by $t_1 = 1$ and $r = 10$.

1 ENTER × 10 ENTER

Press ENTER 18 more times to generate the rest of the terms.

Use a similar keystroke sequence for parts **b–f**.

LESSON 11.5

EXAMPLE 1 Page 721

Find S_{10} to the nearest tenth for the series defined by $t_1 = 3$ and $r = 1.5$.

Set the calculator to round answers to the nearest tenth:

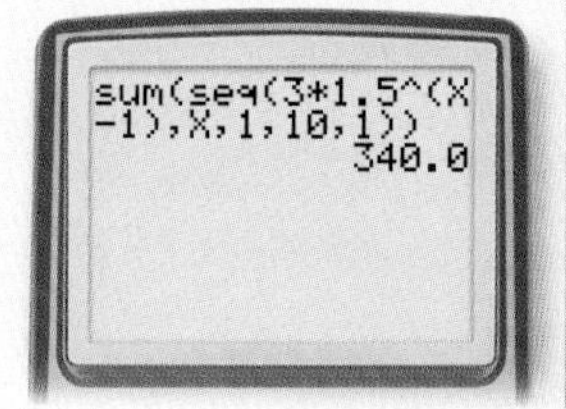

Find the sum:

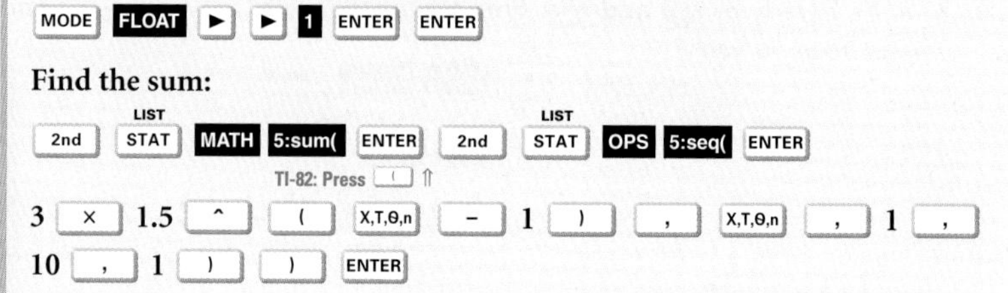

EXAMPLE 3 Page 722

Be sure that your calculator is returned to the floating decimal mode setting.

Evaluate $\sum_{k=1}^{6} 2(3^{k-1})$.

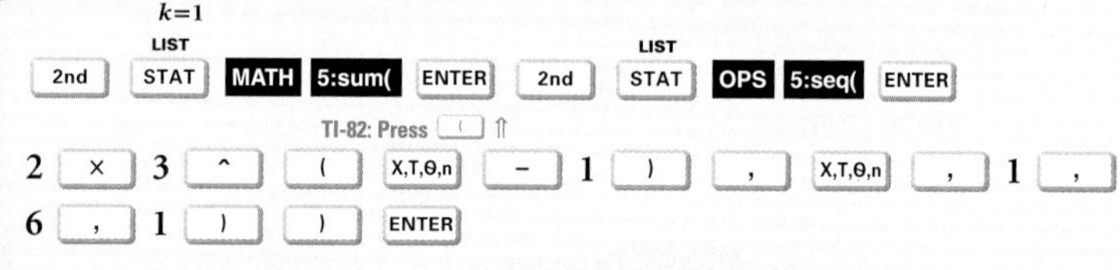

LESSON 11.6

Activity
Page 729

First clear old data from Lists 1 and 2.

For Step 1, part a, create a table of values for n and S_n for the infinite geometric series defined by $t_1 = -5$ and $r = -0.25$. Use n-values of 1, 2, 3, 5, 10, and 100.

STAT EDIT 1:Edit L1 1 ENTER 2 ENTER 3 ENTER 5 ENTER 10

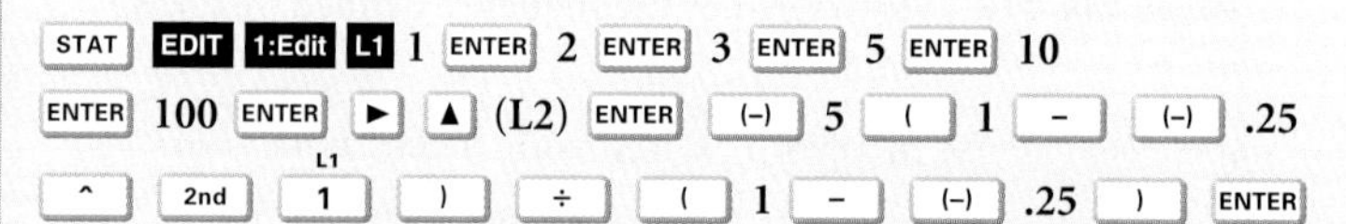

Use a similar keystroke sequence for Step 1, parts **b–d**, and for Step 2.

EXAMPLES 1 and 2 For Example 1, graph the formula for the sum of n terms of the geometric series in which $t_1 = 3$ and $r = 0.4$.

Pages 730 and 731

Be sure that your calculator is returned to function mode.

Use friendly viewing window [0, 28.2] by [0, 10].

Y= 3 (1 - .4 ^ X,T,θ,n) ÷ (1 - .4) GRAPH TRACE ►

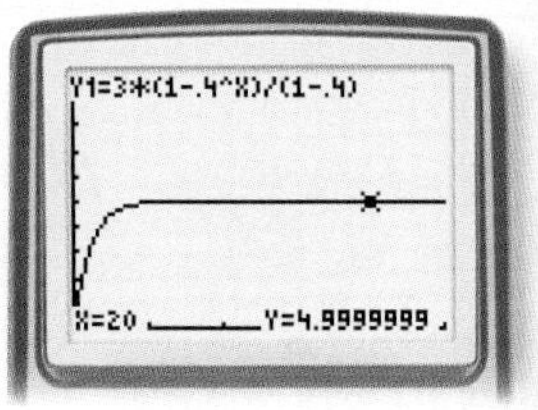

Y1=3*(1-.4^X)/(1-.4)

X=20 Y=4.9999999

Use a similar keystroke sequence for Example 2. Use friendly viewing window [0, 28.2] by [0, 0.5].

LESSON 11.7

EXAMPLE 1 To find the 4th and 10th entries in row 12 of Pascal's triangle, evaluate $_{12}C_3$ and $_{12}C_9$.

Page 737

12 MATH PRB 3:nCr ENTER 3 ENTER

12 MATH PRB 3:nCr ENTER 9 ENTER

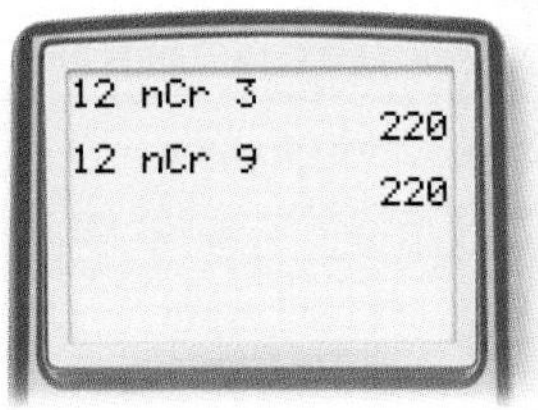

12 nCr 3
220
12 nCr 9
220

EXAMPLE 3 Evaluate $\frac{_7C_3}{2^7} + \frac{_7C_4}{2^7}$.

Page 738

(7 MATH PRB 3:nCr ENTER 3) ÷ (2 ^ 7) + (7 MATH PRB 3:nCr ENTER 4) ÷ (2 ^ 7) ENTER

(7 nCr 3)/(2^7)+
(7 nCr 4)/(2^7)
.546875

LESSON 11.8

EXAMPLE 1 Find the coefficients in the expansion of $(x + y)^7$.

Page 742

2nd STAT (LIST) OPS 5:seq(ENTER 7 MATH PRB 3:nCr ENTER X,T,θ,n , X,T,θ,n , 0 , 7 , 1) ENTER

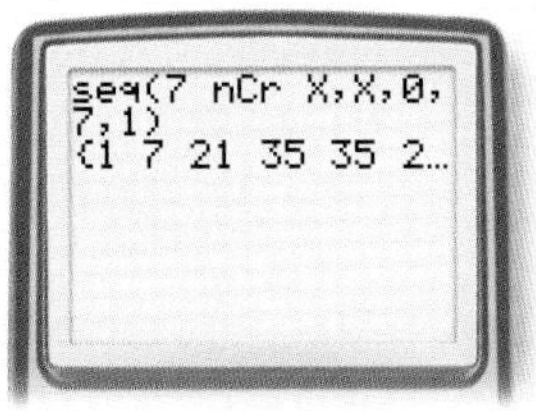

seq(7 nCr X,X,0,
7,1)
{1 7 21 35 35 2…

12 Discrete Mathematics: Statistics

Lesson Presentation CD-ROM
PowerPoint® presentations for each lesson 12.1–12.6

CHAPTER PLANNING GUIDE

Lesson	12.1	12.2	12.3	12.4	12.5	12.6	Project and Review
Pupil's Edition Pages	764–771	772–780	781–789	792–798	799–805	806–813	814–815, 816–821
Practice and Assessment							
Extra Practice (Pupil's Edition)	979	980	980	981	981	982	
Practice Workbook	76	77	78	79	80	81	
Practice Masters Levels A, B, and C	226–228	229–231	232–234	235–237	238–240	241–243	
Standardized Test Practice Masters	87	88	89	90	91	92	93
Assessment Resources	153	154	155	157	158	159	156, 160–165
Visual Resources							
Lesson Presentation Transparencies Vol. 2	116–119	120–123	124–127	128–131	132–135	136–139	
Teaching Transparencies	48	49, 50	51		52	53	
Answer Key Transparencies	422–426	427–445	446–454	455–457	458–459	460–462	463–469
Quiz Transparencies	12.1	12.2	12.3	12.4	12.5	12.6	
Teacher's Tools							
Reteaching Masters	151–152	153–154	155–156	157–158	159–160	161–162	
Make-Up Lesson Planner for Absent Students	76	77	78	79	80	81	
Student Study Guide	76	77	78	79	80	81	
Spanish Resources	76	77	78	79	80	81	
Block Scheduling Handbook							24–25
Activities and Extensions							
Lesson Activities	76	77	78	79	80	81	
Enrichment Masters	76	77	78	79	80	81	
Cooperative-Learning Activities	76	77	78	79	80	81	
Problem Solving/ Critical Thinking	76	77	78	79	80	81	
Student Technology Guide	76	77	78	79	80	81	
Long Term Projects							45–48
Writing Activities for Your Portfolio							34–36
Tech Prep Masters							55–58
Building Success in Mathematics							30–32

LESSON PACING GUIDE

Lesson	12.1	12.2	12.3	12.4	12.5	12.6	Project and Review
Traditional	2 days	2 days	1 day	1 day	1 day	1 day	2 days
Block	1 day	1 day	$\frac{1}{2}$ day	$\frac{1}{2}$ day	$\frac{1}{2}$ day	$\frac{1}{2}$ day	1 day
Two-Year	4 days	4 days	2 days	2 days	2 days	2 days	4 days

CONNECTIONS AND APPLICATIONS

Lesson	12.1	12.2	12.3	12.4	12.5	12.6	Review
Algebra	764–771	772–780	781–789	792–798	799–805	806–813	814–821
Patterns in Data		779	786				
Probability		775, 777					
Transformations		780		795, 797			
Business and Economics	764, 765, 769, 770, 771	772, 774, 777, 778, 779		793, 794, 797		811, 812. 813	818
Life Skills	771	778		796	804, 805	809, 811, 812. 813	
Science	766, 767		781, 784, 785	792	803, 805	806	
Social Studies	770		782, 786, 787, 788	797	804		819
Sports and Leisure	768	773	784	797	800	810	821
Other		776, 778, 779	789		799, 801, 803, 804		

BLOCK SCHEDULING GUIDE

Day	Lesson	Teacher Directed: Lesson Examples, Teaching Transparencies	Student Guided Activity, Try This	Cooperative-Learning Activity, Lesson Activity, Student Technology Guide	Practice: Practice & Apply, Extra Practice, Practice Workbook	Assessment: Quiz, Mid-Chapter Assessment	Problem Solving, Reteaching
1	12.1	10 min	15 min	15 min	65 min	15 min	15 min
2	12.2	10 min	15 min	15 min	65 min	15 min	15 min
3	12.3	5 min	7 min	7 min	30 min	7 min	7 min
	12.4	5 min	8 min	8 min	35 min	8 min	8 min
4	12.5	5 min	7 min	7 min	30 min	7 min	7min
	12.6	5 min	8 min	8 min	35 min	8 min	8 min
5	Assess.	50 min **PE:** Chapter Review	90 min **PE:** Chapter Project, Writing Activities	90 min Tech Prep Masters	65 min **PE:** Chapter Assessment, Test Generator	30 min Chap. Assess. (A or B), Alt. Assess. (A or B), Test Generator	

PE: Pupil's Edition

Alternative Assessment

The following suggest alternative assessments for students who may benefit from a different type of assessment than the regular chapter quizzes and the mid-chapter/end-of-chapter test. Visit the HRW web site to get additional Alternative Assessment material.

internet connect

Alternative Assessment
Go To: **go.hrw.com**
Keyword: **MB1 Alt Assess**

Performance Assessment

1. Consider the data set {5, 4, 8, 4, 5, 4, 2, 8}.

a. Organize the data so that range, median, mode, and mean can be easily found.

b. Find the measures of central tendency and dispersion named in part **a.** Also find the standard deviation.

c. If you were to show the data graphically, which display type(s) would you use? Sketch the display(s).

2. A bag has 5 red marbles and 7 green marbles. A marble is selected and then is replaced. Selecting 3 red marbles in 5 tries wins a prize.

a. Explain why this experiment is a binomial probability experiment.

b. Find the probability of winning.

Portfolio Project

1. a. Create a data set that meets the following specifications:

i. There are 10 members in the data set.

ii. The minimum is 1, and the range is 9.

iii. The mean is 5.75, and the median is 6.5.

b. Investigate whether there is a data set that meets the specified conditions and also has a standard deviation of 1.

c. Experiment with other types of equations.

2. Several binomial experiments are conducted and each consists of 10 trials.

a. Using technology, illustrate the distributions for each of the following probabilities of success: 0.30, 0.35, 0.40, 0.45, 0.50, 0, 0.55, 0.60

b. From the patterns exhibited in your distributions in part **a,** what happens if p falls below 0.30 or p rises above 0.60. Write a report that illustrates and explains your findings. How can you *skew* a distribution left or right of *center*?

internet connect

The table below identifies the pages in this chapter that contain internet and technology information.

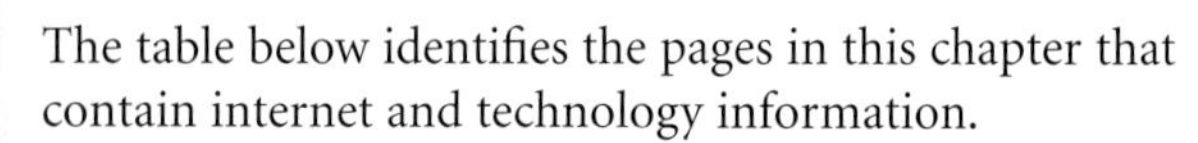

Content Links

Activities Online	**pages 785, 811**
Portfolio Extensions	**pages 771, 798**
Homework Help Online	**pages 769, 778, 786, 796, 804, 812**
Graphic Calculator Support	**page 822**

Resource Links

Parents can go online and find concepts that students are learning—lesson by lesson—and questions that pertain to each lesson, which facilitate parent-student discussion.

Go To: **go.hrw.com**
Keyword: **MB1 Parent Guide**

Technical Support

The following may be used to obtain technical support for any HRW software product.

Online Help: **www.hrwtechsupport.com**
e-mail: **tschrw@hrwtechsupport.com**
HRW Technical Support Center: **(800)323-9239**
7 AM to 10 PM Monday through Friday Central Time

Visit the HRW math web site at: **www.hrw.com/math**

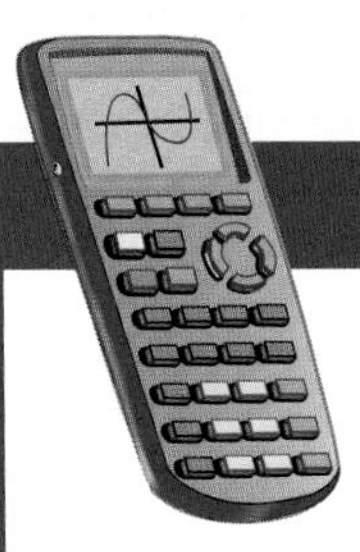

Technology

Lesson Suggestions and Calculator Examples

(Keystrokes are based on a TI-83 calculator.)

Lesson 12.1 Measures of Central Tendency

Have students enter 7, 5, 4, 7, 9, 6, and 10 into **L1**. They can find the mean or median by pressing 2nd STAT, selecting **MATH** **3: mean(** (or **4:median(**), and pressing 2nd 1. They can also find the mean or median when data values are in **L1** and their corresponding frequencies are in **L2** by using the same procedure and entering **L1,L2)** after **mean(** or **median(** as shown below.

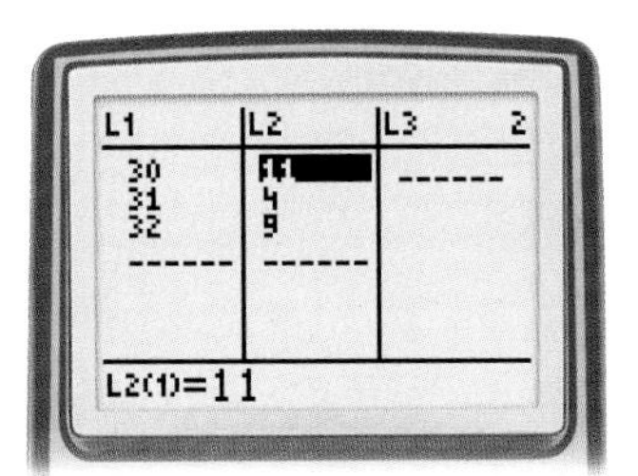

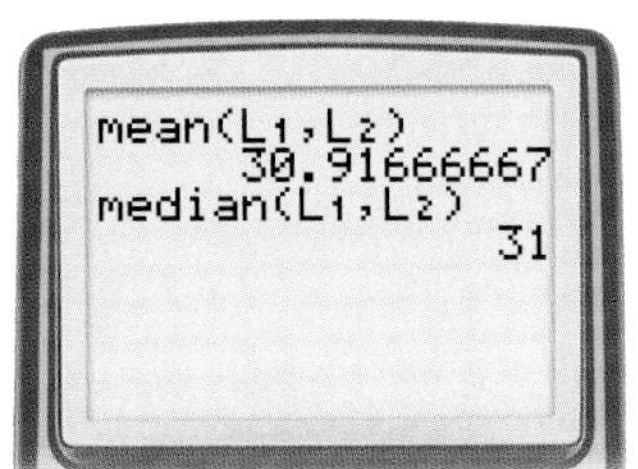

Lesson 12.2 Stem-and-Leaf Plots, Histograms, and Circle Graphs

To create a histogram of the data shown in the frequency table at right, display the **STAT PLOT** menu screen shown below by pressing 2nd Y=, selecting **Plot1**, and pressing ENTER.

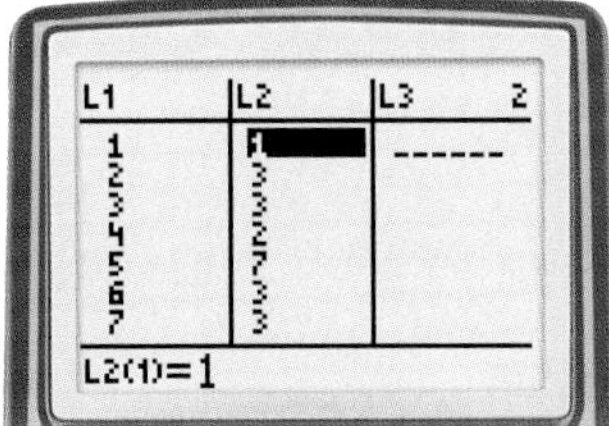

Turn on **Plot1** by selecting **On** and pressing ENTER. By using the down arrow key and the right arrow key, select the histogram icon, and press ENTER. Enter the list name that contains the data after **Xlist:** and the list name that contains the frequency after **Freq:**. Clear any functions in the **Y=** editor.

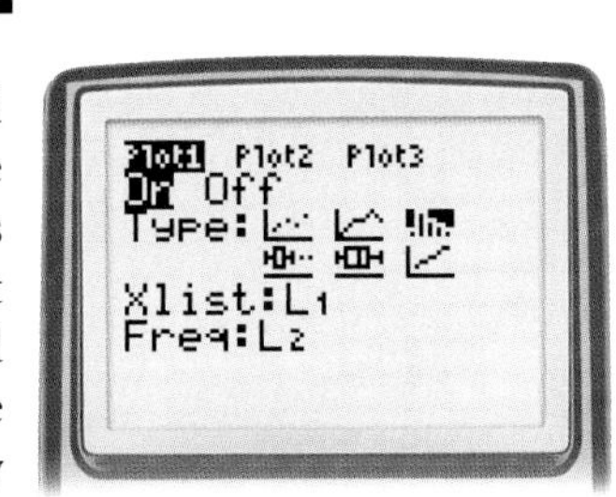

Press WINDOW and use the following settings:

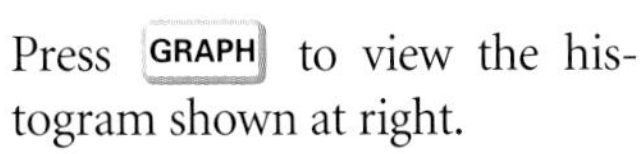

Xmin=0	**Ymin=0**
Xmax=10	**Ymax=10**
Xscl=1	**Yscl=1**

Press GRAPH to view the histogram shown at right.

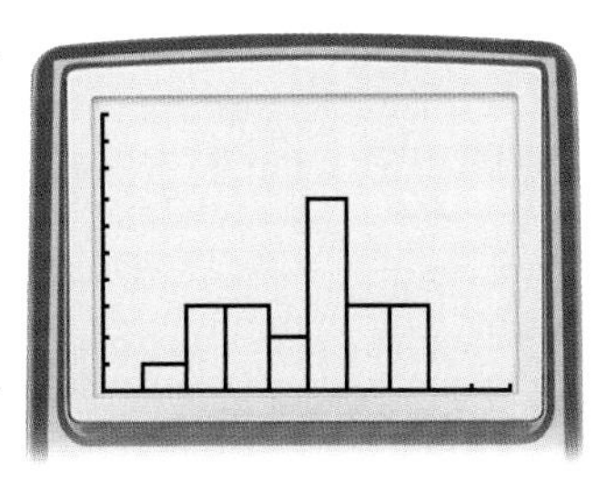

Lesson 12.3 Box-and-Whisker Plots

The procedure for graphing a box-and-whisker plot is the same as that described in Lesson 12.2 above, but students should select the icon of the desired box-and-whisker plot under **TYPE:**. They can press TRACE to view the minimum, first quartile, median, third quartile, and the maximum data values from the plot.

Lesson 12.4 Measure of Dispersion

The one-variable statistical report is obtained by pressing STAT, then selecting **CALC** and **1:1–Var Stats**. Note that both **σx,** the standard deviation of a population, and **Sx,** the standard deviation of a population based on a sample, are given. Tell students to use only the values shown for **σx.**

Lesson 12.5 Binomial Distributions

To solve problems that involve binomial experiments, have students press **DISTR** **0: binomialpdf(** .

In the display at left below, the binomial probability of each success is 0.96. The probability of 6, 7, or 8 successes in 8 trials is shown to be about 0.9969. In the display at right below, the probability of success 0.85. In 20 trials, the probability of 18 successes or 19 successes is shown to be about 0.366.

Lesson 12.6 Normal Distributions

To calculate normal distribution probabilities, students press **DISTR** **2: normalcdf(** . Shown below are the calculations of $P(x \geq 1.6)$ and $P(0.4 \leq x \leq 2.0)$, respectively.

Point out that if students want to calculate a probability such as $P(x \leq 1.6)$, students should use 10^{-99} as the lower bound and 1.6 as the upper bound.

internet connect

For keystrokes of other graphing calculators models, visit the HRW web site at **go.hrw.com** and enter the keyword **MB1 CALC.**

About Chapter 12

Background Information

In this chapter, students learn to represent data with frequency tables, graphs, and plots, to summarize data by using statistics, to analyze and interpret data, and to make predictions. Students organize data sets by using histograms, circle graphs, stem-and-leaf plots, and box-and-whisker plots. Students also find three measures of central tendency—mean, median, and mode—and use measures of dispersion, such as range, interquartile range, variance, and standard deviation. The fifth lesson introduces the binomial probability theorem, which students use to find the probability of an event with repeated trials. In the sixth lesson, students are introduced to normal distribution, and they use tables and calculators to find the percent of data in a given interval of a distribution.

CHAPTER RESOURCES

- Block-Scheduling Handbook
- Writing Activities for Your Portfolio
- Tech Prep Masters
- Long-Term Project
- Assessment Resources: Mid-Chapter Assessment Chapter Assessments Alternative Assessments
- Test and Practice Generator
- Technology Handbook

Chapter Objectives

- Find the mean, median, and mode of a data set. [**12.1**]

12

DISCRETE MATHEMATICS Statistics

STATISTICS IS A BRANCH OF APPLIED MATHEMATICS that involves collecting, organizing, interpreting, and making predictions from data. Statistics is categorized as descriptive or inferential. Descriptive statistics uses tables, graphs, and summary measures to describe data. Inferential statistics consists of analyzing and interpreting data to make predictions.

Lessons

About the Photos

Data can be collected in many ways: by using sales records, by conducting telephone surveys, by using census figures, etc. It is often felt that it is easy to collect good data, but the opposite is usually true. Collecting good data in an efficient and timely manner is the most difficult part of the process of producing accurate statistics and requires careful planning. If the sample is not representative of the entire population, then the conclusions drawn from the sample data will not reflect the population as a whole and misleading statistics may be found.

Before collecting data, a statistician identifies the population to be studied, noting the population's size as well as other characteristics. If the population is large, a sampling method and the size of the sample must be chosen. Random sampling is the most accepted method because its results have been proven to be valid in virtually every situation by both practical experience and mathematical theory.

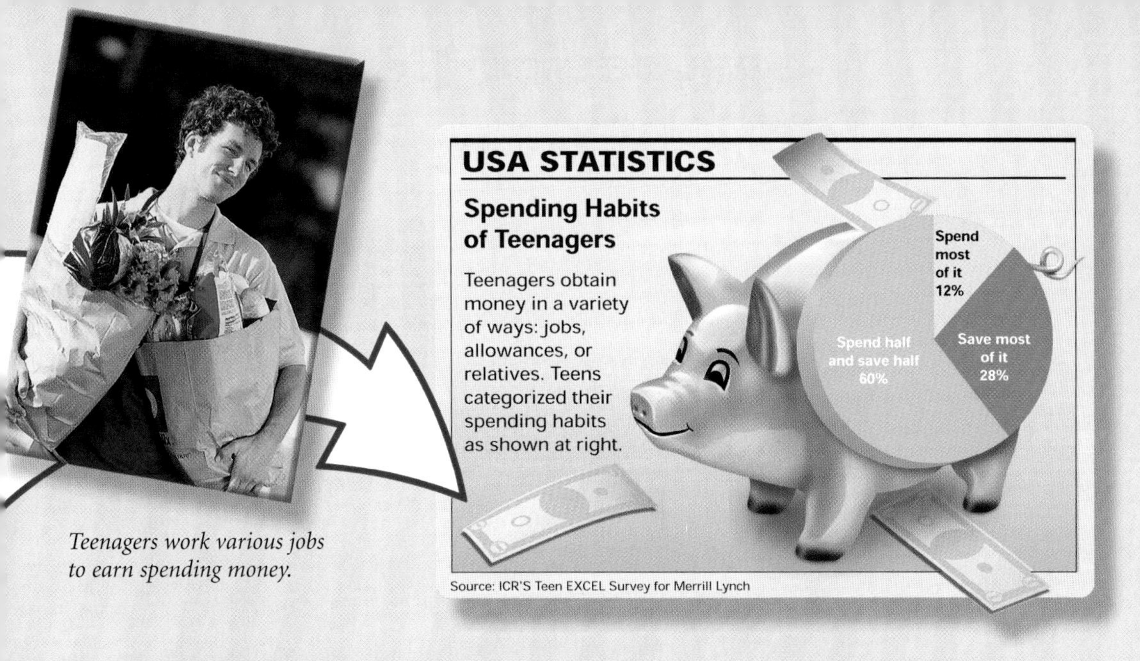

USA STATISTICS
Spending Habits of Teenagers
Teenagers obtain money in a variety of ways: jobs, allowances, or relatives. Teens categorized their spending habits as shown at right.
Spend most of it 12%
Spend half and save half 60%
Save most of it 28%
Source: ICR'S Teen EXCEL Survey for Merrill Lynch

Teenagers work various jobs to earn spending money.

About the Chapter Project

To obtain information about a large group, or population, smaller parts, or samples are studied. A sample is any part of a population. A sampling method is a procedure for selecting a sample to represent the population. A sampling method is *biased* if it deliberately or unintentionally favors particular outcomes. In the Chapter Project, *That's Not Fair!*, you will design a sampling method to provide a reasonable representation of a population and use the method to conduct a survey.

After completing the Chapter Project, you will be able to do the following:

- Determine whether a sampling procedure provides a poor or a reasonable representation of the total population.
- Design a sampling method that provides a reasonable representation of the population.
- Use statistics to support conclusions made from results of a survey.

About the Portfolio Activities

Throughout the chapter, you will be given opportunities to complete Portfolio Activities that are designed to support your work on the Chapter Project.

- Comparing measures of central tendency is included in the Portfolio Activity on page 771.
- Collecting data and choosing a method of visual representation for your data is included in the Portfolio Activity on page 780.
- Collecting data and explaining what some features of a box-and-whisker plot tell you about the data is included in the Portfolio Activity on page 789.
- Explaining what the measures of dispersion tell you about data that you have collected is included in the Portfolio Activity on page 798.

- Find or estimate the mean from a frequency table of data. [**12.1**]
- Make a stem-and-leaf plot, a histogram, or a circle graph for a data set. [**12.2**]
- Find and use relative frequencies to solve probability problems. [**12.2**]
- Find the range, quartiles, and interquartile range of a data set. [**12.3**]
- Make and use a box-and-whisker plot for a data set. [**12.3**]
- Calculate and use measures of dispersion, such as range, mean deviation, variance, and standard deviation. [**12.4**]
- Find the probability of r successes in n trials of a binomial experiment. [**12.5**]
- Find the probability of an event given that the data is normally distributed and its mean and standard deviation are known. [**12.6**]
- Use z-scores to find probabilities. [**12.6**]

Portfolio Activities appear at the end of Lessons 12.1, 12.2, 12.3, and 12.4. Each serves as preparation for the Chapter Project. The Portfolio Activities as well as the Chapter Project Activities are appropriate for inclusion in the student's portfolio. Students should be encouraged to include in their portfolios any other work in which they feel a sense of pride or a sense of accomplishment.

internet connect go.hrw.com

Chapter Internet Features and Online Activities

Lesson	Keyword	Page
12.1	MB1 Homework Help	769
	MB1 Central	771
12.2	MB1 Homework Help	778
12.3	MB1 Homework Help	786
	MB1 Actresses	785
	MB1 Oil Production	785

Lesson	Keyword	Page
12.4	MB1 Homework Help	796
	MB1 Simpson	798
12.5	MB1 Homework Help	804
12.6	MB1 Homework Help	812
	MB1 Distribution	811

Prepare

NCTM PRINCIPLES & STANDARDS 1, 4–10

QUICK WARM-UP

Find the average of each pair of numbers.

1. 10 and 16 — 13
2. 12 and 23 — 17.5
3. 42 and 42 — 42
4. 0 and 208 — 104
5. 1.9 and 3.6 — 2.75
6. 5 and 2.3 — 3.65
7. Arrange the set of numbers below in order from least to greatest.
   6.4, 0.8, 3.7, 1.5, 2.6, 3.9
   0.8, 1.5, 2.6, 3.7, 3.9, 6.4

Also on Quiz Tranparency 12.1

Teach

Why A simple way to summarize a large quantity of numbers is to reduce them to a few special numbers such as the mean, median, and mode. Ask students which of the following statements best describes how well a baseball player is hitting: the player has gone to bat 73 times and has 23 hits, or the player's batting average is 0.315.

Measures of Central Tendency

Objectives

- Find the mean, median, and mode of a data set.
- Find or estimate the mean from a frequency table of data.

Why *You can use measures of central tendency to analyze many real-world situations, such as the number of minutes that a commercial radio station devotes to music.*

APPLICATION
BROADCASTING

Guillermo wondered how many minutes his favorite commercial radio station actually devotes to music. He recorded the broadcast between 3:00 P.M. and 4:00 P.M. for 10 successive weekdays. Guillermo's results, in minutes, were 40, 45, 39, 40, 41, 42, 37, 41, 41, and 40.

Find *measures of central tendency* that summarize this data and compare the measures. *You will solve this problem in Example 1.*

Comparing Central Tendencies

Measures of central tendency are values that are representative of an entire data set. Three commonly used measures of central tendency are the *mean*, *median*, and *mode*.

Measures of Central Tendency

The **mean**, or arithmetic average, denoted $\overline{x}$, of a data set is the sum of all of the values in the data set divided by the number of values.

The **median** of a data set is the numerical middle value when the data values are arranged in ascending or descending order. If there are an even number of values, the median is the mean of the two middle values.

The **mode** of a data set is the value that is repeated most often in the data set. There can be one, more than one, or no mode.

Alternative Teaching Strategy

USING MANIPULATIVES To help students visualize central tendencies, have them cut out 9 one-inch wide paper strips of various lengths. At least two of the strips should be same length. Have students measure the length of the strips, write the length on each strip, and arrange the strips in order from shortest to longest. Ask students to find the lengths of the strips that represent the median and the mode. They should recognize that the median is the length of the strip in the middle, and the mode is the most frequently occurring length among the strips. Now tape all of the strips together end-to-end and fold the resulting strips into 9 equal lengths to find the mean.

EXAMPLE 1 Refer to the radio station data given at the beginning of the lesson.

APPLICATION BROADCASTING

Find the measures of central tendency, and compare them.
a. mean **b.** median **c.** mode

SOLUTION

a. $\bar{x} = \frac{40 + 45 + 39 + 40 + 41 + 42 + 37 + 41 + 41 + 40}{10} = \frac{406}{10} = 40.6$

The mean number of minutes devoted to music is 40.6.

PROBLEM SOLVING

b. Make an organized list. Arrange values in the data set in ascending order.

37, 39, 40, 40, 40, 41, 41, 41, 42, 45

The median is the mean of the fifth and sixth values, 40 and 41, which is $\frac{40 + 41}{2} = 40.5$. Thus, the median number of minutes is 40.5.

c. The most often repeated values, 40 and 41, both appear the same number of times. Because there are two modes, the data set is called *bimodal.*

The mean (40.6), median (40.5), and modes (40 and 41) are all very similar. Thus, the measures of central tendency for the number of minutes devoted to music during this hour are all between 40 and 41, inclusive.

TRY THIS Using the data 88, 74, 98, 76, 68, 74, 89, and 92, find the mean, median, and mode, and compare them.

CRITICAL THINKING What percent of the time does Guillermo's radio station play music? How broadly can this generalization be applied? Explain.

Any of the measures of central tendency can be misleading (not truly representative or typical) for certain data sets, as shown in Example 2.

EXAMPLE 2 **The salaries at a small business with 7 employees are as follows: $255,000, $32,000, $30,000, $28,000, $24,000, $22,000, and $22,000.**

APPLICATION BUSINESS

a. Find the mean, median, and mode of the salaries.
b. Explain which measures best represent a typical employee's salary.

SOLUTION

a. mean: $\bar{x} = \frac{255{,}000 + 32{,}000 + 30{,}000 + 28{,}000 + 24{,}000 + 22{,}000 + 22{,}000}{7}$

$= \frac{413{,}000}{7} = 59{,}000$

median: The middle value is $28,000.

mode: $22,000

b. The mean, $59,000, does not represent a typical salary because all except the top salary are much lower. The mode, $22,000, is a better representation of a typical salary than the mean, but it is still not the best representation because it is the lowest salary. The median, $28,000, is the best representation of the typical salary.

Interdisciplinary Connection

GEOGRAPHY The 10 longest rivers in Europe have the following lengths in miles:

Danube	Rhine	Elbe	Loire	Tagus
1766	850	725	630	627

Meuse	Ebro	Rhone	Guadiana	Seine
590	580	505	500	482

Have students find the mean and median of the river lengths and explain which measure is most representative. **The mean, 725.5, is not typical because 8 of the rivers have lengths less than the mean; the median, 608.5, is the most representative.**

Use Teaching Transparency 48.

ADDITIONAL EXAMPLE 1

Every morning, Janet runs around the lake that is near her house. For one week, Janet recorded the amount of time it took for each run. The times in minutes were 25, 33, 26, 30, 28, 28, and 29. **Find the measures of central tendency and compare them. Give answers to the nearest tenth, if necessary.**

mean: ≈28.4 min
median: 28 min
mode: 28 min

The mean, median, and mode are all very similar. The amount of time it takes Janet to run around the lake is approximately 28 min.

TRY THIS

Mean: 82.375; median: 82; mode: 74; the mean and median are very close, whereas the mode is lower than the others.

CRITICAL THINKING

≈68% of the time; applies only between 3 P.M. and 4 P.M. on weekdays

ADDITIONAL EXAMPLE 2

Last month, 6 houses were sold in Centerville. The sale prices were as follows: $140,000, $165,000, $150,000, $300,000, $146,000, and $125,000.

a. Find the mean, median, and mode of the sale prices.
mean: $171,000
median: $148,000
mode: none

b. Explain which measure best represents the typical sale price of a house in Centerville last month.
median

TRY THIS
Mean: $71,000; median: $85,000; mode: none; the median is the best measure because the mean is below every value except $0.

CRITICAL THINKING
No; the mean is less influenced by the outlier because it uses the sum of all the data values divided by the number of employees.

Activity Notes

Be sure students notice that adding 3 to each value increases the mean, median, and mode by 3. Multiplying each value by 2 multiplies the mean, median, and mode by 2, and taking half of each value results in half of the mean, median, and mode.

Cooperative Learning

Students can work in groups of three, with each group consisting of a leader, a recorder, and a reporter. Members can choose their roles. The leader will roll both number cubes, and the recorder will record the sums. Each member will answer the questions in Steps 1, 2, 3, and 4. The group will prepare a summary, and the reporter will present it to the class.

CHECKPOINT ✔
5. **The mean, median, and mode each change by the amount you add, subtract, multiply, or divide each data value.**

CHECKPOINT ✔
6. **Find the mean of the data values without the zeros. Then add 3 zeros to the mean. If each number is multiplied by 1000, the mean of the numbers is also multiplied by 1000.**

TRY THIS The yearly bonuses for five managers are $90,000, $85,000, $100,000, $0, and $80,000. Find the mean, median, and mode, and explain which measures are most representative.

CHECKPOINT ✔ Suppose that the number of employees in Example 2 increases by 6, and these employees all receive the same salary, $30,000. Does the one large salary of $255,000 have as much influence on the new mean as it did on the original mean? Why?

In the Activity below you can explore how the mean, median, and mode are influenced by various changes to the data set.

Activity
Exploring Measures of Central Tendency

You will need: 2 number cubes

1. Roll both number cubes and record their sum for 20 trials. Find the mean, median, and mode(s) of the sums.
2. Add 3 to each sum. Find the new mean, median, and mode(s), and describe how the measures have changed.
3. Double each sum from Step 1. Find the new mean, median, and mode, and describe how the measures have changed.
4. Take half of each sum from Step 1, and then subtract 1 from each value. Find the new mean, median, and mode(s), and describe how the measures have changed.

CHECKPOINT ✔ 5. Make a conjecture about what happens to the mean, median, and mode when you add, subtract, multiply, or divide each data value.

CHECKPOINT ✔ 6. Each value in Example 2 ends with three zeros. How can you use your conjecture to find the mean for the values in Example 2 by performing calculations without the final three zeros? Explain.

Frequency Tables

APPLICATION
ECOLOGY

The number of chirps that a cricket makes is related to the temperature according to the following relationship:

$$\text{number of chirps in 15 seconds} + 40 = \text{temperature in degrees Fahrenheit}$$

To verify this relationship, a class of 24 students counted cricket chirps for 15 seconds with the following results:

30	32	30	30	30	30	32	31
30	32	32	30	32	30	32	32
30	30	31	32	31	30	32	31

Inclusion Strategies

ENGLISH LANGUAGE DEVELOPMENT Seven salespeople earned the following commissions last week: $150, $150, $210, $215, $245, $370, and $490. The owner of the company is advertising to hire more salespeople. Ask students which measure the owner should use in order to encourage people to apply for the job. Then have them explain which measure gives the fairest picture of the salespeople's weekly commissions.
Mean: $261.43; median: $215; mode: $150; the mean is the most attractive average to use in the ad; the median is the fairest average.

This *raw data* of cricket chirps can be organized into a **frequency table** that lists the number of times, or frequency, that each data value appears.

To make a frequency table, first list each distinct value. Then make a mark for each value in the data set. Finally, count the number of marks to get the respective frequency for each value, as shown below.

Number of chirps	Tally	Frequency
30	~~||||~~ ~~||||~~ |	11
31	||||	4
32	~~||||~~ ||||	9
	Total	24

Example 3 demonstrates how to find the mean of a data set that is organized in a frequency table.

EXAMPLE 3 **Find the mean number of cricket chirps from the frequency table above. Then estimate the temperature at the time the chirps were counted.**

APPLICATION ECOLOGY

SOLUTION

Multiply each number of chirps by its corresponding frequency.

Frequency Table for Cricket Chirps

Number of chirps	Frequency	Product
30	11	330
31	4	124
32	9	288
Total	24	742

Add the products, and divide by the total number of values.

$$\bar{x} = \frac{742}{24} \approx 30.9$$

The mean number of cricket chirps during the 15 seconds was 30.9. Thus, according to the given relationship, the temperature should have been $30.9 + 40 = 70.9$, or about 71°F.

CHECKPOINT ✔ Would you obtain the same mean in Example 3 if you added each of the 24 values and then divided this sum by 24? Explain. Would you obtain the same mean if you calculated $\frac{30 + 31 + 32}{3}$? Explain.

TRY THIS Suppose that 9 students counted 30 chirps, 5 students counted 31 chirps, and 10 students counted 32 chirps. Find the mean number of chirps by using a frequency table. Then find the corresponding temperature.

Enrichment

Give students the following test scores for a class: 5 As, 10 Bs, 10 Cs, and 5 Ds. Point out that the data are not numerical. Ask them to explain which measures of central tendency make sense for this data set. **mode** Be sure they notice that the mean and median of the letter grades cannot be calculated, but that the mode can be determined. Now ask students to develop a grading system that converts the letter grades to numerical values. Have them explain how the measures of central tendency could be found for their system.

Answers may vary. Sample answer: Letting 4 represent A, 3 represent B, 2 represent C, and 1 represent D, the mean is 2.5 and the median is 2.5.

ADDITIONAL EXAMPLE 3

Make a frequency table for the data below on the number of books read by 30 students last month. Then use your frequency table to find the mean.

4, 0, 1, 2, 6, 0, 1, 2, 0, 4, 0, 2, 2, 1, 1, 3, 1, 0, 3, 2, 0, 1, 2, 2, 4, 1, 1, 1, 0, 4

Number of books	Frequency	Product
0	7	0
1	9	9
2	7	14
3	2	6
4	4	16
5	0	0
6	1	6
Total	30	51

mean: 1.7

CHECKPOINT ✔

Yes; the 24 values will add to 742; no; this method implies 30 chirps, 31 chirps, and 32 chirps each occurred 1 time.

TRY THIS

mean ≈ 31.0; temperature ≈ 71.0°F

Teaching Tip

TECHNOLOGY When using a TI-83 or TI-82 graphics calculator, the following are general guidelines for finding the mean and median of data in a single list:

1. Enter the data into a list by pressing STAT, selecting 1:Edit..., and entering the data into one of the lists.
2. To calculate the mean or median of the data listed in a single list, press 2nd STAT, select MATH, and choose 3:mean(or 4:median(. Then press 2nd 1 for **L1**, 2nd 2 for **L2**, etc., and then press) ENTER.

ADDITIONAL

EXAMPLE 4

The grouped frequency table below lists the numbers of CDs bought by 50 students last year. **Estimate the mean number of CDs bought by these students last year.**

Number of CDs	Frequency
0	4
1–5	14
6–10	9
11–15	10
16–20	8
21–25	1
26–30	2
31–35	1
36–40	1

mean: ≈11

CRITICAL THINKING

The mean is based on the class mean, which represents a range of values.

Teaching Tip

TECHNOLOGY When using a TI-83 or TI-82 graphics calculator, the following are guidelines for finding the mean and median of data in a frequency table:

1. Enter a frequency table into two lists, where one list contains the data and the other contains the frequency. In the example below, **L1** contains the data and **L2** contains the frequency.
2. Press 2nd STAT, select MATH, choose 3:mean(or 4:median(, and press 2nd 1 , 2nd 2).

The following are the syntaxes for mean and median:

mean(*list*[,*freqlist*])
median(*list*[,*freqlist*])

When the optional parameter, ***freqlist***, is omitted, the frequency is assumed to be 1.

When there are many different values, a *grouped* frequency table is used. In a **grouped frequency table**, the values are grouped into *classes* that contain a range of values. Example 4 shows the procedure for estimating the mean from a grouped frequency table.

EXAMPLE 4

APPLICATION
MUSIC

The grouped frequency table at right lists the results of a survey of 80 musicians who were asked how many hours per week they spend practicing.

Estimate the mean number of hours that these musicians practice each week.

Hours	Frequency
1–5	13
6–10	9
11–15	9
16–20	14
21–25	16
26–30	8
31–35	8
36–40	3
Total	80

SOLUTION

First find the *class mean*, or midpoint value, for each class. Then multiply each class mean by its corresponding frequency. Add the products, and divide by the total number of musicians surveyed.

$$\bar{x} = \frac{1450}{80} = 18.125$$

Hours	Class mean	Frequency	Product
1–5	3	13	39
6–10	8	9	72
11–15	13	9	117
16–20	18	14	252
21–25	23	16	368
26–30	28	8	224
31–35	33	8	264
36–40	38	3	114
	Total	80	1450

Thus, a reasonable estimate of the mean number of hours that these musicians practice each week is 18 hours.

CRITICAL THINKING Explain why the mean of the values in a grouped frequency table is an estimate.

Exercises

Communicate

1. Which measure is the easiest to determine: mean, median, or mode? Which measure is the most difficult? Why?
2. Suppose that the largest and smallest value in a data set are omitted. Will the median change? Will the mean change? Explain.
3. Describe a data set for which the mean is not as representative as the mode or median.

Reteaching the Lesson

WORKING BACKWARD Ask students to create a data set of five numbers that satisfies the following conditions:

- There is no mode.
- The median is equal to the mean.
- The smallest number is $\frac{1}{3}$ of the largest.
- When arranged in order from least to greatest, each number is 2 more than the preceding number.

4, 6, 8, 10, 12

Guided Skills Practice

4. $\bar{x} = \$5.75$; \$5.625; \$5.00; the mean and median are close, but the mode is the smallest value in the data set.

4. Find the mean, median, and mode for the hourly wages below and then compare the measures. ***(EXAMPLE 1)***

\$5.25, \$5.00, \$6.50, \$6.00, \$5.00, \$6.75, \$6.50, \$5.00

5. The hours worked in one week by 10 cashiers at a grocery store were 36, 40, 34, 38, 33, 0, 40, 32, 35, and 37. ***(EXAMPLE 2)***

a. Find the mean, median, and mode of the hours worked that week.

b. Explain which measures best represent the number of the hours worked by a typical cashier in that week.

6. Find the mean of the data set given in the horizontal frequency table at right.
(EXAMPLE 3) **≈4.23**

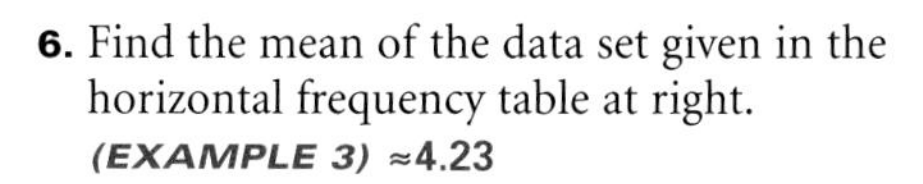

Value	3	4	5	Total
Frequency	6	11	13	30

7. MARKETING Thirty people were asked how many magazines they read in one month. A grouped frequency table for the responses is shown at right. Estimate the mean number of magazines read in one month.
(EXAMPLE 4) **4.2 magazines per month**

Number	Frequency
0–2	10
3–5	12
6–8	4
9–11	4
Total	30

Practice and Apply

internet connect
Homework Help Online
Go To: go.hrw.com
Keyword:
MB1 Homework Help for Exercises 8–21

Find the mean, median, and mode of each data set. Give answers to the nearest thousandth, when necessary.

8. 1, 3, 4, 8, 1, 7, 1, 5 **$\bar{x} = 3.75$; 3.5; 1**

9. 2, 5, 3, 6, 3, 1, 3, 4 **$\bar{x} = 3.375$; 3; 3**

10. 18, 13, 16, 20, 21, 13, 19

11. 14, 16, 19, 14, 12, 15, 13

12. −5, −1, 2, −6, −2,

13. −12, −10, −13, −9, −11

14. 2.1, 3.4, 3.7, 2.2, 2.1, 2.2

15. 1.7, 1.6, 3.8, 5.1, 1.6, 3.8

16. 0.33, 1.24, 2.71, 7.42, 6.21

17. 4.82, 5.22, 8.32, 3.22, 1.56

18. $\frac{1}{2}, \frac{3}{4}, \frac{5}{8}, \frac{7}{8}$

19. $\frac{1}{3}, \frac{1}{9}, \frac{5}{6}, \frac{5}{9}$

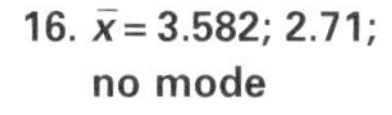

16. $\bar{x} = 3.582$; 2.71; no mode

17. $\bar{x} = 4.628$; 4.82; no mode

18. $\bar{x} = \frac{11}{16} \approx 0.688$; $\frac{11}{16} \approx 0.688$; no mode

19. $\bar{x} = \frac{11}{24} \approx 0.458$; $\frac{4}{9} \approx 0.444$; no mode

20. $\bar{x} = \$12.71$; \$11.25; \$8, \$10, and \$16; mean and median seem most representative

Find the mean, median, and mode of each data set, and compare them.

20. the price of haircuts (rounded to nearest half-dollar):
\$6.50, \$7.50, \$8, \$8, \$10, \$10, \$12.50, \$14, \$16, \$16, \$20, \$24

21. the cost for a gallon of unleaded gasoline (rounded to the nearest cent):
\$1.20, \$1.23, \$1.25, \$1.16, \$1.32, \$1.24, \$1.33, \$1.23, \$1.21, \$1.30, \$1.20, \$1.20, \$1.21, \$1.28 **$\bar{x} = \$1.24$; \$1.23; \$1.20; all 3 measures are similar and seem to be good measures of central tendency.**

Make a frequency table for each data set, and find the mean.

22. the number of days students in 4th period were absent: **1.8 days**
1, 0, 3, 4, 1, 0, 2, 0, 3, 4, 1, 3, 4, 1, 2, 0, 1, 2, 0, 4, 3, 1, 2, 2, 2, 1, 3, 1, 1, 2

23. the number of pets that students in a class have: **1.7 pets**
3, 4, 0, 1, 2, 3, 2, 0, 4, 0, 1, 0, 1, 2, 0, 1, 0, 4, 2, 0, 1, 1, 4, 2, 3, 3, 4, 2, 1, 0

10. $\bar{x} \approx 17.143$; 18; 13

11. $\bar{x} \approx 14.714$; 14; 14

12. $\bar{x} = -2.4$; −2; no mode

13. $\bar{x} = -11$; −11; no mode

14. $\bar{x} \approx 2.617$; 2.2; 2.1 and 2.2

15. $\bar{x} \approx 2.933$; 2.75; 1.6 and 3.8

22.

Number of days	Frequency
0	5
1	9
2	7
3	5
4	4
Total	30

$\bar{x} = 1.8$ days

23.

Number of pets	Frequency
0	8
1	7
2	6
3	4
4	5
Total	30

$\bar{x} = 1.7$ pets

Assess

Selected Answers

Exercises 4–7, 9–47 odd

ASSIGNMENT GUIDE

In Class	1–7
Core	9–37 odd
Core Plus	8–38 even
Review	39–48
Preview	49–51

Extra Practice can be found beginning on page 940.

Error Analysis

Students may not understand why they need to find both the median and mode of a data set. Emphasize that the median is most useful when the data contain outliers. The mode is most useful when the data values are clustered.

5a. $\bar{x} = 32.5$ hr; median = 35.5 hr; mode = 40 hr

b. The median best represents the data because mean is more influenced by the extreme value, 0. The mode is the largest value, which happens to occur twice.

Make a grouped frequency table for each data set, and estimate the mean.

24. test scores of a class: 66, 75, 74, 78, 88, 99, 75, 88, 76, 74, 66, 89, 82, 92, 67, 89, 88, 84, 92, 65, 75, 85, 78, 79, 84, 94, 91, 81, 61, 79 **80**

25. the miles per gallon for cars driven to school: 30, 21, 18, 19, 23, 24, 26, 32, 30, 22, 12, 15, 21, 28, 27, 18, 16, 19, 23, 29, 24, 25, 16 **22 mpg**

For each situation described below, decide whether you would represent the data in a frequency table or a grouped frequency table. Explain your choice.

26. the number of pencils carried by each student to class

27. the dollar value of sales recorded by the school store for one month

28. the number of points scored by a basketball team in each game for a season, which vary from 38 to 75

29. the number of brothers and sisters of each student in a class

Survey about 15 students in your class about the topics below, and find the mean, median, and mode of the responses. Then describe which measures are most representative.

30. the distance that students live from school (to the nearest tenth of a mile)

31. the estimated number of movies that students saw last year

32. the time spent sleeping the previous night

33. the time spent studying the previous day

APPLICATIONS

34. BUSINESS A vending company claims that one of its beverage machines dispenses about 8 fluid ounces into each cup. To verify this claim, they measured the amount dispensed into 40 cups. The results are listed below.

7.8	7.9	7.6	8.0	7.8	8.0	7.9	7.6	8.0	8.0	7.5	7.9	7.8	7.9
7.8	8.0	8.2	7.9	7.6	8.0	7.6	8.1	8.1	7.9	7.5	8.0	7.8	7.8
8.0	8.2	7.9	8.1	8.1	7.9	7.5	8.0	7.8	7.8	8.0	8.2		

a. Find the mean, median, and mode, and compare them.

b. Do the results from part **a** support the company's claim?

35. INVENTORY The manager of the women's shoe department recorded the following sizes of shoes sold in one day:

$$5, 6, 5\tfrac{1}{2}, 7, 9, 6, 5, 7\tfrac{1}{2}, 7, 5, 8, 6\tfrac{1}{2}, 7, 8, 6, 4, 6\tfrac{1}{2}, 10, 7$$

a. Find the mean, median, and mode of the shoe sizes sold.

b. Explain which measure of central tendency is the most helpful for the manager.

36. WORKFORCE The distribution of ages in the workforce of the United States is shown in the table at right.

a. Make three grouped frequency tables, one for each year.

b. Find the estimated mean age of a worker for each year. Use 60 for the class mean of the 55+ class.

c. Compare the estimated means from part **b**. What do they indicate about the workforce?

Percent of the Labor Force by Age

Age	1979	1992	2005
16–24	24%	16%	16%
25–34	27%	28%	21%
35–44	19%	27%	25%
45–54	16%	18%	24%
55+	14%	12%	14%

[Source: Bureau of Labor Statistics]

24–25. Answers may vary due to arrangement of groups. Possible answers:

24.

Scores	Class mean	Freq.
60–64	62	1
65–69	67	4
70–74	72	2
75–79	77	8
80–84	82	4
85–89	87	6
90–94	92	4
95–99	97	1
	Total	30

25.

MPG	Class mean	Freq.
10–14	12	1
15–19	17	7
20–24	22	7
25–29	27	5
30–35	32	3
	Total	23

26. frequency table
27. grouped frequency table
28. grouped frequency table
29. frequency table

34a. $\bar{x} = 7.8875$ fl oz; 7.9 fl oz; 8.0 fl oz; all measures are very similar.

b. Sample answer: The mean and median can support the claim of "about 8 ounces." However, it requires rounding up.

35a. $\bar{x} \approx 6.632$; $6\frac{1}{2}$; 7; the mean and median are similar, and the mode is larger.

b. Mode is the most helpful for maintaining stock.

36b. 1979: $\bar{x} = 36.59$; 1992: $\bar{x} \approx 37.86$; 2005: $\bar{x} = 39.55$

c. The mean age for the work force is increasing.

30–33. Answers will vary. Find the following measures: mean $= \dfrac{\text{sum of data values}}{\text{number of values}}$, median = numerical middle value (or average of the two middle values if there are an even number of values), and mode = the value or values repeated most often. The most representative measures will depend on the data collected.

36a.

1979			
Age	Class mean	Freq.	Product
16–24	20	24	480
25–34	29.5	27	796.5
35–44	39.5	19	750.5
45–54	49.5	16	792
55+	60	14	840
	Total	100	3659

Practice

NAME ______ CLASS ______ DATE ______

Practice
12.1 Measures of Central Tendency

Find the mean, median, and mode of each data set. Round answers to the nearest thousandth, when necessary.

1. 62, 54, 63, 92, 62, 79, 54, 62 ___66; 62; 62___
2. 12, 28, 40, 22, 33, 28, 9 ___24.571; 28; 28___
3. 3.6, 6.3, 1.3, 3.6, 1.0, 5.9 ___3.617; 3.6; 3.6___
4. 277, 725, 920, 835, 255, 725 ___622.833; 725; 725___
5. 1828, 1008, 1600, 7309, 2215, 1600 ___2593.333; 1714; 1600___

Find the mean, median, and mode of the data, and compare them.

6. percent of total social welfare expenditures for education in 1985–1992: 22.8, 23.2, 24.5, 24.8, 25.0, 25.0, 24.0, 23.1 24.05; 24.25; 25; all are reasonable measures because they are so close, but the mean and median are better than the mode because 6 of the 8 numbers are less than the mode.

Make a frequency table for the data, and find the mean.

7. ages (in years) of members of the swim team: 14, 15, 17, 17, 18, 16, 15, 14, 16, 17, 17, 18, 17, 16, 16, 15, 14, 17

mean: ___$16.\overline{05}$___

Age (years)	Tally	Frequency
14	III	3
15	III	3
16	IIII	4
17	卌 I	6
18	II	2

Make a grouped frequency table for the data, and estimate the mean.

8. number of books read by the members of a class in the past year: 5, 4, 12, 22, 30, 5, 7, 3, 1, 10, 12, 16, 26, 15, 10, 2, 5, 3, 10, 21

estimated mean: ___10.75___

Number of books	Class mean	Frequency	Product
1–5	3	8	24
6–10	8	4	32
11–15	13	3	39
16–20	18	1	18
21–25	23	2	46
25–30	28	2	56

APPLICATIONS

37. ACADEMICS A student's test scores are 86, 72, 85, and 90. What is the lowest score that the student can get on the next test and still have a test average (mean) of at least 80? **67**

CHALLENGE

38. ACCOUNTING The record of Jacob's gasoline purchases on a recent vacation is given in the table below. What is the average cost per gallon of gasoline for the entire trip? **$1.16**

Price per gallon	$1.18	$1.04	$1.29	$1.12	$1.21
Number of gallons purchased	21	17	16	19	11

Look Back

39. Given $f(x) = x^3$ and $g(x) = x^3 + 2x - 3$, find $f + g$ and $f - g$. ***(LESSON 2.4)***

Write each equation in logarithmic form. ***(LESSON 6.3)***

40. $16^{\frac{1}{2}} = 4$ **41.** $5^4 = 625$ **42.** $\left(\frac{1}{3}\right)^2 = \frac{1}{9}$ **43.** $2^{-3} = \frac{1}{8}$

Write each equation in exponential form. ***(LESSON 6.3)***

44. $\log_5 125 = 3$ $\mathbf{5^3 = 125}$ **45.** $4 = \log_5 625$ $\mathbf{5^4 = 625}$ **46.** $-4 = \log_3 \frac{1}{81}$ $\mathbf{3^{-4} = \frac{1}{81}}$

Factor each polynomial expression. ***(LESSON 7.3)***

47. $x^3 + 125$ $\mathbf{(x + 5)(x^2 - 5x + 25)}$ **48.** $x^3 - 27$ $\mathbf{(x - 3)(x^2 + 3x + 9)}$

39. $(f + g)(x) = 2x^3 + 2x - 3$; $(f - g)(x) = -2x + 3$

40. $\log_{16} 4 = \frac{1}{2}$

41. $\log_5 625 = 4$

42. $\log_{\frac{1}{3}} \frac{1}{9} = 2$

43. $\log_2 \frac{1}{8} = -3$

Look Beyond

Create a data set with at least 5 values for each description below.

49. The mean, median, and mode are all equal.

50. The mean is 4, the median is 5, and the mode is 4.

51. When one data value is deleted, the mean increases by 10.

Sample answers:

49. 2, 3, 3, 3, 4

50. −6, 4, 4, 5, 6, 7, 8

51. −10, 20, 30, 40, 70; $\bar{x} = 30$
20, 30, 40, 70; $\bar{x} = 40$

47	71	75	70	59	78
88	82	89	72	70	74
95	91	74	85	92	62
93	85	98	73	75	97

The scores in Allison's world history class are given above. Allison's score was 78.

1. Find the mean and the median score for this class.
2. Should Allison use the mean or the median to compare her score with the rest of the class when she reports her grade to her parents? Explain.
3. Choose a score that is greater than the median and raise it by 10 points. Find the new mean and median. Explain what happens to each measure.
4. Choose a score that is less than the median and lower it by 10 points. Find the new mean and median. Explain what happens to each measure.

1992			
Age	Class mean	Freq.	Product
16–24	20	16	320
25–34	29.5	28	826
35–44	39.5	27	1066.5
45–54	49.5	18	891
55+	60	12	720
	Total	101	3823.5

2005			
Age	Class mean	Freq.	Product
16–24	20	16	320
25–34	29.5	21	619.5
35–44	39.5	25	987.5
45–54	49.5	24	1188
55+	60	14	840
	Total	100	3955

Look Beyond

These exercises allow students to make up a data set that satisfies certain conditions related to the measures of central tendency.

ALTERNATIVE
Assessment

Portfolio Activity

The Portfolio Activity can be used as preparation for the Chapter Project or as a separate activity. In the Portfolio Activity on this page, students calculate measures of central tendency and use these measures to compare or contrast data sets.

Answers to Portfolio Activities can be found in Additional Answers of the Teacher's Edition.

Student Technology Guide

NAME ______ CLASS ______ DATE ______

Student Technology Guide
12.1 *Measures of Central Tendency*

A graphics calculator can find the mean and median of a data set.

Example: The number of songs on 13 CDs are 11, 16, 12, 12, 12, 15, 11, 10, 15, 18, 14, 9, and 17. Find the mean and median of this data set.

- Enter the data into a list. Press STAT 1:Edit... ENTER. Type the numbers in the data set into L1. Press ENTER or ▼ after each number. After entering the data, press 2nd MODE.
- Press STAT CALC 1:1-Var Stats ENTER ENTER. The mean is given by $\bar{x}$=, so the mean of this data set is about 13.23, as shown at left below.
- To find the median, press ▼ three times. The median of the data set is 12, as shown at right below.

```
1-Var Stats
 x̄=13.23076923
 Σx=172
 Σx²=2370
 Sx=2.803386231
 σx=2.693406369
↓n=13
```

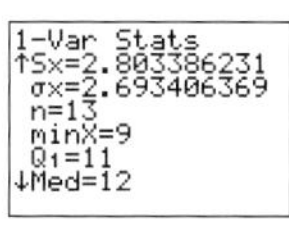

Example: A 14th CD has 22 songs. If this CD is added to the collection in the first example, how is the mean affected?

- To add a number to L1, press STAT 1:Edit... ENTER and use ▼ to move to the end of the list. Type 22 and press ENTER or ▼. Then press 2nd MODE.
- Press STAT CALC 1:1-Var Stats ENTER ENTER.

The mean of the data set increases to about 13.86.

Find the mean and median of each data set. Round answers to the nearest hundredth.

1. 4, 8, 12, 4, 15, 22, 18, 21, 16
   mean ≈ 13.33; median = 15
2. 0.76, 0.84, 0.67, 0.58, 0.91, 0.86
   mean = 0.77; median = 0.8
3. Five small homes in a neighborhood sold for $152,000, $164,000, $140,000, $161,000, and $166,000. The next week, a large home sold for $525,000. Compare the mean and the median prices before and after the sixth home was sold.
   mean price before sixth house: $156,600; median price: $161,000; mean price after sixth house: $218,000; median price: $162,500; both the median and the mean prices increased, but the mean increased by more.

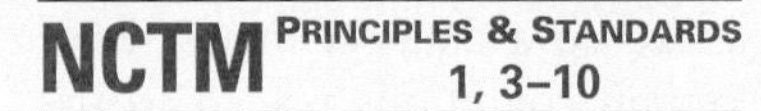

NCTM PRINCIPLES & STANDARDS 1, 3–10

QUICK WARM-UP

Refer to the following data set: 17, 22, 8, 23, 20, 22, 21, 17, 19, 20, 22, and 18.

1. Find the mean. ≈19.1
2. Find the median. 20
3. Find the mode(s). 22
4. Make a grouped frequency table for the data.

Number	Frequency
1–5	0
6–10	1
11–15	0
16–20	6
21–25	5

5. Use your grouped frequency table from Exercise 4 to estimate the mean. ≈19.25

Also on Quiz Tranparency 12.2

Teach

Why Displaying data visually in stem-and-leaf plots, histograms, or circle graphs can help readers see patterns more clearly. Poll students on their eye color. Then have them graph the data.

Stem-and-Leaf Plots, Histograms, and Circle Graphs

Objectives

- Make a stem-and-leaf plot, a histogram, or a circle graph for a data set.
- Find and use relative frequencies to solve probability problems.

APPLICATION
MARKETING

73	24	5	72
64	38	66	70
20	41	55	67
8	25	12	37
21	58	54	42
61	45	19	6
19	36	42	14

Why You can use a variety of visual displays to view the general distribution of real-world data, such as the number of times that an Internet site is visited.

An Internet site recorded the number of "hits" between 4 P.M. and 6 P.M. on 28 randomly selected weekdays. The results are listed above.

Stem-and-Leaf Plots

Stem	Leaf (4|1 = 41)
0	5, 6, 8
1	2, 4, 9, 9
2	0, 1, 4, 5
3	2, 6, 8
4	1, 2, 2, 5
5	4, 5, 8
6	1, 4, 6, 7
7	0, 2, 3

This is an ordered stem-and-leaf plot because the leaves of each stem are listed in order.

A *stem-and-leaf plot* of the Internet data is shown at left. A **stem-and-leaf plot** is a quick way to arrange a set of data and view its shape, or general distribution.

In a stem-and-leaf plot, each data value is split into two parts: a *stem* and a *leaf.* The stems and leaves are chosen so that the data is represented in the most informative way.

A key, shown in the upper right corner of a stem-and-leaf plot, explains what the stems and leaves represent. For example, the key at left tells you that the stem 4 and the leaf 1 represent the data value 41.

From the stem-and-leaf plot above, you can easily see that the maximum number of "hits" is 73 and the minimum is 5. You can also see that the data values are fairly evenly distributed, with 3 or 4 leaves in each stem.

CHECKPOINT ✓ Suppose that you want to make a stem-and-leaf plot for values ranging from 105 to 162. What stems would you use? Why? What stems would you use if the values range from 12.4 to 19.3? Why?

CRITICAL THINKING Describe what the fairly even distribution of data values indicates about the Internet site.

Alternative Teaching Strategy

USING COGNITIVE STRATEGIES Have students rotate the stem-and-leaf plot in Example 1 90° counterclockwise so that the leaves extend vertically upward. Now have them construct a histogram with the following intervals: 0–10, 11–19, 20–29, 30–39 40–49, 50–59, 60–69, 70–79, 80–89, and 90–99. Ask them to describe how the stem-and-leaf plot and histogram are alike. **The histogram allows them to visually compare the frequencies of the data, and a stem-and-leaf plot displays the frequencies as well as the individual data items in an ordered way.**

You can also find the median and the mode(s) of a data set from a stem-and-leaf plot, as shown in Example 1.

EXAMPLE 1

APPLICATION
RECREATION

Rosa is planning the annual Degollado family reunion. She has collected the ages of family members who plan to attend.

32	32	34	91	38
12	17	62	22	51
27	34	43	44	44
8	30	30	31	40
34	37	38	38	78
50	26	54	28	29
19	6	45		

a. Make a stem-and-leaf plot of the ages.
b. Find the median and mode of the ages.
c. How can the stem-and-leaf plot be used to plan the reunion?

SOLUTION

a. Choose digits in the tens place for the stems, as shown at right.
b. Because there are 33 values, the median is the 17th value, 34. The modes are 34 and 38.
c. The stem-and-leaf plot organizes the ages so that events can be planned for different age groups. For example, you can see that 3 members are 12 or younger, 2 members are teenagers, and 3 members are 60 or older. You can also see that the ages *cluster* around the 30s, forming a *mound-shaped* distribution.

Stem	Leaf 5|2 = 52
0	6, 8
1	2, 7, 9
2	2, 6, 7, 8, 9
3	0, 0, 1, 2, 2, 4, 4, 4, 7, 8, 8, 8
4	0, 3, 4, 4, 5
5	0, 1, 4
6	2
7	8
8	
9	1

TRY THIS Make a stem-and-leaf plot for the data at right. Find the median and mode of the data. How many values are between 5.0 and 6.0? Describe the shape of the distribution.

3.3	5.5	5.3	7.7	4.2	2.5	6.5	9.2	5.6
4.2	6.9	2.3	9.1	5.6	4.5	7.0	7.2	4.5
5.1	7.2	5.4	2.3	3.2	6.2	3.2	6.2	2.3

CHECKPOINT ✔ Use the stem-and-leaf plot at right to answer the questions below.
a. How can you find the number of values in the data set?
b. How can you find the maximum, minimum, median, and mode(s)?
c. In what order are the values in the stem-and-leaf plot arranged?

Stem	Leaf 1|0 = 10
0	1, 1, 3, 7, 8
1	0, 0, 5, 6, 9
2	1, 2, 3, 5
3	2
4	2, 5, 6

Interdisciplinary Connection

BUSINESS An inventory was taken on the stock of a bookstore. It showed that 25% of the total stock were history books, 10% were science books, 45% were fiction books, 5% were reference books, and 15% were other types of books. Have students make a circle graph to represent this data. Then have them find the probability that a randomly selected book will be a history book or a science book. 35%

CHECKPOINT ✔
Stems: 10, 11, 12, 13, 14, 15, 16; stems: 12, 13, 14, 15, 16, 17, 18, 19; data can be placed in a reasonable number of categories.

CRITICAL THINKING
Each data value is approximately equally likely.

Use Teaching Transparency 49.

ADDITIONAL EXAMPLE 1

A bakery collected the following data about the number of loaves of fresh bread sold on each of 24 business days:

53	49	27	48	60	52
44	38	47	52	82	46
55	31	39	54	51	47
50	45	50	61	43	64

a. Make a stem-and-leaf plot of the data.

Stems	Leaves 2|7 = 27
2	7
3	1, 8, 9
4	3, 4, 5, 6, 7, 7, 8, 9
5	0, 0, 1, 2, 2, 3, 4, 5
6	0, 1, 4
7	
8	2

b. Find the median and mode(s) of the data.
median: 49.5
modes: 47, 50, 52

c. How can the owner use the stem-and-leaf plot to make plans for baking bread? The owner should plan to bake about 4 dozen loaves of bread.

TRY THIS
median = 5.4; mode = 2.3; 6; slightly mound-shaped

CHECKPOINT ✔
a. Count the number of leaves.
b. Max. is the last stem-leaf combination; min. is the first stem-leaf combination; median is the middle stem-leaf combination; mode is the stem-leaf combination with the leaf that is most repeated.
c. ascending

ADDITIONAL
EXAMPLE 2

The data below show the number of people in 24 vehicles that passed a designated checkpoint.

1 4 1 2 2 1 3 1
3 2 2 6 4 2 1 1
2 4 3 1 2 4 2 3

a. Make a frequency table for these data.

No.	1	2	3	4	5	6
Freq.	7	8	4	4	0	1

b. Make a histogram from the frequency table.

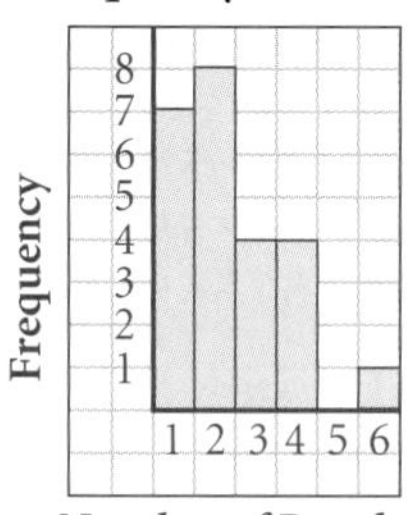

Teaching Tip

TECHNOLOGY When using a TI-83 or TI-82 graphics calculator to create the histogram for Example 2, enter the hours into **L1** and the frequency into **L2**. Press 2nd Y=, select **1:Plot1...**, and press ENTER. Use the following settings:

Use the following viewing window:

Xmin=1 **Ymin=0**
Xmax=9 **Ymax=10**
Xscl=1 **Yscl=1**

Now press GRAPH to view the histogram.

TRY THIS
Check students' tables and histograms.

Histograms

A **histogram** is a bar graph that gives the frequency of each value.

In a histogram, the horizontal axis is like a number line divided into equal widths. Each width represents a data value or range of data values. The height of each bar indicates the frequency of that data value or range of data values.

EXAMPLE 2 Isaac manages a canoe rental business. He recorded the number of hours that each of 30 customers rented canoes.

APPLICATION
BUSINESS

Make a frequency table and a histogram for the canoe rental data below.

1	4	5	2	2	2	3	2	3	3
2	3	7	2	3	3	2	6	4	2
1	1	3	5	5	4	4	2	8	2

SOLUTION

PROBLEM SOLVING

Make a table. Organize the data in a frequency table. Make the histogram by measuring 8 equal widths for the number of hours the canoes were rented. Draw vertical bars to the height of the corresponding frequencies.

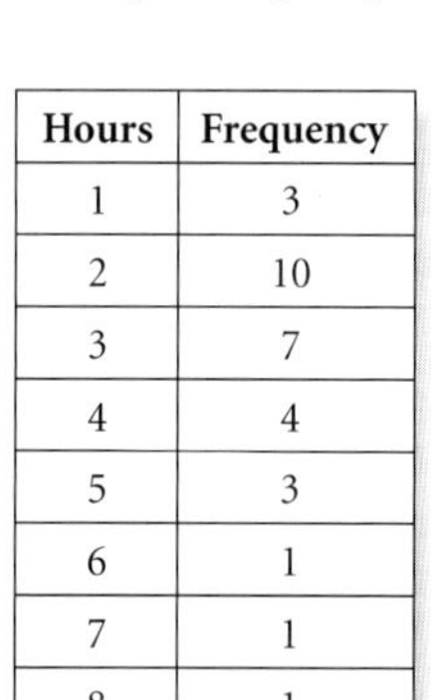

Hours	Frequency
1	3
2	10
3	7
4	4
5	3
6	1
7	1
8	1

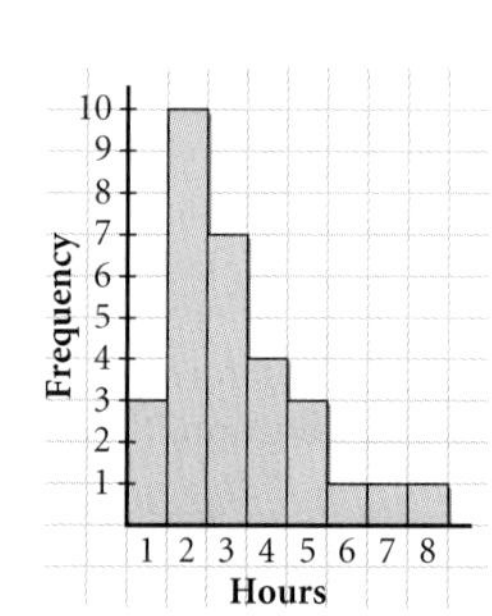

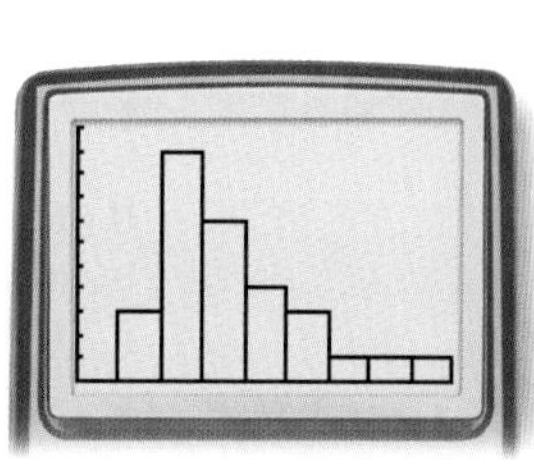

You can also make histograms on your graphics calculator by entering the frequency data into lists.

TRY THIS Make a frequency table and histogram for the data below.

11	12	16	16	12	16	15	11	15	14	13	12
17	15	17	13	14	13	11	13	12	15	13	15

CHECKPOINT ✔ Given a histogram, how can you make a frequency table for the data?

CRITICAL THINKING Describe the histogram for a set of data that has no mode.

Inclusion Strategies

USING DISCUSSION Some types of graphs are better suited to display particular patterns of data than others. To help students see this, give them situations for which they must chose the most appropriate graph. For example, have them explain whether a circle graph or stem-and-leaf plot is the best way for a teacher to represent class scores on an algebra test. **stem-and-leaf plot** Ask them to explain whether a histogram or circle graph is the better way to graph the average monthly rainfall in inches for a city in one year. **histogram**

Relative frequency tables are frequency tables that include a column that displays how frequently a value appears *relative* to the entire data set. The *relative frequency* column is the percent frequency, or probability. A relative frequency table and histogram are shown below for the canoe rental data.

Relative Frequency Table

Hours	Frequency	Relative frequency
1	3	$\frac{3}{30} = 0.10$, or 10%
2	10	$\frac{10}{30} = 0.\overline{3}$, or $33.\overline{3}\%$
3	7	$\frac{7}{30} = 0.2\overline{3}$, or $23.\overline{3}\%$
4	4	$\frac{4}{30} = 0.1\overline{3}$, or $13.\overline{3}\%$
5	3	$\frac{3}{30} = 0.10$, or 10%
6	1	$\frac{1}{30} = 0.0\overline{3}$, or $3.\overline{3}\%$
7	1	$\frac{1}{30} = 0.0\overline{3}$, or $3.\overline{3}\%$
8	1	$\frac{1}{30} = 0.0\overline{3}$, or $3.\overline{3}\%$
Total	30	$\frac{30}{30} = 1.00$, or 100%

Relative Frequency Histogram

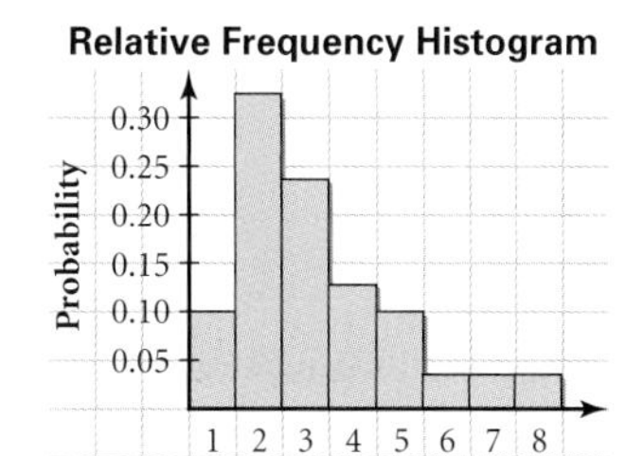

There is a $0.0\overline{3}$ probability that a randomly selected customer will rent a canoe for 8 hours.

EXAMPLE 3 **Use the relative frequencies given above to estimate the probability that a randomly selected customer will rent a canoe for 5 or more hours.**

SOLUTION

Let event $A = 5$ hours, event $B = 6$ hours, event $C = 7$ hours, and event $D = 8$ hours. Assume that events A, B, C and D are mutually exclusive.

$$\begin{aligned} P(A \text{ or } B \text{ or } C \text{ or } D) &= P(A) + P(B) + P(C) + P(D) \\ &= 0.10 + 0.0\overline{3} + 0.0\overline{3} + 0.0\overline{3} \\ &\approx 0.20 \end{aligned}$$

Thus, the probability that a randomly selected customer will rent a canoe for 5 or more hours is approximately 0.20, or about 20%.

TRY THIS Estimate the probability that a randomly selected customer will rent a canoe for less than 4 hours?

Activity: Exploring the Shapes of Histograms

You will need: coins and a pair of number cubes

1. Roll a number cube and record the result. Repeat this procedure for a total of 20 trials. Make a histogram with the number rolled (1–6) on the horizontal axis.
2. Toss a coin 6 times and record the number of heads that appear. Repeat this procedure for a total of 20 trials. Make a histogram with the number of heads out of 6 tosses on the horizontal axis.

CHECKPOINT ✔

3. Compare the shape of the histogram for Step 1 with the one for Step 2. Which is flatter? Which is more mound-shaped?

CHECKPOINT ✔
The horizontal axis gives the data values, and the height of the bar gives the frequency.

CRITICAL THINKING
Bar heights are all the same.

Use Teaching Transparency 50.

ADDITIONAL EXAMPLE 3

Refer to the vehicle data in Additional Example 2. **Find the probability that there are 4 or more people in a randomly selected vehicle.**
about 21%

TRY THIS
≈66.7%

Math CONNECTION

PROBABILITY Be sure students notice that in a frequency distribution, the sum of the frequencies must equal the total number of data values and that in a probability distribution, the sum of the probabilities must equal 1.

Activity Notes

Students should observe that the histogram in Step 2 is more mound shaped than the histogram in Step 1. The histogram in Step 4 is also mound-shaped.

Enrichment

Have students make a stem-and-leaf plot for the following data:

1, 2, 4, 7, 10, 11, 13, 15, 18, 19,
22, 23, 24, 25, 32, 41, 45, 48

Now have them compare the distribution of this stem-and-leaf plot with that of the stem-and-leaf plot for the Checkpoint data on page 773. They can do this by making a scatter plot of ordered pairs from both data sets. If they are using a TI-83 or TI-82 graphics calculator, have them enter the ordered data from the sets into **L1** and **L2**, respectively. Otherwise, have them use the items from one set as x-values and the items from the other set as y-values in order to create the ordered pairs. Now have students find the least-squares line and the correlation. Point out that if the correlation is high, the two data sets have similar distributions. **Letting x represent the data from page 773 and y represent the data in this activity, the least-squares line is $y = 0.983x + 1.111$ and the correlation coefficient, r, is approximately 0.998.**

Cooperative Learning

Have students work in pairs, with each member alternately completing Steps 1 and 2. Then each member should make the required histograms and compare them.

CHECKPOINT ✔

3. **Step 1 is flatter, and Step 2 is more mound-shaped.**

CHECKPOINT ✓

5. Answers may vary.
Sample:

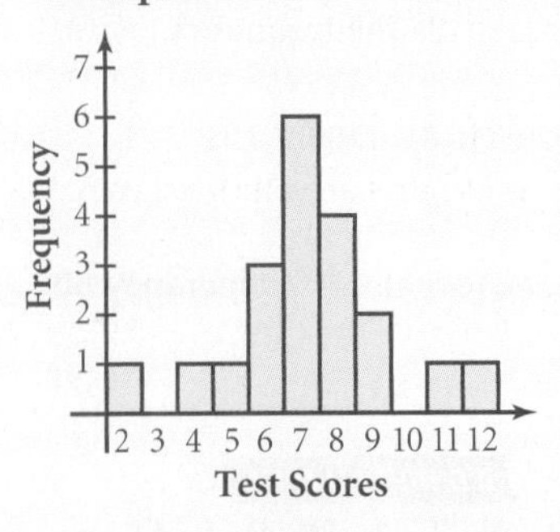

The histogram should be roughly mound-shaped.

ADDITIONAL EXAMPLE 4

The table below shows the distribution by region of the resident population of the United States in 1996.

Resident Population of the United States in 1996	
Region	**Population (in millions)**
Northeast	51.6
Midwest	62.1
South	93.1
West	58.5

[*Source: Statistical Abstract of the United States, 1997*]

a. Make a circle graph to represent the data.

Resident Population of the United States in 1996

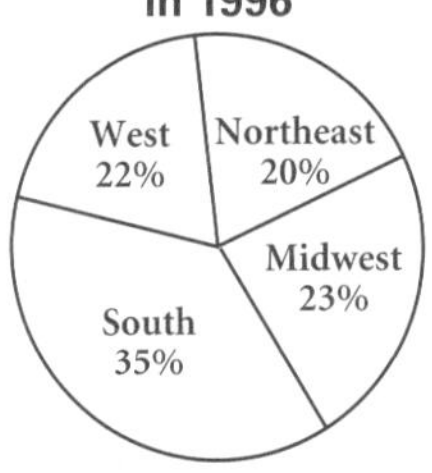

b. Find the probability that a randomly chosen resident of the United States in 1996 was not a resident of the South. about 65%

4. Suppose that a pair of number cubes are rolled and the sum is recorded for 20 trials. What shape would you expect for the histogram of this data? Explain.

CHECKPOINT ✓ **5.** Roll a pair of number cubes as described in Step 4 and record the results. Make a histogram and describe its shape. Is the histogram different from the shape you expected? Explain.

Circle Graphs

You can use a **circle graph** to display the distribution of non-overlapping *parts of a whole*, as shown in Example 4.

EXAMPLE 4 **The causes of 1200 fires are listed at right.**

APPLICATION PUBLIC SAFETY

a. Make a circle graph to represent this data.
b. Find the probability that a given fire was caused by smoking, children, or cooking.

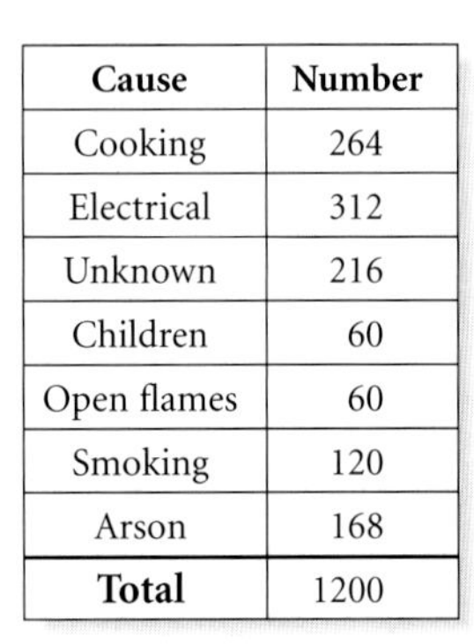

Cause	Number
Cooking	264
Electrical	312
Unknown	216
Children	60
Open flames	60
Smoking	120
Arson	168
Total	1200

SOLUTION

a. Find the relative frequency, or percent, for each cause. Multiply this percent by 360° to obtain the corresponding central angle.

Cause	Relative Frequency	Central Angle
Cooking	$\frac{264}{1200} = 0.22$	$0.22 \times 360^\circ \approx 79^\circ$
Electrical	$\frac{312}{1200} = 0.26$	$0.26 \times 360^\circ \approx 94^\circ$
Unknown	$\frac{216}{1200} = 0.18$	$0.18 \times 360^\circ \approx 65^\circ$
Children	$\frac{60}{1200} = 0.05$	$0.05 \times 360^\circ = 18^\circ$
Open flames	$\frac{60}{1200} = 0.05$	$0.05 \times 360^\circ = 18^\circ$
Smoking	$\frac{120}{1200} = 0.10$	$0.10 \times 360^\circ = 36^\circ$
Arson	$\frac{168}{1200} = 0.14$	$0.14 \times 360^\circ \approx 50^\circ$
Total	$\frac{1200}{1200} = 1.00$	$1.00 \times 360^\circ = 360^\circ$

Use a protractor to measure the central angle for each category.

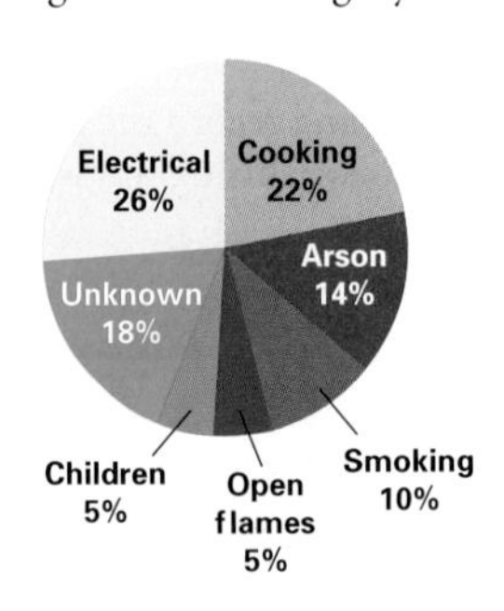

Reteaching the Lesson

INVITING PARTICIPATION Have students survey their class to find the total number of letters in the first and last names of each student. Have them organize the data in a table. Now have them make a stem-and-leaf plot, histogram, and circle graph for the data set. Ask students to compare each graph and to choose the graph that best displays the data in the table. Then have them summarize their findings.

b. Let each event be the cause of a given fire: event A = smoking, event B = children, and event C = cooking. Assume that events A, B, and C are mutually exclusive.

$$\begin{aligned} P(A \text{ or } B \text{ or } C) &= P(A) + P(B) + P(C) \\ &= 0.10 + 0.05 + 0.22 \\ &= 0.37 \end{aligned}$$

Thus, the probability that a fire was caused by smoking, children, or cooking is 0.37, or 37%.

SUMMARY OF DISPLAYS FOR DATA

Name	When useful
stem-and-leaf plot	• to quickly arrange raw data • to show distribution and retain actual values for analysis
histogram	• to show frequency or probability distributions
circle graph	• to show how parts relate to a whole

Exercises

Communicate

1. Describe the different shapes of distributions discussed in the lesson.

2. Which contains more information, a stem-and-leaf plot or a histogram? When is each representation preferred?

3. Explain why the relative frequencies for a data set may also be probabilities.

4. Circle graphs are also called pie charts. Describe the characteristics of a circle graph that are implied by the name *pie chart*.

Guided Skills Practice

APPLICATION

5. BUSINESS The number of calls for customer service during 24 randomly selected days are listed below. ***(EXAMPLE 1)***

22	32	25	42	48	42	36	51	42	53	53	29
31	38	52	51	48	24	39	37	71	51	39	21

a. Make a stem-and-leaf plot of the calls received.
b. Find the median and mode of the number of calls.
c. How could a stem-and-leaf plot of the number of calls be used?

5b. median = 40.5; modes = 42 and 51

c. sample answer: to determine how many people should be available to answer calls

Math CONNECTION

PROBABILITY Remind students that when two events cannot occur at the same time, the events are said to be *mutually exclusive*. If events A, B, and C from part **b** of Example 4 are mutually exclusive, then the probability that the fire was caused by smoking, children playing, or cooking is the sum of the individual probabilities.

Assess

Selected Answers

Exercises 5–8, 9–45 odd

ASSIGNMENT GUIDE

In Class	1–8
Core	9–27 odd
Core Plus	10–28 even, 29
Review	30–45
Preview	46

Extra Practice can be found beginning on page 940.

5a.

Stem	Leaf 2|1 = 21
2	1, 2, 4, 5, 9
3	1, 2, 6, 7, 8, 9, 9
4	2, 2, 2, 8, 8
5	1, 1, 1, 2, 3, 3
6	
7	1

Error Analysis

Some students may be confused by the terms *frequency* and *relative frequency*. Tell them that frequency values are whole numbers and that relative frequency values are values from 0 to 1.

6.

Responses	Frequency
3	2
4	4
5	5
6	5
7	4
8	3
9	1

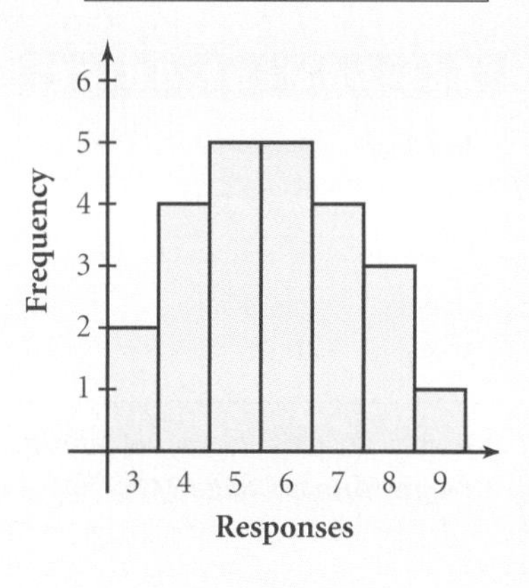

Practice

NAME ______ CLASS ______ DATE ______

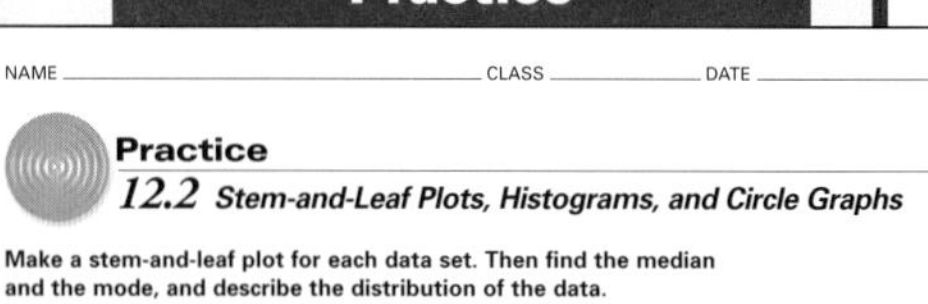

Practice
12.2 **Stem-and-Leaf Plots, Histograms, and Circle Graphs**

Make a stem-and-leaf plot for each data set. Then find the median and the mode, and describe the distribution of the data.

1. 40, 64, 54, 38, 42, 45, 33, 37, 56, 58, 64

Stem	Leaf 3|8 = 38
3	3, 7, 8
4	0, 2, 5
5	4, 6, 8
6	4, 4

45; 64; flat

2. 3.6, 4.8, 3.9, 1.7, 4.3, 2.3, 4.8, 3.1, 4.0, 2.3

Stem	Leaf 3|1 = 3.1
1	7
2	3, 3
3	1, 6, 9
4	0, 3, 8, 8

3.75; 2.3 and 4.8; mound-shaped

Make a frequency table and a histogram for the data.

3. 1.0, 1.3, 1.1, 1.4, 1.4, 1.2, 1.1, 1.0, 1.0, 1.3, 1.3, 1.4, 1.3, 1.2, 1.0, 1.3, 1.4, 1.2, 1.0, 1.3

Number	Frequency
1.0	5
1.1	2
1.2	3
1.3	6
1.4	4

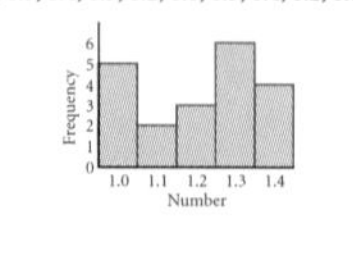

Complete the table, and make a relative frequency histogram for the data.

4.

Number	Frequency	Relative frequency
10	8	16%
11	12	24%
12	7	14%
13	11	22%
14	12	24%

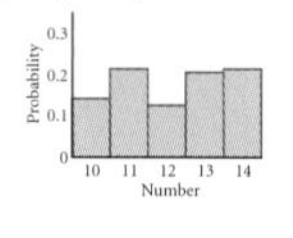

Make a circle graph for the data.

5.

Motor Vehicle Registration by Type, 1994			
Passenger cars	Motorcycles	Buses	Trucks
33.2%	0.9%	15.9%	50%

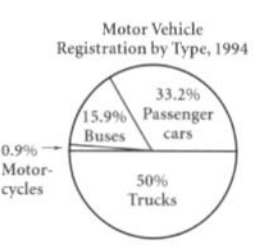

APPLICATIONS

College campus in New York

6. **LAW ENFORCEMENT** The number of calls that a police department responded to during 24 randomly selected days are listed below. ***(EXAMPLE 2)***

5	7	8	4	9	4	5	6	7	4	6	3
6	7	6	8	5	4	6	8	3	7	5	5

Make a frequency table and a histogram for the number of responses.

7. **BUSINESS** Refer to the relative frequency table on page 775. Find the probability that a randomly selected customer rents a canoe for 5 hours or less. ***(EXAMPLE 3)*** **0.9**

8. **EDUCATION** The table at right lists the majors of students at a small college. ***(EXAMPLE 4)***
   a. Make a circle graph to represent this data.
   b. Find the probability that a randomly selected student is a liberal arts major or undecided. **0.6**

Major	Number
Liberal arts	891
Natural sciences	627
Undecided	235
Other	122

9. 8.55; 7.5 and 8.2; mound-shaped
10. 28.3; no mode; mound-shaped
11. 347; 347 and 348; flat-shaped
12. 7.85; 7.85; flat-shaped

Practice and Apply

Make a stem-and-leaf plot for each data set. Then find the median and the mode, and describe the distribution of the data.

9. 8.9, 8.8, 7.2, 7.5, 9.2, 7.9, 8.2, 9.1, 8.7, 8.2, 8.5, 8.6, 9.5, 7.5
10. 27.2, 26.3, 30.1, 26.8, 27.3, 28.7, 28.3, 29.8, 29.4, 28.4, 29.1, 28.1, 27.6
11. 359, 357, 348, 347, 337, 347, 340, 335, 338, 348, 339, 356, 336, 358
12. 6.15, 8.55, 7.85, 9.65, 7.85, 8.45, 7.35, 6.35, 8.45, 9.65, 7.85, 9.75, 6.35

Make a frequency table and histogram for each data set.

13. 5, 4, 6, 7, 9, 2, 3, 9, 6, 9, 3, 2, 8, 10, 10
14. 3, 5, 7, 9, 2, 4, 6, 10, 7, 8, 2, 3, 9, 7, 6
15. 1.0, 0.5, 1.5, 2.0, 2.5, 1.0, 1.5, 2.0, 0.5, 1.5, 2.5, 3.0, 2.0, 1.5, 1.0
16. 0.8, 0.4, 0.6, 0.4, 0.2, 0.2, 0.6, 0.6, 0.4, 0.2, 0.4, 0.4, 0.4, 0.8, 0.6

Make a relative frequency table and relative frequency histogram for each data set.

17. 1, 3, 1, 4, 3, 2, 7, 5, 8, 3, 7, 1, 4, 8, 5, 7, 4, 2, 3, 4, 7, 3, 8, 1
18. 40, 60, 80, 30, 70, 80, 80, 60, 60, 40, 30, 40, 70, 100, 60, 70, 30, 40
19. 0.1, 0.3, 0.1, 0.2, 0.4, 0.2, 0.1, 0.5, 0.3, 0.1, 0.1, 0.2, 0.1, 0.3, 0.2
20. 8.1, 8.5, 7.6, 7.9, 8.0, 7.8, 7.9, 8.1, 8.0, 8.1, 8.3, 8.1, 7.9, 7.5, 8.0

Make a histogram with the following intervals for each data set: $0 \le x < 1$, $1 \le x < 2$, $2 \le x < 3$, $3 \le x < 4$, $4 \le x < 5$, and $5 \le x < 6$.

21. 0.2, 1.3, 5.4, 4.3, 2.2, 4.3, 4.6, 3.5, 5.1, 4.8, 1.5, 3.7, 5.4, 4.0, 4.2, 5.2
22. 2.2, 4.6, 3.2, 1.2, 2.8, 3.8, 4.2, 1.2, 2.2, 1.5, 0.5, 2.9, 3.6, 0.9, 1.0
23. 3.4, 4.8, 1.2, 2.5, 3.6, 5.2, 5.0, 4.1, 3.8, 3.5, 4.2, 5.1, 4.8, 4.4, 4.9
24. 0.2, 1.9, 1.2, 0.7, 2.3, 3.1, 2.5, 1.8, 1.6, 1.4, 0.8, 1.3, 0.9, 2.2, 1.7

8a.

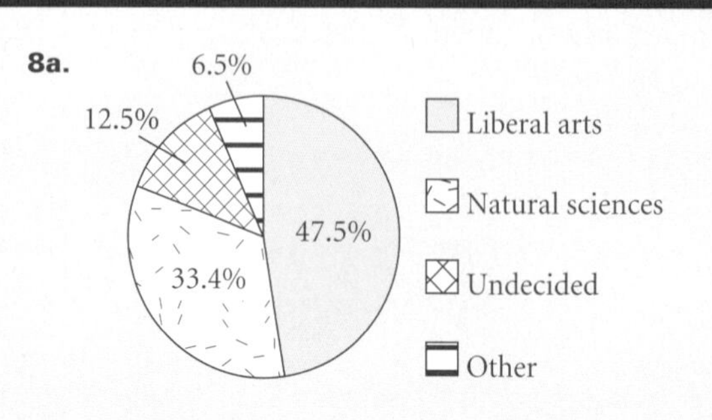

9.

Stem	Leaf 7|2 = 7.2
7	2, 5, 5, 9
8	2, 2, 5, 6, 7, 8, 9
9	1, 2, 5

median = 8.55; modes = 7.5 and 8.2; mound-shaped

The answers to Exercises 10–24 can be found in Additional Answers beginning on page 1002.

CONNECTION

25. PATTERNS IN DATA Survey at least 20 students in your class about how they travel to school (walk, bus, car, bike, subway, . . .). Organize your data into a table and make a circle graph. Explain what your graph illustrates.

APPLICATIONS

26. BUSINESS The dollar amount of the total purchases from a vending machine in an office over a 3-month period is shown below for 21 randomly selected individuals.

$10	$43	$5	$18	$8	$63	$10	$6	$30	$22	
$27	$25	$14	$18	$30	$41	$27	$22	$31	$32	$42

a. Make a stem-and-leaf plot of the dollar amounts. Describe the distribution.
b. Find the median and mode of the dollar amounts.
c. Find the probability that a randomly selected individual spent $20 or less on vending machine purchases.

26a. mound-shaped
b. $25; $10, $18, $22, $27, and $30
c. about 38%

27. INCOME The percent distribution by income of all households in the United States is listed in the table below for 1994.

Under $14,999	$15,000–$24,999	$25,000–$34,999	$35,000–$49,999	$50,000–$74,999	Over $75,000
22.7%	16.7%	14.2%	16.3%	16.5%	13.6%

[*Source: U.S. Bureau of the Census*]

a. Explain whether a histogram or circle graph would best display the data in the table and draw the display.
b. Interpret your display by describing what it illustrates. **fairly flat**
c. Find the probability that a randomly selected household in the United States has an income of $50,000 or more. **30.1%**

28. ARMED FORCES The table below lists the location and number of the active-duty military personnel from the United States worldwide.

Location	Number
U.S., U.S. Territories, and special locations	1,397,083
Western and Southern Europe	166,249
East Asia and Pacific	99,022
North Africa, Middle East, and South Asia	11,490
Sub-Saharan Africa	6864
Other Western Hemisphere	17,758

[*Source: The World Almanac and Book of Facts, 1995*]

a. Display this information in your choice of a stem-and-leaf plot, histogram, or circle graph. Justify your choice.
b. Find the probability that a randomly selected active-duty military person from the United States is not located in the United States. **17.**

CHALLENGE

29. MARKETING A taste test of three sodas, labeled A, B, and C, included 175 participants. In the test, soda B was selected by twice as many people as soda A, and soda C was selected by twice as many people as soda B. Represent these results in a circle graph.

25. Answers may vary.

26a.

Stem	Leaf 2|2 = 22
0	5, 6, 8
1	0, 0, 4, 8, 8
2	2, 2, 5, 7, 7
3	0, 0, 1, 2
4	1, 2, 3
5	
6	3

The answers to Exercises 28a–29 can be found in Additional Answers beginning on page 1002.

27a. Answers may vary. Sample answers: A histogram most effectively shows how the incomes are divided into fairly even percents. Alternatively, a circle graph best illustrates that the incomes represent the entire United States with fairly even parts that represent the percent in each income.

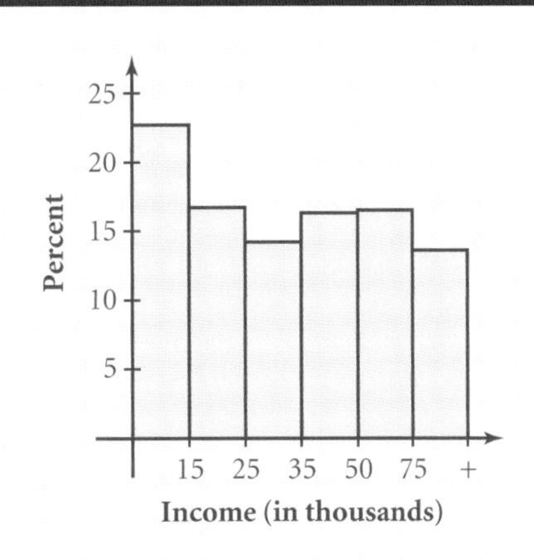

Student Technology Guide

NAME ______________ CLASS ________ DATE ________

Student Technology Guide
12.2 Stem-and-Leaf Plots, Histograms, and Circle Graphs

A histogram shows the frequency of each value in a numerical data set. You can use a graphics calculator to make a histogram.

Example: An ecologist reported the number of sightings of an endangered bird of prey every day for two weeks. Make a histogram for this data.

0, 1, 4, 0, 2, 2, 0, 1, 1, 0, 3, 1, 0, 2

- Press STAT 1:Edit... ENTER. Type the numbers in the data set into L1. Press ENTER or ▼ after each number. After entering the data, press 2nd MODE.
- Press 2nd Y= 1:Plot1... ENTER.
- The cursor is over On. Press ENTER to turn the plot on.
- Press ▼, then press ► twice to move to the picture of a histogram. Press ENTER.
- Be sure Xlist is set to L1 and Freq is 1. Press 2nd MODE.
- Press WINDOW. Set Xmin = 0, Xmax = 5, Xscl = 1, Ymin = 0, Ymax = 6, and Yscl = 1.
- Press GRAPH. To see the frequency represented by each bar, press TRACE and use ► and ◄ to move from bar to bar. The screen at the right shows that one sighting occurred 4 times, because min = 1 and $n = 4$.

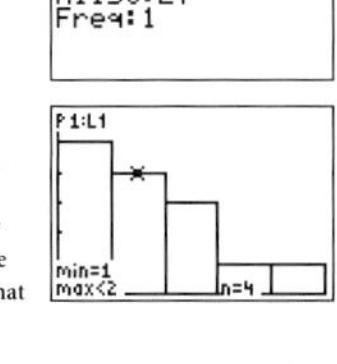

On a separate sheet of paper, make a histogram for each data set.

1. 5, 8, 5, 6, 8, 9, 9, 5, 6, 9, 4, 6, 7, 5, 8
2. 14, 14, 17, 22, 23, 28, 15, 15, 14, 22, 23, 16, 23
3. The list below shows the number of cities whose population exceeded 250,000 in 1992 in each of the 50 states. For instance, the 8 represents the fact that Texas had 8 such cities. Make a histogram for this data set. Describe the distribution of the data.

10, 8, 4, 3, 3, 2, 2, 2, 2, 2, 2, 2, 2, 1, 1, 1, 1, 1, 1, 1, 1, 1, 1, 1, 1, 1, 1, 1, 1, 1, 1, 1, 0, 0, 0, 0, 0, 0, 0, 0, 0, 0, 0, 0, 0, 0, 0, 0

Descriptions may vary. The data values at the lower end of the distribution occur more frequently.

Exercise 46 involves a binomial experiment in which an unfair coin is tossed 3 times. In Lesson 12.5, students will find the theoretical probability that exactly r successes will occur in n trials of a binomial experiment.

ALTERNATIVE

Assessment

Portfolio Activity

The Portfolio Activity can be used as preparation for the Chapter Project or as a separate activity. In the Portfolio Activity on this page, students will collect data and represent the data with a stem-and-leaf plot, a histogram, and a circle graph.

Answers to Portfolio Activities can be found in Additional Answers of the Teacher's Edition.

Solve each proportion for x. ***(LESSON 1.4)***

30. $\frac{3x-2}{4} = \frac{x}{9}$ **$\frac{18}{23}$**

31. $\frac{-4x}{7} = x + 2$ **$-\frac{14}{11}$**

32. $\frac{20x}{-6} = \frac{x-4}{3}$ **$\frac{4}{11}$**

Factor each expression. ***(LESSON 5.3)***

33. $x^2 + 16x + 64$ **$(x+8)^2$**

34. $x^2 - 10x + 25$ **$(x-5)^2$**

35. $3(2x-5) - x(2x-5)$ **$(3-x)(2x-5)$**

36. $25a^2 - 49b^2$ **$(5a+7b)(5a-7b)$**

Divide by using synthetic division. ***(LESSON 7.3)***

37. $(x^3 + x^2 - 20) \div (x+2)$ **$x^2 - x + 2 - \frac{24}{x+2}$**

38. $(2x^3 + 10x^2 - 2x + 8) \div (x-3)$ **$2x^2 + 16x + 46 + \frac{146}{x-3}$**

Use a graph, synthetic division, and factoring to find all of the roots of each equation. ***(LESSON 7.4)***

39. $x^3 - 2x^2 - 4x + 8 = 0$ **−2, 2**

40. $x^3 + x^2 - x - 1 = 0$ **−1, 1**

41. $c^3 + 5c^2 + 8c + 4 = 0$ **−2, −1**

42. $y^3 - 4y^2 + 4y = 0$ **0, 2**

CONNECTION

43. $\frac{(x-4)^2}{4} + \frac{(y-2)^2}{6.25} = 1$

43. **TRANSFORMATIONS** Translate the ellipse defined by the equation $25x^2 + 16y^2 = 100$ down 2 units and to the right 4 units. Write the standard equation of the resulting ellipse. ***(LESSON 9.4)***

A coin is tossed 3 times. Use a tree diagram to find the probability that 2 of the 3 tosses land tails up, given that each event below occurs. ***(LESSON 10.6)***

44. The first coin lands tails up. **$\frac{1}{2}$**

45. The first coin lands heads up. **$\frac{1}{4}$**

46. A trick coin with a 0.75 probability of heads is tossed 3 times. What is the probability of getting 2 heads in 3 tosses? **≈0.42**

Collect data from the National Weather Service about the average number of days per year that different cities have snowfall. Choose one city in each of the 50 states, including the city in which you reside. Represent the data in a stem-and-leaf plot, a histogram, and a circle graph. Then answer the questions below.

1. How does your city compare with the other cities? Which display—a stem-and-leaf plot, a histogram, or a circle graph—best compares this information? Why?
2. How many cities have the same average number of days of snowfall per year as your city? Which display—a stem-and-leaf plot, a histogram, or a circle graph—best illustrates this information? Why?
3. Which display—a stem-and-leaf plot, a histogram, or a circle graph—do you prefer for this data? Why?

WORKING ON THE CHAPTER PROJECT

You should now be able to complete Activity 1 of the Chapter Project.

Box-and-Whisker Plots

Hollywood Hill, Los Angeles, California

Wrigley Field, home of the Chicago Cubs

Objectives

- Find the range, quartiles, and interquartile range for a data set.
- Make a box-and-whisker plot for a data set.

Why *You can use box-and-whisker plots to compare the distributions of two sets of similar data, such as the monthly mean temperatures for two cities.*

APPLICATION
CLIMATE

The mean monthly temperatures for Los Angeles, California, and Chicago, Illinois, are listed at right. Construct a box-and-whisker plot for the temperatures in each city and compare them. *You will solve the problem in Example 2.*

Los Angeles (1961–1990) Monthly Mean Temperatures (°F)

Jan.	55.9
Feb.	57.0
Mar.	58.3
Apr.	60.8
May	63.3
June	66.7
July	70.9
Aug.	71.8
Sep.	70.5
Oct.	66.6
Nov.	62.1
Dec.	57.6

[*Source: U.S. NOAA*]

Chicago (1961–1990) Monthly Mean Temperatures (°F)

Jan.	21.0
Feb.	25.5
Mar.	37.0
Apr.	48.6
May	58.8
June	68.5
July	73.0
Aug.	71.6
Sep.	64.4
Oct.	52.7
Nov.	39.9
Dec.	26.6

[*Source: U.S. NOAA*]

Quartiles

Exploring Quartiles

You will need: a calculator and graph paper

The table below gives the mean monthly precipitation (1961–1990) in inches for Dallas–Fort Worth, Texas. [*Source: U.S. NOAA*]

Jan.	Feb.	Mar.	Apr.	May	Jun.	Jul.	Aug.	Sep.	Oct.	Nov.	Dec.
1.9	2.2	2.8	3.5	4.9	3.0	2.3	2.2	3.4	3.5	2.3	1.8

1. Find the median of the monthly precipitation data. What percent of the data values are *below* the median? *above* the median?

Alternative Teaching Strategy

INVITING PARTICIPATION Randomly choose 9 students from the class and have them stand an equal distance from each other, arranged by height from the shortest to the tallest. The height of the student in the middle represents the median height of the group, or the second quartile. Place a piece of paper on the floor in front of that student labeled "second quartile." Now have the class find the median height of the shorter half of the group and then find the median height of the taller half of the group. These medians will fall between the middle two students in the shorter half and between the middle two students in the taller half of the group. Place a piece of paper labeled "first quartile" on the floor between the middle two students in the shorter half and one labeled "third quartile" between the middle two students in the taller half. To find the interquartile range, have students find the difference between the third quartile value and the first quartile value.

Prepare

NCTM PRINCIPLES & STANDARDS 1, 4–10

QUICK WARM-UP

Find the mean, median, and mode(s) of each data set.

1. 2, 3, 9, 4, 5, 7
   mean: 5; median: 4.5; no mode
2. 8.2, 8.2, 8.2, 8.3, 8.3, 8.3
   mean: 8.25; median: 8.25; modes: 8.2, 8.3
3. 10, 11, 12, 13, 14
   mean: 12; median: 12; no mode
4. 1.6, 1.6, 1.6, 1.6, 1.6, 1.6
   mean: 1.6; median: 1.6; mode: 1.6

Also on Quiz Tranparency 12.3

Teach

Why Data sets often include extreme values that distort some measures of central tendency. Box-and-whisker plots are displays of data that show both the measures of central tendency and extreme values. Ask each student how many hours of television they watched the previous day and tally their responses. Ask students to identify any extreme values.

Activity Notes

Students should find that 25% of the values are below the median of the lower half, 50% of the values are below the median of all the values, 75% of the values are below the median of the upper half, and 25% of the data values are above the median of the upper half.

Cooperative Learning

Divide the class into groups of three. Each member should complete Steps 1, 2, 3, and 4 independently. To check their computations, group members can exchange papers and then work together to write a generalization about the patterns they noticed.

CHECKPOINT ✔

4.

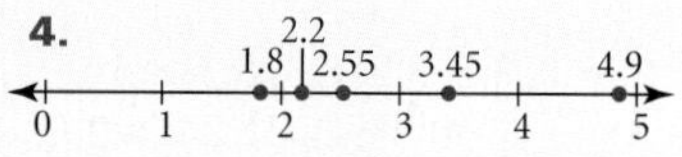

1.9–2.2: 25%; 2.2–2.55: 25%
2.55–3.45: 25%; 3.45–4.9: 25%

ADDITIONAL EXAMPLE 1

The data below shows the age of 43 Florida Marlins baseball players at the 1997 Worlds Series.

29 25 31 19 34 22
32 34 20 31 35 27
35 22 38 28 25 26
21 22 27 26 21 32
25 25 23 35 32 27
28 35 25 30 22 23
38 25 36 30 26 34
26

a. Find the quartiles, range, and interquartile range for the data. quartiles: $Q_1 = 25$, $Q_2 = 27$, $Q_3 = 32$; range: 19; interquartile range: 7

b. Identify any outliers. no outliers

2. Find the median of the lower half of the values. What percent of data values are below this "lower" median?
3. Find the median of the upper half of the values. What percent of data values are above this "upper" median?

CHECKPOINT ✔ 4. Draw a number line and plot each of the medians you calculated in Steps 1–3 along with the highest and lowest values. Write the percent of data values that are between each plotted value.

Related to the median, which divides a data set into halves, are **quartiles** which divide a data set into quarters. The second quartile, Q_2, is the median that divides the lower half of the data values from the upper half. The lower quartile, Q_1, is the median of the lower half of the data values, and the upper quartile, Q_3, is the median of the upper half of the data values.

These five measures are often called the five-point summary of a data set.

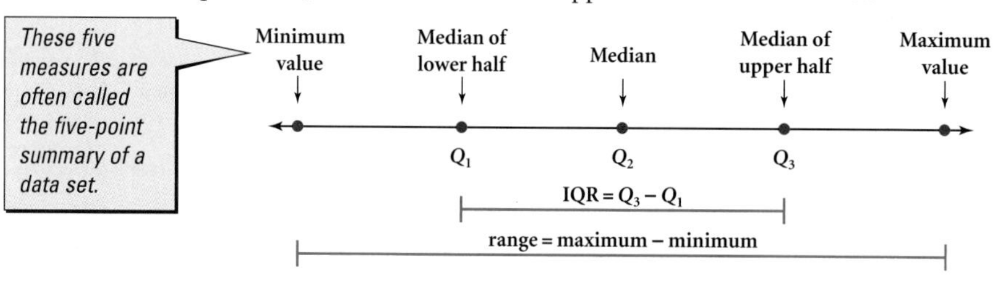

The difference between the maximum value and the minimum value is called the **range**. The difference between the upper and lower quartiles, $Q_3 - Q_1$, is called the **interquartile range**, denoted IQR. When a data value is less than $Q_1 - 1.5(\text{IQR})$ or greater than $Q_3 + 1.5(\text{IQR})$, the data value may be called an **outlier**.

EXAMPLE 1

APPLICATION SOCIAL SERVICES

The number of calls received by a crisis hotline during 17 randomly selected days is listed at right.

50	57	77	66
53	72	51	88
82	70	112	107
69	88	98	65
155			

a. Find the minimum and maximum values, quartiles, range, and interquartile range for the data.
b. Identify any outliers.

SOLUTION

a. Arrange the values in order. The median is the ninth value, 72. Then find the median of the lower half, $Q_1 = \frac{57+65}{2} = 61$, and the median of the upper half, $Q_3 = \frac{88+98}{2} = 93$.

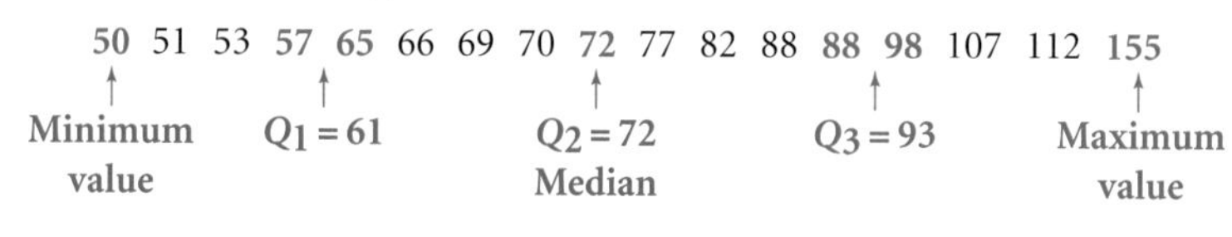

range = maximum − minimum
= 155 − 50 = 105

All data values lie within the range.

IQR = $Q_3 - Q_1$
= 93 − 61 = 32

The middle half of the values lie within the interquartile range.

Interdisciplinary Connection

ASTRONOMY The table below shows the time it would take a jet airplane, traveling at a speed of about 600 miles per hour, to travel from Earth to other planets in our solar system.

	Mercury	Venus	Mars	Jupiter
Time	10 yr 10 mo	5 yr 5 mo	8 yr 10 mo	74 yr 3 mo

	Saturn	Uranus	Neptune	Pluto
Time	150 yr 5 mo	318 yr 6 mo	513 yr 2 mo	690 yr 1 mo

[*Source: NCTM, Mission Mathematics, 1997*]

Have students find the quartiles, range, and interquartile range and identify any outliers for the travel times from Earth.
Q_1 = 9 yr 10 mo, Q_2 = 112 yr 4 mo, Q_3 = 415 yr 10 mo; range = 684 yr 8 mo; IQR = 406 yr; there are no outliers.

b. Find any possible outliers below Q_1.
$Q_1 - 1.5(IQR) = 61 - 1.5(32) = 13$

Find any possible outliers above Q_3.
$Q_3 + 1.5(IQR) = 93 + 1.5(32) = 141$

There are no values less than or equal to 13. Because the data value 155 is greater than 141, 155 is a possible outlier.

TRY THIS Find the minimum and maximum values, quartiles, range, and interquartile range for the set of data below. Identify any possible outliers.

4	7	9	31	34	2	35	37	24	34	31	50
11	33	36	2	8	13	52	57	60	69	78	83

CRITICAL THINKING What could the possible outlier in Example 1 indicate for the crisis hotline?

Box-and-Whisker Plots

A **box-and-whisker plot** displays how data values are distributed. A box-and-whisker plot is shown below for the crisis-hotline data from Example 1.

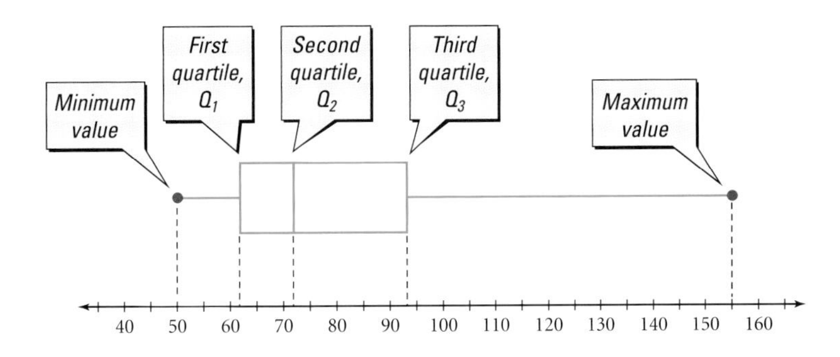

Notice that five measures define a box-and-whisker plot: the minimum value, the maximum value, Q_1, Q_2, and Q_3.

CHECKPOINT ✔ What part of the box-and-whisker plot represents the middle half of the values?

MAKING A BOX-AND-WHISKER PLOT

Step 1. Arrange the values in increasing order and compute Q_1, Q_2, and Q_3.

Step 2. Draw a number line that includes the minimum and maximum values.

Step 3. Make a box whose left end is at Q_1 and whose right end is at Q_3..

Step 4. Draw a vertical line segment to divide the box at, Q_2, the median.

Step 5. Draw a line segment from Q_1 to the minimum value and another line segment from Q_3 to the maximum value for the left and right whiskers.

TRY THIS
minimum = 2; maximum = 83; $Q_1 = 10$; $Q_2 = 33.5$; $Q_3 = 51$; range = 81; IQR = 41; no outliers

CRITICAL THINKING
Maybe there was a single event that disturbed many people.

Teaching Tip

TECHNOLOGY When using a TI-83 or TI-82 graphics calculator to create a box-and-whisker plot for Example 1, enter the data into **L1**, press 2nd Y=, select **1:Plot 1...**, and choose the following:

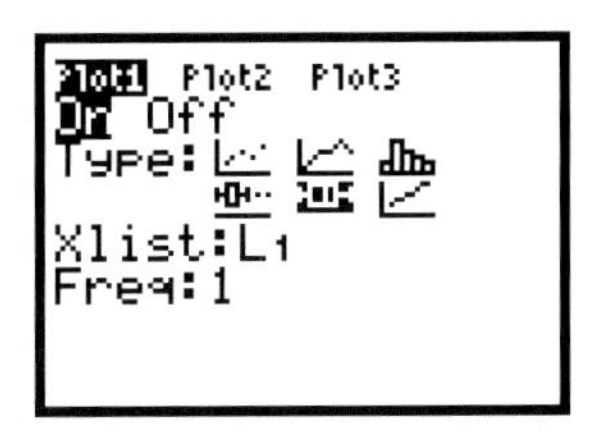

Then press ZOOM and select **9:ZoomStat** to view the box-and-whisker plot. You can use the **TRACE** feature and the arrow keys to display the upper and lower quartiles, the median, and the greatest and least values.

CHECKPOINT ✔
the box, or Q_1 to Q_3

Inclusion Strategies

USING MODELS Set up a 10-unit number line. Have students write an integer from 0 to 10 on a piece of paper. Collect the numbers and plot them on the number line, vertically stacking the points for repeated integers. Students can use the box-and-whisker feature of a graphics calculator to examine how the chosen integers are distributed.

Enrichment

Tell students that the minimum value of a set of data is 10, the maximum value is 90, $Q_1 = 40$, $Q_2 = 50$, and $Q_3 = 60$. Have students make a box-and-whisker plot for the data. Then have students create a stem-and-leaf plot that could model the data in the box-and-whisker plot. Be sure students notice that 50% of the values are between 40 and 60.

ADDITIONAL
EXAMPLE 2

Refer to the precipitation data below.

Normal Monthly Precipitation in Inches (1961–1990)

Month	Chicago	L.A.
Jan.	1.53	2.40
Feb.	1.36	2.51
Mar.	2.69	1.98
Apr.	3.64	0.72
May	3.32	0.14
Jun.	3.78	0.03
Jul.	3.66	0.01
Aug.	4.22	0.15
Sep.	3.82	0.31
Oct.	2.41	0.34
Nov.	2.92	1.76
Dec.	2.47	1.66

[*Source: Statistical Abstract of the United States, 1997*]

a. Make a box-and-whisker plot for the amount of precipitation in each city.

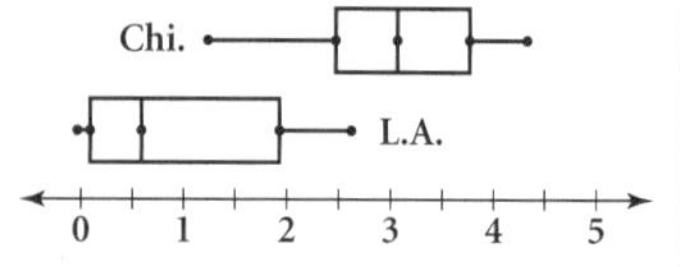

b. Compare the box-and-whisker plots.

The normal amounts of precipitation in L.A. are lower than those in Chicago. The slightly longer box for L.A. indicates that the normal amounts of precipitation vary a bit more than those in Chicago.

CHECKPOINT ✔

IQR: Chi. = 34.65, L.A. = 10.65; The temperature varies more in Chicago. Chicago and New York City have similar temperature variation.

EXAMPLE 2

APPLICATION
CLIMATE

Refer to the temperature data at the beginning of the lesson.

a. Make a box-and-whisker plot for the temperatures in each city.

b. Compare the box-and-whisker plots.

SOLUTION

a. Method 1

Calculate the five measures that define the box-and-whisker plot for each city's temperatures.

TECHNOLOGY
GRAPHICS CALCULATOR
Keystroke Guide, page 822

	Chicago	LA
minimum	21.0	55.9
Q_1	31.8	57.95
median, Q_2	50.65	62.7
Q_3	66.45	68.6
maximum	73.0	71.8

Then draw both plots as shown below.

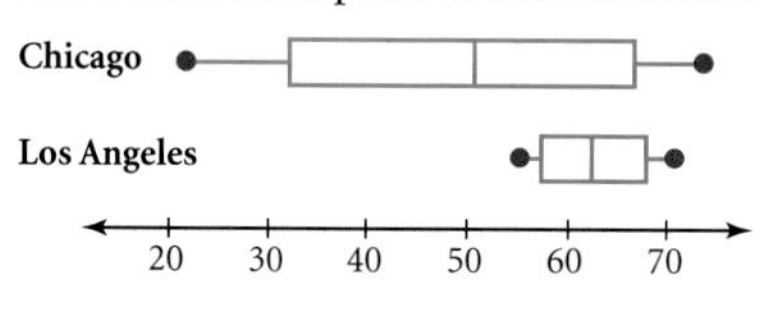

Method 2

Use a graphics calculator. First enter the temperatures for each city into separate lists. Then select box-and-whisker plots for each list of data.

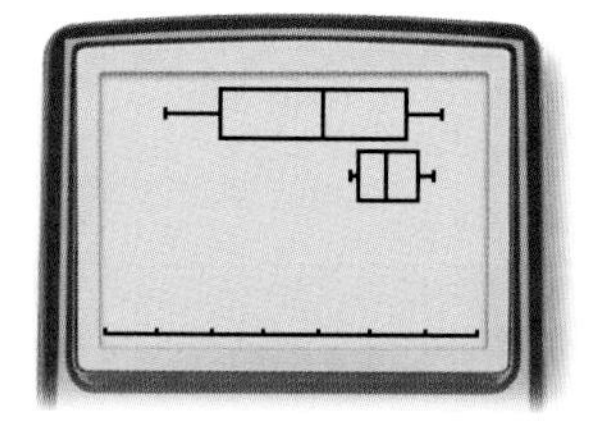

Use the trace feature to identify the maximum, minimum, and quartiles that define each box-and-whisker plot.

b. The longer box and longer whiskers for Chicago indicate that the temperature in Chicago varies much more than the temperature in Los Angeles. Over one-half of Chicago's months had an average temperature less than that of any month in Los Angeles. The maximum average monthly temperature in Chicago was also slightly greater than any average monthly temperature in Los Angeles.

CHECKPOINT ✔ Calculate the IQR for Chicago and for Los Angeles in Example 2. What do your results tell you? What can you conjecture about the temperature variation in New York City, which has an IQR of 35.5?

CRITICAL THINKING Can a box-and-whisker plot have only one whisker? no whiskers? Explain.

APPLICATION
SPORTS

The regular-season batting averages for the top 10 players with at least 175 at bats are displayed at right for the teams in the 1997 World Series.

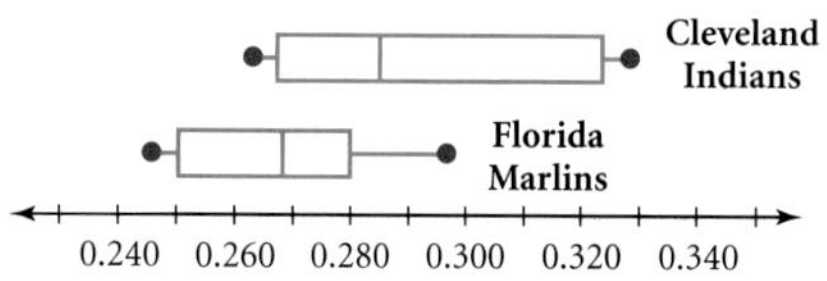

CHECKPOINT ✔ Compare the two box-and-whisker plots above.

Although the graphs show that Cleveland's players had higher batting averages, they lost the 1997 World Series to Florida. Regular-season batting averages are clearly not the only factor to consider when predicting who will win.

Reteaching the Lesson

USING TECHNOLOGY A visual display will help students identify the interquartile range and any outliers. Have students enter the data from Example 1 on page 782 into a TI-82 or TI-83 graphics calculator by entering the numbers 1 through 17 to represent the randomly selected days in **L1** and entering the number of calls received each day in **L2**. Students should put the data in **L2** into ascending order by pressing STAT, selecting **2:SortA(**, and pressing 2nd 2 (to enter **L2**)) ENTER. Then they should create a scatter plot with the following viewing window:

Xmin=0	Ymin=30
Xmax=18	Ymax=180
Xscl=5	Yscl=6

While viewing the scatter plot, students can use the **TRACE** feature to find the ninth value (72), which is the median, the average of the fourth and fifth values (61), which is Q_1, the average of the 13th and 14th values (93), which is Q_3, and the 17th value (155), which is an outlier.

Exercises

Communicate

1. Explain how finding the quartiles for a data set with 20 values is different from finding the quartiles for a data set with 15 values.

2. The box-and-whisker plots at right show the test scores for a history class. Explain what the plots indicate about the class performance on the tests.

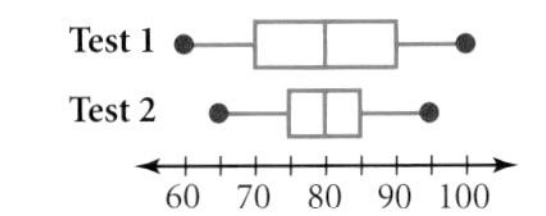

3. What does a box-and-whisker plot tell you about the data set that it represents?

4. Construct two data sets that have the following quartiles: $Q_1 = 5$, $Q_2 = 7$, and $Q_3 = 11$. Explain why you cannot determine from the quartiles what the actual data values are.

internet connect
Activities Online
Go To: **go.hrw.com**
Keyword:
MB1 Actresses

Guided Skills Practice

APPLICATIONS

5. ECOLOGY The lengths (in centimeters) of 24 American burying beetles are given in the table below. ***(EXAMPLE 1)***

Lengths of American Burying Beetles (cm)

2.5	2.8	3.1	3.6	3.4	3.8	3.0	2.8	3.5	3.3	2.6	3.0
2.9	2.7	3.4	3.2	3.7	2.5	3.1	2.9	2.5	3.1	3.8	2.9

a. Find the minimum and maximum values, quartiles, range, and interquartile range for the data. **min = 2.5; Q_1 = 2.8; Q_2 = 3.05; Q_3 = 3.4;**
b. Identify any possible outliers. **max = 3.8; range = 1.3; IQR = 0.6**
no outliers

6. CLIMATE Refer to the temperature data below. ***(EXAMPLE 2)***
a. Make a box-and-whisker plot for the temperatures in each city.
b. Compare the box-and-whisker plots.

Honolulu (1961–1990)
Monthly Mean Temperatures (°F)

Jan.	71.8
Feb.	71.8
Mar.	72.5
Apr.	73.9
May	75.7
June	77.5
July	78.6
Aug.	79.3
Sep.	79.2
Oct.	77.9
Nov.	75.6
Dec.	73.2

[*Source: U.S. NOAA*]

Phoenix (1961–1990)
Monthly Mean Temperatures (°F)

Jan.	51.6
Feb.	55.8
Mar.	61.0
Apr.	68.4
May	76.5
June	85.8
July	91.0
Aug.	89.2
Sep.	83.5
Oct.	72.0
Nov.	59.9
Dec.	52.7

[*Source: U.S. NOAA*]

internet connect
Activities Online
Go To: **go.hrw.com**
Keyword:
MB1 Oil Production

6a.
HI
AZ
50 55 60 65 70 75 80 85 90 95

b. Hawaii's median temperature is higher than Arizona's, but Arizona's temperature is more variable than Hawaii's temperature.

CRITICAL THINKING
Yes; yes; if the minimum value is the same as Q_1 or the maximum value is the same as Q_3, then there will not be two whiskers.

CHECKPOINT ✔
Each of the 5 numbers in the Cleveland Indians' box-plot is higher than those of the corresponding numbers for the Florida Marlins, and the Indians' batting averages are more variable than those of the Marlins.

Assess

Selected Answers
Exercises 5, 6, 7–29 odd

ASSIGNMENT GUIDE

In Class	1–6
Core	7–21 odd, 25–29 odd
Core Plus	8–28 even
Review	30–47
Preview	48

✎ Extra Practice can be found beginning on page 940.

Error Analysis

Some students may think that the median must be one of the values of the data set. Make sure students understand that if a data set has an even number of values, the median is the average of the middle two values.

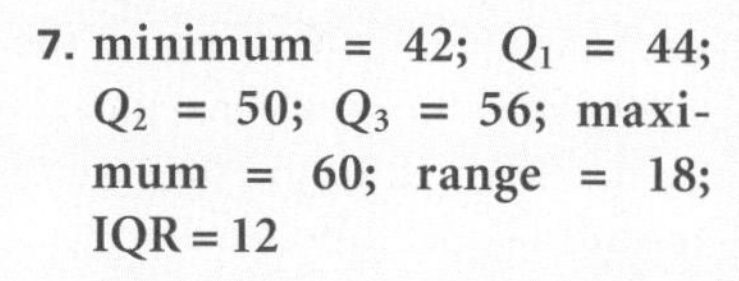

7. minimum = 42; Q_1 = 44; Q_2 = 50; Q_3 = 56; maximum = 60; range = 18; IQR = 12

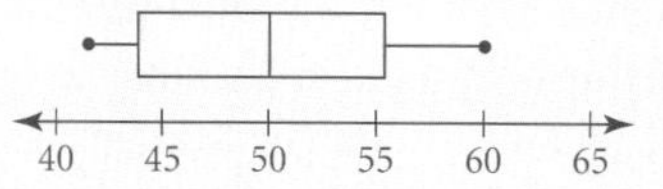

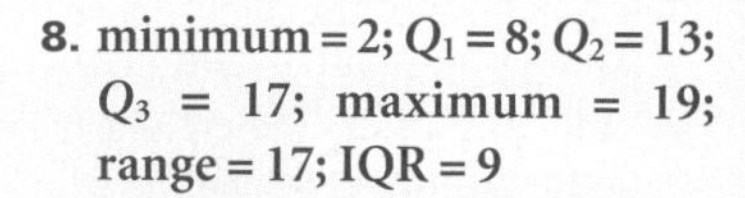

8. minimum = 2; Q_1 = 8; Q_2 = 13; Q_3 = 17; maximum = 19; range = 17; IQR = 9

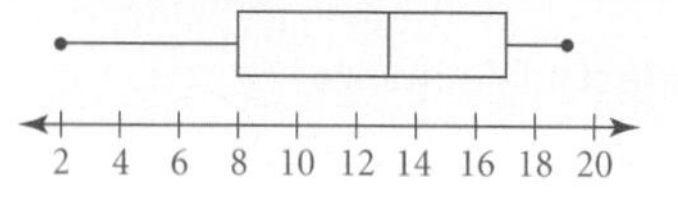

9. minimum = 102; Q_1 = 125; Q_2 = 130; Q_3 = 175; maximum = 190; range = 88; IQR = 50

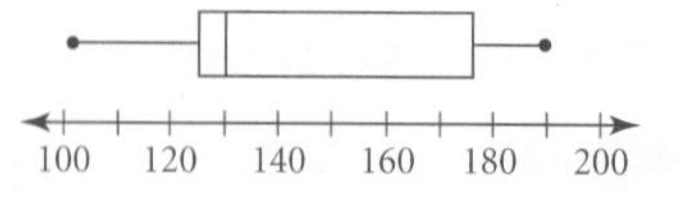

10. minimum = 525; Q_1 = 545; Q_2 = 577.5; Q_3 = 585; maximum = 595; range = 70; IQR = 40

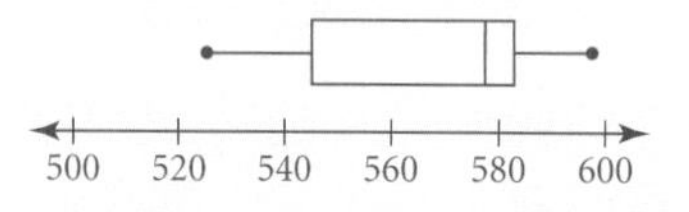

11. minimum = 14; Q_1 = 22; Q_2 = 50; Q_3 = 82; maximum = 93; range = 79; IQR = 60

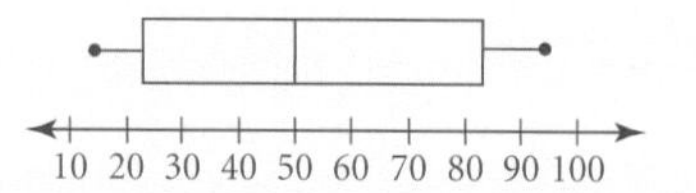

12. minimum = 11; Q_1 = 12.5; Q_2 = 21; Q_3 = 84.5; maximum = 98; range = 87; IQR = 72

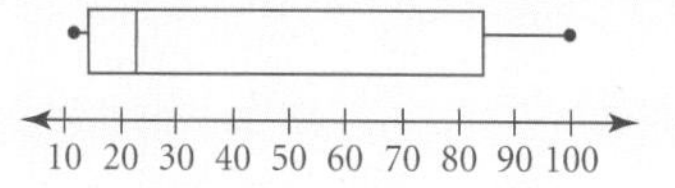

Practice and Apply

internet connect
Homework Help Online
Go To: go.hrw.com
Keyword:
MB1 Homework Help
for Exercises 7–21

Find the minimum and maximum values, quartiles, range, and interquartile range for each data set. Then make a box-and-whisker plot for each data set.

7. 42, 45, 56, 48, 59, 60, 51, 54, 44, 51, 50, 44, 42, 49, 56

8. 2, 16, 4, 11, 14, 8, 17, 19, 13, 19, 9, 15, 8, 13, 17

9. 102, 120, 154, 130, 130, 180, 190, 175, 125, 130

10. 525, 575, 580, 585, 590, 530, 545, 569, 595, 580

11. 22, 50, 78, 22, 77, 93, 27, 86, 14

12. 12, 73, 11, 96, 45, 21, 16, 98, 13

13. 3.2, 4.8, 7.8, 2.2, 7.7, 2.3, 2.7, 8.8, 4.8, 6.5

14. 6.2, 5.1, 4.5, 3.2, 3.5, 5.2, 3.2, 4.8, 8.7, 5.3

Three classes took the same test. Use the box-and-whisker plots below for the scores of each class to answer each question.

Class 2 **15.** Which class had the highest score?

Class 2 **16.** Which class had the greatest range?

Class 3 **17.** Which class had the highest median?

Classes 1 and 2 **18.** Which class had the highest Q_1?

Class 3 **19.** Which class had the greatest IQR?

Class 2 **20.** In which class do the middle half of the scores have the smallest range?

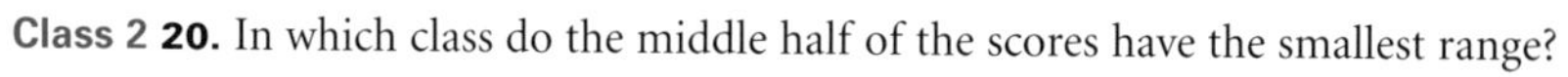

Class 1
Class 2
Class 3
50 60 70 80 90 100

75% **21.** What percent of scores are greater than Q_1 for each class?

CHALLENGE
22. Answers may vary. Sample answer: Students get an idea of how variable the scores of the middle 50% of the class are.

22. Colleges report the SAT I scores of the entering freshman class. Instead of a median score, many schools give a range of scores to represent the middle 50% of the class, such as a range from 520 to 610 for math scores. Explain what this range describes as well as the advantages of using it rather than a median score.

CONNECTION
23. Answers may vary.

23. PATTERNS IN DATA Research the mean monthly temperatures for your city. Find the quartiles and make a box-and-whisker plot. How does the plot for your city compare with those for Chicago and Los Angeles given at the beginning of the lesson?

APPLICATION

A woman in Japan seeds oysters for pearls.

24. DEMOGRAPHICS The table at right gives the percent of women in the labor force of eight major countries for the years 1980 and 1992. [*Source: Information Please: Almanac, Atlas, and Yearbook, 1996*]

a. Find Q_1, Q_2, and Q_3 for both years.

b. Make box-and-whisker plots for the data from 1980 and from 1992.

c. Compare the two box-and-whisker plots.

Females as a Percent of the Total Labor Force

Country	1980	1992
Australia	36.4	42.1
Canada	39.7	45.5
France	39.5	43.8
Germany	38.0	42.0
Japan	38.4	40.5
Sweden	45.2	48.3
United Kingdom	40.4	44.9
United States	42.4	45.7

13. minimum = 2.2; Q_1 = 2.7; Q_2 = 4.8; Q_3 = 7.7; maximum = 8.8; range = 6.6; IQR = 5.0

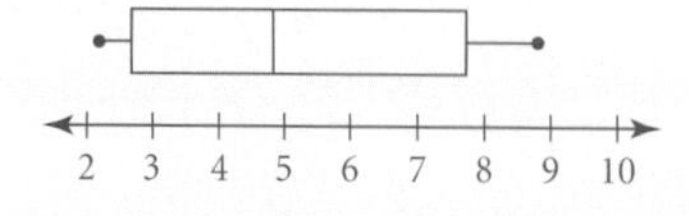

14. minimum = 3.2; Q_1 = 3.5; Q_2 = 4.95; Q_3 = 5.3; maximum = 8.7; range = 5.5; IQR = 1.8

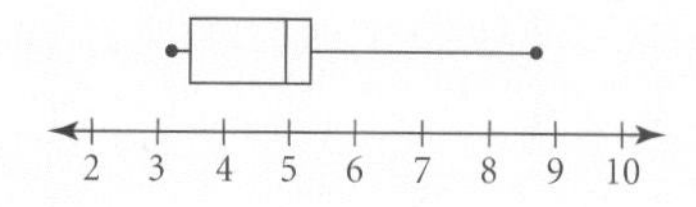

24a. 1992: Q_1 = 42.05, Q_2 = 44.35, Q_3 = 45.6
1980: Q_1 = 38.2, Q_2 = 39.6, Q_3 = 41.4

b.

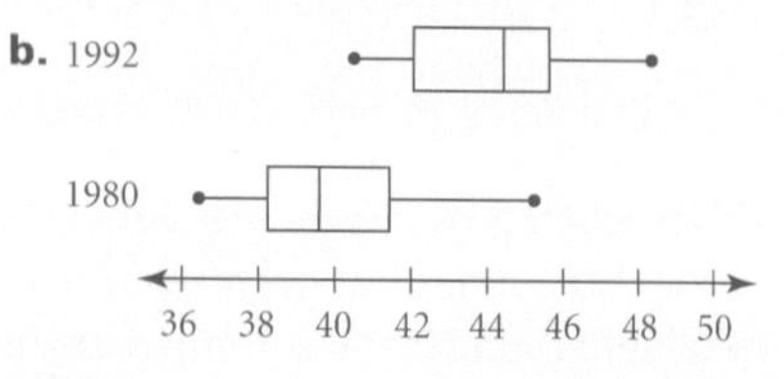

c. The percent of females in the total labor force has increased from 1980 to 1992

APPLICATIONS

DEMOGRAPHICS The table below gives the life expectancy at birth for selected countries in North, Central, and South America for 1994.

Life Expectancy at Birth (years)

Country	Both genders	Male	Female
Brazil	62	57	67
Canada	78	75	82
Chile	75	72	78
Costa Rica	78	76	80
Ecuador	70	67	73
Guatemala	64	62	67
Mexico	73	69	77
Panama	75	72	78
Peru	66	63	68
Trinidad and Tobago	71	68	73
United States	76	73	79
Uruguay	74	71	77
Venezuela	73	70	76

[*Source: Information Please: Almanac, Atlas, and Yearbook, 1996*]

25a. min = 62; max = 78; $Q_1 = 68$; $Q_2 = 73$; $Q_3 = 75.5$

26a. male: min = 57; max = 76; $Q_1 = 65$; $Q_2 = 70$; $Q_3 = 72.5$

female: min = 67; max = 82; $Q_1 = 70.5$; $Q_2 = 77$; $Q_3 = 78.5$

25. a. Find the minimum and maximum values, Q_1, Q_2, and Q_3 for both genders.
b. Make a box-and-whisker plot of the life expectancy for both genders.

26. a. Find the minimum and maximum values, Q_1, Q_2, and Q_3 for males and for females.
b. Using the same number line as in Exercise 25, make a box-and-whisker plot of the life expectancy for males and for females. Then compare them.

27 **GEOGRAPHY** The table below gives the area for each of the 50 states.

Area of States in Square Miles

AL	52,423	HI	10,932	ME	35,387	NJ	8722	SD	77,121
AK	656,424	IA	56,276	MI	96,705	NM	121,598	TN	42,146
AR	53,182	ID	83,574	MN	86,943	NV	110,567	TX	268,601
AZ	114,006	IL	57,918	MO	69,709	NY	54,471	UT	84,904
CA	163,707	IN	36,420	MS	48,434	OH	44,828	VA	42,777
CO	104,100	KS	82,282	MT	147,046	OK	69,903	VT	9615
CT	5544	KY	40,411	NC	53,821	OR	98,386	WA	71,302
DE	2489	LA	51,843	ND	70,704	PA	46,058	WI	65,499
FL	65,756	MA	10,555	NE	77,358	RI	1545	WV	24,231
GA	59,441	MD	12,407	NH	9351	SC	32,008	WY	97,818

[*Source: The World Almanac and Book of Facts, 1996*]

27a. min = 1545; max = 656,424; $Q_1 = 36{,}420$; $Q_2 = 57{,}097$; $Q_3 = 84{,}904$

c. AK, CA, TX

a. Find the minimum and maximum values, Q_1, Q_2, and Q_3 for this data.
b. Make a box-and-whisker plot with the values from part **a**.
c. Identify any possible outliers by naming the corresponding states.

26b.

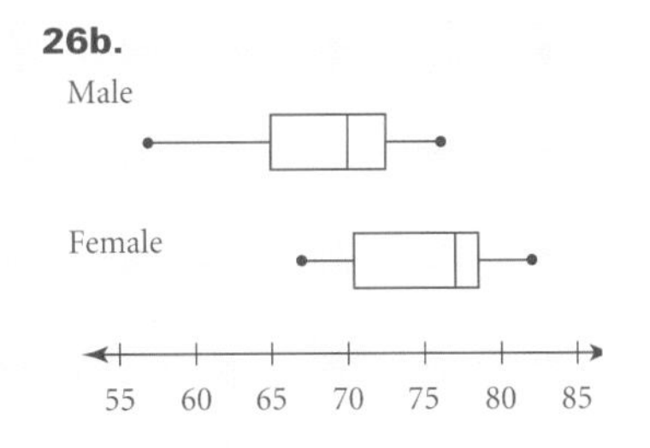

Each of the females' numbers in the 5-number summary is larger than corresponding numbers for males. Women live longer.

27b.

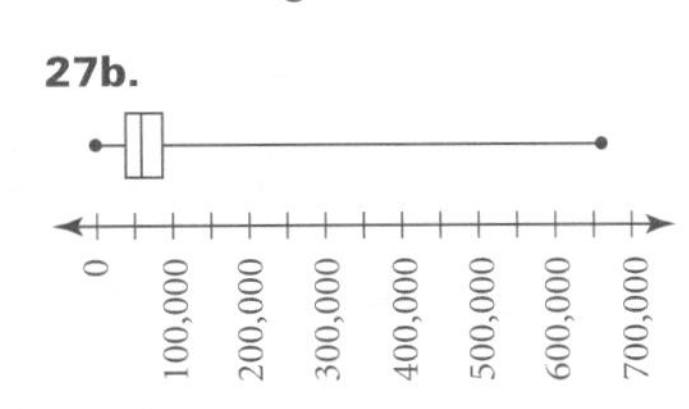

24c. The percent of females in the total labor force increased from 1980 to 1992. Each value in 1992 is about 4% higher than its corresponding value in 1980.

25b.

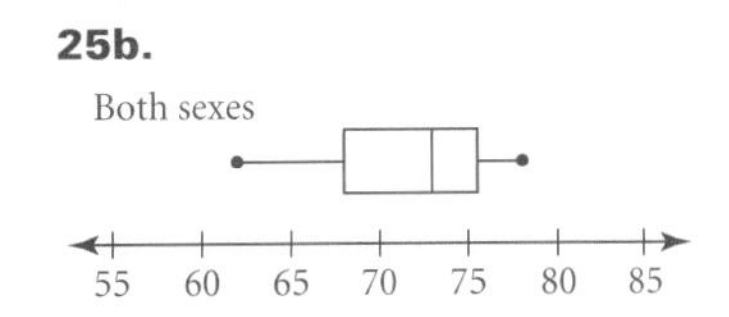

28b.

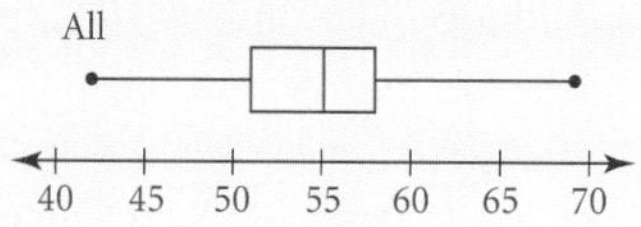

29a.

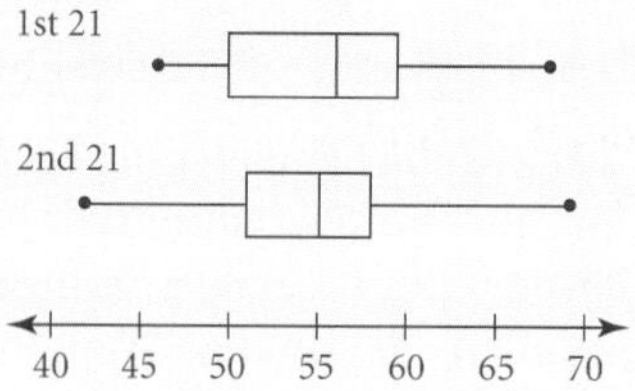

APPLICATION

GOVERNMENT The table below lists the presidents and their ages at inauguration.

John F. Kennedy was 43 years old at inauguration.

Age of the Presidents of the United States at Inauguration

Name	Age	Name	Age
1. Washington	57	**22.** Cleveland	47
2. J. Adams	61	**23.** B. Harrison	55
3. Jefferson	57	**24.** Cleveland	55
4. Madison	57	**25.** McKinley	54
5. Monroe	58	**26.** T. Roosevelt	42
6. J. Q. Adams	57	**27.** Taft	51
7. Jackson	61	**28.** Wilson	56
8. Van Buren	54	**29.** Harding	55
9. W. H. Harrison	68	**30.** Coolidge	51
10. Tyler	51	**31.** Hoover	54
11. Polk	49	**32.** F. D. Roosevelt	51
12. Taylor	64	**33.** Truman	60
13. Filmore	50	**34.** Eisenhower	62
14. Pierce	48	**35.** Kennedy	43
15. Buchanan	65	**36.** L. B. Johnson	55
16. Lincoln	52	**37.** Nixon	56
17. A. Johnson	56	**38.** Ford	61
18. Grant	46	**39.** Carter	52
19. Hayes	54	**40.** Reagan	69
20. Garfield	49	**41.** Bush	64
21. Arthur	50	**42.** Clinton	46

[*Source: Information Please: Almanac, Atlas, and Yearbook, 1996*]

28a. min = 42; max = 69; $Q_1 = 51$; $Q_2 = 55$; $Q_3 = 58$; Reagan

28. **a.** Find the minimum and maximum values, Q_1, Q_2, and Q_3 for the ages of the presidents at inauguration. Are there any possible outliers?
b. Make a box-and-whisker plot for ages of the presidents at inauguration.

29. **a.** Using the same number line, make a box-and-whisker plot for the first 21 presidents and another box-and-whisker plot for the last 21 presidents.
b. Compare the box-and-whisker plots. What do they indicate?
The distribution of ages is fairly even. The perception of how old a president should be has not changed much.

Look Back

Tell whether each function represents exponential growth or decay. ***(LESSON 6.2)***

30. $f(x) = 0.7^x$ decay **31.** $f(x) = 0.7^{-x}$ growth **32.** $f(x) = 7^x$ growth **33.** $f(x) = 7^{-x}$ decay

34. Write an exponential function that models 8% annual growth with a y-value of 1000 at $x = 0$. ***(LESSON 6.2)*** **$f(x) = 1000(0.08)^x$**

Describe the end behavior of each function. ***(LESSON 7.2)***

35. $P(x) = x^3 + 2x^2 - x + 1$ **36.** $P(x) = -2x^3 - 5x + 4$

35. falls to the left and rises to the right

36. rises to the left and falls to the right

Practice

NAME ______ CLASS ______ DATE ______

Practice
12.3 ***Box-and-Whisker Plots***

Find the quartiles, the range, and the interquartile range for each data set.

1. 9, 5, 2, 8, 2, 8, 7, 8, 3, 2, 8, 1, 9, 1, 5, 8, 9, 7, 9, 6
$Q_1 = 2$, $Q_2 = 7$, $Q_3 = 8$; range = 8; $IQR = 6$

2. 13, 14, 15, 19, 16, 19, 8, 17, 10, 7, 5, 18, 10, 16, 17, 12
$Q_1 = 10$, $Q_2 = 14.5$, $Q_3 = 17$; range = 14; $IQR = 7$

3. 35.1, 40.3, 13.8, 15.3, 42.7, 40.8, 15.5, 38.5, 28.4, 11.0, 11.7, 12.1, 23.9, 8.9, 24.0
$Q_1 = 12.1$, $Q_2 = 23.9$, $Q_3 = 38.5$; range = 33.8; $IQR = 26.4$

Find the minimum and maximum values, quartiles, range, and interquartile range for each data set. Then make a box-and-whisker plot for each data set.

4. 17, 14, 5, 8, 15, 4, 11, 6, 13, 17, 17, 13, 7, 9, 5, 3
minimum = 3, maximum = 17, $Q_1 = 5$,
$Q_2 = 10$, $Q_3 = 15$; range = 14; $IQR = 10$

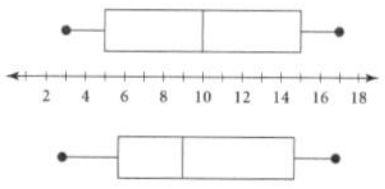

5. 35, 35, 23, 20, 29, 13, 26, 21, 39, 22, 14, 35, 10, 16, 36
minimum = 10, maximum = 39, $Q_1 = 16$,
$Q_2 = 23$, $Q_3 = 35$; range = 29; $IQR = 19$

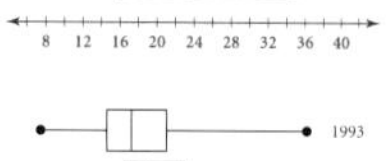

The box-and-whisker plots at right compare birth rates (per 1000 population) for the fifty states for 1992 and 1993.

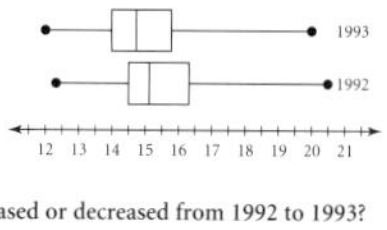

6. Which set of data has the greater median? the data for 1992

7. Does it appear that, in general, the birth rates increased or decreased from 1992 to 1993?
The quartiles and the extremes all decreased, so it appears that the birth rates decreased in general.

8. What percent of the data are less than Q_3 for each year? ≈75%

42. $C(1, -5)$; vertices: $(1, -17)$ and $(1, 7)$; co-vertices: $(-1, -5)$ and $(3, -5)$

43. $C(-2, 0)$; vertices: $(-7, 0)$ and $(3, 0)$; co-vertices: $\left(-2, -\frac{5}{3}\right)$ and $\left(-2, \frac{5}{3}\right)$

APPLICATION

44. 4096 different keys

45. 210 ways

46. 8, 9.5, 11

47. 10.368, 12.4416, 14.92992

Find the constant of variation, and write the equation for the relationship. ***(LESSON 8.1)***

37. y varies directly as x, and $y = 12$ when $x = 3$. $k = 4$; $y = 4x$

38. y varies inversely as x, and $y = 20$ when $x = 3$. $k = 60$; $y = \frac{60}{x}$

39. y varies as the square of x, and $y = 2$ when $x = 3$. $k = \frac{2}{9}$; $y = \frac{2}{9}x^2$

Write the standard equation for each parabola with the given characteristics. ***(LESSON 9.2)***

40. vertex: (0, 0); focus: (3, 0) $x = \frac{1}{12}y^2$ **41.** directrix: $x = 4$; focus: $(-1, 2)$ $x - \frac{3}{2} = -\frac{1}{10}(y - 2)^2$

Write the center, vertices, and co-vertices of each ellipse. ***(LESSON 9.4)***

42. $\frac{(x-1)^2}{4} + \frac{(y+5)^2}{144} = 1$ **43.** $x^2 + 4x + 9y^2 - 21 = 0$

44. SECURITY A certain type of blank key has 6 possible notches, each of which can be cut to 4 different heights (including no cut at all). How many different keys can be made from this type of blank key? ***(LESSON 10.1)***

45. In a group of 7 balloons, 2 are red, 3 are blue, and the rest are white. In how many distinct ways can the balloons be arranged in a row? ***(LESSON 10.2)***

46. Find the next three terms of the arithmetic sequence 2, 3.5, 5, 6.5, . . . ***(LESSON 11.2)***

47. Find the next three terms of the geometric sequence 5, 6, 7.2, 8.64, . . . ***(LESSON 11.4)***

Lewis	Adams
15	20
25	20
30	18
10	22
20	20

48. The table at left shows the points scored by two basketball players in the first five games of the season.

a. Calculate each player's mean points per game.

b. Which player is more consistent, based on this data? Justify your response. Did the player's mean score from part **a** help you to decide which player is more consistent? Why or why not?

48a. Lewis: $\bar{x} = 20$; Adams: $\bar{x} = 20$

b. Adams scores cluster around 20; no; both means were equal.

Obtain the weekly television ratings from the A. C. Nielsen Company. Explain what the ratings represent. Then make a box-and-whisker plot of the data and answer the questions below.

1. What percent of the ratings are below the median, below the lower quartile, above the upper quartile, in the box, and in the whiskers?

2. Is one whisker longer than the other? If so, what does this indicate? If not, what does this indicate?

3. Is the median centered in the box? If so, what does this indicate? If not, what does this indicate?

WORKING ON THE CHAPTER PROJECT

You should now be able to complete Activity 2 of the Chapter Project.

Exercise 48 encourages students to notice the extent to which data values are dispersed around the mean. Measures of dispersion, such as range, variance, and standard deviation, will be studied in depth in Lesson 12.4.

ALTERNATIVE
Assessment

Portfolio Activity

The Portfolio Activity can be used as preparation for the Chapter Project or as a separate activity. In the Portfolio Activity on this page, students collect data and display the data in a box-and-whisker plot.

Answers to Portfolio Activities can be found in Additional Answers of the Teacher's Edition.

Student Technology Guide

NAME ______ CLASS ______ DATE ______

Student Technology Guide
12.3 Box-and-Whisker Plots

Viewing the box-and-whisker plots for two data sets is a good way to see differences in how the data is distributed. You can use a graphics calculator to make side-by-side box-and-whisker plots.

Example: The table below shows the median annual earnings, in thousands of dollars, of workers in ten different occupational categories. Construct one box-and-whisker plot for men's earnings and another for women's earnings. Compare the plots.

Men	42.7	45.1	35.0	32.3	26.7	27.7	23.4	26.5	17.6	20.9
Women	28.9	31.9	26.3	18.7	20.7	21.4	15.4	19.7	14.8	13.1

- Press STAT 1:Edit... ENTER. Enter the earnings for men into L1. Press ENTER or ▼ to enter each number.
- Press ► to move to L2. Enter the earnings for women. Press 2nd MODE.
- Press 2nd Y= 1:Plot1... ENTER.
- The cursor is over On. Press ENTER to turn the plot on.
- Press ▼, then use ► to move to the picture of a box-and-whisker plot. Press ENTER. Be sure Xlist is set to L1 and Freq is 1.
- Use ▲ to move to the top of the screen. Press ► to move to Plot2, then press ENTER.
- Turn Plot2 on and select a box-and-whisker plot. Move to Xlist and press 2nd 2 to select L2. Press 2nd MODE.
- Press ZOOM 9:ZoomStat ENTER to view all data values.

Plot1 Plot2 Plot3
On Off
Type:
Xlist:L1
Freq:1

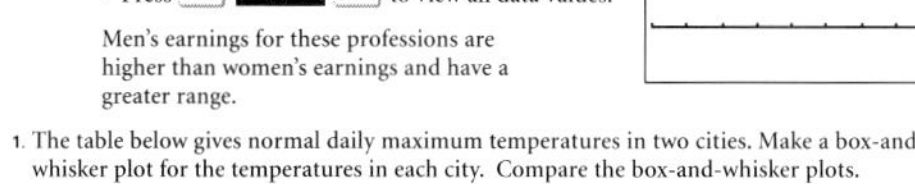

Men's earnings for these professions are higher than women's earnings and have a greater range.

1. The table below gives normal daily maximum temperatures in two cities. Make a box-and-whisker plot for the temperatures in each city. Compare the box-and-whisker plots.

	Jan.	Feb.	Mar.	Apr.	May	June	July	Aug.	Sept.	Oct.	Nov.	Dec.
Bismarck, ND	20.2	26.4	38.5	54.9	67.8	77.1	84.4	82.7	70.8	58.7	39.3	24.5
San Francisco, CA	55.6	59.4	60.8	63.9	66.5	70.3	71.6	72.3	73.6	70.1	62.4	56.1

Comparisons may vary. The box-and-whisker plot for Bismarck is much longer than that for San Francisco, showing that Bismarck's temperatures are more extreme than San Francisco's.

Focus

Students read an article about the value and uses of true random-number sequences for a wide range of scientific and mathematical applications. The article also describes the difficulties and lack of success in artificially generating truly random numbers by using electronic or mechanical devices. Students will use a geometric model in an experiment to see how closely they can approximate a truly random sequence of numbers.

Motivate

Have students assume that they are picking a random sequence of six numbers trying to win a lottery. Ask them what it means to pick a random sequence of numbers and how they would go about choosing numbers that were random. Next ask the students how they think the winning numbers in a lottery are selected. Do they think that this method ensures fairness?

Use Teaching Transparency 51.

EYEWITNESS MATH

Is it RANDOM?

The Quest for True Randomness Finally Appears Successful

by James Gleick

One of the strangest quests of modern computer science seems to be reaching its goal; mathematicians believe they have found a process for making perfectly random strings of numbers.

Sequences of truly patternless, truly unpredictable digits have become a perversely valuable commodity, in demand for a wide variety applications in science and industry. Randomness is a tool for insuring fairness in statistical studies or jury selection, for designing safe cryptographic schemes and for helping scientists simulate complex behavior.

Yet random numbers—as unbiased and disorganized as the result of millions of imaginary coin tosses—have long proved extremely hard to make, either with electronic computers or mechanical devices. Consumers of randomness have had to settle for numbers that fall short, always hiding some subtle pattern.

Random number generators are sold for every kind of computer. Every generator now in use has some kind of flaw, though often the flaw can be hard to detect. Furthermore, in a way, the idea of using a predictable electronic machine to create true randomness is nonsense. No string of numbers is really random if it can be produced by a simple computer process. But in a more practical sense, a string is random if there is no way to distinguish it from a string of coin flips.

Several theorists presented details of the apparent breakthrough in random-number generation. The technique will now be subjected to batteries of statistical tests, meant to see whether it performs as well as the theorists believe it will. The way people perceive randomness in the world around them differs sharply from the way mathematicians understand it and test for it.

The need for randomness in human institutions seems to begin at whatever age "eeny-meeny-miny-moe" becomes a practical decision-making procedure: randomness is meant to insure fairness. Like "eeny-meeny-miny-moe," most such procedures prove far from random. Even the most carefully designed mechanical randomness-makers break down under scrutiny.

One such failure, on a dramatic scale, struck the national draft lottery in 1969, its first year. Military officials wrote all the possible birthdays on 366 pieces of paper and put them into 366 capsules. Then they poured the January capsules into a box and mixed them. Then they added the February capsules and mixed again—and so on.

At a public ceremony, the capsules were drawn from the box by hand. Only later did statisticians establish that the procedure had been far from random; people born toward the end of the year had a far greater chance of being drafted than people born in the early months. In general, the problem of mixing or stirring or shuffling things to insure randomness is more complicated than most experts assume.

An expert in exposing the flaws in pseudorandom-number generators, George Marsaglia of Florida State University, has begun to test the new technique. Dr. Marsaglia judges sequences (of numbers) not just by uniformity—a good distribution of numbers in a sequence —but also by "independence." No number or string of numbers should change the probability of the number or numbers that follow, any more than flipping a coin and getting 10 straight tails changes the likelihood of getting heads on the 11th flip.

[*Source:* New York Times, *April 19, 1988*]

How hard can it be to write a bunch of random numbers? Do you think you can write a sequence of 0s and 1s that looks like it came from coin flips? Do you think you could fool a psychologist or mathematician?

To find out, start by writing down a sequence of 120 zeros and ones in an order that you think looks random.

2. Answers may vary. Sample:
Sequence A is a random sequence.
Sequence B is a sequence made by guessing.
Sequence C is a fake sequence.

	Sequence A	Sequence B	Sequence C
(0, 0, 0)	5	2	3
(1, 1, 1)	6	2	4

3. The probability of all heads is $\frac{1}{8}$, and the probability of all tails is also $\frac{1}{8}$. The events are independent and the basic probability of heads is $\frac{1}{2}$, so $\frac{1}{2} \cdot \frac{1}{2} \cdot \frac{1}{2} = \frac{1}{8}$. The same is true for tails.

Before you put your made-up sequence to the test, you need to find out how to check for randomness. One way is by breaking the sequence into smaller sequences, such as sets of 3 digits. Then you can see whether some sets occur too often or not enough.

You can visualize these sets of digits by plotting them as points. Think of the first 3 digits in your sequence as an *ordered triple*, the *x*, *y*, and *z* coordinates of a point in space. Using the digits 0 and 1 to model heads and tails, respectively, there are 8 possible ordered triples, as shown.

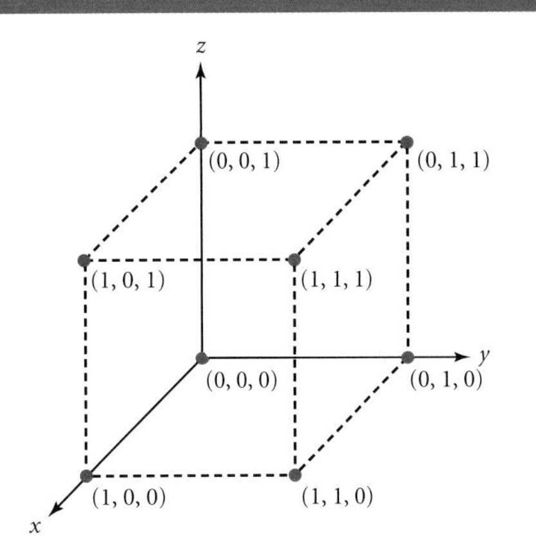

Cooperative Learning

Now you can see whether your classmates are able to distinguish your made-up sequence from an actual sequence.

1. Generate a random sequence of 120 zeros and ones by tossing a coin (heads = 0, tails = 1) or by using a random-number table (even numbers = 0, odd numbers = 1.)

   Exchange this actual sequence and your made-up sequences with another group. (Be sure to mark the sequences so that only you know which one is made up.)

2. Test each sequence you are given by following these steps:
   a. Separate the sequence into 40 ordered triples, as shown here:
   b. Count the number of times the triples **000** and **111** appear. Record your results in a chart.

	Sequence A	Sequence B	Sequence C
000			
111			

3. If you flip a coin 3 times, what is the probability of getting 3 heads? 3 tails? Explain.
4. In each new sequence you test, about how many of the 40 triples would you expect to be 000? 111?
5. How might you use your answer to Step 4 to distinguish real random sequences from fake ones?
6. Use your answer to Step 5 to tell which sequence is the fake one. Then find out whether you are correct.
7. Do you think the test would work better if you used sequences of 1200 zeros and ones instead of 120? Why?
8. Write a definition for *random numbers*.

4. **Out of 40 ordered triples, there should be about 5.**

5. **In actual sequences, the triples 000 and 111 will probably appear about 5 times each. In sequences people make up, these triples are more likely to appear just once or twice each or even not at all.**

6. **Answers may vary.**

7. **The longer the actual sequences, the more likely the results will resemble expected results.**

Cooperative Learning

Form the class into teams of two students in order to perform the experiment. Each team will produce one "fake" sequence and one random sequence. They can decide to work together to create sequences as a team effort or work separately to create the sequences individually and then choose the sequence they want to use. Each team should submit one fake sequence and one random sequence to be tested by another team. Each team should also prepare a written response to the question in Step 7 and a written definition for *random numbers* as outlined in Step 8.

Discuss

Each team will give a presentation to the class. The presentation should give the team's conclusion as to which number sequence was fake and which was random. This should include a discussion of their answers to the questions posed in Steps 3, 4, 5, and 6. The team whose number sequences were tested should disclose which one was fake and which one was random. If the testing team's conclusions were incorrect, their reasoning should be discussed and clarified. Each team should conclude by answering the question in Step 7 and giving the definition of *random numbers*. As an extension of the experiment, have students do library research on real-world applications of random numbers.

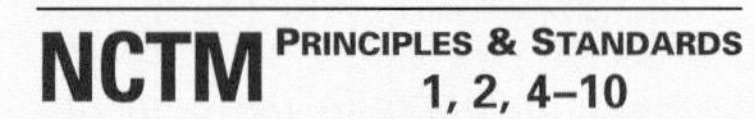

Prepare

NCTM PRINCIPLES & STANDARDS 1, 2, 4–10

QUICK WARM-UP

Find the indicated statistical measure for the data below. Round to the nearest tenth when necessary.
45, 21, 39, 52, 37, 44, 40, 36, 37, 43, 51, 45

1. mean ≈**40.8**
2. median **41.5**
3. mode(s) **37 and 45**
4. range **31**
5. Q_1 **37**
6. Q_2 **41.5**
7. Q_3 **45**
8. interquartile range **8**

Also on Quiz Transparency 12.4

Teach

Why The data involved in most situations are distributed in varying patterns that are centered around some central number. Measures of dispersion describe how a data set is dispersed and are invaluable in evaluating the data. Give students two sets of five test scores: 78, 80, 80, 80, 82 and 100, 80, 90, 70, 60. Ask them to find the mean of each set of data. **80** Then ask them to discuss which set of scores indicates a more consistent performance. **the first set**

Measures of Dispersion

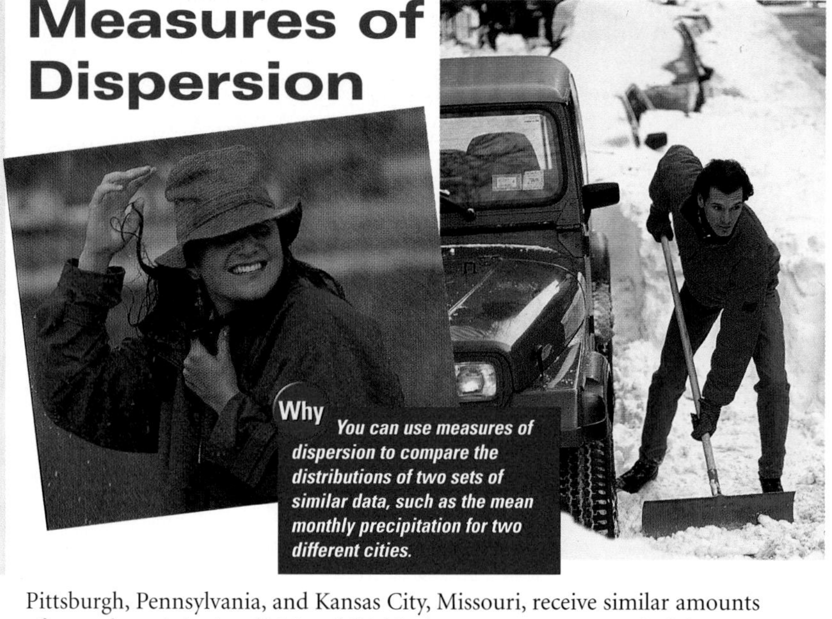

Why You can use measures of dispersion to compare the distributions of two sets of similar data, such as the mean monthly precipitation for two different cities.

Objective
- Calculate and use measures of dispersion, such as range, mean deviation, variance, and standard deviation.

APPLICATION
CLIMATE

Pittsburgh, Pennsylvania, and Kansas City, Missouri, receive similar amounts of annual precipitation (36.5 and 37.6 inches on average, respectively). However, how the precipitation is spread out, or dispersed, over the year is dramatically different. This is illustrated by the box-and-whisker plots below.

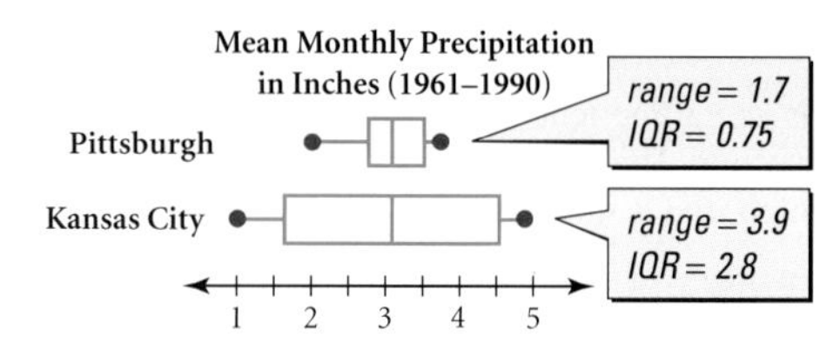

The range and the interquartile range are *measures of dispersion* because they indicate how the monthly precipitation values are *spread* out. **Measures of dispersion** indicate the extent to which values are spread around a central value such as the mean.

As a measure of dispersion, the range is not very reliable because it depends on only two data values, the maximum and the minimum. Likewise, the interquartile range depends on only two values, the first and third quartiles.

Another measure of dispersion, which depends on each data value, is the *mean deviation.* The mean deviation gives the average (mean) amount that the values in a data set differ from the mean.

Mean Deviation

The **mean deviation** of $x_1, x_2, \ldots, x_n$ is the mean of the absolute values of the differences between the data values and the mean, $\overline{x}$.

$$\text{mean deviation} = \frac{1}{n}\sum_{i=1}^{n}|x_i - \overline{x}|$$

Alternative Teaching Strategy

USING ALGORITHMS Show students how to organize their work by giving them the following step-by-step procedure for calculating the variance and standard deviation of a data set:

1. Write the data values in a column.
2. Calculate the mean of the data values.
3. Find the deviation of each value from the mean. Record the result in a second column.
4. Record the square of each deviation in the third column, and add this column to find the variance.
5. Compute the square root of the variance to find the standard deviation.

If available, an electronic spreadsheet is very useful for listing data in columns and for calculating deviations from the mean, variance, and standard deviation. In many spreadsheets, the **VAR** function yields the variance of a population based on the sample, and the **VARP** function yields the variance when the data is viewed as the entire population. Similarly, **STDEV** yields the standard deviation of a population based on the given data sample, while **STDEVP** yields the standard deviation of the data.

EXAMPLE 1 The number of miles, in thousands, obtained in five tests of two different tires is listed in the table below.

APPLICATION
MANUFACTURING

Tire A	66	43	37	50	54
Tire B	54	49	47	48	52

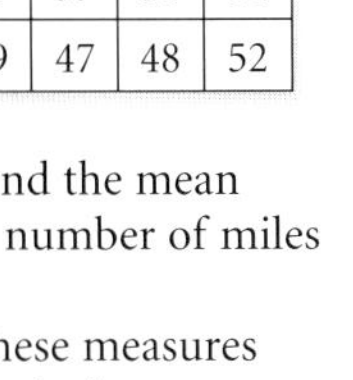

a. Find the range and the mean deviation of the number of miles for each tire.

b. Describe what these measures indicate about each tire.

SOLUTION

a. Tire A

range: $66 - 37 = 29$
The range is 29,000 miles.

To find the mean deviation, first find the mean.

$\overline{x} = \frac{66 + 43 + 37 + 50 + 54}{5} = \frac{250}{5} = 50$

PROBLEM SOLVING

Make a table to find the absolute values of the differences from the mean. Then find and the sum of the absolute values.

x_i	$\lvert x_i - \overline{x} \rvert$
66	16
43	7
37	13
50	0
54	4
Total	40

Divide the total by n.

$n = \frac{40}{5} = 8$

The mean deviation is 8000 miles.

Tire B

range: $54 - 47 = 7$
The range is 7000 miles.

To find the mean deviation, first find the mean.

$\overline{x} = \frac{54 + 49 + 47 + 48 + 52}{5} = \frac{250}{5} = 50$

Make a table to find the absolute values of the differences from the mean. Then find and the sum of the absolute values.

x_i	$\lvert x_i - \overline{x} \rvert$
54	4
49	1
47	3
48	2
52	2
Total	12

Divide the total by n.

$n = \frac{12}{5} = 2.4$

The mean deviation is 2400 miles.

b. Because tire B has a smaller mean deviation than tire A ($2400 < 8000$), the individual values for tire B deviate less from the mean. This indicates that the mean for tire B is a more reliable measure of its central tendency. Thus, the expected mileage for tire B is more predictable.

TRY THIS Find the range and the mean deviation for the data on tire C below. Then compare these measures with those for tires A and B.

Tire C	64	52	50	49	35

CHECKPOINT ✔ Can two data sets have the same range and different mean deviations? Justify your answer with a sample set of data.

ADDITIONAL EXAMPLE 1

The table below lists a student's test scores in two subjects this year.

History	Math
85	82
91	92
96	100
85	77
93	84

a. Find the range and the mean deviation for the test scores.
history: range = 11, mean deviation = 4; math: range = 23, mean deviation = 7.2

b. Describe what these measures indicate about each set of test scores.
The history test scores have a smaller range and a smaller mean deviation. This means that the student performs with greater consistency on the history tests.

TRY THIS

Range: 29,000 mi; mean deviation: 6400 mi; the range for tire C is the same as the range for tire A and is larger than the range for tire B; the mean deviation for tire C is larger than the mean deviation for tire B and smaller than the mean deviation for tire A.

CHECKPOINT ✔

Yes; Answers may vary. Sample answer: The data sets below both have ranges of 29. The mean deviation of *A* is 8, while the mean deviation of *B* is 6.4.

A	66	43	37	50	54
B	64	52	50	49	35

Interdisciplinary Connection

HEALTH The table below lists the ages of 12 rubella patients and the duration (in days) that they were ill.

No.	Age	Days	No.	Age	Days	No.	Age	Days
1	5	4	5	6	4	9	5	3
2	10	5	6	5	3	10	4	8
3	9	4	7	6	5	11	4	5
4	7	4	8	8	1	12	7	7

Have students find the mean, range, variance, and standard deviation for the ages of the patients and for the duration of their illness. Have students describe what these measures indicate about the ages of the patients and the number of days they were ill.

age: mean = 6.3, range = 6, $\sigma^2 = 3.39$, $\sigma = 1.84$; days: mean = 4.4, range = 7, $\sigma^2 = 3.08$, $\sigma = 1.75$; the mean age of an ill child was 6.3, the mean number of days the children were ill was 4.4, and the data indicate that the number of days the patients were ill varies less from its mean than the age of the patients varies from its mean.

ADDITIONAL
EXAMPLE 2

Find the standard deviation for each set of test scores given in Additional Example 1.
history: variance = 19.2, standard deviation ≈ 4.4; math: variance = 65.6, standard deviation ≈ 8.1

TRY THIS
$\sigma \approx 9230$ mi

CHECKPOINT ✔
Tire D has less variation (is more consistent) in mileage than either of the other two tires.

CRITICAL THINKING
Standard deviation; it has the same units as the original data.

Teaching Tip

TECHNOLOGY When using a calculator or spreadsheet with statistical features to calculate the standard deviation of a set of data that represents the entire population, make sure that the calculator or program uses the formula to compute the proper standard deviation. On a TI-83 or TI-82 graphics calculator, enter the data into **L1**, press STAT, select CALC 1:1-Var Stats, and press ENTER.

The display will show both **Sx**, the standard deviation of the population estimated from the data sample, and **σx**, the population standard deviation. The formula given in the book is shown as **σx**.

Note that the **stdDev** feature, which is listed under the **MATH** submenu in the **LIST** menu, produces **Sx**, not **σx**.

The formulas for the different types of standard deviations are as follows:

$$\mathbf{Sx} = \frac{1}{n-1}\sqrt{\sum(x-\bar{x})^2}$$

$$\boldsymbol{\sigma}\mathbf{x} = \frac{1}{n}\sqrt{\sum(x-\bar{x})^2}$$

The variance and standard deviation are two more measures of dispersion that are commonly used in comparing and analyzing data.

Variance and Standard Deviation

If a data set has n data values $x_1, x_2, \ldots, x_n$ and mean $\bar{x}$, then the **variance** and **standard deviation** of the data are defined as follows:

$$\text{Variance: } \sigma^2 = \frac{1}{n}\sum_{i=1}^{n}(x_i - \bar{x})^2$$

$$\text{Standard deviation: } \sigma = \sqrt{\sigma^2}$$

EXAMPLE 2 **Find the standard deviation for the tire data in Example 1.**

APPLICATION
MANUFACTURING

SOLUTION

PROBLEM SOLVING

Tire A

$$\bar{x} = \frac{66 + 43 + 37 + 50 + 54}{5} = 50$$

Make a table to organize your calculations.

x_i	$x_i - \bar{x}$	$(x_i - \bar{x})^2$
66	16	256
43	−7	49
37	−13	169
50	0	0
54	4	16
Total	0	490

variance: $\sigma^2 = \frac{490}{5} = 98$
standard deviation: $\sigma = \sqrt{98} \approx 9.9$

The standard deviation is 9900 miles.

Tire B

$$\bar{x} = \frac{54 + 49 + 47 + 48 + 52}{5} = 50$$

Make a table to organize your calculations.

x_i	$x_i - \bar{x}$	$(x_i - \bar{x})^2$
54	4	16
49	−1	1
47	−3	9
48	−2	4
52	2	4
Total	0	34

variance: $\sigma^2 = \frac{34}{5} = 6.8$
standard deviation: $\sigma = \sqrt{6.8} \approx 2.6$

The standard deviation is 2600 miles.

As expected from the results in Example 1, tire B has a lower standard deviation than tire A, which indicates a greater consistency in its individual test scores.

TRY THIS Find the standard deviation for the data on tire C given in the Try This after Example 1.

CHECKPOINT ✔ Suppose that you are given a standard deviation of 1500 miles for tire D. What does this tell you about tire D relative to tires A and B in Example 2?

CRITICAL THINKING Which measure do you think is used most, the variance or the standard deviation? Why?

Enrichment

Give students the following alternative formula for variance:

$$\sigma^2 = \frac{1}{n}\sum_{i=1}^{n} x_i^2 - \left(\frac{1}{n}\sum_{i=1}^{n} x_i\right)^2$$

Show students how to use this alternative variance formula to find the variance and standard deviation of the data set {3, 5, 18, and 13}. Then ask them to compare this new formula and the original formula. Students should see that the new formula sometimes simplifies computations done by hand or in a spreadsheet. **≈ 3.8**

Inclusion Strategies

USING MANIPULATIVES Have students draw a number line and label each one-inch interval from 1 to 7. Give each student 5 x-tiles. Have them place all 5 x-tiles on the number 4 to model the data set {4, 4, 4, 4, 4}. Tell them to record the data set and find the standard deviation. Now have them move 1 x-tile to another point on the number line. Ask them to write the new data set and find the standard deviation. Have them continue the process of removing 1 x-tile from the original pile until only 1 x-tile remains at 4, finding the standard deviation each time.

Investigating Standard Deviation

You will need: a calculator

1. Find the standard deviation for the following data: 0, 5, 10, 15, 20.
2. Add a constant to each data value in Step 1. Find the new standard deviation, and describe how it changed.
3. Multiply each data value in Step 1 by a positive constant. Find the new standard deviation, and describe how it changed.
4. Divide each data value in Step 1 by 2, and then subtract 5. Find the new standard deviation, and describe how it changed.

CHECKPOINT ✔ 5. Make a conjecture about what happens to the standard deviation when you add, subtract, multiply, or divide each data value by a constant.

CHECKPOINT ✔ 6. Suppose that a basketball player has a mean of 12 points per game with a standard deviation of 2 points. What is a comparable standard deviation for a basketball player who has a mean of 24 points per game? 30 points per game? Explain.

The precipitation data for the box-and-whisker plots from the beginning of the lesson are shown below.

Mean Monthly Precipitation in Inches (1961–1990)

	Jan.	Feb.	Mar.	Apr.	May	Jun.
Pittsburgh	2.0	2.2	2.9	3.1	3.5	3.7
Kansas City	1.1	1.1	2.5	3.1	5.0	4.7

	Jul.	Aug.	Sep.	Oct.	Nov.	Dec.
Pittsburgh	3.5	3.4	3.4	2.5	3.2	3.1
Kansas City	4.4	4.0	4.9	3.3	1.9	1.6

[*Source: U.S. NOAA*]

You can use a graphics calculator to find the standard deviation for the precipitation of each city by entering the data into lists. Then select the appropriate statistical feature.

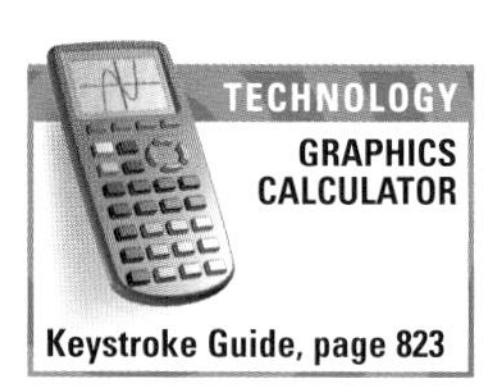

Pittsburgh, Pennsylvania

```
1-Var Stats
x̄=3.041666667
Σx=36.5
Σx²=114.27
Sx=.5434876152
σx=.5203497755
↓n=12
```

Kansas City, Missouri

```
1-Var Stats
x̄=3.133333333
Σx=37.6
Σx²=141.8
Sx=1.47668753
σx=1.413820671
↓n=12
```

standard deviation

The means are within 0.1 of an inch of each other, but the standard deviations are about 0.5 inches for Pittsburgh and 1.4 inches for Kansas City. Thus, Kansas City's monthly precipitation varies much more than Pittsburgh's monthly precipitation.

Activity Notes

Be sure students notice that if each value in a data set differs by a constant from its corresponding value in another set, then the standard deviations of the two sets will be equal. If each value in a data set is a constant multiple of its corresponding value in another set, the standard deviation of the first data set will be the absolute value of that constant times the standard deviation of the second data set.

Math CONNECTION

TRANSLATIONS Point out that when a data set is transformed by a translation, the mean, median, and mode of the original data set are changed, but the range, interquartile range, variance, and standard deviation are the same for both the original data set and the translated data set.

Cooperative Learning

Divide the class into two-member teams. Both members of each team will complete Steps 1, 2, 3, and 4 of the Activity independently. Working as a team, students should verify their calculations and write a conjecture about the patterns they noticed.

Reteaching the Lesson

To show students how standard deviation describes the dispersion of data about the mean, give them the two data sets below.

A: 100, 93, 81, 75, 40, 66, 61, 82, 71 88, 29
B: 75, 72, 105, 78, 76, 34, 78, 70, 70, 72, 56

Ask them to find the mean and range of each set. They should notice that both sets have the same mean, ≈71.45, and the same range, 71. Then have them draw a histogram for each set of data. Be sure they notice that the histograms show that the values for set A are farther from the mean than those for set B. Now ask them to find the standard deviation for both sets. **standard deviation of 20.67 for set A, standard deviation of 16.21 for set B** Point out that the data values of set A are farther from the mean than the data values for set B.

CHECKPOINT ✔

5. **No change when adding or subtracting a constant; the standard deviation is multiplied or divided by the same constant that multiplied or divided the original data**

CHECKPOINT ✔

6. **4 points; 5 points; the standard deviation is one-sixth of the mean.**

Assess

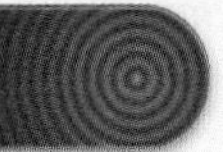

Selected Answers

Exercises 4, 5, 7–47 odd

ASSIGNMENT GUIDE	
In Class	1–5
Core	7–35 odd
Core Plus	6–34 even
Review	36–47
Preview	48

Extra Practice can be found beginning on page 940.

Mid-Chapter Assessment for Lessons 12.1 through 12.3 can be found on page 156 of the *Assessment Resources*.

22. mean deviation = 176.25; standard deviation ≈ 203.64; mean deviation is slightly less affected.

23. mean deviation = 191.25; standard deviation ≈ 220.95; mean deviation is slightly less affected.

Practice

NAME ______ CLASS ______ DATE ______

Practice
12.4 ***Measures of Dispersion***

Find the range and mean deviation for each data set.

1. 24, 20, 38, 36, 52 — 32; 9.6
2. 12, 11, 15, 18, 22, 30 — 19; $5.\overline{3}$
3. 71, 56, 88, 82, 40, 95 — 55; $16.\overline{3}$
4. 120, 142, 167, 188, 167, 200 — 80; 22
5. 5.8, 3.4, 7.2, 10.5, 8.6 — 7.1; 2
6. 38, 52, 40, 61, 53, 90, 100 — 62; ≈18.9

Find the variance and standard deviation for each data set.

7. 13, 13, 17, 11, 22, 20 — 16; 4
8. 82, 44, 67, 52, 120 — 721.6; ≈26.9
9. 1215, 1805, 1715, 2010, 1875 — 74,064; ≈272.1
10. 12, 14.5, 18, 16, 11.5, 15 — 5; ≈2.2
11. 30, 40.2, 40.8, 22.6, 18 — 83.7856; ≈9.2
12. 19.4, 19, 19.2, 19.6, 19.8, 19 — $0.0\overline{8}$; ≈0.3

The table shows the winning scores in the United States Women's Open Golf Championships from 1977 to 1996. Refer to the data in the table for Exercises 13–16.

292	289	284	280	279	283	290	290	280	287
285	277	278	284	283	280	280	277	278	272

13. Find the range. — 20
14. Find the mean deviation. — 4.3
15. Find the variance. — 26.24
16. Find the standard deviation. — about 5.1

Exercises

Communicate

1. Explain why the mean deviation and the standard deviation of a set of data are always nonnegative.
2. Describe the relationship between variance and standard deviation. Is the standard deviation always less than the variance? Explain.
3. Explain why the mean deviation and standard deviation are more reliable measures of dispersion than the range or interquartile range.

Guided Skills Practice

APPLICATION

4a. Tricia: range = 7; mean deviation = 2
Morgan: range = 40; mean deviation = 16.4

b. Tricia's scores are less variable (or more consistent) than Morgan's scores.

4. **EDUCATION** The table at right lists five test scores for two students. ***(EXAMPLE 1)***
   a. Find the range and the mean deviation of the scores for each student.
   b. Describe what these measures indicate about each student's test scores.

Tricia	Morgan
81	98
84	68
88	99
82	59
85	96

5. Find the standard deviation of the scores for each student in Exercise 4. ***(EXAMPLE 2)***
   Tricia: σ ≈ 2.45
   Morgan: σ ≈ 17.01

Practice and Apply

Find the range and mean deviation for each data set.

6. 8, 10, 3, 9, 10 **7; 2**
7. 1, 2, 4, 2, 6 **5; 1.6**
8. 31, 103, 34, 98, 107, 23 **84; ≈36.7**
9. 32, 23, 68, 74, 26, 93 **70; ≈25.7**
10. 13.2, 9.4, 7.3, 12.3, 8.6, 7.6
11. 11.1, 14.2, 8.4, 12.2, 15.2, 10.9
12. −1.22, 4.35, −2.42, 2.33, 4.66
13. 8.72, 7.43, −2.92, −3.56, 5.78

10. 5.9; ≈2.01
11. 6.8; ≈1.87
12. 7.08; ≈2.69
13. 12.28; ≈5.06

internet connect
Homework Help Online
Go To: go.hrw.com
Keyword: MB1 Homework Help for Exercises 14–24

Find the variance and standard deviation for each data set.

14. 9, 10, 10, 8, 7, 11, 12, 9
15. 12, 8, 13, 9, 13, 11, 12, 11, 9
16. 8.1, 10.3, 3.4, 9.8, 10.7
17. 19.2, 12.3, 4.8, 22.4, 26
18. 2.42, 7.46, 4.97, 4.22, 6.44
19. 1.34, 2.56, 4.78, 11.89, 8.92
20. −3, 2, −5, 4, −2, 8, 9, −1
21. 2, 4, −8, 8, 7, −2, −4, 3, 7

Find the mean deviation and standard deviation for each data set. Which measure of variation is less affected by an extreme value?

22. 20, 30, 40, 500
23. 0, 500, 510, 520

CHALLENGES

24. Create two data sets with the same range and the same interquartile range, but different standard deviations.
25. Can the standard deviation of a data set be 0? If so, under what conditions? Use sample data sets in your explanation.

14. 2.25; 1.5
15. ≈2.99; ≈1.73
16. ≈7.19; ≈2.68
17. ≈57.18; ≈7.56
18. ≈3.07; ≈1.75
19. ≈15.64; ≈3.95
20. 23.25; ≈4.82
21. ≈26.99; ≈5.19

24. Answers may vary. sample answer:
data set 1: 1, 2, 3, 4, 5, 6
range = 5, IQR = 3, σ ≈ 1.71
data set 2: 0, 2, 3.2, 4.5, 5, 5
range = 5, IQR = 3, σ ≈ 1.82

25. yes; if all the data values are the same

CONNECTION

26. σ is unchanged; new standard deviation is $c\sigma$.

APPLICATIONS

27. range = 5; mean dev. = 1.02

28. σ ≈ 1.29

31. Location 1: range = 2167; mean dev. ≈ 527.3
Location 2: range = 1815; mean dev. ≈ 442.3
Sales at location 1 are more variable.

32. Location 1: σ ≈ 633.02
Location 2: σ ≈ 540.39
Sales at location 1 are more variable.

26. **TRANSFORMATIONS** What happens to the standard deviation of a set of data if a constant, c, is added to each value in the data set? What happens to the standard deviation for a data set if each value is multiplied by the same constant, c?

SURVEYS In a survey, 30 people were asked to rank a new soda on a scale from 1 to 10. The results are shown in the table at right.

5	7	9	6	8	10
7	8	8	9	7	8
10	8	7	9	6	8
8	10	9	8	10	10
7	9	8	7	7	9

27. Find the range and the mean deviation of the rankings.

28. Find the standard deviation of the rankings.

MANUFACTURING The table below lists the diameters in millimeters of the ball bearings produced by a certain machine.

5.001	4.9998	4.999	5.002	4.999	5.001	5.002	4.998
4.999	5.000	5.001	4.999	5.000	4.998	4.999	5.003
4.998	4.999	5.001	5.002	5.001	4.999	4.997	5.001
5.001	5.002	5.001	4.998	4.999	5.001	5.002	4.997

29. Find the range and mean deviation. **range = 0.006 mm; mean dev. ≈ 0.0014 mm**

30. Find the standard deviation.
σ ≈ 0.0016 mm

BUSINESS The tables below list the number of customers for two fast-food locations.

Location 1

12,375	13,890	13,202
12,825	11,982	12,098
11,829	13,234	12,025
12,502	12,654	11,723

Location 2

13,245	13,543	12,983
12,825	12,925	11,924
12,645	11,982	11,728
12,987	13,125	12,887

31. Find the range and the mean deviation of the data for each location. Describe what these measures indicate about each location.

32. Find the standard deviation of the data for each location. Describe what these measures indicate about each location.

SPORTS The winning times (in minutes:seconds.hundredths of a second) for the men's and women's 1500-meter speed-skating competition in several Olympics are shown below.

	1976	1980	1984	1988	1992	1994	1998
Men	1:59.38	1:55.44	1:58.36	1:52.06	1:54.81	1:51.29	1:47.87
Women	2:16.58	2:10.95	2:03.42	2:00.68	2:05.87	2:02.19	1:57.58

33. Find the mean and median winning times for men and for women.

34. Find the range and the mean deviation for men and for women. Describe what these measures indicate about the men's and women's times.

35. Find the standard deviation for men and for women. Describe what these measures indicate about men's and women's times.

Error Analysis

Some students may incorrectly add the differences and then square the sum when calculating the standard deviation by hand. Suggest that students make a table with the following column headings before calculating the standard deviation:

x_i	$x_i - \overline{x}$	$(x_i - \overline{x})^2$

33. men: $\overline{x}$ = 1:54.17, median = 1:54.81; women: $\overline{x}$ = 2:05.32, median = 2:03.42

34. men: range = 11.51, mean deviation = 3.2386; women: range = 19, mean deviation = 4.9786; The men's times deviate less from the mean than the women's times do.

35. men: σ = 3.7637; women: σ = 6.0194; The men's times are more consistent, and the women's times are more variable.

Student Technology Guide

NAME ____________ CLASS ________ DATE ________

Student Technology Guide
12.4 Measures of Dispersion

The *standard deviation* of a data set is a measure of the spread of the data. You can use a calculator to find the standard deviation of a data set.

Example: The table below shows the average price of gold in the United States for the years 1985 through 1994. Find the standard deviation of the average price of gold.

Year	1985	1986	1987	1988	1989	1990	1991	1992	1993	1994
Price (dollars)	318	368	478	438	383	385	363	345	361	389

- Enter the data into a list. Press STAT 1:Edit... ENTER. Type the prices into L1. Press ENTER or ▼ to enter each number. After entering the last number, press 2nd MODE.
- Press STAT CALC 1:1-Var Stats ENTER ENTER. The standard deviation is given by σx =.

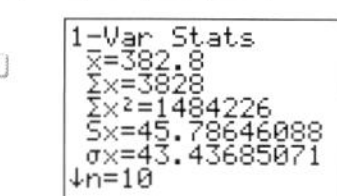

The standard deviation of the average price of gold is about $43.44, as shown.

Find the standard deviation for each set of data. Round answers to the nearest hundredth.

1. 8, 9, 12, 18, 9, 7, 10, 11, 14, 12
3.07

2. 25, 30, 40, 55, 22, 34, 53, 42, 48, 38
10.65

3. 3.14, 2.71, 1.41, 1.73, 6.67, 6.02, 9.81
2.87

4. 284, 312, 325, 298, 297, 268, 302, 295, 310
15.57

5. The table below shows the team batting averages for the National and American Leagues in 1995. Find the standard deviations for the two data sets to the nearest thousandth. Which league had a greater standard deviation in batting averages? Describe what these measures tell you about the team batting averages in the two leagues.

NL	0.281	0.276	0.271	0.270	0.266	0.264	0.264	0.261	0.261	0.259	0.258	0.252	0.249	0.247
AL	0.290	0.280	0.279	0.279	0.276	0.275	0.275	0.265	0.265	0.263	0.261	0.260	0.259	0.247

standard deviation for National League ≈ 0.009; standard deviation for American League ≈ 0.011; the American League had a greater standard deviation. Descriptions may vary. The averages for the American League were more widely scattered than those for the National League.

Look Beyond

In this exercise, students explore the difference between the formulas for the variance and standard deviation of a population and for the variance and standard deviation of a sample.

ALTERNATIVE

Assessment

Portfolio Activity

The Portfolio Activity can be used as preparation for the Chapter Project or as a separate activity. In the Portfolio Activity on this page, students will find the mean and standard deviation of each data set that they collected. Then they will explain what the measures of dispersion tell them about the data.

Answers to Portfolio Activities can be found in Additional Answers of the Teacher's Edition.

Look Back

Solve each system by using a matrix equation. *(LESSON 4.4)*

36 $\begin{cases} 2.3x + 3.2y = 16.1 \\ 4.2x - 4.6y = 12.3 \end{cases}$ **(≈4.72, ≈1.64)**　**37** $\begin{cases} 7.2x + 10.2y = 20.1 \\ 3.8x + 9.5y = 25.6 \end{cases}$ **(≈−2.37, ≈3.64)**

Use the quadratic formula to solve each equation. Give answers to the nearest tenth if their solutions are irrational. *(LESSON 5.5)*

38. $3x^2 + 10x + 1 = 0$ **−0.1, −3.2**　**39.** $2x^2 + 12x - 4 = 0$ **0.3, −6.3**

Find each value. *(LESSONS 10.2 AND 10.3)*

40. ${}_8C_3$ **56**　**41.** ${}_{10}C_3$ **120**　**42.** ${}_{17}P_3$ **4080**　**43.** ${}_{21}P_3$ **7980**

Find the sum of each infinite geometric series, if it exists. *(LESSON 11.6)*

44. $\sum_{k=1}^{\infty} 0.765^k$　**45.** $\sum_{k=1}^{\infty} 0.45^k$　**46.** $\sum_{k=1}^{\infty} \left(\frac{3}{2}\right)^k$　**47.** $\sum_{k=1}^{\infty} 2.7^k$

44. $\frac{153}{47} \approx 3.26$
45. $\frac{9}{11} = 0.\overline{81}$
46. does not exist
47. does not exist

Look Beyond

48. A sample is often used to make predictions about a larger population. To estimate the mean of a larger population, the mean of the sample is used. However, to estimate the variance or standard deviation of a population, the *sample variance*, denoted S^2, is calculated. The formula for the sample variance is $S^2 = \frac{1}{n-1}\sum_{i=1}^{n}(x_i - \overline{x})^2$. This formula differs from the formula for variance, σ^2, in that the sum is divided by $n - 1$ instead of n.

a. Find the sample variance, S^2, and sample standard deviation, S, for the following sample of a larger population: 15, 18, 7, 16, 5, 12.

b. A random survey of 10 households in a certain city revealed the following numbers of automobiles: 2, 3, 2, 1, 1, 4, 2, 1, 3, 4. Use this sample to estimate the mean number of automobiles per household and the sample standard deviation of the data for the entire city.

48a. $S^2 \approx 26.97$; $S \approx 5.19$
b. $\overline{x} \approx 2.3$ autos per household; $S \approx 1.16$ autos per household

Find the mean and the standard deviation for each of the data sets that you collected for the Portfolio Activities on pages 780 and 789. Explain what these measures tell you about the dispersion of each of the data sets.

Binomial Distributions

Why *You can use a binomial experiment to find probabilities of real-world events, such as the probability that enough ambulances will be available for emergencies.*

Objective

- Find the probability of r successes in n trials of a binomial experiment.

APPLICATION
EMERGENCY SERVICES

A medical center has 8 ambulances. Given the ambulance's current condition, regular maintenance, and restocking of medical supplies, the probability of an ambulance being operational is 0.96. Find the probability that at least 6 of the 8 ambulances are operational. *You will solve this problem in Example 2.*

The ambulance problem above is an example of a *binomial experiment*.

Binomial Experiment

A probability experiment is a **binomial experiment** if both of the following conditions are met:

- The experiment consists of n trials whose outcomes are either successes (the outcome *is* the event in question) or failures (the outcome is *not* the event in question).
- The trials are identical and independent with a constant probability of success, p, and a constant probability of failure, $1 - p$.

Activity
Exploring Binomial Probability

You will need: a 6-sided number cube

Let a roll of 1 on the number cube be considered a success, S, and the roll of any other number be considered a failure, F.

1. Write a fraction for the probability of a success, $P(S)$, and for the probability of a failure, $P(F)$, in 1 roll of a number cube.
2. For each arrangement of outcomes in 5 rolls of a number cube, write the probability as a product of fractions. Use exponents in your products.
   a. S F F F F **b.** F F S F F **c.** S S F F F **d.** S F F S F

Prepare

NCTM PRINCIPLES & STANDARDS 1, 3–10

QUICK WARM-UP

Evaluate.

1. ${}_{10}C_4$ 210 **2.** ${}_{8}C_1$ 8 **3.** ${}_{12}C_{12}$ 1

A fair coin is tossed 4 times. Find the probability of tossing

4. 4 heads. $\frac{1}{16}$, or 6.25%

5. fewer than 4 heads. $\frac{15}{16}$, or 93.75%

6. 4 heads *or* 3 heads. $\frac{5}{16}$, or 31.25%

7. 4 heads *or* 4 tails. $\frac{2}{16}$, or 12.5%

Also on Quiz Transparency 12.5

Teach

Why Students are often faced with situations that have only two possible outcomes, such as yes-no, true-false, and success-failure. Binomial experiments are used in statistical evaluations of situations like these. Ask students to discuss how they choose the answer on a true-false test question when they really do not know which answer is correct.

Alternative Teaching Strategy

USING PATTERNS Ask students to write $(0.1 + 0.9)^5$ in expanded form as $(0.1)^0(0.9)^5 + 5(0.1)^1(0.9)^4 + 10(0.1)^2(0.9)^3 + 10(0.1)^3(0.9)^2 + 5(0.1)^4(0.9)^1 + (0.1)^5(0.9)^0$. Tell them that a baseball player's batting average is 0.100. Show them that the last term of the binomial expansion represents the probability of the player getting 5 hits in next 5 at bats. The fifth term represents getting 4 hits in the next 5 at bats. That is, $(0.1)^4$ represents the probability of 4 hits, $(0.9)^1$ represents the probability of 1 out (or no hit), and ${}_5C_1$, or 5, represents the number of ways that 4 hits in 5 at bats can occur. Now ask them what the first, second, third, and fourth terms of the expansion represent. Now define *binomial experiment* and *binomial probability*. Point out that even though the player's batting average changes slightly each time the player is at bat, you can use the binomial probability theorem to approximate the probability of r successful hits in n at bats.

Activity Notes

Students should find that rolling a number cube is a binomial experiment and, in this Activity, that $P(S) = \frac{1}{6}$, $P(S) = \frac{5}{6}$, the number of ways to get r successes in n rolls is given by ${}_nC_r$, and the probability of rolling r successes in n rolls is given by ${}_nC_r\left(\frac{1}{6}\right)^r\left(\frac{5}{6}\right)^{n-r}$.

Cooperative Learning

Divide the class into teams of two. Each member should complete Steps 1, 2, 3, and 4 independently. Working as a team, the students should verify their calculations and complete Step 5.

CHECKPOINT ✔

3. ${}_5C_1$; ${}_5C_2$

CHECKPOINT ✔

5. ${}_5C_3\left(\frac{1}{6}\right)^3\left(\frac{5}{6}\right)^2 \approx 0.032$

ADDITIONAL EXAMPLE 1

Suppose that the probability of a VCR that was manufactured in a certain factory being defective is 2%. **What is the probability that 2 VCRs are defective in a shipment of 12 VCRs from this factory?**

${}_{12}C_2\ (0.02)^2(0.98)^{10} \approx 0.022$

TRY THIS

≈0.13, or 13%

CHECKPOINT ✔ **3.** Using combination notation, how many ways can you obtain exactly 1 success in 5 rolls of a number cube? exactly 2 successes?

4. Using your answers from Steps 2 and 3, write each probability below as a product, using combination notation and fractions with exponents.

a. P(exactly 1 success) **b.** P(exactly 2 successes)

CHECKPOINT ✔ **5.** Find the probability of exactly 3 successes in 5 rolls of a number cube.

In the preceding Activity, *a roll of 1* on a 6-sided number cube is considered a success and *a roll of not 1* is considered a failure. In 5 rolls of the number cube, the probability of 2 successes followed by 3 failures is as follows:

$$\frac{1}{6} \times \frac{1}{6} \times \frac{5}{6} \times \frac{5}{6} \times \frac{5}{6} = \left(\frac{1}{6}\right)^2\left(\frac{5}{6}\right)^3$$

Roll number 1 2 3 4 5

However, this probability accounts for only one way in which 2 successes in 5 rolls can occur. All of the ways in which 2 successes can occur is the combination ${}_5C_2$. Thus, the probability of exactly 2 successes in 5 rolls of a number cube is ${}_5C_2\left(\frac{1}{6}\right)^2\left(\frac{5}{6}\right)^3$, or approximately 0.16.

To find the theoretical probability that exactly r successes will occur in n trials of a binomial experiment, you can use the formula below.

Binomial Probability

In a binomial experiment consisting of n trials, the probability, P, of r successes (where $0 \le r \le n$, p is the probability of success, and $1 - p$ is the probability of failure) is by the following formula:

$$P = {}_nC_r p^r (1 - p)^{n-r}$$

EXAMPLE 1 Suppose that the probability a seed will germinate is 80%.

What is the probability that 7 of these seeds will germinate when 10 are planted?

APPLICATION
GARDENING

SOLUTION

$n = 10$, $r = 7$, $p = 0.8$, and $1 - p = 1 - 0.8$

$$P = {}_nC_r\, p^r\, (1 - p)^{n-r}$$
$$= {}_{10}C_7\, (0.8)^7\, (1 - 0.8)^{10-7}$$
$$\approx 0.201$$

Thus, the probability that 7 of these seeds will germinate when 10 are planted is about 0.201, or 20.1%.

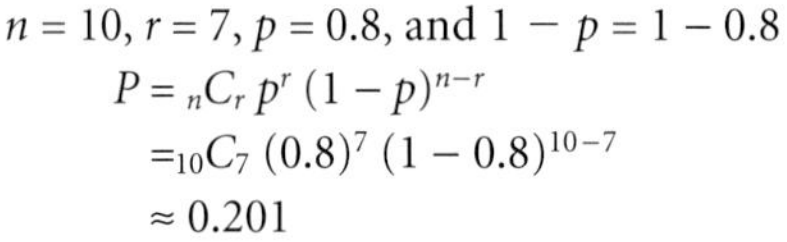

Keystroke Guide, page 823

TRY THIS Suppose that the probability a seed will germinate is 85%. What is the probability that 7 of these seeds will germinate when 10 are planted?

Inclusion Strategies

READING STRATEGY Ask students whether the binomial probability theorem can be applied to the experiments below. If the experiment is binomial, have them describe the properties that make it binomial. If it is not binomial, have them explain why.

1. A multiple-choice exam consists of 20 questions. Each question has 4 possible answers, of which exactly 1 is correct. If a student guesses on every question (with all outcomes equally likely), what is the probability of getting at least 10 questions correct? **binomial**
2. An inspection procedure determines that 90% of newly assembled computers have no defects, 5% have at least one major defect, and 5% have only minor defects. If 5 computers are chosen at random, find the probability that at least 2 of them have only minor defects. **not binomial**
3. A committee of 2 students is chosen from among 5 seniors and 3 juniors. Find the probability that the committee consists of exactly 1 senior and 1 junior. **not binomial**

CHECKPOINT ✔ Refer to Example 1. Find the probability that all 10 of these seeds will germinate when 10 are planted. Then find the probability that none of these seeds will germinate when 10 are planted. What happens to the binomial probability formula when $r = n$? when $r = 0$?

In a binomial experiment, the events *r successes* and *s successes* are mutually exclusive. Thus, you can find the probability of exactly *r* successes or exactly *s* successes by adding the probabilities.

$$P(r \text{ successes } \textit{or } s \text{ successes}) = P(r \text{ successes}) + P(s \text{ successes})$$

EXAMPLE 2

APPLICATION
EMERGENCY SERVICES

Refer to the ambulance problem posed at the beginning of the lesson.

Find the probability that at least 6 of the 8 ambulances are operational. Round to the nearest tenth of a percent.

SOLUTION

At least 6 ambulances are operational when exactly 6, 7, or 8 ambulances are operational.

Find $P(\text{exactly } 6) + P(\text{exactly } 7) + P(\text{exactly } 8)$.

Use $n = 8$, $p = 0.96$, and $1 - p = 1 - 0.96$.

$$P(\text{exactly } 6) = {}_8C_6(0.96)^6(1 - 0.96)^{8-6} \approx 0.0351$$

$$P(\text{exactly } 7) = {}_8C_7(0.96)^7(1 - 0.96)^{8-7} \approx 0.2405$$

$$P(\text{exactly } 8) = {}_8C_8(0.96)^8(1 - 0.96)^{8-8} \approx 0.7214$$

$$P(\text{exactly } 6) + P(\text{exactly } 7) + P(\text{exactly } 8) \approx 0.0351 + 0.2405 + 0.7214 \approx 0.997$$

Thus, the probability is about 99.7%.

TECHNOLOGY
GRAPHICS CALCULATOR
Keystroke Guide, page 823

CHECK

You can check your answer by using a calculator with a built-in binomial probability feature. In the display at right, the command **binompdf** is used for 6, 7, and 8 successes. The sum gives the desired probability.

TRY THIS Find the probability that at least 7 of 8 ambulances are operational, given that the probability of an ambulance being operational is 0.95. Round to the nearest tenth of a percent.

Use Teaching Transparency 52.

CHECKPOINT ✔
≈ 0.107 or 10.7%; ≈ 0 or 0%; Because ${}_nC_n = {}_nC_0 = 1$, when $r = n$ the binomial probability formula becomes p^n, and when $r = 0$ the binomial probability formula becomes $(1 - p)^n$.

ADDITIONAL EXAMPLE 2

A landscaping plan specifies that 10 trees of a certain type are to be planted in front of a building. When this type of tree is planted in the autumn, the probability that it will survive the winter is 85%. **What is the probability that no fewer than 8 of the 10 trees will survive the winter if planted in the autumn?**

$$\sum_{i=8}^{10} {}_{10}C_i\,(0.85)^i(0.15)^{10-i} \approx 82.0\%$$

TRY THIS
$\approx 94.3\%$

Teaching Tip

TECHNOLOGY When using a TI-83 or TI-82 graphics calculator to compute the binomial probability for Example 2, use the following keystrokes:

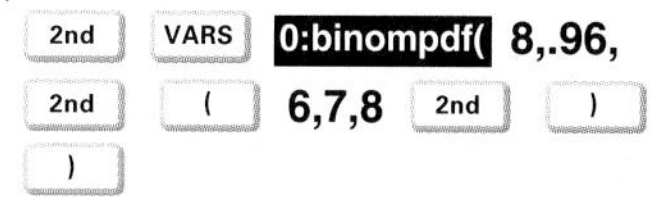

Then press ENTER. The syntax for **binompdf** is (*number of trials, probability of success* [,*number of successes*]), where *number of successes* is given within braces. If *number of successes* is omitted, all terms of the binomial expansion are displayed and can be viewed by using the arrow keys.

Interdisciplinary Connection

DRIVER'S EDUCATION Tell students that $\frac{3}{4}$ of the senior class are licensed drivers. Have them complete the table below and make a relative frequency histogram to determine the probabilities that 0, 1, 2, 3, and 4 licensed drivers are selected when 4 members of the senior class are chosen at random.

No.	4	3	2	1	0
Prob.	0.316	0.422	0.211	0.047	0.004

Enrichment

Tell students that there are two games in which a coin is flipped. They win the first game if between 45% and 55% of the results are heads. They win the second game if at least 75% of the results are heads. For each game, ask students to determine whether they have a better chance of winning each game if they flip the coin 50 times or 500 times. Suggest that they use a graphics calculator to calculate the probabilities. **There is a better chance of winning the first game if they flip the coin 500 times and a better chance of winning the second game if they flip the coin 50 times.**

Teaching Tip

TECHNOLOGY When using a TI-83 or TI-82 graphics calculator to find the sum of the probabilities shown in Example 2, the following is an alternate procedure:

1. Find the terms of the binomial by pressing 2nd VARS, selecting **0:binompdf(**, and entering **8,.96,** 2nd (**6,7,8** 2nd)) ENTER.
2. Then store the list into **L1** by pressing STO▶ 2nd 1.
3. Find the sum of **L1** by pressing 2nd STAT, choosing **MATH 5:sum(**, and then pressing 2nd 1) ENTER.

CHECKPOINT ✔
They are all mound-shaped, but the first distribution is almost symmetric about its center, the middle distribution is clustered about its center, and the last distribution is skewed to the left.

CRITICAL THINKING
For $p > 0.5$, a success is more likely and the peak moves to the right while its tail moves to the left. Therefore, the distribution is skewed to the left. Similarly, when $p < 0.5$, success is less likely, so the distribution is skewed to the right.

In the binomial experiment of tossing a coin 6 times, success is heads, $n = 6$, $p = 0.5$, and r ranges from 0 heads to 6 heads. The probabilities of tossing each possible number of successes are organized in the table and the relative frequency histogram below.

Heads, r	P(heads), $P = {}_6C_r(0.5)^r(0.5)^{6-r}$
0	≈0.016
1	≈0.094
2	≈0.234
3	≈0.312
4	≈0.234
5	≈0.094
6	≈0.016

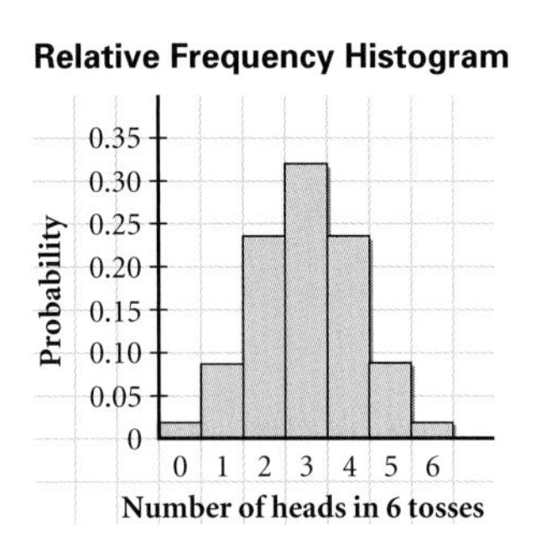

Notice that the distribution of successes is symmetric about the mode because the probabilities of success and failure are equal.

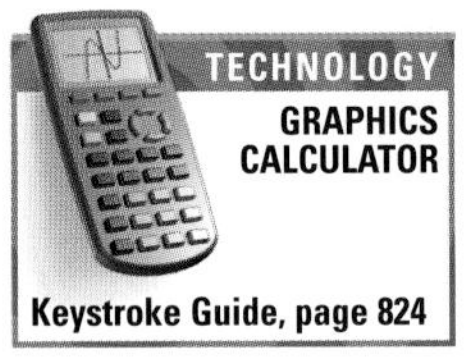

You can use a graphics calculator to generate simulations of tossing a coin 6 times. By graphing the experimental results in a histogram, you can compare the distributions obtained experimentally with the distribution obtained theoretically. Each calculator display below shows the distribution of the results of 100 simulations.

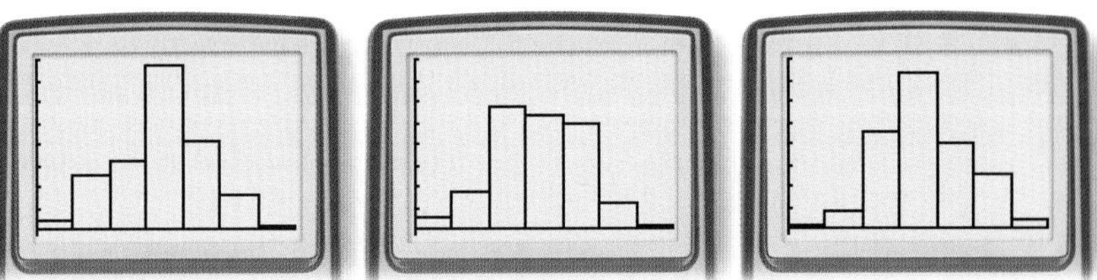

CHECKPOINT ✔ How are the distributions alike and how are they different?

When the probabilities of success and of failure are *not* equal, the binomial distribution obtained by theoretical probability will not be symmetric.

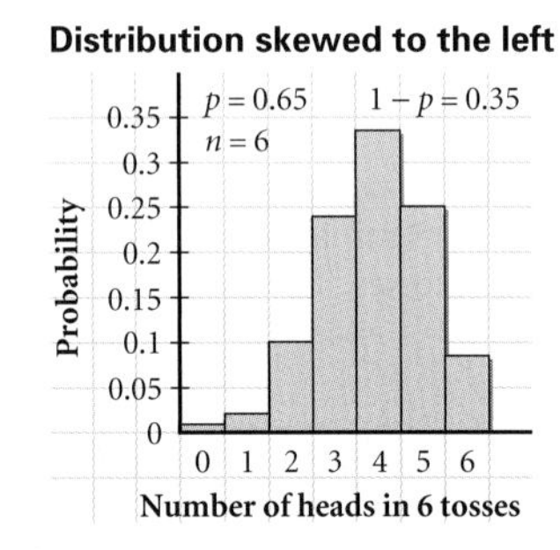

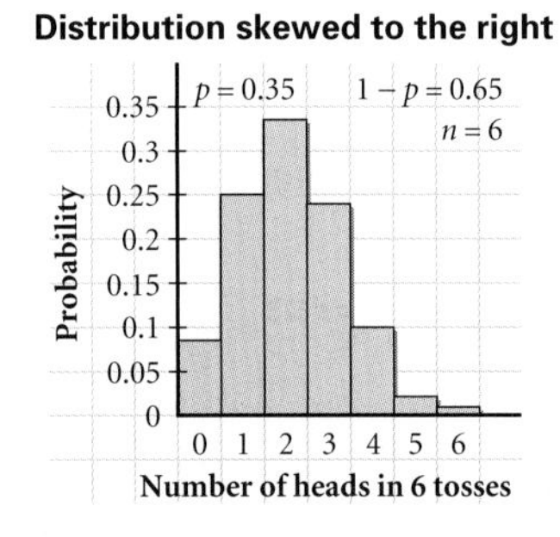

CRITICAL THINKING Explain why the distribution is skewed to the left when $p > 0.5$ and to the right when $p < 0.5$.

Reteaching the Lesson

USING VISUAL MODELS Have students make a tree diagram to visually display all possible outcomes of an event. For example, to find all possible outcomes for 3 coin tosses, have students create a tree diagram in which H represents heads and T represents tails.

H	H	H	HHH
		T	HHT
	T	H	HTH
		T	HTT
T	H	H	THH
		T	THT
	T	H	TTH
		T	TTT

Have students determine the probabilities of getting 0, 1, 2, or 3 heads. **0.125, 0.375, 0.375, 0.125** Then have them draw a relative frequency histogram and find the mean and standard deviation of the number of heads by using the following formulas, where n is the number of trials and p is the probability of success: $\bar{x} = np$ and $\sigma = \sqrt{np(1-p)}$. **$\bar{x} = \frac{3}{2}$, or 1.5; $\sigma = \frac{\sqrt{3}}{2}$, or ≈0.87**

Exercises

Communicate

1. Describe the conditions that an experiment must satisfy in order to be a binomial experiment.
2. Explain why ${}_nC_r$ appears in the formula for the probability of a binomial experiment, $P = {}_nC_r p^r(1 - p)^{n-r}$.
3. A person randomly selects answers to all 12 questions of a multiple-choice test. Each question has 1 correct and 4 incorrect responses. Describe what n, r, and p represent when finding the binomial probability of 9 answers being correct.
4. Explain the conditions under which a binomial distribution is symmetric and the conditions under which it is skewed.

Guided Skills Practice

APPLICATIONS

5. **VETERINARY MEDICINE** Suppose that the probability of a sick animal recovering within a week is 70%. What is the probability that 4 out of 6 sick animals will recover within a week? ***(EXAMPLE 1)*** **≈0.324, or 32.4%**

6. **PUBLIC SAFETY** The probability that a driver is not wearing a seat belt is 0.18. Find the probability that at least 2 of 10 drivers are not wearing seat belts. ***(EXAMPLE 2)*** **≈0.561, or 56.1%**

Practice and Apply

A coin is flipped 8 times. Find the probability of each event.

7. exactly 5 heads **0.219**
8. exactly 3 heads **0.219**
9. exactly 2 heads **0.109**
10. exactly 6 heads **0.109**
11. exactly 0 heads **0.004**
12. exactly 8 heads **0.004**

A family has 4 children. Find the probability of each event, assuming that the probability of a male equals the probability of a female.

13. exactly 2 males **0.375**
14. exactly 1 male **0.25**
15. 1 female and 3 males **0.25**
16. all females **0.0625**

Suppose that 70% of the adults in a certain city are registered voters. In a group of 10 randomly selected adults in the city, find the probability that the indicated number are registered voters.

17. exactly 5 **0.103**
18. exactly 8 **0.233**
19. at least 6 **0.849**
20. at least 8 **0.383**
21. at most 5 **0.151**
22. at most 6 **0.351**
23. 2 are not registered voters **0.233**
24. 4 are not registered voters **0.200**

Assess

Selected Answers

Exercises 5, 6, 7–59 odd

ASSIGNMENT GUIDE

In Class	1–6
Core	7–49 odd
Core Plus	8–50 even
Review	51–59
Preview	60

Extra Practice can be found beginning on page 940.

Error Analysis

Students often mistakenly believe that the probability of 2 successes in 4 trials is the same as that of 4 successes in 8 trials. If the probability of obtaining a 1 on a roll of a number cube is considered a success, have students determine the probability of 2 successes in 4 rolls of a number cube. Then have them determine the probability of 3 successes in 6 rolls and that of 4 successes in 8 rolls. Be sure they notice that the probabilities are not the same.

A person randomly selects answers to all 10 questions on a multiple-choice quiz. Each question has 1 correct answer and 3 incorrect answers. Find the probability that the indicated number of answers are correct.

25. exactly 6 **0.016** **26.** exactly 7 **0.003** **27.** at least 8 **0.0004**

28. at least 5 **0.078** **29.** at most 3 **0.776** **30.** at most 4 **0.922**

Find the probability that a batter with each batting average given below will get at least 3 hits in the next 5 at bats.

31. 0.200 **0.058** **32.** 0.250 **0.104** **33.** 0.300 **0.163** **34.** 0.350 **0.235**

Find the probability that a batter with each batting average given below will get at most 3 hits in the next 5 at bats.

35. 0.200 **0.993** **36.** 0.250 **0.984** **37.** 0.300 **0.969** **38.** 0.350 **0.946**

APPLICATIONS

HOSPITAL STATISTICS Suppose that a hospital found an 8.5% probability that the birth of a baby will require the presence of more than one doctor.

39. Find the probability that 2 of the next 20 babies born at the hospital will require the presence of more than one doctor. **0.277**

40. Find the probability that 3 of the next 20 babies born at the hospital will require the presence of more than one doctor. **0.155**

SURVEYS In a 1998 survey, 54% of U.S. men and 36% of U.S. women consider themselves basketball fans. [*Source: Bruskin-Goldring Research*]

41. Find the probability that 5 men randomly selected from a group of 10 U.S. men consider themselves basketball fans. **0.238**

42. Find the probability that 3 women randomly selected from a group of 10 U.S. women consider themselves basketball fans. **0.246**

internet connect
Homework Help Online
Go To: **go.hrw.com**
Keyword:
MB1 Homework Help for Exercises 25–30, 46–49

SURVEYS In 1997, 40% of U.S. households owned at least one cellular or wireless phone. [*Source: The Wirthin Report*]

43. What is the probability that 2 in 10 U.S. households owned a cellular or wireless phone? **0.121**

44. What is the probability that at least 2 in 10 U.S. households owned a cellular or wireless phone? **0.954**

45. What is the probability that at least 2 in 8 U.S. households owned a cellular or wireless phone? **0.894**

AWARDS Three different prizes are placed in cereal boxes, with no more than 1 prize per box. Prize A is in 10% of the cereal boxes, prize B is in 20%, and prize C is in 30%. Find the probability of each event.

0.6 **46.** 1 prize in 1 box

0.48 **47.** exactly 1 prize in 2 boxes

0.036 **48.** 3 different prizes in 3 boxes

49. at least 2 prizes in 3 boxes **0.648**

Practice

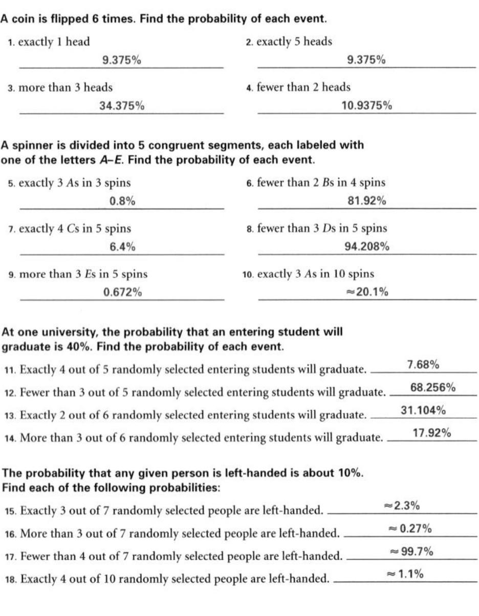

NAME ______ CLASS ______ DATE ______

Practice
12.5 ***Binomial Distributions***

A coin is flipped 6 times. Find the probability of each event.

1. exactly 1 head 9.375%
2. exactly 5 heads 9.375%
3. more than 3 heads 34.375%
4. fewer than 2 heads 10.9375%

A spinner is divided into 5 congruent segments, each labeled with one of the letters *A–E*. Find the probability of each event.

5. exactly 3 *A*s in 3 spins 0.8%
6. fewer than 2 *B*s in 4 spins 81.92%
7. exactly 4 *C*s in 5 spins 6.4%
8. fewer than 3 *D*s in 5 spins 94.208%
9. more than 3 *E*s in 5 spins 0.672%
10. exactly 3 *A*s in 10 spins ≈20.1%

At one university, the probability that an entering student will graduate is 40%. Find the probability of each event.

11. Exactly 4 out of 5 randomly selected entering students will graduate. 7.68%
12. Fewer than 3 out of 5 randomly selected entering students will graduate. 68.256%
13. Exactly 2 out of 6 randomly selected entering students will graduate. 31.104%
14. More than 3 out of 6 randomly selected entering students will graduate. 17.92%

The probability that any given person is left-handed is about 10%. Find each of the following probabilities:

15. Exactly 3 out of 7 randomly selected people are left-handed. ≈2.3%
16. More than 3 out of 7 randomly selected people are left-handed. ≈ 0.27%
17. Fewer than 4 out of 7 randomly selected people are left-handed. ≈ 99.7%
18. Exactly 4 out of 10 randomly selected people are left-handed. ≈ 1.1%

APPLICATION

50. AVIATION A certain twin-engine airplane can fly with only one engine. The probability of engine failure for each of this airplane's engines is 0.002. Determine whether each airplane described below is more likely than this twin-engine airplane to crash due to engine failure.

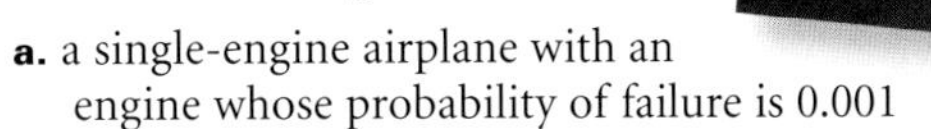

a. a single-engine airplane with an engine whose probability of failure is 0.001

b. an experimental 3-engine airplane that can fly with only 2 engines, equipped with engines whose probability of failure is 0.001

c. an airplane with 4 engines that can fly with only 2 engines, equipped with engines whose probability of failure is 0.0008.

CHALLENGE

d. Find p such that a single-engine airplane with p probability of engine failure is as safe as the 4-engine airplane described in part **c**.

50a. more likely
b. less likely
c. less likely
d. $p = 2 \times 10^{-9}$

Look Back

A number cube is rolled once. Find each probability. ***(LESSON 10.4)***

51. $P(2 \text{ or } 3)$ $\frac{1}{3}$ **52.** $P(1 \text{ or } 2 \text{ or } 6)$ $\frac{1}{2}$ **53.** $P(\text{even or } 6)$ $\frac{1}{2}$

54. $P(\text{not } 6 \text{ or } < 2)$ $\frac{5}{6}$ **55.** $P(< 5)$ $\frac{2}{3}$ **56.** $P(> 2 \text{ or } 4)$ $\frac{2}{3}$

57. Construct an arithmetic sequence with a common difference of 6. ***(LESSON 11.2)*** sample answer: –3, 3, 9, 15, . . .

58. Construct a geometric sequence with a common ratio of 6. ***(LESSON 11.4)*** sample answer: 1, 6, 36, 216, . . .

APPLICATION

59. EDUCATION Refer to the data below. ***(LESSON 12.4)***

Percent of Recent High School Graduates Enrolled in College

	1988	1989	1990	1991	1992	1993	1994
Male	57.0	57.6	57.8	57.6	59.6	59.7	60.6
Female	60.8	61.6	62.0	67.1	63.8	65.4	63.2

[*Source: U.S. Department of Education Statistics*]

a. Find the range and mean deviation for the percent enrollments of males and females.

b. Find the standard deviation for the percent enrollments of males and of females.

c. Describe what the measures indicate about the percent enrollments of males and females.

59a. male: range = 3.6; mean dev. ≈ 1.21
female: range = 6.3; mean dev. ≈ 1.73
b. male: σ ≈ 1.276
female: σ ≈ 2.065
c. The percent of female enrollment is more variable than male enrollment.

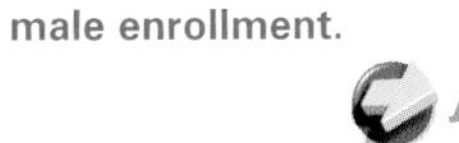

Look Beyond

60. The standard deviation for a binomial distribution is given by $\sigma = \sqrt{np(1-p)}$, where n is the number of trials and p is the probability of a particular event occurring. A number cube is rolled 1000 times. Find the mean ($\bar{x} = np$) and the standard deviation for the event of rolling a 5.

60. $\bar{x} = 166.\bar{6}$; $\sigma \approx 11.79$

Look Beyond

In Exercise 60, students are given the formulas for the mean and standard deviation of a binomial distribution and are asked to find the mean and standard deviation for the event "rolling a 5" on a number cube that is rolled 1000 times.

Student Technology Guide

NAME ____________ CLASS ________ DATE ________

Student Technology Guide
12.5 ***Binomial Distributions***

A binomial distribution is a good model for many real-world situations. You can use a graphics calculator to find probabilities for a binomial distribution.

Example: In 1992, about 48.8% of the babies born in the United States were girls. If this percentage holds, what is the probability that 11 of 20 babies born at a hospital on a particular day are girls?

- The probability of r out of n successes (girls) is: $P = {}_nC_r\, p^r(1-p)^{n-r} = {}_{20}C_{11}(0.488)^{11}(1-0.488)^9$.
- Enter this expression into the calculator. Press 20 MATH PRB 3:nCr ENTER 11 × .488 ^ 11 × (1 − .488) ^ 9 ENTER.

```
20 nCr 11*.488^
11*(1-.488)^9
      .1517936335
```

The probability that 11 out of 20 babies are girls is about 0.15, or 15%.

1. Six babies were born on Douglass Street in one year. Find the probability that exactly two of them were girls. (Assume that 48.8% of babies are girls.) ≈ 0.25, or 25%

2. A multiple-choice test has four choices for each question. Find the probability of answering exactly 6 of 10 questions correctly by randomly guessing. ≈ 0.02, or 2%

Example: In the United States, about 80% of males between ages 18 and 24 are 6 feet tall or shorter. For 5 randomly chosen males in this age group, what is the probability that at least 3 of them are over 6 feet tall?

You can solve this problem by using the method shown above, but there is a faster way if your calculator has a binomial probability feature.

- The probability is $P = P(\text{exactly } 3) + P(\text{exactly } 4) + P(\text{exactly } 5)$.
- Press 2nd VARS 0:binompdf(ENTER 5 , .2 , 3) + 2nd VARS 0:binompdf(ENTER 5 , .2 , 4) + 2nd VARS 0:binompdf(ENTER 5 , .2 , 5) ENTER. (The numbers entered are the total number of trials, the probability of a success, and the number of successes you are interested in.)

```
binompdf(5,.2,3
)+binompdf(5,.2
,4)+binompdf(5,
.2,5)
            .05792
```

The probability that at least 3 out of the 5 males are over 6 feet tall is about 0.06, or 6%.

3. For a group of 5 males between 18 and 24, what is the probability that at least 4 of them are under 6 feet tall? ≈ 0.74, or 74%

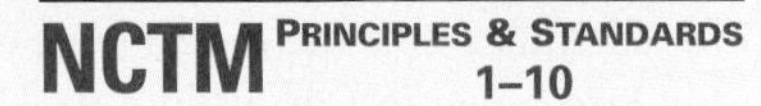

NCTM PRINCIPLES & STANDARDS 1–10

QUICK WARM-UP

Find each statistical measure for the data below. Round to the nearest thousandth when necessary.

2.0 1.3 2.1 2.8
1.8 2.6 1.9 2.3

1. mean 2.1
2. median 2.05
3. mode(s) none
4. quartiles 1.85, 2.05, 2.45
5. range 1.5
6. mean deviation 0.35
7. variance 0.195
8. standard deviation ≈0.442

Also on Quiz Transparency 12.6

Teach

Why Many natural phenomena can be described by data that approximate a bell-shaped or normal distribution, and it is important that students be able to interpret the data in a normal distribution. Ask students to discuss how they interpreted their scores on the SAT or any other standardized test they have taken.

Normal Distributions

Objectives

- Find the probability of an event given that the data is normally distributed and its mean and standard deviation are known.
- Use z-scores to find probabilities.

Why You can use the characteristics of a normal distribution to find probabilities of many real-world events, such as the number of deaths caused by lightning each month.

APPLICATION
METEOROLOGY

The histogram at right shows that deaths caused by lightning occur much more frequently in the summer months. In fact, about 70% of the deaths occurred in June, July, or August. [*Source: NOAA*]

Because this distribution is nearly symmetric about July, you can say that this data represents a *normal distribution.*

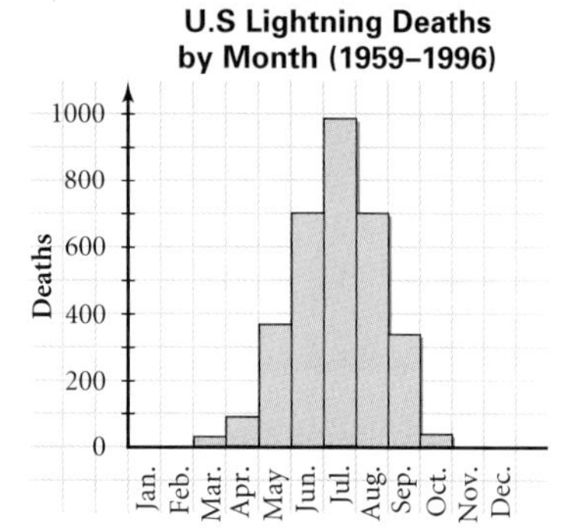

A **normal distribution** of data varies randomly from the mean, creating a mound-shaped pattern that is symmetric about the mean when graphed. Some examples of normally distributed data are human heights and weights.

The model obtained by drawing a curve through the midpoints of the tops of the bars in a histogram of normally distributed data is called a **normal curve.** A normal curve is defined by the mean and the standard deviation. A **standard normal curve** is a normal curve with mean of 0 and standard deviation of 1.

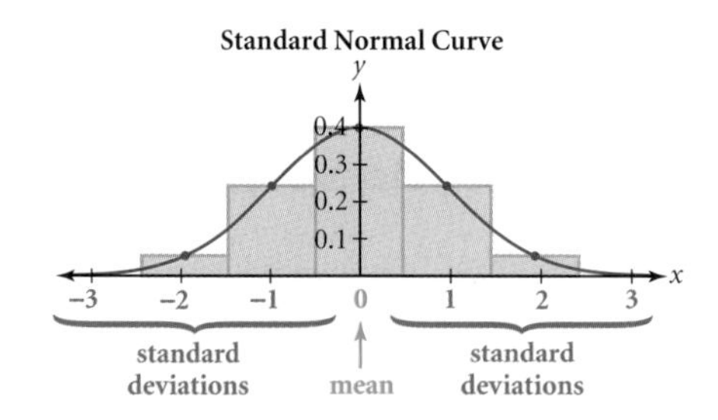

Alternative Teaching Strategy

USING VISUAL MODELS Histograms can be thought of as areas that represent probabilities. Have students refer to the table of probabilities and the frequency histogram at the top of page 802. The base of each bar is 1 unit wide and the height of each is $P(r)$, the probability of r heads. The area of each bar is $1 \cdot P(r)$, and the total area of all the bars is 1. Now tell them that probabilities can be found by using areas. Then ask students to use areas to find the probability of at least 4 heads in 6 tosses of a coin. **0.343** Have students find the mean and standard deviation of the distribution by using the formulas $\bar{x} = np$ and $\sigma = \sqrt{np(1-p)}$, respectively. $\bar{x} = 3$, $\sigma = \frac{\sqrt{6}}{2}$, or ≈ 1.2 Now have them draw a curve that connects the midpoints of the top of the bars and draw a scale with values of $-3\sigma, -2\sigma, -\sigma, 0, \sigma, 2\sigma$, and 3σ marked underneath the horizontal scale such that $\bar{x}$ corresponds to 0. Have them use the table shown on page 807 to approximate the probability of getting at least 4 heads in 6 tosses by using $\sigma = 0.4$. **0.3446** Point out that the table gives the area between the mean and the number of standard deviations that the data point is away from the mean.

Recall from 12.2 that relative frequency histograms give probabilities. You can use the area under a normal curve (and above the x-axis) to approximate probabilities.

The total area under a normal curve is 1, with an area of 0.5 of the total to the right of the mean and an area of 0.5 of the total to the left of the mean.

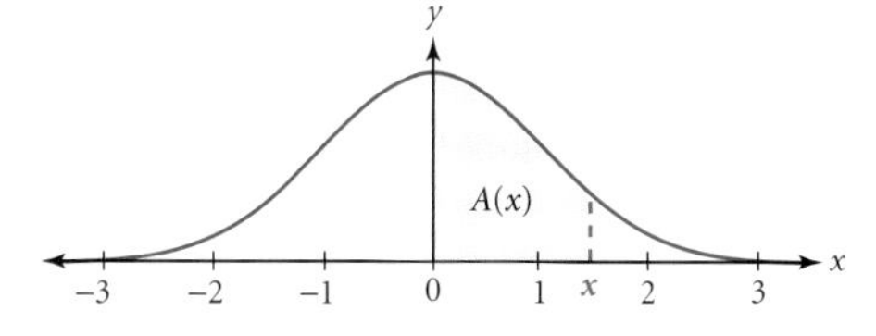

A table of approximate areas, $A(x)$, under the standard normal curve between the mean, 0, and the number of standard deviations, x, is given below.

Standard Normal Curve Areas

x	0	0.2	0.4	0.6	0.8	1.0	1.2	1.4	1.6	1.8	2.0
$A(x)$	0.0000	0.0793	0.1554	0.2257	0.2881	0.3413	0.3849	0.4192	0.4452	0.4641	0.4772

EXAMPLE 1 **Approximate each probability by using the area table for a standard normal curve.**

a. $P(x \geq 1.6)$ **b.** $P(-2.0 \leq x \leq 0.4)$

SOLUTION

a.

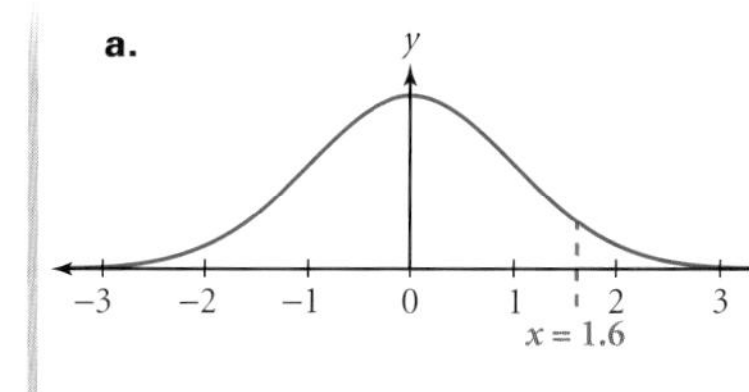

$P(x \geq 1.6) = 0.5 - P(0 \leq x \leq 1.6)$
$\approx 0.5 - 0.4452$
≈ 0.0548

b.

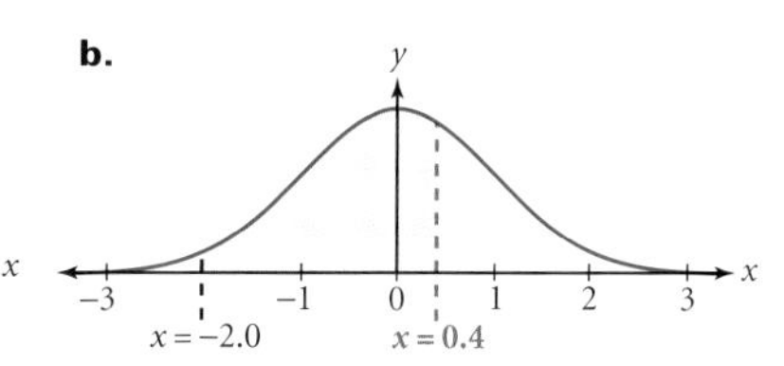

$P(-2.0 \leq x \leq 0.4)$
$= P(-2.0 \leq x \leq 0) + P(0 \leq x \leq 0.4)$
$= P(0 \leq x \leq 2.0) + P(0 \leq x \leq 0.4)$
$\approx 0.4772 + 0.1554$
≈ 0.6326

Keystroke Guide, page 824

CHECK

You can use a graphics calculator to check the probabilities.

TRY THIS Find each probability by using the area table for a standard normal curve.
a. $P(x \leq -0.8)$ **b.** $P(-1.2 \leq x \leq 1.6)$

CHECKPOINT ✔ Using the area table for a standard normal curve, describe the trend for $A(x)$ as x increases.

CRITICAL THINKING Why is 10^99 entered in the graphics calculator for the upper bound in part **a** of Example 1? What would you enter to find $P(x \leq 1.6)$?

Interdisciplinary Connection

TYPING A large group of students took a typing exam. Scores on the exam were normally distributed with a mean of 62 words per minute and a standard deviation of 12 words per minute. A student is selected at random from this group. Have students find the following:

a. the probability that the selected student typed more than 86 words per minute **≈0.02**

b. the probability that the selected student typed between 38 and 50 words per minute **≈0.14**

ADDITIONAL EXAMPLE 1

Find each probability by using the area table for a standard normal curve.

a. $P(x \leq 1.2)$ **≈0.8849**

b. $P(-0.4 \leq x \leq 1.8)$ **≈0.6195**

Teaching Tip

TECHNOLOGY When using a TI-83 or TI-82 graphics calculator to check part **a** of Example 1, press 2nd VARS, choose **2:normalcdf(**, and enter **1.6,10^99**) ENTER.

To check part **b** of Example 1, use the following keystrokes: 2nd VARS **2:normalcdf(** **−2,.4**) ENTER.

The syntax is **2:normalcdf(***lower bound, upper bound* [*,mean,standard deviation*]). The mean and standard deviation are optional parameters and, when omitted, default to 0 and 1, respectively.

Note that **1:normalpdf(** gives the height of the normal distribution for a given value of x, and it is used when graphing the normal distribution function. The syntax is **1:normalpdf(**x,[μ,σ]).

TRY THIS

a. **≈0.2119** **b.** **≈0.8301**

CHECKPOINT ✔

$A(x)$ approaches 0.5.

CRITICAL THINKING

10^{99} is a very large number, and we want the probability of x greater than or equal to 1.6; **2:normalcdf(−10^99,1.6,0,1)**

Activity Notes

Be sure students notice that the graph is symmetric with respect to the y-axis; that is, the area from $x = 0$ to $x = 1$ is about 0.34, and the area from $x = -1$ to $x = 0$ is about 0.34.

Cooperative Learning

Have students work together in pairs to record and interpret the instructions necessary for performing this Activity with a graphics calculator. They should assist each other in the application of these instructions.

Teaching Tip

TECHNOLOGY You can graph the standard normal curve by using the normal probability density feature. If you are using a T1-82 or TI-83 graphics calculator, set the viewing window as follows:

Xmin=–4.7	**Ymin=–.2**
Xmax=4.7	**Ymax=.5**
Xscl=1	**Yscl=.1**

Press Y= 2nd VARS and choose **1:normalpdf(**. To the right of the parenthesis, enter **X,0,1)**. To see the standard normal curve, press GRAPH.

To use the $\int f(x)dx$ feature to display the area under the curve, press 2nd TRACE and choose **7:∫f(x)dx**. For the lower limit, enter **0** and press ENTER. For upper limit, enter **1** and press ENTER. The shaded region represents the area under the curve from $x = 0$ to $x = 1$. The area, 0.34134475, is displayed at the bottom of the screen.

Activity
Exploring the Standard Normal Curve

TECHNOLOGY
GRAPHICS CALCULATOR
Keystroke Guide, page 824

You will need: a graphics calculator

1. Graph $f(x) = \frac{1}{\sqrt{2\pi}} e^{\frac{-x^2}{2}}$.

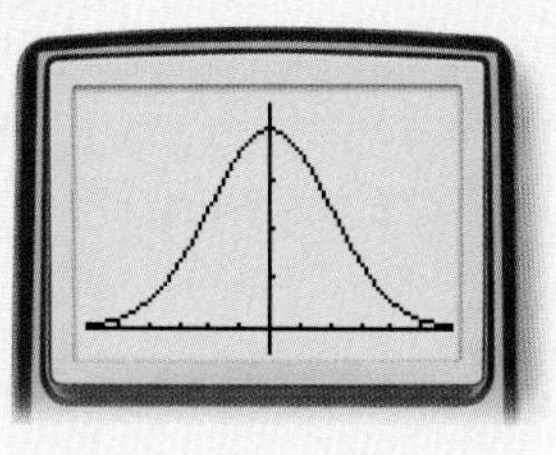

2. Use the trace feature to verify that the graph is symmetric about the y-axis and that the x-axis is a horizontal asymptote.
3. Use the $\int f(x)dx$ feature to find the area under the curve (and above the x-axis) from $x = 0$ to $x = 1$ and from $x = -1$ to $x = 0$.
4. CHECKPOINT ✔ What area represents 1 standard deviation on either side of the mean ($x = 0$)?
5. Repeat Step 3 for values from $x = 0$ to $x = 2$ and from $x = -2$ to $x = 0$.
6. CHECKPOINT ✔ What area represents 2 standard deviations on either side of $x = 0$?
7. Repeat Step 3 for values from $x = 0$ to $x = 3$ and from $x = -3$ to $x = 0$.
8. CHECKPOINT ✔ What area represents 3 standard deviations on either side of $x = 0$?
9. Find the areas corresponding to values less than $x = -3$ and to values greater than $x = 3$.
10. CHECKPOINT ✔ What probability is represented by the total area under the curve?

All normal distributions have the properties listed below.

Properties of Normal Distributions

- The curve is symmetric about the mean, $\bar{x}$.
- The total area under the curve is 1.
- The mean, median, and mode are about equal.
- About 68% of the area is within 1 standard deviation (σ) of the mean.
- About 95% of the area is within 2 standard deviations (2σ) of the mean.
- About 99% of the area is within 3 standard deviations (3σ) of the mean.

2% 14% 34% 34% 14% 2%
$\bar{x}-3\sigma$ $\bar{x}-2\sigma$ $\bar{x}-\sigma$ $\bar{x}$ $\bar{x}+\sigma$ $\bar{x}+2\sigma$ $\bar{x}+3\sigma$ x

CHECKPOINT ✔ Examine the two normal curves graphed below. How are the standard deviations different? How does the size of the standard deviation of a normal distribution affect the graph?

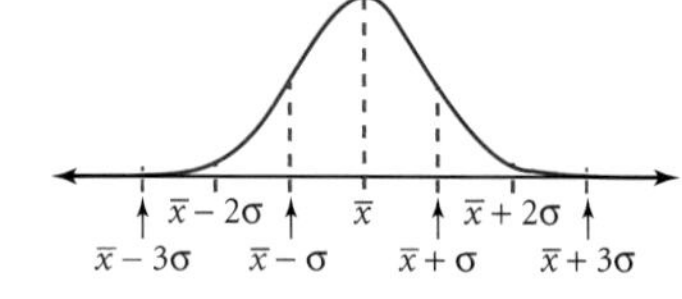

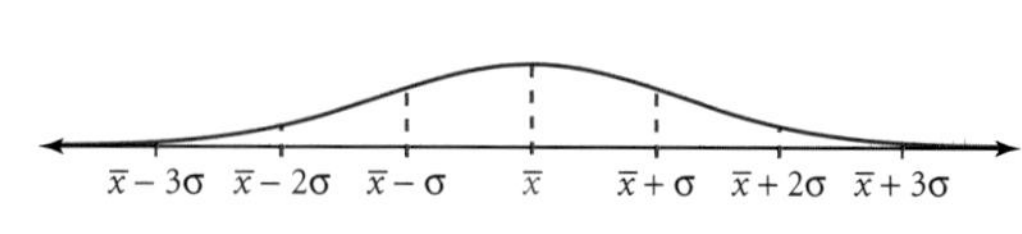

Inclusion Strategies

ENGLISH LANGUAGE DEVELOPMENT Be sure that students can verbally express the mathematical notation associated with normal curves. The result of part **a** of Example 1 is read as "The probability that x is greater than 1.6 is approximately 0.0548." In part **b** of Example 1, $P(-2.0 \le x \le 0.4)$ is read as "the probability that x lies between –2 and 0.4." Tell students that on a graphics calculator, $\int f(x)dx$ can be used to find "the area bounded by the x-axis and the normal curve between two values of x." When the $\int f(x)dx$ feature is used on a TI-83 graphics calculator, the user is asked to enter the lower and upper limits that define the desired region. Have students refer to the Try This on page 810 and express the answer, using both mathematical and verbal notation. Be sure that they translate it as both $P(82 \le x \le 95) \approx 0.5893$ and as the area, ≈ 0.5893, bounded by the x-axis and the normal curve with a mean of 85 and a standard deviation of 7 for values of x between 82 and 95.

Normal distributions occur in many large data sets. For example, standardized test scores are usually normally distributed, as shown in Example 2.

EXAMPLE 2

APPLICATION
ACADEMICS

Scores for a certain professional exam are approximately normally distributed with a mean of 650 and a standard deviation of 100.

a. What is the probability that a randomly selected test score is between 450 and 850?

b. Out of 1000 randomly selected test scores, how many would you expect to be between 450 and 850?

SOLUTION

a. Because $\overline{x} = 650$ and $\sigma = 100$, the interval from 450 to 850 is $650 - 2\sigma \leq \overline{x} \leq 650 + 2\sigma$.

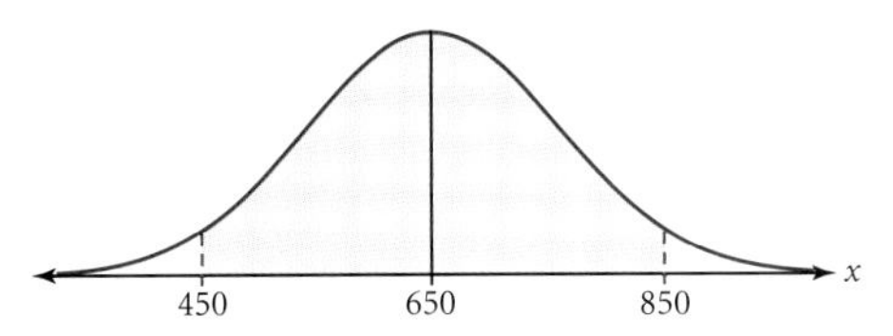

Thus, the probability that a randomly selected test score is between 450 and 850 is about 95%.

b. 95% of 1000 = $0.95 \times 1000 = 950$

Thus, you could expect about 950 out of 1000 randomly selected test scores to be between 450 and 850.

TRY THIS Refer to Example 2 above. Out of 2300 randomly selected scores, how many would you expect to be between 650 and 950?

Using *z*-Scores

A measure called a *z-score* tells how far a data value is from the mean in terms of standard deviations. For example, if a data set has a mean of 50 and a standard deviation of 10, then a data value of 70 has a *z*-score of 2 because it is 2 standard deviations above the mean. Similarly, a score of 20 has a *z*-score of −3 because it is 3 standard deviations below the mean.

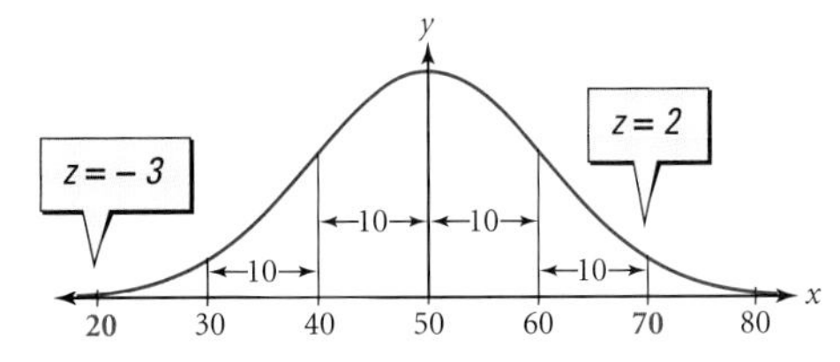

CHECKPOINT ✔
4. ≈0.6826

CHECKPOINT ✔
6. ≈0.9545

CHECKPOINT ✔
8. 0.9973

CHECKPOINT ✔
10. 1

CHECKPOINT ✔
The standard deviation of the first distribution is less than the standard deviation of the second distribution. When the standard deviation is larger, the height of the curve is shorter and the spread of the curve is wider.

ADDITIONAL EXAMPLE 2

On a box of cereal, the weight is listed as 12 ounces. Among all boxes of cereal of this type, however, the actual weights approximate a normal distribution with a mean of 12 ounces and a standard deviation of 0.1 ounce.

a. What is the probability that the contents of a randomly selected box of cereal of this type is within 0.1 ounce of 12 ounces? about 68%

b. How many boxes of cereal out of 5000 can the manufacturer expect to be within this range? about 3400

TRY THIS
≈1147

Enrichment

Have students graph the functions below.

$f(x) = e^{-x^2}$ $\quad g(x) = e^{-\frac{x^2}{2}}$

$h(x) = \frac{1}{\sqrt{2\pi}}e^{-\frac{x^2}{2}}$ $\quad j(x) = \frac{1}{\sqrt{2\pi}}e^{-\frac{(x-2)^2}{2}}$

If students are using TI-82 or TI-83 graphics calculators, have them set the viewing window as follows:

Xmin=−3.5 **Ymin=0**
Xmax=5.5 **Ymax=1.2**

Have them describe the symmetry and the end behavior of each function. Then have them compare the graphs of *f* and *g*, *g* and *h*, and *h* and *j*. Then ask them to guess which graphs have an area of 1 between the graph and the *x*-axis.

Functions *f*, *g*, and *h* are symmetric with respect to the *y*-axis, and *j* is symmetric with respect to $x = 2$. Because the *x*-axis is a horizontal asymptote, as *x* increases in both the positive and negative directions, *f*, *g*, *h*, and *j* approach 0. The graph of *g* is a horizontal stretch of *f*, *h* is a vertical compression of *g*, and *j* is a translation of *h*. The area under *h* and *j* is equal to 1.

ADDITIONAL
EXAMPLE 3

A manufacturer finds that the life expectancy of its best-selling model of clothes dryer is 11 years and the standard deviation is 1.5 years. **Find the probability that this model of clothes dryer will last from 10 to 12 years.**
about 49.5%

Teaching Tip

TECHNOLOGY When using a TI-83 or TI-82 graphics calculator to create the shaded graphs shown in Example 3, use the keystrokes below.

Press 2nd VARS, select **DRAW** **1:ShadeNorm(**, enter the parameters shown below for each method, and press) ENTER.

Method 1:
Use the following viewing window and parameters:

Xmin=–5	**Ymin=–.02**
Xmax=110	**Ymax=.08**
Xscl=10	**Yscl=.01**

parameters: **75,90,85,7**

Method 2:
Use the following viewing window and parameters:

Xmin=–3	**Ymin=–.2**
Xmax=3	**Ymax=.5**
Xscl=1	**Yscl=.1**

parameters: **–1.43,.71,0,1**

CHECKPOINT ✔
78 minutes, 92 minutes; the probability that the airliner will arrive between 78 minutes and 92 minutes, or within 1 standard deviation is 68%, which is close to the probability of 69% found in Example 3.

TRY THIS
≈0.59, or 59%

z-Score

If a data set is normally distributed with a mean of $\bar{x}$ and a standard deviation of σ, then the *z*-score for any data value, *x*, in that data set is given by $z = \frac{x - \bar{x}}{\sigma}$.

EXAMPLE 3

APPLICATION
TRAVEL

An airline finds that the travel times between two cities have a mean of 85 minutes and a standard deviation of 7 minutes. Assume that the travel times are normally distributed.

Find the probability that a flight from the first city to the second city will take between 75 minutes and 90 minutes.

SOLUTION

TECHNOLOGY
GRAPHICS CALCULATOR
Keystroke Guide, page 825

Method 1
Find the area between 75 and 90 under a normal curve with a mean of 85 and a standard deviation of 7.

The area is about 0.686.

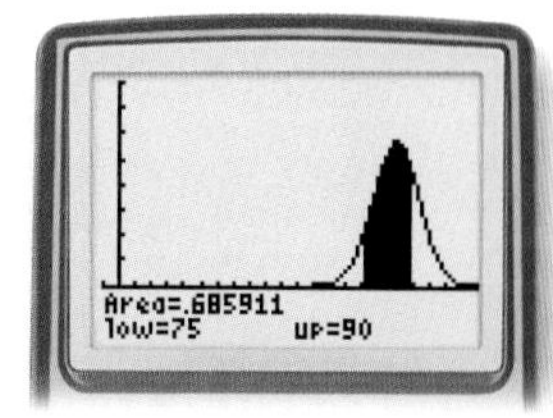

Method 2
Find the *z*-scores for $x_1 = 75$ and $x_2 = 90$.

$$z_1 = \frac{x_1 - \bar{x}}{\sigma} = \frac{75 - 85}{7} \approx -1.43 \qquad z_2 = \frac{x_2 - \bar{x}}{\sigma} = \frac{90 - 85}{7} \approx 0.71$$

Find the area between $z_1 \approx -1.43$ and $z_2 \approx 0.71$ under the standard normal curve.

The area is about 0.685.

Thus, there is about a 69% probability that the flight will take between 75 minutes and 90 minutes.

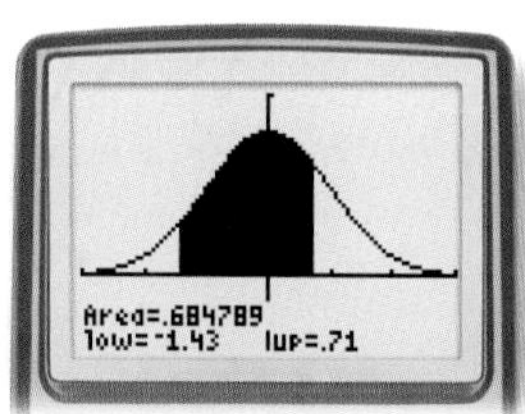

CHECKPOINT ✔ Find the flight times that correspond to $\bar{x} - \sigma$ and $\bar{x} + \sigma$. How do your answers confirm that the solution in Example 3 is reasonable?

TRY THIS Find the probability that a flight from the first city to the second city will take between 82 minutes and 95 minutes.

Reteaching the Lesson

USING VISUAL MODELS Give each student an accurate graph of a normal curve. Explain that the area of an interval on the *x*-axis under the graph represents the probability that a sample data value, taken from a population that varies normally about its mean, will be in that particular interval. Show students the function that describes the curve, and point out that the probability intervals are bordered by multiples of the standard deviation. Have students point to the mean, the interval located 1 standard deviation below the mean, and the interval located 1 standard deviation above the mean. Tell students about two situations in which the means are both 30 but the standard deviations are 2 and 10. Have them draw horizontal scales showing the correspondence between the *z*-scores and the actual data for both situations beneath the *x*-axis, and have them identify the data values that represent -3σ, -2σ, $-\sigma$, 0, σ, 2σ, and 3σ for both situations. Remind them that about 68% of the total area under the curve is between $-\sigma$ and σ, 95% of the total area is between -2σ and 2σ, and 99% of the total areas is between -3σ and 3σ.

Exercises

Communicate

1. Describe the characteristics of the normal curve, including the measures that define it.
2. Explain how to find the area between 25 and 50 under a normal curve with a mean of 30 and a standard deviation of 10.
3. Sketch three normal curves that have the same mean but different standard deviations. Explain how the standard deviation affects the shape of the normal curve.

Guided Skills Practice

Find each probability by using the area table for a standard normal curve given on page 807. ***(EXAMPLE 1)***

4. $P(-0.6 \le x \le 1.4)$ **0.6449**
5. $P(0.2 \le x \le 1.8)$ **0.3848**

APPLICATIONS

6. **MANUFACTURING** The lengths of a certain bolt are normally distributed with a mean of 8 centimeters and a standard deviation of 0.01 centimeter. ***(EXAMPLE 2)***
   a. What is the probability that a randomly selected bolt is within 0.03 centimeter of 8 centimeters? **≈0.99**
   b. Out of 1000 randomly selected bolts, how many can the manufacturer expect to be within 0.03 centimeter of 8 centimeters? **≈990**
7. **TRANSPORTATION** A bus route takes a mean of 40 minutes to complete, with a standard deviation of 5 minutes. Assume that completion times for the route are normally distributed. Find the probability that it takes between 30 minutes and 45 minutes to complete the route. ***(EXAMPLE 3)*** **0.8185**

Practice and Apply

Let x be a random variable with a standard normal distribution. Use the area table for a standard normal curve, given on page 807, to find each probability.

8. $P(x \ge 0.8)$ **0.2119**
9. $P(x \ge 0.4)$ **0.3446**
10. $P(1.0 \le x \le 1.6)$ **0.1039**
11. $P(0.2 \le x \le 1.8)$ **0.3848**
12. $P(x \ge 1)$ **0.1587**
13. $P(x \ge 2)$ **0.0228**
14. $P(x \le -0.2)$ **0.4207**
15. $P(x \le -1.4)$ **0.0808**
16. $P(-0.4 \le x \le 0.4)$ **0.3108**
17. $P(-0.2 \le x \le 0.2)$ **0.1586**
18. $P(x \ge -1.6)$ **0.9452**
19. $P(x \ge -0.8)$ **0.7881**

Let x be a random variable with a standard normal distribution. Use a graphics calculator to find each probability. Round answers to the nearest ten-thousandth.

20. $P(0.653 \le x)$ **0.2569**
21. $P(1.456 \le x)$ **0.0727**
22. $P(1.254 \le x)$ **0.1049**
23. $P(1.457 \le x)$ **0.0726**
24. $P(0.842 \le x \le 1.233)$
25. $P(0.423 \le x \le 1.438)$

24. 0.0911
25. 0.2609

Assess

Selected Answers

Exercises 4–7, 9–71 odd

ASSIGNMENT GUIDE	
In Class	1–7
Core	9–43 odd, 47, 49
Core Plus	8–50 even, 51
Review	52–72
Preview	73

Extra Practice can be found beginning on page 940.

Error Analysis

Encourage students to make a sketch of the conditions for the problem on a normal curve in order to check the results from the calculator. This will help students find errors caused by inaccurate data entry. Remind them that they must use *z*-scores when using a table to find the area under the normal curve between the mean and a given value.

Practice

NAME ______ CLASS ______ DATE ______

Practice
12.6 **Normal Distributions**

Let *x* be a random variable with a standard normal distribution. Use the area table for a standard normal curve, given on page 807 of the textbook, to find each probability.

1. $P(x \geq 0)$ 0.5
2. $P(x \leq 1.2)$ 0.8849
3. $P(x \geq -1.8)$ 0.9641
4. $P(0 \leq x < 0.4)$ 0.1554
5. $P(0 \leq x \leq 2.0)$ 0.4772
6. $P(-0.2 \leq x \leq 0)$ 0.0793
7. $P(1.0 \leq x \leq 2.0)$ 0.1359
8. $P(-0.2 \leq x \leq 0.2)$ 0.1586
9. $P(-0.4 \leq x \leq 1.2)$ 0.5403

The time required to finish a given test is normally distibuted with a mean of 40 minutes and a standard deviation of 8 minutes.

10. What is the probability that a student chosen at randon will finish in less than 32 minutes? 16%
11. What is the probability that a student chosen at random will take more than than 56 minutes to finish? 2%
12. What is the probability that a student chosen at random will take between 24 minutes and 48 minutes? 82%

The owners of a restaurant determine that the number of minutes that a customer waits to be served is normally distributed with a mean of 6 minutes and a standard deviation of 2 minutes.

13. What is the probability that a randomly selected customer will be served in less than 4 minutes? 16%
14. During a survey, 500 customers are served. How many would you expect to be served in less than 8 minutes? ≈420 customers
15. If 1000 customers are served, how many would you expect to wait between 4 minutes and 10 minutes? ≈820 customers

internet connect
Homework Help Online
Go To: **go.hrw.com**
Keyword:
MB1 Homework Help
for Exercises 26–37

A city's annual rainfall is approximately normally distributed with a mean of 40 inches and a standard deviation of 6 inches. Find the probability, to the nearest ten-thousandth, for each amount of annual rainfall in the city.

26. less than 34 inches **0.1587**
27. greater than 46 inches **0.1587**
28. greater than 52 inches **0.0228**
29. less than 28 inches **0.0228**
30. between 34 and 40 inches **0.3413**
31. between 34 and 46 inches **0.6826**

Scores on a professional exam are normally distributed with a mean of 500 and a standard deviation of 50. Out of 28,000 randomly selected exams, find the number of exams that could be expected to have each score.

32. greater than 500 **14,000**
33. greater than 550 **4444**
34. less than 600 **27,362**
35. less than 400 **638**
36. between 450 and 550 **19,113**
37. between 400 and 600 **26,723**

A survey of male shoe sizes is approximately normally distributed with a mean size of 9 and a standard deviation of 1.5. Use *z*-scores to find each probability to the nearest ten-thousandth.

38 $P(9 \leq x \leq 10)$ **0.2486**
39 $P(10 \leq x \leq 11)$ **0.1596**
40 $P(7 \leq x \leq 11)$ **0.8164**
41 $P(7 \leq x \leq 13)$ **0.9044**
42 $P(7 \leq x \leq 12)$ **0.8854**
43 $P(5 \leq x)$ **0.9962**

APPLICATIONS

TRANSPORTATION Tests show that a certain model of a new car averages 36 miles per gallon on the highway with a standard deviation of 3 miles per gallon. Assuming that the distribution is normal, find the probability, to the nearest ten-thousandth, that a car of this model gets the indicated mileage.

44. more than 40 miles per gallon **0.0918**
45. less than 32 miles per gallon **0.0918**
46. between 34 and 38 miles per gallon **0.4972**

MORTGAGE Mortgage statistics indicate that the number of years that a new homeowner will occupy a house before moving or selling is normally distributed with a mean of 6.3 years and a standard deviation of 2.3 years. Find the probability, to the nearest ten-thousandth, of each event.

47. A homeowner will sell or move within 3 years of buying the house. **0.0764**
48. A homeowner will sell or move after 10 years of buying the house. **0.0537**
49. A homeowner will sell or move after 6 to 8 years of buying the house. **0.3221**

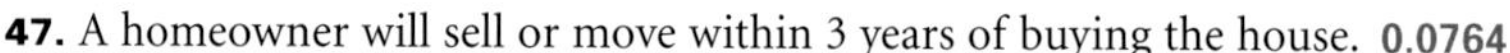

50. QUALITY CONTROL A machine fills containers with perfume. When the machine is adjusted properly, it fills the containers with 4 fluid ounces, with a standard deviation of 0.1 fluid ounce. If 400 randomly selected containers are tested, how many containers with less than 3.8 fluid ounces of perfume will indicate that the machine needs to be adjusted? **10 or more containers**

CHALLENGE

51. QUALITY CONTROL The lengths of a mechanical component are normally distributed with a mean of 30.0 inches and standard deviation of 0.2 inch. All components not within 0.4 inch of 30.0 inches are rejected.

a. What percent of the components are acceptable? **95%**

b. If one of the *acceptable* components is randomly selected, what is the probability, to the nearest percent, that its length is within 0.1 inch of 30.0 inches? **40%**

Look Back

Write an equation in slope-intercept form for the line that contains the given point and is parallel to the given line. *(LESSON 1.3)*

52. $(9, -4)$, $y = -12x + 3$ **53.** $(1, -4)$, $y = 100x + 12$

Write an equation in slope-intercept form for the line that contains the given point and is perpendicular to the given line. *(LESSON 1.3)*

54. $(6, -3)$, $y = -2x + 18$ **55.** $(-2, 7)$, $6x - 7y = 8$

Simplify each expression. *(LESSON 8.4)*

56. $\frac{1}{x} + \frac{1}{x+2}$ **57.** $\frac{1}{x+1} + \frac{1}{x+2}$ **58.** $\frac{x}{1+x} + \frac{x^2+x}{x}$ **59.** $\frac{1}{x} + \frac{x+1}{x+2}$

Write each expression with a rational denominator and in simplest form. *(LESSON 8.7)*

60. $\frac{1}{\sqrt{3}}$ $\frac{\sqrt{3}}{3}$ **61.** $\frac{1}{\sqrt{5}}$ $\frac{\sqrt{5}}{5}$ **62.** $\frac{1}{\sqrt{2}-1}$ $\sqrt{2}+1$ **63.** $\frac{1}{1-\sqrt{3}}$ $-\frac{1+\sqrt{3}}{2}$

52. $y = -12x + 104$
53. $y = 100x - 104$
54. $y = \frac{1}{2}x - 6$
55. $y = -\frac{7}{6}x + \frac{14}{3}$
56. $\frac{2x+2}{x^2+2x}$
57. $\frac{2x+3}{x^2+3x+2}$
58. $\frac{x^2+3x+1}{1+x}$
59. $\frac{x^2+2x+2}{x^2+2x}$

APPLICATION

64. EDUCATION The table at right shows the final exam scores for each student in an economics class. ***(LESSON 10.4)***

72	88	96	75	85
98	80	87	78	90
80	87	82	88	93
92	84	97	98	83

a. Find the range and the mean deviation of the scores.

b. Find the variance and the standard deviation of the scores.

64a. range = 26; mean dev. ≈ 6.09
b. $\sigma^2 = 54.7275$; $\sigma \approx 7.40$

Find the indicated sum of the arithmetic series 5 + 7 + 9 + 11 + · · · *(LESSON 11.3)*

65. S_3 **21** **66.** S_4 **32** **67.** S_{10} **140** **68.** S_{15} **285**

Find the indicated sum of the geometric series $1 + \frac{3}{2} + \frac{9}{4} + \frac{27}{8} + \cdots$ *(LESSON 11.5)*

69. S_3 **70.** S_4 **71.** S_{10} **72.** S_{15}

69. $\frac{19}{4} = 4.75$
70. $\frac{65}{8} = 8.125$
71. $\frac{58{,}025}{512} \approx 113.33$
72. $\frac{14{,}316{,}139}{16{,}384} \approx 873.79$

Look Beyond

73 *Exponential distributions* are widely used to model lifetimes of electronic components which are generally not affected by how long they have already been operated. Suppose that the lifetime (in months) of a certain fuse can be modeled by the exponential density function $f(x) = \frac{1}{100}e^{\frac{-x}{100}}$ for $x \geq 0$. Graph f and describe its general shape.

Look Beyond

In Exercise 73, students are introduced to a function that represents an exponential distribution. This type of distribution is widely used to model lifetimes of mechanical objects and other phenomena. The general form of this type of distribution is given by the following density curve:

$$f(x) = \begin{cases} \frac{1}{\mu} e^{\frac{-x}{\mu}} & x \geq 0 \\ 0 & \text{otherwise} \end{cases}$$

The mean of the distribution is μ, and the area under the curve to the left of x is given by $g(x) = 1 - e^{\frac{-x}{\mu}}$, which is used to compute probabilities.

73.

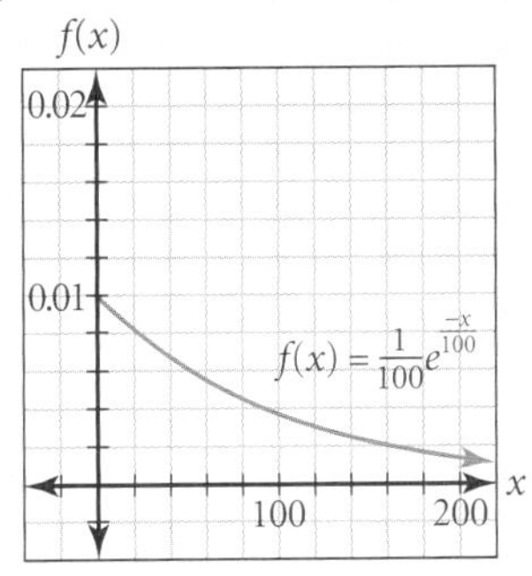

The graph of f is the graph of $g(x) = e^{-x}$, where $x \geq 0$, but f is compressed vertically by a factor of 100 and stretched horizontally by a factor of 100.

Student Technology Guide

NAME ______ CLASS ______ DATE ______

Student Technology Guide
12.6 Normal Distributions

When data is normally distributed, you can use the area under the normal curve to find the probability that an outcome will be less than a particular number, greater than a particular number, or between two numbers. You can use a graphics calculator to find probabilities for normally distributed data.

Example: Suppose that scores on a standardized test are normally distributed with a mean of 50 and a standard deviation of 10. Find the probability that a student scores between 45 and 65 on this test.

Method 1: Find the area under the curve

- Set the viewing window. Press WINDOW. For this situation, set Xmin = 0, Xmax = 100, and Xscl = 10 in order to reflect a distribution whose mean is 50. Set Ymin = −0.02, Ymax = 0.05 and Yscl = 0.01.
- Press 2nd PRGM DRAW 1:ClrDraw ENTER ENTER to clear any previous drawings.
- Press 2nd VARS DRAW 1:ShadeNorm(ENTER to draw a normal curve.
- Press 45 , 65 , 50 , 10) to enter the lower bound, upper bound, mean, and standard deviation. Press ENTER.
- The graph shades the part of the curve between 45 and 65 and displays the area of this portion of the curve.

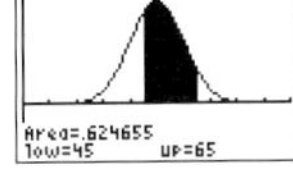

The probability that a student scores between 45 and 65 on this test is about 0.62, or 62%.

Method 2: Use the probability distribution

- Press 2nd VARS 2:normalcdf(ENTER to select cumulative probabilities for a normal distribution.
- Press 45 , 65 , 50 , 10) to enter the lower bound, upper bound, mean, and standard deviation. Press ENTER.

This method confirms the probability found above.

Note: To find the probability that a number is *above* a certain value, use an upper bound of 10^{99}, and to find the probability that a number is *below* a certain value, use a lower bound of -10^{99}.

A bowling league finds that its players' scores are approximately normally distributed with a mean of 142 and a standard deviation of 29. Find the probability that a bowler in this league scores

1. between 122 and 158 **≈ 0.46, or 46%**
2. between 150 and 175 **≈ 0.26, or 26%**
3. more than 200 **≈ 0.02, or 2%**
4. less than 100 **≈ 0.07, or 7%**

Focus

The characteristics of a total population of individuals or objects can often be estimated by using a much smaller subset, or sample, of that population. The accuracy of the estimate is determined in large part by the size of the sample and by the method used to select the sample. Students will review a case study that describes and evaluates a number of sampling procedures and then will make their own evaluation of the sample procedures used in nine other surveys. Then they will design a survey to be conducted in school. Students' designs should include selecting a sample, setting up a sampling procedure, conducting the survey, presenting the data graphically, and drawing conclusions from the results.

Motivate

Ask students to discuss their ideas about how election predictions are made, or how an elected public official's popularity with voters is determined. Ask students how many people they think should be questioned and what selection procedures should be used to obtain accurate results.

That's NOT Fair!

Several weeks prior to an election, a poll can be taken to predict the percent of the votes that each candidate will receive on election day. Since contacting every eligible voter would be impractical, a subset, or *sample*, of all eligible voters is polled. This sample should represent the total population of voters as accurately as possible.

A certain city with 80,000 eligible voters can be divided into four areas that roughly correspond to the annual family incomes shown in the table below.

Area	Number of voters	Family income
Rolling Hills	400	over $100,000
South Park	40,600	$75,000–$100,000
West Side	27,000	$35,000–$74,999
North End	12,000	under $35,000

The sampling methods below will not give a good representation of the total population of voters in this city for the reasons given.

1. **Every home in Rolling Hills** The Rolling Hills area has families with the highest income level. Voters from this area might tend to vote for a candidate who will represent their interests by promising to lower taxes, to provide services based on property and home values, and so on.
2. **100 homes in each of the four areas** The 80,000 eligible voters in the city are not divided equally among the four areas. South Park and West Side would be *under* represented in the sample while Rolling Hills and North End would be *over* represented.
3. **Every 10th person entering the terminal building at the local commercial airport that serves this city** The frequency of airline trips is not a characteristic that is equally distributed among the total population of the city. Business people tend to fly more often than many other people.

Activity 1

1. Poor; the men in the U.S. Naval Academy are usually healthy, fit, and not overweight. This sample would underrepresent the 20-year-old males who are overweight.
2. Poor; not all people receive the newspaper. In addition, the sample may be overrepresented by the people with a particularly strong opinion, which causes them to take the time and money to send in the ballot.
3. Poor; first, it only allows for people to be sampled that have televisions and telephones. Second, since it is a controversial topic, the only people that may take the time to call are those with strong opinions.
4. Reasonable; every 5th cookie should be a representative sample.
5. Poor; the automobiles produced on Friday afternoons are not representative of all the cars manufactured at all other times of the week.

In contrast to the faulty sampling methods discussed above, there are efficient methods that provide a more representative sample of a given population. Two examples are:

- Telephone every nth person listed in the telephone book.
- Mail a questionnaire with a stamped return envelope to every nth person in the telephone book. Notice that even these methods are not truly representative. For example, some people don't have telephones, some have unlisted numbers, some might not return their questionnaires, and so on. Still, for a specific city or area, these methods often give a fairly representative sample.

Activity 1

Tell whether each sampling procedure below provides a poor or a reasonable representation of the total population. Justify your answers.

1. To determine the mean (average) weight of all 20-year-old males, find the weight of all 20-year-old males in the U.S. Naval Academy.
2. To determine the majority opinion on whether a city should treat its water supply with fluoride, publish a ballot in the city's newspaper and ask the public to vote and mail the ballot to the newspaper's office.
3. To determine the majority opinion on a very controversial topic, have television viewers telephone in a "yes" or "no" response.
4. To determine the average number of chocolate chips per cookie in a bag of cookies, find the average number in every fifth cookie.
5. To determine the quality of a given model of automobile, test the last one assembled on every Friday of the model year.
6. To estimate the average brightness of the stars in the summer sky, an astronomer measures the brightness of all the stars that she can see and computes the average brightness.
7. To determine whether the effectiveness of a herbicide was less than advertised, a researcher samples a group of 50 randomly selected users of the herbicide for their opinion about its effectiveness.

Activity 2

1. Design a survey to conduct among a sample of your classmates that represents the student population at your school. Describe what you think the results of your survey will reveal. Explain why you think your sample is a reasonable and fair representation of the entire student population.
2. Conduct your survey. Then represent the data in two of the following formats: a stem-and-leaf plot, histogram, circle graph, or box-and-whisker plot. Explain why you chose these two formats to represent your data.
3. What conclusions can you make about the results of your survey? How do statistical measures and graphs support your conclusions? Are the results what you expected?

Cooperative Learning

Divide the class into survey teams of four members each. Each team will work to together to review and discuss the case study on sampling procedures. Each team will make a written evaluation of the nine sampling procedures presented in **Activity 1** and complete a survey as specified in **Activity 2.**

Discuss

Each survey team should make a report to the class. The report should specify what question was chosen and what results were anticipated, as well as the sample size and sampling procedure used. Then two graphic representations of the data should be presented with an explanation of why those two formats were chosen. Each team should give their conclusions of the survey results, using the data and graphs as support, and specify whether these results were consistent with the anticipated results.

6. Poor; she is only collecting a sample of the brightness of the brightest stars, which are the only ones she can see, not all the stars in the sky.

7. Reasonable; the researcher collected a random sample of users, not just a random sample of users who were disappointed and had complained about the product.

Activity 2

1. Student surveys should include one question, such as "How many siblings do you have?" or "What is your favorite movie?" They should also describe their sample population and explain why it is a good representative sample.

2. Students should choose the format that best represents their data. For example, if there is a large variation in their data, they might choose a stem-and-leaf plot or a histogram.

3. Students should comment on the variation and distribution of their data and draw conclusions based on their results.

12 Chapter Review and Assessment

VOCABULARY

Key Skills & Exercises

LESSON 12.1

Key Skills

Find the mean, median, and mode.

For the data set 8, 7, 4, 5, 9, 4, 5, 4, 2, 3:

mean

$$\bar{x} = \frac{2+3+4+4+4+5+5+7+8+9}{10} = 5.1$$

median

2, 3, 4, 4, 4, 5, 5, 7, 8, 9

The middle numbers are 4 and 5, so the median is 4.5.

mode

The mode is 4.

Exercises

Find the mean, median, and mode of each data set.

1. 7, 9, 2, 9, 0, 2, 8, 9, 1 $\bar{x} = 5.\bar{2}$; 7; 9
2. −3, 8, 2, 3, 2, 4, 3, 2 $\bar{x} = 2.625$; 2.5; 2

Make a frequency table for each data set, and find the mean.

3. 5, 4, 6, 5, 4, 6, 6, 5, 4, 7, 4, 5, 6 5.15
4. 9, 10, 11, 8, 10, 10, 11, 9, 8, 10 9.6

LESSON 12.2

Key Skills

Make a stem-and-leaf plot.

Data: 4, 4, 28, 3, 29, 15, 12, 16, 17, 24, 16, 28, 5, 28, 29

Stems	Leaves (1|2 = 12)
0	3, 4, 4, 5
1	2, 5, 6, 6, 7
2	4, 8, 8, 8, 9, 9

Make a histogram.

Data: 5, 2, 3, 3, 6, 1, 3, 4, 2, 3, 1, 5, 5, 2, 4

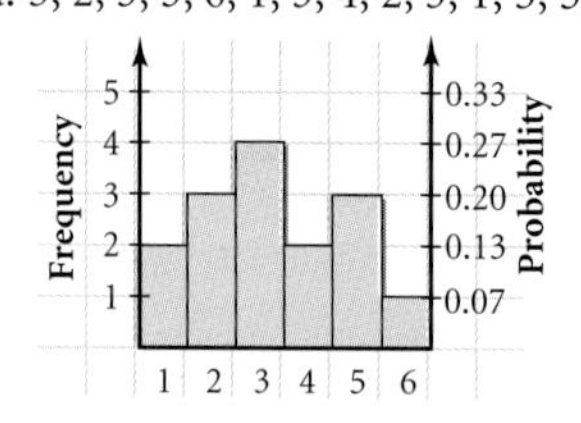

Exercises

Make a stem-and-leaf plot for each data set.

5. 35, 38, 45, 49, 45, 53, 57, 58
6. 67, 87, 82, 73, 81, 78, 79, 69

Make a histogram for each data set.

7. 4, 3, 2, 5, 4, 5, 6, 1, 2, 4, 3, 1
8. 24, 28, 30, 30, 22, 21, 29, 29, 30, 27, 26, 28, 22, 23, 25, 27, 28, 28, 24

Make a relative frequency histogram for each data set.

9. 1, 1, 3, 3, 5, 2, 5, 3
10. 10, 14, 14, 11, 15, 13, 12

3.

Data	Frequency
4	4
5	4
6	4
7	1
Total	13

4.

Data	Frequency
8	2
9	2
10	4
11	2
Total	10

The answers to Exercises 5–10 can be found in Additional Answers beginning on page 1002.

Chapter Test, Form A

NAME ______ CLASS ______ DATE ______

Chapter Assessment

Chapter 12, Form A, page 1

Write the letter that best answers the question or completes the statement.

b 1. Find the median of this set of data: 52, 45, 34, 67, 21, 54, 67, 34, 89, 43, 50, 31.
a. 37.6 b. 47.5 c. 48.9 d. 60.5

d 2. Find the mean of the data shown in the stem-and-leaf plot at right.
a. 4.4 b. 17.8 c. 34.2 d. 51.3

Stem	Leaf
6	0 3 7
5	4 4 6 8 9
4	1 3 5
3	3 4

c 3. A video store surveyed its customers to find the number of movies they rent per month. The results are shown in the grouped frequency table at right. Estimate the mean number of movies per month rented by the customers of this video store.
a. 7.5 b. 8.5 c. 9.5 d. 10.5

Movies Rented	Frequency
1–5	8
6–10	10
11–15	7
16–20	5

a 4. Find Q_1 in this set of data: 17, 21, 24, 32, 45, 46, 51, 52, 63, 72, 81.
a. 24 b. 28 c. 46 d. 63

c 5. The data from a survey on the number of pieces of junk mail people in a town receive is summarized in the box-and-whisker plot at right. Find the IQR for the data.
a. 10 b. 20 c. 30 d. 45

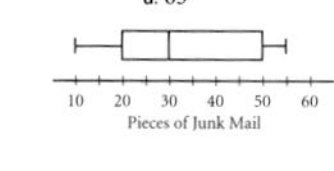

a 6. A teacher surveyed his class to find the number of minutes students spend traveling to school each morning. The results in minutes are shown below.
10 25 15 20 15 5 22 15 20 18
Find σ for the data.
a. 5.6 minutes b. 9.2 minutes c. 16.5 minutes d. 31.1 minutes

d 7. Find the variance of a data set with a standard deviation of 16.
a. 4 b. 32 c. 128 d. 256

a 8. Find the mean deviation of this set of data: 15, 25, 28, 35, 33, 17, 22.
a. 6 b. 7 c. 12 d. 25

NAME ______ CLASS ______ DATE ______

Chapter Assessment

Chapter 12, Form A, page 2

d 9. Find $P(x \geq 1.54)$ using a standard normal curve area table or a graphics calculator.
a. 0.9382 b. 0.7438 c. 0.2754 d. 0.0618

a 10. The scores for a statistics test are shown below.
84 93 67 84 53 73 73 67 95 84 72 85 78 71 87 90 64
Find the range and the mode for the test scores.
a. range = 42; mode = 84 b. range = 40; mode = 73
c. range = 67; mode = 84 d. range = 84; mode = 73

b 11. The results of a survey on the number of television sets owned by households in a town are shown in the histogram at right. If one of the households in the survey were randomly selected, find the probability that the household has 4 or more television sets.
a. 40% b. 50% c. 60% d. 70%

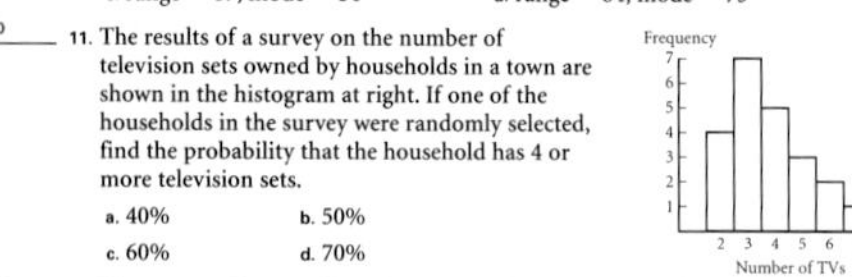

c 12. If a person rolls a number cube 7 times, find the probability that an even number results exactly 5 times.
a. 0.714 b. 0.418 c. 0.164 d. 0.031

b 13. A restaurant has found that 55% of their customers have coffee with their meals. If 12 people eat at this restaurant, find the probability that exactly 7 of them will have coffee with their meals.
a. 0.015 b. 0.222 c. 0.4242 d. 0.577

b 14. A set of data is normally distributed with a mean of 12 and a standard deviation of 4. Find the z-score for 6 in this set of data.
a. 1.5 b. −1.5 c. 0.17 d. −0.17

d 15. The weights of the fish in a lake are normally distributed with a mean of 9.4 pounds and a standard deviation of 3.2 pounds. If a person catches a fish in this lake, find the probability that the fish weighs 15 pounds or less.
a. 0.0401 b. 0.1038 c. 0.8962 d. 0.9599

Make a circle graph.

For	98
Against	67
Undecided	20

For 53%
Undecided 11%
Against 36%

Divide each category in the table by the total, 185, to find its percent. Then multiply its percent by 360° to find the measure of the central angle.

11. The table below lists the number of people who selected each brand. Make a circle graph of the results.

Brand X	53
Brand Y	32
Brand Z	89

LESSON 12.3

Key Skills

Find the minimum and maximum values, range, quartiles, and interquartile range for a data set.

Data: 5, 6, 9, 2, 3, 7, 2, 9, 8
Arrange the data set in ascending order.
2 2 3 5 6 7 8 9 9

$$Q_1 = \frac{2+3}{2} = 2.5 \quad Q_2 = 6 \quad Q_3 = \frac{8+9}{2} = 8.5$$

$$\text{range} = 9 - 2 = 7 \qquad \text{IQR} = 8.5 - 2.5 = 6$$

Make a box-and-whisker plot.

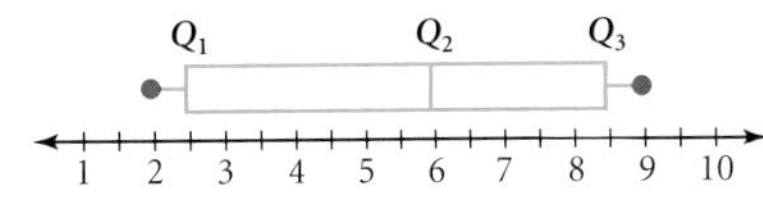

Exercises

Find the quartiles of each data set.

12. 12, 18, 13, 15, 16, 19, 17, 13, 14, 19, 17, 16

13. 23, 28, 34, 36, 34, 29, 35, 31, 45, 22, 23, 25

Make a box-and-whisker plot for each data set.

14. 5, 4, 9, 3, 1, 9, 0, 6, 3, 2

15. 22, 28, 29, 24, 25, 28, 29, 23, 23, 29

12. $Q_1 = 13.5$; $Q_2 = 16$; $Q_3 = 17.5$

13. $Q_1 = 24$; $Q_2 = 30$; $Q_3 = 34.5$

LESSON 12.4

Key Skills

Find the range, variance, mean deviation, and standard deviation of a data set.

Data: 3, 2, 3, 5, 7, 5
In ascending order:
2, 3, 3, 5, 5, 7

Range: 7 − 2 = 5

Mean:
$$\frac{2+3+3+5+5+7}{6} \approx 4.167$$

Mean deviation: $\frac{9}{6} = 1.5$

Variance: $\frac{1}{6}\sum_{i=1}^{6}(x_i - \overline{x})^2 \approx 2.8$

Standard deviation: $\sqrt{2.8} \approx 1.7$

x_i	$\lvert x_i - \overline{x}\rvert$
3	1.167
2	2.167
3	1.167
5	0.833
7	2.833
5	0.833
Total	9

Exercises

Find the range and mean deviation of each data set.

16. 6, 10, 12, 4, 14, 8, 11, 14 **10; ≈2.91**

17. 20, 22, 15, 14, 13, 17 **9; ≈2.83**

18. 3, 6, −7, 9, −3, 2 **16; ≈4.44**

19. 4, −8, 12, 13, −22, 24, 21 **46; ≈12.82**

Find the variance and standard deviation of each data set.

20. 10, 12, 15, 18, 11, 13, 14, 16, 19, 20 **$\sigma^2 \approx 10.56$; $\sigma \approx 3.25$**

21. 100, 140, 130, 180, 80, 160 **$\sigma^2 \approx 1147.22$; $\sigma \approx 33.87$**

22. 8, 9, 12, 14, 7, 9, 11, 13, 14 **$\sigma^2 \approx 6.17$; $\sigma \approx 2.48$**

23. 3, 4, 12, 2, 3, 4, 6, 12, 18, 20, 2 **$\sigma^2 \approx 39.42$; $\sigma \approx 6.28$**

11.

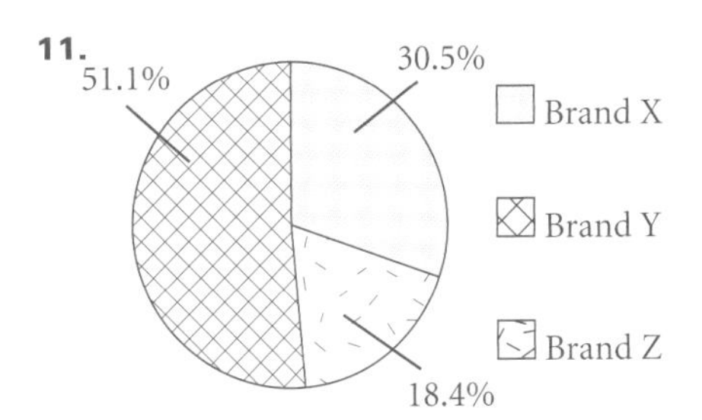

14.

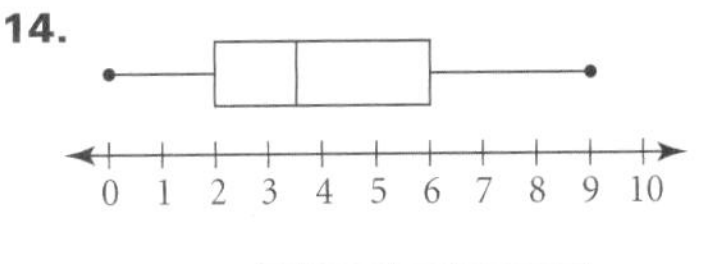

15.

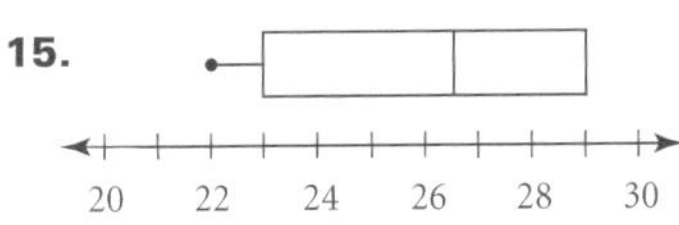

Chapter Test, Form B

NAME ______ CLASS ______ DATE ______

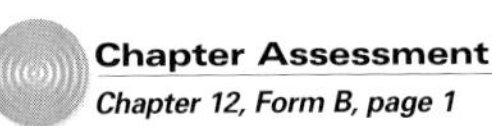

Chapter Assessment
Chapter 12, Form B, page 1

1. A principal takes a random sample of the classes at her school and finds the number of students in each class. The data is shown below.
17 23 25 31 14 19 27 33 20 23
11 30 23 35 16 25 17 28 32 21
a. In the space at right, make a stem-and-leaf plot of the data.
b. Find the mean, median, and mode for the data.
mean: 23.5 students; median: 23 students; mode: 23 students

Stem	Leaf 3|0 = 30
3	0 1 2 3 5
2	0 1 3 3 3 5 5 7 8
1	1 4 6 7 7 9

2. A travel agent is interested in the length of time people in a town go on vacation. The results from a random sample of people in the town are shown below.
5 3 4 5 7 6 5 3 3 4 5 7 3
4 4 6 3 5 7 5 4 5 6 5 4
a. In the space at right, make a histogram of the data.
b. If one person from the sample is randomly selected, find the probability that the person spends at least 5 days on vacation.
56%

Frequency
Number of Days

3. The data below represents the number of people that stopped at a display in a mall during a random sample of days.
45 56 67 32 43 85 67 54 35 62 60
55 74 41 78 63 35 48 65 39 58
a. Find the quartiles and the IQR for the data.
Q_1 = 42 people; Q_2 = 56 people; Q_3 = 66 people; IQR = 24 people
b. In the space at right, make a box-and-whisker plot of the data.

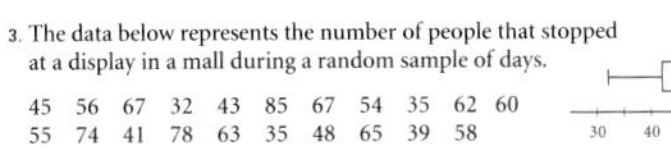

4. The data below represents the number of inches of rain that fell in Knox during 10 consecutive months.
6.3 5.7 7.3 4.7 5.3 3.7 6.8 7.6 4.2 3.4
Find the range, variance, standard deviation, and mean deviation for the data. Round your answers to the nearest hundredth.
range: 4.2 inches; variance: 2.00 square inches; standard deviation: 1.42 inches; mean deviation: 1.24 inches

NAME ______ CLASS ______ DATE ______

Chapter Assessment
Chapter 12, Form B, page 2

5. Make a circle graph of the data below, which shows the favorite fast food of a randomly selected sample of people at a mall.

Food	Frequency
Pizza	15
Hamburgers	18
Tacos	15
Chicken Nuggets	12

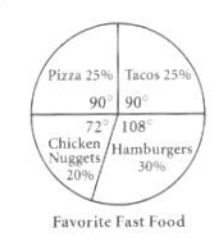

Favorite Fast Food

Find each probability by using a standard normal curve area table or a graphics calculator.

6. $P(x \geq -0.67)$ 0.7486
7. $P(x < 2.33)$ 0.9901
8. $P(-1.65 < x \leq -1.29)$ 0.049

9. If a fair coin is tossed 14 times, find the probability that tails results exactly 10 times. Round your answer to the nearest thousandth. 0.061

10. An insurance agent has found that 7% of drivers in Plymouth have an accident in any given year. If the insurance agent has insured 14 drivers in Plymouth, find the probability that 3 or fewer will have an accident this year. Round your answer to the nearest thousandth. 0.986

11. The scores on a reading test are normally distributed with a mean of 150 and a standard deviation of 20.
a. Find the z-score of a test score of 175. $z = 1.25$
b. If a randomly selected person takes this reading test, find the probability that the person scores 175 or lower on the test. 0.8944

12. The amount of soda a machine dispenses into cups is normally distributed with a mean of 250 milliliters and a standard deviation of 8 milliliters. If one cup of soda dispensed by this machine is randomly selected, find the following probabilities:
a. Find the probability that the cup contains more than 238 milliliters of soda. 0.9332
b. Find the probability that the cup contains between 234 and 262 milliliters of soda. 0.9104

LESSON 12.5

Key Skills

Find probabilities of events in a binomial experiment.

A certain trick coin has a probability of 0.75 for heads and 0.25 for tails.

The probability of 3 heads appearing in 5 tosses of the coin is found by using the formula for binomial probability, as shown below.

$$P(\text{exactly 3 heads}) = {}_5C_3(0.75)^3(0.25^{5-3}) \approx 0.26$$

Exercises

Given 10 trials of a binomial experiment with 0.6 probability of success, find each probability below.

24. $P(\text{exactly 2 successes})$ **0.0106**

25. $P(\text{exactly 3 successes})$ **0.0425**

26. $P(\text{at least 3 successes})$ **0.9877**

27. Find the probability of at least 3 heads in 5 tosses of a fair coin. **0.5**

LESSON 12.6

Key Skills

Find probabilities by using a normal distribution.

Let x be a random variable with a standard normal distribution. To find $P(-1.6 \le x \le 0.4)$, use the area table for a standard normal curve on page 807.

$$P(-1.6 \le x \le 0.4) \approx 0.4452 + 0.1554 \approx 0.6006$$

Use *z*-scores to find probabilities.

If x is a random data value with a normally distributed data set with a mean of 6 and a standard deviation of 3, find $P(8 \le x \le 11)$.

$$z_1 = \frac{8-6}{3} = \frac{2}{3} \approx 0.667$$

$$z_2 = \frac{11-6}{3} = \frac{5}{3} \approx 1.667$$

$$P(8 \le x \le 11) \approx 0.205$$

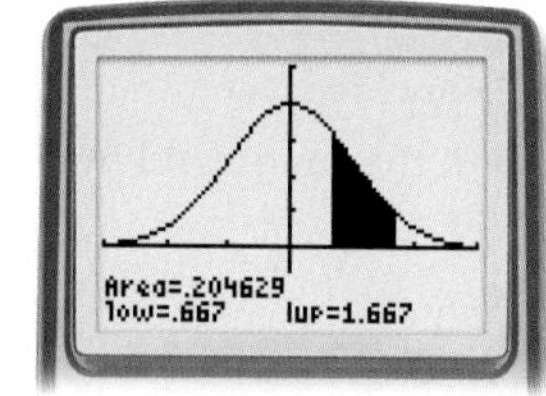

Exercises

Let x be a random variable with a standard normal distribution. Use the area table for a standard normal curve on page 807 to find each probability.

28. $P(x \le 0.4)$ **0.6554**

29. $P(x \ge -1.8)$ **0.9641**

30. $P(-1.6 \le x \le 1.6)$ **0.8904**

31. $P(-0.8 \le x \le 0.8)$ **0.5762**

Let x be a random data value with a normally distributed data set with a mean of 20 and a standard deviation of 2. Use *z*-scores to find each probability to the nearest percent.

32 $P(x \le -21)$ **0.6915**

33 $P(x \le 18.6)$ **0.2420**

34 $P(16.5 \le x \le 23.5)$ **0.9199**

35 $P(17.2 \le x \le 22)$ **0.7606**

Applications

36. AUTOMOBILE DISTRIBUTION The table at right gives the number of each type of vehicle on a car dealer's lot. Make a circle graph to represent this data.

Type	Number
cars (other than sports cars)	50
sports cars	20
trucks	30
sports utility vehicles	28

36.

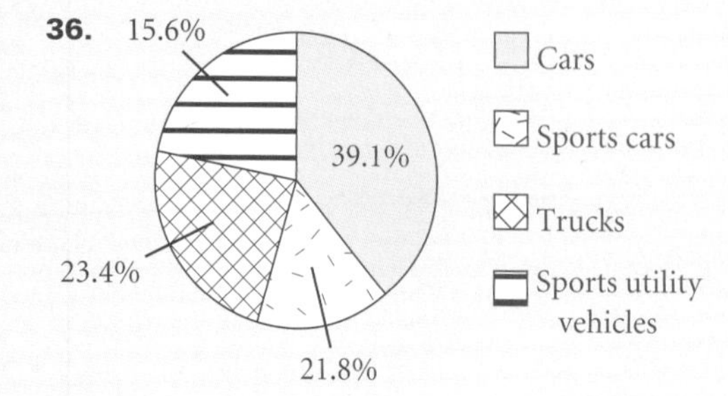

Chapter Test

Find the mean, median, and mode of each data set.

1. 2, 5, 8, 2, 4, 2, 8, 1, 3, 5, 2, 2, 8, 3, 2 $\bar{x} = 3.8$; 3; 2
2. 12, 10, 15, 13, 15, 16, 15, 14, 18, 12 $\bar{x} = 14$; 14.5; 15

Make a frequency table for each data set, and find the mean.

3. the number of snow days missed during the past 20 years: 1, 2, 3, 4, 4, 1, 3, 1, 3, 2, 2, 2, 4, 1, 3, 5, 3, 2, 3, 4
4. the age of students in a tennis class: 14, 18, 15, 17, 16, 16, 14, 15, 17, 15, 16, 15, 17, 16, 15, 15, 14, 15, 16, 14, 15

Make a stem-and-leaf plot for each data set.

5. 15, 20, 100, 50, 45, 37, 23, 75, 12, 25, 40, 64, 85, 20, 57, 44, 28, 60, 24, 10, 36, 30
6. 1.8, 2.4, 5.9, 4.6, 2.3, 2.8, 1.9, 3.2, 6.5, 3.4, 3.9, 2.5

Make a histogram for each data set.

7. 3, 6, 2, 4, 4, 4, 3, 5, 4, 5, 2, 3, 4, 4, 4, 3, 5
8. 13, 9, 10, 12, 11, 9, 10, 11, 8, 10, 11, 9, 8, 12, 10, 11, 11, 10, 10, 9
9. **POLITICS** The table shows the number of people voting in the last election. Represent these data with a circle graph.

Party affiliation	Number of people voting
Democrat	160
Republican	200
Independent	100
Other	40

Find the minimum and maximum values, quartiles, range, and interquartile range for each data set.

10. 18, 24, 59, 46, 23, 28, 19, 32, 65, 34, 39, 25, 27, 42, 54, 23, 34, 19, 41
11. 3, 5, 8, 9, 4, 1, 6, 5, 2, 6, 2, 8, 5, 7

Find the range and mean deviation for each data set.

12. 3, 5, 7, 9, 11, 13 10; 3
13. 10, 12, 15, 18, 23, 25, 30, 33 23; 7

Find the variance and standard deviation for each data set.

14. 3, 5, 7, 9, 11, 13 11.67; 3.42
15. 10, 12, 15, 18, 23, 25, 30, 33 61.44; 7.84
16. What can be said about a set of data if its standard deviation is 0? All the data points are equal to one another.

Use this information for Exercises 17 and 18. The probability that a battery will fail in the cold is 8%. A new digital camcorder uses 4 of these batteries.

17. What is the probability that the camcorder will work with the supplied batteries? 0.72
18. What is the probability that exactly 2 of the batteries will fail when first used? 0.033
19. A fair coin is flipped 7 times. Find the probability of exactly 6 heads. 0.055

Let x be a random variable with a standard normal distribution. Find each probability.

20. $P(x \leq 1.6)$ 0.9452
21. $P(x \geq -1.2)$ 0.8849
22. $P(-1.4 \leq x \leq 0.8)$ 0.7073
23. **PSYCHOLOGY** IQ scores are normally distributed with a mean of 100 and a standard deviation of 15. If a genius has an IQ score above 127, what percentage of people can be classified as a genius? 3.6%
24. **EDUCATION** Grade point averages are normally distributed with a mean of 2.5 and a standard deviation of 0.5. What percentage of students graduate with a grade point average of 3.5 or more? 2.3%

Chapter Test

3. mean = 2.65

Days	Frequency
1	4
2	5
3	6
4	4
5	1

4. mean = 15.5

Age	Frequency
14	4
15	8
16	5
17	3
18	1

5. 1|0 = 10

Stems	Leaf
1	0, 2, 5
2	0, 0, 3, 4, 5, 8
3	0, 6, 7
4	0, 4, 5
5	0, 7
6	0, 4
7	5
8	5
9	
10	0

6. 1|8 = 1.8

Stems	Leaf
1	8, 9
2	3, 4, 5, 8
3	2, 4, 9
4	6
5	9
6	5

7.

8.

9.

10. minimum = 18; Q1 = 23; Q2 = 32; Q3 = 42; maximum = 65; range = 47; IQR = 19

11. minimum = 1; Q1 = 3; Q2 = 5; Q3 = 7; maximum = 9; range = 8; IQR = 4

College Entrance Exam Practice

Multiple-Choice and Quantitative-Comparison Samples

The first half of the Cumulative Assessment contains two types of items found on standardized tests—multiple-choice questions and quantitative-comparison questions. Quantitative-comparison items emphasize the concepts of equality, inequality, and estimation.

Free-Response Grid Samples

The second half of the Cumulative Assessment is a free-response section. This part of the Cumulative Assessment requires student-produced response items like those commonly found on college entrance exams. These questions require the use of machine-scored answer grids. You may wish to have students practice answering these items in preparation for standardized tests.

College Entrance Exam Practice

QUANTITATIVE COMPARISON For Items 1–5, write
A if the quantity in Column A is greater than the quantity in Column B;
B if the quantity in Column B is greater than the quantity in Column A;
C if the quantities are equal; or
D if the relationship cannot be determined from the given information.

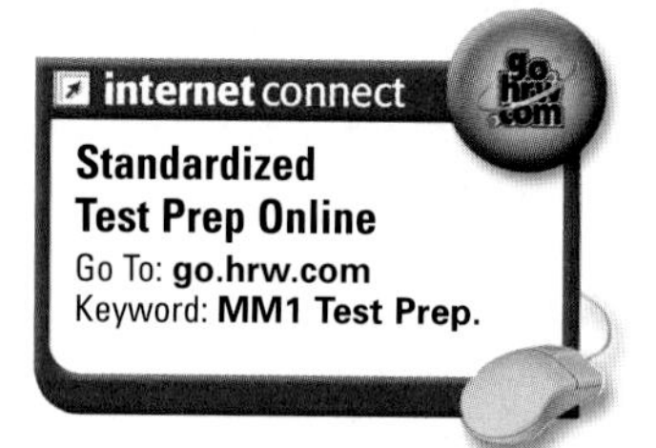

	Column A	Column B	Answers
1. A	The value of x: $\frac{2x}{7} = \frac{3x-1}{4}$	The value of x: $\frac{x}{6} = \frac{2x}{5}$	(A) (B) (C) (D) [Lesson 12.1]
2. B	$\lceil 3.95 \rceil$	$\lceil 4.95 \rceil$	(A) (B) (C) (D) [Lesson 2.6]
3. A	The maximum value of $f(x) = -(x+4)^2 + 3$	The minimum value of $g(x) = 2(x-5)^2 - 1$	(A) (B) (C) (D) [Lesson 10.3]
4. B	${}_5C_3$	${}_5P_3$	(A) (B) (C) (D) [Lesson 5.1]
5. B	The mean of 76, 78, 71, 78, 72, 75	The median of 76, 78, 71, 78, 72, 75	(A) (B) (C) (D) [Lesson 1.4]

6. Solve $4(5 + 2x) = 42 - 3x$. ***(LESSON 1.6)***
c
a. $x = -2$ **b.** $x = 4$
c. $x = 2$ **d.** $x = 12$

7. Which is the axis of symmetry of the graph of $y = 2x^2 - 8x - 1$? ***(LESSON 5.1)***
c
a. $x = 1$ **b.** $x = -1$
c. $x = 2$ **d.** $x = -2$

8. Which expression gives the function rule for $f \circ g$, where $f(x) = 3 - x$ and $g(x) = 2x$? ***(LESSON 2.4)***
a
a. $3 - 2x$ **b.** $3 + 2x$
c. $2(3 - x)$ **d.** $6 - 2x$

9. Simplify $\left(-\frac{2a^{-2}}{b^{-1}}\right)^3$. ***(LESSON 2.2)***
c
a. $\frac{8a^6}{b^3}$ **b.** $\frac{-8b^6}{a^3}$
c. $\frac{-8b^3}{a^6}$ **d.** $-\frac{8a^6}{b^3}$

10. Which property is illustrated by $a + 0 = a$, where a represents a real number? ***(LESSON 2.1)***
a
a. Identity **b.** Inverse
c. Reciprocal **d.** Closure

11. How many solutions does a consistent system of linear equations have? ***(LESSON 3.1)***
c
a. 0 **b.** 1
c. at least 1 **d.** infinite

12. Which expression is equivalent to $|x - 2| = 6$? ***(LESSON 1.8)***
c
a. $x = 4$ **b.** $x = 8$
c. $x = -4$ *or* $x = 8$ **d.** $x = 8$ *and* $x = -8$

13. Which best describes the roots of $3x^2 + 2x - 5 = 0$? ***(LESSON 5.6)***
b
a. 1 real root **b.** 2 rational roots
c. 2 irrational roots **d.** 2 imaginary roots

14. Which function represents exponential decay? *(LESSON 6.2)* c

a. $f(x) = 2.5^x$ b. $f(x) = 2(5)^x$
c. $f(x) = 2(0.5)^x$ d. $f(x) = 2x^2$

15. Evaluate $-\frac{1}{3}\sqrt[3]{27}$. *(LESSON 8.6)* d

a. 3 b. −3 c. 1 d. −1

16. Which is the remainder when $2x^2 - 5x + 8$ is divided by $x + 4$? *(LESSON 7.3)* a

a. 60 b. −44 c. 0 d. 20

17. Which expression is equivalent to $\log_a \frac{xy}{2}$? *(LESSON 6.4)* c

a. $\log_a 2xy$ b. $\log_a y + 2 \log_a a$
c. $\log_a xy - \log_a 2$ d. $\log_a xy + \log_a 2$

18. Solve $\begin{cases} 2x^2 + y^2 = 36 \\ x^2 - y^2 = 12 \end{cases}$. *(LESSON 9.6)* d

a. $(2\sqrt{13}, 8), (-2\sqrt{13}, 8)$
b. $(8, 2\sqrt{13}), (-8, 2\sqrt{13})$
c. $(2, 4), (2, -4), (-2, 4), (-2, -4)$
d. $(4, 2)(-4, 2), (4, -2), (-4, -2)$

19. Write the equation in slope-intercept form for the line that has a slope of $-\frac{5}{3}$ and contains the point $(8, -3)$. *(LESSON 1.3)* $y = -\frac{5}{3}x + \frac{31}{3}$

20. Write the function that represents the graph of $f(x) = |x|$ translated 2 units to the left. *(LESSON 2.7)* $f(x) = |x + 2|$

21. Factor $25x^2 - 9$, if possible. *(LESSON 5.3)* $(5x + 3)(5x - 3)$

22. Graph $-2 < y \leq 3$ in a coordinate plane. *(LESSON 3.4)*

Let $A = \begin{bmatrix} 1 & 4 \\ -3 & -1 \end{bmatrix}$ and $B = \begin{bmatrix} 1 & -11 \\ 6 & 15 \end{bmatrix}$. Perform the indicated operations. *(LESSON 4.1)*

23. AB 24. $3BA$

25. $A + B$ 26. $B - 2A$

27. Use elimination to solve the system. *(LESSON 3.2)* $\begin{cases} 5x + 11y = -7 \\ -3x + 2y = 30 \end{cases}$ (−8, 3)

28. Multiply $(5 - 2i)(5 - 2i)$. *(LESSON 5.6)* $21 - 20i$

29. Find the domain of $g(x) = \frac{x^2 - 1}{x - 5}$. *(LESSON 8.2)* $x \neq 5$

30. Evaluate $e^{\ln 2} + \ln e^{-7}$. *(LESSON 6.5)* −5

31. Describe the end behavior of $P(x) = -2x^3 + x^2 - 11x + 4$. *(LESSON 7.2)*

32. If a coin is tossed 10 times, what is the probability that it will land heads up exactly 4 times? *(LESSON 12.5)* ≈0.205

33. Divide by using synthetic division. $(3x^3 - 18x + 12) \div (x - 3)$ *(LESSON 7.3)*

34. Solve $x^2 + 2x - 1 \leq 0$, and graph its solution. *(LESSON 5.7)*

35. Solve $\frac{3}{x^2 - 4} = \frac{-2}{5x + 10}$. *(LESSON 8.5)* −5.5

36. Find the domain of $g(x) = \sqrt{1 - 2(x + 1)}$. *(LESSON 8.6)* $x \leq -\frac{1}{2}$

37. In how many ways can a committee of 5 be chosen from a group of 8? *(LESSON 10.3)* 56

38. Factor $x^3 + 125$. *(LESSON 7.3)*

FREE-RESPONSE GRID **The following questions may be answered by using a free-response grid such as that commonly used by standardized-test services.**

39. Simplify $\sqrt{27^{\frac{2}{3}}}$. *(LESSON 8.7)* 3

40. Evaluate $10^{\log_{10} 1000} - \log_2 2$. *(LESSON 6.4)* 999

41. Find the value of v in $2 = \log_v 64$. *(LESSON 6.3)* 8

42. Find the sixth term of the sequence −100, 500, −250, . . . *(LESSON 11.4)* 312,500

43. Find the distance between the points $P(-3, 2)$ and $Q(1, 5)$. *(LESSON 9.1)* 5

ENTERTAINMENT **The top 10 movies grossed the dollar amounts (in millions) shown at right in one weekend.**

4.0	14.8
3.7	10.6
3.0	6.1
2.9	5.6
2.5	4.2

44. Find the mean for this data set. *(LESSON 12.1)* 5.74

45. Find the range for this data set. *(LESSON 12.4)* 12.3

22.

23. $\begin{bmatrix} 24 & 49 \\ -6 & 18 \end{bmatrix}$

24. $\begin{bmatrix} 99 & 33 \\ -117 & 27 \end{bmatrix}$

25. $\begin{bmatrix} 1 & -7 \\ 3 & 14 \end{bmatrix}$

26. $\begin{bmatrix} -2 & -19 \\ 12 & 17 \end{bmatrix}$

31. rises on the left and falls on the right

33. $3x^2 + 9x + 9 + \frac{39}{x - 3}$

34. $-1 - \sqrt{2} \leq x \leq -1 + \sqrt{2}$, or approx. $-2.414 \leq x \leq 0.414$

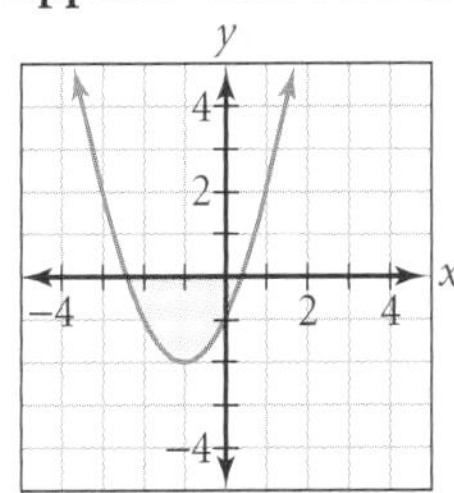

38. $(x + 5)(x^2 - 5x + 25)$

Keystroke Guide for Chapter 12

Essential keystroke sequences (using the model TI-82 or TI-83 graphics calculator) are presented below for all Activities and Examples found in this chapter that require or recommend the use of a graphics calculator.

internet connect

For Keystrokes of other graphing calculator models, visit the HRW web site at **go.hrw.com** and enter the keyword **MB1 CALC**.

LESSON 12.2

EXAMPLE 2 Make a histogram for the canoe rental data given.

Page 774

First clear old data and old equations.

Use viewing window [0, 9] by [0, 11].

Enter the data:

STAT EDIT 1:Edit ENTER L1 1 ENTER 4 ENTER 5 ENTER 2 ENTER 2 ENTER 2 ENTER 3 ENTER 2 ENTER 3 ENTER . . .

Continue until all of the data values are entered into List 1.

Make a histogram:

2nd STAT PLOT (Y=) STAT PLOTS 1:Plot 1 ENTER ON ENTER ▼ (Type:)

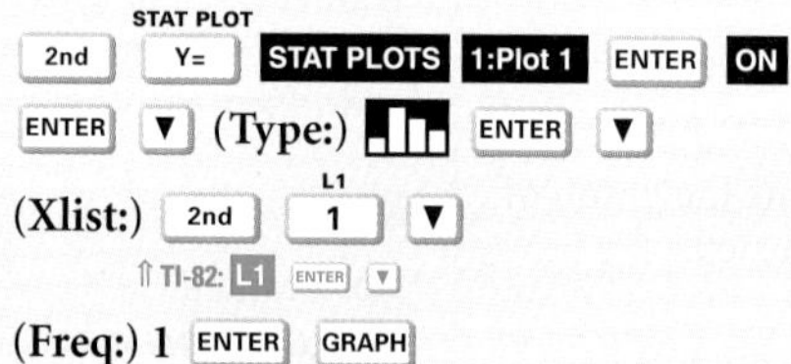

ENTER ▼

(Xlist:) 2nd L1 (1) ▼
⇑ TI-82: L1 ENTER ▼

(Freq:) 1 ENTER GRAPH
⇑ TI-82: 1 ENTER GRAPH

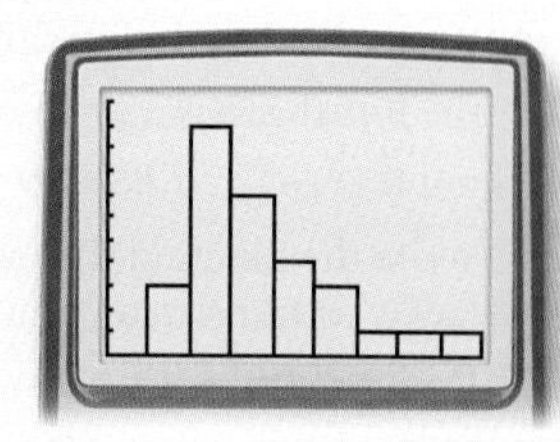

LESSON 12.3

EXAMPLE 2 Make box-and-whisker plots for the temperature data given.

Page 784

Use viewing window [10, 80] by [0, 1].

Enter the data:

STAT EDIT 1:Edit ENTER L1 21 ENTER 25.5 ENTER 37 ENTER 48.6 ENTER . . .

Continue until all of the data values for Chicago are entered into List 1.

► L2 55.9 ENTER 57 ENTER 58.3 ENTER 60.8 ENTER 63.3 ENTER . . .

Continue until all of the data values for Los Angeles are entered into List 2.

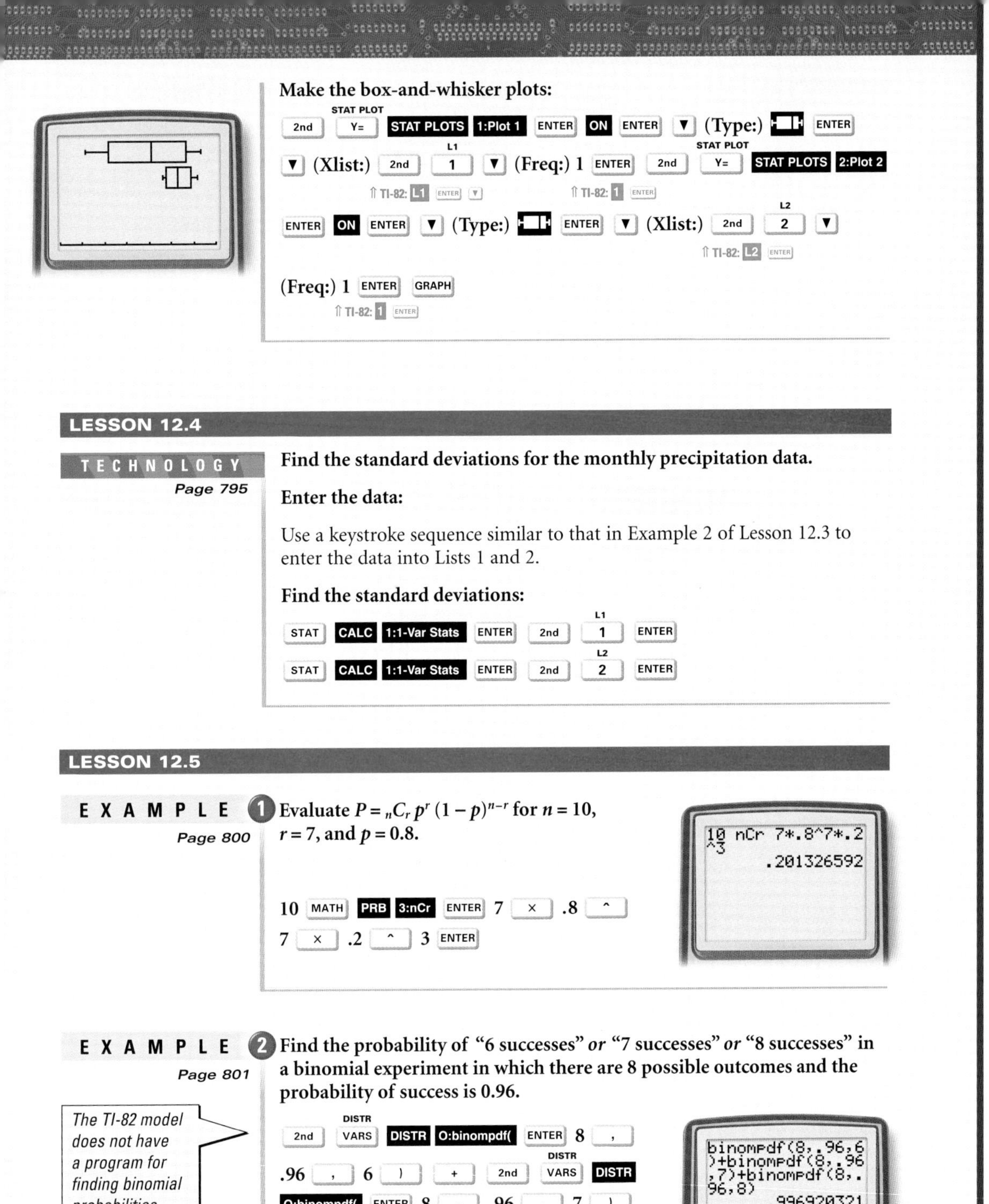

Make the box-and-whisker plots:

2nd [STAT PLOT] Y= | STAT PLOTS | 1:Plot 1 | ENTER | ON | ENTER | ▼ | (Type:) ▭ | ENTER

▼ (Xlist:) 2nd [L1] 1 ▼ (Freq:) 1 ENTER 2nd [STAT PLOT] Y= | STAT PLOTS | 2:Plot 2

⇑ TI-82: L1 ENTER ▼ ⇑ TI-82: 1 ENTER

ENTER | ON | ENTER | ▼ | (Type:) ▭ | ENTER | ▼ | (Xlist:) 2nd [L2] 2 ▼

⇑ TI-82: L2 ENTER

(Freq:) 1 ENTER GRAPH

⇑ TI-82: 1 ENTER

LESSON 12.4

TECHNOLOGY
Page 795

Find the standard deviations for the monthly precipitation data.

Enter the data:

Use a keystroke sequence similar to that in Example 2 of Lesson 12.3 to enter the data into Lists 1 and 2.

Find the standard deviations:

STAT | CALC | 1:1-Var Stats | ENTER | 2nd [L1] 1 | ENTER

STAT | CALC | 1:1-Var Stats | ENTER | 2nd [L2] 2 | ENTER

LESSON 12.5

EXAMPLE 1
Page 800

Evaluate $P = {}_nC_r\, p^r (1-p)^{n-r}$ for $n = 10$, $r = 7$, and $p = 0.8$.

10 MATH | PRB | 3:nCr | ENTER 7 × .8 ^ 7 × .2 ^ 3 ENTER

```
10 nCr 7*.8^7*.2
^3
          .201326592
```

EXAMPLE 2
Page 801

The TI-82 model does not have a program for finding binomial probabilities.

Find the probability of "6 successes" *or* "7 successes" *or* "8 successes" in a binomial experiment in which there are 8 possible outcomes and the probability of success is 0.96.

2nd [DISTR] VARS | DISTR | 0:binompdf(| ENTER 8 , .96 , 6) + 2nd [DISTR] VARS | DISTR | 0:binompdf(| ENTER 8 , .96 , 7) + 2nd [DISTR] VARS | DISTR | 0:binompdf(| ENTER 8 , .96 , 8) ENTER

```
binompdf(8,.96,6
)+binompdf(8,.96
,7)+binompdf(8,.
96,8)
          .996920321
```

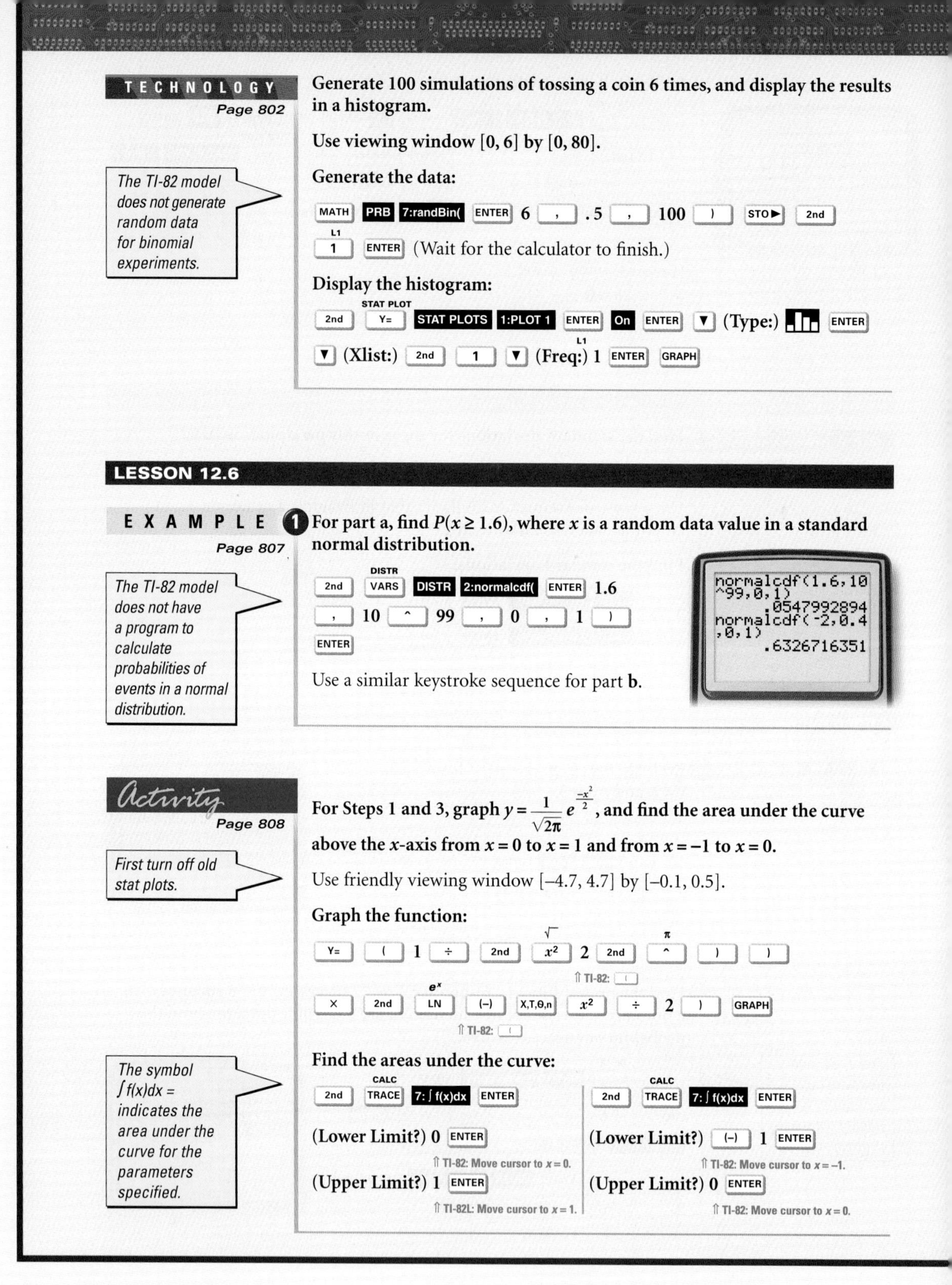

TECHNOLOGY
Page 802

The TI-82 model does not generate random data for binomial experiments.

Generate 100 simulations of tossing a coin 6 times, and display the results in a histogram.

Use viewing window [0, 6] by [0, 80].

Generate the data:

[MATH] [PRB] [7:randBin(] [ENTER] 6 [,] .5 [,] 100 [)] [STO▶] [2nd] [1] (L1) [ENTER] (Wait for the calculator to finish.)

Display the histogram:

[2nd] [Y=] (STAT PLOT) [STAT PLOTS] [1:PLOT 1] [ENTER] [On] [ENTER] [▼] (Type:) [histogram icon] [ENTER] [▼] (Xlist:) [2nd] [1] (L1) [▼] (Freq:) 1 [ENTER] [GRAPH]

LESSON 12.6

EXAMPLE 1
Page 807

The TI-82 model does not have a program to calculate probabilities of events in a normal distribution.

For part a, find $P(x \geq 1.6)$, where x is a random data value in a standard normal distribution.

[2nd] [VARS] (DISTR) [DISTR] [2:normalcdf(] [ENTER] 1.6 [,] 10 [^] 99 [,] 0 [,] 1 [)] [ENTER]

```
normalcdf(1.6,10
^99,0,1)
      .0547992894
normalcdf(-2,0.4
,0,1)
      .6326716351
```

Use a similar keystroke sequence for part **b**.

Activity
Page 808

First turn off old stat plots.

For Steps 1 and 3, graph $y = \frac{1}{\sqrt{2\pi}} e^{\frac{-x^2}{2}}$, and find the area under the curve above the x-axis from $x = 0$ to $x = 1$ and from $x = -1$ to $x = 0$.

Use friendly viewing window [–4.7, 4.7] by [–0.1, 0.5].

Graph the function:

[Y=] [(] 1 [÷] [2nd] [x^2] ($\sqrt{\ }$) 2 [2nd] [^] (π) [)] [)]
⇑ TI-82: [(]

[×] [2nd] [LN] (e^x) [(–)] [X,T,θ,n] [x^2] [÷] 2 [)] [GRAPH]
⇑ TI-82: [(]

The symbol ∫ f(x)dx = indicates the area under the curve for the parameters specified.

Find the areas under the curve:

[2nd] [TRACE] (CALC) [7:∫ f(x)dx] [ENTER]

(Lower Limit?) 0 [ENTER]
⇑ TI-82: Move cursor to $x = 0$.

(Upper Limit?) 1 [ENTER]
⇑ TI-82L: Move cursor to $x = 1$.

[2nd] [TRACE] (CALC) [7:∫ f(x)dx] [ENTER]

(Lower Limit?) [(–)] 1 [ENTER]
⇑ TI-82: Move cursor to $x = -1$.

(Upper Limit?) 0 [ENTER]
⇑ TI-82: Move cursor to $x = 0$.

EXAMPLE 3 For Method 1, find the area between 75 and 90 under a normal curve with a mean of 85 and a standard deviation of 7.

Page 810

To clear anything created through a draw command, press 2nd DRAW PRGM 1:ClrDraw ENTER ENTER.

Use viewing window [–5, 110] by [–0.02, 0.08].

To find these areas with a model TI-82, use a keystroke sequence similar to that used in the Activity for Lesson 12.6.

2nd DISTR VARS DRAW 1:ShadeNorm(ENTER

75 , 90 , 85 , 7) ENTER

Area=.685911
low=75 up=90

For Method 2, find the area between $x \approx -1.43$ and $x \approx 0.71$ under the standard normal curve.

Use viewing window [–3, 3] by [–0.2, 0.5].

2nd DISTR VARS DRAW 1:ShadeNorm(ENTER (–)

1.43 , .71) ENTER

Area=.684789
low=-1.43 up=.71

13 Trigonometric Functions

Lesson Presentation CD-ROM
PowerPoint® presentations for each lesson 13.1–13.6

CHAPTER PLANNING GUIDE

Lesson	13.1	13.2	13.3	13.4	13.5	13.6	Project and Review
Pupil's Edition Pages	828–835	836–842	843–850	851–857	858–866	867–873	874–875, 876–881
Practice and Assessment							
Extra Practice (Pupil's Edition)	982	983	983	984	984	985	
Practice Workbook	82	83	84	85	86	87	
Practice Masters Levels A, B, and C	244–246	247–249	250–252	253–255	256–258	259–261	
Standardized Test Practice Masters	94	95	96	97	98	99	100
Assessment Resources	166	167	168	169	171	172	170, 173–178
Visual Resources							
Lesson Presentation Transparencies Vol. 2	140–143	144–147	148–151	152–155	156–159	160–163	
Teaching Transparencies		54	55, 56	57	58	59, 60	
Answer Key Transparencies	470–472	473–478	479–484	485–489	490–499	500–502	503–509
Quiz Transparencies	13.1	13.2	13.3	13.4	13.5	13.6	
Teacher's Tools							
Reteaching Masters	163–164	165–166	167–168	169–170	171–172	173–174	
Make-Up Lesson Planner for Absent Students	82	83	84	85	86	87	
Student Study Guide	82	83	84	85	86	87	
Spanish Resources	82	83	84	85	86	87	
Block Scheduling Handbook							26–27
Activities and Extensions							
Lesson Activities	82	83	84	85	86	87	
Enrichment Masters	82	83	84	85	86	87	
Cooperative-Learning Activities	82	83	84	85	86	87	
Problem Solving/ Critical Thinking	82	83	84	85	86	87	
Student Technology Guide	82	83	84	85	86	87	
Long Term Projects							49–52
Writing Activities for Your Portfolio							37–39
Tech Prep Masters							61–64
Building Success in Mathematics							33–35

LESSON PACING GUIDE

Lesson	13.1	13.2	13.3	13.4	13.5	13.6	Project and Review
Traditional	2 days	2 days	1 day	1 day	1 day	1 day	2 days
Block	1 day	1 day	$\frac{1}{2}$ day	$\frac{1}{2}$ day	$\frac{1}{2}$ day	$\frac{1}{2}$ day	1 day
Two-Year	4 days	4 days	2 days	2 days	2 days	2 days	4 days

CONNECTIONS AND APPLICATIONS

Lesson	13.1	13.2	13.3	13.4	13.5	13.6	Review
Algebra	828–835	836–842	843–850	851–857	858–866	867–873	876–881
Geometry	832, 834		843, 849, 850	851, 856			
Probability		842					881
Transformations					860, 861, 862, 863		
Life Skills	834				866		
Business and Economics	834		849			872	
Science	828, 831, 833, 834	836, 837, 841, 842	843, 845, 848, 849	851, 854, 856, 857	858, 862, 865	867, 871, 872	879
Social Studies					865	873	878
Sports and Leisure				855, 856		873	
Cultural Connection: Africa	831						

BLOCK SCHEDULING GUIDE

Day	Lesson	Teacher Directed: Lesson Examples, Teaching Transparencies	Student Guided Activity, Try This	Cooperative-Learning Activity, Lesson Activity, Student Technology Guide	Practice: Practice & Apply, Extra Practice, Practice Workbook	Assessment: Quiz, Mid-Chapter Assessment	Problem Solving, Reteaching
1	13.1	10 min	15 min	15 min	65 min	15 min	15 min
2	13.2	10 min	15 min	15 min	65 min	15 min	15 min
3	13.3	10 min	10 min	7 min	25 min	7 min	7 min
	13.4	10 min	10 min	8 min	25 min	8 min	8 min
4	13.5	10 min	10 min	7 min	25 min	7 min	7 min
	13.6	10 min	10 min	8 min	25 min	8 min	8 min
5	Assess.	50 min **PE:** Chapter Review	90 min **PE:** Chapter Project, Writing Activities	90 min Tech Prep Masters	65 min **PE:** Chapter Assessment, Test Generator	30 min Chap. Assess. (A or B), Alt. Assess. (A or B), Test Generator	

PE: Pupil's Edition

Alternative Assessment

The following suggest alternative assessments for students who may benefit from a different type of assessment than the regular chapter quizzes and the mid-chapter/end-of-chapter test. Visit the HRW web site to get additional Alternative Assessment material.

internet connect

Alternative Assessment
Go To: **go.hrw.com**
Keyword: **MB1 Alt Assess**

Performance Assessment

1. Find the values of the trigonometric functions for an angle whose measure is 210° without using a calculator.
   a. Using a diagram, explain how to use a reference angle to find the answers.
   b. Find the required values by using the values of the trigonometric functions of a 30° angle.
2. Two points on a circle with a radius of 4 inches are 15 inches apart when distance is measured along the circle. What angle in degrees do the points and the center of the circle determine? How does radian measure figure into the solution?
3. Consider $y = -2\cos(3\theta - 30°) + 1$.
   a. Find amplitude, period, vertical shift, and phase shift for this function?
   b. Use transformations of the parent function to sketch the graph.

Portfolio Project

1. You can always draw an altitude in a triangle to divide the triangle into two right triangles. What information do you need to know about the original triangle so that right-triangle trigonometry can be used to find other measures of the given triangle? Use illustrations to justify your answer.
2. In the *polar coordinate system*, point P in the plane is represented by the distance between P and the *origin* O and the angle formed by $\overrightarrow{OP}$ and a fixed reference ray from O, called the *polar axis*.
   a. Using a graphics calculator set in polar-equation and radian mode, graph each of the following polar functions over the specified intervals. Do you see any patterns?

   $r = \sin\frac{\theta}{4}$, $0 \le \theta \le 4\pi$ $\quad r = \sin\frac{\theta}{6}$, $0 \le \theta \le 6\pi$ $\quad r = \sin\frac{\theta}{8}$, $0 \le \theta \le 8\pi$

internet connect

The table below identifies the pages in this chapter that contain internet and technology information.

Content Links

Activities Online	**pages 832, 854, 864**
Portfolio Extensions	**pages 835, 850**
Homework Help Online	**pages 833, 841, 848, 855, 864, 872**
Graphic Calculator Support	**page 882**

Resource Links

Parents can go online and find concepts that students are learning—lesson by lesson—and questions that pertain to each lesson, which facilitate parent-student discussion.

Go To: **go.hrw.com**
Keyword: **MB1 Parent Guide**

Technical Support

The following may be used to obtain technical support for any HRW software product.

Online Help: **www.hrwtechsupport.com**
e-mail: **tschrw@hrwtechsupport.com**

HRW Technical Support Center: **(800)323-9239**
7 AM to 10 PM Monday through Friday Central Time

Visit the HRW math web site at: **www.hrw.com/math**

Technology

Lesson Suggestions and Calculator Examples

(Keystrokes are based on a TI-83 calculator.)

Lesson 13.1 Trigonometry of Right Triangles

In this lesson, students will need to express angle measures in degrees. To set their calculators to degree mode, students should press MODE, select **Degree**, and press ENTER.

Consider posing the problem below.

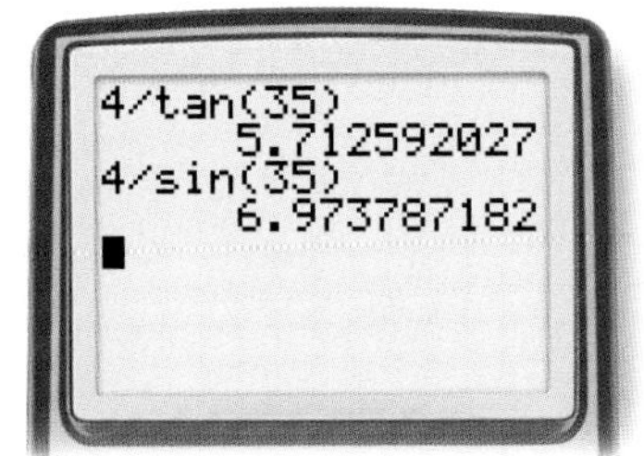

Suppose that in $\triangle ABC$, $m\angle A = 35°$ and $BC = 4.0$. To find AC and AB, students should write $\tan 35° = \frac{4.0}{AC}$ and $\sin 35° = \frac{4.0}{AB}$ and then evaluate the expressions, as shown at right.

Point out to students that the calculator does not have keys for the evaluation of cosecant, secant, and cotangent. There is no need for such keys because of the following argument:

$$\text{cosecant } A = \frac{\text{hyp}}{\text{opp}} = \frac{1}{\frac{\text{opp}}{\text{hyp}}} = \frac{1}{\sin A}$$

Therefore, on the calculator, students can evaluate the reciprocal of sin A to find csc A.

Lesson 13.2 Trigonometric Functions of General Angles

To help students understand that coterminal angles have the same sine, consider an activity based on the display at right. Have students experiment with other multiples of 360° and with angles whose measure is different from 32°.

Lesson 13.3 Common Angles and Periodic Functions

The exploration of coterminal angles suggested above in Lesson 13.2 can be of use in exploring the periodic nature of trigonometric functions. Alternatively, have students graph $y = \sin \theta$, using degree mode and the window settings shown below.

Xmin=–720 **Ymin=–2**
Xmax=720 **Ymax=2**
Xscl=90 **Yscl=1**

Continue the exploration of the periods of functions by changing $\sin \theta$ to $\cos \theta$.

Lesson 13.4 Radian Measure and Arc Length

In this lesson, you will need to introduce students to radian measure, accessed via the **MODE** menu. You will also introduce students to changing an angle measure from one measurement system to another. Pressing 2nd MATRX (or 2nd APPS on the TI-83 Plus) allows you to access the **ANGLE** menu, which will convert between radians and degrees automatically. However, it is suggested that students not be shown this until they are easily able to make the conversions by themselves.

Lesson 13.5 Graphing Trigonometric Functions

Have students explore how changing a, b, c, and d in a function such as $y = a\sin(b\theta - c) + d$ affects the graph. Show students that certain translations of trigonometric graphs have special results. Consider having students graph the functions below.

$$y = \sin \theta \text{ and } y = \sin(180° - \theta)$$

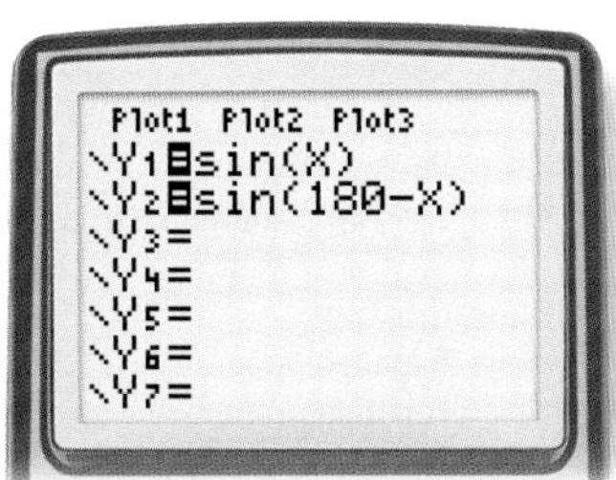

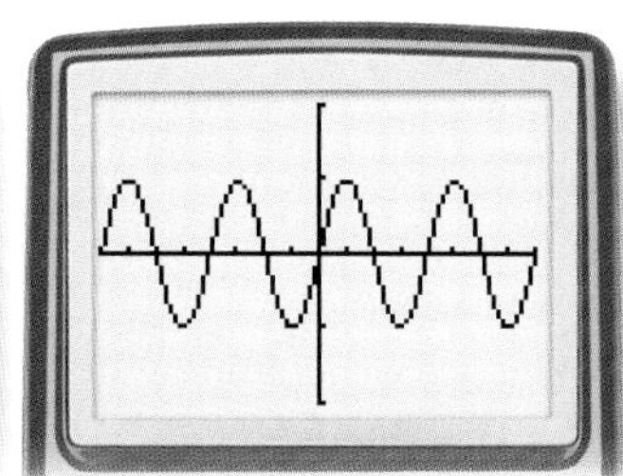

Ask students why the function list contains two functions but the graphical display shows only one graph.

Lesson 13.6 Inverses of Trigonometric Functions

The chapter ends with a discussion of the inverses of trigonometric functions. The inverse of sine, cosine, and tangent are accessed by pressing 2nd and then pressing the key for the desired function whose inverse is required. Students should check that the calculator is set in the proper mode before working with inverse trigonometric functions.

For further information, refer to the
- technology discussions in the lessons.
- lesson-related teacher's commentary in the side columns of this *Teacher's Edition*.
- lesson-related *Student Technology Guide* masters.
- *HRW Technology Handbook*.

internet connect

For keystrokes of other graphing calculators models, visit the HRW web site at **go.hrw.com** and enter the keyword **MB1 CALC**.

About Chapter 13

Background Information

In this chapter, students study trigonometric functions. They begin by exploring trigonometric functions in right triangles and investigating some common angles. They learn about radian measures of an angle and about converting between degrees and radians. Students then investigate the graphs of the trigonometric functions as well as their inverses.

CHAPTER RESOURCES

- Block-Scheduling Handbook
- Writing Activities for Your Portfolio
- Tech Prep Masters
- Long-Term Project
- Assessment Resources:
  Mid-Chapter Assessment
  Chapter Assessments
  Alternative Assessments
- Test and Practice Generator
- Technology Handbook

Chapter Objectives

- Find the trigonometric functions of acute angles. **[13.1]**
- Solve a right triangle by using trigonometric functions. **[13.1]**
- Find coterminal and reference angles. **[13.2]**
- Find the trigonometric function values of angles in standard position. **[13.2]**

Trigonometric Functions

THE WORD TRIGONOMETRY COMES FROM THE Greek words for triangle (*trigonon*) and measure (*metria*). Trigonometry is commonly described as the study of the relationship between the angles and sides of a triangle.

Trigonometry has a wide variety of applications in physics, astronomy, architecture, engineering, and other disciplines. People's understanding and interest in triangles can be observed in the structures shown here.

Lessons

The Pyramids at Giza, Egypt

The Louvre Museum, Paris, France

About the Photos

The Egyptians used primitive trigonometry and similar-triangle theory in the construction of the pyramids. To maintain a uniform slope of the faces, they used a ratio similar to the cotangent of an angle. Today the slope of a line is measured by the ratio of rise over run, but the Egyptians used the reciprocal of this ratio. The Egyptian *seked* was the measure of the horizontal departure of an oblique line from the vertical axis for every unit of change in the height. Their unit of measure for the vertical distance was the *cubit* (distance from the elbow to the tip of the middle finger, or ≈18 in.), and their unit of measure for the horizontal distance was the *hand.* A cubit is equivalent to 7 hands. The great Cheops Pyramid is 3080 hands wide and 280 cubits high and has a seqt of $\frac{1540}{280}$, or $5\frac{1}{2}$ hands per cubit.

Some of the presumed geometrical relationships among the dimensions of the Great Pyramid are false, but the ratio of its perimeter to its height is very close to $\frac{44}{7}$, or twice $\frac{22}{7}$ (a close approximation of π). However, most historians feel that the Egyptians used $3\frac{1}{6}$, not $3\frac{1}{7}$, as the ratio of the circumference of a circle to its diameter.

About the Chapter Project

George W. G. Ferris constructed the first Ferris wheel for the World's Exposition in Chicago in 1893. In the Chapter Project, you will create a model for the altitude of a rider on the Ferris wheel at Chicago's Navy Pier, which is modeled after Ferris's original wheel.

After completing the Chapter Project, you will be able to do the following:

- Model the height of a point on a Ferris wheel as a function of time.
- Interpret the real-world meaning of each parameter in your model.
- Find the speed of a point on a given Ferris wheel.

About the Portfolio Activities

Throughout the chapter, you will be given opportunities to complete Portfolio Activities that are designed to support your work on the Chapter Project.

- Sketching a rough graph for the height of a point on the Ferris wheel (relative to the center of the wheel) as a function of the angle of rotation is included in the Portfolio Activity on page 835.
- Sketching a graph for the altitude of a rider on the Ferris wheel as a function of the angle of rotation is included in the Portfolio Activity on page 850.
- Sketching a graph for the altitude of a rider on the Ferris wheel as a function of time and finding the speed of a rider on the Ferris wheel is included in the Portfolio Activity on page 857.
- Creating and interpreting a model for the altitude of a rider on the Ferris wheel is included in the Portfolio Activity on page 866.

Rock and Roll Hall of Fame, Cleveland, Ohio

internet connect

Chapter Internet Features and Online Activities

Lesson	Keyword	Page	Lesson	Keyword	Page
13.1	MB1 Homework Help	833	13.4	MB1 Homework Help	855
	MB1 Gazebo	832		MB1 RPM	854
	MB1 Ferris	835	13.5	MB1 Homework Help	864
13.2	MB1 Homework Help	841		MB1 Biorhythms	864
13.3	MB1 Homework Help	848	13.6	MB1 Homework Help	872
	MB1 Trig Hist	850			

- Find exact values for trigonometric functions of special angles and their multiples. **[13.3]**
- Find approximate values for trigonometric functions of any angle. **[13.3]**
- Convert from degree measure to radian measure and vice versa. **[13.4]**
- Find arc length. **[13.4]**
- Graph the sine, cosine, and tangent functions and their transformations. **[13.5]**
- Use the graph of the sine function to solve problems. **[13.5]**
- Evaluate trigonometric expressions involving inverses. **[13.6]**

Portfolio Activities appear at the end of Lessons 13.1, 13.3, 13.4, and 13.5. Each serves as preparation for the Chapter Project. The Portfolio Activities as well as the Chapter Project Activities are appropriate for inclusion in the student's portfolio. Students should be encouraged to include in their portfolios any other work in which they feel a sense of pride or a sense of accomplishment.

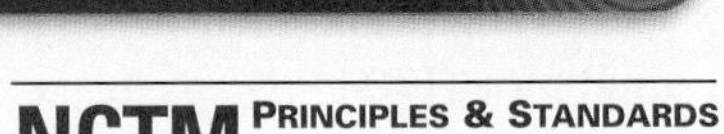

NCTM PRINCIPLES & STANDARDS 1–4, 6–10

QUICK WARM-UP

Rewrite as a decimal rounded to the nearest ten-thousandth.

1. $\frac{13}{24}$ ≈ 0.5417 **2.** $\frac{24}{13}$ ≈ 1.8462

Solve each equation.

3. $0.3 = \frac{m}{8}$ $m = 2.4$ **4.** $1.25 = \frac{7}{y}$ $y = 5.6$

5. Find the unknown length in the right triangle ABC. $a \approx 5.9$

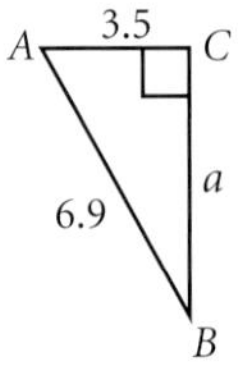

(Figure: right triangle ABC with $AC = 3.5$, $AB = 6.9$, $BC = a$, right angle at C)

Also on Quiz Transparency 13.1

Teach

Why Many problem situations involve angles of elevation and angles of depression. Both of these angles are measured from a horizontal line. Have students discuss situations in which these angles might play a role. **samples: an airplane landing, the viewing angle to a movie screen from a seat, the line of sight to the bottom of a canyon**

CHECKPOINT ✔
$\overline{AC}$ is the leg opposite $\angle B$.
$\overline{BC}$ is the leg adjacent to $\angle B$.

Right-Triangle Trigonometry

Objectives

- Find the trigonometric functions of acute angles.
- Solve a right triangle by using trigonometric functions.

Why ***You can use right-triangle trigonometry to solve real-world problems such as finding the height of the puffins above the water.***

APPLICATION
WILDLIFE

An ornithologist is taking pictures of puffins on the edge of a cliff. To find the height of the puffins above the water, she measures a 28° *angle of elevation* of her line of sight to the puffins. If her position is about 15 feet above the water, and about 40 feet from the cliff, how high above the water are the puffins? *You will solve this problem in Example 3.*

To find the height of the puffins, you can use *trigonometry*. **Trigonometry** can be used to find the measure of an unknown angle or an unknown side length of a right triangle.

Recall from Lesson 5.2 that the *hypotenuse* of a right triangle is the side opposite the right angle. The *legs* of a right triangle are the other two sides. Each leg is opposite one of the two *acute* angles and adjacent to the other. For example, in $\triangle ABC$, $\overline{BC}$ is the leg *opposite* angle A and $\overline{AC}$ is the leg *adjacent* to angle A.

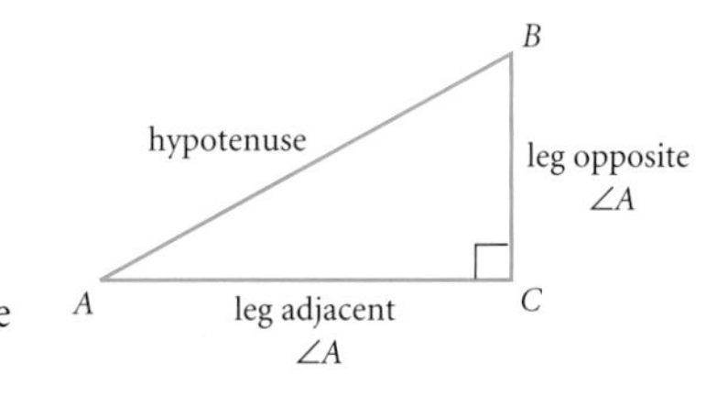

CHECKPOINT ✔ Identify the legs of $\triangle ABC$ above by their relationship to $\angle B$.

Alternative Teaching Strategy

USING MODELS Instead of using right triangles to introduce the trigonometric (trig) functions, you may want to use the unit circle. Begin by placing angle θ in standard position and drawing the unit circle $x^2 + y^2 = 1$. Let (x, y) represent the point where the terminal side of the angle intersects the unit circle. Then define the six trig functions as follows:

$\sin\theta = y$ $\csc\theta = \frac{1}{y}$

$\cos\theta = x$ $\sec\theta = \frac{1}{x}$

$\tan\theta = \frac{y}{x}\ (x \neq 0)$ $\cot\theta = \frac{x}{y}\ (y \neq 0)$

Give students the list below to help them remember the definitions of and relationships among the trig functions.

1. $\cos\theta$ is the first coordinate of the point where the terminal side of the angle intersects the unit circle, $\sin\theta$ is the second coordinate, and $\tan\theta$ is the ratio of the second coordinate to the first.
2. The following are reciprocals of each other:
   $\sin\theta$ and $\csc\theta$
   $\cos\theta$ and $\tan\theta$
   $\sec\theta$ and $\cot\theta$

Triangle ABC at right shows the abbreviations for the lengths of the sides of a right triangle in terms of $\angle A$. The six trigonometric functions are defined below with these abbreviations.

Trigonometric Functions of $\angle A$

sine of $\angle A = \frac{\text{opp.}}{\text{hyp.}}$	cosecant of $\angle A = \frac{\text{hyp.}}{\text{opp.}}$
cosine of $\angle A = \frac{\text{adj.}}{\text{hyp.}}$	secant of $\angle A = \frac{\text{hyp.}}{\text{adj.}}$
tangent of $\angle A = \frac{\text{opp.}}{\text{adj.}}$	cotangent of $\angle A = \frac{\text{adj.}}{\text{opp.}}$

The six trigonometric functions are abbreviated as sin A, cos A, tan A, csc A, sec A, and cot A.

EXAMPLE 1 **Find the values of the six trigonometric functions of $\angle X$ for $\triangle XYZ$ at right. Give exact answers and answers rounded to the nearest ten-thousandth.**

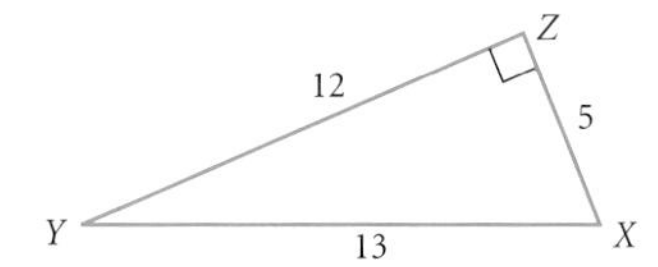

SOLUTION

$\sin X = \frac{\text{opp.}}{\text{hyp.}} = \frac{12}{13} \approx 0.9231$ $\qquad \csc X = \frac{\text{hyp.}}{\text{opp.}} = \frac{13}{12} \approx 1.0833$

$\cos X = \frac{\text{adj.}}{\text{hyp.}} = \frac{5}{13} \approx 0.3846$ $\qquad \sec X = \frac{\text{hyp.}}{\text{adj.}} = \frac{13}{5} = 2.6$

$\tan X = \frac{\text{opp.}}{\text{adj.}} = \frac{12}{5} = 2.4$ $\qquad \cot X = \frac{\text{adj.}}{\text{opp.}} = \frac{5}{12} \approx 0.4167$

TRY THIS Find the values of the six trigonometric functions of $\angle Y$ for $\triangle XYZ$. Give exact answers and answers rounded to the nearest ten-thousandth.

Notice that in the solution to Example 1, the ratios for sin X and csc X are reciprocals. Likewise, the ratios for cos X and sec X are reciprocals, and the ratios for tan X and cot X are reciprocals.

The cosecant, secant, and cotangent ratios can be expressed in terms of the sine, cosine, and tangent ratios, respectively.

$$\csc A = \frac{1}{\sin A} \qquad \sec A = \frac{1}{\cos A} \qquad \cot A = \frac{1}{\tan A}$$

CHECKPOINT ✔ Show that the following statements are also true:

$$\sin A = \frac{1}{\csc A} \qquad \cos A = \frac{1}{\sec A} \qquad \tan A = \frac{1}{\cot A}$$

CRITICAL THINKING Find the number of different possible ratios of the lengths of two sides of a triangle by using permutations. Explain what your result means. Are all possible ratios given by the functions above?

Interdisciplinary Connection

SCIENCE Surveyors use trigonometry to find distances that are not easily measured. A surveyor needs to find the distance across a canyon. One person stands on one side of the caynon at point A, and another stands on the other side at point B, which is directly across from point A. The person at point B walks perpendicularly to $\overline{AB}$ for a distance of 5 meters, stopping at point C. The person at point A measures the angle between points B and C to be 6°. Find the distance, AB, across the canyon. ≈47.57 m

Inclusion Strategies

ENGLISH LANGUAGE DEVELOPEMENT Some students have trouble remembering the trig ratios. The following acronym may help: SOH-CAH-TOA (Sine is Opposite over Hypotenuse-Cosine is Adjacent over Hypotenuse-Tangent is Opposite over Adjacent). Have students write a story about SOH-CAH-TOA that mentions the trig ratios.

ADDITIONAL EXAMPLE 1

Find the values of the six trig functions of θ in $\triangle KLM$ at right. Give exact answers and answers rounded to the nearest ten-thousandth.

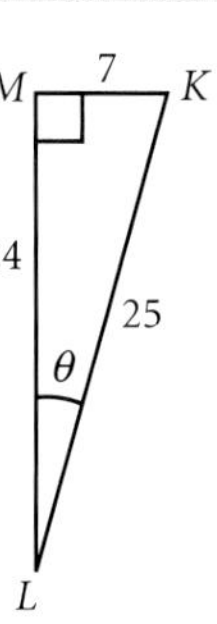

$\sin \theta = \frac{7}{25} = 0.2800$

$\cos \theta = \frac{24}{25} = 0.9600$

$\tan \theta = \frac{7}{24} \approx 0.2917$

$\cot \theta = \frac{24}{7} = 3.4286$

$\sec \theta = \frac{25}{24} \approx 1.0417$

$\csc \theta = \frac{25}{7} \approx 3.5714$

TRY THIS

$\sin Y = \frac{5}{13} \approx 0.3846$

$\cos Y = \frac{12}{13} \approx 0.9231$

$\tan Y = \frac{5}{12} \approx 0.4167$

$\cot Y = \frac{12}{5} = 2.4$

$\sec Y = \frac{13}{12} \approx 1.0833$

$\csc Y = \frac{13}{5} = 2.6$

CHECKPOINT ✔

$\sin A = \frac{\text{opp.}}{\text{hyp.}} = \frac{1}{\frac{\text{hyp.}}{\text{opp.}}} = \frac{1}{\csc A}$

$\cos A = \frac{\text{adj.}}{\text{hyp.}} = \frac{1}{\frac{\text{hyp.}}{\text{adj.}}} = \frac{1}{\sec A}$

$\tan A = \frac{\text{opp.}}{\text{adj.}} = \frac{1}{\frac{\text{adj.}}{\text{opp.}}} = \frac{1}{\cot A}$

CRITICAL THINKING

3 sides taken 2 at a time gives $_3P_2 = 6; \frac{a}{b}, \frac{a}{c}, \frac{b}{a}, \frac{b}{c}, \frac{c}{a}, \frac{c}{b}$; yes.

Activity Notes

In this Activity, students investigate the sine, cosine, and tangent of 20° in various right triangles. Students should see that no matter which triangle is used, the sine, cosine, and tangent of 20° is constant. Remind them that similar triangles have angles that are congruent and proportional sides. That is, if $\triangle ABC \sim \triangle XYZ$, then $\frac{AB}{AC} = \frac{XY}{XZ}$, and the same is true for the other pairs of corresponding sides.

Teaching Tip

TECHNOLOGY Remind students they should make sure that their calculators are in **Degree** mode (not in **Radian** mode) before they find trig ratios unless otherwise directed. On a TI-82 or a TI-83, press MODE, select **Degree**, and press ENTER.

Some calculators require the user to enter the angle first and then press the trig ratio, while others require the user to press the trig ratio first and then enter the angle.

Cooperative Learning

Have students work in groups of four on the Activity. Have each student in the group draw a different set of triangles. Then have each member of the group compare the values for each trig ratio. Have the class discuss their results, and point out that the trig ratios for a given angle are constant.

CHECKPOINT ✔

4. The ratios for the sine, cosine, and tangent of an acute angle in similar triangles do not depend on the lengths of the sides.

Exploring Trigonometric Functions

You will need: a protractor, a centimeter ruler, and a calculator

On a sheet of paper, make a large diagram like the one shown below. Place segments $\overline{X_1Y_1}$, $\overline{X_2Y_2}$, and $\overline{X_3Y_3}$ wherever you wish, as long as they are perpendicular to $\overrightarrow{AP}$.

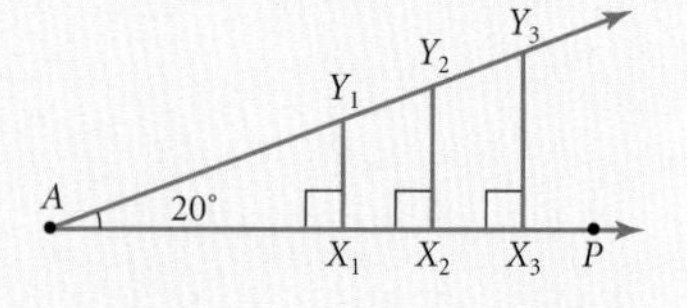

1. Copy and complete the table below by measuring the indicated sides and calculating sin *A*, cos *A*, and tan *A*.
2. Are all of the entries approximately equal in the sin *A* column? in the cos *A* column? in the tan *A* column?

	opp. $\angle A$	adj. $\angle A$	hyp.	$\sin A = \frac{\text{opp.}}{\text{hyp.}}$	$\cos A = \frac{\text{adj.}}{\text{hyp.}}$	$\tan A = \frac{\text{opp.}}{\text{adj.}}$
$\triangle AY_1X_1$						
$\triangle AY_2X_2$						
$\triangle AY_3X_3$						

3. Compare your results from Step 2 with your classmates' results.

CHECKPOINT ✔ 4. What can you conjecture about the sine, cosine, and tangent functions for the measure of $\angle A$?

As the results of the Activity suggest, the ratios for the sine, cosine, and tangent of an acute angle in similar triangles does not depend on the lengths of sides. The trigonometric functions depend only on the measure of the acute angle.

For any given angle measure, you can obtain values for the sine, cosine, and tangent of the angle by using a scientific calculator in degree mode. These trigonometric function values can be used to find the unknown lengths of sides, as shown in Example 2.

EXAMPLE 2

For $\triangle ABC$ shown at right, find each side length to the nearest tenth.

a. *AB* **b.** *BC*

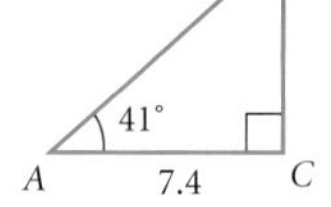

SOLUTION

From the diagram, $m\angle A = 41°$ and the length of the leg adjacent to $\angle A$ is 7.4.

a. To find *AB*, the length of the hypotenuse, use the cosine ratio.

$$\cos A = \frac{\text{adj.}}{\text{hyp.}}$$
$$\cos 41° = \frac{7.4}{AB}$$
$$AB = \frac{7.4}{\cos 41°}$$
$$AB \approx \frac{7.4}{0.7547} \approx 9.8$$

b. To find *BC*, the length of the leg opposite $\angle A$, use the tangent ratio.

$$\tan A = \frac{\text{opp.}}{\text{adj.}}$$
$$\tan 41° = \frac{BC}{7.4}$$
$$7.4 \times \tan 41° = BC$$
$$7.4 \times 0.8693 \approx BC$$
$$BC \approx 6.4$$

Enrichment

The circle below has its center at (0, 0) and a radius of *r*.

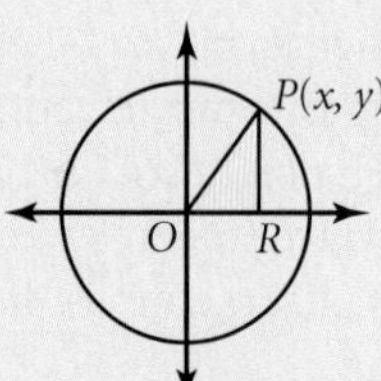

Have students label the sides of the right triangle with *x*, *y*, and *r* and then use the triangle and the equation of the circle, $x^2 + y^2 = r^2$, to show that $\sin^2 \angle ROP + \cos^2 \angle ROP = 1$.

Label sides as $OR = x$, $RP = y$, and $OP = r$; because $x^2 + y^2 = r^2$, $\left(\frac{y}{r}\right)^2 + \left(\frac{x}{r}\right)^2 = 1$ and $\sin^2 \angle ROP + \cos^2 \angle ROP = 1$ by the definitions of sine and cosine.

TRY THIS For $\triangle KLM$ shown at right, find KL and LM to the nearest tenth.

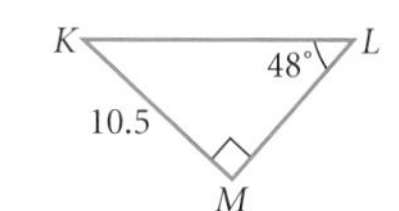

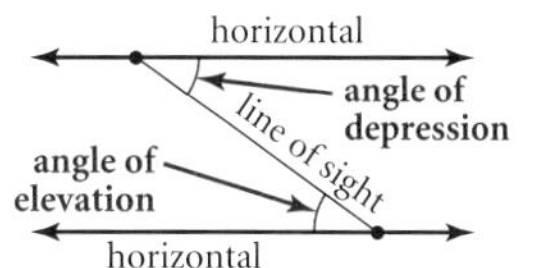

An **angle of elevation** is an angle between a horizontal line and a line of sight to a point above. An **angle of depression** is an angle between a horizontal line and a line of sight to a point below.

EXAMPLE 3 Refer to the ornithologist's problem described at the beginning of the lesson.

APPLICATION
WILDLIFE

How high above the water are the puffins? Give your answer to the nearest foot.

SOLUTION

In the diagram, BE represents the height of the puffins above the water. Because $BE = BA + AE$ and you know that EA is 15, you need to find AB. The angle of elevation is 28°.

$$\tan 28^\circ = \frac{AB}{40}$$
$$40 \tan 28^\circ = AB$$
$$AB \approx 21.3$$

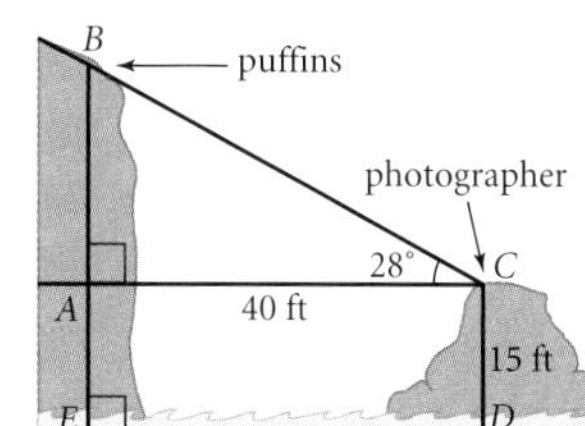

Then find BE.

$$BE = EA + AB$$
$$\approx 15 + 21.3$$
$$\approx 36.3$$

The puffins are about 36 feet above the water.

When you know the trigonometric ratio of an angle, you can find the measure of that angle by using the *inverse relation* of the trigonometric ratio. For example, if $\tan A$ is $\frac{4}{3}$, then the angle whose tangent is $\frac{4}{3}$ is written $\tan^{-1}\frac{4}{3}$.

$$\tan A = \frac{4}{3}$$
$$m\angle A = \tan^{-1}\frac{4}{3}$$
$$m\angle A \approx 53^\circ$$

Most calculators have SIN⁻¹, COS⁻¹, and TAN⁻¹ keys for these inverse trigonometric relations. Note that $\sin^{-1} x$ is *not* the same as $\frac{1}{\sin x}$. These inverse relations are sometimes called the *arcsine*, *arccosine*, and *arctangent* relations.

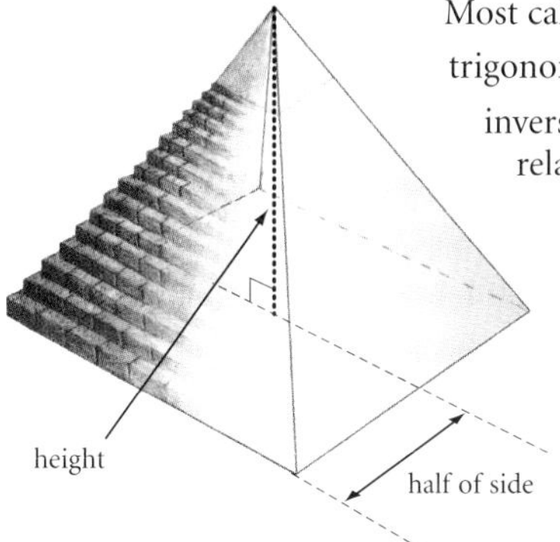

CULTURAL CONNECTION: AFRICA The Egyptians used a triangle relation called the *seked* to denote the slope of the inclined face of a pyramid.

$$seked = \frac{\text{half of the pyramid's side length in palms}}{\text{height of the pyramid in cubits, where 1 cubit equals 7 palms}}$$

Today, we use the tangent ratio, which is similar to the reciprocal of the *seked*.

Reteaching the Lesson

USING MODELS Have students work in groups of three or four students. Have each group draw a right triangle for each of the properties listed below, record the lengths of the three sides, find the measures of the two acute angles, and find the three basic trig functions for both acute angles.

1. The leg opposite the smaller acute angle is half as long as the hypotenuse. **The acute angles are 30° and 60°; sin 30° = 0.5, cos 30° ≈ 0.8660, tan 30° ≈ 0.5774; sin 60° ≈ 0.8660, cos 60° = 0.5, tan 60° ≈ 1.7321.**
2. The legs are the same length. **Both acute angles are 45°; sin 45° ≈ 0.7071, cos 45° ≈ 0.7071, tan 45° = 1.**
3. The length of the three sides have the ratio 3:4:5. **The acute angles are approximately 36.9° and 53.1°; sin 36.9° = 0.6000, cos 36.9° = 0.8000, tan 36.9° = 0.7500; sin 53.1° = 0.8000, cos 53.1° = 0.6000, tan 53.1° ≈ 1.3319.**

Each group should show the triangles they drew for each case and present its findings to the class.

ADDITIONAL EXAMPLE 2

For $\triangle RST$ shown below, find each side length to the nearest tenth.

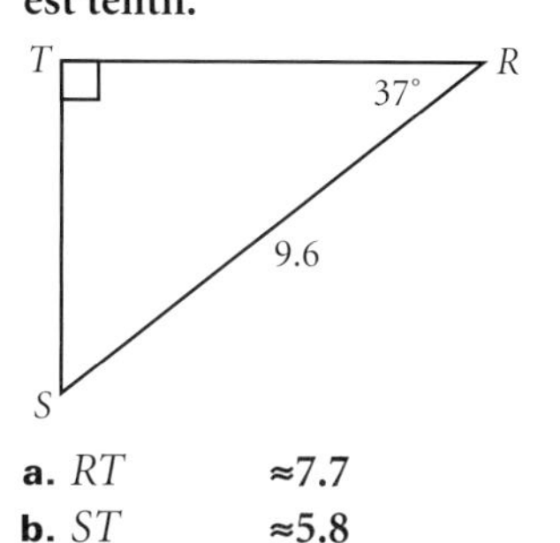

a. RT **≈7.7**
b. ST **≈5.8**

TRY THIS
$KL \approx 14.1$; $LM \approx 9.5$

ADDITIONAL EXAMPLE 3

The height of an observation tower in a state park is 30 feet. A ranger at the top of the tower sees a fire along a line of sight that is at a 1° angle of depression. **How far is the fire from the base of the tower? Round your answer to the nearest foot.**

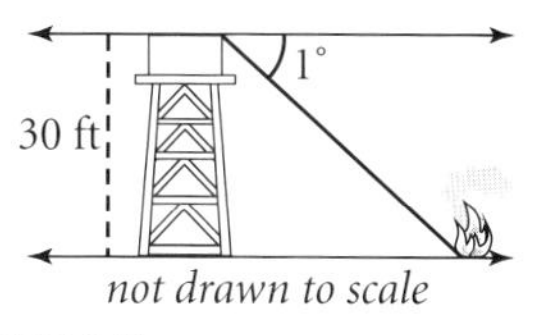

≈1719 ft

Teaching Tip

The inverse reciprocal trig relations are usually written as $\cot^{-1} x$, $\sec^{-1} x$, and $\csc^{-1} x$, and they are sometimes called the *arccotangent*, *arcsecant*, and *arccosecant*. They are rarely included in a calculator's features, so to find the value of $\csc^{-1} x$, where $-1 \le x \le 1$ and $x \ne 0$, students should calculate $\frac{1}{\sin^{-1} x}$.

Math CONNECTION

GEOMETRY Students use the fact that the sum of the angles in a triangle is 180° to recall the relationship between the acute angles of a right triangle. This fact is helpful when finding angle measures.

ADDITIONAL EXAMPLE 4

Solve $\triangle PQR$. Give $m\angle Q$ to the nearest degree and RP and RQ to the nearest tenth of a unit.

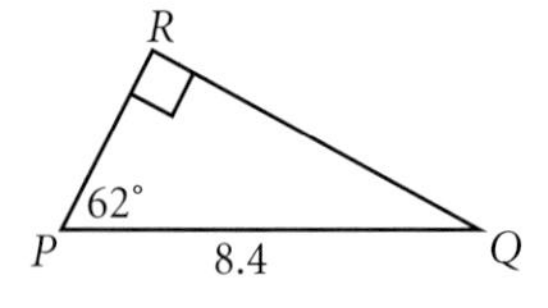

$m\angle Q = 28°$; $RP \approx 3.9$; $RQ \approx 7.4$

TRY THIS

$m\angle K \approx 29°$; $m\angle L \approx 61°$; $LM \approx 4.6$

CHECKPOINT ✔

Use the Pythagorean theorem to find RS, find the measure of $\angle R$ by using the inverse sine function, and then compute $m\angle S = 90 - m\angle R$.

$RS = \sqrt{6.8^2 + 3.8^2} \approx 7.8$

$m\angle R = \sin^{-1}\frac{3.8}{7.8} \approx 29.2°$

$m\angle S \approx 90° - 29.2° = 60.8°$

Assess

Selected Answers

Exercises 4–7, 9–59 odd

ASSIGNMENT GUIDE	
In Class	1–7
Core	9–47 odd
Core Plus	8–48 even, 49
Review	50–59
Preview	60

Extra Practice can be found beginning on page 940.

Inverse trigonometric relations are often used to solve a triangle. **Solving a triangle** involves finding the measures of all of the unknown sides and angles of the triangle. A geometry fact used in solving triangles is that the sum of the measures of all the angles in a triangle is 180°. For right triangles, the sum of the measures of the two acute angles is 90°.

EXAMPLE 4 **Solve $\triangle RST$. Give $m\angle R$ and $m\angle S$ to the nearest degree, and give RS to the nearest tenth of a unit.**

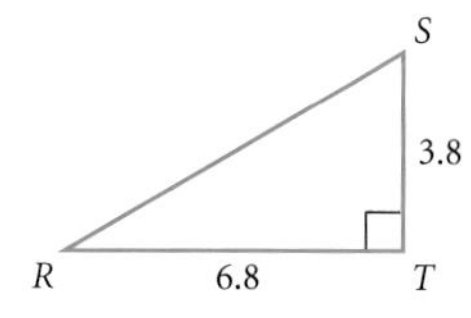

SOLUTION

1. First find $m\angle R$.

$$\tan R = \frac{3.8}{6.8}$$

$$R = \tan^{-1}\frac{3.8}{6.8} \approx 29°$$

2. Then find $m\angle S$.

$$R + S + T = 180° \qquad \text{or} \qquad R + S = 90°$$

$$29° + S + 90° \approx 180° \qquad 29° + S \approx 90°$$

$$S \approx 61° \qquad S \approx 61°$$

3. Use the Pythagorean Theorem to find RS.

$$(RS)^2 = (6.8)^2 + (3.8)^2$$

$$RS = \sqrt{(6.8)^2 + (3.8)^2}$$

$$RS \approx 7.8$$

Thus, $m\angle R \approx 29°$, $m\angle S \approx 61°$, and $RS \approx 7.8$.

TRY THIS Solve $\triangle KLM$. Give $m\angle K$ and $m\angle L$ to the nearest degree and LM to the nearest tenth of a unit.

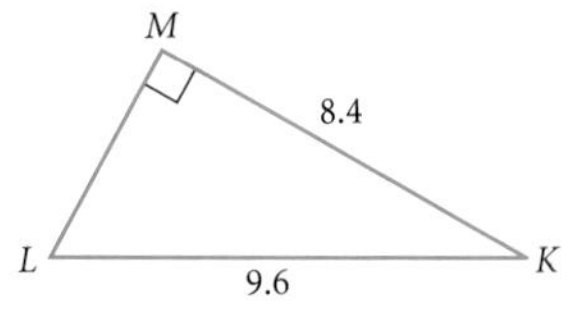

CHECKPOINT ✔ Explain how to solve $\triangle RST$ in Example 4 by finding RS first and then using the sine or cosine to find $m\angle R$.

Exercises

Communicate

1. Explain how to find the values of the six trigonometric functions of $\angle A$ at right.
2. Explain how to find the measures of $\angle A$ and $\angle B$ in $\triangle ABC$ at right.
3. Explain how the expressions $\frac{1}{\sin A}$ and $\sin^{-1} A$ are different.

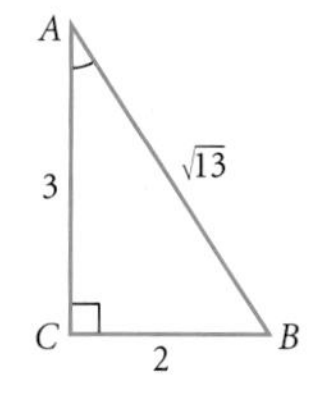

Guided Skills Practice

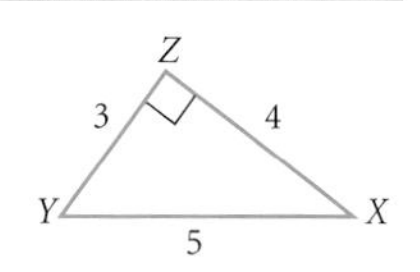

4. Find the values of the six trigonometric functions of $\angle X$ in $\triangle XYZ$ at right. Give exact answers and answers rounded to the nearest ten-thousandth. ***(EXAMPLE 1)***

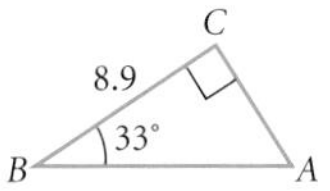

5. For $\triangle ABC$ at left, find AC and BA to the nearest tenth. ***(EXAMPLE 2)*** **$AC \approx 5.8$; $BA \approx 10.6$**

APPLICATION

6. SURVEYING An engineer stands 50 feet away from a building and sights the top of the building with a surveying device mounted on a tripod. If the surveying device is 5 feet above the ground and the angle of elevation is 50°, how tall is the building? ***(EXAMPLE 3)*** **≈64.6 feet**

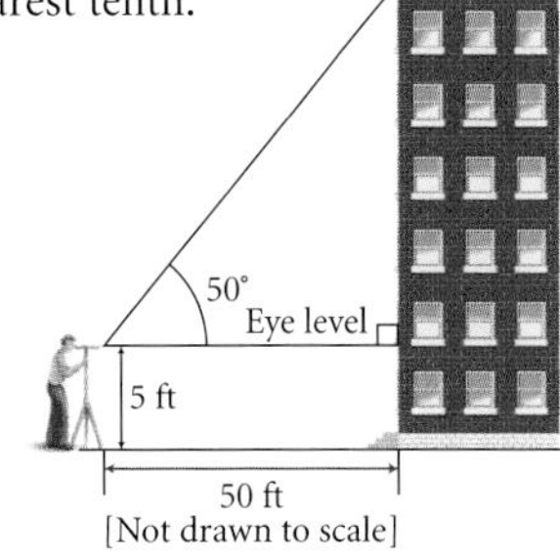

7. Solve $\triangle ABC$ shown below. Give $m\angle A$ and $m\angle B$ to the nearest degree, and give AB to the nearest tenth of a unit. ***(EXAMPLE 4)*** **$m\angle A \approx 47°$**
$m\angle B \approx 43°$
$AB \approx 7.3°$

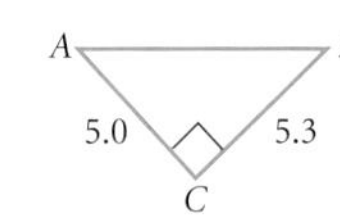

8. $\frac{8}{17} \approx 0.4706$
9. $\frac{15}{17} \approx 0.8824$
10. $\frac{8}{17} \approx 0.4706$
11. $\frac{15}{17} \approx 0.8824$
12. $\frac{8}{15} \approx 0.5333$
13. $\frac{15}{8} \approx 1.875$
14. $\frac{17}{15} \approx 1.1333$
15. $\frac{17}{8} = 2.125$
16. $\frac{17}{15} \approx 1.1333$
17. $\frac{17}{8} = 2.125$
18. $\frac{8}{15} \approx 0.5333$
19. $\frac{15}{8} = 1.875$
20. $\frac{3}{\sqrt{13}} \approx 0.8321$

Practice and Apply

Refer to $\triangle JKL$ below to find each value listed. Give exact answers and answers rounded to the nearest ten-thousandth.

8. $\sin K$	**9.** $\sin J$	**10.** $\cos J$
11. $\cos K$	**12.** $\tan K$	**13.** $\tan J$
14. $\csc J$	**15.** $\csc K$	**16.** $\sec K$
17. $\sec J$	**18.** $\cot J$	**19.** $\cot K$

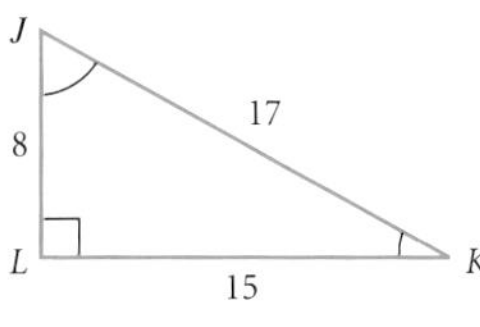

Refer to $\triangle FGH$ below to find each value listed. Give exact answers and answers rounded to the nearest ten-thousandth.

20. $\sin G$	**21.** $\sin F$	**22.** $\cos G$ **$\frac{2}{\sqrt{13}} \approx 0.5547$**
23. $\cos F$	**24.** $\tan G$ **$\frac{3}{2} = 1.5$**	**25.** $\tan F$ **$\frac{2}{3} \approx 0.6667$**
26. $\csc G$	**27.** $\csc F$	**28.** $\sec G$ **$\frac{\sqrt{13}}{2} = 1.8028$**
29. $\sec F$	**30.** $\cot G$	**31.** $\cot F$ **$\frac{3}{2} = 1.5$**

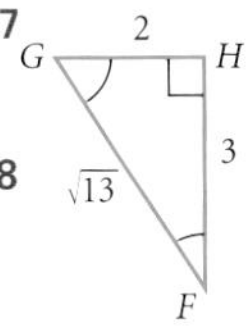

Find $m\angle A$ by using inverse trigonometric functions.

32. **35.5°**

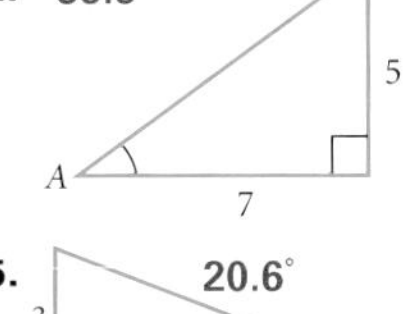
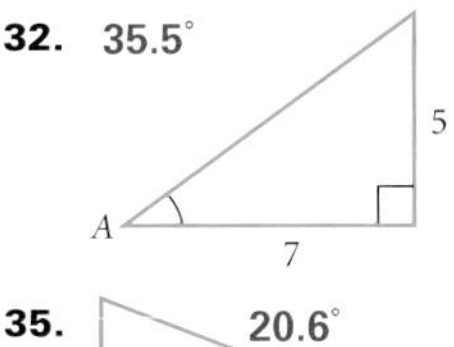

33. **30°**

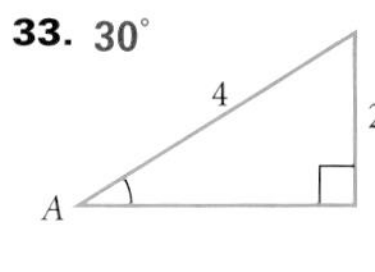

34. **48.2°**

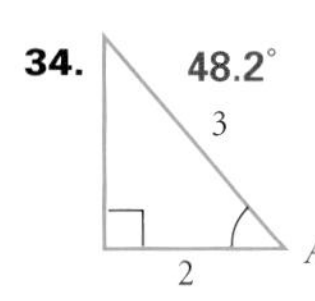

35. **20.6°**

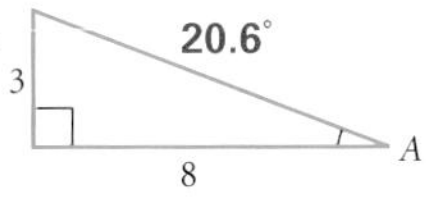

36. **48.6°**

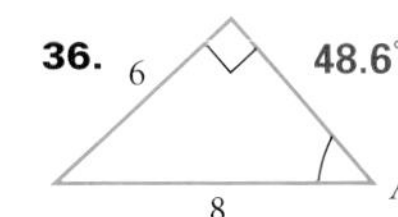

37. **41.4°**

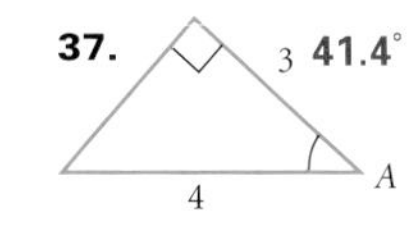

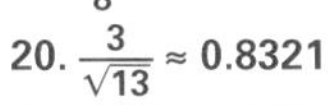

internet connect

Homework Help Online
Go To: **go.hrw.com**
Keyword:
MB1 Homework Help for Exercises 32–37

Error Analysis

Students often have trouble using inverse trig functions to solve for an angle measure in a given triangle. Remind students that the sine of an angle is calculated in order to get a ratio and that the inverse sine of a ratio is calculated in order to get an angle measure. Therefore, students should first write the expression without using inverse trig functions and then should use the related inverse trig function to find the measure of the angle. For example, first write $\sin A = 0.75$ and then write $A = \sin^{-1} 0.75 \approx 48.6°$.

If students are getting incorrect values for trig ratios or inverse trig functions from their calculators, make sure that their calculators are in **Degree** mode, not **Radian** mode.

4. $\sin x = \frac{3}{5} = 0.6$;
$\csc x = \frac{5}{3} \approx 1.6667$;
$\cos x = \frac{4}{5} = 0.8$;
$\sec x = \frac{5}{4} = 1.25$;
$\tan x = \frac{3}{4} = 0.75$;
$\cot x = \frac{4}{3} \approx 1.3333$

21. $\frac{2}{\sqrt{13}} \approx 0.5547$

23. $\frac{3}{\sqrt{13}} \approx 0.8321$

26. $\frac{\sqrt{13}}{3} \approx 1.2019$

27. $\frac{\sqrt{13}}{2} \approx 1.8028$

29. $\frac{\sqrt{13}}{3} \approx 1.2019$

30. $\frac{2}{3} \approx 0.6667$

In Exercise 60, students are asked to find the angles of rotation that create a semicircle and a quarter circle. In Lesson 13.2, students will study angles of rotation.

38. $m\angle A \approx 39^\circ$
$m\angle B \approx 51^\circ$
$AB \approx 6.4$

39. $m\angle R \approx 32^\circ$
$m\angle S \approx 58^\circ$
$ST \approx 2.1$

40. $m\angle A = 40^\circ$
$AB \approx 9.5$
$AC \approx 7.3$

41. $m\angle S = 50^\circ$
$RS \approx 11.4$
$RT \approx 8.7$

CONNECTION

42. $m\angle B = 55^\circ$
$AC \approx 10.2$
$BC \approx 7.2$

APPLICATIONS

43. $m\angle B = 48^\circ$
$BC \approx 3.8$
$AC \approx 4.2$

Solve each triangle. Give angle measures to the nearest degree and side lengths to the nearest tenth.

38. 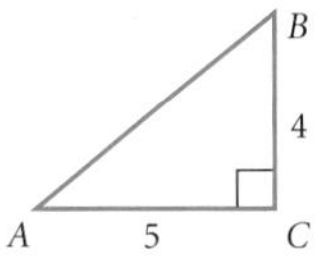 (triangle ABC: $AC = 5$, $BC = 4$, right angle at C)

39. 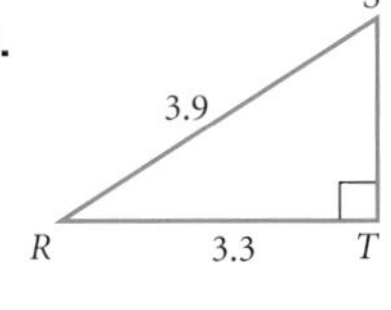

40. 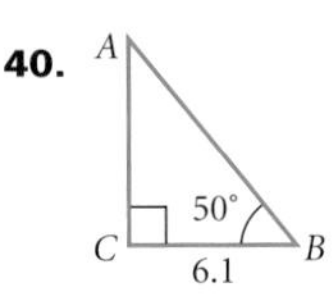

41. 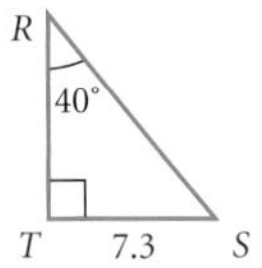 (triangle RTS: $m\angle R = 40^\circ$, $TS = 7.3$, right angle at T)

42. 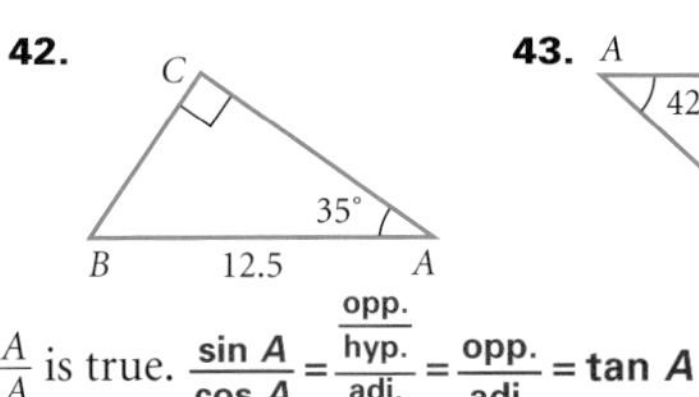

43. 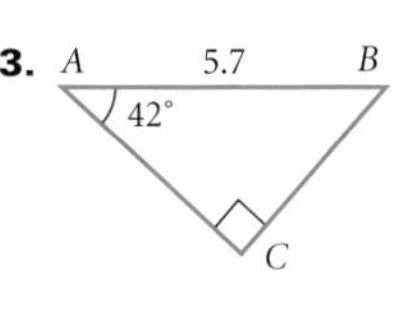

44. Show that $\tan A = \frac{\sin A}{\cos A}$ is true. $\frac{\sin A}{\cos A} = \frac{\frac{\text{opp.}}{\text{hyp.}}}{\frac{\text{adj.}}{\text{hyp.}}} = \frac{\text{opp.}}{\text{adj.}} = \tan A$

45. GEOMETRY Quadrilateral $ABCD$ at right is a rectangle. Find AD and AC to the nearest tenth of a foot. **$AD \approx 687.9$ ft; $AC \approx 829.8$ ft**

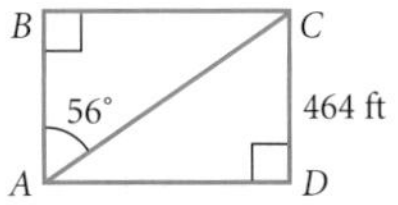

(rectangle $ABCD$: $m\angle BAC = 56^\circ$, $CD = 464$ ft)

46. HOME IMPROVEMENT Mary and Chris want to build a right-triangular deck behind their house. They would like the hypotenuse of the deck to be 20 feet long, and they would like the other two sides of the deck to be equal in length.

a. Find the length of the sides of the deck. **14.1 ft**

b. Find the area of the deck. **99.4 sq ft**

AVIATION A commercial airline pilot is flying at an altitude of 6.5 miles. To make a gentle descent for landing, the pilot begins descending toward the airport when still fairly far away.

47. If the pilot begins descending 186 miles from the airport (measured on the ground), what angle will the plane's path make with the runway (without further adjustment)? **2°**

48. If the plane's path is to make an angle of 5° with the runway (without further adjustment), how far from the airport (measured on the ground) must the pilot begin descending? **≈74 miles**

CHALLENGE

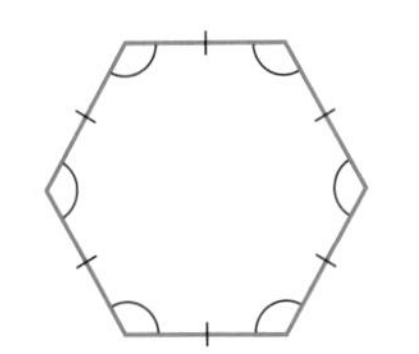

49. CONSTRUCTION The city park manager would like to build a gazebo in the shape of a regular hexagon with sides 10 feet long. (A regular hexagon is a 6-sided polygon with all sides equal in length and all angles equal in measure.) Paving costs $15 per square foot. Use trigonometric ratios to find the cost of paving the hexagonal area. **$3897**

Practice

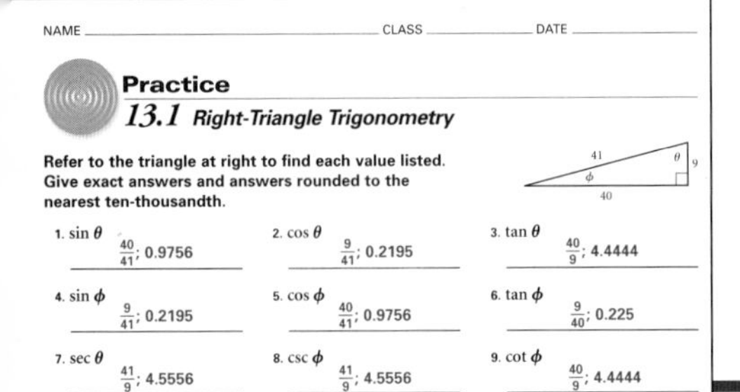

NAME ____ CLASS ____ DATE ____

Practice
13.1 **Right-Triangle Trigonometry**

Refer to the triangle at right to find each value listed. Give exact answers and answers rounded to the nearest ten-thousandth.

1. $\sin\theta$ $\frac{40}{41}$; 0.9756
2. $\cos\theta$ $\frac{9}{41}$; 0.2195
3. $\tan\theta$ $\frac{40}{9}$; 4.4444
4. $\sin\phi$ $\frac{9}{41}$; 0.2195
5. $\cos\phi$ $\frac{40}{41}$; 0.9756
6. $\tan\phi$ $\frac{9}{40}$; 0.225
7. $\sec\theta$ $\frac{41}{9}$; 4.5556
8. $\csc\phi$ $\frac{41}{9}$; 4.5556
9. $\cot\phi$ $\frac{40}{9}$; 4.4444

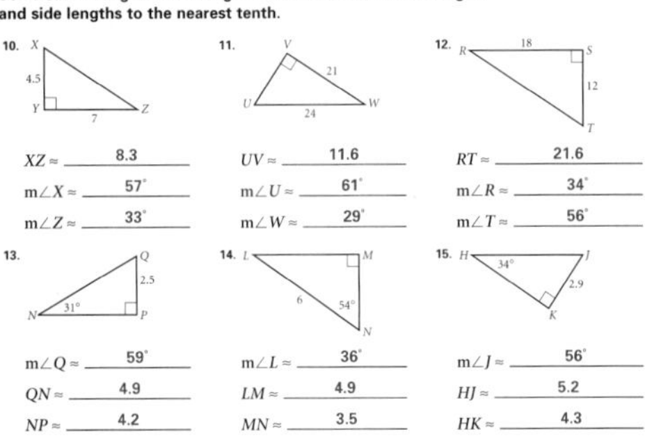

Solve each triangle. Round angle measures to the nearest degree and side lengths to the nearest tenth.

10. $XZ \approx$ 8.3; $m\angle X \approx$ 57°; $m\angle Z \approx$ 33°
11. $UV \approx$ 11.6; $m\angle U \approx$ 61°; $m\angle W \approx$ 29°
12. $RT \approx$ 21.6; $m\angle R \approx$ 34°; $m\angle T \approx$ 56°
13. $m\angle Q \approx$ 59°; $QN \approx$ 4.9; $NP \approx$ 4.2
14. $m\angle L \approx$ 36°; $LM \approx$ 4.9; $MN \approx$ 3.5
15. $m\angle J \approx$ 56°; $HJ \approx$ 5.2; $HK \approx$ 4.3

Look Back

Determine the degree of each polynomial function. ***(LESSON 7.1)***

50. $f(x) = 3x^5 - 5x^8 + 4x^3 + 2$ **8**

51. $f(x) = (x^2 - 9)(x^3 + 4)$ **5**

Write each polynomial in factored form. ***(LESSON 7.3)***

52. $2x^3 - 18x$ $\mathbf{2x(x+3)(x-3)}$

53. $3x^3 - 7x^2 + 2x$ $\mathbf{x(3x-1)(x-2)}$

Write each expression with a rational denominator in simplest form. ***(LESSON 8.7)***

54. $\frac{3}{\sqrt{2}}$ $\mathbf{\frac{3\sqrt{2}}{2}}$

55. $\frac{1}{\sqrt{3}}$ $\mathbf{\frac{\sqrt{3}}{3}}$

56. $\frac{5}{1-\sqrt{2}}$ $\mathbf{-5-5\sqrt{2}}$

57. $\frac{-2}{\sqrt{2}+\sqrt{3}}$ $\mathbf{2\sqrt{2}-2\sqrt{3}}$

Find the minimum and maximum values, Q_1, Q_2, and Q_3, and make a box-and-whisker plot for each set of data. ***(LESSON 12.3)***

58. 102, 107, 122, 99, 103, 121, 113, 100, 78, 130, 125, 119, 110

59. 12, 34, 18, 25, 53, 46, 17, 14, 25, 36, 24, 19, 17, 28, 26, 22

58. min = 78; max = 130
$Q_1 = 101$; $Q_2 = 110$;
$Q_3 = 121.5$

59. min = 12; max = 53;
$Q_1 = 17.5$; $Q_2 = 24.5$;
$Q_3 = 31$

internet connect
Portfolio Extension
Go To: **go.hrw.com**
Keyword:
MB1 Ferris

Look Beyond

60. **GEOMETRY** A 360° angle of rotation creates a circle. What angle of rotation creates a semicircle? a quarter circle? **180°; 90°**

PORTFOLIO ACTIVITY

The world's largest Ferris wheel, as of 1998, is the Cosmoclock 21 in Yokohama City, Japan. Its center is 344.5 feet above the ground and it has a diameter of 328 feet.
[*Source: Guiness Book of World Records, 1998*]

The center of the Cosmoclock 21 is located at the origin of the coordinate plane at right. Assume that a point, *P*, begins its rotation at (164, 0) and that it rotates in a counterclockwise direction.

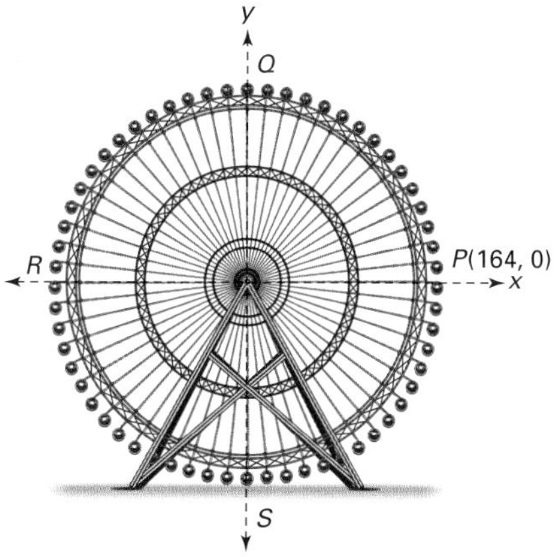

1. Identify the coordinates of *Q*, *R*, and *S*.

2. Copy and complete the table below.

Rotation (degrees)	0°	90°	180°	270°	...	810°
Height of *P* relative to the *x*-axis (feet)	0	164		−164	...	

3. Plot the points from your table in a new coordinate plane. Sketch a *smooth curve* through the points.

4. Does your graph appear to be linear, quadratic, exponential, logarithmic, or none of these?

WORKING ON THE CHAPTER PROJECT

You should now be able to complete Activity 1 of the Chapter Project.

ALTERNATIVE Assessment

Portfolio Activity

The Portfolio Activity can be used as preparation for the Chapter Project or as a separate activity. In the Portfolio Activity on this page, students examine a Ferris wheel. They find the height of a point on the Ferris wheel at various degrees of rotations and then plot this data. The student should see that a sinusoidal curve appears if the points are connected with a smooth curve.

Answers to Portfolio Activities can be found in Additional Answers of the Teacher's Edition.

58.

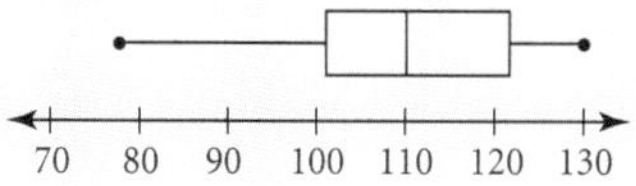

59.

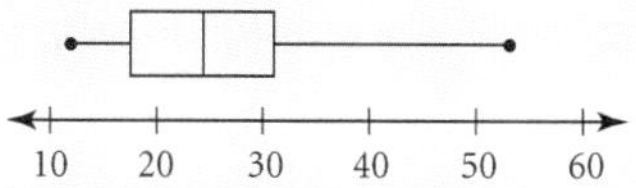

Student Technology Guide

NAME ______ CLASS ______ DATE ______

Student Technology Guide
13.1 ***Right-Triangle Trigonometry***

When you *solve* a triangle, you find all of the unknown side lengths and angle measures. A calculator can help you solve right triangles.

Before using trigonometric functions of angles, press MODE, use the arrow keys to move to Degree, and then press ENTER. Press 2nd MODE.

Example: Solve $\triangle ABC$ shown at right. Give $m\angle A$ and $m\angle C$ to the nearest degree and AB to the nearest tenth of a unit.

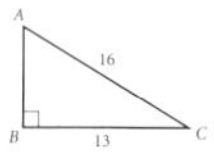

- Since you know the length of the side adjacent to $\angle C$ and the length of the hypotenuse, use cosine to find $m\angle C$. $\cos C = \frac{13}{16}$, so $C = \cos^{-1}\left(\frac{13}{16}\right)$. Press 2nd COS 13 ÷ 16) ENTER. To the nearest degree, $m\angle C = 36°$.
- To find $m\angle A$, subtract $m\angle B + m\angle C$ from 180°. Press 180 − 90 − 36 ENTER; $m\angle A = 54°$.
- Use the Pythagorean Theorem to find AB. $(AB)^2 = 16^2 - 13^2$, so $AB = \sqrt{16^2 - 13^2}$. Press 2nd x^2 16 x^2 − 13 x^2) ENTER. To the nearest tenth of a unit, $AB = 9.3$.

```
cos⁻¹(13/16)
       35.6590877
180-90-36
               54
√(16²-13²)
      9.327379053
```

For $\triangle ABC$, $m\angle A = 54°$, $m\angle C = 36°$, and $AB = 9.3$.

Solve each triangle. Give angle measures to the nearest degree and side lengths to the nearest tenth.

1. 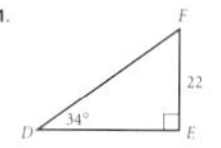

$m\angle F = 56°$, $DE = 32.6$, $FD = 39.3$

2. 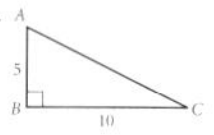

$m\angle A = 63°$, $m\angle C = 27°$, $AC = 11.2$

3. 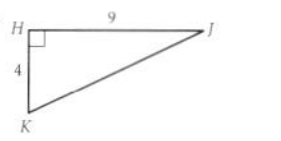

$m\angle J = 24°$, $m\angle K = 66°$, $JK = 9.8$

4. 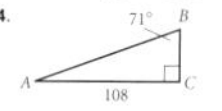

$m\angle A = 19°$, $AB = 114.2$, $BC = 37.2$

5. 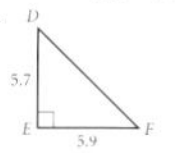

$m\angle D = 46°$, $m\angle F = 44°$, $DF = 8.2$

6. 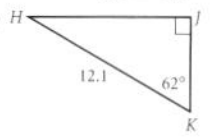

$m\angle H = 28°$, $JK = 5.7$, $HJ = 10.7$

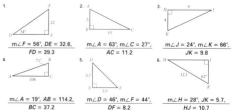

Prepare

NCTM PRINCIPLES & STANDARDS 1–4, 6–10

QUICK WARM-UP

Find the values of each trigonometric function for angle θ. Give exact answers and answers rounded to the nearest ten-thousandth.

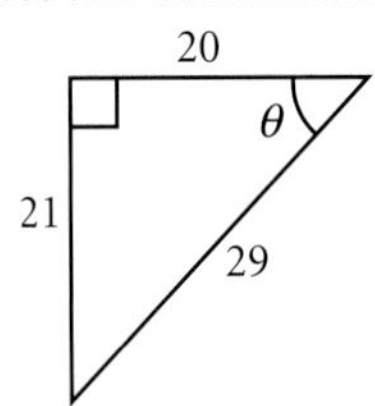

1. $\sin\theta$ $\quad \frac{21}{29} \approx 0.7241$
2. $\cos\theta$ $\quad \frac{20}{29} \approx 0.6897$
3. $\tan\theta$ $\quad \frac{21}{20} = 1.05$
4. $\csc\theta$ $\quad \frac{29}{21} \approx 1.3810$
5. $\sec\theta$ $\quad \frac{29}{20} = 1.45$
6. $\cot\theta$ $\quad \frac{20}{21} \approx 0.9524$

Also on Quiz Transparency 13.2

Teach

Why Have students list objects that make revolutions around a fixed point. Ask them to consider how many degrees the object passes through in one revolution. 360° in two revolutions? 720°

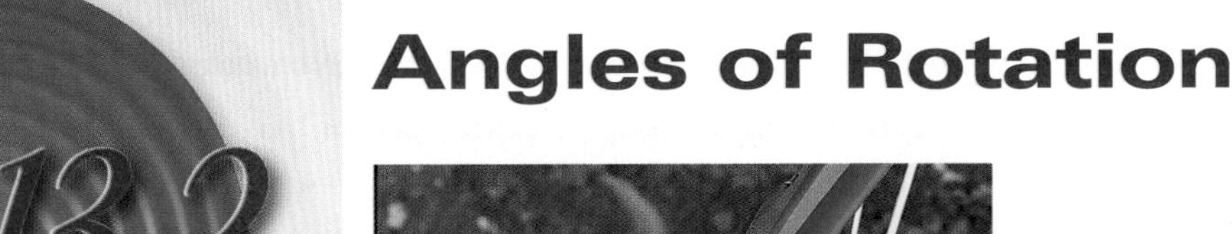

Angles of Rotation

Why *You can use angles of rotation to describe the rate at which an airplane propeller rotates.*

Objectives

- Find coterminal and reference angles.
- Find the trigonometric function values of angles in standard position.

APPLICATION
AVIATION

The propeller of an airplane rotates 1100 times per minute. Through how many degrees will a point on the propeller rotate in 1 second? *You will solve this problem in Example 1.*

In geometry, an angle is defined by two rays that have a common endpoint. In trigonometry, an angle is defined by a ray that is rotated around its endpoint. Each position of the rotated ray, relative to its starting position, creates an **angle of rotation.** The Greek letter *theta*, θ, is commonly used to name an angle of rotation.

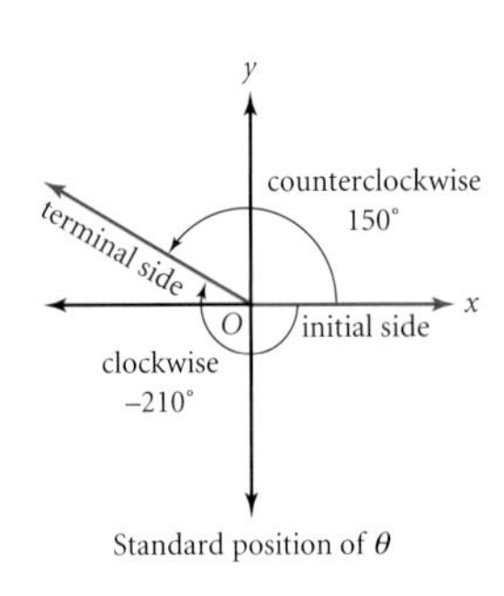

Standard position of θ

The initial position of the ray is called the **initial side** of the angle, and the final position is called the **terminal side** of the angle. When the initial side lies along the positive x-axis and its endpoint is at the origin, the angle is said to be in **standard position**.

If the direction of rotation is counterclockwise, the angle has a **positive measure**. If the direction of rotation is clockwise, the angle has a **negative measure**.

The most common unit for angle measure is the **degree.** A complete rotation of a ray is assigned a measure of 360°. Thus, a measure of 1° is $\frac{1}{360}$ of a complete rotation.

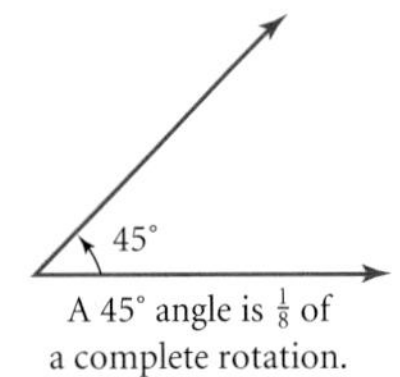

A 45° angle is $\frac{1}{8}$ of a complete rotation.

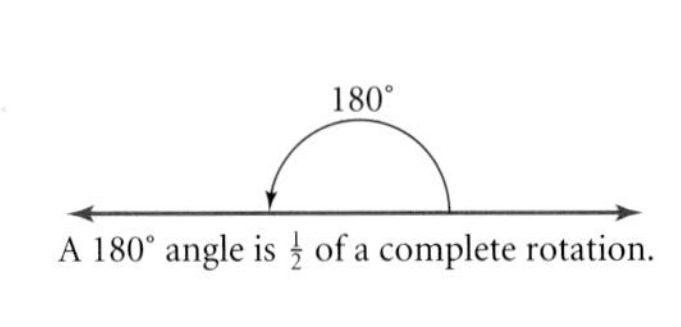

A 180° angle is $\frac{1}{2}$ of a complete rotation.

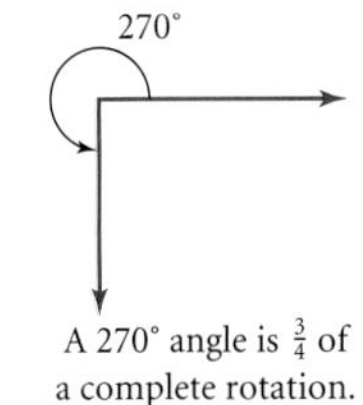

A 270° angle is $\frac{3}{4}$ of a complete rotation.

CHECKPOINT ✔ What direction of rotation generates an angle with a measure of −90°? of 120°? What portion of a complete rotation is −90°? 120°?

Alternative Teaching Strategy

HANDS-ON STRATEGIES Have each student make an angle dial by cutting a large circle out of a piece of paper. Draw the coordinate axes on the circle, with the origin at the center of the circle. Using a different color of paper, cut out a smaller circle. Cut a slit along any radius of the smaller circle and a slit along the positive x-axis of the larger circle. Place the smaller circle behind the larger circle, align the centers, and insert the lower flap of the smaller circle through the slit in the larger circle. Holding the two circles at their common center, the smaller circle can be rotated so that a portion of the larger circle appears, as shown in the diagram. Students can use this dial to visualize different angles as the terminal side is rotated counterclockwise. Students can also use the dial to help them draw the reference angle of a given angle.

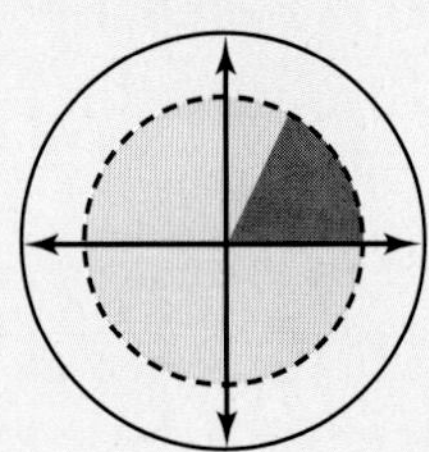

EXAMPLE 1 Refer to the propeller problem at the beginning of the lesson.

APPLICATION
AVIATION

Find the number of degrees through which a point on the propeller rotates in 1 second.

SOLUTION

The propeller rotates 1100 times per minute. The number of degrees through which a point rotates in 1 minute is $1100 \times 360° = 396{,}000°$. The number of degrees through which a point rotates in 1 second is $\frac{396{,}000°}{60} = 6600°$.

TRY THIS A record player makes 33.3 revolutions in 1 minute. Find the number of degrees through which a point on the record rotates in 1 second.

Angles in standard position are **coterminal** if they have the same terminal side.

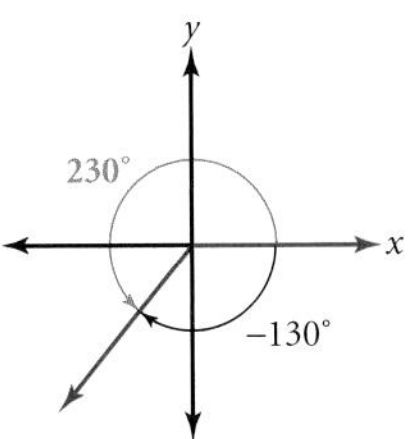

A 230° angle and a −130° angle are coterminal.

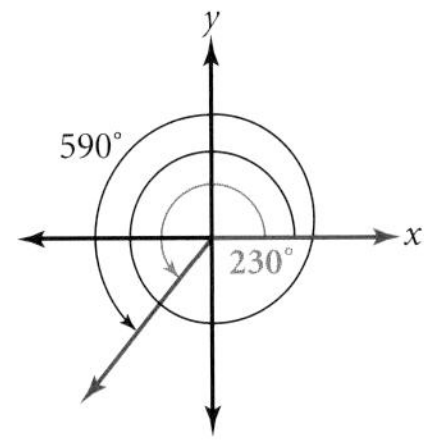

A 230° angle and a 590° angle are coterminal.

You can find coterminal angles by adding or subtracting integer multiples of 360°. This is shown in Example 2.

EXAMPLE 2 **Find the coterminal angle, θ, for each angle below such that $-360° < \theta < 360°$.**

a. 180° **b.** −27°

SOLUTION

Add and subtract 360° from each given angle. Discard answers that are not in the given range, $-360° < \theta < 360°$.

a. ~~$\theta = 180° + 360° = 540°$~~
$\theta = 180° - 360° = -180°$
The coterminal angle is −180°.

b. $\theta = -27 + 360° = 333°$
~~$\theta = -27 - 360° = 387°$~~
The coterminal angle is 333°.

TRY THIS Find the coterminal angle, θ, for 123° and for −185° such that $-360° < \theta < 360°$.

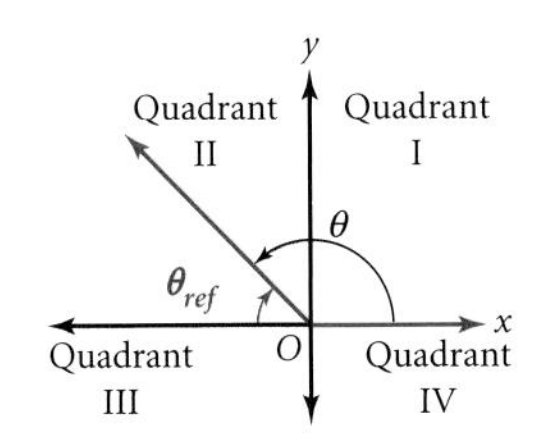

In Lesson 13.3 you will learn how to find trigonometric values for angles in standard position that are larger than 90° (or smaller than 0°). In order to do this, you will need to know how to find the measures of *reference angles.*

For an angle θ in standard position, the **reference angle,** θ_{ref}, is the positive acute angle formed by the terminal side of θ and the nearest part (positive or negative) of the x-axis. Use the positive x-axis for angles in Quadrants I and IV, and use the negative x-axis for angles in Quadrants II and III.

Inclusion Strategies

KINESTHETIC LEARNERS Use masking tape to create a large coordinate plane on the floor of the classroom. (A tile floor provides a wonderful grid system.) On the tape, label the intervals for the axes. Tape one end of a piece of string to the origin. Choose an angle between 0° and 360°. Have a student start on the positive x-axis, holding the string taut, and map out the angle. Another student can use a board protractor held at the origin of the axes to help the student decide how far to walk. Let several students map out various angles, both positive and negative, while the class gives them guidance. Then give students a point on the terminal side of an angle and have them use the string to model the angle. Using the coordinates of the point, the class should find the trig ratios of the angle modeled.

CHECKPOINT ✔
clockwise; counterclockwise; $\frac{1}{4}$; $\frac{1}{3}$

ADDITIONAL EXAMPLE 1

The propeller of an experimental aircraft rotates 900 times per minute. **Find the number of degrees through which a point on the propeller rotates in 1 second.**
5400°

TRY THIS
199.8°

ADDITIONAL EXAMPLE 2

Find the coterminal angle, θ, for each angle below such that $-360° < \theta < 360°$.

a. 90° −270°
b. −210° 150°

Teaching Tip

For any given angle, there are an infinite number of coterminal angles. However, every angle except 0° has two coterminal angles in the open interval (−360°, 360°). An angle of 0° has one coterminal angle in that interval, namely 0°.

TRY THIS
−237°; 175°

Use Teaching Transparency 54.

Teaching Tip

For a given angle θ in standard position and measured in degrees, $0° \le \theta_{ref} \le 90°$. When radian measure is used, the reference angle is sometimes referred to as the reference number because radian angle measures are real numbers.

ADDITIONAL EXAMPLE 3

Find the reference angle, θ_{ref}, for each angle.

a. $\theta = 212°$ **$\theta_{ref} = 32°$**

b. $\theta = -340°$ **$\theta_{ref} = 20°$**

c. $\theta = 40°$ **$\theta_{ref} = 40°$**

d. $\theta = 124°$ **$\theta_{ref} = 56°$**

TRY THIS
45°; 55°

CRITICAL THINKING
4

Teaching Tip

Some commonly used Greek letters and their names are listed below.

α	alpha
β	beta
γ	gamma
δ	delta
θ	theta
λ	lambda
ϕ or φ	phi
τ	tau
ω	omega

EXAMPLE 3 **Find the reference angle, θ_{ref}, for each angle.**

a. $\theta = 94°$ **b.** $\theta = 245°$ **c.** $\theta = 290°$ **d.** $\theta = -110°$

SOLUTION

a. $\theta = 94°$ is in Quadrant II.
Use the negative x-axis.

$\theta_{ref} = |180° - \theta|$

$\theta_{ref} = |180° - 94°| = 86°$

b. $\theta = 245°$ is in Quadrant III.
Use the negative x-axis.

$\theta_{ref} = |180° - \theta|$

$\theta_{ref} = |180° - 245°| = 65°$

c. $\theta = 290°$ is in Quadrant IV.
Use the positive x-axis.

$\theta_{ref} = |360° - \theta|$

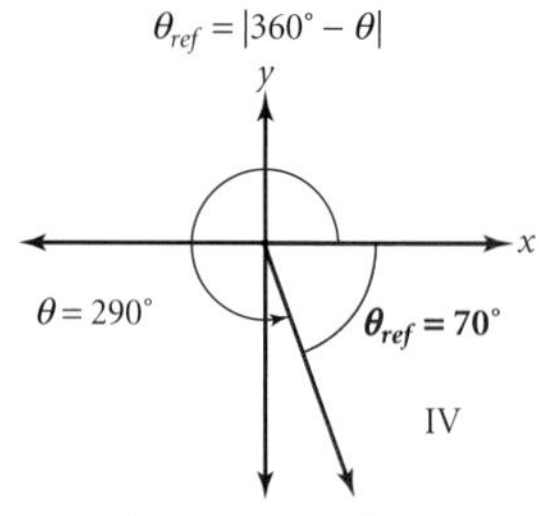

$\theta_{ref} = |360° - 290°| = 70°$

d. $\theta = -110°$ is in Quadrant III.
Use the negative x-axis.

$\theta_{ref} = |180° - \theta|$

The positive coterminal angle for $-110°$ is $250°$.
$\theta_{ref} = |180° - 250°| = 70°$

TRY THIS Find the reference angle, θ_{ref}, for $\theta = 315°$ and $\theta = -235°$.

CRITICAL THINKING How many angles in standard position between 0° and 360° have the same reference angle?

If you think of x and y as the coordinates of a point on the terminal side of an angle in standard position, you will be able to determine the correct sign of the values for the trigonometric functions.

Trigonometric Functions of θ

Let $P(x, y)$ be a point on the terminal side of θ in standard position. The distance from the origin to P is given by $r = \sqrt{x^2 + y^2}$.

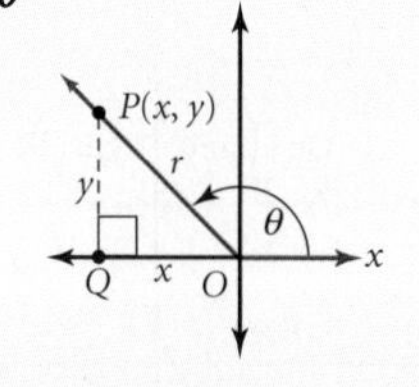

$\sin\theta = \frac{y}{r}$ $\cos\theta = \frac{x}{r}$ $\tan\theta = \frac{y}{x}, x \ne 0$

$\csc\theta = \frac{r}{y}, y \ne 0$ $\sec\theta = \frac{r}{x}, x \ne 0$ $\cot\theta = \frac{x}{y}, y \ne 0$

Enrichment

Suppose that a circle on a coordinate plane has a radius of 1. Label the coordinates of the circle and the terminal side of 0°, 90°, 180°, and 270°. **(1, 0), (0, 1), (−1, 0), (0, −1)** Using these coordinates and the trig functions of θ found on page 838 of the text, find the sine, cosine, and tangent of each of these angles. Use a calculator to check your answers.

EXAMPLE 4 Let $P(-2, -3)$ be a point on the terminal side of θ in standard position. Find the exact values of the six trigonometric functions of θ.

SOLUTION

PROBLEM SOLVING

Draw a diagram. You know that $x = -2$ and $y = -3$. Find r.

$$r = \sqrt{x^2 + y^2}$$

$$r = \sqrt{(-2)^2 + (-3)^2} = \sqrt{13}$$

$$\sin \theta = \frac{y}{r} = \frac{-3}{\sqrt{13}} = -\frac{3\sqrt{13}}{13} \qquad \cos \theta = \frac{x}{r} = \frac{-2}{\sqrt{13}} = -\frac{2\sqrt{13}}{13} \qquad \tan \theta = \frac{y}{x} = \frac{-3}{-2} = \frac{3}{2}$$

To find csc θ, sec θ, and cot θ, use reciprocals.

$$\csc \theta = -\frac{\sqrt{13}}{3} \qquad \sec \theta = -\frac{\sqrt{13}}{2} \qquad \cot \theta = \frac{2}{3}$$

TRY THIS Let $P(3, -5)$ be a point on the terminal side of θ in standard position. Find the exact values of the six trigonometric functions of θ.

Activity
Investigating the Signs in Each Quadrant

You will need: no special materials

1. Copy and complete the table below with the signs of the trigonometric functions of θ in standard position for each quadrant.

	Quadrant			
Trig value	**I**	**II**	**III**	**IV**
sin θ and csc θ	+			
cos θ and sec θ				
tan θ and cot θ				

Quadrant II (−, +) | Quadrant I (+, +) | Quadrant III (−, −) | Quadrant IV (+, −)

2. In what quadrant is the terminal side of θ if $\sin \theta = -\frac{2}{7}$? if $\cos \theta = -\frac{2}{7}$? if $\tan \theta = -\frac{1}{5}$? Give all possible answers.

CHECKPOINT ✔ 3. Does the value of r affect the sign of any of the trigonometric values? Explain.

CHECKPOINT ✔ 4. Which coordinate, x or y, determines the sign of sin θ and csc θ? of cos θ and sec θ? of tan θ and cot θ?

In each quadrant at right are listed the trigonometric function values that are positive for any angle θ in that quadrant. For example, for any angle θ in Quadrant II, sin θ and csc θ are positive while all other trigonometric function values are negative. This occurs because the sign of sin θ and csc θ both depend on y, which is positive in Quadrant II.

Quadrant II: sin θ, csc θ | Quadrant I: all | Quadrant III: tan θ, cot θ | Quadrant IV: cos θ, sec θ

Interdisciplinary Connection

SPORTS A mountain bike uses two sets of gears. Pedaling turns the front gears. The front gears turn the rear gears, which turn the rear wheel of the bike, causing it to move forward. Suppose that a bicycle has 42 equally spaced teeth on the front gear and 14 equally spaced teeth on the rear gear. For every revolution of the front gear, how many revolutions does the rear gear make? **3** Use this example to describe how mountain bikers might shift gears when dealing with varying terrain. **A greater ratio between the radius of the front gear and the radius of the rear gear means that fewer revolutions of the pedals would be required to move the bicycle.**

ADDITIONAL
EXAMPLE 4

Let $P(-1, 3)$ be a point on the terminal side of θ in standard position. **Find the exact values of the six trigonometric functions of θ.**

$\sin \theta = \frac{3\sqrt{10}}{10}$ $\quad \csc \theta = \frac{\sqrt{10}}{3}$

$\cos \theta = -\frac{\sqrt{10}}{10}$ $\quad \sec \theta = -\sqrt{10}$

$\tan \theta = -3$ $\quad \cot \theta = -\frac{1}{3}$

TRY THIS

$\sin \theta = -\frac{5\sqrt{34}}{34}$ $\quad \csc \theta = -\frac{\sqrt{34}}{5}$

$\cos \theta = \frac{3\sqrt{34}}{34}$ $\quad \sec \theta = \frac{\sqrt{34}}{3}$

$\tan \theta = -\frac{5}{3}$ $\quad \cot \theta = -\frac{3}{5}$

Activity Notes

In this Activity, students should find that all six trig ratios are positive in Quadrant I, only the sine and cosecant are positive in Quadrant II, only the tangent and cotangent are positive in Quadrant III, and only the cosine and secant are positive in Quadrant IV. *(See Error Analysis on page 842.)*

Cooperative Learning

Put students in groups of three for this Activity, and have them discuss what determines the sign of each trig ratio. The group should select an angle in each quadrant and assign one of the three ratio categories in the table to each member. Each group member should find the sign of the trig ratios for the chosen angle and share the result with the group. The group should summarize its findings and report to the class.

CHECKPOINT ✔
3. No; r is always positive.

CHECKPOINT ✔
4. y; x; both x and y

ADDITIONAL EXAMPLE 5

The terminal side of θ lies in Quadrant IV, and $\cos\theta = \frac{5}{13}$. Find the values of all six trigonometric functions of θ.

$\sin\theta = -\frac{12}{13}$ $\quad\csc\theta = -\frac{13}{12}$

$\cos\theta = \frac{5}{13}$ $\quad\sec\theta = \frac{13}{5}$

$\tan\theta = -\frac{12}{5}$ $\quad\cot\theta = -\frac{5}{12}$

TRY THIS

$\sin\theta = -\frac{4}{5}$ $\quad\csc\theta = -\frac{5}{4}$

$\cos\theta = -\frac{3}{5}$ $\quad\sec\theta = -\frac{5}{3}$

$\tan\theta = \frac{4}{3}$ $\quad\cot\theta = \frac{3}{4}$

CHECKPOINT ✔

$\sin\theta = 1$

$\cos\theta = 0$

$\tan\theta$ is undefined.

$\cot\theta = 0$

$\sec\theta =$ is undefined.

$\csc\theta = 1$

Practice

NAME ______________ CLASS ________ DATE ________

Practice
13.2 ***Angles of Rotation***

For each angle below, find all coterminal angles, θ, such that $-360° < \theta < 360°$. Then find the corresponding reference angle.

1. 47° ___−313°; 47°___ 2. −123° ___237°; 57°___ 3. 218° ___−142°; 38°___
4. 512° ___152°; 28°___ 5. −222° ___138°; 42°___ 6. 307° ___−53°; 53°___
7. 1122° ___42°; 42°___ 8. −185° ___175°; 5°___ 9. 645° ___285°; 75°___

Find the reference angle.

10. 105° ___75°___ 11. −213° ___33°___ 12. 715° ___5°___
13. −144° ___36°___ 14. 860° ___40°___ 15. −72° ___72°___
16. −2° ___2°___ 17. 1000° ___80°___ 18. −420° ___60°___

Find the exact values of the six trigonometric functions of θ, given each point on the terminal side of θ in standard position.

19. (12, 8)	20. (−5, 10)	21. (4, 9)
$\sin\theta = \frac{2\sqrt{13}}{13}$	$\sin\theta = \frac{2\sqrt{5}}{5}$	$\sin\theta = \frac{9\sqrt{97}}{97}$
$\cos\theta = \frac{3\sqrt{13}}{13}$	$\cos\theta = -\frac{\sqrt{5}}{5}$	$\cos\theta = \frac{4\sqrt{97}}{97}$
$\tan\theta = \frac{2}{3}$	$\tan\theta = -2$	$\tan\theta = 2.25$
$\csc\theta = \frac{\sqrt{13}}{2}$	$\csc\theta = \frac{\sqrt{5}}{5}$	$\csc\theta = \frac{\sqrt{97}}{9}$
$\sec\theta = \frac{\sqrt{13}}{3}$	$\sec\theta = -\sqrt{5}$	$\sec\theta = \frac{\sqrt{97}}{4}$
$\cot\theta = \frac{3}{2}$	$\cot\theta = -\frac{1}{2}$	$\cot\theta = \frac{4}{9}$

Given the quadrant of θ in standard position and a trigonometric function value of θ, find exact values for the indicated trigonometric function.

22. IV, $\sin\theta = -\frac{3}{5}$; $\tan\theta$ ___$-\frac{3}{4}$___
23. I, $\tan\theta = \frac{5}{8}$; $\csc\theta$ ___$\frac{\sqrt{89}}{5}$___
24. II, $\cos\theta = -\frac{5}{8}$; $\sin\theta$ ___$\frac{\sqrt{39}}{8}$___
25. III, $\csc\theta = -1.25$; $\tan\theta$ ___$\frac{4}{3}$___
26. II, $\cot\theta = -2.4$; $\sin\theta$ ___$\frac{5}{13}$___
27. IV, $\sec\theta = \frac{4}{3}$; $\cot\theta$ ___$\frac{3\sqrt{7}}{7}$___

If you know which quadrant contains the terminal side of θ in standard position and the exact value of one trigonometric function of θ, you can find the values of the other trigonometric functions of θ. This is shown in Example 5.

EXAMPLE 5 **The terminal side of θ in standard position is in Quadrant II, and $\cos\theta = -\frac{3}{5}$. Find the exact values of the six trigonometric functions of θ.**

SOLUTION

PROBLEM SOLVING

Draw a diagram and find the x- and y-coordinates of P.

$$\cos\theta = \frac{x}{r} = -\frac{3}{5}$$

In Quadrant II, x is negative. Thus, $x = -3$ and $r = 5$.

Use the Pythagorean Theorem to find y.

$$5^2 = (-3)^2 + y^2$$
$$y^2 = 25 - 9$$
$$y = \pm\sqrt{16}$$
$$y = 4 \quad \textit{Because P is in Quadrant II, y is positive.}$$

$\sin\theta = \frac{y}{r} = \frac{4}{5}$ $\qquad\cos\theta = \frac{x}{r} = \frac{-3}{5}$ $\qquad\tan\theta = \frac{y}{x} = \frac{4}{-3}$

$\csc\theta = \frac{r}{y} = \frac{5}{4}$ $\qquad\sec\theta = \frac{r}{x} = \frac{5}{-3}$ $\qquad\cot\theta = \frac{x}{y} = \frac{-3}{4}$

TRY THIS The terminal side of θ in standard position is in Quadrant III, and $\sin\theta = -\frac{4}{5}$. Find the exact values of the six trigonometric functions of θ.

If the terminal side of an angle, θ, in standard position coincides with a coordinate axis (such that x or y is 0), some trigonometric functions of θ will be undefined. For example csc 0°, sec 90°, tan 180°, and cot 270° are all undefined because they involve division by zero. These angle measures are excluded values in the domain of the respective functions.

CHECKPOINT ✔ Find the exact values of the six trigonometric functions for $\theta = 90°$.

Exercises

Communicate

1. Describe the differences between angles in right triangles and angles of rotation.
2. Describe the difference that may exist between the trigonometric functions of an angle and those of its reference angle. Explain the reason for this difference.
3. Do you need to know the measure of an angle in order to find the exact values of its trigonometric functions? Explain.

Reteaching the Lesson

USING TECHNOLOGY Have groups of students choose an angle measure, θ, greater than 90° and use a scientific calculator to fill in the table below.

	θ	$\theta + 360°$	$\theta + 720°$	$\theta + 360n$
$\sin\theta$				
$\cos\theta$				
$\tan\theta$				

Have each group share its observations with the class. **The values are constant in each row of the table.** Have students discuss why this is true. Have students find the reference angle for the angle and discuss the entries of the table if multiples of 90° or 180° were added to θ rather than multiples of 360°. **The absolute value of the entries in each row would be the same.**

Guided Skills Practice

APPLICATION

4. AVIATION The main rotor of a helicopter rotates 430 times per minute. Find the number of degrees through which a point on the main rotor rotates in 1 second. ***(EXAMPLE 1)*** 2580°/s

5. Find the coterminal angle, θ, for 271° such that $-360° < \theta < 360°$. ***(EXAMPLE 2)*** −89°

6. Find the reference angle for 93°, 280°, and −36°. ***(EXAMPLE 3)*** 87°; 80°; 36°

7. Let $P(3, -2)$ be a point on the terminal side of θ in standard position. Find the exact values of the six trigonometric functions of θ. ***(EXAMPLE 4)***

8. The terminal side of θ in standard position is in Quadrant III, and $\sin\theta = -\frac{12}{13}$. Find the exact values of the six trigonometric functions of θ. ***(EXAMPLE 5)***

Practice and Apply

17. 252°, −108°; 72°
18. 118°, −242°; 62°
23. −90°, 270°; 90°
24. −125°, 235°; 55°
25. 180°, −180°; 0°
29. 50°, −310°; 50°
30. 200°, −160°; 20°
35. −35°, 325°; 35°
36. −180°, 180°; 0°

Sketch each angle in standard position.

9. 115° **10.** 280° **11.** −300° **12.** −130°

For each angle below, find all coterminal angles such that −360° < θ < 360°. Then find the corresponding reference angle, if it exists.

13. 35° −325°; 35°	**14.** 23° −337°; 23°	**15.** 112° −248°; 68°	**16.** 160° −200°; 20°
17. 612°	**18.** 478°	**19.** −135° 225°; 45°	**20.** −315° 45°; 45°
21. 90° −270°; 90°	**22.** −180° 180°; 0°	**23.** −450°	**24.** −485°
25. 540°	**26.** 270° −90°; 90°	**27.** 225° −135°; 45°	**28.** 195° −165°; 15°
29. 410°	**30.** 560°	**31.** −120° 240°; 60°	**32.** −280° 80°; 80°
33. −175° 185°; 5°	**34.** −295° 65°; 65°	**35.** −395°	**36.** −540°

Find the exact values of the six trigonometric functions of θ.

37.

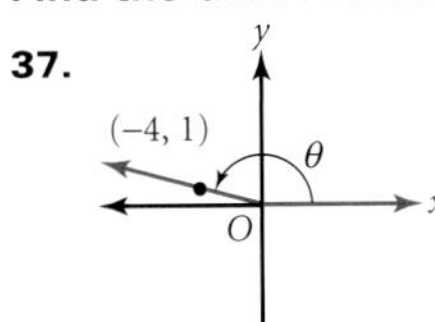

38.

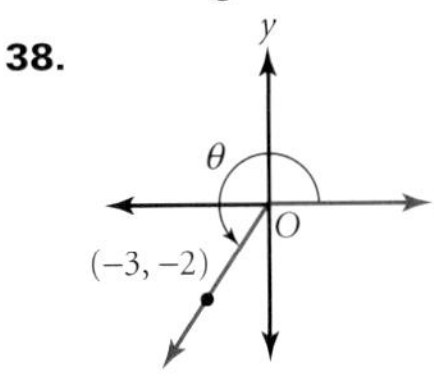

39.

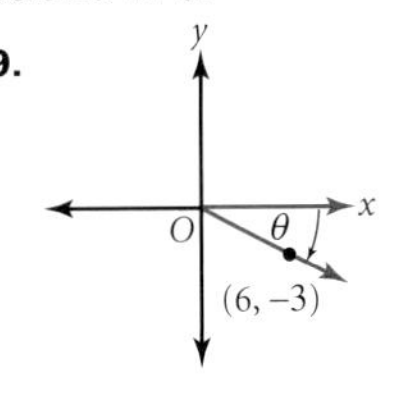

Find the exact values of the six trigonometric functions of θ given each point on the terminal side of θ in standard position.

40. (3, 4) **41.** (5, 2) **42.** (−4, 2) **43.** (−4, 6)

44. $(\sqrt{3}, -3)$ **45.** $(2\sqrt{5}, -1)$ **46.** (−4, −3) **47.** (−1, −8)

internet connect
Homework Help Online
Go To: **go.hrw.com**
Keyword:
MB1 Homework Help for Exercises 48–55

Given the quadrant of θ in standard position and a trigonometric function value of θ, find exact values for the indicated functions.

48. I, $\cos\theta = 0.25$; $\tan\theta$ $\sqrt{15}$

49. III, $\cos\theta = -\frac{1}{2}$; $\tan\theta$ $\sqrt{3}$

50. IV, $\tan\theta = -1$; $\csc\theta$ $-\sqrt{2}$

51. I, $\tan\theta = 2$; $\csc\theta$ $\frac{\sqrt{5}}{2}$

52. III, $\sin\theta = -\frac{1}{2}$; $\sec\theta$ $-\frac{2\sqrt{3}}{3}$

53. II, $\sin\theta = 0.4$; $\sec\theta$ $-\frac{5\sqrt{21}}{21}$

54. IV, $\cot\theta = -1.2$; $\cos\theta$ $\frac{6\sqrt{61}}{61}$

55. II, $\cot\theta = -1.75$; $\cos\theta$ $-\frac{7\sqrt{65}}{65}$

9.

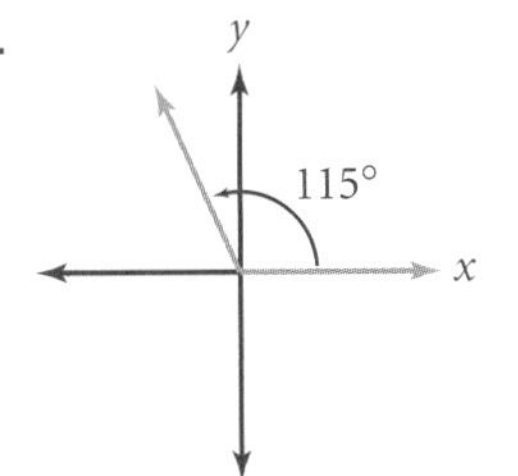

10.

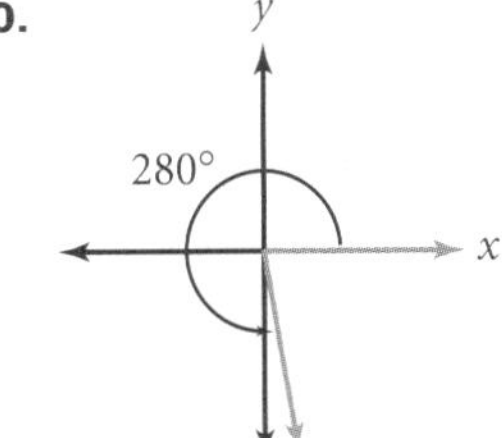

11.

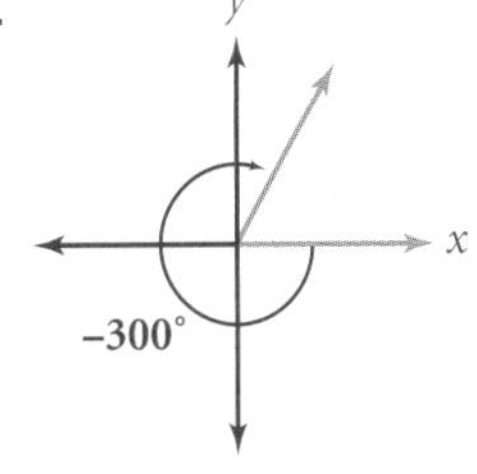

The answers to Exercises 12 and 37–47 can be found in Additional Answers beginning on page 1002.

Assess

Selected Answers

Exercises 4–8, 9–79 odd

ASSIGNMENT GUIDE

In Class	1–8
Core	9–67 odd
Core Plus	10–68 even
Review	69–79
Preview	80–82

Extra Practice can be found beginning on page 940.

7. $\sin\theta = -\frac{2\sqrt{13}}{13}$; $\csc\theta = -\frac{\sqrt{13}}{2}$;
$\cos\theta = \frac{3\sqrt{13}}{13}$; $\sec\theta = \frac{\sqrt{13}}{3}$;
$\tan\theta = -\frac{2}{3}$; $\cot\theta = -\frac{3}{2}$

8. $\sin\theta = -\frac{12}{13}$; $\csc\theta = -\frac{13}{12}$;
$\cos\theta = -\frac{5}{13}$; $\sec\theta = -\frac{13}{5}$;
$\tan\theta = \frac{12}{5}$; $\cot\theta = \frac{5}{12}$

NAME ____________ CLASS ________ DATE ________

Student Technology Guide
13.2 ***Angles of Rotation***

Two angles are *coterminal* if their measures differ by a multiple of 360°. A calculator can quickly find measures of coterminal angles.

Press MODE, use the arrow keys to move to **Degree**, then press ENTER.
Press 2nd MODE.

Example: Find the coterminal angle, θ, for each angle below such that $-360° < \theta < 360°$.
a. −74° **b.** 227°

Add and subtract 360° from each given angle measure. Discard answers that are not in the given range.

a. To add 360°, press (-) 74 + 360 ENTER. To subtract 360°, press 2nd ENTER and use ◄ to move to the plus sign. Press − ENTER.

The coterminal angle is 286°.

b. To add 360°, press 227 + 360 ENTER. To subtract 360°, press 2nd ENTER and use ◄ to move to the plus sign. Press − ENTER.

The coterminal angle is −133°.

The angles in the example above involve less than one rotation. There is an efficient way to find a coterminal angle between −360° and 360° for an angle that involves several rotations.

Example: Find the coterminal angle, θ, for −2226° such that $-360° < \theta < 360°$.

- Estimate the multiple of 360° you need to add to −2226 so that the sum will be between −360° and 0°.
  Press (-) 2226 + 5 × 360 ENTER.
- Modify your estimate by changing 5 to 6 and pressing ENTER.
- Press 2nd (-) + 360 ENTER.

```
-2226+5*360
           -426
-2226+6*360
            -66
Ans+360
            294
```

The coterminal angles are −66° and 294°.

Find the coterminal angle, θ, for each angle below such that −360° < θ < 360°.

1. −42° 318°
2. 302° −58°
3. 1743° 303°
4. −117,281° −281°

Error Analysis

Students often have trouble determining the signs of the trig ratios in various quadrants. Have students determine which of the three basic trig functions are positive in each quadrant. Then have them make a mnemonic from the first letter of each function. For example, starting in Quadrant I and going counterclockwise, "All Students Take Calculus" means that All three basic trig ratios are positive in Quadrant I, only Sine is positive in Quadrant II, only Tangent is positive in Quadrant III, and only Cosine is positive in Quadrant IV. Remind students that each reciprocal function has the same sign as its parent function.

QII	QI
Students	*All*
Take	*Calculus*
QIII	QIV

In Exercises 80–82, students explore the trig ratios of three special angles. Students are given the coordinates of points on the terminal sides of angles with measures of 30°, 45°, and 60° and are asked to find the six trig ratios for each. Common angles such as these are studied further in Lesson 13.3.

Find the number of rotations or the fraction of a rotation represented by each angle below. Indicate whether the rotation is clockwise or counterclockwise. clockwise = CW and counterclockwise = CCW

56. 45° $\frac{1}{8}$ **CCW** **57.** 90° $\frac{1}{4}$ **CCW** **58.** −180° $\frac{1}{2}$ **CW** **59.** −270° $\frac{3}{4}$ **CW**

60. 450° $1\frac{1}{4}$ **CCW** **61.** 720° **2 CCW** **62.** −420° $1\frac{1}{6}$ **CW** **63.** −640° $1\frac{7}{9}$ **CW**

CHALLENGES

64. Find $\cos\theta$ if $\sin\theta = 0.375$ and $\tan\theta$ is less than 0. **≈−0.9270**

65. Find $\tan\theta$ if $\cos\theta = 0.809$ and $\sin\theta$ is less than 0. **≈−0.7266**

CONNECTION

66. PROBABILITY An angle of rotation of 120° is colored red on a circular spinner at a school fair. If the spinner lands anywhere in the red space, the contestant wins. What is the probability of a contestant winning? $\frac{1}{3}$, **or ≈33%**

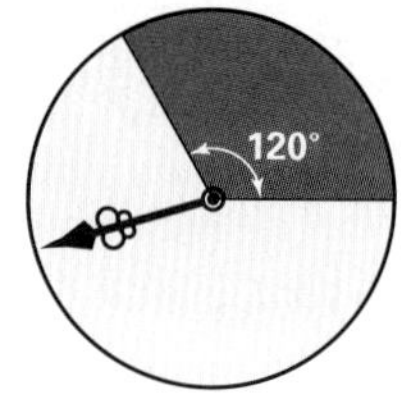

APPLICATIONS

67. ENGINEERING The flywheel of an engine rotates 900 times per minute. Through how many degrees does a point on the flywheel rotate in 1 second? **5400°/s**

68. NAVIGATION Airline pilots and sea captains both use *nautical miles* to measure distance. A nautical mile is approximately equal to the arc length intercepted on the surface of the Earth by a central angle measure of 1 *minute* (there are 60 minutes in 1 degree). The diameter of the Earth at the equator is approximately 7926.41 miles.

a. How many minutes are there in the circumference of the Earth? **21,600 min**

b. Find the approximate circumference of the Earth in miles. **24,902 mi**

c. Approximately how many miles are equal to one nautical mile? **1.15 mi**

69. Solve $x^2 - 8 = 188$ for x. ***(LESSONS 5.5 AND 5.6)*** **±14**

Graph each number and its conjugate in the complex plane. *(LESSON 5.6)*

70. $-6 + 4i$ **71.** $5i$ **72.** -1 **73.** $-3 - 4i$

Evaluate. *(LESSON 5.6)*

74. $|1 + i|$ $\sqrt{2}$ **75.** $|2 + 3i|$ $\sqrt{13}$ **76.** $\left|\frac{\sqrt{2}}{2} + \frac{\sqrt{2}}{2}i\right|$ **1** **77.** $\left|\frac{\sqrt{3}}{3} + \frac{\sqrt{6}}{3}i\right|$ **1**

78. Find the standard equation for the hyperbola centered at (1, 4) with vertices at (−4, 4) and (6, 4) and co-vertices at (1, −5) and (1, 13). Graph the equation. ***(LESSON 9.5)*** $\frac{(x-1)^2}{25} - \frac{(y-4)^2}{81} = 1$

79. How many ways are there to choose a committee of 4 from a group of 10 people? ***(LESSON 10.3)*** **210**

Look Beyond

Find the exact values of the six trigonometric functions of θ given each point on the terminal side of θ in standard position.

80. $\left(\frac{\sqrt{3}}{2}, \frac{1}{2}\right)$ **81.** $\left(\frac{\sqrt{2}}{2}, \frac{\sqrt{2}}{2}\right)$ **82.** $\left(\frac{1}{2}, \frac{\sqrt{3}}{2}\right)$

70.

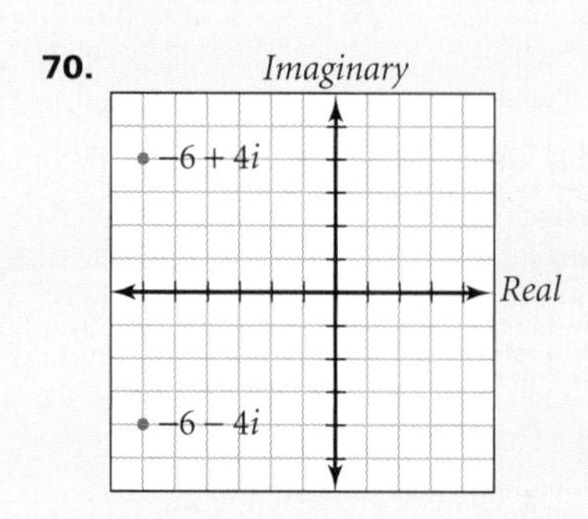

71.

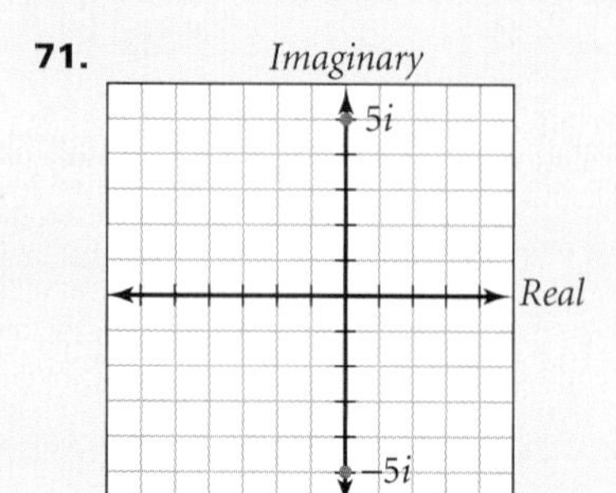

72.

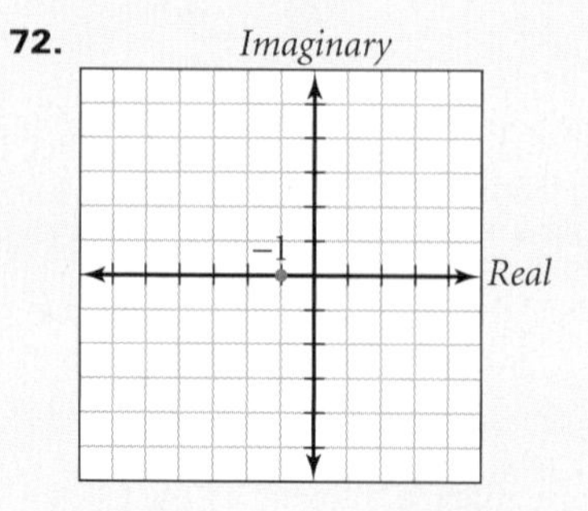

73.

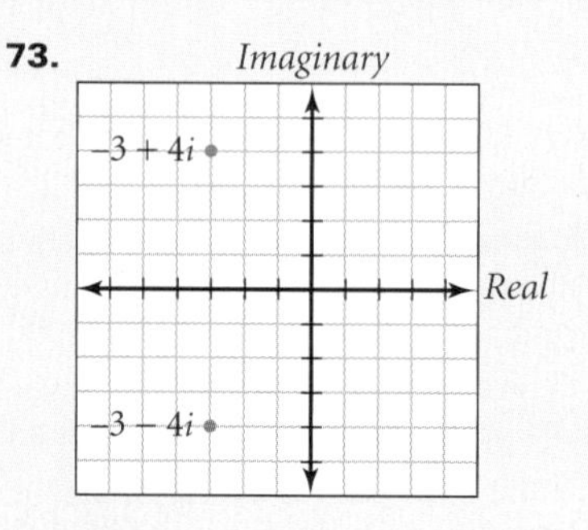

The answers to Exercises 80–82 can be found in Additional Answers beginning on page 1002.

Trigonometric Functions of Any Angle

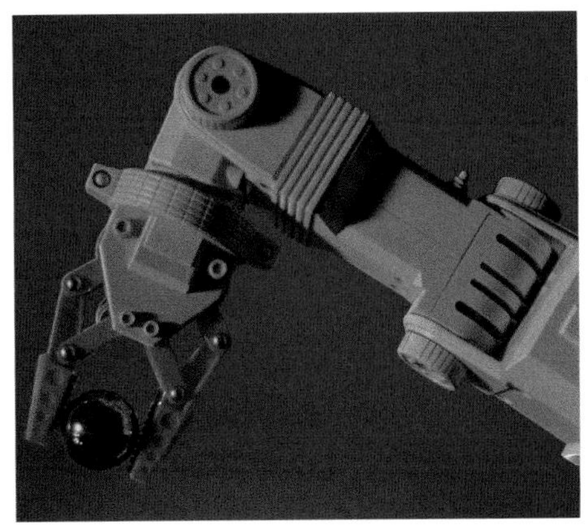

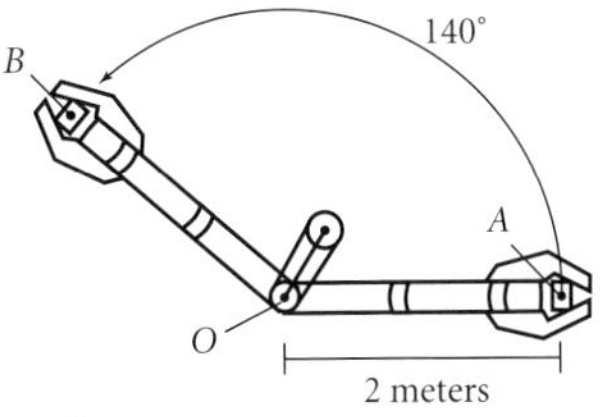

Why *You can use trigonometric functions of any angle to solve real-world problems such as designing the motion of a robotic arm.*

Objective

- Find exact values for trigonometric functions of special angles and their multiples.
- Find approximate values for trigonometric functions of any angle.

APPLICATION
ROBOTICS

Steve is programming a 2-meter long robotic arm. The arm grasps an object at point *A*, located directly to the right of the pivot point, *O*. The arm swings through an angle of 140° and releases the object at point *B*. What is the new position of the object relative to the pivot point? *You will solve this problem in Example 3.*

There are certain angles whose exact trigonometric function values can be found without a calculator. You will explore these angles in the Activity below.

Activity Exploring Special Triangles

CONNECTION
GEOMETRY

You will need: no special materials

1. When a square is bisected along a diagonal, two 45-45-90 triangles are formed, as shown below.

   Find the length of c in terms of a by using the Pythagorean Theorem.

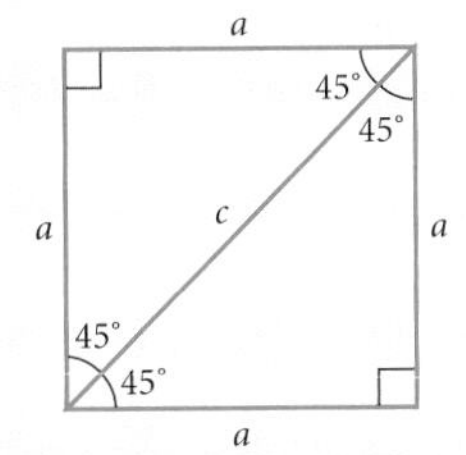

2. An equilateral triangle has three 60° angles. When one angle is bisected, two 30-60-90 triangles are formed, as shown below.

   Find the length of c in terms of a by using the Pythagorean Theorem.

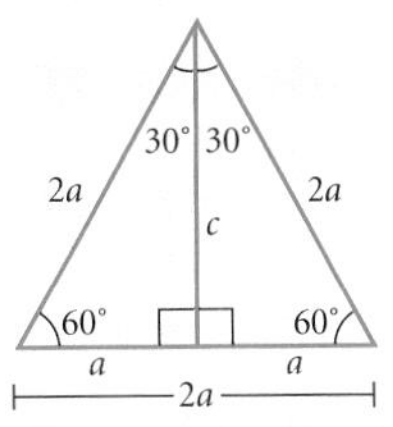

CHECKPOINT ✔ 3. State the relationship between the sides of a 45-45-90 triangle and the relationship between the sides of a 30-60-90 triangle.

Prepare

NCTM PRINCIPLES & STANDARDS 1–4, 6–10

QUICK WARM-UP

Find the reference angle, θ_{ref}, for each angle.

1. $\theta = 163°$ $\theta_{ref} = 17°$
2. $\theta = -163°$ $\theta_{ref} = 17°$
3. $\theta = -280°$ $\theta_{ref} = 80°$
4. $\theta = 311°$ $\theta_{ref} = 49°$

The point (1, −2) is located on the terminal side of θ. **Find the exact value of each trigonometric function of θ.**

5. $\sin\theta$ $-\frac{2\sqrt{5}}{5}$
6. $\cos\theta$ $\frac{\sqrt{5}}{5}$
7. $\tan\theta$ -2
8. $\csc\theta$ $-\frac{\sqrt{5}}{2}$
9. $\sec\theta$ $\sqrt{5}$
10. $\cot\theta$ $-\frac{1}{2}$

Also on Quiz Transparency 13.3

Teach

Why Use a board compass to model the robotic arm described in the text. Create a coordinate plane on the floor of the classroom and place the point of the compass at the origin. Have students decide how to swing the compass arm to model the motion of the robotic arm.

Activity Notes

In this Activity, students use properties of special triangles to find the side lengths of 45-45-90 and 30-60-90 triangles. Students should find that the sides of a 45-45-90 triangle are in the ratio $1 : 1 : \sqrt{2}$ and that the sides of a 30-60-90 triangle are in the ratio $1 : \sqrt{3} : 2$. The ratios of these lengths can then be used for any similar triangle.

CHECKPOINT ✔

3. 45-45-90: $1 : 1 : \sqrt{2}$; 30-60-90: $1 : \sqrt{3} : 2$

Alternative Teaching Strategy

USING MANIPULATIVES Have students use a compass, protractor, and ruler to construct an isosceles right triangle with 1-inch legs and a 30-60-90 triangle with a 2-inch hypotenuse. Have students cut these "reference triangles" out of poster board. Along each side of the triangle, have students write the length and then label the angle measures. Have students place these triangles along the x- and y-axes in appropriate quadrants to help them solve problems involving these angles. For example, as students work through Example 1 in the text, they will draw a coordinate system and use a protractor to mark a 120° angle. Once the terminal side of the angle is drawn, students can find the reference angle and check whether the 60° angle on their reference triangle fits between the x-axis and the terminal side of the angle. Then students can use similar triangles to solve the problem. Having the reference triangles will help students use the correct sides when setting up their trig ratios.

Math CONNECTION

GEOMETRY When solving the 45-45-90 triangle, students use properties of a square and the Pythagorean theorem to find that the legs of the triangle are equal in length and that the hypotenuse is $\sqrt{2}$ times the length of a leg. When solving the 30-60-90 triangle, students must use the fact that the altitude of an equilateral triangle bisects the base of the triangle. Then students can use the Pythagorean theorem to show the lengths of the legs and hypotenuse have a ratio of $1 : \sqrt{3} : 2$.

CHECKPOINT ✔

θ	$\sin\theta$	$\cos\theta$	$\tan\theta$
30°	$\frac{1}{2}$ 0.5	$\frac{\sqrt{3}}{2}$ 0.8660	$\frac{\sqrt{3}}{3}$ 0.5774
45°	$\frac{\sqrt{2}}{2}$ 0.7071	$\frac{\sqrt{2}}{2}$ 0.7071	1
60°	$\frac{\sqrt{3}}{2}$ 0.8660	$\frac{1}{2}$ 0.5	$\sqrt{3}$ 1.7321

Use Teaching Transparency 55.

In the Activity, you found that the lengths of the sides of a 45-45-90 triangle have a ratio of 1 to 1 to $\sqrt{2}$, and the lengths of the sides of a 30-60-90 triangle have a ratio of 1 to $\sqrt{3}$ to 2. You can use these relationships to find the exact values of the sine, cosine, and tangent of 30°, 45°, and 60° angles.

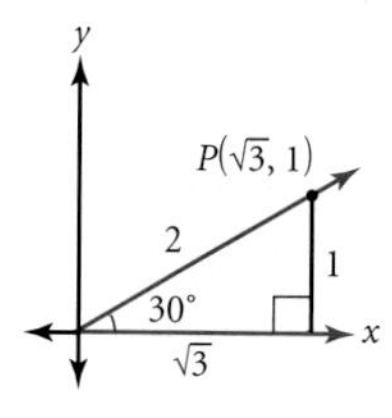

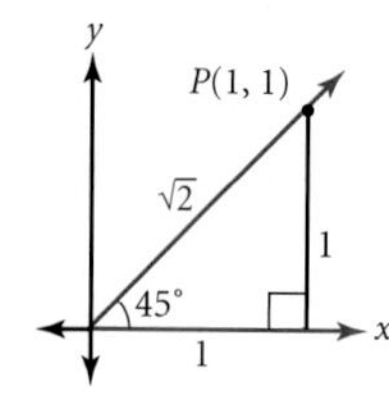

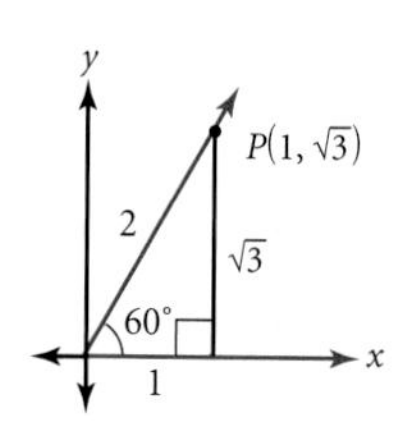

$\sin 30° = \frac{1}{2}$ $\quad$ $\cos 30° = \frac{\sqrt{3}}{2}$ $\quad$ $\tan 30° = \frac{1}{\sqrt{3}}$, or $\frac{\sqrt{3}}{3}$

$\sin 45° = \frac{1}{\sqrt{2}}$, or $\frac{\sqrt{2}}{2}$ $\quad$ $\cos 45° = \frac{1}{\sqrt{2}}$, or $\frac{\sqrt{2}}{2}$ $\quad$ $\tan 45° = 1$

$\sin 60° = \frac{\sqrt{3}}{2}$ $\quad$ $\cos 60° = \frac{1}{2}$ $\quad$ $\tan 60° = \sqrt{3}$

Throughout this chapter it will be helpful to be familiar with the exact values of the sine, cosine, and tangent of 30°, 45°, and 60° angles, given above.

CHECKPOINT ✔ Make a table of the exact values and the decimal approximations of the sine, cosine, and tangent of 30°, 45°, and 60°.

You can use the exact values of the sine, cosine, and tangent given above to evaluate any angle whose reference angle is 30°, 45°, or 60°. This is shown in Example 1.

EXAMPLE 1 **Find exact values of sin 315°, cos 315°, and tan 315°.**

SOLUTION

PROBLEM SOLVING **Draw a diagram** and find the reference angle.

$$\theta_{ref} = |360° - 315°| = 45°$$

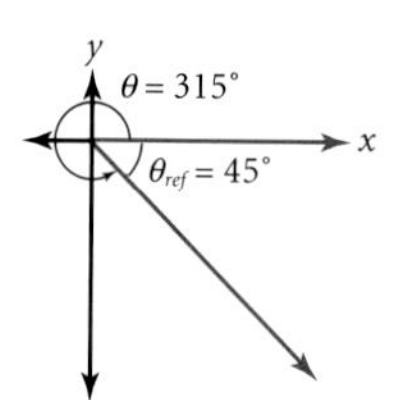

Because 315° is in Quadrant IV, where y is negative, the sine and tangent are negative.

$\sin 315° = -\sin 45° = -\frac{1}{\sqrt{2}}$, or $-\frac{\sqrt{2}}{2}$

$\cos 315° = \cos 45° = \frac{1}{\sqrt{2}}$, or $\frac{\sqrt{2}}{2}$

$\tan 315° = -\tan 45° = -1$

TRY THIS Find exact values of $\sin(-150°)$, $\cos(-150°)$, and $\tan(-150°)$.

In Lesson 13.2, a point on the terminal side of an angle was used to find trigonometric function values. In Example 2 on the next page, trigonometric function values are used to find the exact coordinates of a point on the terminal side of an angle.

Interdisciplinary Connection

HISTORY Thales, an ancient Greek mathematician and philosopher, used special right triangles to find the height of unknown objects. He waited for a time of day when the shadow of a meter stick was exactly one meter long. At this exact time, Thales said the height of the unknown object would also be the same as the length of its shadow. Explain why this theory works. **because isosceles right triangles are created**

Inclusion Strategies

USING VISUAL MODELS Have students draw a unit circle and indicate the point, $P(x, y)$, where the terminal side of a 60° angle intersects the circle. Have students find the coordinates of P by using the relationship $(x, y) = (\cos\theta, \sin\theta)$. **(0.5, 0.8660)** Have students write a paragraph explaining why cos 60° is the x-coordinate of P and sin 60° is the y-coordinate.

EXAMPLE 2 **Find the exact coordinates of point P, located at the intersection of a circle with a radius of 5 and the terminal side of a 150° angle in standard position.**

SOLUTION

PROBLEM SOLVING

Draw a diagram, and find the reference angle.

$$\theta_{ref} = |180° - 150°| = 30°$$

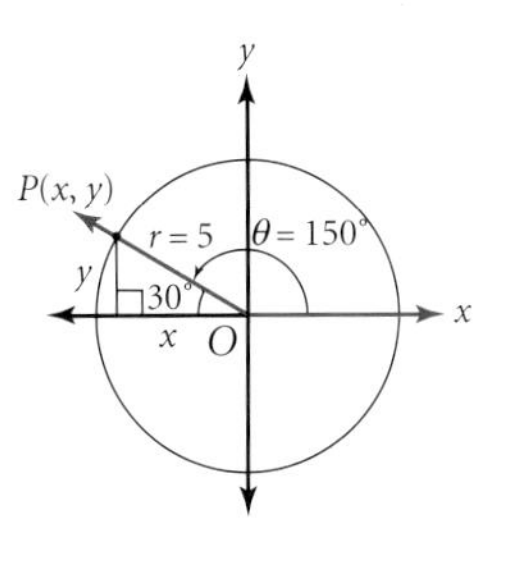

Because P is in Quadrant II, where x is negative, the cosine is negative.

$$\cos 150° = \frac{x}{r} \qquad \sin 150° = \frac{y}{r}$$

$$-\cos 30° = \frac{x}{5} \qquad \sin 30° = \frac{y}{5}$$

$$-\frac{\sqrt{3}}{2} = \frac{x}{5} \qquad \frac{1}{2} = \frac{y}{5}$$

$$-\frac{5\sqrt{3}}{2} = x \qquad \frac{5}{2} = y$$

The exact coordinates of P are $\left(-\frac{5\sqrt{3}}{2}, \frac{5}{2}\right)$.

TRY THIS

Find the exact coordinates of point P, located at the intersection of a circle with a radius of 12 and the terminal side of a 300° angle in standard position.

For any point on the terminal side of an angle in Quadrant I, $\cos \theta = \frac{x}{r}$, so $r \cos \theta = x$, and $\sin \theta = \frac{y}{r}$, so $r \sin \theta = y$. This allows you to find the coordinates of any point on a circle centered at the origin with radius r.

Coordinates of a Point on a Circle

If P lies at the intersection of the terminal side of θ in standard position and a circle with a radius of r centered at the origin, then the coordinates of P are $(r \cos \theta, r \sin \theta)$.

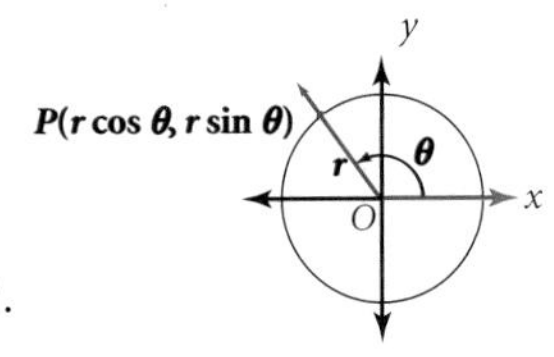

EXAMPLE 3 Refer to the robotic arm described at the beginning of the lesson.

APPLICATION
ROBOTICS

What is the new position of the object relative to the pivot point?

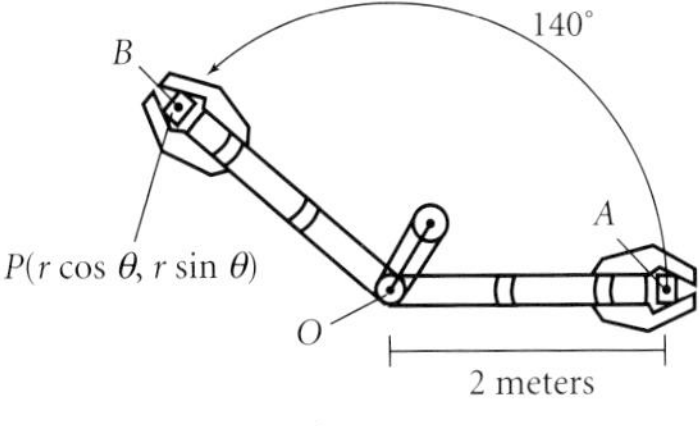

SOLUTION

TECHNOLOGY
SCIENTIFIC CALCULATOR

Place the pivot point at the origin. The new position at point B has the coordinates $(r \cos \theta, r \sin \theta)$. Substitute 2 for r and 140° for θ.

$$B(r \cos \theta, r \sin \theta) = B(2 \cos 140°, 2 \sin 140°)$$
$$\approx B(-1.53, 1.29)$$ *Use a scientific calculator in degree mode.*

The object is about 1.53 meters to the left of the pivot point and about 1.29 meters above the pivot point.

Enrichment

Make transparencies of polar graph paper and give one to each student along with some Cartesian graph paper. Explain to students that any point in a plane can be located by giving the distance to the origin, r, and the angle of rotation from the positive x-axis, θ. Like points in the Cartesian coordinate system, (x, y), points in the polar coordinate system are named with coordinates, (r, θ). Have students discuss similarities and differences between the two coordinate systems. **Both have an origin and axes, but the polar coordinate system is defined by angle measures and the radii of concentric circles.**

Have students plot a point and label it with Cartesian coordinates. Then have them place the polar transparency over the Cartesian graph paper, align the origins, and find the polar coordinates of the point. Have students compare the two representations of the point's coordinates. **A point represented by (x, y) in Cartesian coordinates is represented by (r, θ) in polar coordinates, where $r = x^2 + y^2$ and $\theta = \tan^{-1} \frac{y}{x}$. A point represented by (r, θ) is represented by (x, y), where $x = r \cos \theta$ and $y = r \sin \theta$.**

ADDITIONAL **EXAMPLE 1**

Find exact values of sin 240°, cos 240°, and tan 240°.

$\sin 240° = -\frac{\sqrt{3}}{2}$

$\cos 240° = -\frac{1}{2}$

$\tan 240° = \sqrt{3}$

TRY THIS

$\sin(-150°) = -\frac{1}{2}$

$\cos(-150°) = -\frac{\sqrt{3}}{2}$

$\tan(-150°) = \frac{\sqrt{3}}{3}$

ADDITIONAL **EXAMPLE 2**

Find the exact coordinates of a point, R, that is located at the intersection of a circle with a radius of 7 and the terminal side of a 135° angle in standard position.

$\left(-\frac{7\sqrt{2}}{2}, \frac{7\sqrt{2}}{2}\right)$

TRY THIS

$(6, -6\sqrt{3})$

ADDITIONAL **EXAMPLE 3**

Refer to the robotic arm described at the beginning of the lesson. Suppose that the length of the robotic arm is 3 meters. **Find the new position of the object after the arm swings through an angle of 200°.**

The object is about 2.82 m to the left of the pivot point and about 1.03 m below it.

Unit Circle

When a circle centered at the origin has a radius of 1, it is called a **unit circle.** Because r is 1, the coordinates of P are $(\cos \theta, \sin \theta)$.

Unit circles are helpful in demonstrating the behavior of trigonometric functions. The unit circle below shows the x- and y-coordinates of P for special angles between 0° and 360°.

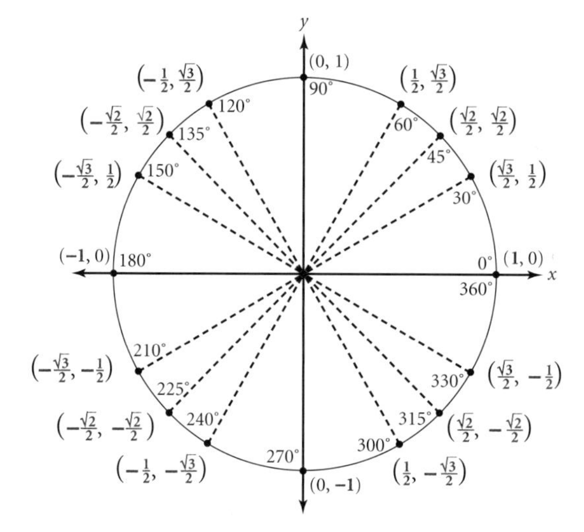

CHECKPOINT ✔ Using the coordinates on the unit circle above, find the following trigonometric function values: sin 90°, cos 60°, sin −45°, cos −30°, sin 360°, and cos 180°.

Recall from Lesson 13.2 that an angle of 90° is coterminal with angles of 450° and 810°. In fact, an angle of 90° is coterminal with any angle whose measure can be represented by $90° + n360°$, where n is an integer. All coterminal angles share the same reference angle and have the same trigonometric function values. Because the values of sine and cosine repeat every 360°, they are called *periodic functions* and their *period* is 360°.

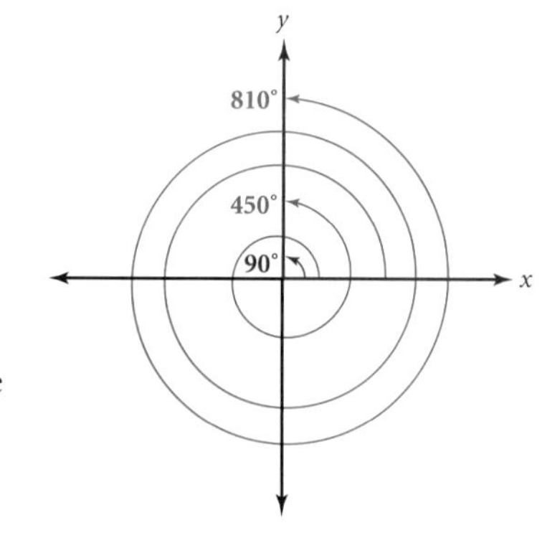

Periodic Functions

A function, f, is **periodic** if there is a number p such that $f(x + p) = f(x)$ for every x in the domain of f.

The smallest positive number p that satisfies the equation above is called the **period** of the function.

CRITICAL THINKING What is the period of the tangent function?

CHECKPOINT ✔

$\sin 90° = 1$ $\cos 60° = \frac{1}{2}$

$\sin(-45°) = -\frac{\sqrt{2}}{2}$ $\cos(-30°) = \frac{\sqrt{3}}{2}$

$\sin 360° = 0$ $\cos 180° = -1$

Use Teaching Transparency 56.

Teaching Tip

Have students draw the unit circle, $x^2 + y^2 = 1$, and place an index finger at the point (1, 0). Have them move their finger counterclockwise around the circle and keep track of what happens to the x-coordinate of the point marked by their fingertip. They should see that the x-value begins at 1 and decreases through 0 to −1. The x-value then increases back through 0 until it reaches 1 again. Tell student that the x-value, or cosine function, repeats and the *period* of repetition is 360°. Have them demonstrate the period of the sine function by using a similar procedure and noticing the change in the y-values. Then have them demonstrate the period of the tangent functions by noting the value of the ratio $\frac{y}{x}$. Have them record the values of θ where $\tan \theta$ is undefined.

CRITICAL THINKING
180°

Reteaching the Lesson

USING TABLES Have students fill in the table below.

Reference triangle	θ	$\sin \theta$	$\cos \theta$	$\tan \theta$
30°, 60°, sides $\sqrt{3}$, 2, 1	30°	0.50	0.87	0.58
	150°	0.50	−0.87	−0.58
	210°	−0.50	−0.87	0.58
	330°	−0.50	0.87	−0.58

Students should draw the reference triangle in the first column and label the sides. The second column should consist of all angles between 0° and 360° that have a 30° reference angle. Have students create similar tables for all angles with a reference angle of 60° and of 45°. Students may use the triangles and quadrants to fill in the rest of the tables and should check their answers with a scientific calculator. Discuss each table with the class, pointing out that the absolute values for the trig functions are the same in each column. Have students make observations about each table and compare all three charts.

EXAMPLE 4 **Find the exact values of the sine, cosine, and tangent of each angle.**

a. 900°　　　　**b.** −930°

SOLUTION

Find a coterminal angle with a positive measure between 0° and 360°.

a.

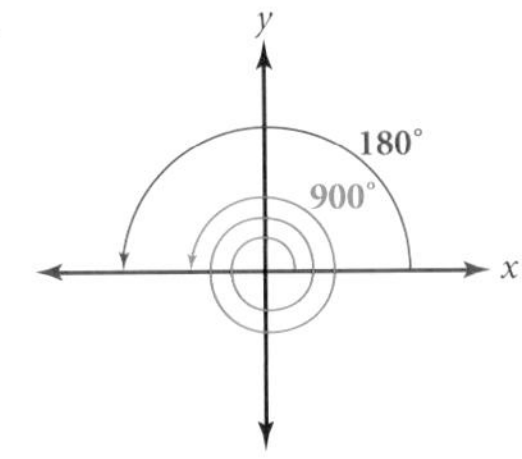

b.

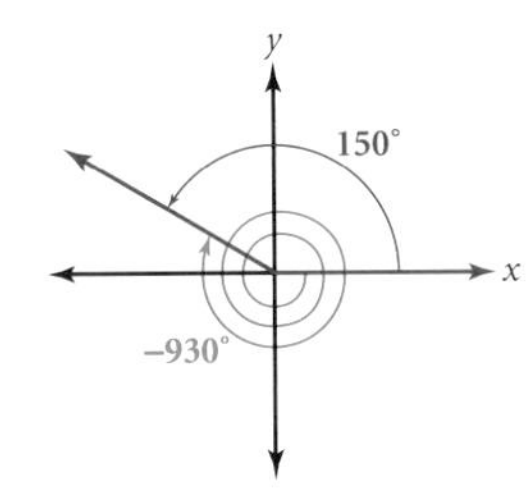

a.	b.
$900° - 2 \cdot 360° = \mathbf{180°}$	$-930° + 3 \cdot 360° = \mathbf{150°}$
$\sin 900° = \sin 180° = 0$	$\sin 930° = \sin 150° = \frac{1}{2}$
$\cos 900° = \cos 180° = -1$	$\cos 930° = \cos 150° = -\frac{\sqrt{3}}{2}$
$\tan 900° = \tan 180° = 0$	$\tan 930° = \tan 150° = -\frac{\sqrt{3}}{3}$

TRY THIS Find the exact values of the sine, cosine, and tangent of each angle.

a. 1110°　　　　**b.** −1110°

Exercises

Communicate

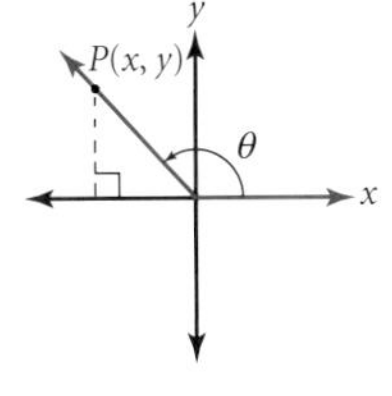

1. Explain when it is possible to use a reference angle rather than a scientific calculator to solve a trigonometric problem.
2. Explain the relationship of the sine and cosine of θ to the coordinates of point P in the diagram at right.
3. Describe how to find the exact value of sin 420°.

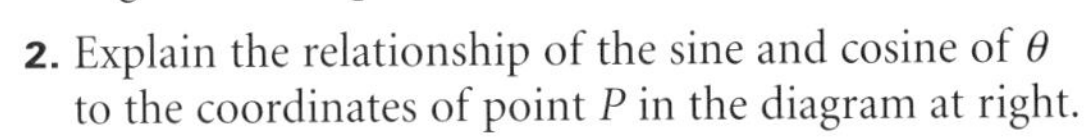

Guided Skills Practice

4. Find the exact values of sin 150°, cos 150°, and tan 150°. ***(EXAMPLE 1)***
5. Find the exact coordinates of a point, P, that is located at the intersection of a circle with a radius of 9 and the terminal side of a 315° angle in standard position. ***(EXAMPLE 2)***

4. $\sin 150° = \frac{1}{2}$
   $\cos 150° = -\frac{\sqrt{3}}{2}$
   $\tan 150° = -\frac{\sqrt{3}}{3}$

5. $\left(\frac{9\sqrt{2}}{2}, -\frac{9\sqrt{2}}{2}\right)$

ADDITIONAL EXAMPLE 4

Find the exact value of the sine, cosine, and tangent of each angle.

a. $\theta = 585°$

$\sin 585° = -\frac{\sqrt{2}}{2}$

$\cos 585° = -\frac{\sqrt{2}}{2}$

$\tan 585° = 1$

b. $\theta = -240°$

$\sin(-240°) = \frac{\sqrt{3}}{2}$

$\cos(-240°) = -\frac{1}{2}$

$\tan(-240°) = -\sqrt{3}$

TRY THIS

$\sin(1110°) = \frac{1}{2}$

$\cos(1110°) = \frac{\sqrt{3}}{2}$

$\tan(1110°) = \frac{\sqrt{3}}{3}$

$\sin(-1110°) = -\frac{1}{2}$

$\cos(-1110°) = \frac{\sqrt{3}}{2}$

$\tan(-1110°) = -\frac{\sqrt{3}}{3}$

Assess

Selected Answers

Exercises 4–7, 9–111 odd

ASSIGNMENT GUIDE

In Class	1–7
Core	9–91 odd, 95–99 odd
Core Plus	8–100 even
Review	101–111
Preview	112

Extra Practice can be found beginning on page 940.

Mid-Chapter Assessment for Lessons 13.1 through 13.3 can be found on page 170 of the *Assessment Resources.*

Error Analysis

When students find the reference angle for a given angle, they sometimes use the y-axis instead of the x-axis. For example, a 300° angle is 60° from the positive x-axis and 30° from the negative y-axis. Remind students that the reference angle is always measured from the x-axis, and encourage them to draw a *vertical* line to the x-axis each time they look for the reference angle. This will always create a reference triangle.

39. $\frac{1}{2}; -\frac{\sqrt{3}}{2}; -\frac{\sqrt{3}}{3}$

40. $\frac{\sqrt{2}}{2}; -\frac{\sqrt{2}}{2}; -1$

41. $-\frac{\sqrt{3}}{2}; \frac{1}{2}; -\sqrt{3}$

42. $-\frac{\sqrt{2}}{2}; -\frac{\sqrt{2}}{2}; 1$

43. $\frac{\sqrt{3}}{2}; -\frac{1}{2}; -\sqrt{3}$

Practice

NAME ______ CLASS ______ DATE ______

Practice
13.3 *Trigonometric Functions of Any Angle*

Point P is located at the intersection of a circle with a radius of r and the terminal side of angle θ. Find the exact coordinates of P.

1. $\theta = 45°; r = 5$ $\left(\frac{5\sqrt{2}}{2}, \frac{5\sqrt{2}}{2}\right)$
2. $\theta = 60°; r = 12$ $(6, 6\sqrt{3})$
3. $\theta = -120°; r = 15$ $\left(-\frac{15}{2}, -\frac{15\sqrt{3}}{2}\right)$
4. $\theta = 330°; r = 40$ $(20\sqrt{3}, -20)$
5. $\theta = 135°; r = 10$ $(-5\sqrt{2}, 5\sqrt{2})$
6. $\theta = 750°; r = 4$ $(2\sqrt{3}, 2)$

Point P is located at the intersection of the unit circle and the terminal side of angle θ in standard position. Find the coordinates of P to the nearest thousandth.

7. $\theta = 42°$ (0.743, 0.669)
8. $\theta = 129°$ (−0.629, 0.777)
9. $\theta = 244°$ (−0.438, −0.899)
10. $\theta = 305°$ (0.574, −0.819)
11. $\theta = -41°$ (0.755, −0.656)
12. $\theta = -105°$ (−0.259, −0.966)

Find the exact values of the sine, cosine, and tangent of each angle.

13. 2160° sin: 0 cos: 1 tan: 0
14. 315° sin: $-\frac{\sqrt{2}}{2}$ cos: $\frac{\sqrt{2}}{2}$ tan: −1
15. −240° sin: $\frac{\sqrt{3}}{2}$ cos: $-\frac{1}{2}$ tan: $-\sqrt{3}$
16. 1770° sin: $-\frac{1}{2}$ cos: $\frac{\sqrt{3}}{2}$ tan: $-\frac{\sqrt{3}}{3}$

Find each trigonometric function value. Give exact answers.

17. sin 420° $\frac{\sqrt{3}}{2}$
18. cos(−150°) $-\frac{\sqrt{3}}{2}$
19. csc(−480°) $-\frac{2\sqrt{3}}{3}$
20. tan 300° $-\sqrt{3}$
21. cos 1035° $\frac{\sqrt{2}}{2}$
22. sin 1470° $\frac{1}{2}$
23. cot(−120°) $\frac{\sqrt{3}}{3}$
24. tan 495° −1
25. csc 210° −2

APPLICATION

8. $-\frac{\sqrt{3}}{2}; \frac{1}{2}; -\sqrt{3}$
9. $-\frac{1}{2}; -\frac{\sqrt{3}}{2}; \frac{\sqrt{3}}{3}$
10. $-\frac{\sqrt{2}}{2}; -\frac{\sqrt{2}}{2}; 1$
11. $\frac{\sqrt{3}}{2}; -\frac{1}{2}; -\sqrt{3}$
12. $\frac{\sqrt{2}}{2}; -\frac{\sqrt{2}}{2}; -1$
13. $-\frac{\sqrt{3}}{2}; -\frac{1}{2}; \sqrt{3}$
14. $\frac{\sqrt{2}}{2}; -\frac{\sqrt{2}}{2}; 1$
15. $-\frac{\sqrt{2}}{2}; -\frac{\sqrt{2}}{2}; 1$
16. $-\frac{1}{2}; -\frac{\sqrt{3}}{2}; \frac{\sqrt{3}}{3}$
17. $\frac{1}{2}; \frac{\sqrt{3}}{2}; \frac{\sqrt{3}}{3}$
18. $-\frac{\sqrt{3}}{2}; -\frac{1}{2}; \sqrt{3}$
19. $\frac{1}{2}; -\frac{\sqrt{3}}{2}; -\frac{\sqrt{3}}{3}$
23. $\left(-\frac{9\sqrt{2}}{2}, \frac{9\sqrt{2}}{2}\right)$
24. $(-25, -25\sqrt{3})$
25. $(-45, 0)$
26. $(3.1\sqrt{3}, -3.1)$
27. $(-3.8\sqrt{2}, -3.8\sqrt{2})$
28. $(0.45\sqrt{3}, 0.45)$

internet connect
Homework Help Online
Go To: go.hrw.com
Keyword:
MB1 Homework Help for Exercises 8–19, 37–92

29. (0.26, 0.97)
30. (−0.91, −0.42)
31. (−0.64, −0.77)
32. (−0.91, 0.42)
33. (1.00, 0.07)
34. (0.05, 1.00)
35. (0.03, −1.00)
36. (−1.00, 0.05)
37. $\frac{\sqrt{2}}{2}; \frac{\sqrt{2}}{2}; 1$
38. $-\frac{1}{2}; \frac{\sqrt{3}}{2}; -\frac{\sqrt{3}}{3}$

6. ROBOTICS Refer to the robotic arm described at the beginning of the lesson. Find the new position of the object relative to the pivot point after the 2-meter robotic arm swings through an angle of 110° degrees. ***(EXAMPLE 3)*** **(−0.68, 1.88)**

7. Find the exact values of the sine, cosine, and tangent of −1200°. ***(EXAMPLE 4)***
$\sin(-1200°) = -\frac{\sqrt{3}}{2}$, $\cos(-1200°) = -\frac{1}{2}$, $\tan(-1200°) = \sqrt{3}$

Practice and Apply

Find the exact values of the sine, cosine, and tangent of each angle.

8. 300° **9.** 210° **10.** 225° **11.** 120°
12. 135° **13.** 240° **14.** −225° **15.** −135°
16. −150° **17.** −330° **18.** −120° **19.** −210°

Point P is located at the intersection of a circle with a radius of r and the terminal side of angle θ. Find the exact coordinates of P.

20. $\theta = 60°, r = 3$ $\left(\frac{3}{2}, \frac{3\sqrt{3}}{2}\right)$ **21.** $\theta = 30°, r = 5$ $\left(\frac{5\sqrt{3}}{2}, \frac{5}{2}\right)$ **22.** $\theta = 120°, r = 8$ $(-4, 4\sqrt{3})$
23. $\theta = 135°, r = 9$ **24.** $\theta = 240°, r = 50$ **25.** $\theta = 180°, r = 45$
26. $\theta = -30°, r = 6.2$ **27.** $\theta = -135°, r = 7.6$ **28.** $\theta = -330°, r = 0.9$

Point P is located at the intersection of the unit circle and the terminal side of angle θ in standard position. Find the coordinates of P to the nearest hundredth.

29. $\theta = 75°$ **30.** $\theta = 205°$ **31.** $\theta = -130°$ **32.** $\theta = -205°$
33. $\theta = 4°$ **34.** $\theta = 87°$ **35.** $\theta = -88°$ **36.** $\theta = -183°$

Find the exact values of the sine, cosine, and tangent of each angle.

37. 405° **38.** 690° **39.** 870° **40.** 855°
41. 1380° **42.** 1305° **43.** −600° **44.** −510°
45. −495° **46.** −480° **47.** −840° **48.** −1020°

Find each trigonometric function value. Give exact answers.

49. sin 135° $\frac{\sqrt{2}}{2}$ **50.** cos 120° $-\frac{1}{2}$ **51.** tan 150° $-\frac{\sqrt{3}}{3}$ **52.** sin 240° $-\frac{\sqrt{3}}{2}$
53. cos 210° $-\frac{\sqrt{3}}{2}$ **54.** tan 225° **1** **55.** sin 300° $-\frac{\sqrt{3}}{2}$ **56.** cos 315° $\frac{\sqrt{2}}{2}$
57. tan 330° $-\frac{\sqrt{3}}{3}$ **58.** sin 0° **0** **59.** cos 0° **1** **60.** tan 180° **0**
61. sin 90° **1** **62.** cos 90° **0** **63.** tan 270° **undef.** **64.** sin 180° **0**
65. cos 180° **−1** **66.** tan 90° **undef.** **67.** sin(−90°) **−1** **68.** cos(−90°) **0**
69. tan(−180°) **0** **70.** sin 720° **0** **71.** cos 1080° **1** **72.** cos 450° **0**
73. sin 495° $\frac{\sqrt{2}}{2}$ **74.** sin(−45°) $-\frac{\sqrt{2}}{2}$ **75.** cos(−135°) $-\frac{\sqrt{2}}{2}$ **76.** cos(−270°) **0**
77. sin(−405°) $-\frac{\sqrt{2}}{2}$ **78.** tan(−150°) $\frac{\sqrt{3}}{3}$ **79.** tan(−30°) $-\frac{\sqrt{3}}{3}$ **80.** sin 1125° $\frac{\sqrt{2}}{2}$
81. cos 810° **0** **82.** tan 390° $\frac{\sqrt{3}}{3}$ **83.** tan 780° $\sqrt{3}$ **84.** csc 135° $\sqrt{2}$
85. sec 120° **−2** **86.** cot 150° $-\sqrt{3}$ **87.** csc(−660°) $\frac{2\sqrt{3}}{3}$ **88.** sec(−990°) **undef.**
89. cot(−765°) **−1** **90.** sec 405° $\sqrt{2}$ **91.** csc 1140° $\frac{2\sqrt{3}}{3}$ **92.** cot 1500° $\frac{\sqrt{3}}{3}$

CHALLENGE

93. Use the definition of a periodic function to show that the function $f(x) = x$ is not periodic.

44. $\sin(-510°) = -\frac{1}{2}$
$\cos(-510°) = -\frac{\sqrt{3}}{2}$
$\tan(-510°) = \frac{\sqrt{3}}{3}$

45. $\sin(-495°) = -\frac{\sqrt{2}}{2}$
$\cos(-495°) = -\frac{\sqrt{2}}{2}$
$\tan(-495°) = 1$

46. $\sin(-480°) = -\frac{\sqrt{3}}{2}$
$\cos(-480°) = -\frac{1}{2}$
$\tan(-480°) = \sqrt{3}$

47. $\sin(-840°) = -\frac{\sqrt{3}}{2}$
$\cos(-840°) = -\frac{1}{2}$
$\tan(-840°) = \sqrt{3}$

48. $\sin(-1020°) = \frac{\sqrt{3}}{2}$
$\cos(-1020°) = \frac{1}{2}$
$\tan(-1020°) = \sqrt{3}$

CONNECTION

94. $AB = 12$ cm
$AC = 6\sqrt{3}$ cm
$m\angle B = 60^\circ$

97. $PR = 12$ in.
$PQ = 12\sqrt{2}$ in.
$m\angle Q = 45^\circ$

GEOMETRY **Solve each triangle.**

94.

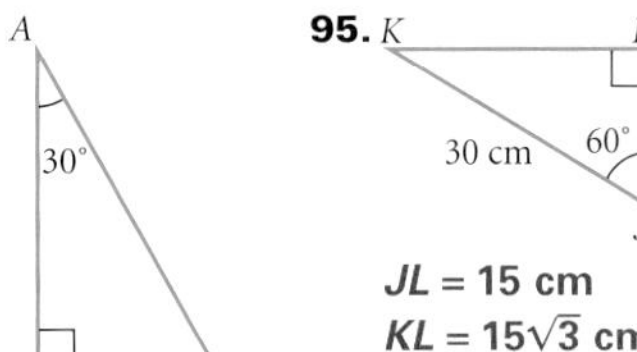

95.

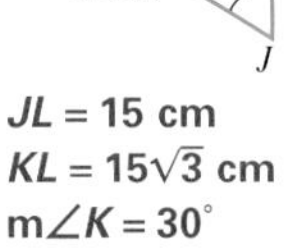

$JL = 15$ cm
$KL = 15\sqrt{3}$ cm
$m\angle K = 30^\circ$

96.

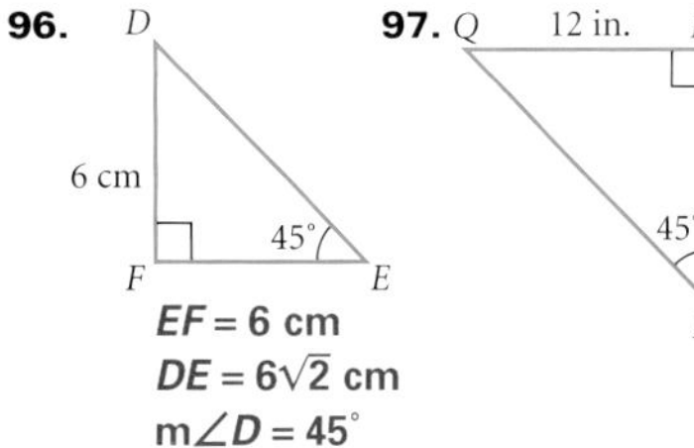

$EF = 6$ cm
$DE = 6\sqrt{2}$ cm
$m\angle D = 45^\circ$

97. Q 12 in. R 45° P

APPLICATIONS

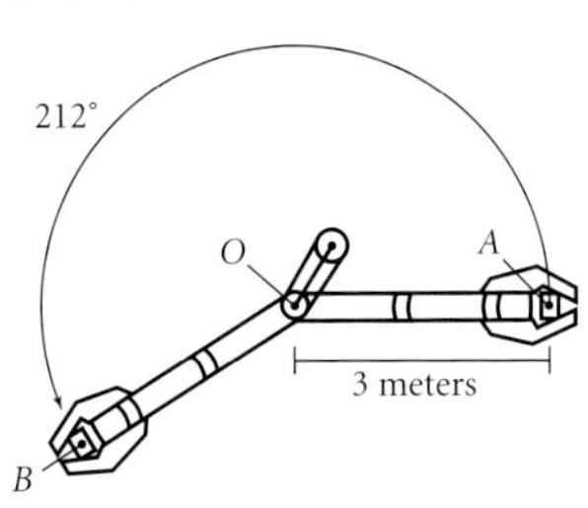

98a. 2.54 m to the left of O and 1.59 m below O
b. 1.37 m to the left of O and 3.76 m below O
c. extend 2.24 m and swing 63.43°

98. ROBOTICS A robotic arm attached at point O picks up an object at point A, which is 3 meters to the right of O.

a. The arm rotates through an angle of 212° and releases the object at point B. What is the location of the object at point B relative to O?

b. If the arm lifts an object at A, rotates through an angle of 250°, extends to a length of 4 meters, and then places the object at point C, what is the location of the object at point C relative to O?

c. How far must the arm be extended and through what angle must it swing in order for it to move an object from point A to a point that is located 1 meter to the right of and 2 meters below point O?

99. BICYCLE DESIGN The tires of a bicycle have a diameter of 26 inches. In the lowest gear, one complete revolution of the pedals causes the back wheel to rotate through an angle of 106°. How far, in inches, does this cause the bike to move? **≈24 in.**

100. CONSTRUCTION The Williams are building a new fence for their horse pen. The fence will have square sections with diagonal supports, as shown at right. If the height of a square fence section is 3.5 feet, find the length of a diagonal support. **5.0 ft**

Look Back

Let $f(x) = 3x + 2$ and $g(x) = 4x - 1$. Find each composite function. ***(LESSON 2.4)***

101. $f \circ g$ **$12x - 1$** **102.** $g \circ f$ **$12x + 7$** **103.** $g \circ g$ **$16x - 5$**

Matrix J represents the amounts of money that Sheree and her brother Donnell had in their savings and checking accounts at the end of January. ***(LESSON 4.1)***

	Savings	Checking
Sheree	325	512
Donnell	408	275

$= J$

104. What are the dimensions of matrix J? **2 × 2**

105. Describe the data in location j_{21}. **\$408: amt. Donnell has in savings**

106. Find the total amount Sheree had in the bank at the end of January. **\$837**

93. If f is periodic, then there is a value, p, such that $f(x + p) = f(x)$. Then for the value, p, $f(x) = x$, $f(x + p) = x + p$ and $x + p = x$ for some p, which is not possible unless $p = 0$.

Student Technology Guide

NAME ______ CLASS ______ DATE ______

Student Technology Guide
13.3 Trigonometric Functions of Any Angle

You can use the cosine and sine of an angle to find the coordinates of a point located a specified distance from the origin. You can find the cosine and sine values with a calculator. Be sure that your calculator is in degree mode for this lesson.

Example: In the figure at right below, point P is located at the intersection of a circle with a radius of 6 and the terminal side of a 122° angle. Find the coordinates of P. Round your answers to the nearest thousandth.

- The coordinates of P are (x, y). Since $\cos\theta = \frac{x}{r}$, $x = r \cdot \cos\theta$. Press 6 × COS 122) ENTER. To the nearest thousandth, the x-coordinate is −3.180.
- Since $\sin\theta = \frac{y}{r}$, $y = r \cdot \sin\theta$. Press 6 × SIN 122) ENTER. To the nearest thousandth, the y-coordinate is 5.088.

The coordinates of P are (−3.180, 5.088).

Point P is located at the intersection of a circle with a radius of r and the terminal side of angle θ. Find the coordinates of P to the nearest thousandth.

1. $\theta = 18^\circ$, $r = 10$ (9.511, 3.090)
2. $\theta = 275^\circ$, $r = 2.9$ (0.253, −2.889)
3. $\theta = 350^\circ$, $r = 8.12$ (7.997, −1.410)

Some calculators, such as the TI-92, can give exact values of trigonometric functions.

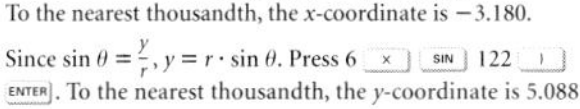

To set the calculator for exact values, press MODE. Use the down arrow to move to Exact/Approx. Press the right arrow to move to 2:EXACT. Press ENTER ENTER.

To find the exact value of csc 405°, use the fact that cosecant and sine are reciprocals. Press 1 ÷ SIN 405) ENTER.

As shown, csc 405° = $\sqrt{2}$.

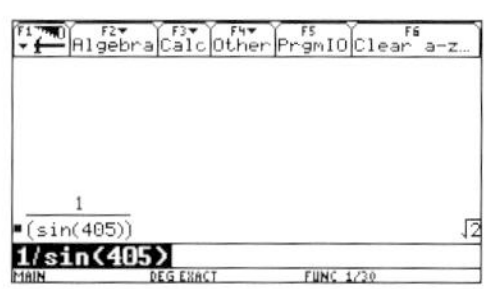

Find each trigonometric value. Give exact answers.

4. sec 840° −2
5. csc(−225°) $\sqrt{2}$
6. cot 1440° undefined
7. tan(−2010°) $-\frac{\sqrt{3}}{3}$

In Lesson 13.4, students learn about angles measured in radians and their relationship to the arc length of a circle. In Exercise 112, students begin to explore this concept with special angles.

ALTERNATIVE
Assessment

Portfolio Activity

The Portfolio Activity can be used as preparation for the Chapter Project or as a separate activity. In the Portfolio Activity on this page, students convert various points on the Ferris wheel to heights above the ground. This is different from the Activity on page 835, in which students found the height relative to the origin, (0, 0). For this activity, students must add 50 to each height value in order to get the height above the ground. The resulting graph should have a shape that is similar to the graph on page 835 but is shifted up 50 meters.

Answers to Portfolio Activities can be found in Additional Answers of the Teacher's Edition.

CONNECTION

107. $AB = BC = 10$

108. $\bar{x} = 6\frac{1}{3}$; 7; 4, 8

109. $\bar{x} = 8\frac{1}{3}$; 8; 3, 8

internet connect

Portfolio Extension

Go To: **go.hrw.com**
Keyword:
MB1 TrigHist

107. COORDINATE GEOMETRY Show that the triangle with vertices at $A(-2, 5)$, $B(4, 13)$, and $C(10, 5)$ is an isosceles triangle. ***(LESSON 9.1)***

Find the mean, median, and mode of each data set. Give answers to the nearest hundredth. ***(LESSON 12.1)***

108. 2, 8, 4, 11, 13, 4, 7, 8, 0

109. 5, 8, 3, 8, 12, 3, 16, 9, 11

Find the mean and standard deviation of each data set. Give answers to the nearest hundredth. ***(LESSON 12.4)***

110. 5, 7, 38, 4, 9, 10, 11, 9, 3

$\bar{x} = 10\frac{2}{3}$; $\sigma \approx 10.01$

111. 44, 43, 0, 47, 53, 54, 45, 48

$\bar{x} = 41.75$; $\sigma \approx 16.22$

Look Beyond

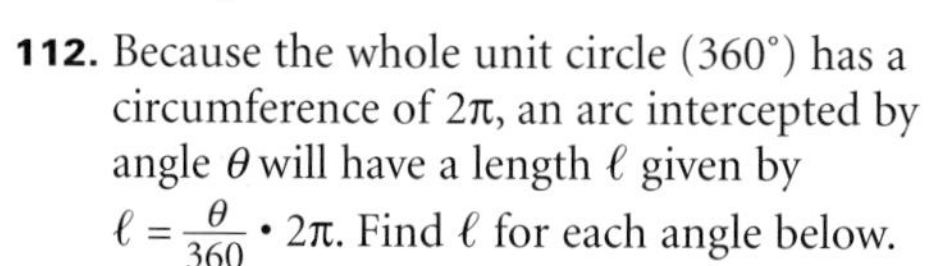

112. Because the whole unit circle (360°) has a circumference of 2π, an arc intercepted by angle θ will have a length ℓ given by $\ell = \frac{\theta}{360} \cdot 2\pi$. Find ℓ for each angle below.

a. $\theta = 180°$ π **b.** $\theta = 90°$ $\frac{\pi}{2}$ **c.** $\theta = 360°$ 2π **d.** $\theta = 45°$ $\frac{\pi}{4}$

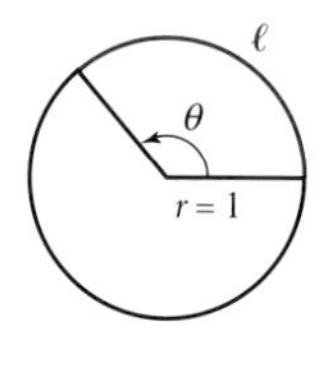

Refer to the Cosmoclock 21 described in the Portfolio Activity on page 835. The center of the Ferris wheel is located at the origin of the coordinate plane at right. Assume that point P begins its rotation at (164, 0) and that it rotates counterclockwise.

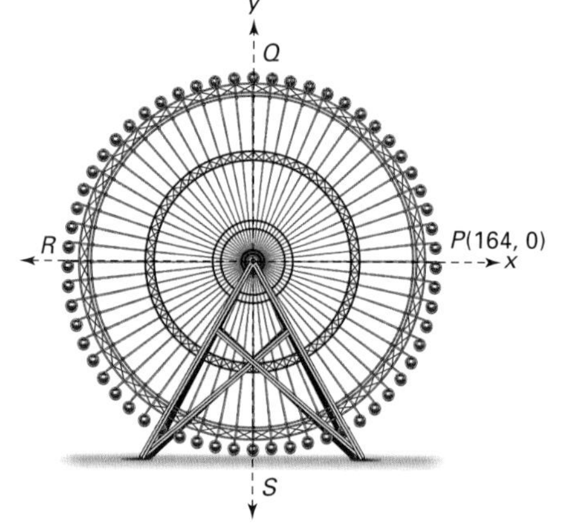

1. Find the height of P relative to the x-axis for each angle of rotation.
   a. 30° **b.** 45° **c.** 60°
2. Find the altitude of P, or height of P relative to the ground, for each angle of rotation given in Step 1.
3. Describe a general rule for converting from the height of P relative to the x-axis to the altitude of P.
4. Create a table of values for the altitude of P. Include all special angle measures for θ such that $0° \le \theta \le 360°$.
5. Plot the points from your table on graph paper. Sketch a *smooth curve* through the points.
6. Describe how the graph that you sketched in Step 5 compares with the graph that you sketched in Step 3 of the Portfolio Activity on page 835.

WORKING ON THE CHAPTER PROJECT

You should now be able to complete Activity 2 of the Chapter Project.

Radian Measure and Arc Length

Objectives

- Convert from degree measure to radian measure and vice versa.
- Find arc length.

Why *Radian measure is used to describe periodic phenomena such as seismic waves, climatic and population cycles, and the motion of circular orbiting objects such as satellites.*

APPLICATION
METEOROLOGY

A weather satellite orbits the Earth at an altitude of approximately 22,200 miles above Earth's surface. If the satellite observes a fixed region on Earth and has a period of revolution of 24 hours, what is the linear speed of the satellite? What is its angular speed? *You will answer these questions in Example 4.*

A useful angle measure other than degrees is *radian* measure. In the Activity below, you can investigate the relationship between measures in a circle, which is fundamental to the definition of radian measure.

Investigating Circle Ratios

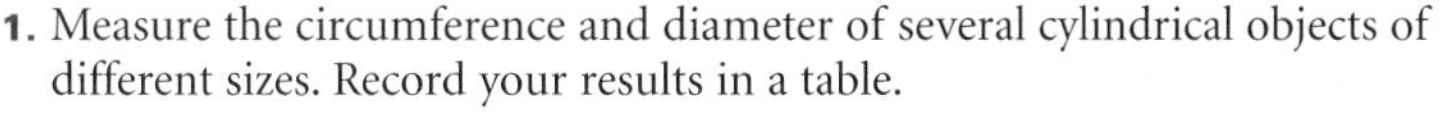

You will need: centimeter measuring tape and various cylindrical cans

1. Measure the circumference and diameter of several cylindrical objects of different sizes. Record your results in a table.
2. Plot your values as ordered pairs, with diameter on the x-axis.
3. Find the least-squares line for this data, and determine its slope.

CHECKPOINT ✔ 4. The slope should be approximately equal to a famous number that relates circumference and diameter. What is it? How close were you?

Prepare

NCTM PRINCIPLES & STANDARDS 1–4, 6–10

QUICK WARM-UP

Find each trigonometric value. Give exact answers.

1. sin 60° $\frac{\sqrt{3}}{2}$	**2.** cos 135° $-\frac{\sqrt{2}}{2}$
3. tan(−30°) $-\frac{\sqrt{3}}{3}$	**4.** sin(−150°) $-\frac{1}{2}$
5. cos 360° 1	**6.** sin(−90°) −1
7. tan(−180°) 0	**8.** tan(270°) undefined

Find the number of rotations represented by each angle.

9. 720° 2	**10.** 90° $\frac{1}{4}$
11. −60 $\frac{1}{6}$	**12.** −540° $\frac{3}{2}$

Also on Quiz Transparency 13.4

Teach

Why Show students a map of airline flight routes. Have the class discuss why the routes are curved and what they actually represent on Earth. **The routes represent arcs just above the surface of Earth.**

Activity Notes

In this Activity, students explore the relationship between the circumference and diameter of various circles. Students should find that for any circle, the circumference is slightly more than 3 times the diameter.

CHECKPOINT ✔

4. Students should recognize that the slope is π.

Alternative Teaching Strategy

HANDS-ON STRATEGIES Have students draw a small circle and a ray containing a radius of the circle. Using a piece of string, have students measure the radius and place this length of string along the circle with one end at the intersection of the ray and the circle. Have students draw another ray from the center of the circle through the point at the other end of the string. The angle created has a defined measure of 1 radian. Have students draw a larger circle with the same center and repeat the string exercise. The string should fit exactly between the two rays already drawn. Have students repeat this several times to see that 1 radian is a consistent angle measure defined by an intercepted arc that is equal to the radius.

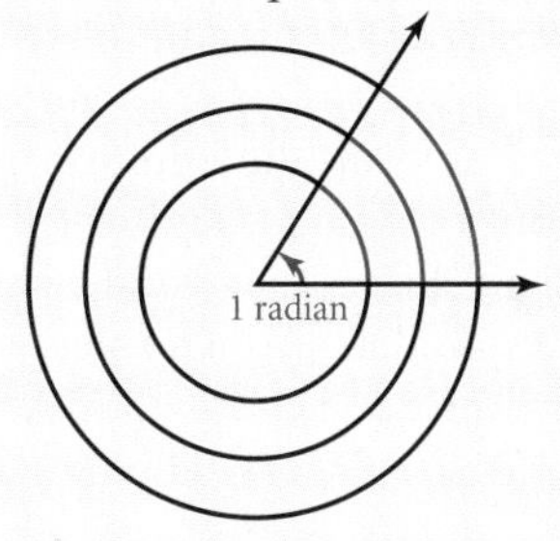

Cooperative Learning

Have students work in pairs for this Activity. Students will need to work together to physically measure the distance around the cans by using the strips. One student should hold the strip while the other reads the distance.

ADDITIONAL EXAMPLE 1

Convert from degrees to radians and from radians to degrees.

a. $-150°$ **b.** $\frac{4\pi}{3}$ radians

$-\frac{5\pi}{6}$ radians 240°

TRY THIS

$-120° \times \frac{\pi}{180°} = -\frac{2\pi}{3}$ radians;

$-\frac{2\pi}{3} \times \frac{180}{\pi} = -120°$

CHECKPOINT ✔

$1° = \frac{\pi}{180}$ radians

Use Teaching Transparency 57.

Teaching Tip

TECHNOLOGY Make sure that students have their calculators set in radian mode throughout this lesson.

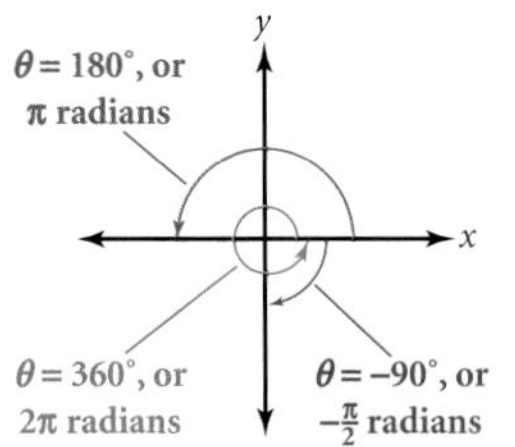

The circumference of a circle with a radius of r is $2\pi r$. Because the radius of a unit circle is 1, its circumference is 2π. The **radian** measure of an angle is equal to the length of the arc on the unit circle that is intercepted by the angle in standard position. Thus, an angle of rotation of 360° has a measure of 2π radians, an angle of 180° has a measure of π radians, and an angle of −90° has a measure of $-\frac{\pi}{2}$ radians.

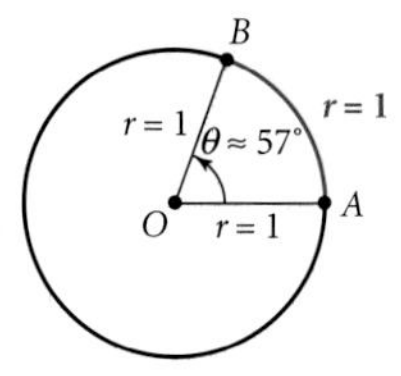

When the length of the arc determined by an angle of rotation equals the radius, the measure of the angle is 1 radian. Because an arc length of 1 radian represents $\frac{1}{2\pi}$ of the entire circumference, 1 radian $= \frac{1}{2\pi} \cdot 360°$, or about 57°.

You can convert from degrees to radians, and vice versa, by using the relationship below to *multiply by 1.*

$$1 = \frac{1 \text{ rotation}}{1 \text{ rotation}} = \frac{2\pi \text{ radians}}{360° \text{ degrees}} = \frac{\pi \text{ radians}}{180° \text{ degrees}}$$

CONVERTING ANGLE MEASURES	
Degrees to radians	**Radians to degrees**
Multiply by $\frac{\pi \text{ radians}}{180°}$.	Multiply by $\frac{180°}{\pi \text{ radians}}$.

EXAMPLE 1 **Convert from degrees to radians and from radians to degrees.**

a. 40° **b.** 3π radians

SOLUTION

a. $40° \cdot \frac{\pi \text{ radians}}{180°} = \frac{2\pi}{9}$ radians **b.** $3\pi \cdot \frac{180°}{\pi \text{ radians}} = 540°$

TRY THIS Convert −120° to radians and $-\frac{2}{3}\pi$ radians to degrees.

CHECKPOINT ✔ How many radians correspond to 1°?

EXAMPLE 2 **Evaluate. Give exact values.**

a. $\sin \frac{\pi}{3}$ **b.** $\cos \frac{3\pi}{4}$ **c.** $\tan \frac{4\pi}{3}$

SOLUTION

Convert from radians to degrees. Then evaluate.

a. $\frac{\pi}{3} \times \frac{180°}{\pi} = 60°$

$\sin \frac{\pi}{3} = \sin 60° = \frac{\sqrt{3}}{2}$

b. $\frac{3\pi}{4} \times \frac{180°}{\pi} = 135°$

$\cos \frac{3\pi}{4} = \cos 135° = -\frac{\sqrt{2}}{2}$

c. $\frac{4\pi}{3} \times \frac{180°}{\pi} = 240°$

$\tan \frac{4\pi}{3} = \tan 240° = \sqrt{3}$

TRY THIS Evaluate $\sin \frac{3\pi}{2}$, $\cos \frac{2\pi}{3}$, and $\tan \frac{5\pi}{4}$. Give exact values.

CHECKPOINT ✔ Draw a unit circle and label all of the special angles in radians from 0 to 2π.

Interdisciplinary Connection

ASTRONOMY Planets travel around the Sun in elliptical orbits. As they travel closer to the Sun, their velocity and the distance they can travel in one unit of time increase. However, the area of the sector it sweeps out in one unit of time remains constant. This is *Kepler's third law of planetary motion.* Have students find information in an almanac about the velocity and distance traveled by one of the planets. They will need at least 10 pairs of data spread along the planet's orbit. Have students demonstrate Kepler's third law by using the data.

Inclusion Strategies

USING ALGORITHMS Some students have difficulty remembering the order in which to multiply and divide the conversion factors for converting angle measurements. Give students the following proportion: $\frac{\text{radians}}{2\pi} = \frac{\text{degrees}}{360}$. This allows students to use the same proportion whether they are converting from radians to degrees or from degrees to radians.

Arc Length

A circle with a radius of r and a central angle of θ, whose vertex at the center of the circle, is shown at right. You can use proportions to find a formula for the length of the intercepted arc, s, as follows:

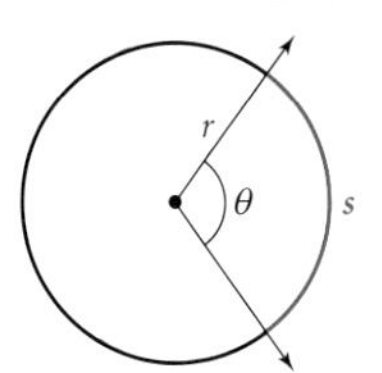

$$\frac{\text{radian measure of } \theta}{\text{radian measure of circle}} \rightarrow \frac{\theta}{2\pi} = \frac{s}{2\pi r} \leftarrow \frac{\text{arc length of } \theta}{\text{arc length of circle}}$$

$$\theta = \frac{s}{r} \quad \textit{Multiply each side by } 2\pi.$$

$$s = r\theta$$

Arc Length

If θ is the radian measure of a central angle in a circle with a radius of r, then the length, s, of the arc intercepted by θ is $s = r\theta$.

CRITICAL THINKING Define radian measure by using the definition of arc length. What does your definition tell you about the units for radians?

EXAMPLE 3 **A central angle in a circle with a diameter of 30 meters measures $\frac{\pi}{3}$ radians. Find the length of the arc intercepted by this angle.**

SOLUTION

Because the diameter is 30 meters, the radius is 15 meters.

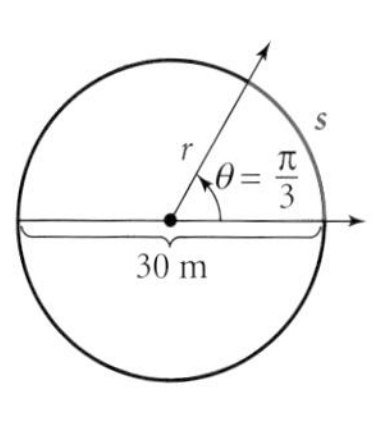

$$s = r\theta$$
$$s = 15\left(\frac{\pi}{3}\right)$$
$$s = 5\pi$$

The arc length is 5π meters, or about 15.7 meters.

TRY THIS A central angle in a circle with a radius of 1.25 feet measures 0.6 radian. Find the length of the arc intercepted by this angle.

Merry-go-round

When an object is moving at a constant speed in a circular path with a radius of r, the **linear speed** of the object is a measure of how fast the position of the object changes and is given by $\frac{s}{t}$, or $\frac{r\theta}{t}$, where t is time and θ is an angle measure in radians. This is a form of the ratio $\frac{\text{distance}}{\text{time}}$.

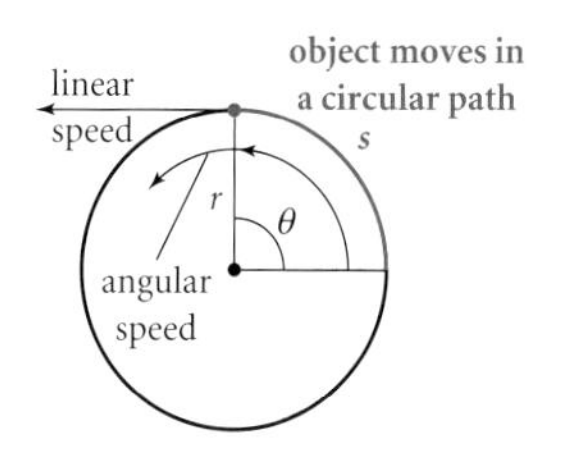

The **angular speed** of the object is a measure of how fast the angle of rotation for the object changes and is given by $\frac{\theta}{t}$, where θ is an angle measure in radians and t is time.

ADDITIONAL EXAMPLE 2

Evaluate. Give exact values.

a. $\sin\frac{\pi}{4}$ $\quad \frac{\sqrt{2}}{2}$

b. $\cos\frac{2\pi}{3}$ $\quad -\frac{1}{2}$

c. $\tan\frac{3\pi}{4}$ $\quad -1$

Math CONNECTION

GEOMETRY An *arc* is an unbroken part of a circle. A *central angle* is an angle in the plane of the circle whose vertex is the center of the circle. If the endpoints of an arc lie on the sides of a central angle and all other points of the arc are in the interior of the angle, the arc is said to be *intercepted* by the angle.

TRY THIS

$\sin\frac{3\pi}{2} = -1$; $\cos\frac{2\pi}{3} = -\frac{1}{2}$; $\tan\frac{5\pi}{4} = 1$

CHECKPOINT ✔

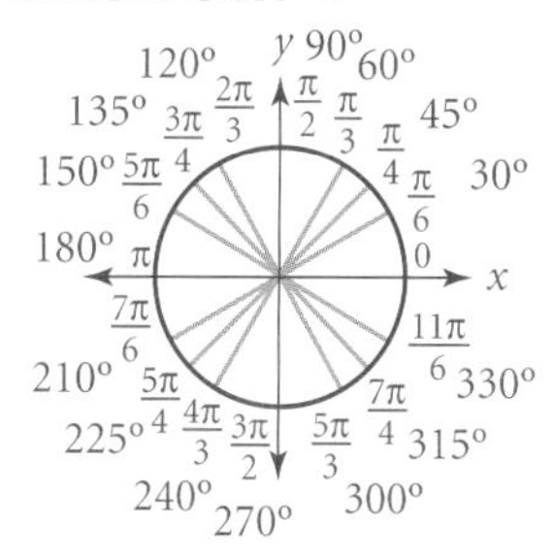

CRITICAL THINKING

Given an arc of length s on a circle of radius r, the radian measure of the central angle intercepted by the arc is $\theta = \frac{s}{r}$; radians have no units.

ADDITIONAL EXAMPLE 3

A central angle in a circle with a radius of 24 inches measures $\frac{5\pi}{4}$ radians. **Find the length of the arc intercepted by this angle.**

30π in., or about 94.2 in.

TRY THIS

0.75 ft

Enrichment

Have students design a Ferris wheel with a radius of 40 feet. Students should decide how many seats to put on the Ferris wheel. Suppose that there are 12 evenly spaced seats. Have students find the distance along the curve of the Ferris wheel between each seat. **≈20.9 ft** Have students repeat the problem for 13 seats, for 14 seats, and for 15 seats. **13: ≈19.3 ft; 14: ≈18.0 ft; 15: ≈16.8 ft** Have students determine the maximum number of seats that the Ferris wheel can reasonably have. Students should justify their response. **between 30 seats (8.4 ft) and 40 seats (6.3 ft)**

Reteaching the Lesson

HANDS-ON STRATEGIES Draw several central angles in standard position within a unit circle. For each angle, have students use a protractor to find its degree measure and then use a piece of string to measure the intercepted arc. Have students convert the angle measure to radians and compare it with the length of string. **The length of the arc is the same as the radian measure of the angle.**

ADDITIONAL EXAMPLE 4

A bicycle wheel with a radius of 12 inches is rotating at a constant rate of 3 revolutions every 4 seconds.

a. What is the linear speed of a point on the rim of this wheel?
18π, or ≈ 56.5, in./s

b. What is the angular speed of a point on the rim of this wheel?
$\frac{3\pi}{2}$ radians/s

TRY THIS
linear speed $\approx$ 1037 mph;
angular speed $= \frac{\pi}{12}$ radians/hr, or ≈ 0.26 radians/hr

Teaching Tip

For an object traveling at a constant rate along a circular path with a radius of r, the linear speed v and the angular speed ω are related by the equation $v = r\omega$.

EXAMPLE 4 **Refer to the weather satellite described at the beginning of the lesson. Assume that the radius of the Earth is 3960 miles.**

APPLICATION
METEOROLOGY

a. What is the linear speed of the satellite?
b. What is the angular speed of the satellite?

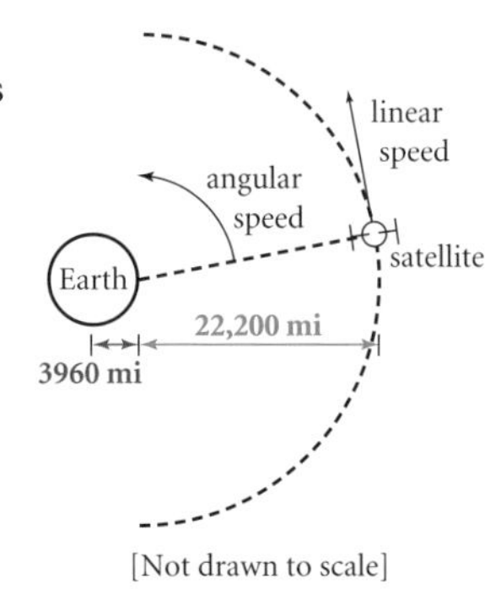

[Not drawn to scale]

SOLUTION

a. Find the radius of the satellite's orbit.

$$\begin{aligned} \textbf{radius of orbit} &= \textbf{Earth's radius} + \textbf{satellite's altitude} \\ &= \textbf{3960} + \textbf{22,200} \\ &= \textbf{26,160} \end{aligned}$$

Find the linear speed of the satellite if it makes one complete revolution (2π radians) in 24 hours.

$$\begin{aligned} \text{linear speed} &= \frac{r\theta}{t} \\ &= \frac{26{,}160 \cdot 2\pi}{24} \\ &\approx 6848 \end{aligned}$$

The linear speed of the satellite is about 6848 miles per hour.

Computer enhanced image from a weather satellite

b. Use the formula for angular speed.

$$\begin{aligned} \text{angular speed} &= \frac{\theta}{t} \\ &= \frac{2\pi}{24} \\ &= \frac{\pi}{12} \end{aligned}$$

The angular speed of the satellite is $\frac{\pi}{12}$ radians per hour.

TRY THIS Find the linear and angular speeds of a person standing on Earth, 3960 miles from its center.

Exercises

Communicate

1. Explain what the radian measure of an angle is and how it differs from degree measure.
2. Describe how to convert from radians to degrees and vice versa.
3. What happens to the length of an arc intercepted by a given central angle of a circle if the radius of the circle is doubled? Why?
4. Describe the linear and angular speeds associated with circular motion. How do they differ?

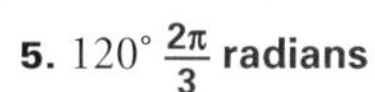

Guided Skills Practice

Convert from degrees to radians and from radians to degrees. ***(EXAMPLE 1)***

5. 120° **$\frac{2\pi}{3}$ radians** **6.** $\frac{\pi}{4}$ radians **45°**

Evaluate. Give exact values. ***(EXAMPLE 2)***

7. $\sin \frac{2\pi}{3}$ **$\frac{\sqrt{3}}{2}$** **8.** $\cos \frac{5\pi}{4}$ **$-\frac{\sqrt{2}}{2}$** **9.** $\tan \frac{5\pi}{3}$ **$-\sqrt{3}$**

10. A central angle in a circle with a diameter of 90 centimeters measures $\frac{4\pi}{3}$ radians. Find the length of the arc intercepted by this angle. ***(EXAMPLE 3)*** **60π, or 188.5 cm**

11. ENTERTAINMENT The outer 14 feet of the Space Needle Restaurant in Seattle rotates once every 58 minutes. Find the linear speed in feet per minute of a person sitting by the window of this restaurant if the diameter of the restaurant is 194.5 feet. How fast is this in miles per hour? ***(EXAMPLE 4)*** **10.54 ft/min; 0.12 mi/hr**

Seattle's Space Needle

Practice and Apply

internet connect

Homework Help Online
Go To: **go.hrw.com**
Keyword:
MB1 Homework Help for Exercises 12–35

Convert each degree measure to radian measure. Give exact answers.

12. 180° **π** **13.** 90° **$\frac{\pi}{2}$** **14.** 360° **2π** **15.** 270° **$\frac{3\pi}{2}$**
16. −30° **$-\frac{\pi}{6}$** **17.** −120° **$-\frac{2\pi}{3}$** **18.** −210° **$-\frac{7\pi}{6}$** **19.** −240° **$-\frac{4\pi}{3}$**
20. 720° **4π** **21.** 930° **$\frac{31\pi}{6}$** **22.** 80° **$\frac{4\pi}{9}$** **23.** 160° **$\frac{8\pi}{9}$**

Convert each radian measure to degree measure. Round answers to the nearest tenth of a degree.

24. 2π **360°** **25.** π **180°** **26.** $\frac{\pi}{2}$ **90°** **27.** $\frac{\pi}{4}$ **45°**
28. $\frac{\pi}{3}$ **60°** **29.** $\frac{\pi}{6}$ **30°** **30.** $-\frac{\pi}{2}$ **−90°** **31.** $-\frac{\pi}{4}$ **−45°**
32. −3.91 **−224.0°** **33.** −9.799 **−561.4°** **34.** 9.27 **531.1°** **35.** 4.96 **284.2°**

Evaluate each expression. Give exact values.

36. $\sin \pi$ **0** **37.** $\cos \pi$ **−1** **38.** $\cos \frac{\pi}{3}$ **$\frac{1}{2}$** **39.** $\sin \frac{7\pi}{6}$ **$-\frac{1}{2}$**
40. $\sin\left(-\frac{\pi}{6}\right)$ **$-\frac{1}{2}$** **41.** $\cos\left(-\frac{5\pi}{3}\right)$ **$\frac{1}{2}$** **42.** $\tan \pi$ **0** **43.** $\tan \frac{\pi}{4}$ **1**
44. $\cos \frac{2\pi}{3}$ **$-\frac{1}{2}$** **45.** $\cos\left(-\frac{7\pi}{4}\right)$ **$\frac{\sqrt{2}}{2}$** **46.** $\sin \frac{11\pi}{2}$ **−1** **47.** $\cos 5\pi$ **−1**
48. $\tan \frac{9\pi}{4}$ **1** **49.** $\sec \frac{\pi}{4}$ **$\sqrt{2}$** **50.** $\cot \frac{\pi}{6}$ **$\sqrt{3}$** **51.** $\csc\left(-\frac{\pi}{3}\right)$ **$-\frac{2\sqrt{3}}{3}$**

A circle has a diameter of 10 meters. For each central angle measure below, find the length in meters of the arc intercepted by the angle.

52. 3.8 radians **19 m** **53.** 2.4 radians **12 m** **54.** 45 radians **225 m**
55. 72 radians **360 m** **56.** 4.28 radians **21.4 m** **57.** 0.67 radians **3.35 m**
58. $\frac{\pi}{3}$ radians **$\frac{5\pi}{3}$ m** **59.** $\frac{2\pi}{3}$ radians **$\frac{10\pi}{3}$ m** **60.** $\frac{\pi}{4}$ radians **$\frac{5\pi}{4}$ m**
61. $\frac{\pi}{2}$ radians **$\frac{5\pi}{2}$ m** **62.** $\frac{7\pi}{4}$ radians **$\frac{35\pi}{4}$ m** **63.** $\frac{7\pi}{6}$ radians **$\frac{35\pi}{6}$ m**

Assess

Selected Answers

Exercises 5–11, 13–85 odd

ASSIGNMENT GUIDE	
In Class	1–11
Core	13–71 odd
Core Plus	12–72 even
Review	73–86
Preview	87

Extra Practice can be found beginning on page 940.

Error Analysis

Students often find it counterintuitive that negative angles are measured in a clockwise direction and positive angles are measured in a counterclockwise direction. Have them draw several angles with both negative and positive measures to reinforce the concept that positive angles are measured counterclockwise.

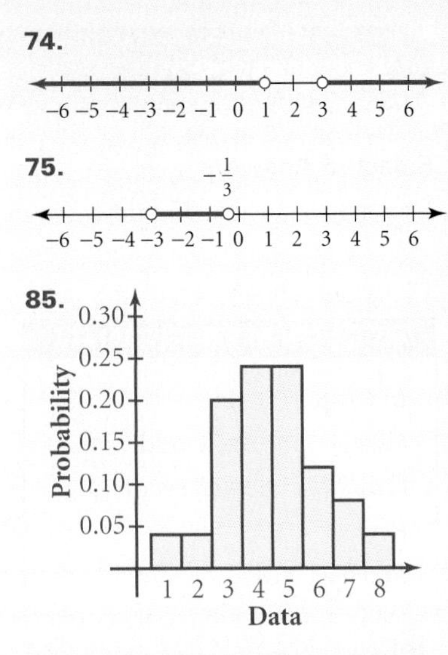

Practice

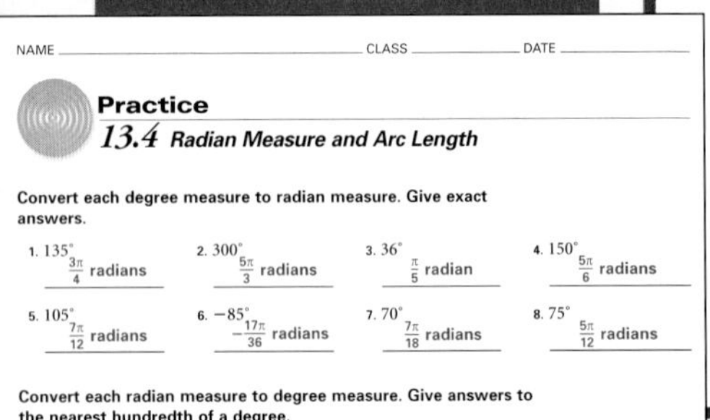

NAME ______ CLASS ______ DATE ______

Practice
13.4 **Radian Measure and Arc Length**

Convert each degree measure to radian measure. Give exact answers.

1. 135° $\frac{3\pi}{4}$ radians	2. 300° $\frac{5\pi}{3}$ radians	3. 36° $\frac{\pi}{5}$ radian	4. 150° $\frac{5\pi}{6}$ radians
5. 105° $\frac{7\pi}{12}$ radians	6. −85° $-\frac{17\pi}{36}$ radians	7. 70° $\frac{7\pi}{18}$ radians	8. 75° $\frac{5\pi}{12}$ radians

Convert each radian measure to degree measure. Give answers to the nearest hundredth of a degree.

9. $\frac{5\pi}{2}$ radians 450°	10. $\frac{11\pi}{12}$ radians 165°	11. $\frac{7\pi}{9}$ radians 140°	12. $\frac{13\pi}{12}$ radians 195°
13. 8.25 radians 472.69°	14. 1.8 radians 103.13°	15. 3 radians 171.89°	16. 0.5 radian 28.65°

A circle has a diameter of 20 feet. For each central angle measure below, find the length in feet of the arc intercepted by the angle.

17. $\frac{3\pi}{4}$ radians 47.1 feet	18. $\frac{\pi}{12}$ radian 5.2 feet	19. $\frac{2\pi}{3}$ radians 41.9 feet	20. $\frac{\pi}{6}$ radian 10.5 feet
21. 2.5 radians 50 feet	22. 4 radians 80 feet	23. 7.3 radians 146 feet	24. 10 radians 200 feet

Evaluate each expression. Give exact values.

25. $\sin 3\pi$ 0	26. $\cos \frac{2\pi}{3}$ $-\frac{1}{2}$	27. $\tan \frac{5\pi}{3}$ $-\sqrt{3}$	28. $\csc\left(-\frac{\pi}{2}\right)$ −1
29. $\tan \frac{13\pi}{6}$ $\frac{\sqrt{3}}{3}$	30. $\sin \frac{7\pi}{4}$ $-\frac{\sqrt{2}}{2}$	31. $\cos \frac{5\pi}{2}$ 0	32. $\sec \frac{5\pi}{3}$ 2

CONNECTION

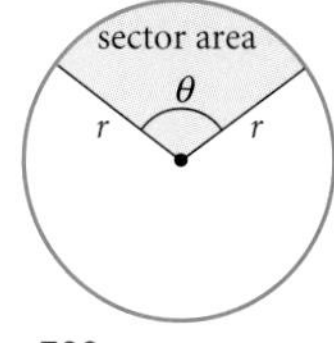

GEOMETRY The **area of a sector**, A, which resembles the slice of a pie, is a fraction $\left(\frac{\theta}{2\pi}\right)$ of the area of a complete circle (πr^2), so $A = \frac{\theta}{2\pi} \cdot \pi r^2 = \frac{1}{2}r^2\theta$, where θ is the measure of the central angle in radians.

64. Find the area of a sector with a central angle of $\frac{7\pi}{6}$ radians in a circle with a radius of 20 meters. **$\approx\frac{700\pi}{3}$, or 733 m^2**

65. Find the central angle for a sector with an area of 55.5 square inches in a circle with a radius of 12 inches. **$\frac{37}{48}$ radian**

APPLICATIONS

ENGINEERING The rear windshield wiper shown below moves through an angle of $\frac{3\pi}{4}$ radians in 0.9 second at normal speed.

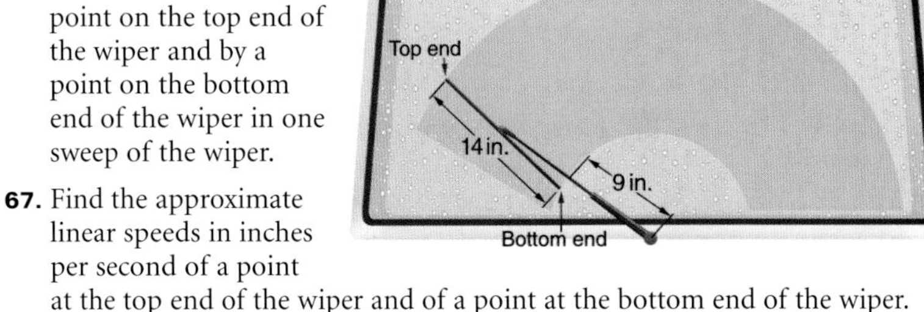

66. Find the approximate distances traveled by a point on the top end of the wiper and by a point on the bottom end of the wiper in one sweep of the wiper.

66. top: 54.2 in. bottom: 21.2 in.

67. Find the approximate linear speeds in inches per second of a point at the top end of the wiper and of a point at the bottom end of the wiper. What are these speeds in miles per hour?

67. top: 60.2 in./s, or 3.42 mi/hr bottom: 23.6 in./s, or 1.34 mi/hr

TECHNOLOGY A CD player rotates a CD at different speeds depending on where the laser is reading the disc. Assume that information is stored within a 6-centimeter diameter on the disc.

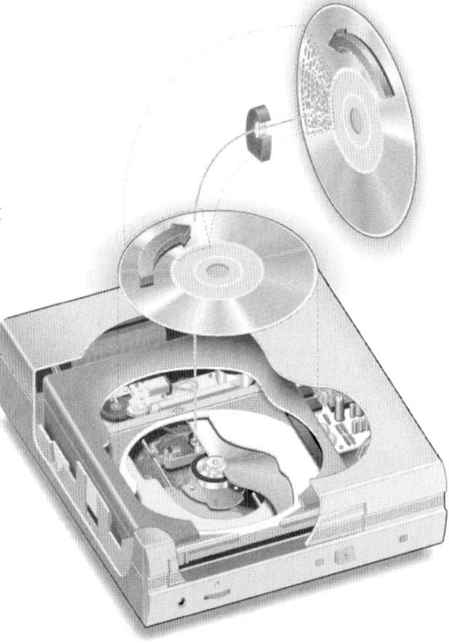

68. Find the linear speed of a point on the outer edge of the CD when the CD player is rotating at 200 revolutions per minute.

68. 1200π, or ≈3770 cm/min

69. Find the linear speed of a point 2 centimeters from the outer edge of the CD when the CD player is rotating at 240 revolutions per minute.

69. 480π, or ≈1508 cm/min

AUTO RACING In 17.5 seconds, a car covers an arc intercepted by a central angle of 120° on a circular track with a radius of 300 meters.

70. What is the car's linear speed in meters per second? **35.9 m/s**

71. What is the car's angular speed in radians per second? **0.12 rad/s**

86.

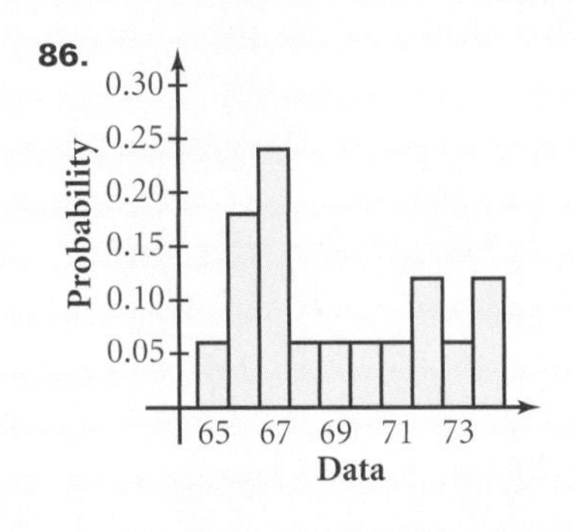

Frequency tables for Exercises 85 and 86 can be found in Additional Answers beginning on page 1002.

CHALLENGE

72. MACHINERY The large gear shown at right rotates through angle θ_1 (measured in radians), causing the small gear to rotate through angle θ_2. Find an expression for θ_2 in terms of θ_1, r_1, and r_2. $\theta_2 = \frac{\theta_1 r_1}{r_2}$

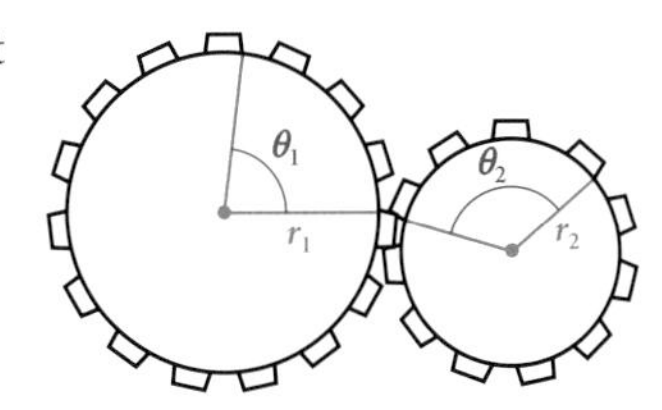

Solve each inequality. Graph the solution on a number line. *(LESSON 1.8)*

73. $|x-4| \leq -2$ no solution

74. $|2-x| > 1$ $x < 1$ or $x > 3$

75. $|3x+5| < 4$ $-3 < x < -\frac{1}{3}$

Multiply. *(LESSON 5.6)*

76. $(1+i)(2+3i)$ $-1+5i$

77. $(1-2i)(-2+i)$ $5i$

78. $(3+4i)(2-3i)$ $18-i$

Solve each equation. Round your answers to the nearest hundredth. *(LESSON 6.3)*

79. $4^x = 35$ 2.56

80. $\log x^2 = 4$ 100

81. $3\log(x+1) = 5$ 45.42

Solve each rational equation algebraically. Check your solutions by any method. *(LESSON 8.5)*

82. $\frac{x-3}{x+5} = \frac{x}{x+1}$ $-\frac{3}{7}$

83. $\frac{x-8}{2x} = \frac{x}{6}$ $\frac{3 \pm i\sqrt{87}}{2}$

84. $\frac{y}{y-4} - \frac{y}{y+2} = \frac{5}{y^2-2y-8}$ $\frac{5}{6}$

Make relative frequency table and histogram of probabilities for each set of data. *(LESSON 12.2)*

85. 3, 4, 5, 4, 5, 6, 5, 7, 4, 5, 2, 1, 3, 4, 7, 5, 3, 4, 6, 5, 4, 8, 6, 3, 3

86. 69, 74, 66, 68, 72, 74, 67, 71, 73, 67, 67, 65, 66, 67, 66, 70, 72

87 Graph $y = \sin x$ and $y = \cos x$ over the interval $-4\pi \leq x \leq 4\pi$, and compare the graphs.

Refer to the Cosmoclock 21 described in the Portfolio Activity on page 835.

1. If the Cosmoclock 21 completes 1.5 revolutions per minute, how many seconds does it take to make 1 complete revolution? $\frac{1}{4}$ of a revolution?
2. Refer to the table of values that you created in Step 4 of the Portfolio Activity on page 850. Find the corresponding time in seconds for each angle of rotation in the table.
3. Plot the altitude of P versus time on graph paper. Sketch a *smooth curve* through the points.
4. What is the period of the graph? What does the period represent?
5. Find the linear speed of a rider on the Cosmoclock 21.

WORKING ON THE CHAPTER PROJECT

You should now be able to complete Activity 3 of the Chapter Project.

In Exercise 87, students use technology to see the graph of $y = \sin x$. Students should be sure that the graphics calculator is set in radian mode. Students will study the graphs of trigonometric functions in Lesson 13.5.

ALTERNATIVE

Assessment

Portfolio Activity

The Portfolio Activity can be used as preparation for the Chapter Project or as a separate activity. In the Portfolio Activity on this page, students graph the height of the Ferris wheel versus the time it takes to reach that height. Students should sketch a sinusoidal curve through their data points. They are also asked to find the period of the graph, which is covered in Lesson 13.3.

Answers to Portfolio Activities can be found in Additional Answers of the Teacher's Edition.

Student Technology Guide

NAME ______ CLASS ______ DATE ______

Student Technology Guide

13.4 ***Radian Measure and Arc Length***

In Lesson 13.4 of the textbook, you saw how radians are used to measure angles. You can use a graphics calculator to convert angle measures from radians to degrees and vice versa.

Example: Convert 198° to radians. Round the answer to the nearest hundredth of a radian.

- One way to convert degrees to radians is to multiply by $\frac{\pi \text{ radians}}{180°}$. Press 198 [×] [2nd] [^] [÷] 180 [ENTER].
- The second method uses the radian mode of your calculator. Press [MODE], use the arrow keys to move to **Radian**, and then press [ENTER]. Press 198 [2nd] [MATRX] **1:°** [ENTER] [ENTER].

```
198*π/180
      3.455751919
198°
      3.455751919
```

To the nearest hundredth, 198° = 3.46 radians.

Convert each degree measure to radian measure. Round answers to the nearest hundredth of a radian.

1. 114° 1.99
2. 245° 4.28
3. −897° −15.66
4. −219° −3.82

Example: Convert 7.2 radians to degrees. Round the answer to the nearest hundredth of a degree.

- One way to convert radians to degrees is to multiply by $\frac{180°}{\pi \text{ radians}}$. Press 7.2 [×] 180 [÷] [2nd] [^] [ENTER].
- The second method uses the degree mode of your calculator. Press [MODE]. Use the arrow keys to move to **Degree** and then press [ENTER]. Press 7.2 [2nd] [MATRX] **3:ʳ** [ENTER] [ENTER].

```
7.2*180/π
      412.5296125
7.2ʳ
      412.5296125
```

To the nearest hundredth, 7.2 radians = 412.53°.

Convert each radian measure to degree measure. Round answers to the nearest hundredth of a degree.

5. −1.55 −88.81°
6. 12 687.55°
7. 6.44 368.98°
8. −19.991 −1145.40°

87.

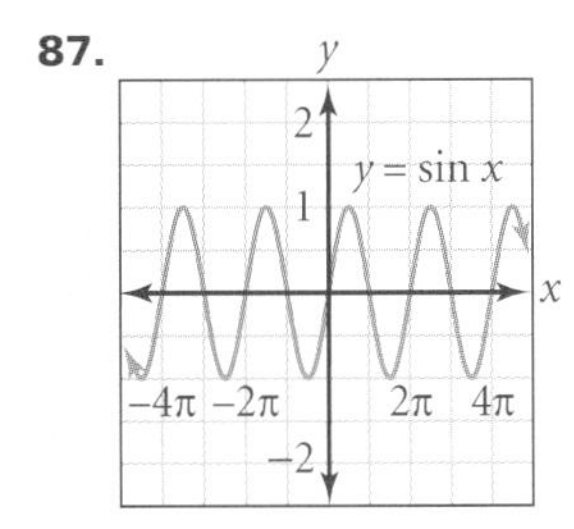

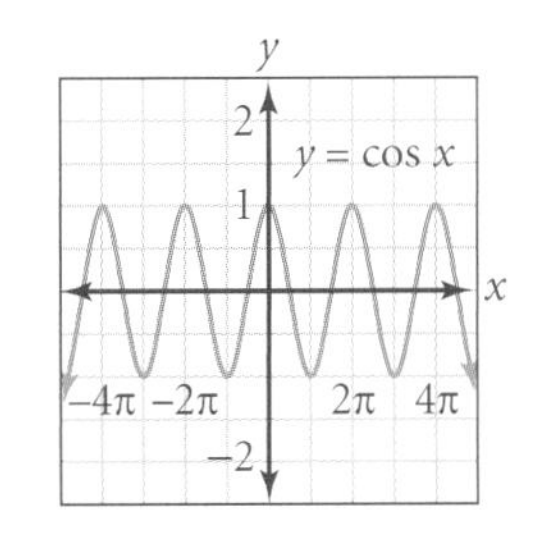

Both graphs have the same shape, both have the same domain and range, and both have a maximum value 1 and a minimum value −1. The y-intercept for $y = \sin x$ is 0, whereas the y-intercept for $y = \cos x$ is 1. The graph of $y = \sin x$ appears to be the same shape as the graph of $y = \cos x$ translated $\frac{\pi}{2}$ to the right.

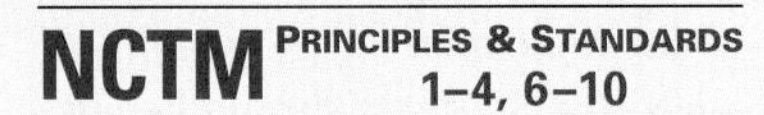

Prepare

NCTM PRINCIPLES & STANDARDS 1–4, 6–10

QUICK WARM-UP

Identify the transformation(s) applied to the parent function $f(x) = \sqrt{x}$.

1. $f(x) = \sqrt{x-2}$ horizontal translation 2 units right
2. $f(x) = \sqrt{x} - 2$ vertical translation 2 units down
3. $f(x) = -3\sqrt{x}$ vertical stretch by a factor of 3, reflection across the x-axis
4. $f(x) = \sqrt{-x}$ reflection across the y-axis
5. $f(x) = \sqrt{6x+1}$ horizontal compression by a factor of $\frac{1}{6}$, horizontal translation $\frac{1}{6}$ unit left

Also on Quiz Transparency 13.5

Teach

Why Have students discuss naturally occurring periodic functions, such as the height of tides versus time. For each example given, sketch the graph and have the students discuss what changes in the graph indicate.

Use Teaching Transparency 58.

Graphing Trigonometric Functions

Why *The graphs of trigonometric functions can be used to model real-world events such as the changes in air pressure that create sounds.*

Objectives

- Graph the sine, cosine, and tangent functions and their transformations.
- Use the sine function to solve problems.

APPLICATION
ACOUSTICS

Sound occurs when an object, such as a speaker, vibrates. This vibration causes small changes in air pressure, which travel away from the object in waves. The sound from an electric keyboard can be modeled by a transformed graph of a trigonometric function. *You will do this in Example 3.*

Activity: Exploring Trigonometric Graphs

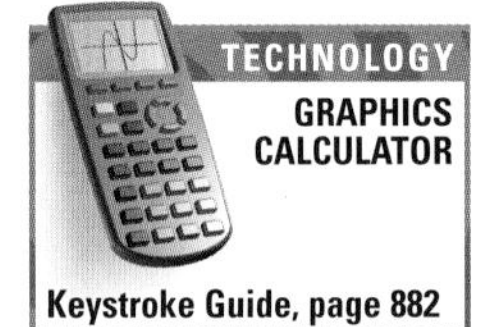

Keystroke Guide, page 882

You will need: a graphics calculator

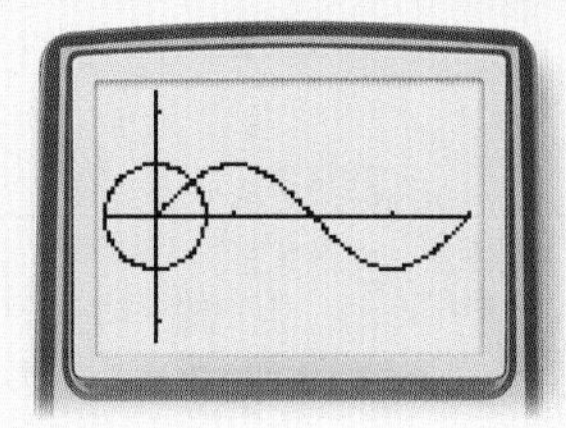

Using radian, parametric, and simultaneous modes, enter $x_{1t} = \cos t$ and $y_{1t} = \sin t$ to generate the unit circle, and enter $x_{2t} = t$ and $y_{2t} = \sin t$ for the function. Set a square viewing window such that t takes on values from 0 to 2π inclusive.

1. Graph the equations and watch as both curves emerge simultaneously. What is the function that you have just graphed?

CHECKPOINT ✔

2. Describe any relationships that you see between the two curves during the graphing process. You can also use the trace feature to compare x- and y-values with t-values.

CHECKPOINT ✔

3. **a.** What is the period of the function?
   b. Predict how the curves will change for $x_{1t} = 2\cos t$, $y_{1t} = 2\sin t$, $x_{2t} = t$, and $y_{2t} = 2\sin t$. Check your prediction by graphing.
   c. Predict how the curves will differ from the original curves for $x_{1t} = \cos 3t$, $y_{1t} = \sin 3t$, $x_{2t} = t$, and $y_{2t} = \sin 3t$. Check your prediction by graphing. (Watch the graphing of the circle carefully.)
4. Repeat Steps 1–3 for $y_{2t} = \cos t$ and $y_{2t} = \tan t$.

Alternative Teaching Strategy

USING VISUAL MODELS Have students construct a unit circle and mark several common angles around the circle. Recall that every point on the unit circle can be described by $P(\cos\theta, \sin\theta)$, where θ is the central angle in standard position. The function values of $f(\theta) = \sin\theta$ are equal to the y-coordinates of the point as it moves around the circle. Have students create a separate set of coordinate axes and label the radian angle measures along the x-axis. Have students mark the angle measure as $\frac{\pi}{6}$. Then have them cut a piece of string whose length is equal to the distance from the x-axis to the intersection of this ray and the unit circle. Then have them tape the string perpendicular to the x-axis at $x = \frac{\pi}{6}$. Repeating this process for each angle marked on the unit circle will map out the graph of $f(\theta) = \sin\theta$.

For the graph of $f(\theta) = \cos\theta$, students should use string to represent the x-coordinate of special angles marked around the unit circle by measuring the distance from the y-axis to the intersection of the ray and the circle. Have them tape the string perpendicular to the x-axis.

The Sine and Cosine Functions

Recall from Lesson 13.3 that one period of the parent function $y = \sin\theta$ or $y = \cos\theta$ is 360°

The table of values at right and the graphs below show one period of the sine and cosine function over the interval $0° \le \theta \le 360°$.

The vertical line segments in the unit circle next to the graph below of $y = \sin\theta$ represent selected values of $\sin\theta$ that can be used to construct the graph.

The horizontal line segments in the unit circle next to the graph below of $y = \cos\theta$ represent selected values of $\cos\theta$ that can be used to construct the graph.

	$y = \sin\theta$		$y = \cos\theta$	
θ	Exact	Approx.	Exact	Approx.
0°	0	0.00	1	1.00
30°	$\frac{1}{2}$	0.50	$\frac{\sqrt{3}}{2}$	0.87
45°	$\frac{\sqrt{2}}{2}$	0.71	$\frac{\sqrt{2}}{2}$	0.71
60°	$\frac{\sqrt{3}}{2}$	0.87	$\frac{1}{2}$	0.50
90°	1	1.00	0	0.00
120°	$\frac{\sqrt{3}}{2}$	0.87	$-\frac{1}{2}$	−0.50
135°	$\frac{\sqrt{2}}{2}$	0.71	$-\frac{\sqrt{2}}{2}$	−0.71
150°	$\frac{1}{2}$	0.50	$-\frac{\sqrt{3}}{2}$	−0.87
180°	0	0.00	−1	−1.00
210°	$-\frac{1}{2}$	−0.50	$-\frac{\sqrt{3}}{2}$	−0.87
225°	$-\frac{\sqrt{2}}{2}$	−0.71	$-\frac{\sqrt{2}}{2}$	−0.71
240°	$-\frac{\sqrt{3}}{2}$	−0.87	$-\frac{1}{2}$	−0.50
270°	−1	−1.00	0	0.00
300°	$-\frac{\sqrt{3}}{2}$	−0.87	$\frac{1}{2}$	0.50
315°	$-\frac{\sqrt{2}}{2}$	−0.71	$\frac{\sqrt{2}}{2}$	0.71
330°	$-\frac{1}{2}$	−0.50	$\frac{\sqrt{3}}{2}$	0.87
360°	0	0.00	1	1.00

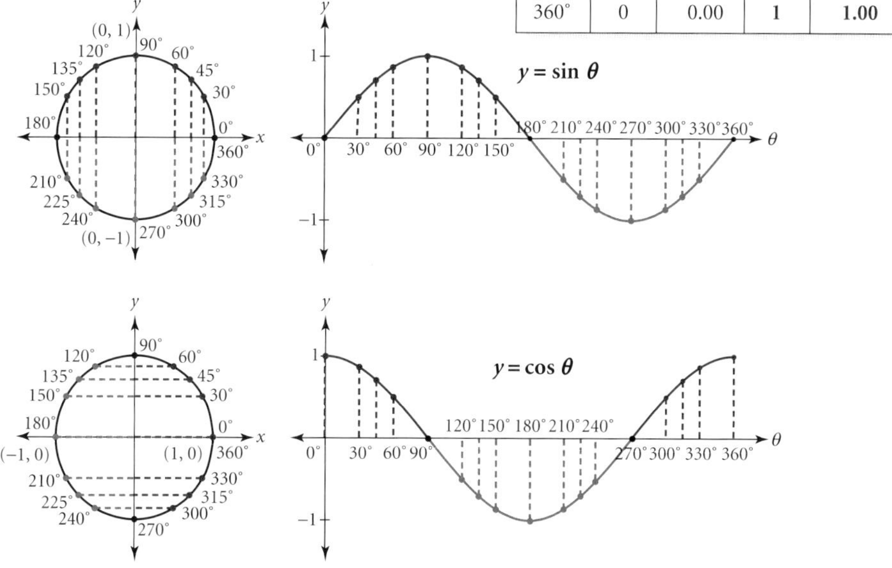

Activity Notes

This Activity leads students through a real-time simulation of "unwrapping" the unit circle in order to create the graph of $y = \sin x$. Using the parametric mode on a graphics calculator, students can see where the values of the function $y = \sin x$ originate on the unit circle. Students see changes in the amplitude by graphing $y = 2 \sin x$ and see changes in the period by graphing $y = \sin 3x$. Students then follow a similar procedure to investigate the graphs of $y = \cos x$ and $y = \tan x$.

Teaching Tip

TECHNOLOGY When using a TI-83 or TI-82 graphics calculator in the Activity, press MODE and choose radian, parametric, and simultaneous modes by using the arrow keys and pressing ENTER after each selection. The value of **Tstep** can be used to slow down the drawing.

Press WINDOW and set the viewing window as follows:

Tmin=0
Tmax=2π
Tstep=.5 (fast) or **.01** (slow)
Xmin=−1 **Ymin=−2**
Xmax=2π **Ymax=2**
Xscl=π/2 **Yscl=1**

CHECKPOINT ✔

2. Answers may vary. For each value of t, the y-coordinates of each curve are the same.

CHECKPOINT ✔

3a. 2π

b. The circle will have a radius of 2, so it will be larger. The height of the sine curve will range from −2 to 2.

c. The circle will have a radius of 1, and the period of the sine function will be compressed by $\frac{1}{3}$.

Interdisciplinary Connection

MUSICAL SOUNDS The sound wave generated by playing middle C and high C at the same time on a piano can be represented by the function $f(x) = \cos x + \sin 2x$. To sketch the graph of f quickly, first sketch the graph of $g(x) = \cos x$ and $h(x) = \sin 2x$ on the same coordinate axes. Mark critical points on the x-axis, such as $x = 0, \frac{\pi}{3}, \frac{\pi}{4}, \frac{\pi}{6}$, and $\frac{\pi}{2}$, and plot points for the graph of f by adding the y-values of points from the graphs of g (shown in blue) and h (shown in green). Then connect the points to the get the graph of f (shown in red).

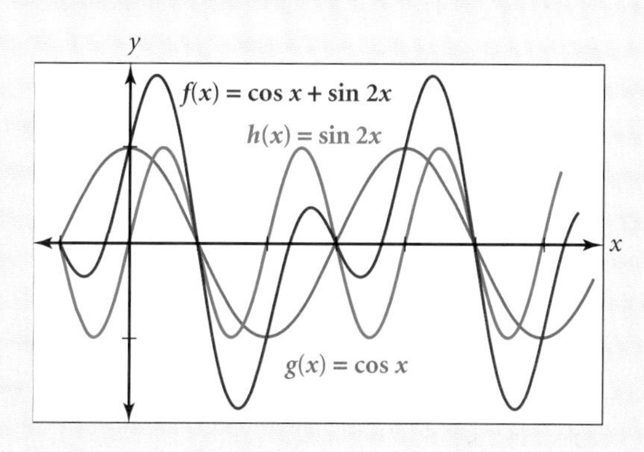

ADDITIONAL
EXAMPLE 1

Graph one period of each trig function along with its parent function.

a. $y = \frac{1}{2}\cos\theta$

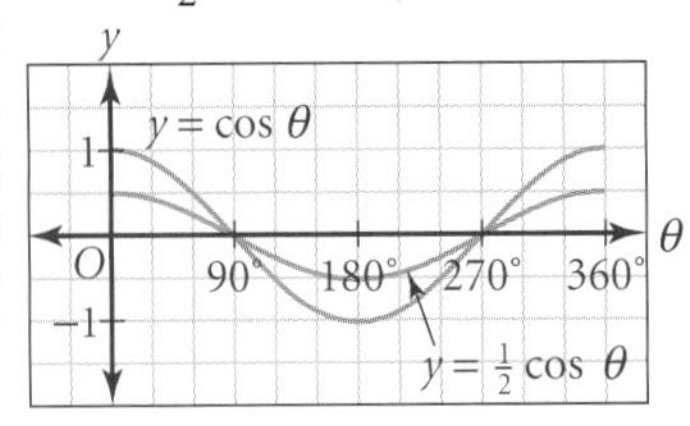

b. $y = \sin 2\theta$

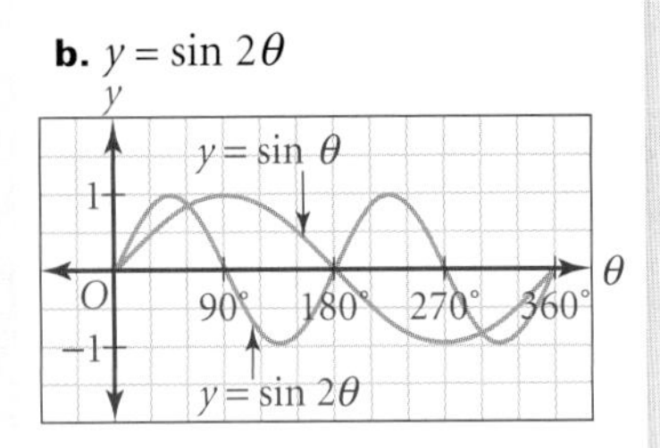

Math CONNECTION

TRANSFORMATIONS Students apply translations, reflections, stretches, and compressions to the graphs of the trigonometric functions.

Teaching Tip

TECHNOLOGY When using a graphics calculator to produce trig graphs, make sure that if the calculator is in **Degree** mode, then the values for **Xmin, Xmax,** and **Xscl** are measured in degrees, such as $-360° \le x \le 360°$ with an x-scale of 45. If the calculator is in **Radian** mode, then use radian measures for **Xmin, Xmax,** and **Xscl,** such as $-2\pi \le x \le 2\pi$ with an x-scale of $\frac{\pi}{4}$.

TRY THIS

a.

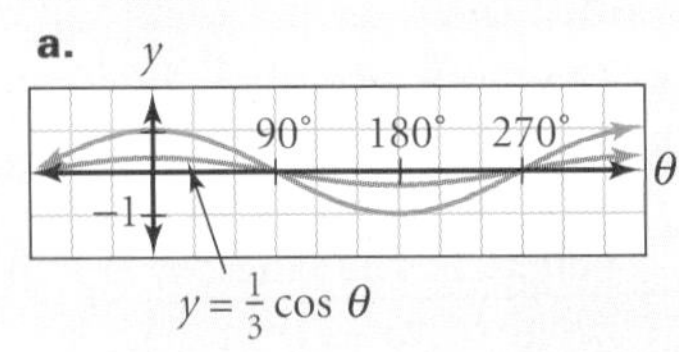

b.

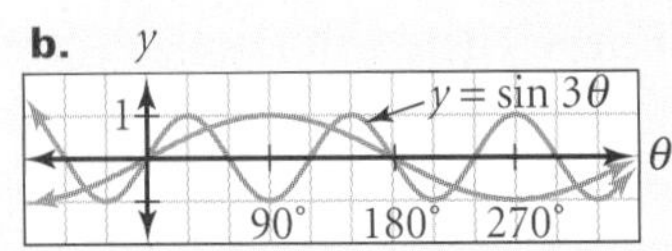

CHECKPOINT ✔

$2; \frac{1}{4}$

Stretches and Compressions

Transformations of the sine and cosine functions are quite common in real-world applications. Recall from Lesson 2.7 that the graph of the function $y = af(x)$ is a vertical stretch or compression of the graph of the parent function $y = f(x)$ by a factor of a. Similarly, the graph of $y = f(bx)$ is a horizontal stretch or compression of the graph of $y = f(x)$ by a factor of $\frac{1}{|b|}$.

EXAMPLE 1 **Graph at least one period of each trigonometric function along with its parent function.**

a. $y = 2\sin\theta$ **b.** $y = \cos\frac{1}{2}\theta$

SOLUTION

a. The 2 causes the function values to be twice as big, so the graph of $y = 2\sin\theta$ is a vertical stretch of the graph of $y = \sin\theta$ by a factor of 2. Because the zeros are the same as in the parent function, this kind of transformation is relatively easy to graph without creating a table of values.

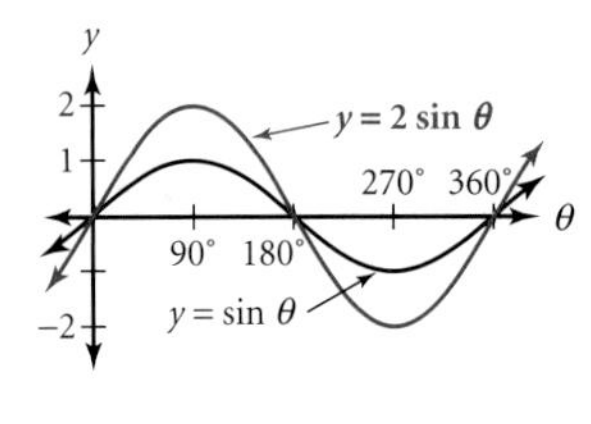

b. The values of θ are multiplied by $\frac{1}{2}$, so the curve "unfolds" half as fast, or takes twice as long to "unfold." The graph of $y = \cos\frac{1}{2}\theta$ is a horizontal stretch of the graph of $y = \cos\theta$ by a factor of 2. Note that the graph of $y = \cos\theta$ completes one period over the interval $0° \le \theta \le 360°$, while the graph of $y = \cos\frac{1}{2}\theta$ completes one period over the interval $0° \le \theta \le 720°$.

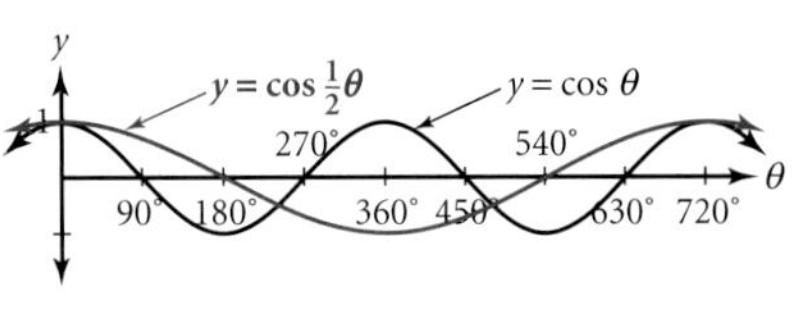

TRY THIS Graph at least one period of each trigonometric function along with its parent function.

a. $y = \frac{1}{3}\cos\theta$ **b.** $y = \sin 3\theta$

The **amplitude** of a periodic function is defined as follows:

$$\text{amplitude} = \frac{1}{2}(\text{maximum value} - \text{minimum value})$$

Because $y = \sin\theta$ and $y = \cos\theta$ each have a minimum of -1 and a maximum of 1, the amplitude of each function is $\frac{1}{2}[1 - (-1)] = 1$.

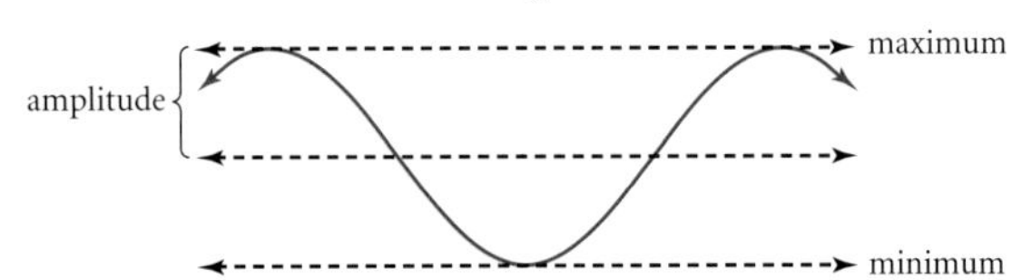

CHECKPOINT ✔ What is the amplitude of $y = 2\sin\theta$ in Example 1? of $y = \frac{1}{4}\cos 3\theta$?

Inclusion Strategies

USING VISUAL MODELS Have students use a graphics calculator set in degree mode to graph $y = a + b\sin c(x - d)$ for $a = 0$, $b = 1$, $c = 1$, and $d = 0$. Then graph the function for each of the following values, changing the value of one variable at a time:

a: 0, 1 and −1 b: 1, 0.5 and 2
c: 1, 0.5 and 2 d: 0, π and $-\pi$

Have students produce some of the graphs in both degree and radian modes. They should note that **Xmin, Xmax, Xscl,** and the value of d (which represents a horizontal shift) must be changed to produce equivalent curves in different modes.

Students should draw and label each of these functions on 3" × 5" index cards. Then have them organize these cards in a manner that helps them see how each variable and the different modes affect the graph.

Translations

Recall from Lesson 2.7 that the graph of $y = f(x - h)$ is a horizontal translation of the graph of $y = f(x)$. The translation is h units to the right for $h > 0$ and $|h|$ units to the left for $h < 0$. A horizontal translation of a sine or cosine function is also called a **phase shift.** The graph of $y = f(x) + k$ is a vertical translation of the graph of $y = f(x)$. The translation is k units up for $k > 0$ and $|k|$ units down for $k < 0$.

EXAMPLE 2 **Graph at least one period of each trigonometric function along with its parent function.**

a. $y = \sin(\theta + 45^\circ)$ **b.** $y = \cos\theta + 1$

CONNECTION
TRANSFORMATIONS

SOLUTION

a. Rewrite $y = \sin(\theta + 45^\circ)$ in the form $y = \sin(\theta - h)$.

$$y = \sin[\theta - (-45^\circ)]$$

The graph of $y = \sin(\theta + 45^\circ)$ is a horizontal translation 45° to the left of the graph of $y = \sin\theta$. This is reasonable because when 45° is added to the values of θ, the graph unfolds 45° *sooner.*

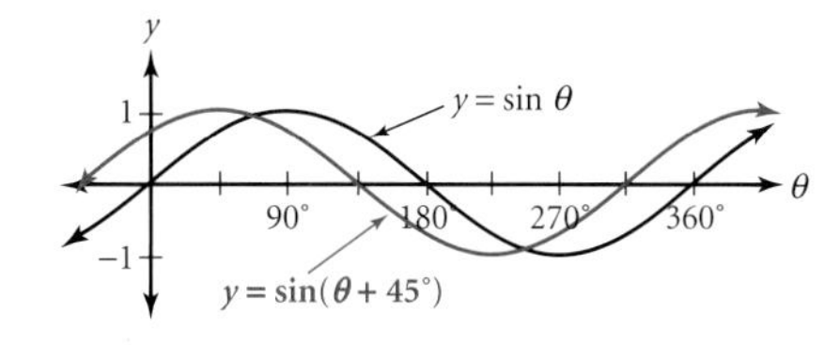

b. The graph of $y = \cos\theta + 1$ is a vertical translation of the graph of $y = \cos\theta$ 1 unit up. This is reasonable because 1 is added to each function value.

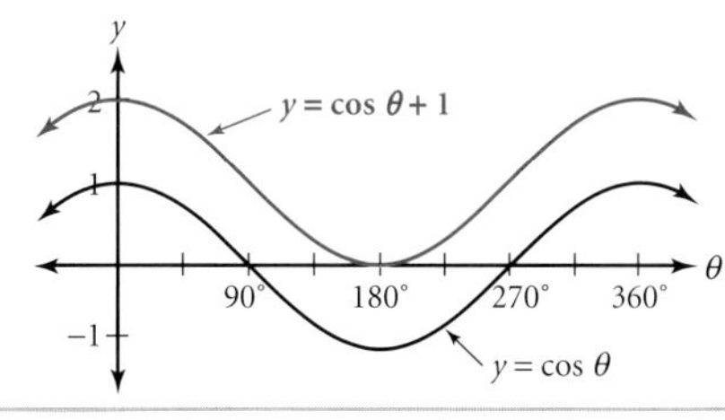

TRY THIS

Graph at least one period of each trigonometric function along with its parent function.

a. $y = \cos(\theta - 45^\circ)$ **b.** $y = \sin\theta - 1$

CRITICAL THINKING

Write the translated sine function and the translated cosine function represented by the graph below.

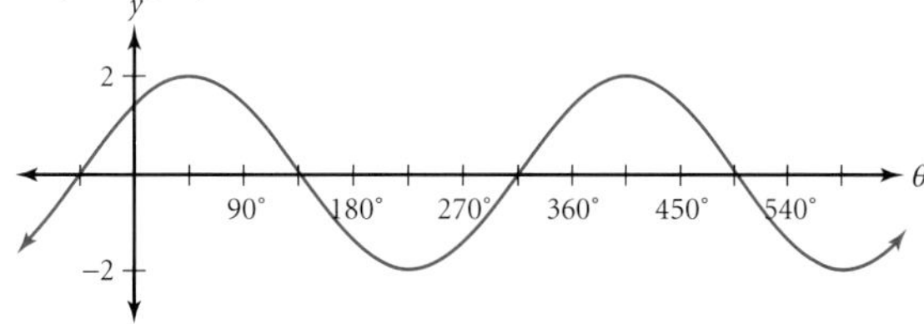

Enrichment

Have students use a graphics calculators in radian mode to compare the graphs of $y = \sin x$ and $y = \cos(x + \pi)$ and the graphs of $y = \sin(x + \pi)$ and $y = \cos x$. Encourage students to speculate about the possibility of expressing all cosine functions in terms of sine functions.

Have students write a rule that converts sine functions to cosine functions. $\cos x = \sin\left(x + \frac{\pi}{2}\right)$ Then have students test their rule with a graphics calculator.

ADDITIONAL
EXAMPLE 2

Graph one period of each trigonometric function along with its parent function.

a. $y = \cos(\theta - 45^\circ)$

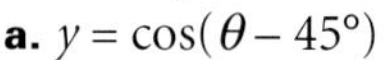

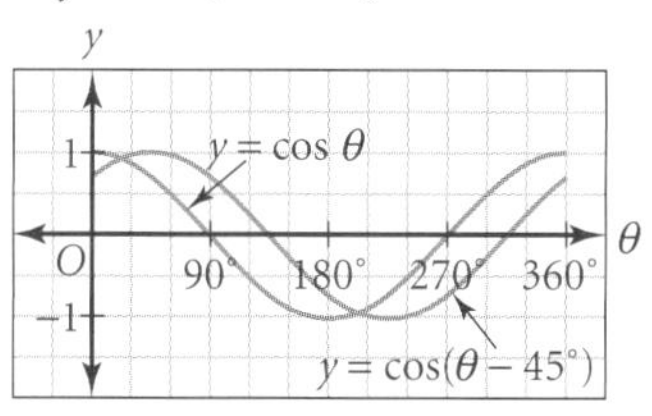

b. $y = \sin\theta - 1.5$

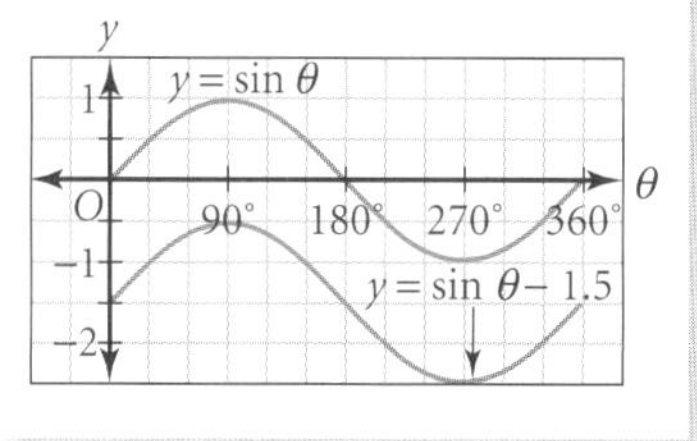

Math CONNECTION

TRANSFORMATIONS In Example 2, students examine vertical and horizontal translations of the trigonometric functions.

TRY THIS

a.

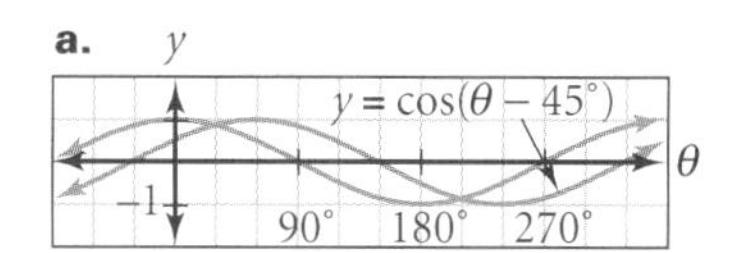

b.

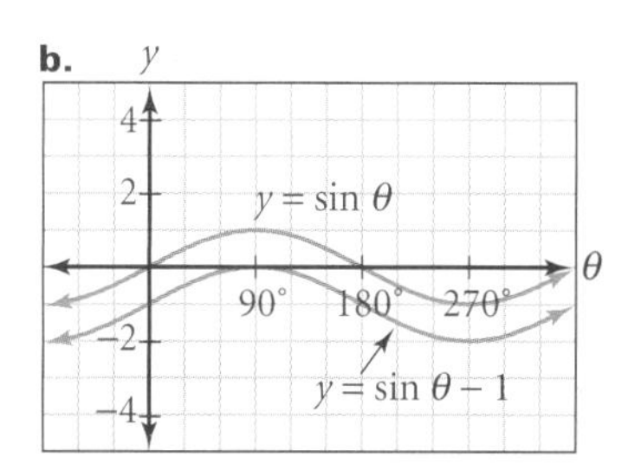

CRITICAL THINKING

For a sine function, the graph has been translated 45° to the left: $y = 2\sin(\theta + 45^\circ)$. For a cosine function, the graph has been translated 45° to the right: $y = 2\cos(\theta - 45^\circ)$.

Use Teaching Transparency 59.

Math CONNECTION

TRANSFORMATIONS Each parameter of the transformations applied to the general sine function is related to a specific attribute of a sound.

Teaching Tip

TECHNOLOGY When using a TI-83 or TI-82 graphics calculator for Example 3, set the calculator to radian, function, and sequential modes, and use the following viewing window:

Xmin=–1/110	Ymin=–4
Xmax=2/110	Ymax=4
Xscl=1/440	Yscl=1

ADDITIONAL EXAMPLE 3

A particular sound has a frequency of 60 hertz and an amplitude of 2.

a. **Write a transformed sine function to represent this sound.**

$y = 2 \sin 120\pi t$

b. **Write a new function that represents a phase shift of $\frac{1}{2}$ of a period to the right. Then graph one period of both functions on the same coordinate plane.**

$y = 2 \sin 120\pi\left(t - \frac{1}{120}\right)$

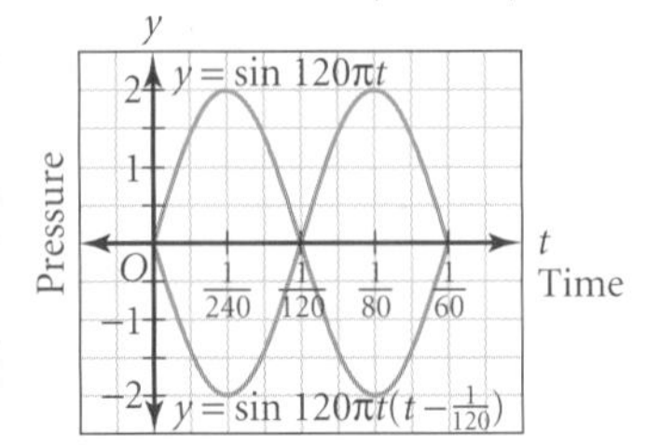

TRY THIS

$y = 1.5 \sin 240\pi\left(t + \frac{1}{360}\right)$

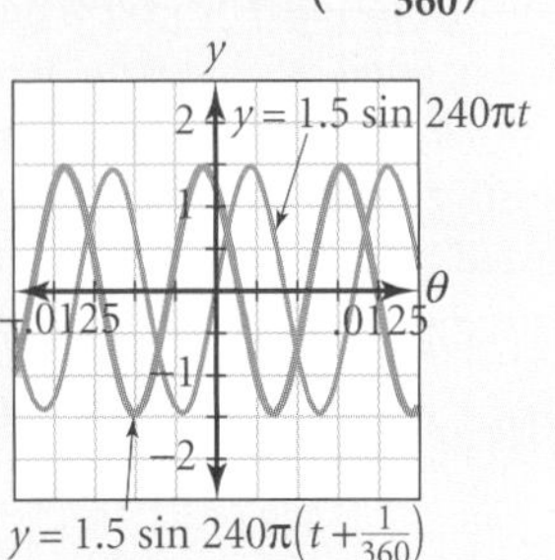

CONNECTION
TRANSFORMATIONS

The function $y = a \sin b(t - c)$, where t is in radians and y represents relative air pressure, can represent a particular sound as follows:

- The amplitude of the graph, a, represents the relative *intensity* of the changes in air pressure. A higher intensity results in a louder sound.
- The frequency of the sound wave, measured in units called hertz (Hz), or cycles per second, determines the pitch of the sound. The value of b in the function is equal to the frequency multiplied by 2π.
- The phase shift, c, represents the change in the position of the sound wave over time.
- The period of the sound wave is the reciprocal of the frequency.

EXAMPLE 3

APPLICATION
ACOUSTICS

A particular sound has a frequency of 55 hertz and an amplitude of 3.

a. Write a transformed sine function to represent this sound.

b. Write a new function that represents a phase shift of $\frac{1}{2}$ of a period to the right of the function from part **a**. Then use a graphics calculator to graph at least one period of both functions on the same coordinate plane.

Denyce Graves singing in the opera Carmen

SOLUTION

a. The parent function is $y = \sin t$. Write the transformed function in the form $y = a \sin b(t - c)$.

The amplitude is 3.	$a = 3$
The frequency is related to b.	$b = 2\pi \times 55 = 110\pi$
There is no phase shift.	$c = 0$

Thus, the transformed function is $y = 3 \sin 110\pi t$.

b. The period is the reciprocal of the frequency: $\frac{1}{55}$. One-half of the period is $\frac{1}{2} \times \frac{1}{55} = \frac{1}{110}$. Thus, a phase shift of $\frac{1}{110}$ radian to the right of the function $y = 3 \sin 110\pi t$ is given by $y = 3 \sin 110\pi\left(t - \frac{1}{110}\right)$. Graph both functions in radian mode.

TECHNOLOGY
GRAPHICS CALCULATOR

Keystroke Guide, page 883

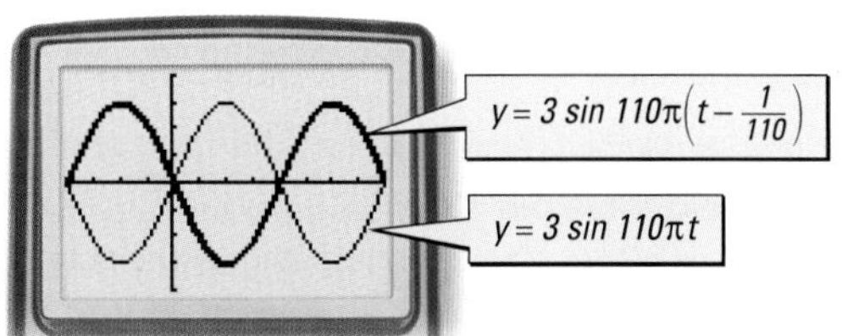

TRY THIS Write the function for a sound with a frequency of 120 hertz, an amplitude of 1.5, and a phase shift of $\frac{1}{3}$ of a period to the left. Graph at least one period of the function along with its parent function.

Reteaching the Lesson

USING TECHNOLOGY Have students put their graphics calculator in degree mode, enter **Y1=sin(X)**, press 2nd WINDOW, and set **TblMin=0** and **ΔTbl=15**. Have students press 2nd GRAPH, examine the table of values, and compare it with the sine of the common angles studied in the previous lesson. Have them discuss their observations and predict the shape of the graph. Have students graph the function to verify their observations. On the same screen, have students graph $y = 2 \sin x$, $y = 3 \sin x$, and $y = 4 \sin x$. Students should notice that these graphs differ only in amplitude. Have students clear the old graphs and graph $y = \sin x$ and $y = -\sin x$ on the same screen. Students should notice that these graphs are reflections of each other. Finally, have students graph $y = \sin x$, $y = \sin(x + 45)$, $y = \sin(x + 90)$, and $y = \sin(x + 180)$ and discuss the relationships among these graphs. Students should notice that the graphs are horizontal shifts of the parent graph. Have students continue this activity to investigate the graphs of $y = \cos x$ and $y = \tan x$ and the horizontal shifts of these graphs.

The Tangent Function

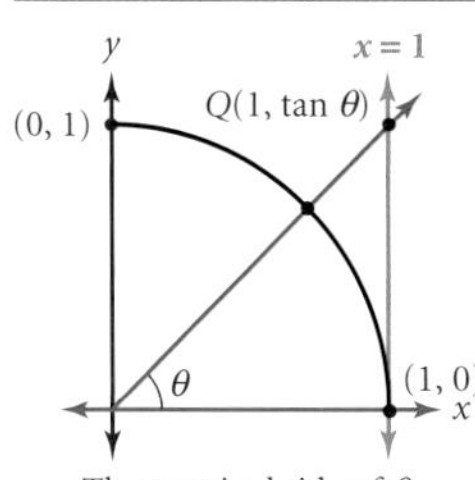

The terminal side of θ intersects the tangent line $x = 1$ at Q.

The sine and cosine functions are related to the coordinates of the points on a unit circle. The tangent function, however, relates to the points on a line *tangent* to the unit circle. Recall from geometry that a **tangent line** is perpendicular to a radius of the circle and touches the circle at only one point.

The table of values and graph for $y = \tan\theta$ below show that its period is 180°. The graph of $y = \tan\theta$ has vertical asymptotes where the function is undefined and has no amplitude because its range is all real numbers.

	θ	−90° 90°	−60° 120°	−45° 135°	−30° 150°	0° 180°	30° 210°	45° 225°	60° 240°
$\tan\theta$	exact	not defined	$-\sqrt{3}$	−1	$-\frac{\sqrt{3}}{3}$	0	$\frac{\sqrt{3}}{3}$	1	$\sqrt{3}$
	approx.	not defined	−1.73	−1	−0.58	0	0.58	1	1.73

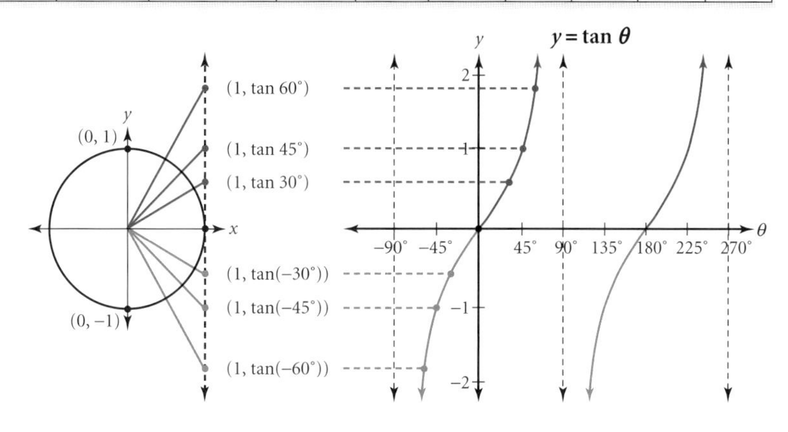

EXAMPLE 4 **Graph at least one period of $y = 3\tan\frac{\theta}{2}$ along with its parent function.**

SOLUTION

Make a table of approximate values. Then graph.

θ	$\frac{\theta}{2}$	$\tan\frac{\theta}{2}$	$3\tan\frac{\theta}{2}$
−180°	−90°	not defined	not defined
−120°	−60°	−1.73	−5.20
−90°	−45°	−1	−3
0°	0°	0	0
90°	45°	1	3
120°	60°	1.73	5.20
180°	90°	not defined	not defined

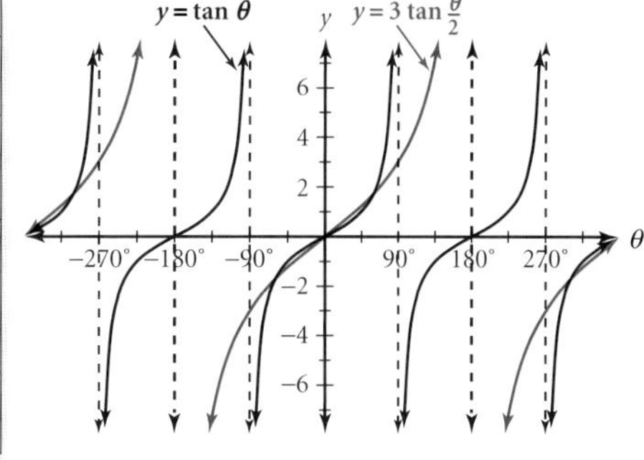

Notice that the graph of $y = 3\tan\frac{\theta}{2}$ is a horizontal stretch of the graph of $y = \tan\theta$ by a factor of 2 and a vertical stretch by a factor of 3.

TRY THIS Graph at least one period of $y = \frac{1}{2}\tan\theta - 3$ along with its parent function.

ADDITIONAL EXAMPLE 4

Graph at least one period of $y = \tan 2\theta + 1$.

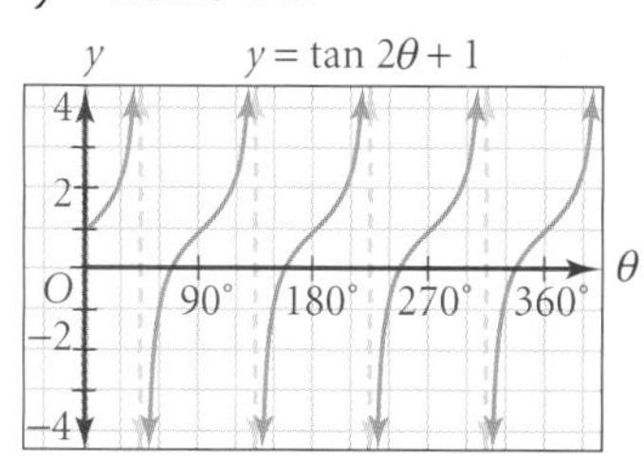

Math CONNECTION

TRANSFORMATIONS Students use stretches and compressions to transform the graph of $y = \tan x$.

TRY THIS

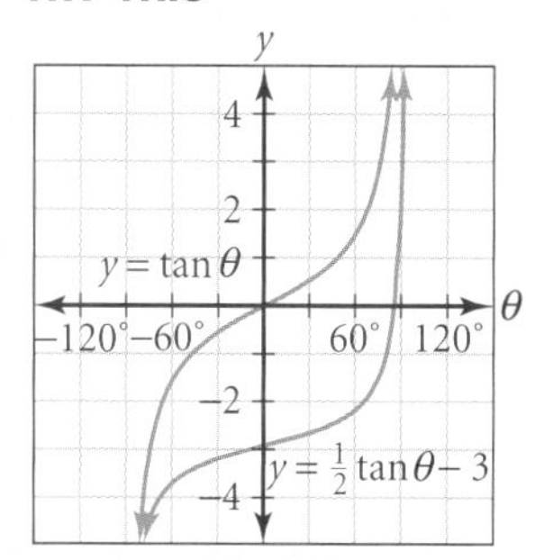

Assess

Selected Answers

Exercises 5–10, 11–67 odd

ASSIGNMENT GUIDE	
In Class	1–10
Core	11–57 odd
Core Plus	12–56 even
Review	58–67
Preview	68–72

Extra Practice can be found beginning on page 940.

Error Analysis

Students often have trouble finding the period of functions written in the form $y = \sin Bx$. Have students look at the graphs of $y = \sin x$ and $y = \sin 2x$ on a graphics calculator. Point out to students that when $x = \pi$, the graph of $y = \sin 2x$ has a y-value of $\sin 2\pi$, or 0, so the cycle of the function is completed in half the distance or time. Therefore, the period of $y = \sin 2x$ is π.

Practice

NAME ______ CLASS ______ DATE ______

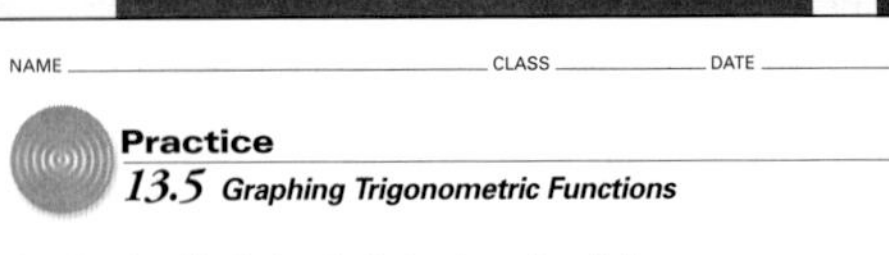

Practice
13.5 Graphing Trigonometric Functions

Complete the table of values for the function and graph the function along with its parent function.

1.

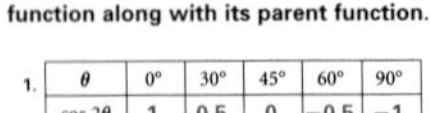

θ	0°	30°	45°	60°	90°
$\cos 2\theta$	1	0.5	0	−0.5	−1

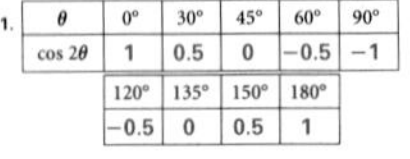

120°	135°	150°	180°
−0.5	0	0.5	1

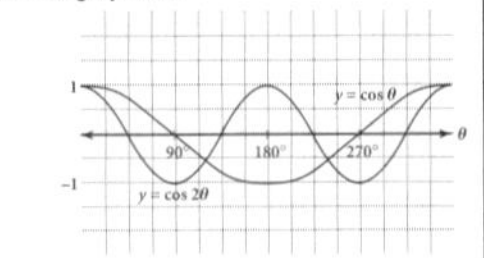

Identify the amplitude, if it exists, and the period of each function.

2. $y = 4.5 \cos 2\theta$ — 4.5; π radians

3. $y = 3\tan\left(x - \frac{\pi}{2}\right) + 1$ — does not exist; π radians

4. $y = 1.2\cos(x + \pi)$ — 1.2; 2π radians

Identify the phase shift and vertical translation of each function from its parent function. Then graph at least one period of the function for $0° \le \theta \le 360°$, or $0 \le x \le 2\pi$.

5. $y = 2\cos(\theta - 45°) + 1.5$ — shift of 45° right; translation of 1.5 units up

6. $y = \sin 2(x + \pi) - 1$ — shift of π units left; translation of 1 unit down

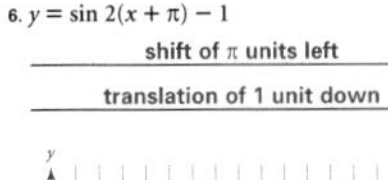

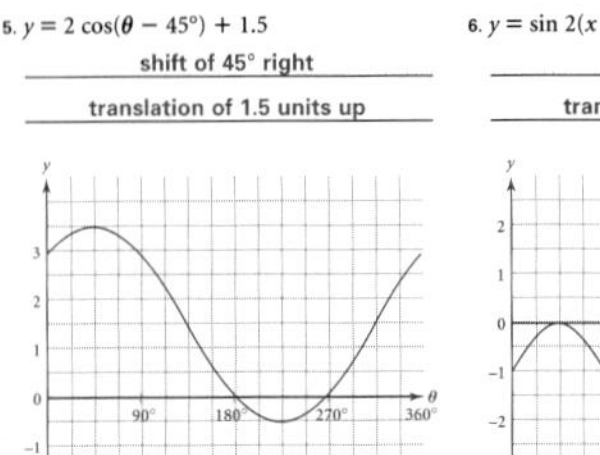

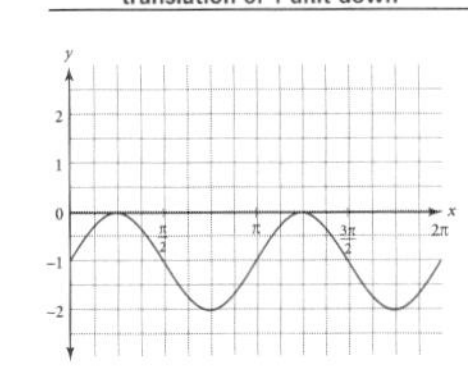

Exercises

internet connect
Activities Online
Go To: go.hrw.com
Keyword: MB1 Biorhythms

Communicate

1. Compare and contrast the graph of $y = \sin\theta$ with the graph of $y = \cos\theta$.
2. Describe the shape of the graph of the tangent function.
3. Explain why the amplitude of $y = -4\sin\theta$ is larger than that of $y = 3\sin\theta + 2$.
4. Describe at least four ways in which the graph of $y = \tan\theta$ differs from the graph of $y = \sin\theta$.

Guided Skills Practice

Graph at least one period of each trigonometric function along with its parent function. ***(EXAMPLE 1)***

5. $y = \frac{1}{3}\cos\theta$
6. $y = \sin\frac{3}{2}\theta$

Graph at least one period of each trigonometric function along with its parent function. ***(EXAMPLE 2)***

7. $y = \cos(\theta - 90°)$
8. $y = \sin\theta - 1.5$
9. Write the function for a sound with a frequency of 30 hertz, an amplitude of 2, and a phase shift of $\frac{1}{4}$ of a period to the left. Graph at least one period of the function. ***(EXAMPLE 3)*** $y = 2\sin 60\pi\left(t + \frac{1}{120}\right)$
10. Graph at least one period of the function $y = \frac{3}{2}\tan 3\theta$ along with its parent function. ***(EXAMPLE 4)***

11. 2.5; π
12. 1.5; $\frac{\pi}{2}$
13. none; $\frac{\pi}{3}$
14. none; $\frac{\pi}{3}$
15. 5; 4π
16. 6; 8π
17. 3; 2π
18. 2; 2π
19. 1; 2π
20. 90° right; up 3
21. 45° right; down 2
22. 30° left; down 2

Practice and Apply

internet connect
Homework Help Online
Go To: go.hrw.com
Keyword: MB1 Homework Help for Exercises 11–19

Identify the amplitude, if it exists, and the period of each function.

11. $y = 2.5\sin 2\theta$
12. $y = 1.5\sin 4\theta$
13. $y = 4.5\tan 3\theta$
14. $y = -4\tan 3\theta$
15. $y = -5\cos\frac{1}{2}\theta$
16. $y = -6\sin\frac{1}{4}\theta$
17. $y = 3\cos(\theta + 90°)$
18. $y = -2\sin(\theta - 30°)$
19. $y = -\sin(\theta + 45°)$

Identify the phase shift and the vertical translation of each function from its parent function.

20. $y = \sin(\theta - 90°) + 3$
21. $y = \cos(\theta - 45°) - 2$
22. $y = \cos(\theta + 30°) - 2$
23. $y = \sin(\theta + 60°) + 1$
24. $y = 3 - \sin(\theta - 45°)$
25. $y = 2 + \cos(\theta + 30°)$
26. $y = 4\cos[3(\theta + 180°)] + 1$
27. $y = 3\sin[2(\theta - 135°)] - 3$

23. 60° left; up 1
24. 45° right; up 3
25. 30° left; up 2
26. 180° left; up 1
27. 135° right; down 3

5.

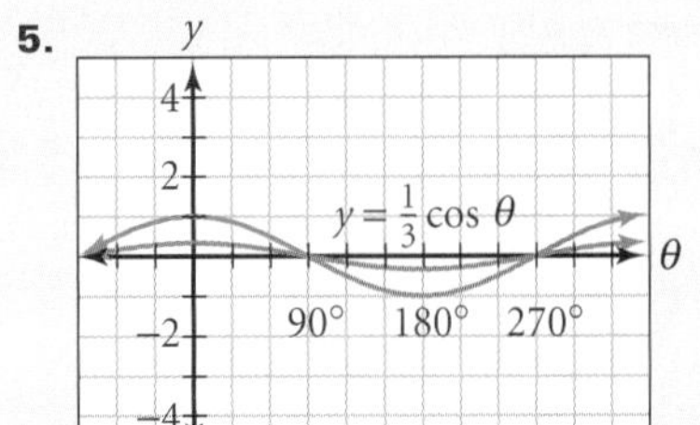

6.

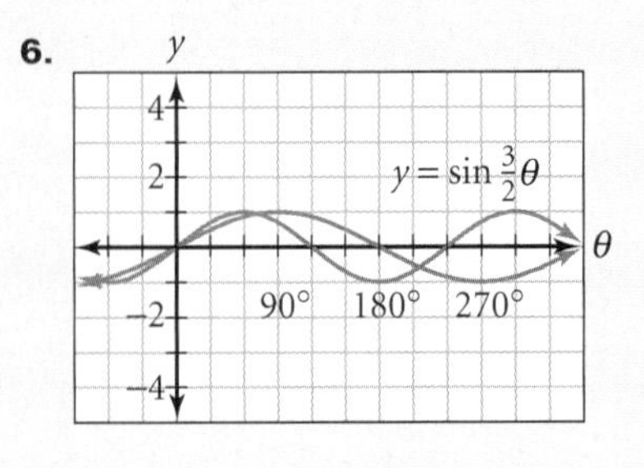

9. Check students' graphs.

The answers to Exercises 7–8 and 10 can be found in Additional Answers beginning on page 1002.

Describe the transformation of each function from its parent function. Then graph at least one period of the function along with its parent function.

28. $y = 2\cos\theta$
29. $y = 4\sin\theta$
30. $y = -2\sin\theta$
31. $y = -3\cos\theta$
32. $y = \sin(\theta - 90°)$
33. $y = \cos(\theta + 90°)$
34. $y = 3\cos(\theta + 90°)$
35. $y = 2\sin(\theta - 90°)$
36. $y = -4\cos\frac{1}{2}\theta$
37. $y = -2\cos\frac{1}{3}\theta$
38. $y = \frac{1}{2}\sin 3\theta$
39. $y = \frac{1}{3}\sin 2\theta$
40. $y = 3\tan\theta$
41. $y = 2\tan\theta$
42. $y = \tan\theta + 3$
43. $y = \tan\theta - 2$
44. $y = \tan 2\theta$
45. $y = \tan 3\theta$
46. $y = 2\tan\frac{1}{3}\theta$
47. $y = 3\tan\frac{1}{2}\theta$

CHALLENGE

48. Write a function of the form $f(\theta) = a\cos[b(\theta - c)] + d$ for the graph below.
$y = 3\cos 2\theta + 1$

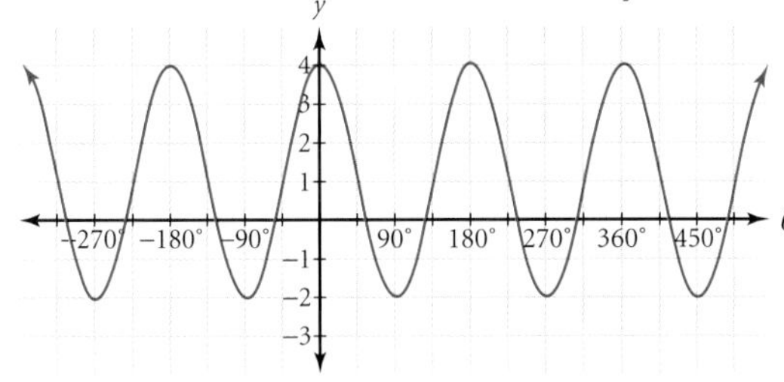

APPLICATIONS

EMPLOYMENT The number of people employed in a resort town can be modeled by the function $g(x) = 1.5\sin\left(\frac{\pi x}{6} + 1\right) + 5.2$, where x is the month of the year (beginning with 1 for January) and $g(x)$ is the number of people (in thousands) employed in the town that month.

49. What type of resort might this be? Explain. **ski resort or tropical resort because population highest in January**

50. About how many people are permanently employed in the town? **≈3700**

51. About how many people are employed in February? **≈6533**

52. Find two months when there are about 4500 people employed in the town. **May and September**

53. If a major year-round business in the town were to close, which one of the constants in the function model would decrease? **5.2**

TEMPERATURE The temperature in an air-conditioned office on a hot day can be modeled by the function $t(x) = 1.5\cos\left(\frac{\pi x}{12}\right) + 67$, where x is the time in minutes after the air conditioner is turned on and $t(x)$ is the temperature in degrees Fahrenheit after x minutes.

54. How long does the air conditioner run after being turned on? **12 min**

55. Find the maximum and minimum temperatures in the office building.

56. Find the temperature 10 minutes after the air conditioner is turned on.

57. Adjust the function to model the temperature in the office when the thermostat is set to a higher temperature.

55. max: 68.5°F; min: 65.5°F

56. ≈65.7°F

57. Answers may vary, but 67 should be increased.

28. stretched vertically by a factor of 2

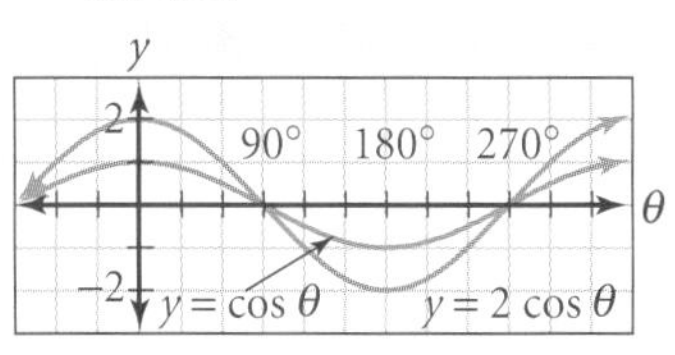

29. stretched vertically by a factor of 4

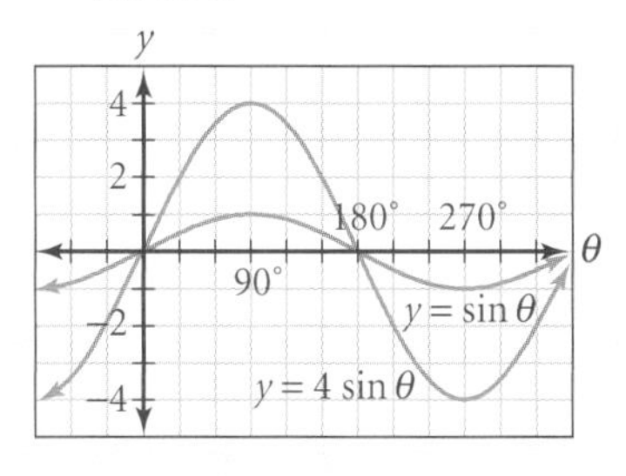

30. reflected across θ-axis and stretched vertically by a factor of 2

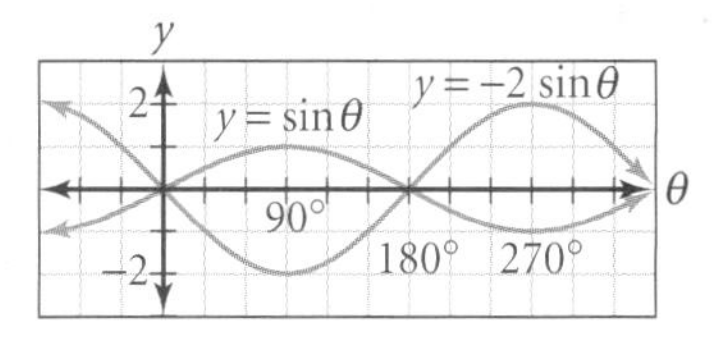

The answers to Exercises 34–47 can be found in Additional Answers beginning on page 1002.

31. reflected across θ-axis and stretched vertically by a factor of 3

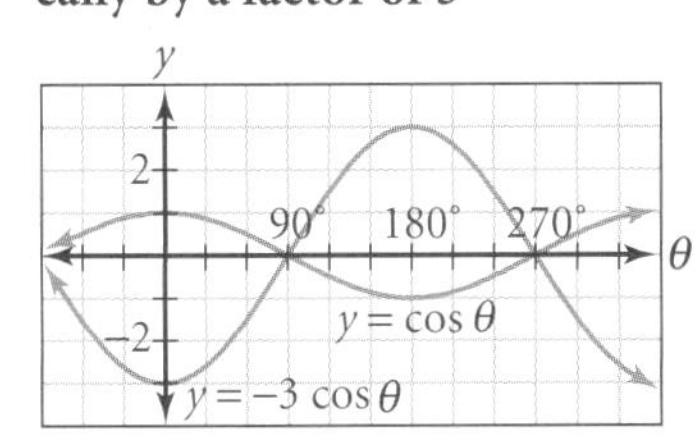

32. translated 90° to the right

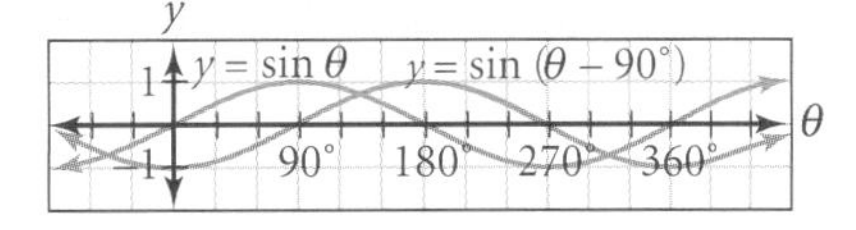

33. translated 90° to the left

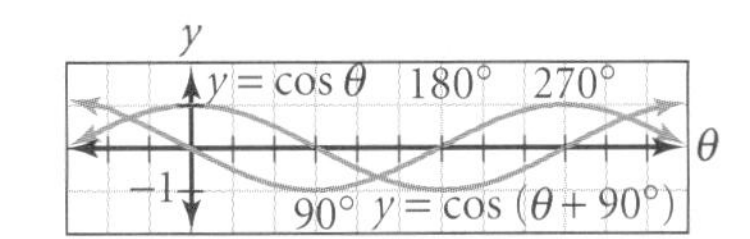

Student Technology Guide

NAME ______ CLASS ______ DATE ______

Student Technology Guide
13.5 Graphing Trigonometric Functions

Transformations of the parent trigonometric functions are important mathematical models. A graphics calculator is an excellent tool for graphing transformations of trigonometric functions.

Example: Graph one period of $y = 4\cos 3\theta + 2$ along with its parent function.

- First, press MODE, use the arrow keys to move to Degree, and then press ENTER. Press 2nd MODE.
- Analyze the period and phase shift to determine a good scale for the horizontal axis. Since $b = 3$, the period is $\frac{360°}{3} = 120°$. Since $c = 0$, there is no phase shift. A complete period of the graph begins at 0° and ends at 120°.
- Analyze the amplitude and vertical shift to determine a good scale for the vertical axis. Since $a = 4$, the amplitude is 4. Since $d = 2$, the vertical shift is 2. The y-values will vary between $2 - 4 = -2$ and $2 + 4 = 6$.
- To set the viewing window to the desired values, press WINDOW. Change Xmin to 0, Xmax to 120, Xscl to 30, Ymin to −2, Ymax to 6, and Yscl to 1.
- Press Y= 4 COS 3 X,T,θ,n) + 2.
- To enter the parent function, $y = \cos\theta$, move to Y2 and press COS X,T,θ,n).
- Press GRAPH.

```
WINDOW
 Xmin=0
 Xmax=120
 Xscl=30
 Ymin=-2
 Ymax=6
 Yscl=1
 Xres=1
```

The graph of $y = 4\cos 3\theta + 2$ and its parent function $y = \cos\theta$ are shown at right. Notice that only one-third of one period of the parent function is seen in this window.

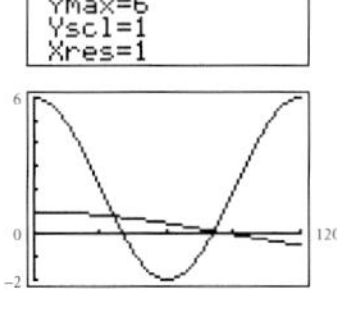

Graph one period of each trigonometric function along with its parent function. State the smallest interval over which you can see one period of the graph.

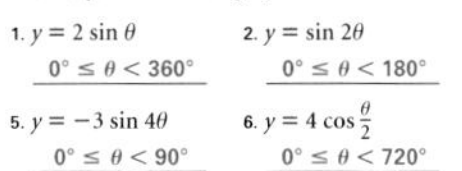

1. $y = 2\sin\theta$ **$0° \le \theta < 360°$**
2. $y = \sin 2\theta$ **$0° \le \theta < 180°$**
3. $y = \cos\theta + 4$ **$0° \le \theta < 360°$**
4. $y = \cos(\theta - 30°)$ **$30° \le \theta < 390°$**
5. $y = -3\sin 4\theta$ **$0° \le \theta < 90°$**
6. $y = 4\cos\frac{\theta}{2}$ **$0° \le \theta < 720°$**
7. $y = 3\sin 6\theta - 2$ **$0° \le \theta < 60°$**
8. $y = 4\tan\theta - 1$ **$-90° \le \theta < 90°$**

Use Teaching Transparency 60.

Students examine the graphs of $y = \csc x$ and $y = \sec x$, which are the reciprocal functions of sine and cosine.

ALTERNATIVE
Assessment

Portfolio Activity

The Portfolio Activity can be used as preparation for the Chapter Project or as a separate activity. In the Portfolio Activity on this page, students find the period and amplitude of graphs created in previous Portfolio Activities. Using this information, students write a function to model the data from the graph.

Answers to Portfolio Activities can be found in Additional Answers of the Teacher's Edition.

Look Back

APPLICATION

58. INCOME When Julie baby-sat for 4 hours and did yard work for 4 hours, she made a total of \$47. When she baby-sat for 7 hours and did yard work for 2 hours, she made a total of \$51. How much does Julie get paid for each type of work? ***(LESSON 3.2)*** **babysitting: \$5.50; yard work: \$6.25**

Write each pair of parametric equations as a single equation in *x* and *y*. ***(LESSON 3.6)***

59. $\begin{cases} x(t) = 2t + 1 \\ y(t) = 2t - 3 \end{cases}$ **60.** $\begin{cases} x(t) = t - 2 \\ y(t) = 4t + 1 \end{cases}$ **61.** $\begin{cases} x(t) = 2t + 1 \\ y(t) = 8t + 13 \end{cases}$

59. $x - y = 4$
60. $4x - y = -9$
61. $4x - y = -9$
62. $(-8.58, 8.22)$
63. $(0.86, 0.49)$

Solve each system of equations by using a matrix equation. Give answers to the nearest hundredth. ***(LESSON 4.4)***

62 $\begin{cases} 2.25x + 3.78y = 11.78 \\ 3.56x + 4.89y = 9.67 \end{cases}$ **63** $\begin{cases} 3.87x + 8.45y = 7.48 \\ 6.67x + 8.36y = 9.85 \end{cases}$

Find *x* to the nearest ten-thousandth. ***(LESSONS 6.3 AND 6.5)***

64. $e^{2x+1} = 13$ **≈0.7825** **65.** $e^{3x-2} = 25$ **≈1.7396** **66.** $2^{2x+4} = 20$ **≈0.1610** **67.** $2^{3x-21} = 31$ **≈8.6514**

Look Beyond

68 Graph the functions $f(x) = \sin x$ and $f(x) = \frac{1}{\sin x}$ on the same axes. Compare their graphs. How are they alike? How are they different?

69 Graph the functions $f(x) = \cos x$ and $f(x) = \frac{1}{\cos x}$ on the same axes. Compare their graphs. How are they alike? How are they different?

Using the definitions from Lesson 13.1, make a table of values and graph each function.

70. $y = \sec \theta$ **71.** $y = \csc \theta$ **72.** $y = \cot \theta$

Refer to the data and the graph you created in the Portfolio Activity on page 857.

1. Find the amplitude of your graph. What does the amplitude represent in terms of the Cosmoclock 21?
2. Write an equation of the form $y = a \sin bt$ to model the data that was graphed.
3. Create a scatter plot of the data. Then graph your equation on the same coordinate axes as your scatter plot. Is the equation a good model for the data? Explain.

WORKING ON THE CHAPTER PROJECT

You should now be able to complete Activity 4 of the Chapter Project.

68.

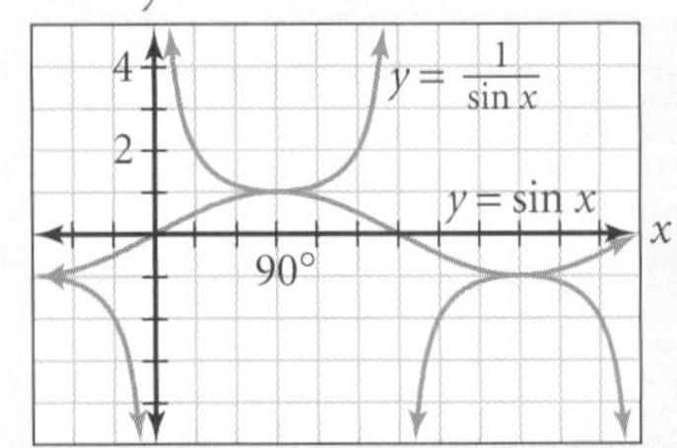

The graphs are alike in that they share common points at odd-number multiples of 90°, such as (90°, 1) and (270°, −1). The graphs are different in that the sine curve has one piece, but $\frac{1}{\sin x}$ has disconnected pieces.

69. The graphs, shown at right, are alike in that they share common points at multiples of 180°, such as (0°, 1) and (180°, −1). The graphs are different in that the cosine curve has one piece, but $\frac{1}{\cos x}$ has disconnected pieces.

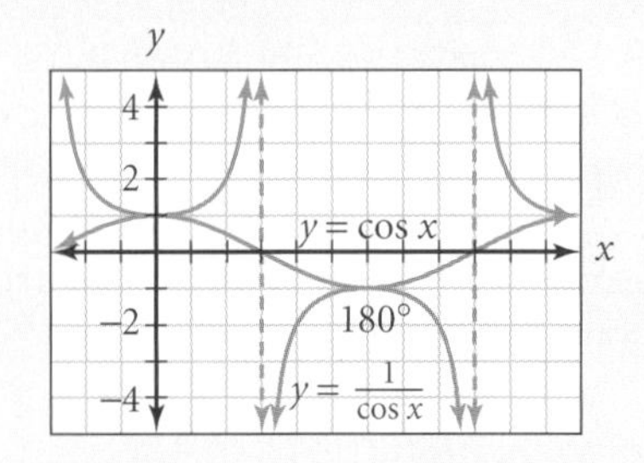

The answers to Exercises 70–72 can be found in Additional Answers beginning on page 1002.

Inverses of Trigonometric Functions

Why *You can use inverses of trigonometric functions to solve real-world problems, such as finding the angle of depression between a supersonic aircraft and the ground.*

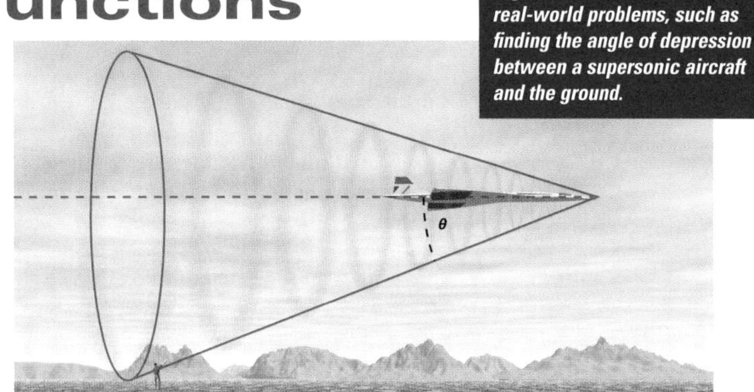

Objective

- Evaluate trigonometric expressions involving inverses.

APPLICATION
AVIATION

When an aircraft flies faster than the speed of sound, which is about 730 miles per hour, shock waves in the shape of a cone are created. When the cone passes a person on the ground, a sonic boom is heard.

The speed of an aircraft can be expressed as a *Mach number*, which gives the aircraft's speed relative to the speed of sound. For example, an aircraft flying at 1000 miles per hour has a speed of $\frac{1000}{730}$, or about Mach 1.4. If θ is the angle of depression between the aircraft and the ground, then θ depends on the speed of the aircraft in Mach numbers as follows:

$$\sin \theta = \frac{1}{\text{speed of aircraft}}$$

What is θ if the speed of the aircraft is Mach 1.3? Mach 1.8? *You will solve this problem in Example 4.*

In Lesson 2.5 you found inverse relations by interchanging the domain and range of the given relation or function. This procedure is used in the following Activity involving trigonometric functions.

Activity Exploring the Inverse Relation of $y = \sin x$

You will need: graph paper

1. Create a table of values and sketch the graph of $y = \sin x$ over the interval $-2\pi \le x \le 2\pi$. Use x-values of $0, \pm\frac{\pi}{2}, \pm\pi, \pm\frac{3\pi}{2}$, and $\pm 2\pi$.
2. Create a new table of ordered pairs by interchanging the x and y values in your table from Step 1. Plot the new ordered pairs on the same axes, and sketch the resulting curve, which represents $y = \sin^{-1} x$.
3. Add the line $y = x$ to your graph.
4. CHECKPOINT ✔ Fold the graph paper along the line $y = x$. Describe what happens to the graphs of $y = \sin x$ and $y = \sin^{-1} x$ and what this result means.
5. CHECKPOINT ✔ Describe what $y = \sin^{-1} x$ represents. Is $y = \sin^{-1} x$ a function? Explain.

Alternative Teaching Strategy

USING TECHNOLOGY Remind students of the definition of an inverse. Have them set their graphics calculator to radian mode. Using the **STAT** menu of a TI-82 or TI-83 graphics calculator, enter the following values into **L1**:

$-\frac{\pi}{2} \quad -\frac{\pi}{3} \quad -\frac{\pi}{4} \quad -\frac{\pi}{6} \quad 0 \quad \frac{\pi}{6} \quad \frac{\pi}{4} \quad \frac{\pi}{3} \quad \frac{\pi}{2}$

In **L2**, have the calculator compute the sine of the values in **L1**. Set the viewing window as follows:

Xmin=−2	**Ymin=−2**
Xmax=2	**Ymax=2**

Make a scatterplot of **L1** versus **L2** and of **L2** versus **L1** on the same axes. Have students discuss these graphs and realize that they are inverses of each other. Have students discuss what would happen to the graph of the inverse if more values were entered, such as extending the values in **L1** through π. **The inverse is no longer a function.** Have students repeat this activity for cosine and tangent, restricting the values in **L1** so that the inverse is a function.

Prepare

NCTM PRINCIPLES & STANDARDS 1–4, 6–10

QUICK WARM-UP

Find the inverse of each function.

1. $f(x) = x - 4$
   $f^{-1}(x) = x + 4$
2. $g(x) = -5x$
   $g^{-1}(x) = \frac{x}{-5}$
3. $h(x) = 2x + 6$
   $h^{-1}(x) = 0.5x - 3$
4. $j(x) = 3^x$
   $j^{-1}(x) = \log_3 x$
5. $k(x) = \ln x$
   $k^{-1}(x) = e^x$

Also on Quiz Transparency 13.6

Teach

Why Have students discuss situations in which the lengths of two sides of a right triangle are known but the measure of an angle is needed. **sample: angle of depression of a landing airplane** Inverse trigonometric functions allow you to find an angle given the ratio of the sides.

Activity Notes

In this Activity, students create a table of values for $y = \sin x$. By interchanging the x- and y-values in the table, the students graph the inverse of $y = \sin x$. Students should see that $y = \sin x$ and $y = \sin^{-1} x$ are reflections of each other across the line $y = x$.

☞For Checkpoint answers, see page 868.

Cooperative Learning

Have students work in pairs for the Activity. One student should find the values and record them in the table, and the other student should plot the values on graph paper. The pairs should discuss the meaning of interchanging the x- and y-values and how the change relates to the line $y = x$.

CHECKPOINT ✔

4. The graphs match up with each other; $y = \sin x$ and $y = \sin^{-1} x$ are inverses of each other.

CHECKPOINT ✔

5. $y = \sin^{-1} x$ is the inverse of $y = \sin x$; $y = \sin^{-1}$ is not a function because some x-values correspond to more than one y-value.

ADDITIONAL EXAMPLE 1

Find all possible values of $\tan^{-1}(-1)$. $135° + n(360°)$ or $315° + n(360°)$, where n is an integer

TRY THIS

$45° + 360n$ and $135° + 360n$, where n is an integer

Because the trigonometric functions are periodic, many values in their domains have the same function values. For example, examine the sine, cosine, and tangent of 30°, 390°, and −330° shown below.

$\sin 30° = \frac{1}{2}$ $\quad\sin 390° = \frac{1}{2}$ $\quad\sin(-330°) = \frac{1}{2}$

$\cos 30° = \frac{\sqrt{3}}{2}$ $\quad\cos 390° = \frac{\sqrt{3}}{2}$ $\quad\cos(-330°) = \frac{\sqrt{3}}{2}$

$\tan 30° = \frac{1}{\sqrt{3}}$, or $\frac{\sqrt{3}}{3}$ $\quad\tan 390° = \frac{1}{\sqrt{3}}$, or $\frac{\sqrt{3}}{3}$ $\quad\tan(-330°) = \frac{1}{\sqrt{3}}$, or $\frac{\sqrt{3}}{3}$

These repeated function values are readily apparent in the graph of a trigonometric function. For example, examine the graph of $y = \sin \theta$ shown below.

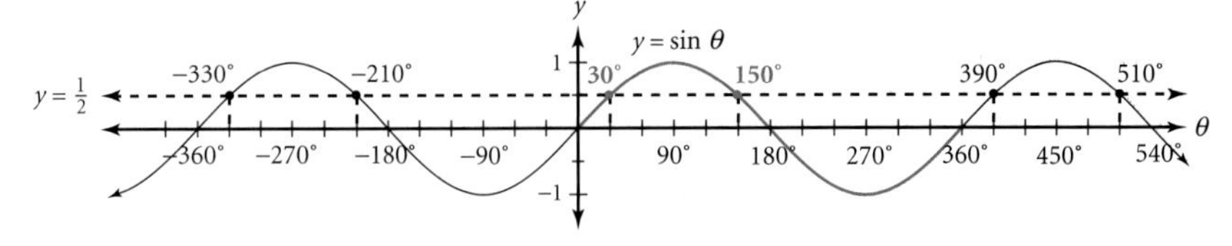

Recall from Lesson 13.1 that the inverse relations of the sine, cosine, and tangent functions are denoted as $y = \sin^{-1} x$, $y = \cos^{-1} x$, and $y = \tan^{-1} x$. To find all possible values of an inverse trigonometric function, such as $y = \sin^{-1} \frac{1}{2}$, first find all values that occur within one period. In the period between 0° and 360°, as shown above in blue, $\sin 30° = \frac{1}{2}$ and $\sin 150° = \frac{1}{2}$. Therefore, $\sin^{-1} \frac{1}{2} = 30°$ and $\sin^{-1} \frac{1}{2} = 150°$. The other values of $\sin^{-1} \frac{1}{2}$ occur at values every period before or after 30° and 150°. Thus, if n is an integer, then all possible values of $\sin^{-1} \frac{1}{2}$ occur at $30° + n360°$ and $150° + 360°$.

EXAMPLE 1 **Find all possible values of $\cos^{-1} \frac{1}{2}$.**

SOLUTION

Find all possible values of $\cos^{-1} \frac{1}{2}$ within one period. The cosine function is positive in Quadrants I and IV.

$\cos 60° = \frac{1}{2}$ *Quadrant I* $\qquad \cos 300° = \frac{1}{2}$ *Quadrant IV*

Thus, all possible values of $\cos^{-1} \frac{1}{2}$ are $60° + n360°$ and $300° + n360°$.

TRY THIS Find all possible values of $\sin^{-1} \frac{\sqrt{2}}{2}$.

The graph of $y = \sin \theta$, shown above, clearly fails the horizontal-line test. Therefore, $y = \sin \theta$ is not a one-to-one function and its inverse cannot be a function. This is true for all trigonometric functions unless their domains are restricted in such a way that their inverses can be functions. The functions *Sine, Cosine,* and *Tangent* (denoted by capital letters) are defined as the sine, cosine, and tangent functions with the restricted domains defined on the next page. The restricted domains are called **principal values.**

Interdisciplinary Connection

PHYSICAL SCIENCE As an aircraft approaches the airport where it will land, the pilot must calculate the angle of descent. Suppose that the aircraft is flying at an altitude of 15,000 feet and is 70,000 feet from the airport, as shown in the diagram below. Find the angle of descent, θ. **$\theta \approx 12.09°$**

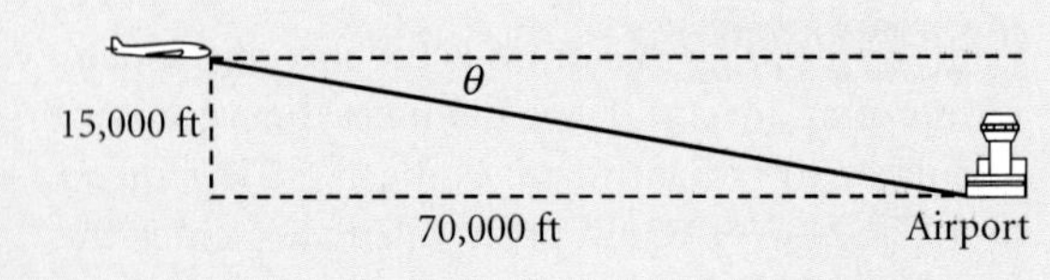

Principal Values of Sin θ, Cos θ, and Tan θ

$\text{Sin } \theta = \sin \theta$ for $-90° \le \theta \le 90°$

$\text{Cos } \theta = \cos \theta$ for $0° \le \theta \le 180°$

$\text{Tan } \theta = \tan \theta$ for $-90° < \theta < 90°$

The graph of $y = \sin \theta$ and the portion of the curve that represents $y = \text{Sin } \theta$ are shown below.

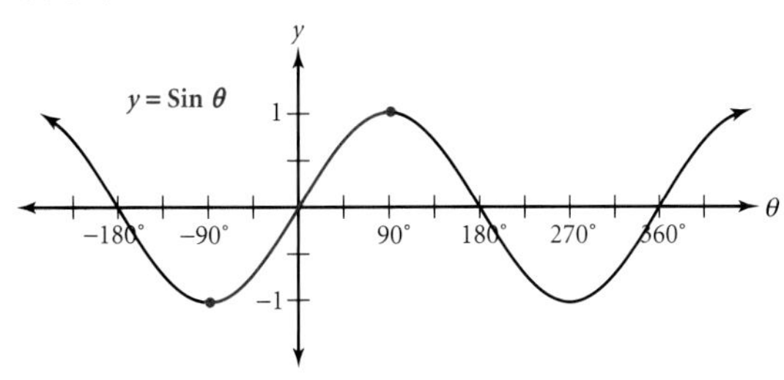

CHECKPOINT ✔ Graph $y = \cos \theta$ and $y = \tan \theta$ and indicate the portion of these curves that represent $y = \text{Cos } \theta$ and $y = \text{Tan } \theta$.

CHECKPOINT ✔

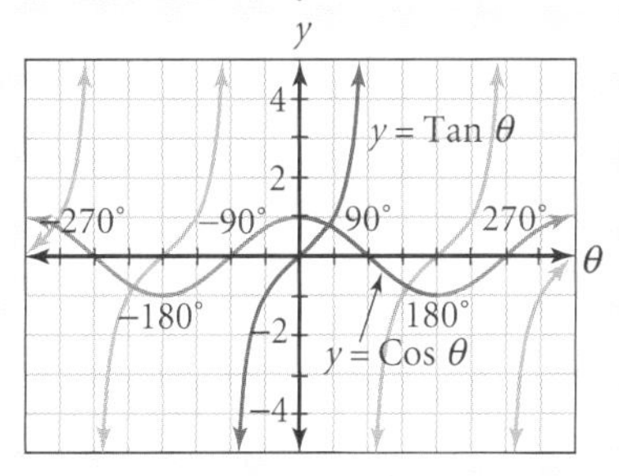

The functions $y = \text{Sin } \theta$, $y = \text{Cos } \theta$, and $y = \text{Tan } \theta$ are one-to-one functions and have inverses that are also functions. The inverse functions are denoted by $y = \text{Sin}^{-1} x$, $y = \text{Cos}^{-1} x$, and $y = \text{Tan}^{-1} x$, respectively.

Inverse Trigonometric Functions

If $y = \text{Sin } x$, then its inverse function is $y = \text{Sin}^{-1} x$.

If $y = \text{Cos } x$, then its inverse function is $y = \text{Cos}^{-1} x$.

If $y = \text{Tan } x$, then its inverse function is $y = \text{Tan}^{-1} x$.

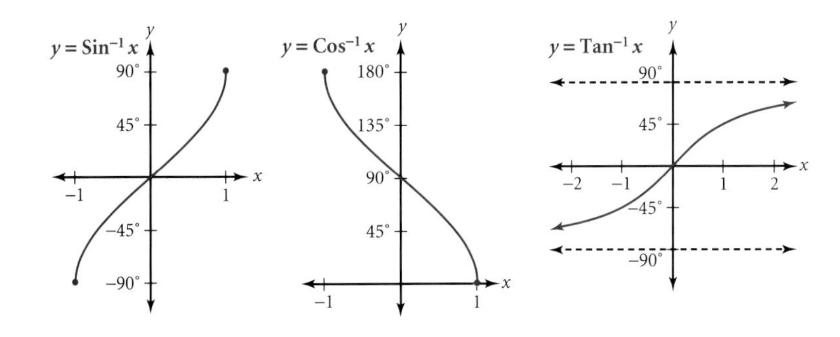

CHECKPOINT ✔ What is the range of $y = \text{Sin}^{-1} x$, of $y = \text{Cos}^{-1} x$, and of $y = \text{Tan}^{-1} x$?

Inclusion Strategies

TACTILE LEARNERS Have students graph each trig function and the line $y = x$ on graph paper, and have them fold the paper along the line $y = x$ so that the y-axis is above the fold. Then have students use a different color to graph the inverse of each trig function as the reflection of the function's graph across the line $y = x$. Students should label each of the trig functions and their inverses. Encourage students to keep these graphs for reference as they work through the examples in the text or homework.

Enrichment

One way that an inverse function is defined is $f(f^{-1}(x)) = x$. Have students discuss this definition and then have them apply it to the trig functions for a few numbers. For example, $\sin(\sin^{-1} 0.5) = 0.5$ and $\sin^{-1}(\sin 30°) = 30°$. Have students write a paragraph explaining this definition of an inverse function.

CHECKPOINT ✔

range of $\text{Sin}^{-1} x$: $-90° \le x \le 90°$
range of $\text{Cos}^{-1} x$: $0° \le x \le 180°$
range of $\text{Tan}^{-1} x$: $-90° \le x \le 90°$

EXAMPLE 2 **Evaluate each inverse trigonometric expression.**

a. $\text{Sin}^{-1}\frac{1}{2}$ **b.** $\text{Cos}^{-1}\left(-\frac{\sqrt{3}}{2}\right)$ **c.** $\text{Tan}^{-1}\left(-\frac{\sqrt{3}}{3}\right)$

SOLUTION

a. Find the angle between the principal values –90° and 90°, inclusive, whose sine is $\frac{1}{2}$. Because $\sin 30° = \frac{1}{2}$, $\text{Sin}^{-1}\frac{1}{2} = 30°$.

b. Find the angle between the principal values 0° and 180°, inclusive, whose cosine is $-\frac{\sqrt{3}}{2}$. Because $\cos 150° = -\frac{\sqrt{3}}{2}$, $\text{Cos}^{-1}\left(-\frac{\sqrt{3}}{2}\right) = 150°$.

c. Find the angle between the principal values –90° and 90°, inclusive, whose tangent is $-\frac{\sqrt{3}}{3}$. Because $\tan(-30°) = -\frac{\sqrt{3}}{3}$, $\text{Tan}^{-1}\left(-\frac{\sqrt{3}}{3}\right) = -30°$.

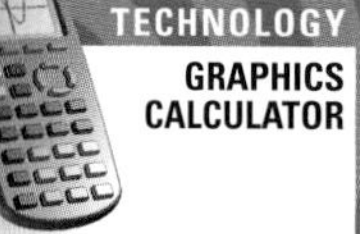

Keystroke Guide, page 883

CHECK

A scientific or graphics calculator returns principal values for inverse trigonometric functions.

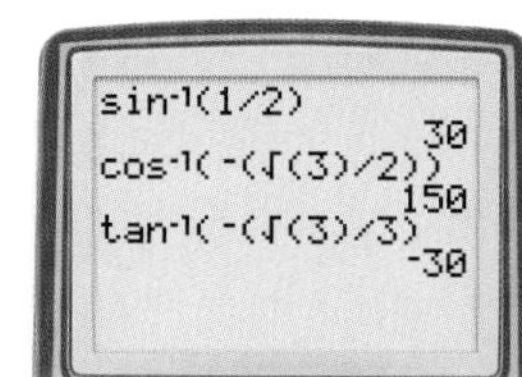

TRY THIS Evaluate each inverse trigonometric expression.

a. $\text{Sin}^{-1}\frac{\sqrt{3}}{2}$ **b.** $\text{Cos}^{-1}\left(-\frac{\sqrt{2}}{2}\right)$ **c.** $\text{Tan}^{-1}\sqrt{3}$

To evaluate an expression such as $\sin\left(\text{Cos}^{-1}\frac{\sqrt{3}}{2}\right)$, first evaluate the expression inside the parentheses. Then evaluate the resulting expression.

EXAMPLE 3 **Evaluate each trigonometric expression.**

a. $\sin\left(\text{Cos}^{-1}\frac{\sqrt{3}}{2}\right)$ **b.** $\text{Tan}^{-1}(\cos 180°)$

SOLUTION

a. Because $\cos 30° = \frac{\sqrt{3}}{2}$, substitute 30° for $\text{Cos}^{-1}\frac{\sqrt{3}}{2}$.

$$\sin\left(\text{Cos}^{-1}\frac{\sqrt{3}}{2}\right) = \sin 30° = \frac{1}{2}$$

b. First evaluate cos 180°. Then use the fact that $\tan(-45°) = -1$.

$$\text{Tan}^{-1}(\cos 180°) = \text{Tan}^{-1}(-1) = -45°$$

TRY THIS Evaluate each trigonometric expression.

a. $\text{Cos}^{-1}(\sin 315°)$ **b.** $\tan\left[\text{Sin}^{-1}\left(-\frac{\sqrt{3}}{2}\right)\right]$

CRITICAL THINKING Explain why $\text{Cos}^{-1}(\cos x) \neq x$, $\text{Sin}^{-1}(\sin x) \neq x$, and $\text{Tan}^{-1}(\tan x) \neq x$ for all values of x.

ADDITIONAL EXAMPLE 2

Evaluate each inverse trigonometric function.

a. $\text{Sin}^{-1}\left(-\frac{1}{2}\right)$ **–30°**

b. $\text{Cos}^{-1}\frac{\sqrt{3}}{2}$ **30°**

c. $\text{Tan}^{-1}(-\sqrt{3})$ **–60°**

TRY THIS

a. $\text{Sin}^{-1}\frac{\sqrt{3}}{2} = 60°$

b. $\text{Cos}^{-1}\left(-\frac{\sqrt{2}}{2}\right) = 135°$

c. $\text{Tan}^{-1}\sqrt{3} = 60°$

ADDITIONAL EXAMPLE 3

Evaluate each trigonometric expression.

a. $\cos\left(\text{Tan}\frac{-1\sqrt{3}}{3}\right)$ $\frac{\sqrt{3}}{2}$

b. $\text{Tan}^{-1}(\sin 90°)$ **45°**

TRY THIS

a. $\text{Cos}^{-1}(\sin 315°) = 135°$

b. $\tan\left[\text{Sin}^{-1}\left(-\frac{\sqrt{3}}{2}\right)\right] = -\sqrt{3}$

CRITICAL THINKING

Because cos x and sin x are not one-to-one functions, inverse functions do not exist for these functions for all values of x.

Reteaching the Lesson

USING VISUAL MODELS The inverse of the sine can be determined by drawing a triangle. For example, to find $\sin^{-1} 0.7$, construct a right triangle with a hypotenuse of 1 and one leg of 0.7. Then use a protractor to measure the angle opposite the leg that has a length of 0.7. Have students verify their answer with a scientific calculator. Repeat this activity for inverse trig functions. This exercise will help students see that the output of an inverse trig function is an angle measure.

Then show students how to find $\sin\left(\text{Cos}^{-1}\frac{\sqrt{3}}{2}\right)$ by using the associated triangle. Because the cosine is defined by adjacent side over hypotenuse, the length of the third side is $2^2 - \sqrt{3}^2 = 1$, so they should draw a triangle with side lengths of 1 and $\sqrt{3}$ units and a hypotenuse of 2 units. Identify the angle that produces the cosine of $\frac{\sqrt{3}}{2}$ and find the sine of that angle. $\frac{1}{2}$ Have students find other composite trig ratios by using the special triangles studied earlier.

EXAMPLE 4 Refer to the aviation problem at the beginning of the lesson.

Find the angle of depression, θ, for an airplane flying at each speed.

a. Mach 1.3 **b.** Mach 1.8

APPLICATION AVIATION

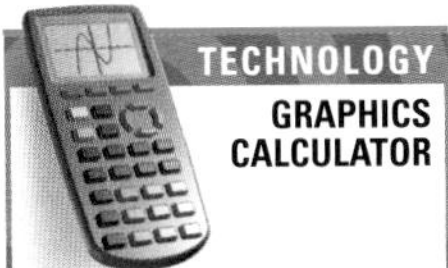

Keystroke Guide, page 883

SOLUTION

a. $\sin\theta = \frac{1}{1.3}$

$\theta = \mathrm{Sin}^{-1}\frac{1}{1.3} \approx 50.28°$

The angle of depression is about 50.28°.

b. $\sin\theta = \frac{1}{1.8}$

$\theta = \mathrm{Sin}^{-1}\frac{1}{1.8} \approx 33.75°$

The angle of depression is about 33.75°.

ADDITIONAL EXAMPLE 4

Refer to the problem at the beginning of the lesson. **Find the angle of depression, θ, for an airplane flying at each speed.**

a. Mach 1.4 ≈45.58°

b. Mach 2.0 30°

Exercises

Communicate

1. Describe how Sin θ, Cos θ, and Tan θ are related to sin θ, cos θ, and tan θ, respectively.
2. Describe how $\mathrm{Sin}^{-1} x$, $\mathrm{Cos}^{-1} x$, and $\mathrm{Tan}^{-1} x$ are related to Sin x, Cos x, and Tan x, respectively.
3. Explain why $\mathrm{Sin}^{-1} 5$ is not defined. Is $\mathrm{Tan}^{-1} 5$ defined? Explain.

Guided Skills Practice

4. Find all possible values of $\cos^{-1}\left(\frac{\sqrt{3}}{2}\right)$. *(EXAMPLE 1)*

Evaluate each inverse trigonometric expression. *(EXAMPLE 2)*

5. $\mathrm{Sin}^{-1}\left(-\frac{\sqrt{3}}{2}\right)$ −60°
6. $\mathrm{Cos}^{-1}\left(\frac{1}{2}\right)$ 60°
7. $\mathrm{Tan}^{-1}(-\sqrt{3})$ −60°

Evaluate each trigonometric expression. *(EXAMPLE 3)*

8. $\sin\left(\mathrm{Cos}^{-1}\frac{1}{2}\right)$ $\frac{\sqrt{3}}{2}$
9. $\mathrm{Cos}^{-1}(\sin 30°)$ 60°
10. $\mathrm{Tan}^{-1}(\tan 150°)$ −30°

11. Find the angle of depression, θ, for an airplane flying at Mach 1.5. *(EXAMPLE 4)* 41.81°

Practice and Apply

Find all possible values for each expression.

12. $\cos^{-1}\left(-\frac{1}{2}\right)$
13. $\sin^{-1}\left(-\frac{\sqrt{2}}{2}\right)$
14. $\tan^{-1}\left(-\frac{\sqrt{3}}{3}\right)$
15. $\tan^{-1}\sqrt{3}$
16. $\sin^{-1} 1$
17. $\cos^{-1} 0$
18. $\sin^{-1}\frac{\sqrt{3}}{2}$
19. $\cos^{-1}\frac{\sqrt{2}}{2}$
20. $\tan^{-1} 0$

Answers:

4. $30° + n360°$ and $330° + n360°$, where n is an integer
12. $120° + n360°$ and $240° + n360°$
13. $225° + n360°$ and $315° + n360°$
14. $150° + n360°$ and $330° + n360°$
15. $60° + n360°$ and $240° + n360°$
16. $90° + n360°$
17. $90° + n360°$ and $270° + n360°$
18. $60° + n360°$ and $120° + n360°$
19. $45° + n360°$ and $315° + n360°$
20. $n360°$ and $180° + n360°$

Assess

Selected Answers

Exercises 4–11, 13–63 odd

ASSIGNMENT GUIDE

In Class	1–11
Core	13–43 odd, 47–51 odd
Core Plus	12–44 even, 45, 46–52 even
Review	53–64
Preview	65–66

Extra Practice can be found beginning on page 940.

Error Analysis

When solving problems involving inverse trig functions, students often try to find the inverse sine of an angle instead of the ratio whose value is between 0 and 1. Encourage students to write the trig relationship first and then calculate an inverse if the angle is the unknown value. For example, suppose that a student wants to find θ if $\cos\theta = 0.788$. Since the angle is the unknown value, isolate θ by taking the inverse cosine of both sides of the equation. This results in $\cos^{-1}(\cos\theta) = \cos^{-1}0.788$, which gives $\theta \approx 38°$.

45. Let $f(\theta) = \text{Sin}\,\theta$.
$(f^{-1} \circ f)(\theta) = (f \circ f^{-1})(\theta) = \theta$
by definition of inverse functions
$(\text{Sin}^{-1} \circ \text{Sin}\,\theta)(\theta)$
$= (\text{Sin} \circ \text{Sin}\,\theta^{-1})(\theta) = \theta$
Substitute Sin for f.
$(\text{Sin}^{-1} \circ \text{Sin}\,\theta)(\theta) = \theta$

46. Let $f^{-1}(x) = \text{Cos}^{-1}x$
$(f^{-1} \circ f)(\theta) = (f \circ f^{-1})(\theta) = \theta$
by definition of inverse functions
$(\text{Cos} \circ \text{Cos}\,\theta^{-1})(\theta)$
$= (\text{Cos}^{-1} \circ \text{Cos}\,\theta)(\theta) = \theta$
Substitute Cos for f.
$(\cos \circ \text{Cos}^{-1})(x) = x$
$\text{Cos}\,x = \cos x$ for $0° \le x \le 180°$

internet connect
Homework Help Online
Go To: **go.hrw.com**
Keyword:
MB1 Homework Help
for Exercises 21–24

Evaluate each trigonometric expression.

21. $\text{Sin}^{-1}\left(-\frac{1}{2}\right)$ **−30°**   **22.** $\text{Sin}^{-1}\frac{\sqrt{2}}{2}$ **45°**   **23.** $\text{Cos}^{-1}\frac{\sqrt{2}}{2}$ **45°**

24. $\text{Cos}^{-1}\left(-\frac{\sqrt{3}}{2}\right)$ **150°**   **25.** $\text{Tan}^{-1}\sqrt{3}$ **60°**   **26.** $\text{Tan}^{-1}\frac{\sqrt{3}}{3}$ **30°**

27. $\text{Sin}^{-1}\frac{\sqrt{3}}{2}$ **60°**   **28.** $\text{Sin}^{-1}1$ **90°**   **29.** $\text{Cos}^{-1}(-1)$ **180°**

30. $\text{Cos}^{-1}\frac{1}{2}$ **60°**   **31.** $\text{Tan}^{-1}1$ **45°**   **32.** $\text{Tan}^{-1}(-1)$ **−45°**

Evaluate each trigonometric expression.

33. $\tan\left(\text{Sin}^{-1}\frac{\sqrt{2}}{2}\right)$ **1**   **34.** $\cos\left(\text{Sin}^{-1}\frac{1}{2}\right)$ $\frac{\sqrt{3}}{2}$ **35.** $\sin(\text{Tan}^{-1}1)$ $\frac{\sqrt{2}}{2}$

36. $\cos\left(\text{Tan}^{-1}\frac{\sqrt{3}}{2}\right)$ **0.7560**   **37.** $\tan\left(\text{Cos}^{-1}\frac{1}{2}\right)$ $\sqrt{3}$ **38.** $\tan\left(\text{Cos}^{-1}\frac{\sqrt{2}}{2}\right)$ **1**

39. $\text{Tan}^{-1}(\sin 30°)$ **26.57°**   **40.** $\text{Tan}^{-1}(\cos 135°)$ **−35.26°**   **41.** $\text{Cos}^{-1}(\tan 225°)$ **0°**

42. $\text{Cos}^{-1}(\sin 60°)$ **30°**   **43.** $\text{Sin}^{-1}(\cos 120°)$ **−30°**   **44.** $\text{Sin}^{-1}(\cos 300°)$ **30°**

CHALLENGE

Prove each statement.

45. $\text{Sin}^{-1}(\text{Sin}\,\theta) = \theta$ **46.** $\cos(\text{Cos}^{-1}x) = x$

APPLICATIONS

47. FORESTRY A tree casts a 35-foot shadow on the ground when the angle of elevation from the edge of the shadow to the sun is about 40°. How tall is the tree? Draw a diagram to illustrate this situation. **29.4 feet**

48. ASTRONOMY Katie is setting up a new telescope in her backyard. Her neighbor's house, which is 50 feet away from the telescope, is 30 feet tall, and the eyepiece of the telescope is 5 feet above the ground. What is the minimum angle that the telescope must make with the horizon in the direction of her neighbor's house in order to see over the house? **≈26.57°**

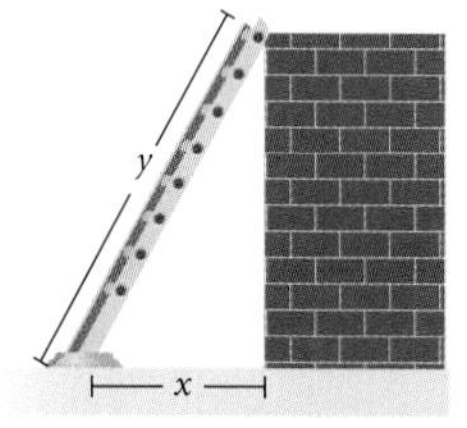

49. CARPENTRY When using a ladder, it is recommended that the distance from the base of the ladder to the structure, x, be $\frac{1}{4}$ of the distance from the base of the ladder to the support for the top of the ladder, y. Find the measure of the angle formed by the ladder and the ground. **≈75.5°**

50. ARCHITECTURE A person's eyes are 6 feet above the ground and 15 feet from a building. The angle of elevation from the person's line of sight to the top of the building is 75°.

a. How tall is the building? **≈62 feet**

b. What would be the angle of elevation if the person were standing 50 feet from the building? 100 feet? **≈48°; ≈29°**

47. Let x be the height of the tree.

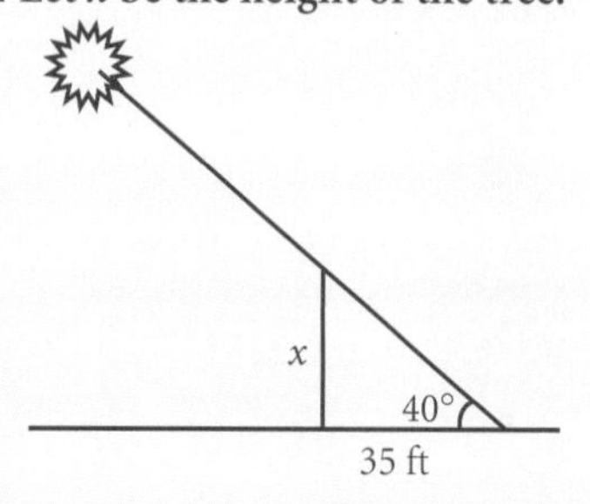

Practice

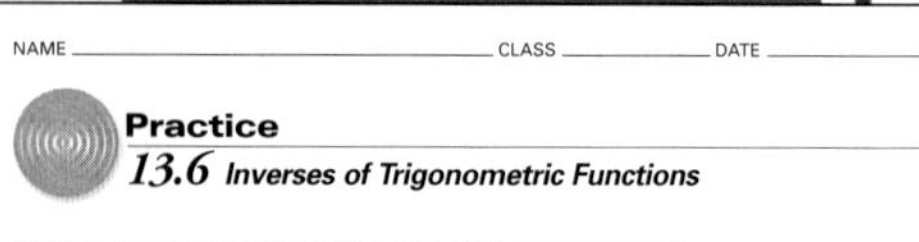

NAME ______ CLASS ______ DATE ______

Practice
13.6 Inverses of Trigonometric Functions

Evaluate each trigonometric expression. Give exact answers in degrees.

1. $\text{Sin}^{-1}\left(-\frac{\sqrt{3}}{2}\right)$ −60°	2. $\text{Cos}^{-1}\left(-\frac{1}{2}\right)$ 120°	3. $\text{Tan}^{-1}0$ 0°
4. $\text{Cos}^{-1}1$ 0°	5. $\text{Sin}^{-1}\left(-\frac{\sqrt{2}}{2}\right)$ −45°	6. $\text{Tan}^{-1}(-\sqrt{3})$ −60°

Evaluate each trigonometric expression.

7. $\cos(\text{Sin}^{-1}1)$ 0	8. $\tan\left(\text{Cos}^{-1}\left(-\frac{\sqrt{2}}{2}\right)\right)$ −1	9. $\sin(\text{Tan}^{-1}\sqrt{3})$ $\frac{\sqrt{3}}{2}$
10. $\text{Cos}^{-1}\left(\sin\frac{\pi}{3}\right)$ $\frac{\pi}{6}$ radians	11. $\text{Tan}^{-1}(\cos(-720°))$ −45°	12. $\text{Sin}^{-1}\left(\tan\frac{5\pi}{4}\right)$ $\frac{\pi}{2}$ radians

Find each value. Give answers in radians, rounded to the nearest ten-thousandth.

13. $\text{Tan}^{-1}40.2356$ 1.5459 radians	14. $\text{Sin}^{-1}0.0345$ 0.0345 radians	15. $\text{Cos}^{-1}(-0.8114)$ 2.5173 radians
16. $\text{Cos}^{-1}0.7756$ 0.6831 radians	17. $\text{Tan}^{-1}(-38.2004)$ −1.5446 radians	18. $\text{Sin}^{-1}(-0.5454)$ −0.5769 radians

Use inverse trigonometric functions to solve each problem.

19. A ramp that is 18 feet long rises to a loading platform that is 3 feet above the ground. Find, to the nearest tenth of a degree, the angle that the ramp makes with the ground. 9.6°

20. At one point in the day, a tower that is 150 feet high casts a shadow that is 210 feet long. Find, to the nearest tenth of a degree, the angle of elevation of the sun at that point. 35.5°

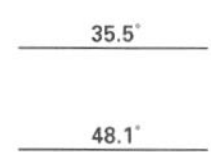

21. A kite is flying 67 meters above the ground, and its string is 90 meters long. Find the angle, to the nearest tenth of a degree, that the kite string makes with the horizontal. 48.1°

APPLICATIONS

51. HIKING Sandra wants to find the height of a mountain. From her first location on the ground, she finds the angle of elevation to the top of the mountain to be about 35°. After moving 1000 meters closer to the mountain on level ground, she finds the angle of elevation to be 50°. Find the height of the mountain to the nearest meter. (Hint: Find two equations with two unknowns, and solve them for each unknown.) **≈1698 feet**

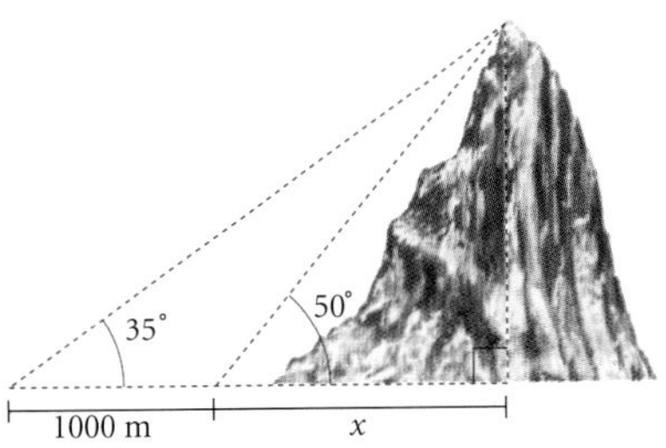

52. PUBLIC SAFETY A driver approaching an intersection sees the word *STOP* written on the road. The driver's eyes are 3.5 feet above the road and 50 feet from the nearest edge of the 8-foot long letters. What angle, θ, does the word make with the driver's eyes? **≈0.55°**

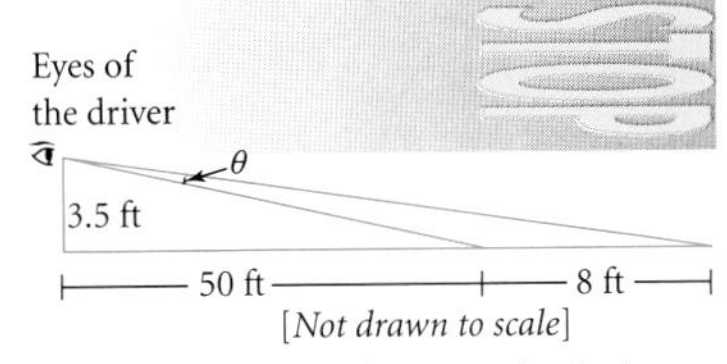

Eyes of the driver

θ

3.5 ft

50 ft — 8 ft

[Not drawn to scale]

To simulate the driver's perspective, look at the page from its left edge.

Look Back

53 Solve the system of equations by using a matrix equation. ***(LESSON 4.4)*** **(3, 5, −1)**

$$\begin{cases} 2x + 3y + 5z = 16 \\ -4x + 2y + 3z = -5 \\ 3x - y - z = 5 \end{cases}$$

Solve each equation for *x*. ***(LESSON 6.7)***

54. $5e^{2x-1} = 60$ **≈1.74**

55. $2 \log_3 x + \log_3 9 = 4$ **3**

56. Find the zeros of the function $f(x) = x^3 + 5x^2 - 8x - 12$. ***(LESSON 7.5)*** **−6, −1, and 2**

Find the reference angle for each angle below. ***(LESSON 13.2)***

57. 337° **23°**

58. −118° **62°**

59. −23° **23°**

60. 520° **20°**

Convert from radians to degrees. ***(LESSON 13.4)***

61. $\frac{5\pi}{12}$ radians **75°**

62. $\frac{7\pi}{8}$ radians **157.5°**

63. 2.38 radians **≈136.4°**

64. 4.72 radians **≈270.4°**

Look Beyond

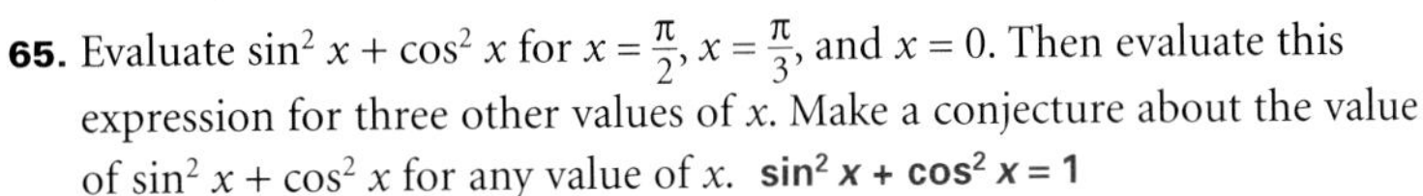

65. Evaluate $\sin^2 x + \cos^2 x$ for $x = \frac{\pi}{2}$, $x = \frac{\pi}{3}$, and $x = 0$. Then evaluate this expression for three other values of x. Make a conjecture about the value of $\sin^2 x + \cos^2 x$ for any value of x. $\sin^2 x + \cos^2 x = 1$

66. Verify that $\frac{\sin A}{4} = \frac{\sin B}{5} = \frac{\sin C}{3}$ is true for the triangle at left.

(Triangle: A, B, C; AB = 3, BC = 4, AC = 5)

Look Beyond

In these exercises, students look at numerical examples of some common trig expressions and make conjectures about their values. Exercise 65 gives students the expression $\sin^2 x + \cos^2 x$. The related identity will be studied in Lesson 14.3. In Exercise 66, students are introduced to the law of sines, which will be studied in Lesson 14.2.

66. $\frac{\sin A}{4} = \frac{\sin B}{5} = \frac{\sin C}{3} = \frac{1}{5}$

Student Technology Guide

NAME ____ CLASS ____ DATE ____

Student Technology Guide
13.6 Inverses of Trigonometric Functions

When you know the value of a trigonometric ratio and you need to find an angle, use a trigonometric inverse function. A calculator can find values of trigonometric inverses.

Example: Find Cos^{-1} 0.1238 to the nearest degree.

- Press MODE, use the arrow keys to move to Degree, and then press ENTER. Press 2nd MODE.
- Press 2nd COS .1238) ENTER.

To the nearest degree, Cos^{-1} 0.1238 = 83°.

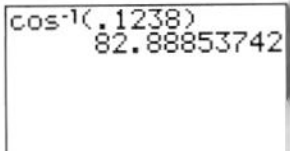

Find each value. Give answers to the nearest degree.

1. Cos^{-1} 0.5 — 60°
2. $\text{Sin}^{-1}(-0.89)$ — −63°
3. Tan^{-1} 15.3 — 86°
4. Cos^{-1} 15.3 — undefined

Example: Find $\text{Tan}^{-1}(-7.58)$ to the nearest ten thousandth of a radian.

- Press MODE, use the arrow keys to move to Radian, and then press ENTER. Press 2nd MODE.
- Press 2nd TAN (-) 7.58) ENTER.

To the nearest ten-thousandth, $\text{Tan}^{-1}(-7.58) = -1.4396$.

Find each value. Give answers to the nearest ten-thousandth of a radian.

5. Cos^{-1} 0.125 — 1.4455
6. $\text{Sin}^{-1}(-0.7071)$ — −0.7854
7. $\text{Tan}^{-1}(-31.7)$ — −1.5393
8. Sin^{-1} 0.89 — 1.0973

Example: The height of a building is 108 feet. To the nearest degree, what is the angle of elevation to the top of the building from a distance of 150 feet?

- Since $\tan A = \frac{108}{150}$, $m\angle A = \tan^{-1}\frac{108}{150}$. Make sure that the calculator is in degree mode and press 2nd TAN 108 ÷ 150) ENTER. You will find that $m\angle A$ is about 36°.

9. In the example above, find the angle of elevation if the building is 80 feet tall. $m\angle A \approx 28°$

Focus

As a person rides a Ferris wheel that is rotating at a constant rate, the distance between a person and the ground can be modeled by a sine function. The Ferris wheel presents a nice model of a unit circle, and the sine function represents the seat's distance from the ground as it rotates around the Ferris wheel. Students are given the dimensions of a Ferris wheel in Chicago and are asked to create a sinusoidal model for its motion.

Motivate

Have students discuss the similarities between a Ferris wheel ride and a sine function. The discussion should involve the concept of a periodic function returning to the same position at fixed intervals. Have students discuss the Ferris wheel as a model of the unit circle.

CHAPTER PROJECT THIRTEEN

Reinventing the Wheel

Navy Pier's most visible attraction, a 150-foot Ferris wheel, is modeled after the very first Ferris wheel, which was built for Chicago's 1893 World Colombian Exposition. The Navy Pier Ferris wheel offers unparalleled views of Chicago's skyline and lakefront. With 40 gondolas that seat 6 passengers each, the Ferris wheel can accommodate up to 240 passengers at a time. In the evening, the Ferris wheel's 40 spokes, spanning a diameter of 140 feet, are illuminated by thousands of lights. The Ferris wheel takes approximately 440 seconds to make one complete revolution.

Activity 1

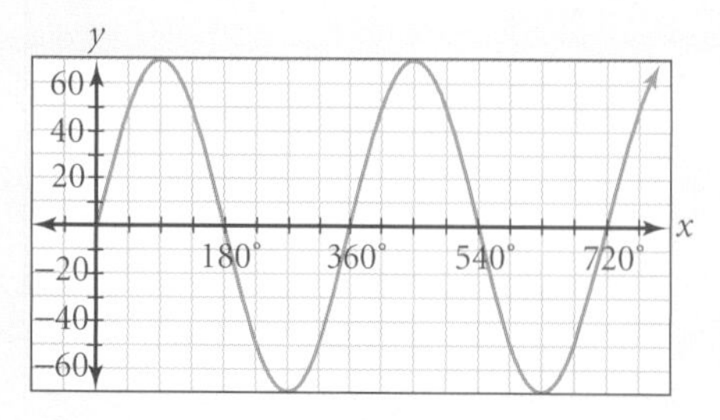

Activity 2

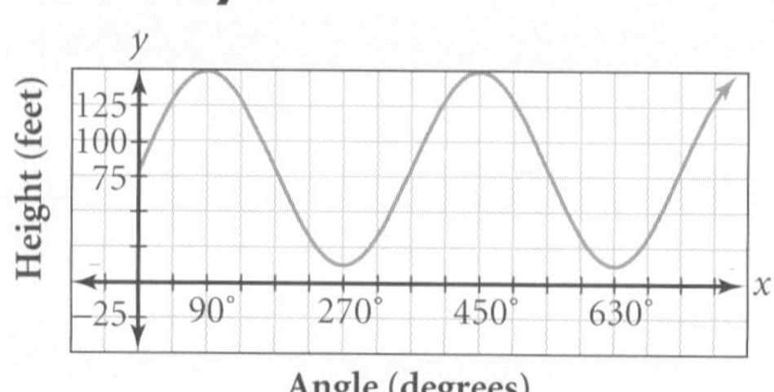

George W. G. Ferris (1859–1896)

Activity 1

Graph a representation of the Navy Pier Ferris wheel on a coordinate plane with its center at the origin. Create a table of values for the distance from a point on the Ferris wheel to the x-axis. Use the following angles of counterclockwise rotation: $0°$, $90°$, $180°$, $270°$, . . . , $810°$. Graph the resulting ordered pairs, and sketch a *smooth curve* through the points.

Activity 2

Create a table of values for the altitude (height above the ground) of a rider on the Ferris wheel. Let the independent variable include all common angle measures for θ such that $0° \le \theta \le 810°$. Graph the points, and sketch a *smooth curve* through the points.

Activity 3

1. Using the table of values that you created in Activity 2 and the fact that one complete revolution takes 440 seconds, convert the units of the independent variable from degrees to time in seconds. Graph the points, and sketch a *smooth curve* through the points.
2. Find the linear speed, in miles per hour, of a rider on the Ferris wheel.

Activity 4

Write an equation of the form $y = a \sin bt$ to model the altitude of a rider on the Ferris wheel as a function of time. Describe what each variable in your model represents.

Cooperative Learning

Students should work in groups of three or four for this activity. Students should alternate in the roles of following the point around the Ferris wheel, calculating positions, and plotting the data points on graph paper.

Discuss

After each step, have the class come together to discuss their findings. Have each group draw their graphs on a large piece of graph paper and display the graphs around the room. Then students can quickly see the graphs of each group and discuss any differences. As students progress through the steps, have them compare their graphs with those from the previous Portfolio Activities in this chapter. Students might extend this project to calculate the amount of possible income from the Ferris wheel. Students should decide on a price and length of time per ride. Using time as the independent variable and price per ride as the dependent variable in the model, students can estimate the revenue earned on a given day.

Activity 3

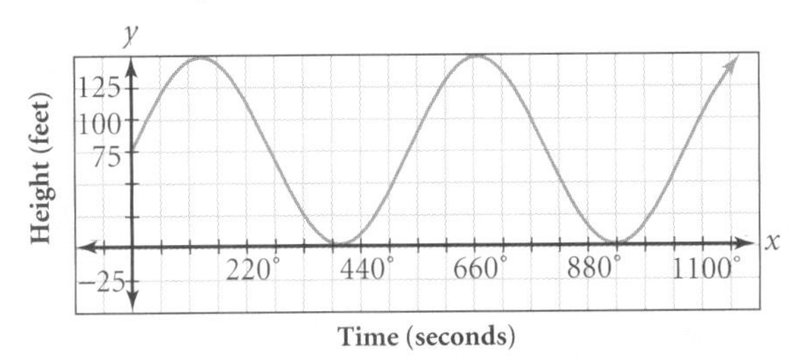

2. **0.68 mph**

Activity 4

$y = 70 \sin \frac{9}{11}t + 80$

$a = 70$: radius of the Ferris wheel

$b = \frac{9}{11}$: conversion from seconds to degrees

$d = 80$: translation of center of Ferris wheel above the ground

The tables of values for Activities 1–3 can be found in Additional Answers beginning on page 1002.

Chapter Review

13 Chapter Review and Assessment

VOCABULARY

Chapter Test, Form A

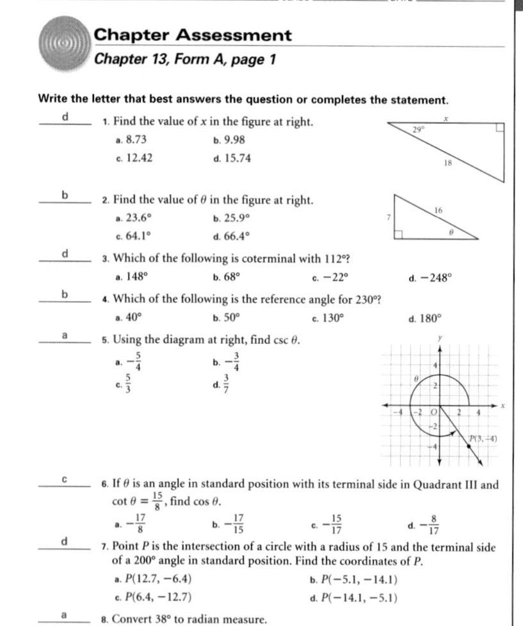

NAME ______ CLASS ______ DATE ______

Chapter Assessment
Chapter 13, Form A, page 1

Write the letter that best answers the question or completes the statement.

d 1. Find the value of x in the figure at right.
a. 8.73 b. 9.98
c. 12.42 d. 15.74

b 2. Find the value of θ in the figure at right.
a. 23.6° b. 25.9°
c. 64.1° d. 66.4°

d 3. Which of the following is coterminal with 112°?
a. 148° b. 68° c. −22° d. −248°

b 4. Which of the following is the reference angle for 230°?
a. 40° b. 50° c. 130° d. 180°

a 5. Using the diagram at right, find csc θ.
a. $-\frac{5}{4}$ b. $-\frac{3}{4}$
c. $\frac{5}{3}$ d. $\frac{3}{7}$

c 6. If θ is an angle in standard position with its terminal side in Quadrant III and $\cot\theta = \frac{15}{8}$, find $\cos\theta$.
a. $-\frac{17}{8}$ b. $-\frac{17}{15}$ c. $-\frac{15}{17}$ d. $-\frac{8}{17}$

d 7. Point P is the intersection of a circle with a radius of 15 and the terminal side of a 200° angle in standard position. Find the coordinates of P.
a. $P(12.7, -6.4)$ b. $P(-5.1, -14.1)$
c. $P(6.4, -12.7)$ d. $P(-14.1, -5.1)$

a 8. Convert 38° to radian measure.
a. 0.66 b. 12.10 c. 320.86 d. 2177.24

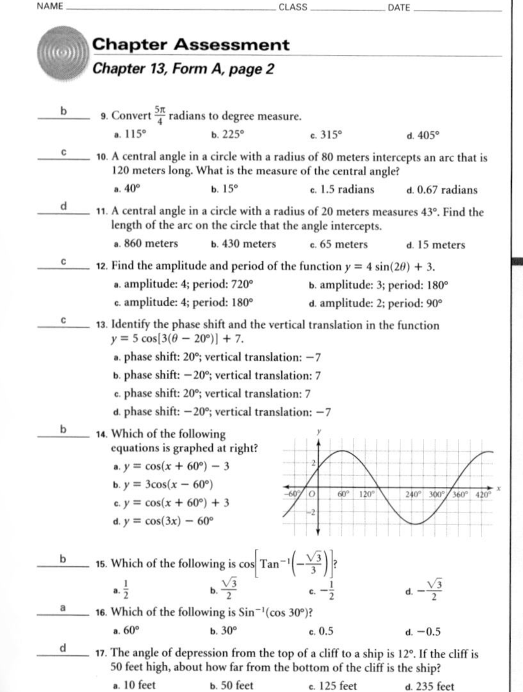

NAME ______ CLASS ______ DATE ______

Chapter Assessment
Chapter 13, Form A, page 2

b 9. Convert $\frac{5\pi}{4}$ radians to degree measure.
a. 115° b. 225° c. 315° d. 405°

c 10. A central angle in a circle with a radius of 80 meters intercepts an arc that is 120 meters long. What is the measure of the central angle?
a. 40° b. 15° c. 1.5 radians d. 0.67 radians

d 11. A central angle in a circle with a radius of 20 meters measures 43°. Find the length of the arc on the circle that the angle intercepts.
a. 860 meters b. 430 meters c. 65 meters d. 15 meters

c 12. Find the amplitude and period of the function $y = 4\sin(2\theta) + 3$.
a. amplitude: 4; period: 720° b. amplitude: 3; period: 180°
c. amplitude: 4; period: 180° d. amplitude: 2; period: 90°

c 13. Identify the phase shift and the vertical translation in the function $y = 5\cos[3(\theta - 20°)] + 7$.
a. phase shift: 20°; vertical translation: −7
b. phase shift: −20°; vertical translation: 7
c. phase shift: 20°; vertical translation: 7
d. phase shift: −20°; vertical translation: −7

b 14. Which of the following equations is graphed at right?
a. $y = \cos(x + 60°) - 3$
b. $y = 3\cos(x - 60°)$
c. $y = \cos(x + 60°) + 3$
d. $y = \cos(3x) - 60°$

b 15. Which of the following is $\cos\left[\text{Tan}^{-1}\left(-\frac{\sqrt{3}}{3}\right)\right]$?
a. $\frac{1}{2}$ b. $\frac{\sqrt{3}}{2}$ c. $-\frac{1}{2}$ d. $-\frac{\sqrt{3}}{2}$

a 16. Which of the following is $\text{Sin}^{-1}(\cos 30°)$?
a. 60° b. 30° c. 0.5 d. −0.5

d 17. The angle of depression from the top of a cliff to a ship is 12°. If the cliff is 50 feet high, about how far from the bottom of the cliff is the ship?
a. 10 feet b. 50 feet c. 125 feet d. 235 feet

Key Skills & Exercises

LESSON 13.1

Key Skills

Solve a right triangle by using trigonometric functions.

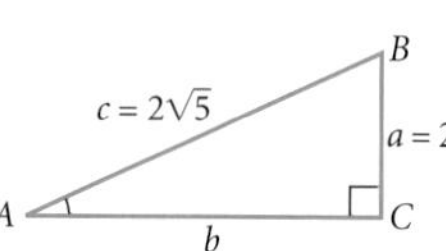

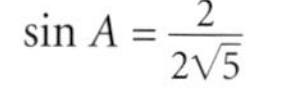

$\sin A = \frac{2}{2\sqrt{5}}$

$A = \sin^{-1}\frac{2}{2\sqrt{5}}$

$A \approx 26.6°$

$A + B = 90°$

$B \approx 90° - 26.6° \approx 63.4°$

$2^2 + b^2 = (2\sqrt{5})^2$

$b = \sqrt{(2\sqrt{5})^2 - 2^2}$

$b = 4$

Exercises

Solve each triangle.

1. $a = \sqrt{5}$, $A \approx 48.2°$, $B \approx 41.8°$

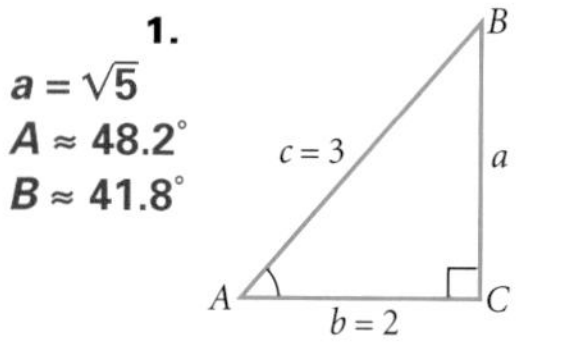

2. $c \approx 7.8$, $a \approx 4.5$, $B = 55°$

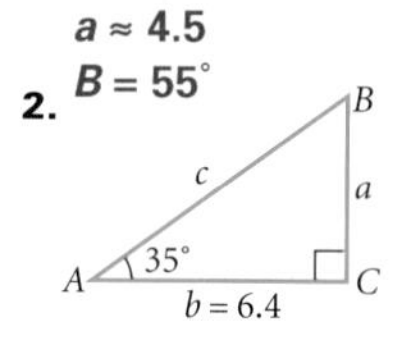

3.

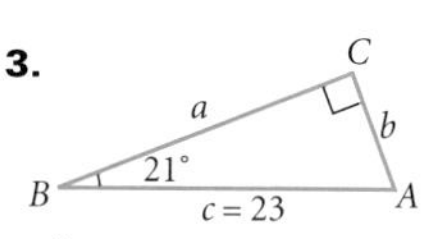

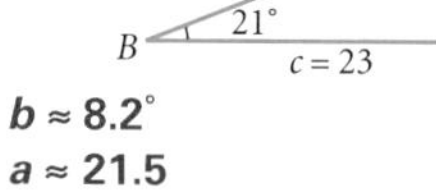

$b \approx 8.2°$, $a \approx 21.5$, $A = 69°$

4. $c = 0.5$, $A \approx 36.9°$, $B \approx 53.1°$

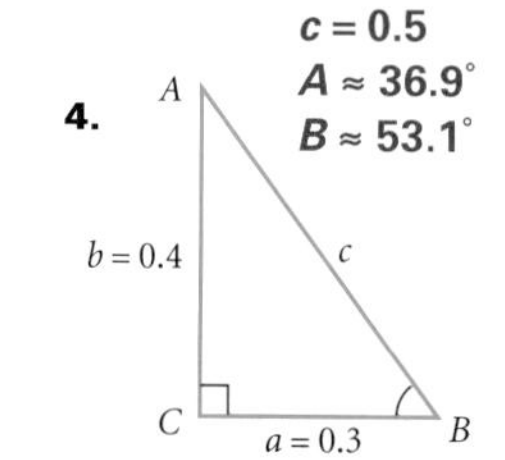

LESSON 13.2

Key Skills

Find coterminal angles and reference angles.

Coterminal angles of 480°, such that $-360° < \theta < 360°$, are found as follows:

$$480° - 360° = 120°$$
$$480° - (2)(360°) = -240°$$

$$\theta_{ref} = |180° - 120°| = 60°$$

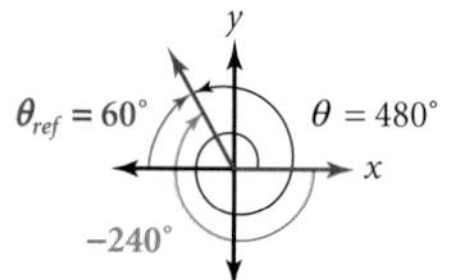

Exercises

For each angle below, find all coterminal angles such that $-360° < \theta < 360°$. Then find their corresponding reference angles.

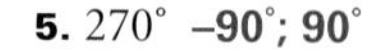

5. 270° −90°; 90°
6. 150° −210°; 30°
7. −135° 225°; 45°
8. −225° 135°; 45°
9. 380° 20°, −340°; 20°
10. 440° 80°, −280°; 80°
11. 1028° 308°, −52°; 52°
12. 973° 253°, −107°; 73°
13. −515° 205°, −155°; 25°
14. −612° 108°, −252°; 72°

Find the trigonometric function values of angles in standard position.

The terminal side of θ in standard position is in Quadrant IV, and $\cos\theta = \frac{5}{13}$. To find $\sin\theta$, find the length of the longer leg. Use the Pythagorean Theorem.

$$y = \pm\sqrt{13^2 - 5^2} = -12$$

Because P is in Quadrant IV, y is negative.

Thus, $\sin\theta = \frac{y}{r} = -\frac{12}{13}$.

Find the exact values of the six trigonometric functions of θ, given each point on the terminal side of θ in standard position.

15. $P(3, -4)$ **16.** $P(-2, 5)$

17. $P(-1, -8)$ **18.** $P(6, 2)$

Given the quadrant of θ in standard position and a trigonometric function value of θ, find exact values for the indicated functions.

19. III, $\sin\theta = -\frac{2}{7}$; $\tan\theta$ $\frac{2\sqrt{5}}{15}$

20. IV, $\cos\theta = \frac{1}{3}$; $\sin\theta$ $-\frac{2\sqrt{2}}{3}$

21. I, $\tan\theta = 1$; $\cos\theta$ $\frac{\sqrt{2}}{2}$

22. II, $\tan\theta = -\sqrt{3}$; $\sin\theta$ $\frac{\sqrt{3}}{2}$

LESSON 13.3

Key Skills

Find exact values for trigonometric functions of special angles and their multiples.

$$\cos 390° = \cos(390° - 360°) = \cos 30° = \frac{\sqrt{3}}{2}$$

Find the coordinates of a point on a circle given an angle of rotation and the radius of the circle.

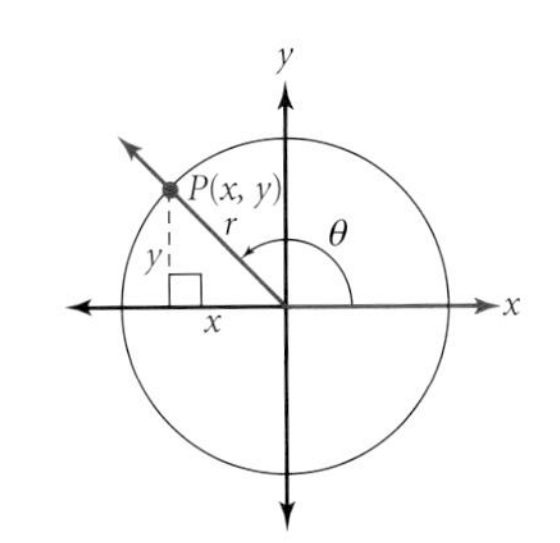

The coordinates of $P(x, y)$ shown at right are $P(r\cos\theta, r\sin\theta)$.

Exercises

Find each trigonometric function value. Give exact answers.

23. $\cos 135°$ $-\frac{\sqrt{2}}{2}$ **24.** $\sin 315°$ $-\frac{\sqrt{2}}{2}$

25. $\tan 225°$ **1** **26.** $\cos 0°$ **1**

27. $\sin(-270°)$ **1** **28.** $\cos(-180°)$ **−1**

29. $\tan(-90°)$ **undefined** **30.** $\cos 675°$ $\frac{\sqrt{2}}{2}$

31. $\sin 600°$ $-\frac{\sqrt{3}}{2}$ **32.** $\tan 765°$ **−1**

Point P is located at the intersection of a circle with a radius of r and the terminal side of angle θ in standard position. Find the exact coordinates of P.

33. $\theta = 60°, r = 1$ $\left(\frac{1}{2}, \frac{\sqrt{3}}{2}\right)$ **34.** $\theta = -30°, r = 2$ $(\sqrt{3}, -1)$

35. $\theta = 240°, r = 5$ $\left(-\frac{5}{2}, -\frac{5\sqrt{3}}{2}\right)$ **36.** $\theta = -240°, r = 3$ $\left(-\frac{3}{2}, \frac{3\sqrt{3}}{2}\right)$

LESSON 13.4

Key Skills

Convert from degrees to radians and vice versa.

Multiply degrees by $\frac{\pi \text{ radians}}{180°}$.

Multiply radians by $\frac{180°}{\pi \text{ radians}}$.

Find arc length.

Use $s = r\theta$, where s is the arc length, r is the radius, and θ is the angle measure in radians.

Exercises

Convert each degree measure to radian measure, giving exact answers. Convert each radian measure to degree measure, rounding answers to the nearest tenth of a degree.

37. $78°$ $\frac{13\pi}{30}$ **38.** $334.61°$ $\frac{33{,}461\pi}{18{,}000}$ **39.** $-23°$ $-\frac{23\pi}{180}$

40. $\frac{\pi}{7}$ radians **25.7°** **41.** $-\frac{15\pi}{16}$ radians **−450°** **42.** 8.87 radians **508.2°**

43. Find the length of the arc intercepted by a central angle of 30° in a circle with a radius of 4.5 meters. **≈2.36 meters**

---

17. $\sin\theta = -\frac{8\sqrt{65}}{65}$ $\csc\theta = -\frac{\sqrt{65}}{8}$

$\cos\theta = -\frac{\sqrt{65}}{65}$ $\sec\theta = -\sqrt{65}$

$\tan\theta = 8$ $\cot\theta = \frac{1}{8}$

18. $\sin\theta = \frac{\sqrt{10}}{10}$ $\csc\theta = \sqrt{10}$

$\cos\theta = \frac{3\sqrt{10}}{10}$ $\sec\theta = \frac{\sqrt{10}}{3}$

$\tan\theta = \frac{1}{3}$ $\cot\theta = 3$

15. $\sin\theta = -\frac{4}{5}$ $\csc\theta = -\frac{5}{4}$

$\cos\theta = \frac{3}{5}$ $\sec\theta = \frac{5}{3}$

$\tan\theta = -\frac{4}{3}$ $\cot\theta = -\frac{3}{4}$

16. $\sin\theta = \frac{5\sqrt{29}}{29}$ $\csc\theta = \frac{\sqrt{29}}{5}$

$\cos\theta = -\frac{2\sqrt{29}}{29}$ $\sec\theta = -\frac{\sqrt{29}}{2}$

$\tan\theta = -\frac{5}{2}$ $\cot\theta = -\frac{2}{5}$

Chapter Test, Form B

NAME ____ CLASS ____ DATE ____

Chapter Assessment

Chapter 13, Form B, page 1

1. Find x and y to the nearest tenth in the triangle at right. $x \approx 64.3$; $y \approx 71.0$

2. Find θ to the nearest tenth of a degree in the triangle at right. $\theta \approx 36.9°$

Find the reference angle, θ_{ref}, for each angle.

3. −74° 74° 4. 105° 75° 5. 275° 85°

Evaluate each expression. Round answers to the nearest ten-thousandth.

6. $\tan\left(-\frac{2\pi}{5}\right)$ −3.0777 7. $\cos 274°$ 0.0698 8. $\sin 6.2$ −0.0831

9. The point P is the intersection of a circle with a radius of 30 and the terminal side of a 310° central angle. Find the coordinates of P to the nearest tenth. $P(19.3, -23.0)$

10. The terminal side of θ lies in Quadrant II, and $\sin\theta = \frac{20}{29}$. Find exact values for $\tan\theta$ and $\sec\theta$. $\tan\theta = -\frac{20}{21}$; $\sec\theta = -\frac{29}{21}$

11. Find exact values for $\csc\theta$ and $\cot\theta$ if θ is in standard position and $(-12, 5)$ is on the terminal side of θ. $\csc\theta = \frac{13}{5}$; $\cot\theta = -\frac{12}{5}$

12. Find the exact values for the sine and cosine of 1650°. $\sin 1650° = -\frac{1}{2}$; $\cos 1650° = -\frac{\sqrt{3}}{2}$

13. Find an angle between −360° and 360° that is coterminal with 250°. −110°

Convert to radian measure. Give exact answers.

14. 240° $\frac{4\pi}{3}$ 15. 315° $\frac{7\pi}{4}$

Convert to degree measure. Round answers to the nearest degree.

16. $\frac{7\pi}{2}$ 630° 17. 2.6 149°

NAME ____ CLASS ____ DATE ____

Chapter Assessment

Chapter 13, Form B, page 2

18. A central angle in a circle with a radius of 60 centimeters intercepts an arc that is 75 centimeters long. Find the measure of the central angle in radians. 1.25 radians

19. A central angle in a circle with a radius of 25 meters measures 200°. Find the length of the arc on the circle that the angle intercepts to the nearest tenth of a meter. 87.3 meters

20. Graph one period of $y = 5\cos(x + 30°)$.

21. Graph one period of $y = \sin\left(\frac{3}{4}x\right) - 2$.

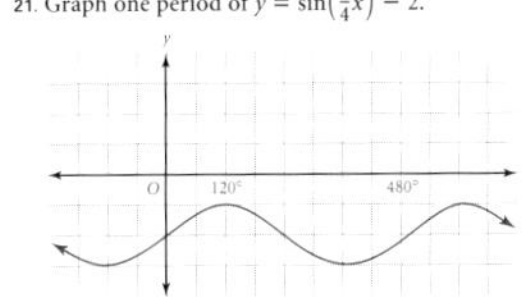

22. Find the phase shift and amplitude of $y = 7.2\sin[4(x - 10°)] - 9$. phase shift: 10°; amplitude: 7.2

Evaluate each composite trigonometric function. Give exact answers.

23. $\tan\left[\left(\text{Cos}^{-1}\left(-\frac{1}{2}\right)\right)\right]$ $-\sqrt{3}$ 24. $\text{Sin}^{-1}(\cos 25°)$ 65°

25. The angle of depression from the top of a building to a street vendor is 34°. If the building is 30 feet high, how far, to the nearest tenth of a foot, from the bottom of the building is the vendor? 44.5 feet

48. 90° to the right; no vertical translation

49. 45° to the left; no vertical translation

50. 215° to the right; 3 up

51. 120° to the left; 2 down

52. $y = \cos 4\theta$ is $y = \cos \theta$ compressed horizontally by a factor of $\frac{1}{4}$.

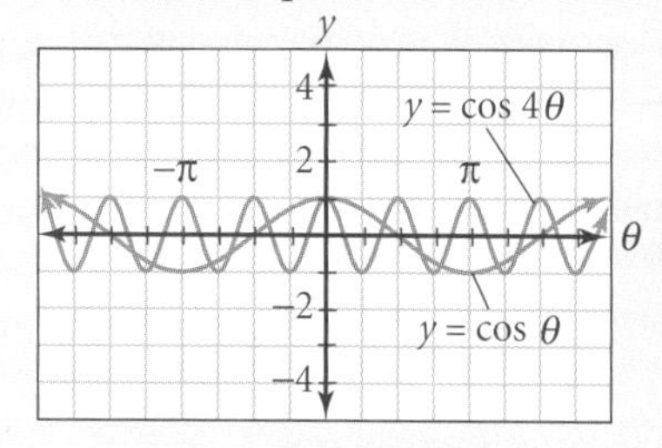

53. $y = \tan \frac{\theta}{2}$ is $\tan \theta$ stretched horizontally by a factor of 2.

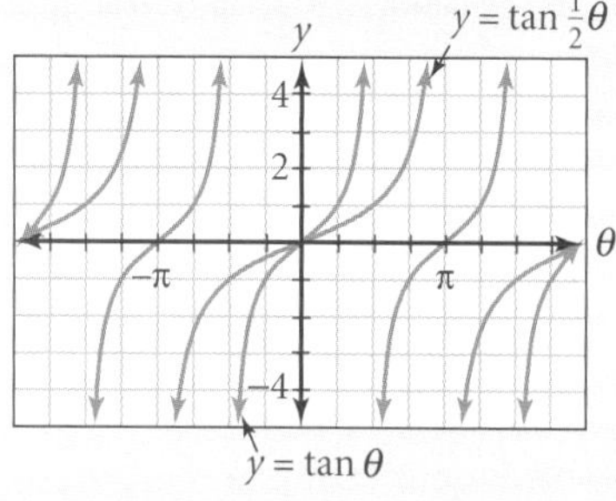

54. $y = 2 \sin(\theta - 45°)$ is $y = \sin \theta$ stretched vertically by a factor of 2 and translated 45° to the right.

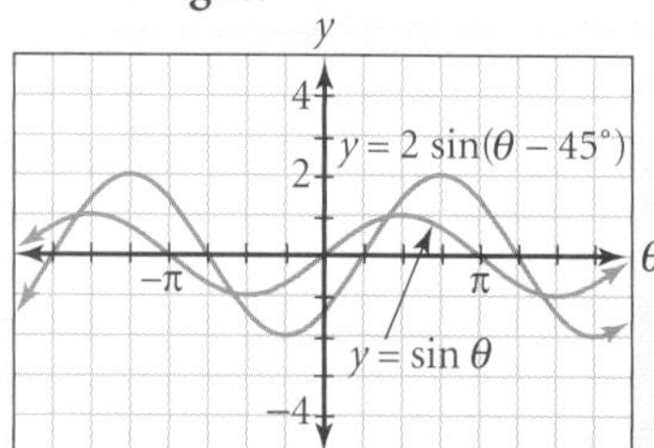

LESSON 13.5

Key Skills

Graph transformations of trigonometric functions.

For $y = a \sin(b\theta - c) + d$ and $y = a \cos(b\theta - c) + d$, $|a|$ is the amplitude; $\frac{2\pi}{|b|}$, where $b \neq 0$, is the period; $|c|$ is the phase shift (to the right for $c > 0$ and to the left for $c < 0$); and $|d|$ is the vertical shift (upward for $d > 0$ and downward for $d < 0$).

For $y = a \tan(b\theta - c) + d$, there is no amplitude, but a represents a vertical stretch or compression, and $\frac{\pi}{|b|}$, where $b \neq 0$, is the period. The phase shift and vertical shift are the same as those of sine and cosine.

Exercises

Identify the amplitude, if it exists, and the period of each trigonometric function.

44. $y = -3 \sin \theta$ **3; 2π**

45. $y = -4 \sin \theta$ **4; 2π**

46. $y = \frac{1}{2} \sin 3\theta$ **2; $\frac{2\pi}{3}$**

47. $y = 2 \tan \frac{1}{3}\theta$ **no amplitude; 3π**

Identify the phase shift and the vertical shift of each function from its parent function.

48. $y = 2 \sin(\theta - 90°)$

49. $y = -4 \cos(\theta + 45°)$

50. $y = \sin(\theta - 215°) + 3$

51. $y = \cos(\theta + 120°) - 2$

Describe the transformation of each function from its parent function. Then graph at least one period of the function along with its parent function.

52. $y = \cos 4\theta$

53. $y = \tan \frac{1}{2}\theta$

54. $2 \sin(\theta - 45°)$

55. $y = 3 \cos(\theta + 45°)$

LESSON 13.6

Key Skills

Evaluate trigonometric expressions involving inverses.

Find all possible values of $\sin^{-1} \frac{\sqrt{2}}{2}$.

For $0° \leq \theta < 360°$, $\sin^{-1} \frac{\sqrt{2}}{2} = 45°$ and $\sin^{-1} \frac{\sqrt{2}}{2} = 135°$.

Thus, all possible values are $45° + n360°$ and $135° + n360°$, where n is an integer.

Evaluate composite trigonometric functions.

$$\cos\left(\text{Sin}^{-1} \frac{\sqrt{3}}{2}\right) = \cos 60° = \frac{1}{2}$$

Exercises

Find all possible values for each expression.

56. $\sin^{-1} 0$

57. $\cos^{-1} 1$

58. $\tan^{-1}\left(-\frac{\sqrt{3}}{3}\right)$

59. $\cos^{-1}\left(-\frac{\sqrt{2}}{2}\right)$

60. $\sin^{-1}\left(-\frac{\sqrt{3}}{2}\right)$

61. $\tan^{-1} 1$

Evaluate each trigonometric expression.

62. $\text{Sin}^{-1} \frac{1}{2}$ **30°**

63. $\text{Cos}^{-1}\left(-\frac{\sqrt{3}}{2}\right)$ **150°**

64. $\text{Tan}^{-1} \sqrt{3}$ **60°**

65. $\text{Sin}^{-1}\left(-\frac{\sqrt{2}}{2}\right)$ **−45°**

66. $\tan\left(\text{Sin}^{-1} \frac{1}{2}\right)$ **$\frac{\sqrt{3}}{3}$**

67. $\cos[\text{Tan}^{-1}(-\sqrt{3})]$ **$\frac{1}{2}$**

68. $\text{Tan}^{-1}(\cos 135°)$ **≈−35.26°**

69. $\text{Sin}^{-1}[\sin(-120°)]$ **−60°**

Applications

70. MAPMAKING A map maker is making a map of a park. She would like to find the distance across a river. She marks a point on the bank of the river at C and then walks 200 meters up the river to point A, where she measures an angle of 50° between points B and C. What is the distance from point B to point C across the river? **≈238.4 meters**

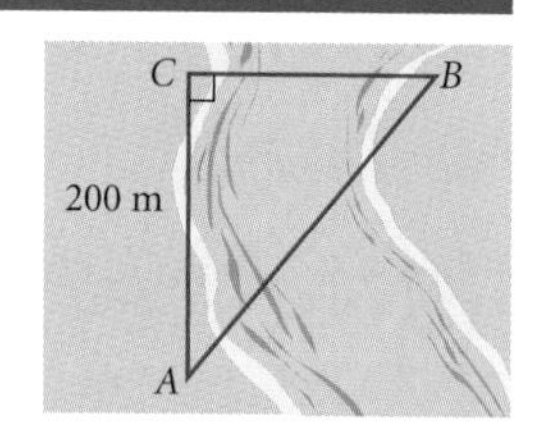

55. $y = 3 \cos(\theta + 45°)$ is $y = \cos \theta$ stretched vertically by a factor of 3 and translated 40° to the left.

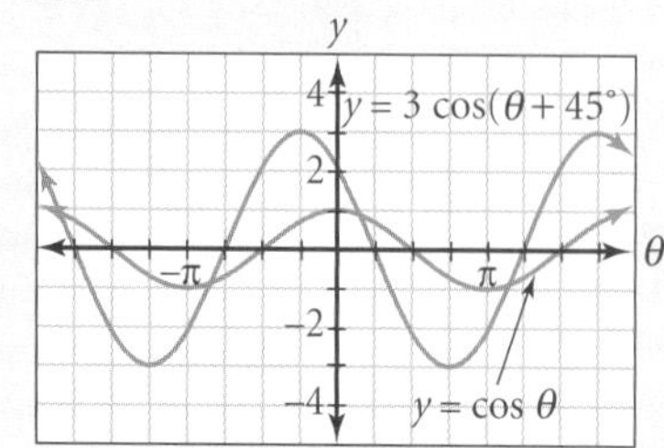

56. $n180°$ (or $0° + n360°$ and $180° + n360°$), where n is an integer

57. $n360°$, where n is an integer

58. $150° + n360°$ and $330° + n360°$, where n is an integer

59. $135° + n360°$ and $225° + n360°$, where n is an integer

60. $240° + n360°$ and $300° + n360°$, where n is an integer

61. $45° + n360°$ and $225° + n360°$, where n is an integer

Chapter Test

13

Solve each triangle. Give angle measures to the nearest degree and side lengths to the nearest tenth.

1.

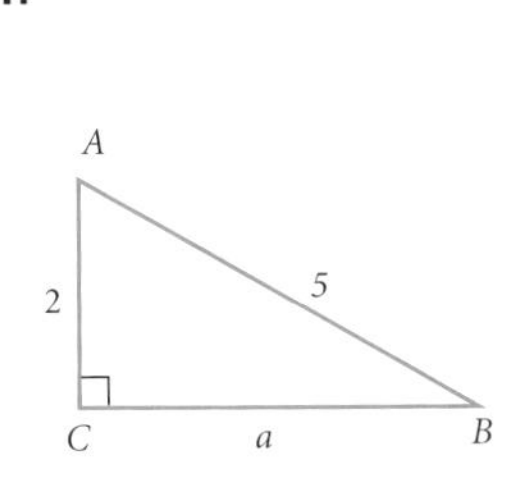

2.

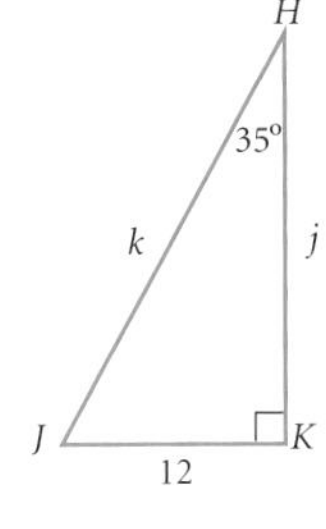

3. GEOMETRY A regular octagon is made by cutting off the corners of a square so that each triangle cut off is isosceles. What is the length of the side of an octagon made from a square of side length 10? **4.14**

For each angle below, find all coterminal angles such that $-360° < \theta < 360°$. Then find the corresponding reference angle, if it exists.

4. 137° **−223°; 43°**

5. 515° **155°; −205°; 25°**

6. 38° **−322°; 38°**

7. 1729° **289°; −71°; 71°**

Given the quadrant of θ in standard position and a trigonometric function value of θ, find exact values for the indicated functions.

8. IV, $\cos\theta = \frac{5}{13}$; $\sin\theta$ $-\frac{12}{13}$

9. II, $\tan\theta = -\frac{1}{2}$; $\sin\theta$ $\frac{1}{\sqrt{5}} = \frac{\sqrt{5}}{5}$

Find each trigonometric function value. Give exact answers.

10. sin 330° $-\frac{1}{2}$

11. cos (−150°) $-\frac{\sqrt{3}}{2}$

12. sin 720° **0**

13. tan (−765°) **−1**

14. cos 300° $\frac{1}{2}$

15. tan 270° **undefined**

Point P is located at the intersection of a circle with a radius of r and the terminal side of an angle θ in standard position. Find the exact coordinates of P.

16. $\theta = 30°, r = 5$ $\left(\frac{5\sqrt{3}}{2}, \frac{5}{2}\right)$

17. $\theta = 225°, r = 12$ $(-6\sqrt{2}, -6\sqrt{2})$

18. $\theta = -150°, r = 4$ $(-2\sqrt{3}, -2)$

19. $\theta = 300°, r = 8$ $(4, -4\sqrt{3})$

20. CONSTRUCTION A roofline has a $\frac{7}{12}$ pitch, meaning that the roof rises 7 feet for each 12 feet of horizontal distance. What angle does the roof make with the horizontal? **30.3°**

Convert each degree measure to radian measure or radian measure to degree measure. Give exact answers.

21. 315° $\frac{7\pi}{4}$

22. −150° $-\frac{5\pi}{6}$

23. 495° $\frac{11\pi}{4}$

24. $\frac{\pi}{12}$ **15°**

25. $\frac{5\pi}{4}$ **225°**

26. $\frac{-5\pi}{3}$ **−300°**

27. TECHNOLOGY A 12-inch LP vinyl record rotates at $33\frac{1}{3}$ rpm. When the record needle is placed 4 inches from the center of the record, what is the linear speed of the record at that point? **837.8 in./min or 69.8 ft./min.**

Identify the amplitude, if it exists, and the period of each function.

28. $y = 3 \sin 2\theta$ **3; π**

29. $y = \frac{2}{3} \tan 5\theta$ **5; 6π**

30. $y = 5 \sin \frac{1}{3}\theta$ **none; $\frac{\pi}{5}$**

31. $y = -3 \cos\left(\theta - \frac{\pi}{4}\right)$ **3; 2π**

Describe the transformations of each function from its parent function.

32. $y = \sin(\theta + 60°) + 2$

33. $y = -3 \cos(\theta - 300°)$

Find all possible values for each expression.

34. $\sin^{-1}\left(\frac{\sqrt{2}}{2}\right)$

35. $\cos^{-1}\left(\frac{1}{2}\right)$

36. $\tan^{-1}(1)$

37. $\tan^{-1}(\sqrt{3})$

Evaluate each trigonometric expression.

38. $\sin\left(\text{Cos}^{-1}\frac{1}{2}\right)$ $\frac{\sqrt{3}}{2}$

39. $\text{Tan}^{-1}(\sin 90°)$ **45°**

Chapter Test

1. $A = 66°; B = 24°; a = 4.6$

2. $J = 55°; k = 20.9; j = 17.1$

32. translated to the left 60° and up 2
33. reflected across the x-axis, stretched vertically by a factor of 3, and translated 300° to the right.
34. $45° + 360°n$ or $135° + 360°n$
35. $60° + 360°n$ or $300° + 360°n$
36. $45° + 360°n$ or $225° + 360°n$
37. $60° + 360°n$ or $240° + 360°n$

Cumulative Assessment

College Entrance Exam Practice

Multiple-Choice and Quantitative-Comparison Samples

The first half of the Cumulative Assessment contains two types of items found on standardized tests—multiple-choice questions and quantitative-comparison questions. Quantitative-comparison items emphasize the concepts of equality, inequality, and estimation.

Free-Response Grid Samples

The second half of the Cumulative Assessment is a free-response section. This part of the Cumulative Assessment requires student-produced response items like those commonly found on college entrance exams. These questions require the use of machine-scored answer grids. You may wish to have students practice answering these items in preparation for standardized tests.

CHAPTERS 1–13 CUMULATIVE ASSESSMENT

College Entrance Exam Practice

QUANTITATIVE COMPARISON For Items 1–6, write
A if the quantity in Column A is greater than the quantity in Column B;
B if the quantity in Column B is greater than the quantity in Column A;
C if the quantities are equal; or
D if the relationship cannot be determined from the given information.

internet connect
Standardized Test Prep Online
Go To: **go.hrw.com**
Keyword: **MM1 Test Prep.**

	Column A	Column B	Answers
1. A	6^{-2}	2^{-6}	Ⓐ Ⓑ Ⓒ Ⓓ [Lesson 2.2]
2. B	The x-coordinate of the center of the circle: $(x+2)^2+y^2=1$	$x^2+y^2-4x+4=1$	Ⓐ Ⓑ Ⓒ Ⓓ [Lesson 9.3]
3. A	The degree of the polynomial: $3x^3-2x^4+1$	$(3x-2)(2x+1)^2$	Ⓐ Ⓑ Ⓒ Ⓓ [Lesson 7.1]
4. B	$\frac{5\pi}{3}$ radians	$305°$	Ⓐ Ⓑ Ⓒ Ⓓ [Lesson 13.4]
5. C	${}_{10}C_2$	${}_{10}C_8$	Ⓐ Ⓑ Ⓒ Ⓓ [Lesson 10.3]
6. B	$\lvert -1\rvert - \lvert -3\rvert$	$[-2.01]-[-1.33]$	Ⓐ Ⓑ Ⓒ Ⓓ [Lesson 2.6]

7. The domain of $f(x)=\frac{2x-3}{x+1}$ includes all real numbers except which value? ***(LESSON 8.2)***
b
a. 1 **b.** -1 **c.** $-\frac{3}{2}$ **d.** $\frac{3}{2}$

8. Which expression is not equivalent to the others? ***(LESSON 6.4)***
d
a. $\log_3 (2x)^{\frac{1}{4}}$ **b.** $\frac{1}{4}\log_3 2x$
c. $\log_3 \sqrt[4]{2x}$ **d.** $3^{\frac{1}{4}\log_3 2x}$

9. Which represents $(2x^3-x^4)+(3x^2-5)-(x^2-x^4+1)$ as a polynomial in standard form? ***(LESSON 7.1)***
b
a. $-2x^4+2x^3+3x^2-6$
b. $2x^3+2x^2-6$
c. $2x^3+2x^2-4$
d. $2x^3+3x^2+4$

10. Which is the value of x at the vertex of the graph of $f(x)=2x^2-4x+1$? ***(LESSON 5.1)***
b
a. -1 **b.** 1 **c.** 2 **d.** $-\frac{1}{2}$

11. Which describes a number that cannot be written as the ratio of two integers? ***(LESSON 2.1)***
d
a. prime **b.** integer
c. rational **d.** irrational

12. Which of the following is the complex conjugate of $3-2i$? ***(LESSON 5.6)***
b
a. $-3+2i$ **b.** $3+2i$
c. $2i-3$ **d.** $-3-2i$

13. Solve $\lvert 2x+5\rvert = 11$. ***(LESSON 1.8)***
c
a. $3, -3$ **b.** $8, -8$ **c.** $3, -8$ **d.** $8, -3$

14. If y varies directly as x and y is 8 when x is 4,
b which of the following is the constant of variation? ***(LESSON 1.4)***

a. $\frac{1}{2}$ **b.** 2 **c.** 32 **d.** −2

15. If an entire population of 100 bacteria doubles
d every hour, how many bacteria are in the population after 3 hours? ***(LESSON 6.1)***

a. 200 **b.** 300 **c.** 400 **d.** 800

16. Which of the following is true of the ellipse
a given by $\frac{x^2}{36} + \frac{y^2}{4} = 1$? ***(LESSON 9.4)***

a. The major axis is horizontal.
b. The major axis is vertical.
c. The foci are $(0, \pm 4\sqrt{2})$.
d. The length of the major axis is 6.

17. Which equation defines the inverse of
c $f(x) = \frac{1}{3}x + 1$? ***(LESSON 2.5)***

a. $y = -\frac{1}{3}x + 1$ **b.** $y + 1 = \frac{1}{3}x$
c. $y + 3 = 3x$ **d.** $y + 3x = 3$

18. Graph $y \le \frac{1}{4}x - 6$. ***(LESSON 3.3)***

19. Evaluate $3(\sqrt{45})^2$. ***(LESSON 8.6)*** **135**

20. Write the function represented by the graph of $f(x) = x^2$ translated 3 units to the left. ***(LESSON 2.7)*** **$g(x) = (x + 3)^2$**

Use the diagram below to find each value. Give exact answers. ***(LESSON 13.1)***

21. $\sin A$ $\frac{4}{5}$

22. $\cos A$ $\frac{3}{5}$

23. $\tan B$ $\frac{3}{4}$

24. $\cot A$ $\frac{3}{4}$

25. $\sec B$ $\frac{5}{4}$

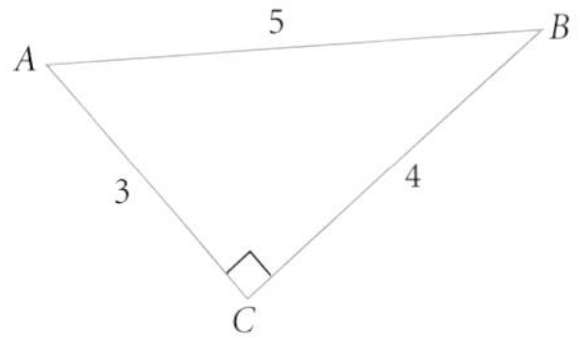

26. Use the quadratic formula to solve $5x^2 + x - 2 = 0$. ***(LESSON 5.5)*** $\frac{-1 \pm \sqrt{41}}{10}$

27. Write the pair of parametric equations as a single equation in only x and y. ***(LESSON 3.6)***

$$\begin{cases} x(t) = 3t \\ y(t) = 5 - 2t \end{cases}$$ **$2x + 3y = 15$**

28. Solve $\frac{6}{x-2} > \frac{5}{x-3}$. ***(LESSON 8.5)*** **$x > 8$**

29. Write an equation in standard form for the line that contains (−3, 4) and is perpendicular to $y = 3x - 5$. ***(LESSON 1.3)*** **$x + 3y = 9$**

30. Simplify $\frac{x}{x+4} \div \frac{6x^2}{3x+12}$. ***(LESSON 8.3)*** $\frac{1}{2x}$

31. State the domain of $f(x) = \sqrt{2 - 3x}$. $x \le \frac{2}{3}$ ***(LESSON 8.6)***

32. Factor $3y(5x + 2) - 4(5x + 2)$, if possible. ***(LESSON 5.3)*** **$(3y - 4)(5x + 2)$**

33. Find the product $\begin{bmatrix} 2 & -2 & 4 \\ -1 & 3 & 5 \end{bmatrix} \begin{bmatrix} 3 & 5 \\ 2 & -2 \\ -1 & 4 \end{bmatrix}$, if it exists. ***(LESSON 4.2)***

FREE-RESPONSE GRID The following questions may be answered by using a free-response grid such as that commonly used by standardized-test services.

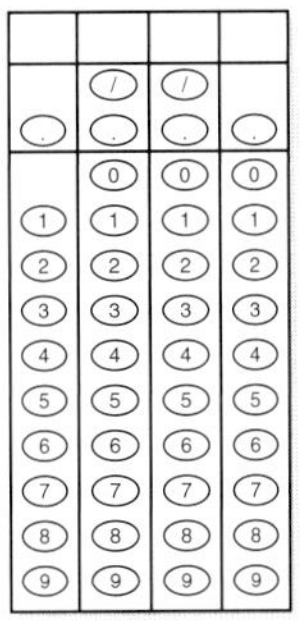

34. Solve $\ln(2x - 7) = \ln 13$. ***(LESSON 6.7)*** **10**

35. Evaluate $\sum_{n=1}^{3} 8$. **24** ***(LESSON 11.1)***

36. Evaluate $\sin\left(\text{Cos}^{-1}\frac{1}{2}\right)$. ***(LESSON 13.6)*** **0.866**

37. Give the exact value of $\sin\left(\frac{\pi}{2}\right)$. **1** ***(LESSON 13.4)***

38. Evaluate $\log_{10} 10^5 + 7^{\log_7 6}$. ***(LESSON 6.4)*** **11**

39. Find the reference angle, θ_{ref}, for 640°. **80°** ***(LESSON 13.2)***

40. Find the value of v in $\frac{1}{2} = \log_v 4$. **16** ***(LESSON 6.3)***

41. **PROBABILITY** In how many different ways can a committee chairman and vice-chairman be selected from 15 members? ***(LESSON 10.2)*** **210**

42. What is the minimum value of $f(x) = 3 + 2\cos\left(x + \frac{\pi}{2}\right)$? ***(LESSON 13.5)*** **1**

43. What is the 13th term of the arithmetic sequence 50, 46, 42, . . . ? ***(LESSON 11.2)*** **2**

18.

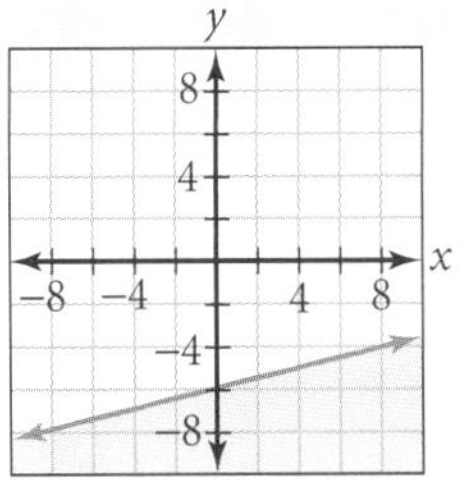

33. $\begin{bmatrix} -2 & 30 \\ -2 & 9 \end{bmatrix}$

Keystroke Guide for Chapter 13

Essential keystroke sequences (using the model TI-82 or TI-83 graphics calculator) are presented below for all Activities and Examples found in this chapter that require or recommend the use of a graphics calculator.

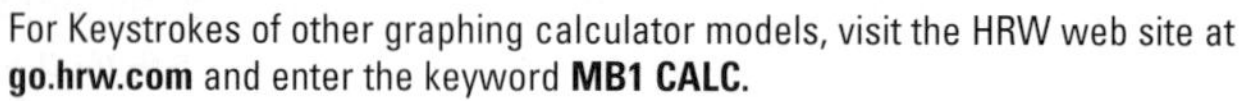

For Keystrokes of other graphing calculator models, visit the HRW web site at **go.hrw.com** and enter the keyword **MB1 CALC**.

LESSON 13.5

Page 858

For Step 1, use radian, parametric, and simultaneous modes to graph the parametric equations $x_1(t) = \cos t$ and $y_1(t) = \sin t$ in order to generate the unit circle and $x_2(t) = t$ and $y_2(t) = \sin t$ in order to generate the basic sine function.

Set the modes:

Set the viewing window:

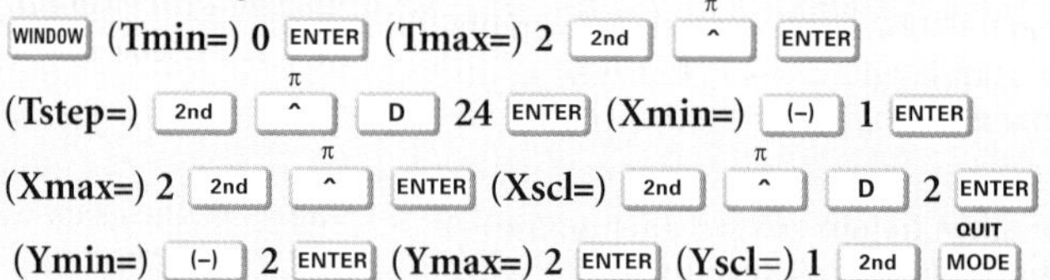

WINDOW (Tmin=) 0 ENTER (Tmax=) 2 2nd π ^ ENTER
(Tstep=) 2nd π ^ D 24 ENTER (Xmin=) (-) 1 ENTER
(Xmax=) 2 2nd π ^ ENTER (Xscl=) 2nd π ^ D 2 ENTER
(Ymin=) (-) 2 ENTER (Ymax=) 2 ENTER (Yscl=) 1 2nd QUIT MODE

Graph the parametric equations:

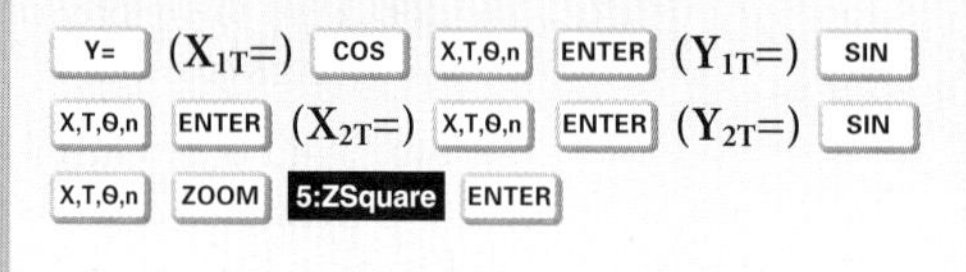

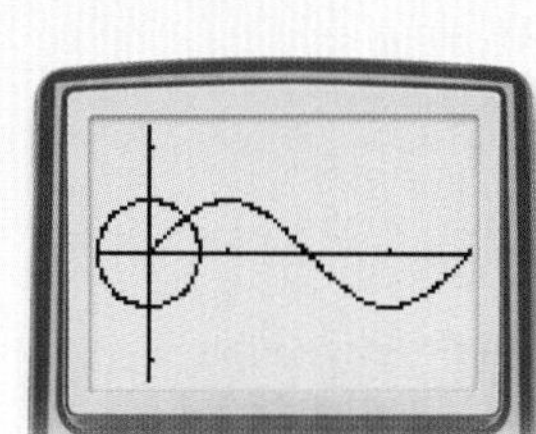

For Step 2, use the trace feature to compare the *x*-values and *y*-values for various *t*-values.

Use TRACE and the cursor keys. The ◄ and ► keys move the cursor along the graph. The ▲ and ▼ keys move the cursor to and from corresponding points on the two curves.

For part b of Step 3, graph the parametric equations $x_1(t) = 2 \cos t$, $y_1(t) = 2 \sin t$, $x_2(t) = t$, and $y_2(t) = 2 \sin t$.
Change Xmin to −2 in the viewing window, and use a keystroke sequence similar to that for Step 1.

For part c of Step 3, graph the parametric equations $x_1(t) = 3 \cos t$, $y_1(t) = 3 \sin t$, $x_2(t) = t$, and $y_2(t) = 3 \sin t$.
Change Xmin to −3 in the viewing window, and use a keystroke sequence similar to that for Step 1.

EXAMPLE 3 **In radian, function, and sequential modes, graph $y = 3 \sin 110\pi x$ and $y = 3 \sin 110\pi\left(x - \frac{1}{110}\right)$ on the same screen.**

Page 862

Use a viewing window $\left[-\frac{1}{110}, \frac{2}{110}\right]$ by [−4, 4] and an x-scale of $\frac{1}{440}$.

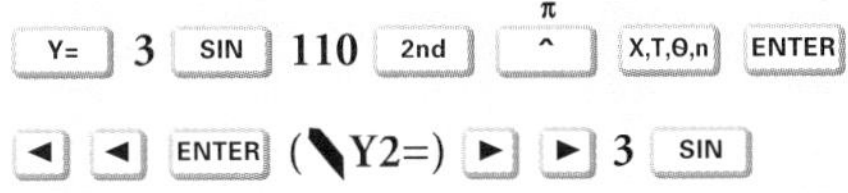

110 2nd ^ (π) (X,T,θ,n − (1
⇑ TI-82: (

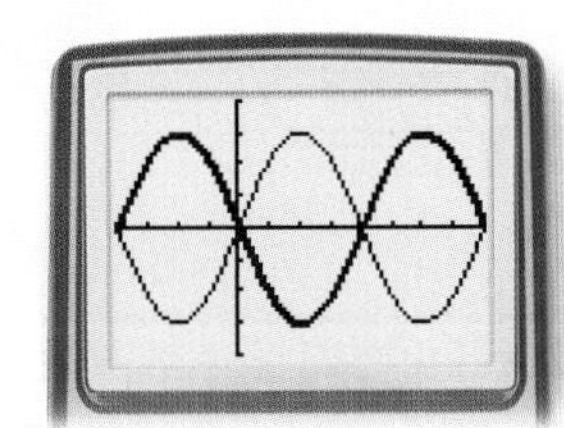

LESSON 13.6

EXAMPLE 2 **Evaluate $\text{Sin}^{-1} \frac{1}{2}$, $\text{Cos}^{-1}\left(-\frac{\sqrt{3}}{2}\right)$, and $\text{Tan}^{-1}\left(-\frac{\sqrt{3}}{3}\right)$.**

Page 870

Set the mode:

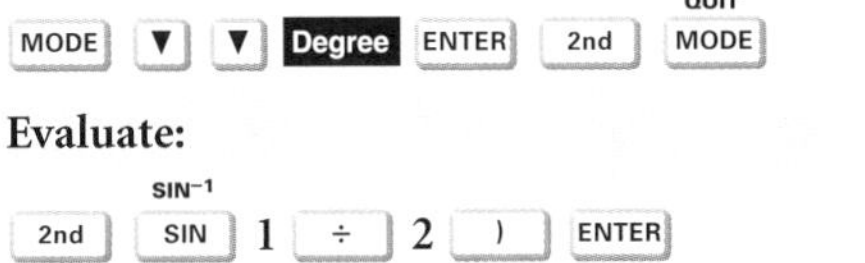

Evaluate:

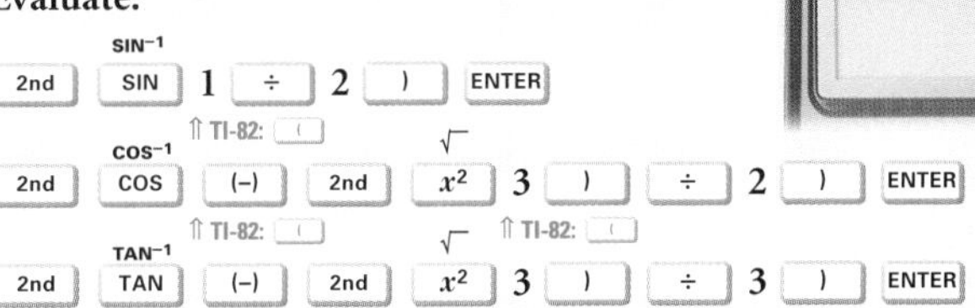

sin⁻¹(1/2)
30
cos⁻¹(-(√(3)/2))
150
tan⁻¹(-(√(3)/3))
-30

EXAMPLE 4 **For part a, find $\text{Sin}^{-1} \frac{1}{1.3}$ in degrees.**

Page 871

Set the mode:

MODE ▼ ▼ Degree ENTER 2nd MODE (QUIT)

Evaluate:

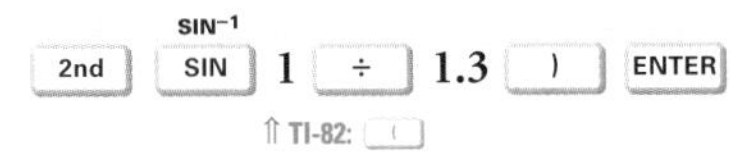

Further Topics in Trigonometry

Lesson Presentation CD-ROM
PowerPoint® presentations for each lesson 14.1–14.6

CHAPTER PLANNING GUIDE

Lesson	14.1	14.2	14.3	14.4	14.5	14.6	Project and Review
Pupil's Edition Pages	886–893	894–901	902–908	909–916	917–921	922–927	928–929, 930–935
Practice and Assessment							
Extra Practice (Pupil's Edition)	985	986	986	987	987	987	
Practice Workbook	88	89	90	91	92	93	
Practice Masters Levels A, B, and C	262–264	265–267	268–270	271–273	274–276	277–279	
Standardized Test Practice Masters	101	102	103	104	105	106	107
Assessment Resources	179	180	181	183	184	185	182, 186–191
Visual Resources							
Lesson Presentation Transparencies Vol. 2	164–167	168–171	172–175	176–179	180–183	184–187	
Teaching Transparencies	61, 62	63					
Answer Key Transparencies	510–512	513–515	516–521	522–529	530–532	533–534	535–538
Quiz Transparencies	14.1	14.2	14.3	14.4	14.5	14.6	
Teacher's Tools							
Reteaching Masters	175–176	177–178	179–180	181–182	183–184	185–186	
Make-Up Lesson Planner for Absent Students	88	89	90	91	92	93	
Student Study Guide	88	89	90	91	92	93	
Spanish Resources	88	89	90	91	92	93	
Block Scheduling Handbook							28–29
Activities and Extensions							
Lesson Activities	88	89	90	91	92	93	
Enrichment Masters	88	89	90	91	92	93	
Cooperative-Learning Activities	88	89	90	91	92	93	
Problem Solving/ Critical Thinking	88	89	90	91	92	93	
Student Technology Guide	88	89	90	91	92	93	
Long Term Projects							53–56
Writing Activities for Your Portfolio							40–42
Tech Prep Masters							65–68
Building Success in Mathematics							37–38

LESSON PACING GUIDE

Lesson	14.1	14.2	14.3	14.4	14.5	14.6	Project and Review
Traditional	1 day	1 day	2 days	2 days	2 days	2 days	2 days
Block	$\frac{1}{2}$ day	$\frac{1}{2}$ day	1 day	1 day	1 day	1 day	1 day
Two-Year	2 days	2 days	4 days	4 days	4 days	4 days	4 days

CONNECTIONS AND APPLICATIONS

Lesson	14.1	14.2	14.3	14.4	14.5	14.6	Review
Algebra	886–893	894–901	902–908	989–916	917–921	922–927	928–935
Geometry	886, 887, 892	898, 900	906			926	933
Probability							935
Transformations			904	912, 914, 915	921		
Maximum/Minimum		900					
Business and Economics		901		916			932
Science	888, 892, 893	894, 896, 899, 901	902, 905, 906, 907, 908	915	917, 919	926, 927	932
Sports and Leisure					921	922, 925	932
Other	892, 893			909, 913, 914, 916			
Cultural Connection: Africa					920		

BLOCK SCHEDULING GUIDE

Day	Lesson	Teacher Directed: Lesson Examples, Teaching Transparencies	Student Guided Activity, Try This	Cooperative-Learning Activity, Lesson Activity, Student Technology Guide	Practice: Practice & Apply, Extra Practice, Practice Workbook	Assessment: Quiz, Mid-Chapter Assessment	Problem Solving, Reteaching
1	14.1	20 min	20 min	15 min	50 min	15 min	15 min
	14.2	10 min	15 min	15 min	65 min	15 min	15 min
2	14.3	10 min	15 min	15 min	65 min	15 min	15 min
3	14.4	10 min	15 min	15 min	65 min	15 min	15 min
4	14.5	10 min	15 min	15 min	65 min	15 min	15 min
5	14.6	10 min	15 min	15 min	65 min	15 min	15 min
6	Assess.	50 min **PE:** Chapter Review	90 min **PE:** Chapter Project, Writing Activities	90 min Tech Prep Masters	65 min **PE:** Chapter Assessment, Test Generator	30 min Chap. Assess. (A or B), Alt. Assess. (A or B), Test Generator	

PE: Pupil's Edition

Alternative Assessment

The following suggest alternative assessments for students who may benefit from a different type of assessment than the regular chapter quizzes and the mid-chapter/end-of-chapter test. Visit the HRW web site to get additional Alternative Assessment material.

internet connect

Alternative Assessment
Go To: **go.hrw.com**
Keyword: **MB1 Alt Assess**

Performance Assessment

1. In $\triangle ABC$, $AB = 10$, $AC = 12$, and $m\angle A = 42°$.
  a. To the nearest tenth of a square unit, find the area of the triangle.
  b. Outline and apply two different strategies to find the measures of the other parts of the triangle.

2. How is $\sin(270° + \theta)$ related to $\sin \theta$? Justify your response by using a trigonometric identity.

3. Consider $\cos^2 x - 11 \cos x + 28 = 0$. Show that this equation has no solution.

Portfolio Project

Suggest that students choose one of the following projects for inclusion in their portfolios

1. a. Find the area of polygon $ABCDE$ by subdividing it into trianglular regions.

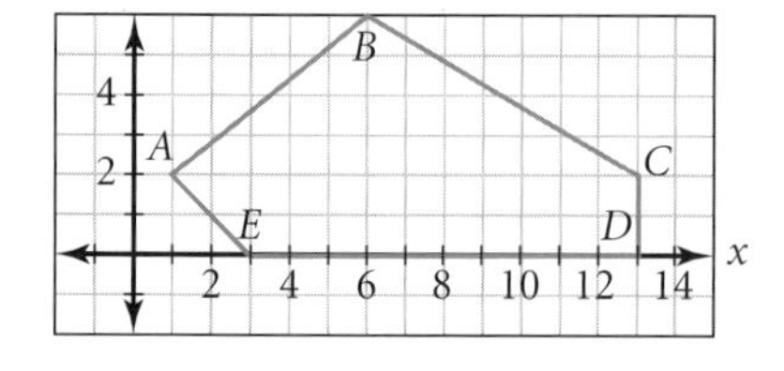

  b. Devise and describe a method for finding the area of a polygon whose vertices have given coordinates. Illustrate your method(s) by using polygons with 6 or 7 sides. Indicate how many triangles must be considered.

2. a. Solve $\begin{cases} a \cos 15° + b \sin 30° = 1 \\ a \sin 30° + b \cos 15° = 1 \end{cases}$ by using the elimination method and sum or difference identities.
  b. Generalize your results to solve the system $\begin{cases} a \cos x + b \sin y = 1 \\ a \sin y + b \cos x = 0 \end{cases}$ for a and b in terms of x and y.

internet connect

The table below identifies the pages in this chapter that contain internet and technology information.

Content Links

Activities Online	**pages 890, 920**
Portfolio Extensions	**page 901**
Homework Help Online	**pages 891, 899, 907, 914, 921, 926**
Graphic Calculator Support	**page 936**

Resource Links

Parents can go online and find concepts that students are learning—lesson by lesson—and questions that pertain to each lesson, which facilitate parent-student discussion.

Go To: **go.hrw.com**
Keyword: **MB1 Parent Guide**

Technical Support

The following may be used to obtain technical support for any HRW software product.

Online Help: **www.hrwtechsupport.com**
e-mail: **tschrw@hrwtechsupport.com**
HRW Technical Support Center: **(800)323-9239**
7 AM to 10 PM Monday through Friday Central Time

Visit the HRW math web site at: **www.hrw.com/math**

Technology

Lesson Suggestions and Calculator Examples

(Keystrokes are based on a TI-83 calculator.)

Lesson 14.1 The Law of Sines

Example 4 illustrates two situations that can arise when students try to solve a triangle using the law of sines. In part **a,** no triangle is formed. Students can read the calculator display to see this. In part **b,** two triangles are formed but the display does not immediately reveal this. Students must use reason to find out that $\angle C$ can have two measures. The display only shows that one angle measure is possible.

Lesson 14.2 The Law of Cosines

The strategy shown in Example 4 of this lesson involves both the law of cosines and the law of sines. The display at the right shows the steps in finding the measure of $\angle C$. When students press 2nd ENTER, the expression will reappear on the display. Editing the numbers according to the law of cosines can give the measure of $\angle B$ and then, after further editing, the measure of $\angle A$.

Lesson 14.3 Fundamental Trigonometric Identities

The graph related to Example 3 is shown below. Notice that the graph looks suspiciously like the graph of $y = \cos\theta$, but it is raised 1 unit up. This observation can guide students to the simplification of $\frac{\sin^2\theta}{1-\cos\theta}$. In Example 4, entering the expression $\sec\theta - \tan\theta\sin\theta$ into the calculator provides some insight into how the student can simplify the expression algebraically.

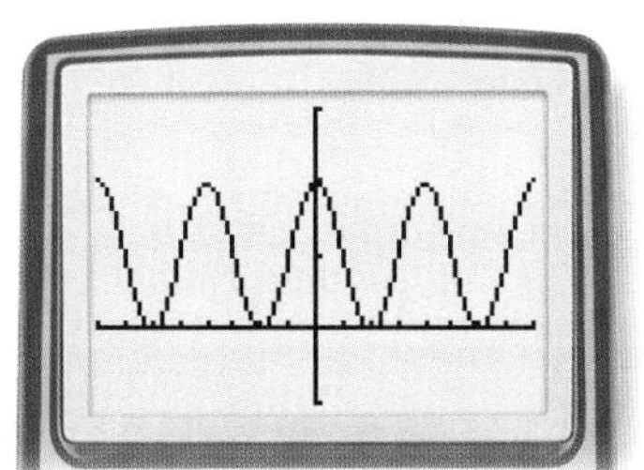

To enter the expression, students must transform it into the expression shown below.

$$\frac{1}{\cos\theta} - \tan\theta\sin\theta$$

Lesson 14.4 Sum and Difference Identities

The calculator display at the left below shows the graphs of $f(x) = x^4 - x^2 + 2$ and $g(x) = x^3 - 2x$. f is an even function and g is an odd function. The graphs of $y = \sin x$ and $y = \cos x$ are shown below at right. Ask students how they can use the display at left below to help tell which trigonometric function at right below is even and which is odd.

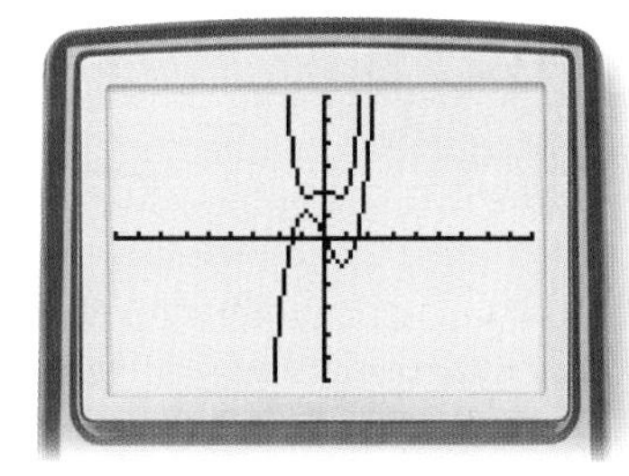

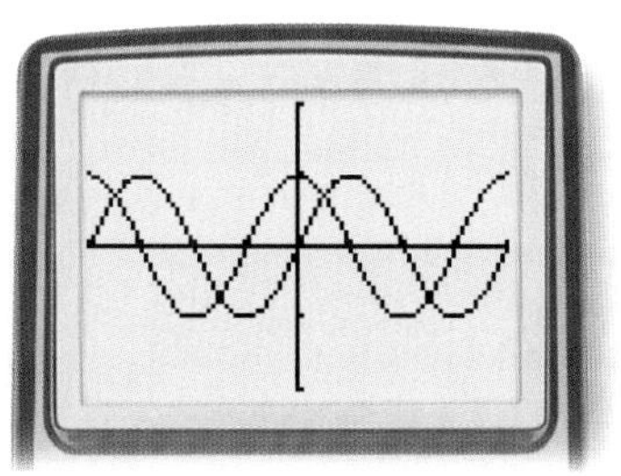

Lesson 14.5 Double- and Half-Angle Identities

In this lesson, one technology objective is to use the graphics calculator to dispel common false beliefs.

To find the sin of 2θ, find $\sin\theta$, and then multiply it by 2.

A graphics calculator display quickly shows that this belief is untrue. The graphs of $y = \sin 2\theta$ and $y = 2\sin\theta$ are shown at right. Clearly, they are not the same.

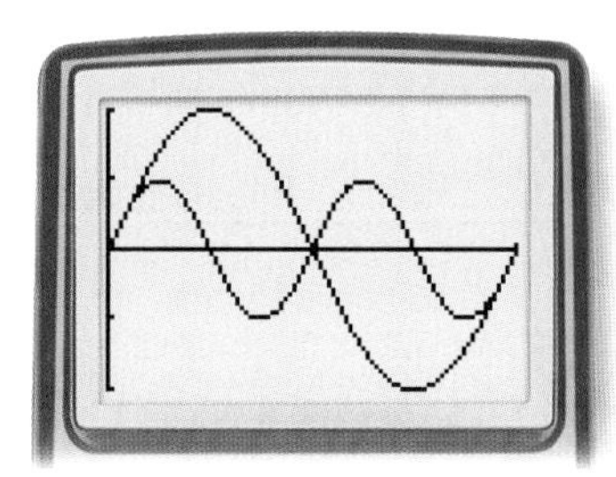

Lesson 14.6 Solving Trigonometric Equations

Equations in this lesson can be explored graphically and algebraically. Consider posing an equation to students that requires a graphical approach. Consider, for example, $\tan x = x$.

For further information, refer to the

- technology discussions in the lessons.
- lesson-related teacher's commentary in the side columns of this *Teacher's Edition.*
- lesson-related *Student Technology Guide* masters.
- *HRW Technology Handbook.*

internet connect

For keystrokes of other graphing calculators models, visit the HRW web site at **go.hrw.com** and enter the keyword **MB1 CALC**.

About Chapter 14

Background Information

Trigonometric (trig) functions are used to model many phenomena in the world around us. In this chapter, students will learn more about working with trig functions. Students will learn to find sides and angles of triangles by applying the laws of sines and cosines. They learn to simplify trigonometric expressions by using identities. Finally, students learn to solve trig equations.

CHAPTER RESOURCES

- Block-Scheduling Handbook
- Writing Activities for Your Portfolio
- Tech Prep Masters
- Long-Term Project
- Assessment Resources:
  - Mid-Chapter Assessment
  - Chapter Assessments
  - Alternative Assessments
- Test and Practice Generator
- Technology Handbook

Chapter Objectives

- Solve mathematical and real-world problems by using the law of sines. [**14.1**]
- Use the law of cosines to solve triangles. [**14.2**]
- Prove fundamental trigonometric identities. [**14.3**]
- Use fundamental trigonometric identities to rewrite expressions. [**14.3**]
- Evaluate expressions by using the sum and difference identities. [**14.4**]

14 Further Topics in Trigonometry

IN THIS SECOND CHAPTER ON TRIGONOMETRY, YOU will study additional uses of trigonometry and several relationships among the trigonometric functions. For example, the law of sines and the law of cosines will enable you to solve more general triangles. A classic application of trigonometry is ship navigation, which involves the instruments shown here.

Lessons

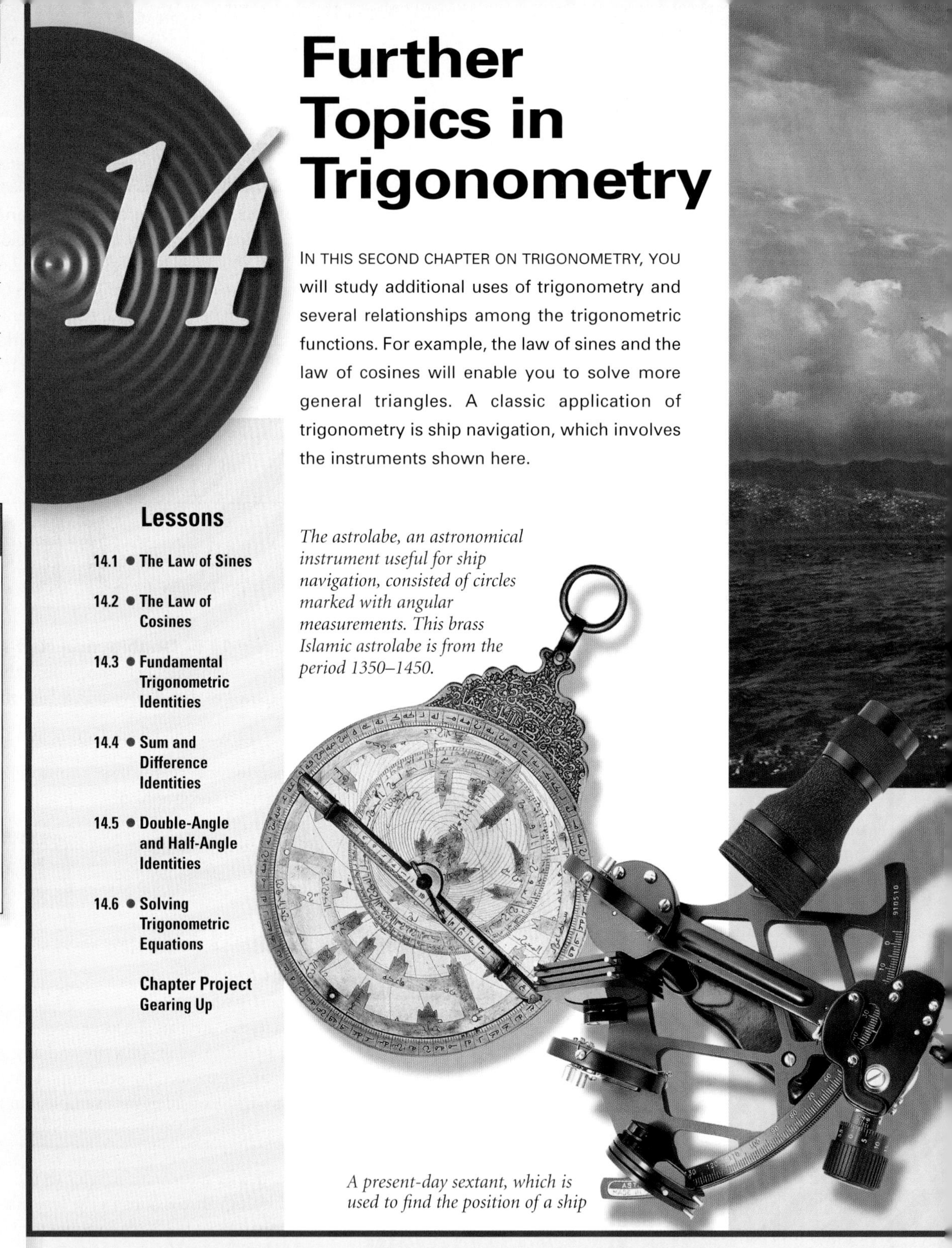

The astrolabe, an astronomical instrument useful for ship navigation, consisted of circles marked with angular measurements. This brass Islamic astrolabe is from the period 1350–1450.

A present-day sextant, which is used to find the position of a ship

About the Photos

One of the first instruments for measuring the altitude of the Sun and stars in order to in locate a position on Earth was the *astrolabe,* which is shown above at left. The principles of an astrolabe projection were known before 150 B.C.E., and the Islamic world improved it by 800 C.E. and introduced it to Europe by the early twelfth century. The astrolabe was the most popular navigational instrument until the mid-seventeenth century, when it was replaced with the *sextant.*

A sextant, which is shown above at right, is a double-reflecting optical instrument that locates latitude and longitude by measuring the altitudes of the Sun, Moon, and stars. Until 1998, the U.S. Naval Academy required every midshipman to learn how to use a sextant. The Naval Academy now uses the Global Positioning System (GPS), which determines positions by using computers and satellites.

About the Chapter Project

Gear design has evolved over hundreds of years and can involve some complex mechanical engineering. The Chapter Project, *Gearing Up*, will give you some insight into the mathematics that allow gears to mesh smoothly.

After completing the Chapter Project, you will be able to:

- Determine a gear tooth profile and the spacing of gear teeth around a base circle.
- Design a gear template for a set of gears.
- Make a working model of a set of gears that mesh together smoothly.

About the Portfolio Activities

Throughout the chapter, you will be given opportunities to complete the Portfolio Activities that are designed to support your work on the Chapter Project.

- Sketching an involute gear profile and determining the radius of the curved edge of a gear tooth is included in the Portfolio Activity on page 901.
- Using rotation matrices to find the positions of gear teeth on a gear's base circle is included in the Portfolio Activity on page 916.

internet connect

Chapter Internet Features and Online Activities

Lesson	Keyword	Page
14.1	MB1 Homework Help	891
	MB1 Surveying	890
14.2	MB1 Homework Help	899
	MB1 Trig Apps	901
14.3	MB1 Homework Help	907
14.4	MB1 Homework Help	914
14.5	MB1 Homework Help	921
	MB1 Archimedes	920
14.6	MB1 Homework Help	926

- Use matrix multiplication with sum and difference identities to perform rotations. [**14.4**]
- Evaluate and simplify expressions by using double-angle and half-angle identities. [**14.5**]
- Solve trigonometric equations algebraically and graphically. [**14.6**]
- Solve real-world problems by using trigonometric equations. [**14.6**]

Portfolio Activities appear at the end of Lessons 14.2 and 14.4. Each serves as preparation for the Chapter Project. The Portfolio Activities as well as the Chapter Project Activities are appropriate for inclusion in the student's portfolio. Students should be encouraged to include in their portfolios any other work in which they feel a sense of pride or a sense of accomplishment.

Prepare

NCTM PRINCIPLES & STANDARDS 1–4, 6–10

QUICK WARM-UP

Solve each of the following:

1. $\frac{x}{6} = \frac{3}{21}$ $x \approx 0.857$
2. $\frac{51}{90} = \frac{30}{x}$ $x \approx 52.94$
3. $\frac{0.7071}{45} = \frac{0.7660}{x}$ $x \approx 48.75$
4. $\frac{0.5}{30} = \frac{x}{45}$ $x = 0.75$
5. $\frac{300}{x} = 0.0152$ $x \approx 9736.84$

Also on Quiz Transparency 14.1

Teach

Why Ask students to describe situations in which it is either difficult or impossible to measure a distance directly. **sample: measuring the distance across a canyon or lake** Have students discuss strategies they might use to find these distances.

Math CONNECTION

GEOMETRY The formula of the area of a triangle that uses its base and height is used to derive the formula of the area of a triangle that uses two sides and the included angle.

The Law of Sines

Why *The law of sines can be derived from an area formula that contains the sine function. This area formula can be used to find the area of the triangular-shaped park represented below.*

Objective

- Solve mathematical and real-world problems by using the law of sines.

The triangular piece of land represented at right will be used for a new park. What is the approximate area of the land?

Let K represent the area of $\triangle ABC$.

$$K = \frac{1}{2} \times \text{base} \times \text{height}$$

$$K = \frac{1}{2} \times AC \times BD$$

$$K = \frac{1}{2} \times 2.5 \times (2 \sin 32^\circ)$$ *Because $\sin 32^\circ = \frac{BD}{2}$, substitute $2 \sin 32^\circ$ for BD.*

$$K \approx 1.3$$

The area of the triangular piece of land is about 1.3 square miles.

The information given in the park problem above includes AC, BC, and the measure of the included angle, C. This information is known as side-angle-side, or SAS, information. Given SAS information for a triangle, you can always find its area with one of the formulas below.

Area of a Triangle

The area, K, of $\triangle ABC$ is given by the equations below.

$$K = \frac{1}{2}bc \sin A \qquad K = \frac{1}{2}ac \sin B$$

$$K = \frac{1}{2}ab \sin C$$

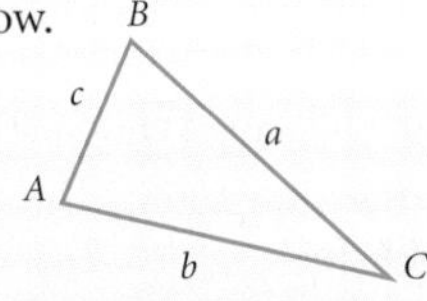

When labeling triangles, it is customary to use capital letters for the angles or angle measures, and lowercase letters for the sides or side lengths. Furthermore, an angle and the side opposite that angle are labeled with the same letter.

Alternative Teaching Strategy

USING SYMBOLS

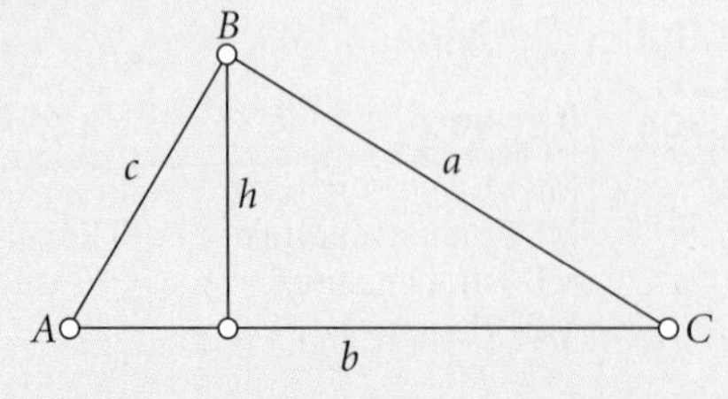

In the triangle above, have students use the two right triangles formed by the altitude, h, to write two trig expressions for h. $\mathbf{h = c \sin A}$; $\mathbf{h = a \sin C}$

Have students use the transitive property to get $\frac{\sin A}{a} = \frac{\sin C}{c}$. Then have them draw the altitude to side $\overline{BC}$, name it h_2, and find two expressions for h_2 involving b, c, $\sin B$, and $\sin C$. $\mathbf{h_2 = b \sin C}$; $\mathbf{h_2 = c \sin B}$ They can then use the transitive property to derive the complete law of sines.

EXAMPLE 1 **Find the area of $\triangle RST$ to the nearest tenth of a square unit.**

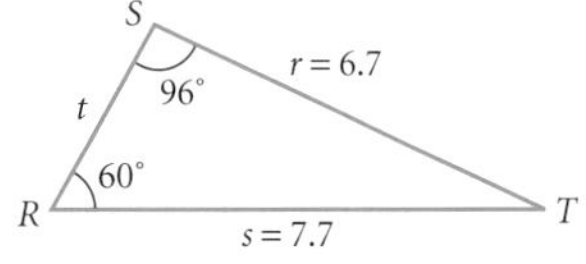

SOLUTION

In order to have SAS information, you need to know T. Recall that the sum of the angle measures of any triangle is 180°.

$$R + S + T = 180°$$
$$60° + 96° + T = 180°$$
$$T = 24°$$

With SAS information, you can use one of the area formulas from the previous page to find the area of $\triangle RST$.

$$K = \frac{1}{2}sr \sin T$$
$$K = \frac{1}{2}(7.7)(6.7) \sin 24°$$
$$K \approx 10.5$$

Thus, the area of $\triangle RST$ is about 10.5 square units.

TRY THIS Find the area of $\triangle XYZ$ to the nearest tenth of a square unit.

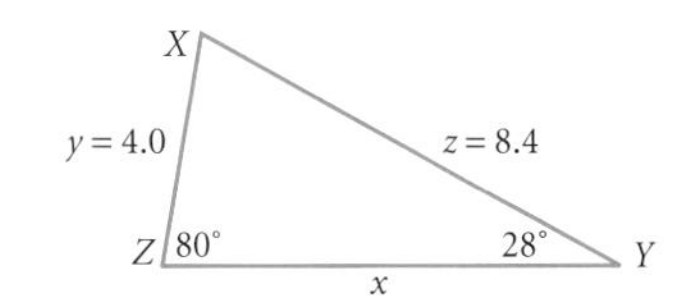

Law of Sines

Using the area formulas, you can derive the *law of sines*. Because the formulas $K = \frac{1}{2}bc \sin A$, $K = \frac{1}{2}ac \sin B$, and $K = \frac{1}{2}ab \sin C$ all represent the area of $\triangle ABC$, they are equal.

$$\frac{1}{2}bc \sin A = \frac{1}{2}ac \sin B = \frac{1}{2}ab \sin C$$
$$bc \sin A = ac \sin B = ab \sin C$$
$$\frac{bc \sin A}{abc} = \frac{ac \sin B}{abc} = \frac{ab \sin C}{abc}$$
$$\frac{\sin A}{a} = \frac{\sin B}{b} = \frac{\sin C}{c}$$

Law of Sines

For $\triangle ABC$, the **law of sines** states the following:

$$\frac{\sin A}{a} = \frac{\sin B}{b} = \frac{\sin C}{c}$$

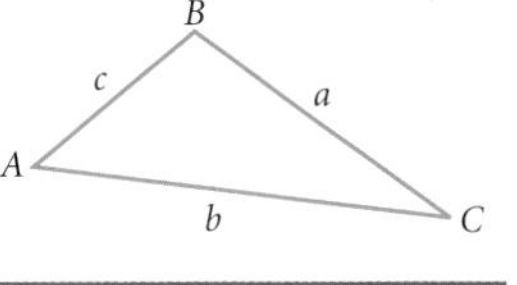

CRITICAL THINKING Show that $\frac{\sin A + \sin B}{\sin B} = \frac{a + b}{b}$ is true for any $\triangle ABC$.

Interdisciplinary Connection

BIOLOGY Biologists have found a very large, old, leaning tree, and they are interested in its length. The biologists find that the angle between the tree and the ground is 83°. They walk 25 feet away and find that the angle from that point on the ground to the top of the tree is 35°. Explain how these measurements help the biologist find the length of the tree. **They are used in the equations for the law of sines.** Find the length of the tree to the nearest tenth of a foot.

$\frac{\sin 35°}{x} = \frac{\sin 62°}{25}$, **so $x \approx 16.2$ ft**

Inclusion Strategies

KINESTHETIC LEARNERS Have students go to a large open space near the school, such as a soccer field or a parking lot. Using tape measures and protractors, have the students create triangles like the ones in Exercises 24–43 on page 891. Students should use the law of sines to solve the triangle and then physically measure the unknowns to check their work. Remind students to be precise when measuring.

Teaching Tip

In this chapter, there will be no notational distinction between an angle and its measure, and the angle symbol, ∠, will no longer be used. That is, the angle whose vertex is the point A and its measure will both be denoted as A. The meaning of the symbol should be clear from its context.

ADDITIONAL EXAMPLE 1

Find the area of $\triangle JKL$ to the nearest tenth of a square unit.

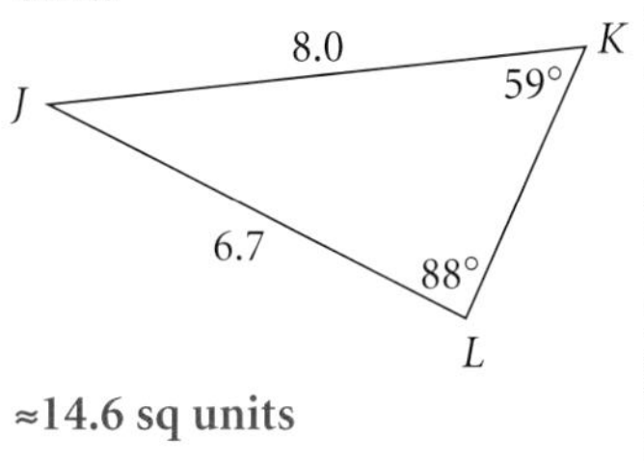

≈14.6 sq units

Math CONNECTION

GEOMETRY In Example 1, students must use the formula for the sum of the angles of a triangle and the trig formula for the area of a triangle given on page 886.

TRY THIS

$K \approx 16.0$ sq units

Teaching Tip

An additional method for finding the area of a triangle is Heron's formula, given by the equation

$$\text{area} = \sqrt{s(s-a)(s-b)(s-c)},$$

where $s = \frac{a+b+c}{2}$.

Use Teaching Transparency 61.

CRITICAL THINKING

$$\frac{\sin A}{a} = \frac{\sin B}{b}$$
$$\frac{\sin A}{\sin B} = \frac{a}{b}$$
$$\frac{\sin A}{\sin B} + \frac{\sin B}{\sin B} = \frac{a}{b} + \frac{b}{b}$$
$$\frac{\sin A + \sin B}{\sin B} = \frac{a+b}{b}$$

ADDITIONAL
EXAMPLE 2

Solve $\triangle PQR$. Give answers to the nearest tenth, if necessary.

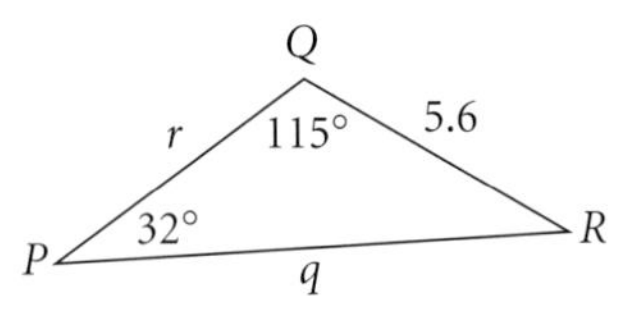

$R = 33°$; $q \approx 9.6$; $r \approx 5.8$

TRY THIS
$F = 52°$, $d \approx 8.5$, $e \approx 5.4$

ADDITIONAL
EXAMPLE 3

In $\triangle XYZ$ below, find z to the nearest whole number.

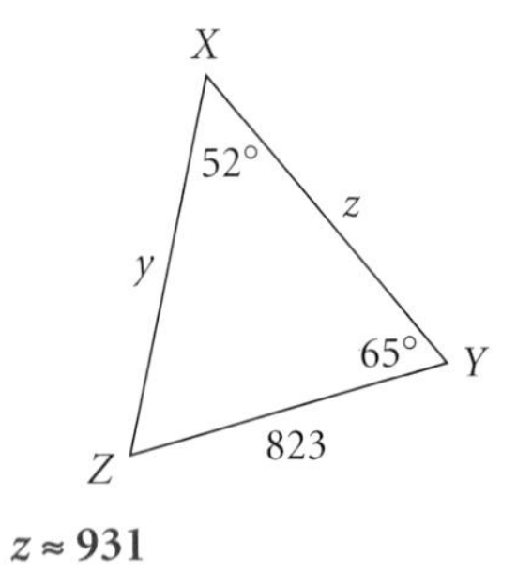

$z \approx 931$

TRY THIS
$m \approx 638$

When you are given the measures of two angles and the length of the included side in a triangle, this is called angle-side-angle (ASA) information. Given ASA information, you can use the law of sines to solve triangles, as shown in Example 2.

EXAMPLE 2 **Solve $\triangle ABC$. Give answers to the nearest tenth, if necessary.**

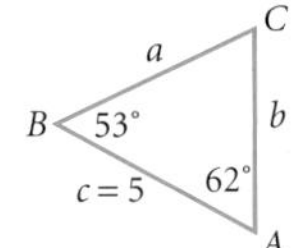

SOLUTION

First find C.

$$A + B + C = 180°$$
$$62° + 53° + C = 180°$$
$$C = 65°$$

Now apply the law of sines to find sides a and b.

$$\frac{\sin A}{a} = \frac{\sin C}{c} \qquad \frac{\sin B}{b} = \frac{\sin C}{c}$$
$$\frac{\sin 62°}{a} = \frac{\sin 65°}{5} \qquad \frac{\sin 53°}{b} = \frac{\sin 65°}{5}$$
$$a = \frac{5 \sin 62°}{\sin 65°} \qquad b = \frac{5 \sin 53°}{\sin 65°}$$
$$a \approx 4.9 \qquad b \approx 4.4$$

TRY THIS Solve $\triangle DEF$. Give answers to the nearest tenth, if necessary.

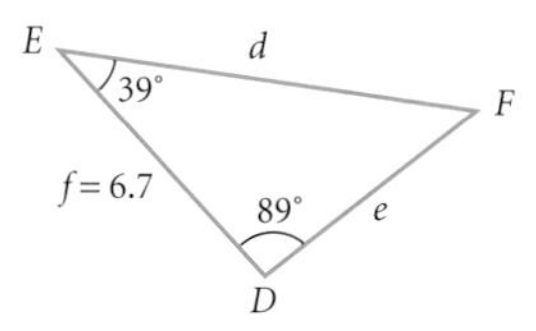

Example 3 shows how you can use the law of sines to solve a triangle for which you are given side-angle-angle (SAA) information.

EXAMPLE 3

APPLICATION
SURVEYING

A surveying crew needs to find the distance between two points, A and B. They cannot measure the distance directly because there is a hill between the two points. The surveyors obtain the information shown in the diagram at right.

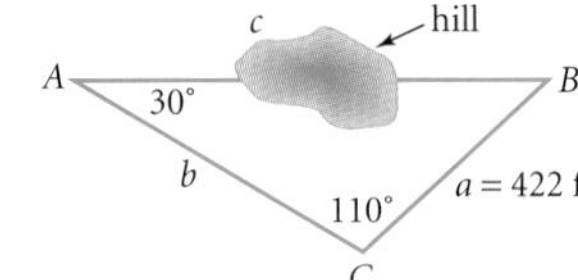

Find c to the nearest foot.

SOLUTION

$$\frac{\sin C}{c} = \frac{\sin A}{a}$$
$$\frac{\sin 110°}{c} = \frac{\sin 30°}{422} \qquad \textit{Substitute } A = 30°,\ a = 422, \textit{ and } C = 110°.$$
$$c = \frac{422 \sin 110°}{\sin 30°}$$
$$c \approx 793$$

Thus, the distance between points A and B is approximately 793 feet.

TRY THIS In $\triangle KLM$ at right, find m to the nearest whole number.

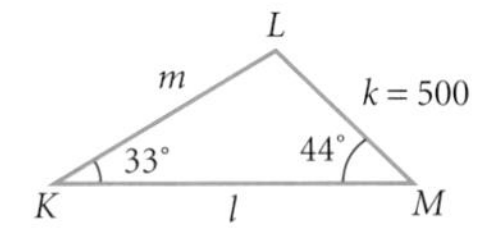

Enrichment

The orbits of Earth and Venus are approximately circular, with the Sun at the center. A sighting of Venus is made from Earth and the angle between this line of sight and a line of sight to the Sun is found to be 18°. If the diameter of the orbit of the Earth is 18.6×10^7 miles and the diameter of the orbit of Venus is 13.4×10^7 miles, what are the possible distances from Earth to Venus?
14.9×10^7 mi or 27.9×10^6 mi

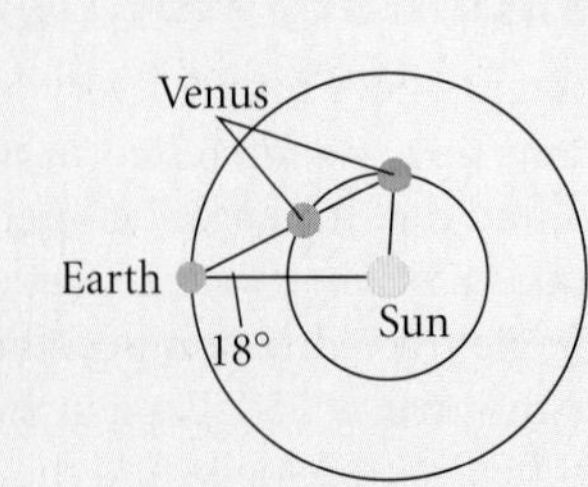

The Ambiguous Case

When you are given two side lengths and the measure of an angle that is not between the sides, the given information is called side-side-angle (SSA) information.

Exploring SSA Information

You will need: no special materials

1. The figure below illustrates SSA triangle information. Side a is free to pivot about C. How many triangles can be formed when $a < h$, where h is the altitude of the triangle? Use the illustration to explain your answer.

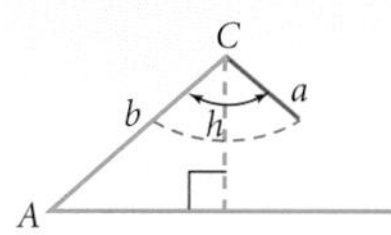

2. How many triangles can be formed when $a = h$? Use the illustration to explain your answer.

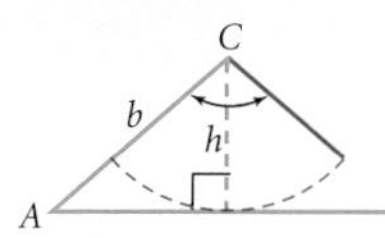

3. How many triangles can be formed when $a > h$ and $a < b$? Use the illustration to explain your answer.

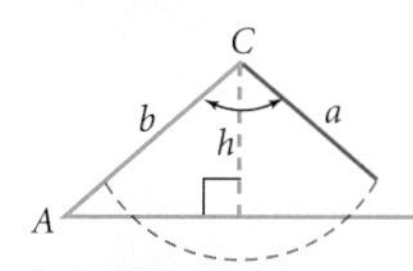

4. How many triangles may be formed when $a > h$ and $a > b$? Use the illustration to explain your answer.

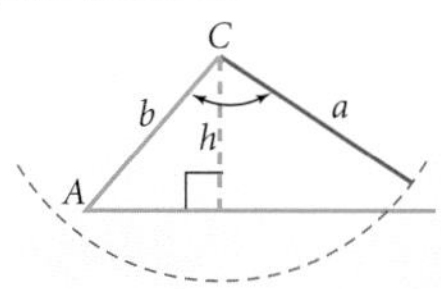

CHECKPOINT ✔ 5. Explain how you may find 0, 1, or 2 triangles, given SSA information.

Recall from geometry that SSA information is not sufficient to prove triangle congruence. With SSA information, 0, 1, or 2 triangles may be possible.

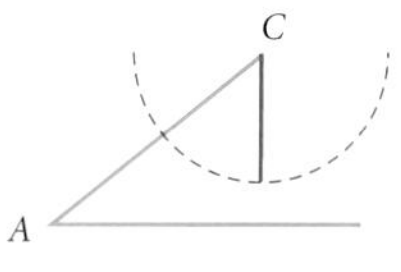
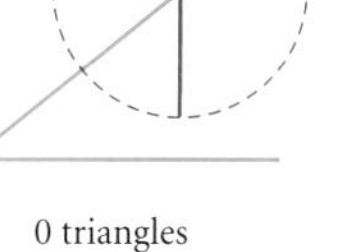

0 triangles

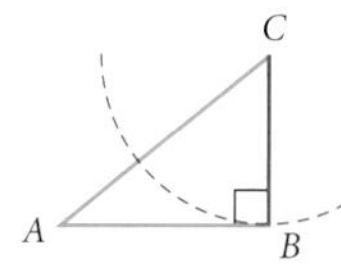
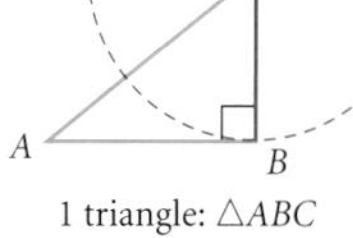

1 triangle: $\triangle ABC$

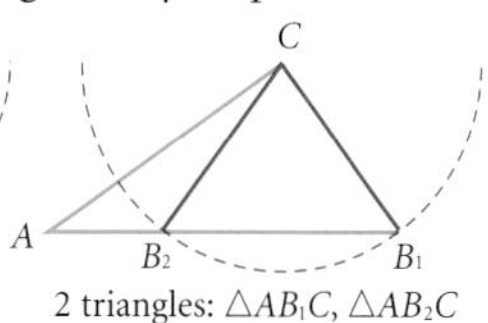

2 triangles: $\triangle AB_1C$, $\triangle AB_2C$

Reteaching the Lesson

USING COGNITIVE STRATEGIES Have students recall how to solve a proportion for one unknown. Discuss with students the law of sines as a proportional relationship. When solving a triangle, have students list all known values and unknown values. To use the law of sines, the set of values must include at least two angle measures and two side lengths, any one of which may be unknown. Once these values are identified, students need to match each angle with the opposite side and then write the ratio given by the law of sines. If SSA information is given, then there is the possibility of 0, 1, or 2 triangles. Suppose that angle A and the lengths of sides a and b are given. The four possible cases are given below. The h represents the altitude of the triangle drawn to side $\overline{AB}$, and $h = b \sin A$.

1. If $a < h$, then 0 triangles can be formed.
2. If $a = h$, then 1 triangle can be formed.
3. If $h < a < b$, then 2 triangles can be formed.
4. If $a > b$, then 1 triangle can be formed.

Have students illustrate each of the four possible cases.

Use Teaching Transparency 62.

Math CONNECTION

GEOMETRY In geometry, students study conditions under which triangles are congruent, such as SSS, SAS, AAS, and ASA. Students also learn that SSA does not provide sufficient information to obtain a unique triangle.

Activity Notes

In this Activity, students explore the triangles formed by using a variation of SSA. When the side opposite the given angle, a, is less than the altitude, h, no triangle can be formed. When $a = h$ or $a > b$, where b is the length of a side adjacent to the given angle, a unique right triangle can be formed. When $h < a < b$, two different triangles can be formed.

Cooperative Learning

You may wish to have students do the Activity in groups of three. Students should discuss the strategies for each construction with team members, construct their own triangles, and then compare their results with the group.

CHECKPOINT ✔

5. If $a < h$, then 0 triangles can be formed. If $a = h$ or if $a > h$ and $a > b$, then 1 triangle can be formed. If $a > h$ and $a < b$, then 2 triangles can be formed.

ADDITIONAL
EXAMPLE 4

Determine whether the given SSA information defines 0, 1, or 2 triangles.

a. $a = 6.5$, $b = 7.4$, $B = 79°$
One triangle is defined.

b. $a = 10.9$, $c = 3.2$, $C = 42°$
No triangle is defined.

TRY THIS

a. 0 triangles

b. 2 triangles

Assess

Selected Answers

Exercises 5–9, 11–67 odd

ASSIGNMENT GUIDE	
In Class	1–9
Core	11–55 odd
Core Plus	10–56 even
Review	57–68
Preview	69

Extra Practice can be found beginning on page 940.

EXAMPLE 4 **Determine whether the given SSA information defines 0, 1, or 2 triangles.**

a. $b = 2$, $c = 8$, and $B = 120°$

b. $c = 10$, $a = 6$, and $A = 28°$

SOLUTION

a.

$$\frac{\sin C}{c} = \frac{\sin B}{b}$$

$$\frac{\sin C}{8} = \frac{\sin 120°}{2}$$

$$\sin C = \frac{8 \sin 120°}{2}$$

$$\sin C \approx 3.4641$$

The range of $y = \sin \theta$ is $-1 \le y \le 1$. Because there is no angle whose sine is 3.4641, no triangle can be formed.

b.

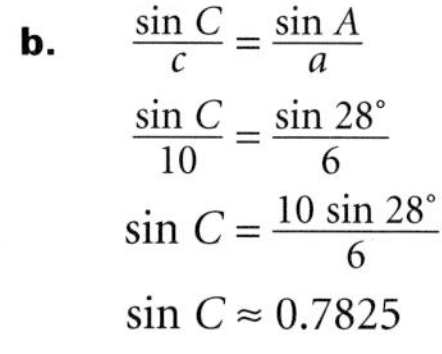

$$\frac{\sin C}{c} = \frac{\sin A}{a}$$

$$\frac{\sin C}{10} = \frac{\sin 28°}{6}$$

$$\sin C = \frac{10 \sin 28°}{6}$$

$$\sin C \approx 0.7825$$

Recall from Lesson 13.3 that there are two possible values of θ between 0° and 180° for which $\sin \theta \approx 0.7825$. Thus, there are two possible triangles.

$C \approx 51.5°$ *or* $C \approx 180° - 51.5°$
$C \approx 128.5°$

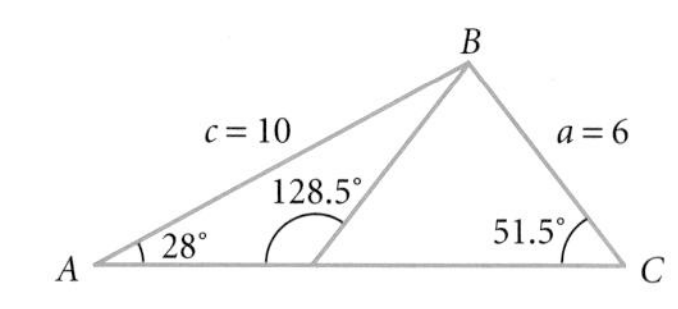

TRY THIS Determine whether the given SSA information defines 0, 1, or 2 triangles.

a. $a = 10$, $c = 4$, and $C = 148°$

b. $a = 2.4$, $b = 3.1$, and $A = 24°$

LAW OF SINES			
Given	**You can**	**Given**	**You can**
SAS	find the area of a triangle	SAA	solve a triangle
ASA	solve a triangle	SSA	define 0, 1, or 2 triangles

Exercises

Communicate

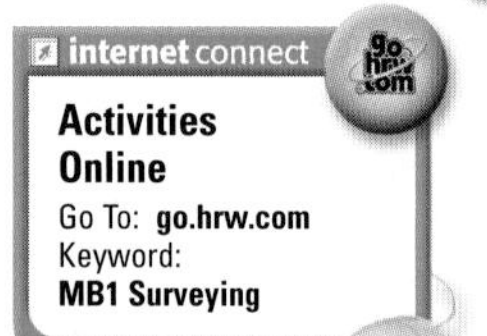

1. Explain how to solve a triangle when ASA information is known.

2. Explain how to solve a triangle when SAA information is known.

3. Explain under what circumstances SSA information does *not* determine a triangle.

4. Explain how information about sides and angles may determine two different triangles.

Guided Skills Practice

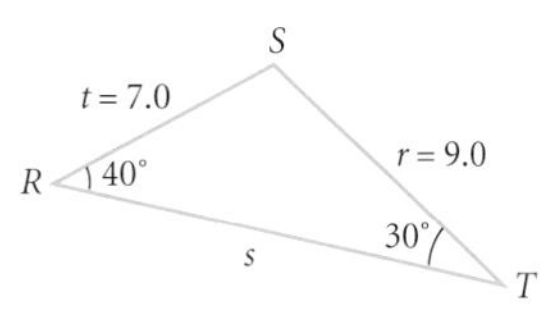

5. Find the area of $\triangle RST$ to the nearest tenth of a square unit. ***(EXAMPLE 1)*** **29.6 square units**

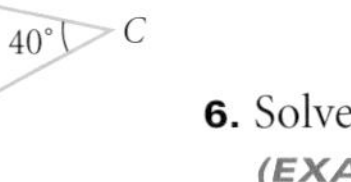

6. Solve $\triangle ABC$ at left. Give answers to the nearest tenth, if necessary. ***(EXAMPLE 2)*** **$B = 115°$; $a \approx 3.7$; $c \approx 5.7$**

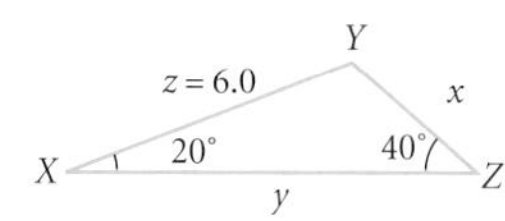

7. In $\triangle XYZ$ at right, find x to the nearest tenth. ***(EXAMPLE 3)*** **$x \approx 3.2$**

Determine whether the given SSA information defines 0, 1, or 2 triangles. ***(EXAMPLE 4)***

8. $a = 12$, $b = 15$, and $A = 30°$
2 triangles

9. $c = 2$, $b = 20$, and $C = 150°$
0 triangles

Practice and Apply

Find the area of $\triangle ABC$ to the nearest tenth of a square unit.

10. $b = 5$ in., $c = 8$ in., $A = 45°$
11. $a = 10$ ft, $c = 12$ ft, $B = 30°$
12. $a = 9$ in., $b = 11$ in., $C = 60°$
13. $b = 7$ cm, $c = 10$ cm, $A = 45°$
14. $a = 6$ km, $c = 10$ km, $B = 57°$
15. $a = 25$ ft, $b = 32$ ft, $C = 67°$
16. $b = 13$ in., $c = 16$ in., $A = 110°$
17. $a = 17$ m, $c = 22$ m, $B = 121°$
18. $a = 19$ ft, $b = 8$ ft, $C = 102.32°$
19. $b = 5$ cm, $c = 8$ cm, $A = 94.75°$
20. $a = 5$ km, $c = 9$ km, $B = 67.23°$
21. $a = 87$ ft, $b = 42$ ft, $C = 73.97°$
22. $a = 7$ m, $c = 10$ m, $C = 23°$
23. $a = 33$ ft, $c = 49$ ft, $B = 63.34°$

Use the given information to find the indicated side length in $\triangle ABC$. Give answers to the nearest tenth.

24. Given $A = 42°$, $B = 35°$, and $a = 10$, find b. **8.6**
25. Given $A = 50°$, $C = 25°$, and $a = 15$, find c. **8.3**
26. Given $B = 60°$, $C = 70°$, and $c = 15$, find b. **13.8**
27. Given $C = 55°$, $A = 100°$, and $c = 8$, find a. **9.6**
28. Given $A = 115°$, $B = 30°$, and $c = 10$, find a. **15.8**
29. Given $A = 40°$, $C = 80°$, and $b = 15$, find c. **17.1**

Solve each triangle. Give answers to the nearest tenth, if necessary.

30. $A = 43°$, $B = 52°$, $b = 20$
31. $B = 27°$, $C = 52°$, $c = 6$
32. $A = 20°$, $C = 60°$, $a = 10$
33. $A = 35°$, $B = 62°$, $a = 8$
34. $A = 23°$, $B = 62°$, $c = 15$
35. $B = 80°$, $C = 20°$, $a = 10$
36. $A = 60°$, $B = 40°$, $a = 10$
37. $B = 35°$, $C = 48°$, $b = 12$
38. $B = 40°$, $C = 60°$, $b = 8$
39. $A = 37°$, $C = 42°$, $b = 20$
40. $A = 40°$, $B = 45°$, $c = 16$
41. $C = 42°$, $B = 58°$, $c = 9$
42. $B = 30°$, $C = 45°$, $a = 9$
43. $A = 45°$, $C = 23°$, $b = 11$

10. 14.1 in.^{2}
11. 30 ft^2
12. 42.9 in.^{2}
13. 24.7 cm^2
14. 25.2 km^2
15. 368.2 ft^2
16. 97.7 in.^{2}
17. 160.3 m^2
18. 74.2 ft^2
19. 19.9 cm^2
20. 20.7 km^2
21. 1756.0 ft^2
22. 13.7 m^2
23. 722.5 ft

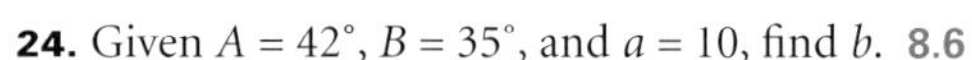

internet connect
Homework Help Online
Go To: **go.hrw.com**
Keyword:
MB1 Homework Help for Exercises 24–43

30. $C = 85°$; $a \approx 17.3$; $c = 25.3$
31. $A = 101°$; $b \approx 3.5$; $a \approx 7.5$
32. $B = 100°$; $b \approx 28.8$; $c \approx 25.3$
33. $C = 83°$; $b \approx 12.3$; $c \approx 13.8$
34. $C = 95°$; $a \approx 5.9$; $b \approx 13.3$

Error Analysis

Students often make mistakes when relating an angle to its opposite side. When finding the side opposite an angle, have students lightly draw a ray starting at the vertex of the angle, passing through the middle of the angle, and ending at the opposite side.

Students may check whether a triangle has been solved correctly by using *Mollweide's equation*, which is given below.

$$(a - b)\cos\frac{C}{2} \approx c\sin\frac{A - B}{2}$$

After substituting all values of the triangle into the equation, if the left side is not *approximately* equal to the right side, then an error has been made.

35. $A = 80°$; $b \approx 10$; $c \approx 3.5$
36. $C = 80°$; $b \approx 7.4$; $c \approx 11.4$
37. $A = 97°$; $a \approx 20.8$; $c \approx 15.5$
38. $A = 80°$; $a \approx 12.3$; $c \approx 10.8$
39. $B = 101°$; $a \approx 12.3$; $c \approx 13.6$
40. $C = 95°$; $a \approx 10.3$; $b \approx 11.4$
41. $A = 80°$; $a \approx 13.2$; $b \approx 11.4$
42. $A = 105°$; $b \approx 4.7$; $c \approx 6.6$
43. $B = 112°$; $a \approx 8.4$; $c \approx 4.6$

45. 2 possible triangles
$A = 30°, B = 108.2°, C = 41.8°,$
$a = 1.5, b = 2.8, c = 2$ and
$A = 30°, B = 11.8°, C = 138.2°,$
$a = 1.5, b = 0.6, c = 2$

46. 1 possible triangle
$A = 45°, B = 100.6°, C = 34.4°,$
$a = 5, b = 7.0, c = 4$

48. 1 possible triangle
$A = 30°, B = 60°, C = 90°,$
$a = 1, b = 1.7, c = 2$

49. 1 possible triangle
$A = 45°, B = 45°, C = 90°,$
$a = \frac{5\sqrt{2}}{2}, b = 3.5, c = 5$

State the number of triangles determined by the given information. If 1 or 2 triangles are formed, solve the triangle(s). Give answers to the nearest tenth, if necessary.

44. $A = 45°, c = 10, a = 2$ **0**

45. $A = 30°, c = 2, a = 1.5$ **2**

46. $A = 45°, c = 4, a = 5$ **2**

47. $A = 60°, c = 8, a = 2$ **0**

48. $A = 30°, c = 2, a = 1$ **1**

49. $A = 45°, c = 5, a = \frac{5\sqrt{2}}{2}$ **1**

CHALLENGE

50. Find the length of side x in the figure at right to the nearest tenth. **2.53**

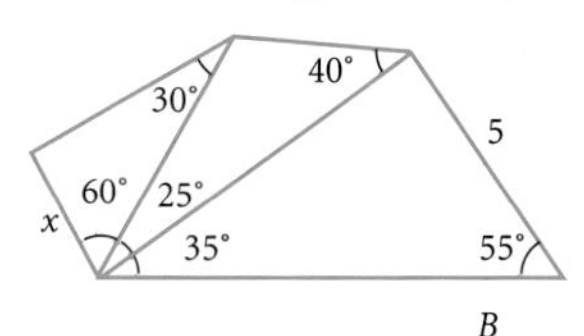

CONNECTION

51. GEOMETRY In the figure at right, $CD = 100$ centimeters, $m\angle 1 = 33°$, $m\angle 2 = 42°$, $m\angle 3 = m\angle 1 + m\angle 2$, $m\angle 4 = 37°$, $m\angle 5 = 78°$, and $m\angle 6 = 50°$. Find AB to the nearest centimeter. **73 cm**

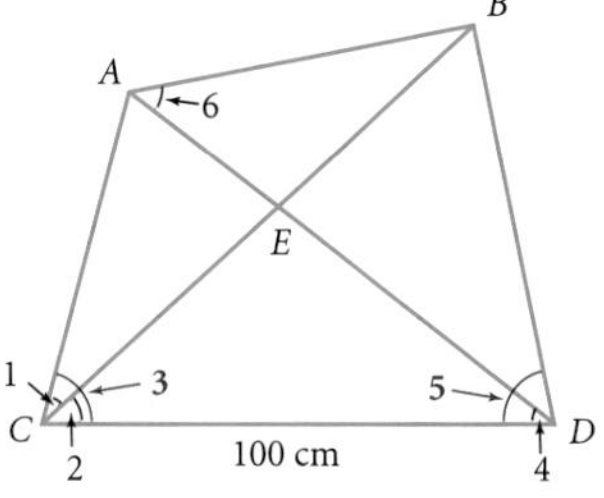

APPLICATIONS

52. FIRE FIGHTING Two rangers, one at station A and one at station B, observe a fire in the forest. The angle at station A formed by the lines of sight to station B and to the fire is 65.23°. The angle at station B formed by the lines of sight to station A and to the fire is 56.47°. The stations are 10 kilometers apart.

a. How far from station A is the fire? **9.8 km**

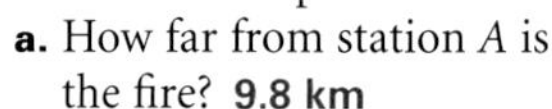

b. How far from station B is the fire? **10.7 km**

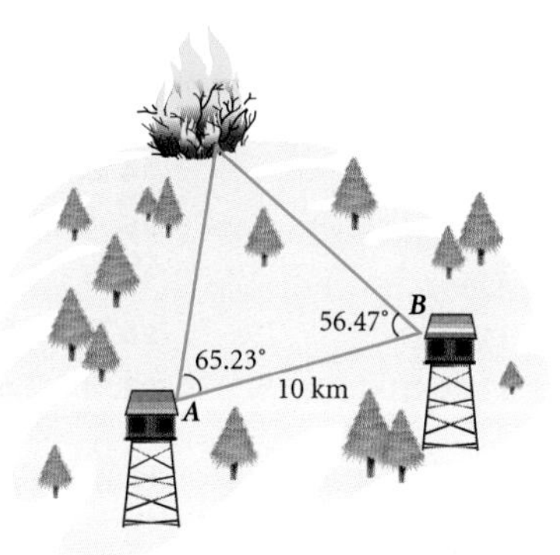

53. SURVEYING Refer to the diagram below. Find the distance from point A to point B across the river. Give your answer to the nearest meter. **691 m**

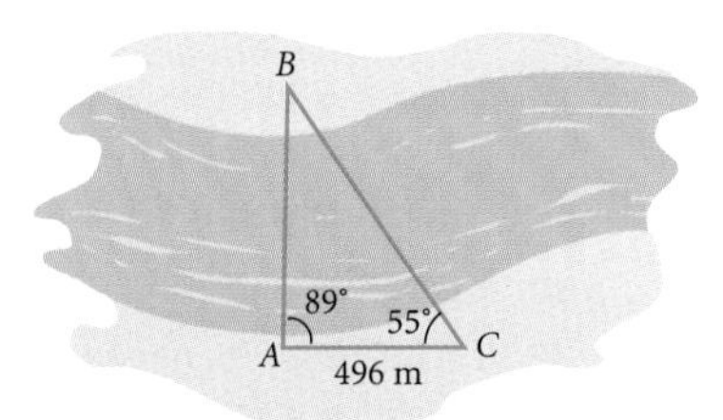

54. FORESTRY The angle of elevation between a straight path and a horizontal is 6°. A tree at the higher end of the path casts a 6.5-meter shadow down the path. The angle of elevation from the end of the shadow to the top of the tree is 32°. How tall is the tree? **3.36 m**

Practice

NAME ______________________ CLASS ____________ DATE ____________

Practice
14.1 **The Law of Sines**

Use the given information to find the indicated side length in ΔABC. Round answers to the nearest tenth.

1. Given $m\angle A = 28°$, $m\angle B = 95°$, and $a = 12$, find b. 25.5
2. Given $m\angle B = 51°$, $m\angle C = 70°$, and $c = 30$, find b. 24.8
3. Given $m\angle A = 105°$, $m\angle B = 64°$, and $a = 18$, find b. 16.7
4. Given $m\angle B = 48°$, $m\angle C = 62°$, and $b = 25$, find c. 29.7
5. Given $m\angle C = 100°$, $m\angle A = 82°$, and $a = 5.6$, find c. 5.6
6. Given $m\angle A = 75°$, $m\angle B = 55°$, and $b = 24.5$, find a. 28.9

Solve each triangle. Round answers to the nearest tenth.

7. $m\angle A = 82°, m\angle B = 60°, a = 5$
   $m\angle C = 38°, b = 4.4, c = 3.1$
8. $m\angle B = 65°, m\angle C = 80°, b = 20$
   $m\angle A = 35°, c = 21.7, a = 12.7$
9. $m\angle A = 100°, m\angle B = 35°, b = 15$
   $m\angle C = 45°, a = 25.8, c = 18.5$
10. $m\angle A = 72°, m\angle C = 64°, c = 5.2$
   $m\angle B = 44°, a = 5.5, b = 4.0$
11. $m\angle A = 46°, m\angle B = 52°, b = 17$
   $m\angle C = 82°, a = 15.5, c = 21.4$
12. $m\angle B = 39°, m\angle C = 66°, b = 54$
   $m\angle A = 75°, a = 82.9, c = 78.4$

State the number of triangles determined by the given information. If 1 or 2 triangles are formed, solve the triangle(s). Round answers to the nearest tenth, if necessary.

13. $m\angle A = 64°, b = 16, a = 20$ one; $m\angle B = 46.0°, m\angle C = 70.0°, c = 20.9$
14. $m\angle B = 98°, a = 10.5, b = 8.8$ none
15. $m\angle B = 28°, a = 40, b = 26$ two; (1) $m\angle A = 46.2°, m\angle C = 105.8°, c = 53.3$; (2) $m\angle A = 133.8°, m\angle C = 18.2°, c = 17.3$
16. Find, to the nearest tenth of a foot, the length of fence needed to enclose the triangular piece of land shown in the diagram.
   1532.7 feet

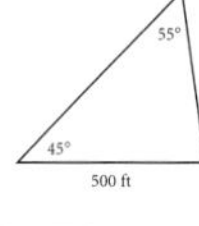

APPLICATIONS

The U.S. Coast Guard aids vessels in distress.

55. RESCUE A boat in distress at sea is sighted from two coast guard observation posts, A and B, on the shore. The angle at post A formed by the lines of sight to post B and to the boat is 41.67°. The angle at post B formed by the lines of sight to post A and to the boat is 36.17°. Find the distance, to the nearest tenth of a kilometer, from observation post A to the boat. **14.5 km**

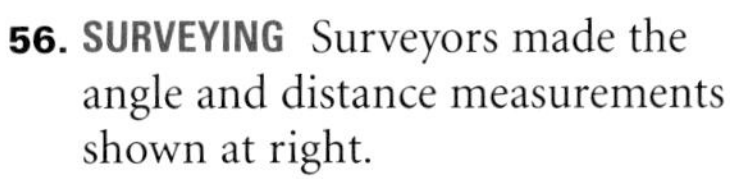

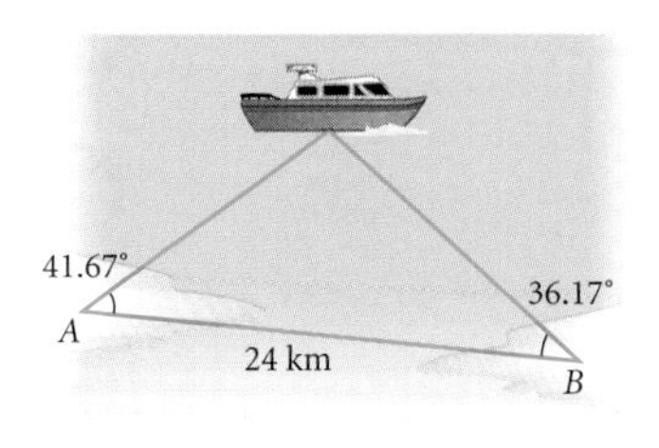

56. SURVEYING Surveyors made the angle and distance measurements shown at right.

a. Find distance c to the nearest meter. **≈156 m**

b. Find distance a to the nearest meter. **≈184 m**

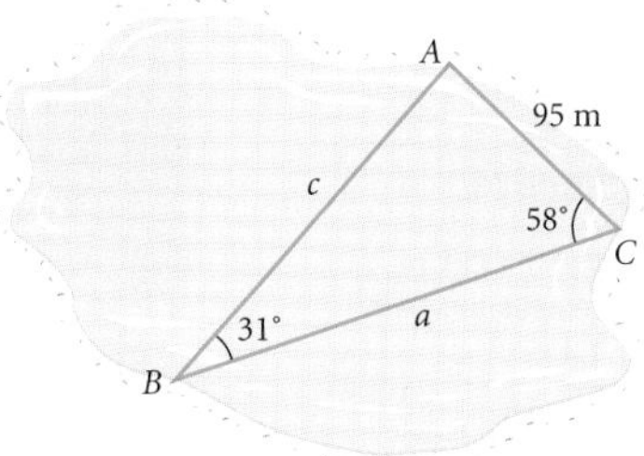

Look Back

APPLICATION

57. INVESTMENTS How long does it take for an investment to double at an annual interest rate of 5% compounded continuously? ***(LESSON 6.6)*** **about 14 years**

Factor each polynomial. ***(LESSON 7.3)***

58. $3x^3 - 12x$ **$3x(x-2)(x+2)$**

59. $2x^4 - 12x^3 + 18x^2$ **$2x^2(x-3)^2$**

60. Identify all asymptotes and holes in the graph of $f(x) = \frac{2x^2 + 10x}{x^2 + 2x - 15}$. ***(LESSON 8.2)*** **$x = 3$; hole $(x = -5)$; $y = 2$**

Convert each degree measure to radian measure. Give exact answers. ***(LESSON 13.4)***

61. 90° $\frac{\pi}{2}$

62. −180° $-\pi$

63. 135° $\frac{3\pi}{2}$

64. 120° $\frac{2\pi}{3}$

Convert each radian measure to degree measure. Round answers to the nearest tenth of a degree. ***(LESSON 13.4)***

65. $-\frac{\pi}{5}$ **−36°**

66. $\frac{3\pi}{7}$ **77.1°**

67. 4.1802 **239.5°**

68. −2.3221 **−133.1°**

Look Beyond

69. For each triangle, verify that $c^2 = a^2 + b^2 - 2ab \cos C$.

a.

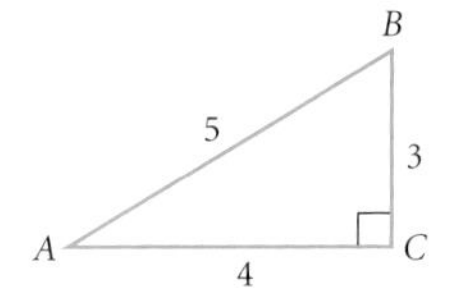

b.

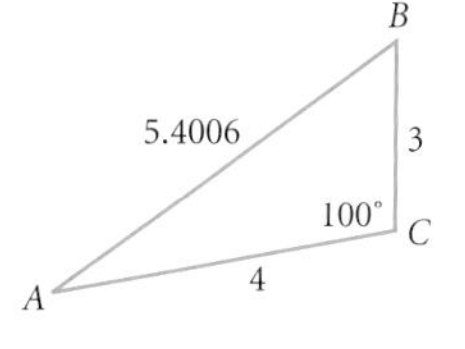

In Exercise 69, students ve[rify the] law of cosines. This law is [...] in Lesson 14.2.

69a. $c^2 = a^2 + b^2 - 2ab \cos$
$5^2 = 3^2 + 4^2 - 2(3)(4)$
$25 = 25$

b. $c^2 = a^2 + b^2 - 2ab \cos$
$5.4006^2 = 3^2 + 4^2 -$
$2(3)(4)\cos 100°$
$29.17 = 29.17$

Student Techno[logy] Guide

NAME ______ CLASS ______

Student Technology Guide
14.1 **The Law of Sines**

You can use a calculator to solve problems involving the law o[f sines.]

Before doing this lesson, press MODE, use the arrow keys to mo[ve to] Degree, and then press ENTER. Press 2nd MODE.

Example: Solve $\triangle ABC$ if $m\angle C = 31°$, $c = 5$, and $a = 9$. Round [...]

This is SSA information, so there may be 0, 1, or 2 tri[angles.]

- You know $m\angle C$ and the lengths of c and a. By the law of sines, $\frac{\sin 31°}{5} = \frac{\sin A}{9}$. Solving for $\sin A$ give[s] $\sin A = \frac{9 \sin 31°}{5}$.
- Press 9 × SIN 31) ÷ 5 ENTER. $\sin A \approx 0.927$
- Because $\sin A < 1$, consider two possibilities for $m\angle A$. To find the acute possibility, press 2nd SIN 2nd (-)) ENTER. To find the obtuse possibility, press 180 − 2nd (-) ENTER. To t[he] nearest tenth of a degree, $m\angle A = 68.0°$ or $112.0°$.
- Find both possibilities for $m\angle B$. Press 180 − 31 − 68 ENTER. Then press 2nd ENTER, change the 68 to 112, and press ENTER again. To the near[est] tenth of a degree, $m\angle B = 81.0°$ or $37.0°$.
- Find the possibilities for b. One possibility is $b = \frac{5 \sin 81°}{\sin 31°}$; the other is $b = \frac{5 \sin 37°}{\sin 31°}$. Press 5 × SIN 81) ÷ SIN 31) ENTER. Then press 2nd ENTER, change 81 to 37, and pre[ss] ENTER again.

To the nearest tenth, one triangle has $m\angle A = 68$ [...] $b = 9.6$; the other has $m\angle A = 112.0°$, $m\angle B = 3$[...]

State the number of triangles determined by the given information. Then solve the triangle(s). Round your ans[wers to] the nearest tenth.

1. $m\angle A = 41°$, $m\angle B = 22°$, $b = 10.1$
one triangle; $a = 17.7$, $m\angle C = 117.0°$, $c = 24.0$

2. $m\angle A =$ [...]
or

The Law of Cosines

Why *The law of cosines can be used to find distances that cannot be measured easily. For example, the law of cosines can be used to find the distance between two boats when a distress signal is sent.*

Objective

- Use the law of cosines to solve triangles.

APPLICATION
NAVIGATION

Two fishing boats, the *Tina Anna* and the *Melissa Jane*, leave the same dock at the same time. The *Tina Anna* sails at 15 nautical miles per hour and the *Melissa Jane* sails at 17 nautical miles per hour in directions that create a 115° angle between their paths.

After 3 hours, the *Tina Anna* sends a distress signal to the *Melissa Jane*. How far apart are the two boats when the distress signal is sent? As soon as the signal is sent, the *Tina Anna* stops and the *Melissa Jane* sails at 17 nautical miles per hour directly toward the *Tina Anna*. How long will it take for the *Melissa Jane* to reach the *Tina Anna*? To answer these questions, you can use the *law of cosines*. *You will solve this problem in Example 2.*

The law of cosines is used in solving triangles for which side-side-side (SSS) or side-angle-side (SAS) information is given. In these cases, the law of sines can be used only after more information is found by using the law of cosines.

To derive the law of cosines, consider $\triangle ABC$ with altitude $\overline{BD}$ whose length is h.

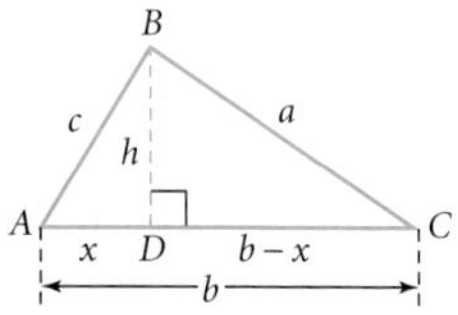

In $\triangle ABD$:

$c^2 = x^2 + h^2$ and $\cos A = \frac{x}{c}$, or $x = c \cos A$

In $\triangle CBD$:

$$a^2 = (b - x)^2 + h^2$$
$$a^2 = b^2 - 2bx + x^2 + h^2$$
$$a^2 = b^2 - 2bx + c^2 \quad \text{Substitute } c^2 \text{ for } x^2 + h^2.$$
$$a^2 = b^2 - 2b(c \cos A) + c^2 \quad \text{Substitute } c \cos A \text{ for } x.$$
$$a^2 = b^2 + c^2 - 2bc \cos A \quad \text{Simplify.}$$

Alternative Teaching Strategy

Tell students that in *any* triangle, the square of one side is equal to the sum of the squares of the other two sides minus twice the product of the other two sides and the cosine of the angle between them. Show them the triangle and diagram at right. Emphasize how the quantities are related by pointing to the different items in the triangle that are in the equation. Have students draw a similar diagram for the two other expressions in the law of cosines. Remind students to always sketch the triangle and label the given quantities before writing the equation.

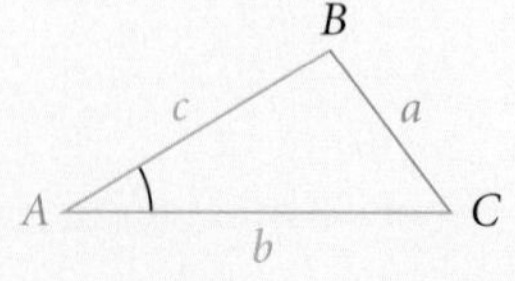

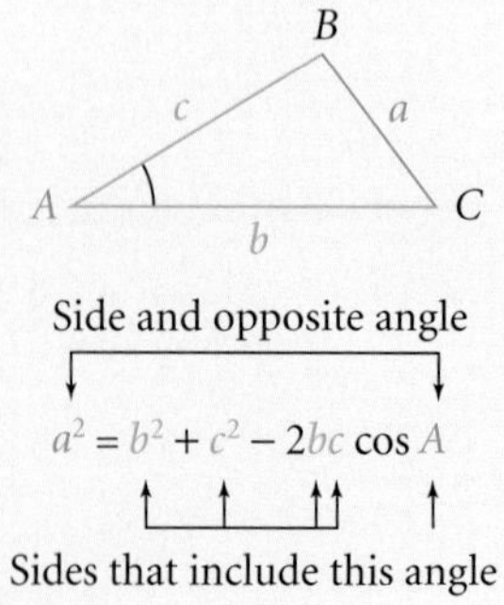

Side and opposite angle

$a^2 = b^2 + c^2 - 2bc \cos A$

Sides that include this angle

The two other formulas for the law of cosines can be derived in a similar fashion.

Law of Cosines

In any triangle $\triangle ABC$, the law of cosines states the following:

$$a^2 = b^2 + c^2 - 2bc \cos A$$
$$b^2 = a^2 + c^2 - 2ac \cos B$$
$$c^2 = a^2 + b^2 - 2ab \cos C$$

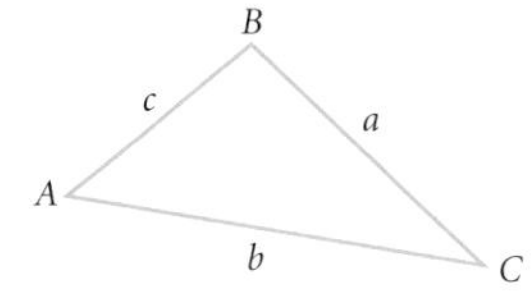

CRITICAL THINKING Show that the Pythagorean Theorem is a special case of the law of cosines. Then show that if $c^2 = a^2 + b^2$, then $\triangle ABC$ is a right triangle in which C is the right angle.

Example 1 below shows you how to use the law of cosines in two situations to find the unknown length of a side of a triangle when given SAS information and the unknown measure of an angle when given SSS information.

EXAMPLE 1 **Find the indicated measure to the nearest tenth for $\triangle ABC$.**

a. Given $a = 123$, $c = 97$, and $B = 22°$, find b.

b. Given $a = 11.3$, $b = 7.2$, and $c = 14.8$, find A.

SOLUTION

PROBLEM SOLVING

a. Draw a diagram.

A, C, B; b; $c = 97$; $22°$; $a = 123$

You are given SAS information. Use the law of cosines to find b.

$b^2 = a^2 + c^2 - 2ac \cos B$

$b^2 = 123^2 + 97^2 - 2(123)(97) \cos 22°$

$b = \sqrt{123^2 + 97^2 - 2(123)(97) \cos 22°}$

$b \approx 49.1$

CHECK

Note that the side opposite the smallest angle has the shortest length.

b. Draw a diagram.

C, A, B; $b = 7.2$; $a = 11.3$; $c = 14.8$

You are given SSS information. Use the law of cosines to find A.

$a^2 = b^2 + c^2 - 2bc \cos A$

$11.3^2 = 7.2^2 + 14.8^2 - 2(7.2)(14.8) \cos A$

$\cos A = \dfrac{11.3^2 - 7.2^2 - 14.8^2}{-2(7.2)(14.8)}$

$\cos A \approx 0.6719$

$A \approx \cos^{-1}(0.6719)$

$A \approx 47.8°$

TRY THIS Find the indicated measure, to the nearest tenth, for $\triangle XYZ$.

a. Given $x = 82$, $z = 63.2$, and $Y = 114°$, find y.

b. Given $x = 2.47$, $y = 3.80$, and $z = 4.24$, find X.

CRITICAL THINKING

If $C = 90°$, then $\cos 90° = 0$ and $c^2 = a^2 + b^2 - 2ab \cos C$ becomes $c^2 = a^2 + b^2$. If $c^2 = a^2 + b^2$, then $-2ab \cos C = 0$. Because a and b cannot be 0, $\cos C$ must equal 0. Since $\cos 90° = 0$, the triangle is a right triangle.

ADDITIONAL EXAMPLE 1

Find the indicated measures for $\triangle PQR$.

a. Given $q = 46$, $r = 72$, and $P = 126°$, find p.
$p \approx 106$

b. Given $p = 4.1$, $q = 6.7$, and $r = 3.3$, find Q.
$Q \approx 129.4°$

Teaching Tip

When given three sides of a triangle and asked to find the measure of the angles, the following alternative form of the law of cosines is useful:

$$A = \cos^{-1}\left(\frac{b^2 + c^2 - a^2}{2bc}\right)$$

In $\triangle ABC$:

1. If $a^2 = b^2 + c^2$, then A is a right angle.
2. If $a^2 < b^2 + c^2$, then A is an acute angle.
3. If $a^2 > b^2 + c^2$, then A is an obtuse angle.

TRY THIS

a. $y \approx 122.2$

b. $X \approx 35.2°$

Interdisciplinary Connection

ENVIRONMENTAL SCIENCE An oil tanker runs aground and begins leaking oil. By the time the leak is contained, a large circular spill has formed. An observer on shore sights the angle between her position and the endpoints of a diameter, which is parallel to the shore, of the circular region and reads a measure of 102°. The observer gets in a boat, measures the distance from the point where she was on shore to each endpoint of the diameter, and records distances of 1.2 miles and 3.05 miles. Use the law of cosines to find the diameter of the oil spill. **≈3.5 mi** Approximate the area of the spill. **≈9.62 sq mi**

Use Teaching Transparency 63.

ADDITIONAL EXAMPLE 2

Leaving from the same airport, airplane A flies west at 300 mph and airplane B flies southeast at 170 mph.

a. **How far apart will the planes be in 1 hour? Round your answer to the nearest mile.**

≈437 mi

b. **If both planes land and airplane A then flies at 290 mph to the airport where airplane B is located, how long will it take for airplane A to arrive?**

≈1 hr and 30 min

Activity Notes

In this Activity, students should find that there are two possible solutions in the interval (0, 180) when using the law of sines to find the measure of an angle and that there is one possible solution when using the law of cosines.

CHECKPOINT ✔

3. Two solutions are possible because $\sin \theta = x$ has two solutions for any θ in the interval [0, 180).

CHECKPOINT ✔

6. One solution is possible because $\cos \theta = x$ has one solution for any θ in the interval [0, 180).

EXAMPLE 2 Refer to the two fishing boats described at the beginning of the lesson. The boats leave the dock at the same time, and after 3 hours the *Tina Anna* sends a distress signal to the *Melissa Jane.*

APPLICATION NAVIGATION

a. How far apart are the two boats when the distress signal is sent? Give your answer to the nearest tenth of a nautical mile.

b. If the *Tina Anna* stops and the *Melissa Jane* sails at 17 nautical miles per hour toward the *Tina Anna*, about how long will it take for the *Melissa Jane* to reach the *Tina Anna*?

Tina Anna
T
15 nautical miles per hour
115°
D
17 nautical miles per hour
M
Melissa Jane

SOLUTION

PROBLEM SOLVING

a. **Draw a diagram.** Use the formula *distance* = *rate* × *time* to find *TD* and *MD*.

T
45
115°
D
51
M

$TD = 15(3) = 45$

$MD = 17(3) = 51$

Use the law of cosines to find *MT*.

$$(MT)^2 = (TD)^2 + (MD)^2 - 2(TD)(MD) \cos D$$
$$(MT)^2 = (45)^2 + (51)^2 - 2(45)(51) \cos 115°$$
$$MT = \sqrt{(45)^2 + (51)^2 - 2(45)(51) \cos 115°}$$
$$MT \approx 81.0$$

The boats are about 81.0 nautical miles apart.

b. To find the time, use the formula *distance* = *rate* × *time.*

$$d = rt$$
$$81.0 = 17t$$
$$t = \frac{81.0}{17}$$
$$t \approx 4.8$$

It will take the *Melissa Jane* about 4.8 hours, or 4 hours and 48 minutes, to reach the *Tina Anna.*

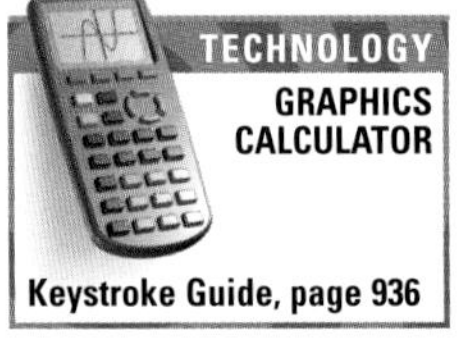

TECHNOLOGY GRAPHICS CALCULATOR

Keystroke Guide, page 936

Activity: Using Graphs to Explore Solutions

You will need: a graphics calculator or graph paper

1. Graph $y = \sin \theta$ for $0° \leq \theta \leq 180°$.
2. How many times does $\sin \theta = \frac{1}{2}$ on this interval?
3. What is the maximum number of solutions that are possible when finding an angle of a triangle by using the law of sines? Explain. (CHECKPOINT ✔)
4. Graph $y = \cos \theta$ for $0° \leq \theta \leq 180°$.
5. How many times does $\cos \theta = \frac{1}{2}$ on this interval?
6. What is the maximum number of solutions that are possible when finding an angle of a triangle by using the law of cosines? Explain. (CHECKPOINT ✔)

Inclusion Strategies

VISUAL LEARNERS The law of cosines can be difficult to apply because of the many variables and correspondences that need to be made. In the form $c^2 = a^2 + b^2 - 2ab \cos C$, the students must identify angle *C* and the side opposite that angle. Have students use colored pencils to label the angle and opposite side of a triangle with the same color. The order of the other two sides is unimportant because the commutative property allows *a* and *b* to be interchanged. The critical pieces of information to identify are angle *C* and side *c*.

Example 3 below shows you how to use the law of cosines and the law of sines with SAS information to solve a triangle.

EXAMPLE 3 **Solve $\triangle DFG$ at right. Give answers to the nearest tenth.**

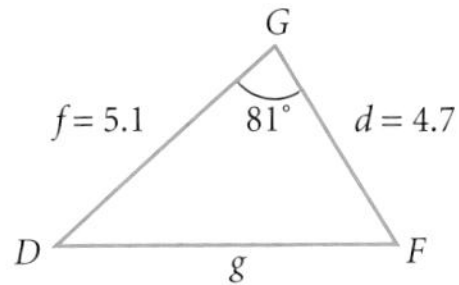

SOLUTION

1. Use the law of cosines to find g.

$$g^2 = d^2 + f^2 - 2df \cos G$$
$$g^2 = (4.7)^2 + (5.1)^2 - 2(4.7)(5.1) \cos 81^\circ$$
$$g = \sqrt{(4.7)^2 + (5.1)^2 - 2(4.7)(5.1) \cos 81^\circ}$$
$$g \approx 6.4$$

2. Then use the law of sines to find a second angle. Find D.

$$\frac{\sin D}{d} = \frac{\sin G}{g}$$
$$\frac{\sin D}{4.7} = \frac{\sin 81^\circ}{6.4}$$
$$\sin D = \frac{4.7 \sin 81^\circ}{6.4}$$
$$\sin D = 0.7253$$
$$D \approx 46.5^\circ \text{ or } D \approx 133.5^\circ$$

If $d < g$, then $D < G$. Therefore, $D \approx 46.5^\circ$.

Remember to consider both possible angle measures when using the law of sines.

3. Find G.

$$D + G + F = 180^\circ$$
$$F \approx 180^\circ - 81^\circ - 46.5^\circ$$
$$F \approx 52.5^\circ$$

Thus, $g \approx 6.4$, $D \approx 46^\circ$, and $F \approx 53^\circ$.

TRY THIS Solve $\triangle XYZ$ at right. Give answers to the nearest tenth.

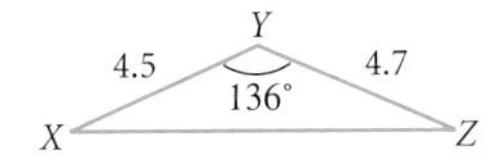

Example 4 below shows you how to use the law of cosines and the law of sines with SSS information to solve a triangle.

EXAMPLE 4 **Solve $\triangle ABC$ at right. Give answers to the nearest tenth.**

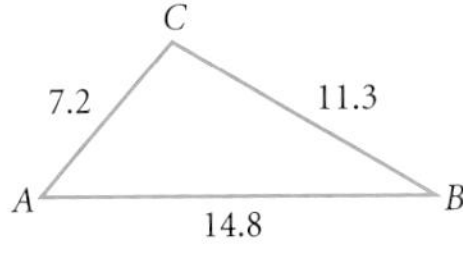

SOLUTION

The angle opposite the largest side can be either obtuse or acute. If you begin by using the law of cosines to find this angle, then you can use the law of sines to find the next angle without having to consider the ambiguous case.

1.
$$c^2 = a^2 + b^2 - 2ab \cos C$$
$$(14.8)^2 = (11.3)^2 + (7.2)^2 - 2(11.3)(7.2) \cos C$$
$$(14.8)^2 - (11.3)^2 - (7.2)^2 = -2(11.3)(7.2) \cos C$$
$$\cos C = \frac{(14.8)^2 - (11.3)^2 - (7.2)^2}{-2(11.3)(7.2)}$$
$$\cos C \approx -0.2428$$
$$C \approx 104.1^\circ$$

ADDITIONAL EXAMPLE 3

Solve $\triangle JKL$ below. Give k to the nearest tenth of a unit and J and L to the nearest degree.

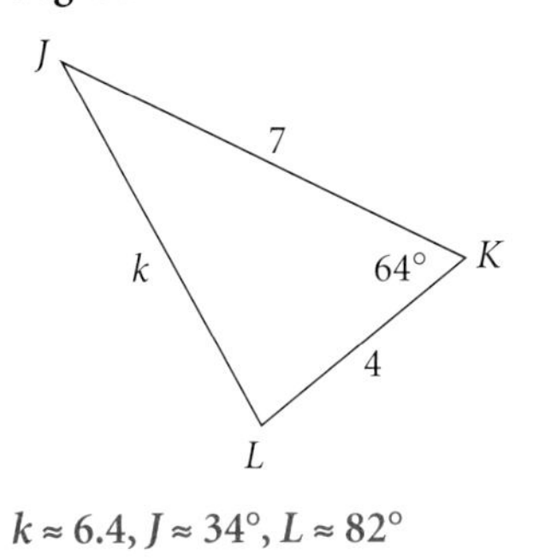

$k \approx 6.4, J \approx 34^\circ, L \approx 82^\circ$

TRY THIS

$X \approx 22.5^\circ, y \approx 8.5, Z \approx 21.5^\circ$

ADDITIONAL EXAMPLE 4

Solve $\triangle RST$ below. Give your answers to the nearest degree.

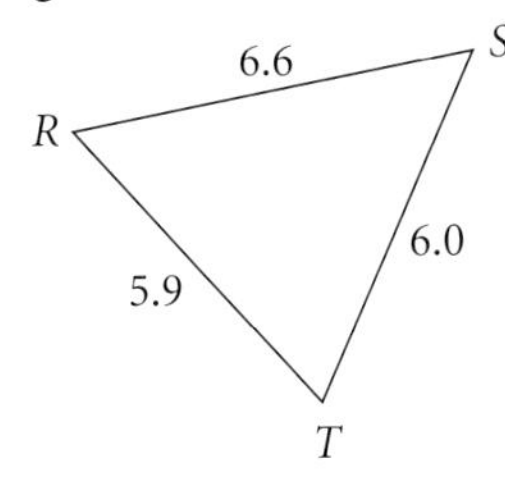

$R \approx 57^\circ; S \approx 56^\circ; T \approx 67^\circ$

Enrichment

Three circles with radii of 3 inches, 5 inches, and 9 inches are tangent to each other as shown below. To the nearest tenth of a degree, find the three angles formed by the segments joining their centers. **approximately 34.8°, 58.8°, 86.4°**

Assess

Selected Answers

Exercises 5–8, 9–55 odd

ASSIGNMENT GUIDE	
In Class	1–8
Core	9–41 odd, 45–49 odd
Core Plus	10–50 even
Review	51–56
Preview	57

Extra Practice can be found beginning on page 940.

PROBLEM SOLVING

Identify the wanted, given, and needed information. You could continue to use the law of cosines to find the measure of the second angle, but the law of sines is the easier method to use. Because C is obtuse, you know that the other angles are acute, so there will be no guessing involved when using the law of sines.

2. Find the measure of one of the other angles.

$$\frac{\sin B}{b} = \frac{\sin C}{c}$$

$$\frac{\sin B}{7.2} = \frac{\sin 104.1^\circ}{14.8}$$

$$\sin B = \frac{7.2 \sin 104.1^\circ}{14.8}$$

$$\sin B = 0.4718$$

$$B \approx 28.2^\circ$$

C, A, B; 7.2; 104.1°; 11.3; 14.8

3. Find the measure of the remaining angle.

$$A + B + C = 180^\circ$$

$$A + 28.2 + 104.1 \approx 180^\circ$$

$$A \approx 180^\circ - 28.2^\circ - 104.1^\circ$$

$$A \approx 47.7^\circ$$

SOLVING A TRIANGLE	
Given:	**Use:**
SSS	law of cosines, then law of sines
SSA	law of sines (ambiguous)
SAA	law of sines
ASA	law of sines
SAS	law of cosines, then law of sines
AAA	not possible

Exercises

Communicate

1. What variables appear in each of the three versions of the law of cosines, and what do these variables represent?

2. Explain how to solve a triangle by using the law of cosines if the lengths of the three sides of the triangle are known.

3. Explain how to solve a triangle by using the law of cosines if the lengths of two sides and the measure of the angle between them are known.

CONNECTION

4. GEOMETRY Explain why it is not possible to solve a triangle by using AAA information.

Reteaching the Lesson

INVITING PARTICIPATION Put students into groups of five and assign each team member one of the following duties: supervise other members, find side a, find side b, find side c, find angle C. Each group will solve the same triangle but will be given different pieces of information. Give a triangle with all of the sides and angles labeled to each supervisor. The supervisor will then give each group member the three pieces of information needed to solve his or her triangle. Once all group members have their solutions, the group should compare their results with the triangle given to the supervisor and discuss any discrepancies. Students should rotate until each member has assumed each duty.

Guided Skills Practice

5. Find the indicated measure to the nearest tenth for $\triangle ABC$. *(EXAMPLE 1)*

a. Given $a = 65$, $c = 52$, and $B = 31°$, find b. 33.7

b. Given $a = 8$, $b = 12.1$, and $c = 9.4$, find A. 41.3°

APPLICATION

6. NAVIGATION Refer to the two boats described at the beginning of the lesson and continued in Example 2. Assume that $D = 128°$, the *Tina Anna* sails at 21 nautical miles per hour and the *Melissa Jane* sails at 18.5 nautical miles per hour. How far apart are the boats when the distress signal is sent? How long will it take the *Melissa Jane* to reach the *Tina Anna*? *(EXAMPLE 2)*

6. ≈106.6 nautical miles; ≈5.8 hours, or 5 hours and 48 minutes

7. Solve $\triangle DEF$. Give answers to the nearest tenth. *(EXAMPLE 3)* $e \approx 8.0$; $D \approx 59°$; $F \approx 49°$

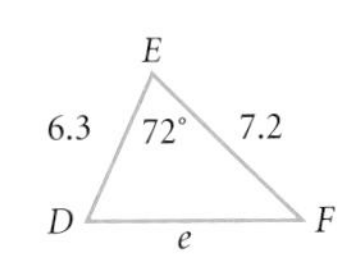

8. Solve $\triangle ABC$ at right. Give answers to the nearest tenth. *(EXAMPLE 4)* $A = 20.2°$; $B \approx 43.1°$; $C \approx 116.7°$

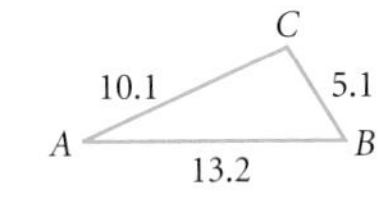

Practice and Apply

Classify the type of information given, and then find the measure of A in each triangle. Give answers to the nearest tenth.

9. SAS; $A \approx 52.5°$

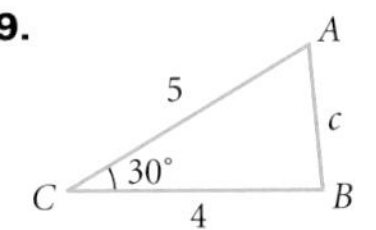

10. SAS; $A \approx 20.4°$

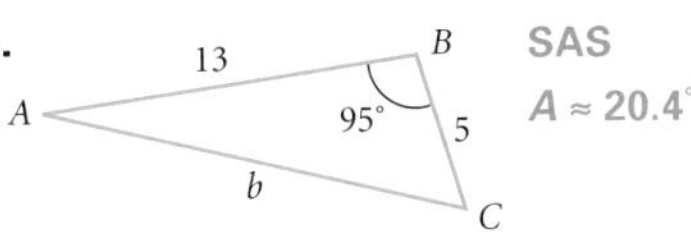

11. SSS; $A \approx 88.0°$

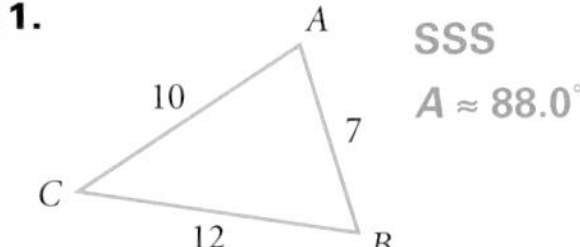

12. SSS; $A \approx 48.2°$

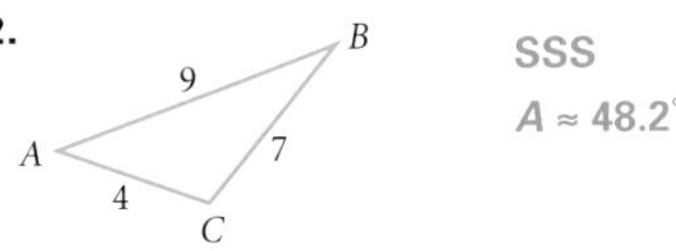

13. SAS; $c \approx 7.1$
14. SAS; $a \approx 8.6$
15. SAS; $b \approx 34.0$
16. SAS; $c \approx 64.0$
17. SAS; $a \approx 4.1$
18. SSA; $c \approx 5.3$

internet connect
Homework Help Online
Go To: **go.hrw.com**
Keyword:
MB1 Homework Help for Exercises 13–42

Classify the type of information given, and then use the law of cosines to find the missing side length of $\triangle ABC$ to the nearest tenth.

13. $a = 10$, $b = 15$, $C = 24°$

14. $b = 20$, $c = 14$, $A = 21°$

15. $a = 24.4$, $c = 16.2$, $B = 112°$

16. $a = 47.5$, $b = 58.0$, $C = 74°$

17. $A = 78°$, $b = 2$, $c = 4$

18. $B = 108°$, $a = 7$, $b = 10$

Solve each triangle. Give answers to the nearest tenth.

19. $a = 35$, $b = 49$, $c = 45$

20. $a = 8$, $b = 9$, $c = 33$

21. $a = 12.3$, $b = 14.0$, $c = 15.7$

22. $a = 18.1$, $b = 21.0$, $c = 23.7$

23. $a = 0.7$, $b = 0.9$, $c = 1.2$

24. $a = 8.4$, $b = 9.6$, $c = 11.4$

Error Analysis

When solving for an angle measure by using the law of cosines, students often calculate $a^2 + b^2 - 2ab$, forgetting the multiplication of the last term by cos C. When the unknown is an angle measure, remind students to subtract a^2 and b^2 from the left side of the equation and then divide by $-2ab$ before finding $\cos^{-1} \frac{c^2 - a^2 - b^2}{-2ab}$. Remind students to place parentheses around the denominator, $-2ab$, when entering it into a scientific calculator.

19. $A \approx 43.5°$; $B \approx 74.4°$; $C \approx 62.1°$
20. not possible
21. $A \approx 48.5°$; $B \approx 58.5°$; $C \approx 73.0°$
22. $A \approx 47.3°$; $B \approx 58.5°$; $C \approx 74.2°$
23. $A \approx 35.4°$; $B \approx 48.2°$; $C \approx 96.4°$
24. $A \approx 46.2°$; $B \approx 55.5°$; $C \approx 78.3°$

In Exercise 57, students explore the graph of the trig identity $\sin^2 x + \cos^2 x = 1$. The graph that students see on their graphics calculator should be the horizontal line $y = 1$.

35. SSS; $A \approx 93.8°, B \approx 29.9°, C \approx 56.3°$

36. SSS; $A \approx 75.5°, B \approx 57.9°, C \approx 46.6°$

37. SSA; $B = 42.7°, C \approx 79.3°, c \approx 11.6°$

38. SSA; $B \approx 63.1°, C \approx 74.9°, c \approx 13.0$ and $B \approx 116.9°, C \approx 21.1°, c \approx 4.8$

39. SSS; not possible

40. SSS; not possible

41. SSS; $A \approx 52.0°, B \approx 68.6°, C \approx 59.4°$

42. SSS; $A \approx 74.2°, B \approx 56.0°, C \approx 49.8°$

Practice

NAME ______ CLASS ______ DATE ______

Practice
14.2 The Law of Cosines

Classify the type of information given, and then use the law of cosines to find the missing side length of ΔABC. Round answers to the nearest tenth.

1. $m\angle A = 46°, b = 24, c = 18$ — SAS; $a = 17.3$
2. $a = 25, b = 28, m\angle C = 64°$ — SAS; $c = 28.2$
3. $m\angle A = 32°, b = 8, c = 10$ — SAS; $a = 5.3$
4. $a = 10, c = 12, m\angle A = 52°$ — SSA; $b = 9.8$
5. $m\angle C = 110°, a = 16, b = 22$ — SAS; $c = 31.3$
6. $b = 8, c = 14, m\angle A = 73°$ — SAS; $a = 13.9$

Solve each triangle by using the law of cosines and, where appropriate, the law of sines. Round answers to the nearest tenth.

7. $a = 17, b = 15, c = 24$ — $m\angle A = 44.7°, m\angle B = 38.3°, m\angle C = 97.0°$
8. $a = 50, b = 31, c = 46$ — $m\angle A = 78.3°, m\angle B = 37.4°, m\angle C = 64.3°$
9. $a = 12, b = 11, c = 9$ — $m\angle A = 73.0°, m\angle B = 61.2°, m\angle C = 45.8°$
10. $a = 4.5, b = 3.2, c = 6.1$ — $m\angle A = 45.8°, m\angle B = 30.7°, m\angle C = 103.5°$
11. $a = 10, b = 15, c = 13$ — $m\angle A = 41.1°, m\angle B = 80.3°, m\angle C = 58.6°$

Classify the type of information given, and then solve ΔABC. Round answers to the nearest tenth. If no such triangle exists, write *not possible*.

12. $m\angle B = 110°, a = 75, c = 85$ — SAS; $b = 131.2, m\angle A = 32.5°, m\angle C = 37.5°$
13. $a = 6.2, b = 8, c = 4.2$ — SSS; $m\angle A = 50.0°, m\angle B = 98.7°, m\angle C = 31.3°$
14. $a = 200, b = 100, c = 150$ — SSS; $m\angle A = 104.5°, m\angle B = 29.0°, m\angle C = 46.5°$
15. $m\angle A = 50°, b = 3, c = 8$ — SAS; $a = 6.5, m\angle B = 20.7°, m\angle C = 109.3°$
16. $m\angle C = 95°, a = 8, c = 6$ — SSA; not possible
17. Find, to the nearest tenth of a degree, the measures of the angles of an isosceles triangle in which the base is half as long as each side. — 29.0°, 75.5°, and 75.5°

25. SSS; $A \approx 79.0°, B \approx 54.9°, C \approx 46.1°$
26. SSS; $A \approx 29.0°, B \approx 104.5°, C \approx 46.5°$
27. SAS; $b \approx 49.1, A \approx 110.3°, C \approx 47.7°$
28. SAS; $a \approx 159.5, B \approx 44.8°, C \approx 21.2°$
29. SAS; $b \approx 3.2, A \approx 38.7°, C \approx 111.3°$
30. SAS; $b \approx 3.6, A \approx 36.1°, C \approx 98.9°$
31. SSS; not possible
32. SSS; not possible
33. SAS; $c \approx 6.2, A \approx 77.9°, B \approx 42.1°$
34. SAS; $c \approx 6.8, A \approx 17.1°, B \approx 132.9°$

Classify the type of information given, and then solve $\triangle ABC$. Give answers to the nearest tenth. If no such triangle exists, write *not possible*.

25. $a = 30, b = 25, c = 22$

26. $a = 10, b = 20, c = 15$

27. $a = 123, c = 97, B = 22°$

28. $b = 123, c = 63.2, A = 114°$

29. $B = 30°, a = 4, c = 6$

30. $B = 45°, a = 3, c = 5$

31. $a = 7, b = 9, c = 18$

32. $a = 8, b = 12, c = 21$

33. $C = 60°, a = 7, b = 5$

34. $C = 30°, a = 4, b = 10$

35. $a = 6, b = 3, c = 5$

36. $a = 8, b = 7, c = 6$

37. $A = 58°, a = 10, b = 8$

38. $A = 42°, a = 9, b = 12$

39. $a = 9, b = 15, c = 5$

40. $a = 4, b = 8, c = 13$

41. $a = 11, b = 13, c = 12$

42. $a = 29, b = 25, c = 23$

CHALLENGE

43. Find x in the figure at right. $x \approx 6.8$

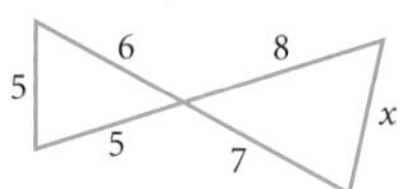

CONNECTIONS

44. GEOMETRY In parallelogram $ABCD$, $AC = 8.4$, $BD = 5.6$, and $m\angle CED = 80°$. Find the length of the sides of parallelogram $ABCD$ to the nearest tenth. $CD = AB \approx 4.6$; $AD = BC \approx 5.4$

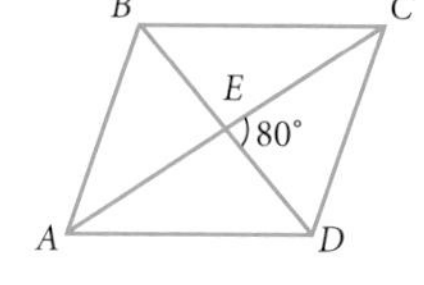

45. GEOMETRY Find all of the angle measures in an isosceles triangle whose base is $\frac{1}{3}$ as long as its legs. 19.2°; 80.4°; 80.4°

MAXIMUM/MINIMUM A TV antenna is to be installed on a roof that has a pitch of 5 in 12, or a rise of 5 feet for every 12 feet of run, as shown below. The manufacturer's instructions state that the angle that each of the two guy wires make with the pole should be no less than 30°.

46. 18 feet

46. What is the minimum length of each guy wire that could be used? Round your answer to the nearest foot.

47. longest pole: 15.8 feet; guy wire: 24.0 feet

47. What is the height to the nearest tenth of a foot of the longest antenna pole that could fit on the roof shown at right if the roof attachment point can vary? What is the length to the nearest foot of the guy wire that will be required for an antenna pole of this length?

48. no

48. Is it possible for the guy wire to make a 40° angle with a 13-foot antenna pole?

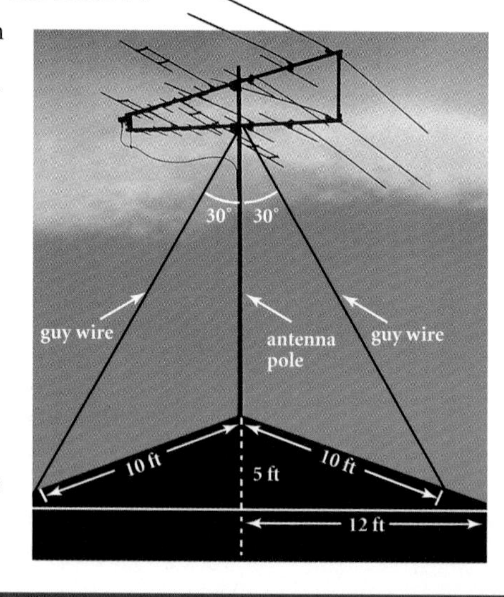

APPLICATIONS

49. SURVEYING A surveying crew needs to find the distance between two points, A and B, but a boulder blocks the path. The surveyors obtain the information shown in the diagram at right. Find AB. Give your answer to the nearest foot. **239 ft**

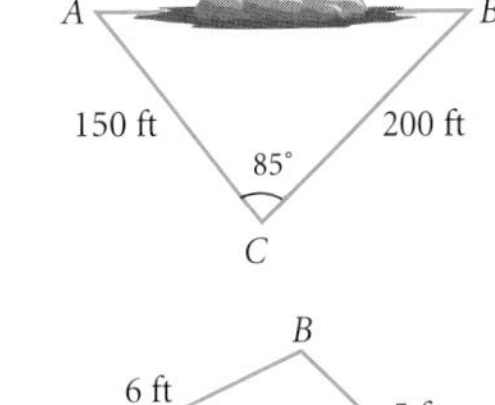

50. MANUFACTURING A piece of sheet metal is to be cut using a blowtorch so that it forms a triangle with the side lengths shown at right. Find the measures of angles A, B, and C. **$A \approx 31.6°$, $B \approx 109.5°$, $C \approx 38.9°$**

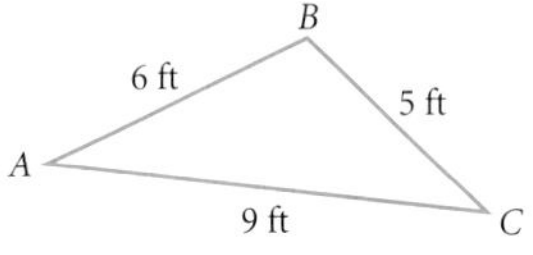

Graph each pair of parametric equations for the given interval of t. ***(LESSON 3.6)***

51. $\begin{cases} x(t) = 4t \\ y(t) = 2 - t \end{cases}$ for $2 \le t \le 6$

52. $\begin{cases} x(t) = 2t - 1 \\ y(t) = \frac{1}{2}t \end{cases}$ for $-4 \le t \le 4$

Use the quadratic formula to solve each equation. Give exact answers. ***(LESSON 5.5)***

53. $y = 6x^2 - x - 12$ **$-\frac{4}{3}, \frac{3}{2}$**

54. $y = 2x^2 + 5x + 2$ **$-2, -\frac{1}{2}$**

55. $y = x^2 + 3x - 2$ **$\frac{-3 \pm \sqrt{17}}{2}$**

56. $y = 2x^2 + 3x$ **$-\frac{3}{2}, 0$**

internet connect

Portfolio Extension

Go To: **go.hrw.com**
Keyword:
MB1 Trig Apps

57. Graph the function $y = \sin^2 x + \cos^2 x$ for x-values from 0 to 2π. Describe the graph. **The graph is $y = 1$.**

Gear-Tooth Design Gear makers often shape the sides of gear teeth so that they curve toward each other. This gear design allows gears to mesh smoothly without seizing up or jamming.

On gear tooth $ABCD$, $\overset{\frown}{BC}$ is the arc of a circle with its center at A and radius AC. In the diagram, $\theta = 13.5°$. The radius of the gear's base circle is 6 centimeters and the radius of the outer circle is 7 centimeters.

1. Make a sketch like the gear tooth above by using the given lengths and angle measures.

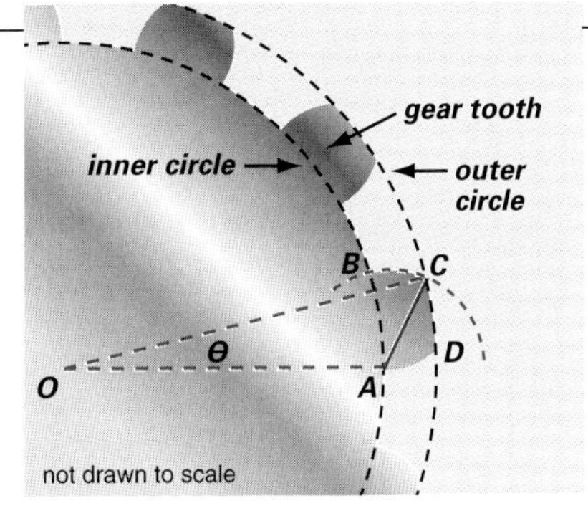

2. Use the law of cosines to find AC, the radius of the circle that generates $\overset{\frown}{BC}$.

WORKING ON THE CHAPTER PROJECT

You should now be able to complete Activity 1 of the Chapter Project.

ALTERNATIVE Assessment

Portfolio Activity

The Portfolio Activity can be used as preparation for the Chapter Project or as a separate activity. In the Portfolio Activity on this page, students use the law of cosines to find the lengths of a gear tooth.

Answers to Portfolio Activities can be found in Additional Answers of the Teacher's Edition.

Student Technology Guide

NAME ______ CLASS ______ DATE ______

Student Technology Guide
14.2 ***The Law of Cosines***

You can use a calculator to solve problems involving the law of cosines.

Before doing this lesson, press MODE, use the arrow keys to move to **Degree**, and then press ENTER. Press 2nd MODE.

Example: Find x if $y = 7.5$, $z = 10.4$, and $m\angle X = 58°$. Round your answer to the nearest tenth.

- By the law of cosines, $x^2 = y^2 + z^2 - 2yz \cos X$. Substituting and solving for x gives $x = \sqrt{7.5^2 + 10.4^2 - 2(7.5)(10.4) \cos 58°}$.
- Press 2nd x^2 7.5 x^2 + 10.4 x^2 - (2 × 7.5 × 10.4 × COS 58)) ENTER.

```
√(7.5²+10.4²-(2*
7.5*10.4*cos(58)
)
         9.041161141
```

To the nearest tenth, $x = 9.0$.

Example: Find $m\angle A$ if $a = 23.3$, $b = 18.5$, and $c = 12.9$. Round your answer to the nearest tenth of a degree.

- By the law of cosines, $a^2 = b^2 + c^2 - 2bc \cos A$. Substituting for a, b, and c gives $23.3^2 = 18.5^2 + 12.9^2 - 2(18.5)(12.9) \cos A$. Solving for $m\angle A$ gives $m\angle A = \cos^{-1}\left(\frac{18.5^2 + 12.9^2 - 23.3^2}{2(18.5)(12.9)}\right)$.
- Press 2nd COS (18.5 x^2 + 12.9 x^2 - 23.3 x^2) ÷ (2 × 18.5 × 12.9)) ENTER.

```
cos⁻¹((18.5²+12.9
²-23.3²)/(2*18.5
*12.9))
         94.11254891
```

To the nearest tenth of a degree, $m\angle A = 94.1°$.

(Note: In both examples, the simplest next step in solving the triangle is to use the law of sines to find the measure of a second angle. Then you can subtract the measures of the two known angles from 180° to find the measure of the third angle.)

Use the law of cosines to find the indicated side length or angle measure. Round your answers to the nearest tenth.

1. If $a = 2.7$, $b = 3.5$, and $c = 5.1$, find $m\angle C$. $m\angle C = 110.0°$
2. If $t = 28$, $u = 41$, and $m\angle V = 39°$, find v. $v = 26.1$

51.

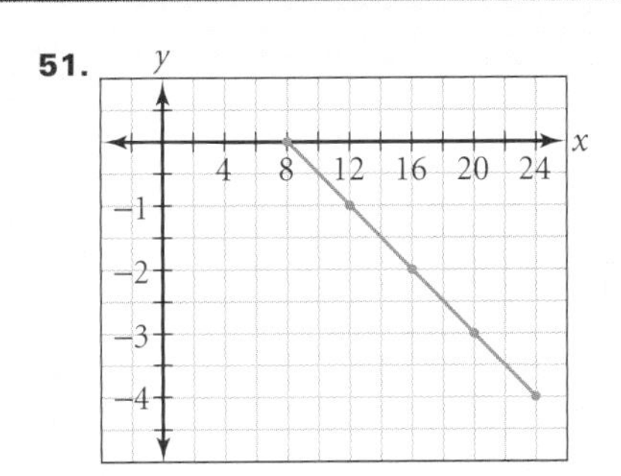

52.

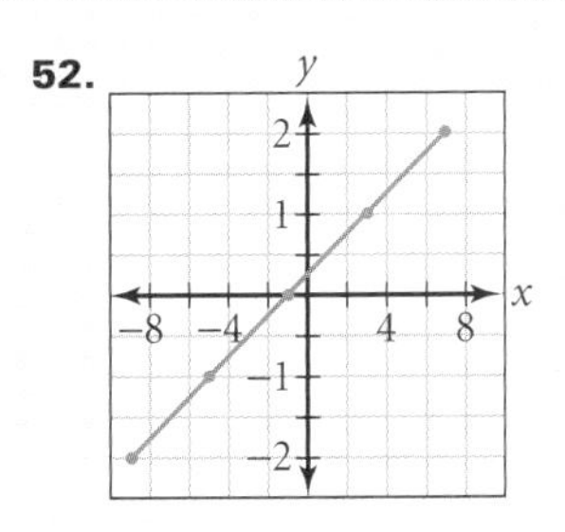

57.

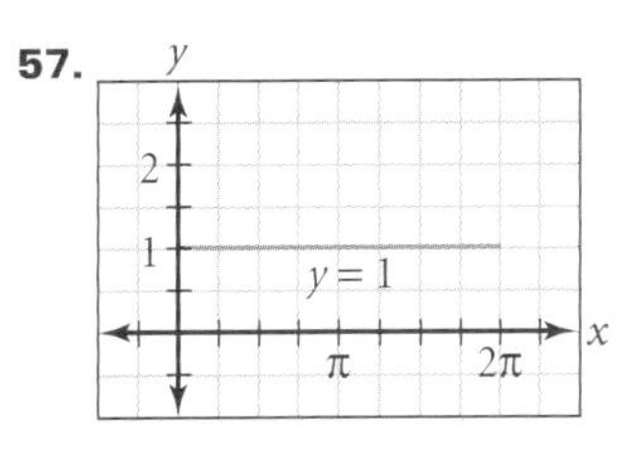

The graph is $y = 1$.

Prepare

NCTM PRINCIPLES & STANDARDS 1, 2, 4, 6–10

LESSON RESOURCES

- Practice 14.3
- Reteaching Master . . . 14.3
- Enrichment Master . . 14.3
- Cooperative-Learning Activity 14.3
- Lesson Activity 14.3
- Problem Solving/Critical Thinking Master 14.3

QUICK WARM-UP

Given the information below, solve for every possible $\triangle ABC$. Round angle measures to the nearest degree and lengths to the nearest tenth.

1. $a = 3, b = 4, c = 5$
   $A \approx 37°, B \approx 53°, C = 90°$
2. $A = 45°, B = 30°, c = 10$
   $C = 105°, a \approx 7.3, b \approx 5.2$
3. $B = 120°, a = 12, c = 10$
   $A \approx 33°, C \approx 27°, b \approx 19.1$
4. $A = 45°, a = 10, b = 12$
   two possible triangles:
   $B \approx 58°, C \approx 77°, c \approx 13.8;$
   or $B \approx 122°, C \approx 13°, c \approx 3.2$
5. $B = 30°, a = 12, b = 4$
   no possible triangles

Also on Quiz Transparency 14.3

Teach

Why Have students discuss why the sine function is used in the formula for the force causing the block to slide. In the right triangle shown, the sine function represents the dotted vertical line. Have students discuss the coefficient of friction for different types of surfaces.

14.3 Fundamental Trigonometric Identities

The coefficients of friction between rubber and concrete keeps this runner from slipping.

Objectives

- Prove fundamental trigonometric identities.
- Use fundamental trigonometric identities to rewrite expressions.

Why *Fundamental trigonometric identities can be used to rewrite a trigonometric expression as a single trigonometric function. This lets you solve real-world problems such as finding the angle of incline where rubber on concrete starts to slip.*

APPLICATION PHYSICS

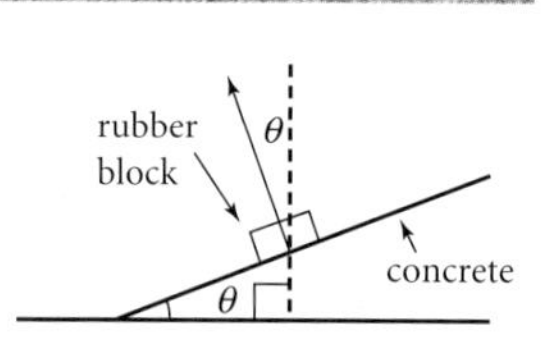

A block of rubber rests on a concrete platform. One end of the platform is slowly elevated. At what angle, θ, will the block of rubber begin to slide down the concrete platform? This angle is used to find the *coefficient of static friction*, μ_s (read "*mu* sub *s*"), between the rubber block and the cement. For a rubber block on concrete, this number is $\mu_s = 1.4$.

The force of friction that prevents the block from sliding is equal to $\mu_s mg \cos \theta$, where m is the mass of the block and g is the acceleration due to gravity. The force that causes the block to slide is $mg \sin \theta$. At the instant that the block begins to slide, both forces are equal, as shown below.

$$mg \sin \theta = 1.4mg \cos \theta$$

Use this equation to find the angle, θ, at which the block begins to slide. *You will solve this problem in Example 5.*

Trigonometric identities are equations that are true for all values of the variables for which the expressions on each side of the equation are defined. Recall from Chapter 13 that if $P(x, y)$ is a point on the terminal side of θ in standard position, then $\tan \theta = \frac{x}{y}$ and $P(x, y) = P(r \cos \theta, r \sin \theta)$. You can use these definitions to prove the identity $\tan \theta = \frac{\sin \theta}{\cos \theta}$.

Alternative Teaching Strategy

USING VISUAL MODELS Draw the diagram at right of a unit circle on the board or overhead. The blue legs of the smaller right triangle have lengths of $\sin \alpha$ and $\cos \alpha$. The larger right triangle is similar to the smaller right triangle. The green leg of the larger right triangle is a tangent of the circle and the other leg of this right triangle is a radius. Using similar triangles, we can set up a proportion between the sides: $\frac{\tan \alpha}{1} = \frac{\sin \alpha}{\cos \alpha}$. This gives the tangent identity. The values for the smaller right triangle and the Pythagorean theorem give the identity $\sin^2 \alpha + \cos^2 \alpha = 1$. The other two Pythagorean identities are derived by dividing each term in the first identity by either $\sin^2 \alpha$ or $\cos^2 \alpha$.

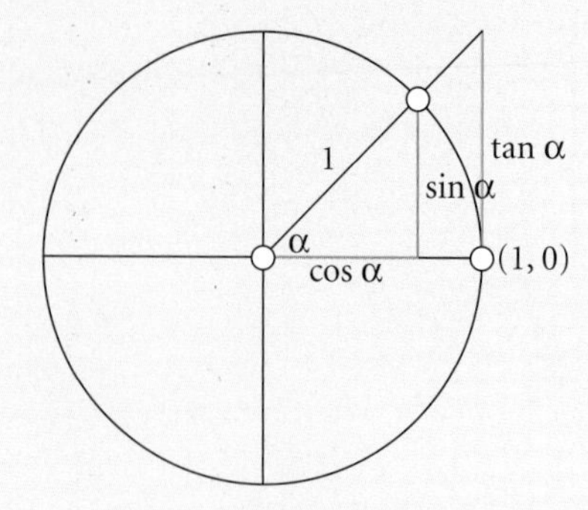

EXAMPLE 1 **Prove the identity $\tan\theta = \frac{\sin\theta}{\cos\theta}$.**

SOLUTION

$\tan\theta = \frac{y}{x}$ — Use the definition of tan θ.

$\tan\theta = \frac{r\sin\theta}{r\cos\theta}$ — Use substitution.

$\tan\theta = \frac{\sin\theta}{\cos\theta}$

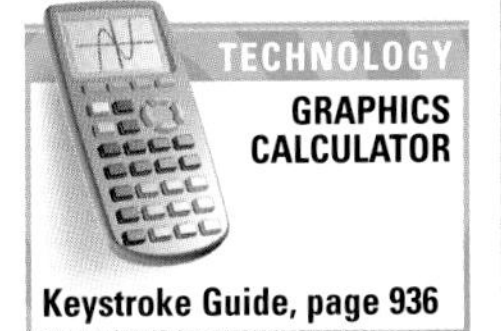

CHECK

In degree mode, graph $y = \tan x$ and $y = \frac{\sin x}{\cos x}$ on the same screen. The graphs appear to be the same. Note that this does not prove the identity, it only verifies it.

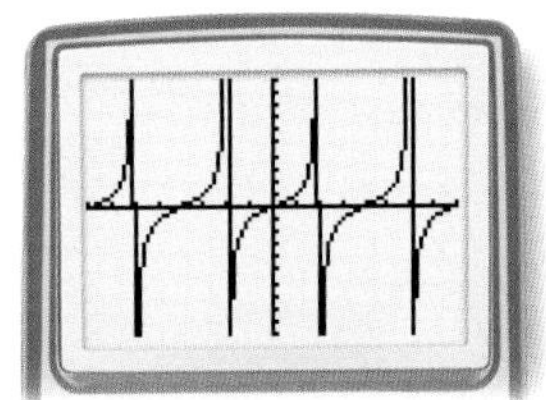

TRY THIS Prove the identity $\cot\theta = \frac{\cos\theta}{\sin\theta}$.

You can use a procedure similar to that shown in Example 1 to prove *ratio identities* and *reciprocal identities*. You can use the Pythagorean Theorem and the definitions of the trigonometric functions to prove the *Pythagorean identities.*

Fundamental Identities

Ratio identities	Reciprocal identities	Pythagorean identities
$\tan\theta = \frac{\sin\theta}{\cos\theta}$	$\csc\theta = \frac{1}{\sin\theta}$	$\cos^2\theta + \sin^2\theta = 1$
		$\sin^2\theta = 1 - \cos^2\theta$
$\cot\theta = \frac{\cos\theta}{\sin\theta}$	$\sec\theta = \frac{1}{\cos\theta}$	$\cos^2\theta = 1 - \sin^2\theta$
		$\tan^2\theta + 1 = \sec^2\theta$
	$\cot\theta = \frac{1}{\tan\theta}$	$1 + \cot^2\theta = \csc^2\theta$

Note that the square of $\sin\theta$ is written as $\sin^2\theta$. This form is used for all trigonometric functions.

EXAMPLE 2 **Prove the identity $\cos^2\theta + \sin^2\theta = 1$.**

SOLUTION

$\cos^2\theta + \sin^2\theta = \left(\frac{x}{r}\right)^2 + \left(\frac{y}{r}\right)^2$ — Use definitions of cos θ and sin θ.

$\cos^2\theta + \sin^2\theta = \frac{x^2+y^2}{r^2}$

$\cos^2\theta + \sin^2\theta = \frac{r^2}{r^2}$ — By the Pythagorean Theorem, $x^2 + y^2 = r^2$.

$\cos^2\theta + \sin^2\theta = 1$

TRY THIS Prove the identity $\tan^2\theta + 1 = \sec^2\theta$.

Inclusion Strategies

USING MODELS Have students set their calculators to radian mode and complete the table below, rounding decimals to three places.

θ	$\cos\theta$	$1 - \frac{\theta^2}{2}$	$\sin\theta$	$\theta - \frac{\theta^3}{6}$
0.05	0.999	0.999	0.050	0.050
0.1	0.995	0.995	0.998	0.998
0.2	0.980	0.980	0.199	0.199
0.3	0.955	0.955	0.300	0.300

Point out that trig functions can be closely approximated by very simple polynomials known as *Taylor polynomials.* Two other trig functions and the Taylor polynomials that approximate them are given below.

$$\tan\theta \approx \theta + \tfrac{1}{3}\theta^3 + \tfrac{2}{15}\theta^5$$

$$e^x\sin\theta \approx \theta + \theta^2 + \tfrac{1}{3}\theta^3$$

ADDITIONAL EXAMPLE 1

Prove the identity $\sec\theta = \frac{1}{\cos\theta}$.

$\sec\theta = \frac{r}{x}$ — Def. of secant

$\sec\theta = \frac{r}{r\cos\theta}$ — Substitution

$\sec\theta = \frac{1}{\cos\theta}$ — Simplify

TRY THIS

$\cot\theta = \frac{x}{y}$ — Def. cot θ

$= \frac{r\cos\theta}{r\sin\theta}$ — Substitution

$= \frac{\cos\theta}{\sin\theta}$ — Simplify

Teaching Tip

TECHNOLOGY Make sure that students have their calculators set in degree mode. Students should use the following window to view the graphs in Example 1:

Xmin=–360	Ymin=–10
Xmax=360	Ymax=10
Xscl=90	Yscl=1

ADDITIONAL EXAMPLE 2

Prove the identity $1 + \cot^2\theta = \csc^2\theta$.

$1 + \cot^2\theta$

$= 1 + \left(\frac{x}{y}\right)^2$ — Def. cot θ

$= \frac{y^2+x^2}{y^2}$ — Simplify

$= \frac{r^2}{y^2}$ — $r^2 = x^2 + y^2$

$= \csc^2\theta$ — Def. csc θ

TRY THIS

$\tan^2\theta + 1$

$= \left(\frac{y}{x}\right)^2 + 1$ — Def. tan θ

$= \frac{y^2+x^2}{x^2}$ — Simplify

$= \frac{r^2}{x^2}$ — $r^2 = x^2 + y^2$

$= \sec^2\theta$ — Def. sec θ

You can use the fundamental identities to rewrite trigonometric expressions in terms of a single trigonometric function.

EXAMPLE 3 **Write $\frac{\sin^2\theta}{1-\cos\theta}$ in terms of a single trigonometric function.**

SOLUTION

$$\frac{\sin^2\theta}{1-\cos\theta} = \frac{1-\cos^2\theta}{1-\cos\theta}$$ *Use $\sin^2\theta = 1-\cos^2\theta$.*

$$= \frac{(1+\cos\theta)(1-\cos\theta)}{1-\cos\theta}$$ *Factor the difference of two squares.*

$$= \frac{(1+\cos\theta)\cancel{(1-\cos\theta)}}{\cancel{1-\cos\theta}}$$

$$= 1+\cos\theta$$

TECHNOLOGY GRAPHICS CALCULATOR

Keystroke Guide, page 936

CHECK

Graph $y = \frac{\sin^2\theta}{1-\cos\theta}$ and $y = 1+\cos\theta$ on the same screen. The graphs appear to coincide.

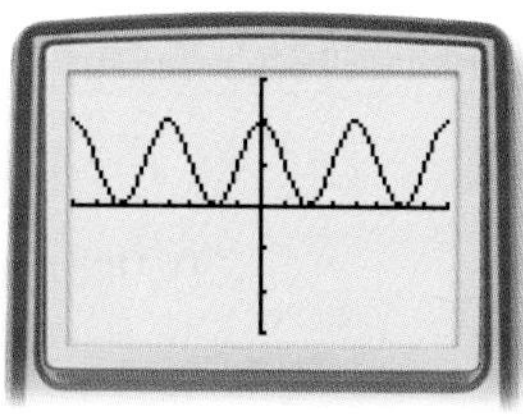

TRY THIS

Write $\frac{\cos^2\theta}{1-\sin\theta}$ in terms of a single trigonometric function.

Activity Exploring Graphing Methods

TECHNOLOGY GRAPHICS CALCULATOR

Keystroke Guide, page 937

You will need: a graphics calculator

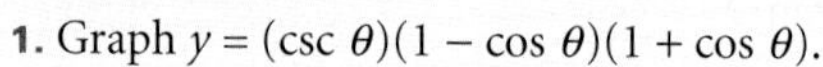

1. Graph $y = (\csc\theta)(1-\cos\theta)(1+\cos\theta)$.

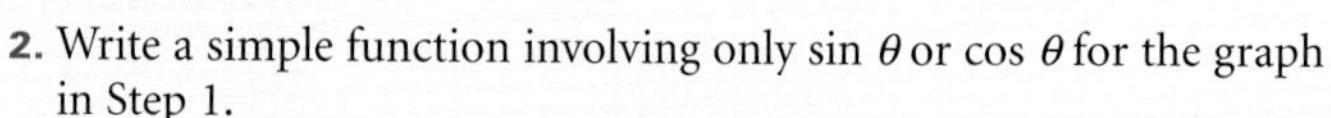

2. Write a simple function involving only $\sin\theta$ or $\cos\theta$ for the graph in Step 1.

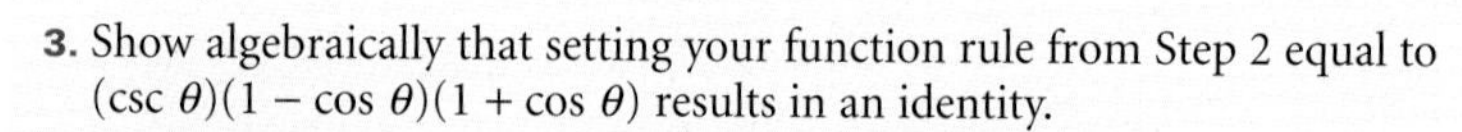

3. Show algebraically that setting your function rule from Step 2 equal to $(\csc\theta)(1-\cos\theta)(1+\cos\theta)$ results in an identity.
4. Repeat Steps 1–3, using $y = \tan\theta(\csc\theta - \tan\theta\cos\theta)$.

CHECKPOINT ✔

5. Describe one advantage to graphing the related function for a trigonometric expression to help simplify the expression.
6. Use your own example to illustrate how a graph can help you simplify a trigonometric expression.

CONNECTION TRANSFORMATIONS

You can use a graphics calculator to get hints on the outcome of rewriting a trigonometric expression. For example, you can graph $y = \frac{\sin^2\theta}{1-\cos\theta}$ from Example 3 before rewriting the expression to see that it appears to be the graph of $y = \cos\theta$ translated 1 unit up.

CHECKPOINT ✔ Graph $y = \tan^2\theta - \sec^2\theta$. What does it suggest to you about the result of rewriting the expression $\tan^2\theta - \sec^2\theta$?

Teaching Tip

Parentheses are often omitted in multiplication. For example, $(\sin\theta)(\cos\theta)$ is usually written as $\sin\theta\cos\theta$ and $2(\sin\theta)$ is usually written as $2\sin\theta$. Parentheses are required in expressions such as $\sin(\theta+\phi)$, which represents the sine of the sum of θ and ϕ, while the expression $\sin\theta + \phi$ represents the sum of $\sin\theta$ and ϕ.

ADDITIONAL EXAMPLE 3

Write $\frac{\sin\theta}{\tan\theta}$ in terms of a single trig function. $\cos\theta$

Teaching Tip

TECHNOLOGY For the graph in Example 3, make sure that students have their calculators set in degree mode. Students should use the following viewing window:

Xmin=–720	Ymin=–3
Xmax=720	Ymax=3
Xscl=90	Yscl=1

TRY THIS

$1+\sin\theta$

Activity Notes

In this Activity, students simplify trig expressions by using the graph of the expression as a guide. In Step 1, the function simplifies to $y = \sin\theta$. In Step 4, the function simplifies to $y = \cos\theta$.

CHECKPOINT ✔

5. **Graphing the related function can help identify the simplified expression that is the desired result.**

CHECKPOINT ✔

$\tan^2\theta - \sec^2\theta = -1$

Enrichment

Because sine and cosine ratios sometimes equal zero, the other trigonometric functions will be undefined for certain values. For example, if $\sin\theta$ is 0, then $\csc\theta$ is undefined for that value of θ. Have students explore the values between 0° and 360° for which each basic function is zero. For each value found, students should determine which other trig functions are also affected. Finally, have students graph each pair of functions at right on the same coordinate plane.

- $y = \sin x$ and $y = \csc x$
- $y = \cos x$ and $y = \sec x$
- $y = \cos x$ and $y = \tan x$
- $y = \sin x$ and $y = \cot x$

For each pair of graphs, the values for which the reciprocal function is undefined are indicated by vertical asymptotes.

EXAMPLE 4 **Write $\sec\theta - \tan\theta\sin\theta$ in terms of $\cos\theta$.**

TECHNOLOGY GRAPHICS CALCULATOR
Keystroke Guide, page 937

SOLUTION

Graph $y = \sec\theta - \tan\theta\sin\theta$.

Notice that the graph appears to be the same as $y = \cos\theta$.

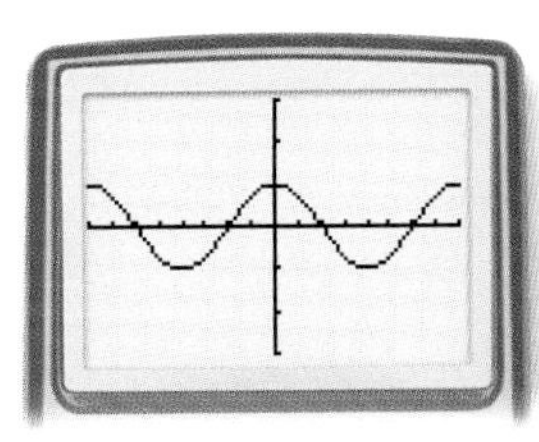

Use algebra to verify this.

$$\begin{aligned}\sec\theta - \tan\theta\sin\theta &= \frac{1}{\cos\theta} - \left(\frac{\sin\theta}{\cos\theta}\right)(\sin\theta) && \text{Use reciprocal and ratio identities.}\\ &= \frac{1}{\cos\theta} - \frac{\sin^2\theta}{\cos\theta}\\ &= \frac{1-\sin^2\theta}{\cos\theta}\\ &= \frac{\cos^2\theta}{\cos\theta} && \text{Use a Pythagorean identity.}\\ &= \cos\theta\end{aligned}$$

TRY THIS Write $\frac{1}{\sec^2\theta}$ in terms of $\sin\theta$.

CRITICAL THINKING Write $\tan\theta$ in terms of $\sin\theta$.

Example 5 uses substitution of equivalent trigonometric expressions to solve problems.

EXAMPLE 5 Refer to the friction problem at the beginning of the lesson.

APPLICATION PHYSICS

Use the equation $mg\sin\theta = \mu_s mg\cos\theta$ to determine the angle at which each material begins to slide.

a. rubber block on cement: $\mu_s = 1.4$

b. glass block on lubricated metal: $\mu_s = 0.25$

SOLUTION

a.
$$\begin{aligned}mg\sin\theta &= \mu_s mg\cos\theta\\ mg\sin\theta &= 1.4mg\cos\theta\\ \sin\theta &= 1.4\cos\theta\\ \frac{\sin\theta}{\cos\theta} &= 1.4\\ \tan\theta &= 1.4\\ \theta &\approx 54.5^\circ\end{aligned}$$

Thus, rubber will begin to slide on cement at an angle of about 54.5°.

b.
$$\begin{aligned}mg\sin\theta &= \mu_s mg\cos\theta\\ mg\sin\theta &= 0.25mg\cos\theta\\ \sin\theta &= 0.25\cos\theta\\ \frac{\sin\theta}{\cos\theta} &= 0.25\\ \tan\theta &= 0.25\\ \theta &\approx 14.4^\circ\end{aligned}$$

Thus, glass will begin to slide on lubricated metal at an angle of about 14.4°.

TRY THIS The coefficient of static friction for a certain type of leather on metal is $\mu_s = 0.8$. At what angle will a block of this type of leather begin to slide on a metal platform?

Reteaching the Lesson

USING COGNITIVE STRATEGIES Give students some identities and have them substitute values for the variable to illustrate that one side of the identity has the same value as the other side. Have students note the values of the variable for which each expression is undefined. Then give them the following list to help them simplify trig identities:

1. Simplify the more complicated expression first.
2. Perform all algebraic operations, such as squaring, factoring, adding fractions, and multiplying the numerator and denominator by a nonzero factor.
3. If other approaches fail, rewrite all trig expressions in terms of sine and cosine.
4. Rewrite one side of the identity in terms of a single trig expression.
5. Use previously proven identities.
6. Substitute values into both the original and the simplified identities to help identify possible errors.
7. Graph both the original and simplified identities on the same screen.

Math CONNECTION

TRANSFORMATIONS In Example 3, students use the graph of $y = \cos\theta$ and their understanding of the translation of functions to help them simplify the trig expression $\frac{\sin^2\theta}{1-\cos\theta}$ to $1 + \cos\theta$.

Teaching Tip

TECHNOLOGY For the Activity on page 904, have students set their viewing window as follows:

Xmin=–360 Ymin=–5
Xmax=360 Ymax=5

When graphing the trig functions, students should use **X** in place of θ. Remind students that the cosecant ratio is the reciprocal of the sine ratio, so they must enter **1/sin(X)** for csc θ.

ADDITIONAL EXAMPLE 4

Write $2\cos^2\theta - \sin^2\theta + 1$ in terms of $\cos\theta$. $3\cos^2\theta$

TRY THIS
$\frac{1}{\sec^2\theta} = 1 - \sin^2\theta$

CRITICAL THINKING
$\tan\theta = \frac{\sin\theta}{\sqrt{1-\sin^2\theta}}$, or $\frac{\sin\theta\sqrt{1-\sin^2\theta}}{1-\sin^2\theta}$

ADDITIONAL EXAMPLE 5

At what angle will a block begin to slide on a metal platform for the following coefficients of static friction?

a. $\mu_s = 0.10$ $\theta \approx 5.7^\circ$

b. $\mu_s = 2$ $\theta \approx 63.4^\circ$

TRY THIS
≈38.7°

Teaching Tip

The final answers to Exercises 13–26 should be in *simplest* form. For example, $\frac{1}{\cot \theta}$ should be further simplified to $\tan \theta$.

Assess

Selected Answers

Exercises 4–8, 9–71 odd

ASSIGNMENT GUIDE	
In Class	1–8
Core	9–53 odd, 59, 61
Core Plus	10–62 even
Review	63–72
Preview	73

Extra Practice can be found beginning on page 940.

Mid-Chapter Assessment for Lessons 14.1 through 14.3 can be found on page 182 of the *Assessment Resources*.

Exercises

Communicate

1. How is the tangent function related to the sine and cosine functions?
2. Describe two strategies that can be used to rewrite trigonometric expressions.

CONNECTION

3. GEOMETRY Explain how the trigonometric Pythagorean identity $\sin^2 \theta + \cos^2 \theta = 1$ is related to the Pythagorean Theorem.

Guided Skills Practice

4. Prove the identity $\sec \theta = \frac{1}{\cos \theta}$, for $\cos \theta \neq 0$. *(EXAMPLE 1)*
5. Prove the identity $\sin^2 \theta = 1 - \cos^2 \theta$. *(EXAMPLE 2)*
6. Write $\frac{\cos^2 \theta}{1 + \sin \theta}$ in terms of a single trigonometric function. *(EXAMPLE 3)*
7. Write $\cot^2 \theta$ in terms of $\sin \theta$. *(EXAMPLE 4)*

APPLICATION

8. PHYSICS The coefficient of static friction for a certain type of rubber on concrete is $\mu_s = 1.2$. At what angle will a block of this type of rubber begin to slide on a concrete platform? *(EXAMPLE 5)*

6. $1 - \sin \theta$

7. $\frac{1 - \sin^2 \theta}{\sin^2 \theta}$

8. 50.2°

Practice and Apply

Use definitions to prove each identity.

9. $\cot \theta = \frac{1}{\tan \theta}$, $\tan \theta \neq 0$
10. $\csc \theta = \frac{1}{\sin \theta}$, $\sin \theta \neq 0$
11. $\cos^2 \theta = 1 - \sin^2 \theta$
12. $1 + \cot^2 \theta = \csc^2 \theta$

Write each expression in terms of a single trigonometric function.

13. $\cot \theta \sin \theta$ — $\cos \theta$
14. $\tan \theta \cos \theta$ — $\sin \theta$
15. $\tan \theta \csc \theta$ — $\sec \theta$
16. $\tan \theta \sec \theta \sin \theta$ — $\tan^2 \theta$
17. $\csc \theta \sin^2 \theta$ — $\sin \theta$
18. $\sec \theta \cos^2 \theta$ — $\cos \theta$
19. $\left(\frac{\sin^2 \theta}{\cos \theta}\right)(\csc \theta)$ — $\tan \theta$
20. $\left(\frac{\cos^3 \theta}{\sin \theta}\right)(\sec^2 \theta)$ — $\cot \theta$
21. $\frac{\sin \theta}{\tan \theta}$ — $\cos \theta$
22. $\frac{\cos \theta}{\cot \theta}$ — $\sin \theta$
23. $\frac{\csc^2 \theta}{\cot^2 \theta}$ — $\sec^2 \theta$
24. $\frac{\sec^2 \theta}{\tan^2 \theta}$ — $\csc^2 \theta$
25. $\left(\frac{\sin \theta}{\cot \theta}\right)(\cos \theta)$ — $\sin^2 \theta$
26. $\left(\frac{\cos \theta}{\tan \theta}\right)(\sin \theta)$ — $\cos^2 \theta$

Practice

NAME ________ CLASS ________ DATE ________

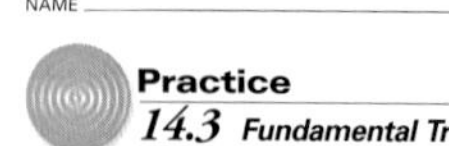

Practice
14.3 Fundamental Trigonometric Identities

Use definitions to prove each identity.

1. $\frac{\cot \theta}{\cos \theta} = \csc \theta$

$\frac{\cot \theta}{\cos \theta} = \frac{x}{y} \div \frac{x}{r}$
$= \frac{x}{y} \cdot \frac{r}{x}$
$= \frac{r}{y} = \csc \theta$

2. $(\sin \theta - 1)(\sin \theta + 1) = -\cos^2 \theta$

$(\sin \theta - 1)(\sin \theta + 1) = \left(\frac{y}{r} - 1\right)\left(\frac{y}{r} + 1\right)$
$= \frac{y^2}{r^2} - 1$
$= \frac{y^2 - r^2}{r^2}$
$= -\frac{x^2}{r^2} = -\cos^2 \theta$

3. $1 + \cot^2 \theta = \csc^2 \theta$

$1 + \cot^2 \theta = 1 + \frac{x^2}{y^2}$
$= \frac{x^2 + y^2}{y^2}$
$= \frac{r^2}{y^2} = \csc^2 \theta$

4. $1 + \tan^2 \theta = \sec^2 \theta$

$1 + \tan^2 \theta = 1 + \frac{y^2}{x^2}$
$= \frac{x^2 + y^2}{x^2}$
$= \frac{r^2}{x^2} = \sec^2 \theta$

Write each expression in terms of a single trigonometric function.

5. $\frac{1 - \cos^2 \theta}{\sin \theta}$ — $\sin \theta$
6. $(\sec \theta - 1)(\sec \theta + 1)$ — $\tan^2 \theta$
7. $\cot^2 \theta - \csc^2 \theta$ — -1
8. $\frac{(\sin \theta)(\cos \theta)}{1 - \sin^2 \theta}$ — $\tan \theta$
9. $\csc \theta - (\cos \theta)(\cot \theta)$ — $\sin \theta$
10. $(\sin \theta)(\tan \theta) + \cos \theta$ — $\sec \theta$

Write each expression in terms of $\sin \theta$.

11. $\frac{\csc \theta - \sin \theta}{\cot^2 \theta}$ — $\frac{\frac{1}{\sin \theta} - \sin \theta}{\frac{\cos^2 \theta}{\sin^2 \theta}} = \sin \theta$

12. $(\csc \theta - \cot \theta)(1 + \cos \theta)$ — $\left(\frac{1}{\sin \theta} - \frac{\cos \theta}{\sin \theta}\right)(1 + \cos \theta) = \sin \theta$

13. $\sin \theta + (\cos \theta)(\cot \theta)$ — $\sin \theta + \cos \theta \left(\frac{\cos \theta}{\sin \theta}\right) = \csc \theta = \frac{1}{\sin \theta}$

4. $\sec \theta = \frac{r}{x}$
$= \frac{1}{\frac{x}{r}}$
$= \frac{1}{\cos \theta}$

5. $\sin^2 \theta = \left(\frac{y}{r}\right)^2$
$= \frac{y^2}{r^2}$
$= \frac{r^2 - x^2}{r^2}$
$= \frac{r^2}{r^2} - \frac{x^2}{r^2}$
$= 1 - \left(\frac{x}{r}\right)^2$
$= 1 - \cos^2 \theta$

9. $\cot \theta = \frac{x}{y}$
$= \frac{1}{\frac{y}{x}}$
$= \frac{1}{\tan \theta}$, $\tan \theta \neq 0$

The answers to Exercises 10–12 can be found in Additional Answers beginning on page 1002.

27. $1 - 2\cos^2\theta$

28. $\cos^2\theta + 1$

29. $\frac{1 - 2\cos^2\theta}{(1 - \cos^2\theta)^2}$

30. $\frac{1}{\cos\theta}$

31. $\frac{1 - \sin^2\theta}{\sin\theta}$

32. $\frac{1}{\sin^2\theta} - 1$ or $\frac{1 - \sin^2\theta}{\sin^2\theta}$

Homework Help Online
Go To: **go.hrw.com**
Keyword:
MB1 Homework Help
for Exercises 35–50

Write each expression in terms of cos θ.

27. $2\sin^2\theta - 1$

28. $(1 - \sin^2\theta)(1 + \sec^2\theta)$

29. $(1 - \cot^2\theta)(\cot^2\theta + 1)$

30. $\frac{\tan\theta}{\sin\theta}$

Write each expression in terms of sin θ.

31. $\cot\theta\cos\theta$

32. $\cot^2\theta$

33. $\tan^2\theta\cos^2\theta + \csc\theta\sin^2\theta + \frac{1}{\sin\theta}$

34. $\frac{1}{\sec^2\theta}$ $1 - \sin^2\theta$

Use identities to verify that each statement is true.

35. $\frac{\sec\theta}{\csc\theta} = \tan\theta$

36. $\frac{\csc\theta}{\sec\theta} = \cot\theta$

37. $\frac{\tan^2\theta}{\sec^2\theta} = \sin^2\theta$

38. $\frac{\cot^2\theta}{\csc^2\theta} = \cos^2\theta$

39. $\cot^2\theta = \cos^2\theta\csc^2\theta$

40. $\tan^2\theta = \sin^2\theta\sec^2\theta$

41. $\frac{\sec\theta}{\cos\theta} = \sec^2\theta$

42. $\frac{\csc\theta}{\sin\theta} = \csc^2\theta$

43. $\frac{\cos\theta}{1 - \sin^2\theta} = \sec\theta$

44. $\frac{\sin\theta}{1 - \cos^2\theta} = \csc\theta$

45. $(\sec\theta)(1 - \sin^2\theta) = \cos\theta$

46. $(\csc\theta)(1 - \cos^2\theta) = \sin\theta$

47. $(\tan\theta)(\csc\theta)(\sec\theta) = \sec^2\theta$

48. $(\cot\theta)(\csc\theta)(\sec\theta) = \csc^2\theta$

49. Write $\tan^2\theta - 2\sec\theta\sin\theta$ in terms of $\sin\theta$ and $\cos\theta$.

50. Write $\tan^2\theta - 2\sec\theta\sin\theta$ in terms of $\tan\theta$.

51. $\pm\sqrt{1 - \cos^2\theta}$

52. $\pm\frac{1}{\sqrt{1 - \cos^2\theta}}$

53. $\pm\frac{\sqrt{1 - \cos^2\theta}}{\cos\theta}$

54. $\pm\frac{\cos\theta}{\sqrt{1 - \cos^2\theta}}$

Write each expression in terms of cos θ.

51. $\sin\theta$

52. $\csc\theta$

53. $\tan\theta$

54. $\cot\theta$

CHALLENGE

Write all of the trigonometric functions in terms of each given function.

55. $\sin\theta$

56. $\tan\theta$

57. $\cot\theta$

58. $\sec\theta$

APPLICATION

PHYSICS Refer to the friction problem described at the beginning of the lesson and continued in Example 5. Use the equation $mg\sin\theta = \mu_s mg\cos\theta$ to determine the angle at which each material begins to slide.

8.0° **59.** waxed wood on wet snow: $\mu_s = 0.14$

21.8° **60.** wood on wood: $\mu_s = 0.4$

31.0° **61.** wood on brick: $\mu_s = 0.6$

14.0° **62.** silk on silk: $\mu_s = 0.25$

Friction slows the motion of a skier, allowing them to turn and stop.

Error Analysis

Students often confuse the reciprocals of sine and cosine. Point out to students that the reciprocal functions do not begin with the same prefix. The reciprocal of cosine is secant, *not* cosecant, and the reciprocal of sine is cosecant, *not* secant. Also remind them that the reciprocal trig functions are not the same as the inverse functions.

35. $\tan\theta = \frac{\sec\theta}{\csc\theta} = \frac{\frac{1}{\cos\theta}}{\frac{1}{\sin\theta}} = \frac{\sin\theta}{\cos\theta} = \tan\theta$

36. $\cot\theta = \frac{\csc\theta}{\sec\theta} = \frac{\frac{1}{\sin\theta}}{\frac{1}{\cos\theta}} = \frac{\cos\theta}{\sin\theta} = \cot\theta$

37. $\sin^2\theta = \frac{\tan^2\theta}{\sec^2\theta} = \frac{\frac{\sin^2\theta}{\cos^2\theta}}{\frac{1}{\cos^2\theta}} = \sin^2\theta$

Student Technology Guide

NAME ______ CLASS ______ DATE ______

Student Technology Guide
14.3 Fundamental Trigonometric Identities

A calculator cannot prove a trigonometric identity. However, you can use a graphics calculator to check simplifications of trigonometric expressions.

Before doing this lesson, press MODE, use the arrow keys to move to Degree, and then press ENTER. Press 2nd MODE.

Example: Use graphing to confirm that $\frac{1}{\sin\theta\cos\theta} - \tan\theta = \cot\theta$.

To check that the two sides of the equation are equal, graph $y = \frac{1}{\sin\theta\cos\theta} - \tan\theta$ and $y = \cot\theta$ on the same screen and see whether the graphs coincide.

- First set the viewing window. Press WINDOW and change Xmin to −360, Xmax to 360, Xscl to 90, Ymin to −10, Ymax to 10, and Yscl to 1.
- Press Y= 1 ÷ (SIN X,T,θ,n) × COS X,T,θ,n)) − TAN X,T,θ,n) to enter $\frac{1}{\sin\theta\cos\theta} - \tan\theta$ as Y1.
- Use ▼ to move to Y2. Most calculators do not have a cotangent key, so you will need to use the identity $\frac{1}{\tan\theta} = \cot\theta$. To do this, press TAN X,T,θ,n) x^{-1}.
- Press GRAPH.

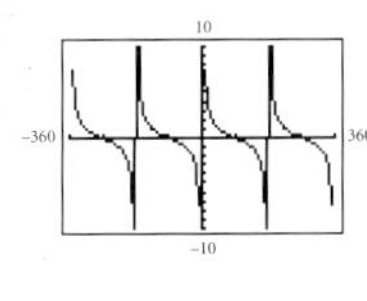

The graphs coincide, indicating that $\frac{1}{\sin\theta\cos\theta} - \tan\theta = \cot\theta$.

Simplify. Check your answers by graphing.

1. $\frac{\cos\theta}{\sin\theta}\cdot\tan\theta$ ___1___

2. $\frac{1 - \cos^2\theta}{\sin\theta}$ ___$\sin\theta$___

3. $\sin\theta\sec\theta$ ___$\tan\theta$___

Write each expression in terms of a single trigonometric function. Use a graphics calculator to check your answers.

4. $\frac{1 - \cos^2\theta}{1 - \sin^2\theta}$ ___$\tan^2\theta$___

5. $\tan\theta\cos\theta\sin\theta$ ___$\sin^2\theta$___

6. $(\csc^2\theta - 1)\sin^2\theta$ ___$\cos^2\theta$___

38. $\sin^2\theta = \frac{\cot^2\theta}{\csc^2\theta} = \frac{\frac{\cos^2\theta}{\sin^2\theta}}{\frac{1}{\sin^2\theta}} = \cos^2\theta$

39. $\cot^2\theta = \cos^2\theta\csc^2\theta = \cos^2\theta\left(\frac{1}{\sin^2\theta}\right) = \frac{\cos^2\theta}{\sin^2\theta} = \cot^2\theta$

40. $\tan^2\theta = \sin^2\theta\sec^2\theta = \sin^2\theta\frac{1}{\cos^2\theta} = \frac{\sin^2\theta}{\cos^2\theta} = \tan^2\theta$

The answers to Exercises 41–50 and 55–58 can be found in Additional Answers beginning on page 1002.

In calculus, students will use algebraic graphs to approximate trig functions. In Exercise 75, students see that for a very small window, $\sin x \approx x$ but that in a larger window this equation is obviously not an identity. Make sure that they have their calculators set to radian mode.

70.

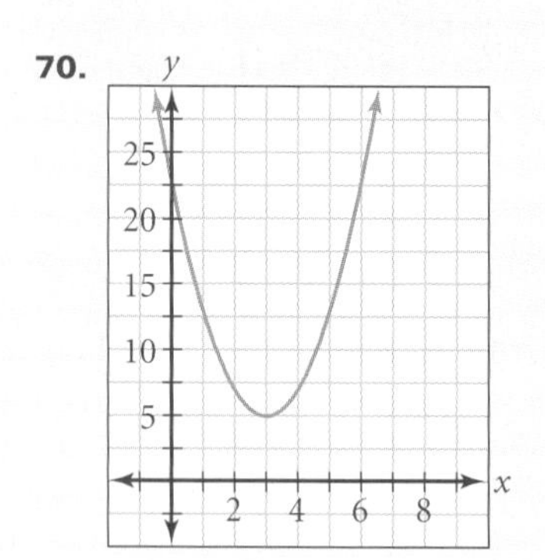

vertex: $(3, 5)$

focus: $\left(3, \frac{41}{8}\right)$

directrix: $y = \frac{39}{8}$

71.

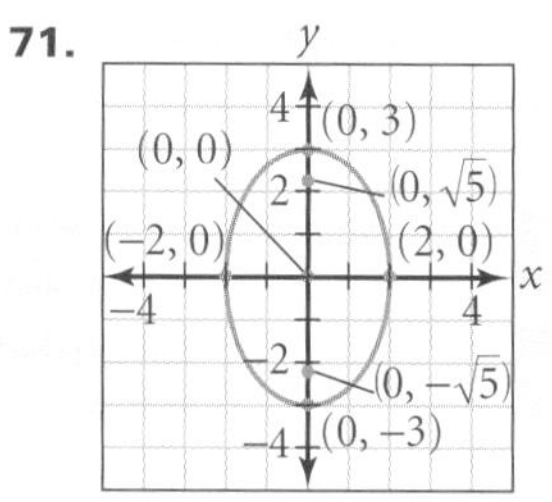

center: $(0, 0)$

foci: $(0, -\sqrt{5})$ and $(0, \sqrt{5})$

vertices: $(0, -3)$ and $(0, 3)$

co-vertices: $(-2, 0)$ and $(2, 0)$

73a. Yes; the graphs look the same.

b. No; the graphs look different.

c. Graphs can sometimes be misleading about identities, especially if the domain is limited.

63. $\begin{bmatrix} 2 & 1 & -6 \\ -1 & 1 & 1 \\ 5 & -3 & 7 \end{bmatrix}\begin{bmatrix} x \\ y \\ z \end{bmatrix} = \begin{bmatrix} -12 \\ -7 \\ 11 \end{bmatrix}$; $\left(-\frac{30}{11}, -\frac{102}{11}, -\frac{5}{11}\right)$, or about $(-2.7, -9.3, -0.5)$

Look Back

63 Write the matrix equation that represents the system at right and solve the system, if possible. *(LESSON 4.4)*

$$\begin{cases} 2x + y - 6z = -12 \\ -x + y + z = -7 \\ 5x - 3y + 7z = 11 \end{cases}$$

APPLICATION

GEOLOGY Recall from Lesson 6.7 that on the Richter scale, the magnitude, M, of an earthquake depends on the amount of energy, E, in ergs released by the earthquake as given by the equation $M = \frac{2}{3} \log \frac{E}{10^{11.8}}$. The list below gives information about some earthquakes that have occurred in the recent past.

Central square of Leninakan, Armenia, in 1988

Year	Location	Richter magnitude
1976	Tangshan, China	8.2
1978	Northeast Iran	7.7
1985	Mexico City, Mexico	8.1
1988	Northwest Armenia	6.8
1989	San Francisco, CA	7.1
1990	Northwest Iran	7.7
1993	South India	6.4
1994	Northridge, CA	6.8
1995	Kobe, Japan	7.2

Compare the amounts of energy released by the earthquakes listed for the indicated years. How much more energy was released by the greater earthquake? *(LESSON 6.7)*

64. 1976 and 1989 **44.7**

65. 1976 and 1985 **1.4**

66. 1978 and 1993 **89.1**

67. 1990 and 1993 **89.1**

68. 1976 and 1995 **31.6**

69. 1985 and 1994 **89.1**

70. Graph $y = 2(x - 3)^2 + 5$. Label the vertex, focus, and directrix. *(LESSON 9.2)*

71. Graph $\frac{x^2}{4} + \frac{y^2}{9} = 1$. Label the center, vertices, co-vertices, and foci. *(LESSON 9.4)*

72. In the diagram at right, $\angle BAC$ is the angle of depression. Point A represents the eyes of a person who is standing on top of a building and sees a traffic accident at C. How far from the base of the building, D, is the accident? Give your answer to the nearest foot. *(LESSON 13.1)* **255 feet**

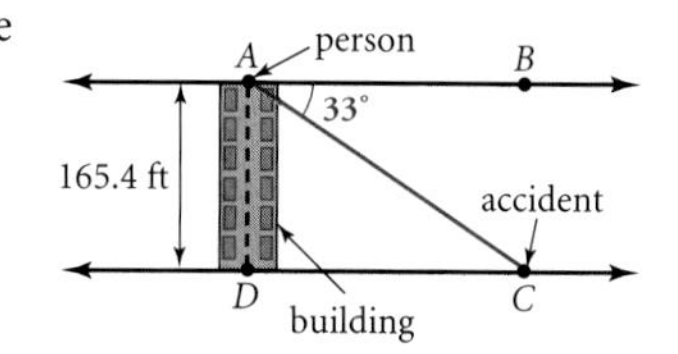

Look Beyond

73. Carry out the procedure below by using radian measure.

a. On the same axes, graph $y = \sin x$ and $y = x$ for $-0.3 \le x \le 0.3$. Does this suggest that $\sin x = x$?

b. Repeat part **a** for $-2 \le x \le 2$.

c. Draw a conclusion about using a graph to verify identities.

Sum and Difference Identities

Objectives

- Evaluate expressions by using the sum and difference identities.
- Use matrix multiplication with sum and difference identities to perform rotations.

Why *You can use the sum and difference identities and matrix multiplication to create designs that are composed of multiple rotations of an image.*

APPLICATION DESIGN

A design is made by rotating a rectangular figure as shown at left below. The figure on the positive x-axis has the vertices A, B, C, and D, shown at right below.

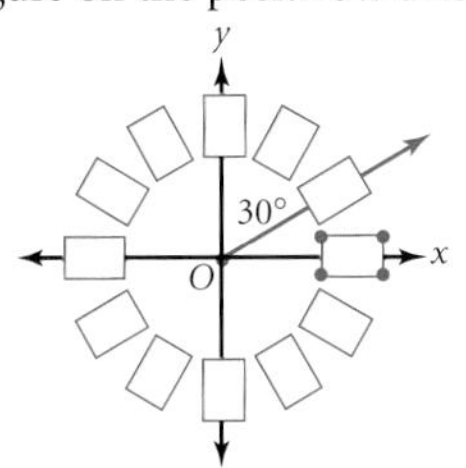

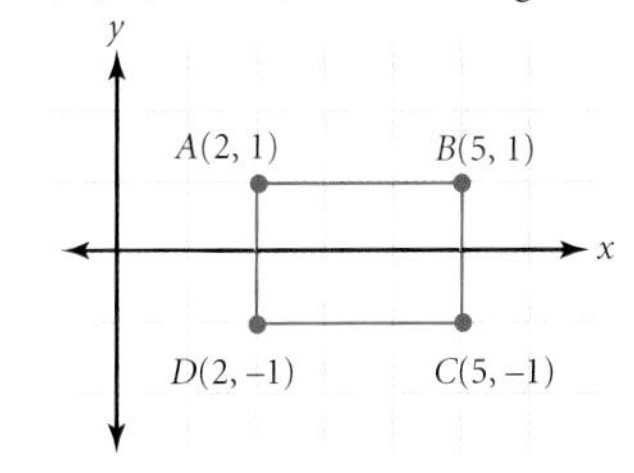

Find the coordinates of the vertices after a 30° rotation about the origin.
You will solve this problem in Example 5.

To solve the problem above, you can use a rotation matrix. The entries in the matrix are found by using the trigonometric *sum and difference identities.*

Activity: Proving the Difference Identity for Cosine

You will need: no special materials

In the diagram at left, $m\angle POQ = \alpha - \beta$.

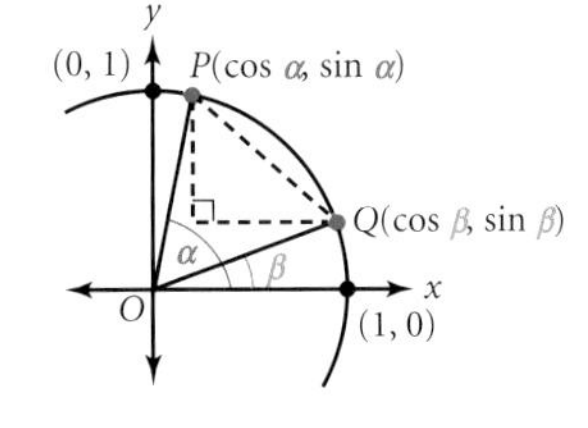

1. Using the distance formula, $d = \sqrt{(x_2 - x_1)^2 + (y_2 - y_1)^2}$, you can write the equation below.

$$(PQ)^2 = (\cos \alpha - \cos \beta)^2 + (\sin \alpha - \sin \beta)^2$$

   a. Show that the right side of the equation can be rewritten as $2 - 2 \cos \alpha \cos \beta - 2 \sin \alpha \sin \beta$.

   b. What identity did you use in part **a**?

Prepare

NCTM PRINCIPLES & STANDARDS 1–4, 6–10

QUICK WARM-UP

Find the distance between each pair of points below. Give approximate values rounded to the nearest tenth.

1. (4, 1) and (4, 0) — 1
2. (3, 2) and (–2, –5) — ≈8.6
3. (–5, 6) and (5, 1) — ≈11.2
4. $\left(\frac{\sqrt{2}}{2}, \frac{\sqrt{2}}{2}\right)$ and $\left(-\frac{1}{2}, \frac{\sqrt{3}}{2}\right)$ — ≈1.2
5. (cos 30°, sin 30°) and (cos 45°, sin 45°) — ≈0.26

Also on Quiz Transparency 14.4

Teach

Why The exact value of many trig functions can be found by using special angles and the sum and difference identities. Have students recall the special angles and the definitions of the basic trig functions. Then have them give the value of the trig functions for several other angles by using the special angles and the sum and difference identities.

Activity Notes

In this Activity, students use the distance formula to prove the difference identity for cosine. Students find two different expressions for the lengths PQ and RS. They should simplify each expression by using the Pythagorean identity, $\sin^2 \theta + \cos^2 \theta = 1$, and should see that the two expressions are equal because $\overline{RS}$ and $\overline{PQ}$ are corresponding parts of congruent triangles.

Alternative Teaching Strategy

USING TECHNOLOGY Give students a mixed list of equations that includes the sum and difference identities as well as untrue equations like the following:

- $\cos(A + B) = \cos A + \cos B$
- $\cos(A - B) = \cos A - \cos B$
- $\sin(A + B) = \sin A + \sin B$
- $\sin(A - B) = \sin A - \sin B$

Have students work in groups with a scientific calculator to determine which of the equations are identities. Students should substitute real numbers for A and B to test each equation. Make sure that students test enough values for A and B to decide whether the equation is an identity. Remind students that providing one value for which one side does not equal the other is enough to prove an equation is *not* an identity. On the other hand, providing many values for which the sides are equal is not enough to *prove* that the equation is an identity, but it suggests that the equation *may* be an identity. Have students create a list that summarizes all the identities and have them share their list with the class.

CHECKPOINT ✔

3. yes; $OP = OQ = OR = OS = 1$ by radii of the unit circle, $\triangle POQ \cong \triangle ROS$ by SAS, $PQ \cong RS$ by corr. parts; yes; by principle of powers

CHECKPOINT ✔

4. $2 - 2\cos(\alpha - \beta) = 2 - 2\cos\alpha\cos\beta - 2\sin\alpha\sin\beta$

$-2\cos(\alpha - \beta) = -2\cos\alpha\cos\beta - 2\sin\alpha\sin\beta$

$-2\cos(\alpha - \beta) = -2(\cos\alpha\cos\beta + \sin\alpha\sin\beta)$

$\cos(\alpha - \beta) = \cos\alpha\cos\beta + \sin\alpha\sin\beta$

ADDITIONAL EXAMPLE 1

Find the exact value of each expression.

a. $\sin(210° + 30°)$ $\quad -\frac{\sqrt{3}}{2}$

b. $\cos(210° + 30°)$ $\quad -\frac{1}{2}$

TRY THIS

a. -1

b. $\frac{\sqrt{2} - \sqrt{6}}{4}$

Teaching Tip

Encourage students to check each exact value with a calculator's approximate value.

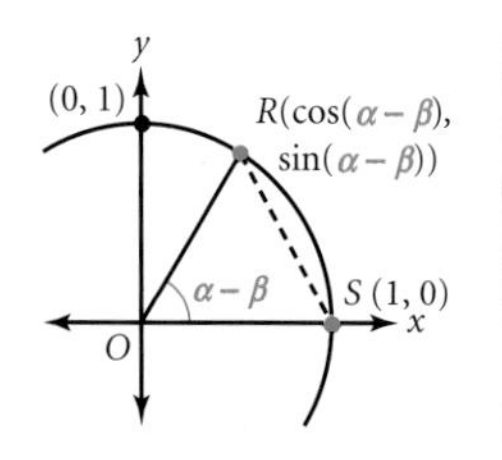

In the diagram at left, $m\angle ROS = \alpha - \beta$.

2. Using the distance formula, you can write the equation below.

$$(RS)^2 = [\cos(\alpha - \beta) - 1]^2 + [\sin(\alpha - \beta) - 0]^2$$

a. Show that the right side of the equation can be rewritten as $2 - 2\cos(\alpha - \beta)$.

b. What identity did you use in part **a**?

CHECKPOINT ✔ 3. Does $PQ = RS$? Explain. Does $(PQ)^2 = (RS)^2$? Explain.

CHECKPOINT ✔ 4. Using $(PQ)^2 = (RS)^2$ and the expressions for $(PQ)^2$ and $(RS)^2$ given in Step 1**a** and 2**a**, derive an expression for $\cos(\alpha - \beta)$ that includes $\sin\alpha$, $\sin\beta$, $\cos\alpha$, and $\cos\beta$.

All of the sum and difference identities can be proved in a manner similar to that explored in the Activity.

Sum and Difference Identities

$\sin(A + B) = \sin A\cos B + \cos A\sin B$
$\sin(A - B) = \sin A\cos B - \cos A\sin B$

$\cos(A + B) = \cos A\cos B - \sin A\sin B$
$\cos(A - B) = \cos A\cos B + \sin A\sin B$

EXAMPLE 1 **Find the exact value of each expression.**

a. $\sin(120° + 45°)$ **b.** $\cos(120° + 45°)$

SOLUTION

a. $\sin(120° + 45°) = (\sin 120°)(\cos 45°) + (\cos 120°)(\sin 45°)$

$$= \left(\frac{\sqrt{3}}{2}\right)\left(\frac{\sqrt{2}}{2}\right) + \left(-\frac{1}{2}\right)\left(\frac{\sqrt{2}}{2}\right)$$
$$= \frac{\sqrt{6}}{4} - \frac{\sqrt{2}}{4}$$
$$= \frac{\sqrt{6} - \sqrt{2}}{4}$$

b. $\cos(120° + 45°) = (\cos 120°)(\cos 45°) - (\sin 120°)(\sin 45°)$

$$= \left(-\frac{1}{2}\right)\left(\frac{\sqrt{2}}{2}\right) - \left(\frac{\sqrt{3}}{2}\right)\left(\frac{\sqrt{2}}{2}\right)$$
$$= \frac{-\sqrt{2}}{4} - \frac{\sqrt{6}}{4}$$
$$= \frac{-\sqrt{2} - \sqrt{6}}{4}$$

TRY THIS Find the exact value of each expression.

a. $\cos(210° - 30°)$ **b.** $\sin(330° - 135°)$

Interdisciplinary Connection

PHYSICS The motion of the prongs of a tuning fork and particles of air during the passage of a simple sound wave are examples of simple harmonic motion. The position of a point on such items at a certain time, t, measured in seconds can be described by the function $y = A\cos(2\pi ft + P_0)$, where A is the amplitude, $2\pi f$ is the angular frequency measured in radians per second, and P_0 is the initial phase shift measured in radians. The velocity of the point at any time, t, is given by $v = 2A\pi f\sin(2\pi ft + P_0)$. Given that the simple harmonic motion of a point is described by the equation $y = 3\cos(6\pi t + 4)$, have students rewrite the position and velocity functions by using the sum identities.

Inclusion Strategies

USING COGNITIVE STRATEGIES Have students make a list of the sum and difference identities. Put students in groups of 3 or 4, and have each group make up a mnemonic device to help them remember the identities. This might be in the form of a chant, a song, or some other device. Have each group share their mnemonic device with the class.

You can use the difference identities to derive other identities.

EXAMPLE 2 **Prove the identity $\sin(180° - \theta) = \sin \theta$.**

SOLUTION

$$\begin{aligned}\sin(180° - \theta) &= (\sin 180°)(\cos \theta) - (\cos 180°)(\sin \theta)\\ &= (0)(\cos \theta) - (-1)(\sin \theta)\\ &= \sin \theta\end{aligned}$$

TRY THIS Prove the identity $\sin(90° - \theta) = \cos \theta$.

A function, f, is **even** if $f(-x) = f(x)$ for all values of x in its domain. A function f is **odd** if $f(-x) = -f(x)$ for all x in its domain. You can use the difference identities to show that cosine is an even function and sine is an odd function.

$$\begin{aligned}\cos(-\theta) &= \cos(0° - \theta)\\ &= \cos 0° \cos \theta + \sin 0° \sin \theta\\ &= (1)(\cos \theta) + (0)\sin \theta\\ &= \cos \theta\end{aligned}$$

$$\begin{aligned}\sin(-\theta) &= \sin(0° - \theta)\\ &= \sin 0° \cos \theta - \cos 0° \sin \theta\\ &= (0)(\cos \theta) - (1)(\sin \theta)\\ &= -\sin \theta\end{aligned}$$

Thus, the cosine function is an even function and the sine function is an odd function.

CHECKPOINT ✔ Use the reciprocal identities to determine whether the secant and cosecant functions are odd or even.

EXAMPLE 3 **Find the exact value of each expression.**

a. $\sin(-165°)$ **b.** $\sin 195°$

SOLUTION

a.
$$\begin{aligned}\sin(-165°) &= -\sin 165°\\ &= -\sin(120° + 45°)\\ &= -[(\sin 120°)(\cos 45°) + (\cos 120°)(\sin 45°)]\\ &= -\left[\left(\frac{\sqrt{3}}{2}\right)\left(\frac{\sqrt{2}}{2}\right) + \left(-\frac{1}{2}\right)\left(\frac{\sqrt{2}}{2}\right)\right]\\ &= -\left(\frac{\sqrt{6} - \sqrt{2}}{4}\right)\\ &= \frac{\sqrt{2} - \sqrt{6}}{4}\end{aligned}$$

b.
$$\begin{aligned}\cos 195° &= \cos(150° + 45°)\\ &= (\cos 150°)(\cos 45°) - (\sin 150°)(\sin 45°)\\ &= \left(-\frac{\sqrt{3}}{2}\right)\left(\frac{\sqrt{2}}{2}\right) - \left(\frac{1}{2}\right)\left(\frac{\sqrt{2}}{2}\right)\\ &= \frac{-\sqrt{6} - \sqrt{2}}{4}\end{aligned}$$

TRY THIS Find the exact value of each expression.

a. $\cos(-105°)$ **b.** $\sin 285°$

Enrichment

Have students prove the following:

1. The tangent is an odd function.

$$\tan(-\theta) = \frac{\sin(-\theta)}{\cos(-\theta)} = \frac{-\sin \theta}{\cos \theta} = -\frac{\sin \theta}{\cos \theta} = -\tan \theta$$

Because $\tan(-\theta) = -\tan \theta$, the tangent is an odd function.

2. $\frac{\cos 2\theta}{\sin \theta} - \frac{\sin 2\theta}{\cos \theta} = \frac{\cos 3\theta}{\sin \theta \cos \theta}$

$$\begin{aligned}\frac{\cos 2\theta}{\sin \theta} - \frac{\sin 2\theta}{\cos \theta} &= \frac{\cos 2\theta \cos \theta - \sin 2\theta \sin \theta}{\sin \theta \cos \theta}\\ &= \frac{\cos(2\theta + \theta)}{\sin \theta \cos \theta}\\ &= \frac{\cos 3\theta}{\sin \theta \cos \theta}\end{aligned}$$

ADDITIONAL EXAMPLE 2

Prove $\tan(\alpha + \beta) = \frac{\tan \alpha + \tan \beta}{1 - \tan \alpha \tan \beta}$.

$$\begin{aligned}\tan(\alpha + \beta) &= \frac{\sin(\alpha + \beta)}{\cos(\alpha + \beta)}\\ &= \frac{\sin \alpha \cos \beta + \cos \alpha \sin \beta}{\cos \alpha \cos \beta - \sin \alpha \sin \beta}\\ &= \frac{\frac{\sin \alpha \cos \beta}{\cos \alpha \cos \beta} + \frac{\cos \alpha \sin \beta}{\cos \alpha \cos \beta}}{\frac{\cos \alpha \cos \beta}{\cos \alpha \cos \beta} - \frac{\sin \alpha \sin \beta}{\cos \alpha \cos \beta}}\\ &= \frac{\frac{\sin \alpha}{\cos \alpha} + \frac{\sin \beta}{\cos \beta}}{1 - \frac{\sin \alpha \sin \beta}{\cos \alpha \cos \beta}}\\ &= \frac{\tan \alpha + \tan \beta}{1 - \tan \alpha \tan \beta}\end{aligned}$$

TRY THIS

$$\begin{aligned}\sin(90° - \theta) &= \sin 90° \cos \theta - \cos 90° \sec \theta\\ &= 1 \cos \theta - 0 \bullet \sin \theta\\ &= \cos \theta\end{aligned}$$

CHECKPOINT ✔

$\sec(-\theta) = \frac{1}{\cos(-\theta)} = \frac{1}{\cos \theta} = \sec \theta$, secant is even;

$\csc(-\theta) = \frac{1}{\sin(-\theta)} = \frac{1}{-\sin \theta} = -\csc \theta$, cosecant is odd.

ADDITIONAL EXAMPLE 3

Find the exact value of each expression.

a. $\cos(-165°)$ $\quad \frac{-\sqrt{6} - \sqrt{2}}{4}$

b. $\cos 255°$ $\quad \frac{\sqrt{2} - \sqrt{6}}{4}$

TRY THIS

a. $\frac{\sqrt{2} - \sqrt{6}}{4}$

b. $\frac{-\sqrt{2} - \sqrt{6}}{4}$

ADDITIONAL
EXAMPLE 4

Graph $y = \tan(30° - \theta)$.

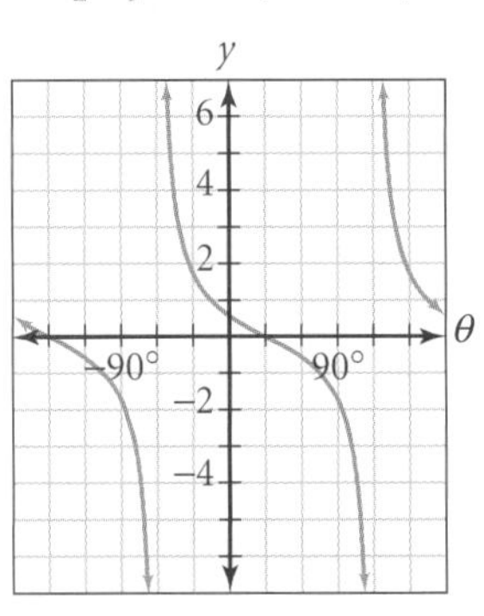

TRY THIS

$y = \cos(45° - \theta)$

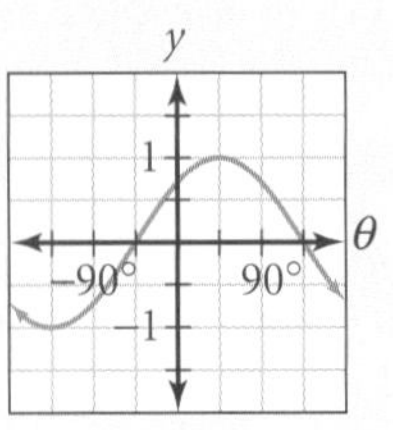

EXAMPLE 4 Graph $y = \sin(30° - \theta)$.

CONNECTION
TRANSFORMATIONS

SOLUTION

$y = \sin(30° - \theta)$
$= \sin[-(\theta - 30°)]$
$= -\sin(\theta - 30°)$ *The sine function is odd.*

Graph $y = \sin\theta$.

Then translate the graph 30° to the right, and reflect the graph across the x-axis.

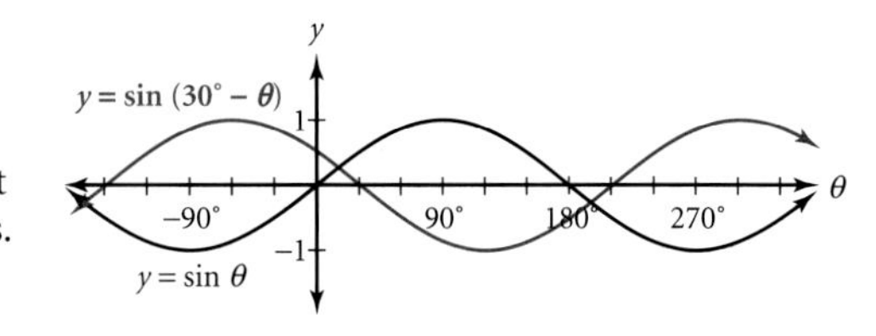

TRY THIS Graph $y = \cos(45° - \theta)$.

Rotation Matrices

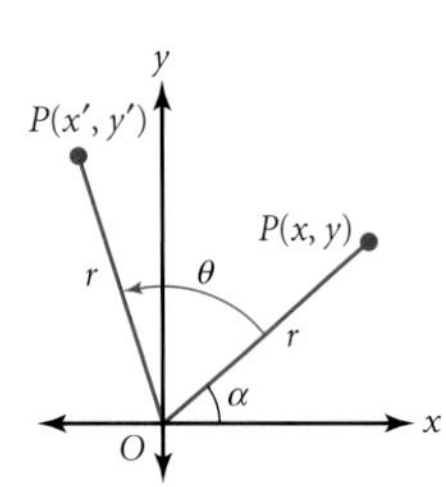

Matrix multiplication can be used in combination with sum and difference identities to determine the coordinates of points rotated on a plane about the origin. In the diagram at left, $P'(x', y')$ is the image of $P(x, y)$ after a rotation of θ degrees.

Recall from Lesson 13.3 that the coordinates of P are $(r\cos\alpha, r\sin\alpha)$. The coordinates of P' are found below.

$x' = r\cos(\alpha + \theta)$
$= r[(\cos\alpha)(\cos\theta) - (\sin\alpha)(\sin\theta)]$
$= (r\cos\alpha)(\cos\theta) - (r\sin\alpha)(\sin\theta)$
$= x(\cos\theta) - y(\sin\theta)$
$= x\cos\theta - y\sin\theta$

$y' = r\sin(\alpha + \theta)$
$= r[(\sin\alpha)(\cos\theta) + (\cos\alpha)(\sin\theta)]$
$= (r\sin\alpha)(\cos\theta) + (r\cos\alpha)(\sin\theta)$
$= y(\cos\theta) + x(\sin\theta)$
$= x\sin\theta + y\cos\theta$

Thus, you can find the coordinates of the image point, $P'(x', y')$, by using a *rotation matrix.*

$$\begin{bmatrix} x' \\ y' \end{bmatrix} = \begin{bmatrix} x\cos\theta - y\sin\theta \\ x\sin\theta + y\cos\theta \end{bmatrix} = \begin{bmatrix} \cos\theta & -\sin\theta \\ \sin\theta & \cos\theta \end{bmatrix}\begin{bmatrix} x \\ y \end{bmatrix}$$

Rotation Matrix

If $P(x, y)$ is any point in a plane, then the coordinates of the image of point $P'(x', y')$ after a rotation of θ degrees about the origin can be found by using a *rotation matrix* as follows:

$$\begin{bmatrix} \cos\theta & -\sin\theta \\ \sin\theta & \cos\theta \end{bmatrix}\begin{bmatrix} x \\ y \end{bmatrix} = \begin{bmatrix} x' \\ y' \end{bmatrix}$$

Reteaching the Lesson

USING COGNITIVE STRATEGIES Have students verify each identity by computing both sides for given values of A and B. Show students that if they rewrite 75° as 50° + 25°, the identities do not help them to compute sin 75° exactly. However, if students write 75° as 30° + 45°, the identities for the sum will give exact values for both sin 75° and cos 75°. For each identity, show students a sample decomposition that does not give exact values and one that does. Discuss techniques for finding decompositions that work well.

USING MODELS Have students cut out a polygonal figure from poster board and draw a set of axes on a piece of paper. Have them mark an angle of 120° on the axes, place one vertex of the polygonal figure at the origin of the axes, and record the coordinates of each vertex. Then have them rotate the figure 120°, keeping the vertex at the origin fixed. Then have them record the new coordinates of each vertex. Have them compare the new coordinates with those produced by multiplying a rotation matrix by a matrix of the original coordinates.

CHECKPOINT ✔ Write the rotation matrix for each angle of rotation.
a. 90° **b.** 180° **c.** 270°

EXAMPLE 5 Refer to the rectangular figure described at the beginning of the lesson.

APPLICATION
DESIGN

Find the coordinates to the nearest hundredth of the vertices after a 30° rotation about the origin.

SOLUTION

The rectangular figure has vertices at $A(2, 1)$, $B(5, 1)$, $C(5, -1)$, and $D(2, -1)$.

Write matrices for a 30° rotation and for the vertices of figure $ABCD$.

$$R_{30°} = \begin{bmatrix} \cos 30° & -\sin 30° \\ \sin 30° & \cos 30° \end{bmatrix} \qquad S = \begin{bmatrix} 2 & 5 & 5 & 2 \\ 1 & 1 & -1 & -1 \end{bmatrix}$$

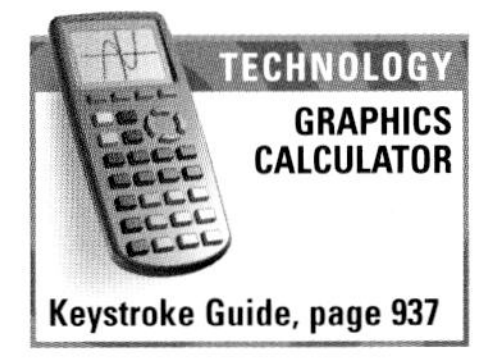

Find the matrix product.

$$R_{30°} \times S = \begin{bmatrix} \cos 30° & -\sin 30° \\ \sin 30° & \cos 30° \end{bmatrix} \begin{bmatrix} 2 & 5 & 5 & 2 \\ 1 & 1 & -1 & -1 \end{bmatrix}$$
$$\approx \begin{bmatrix} 1.23 & 3.83 & 4.83 & 2.23 \\ 1.87 & 3.37 & 1.63 & 0.13 \end{bmatrix}$$

The approximate coordinates of the vertices for the image of figure $ABCD$ are $A'(1.23, 1.87)$, $B'(3.83, 3.37)$, $C'(4.83, 1.63)$, and $D'(2.23, 0.13)$.

TRY THIS Given the figure described in Example 5, find the coordinates of the vertices for the image of this figure after a 60° rotation about the origin.

CRITICAL THINKING Show that a 90° rotation about the origin is the same as a 60° rotation about the origin followed by a 30° rotation.

Exercises

Communicate

1. Explain how to use sum or difference identities to find the exact value of cos 75°.
2. Explain how to use sum or difference identities to find the exact value of sin(−15°).
3. Explain how a matrix can be used to rotate the point $P(2, 3)$ at right 45° about the origin.

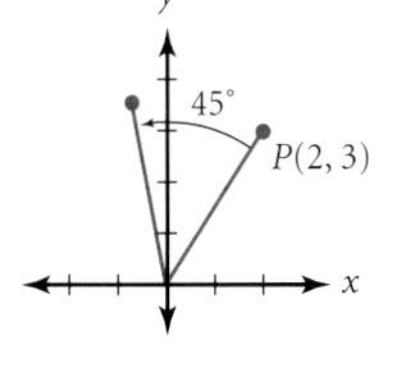

CHECKPOINT ✔

a. $\begin{bmatrix} 0 & -1 \\ 1 & 0 \end{bmatrix}$ **b.** $\begin{bmatrix} -1 & 0 \\ 0 & -1 \end{bmatrix}$

c. $\begin{bmatrix} 0 & 1 \\ -1 & 0 \end{bmatrix}$

ADDITIONAL EXAMPLE 5

Refer to the design problem described at the beginning of the lesson. **Find the coordinates to the nearest hundredth of the vertices after a 150° rotation about the origin.**

$A'(-2.23, 0.13)$,
$B'(-4.83, 1.63)$,
$C'(-3.83, 3.37)$,
$D'(-1.23, 1.87)$

Teaching Tip

TECHNOLOGY Make sure that students have their calculators in degree mode. To enter the matrices, access the **MATRIX** menu and then select **EDIT** to enter values into the desired matrix. Enter matrix $R_{30°}$ into the calculator's matrix **[A]** as a **2 x 2** matrix, and enter matrix S into the calculator's matrix **[B]** as a **2 x 4** matrix. To multiply the matrices, return to the general screen, access the **MATRIX** menu, select **1:A** under **NAMES**, press ×, return to the **MATRIX** menu, and select **2:B** under **NAMES**. Press ENTER to view the matrix product. You may enter the coordinates of the vertices in any order, but make sure that the coordinates of each vertex are aligned in a single column. The product matrix will display the new coordinates in the same order.

TRY THIS
$A'(0.13, 2.23)$, $B'(1.63, 4.83)$
$C'(3.37, 3.83)$, $D'(1.87, 1.23)$

For the answer to Critical Thinking, see page 914.

CRITICAL THINKING

$$\begin{bmatrix}\cos 30^\circ & -\sin 30^\circ\\ \sin 30^\circ & \cos 30^\circ\end{bmatrix}\cdot\begin{bmatrix}\cos 60^\circ & -\sin 60^\circ\\ \sin 60^\circ & \cos 60^\circ\end{bmatrix}\begin{bmatrix}x\\ y\end{bmatrix}$$

$$=\begin{bmatrix}\frac{\sqrt{3}}{2} & -\frac{1}{2}\\ \frac{1}{2} & \frac{\sqrt{3}}{2}\end{bmatrix}\cdot\begin{bmatrix}\frac{1}{2} & -\frac{\sqrt{3}}{2}\\ \frac{\sqrt{3}}{2} & \frac{1}{2}\end{bmatrix}\begin{bmatrix}x\\ y\end{bmatrix}$$

$$=\begin{bmatrix}\frac{\sqrt{3}}{2} & -\frac{1}{2}\\ \frac{1}{2} & \frac{\sqrt{3}}{2}\end{bmatrix}\cdot\begin{bmatrix}\frac{1}{2}x-\frac{\sqrt{3}}{2}y\\ \frac{\sqrt{3}}{2}x+\frac{1}{2}y\end{bmatrix}$$

$$=\begin{bmatrix}-y\\ x\end{bmatrix}$$

$$=\begin{bmatrix}0 & -1\\ 1 & 0\end{bmatrix}\begin{bmatrix}x\\ y\end{bmatrix}$$

$$=\begin{bmatrix}\cos 90^\circ & -\sin 90^\circ\\ \sin 90^\circ & \cos 90^\circ\end{bmatrix}\begin{bmatrix}x\\ y\end{bmatrix}$$

6. $-\cos(\theta+180^\circ)$
$= [\cos\theta\cos 180^\circ - \sin\theta\sin 180^\circ]$
$= [\cos\theta(-1) - \sin\theta(0)]$
$= -(-\cos\theta)$
$= \cos\theta$

9.

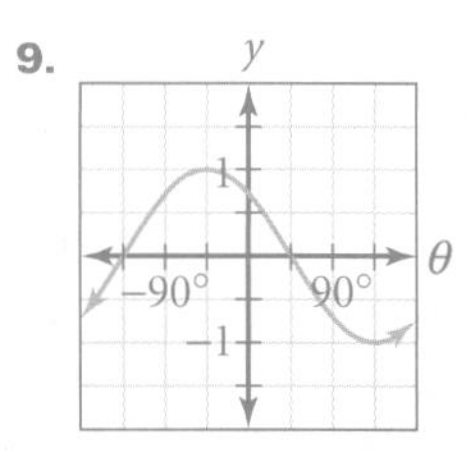

11. $\frac{\sqrt{2}+\sqrt{6}}{4}$
12. $\frac{-\sqrt{2}+\sqrt{6}}{4}$
13. $\frac{-\sqrt{6}-\sqrt{2}}{4}$
14. $\frac{\sqrt{6}-\sqrt{2}}{4}$
15. $-\frac{\sqrt{2}}{2}$
16. $-\frac{\sqrt{2}}{2}$
17. $\frac{-\sqrt{2}+\sqrt{6}}{4}$
18. $\frac{-\sqrt{6}+\sqrt{2}}{4}$
19. $\frac{-\sqrt{2}-\sqrt{6}}{4}$
20. $\frac{-\sqrt{6}-\sqrt{2}}{4}$
21. $\frac{-\sqrt{6}+\sqrt{2}}{4}$
22. $\frac{-\sqrt{6}-\sqrt{2}}{4}$

Guided Skills Practice

Find the exact value of each expression. ***(EXAMPLE 1)***

4. $\sin(145^\circ + 60^\circ)$ $\frac{\sqrt{2}-\sqrt{6}}{4}$ **5.** $\cos(60^\circ - 45^\circ)$ $\frac{\sqrt{2}+\sqrt{6}}{4}$

6. Prove the identity $-\cos(\theta + 180^\circ) = \cos\theta$. ***(EXAMPLE 2)***

Find the exact value of each expression. ***(EXAMPLE 3)***

7. $\cos(-270^\circ)$ 0 **8.** $\sin(-240^\circ)$ $\frac{\sqrt{3}}{2}$

CONNECTION

9. TRANSFORMATIONS Graph $y = \sin(45^\circ - \theta)$. ***(EXAMPLE 4)***

APPLICATION

10. DESIGN Refer to the design problem described at the beginning of the lesson. Find the coordinates to the nearest hundredth of the vertices after a 120° rotation about the origin. ***(EXAMPLE 5)***
A′(−1.87, 1.23), *B′*(−3.37, 3.83), *C′*(−1.63, 4.83), *D′*(−0.13, 2.23)

Practice and Apply

Find the exact value of each expression.

11. $\sin(30^\circ + 45^\circ)$ **12.** $\sin(30^\circ + 135^\circ)$ **13.** $\cos(30^\circ + 135^\circ)$
14. $\cos(30^\circ + 45^\circ)$ **15.** $\sin(135^\circ + 180^\circ)$ **16.** $\sin(135^\circ + 180^\circ)$
17. $\cos(120^\circ - 45^\circ)$ **18.** $\cos(150^\circ - 45^\circ)$ **19.** $\sin(210^\circ - 315^\circ)$
20. $\sin(240^\circ - 315^\circ)$ **21.** $\cos(225^\circ - 330^\circ)$ **22.** $\cos(135^\circ - 330^\circ)$

internet connect
Homework Help Online
Go To: **go.hrw.com**
Keyword:
MB1 Homework Help for Exercises 23–28

Prove each identity.

23. $\sin(90^\circ - \theta) = \cos\theta$ **24.** $\cos(90^\circ - \theta) = \sin\theta$
25. $\cos(90^\circ + \theta) = -\sin\theta$ **26.** $\sin(270^\circ + \theta) = -\cos\theta$
27. $\sin(180^\circ - \theta) = \sin\theta$ **28.** $\cos(180^\circ - \theta) = -\cos\theta$

Use substitution to verify each statement.

29. $\sin(A + B) \neq \sin A + \sin B$ **30.** $\cos(A + B) \neq \cos A + \cos B$
31. $\sin(A - B) \neq \sin A - \sin B$ **32.** $\sin(A - B) \neq \cos A - \cos B$

Find the exact value of each expression.

33. $\sin 105^\circ$ $\frac{\sqrt{2}+\sqrt{6}}{4}$ **34.** $\sin 165^\circ$ $\frac{-\sqrt{2}+\sqrt{6}}{4}$ **35.** $\cos 195^\circ$ $\frac{-\sqrt{2}-\sqrt{6}}{4}$
36. $\cos 225^\circ$ $-\frac{\sqrt{2}}{2}$ **37.** $\sin 15^\circ$ $\frac{\sqrt{6}-\sqrt{2}}{4}$ **38.** $\sin 75^\circ$ $\frac{\sqrt{6}+\sqrt{2}}{4}$
39. $\cos 165^\circ$ $\frac{-\sqrt{2}-\sqrt{6}}{4}$ **40.** $\cos 285^\circ$ $\frac{\sqrt{6}-\sqrt{2}}{4}$ **41.** $\sin(-135^\circ)$ $-\frac{\sqrt{2}}{2}$
42. $\sin(-210^\circ)$ $\frac{1}{2}$ **43.** $\cos(-235^\circ)$ $-\frac{\sqrt{2}}{2}$ **44.** $\cos(-15^\circ)$ $\frac{\sqrt{6}+\sqrt{2}}{4}$

Find the rotation matrix for each angle of rotation. Round entries to the nearest hundredth, if necessary.

45. 45° **46.** 60° **47.** 320° **48.** 224°
49. −120° **50.** −200° **51.** −135° **52.** −320°

Graph each function.

53. $y = \sin(\theta - 60^\circ)$ **54.** $y = \sin(\theta - 45^\circ)$ **55.** $y = \cos(30^\circ - \theta)$
56. $y = \cos(180^\circ - \theta)$ **57.** $y = \sin(120^\circ - \theta)$ **58.** $y = \cos(135^\circ - \theta)$

23. $\sin(90^\circ - \theta) = \sin 90^\circ\cos\theta - \cos 90^\circ\sin\theta$
$= 1\cdot\cos\theta - 0\cdot\sin\theta$
$= \cos\theta$

24. $\cos(90^\circ - \theta) = \cos 90^\circ\cos\theta + \sin 90^\circ\sin\theta$
$= 0\cdot\cos\theta + 1\cdot\sin\theta$
$= \sin\theta$

25. $\cos(90^\circ + \theta) = \cos 90^\circ\cos\theta - \sin 90^\circ\sin\theta$
$= 0\cdot\cos\theta - 1\cdot\sin\theta$
$= -\sin\theta$

26. $\sin(270^\circ + \theta) = \sin 270^\circ\cos\theta + \cos 270^\circ\sin\theta$
$= -1\cdot\cos\theta + 0\cdot\sin\theta$
$= -\cos\theta$

27. $\sin(180^\circ - \theta) = \sin 180^\circ\cos\theta - \cos 180^\circ\sin\theta$
$= 0\cdot\cos\theta - (-1)\sin\theta$
$= \sin\theta$

28. $\cos(180^\circ - \theta) = \cos 180^\circ\cos\theta + \sin 180^\circ\sin\theta$
$= -1\cdot\cos\theta + 0\cdot\sin\theta$
$= -\cos\theta$

The answers to Exercises 29–32 and 45–58 can be found in Additional Answers beginning on page 1002.

Practice

NAME ______ CLASS ______ DATE ______

Practice
14.4 ***Sum and Difference Identities***

Find the exact value of each expression.

1. $\sin\left(\frac{3\pi}{4}+\frac{\pi}{3}\right)$ $\frac{\sqrt{2}}{4}-\frac{\sqrt{6}}{4}$
2. $\sin\left(\frac{3\pi}{4}-\frac{\pi}{3}\right)$ $\frac{\sqrt{2}}{4}+\frac{\sqrt{6}}{4}$
3. $\cos\left(\frac{3\pi}{4}+\frac{\pi}{3}\right)$ $-\frac{\sqrt{2}}{4}-\frac{\sqrt{6}}{4}$
4. $\cos\left(\frac{3\pi}{4}-\frac{\pi}{3}\right)$ $-\frac{\sqrt{2}}{4}+\frac{\sqrt{6}}{4}$
5. $\sin\left(\frac{3\pi}{4}+\frac{2\pi}{3}\right)$ $-\frac{\sqrt{2}}{4}-\frac{\sqrt{6}}{4}$
6. $\cos\left(\frac{3\pi}{4}+\frac{2\pi}{3}\right)$ $\frac{\sqrt{2}}{4}-\frac{\sqrt{6}}{4}$
7. $\sin\left(\frac{\pi}{6}+\frac{\pi}{4}\right)$ $\frac{\sqrt{2}}{4}+\frac{\sqrt{6}}{4}$
8. $\sin\left(\frac{\pi}{6}-\frac{\pi}{4}\right)$ $\frac{\sqrt{2}}{4}-\frac{\sqrt{6}}{4}$
9. $\cos\left(\frac{\pi}{6}+\frac{\pi}{4}\right)$ $\frac{\sqrt{6}}{4}-\frac{\sqrt{2}}{4}$

Find the exact value of each expression.

10. $\sin(-285^\circ)$ $\frac{\sqrt{2}}{4}+\frac{\sqrt{6}}{4}$
11. $\cos(-285^\circ)$ $-\frac{\sqrt{2}}{4}+\frac{\sqrt{6}}{4}$
12. $\sin 135^\circ$ $\frac{\sqrt{2}}{2}$
13. $\cos 210^\circ$ $-\frac{\sqrt{3}}{2}$
14. $\cos(-75^\circ)$ $-\frac{\sqrt{2}}{4}+\frac{\sqrt{6}}{4}$
15. $\sin 345^\circ$ $\frac{\sqrt{2}}{4}-\frac{\sqrt{6}}{4}$
16. $\cos 345^\circ$ $\frac{\sqrt{2}}{4}+\frac{\sqrt{6}}{4}$
17. $\sin 240^\circ$ $-\frac{\sqrt{3}}{2}$
18. $\sin(-75^\circ)$ $-\frac{\sqrt{2}}{4}-\frac{\sqrt{6}}{4}$

Find the rotation matrix for each angle. Round entries to the nearest hundredth, if necessary.

19. 120° $\begin{bmatrix}-0.5 & -0.87\\ 0.87 & -0.5\end{bmatrix}$
20. 135° $\begin{bmatrix}-0.71 & -0.71\\ 0.71 & -0.71\end{bmatrix}$
21. 225° $\begin{bmatrix}-0.71 & 0.71\\ -0.71 & -0.71\end{bmatrix}$
22. 65° $\begin{bmatrix}0.42 & -0.91\\ 0.91 & 0.42\end{bmatrix}$
23. −40° $\begin{bmatrix}0.77 & 0.64\\ -0.64 & 0.77\end{bmatrix}$
24. 112° $\begin{bmatrix}-0.37 & -0.93\\ 0.93 & -0.37\end{bmatrix}$

25. A rectangle has vertices at (3, 5), (3, 10), (7, 10), and (7, 5). Find the coordinates of the vertices after a 135° counterclockwise rotation about the origin. Round coordinates to the nearest hundredth.
(−5.66, −1.41), (−9.19, −4.95), (−12.02, −2.12), (−849, 1.41)

59. $P'\left(-\frac{5\sqrt{2}}{2}, -\frac{\sqrt{2}}{2}\right)$

60. $P'(-3\sqrt{2}, -2\sqrt{2})$

61. $P'\left(\frac{\sqrt{2}}{2}, -\frac{5\sqrt{2}}{2}\right)$

Find the coordinates of the image of each point after a 135° rotation.

59. $P(2, 3)$ **60.** $P(1, 5)$ **61.** $P(-3, 2)$ **62.** $P(4, -5)$

Find the coordinates of the image of each point after a –30° rotation.

63. $P(-1, 2)$ **64.** $P(2, -3)$ **65.** $P(10, 23)$ **66.** $P(7, 35)$

CHALLENGES

62. $P'\left(\frac{\sqrt{2}}{2}, \frac{9\sqrt{2}}{2}\right)$

67. Use the sum and difference identities for the sine function to show that $\sin(A + B) + \sin(A - B) = 2 \sin A \cos B$ is true.

68. Use the definition of an inverse matrix to verify that the rotation matrix for θ degrees and the rotation matrix for $-\theta$ degrees are inverse matrices.

CONNECTION

63. $P'\left(\frac{-\sqrt{3}+2}{2}, \frac{1+2\sqrt{3}}{2}\right)$

64. $P'\left(\frac{2\sqrt{3}-3}{2}, \frac{-2-3\sqrt{3}}{2}\right)$

65. $P'\left(\frac{10\sqrt{3}+23}{2}, \frac{-10+23\sqrt{3}}{2}\right)$

TRANSFORMATIONS A rectangular figure has vertices at $W(3, 0)$, $X(3, 2)$, $Y(6, 2)$, and $Z(6, 0)$. Find the coordinates, to the nearest hundredth, of the vertices after the indicated rotation.

69. 60° counterclockwise **70.** 120° counterclockwise
71. 30° clockwise **72.** 150° clockwise
73. 225° clockwise **74.** 330° clockwise
75. 270° counterclockwise **76.** 240° counterclockwise

APPLICATIONS

The up-and-down motion of a pogo stick spring can illustrate simple harmonic motion.

77. PHYSICS The function $f(t) = a \cos(At + B)$ represents simple harmonic motion, where t is time, a is the amplitude of the motion, A is the angular frequency in radians, and B is the phase shift in radians. Show that f can be expressed in terms of a difference of the cosine and sine functions when $a = 3$, $A = 1$, and $B = \frac{\pi}{4}$. $\frac{3\sqrt{2}}{2}(\cos t - \sin t)$

78 PHYSICS The *superposition principle* states that if two or more waves are traveling in the same medium (air, water, and so on) the resulting wave is found by adding together the displacements of the individual waves. For instance, for waves f and g, the resulting wave is $y = f(x) + g(x)$.

a. Graph $f(x) = \cos 2x$ and $g(x) = \cos(2x - 1)$ over the interval $0 \le x \le 2\pi$. How do the graphs of f and g differ?

b. Graph $h(x) = f(x) + g(x)$ in the same screen as f and g. Describe the behavior of the graph of h.

c. Compare the periods of f, g, and h. They are all the same.

d. The identity $\cos x + \cos y = 2 \cos \frac{1}{2}(x + y) \cos \frac{1}{2}(x - y)$ can be derived from the identities of this lesson. Use this identity to simplify h.

Ocean waves generally pass through one another without being altered. However, a momentary combination of waves can result in an unusually tall wave.

Assess

Selected Answers

Exercises 4–10, 11–81 odd

ASSIGNMENT GUIDE

In Class	1–10
Core	11–65 odd, 69–77 odd
Core Plus	12–78 even
Review	79–81
Preview	82–83

Extra Practice can be found beginning on page 940.

Error Analysis

Students often use the wrong sign in the sum and difference identities for cosine. Point out to students that the two cosine identities contain the sign that is the opposite of their name. For example, the *sum* identity for cosine consists of the *difference* of the two products.

66. $P'\left(\frac{7\sqrt{3}+35}{2}, \frac{-7+35\sqrt{3}}{2}\right)$

Student Technology Guide

NAME ______ CLASS ______ DATE ______

Student Technology Guide
14.4 Sum and Difference Identities

You can rotate a set of points θ degrees about the origin by multiplying the matrix $\begin{bmatrix} \cos\theta & -\sin\theta \\ \sin\theta & \cos\theta \end{bmatrix}$ by a matrix containing the coordinates of the points. A graphics calculator can do the matrix multiplication quickly.

Before doing this lesson, press MODE, use the arrow keys to move to Degree, and then press ENTER. Press 2nd MODE.

Example: Find the image, P', of point $P(-2, 4)$ after a 60° rotation about the origin. Round coordinates to the nearest hundredth.

- Write the rotation matrix, $R_{60°}$, and the matrix of coordinates I. For this problem, $R_{60°} = \begin{bmatrix} \cos 60° & -\sin 60° \\ \sin 60° & \cos 60° \end{bmatrix}$ and $I = \begin{bmatrix} -2 \\ 4 \end{bmatrix}$.
- Enter $R_{60°}$ as matrix A. Press MATRX EDIT 1:[A] ENTER. A flashing cursor appears over the first dimension of matrix A. To set up a 2 × 2 matrix, press 2 ENTER 2 ENTER. To store the entries in the matrix itself, press COS 60 ENTER (-) SIN 60 ENTER SIN 60 ENTER COS 60 ENTER.
- Enter I as matrix B. Press MATRX EDIT 2:[B] ENTER. Then set up a 2 × 1 matrix and enter the coordinates (−2, 4).
- Press 2nd MODE. Then press MATRX 1:[A] ENTER × MATRX 2:[B] ENTER ENTER.

```
[A]*[B]
[[-4.464101615]
 [.2679491924 ]]
```

To the nearest hundredth, the coordinates of P' are (−4.46, 0.27).

Find the image of each point after a rotation of −45° about the origin. Round coordinates to the nearest hundredth.

1. $P(3, 0)$ (2.12, −2.12)
2. $P(-5, 2)$ (−2.12, 4.95)
3. $P(11, -4)$ (4.95, −10.61)
4. $P(-13, 18)$ (3.54, 21.92)

5. Rectangle $ABCD$ has vertices at $A(-1, 4)$, $B(-1, -2)$, $C(3, 4)$, and $D(3, -2)$. Find the coordinates of the vertices of the image, $A'B'C'D'$, of $ABCD$ after a 90° rotation about the origin. What do you notice? (Hint: In this problem, I is a 2 × 4 matrix.)
$A'(-4, -1)$, $B'(2, -1)$, $C'(-4, 3)$, $D'(2, 3)$; the x-coordinate of each image point is the opposite of the y-value of its pre-image, and the y-coordinate of each image point is the x-coordinate of its pre-image.

67.
$$\begin{aligned} \sin(A + B) + \sin(A - B) &= \sin A \cos B + \cos A \sin B \\ &\quad + \sin A \cos B - \cos A \sin B \\ &= (\sin A \cos B + \sin A \cos B) \\ &\quad + (\cos A \sin B - \cos A \sin B) \\ &= 2 \sin A \cos B + 0 \\ &= 2 \sin A \cos B \end{aligned}$$

68.
$$R_\theta = \begin{bmatrix} \cos\theta & -\sin\theta \\ \sin\theta & \cos\theta \end{bmatrix}$$
$$R_{-\theta} = \begin{bmatrix} \cos(-\theta) & -\sin(-\theta) \\ \sin(-\theta) & \cos(-\theta) \end{bmatrix} = \begin{bmatrix} \cos\theta & \sin\theta \\ -\sin\theta & \cos\theta \end{bmatrix}$$
$$R_\theta \cdot R_{-\theta} = \begin{bmatrix} \cos\theta & -\sin\theta \\ \sin\theta & \cos\theta \end{bmatrix}\begin{bmatrix} \cos\theta & \sin\theta \\ -\sin\theta & \cos\theta \end{bmatrix} = \begin{bmatrix} 1 & 0 \\ 0 & 1 \end{bmatrix}$$

The answers to Exercises 69–78 can be found in Additional Answers beginning on page 1002.

In Exercises 82–83, students are asked to justify whether the sine of half of any angle is half the sine of that angle and whether or not for all angles the sine of twice of any angle is twice the sine of that angle.

ALTERNATIVE
Assessment

Portfolio Activity

The Portfolio Activity can be used as preparation for the Chapter Project or as a separate activity. In the Portfolio Activity on this page, students find the coordinates of the vertices of a gear by using a rotation matrix.

Answers to Portfolio Activities can be found in Additional Answers of the Teacher's Edition.

79. $f^{-1}(x) = \frac{7-x}{3}$

APPLICATION

80. Assuming the offices are located at (−1000, 0) and (1000, 0), the equation is $\frac{x^2}{93,636} - \frac{y^2}{906,364} = 1$.

79. Find an equation for the inverse of $f(x) = -3x + 7$. Then use composition to verify that the equation you wrote is the inverse. *(LESSON 2.5)*

80. **LAW ENFORCEMENT** An explosion is heard by two law enforcement officers who are 2000 meters apart. Electronic equipment allows them to determine that one officer heard the explosion 1.8 seconds after the other officer. The speed of sound in air (at 20°C) is approximately 340 meters per second. Write an equation for the possible locations of the explosions relative to the two law enforcement officers. *(LESSON 9.5)*

81. Solve $\triangle ABC$ given that $a = 2.96$, $b = 3.78$, and $c = 4.54$. *(LESSON 14.2)*
$A \approx 40.4°$; $B \approx 55.9°$; $C \approx 83.7°$

82. Is $\sin \frac{A}{2} = \frac{\sin A}{2}$ true for all angle measures A? Justify your response. **no**

83. Is $\sin 2A = 2 \sin A$ true for all angle measures A? Justify your response. **no**

PROGRAMMING A machine-tool operator needs to program a gear-cutting machine to cut a gear with 12 teeth. The teeth are to be positioned at 30° intervals. The coordinates for the vertices of the gear-tooth profile are $A(8.00, 0.00)$, $B(7.52, 2.74)$, $C(9.72, 2.33)$, and $D(9.94, 1.13)$.

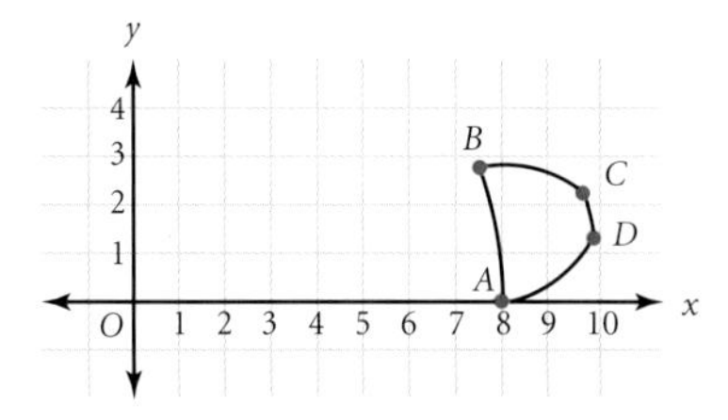

1. Find the coordinates for the image of the gear-tooth profile after a 30° rotation about the origin.
   a. Write matrices for a 30° rotation and for the vertices of the gear-tooth profile.
   b. Find the matrix product.

2. Find the coordinates to the nearest hundredth for the image of the gear-tooth profile after a 60° rotation about the origin.

WORKING ON THE CHAPTER PROJECT

You should now be able to complete Activity 2 of the Chapter project.

Double-Angle and Half-Angle Identities

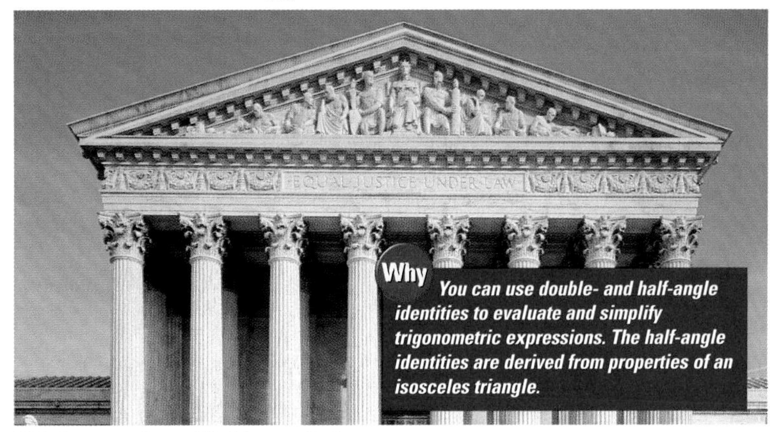

Why *You can use double- and half-angle identities to evaluate and simplify trigonometric expressions. The half-angle identities are derived from properties of an isosceles triangle.*

U.S. Supreme Court, Washington, D.C.

Objective

- Evaluate and simplify expressions by using double-angle and half-angle identities.

APPLICATION
ARCHITECTURE

An isosceles roof system, such as the one shown above, is represented at right. The width, s, can be written in terms of a and θ as follows:

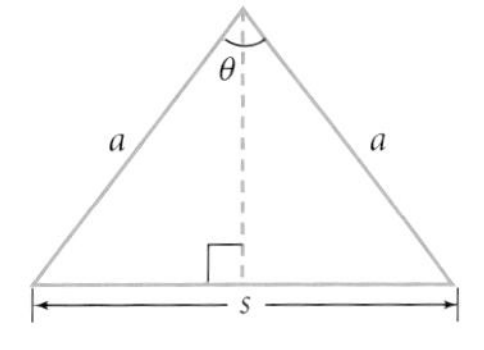

$$\sin\frac{\theta}{2} = \frac{\left(\frac{s}{2}\right)}{a}$$

$$s = 2a\sin\frac{\theta}{2}$$

Another expression for s can be derived from the law of cosines.

$$s^2 = a^2 + a^2 - 2aa\cos\theta$$
$$s^2 = 2a^2 - 2a^2\cos\theta$$
$$s^2 = a^2(2 - 2\cos\theta)$$
$$s = a\sqrt{2 - 2\cos\theta}$$

By equating these two expressions for s, you can write what is called a *half-angle identity* for the sine function. This is shown on page 919.

Double-Angle Identities

You can use sum identities to prove the *double-angle identities* for the sine and cosine functions.

$$\sin 2\theta = \sin(\theta + \theta) = \sin\theta\cos\theta + \cos\theta\sin\theta = 2\sin\theta\cos\theta$$

$$\cos 2\theta = \cos(\theta + \theta) = \cos\theta\cos\theta - \sin\theta\sin\theta = \cos^2\theta - \sin^2\theta$$

Double-Angle Identities

$$\sin 2\theta = 2\sin\theta\cos\theta \qquad \cos 2\theta = \cos^2\theta - \sin^2\theta$$

Prepare

NCTM PRINCIPLES & STANDARDS 1–4, 6–10

QUICK WARM-UP

Find the exact value of each expression.

1. $\sin 15°$	$\frac{\sqrt{6}-\sqrt{2}}{4}$
2. $\cos 15°$	$\frac{\sqrt{6}+\sqrt{2}}{4}$
3. $\sin 255°$	$-\frac{\sqrt{6}+\sqrt{2}}{4}$
4. $\cos 255°$	$\frac{\sqrt{2}-\sqrt{6}}{4}$
5. $\tan(-255°)$	$-2-\sqrt{3}$

Also on Quiz Transparency 14.5

Teach

Why Ask students to state the angles whose exact trig ratios can be calculated directly. **0°, 30°, 45°, 60°, 90°, sums and differences of these angles, and the related angles in other quadrants** Now ask students to state the angles whose exact trig ratios can be calculated by using double-angle and half-angle identities. **22.5°, 11.25°, 15°, 7.5°, and any angles equal to the sums of these and the previous angles**

Alternative Teaching Strategy

USING TECHNOLOGY Have students use a graphics calculator to graph both sides of each double-angle and half-angle identity. The graphs should coincide, showing that the equation is in fact an identity. Have students evaluate each side of the identities for random values in order to see numeric verification that the identity holds true.

You can use double-angle identities to simplify a trigonometric expression, as shown in Example 1.

EXAMPLE 1 **Simplify $(\cos\theta + \sin\theta)^2$.**

SOLUTION

$$(\cos\theta+\sin\theta)^2 = \cos^2\theta + 2\cos\theta\sin\theta + \sin^2\theta \quad \textit{Expand.}$$
$$= \cos^2\theta + \sin^2\theta + 2\sin\theta\cos\theta \quad \textit{Rearrange terms.}$$
$$= 1 + \sin 2\theta \quad \textit{Use substitution.}$$

CRITICAL THINKING Use a double-angle identity to write $\sin 3\theta$ in terms of $\sin\theta$ and $\cos\theta$.

You can use double-angle identities to find the exact value of a double-angle given certain information.

EXAMPLE 2 **Given $90° \le \theta \le 180°$ and $\cos\theta = -\frac{3}{4}$, find the exact value of $\cos 2\theta$.**

SOLUTION

PROBLEM SOLVING **Draw a diagram** and find the exact value of $\sin\theta$.

$$x^2 + y^2 = r^2$$
$$(-3)^2 + y^2 = 4^2$$
$$y = \pm\sqrt{4^2 - (-3)^2}$$
$$y = \sqrt{7} \quad \textit{y is positive in Quadrant II.}$$

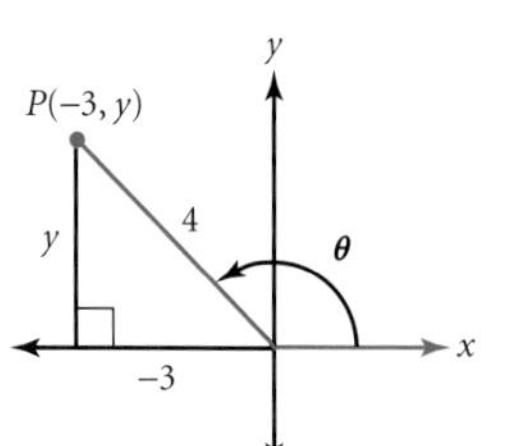

Thus, $\sin\theta = \frac{y}{r} = \frac{\sqrt{7}}{4}$.

Use the double-angle identity.

$$\cos 2\theta = \cos^2\theta - \sin^2\theta$$
$$= \left(-\frac{3}{4}\right)^2 - \left(\frac{\sqrt{7}}{4}\right)^2$$
$$= \frac{2}{16}, \text{ or } \frac{1}{8}$$

TRY THIS Given $270° \le \theta \le 360°$ and $\cos\theta = \frac{1}{4}$, find the exact value of $\sin 2\theta$.

You can use the identity $\cos^2\theta = 1 - \sin^2\theta$ or $\sin^2\theta = 1 - \cos^2\theta$ to write alternative identities for $\cos 2\theta$.

$$\cos 2\theta = \cos^2\theta - \sin^2\theta = (1 - \sin^2\theta) - \sin^2\theta = 1 - 2\sin^2\theta$$

$$\cos 2\theta = \cos^2\theta - \sin^2\theta = \cos^2\theta - (1 - \cos^2\theta) = 2\cos^2\theta - 1$$

Alternative Double-Angle Identities for Cosine

$\cos 2\theta = 1 - 2\sin^2\theta$ $\qquad$ $\cos 2\theta = 2\cos^2\theta - 1$

CHECKPOINT ✔ Solve the problem in Example 2 by using each of the alternative double-angle identities for cosine.

Inclusion Strategies

USING VISUAL MODELS Have students make an index card for each identity. On one side of the card, have students write the identity. On the other side, have students create a visual model that will help them remember the identity. For example, students might use an isosceles triangle with an altitude drawn to the unequal side for the half-angle identities. Have students work in pairs, with one student holding up the visual model and the other student recalling the identity.

Enrichment

Have students derive an identity for $\tan 2\theta$. Students may choose to rewrite the tangent ratio as a quotient of sine and cosine, or they may choose to use the sum identity from Additional Example 2 in Lesson 14.4, shown on page 911. There are many ways to derive this identity, so students should justify their identity algebraically, graphically, and numerically.

$$\tan 2\theta = \tan(\theta+\theta) = \frac{\tan\theta + \tan\theta}{1 - \tan\theta\tan\theta} = \frac{2\tan\theta}{1 - \tan^2\theta}$$

ADDITIONAL EXAMPLE 1

Simplify $\cos^4\theta - \sin^4\theta$.

$\cos^4\theta - \sin^4\theta$
$= (\cos^2\theta - \sin^2\theta)(\cos^2\theta + \sin^2\theta)$
$= (\cos 2\theta)(1)$
$= \cos 2\theta$

CRITICAL THINKING

$\sin 3\theta$
$= \sin(2\theta + \theta)$
$= \sin 2\theta\cos\theta + \cos 2\theta\sin\theta$
$= 2(\sin\theta\cos\theta)\cos\theta + (\cos^2\theta - \sin^2\theta)\sin\theta$
$= 2\sin\theta\cos^2\theta + \sin\theta\cos^2\theta - \sin^3\theta$
$= 3\sin\theta\cos^3\theta - \sin^3\theta$

ADDITIONAL EXAMPLE 2

Given $180° \le \theta \le 270°$ and $\cos\theta = -\frac{2}{3}$, find the exact values of $\sin\theta$ and $\sin 2\theta$.

$\sin\theta = -\frac{\sqrt{5}}{3}$;
$\sin 2\theta = 2\sin\theta\cos\theta$
$= 2\left(-\frac{5}{3}\right)\left(-\frac{2}{3}\right)$
$= \frac{4\sqrt{5}}{9}$

TRY THIS

$\sin 2\theta = -\frac{\sqrt{15}}{8}$

CHECKPOINT ✔

$\sin\theta = -\frac{\sqrt{7}}{4}$
$\cos 2\theta = 1 - 2\sin^2\theta$
$= 1 - 2\left(\frac{\sqrt{7}}{4}\right)^2$
$= 1 - \frac{7}{8}$
$= \frac{1}{8}$
$\cos 2\theta = 2\cos^2\theta - 1$
$= 2\left(\frac{3}{4}\right)^2 - 1$
$= \frac{9}{8} - 1$
$= \frac{1}{8}$

Half-Angle Identities

APPLICATION
ARCHITECTURE

Refer to the isosceles triangle described at the beginning of the lesson. A half-angle identity for the sine function can be found by solving the two expressions for s that relate $\sin\frac{\theta}{2}$ and $\cos\theta$.

$$2a\sin\frac{\theta}{2} = a\sqrt{2-2\cos\theta} \quad (0^\circ < \theta < 180^\circ \text{ and } a > 0)$$
$$\sin\frac{\theta}{2} = \frac{\sqrt{2-2\cos\theta}}{2}$$
$$\sin\frac{\theta}{2} = \sqrt{\frac{2-2\cos\theta}{4}}$$
$$\sin\frac{\theta}{2} = \sqrt{\frac{1-\cos\theta}{2}}$$

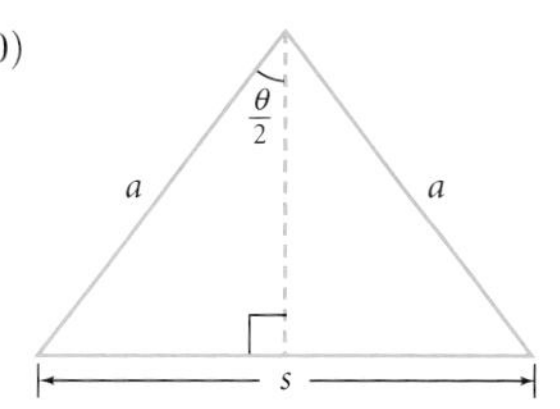

The *half-angle identities* for the sine and cosine of any angle are given below.

Half-Angle Identities

$$\sin\frac{\theta}{2} = \pm\sqrt{\frac{1-\cos\theta}{2}} \qquad \cos\frac{\theta}{2} = \pm\sqrt{\frac{1+\cos\theta}{2}}$$

Choose + or – depending on the sign of the value for $\sin\frac{\theta}{2}$ or $\cos\frac{\theta}{2}$.

EXAMPLE 3 **Given $180^\circ \le \theta \le 270^\circ$ and $\sin\theta = -\frac{2}{3}$, find the exact value of $\cos\frac{\theta}{2}$.**

SOLUTION

PROBLEM SOLVING

Draw a diagram and find the exact value of $\cos\theta$.

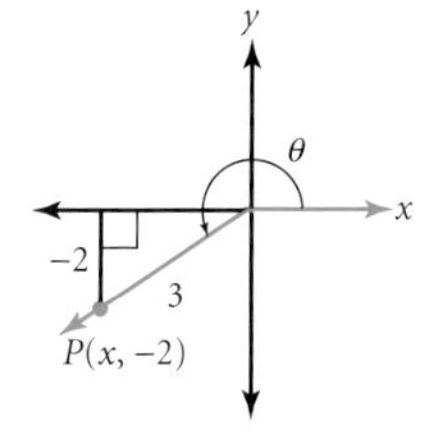

$$x^2 + y^2 = r^2$$
$$x^2 + (-2)^2 = 3^2$$
$$x = \pm\sqrt{3^2 - (-2)^2}$$
$$x = -\sqrt{5} \qquad \textit{x is negative in Quadrant III.}$$

Thus, $\cos\theta = \frac{x}{r} = \frac{-\sqrt{5}}{3}$.

Use the half-angle identity for cosine. If $180^\circ \le \theta \le 270^\circ$, then $\frac{180^\circ}{2} \le \frac{\theta}{2} \le \frac{270^\circ}{2}$, or $90^\circ \le \theta \le 135^\circ$. Therefore, the sign of the value for $\cos\frac{\theta}{2}$ will be negative.

$$\cos\frac{\theta}{2} = -\sqrt{\frac{1+\cos\theta}{2}}$$
$$= -\sqrt{\frac{1+\left(\frac{-\sqrt{5}}{3}\right)}{2}}$$
$$= -\sqrt{\frac{1}{2}\left(1-\frac{\sqrt{5}}{3}\right)}$$
$$= -\sqrt{\frac{1}{2}-\frac{\sqrt{5}}{6}}$$

TRY THIS Given $90^\circ \le \theta \le 180^\circ$ and $\cos\theta = -\frac{1}{3}$, find the exact value of $\sin\frac{\theta}{2}$.

Teaching Tip

Remind students that the identities $\sin^2 2\theta + \cos^2 2\theta = 1$ and $\sin^2\frac{\theta}{2} + \cos^2\frac{\theta}{2} = 1$ can be used to check the values of $\sin 2\theta$ and $\cos 2\theta$ and of $\sin\frac{\theta}{2}$ and $\cos\frac{\theta}{2}$, respectively.

ADDITIONAL EXAMPLE 3

For $\sin\theta = \frac{1}{3}$ and $0^\circ < \theta < 90^\circ$, find the exact value of $\cos\frac{\theta}{2}$.

$\pm\sqrt{\frac{1+\frac{2\sqrt{2}}{3}}{2}}$, or $\pm\sqrt{\frac{1}{2}+\frac{\sqrt{2}}{3}}$

TRY THIS

$\frac{\sqrt{6}}{3}$

Reteaching the Lesson

USING TECHNOLOGY Have students input each side of the double-angle identity for the sine ratio into the **Y=** menu of a graphics calculator. That is, have them enter **Y1=sin(2X)** and **Y2=2sin(x)cos(x)**. Have students graph both equations in the same window and see that the graphs coincide. Have students repeat this activity for the other identities in this lesson. When students use a graphics calculator to verify the half-angle identities, they will need to further separate the right side into positive and negative square roots. If the students have a calculator that will graph in different line styles, it will distinguish the various graphs. If the calculator does not have this feature, have students use the trace feature to distinguish each graph. Have students discuss why the identity is true for only certain domains. Students may need to recall the quadrants in which the sine and cosine are positive and the quadrants in which they are negative. Sine is positive in Quadrants I and II; cosine is positive in Quadrants I and IV.

CRITICAL THINKING

$$\cos 2\left(\frac{\theta}{2}\right) = 2\cos^2\frac{\theta}{2} - 1$$
$$\cos\theta = 2\cos^2\frac{\theta}{2} - 1$$
$$1 + \cos\theta = 2\cos^2\frac{\theta}{2}$$
$$\frac{1+\cos\theta}{2} = \cos^2\frac{\theta}{2}$$
$$\pm\sqrt{\frac{1+\cos\theta}{2}} = \cos\frac{\theta}{2}$$

Assess

Selected Answers

Exercises 3–7, 9–51 odd

ASSIGNMENT GUIDE	
In Class	1–7
Core	9–21 odd, 25–41 odd
Core Plus	8–40 even
Review	42–52
Preview	53

Extra Practice can be found beginning on page 940.

CULTURAL CONNECTION: AFRICA Ptolemy was an astronomer who lived and worked in the North African city of Alexandria during the second century C.E. Ptolemy wrote the most authoritative work on trigonometry of that time, called the *Almagest*, which included the double-angle identities for the sine of an angle.

Ptolemy, a second century astronomer

CRITICAL THINKING Use identities to show that $\cos 2\left(\frac{\theta}{2}\right) = 2\cos^2\frac{\theta}{2} - 1$ can be rewritten as the identity $\cos\frac{\theta}{2} = \pm\sqrt{\frac{1+\cos\theta}{2}}$.

Exercises

internet connect — Activities Online. Go To: go.hrw.com Keyword: MB1 Archimedes

Communicate

1. Use the range of $y = \sin\theta$ to explain why $\sin 2\theta$ is not, in general, equivalent to $2\sin\theta$.
2. Describe how to determine the sign of the value of $\sin\frac{\theta}{2}$ or $\cos\frac{\theta}{2}$ if you are given $0° < \theta < 360°$ and the quadrant in which θ terminates.

Guided Skills Practice

3. Simplify $\cos^4\theta - \sin^4\theta$. *(EXAMPLE 1)* **$\cos 2\theta$**

Given $0° \le \theta \le 90°$ and $\sin\theta = \frac{2}{5}$, find the exact value of each expression. *(EXAMPLES 2 AND 3)*

4. $\cos 2\theta$ **$\frac{17}{25}$**
5. $\sin 2\theta$ **$\frac{4\sqrt{21}}{25}$**
6. $\cos\frac{\theta}{2}$ **$\sqrt{\frac{1}{2}+\frac{\sqrt{21}}{10}}$**
7. $\sin\frac{\theta}{2}$ **$\sqrt{\frac{1}{2}-\frac{\sqrt{21}}{10}}$**

Practice and Apply

Simplify.

8. $\frac{\sin 2\theta}{\cos\theta}$ **$2\sin\theta$**
9. $\cos 2\theta + 1$ **$2\cos^2\theta$**
10. $\cos 2\theta + 2\sin^2\theta$ **1**
11. $\frac{\cos\theta\sin 2\theta}{1+\cos 2\theta}$ **$\sin\theta$**
12. $\frac{\cos 2\theta}{\cos\theta + \sin\theta}$ **$\cos\theta - \sin\theta$**
13. $\frac{\cos 2\theta}{\cos\theta - \sin\theta} - \sin\theta$ **$\cos\theta$**

Write each expression in terms of trigonometric functions of θ rather than multiples of θ.

14. $\sin^2 2\theta$
15. $\sin 4\theta$
16. $\cos^2 2\theta$
17. $\cos 4\theta$
18. $\sin 2\theta - 1 + \cos^2\theta$ **$2\sin\theta\cos\theta - \sin^2\theta$**
19. $\frac{\sin^2\theta}{\sin 2\theta}$ **$\frac{\sin\theta}{2\cos\theta}$**

Answers: 14. **$4\sin^2\theta\cos^2\theta$** 15. **$4\sin\theta\cos^3\theta - 4\sin^3\theta\cos\theta$** 16. **$\cos^4\theta - 2\cos^2\theta\sin^2\theta + \sin^4\theta$** 17. **$\cos^4\theta - 6\sin^2\theta\cos^2\theta + \sin^4\theta$**

Write each expression in terms of a single trigonometric function.

20. $\sin\theta\cos\theta$ **$\frac{\sin 2\theta}{2}$**
21. $2\cos^2\theta - 2\sin^2\theta$ **$2\cos 2\theta$**
22. $2\cos^2\theta - \sin^2\theta$ **$\cos^2\theta + \cos 2\theta$**

CHALLENGE

23. Find the exact value of $\sin 7.5°$. **$\sqrt{\frac{2-\sqrt{2+\sqrt{3}}}{2}}$**

Practice

NAME ________ CLASS ________ DATE ________

Practice 14.5 *Double-Angle and Half-Angle Identities*

Verify that the double-angle identities and the half-angle identities are true for the sine and cosine of each angle. 1–4. Check students' work.

1. 90°
2. 120°
3. $\frac{\pi}{3}$
4. $\frac{2\pi}{3}$

Write each expression in terms of trigonometric functions of θ rather than multiples of θ. 5–12. Answers may vary.

5. $\sin^2\left(\frac{\theta}{2}\right)$ $\frac{1-\cos\theta}{2}$
6. $\cos^2\left(\frac{\theta}{2}\right)$ $\frac{1+\cos\theta}{2}$
7. $\frac{\sin 2\theta}{\tan\theta}$ $2\cos^2\theta$

Simplify.

8. $\frac{1-\cos 2\theta}{1+\cos 2\theta}$ $\tan^2\theta$
9. $\frac{1+\sin\theta-\cos 2\theta}{\cos\theta+\sin 2\theta}$ $\tan\theta$
10. $\frac{1+\cos 2\theta}{\sin 2\theta}$ $\cot\theta$
11. $\left(\sin\left(\frac{\theta}{2}\right)+\cos\left(\frac{\theta}{2}\right)\right)^2$ $1+\sin\theta$
12. The angle of elevation of a flagpole was measured at distances of 45 feet and 14.4 feet from the flagpole. The second measure of the angle of elevation was twice the first. Find the height of the flagpole. 27 feet

internet connect
Homework Help Online
Go To: go.hrw.com
Keyword:
MB1 Homework Help for Exercises 24–39

Use the information given to find the exact value of sin 2θ and cos 2θ.

24. $90° \le \theta \le 180°$; $\cos \theta = -\frac{3}{5}$
25. $90° \le \theta \le 180°$; $\sin \theta = \frac{3}{5}$
26. $270° \le \theta \le 360°$; $\cos \theta = \frac{2}{5}$
27. $270° \le \theta \le 360°$; $\sin \theta = -\frac{2}{5}$
28. $0° \le \theta \le 90°$; $\sin \theta = \frac{1}{4}$
29. $0° \le \theta \le 90°$; $\cos \theta = \frac{1}{4}$
30. $180° \le \theta \le 270°$; $\sin \theta = -\frac{\sqrt{5}}{4}$
31. $180° \le \theta \le 270°$; $\cos \theta = -\frac{\sqrt{5}}{4}$

Use the information given to find the exact value of $\sin \frac{\theta}{2}$ and $\cos \frac{\theta}{2}$.

32. $0° \le \theta \le 90°$; $\sin \theta = \frac{1}{5}$
33. $0° \le \theta \le 90°$; $\cos \theta = \frac{1}{5}$
34. $90° \le \theta \le 180°$; $\cos \theta = -\frac{5}{6}$
35. $90° \le \theta \le 180°$; $\sin \theta = \frac{5}{6}$
36. $180° \le \theta \le 270°$; $\sin \theta = -\frac{\sqrt{5}}{3}$
37. $180° \le \theta \le 270°$; $\cos \theta = -\frac{\sqrt{5}}{3}$
38. $270° \le \theta \le 360°$; $\cos \theta = \frac{3}{8}$
39. $270° \le \theta \le 360°$; $\sin \theta = -\frac{3}{8}$

30. $\frac{\sqrt{55}}{8}; \frac{3}{8}$
31. $\frac{\sqrt{55}}{8}; -\frac{3}{8}$
32. $\sqrt{\frac{1}{2} - \frac{\sqrt{6}}{5}}; \sqrt{\frac{1}{2} + \frac{\sqrt{6}}{5}}$
33. $\sqrt{\frac{2}{5}}; \sqrt{\frac{3}{5}}$

APPLICATION

SPORTS A golf ball is struck with an initial velocity of v_0, in feet per second, and leaves the ground at angle x. The distance that the ball travels is given by the function $d(x) = \frac{(v_0)^2 \sin x \cos x}{16}$.

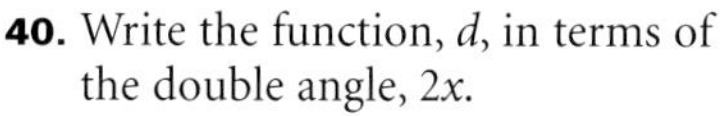

40. Write the function, d, in terms of the double angle, $2x$.

41. At what angle must a golf ball be hit in order to achieve the maximum possible distance for a given initial velocity? Explain.

Tiger Woods, 1997

34. $\sqrt{\frac{11}{12}}; \sqrt{\frac{1}{12}}$
35. $\sqrt{\frac{1}{2} + \frac{11}{12}}; \sqrt{\frac{1}{2} - \frac{11}{12}}$
36. $\sqrt{\frac{5}{6}}; -\sqrt{\frac{1}{6}}$
37. $\sqrt{\frac{1}{2} + \frac{\sqrt{5}}{6}}; -\sqrt{\frac{1}{2} - \frac{\sqrt{5}}{6}}$
38. $\frac{\sqrt{5}}{4}; -\frac{\sqrt{11}}{4}$
39. $\sqrt{\frac{1}{2} - \frac{\sqrt{55}}{16}}; -\sqrt{\frac{1}{2} + \frac{\sqrt{55}}{16}}$
40. $d(x) = \frac{(v_0)^2 \sin 2x}{32}$
41. 45°; sample answer: Graph reaches its maximum at 45°.

Look Back

State whether each relation represents a function. ***(LESSON 2.3)***

42. $\{(-1, 6), (0, 3), (1, 3), (2, 6)\}$ function
43. $\{(1, 2), (2, 3), (2, 4), (3, 5)\}$ not a function

Evaluate. ***(LESSONS 10.2 AND 10.3)***

44. $\frac{5!}{2!}$ 60
45. $0!$ 1
46. ${}_{12}C_4$ 495
47. ${}_{13}C_3$ 286
48. ${}_{10}P_4$ 5040
49. ${}_{15}P_3$ 2730
50. $\binom{7}{3}$ 35
51. $\binom{10}{2}$ 45

CONNECTION

52. TRANSFORMATIONS For the function $f(\theta) = 3 \sin(2\theta - 60°)$, describe the transformation from its parent function. Then graph at least one period of the function along with the parent function. ***(LESSON 13.5)***

vertical stretch of 3, horizontal compression of $\frac{1}{2}$, horizontal translation of 30° to the right

Look Beyond

$\theta = 90°$ **53.** Consider the equation $\sin^2 \theta + 2 \sin \theta - 3 = 0$. Substitute x for $\sin \theta$, and solve the resulting equation for x. Then solve for θ given that $0° \le \theta \le 360°$.

24. $-\frac{24}{25}; -\frac{7}{25}$
25. $-\frac{24}{25}; \frac{7}{25}$
26. $-\frac{4\sqrt{21}}{25}; -\frac{17}{25}$
27. $-\frac{4\sqrt{21}}{25}; \frac{17}{25}$
28. $\frac{\sqrt{15}}{8}; \frac{7}{8}$
29. $\frac{\sqrt{15}}{8}; -\frac{7}{8}$

Error Analysis

When using the double-angle identities for cosine, students may not pay close attention to the order of the terms in the subtraction. Remind them that subtraction is not commutative, so $2 \sin^2 \alpha - 1$ is equal to $-\cos 2\alpha$, not $\cos 2\alpha$.

Look Beyond

In Exercise 53, students substitute a variable into a trig equation in order to get a quadratic equation that can be solved algebraically. Students will learn more about solving trig equations in Lesson 14.6.

52. vertical stretch of 3; horizontal compression of $\frac{1}{2}$; horizontal translation of 30° right;

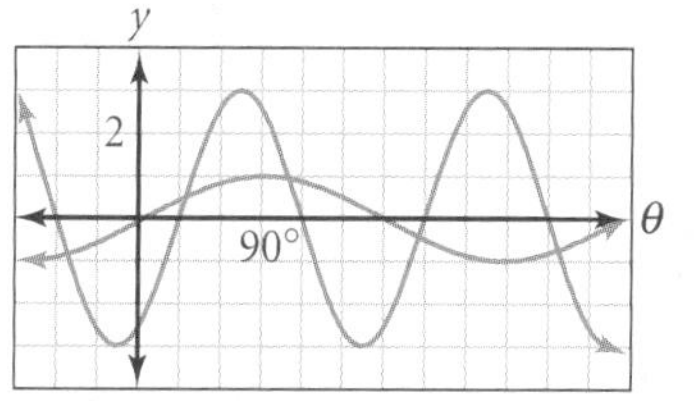

(graph: axes y and θ; labels 2, 90°)

Student Technology Guide

NAME ______ CLASS ______ DATE ______

Student Technology Guide
14.5 Double-Angle and Half-Angle Identities

Double-angle and half-angle identities allow you to simplify trigonometric expressions by reducing the number of trigonometric functions involved. You can use a graphics calculator to help verify these identities.

Before doing this lesson, press MODE, use the arrow keys to move to Degree, and then press ENTER. Press 2nd MODE.

Example: Verify that the double-angle identity for sine is true for $\theta = 88°$.

To verify $\sin 2\theta = 2 \sin \theta \cos \theta$ for $\theta = 88°$, compare the values of $\sin 2(88°)$ and $2 \sin 88° \cos 88°$.

- Press SIN 2 × 88) ENTER.
- Press 2 × SIN 88) × COS 88) ENTER.

```
sin(2*88)
      .0697564737
2*sin(88)*cos(88
)
      .0697564737
```

The answers are equal, verifying the identity.

You can also verify an identity by graphing.

Example: Use graphing to confirm the double-angle identity $\cos 2\theta = \cos^2 \theta - \sin^2 \theta$.

To check that the two sides of the equation are equal, graph $y = \cos 2\theta$ and $y = \cos^2 \theta - \sin^2 \theta$ on the same screen.

- Press WINDOW. Change Xmin to −360, Xmax to 360, Xscl to 90, Ymin to −2, Ymax to 2, and Yscl to 1.
- Press Y= COS 2 × X,T,θ,n) to enter $y = \cos 2\theta$ as Y1.
- Use ▼ to move to Y2. Press COS X,T,θ,n) x² − SIN X,T,θ,n) x².
- Press GRAPH.

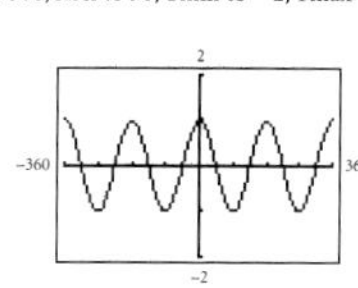

The graphs coincide, indicating that $\cos 2\theta = \cos^2 \theta - \sin^2 \theta$.

Use a graph to determine if each of the following is an identity.

1. $\sin^2 \theta = \frac{1}{2}(1 - \cos 2\theta)$ yes
2. $\frac{1}{(\sin \theta)(\cos \theta)} - \frac{\cos \theta}{\sin \theta} = \cot \theta$ no
3. $\sin 2\theta = 2(\sin^2 \theta)(\tan \theta)$ no
4. $(\sin \theta + \cos \theta)^2 = 1 + \sin 2\theta$ yes

Prepare

NCTM PRINCIPLES & STANDARDS 1, 2, 4, 6–10

QUICK WARM-UP

Solve for x.

1. $2x^2 + x - 1 = 0$
   $x = -1$ *or* $x = \frac{1}{2}$
2. $x^2 - 2x - 3 = 0$
   $x = -1$ *or* $x = 3$

Solve for $0° \leq \theta \leq 360°$.

3. $\sin \theta = -1$ $\theta = 270°$
4. $\cos \theta = -\frac{\sqrt{3}}{2}$
   $\theta = 150°$ *or* $\theta = 210°$
5. $\sin \theta = \frac{\sqrt{6} - \sqrt{2}}{4}$
   $\theta = 15°$ *or* $\theta = 165°$

Also on Quiz Transparency 14.6

Teach

Why Have students discuss phenomena that are periodic in nature. **samples: sunrise times, waves, music** Have them discuss when equations are used to describe these phenomena.

Solving Trigonometric Equations

Objectives

- Solve trigonometric equations algebraically and graphically.
- Solve real-world problems by using trigonometric equations.

Why ***Trigonometric equations can be used to solve real-world problems such as finding the angle at which a batter hits a ball.***

APPLICATION
SPORTS

When a batter hits a baseball, the ball travels in a parabolic path. For a certain hit, the path of the ball in terms of time, t, in seconds is represented by the parametric equations below.

$$\begin{cases} x(t) = 122t \cos \theta & x(t) \text{ is the distance in feet.} \\ y(t) = 122t \sin \theta - 16t^2 & y(t) \text{ is the height in feet.} \end{cases}$$

At what angle is the ball hit if it has a height of 15 feet after 3 seconds? *You will answer this question in Example 4.*

Notice that each of the parametric equations is a trigonometric equation. A **trigonometric equation** is an equation that contains at least one trigonometric function. Each of the equations below are trigonometric equations.

$$\cos \theta = 0.5 \qquad \sin^2\left(2x - \frac{\pi}{3}\right) = \frac{\sqrt{3}}{2} \qquad \cos^2 x - 2 \cos x + 3 = 0$$

A solution to a trigonometric equation is any value of the variable for which the equation is true.

Alternative Teaching Strategy

USING COGNITIVE STRATEGIES Give students a trig equation to solve. The general method of solution is to use identities to simplify the equation to one trig function and then substitute x for that trig function. For example, x might be substituted for $\sin \alpha$. This transforms the problem into an algebraic equation. Show students that in order to solve the algebraic equation for x, they should use the techniques they have learned throughout this text. Once the algebraic equation has been solved, students should solve the trig equation by replacing the substituted function for x, that is, by substituting $\sin \alpha$ for x. Once this is done, the complete set of solutions for the original trig equation should be found and checked.

Exploring Trigonometric Equations

You will need: a graphics calculator in degree mode

1. Graph $y = \sin\theta$ for the interval $0^\circ \leq \theta < 360^\circ$. Over this range, how many values of θ satisfy $\sin\theta = 1$? What are these values?
2. Graph $y = \sin\theta$ for the interval $0^\circ \leq \theta < 720^\circ$. Over this range, how many values of θ satisfy $\sin\theta = 0.5$? What are these values?
3. Copy and complete the table below. Based on the table, write a complete solution to $\sin\theta = 1$ in general terms.

Interval	$0^\circ \leq \theta < 360^\circ$	$0^\circ \leq \theta < 720^\circ$	$0^\circ \leq \theta < 1080^\circ$	$0^\circ \leq \theta < 1440^\circ$
Number of solutions				
Solutions				

CHECKPOINT ✔

4. Let $\sin\theta = a$, where θ is any real number and a is a fixed real number. Find one value of a for which $\sin\theta = a$ has no solution.
5. Is there a value of a for which there is exactly one solution to $\sin\theta = a$? Justify your response.
6. If $\sin\theta = a$ has at least one solution, must it have infinitely many solutions? Explain your response.

Trigonometric equations are true only for certain values of the variables, unlike trigonometric identities, which are true for all values of the variables. Example 1 shows you how to find all possible solutions of a trigonometric equation.

EXAMPLE 1 **Find all solutions of $\cos\theta = \sqrt{3} - \cos\theta$.**

SOLUTION

Method 1 Use algebra.

First solve for $0^\circ \leq \theta < 360^\circ$.

$$\cos\theta = \sqrt{3} - \cos\theta$$
$$2\cos\theta = \sqrt{3}$$
$$\cos\theta = \frac{\sqrt{3}}{2}$$
$$\theta = 30^\circ \text{ or } \theta = 330^\circ$$

Method 2 Use a graph.

Graph $y = \cos\theta$ and $y = \sqrt{3} - \cos\theta$ on the same screen over the interval $0^\circ \leq \theta < 360^\circ$, and find any points of intersection.

The graphs intersect at $\theta = 30^\circ$ and at $\theta = 330^\circ$.

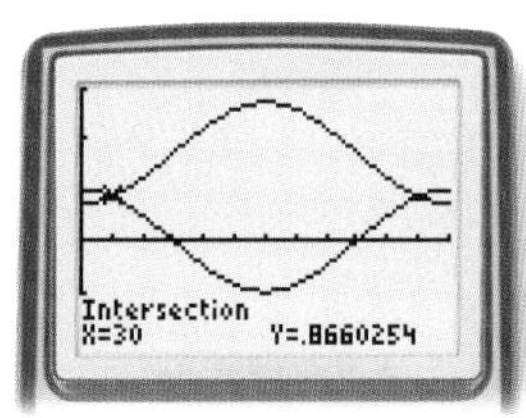

Thus, $\theta = 30^\circ + n360^\circ$ *or* $\theta = 330^\circ + n360^\circ$.

TRY THIS Find all solutions of $1 - 2\sin\theta = 0$.

Interdisciplinary Connection

ASTRONOMY Have students find a function that models the sunrise times for each day of a certain year by using data found on the Internet or in an almanac. Using this model, have students find the dates that the sunrise will be at 6:48 AM. (Note: Some sources of data may not account for daylight savings time, so students may have to adjust some of the data.)

Inclusion Strategies

VISUAL LEARNERS Have students work in pairs or individually to make a poster showing the graph of some of the equations from Exercises 20–43. Each side of the equations should be graphed in a different color, and the intersection should be clearly labeled. Graphics calculators with a table feature can be used to assist students in determining the coordinates of points.

Activity Notes

In this Activity, students investigate the multiple solutions of trig functions over four different intervals. Students should find that over the interval $0^\circ \leq \theta < 360n$ there are n solutions to $\sin\theta = a$ when $a = 1$ or -1, that there are $2n$ solutions when $-1 < a < 1$, and that there are no solutions when $|a| > 0$.

Teaching Tip

TECHNOLOGY For each step of this Activity, have students set the value for **Xmax** to the high end of the interval given for that step.

CHECKPOINT ✔

4. Any $a > 1$ or any $a < -1$; possible answer: -2

ADDITIONAL EXAMPLE 1

Find all solutions of $\sqrt{2}\cos\theta + 1 = 0$.

$135^\circ + n360^\circ$ *or* $225^\circ + n360^\circ$

Teaching Tip

TECHNOLOGY Students should use the following viewing window:

Xmin=0	**Ymin=–1**
Xmax=360	**Ymax=3**

Make sure that they have their calculators in degree mode.

TRY THIS

$\theta = 30^\circ + n360^\circ$ *or*
$\theta = 150^\circ + n360^\circ$

A trigonometric equation may also be solved by using methods for solving quadratic equations.

EXAMPLE 2 **Find the exact solutions of $\sin^2\theta - 2\sin\theta - 3 = 0$ for $0° \le \theta < 360°$.**

SOLUTION

$$\sin^2\theta - 2\sin\theta - 3 = 0$$
$$(\sin\theta)^2 - 2(\sin\theta) - 3 = 0$$
$$u^2 - 2u - 3 = 0 \quad \textit{Substitute u for sin } \theta.$$
$$(u+1)(u-3) = 0 \quad \textit{Factor the quadratic expression.}$$
$$(\sin\theta + 1)(\sin\theta - 3) = 0 \quad \textit{Substitute sin } \theta \textit{ for u.}$$
$$\sin\theta = -1 \text{ or } \sin\theta = 3 \quad \textit{Apply the Zero-Product Property.}$$

For $\sin\theta = -1$, $\theta = 270°$. The equation $\sin\theta = 3$ has no solution.

Thus, the solution for $0° \le \theta < 360°$ is $\theta = 270°$.

TRY THIS Find the exact solutions of $\cos^2\theta - \sqrt{2}\cos\theta + \frac{1}{2} = 0$ for $0° \le \theta < 360°$.

CRITICAL THINKING Use substitution to find the exact solutions of $\sin 3\theta = \frac{1}{2}$ for $0° \le \theta < 360°$.

A trigonometric equation may contain two trigonometric functions. You can often use trigonometric identities to write the equation in terms of only one of the functions. This is shown in Example 3.

EXAMPLE 3 **Solve $2\cos^2\theta = \sin\theta + 1$ for $0° \le \theta < 360°$.**

SOLUTION

$$2\cos^2\theta = \sin\theta + 1$$
$$2(1 - \sin^2\theta) = \sin\theta + 1 \quad \textit{Substitute } 1 - \sin^2\theta \textit{ for } \cos^2\theta = 1.$$
$$2 - 2\sin^2\theta = \sin\theta + 1$$
$$2\sin^2\theta + \sin\theta - 1 = 0$$
$$2u^2 + u - 1 = 0 \quad \textit{Substitute u for sin } \theta.$$
$$(u+1)(2u-1) = 0 \quad \textit{Factor.}$$
$$(\sin\theta + 1)(2\sin\theta - 1) = 0$$

$$\sin\theta = -1 \quad or \quad \sin\theta = \frac{1}{2}$$
$$\theta = 270° \quad or \quad \theta = 30° \quad or \quad \theta = 150°$$

CHECK

Graph $y = 2\cos^2 x$ and $y = \sin x + 1$ on the same screen for $0° \le x < 360°$, and find any points of intersection.

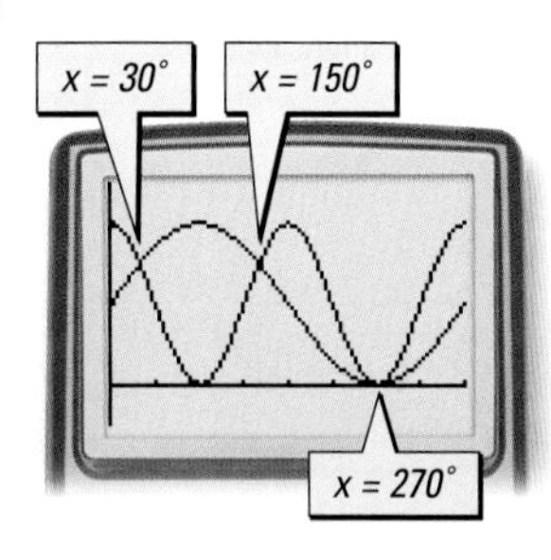

The graph shows intersections at $x = 30°$, $x = 150°$, and $x = 270°$.

TRY THIS Solve $1 + \tan^2\theta + \sec\theta = 0$ for $0° \le \theta < 360°$.

ADDITIONAL EXAMPLE 2

Find the exact solutions of $2\cos^2\theta + 3\cos\theta + 1 = 0$ for $0° \le \theta < 360°$.

$\theta = 120°$, $\theta = 180°$, *or* $\theta = 240°$

TRY THIS

$\theta = 45°$ *or* $\theta = 315°$

CRITICAL THINKING

$\theta = 10°, 50°, 130°, 170°, 250°, 290°$

ADDITIONAL EXAMPLE 3

Solve $\sin\theta + \cos\theta = 1$ for $0° \le \theta < 360°$.

$\sin\theta + \cos\theta = 1$
$\sin\theta = 1 - \cos\theta$
$\sin^2\theta = (1 - \cos\theta)^2$
$\sin^2\theta = 1 - 2\cos\theta + \cos^2\theta$
$1 - \cos^2\theta = 1 - 2\cos\theta + \cos^2\theta$
$2\cos^2\theta - 2\cos\theta = 0$
$2\cos\theta(\cos\theta - 1) = 0$
$2\cos\theta = 0$ *or* $\cos\theta = 1$
$\theta = 90°$ *or* $\theta = 270°$ *or* $\theta = 0°$

Teaching Tip

TECHNOLOGY Students should use the following viewing window:

Xmin=0	Ymin=–.5
Xmax=360	Ymax=2.5

Make sure that students' calculators are in degree mode.

TRY THIS

$\theta = 180°$

Enrichment

In each example in this lesson, another trig function can be used in place of sin x or cos x. For instance, in Example 2, sec x can be used in place of sin x. Ask students to explain whether this changes the solution or the method of solution. **It changes the solution but not the method.**

EXAMPLE 4 Refer to the baseball problem at the beginning of the lesson.

APPLICATION
SPORTS

At what angle is the ball hit if it has a height of 15 feet after 3 seconds?

SOLUTION

Substitute 15 for $y(t)$ and 3 for t in the equation for the height.

$$y(t) = 122t \sin \theta - 16t^2$$
$$15 = 122(3) \sin \theta - 16(3)^2$$
$$15 = 366 \sin \theta - 144$$
$$\sin \theta = \frac{15 + 144}{366}$$
$$\theta \approx \sin^{-1}(0.4344)$$
$$\approx 25.7^\circ$$

TECHNOLOGY
GRAPHICS CALCULATOR
Keystroke Guide, page 939

CHECK

Graph $y = 366 \sin x - 144$ and $y = 15$ on the same screen, and look for any points of intersection.

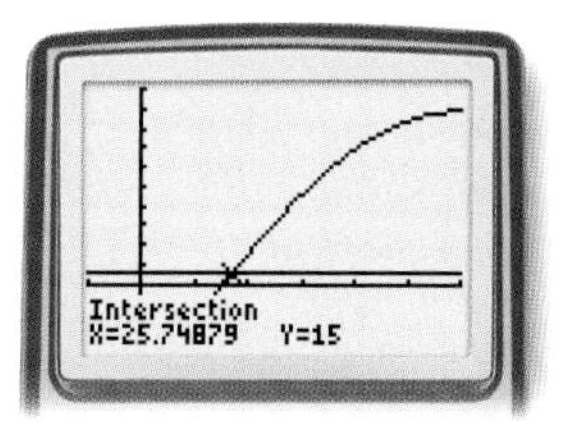

The graph shows an intersection at $\theta \approx 25.7^\circ$.

CHECKPOINT ✔ What distance has the ball traveled if it has a height of 15 feet after 3 seconds? Is the hit a home run if the fence is 325 feet from home plate and 15 feet tall?

Exercises

Communicate

1. Describe how trigonometric equations differ from trigonometric identities.
2. Give the number of solutions to $\sin x = -1$ and $\cos x = 2$, and explain why they differ.
3. Summarize the different methods used to solve trigonometric equations in Examples 1, 2, and 3.

Guided Skills Practice

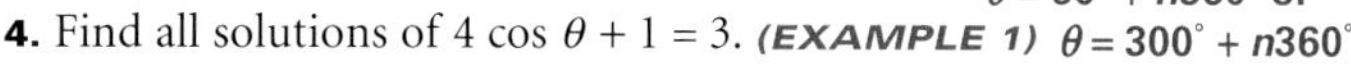

4. Find all solutions of $4 \cos \theta + 1 = 3$. ***(EXAMPLE 1)*** $\theta = 60^\circ + n360^\circ$ or $\theta = 300^\circ + n360^\circ$

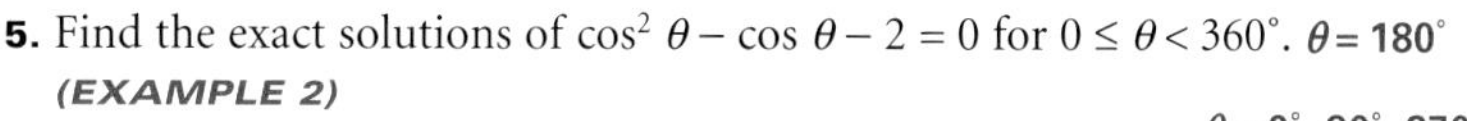

5. Find the exact solutions of $\cos^2 \theta - \cos \theta - 2 = 0$ for $0 \leq \theta < 360^\circ$. $\theta = 180^\circ$ ***(EXAMPLE 2)***
6. Solve $1 - 2 \cos \theta + \cos^2 \theta = \sin^2 \theta$ for $0 \leq \theta < 360^\circ$. ***(EXAMPLE 3)*** $\theta = 0^\circ, 90^\circ, 270^\circ$
7. **SPORTS** Refer to the baseball problem at the beginning of the lesson. At what angle is the ball hit if it has a height of 20 feet after 3 seconds? ***(EXAMPLE 4)*** $\theta \approx 26.6^\circ$

ADDITIONAL EXAMPLE 4

Refer to the baseball problem at the beginning of the lesson. **At what angle is the ball hit if it has a height of 30 feet after 3 seconds?** $\approx 28.4^\circ$

Teaching Tip

TECHNOLOGY Students should use the following viewing window:

Xmin=–15	**Ymin=–75**
Xmax=90	**Ymax=250**

Make sure that students have their calculators in degree mode.

You may want to note that in the parametric equations given for the motion of the ball at the beginning of this lesson, 122 represents the initial speed of the ball in feet per second, θ represents the angle of the ball's path with the horizontal, y represents the vertical position of the ball at time t, and x represents its horizontal position at time t. The given equations assume that the initial height is 0 and that there is no air resistance or spin.

CHECKPOINT ✔

$x(3) = 122(3) \cos(25.7)$
≈ 329.8 ft; yes

Reteaching the Lesson

INVITING PARTICIPATION Have students work in groups of three or four at the board. Write a trig equation on a transparency, and have each group work to solve the equation on the board. Once the groups finish solving the equation, ask them to walk around the room to look at other groups' solutions. Have at least one group describe their solution method. If another group used a different method, have them explain their work. Call on various members of the group to ensure that each person in the group can explain the solution.

Assess

Selected Answers

Exercises 4–7, 9–61 odd

ASSIGNMENT GUIDE	
In Class	1–7
Core	9–57 odd
Core Plus	8–56 even
Review	58–61
Preview	62

Extra Practice can be found beginning on page 940.

Error Analysis

When students solve trig equations by substituting a variable for the trig function, they sometimes stop after obtaining the algebraic solution. Remind students that they must put the trig function back into the solution and solve for the angle. Also, remind them that trig equations have multiple solutions most of the time, depending on the domain.

44. 30°, 150°, 270°

Practice

NAME ______ CLASS ______ DATE ______

Practice
14.6 Solving Trigonometric Equations

Find all solutions of each equation.

1. $\sin^2 \theta = \frac{1}{4}$ 30° + n(360°), 150° + n(360°), 210° + n(360°), and 330° + n(360°)
2. $\tan \theta - 1 = 0$ 45° + n(360°)
3. $\sec \frac{\theta}{2} = 2$ 120° + n(360°)
4. $2 \cos \frac{\theta}{2} - 1 = 0$ 120° + n(360°)

Find the exact solutions of each equation for $0° \le \theta < 360°$.

5. $2 \cos^2 \theta - 3 \cos \theta = 2$ 120° and 240°
6. $2 \sin^2 \theta + 3 \sin \theta + 1 = 0$ 210°, 270°, and 330°
7. $4 \cos^2 \theta - 2 = 0$ 45°, 135°, 225°, and 315°
8. $2 \sin^2 \theta = \cos 2\theta$ 30°, 150°, 210°, and 330°

Find the exact solutions of each equation for $0 \le x < 2\pi$.

9. $2 \sin \frac{x}{2} - 1 = 0$ $\frac{\pi}{3}$ and $\frac{5\pi}{3}$
10. $\cos x - \sin x = 0$ $\frac{\pi}{4}$ and $\frac{5\pi}{4}$
11. $2 \cos 3x - 1 = 0$ $\frac{\pi}{9}, \frac{5\pi}{9}, \frac{7\pi}{9}, \frac{11\pi}{9}, \frac{13\pi}{9}$, and $\frac{17\pi}{9}$
12. $2 \sin^2 x - \sin x - 1 = 0$ $\frac{\pi}{2}, \frac{7\pi}{6}$, and $\frac{11\pi}{6}$

Solve each equation to the nearest hundredth of a radian for $0 \le x < 2\pi$.

13. $9 \cos^2 x - 1 = 0$ 1.23, 1.91, 4.37, and 5.05
14. $6 \sin^2 x - 5 \sin x + 1 = 0$ 0.34, 0.52, 2.62, and 2.80
15. The equation $y(t) = 122t \sin \theta - 16t^2$ describes the altitude of a ball t seconds after it was hit at an angle of θ degrees. Determine, to the nearest tenth of a degree, the measure of the angle at which the ball was hit if it had an altitude of 20 feet after 2.8 seconds. 25.2°

internet connect

Homework Help Online
Go To: **go.hrw.com**
Keyword: **MB1 Homework Help** for Exercises 8–43

8. 30° + n360° *or* 150° + n360°
9. 120° + n360° *or* 240° + n360°
10. 240° + n360° *or* 300° + n360°
11. 60° + n360° *or* 300° + n360°
12. 45° + n360° *or* 135° + n360°
13. 30° + n360° *or* 150° + n360°
14. 90° + n360° *or* 270° + n360°
15. 180° + n360°
16. 60° + n360° *or* 240° + n360°
17. 135° + n360° *or* 315° + n360°

28. 0°, 30°, 150°, 180°
29. 90°, 150°, 210°, 270°
36. 45°, 135°, 225°, 315°

Practice and Apply

Find all solutions of each equation.

8. $2 \sin \theta - 1 = 0$

9. $2 \cos \theta + 1 = 0$

10. $4 \sin \theta + 2\sqrt{3} = 0$

11. $4 \cos \theta - 2 = 0$

12. $\sin \theta = \sqrt{2} - \sin \theta$

13. $1 - \sin \theta = \sin \theta$

14. $2 - 3 \cos \theta = \cos \theta + 2$

15. $1 + 5 \cos \theta = 2 \cos \theta - 2$

16. $\tan \theta - \sqrt{3} = 0$

17. $\cot \theta + 1 = 0$

18. $6 \cos \theta - 1 = 3 + 4 \cos \theta$ **no solution**

19. $1 - 2 \sin \theta = \sin \theta - \sqrt{3}$ **≈65.6° + n360° *or* ≈114.4° + n360°**

Find the exact solutions of each equation for $0° \le \theta < 360°$.

20. $2 \cos^2 \theta - \cos \theta = 1$ **0°, 120°, 240°**

21. $2 \sin^2 \theta = 1 - \sin \theta$ **30°, 150°, 270°**

22. $2 \sin^2 \theta - 5 \sin \theta = -2$ **30°, 150°**

23. $2 \cos^2 \theta - 3 \cos \theta = 2$ **120°, 240°**

24. $3 \cos \theta + 2 = -\cos^2 \theta$ **180°**

25. $3 - \sin \theta = -\sin^2 \theta$ **no solution**

26. $\sin \theta + \sin \theta \cos \theta = 0$ **0°, 180°**

27. $\cos^2 \theta + \cos \theta = 0$ **90°, 180°, 270°**

28. $6 \sin^2 \theta - 3 \sin \theta = 0$

29. $2 \cos^2 \theta + \sqrt{3} \cos \theta = 0$

30. $\cos^2 \theta + 2 \cos \theta = -2$ **no solution**

31. $25 \sin^2 \theta + 8 \sin \theta = -8$ **no solution**

32. $2 \cos^2 \theta = \sin^2 \theta + 2$ **0°, 180°**

33. $\cos^2 \theta = \sin^2 \theta + 1$ **0°, 180°**

34. $\cos \theta \tan \theta = -1$ **270°**

35. $\sec \theta \cos^2 \theta - 1 = 0$ **0°**

36. $2 \tan^2 \theta = \sec^2 \theta$

37. $\sec^2 \theta + 2 \cot^2 \theta = 3$ **no solution**

38. $\cos \theta - \sin^2 \theta = 1$ **0°**

39. $3 + 3 \sin \theta = \cos^2 \theta$ **270°**

40. $2 \cos^2 \theta = \sin \theta + 1$ **30°, 150°, 270°**

41. $-\cos \theta - \sin^2 \theta = 1$ **180°**

42. $2 \cos^2 \theta - 2 \cos \theta = -\sin^2 \theta$ **0°**

43. $1 - \cos^2 \theta = \cos^2 \theta + 2 \cos \theta + 1$ **90°, 180°, 270°**

Solve each equation to the nearest tenth of a degree for $0° \le \theta < 360°$.

44 $2 \cos^2 \theta - \sin \theta - 1 = 0$

45 $\tan^2 \theta + \cot^2 \theta + 2 = 0$ **no solution**

46 $\cos 2\theta - \sin \theta = 0$ **30°, 150°, 270°**

47 $\sin 2\theta + \sin \theta = 0$ **0°, 120°, 180°, 240°**

48 $\cos^2 \theta + 5 \cos \theta + 2 = 0$ **≈116.0°, ≈244.0°**

49 $4 \sin^2 \theta - 3 \sin \theta - 2 = 0$ **≈205.2°, ≈334.8°**

CHALLENGE

50. Solve $|\sin \theta| = \sin \theta$ and $|\cos \theta| = \cos \theta$ over the interval $-360° \le \theta < 360°$.

50. $-360° \le \theta \le -180°$ *or* $0° \le \theta \le 180°$; $-360° \le \theta \le 270°$ *or* $90° \le \theta \le 90°$ *or* $270° \le \theta \le 360°$

CONNECTION

51. GEOMETRY A circular cap is formed by subtracting the area of the triangle (with legs of length r and angle θ in radians) from the area of the sector, as shown in the figure at right. The area of a circular cap is given by the formula $A = \frac{1}{2}r^2(\theta - \sin \theta)$. A circular sector with a radius of 5 and angle θ has a circular cap whose area is 20. What is the measure of angle θ in radians? **$\theta \approx$ 1.6 radians, or 91.7°**

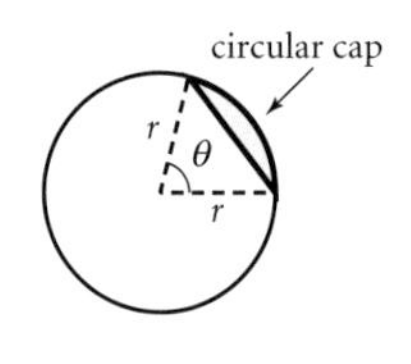

APPLICATION

PHYSICS The position of a weight attached to an oscillating spring is given by $y = 5 \cos \pi t$, where t is time in seconds and y is vertical distance in centimeters. Rest position is at the point $y = 0$.

52. Find the times at which the weight is 5 centimeters above its rest position.

53. Find the times at which the weight is 4 centimeters above its rest position.

52. $t = 0 + 2n$ seconds
53. $t \approx 0.2 + 2n$ *or* $t \approx 1.8 + 2n$ seconds

APPLICATION

PHYSICS When light travels from one medium to another, the path of the light ray changes direction. Snell's law states that a light ray traveling from air to water changes direction according to the equation $\frac{\sin \theta_{air}}{\sin \theta_{water}} = n_{water}$, where θ_{air} and θ_{water} are the angles shown in the diagram below, and n_{water} is a constant called the *index of refraction.*

For example, when looking at a fish underwater, a bird hovering in the air perceives the fish to be nearer to the water's surface than it actually is. Conversely, the fish perceives the bird in the air to be farther away from the water's surface than it actually is.

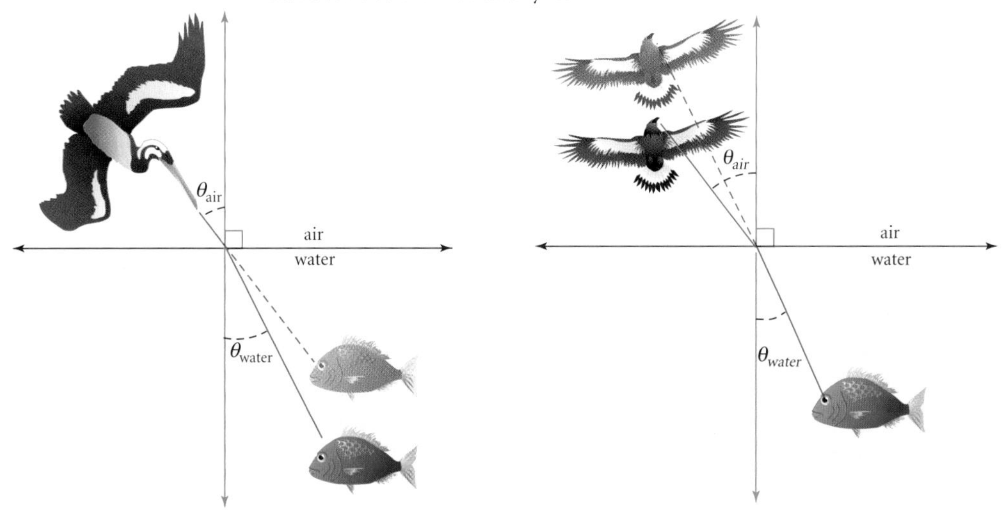

Given the index of refraction for water, n_{water}, is approximately 1.33, find each indicated angle.

54. θ_{air} if θ_{water} is 30°. **41.7°**

55. θ_{water} if θ_{air} is 60°. **40.6°**

56. θ_{water} if θ_{air} is 42°. **30.2°**

57. θ_{air} if θ_{water} is 25°. **34.2°**

Look Back

Let $A = \begin{bmatrix} 5 & -4 \\ 7 & 3 \end{bmatrix}$, $B = \begin{bmatrix} 3 \\ -5 \end{bmatrix}$, and $C = \begin{bmatrix} 7 \\ -3 \end{bmatrix}$. Find each product. If the product does not exist, write *none*. *(LESSON 4.2)*

58. AC $\begin{bmatrix} 47 \\ 40 \end{bmatrix}$

59. BA none

60. BC none

61. Solve $\triangle ABC$ given that $a = 12.7$, $c = 10.4$, and $B = 26.5°$. Give answers to the nearest tenth. ***(LESSON 14.2)*** **$b \approx 5.7$; $A \approx 83.8°$, $c \approx 69.7°$**

A pencil in a glass of water appears bent or broken because of the bending of light, or refraction.

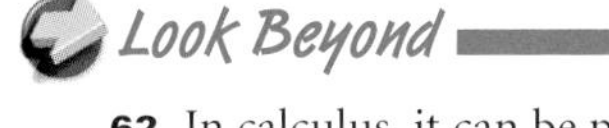

Look Beyond

62. In calculus, it can be proved that $\sin x = x - \frac{x^3}{3!} + \frac{x^5}{5!} - \frac{x^7}{7!} + \cdots$ and $\cos x = 1 - \frac{x^2}{2!} + \frac{x^4}{4!} - \frac{x^6}{6!} + \cdots$ (where x is expressed in radians) are true. Use the first five terms of these series, called the *Taylor series*, to approximate the value of $\sin \frac{\pi}{6}$ and $\cos \pi$. Compare your answers with the exact values.

$\sin \frac{\pi}{6} \approx 0.5$; $\frac{1}{2}$

$\cos \pi \approx -0.976$; 1

Look Beyond

Students use graphs to explore algebraic approximations of trig functions. These approximations, known as *Taylor series*, are studied in calculus.

Student Technology Guide

NAME ________ CLASS ________ DATE ________

Student Technology Guide
14.6 ***Solving Trigonometric Equations***

The intersect feature of a graphics calculator can help you solve a trigonometric equation.

Before doing this lesson, press MODE, use the arrow keys to move to Radian, and then press ENTER. Press 2nd MODE.

Example: Find the approximate solutions of $4\cos^2 x = \sin x + 3$ over the interval $0 \le x \le 2\pi$. Round solutions to the nearest hundredth of a radian.

To solve this equation by graphing, graph $y = 4\cos^2 x$ and $y = \sin x + 3$ on the same screen and find the points of intersection.

- First set the viewing window. Press WINDOW and change Xmin to 0, Xmax to 2π (press 2 × 2nd ^ ENTER), Xscl to 1, Ymin to −5, Ymax to 5, and Yscl to 1.
- Press Y= 4 × COS X,T,θ,n) x^2 to enter $4\cos^2 x$ as Y1.
- Use ▼ to move to Y2. Press SIN X,T,θ,n) + 3 to enter $\sin x + 3$ as Y2.
- Press GRAPH. Notice that you can see four intersection points in this viewing window.
- Press TRACE. Use ◄ to move as close as you can to the leftmost intersection point. This point appears to be near (0.4, 3.4), as shown.
- You can use the intersect feature to get a better estimate. Press 2nd TRACE 5:intersect ENTER ENTER ENTER ENTER. The intersection point shows that, to the nearest hundredth, the first solution is $x = 0.40$.
- Use the trace and intersect features as described above to find the other three solutions in the interval.

Y1=4cos(X)^2
X=.40105438 Y=3.3903849

Intersection
X=.40105322 Y=3.3903882

To the nearest hundredth, the solutions of $4\cos^2 x = \sin x + 3$ over the interval $0 \le x \le 2\pi$ are 0.40, 2.74, 3.84, and 5.59.

Solve each equation to the nearest hundredth of a radian for $0 \le x \le 2\pi$.

1. $\sin x + \cos x = 0.9$ — 1.67 and 6.19
2. $\tan^2 x + \sin x = \cos x$ — 0.53 and 5.41
3. $2\sin^2 x + 3\sin x - 1 = 0$ — 0.28 and 2.86

Focus

Students will use the law of cosines and rotation matrices to construct a gear. They will cut out two copies of the gear from cardboard to see how the gears work in tandem. Their calculations and construction techniques must be precise in order for the gears to work together properly.

Motivate

Bring in examples of gears. Many infant and toddler toys provide nice examples of large gears. Have students think of other objects that contain gears, such as a bicycle, can opener, or another mechanical device.

Activity 1

1–2.

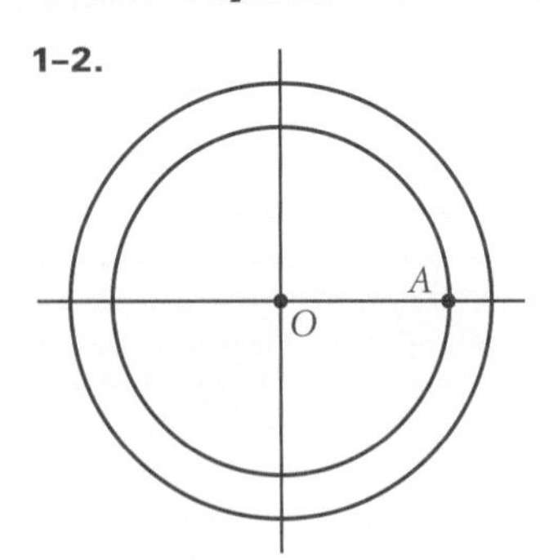

3–4.

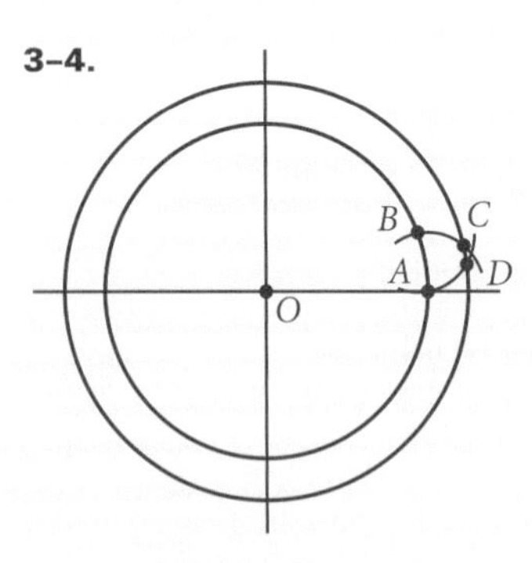

5. arc $AB \approx 20°$

Activity 2

1. $R_{0°} = \begin{bmatrix} 8.00 & 7.52 & 9.72 & 9.94 \\ 0.00 & 2.74 & 2.33 & 1.13 \end{bmatrix}$

$R_{30°} = \begin{bmatrix} 6.93 & 5.14 & 7.25 & 8.04 \\ 4.00 & 6.13 & 6.88 & 5.95 \end{bmatrix}$

$R_{60°} = \begin{bmatrix} 4.00 & 1.39 & 2.84 & 3.99 \\ 6.93 & 7.88 & 9.58 & 9.17 \end{bmatrix}$

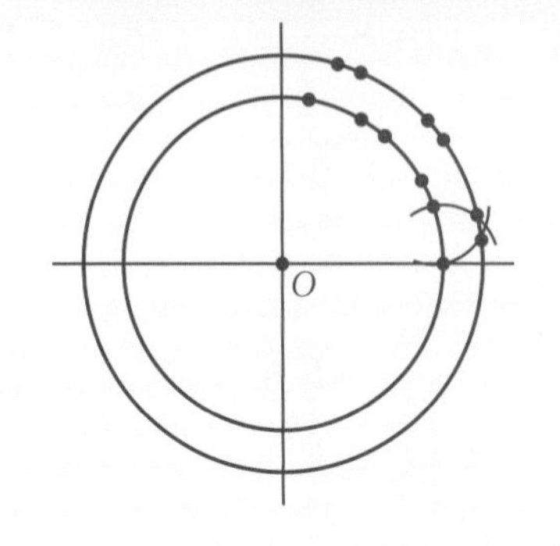

2. $R_{90°} = \begin{bmatrix} 0.00 & -2.74 & -2.33 & -1.13 \\ 8.00 & 7.52 & 9.72 & 9.94 \end{bmatrix}$

$R_{120°} = \begin{bmatrix} -4.00 & -6.13 & -6.88 & -5.95 \\ 6.93 & 5.14 & 7.25 & 8.04 \end{bmatrix}$

$R_{150°} = \begin{bmatrix} -6.93 & -7.88 & -9.58 & -9.17 \\ 4.00 & 1.39 & 2.84 & 3.99 \end{bmatrix}$

$R_{180°} = \begin{bmatrix} -8.00 & -7.52 & -9.72 & -9.94 \\ 0.00 & -2.74 & -2.33 & -1.13 \end{bmatrix}$

$R_{210°} = \begin{bmatrix} -6.93 & -5.14 & -7.25 & -8.04 \\ -4.00 & -6.13 & -6.88 & -5.95 \end{bmatrix}$

CHAPTER FOURTEEN PROJECT

GEARING UP

Much of the activity in today's world depends on electricity. This modern technology is in turn equally dependent on a much older technology that you will study in this Chapter Project. Electricity flows into our homes because turbines turn, and the smooth and efficient transfer of this mechanical energy to electricity often depends on systems of gears.

A gear is usually classified as a "simple machine"—a wheel with teeth. But what size and shape should the teeth be so that a system of gears will mesh together properly? It turns out that this is not a simple question to answer.

In this project, you will design a template for a set of gears that will mesh together smoothly. Then you will make the gears to see if they work properly.

Activity 1

You may lay out your gear template on a sheet of centimeter graph paper or on an unlined sheet of white paper. In both cases, orient the paper horizontally rather than vertically.

1. Mark and label point O in the center of the graph paper, and draw the x-axis and y-axis.
2. Use a compass to draw a *base circle* centered at point O with a radius of 8 centimeters and an *outer circle* centered at point O with a radius of 10 centimeters. Mark and label point $A(8, 0)$. If you are using unlined paper, place point A anywhere on the base circle.
3. Set a compass to draw a radius of 2.8 centimeters. Put the compass point on A and draw an arc that passes through the base circle and the outer circle. Mark and label point B, the arc's intersection with the base circle, and point C, the arc's intersection with the outer circle.
4. With the compass point on B and the pencil point on A, draw an arc that passes through the outer circle, and label the point of intersection D. Quadrilateral $ABCD$ constitutes the *gear tooth profile.*
5. Use the Law of Cosines to determine the measure of arc AB on the base circle to the nearest degree.

Activity 2

In this activity you will complete the gear template you began in Activity 1.

1. Use the coordinates from the Portfolio Activity on page 916 for the images of the original gear tooth profile under a 30° and a 60° rotation about the origin. Carefully plot these coordinates on the base circle and on the outer circle.
2. Use matrix multiplication to determine the coordinates for the images of the original gear tooth profile under rotations of 90°, 120°, 150°, 180°, 210°, 240°, 270°, 300°, and 330° about the origin. Carefully plot these coordinates on the base circle and on the outer circle. Note: Plot the points as carefully as you can. Rounding error and the precision of the centimeter grid will often make your vertices fall just above or below the base circle and the outer circle. Use these circles as guides for positioning the vertices of the gear teeth.
3. Use a compass to draw arcs for the curved sides of each gear tooth. (See Steps 3 and 4 in Activity 1.)

Activity 3

Make working models of the gears.

1. Cut out the gear template that you completed in Activity 3.
2. Pin the gear template to cardboard or fiberboard, and carefully trace around it. Then outline a second gear.
3. Cut out the gears. Position them on a piece of cardboard so that their teeth mesh, and pin them at their centers to keep them in place. The gears should turn smoothly in opposite directions. Observe how the curved surfaces touch each other when the gear teeth engage.

Cooperative Learning

Have students work in groups of three on this project. Students should take the roles of calculations manager, construction manager, and supervisor. The calculations manager will perform all mathematical calculations. The construction manager will be responsible for the physical construction of the gears. The supervisor will oversee the two managers and check for accuracy.

Discuss

After each activity, have groups share their results and discuss any problems they encountered. Once the gears are actually constructed, have each group model their working gears. As an extension, have students consider how two gears of different sizes would work together. Have them analyze models (for example, infant toys) and take measurements to discover the necessary proportions. Then have them try to construct a different gear that will work in tandem with the gear they have already constructed.

Activity 2 Step 2 (cont.)

$$R_{240^\circ} = \begin{bmatrix} -4.00 & -1.39 & -2.84 & -3.99 \\ -6.93 & -7.88 & -9.58 & -9.17 \end{bmatrix}$$

$$R_{270^\circ} = \begin{bmatrix} 0.00 & 2.74 & 2.33 & 1.13 \\ -8.00 & -7.52 & -9.72 & -9.94 \end{bmatrix}$$

$$R_{300^\circ} = \begin{bmatrix} 4.00 & 6.13 & 6.88 & 5.95 \\ -6.93 & -5.14 & -7.25 & -8.04 \end{bmatrix}$$

$$R_{330^\circ} = \begin{bmatrix} 6.93 & 7.88 & 9.58 & 9.17 \\ -4.00 & -1.39 & -2.84 & -3.99 \end{bmatrix}$$

3.

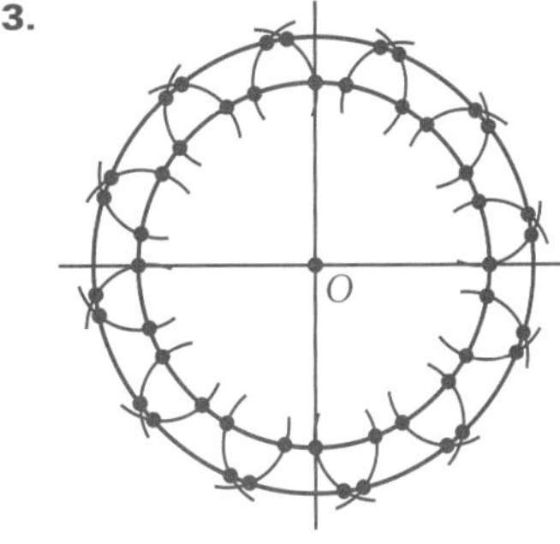

Activity 3

Students should follow the directions given in Steps 1, 2, and 3 to create a working gear model.

Chapter Review and Assessment

VOCABULARY

Key Skills & Exercises

LESSON 14.1

Key Skills

Use the law of sines to solve triangles.

Solve triangle $\triangle ABC$ given that $A = 42°$, $B = 52°$, and $a = 10$.

$$\frac{\sin A}{a} = \frac{\sin B}{b}$$
$$\frac{\sin 42°}{10} = \frac{\sin 52°}{b}$$
$$b = \frac{10 \sin 52°}{\sin 42°} \approx 11.8$$

Determine whether the given SSA information defines 0, 1, or 2 triangles.

SSA information: $c = 12$, $a = 8$, and $A = 30°$

$$\frac{\sin A}{a} = \frac{\sin C}{c}$$
$$\frac{\sin 30°}{8} = \frac{\sin C}{12}$$
$$\sin C \approx 0.75$$
$$C \approx 48.6° \text{ or } C \approx 131.4°$$

There are 2 possible triangles.

Exercises

Solve each triangle. Give answers to the nearest tenth, if necessary.

1. $A = 35°$, $B = 45°$, $a = 12$ **$b \approx 14.8$, $c \approx 20.6$, $C = 100°$**
2. $B = 27°$, $C = 40°$, $b = 20$ **$a \approx 40.6$, $c \approx 28.3$, $A = 113°$**
3. $A = 30°$, $c = 10$, $B = 50°$ **$a \approx 5.1$, $b \approx 7.8$, $C = 100°$**
4. $B = 42°$, $a = 14$, $C = 57°$ **$b \approx 9.5$, $c \approx 11.9$, $A = 81°$**

State the number of triangles determined by the given information. If 1 or 2 triangles are formed, solve the triangle(s). Give answers to the nearest tenth, if necessary.

5. $A = 40°$, $c = 20$, $a = 5$ **0 triangles**
6. $A = 60°$, $c = 10$, $a = 5\sqrt{3}$
7. $B = 30°$, $b = 3$, $c = 5$
8. $B = 75°$, $c = 9$, $b = 3$ **0 triangles**

6. 1 triangle; $b = 5$, $B = 30°$, $C = 90°$
7. 2 triangles; $C \approx 56.4°$, $A \approx 93.6°$, $a \approx 6.0$ *or* $C \approx 123.6°$, $A \approx 26.4°$, $a \approx 2.7$

LESSON 14.2

Key Skills

Use the law of cosines to solve triangles.

$$a^2 = b^2 + c^2 - 2bc \cos A$$
$$b^2 = a^2 + c^2 - 2ac \cos B$$
$$c^2 = a^2 + b^2 - 2ab \cos C$$

Given:	Use:
SSS	law of cosines, then law of sines
SSA	law of sines (ambiguous)
SAA	law of sines
ASA	law of sines
SAS	law of cosines, then law of sines
AAA	not possible

Exercises

Classify the type of information given, and then solve $\triangle ABC$ to the nearest tenth.

9. $A = 37°$, $b = 10$, $c = 14$
10. $B = 63°$, $a = 12$, $c = 15$
11. $a = 6$, $b = 3$, $c = 5$
12. $a = 9$, $b = 4$, $c = 12$
13. $A = 35°$, $c = 30$, $a = 20$

9. SAS; $a \approx 8.5$, $B \approx 45.0°$, $C \approx 98.0°$
10. SAS; $b \approx 14.3$, $A \approx 48.4°$, $C \approx 68.6°$
11. SSS; $A \approx 93.8°$, $B \approx 29.9°$, $C \approx 56.3°$
12. SSS; $A \approx 34.6°$, $B \approx 14.6°$, $C \approx 130.8°$
13. SSA; $B \approx 85.6°$, $C \approx 59.4°$, $b \approx 34.8$ *or* $B \approx 24.4°$, $C \approx 120.6°$, $b \approx 14.4$

Chapter Test, Form A

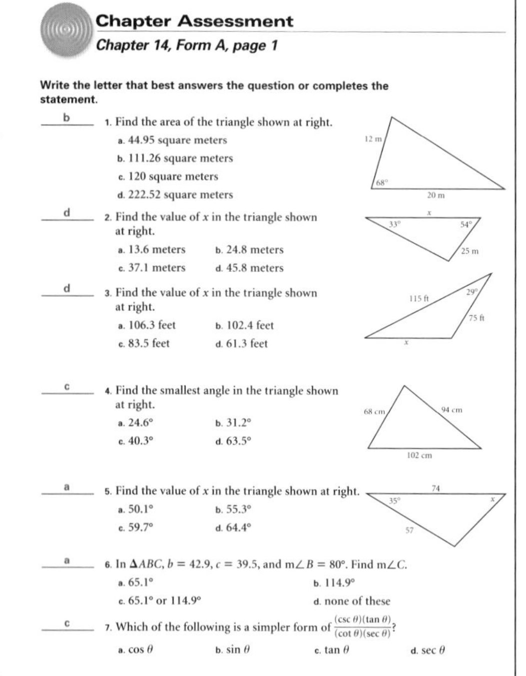

NAME ______ CLASS ______ DATE ______

Chapter Assessment
Chapter 14, Form A, page 1

Write the letter that best answers the question or completes the statement.

b 1. Find the area of the triangle shown at right.
a. 44.95 square meters
b. 111.26 square meters
c. 120 square meters
d. 222.52 square meters

d 2. Find the value of x in the triangle shown at right.
a. 13.6 meters b. 24.8 meters
c. 37.1 meters d. 45.8 meters

d 3. Find the value of x in the triangle shown at right.
a. 106.3 feet b. 102.4 feet
c. 83.5 feet d. 61.3 feet

c 4. Find the smallest angle in the triangle shown at right.
a. 24.6° b. 31.2°
c. 40.3° d. 63.5°

a 5. Find the value of x in the triangle shown at right.
a. 50.1° b. 55.3°
c. 59.7° d. 64.4°

a 6. In $\triangle ABC$, $b = 42.9$, $c = 39.5$, and $m\angle B = 80°$. Find $m\angle C$.
a. 65.1° b. 114.9°
c. 65.1° or 114.9° d. none of these

c 7. Which of the following is a simpler form of $\frac{(\csc\theta)(\tan\theta)}{(\cot\theta)(\sec\theta)}$?
a. $\cos\theta$ b. $\sin\theta$ c. $\tan\theta$ d. $\sec\theta$

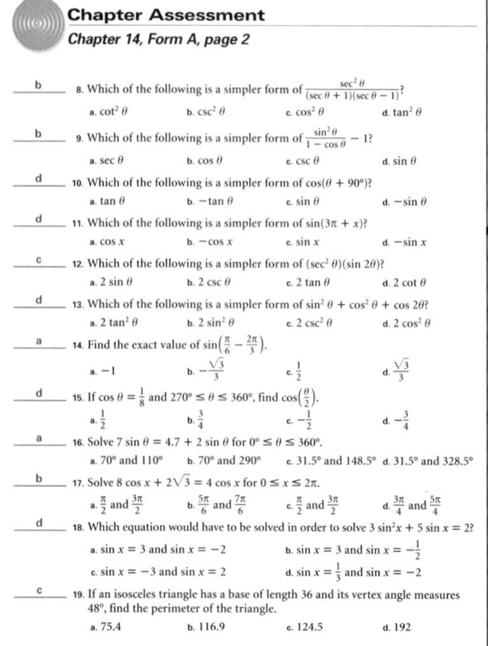

NAME ______ CLASS ______ DATE ______

Chapter Assessment
Chapter 14, Form A, page 2

b 8. Which of the following is a simpler form of $\frac{\sec^2\theta}{(\sec\theta + 1)(\sec\theta - 1)}$?
a. $\cot^2\theta$ b. $\csc^2\theta$ c. $\cos^2\theta$ d. $\tan^2\theta$

b 9. Which of the following is a simpler form of $\frac{\sin^2\theta}{1 - \cos\theta} - 1$?
a. $\sec\theta$ b. $\cos\theta$ c. $\csc\theta$ d. $\sin\theta$

d 10. Which of the following is a simpler form of $\cos(\theta + 90°)$?
a. $\tan\theta$ b. $-\tan\theta$ c. $\sin\theta$ d. $-\sin\theta$

d 11. Which of the following is a simpler form of $\sin(3\pi + x)$?
a. $\cos x$ b. $-\cos x$ c. $\sin x$ d. $-\sin x$

c 12. Which of the following is a simpler form of $(\sec^2\theta)(\sin 2\theta)$?
a. $2\sin\theta$ b. $2\csc\theta$ c. $2\tan\theta$ d. $2\cot\theta$

d 13. Which of the following is a simpler form of $\sin^2\theta + \cos^2\theta + \cos 2\theta$?
a. $2\tan^2\theta$ b. $2\sin^2\theta$ c. $2\csc^2\theta$ d. $2\cos^2\theta$

a 14. Find the exact value of $\sin\left(\frac{\pi}{6} - \frac{2\pi}{3}\right)$.
a. -1 b. $-\frac{\sqrt{3}}{3}$ c. $\frac{1}{2}$ d. $\frac{\sqrt{3}}{3}$

d 15. If $\cos\theta = \frac{1}{8}$ and $270° \le \theta \le 360°$, find $\cos\left(\frac{\theta}{2}\right)$.
a. $\frac{1}{2}$ b. $\frac{3}{4}$ c. $-\frac{1}{2}$ d. $-\frac{3}{4}$

a 16. Solve $7\sin\theta = 4.7 + 2\sin\theta$ for $0° \le \theta \le 360°$.
a. 70° and 110° b. 70° and 290° c. 31.5° and 148.5° d. 31.5° and 328.5°

b 17. Solve $8\cos x + 2\sqrt{3} = 4\cos x$ for $0 \le x \le 2\pi$.
a. $\frac{\pi}{2}$ and $\frac{3\pi}{2}$ b. $\frac{5\pi}{6}$ and $\frac{7\pi}{6}$ c. $\frac{\pi}{2}$ and $\frac{3\pi}{2}$ d. $\frac{3\pi}{4}$ and $\frac{5\pi}{4}$

d 18. Which equation would have to be solved in order to solve $3\sin^2 x + 5\sin x = 2$?
a. $\sin x = 3$ and $\sin x = -2$ b. $\sin x = 3$ and $\sin x = -\frac{1}{2}$
c. $\sin x = -3$ and $\sin x = 2$ d. $\sin x = \frac{1}{3}$ and $\sin x = -2$

c 19. If an isosceles triangle has a base of length 36 and its vertex angle measures 48°, find the perimeter of the triangle.
a. 75.4 b. 116.9 c. 124.5 d. 192

LESSON 14.3

Key Skills

Use fundamental trigonometric identities to rewrite expressions.

Verify the identity $(\tan\theta)\frac{\cos\theta}{\sin^2\theta} = \csc\theta$.

$$(\tan\theta)\frac{\cos\theta}{\sin^2\theta} = \frac{\sin\theta}{\cos\theta} \cdot \frac{\cos\theta}{\sin^2\theta} = \frac{1}{\sin\theta} = \csc\theta$$

Exercises

Write each expression in terms of a single trigonometric function.

14. $(\sec\theta)(\cos^2\theta)$ **$\cos\theta$**

15. $(\csc\theta)(\tan\theta)$ **$\sec\theta$**

16. $\frac{2\cos^2\theta}{1-\sin^2\theta}$ **2**

17. $\frac{-3\tan^2\theta}{1+\sec^2\theta}$ **$\frac{-3\tan^2\theta}{2+\tan^2\theta}$**

LESSON 14.4

Key Skills

Use the sum and difference identities.

Find the exact value of $\sin(30^\circ - 45^\circ)$.

$$\begin{aligned}\sin(30^\circ - 45^\circ) &= (\sin 30^\circ)(\cos 45^\circ) - (\cos 30^\circ)(\sin 45^\circ)\\ &= \left(\frac{1}{2}\right)\left(\frac{\sqrt{2}}{2}\right) - \left(\frac{\sqrt{3}}{2}\right)\left(\frac{\sqrt{2}}{2}\right) = \frac{\sqrt{2}-\sqrt{6}}{4}\end{aligned}$$

Use rotation matrices.

Find the coordinates of the image of $P(-3, 4)$ after a 20° rotation about the origin.

$$\begin{bmatrix}\cos 20^\circ & -\sin 20^\circ\\ \sin 20^\circ & \cos 20^\circ\end{bmatrix}\begin{bmatrix}-3\\ 4\end{bmatrix} \approx \begin{bmatrix}-4.2\\ 2.7\end{bmatrix}$$

Exercises

Find the exact value of each expression.

18. $\sin(45^\circ - 210^\circ)$ **$-\frac{\sqrt{6}+\sqrt{2}}{4}$**

19. $\sin(60^\circ + 270^\circ)$ **$-\frac{1}{2}$**

20. $\cos(90^\circ + 60^\circ)$ **$-\frac{\sqrt{3}}{2}$**

21. $\cos(120^\circ - 135^\circ)$

22. $\sin 195^\circ$ **$\frac{\sqrt{2}-\sqrt{6}}{4}$**

23. $\cos 75^\circ$ **$\frac{\sqrt{6}-\sqrt{2}}{4}$**

24. $\cos(-210^\circ)$ **$-\frac{\sqrt{3}}{2}$**

25. $\sin(-15^\circ)$ **$\frac{\sqrt{2}-\sqrt{6}}{4}$**

Find the coordinates of the image of each point after a 120° rotation.

26. $(3, -5)$ **(2.83, 5.10)**

27. $(-2, 7)$ **(−5.06, −5.23)**

21. **$\frac{\sqrt{2}+\sqrt{6}}{4}$**

LESSON 14.5

Key Skills

Use the double- and half-angle identities.

$\sin 2\theta = 2\sin\theta\cos\theta \qquad \sin\frac{\theta}{2} = \pm\sqrt{\frac{1-\cos\theta}{2}}$

$\cos 2\theta = \cos^2\theta - \sin^2\theta \qquad \cos\frac{\theta}{2} = \pm\sqrt{\frac{1+\cos\theta}{2}}$

Given $180^\circ \le \theta \le 270^\circ$ and $\sin\theta = -\frac{4}{7}$, find the exact value of $\cos\frac{\theta}{2}$.

Find $\cos\theta$: $\quad x^2 + y^2 = r^2$

$x^2 + (-4)^2 = 7^2$

$x = -\sqrt{33} \leftarrow$ *Quadrant III*

$\cos\theta = \frac{-\sqrt{33}}{7}$

$$\begin{aligned}\cos\frac{\theta}{2} &= \pm\sqrt{\frac{1+\cos\theta}{2}}\\ &= -\sqrt{\frac{1+\left(-\frac{\sqrt{33}}{7}\right)}{2}} \leftarrow 90^\circ \le \frac{\theta}{2} \le 135^\circ\\ &= -\sqrt{\frac{1}{2} - \frac{\sqrt{33}}{14}}\end{aligned}$$

Exercises

Given $0^\circ \le \theta \le 90^\circ$ and $\cos\theta = \frac{1}{8}$, find the exact value of each expression.

28. $\cos 2\theta$ **$-\frac{31}{32}$**

29. $\sin 2\theta$ **$\frac{\sqrt{63}}{32}$**

30. $\cos\frac{\theta}{2}$ **$\frac{3}{4}$**

31. $\sin\frac{\theta}{2}$ **$\frac{\sqrt{7}}{4}$**

Given $270^\circ \le \theta \le 360^\circ$ and $\sin\theta = -\frac{5}{8}$, find the exact value of each expression.

32. $\sin 2\theta$ **$-\frac{5\sqrt{39}}{32}$**

33. $\cos 2\theta$ **$\frac{7}{32}$**

34. $\sin\frac{\theta}{2}$ **$\frac{\sqrt{8-\sqrt{39}}}{4}$**

35. $\cos\frac{\theta}{2}$ **$-\frac{\sqrt{8+\sqrt{39}}}{4}$**

Write each expression in terms of trigonometric functions of θ rather than multiples of θ.

36. $\sin 2\theta + \cos 2\theta$

37. $\cos 4\theta$

36. **$\cos^2\theta + 2\sin\theta\cos\theta - \sin^2\theta$**

37. **$\cos^4\theta - 6\cos^2\theta\sin^2\theta + \sin^4\theta$**

Chapter Test, Form B

NAME ________ CLASS ________ DATE ________

Chapter Assessment

Chapter 14, Form B, page 1

1. Find x in the triangle at right to the nearest tenth of a foot.

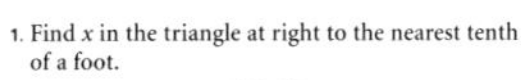

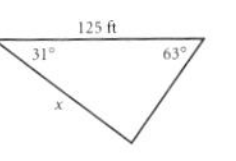

111.6 feet

2. Find x in the triangle at right to the nearest tenth of a centimeter.

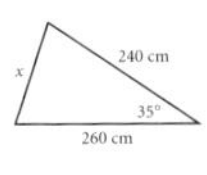

151.6 centimeters

3. Find the area of the triangle at right to the nearest tenth of a square meter.

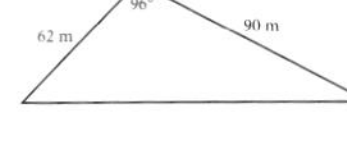

2774.7 square meters

4. Solve ΔABC. Give the measure of each angle to the nearest tenth of a degree.

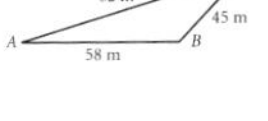

$m\angle A \approx 32.0^\circ$; $m\angle B \approx 104.8^\circ$; $m\angle C \approx 43.2^\circ$

5. Solve ΔABC. Give c to the nearest tenth of a foot and $m\angle A$ and $m\angle C$ to the nearest tenth of a degree.

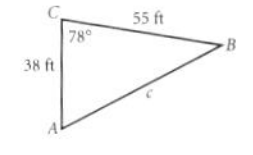

$c \approx 60.0$ feet; $m\angle A \approx 63.7^\circ$; $m\angle B \approx 38.3^\circ$

6. In ΔABC, $m\angle A = 67^\circ$, $a = 20$, and $c = 21.5$. Find all possible values for $m\angle C$ to the nearest tenth of a degree.

$m\angle C \approx 81.7^\circ$ or $m\angle C \approx 98.3^\circ$

7. If $\cos\theta = \frac{7}{8}$ and $270^\circ \le \theta \le 360^\circ$, find $\sin\left(\frac{\theta}{2}\right)$. **$-\frac{1}{4}$**

8. If $\sin\theta = -\frac{2}{3}$ and $180^\circ \le \theta \le 270^\circ$, find $\cos 2\theta$. **$\frac{1}{9}$**

9. Find the exact value of $\sin\left(\frac{\pi}{3} - \frac{\pi}{2}\right)$. **$-\frac{1}{2}$**

NAME ________ CLASS ________ DATE ________

Chapter Assessment

Chapter 14, Form B, page 2

10. Find the exact value of $\cos\left(\frac{3\pi}{2} + \frac{\pi}{3}\right)$. **$\frac{\sqrt{3}}{2}$**

11. Find the image of $(-2, 6)$ after a rotation of 90° about the origin. **(−6, −2)**

Prove each identity.

12. $\cos(270^\circ - \theta) = -\sin\theta$

$\cos(270^\circ - \theta)$
$= (\cos 270^\circ)(\cos\theta) + (\sin 270^\circ)(\sin\theta)$
$= (0)(\cos\theta) + (-1)(\sin\theta)$
$= -\sin\theta$

13. $\sin\left(x + \frac{\pi}{2}\right) = \cos x$

$\sin\left(x + \frac{\pi}{2}\right) = (\sin x)\left(\cos\frac{\pi}{2}\right) + (\cos x)\left(\sin\frac{\pi}{2}\right)$
$= (\sin x)(0) + (\cos x)(1)$
$= \cos x$

Write each expression in terms of a single trigonometric function.

14. $(\sec^2\theta)(\cot\theta)(\sin^2\theta)$ **$\tan\theta$**

15. $(\sin\theta)(\cot^2\theta + 1)$ **$\frac{1}{\sin\theta} = \csc\theta$**

16. $\frac{\tan^2\theta}{\sec\theta + 1}$ **$\sec\theta - 1$**

17. $\frac{\cos 2\theta + \sin^2\theta}{\sin 2\theta}$ **$\frac{\cot\theta}{2}$**

Solve each equation to the nearest tenth of a degree for $0^\circ \le \theta \le 360^\circ$.

18. $6\sin\theta + 7 = 9 - 4\sin\theta$ **11.5° and 168.5°**

19. $4\cos^2\theta - 5\cos\theta + 1 = 0$ **75.5°, 284.5°, 0°, and 360°**

Solve each equation to the nearest hundredth of a radian for $0 \le x \le 2\pi$.

20. $5(\sin x)(\cot x) + 2 = 6$ **0.64 and 5.64**

21. $5\sin x + 1 = 3\cos^2 x$ **0.34 and 2.80**

LESSON 14.6

Key Skills

Solve trigonometric equations.

Find the exact solutions of $-\cos^2 \theta = 1 + 5 \sin \theta$ for $0° \le \theta < 360°$.

$$-2\cos^2 \theta = 1 + 5\sin \theta$$
$$-2(1 - \sin^2 \theta) = 1 + 5 \sin \theta$$
$$2\sin^2 \theta - 5 \sin \theta - 3 = 0$$
$$(2\sin \theta + 1)(\sin \theta - 3) = 0$$
$$\sin \theta = -\frac{1}{2} \quad or \quad \sin \theta = 3$$

For $\sin \theta = -\frac{1}{2}$, $\theta = 225°$ *or* $\theta = 315°$. For $\sin \theta = 3$, there is no solution.

Exercises

Find the exact solutions of each equation for $0° \le \theta < 360°$.

38. $2\cos \theta - \sqrt{2} = 0$ **45°, 315°**

39. $\sin^2 \theta + \sin \theta = 2$ **90°**

40. $2\cos^2 \theta - \cos \theta - 1 = 0$ **0°, 120°, 240°**

41. $\cos^2 \theta - \sin^2 \theta + 1 = 0$ **90°, 270°**

42. $4\sin^2 \theta + 4\cos \theta - 1 = 0$ **120°, 240°**

43. $4\sin^2 \theta - 2(\sqrt{2} + 1)\sin \theta + \sqrt{2} = 0$

30°, 45°, 135°, 150°

Applications

44. SURVEYING A map maker makes the measurements shown in the diagram at right. Find the distance across the river from *B* to *C*. **175.3 m**

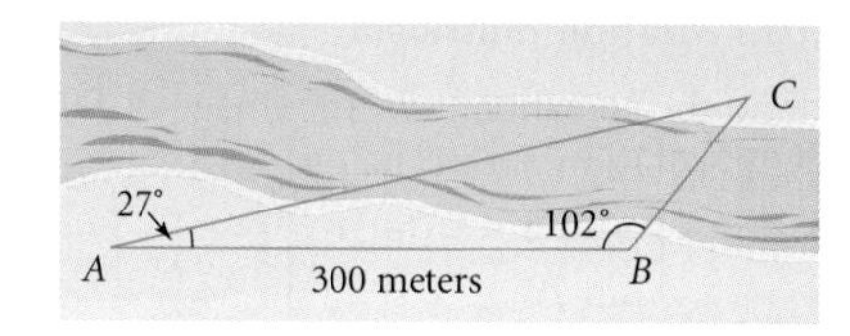

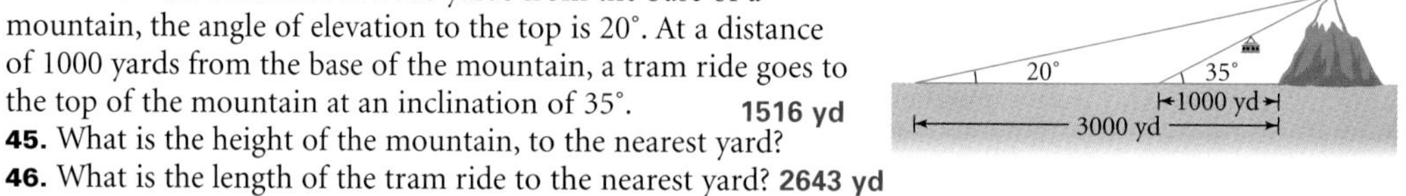

RECREATION At a distance of 3000 yards from the base of a mountain, the angle of elevation to the top is 20°. At a distance of 1000 yards from the base of the mountain, a tram ride goes to the top of the mountain at an inclination of 35°.

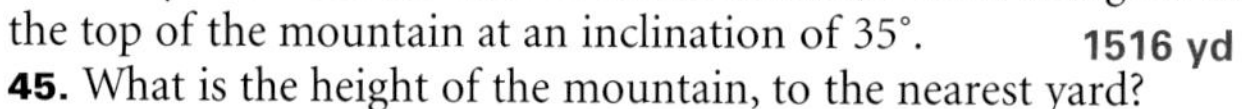

45. What is the height of the mountain, to the nearest yard? **1516 yd**

46. What is the length of the tram ride to the nearest yard? **2643 yd**

47. REAL ESTATE A triangular lot has sides of 215 feet, 185 feet, and 125 feet. Find the measures of the angles at its corners. **85.5°, 59.1°, 35.4°**

48. SURVEYING A tunnel from point *A* to point *B* runs through a mountain. From point *C*, both ends of the tunnel can be observed. If $AC = 165$ meters, $BC = 115$ meters, and $C = 74°$, find *AB*, the length of the tunnel. **173.2 m**

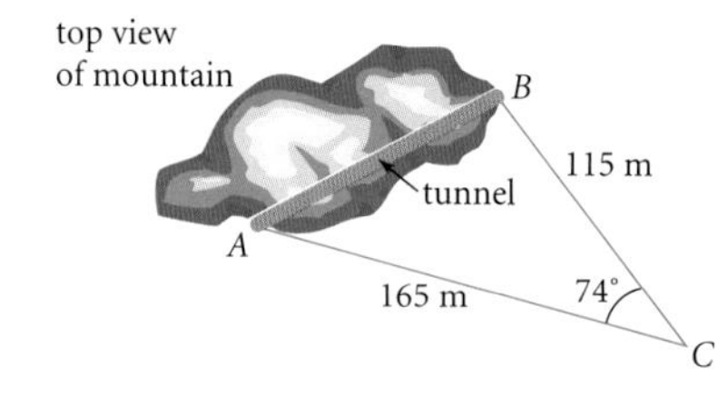

49. SPACE SCIENCE Two space-monitoring stations are located at two different points, *A* and *B*, on Earth's surface. The inscribed angle at the Earth's center is $C = 110°$. The radius of the Earth is about 3960 miles. Find the straight-line distance, *AB*, between the two stations. **6487.7 mi**

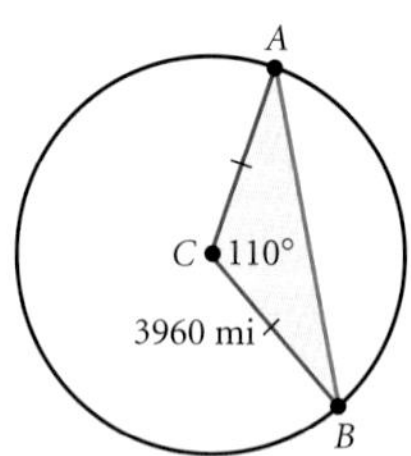

Chapter Test

Solve each triangle. Give answers to the nearest tenth.

1. $B = 18°, C = 65°, b = 9$ **A = 97°, a = 28.9, c = 26.4**
2. $A = 23°, C = 57°, c = 45$ **B = 100°, a = 21.0, b = 52.8**
3. PIONEERING Two hikers leave a campsite, walking at a constant rate of 2.4 miles per hour. One follows a northeast path 37° north of the east-west line. The other is on a southeast path 57° south of the east-west line. How far apart are they after 3 hours? **10.5 miles**

State the number of triangles determined by the given information. If one or two are formed, solve the triangle(s). Give answers to the nearest tenth.

4. $A = 60°, c = 12, a = 3$ **0**
5. $A = 30°, a = 24, c = 36$

Find the area of ΔABC to the nearest tenth of a square unit.

6. $b = 15, c = 8, A = 35°$ **34.4 u^2**
7. $a = 15, c = 13, B = 70°$ **91.6 u^2**

Classify the type of information given, then solve each triangle to the nearest tenth.

8. $A = 35°, b = 8, c = 7$ **SAS; B = 84.4°, C = 60.6°, a = 4.6**
9. $a = 15, b = 18, C = 105°$ **SAS; A = 33.5°, B = 41.5°, c = 26.2**
10. AIR TRAFFIC CONTROL An airplane is spotted on radar at a line of sight distance of 75 miles east at an angle 16° above the horizon. A second airplane is spotted west at an angle of 24° above the horizon with a line of sight distance of 36 miles. Find the line of sight distance between the two planes. **105.2 miles**

Write each expression in terms of a single trigonometric function.

11. $\frac{\sec\theta}{\cos\theta} - \sec\theta$ **$\frac{1-\cos\theta}{\cos^2\theta}$**
12. $\csc\theta - \cot\theta\cos\theta$ **$\sin\theta$**
13. $\frac{1}{\cos^2\theta} - \tan^2\theta - \sin^2\theta$ **$\cos^2\theta$**

Find the exact value of each expression.

14. $\sin(330° - 225°)$ **$\frac{\sqrt{6}+\sqrt{2}}{4}$**
15. $\cos(45° - 60°)$ **$\frac{\sqrt{6}+\sqrt{2}}{4}$**
16. $\cos 105°$ **$\frac{\sqrt{2}-\sqrt{6}}{4}$**

Find the coordinates of the image of each point after the given rotation.

17. $(5, 8); 30°$ **$\left(\frac{5\sqrt{3}-8}{2}, \frac{8\sqrt{3}+5}{2}\right)$**
18. $(-2, 6); -60°$ **$\left(3\sqrt{3}-1, \sqrt{3}+3\right)$**
19. $(8, -1); 60°$ **$\left(4+\frac{\sqrt{3}}{2}, -\frac{1}{2}+4\sqrt{3}\right)$**

Given $0° \leq \theta \leq 90°$ and $\cos\theta = \frac{2}{3}$, find the exact value of each expression.

20. $\cos 2\theta$ **$-\frac{1}{9}$**
21. $\sin 2\theta$ **$\frac{4\sqrt{5}}{9}$**
22. $\sin\frac{\theta}{2}$ **$\frac{\sqrt{6}}{6}$**
23. $\cos\frac{\theta}{2}$ **$\frac{\sqrt{30}}{6}$**

Write each expression below in terms of trigonometric functions of θ rather than multiples of θ.

24. $\frac{2\sin^2\theta}{\sin 2\theta}$ **$\frac{\sin\theta}{\cos\theta}$**
25. $\frac{\cos 2\theta}{\cos\theta - \sin\theta}$ **$\cos\theta + \sin\theta$**
26. $\frac{1 + \sin 2\theta}{\sin\theta + \cos\theta}$ **$\sin\theta + \cos\theta$**
27. The minute hand on a clock face is 3 inches long. How far does the minute hand travel in 5 minutes? **1.57 in.**

Find the exact solution of each equation for $0° \leq \theta < 360°$.

28. $2\sin^2\theta = -\sqrt{3}\sin\theta$ **0°, 180°, 240°, 300°**
29. $2\sin^2\theta = -\sin\theta + 4$ **no solution**
30. $2\cos^2\theta = -\sqrt{2}\cos\theta$ **90°, 135°, 225°, 270°**
31. $\cos 2\theta - \sin\theta + 2\sin^2\theta = 0$ **90°**

Chapter Test

5. two; $C = 48.6°, B = 101.4°, b = 47.1, C = 131.4°, B = 18.6°, b = 15.3$

College Entrance Exam Practice

Multiple-Choice and Quantitative-Comparison Samples

The first half of the Cumulative Assessment contains two types of items found on standardized tests—multiple-choice questions and quantitative-comparison questions. Quantitative-comparison items emphasize the concepts of equality, inequality, and estimation.

Free-Response Grid Samples

The second half of the Cumulative Assessment is a free-response section. This part of the Cumulative Assessment requires student-produced response items like those commonly found on college entrance exams. These questions require the use of machine-scored answer grids. You may wish to have students practice answering these items in preparation for standardized tests.

CHAPTERS CUMULATIVE ASSESSMENT

1–14 College Entrance Exam Practice

QUANTITATIVE COMPARISON For Items 1–6, write
A if the quantity in Column A is greater than the quantity in Column B;
B if the quantity in Column B is greater than the quantity in Column A;
C if the quantities are equal; or
D if the relationship cannot be determined from the given information.

internet connect
Standardized Test Prep Online
Go To: **go.hrw.com**
Keyword: **MM1 Test Prep.**

	Column A	Column B	Answers
1. B	The period of f $f(\theta) = 2 + 2\cos 5\theta$	$f(\theta) = -3\sin\frac{\theta}{2}$	(A) (B) (C) (D) [Lesson 13.5]
2. A	$\sum_{n=1}^{5} 4n$	$\sum_{n=1}^{4} 5n$	(A) (B) (C) (D) [Lesson 11.1]
3. A	The number of real-number solutions $2x^2 - 3x - 3 = 0$	$4x^2 - 4x + 1 = 0$	(A) (B) (C) (D) [Lesson 5.6]
4. D	$\ln x$	$\log x$	(A) (B) (C) (D) [Lesson 6.6]
5. B	The given measure for 76, 78, 71, 78, 72, and 75 mean	median	(A) (B) (C) (D) [Lesson 12.1]
6. A	$3x - 5y = 6$ slope	y-intercept	(A) (B) (C) (D) [Lesson 1.2]

7. Which word describes the number $\frac{1}{12}$? *(LESSON 2.1)* c
- **a.** prime
- **b.** integer
- **c.** rational
- **d.** irrational

8. Simplify $(x^3 - 8x^2 + 17x - 6) \div (x^2 - 5x + 2)$. *(LESSON 7.3)* b
- **a.** $x + 3$
- **b.** $x - 3$
- **c.** $x + 1$
- **d.** $x - 1$

9. Which equation below has a graph that opens downward? *(LESSON 5.1)* b
- **a.** $y = 5x^2$
- **b.** $y = 5 - x^2$
- **c.** $y - x^2 = 5$
- **d.** $y = x^2 - 5$

10. Which equation represents the line that has a slope of $\frac{1}{2}$ and contains the point $P(2, -2)$? *(LESSON 1.3)* a
- **a.** $y + 2 = \frac{1}{2}(x - 2)$
- **b.** $y - 2 = \frac{1}{2}(x - 2)$
- **c.** $y - 2 = \frac{1}{2}(x + 2)$
- **d.** $y + 2 = \frac{1}{2}(x + 2)$

11. Which expression represents $(a^3 - 1) - (a^3 - a^2 + 5)$ in standard form? *(LESSON 7.1)* c
- **a.** $2a^3 + a^2 - 6$
- **b.** $-a^2 + 6$
- **c.** $a^2 - 6$
- **d.** $-2a^3 + 4$

12. What is the amplitude of $y = 3\sin(2\theta - 5)$? **(LESSON 13.5)**
b
a. 1 **b.** 3 **c.** 2 **d.** 5

13. Simplify $(4 - 5i) + (-7 - 5i)$. **(LESSON 5.6)**
b
a. $-(7 + i)$ **b.** $-3 - 10i$
c. $-3 + i$ **d.** -3

14. Which ellipse has a horizontal major axis that is 18 units long? **(LESSON 9.4)**
c
a. $\frac{x^2}{4} + \frac{y^2}{81} = 1$ **b.** $\frac{x^2}{9} + \frac{y^2}{4} = 1$
c. $4x^2 + 81y^2 = 324$ **d.** $81x^2 + 4y^2 = 324$

15. A committee of 3 is selected from 8 eligible members. How many different committees are possible? **(LESSON 10.3)**
d
a. $8!$ **b.** $3!$ **c.** $\frac{8!}{5!}$ **d.** $\frac{8!}{3!5!}$

16. Find A in $\triangle ABC$ if $a = 6$, $b = 9$, and $C = 30°$. **(LESSON 14.2)**
b
a. $\approx 68.3°$ **b.** $\approx 38.3°$
c. $\approx 111.7°$ **d.** $60°$

17. Which expression is not equivalent to the others? **(LESSON 6.4)**
b
a. $3\log_5 2x$ **b.** $\log_5 2x^3$
c. $\log_5(2x)^3$ **d.** $\log_5 8x^3$

For Items 18–19, state the property that is illustrated in each statement. All variables represent real numbers. *(LESSON 2.1)*

18. $3(x^2 - 1) = 3x^2 - 3$ **Distributive**

19. $-4 + 4 = 0$ **Inverse Prop of +**

20. Simplify $\frac{-9x - 3}{x^2 - 11x + 18} + \frac{x + 3}{x - 9}$. **(LESSON 8.4)**

21. Find the area of $\triangle ABC$ to the nearest tenth of a square unit if $a = 4.5$, $c = 8.3$, and $B = 55°$. **(LESSON 14.1)** ≈ 15.3

22. Find the inverse of $f(x) = -\frac{1}{4}x + 2$. **(LESSON 2.5)** $f^{-1}(x) = -4x + 8$

23. Solve the literal equation $R = \frac{S + F + P}{S + P}$ for S. **(LESSON 1.6)** $S = \frac{F + P - RP}{R - 1}$

24. Solve $\begin{cases} 3x - 4y = -14 \\ 3x + 2y = 16 \end{cases}$. **(LESSON 3.2)** (2, 5)

25. Use the quadratic formula to solve $x^2 - 2x + 4 = 0$. **(LESSON 5.6)** $x = 1 \pm i\sqrt{3}$

26. Graph the solution of $\begin{cases} x < 2 \\ y \le 3x - 4 \end{cases}$. **(LESSON 3.4)**

27. Simplify the complex fraction $\frac{\frac{3x}{5}}{\frac{x}{2}}$. **(LESSON 8.3)** $\frac{6}{5}$

28. Factor $4x^4 - 17x^2 + 4$ completely. **(LESSON 7.3)** $(2x - 1)(2x + 1)(x - 2)(x + 2)$

29. Find the domain of $f(x) = \frac{x^2 - 4}{x - 2}$. **(LESSON 8.2)** $x \ne 2$

FREE-RESPONSE GRID **The following questions may be answered by using a free-response grid such as that commonly used by standardized-test services.**

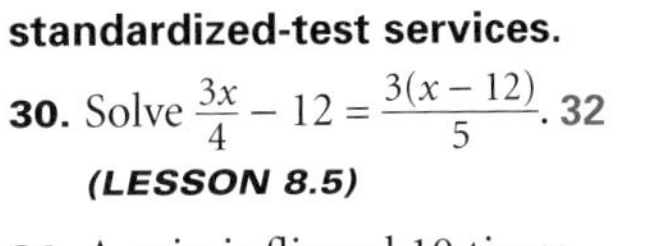

30. Solve $\frac{3x}{4} - 12 = \frac{3(x - 12)}{5}$. **(LESSON 8.5)** 32

31. A coin is flipped 10 times. Find the probability of exactly 4 heads appearing. **0.205** **(LESSON 11.7)**

32. Find the coterminal angle, θ, for 255° such that $-360° < \theta < 360°$. **(LESSON 13.2)** −105

33. Find the value of v if $v = \log_3 1$. **(LESSON 6.3)** 0

34. Convert $\frac{3\pi}{5}$ radians to degrees. **(LESSON 13.4)** 108°

35. Find the final amount of a \$1000 investment earning 5% interest compounded annually for 10 years. Round answer to the nearest dollar. **(LESSON 6.2)** \$1628.89

36. Evaluate $\text{Cos}^{-1}(\sin 30°)$. **(LESSON 13.6)** 60°

37. Evaluate $\log_5 73.25$ to the nearest hundredth. **(LESSON 6.4)** 2.67

38. Evaluate $\sum_{n=1}^{4}(8 - 5n)$. **(LESSON 11.3)** −18

39. **PROBABILITY** A 6-sided number cube is rolled once. Find the probability of getting an even number *or* 1. **(LESSON 10.4)** $\frac{2}{3}$

40. **PROBABILITY** A coin is flipped 5 times. Find the probability of exactly 3 heads appearing. **(LESSON 12.5)** 0.3125

20. 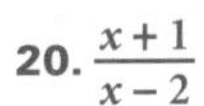

26.

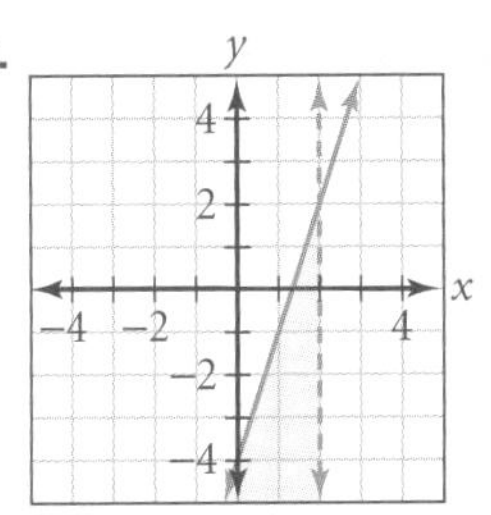

Keystroke Guide for Chapter 14

Essential keystroke sequences (using the model TI-82 or TI-83 graphics calculator) are presented below for all Activities and Examples found in this chapter that require or recommend the use of a graphics calculator.

internet connect

For Keystrokes of other graphing calculator models, visit the HRW web site at **go.hrw.com** and enter the keyword **MB1 CALC**.

LESSON 14.2

Activity
Page 896

For Steps 1 and 2, graph $y = \sin x$ and $y = \frac{1}{2}$ on the same screen for $0° \leq x \leq 180°$.

Use viewing window [0, 180] by [−2, 2] in degree mode.

Y= SIN X,T,θ,n ENTER (Y2=) 1 ÷ 2 GRAPH

Use a similar keystroke sequence for Steps 4 and 5.

LESSON 14.3

EXAMPLE 1
Page 903

Graph $y = \tan x$ and $y = \frac{\sin x}{\cos x}$ on the same screen.

Use viewing window [−360, 360] by [−10, 10] in degree mode.

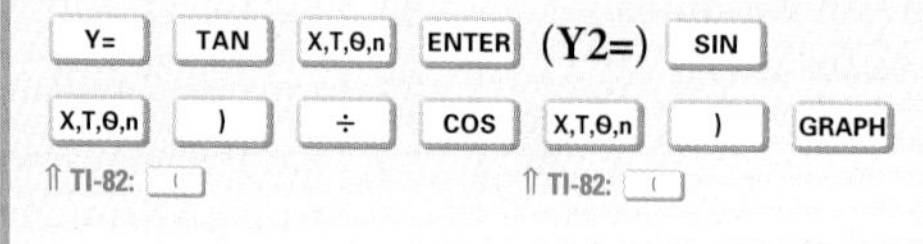

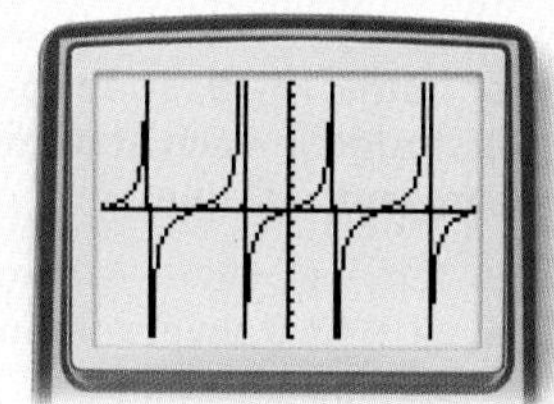

EXAMPLE 3
Page 904

Graph $y = \frac{\sin^2 x}{1 - \cos x}$ and $y = 1 + \cos x$ on the same screen.

Use viewing window [−720, 720] by [−3, 3] in degree mode.

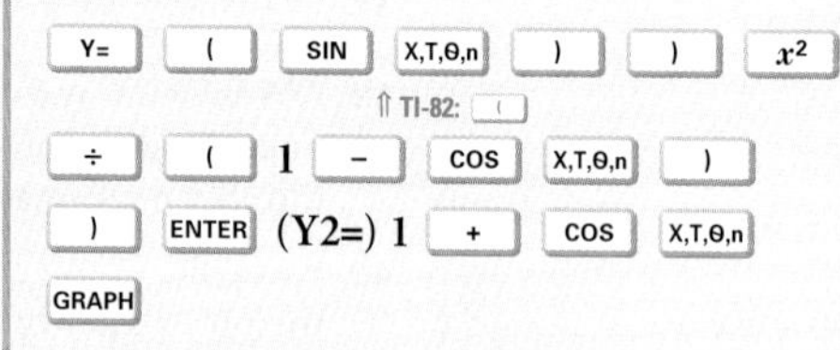

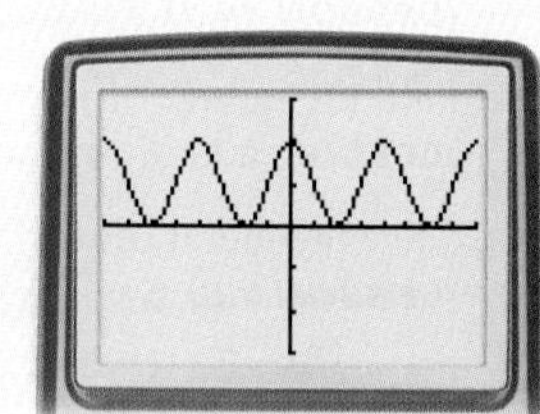

Activity Page 904

For Step 1, graph $y = (\csc x)(1 - \cos x)(1 + \cos x)$.

Use viewing window [−360, 360] by [−3, 3] in degree mode.

Y= (1 ÷ SIN X,T,θ,n))) (1 − COS X,T,θ,n
⇑ TI-82: (⇑ TI-82: (

)) (1 + COS X,T,θ,n)) GRAPH
⇑ TI-82: (

For Step 3, use the same viewing window and a similar keystroke sequence.

EXAMPLE 4 Page 905

Graph $y = \sec x - \tan x \sin x$.

Use viewing window [−360, 360] by [−3, 3] in degree mode.

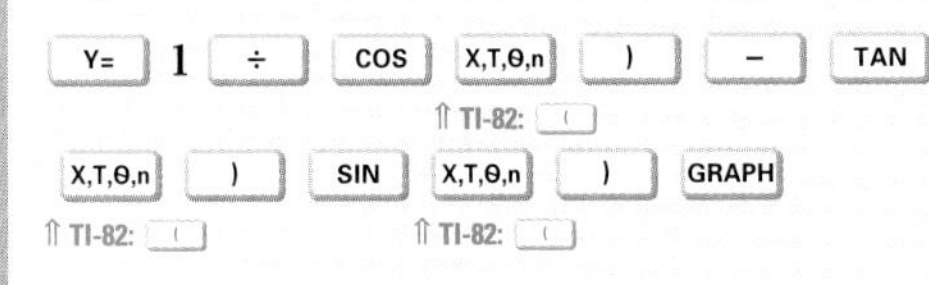

Y= 1 ÷ COS X,T,θ,n) − TAN
⇑ TI-82: (

X,T,θ,n) SIN X,T,θ,n) GRAPH
⇑ TI-82: (⇑ TI-82: (

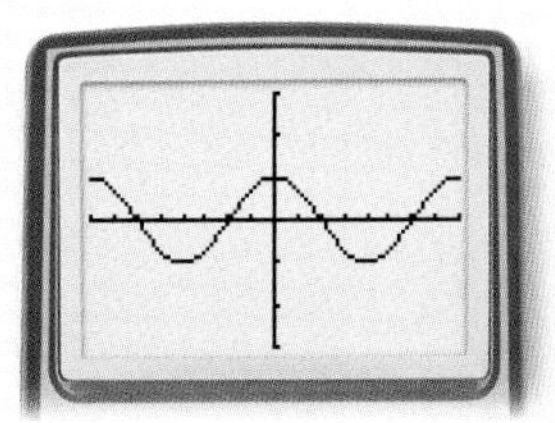

LESSON 14.4

EXAMPLE 5 Page 913

Find the product $\begin{bmatrix} \cos 30^\circ & -\sin 30^\circ \\ \sin 30^\circ & \cos 30^\circ \end{bmatrix} \begin{bmatrix} 2 & 5 & 5 & 2 \\ 1 & 1 & -1 & -1 \end{bmatrix}$.

Set the mode:

MODE ▼ 2 ENTER 2nd QUIT MODE

Enter the matrices:

For TI-83 Plus, press 2nd MATRX x^{-1} to access the matrix menu.

MATRX EDIT 1:[A] ENTER (Matrix[A]) 2 ENTER 2 ENTER COS 30) ENTER
⇑ TI-82: (

(−) SIN 30) ENTER SIN 30) ENTER COS 30) ENTER
⇑ TI-82: (⇑ TI-82: (⇑ TI-82: (

MATRX EDIT 2:[B] ENTER (Matrix[B]) 2 ENTER 4 ENTER 2 ENTER 5 ENTER 5 ENTER

2 ENTER 1 ENTER 1 ENTER (−) 1 ENTER (−) 1 ENTER 2nd QUIT MODE

Multiply the matrices:

MATRX NAMES 1:[A] ENTER × MATRX

NAMES 2:[B] ENTER ENTER

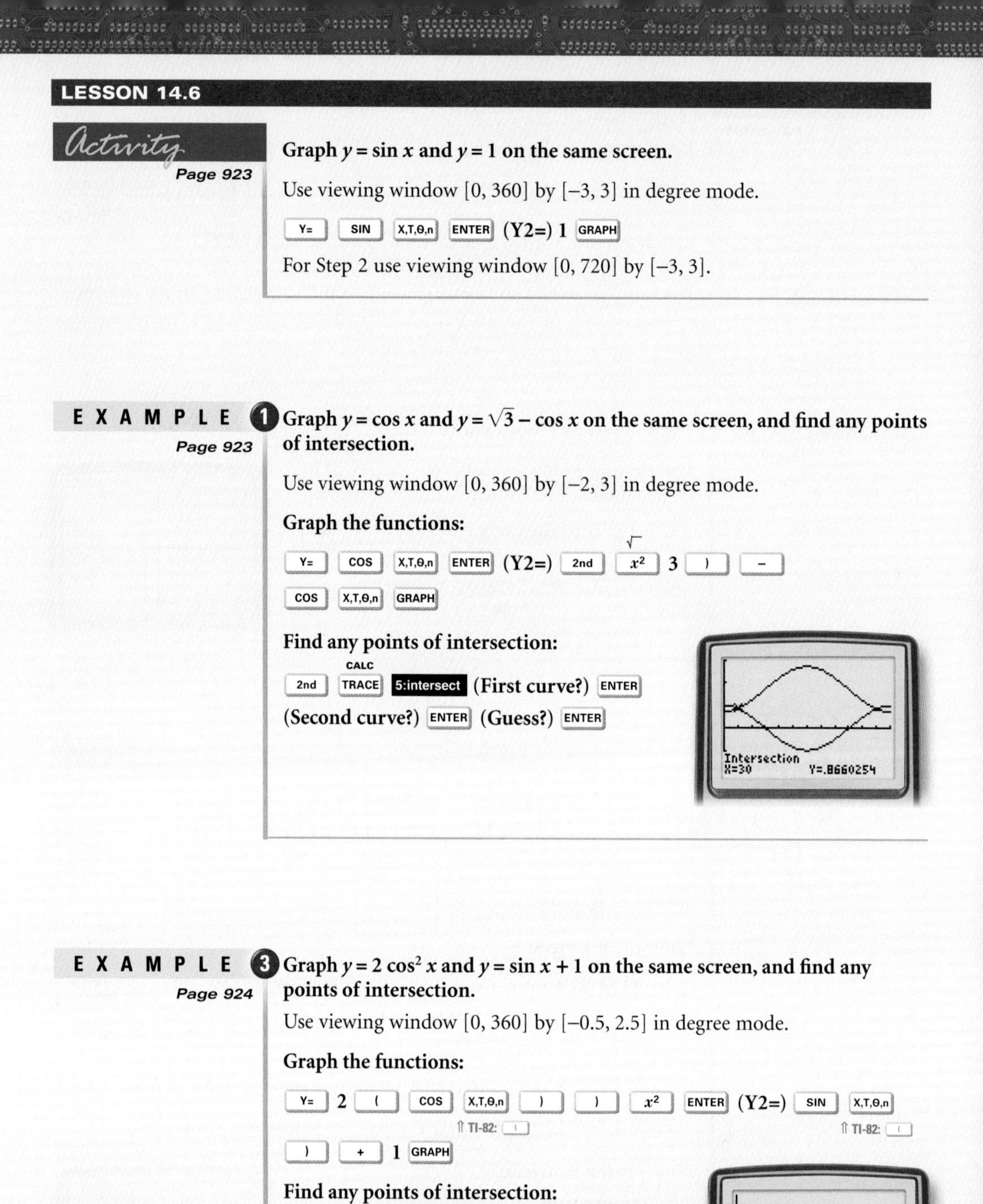

LESSON 14.6

Activity
Page 923

Graph $y = \sin x$ and $y = 1$ on the same screen.

Use viewing window [0, 360] by [−3, 3] in degree mode.

[Y=] [SIN] [X,T,θ,n] [ENTER] (Y2=) 1 [GRAPH]

For Step 2 use viewing window [0, 720] by [−3, 3].

EXAMPLE 1
Page 923

Graph $y = \cos x$ and $y = \sqrt{3} - \cos x$ on the same screen, and find any points of intersection.

Use viewing window [0, 360] by [−2, 3] in degree mode.

Graph the functions:

[Y=] [COS] [X,T,θ,n] [ENTER] (Y2=) [2nd] [x^2] ($\sqrt{\ }$) 3 [ )] [−]
[COS] [X,T,θ,n] [GRAPH]

Find any points of intersection:

[2nd] [TRACE] (CALC) **5:intersect** (First curve?) [ENTER]
(Second curve?) [ENTER] (Guess?) [ENTER]

EXAMPLE 3
Page 924

Graph $y = 2\cos^2 x$ and $y = \sin x + 1$ on the same screen, and find any points of intersection.

Use viewing window [0, 360] by [−0.5, 2.5] in degree mode.

Graph the functions:

[Y=] 2 [( ] [COS] [X,T,θ,n] [ )] [ )] [x^2] [ENTER] (Y2=) [SIN] [X,T,θ,n]
⇑ TI-82: [( ] ⇑ TI-82: [( ]
[ )] [+] 1 [GRAPH]

Find any points of intersection:

[2nd] [TRACE] (CALC) **5:intersect** (First curve?) [ENTER]
(Second curve?) [ENTER] (Guess?) [ENTER]

The calculator may return an error message when looking for the intersection point (270°, 0).

EXAMPLE 4 Page 925

Graph $y = 366 \sin x - 144$ and $y = 15$ on the same screen, and find any points of intersection.

Use viewing window [−15, 90] by [−75, 250] in degree mode.

Graph the functions:

Y= 366 SIN X,T,θ,n − 144 ENTER (Y2=) 15 GRAPH

Find any points of intersection:

2nd TRACE (CALC) 5:intersect (First curve?) ENTER (Second curve?) ENTER (Guess?) ENTER

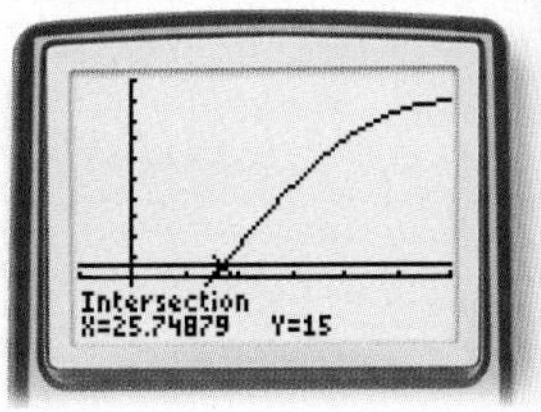

Intersection
X=25.74879 Y=15

Extra Practice
Tables
ry
Parent Functions
Additional Answers
Index

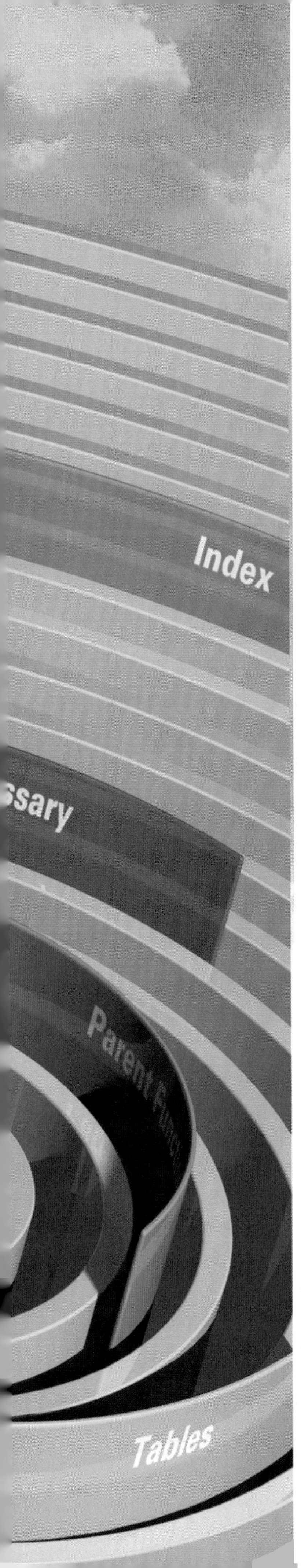

Info Bank

Lesson 1.1

12.

x	–1	0	1	2
y	–4	–1	2	5

Yes; there is a constant difference in the x-values, 1, and a constant difference in the y-values, 3.

13.

x	–6	–3	–1	2
y	–3	–4	–5	–6

No; there is no constant difference in the x-values.

14.

x	1	3	5	7
y	–6	–3	0	3

Yes; there is a constant difference in the x-values, 2, and a constant difference in the y-values, 3.

Lesson 1.2

8. undefined slope; no y-intercept;

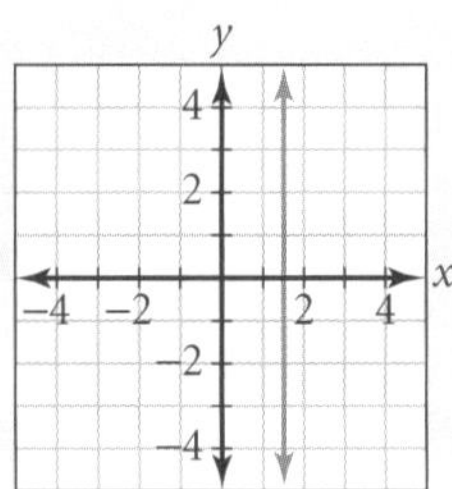

9. $m = 2, b = 1$;

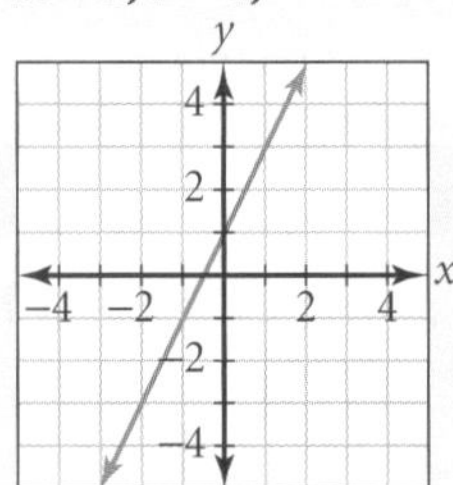

10. $m = 3, b = -2$;

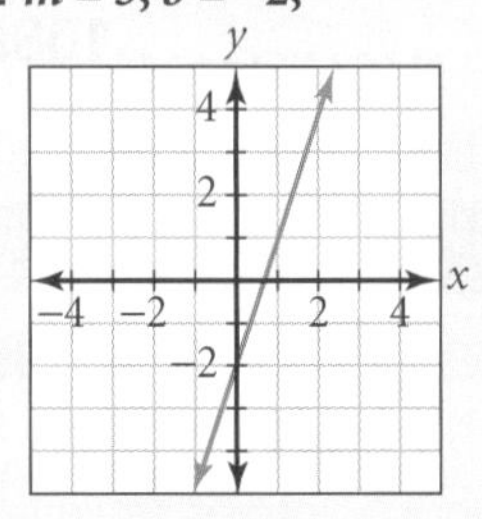

11. $m = 0, b = -3$;

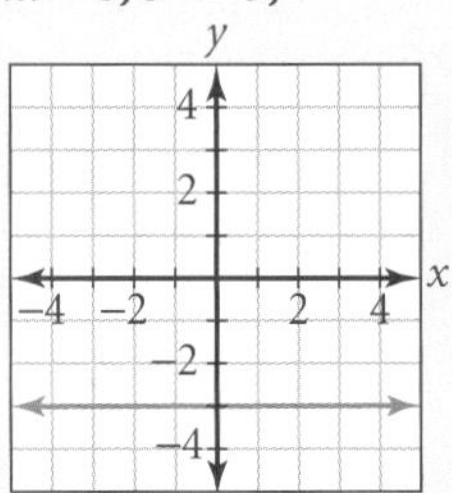

Extra Practice

Chapter 1

LESSON 1.1

State whether each equation is a linear equation.

1. $y = 2x$ linear **2.** $y = -3x - 1$ linear **3.** $y = x + 3$ linear **4.** $y = \frac{3}{4}x + 1$ linear

5. $y = 10 - x$ linear **6.** $y = \frac{4x}{5}$ linear **7.** $y = 7 + x^2$ not linear **8.** $y = 2 - x^3$ not linear

Determine whether each table represents a linear relationship between *x* and *y*. If the relationship is linear, write the next ordered pair that would appear in the table.

9.

x	0	1	2	3
y	1	6	11	16

linear; (4, 21)

10.

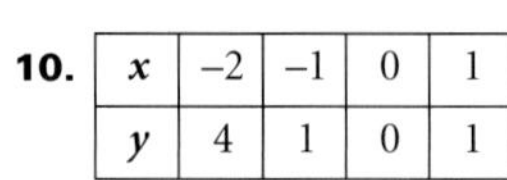

x	–2	–1	0	1
y	4	1	0	1

not linear

11.

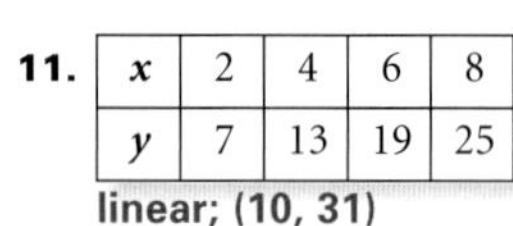

x	2	4	6	8
y	7	13	19	25

linear; (10, 31)

For each graph, make a table of values to represent the points. Does the table represent a linear relationship? Explain.

12. yes

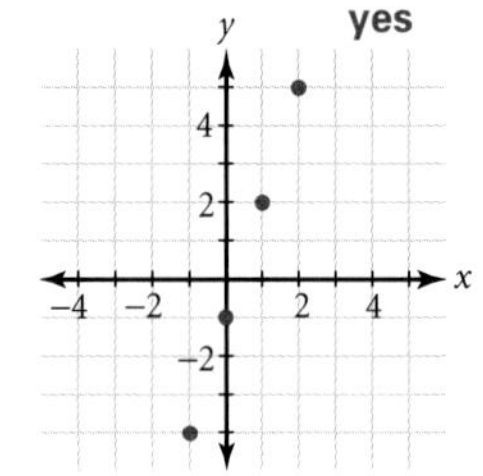

13. no

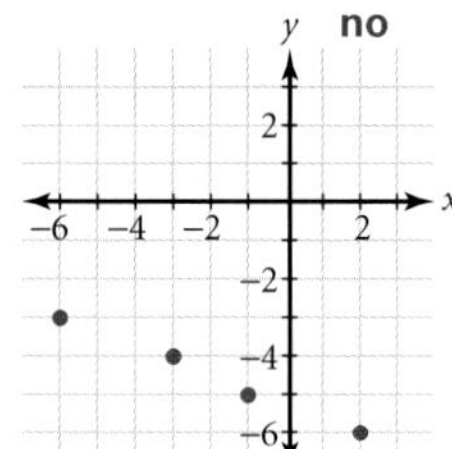

14. yes

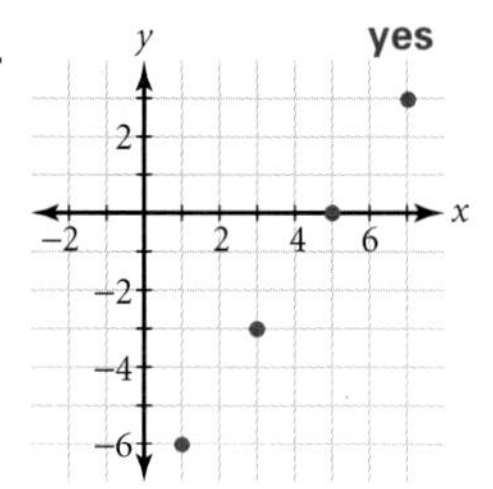

LESSON 1.2

Write an equation in slope-intercept form for the line that has the indicated slope, *m*, and *y*-intercept, *b*.

1. $m = -3, b = 6$ $y = -3x + 6$ **2.** $m = -2, b = 0$ $y = -2x$ **3.** $m = 0, b = \frac{1}{2}$ $y = \frac{1}{2}$ **4.** $m = \frac{4}{5}, b = 1$ $y = \frac{4}{5}x + 1$

Find the slope of the line containing the indicated points.

5. (0, 0) and (–2, –8) 4 **6.** (1, 2) and (2, 6) 4 **7.** (–4, 5) and (1, 5) 0

Identify the slope, *m*, and the *y*-intercept, *b*, for each line. Then graph.

8. $x = 1.5$ no slope **9.** $-2x + y = 1$ $m = 2$; $b = 1$ **10.** $3x - y = 2$ $m = 3$; $b = -2$ **11.** $y = -3$ $m = 0$; $b = -3$

Write an equation in slope-intercept form for each line.

12.

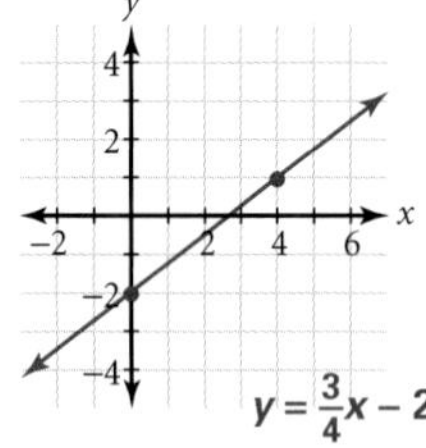

$y = \frac{3}{4}x - 2$

13.

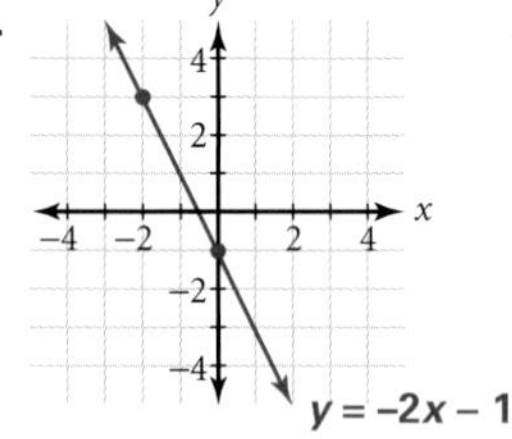

$y = -2x - 1$

14.

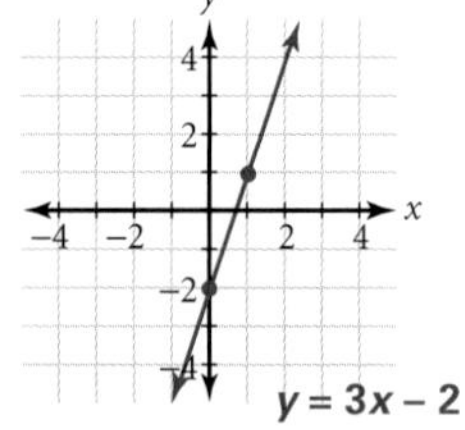

$y = 3x - 2$

LESSON 1.3

Write an equation in slope-intercept form for the line containing the indicated points.

1. $(4, -7)$ and $(9, 3)$ $y = 2x - 15$
2. $(3, 10)$ and $(-8, -12)$ $y = 2x + 4$
3. $(8, -11)$ and $(13, -11)$ $y = -11$
4. $(3, -17)$ and $(-4, 18)$ $y = -5x - 2$
5. $(-7, -1)$ and $(5, 7)$ $y = \frac{2}{3}x + \frac{11}{3}$
6. $(0, 13)$ and $(4, -11)$ $y = -6x + 13$

Write an equation in slope-intercept form for the line that has the indicated slope, *m*, and that contains the given point.

7. $m = -8$; $(-1, -2)$ $y = -8x - 10$
8. $m = -\frac{1}{2}$; $(4, 6)$ $y = -\frac{1}{2}x + 8$
9. $m = -13$; $(5, 1)$ $y = -13x + 66$
10. $m = \frac{5}{6}$; $(-12, 8)$ $y = \frac{5}{6}x + 18$

Write an equation in slope-intercept form for the line that contains the given point and is parallel to the given line.

11. $(0, 7)$; $y = 5x - 3$ $y = 5x + 7$
12. $(-11, 8)$; $y = -x + 5$ $y = -x - 3$
13. $\left(4\frac{1}{2}, 9\right)$; $y = -2x + 15$ $y = -2x + 18$
14. $(9, 7)$; $y = \frac{2}{3}x - 1$ $y = \frac{2}{3}x + 1$
15. $(-2, -3)$; $3x + 4y = 9$ $y = -\frac{3}{4}x - \frac{9}{2}$
16. $(-7, 2)$; $-x + 3y = 1$ $y = \frac{1}{3}x + \frac{13}{3}$

Write an equation in slope-intercept form for the line that contains the given point and is perpendicular to the given line.

17. $(3, -6)$; $y = \frac{1}{2}x + 7$ $y = -2x$
18. $(4, 5)$; $y = -\frac{1}{8}x + 17$ $y = 8x - 27$
19. $(-3, 9)$; $y = 5x + \frac{1}{4}$ $y = -\frac{1}{5}x + \frac{42}{5}$
20. $(0, 7)$; $x - 4y = 3$ $y = -4x + 7$
21. $(3, -3)$; $5x + 2y = 1$ $y = \frac{2}{5}x - \frac{21}{5}$
22. $(6, -4)$; $3x - y = 0.5$ $y = -\frac{1}{3}x - 2$

LESSON 1.4

In Exercises 1–6, *y* varies directly as *x*. Find the constant of variation, and write an equation of direct variation that relates the two variables.

1. $y = 32$ when $x = 8$ 4; $y = 4x$
2. $y = -12$ when $x = -4$ 3; $y = 3x$
3. $y = 4.2$ when $x = 0.7$ 6; $y = 6x$
4. $y = 65$ when $x = 13$ 5; $y = 5x$
5. $y = -3$ when $x = 11$ $-\frac{3}{11}$; $y = -\frac{3}{11}x$
6. $y = \frac{1}{4}$ when $x = \frac{1}{2}$ $\frac{1}{2}$; $y = \frac{1}{2}x$

Solve each proportion for the indicated variable. Check your answers.

7. $\frac{5}{32} = \frac{10}{x}$ $x = 64$
8. $\frac{w}{14} = \frac{14}{49}$ $w = 4$
9. $\frac{0.7}{10} = \frac{21}{k}$ $k = 300$
10. $\frac{4}{5} = \frac{t}{8}$ $t = 6.4$
11. $\frac{n+1}{7} = \frac{3}{5}$ $n = \frac{16}{5}$, or 3.2
12. $\frac{3y+4}{10} = \frac{y}{5}$ $y = -4$
13. $\frac{z+2}{7} = \frac{z}{6}$ $z = 12$
14. $\frac{m-2}{15} = \frac{m+2}{6}$ $m = -\frac{14}{3}$, or $-4\frac{2}{3}$
15. $\frac{5x-3}{7} = \frac{2x}{3}$ $x = 9$

Determine whether the values in each table are related by a direct variation. If so, write an equation for the variation. If not, explain.

16.

x	1	2	3	4	5
y	1.5	3	4.5	6	7.5

yes; $y = 1.5x$

17.

x	1	2	3	4	6
y	24	12	8	6	4

no; $\frac{24}{1} \neq \frac{12}{2}$

18.

x	5	6	7	8	9
y	6	7.2	8.4	9.6	10.8

yes; $y = 1.2x$

19.

x	−2	−1	1	2
y	14	7	−7	−14

yes; $y = -7x$

Extra Practice

Lesson 1.5

1.

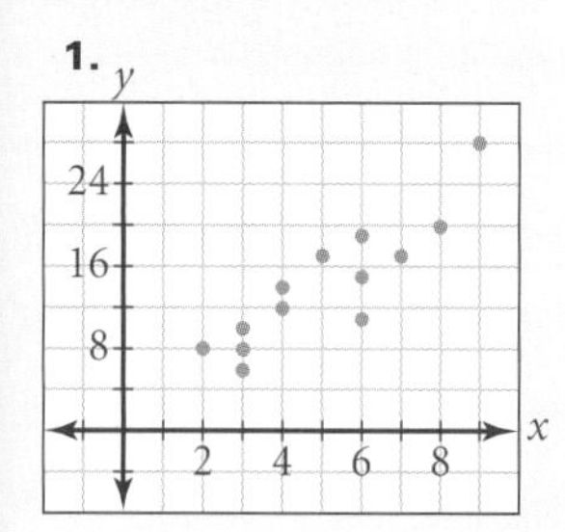

The correlation is positive; $y \approx 2.51x + 1.41$.

2.

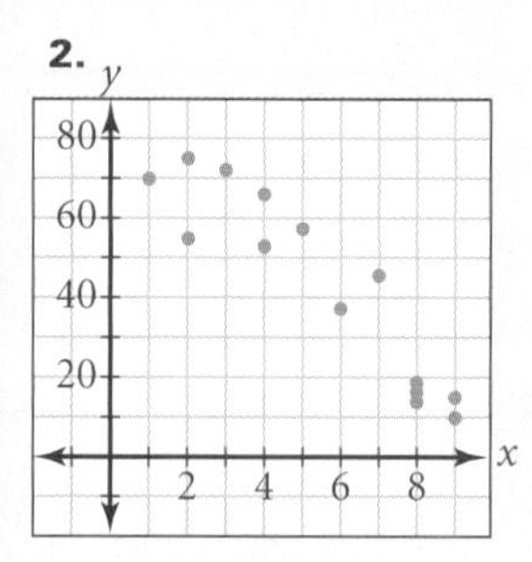

The correlation is negative; $y \approx -8.00x + 86.77$.

LESSON 1.5

Create a scatter plot of the data in each table. Describe the correlation. Then find an equation for the least-squares line.

1.

x	5	8	4	9	7	6	3	2	6	6	4	3	3	9
y	17	20	12	28	17	15	6	8	11	18	14	10	8	24

positive; $y \approx 2.51x + 1.41$

2.

x	1	2	2	3	4	5	6	7	8	8	4	8	9	9
y	70	55	75	72	66	58	38	45	17	19	52	15	15	10

negative; $y \approx -8.00x + 86.77$

The table lists the total weight lifted by the winners in eight weight classes of the 1996 Women's National Weightlifting Championship.

Weight class (kg)	46	50	54	59	64	70	76	83
Total lifted (kg)	140.0	127.5	167.5	167.5	192.5	185.0	197.5	200.0

3 Let x represent weight class, and let y represent the total weight lifted. Enter the data into a graphics calculator, and find an equation of the least-squares line. **$y \approx 1.86x + 55.38$**

4 Find the correlation coefficient, r, to the nearest tenth. **$r \approx 0.90$**

5 Suppose that there were a 68-kilogram weight class. Predict the total weight lifted by the winner. **≈182 kg**

LESSON 1.6

Solve each equation.

1. $x - 9 = -23$ **−14**

2. $\frac{1}{2}x - 3 = 11$ **28**

3. $3x - 2 = 13$ **5**

4. $-3x + 5 = 19$ **$-\frac{14}{3}$, or $-4\frac{2}{3}$**

5. $12 - 5x = -8$ **4**

6. $\frac{3}{4}x + \frac{1}{4} = 9$ **$\frac{35}{3}$, or $11\frac{2}{3}$**

7. $\frac{9}{10}x - 17 = 19$ **40**

8. $\frac{16}{15}x + 78 = 14$ **−60**

9. $\frac{5}{12}x - 12 = 48$ **144**

10. $7x + 8 = 11x$ **2**

11. $4x + 12 = 7x$ **4**

12. $9x - 42 = 3x$ **7**

13. $4x + 5 = x - 3$ **$-\frac{8}{3}$, or $-2\frac{2}{3}$**

14. $7x - 22 = 3x + 18$ **10**

15. $\frac{x+3}{2} = x - 4$ **11**

16. $8\left(\frac{3}{4}x + \frac{1}{2}\right) = 5x$ **−4**

17. $8x = 6x - 11$ **$-\frac{11}{2}$, or $-5\frac{1}{2}$**

18. $\frac{2}{3}x - 7 = 3$ **15**

19. $-3x + 5 = 5x - 3$ **1**

20. $18 = 6x + 8$ **$\frac{5}{3}$, or $1\frac{2}{3}$**

21. $x + \frac{15}{8} = \frac{3x}{2}$ **$\frac{15}{4}$, or $3\frac{3}{4}$**

22. $0.7x + 0.3x = 2x - 4$ **4**

23. $\frac{1}{2}x + 6 = x - 4$ **20**

24. $x - 7 = 3\frac{1}{2} + 2x$ **$-\frac{21}{2}$, or $-10\frac{1}{2}$**

Solve each literal equation for the indicated variable.

25. $S = 180(n - 2)$ for n **$n = \frac{S}{180} + 2$**

26. $A = \frac{1}{2}pr$ for p **$p = \frac{2A}{r}$**

27. $V = \frac{1}{3}Bh$ for h **$h = \frac{3V}{B}$**

28. $m = \frac{1}{2}(a - b)$ for a **$a = 2m + b$**

29. $m = \frac{1}{2}(a - b)$ for b **$b = a - 2m$**

30. $\frac{x}{a} + \frac{y}{b} = 1$ for a **$a = \frac{bx}{b - y}$**

31. $m = \frac{y_2 - y_1}{x_2 - x_1}$ for x_1 **$x_1 = \frac{y_1 - y_2}{m} + x_2$**

32. $F = \frac{W}{d}$ for d **$d = \frac{W}{F}$**

33. $E = IR$ for R **$R = \frac{E}{I}$**

34. $t = -0.55\left(\frac{a}{1000}\right)$ for a **$a = -\frac{1000t}{0.55}$**

35. $R = \frac{s^2}{A}$ for A **$A = \frac{s^2}{R}$**

36. $S = \frac{1}{2}(a + b + c)$ for b **$b = 2S - a - c$**

37. Given the equation $x = 4 + y$, use substitution to solve $3y - x = -14$ for y. **$y = -5$**

38. Given the equation $x - 2y = 8$, use substitution to solve $6y = -3x$ for x. **$x = 4$**

LESSON 1.7

Write an inequality that describes each graph.

1. −6 −4 −2 0 2 4 6 $x < -2$

2. −2 0 2 4 6 8 $x \geq 7$

3. −10 −8 −6 −4 −2 0 2 $x > -7$

4. 0 1 2 3 4 5 6 $x < 3.5$

5. −20 −18 −16 −14 −12 −10 −8 $x \geq -14$

6. 30 32 34 36 38 40 42 $x \geq 33$

Solve each inequality, and graph the solution on a number line.

7. $8x < 64$ **$x < 8$**
8. $14x \leq -42$ **$x \leq -3$**
9. $x + 15 > 7$ **$x > -8$**
10. $-3x > -21$ **$x < 7$**
11. $x - 9 \geq 1$ **$x \geq 10$**
12. $3x - 2 \leq 13$ **$x \leq 5$**
13. $11 - x < 7$ **$x > 4$**
14. $6 - 7x \geq -8$ **$x \leq 2$**
15. $-\frac{x}{4} \geq 3$ **$x \leq -12$**
16. $7x - 2 < 3x + 4$ **$x < 1\frac{1}{2}$**
17. $4 - 2(x + 1) \leq -3$ **$x \geq 2\frac{1}{2}$**
18. $3(x - 2) + 5 \geq 7 + x$ **$x \geq 4$**

Graph the solution of each compound inequality on a number line.

19. $x \geq -5$ *and* $x < 3$ **$-5 \leq x < 3$**
20. $2x + 1 \leq 7$ *and* $-3x + 1 < -5$ **$2 < x \leq 3$**
21. $x + 5 \geq 2$ *and* $x - 3 < 2$ **$-3 \leq x < 5$**
22. $7 - x < 4$ *or* $2x + 1 < -2$ **$x < -\frac{3}{2}$ or $x > 3$**
23. $4x < -20$ *or* $5x - 2 \geq 3$ **$x < -5$ or $x \geq 1$**
24. $3x - 6 < 12$ *and* $1 - 2x \leq 17$ **$-8 \leq x < 6$**
25. $\frac{1}{3}(x + 9) \geq 4$ *or* $3 < -2x - 5$ **$x < -4$ or $x \geq 3$**
26. $\frac{3}{4}(12 - 2x) \leq 0$ *or* $8 - x > 3$ **$x < 5$ or $x \geq 6$**

LESSON 1.8

Match each statement on the left with a statement on the right.

1. $|x - 3| = -4$ **d**
2. $|x - 3| \leq 4$ **b**
3. $|x - 3| \geq -4$ **e**
4. $|x - 3| = 4$ **a**
5. $|x - 3| \geq 4$ **c**
6. $|x - 3| \leq -4$ **d**

a. $x = -1$ *or* $x = 7$
b. $x \geq -1$ *and* $x \leq 7$
c. $x \leq -1$ *or* $x \geq 7$
d. There is no solution.
e. All real numbers are solutions.

Solve each absolute-value equation. If the equation has no solution, write *no solution*.

7. $|x - 1| = 3$ **−2, 4**
8. $|x + 9| = 13$ **−22, 4**
9. $|x - 6| = 9$ **−3, 15**
10. $|7 + x| = 13$ **−20, 6**
11. $|2x - 2| = 5$ **$-1\frac{1}{2}$, $3\frac{1}{2}$**
12. $|4x + 1| = 10$ **$-2\frac{3}{4}$, $2\frac{1}{4}$**
13. $|3x| = -14$ **no solution**
14. $|3x - 5| + 7 = 4$ **no solution**
15. $|10x + 5| - 7 = 8$ **−2, 1**
16. $3 = \left|\frac{1}{5}(2 - x)\right|$ **−13, 17**
17. $\left|\frac{2}{3}x\right| = \frac{1}{12}$ **$\frac{1}{8}$, $-\frac{1}{8}$**
18. $|1 - x| = 8$ **−7, 9**

Solve each absolute-value inequality. Graph the solution on a number line.

19. $|2x| < 8$ **$-4 < x < 4$**
20. $|x - 1| \geq 4$ **$x \leq -3$ or $x \geq 5$**
21. $|2 - x| \leq 6$ **$-4 \leq x \leq 8$**
22. $|3x + 5| > 2$ **$x < -2\frac{1}{3}$ or $x > -1$**
23. $|5x - 10| > 0$ **$x < 2$ or $x > 2$**
24. $|6x + 3| \leq -7$ **no solution**
25. $|6 - 12x| \leq -7$ **no solution**
26. $|4x + 3| > 5$ **$x < -2$ or $x > \frac{1}{2}$**
27. $|6 - 3x| \leq 2$ **$\frac{4}{3} \leq x \leq \frac{8}{3}$**
28. $|2x - 6| < 4$ **$1 < x < 5$**
29. $|3x - 7| \geq 2$ **$x \leq \frac{5}{3}$ or $x \geq 3$**
30. $|4x + 3| - 2 > 4$ **$x < -\frac{9}{4}$ or $x > \frac{3}{4}$**

Lesson 1.7

7. $x < 8$
0 2 4 6 8 10 12

8. $x \leq -3$
−6 −4 −2 0 2 4 6

9. $x > -8$
−10 −8 −6 −4 −2 0 2

10. $x < 7$
0 2 4 6 8 10 12

11. $x \geq 10$
0 2 4 6 8 10 12

12. $x \leq 5$
0 2 4 6 8 10 12

13. $x > 4$
0 2 4 6 8 10 12

14. $x \leq 2$
−6 −4 −2 0 2 4 6

15. $x \leq -12$
−14 −12 −10 −8 −6 −4 −2

16. $x < 1\frac{1}{2}$
−6 −4 −2 0 2 4 6

17. $x \geq 2\frac{1}{2}$
−6 −4 −2 0 2 4 6

18. $x \geq 4$
−6 −4 −2 0 2 4 6

19. $x \geq -5$ *and* $x < 3$
−6 −4 −2 0 2 4 6

20. $x \leq 3$ *and* $x > 2$
−6 −4 −2 0 2 4 6

21. $x \geq -3$ *and* $x < 5$
−6 −4 −2 0 2 4 6

22. $x > 3$ *or* $x < -\frac{3}{2}$
−6 −4 −2 0 2 4 6

23. $x < -5$ *or* $x \geq 1$
−6 −4 −2 0 2 4 6

24. $x < 6$ *and* $x \geq -8$
−12 −8 −4 0 4 8 12

25. $x \geq 3$ *or* $x < -4$
−6 −4 −2 0 2 4 6

26. $x \geq 6$ *or* $x < 5$
−6 −4 −2 0 2 4 6

Lesson 1.8

19. $x > -4$ *and* $x < 4$
−6 −4 −2 0 2 4 6

20. $x \leq -3$ *or* $x \geq 5$
−6 −4 −2 0 2 4 6

21. $x \geq -4$ *and* $x \leq 8$
−4 −2 0 2 4 6 8

22. $x < -2\frac{1}{3}$ *or* $x > -1$
−4 −3 −2 −1 0

23. $x < 2$ *or* $x > 2$
−6 −4 −2 0 2 4 6

26. $x < -2$ *or* $x > \frac{1}{2}$
−3 −2 −1 0 1 2 3

27. $\frac{4}{3} \leq x \leq \frac{8}{3}$
0 1 2 3 4

28. $1 < x < 5$
0 1 2 3 4 5 6

29. $x \leq \frac{5}{3}$ *or* $x \geq 3$
0 1 2 3 4 5 6

30. $x < -\frac{9}{4}$ *or* $x > \frac{3}{4}$
−3 −2 −1 0 1 2 3

Lesson 2.1

1. natural, whole, integer, rational, real
2. rational, real
3. irrational, real
4. integer, rational, real
5. rational, real
6. rational, real
7. irrational, real
8. rational, real

Chapter 2

LESSON 2.1

Classify each number in as many ways as possible.

1. 47 **2.** 12.86 **3.** $\sqrt{7}$ **4.** $-\sqrt{100}$

5. $\frac{7}{9}$ **6.** $0.\overline{456}$ **7.** $123.45678\ldots$ **8.** $12.888888\ldots$

State the property that is illustrated in each statement. All variables represent real numbers.

9. $z + 1.09 = 1.09 + z$ **Comm. (+)**

10. $152 + 0 = 152$ **Identity (+)**

11. $23(x + 34) = 23x + 23(34)$ **Distrib.**

12. $-7 + (19 + 2) = (-7 + 19) + 2$ **Assoc. (+)**

13. $-42y = y(-42)$ **Commutative (×)**

14. $\frac{12}{y} \cdot \frac{y}{12} = 1$, where $y \neq 0$ **Inverse (×)**

15. $1 \cdot 77 = 77$ **Identity (×)**

16. $422 + (-422) = 0$ **Inverse (+)**

Evaluate each expression by using the order of operations.

17. $24 \div 8 + 4$ **7**

18. $2 + 7^2$ **51**

19. $3(11 + 2^2) + 1$ **46**

20. $(29 + 7) + 12 \div 6$ **38**

21. $8 + 2 \times 5 + 7$ **25**

22. $1 + 2 \times 7 - 5$ **10**

23. $\frac{15 + 10}{5} - 16 \div 4$ **1**

24. $\frac{7^2 - 1}{3 + 5}$ **6**

25. $16 \times 2 \div (1 - 5)$ **−8**

26. $1 - \frac{4}{2 \cdot 7 + 4}$ $\mathbf{\frac{7}{9}}$

27. $\frac{13 - 2 \cdot 6}{5 + 2 \cdot 3}$ $\mathbf{\frac{1}{11}}$

28. $3 + \frac{4(3 - 1)}{8}$ **4**

29. $\frac{5}{8} + \frac{6}{2(7 + 1)}$ **1**

30. $\frac{2(7 - 3)}{3(4 + 5)}$ $\mathbf{\frac{8}{27}}$

31. $20 \cdot 9 + \frac{8 - 3}{5}$ **181**

LESSON 2.2

Evaluate each expression.

1. 12^1 **12**

2. $(-27)^0$ **1**

3. 5^{-1} $\mathbf{\frac{1}{5}}$

4. $\left(\frac{4}{5}\right)^{-2}$ $\mathbf{\frac{25}{16}}$**, or** $\mathbf{1\frac{9}{16}}$

5. $\left(\frac{1}{3}\right)^{-2}$ **9**

6. $\left(\frac{2}{3}\right)^{-4}$ $\mathbf{\frac{81}{16}}$**, or** $\mathbf{5\frac{1}{16}}$

7. $8^{\frac{1}{3}}$ **2**

8. $\left(\frac{1}{8}\right)^{-3}$ **512**

9. $81^{\frac{3}{4}}$ **27**

10. $27^{\frac{2}{3}}$ **9**

11. $\left(100^{\frac{2}{3}}\right)^{\frac{3}{4}}$ **10**

12. $125^{\frac{1}{3}}$ **5**

Simplify each expression, assuming that no variable equals zero. Write your answer with positive exponents only.

13. z^5z^{-3} $\mathbf{z^2}$

14. $\left(\frac{a^{\frac{-2}{3}}}{b^8}\right)^{-2}$ $\mathbf{a^{\frac{4}{3}}b^{16}}$

15. $3b^3 \cdot b^4 \cdot b^{-2}$ $\mathbf{3b^5}$

16. $\left(\frac{9z^4}{16w^8}\right)^{-2}$ $\mathbf{\frac{256w^{16}}{81z^8}}$

17. $2xy(-3x^2y^3)$ $\mathbf{-6x^3y^4}$

18. $(-6ab^2c^5)^2$ $\mathbf{36a^2b^4c^{10}}$

19. $\frac{k^7}{k^5}$ $\mathbf{k^2}$

20. $\frac{n^{-3}}{n^4}$ $\mathbf{\frac{1}{n^7}}$

21. $\left(\frac{3x^{-4}}{y^3}\right)^2$ $\mathbf{\frac{9}{x^8y^6}}$

22. $(x^{-1}y^{-2}z^3)^{-2}(x^2y^{-4}z^6)$ $\mathbf{x^4}$

23. $(a^{-3}b^{-4})^{-2}(a^9b^{-7})^0$ $\mathbf{a^6b^8}$

24. $\left(\frac{24x^4y^{-5}}{-4x^{-2}y^{-1}}\right)^{-3}$ $\mathbf{-\frac{y^{12}}{216x^{18}}}$

25. $\frac{(x^3y^2)^{-2}}{x^2y^4}$ $\mathbf{\frac{1}{x^8y^8}}$

26. $\left(\frac{a^{-3}}{b^{-5}}\right)^{-2}\left(\frac{b^3}{a^2}\right)^{-1}$ $\mathbf{\frac{a^8}{b^{13}}}$

27. $\left(\frac{5c^{-2}}{z^3}\right)^2\left(\frac{c^3z^3}{x}\right)^{-2}$ $\mathbf{\frac{25x^2}{c^{10}z^{12}}}$

LESSON 2.3

State whether each relation represents a function. Explain.

1. yes

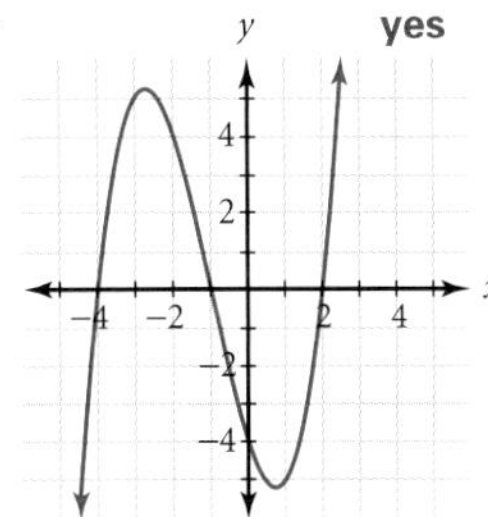

2. no

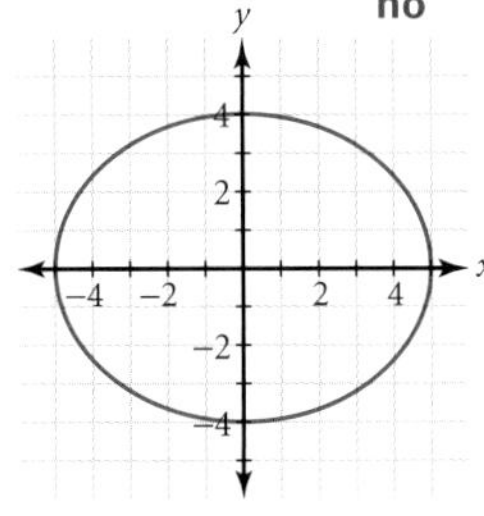

3. no

x	y
−2	8
−2	7
−1	6
0	5

4. $\{(32, 1), (48, 15), (56, 19)\}$ yes

5. $\{(-8, 25), (-8, 24), (-8, 23)\}$ no

6. $\{(7, 8), (9, 10), (11, 12)\}$ yes

State the domain and range of each function.

7. $\{(-2, 8), (3, 13)\}$ D: {−2, 3}; R: {8, 13}

8. $\{(0, 1), (1, 4), (2, 7), (3, 10)\}$ D: {0, 1, 2, 3}; R: {1, 4, 7, 10}

9. $\{(-4, 16), (0, 0), (2, 4)(4, 16)\}$ D: {−4, 0, 2, 4}; R: {0, 4, 16}

10. $\{(1.1, 5), (2.2, 10), (3.3, 15), (4.4, 20)\}$ D: {1.1, 2.2, 3.3, 4.4}; R: {5, 10, 15, 20}

Evaluate each function for the given values of *x*.

11. $f(x) = 1 - 2x$ for $x = -3$ and $x = 1$ 7, −1

12. $f(x) = \frac{x-1}{4}$ for $x = 7$ and $x = -3$ $\frac{3}{2}$, −1

13. $f(x) = 10x - 7$ for $x = -1$ and $x = 0$ −17, −7

14. $f(x) = x^2 + 2x + 1$ for $x = 3$ and $x = 4$ 16, 25

15. $f(x) = -\frac{3x}{5}$ for $x = 10$ and $x = -2$ −6, $\frac{6}{5}$

16. $f(x) = -2x^2 - x$ for $x = -3$ and $x = \frac{1}{4}$ −15, $-\frac{3}{8}$

LESSON 2.4

Let $f(x) = x^2$ and $g(x) = 2x - 1$. Find each new function, and state any domain restrictions.

1. $f + g = x^2 + 2x - 1$

2. $f - g = x^2 - 2x + 1$

3. $g - f = -x^2 + 2x - 1$

4. $f \cdot g = 2x^3 - x^2$

5. $\frac{f}{g} = \frac{x^2}{2x-1}, x \neq \frac{1}{2}$

6. $\frac{g}{f} = \frac{2x-1}{x^2}, x \neq 0$

Let $f(x) = -2x^2$ and $g(x) = x + 1$. Find each new function, and state any domain restrictions.

7. $f + g = -2x^2 + x + 1$

8. $f - g = -2x^2 - x - 1$

9. $g - f = 2x^2 + x + 1$

10. $f \cdot g = -2x^3 - 2x^2$

11. $\frac{f}{g} = -\frac{2x^2}{x+1}, x \neq -1$

12. $\frac{g}{f} = -\frac{x+1}{2x^2}, x \neq 0$

Find $f \circ g$ and $g \circ f$.

13. $f(x) = 4x$ and $g(x) = x^2 - 2$ $(f \circ g)(x) = 4x^2 - 8$; $(g \circ f)(x) = 16x^2 - 2$

14. $f(x) = x^2 - 3x$ and $g(x) = 2x$ $(f \circ g)(x) = 4x^2 - 6x$; $(g \circ f)(x) = 2x^2 - 6x$

15. $f(x) = -3x$ and $g(x) = 2x^2 - 3x$ $(f \circ g)(x) = -6x^2 + 9x$; $(g \circ f)(x) = 18x^2 + 9x$

16. $f(x) = \frac{1}{3}x$ and $g(x) = -9x^2$ $(f \circ g)(x) = -3x^2$; $(g \circ f)(x) = -x^2$

Let $f(x) = 3x$, $g(x) = -2x^2$, and $h(x) = x^2 - 1$. Evaluate each composite function.

17. $(f \circ g)(3)$ −54

18. $(f \circ h)(2)$ 9

19. $(g \circ f)(1)$ −18

20. $(h \circ f)(0)$ −1

21. $(f \circ f)(1)$ 9

22. $(h \circ g)(-2)$ 63

Lesson 2.3

1. Yes; no vertical line intersects more than one point.

2. No; many vertical lines intersect more than one point.

3. No; two different values of y are paired with the x-value −2.

4. Yes; for each value of x, there is exactly one value of y.

5. No; three different y-values are paired with the x-value −8.

6. Yes; for each value of x, there is exactly one value of y.

LESSON 2.5

Find the inverse of each relation. State whether the relation is a function. State whether the inverse is a function.

1. $\{(1, 2), (2, 2), (3, 2), (4, 2)\}$
2. $\{(1, 0), (1, 2), (4, 3), (4, -2)\}$
3. $\{(1, 6), (2, 9), (3, 12), (4, 19)\}$
4. $\{(-2, 9), (-1, 8), (0, 7), (1, 6)\}$

For each function, find an equation for the inverse. Then use composition to verify that the equation you wrote is the inverse.

5. $f(x) = 10x - 6$ $y = \frac{x+6}{10}$
6. $g(x) = -6x + 5$ $y = \frac{5-x}{6}$
7. $h(x) = 0.5x + 2.5$ $y = 2x - 5$
8. $g(x) = 9.5 - x$ $y = 9.5 - x$
9. $h(x) = \frac{x-2}{6}$ $y = 6x + 2$
10. $f(x) = 14.4x$ $y = \frac{x}{14.4}$

Graph each function, and use the horizontal-line test to determine whether the inverse is a function.

11. $f(x) = 10x - 6$ yes
12. $h(x) = 1 - 2x^2$ no
13. $g(x) = \frac{1}{2}x + 4$ yes
14. $g(x) = x^3 + 1$ yes
15. $f(x) = \frac{1}{x}$ yes
16. $f(x) = \frac{1}{x^2}$ no

LESSON 2.6

Graph each function.

1. $g(x) = \begin{cases} x & \text{if } x \le 0 \\ -x & \text{if } x > 0 \end{cases}$
2. $f(x) = \begin{cases} -2x + 1 & \text{if } x \le 2 \\ 4x + 1 & \text{if } x > 2 \end{cases}$
3. $h(x) = \begin{cases} 9 & \text{if } x < -3 \\ x^2 & \text{if } -3 \le x \le 3 \\ 9 & \text{if } x > 3 \end{cases}$
4. $g(x) = |x| - 2$
5. $f(x) = \lceil x \rceil + 1$
6. $g(x) = -[x]$

Write the piecewise function represented by each graph.

7.
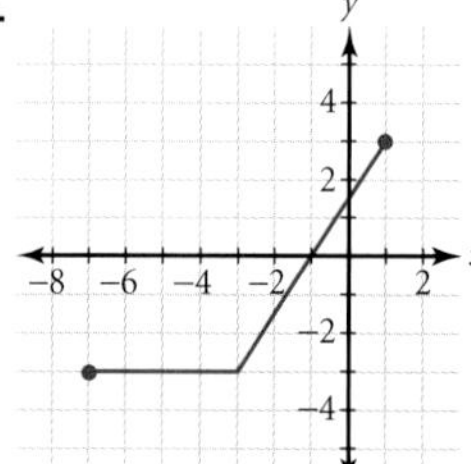

8.
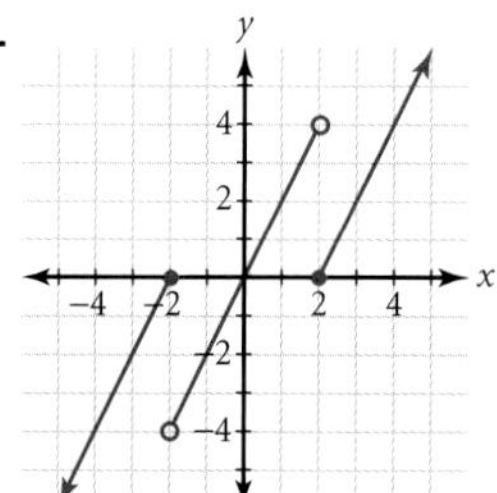

9.
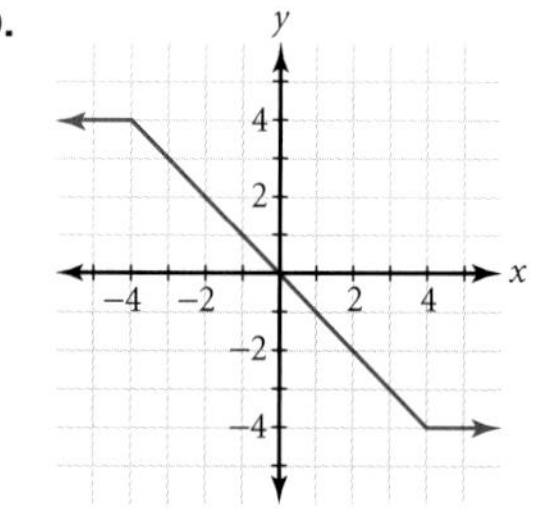

10.
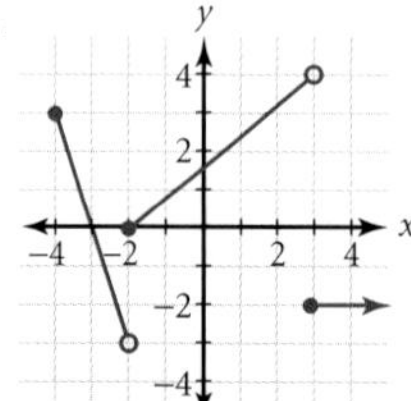

11.
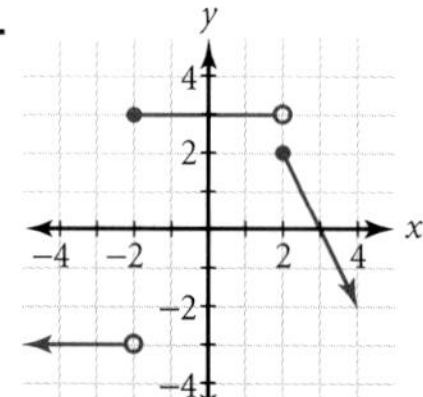

12.
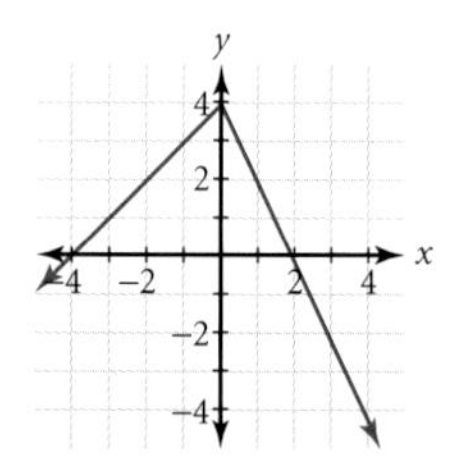

Evaluate.

13. $\lceil 14.07 \rceil$ 15
14. $[14.07]$ 14
15. $\lceil -8.65 \rceil$ −8
16. $[72.4] + |-6|$ 78
17. $\lceil 23.4 \rceil + [23.4]$ 47
18. $[12.6] - [9.2]$ 3
19. $|-12| + |12|$ 24
20. $\lceil -7 \rceil - |-7|$ −14

Extra Practice

Lesson 2.5

1. $\{(2, 1), (2, 2), (2, 3), (2, 4)\}$; yes; no
2. $\{(0, 1), (2, 1), (3, 4), (-2, 4)\}$; no; yes
3. $\{(6, 1), (9, 2), (12, 3), (19, 4)\}$; yes; yes
4. $\{(9, -2), (8, -1), (7, 0), (6, 1)\}$; yes; yes

Lesson 2.6

1.

2.
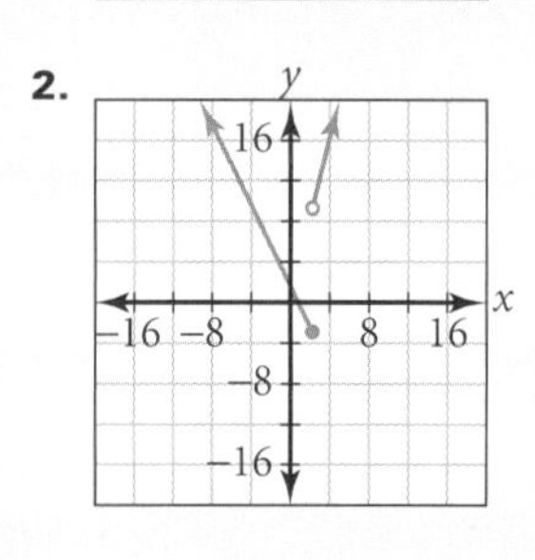

3.
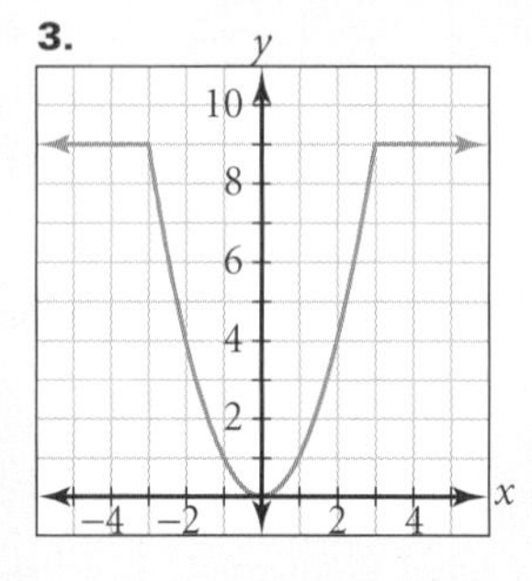

4.
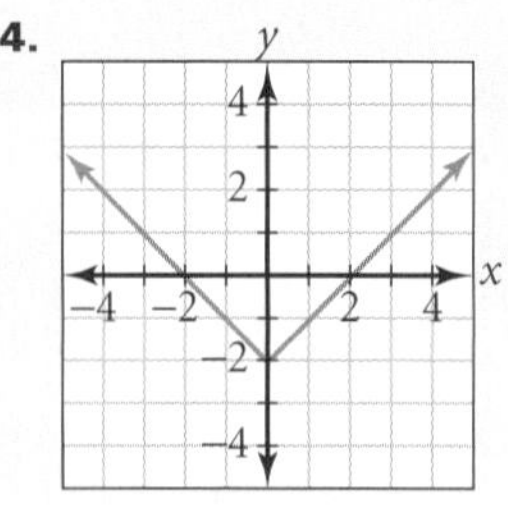

5.
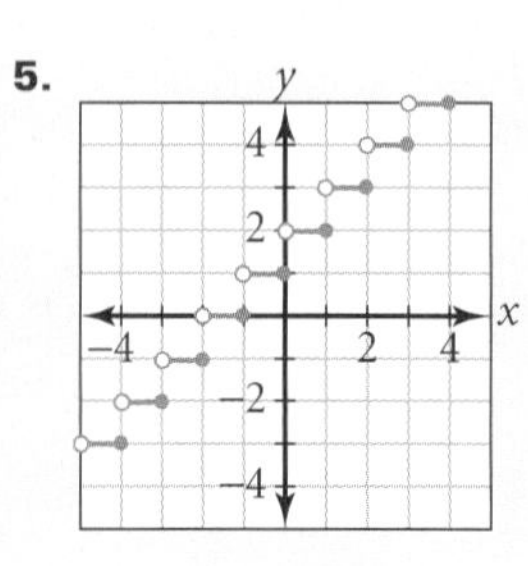

6.

7. $f(x) = \begin{cases} -3 & \text{if } -7 \le x \le -3 \\ \frac{3}{2}x + \frac{3}{2} & \text{if } -3 < x \le 1 \end{cases}$

8. $f(x) = \begin{cases} 2x + 4 & \text{if } x \le -2 \\ 2x & \text{if } -2 < x < 2 \\ 2x - 4 & \text{if } x \ge 2 \end{cases}$

9. $f(x) = \begin{cases} 4 & \text{if } x < -4 \\ -x & \text{if } -4 \le x \le 4 \\ -4 & \text{if } x > 4 \end{cases}$

10. $f(x) = \begin{cases} -3x - 9 & \text{if } -4 \le x < -2 \\ \frac{4}{5}x + \frac{8}{5} & \text{if } -2 \le x < 3 \\ -2 & \text{if } x \ge 3 \end{cases}$

11. $f(x) = \begin{cases} -3 & \text{if } x < -2 \\ 3 & \text{if } -2 \le x < 2 \\ -2x + 6 & \text{if } x \ge 2 \end{cases}$

12. $f(x) = \begin{cases} x + 4 & \text{if } x < 0 \\ -2x + 4 & \text{if } x \ge 0 \end{cases}$

LESSON 2.7

Identify each transformation from the parent function $f(x) = x^4$ to g.

1. $g(x) = 8x^4$ **2.** $g(x) = (x + 11)^4$ **3.** $g(x) = -x^4$

4. $g(x) = (-6x)^4$ **5.** $g(x) = x^4 - 16$ **6.** $g(x) = x^4 + 5$

7. $g(x) = (-x)^4 + 13$ **8.** $g(x) = 7(x - 4)^4$ **9.** $g(x) = 2(x + 3)^4$

Identify each transformation from the parent function $f(x) = |x|$ to g.

10. $g(x) = |34x|$ **11.** $g(x) = |x - 19|$ **12.** $g(x) = |x| - 19$

13. $g(x) = 2|3x|$ **14.** $g(x) = \left|\frac{1}{3}x\right|$ **15.** $g(x) = 24|x| + 9$

Write a function, g, for the graph described.

16. the graph of $f(x) = x^2$ translated 18 units down **$g(x) = x^2 - 18$**

17. the graph of $f(x) = |x|$ vertically stretched by a factor of 10 **$g(x) = 10|x|$**

18. the graph of $f(x) = \sqrt{x}$ reflected across the x-axis **$g(x) = -\sqrt{x}$**

19. the graph of $f(x) = 21x + 17$ reflected across the y-axis **$g(x) = 21(-x) + 17$**

20. the graph of $f(x) = x^2$ horizontally stretched by a factor of 6 **$g(x) = \left(\frac{1}{6}x\right)^2$**

21. the graph of $f(x) = x^2$ translated 3.5 units to the left **$g(x) = (x + 3.5)^2$**

22. the graph of $f(x) = x^3$ translated 11 units up **$g(x) = x^3 + 11$**

23. the graph of $f(x) = |x|$ horizontally compressed by a factor of $\frac{1}{3}$ **$g(x) = |3x|$**

Chapter 3

LESSON 3.1

Graph and classify each system. Then find the solution from the graph.

1 $\begin{cases} y = 7x + 8 \\ y = 3x \end{cases}$ **independent; (−2, −6)**

2 $\begin{cases} 3x - 2y = 9 \\ 2y - 3x = 6 \end{cases}$ **inconsistent; no solution**

3 $\begin{cases} 2x + y = -1 \\ 2y = -4x - 2 \end{cases}$ **dependent; infinitely many**

4 $\begin{cases} 4x + 5y = 2 \\ x + y = 1 \end{cases}$ **independent; (3, −2)**

5 $\begin{cases} x - 2y = 1 \\ 2x - 5y = -1 \end{cases}$ **independent; (7, 3)**

6 $\begin{cases} 4x - y = 5 \\ y = 4x + 3 \end{cases}$ **inconsistent; no solution**

7 $\begin{cases} 5x + 2y = 6 \\ x - y = -3 \end{cases}$ **independent; (0, 3)**

8 $\begin{cases} 3x - 2y = -2 \\ x + 2y = -6 \end{cases}$ **independent; (−2, −2)**

9 $\begin{cases} 2x - y = 5 \\ 3x + 2y = 4 \end{cases}$ **independent; (2, −1)**

10 $\begin{cases} y - x = 1 \\ 2x + y = -5 \end{cases}$ **independent; (−2, −1)**

11 $\begin{cases} y - 2x = -1 \\ x + 3y = 4 \end{cases}$ **independent; (1, 1)**

12 $\begin{cases} 2x + 3y = -4 \\ y = \frac{1}{2}x + 1 \end{cases}$ **independent; (−2, 0)**

Use substitution to solve each system.

13. $\begin{cases} y = x - 13 \\ 2x + 3y = 1 \end{cases}$ **(8, −5)**

14. $\begin{cases} x + y = 12 \\ x - y = 8 \end{cases}$ **(10, 2)**

15. $\begin{cases} y = 2x - 7 \\ x + y = 5 \end{cases}$ **(4, 1)**

16. $\begin{cases} y = \frac{1}{2}x - 3 \\ x = y + 1 \end{cases}$ **(−4, −5)**

17. $\begin{cases} 6x - 2y = 0 \\ x - y = 12 \end{cases}$ **(−6, −18)**

18. $\begin{cases} \frac{2}{3}x + \frac{1}{2}y = 2 \\ x - y = 10 \end{cases}$ **(6, −4)**

19. $\begin{cases} y = 4x - 9 \\ y = 3x + 5 \end{cases}$ **(14, 47)**

20. $\begin{cases} 2x - y = -1 \\ x + y = -17 \end{cases}$ **(−6, −11)**

21. $\begin{cases} x + 3y = 12 \\ x - 2y = -8 \end{cases}$ **(0, 4)**

22. $\begin{cases} 2x + 5y + z = -4 \\ 4y + z = 0 \\ z = 8 \end{cases}$ **(−1, −2, 8)**

23. $\begin{cases} x - 2y + 3z = 9 \\ x + y = -3 \\ x = -2 \end{cases}$ **(−2, −1, 3)**

24. $\begin{cases} x + y + z = 10 \\ x + z = 8 \\ z = 5 \end{cases}$ **(3, 2, 5)**

Lesson 2.7

1. a vertical stretch by a factor of 8

2. a translation 11 units to the left

3. a reflection across the x-axis

4. a horizontal compression by a factor of $\frac{1}{6}$

5. a translation 16 units down

6. a translation 5 units up

7. a translation 13 units up

8. a vertical stretch by a factor of 7 and a translation 4 units to the right

9. a vertical stretch by a factor of 2 and a translation 3 units to the left

10. a horizontal compression by a factor of $\frac{1}{34}$

11. a translation 19 units to the right

12. a translation 19 units down

13. a horizontal compression by a factor of $\frac{1}{3}$ and a vertical stretch by a factor of 2

14. a horizontal stretch by a factor of 3

15. a vertical stretch by a factor of 24 and a translation 9 units up

Extra Practice

Extra Practice

LESSON 3.2

Use elimination to solve each system. Check your solution.

1. $\begin{cases} 13x - 2y = 10 \\ 8x + y = 24 \end{cases}$ **(2, 8)**

2. $\begin{cases} 7x - y = 5 \\ 3x + y = 15 \end{cases}$ **(2, 9)**

3. $\begin{cases} -2x - y = 7 \\ x - 3y = 14 \end{cases}$ **(–1, –5)**

4. $\begin{cases} 3x - 4y = 12 \\ 8y - 6x = -24 \end{cases}$ **infinitely many**

5. $\begin{cases} \frac{1}{2}x - y = 5 \\ x + 2y = -2 \end{cases}$ **(4, –3)**

6. $\begin{cases} 7x - 4y = -9 \\ 3x + 2y = -15 \end{cases}$ **(–3, –3)**

7. $\begin{cases} 7x - 3y = 5 \\ 3y - 7x = 8 \end{cases}$ **no solution**

8. $\begin{cases} \frac{3}{4}x + \frac{1}{2}y = 11 \\ 3x - y = 14 \end{cases}$ **(8, 10)**

9. $\begin{cases} 2x + 11y = 18 \\ 5x + 3y = -4 \end{cases}$ **(–2, 2)**

Use any method to solve each system. Check your solution.

10. $\begin{cases} y = 12x - 3 \\ 4x - y = -1 \end{cases}$ $\left(\frac{1}{2}, 3\right)$

11. $\begin{cases} 2x + 5y = 43 \\ 7x - y = -16 \end{cases}$ **(–1, 9)**

12. $\begin{cases} 2x + y = -3 \\ 2x - y = -17 \end{cases}$ **(–5, 7)**

13. $\begin{cases} -3x + 2y = 1 \\ 4y - 6x = 2 \end{cases}$ **infinitely many**

14. $\begin{cases} y = 4x - 2 \\ y = 2x + 8 \end{cases}$ **(5, 18)**

15. $\begin{cases} \frac{3}{4}x - y = 2 \\ 3x - 4y = 3 \end{cases}$ **no solution**

16. $\begin{cases} 9x + 3y = -3 \\ y - x = 11 \end{cases}$ **(–3, 8)**

17. $\begin{cases} 5y - 7x = -13 \\ 2x + 3y = 17 \end{cases}$ **(4, 3)**

18. $\begin{cases} 17 - 5y = 3x \\ x + y = 5 \end{cases}$ **(4, 1)**

LESSON 3.3

Graph each linear inequality.

1. $y < 2x - 4$ **2.** $y \geq -x + 3$ **3.** $y \leq 4x - 1$

4. $y > 4x + 4$ **5.** $-y \geq -3x$ **6.** $-y < 3x$

7. $x + 2y \leq 5$ **8.** $2x + y > -1$ **9.** $x + 4y < 8$

10. $3x + 2y \leq 6$ **11.** $5x - y > 6$ **12.** $\frac{2}{3}x + y \geq -2$

13. $x \geq -3$ **14.** $y < 0$ **15.** $x > 2$

16. $x + 5 < 0$ **17.** $-\frac{1}{2}y \leq 3$ **18.** $-\frac{2}{3}y \leq -4$

Write an inequality for each graph.

19. $y \leq -3x + 5$

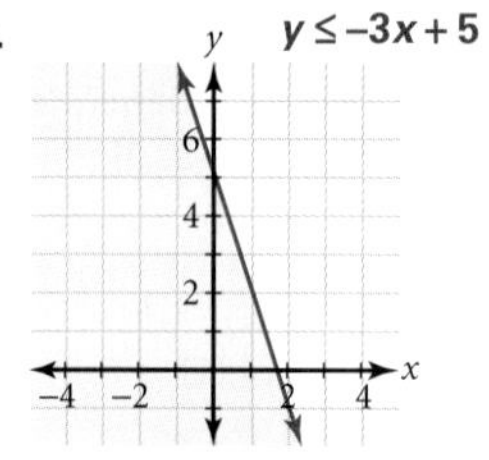

20. $y > \frac{1}{2}x - 3$

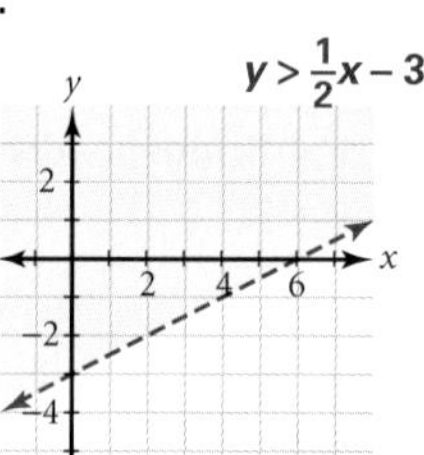

21. $x < 4$

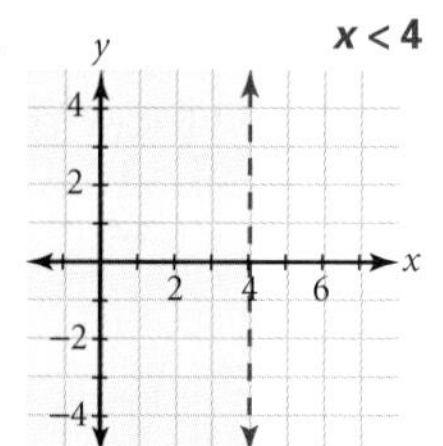

22. $y \geq -2x - 2$

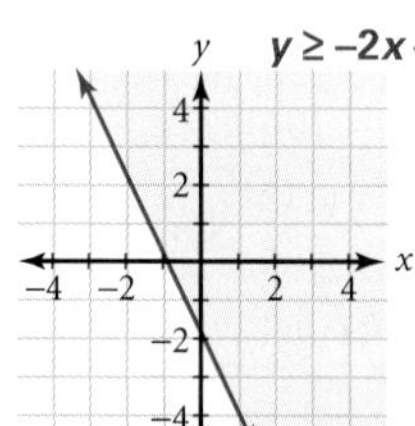

23. $y < \frac{1}{2}x + \frac{3}{2}$

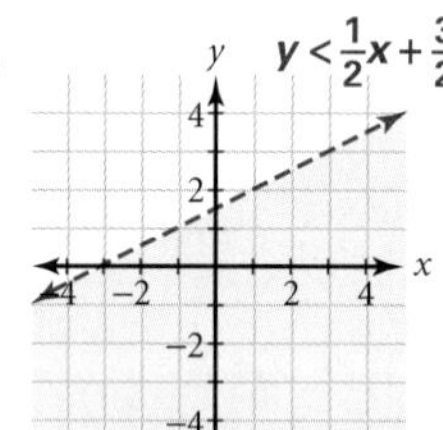

24. $y < -2$

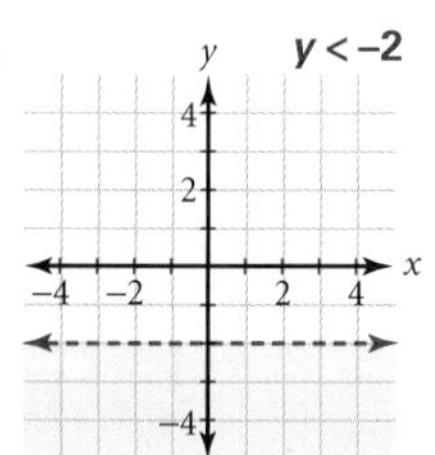

Lesson 3.3

1.

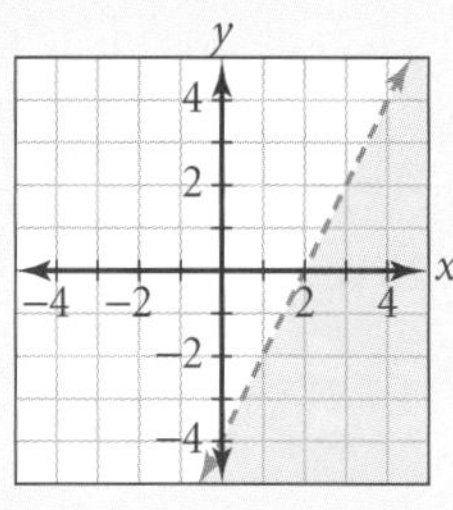

2.

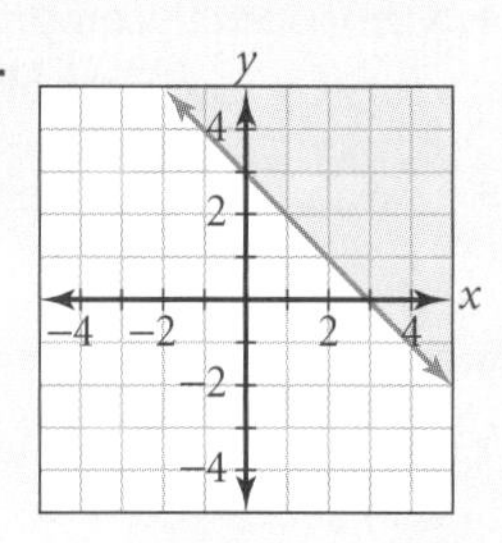

3.

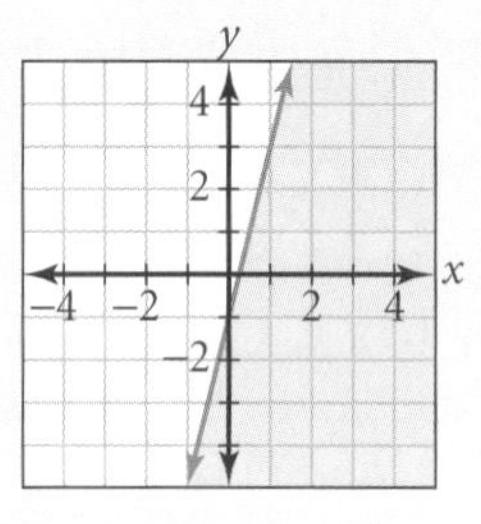

4.

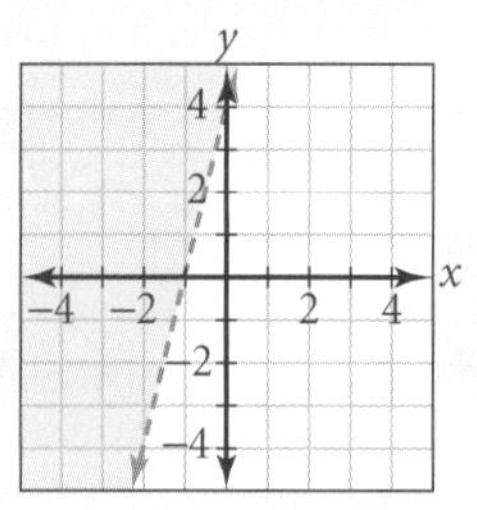

5.

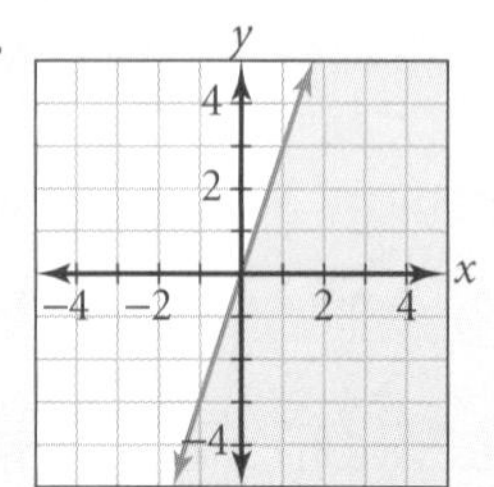

6.

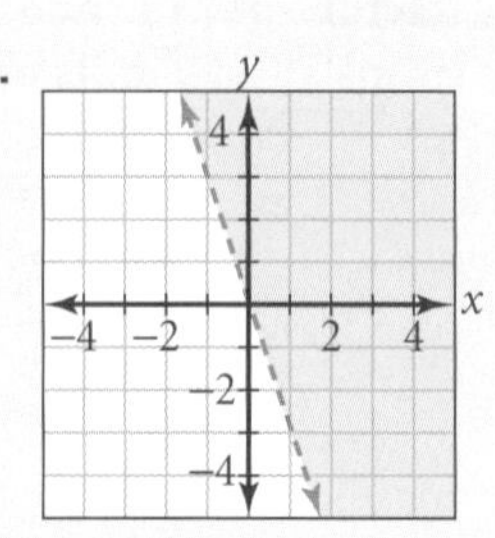

7.

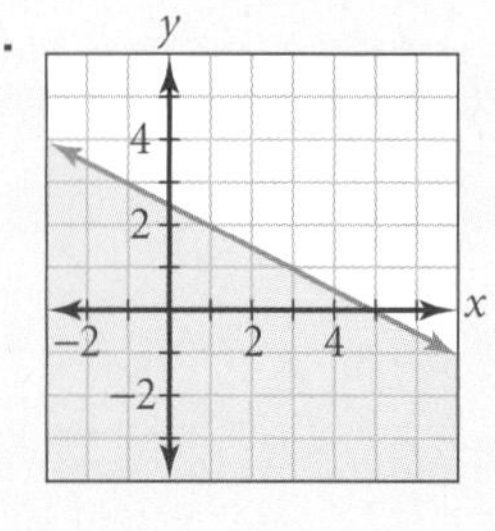

8.

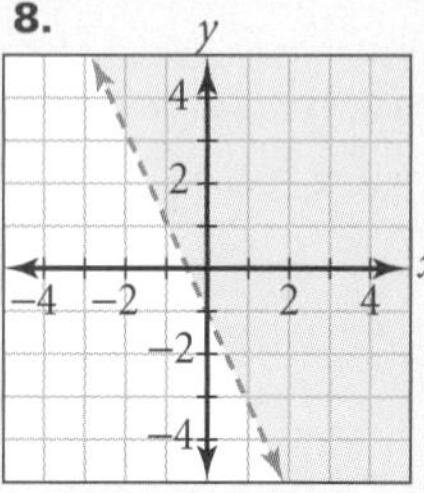

9.

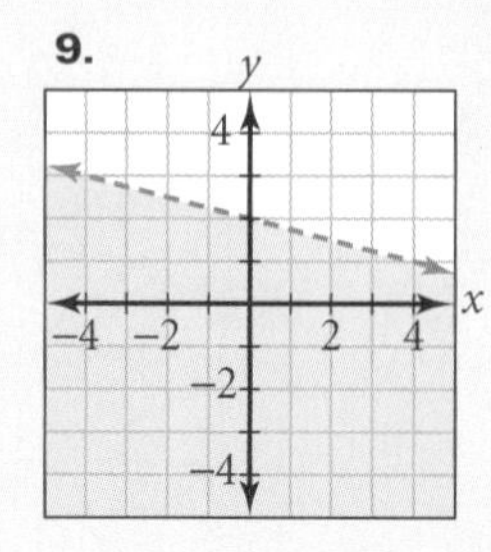

10.

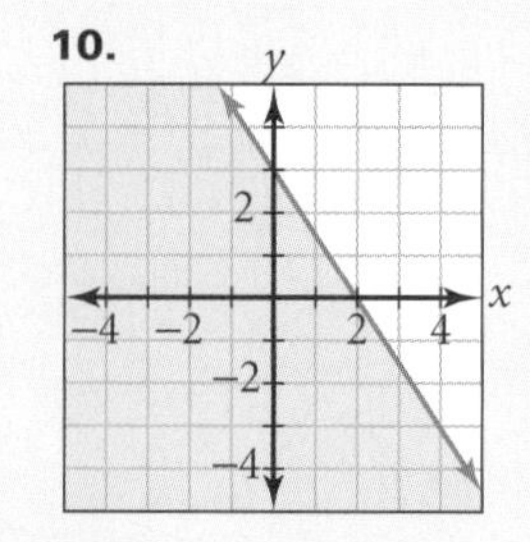

11.

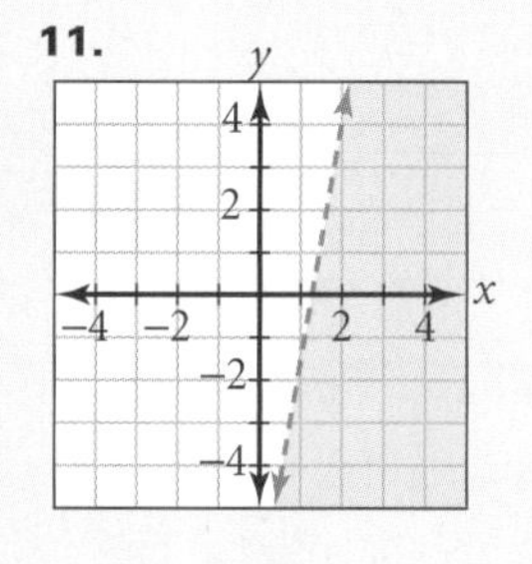

LESSON 3.4

Graph each compound inequality in a coordinate plane.

1. $-3 \le x \le 0$ **2.** $-4 < x < -1$ **3.** $-6 < y < -1$

4. $-2 \le y \le 4$ **5.** $2 \le x < 5$ **6.** $1 < y \le 4$

Graph each system of linear inequalities.

7. $\begin{cases} y < 3 \\ y \ge 2x - 1 \end{cases}$ **8.** $\begin{cases} x \le 2 \\ y > x \end{cases}$ **9.** $\begin{cases} y < 5 - x \\ y > 2x + 1 \end{cases}$

10. $\begin{cases} x > 1 \\ y \le 4 \\ y < 2x + 1 \end{cases}$ **11.** $\begin{cases} x > 0 \\ y < -x + 5 \\ y > x - 5 \end{cases}$ **12.** $\begin{cases} x \ge 2 \\ y \ge 0 \\ y \ge -x + 3 \\ y \le x + 1 \end{cases}$

Write the system of inequalities whose solution is graphed. Assume that each vertex has integer coordinates.

13.

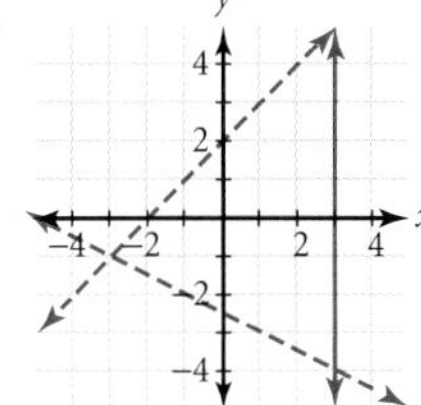

14.

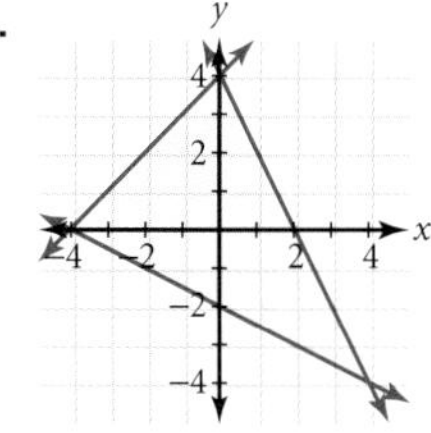

15.

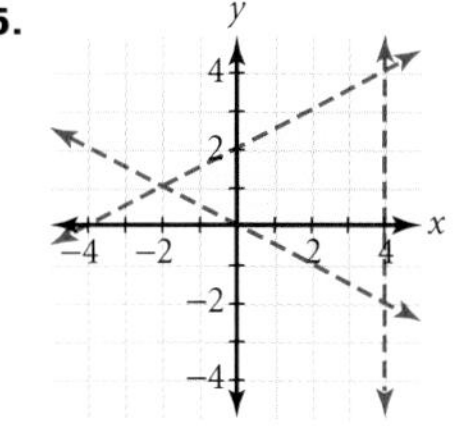

LESSON 3.5

Graph the feasible region for each set of constraints.

1. $\begin{cases} x + y \le 5 \\ 2x + y \le 9 \\ x \ge 0 \\ y \ge 0 \end{cases}$ **2.** $\begin{cases} 2x + 3y \le 15 \\ x - y \le 4 \\ x \ge 0 \\ y \ge 0 \end{cases}$ **3.** $\begin{cases} y \le \frac{1}{2}x + 4 \\ y \ge \frac{1}{2}x - 2 \\ 0 \le x < 6 \end{cases}$

The feasible region for a set of constraints has vertices at (0, 0), (60, 0), (60, 30), and (10, 50). Given this feasible region, find the maximum and minimum values of each objective function.

4. $C = 100x + 25y$ max = 6750 min = 0

5. $P = 40x + 65y$ max = 4350 min = 0

Find the maximum and minimum values, if they exist, of each objective function for the given constraints.

6. $P = 5x + 2y$
Constraints:
$\begin{cases} x + y \le 7 \\ x - y \le 5 \\ x \ge 0 \\ y \ge 0 \end{cases}$ max = 32 min = 0

7. $P = 10x + y$
Constraints:
$\begin{cases} 5x + 2y \le 20 \\ x - y \le 4 \\ x \ge 0 \\ y \ge 0 \end{cases}$ max = 40 min = 0

8. $E = 4x + 5y$
Constraints:
$\begin{cases} x + 2y \ge 6 \\ x + 2y \le 12 \\ x \ge 0 \\ y \ge 0 \end{cases}$ max = 48 min = 15

12.

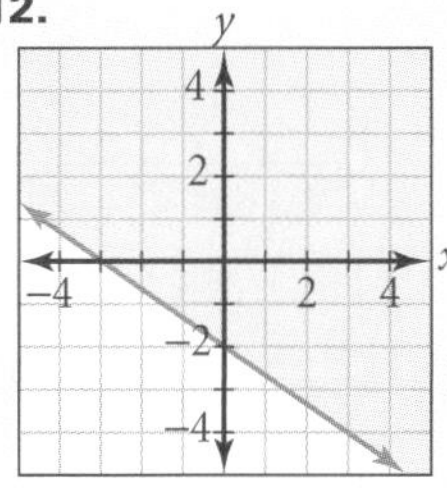

13.

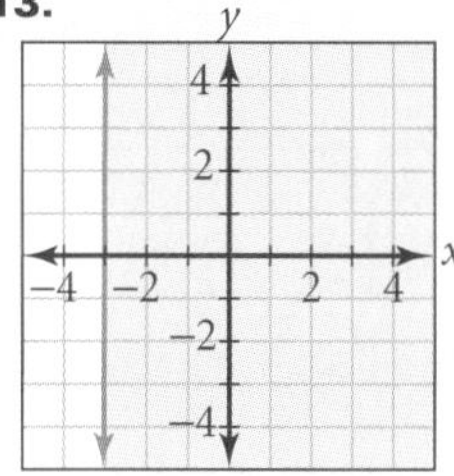

14.

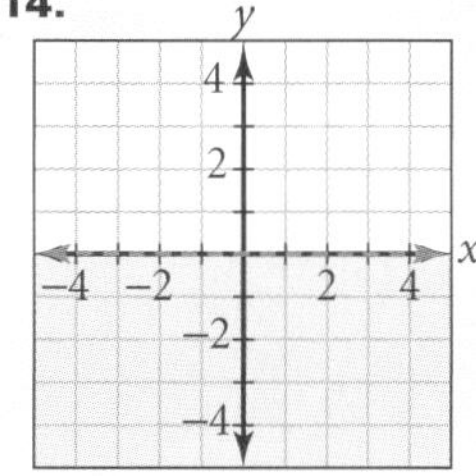

15.

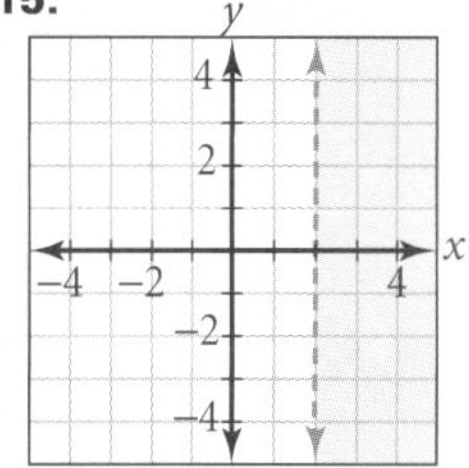

16.

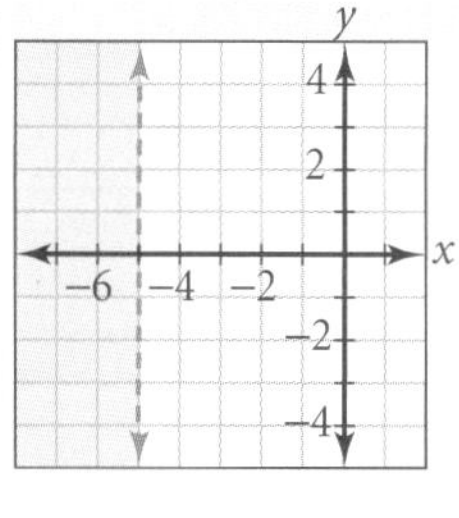

17.

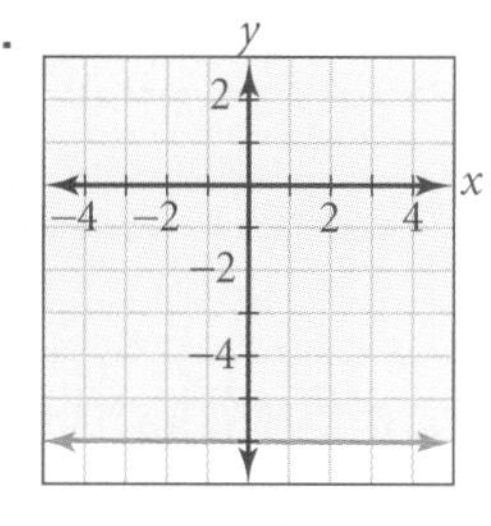

18.

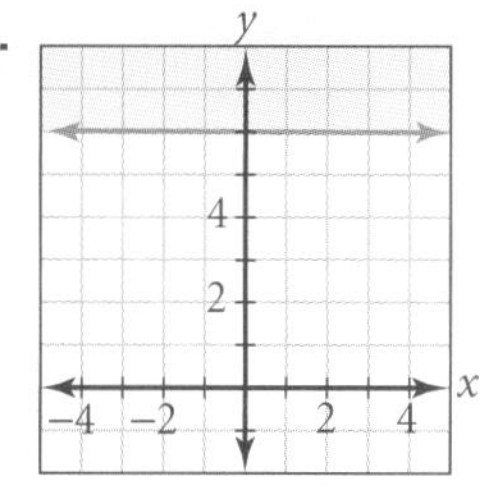

Lesson 3.4

1.

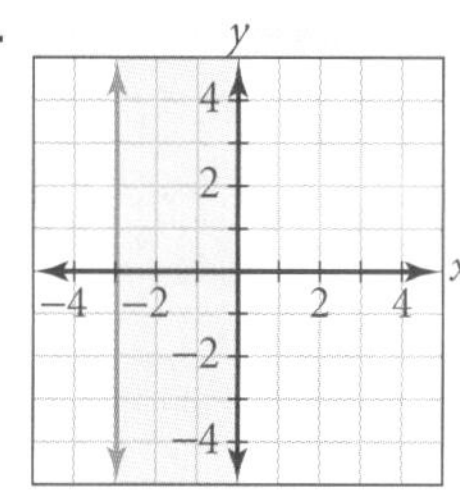

2.

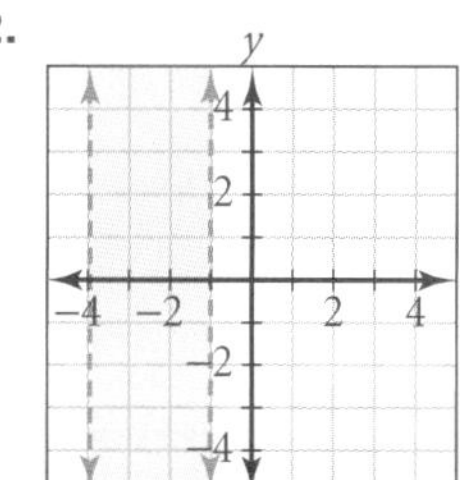

3.

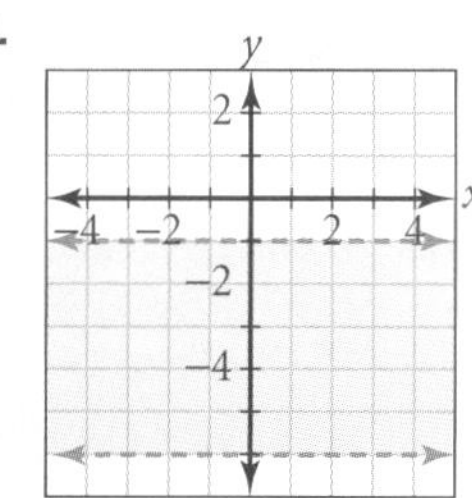

4.

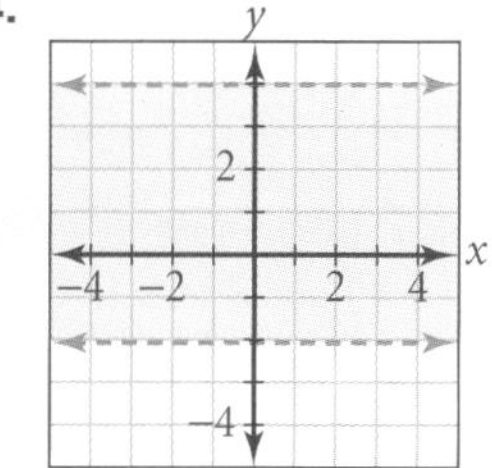

☞ For answers to Lesson 3.4 Exercises 5–15, see page 952; for answers to Lesson 3.5 Exercises 1–3, see page 953.

Lesson 3.4 (cont.)

5.

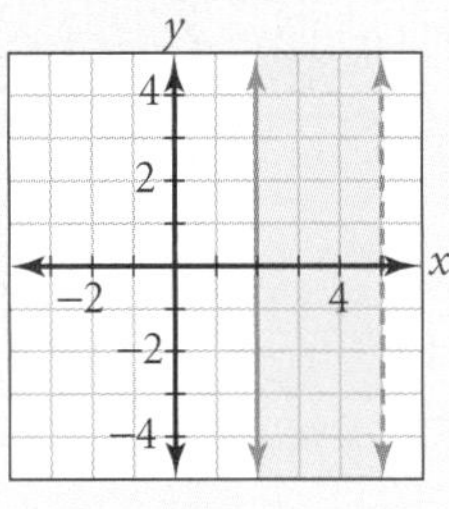

6.

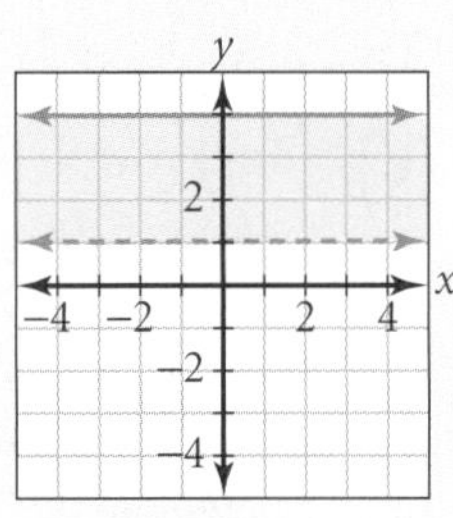

7.

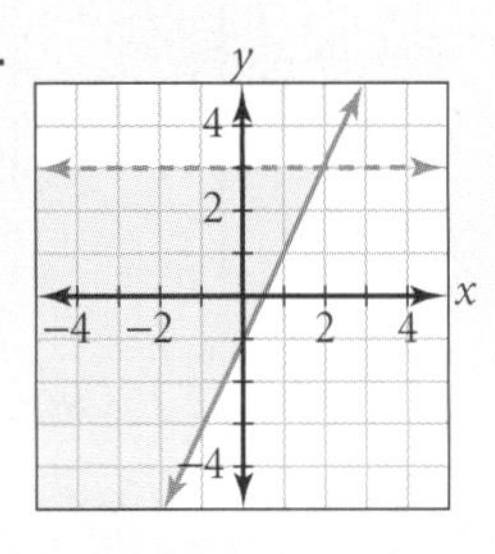

8.

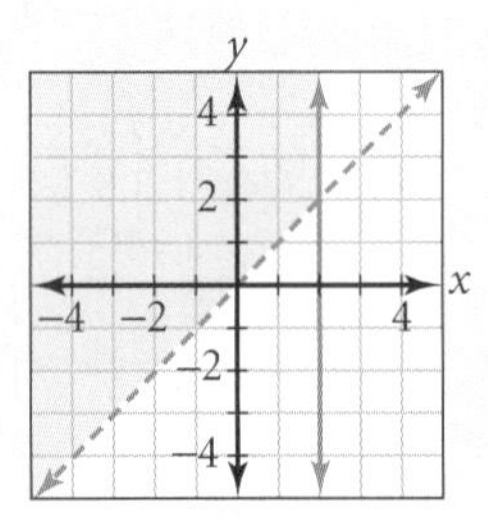

9.

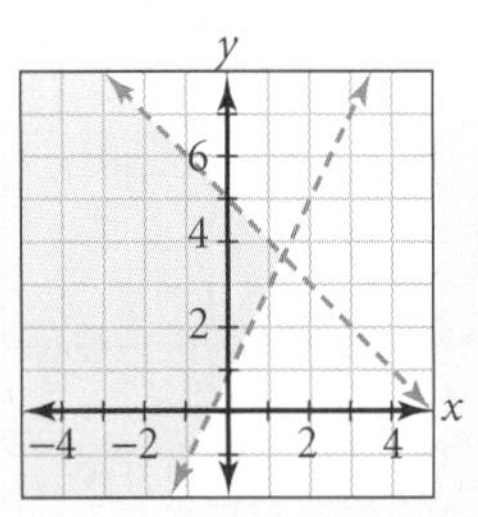

10.

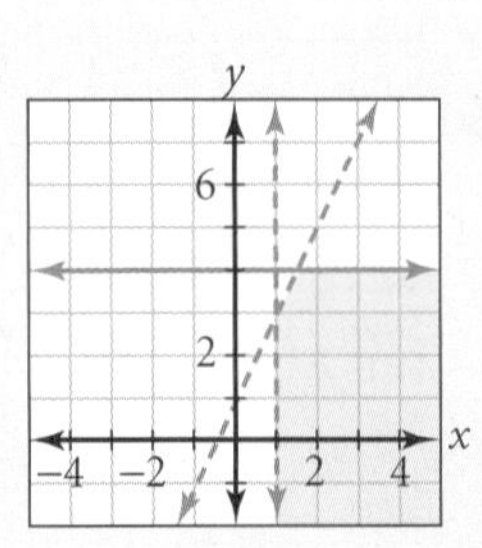

LESSON 3.6

Graph each pair of parametric equations for the given interval of *t*.

1. $\begin{cases} x(t) = t + 4 \\ y(t) = t - 3 \end{cases}$ for $-4 \le t \le 4$

2. $\begin{cases} x(t) = -4t \\ y(t) = t + 2 \end{cases}$ for $-3 \le t \le 3$

3. $\begin{cases} x(t) = 2t + 1 \\ y(t) = t - 4 \end{cases}$ for $-3 \le t \le 3$

4. $\begin{cases} x(t) = 2 - t \\ y(t) = \frac{1}{2}t + 1 \end{cases}$ for $-4 \le t \le 4$

Write each pair of parametric equations as a single equation in *x* and *y*.

5. $\begin{cases} x(t) = 2 - t \\ y(t) = 3 + t \end{cases}$ $y = 5 - x$

6. $\begin{cases} x(t) = t + 4 \\ y(t) = 1 - 2t \end{cases}$ $y = 9 - 2x$

7. $\begin{cases} x(t) = 4 - t \\ y(t) = t + 1 \end{cases}$ $y = 5 - x$

8. $\begin{cases} x(t) = 4t \\ y(t) = 3t - 1 \end{cases}$ $y = \frac{3}{4}x - 1$

9. $\begin{cases} x(t) = 2t \\ y(t) = \frac{t}{3} \end{cases}$ $y = \frac{x}{6}$

10. $\begin{cases} x(t) = 3t \\ y(t) = t^2 \end{cases}$ $y = \frac{x^2}{9}$

Graph the function represented by each pair of parametric equations. Then graph its inverse in the same coordinate plane.

11. $\begin{cases} x(t) = t^2 - 4 \\ y(t) = t \end{cases}$

12. $\begin{cases} x(t) = t^2 \\ y(t) = t - 4 \end{cases}$

13. $\begin{cases} x(t) = t^2 + 2 \\ y(t) = 1 - t \end{cases}$

14. $\begin{cases} x(t) = 2 - t^2 \\ y(t) = t + 1 \end{cases}$

15. $\begin{cases} x(t) = t^2 + 3t - 1 \\ y(t) = t + 2 \end{cases}$

16. $\begin{cases} x(t) = t^2 - 4t + 2 \\ y(t) = 1 - t \end{cases}$

Chapter 4

LESSON 4.1

For Exercises 1–14, let $A = \begin{bmatrix} 2 & 3 & -2 \\ 1 & 4 & 5 \\ 0 & 1 & 7 \end{bmatrix}$, $B = \begin{bmatrix} -3 & -2 & 1 \\ 0 & 5 & 5 \\ 1 & 2 & -1 \end{bmatrix}$, **and** $C = \begin{bmatrix} 8 & 0 & 1 \\ -6 & 4 & 3 \end{bmatrix}$.

Give the dimensions of each matrix.

1. A 3×3 **2.** B 3×3 **3.** C 2×3

Give the entry at the indicated address in matrix *A*, *B*, or *C*.

4. a_{13} -2 **5.** c_{21} -6 **6.** b_{31} 1

Perform the indicated matrix operations. If it is not possible, explain why.

7. $-A$ **8.** $A + B$ **9.** $-2C$ **10.** $B - A$

11. $A + C$ **12.** $2A - B$ **13.** $2B + A$ **14.** $4C$

Solve for *x* and *y*.

15. $\begin{bmatrix} x + 3 & -3 \\ 2 & 3y + 1 \end{bmatrix} = \begin{bmatrix} 10 & -3 \\ 2 & 10 \end{bmatrix}$ $x = 7$, $y = 3$

16. $\begin{bmatrix} -4 & 21 \\ -4y - 3 & 1 \end{bmatrix} = \begin{bmatrix} -4 & -2x + 5 \\ -19 & 1 \end{bmatrix}$ $x = -8$, $y = 4$

17. Quadrilateral *ABCD* has vertices at $A(0, 0)$, $B(1, 4)$, $C(4, 2)$, and $D(2, 0)$.

a. Represent quadrilateral *ABCD* in a matrix called *Q*.

b. Find $-2Q$.

c. Sketch quadrilateral *ABCD* and its image, *A'B'C'D'*, represented by $-2Q$. Describe the transformation.

11.

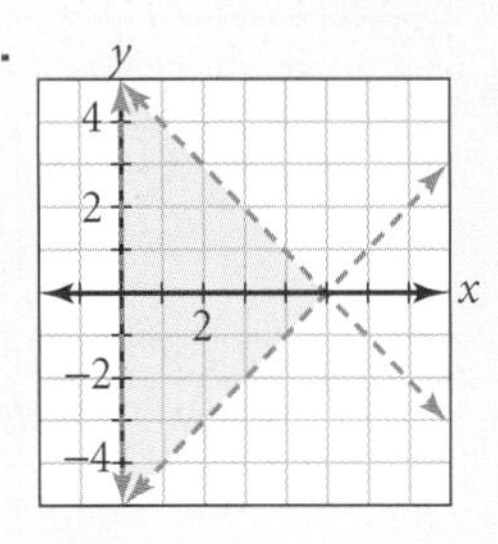

12.

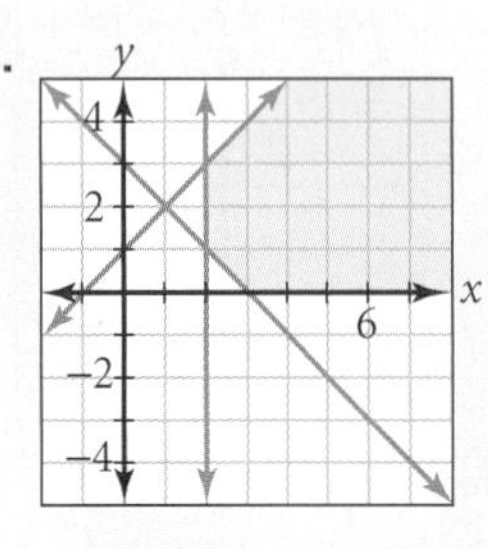

13. $\begin{cases} x \le 3 \\ y < x + 2 \\ y > -\frac{1}{2}x - \frac{5}{2} \end{cases}$

14. $\begin{cases} y \le x + 4 \\ y \le -2x + 4 \\ y \ge -\frac{1}{2}x - 2 \end{cases}$

15. $\begin{cases} y \le \frac{1}{2}x + 2 \\ y > -\frac{1}{2}x \\ x < 4 \end{cases}$

Lesson 3.5

1.

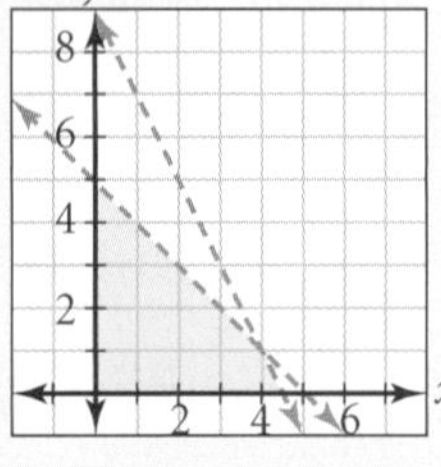

2.

LESSON 4.2

Find each product, if it exists.

1. $\begin{bmatrix} 2 & 4 & 3 \end{bmatrix}\begin{bmatrix} -4 \\ 0 \\ 6 \end{bmatrix}$ **[10]**

2. $\begin{bmatrix} -2 & -1 \\ 0 & 4 \\ 0 & -4 \end{bmatrix}\begin{bmatrix} 8 \\ 0 \\ 12 \end{bmatrix}$ **does not exist**

3. $\begin{bmatrix} -6 & 2 \\ 4 & 0 \end{bmatrix}\begin{bmatrix} 3 & 2 \\ 1 & -1 \end{bmatrix}$ $\begin{bmatrix} -16 & -14 \\ 12 & 8 \end{bmatrix}$

4. $\begin{bmatrix} -7 & 2 \\ 2 & 3 \end{bmatrix}\begin{bmatrix} 8 & -1 & 2 \\ 3 & 2 & 4 \\ -3 & 5 & 1 \end{bmatrix}$ **does not exist**

5. $\begin{bmatrix} -3 & -3 & 7 \end{bmatrix}\begin{bmatrix} 9 & 1 \\ 0 & 3 \\ 3 & 5 \end{bmatrix}$ **[−6 23]**

6. $\begin{bmatrix} 2 & 1 & 0 & 3 \end{bmatrix}\begin{bmatrix} -5 & 4 \\ 4 & 9 \\ 2 & 7 \\ -3 & -6 \end{bmatrix}$ **[−15 −1]**

7. $\begin{bmatrix} 1 & 4 \\ 2 & 3 \\ -3 & -3 \end{bmatrix}\begin{bmatrix} -5 & 3 & 4 \\ 1 & 1 & 0 \end{bmatrix}$

8. $\begin{bmatrix} 1 & 0 & 4 \\ -2 & 2 & 3 \end{bmatrix}\begin{bmatrix} 6 \\ 2 \\ 5 \end{bmatrix}$ $\begin{bmatrix} 26 \\ 7 \end{bmatrix}$

9. $\begin{bmatrix} 9 & 4 & 1 \end{bmatrix}\begin{bmatrix} -5 & 1 \\ 0.5 & 4 \\ 2 & 1 \end{bmatrix}$ **[−41 26]**

10. Triangle DEF has vertices at $D(-2, 4)$, $E(4, 0)$, and $F(0, -4)$. $T = \begin{bmatrix} -2 & 4 & 0 \\ 4 & 0 & -4 \end{bmatrix}$

a. Represent $\triangle DEF$ in a matrix called T.

b. Multiply T by the transformation matrix $S = \begin{bmatrix} \frac{3}{4} & 0 \\ 0 & \frac{3}{4} \end{bmatrix}$. $ST = \begin{bmatrix} -1.5 & 3 & 0 \\ 3 & 0 & -3 \end{bmatrix}$

c. Sketch $\triangle DEF$ and its image, $\triangle D'E'F'$, represented by ST. Describe the transformation. **compression by a scale factor of $\frac{3}{4}$**

LESSON 4.3

Determine whether each pair of matrices are inverses of each other.

1. $\begin{bmatrix} 5 & -4 \\ 1 & 6 \end{bmatrix}$ and $\begin{bmatrix} -5 & 4 \\ -1 & 6 \end{bmatrix}$ **no**

2. $\begin{bmatrix} 2 & 4 \\ 0 & -3 \end{bmatrix}$ and $\begin{bmatrix} \frac{1}{2} & \frac{2}{3} \\ 0 & -\frac{1}{3} \end{bmatrix}$ **yes**

3. $\begin{bmatrix} 2 & -3 \\ -1 & 2 \end{bmatrix}$ and $\begin{bmatrix} 2 & 3 \\ 1 & 2 \end{bmatrix}$ **yes**

4. $\begin{bmatrix} 3 & 2 \\ 5 & 4 \end{bmatrix}$ and $\begin{bmatrix} 2 & -1 \\ -2.5 & 1.5 \end{bmatrix}$ **yes**

Find the determinant, and state whether each matrix has an inverse.

5. $\begin{bmatrix} 5 & 3 \\ 2 & 1 \end{bmatrix}$ **−1; yes**

6. $\begin{bmatrix} 4 & 2 \\ 3 & 2 \end{bmatrix}$ **2; yes**

7. $\begin{bmatrix} 4 & 2 \\ 6 & 3 \end{bmatrix}$ **0; no**

8. $\begin{bmatrix} -3 & 1 \\ 7 & 4 \end{bmatrix}$ **−19; yes**

9. $\begin{bmatrix} \frac{1}{2} & -\frac{1}{4} \\ \frac{3}{4} & \frac{5}{2} \end{bmatrix}$ **1.4375; yes**

10. $\begin{bmatrix} \frac{2}{3} & -\frac{3}{4} \\ \frac{1}{3} & -5 \end{bmatrix}$ **≈−3.083; yes**

Find the inverse matrix, if it exists. Round entries to the nearest hundredth, if necessary. If the inverse matrix does not exist, write *no inverse*.

11 $\begin{bmatrix} 5 & 8 \\ 2 & 3 \end{bmatrix}$ $\begin{bmatrix} -3 & 8 \\ 2 & -5 \end{bmatrix}$

12 $\begin{bmatrix} 2.5 & 5 \\ 2 & 4 \end{bmatrix}$ **no inverse**

13 $\begin{bmatrix} 8 & 3 \\ 4 & 2 \end{bmatrix}$ $\begin{bmatrix} 0.5 & -0.75 \\ -1 & 2 \end{bmatrix}$

14 $\begin{bmatrix} 12 & 9 \\ 4 & 3 \end{bmatrix}$ **no inverse**

15 $\begin{bmatrix} 3 & 2 \\ 13 & 8 \end{bmatrix}$ $\begin{bmatrix} -4 & 1 \\ 6.5 & -1.5 \end{bmatrix}$

16 $\begin{bmatrix} 3 & 8 \\ 4 & 12 \end{bmatrix}$ $\begin{bmatrix} 3 & -2 \\ -1 & 0.75 \end{bmatrix}$

17 $\begin{bmatrix} 4 & -2 & 1 \\ 1 & 1 & 3 \\ 1 & -1 & 1 \end{bmatrix}$

18 $\begin{bmatrix} 6 & 1 & -7 \\ -2 & 0 & 4 \\ 4 & 5 & -3 \end{bmatrix}$

19 $\begin{bmatrix} \frac{1}{3} & \frac{2}{3} & 0 \\ -\frac{1}{2} & 3 & -1 \\ 4 & -\frac{1}{4} & 3 \end{bmatrix}$

11.

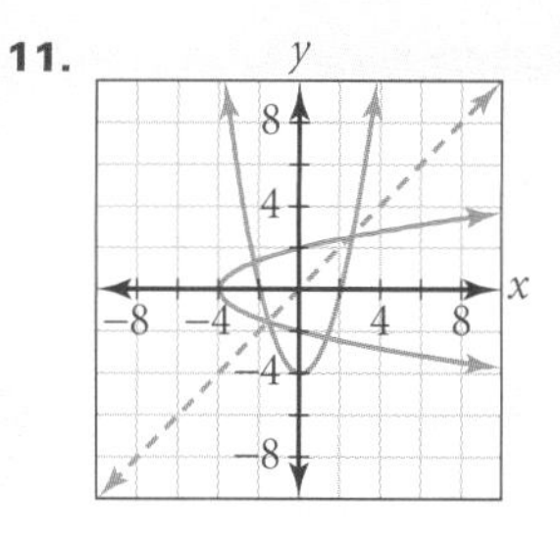

12.

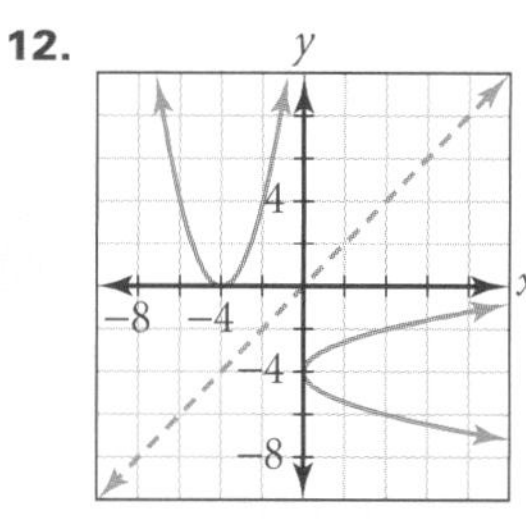

13.

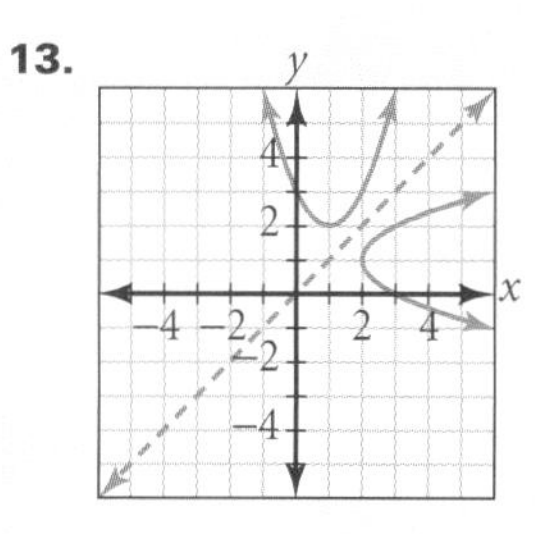

14.

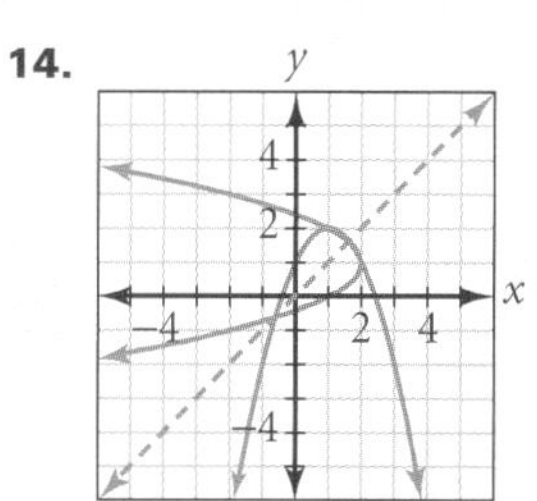

15.

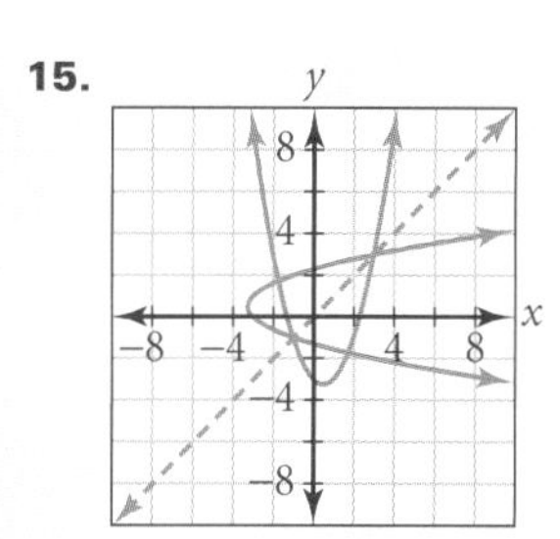

16.

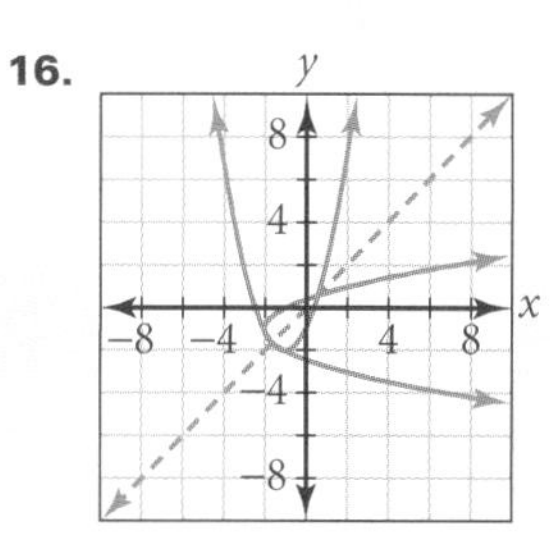

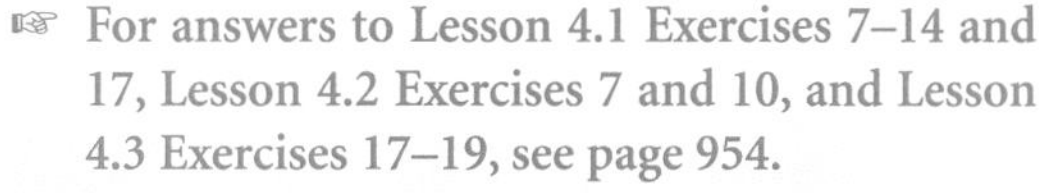

☞ For answers to Lesson 4.1 Exercises 7–14 and 17, Lesson 4.2 Exercises 7 and 10, and Lesson 4.3 Exercises 17–19, see page 954.

3.

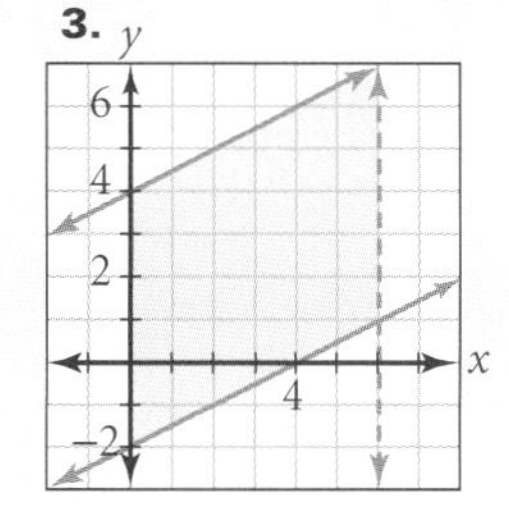

Lesson 3.6

1.

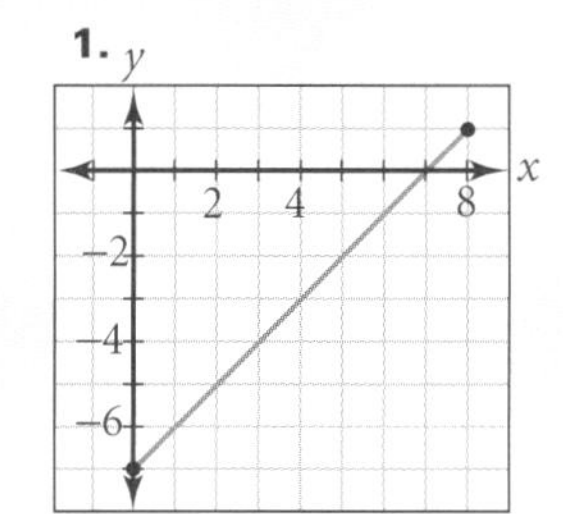

2.

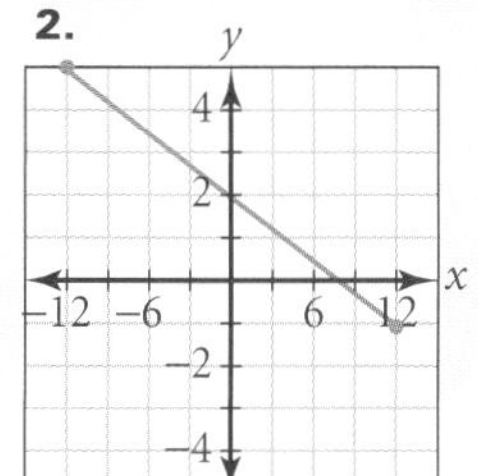

3.

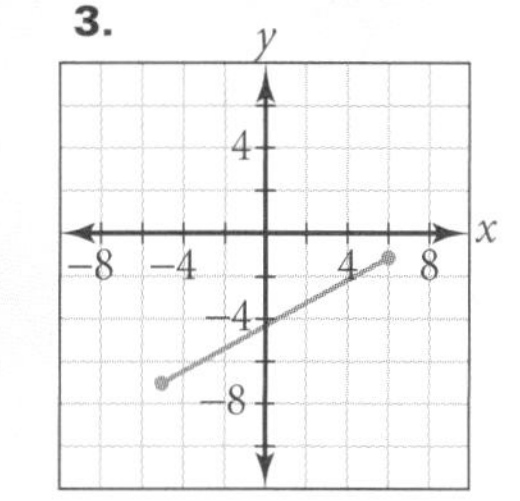

4.

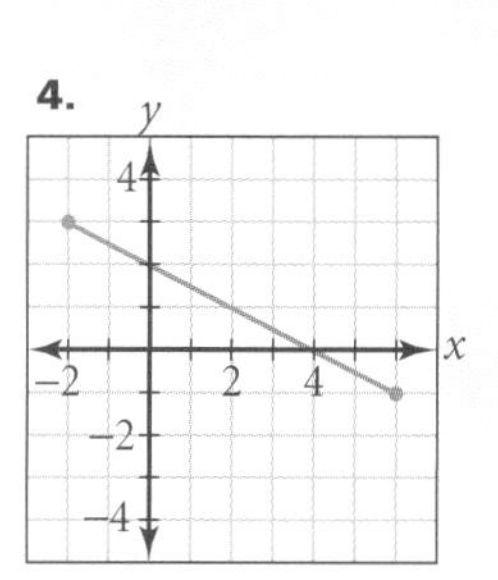

Extra Practice

Lesson 4.1

7. $\begin{bmatrix} -2 & -3 & 2 \\ -1 & -4 & -5 \\ 0 & -1 & -7 \end{bmatrix}$

8. $\begin{bmatrix} -1 & 1 & -1 \\ 1 & 9 & 10 \\ 1 & 3 & 6 \end{bmatrix}$

9. $\begin{bmatrix} -16 & 0 & -2 \\ 12 & -8 & -6 \end{bmatrix}$

10. $\begin{bmatrix} -5 & -5 & 3 \\ -1 & 1 & 0 \\ 1 & 1 & -8 \end{bmatrix}$

11. Not possible; the matrices do not have the same dimensions.

12. $\begin{bmatrix} 7 & 8 & -5 \\ 2 & 3 & 5 \\ -1 & 0 & 15 \end{bmatrix}$

13. $\begin{bmatrix} -4 & -1 & 0 \\ 1 & 14 & 15 \\ 2 & 5 & 5 \end{bmatrix}$

14. $\begin{bmatrix} 32 & 0 & 4 \\ -24 & 16 & 12 \end{bmatrix}$

17a. $Q = \begin{bmatrix} 0 & 1 & 4 & 2 \\ 0 & 4 & 2 & 0 \end{bmatrix}$

b. $-2Q = \begin{bmatrix} 0 & -2 & -8 & -4 \\ 0 & -8 & -4 & 0 \end{bmatrix}$

c.

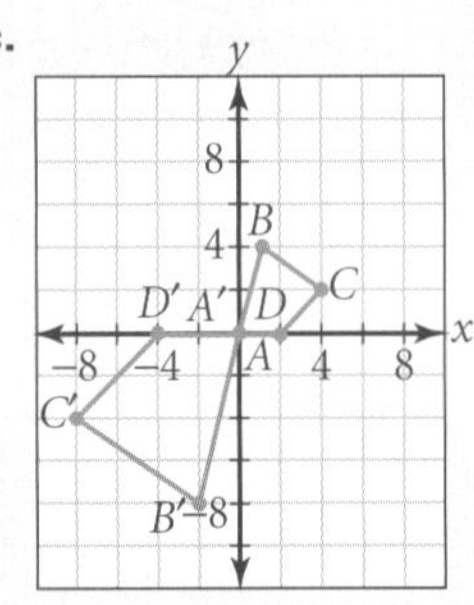

Lesson 4.2

7. $\begin{bmatrix} -1 & 7 & 4 \\ -7 & 9 & 8 \\ 12 & -12 & -12 \end{bmatrix}$

10c.

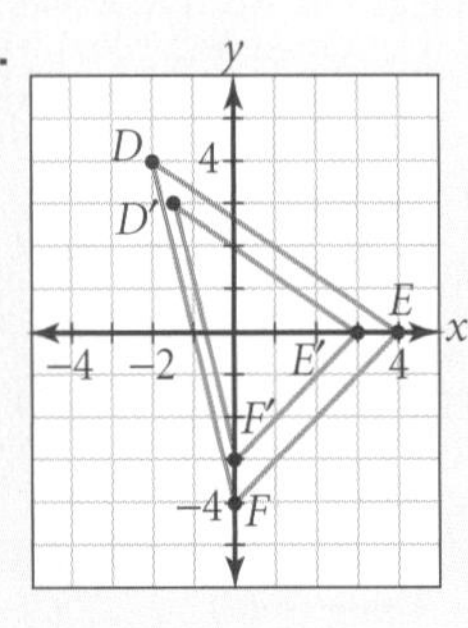

Extra Practice

LESSON 4.4

Write the matrix equation that represents each system.

1. $\begin{cases} 2x + y = 2 \\ 5x - 3y = -17 \end{cases}$

2. $\begin{cases} 8x + 2y = 10 \\ 5x + y = 7 \end{cases}$

3. $\begin{cases} 3x - 2y = 4 \\ x - 4y = 15 \end{cases}$

4. $\begin{cases} 2x + y + z = 1 \\ x + 3y - 4z = 19 \\ 4x - 2y + 3z = -9 \end{cases}$

5. $\begin{cases} 2x - 3y + z = 3 \\ x - 5y + 2z = 4 \\ 4x - y - z = -1 \end{cases}$

6. $\begin{cases} 3x - y + 2z = 3 \\ 2x + 5y - 3z = -12 \\ x - 3y + 4z = 8 \end{cases}$

Write the system of equations represented by each matrix equation.

7. $\begin{bmatrix} -2 & 1 & -1 \\ 5 & -1 & 2 \\ 3 & 4 & 1 \end{bmatrix}\begin{bmatrix} x \\ y \\ z \end{bmatrix} = \begin{bmatrix} 4 \\ -6 \\ 10 \end{bmatrix}$

8. $\begin{bmatrix} 0.5 & -2 & -1 \\ 3 & -6 & 2 \\ 1 & 1 & -5 \end{bmatrix}\begin{bmatrix} x \\ y \\ z \end{bmatrix} = \begin{bmatrix} -1 \\ 8 \\ 0 \end{bmatrix}$

Write the matrix equation that represents each system, and solve the system, if possible, by using a matrix equation.

9. $\begin{cases} 15x - 7y = 9 \\ 11x + 5y = 37 \end{cases}$ (2, 3)

10. $\begin{cases} x + y - z = 2 \\ 2x - y + z = 7 \\ x + y = 10 \end{cases}$ (3, 7, 8)

11. $\begin{cases} 2x + y + z = 5 \\ x - y + 3z = -11 \\ y + z = 1 \end{cases}$ (2, 4, −3)

12. $\begin{cases} x - 2y + 3z = -6 \\ 2x + y - 4z = -7 \\ 5x + 3y - 2z = 10 \end{cases}$ (−1, 7, 3)

13. $\begin{cases} 4x - 3y + z = 9 \\ 2x + y - 3z = -7 \\ 3x + 2y + z = 12 \end{cases}$ (2, 1, 4)

14. $\begin{cases} x + y - z = 15 \\ 2x - y + z = 0 \\ 3x + 2y - 3z = 38 \end{cases}$ (5, 7, −3)

LESSON 4.5

Write the augmented matrix for each system of equations.

1. $\begin{cases} 4x - 3y = 7 \\ 2x - y = 5 \end{cases}$

2. $\begin{cases} 11x - 5y + z = 9 \\ 3x + 7y - z = 3 \\ x - y + 4z = 8 \end{cases}$

3. $\begin{cases} 6x - y + z = 6 \\ 3x + 4y - z = 3 \\ 9x - 3y + 2z = 9 \end{cases}$

Find the reduced row-echelon form of each matrix.

4. $\left[\begin{array}{ccc|c} 2 & 1 & -1 & -1 \\ 1 & -1 & 3 & 8 \\ 1 & 1 & 1 & 2 \end{array}\right]$

5. $\left[\begin{array}{ccc|c} 1 & 1 & 2 & 2 \\ 1 & 0 & 2 & -3 \\ -1 & 0 & 3 & 5 \end{array}\right]$

6. $\left[\begin{array}{ccc|c} -1 & -2 & 0 & -4 \\ 1 & 2 & 1 & 7 \\ 3 & 6 & 3 & 21 \end{array}\right]$

Solve each system of equations by using the row-reduction method. Show each step.

7. $\begin{cases} 8x - 5y = -6 \\ 4x + 3y = 8 \end{cases}$ $\left(\frac{1}{2}, 2\right)$

8. $\begin{cases} 2x - 5y = 4 \\ 5x - 7y = -1 \end{cases}$ (−3, −2)

9. $\begin{cases} 7x - 3y = 29 \\ 10x + 3y = 5 \end{cases}$ (2, −5)

10. $\begin{cases} 2x - y + 3z = -7 \\ x + 4y - 2z = 17 \\ 3x + y + 2z = 2 \end{cases}$ (1, 3, −2)

11. $\begin{cases} x + 5y - 3z = 14 \\ 2x + y - z = 10 \\ x - 2y + z = 0 \end{cases}$ (4, 2, 0)

12. $\begin{cases} -x - 3y + 2z = -10 \\ 2x + y + z = 5 \\ 3x - 2y + 3z = -5 \end{cases}$ (0, 4, 1)

Classify each system as inconsistent, dependent, or independent.

13. $\begin{cases} -y - z = -1 \\ x + y + z = -2 \\ 2x - y - z = -7 \end{cases}$ dependent

14. $\begin{cases} x - 2y + 3z = 13 \\ 2x + 5y - 3z = -19 \\ x + 4y - 5z = -21 \end{cases}$ indep.

15. $\begin{cases} x - y + z = 4 \\ -2x + y - 2z = -7 \\ x + z = 6 \end{cases}$ inconsistent

Lesson 4.3

17. $\begin{bmatrix} 0.4 & 0.1 & -0.7 \\ 0.2 & 0.3 & -1.1 \\ -0.2 & 0.2 & 0.6 \end{bmatrix}$

18. $\begin{bmatrix} 0.5 & 0.8 & -0.1 \\ -0.25 & -0.25 & 0.25 \\ 0.25 & 0.65 & -0.05 \end{bmatrix}$

19. $\begin{bmatrix} 7 & -1.6 & -0.53 \\ -2 & 0.8 & 0.27 \\ -9.5 & 2.2 & 1.07 \end{bmatrix}$

Chapter 5

LESSON 5.1

Show that each function is a quadratic function by writing it in the form $f(x) = ax^2 + bx + c$ and identifying a, b, and c.

1. $f(x) = (x+1)(x-7)$ $= x^2 - 6x - 7$

2. $f(x) = (x-11)(x+2)$ $= x^2 - 9x - 22$

3. $f(x) = (2x+1)(x-4)$ $= 2x^2 - 7x - 4$

4. $f(x) = (12-x)(1+x)$ $= -x^2 + 11x + 12$

5. $f(x) = (2x+7)(x-5)$ $= 2x^2 - 3x - 35$

6. $f(x) = (x-2)^2 + 9$ $= x^2 - 4x + 13$

Identify whether each function is a quadratic function. Use a graph to check your answers.

7. $g(x) = 5 - 2x + x^2$ yes

8. $h(x) = (x-1)(x^2+1)$ no

9. $f(x) = x^2 - (x+1)^2$ no

10. $g(x) = 3x(-x+4)$ yes

State whether each parabola opens up or down and whether the y-coordinate of the vertex is the minimum value or the maximum value of the function.

11. $f(x) = 16 - x^2$ down; max

12. $h(x) = x^2 - x - 12$ up; min

13. $g(x) = (4-x)(6-x)$ up; min

14. $f(x) = (x-4)^2 - 2x^2$ down; max

Graph each function and give the approximate coordinates of the vertex.

15 $f(x) = 4 + 2x + x^2$ (–1, 3)

16 $h(x) = 3x^2 + x - 3$ (–0.167, –3.083)

17 $k(x) = \frac{1}{2}x^2 + 5$ (0, 5)

18 $g(x) = (x-3)^2$ (3, 0)

19 $f(x) = 0.5x + x^2$ (–0.25, –0.0625)

20 $h(x) = -(x-2)(x-3)$ (2.5, 0.25)

LESSON 5.2

Solve each equation. Give exact solutions. Then approximate each solution to the nearest hundredth, if necessary.

1. $x^2 = 51$ $\pm\sqrt{51} \approx \pm 7.14$

2. $16x^2 = 49$ ± 1.75

3. $3x^2 = 39$ $\pm\sqrt{13} \approx \pm 3.61$

4. $x^2 - 44 = 0$ $\pm 2\sqrt{11} \approx \pm 6.63$

5. $7x^2 + 3 = 29$ $\pm\frac{\sqrt{182}}{7} \approx \pm 1.93$

6. $(x-4)^2 = 19$ $4 - \sqrt{19} \approx -0.36$ *or* $4 + \sqrt{19} \approx 8.36$

Find the unknown length in each triangle. Give answers to the nearest tenth.

7.

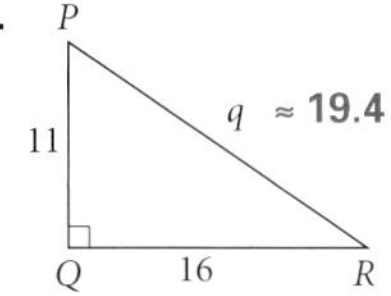

8.

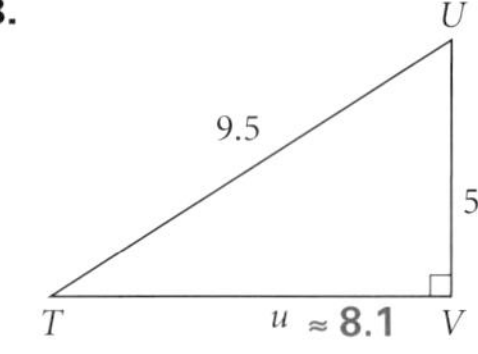

9.

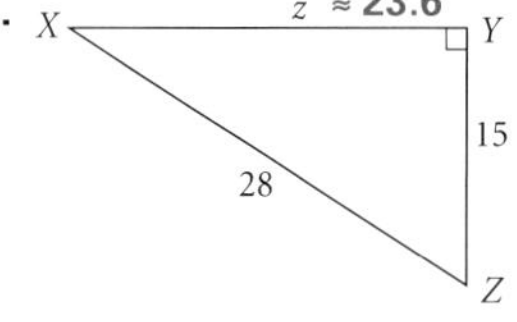

Find the missing side length in right triangle *ABC*. Round answers to the nearest tenth, if necessary.

10. $a = 17$ and $b = 8$ $c \approx 18.8$

11. $a = 20$ and $c = 32$ $b \approx 25.0$

12. $b = 7.2$ and $c = 13$ $a \approx 10.8$

13. $b = 4.5$ and $c = 16$ $a \approx 15.4$

14. $c = \sqrt{102}$ and $a = 6$ $b \approx 8.1$

15. $a = 9$ and $b = 11$ $c \approx 14.2$

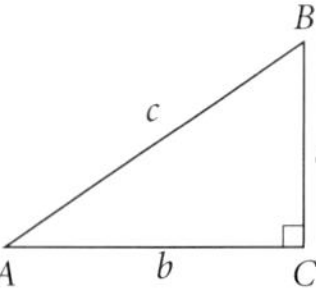

Extra Practice

Lesson 4.5

1. $\left[\begin{array}{cc|c} 4 & -3 & 7 \\ 2 & -1 & 5 \end{array}\right]$

2. $\left[\begin{array}{ccc|c} 11 & -5 & 1 & 9 \\ 3 & 7 & -1 & 3 \\ 1 & -1 & 4 & 8 \end{array}\right]$

3. $\left[\begin{array}{ccc|c} 6 & -1 & 1 & 6 \\ 3 & 4 & -1 & 3 \\ 9 & -3 & 2 & 9 \end{array}\right]$

4. $\left[\begin{array}{ccc|c} 1 & 0 & 0 & 1 \\ 0 & 1 & 0 & -1 \\ 0 & 0 & 1 & 2 \end{array}\right]$

5. $\left[\begin{array}{ccc|c} 1 & 0 & 0 & -3.8 \\ 0 & 1 & 0 & 5 \\ 0 & 0 & 1 & 0.4 \end{array}\right]$

6. $\left[\begin{array}{ccc|c} 1 & 2 & 0 & 4 \\ 0 & 0 & 1 & 3 \\ 0 & 0 & 0 & 0 \end{array}\right]$

Extra Practice

Lesson 4.4

1. $\begin{bmatrix} 2 & 1 \\ 5 & -3 \end{bmatrix}\begin{bmatrix} x \\ y \end{bmatrix} = \begin{bmatrix} 2 \\ -17 \end{bmatrix}$

2. $\begin{bmatrix} 8 & 2 \\ 5 & 1 \end{bmatrix}\begin{bmatrix} x \\ y \end{bmatrix} = \begin{bmatrix} 10 \\ 7 \end{bmatrix}$

3. $\begin{bmatrix} 3 & -2 \\ 1 & -4 \end{bmatrix}\begin{bmatrix} x \\ y \end{bmatrix} = \begin{bmatrix} 4 \\ 15 \end{bmatrix}$

4. $\begin{bmatrix} 2 & 1 & 1 \\ 1 & 3 & -4 \\ 4 & -2 & 3 \end{bmatrix}\begin{bmatrix} x \\ y \\ z \end{bmatrix} = \begin{bmatrix} 1 \\ 19 \\ -9 \end{bmatrix}$

5. $\begin{bmatrix} 2 & -3 & 1 \\ 1 & -5 & 2 \\ 4 & -1 & -1 \end{bmatrix}\begin{bmatrix} x \\ y \\ z \end{bmatrix} = \begin{bmatrix} 3 \\ 4 \\ -1 \end{bmatrix}$

6. $\begin{bmatrix} 3 & -1 & 2 \\ 2 & 5 & -3 \\ 1 & -3 & 4 \end{bmatrix}\begin{bmatrix} x \\ y \\ z \end{bmatrix} = \begin{bmatrix} 3 \\ -12 \\ 8 \end{bmatrix}$

7. $\begin{cases} -2x + y - z = 4 \\ 5x - y + 2z = -6 \\ 3x + 4y + z = 10 \end{cases}$

8. $\begin{cases} 0.5x - 2y - z = -1 \\ 3x - 6y + 2z = 8 \\ x + y - 5z = 0 \end{cases}$

9. $\begin{bmatrix} 15 & -7 \\ 11 & 5 \end{bmatrix}\begin{bmatrix} x \\ y \end{bmatrix} = \begin{bmatrix} 9 \\ 37 \end{bmatrix}$

10. $\begin{bmatrix} 1 & 1 & -1 \\ 2 & -1 & 1 \\ 1 & 1 & 0 \end{bmatrix}\begin{bmatrix} x \\ y \\ z \end{bmatrix} = \begin{bmatrix} 2 \\ 7 \\ 10 \end{bmatrix}$

11. $\begin{bmatrix} 2 & 1 & 1 \\ 1 & -1 & 3 \\ 0 & 1 & 1 \end{bmatrix}\begin{bmatrix} x \\ y \\ z \end{bmatrix} = \begin{bmatrix} 5 \\ -11 \\ 1 \end{bmatrix}$

12. $\begin{bmatrix} 1 & -2 & 3 \\ 2 & 1 & -4 \\ 5 & 3 & -2 \end{bmatrix}\begin{bmatrix} x \\ y \\ z \end{bmatrix} = \begin{bmatrix} -6 \\ -7 \\ 10 \end{bmatrix}$

13. $\begin{bmatrix} 4 & -3 & 1 \\ 2 & 1 & -3 \\ 3 & 2 & 1 \end{bmatrix}\begin{bmatrix} x \\ y \\ z \end{bmatrix} = \begin{bmatrix} 9 \\ -7 \\ 12 \end{bmatrix}$

14. $\begin{bmatrix} 1 & 1 & -1 \\ 2 & -1 & 1 \\ 3 & 2 & -3 \end{bmatrix}\begin{bmatrix} x \\ y \\ z \end{bmatrix} = \begin{bmatrix} 15 \\ 0 \\ 38 \end{bmatrix}$

Lesson 5.1

1. $f(x) = x^2 - 6x - 7;$ $a = 1, b = -6, c = -7$

2. $f(x) = x^2 - 9x - 22;$ $a = 1, b = -9, c = -22$

3. $f(x) = 2x^2 - 7x - 4;$ $a = 2, b = -7, c = -4$

4. $f(x) = -x^2 + 11x + 12;$ $a = -1, b = 11, c = 12$

5. $f(x) = 2x^2 - 3x - 35;$ $a = 2, b = -3, c = -35$

6. $f(x) = x^2 - 4x + 13;$ $a = 1, b = -4, c = 13$

Lesson 5.4

22. $f(x) = 9x^2; (0, 0); x = 0$

23. $f(x) = x^2 + (-3);$ $(0, -3); x = 0$

24. $f(x) = -x^2 + 12; (0, 12);$ $x = 0$

25. $f(x) = [x - (-2)]^2 + (-4);$ $(-2, 4); x = -2$

26. $f(x) = -\left(x - \frac{3}{2}\right)^2 + \frac{13}{4};$ $\left(\frac{3}{2}, \frac{13}{4}\right); x = \frac{3}{2}$

27. $f(x) = [x - (-3)]^2 + (-4);$ $(-3, -4); x = -3$

28. $f(x) = [x - (-6)]^2 + 6;$ $(-6, 6); x = -6$

29. $f(x) = 2[x - (-1)]^2 + (-5);$ $(-1, -5); x = -1$

30. $f(x) = -\left[x - \left(-\frac{3}{2}\right)\right]^2 + \frac{37}{4};$ $\left(-\frac{3}{2}, \frac{37}{4}\right); x = -\frac{3}{2}$

LESSON 5.3

Factor each expression.

1. $7x + 49$ **$7(x + 7)$**
2. $36x + 108x^2$ **$36x(1 + 3x)$**
3. $4x^2 - 28x$ **$4x(x - 7)$**
4. $-9x^2 + 3x$ **$3x(-3x + 1)$**
5. $14(3 - x^2) + x(3 - x^2)$ **$(14 + x)(3 - x^2)$**
6. $3x(x - 8) + 7(x - 8)$ **$(3x + 7)(x - 8)$**
7. $-3x^3 - 12x^2$ **$-3x^2(x + 4)$**
8. $(9 - 2x)3x - 8(9 - 2x)$ **$(3x - 8)(9 - 2x)$**

Factor each quadratic expression.

9. $x^2 + 14x + 49$ **$(x + 7)^2$**
10. $x^2 - 13x + 30$ **$(x - 10)(x - 3)$**
11. $x^2 + 17x + 42$ **$(x + 3)(x + 14)$**
12. $x^2 - 7x - 60$ **$(x - 12)(x + 5)$**
13. $3x^2 + 8x + 4$ **$(3x + 2)(x + 2)$**
14. $2x^2 - 7x + 6$ **$(2x - 3)(x - 2)$**
15. $5x - 2x^2 - 3$ **$(3 - 2x)(x - 1)$**
16. $x + 2 - 6x^2$ **$(1 + 2x)(2 - 3x)$**

Solve each equation by factoring and applying the Zero-Product Property.

17. $x^2 - 144 = 0$ **± 12**
18. $x^2 - 18x + 81 = 0$ **9**
19. $16x^2 - 25 = 0$ **$\pm\frac{5}{4}$**
20. $x^2 + 3x - 4 = 0$ **1 *or* -4**
21. $9x^2 - 6x + 1 = 0$ **$\frac{1}{3}$**
22. $x^2 + 10x + 25 = 0$ **-5**

Use factoring and the Zero-Product Property to find the zeros of each quadratic function.

23. $h(x) = x^2 - 12x$ **$0, 12$**
24. $k(x) = x^2 - 4x - 21$ **$7, -3$**
25. $g(x) = 2x^2 + 17x - 9$ **$-9, 0.5$**
26. $f(x) = x^2 + 16x + 55$ **$-5, -11$**
27. $h(x) = x^2 - 12x + 35$ **$5, 7$**
28. $a(x) = 2x^2 + 9x - 5$ **$\frac{1}{2}, -5$**

LESSON 5.4

Complete the square for each quadratic expression in order to form a perfect-square trinomial. Then write the new expression as a binomial squared.

1. $x^2 + x$ **$x^2 + x + \frac{1}{4}; \left(x + \frac{1}{2}\right)^2$**
2. $x^2 + 18x$ **$x^2 + 18x + 81; (x + 9)^2$**
3. $x^2 - 10x$ **$x^2 - 10x + 25; (x - 5)^2$**
4. $x^2 + 26x$ **$x^2 + 26x + 169; (x + 13)^2$**
5. $x^2 - 9x$ **$x^2 - 9x + \frac{81}{4}; \left(x - \frac{9}{2}\right)^2$**
6. $x^2 + 3x$ **$x^2 + 3x + \frac{9}{4}; \left(x + \frac{3}{2}\right)^2$**
7. $x^2 - 22x$ **$x^2 - 22x + 121; (x - 11)^2$**
8. $x^2 + 17x$ **$x^2 + 17x + \frac{289}{4}; \left(x + \frac{17}{2}\right)^2$**
9. $x^2 - 0.5x$ **$x^2 - 0.5x + 0.0625; (x - 0.25)^2$**

Solve each equation by completing the square. Give exact solutions.

10. $x^2 - 10x + 4 = 0$ **$5 \pm \sqrt{21}$**
11. $x^2 + 2x - 7 = 0$ **$-1 \pm 2\sqrt{2}$**
12. $x^2 - 12x = 20$ **$6 \pm 2\sqrt{14}$**
13. $x^2 = 8x + 12$ **$4 \pm 2\sqrt{7}$**
14. $0 = x^2 - 14x + 2$ **$7 \pm \sqrt{47}$**
15. $21 = x^2 - 16x$ **$8 \pm \sqrt{85}$**
16. $x^2 + 5x = 7$ **$-\frac{5}{2} \pm \frac{\sqrt{53}}{2}$**
17. $x^2 + 3x + 6 = 17$ **$-\frac{3}{2} \pm \frac{\sqrt{53}}{2}$**
18. $x^2 - 7x = 15 - x$ **$3 \pm 2\sqrt{6}$**
19. $2x^2 - 15 = 8x$ **$2 \pm \frac{\sqrt{46}}{2}$**
20. $4x^2 + 12x = 14$ **$-\frac{3}{2} \pm \frac{\sqrt{23}}{2}$**
21. $x^2 + x - 3 = 0$ **$-\frac{1}{2} \pm \frac{\sqrt{13}}{2}$**

Write each quadratic function in vertex form. Give the coordinates of the vertex and the equation of the axis of symmetry. coordinates of vertex:

22. $f(x) = 9x^2$ **$(0, 0)$**
23. $f(x) = x^2 - 3$ **$(0, -3)$**
24. $f(x) = -x^2 + 12$ **$(0, 12)$**
25. $f(x) = x^2 + 4x$ **$(-2, -4)$**
26. $f(x) = -x^2 + 3x + 1$ **$\left(\frac{3}{2}, \frac{13}{4}\right)$**
27. $f(x) = x^2 + 6x + 5$ **$(-3, -4)$**
28. $f(x) = x^2 + 12x + 42$ **$(-6, 6)$**
29. $f(x) = 2x^2 + 4x - 3$ **$(-1, -5)$**
30. $f(x) = -x^2 - 3x + 7$ **$\left(-\frac{3}{2}, \frac{37}{4}\right)$**

LESSON 5.5

Use the quadratic formula to solve each equation. Give exact solutions.

1. $x^2 - 8x + 15 = 0$ **3, 5**
2. $x^2 + 14x = 0$ **0, −14**
3. $x^2 - 5x + 1 = 0$ $\frac{5 \pm \sqrt{21}}{2}$
4. $x^2 - x = 12$ **−3, 4**
5. $x^2 - 10x + 14 = 5$ **1, 9**
6. $x^2 + 5x + 3 = 0$ $\frac{-5 \pm \sqrt{13}}{2}$
7. $x^2 - 7 = 2x$ $1 \pm 2\sqrt{2}$
8. $2x^2 + 3x - 1 = 0$ $\frac{-3 \pm \sqrt{17}}{4}$
9. $(x-3)(x+4) = 5$ $\frac{-1 \pm \sqrt{69}}{2}$
10. $3x^2 - 2x - 5 = 0$ $-1, \frac{5}{3}$
11. $7x^2 - 2x - 8 = 0$ $\frac{1 \pm \sqrt{57}}{7}$
12. $-2x^2 + 3x + 1 = 0$ $\frac{-3 \pm \sqrt{17}}{-4}$
13. $4x + x^2 - 9 = 0$ $-2 \pm \sqrt{13}$
14. $4x^2 = x + 6$ $\frac{1 \pm \sqrt{97}}{8}$
15. $x^2 - 14 = x$ $\frac{1 \pm \sqrt{57}}{2}$

For each quadratic function, write the equation for the axis of symmetry and find the coordinates of the vertex. coordinates of the vertex:

16. $y = 3x^2 + 6x - 2$ $(-1, -5)$
17. $y = x^2 + 4x - 11$ $(-2, -15)$
18. $y = -x^2 + 8x + 12$ $(4, 28)$
19. $y = 2x^2 + 3x - 5$ $\left(-\frac{3}{4}, -6\frac{1}{8}\right)$
20. $y = 6 + 2x - x^2$ $(1, 7)$
21. $y = x^2 + x - 9$ $\left(-\frac{1}{2}, -9\frac{1}{4}\right)$
22. $y = -3x^2 + 2x + 8$ $\left(\frac{1}{3}, 8\frac{1}{3}\right)$
23. $y = -3x^2 - 4x + 2$ $\left(-\frac{2}{3}, 3\frac{1}{3}\right)$
24. $y = 5x^2 + 10x + 3$ $(-1, -2)$
25. $y = 4x^2 + 2x - 1$ $\left(-\frac{1}{4}, -1\frac{1}{4}\right)$
26. $y = -7 + 6x + x^2$ $(-3, -16)$
27. $y = 3x^2 - 9x + 5$ $\left(\frac{3}{2}, -1\frac{3}{4}\right)$
28. $y = -x^2 + 8x - 16$ $(4, 0)$
29. $y = 4x^2 - 3x - 5$ $\left(\frac{3}{8}, -5\frac{9}{16}\right)$
30. $y = -5x^2 - 4x + 7$ $\left(-\frac{2}{5}, 7\frac{4}{5}\right)$

LESSON 5.6

Find the discriminant and determine the number of real solutions. Then solve.

1. $x^2 - 3x - 5 = 0$ $29; 2; \frac{3}{2} \pm \frac{\sqrt{29}}{2}$
2. $2x^2 - 4x + 3 = 0$ $-8; 0; 1 \pm \frac{i\sqrt{2}}{2}$
3. $x^2 - 7x + 17 = 0$ $-19; 0; \frac{7}{2} \pm \frac{i\sqrt{19}}{2}$
4. $3x^2 - 5x + 8 = 0$ $-71; 0; \frac{5}{6} \pm \frac{i\sqrt{71}}{6}$
5. $4x - 8x^2 = 3$ $-80; 0; \frac{1}{4} \pm \frac{i\sqrt{5}}{4}$
6. $x^2 - 5x + 10 = 0$ $-15; 0; \frac{5}{2} \pm \frac{i\sqrt{15}}{2}$
7. $4x^2 + 9 = 2x$ $-140; 0; \frac{1}{4} \pm \frac{i\sqrt{35}}{4}$
8. $x - x^2 = 11$ $-43; 0; \frac{1}{2} \pm \frac{i\sqrt{43}}{2}$
9. $-3x^2 = 6x - 2$ $60; 2; -1 \pm \frac{\sqrt{15}}{3}$

Write the conjugate of each complex number.

10. -2 -2
11. $13 - 5i$ $13 + 5i$
12. $7 + 2i$ $7 - 2i$
13. $-3i + 4$ $4 + 3i$
14. $i - 1$ $-1 - i$
15. $3i$ $-3i$

Simplify.

16. $(6 + 5i) + (13 - 4i)$ $19 + i$
17. $(-2 + 14i) - (-7 + 12i)$ $5 + 2i$
18. $(17 + 8i) + (-4 + i)$ $13 + 9i$
19. $(6i + 2) + (3i - 1)$ $1 + 9i$
20. $(3i - 1) - (-2i + 1)$ $-2 + 5i$
21. $(-5 - 3i) + (2i - 6)$ $-11 - i$
22. $(20 - 16i) - (9 + 2i)$ $11 - 18i$
23. $(15 - 6i) + (15 + 6i)$ 30
24. $(14 + 3i) - (3 + i)$ $11 + 2i$
25. $3(-9 + 12i)$ $-27 + 36i$
26. $2i(-4 - 8i)$ $16 - 8i$
27. $(7 + 3i)(2 - 5i)$ $29 - 29i$
28. $(2 - 4i)(7 + i)$ $18 - 26i$
29. $(-3i + 2)(4 - i)$ $5 - 14i$
30. $(3 + 5i)(-2i - 4)$ $-2 - 26i$
31. $\frac{5 - i}{3 + i}$ $\frac{7}{5} - \frac{4i}{5}$
32. $\frac{4 - 7i}{-1 + i}$ $-\frac{11 - 3i}{2}$
33. $\frac{6 - 2i}{-4 + i}$ $-\frac{26 - 2i}{17}$
34. $\frac{5 + i}{i}$ $1 - 5i$
35. $\frac{4 + 3i}{2i}$ $\frac{3}{2} - 2i$
36. $\frac{2i + 5}{-3i}$ $\frac{-2 + 5i}{3}$
37. $(2i - 9)^2$ $77 - 36i$
38. $(-3 + 10i)^2$ $-91 - 60i$
39. $(4 - 5i)^2$ $-9 - 40i$

Graph each number and its conjugate in the complex plane.

40. $4 + i$ $(4, 1), (4, -1)$
41. $4i$ $(0, 4), (0, -4)$
42. 4 $(4, 0), (4, 0)$
43. $i + 2$ $(2, 1), (2, -1)$
44. $-i - 3$ $(-3, -1), (-3, 1)$
45. -6 $(-6, 0), (-6, 0)$

Lesson 5.5

16. $x = -1; (-1, -5)$

17. $x = -2; (-2, -15)$

18. $x = 4; (4, 28)$

19. $x = -\frac{3}{4}; \left(-\frac{3}{4}, -6\frac{1}{8}\right)$

20. $x = 1; (1, 7)$

21. $x = -\frac{1}{2}; \left(-\frac{1}{2}, -9\frac{1}{4}\right)$

22. $x = \frac{1}{3}; \left(\frac{1}{3}, 8\frac{1}{3}\right)$

23. $x = -\frac{2}{3}; \left(-\frac{2}{3}, 3\frac{1}{3}\right)$

24. $x = -1; (-1, -2)$

25. $x = -\frac{1}{4}; \left(-\frac{1}{4}, -1\frac{1}{4}\right)$

26. $x = -3; (-3, -16)$

27. $x = \frac{3}{2}; \left(\frac{3}{2}, -1\frac{3}{4}\right)$

28. $x = 4; (4, 0)$

29. $x = \frac{3}{8}; \left(\frac{3}{8}, 5\right)$

30. $x = -\frac{2}{5}; \left(-\frac{2}{5}, 7\frac{4}{5}\right)$

Lesson 5.6

40.

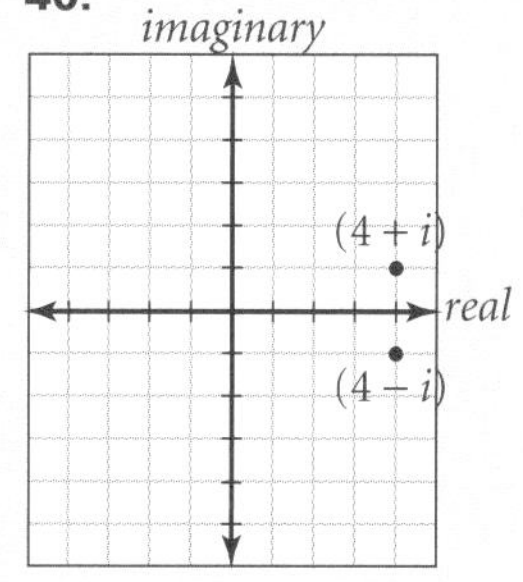

41.

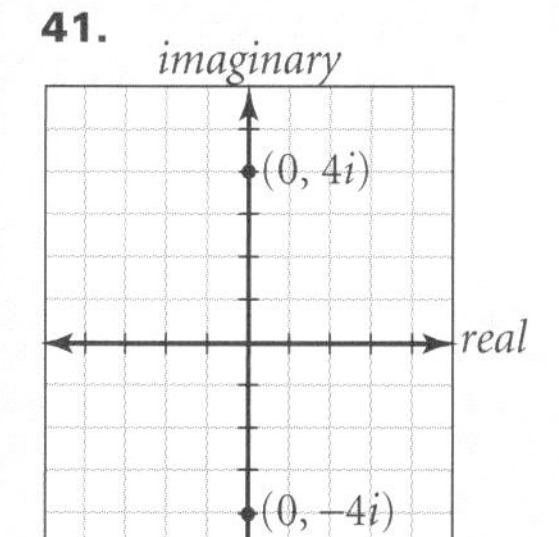

42.

imaginary
(4, 0)
real

43.

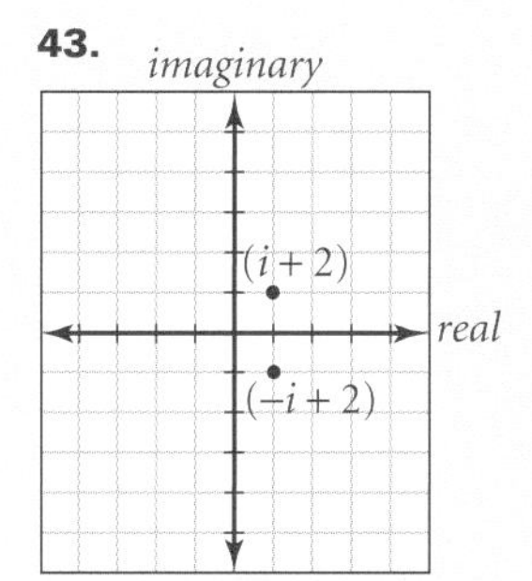

44.

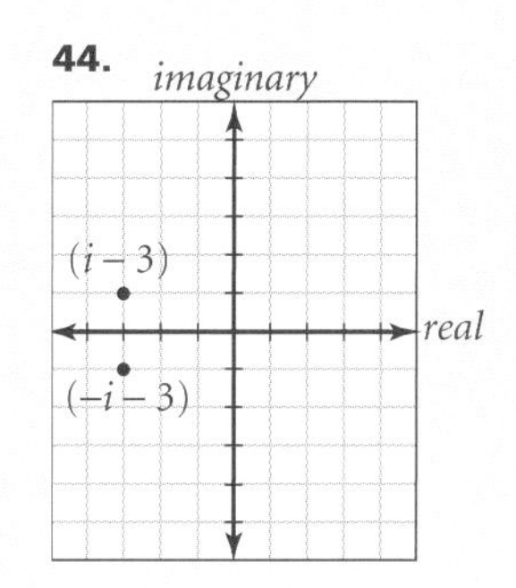

45.

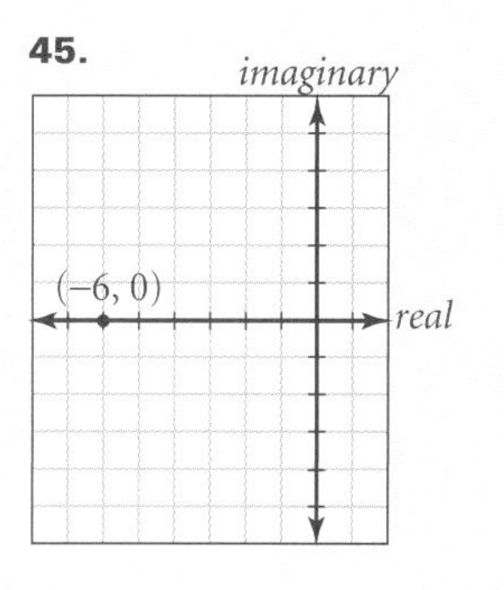

Extra Practice

Lesson 5.7

3. $f(x) = x^2 + 7x - 1$

7. $f(x) = x^2 - 18x - 59$

9. $f(x) = 0.5x^2 + 3x - 1$

13. $f(x) = 0.25x^2 - 0.5x + 1$

Lesson 5.8

1. $-3 < x < 3$

2. $x \le -5$ *or* $x \ge 5$

3. $-9 \le x \le -2$

4. $5 < x < 6$

5. $x < -4$ *or* $x > 3$

6. $x \le 1$ *or* $x \ge 4$

7. $x = 4$

8. $x = 3$

9. $x < -1$ *or* $x > 7$

10. $x < -2 - \sqrt{5}$ *or* $x > -2 + \sqrt{5}$

11. $\frac{3}{2} - \frac{\sqrt{29}}{2} \le x \le \frac{3}{2} + \frac{\sqrt{29}}{2}$

12. $-1 - \frac{1}{2}\sqrt{2} < x < -1 + \frac{1}{2}\sqrt{2}$

Extra Practice

LESSON 5.7

Solve a system of equations in order to find a quadratic function that fits each set of data points exactly.

1. $(-7, 51), (5, 27), (-4, 18)$ $f(x) = x^2 + 2$
2. $(-2, 9), (6, -47), (-1, 9)$ $f(x) = -x^2 - 3x + 7$
3. $(-4, -13), (-2, -11), (2, 17)$
4. $(-3, 14), (2, 19), (4, 49)$ $f(x) = 2x^2 + 3x + 5$
5. $(0, 6), (1, 8), (3, 18)$ $f(x) = x^2 + x + 6$
6. $(-5, 19), (-3, 7), (6, 52)$ $f(x) = x^2 + 2x + 4$
7. $(-3, 4), (-2, -19), (9, -140)$
8. $(-6, -114), (-5, -81), (4, -54)$ $f(x) = -3x^2 - 6$
9. $(2, 7), (4, 19), (-6, -1)$
10. $(3, 47), (-1, 13), (0, 1)$ $f(x) \approx 6.83x^2 - 5.17x + 1$
11. $(4, 0), (6, -12), (-2, -12)$ $f(x) = -x^2 + 4x$
12. $(1, 4), (2, -20), (-1, -8)$ $f(x) = -10x^2 + 6x + 8$
13. $(4, 3), (-4, 7), (2, 1)$
14. $(-3, 51), (3, 27), (0, 3)$ $f(x) = 4x^2 - 4x + 3$
15. $(-8, -28), (6, 0), (3, 10.5)$ $f(x) = -0.5x^2 + x + 12$
16. $(-5, 17), (5, 47), (3, 25)$ $f(x) = x^2 + 3x + 7$

Randall plays baseball for his high school team. The table shows the height, *y*, of the ball *x* seconds after Randall hit it.

Time (seconds)	Height (feet)
1	48
2	60
3	40

17. Find a quadratic function that fits the data by solving a system. $f(x) = -16x^2 + 60x + 4$
18. Find the height of the ball 2.5 seconds after it was hit. **54 ft**
19. After how many seconds did the ball hit the ground? **about 3.8 seconds**

LESSON 5.8

Solve each inequality. Graph the solution on a number line.

1. $x^2 - 9 < 0$ $-3 < x < 3$
2. $x^2 - 25 \ge 0$ $x \le -5$ *or* $x \ge 5$
3. $x^2 + 11x + 18 \le 0$ $-9 \le x \le -2$
4. $x^2 - 11x + 30 < 0$ $5 < x < 6$
5. $x^2 + x - 12 > 0$ $x < -4$ *or* $x > 3$
6. $x^2 - 5x + 4 \ge 0$ $x \le 1$ *or* $x \ge 4$
7. $x^2 - 8x + 16 \le 0$ $x = 4$
8. $x^2 - 6x + 9 \le 0$ $x = 3$
9. $x^2 - 6x - 7 > 0$ $x < -1$ *or* $x > 7$
10. $x^2 + 4x - 1 > 0$ $x < -2 - \sqrt{5}$ *or* $x > -2 + \sqrt{5}$
11. $x^2 - 3x - 5 \le 0$ $\frac{3}{2} - \frac{\sqrt{29}}{2} \le x \le \frac{3}{2} + \frac{\sqrt{29}}{2}$
12. $x^2 + 2x + \frac{1}{2} < 0$ $-1 - \frac{1}{2}\sqrt{2} < x < -1 + \frac{1}{2}\sqrt{2}$

Sketch the graph of each inequality. Then decide which of the given points are in the solution region.

13. $y < (x + 2)^2 - 3$ **C** $A(-3, -1)$ $B(-3, -2)$ $C(-3, -4)$
14. $y \le -(x + 1)^2 + 4$ **B, C** $A(1, 2)$ $B(1, -2)$ $C(1, -5)$
15. $y > -(x - 4)^2 + 3$ **C** $A(3, 2)$ $B(4, -3)$ $C(7, -5)$

Graph each inequality and shade the solution region.

16. $y \ge (x - 3)^2$
17. $y \le (x - 1)^2$
18. $y > -x^2 + 2x - 1$
19. $y \le x^2 + 4x + 3$
20. $y \ge x^2 + 5x + 4$
21. $y \le (x + 2)^2 + 2$
22. $y > x^2 + 2x - 4$
23. $y < 2x^2 + 2x - 2$
24. $y > -(x + 4)^2$
25. $y < (x - 4)^2 - 1$
26. $y \ge 2x^2 + 6x + 9$
27. $y < x^2 + 4x + 1$

13.

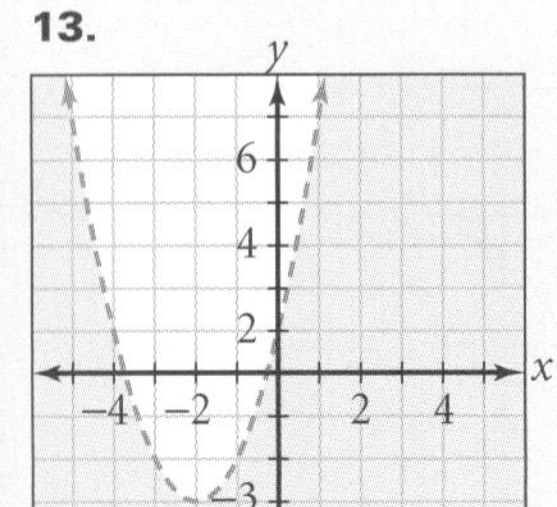

A

14.

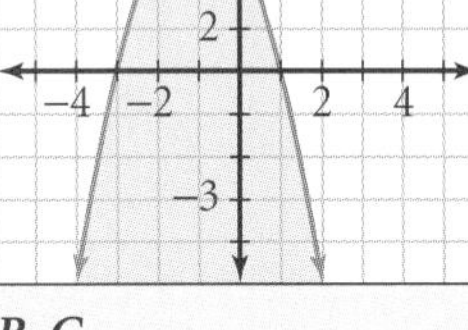
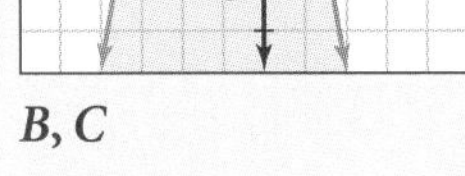

B, C

15.

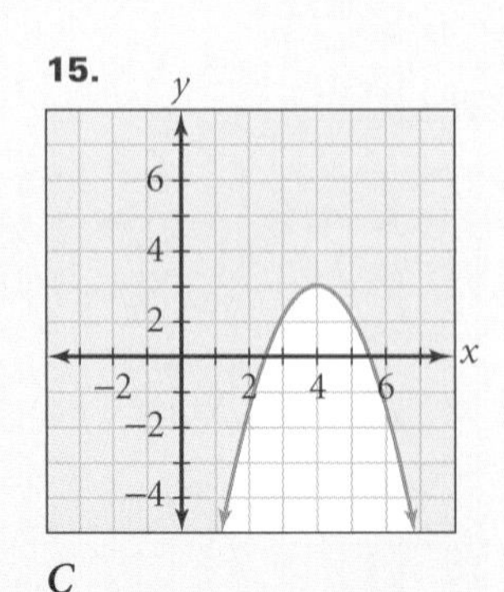

C

16.

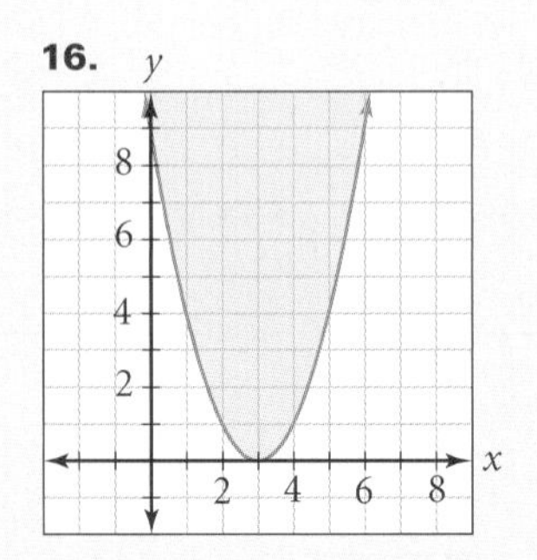

Chapter 6

LESSON 6.1

Find the multiplier for each rate of exponential growth or decay.

1. 3% growth **1.03**
2. 2.4% growth **1.024**
3. 10% decay **0.9**
4. 4% decay **0.96**
5. 0.7% growth **1.007**
6. 1.4% growth **1.014**
7. 18% growth **1.18**
8. 9% decay **0.91**
9. 0.04% growth **1.0004**
10. 7.2% decay **0.928**
11. 1.15% growth **1.0115**
12. 0.7% decay **0.993**

Evaluate each expression to the nearest thousandth for the given value of *x*.

13. 2^x for $x = 1.5$ **2.828**
14. $40(2)^{2x}$ for $x = 2.5$ **1280**
15. $30(0.5)^x$ for $x = 3$ **3.750**
16. $20(2)^{x+1}$ for $x = 0.5$ **56.569**
17. 2^{3x} for $x = 0.8$ **5.278**
18. $50(0.5)^{2x}$ for $x = 0.75$ **17.678**
19. $3(0.5)^x$ for $x = 5$ **0.094**
20. $10(2)^x$ for $x = 2.4$ **52.780**
21. A physician gives a patient 250 milligrams of an antibiotic that is eliminated from the bloodstream at a rate of 15% per hour. Predict the number of milligrams remaining after 3 hours. **153.53 mg**
22. A lab sample contains 400 bacteria that double every 15 minutes. Predict the number of bacteria after 3 hours. **1,638,400**
23. The population of a city was approximately 450,000 in the year 2000 and was projected to grow at an annual rate of 2.3%. Predict the population, to the nearest ten thousand, for the year 2006. **520,000**

LESSON 6.2

Identify each function as linear, quadratic, or exponential.

1. $f(x) = x^2 - 12$ **quadratic**
2. $g(x) = 2x - 12$ **linear**
3. $h(x) = (x + 3)^2$ **quadratic**
4. $k(x) = \left(\frac{5}{4}\right)^{2x}$ **exponential**
5. $p(x) = 3x + 2^2$ **linear**
6. $q(x) = 3^{x+2}$ **exponential**

Tell whether each function represents exponential growth or decay.

7. $b(x) = 40(3.8)^x$ **growth**
8. $f(x) = 100(0.18)^x$ **decay**
9. $g(x) = \left(\frac{1}{5}\right)^x$ **decay**
10. $w(x) = 3.5(1.01)^x$ **growth**
11. $z(x) = 0.4^x$ **decay**
12. $m(x) = 450(2.04)^x$ **growth**
13. $k(x) = 500(0.99)^x$ **decay**
14. $h(x) = 20(1.75)^x$ **growth**
15. $f(x) = 17(4)^{-x}$ **decay**

Find the final amount for each investment.

16. $1200 earning 5% interest compounded annually for 10 years **$1954.67**
17. $900 earning 6% interest compounded annually for 15 years **$2156.90**
18. $5000 earning 6.5% interest compounded semiannually for 12 years **$10,772.87**
19. $500 earning 5.5% interest compounded semiannually for 3 years **$588.38**
20. $8000 earning 8% interest compounded quarterly for 5 years **$11,887.58**
21. $600 earning 7.5% interest compounded quarterly for 2 years **$696.13**
22. $10,000 earning 8% interest compounded daily for 1 year **$10,832.78**
23. $4000 earning 5.25% interest compounded daily for 2 years **$4442.81**

17.

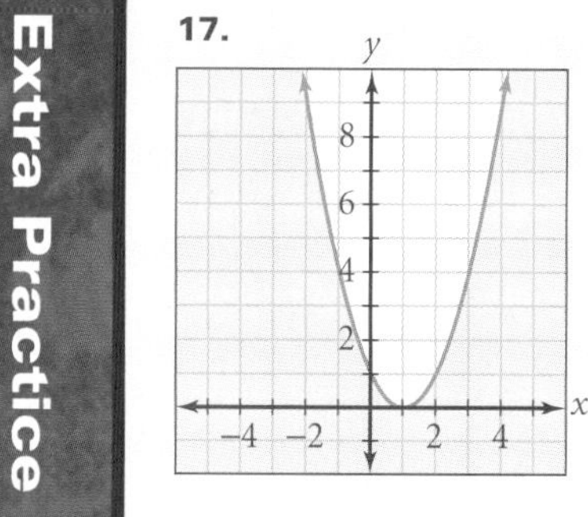

18.

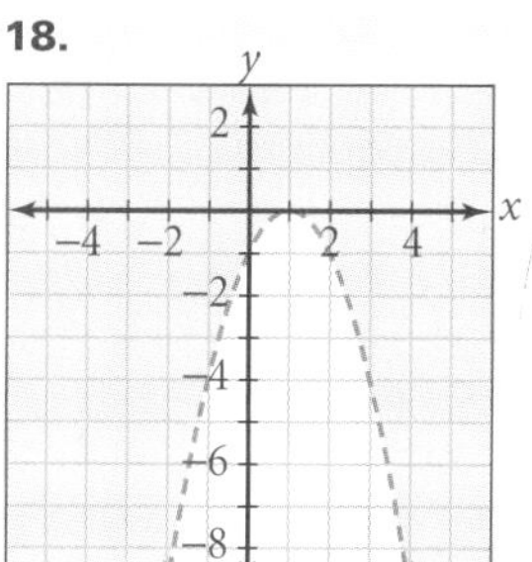

19.

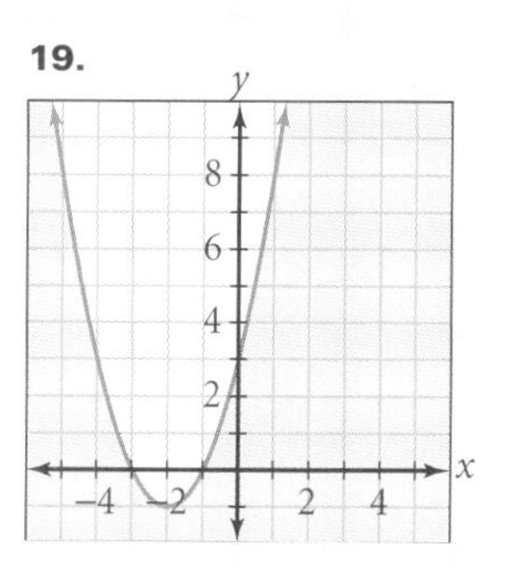

20.

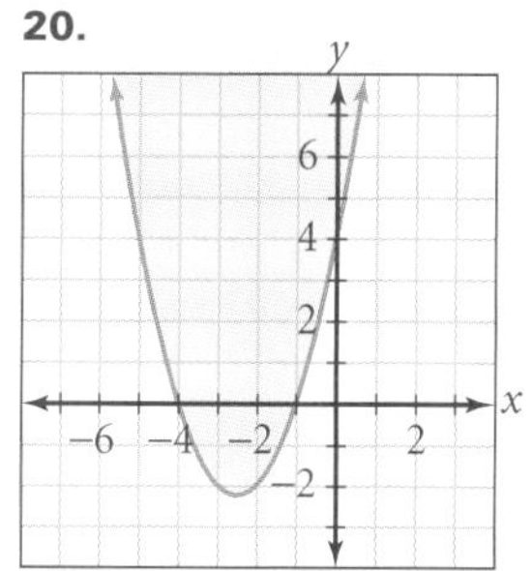

21.

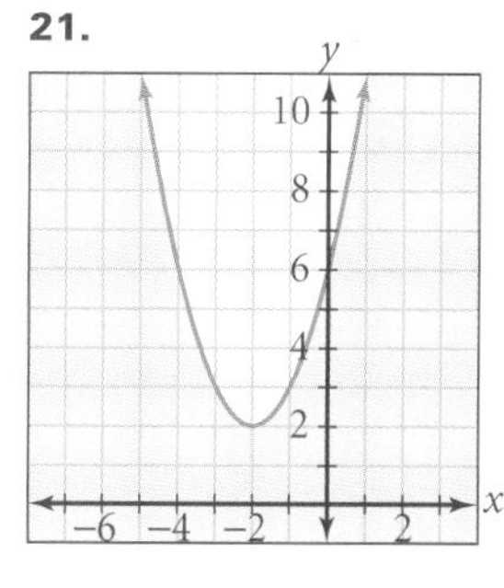

22.

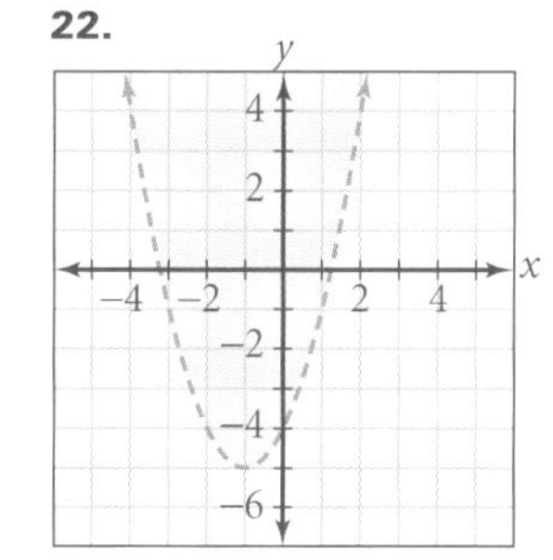

23.

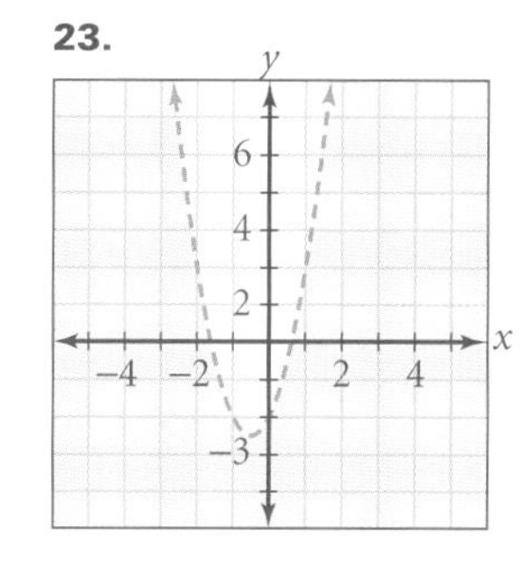

24.

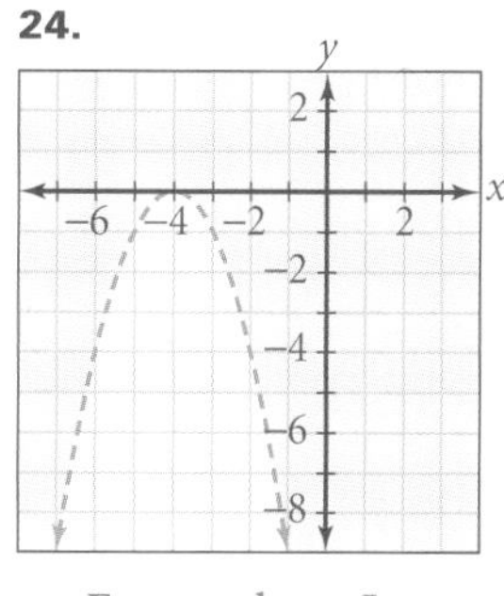

25.

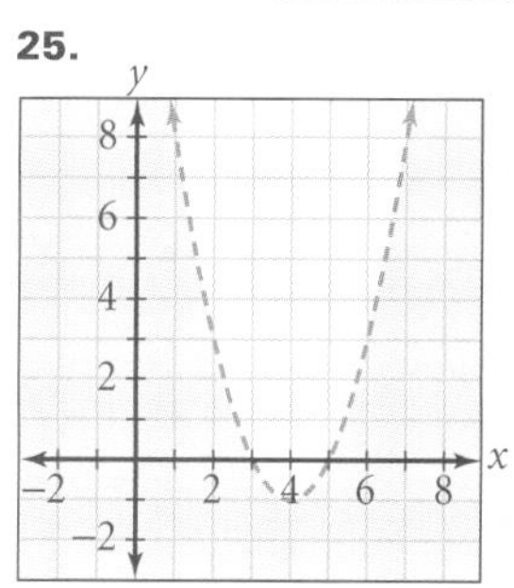

For graphs to Lesson 5.8 Exercises 26 and 27, see page 960.

Extra Practice

Extra Practice

LESSON 6.3

Write each equation in logarithmic form.

1. $3^4 = 81$ $\log_3 81 = 4$
2. $4^3 = 64$ $\log_4 64 = 3$
3. $\left(\frac{1}{2}\right)^7 = \frac{1}{128}$ $\log_{\left(\frac{1}{2}\right)} \frac{1}{128} = 7$
4. $\left(\frac{1}{4}\right)^3 = \frac{1}{64}$ $\log_{\left(\frac{1}{4}\right)} \frac{1}{64} = 3$
5. $\left(\frac{1}{3}\right)^{-4} = 81$ $\log_{\left(\frac{1}{3}\right)} 81 = -4$
6. $\left(\frac{1}{15}\right)^{-2} = 225$ $\log_{\left(\frac{1}{15}\right)} 225 = -2$
7. $5^{-3} = \frac{1}{125}$ $\log_5 \frac{1}{125} = -3$
8. $10^{-2} = 0.01$ $\log_{10} 0.01 = -2$
9. $9^{-2} = \frac{1}{81}$ $\log_9 \frac{1}{81} = -2$

Write each equation in exponential form.

10. $\log_{14} 196 = 2$ $14^2 = 196$
11. $\log_7 2401 = 4$ $7^4 = 2401$
12. $\log_8 \frac{1}{512} = -3$ $8^{-3} = \frac{1}{512}$
13. $\log_6 \frac{1}{1296} = -4$ $6^{-4} = \frac{1}{1296}$
14. $\log_3 81 = 4$ $3^4 = 81$
15. $\log_2 256 = 8$ $2^8 = 256$
16. $\log_{17} 289 = 2$ $17^2 = 289$
17. $\log_{10} 0.0001 = -4$ $10^{-4} = 0.0001$
18. $\log_{10} 10{,}000 = 4$ $10^4 = 10{,}000$

Solve each equation for *x*. Round your answers to the nearest hundredth.

19. $10^x = 15$ ≈ 1.18
20. $10^x = 72$ ≈ 1.86
21. $10^x = 4.5$ ≈ 0.65
22. $10^x = 7.8$ ≈ 0.89
23. $10^x = 1042$ ≈ 3.02
24. $10^x = 2509$ ≈ 3.40
25. $10^x = 0.835$ ≈ -0.08
26. $10^x = 0.007$ ≈ -2.15
27. $10^x = 14.2$ ≈ 1.15

Find the value of *v* in each equation.

28. $v = \log_4 1024$ **5**
29. $v = \log_{13} 1$ **0**
30. $\log_6 \frac{1}{36} = v$ **−2**
31. $4 = \log_5 v$ **625**
32. $\log_4 v = -3$ $\frac{1}{64}$
33. $-6 = \log_2 v$ $\frac{1}{64}$
34. $-3 = \log_v \frac{1}{27}$ **3**
35. $\log_v \frac{1}{625} = -4$ **5**
36. $7 = \log_v 128$ **2**

LESSON 6.4

Write each expression as a sum or difference of logarithms. Then simplify, if possible.

1. $\log_3 9x$ $2 + \log_3 x$
2. $\log_3 27x$ $3 + \log_3 x$
3. $\log_4(2 \cdot 3 \cdot 4)$ $1.5 + \log_4 3$
4. $\log_2 \frac{16}{y}$ $4 - \log_2 y$
5. $\log_5 \frac{4}{5}$ $\log_5 4 - 1$
6. $\log_{10} \frac{xy}{10}$ $\log_{10} x + \log_{10} y - 1$

Write each expression as a single logarithm. Then simplify, if possible.

7. $\log_2 3 + \log_2 7$ $\log_2 21$
8. $\log_6 12 + \log_6 15 - \log_6 5$ $\log_6 36 = 2$
9. $2 \log_4 5 - \log_4 6$ $\log_4 \frac{25}{6}$
10. $\log_9 x - 3 \log_9 y$ $\log_9 \frac{x}{y^3}$
11. $3 \log_5 3 - \log_5 5.4$ $\log_5 5 = 1$
12. $\frac{1}{2} \log_b 25 + 3 \log_b z$ $\log_b 5z^3$

Evaluate each expression.

13. $\log_3 3^4 - \log_8 8^4$ **0**
14. $\log_7 7^5 + \log_6 6^3$ **8**
15. $4^{\log_4 87} + \log_5 25$ **89**
16. $8^{\log_8 9} - \log_4 16$ **7**
17. $\log_3 \frac{1}{81} + \log_4 64$ **−1**
18. $\log_2 64 - 7^{\log_7 1}$ **5**

Solve for *x*, and check your answers. If the equation has no solution, write *no solution*.

19. $\log_8(x + 1) = \log_8(2x - 2)$ **3**
20. $\log_3(3x - 4) = \log_3(8 - 5x)$ **1.5**
21. $\log_7(6x + 4) = \log_7(-3x - 5)$ **no sol.**
22. $\log_{10}(6x + 3) = \log_{10} 3x$ **no solution**
23. $\log_2 x + \log_2(x - 4) = 5$ **8**
24. $\log_8(3x + 1) + \log_8(x - 1) = 2$ **5**
25. $2 \log_b x = \log_b 2 + \log_b(2x - 2)$ **2**
26. $2 \log_b x = \log_b(x - 1) + \log_b 4$ **2**

26.

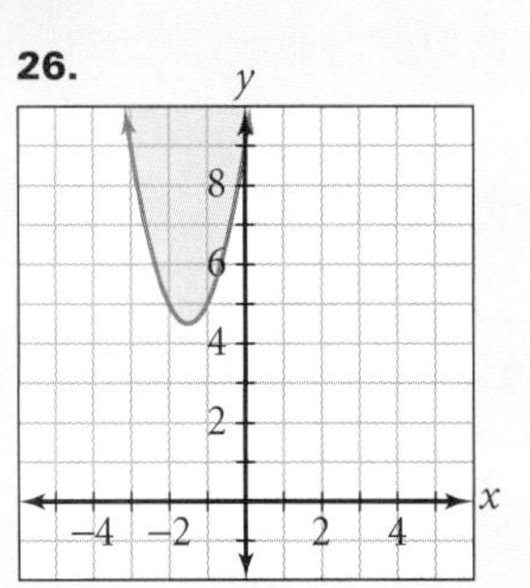

27.

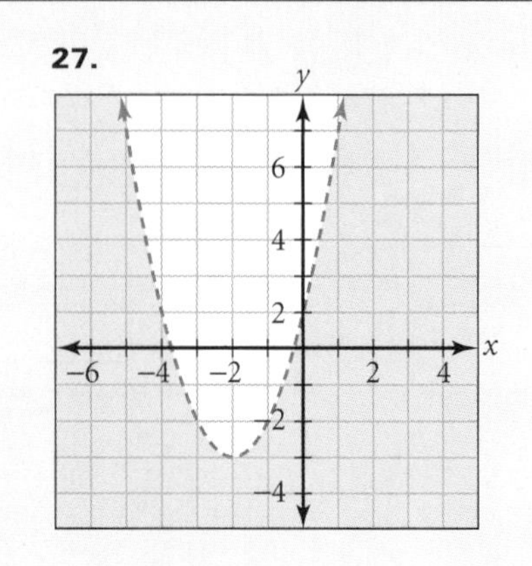

Extra Practice

LESSON 6.5

Evaluate each logarithmic expression to the nearest hundredth.

1. $\log_2 51$ **5.67**
2. $\log_5 64$ **2.58**
3. $\log_6 0.5$ **−0.39**
4. $\log_4 9$ **1.58**
5. $\log_9 14$ **1.20**
6. $\log_7 32$ **1.78**
7. $\log_8 0.23$ **−0.71**
8. $\log_{\frac{1}{2}} 15$ **−3.91**
9. $\log_2 0.72$ **−0.47**
10. $\log_{\frac{1}{4}} 16$ **−2**
11. $2 - \log_5 7$ **0.79**
12. $\log_9 10$ **1.05**
13. $\log_6 \frac{2}{3}$ **−0.23**
14. $\log_8 50$ **1.88**
15. $3 + \log_3 22$ **5.81**
16. $\log_7 \frac{3}{4}$ **−0.15**
17. $\log_8 \frac{1}{3}$ **−0.53**
18. $\log_7 8$ **1.07**
19. $10 + \log_4 25$ **12.32**
20. $\log_{15} 40$ **1.36**
21. $\log_9 \frac{3}{4}$ **−0.13**

Solve each equation. Round your answers to the nearest hundredth.

22. $5^x = 24$ **≈1.97**
23. $6^x = 44$ **≈2.11**
24. $8^x = 0.9$ **≈−0.05**
25. $2^x = 3.5$ **≈1.81**
26. $9^x = 17$ **≈1.29**
27. $3^x = 41$ **≈3.38**
28. $8^{-x} = 0.25$ **≈0.67**
29. $4^x = 22$ **≈2.23**
30. $9^x = 2$ **≈0.32**
31. $2.5^x = 17$ **≈3.09**
32. $7^x = 3$ **≈0.56**
33. $12^x = 140$ **≈1.99**
34. $1 + 3^x = 14$ **≈2.33**
35. $3^{-x} = 0.9$ **≈0.10**
36. $4^{x+1} = 64$ **2**
37. $5^{2x} = 114$ **≈1.47**
38. $7 - 2^x = 1$ **≈2.58**
39. $4 + 4^x = 14$ **≈1.66**
40. $5^{x-2} = 70$ **≈4.64**
41. $5^x = 20.5$ **≈1.88**
42. $7^x = 22$ **≈1.59**

LESSON 6.6

Evaluate each expression, if possible, to the nearest thousandth.

1. e^3 **20.086**
2. e^{-2} **0.135**
3. $e^{4.5}$ **90.017**
4. $e^{0.6}$ **1.822**
5. $e^{\sqrt{3}}$ **5.652**
6. $\ln 17$ **2.833**
7. $\ln \sqrt{7}$ **0.973**
8. $\ln 45$ **3.807**
9. $\ln(-12)$ **not defined**
10. $\ln(-5)$ **not defined**
11. $\ln 0.8$ **−0.223**
12. $\ln \sqrt{3}$ **0.549**

Write an equivalent logarithmic or exponential equation.

13. $e^{3.22} \approx 25.03$ **$\ln 25.03 \approx 3.22$**
14. $e^5 \approx 148.41$ **$\ln 148.41 \approx 5$**
15. $\ln 50 \approx 3.91$ **$e^{3.91} \approx 50$**
16. $\ln 3.6 \approx 1.28$ **$e^{1.28} \approx 3.6$**
17. $e^{3.4} \approx 29.96$ **$\ln 29.96 \approx 3.4$**
18. $\ln 5 \approx 1.61$ **$e^{1.61} \approx 5$**
19. $\ln 25 \approx 3.22$ **$e^{3.22} \approx 25$**
20. $e^{\frac{1}{2}} \approx 1.65$ **$\ln 1.65 \approx \frac{1}{2}$**
21. $e^{\frac{2}{3}} \approx 1.95$ **$\ln 1.95 \approx \frac{2}{3}$**
22. $e^{-7} \approx 0.000912$ **$\ln 0.000912 \approx -7$**
23. $\ln\left(\frac{1}{4}\right) \approx -1.39$ **$e^{-1.39} \approx \frac{1}{4}$**
24. $\ln\left(\frac{3}{4}\right) \approx -0.29$ **$e^{-0.29} \approx \frac{3}{4}$**

Solve each equation for *x* by using the natural logarithm function. Round your answers to the nearest hundredth.

25. $15^x = 27$ **≈1.22**
26. $4.2^x = 15$ **≈1.89**
27. $7^{-x} = 120$ **≈−2.46**
28. $0.5^x = 11$ **≈−3.46**
29. $8^{\frac{-x}{2}} = 21$ **≈−2.93**
30. $9^{-x} = 0.2$ **≈0.73**
31. $\left(\frac{1}{3}\right)^{-2x} = 125$ **≈2.20**
32. $\left(\frac{2}{3}\right)^x = 12$ **≈−6.13**
33. $1.5^{3x} = 1500$ **≈6.01**
34. $2.3^x = 15$ **≈3.25**
35. $7^x = 14{,}000$ **≈4.91**
36. $11^{-2x} = 15{,}000$ **≈−2.01**

37. An investor puts \$5000 in an account that earns 6.5% annual interest which is compounded continuously. Find the amount that will be in the account at the end of 5 years if no deposits or withdrawals are made. **\$6920.15**

Lesson 7.1

1. yes; cubic trinomial
2. yes; cubic binomial
3. no
4. no
5. yes; quartic binomial
6. yes; seventh-degree trinomial

LESSON 6.7

Solve each equation for *x*. Write the exact solution and the approximate solution to the nearest hundredth, when appropriate.

1. $2^x = 2^5$ **5**
2. $\log_6 216 = x$ **3**
3. $3^{x-1} = 3^4$ **5**
4. $\log x = 2.1$ **$10^{2.1} \approx 125.89$**
5. $x = \log_3 27$ **3**
6. $\log_9 x = 2$ **81**
7. $5 = \log_x 32$ **2**
8. $\log_x \frac{1}{4} = -1$ **4**
9. $3^x = 4$ **$\frac{\ln 4}{\ln 3} \approx 1.26$**
10. $\log_4(x - 2) = 2$ **18**
11. $10^{x-1} = 121$ **$1 + \log 121 \approx 3.08$**
12. $e^{x+1} = 14$ **$-1 + \ln 14 \approx 1.64$**
13. $e^{2x-1} = 9$ **$\frac{1}{2}(1 + \ln 9) \approx 1.60$**
14. $\ln(x + 1) = \ln 7$ **6**
15. $\ln(x - 5) = \ln(3x + 1)$ **no solution**
16. $2 \ln\left(x + \frac{1}{2}\right) = \ln \frac{1}{4}$ **0**
17. $e^{-3x+4} = 22$ **$\frac{1}{3}(4 - \ln 22) \approx 0.30$**
18. $2 \ln x = \ln(2x - 1)$ **1**

In Exercises 19 and 20, use the equation $M = \frac{2}{3} \log \frac{E}{10^{11.8}}$.

19. On October 15, 1997, an earthquake with a magnitude of 6.8 struck parts of Chile and Argentina. Find the amount of energy released by the earthquake. **10^{22} ergs**
20. From December 1811 to early 1812, a series of earthquakes shook the Mississippi Valley near New Madrid, Missouri. One of the earthquakes released about 2.5×10^{24} ergs of energy. Find the earthquake's magnitude on the Richter scale. Round your answer to the nearest tenth. **8.4**

Chapter 7

LESSON 7.1

Determine whether each expression is a polynomial. If so, classify the polynomial by degree and by number of terms.

1. $12x^3 - 2x^2 + \frac{1}{2}$
2. $\frac{x^3}{11} + \frac{x^2}{8}$
3. $\frac{11}{x^3} + \frac{8}{x^2}$
4. $4x - 2^{2x} + 3^{5x}$
5. $0.66x^4 - 1$
6. $10x^3 + 6x^7 - 15x$

Evaluate each polynomial expression for the indicated value of *x*.

7. $x^3 - 3x^2 + 4x$ for $x = -2$ **−28**
8. $4 - 2x + 3x^2 - x^4$ for $x = -1$ **8**
9. $\frac{3}{4}x^4 + \frac{1}{2}x^2 - \frac{1}{4}x + \frac{1}{2}$ for $x = 2$ **14**
10. $-x^4 - x^3 - x^2 + 12$ for $x = 5$ **−763**
11. $0.5x^4 + 2.5x^3 - x^2$ for $x = 4$ **272**
12. $x^4 - 3x^3 + 3x^2 - 9$ for $x = 3$ **18**

Write each sum or difference as a polynomial expression in standard form. Then classify the polynomial by degree and by number of terms.

13. $(-2x^3 + 5x^2 - 3x + 7) + (5x^3 + x^2 + 9)$ **$3x^3 + 6x^2 - 3x + 16$; cubic polynomial**
14. $(7x^4 - 3x^3 + 5x) - (2x^4 + x^3 + x^2 + 3x - 2)$ **$5x^4 - 4x^3 - x^2 + 2x + 2$; quartic polynomial**
15. $(4.1x^3 + 3.5x - 6x^2 - 11) - (3x^2 - 4x^3 + 9)$ **$8.1x^3 - 9x^2 + 3.5x - 20$; cubic polynomial**
16. $(5x^3 - 2x^4 + 3x - 6) + (x^4 - 7x^2 + 3x - 9)$ **$-x^4 + 5x^3 - 7x^2 + 6x - 15$; quartic polynomial**
17. $(3x^5 - 4x^2 + 2x^3) - (4x^4 + 3x^3 - 9x^2 - 7)$ **$3x^5 - 4x^4 - x^3 + 5x^2 + 7$; quintic polynomial**
18. $(7.5x^3 + 3.2x^4 + 5.1x^2 + x) + (x^4 - 7x^2 - 3x)$ **$4.2x^4 + 7.5x^3 - 1.9x^2 - 2x$; quartic polynomial**

Graph each function. Describe the general shape of the graph.

19. $f(x) = x^3 - x^2 - x + 1$ **S-shape**
20. $g(x) = -2x^3 + 3x - 1$ **S-shape**
21. $k(x) = x^4 - x^3 - 3x^2 + 1$ **W-shape**
22. $h(x) = 0.5x^4 - 3x^2$ **W-shape**

LESSON 7.2

Graph each function. Approximate any local maxima or minima to the nearest tenth.

1 $P(x) = 10x - 8x^2$ **max: 3.1**

2 $P(x) = x^2 - 3x - 2$ **min: -4.3**

3 $P(x) = x^4 - x^3 - x^2$ **max: 0; min: -1.1, -0.1**

4 $P(x) = 5x - x^3 + 1$ **max: 5.3; min: -3.3**

Graph each function. Approximate any local maxima or minima to the nearest tenth. Find the intervals over which the function is increasing and decreasing.

5 $P(x) = x^3 + x^2; -6 \leq x \leq 6$

6 $P(x) = 0.5x^4 + x^2 - 3; -5 \leq x \leq 5$

7 $P(x) = -2x^4 + 3x^3 + 2x^2; -4 \leq x \leq 4$

8 $P(x) = x^3 - 4x^2 + 1; -5 \leq x \leq 5$

9 $P(x) = 0.2x^3 - 6x + 4; -6 \leq x \leq 6$

10 $P(x) = 3x^4 - 8x^2 - 2; -2 \leq x \leq 2$

Describe the end behavior of each function.

11. $P(x) = 17x^2 - 8x^3 - 6$

12. $P(x) = 9 - 2x - x^2 + 5x^3$

13. $P(x) = 4x^4 - 6x^3 + 2x^2 - x$

14. $P(x) = -8 + x^3 - 5x^4$

11. rises on left, falls on right
12. falls on left, rises on right
13. rises on left and right
14. falls on left and right

15 The number (in thousands) of federal employees in the United States is given in the table below. Find a quartic regression model for the data by using $x = 0$ for 1982. [*Source: U.S. Bureau of the Census*]

1982	1983	1984	1985	1986	1987	1988	1989
15,841	16,034	16,436	16,690	16,933	17,212	17,588	18,369

$f(x) = 3.83x^4 - 44.15x^3 + 156.86x^2 + 112.93x + 15{,}833.87$

LESSON 7.3

Write each product as a polynomial in standard form.

1. $2x^3(-5x^4 + 3x^3 - 2x - 6)$

2. $(4x - 7)(3x + 4)$

3. $(x - 6)(x^2 + 3x - 5)$

4. $(2x - 3)(x + 4)^2$

5. $(2x - 1)^3$

6. $(x + 7)(2x^2 - 3x - 4)$

Use substitution to determine whether the given linear expression is a factor of the polynomial.

7. $2x^3 + 7x^2 - 15x; x + 5$ **yes**

8. $x^3 - 4x^2 - 20x - 7; x - 7$ **yes**

9. $2x^3 + 15x^2 - 9x - 10; x + 8$ **no**

Divide by using long division.

10. $(2x^3 + 3x^2 - 6x - 3) \div (2x - 3)$ **10.** $x^2 + 3x + 1.5 + \frac{1.5}{2x - 3}$

11. $(x^3 + 3x + 4) \div (x + 1)$ $x^2 - x + 4$

12. $(x^2 - 27x + x^3 + 28) \div (x - 4)$ $x^2 + 5x - 7$

13. $\left(\frac{1}{2}x^2 - 5x + 3x^3 + 2\right) \div \left(x - \frac{1}{2}\right)$ $3x^2 + 2x - 4$

Divide by using synthetic division.

14. $(x^2 + 9x - 36) \div (x - 3)$ $x + 12$

15. $(x^3 - 5x^2 - 2x + 24) \div (x - 3)$ $x^2 - 2x - 8$

16. $(x^3 + 8) \div (x - 4)$ $x^2 + 4x + 16 + \frac{72}{x - 4}$

17. $(9x - 17x^2 - 9 + 5x^3) \div (x - 3)$ $5x^2 - 2x + 3$

For each function, use both synthetic division and substitution to find the indicated value.

18. $P(x) = x^4 + 3x^3 - 2x + 1; P(2)$ **37**

19. $P(x) = x^3 - 8; P(-2)$ **-16**

20. $P(x) = 5x^2 - 4x + 3; P(-1)$ **12**

21. $P(x) = -3x^3 + 4 - x^2; P(3)$ **-86**

Lesson 7.2

5. local maximum of 0.1; local minimum of 0; increasing for $-6 < x < -0.7$ and $0 < x < 6$, decreasing for $-0.7 < x < 0$

6. local minimum of -3; decreasing for $-5 < x < 0$ and increasing for $0 < x < 5$

7. local maxima of 0.1 and 4.5; local minimum of 0; increasing for $-4 < x < -0.3$ and $0 < x < 1.5$, decreasing for $-0.3 < x < 0$ and $1.5 < x < 4$

8. local maximum of 1; local minimum of -8.5; increasing for $-5 < x < 0$ and $2.7 < x < 5$, decreasing for $0 < x < 2.7$

9. local maximum of 16.6; local minimum of -8.6; increasing for $-6 < x < -3.2$ and $3.2 < x < 6$, decreasing for $-3.2 < x < 3.2$

10. local maximum of -2; local minima of -7.3 and 7.3; increasing for $-1.2 < x < 0$ and $1.2 < x < 2$, decreasing for $-2 < x < -1.2$ and $0 < x < 1.2$

Lesson 7.3

1. $-10x^7 + 6x^6 - 4x^4 - 12x^3$

2. $12x^2 - 5x - 28$

3. $x^3 - 3x^2 - 23x + 30$

4. $2x^3 + 13x^2 + 8x - 48$

5. $8x^3 - 12x^2 + 6x - 1$

6. $2x^3 + 11x^2 - 25x - 28$

Extra Practice

LESSON 7.4

Use factoring to solve each equation.

1. $x^3 + 3x^2 - 10x = 0$ **0, –5, 2**
2. $x^3 - 64x = 0$ **–8, 0, 8**
3. $x^3 + 49x = 14x^2$ **0, 7 (mult. 2)**
4. $2x^3 - 22x^2 + 56x = 0$ **0, 4, 7**
5. $3x^3 + 3x^2 = 6x$ **–2, 0, 1**
6. $x^3 - 77x + 4x^2 = 0$ **1, 7, –11**

Use a graph, synthetic division, and factoring to find all of the roots of each equation.

7. $x^3 - 2x^2 - 4x + 8 = 0$ **–2, 2 (mult. 2)**
8. $x^3 - 3x^2 - x + 3 = 0$ **–1, 1, 3**
9. $x^3 - 2x^2 - 13x - 10 = 0$ **–1, –2, 5**
10. $x^3 + 5x^2 = x + 5$ **–5, –1, 1**
11. $x^3 + x + 6 = 4x^2$ **–1, 2, 3**
12. $x^3 - 9x^2 + 15x - 7 = 0$ **1 (mult. 2), 7**

Use variable substitution and factoring to find all of the roots of each equation.

13. $x^4 - 14x^2 + 45 = 0$ **$\pm 3, \pm\sqrt{5}$**
14. $x^4 - 16x^2 + 15 = 0$ **$\pm 1, \pm\sqrt{15}$**
15. $x^4 + 12 = 13x^2$ **$\pm 1, \pm 2\sqrt{3}$**
16. $x^4 - 25x^2 + 144 = 0$ **$\pm 4, \pm 3$**
17. $x^4 + 33 = 14x^2$ **$\pm\sqrt{11}, \pm\sqrt{3}$**
18. $x^4 - 8x^2 + 7 = 0$ **$\pm 1, \pm\sqrt{7}$**

Use a graph and the Location Principle to find the real zeros of each function. Give approximate values to the nearest hundredth, if necessary.

19. $g(x) = x^3 - 5x^2 + 7x$ **0**
20. $b(x) = 0.8x^4 - 2x^2 + 1$ **±1.35, ±0.83**
21. $f(x) = x^3 - 4x + 2$ **–2.21, 0.54, 1.68**

LESSON 7.5

Find all of the rational roots of each polynomial equation.

1. $3x^2 - 14x + 8 = 0$ **$\frac{2}{3}, 4$**
2. $2x^2 - 5x - 3 = 0$ **$-\frac{1}{2}, 3$**
3. $10x^3 - 31x^2 + 25x - 6 = 0$ **$\frac{1}{2}, \frac{3}{5}, 2$**
4. $2x^3 - 9x^2 + 7x + 6 = 0$ **$-\frac{1}{2}, 2, 3$**
5. $4x^3 - 9x^2 - x + 6 = 0$ **$-\frac{3}{4}, 1, 2$**
6. $15x^3 - 47x^2 + 38x - 8 = 0$ **$\frac{1}{3}, \frac{4}{5}, 2$**

Find all zeros of each polynomial function.

7. $M(x) = x^3 - x^2 - 7x + 3$ **$-1 \pm \sqrt{2}, 3$**
8. $H(x) = x^3 - 3x^2 - 5x + 15$ **$\pm\sqrt{5}, 3$**
9. $J(x) = x^3 - 2x^2 + 7x - 14$ **$2, \pm i\sqrt{7}$**
10. $R(x) = x^4 - 3x^3 - x^2 - 9x - 12$ **$-1, 4, \pm i\sqrt{3}$**
11. $F(x) = x^4 - 5x^2 - 24$ **$\pm 2\sqrt{2}, \pm i\sqrt{3}$**
12. $H(x) = x^4 + 5x^3 + x^2 - 20x - 20$ **$\pm 2, -\frac{5}{2} \pm \frac{\sqrt{5}}{2}$**

Find all real values of *x* for which the functions are equal. Give your answers to the nearest hundredth.

13. $P(x) = x^4 + 3x^2 + 2x$ and $Q(x) = x + 2$ **–0.87, 0.63**
14. $P(x) = -x^2 + 6x - 2$ and $Q(x) = 0.2x^4 + x^3 - 3$ **–0.16, 1.85**
15. $P(x) = x^4 - 3x^2$ and $Q(x) = x^2 + 3x - 5$ **0.86, 2.07**
16. $P(x) = x^4 - x^3 + 1$ and $Q(x) = x^2 - 2x + 3$ **±1.41**

Write a polynomial function, *P*, in factored form and in standard form by using the given information.

17. P is of degree 3; $P(0) = 2$; zeros: $-1, 1, 2$
18. P is of degree 4; $P(0) = 6$; zeros: $2, -3, \frac{1}{2}, -\frac{1}{2}$
19. P is of degree 4; $P(0) = -9$; zeros: $\frac{1}{2}$ (multiplicity 2), $\frac{3}{2}, -\frac{3}{2}$
20. P is of degree 3; $P(0) = -1$; zeros: $\frac{1}{2}, \frac{1}{3}$ (multiplicity 2)
21. P is of degree 3; $P(0) = -36$; zeros: $2, 3i$
22. P is of degree 4; $P(0) = 24$; zeros: $2, 3, i$

17. $P(x) = (x+1)(x-1)(x-2) = x^3 - 2x^2 - x + 2$

18. $P(x) = 4(x-2)(x+3)\left(x-\frac{1}{2}\right)\left(x+\frac{1}{2}\right) = 4x^4 + 4x^3 - 25x^2 - x + 6$

19. $P(x) = 16\left(x-\frac{1}{2}\right)^2\left(x-\frac{3}{2}\right)\left(x+\frac{3}{2}\right) = 16x^4 - 16x^3 - 32x^2 + 36x - 9$

20. $P(x) = 18\left(x-\frac{1}{2}\right)\left(x-\frac{1}{3}\right)^2 = 18x^3 - 21x^2 + 8x - 1$

21. $P(x) = 2(x-2)(x-3i)(x+3i) = 2x^3 - 4x^2 + 18x - 36$

22. $P(x) = 4(x-2)(x-3)(x+i)(x-i) = 4x^4 - 20x^3 + 28x^2 - 20x + 24$

Chapter 8

LESSON 8.1

For Exercises 1–4, y varies inversely as x. Write the appropriate inverse-variation equation, and find y for the given values of x.

1. $y = 18$ when $x = 8$; $x = 4, 9, 15,$ and 20 **$y = \frac{144}{x}$; 36, 16, 9.6, 7.2**
2. $y = 7.5$ when $x = 8$; $x = 4, 5, 6,$ and 18 **$y = \frac{60}{x}$; 15, 12, 10, $3\frac{1}{3}$**
3. $y = 0.32$ when $x = 5$; $x = 0.1, 0.2, 4,$ and 8 **$y = \frac{1.6}{x}$; 16, 8, 0.4, 0.2**
4. $y = 9.5$ when $x = 4$; $x = 1.9, 5, 6,$ and 20 **$y = \frac{38}{x}$; 20, 7.6, $6\frac{1}{3}$, 1.9**

For Exercises 5–8, y varies jointly as x and z. Write the appropriate joint-variation equation, and find y for the given values of x and z.

5. $y = 18$ when $x = 4$ and $z = 3$; $x = 4.5$ and $z = 6$ **$y = 1.5xz$; 40.5**
6. $y = 375$ when $x = 6$ and $z = 5$; $x = 0.4$ and $z = 30$ **$y = 12.5xz$; 150**
7. $y = 24$ when $x = 2$ and $z = -4$; $x = -3$ and $z = 0.8$ **$y = -3xz$; 7.2**
8. $y = 1.5$ when $x = 1.5$ and $z = 6$; $x = 8$ and $z = 0.4$ **$y = \frac{1}{6}xz$; ≈0.53**

For Exercises 9–12, z varies jointly as x and y and inversely as w. Write the appropriate inverse-variation equation, and find z for the given values of x, y, and w.

9. $z = 82.5$ when $x = 12$, $y = 5$, and $w = 4$; $x = 8$, $y = 4.5$, and $w = 10$ **$z = \frac{5.5xy}{w}$; 19.8**
10. $z = 19.2$ when $x = 20$, $y = 4.2$, and $w = 3.5$; $x = 16$, $y = 9$, and $w = 5$ **$z = \frac{0.8xy}{w}$; 23.04**
11. $z = 6$ when $x = 12$, $y = -2$, and $w = 5$; $x = 7$, $y = 0.2$, and $w = 14$ **$z = \frac{-1.25xy}{w}$; −0.125**
12. $z = 2.5$ when $x = -6$, $y = 5$, and $w = 4$; $x = 3$, $y = -0.6$, and $w = 2$ **$z = -\frac{xy}{3w}$; 0.3**
13. The time, t, that it takes to travel a given distance, d, varies inversely as r, the rate of speed. A certain trip can be made in 7.5 hours at a rate of 60 miles per hour. Find the constant of variation, and write an inverse-variation equation. Find t to the nearest tenth when r is 40, 45, 50, and 55. **450; $t = \frac{450}{r}$; 11.3, 10, 9, and 8.2 hours**

LESSON 8.2

Determine whether each function is a rational function. If so, find the domain. If not, explain.

1. $f(x) = \frac{x - 0.5}{(x+2)(x+5)}$ **yes; $x \neq -2, -5$**
2. $g(x) = \frac{2^{x+1}}{x^2 + 1}$ **no; 2^{x+1} not poly.**
3. $h(x) = \frac{x^2 + 3x + 1}{x^2 - 9}$ **yes; $x \neq -3, 3$**

Identify all asymptotes and holes in the graph of each rational function.

4. $b(x) = \frac{x+6}{2x-1}$ **$x = \frac{1}{2}$, $y = \frac{1}{2}$**
5. $m(x) = \frac{x-5}{x^2-3}$ **$x = \pm\sqrt{3}$, $y = 0$**
6. $k(x) = \frac{2x}{x^2 - x - 2}$ **$x = -1$, $x = 2$; $y = 0$**
7. $f(x) = \frac{2x^3 + 6x^2}{x^2 - x - 12}$ **$x = 4$; hole $(x = -3)$**
8. $f(x) = \frac{-x^3 + x^2}{x^2 + x - 2}$ **$x = -2$; hole $(x = 1)$**
9. $f(x) = \frac{x^2 + 2x - 15}{x - 3}$ **hole $(x = 3)$**
10. $f(x) = \frac{x^2 - 2x - 3}{x + 1}$ **hole $(x = -1)$**
11. $f(x) = \frac{x^2 - 4}{2x^2 - 5x + 2}$ **$x = \frac{1}{2}$, $y = \frac{1}{2}$; hole $(x = 2)$**
12. $f(x) = \frac{3x^2 + x - 4}{x^2 + 2x - 3}$ **$x = -3$, $y = 3$; hole $(x = 1)$**

Find the domain of each rational function. Identify all asymptotes and holes in the graph of each function. Then graph.

13. $h(x) = \frac{x+4}{x-4}$ **D: $x \neq 4$; $x = 4$, $y = 1$**
14. $f(x) = \frac{x}{3x(x-1)}$ **D: $x \neq 0, 1$; $x = 1$, $y = 0$; hole $(x = 0)$**
15. $g(x) = \frac{x+3}{x^2 + 8x + 15}$ **D: $x \neq -3, -5$; $x = -5$, $y = 0$; hole $(x = -3)$**
16. $f(x) = \frac{1}{x+2}$ **D: $x \neq -2$; $x = -2$, $y = 0$**
17. $b(x) = \frac{x-1}{x+2}$ **D: $x \neq -2$; $x = -2$, $y = 1$**
18. $d(x) = \frac{x-1}{x^2 - 5x + 4}$ **D: $x \neq 4, 1$; $x = 4$, $y = 0$; hole $(x = 1)$**

Lesson 8.2

13. vertical asymptote: $x = 2$; horizontal asymptote: $y = 3$; hole when $x = -\frac{1}{2}$

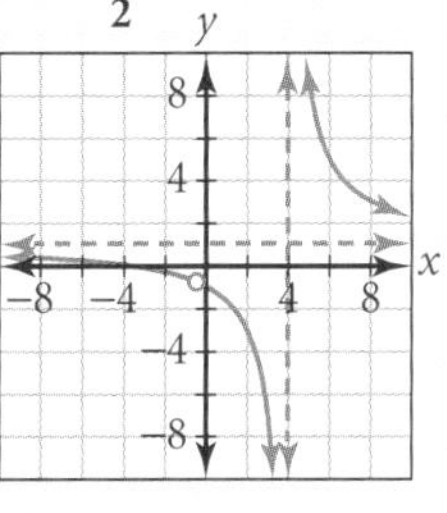

14. vertical asymptote: $x = 1$; horizontal asymptote: $y = 0$; hole when $x = 0$

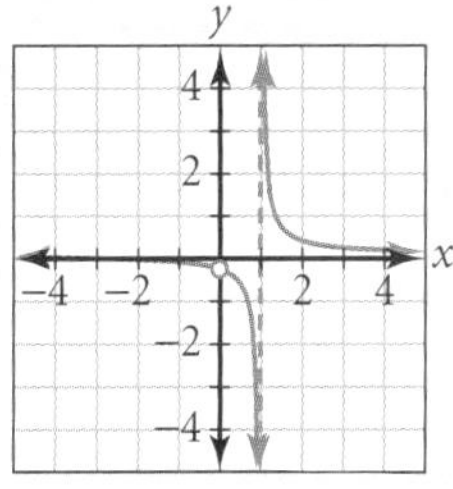

15. vertical asymptote: $x = -5$; horizontal asymptote: $y = 0$; hole when $x = -3$

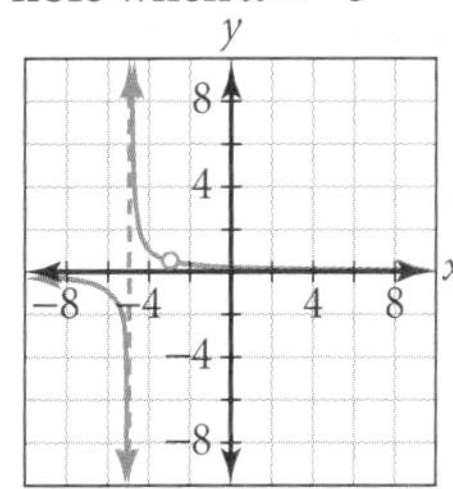

16. vertical asymptote: $x = -2$; horizontal asymptote: $y = 0$; no holes

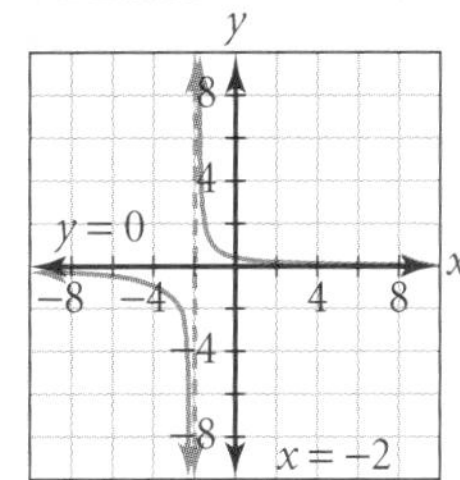

17. vertical asymptote: $x = -2$; horizontal asymptote: $y = 1$; no holes

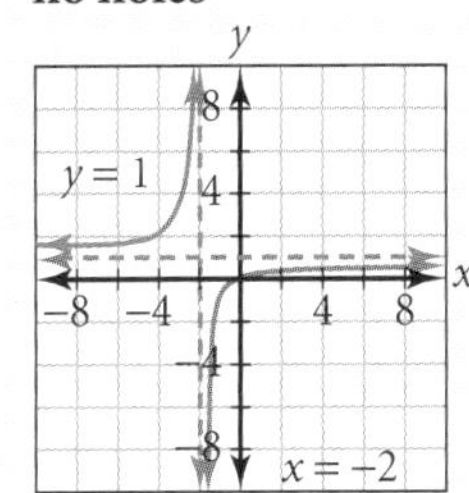

18. vertical asymptote: $x = 4$; horizontal asymptote: $y = 0$; hole when $x = 1$

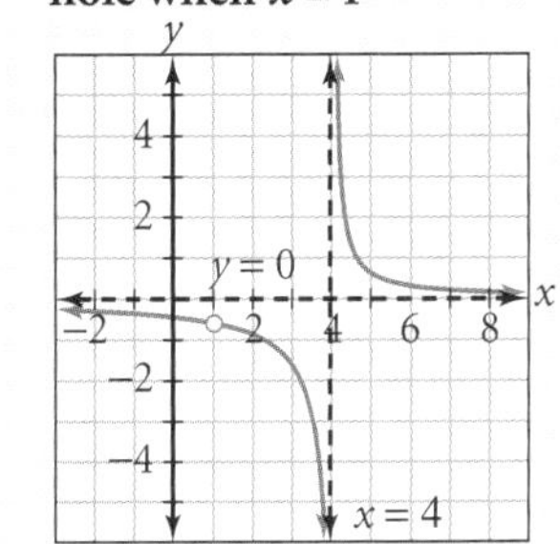

LESSON 8.3

Simplify each rational expression.

1. $\frac{7x(12x^5)}{3x^2(28x^2)}$ $\quad \mathbf{x^2}$
2. $\frac{x^2+8x-9}{x^2-81}$ $\quad \mathbf{\frac{x-1}{x-9}}$
3. $\frac{x^2+10x+24}{x^2+x-12}$ $\quad \mathbf{\frac{x+6}{x-3}}$
4. $\frac{x^2+4x-12}{3x^2-12x+12}$ $\quad \mathbf{\frac{x+6}{3x-6}}$
5. $\frac{x^2-8x-20}{12x-x^2-20}$ $\quad \mathbf{-\frac{x+2}{x-2}}$
6. $\frac{x^2-25}{2x^2-7x-15}$ $\quad \mathbf{\frac{x+5}{2x+3}}$

Simplify each product or quotient.

7. $\frac{2x^2-x-1}{3x^2-2x-1} \cdot \frac{15x^3+5x^2}{4x^2-1}$ $\quad \mathbf{\frac{5x^2}{2x-1}}$
8. $\frac{x^2-2x-3}{x^2+x-20} \div \frac{x^2+2x+1}{x^2+6x+5}$ $\quad \mathbf{\frac{x-3}{x-4}}$
9. $\frac{4x+8}{5x-20} \div \frac{10+3x-x^2}{x^2-4x}$ $\quad \mathbf{-\frac{4x}{5x-25}}$
10. $\frac{2x^2-12x-14}{x^3-16x} \cdot \frac{-16-4x}{6x-42}$ $\quad \mathbf{-\frac{4x+4}{3x^2-12x}}$
11. $\frac{3x^2+10x-8}{3x^2-17x+10} \cdot \frac{5+9x-2x^2}{x^2+3x-4}$ $\quad \mathbf{-\frac{2x+1}{x-1}}$
12. $\frac{3x^2+14x-5}{x^2+2x-15} \div \frac{3x^2-25x+8}{8+15x-2x^2}$ $\quad \mathbf{-\frac{2x+1}{x-3}}$

Simplify each complex fraction.

13. $\frac{\frac{x^2-49}{x^2-100}}{\frac{x-7}{x+10}}$ $\quad \mathbf{\frac{x+7}{x-10}}$
14. $\frac{\frac{(x+7)^2}{(2x-3)^2}}{\frac{x^2-49}{2x^2-17x+21}}$ $\quad \mathbf{\frac{x+7}{2x-3}}$
15. $\frac{\frac{x^2+8x-33}{x^2-x-6}}{\frac{x^2+10x-11}{x^2+9x+14}}$ $\quad \mathbf{\frac{x+7}{x-1}}$

Simplify each product or quotient involving complex fractions.

16. $\frac{\frac{1}{x+1}}{\frac{x}{x^2-1}} \cdot \frac{x}{x-1}$ $\quad \mathbf{1}$
17. $\frac{x-3}{x-4} \cdot \frac{\frac{2x-8}{x^2-9}}{\frac{x+5}{x+3}}$ $\quad \mathbf{\frac{2}{x+5}}$
18. $\frac{x+4}{x^2-9} \div \frac{\frac{x^2+4x}{x+3}}{\frac{x-3}{x}}$ $\quad \mathbf{\frac{1}{x^2}}$
19. $\frac{\frac{-2x-5}{x^2-1}}{\frac{x+1}{x-3}} \cdot \frac{x^2-1}{2x^2-x-15}$ $\quad \mathbf{-\frac{1}{x+1}}$
20. $\frac{x}{x+4} \cdot \frac{\frac{x^2+6x+8}{x+10}}{\frac{x^2}{2x+20}}$ $\quad \mathbf{\frac{2x+4}{x}}$
21. $\frac{\frac{x^2+4x-32}{x^2-12x+35}}{\frac{16x-4x^2}{x^2-4x-21}} \cdot \frac{x^2-10x}{x^2+11x+24}$ $\quad \mathbf{-\frac{x-10}{4x-20}}$

LESSON 8.4

Simplify each sum or difference.

1. $\frac{3x}{2x+5} - \frac{2x}{2x+5}$ $\quad \mathbf{\frac{x}{2x+5}}$
2. $\frac{2x+1}{3x-4} + \frac{x-1}{3x-4}$ $\quad \mathbf{\frac{3x}{3x-4}}$
3. $\frac{3x^2+x}{12} - \frac{x^2+1}{4}$ $\quad \mathbf{\frac{x-3}{12}}$
4. $\frac{5x-15}{x^2-9} - \frac{2}{x+3}$ $\quad \mathbf{\frac{3}{x+3}}$
5. $\frac{2x-1}{x+8} + \frac{34x}{x^2-64}$ $\quad \mathbf{\frac{2x+1}{x-8}}$
6. $\frac{2}{x-8} + \frac{1}{x+2}$ $\quad \mathbf{\frac{3x-4}{x^2-6x-16}}$
7. $\frac{3x-10}{x^2+4x-12} - \frac{2}{x+6}$ $\quad \mathbf{\frac{x-6}{x^2+4x-12}}$
8. $\frac{-2x-3}{x^2-3x} - \frac{-x}{x-3}$ $\quad \mathbf{\frac{x+1}{x}}$
9. $\frac{2x+1}{5-x} + \frac{1}{3x+2}$ $\quad \mathbf{-\frac{6x^2+6x+7}{3x^2-13x-10}}$

Simplify each sum or difference involving complex fractions.

10. $\frac{5}{\frac{x}{3x+1}} - \frac{7x^2+3x}{x}$ $\quad \mathbf{-\frac{7x^2-12x-5}{x}}$
11. $\frac{\frac{3}{2x-1}}{\frac{6x}{2x-1}} + \frac{3}{x}$ $\quad \mathbf{\frac{7}{2x}}$
12. $\frac{\frac{x+1}{x-2}}{\frac{x+2}{2}} - \frac{x}{x^2-4}$ $\quad \mathbf{\frac{1}{x-2}}$
13. $\frac{\frac{7x}{x^2-4}}{\frac{6}{x-2}} - \frac{\frac{2x-6}{x+2}}{\frac{x-3}{2x}}$ $\quad \mathbf{-\frac{17x}{6x+12}}$
14. $\frac{\frac{2}{x+5}}{\frac{x-5}{x}} + \frac{\frac{5x-10}{x-5}}{\frac{x^2+3x-10}{2}}$ $\quad \mathbf{\frac{2}{x-5}}$
15. $\frac{\frac{2}{x^2+2x-3}}{\frac{x-4}{x^3-x^2}} + \frac{\frac{6}{x^2+5x+6}}{\frac{x-4}{x+2}}$ $\quad \mathbf{\frac{2x^2+6}{x^2-x-12}}$

Write each expression as a single rational expression in simplest form.

16. $\frac{3}{x+7} - \frac{2x+8}{x+7} + \frac{4x+19}{x+7}$ $\quad \mathbf{2}$
17. $\frac{8x-5}{2x+3} + \frac{x+4}{2x+3} - \frac{3x-10}{2x+3}$ $\quad \mathbf{3}$
18. $\frac{3x}{x+2} - \frac{3x}{x+5} + \frac{18}{x^2+7x+10}$ $\quad \mathbf{\frac{9}{x+5}}$
19. $\frac{2x-1}{x+5} + \frac{x}{x-2} - \frac{5x+4}{x^2+3x-10}$ $\quad \mathbf{\frac{3x+1}{x+5}}$

LESSON 8.5

Solve each equation. Check your solution.

1. $\frac{x-3}{x-1} = \frac{x}{x+4}$ **6**
2. $\frac{x+1}{x-2} = \frac{x+3}{x-1}$ **5**
3. $\frac{x-7}{x+3} = \frac{x-9}{x-3}$ **12**
4. $\frac{1}{x} - \frac{5}{6x} = \frac{2}{3}$ $\frac{1}{4}$
5. $\frac{3}{2} - \frac{3}{x} = \frac{9}{2x}$ **5**
6. $\frac{x-7}{x+1} - \frac{x-4}{3x-2} = 0$ **1, 9**
7. $\frac{3}{x-2} + \frac{5}{x+2} = \frac{4x^2}{x^2-4}$ **1**
8. $\frac{3}{x-1} - \frac{1}{x+1} = \frac{3}{x^2-1}$ $-\frac{1}{2}$
9. $5 - \frac{26}{x+2} = \frac{27}{x^2-4}$ $\frac{1}{5}$, **5**
10. $\frac{2x-3}{4} + 2 = \frac{2x+1}{3}$ $\frac{11}{2}$
11. $\frac{3x}{4} - \frac{2x-1}{2} = \frac{x-7}{6}$ **4**
12. $\frac{x+1}{x-3} = \frac{3}{x} + \frac{12}{x^2-3x}$ **−1**
13. $\frac{4}{x^2-8x+12} = \frac{x}{x-2} + \frac{1}{x-6}$ **−1**
14. $\frac{2x-3}{x-5} = \frac{x}{x+4} + \frac{20x-37}{x^2-x-20}$ **no solution**
15. $\frac{x-2}{x+1} = \frac{x-3}{x^2-5x-6} - \frac{2x-7}{x-6}$ **4,** $\frac{2}{3}$

Solve each inequality. Check your solution.

16. $\frac{x-2}{x+6} > 4$ $-8\frac{2}{3} < x < -6$
17. $\frac{3x+2}{2x} < 1$ $-2 < x < 0$
18. $\frac{3x+3}{2x} > 1$ $x < -3$ *or* $x > 0$
19. $\frac{4x}{3x-2} > \frac{1}{2}$ $x < -\frac{2}{5}$ *or* $x > \frac{2}{3}$
20. $\frac{x-2}{x+2} < 3$ $x < -4$ *or* $x > -2$
21. $\frac{3x+5}{2x-3} < 6$ $x < \frac{3}{2}$ *or* $x > \frac{23}{9}$

Use a graphics calculator to solve each rational inequality. Round answers to the nearest tenth.

22. $\frac{3x+5}{2x-3} < 0$ $-\frac{5}{3} < x < \frac{3}{2}$
23. $\frac{2}{x} > x^2 + 1$ $0 < x < 1$
24. $\frac{3}{2x} < x^2 + 2$ $x < 0$ *or* $x > 0.6$
25. $\frac{x+4}{x-2} < x$ $-1 < x < 2$ *or* $x > 4$
26. $\frac{x+3}{x+1} > 2x$ $-1 < x < 1$ *or* $x < -1.5$
27. $x^2 - 3 \geq \frac{1}{x^2}$ $x \leq -1.8$ *or* $x \geq 1.8$
28. $\frac{x+5}{x-2} < \frac{36}{x^2-4}$ $-9.7 < x < -2$ *or* $2 < x < 2.7$
29. $\frac{x+1}{6-x} \leq \frac{2-x}{x+1}$ $-1 < x \leq 1.1$ *or* $x > 6$
30. $\frac{x}{x+1} + \frac{2x}{x-1} > \frac{2}{x^2-1}$ $x < -1, -1 < x < \frac{2}{3}, x > 1$

LESSON 8.6

Evaluate each expression.

1. $\frac{3}{5}\sqrt[3]{-27}$ $-\frac{9}{5}$
2. $0.25\sqrt[4]{16}$ **0.5**
3. $3\left(\sqrt[3]{-512}\right)^2$ **192**
4. $\frac{3}{2}\left(\sqrt[3]{-1000}\right)^2$ **150**
5. $\frac{1}{3}\left(\sqrt[3]{64}\right)^2$ $\frac{16}{3}$
6. $5\left(\sqrt{81}\right)^{-2}$ $\frac{5}{81}$
7. $\frac{1}{2}\left(\sqrt[3]{-512}\right)^{-1}$ $-\frac{1}{16}$
8. $\left(\sqrt[4]{1296} - 2\right)^{\frac{1}{2}}$ **2**
9. $3\left(\sqrt[4]{625} + 3\right)^{\frac{1}{3}}$ **6**
10. $\frac{1}{3}\left(\sqrt[5]{-243}\right)^2 - 3$ **0**
11. $\frac{2}{3}\left(\sqrt[4]{\frac{21}{8}}\right)^4 + \frac{1}{4}$ **2**
12. $\frac{3}{5}\left(\sqrt[4]{\frac{35}{9}}\right)^4 + \frac{2}{3}$ **3**

Find the domain of each radical function.

13. $f(x) = \sqrt{x^2 - 16}$ $x \leq -4$ *or* $x \geq 4$
14. $f(x) = \sqrt{3x + 6}$ $x \geq -2$
15. $f(x) = \sqrt{4(x-1)}$ $x \geq 1$
16. $f(x) = \sqrt{4x^2 - 9}$ $x \leq -\frac{3}{2}$ *or* $x \geq \frac{3}{2}$
17. $f(x) = \sqrt{x^2 + 2x + 1}$ **all reals**
18. $f(x) = \sqrt{x^2 + 7x + 12}$ $x \leq -4$ *or* $x \geq -3$

Find the inverse of each quadratic function. Then graph the function and its inverse in the same coordinate plane.

19. $y = x^2 + 4$ $y = \pm\sqrt{x-4}$
20. $y = x^2 - 3$ $y = \pm\sqrt{x+3}$
21. $y = x^2 + 4x$ $y = -2 \pm \sqrt{x+4}$
22. $y = x^2 - 8x + 16$ $y = 4 \pm \sqrt{x}$
23. $y = x^2 - 6x + 9$ $y = 3 \pm \sqrt{x}$
24. $y = x^2 - 4x + 1$ $y = 2 \pm \sqrt{x+3}$

25. The speed of an ocean wave depends on the depth of the water in which it travels. A wave's speed, in miles per hour, in water that is x feet deep is given by the function $f(x) = \sqrt{21.92x}$. Find the speed of a wave in water that is 25, 50, and 100 feet deep. Round your answers to the nearest tenth. **23.4 mph; 33.1 mph; 46.8 mph**

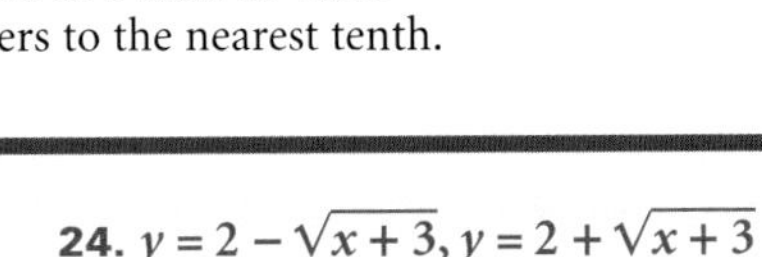

Extra Practice

Lesson 8.6

19. $y = -\sqrt{x-4}, y = \sqrt{x-4}$

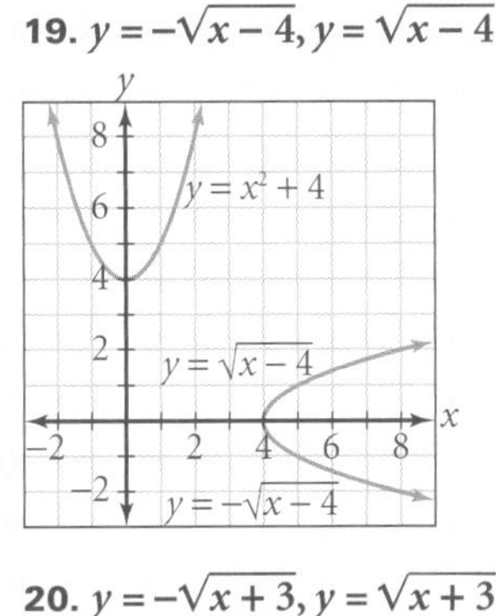

20. $y = -\sqrt{x+3}, y = \sqrt{x+3}$

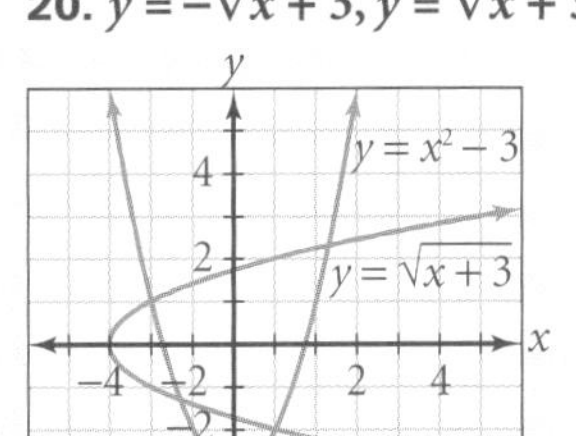
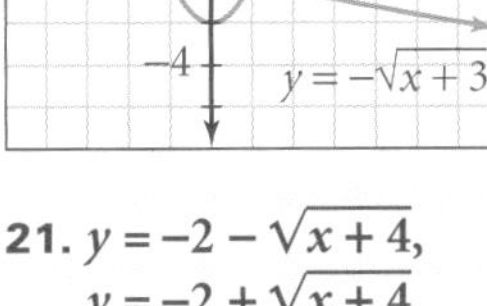

21. $y = -2 - \sqrt{x+4}$, $y = -2 + \sqrt{x+4}$

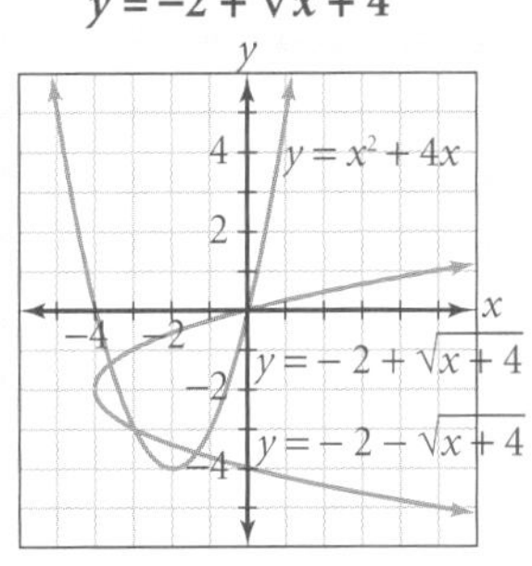

22. $y = 4 - \sqrt{x}, y = 4 + \sqrt{x}$

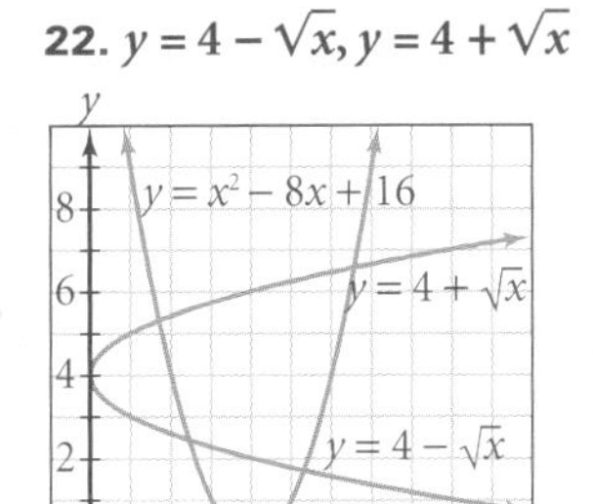

23. $y = 3 - \sqrt{x}, y = 3 + \sqrt{x}$

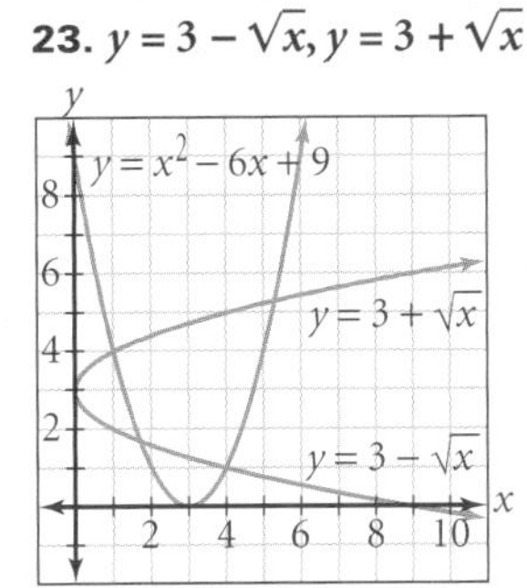

24. $y = 2 - \sqrt{x+3}, y = 2 + \sqrt{x+3}$

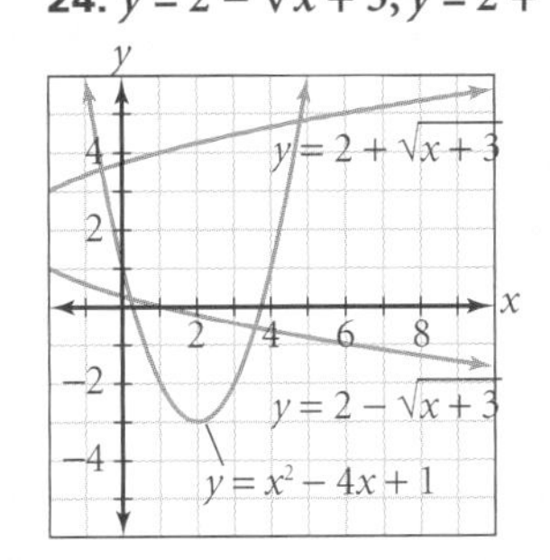

LESSON 8.7

Simplify each radical expression by using the Properties of *n*th Roots.

1. $\sqrt{125}$ $\mathbf{5\sqrt{5}}$ **2.** $\sqrt[3]{162x^6y^3}$ $\mathbf{3x^2y\sqrt[3]{6}}$ **3.** $\sqrt[4]{80x^8z^{10}}$ $\mathbf{2x^2z^2\sqrt[4]{5z^2}}$

4. $\sqrt[3]{-56x^4y^4z^3}$ $\mathbf{-2xyz\sqrt[3]{7xy}}$ **5.** $(75x^2y^3z)^{\frac{1}{2}}$ $\mathbf{5|xy|\sqrt{3yz}}$ **6.** $(54x^5)^{\frac{1}{3}}$ $\mathbf{3x\sqrt[3]{2x^2}}$

Simplify each product or quotient. Assume that the value of each variable is positive.

7. $\sqrt[3]{9x^2} \cdot \sqrt[3]{3x}$ $\mathbf{3x}$ **8.** $\sqrt[3]{4x^5} \cdot \sqrt[3]{54xy^2}$ $\mathbf{6x^2\sqrt[3]{y^2}}$ **9.** $\sqrt{8x^3} \cdot (2xz^5)^{\frac{1}{2}} \cdot \sqrt{4x^3z^4}$ $\mathbf{8x^3z^4\sqrt{xz}}$

10. $\dfrac{(81y^5)^{\frac{1}{4}}}{\sqrt[4]{x^4y}}$ $\mathbf{\dfrac{3y}{x}}$ **11.** $\dfrac{\sqrt[3]{48x^2y^4z^4}}{\sqrt[3]{6x}}$ $\mathbf{2yz\sqrt[3]{xyz}}$ **12.** $\dfrac{\sqrt{15x^9y^3}}{\sqrt{5x^5y}}$ $\mathbf{x^2y\sqrt{3}}$

13. $\sqrt[4]{8x^5} \cdot \sqrt[4]{4x^7}$ $\mathbf{2x^3\sqrt[4]{2}}$ **14.** $\dfrac{\sqrt{9b^7}}{(12b^5)^{\frac{1}{2}}}$ $\mathbf{\dfrac{3b}{2\sqrt{3}}}$, or $\mathbf{\dfrac{b\sqrt{3}}{2}}$ **15.** $\dfrac{\sqrt[4]{8x^5}}{(20x^2)^{-\frac{1}{4}}}$ $\mathbf{2x\sqrt[4]{10x^3}}$

Find each sum, difference, or product. Give your answer in simplest radical form.

16. $(12-\sqrt{2})+(15+\sqrt{2})$ $\mathbf{27}$ **17.** $(9+2\sqrt{5})-(1+\sqrt{45})$ $\mathbf{8-\sqrt{5}}$

18. $(7-2\sqrt{6})(7+2\sqrt{6})$ $\mathbf{25}$ **19.** $(3-\sqrt{8})(5+\sqrt{2})$ $\mathbf{11-7\sqrt{2}}$

20. $(4+\sqrt{3})(-2+\sqrt{2})$ $\mathbf{-8-2\sqrt{3}+4\sqrt{2}+\sqrt{6}}$ **21.** $6\sqrt{3}(2\sqrt{5}+4\sqrt{6})$ $\mathbf{12\sqrt{15}+72\sqrt{2}}$

22. $7\sqrt{20}+8\sqrt{5}-2\sqrt{45}$ $\mathbf{16\sqrt{5}}$ **23.** $6\sqrt{8}-(\sqrt{24}-3\sqrt{72}+\sqrt{54})$ $\mathbf{30\sqrt{2}-5\sqrt{6}}$

24. $4\sqrt{2}(\sqrt{12}-3\sqrt{2}+4\sqrt{8})$ $\mathbf{40+8\sqrt{6}}$ **25.** $(4\sqrt{2}-2\sqrt{3})(5\sqrt{2}-\sqrt{3})$ $\mathbf{46-14\sqrt{6}}$

Write each expression with a rational denominator and in simplest form.

26. $\dfrac{3}{\sqrt{15}}$ $\mathbf{\dfrac{\sqrt{15}}{5}}$ **27.** $\dfrac{\sqrt{135}}{\sqrt{15}}$ $\mathbf{3}$ **28.** $\dfrac{5}{1-\sqrt{6}}$ $\mathbf{-1-\sqrt{6}}$ **29.** $\dfrac{-3}{\sqrt{6}-\sqrt{2}}$ $\mathbf{\dfrac{-3\sqrt{6}-3\sqrt{2}}{4}}$

30. $\dfrac{14}{\sqrt{5}+\sqrt{3}}$ $\mathbf{7\sqrt{5}-7\sqrt{3}}$ **31.** $\dfrac{\sqrt{3}-\sqrt{2}}{\sqrt{3}+\sqrt{2}}$ $\mathbf{5-2\sqrt{6}}$ **32.** $\dfrac{2\sqrt{5}-\sqrt{3}}{\sqrt{5}+\sqrt{3}}$ $\mathbf{\dfrac{13-3\sqrt{15}}{2}}$ **33.** $\dfrac{2\sqrt{x}}{3\sqrt{x}-4\sqrt{y}}$ $\mathbf{\dfrac{6x+8\sqrt{xy}}{9x-16y}}$

LESSON 8.8

Solve each radical equation by using algebra. If the equation has no real solution, write *no solution*. Check your solutions.

1. $\sqrt{x-5}=3$ $\mathbf{14}$ **2.** $\sqrt{x^2-15}=7$ $\mathbf{\pm 8}$ **3.** $\sqrt{x-4}=\sqrt{x+4}$ **no solution**

4. $\sqrt{2x-5}+4=3$ **no solution** **5.** $\sqrt{3x-5}=5$ $\mathbf{10}$ **6.** $\sqrt{5x-11}=x-1$ $\mathbf{3, 4}$

7. $\sqrt{2x-1}=x$ $\mathbf{1}$ **8.** $\sqrt[3]{x+5}=\sqrt[3]{3x-2}$ $\mathbf{3\frac{1}{2}}$ **9.** $\sqrt{x^2-4x-5}=\sqrt{5x-x^2}$ $\mathbf{5}$

Solve each radical inequality by using algebra. If the inequality has no real solution, write *no solution*. Check your solution.

10. $\sqrt{x-3}\ge 2$ $\mathbf{x\ge 7}$ **11.** $3>\sqrt{2x}$ $\mathbf{0<x<4.5}$ **12.** $\sqrt{4x-1}>2$ $\mathbf{x>\frac{5}{4}}$

13. $3\ge\sqrt{x^2-4x+4}$ $\mathbf{2\le x\le 5}$ **14.** $\sqrt{1-x}>3$ $\mathbf{x<-8}$ **15.** $\sqrt{3x-2}\le 2$ $\mathbf{\frac{2}{3}\le x\le 2}$

16. $4\le\sqrt{7-x}$ $\mathbf{x\le -9}$ **17.** $\sqrt{5x-6}>12$ $\mathbf{x>30}$ **18.** $\sqrt{4x+1}\ge 5$ $\mathbf{x\ge 6}$

***Solve* each radical equation or inequality *by graphing.* Round solutions to the nearest tenth. Check your solutions by any method.**

19 $2\sqrt{x}\le 3x-4$ $\mathbf{x\ge 2.4}$ **20** $3\sqrt{x+2}\ge\sqrt{x^2+4}$ $\mathbf{-1.4\le x\le 10.4}$ **21** $0.25\sqrt{3x-1}<x+2$ $\mathbf{x\ge\frac{1}{3}}$

22 $\sqrt[3]{x^2+1}=x$ $\mathbf{x=1.5}$ **23** $\sqrt[3]{x+2}=\sqrt{x}$ $\mathbf{x=2.9}$ **24** $\sqrt[3]{2x-1}>2\sqrt{x-4}$ $\mathbf{4<x<5.1}$

Chapter 9

LESSON 9.1

Graph each equation and identify the conic section.

1. $x^2 + y^2 = 4$ **circle**
2. $y^2 - 9x = 0$ **parabola**
3. $9x^2 - 4y^2 = 25$ **hyperbola**
4. $x^2 + 4y^2 = 16$ **ellipse**
5. $x^2 - y^2 = 16$ **hyperbola**
6. $4x^2 - y = 0$ **parabola**
7. $16x^2 + y^2 = 81$ **ellipse**
8. $x^2 + y^2 = 144$ **circle**
9. $9x^2 - 16y^2 = 144$ **hyperbola**

Find the distance between *P* and *Q* and the coordinates of *M*, the midpoint of $\overline{PQ}$. Give exact answers and approximate answers to the nearest hundredth when appropriate.

10. $P(-6, 4)$ and $Q(2, -2)$ **10; $M(-2, 1)$**
11. $P(6, -2)$ and $Q(2, 4)$ **$2\sqrt{13} \approx 7.21$; $M(4, 1)$**
12. $P(8, -4)$ and $Q(6, 0)$ **$2\sqrt{5} \approx 4.47$; $M(7, -2)$**
13. $P(0, 1)$ and $Q(4, 7)$ **$2\sqrt{13} \approx 7.21$; $M(2, 4)$**
14. $P(-2, 3)$ and $Q(5, 4)$ **$5\sqrt{2} \approx 7.07$; $M\left(\frac{3}{2}, \frac{7}{2}\right)$**
15. $P(\sqrt{2}, 4)$ and $Q(3\sqrt{2}, 0)$ **$2\sqrt{6} \approx 4.90$; $M(2\sqrt{2}, 2) \approx (2.83, 2)$**

Find the center, circumference, and area of a circle whose diameter has the given endpoints.

16. $P(12, -8)$ and $Q(6, 0)$ **$(9, -4)$; 10π; 25π**
17. $P(3, 4)$ and $Q(10, -20)$ **$\left(\frac{13}{2}, -8\right)$; 25π; 156.25π**
18. $P(0, 10)$ and $Q(2, -6)$ **$(1, 2)$; $2\pi\sqrt{65}$; 65π**
19. $P(14, 8)$ and $Q(-2, -8)$ **$(6, 0)$; $16\pi\sqrt{2}$; 128π**
20. $P(4, -5)$ and $Q(8, 3)$ **$(6, -1)$; $4\pi\sqrt{5}$; 20π**
21. $P(24, 16)$ and $Q(-2, 18)$ **$(11, 17)$; $2\pi\sqrt{170}$; 170π**

LESSON 9.2

Write the standard equation for each parabola below.

1. **$y = \frac{1}{2}x^2$**

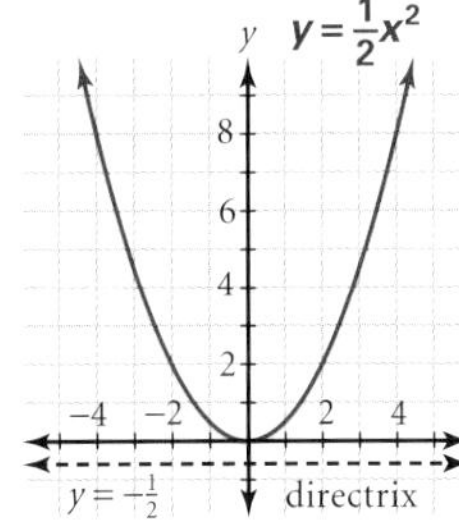

2. **$x - 2 = \frac{1}{4}(y + 5)^2$**

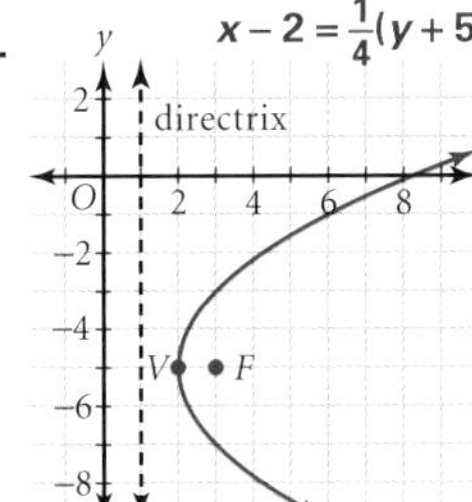

3. **$y - 2 = -\frac{1}{12}(x - 3)^2$**

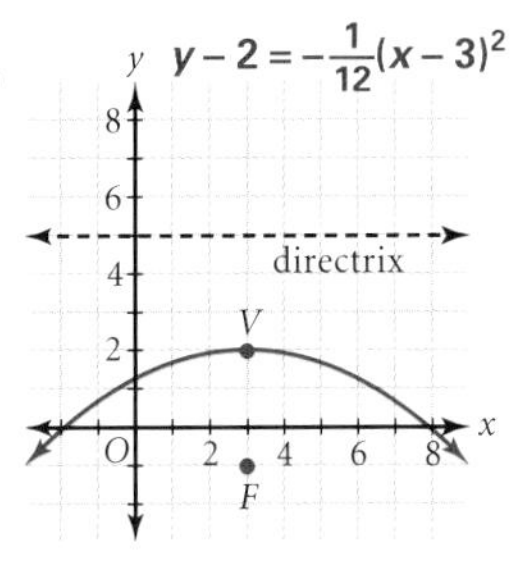

Graph each equation. Label the vertex, focus, and directrix.

4. $y = \frac{1}{2}x^2$
5. $x = -\frac{1}{4}y^2$
6. $y = -\frac{1}{12}x^2$
7. $y - 2 = (x - 2)^2$
8. $x + 3 = \frac{1}{8}(y - 1)^2$
9. $y - x^2 + 6x = 0$
10. $y - x^2 + 2x = 0$
11. $y - x^2 - 10x = 27$
12. $x^2 + 2x - 3y = 5$

Write the standard equation for a parabola with the given characteristics.

13. vertex: $(0, 0)$, directrix: $x = -15$ **$x = \frac{1}{60}y^2$**
14. focus: $(-2, 3)$, directrix: $x = 2$ **$x = -\frac{1}{8}(y - 3)^2$**
15. vertex: $(0, 2)$, directrix: $y = -4$ **$y - 2 = \frac{1}{24}x^2$**
16. vertex: $(2, 0)$, focus: $(6, 0)$ **$x - 2 = \frac{1}{16}y^2$**
17. focus: $(1, 1)$, directrix: $y = 0$ **$y - \frac{1}{2} = \frac{1}{2}(x - 1)^2$**
18. vertex: $(3, 2)$, focus: $(6, 2)$ **$x - 3 = \frac{1}{12}(y - 2)^2$**

19. The parabola defined by the equation $y = -2x^2 + 12x - 13$ is translated 3 units up and 2 units to the left. Write the standard equation of the resulting parabola. **$y - 8 = -2(x - 1)^2$**

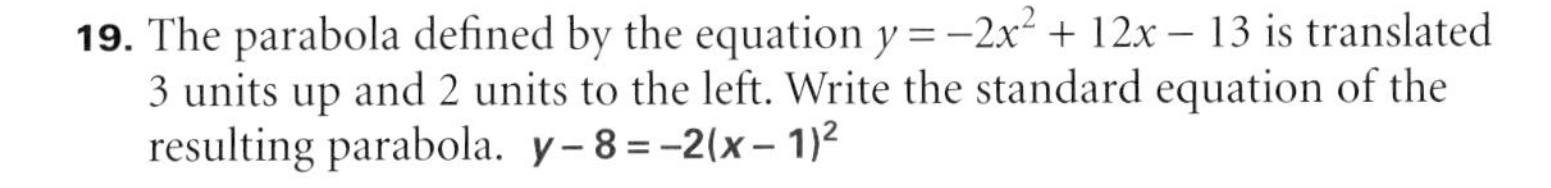

Lesson 9.2

4.

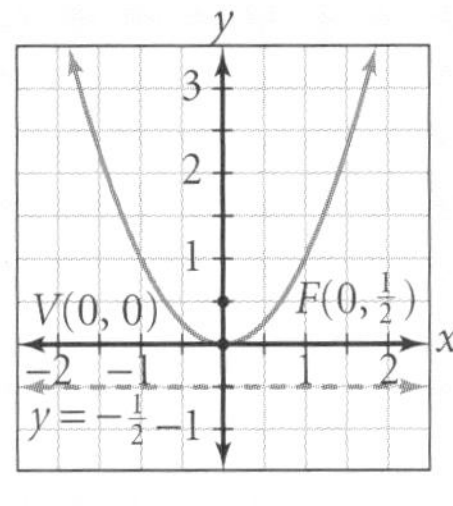

5.

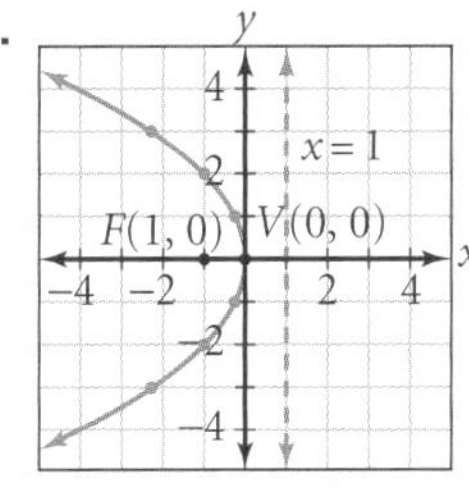

6.

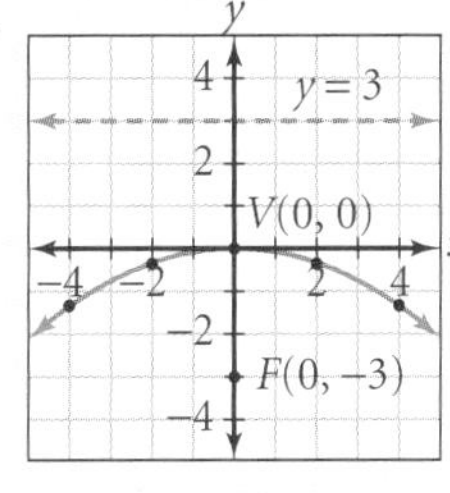

7.

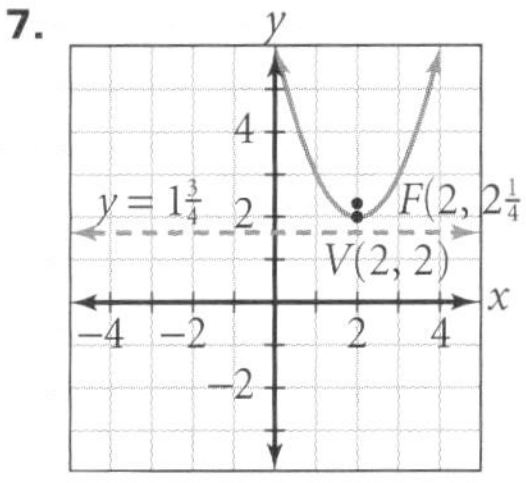

8.

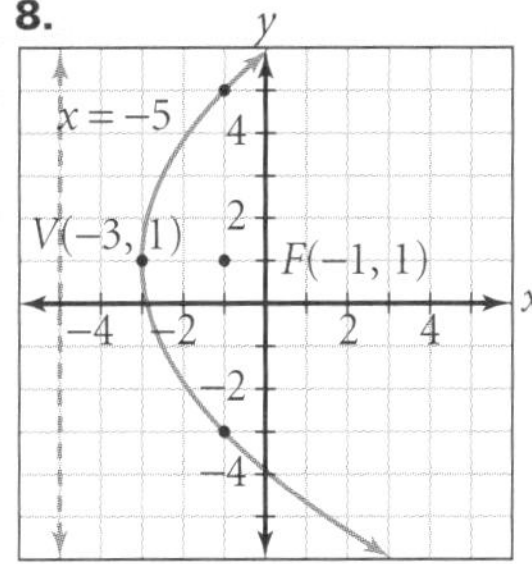

9.

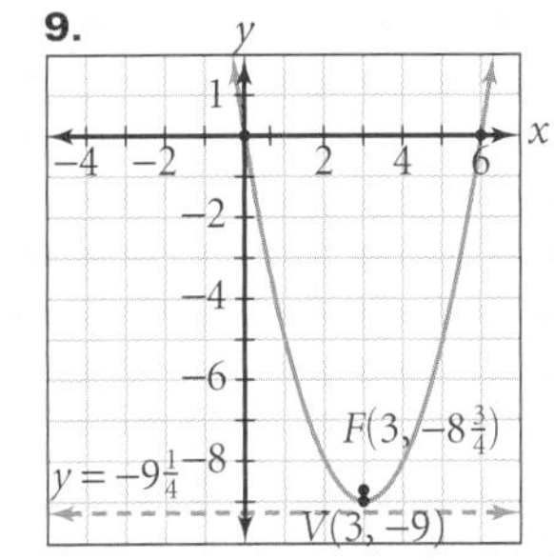

10.

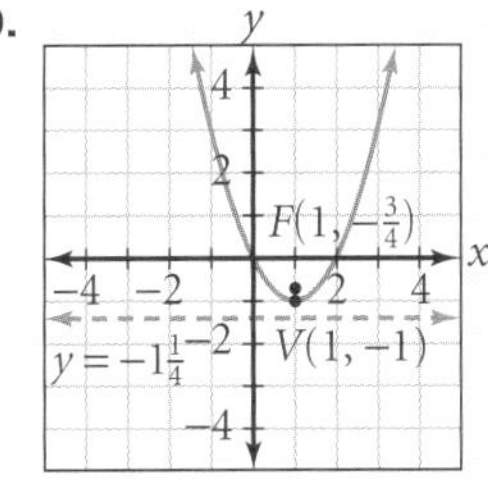

11.

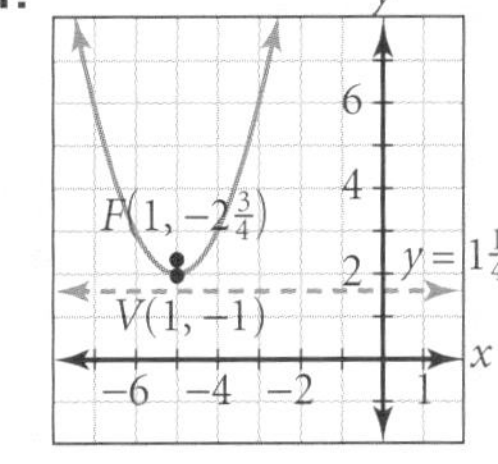

12.

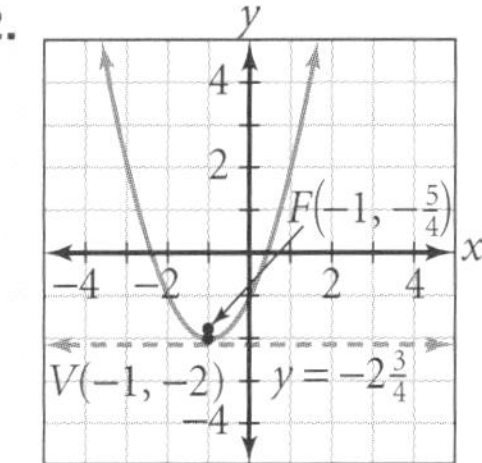

Lesson 9.3

10.

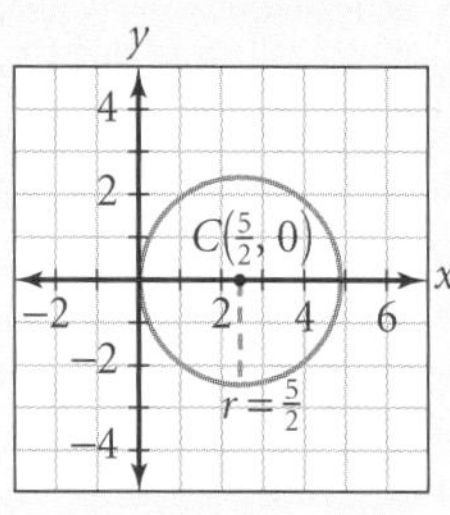

11.

C(5, 9)
r = 10

12.

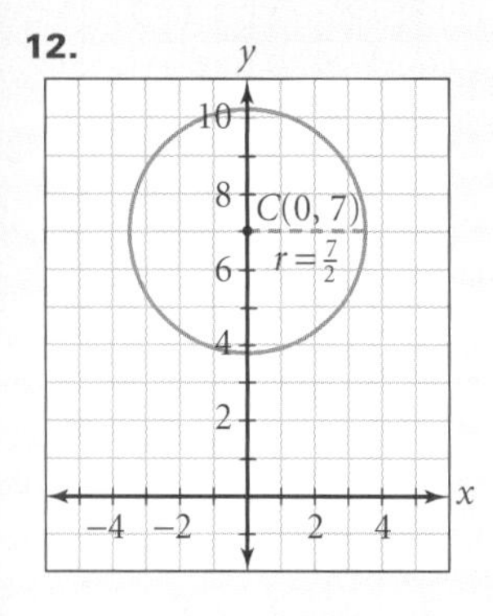

Lesson 9.4

4. $\frac{x^2}{9}+\frac{y^2}{25}=1$; center: (0, 0); vertices: (0, −5) and (0, 5); co-vertices: (−3, 0) and (3, 0); foci: (0, −4) and (0, 4)

5. $x^2 + \frac{y^2}{49} = 1$; center: (0, 0); vertices: (0, −7) and (0, 7); co-vertices: (−1, 0) and (1, 0); foci: $(0, -4\sqrt{3})$ and $(0, 4\sqrt{3})$

6. $\frac{x^2}{64} + \frac{y^2}{16} = 1$; center: (0, 0); vertices: (−8, 0) and (8, 0); co-vertices: (0, −4) and (0, 4); foci: $(-4\sqrt{3}, 0)$ and $(4\sqrt{3}, 0)$

7. $\frac{(x-1)^2}{9} + y^2 = 1$; center: (1, 0); vertices: (−2, 0) and (4, 0); co-vertices: (1, 1) and (1, −1); foci: $(1 - 2\sqrt{2}, 0)$ and $(1 + 2\sqrt{2}, 0)$

LESSON 9.3

Write the standard equation for each circle below.

1. $(x-4)^2 + (y-2)^2 = 4$

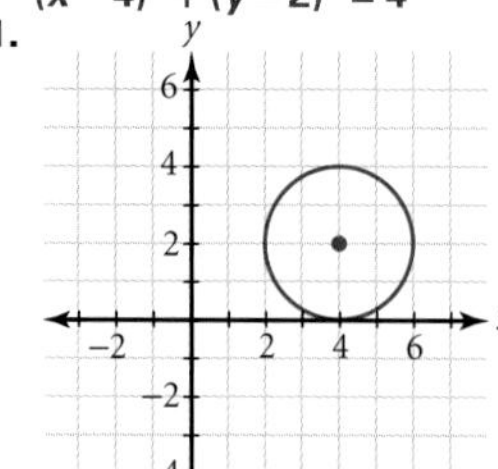

2. $(x+3)^2 + (y+3)^2 = 9$

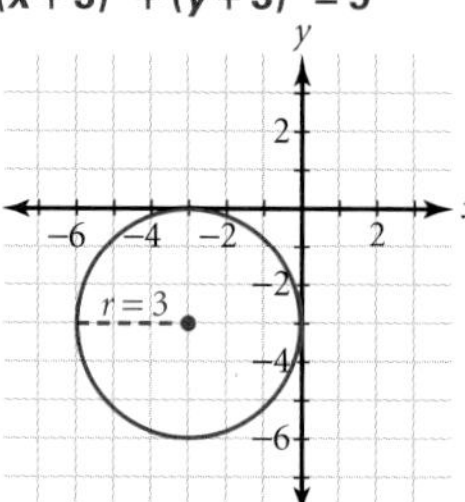

3. $(x+1)^2 + (y-3)^2 = \frac{49}{4}$

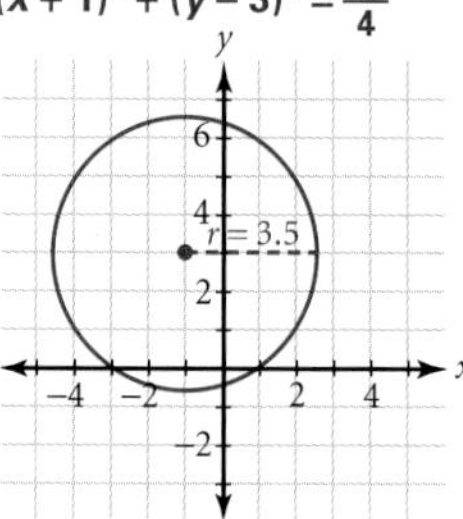

Write the standard equation of a circle with each given radius and center.

4. $r = 5$; $C(0, 4)$ $x^2 + (y-4)^2 = 25$
5. $r = 4.5$; $C(0, 0)$ $x^2 + y^2 = 20.25$
6. $r = 1.5$; $C(-3, -5)$ $(x+3)^2 + (y+5)^2 = 2.25$

Write the standard equation for each circle. Then state the coordinates of its center and give its radius.

7. $x^2 + y^2 - 16x + 4y = -43$ $(x-8)^2 + (y+2)^2 = 25$; $r = 5$; $C(8, -2)$
8. $x^2 - 8x + y^2 = 33$ $(x-4)^2 + y^2 = 49$; $r = 7$; $C(4, 0)$
9. $x^2 + y^2 - 20x - 10y + 61 = 0$ $(x-10)^2 + (y-5)^2 = 64$; $r = 8$; $C(10, 5)$

Graph each equation. Label the center and the radius.

10. $\left(x - \frac{5}{2}\right)^2 + y^2 = \frac{25}{4}$
11. $(x-5)^2 + (y-9)^2 = 100$
12. $x^2 + (y-7)^2 = \frac{49}{4}$

13. State whether $C(-1, 3)$ is inside, outside, or on the circle whose equation is $x^2 + y^2 - 12x - 2y = 8$. outside

LESSON 9.4

Write the standard equation for each ellipse below.

1. $\frac{x^2}{4} + \frac{y^2}{16} = 1$

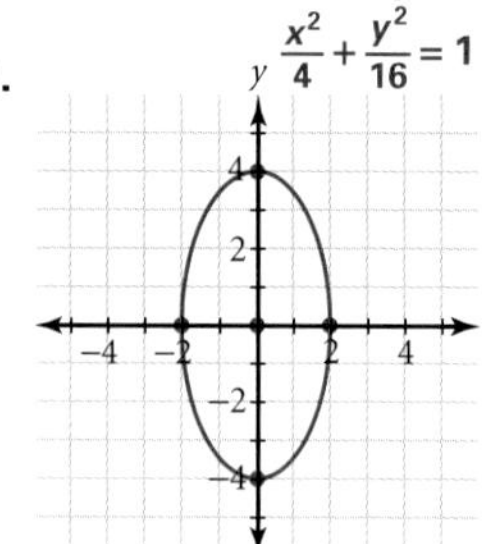

2. $\frac{(x-1)^2}{9} + (y-1)^2 = 1$

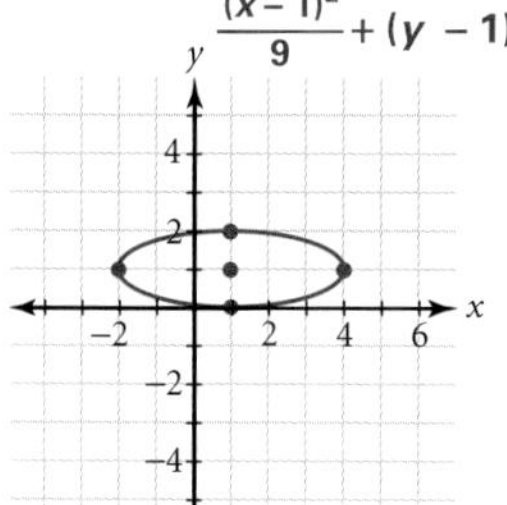

3. $\frac{(x+1)^2}{4} + \frac{(y-1)^2}{9} = 1$

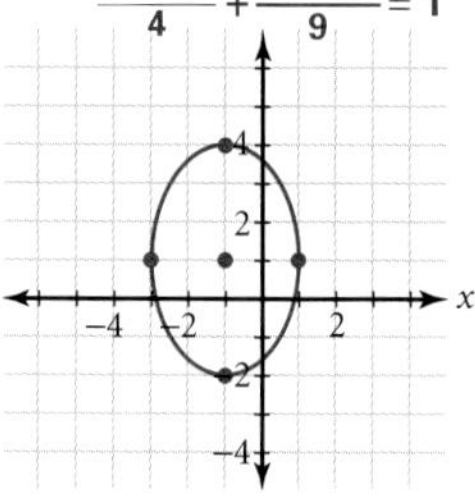

Write the standard equation of each ellipse. Find the coordinates of its center, vertices, co-vertices, and foci.

4. $25x^2 + 9y^2 = 225$
5. $49x^2 + y^2 = 49$
6. $x^2 + 4y^2 = 64$
7. $x^2 - 2x + 9y^2 - 8 = 0$
8. $x^2 + 4y^2 - 18x - 8y + 81 = 0$
9. $9x^2 + 4y^2 - 144x - 8y = -544$

Write the standard equation for an ellipse with the given characteristics.

10. foci: (−2, 0), (2, 0); vertices: (−6, 0), (6, 0) $\frac{x^2}{36} + \frac{y^2}{32} = 1$
11. foci: (0, −3), (0, 3); vertices: (0, −4), (0, 4) $\frac{x^2}{16} + \frac{y^2}{7} = 1$

Graph each ellipse. Label the center, foci, vertices, and co-vertices.

12. $\frac{x^2}{100} + \frac{y^2}{25} = 1$

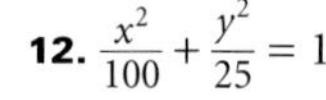

13. $\frac{x^2}{4} + \frac{y^2}{25} = 1$
14. $\frac{(x-2)^2}{16} + \frac{(y-1)^2}{25} = 1$

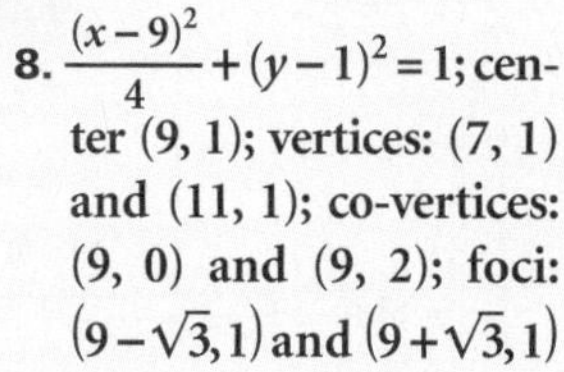

8. $\frac{(x-9)^2}{4} + (y-1)^2 = 1$; center (9, 1); vertices: (7, 1) and (11, 1); co-vertices: (9, 0) and (9, 2); foci: $(9 - \sqrt{3}, 1)$ and $(9 + \sqrt{3}, 1)$

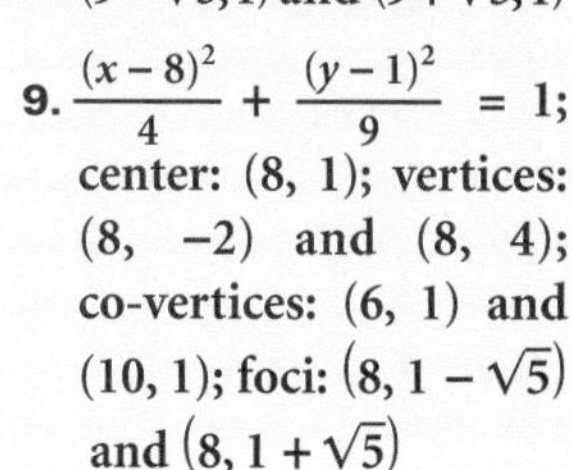

9. $\frac{(x-8)^2}{4} + \frac{(y-1)^2}{9} = 1$; center: (8, 1); vertices: (8, −2) and (8, 4); co-vertices: (6, 1) and (10, 1); foci: $(8, 1 - \sqrt{5})$ and $(8, 1 + \sqrt{5})$

12.

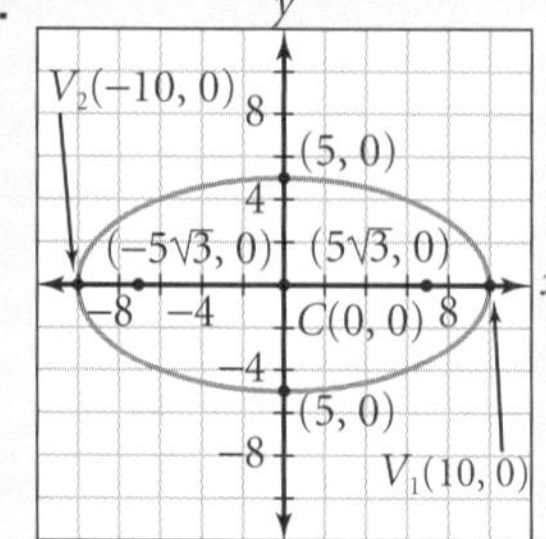

13.

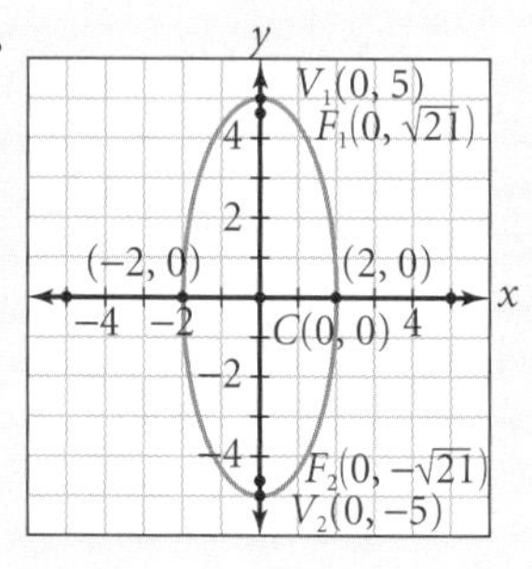

Write the standard equation for each hyperbola below.

1. $\frac{x^2}{16} - \frac{y^2}{4} = 1$

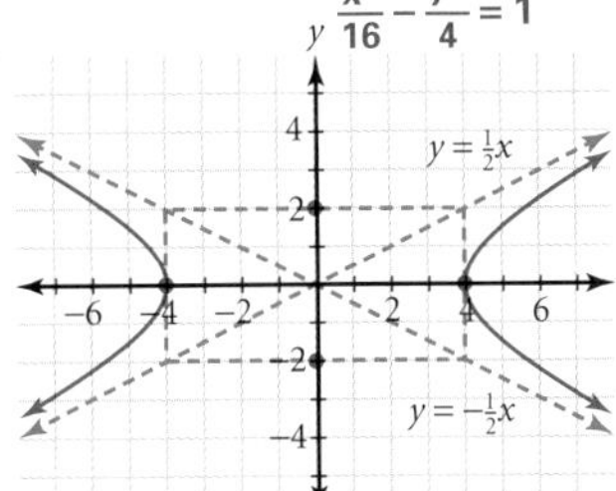

2. $\frac{y^2}{9} - \frac{x^2}{16} = 1$

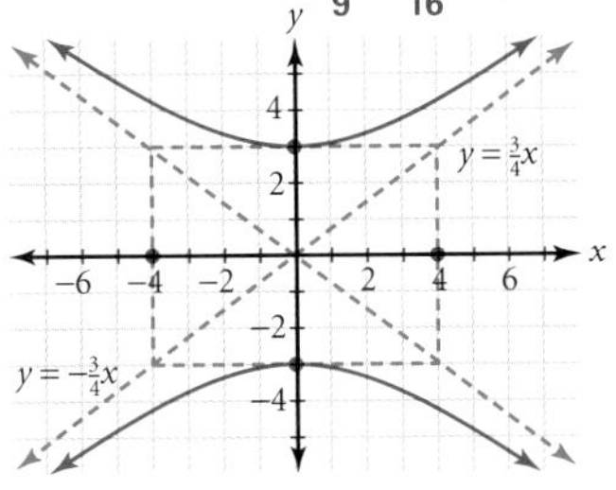

Write the standard equation for each hyperbola. Give the coordinates of the center, vertices, co-vertices, and foci.

3. $x^2 - 100y^2 = 100$
4. $25y^2 - 100y + 100 - x^2 = 25$
5. $9x^2 - 4y^2 - 18x + 8y = 31$

Write the standard equation for a hyperbola with the given characteristics.

6. vertices: $(-6, 0)$ and $(6, 0)$; foci: $(-8, 0)$ and $(8, 0)$ $\frac{x^2}{36} - \frac{y^2}{28} = 1$
7. vertices: $(0, -12)$ and $(0, 12)$; co-vertices: $(-11, 0)$ and $(11, 0)$ $\frac{y^2}{144} - \frac{x^2}{121} = 1$

Graph each hyperbola. Label the center, vertices, co-vertices, foci, and asymptotes.

8. $\frac{x^2}{9} - \frac{y^2}{4} = 1$
9. $\frac{y^2}{36} - \frac{x^2}{9} = 1$
10. $\frac{(x-2)^2}{4} - (y-1)^2 = 1$

11. The hyperbola defined by the equation $25x^2 - 36y^2 = 900$ is translated 4 units up and 3 units to the right. Write the standard equation of the resulting hyperbola. $\frac{(x-3)^2}{36} - \frac{(y-4)^2}{25} = 1$

LESSON 9.6

Use the substitution method to solve each system. If there are no real solutions, write *none*.

1. $\begin{cases} y = 7 - x \\ y = x^2 + 1 \end{cases}$ $(-3, 10)$ and $(2, 5)$
2. $\begin{cases} y = 12 - 6x \\ x^2 + y = 4 \end{cases}$ $(2, 0)$ and $(4, -12)$
3. $\begin{cases} x = 12y - 4 \\ x^2 - 16y^2 = 16 \end{cases}$ $(-4, 0)$ and $\left(5, \frac{3}{4}\right)$

Use the elimination method to solve each system. If there are no real solutions, write *none*.

4. $\begin{cases} 4x^2 + y^2 = 20 \\ x^2 + 4y^2 = 20 \end{cases}$ 4 solutions: $(\pm 2, \pm 2)$
5. $\begin{cases} x^2 + y^2 = 36 \\ 9x^2 + 4y^2 = 16 \end{cases}$ none
6. $\begin{cases} 16x^2 + 9y^2 = 144 \\ -48x^2 + y^2 = 144 \end{cases}$ none

Solve each system by graphing. Round answers to the nearest hundredth, if necessary. If there are no real solutions, write *none*.

7. $\begin{cases} x^2 + y^2 = 16 \\ 2x^2 - y^2 = -4 \end{cases}$ 4 solutions: $(\pm 2, \pm 3.46)$
8. $\begin{cases} 3y^2 - 4x^2 = 1 \\ x^2 + y^2 = 9 \end{cases}$ 4 solutions: $(\pm 1.93, \pm 2.30)$
9. $\begin{cases} 9x^2 + 16y^2 = 144 \\ 3x^2 - 2y^2 = 6 \end{cases}$ 4 solutions: $(\pm 2.41, \pm 2.39)$

Classify the conic section defined by each equation. Write the standard equation of the conic section, and sketch the graph.

10. $x^2 - 8x + y = -2$ parabola
11. $x^2 + y^2 - 6x - 10y = 2$ circle
12. $x^2 + 10x - 8y = -73$ parabola
13. $9x^2 + 4y^2 - 72x - 32y = -172$ ellipse
14. $9x^2 - 4y^2 + 54x + 8y + 41 = 0$ hyperbola
15. $4x^2 + 25y^2 + 16x + 50y - 59 = 0$ ellipse

Lesson 9.5

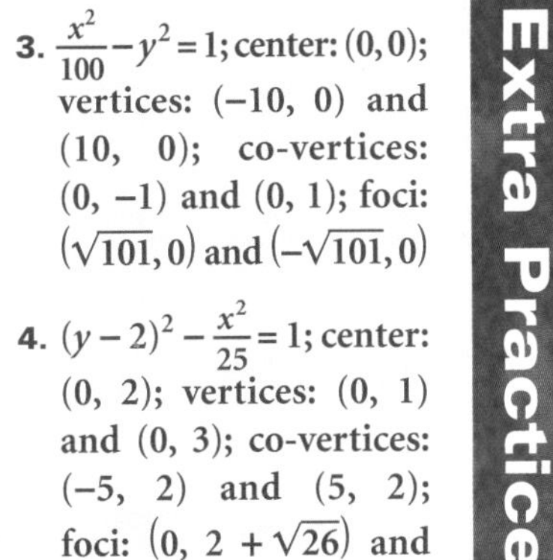

3. $\frac{x^2}{100} - y^2 = 1$; center: $(0, 0)$; vertices: $(-10, 0)$ and $(10, 0)$; co-vertices: $(0, -1)$ and $(0, 1)$; foci: $(\sqrt{101}, 0)$ and $(-\sqrt{101}, 0)$

4. $(y-2)^2 - \frac{x^2}{25} = 1$; center: $(0, 2)$; vertices: $(0, 1)$ and $(0, 3)$; co-vertices: $(-5, 2)$ and $(5, 2)$; foci: $(0, 2 + \sqrt{26})$ and $(0, 2 - \sqrt{26})$

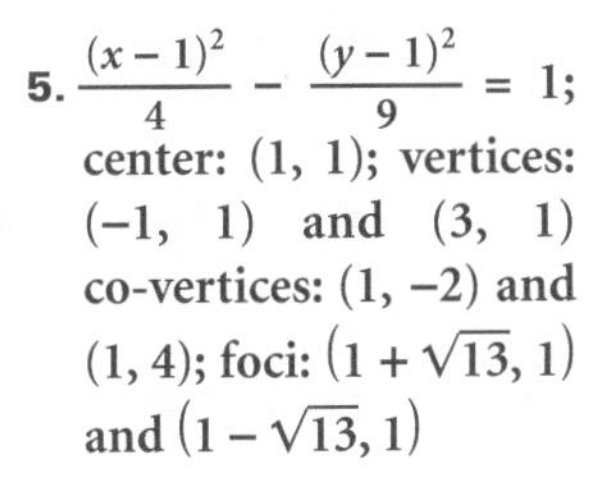

5. $\frac{(x-1)^2}{4} - \frac{(y-1)^2}{9} = 1$; center: $(1, 1)$; vertices: $(-1, 1)$ and $(3, 1)$ co-vertices: $(1, -2)$ and $(1, 4)$; foci: $(1 + \sqrt{13}, 1)$ and $(1 - \sqrt{13}, 1)$

8.

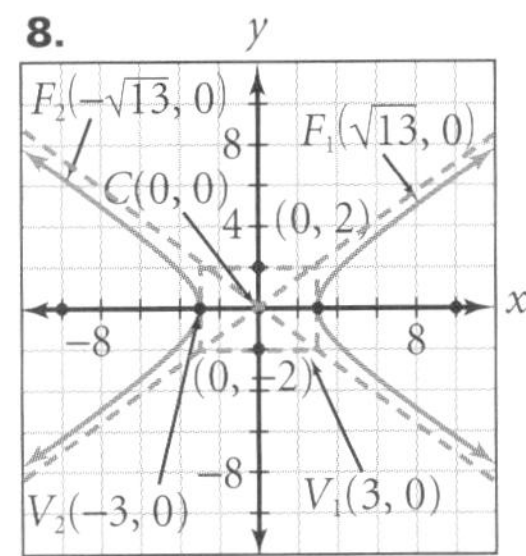

9.

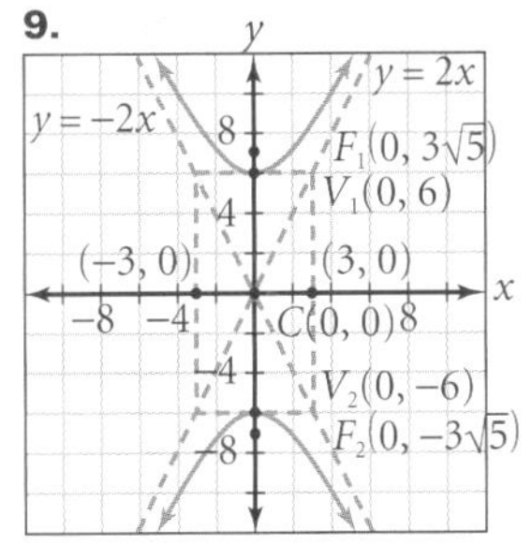

10.

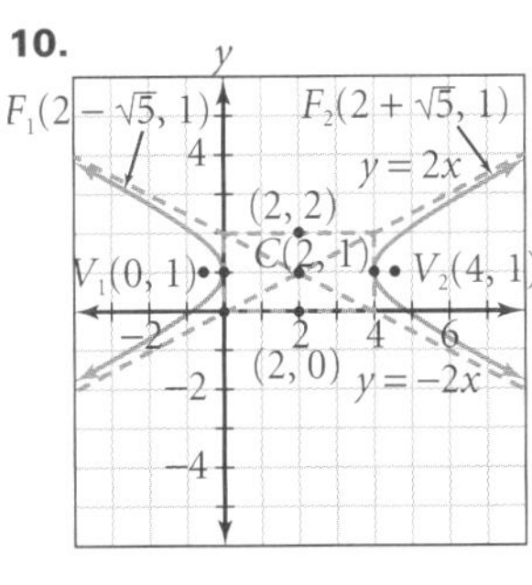

14.

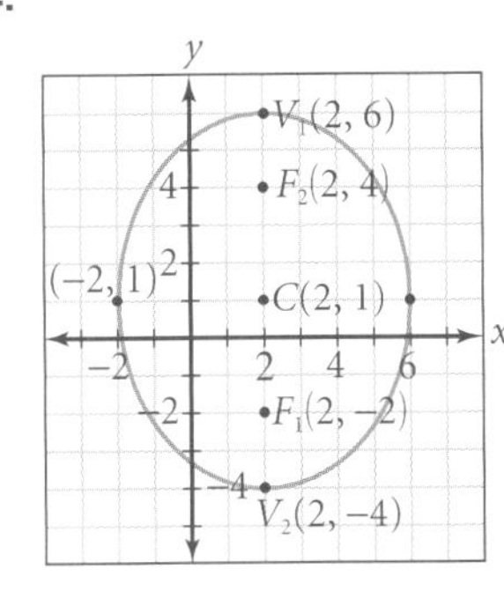

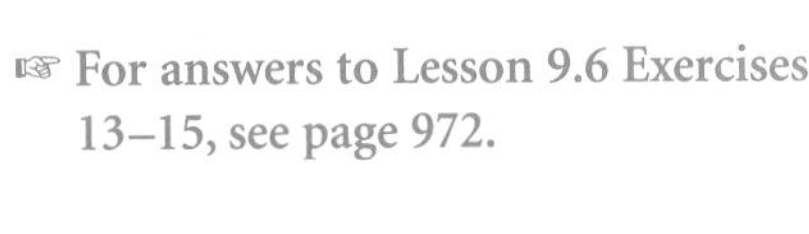

☞ For answers to Lesson 9.6 Exercises 13–15, see page 972.

Lesson 9.6

10. parabola; $y - 14 = -(x-4)^2$

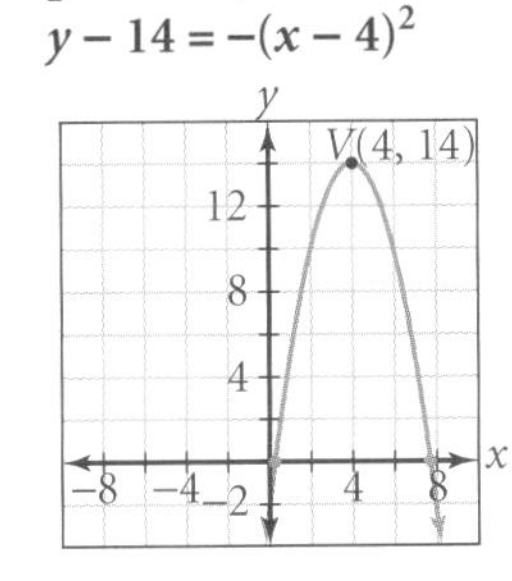

11. circle; $(x-3)^2 + (y-5)^2 = 36$

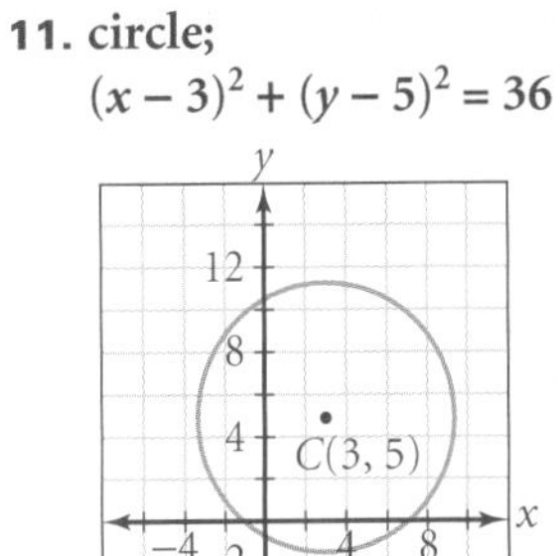

12. parabola; $y - 6 = \frac{1}{8}(x+5)^2$

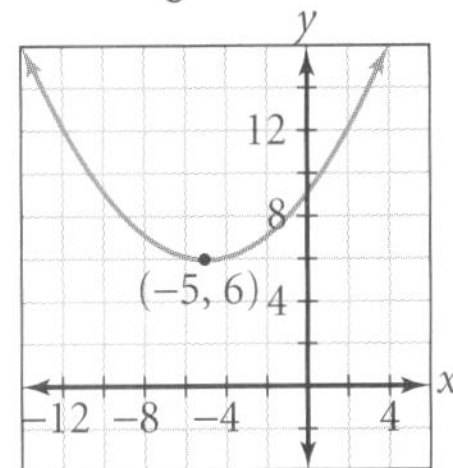

13. ellipse; $\frac{(x-4)^2}{4}+\frac{(y-4)^2}{9}=1$

14. hyperbola; $\frac{(x-3)^2}{4}-\frac{(y-1)^2}{9}=1$

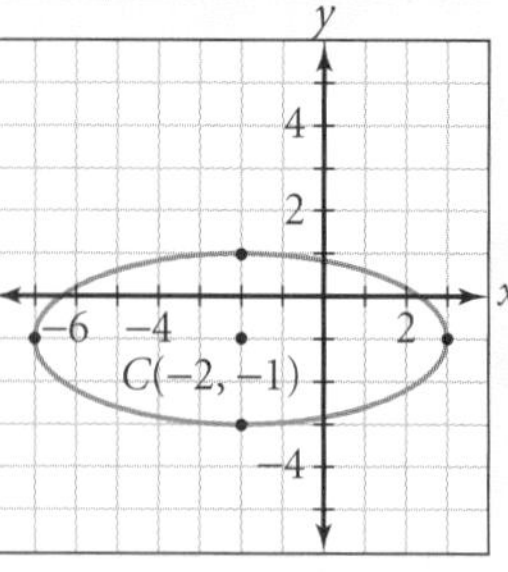

15. ellipse; $\frac{(x+2)^2}{25}+\frac{(y+1)^2}{4}=1$

Chapter 10

LESSON 10.1

Find the probability of each event.

1. A yellow marble is drawn at random from a bag containing 3 white, 2 yellow, 1 green, and 4 blue marbles. $\frac{1}{5}$
2. Arden arrives at home at 10:07 P.M. and is there to receive a package that was expected at any time between 10:00 P.M. and 10:15 P.M. $\frac{8}{15}$
3. When a number cube is rolled, a number less than 5 appears. $\frac{2}{3}$
4. A card chosen at random from a standard deck is red. $\frac{1}{2}$
5. When an 8-sided number die is rolled, a multiple of 3 appears. $\frac{1}{4}$
6. A card chosen at random from a standard deck is a jack, queen, king, or ace. $\frac{4}{13}$

A spinner is divided into three colored regions. The results of 125 spins are recorded in the table at right. Find the probability of each event.

red	53
blue	32
yellow	40

7. red $\frac{53}{125}=42.4\%$ **8.** blue $\frac{32}{125}=25.6\%$ **9.** yellow $\frac{8}{25}=32\%$

Find the number of possible passwords (with no letters or digits excluded) for each of the following conditions:

10. 2 letters followed by 3 digits, with no letters or digits excluded **676,000**
11. 3 letters followed by 3 digits, with only vowels and even digits allowed (Consider *y* a consonant.) **15,625**
12. 2 letters followed by 2 digits followed by 3 letters, with no letters or digits excluded **1,188,137,600**
13. In a 5-digit U.S. zip code, the last 2 digits identify the local delivery area. How many local delivery areas can be designated if any digit can be used? **100**
14. For a given telephone area code, how many 7-digit telephone numbers are possible if the first digit cannot be 0? 9×10^6 ***or*** **9,000,000**

LESSON 10.2

A city recreation program has one summer job available at each of its 6 parks. In how many ways can the 6 jobs be assigned if the given number of employees are available?

1. 6 **720** **2.** 10 **151,200** **3.** 12 **665,280** **4.** 15 **3,603,600**

Find the number of permutations of the digits 1–5 for each situation.

5. using all 5 digits **120** **6.** taking 3 digits at a time **60**

7. taking 4 digits at a time **120** **8.** taking 2 digits at a time **20**

Find the number of permutations of the letters in each word.

9. *heptagon* **40,320** **10.** *pentagon* **20,160** **11.** *circle* **360** **12.** *textbook* **10,080**

13. The United Nations Security Council has 15 member nations. In how many ways can the representatives be seated around a circular table if each nation has one representative? **about** 8.7×10^{10}
14. The 8 winning entries in a school art contest are to be displayed in a row. In how many different orders can the entries be displayed? **40,320**
15. A shoe store has 6 different models of running shoes on sale. In how many ways can a rotating display of the models be arranged? **120**

LESSON 10.3

Find the number of ways in which each committee can be selected.

1. 2 people from a group of 5 people **10**
2. 4 people from a group of 7 people **35**
3. 3 people from a group of 8 people **56**
4. 1 person from a group of 9 people **9**

A take-out restaurant offers a selection of 5 main dishes, 4 vegetables, and 3 desserts. In how many ways can a family choose a meal consisting of the following?

5. 2 different main dishes, 3 different vegetables, and 1 dessert **120**
6. 3 different main dishes, 2 different vegetables, and 3 different desserts **60**
7. 5 different main dishes, 4 different vegetables, and 3 different desserts **1**

Four marbles are chosen at random (without replacement) from a bag containing 4 white marbles and 6 green marbles. Find the probability of selecting each combination.

8. 4 green **≈7%**
9. 2 white *and* 2 green **≈43%**
10. 1 white *and* 3 green **≈38%**

Fifteen students are entered in a public-speaking contest. Determine whether each situation involves a permutation or a combination.

11. The order in which the contestants speak must be chosen. **permutation**
12. First, second, and third prizes are awarded. **permutation**
13. Two of the 15 represent the school in a regional contest. **combination**

14. In a survey of 50 voters, 33 favor a policy change, and 17 oppose it or have no opinion. Find the probability that in a random sample of 10 respondents from this survey, exactly 8 favor the proposed regulation and 2 oppose it or have no opinion. **about 0.18, or 18%**

LESSON 10.4

A card is drawn at random from a standard deck of playing cards. State whether the events *A* and *B* are inclusive or mutually exclusive. Then find *P*(*A or B*). (Note: A standard deck of 52 cards has 12 face cards, 4 of which are kings. Also, exactly half of the cards are red, including 6 of the face cards and 2 of the kings.)

1. *A*: The card is a queen. *B*: The card is a king. **mutually exclusive; $\frac{2}{13} \approx 15\%$**
2. *A*: The card is red. *B*: The card is a king. **inclusive; $\frac{7}{13} \approx 54\%$**
3. *A*: The card is red. *B*: The card is not a king. **inclusive; $\frac{25}{26} \approx 96\%$**
4. *A*: The card is a face card. *B*: The card is not a king. **inclusive; 1 = 100%**
5. *A*: The card is red. *B*: The card is not red. **mutually exclusive; 1 = 100%**
6. *A*: The card is a face card. *B*: The card is a king. **inclusive; $\frac{3}{13} \approx 23\%$**

A number cube is rolled once, and the number on the top face is recorded. Find the probability of each pair of events.

7. The number is even *or* greater than 5.
8. The number is 5 *or* a multiple of 3.
9. The number is odd *or* greater than 4.
10. The number is greater than 2 *or* less than 5.

7. $\frac{1}{2} = 50\%$ 8. $\frac{1}{2} = 50\%$ 9. $\frac{2}{3} \approx 67\%$ 10. 1 = 100%

Each of the digits from 0 to 9 is written on a card. The cards are placed in a sack, and one is drawn at random. Find the probability of each pair of events.

11. digit is odd *or* a multiple of 3 **$\frac{3}{5} = 60\%$**
12. digit is less than 2 *or* greater than 8 **$\frac{3}{10} = 30\%$**
13. digit is odd *or* less than 5 **$\frac{4}{5} = 80\%$**
14. digit is greater than 7 *or* even **$\frac{3}{5} = 60\%$**

Extra Practice

Extra Practice

LESSON 10.5

A coin is tossed 3 times. Find the probability of each event.

1. All 3 tosses are heads. $\frac{1}{8} = 12.5\%$

2. The first toss is heads, but the second and the third are tails. $\frac{1}{8} = 12.5\%$

Events *Q*, *R*, and *S* are independent, and $P(Q) = 0.2$, $P(R) = 0.4$, and $P(S) = 0.1$. Find each probability.

3. $P(Q \text{ and } R)$ $0.08 = 8\%$ **4.** $P(Q \text{ and } S)$ $0.02 = 2\%$ **5.** $P(R \text{ and } S)$ $0.04 = 4\%$

A red number cube and a green number cube are rolled. Find the probability of each event.

6. The red cube is a 3, *and* the green cube is greater than 3. $\frac{1}{12} \approx 8.3\%$

7. The red cube is greater than 1, *and* the green cube is less than 6. $\frac{25}{36} \approx 69.4\%$

8. The red cube is less than or equal to 4, *and* the green cube is greater than or equal to 5. $\frac{2}{9} \approx 22.2\%$

9. The green cube is less than or equal to 6, *and* the red cube is greater than or equal to 1. $1 = 100\%$

A bag contains 3 red, 2 green, and 5 blue marbles. A marble is picked at random and is replaced. A second marble is picked at random. Find the probability of each event.

10. Both marbles are red. $0.09 = 9\%$

11. The first marble is green, *and* the second is blue. $0.1 = 10\%$

12. Neither marble is green. $0.64 = 64\%$

13. The first marble is blue, *and* the second is not blue. $0.25 = 25\%$

A number cube is rolled twice. On each roll, the number on the top face of the cube is recorded. Find the probability of each event.

14. The first number is even, *and* the second number is greater than 3. $0.25 = 25\%$

15. The first number is greater than 4, *and* the second number is less than 3. $\frac{1}{9} \approx 11\%$

LESSON 10.6

A bag contains 5 red, 7 blue, and 4 white marbles. Two consecutive draws are made from the bag without replacement of the first draw. Find each probability.

1. red first *and* blue second $\frac{7}{48}$

2. red first *and* white second $\frac{1}{12}$

3. blue first *and* red second $\frac{7}{48}$

4. white first *and* white second $\frac{1}{20}$

Two number cubes are rolled, and the first cube shows a 3. Find the probability of each event below.

5. a sum of 8 $\frac{1}{6}$

6. one even number *and* one odd number $\frac{1}{2}$

7. a sum of less than 6 $\frac{1}{3}$

8. a sum of greater than 5 *and* less than 9 $\frac{1}{2}$

For one roll of a number cube, let *A* be the event "even" and let *B* be the event "4". Find each probability.

9. $P(A)$ $\frac{1}{2}$ **10.** $P(B)$ $\frac{1}{6}$ **11.** $P(A \text{ and } B)$ $\frac{1}{6}$ **12.** $P(A \text{ or } B)$ $\frac{1}{2}$ **13.** $P(A \mid B)$ 1 **14.** $P(B \mid A)$ $\frac{1}{3}$

For one roll of a number cube, let *A* be the event "less than 4" and let *B* be the event "1 *or* 2". Find each probability.

15. $P(A)$ $\frac{1}{2}$ **16.** $P(B)$ $\frac{1}{3}$ **17.** $P(A \text{ and } B)$ $\frac{1}{3}$ **18.** $P(A \text{ or } B)$ $\frac{1}{2}$ **19.** $P(A \mid B)$ 1 **20.** $P(B \mid A)$ $\frac{2}{3}$

21. Given that $P(A \text{ and } B) = 0.2$ and $P(A) = 0.5$, find $P(B \mid A)$. 0.4

22. Given that $P(B \mid A) = 0.8$ and $P(A \text{ and } B) = 0.6$, find $P(A)$. 0.75

LESSON 10.7

Use a simulation with 20 trials to estimate each probability. Simulation results may vary.

1. In 3 tosses of a coin, 2 consecutive heads will appear. **≈38%**
2. In 4 tosses of a coin, tails will appear exactly once. **≈25%**
3. In 3 rolls of a number cube, the number 3 will appear exactly twice. **≈7%**
4. In 5 rolls of a number cube, the number 4 will appear exactly 3 times. **≈3%**
5. In 6 rolls of a number cube, the number 1 will not appear. **≈33%**

Of 150 motorists observed at an intersection, 47 turned left, 72 went straight, and 31 turned right. Use a simulation with 10 trials to estimate the probability of each event.

6. At least 2 out of every 5 consecutive motorists go straight. **≈79%**
7. More than 1 out of every 5 consecutive motorists turn left. **≈50%**
8. Less than 3 out of every 5 consecutive motorists turn right. **≈94%**
9. At least 3 out of every 5 consecutive motorists do not go straight. **≈54%**
10. No more than 1 out of every 5 consecutive motorists goes left. **≈50%**

11. Assume that a person who is learning to play darts has acquired enough skill to hit the target but is just as likely to hit any one spot on the target as any other. Use a simulation with 10 trials to estimate the probability that exactly 3 out of 5 darts that land on the square target shown at right land inside the circle. **≈22%**

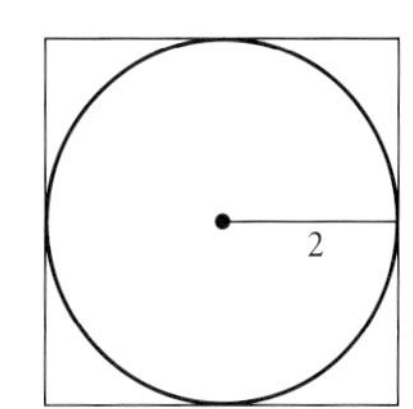

Chapter 11

LESSON 11.1

Write the first five terms of each sequence.

1. $t_n = -3n + 8$ **5, 2, –1, –4, –7**
2. $t_n = 4n - 12$ **–8, –4, 0, 4, 8**
3. $t_n = 2n^2$ **2, 8, 18, 32, 50**
4. $t_1 = 1$, $t_n = t_{n-1} + 5$ **1, 6, 11, 16, 21**
5. $t_1 = 16$, $t_n = t_{n-1} - 6$ **16, 10, 4, –2, –8**
6. $t_1 = 3$, $t_n = 2t_{n-1}$ **3, 6, 12, 24, 48**

Write a recursive formula for each sequence and find the next three terms.

7. 1, 10, 19, 28, . . . **$t_1 = 1$, $t_n = t_{n-1} + 9$; 37, 46, 55**
8. 2, 7, 12, 17, . . . **$t_1 = 2$, $t_n = t_{n-1} + 5$; 22, 27, 32**
9. 4, 9, 19, 34, . . . **$t_1 = 4$, $t_n = t_{n-1} + 5(n-1)$; 54, 79, 109**

Write the terms of each series and then evaluate the sum.

10. $\sum_{n=1}^{6}(12n + 1)$ **13, 25, 37, 49, 61, 73; 258**
11. $\sum_{j=1}^{5}(j - 1)^2$ **0, 1, 4, 9, 16; 30**
12. $\sum_{k=1}^{4}(3k^2 + 4k)$ **7, 20, 39, 64; 130**

Evaluate.

13. $\sum_{m=1}^{6} 5m$ **105**
14. $\sum_{n=1}^{5}(10n - 3)$ **135**
15. $\sum_{j=1}^{4}(2j^2 - 3j + 1)$ **34**

Extra Practice

Extra Practice

LESSON 11.2

Based on the terms given, state whether each sequence is arithmetic. If it is, identify the common difference, d.

1. 5, 7, 10, 14, 19, . . . **no**

2. –8, –5, –2, 1, 4, . . . **yes; 3**

3. 1, 8, 27, 64, 125, . . . **no**

4. 0.1, 0.2, 0.3, 0.4, 0.5, . . . **yes; 0.1**

5. 4.4, 5.5, 6.6, 7.7, . . . **yes; 1.1**

6. 1.2, 3.4, 5.6, 7.8, . . . **yes; 2.2**

Write an explicit formula for the nth term of each arithmetic sequence.

7. 5, 8, 11, 14, 17, . . . $\mathbf{t_n = 3n + 2}$

8. 7, 3, –1, –5, –9, . . . $\mathbf{t_n = -4n + 11}$

9. 10, 16, 22, 28, 34, . . . $\mathbf{t_n = 6n + 4}$

10. 30, 37, 44, 51, 58, . . . $\mathbf{t_n = 7n + 23}$

11. –12, –8, –4, 0, 4, . . . $\mathbf{t_n = 4n - 16}$

12. 40, 33, 26, 19, 12, . . . $\mathbf{t_n = -7n + 47}$

List the first four terms of each arithmetic sequence.

13. $t_1 = 5, t_n = t_{n-1} + 11$ **5, 16, 27, 38**

14. $t_1 = -13, t_n = t_{n-1} + 10$ **–13, –3, 7, 17**

15. $t_1 = 0, t_n = t_{n-1} + 20$ **0, 20, 40, 60**

16. $t_n = 5n - 8$ **–3, 2, 7, 12**

17. $t_n = -8n - 4$ **–12, –20, –28, –36**

18. $t_n = 6n - 10$ **–4, 2, 8, 14**

Find the indicated number of arithmetic means between the two given numbers.

19. three arithmetic means between 5 and 29 **11, 17, and 23**

20. four arithmetic means between –12 and 28 **–4, 4, 12, and 20**

21. two arithmetic means between 6.5 and 15.8 **9.6 and 12.7**

22. three arithmetic means between –6 and 4 **–3.5, –1, and 1.5**

LESSON 11.3

Use the formula for an arithmetic series to find each sum.

1. 3 + 6 + 9 + 12 + 15 **45**

2. –8 + (–15) + (–22) + (–29) + (–36) **–110**

3. 40 + 42 + 44 + · · · + 68 **810**

4. –45 + (–40) + (–35) + (–30) + · · · + 25 **–150**

5. Find the sum of the first 175 natural numbers. **15,400**

6. Find the sum of the multiples of 6 from 18 to 120 inclusive. **1242**

7. Find the sum of the multiples of 8 from 40 to 480 inclusive. **14,560**

For each arithmetic series, find S_{20}.

8. 4, 8, 12, 16, 20, 24, . . . **840**

9. 3, 8, 13, 18, 23, . . . **1010**

10. –15, –12, –9, –6, –3, . . . **270**

11. –20, –40, –60, –80, –100, . . . **–4200**

12. $\pi, 3\pi, 5\pi, 7\pi, 9\pi, \ldots$ $\mathbf{400\pi \approx 1257}$

13. $\sqrt{7}, 4\sqrt{7}, 7\sqrt{7}, 10\sqrt{7}, \ldots$ $\mathbf{590\sqrt{7} \approx 1561}$

Evaluate.

14. $\sum_{j=1}^{5}(24 - 3j)$ **75**

15. $\sum_{n=1}^{8}(15n - 1)$ **532**

16. $\sum_{k=1}^{10}(3k + 50)$ **665**

17. $\sum_{m=1}^{20}(120 - 10m)$ **300**

18. $\sum_{n=1}^{12}(25n + 4)$ **1998**

19. $\sum_{k=1}^{15}(10 + 20k)$ **2550**

LESSON 11.4

Determine whether each sequence is a geometric sequence. If so, identify the common ratio, *r*, and give the next three terms.

1. $12, 6, 3, \frac{3}{2}, \ldots$ **yes;** $\frac{1}{2}$; $\frac{3}{4}, \frac{3}{8}, \frac{3}{16}$

2. $11, 22, 44, 88, \ldots$ **yes; 2; 176, 352, 704**

3. $-2, -6, -18, -54, \ldots$ **no**

4. $27, 9, 3, 1, \ldots$ **yes;** $\frac{1}{3}$; $\frac{1}{3}, \frac{1}{9}, \frac{1}{27}$

5. $25, 36, 49, 64, \ldots$ **no**

6. $25, 2.5, 0.25, 0.025, \ldots$ **yes; 0.1; 0.0025, 0.00025, 0.000025**

List the first four terms of each geometric sequence.

7. $t_1 = 5$, $t_n = 0.2t_{n-1}$ **5, 1, 0.2, 0.04**

8. $t_1 = 4$, $t_n = 10t_{n-1}$ **4, 40, 400, 4000**

9. $t_1 = -2$, $t_n = -4.5t_{n-1}$ **−2, 9, −40.5, 182.25**

Find the fifth term in the geometric sequence that includes the given terms.

10. $t_2 = 48$; $t_3 = 144$ **1296**

11. $t_2 = 224$; $t_4 = 14$ **3.5**

12. $t_3 = 75$; $t_8 = 234{,}375$ **1875**

Write an explicit formula for the *n*th term of each geometric sequence.

13. $0.04, 0.2, 1, 5, \ldots$ $t_n = 0.04 \cdot 5^{n-1}$

14. $16, 8, 4, 2, \ldots$ $t_n = 16\left(\frac{1}{2}\right)^{n-1}$

15. $\sqrt{6}, 6, 6\sqrt{6}, 36, \ldots$ $t_n = 6^{\frac{n}{2}}$ *or* $(\sqrt{6})^n$

Find the indicated number of geometric means between the two given numbers.

16. two geometric means between 12 and 324 **36 and 108**

17. two geometric means between 6.4 and 21.6 **9.6 and 14.4**

18. three geometric means between 16 and 81 **24, 36, and 54 *or* −24, 36, and −54**

19. three geometric means between 8 and 312.5 **20, 50, and 125 *or* −20, 150, and −125**

LESSON 11.5

Find each sum. Round answers to the nearest tenth, if necessary.

1. S_{10} for the geometric series $3 + 6 + 12 + 24 + \cdots$ **3069**

2. S_8 for the geometric series $-32 + 16 + (-8) + 4 + (-2) + \cdots$ **−21.3**

3. $\frac{3}{4} + \frac{3}{8} + \frac{3}{16} + \frac{3}{32} + \frac{3}{64} + \frac{3}{128}$ $\frac{189}{128} \approx 1.5$

4. $-0.48 + 2.4 - 12 + 60 - 300$ **−250.1**

For Exercises 5–8, refer to the series 0.2 + 0.6 + 1.8 + 5.4 + · · ·

5. Find t_8. **437.4**

6. Find t_{16}. **2,869,781.4**

7. Find S_8. **656**

8. Find S_{16}. **4,304,672**

Evaluate. Round answers to the nearest tenth, if necessary.

9. $\sum_{k=1}^{8} 2(3^k - 1)$ **19,664**

10. $\sum_{n=1}^{6} 4(0.5^n)$ **3.9**

11. $\sum_{k=1}^{10} 3.5^{k-1}$ **110,341.5**

12. $\sum_{k=1}^{20} 0.5(2^{k-1})$ **524,287.5**

Use mathematical induction to prove that each statement is true for every natural number, *n*.

13. $\frac{1}{2} + \frac{1}{2^2} + \frac{1}{2^3} + \cdots + \frac{1}{2^n} = 1 - \frac{1}{2^n}$

14. $1^3 + 3^3 + 5^3 + \cdots + (2n-1)^3 = n^2(2n^2 - 1)$

Extra Practice

Lesson 11.5

13. $\frac{1}{2} = 1 - \frac{1}{2}$, so the statement is true for $n = 1$. Assume that the statement is true for an integer k. Then $\frac{1}{2} + \frac{1}{2^2} + \frac{1}{2^3} + \cdots + \frac{1}{2^k} = 1 - \frac{1}{2^k}$, and $\frac{1}{2} + \frac{1}{2^2} + \frac{1}{2^3} + \cdots + \frac{1}{2^k} + \frac{1}{2^{k+1}} = 1 - \frac{1}{2^k} + \frac{1}{2^{k+1}} = 1 - \left(\frac{1}{2^k} - \frac{1}{2^{k+1}}\right) = 1 - \left(\frac{2}{2^{k+1}} - \frac{1}{2^{k+1}}\right) = 1 - \frac{1}{2^{k+1}}$, and the statement is true for $k + 1$.

14. $1^3 = 1$ and $1^2(2 - 1) = 1$, so the statement is true for $n = 1$. Assume that the statement is true for an integer k. Then $1^3 + 3^3 + 5^3 + \cdots + (2k-1)^3 = k^2(2k^2 - 1)$, and $1^3 + 3^3 + 5^3 + \cdots + (2k-1)^3 + [2(k+1) - 1]^3 = k^2(2k^2 - 1) + (2k+1)^3 = 2k^4 + 8k^3 + 11k^2 + 6k + 1$. But $(k+1)^2[2(k+1)^2 - 1] = (k+1)^2(2k^2 + 4k + 1) = (k^2 + 2k + 1)(2k^2 + 4k + 1) = 2k^4 + 8k^3 + 11k^2 + 6k + 1$. Since the two expressions are equal, the statement is true for $k + 1$.

Extra Practice

Lesson 11.8

1. $a^4 + 4a^3b + 6a^2b^2 + 4ab^3 + b^4$
2. $w^5 + 5w^4z + 10w^3z^2 + 10w^2z^3 + 5wz^4 + z^5$
3. $c^6 + 6c^5d + 15c^4d^2 + 20c^3d^3 + 15c^2d^4 + 6cd^5 + d^6$
4. $1 + 3x + 3x^2 + x^3$
5. $y^5 + 10y^4 + 40y^3 + 80y^2 + 80y + 32$
6. $z^4 + 4z^3 + 6z^2 + 4z + 1$
9. $32x^5 + 80x^4y + 80x^3y^2 + 40x^2y^3 + 10xy^4 + y^5$
10. $81z^4 - 216z^3 + 216z^2 - 96z + 16$
11. $x^4 - 8x^3y + 24x^2y^2 - 32xy^3 + 16y^4$
12. $243y^5 - 810y^4z + 1080y^3z^2 - 720y^2z^3 + 240yz^4 - 32z^5$
13. $\frac{1}{8}x^3 + \frac{1}{4}x^2y + 6xy^2 + 8y^3$
14. $\frac{8}{27}x^3 - \frac{4}{3}x^2y + 2xy^2 - y^3$

Extra Practice

LESSON 11.6

Find the sum of each infinite geometric series, if it exists.

1. $\frac{1}{5} + \frac{1}{25} + \frac{1}{125} + \frac{1}{625} + \cdots$ **$\frac{1}{4}$**
2. $20 + 12 + 7.2 + 4.32 + \cdots$ **50**
3. $0.2 + 0.4 + 0.8 + 1 + \cdots$ **none**
4. $5 + \frac{5}{7} + \frac{5}{49} + \frac{5}{343} + \cdots$ **$5\frac{5}{6} \approx 5.83$**

Find the sum of each infinite geometric series, if it exists.

5. $\sum_{n=0}^{\infty} 4^n$ **none**
6. $\sum_{k=1}^{\infty} \frac{3}{5^k}$ **0.75**
7. $\sum_{n=1}^{\infty} 0.4^n - 1$ **$-\frac{1}{3}$**
8. $\sum_{k=0}^{\infty} (-0.25)^k$ **0.8**

Write each decimal as a fraction in simplest form.

9. $0.\overline{7}$ **$\frac{7}{9}$**
10. $0.\overline{23}$ **$\frac{23}{99}$**
11. $0.\overline{321}$ **$\frac{107}{333}$**
12. $0.\overline{726}$ **$\frac{242}{333}$**

Write an infinite geometric series that converges to the given number.

13. $0.3131313131\ldots$ **$\sum_{n=1}^{\infty} \frac{31}{100^n}$**
14. $0.4747474747\ldots$ **$\sum_{n=1}^{\infty} \frac{47}{100^n}$**
15. $0.357357357\ldots$ **$\sum_{n=1}^{\infty} \frac{357}{1000^n}$**

LESSON 11.7

State the location of each entry in Pascal's triangle. Then give the value of each expression.

1. ${}_4C_2$ **third entry, row 4; 6**
2. ${}_9C_5$ **sixth entry, row 9; 126**
3. ${}_8C_3$ **fourth entry, row 8; 56**
4. ${}_6C_5$ **sixth entry, row 6; 6**
5. ${}_{10}C_7$ **eighth entry, row 10; 120**
6. ${}_{15}C_7$ **eighth entry, row 15; 6435**
7. ${}_{11}C_8$ **ninth entry, row 11; 165**
8. ${}_{20}C_{10}$ **eleventh entry, row 20; 184,756**

Find the fifth and eighth entries in the indicated row of Pascal's triangle.

9. row 8 **70; 8**
10. row 10 **210; 120**
11. row 13 **715; 1716**
12. row 16 **1820; 11,440**

Find the probability of each event.

13. exactly 3 heads in 4 tosses of a fair coin **0.25**
14. 3 *or* 4 heads in 8 tosses of a fair coin **≈0.49**
15. no more than 3 heads in 7 tosses of a fair coin **0.5**
16. no fewer than 4 heads in 9 tosses of a fair coin **≈0.75**
17. 3 *or* 4 *or* 5 heads in 10 tosses of a fair coin **≈0.57**

A student guesses the answers for 8 items on a true-false quiz. Find the probability that the indicated number of answers is correct.

18. exactly 6 **≈0.11**
19. at least 5 **≈0.36**
20. at most 4 **≈0.64**

LESSON 11.8

Expand each binomial.

1. $(a + b)^4$
2. $(w + z)^5$
3. $(c + d)^6$
4. $(1 + x)^3$
5. $(y + 2)^5$
6. $(z + 1)^4$

For Exercises 7–8, refer to the expansion of $(x + y)^{12}$.

7. How many terms are in the expansion? **13 terms**
8. What is the exponent of x in the term containing y^7? What is the term? **5; $792x^5y^7$**

Expand each binomial.

9. $(2x + y)^5$
10. $(3z - 2)^4$
11. $(x - 2y)^4$
12. $(3y - 2z)^5$
13. $\left(\frac{1}{2}x + 2y\right)^3$
14. $\left(\frac{2}{3}x - y\right)^3$

Use the Binomial Theorem to find each theoretical probability for a baseball player with a batting average of 0.250.

15. exactly 4 hits in 5 at bats **about 0.01**
16. no more than 3 hits in 5 at bats **about 0.98**
17. exactly 2 hits in 6 at bats **about 0.30**
18. no more than 3 hits in 6 at bats **about 0.96**

Chapter 12

LESSON 12.1

Find the mean, median, and mode of each data set. Round answers to the nearest thousandth, if necessary.

1. 12, 16, 22, 45, 30, 58, 11, 21, 29, 37 **$\bar{x} = 28.1$; 25.5; no mode**
2. 92, 90, 88, 88, 99, 70, 55, 85, 92, 93, 90 **$\bar{x} = 85.636$; 90; 88, 90, and 92**
3. 12.6, 8.5, 7.7, 9.9, 12.8, 12.6, 12.5, 13.2 **$\bar{x} = 11.225$; 12.55; 12.6**
4. 5, 5, 6, 16, 24, 32, 5, 66, 7, 10, 22, 6 **$\bar{x} = 17$; 8.5; 5**

Find the mean, median, and mode of each data set and compare the three measures. **$\bar{x} = 30{,}100$; 29,300; 35,700**

5. minimum starting salaries (in dollars per year) in selected professional specialties: 35,700; 24,700; 34,100; 35,700; 22,900; 29,300; 28,300
6. number of people (in millions) viewing prime-time television in a given week: 94.5, 93.2, 85.2, 88.8, 79.2, 77.6, 92.8 **$\bar{x} \approx 87.3$; 88.8; no mode**

Make a frequency table for each data set and find the mean.

7. workers' sick days in one year: 0, 5, 2, 1, 3, 5, 5, 10, 22, 0, 0, 4, 3, 2, 0, 0, 1, 1, 1, 8 **3.65**
8. the number of bicycles in students' families: 1, 4, 3, 3, 6, 2, 4, 4, 3, 1, 1, 1, 2, 2, 2, 3, 5, 4, 4, 3 **2.9**

Make a grouped frequency table for each data set and estimate the mean.

9. class test scores: 88, 72, 65, 58, 90, 71, 66, 82, 76, 75, 77, 91, 56, 70, 92, 80, 66, 86, 84, 75 **76.5**
10. the number of hours worked per week: 40, 32, 30, 44, 40, 52, 30, 25, 20, 42, 46, 38, 35, 27, 55, 51 **≈38.9**

9.

Score	Class mean	Freq.	Product
50–59	54.5	2	109
60–69	64.5	3	193.5
70–79	74.5	7	521.5
80–89	84.5	5	422.5
90–99	94.5	3	283.5

estimated mean: 76.5

10.

Score	Class mean	Freq.	Product
20–24	22	1	22
25–29	27	2	54
30–34	32	3	96
35–39	37	2	74
40–44	42	4	168
45–49	47	1	47
50–54	52	2	104
55–59	57	1	57

estimated mean: ≈38.9

Lesson 12.1

5. 30,100; 29,300; 35,700; sample answer: The mode is too high, but the mean and median are reasonable values to describe the data.

6. about 87.3; 88.8; no mode; sample answer: The mean and median are reasonable, and there is no mode.

7.

Number of days	Tally	Frequency
0	𝍸	5
1	IIII	4
2	II	2
3	II	2
4	I	1
5	III	3
8	I	1
10	I	1
22	I	1

mean: 3.65

8.

No. of bicycles	Tally	Frequency
1	IIII	4
2	IIII	4
3	𝍸	5
4	𝍸	5
5	I	1
6	I	1

mean: 2.9

Lesson 12.2

1.

Stem	Leaf 3|0 = 30
1	2, 2, 5, 6
2	2, 5, 6
3	0, 2, 3
4	0, 3, 6, 7

28; 12; flat-shaped

2.

Stem	Leaf 7|2 = 72
5	0, 4, 5, 5, 7, 8
6	0, 1
7	2, 4, 6, 8
8	7, 8
9	0, 2, 6

72; 55; mound-shaped

3.

Number	Frequency
1	2
2	5
3	4
4	2
5	6
6	4
7	5

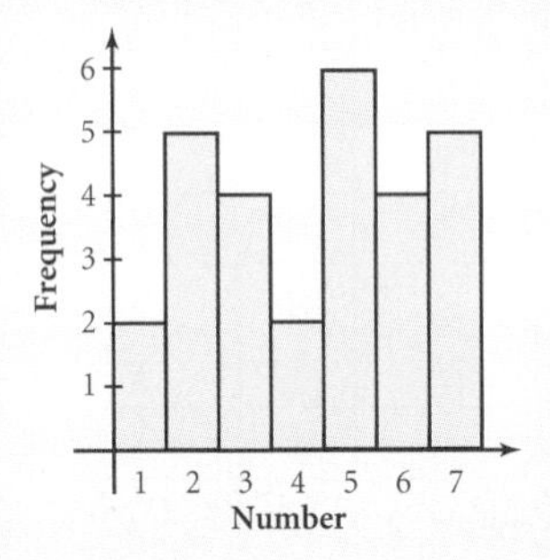

LESSON 12.2

Make a stem-and-leaf plot for each data set. Then find the median and the mode, and describe the distribution of the data.

1. 16, 22, 26, 43, 30, 12, 15, 40, 47, 25, 46, 33, 32, 12 **28; 12; flat-shaped**
2. 55, 87, 92, 50, 54, 58, 57, 72, 88, 96, 90, 78, 74, 55, 60, 61, 76 **72; 55; mound-shaped**

Make a frequency table and a histogram for each data set.

3. 3, 5, 7, 2, 2, 3, 5, 5, 6, 6, 6, 2, 3, 4, 1, 5, 7, 5, 4, 6, 7, 7, 2, 1, 2, 7, 5, 3
4. 13, 16, 15, 12, 12, 15, 18, 20, 17, 16, 15, 16, 12, 13, 17, 19, 20, 15, 15, 14

Make a relative frequency table and a histogram of probabilities for each data set.

5. 20, 22, 25, 21, 24, 22, 25, 23, 23, 20, 21, 25, 22, 21, 24, 21, 22, 23, 25, 22
6. 5, 4, 8, 5, 8, 4, 6, 8, 4, 5, 8, 8, 4, 8, 4, 8, 6, 5, 5, 5, 7, 6, 5, 7, 5

The table below lists the number of United States military personnel by branch.

United States Military Personnel, 1996

Army	Air Force	Navy	Marine Corps
493,330	389,400	436,608	172,287

7a. Make a circle graph to represent this data set.

b. Find the probability that a randomly selected person in the U.S. military is in the marines or the navy. **≈0.41**

LESSON 12.3

Find the minimum and maximum values, quartiles, range, and interquartile range for each data set. Then make a box-and-whisker plot for each data set.

1. 12, 35, 22, 18, 16, 21, 19, 33, 7, 10, 14, 28, 27, 16, 13
2. 5.4, 7.8, 1.1, 9.2, 12.6, 15.5, 18.0, 16.2, 18.8, 12.1, 13.2, 13.2, 15.0, 16.3, 20.2
3. 210, 185, 340, 715, 224, 290, 168, 312, 272, 300
4. 47, 40, 31, 22, 62, 50, 43, 28, 47, 35, 32, 44, 29, 28, 56, 50, 52, 54, 36, 20, 22
5. 5.0, 6.5, 8.0, 3.2, 8.1, 7.4, 6.7, 6.2, 5.0, 12.3, 5.7, 6.3, 6.8, 5.7, 7.2, 8.4

The box-and-whisker plots below compare the state per capita personal incomes in dollars for 1990 and 1995. Refer to the box-and-whisker plots for Exercises 6–8.

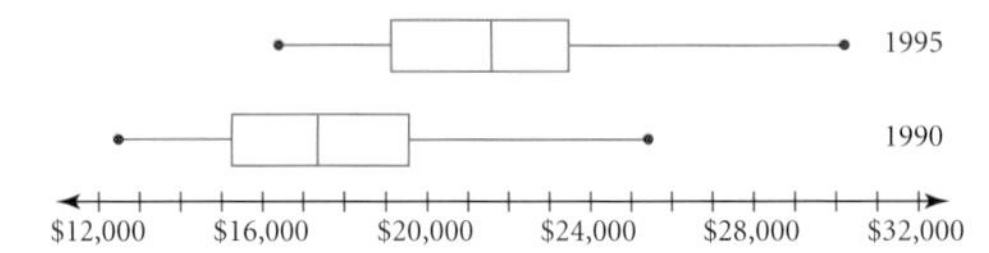

6. Which data set has the greater range? **1995 data**
7. Compare the interquartile ranges for the two data sets. **They are about the same.**
8. Q_1 for the 1995 data and Q_3 for the 1990 data are about the same. Describe what this means in terms of the distribution of the data.

4.

Number	Frequency
12	3
13	2
14	1
15	5
16	3
17	2
18	1
19	1
20	2

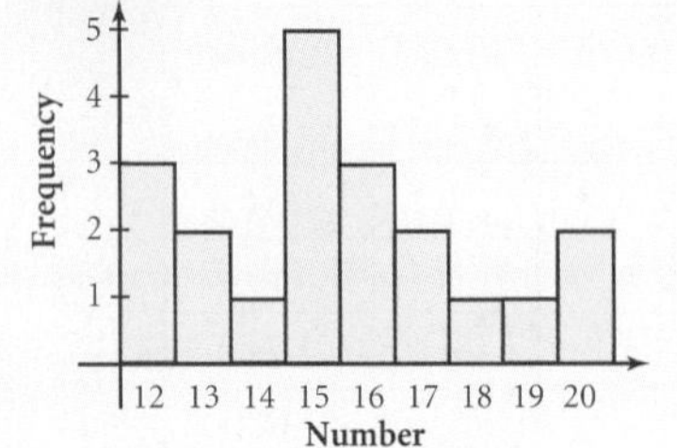

5.

Number	Frequency	Rel. Freq.
20	2	10%
21	4	20%
22	5	25%
23	3	15%
24	2	10%
25	4	20%

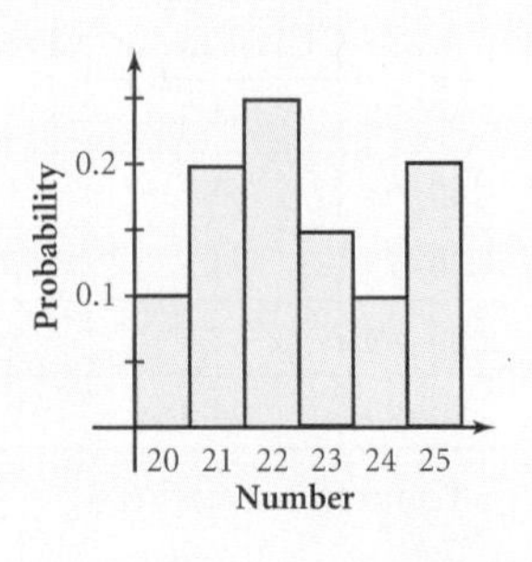

LESSON 12.4

Find the range and the mean deviation for each data set.

1. 6, 5, 3, 2, 4, 6, 8, 6 **6; 1.5**

2. 20, 22, 20, 21, 23, 26 **6; 1.7**

3. 5, 8, 10, 16, 9, 12 **11; 2.7**

4. 28, 40, 20, 32, 30, 54 **34; 8.7**

5. −12, 25, 17, −8, 15 **37; 13.9**

6. 3.5, 4.0, 2.8, 3.8, 7.2, 7.5 **4.7; 1.7**

Find the variance and the standard deviation for each data set.

7. 52, 61, 54, 48, 72 **71.04; ≈8.4**

8. 115, 120, 132, 140, 113 **107.6; ≈10.4**

9. 61, 20, 93, 72, 30, 24 **$738.\overline{3}$; ≈27.2**

10. 0.4, 1.1, 6.9, 9.8, 6.3 **12.9; ≈3.6**

11. 9, 7, 3, 1, 2, 8 **9.7; 3.1**

12. 59, 60, 7, 37, 69 **499; ≈22.3**

The table below shows the number of free throws made by the 15 best free-throw shooters in the National Basketball Association for the 1995–1996 season. Refer to the table for Exercises 13–14.

146	259	338	130	167	425	430	132
125	342	137	146	135	247	172	

13. Find the range and the mean deviation of the data. **305; ≈94.5**

14. Find the standard deviation of the data. **≈107.1**

LESSON 12.5

A coin is flipped 5 times. Find the probability of each event.

1. exactly 3 heads **31.25%**

2. exactly 2 heads **31.25%**

3. at least 4 heads **18.75%**

4. at most 4 heads **96.875%**

5. at most 1 heads **18.75%**

6. at least 2 heads **81.25%**

A spinner is divided into 6 congruent segments, each labeled with one of the letters *A–F*. Find each of the following probabilities:

7. exactly 3 *A*s in 5 spins **≈3.2%**

8. exactly 2 *D*s in 4 spins **≈11.6%**

9. more than 2 *B*s in 5 spins **≈3.5%**

10. no more than 2 *C*s in 4 spins **≈98.4%**

Find the probability that a batter will get *exactly* 3 hits in her next 6 at bats given each batting average below.

11. 0.300 **≈18.5%**

12. 0.280 **≈16.4%**

13. 0.312 **≈19.8%**

Find the probability that a batter will get *at least* 3 hits in his next 6 at bats given each batting average below.

14. 0.290 **23.7%**

15. 0.285 **22.8%**

16. 0.315 **28.4%**

Lesson 12.3

1. min = 7; Q_1 = 13; Q_2 = 18; Q_3 = 27; max = 35; range = 28; IQR = 14

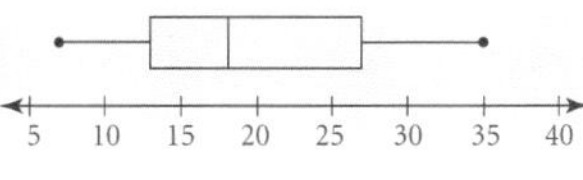

2. min = 1.1; Q_1 = 9.2; Q_2 = 13.2; Q_3 = 16.3; max = 20.2; range = 19.1; IQR = 7.1

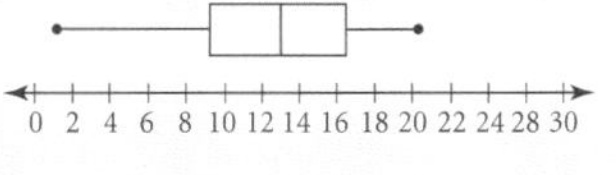

3. min = 168; Q_1 = 210; Q_2 = 281; Q_3 = 312; max = 715; range = 547; IQR = 102

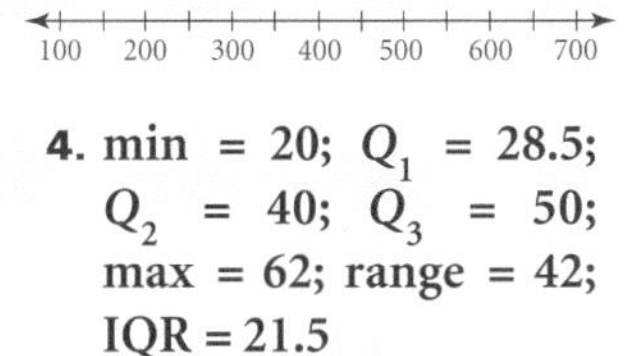

4. min = 20; Q_1 = 28.5; Q_2 = 40; Q_3 = 50; max = 62; range = 42; IQR = 21.5

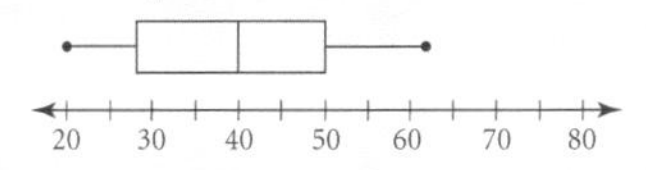

5. min = 3.2; Q_1 = 5.7; Q_2 = 6.6; Q_3 = 7.7; max = 12.3; range = 9.1; IQR = 2

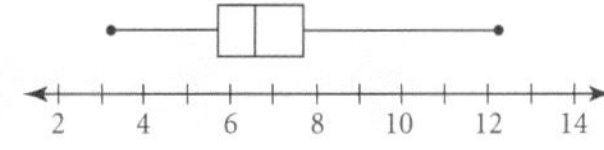

8. Both numbers appear to be slightly greater than \$19,000. About 75% of the data for 1995 are greater than that number, while only about 25% of the data for 1990 are greater than that number.

6.

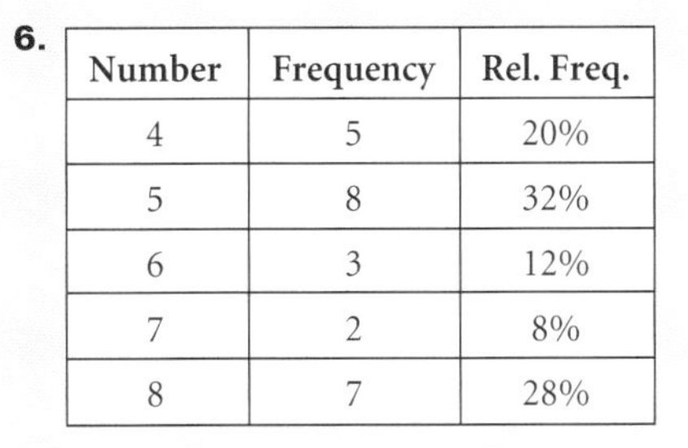

Number	Frequency	Rel. Freq.
4	5	20%
5	8	32%
6	3	12%
7	2	8%
8	7	28%

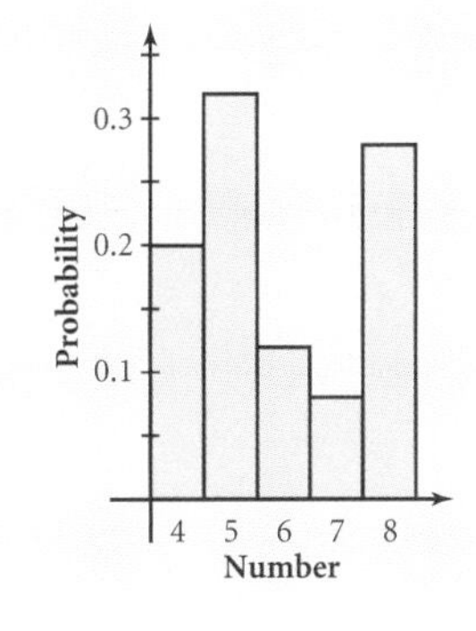

7a.

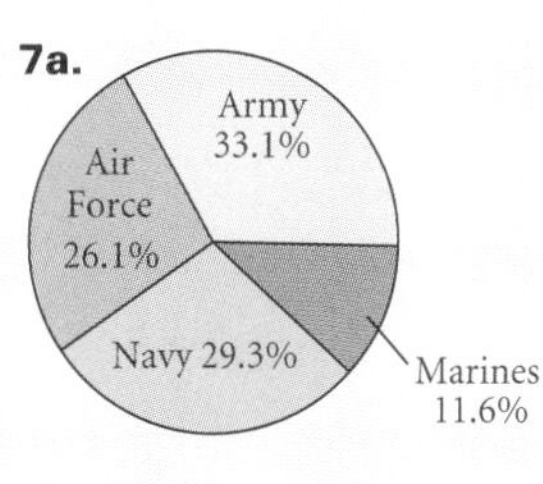

b. ≈ 0.41

LESSON 12.6

Let x be a random variable with a standard normal distribution. Use the area table for a standard normal curve, given on page 807, to find each probability.

1. $P(x \le 0.4)$ **0.6554**

2. $P(x \ge 0.6)$ **0.2743**

3. $P(1.4 \le x \le 1.8)$ **0.0449**

4. $P(x \le -0.8)$ **0.2119**

5. $P(-0.2 \le x \le 1.4)$ **0.4985**

6. $P(x \ge -0.6)$ **0.7257**

At one university, the ages of first-year students are approximately normally distributed with a mean of 19 and a standard deviation of 1. A first-year student is chosen at random. Find the probability that the student is

7. not less than 21 years old. **≈0.02**

8. 19 or younger. **0.5**

9. between 18 and 20 inclusive. **≈0.68**

10. between 19 and 21 inclusive. **≈0.48**

On an assembly line, the time required to perform a certain task is approximately normally distributed with a mean of 140 seconds and a standard deviation of 10 seconds. Of 1000 separate tasks, how many can be expected to take the given number of seconds?

11. more than 150 seconds **≈160**

12. less than 150 seconds **≈840**

13. between 120 and 150 seconds **≈820**

14. more than 160 seconds **≈20**

Chapter 13

LESSON 13.1

Refer to $\triangle XYZ$ at right in order to find each value below. Give exact answers and answers rounded to the nearest ten-thousandth.

Triangle XYZ: $YZ = 21$, $ZX = 20$, $YX = 29$, right angle at Z.

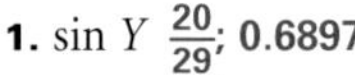

1. $\sin Y$ **$\frac{20}{29}$; 0.6897**

2. $\sin X$ **$\frac{21}{29}$; 0.7241**

3. $\cos Y$ **$\frac{21}{29}$; 0.7241**

4. $\cos X$ **$\frac{20}{29}$; 0.6897**

5. $\tan Y$ **$\frac{20}{21}$; 0.9524**

6. $\tan X$ **$\frac{21}{20}$; 1.05**

7. $\csc Y$ **$\frac{29}{20}$; 1.45**

8. $\csc X$ **$\frac{29}{21}$; 1.3810**

9. $\sec Y$ **$\frac{29}{21}$; 1.3810**

10. $\sec X$ **$\frac{29}{20}$; 1.45**

11. $\cot Y$ **$\frac{21}{20}$; 1.05**

12. $\cot X$ **$\frac{20}{21}$; 0.9524**

Solve each triangle. Round angle measures to the nearest degree and side lengths to the nearest tenth.

13.

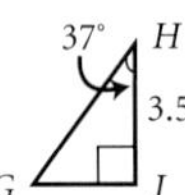

m∠G = 53°; GI ≈ 2.6; GH ≈ 4.4

14.

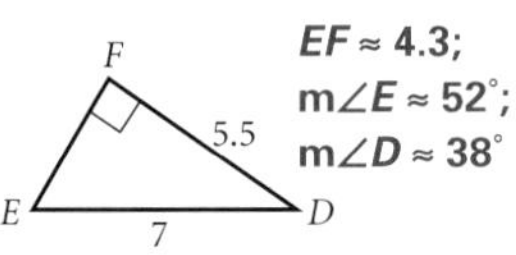

EF ≈ 4.3; m∠E ≈ 52°; m∠D ≈ 38°

15. Triangle JLK: $JL = 8$, $LK = 12$, right angle at L.

JK ≈ 14.4; m∠J ≈ 56°; m∠K ≈ 34°

16.

AB ≈ 6.7; m∠A ≈ 27°; m∠B ≈ 63°

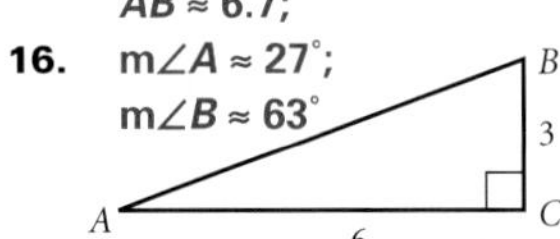

17. Triangle NOM: $m\angle N = 26°$, $OM = 9.7$, right angle at O.

m∠M = 64°; MN ≈ 22.1; NO ≈ 19.9

18.

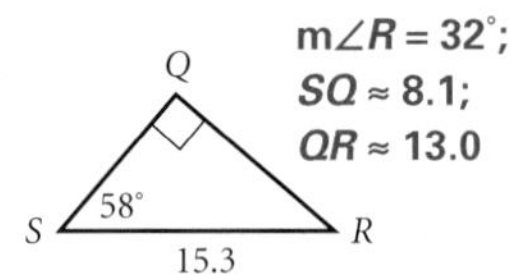

m∠R = 32°; SQ ≈ 8.1; QR ≈ 13.0

LESSON 13.2

For each angle below, find all coterminal angles such that $-360° < \theta < 360°$.

1. 114° **−246°**
2. 22° **−338°**
3. −53° **307°**
4. −272° **88°**
5. 512° **152°, −208°**
6. −495° **−135°, 225°**

Find the reference angle for each angle below.

7. 117° **63°**
8. −78° **78°**
9. 1024° **56°**
10. −512° **28°**
11. 245° **65°**
12. 311° **49°**

Given each point on the terminal side of θ in standard position, find the exact values of the six trigonometric functions of θ.

13. $(3, 6)$
14. $(-4, 5)$
15. $(3, -8)$
16. $(2, -\sqrt{2})$
17. $(-4, -3)$
18. $(6, 5)$

Given the quadrant of θ in standard position and a trigonometric function value of θ, find the exact values for the indicated function.

19. I, $\tan\theta = \sqrt{3}$; $\cos\theta$ $\frac{1}{2}$
20. III, $\sin\theta = -\frac{5}{13}$; $\cos\theta$ $-\frac{12}{13}$
21. IV, $\csc\theta = -\frac{4}{3}$; $\tan\theta$ $-\frac{3\sqrt{7}}{7}$

LESSON 13.3

Point P is located at the intersection of a circle with a radius of r and the terminal side of angle θ in standard position. Find the exact coordinates of P.

1. $\theta = -60°$, $r = 10$ $(5, -5\sqrt{3})$
2. $\theta = 120°$, $r = 6$ $(-3, 3\sqrt{3})$
3. $\theta = 45°$, $r = 100$ $(50\sqrt{2}, 50\sqrt{2})$
4. $\theta = 300°$, $r = 2$ $(1, -\sqrt{3})$
5. $\theta = 135°$, $r = 20$ $(-10\sqrt{2}, 10\sqrt{2})$
6. $\theta = 330°$, $r = 50$ $(25\sqrt{3}, -25)$

Point P is located at the intersection of a unit circle and the terminal side of angle θ in standard position. Find the coordinates of P to the nearest hundredth.

7. $\theta = 50°$ **(0.64, 0.77)**
8. $\theta = -95°$ **(−0.09, −1.00)**
9. $\theta = 345°$ **(0.97, −0.26)**

Find the exact values of the sine, cosine, and tangent of each angle.

10. −300°
11. 405°
12. −420°
13. 1380°
14. 495°
15. 330°
16. −1800°
17. 930°

Find each trigonometric value. Give exact answers.

18. cos 495° $-\frac{\sqrt{2}}{2}$
19. tan 870° $-\frac{\sqrt{3}}{3}$
20. sin 780° $\frac{\sqrt{3}}{2}$
21. sin 330° $-\frac{1}{2}$
22. cos 405° $\frac{\sqrt{2}}{2}$
23. sin 570° $-\frac{1}{2}$
24. tan 405° **1**
25. cos 660° $\frac{1}{2}$
26. tan 420° $\sqrt{3}$
27. tan 1395° **−1**
28. sin(−1485°) $-\frac{\sqrt{2}}{2}$
29. cos 660° $\frac{1}{2}$
30. sin 1110° $\frac{1}{2}$
31. sec 780° **2**
32. cot 300° $-\frac{\sqrt{3}}{3}$
33. sin 390° $\frac{1}{2}$
34. csc(−765°) $-\sqrt{2}$
35. cot(−210°) $-\sqrt{3}$

Lesson 13.2

13–18. Answers are given in order as follows: sin, cos, tan, csc, sec, cot.

13. $\frac{2\sqrt{5}}{5}; \frac{\sqrt{5}}{5}; 2; \frac{\sqrt{5}}{2}; \sqrt{5}; \frac{1}{2}$

14. $\frac{5\sqrt{41}}{41}; -\frac{4\sqrt{41}}{41}; -1.25; \frac{\sqrt{41}}{5}; -\frac{\sqrt{41}}{4}; -\frac{4}{5}$

15. $-\frac{8\sqrt{73}}{73}; \frac{3\sqrt{73}}{73}; -\frac{8}{3}; -\frac{\sqrt{73}}{8}; \frac{\sqrt{73}}{3}; -\frac{3}{8}$

16. $-\frac{\sqrt{3}}{3}; \frac{\sqrt{6}}{3}; -\frac{\sqrt{2}}{2}; -\sqrt{3}; \frac{\sqrt{6}}{2}; -\sqrt{2}$

17. $-\frac{3}{5}; -\frac{4}{5}; \frac{3}{4}; -\frac{5}{3}; -\frac{5}{4}; \frac{4}{3}$

18. $\frac{5\sqrt{61}}{61}; \frac{6\sqrt{61}}{61}; \frac{5}{6}; \frac{\sqrt{61}}{5}; \frac{\sqrt{61}}{6}; \frac{6}{5}$

Lesson 13.3

10–17. Answers are given in order as follows: sin, cos, tan.

10. $\frac{\sqrt{3}}{2}; \frac{1}{2}; \sqrt{3}$

11. $\frac{\sqrt{2}}{2}; \frac{\sqrt{2}}{2}; 1$

12. $-\frac{\sqrt{3}}{2}; \frac{1}{2}; -\sqrt{3}$

13. $-\frac{\sqrt{3}}{2}; \frac{1}{2}; -\sqrt{3}$

14. $\frac{\sqrt{2}}{2}; -\frac{\sqrt{2}}{2}; -1$

15. $-\frac{1}{2}; \frac{\sqrt{3}}{2}; -\frac{\sqrt{3}}{3}$

16. 0; 1; 0

17. $-\frac{1}{2}; -\frac{\sqrt{3}}{2}; \frac{\sqrt{3}}{3}$

LESSON 13.4

Convert each degree measure to radian measure. Give exact answers.

1. $30°$ $\frac{\pi}{6}$ **2.** $-90°$ $-\frac{\pi}{2}$ **3.** $20°$ $\frac{\pi}{9}$

4. $400°$ $\frac{20\pi}{9}$ **5.** $1080°$ 6π **6.** $50°$ $\frac{5\pi}{18}$

Convert each radian measure to degree measure. Round answers to the nearest tenth of a degree, if necessary.

7. $\frac{2\pi}{3}$ radians **120°** **8.** $-\frac{\pi}{9}$ radian **−20°** **9.** 3π radians **540°**

10. $-\frac{3\pi}{4}$ radians **−135°** **11.** 3.245 radians **≈185.9°** **12.** −6.122 radians **≈−350.8°**

Evaluate each expression. Give the values.

13. $\sin\frac{3\pi}{4}$ $\frac{\sqrt{2}}{2}$ **14.** $\csc\frac{\pi}{6}$ **2** **15.** $\tan(-2\pi)$ **0** **16.** $\cos\frac{5\pi}{6}$ $-\frac{\sqrt{3}}{2}$

17. $\cos\frac{\pi}{3}$ $\frac{1}{2}$ **18.** $\tan\left(-\frac{3\pi}{4}\right)$ **1** **19.** $\cot\frac{\pi}{4}$ **1** **20.** $\cos\left(-\frac{11\pi}{6}\right)$ $\frac{\sqrt{3}}{2}$

A circle has a diameter of 8 meters. For each central angle measure below, find the length in meters of the arc intercepted by the angle.

21. $\frac{5\pi}{6}$ radians **≈10.5 m** **22.** $\frac{\pi}{12}$ radians **≈1.05 m** **23.** 2.5 radians **10 m** **24.** 1.2 radians **4.8 m**

LESSON 13.5

Identify the amplitude, if it exists, and the period of each function.

1. $y = 3.5\sin 4\theta$ **3.5; 90°** **2.** $y = 8\tan x$ **does not exist; 180°** **3.** $y = -6\cos(-x)$ **6; 360°**

4. $y = 7\cos\frac{1}{2}\theta$ **7; 720°** **5.** $y = \frac{2}{3}\sin 3\theta$ $\frac{2}{3}$**; 120°** **6.** $y = -\frac{1}{2}\tan 6\theta$ **does not exist; 30°**

Identify the phase shift and vertical translation of each function from its parent function.

7. $y = \cos(\theta + 45°) + 1$ **8.** $y = \sin(\theta + 180°) - 3$ **9.** $y = \tan(\theta - 90°)$

10. $y = 3\cos(\theta - 30°) - 1$ **11.** $y = 3 + \sin(\theta - 45°)$ **12.** $y = 2 - \tan(\theta + 30°)$

Describe the transformation of each function from its parent function. Then graph at least one period of the given function and its parent function.

13. $y = 5\sin\theta$ **14.** $y = \cos(\theta + 90°)$ **15.** $y = \tan 2\theta$

16. The sales of a seasonal product are modeled by the function $s(x) = 40\sin\frac{\pi}{6}x + 74$, where s is thousands of units and x is time in months (beginning with 1 for January). Identify the amplitude, period, and phase shift of the function. Sketch a graph of the function for at least one period. What month shows the greatest number of units sold?

Lesson 13.5

7. 45° left; 1 unit up

8. 180° left; 3 units down

9. 90° right; no vertical translation

10. 30° right; 1 unit down

11. 45° right; 3 units up

12. 30° left; 2 units up

13. vertical stretch by factor of 5

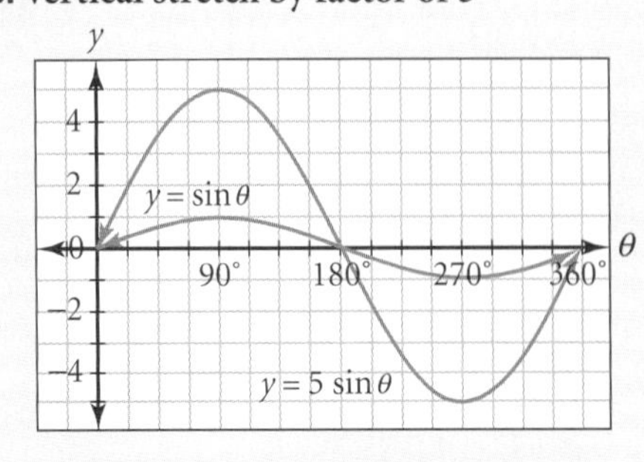

14. horizontal translation 90° to the left

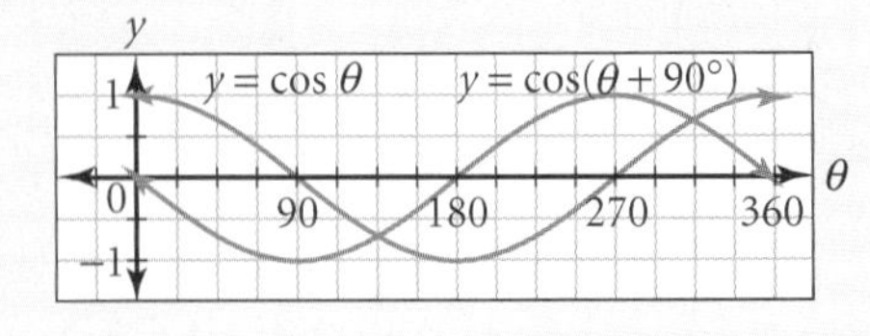

LESSON 13.6

Find all possible values for each expression.

1. $\sin^{-1}\frac{\sqrt{3}}{2}$ **$60^\circ + n360^\circ$ and $120^\circ + n360^\circ$**
2. $\cos^{-1}\frac{1}{2}$ **$60^\circ + n360^\circ$ and $300^\circ + n360^\circ$**
3. $\cos^{-1}\left(-\frac{\sqrt{2}}{2}\right)$ **$135^\circ + n360^\circ$ and $225^\circ + n360^\circ$**
4. $\tan^{-1} 1$ **$45^\circ + n360^\circ$ and $225^\circ + n360^\circ$**
5. $\tan^{-1} -1$ **$135^\circ + n360^\circ$ and $315^\circ + n360^\circ$**
6. $\sin^{-1} 0$ **$0^\circ + n360^\circ$ and $180^\circ + n360^\circ$**

Evaluate each trigonometric expression.

7. $\text{Cos}^{-1} 0$ **90°**
8. $\text{Sin}^{-1}\left(-\frac{1}{2}\right)$ **-30°**
9. $\text{Cos}^{-1}\frac{\sqrt{3}}{2}$ **30°**
10. $\text{Tan}^{-1}(-\sqrt{3})$ **-60°**
11. $\text{Sin}^{-1}\left(-\frac{\sqrt{2}}{2}\right)$ **-45°**
12. $\text{Tan}^{-1} 0$ **0°**

Evaluate each trigonometric expression.

13. $\tan\left(\text{Sin}^{-1}\frac{\sqrt{3}}{2}\right)$ **$\sqrt{3}$**
14. $\cos(\text{Tan}^{-1} 1)$ **$\frac{\sqrt{2}}{2}$**
15. $\sin\left[\text{Cos}^{-1}\left(-\frac{\sqrt{2}}{2}\right)\right]$ **$\frac{\sqrt{2}}{2}$**
16. $\text{Sin}^{-1}(\cos 45^\circ)$ **45°**
17. $\text{Tan}^{-1}(\cos 90^\circ)$ **0°**
18. $\text{Sin}^{-1}(\cos 30^\circ)$ **60°**
19. At one point in the day, a 15-foot flagpole casts a 19.5-foot shadow. Find the angle of elevation between the sun and the far end of the shadow. **$\approx 38^\circ$**

Chapter 14

LESSON 14.1

Find the area of $\triangle ABC$ to the nearest tenth of a square unit.

1. $b = 18$ in., $c = 24$ in., $A = 42^\circ$ **144.5 in.^{2}**
2. $a = 4$ m, $c = 9$ m, $B = 67^\circ$ **16.6 m^2**
3. $c = 20$ cm, $a = 10$ cm, $B = 110^\circ$ **94.0 cm^2**
4. $b = 5$ in., $c = 4$ in., $A = 120^\circ$ **8.7 in.^{2}**

Use the given information to find the indicated side length of $\triangle ABC$. Round answers to the nearest tenth, if necessary.

5. Given $A = 38^\circ$, $B = 50^\circ$, and $a = 7$, find b. **8.7**
6. Given $B = 120^\circ$, $C = 42^\circ$, and $b = 70$, find c. **54.1**
7. Given $A = 52^\circ$, $C = 88^\circ$, and $a = 6$, find c. **7.6**
8. Given $C = 59^\circ$, $B = 63^\circ$, and $c = 15$, find b. **15.6**

Solve each triangle. Round answers to the nearest tenth, if necessary.

9. $A = 65^\circ$, $C = 20^\circ$, $b = 9$
10. $C = 37^\circ$, $B = 20^\circ$, $a = 40$
11. $A = 21^\circ$, $C = 104^\circ$, $b = 10$
12. $B = 50^\circ$, $C = 32^\circ$, $a = 35$
13. $A = 50^\circ$, $C = 44^\circ$, $a = 12$
14. $A = 65^\circ$, $C = 70^\circ$, $b = 15$

State the number of triangles determined by the given information. If 1 or 2 triangles are fomed, solve the triangle(s). Round the angle measures and sides lengths to the nearest tenth.

15. $A = 117^\circ$, $b = 40$, $a = 28$ **0**
16. $B = 39^\circ$, $a = 4$, $b = 3$ **2**
17. A surveyor marks the corners of a triangular lot and labels them as A, B, and C. If $\overline{AC} = 110$ feet, $BC = 158$ feet, and the measure of the angle between $\overline{AC}$ and $\overline{BC}$ is 65°, find the area of the park to the nearest hundredth of a square foot. **7875.81 ft^2**

15. **horizontal compression by a factor of $\frac{1}{2}$**

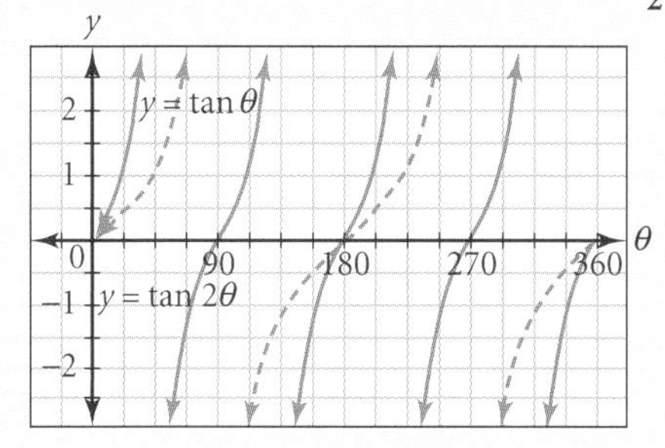

16. **amplitude = 40, period = 12 months, phase shift = 0; March**

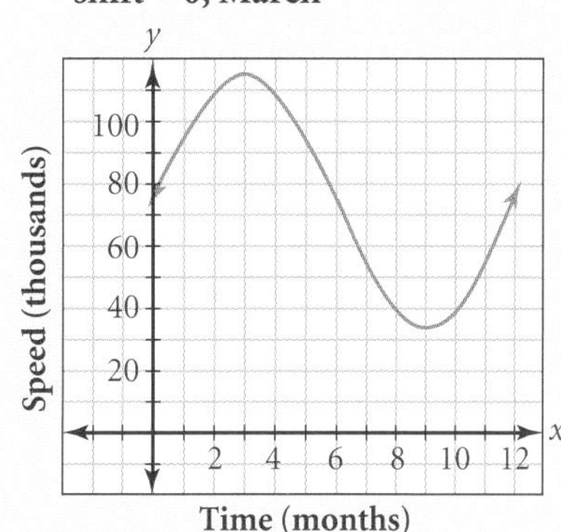

Note: **Throughout, answers may vary slightly due to rounding, method of calculation, or order in which solutions were found.**

Lesson 14.1

9. $B = 95°, a = 8.2, c = 3.1$
10. $A = 123°, b = 16.3, c = 28.7$
11. $B = 55°, a = 4.4, c = 11.8$
12. $A = 98°, b = 27.1, c = 18.7$
13. $B = 86°, b = 15.6, c = 10.9$
14. $B = 45°, a = 19.2, c = 19.9$
16. 2;
(1) $A = 57°, C = 84°$, and $c = 4.7$,
(2) $A = 123°$, $C = 18°$, and $c = 1.5$

Extra Practice

Lesson 14.2

7. $A = 59.5°$, $B = 83.9°$, $C = 36.6°$

8. $A = 39.8°$, $B = 106.1°$, $C = 34.1°$

9. $A = 38.6°$, $B = 48.5°$, $C = 92.9°$

10. $A = 52.3°$, $B = 46.5°$, $C = 81.2°$

11. $A = 24.2°$, $B = 30.8°$, $C = 125.0°$

12. $A = 22.2°$, $B = 27.1°$, $C = 130.7°$

13. SSS; $A = 48.5°$, $B = 55.8°$, $C = 75.7°$

14. SSA; not possible

15. SSA; $A = 51.2°$, $C = 56.8°$, $c = 8.9$

16. SAS; $c = 83.5$, $A = 69.4°$, $B = 60.6°$

17. SSA; not possible

18. SSA; $a = 6.4$, $C = 91.7°$, $B = 18.3°$

Lesson 14.3

1. $(\sec \theta)(\sin \theta) = \frac{r}{x} \cdot \frac{y}{r} = \frac{y}{x} = \tan \theta$

2. $\sin^2 \theta + \cos^2 \theta = \frac{y^2}{r^2} + \frac{x^2}{r^2} = \frac{x^2 + y^2}{r^2} = 1$ and $\sec^2 \theta - \tan^2 \theta = \frac{r^2}{x^2} - \frac{y^2}{x^2} = \frac{r^2 - y^2}{x^2} = \frac{x^2 + y^2 - y^2}{x^2} = 1$

3. $\tan \theta = \frac{y}{x} = \frac{1}{\frac{x}{y}} = \frac{1}{\cot \theta}$

4. $1 + \cot^2 \theta = 1 + \frac{x^2}{y^2} = \frac{x^2 + y^2}{y^2} = \frac{r^2}{y^2} = \csc^2 \theta$

11. $\dfrac{\frac{1}{\cos \theta} - \cos \theta}{\frac{\sin \theta}{\cos \theta}} = \sin \theta$

12. $\frac{1}{\cos^2 \theta} - \frac{\sin^2 \theta}{\cos^2 \theta} = 1$

13. $\frac{\cos \theta}{\sin \theta}(1 - \sin \theta) + \cos \theta = \frac{\cos \theta}{\sin \theta} = \cot^2 \theta$

LESSON 14.2

Classify the type of information given, and then use the law of cosines to find the missing side length of $\triangle ABC$ to the nearest tenth.

1. $b = 64, c = 80, A = 80°$ **SAS; $a = 93.4$**

2. $a = 5, b = 12, C = 46°$ **SAS; $c = 9.3$**

3. $a = 7, c = 9, B = 100°$ **SAS; $b = 12.3$**

4. $b = 42, c = 30, A = 56°$ **SAS; $a = 35.4$**

5. $b = 90, c = 120, A = 55°$ **SAS; $a = 100.6$**

6. $a = 4, b = 6, C = 130°$ **SAS; $c = 9.1$**

Solve each triangle. Round answers to the nearest tenth.

7. $a = 13, b = 15, c = 9$

8. $a = 8, b = 12, c = 7$

9. $a = 20, b = 24, c = 32$

10. $a = 24, b = 22, c = 30$

11. $a = 40, b = 50, c = 80$

12. $a = 10, b = 12, c = 20$

Classify the type of information given, and then solve $\triangle ABC$, if possible. Round answers to the nearest tenth. If no such triangle exists, write *not possible*.

13. $a = 5.8, b = 6.4, c = 7.5$ **SSS**

14. $A = 39°, b = 42, a = 16$ **SSA**

15. $B = 72°, a = 8.2, b = 10$ **SSA**

16. $C = 50°, a = 102, b = 95$ **SAS**

17. $A = 40°, a = 5, b = 20$ **SSA**

18. $A = 70°, a = 6, b = 2$ **SSA**

19. A piece of gold cord 42 inches long will be used to trim the edges of a banner in the shape of an isosceles triangle. If the base of the triangle is 20 inches, find the measure of each angle of the triangle to the nearest tenth of a degree. **24.6°, 24.6°, and 130.8°**

LESSON 14.3

Use definitions to prove each identity.

1. $(\sec \theta)(\sin \theta) = \tan \theta$

2. $\sin^2 \theta + \cos^2 \theta = \sec^2 \theta - \tan^2 \theta$

3. $\tan \theta = \frac{1}{\cot \theta}$

4. $1 + \cot^2 \theta = \csc^2 \theta$

Write each expression in terms of a single trigonometric function, if possible.

5. $(\cos \theta)(\sec \theta) - \cos^2 \theta$ **$\sin^2 \theta$**

6. $\frac{\tan \theta}{\sin \theta}$ **$\sec \theta$**

7. $\frac{(\sin \theta)(\sec \theta)}{\tan \theta}$ **1**

8. $(\csc \theta)(\cos \theta)(\sin \theta)$ **$\cos \theta$**

9. $\sec^2 \theta - \tan^2 \theta + \cot^2 \theta$ **$\cot^2 \theta$**

10. $\frac{\tan \theta + 1}{\tan \theta}$ **$1 + \cot \theta$**

Write each expression in terms of only one trigonometric function. Then simplify, if possible.

11. $\frac{\sec \theta - \cos \theta}{\tan \theta}$

12. $\sec^2 \theta - \tan^2 \theta$

13. $(\cot \theta)(1 - \sin \theta) + \cos \theta$

14. $(\sec \theta)(\sec \theta - \cos \theta)$

15. $\frac{\sin \theta + \tan \theta}{1 + \sec \theta}$

16. $\frac{1 + \cot^2 \theta}{1 + \tan^2 \theta}$

14. $\frac{1}{\cos \theta}\left(\frac{1}{\cos \theta} - \cos \theta\right) = \frac{1}{\cos^2 \theta} - 1 = \sec^2 \theta - 1 = \tan^2 \theta$

15. $\dfrac{\sin \theta + \frac{\sin \theta}{\cos \theta}}{1 + \frac{1}{\cos \theta}} = \sin \theta$

16. $\dfrac{1 + \frac{\cos^2 \theta}{\sin^2 \theta}}{1 + \frac{\sin^2 \theta}{\cos^2 \theta}} = \frac{\cos^2 \theta}{\sin^2 \theta} = \cot^2 \theta$

LESSON 14.4

Find the exact value of each expression.

1. $\sin(45° - 30°)$ $\frac{\sqrt{6}}{4} - \frac{\sqrt{2}}{4}$

2. $\sin(225° + 60°)$ $-\frac{\sqrt{2}}{4} - \frac{\sqrt{6}}{4}$

3. $\cos(120° + 45°)$ $-\frac{\sqrt{2}}{4} - \frac{\sqrt{6}}{4}$

4. $\cos(150° - 135°)$ $\frac{\sqrt{6}}{4} + \frac{\sqrt{2}}{4}$

5. $\sin(90° - 120°)$ $-\frac{1}{2}$

6. $\cos(30° + 120°)$ $-\frac{\sqrt{3}}{2}$

7. $\sin 210°$ $-\frac{1}{2}$

8. $\cos 105°$ $\frac{\sqrt{2}}{4} - \frac{\sqrt{6}}{4}$

9. $\sin 285°$ $-\frac{\sqrt{6}}{4} - \frac{\sqrt{2}}{4}$

10. $\cos 75°$ $\frac{\sqrt{6}}{4} - \frac{\sqrt{2}}{4}$

11. $\cos 15°$ $\frac{\sqrt{6}}{4} + \frac{\sqrt{2}}{4}$

12. $\sin 195°$ $\frac{\sqrt{2}}{4} - \frac{\sqrt{6}}{4}$

Find the rotation matrix for each angle of rotation.

13. 30°

14. −45°

15. 80°

16. A rectangle has vertices at (4, 2), (4, 8), (10, 8), and (10, 2). Find the coordinates, to the nearest hundredth, of the vertices for the image of the rectangle after a 45° rotation about the origin. **(1.41, 4.24), (−2.83, 8.49), (1.41, 12.73), (5.66, 8.49)**

LESSON 14.5

Write each expression in terms of trigonometric functions of θ rather than multiples of θ.

1. $\cos 4\theta$ $\cos^4\theta - 6\sin^2\theta\cos^2\theta + \sin^4\theta$

2. $\frac{1 - \cos 2\theta}{2}$ $\sin^2\theta$

3. $\cos 2\theta + \sin^2\theta$ $\cos^2\theta$

Write each expression in terms of a single trigonometric function.

4. $\frac{\sin^2\theta}{\sin 2\theta}$ $\frac{1}{2}\tan\theta$

5. $\frac{\cos 2\theta}{\cos\theta + \sin\theta} + \sin\theta$ $\cos\theta$

6. $\frac{\sin 2\theta}{1 - \cos 2\theta}$ $\cot\theta$

Use the information given to find the exact value of $\sin 2\theta$ and $\cos 2\theta$.

7. $0° \le \theta \le 360°$; $\cos\theta = \frac{4}{5}$ $\frac{24}{25}, \frac{7}{25}$

8. $90° \le \theta \le 180°$; $\sin\theta = \frac{1}{4}$ $\frac{\sqrt{15}}{8}; \frac{7}{8}$

Use the information given to find the exact value of $\sin\frac{1}{2}\theta$ and $\cos\frac{1}{2}\theta$.

9. $0° \le \theta \le 180°$; $\tan\theta = \frac{4}{3}$ $\frac{\sqrt{5}}{5}; \frac{2\sqrt{5}}{5}$

10. $180° \le \theta \le 270°$; $\cos\theta = -\frac{3}{8}$ $\frac{\sqrt{11}}{4}; -\frac{\sqrt{5}}{4}$

LESSON 14.6

Find all solutions of each equation.

1. $2\sin\theta + \sqrt{2} = 0$

2. $6\sin\theta + 3 = 0$

3. $\sqrt{3} - \sin\theta = \sin\theta$

Find the exact solutions of each equation for $0° \le \theta < 360°$.

4. $\sec^2\theta - 4 = 0$ **60°, 120°, 240°, and 300°**

5. $\sin 2\theta = \sin\theta$ **0°, 60°, 180°, and 300°**

6. $\sec^2\theta + 2\sec\theta = 0$ **120° and 240°**

7. $\cos 2\theta = 3\cos\theta + 1$ **120° and 240°**

8. $2\cos^2\theta - \cos\theta - 1 = 0$ **0°, 120°, and 240°**

9. $2\sin\theta\cos\theta = \tan\theta$ **0°, 45°, 135°, 180°, 225°, and 315°**

Solve each equation to the nearest tenth of a degree for $0° \le \theta < 360°$.

10 $4\cos\theta - 5\sin\theta = 0$ **38.7°, 218.7°**

11 $3\sin^2\theta - 2\sin\theta - 1 = 0$ **90.0°, 199.5°, 340.5°**

Extra Practice

Lesson 14.4

13. $\begin{bmatrix} 0.87 & -0.5 \\ 0.5 & 0.87 \end{bmatrix}$

14. $\begin{bmatrix} 0.71 & 0.71 \\ -0.71 & 0.71 \end{bmatrix}$

15. $\begin{bmatrix} 0.17 & -0.98 \\ 0.98 & 0.17 \end{bmatrix}$

Lesson 14.6

1. $225° + n360°$ and $315° + n360°$

2. $210° + n360°$ and $330° + n360°$

3. $60° + n360°$ and $120° + n360°$

Parent Functions and Their Graphs

The simplest form of any function is called the parent function. Each parent function has a distinctive graph. These two pages summarize the basic graphs of some parent functions.

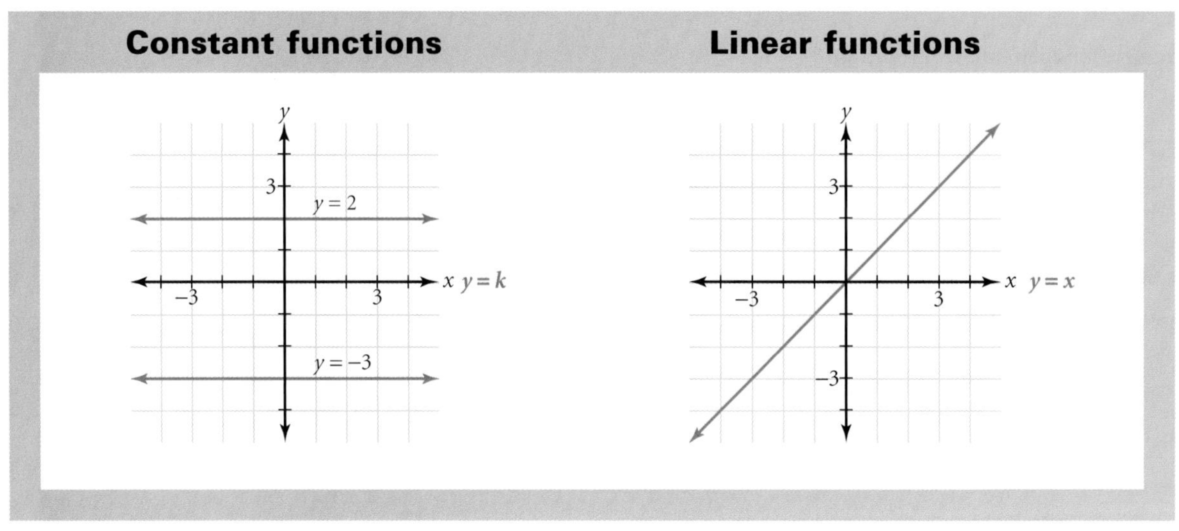

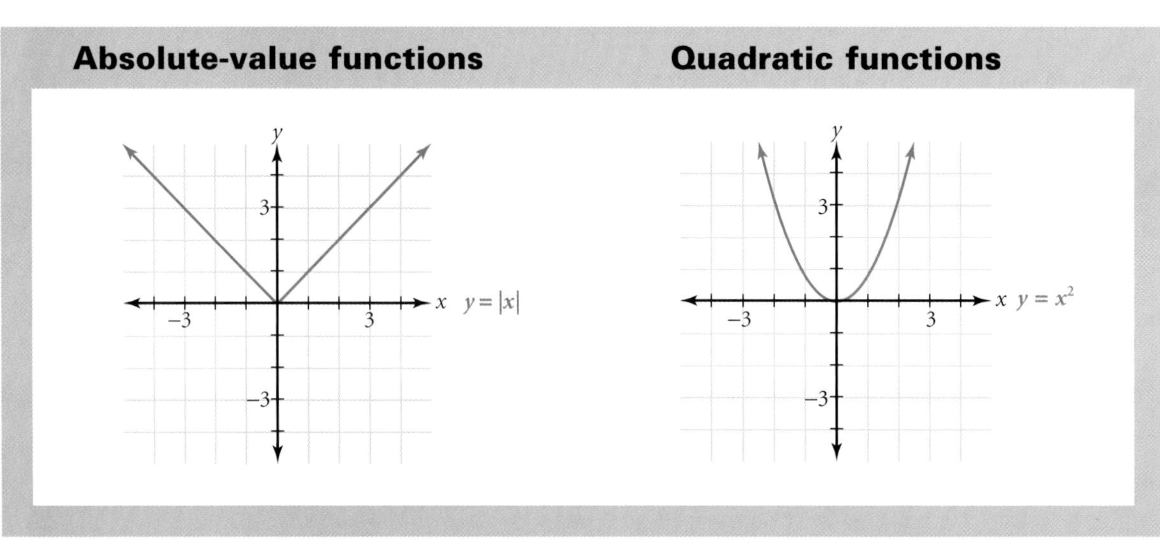

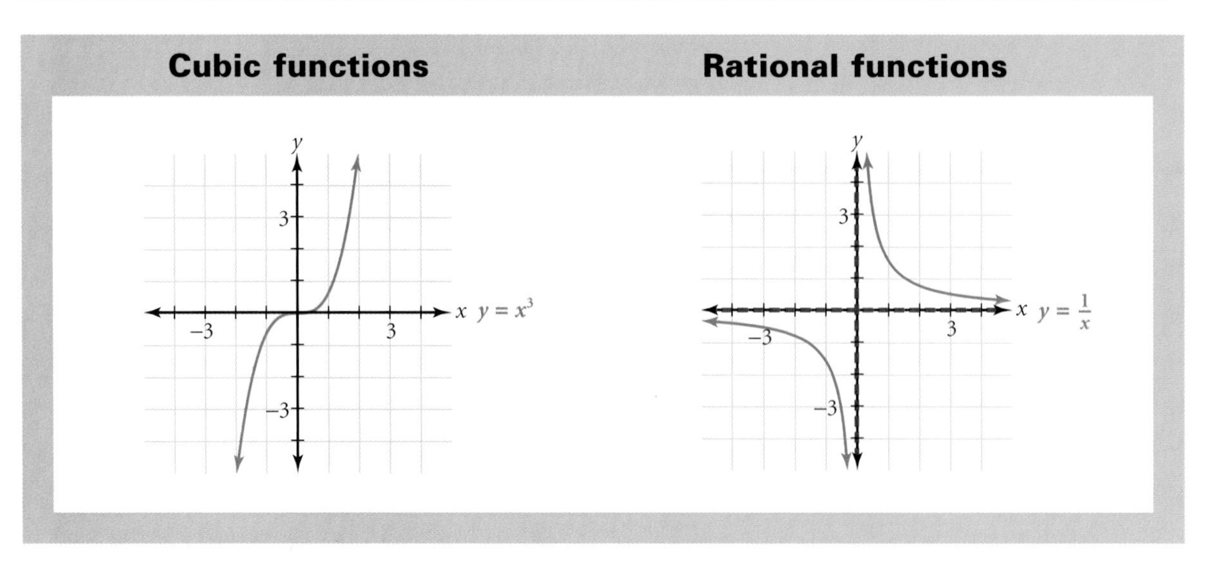

Radical functions

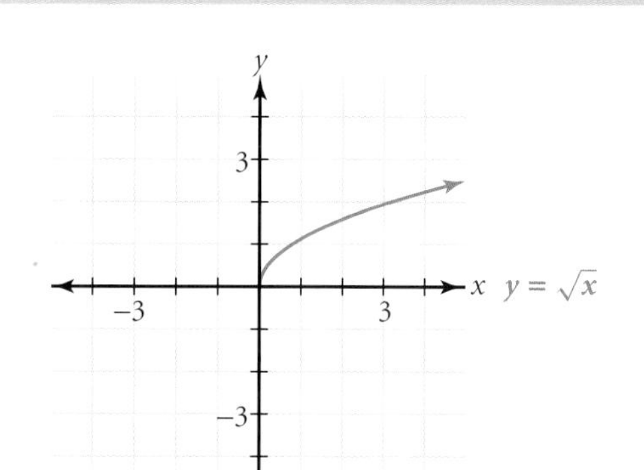

Exponential functions

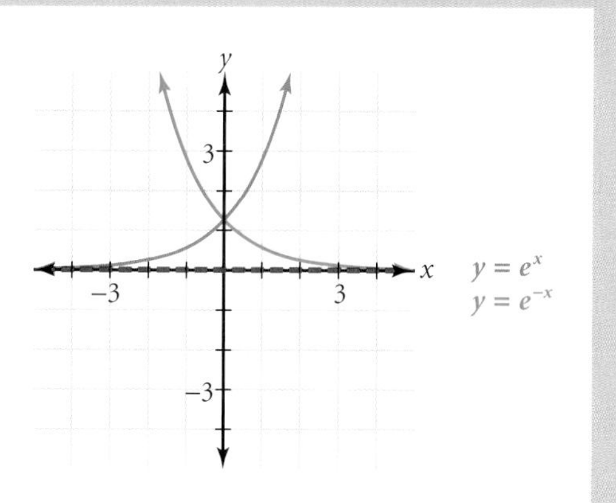

Logarithmic functions

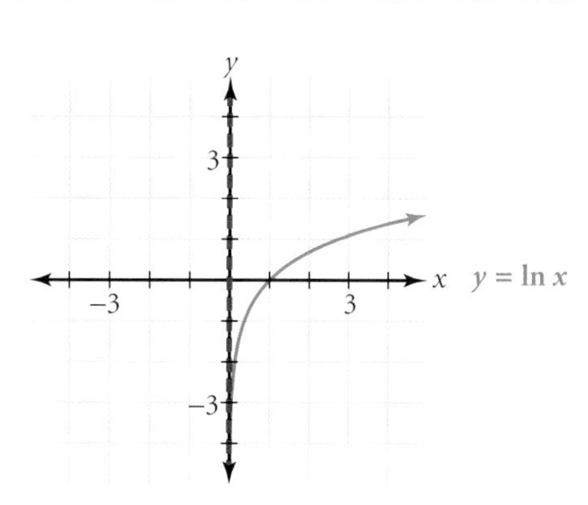

Sine functions

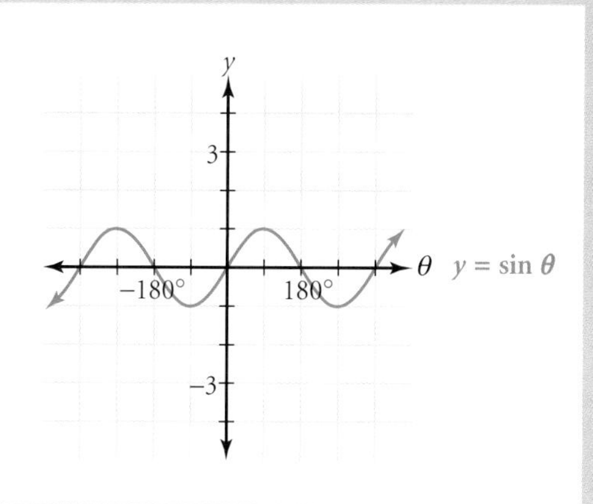

Cosine functions

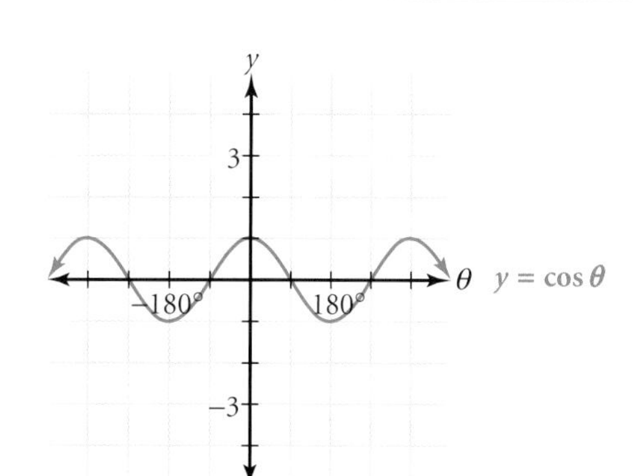

Tangent functions

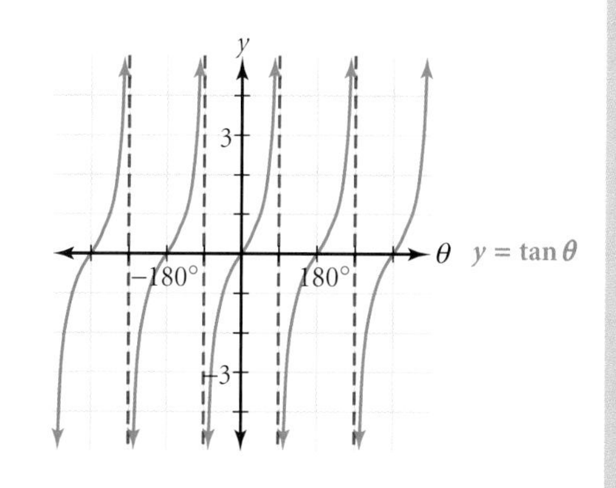

Transformations of Parent Functions

A transformation of a parent function is an alteration of its function rule that results in an alteration of its graph. The new graph retains the distinctive features of the graph of the parent function. These two pages summarize transformations.

Translations

Vertical

If $y = f(x)$, then $y = f(x) + k$ gives a vertical translation of the graph of f. The translation is k units up for $k > 0$ and $|k|$ units down for $k < 0$.

Horizontal

If $y = f(x)$, then $y = f(x - h)$ gives a horizontal translation of the graph of f. The translation is h units to the right for $h > 0$ and $|h|$ units to the left for $h < 0$.

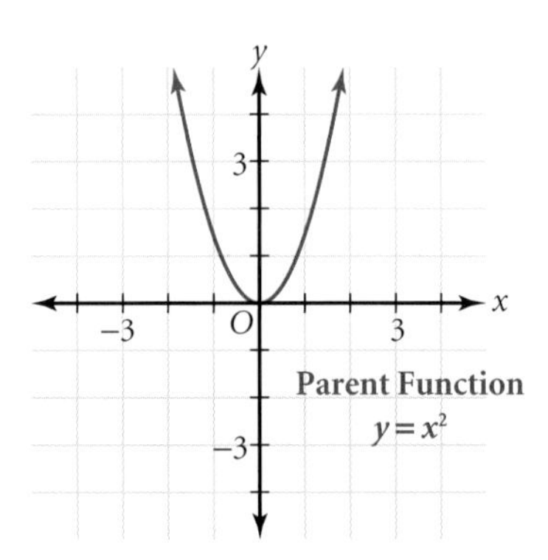

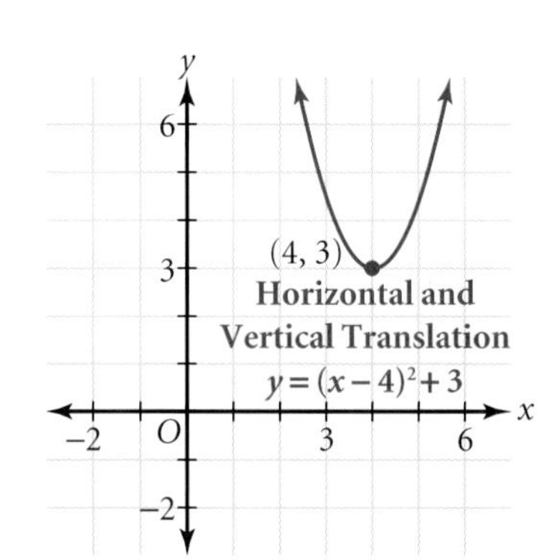

Vertical Stretches and Compressions

If $y = f(x)$, then $y = af(x)$ gives a vertical stretch or vertical compression of the graph of f.

If $a > 1$, the graph is stretched vertically by a factor of a.
If $a < 1$, the graph is compressed vertically by a factor of a.

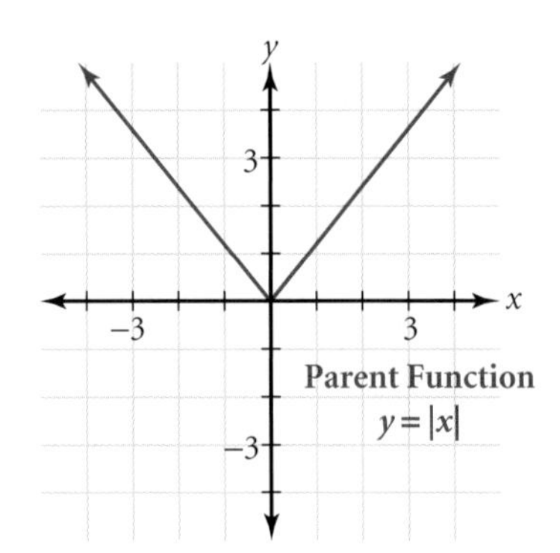

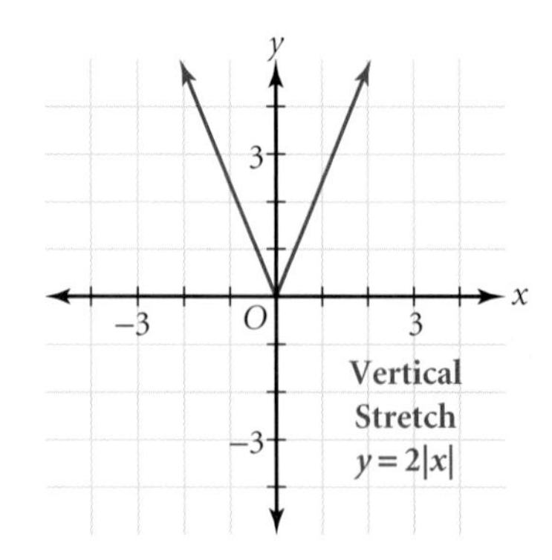

Horizontal Stretches and Compressions

If $y = f(x)$, then $y = f(bx)$ gives a horizontal stretch or horizontal compression of the graph of f.

If $b > 1$, the graph is compressed horizontally by a factor of $\frac{1}{b}$.

If $0 < b < 1$, the graph is stretched horizontally by a factor of $\frac{1}{b}$.

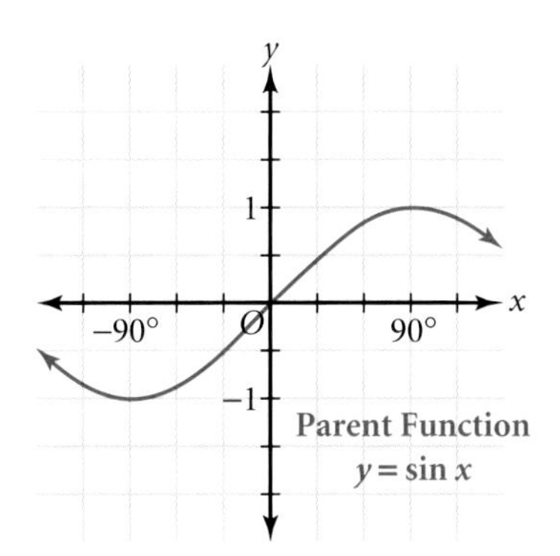

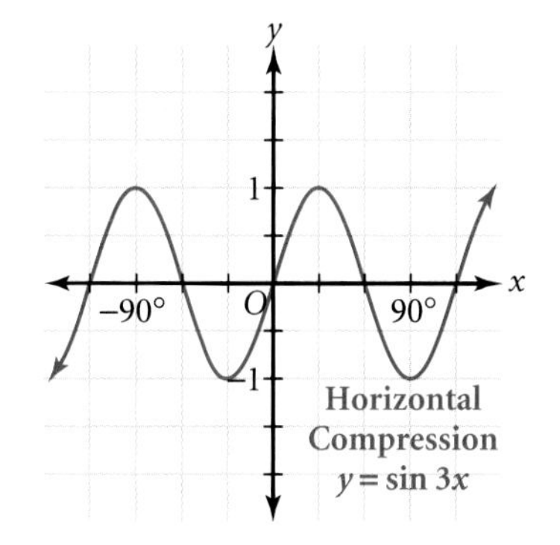

Reflections

If $y = f(x)$, then $y = -f(x)$ gives a reflection of the graph of f across the x-axis.

If $y = f(x)$, then $y = f(-x)$ gives a reflection of the graph of f across the y-axis.

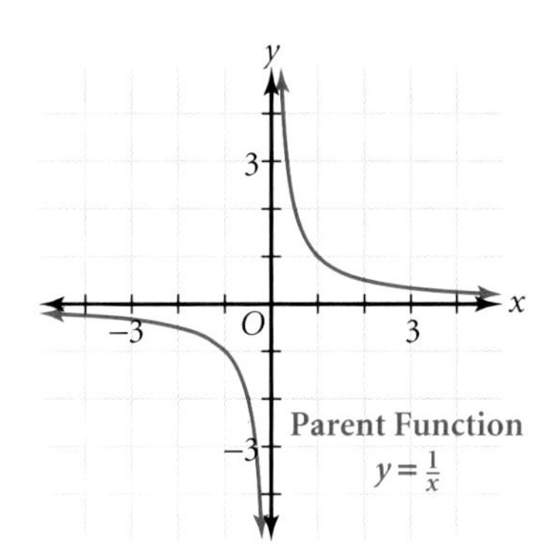

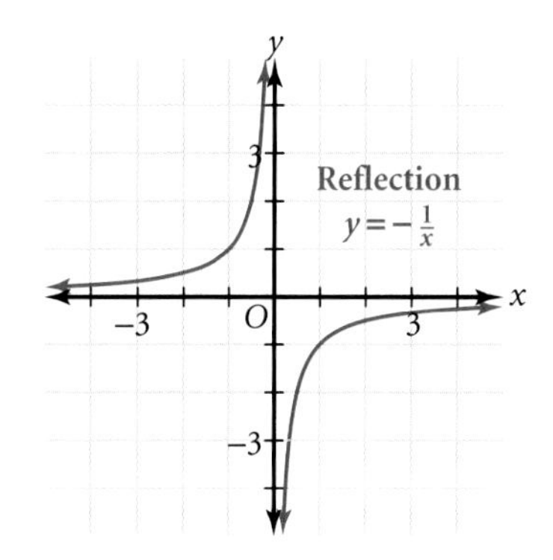

Combining Transformations

Any number of the above transformations can be combined. For example, the graph at right represents $y = -2(x - 4)^2 + 3$. It is a vertical stretch, a horizontal and vertical translation, and a reflection of the graph of the parent function, $y = x^2$.

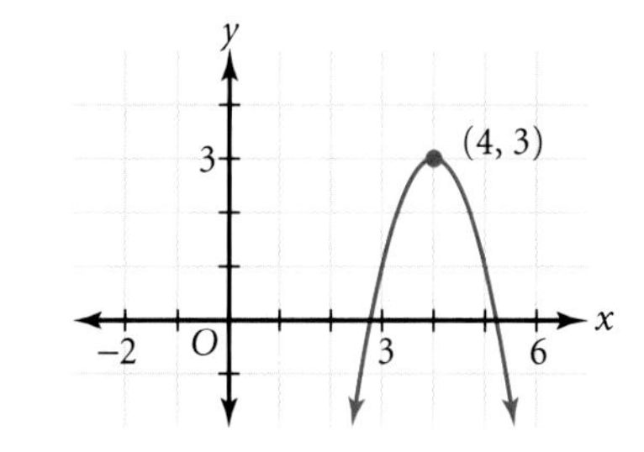

Table of Random Digits

Column / Line	(1)	(2)	(3)	(4)	(5)	(6)	(7)	(8)	(9)	(10)	(11)	(12)	(13)	(14)
1	10480	15011	01536	02011	81647	91646	69179	14194	62590	36207	20969	99570	91291	90700
2	22368	46573	25595	85393	30995	89198	27982	53402	93965	34095	52666	19174	39615	99505
3	24130	48360	22527	97265	76393	64809	15179	24830	49340	32081	30680	19655	63348	58629
4	42167	93093	06243	61680	07856	16376	39440	53537	71341	57004	00849	74917	97758	16379
5	37570	39975	81837	16656	06121	91782	60468	81305	49684	60672	14110	06927	01263	54613
6	77921	06907	11008	42751	27756	53498	18602	70659	90655	15053	21916	81825	44394	42880
7	99562	72905	56420	69994	98872	31016	71194	18738	44013	48840	63213	21069	10634	12952
8	96301	91977	05463	07972	18876	20922	94595	56869	69014	60045	18425	84903	42508	32307
9	89579	14342	63661	10281	17453	18103	57740	84378	25331	12566	58678	44947	05585	56941
10	85475	36857	53342	53988	53060	59533	38867	62300	08158	17983	16439	11458	18593	64952
11	28918	69578	88231	33276	70997	79936	56865	05859	90106	31595	01547	85590	91610	78188
12	63553	40961	48235	03427	49626	69445	18663	72695	52180	20847	12234	90511	33703	90322
13	09429	93969	52636	92737	88974	33488	36320	17617	30015	08272	84115	27156	30613	74952
14	10365	61129	87529	85689	48237	52267	67689	93394	01511	26358	85104	20285	29975	89868
15	07119	97336	71048	08178	77233	13916	47564	81056	97735	85977	29372	74461	28551	90707
16	51085	12765	51821	51259	77452	16308	60756	92144	49442	53900	70960	63990	75601	40719
17	02368	21382	52404	60268	89368	19885	55322	44819	01188	65225	64835	44919	05944	55157
18	01011	54092	33362	94904	31273	04146	18594	29852	71585	85030	51132	01915	92747	64951
19	52162	53916	46369	58586	23216	14513	83149	98736	23495	64350	94738	17752	35156	35749
20	07056	97628	33787	09998	42698	06691	76988	13602	51851	46104	88916	19509	25625	58104
21	48663	91245	85828	14346	09172	30168	90229	04734	59193	22178	30421	61666	99904	32812
22	54164	58492	22421	74103	47070	25306	76468	26384	58151	06646	21524	15227	96909	44592
23	32639	32363	05597	24200	13363	38005	94342	28728	35806	06912	17012	64161	18296	22851
24	29334	27001	87637	87308	58731	00256	45834	15398	46557	41135	10367	07684	36188	18510
25	02488	33062	28834	07351	19731	92420	60952	61280	50001	67658	32586	86679	50720	94953
26	81525	72295	04839	96423	24878	82651	66566	14778	76797	14780	13300	87074	79666	95725
27	29676	20591	68086	26432	46901	20849	89768	81536	86645	12659	92259	57102	80428	25280
28	00742	57392	39064	66432	84673	40027	32832	61362	98947	96067	64760	64584	96096	98253
29	05366	04213	25669	26422	44407	44048	37937	63904	45766	66134	75470	66520	34693	90449
30	91921	26418	64117	94305	26766	25940	39972	22209	71500	64568	91402	42416	07844	69618
31	00582	04711	87917	77341	42206	35126	74087	99547	81817	42607	43808	76655	62028	76630
32	00725	69884	62797	56170	86324	88072	76222	36086	84637	93161	76038	65855	77919	88006
33	69011	65795	95876	55293	18988	27354	26575	08625	40801	59920	29841	80150	12777	48501
34	25976	57948	29888	88604	67917	48708	18912	82271	65424	69774	33611	54262	85963	03547
35	09763	83473	73577	12908	30883	18317	28290	35797	05998	41688	34952	37888	38917	88050
36	91567	42595	27958	30134	04024	86385	29880	99730	55536	84855	29080	09250	79656	73211
37	17955	56349	90999	49127	20044	59931	06115	20542	18059	02008	73708	83517	36103	42791
38	46503	18584	18845	49618	02304	51038	20655	58727	28168	15475	56942	53389	20562	87338
39	92157	89634	94824	78171	84610	82834	09922	25417	44137	48413	25555	21246	35509	20468
40	14577	62765	35605	81263	39667	47358	56873	56307	61607	49518	89656	20103	77490	18062
41	98427	07523	33362	64270	01638	92477	66969	98420	04880	45585	46565	04102	46880	45709
42	34914	63976	88720	82765	34476	17032	87589	40836	32427	70002	70663	88863	77775	69348
43	70060	28277	39475	46473	23219	53416	94970	25832	69975	94884	19661	72828	00102	66794
44	53976	54914	06990	67245	68350	82948	11398	42878	80287	88267	47363	46634	06541	97809
45	76072	29515	40980	07391	58745	25774	22987	80059	39911	96189	41151	14222	60697	59583
46	90725	52210	83974	29992	65831	38857	50490	83765	55657	14361	31720	57375	56228	41546
47	64364	67412	33339	31926	14883	24413	59744	92351	97473	89286	35931	04110	23726	51900
48	08962	00358	31662	25388	61642	34072	81249	35648	56891	69352	48373	45578	78547	81788
49	95012	68379	93526	70765	10592	04542	76463	54328	02349	17247	28865	14777	62730	92277
50	15664	10493	20492	38391	91132	21999	59516	81652	27195	48223	46751	22923	32261	85653

Standard Normal Curve Areas

Tables

The table below gives the area under the standard normal curve between the mean, 0, and the desired number of standard deviations, z.

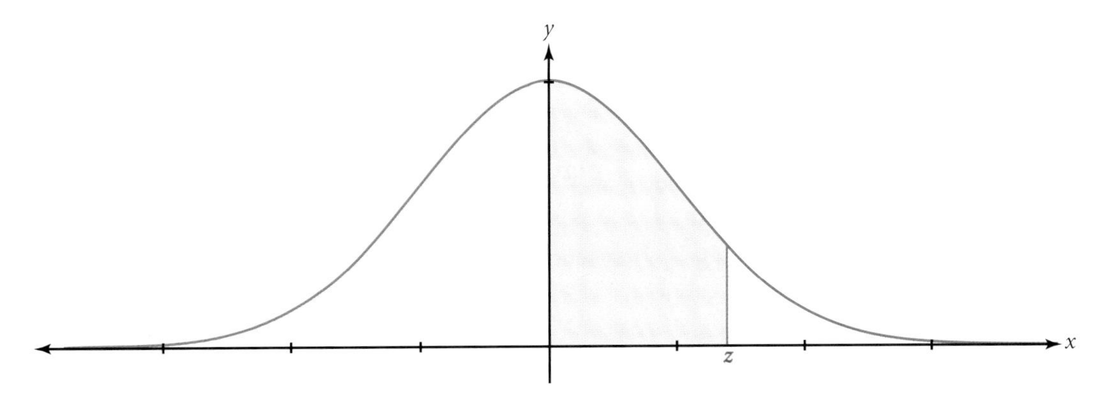

$P(x < z)$ for $z \geq 0$

z	0	1	2	3	4	5	6	7	8	9
0.0	.0000	.0040	.0080	.0120	.0160	.0199	.0239	.0279	.0319	.0359
0.1	.0398	.0438	.0478	.0517	.0557	.0596	.0636	.0675	.0714	.0753
0.2	.0793	.0832	.0871	.0910	.0948	.0987	.1026	.1064	.1103	.1141
0.3	.1179	.1217	.1255	.1293	.1331	.1368	.1406	.1443	.1480	.1517
0.4	.1554	.1591	.1628	.1664	.1700	.1736	.1772	.1808	.1844	.1879
0.5	.1915	.1950	.1985	.2019	.2054	.2088	.2123	.2157	.2190	.2224
0.6	.2257	.2291	.2324	.2357	.2389	.2422	.2454	.2486	.2518	.2549
0.7	.2580	.2612	.2642	.2673	.2704	.2734	.2764	.2794	.2823	.2852
0.8	.2881	.2910	.2939	.2967	.2995	.3023	.3051	.3078	.3106	.3133
0.9	.3159	.3186	.3212	.3238	.3264	.3289	.3315	.3340	.3365	.3389
1.0	.3413	.3438	.3461	.3485	.3508	.3531	.3554	.3577	.3599	.3621
1.1	.3643	.3665	.3686	.3708	.3729	.3749	.3770	.3790	.3810	.3830
1.2	.3849	.3869	.3888	.3907	.3925	.3944	.3962	.3980	.3997	.4015
1.3	.4032	.4049	.4066	.4082	.4099	.4115	.4131	.4147	.4162	.4177
1.4	.4192	.4207	.4222	.4236	.4251	.4265	.4279	.4292	.4306	.4319
1.5	.4332	.4345	.4357	.4370	.4382	.4394	.4406	.4418	.4429	.4441
1.6	.4452	.4463	.4474	.4484	.4495	.4505	.4515	.4525	.4535	.4545
1.7	.4554	.4564	.4573	.4582	.4591	.4599	.4608	.4616	.4625	.4633
1.8	.4641	.4649	.4656	.4664	.4671	.4678	.4686	.4693	.4699	.4706
1.9	.4713	.4719	.4726	.4732	.4738	.4744	.4750	.4756	.4761	.4767
2.0	.4772	.4778	.4783	.4788	.4793	.4798	.4803	.4808	.4812	.4817
2.1	.4821	.4826	.4830	.4834	.4838	.4842	.4846	.4850	.4854	.4857
2.2	.4861	.4864	.4868	.4871	.4875	.4878	.4881	.4884	.4887	.4890
2.3	.4893	.4896	.4898	.4901	.4904	.4906	.4909	.4911	.4913	.4916
2.4	.4918	.4920	.4922	.4925	.4927	.4929	.4931	.4932	.4934	.4936
2.5	.4938	.4940	.4941	.4943	.4945	.4946	.4948	.4949	.4951	.4952
2.6	.4953	.4955	.4956	.4957	.4959	.4960	.4961	.4962	.4963	.4964
2.7	.4965	.4966	.4967	.4968	.4969	.4970	.4971	.4972	.4973	.4974
2.8	.4974	.4975	.4976	.4977	.4977	.4978	.4979	.4979	.4980	.4981
2.9	.4981	.4982	.4982	.4983	.4984	.4984	.4985	.4985	.4986	.4986
3.0	.4986	.4987	.4987	.4988	.4988	.4989	.4989	.4989	.4990	.4990

Glossary

absolute value For any real number x, $|x| = x$ if $x \geq 0$ and $|x| = -x$ if $x < 0$. On a number line, $|x|$ is the distance from x to 0. (62)

absolute value of a complex number The distance of the complex number $a + bi$ from the origin in the complex plane, denoted $|a + bi| = \sqrt{a^2 + b^2}$. (318)

absolute-value function A function described by $f(x) = |x|$. (127)

additive inverse matrix The scalar product of a matrix and −1. (219)

adjacency matrix A representation of a network that indicates how many one-stage (direct) paths are possible from one vertex to another. (228)

amplitude The amplitude of a periodic function is one-half of the difference between the maximum and minimum function values and is always positive. (860)

angle of depression The angle formed by a horizontal line and a line of sight to a point below. (831)

angle of elevation The angle formed by a horizontal line and a line of sight to a point above. (831)

angle of rotation The angle formed by a ray that is rotated around its endpoint. (836)

arc length The length of the arc intercepted by a central angle with a radian measure of θ in a circle with a radius of r is given by the equation $s = r\theta$. (853)

arithmetic means The terms between any two nonconsecutive terms of an arithmetic sequence. (702)

arithmetic sequence A sequence whose successive terms differ by the same number, d, called the common difference. (700)

arithmetic series The indicated sum of the terms of an arithmetic sequence. (707)

asymptote A line that a curve approaches (but does not reach) as its x- or y-values become very large or very small. (362)

asymptotes of a hyperbola The diagonals of the rectangle that is determined by the vertices and co-vertices. (597)

augmented matrix A matrix that consists of the coefficients and the constant terms in a system of linear equations. (251)

axis of symmetry of a parabola A line that divides the parabola into two parts that are mirror images of each other. (276, 310)

base In an exponential expression of the form b^x, b is the base. (94, 362)

binomial A polynomial with exactly two terms. (425)

binomial experiment A probability experiment that meets the following conditions: The experiment consists of n trials whose outcomes are either successes or failures, and the trials are identical and independent with a constant probability of success, p, and a constant probability of failure, $1 - p$. (799)

binomial probability In a binomial experiment consisting of n trials, the probability, P, of r successes (where $0 \leq r \leq n$, p is the probability is of success, and $1 - p$ is the probability of failure) is given by the equation $P = {}_nC_r\, p^5(1 - p)^{n-r}$. (800)

Binomial Theorem A theorem that tells how to expand a positive integer power of a binomial. (742)

box-and-whisker plot A summary display of how values are distributed within a data set. (783)

center A fixed point that is used to define a circle, ellipse, or hyperbola. (579, 587, 596)

change-of-base formula For any positive real numbers $a \neq 1$, $b \neq 1$, and $x > 0$, $\log_b x = \frac{\log_a x}{\log_a b}$. (388)

circle graph A display of the distribution of non-overlapping parts of a whole by using sectors of a circle. (776)

circle The set of all points in a plane that are at a constant distance, called the radius, from a fixed point, called the center. (579)

circular permutation An arrangement of distinct objects in a specified order around a circle. (639)

coefficient The numerical factor of a monomial. (425)

combination An arrangement of a group of objects in which order is *not* important. (643)

combined variation A relationship containing both direct and inverse variation. (484)

common difference The number by which successive terms of an arithmetic sequence differ. (700)

common logarithm A logarithm whose base is 10. (385)

common ratio The ratio by which successive terms of a geometric sequence differ. (713)

complement The complement of event A consists of all outcomes in the sample space that are not in A, denoted A^c. (654)

completing the square A process used to form a perfect-square trinomial. (300)

Complex Conjugate Root Theorem If P is a polynomial function with real-number coefficients and $a + bi$ (where $b \neq 0$) is a root of $P(x) = 0$, then $a - bi$ is also a root of $P(x) = 0$. (461)

complex fraction A quotient that contains one or more fractions in the numerator, the denominator, or both. (500)

complex number Any number that can be written as $a + bi$, where a and b are real numbers and $i = \sqrt{-1}$. (316)

complex plane A set of coordinates axes in which the horizontal axis is the real axis and the vertical axis is the imaginary axis; used to graph complex numbers. (318)

composition of functions The composition of functions f and g, $(f \circ g)(x)$, is defined as $(f(g(x))$. The domain of f must include the range of g. (113)

compound inequalities A pair of inequalities combined by the words *and* or *or*. (56)

conditional probability The probability of event B, given that event A has happened (or will occur), denoted $P(B|A)$. (665)

conic section A plane figure formed by the intersection of a double cone and a plane. (562)

conjugate axis The axis of symmetry of a hyperbola that is perpendicular to the transverse axis. (596)

conjugate of a complex number The conjugate of a complex number $a + bi$ is $a - bi$, denoted $\overline{a + bi}$. (318)

consistent system A system of equations or inequalities that has at least one solution. (157)

constant A monomial with no variables. (425)

constant function A constant function is a function of the form $f(x) = k$. (125)

constant of variation The constant k in an inverse-, joint-, or combined-variation equation. (480)

constraints The inequalities that form the feasible region in a linear-programming problem. (187)

continuous compounding formula If P dollars are invested at an interest rate, r, that is compounded continuously, then the amount, A, of the investment at time t is given by $A = Pe^{rt}$. (393)

continuous function A function whose graph is an unbroken line or smooth curve. (105, 434)

converge Describes a infinite series whose partial sums approach a fixed number as n increases. (729)

Corner-Point Principle A principle in linear programming that identifies the maximum and minimum values of the objective function as occurring at one of the vertices of the feasible region. (189)

correlation coefficient A number represented by the variable r, where $-1 \leq r \leq 1$, that describes how closely points in a scatter plot cluster around the least-squares line. (39)

coterminal angle Describes angles that have the same terminal side when in standard position. (837)

co-vertices The endpoints of the minor axis of an ellipse; the endpoints of the conjugate axis of a hyperbola. (587, 596)

cube root A number, $\sqrt[3]{x}$, that when multiplied by itself three times produces the given number, x. (523)

decreasing function For a function f and any numbers x_1 and x_2 in the domain of f, the function f is decreasing over an open interval if for every $x_1 < x_2$ in the interval, $f(x_1) > f(x_2)$. (433)

degree of a monomial The sum of the exponents of the variables in the monomial. (425)

degree of a polynomial The degree of the monomial with the highest degree after simplification. (425)

degree The most common unit for angle measure; one degree, 1°, is defined as $\frac{1}{360}$ of a complete rotation of a ray. (836)

dependent events Two events are dependent if the occurrence of one event affects the occurrence of the other, or if the events are not independent. (660)

dependent system A system of equations that has infinitely many solutions. (157)

dependent variable The output of a function. For $y = f(x)$, $f(x)$ is the dependent variable. (106)

determinant A real number associated with a square matrix. (238)

diameter A chord of a circle that contains the center of the circle. (566)

dimensions of a matrix A matrix of m horizontal rows and n vertical columns has the dimension $m \times n$. (216)

direct variation The equation $y = kx$ describes a direct variation, where y varies directly as x, k is the constant of variation, and $k \neq 0$. (29)

directrix A fixed line used to define a parabola. (570)

discontinuous function A function whose graph has breaks or holes in it. (434)

discrete function A function whose graph consists of points that are not connected. (105)

discriminant The discriminant of a quadratic equation $ax^2 + bx + c = 0$ is $b^2 - 4ac$. (314)

distance formula The distance, d, between $P(x_1, y_1)$ and $Q(x_2, y_2)$ is $d = \sqrt{(x_2 - x_1)^2 + (y_2 - y_1)^2}$. (563)

diverge Describes a infinite series whose partial sums do not approach a fixed number as n increases. (729)

domain The set of possible values for the first coordinate of a function. (102, 104)

double root For a quadratic equation, if $b^2 - 4ac = 0$, the equation has only one solution, called a double root. (314)

effective yield The annually compounded interest rate that yields the final amount of an investment. (365)

elementary row operations Operations performed on a matrix that result in an equivalent matrix. (252)

elimination method A method of solving a system of equations by multiplying and combining the equations in the system in order to eliminate a variable. (164)

ellipse The set of all points P in a plane such that the sum of the distances from P to two fixed points, F_1 and F_2, called the foci, is a constant. (587)

end behavior What happens to a polynomial function as its domain values get very small and very large. (435)

entry Each value in a matrix; also called an element. (216)

equation A statement of equality between two expressions that may be true or false. (45)

equivalent equations Equations that have the same solution set. (48)

even function A function f for which $f(-x) = f(x)$ for all values of x in its domain. (911)

event An individual outcome or any specified combination of outcomes. (628)

excluded values Real numbers for which a rational function is not defined. (491)

experimental probability A probability approximated by performing trials and recording the ratio of the number of occurrences of the event to the number of trials. (629)

explicit formula A formula that defines the nth term, or general term, of a sequence. (691)

exponential expression An algebraic expression in which the exponent is a variable and the base is a fixed number. (355)

exponential function A function of the form $f(x) = b^x$, where b is a positive real number other than 1 and x is any real number. (362)

exponential growth and decay Represented by a function of the form $f(x) = b^x$, where $b > 1$ or $0 < b < 1$, respectively. (363)

Exponential-Logarithmic Inverse Property For $b > 0$ and $b \neq 0$, $\log_b b^x = x$ and $b^{\log_b x} = x$ for $x > 0$. (380)

extraneous solution A solution to a derived equation that is not a solution to the original equation. (514)

Factor Theorem For a polynomial $P(x)$, if and only if $P(r) = 0$, then $x - r$ is a factor $P(x)$. (442)

factorial If n is a positive integer, then n factorial, written $n!$, is given by $n \times (n-1) \times (n-2) \times \cdots \times 2 \times 1$. (636)

factoring The process that allows a sum to be written as a product. (290)

feasible region The solution set of a linear-programming problem. (187)

finite sequence A sequence that ends and therefore has a last term. (691)

foci Fixed points that are used to define an ellipse or hyperbola. (587, 595)

focus A fixed point used to define a parabola. (570)

frequency table A table that lists the number of times, or frequency, that each data value appears. (767)

function A relation in which, for each first coordinate, there is exactly *one* corresponding second coordinate. (102)

function notation A function is usually defined in terms of x and y, where $y = f(x)$, x is the independent variable, and $f(x)$ is the dependent variable. (106)

Fundamental Counting Principle If there are m ways that one event can occur and n ways that another event can occur, then there are $m \times n$ ways that both events can occur. (631)

Fundamental Theorem of Algebra Every polynomial function with degree $n \geq 1$ has at least one complex zero. Corollary: Every polynomial function with degree $n \geq 1$ has exactly n complex zeros, counting multiplicities. (462)

geometric means The terms between any two nonconsecutive terms of a geometric sequence. (716)

geometric sequence A sequence in which the ratio of successive terms is the same number, r, called the common ratio. (713)

geometric series The indicated sum of the terms of a geometric sequence. (720)

greatest-integer function A function denoted by $f(x) = [x]$ that converts a real number, x, into the largest integer that is less than or equal to x. (125)

grouped frequency table A frequency table in which the values are grouped into classes that contain a range of data values. (767)

histogram A bar graph that gives the frequency of each value in a data set. (774)

hole in the graph If the factor $x - b$ is a factor of both the numerator and denominator of a rational function, then a hole occurs in the graph of the rational function when $x = b$. (494)

horizontal line A line with slope of 0. (16)

horizontal-line test If a horizontal line crosses the graph of a function in more than one point, the inverse of the function is not a function. (120)

hyperbola The set of all points P in a plane such that the absolute value of the difference between the distances from P to two fixed points in the plane, F_1 and F_2, called the foci, is a constant. (595)

identity function A linear function defined by $I(x) = x$. (120)

identity matrix for multiplication An $n \times n$ matrix with 1s along the main diagonal (upper left entry to lower right entry) and 0s elsewhere. (235)

imaginary axis The vertical axis in the complex plane. (318)

imaginary part of a complex number For a complex number $a + bi$, b is the imaginary part. (316)

imaginary unit The imaginary unit i is defined as $i = \sqrt{-1}$ and $i^2 = -1$. (315)

inclusive events Events which can occur at the same time. (652)

inconsistent system A system of equations or inequalities that has no solution. (157)

increasing function For a function f and any numbers x_1 and x_2 in the domain of f, the function f is increasing over an open interval if for every $x_1 < x_2$ in the interval, $f(x_1) < f(x_2)$. (433)

independent events Two events are independent if the occurrence (or non-occurrence) of one event has no effect on the likelihood of the occurrence of the other event. (660)

independent system A system of equations that has exactly one solution. (157)

independent variable The input of a function. For $y = f(x)$, x is the independent variable. (106)

inequality A mathematical sentence that contains $>, <, \geq, \leq$, or $\neq$. (54)

infinite geometric series A geometric series with infinitely many terms. (729)

infinite sequence A sequence that continues without end. (691)

initial side The initial position of a rotated ray. (836)

interquartile range (IQR) The difference between the upper and lower quartiles of a data set. (782)

inverse of a matrix If A is an $n \times n$ matrix with an inverse, then A^{-1} is its inverse matrix, and $AA^{-1} = A^{-1}A = I$. (235)

inverse of a relation The inverse of a relation consisting of the ordered pairs (x, y) is the set of all ordered pairs (y, x). (118)

inverse variation Two variables, x and y, have an inverse-variation relationship if there is a nonzero number k such that $xy = k$, or $y = \frac{k}{x}$. (480)

irrational number A number whose decimal part does not terminate or repeat. (86)

joint variation If $y = kxz$ where k is a nonzero constant, then y varies jointly as x and z ($x \neq 0$ and $z \neq 0$). (482)

law of cosines For $\triangle ABC$, $a^2 = b^2 + c^2 - 2bc \cos A$, $b^2 = a^2 + c^2 - 2ac \cos B$, $c^2 = a^2 + b^2 - 2ab \cos C$. (895)

law of sines For $\triangle ABC$, $\frac{\sin A}{a} = \frac{\sin B}{b} = \frac{\sin C}{c}$. (887)

leading coefficient The coefficient of the term with the highest degree. (435)

least-squares line A linear model that fits a data set. (38)

like terms Two or more monomials that can only differ in their coefficients. (46)

linear equation An equation whose graph is a line. (5)

linear permutation A arrangement of objects in a specified order in a straight line. (636)

linear programming A method of finding a maximum or a minimum value that satisfies all of the given conditions of a particular situation. (187)

linearly related A relationship in which a constant difference in consecutive x-values results in a constant difference in consecutive y-values. (5)

literal equation An equation that contains two or more variables. (47)

local maximum For a function f, $f(a)$ is a local maximum if there is an interval around a such that $f(a) > f(x)$ for all values of x in the interval, where $x \neq a$. (433)

local minimum For a function f, $f(a)$ is a local minimum if there is an interval around a such that $f(a) < f(x)$ for all values of x in the interval, where $x \neq a$. (433)

Location Principle If P is a polynomial function and $P(x_1)$ and $P(x_2)$ have opposite signs, then there is a real number r between x_1 and x_2 that is a zero of P, that is $P(r) = 0$. (450)

logarithmic function A function of the form $y = \log_b x$ with base b, or $x = b^y$, which is the inverse of the exponential function $y = b^x$, where $b \neq 1$ and $b > 0$. (372)

major axis The longer axis of an ellipse. (587)

mathematical induction A type of mathematical proof that uses the following two steps to prove a statement for all natural numbers n: the basis step, which shows that the statement is true for $n = 1$, and the induction step, which assumes that the statement is true for a natural number, k, and proves that the statement is true for the natural number $k + 1$. (722)

matrix Any rectangular array of numbers enclosed in a single set of brackets. (216)

matrix equation An equation of the form $AX = B$, where A is the coefficient matrix, X is the variable matrix, and B is the constant matrix. (244)

matrix multiplication If matrix A has dimension $m \times n$ and matrix B has dimensions $n \times r$, then the product AB has dimensions $m \times r$. (226)

mean The sum of all of the values in a data set divided by the number of values; also called arithmetic average. (764)

mean deviation The average amount that the values in a data set differ from the mean. (792)

median The middle value, denoted Q_2, in a data set that is arranged in ascending or descending order. If there are an even number of data values, the median is the mean of the two middle values. (764)

midpoint formula The coordinates of the midpoint, M, between two points $P(x_1, y_1)$ and $Q(x_2, y_2)$ are $M\left(\frac{x_1 + x_2}{2}, \frac{y_1 + y_2}{2}\right)$. (565)

minor axis The shorter axis of an ellipse. (587)

mode The value in a data set that occurs most often. There can be one, more than one, or no mode. (764)

monomial A numeral, variable, or product of a numeral and one or more variables. (425)

multiplicity The number of times that a factor is repeated in the factorization of a polynomial expression. (449)

multiplier The base of an exponential expression. (355)

mutually exclusive events Events that cannot occur at the same time. (652)

natural base The irrational number e, which is approximately equal to 2.71828... (393)

natural exponential function An exponential function with base e; $f(x) = e^x$. (393)

natural logarithmic function The function $y = \log_e x$, the inverse of the natural exponential function. (394)

normal distribution Data that varies randomly from the mean, creating a bell-shaped pattern that is symmetric about the mean when graphed. (806)

objective function The function to be maximized or minimized in a linear-programming problem. (187)

odd function A function f for which $f(-x) = -f(x)$ for all values of x in its domain. (911)

one-to-one A one-to-one function can be intersected by a horizontal line at no more than one point. The inverse of a one-to-one function is also a function. (120)

One-to-One Property of Exponents If $b^x = b^y$, then $x = y$. (372)

One-to-One Property of Logarithms If $\log_b x = \log_b y$, then $x = y$. (380)

outlier A data value that is less than $Q_1 - 1.5(\text{IQR})$ or greater than $Q_3 + 1.5(\text{IQR})$. (782)

parabola The graph of a quadratic function. (276) The set of all points $P(x, y)$ in the plane whose distance to a point, called the focus, equals the distance to a fixed line, called the directrix. (570)

parallel lines Two lines (in the same plane) that have the same slope. All vertical lines are parallel and all horizontal lines are parallel. (23)

parametric equations A pair of continuous functions that define the x- and y-coordinates of a point in a coordinate plane in terms of a third variable. (196)

partial sum The sum of a specified number of terms of an infinite geometric series. (728)

Pascal's triangle A triangular pattern formed by the coefficients of binomial expansion. (735)

period The smallest positive number p that satisfies the equation in the definition of a periodic function. (846)

periodic function Describes functions for which there is a number p such that $f(x + p) = f(x)$ for every x in the domain of f. (846)

permutation An arrangement of objects in a specified order. (636)

perpendicular lines Two lines whose slopes are negative reciprocals of one another. All vertical and horizontal lines are perpendicular. (24)

phase shift A horizontal translation of a sine or cosine function. (861)

piecewise function A function that consists of different function rules for different parts of the domain. (124)

point-slope form The point-slope form of a line is $y - y_1 = m(x - x_1)$, where m is the slope and (x_1, y_1) is the coordinates of a point on the line. (22)

polynomial A monomial or a sum of terms that are monomials. (425)

polynomial function A function that is defined by a polynomial. (427)

power An expression of the form a^n. (94)

Power Property of Logarithms For $m > 0$, $b > 0$, $b \neq 1$, and any real number p, $\log_b m^p = p \log_b m$. (379)

Principle of Powers If $a = b$ and n is a positive integer, then $a^n = b^n$. (536)

principal square root The positive square root of a number a, denoted $\sqrt{a}$. (281)

principal values The restricted domains of the Sine, Cosine, and Tangent functions. (868)

probability The overall likelihood of the occurrence of an event. (628)

Product Property of Logarithms For $m > 0$, $n > 0$, and $b \neq 1$, $\log_b(mn) = \log_b m + \log_b n$. (378)

Product Property of Radicals For $a \geq 0$. $b \geq 0$, and a positive integer n, $\sqrt[n]{ab} = \sqrt[n]{a} \cdot \sqrt[n]{b}$ (529)

Product Property of Square Roots If $a \geq 0$ and $b \geq 0$, then $\sqrt{ab} = \sqrt{a} \cdot \sqrt{b}$. (281)

Properties of *n*th Roots For any real number a, $\sqrt[n]{a^n} = |a|$ if n is a positive even integer, and $\sqrt[n]{a^n} = a$ if n is a positive odd integer. (529)

proportion An equation that states that two ratios are equal. (31)

Pythagorean Theorem If $\triangle ABC$ is a right triangle with the right angle at C, the $a^2 + b^2 = c^2$. (284)

quadratic expression An expression of the form $ax^2 + bx + c$, where $a \neq 0$. (275)

quadratic formula The quadratic formula, $\frac{-b \pm \sqrt{b^2 - 4ac}}{2a}$, gives the solutions of the quadratic equation $ax^2 + bx + c = 0$ and $a \neq 0$. (308)

quadratic function Any function that can be written in the form $f(x) = ax^2 + bx + c$, where $a \neq 0$. (275)

quadratic inequality in two variables An inequality that can be written in one of the following forms, where a, b, and c are real numbers and $a \neq 0$: $y \geq ax^2 + bx + c$, $y > ax^2 + bx + c$, $y \leq ax^2 + bx + c$, and $y < ax^2 + bx + c$. (333)

quartiles Two values that, along with the median (Q_2), divide a data set into quarters; there is a lower quartile, Q_1, and an upper quartile, Q_3. (782)

Quotient Property of Logarithms For $m > 0$, $n > 0$, and $b \neq 1$, $\log_b \frac{m}{n} = \log_b m - \log_b n$. (378)

Quotient Property of Radicals For $a \geq 0$. $b \geq 0$, and a positive integer n, $\sqrt[n]{\frac{a}{b}} = \frac{\sqrt[n]{a}}{\sqrt[n]{b}}$, where $b \neq 0$. (529)

Quotient Property of Square Roots If $a \geq 0$ and $b \geq 0$, then $\sqrt{\frac{a}{b}} = \frac{\sqrt{a}}{\sqrt{b}}$. (281)

radian A unit of angle measure that is equal to $\frac{1}{2\pi}$ of the circumference of the unit circle; 1 radian is equal to approximately 57°. (852)

radical equation An equation that contains at least one radical expression with a variable in the radicand. (536)

radical expression An expression that contains at least one radical symbol. (524)

radical function A function that contains at least one radical expression. (524)

radical inequality An inequality that contains at least one radical expression. (540)

radical symbol The symbol $\sqrt{}$ in a radical expression. (524)

radicand The number or expression under a radical symbol. (524)

radius A segment with one endpoint at the center of the circle and the other endpoint on the circle. (566)

random Describes outcomes whose occurrences are all equally likely. (628)

range The set of possible values for the second coordinate of a function. (102, 104) The absolute value of the difference between the largest value and the smallest value of a data set. (782)

rational equation An equation that contains at least one rational expression. (512)

rational expression The quotient of two polynomials. (489)

rational function A function defined by a rational expression. (489)

rational inequality An inequality that contains at least one rational expression. (515)

rational number A number that can be expressed as the quotient of two integers, where the denominator is not equal to zero. (86)

Rational Root Theorem Let P be a polynomial function with integers coefficients in standard form. If $\frac{p}{q}$ (in lowest terms) is a root of $P(x) = 0$, then p is a factor of the constant term of P and q is a factor of the leading coefficient of P. (458)

rationalizing the denominator The process of removing an imaginary number from the denominator of a quotient. (318) A procedure that involves transforming a quotient with a radical in the denominator into an expression with no radical in the denominator. (532)

real axis The horizontal axis in the complex plane. (318)

real number Any rational or irrational number. (86)

real part of a complex number For a complex number $a + bi$, a is the real part. (316)

recursive formula A formula for a sequence in which one or more previous terms are used to generate the next term. (691)

reduced row-echelon form An augmented matrix is in this form if the coefficient columns form an identity matrix. (252)

reference angle For an angle in standard position, the reference angle, θ_{ref}, is the positive acute angle formed by the terminal side of θ and the nearest part (positive or negative) of the x-axis. (837)

relation Any set of ordered pairs. (104)

relative frequency table A frequency table that includes a column showing how frequently each value appears relative to the entire data set. (775)

Remainder Theorem If the polynomial expression that defines the function P is divided by $x - a$, then the remainder is the number $P(a)$. (444)

roots Solutions to an equation. (449)

rounding-up function The function, denoted $f(x) = \lceil x \rceil$, that converts a real number, x, into the smallest integer greater than or equal to x. (125)

row-reduction method The process of performing elementary row operations on an augmented matrix to solve a system of equations and determine whether the system is independent, dependent, or inconsistent. (251)

sample space The set of all possible outcomes of an event. (628)

scalar multiplication Multiplication of each entry in a matrix by the same real number. (218)

scatter plot The graph of the ordered pairs that describe a relationship between two sets of data. (37)

sequence An ordered list of numbers. (691)

series The indicated sum of the terms of a sequence. (693)

sigma The Greek letter Σ, used to denote a series. (693)

simplest radical form The expression $\sqrt{a}$ is in simplest radical form if no factor of a is a perfect square. (530)

simulation A representation of events that are likely to occur in the real world that can be used to find experimental probabilities. (672)

slope-intercept form A linear equation in the form $y = mx + b$, where m represents the slope and b represents the y-intercept. (14)

slope of a line The ratio of the change in vertical direction to the corresponding change in the horizontal direction. (13)

solution A value that can replace a variable that makes an equation or inequality true. (45, 55)

solving a triangle Finding the measures of all of the unknown sides and angles of the triangle. (832)

square matrix A matrix that has the same number of columns and rows. (234)

square root A number, $\sqrt{x}$, that when multiplied by itself produces the given number, x. (520)

standard deviation A measure of dispersion for a data set, given by the formula $\sigma = \sqrt{\sigma^2}$. (794)

standard form The standard form of a linear equation is $Ax + By = C$, where A, B, and C are not both 0. (15)

standard form of a quadratic equation A quadratic equation of the form $ax^2 + bx + c = 0$. (294)

standard normal curve A normal curve with a mean of 0 and a standard deviation of 1. (806)

standard position An angle is in standard position when its initial side lies along the x-axis and its endpoint is at the origin. (836)

stem-and-leaf plot A way of displaying a data set in which each data value is split into two parts, a stem and a leaf. (772)

step function A function whose graph looks like a series of steps. (125)

Substitution Property If $a = b$, then a may replace b in any statement containing a and the resulting statement will be true. (46)

summation notation A way to express a series in an abbreviated form by using the Greek letter sigma, Σ. (693)

synthetic division A method of division of a polynomial by a binomial in which only coefficients are used. (442)

system of equations A set of equations in the same variables. (156)

system of linear inequalities A set of linear inequalities in the same variables. (179)

system of nonlinear equations A set of equations in which at least one equation is nonlinear. (606)

tangent line A line that is perpendicular to a radius of a circle and that touches the circle at only one point. (863)

terminal side The final position of a rotated ray. (836)

terms Parts of an algebraic expression separated by addition or subtraction signs. (46) The numbers in a sequence. (691)

theoretical probability The theoretical probability of event A is defined by $P(A) = \frac{\text{number of outcomes in event } A}{\text{number of outcomes in the sample space}}$. (629)

transformation An alteration in the function rule and its graph. (133)

transverse axis The axis of symmetry of a hyperbola that contains vertices and foci. (596)

trial A systematic opportunity for an event to occur. (628)

trigonometric equation An equation that includes at least one trigonometric function. (922)

trigonometric functions A function that uses one of the six trigonometric ratios to assign values to the measures of the acute angles of a right triangle, or angles of rotation. (829, 838)

trigonometric identity An equation that includes trigonometric functions and that is true for all values of the variables for which the expressions on each side of the equation are defined. (902)

trinomial A polynomial with three terms. (425)

turning points The points on the graph of a polynomial function that correspond to local maxima and minima. (433)

unit circle A circle centered at the origin with a radius of 1. (846)

variable A symbol used to represent one or many different numbers. (45)

variance A measure of dispersion for a data set, given by the formula $\sigma^2 = \frac{1}{n}\sum_{i=1}^{n}(x_i - \overline{x})^2$ where n is the number of values in the data set and $\overline{x}$ is the mean. (794)

vertex A point in a finite set of connected points called a network. (228)

vertex form of a parabola If the coordinates of the vertex of the graph of $y = ax^2 + bx + c$, where $a \neq 0$, are (h, k), then the parabola can be represented in vertex form as $y = a(x - h)^2 + k$. (302)

vertex of a parabola Either the lowest point on the graph or the highest point on the graph. (276) The midpoint between the focus and directrix. (571)

vertical-line test If a vertical line crosses the graph of a relation in more than one point, the relation is not a function. (103)

vertical line A line that has an undefined slope. (16)

vertices The endpoints of the major axis of an ellipse; the endpoints of the transverse axis of a hyperbola. (587, 596)

***x*-intercept** The x-coordinate of the point where the graph crosses the x-axis. (15)

***y*-intercept** The y-coordinate of the point where the graph of a line crosses the y-axis. (14)

zero of a function Any number r such that $f(r) = 0$. (294, 434)

Zero-Product Property If $pq = 0$, then $p = 0$ or $q = 0$. (294)

***z*-score** A measure of how far a value is from the mean in terms of the standard deviation. (809)

Additional Answers

Lesson 1.1, pages 4–11

Activity

1. Answers may vary. Sample answer: Usually a commission is payment to the salesperson of a certain percentage of their gross sales over a certain period of time.

2.

Weekly sales, s (dollars)	Weekly wages, w (in dollars)
100	$40 + 0.10(100) = 50$
200	$40 + 0.10(200) = 60$
300	$40 + 0.10(300) = 70$
400	$40 + 0.10(400) = 80$
500	$40 + 0.10(500) = 90$

3. Successive entries increase by a constant amount of \$100; successive entries increase by a constant amount of \$10.

4.

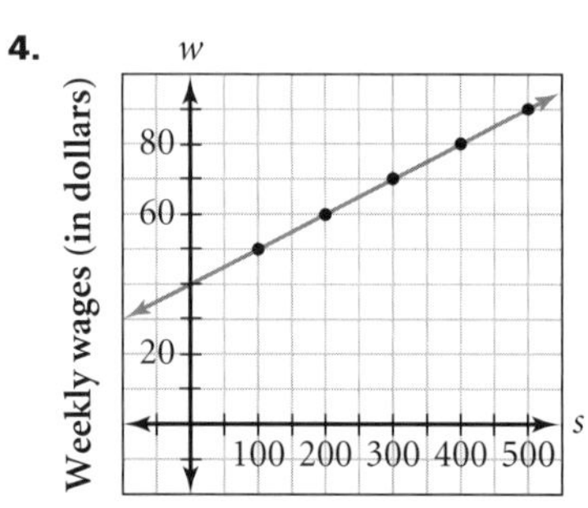

Yes; each point is on the line segment.

CHECKPOINT ✔

5. $w = 0.10s + 40$; the weekly wage is 40 dollars plus 10 percent of weekly sales.

Exercises

Communicate

1. Each pair of x- and y-values in the table is a solution to the equation and represents a point on the graph.

2.

Price	Tax
\$6	0.42
\$8	0.56
\$10	0.70
\$12	0.84

Find the difference in consecutive price values and consecutive tax values. If there is a constant difference in the price values and a constant difference in the tax values, the relationship between price and tax is linear.

3. Plot the three points on a graph, and try to draw a line through all three points. The points are all on the same line.

Practice and Apply

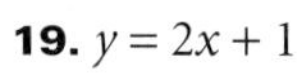

19. $y = 2x + 1$

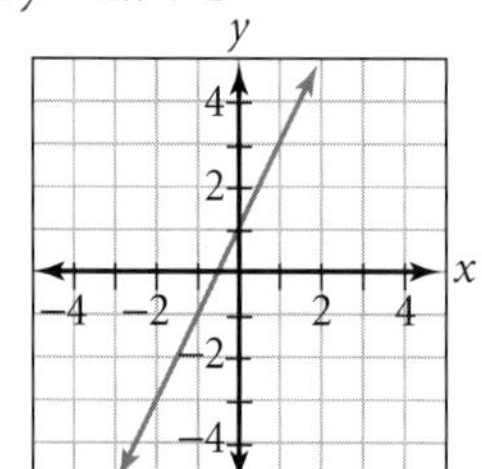

20. $y = 4x + 3$

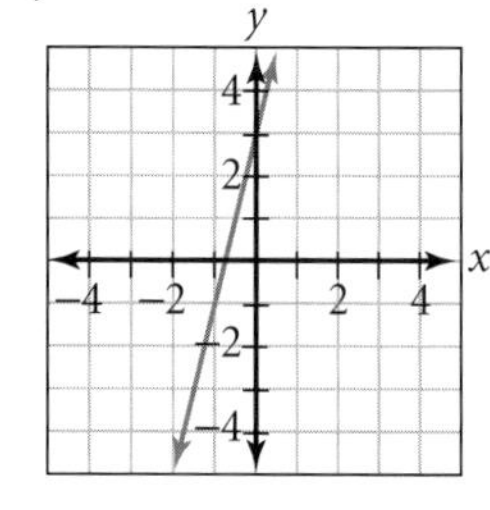

21. $y = 3x - 6$

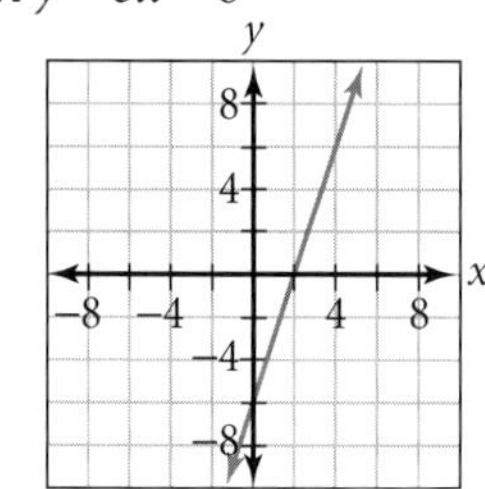

22. $y = 6x - 3$

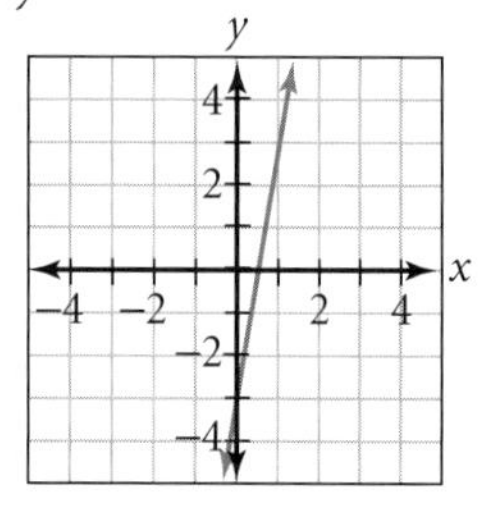

23. $y = 5 - 2x$

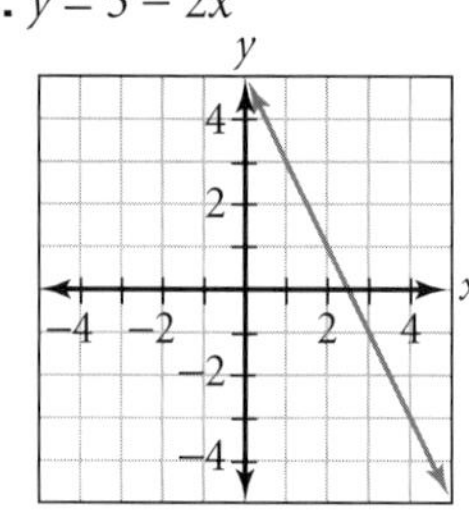

24. $y = 3 - 5x$

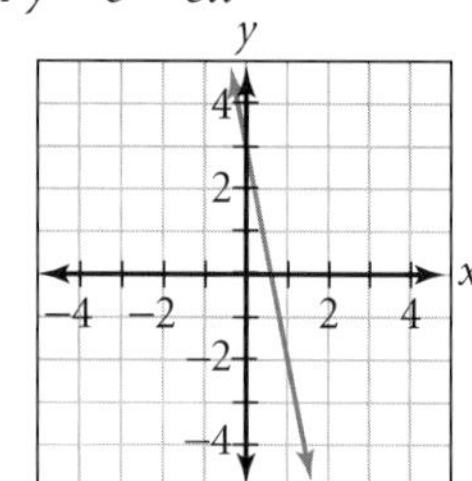

25. $y = -x + 5$

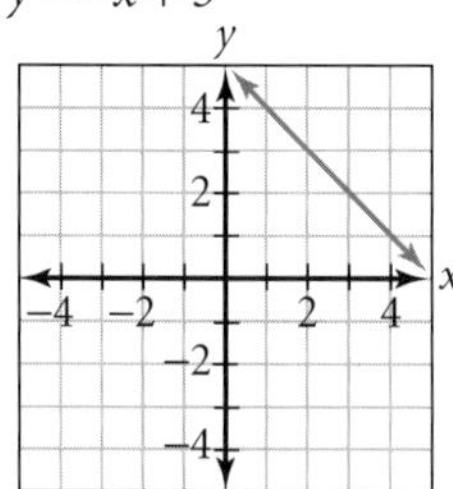

26. $y = -x - 2$

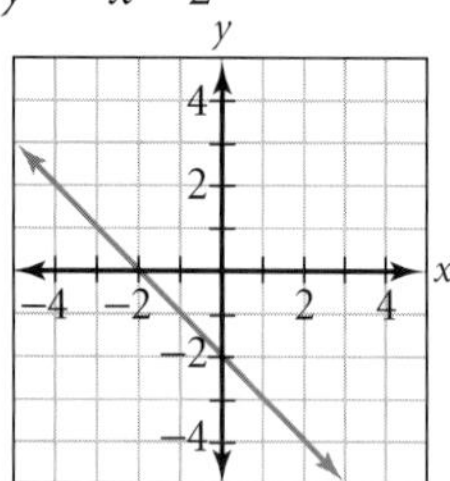

27. $y = \frac{2}{3}x + 4$

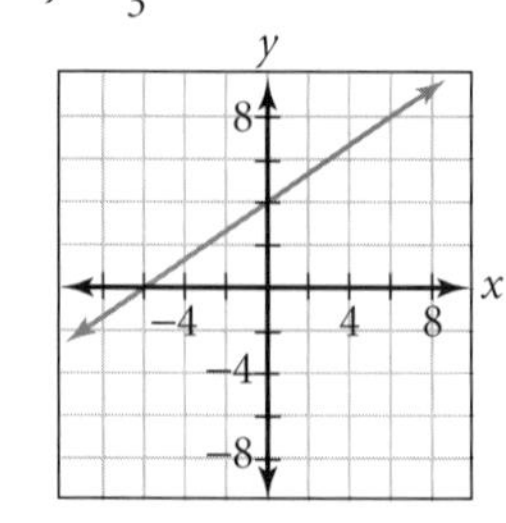

28. $y = \frac{1}{3}x - 5$

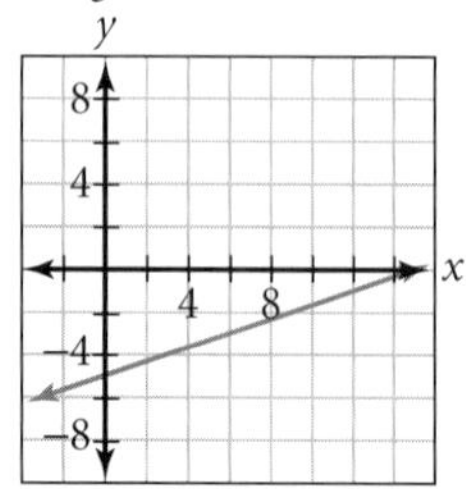

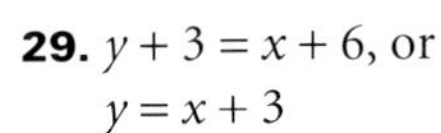

29. $y + 3 = x + 6$, or $y = x + 3$

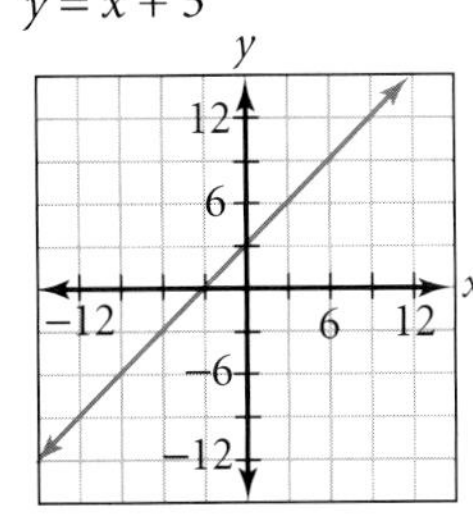

30. $y + 4 = x - 3$, or $y = x - 7$

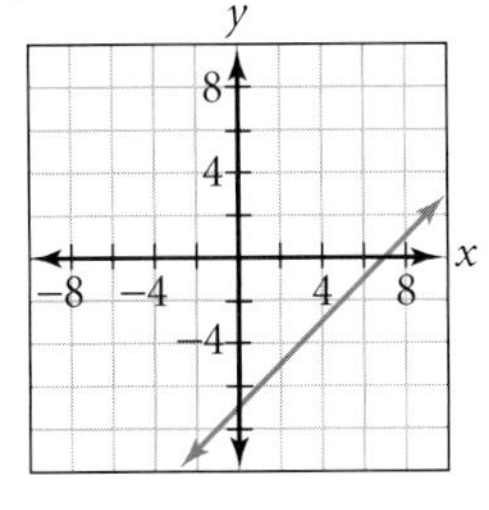

44. $y = 12 - 2.5x$

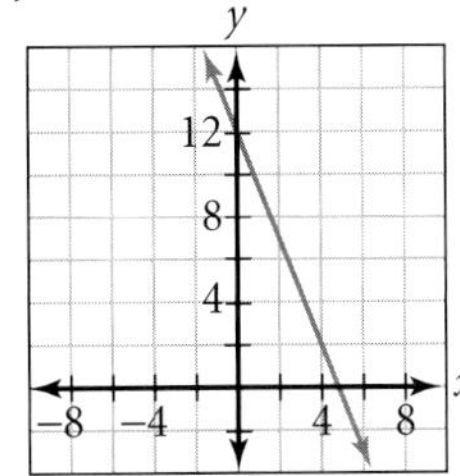

45. $y = 4 - 0.5x$

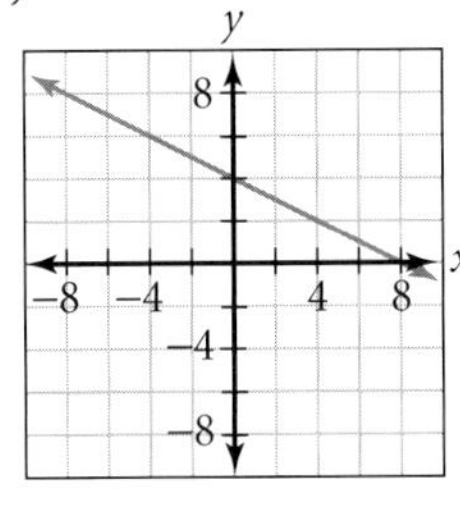

46. $y = \frac{1}{2}x - \frac{3}{5}$

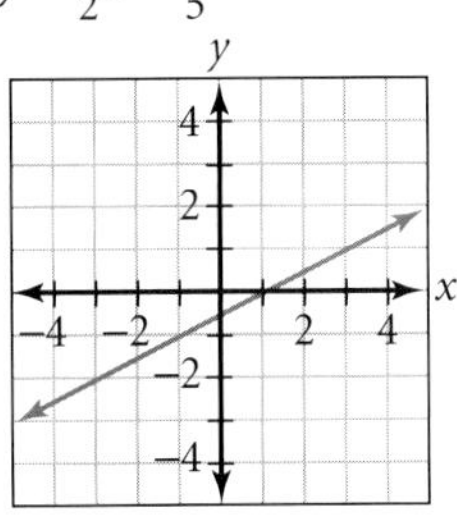

47. $y = -\frac{2}{3}x + \frac{1}{2}$

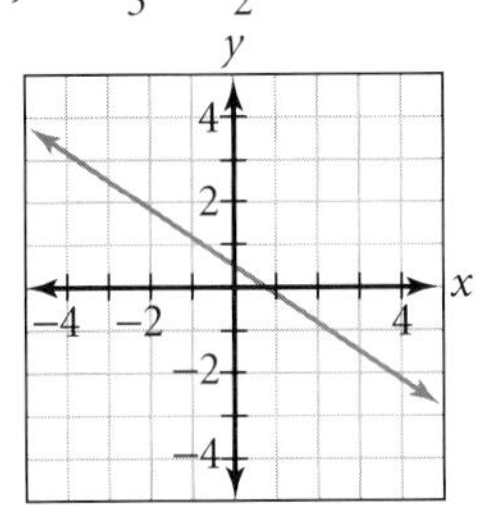

48. When $x = 0$, the graph is a horizontal line that crosses the y-axis at $y = b$. When $y = 0$, the graph is a vertical line that crosses the x-axis at $x = -\frac{b}{m}$.

49. a.

x	y
0	6
1	9
2	12
3	15

d. The answers to parts **b** and **c** could also be found by continuing the table and graph until the desired ordered pairs are found.

3. Answers will vary. Sample data table: This data was recorded during a one hour workout on an exercise bicycle that recorded time in minutes and total distance in tenths of a mile.

Time (minutes)	Distance (miles)
0	0
5	1.6
10	3.5
15	5.2
20	7.1
25	8.9
30	10.5
35	12.3
40	14.0
45	15.6
50	17.3
55	18.8
60	20.4

4. Answers will vary. Sample ordered pairs from data in Step 3: (0, 0), (5, 1.6), (10, 3.5), (15, 5.2), (20, 7.1), (25, 8.9), (30, 10.5), (35, 12.3), (40, 14.0), (45, 15.6), (50, 17.3), (55, 18.8), (60, 20.4)

Bicycle ride

y; Distance (miles): 0, 2, 4, 6, 8, 10, 12, 14, 16, 18, 20

x; Time (minutes): 10, 20, 30, 40, 50, 60

Portfolio Activity

1. Answers will vary. Sample: Three real-world situations in which a distance changes at a fairly constant rate over time are: the distance recorded by an exercise bicycle during a one-hour workout, the distance flown by a glider during a three-hour flight, and the distance traveled by a space probe during a flight to Mars.
2. Answers will vary. Sample: You could collect data by using an exercise bicycle that records time and distance. Ride the bicycle for one hour recording the total distance ridden every 5 minutes.

Lesson 1.2, pages 12–20

Activity

1. The slopes for each pair have the same absolute value but opposite signs.
2. When $m < 0$, the line falls from left to right, and when $m > 0$, the line rises from left to right. The graph of each line of the form $y = mx$ passes through the origin, (0, 0).

CHECKPOINT ✔

3. Answers may vary. Sample answer: $y = 7x$, $y = -7x$

Exercises

Communicate

1. Arrange the absolute values of the slopes from smallest to largest: e, b, c, f, a, d or e, b, c, a, f, d. Notice that f and a are equally steep.

2. Here are 2 possible ways: (1) Write the equation in slope-intercept form, $y = -\frac{3}{2}x + 2$. Plot the y-intercept, (0, 2), and then use the slope, $-\frac{3}{2}$, to find another point by counting 3 units down and 2 units to the right or by counting 3 units up and 2 units to the left. Connect the two points with a line. (2) Use the intercepts. Substitute 0 for x to get a y-intercept of 2, and substitute 0 for y to get an x-intercept of $\frac{4}{3}$. Connect the two intercepts with a line.

3. Two ways are possible: (1) Substitute 0 for x, and solve for y. (2) Write the equation in slope-intercept form. The y-intercept is the number b when the equation is in the form $y = mx + b$. The y-intercept is −2.

4. Solve the equation for y, and write it in the form $y = mx + b$. $y = x - 2$ ($m = 1$ and $b = -2$)

Practice and Apply

32. $m = -\frac{1}{2}, b = -2$

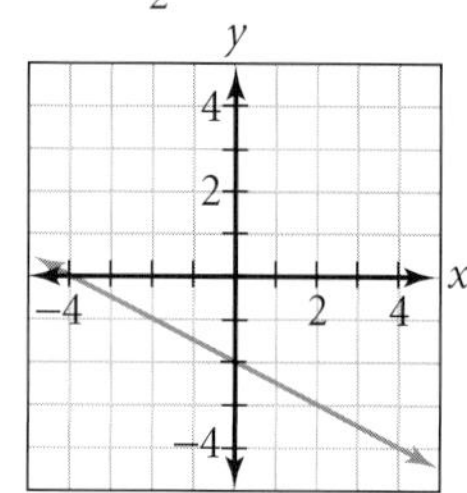

33. $m = 0.6, b = -4$

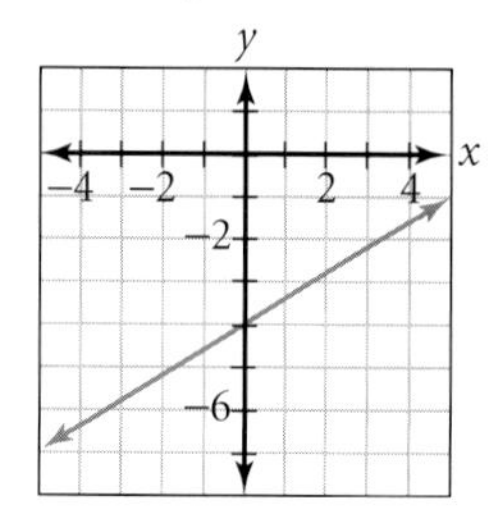

34. $m = -2, b = 1$

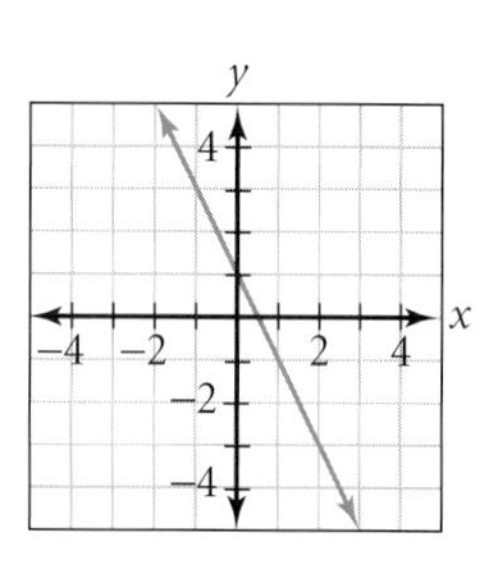

35. m is undefined, no y-intercept

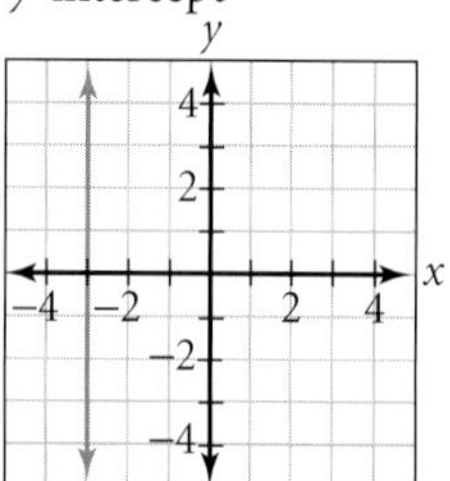

40. $4x + y = -4$

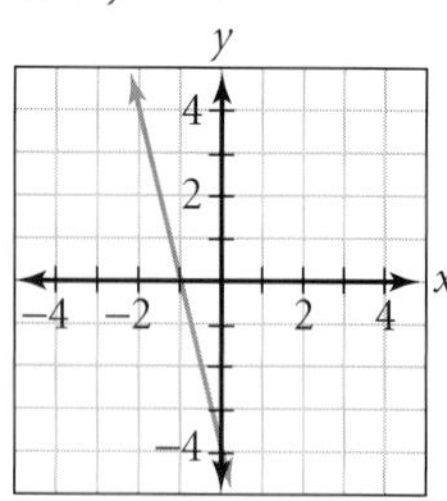

41. $x + 3y = 12$

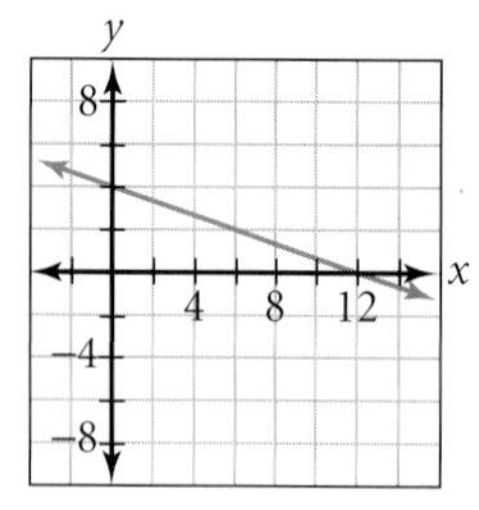

42. $2x - y = 8$

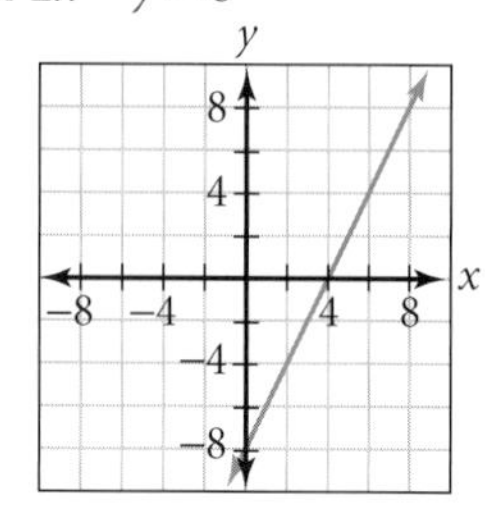

43. $-x + 2y = 5$

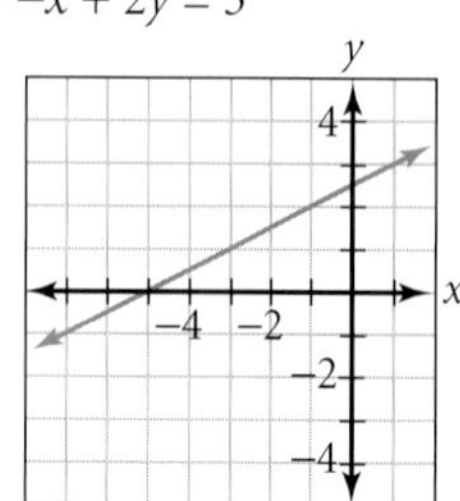

44. $7x + 3y = 2$

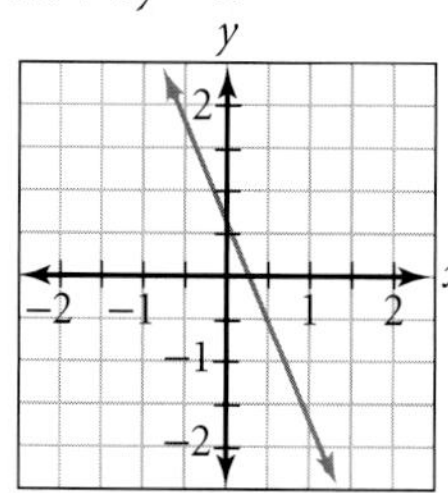

45. $-x + 8y = -6$

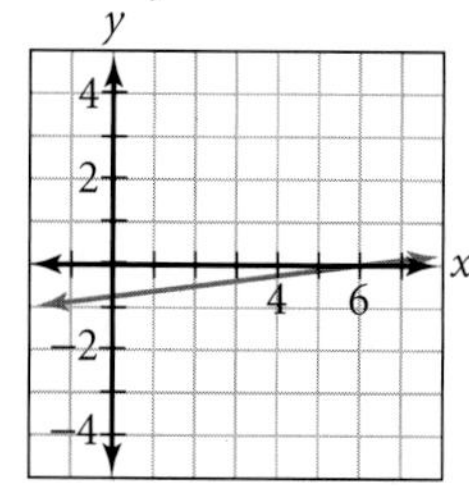

46. $-3x + y = -9$

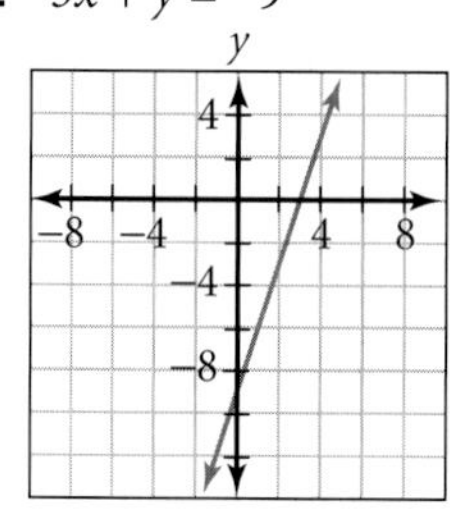

47. $-x - 7y = 3$

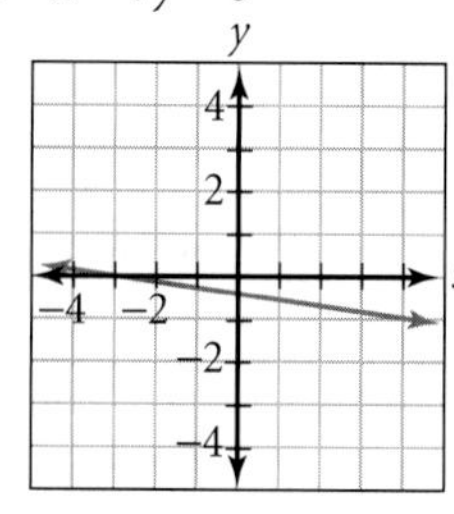

48. $x - y = -1$

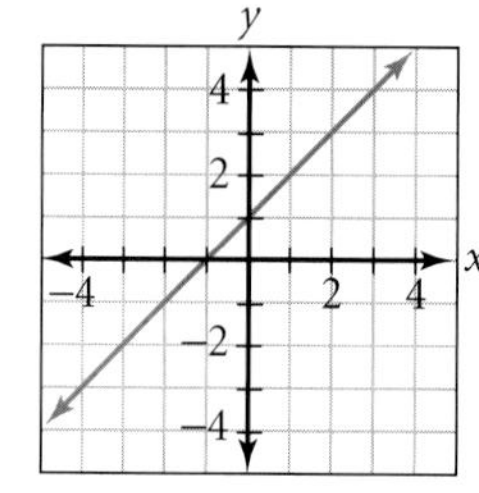

49. $-\frac{1}{2}x + 3y = 7$

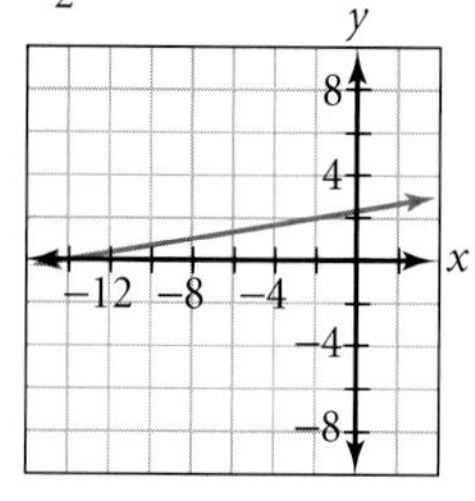

50. $5x - 8y = 16$

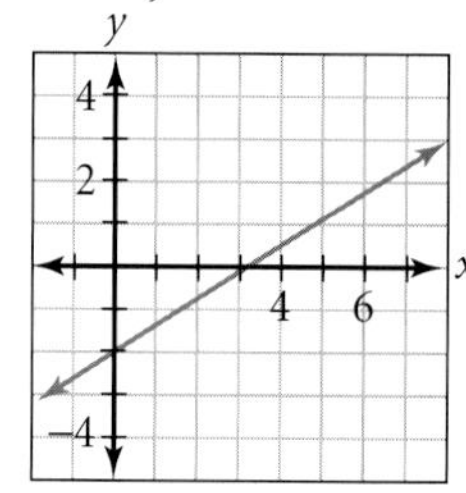

51. $x + \frac{1}{2}y = -2$

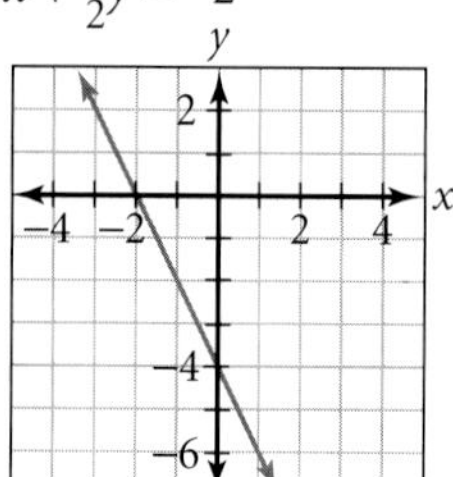

52. Slope is undefined; $x = 5$

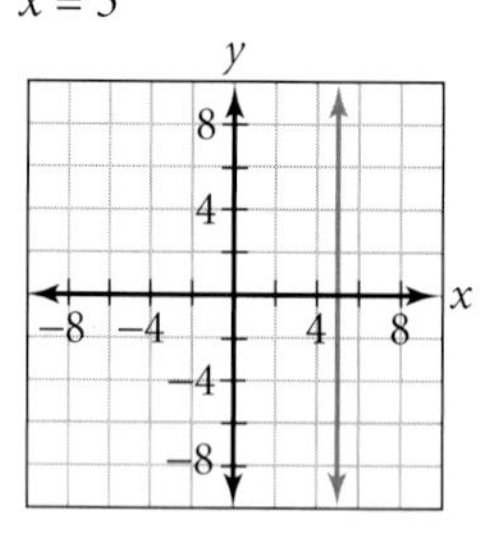

53. Slope is undefined; $x = -2$

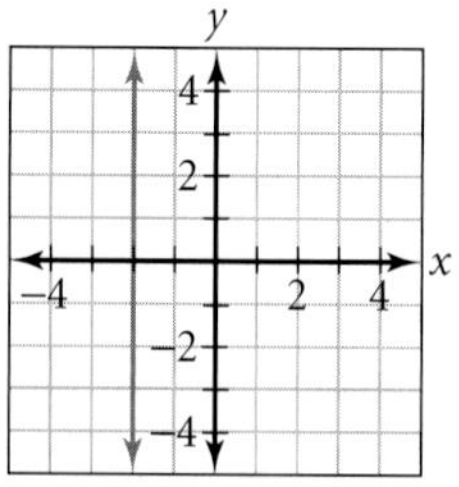

54. Slope = 0; $y = 8$

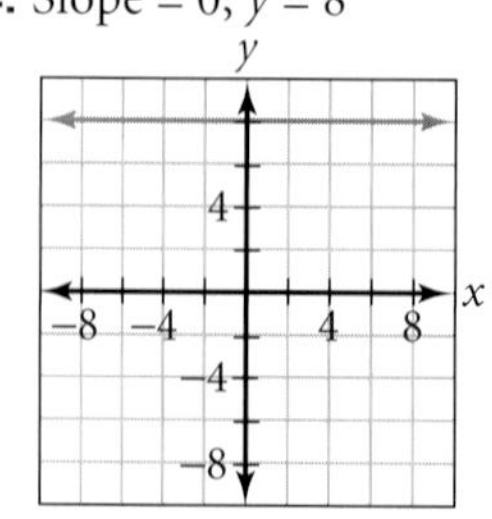

55. Slope = 0; $y = -5$

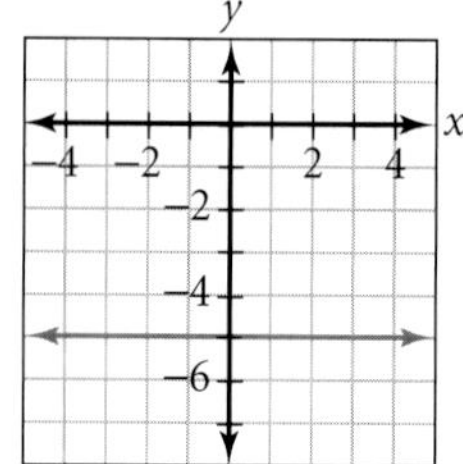

56. Slope is undefined; $x = -1$

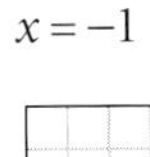

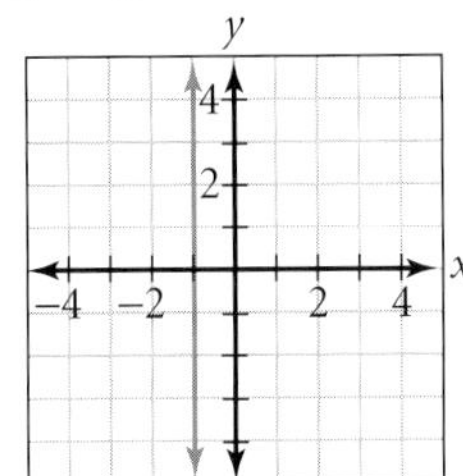

57. Slope is undefined; $x = 9$

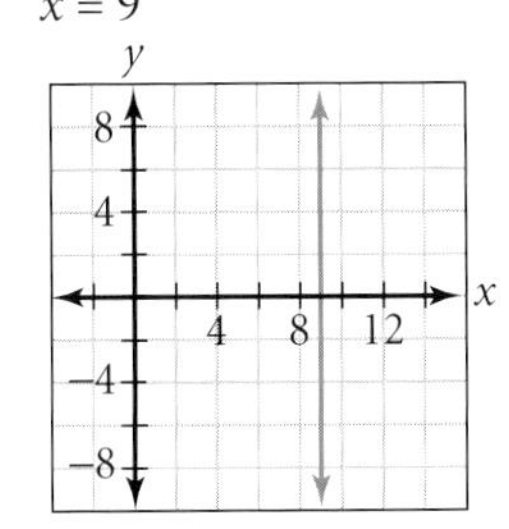

58. Slope = 0; $y = -8$

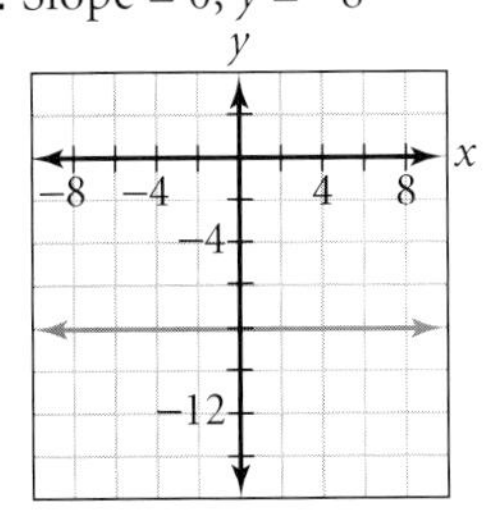

59. Slope = 0; $y = 7$

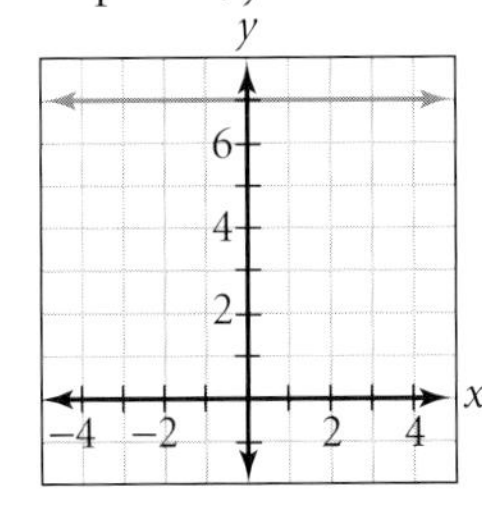

60. Slope is undefined; $x = -\frac{1}{3}$

61. Slope is undefined; $x = -\frac{1}{4}$

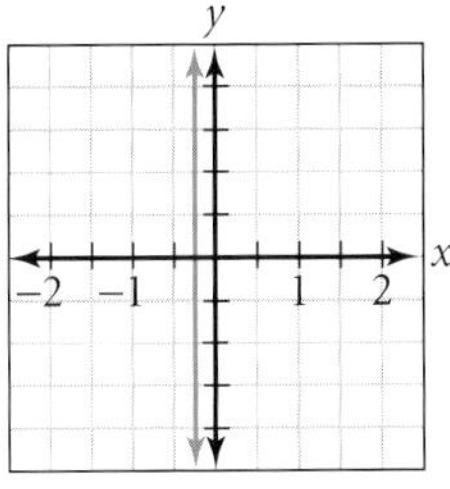

62. Slope = 0; $y = \frac{3}{4}$

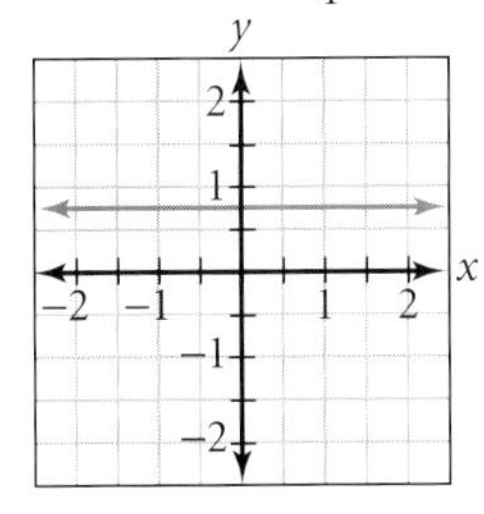

63. Slope = 0; $y = \frac{2}{3}$

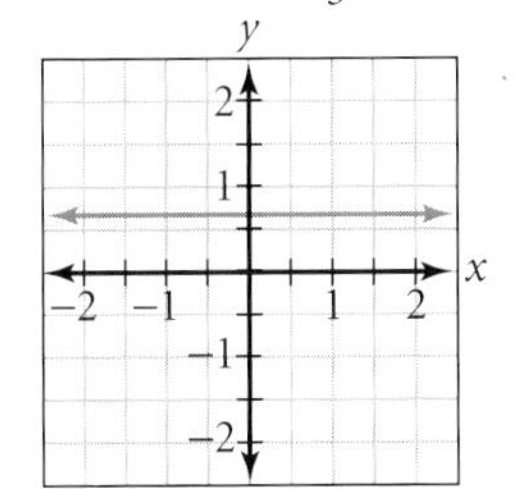

Look Back

74.

x	−2	−1	0	1	2	3	4
y	13	10	7	4	1	−2	−5

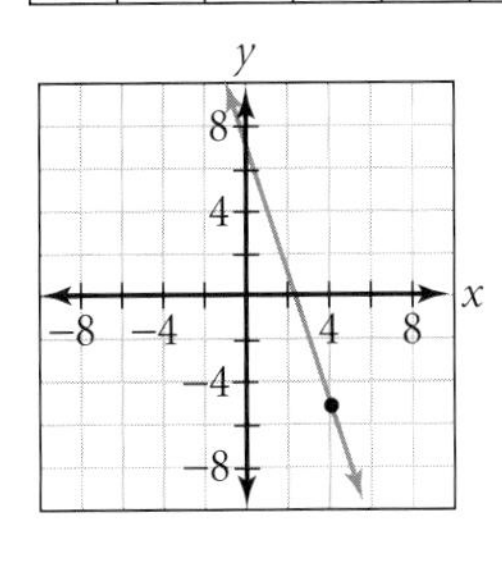

No; the value corresponding to $x = 4$ in the table is $y = -5$. The point $(4, -4)$ is not on the line. Substituting the x-value 4 results in a y-value of −5.

Portfolio Activity

1. Answers will vary. Sample: Two points on the graph that seem to fit the trend of the data are (15, 5.2) and (40, 14.0).

2. Answers will vary. Sample:

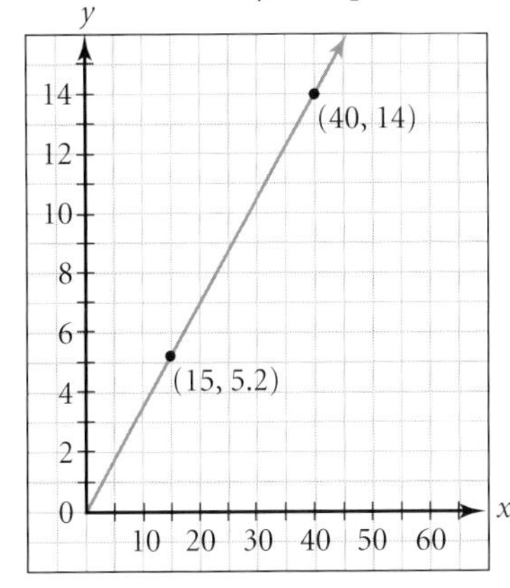

Find the slope of the line by using (15, 5.2) and (40, 14). Let $(x_1, y_1) = (15, 5.2)$ and $(x_2, y_2) = (40, 14)$.

$$m = \frac{y_2 - y_1}{x_2 - x_1} = \frac{14 - 5.2}{40 - 15} = \frac{8.8}{25} = 0.352$$

The y-intercept looks like it could be (0, 0), but this is not certain. To find the y-intercept, substitute (40, 14) into an equation of the form $y = mx + b$, where $m = 0.352$, and solve for b.

$$14 = 0.352(40) + b$$
$$14 = 14.08 + b$$
$$b = 14 - 14.08 = -0.08$$

The equation in slope-intercept form is $y = 0.352x - 0.08$.

3. Answers will vary. Sample: The rate of change indicated by this linear model is miles per minute, or specifically, 0.352 miles per minute.

Lesson 1.3, pages 21–28

Activity

1. The lines $y = 2x$ and $y = 2x - 2.5$ are parallel to the line $y = 2x + 1$. They each have a coefficient of 2 for the x-term.

CHECKPOINT ✔

2. Answers may vary. Sample answer: $y = 2x + 3$

3. The lines $y = -\frac{1}{2}x + 2$ and $y = -\frac{1}{2}x + 3$ are perpendicular to the line $y = 2x + 1$. The coefficient of the x-term in these two equations is the negative of the reciprocal of the coefficient of the x-term in $y = 2x + 1$.

CHECKPOINT ✔

4. Answers may vary. Sample answer: $y = -\frac{1}{2}x - 3$

Exercises

Communicate

1. First find the slope of the line by using the slope formula: $m = \frac{y_2 - y_1}{x_2 - x_1} = \frac{-2 - 3}{4 - 1} = -\frac{5}{3}$. Next find the y-intercept of the line by substituting $-\frac{5}{3}$ for m and the coordinates of either point into $y = mx + b$: $b = \frac{14}{3}$. Now write the equation in $y = mx + b$ form: $y = \frac{5}{3}x + \frac{14}{3}$. You can also use the point-slope form after finding the slope.
2. Pick two points on the line, such as (0, 3) and (4, 0). Find the slope between them, $m = -\frac{3}{4}$. Now choose one of those points, such as (0, 3). Substitute $-\frac{3}{4}$ for m, 0 for x_1 and 3 for y_1 in the point-slope formula $y - y_1 = m(x - x_1)$. Then use algebra to write the other forms of the equation.
3. Solve each equation for y and compare the slopes. If the lines have the same slopes, then they are parallel. If the lines have slopes which are negative reciprocals of each other, then they are perpendicular. The slopes are $-\frac{5}{6}$ and $\frac{6}{5}$ so the lines are perpendicular.
4. Find the slope of $x + 2y = 4$ by rewriting the equation in slope-intercept form, $m = -\frac{1}{2}$. The new slope is the negative reciprocal, 2. Now substitute the new slope, 2, and the point (3, −1) into the point-slope formula to find the y-intercept. Write the equation in $y = mx + b$ form: $y = 2x - 7$.

Lesson 1.4, pages 29–36

Activity

1. $\frac{A'B'}{AB} = \frac{3}{2}$; $\frac{B'C'}{BC} = \frac{3}{2}$; $\frac{A'C'}{AC} = \frac{3}{2}$

CHECKPOINT ✔

2. There is a direct-variation relationship. If the lengths of $\triangle ABC$ are represented by x and the lengths of $\triangle A'B'C'$ are represented by y, then the direct-variation equation is $y = \frac{3}{2}x$.

Exercises

Communicate

1. We know that $y = kx$, so substitute the given values for x and y and solve for k. $k = 2$, so $y = 2x$.
2. Linear equations are not direct variations when there is some nonzero constant added to the x-term, $y = kx + c$. The graphs of linear equations that are direct variations are lines that pass through the origin, but the graphs of linear equations that are not direct variations do not pass through the origin.
3. Use a proportion, such as $\frac{8}{-2} = \frac{12}{x}$, or a direct-variation equation, such as $y = -4x$.
4. No, not in the form $y = kx$.
5. No, not in the form $y = kx$.
6. Yes, has the form $y = kx$.
7. Yes, has the form $y = kx$.

Portfolio Activity

1. Answers will vary. Sample: For the linear model, the variable y, which represents total distance, does not quite vary directly as x, which represents the time. There is a constant, 0.352, by which to multiply the time but then you must add −0.08 to find the distance. For the actual data set, distance does not vary directly as time for each pair of data points. The ratio of distance to time changes so there is not one constant k to use as the constant of variation.
2. Answers will vary. From the data set, $\frac{y_1}{x_1} = \frac{y_2}{x_2}$ is approximately true for some data values such as $\frac{5.2}{15} = 0.34\overline{6}$ and $\frac{14}{40} = 0.35$, but for other values such as $\frac{1.6}{5} = 0.32$ and $\frac{3.5}{10} = 0.35$ it is less accurate. For the real data set, the rate of change is not constant because the rider probably slowed down and sped up during the one-hour ride.
3. Answers will vary. Sample answer: Using the points (15, 5.2) and (40, 14.0) on the linear model, a proportion is $\frac{5.2}{15} = \frac{14}{40}$. This proportion is approximately true. All ratios of the y-value to the x-value should be approximately equal since the line was drawn to have a constant rate of change. For all values in the linear model, the proportions are not exactly equal because the y-intercept is not 0 and this linear model is not a direct variation.

Lesson 1.5, pages 37–44

Activity

1. Graph is shown on page 37.
2. Answers may vary. Sample answer: $y = -6x + 175$ or $y = -7x + 180$
3. Answers may vary. Sample answer: $y = -8x + 185$

CHECKPOINT ✔

4. Slope = −8 cases per year
   The slope indicates that each year there are 8 thousand fewer cases of chicken pox reported in the United States.

Exercises

Communicate

1. False. The range of the correlation coefficient is −1 to 1.
2. True.

3. False. The closer the correlation coefficient is to zero, the less linear the relationship is between the data points.

4. Answers may vary. Sample answer: as population increases, available water supply per person decreases.

5. Answers may vary. Sample answer: as population increases, the volume of garbage produced increases.

6. The points in the scatter plot are widely dispersed, with no apparent line of best fit.

7. The points in the scatter plot lie roughly around a line that slopes downward to the right. However, since r is not close to -1, there will be many points away from the line.

8. The points in the scatter plot lie very close to a line that slopes upward to the right. Since r is very close to 1, most of the points will be very close to the line of best fit.

Portfolio Activity

1. Answers will vary. Sample: The least-squares regression line is $y = 0.3428571429x + 0.1142857143$.

2. Answers will vary.

3. Answers will vary. Sample: The slope of the least-squares regression line is 0.3428571429 and the slope of the linear model is 0.352. These two slopes are very close.

4. Answers will vary. Sample: The correlation coefficient of the least-squares line is 0.9996983203.

Lesson 1.6, pages 45–51

Activity

1. The expression $x + 3$ is equal to the expression $9 - 2x$.

2. The expressions have the same value for $x = 2$.

3. True

CHECKPOINT ✔

4. Graph $y = 2x - 1$ and $y = 2 - x$ on the same screen, and find the x-coordinate of the point where the graphs intersect: $x = 1$.

Exercises

Communicate

1. Addition, Division, and Symmetric Properties

2. Multiplication, Subtraction, and Division Properties

3. Addition and Division Properties

4. Answers may vary. Sample answer: By using the Addition Property and adding 7 to each side of the equation, an equivalent equation of $4x = 21$ would be obtained.

5. Graph $y = \frac{2(x+3)}{7}$ and $y = \frac{9(x-3)}{5}$ on the same coordinate plane. The solution is the x-coordinate of the point of intersection of these two lines: $x = 4.13$.

Look Back

85. $y > -5$ means the values of y are greater than -5.

86. $-3 < x < 3$ means x is greater than -3 and less than 3.

87. $-1 \le y \le 1$ means the values of y are greater than or equal to -1 and also less than or equal to 1.

88. $x \le -3$ means the values of x are less than or equal to -3.

Portfolio Activity

1. Answers will vary. Sample answer: Choose the y-value 6.0 and the equation $y = 0.3428571429x + 0.1142857143$. A reasonable prediction for the time to ride 6 miles is 17.2 minutes.

2. Answers may vary. On the graph of the scatter plot, linear model, and least-squares regression line, the point should lie on the least-squares regression line.

Lesson 1.7, pages 54–60

Activity

1. $x < 3$

2.

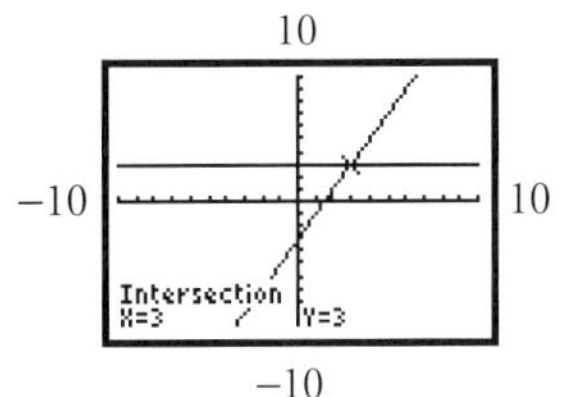

3. All values of x less than 3

4. The answer to Step 3 is the same as the answer obtained in Step 1.

CHECKPOINT ✔

5. Graph $y = 3x + 2$ and $y = 5$ on the same screen. Find the values of x for which the graph of $y = 3x + 2$ is above the graph of $y = 5$.

Exercises

Communicate

1. Solve $7x - 7 > 0$ for x to get $x > 1$. Draw a number line. Put an open circle at $x = 1$. Draw an arrow to the right from the open circle.

2. The graph of $7x - 7 \ge 0$ is the same as the graph of $7x - 7 > 0$ except the open circle at $x = 1$ is a closed circle. The graph of $7x - 7 < 0$ is an open circle at $x = 1$, but the arrow goes to the left on the number line.

3. No. Multiplying each side of an inequality by a negative changes the direction of the inequality sign. $x < 16$ is equivalent to $-x > -16$.

4. *Nonnegative* means greater than or equal to zero, so "x is nonnegative" is expressed as $x \geq 0$.

Practice and Apply

27. $y < -8$

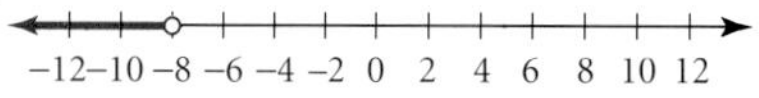

−12 −10 −8 −6 −4 −2 0 2 4 6 8 10 12

28. $x < 8$

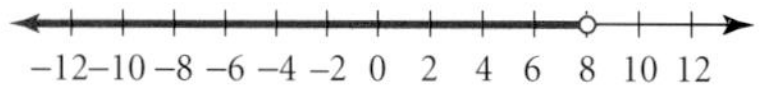

−12 −10 −8 −6 −4 −2 0 2 4 6 8 10 12

29. $x \geq 11$

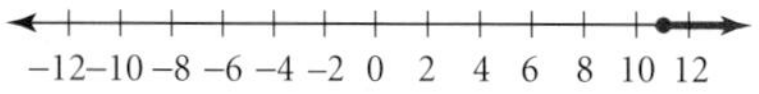

−12 −10 −8 −6 −4 −2 0 2 4 6 8 10 12

30. $d \geq -28$

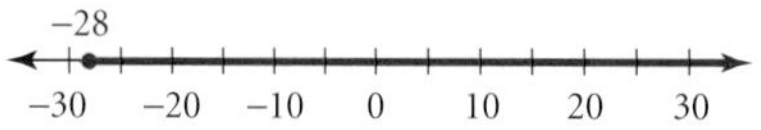

−28

−30 −20 −10 0 10 20 30

31. $x \leq 150$

−300 −200 −100 0 100 200 300

32. $x < -7$

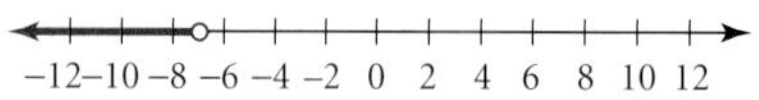

−12 −10 −8 −6 −4 −2 0 2 4 6 8 10 12

33. $x \geq -6$

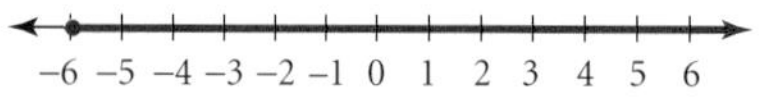

−6 −5 −4 −3 −2 −1 0 1 2 3 4 5 6

34. $x > -33$

−33

−60 −40 −20 0 20 40 60

35. $x < -15$

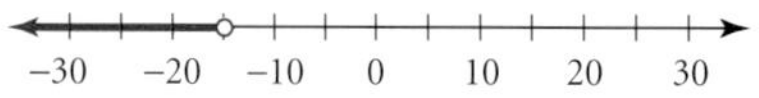

−30 −20 −10 0 10 20 30

36. $x < 2$

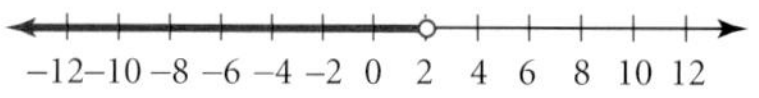

−12 −10 −8 −6 −4 −2 0 2 4 6 8 10 12

37. $x \leq 9$

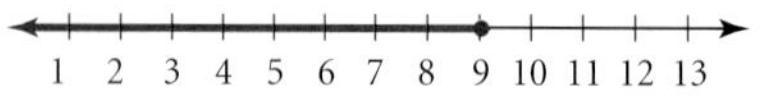

1 2 3 4 5 6 7 8 9 10 11 12 13

38. $x \leq -30$

−60 −50 −40 −30 −20 −10 0

39. $t > -64$

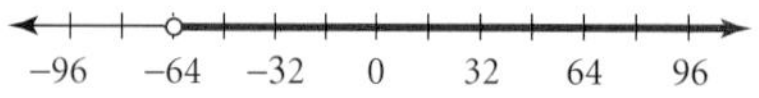

−96 −64 −32 0 32 64 96

40. $p > -6$

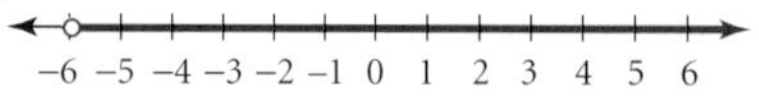

−6 −5 −4 −3 −2 −1 0 1 2 3 4 5 6

41. $x \leq \frac{1}{4}$

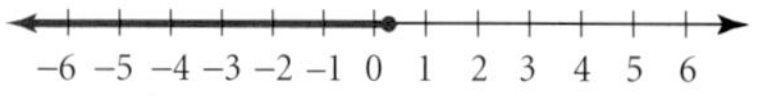

−6 −5 −4 −3 −2 −1 0 1 2 3 4 5 6

42. $y < 1$

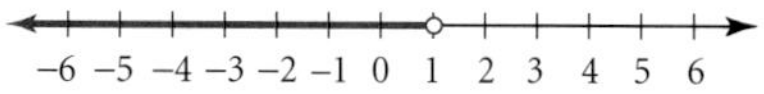

−6 −5 −4 −3 −2 −1 0 1 2 3 4 5 6

43. $a < 5$

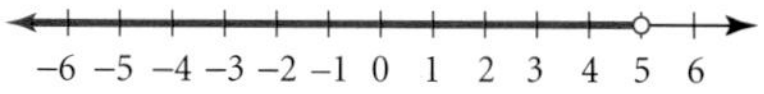

−6 −5 −4 −3 −2 −1 0 1 2 3 4 5 6

44. $x < 4.5$

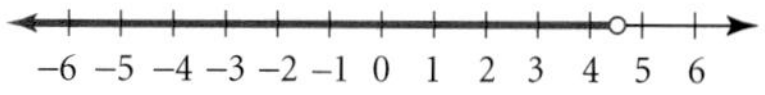

−6 −5 −4 −3 −2 −1 0 1 2 3 4 5 6

45. $x \geq 15$

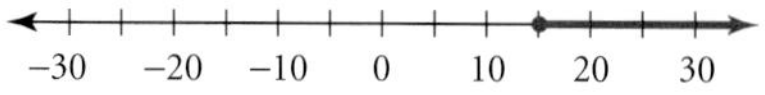

−30 −20 −10 0 10 20 30

46. $x < -7$

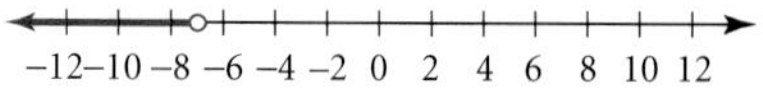

−12 −10 −8 −6 −4 −2 0 2 4 6 8 10 12

47. $x \geq -4$

−6 −5 −4 −3 −2 −1 0 1 2 3 4 5 6

48. $x < \frac{1}{7}$

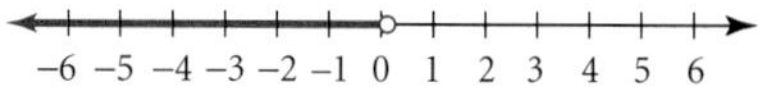

−6 −5 −4 −3 −2 −1 0 1 2 3 4 5 6

49. $x \leq -\frac{6}{19}$

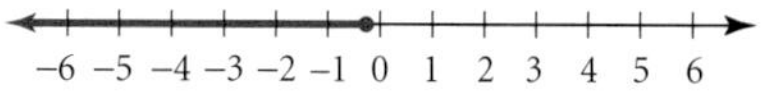

−6 −5 −4 −3 −2 −1 0 1 2 3 4 5 6

50. a. $-4 < x < 2$

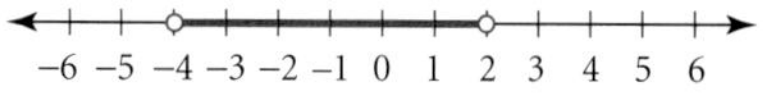

−6 −5 −4 −3 −2 −1 0 1 2 3 4 5 6

b. $x > 2$

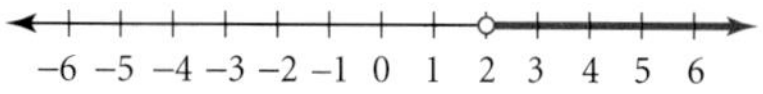

−6 −5 −4 −3 −2 −1 0 1 2 3 4 5 6

c. all real numbers

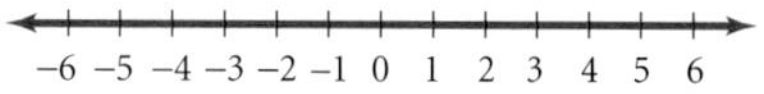

−6 −5 −4 −3 −2 −1 0 1 2 3 4 5 6

d. $x > -4$

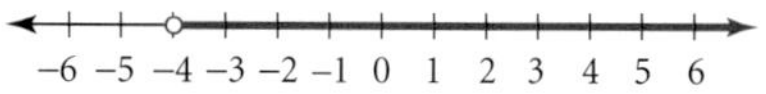

−6 −5 −4 −3 −2 −1 0 1 2 3 4 5 6

51. a. $x < -4$

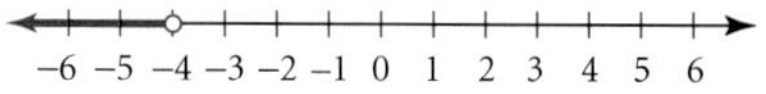

−6 −5 −4 −3 −2 −1 0 1 2 3 4 5 6

b. no solution

c. $x < 2$

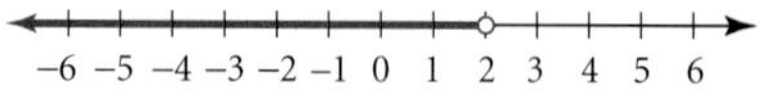

−6 −5 −4 −3 −2 −1 0 1 2 3 4 5 6

d. $x < -4$ *or* $x > 2$

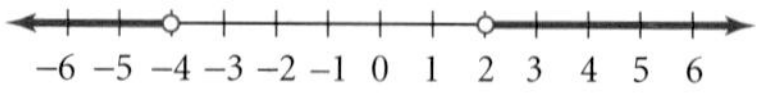

−6 −5 −4 −3 −2 −1 0 1 2 3 4 5 6

66. $b < 4$ *or* $b > 5$

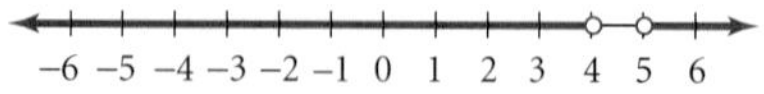

−6 −5 −4 −3 −2 −1 0 1 2 3 4 5 6

67. $x > -4$ *or* $x \leq -5$

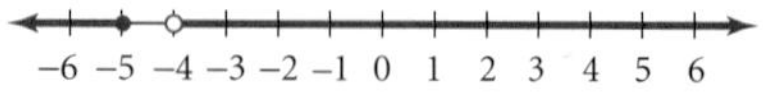

−6 −5 −4 −3 −2 −1 0 1 2 3 4 5 6

68. $x \leq -15$ *and* $x < 2$

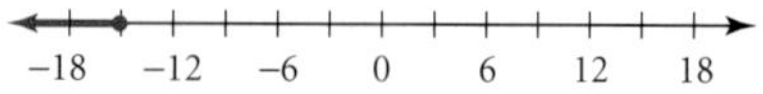

−18 −12 −6 0 6 12 18

69. $m > -\frac{9}{4}$ *and* $m \leq -\frac{5}{2}$; no solution

70. $x > 2$ *or* $x \leq -3$

71. $x > -5$ *or* $x \leq 4$; all real numbers

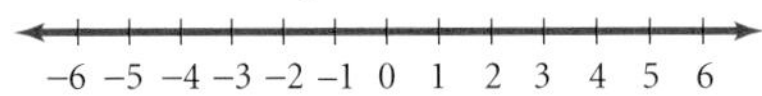

Lesson 1.8, pages 61–69

Activity

1. The graphs intersect at one point, at $x = 2$.
2. Answers may vary. Sample answer: try $m = \frac{1}{2}$, $b = 2$; $x = -\frac{4}{3}$ *or* $x = 4$
3. Answers may vary. Sample answer: try $m = 1$, $b = -3$; no solution to $|x| = x - 3$
4. Answers may vary. Sample answer: try $m = 1$, $b = 0$; infinitely many solutions to $|x| = x$

CHECKPOINT ✔

5. There are either 0, 1, 2, or infinitely many possible solutions.

Exercises

Communicate

1. $|3x - 5|$ is always nonnegative, and a nonnegative number plus 4 can never equal 3. Thus, there is no solution to $|3x - 5| + 4 = 3$.
2. Sometimes a false solution can be introduced.
3. Geometrically, the absolute value of x is the distance between x and 0 on the number line. Since that distance can be measured from the left or right of zero, there are two values of x that satisfy the definition.
4. Mathematically, the word *and* indicates when two or more statements must all be true at the same time. The word *or* indicates that at least one statement must be true.
5. $|x| > a$, where $a < 0$

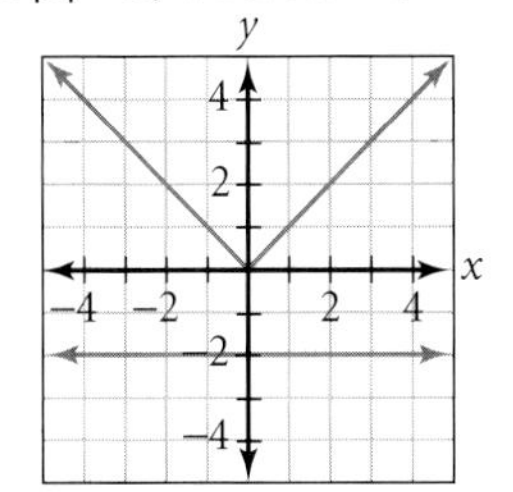

Practice and Apply

40. $x < 3$ *or* $x > 5$

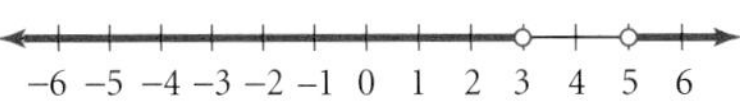

41. $x \geq -12$ *and* $x \leq 2$

42. $x < -5$ *or* $x > 5$

43. $x \geq -6$ *and* $x \leq 6$

44. No solution

45. All real numbers

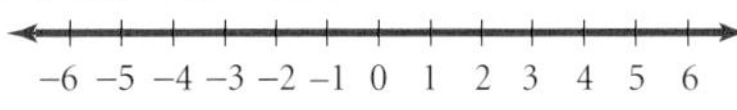

46. No solution

47. $x > -4$ *and* $x < 7$

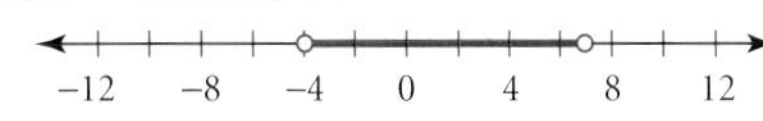

48. $x \geq -5$ *and* $x \leq 2$

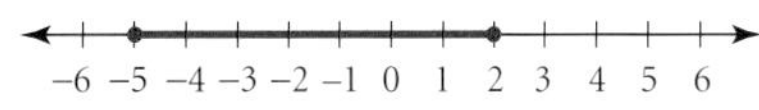

49. $x < \frac{12}{7}$

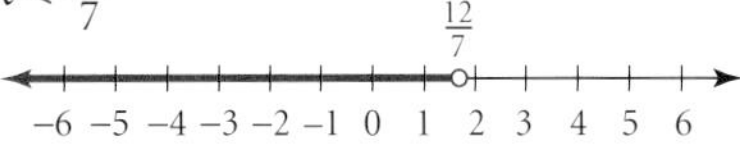

50. $x \leq -\frac{5}{2}$ *or* $x \geq 5$

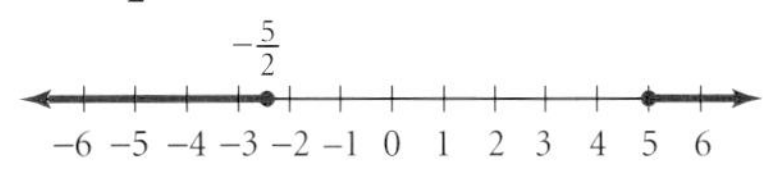

51. All real numbers

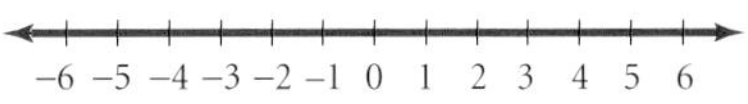

52. All real numbers

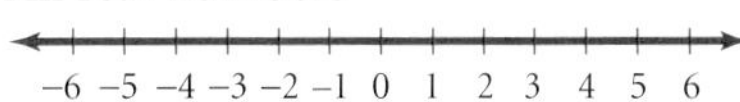

53. No solution

54. No solution

55. $x \leq -\frac{3}{4}$ *or* $x \geq \frac{1}{4}$

56. $-\frac{3}{4} \leq x \leq \frac{1}{4}$

57. No solution

58. $x > 3$ *and* $x < 11$

59. The distance between x and 4 is less than 1.

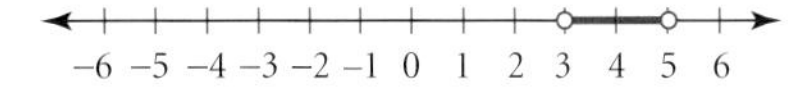

Look Back

66. a. Answers may vary. Sample answer:

Number of years, t	Value (in dollars), v
0	$90{,}000 - 4500(0) = 90{,}000$
1	$90{,}000 - 4500(1) = 85{,}500$
2	$90{,}000 - 4500(2) = 81{,}000$
5	$90{,}000 - 4500(5) = 67{,}500$
10	$90{,}000 - 4500(10) = 45{,}000$

Chapter 1 Project

Activity 2

5. Answers will vary. Samples:
- **a.** For height and arm span, the y-intercept is 0, which tells you that arm span varies directly as height.
- **b.** For height and hand span, the y-intercept is not 0, which tells you that to find a new y-value, you need to multiply the x-value by the slope and then add the y-intercept value. The variation is not direct.
- **c.** For height and distance from head to ceiling, the y-intercept is not 0, which tells you that to find a new y-value, you need to multiply the x-value by the slope and then add the y-intercept value. The variation is not direct.
- **d.** For height and coin value, the y-intercept is not 0 which tells you that to find a new y-value you need to multiply the x-value by the slope and then add the y-intercept value. The variation is not direct.

Activity 3

1. Answers will vary. Samples:
- **a.** For height and arm span, there appears to be a positive correlation between the variables because the points cluster around a line that has a positive slope.
- **b.** For height and hand span, there appears to be a positive correlation between the variables because the points cluster around a line that has a positive slope.
- **c.** For height and distance from head to ceiling, there is a negative correlation between the variables because the points lie on a straight line that has a negative slope.
- **d.** For height and coin value, there appears to be no correlation because the points do not cluster around a line.

2. Answers will vary.
- **a.** The correlation coefficient is about 0.9.
- **b.** The correlation coefficient is about 0.5.
- **c.** The correlation coefficient is about −1.0.
- **d.** The correlation coefficient is about −0.4.

3. Answers will vary. Check students' graphs.
- **a.** $y = 1.05x - 3.8$
- **b.** $y = 0.06x + 4.3$
- **c.** $y = -x + 120$
- **d.** $y = -6.80x + 541.6$

4. Answers will vary. Samples:
- **a.** For height and arm span, the lines for the least-squares line and the linear model are very close. Their slopes are almost equal, but the y-intercepts are slightly different.
- **b.** For height and hand span, the lines for the least-squares line and the linear model are very close. Their slopes are almost equal, but the y-intercepts are slightly different.
- **c.** For height and distance from head to ceiling, the lines for the least-squares line and the linear model are identical.
- **d.** For height and coin value, the lines for the least-squares line and the linear model are not very close. Their slopes are different and their y-intercepts are about 500 units apart.

5. a. Answers will vary. Sample: A prediction for the arm span of a person who is 5 feet tall is 60 inches.
- **b.** Answers will vary. Sample: A prediction for the hand span of a person who is 5 feet tall is $6\frac{3}{4}$ inches.
- **c.** Answers will vary. Sample: A prediction for the distance from the ceiling (10 feet) to the head of a person who is 5 feet tall is 60 inches.
- **d.** Answers will vary. Sample: A prediction for the value of pocket change of a person who is 5 feet tall is 120 cents.

Activity 4

1. Answers will vary. Sample: Use the least-squares regression equation for height and arm span.

$y = 1.05x - 3.8$	
$62 = 1.05x - 3.8$	Substitute 62 for y.
$62 + 3.8 = 1.05x - 3.8 + 3.8$	Use the Addition Property.
$65.8 = 1.05x$	Simplify.
$\frac{65.8}{1.05} = x$	Use the Division Property.
$x \approx 62.7$	Simplify.

A prediction for the height is 62.7 inches.

2. Answers will vary. Sample: Use the least-squares regression equation for height and hand span.

$y = 0.06x + 4.3$	
$8.5 = 0.06x + 4.3$	Substitute 8.5 for y.
$8.5 - 4.3 = 0.06x + 4.3 - 4.3$	Use the Subtraction Property.
$4.2 = 0.06x$	Simplify.
$\frac{4.2}{0.06} = x$	Use the Division Property.
$x \approx 70$	Simplify.

A prediction for the height is 70 inches.

3. Answers will vary. Sample: Use the least-squares regression equation for height and distance of head from ceiling.

$y = -x + 120$	
$28 = -x + 120$	Substitute 28 for y.
$28 - 120 = -x + 120 - 120$	Use the Subtraction Property.
$-92 = -x$	Simplify.
$(-1) - 92 = (-1) - x$	Use the Multiplication Property.
$x = 92$	Simplify.

A prediction for the height (using a 10-foot ceiling) is 92 inches.

4. Answers will vary. Sample: Use the least-squares regression equation for height and coin value.

$y = -6.80x + 541.6$	
$126 = -6.80x + 541.6$	Substitute 126 for y.
$126 - 541.6 = -6.80x + 541.6 - 541.6$	Use the Subtraction Property.
$-415.6 = -6.80x$	Simplify.
$\frac{-415.6}{-6.80} = x$	Use the Division Property.
$x \approx 61$	Simplify.

A prediction for the height is 61 inches.

Chapter 1 Review and Assessment

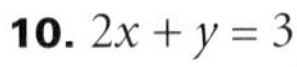
10. $2x + y = 3$

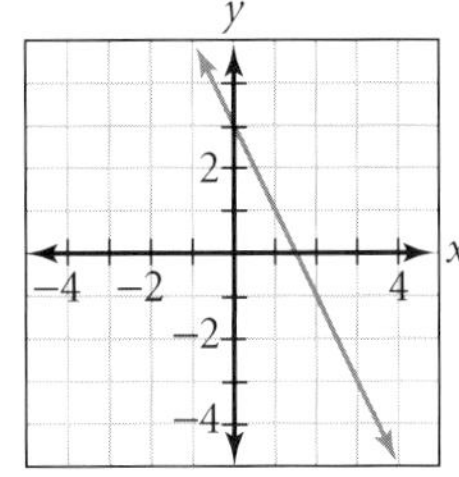

11. $3y - x = 1$

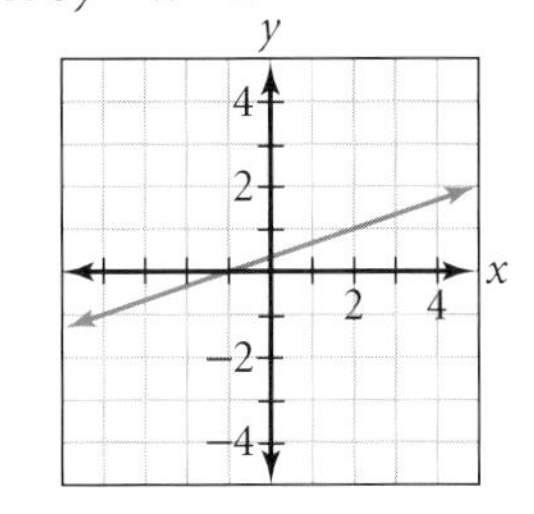

12. $y = 1$

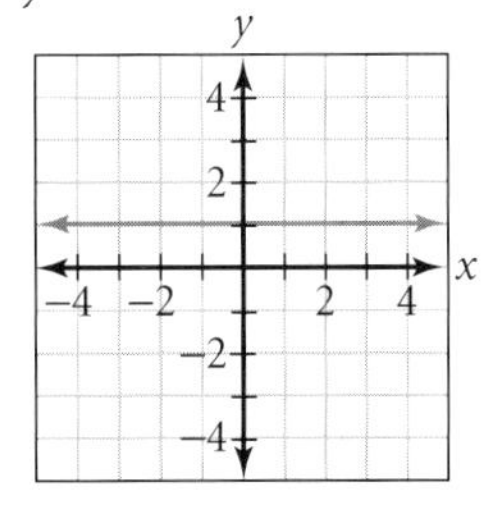

13. $x = -2$

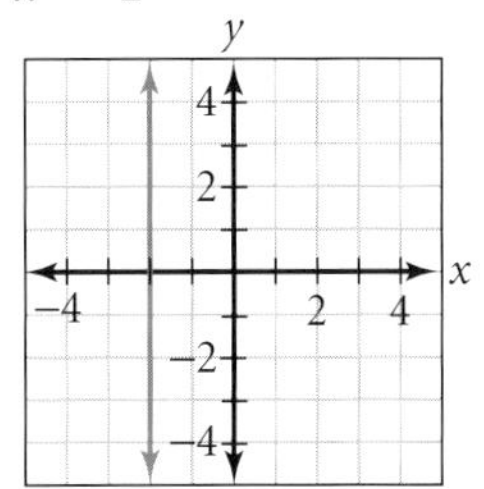

Lesson 2.1, pages 86–93

Activity

1. **a.** $12 + 8 \div 5 = 13.6$
   b. $(12 + 8) \div 5 = 4$

2. In part **a,** division is performed before addition. In part **b,** the parentheses change the order of operations, so addition is performed before division. The result from part **b** is correct.

3. **a.** 5.4; **b.** −1; **c.** 1.666666667

4. In part **a,** the division, $3 \div 5 = 0.6$, is performed first, yielding $8 - 0.6 - 2$. The result is 5.4. In part **b,** the subtraction within the parentheses, $8 - 3$, is performed first and then the division is done before the last subtraction. The result is −1. In part **c,** both subtractions, $8 - 3$ and $5 - 2$, should be performed before division. The result is 1.666666667. The result from part **c** is correct.

Exercises

Communicate

1. Answers may vary. Possible answers may include adding points in a football game (whole numbers), adding the cost of items in a grocery store (rational numbers), or finding the circumference of a circle (irrational numbers).

2. Answers may vary. Possible answer: The Commutative Properties of Addition and Multiplication state that the sum or product of two numbers does not change if the order of the two numbers is changed. The word *commutative* is appropriate because *commute* means "to move."

3. Answers may vary. Possible answer: The Associative Properties of Addition and Multiplication state that the sum or product of three or more numbers does not change if the last two numbers are combined before the first. The word *associative* is appropriate because *associate* means "to group or join."

Portfolio Activity

Space Debris Table	
Year	Number of Debris Objects
1993	7000
1994	7210
1995	7426
1996	7649
1997	7879
1998	8115
1999	8358
2000	8609
2001	8867
2002	9133
2003	9407
2004	9690
2005	9980
2006	10,280
2007	10,588
2008	10,906
2009	11,233
2010	11,570

Lesson 2.2, pages 94–101

Activity

1. $a^3 \cdot a^5 = aaa \cdot aaaaa = aaaaaaaa = a^8$
   Add 3 and the exponents.

2. $(a^3)^5 = a^3 \cdot a^3 \cdot a^3 \cdot a^3 \cdot a^3 = aaa \cdot aaa \cdot aaa \cdot aaa \cdot aaa$
   $= aaaaaaaaaaaaaaa = a^{15}$
   Multiply 3 and the exponents.

CHECKPOINT ✔

3. First add the exponents inside the parentheses: $a^7 \cdot a^3 = a^{7+3} = a^{10}$. Then multiply the resulting exponent by 2: $(a^{10})^2 = a^{10 \cdot 2} = a^{20}$.

Exercises

Communicate

1. $x^5x^3 = x^{5+3} = x^8$ and $(x^5)^3 = x^{5 \cdot 3} = x^{15}$, so x^5x^3 and $(x^5)^3$ are not equivalent.

2. $ax^2 = a \cdot x \cdot x$ while $(ax)^2 = a^2x^2 = a \cdot a \cdot x \cdot x$, so ax^2 and $(ax)^2$ are not equivalent.

3. Use $a^{-n} = \frac{1}{a^n}$: $5^{-2} = \frac{1}{5^2} = \frac{1}{25}$.

4. $4^{\frac{3}{2}} = \left(4^{\frac{1}{2}}\right)^3 = (\sqrt{4})^3 = 2^3 = 8$

Portfolio Activity

1. 1993 or $t = 0$: $d = 7000(1.03)^0$; $d = 7000$
   1994 or $t = 1$: $d = 7000(1.03)^1$; $d \approx 7210$
   1995 or $t = 2$: $d = 7000(1.03)^2$; $d \approx 7426$
   1996 or $t = 3$: $d = 7000(1.03)^3$; $d \approx 7649$
   1997 or $t = 4$: $d = 7000(1.03)^4$; $d \approx 7879$
   1998 or $t = 5$: $d = 7000(1.03)^5$; $d \approx 8115$
   1999 or $t = 6$: $d = 7000(1.03)^6$; $d \approx 8358$
   2000 or $t = 7$: $d = 7000(1.03)^7$; $d \approx 8609$
   2001 or $t = 8$: $d = 7000(1.03)^8$; $d \approx 8867$
   2002 or $t = 9$: $d = 7000(1.03)^9$; $d \approx 9133$
   2003 or $t = 10$: $d = 7000(1.03)^{10}$; $d \approx 9407$
   2004 or $t = 11$: $d = 7000(1.03)^{11}$; $d \approx 9690$
   2005 or $t = 12$: $d = 7000(1.03)^{12}$; $d \approx 9980$
   2006 or $t = 13$: $d = 7000(1.03)^{13}$; $d \approx 10{,}280$
   2007 or $t = 14$: $d = 7000(1.03)^{14}$; $d \approx 10{,}588$
   2008 or $t = 15$: $d = 7000(1.03)^{15}$; $d \approx 10{,}906$
   2009 or $t = 16$: $d = 7000(1.03)^{16}$; $d \approx 11{,}233$
   2010 or $t = 17$: $d = 7000(1.03)^{17}$; $d \approx 11{,}570$

2. 2020 or $t = 27$: $d = 7000(1.03)^{27}$; $d \approx 15{,}549$; according to this model, at the end of the year 2020, there will be about 15,549 space debris objects.

Lesson 2.3, pages 102–110

Activity

1. The height and radius of a cylindrical container will influence how much water it can hold. Consider two containers that differ only in height: the taller will hold more, so height affects the capacity. Now consider two containers that differ only in their radii: the one with the larger radius will hold more, so radius affects the holding capacity.

2. Answers may vary. Sample answer: The cost of produce in a supermarket usually depends on the weight and a price per pound. The domain and range of this function is the set of positive real numbers.

3. From 0 minutes to 4 minutes, the volume of water increases from 0 gallons to 40 gallons. From 4 minutes to 10 minutes, the volume remains constant at 40 gallons. From 10 minutes to 16 minutes, the volume decreases from 40 gallons to 0 gallons.

CHECKPOINT ✔

4. Sample: H is the height in feet of a ball thrown in the air as a function of time, t, in seconds.

Exercises

Communicate

1. In a relation, a value of the first variable can be matched with more than one value of the second variable, but in a function, a value of the first variable must be matched with exactly one value of the second variable.

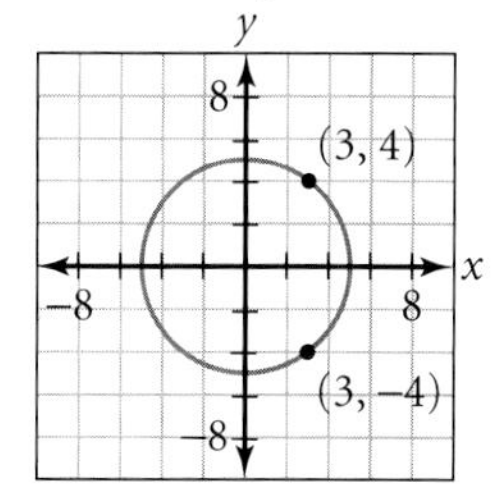

In the figure, notice (3, 4) and (3, −4) are both elements of the relation. This would not be permissible for a function since 3 is paired with two range members, 4 and −4.

2. A *function* may be represented as a table, as a graph, or as a set of ordered pairs.

3. The *domain* of a set of ordered pairs is the set of first coordinates, and the *range* is the set of second coordinates. For the given example, domain: $\{-2, 2, 3, 4\}$, range: $\{0, 2, 5\}$.

4. Since the dollar value of the meals, x, and Cleo's tips, y, are nonnegative values, we have a domain of $x \geq 0$ and a range of $y \geq 0$. Assuming that Cleo's customers give her 15% of the dollar value of their meals, the relationship can be represented by the graph below.

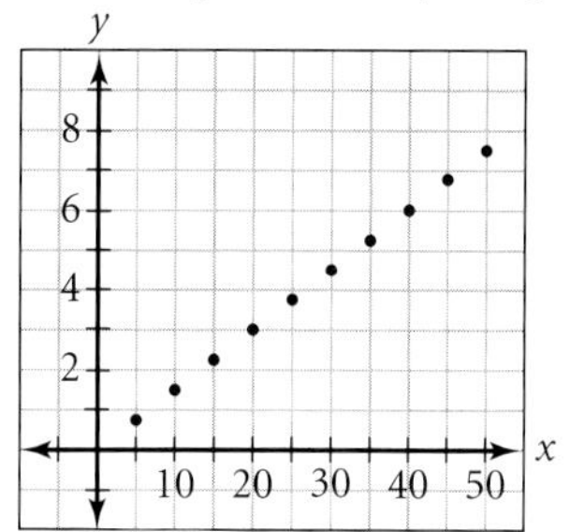

Portfolio Activity

1.

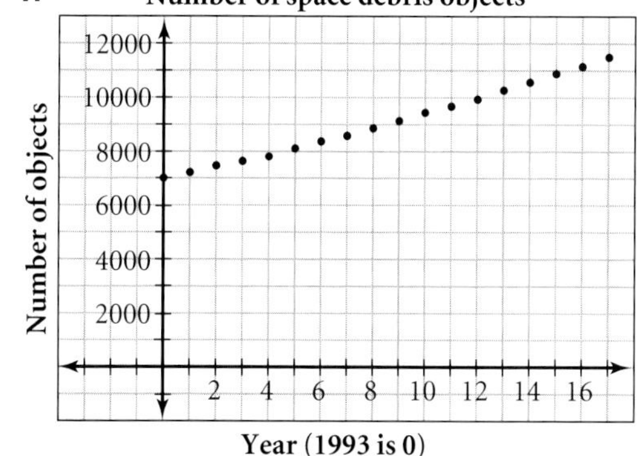

This graph shows that the total number of space debris objects, y, is a function of time, x, because each year is paired with exactly one value for the number of space debris objects.

2. The domain is the set of all years defined for the function. The range is the number of space debris objects for each year defined for the function. The independent variable is the year, x. The dependent variable is the number of space debris objects, y.

3. Using a graphics calculator, a linear function that models the data is approximately $y = 2.67.9 + 6828.5$ and the correlation coefficient, r, is greater than 0.99767. Using a graphics calculator, an exponential function that models the data is approximately $7000(1.03)^x$ and the correlation coefficient, r, is greater than 0.9999.

4. Using a graphics calculator, the linear model fits the data fairly well, but the exponential model fits the data better.

Lesson 2.4, pages 111–117

Activity

1. The function is a linear function.

2. As speed increases, the braking distance values appear to increase by a greater amount.

3. The stopping distance is the sum of the reaction distance and the braking distance. The scatter plot of the initial speed and stopping distance data will be the sum of the two previous scatter plots.

CHECKPOINT ✔

4.

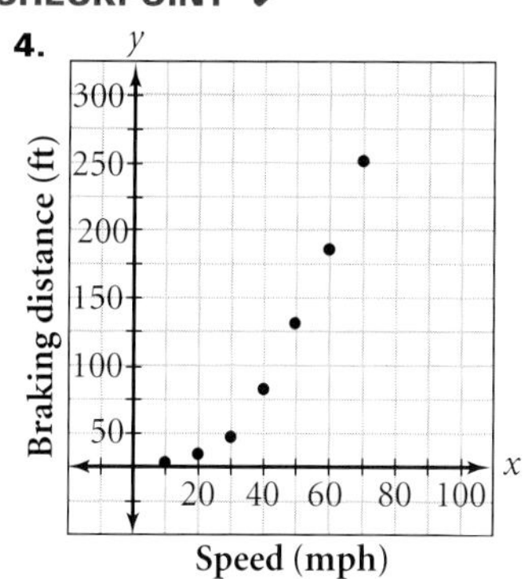

As speed increases, the stopping distance increases and the rate of change of the stopping distance increases.

Exercises

Communicate

1. Since stopping distance = reaction distance + braking distance, or $s(x) = r(x) + b(x)$, then $r(x) = s(x) - b(x)$.

2. To find $f \circ g$, first apply the function g to x and then apply the function f to the result. In other words:

$$\begin{aligned}(f \circ g)(x) &= f(g(x)) \\ &= f(2x^2 + 4) \\ &= 4(2x^2 + 4) - 7 \\ &= 8x^2 + 9\end{aligned}$$

3. First: $(f \circ g)(x) = f(g(x))$
$= f(2x)$
$= 3(2x) + 1 = 6x + 1$
Second: $(g \circ f)(x) = g(f(x))$
$= g(3x + 1)$
$= 2(3x + 1)$
$= 6x + 2$
So, $f \circ g$ and $g \circ f$ are not equivalent.

Portfolio Activity

1. $D(t) = 1550 + 42t$

2. $R(t) = -15t$

3. $(D - R)(t) = D(t) - R(t) = 1550 + 42t - 15t$
$= 1550 + 27t$

4.

Year	t	$d(t) - r(t)$	Difference in y-values
1993	0	1550	
1994	1	1577	27
1995	2	1604	27
1996	3	1631	27
1997	4	1658	27
1998	5	1685	27
1999	6	1712	27
2000	7	1739	27
2001	8	1766	27
2002	9	1793	27
2003	10	1820	27
2004	11	1847	27
2005	12	1874	27
2006	13	1901	27
2007	14	1928	27
2008	15	1955	27
2009	16	1982	27
2010	17	2009	27

In the table, there is a constant difference of 1 for the x-values and a constant difference of 27 for the y-values.

5. This function is a linear function because the equation is of the form $y = ax + b$ and the graph is a straight line.

Practice and Apply

25. $(f + g)(x) = f(x) + g(x)$ — Def. of $f + g$
$= (x^2 - 1) + (2x - 3)$ — Subst.
$= x^2 + 2x - 4$ — Simplify

26. $(f - g)(x) = f(x) - g(x)$ — Def. of $f - g$
$= (x^2 - 1) - (2x - 3)$ — Subst.
$= x^2 - 2x + 2$ — Dist. Prop.

27. $(g - f)(x) = g(x) - f(x)$ — Def. of $f - g$
$= (2x - 3) - (x^2 - 1)$ — Subst.
$= -x^2 + 2x - 2$ — Simplify

28. $(f \cdot g)(x) = f(x) \cdot g(x)$ — Def. of $f \cdot g$
$= (x^2 - 1)(2x - 3)$ — Subst.
$= x^2(2x - 3) - 1(2x - 3)$ — Dist. Prop.
$= 2x^3 - 3x^2 - 2x + 3$ — Simplify

29. $\left(\frac{f}{g}\right)(x) = \frac{f(x)}{g(x)}, g(x) \neq 0$ — Def. of $\frac{f}{g}$
$= \frac{x^2 - 1}{2x - 3}, x \neq \frac{3}{2}$ — Subst.

30. $(f + g)(x) = f(x) + g(x)$ — Def. of $f + g$
$= (x - 3) + (x^2 - 9)$ — Subst.
$= x^2 + x - 12$ — Simplify

31. $(f - g)(x) = f(x) - g(x)$ — Def. of $f - g$
$= (x - 3) - (x^2 - 9)$ — Subst.
$= -x^2 + x + 6$ — Simplify

32. $(g - f)(x) = g(x) - f(x)$ — Def. of $g - f$
$= (x^2 - 9) - (x - 3)$ — Subst.
$= x^2 - x - 6$ — Simplify

33. $(f \cdot g)(x) = f(x) \cdot g(x)$ — Def. of $f \cdot g$
$= (x - 3)(x^2 - 9)$ — Subst.
$= x(x^2 - 9) - 3(x^2 - 9)$ — Dist. Prop.
$= x^3 - 9x - 3x^2 + 27$ — Dist. Prop.
$= x^3 - 3x^2 - 9x + 27$ — Simplify

34. $\left(\frac{g}{f}\right)(x) = \frac{g(x)}{f(x)}, f(x) \neq 0$ — Def. of $\frac{g}{f}$
$= \frac{x^2 - 9}{x - 3}, x \neq 3$ — Subst.
$= x + 3, x \neq 3$ — Simplify

Lesson 2.5, pages 118–123

Activity

1 and 2. The graphs of each function and its inverse are symmetric across the line $y = x$.

CHECKPOINT ✔

3. The graph of the inverse of each function is the reflection of the graph of the function across the line $y = x$.

Exercises

Communicate

1. A function is *one-to-one* if it has an inverse that is also a function.

2. The *vertical-line test* is used to determine if the graph of a relation represents a function. The *horizontal-line test* is used to determine if the graph of an inverse represents a function.

3. In $y = 4x - 1$, first interchange the variables. Then solve for y.
$x = 4y - 1$
$x + 1 = 4y$
$y = \frac{x + 1}{4}$
$y = \frac{1}{4}x + \frac{1}{4}$
The resulting equation is the inverse.

4. The graphs of a function and its inverse are reflections of each other across the line $y = x$.

Lesson 2.6, pages 124–132

Activity

1.

Hours worked, h	10.0	30.0	35.0	40.00	52.5
Wage, $w(h)$	210.00	630.00	735.00	840.00	1233.75

2. Answers may vary. Sample answer:

Hours worked, h	15.0	38.0	42.0	56.0	60.0
Wage, $w(h)$	315.00	798.00	903.00	1344.00	1470.00

3.

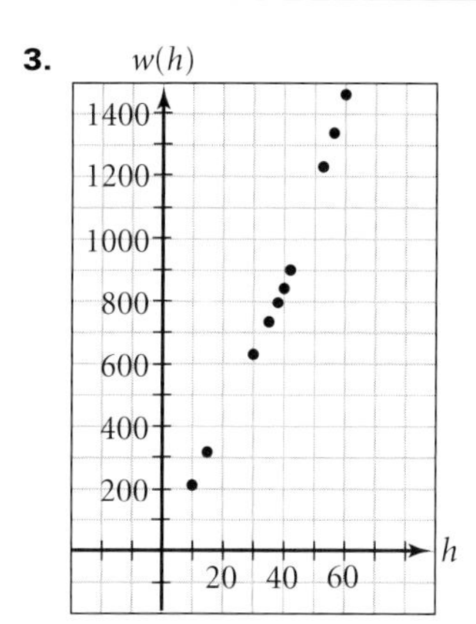

CHECKPOINT ✔

4. The graph appears to be made up of two straight-line segments.

Exercises

Communicate

1. A piecewise function consists of two or more different functions defined on nonoverlapping parts of the domain of the function.

2. The greatest-integer function applied to a number gives the greatest integer less than or equal to the number. Thus, the greatest integer less than or equal to 1.5 is 1, so $[1.5] = 1$. And the greatest integer less than or equal to -1.5 is -2, not -1, so $[-1.5] = 2$.

3. The greatest-integer and round-up functions are both step functions; their domains are the same, the set of all real numbers; and their ranges are the same, the set of integers. But for any noninteger number, the round-up function is one greater than the value of the greatest-integer function. For any integer, the two function values are the same.

4. No, the inverse of $y = |x|$ is not a function. The horizontal line $y = 2$ intersects the graph at $x = -2$ and $x = 2$.

Practice and Apply

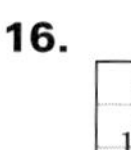

16.

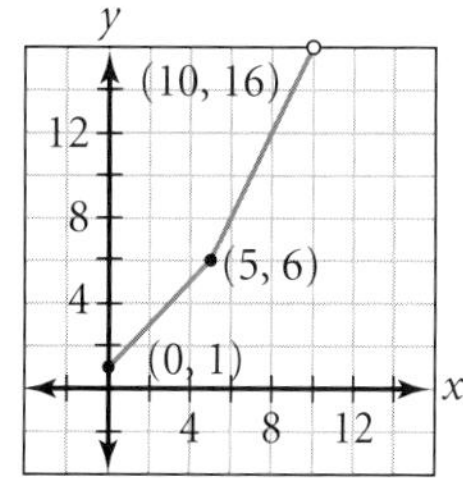

17.

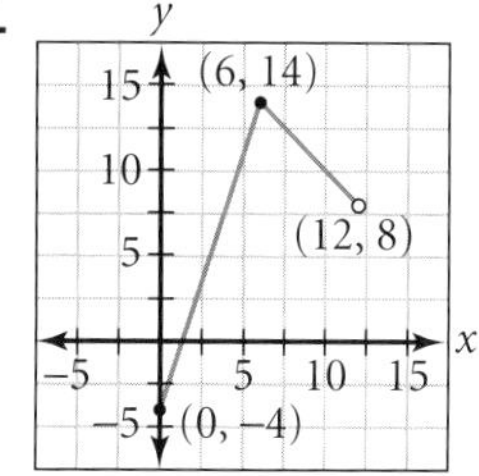

18.

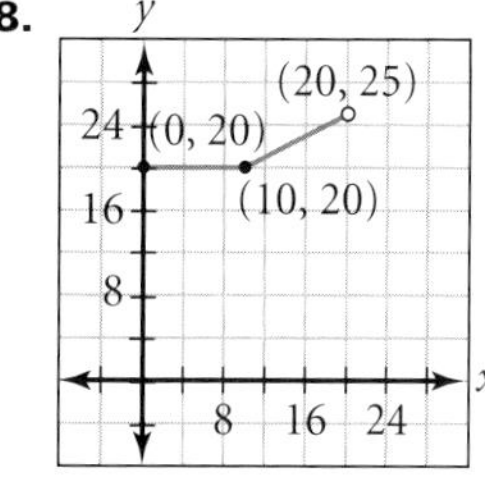

19.

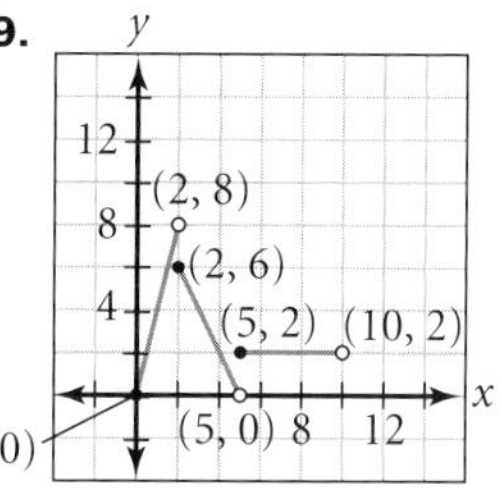

20.

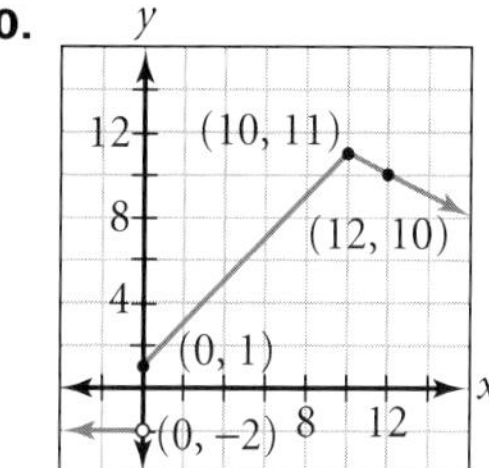

21.

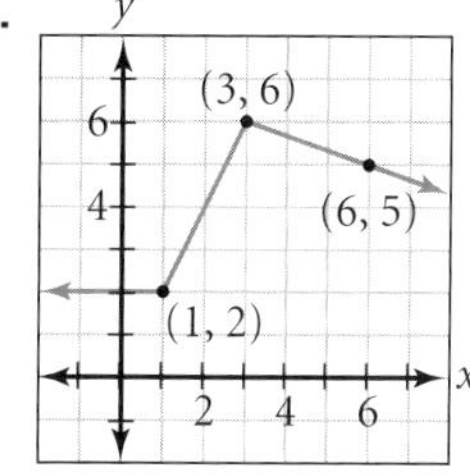

22.

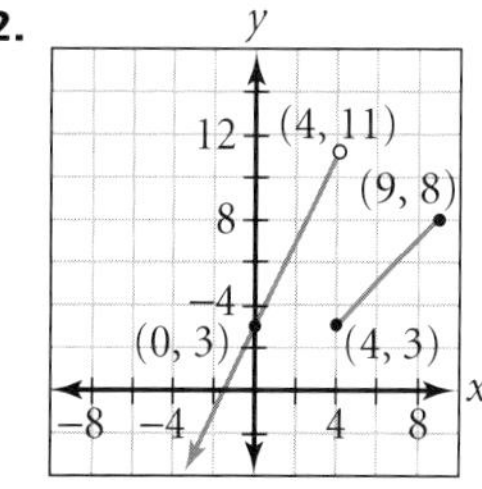

23.

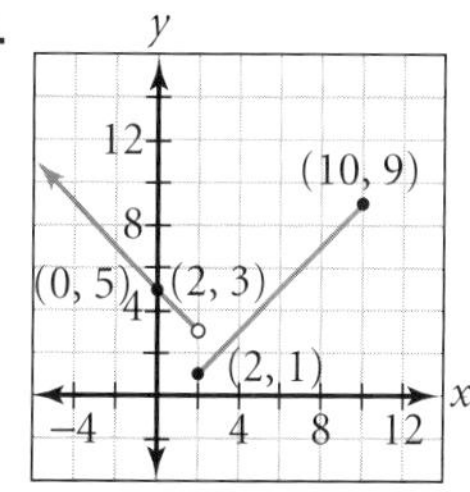

66.

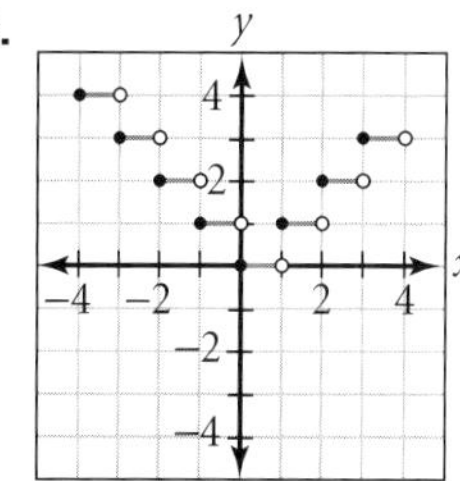

For $x \geq 0$ and all integer values of x, the functions are identical. They differ for negative noninteger values of x. For example, $|[-2.5]| = |-3| = 3$ $[|-2.5|] = [2.5] = 2$, so $|[x]| = [|x|] + 1$ if $x < 0$.

67. $f(x) = \frac{1}{10}\lceil 10x \rceil$ will round up x to the nearest tenth.

$g(x) = \frac{1}{100}[100x]$ will round down to the nearest hundredth.

68. $f(g(x)) = f(x^2 - 2)$
$= |x^2 - 2|$

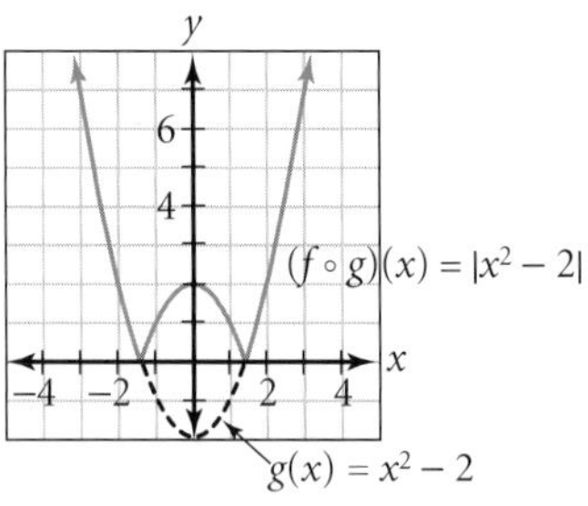

73. a.

Lbs of coffee	Cost
1	9.89
2	19.78
3	29.67
4	36.56
5	49.45
6	57.43
7	65.41
8	73.39
9	81.37
10	89.35

b.

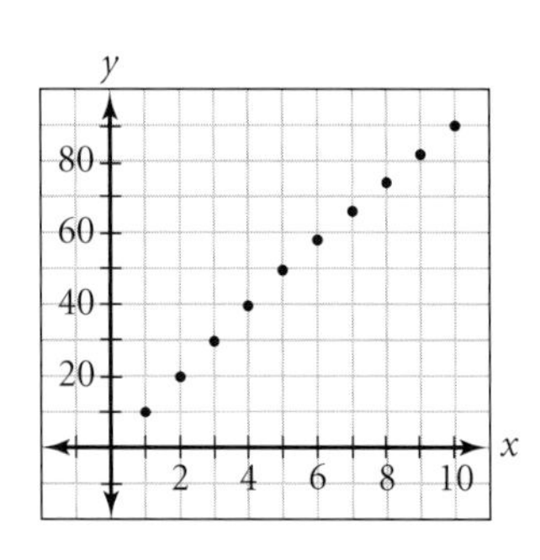

74. a.

Monthly sales	Monthly income
5000	2950
10,000	3050
15,000	3150
20,000	3250
25,000	3350
30,000	3600
35,000	3850
40,000	4100
45,000	4350
50,000	4600

b.

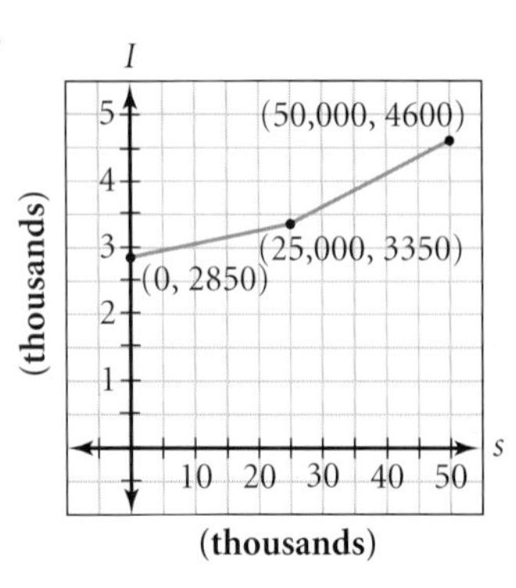

75. a.

Labor hours	Repair charges
0.5	45
1.0	45
1.5	65
2.0	85
2.5	105
3.0	125
3.5	145
4.0	165
4.5	185
5.0	205

b.

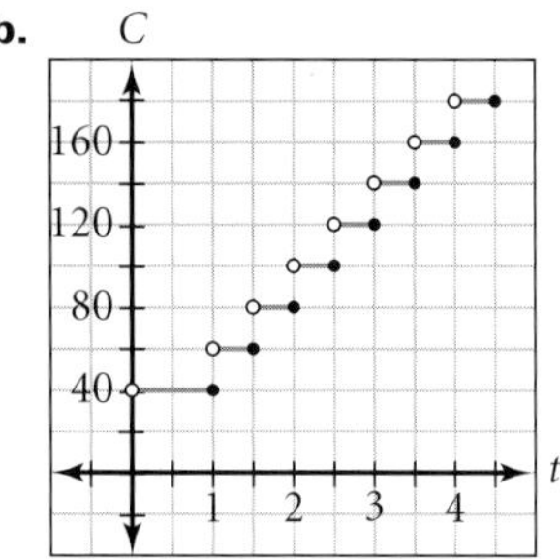

76.

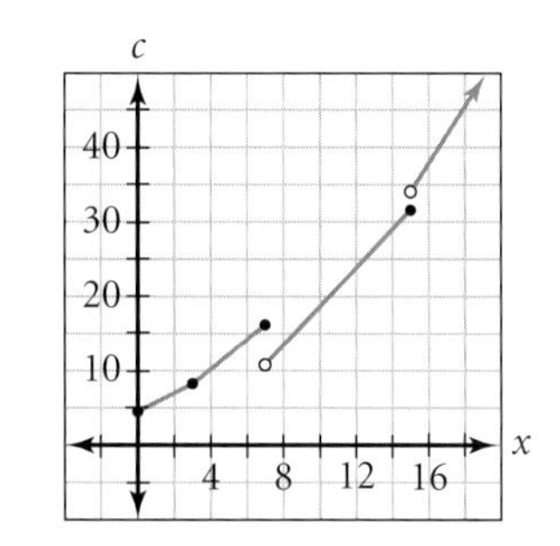

Portfolio Activity

1. $n(t) = \begin{cases} 7000(1.03)^t \text{ if } 0 \le x < 8 \\ 7209 + 200t \text{ if } 8 \le t \le 17 \end{cases}$, where 2001 means $t = 8$
2. 2005 means $t = 12$, so use $7209 + 200t$ as the function: $7209 + 200(12) = 9609$. The total number of space debris objects in 2005 will be 9609.
3. The number of objects in 1993 was 7000. Double that number is 14,000. Using the piecewise function, $n(t) = 7209 + 200t$, solve $14{,}000 = 7209 + 200t$: $14{,}000 = 7209 + 200t$; $6791 = 200t$; $t \approx 34$. Using the function from the Portfolio Activity on page 93, $7000(1.03)^{24} \approx 14{,}230$. Comparing the two functions, it will take about 34 years for the number of space debris objects to double using the piecewise function and about 24 years using the exponential function.
4. The year 2010 is represented by $t = 17$. Using the piecewise function, $7209 + 200t = 339 + 200(17) = 7209 + 3400 = 10{,}609$. The Space Debris Table shows 11,570. The difference in the number of objects from one function to the other is 961.

Lesson 2.7, pages 133–141

Activity

1.

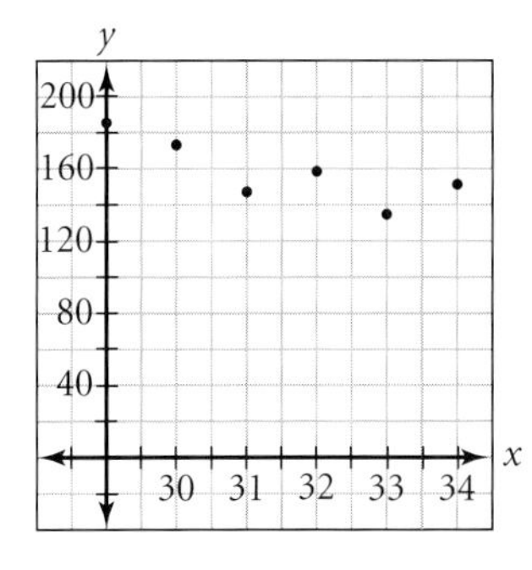

linear regression model: $y \approx -7.85x + 405.73$

2.

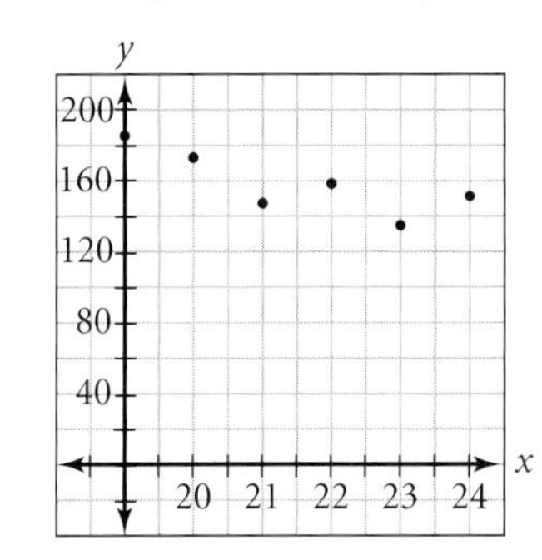

linear regression model: $y \approx -7.85x + 327.18$

CHECKPOINT ✔

3. The equations for the two regression lines have different y-intercepts but have the same slope.

Exercises

Communicate

1. a. The graphed function is a vertical translation 3 units down.

b. The graphed function is a horizontal translation 2 units to the left.

2. $\frac{1}{b} > 1$ if $0 < b < 1$

3. *Reflections* and *translations* are both "rigid" transformations in the sense that the graph is not distorted or twisted but is shifted (translations) or inverted (reflections).

4. If $h > 0$, the graph of $f(x - h)$ will be shifted to the right. If $h < 0$, the graph of $f(x - h)$ will be shifted to the left.

5. Both stretches are "nonrigid" transformations resulting in distortions of the original graph. A *vertical stretch* results from a change affecting the range, or a change after the parent function is evaluated. The graph moves away from the x-axis. A *horizontal stretch* results from a change affecting the domain, or a change before the parent function is evaluated. The graph moves away from the y-axis.

Practice and Apply

61.

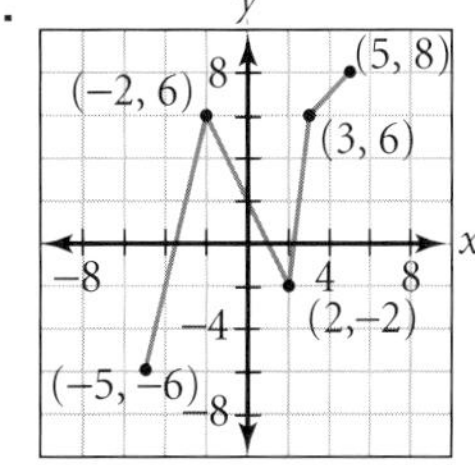

62.

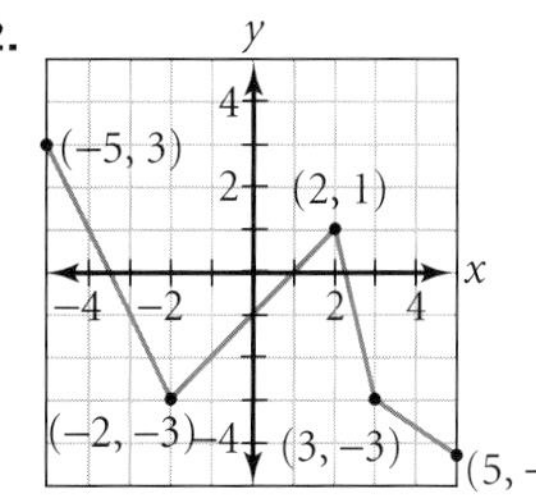

63.

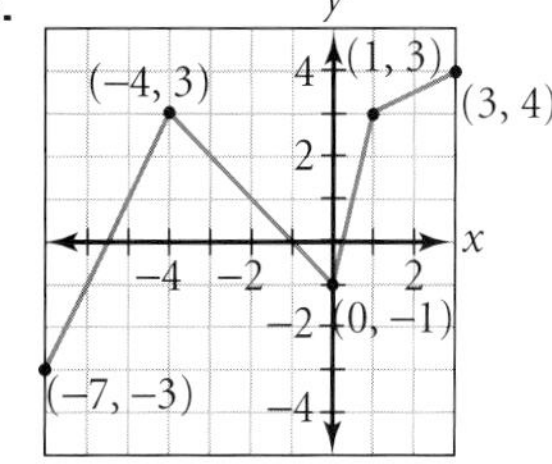

64.

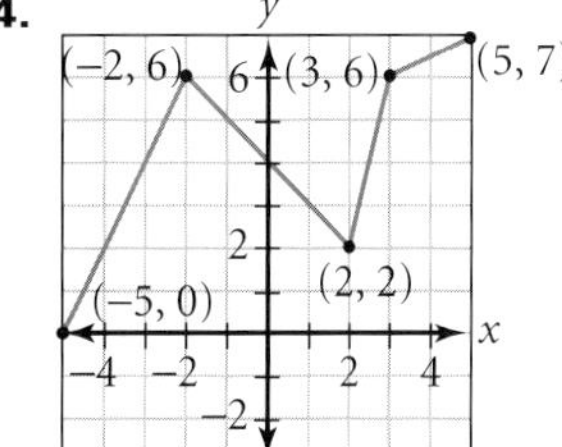

65.

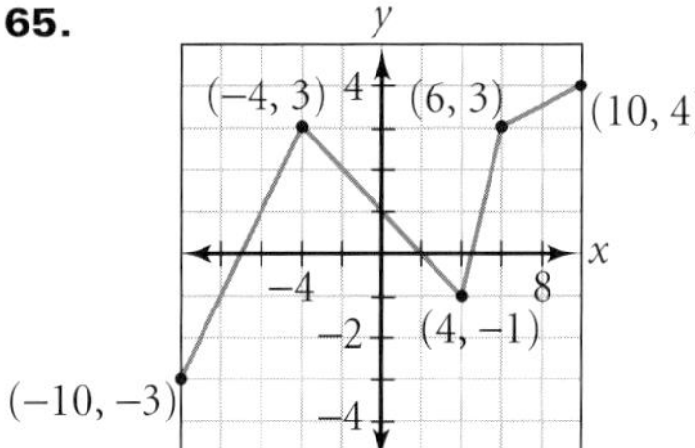

66. a. If $f(t) = t^2$, then $h(t) = -16t^2 + 100$ may be seen as three transformations applied to $f(t)$: a vertical stretch by a factor of 16, a vertical reflection across the x-axis, and a vertical shift of 100 units up.

b. $h(t) = -16t^2 + 25$

c.

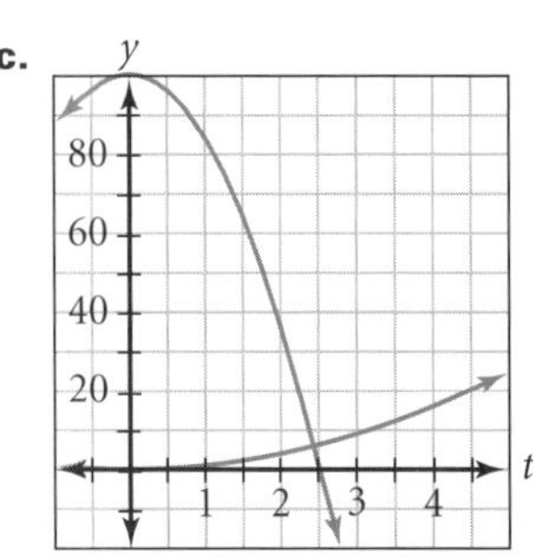

Consider what happens to the point where $x = 1$. On the graph of f, $y = f(1) = (1)^2 = 1$. Following the transformations listed in part **a,** $y = 1$ is first multiplied by 16 ($y = 16$), and then reflected across the x-axis ($y = -16$), and then shifted vertically 100 units up ($y = -16 + 100 = 84$). This is algebraically confirmed by $h(1) = -16(1)^2 + 100 = 84$.

Look Back

73. The relation is a function; the inverse, {(100, 1), (200, 2), (300, 3), (400, 4)}, is a function because each domain member is paired with exactly one range member.

74. The relation is a function; the inverse, {(5, 1), (10, 2), (10, 3) (15, 4)}, is not a function because 10 is paired with both 2 and 3.

75.

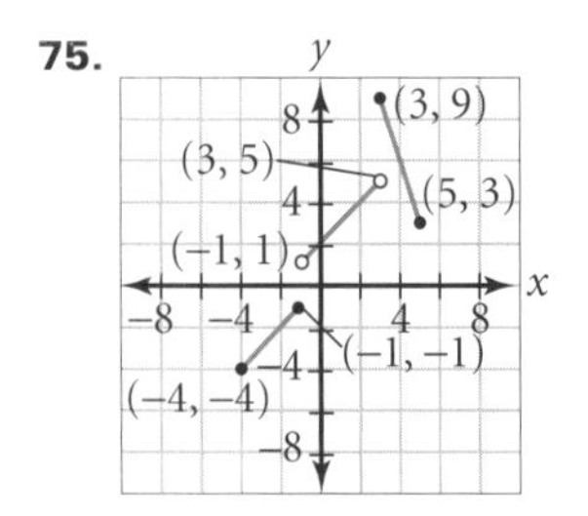

Look Beyond

76. The coordinates of the point of intersection are (−2, 1). The resulting x-value is the same as the x-value of the point of intersection.

Chapter 2 Project

Activity 3

1. Answers will vary. Sample: As the altitude increases, the number of debris objects increases until the altitude reaches 1200 km. Then the number of objects decreases until the altitude reaches 1800 km. The number remains about the same until 2000 km. It then increases from 2000 to 2200 km. After 2200 km, the number decreases again. A function that models the data in the graph is a piecewise function with five parts: $y = 1.075x - 290$ for $400 \le x \le 1200$, $y = -1.4x + 2650$ for $1200 \le x \le 1800$, $y = 200$ for $1800 \le x \le 2000$, $y = 1.5x - 2800$ for $2000 \le x \le 2200$, and $y = -0.41x + 1316$ for $2200 \le x \le 3000$.

2. Answers will vary. Sample using the functions in Step 1: Estimate for 725 km: $y = 1.075x - 290$; $y = 1.075(725) - 290$; $y \approx 489$ objects. Estimate for 725 km: $y = -1.4x + 2650$; $y = -1.4(1450) + 2650 = 620$ objects. Estimate for 1900 km: $y = 200$ objects. The model is complicated and not particularly accurate.

Lesson 3.1, pages 156–163

Activity

1. System I: $\begin{cases} y = 2x - 1 \\ y = -x + 5 \end{cases}$

a. Yes

b. Yes, $x = 2$ and $y = 3$.

2. System II: $\begin{cases} y = 2x - 1 \\ y = 2x + 1 \end{cases}$

a. No

b. No. Answers may vary. Sample answer: $\begin{cases} y = 2x - 1 \\ y = x + 1 \end{cases}$ has the unique solution $x = 2$ and $y = 3$.

System III: $\begin{cases} y = \frac{8 - 3x}{4} \\ y = -\frac{3}{4}x + 2 \end{cases}$

a. Yes

b. No. Answers may vary. Sample answer: $\begin{cases} y = \frac{8 - 3x}{4} \\ y = \frac{3}{4}x + 2 \end{cases}$ has the unique solution $x = 0$ and $y = 2$.

CHECKPOINT ✔

3. If the slopes are the same, the system has no solution or infinitely many solutions. If the slopes are different, then the system has exactly one solution.

Exercises

Communicate

1. An inconsistent system has lines that are parallel and do not coincide. A consistent, dependent system has lines that coincide. A consistent, independent system has lines that have different slopes, and therefore intersect at exactly one point.

2. Answers may vary. Sample answer: $\begin{cases} 3x - 2y = 1 \\ -6x + 4y = 1 \end{cases}$ is an inconsistent system. $\begin{cases} 3x - 2y = 1 \\ -6x + 4y = -2 \end{cases}$ is a dependent system. $\begin{cases} 3x - 2y = 1 \\ 4x - 6y = -2 \end{cases}$ is an independent system.

3. Answers may vary. Sample answer: To solve $\begin{cases} y = 2x - 5 \\ x + y = 13 \end{cases}$ by graphing, graph each equation on the same coordinate plane. The coordinates of the point of intersection (provided one exists) are the solution to the system.

4. Answers may vary. Sample answer: To solve $\begin{cases} p - q = -7 \\ 2p + q = 16 \end{cases}$ by substitution, solve $p - q = -7$ for p: $p = q - 7$; substitute $q - 7$ for p in $2p + q = 16$ and solve for q: $2(q - 7) + q = 16 \rightarrow q = 10$; substitute 10 for q in $p - q = -7$ and solve for p: $p - 10 = -7 \rightarrow p = 3$. Thus, the solution (p, q) is $(3, 10)$.

Lesson 3.2, pages 164–171

Activity

1. **a.** System I: $\begin{cases} x - y = -2 \\ -5x + 5y = 10 \end{cases}$ The system is dependent. There are infinitely many solutions.
   b. The elimination process gives the equation $0 = 0$. The statement is true.
2. **a.** System II: $\begin{cases} 3x + 3y = -5 \\ 2x + 2y = 7 \end{cases}$ The system is inconsistent. There are no solutions.
   b. The elimination process gives $0 = -31$. The statement is false.

CHECKPOINT ✔

3. The statement is a false statement, so the system is inconsistent and there is no solution.

CHECKPOINT ✔

4. The statement is true, so the system is dependent and there are infinitely many solutions.

Exercises

Communicate

1. To solve $\begin{cases} 3x - 4y = 3 \\ 2x + y = -5 \end{cases}$ by elimination, multiply the second equation by 4 and combine (add) the two resulting equations. The y will be eliminated and the sum will contain the variable x only.
2. In attempting to solve by elimination, an inconsistent system will result in a false statement, such as $1 = 0$. A dependent system will result in a true statement, usually $0 = 0$.
3. the Addition Property of Equality
4. Answers may vary. Sample answer: If one of the equations in the system has a variable with coefficient 1 or −1, the substitution method is easier to use. In most all other cases, the elimination method is the easier one to use.

Portfolio Activity

Let x represent the number of regular models and y represent the number of sport models. Write and solve this system of linear equations, or $\begin{cases} 2x + y = 18 \\ x + y = 10 \end{cases}$. The solution is $x = 8$ and $y = 2$, so 8 regular models and 2 sport models are produced per hour.

Lesson 3.3, pages 172–178

Activity

1. $8x + 15y \le 90$

CHECKPOINT ✔

2. Daryll cannot purchase a fraction of a tape or compact disc. He must purchase them in whole-number units.
3. $(0, 0)$, $(10, 0)$, $(0, 6)$, and $(5, 3)$ each satisfy the inequality.
4. 11 tapes and 0 CDs or 0 tapes and 6 CDs
5. 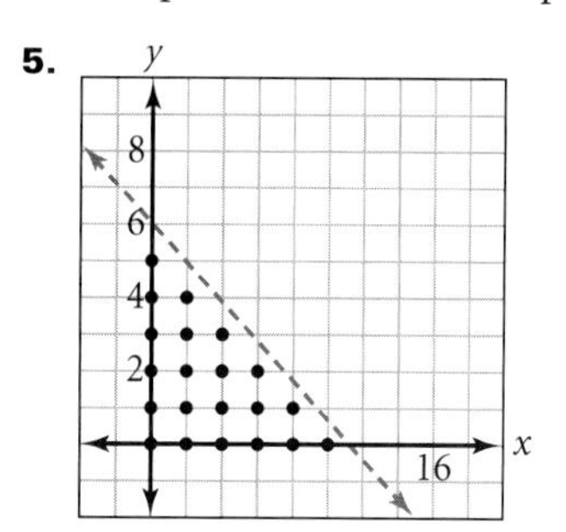

CHECKPOINT ✔

6. any application in which x and y represent quantities measured in whole units

Exercises

Communicate

1. To graph $7x - 5y > 0$, first graph the boundary line $7x - 5y = 0$, or $y = \frac{7}{5}x$. Use a dashed line to show that these values are not included in the solution. Choose a test point. (Usually $(0, 0)$ is a good choice, but it cannot be used here since it is on the boundary line.) If the coordinates of the test point satisfy the inequality, shade the region (the half-plane), that contains the point; otherwise, shade the other region.
2. If the inequality contains a $\le$ or $\ge$ symbol, the boundary line is solid. If the inequality contains a $<$ or $>$ symbol, the boundary line is dashed.
3. Once the boundary line is drawn, select a test point that does not lie on the line. If the coordinates of the point satisfy the inequality, the solution is the region, or half-plane, containing the point. Alternatively, for inequalities of the form $y \le mx + b$ or $y < mx + b$, shade the region that lies below the boundary line; if the inequality is of the form $y \ge mx + b$ or $y > mx + b$, shade the region that lies above the boundary line.

Practice and Apply

13.

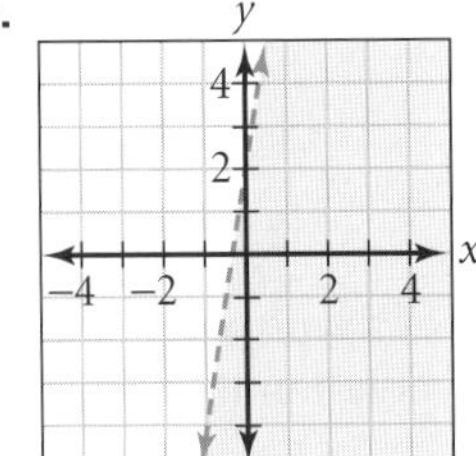

14.

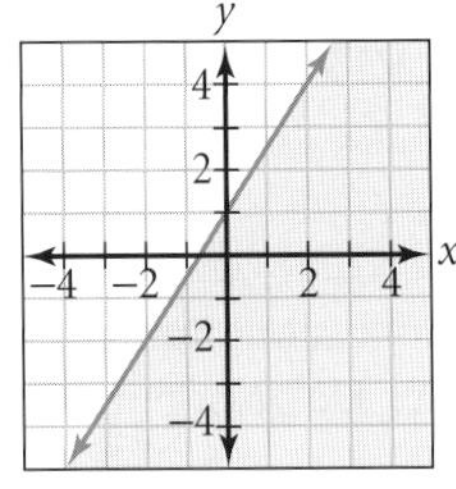

15.

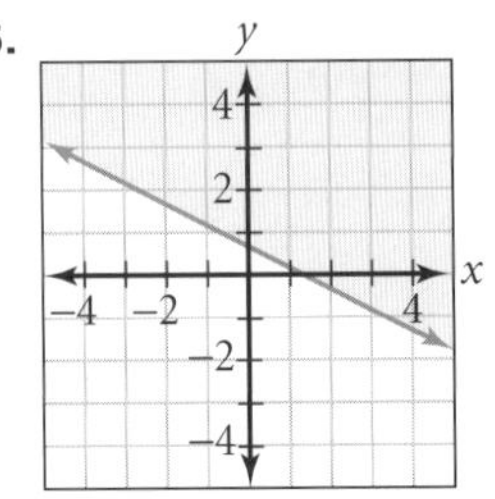

16.

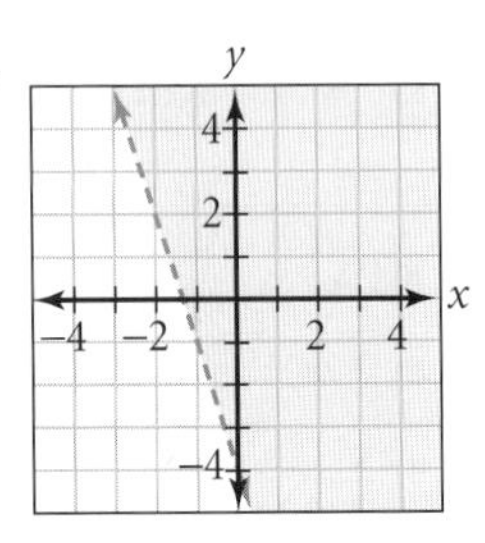

17.

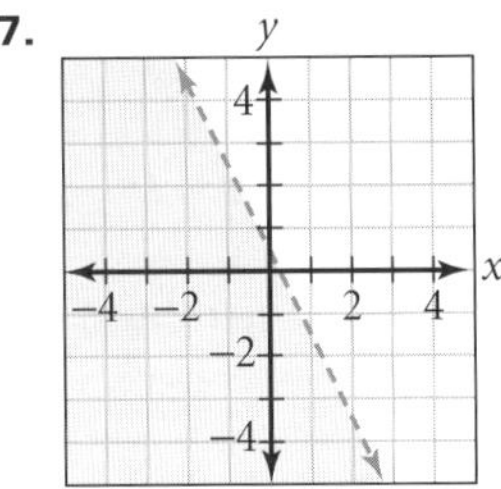

18.

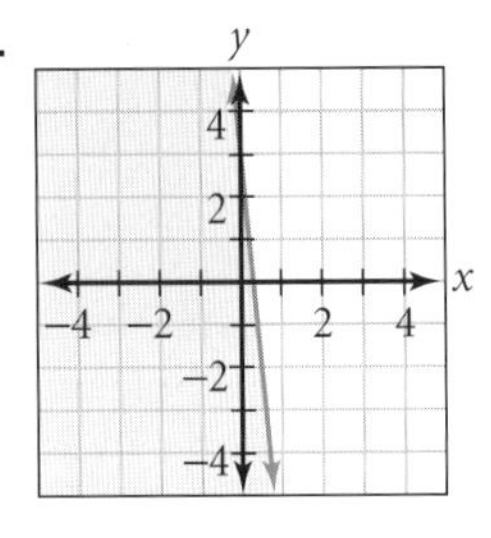

19.

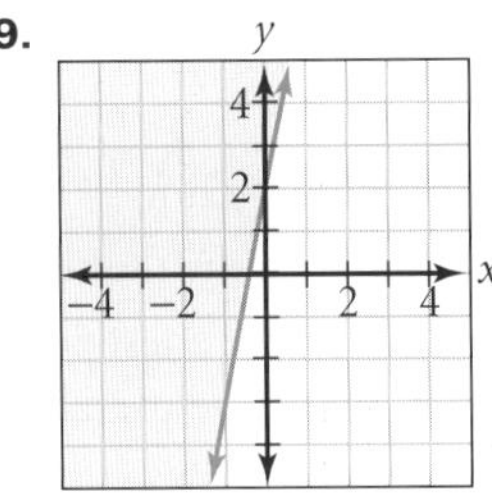

20.

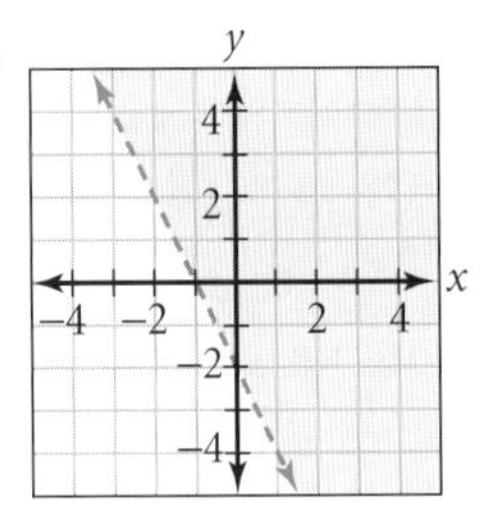

21.

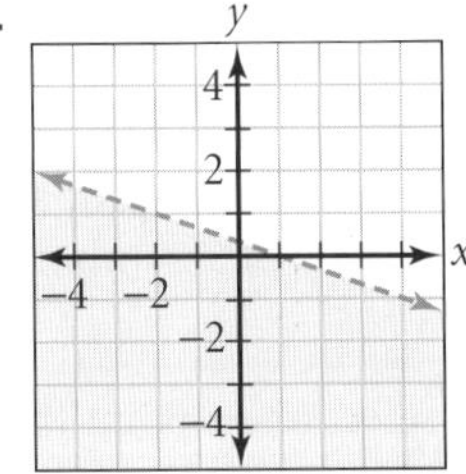

22.

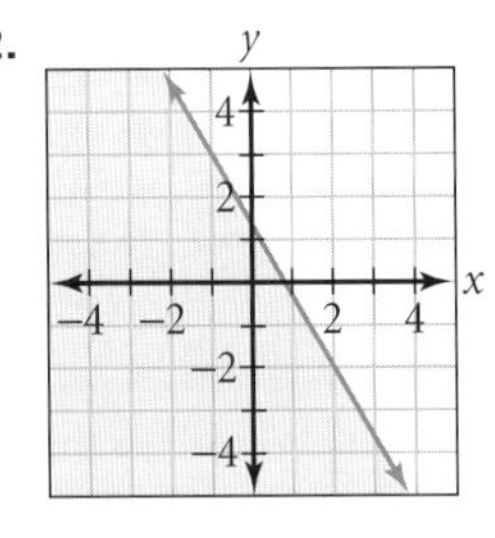

23.

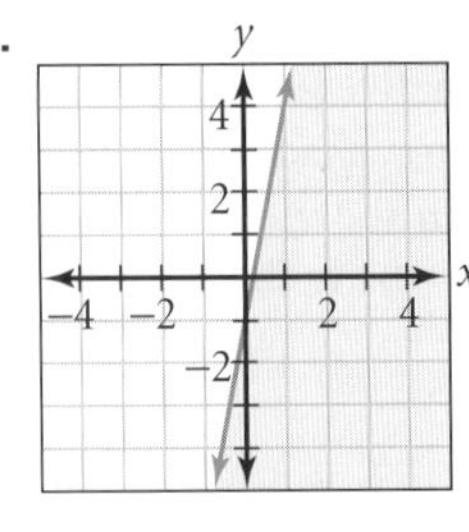

24.

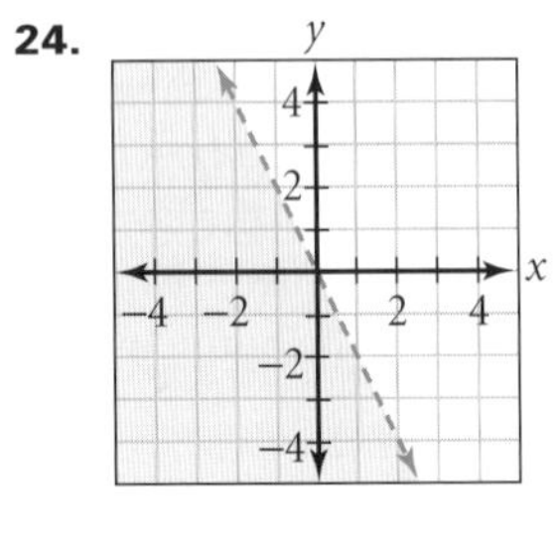

25.

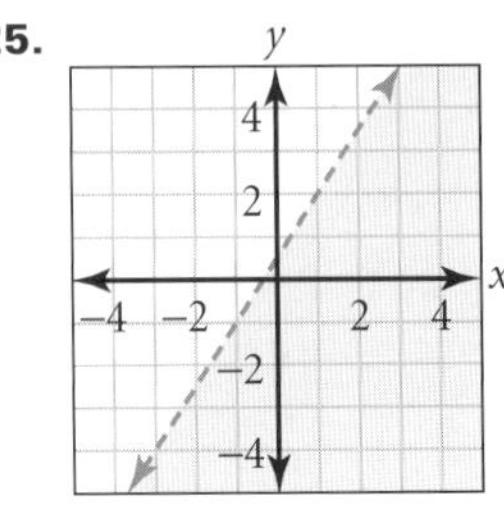

26.

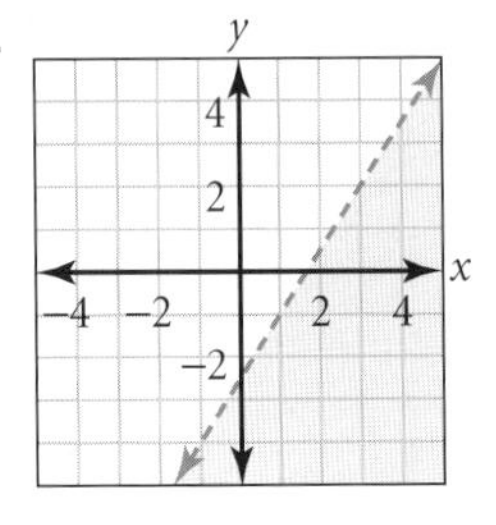

27.

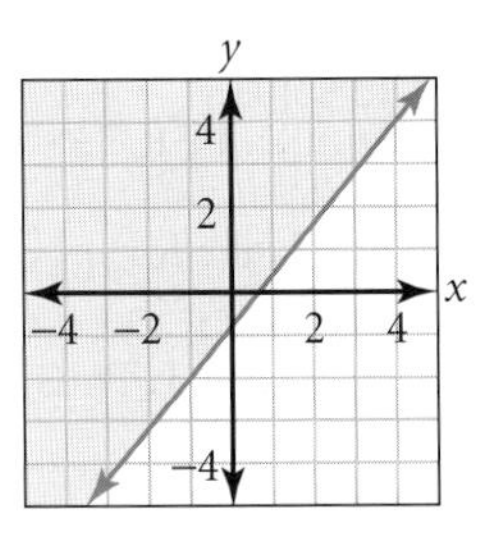

28.

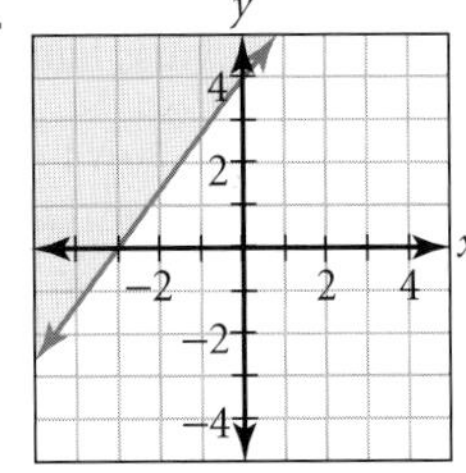

32.

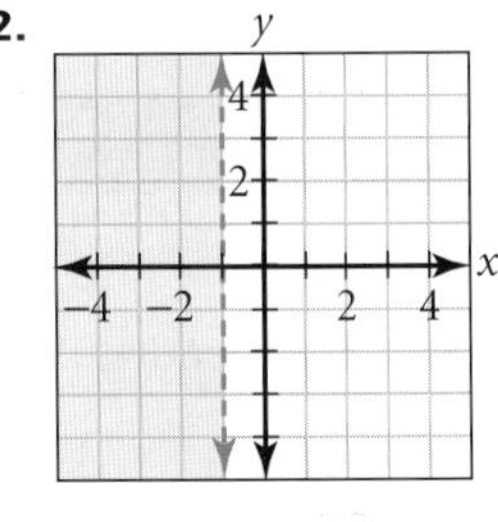

33.

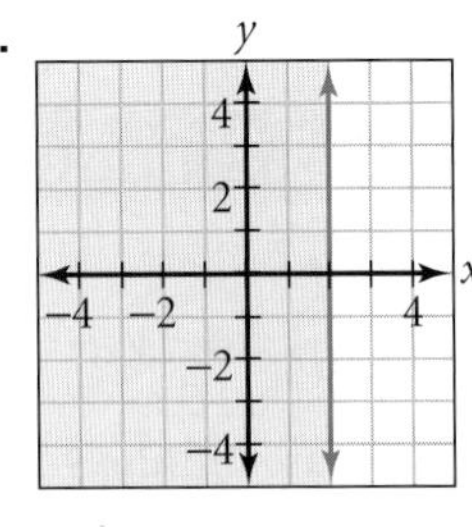

34.

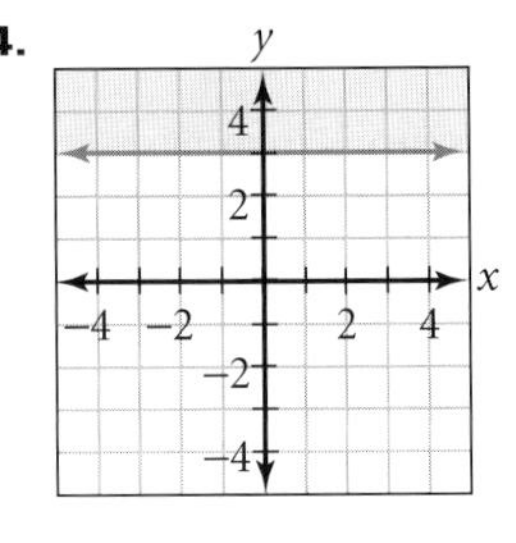

35.

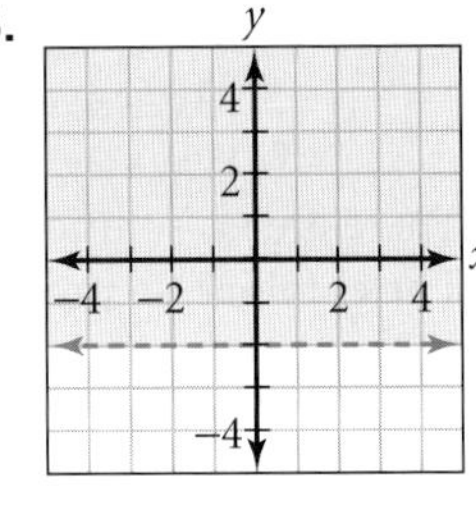

36.

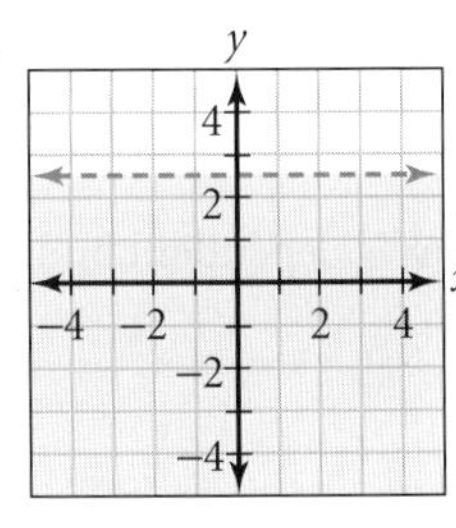

37.

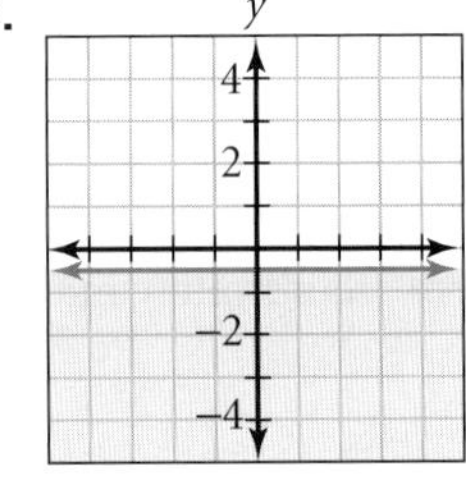

38.

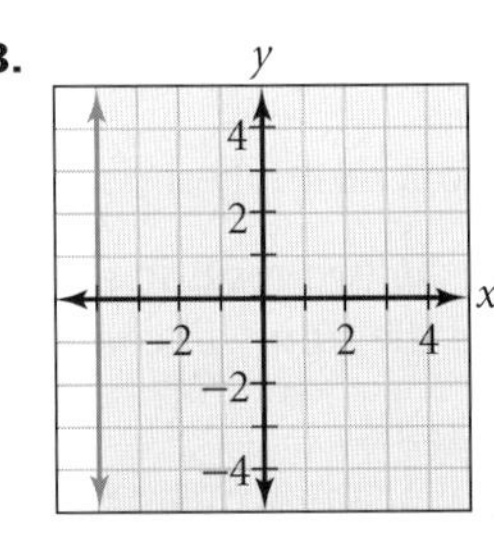

39.

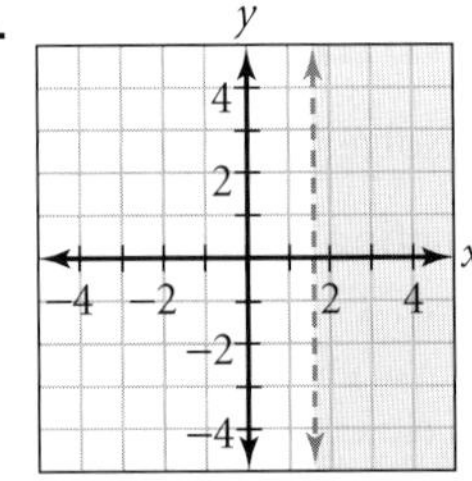

40.

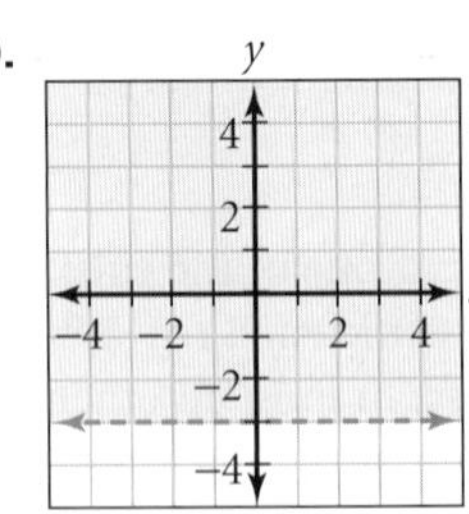

41.

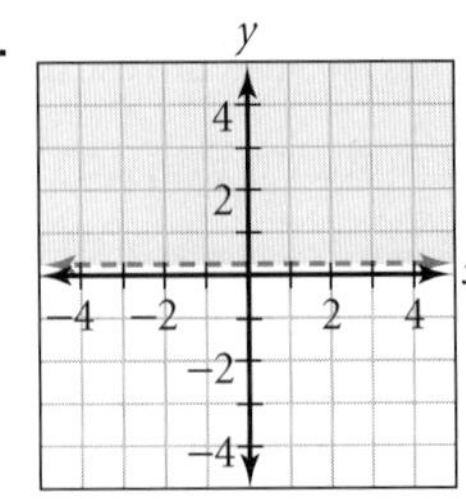

42.

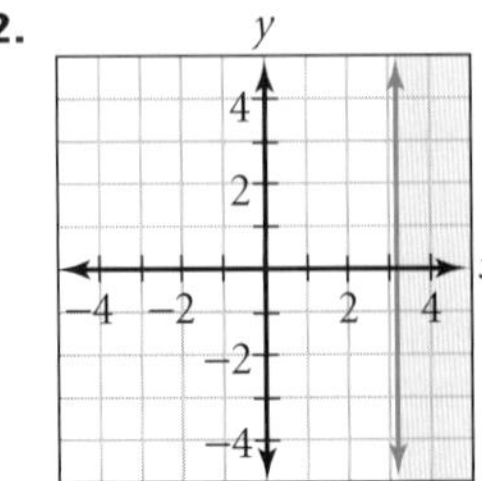

43.

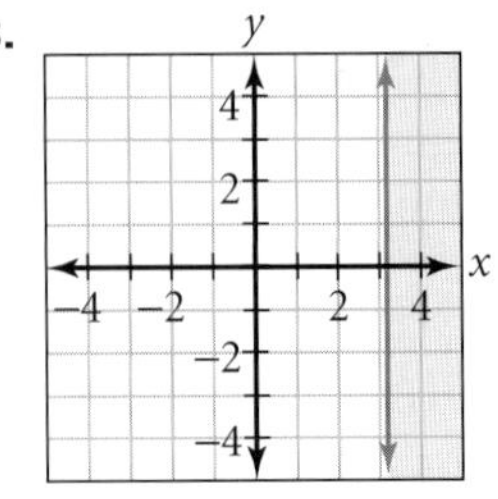

44. a. No, (4, 1) is not in the solution region, and $1 \geq \frac{1}{2} \cdot 4 + 5$ is a false statement.

b. Answers may vary. Sample answer: (6, 20), (6, 10), (6, 15)

c. Answers may vary. Sample answer: (0, 8), (1, 8), (2, 8)

45. $x < n$ for $n = 1, 2, 3,$ and 4 is a series of half-planes which lie to the left of the dashed vertical lines $x = 1$, $x = 2$, $x = 3$, and $x = 4$.

46. $y > (-1)^n x$ for $n = 1, 2, 3,$ and 4 is either the half-plane lying above the dashed line $y = -x$ if n is 1 or 3 or the half-plane lying above the line $y = x$ if n is 2 or 4.

47. $ny \leq 2x$ for $n = 1, 2, 3, 4$ or, equivalently, $y \leq \frac{2}{n}x$ for $n = 1, 2, 3, 4$, is the series of half-planes below the successive solid lines $y = 2x$, $y = x$, $y = \frac{2}{3}x$, and $y = \frac{1}{2}x$.

48. b.

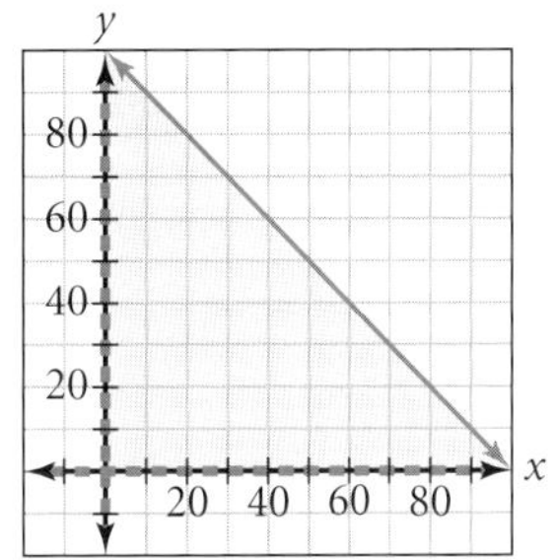

52. b.

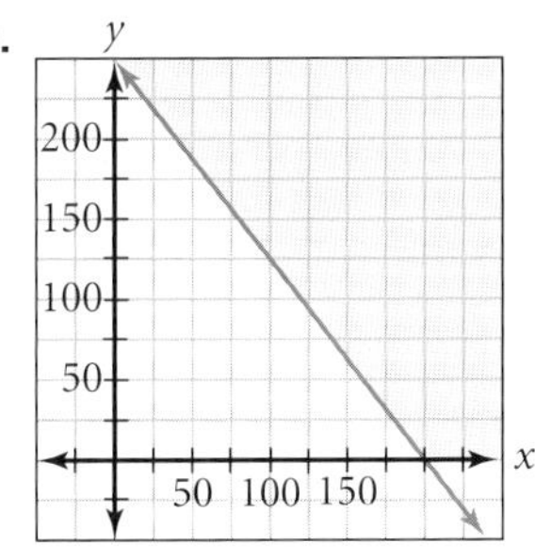

c. Answers may vary. Sample answer: (200, 0), (0, 250), (100, 125), (4, 245)

d. Answers may vary. Sample answer: (100, 150), (100, 175), (100, 200)

e. Answers may vary. Sample answer: (100, 25), (100, 50), (100, 75)

53. b.

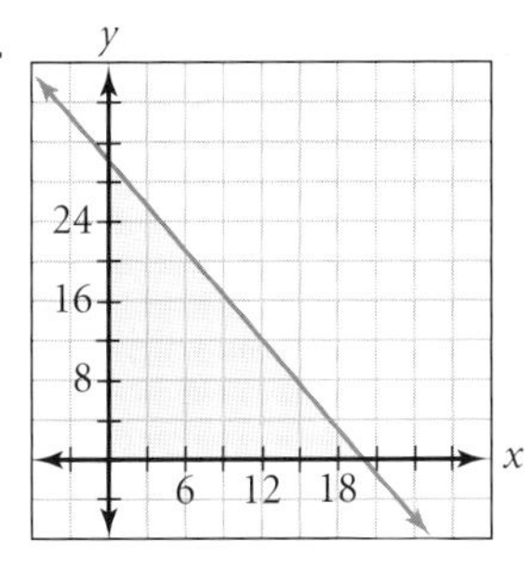

Lesson 3.4, pages 179–186

Activity

1. $\overleftrightarrow{WX}$ represents the boundary at which people begin to feel warm, and $\overleftrightarrow{ZY}$ represents the boundary at which people begin to feel cool. $\overleftrightarrow{WZ}$ and $\overleftrightarrow{XY}$ represent the boundaries at 20% and 60% relative humidity, respectively.

2. W: (20, 79.5); X: (60, 77.5); Y: (60, 72.5); Z: (20, 74.5)

3. $20 \leq x \leq -60$; $y \geq -\frac{1}{20}x + 75.5$; $y \leq -\frac{1}{20}x + 80.5$

CHECKPOINT ✔

4. As relative humidity increases, the comfort zone temperature boundaries decrease. This inverse relationship is why the slopes are negative.

Exercises

Communicate

1. A solid line is used to indicate that points on the boundary line are included in the solution; this occurs if the inequality is of the type ≤ or ≥. A dashed line is used to indicate that points on the boundary line are not included in the solution; this occurs if the inequality is of the type < or >.

2. Shade the region lying below the boundary line.

3. The solution for each inequality contains no points in common. For example: $\begin{cases} y > x + 4 \\ y < x + 3 \end{cases}$

Practice and Apply

12.

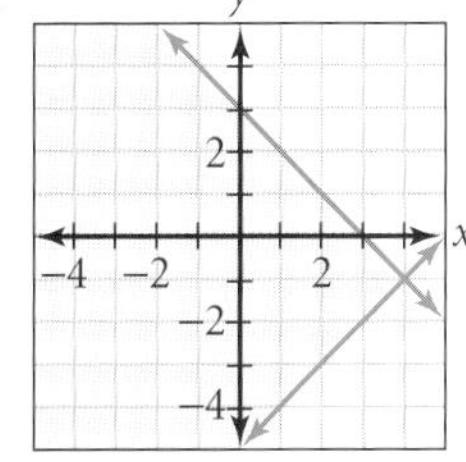

13.

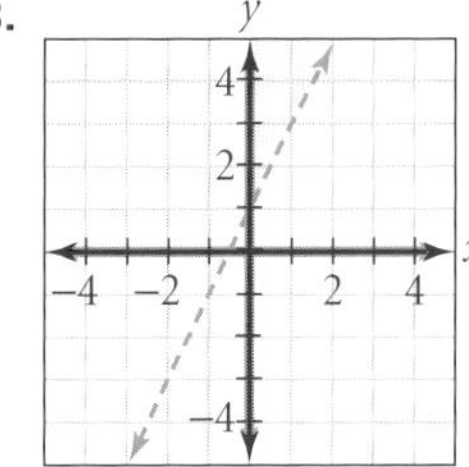

14.

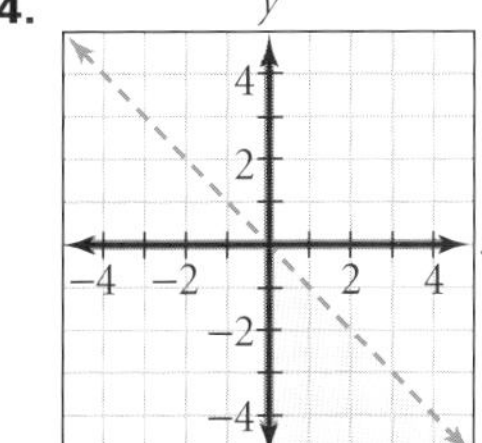

15.

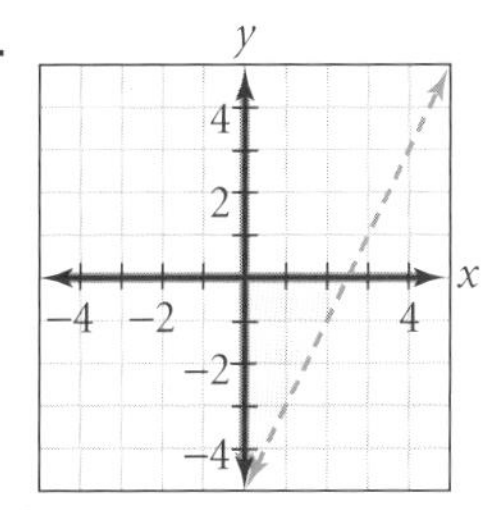

16.

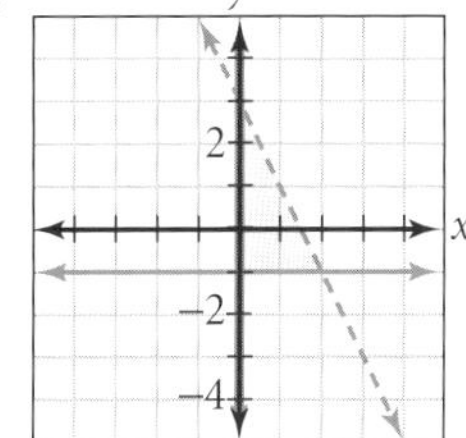

17.

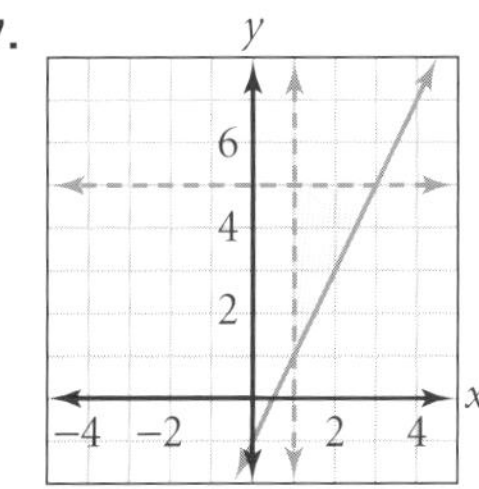

18.

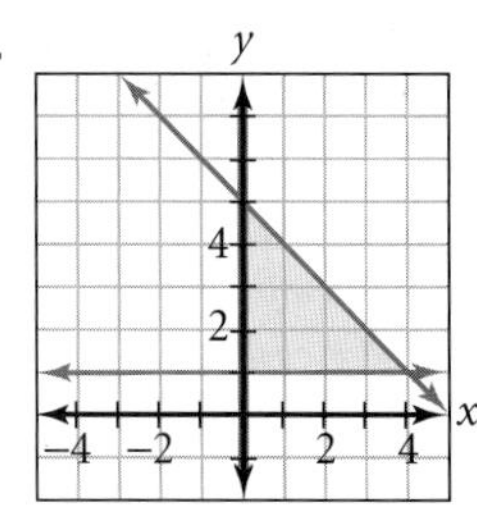

19.

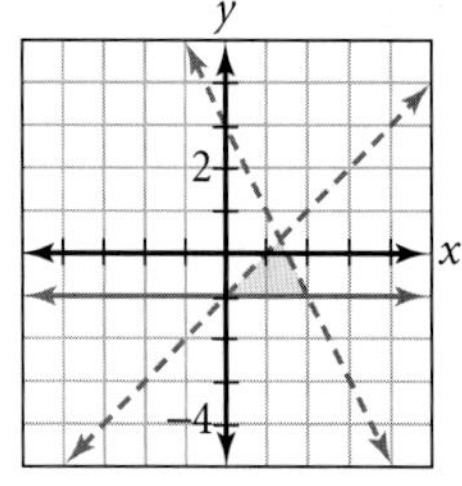

20.

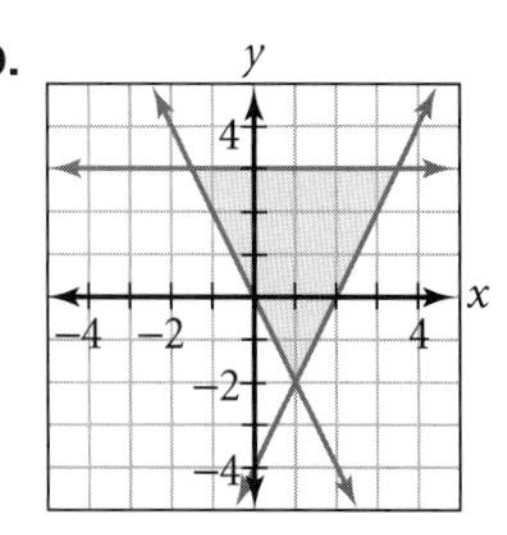

21.

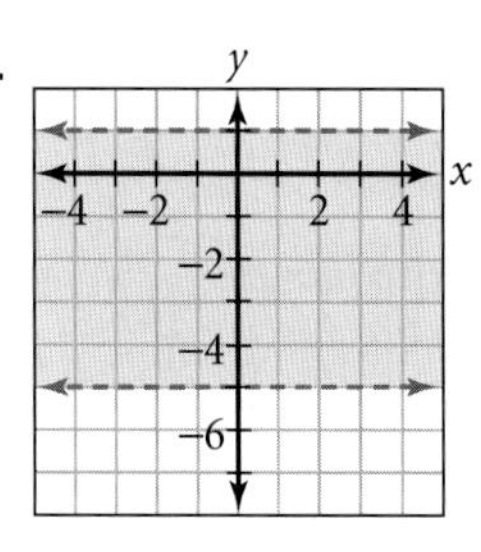

22.

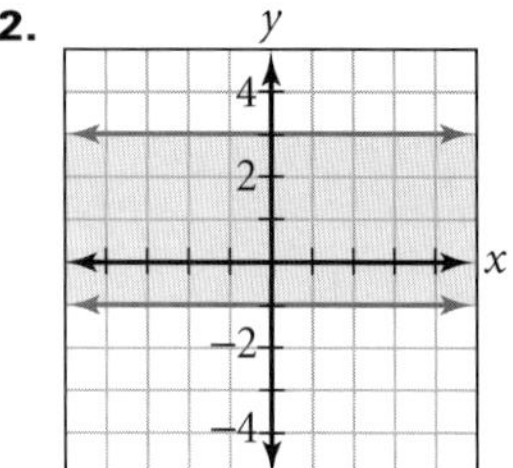

23.

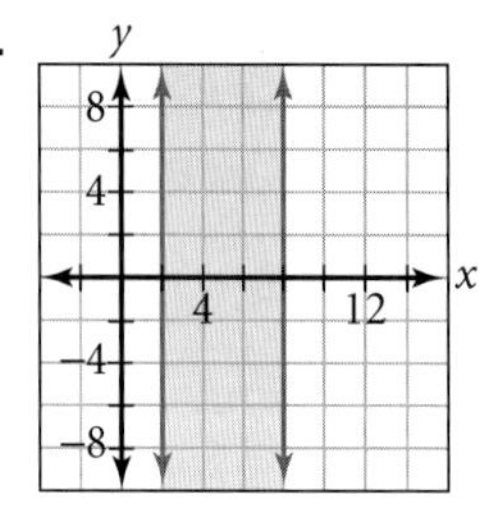

24.

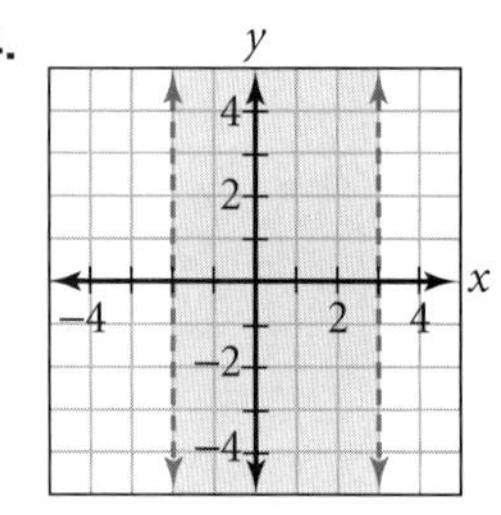

25.

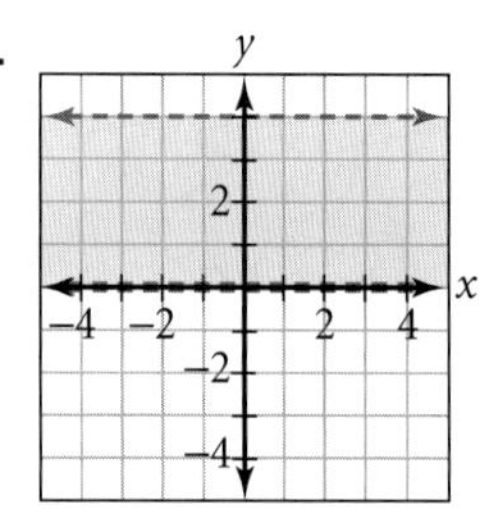

26.

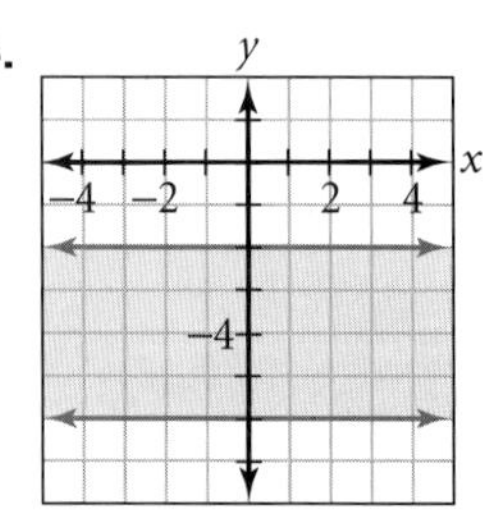

27.

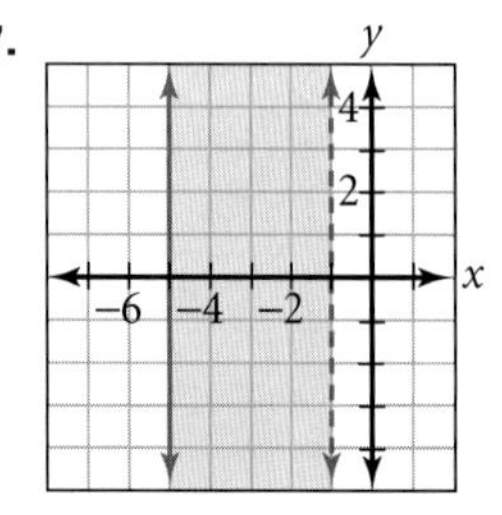

28.

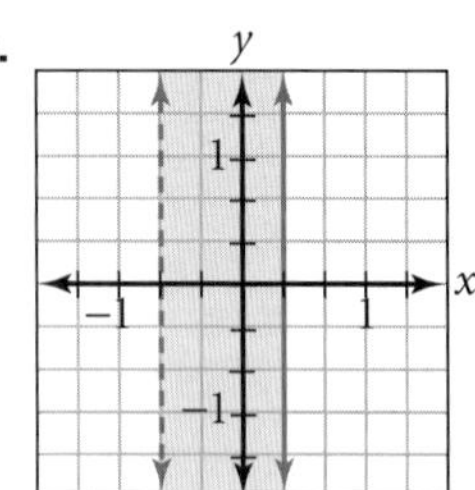

29.

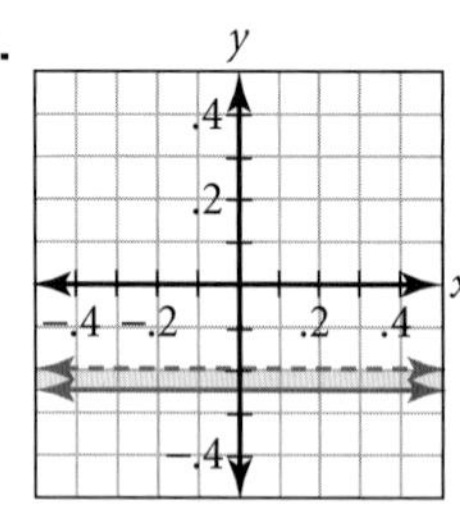

30.

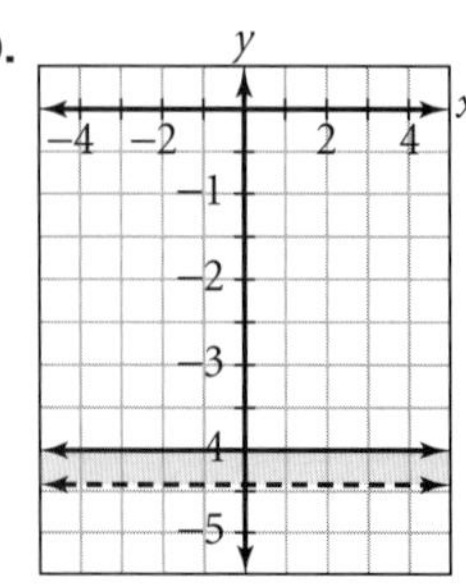

31.

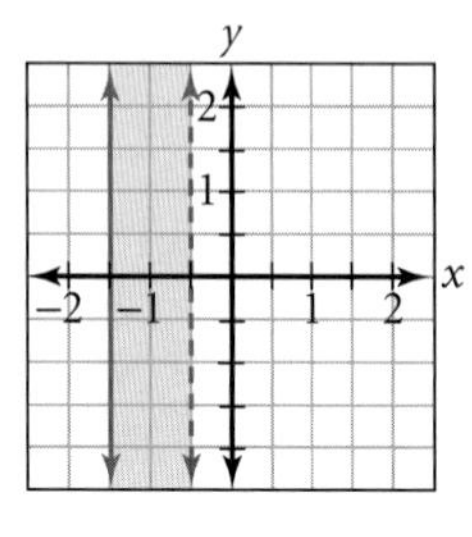

32.

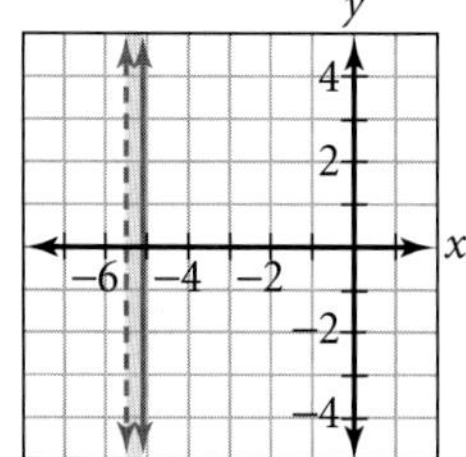

50. b.

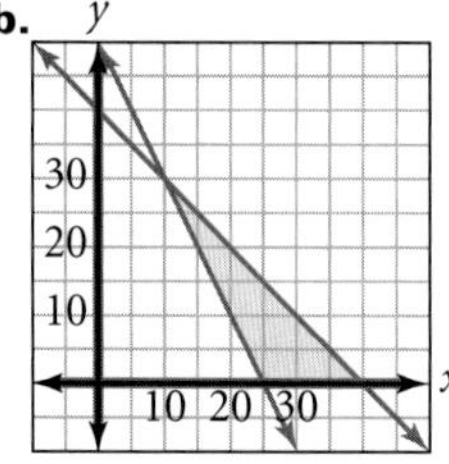

Yes, it is a polygon.

c. Answers may vary. Sample answer: (25, 10) means 25 hours per week of programming and 10 hours per week tutoring.

d. Answers may vary. To minimize work, (25, 0) is best; to maximize income, (40, 0) is best.

51. b.

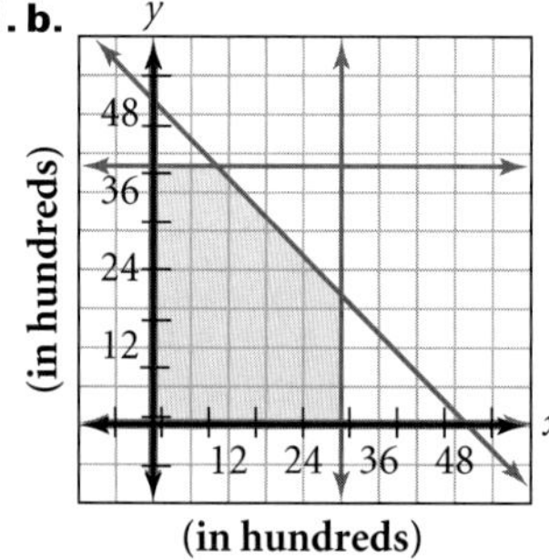

c. The change increases the number of possible combinations.

52. a. $\begin{cases} x + y \geq 45 \\ 50x + 40y \geq 2000 \\ x \geq 0, y \geq 0 \end{cases}$

b.

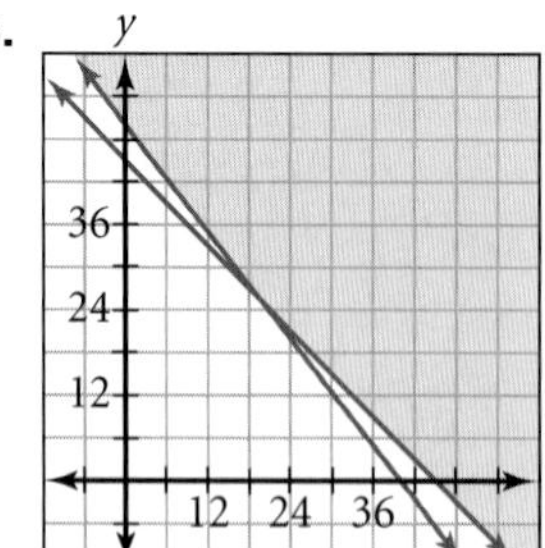

c. 30 gas-powered hedge clippers must be sold.

Look Back

53.

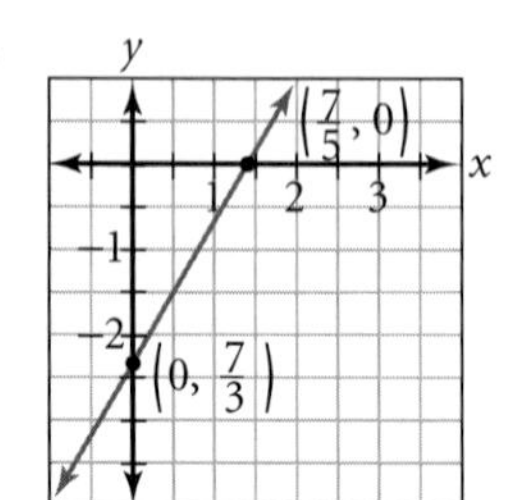

Portfolio Activity

1. Yes; machine A would be used for $4(2) + 5(1) = 13$, which is less than 18 minutes; machine B would be used for $4(1) + 5(3) = 19$, which is less than 24 minutes; and machine C would be used for $4(1) + 5(1) = 9$, which is less than 10 minutes.

2. No; for machine C, $6(1) + 5(1) = 11$, which is greater than the maximum of 10 minutes.

Lesson 3.5, pages 187–194

Activity

1.

Vertex	Objective function
$A(20, 30)$	$R = 357.525(20) + 159.31(30) = 11{,}929.80$
$B(20, 100)$	$R = 357.525(20) + 159.31(100) = 23{,}081.50$
$C(80, 40)$	$R = 357.525(80) + 159.31(40) = 34{,}974.40$
$D(80, 30)$	$R = 357.525(80) + 159.31(30) = 33{,}381.30$

2. $C(80, 40)$; 80 acres of corn and 40 acres of wheat should be grown to maximize the revenue.

3. No

4. No

CHECKPOINT ✔

5. Yes, it appears that the maximum value of the objective function will occur at a vertex.

6. For the minimum value of the object function:

   Step 2
   $A(20, 30)$; 20 acres of corn and 30 acres of wheat should be grown to minimize the revenue.

   Step 3
   None of the other boundary lines points gives a smaller revenue than $A(20, 30)$.

   Step 4
   None of the other interior points gives a smaller revenue than $A(20, 30)$.

   Step 5
   Yes, it appears that the minimum value of the objective function will occur at a vertex.

Exercises

Communicate

1. A *constraint* on a variable is a restriction of the variable to particular values.

2. *Feasible* means suitable and refers to the possible values that a variable can assume in a problem such that all constraints are satisfied simultaneously.

3. Assign variables to the quantities of the problem. Obtain the series of constraints from the information given in the problem and determine the feasible region by graphing. Determine the coordinates of the corner points of this region and then evaluate the objective function at each of the corner points. Write the objective function to be optimized. Evaluate the objective function at each of the corner points and identify the coordinates that yield the maximum or minimum value of the function.

Guided Skills Practice

4. Letting x be the number of acres of corn and y be the number of acres of soybeans:

$$\begin{cases} x + y \le 150 \\ 40 \le x \le 120 \\ 0 \le y \le 100 \end{cases}$$

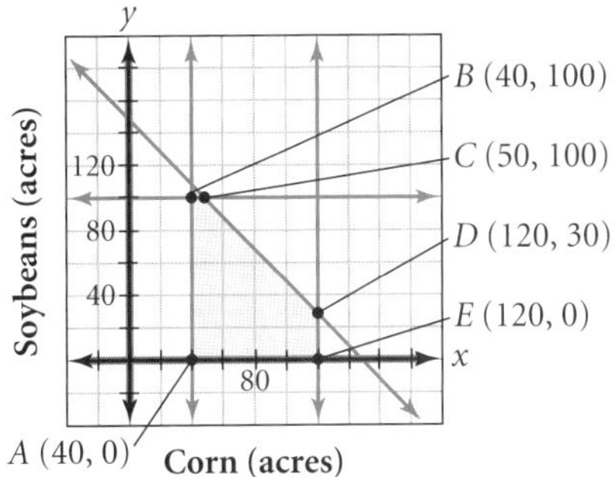

The objective function is $R = 357.525x + 237.32y$.

5. Letting x be the number of acres of wheat and y be the number of acres of soybeans:

$$\begin{cases} x + y \le 220 \\ 100 \le x \le 200 \\ 0 \le y \le 75 \end{cases}$$

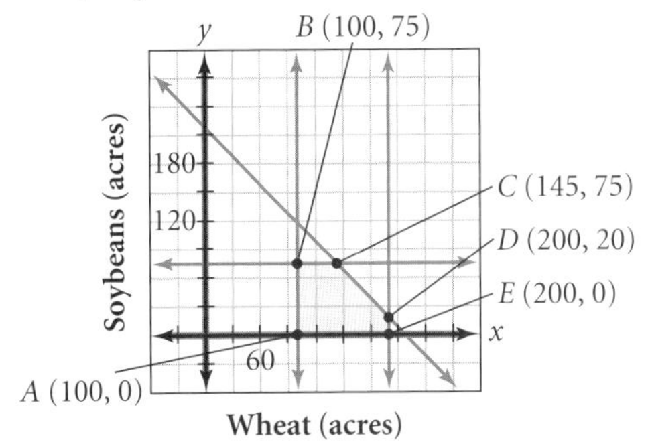

The objective function is $R = 159.31x + 237.32y$.

Portfolio Activity

1. To determine the coordinates of the feasible region, graph the intersection of $x \geq 0$ and $y \geq 0$ with the system of inequalities $\begin{cases} 2x + y \leq 18 \\ x + 3y \leq 24. \\ x + y \leq 10 \end{cases}$

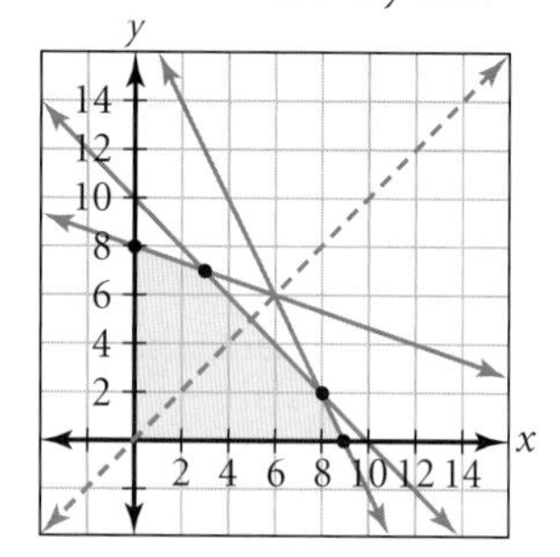

The coordinates of the vertices of the feasible region are (0, 8), (7, 3), (8, 2), and (9, 0).

2. The objective function for the profit, P, is $P = 20x + 30y$.
3. The company should produce 0 regular and 8 sport models to maximize their profit. The profit would be \$240.

Lesson 3.6, pages 195–201

Activity

1.

t	0	1	2	3	4
x	0	120	240	360	480

t	0	1	2	3	4
y	0	13.4	26.8	40.2	53.6

2. The relationship between x and t is linear if there is a constant difference between successive values, that is, $120 - 0 = 120$, $240 - 120 = 120$, etc. The relationship between y and t is linear if there is a constant difference between successive values, that is, $13.4 - 0 = 13.4$, $26.8 - 13.4 = 13.4$, etc. The horizontal and vertical distances in feet after t seconds are given by the parametric equations $\begin{cases} x(t) = 120t \\ y(t) = 13.4t \end{cases}$. Therefore, $y = \frac{67}{600}x$.
3. about 112 seconds; about 13,440 feet

CHECKPOINT ✔

4. No; the parametric equations give the time it takes to reach a certain altitude or horizontal distance.

Exercises

Communicate

1. Substitute x for $x(t)$ and y for $y(t)$. Solve $x = 2t + 3$ for t, and substitute the resulting expression for t in $y = -t + 1$.
2. The relationship of x to t and y to t is lost. You can solve for x and y but not t.
3. $x(t)$ is the horizontal distance in feet that the baseball has traveled in t seconds since being thrown. $y(t)$ is the vertical height in feet of the ball t seconds after being thrown. t is the time in seconds since the ball was thrown.

Chapter 3 Project

Activity 2

3. The coordinates of the vertices are (0, 0), (0, 13), (8, 9), (16, 3), and (18, 0).

Chapter 3 Review and Assessment

25.

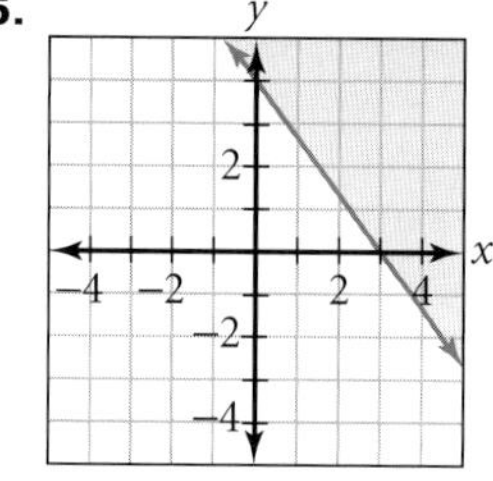

26.

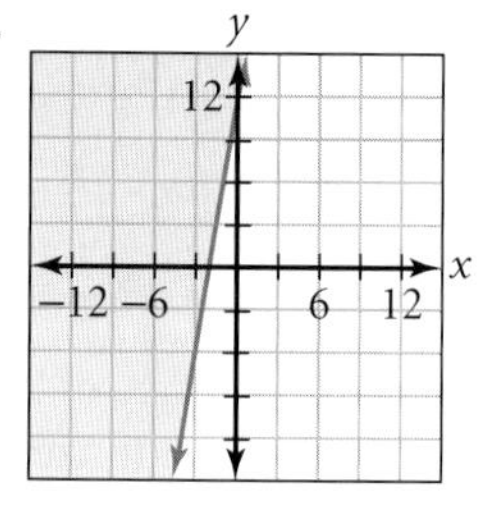

27.

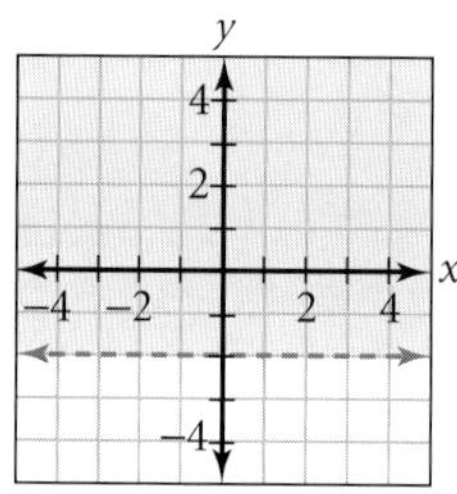

28.

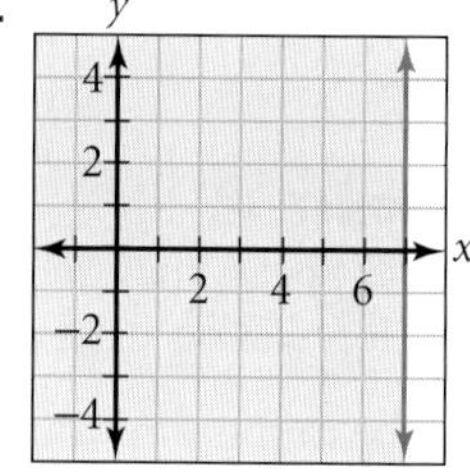

29.

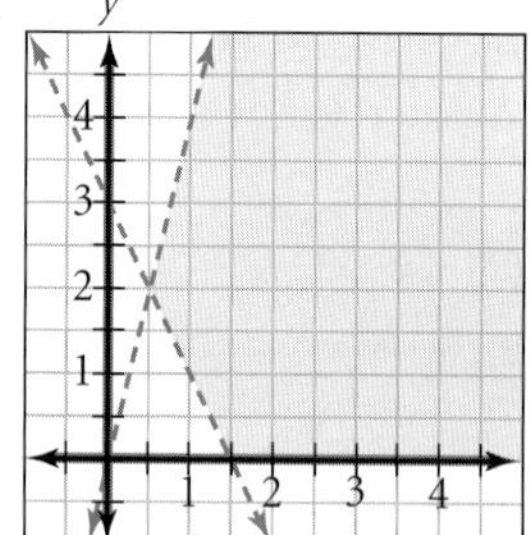

30.

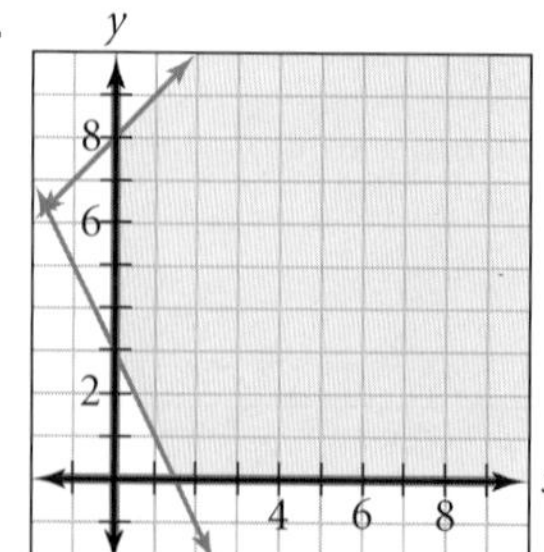

31.

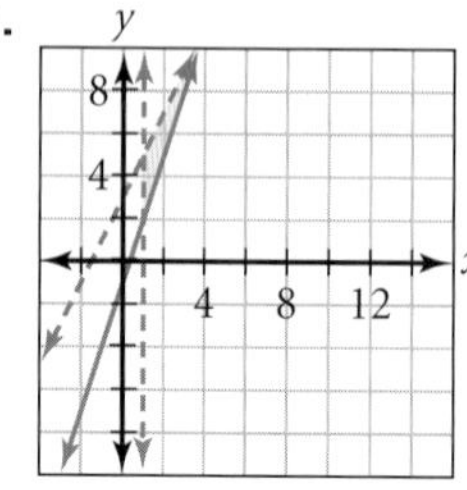

32.

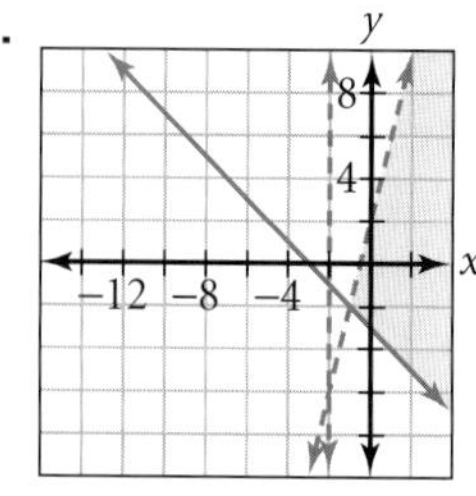

Lesson 4.1, pages 216–224

Exercises

Communicate

1. the entry in the fifth row and the second column
2. the entries in the second row, and in the first, second, third, and fourth columns; at least 2×4
3. Each column represents a point in the polygon. The first row contains the x-coordinates, and the second row contains the y-coordinates.
4. Multiply the matrix representing the polygon by a scalar to find the coordinates of the transformed polygon.

Practice and Apply

48. d.

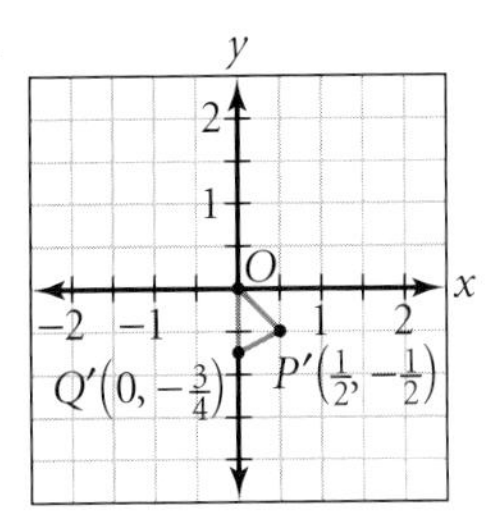

49. a. $\begin{bmatrix} 0 & -3 & -3 \\ 0 & -3 & 0 \end{bmatrix}$

b.

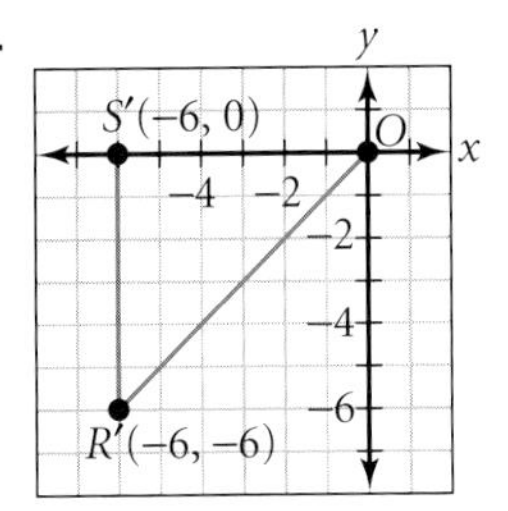

c.

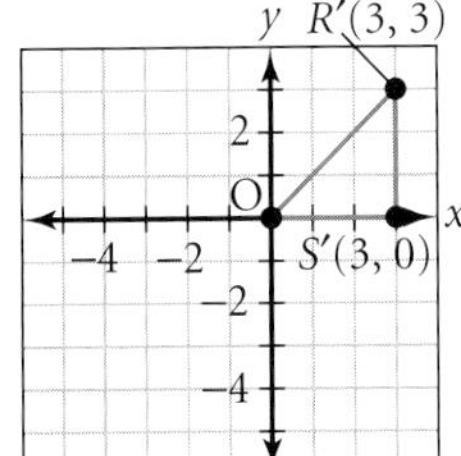

d.

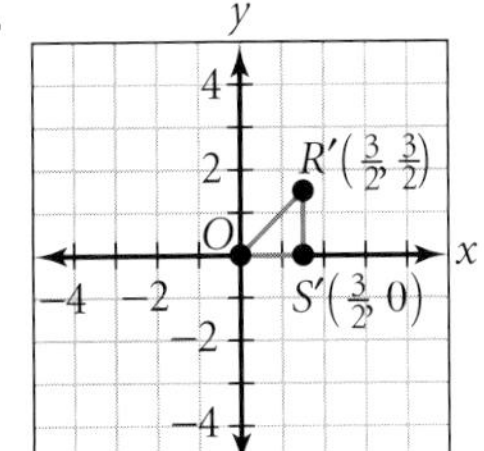

Portfolio Activity

1. Location n_{34} is in the third row and fourth column. This entry represents the number of paths from D to T. Location n_{43} is in the fourth row and third column. This entry represents the number of paths from T to D.
2. The 1 at n_{54} means that there is one path from G to T. The 0 at n_{14} means that there are no paths from M to T.
3. Find the entry that represents the path from G to T. If there is a path, there will be a 1 in location n_{54}.
4. There are no paths from an exhibit to itself.
5. To find the number of paths leading to each exhibit, add the entries in the column representing that exhibit. For example, there are 3 paths to exhibit D, which can be found by adding the entries in the third column.
6. To find the number of paths going from each exhibit, add the entries in the row representing that exhibit. For example, there are 2 paths from exhibit G, which can be found by adding the entries in the fifth row.

Lesson 4.2, pages 225–233

Activity

1.

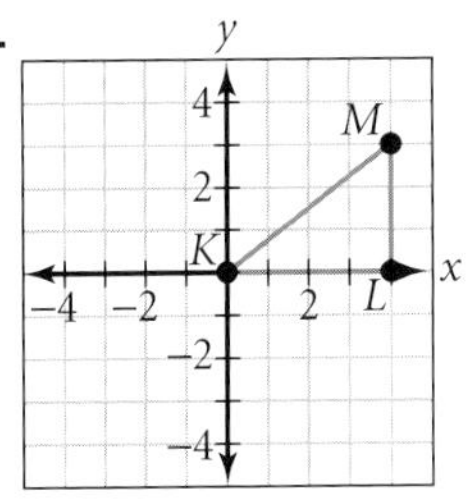

$C = \begin{bmatrix} 0 & 4 & 4 \\ 0 & 0 & 3 \end{bmatrix}$

2. $AC = \begin{bmatrix} 0 & 0 & -3 \\ 0 & 4 & 4 \end{bmatrix}$

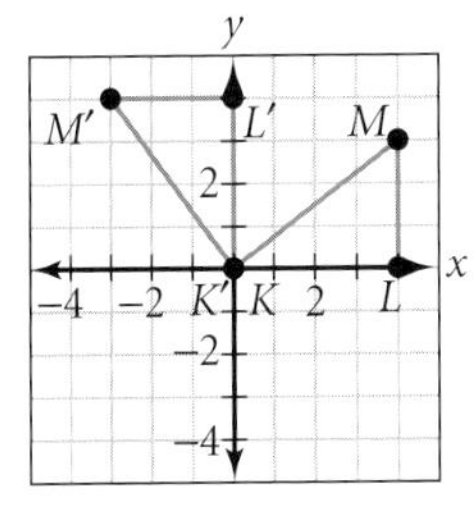

$\triangle K'L'M'$ is $\triangle KLM$ rotated 90° counterclockwise about the origin.

3. $BC = \begin{bmatrix} 0 & 0 & 3 \\ 0 & -4 & -4 \end{bmatrix}$

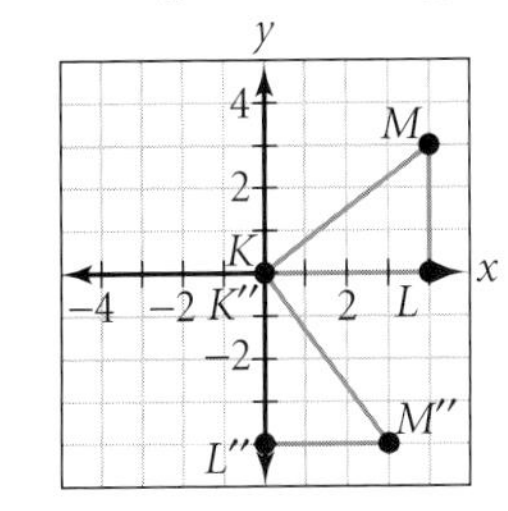

$\triangle K''L''M''$ is $\triangle KLM$ rotated 90° clockwise about the origin.

CHECKPOINT ✔

4. Answers may vary. Multiplying by matrix A rotates a geometric figure 90° counterclockwise about the origin.

CHECKPOINT ✔

5. Answers may vary. Multiplying by matrix B rotates a geometric figure 90° clockwise about the origin.

Exercises

Communicate

1. The inner dimensions must be the same.

2. Find the entry in row i and column j of the product by finding the sum of the products of the corresponding entries in row i of $\begin{bmatrix} 3 & -2 \\ 5 & 7 \end{bmatrix}$ and column j of $\begin{bmatrix} 5 & -3 & 1 \\ -2 & -1 & 4 \end{bmatrix}$.

3. Answers may vary. The entry in row i and column j of the path matrix represents how many one-stage paths are possible from vertex i to vertex j.

Portfolio Activity

1. $N^3 = \begin{bmatrix} 0 & 2 & 0 & 1 & 1 \\ 2 & 0 & 4 & 1 & 1 \\ 0 & 4 & 2 & 4 & 4 \\ 1 & 1 & 4 & 2 & 3 \\ 1 & 1 & 4 & 3 & 2 \end{bmatrix}$

   n_{54} represents the number of three-stage paths from G to T, which is 3. The paths are G → D → G → T, G → T → D → T, and G → T → G → T.

2. n_{43} represents the number of three-stage paths from T to D, which is 4. The paths are T → D → T → D, T → G → T → D, T → D → G → D, and T → D → H → D.

3. The sums of the rows of N^3 are 4, 8, 14, 11, and 11. Exhibit D has the greatest number of three-stage paths.

4. $N^4 = \begin{bmatrix} 2 & 0 & 4 & 1 & 1 \\ 0 & 6 & 2 & 5 & 5 \\ 4 & 2 & 12 & 6 & 6 \\ 1 & 5 & 6 & 7 & 6 \\ 1 & 5 & 6 & 6 & 7 \end{bmatrix}$

   n_{13} is the number of four-stage paths from M to D.

5. The number of four-stage paths from D to itself is found as entry n_{33} which is 12. The paths are: D → G → D → G → D, D → T → D → T → D, D → H → D → H → D, D → H → M → H → D, D → H → D → G → D, D → G → D → T → D, D → T → G → T → D, D → G → D → H → D, D → T → D → H → D, D → G → D → T → D, D → G → T → G → D, and D → T → D → G → D.

Lesson 4.3, pages 234–241

Activity

1. Answers may vary. One example is $A = \begin{bmatrix} 1 & 2 \\ 3 & 7 \end{bmatrix}$.

2. $A^{-1} = \begin{bmatrix} 7 & -2 \\ -3 & 1 \end{bmatrix}$

3. Answers may vary. One example is "GO TEAM" → 7 | 15 | 0 | 20 | 5 | 1 | 13 → $\begin{bmatrix} 7 & 15 & 0 & 20 \\ 5 & 1 & 13 & 0 \end{bmatrix}$

4. $\begin{bmatrix} 1 & 2 \\ 3 & 7 \end{bmatrix}\begin{bmatrix} 7 & 15 & 0 & 20 \\ 5 & 1 & 13 & 0 \end{bmatrix} = \begin{bmatrix} 17 & 17 & 26 & 20 \\ 56 & 52 & 91 & 60 \end{bmatrix} \rightarrow$
   17 | 17 | 26 | 20 | 56 | 52 | 91 | 60

5. $\begin{bmatrix} 7 & -2 \\ -3 & 1 \end{bmatrix}\begin{bmatrix} 17 & 17 & 26 & 20 \\ 56 & 52 & 91 & 60 \end{bmatrix} = \begin{bmatrix} 7 & 15 & 0 & 20 \\ 5 & 1 & 13 & 0 \end{bmatrix} \rightarrow$
   7 | 15 | 0 | 20 | 5 | 1 | 13 | 0

CHECKPOINT ✔

6. Answers may vary. A matrix must be square to have an inverse. An encoding matrix must be square and invertible to be used as the coding matrix.

Exercises

Communicate

1. Answers may vary. *Encryption* is translating a message into coded form, and *decryption* is translating a coded form into a message. A matrix can be used to encrypt a message by multiplying, and its inverse can be used to decrypt the coded form, also by multiplying.

2. **a.** The product is the first matrix because the second matrix is the identity matrix.
   b. The product is the second matrix because the first matrix is the identity matrix.

3. Enter the matrix and then use the x^{-1} key, or reciprocal key.

4. If a matrix has a nonzero determinant, then the matrix has an inverse.

Practice and Apply

47. $A^{-1} = \begin{bmatrix} \frac{1}{2} & 0 \\ 0 & 1 \end{bmatrix}$;

$\begin{bmatrix} \frac{1}{2} & 0 \\ 0 & 1 \end{bmatrix}\begin{bmatrix} 2 & 2 & -2 & -2 \\ 2 & -2 & -2 & 2 \end{bmatrix} = \begin{bmatrix} 1 & 1 & -1 & -1 \\ 2 & -2 & -2 & 2 \end{bmatrix}$

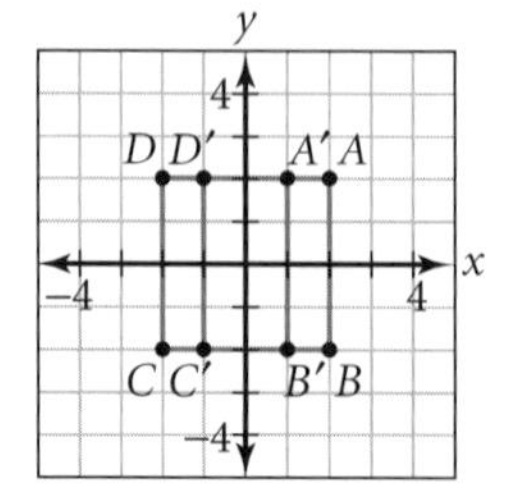

Look Back

66.

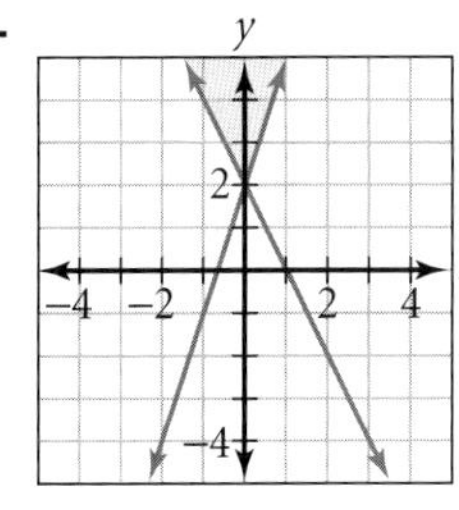

67.

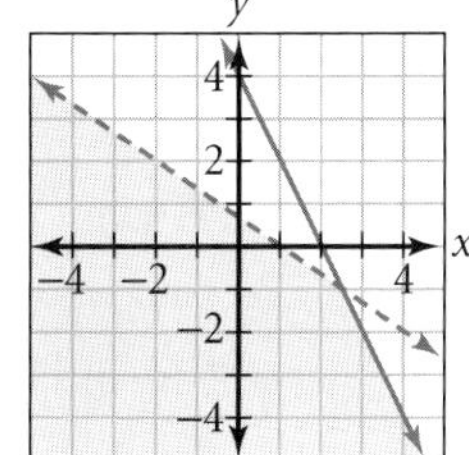

68.

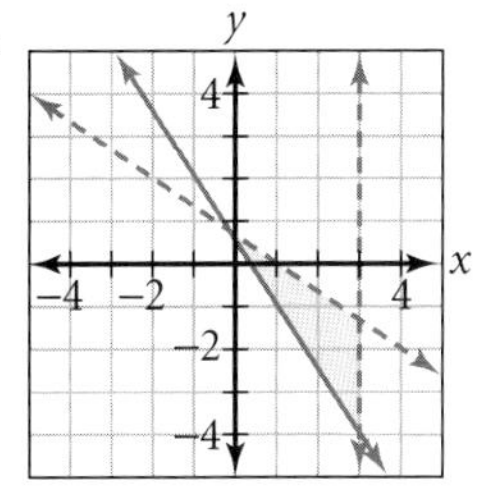

Lesson 4.4, pages 244–250

Activity

1. 2; yes; because the determinant is not zero
2. Answers may vary. Sample answer: $r = 9$ and $s = 5$, so $x = 0$ and $y = 1$.
3. $-\frac{4}{9}, -\frac{2}{5}$; no
4. 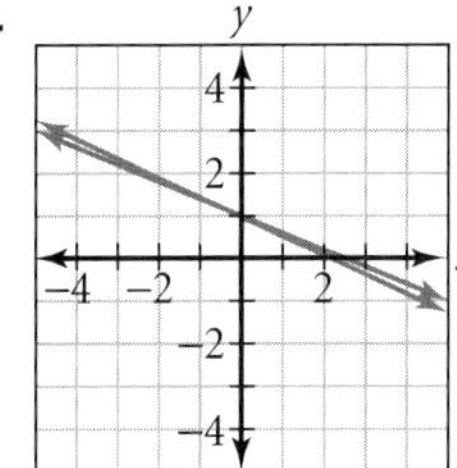

   Answers may vary. Student answers should include that if the slopes are different, then the lines intersect and there is a unique solution.
5. Answers may vary. Students answers should note that as long as the slopes are different, there is a unique solution.

CHECKPOINT ✔

6. Answers may vary. The system has a unique solution. The solution depends on the values of r and s.

Exercises

Communicate

1. A coefficient matrix, A, times a variable matrix, X, equals the constant matrix, B, or $AX = B$.
2. Write each equation with the variables in the order x, y, and z:

   $$\begin{cases} x - y = 5 \\ y - z = -6 \\ 2x - z = 2 \end{cases} \rightarrow \begin{cases} 1x - 1y + 0z = 5 \\ 0x + 1y - 1z = -6 \\ 2x + 0y - 1z = 2 \end{cases}$$

   Place the coefficients of the variables in a matrix:

   $$\begin{bmatrix} 1 & -1 & 0 \\ 0 & 1 & -1 \\ 2 & 0 & -1 \end{bmatrix}$$

   Multiply the coefficient matrix by a matrix of the variable names, and equate the result to a matrix of constants from the right side of the system: $\begin{bmatrix} 1 & -1 & 0 \\ 0 & 1 & -1 \\ 2 & 0 & -1 \end{bmatrix}\begin{bmatrix} x \\ y \\ z \end{bmatrix} = \begin{bmatrix} 5 \\ -6 \\ 2 \end{bmatrix}$
3. Answers may vary. Solve $AX = B$ by multiplying both sides by A^{-1}. Solve $ax = b$ by multiplying both sides by the multiplicative inverse a^{-1}, or $\frac{1}{a}$.
4. Represent the system of equations by the matrix equation, $AX = B$. Find the inverse of the coefficient matrix, A. Multiply both sides of the matrix equation $AX = B$ by A^{-1}, which results in the equation $X = A^{-1}B$. X will contain the value of each variable.
5. Substitute $\begin{bmatrix} -1 \\ -3 \\ 2 \end{bmatrix}$ for $\begin{bmatrix} x \\ y \\ z \end{bmatrix}$ and multiply out the left side to see if it equals the right side of the equation.
6. It does not have a unique solution.

Lesson 4.5, pages 251–259

Activity

1. They have the same slope.
2. Answers may vary. Sample answer: Change the third equation to $6x + 2y = 8$.

   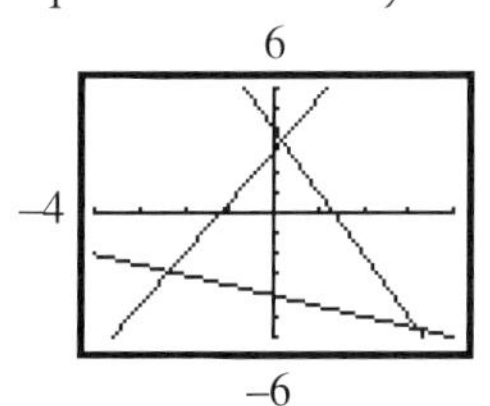

   Each combination of two equations must form a consistent and independent system.

CHECKPOINT ✔

3. Answers may vary. Sample answer: Change the third equation to $-4x + 2y = 8$.

   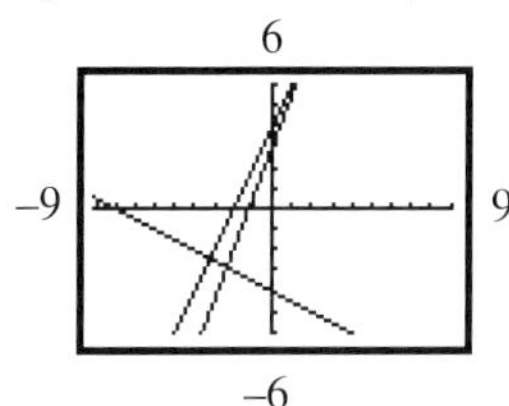

   You need two of the lines to be perpendicular, so their slopes must be negative reciprocals of each other.

Exercises

Communicate

1. Multiplying all entries in a row by a nonzero number corresponds to multiplying both sides of an equation by a nonzero number. Adding a multiple of one row to another corresponds to adding a multiple of one equation to another to eliminate one of the variables.

2. Rewrite each equation so that it contains each variable, adding zeros when necessary.
$$\begin{cases} x - 4y + 7z = 17 \\ 2x + y - z = -5 \\ x + 0y + 4z = 13 \end{cases}$$
Use the coefficients of the terms and the constants to form the augmented matrix: $\begin{bmatrix} 1 & -4 & 7 & \vdots & 17 \\ 2 & 1 & -1 & \vdots & -5 \\ 1 & 0 & 4 & \vdots & 13 \end{bmatrix}$

3. $\begin{cases} -3x - 4y = 2 \\ 4x + 2y - 3z = 6 \\ -2x + z = -6 \end{cases}$

4. $RR_1 \to R_1$
$R_2 + R_3 \to R_3$

Lesson 5.1, pages 274–280

Activity

1. f and g are linear functions; f and g have the same slope, so their graphs are parallel lines.

2. $f \cdot g$ has an x^2-term; its graph is U-shaped, not a straight line.

3. **2nd row:** f and g are linear functions; in fact, f and g are identical, so their graphs coincide; $f \cdot g$ has an x^2-term; its graph is U-shaped, not a straight line.

   3rd row: f and g are linear functions; f and g intersect; $f \cdot g$ has an x^2-term; its graph is U-shaped, not a straight line.

   4th row: f and g are linear functions; f and g intersect; $f \cdot g$ has an x^2-term; its graph is U-shaped, not a straight line.

CHECKPOINT ✔

4. The graphs of f and g are straight lines. The graph of $f \cdot g$ is not a straight line but is a U-shaped curve.

CHECKPOINT ✔

5. The x-intercepts of the U-shaped graph are the same as the x-intercepts of the corresponding lines.

Exercises

Communicate

1. The graph of a linear function is a straight line. It has at most one x-intercept and one y-intercept. The graph of a quadratic function is a U-shaped curve called a parabola. It has one y-intercept but can have 0, 1, or 2 x-intercepts.

2. A linear function is given by an expression in which the x-term has an exponent of 0 or 1 and the y-term has an exponent of one. It may be written in either the form $Ax + By = C$ or the form $y = mx + b$. A quadratic function is given by an expression with an x^2-term and can be written in the form $f(x) = ax^2 + bx + c$, where $a \neq 0$.

3. The quadratic function $f(x) = ax^2 + bx + c$ has a minimum value if $a > 0$ and a maximum value if $a < 0$.

Portfolio Activity

1.
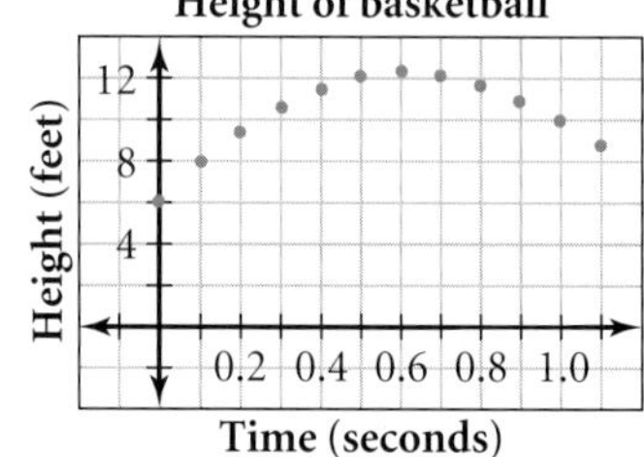

2. Answers will vary. Sample function: $y = -16x^2 + 20x + 6$

3. Answers will vary. Sample using the function in Step 2:
   a. The value of v_0 is 20.
   b. The maximum height is 12.30 ft.
   c. The basketball reaches its maximum height about 0.6 seconds after release.

4. Answers will vary. Sample: Substitute (0.2, 9.36) into the function and solve for v_0.
$$h(t) = -16t^2 + v_0t + h_0$$
$$9.36 = -16(0.2)^2 + v_0(0.2) + 6$$
$$9.36 = -0.64 + 0.2v_0 + 6$$
$$4 = 0.2v_0$$
$$v_0 = 20$$
   a. The value of v_0 is 20.
   b. The maximum height is 12.3 ft.
   c. The basketball reaches its maximum height about 0.6 seconds after release.

Lesson 5.2, pages 281–289

Activity

1 and 2.

Exact solution(s)	Number of solutions	Number of x-intercepts
$x = \pm\sqrt{7}$	2	2
$x = \pm\sqrt{2}$	2	2
$x = 0$	1	1
$x = \pm\sqrt{2}$	2	2
$x = \pm\sqrt{7}$	2	2
$x = 0$	1	1

CHECKPOINT ✔

3. The number of solutions to a quadratic equation and the number of x-intercepts of the related functions are equal.

Exercises

Communicate

1. $5(x+3)^2 = 12$

$(x+3)^2 = \frac{12}{5}$ Divide each side by 5.

$x + 3 = \pm\sqrt{\frac{12}{5}}$ Take the square root of each side.

$x = -3 \pm \sqrt{\frac{12}{5}}$ Subtract 3 from each side.

2. Any application in which only positive measures are possible would require only the principal root. From the examples in this section, finding the length of the side of a triangle (by using the Pythagorean Theorem, or $a^2 + b^2 = c^2$) and finding the time it takes for the life raft to hit the water (by using $h(t) = -16t^2 + 68$) are such applications. Also, finding the radius, r, of a circle given the area A (by using $A = \pi r^2$) is an example.

3. $a^2 + b^2 = c^2$ Use the Pythagorean Theorem.

$3^2 + 4^2 = c^2$

$c^2 = 25$

$c = \sqrt{25}$

$c = 5$ Find the principal root only.

Lesson 5.3, pages 290–298

Activity

1. $x^2 + 4x + 4$ may be represented as the product of the two linear factors, $(x+2)$ and $(x+2)$.

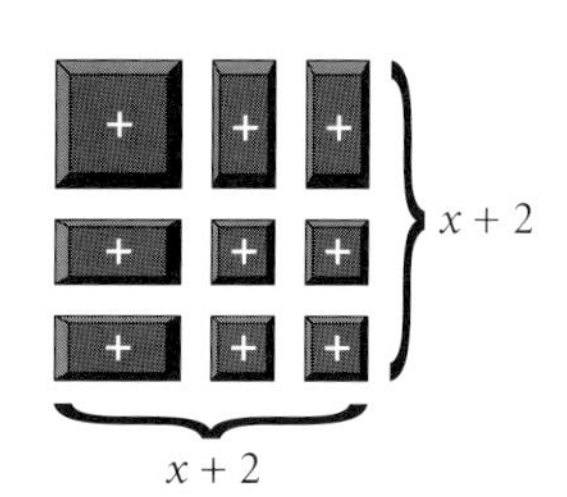

2. $x^2 + 6x + 8$ may be represented as the product of the two linear factors, $(x+2)$ and $(x+4)$.

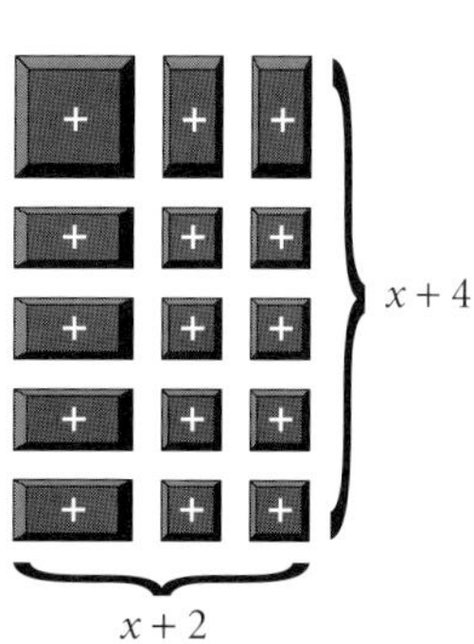

3. $x^2 + 7x + 12$ may be represented as the product of the two linear factors, $(x+3)$ and $(x+4)$.

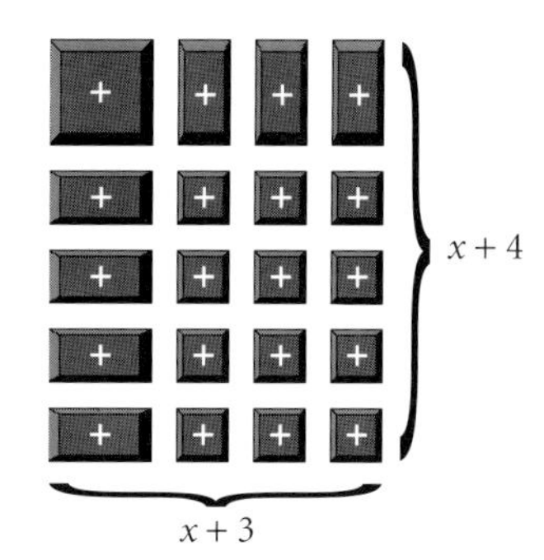

CHECKPOINT ✔

4. After arranging the tiles in the pattern for the product polynomial, the factors should appear as the tiles in the right column and bottom row.

Exercises

Communicate

1. q and s are two numbers whose sum is 34 and whose product is 285. In this case, q is 15 and s is 19 since 15 + 19 = 34 and 15 • 19 = 285.

2. Let the factorization of $x^2 + bx + c$ be $(x+q)(x+s)$. If c is positive, then q and s are both positive or both negative. If c is negative, then q and s have opposite signs—i.e., one is positive and one is negative. If c is positive, the sign of b is the same as the sign of q and s. If c is negative, q is positive, and s is negative, then b is positive when $|q| > |s|$ and b is negative when $|s| > |q|$.

3. If $pq = 0$, then either $p = 0$ or $q = 0$.

Portfolio Activity

1. $h(t) = 6$

$-16t^2 + 20t + 6 = 6$

$-16t^2 + 20t = 0$

$-4t(4t - 5) = 0$

$-4t = 0$ *or* $4t - 5 = 0$

$t = 0$ *or* $t = 1.25$

The basketball will return to a height of 6 feet 1.25 seconds after release.

2. The axis of symmetry is about $x = 0.6$. The basketball will return to a height of 6 feet 1.2 seconds after release.

3. The two answers are 0.05 seconds apart. This is because one is an estimate from a graph and the other was found by solving an equation for an exact answer.

Lesson 5.4, pages 299–306

Activity

1. No, you need unit tiles, too.

2. 4

3. $x^2 + 4x + 4 = (x+2)^2$

4. **Step 1.** no

Step 2. 9

Step 3. $x^2 + 6x + 9 = (x+3)^2$

5. **Step 1.** no

Step 2. 16

Step 3. $x^2 + 8x + 16 = (x+4)^2$

CHECKPOINT ✔

6. Divide the number of x-tiles by two and square the result.
$\left(\frac{8}{2}\right)^2 = 4^2 = 16$

Exercises

Communicate

1.
$$x^2 - 4x - 13 = 0$$
$$x^2 - 4x = 13$$
Add 13 to each side of the equation.
$$x^2 - 4x + \left(\frac{-4}{2}\right)^2 = 13 + \left(\frac{-4}{2}\right)^2$$
Add $\left(\frac{-4}{2}\right)^2$ to each side of the equation.
$$(x-2)^2 = 17$$
Simplify.
$$x - 1 = \pm\sqrt{17}$$
$$x = 1 \pm\sqrt{17}$$

2.
$$2x^2 + 4x = 15$$
$$\frac{1}{2}(2x^2 + 4x) = \frac{1}{2}(15)$$
Divide each side of the equation by 2.
$$x^2 + 2x = \frac{15}{2}$$
Simplify.
$$x^2 + 2x + \left(\frac{2}{2}\right)^2 = \frac{15}{2} + \left(\frac{2}{2}\right)^2$$
Add $\left(\frac{2}{2}\right)^2$ to each side of the equation.
$$(x+1)^2 = \frac{17}{2}$$
Simplify.
$$x + 1 = \pm\sqrt{\frac{17}{2}}$$
$$x = -1 \pm\sqrt{\frac{17}{2}}$$

3. First, graph $y = 2x^2 + 4x$ and $y = 15$ and find the x-coordinates of any points of intersection. Second, graph $y = 2x^2 + 4x - 15$ and locate any x-intercepts on the graph.

4. In the vertex form $y = a(x-h)^2 + k$, h and k are the x- and y-coordinates of the vertex, respectively.

Portfolio Activity

1. Completing the square:
$$h(t) = -16t^2 + 20t + 6$$
$$= -16(t^2 - 1.25t) + 6$$
$$= -16(t^2 - 1.25t + 0.390625) + 6 - (-16)(0.390625)$$
$$= -16(t - 0.625)^2 + 6 + 6.25$$
$$= -16(t - 0.625)^2 + 12.25$$
$$= -16[t - (0.625)]^2 + 12.25$$
The vertex of this parabola is (0.625, 12.25). The maximum height is 12.25 feet.

2. The length of time it takes for the basketball to reach its maximum height is the x-coordinate of the vertex of the parabola. It takes 0.625 seconds.

3. The equation of the axis of symmetry for the height function is $x = 0.625$.

4. To find the length of time for the basketball to drop to the ground, solve the equation $-16t^2 + 20t + 6 = 0$.
$$-16t^2 + 20t + 6 = 0$$
$$-16t^2 + 20t = -6$$
$$-16(t^2 - 1.25t) = -6$$
$$(t^2 - 1.25t) = \frac{6}{16}$$
$$(t^2 - 1.25t + 0.390625) = \frac{6}{16} + 0.390625$$
$$(t - 0.625)^2 = \frac{6}{16} + 0.390625$$
$$t - 0.625 = \pm\sqrt{\frac{6}{16} + 0.390625}$$
$$t = \pm\sqrt{\frac{6}{16} + 0.390625} + 0.625$$
$$t = 1.5 \text{ or } -0.25$$
Reject the negative value for time. The basketball would drop to the ground in 1.5 seconds.

5. The maximum height of the basketball and the time taken to reach that height found in the previous Portfolio Activities are very close to the height and time found in this Portfolio Activity.

Lesson 5.5, pages 307–313

Activity

1 and 2.

Equation	Roots	Average of roots	Related function	x-coord. of vertex
$x^2 + 2x = 0$	$0, -2$	-1	$f(x) = x^2 + 2x$	-1
$-x^2 + 4 = 0$	$-2, 2$	0	$f(x) = -x^2 + 4$	0
$x^2 + 4x + 4 = 0$	-2	-2	$f(x) = x^2 + 4x + 4$	-2
$2x^2 + 5x - 3 = 0$	$-3, \frac{1}{2}$	$-\frac{5}{4}$	$f(x) = 2x^2 + 5x - 3$	$-\frac{5}{4}$
$-x^2 - 3x + 4 = 0$	$-4, 1$	$-\frac{3}{2}$	$f(x) = -x^2 - 3x + 4$	$-\frac{3}{2}$

CHECKPOINT ✔

3. Find the roots of the related equation and average them to get the x-coordinate of the vertex of the graph of a quadratic function.

Exercises

Communicate

1. Graph the function and determine the intercepts by inspection. Find the zeros of function f by factoring then using the Zero Product Principle. Use a calculator's trace feature to locate the x-intercepts.

2. Locate the x-intercepts on the graph and calculate the average of the values. Solve for the x-coordinate by using $x = -\frac{b}{2a}$ and then use that value to find the y-coordinate.

3. The axis of symmetry is the vertical line through the vertex of the parabola.

Portfolio Activity

1. The height function for the basketball is $h(t) = -16t^2 + 20t + 6$. For this function, $a = -16$, $b = 20$, and $c = 6$. The axis of symmetry is $x = -\frac{b}{2a} = -\frac{20}{2(-16)} = -\frac{20}{(-32)} = 0.625$.

2. The maximum height of the basketball occurs at the vertex of the parabola, where $x = 0.625$. Substituting 0.625 for x into the equation results in $y = -16(0.625)^2 + 20(0.625) + 6 = 12.25$. The maximum height is 12.25 feet.

3. The length of time it takes the basketball to reach its maximum height is the x-coordinate of the vertex. It takes 0.625 seconds.

4. To find the length of time for the basketball to drop to the ground, solve the equation $-16t^2 + 20t + 6 = 0$. Using the quadratic formula, $a = -16$, $b = 20$, and $c = 6$.

$$x = \frac{-b \pm \sqrt{b^2 - 4ac}}{2a}$$
$$x = \frac{-20 \pm \sqrt{20^2 - 4(16)(6)}}{2(-16)}$$
$$x = \frac{-20 \pm \sqrt{400 + 384}}{-32}$$
$$x = \frac{-20 \pm \sqrt{784}}{-32}$$
$$x = \frac{-20 \pm 28}{-32}$$
$$x = \frac{-20 + 28}{-32} \text{ or } x = \frac{-20 - 28}{-32}$$
$$x = \frac{8}{-32} = -0.25 \text{ or } x = \frac{-48}{-32} = 1.5$$

Reject the negative value for time. The basketball would drop to the ground in 1.5 seconds.

5. The maximum height of the basketball and the time taken to reach that height estimated in the first two Portfolio Activities are very close to the actual height and time. The answers found in this Portfolio Activity are exactly the same as those found in the Portfolio Activity on page 306.

Lesson 5.6, pages 314–321

Activity

1.

i	$i^2 = -1$	$i^3 = -i$	$i^4 = 1$
$i^5 = i$	$i^6 = -1$	$i^7 = -i$	$i^8 = 1$

2.

$i^9 = i$	$i^{10} = -1$	$i^{11} = -i$	$i^{12} = 1$
$i^{13} = i$	$i^{14} = -1$	$i^{15} = -i$	$i^{16} = 1$

CHECKPOINT ✔

3. The powers of i are cyclic and repeat in a pattern of four numbers: i, -1, $-i$, and 1. To find i^n, divide n by 4 and match the remainder to the power of i in the table below.

$i^1 = i$	$i^2 = -1$
$i^3 = -i$	$i^4 = i^0 = 1$

$$i^{41} = i^{4(10)+1} = i^1 = i$$
$$i^{66} = i^{4(16)+2} = i^2 = -1$$
$$i^{75} = i^{4(18)+3} = i^3 = -i$$
$$i^{100} = i^{4(25)+0} = i^0 = 1$$

Exercises

Communicate

1. The discriminant tells you how many real solutions there are to a quadratic equation.

$$b^2 - 4ac \begin{cases} < 0 & \text{no real solutions} \\ = 0 & \text{one real solution, a double root} \\ > 0 & \text{two real solutions} \end{cases}$$

2. Multiply both the numerator and denominator by the complex conjugate of the denominator.

3. Since a real number has no imaginary part, it would be graphed on the horizontal, or *real* axis. A pure imaginary number has no real part, so it would be graphed on the vertical, or *imaginary* axis.

Practice and Apply

80.

Imaginary
4
2i
2
Conjugate
Real
−4 −2 2 4
−2 −2i
−4

81.

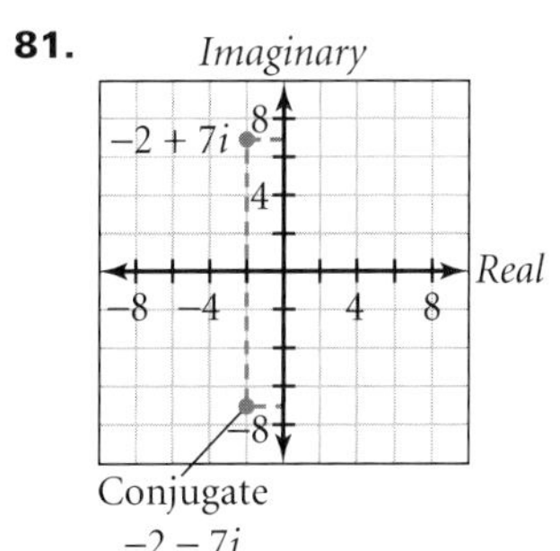

82.

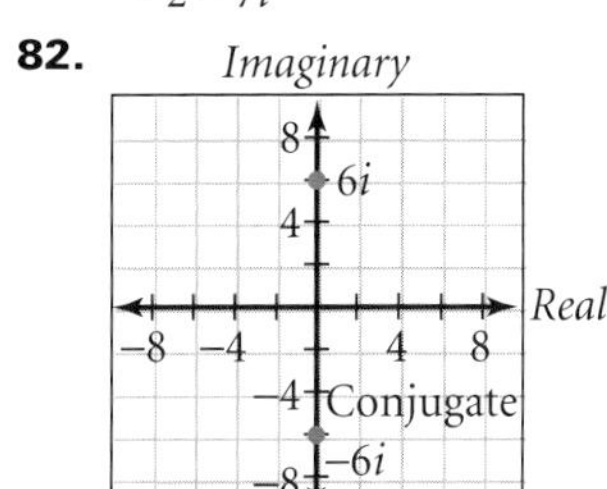

83.

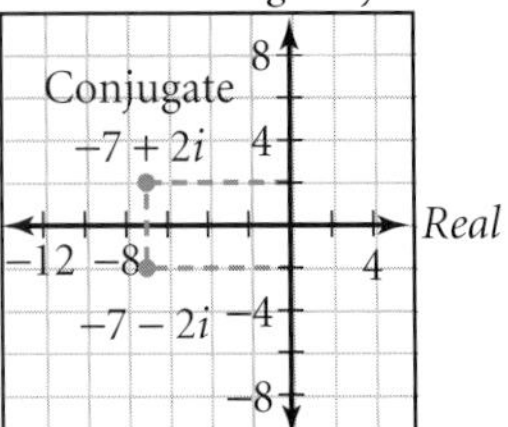

84.

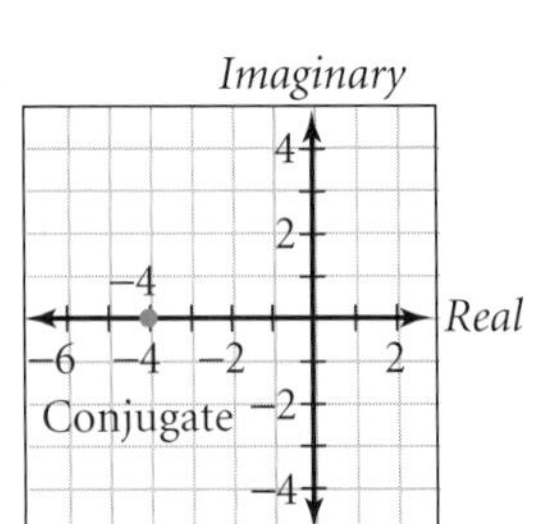

85.

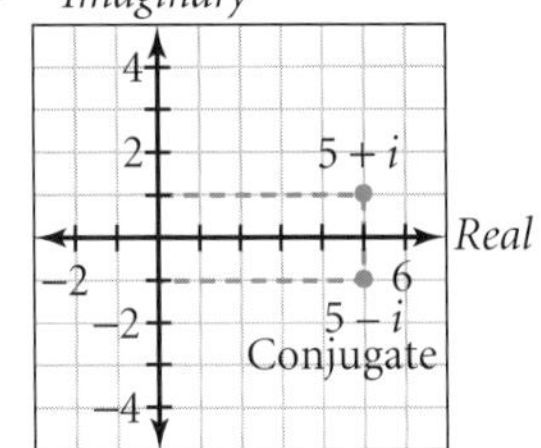

86.

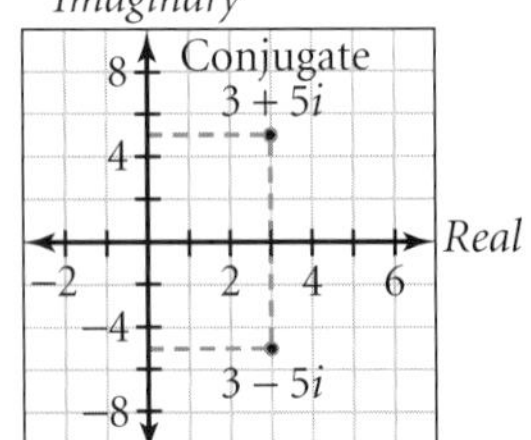

87.

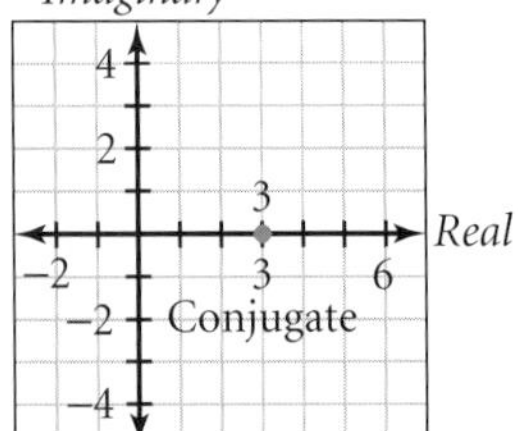

92.

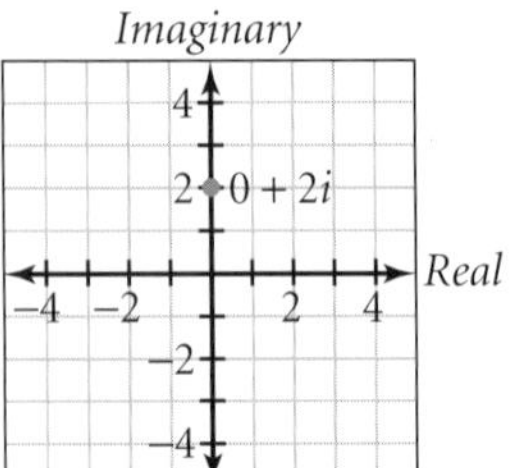

93.

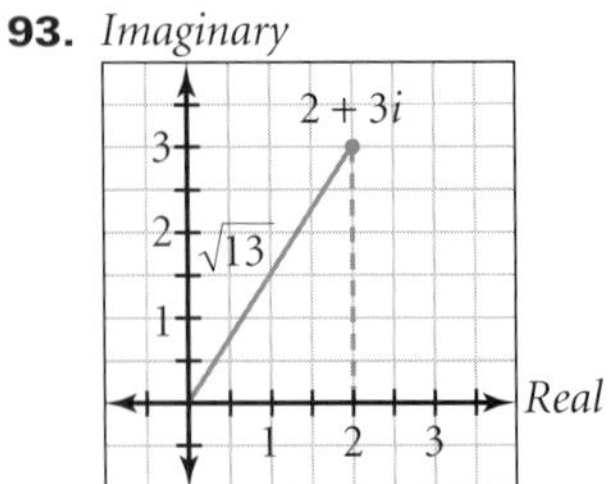

94.

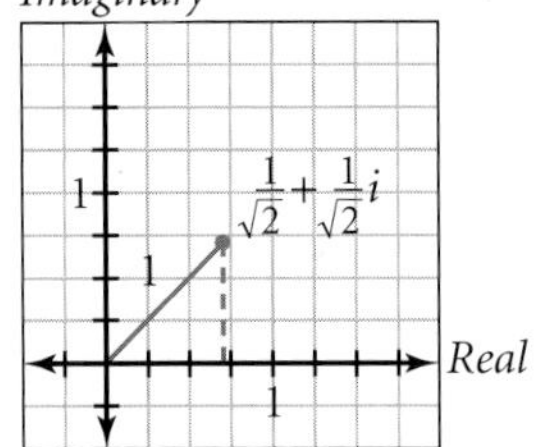

95.

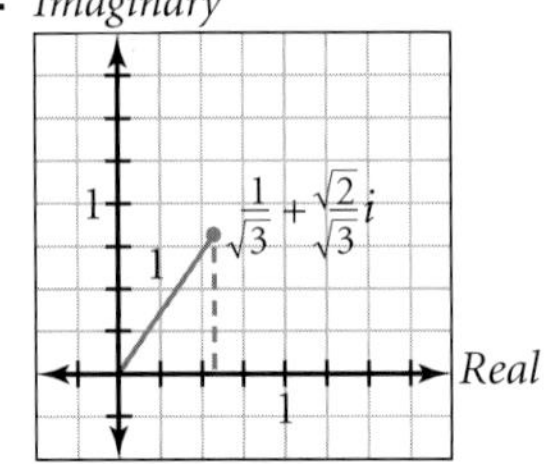

Look Back

101.

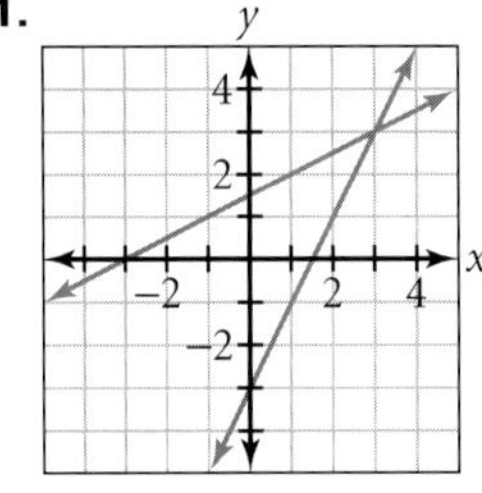

102.

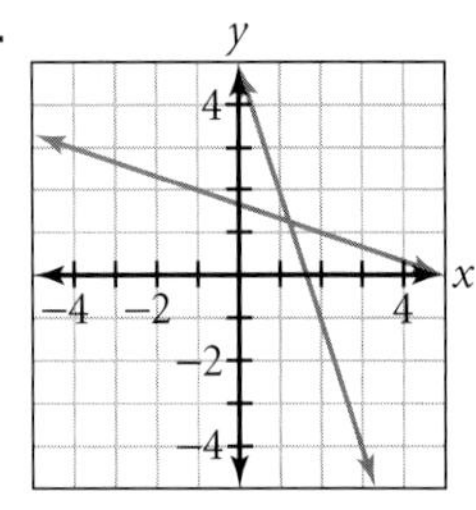

103.

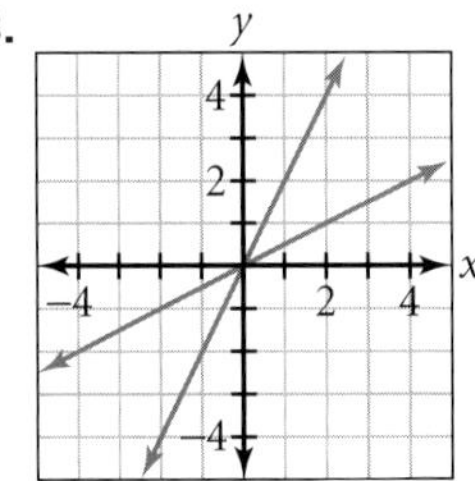

Lesson 5.7, pages 322–329

Activity

Answers may vary. Sample answers:

1.

Elapsed time (in seconds)	Altitude (in feet)
0.5	25
1.0	42
1.5	51
2.0	52
2.5	45
3.0	30
3.5	7

2. 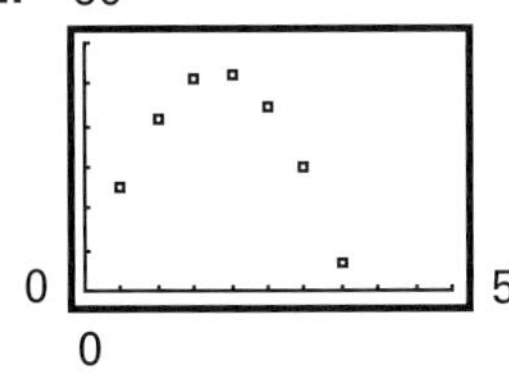

The data indicate a quadratic model.

CHECKPOINT ✔

3. Answers may vary. Sample answer: $h(t) = -16t^2 + 58t$

4. The value in this model agrees with the acceleration due to gravity.

Exercises

Communicate

1. a. Use the general equation for a parabola, $y = ax^2 + bx + c$, and substitute the x- and y-values from the given data, one point at a time, to obtain three equations in three unknowns, a, b, and c.

$$\begin{cases} a + b + c = 6 \\ 9a - 3b + c = 8 \\ 25a + 5b + c = -2 \end{cases}$$

b. Solve the system of three linear equations for a, b, and c, and then substitute these values into the general equation for a parabola to get $y = -0.1875x^2 - 0.875x + 7.0625$.

2. For evenly-spaced x-values, the first differences are the differences between successive y-values; i.e., are found by subtracting the preceding y-value. The second differences are the corresponding differences in first differences: i.e., are found by subtracting the preceding first difference. A constant first difference indicates a linear function. A constant second difference indicates a quadratic function.

Portfolio Activity

1. Answers will vary. Sample using points (0, 6), (0.5, 12), and (1.0, 9.98):
Set up the system of equations.

Point	Substitution	Equation
(0, 6)	$a(0)^2 + b(0) + c = 6$	$c = 6$
(0.5, 12)	$a(0.5)^2 + b(0.5) + c = 12$	$0.25a + 0.5b = 6$
(1.0, 9.98)	$a(1.0)^2 + b(1.0) + c = 9.98$	$a + b = 3.98$

The solution is $a = -16.04$, $b = 20.02$, and $c = 6$.
Write the quadratic function by using these values of a, b, and c.
$f(x) = ax^2 + bx + c$
$f(x) = -16.04x^2 + 20.02x + 6$

2. Answers will vary. Sample using points (0, 6), (0.5, 12), and (1.0, 9.98): Using a calculator, a quadratic regression model is $y = -16.1x^2 + 20.05x + 6$.

3. Answers will vary. Sample: The quadratic models from Steps 1 and 2 are not identical but are fairly close. These models are not the same as the model used in previous Portfolio Activities, but the coefficients are very close.

Lesson 5.8, pages 330–337

Activity

1. Both functions have 2 x-intercepts.

Function	$f(x) = 0$	$f(x) > 0$	$f(x) < 0$
$f(x) = x^2 - 4$	$x = -2$ or $x = 2$	$x < -2$ or $x > 2$	$-2 < x < 2$
$f(x) = -x^2 + 2x + 3$	$x = -1$ or $x = 3$	$-1 < x < 3$	$x < -1$ or $x > 3$

2.

Function	Number of x-intercepts	$f(x) = 0$	$f(x) > 0$	$f(x) < 0$
$f(x) = x^2$	1	$x = 0$	$x < 0$ or $x > 0$	no values
$f(x) = -x^2$	1	$x = 0$	no values	$x < 0$ or $x > 0$

3.

Function	Number of x-intercepts	$f(x) = 0$	$f(x) > 0$	$f(x) < 0$
$f(x) = x^2 + x - 1$	0	no values	no values	all x
$f(x) = x^2 + x + 3$	0	no values	all x	no values

CHECKPOINT ✔

4. a. 3 **b.** 2 **c.** 1

Exercises

Communicate

1. Solve the related equation $x^2 - 2x - 8 = 0$: $x = -2$ or $x = 4$. The roots divide the number line into three intervals and the solution will be of the form $x < -2$ and $x > 4$ or of the form $-2 < x < 4$. Use $x = 0$ as a test point; it lies in the second possible solution set. $0^2 - 2(0) - 8 \stackrel{?}{>} 0$ is false, so the first solution set is the correct one. The solution is $x < -2$ or $x > 4$.

2. Graph the related equation $y = x^2 - 2x - 8$ and locate the x-intercepts, $x = -2$ and $x = 4$. We are interested in finding those values of x for which $y > 0$. y is above the x-axis for $x < -2$ or $x > 4$.

3. First, graph the related equation $y = (x - 2)^2 + 2$. Use a point not on the graph, such as (0, 0), as a test point. $0 \stackrel{?}{<} (0 - 2)^2 + 2$ is true. So, shade the region that contains (0, 0). The region below the curve is the solution.

4. Select a point that lies within the shaded region. Use the point as a test point in the original inequality. If the result is a true statement, the shaded region is the solution; if the statement is false, the shaded region is not the solution.

5. The square of any real number is always greater than or equal to zero. So $(x + 7)^2 < 0$ will have no solution.

Guided Skills Practice

6. $x \le 3$ or $x \ge 4$

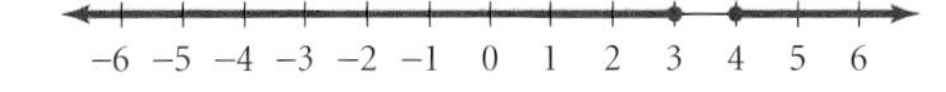

Practice and Apply

18. $-\frac{1}{2} \le x \le \frac{3}{2}$

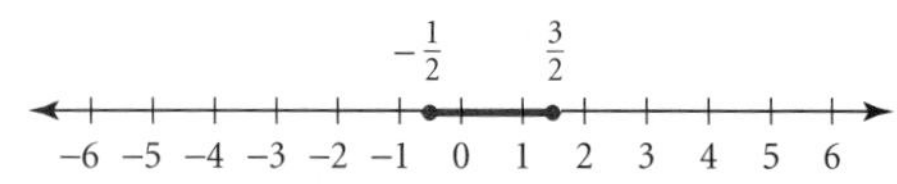

19. $-3 \le x \le 4$

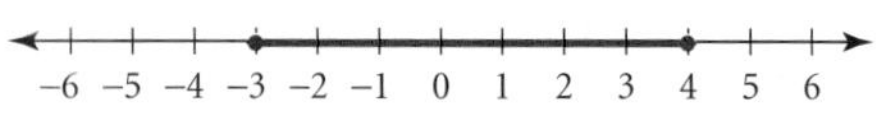

20. no solution

21. $x < -2$ *or* $x > 6$

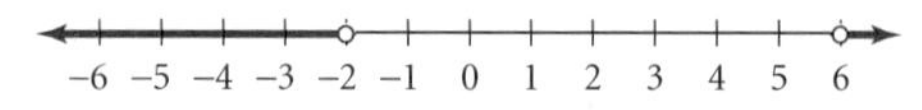

22. $x < -9$ *or* $x > 11$

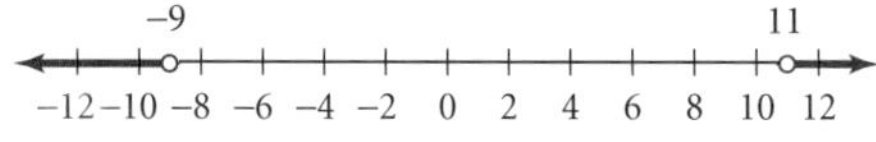

23. $-3 \le x \le 2$

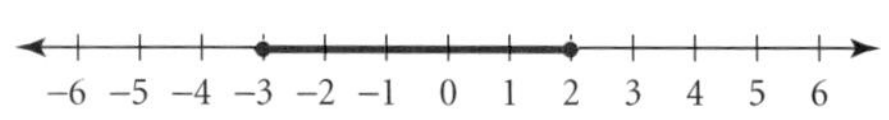

24. $x < -5$ *or* $x > 4$

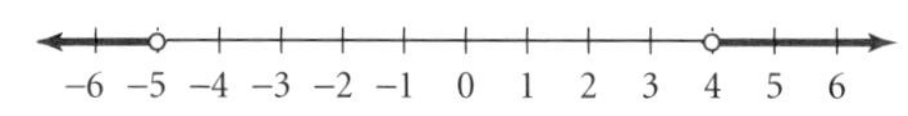

25. $1 \le x \le 6$

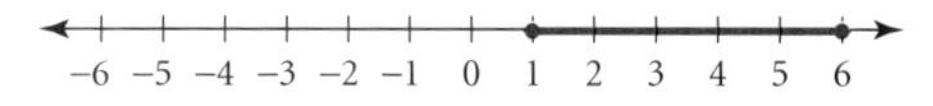

26. $x < -7$ *or* $x > -5$

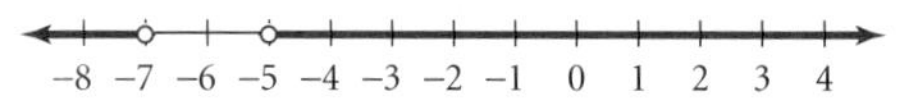

27. $-10 \le x \le 1$

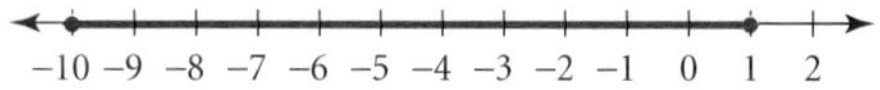

28. $x < -5$ *or* $x > -5$

29. $x < -6$ *or* $x > 3$

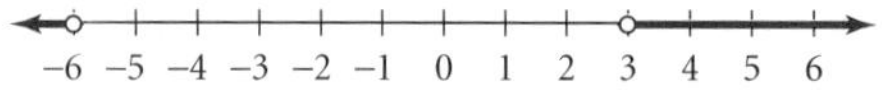

30. $x < -1$ *or* $x > 2$

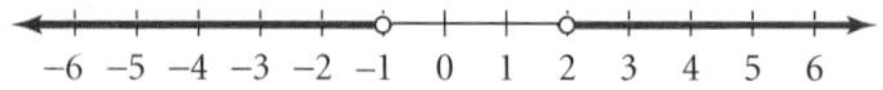

31. $x \le -7$ *or* $x \ge 1$

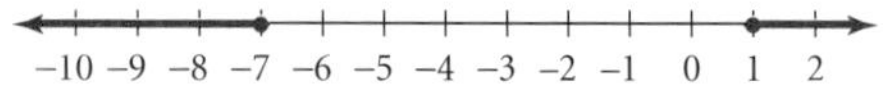

32. $3 \le x \le 5$

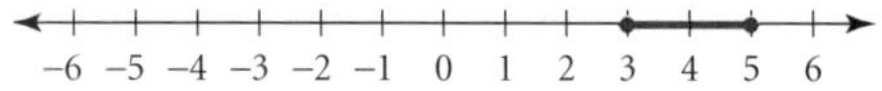

33. $x < \frac{3}{2} - \frac{\sqrt{33}}{2}$ *or* $x > \frac{3}{2} + \frac{\sqrt{33}}{2}$

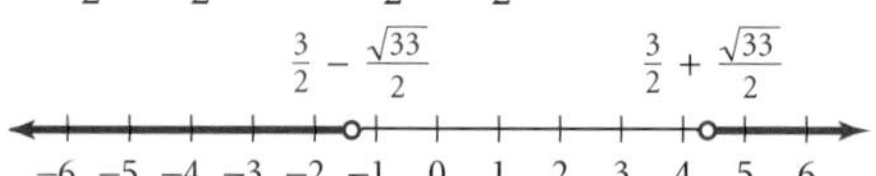

34. $x < -2 - \sqrt{13}$ *or* $x > -2 + \sqrt{13}$

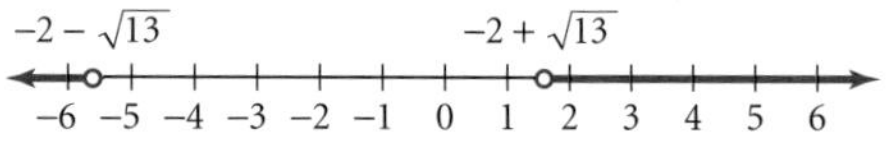

35. $-\frac{1}{2} - \frac{\sqrt{29}}{2} < x < -\frac{1}{2} + \frac{\sqrt{29}}{2}$

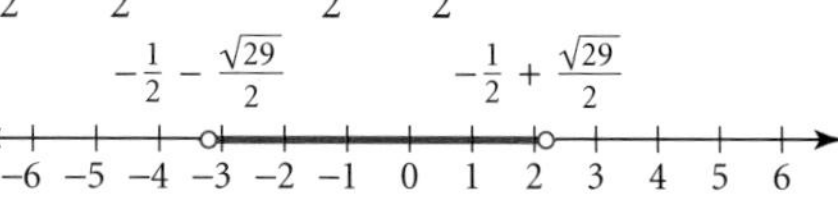

36.

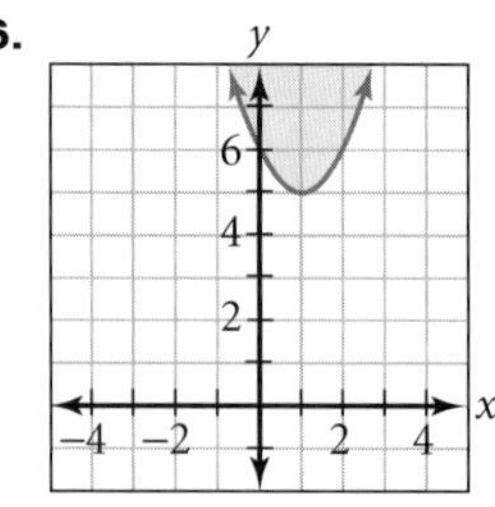

37.

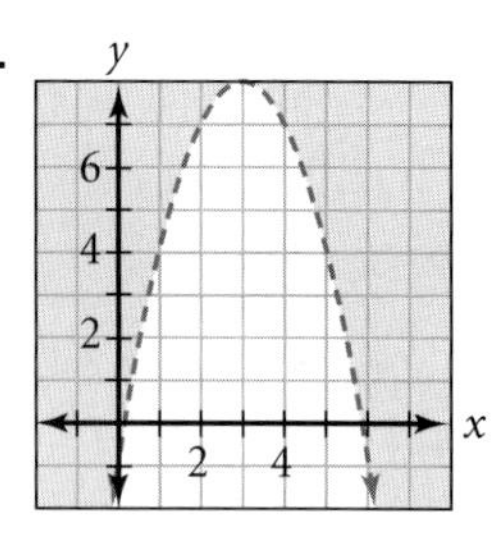

38.

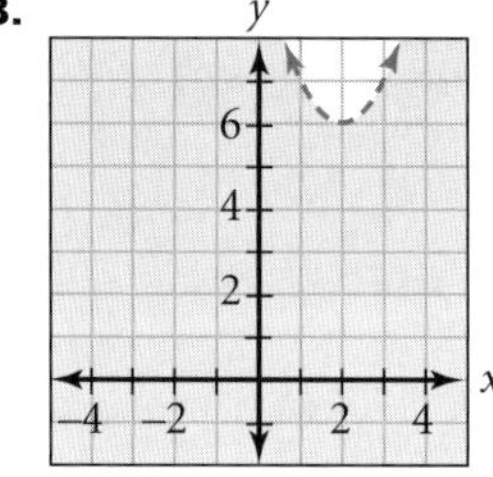

39.

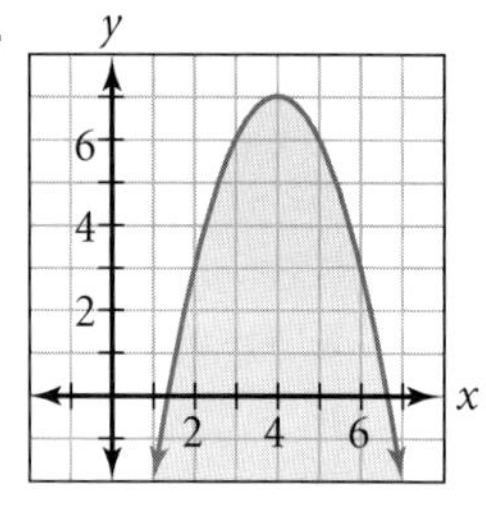

46.

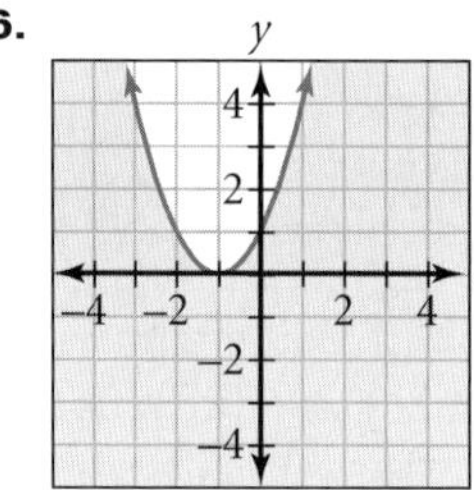

47.

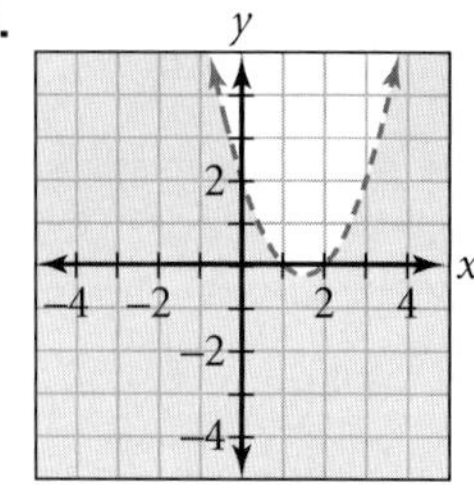

48.

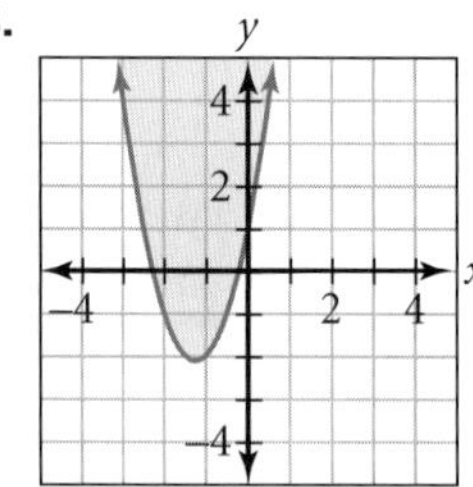

49.

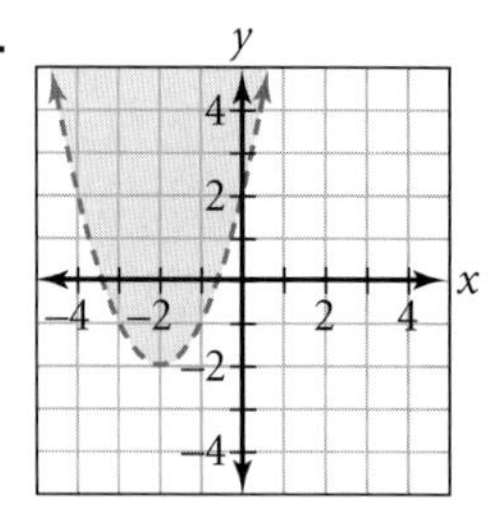

50.

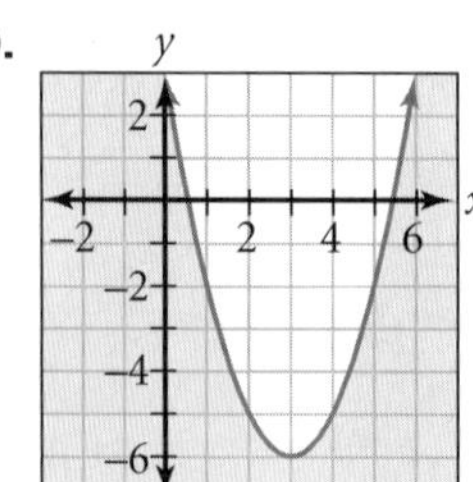

51.

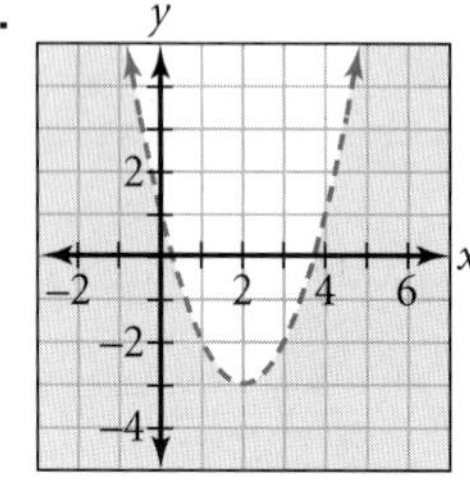

52.

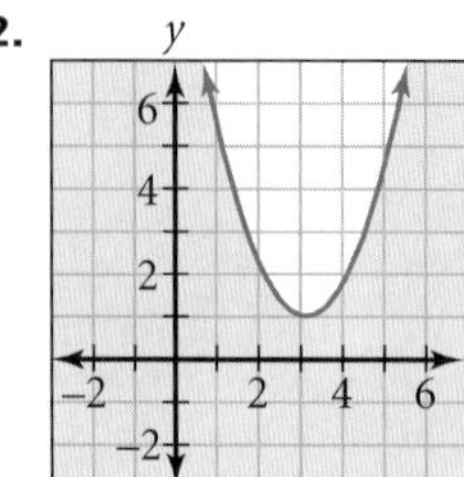

53.

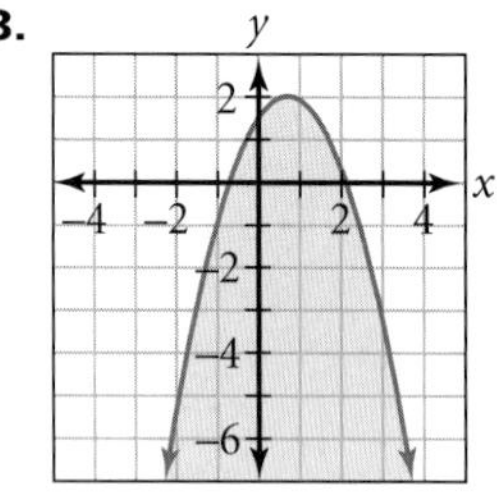

54.

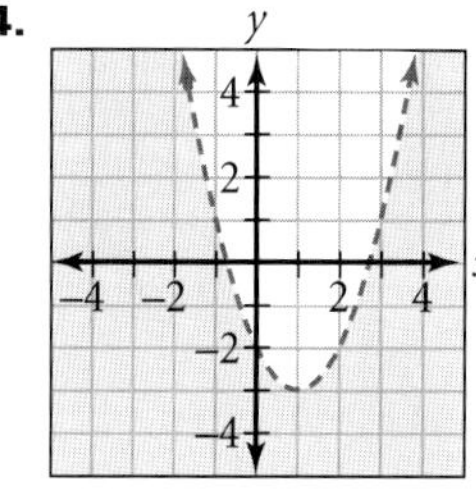

55.

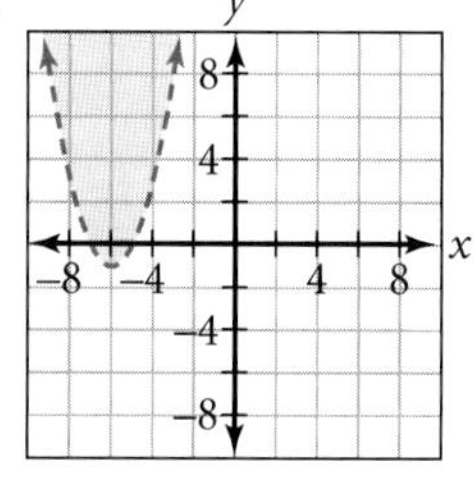

56.

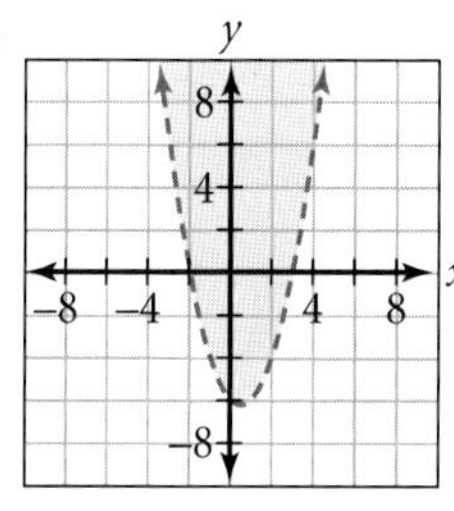

57.

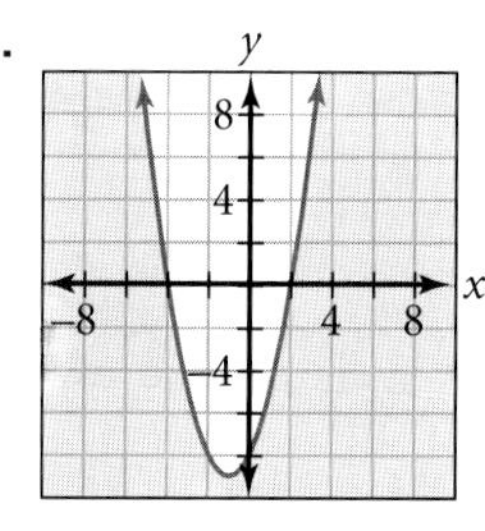

Portfolio Activity

1. To find the time that the basketball is at a height of 10 feet, solve the equation $-16t^2 + 20t + 6 = 10$, or $-16t^2 + 20t - 4 = 0$. Using the quadratic formula, $a = -16$, $b = 20$, and $c = -4$.

$$x = \frac{-b \pm \sqrt{b^2 - 4ac}}{2a}$$

$$x = \frac{-20 \pm \sqrt{20^2 - 4(-16)(-4)}}{2(-16)}$$

$$x = \frac{-20 \pm \sqrt{400 - 256}}{-32}$$

$$x = \frac{-20 \pm \sqrt{144}}{-32}$$

$$x = \frac{-20 \pm 12}{-32}$$

$$x = \frac{-20 + 12}{-32} \text{ or } x = \frac{-20 - 12}{-32}$$

$$x = \frac{-8}{-32} = 0.25 \text{ or } x = \frac{-32}{-32} = 1$$

The basketball is at a height of 10 feet at 0.25 and 1.00 seconds.

2. The basketball is above 10 feet for times greater than 0.25 seconds but less than 1.00 seconds, or $0.25 < t < 1.00$ (t is the time in seconds).

3. The basketball is below 10 feet for times greater than 0 seconds but less than 0.25 seconds and for times greater than 1.00 seconds but less than 1.5 seconds, when it would hit the ground, or $0 < t < 0.25$ and $1.00 < t < 1.5$ (t is the time in seconds).

4.

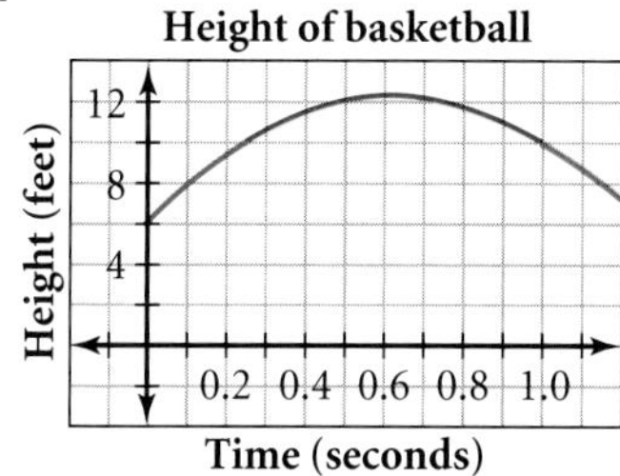

The red area on the graph shows where the basketball is above 10 feet. The blue area on the graph shows where the basketball is below 10 feet.

Lesson 6.1, pages 354–361

Activity

1.

Time (hr)	0	1	2	3	4	5	6
Population	25	50	100	200	400	800	1600

2. $p(n) = 25 \cdot 2^n$

3. 25,600 bacteria; 26,214,400 bacteria

CHECKPOINT ✔

4. 76,800 bacteria; 78,643,200 bacteria

Exercises

Communicate

1. The values of n cannot be negative in either expression because n represents time (hours of bacterial growth or decades of U.S. population growth).

2. The multiplier is found by adding 1 to the growth rate of 8%, or $1 + 0.8 = 1.08$.

3. You assume that the growth rate is constant and that it is proportional to the number present.

4. To find the multiplier for a growth rate of 5%, add 1 to 0.05, so the multiplier for a growth rate of 5% is 1.05. To find the multiplier for a decay rate of 5%, subtract 0.05 from 1, so the multiplier for a decay rate of 5% is 0.95.

Portfolio Activity

1. Answers will vary. Sample data:
Initial room temperature was 24.8.

Elapsed time (seconds)	Temperature (°C)
0	24.8
2	23.9
4	19.1
6	17.0
8	15.2
10	12.0
12	10.8
14	9.02
16	8.42
18	7.34
20	6.72
22	6.25
24	5.77
26	5.44
28	5.29
30	5.13

Elapsed time (seconds)	Temperature (°C)
32	5.13
34	4.97
36	4.97
38	4.97
40	4.81
42	4.81
44	4.61
46	4.81
48	4.61
50	4.81
52	4.45
54	4.32
56	4.15
58	3.96
60	3.8
120	0.5

2. Answers will vary.

a. Sample: A linear regression equation for the data is $y = -0.258t + 15.855$.

b. Using the linear regression model, the temperature after 120 seconds would be $y = -0.258(120) + 15.855 \approx -15.13$°C. (This result will be quite different from the students' actual reading.)

c. A linear model does not model the process well. The linear graph decreases steadily, while the data plot levels off. Logically, the probe temperature cannot fall below the temperature of the water (about 0°C) in the real-life situation.

Lesson 6.2, pages 362–369

Activity

1. The graph is shown in the text on page 363.

2. $x = 0$; $x > 0$; $x < 0$

3.

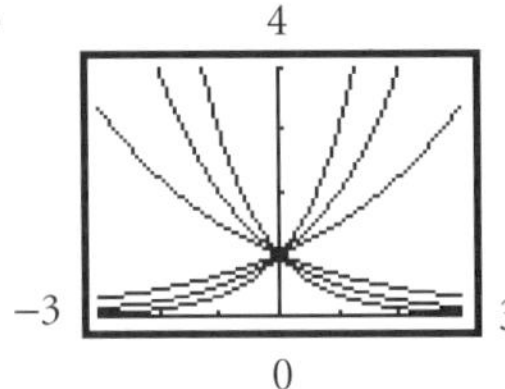

4. reflections of one another across the y-axis; reciprocals of one another

CHECKPOINT ✓

5. $b > 1$; $0 < b < 1$

Exercises

Communicate

1. $0 < b < 1$

2. The domains of both equations are all real numbers, and the ranges of both equations are all real numbers greater than 0.

3. The y-intercept is 2 or a.

4. Both functions have a base and an exponent. However, $y = x^2$ is a quadratic function with a variable base, and $y = 2^x$ is an exponential function with a variable exponent.

Portfolio Activity

1. Answers will vary. Sample:

a. The quadratic regression function for the sample data is $y = 0.0105t^2 - 0.8896t + 21.9589$, where t represents time in seconds.

b. Using this function, the predicted temperature after 120 seconds would be $y = 0.0105(120)^2 - 0.8896(120) + 21.9589 \approx 66.7$°C.

c. A quadratic function does not model the cooling process; the data points descend and then level off, whereas the quadratic curve descends and then rises without bound.

2. Answers will vary. Sample:

a. An exponential regression function for the data is $y = 15.398 \cdot (0.973)^t$.

b. Using this function, the predicted temperature after 120 seconds would be $y = 15.398 \cdot (0.973)^{120} \approx 0.59$°C.

c. An exponential model seems to fit the data fairly well, except for the first few data points. The temperature continues to drop and approaches but does not go below 0°C.

Lesson 6.3, pages 370–376

Activity

1. The y-values are 10 raised to the power of the x-values.

CHECKPOINT ✓

2. a. $x = 3$ **b.** $x = -2$ **c.** $x = -3$ **d.** $x = 0$

3. $x \approx 0.85$

CHECKPOINT ✓

4. $x \approx 1.93$

Exercises

Communicate

1. For any base b, where $b > 0$ and $b \neq 1$, the exponential and logarithmic functions, $f(x) = b^x$ and $g(x) = \log_b x$, are inverse functions.

2. The domain is all positive real numbers. The range is all real numbers. The domain of logarithmic functions is the range of exponential functions. The range of logarithmic functions is the domain of exponential functions.

3. Enter the function $y = 2^x$. Knowing that $2^5 = 32$ and $2^6 = 64$, try values of x between 5 and 6. Continue to narrow the search until the desired number of decimal places is reached.

86. Tables may vary. Sample tables provided.

x	$f(x) = 3^x$
−3	$\frac{1}{27}$
−2	$\frac{1}{9}$
−1	$\frac{1}{3}$
0	1
1	3
2	9
3	27

x	$f^{-1}(x) = \log_3 x$
$\frac{1}{27}$	−3
$\frac{1}{9}$	−2
$\frac{1}{3}$	−1
1	0
3	1
9	2
27	3

87. Tables may vary. Sample tables provided.

x	$f(x) = 3^{-x}$
−3	27
−2	9
−1	3
0	1
1	$\frac{1}{3}$
2	$\frac{1}{9}$
3	$\frac{1}{27}$

x	$f^{-1}(x) = \log_{\frac{1}{3}} x$
27	−3
9	−2
3	−1
1	0
$\frac{1}{3}$	1
$\frac{1}{9}$	2
$\frac{1}{27}$	3

Lesson 6.4, pages 377–384

Activity

1. a. 3; 3 **b.** 4; 4 **c.** 5; 5 **d.** 6; 6

CHECKPOINT ✔

2. $\log_2(a \cdot b) = \log_2 a + \log_2 b$

3. a. 3; 3 **b.** 1; 1 **c.** 2; 2 **d.** 1; 1

CHECKPOINT ✔

4. $\log_2 \frac{a}{b} = \log_2 a - \log_2 b$

Exercises

Communicate

1. Rewrite $\log_{10} 0.005$ as

$$\log_{10}(5 \cdot 0.001) = \log_{10} 5 + \log_{10} 0.001 \approx 0.6990 + (-3) \approx -2.3010$$

Rewrite $\log_{10} 500$ as

$$\log_{10}(5 \cdot 100) = \log_{10} 5 + \log_{10} 100 \approx 0.6990 + 2 \approx 2.6990$$

2. Since the base is the same, $\log_7 32 - \log_7 4 = \log_7 \frac{32}{4} = \log_7 8$

3. Using the Exponential Inverse Property, $4^{\log_4 8} = 8$. Using the Logarithmic Inverse Property, $\log_2 2^7 = 7$.

4. You may get an x-value that leads to finding the logarithm of a negative number, which is undefined.

Portfolio Activity

1. Answers will vary. Sample data using an initial room temperature of 24.8°C:

Elapsed time (seconds)	Temperature (°C)
0	60
2	59.7
4	57.4
6	55
8	53
10	51.1
12	49.3
14	47.5
16	45.1
18	43.6
20	42.3
22	41.1
24	40
26	39
28	38
30	37.2

Elapsed time (seconds)	Temperature (°C)
32	36.3
34	35.5
36	34.7
38	34
40	33.2
42	32.5
44	31.9
46	31.2
48	30.4
50	29.7
52	28.7
54	28.3
56	27.6
58	26.4
60	25.6
120	24.9

2. Answers will vary. Sample:

a. The linear regression function for the data is $y = -0.550t + 56.034$. The quadratic regression function for the data is $y = 0.007t^2 - 0.984t + 60.227$. The exponential regression function for the data is $y = 57.936 \cdot (0.986)^t$.

b. Using the linear regression model, the predicted temperature after 120 seconds would be $y = -0.550(120) + 56.034 \approx -9.99$°C (much colder than the actual temperature). Using the quadratic regression model, the predicted temperature after 120 seconds would be $y = 0.007(120)^2 - 0.984(120) + 60.227 \approx 46.3$°C (much warmer than the actual temperature). Using the exponential regression model, the predicted temperature after 120 seconds would be $y = 57.936 \cdot (0.986)^{120} \approx 11.07$°C (somewhat colder than the actual temperature).

c. Although a linear model fits the 60-second data quite well, it projects temperatures that continue to decrease steadily rather than level off at air temperature. A quadratic model fits the 60-second data well, but it projects temperatures that rise steadily after about 70 seconds, whereas the actual data levels off. Of the three models, the exponential model is the most representative of the cooling process. However, the y-values given by the model should never drop below the air temperature (24.8°C with this data); the y-values of the exponential regression equation descended below (and remained below) this level after about 61 seconds.

3. Answers will vary. Sample data modifying the data from Step 1 and using an initial room temperature of 24.8°C:

a.

Elapsed time (seconds)	Temperature – 24.8°C
0	35.2
2	34.9
4	32.6
6	30.2
8	28.2
10	26.3
12	24.5
14	22.7
16	20.3
18	18.8
20	17.5
22	16.3
24	15.2
26	14.2
28	13.2
30	12.4

Elapsed time (seconds)	Temperature – 24.8°C
32	11.5
34	10.7
36	9.9
38	9.2
40	8.4
42	7.7
44	7.1
46	6.4
48	5.6
50	4.9
52	3.9
54	3.5
56	2.8
58	1.6
60	0.8
120	0.1

b. The exponential regression function for the above data would be $y = 45.8387 \cdot (0.9527)^t$.

c. After adding the sample room temperature of 24.8°C, the resulting approximating function is $y = 45.8387 \cdot (0.9527)^t + 24.8$.

d. Answers will vary. Using the approximating function from part **c**, the predicted temperature after 120 seconds would be $y = 45.839 \cdot (0.953)^{120} + 24.8 \approx 24.9$°C. The graph of this model has the same shape and appears to level off at about the same rate as the plot of the original data. It also predicted the 2-minute reading accurately.

Lesson 6.5, pages 385–391

Activity

1. $\log_3 81 = x$ **2.** $x = \frac{\log 81}{\log 3}$ **3.** $\log_3 81 = \frac{\log 81}{\log 3}$

CHECKPOINT ✓

4. $\log_b y = x$; $b^x = y$; $\log b^x = \log y$; $x \log b = \log y$; $x = \frac{\log y}{\log b}$; $\log_b y = \frac{\log y}{\log b}$

Exercises

Communicate

1. The ratio of sound intensities, $\frac{I}{I_0}$, ranges from 1 to 10^{12}. Such a large range is easier to represent on a logarithmic scale.

2. Take the common log of each side of the equation. Apply the Power Property of Logarithms to move x outside the logarithm. Solve for x in terms of common logarithms, which can be computed on a calculator.
$\log 6^x = \log 39$; $x \log 6 = \log 39$; $x = \frac{\log 39}{\log 6} \approx 2.04$

3. Convert to common logarithms and use the LOG key:
$\log_4 29 = \frac{\log 29}{\log 4} \approx 2.43$

Lesson 6.6, pages 392–399

Activity

1.

Compounding schedule	n	$1\left(1+\frac{1}{n}\right)^n$	Value, A
annually	1	$1\left(1+\frac{1}{1}\right)^1$	2.00000
semiannually	2	$1\left(1+\frac{1}{2}\right)^2$	2.25000
quarterly	4	$1\left(1+\frac{1}{4}\right)^4$	2.44140
monthly	12	$1\left(1+\frac{1}{12}\right)^{12}$	2.61303
daily	365	$1\left(1+\frac{1}{365}\right)^{365}$	2.71456
hourly	8760	$1\left(1+\frac{1}{8760}\right)^{8760}$	2.71812
every minute	525,600	$1\left(1+\frac{1}{525{,}600}\right)^{525{,}600}$	2.71827
every second	31,536,000	$1\left(1+\frac{1}{31{,}536{,}000}\right)^{31{,}536{,}000}$	2.71828

CHECKPOINT ✓

2. The A-values become larger as n becomes larger, but the rate of increase becomes smaller as the values approach 2.71828.

Exercises

Communicate

1. Answers may vary. Similarities: The domain and range of both exponential functions and of both logarithmic functions are the same. Differences: The graph of the exponential base-10 function rises faster than the graph of the natural exponential function. The graph of the natural logarithmic function rises faster than the graph of the common logarithmic function.

2. Answers may vary. Sample answer: Continuous compounding of interest is an example of exponential growth with e as a base. Radioactive decay is an example of exponential decay with e as a base.

3. $A = Pe^{rt}$, where A = amount at time t, P = original investment, r = interest rate, and t = time that has passed

4. If the exponent of e is positive, then the formula represents continuous growth. If the exponent of e is negative, then the formula represents continuous decay.

Practice and Apply

71. The only changes are the following:
a. no changes **b.** x-intercept: changes
c. domain: real numbers greater than the value of the translation; asymptotes: x is the value of the translation; x-intercept: value of the translation plus 1
d. x-intercept: changes

Look Back

87.

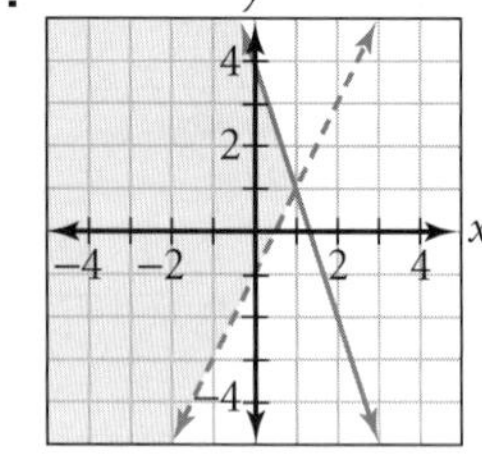

88.

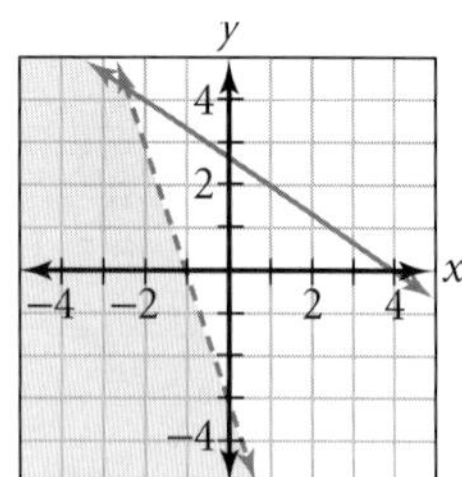

89.

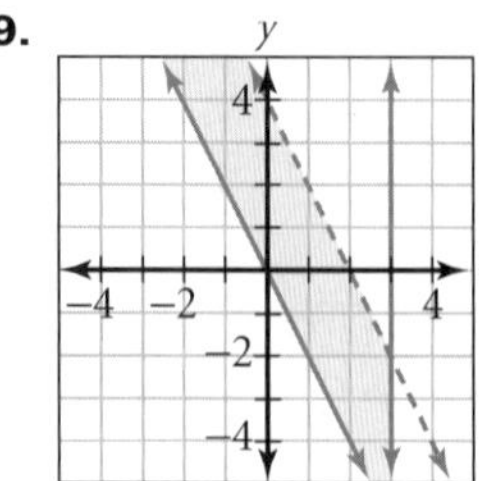

Lesson 6.7, pages 402–409

Activity

1.

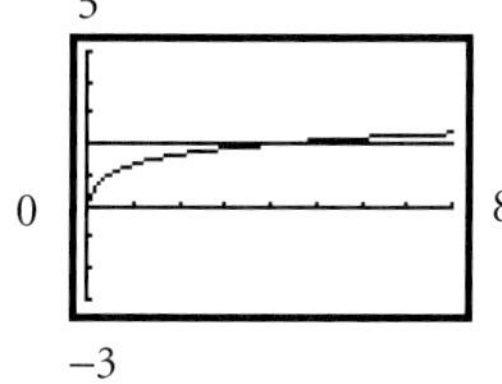

2. $x = 4$; $x < 4$; $x > 4$

CHECKPOINT ✔

3. Find the point of intersection of $y_1 = \log x + \log(x + 21)$ and $y_2 = 2$. The solution to $\log x + \log(x + 21) > 2$ is all values of x such that y_1 is above y_2.

4.

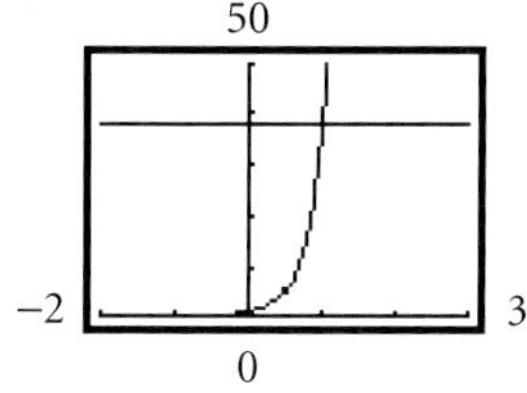

5. $x \approx 0.99$; $x < 0.99$, $x > 0.99$

CHECKPOINT ✔

6. Find the point of intersection of $y_1 = 2e^{4x-1}$ and $y_2 = 38$. The solution to $2e^{4x-1} < 38$ is all values of x such that y_1 is below y_2.

Exercises

Communicate

1. Take the natural logarithm of both sides; evaluate ln 98 and then solve the equation for x.

2.

$\log_2 x + \log_2(x + 3) = 2$	
$\log_2[x(x + 3)] = 2$	Product Property
$x(x + 3) = 2^2$	Raise both sides to a power of 2.
$x^2 + 3x = 4$	Simplify.
$x^2 + 3x - 4 = 0$	Rearrange.
$(x + 4)(x - 1) = 0$	Factor.
$x = -4$ *or* $x = 1$	
x cannot be negative.	Definition of logarithm
$x = 1$	

3. To solve an exponential equation by graphing, put the equation in the form of an exponential expression equal to a constant. Then graph the expression, and trace or solve to find the value of the variable when the expression equals the constant. To solve a logarithmic equation by graphing, put the equation in the form of a logarithmic expression equal to a constant. Then graph the expression, and trace or solve to find the value of the variable when the expression equals the constant.

Portfolio Activity

1. Answers will vary. Sample uses initial air temperature of 24.8°C.

a. $T(t) = 24.8 \cdot (0.969)^t$

b. Answers will vary. The function is $T(t) = 24.8 \cdot (0.969)^t$, where t is time in seconds. The exponential regression function from the Portfolio Activity on page 369 was $y = 15.398 \cdot (0.973)^t$. The difference in the initial y-values of these two functions is about 9°C, but at 90 seconds, there is less than 1°C difference. At 120 seconds, there is less than 0.005°C difference. In comparison with the data values, the Newton function described the data points from 0–8 seconds and 46+ seconds very closely. Between those values, the exponential regression function was closer. Both functions are very good predictors of future temperatures.

2. Answers will vary. Sample used initial room temperature of 24.8°C.

a. $T(t) = 24.8 + 35.2 \cdot (0.939)^t$

b. The function is $T(t) = 24.8 + 35.2 \cdot (0.939)^t$. The approximating function from the Portfolio Activity on page 384 was $y = 24.8 + 45.839 \cdot (0.953)^t$. The approximating function and the function based upon Newton's law produced quite similar answers for the cooling process.

Chapter 6 Project

Sample data using an initial room temperature of 20.7°C:

Elapsed time (seconds)	Temperature (°C)	Elapsed time (seconds)	Temperature (°C)
0	0	32	8.5
2	2.9	34	8.8
4	3.6	36	9.1
6	4.1	38	9.4
8	4.4	40	9.7
10	4.9	42	10.0
12	5.2	44	10.3
14	5.6	46	10.8
16	5.9	48	10.9
18	6.4	50	11.2
20	6.7	52	11.3
22	7.0	54	11.5
24	7.1	56	11.6
26	7.4	58	11.8
28	7.7	60	11.9
30	8.1	120	16.9

Activity 3

1. Answers will vary. Sample:s

Elapsed time (seconds)	Transformed temperatures	Elapsed time (seconds)	Transformed temperatures
0	20.7	32	12.2
2	17.8	34	11.9
4	17.1	36	11.6
6	16.6	38	11.3
8	16.3	40	11.0
10	15.8	42	10.7
12	15.5	44	10.4
14	15.1	46	9.9
16	14.8	48	9.8
18	14.3	50	9.5
20	14.0	52	9.4
22	13.7	54	9.2
24	13.6	56	9.1
26	13.3	58	8.9
28	13.0	60	8.8
30	12.6		

2. Answers will vary. Sample: The exponential regression equation for the data is $y = 18.031 \cdot (0.988)^t$ (t represents the time in seconds). This regression function will not allow the temperature to drop below 0°C, but the temperature will still drop below room temperature.

3. Answers will vary. Sample:
The approximating function is
$y = 20.7 - 18.031 \cdot (0.988)^x$.

4. Answers will vary. Sample: Using the approximating function, the temperature after 120 seconds would be $y = 20.7 - 18.031 \cdot (0.988)^{120} \approx 16.6$. This value is very close to the actual reading, and the temperature will never rise above the air temperature with this model. It seems to represent the process quite well.

Activity 4

1. Answers will vary. Sample uses initial room temperature of 20.7°C. The function is $T(t) = 20.7 - 20.7(0.986)^t$ (t is time in seconds). The temperature after 120 seconds should be 17.0°C.

2. Answers will vary. Sample: The approximating function found in Activity 3 should be more accurate.

Lesson 7.1, pages 424–431

Activity

1. 2, 1 **2.** 3, 2 **3.** 3, 2 **4.** 4, 3 **5.** 4, 3

6. The number of U-turns is one less than the degree of the function.

7. 3, 0 **8.** 3, 0 **9.** 4, 1

CHECKPOINT ✔

10. The number of U-turns in a graph is no greater than one less than the degree of the function.

Exercises

Communicate

1. A polynomial expression is the sum of monomials.

2. The degree of a polynomial function is the greatest exponent of x or the exponent of the first term when the polynomial is written in standard form.

3. A quadratic function can be expressed in the form $a_2x^2 + a_1x + a_0$, where $a_2 \neq 0$. Therefore, a quadratic function is a polynomial function.

Portfolio Activity

1. Answers will vary. Sample: The diameter of the circular base is 6.5 cm. The radius is $6.5 \div 2 = 3.25$ cm. The area of the circular base is $A = \pi r^2 = \pi(3.25)^2 \approx 33.2$ cm^2.

2. Answers will vary. Sample: The height is 6.0 cm. The approximate volume of this part of the bottle is $W(r) = \pi r^2 h_1 = \pi(3.25)^2(6.0) \approx 199.1$ cm^3.

3. Answers will vary. Sample: The height is 6.5 cm. The approximate volume of the air space is $A(r) = \pi r^2 h_2 = \pi(3.25)^2(6.5) \approx 215.7$ cm^3.

4. Answers will vary. Sample: The total volume of the bottle is $V(r) = W(r) + A(r) = \pi r^2 h_1 + \pi r^2 h_2 \approx 199.1 + 215.7 = 414.8$ cm^3.

Lesson 7.2, pages 432–439

Activity

1. a. even **b.** positive **c.** rise, rise
2. a. even **b.** positive **c.** rise, rise
3. a. even **b.** positive **c.** rise, rise
4. a. even **b.** positive **c.** rise, rise
5. a. odd **b.** positive **c.** fall, rise
6. a. odd **b.** positive **c.** fall, rise
7. a. odd **b.** positive **c.** fall, rise
8. a. odd **b.** positive **c.** fall, rise
9. a. even **b.** negative **c.** fall, fall
10. a. even **b.** negative **c.** fall, fall
11. a. even **b.** negative **c.** fall, fall
12. a. even **b.** negative **c.** fall, fall
13. a. odd **b.** negative **c.** rise, fall
14. a. odd **b.** negative **c.** rise, fall
15. a. odd **b.** negative **c.** rise, fall
16. a. odd **b.** negative **c.** rise, fall

CHECKPOINT ✔

17. **a.** rises on left, rises on right
b. falls on left, falls on right
c. falls on left, rises on right
d. rises on left, falls on right

Exercises

Communicate

1. continuous graph, no turning points, falls on left, rises on right
2. Answers may vary. Sample answer: When a function rises and then falls over an interval from left to right, the peak of the curve is called a local maximum. When a function falls and then rises over an interval from left to right, the bottom of the curve is called a local minimum.
3. These answers may be in any order.
   (1) rises on left, rises on right
   (2) rises on left, falls on right
   (3) falls on left, falls on the right
   (4) falls on left, rises on right
4. Answers may vary. Sample answer: When the graph of a function rises from left to right over an interval, it is increasing. When a function falls from left to right over an interval, it is decreasing.

Practice and Apply

37. The end behavior of the graph should be rising on the left and the right. The graph shown falls to the left. A more appropriate window may be [−20, 10] by [−3000, 1000].

38. $P(x) = \frac{5}{6}x^3 - \frac{7}{2}x^2 + \frac{11}{3}x$

39. $f(x) \approx 1.17x^4 - 14.33x^3 + 60.83x^2 - 102.67x + 56$

40. $f(x) \approx 0.083x^4 - 0.167x^3 - 3.583x^2 + 14.667x - 11$

41. $f(x) \approx 0.12x^4 - 2.89x^3 + 23.51x^2 - 76.96x + 84$

42. $f(x) \approx -0.09x^4 + 0.52x^3 + 0.09x^2 - 1.52x + 1$

43. $f(x) \approx 0.176x^4 - 8.677x^3 + 156.139x^2 - 1199.023x + 3662.963$

Portfolio Activity

1. For $f(x) = x^2 - 2x + 1$:
a. The graph of this quadratic function has one U-turn. It opens upward and stays above the x-axis. It has one local minimum and no local maxima.
b. The function has one zero, which is also the local minimum, at (1, 0).

For $g(x) = x^3 - 3x + 2$:
a. The graph of this cubic function has an S-shape with two U-turns. It has one local minimum and one local maximum.
b. The function has one zero, which is also the local minimum, at approximately (0.96, 1).

For $h(x) = x^4 - 4x + 3$:
a. The graph of this quartic function has one U-turn. It opens upward and stays above the x-axis. It has one local minimum and no local maxima.
b. The function has one zero, which is also the local minimum, at (1, 0).

For $j(x) = x^5 - 5x + 4$:
a. The graph of this quintic function has an S-shape with two U-turns. It has one local minimum and one local maximum.
b. The function has one zero, which is also the local minimum, at approximately (0.96, 0).

2. Answers will vary. Sample: f and h have the same local minimum, and each has one U-turn. They both open upward. g and j have approximately the same local minimum, which is also a zero, at about (0.96, 0). Each has an S-shape and two U-turns. All four functions have only one zero each.

3. The function is $V(r) = W(r) + A(r) = \pi r^2 h_1 + \pi r^2 h_2 = \pi r^2(r-1) + \pi r^2(r-2) = \pi r^3 - \pi r^2 + \pi r^3 - 2\pi r^2 = 2\pi r^3 - 3\pi r^2$.

4. Graph the function as $f(x) = 1 - 2\pi x^3 - 3\pi x^2$, where x is the radius. The graph has an S-shape with two U-turns. The local maximum occurs when $x = 0$, which does not make sense as a measurement for the radius of a bottle. The local minimum is about $x = 1$ and produces a negative y-value. The volume cannot be negative, so the function applies only to real-world situations in which the y-values are positive. This occurs when the function crosses the x-axis, at about 1.4. If the height of the water is to be 1 unit less than the radius, then the radius must be greater than 1 in order for a volume to exist.

Lesson 7.3, pages 440–447

Exercises

Communicate

1. The Factor Theorem can be used as a shortcut to test possible linear factors of a polynomial. For example, $x + 1$ is a factor of $x^3 - 2x^2 - 8x - 5$ if and only if −1 is a solution of $x^3 - 2x^2 - 8x - 5 = 0$, that is, if and only if $(-1)^3 - 2(-1)^2 - 8(-1) - 5 = 0$.
2. Synthetic division can be used to divide a polynomial only by a linear binomial of the form $x - r$. When dividing by nonlinear divisors, long division must be used.
3. If you divide $P(x)$ by $x - 5$, then the remainder is the value of $P(5)$.

Lesson 7.4, pages 448–455

Activity

1.

x	y
−3	84
−2	0
−1	−8
0	6
1	12
2	4
3	0

2. $x = -2$ and $x = 3$
3. There is another zero near −0.41.
4. There is a fourth zero at approximately $x = 2.4$.

CHECKPOINT ✔

5. A table can be used to locate whole-number zeros and sign changes. After the sign changes are identified, a graph can approximate the zeros more closely. The zeros of the function are the real roots of the related equation.

Exercises

Communicate

1. For the factor $x - r$, the graph touches the x-axis at $(r, 0)$ but does not cross.
2. The Location Principle states that a zero is located between two x-values whose corresponding y-values have different signs.
3. All of these terms refer to the x-value of the point where the graph of the equation (or related function) crosses the x-axis. Thus, when r is the root or solution of the equation $P(x) = 0$, r is the x-intercept, r is the zero of P, and $x - r$ is a factor of the polynomial that defines P.

Look Beyond

80. a.
$$\begin{aligned}P(2+3i) &= (2+3i)^3 - 2(2+3i)^2 + 5(2+3i) + 26\\ &= (2+3i)(2+3i)^2 - 2(4+12i-9) + 10 + 15i + 26\\ &= (2+3i)(-5+12i) - 8 - 24i + 18 + 10 + 15i + 26\\ &= -10 + 24i - 15i - 36 - 8 - 24i + 18 + 10 + 15i + 26\\ &= 0\end{aligned}$$

$$\begin{aligned}P(2-3i) &= (2-3i)^3 - 2(2-3i)^2 + 5(2-3i) + 26\\ &= (2-3i)(2-3i)^2 - 2(4-12i-9) + 10 - 15i + 26\\ &= (2-3i)(-5-12i) - 8 + 24i + 18 + 10 - 15i + 26\\ &= -10 - 24i + 15i - 36 - 8 + 24i + 18 + 10 - 15i + 26\\ &= 0\end{aligned}$$

b. $[x - (2+3i)][x - (2-3i)] = x^2 - 4x + 13$
Use long division.
$P(x) = (x+2)(x^2 - 4x + 13)$
$= (x+2)[x - (2+3i)][x - (2-3i)]$

Portfolio Activity

1. Answers will vary. Sample: The side length of the square base is 14 cm. The area of the square base is $A = s^2 = (14)^2 \approx 196 \text{ cm}^2$.
2. Answers will vary. Sample: The approximate volume of this part of the bottle is $W(r) = s^2h_1 = (14)^2(12) = 2352 \text{ cm}^3$.
3. Answers will vary. Sample: The approximate volume of the air space is $A(r) = s^2h_2 = (14)^2(13) = 2548 \text{ cm}^3$.
4. Answers will vary. Sample: The total volume of the bottle is $V(r) = W(r) + A(r) = 2352 + 2548 = 4900 \text{ cm}^3$.
5. A function for the volume is $V(r) = W(r) + A(r) = s^2h_1 + s^2h_2 = s^2(s) + s^2(2) = s^3 + 2s^2$. Graph the function $y = x^3 + 2x^2$. Use the trace function to find the x-value for $y = 96$: $x = 4$. The side length of the base is 4 cm.

Lesson 7.5, pages 458–465

Activity

1. three; one
2. R has three real zeros.

CHECKPOINT ✔

3. S has two real zeros.
4. T has one real zero.
5. A cubic function can have either one or three real zeros. By the complex conjugate root theorem, if the cubic function does not have three real zeros, then it has one real zero and two complex zeros which are conjugates.

Exercises

Communicate

1. By the Rational Root Theorem, the only possible rational roots are $\pm\frac{3}{2}$, $\pm\frac{1}{2}$, ±3, or ±1.
2. By the Complex Conjugate Root Theorem, $3 + 2i$ is also a root of the equation.
3. $P(x)$ has five zeros. There is one real zero and four complex zeros, which could be two distinct complex zeros, each with multiplicity 2.
4. $P(x) = x^3 - (1+2i)x^2 + 2ix$ does not have real coefficients. $Q(x) = x^4 - x^3 + 4x^2 - 4x$ does have real coefficients. If P had real coefficients, then by the Complex Conjugate Root Theorem, $1 - 2i$ would also be one of its roots.

Lesson 8.1, pages 480–488

Activity

1. $y = \frac{1}{x}$

x	$\frac{1}{10}$	$\frac{1}{4}$	$\frac{1}{2}$	1	2	3	4	5	6
y	10	4	2	1	$\frac{1}{2} = 0.5$	$\frac{1}{3} \approx 0.33$	$\frac{1}{4} = 0.25$	$\frac{1}{5} = 0.2$	$\frac{1}{6} \approx 0.17$
xy	1	1	1	1	1	1	1	1	1

2. As the values of x increase, the values of y decrease. As the values of x decrease, the values of y increase.

3. $y = \frac{2}{x}$

x	$\frac{1}{10}$	$\frac{1}{4}$	$\frac{1}{2}$	1	2	3	4	5	6
y	20	8	4	2	1	$\frac{2}{3} \approx 0.67$	$\frac{1}{2} = 0.5$	$\frac{2}{5} = 0.4$	$\frac{1}{3} \approx 0.33$
xy	2	2	2	2	2	2	2	2	2

$y = \frac{4}{x}$

x	$\frac{1}{10}$	$\frac{1}{4}$	$\frac{1}{2}$	1	2	3	4	5	6
y	40	16	8	4	2	$\frac{4}{3} \approx 1.33$	1	$\frac{4}{5} = 0.8$	$\frac{2}{3} \approx 0.67$
xy	4	4	4	4	4	4	4	4	4

yes

$y = \frac{3}{x}$

x	$\frac{1}{10}$	$\frac{1}{4}$	$\frac{1}{2}$	1	2	3	4	5	6
y	30	12	6	3	$\frac{3}{2} = 1.5$	1	$\frac{3}{4} = 0.75$	$\frac{3}{5} \approx 0.6$	$\frac{1}{2} \approx 0.5$
xy	3	3	3	3	3	3	3	3	3

CHECKPOINT ✔

4. For $k > 0$, as x increases, $\frac{k}{x}$ decreases; as x decreases, $\frac{k}{x}$ increases.

5. When $x = 0$, $y = \frac{k}{x}$ is not defined.

Exercises

Communicate

1. In direct variation, the variable y varies as a single variable; in joint variation, y varies as more than one variable. For example: Direct variation: Let y be the perimeter of a square with side of length x. Then $y = 4x$. Joint variation: Let y be the area of a rectangle with sides x and z. Then $y = xz$.

2. If $xy = k$, where $k > 0$, then $y = \frac{k}{x}$. As x increases, y must decrease since the relationship is reciprocal.

3. A varies jointly as p and q.

4. A varies directly as the square of s.

5. In $z = \frac{2xy}{d^2}$, z varies jointly as x and y and inversely as the square of d.

6. $a = \frac{bc}{d}$; by dividing both sides by c, we get $\frac{a}{c} = \frac{b}{d}$.

Portfolio Activity

1. 4; 3 **2.** $\overline{OD}$ **3.** $\overline{FD}$

Lesson 8.2, pages 489–497

Activity

1. a.

x	1.0	1.1	1.2	1.3	1.4	1.5	1.6	1.7	1.8	1.9
y	−1	−1.111	−1.25	−1.429	−1.667	−2	−2.5	−3.333	−5	−10

b.

x	3.0	2.9	2.8	2.7	2.6	2.5	2.4	2.3	2.2	2.1
y	1	1.111	1.25	1.4286	1.6667	2	2.5	3.333	5	10

CHECKPOINT ✔

2. y decreases; y increases; y is undefined.

CHECKPOINT ✔

3.

x	−3.8	−3.6	−3.4	−3.2	−3.1
y	−1.25	−1.67	−2.5	−5	−10

x	−2.2	−2.4	−2.6	−2.8	−2.9
y	1.25	1.67	2.5	5	10

y decreases; y increases; y is undefined.

Exercises

Communicate

1. A quotient is a ratio.

2. To find the excluded values, find the zeros of the denominator.

3. If the linear factor in the denominator is also a factor of the numerator, there is a hole in the graph where the factor is equal to zero; if it is not a factor of the numerator, there will be a vertical asymptote where the factor is equal to zero.

4. Write the equations for the asymptotes and graph them as dotted lines. Plot some points on each branch of the graph—on each side of the asymptotes—and sketch the curves through these points that approach but do not reach the asymptotes.

Practice and Apply

28. domain of all real numbers except $x = 5$; vertical asymptote: $x = 5$; horizontal asymptote: $y = 0$; no holes

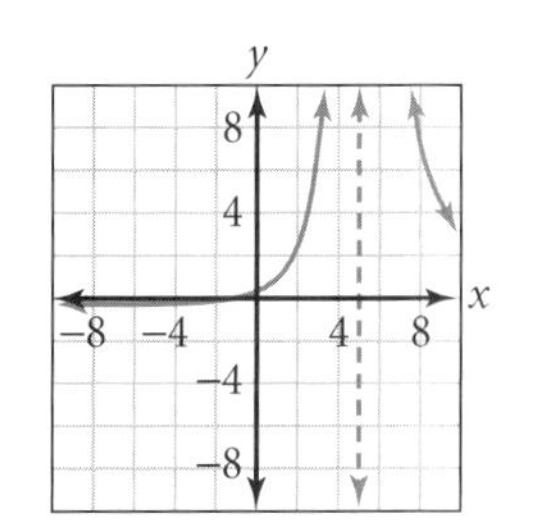

29. domain of all real numbers; no vertical asymptotes; horizontal asymptote: $y = 0$; no holes

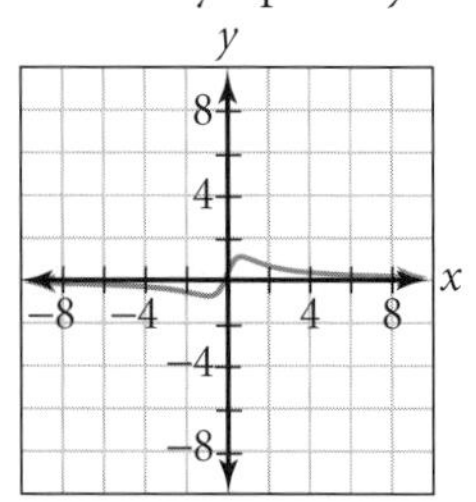

30. domain of all real numbers; no vertical asymptotes; horizontal asymptote: $y = 1$; no holes

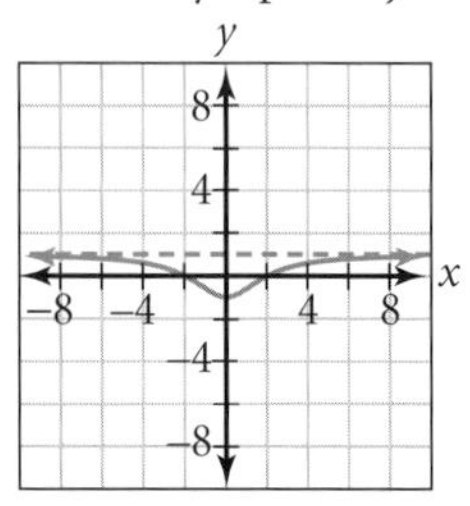

31. domain of all real numbers except $x = 1$ and $x = 6$; vertical asymptotes: $x = 1$ and $x = 6$; no horizontal asymptote; no holes

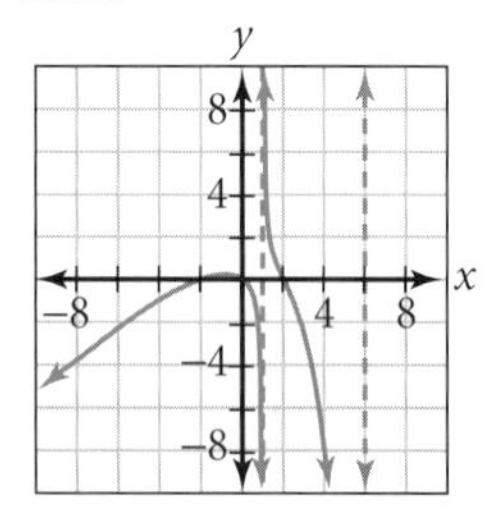

32. domain of all real numbers except $x = 0$, $x = -4$, and $x = 5$; vertical asymptotes: $x = -4$ and $x = 5$; horizontal asymptote: $y = 0$; hole when $x = 0$

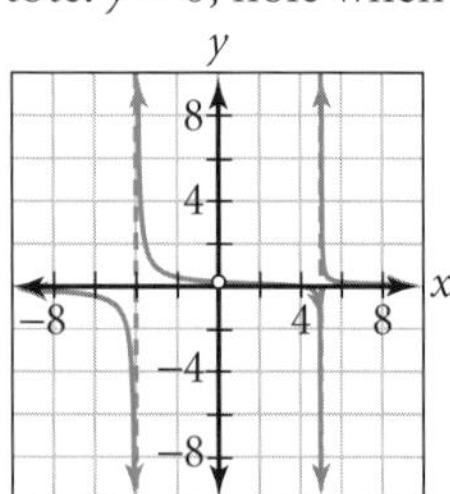

33. domain of all real numbers except $x = 0$ and $x = 4$; vertical asymptote: $x = 4$; horizontal asymptote: $y = 0$; hole when $x = 0$

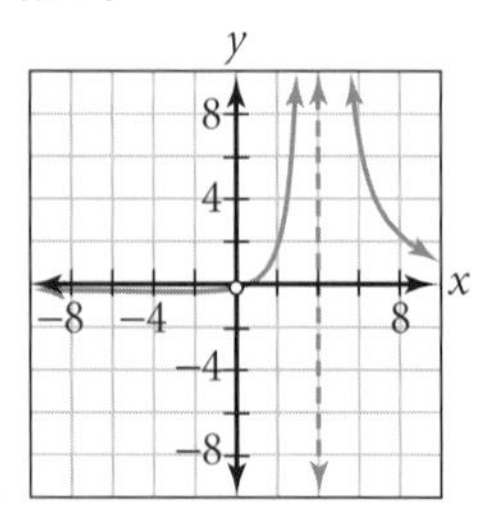

34. domain of all real numbers except $x = 3$; vertical asymptote: $x = 3$; horizontal asymptote: $y = 0$; no holes

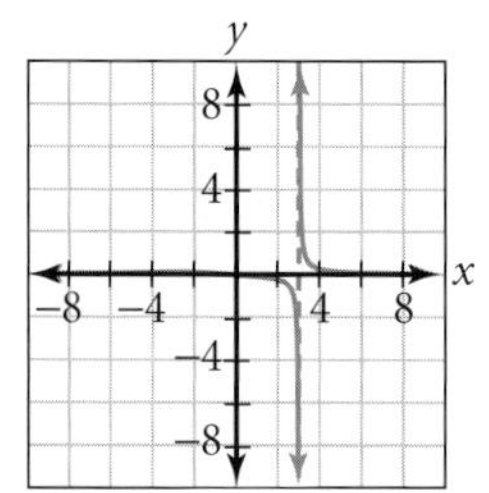

41. a. $c > \frac{9}{4}$; if $b^2 - 4ac = 9 - 4c < 0$, then $x^2 - 3x + c = 0$ has no solutions and there are no vertical asymptotes.

b. $c = \frac{9}{4}$; if $x^2 - 3x + c = 0 - 4c = 0$, then $x^2 - 3x + c = 0$ has 1 solution and there is 1 vertical asymptote.

c. $c < \frac{9}{4}$; if $x^2 - 3x + c = 0 - 4c > 0$, then $x^2 - 3x + c = 0$ has 2 solutions and there are 2 vertical asymptotes.

Portfolio Activity

1. $R(x) = \frac{x^2 + 6x}{4x + 12}$

2. $x = 6$

3. The width is 6 and the length is 12.

4. $S(x) = \frac{x^2 + 4x + 4}{4x + 8}$

5. $x = 6$

6. 8

7. 8

Lesson 8.3, pages 498–504

Activity

1. The graphs appear to be the same.

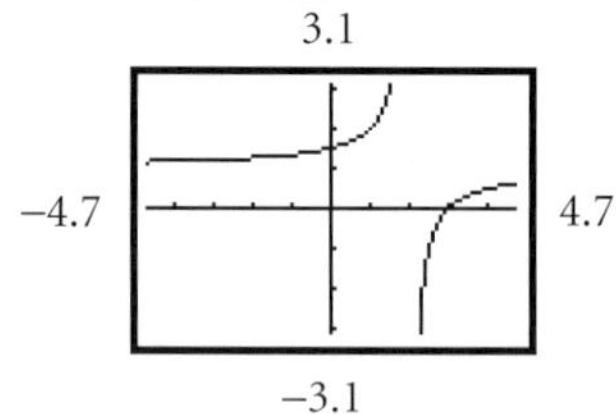

2.

x	−3	−2	−1	0	1	2	3
$f(x)$	1.2	undefined	≈1.33	1.5	2	undefined	0
$g(x)$	1.2	1.25	≈1.33	1.5	2	undefined	0

The entries are the same except at $x = -2$, where f is undefined and $g(-2) = 1.25$.

3. The graphs appear to be the same.

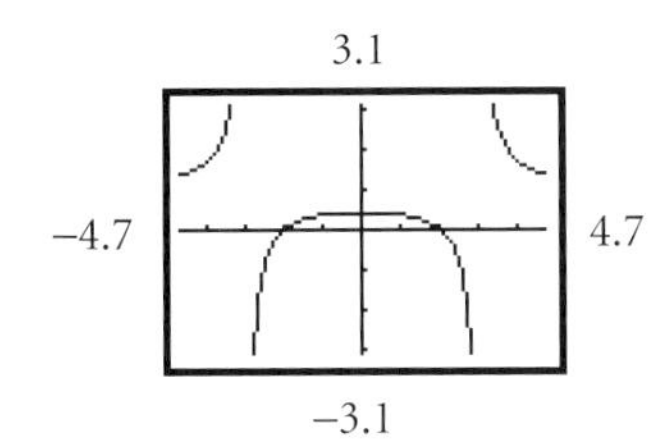

x	−3	−2	−1	0	1	2	3
$f(x)$	undefined	0	0.375	0.44	undefined	0	undefined
$g(x)$	undefined	0	0.375	0.44	0.375	0	undefined

The entries are the same except at $x = 1$, where f is undefined and $g(1) = 0.375$.

CHECKPOINT ✔

4. Find the values of x that make $Q = 0$ and the values that make $R = 0$. If the fraction is reduced by cancellation, some excluded factors will be lost.

Exercises

Communicate

1. In either case, you simplify the product by dividing out common factors in the numerator and denominator before or after you multiply.

2. In either case, in order to divide, you should multiply by the reciprocal of the divisor and simplify by dividing out common factors in the numerator and denominator.

3. First multiply by the reciprocal of the divisor (denominator). Then factor all expressions and simplify by dividing out common factors.

$$\frac{\frac{x^2}{x^2-1}}{\frac{x}{x^2+2x-3}} = \frac{x^2}{x^2-1} \cdot \frac{x^2+2x-3}{x} \quad \text{Multiply by the reciprocal.}$$

$$= \frac{x^{2^1}}{(x+1)(\cancel{x-1})_1} \cdot \frac{(x+3)(\cancel{x-1})^1}{\cancel{x}_1} \quad \text{Factor.}$$

$$= \frac{x^2+3x}{x+1} \quad \text{Divide out the common factor.}$$

The excluded values in the complex form are $x = -1$, $x = 1$, $x = 0$, and $x = -3$. In the simplified form, the only excluded value is $x = -1$.

Lesson 8.4, pages 505–511

Exercises

Communicate

1. The least common denominator (LCD) is the polynomial with lowest degree that is divisible by each polynomial in the denominators. First factor each denominator. The LCD contains each factor as many times as it appears in any one denominator.

$$\frac{x+5}{x^2-7x+6} + \frac{x-1}{x^2-36} = \frac{x+5}{(x-6)(x-1)} + \frac{x-1}{(x-6)(x+6)}$$

Thus, the LCD is $(x-6)(x-1)(x+6)$.

2. Graph the sum as it originally appears and graph your solution. If the solution is correct, the graphs will be the same.

3. **a** and **d** are equivalent. In **d**, the denominators are the same and adding yields $\frac{3}{x^2+4} + \frac{7}{x^2+4} = \frac{3+7}{x^2+4}$, which is equivalent to **a**.

Portfolio Activity

1. $\frac{3abc}{bc+ac+ab}$

2. $\frac{180}{47}$

3. $\frac{4abcd}{bcd+acd+abd+abc}$

4. $\frac{96}{25} = 3.84$

Lesson 8.5, pages 512–519

Activity

1.

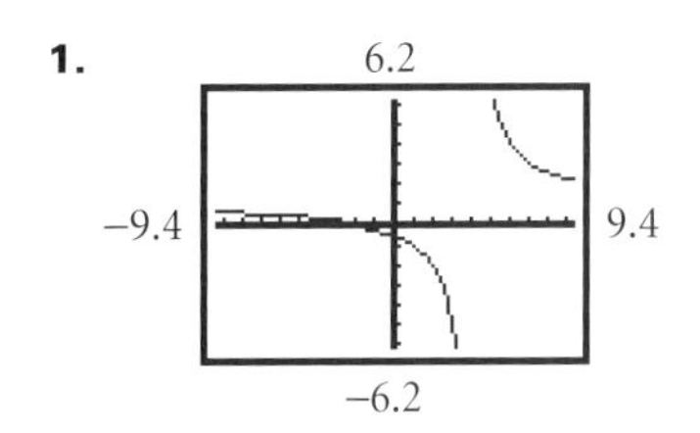

2. If $x = -2$, $y_1 = 0$. If $x = 4$, y_1 is undefined. If $x < -2$ *or* $x > 4$, y_1 is positive. If $-2 < x < 4$, y_1 is negative.

3.

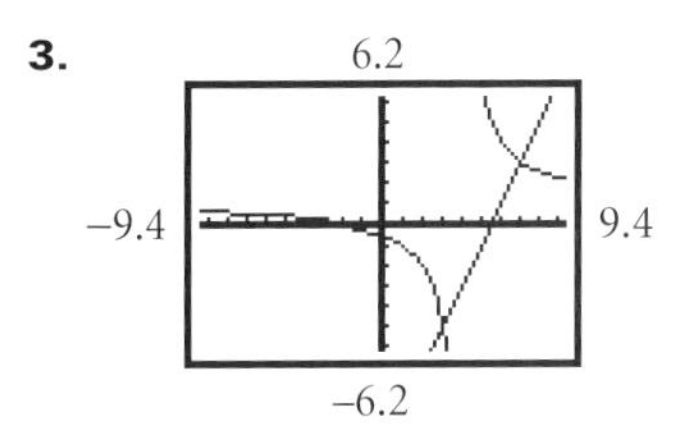

4. $y_1 = y_2$ when $x = 3$ or $x = 7$
$y_1 > y_2$ when $x < 3$ or $4 < x < 7$
$y < y_2$ when $3 < x < 4$ or $x > 7$

CHECKPOINT ✔

5. Graph each function (Steps 1 and 3), and then compare the graphs or the y-values in the table for various values of x.

Exercises

Communicate

1. An extraneous solution is any solution found in the solving process that does not make the original equation true when substituted into the original equation. You can tell if a solution is extraneous if it is an excluded value in the original equation.

2. Graph each side of the equation and find where the graphs intersect. Very often, solutions obtained by using a graph are approximate.

3. Graph the functions $y = \frac{x-1}{x+2}$ and $y = 3$. Those portions of the graph of $y = \frac{x-1}{x+2}$ that lie above the graph of $y = 3$ will determine intervals on the x-axis for which $\frac{x-1}{x+2} > 3$; those portions of the graph of $y = \frac{x-1}{x+2}$ that lie below the graph of $y = 3$ will determine intervals on the x-axis for which $\frac{x-1}{x+2} < 3$.

Portfolio Activity

1. 49.5
2. The answers are the same.
3. The average speed is the total distance over the total time.
4. about 21.3 mph

Lesson 8.6, pages 520–527

Activity

1.

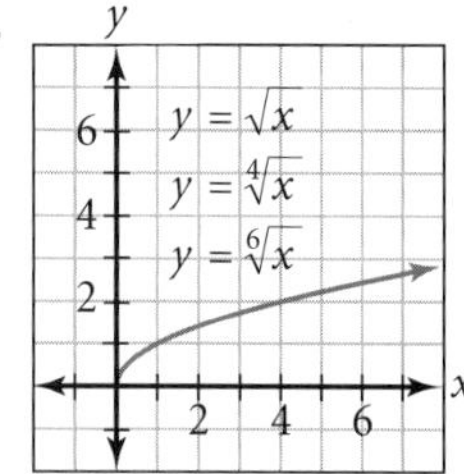

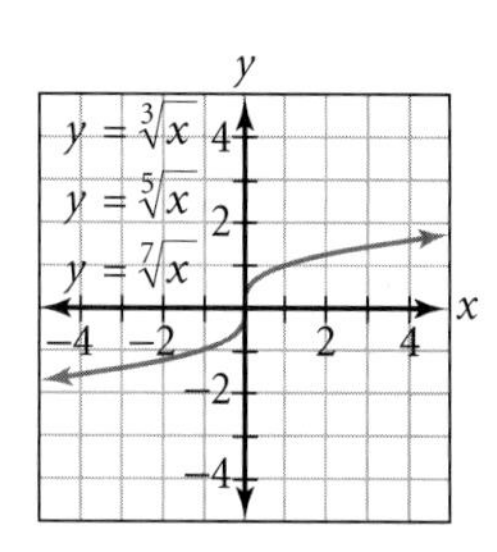

2. $y = \sqrt{x}$, $y = \sqrt[4]{x}$, and $y = \sqrt[6]{x}$ are similar. Also, $y = \sqrt[3]{x}$, $y = \sqrt[5]{x}$, and $y = \sqrt[7]{x}$ are similar.
3. $y = \sqrt{x}$, $y = \sqrt[4]{x}$, and $y = \sqrt[6]{x}$
4. $y = \sqrt[3]{x}$, $y = \sqrt[5]{x}$, and $y = \sqrt[7]{x}$

CHECKPOINT ✔

5. If n is a positive even integer, the domain of $y = \sqrt[n]{x}$ is $x \geq 0$ and the range is $y > 0$. If n is a positive odd integer, the domain of $y = \sqrt[n]{x}$ is all real numbers and the range is all real numbers.

Exercises

Communicate

1. a horizontal translation 4 units to the right, and a vertical stretch by a factor of 3
2. Even roots of negative numbers are undefined in the real numbers, so for even roots, set the radicand (the expression appearing under the radical) greater than or equal to zero and solve. The solution to the inequality will be the domain. For odd roots, the domain is all real numbers.

3. First interchange the roles of x and y.

$y = 4x^3 + 2 \rightarrow x = 4y^3 + 2$

Then solve $x = 4y^3 + 2$ for y.

$$x = 4y^3 + 2$$
$$x - 2 = 4y^3$$
$$\frac{x-2}{4} = y^3$$
$$y = \sqrt[3]{\frac{x-2}{4}}$$

Thus, the inverse of $y = 4x^3 + 2$ is $y = \sqrt[3]{\frac{x-2}{4}}$.

Practice and Apply

25. $y = \pm\sqrt{x+1}$

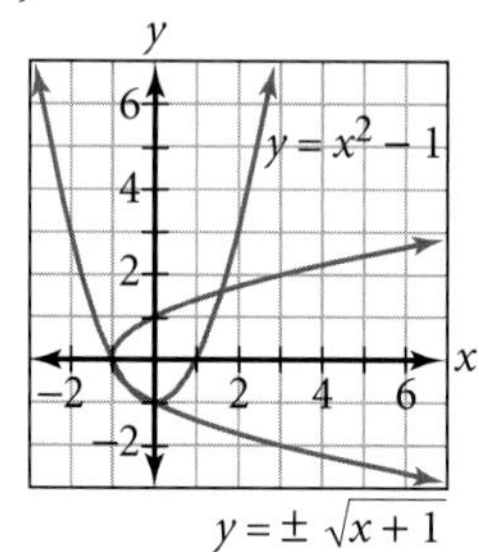

26. $y = \pm\sqrt{x-3}$

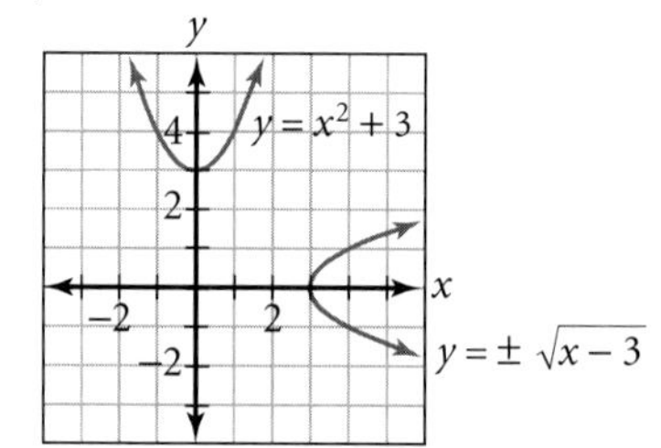

27. $y = \frac{-1 \pm \sqrt{1+12x}}{6}$

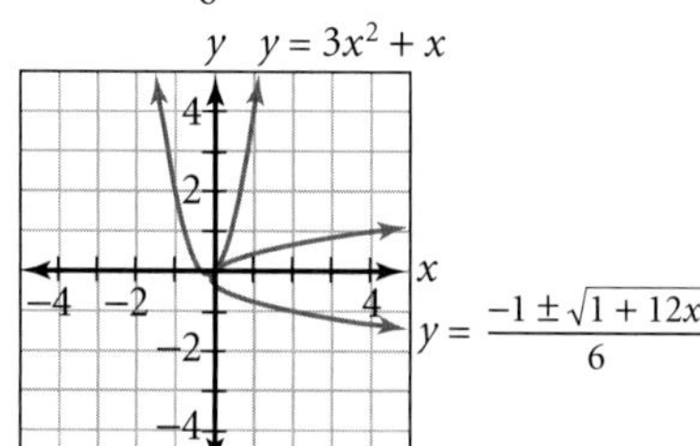

28. $y = -1 \pm \sqrt{1+x}$

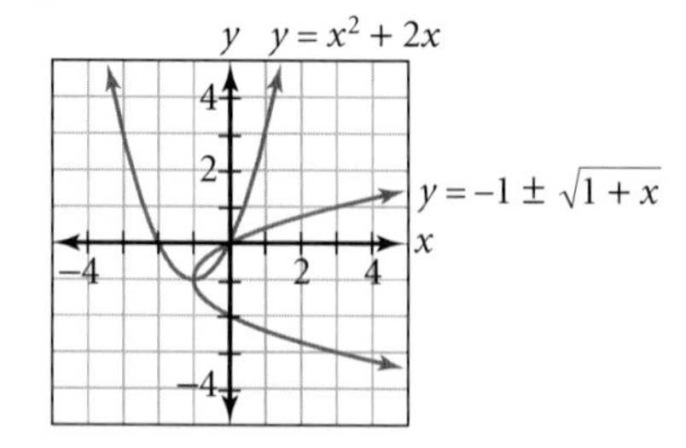

29. $y = -2 \pm \sqrt{x}$

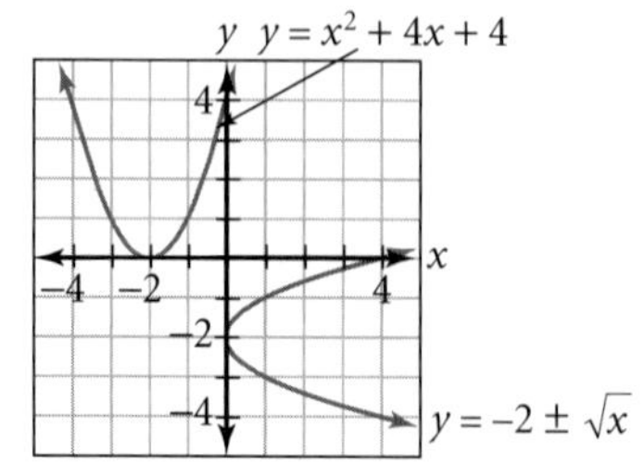

30. 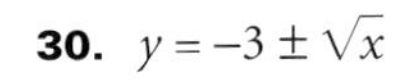 $y = -3 \pm \sqrt{x}$

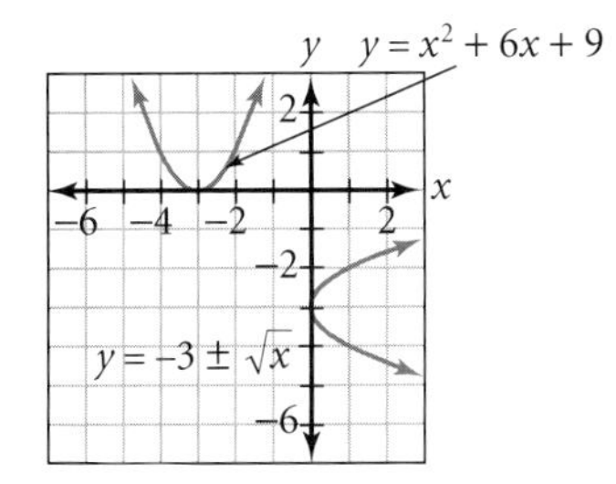

31. $y = 1 \pm \sqrt{x}$

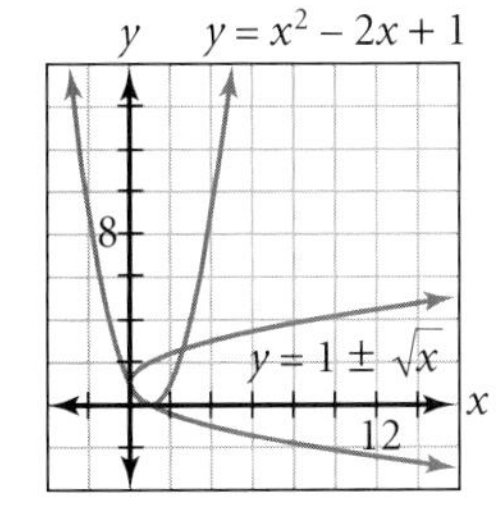

32. $y = \frac{7 \pm \sqrt{1 + 4x}}{2}$

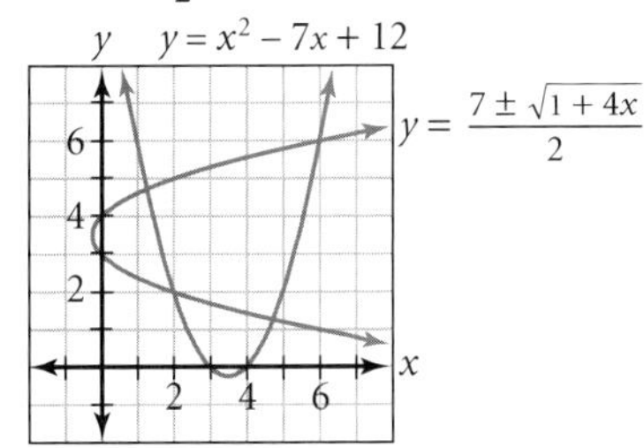

33. $y = \frac{3 \pm \sqrt{25 - 4x}}{2}$

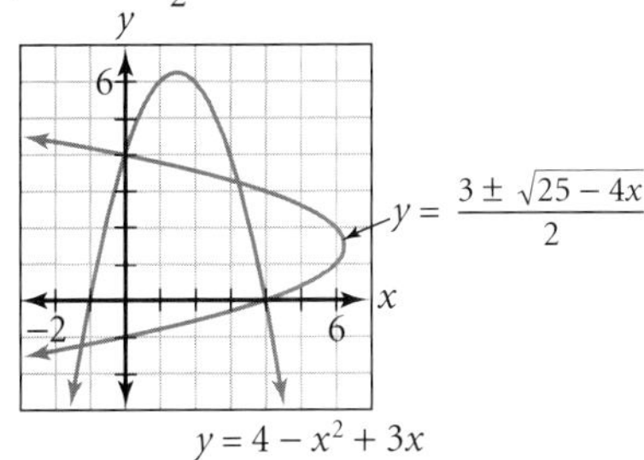

34. $y = -1 \pm \sqrt{9 + x}$

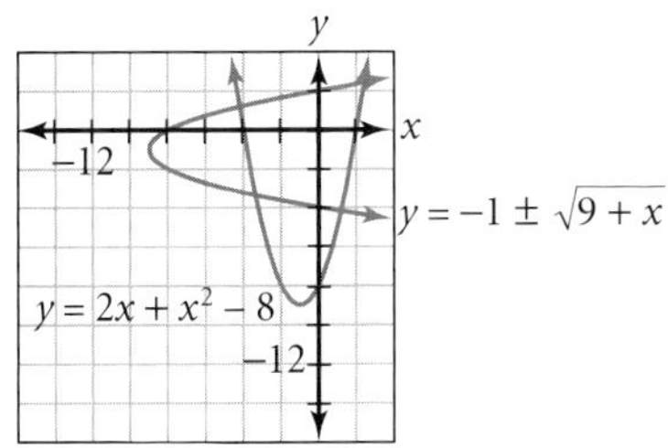

35. $y = \frac{-1 \pm \sqrt{9 + 8x}}{4}$

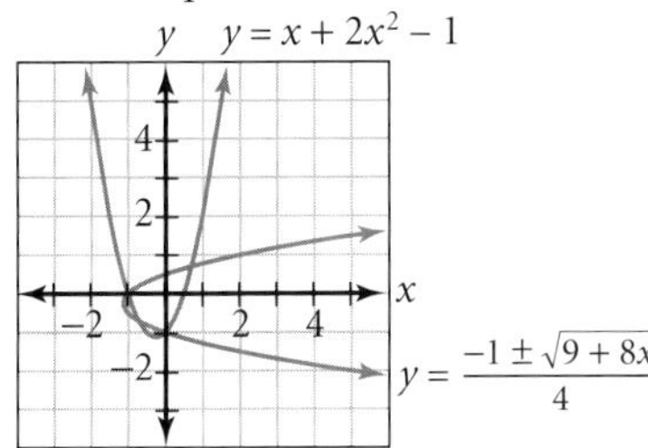

36. $y = \frac{1 \pm \sqrt{25 - 8x}}{4}$

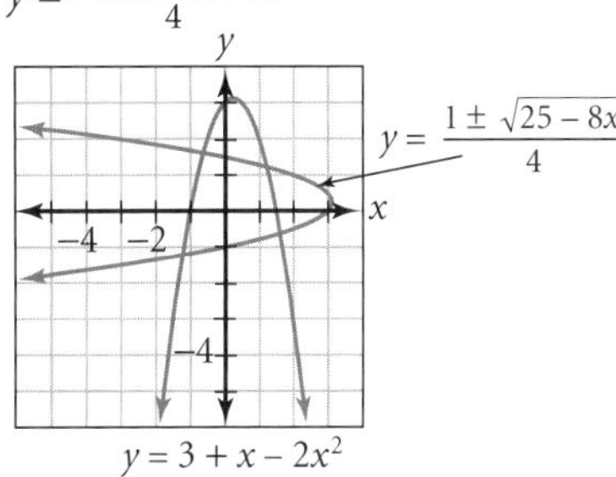

Practice and Apply

62. a horizontal translation 1 unit to the right, a vertical compression by a factor of $\frac{1}{2}$, and a vertical translation 4 units up

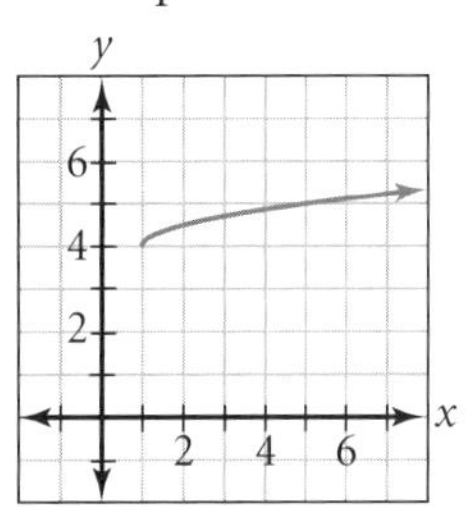

63. a horizontal translation 2 units to the left, a vertical compression by a factor of $\frac{1}{3}$, and a vertical translation 3 units up

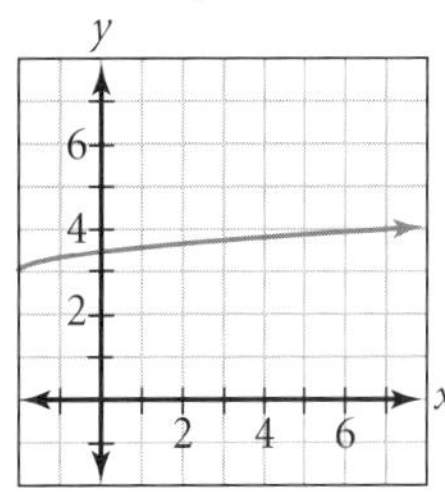

Look Beyond

80. $y = \pm\sqrt{1 - x^2}$; a circle with a radius of 1

Lesson 8.7, pages 528–535

Activity

1.

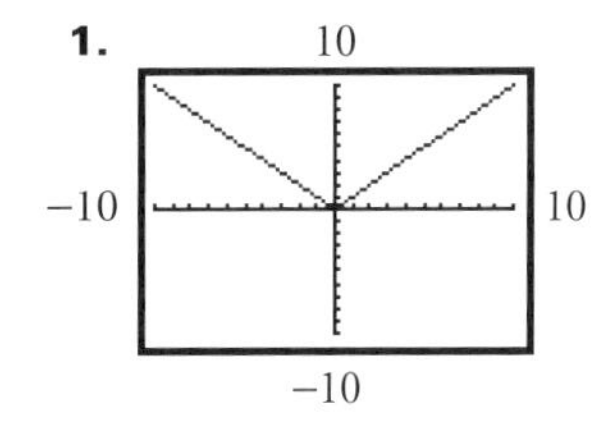

$y = |x|$

2. $y = |x|$; for nonnegative values of x, $\sqrt[4]{x^4} = (x^4)^{\frac{1}{4}} = x$; for negative values of x, x^4 is positive and the fourth root of a positive number is positive. Thus, $\sqrt[4]{x^4} = |x|$.

CHECKPOINT ✔

3. The graph of $y = |x|$; for nonnegative values of x, $\sqrt[n]{x^n} = (x^n)^{\frac{1}{4}}$; for negative values of x, x^4 is positive and the 4th root of a positive number is positive. Thus, $\sqrt[n]{x^n} = |x|$.

4. the graph of $y = x$

5. $y = x$; for all values of x, $(x^5)^{\frac{1}{5}} = x$. Thus, $\sqrt[5]{x^5} = x$.

CHECKPOINT ✔

6. $y = x$; for all values of x, $(x^n)^{\frac{1}{n}} = x$. Thus, $\sqrt[n]{x^n} = x$.

Exercises

Communicate

1. Use the Distributive Property twice to simplify the expression. First distribute $(1 + 2\sqrt{2})$.
$(1 + 2\sqrt{2})(2 - 3\sqrt{2})$
$= (1 + 2\sqrt{2})(2) - (1 + 2\sqrt{2})(3\sqrt{2})$
Then distribute again.
$= 2 + 4\sqrt{2} - 3\sqrt{2} - 12$
$= -10 + \sqrt{2}$

2. To simplify $\sqrt{4x^3}$, factor all perfect squares out of the radical: $\sqrt{4x^3} = \sqrt{2^2x^2x} = 2|x|\sqrt{x}$.

3. To rationalize the denominator, multiply the numerator and the denominator by the conjugate of the denominator.

$$\frac{5 + 3\sqrt{2}}{4 + 7\sqrt{2}} = \frac{5 + 3\sqrt{2}}{4 + 7\sqrt{2}} \cdot \frac{4 - 7\sqrt{2}}{4 - 7\sqrt{2}}$$
$$= \frac{(5)(4) + (5)(-7\sqrt{2}) + (3\sqrt{2})(4) + (3\sqrt{2})(-7\sqrt{2})}{(4)(4) + (4)(-7\sqrt{2}) + (4)(7\sqrt{2}) + (7\sqrt{2})(-7\sqrt{2})}$$
$$= \frac{20 - 35\sqrt{2} + 12\sqrt{2} - 42}{16 - 28\sqrt{2} + 28\sqrt{2} - 98}$$
$$= \frac{-22 - 23\sqrt{2}}{-82}$$
$$= \frac{22 + 23\sqrt{2}}{82}$$

Lesson 8.8, pages 536–543

Activity

1. a. 3

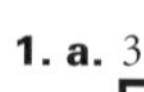

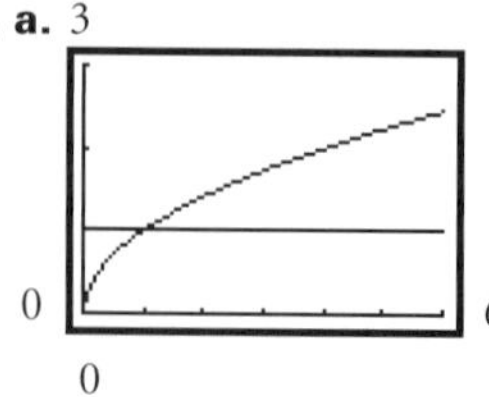

0   6

0

b. $x \geq 1$

c. $0 \leq x \leq 1$

2. For $x \leq 0$, $f(x)$ is undefined.

CHECKPOINT ✔

3. Yes, they will always have solutions. When $a > 0$, $\sqrt{x} \geq a$ always has solutions of x such that $x \geq a^2$, and $\sqrt{x} \leq a$ always has solutions of x such that $x \leq a^2$.

CHECKPOINT ✔

4. Graph $f(x) = \sqrt{x}$ and $g(x) = x - 2$ and determine the points of intersection. The values of x for which f lies on or above g are solutions to $\sqrt{x} \geq x - 2$. The values of x for which f lie on or below g are solutions to $\sqrt{x} \leq x - 2$. The solutions of $\sqrt{x} \geq x - 2$ are $0 \leq x \leq 4$, and the solutions of $\sqrt{x} \leq x - 2$ are $x \geq 4$.

Exercises

Communicate

1. Algebraic: Square both sides of the equation or inequality and then simplify. Graphical: Graph $y_1 = \sqrt{x}$ and $y_2 = 3\sqrt{x - 4}$ and locate their points of intersection. The values of x for which y_0 lies on or below y_2 are solutions for $\sqrt{x} \leq 3\sqrt{x - 4}$.

2. Raising each side of the equation to a power may introduce extraneous solutions.

3. Algebraic: Square both sides of the equation to yield $x = x + 1$. There is no solution to this equation. Graphical: The curve $y = \sqrt{x + 1}$ is a horizontal translation of $y = \sqrt{x}$, so the curves do not intersect.

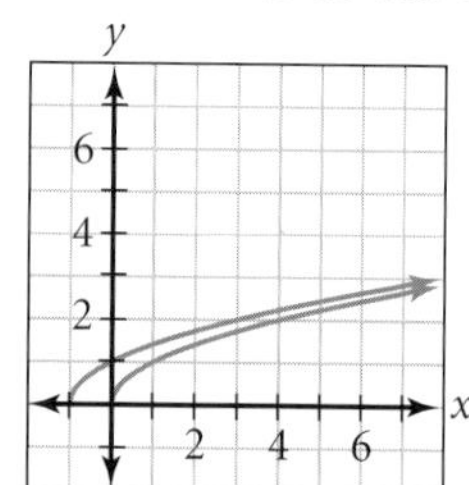

Lesson 9.1, pages 562–569

Activity

1. $x_1 = -2$; $x_2 = 4$; $(x_2 - x_1) = 6$; $(x_2 - x_1)^2 = 36$; $PQ = 2\sqrt{10}$

2. $x_1 = 4$; $x_2 = -2$; $(x_2 - x_1) = -6$; $(x_2 - x_1)^2 = 36$; $PQ = 2\sqrt{10}$

3. no; yes; yes

CHECKPOINT ✔

4. Answers may vary. Sample answer: The distance stays the same because $(x_2 - x_1)^2 = (x_1 - x_2)^2$ and $(y_2 - y_1)^2 = (y_1 - y_2)^2$.

Exercises

Communicate

1. Answers may vary. Sample answer: A plane perpendicular to the axis of a cone at a point other than the cone's vertex forms a circle. A plane parallel to the slant height of a cone forms a parabola or a line. A plane that intersects a cone at an acute angle with its axis forms an ellipse. A plane that intersects the cone at an angle less than the angle formed by the cone's axis and slant line forms a hyperbola.

2. Answers may vary. Sample answer: When you subtract one number from another and square the answer, the answer will be the same regardless of which number was subtracted from which. For example, $(-5 - 2)^2 = [2 - (-5)]^2$ and $[-1 - (-4)]^2 = [-4 - (-1)]^2$.

3. Answers may vary. Sample answer: The radius of the circle is the distance between the center and any point on the circle. Use the distance formula to find the radius, r. The circumference is $C = 2\pi r$, and the area is $A = \pi r^2$.

Lesson 9.2, pages 570–578

Activity

1.

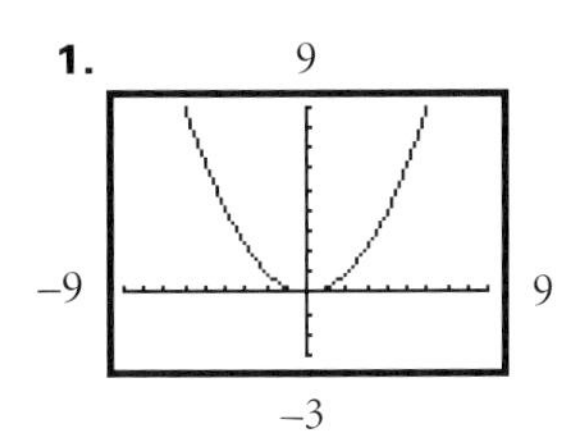

2. $y = \frac{1}{4}(x - 2)^2 + 3$; vertex: $(2, 3)$

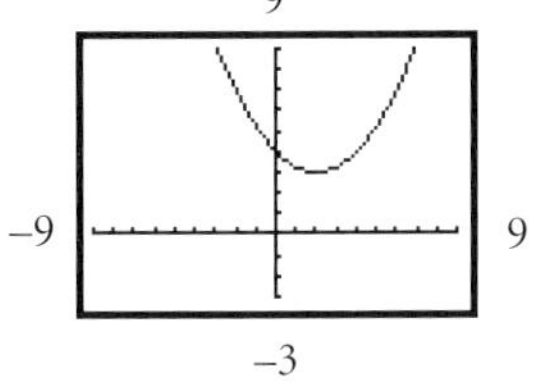

CHECKPOINT ✔

3. 2 units to the right and 3 units up

4. Answers may vary. Sample answer: The x-coordinate of the vertex is the number that is subtracted from x in the equation. The y-coordinate of the vertex is the number that is subtracted from y in the equation.

CHECKPOINT ✔

5. vertex: $(1, -2)$

Exercises

Communicate

1. If the equation has the form $y - k = \frac{1}{4p}(x - h)^2$, the graph opens upward for $p > 0$ and downward for $p < 0$. If the equation has the form $x - h = \frac{1}{4p}(y - k)^2$, the graph opens to the right for $p > 0$ and to the left for $p < 0$.

2. If the standard equation of the parabola is $y - k = \frac{1}{4p}(x - h)^2$, then the vertex is at (h, k), the focus is at $(h, k + p)$, and the directrix is the line $y = k - p$. If the standard equation of the parabola is $x - h = \frac{1}{4p}(y - k)^2$, then the vertex is at (h, k), the focus is at $(h + p, k)$, and the directrix is the line $x = h - p$.

3. First solve the equation for y.

$$\frac{3}{4}y^2 = x$$

$$y^2 = \frac{4}{3}x$$

$$y = \pm\sqrt{\frac{4}{3}x}$$

Then let $y_1 = \sqrt{\frac{4}{3}x}$ and $y_2 = -\sqrt{\frac{4}{3}x}$, or $y_2 = -y_1$.

Practice and Apply

20. vertex: $(0, 0)$; focus: $\left(0, \frac{1}{8}\right)$; directrix: $y = -\frac{1}{8}$

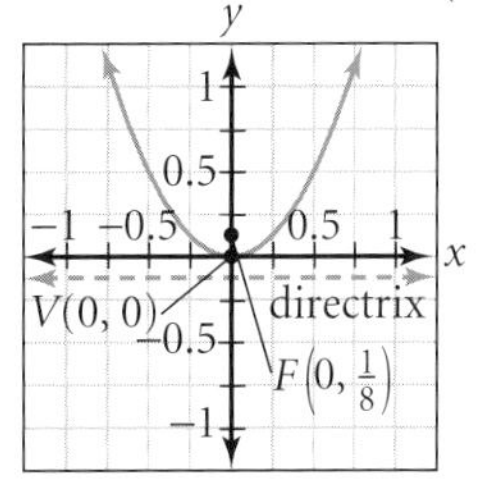

21. vertex: $(-2, -3)$; focus: $(-2, 1)$; directrix: $y = -7$

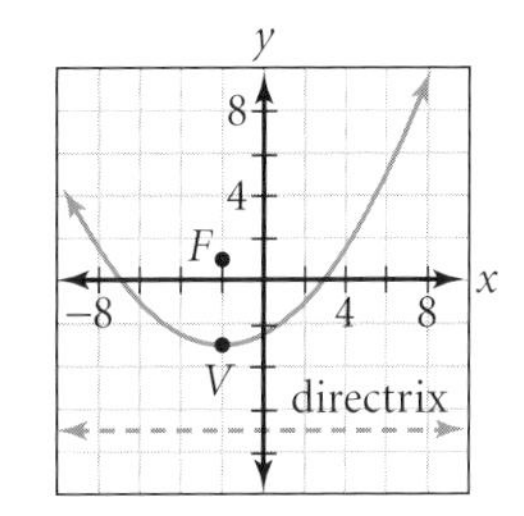

22. vertex: $(1, -2)$; focus: $(4, -2)$; directrix: $x = -2$

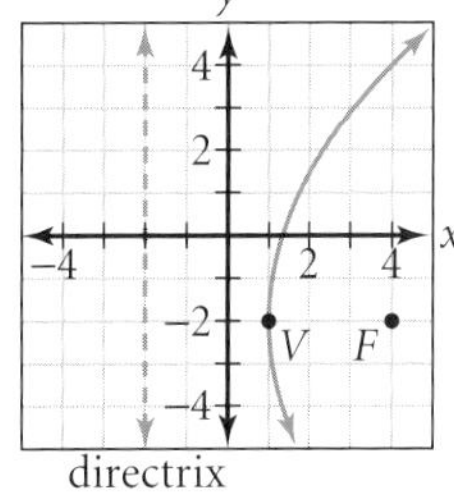

23. vertex: $(1, 4)$; focus: $\left(1, \frac{15}{4}\right)$; directrix: $y = \frac{17}{4}$

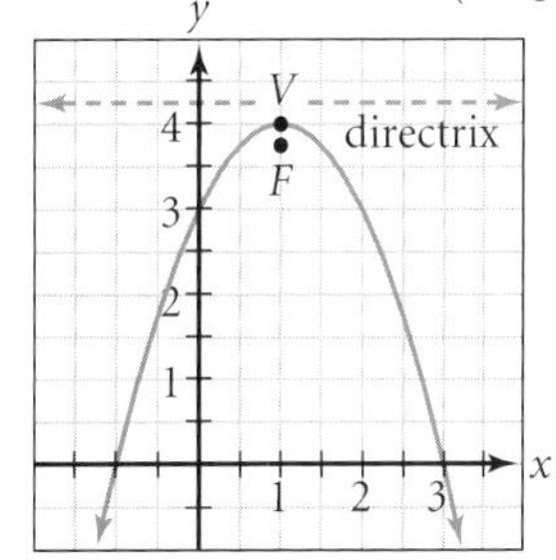

24. vertex: $(1, 1)$; focus: $(1, 2)$; directrix: $y = 0$

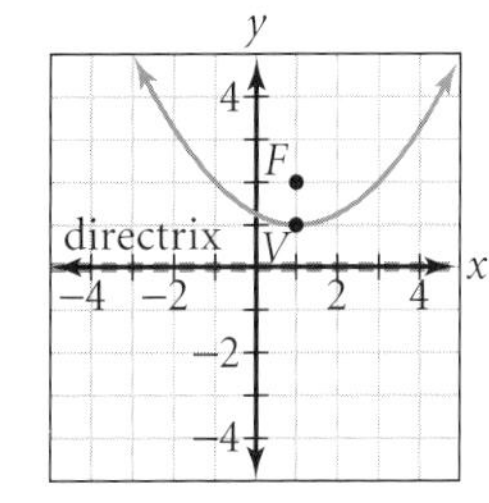

25. vertex: $(1, 0)$; focus: $(1, 2)$; directrix: $y = -2$

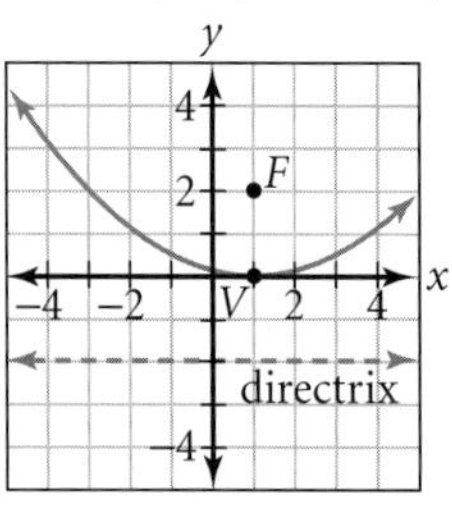

26. vertex: $(3, -1)$; focus: $(1, -1)$; directrix: $x = 5$

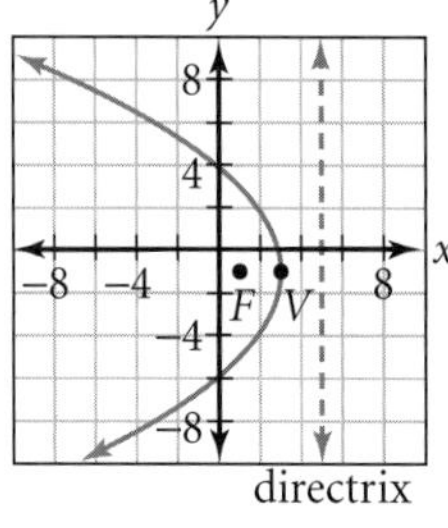

27. vertex: $(0, -3)$; focus: $(0, 0)$; directrix: $y = -6$

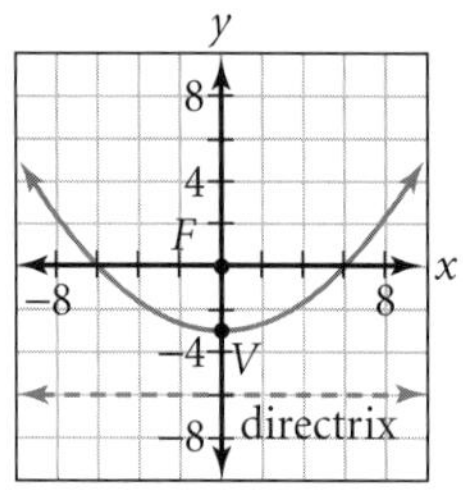

28. vertex: $(-2, 0)$; focus: $(-2, -3)$; directrix: $y = 3$

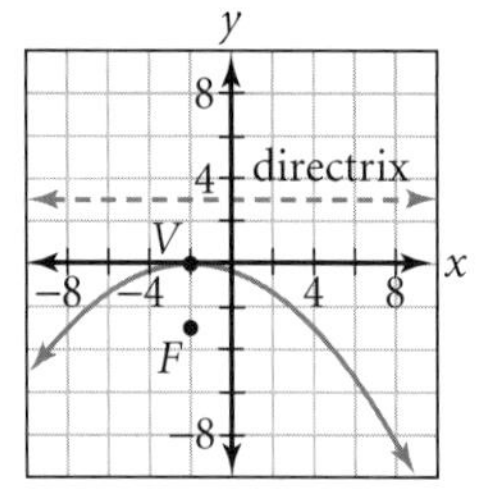

29. vertex: $(1, -2)$; focus: $\left(\frac{3}{2}, -2\right)$; directrix: $x = \frac{1}{2}$

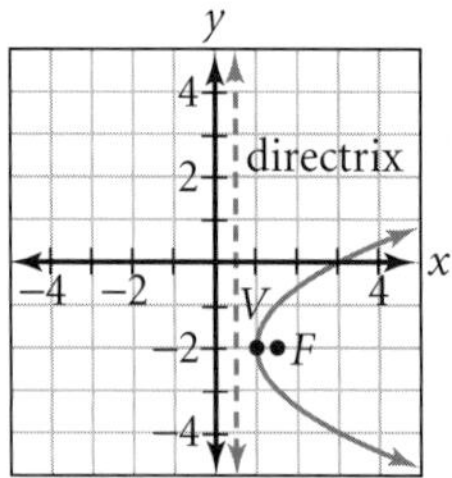

30. vertex: $(-2, 1)$; focus: $\left(-2, \frac{5}{2}\right)$; directrix: $y = -\frac{1}{2}$

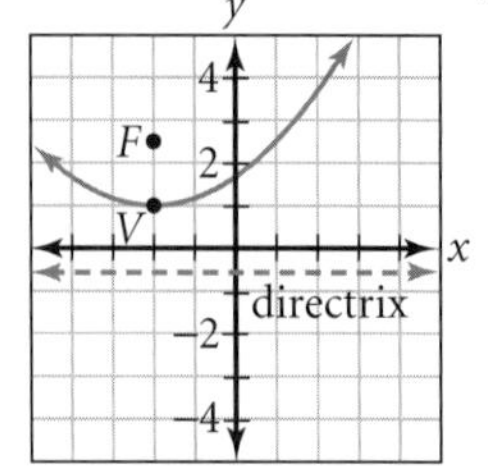

31. vertex: $(3, 1)$; focus: $\left(3, -\frac{3}{2}\right)$; directrix: $y = \frac{7}{2}$

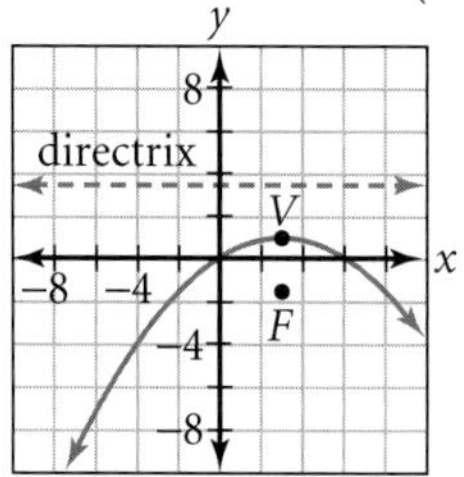

32. vertex: $(4, 4)$; focus: $\left(4, \frac{17}{4}\right)$; directrix: $y = \frac{15}{4}$

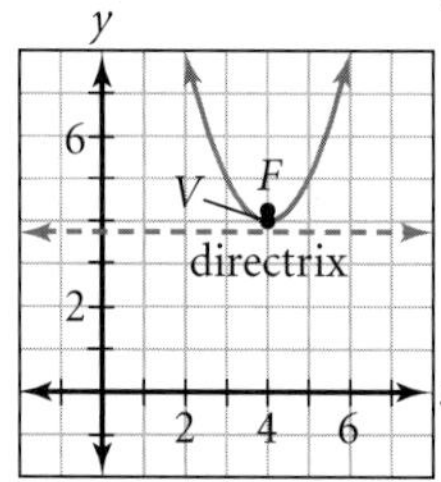

33. vertex: $\left(-\frac{11}{16}, -\frac{3}{2}\right)$; focus: $\left(-\frac{27}{16}, -\frac{3}{2}\right)$; directrix: $x = \frac{5}{16}$

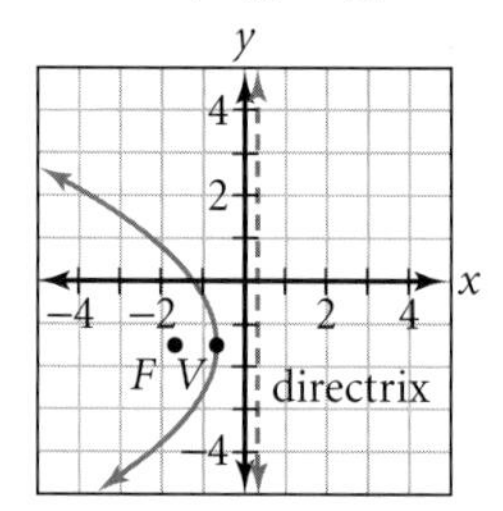

34. vertex: $\left(\frac{9}{2}, 3\right)$; focus: $\left(\frac{7}{2}, 3\right)$; directrix: $x = \frac{11}{2}$,

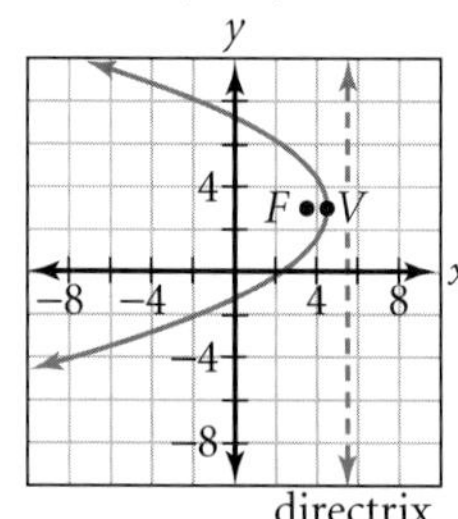

35. vertex: $(-2, 2)$; focus: $\left(-\frac{1}{4}, 2\right)$; directrix: $x = -\frac{15}{4}$

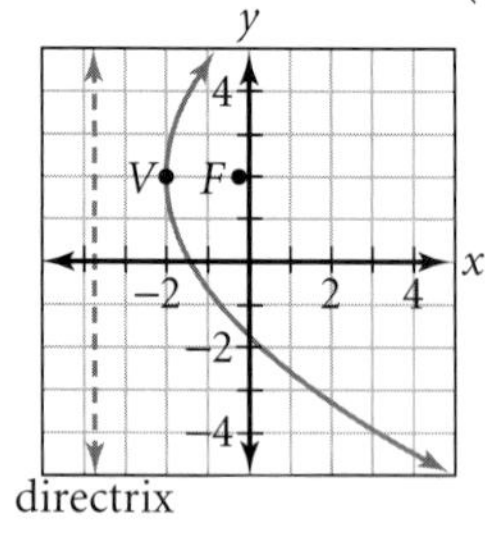

51. If the parabola opens to the right and has its vertex at the origin, the equation is $x = \frac{1}{4}y^2$.

52. $y - 200 = -\frac{1}{450}(x - 300)^2$

Portfolio Activity

1. Check students' work.
2. Check students' work.
3. Check students' work.
4. Check students' work.

5. To make a narrower parabola, move the focus closer to the directrix. To make a wider parabola, move the focus farther away from the directrix.

6. Check students' work.

7. According to the definition of a parabola, each point on the parabola is equidistant.

Lesson 9.3, pages 579–585

Activity

1. $r = 4$

2. $r = 4$; $C(-3, 2)$

CHECKPOINT ✔

3. It is translated 3 units to the left and 2 units up.

4. The center is (h, k), where h is the number subtracted from x and k is the number subtracted from y.

CHECKPOINT ✔

5. $r = 3$; $C(2, -1)$

Exercises

Communicate

1. The coordinates are the numbers subtracted from x and from y, respectively, and the radius is the number whose square appears on the right side of the equation. Because the equation can be written as $(x - 4)^2 + (y - (-1))^2 = 1^2$, the center is at $(4, -1)$ and the radius is 1.

2. You could determine whether the quantity $(x - h)^2 + (y - k)^2$ is less than r^2, greater than r^2, or equal to r^2. For the given point and equation, you need to compare the quantities $(-1 - 2)^2 + (-3 + 1)^2$ and $r^2 = 16$. Because $(-1 - 2)^2 + (-3 + 1)^2 = 9 + 4 = 13$, which is less than 16, the point is inside the circle.

3. You could complete the square in order to see if the equation can be written in the form $(x - h)^2 + (y - k)^2 = r^2$.

$$x^2 + 8x + y^2 + 16 = 0$$
$$x^2 + 8x + y^2 = -16$$
$$x^2 + 8x + 16 + y^2 = -16 + 16$$
$$(x + 4)^2 + y^2 = 0$$

This answer is of the form $(x - h)^2 + (y - k)^2 = r^2$, where $h = -4$, $k = 0$, and $r = 0$. But the radius, r, of a circle must be a positive number, so the equation does not represent a circle.

Guided Skills Practice

4. $x^2 + y^2 = 36$

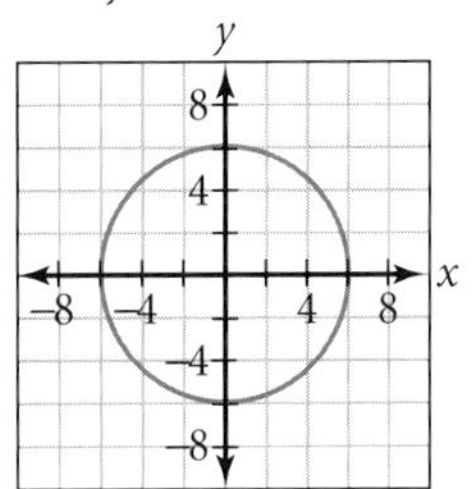

Practice and Apply

35.

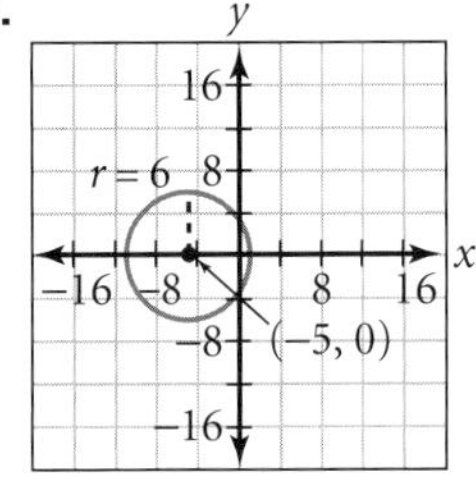

36.

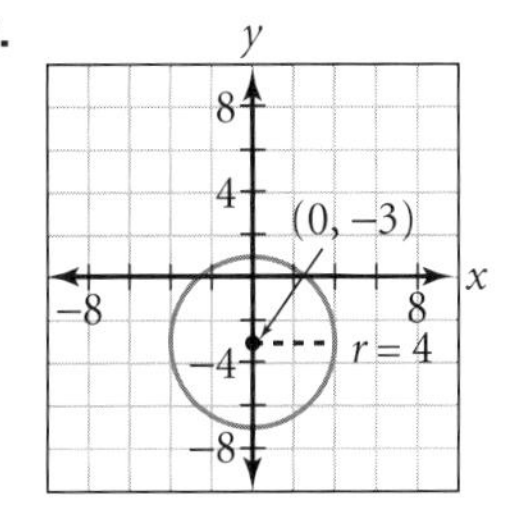

37.

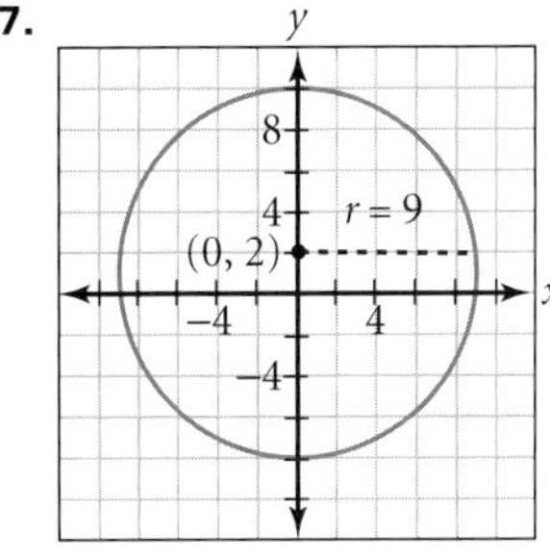

38.

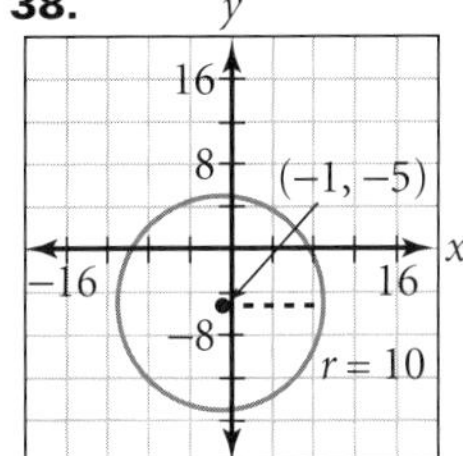

39.

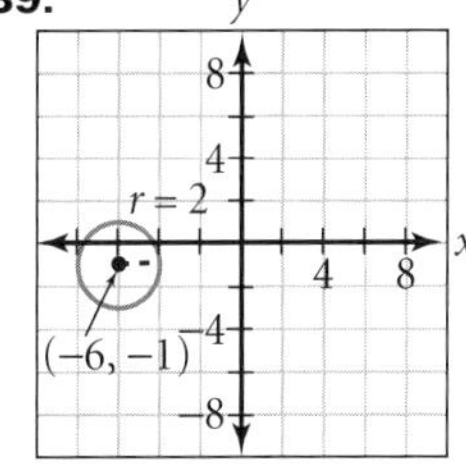

40.

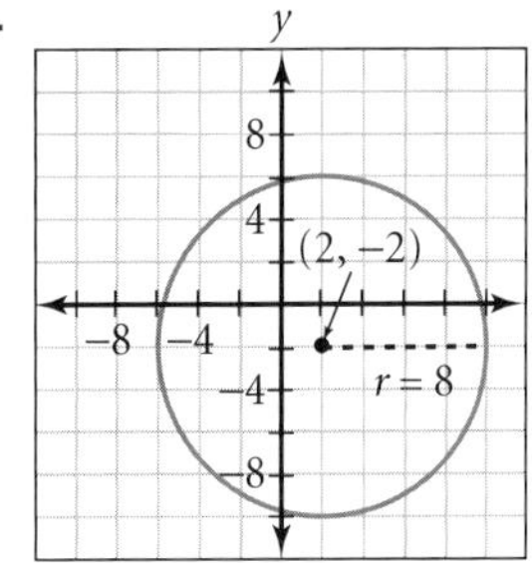

41.

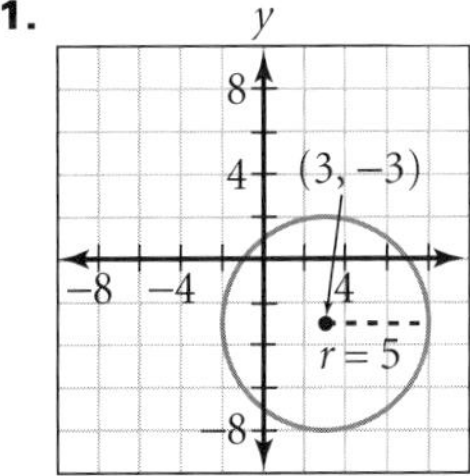

42.

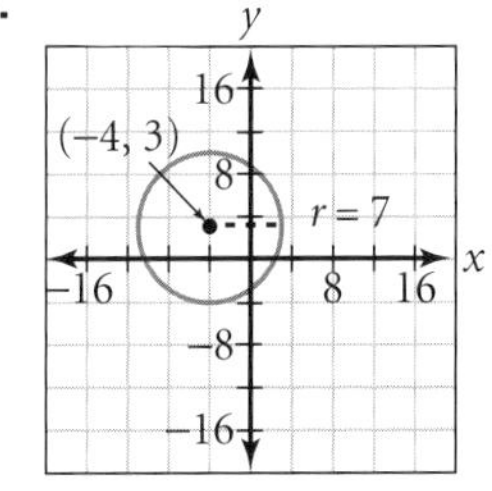

43.

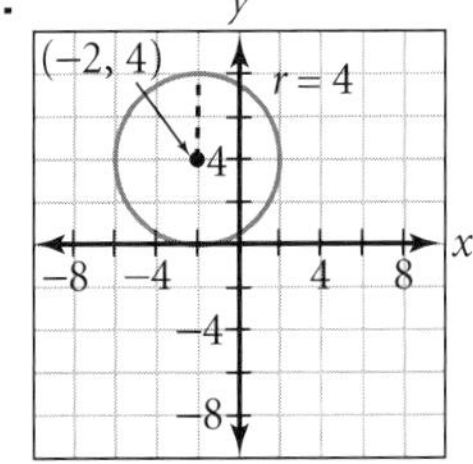

77. a. Answers may vary. Sample answer: The centers of the circles are at (3, 3) and (3, 1), which are 2 units apart. C_1 will be completely enclosed by C_2 if $r_2 - r_1 > 2$. For example, let $r_1 = 4$ and $r_2 = 1$.

b. Answers may vary. Sample answer: If $r_1 = \sqrt{2}$ and $r_2 = \sqrt{2}$, the circles intersect at (2, 2) and (4, 2).

c. If $r_1 + r_2 = 2$, then the circles will intersect at one point. For example, if $r_1 = 1$ and $r_2 = 1$, then the circles will intersect at (3, 2).

Lesson 9.4, pages 586–594

Activity

CHECKPOINT ✔

4. As the length of the string increases, the ellipse becomes larger and more closely resembles a circle.

Exercises

Communicate

1. You can obtain the location of the center, the direction and length of the major axis, and the length of the minor axis. Using this information, you can also find the coordinates of the vertices, co-vertices, and foci. For the given equation, the center is at (0, 2), the vertical major axis has a length of 10, and the minor axis has a length of 8. Therefore, the vertices are at (0, −3) and (0, 7), the co-vertices are at (−4, 2) and (4, 2), and the foci are at (0, −1) and (0, 5).

2. Complete the square and then divide to obtain a 1 on the right-hand side, as follows:

$$9x^2 + 4y^2 + 18x - 40y + 73 = 0$$
$$9(x^2 + 2x) + 4(y^2 - 10y) = -73$$
$$9(x^2 + 2x + 1) + 4(y^2 - 10y + 25) = -73 + 9 + 100$$
$$9(x+1)^2 + 4(y-5)^2 = 36$$
$$\frac{9(x+1)^2}{36} + \frac{4(y-5)^2}{36} = 1$$
$$\frac{(x+1)^2}{4} + \frac{(y-5)^2}{9} = 1$$

3. The center of the graph is at (4, 5). The major axis is horizontal with $a = 3$ and $b = 2$. Plot the vertices, (1, 5) and (7, 5), and the co-vertices, (4, 3) and (4, 7). Connect them in order to form an ellipse.

4. The equation can be written as $x^2 + y^2 = 25$, so the graph is a circle centered at the origin with a radius of 5.

Practice and Apply

15. $\frac{x^2}{4} + \frac{y^2}{1} = 1$; center: (0, 0); vertices: (−2, 0) and (2, 0); co-vertices: (0, −1) and (0, 1); foci: $(-\sqrt{3}, 0)$ and $(\sqrt{3}, 0)$

16. $\frac{x^2}{1} + \frac{y^2}{25} = 1$; center: (0, 0); vertices: (0, −5) and (0, 5); co-vertices: (−1, 0) and (1, 0); foci: $(0, -2\sqrt{6})$ and $(0, 2\sqrt{6})$

17. $\frac{x^2}{\frac{28}{3}} + \frac{y^2}{4} = 1$; center: (0, 0); vertices: $\left(-\frac{2\sqrt{21}}{3}, 0\right)$ and $\left(\frac{2\sqrt{21}}{3}, 0\right)$; co-vertices: (0, −2) and (0, 2); foci: $\left(-\frac{4\sqrt{3}}{3}, 0\right)$ and $\left(\frac{4\sqrt{3}}{3}, 0\right)$

18. $\frac{x^2}{16} + \frac{y^2}{4} = 1$; center: (0, 0); vertices: (−4, 0) and (4, 0); co-vertices: (0, −2) and (0, 2); foci: $(-2\sqrt{3}, 0)$ and $(2\sqrt{3}, 0)$

19. $\frac{x^2}{16} + \frac{y^2}{36} = 1$; center: (0, 0); vertices: (0, −6) and (0, 6); co-vertices: (−4, 0) and (4, 0); foci: $(0, -2\sqrt{5})$ and $(0, 2\sqrt{5})$

20. $\frac{x^2}{9} + \frac{y^2}{36} = 1$; center: (0, 0); vertices: (0, −6) and (0, 6); co-vertices: (−3, 0) and (3, 0); foci: $(0, -3\sqrt{3})$ and $(0, 3\sqrt{3})$

27.

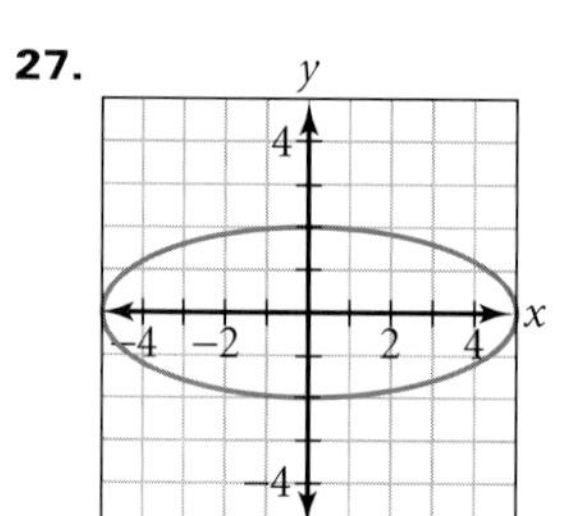

28.

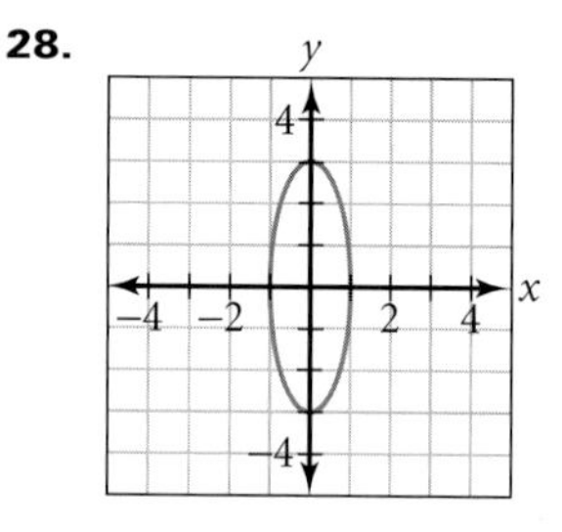

29.

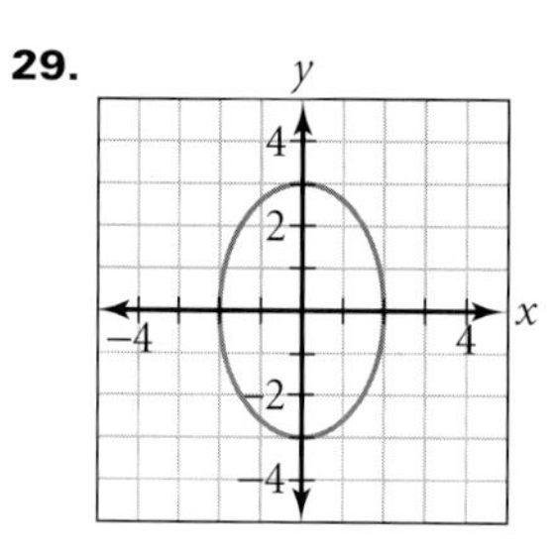

30.

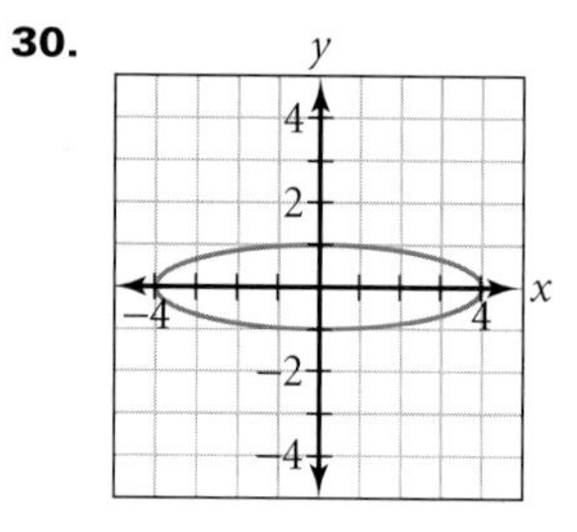

31.

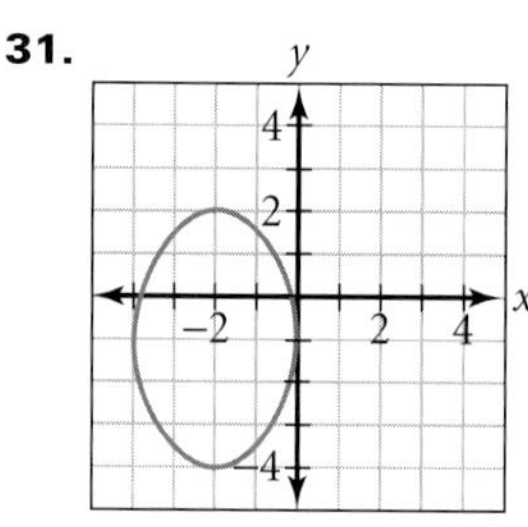

32.

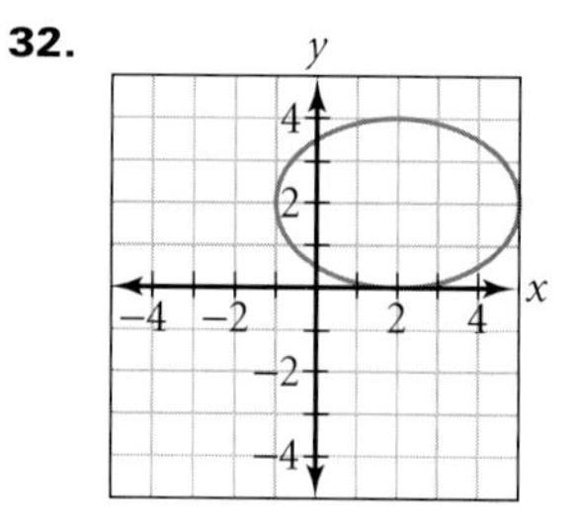

33.

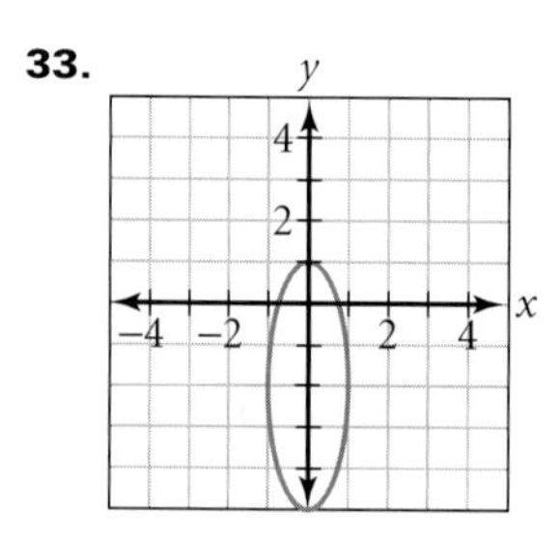

34.

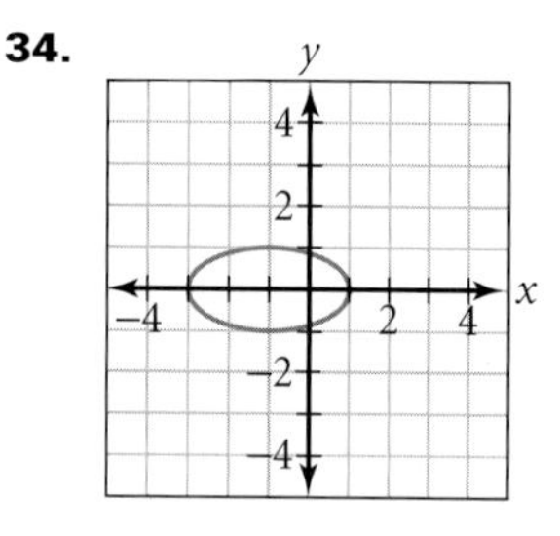

35.

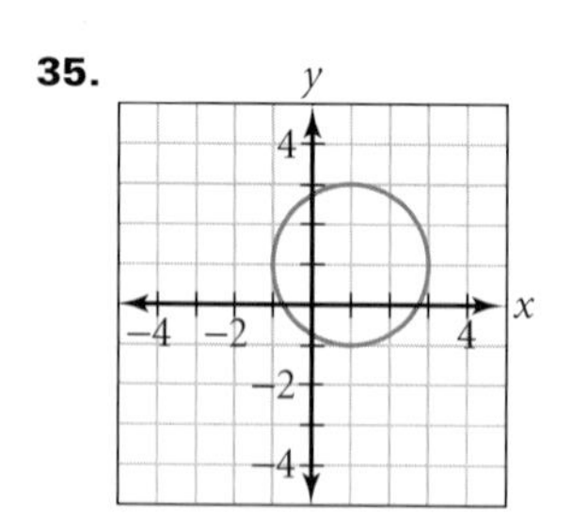

36.

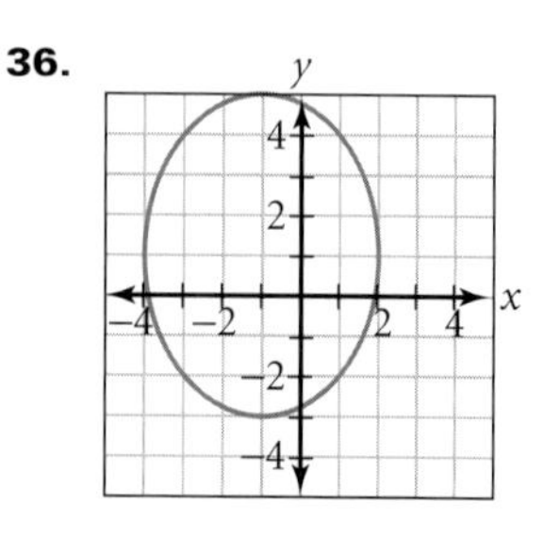

37.

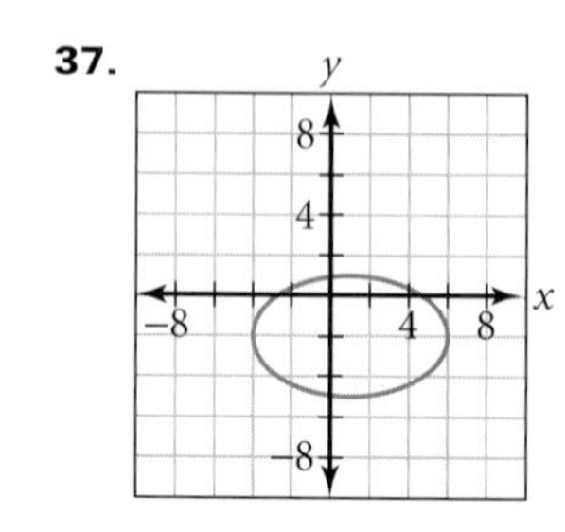

38.

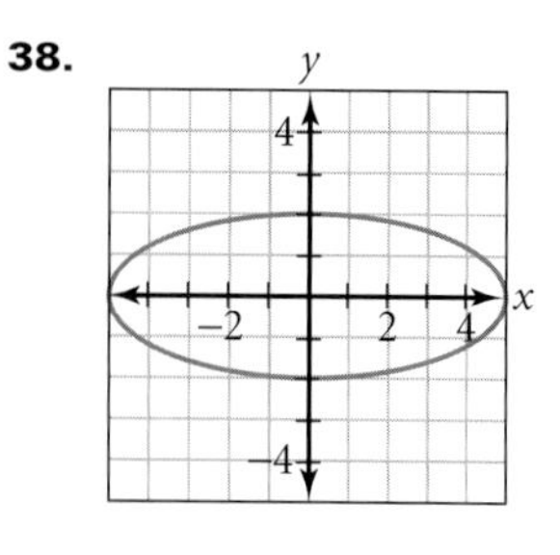

39.

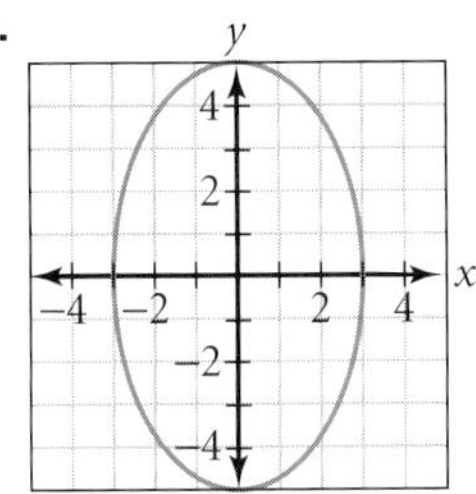

60. If the eccentricity, $E = \frac{c}{a}$, is 1, then the value of c is equivalent to the value of a. We know that $a^2 - b^2 = c^2$, so b has to be zero if a and c are equivalent. So the equation of the ellipse would be $\frac{x^2}{a^2} + \frac{y^2}{0} = 1$. Since we cannot have a 0 in the denominator, $a \neq c$ and the eccentricity, E, cannot be equal to 1.

64. a. $\frac{x^2}{144} + \frac{y^2}{81} = 1$

b. $\frac{x^2}{144} + \frac{y^2}{81} \leq 1$

c. a circle centered at point O

Portfolio Activity

1. Check students' work.
2. Check students' work.
3. Check students' work.
4. Check students' work.
5. Some of the figures are fairly round, and some are rather flat. The figures appear to be ellipses.
6. To make a flatter-shaped ellipse, move the point from Step 3 farther away from the center. To make a rounder-shaped ellipse, move the point from Step 3 closer to the center.
7. The point in Step 3 and the center of the circle are the two foci of the ellipse. The sum of the distances from any point on the ellipse to the two foci is a constant distance.

Lesson 9.5, pages 595–603

1. Yes, the curves get closer to the lines near the edges of the viewing window.

x	$y = \sqrt{9\left(\frac{x^2}{4} - 1\right)}$	$y = -\sqrt{9\left(\frac{x^2}{4} - 1\right)}$	$y = \frac{3}{2}x$	$y = -\frac{3}{2}x$
10	14.70	−14.70	15.00	−15.00
20	29.85	−29.85	30.00	−30.00
30	44.90	−44.90	45.00	−45.00
40	59.93	−59.92	60.00	−60.00
50	74.94	−74.94	75.00	−75.00
60	89.95	−89.95	90.00	−90.00

3. Yes, as x increases, the y-values of $y = \sqrt{9\left(\frac{x^2}{4} - 1\right)}$ and Ï $y = -\sqrt{9\left(\frac{x^2}{4} - 1\right)}$ become closer to the y-values of the lines $y = \frac{3}{2}x$ and $y = -\frac{3}{2}x$, respectively.

CHECKPOINT ✔

4. $y = \frac{4}{5}x$, $y = -\frac{4}{5}x$

x	$y = \sqrt{16\left(\frac{x^2}{25} - 1\right)}$	$y = -\sqrt{16\left(\frac{x^2}{25} - 1\right)}$	$y = \frac{4}{5}x$	$y = -\frac{4}{5}x$
10	6.93	−6.93	8.00	−8.00
20	15.49	−15.49	16.00	−16.00
30	23.66	−23.66	24.00	−24.00
40	31.75	−31.75	32.00	−32.00
50	39.80	−39.80	40.00	−40.00
60	47.83	−47.83	48.00	−48.00

Exercises

Communicate

1. If the equation has the form $\frac{(x-h)^2}{a^2} - \frac{(y-k)^2}{b^2} = 1$, the center is (h, k), the vertices are $(h - a, k)$ and $(h + a, k)$, and the co-vertices are $(h, k - b)$ and $(h, k + b)$. To find the foci, solve $a^2 + b^2 = c^2$ for c. The foci are $(h - c, k)$ and $(h + c, k)$.

   If the equation has the form $\frac{(y-k)^2}{a^2} - \frac{(x-h)^2}{b^2} = 1$, the center is (h, k), the vertices are $(h, k - a)$ and $(h, k + a)$, and the co-vertices are $(h - b, k)$ and $(h + b, k)$. To find the foci, solve $a^2 + b^2 = c^2$ for c. The foci are $(h, k - c)$ and $(h, k + c)$.

2. Sketch the asymptotes, and plot the vertices. Then for each of the two vertices, draw a curve that passes through the vertex and approaches the asymptotes. These curves approximate the hyperbola.

Practice and Apply

17. center: (0, 0); vertices: (−1, 0) and (1, 0); co-vertices: (0, −2) and (0, 2); foci: $(-\sqrt{5}, 0)$ and $(\sqrt{5}, 0)$; asymptotes: $y = -2x$ and $y = 2x$

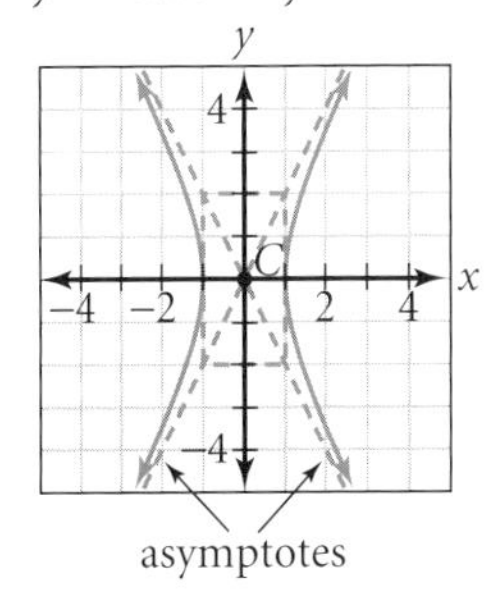

18. center: (0, 0); vertices: (0, −1) and (0, 1); co-vertices: (−3, 0) and (3, 0); foci: $(0, -\sqrt{10})$ and $(0, \sqrt{10})$; asymptotes: $y = -\frac{x}{3}$ and $y = \frac{x}{3}$

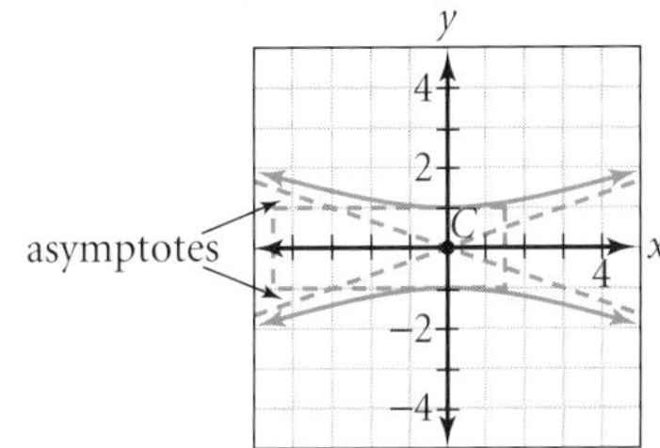

19. center: $(0, 0)$; vertices: $(0, -10)$ and $(0, 10)$; co-vertices: $(-8, 0)$ and $(8, 0)$; foci: $(0, -2\sqrt{41})$ and $(0, 2\sqrt{41})$; asymptotes: $y = -\frac{5x}{4}$ and $y = \frac{5x}{4}$

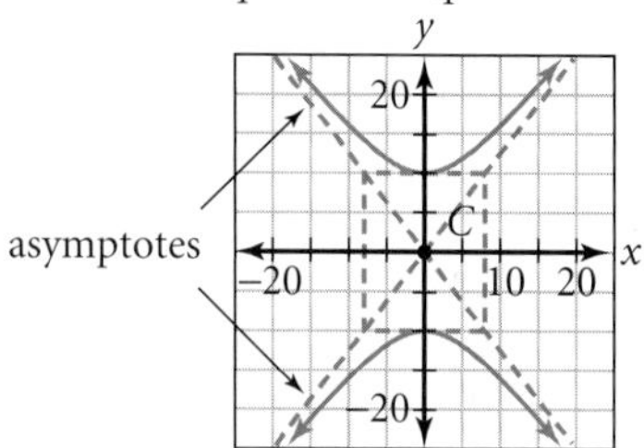

20. center: $(0, 0)$; vertices: $(-5, 0)$ and $(5, 0)$; co-vertices: $(0, -6)$ and $(0, 6)$; foci: $(-\sqrt{61}, 0)$ and $(\sqrt{61}, 0)$; asymptotes: $y = -\frac{6x}{5}$ and $y = \frac{6x}{5}$

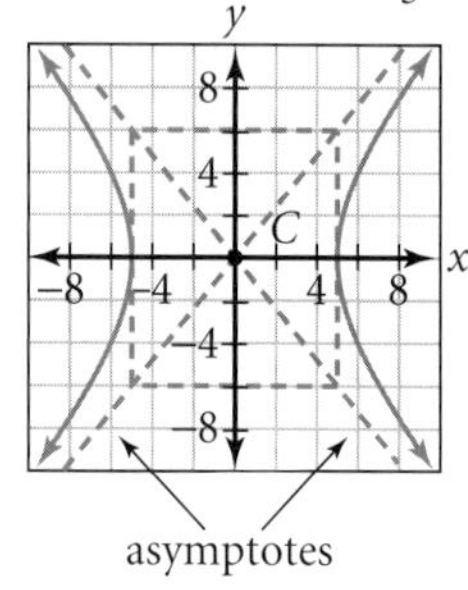

21. center: $(0, 0)$; vertices: $(-5, 0)$ and $(5, 0)$; co-vertices: $(0, -2)$ and $(0, 2)$; foci: $(-\sqrt{29}, 0)$ and $(\sqrt{29}, 0)$; asymptotes: $y = -\frac{2x}{5}$ and $y = \frac{2x}{5}$

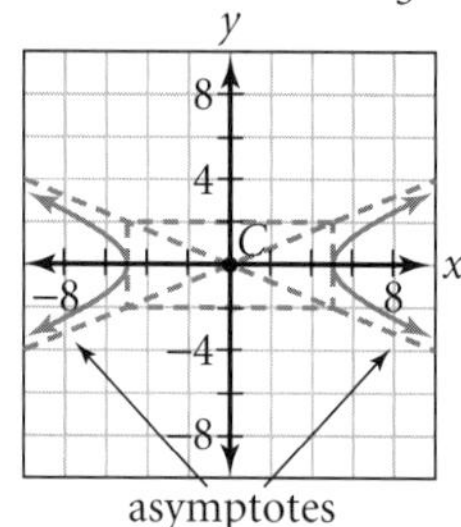

22. center: $(0, 0)$; vertices: $(0, -2)$ and $(0, 2)$; co-vertices: $(-6, 0)$ and $(6, 0)$; foci: $(0, -2\sqrt{10})$ and $(0, 2\sqrt{10})$; asymptotes: $y = -\frac{x}{3}$ and $y = \frac{x}{3}$

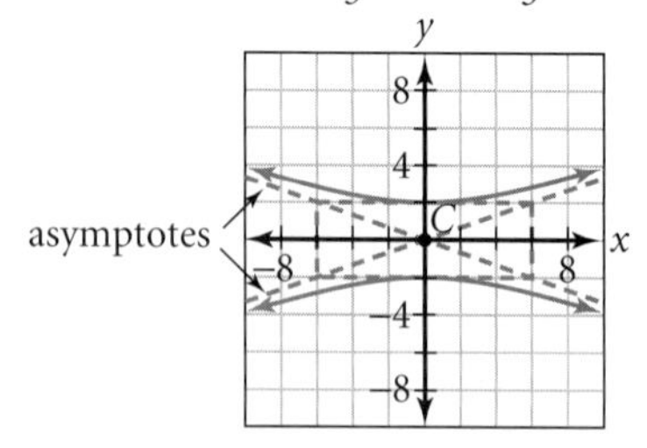

23. center: $(1, -2)$; vertices: $(-1, -2)$ and $(3, -2)$; co-vertices: $(1, -5)$ and $(1, 1)$; foci: $(1 - \sqrt{13}, -2)$ and $(1 - \sqrt{13}, -2)$; asymptotes: $y = -\frac{3x}{2} - \frac{1}{2}$ and $y = \frac{3x}{2} - \frac{7}{2}$

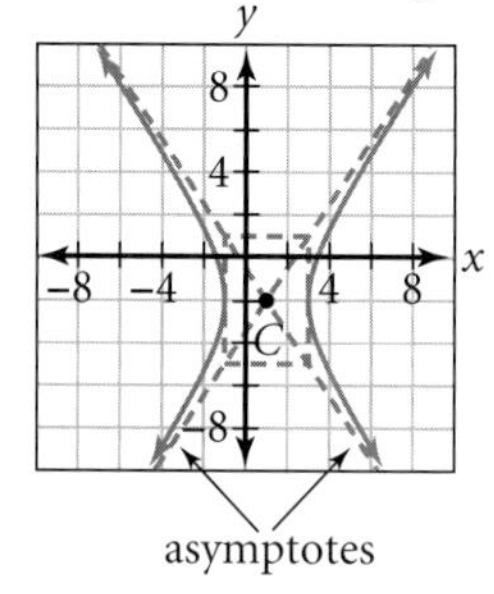

24. center: $(-2, 2)$; vertices: $(-5, 2)$ and $(1, 2)$; co-vertices: $(-2, -2)$ and $(-2, 6)$; foci: $(-7, 2)$ and $(3, 2)$; asymptotes: $y = -\frac{4x}{3} - \frac{2}{3}$ and $y = \frac{4x}{3} + \frac{14}{3}$

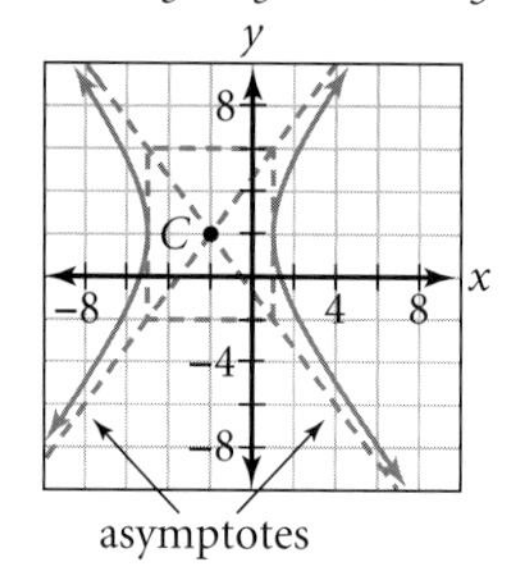

47. a. vertices: $V_1(-a, 0)$ and $V_2(a, 0)$; co-vertices: $C_1(0, -a)$ and $C_2(0, a)$

b. square

c. Answers may vary. Sample answer:

Side	Length	Slope
$\overline{V_1C_1}$	$\sqrt{[0-(-a)]^2 + (-a-0)^2} = \sqrt{2a^2} = a\sqrt{2}$	$\frac{-a-0}{0-(-a)} = -1$
$\overline{C_1V_2}$	$\sqrt{(a-0)^2 + [0-(-a)]^2} = \sqrt{2a^2} = a\sqrt{2}$	$\frac{0-(-a)}{a-0} = 1$
$\overline{V_2C_2}$	$\sqrt{(0-a)^2 + (a-0)^2} = \sqrt{2a^2} = a\sqrt{2}$	$\frac{a-0}{0-a} = -1$
$\overline{C_2V_1}$	$\sqrt{(-a-0)^2 + (0-a)^2} = \sqrt{2a^2} = a\sqrt{2}$	$\frac{0-a}{-a-0} = 1$

The table shows that the sides all have the same length and that the sides meet at right angles because the slopes of adjacent sides are negative reciprocals. Therefore, quadrilateral $V_1C_1V_2C_2$ is a square.

Portfolio Activity

1. Check students' work.

2. Check students' work.

3. Check students' work.

4. Check students' work.

5. Some of the figures are wider than others. The figures are hyperbolas.

6. To make a flatter shape, move the point from Step 3 closer to the circle. To make a more open shape, move the point from Step 3 farther away from the circle.

7. The point in Step 3 is one foci of the hyperbola. The other foci is the center of the circle. The absolute value of the difference between the distance from the point in Step 3 to any point on the hyperbola and the distance from this point on the hyberbola to the center of the circle is a constant.

Lesson 9.6, pages 606–613

Activity

1. **a.** 2 intersection points
   b. 4 intersection points
   c. 2 intersection points
   d. 2 intersection points

CHECKPOINT ✔

2. A circle and a line can intersect at 0, 1, or 2 points.

CHECKPOINT ✔

3. A circle and a hyperbola can intersect at 0, 1, 2, 3, or 4 points.

CHECKPOINT ✔

4. A parabola and a line can intersect at 0, 1, or 2 points.

CHECKPOINT ✔

5. A circle and a parabola can intersect at 0, 1, 2, 3, or 4 points.

Exercises

Communicate

1. Substitute $(2x)^2$, or $4x^2$, for y^2 in the second equation. Then solve for x.

$$x^2 + y^2 = 10$$
$$x^2 + 4x^2 = 10$$
$$5x^2 = 10$$
$$x = \pm\sqrt{2}$$

To find the corresponding y-values, substitute $\sqrt{2}$ and $-\sqrt{2}$ for x in the first equation.

$y = 2x$ $\qquad$ $y = 2x$
$y = 2(\sqrt{2})$ $\qquad$ $y = 2(-\sqrt{2})$
$y = 2\sqrt{2}$ $\qquad$ $y = -2\sqrt{2}$

The solutions are $(\sqrt{2}, 2\sqrt{2})$ and $(-\sqrt{2}, -2\sqrt{2})$.

2. Add the corresponding sides of each equation.

$$2x^2 - y^2 = 1$$
$$\underline{x^2 + y^2 = 5}$$
$$3x^2 = 6$$
$$x^2 = 2$$
$$x = \pm\sqrt{2}$$

Substitute 2 for x^2 in $x^2 + y^2 = 5$.

$$x^2 + y^2 = 5$$
$$2 + y^2 = 5$$
$$y^2 = 3$$
$$y = \pm\sqrt{3}$$

There are 4 solutions, $(\pm\sqrt{2}, \pm\sqrt{3})$, or about $(\pm1.41, \pm1.73)$.

3. no intersection, 1, 2, 3, 4, or infinitely many intersection points

4. If $AC > 0$, then A and C have the same sign (both positive or both negative). If $AC < 0$, then A and C have opposite signs. If $AC = 0$, then $A = 0$, $C = 0$, or both.

Chapter 9 Project

Activity 1

1. **a.**

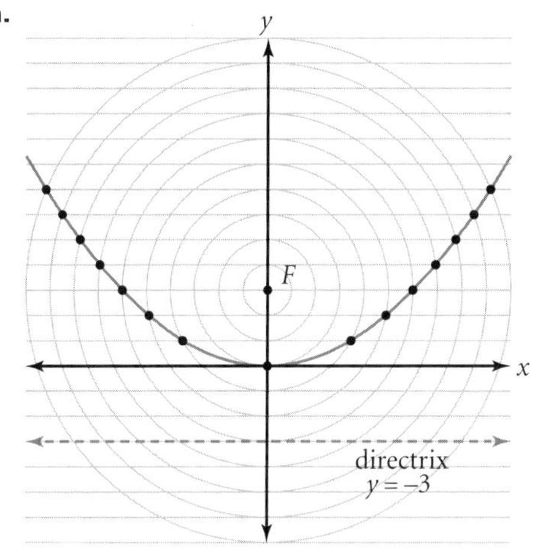

2. **a.**

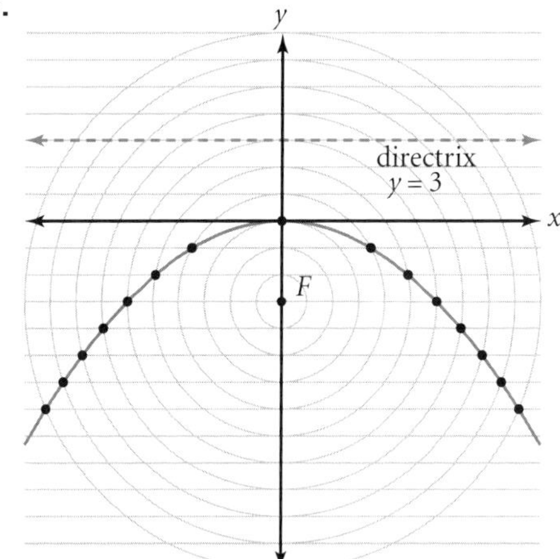

3. **a.**

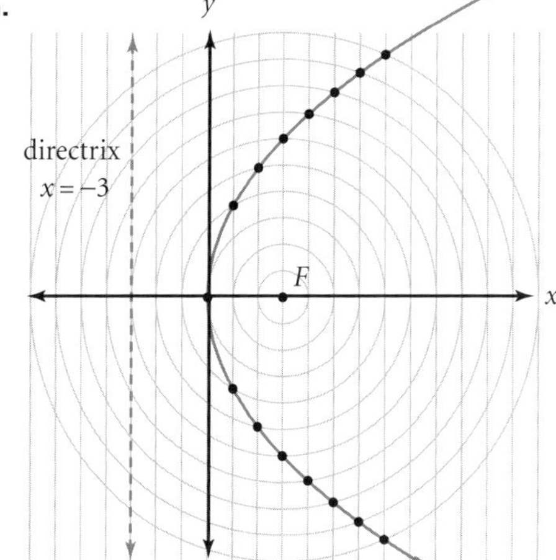

4. a.

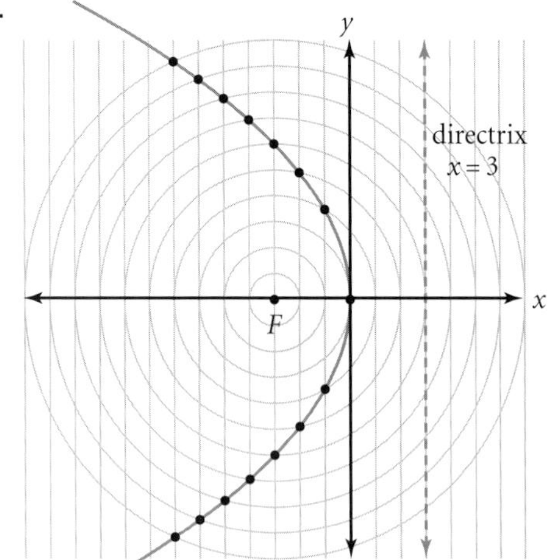

Activity 2

1. a.

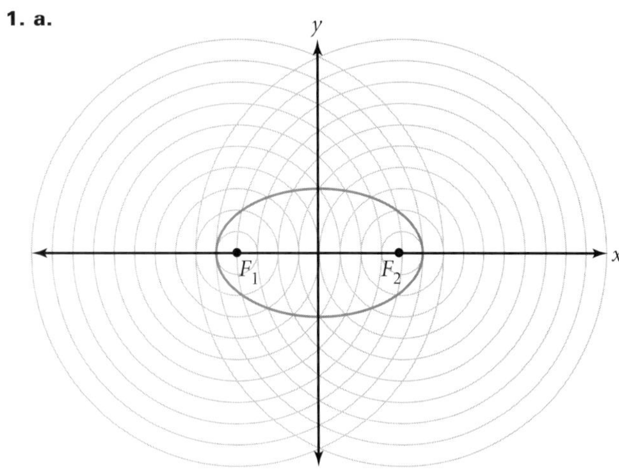

2. a.

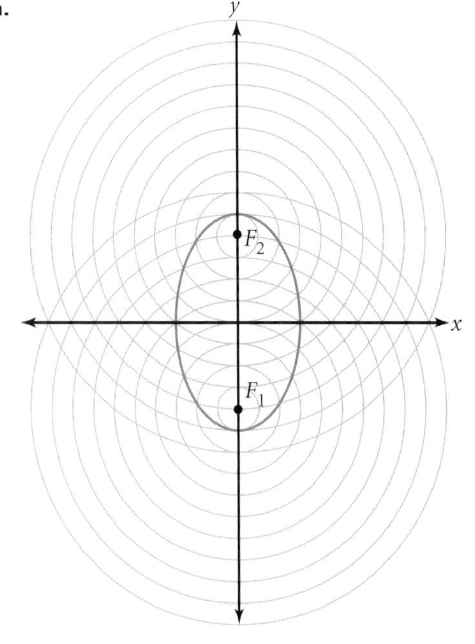

Activity 3

1. d. $y = \frac{4}{3}x$, $y = -\frac{4}{3}x$

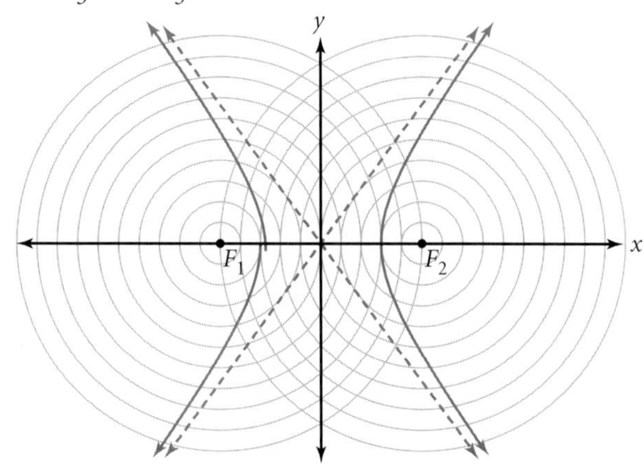

Chapter 9 Review and Assessment

6.

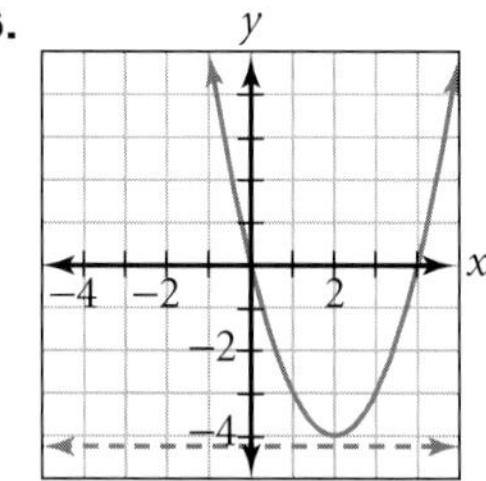

7.

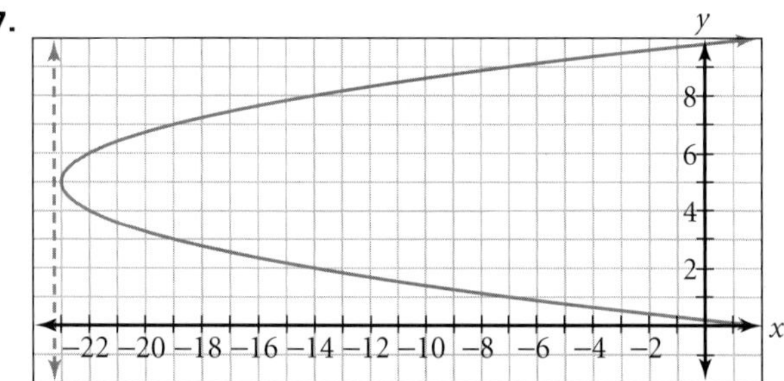

8.

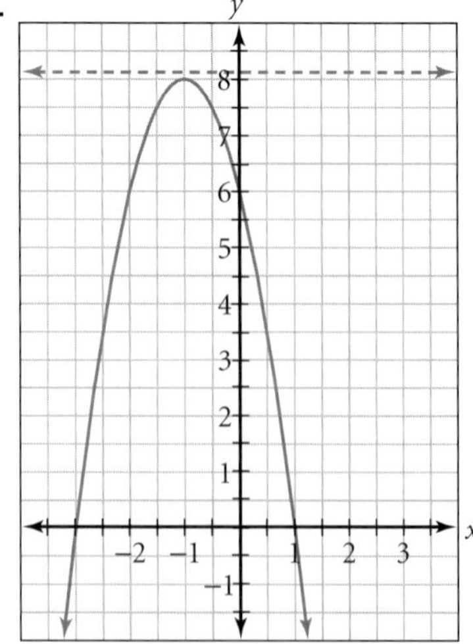

9.

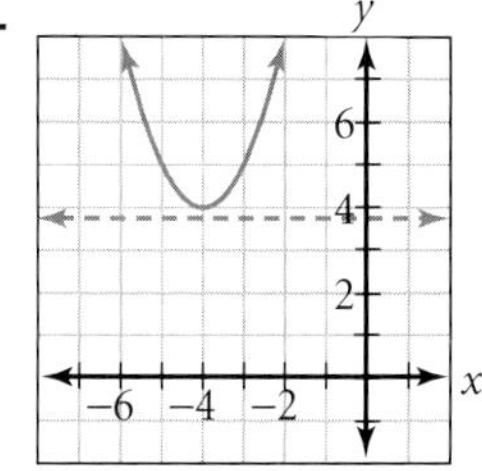

10.

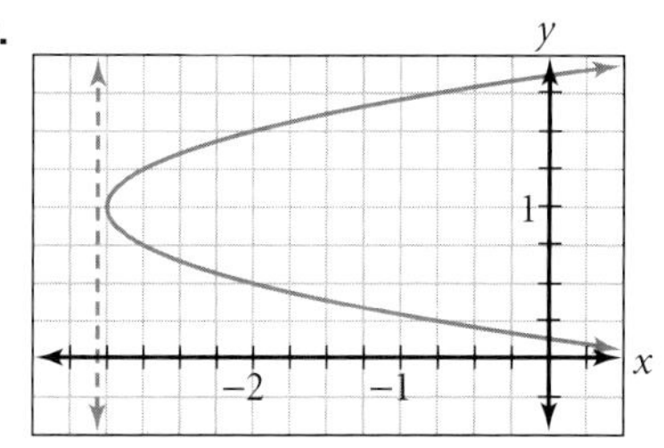

11.

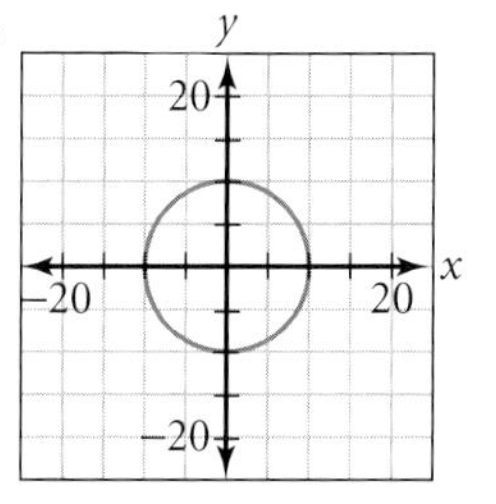

12.

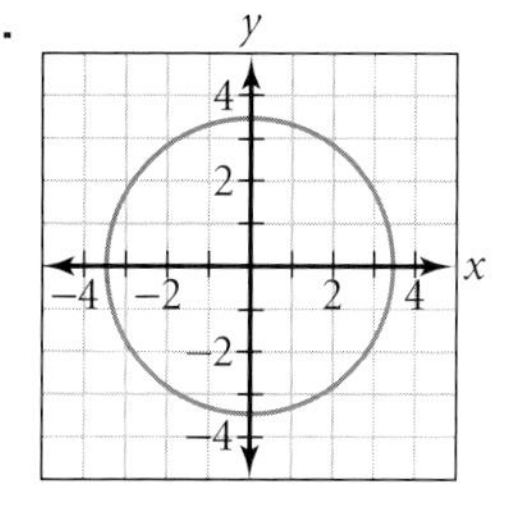

13.

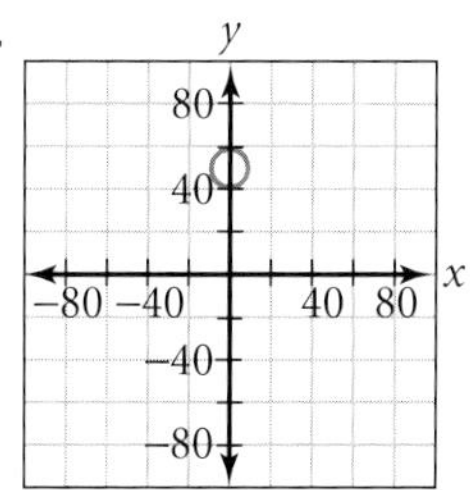

14.

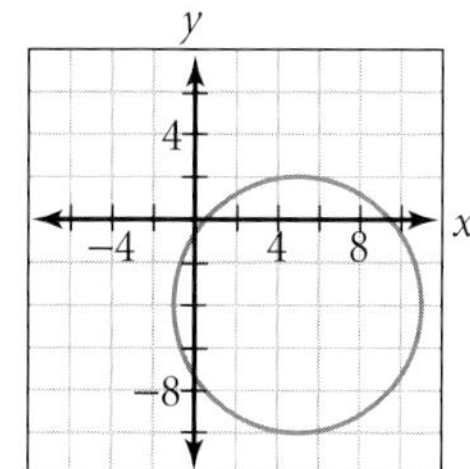

15.

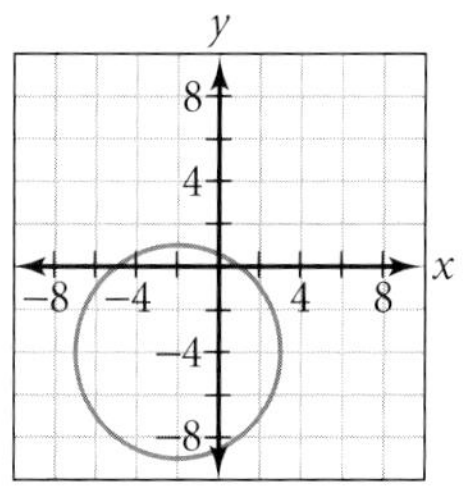

16.

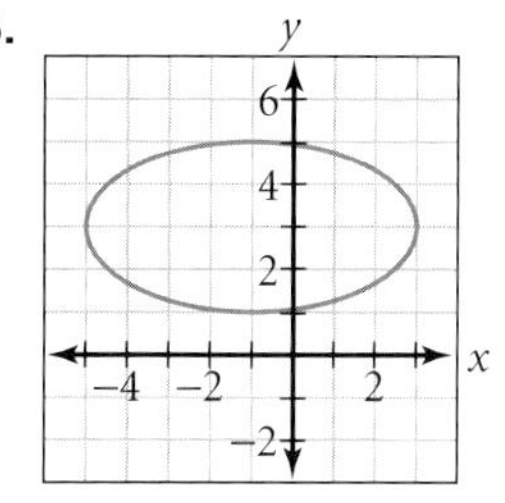

17.

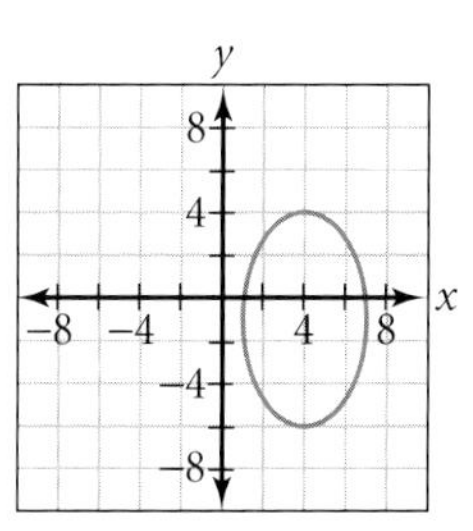

18.

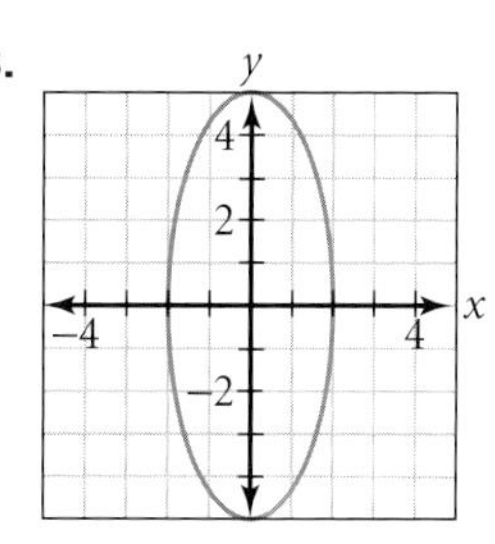

19.

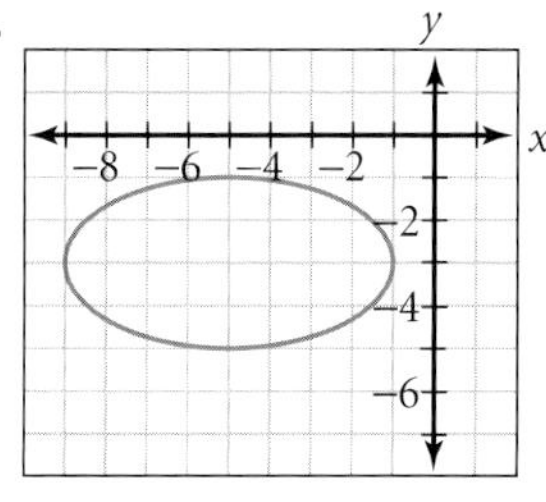

20. $\frac{(x-3)^2}{4} + \frac{(y+1)^2}{16} = 1$; center: $(3, -1)$; vertices: $(3, -5)$ and $(3, 3)$; co-vertices: $(1, -1)$ and $(5, -1)$; foci: $\left(3, -1 - 2\sqrt{3}\right)$ and $\left(3, -1 + 2\sqrt{3}\right)$

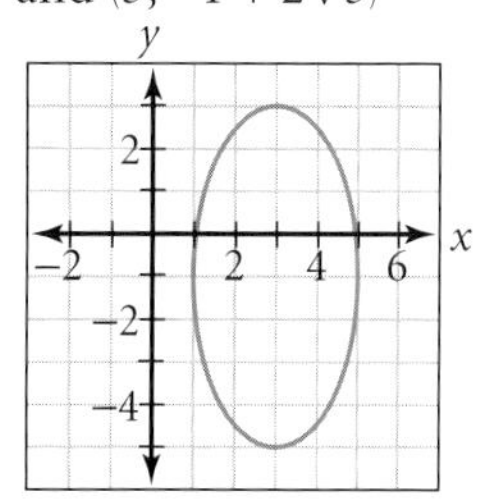

21. $\frac{(x+5)^2}{36} - \frac{(y-1)^2}{64} = 1$; center: $(-5, 1)$; vertices: $(-11, 1)$ and $(1, 1)$; co-vertices: $(-5, -7)$ and $(-5, 9)$; foci: $(-15, 1)$ and $(5, 1)$

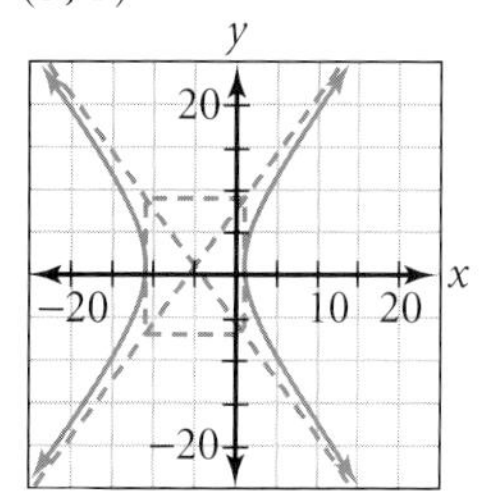

22. $\frac{(y+5)^2}{4} - \frac{(x-4)^2}{9} = 1$; center: $(4, -5)$; vertices: $(4, -7)$ and $(4, -3)$; co-vertices: $(1, -5)$ and $(7, -5)$; foci: $\left(4, -5 - \sqrt{13}\right)$ and $\left(4, -5 + \sqrt{13}\right)$

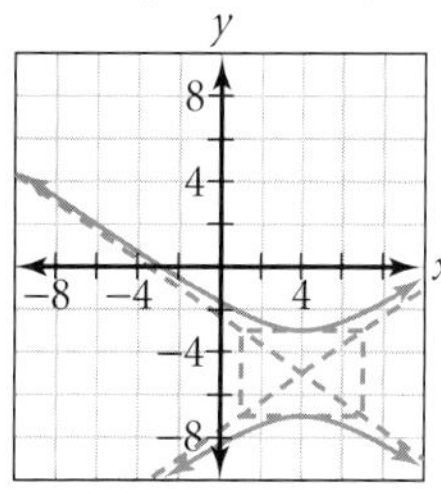

23. $\frac{y^2}{25} - \frac{x^2}{4} = 1$; center: $(0, 0)$; vertices: $(0, -5)$ and $(0, 5)$; co-vertices: $(-2, 0)$ and $(2, 0)$; foci: $\left(0, -\sqrt{29}\right)$ and $\left(0, \sqrt{29}\right)$

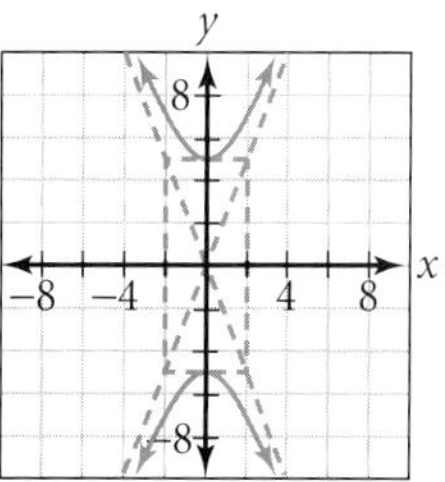

24. $\frac{(x-5)^2}{16} - \frac{(y-1)^2}{9} = 1$; center: $(5, 1)$; vertices: $(1, 1)$ and $(9, 1)$; co-vertices: $(5, -2)$ and $(5, 4)$; foci: $(0, 1)$ and $(10, 1)$

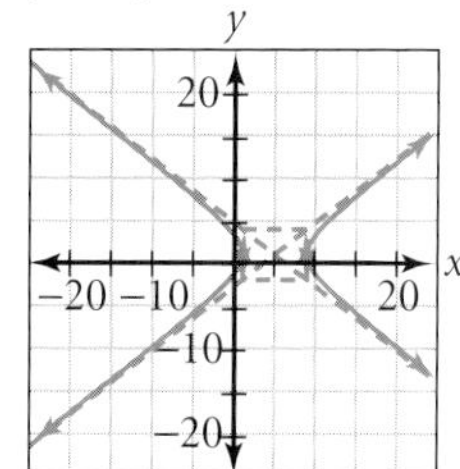

25. $\frac{(y+3)^2}{4} - \frac{(x+5)^2}{36} = 1$; center: $(-5, -3)$; vertices: $(-5, -5)$ and $(-5, -1)$; co-vertices: $(-11, -3)$ and $(1, -3)$; foci: $\left(-5, -3 - 2\sqrt{10}\right)$ and $\left(-5, -3 + 2\sqrt{10}\right)$

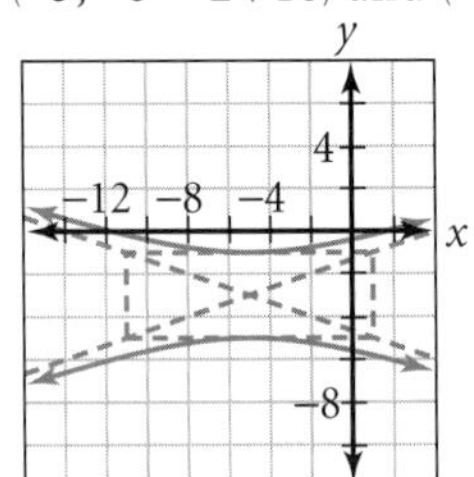

Lesson 10.1, pages 628–635

Activity

2. The order of the items may vary. Sample answer:

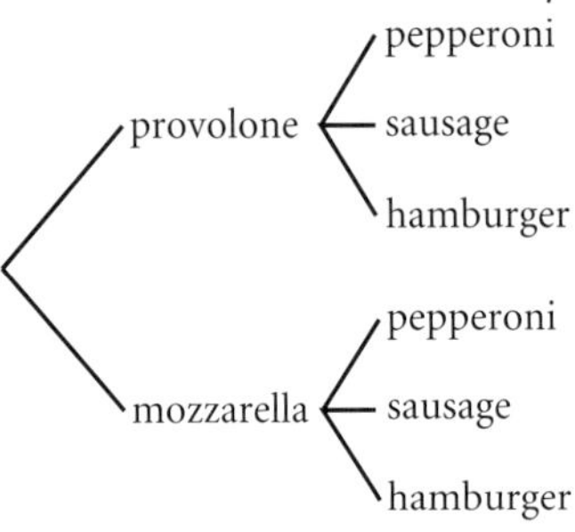

3. 6 combinations are possible.

Exercises

Communicate

1. Answers may vary. Sample answer: rolling a number cube and having an odd number show; throwing a dart at a dartboard and having it hit a dark area of the second ring.

2. Possible answer: Theoretical probability is based on the assumption that all outcomes are equally likely. Experimental probability is based on actual trials. Both theoretical and experimental probabilities are ratios of the number of outcomes in an event and the total number of outcomes (trials).

3. Answers may vary. Sample answer: Using a ratio, the area corresponding to an event can be compared with the area corresponding to the sample space in order to determine the probability of the event.

CHECKPOINT ✔

4. Add a branch for onions and a branch for green peppers after each meat. Twelve combinations are possible.

Practice and Apply

24.

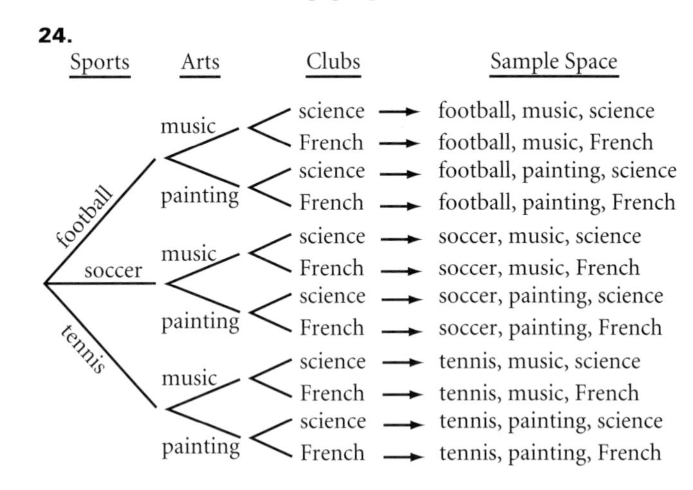

25.

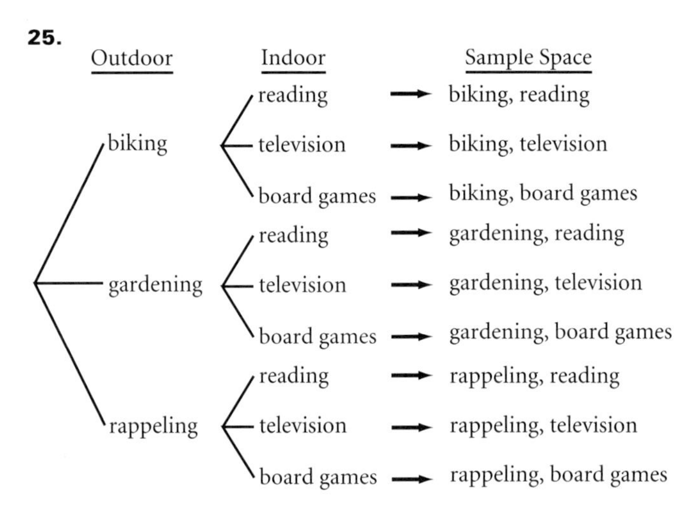

Portfolio Activity

1 and 2. Answers will vary. Sample:

Section	Number of times
A	9
B	5
C	9
D	5
E	4
F	8

The spinner landed in region D 5 times.

3. Answers will vary. Sample for the data collected in Step 1: An estimate for the probability that Eduardo logs onto his mail from 1:30 P.M. to 1:40 P.M. is $\frac{1}{8}$, or 12.5%.

4. The theoretical probability was about 16.7%, so the experiment had a lower probability. This could be due to having only 40 trials and also to the way the spinner was spun.

Lesson 10.2, pages 636–642

Activity

1. **a.** Both equal 210.
   b. Both equal 840.
   c. Both equal 2520.

2. **a.** $7 \times 6 \times \cdots \times (8 - r)$
   b. $n \times (n - 1) \times \cdots \times (n + 1 - r)$

CHECKPOINT ✔

3. ${}_8P_4 = 8 \times 7 \times 6 \times 5 = 1680$
   ${}_8P_5 = 8 \times 7 \times 6 \times 5 \times 4 = 7620$
   ${}_{10}P_8 = 10 \times 9 \times 8 \times 7 \times 6 \times 5 \times 4 \times 3 = 1{,}814{,}400$

Exercises

Communicate

1. Answers may vary. Sample answer: Think of selecting the first object as one event, selecting the second object as another event, and so on. Then there are 4 ways that the first event can occur (4 choices), 3 ways that the second event can occur, 2 ways that the third event can occur, and 1 way that the fourth event can occur. By the Fundamental Counting Principle, there are 4 • 3 • 2 • 1, or 24, permutations of 4 objects.

2. There are more permutations when the letters are taken 3 at a time. Explanations may vary. Sample answer: ${}_4P_2 = \frac{4!}{(4-2)!} = \frac{4}{2} = 12$, while ${}_4P_3 = \frac{4!}{(4-3)!} = \frac{4}{1} = 24$. There are 6 distinct pairs that can be chosen from 4 letters, but each pair can be arranged in only 2!, or 2, ways. There are only 4 distinct triples that can be chosen from 4 letters, but each triple can be arranged in 3!, or 6, ways.

3. ${}_4P_4 = \frac{4!}{(4-4)!} = \frac{4}{0} = 4!$
   ${}_4P_3 = \frac{4!}{(4-3)!} = \frac{4}{1} = 4!$
   ${}_4P_4 = {}_4P_3$ because $0! = 1 = 1!$
   Since ${}_nP_n = \frac{n!}{(n-n)!} = \frac{n!}{0!} = n!$ and ${}_nP_{n-1} = \frac{n!}{(n-(n-1))!} = \frac{n!}{1!}$
   $= n!$, ${}_nP_n = {}_nP_{n-1}$ is always true.

4. Yes, because one distinct circular permutation of 5 objects corresponds to 5 distinct linear permutations of the objects.

Lesson 10.3, pages 643–649

Activity

1. 8-4-1

2. 8-4-1, 8-1-4,
   4-8-1, 4-1-8,
   1-8-4, 1-4-8

3. any-order match

CHECKPOINT ✔

4. There are fewer ways to win with an exact match, so it is harder to do.

Exercises

Communicate

1. In a combination, order does not matter, but in a permutation order does matter. Examples may vary.

2. The number of permutations includes all of the orderings of the combinations, so ${}_5P_3 = 3!({}_5C_3)$, or $6 \times {}_5C_3$. The formula for combinations divides out all of the orderings from the permutation formula, so ${}_nC_r = \frac{n!}{r!(n-r)!} = \frac{1}{r!}({}_nP_r)$.

3. ${}_8C_8 = \frac{8!}{8!0!} = 1$; ${}_nC_n = 1$ because there is only one way to choose the entire set of objects.
   ${}_8C_1 = \frac{8!}{1!7!} = 8$; ${}_nC_1 = n$ because there are n ways to choose 1 at a time.

Look Back

47.

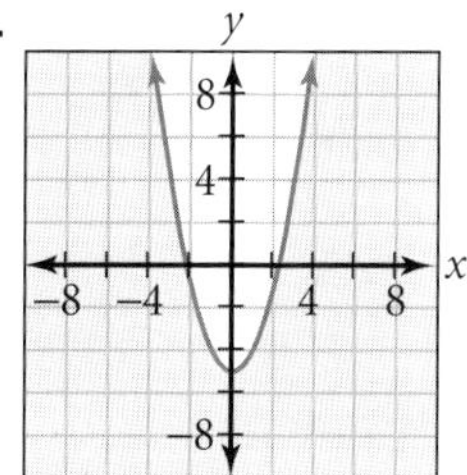

48.

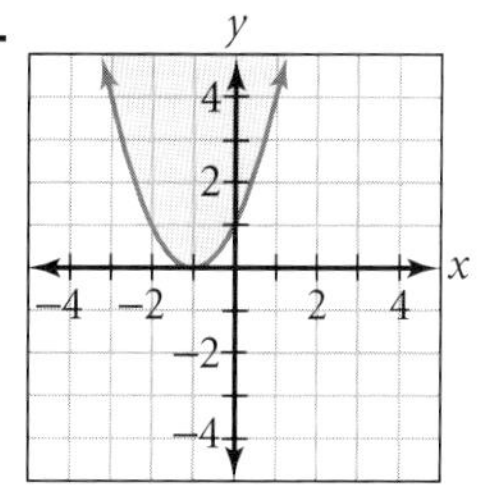

49.

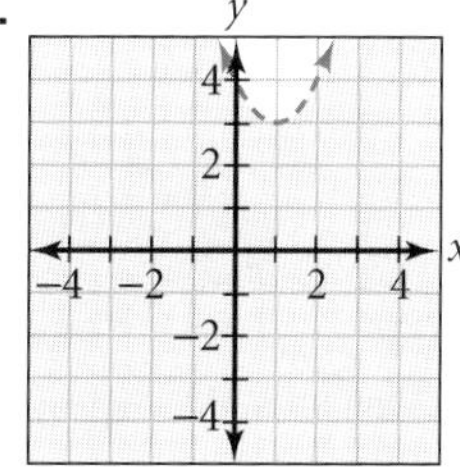

Portfolio Activity

1. Answers will vary. Sample answer:

Trial number	1st #	2nd #	3rd #	4th #	5th #	6th #
1	9	23	3	16	25	2
2	2	2	24	5	19	3
3	16	1	3	1	10	20
4	25	11	25	3	14	6
5	14	21	21	20	25	6
6	7	15	21	19	4	24
7	21	11	16	25	20	22
8	12	11	17	6	24	5
9	15	4	17	20	6	5
10	19	4	1	22	2	22
11	9	8	3	5	6	2
12	25	2	11	13	20	14
13	13	2	14	3	21	18
14	24	17	9	1	5	16
15	4	22	9	15	3	18
16	9	24	12	19	6	9
17	1	10	10	19	23	17
18	2	24	10	2	18	16
19	3	2	1	10	24	17
20	14	12	23	2	24	10
21	1	22	18	17	7	20
22	15	10	13	6	13	25
23	22	7	1	18	6	12
24	5	7	22	9	23	23
25	5	25	18	11	15	22
26	24	15	24	9	7	22
27	6	25	23	25	6	21
28	10	5	24	8	6	12
29	3	22	1	24	24	24
30	15	3	12	8	9	22

2. Answers will vary. Sample answer for the data shown in the answer to Step 1: The total number of times that exactly 2 of the 6 random integers were between 1 and 17 was 4 (in trials 5, 7, 27, and 29). Using these outcomes, the probability that exactly 2 out of 6 respondents chosen favor the proposed regulation is $\frac{2}{15}$, or about 13%.

3. Answers will vary. Sample answer for the data shown in the answer to Step 1: The theoretical probability was about 5%, so the random-number generator produced an 8% higher probability. The model did not produce very good results. The results would probably have been closer if more than 30 trials had been performed.

Lesson 10.4, pages 652–658

Activity

1. The table should show reasonable and correct values.

2. The table should reflect the values in Step 1.

CHECKPOINT ✔

3. $P(A \text{ or } B) = P(A) + P(B)$ is true when A and B are mutually exclusive.

Exercises

Communicate

1. *Mutually exclusive* means that they cannot occur at the same time. For example, if you flip a coin, you cannot get both heads and tails. *Inclusive* means that they can occur at the same time. For example, a student can belong to more than one club at a time.

2. rolling any other number: 3, 4, 5, or 6

3. Because "rolling a 3" is included in "rolling an odd number," the events are inclusive. The probability is $P(\text{odd number}) + P(3) - P(\text{odd number } and \text{ } 3)$, which in this case, is simply $P(\text{odd number})$.

Portfolio Activity

1. Answers will vary. Check students' tables.

2. Answers will vary. An estimate of the probability of rolling "a sum of 7" or "a sum of 11" may be $\frac{5}{40} + \frac{2}{40} = \frac{7}{40}$, or 17.5%. The theoretical probability of rolling "a sum of 7" or "a sum of 11" is $\frac{6}{36} + \frac{2}{36} = \frac{8}{36}$, or about 22%, which is 5% greater than the experimental probability.

3. Answers will vary. Check students' tables.

4. Answers will vary. An estimate of the probability of rolling "a sum of 7" or a "sum of 11" is $\frac{12}{80} + \frac{4}{80} = \frac{8}{40}$, or 20%. The theoretical probability of rolling "a sum of 7" or "a sum of 11" is $\frac{8}{36}$, or about 22%. This is a smaller difference, about 2%. As the number of trials increases, the theoretical and experimental probabilities should become closer to the same value. (Note: Students may need to complete at least 100 trials to obtain improved results.)

Lesson 10.5, pages 659–663

Activity

1. $P(A) = \frac{1}{2}$

2. $P(B) = \frac{1}{6}$

3. No; no; tossing a coin and rolling a number cube have no effect on one another.

4. $\{(H, 1), (H, 2), (H, 3), (H, 4), (H, 5), (H, 6), (T, 1), (T, 2), (T, 3), (T, 4), (T, 5), (T, 6)\}$; $P(A \text{ and } B) = \frac{1}{12}$

5. $P(A) \times P(B) = \frac{1}{2} \times \frac{1}{6} = \frac{1}{12}$

6. $P(C) = \frac{1}{2}$

7. $P(A \text{ and } C) = \frac{3}{12} = \frac{1}{4}$; $P(A) \times P(C) = \frac{1}{2} \times \frac{1}{2} = \frac{1}{4}$; yes, $P(A \text{ and } C) = P(A) \times P(C)$

CHECKPOINT ✔

8. If two events have no effect on one another, the probability that they both occur is the product of their individual probabilities.

Exercises

Communicate

1. Answers may vary. Sample answer: Given two bags, each containing 2 red, 1 yellow, 1 green, and 2 blue marbles, the events of drawing one red marble from each bag are independent. The events of drawing a marble from one bag and then, without replacing the first marble, drawing another marble from the same bag are dependent.

2. For independent events A and B, $P(A \text{ and } B) = P(A) \times P(B)$.

3. If events A and B are mutually exclusive, the occurrence of one makes the other impossible. If events A and B are independent, the occurrence of one does not affect the occurrence of the other. Examples may vary. Sample answer: A coin showing heads and a coin showing tails for a single toss are mutually exclusive events. A coin showing heads on one toss and showing tails on a second toss are independent events.

Lesson 10.6, pages 664–670

Activity

1. $\frac{1}{13}$ 2. $\frac{1}{3}$ 3. $\frac{2}{25}$ 4. $\frac{1}{25}$

CHECKPOINT ✔

5. The sample space is decreased so that the corresponding probabilities are increased; the number of outcomes in the event is decreased so that the probability is decreased; the number of outcomes in the event and/or the sample space may change.

Exercises

Communicate

1. $P(B|A)$ means "the probability of B given A."

2. $P(A \text{ and } B)$ is the probability that both A and B occur. $P(B|A)$ is the probability that B will occur given that A has occurred (or will occur). They are related by the equation $P(B|A) = \frac{P(A \text{ and } B)}{P(A)}$.

3. If A and B are mutually exclusive, $P(B|A)$ must be 0 because B cannot occur if A has occurred. Mathematically, $P(A \text{ and } B) = 0$, so $P(B|A) = \frac{P(A \text{ and } B)}{P(A)} = \frac{0}{P(A)} = 0$.

Lesson 10.7, pages 671–677

Activity

1–3. Answers may vary. The estimated probability should be close to $\frac{1}{4}$.

4. $\frac{1}{4}$

5. Check students' work.

CHECKPOINT ✔

6. Answers may vary. In general, the average ratio should be closest to the theoretical probability. As the number of trials increases, the estimated probability should approach the theoretical probability.

Exercises

Communicate

1. Use **randInt(1,6).**

2. For a situation involving two equally likely outcomes; explanations may vary.

3. 100 trials; as the number of trials in the experiment increases, the results will more closely approximate the theoretical probability.

Chapter 10 Project

Activity 1

t	$P(t)$	N
0	0	000
10	0.052	001–052
20	0.132	053–184
30	0.158	185–342
40	0.135	343–477
50	0.123	478–600
60	0.104	601–704
70	0.058	705–762
80	0.034	763–796
90	0.116	797–912
100	0.050	913–962
110	0.026	963–988
120	0.012	989–1000

Additional Answers

Activity 2

1–3. Students should record simulation results. Sample partial table:

Customer	Number, N	Time, t
1	672	60
2	500	50
3	703	60
4	264	30
5	773	80
6	983	110

Lesson 11.1, pages 690–698

Activity

1. 50; 50

CHECKPOINT ✔

2. Yes, they both equal 50.

3. 10, 15, 20

CHECKPOINT ✔

4. $\sum_{k=1}^{n} c = nc$

Exercises

Communicate

1. A sequence is an ordered list of terms, and a series is the expression of the sum of the terms of a sequence. An example of a sequence is 1, 3, 5, 7, and the corresponding series is 1 + 3 + 5 + 7.

2. Answers may vary. With an explicit formula, any term may be found by substituting a value for n; for $t_n = 2n + 1$, $t_{50} = 101$. With a recursive formula, a term is defined by one or more given terms; for $t_n = 2t_{n-1} - 1$, we cannot find t_{50} without knowing t_{49}.

3. In $\sum_{1=1}^{3} (i + 10)$, 10 is added three times; in $\sum_{i=1}^{3} i + 10$, 10 is added once.

Lesson 11.2, pages 699–706

Activity

1. a.

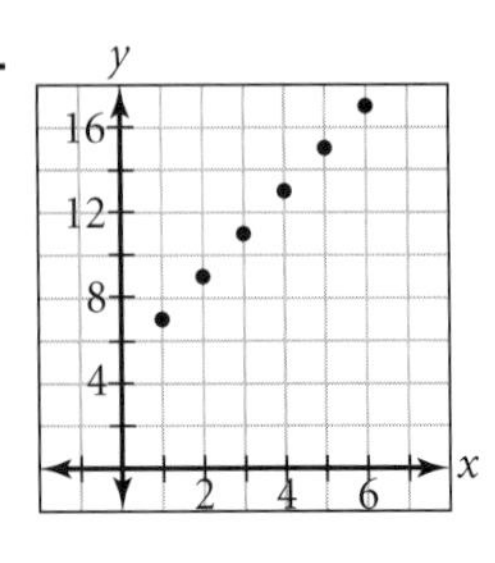

b.

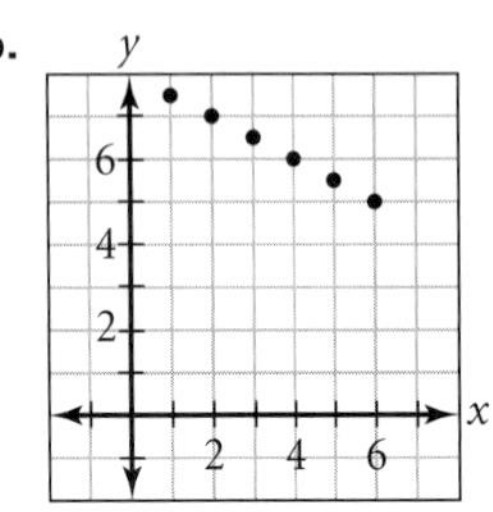

c.

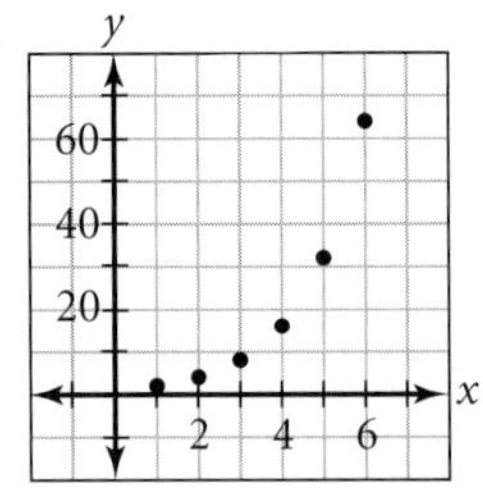

d.

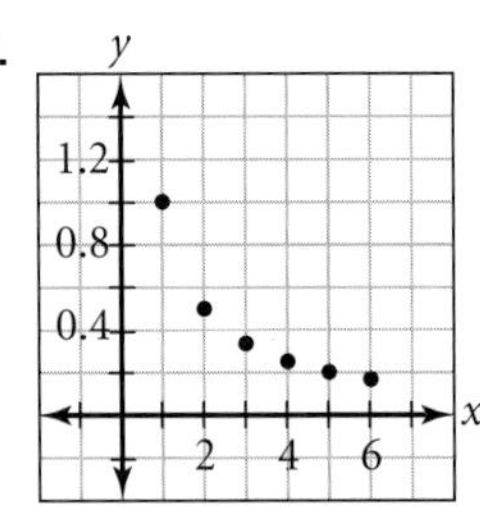

e.

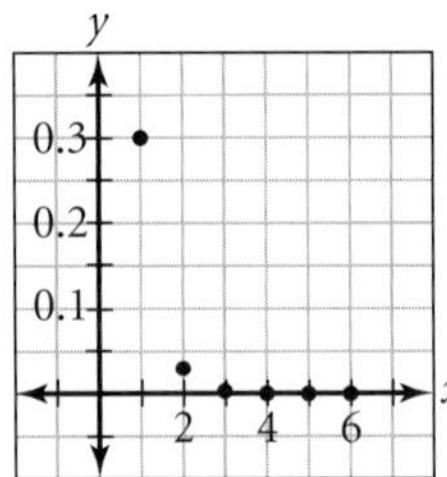

f.

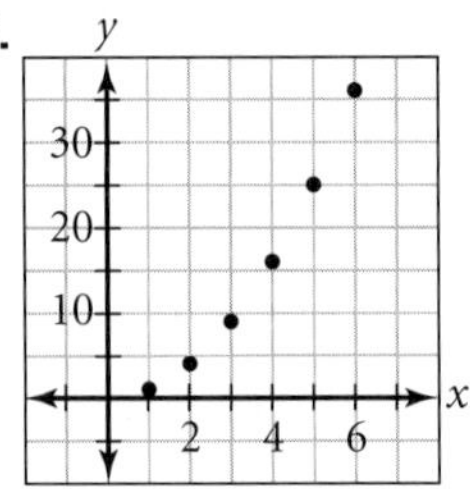

2. **a** and **b**

CHECKPOINT ✔

3. **a** and **b** are linear.
c and **e** are exponential.
d is reciprocal.
f is quadratic.

Exercises

Communicate

1. Find d: $2 - (-4) = 6$ or $8 - 2 = 6$, so $d = 6$.
Find t_1: $t_1 = -4$.
Write and simplify: $t_n = t_1 + (n - 1)d = (-4) + (n - 1)(6) = 6n - 10$

2. The pattern of an arithmetic sequence begins with the first term, t_1. Subsequent terms are found by adding multiples of the differences, d. If nd were used in the explicit formula for the sequence, t_2 would be the first term.

3. They subdivide the interval into 2 and 4 equal parts, and the arithmetic mean of 4 and 20 is the middle of the three arithmetic means of 4 and 20.

Portfolio Activity

The ruler's center of gravity is at the six-inch marking.

Lesson 11.3, pages 707–712

CHECKPOINT ✔

1. Duplicate Figure 1, turn the copy upside-down, and put the pieces together. They form a rectangle with side lengths of 3 and 4 units and an area of 12 square units. So, the area of Figure 1 is half that of the rectangle, or $\frac{3(3+1)}{2}$.

2.

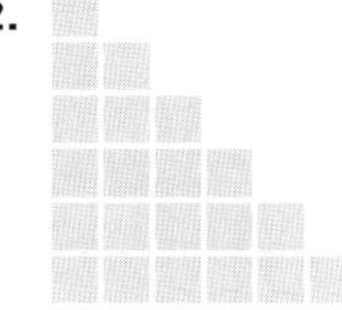

$$1 + 2 + 3 + 4 + 5 + 6 = \frac{6(7)}{2} = \frac{6(1 + 6)}{2} = 21$$

3.

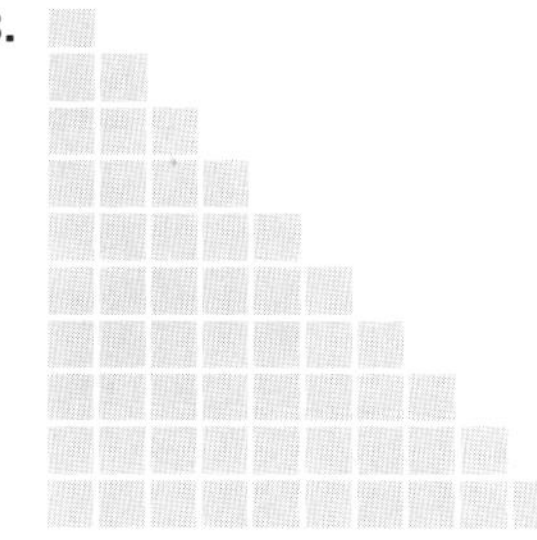

$$1 + 2 + 3 + \cdots + 10 = \frac{10(11)}{2} = \frac{10(1 + 10)}{2} = 55$$

Exercises

Communicate

1. We need to know the first term, t_1, and the nth term, t_n, to find S_n by using the formula $S_n = n\left(\frac{t_1 + t_n}{2}\right)$. Another option is to use $S_n = n\left(\frac{2t_1 + (n - 1)d}{2}\right)$ in which we need to know the first term, t_1, and the difference, d.

2. $\sum_{k=1}^{6} 2k$, or $\sum_{k=1}^{5} 2 + (k - 1)2$

3. many; sample answers: $\sum_{k=1}^{4} 2k = 2\sum_{k=1}^{4} k = \sum_{k=1}^{4} 2 + (k - 1)2 = \sum_{i=0}^{3} 2(i + 1) = \sum_{j=2}^{5} 2(j - 1)$, etc.

Lesson 11.4, pages 713–719

Activity

1. a. 1, 10, 100, 1000, 10,000
b. 1, 2, 4, 8, 16
c. 0.1, 0.11, 0.121, 0.1331, 0.14641
d. 1, 1.01, 1.0201, 1.030301, 1.04060401
e. 1, 0.1, 0.01, 0.001, 0.0001
f. 1, 0.9, 0.81, 0.729, 0.6561

2. largest: **a**
smallest: **e**

CHECKPOINT ✔

3. If $0 < r < 1$, successive terms get smaller. If $r > 1$, successive terms get larger.

Exercises

Communicate

1. First find the common ratio by using $\frac{t_2}{t_1} = 3$. Then substitute into the formula for the nth term, $t_n = t_1 r^{n-1} = (1)(3)^{n-1}$.

2. Let $t_n = t_1 r^{n-1}$ be the general term of the initial sequence. If the common ratio is doubled, $t_n = t_1(2r)^{n-1}$, or $t_n = t_1 2^{n-1} r^{n-1}$. Therefore, each term is multiplied by 2^{n-1}.

3. Let $t_n = t_1 r^{n-1}$ be the general term of the initial sequence. If the first term is doubled, $t_n = (2t_1)r^{n-1}$, or $t_n = 2t_1 r^{n-1}$. Therefore, each term is multiplied by 2.

Lesson 11.5, pages 720–727

Activity

1.

n	1	2	3	4	5	6	7	8
S_n	1	3	7	15	31	63	127	255

2. $S_n = 2^n - 1$

3. 1023

CHECKPOINT ✔

4. They are the same.

Exercises

Communicate

1. A geometric series is the sum of the terms of a geometric sequence; an arithmetic series is the sum of the terms of an arithmetic sequence.

2. $\sum_{k=1}^{5} 2^k = \sum_{j=1}^{5} 2(2^{j-1}) = 2\sum_{i=1}^{5} 2^{i-1}$

3. Substitute any two consecutive numbers for the natural numbers n and $n + 1$ in the statement. The result must be a true statement.

4. No; induction holds only for the natural numbers that come after the number chosen in the basis step. If $n = 3$, then the proof holds for $n = 4, 5, 6, \ldots$

Practice and Apply

62. $2 + 4 + 6 + \cdots + 2n = n(n + 1)$

1. Basis Step

Show that $2 + 4 + 6 + \cdots + 2n = n(n + 1)$ is true for $n = 1$.
$2 = (1)(2)$ **True**

2. Induction Step

Assume that the statement is true for a natural number k.
$2 + 4 + 6 + \cdots + 2k = k(k + 1)$
Determine the statement to be proved by adding $2(k + 1)$ to the left side and substituting $k + 1$ for k on the right.
$2 + 4 + 6 + \cdots + 2k + 2(k + 1) = (k + 1)[(k + 1) + 1]$
$2 + 4 + 6 + \cdots + 2k + 2(k + 1) = (k + 1)(k + 2)$
Rewrite the left side by using the statement assumed to be true in order to obtain the right side.
$2 + 4 + 6 + \cdots + 2k + 2(k + 1) = k(k + 1) + 2(k + 1)$
$= (k + 1)(k + 2)$ **True**

Portfolio Activity

The first tile can be extended to about one-half of its length. The next tile is about one-fourth of its length. This pattern continues. The distance each successive tile can be extended decreases.

Lesson 11.6, pages 728–734

Activity

1. a.

n	S_n
1	−5
2	−3.75
3	−4.0625
5	−4.00390625
10	−3.99999619
100	−4

b.

n	S_n
1	5
2	3.75
3	4.0625
5	4.00390625
10	3.99999619
100	4

c.

n	S_n
1	2
2	$2.\overline{6}$
3	$2.\overline{8}$
5	2.98765432
10	2.99994919
100	3

d.

n	S_n
1	2
2	$1.\overline{3}$
3	$1.\overline{5}$
5	1.50617284
10	1.49997460
100	1.5

2. a.

n	S_n
1	−5
2	$-36.\overline{1}$
3	−285
5	−18,205
10	−596,523,236.1
100	-1.13169×10^{90}

b.

n	S_n
1	5
2	$36.\overline{1}$
3	285
5	18,205
10	596,523,236.1
100	1.13169×10^{90}

c.

n	S_n
1	2
2	8
3	26
5	242
10	59,048
100	5.1538×10^{47}

d.

n	S_n
1	2
2	5
3	14
5	122
10	29,524
100	2.5769×10^{47}

3. All in Step 1; the ratios have an absolute value of less than 1.

4. All in Step 2; the ratios have an absolute value greater than 1.

CHECKPOINT ✔

5. If $|r| < 1$, the series converges. If $|r| \geq 1$, the series diverges.

Exercises

Communicate

1. Find the common ratio of successive terms by using $r = \frac{t_n + 1}{t_n}$. If $|r| < 1$, the series converges and has a sum.

2. Find the common ratios of each series. If $|r| < 1$, the series converges. Because $\left|\frac{1}{2}\right| < 1$, $\sum_{k=1}^{\infty}\left(\frac{1}{2}\right)^k$ converges. Because $|2| > 1$, $\sum_{k=1}^{\infty} 2^k$ diverges.

3. Write the repeating decimal as an infinite geometric series, where $r = \frac{1}{10}$. Compute the sum by using $S = \frac{t_1}{1 - r}$.

Portfolio Activity

2. The maximum distance for tile C is about $\frac{1}{8}$ of a tile.

3. The maximum distance for tile D is about $\frac{1}{16}$ of a tile.

Lesson 11.7, pages 735–740

Activity

1.

Row 0 1
Row 1 1 1
Row 2 1 2 1
Row 3 1 3 3 1
Row 4 1 4 6 4 1
Row 5 1 5 10 10 5 1
Row 6 1 6 15 20 15 6 1
Row 7 1 7 21 35 35 21 7 1
Row 8 1 8 28 56 70 56 28 8 1
Row 9 1 9 36 84 126 126 84 36 9 1
Row 10 1 10 45 120 210 252 210 120 45 10 1
Row 11 1 11 55 165 330 462 462 330 165 55 11 1
Row 12 1 12 66 220 495 792 924 792 495 220 66 12 1
Row 13 1 13 78 286 715 1287 1716 1716 1287 715 286 78 13 1
Row 14 1 14 91 364 1001 2002 3003 3432 3003 2002 1001 364 91 14 1
Row 15 1 15 105 455 1365 3003 5005 6435 6435 5005 3003 1365 455 105 15 1

2. 1, 2, 4, 8, 16, 32, 64, 128, 256, 512, 1024, 2048, 4096, 8192, 16,384, 32,768

CHECKPOINT ✔

3. The sum of row n is 2^n.

CHECKPOINT ✔

4. Rows 2, 3, 7, 11 and 13 have this property too. Row n, where n is a prime number, will have this property.

5.

Row 0 1
Row 1 1 1
Row 2 1 2 1
Row 3 1 3 3 1
Row 4 1 4 6 4 1
Row 5 1 5 10 10 5 1
Row 6 1 6 15 20 15 6 1
Row 7 1 7 21 35 35 21 7 1
Row 8 1 8 28 56 70 56 28 8 1
Row 9 1 9 36 84 126 126 84 36 9 1
Row 10 1 10 45 120 210 252 210 120 45 10 1
Row 11 1 11 55 165 330 462 462 330 165 55 11 1
Row 12 1 12 66 220 495 792 924 792 495 220 66 12 1
Row 13 1 13 78 286 715 1287 1716 1716 1287 715 286 78 13 1
Row 14 1 14 91 364 1001 2002 3003 3432 3003 2002 1001 364 91 14 1
Row 15 1 15 105 455 1365 3003 5005 6435 6435 5005 3003 1365 455 105 15 1

CHECKPOINT ✔

6. The shaded numbers form inverted triangles.

Exercises

Communicate

1. Answers may vary but should include the following: Row n has $n + 1$ entries. The kth entry in row n is ${}_nC_{k-1}$. The sum of the entries in row n is 2^n. ${}_{n+1}C_k + {}_nC_k = {}_{n+1}C_k$, for $0 < k \le n$.

2. Answers may vary but should include the following: If n is a prime number, the entries in row n, except for the first and last, are divisible by n. Even numbers in the triangle form triangular patterns. Each entry, except for the first and last, is the sum of the two closest entries in the row above. The entries in each row are symmetrical, with odd rows containing only a single center item.

3. The probability that exactly 4 heads will result when a coin is tossed 6 times is ${}_6C_4$, which is the fifth entry in row 6 of Pascal's triangle. The probability is given by $P(A) = \frac{{}_6C_4}{2^6} \approx 0.23$.

Practice and Apply

20. ${}_{11}C_4$ is the 5th entry in the 11th row, ${}_{11}C_4 = 330$.

21. ${}_{13}C_7$ is the 8th entry in the 13th row, ${}_{13}C_7 = 1716$.

22. ${}_{12}C_8$ is the 9th entry in the 12th row, ${}_{12}C_8 = 495$.

Lesson 11.8, pages 741–747

Activity

	Product	Expansion
1.	$(x+y)^0 =$	1
	$(x+y)^1 =$	$1x + 1y$
	$(x+y)^2 =$	$1x^2 + 2xy + 1y^2$
2.	$(x+y)^3 =$	$1x^3 + 3x^2y + 3xy^2 + 1y^3$
3.	$(x+y)^4 =$	$1x^4 + 4x^3y + 6x^2y^2 + 4xy^3 + 1y^4$
4.	$(x+y)^5 =$	$1x^5 + 5x^4y + 10x^3y^2 + 10x^2y^3 + 5xy^4 + 1y^5$

CHECKPOINT ✔

5. There are $n + 1$ terms in the expansion of $(x + y)n$, and the coefficients of the terms are symmetric about the center terms.
$(x + y)^6 = 1x^6 + 6x^5y + 15x^4y^2 + 20x^3y^3 + 15x^2y^4 + 6xy^5 + 1y^6$

Exercises

Communicate

1. The entries of row n of Pascal's triangle are the coefficients of the terms in the expansion of $(x + y)^n$.

2. $\binom{6}{3}x^{6-3}y^3 = 20x^3y^3$

3. $\binom{10}{7}\left(\frac{1}{2}\right)^3\left(\frac{1}{2}\right)^7$

Chapter 11 Project

Activity 1

1.

Location of the quarter	none	1 in.	2 in.	3 in.	4 in.	5 in.	6 in.
Location of the center of gravity	6 in.	$5\frac{3}{8}$ in.	$5\frac{1}{2}$ in.	$5\frac{5}{8}$ in.	$5\frac{6}{8}$ in.	$5\frac{7}{8}$ in.	6 in.

Lesson 12.1, pages 764–771

Activity

Answers may vary. Sample answers:

1. $\bar{x} = 6.6$, median = 7, mode = 7
2. $\bar{x} = 9.6$, median = 10, mode = 10; each is increased by 3.
3. $\bar{x} = 13.2$, median = 14, mode = 14; each is doubled.
4. $\bar{x} = 2.3$, median = 2.5, mode = 2.5; each is 1 less than half of the original value.

CHECKPOINT ✔

5. The mean, median, and mode each changed by the amount you added, subtracted, multiplied, or divided.

CHECKPOINT ✔

6. Find the mean of the data values without the zeros. Then add 3 zeros to the mean. If each number is multiplied by 1000, the mean of the numbers is also multiplied by 1000.

Exercises

Communicate

1. Mode is the easiest because it is simply the most repeated value(s). Median is the most difficult because the data values must first be sorted from least to greatest.
2. No; yes; the median will not change because it will still be the middle value, whereas the mean will change unless the difference between the largest number and the mean is the same as the difference between the mean and the smallest number.
3. Answers may vary. Possible answer: If the values in a data set are clustered around one end of the range, the mode or median may be more representative than the mean.

Portfolio Activity

1. $\bar{x} \approx 79$; median = 76.5
2. Allison should use the median of 76.5 because her score of 78 is higher than the median.
3. Answers will vary. Sample: Choose 92. The replacement score would be 102. $\bar{x} \approx 79.4$; median = 76.5; the median did not change, but the mean increased by ≈ 0.4.
4. Answers will vary. Sample: Choose 70. The replacement score would be 60. $\bar{x} \approx 78.5$; median = 76.5; the median did not change, but the mean decreased by ≈ 0.4.

Lesson 12.2, pages 772–780

Activity

Answers may vary. Possible answers:

1.

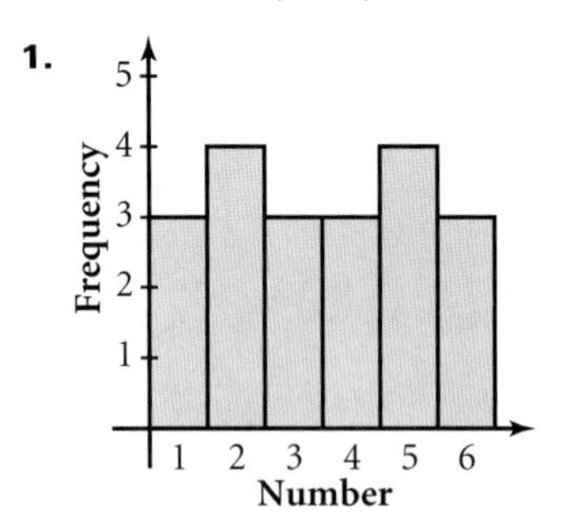

2.

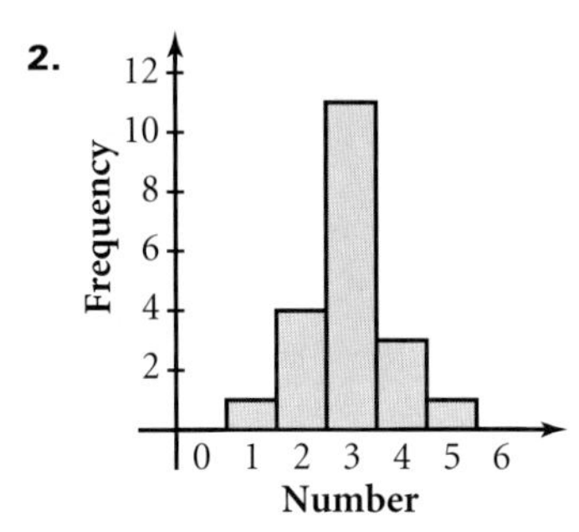

CHECKPOINT ✔

3. Step 1 is flatter, and Step 2 is more mound shaped.
4. Mound-shaped; there should be few sums of 2 or 12 and many sums around 7.

CHECKPOINT ✔

5.

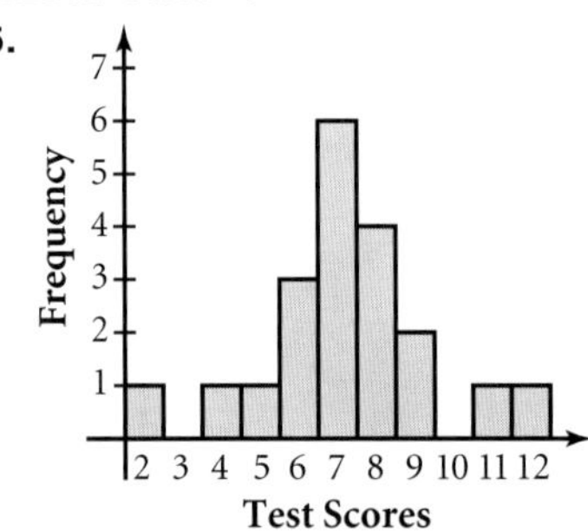

The histogram is roughly mound-shaped.

Exercises

Communicate

1. Even- or uniform-shaped distributions are those where approximately the same number of data values are in each category. Mound-shaped distributions are those where the data values cluster around the middle value or values.

2. Stem-and-leaf plots; the stem-and-leaf plot is preferred when the actual data values must be retained and/or when there are not a lot of data values. The histogram is preferred when frequencies or probability distributions are of interest and/or there are a lot of data values.

3. The relative frequency is the percent frequency, or probability, i.e., a relative frequency of 0.10 is also a probability of 0.10.

4. A circle graph starts with a whole and is divided into pieces to represent the parts of the whole, just as a pie that is cut into pieces that represent individual portions.

Practice and Apply

10.

Stem	Leaf 26|3 = 26.3
26	3, 8
27	2, 3, 6
28	1, 3, 4, 7
29	1, 4, 8
30	1

median = 28.3; no mode; mound-shaped

11.

Stem	Leaf 33|5 = 335
33	5, 6, 7, 8, 9
34	0, 7, 7, 8, 8
35	6, 7, 8, 9

median = 347; modes = 347 and 348; even-shaped

12.

Stem	Leaf 6|15 = 6.15
6	15, 35, 35
7	35, 85, 85, 85
8	45, 45, 55
9	65, 65, 75

median = 7.85, mode = 7.85; even-shaped

13.

Data	Frequency
2	2
3	2
4	1
5	1
6	2
7	1
8	1
9	3
10	2

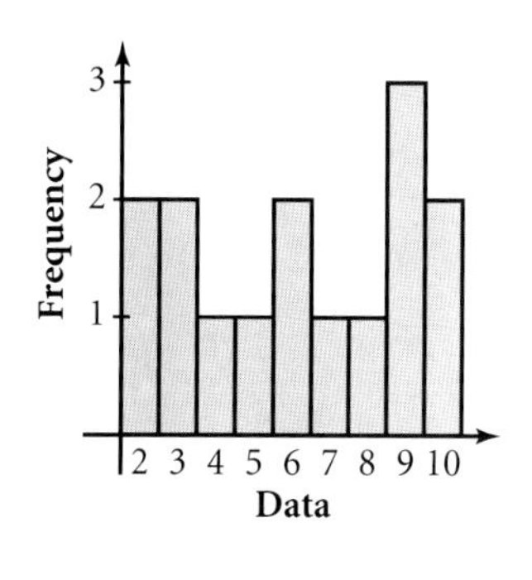

14.

Data	Frequency
2	2
3	2
4	1
5	1
6	2
7	3
8	1
9	2
10	1

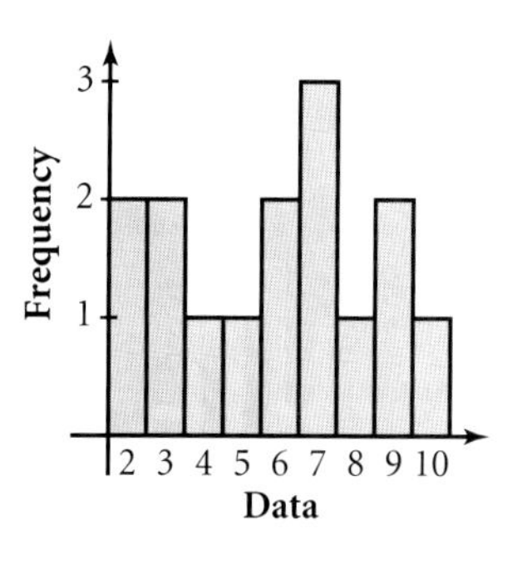

15.

Data	Frequency
0.5	2
1.0	3
1.5	4
2.0	3
2.5	2
3.0	1

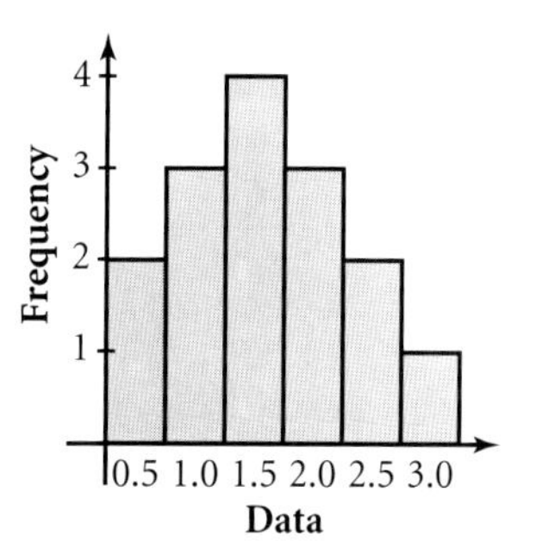

16.

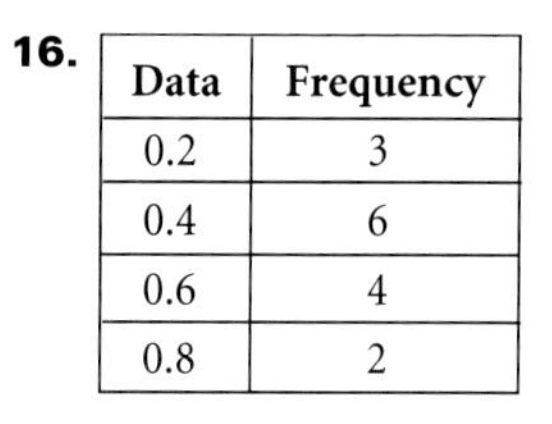

Data	Frequency
0.2	3
0.4	6
0.6	4
0.8	2

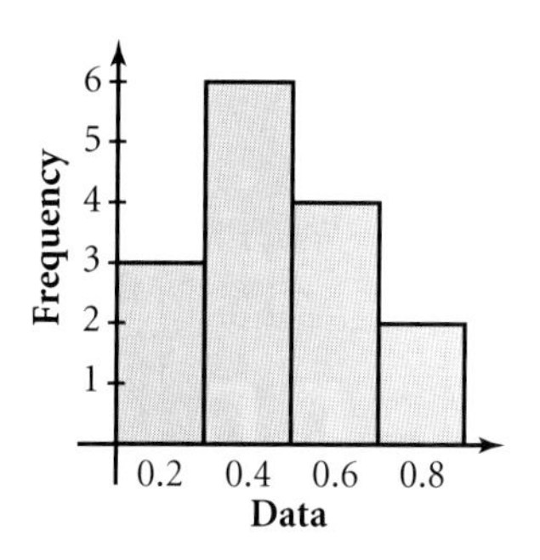

17.

Data	Frequency	Relative frequency
1	4	$\frac{4}{24} = 0.1\overline{6}$
2	2	$\frac{4}{24} = 0.08\overline{3}$
3	5	$\frac{5}{24} = 0.208\overline{3}$
4	4	$\frac{4}{24} = 0.1\overline{6}$
5	2	$\frac{2}{24} = 0.08\overline{3}$
6	0	$\frac{0}{24} = 0$
7	4	$\frac{4}{24} = 0.1\overline{6}$
8	3	$\frac{3}{24} = 0.125$
Total	24	1

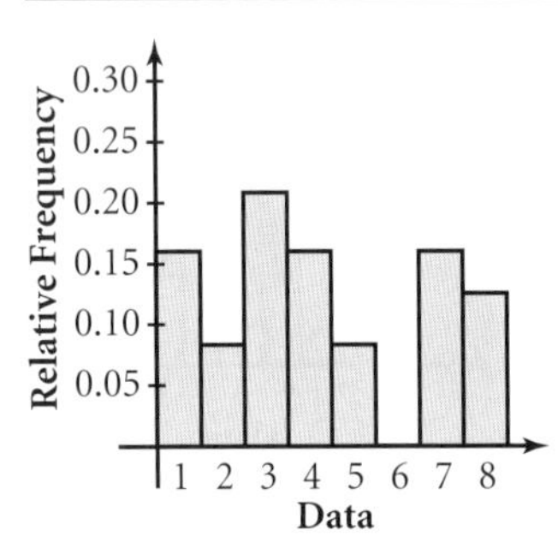

18.

Data	Frequency	Relative frequency
30	3	$\frac{3}{18} = 0.1\overline{6}$
40	4	$\frac{4}{18} = 0.2\overline{2}$
50	0	$\frac{0}{18} = 0$
60	4	$\frac{4}{18} = 0.2\overline{2}$
70	3	$\frac{3}{18} = 0.1\overline{6}$
80	3	$\frac{3}{18} = 0.1\overline{6}$
90	0	$\frac{0}{18} = 0$
100	1	$\frac{1}{18} = 0.0\overline{5}$
Total	18	1

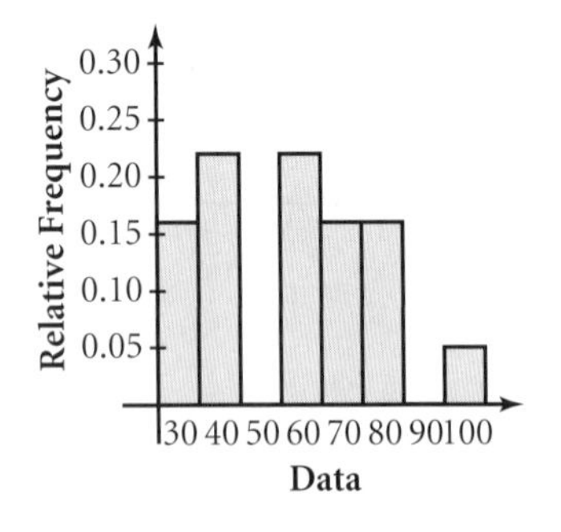

19.

Data	Frequency	Relative frequency
0.1	6	$\frac{6}{15} = 0.4$
0.2	4	$\frac{4}{15} = 0.2\overline{6}$
0.3	3	$\frac{3}{15} = 0.2$
0.4	1	$\frac{1}{15} = 0.0\overline{6}$
0.5	1	$\frac{1}{15} = 0.0\overline{6}$
Total	15	1

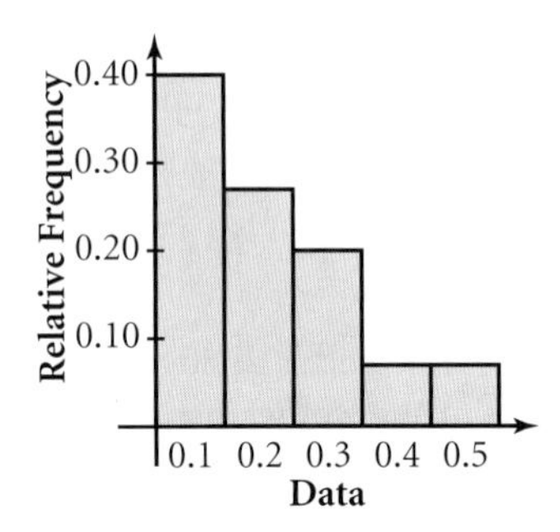

20.

Data	Frequency	Relative frequency
7.5	1	$\frac{1}{15} = 0.0\overline{6}$
7.6	1	$\frac{1}{15} = 0.0\overline{6}$
7.7	0	$\frac{0}{15} = 0$
7.8	1	$\frac{1}{15} = 0.0\overline{6}$
7.9	3	$\frac{3}{15} = 0.2$
8.0	3	$\frac{3}{15} = 0.2$
8.1	4	$\frac{4}{15} = 0.2\overline{6}$
8.2	0	$\frac{0}{15} = 0$
8.3	1	$\frac{1}{15} = 0.0\overline{6}$
8.4	0	$\frac{0}{15} = 0$
8.5	1	$\frac{1}{15} = 0.0\overline{6}$
Total	15	1

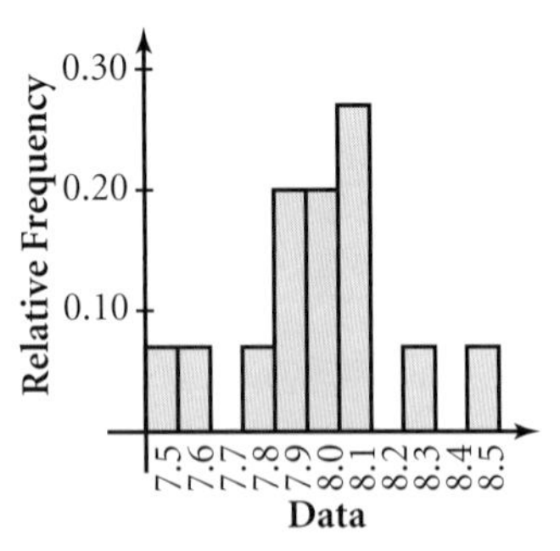

21.

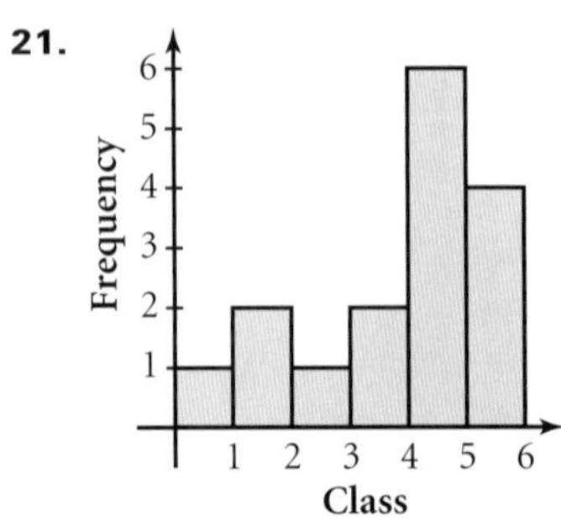

22.

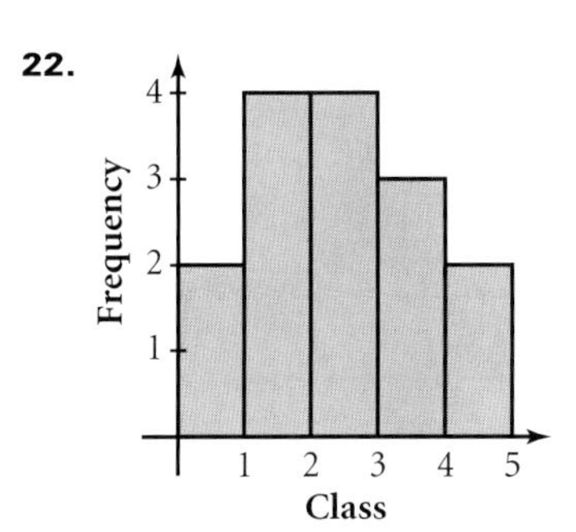

23.

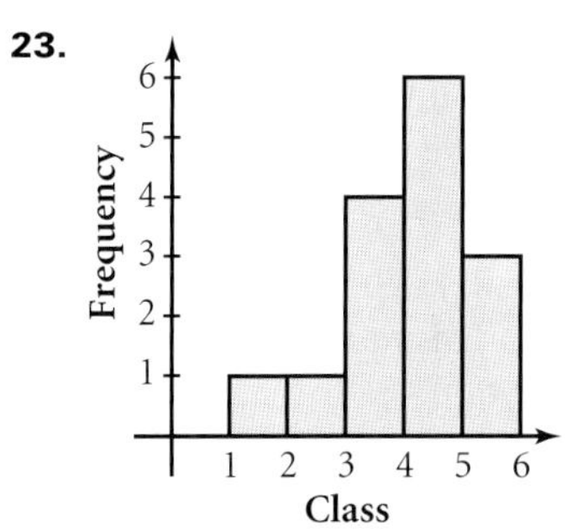

24.

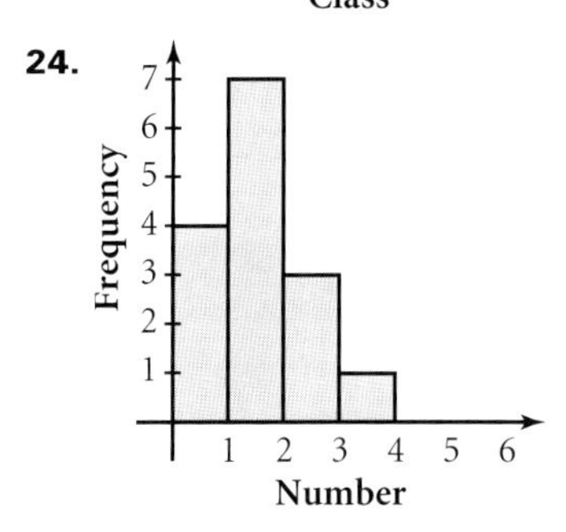

28. a.

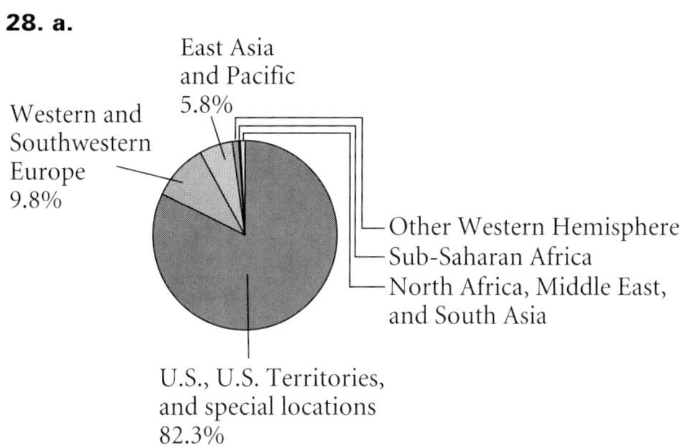

The circle graph reveals at a glance that the vast majority of military personnel are in the U.S., U.S. territories, and special locations.

b. about 17.7%

29.

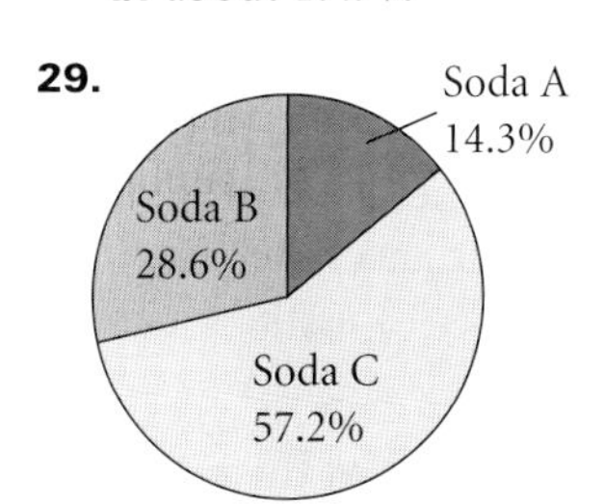

Portfolio Activity

Sample data from the *Statistical Abstract of the U.S.* for average number of days of precipitation per year:

City	Days/year of precip.	City	Days/year of precip.
Mobile, AL	122	Great Falls, MT	101
Juneau, AK	221	Omaha, NE	99
Phoenix, AZ	36	Reno, NV	50
Little Rock, AR	105	Concord, NH	126
San Francisco, CA	62	Atlantic, NJ	112
Denver, CO	89	Albuquerque, NM	61
Hartford, CT	127	Albany, NY	135
Wilmington, DE	116	Charlotte, NC	111
Jacksonville, FL	116	Bismarck, ND	96
Atlanta, GA	115	Cleveland, OH	156
Honolulu, HI	98	Oklahoma City, OK	83
Boise, ID	90	Portland, OR	151
Chicago, IL	126	Philadelphia, PA	117
Indianapolis, IN	126	Providence, RI	124
Des Moines, IA	108	Columbia, SC	109
Wichita, KS	86	Sioux Falls, SD	98
Louisville, KY	125	Memphis, TN	107
New Orleans, LA	114	El Paso, TX	49
Portland, ME	129	Salt Lake City, UT	91
Baltimore, MD	113	Burlington, VT	154
Boston, MA	126	Norfolk, VA	115
Detroit, MI	136	Spokane, WA	113
Minneapolis, MN	115	Charleston, WV	151
Jackson, MS	110	Milwaukee, WI	125
Kansas City, MO	106	Cheyenne, WY	100

1. Answers may vary. Sample answer: The stem-and-leaf plot is good for comparing the amount of precipitation for your city with that of other cities. You can tell how many cities are drier or wetter than your city. The histogram is appropriate for showing categories of amounts of precipitation because you can see whether your city is in a category with a lot of other cities or if it is alone. The circle graph is not appropriate because it shows the part of the total precipitation for all cities that is represented by your city's precipitation.

2. Answers may vary. Sample answer: There were no cities with the same amount of precipitation as my city. The only way to see this was with the stem-and-leaf plot. This is the only plot of the three that shows the individual data points.

3. Answers may vary. Sample answer: The stem-and-leaf plot was best since all the various amounts of precipitation could be seen.

Lesson 12.3, pages 781–789

Activity

1. 2.55 inches; 50%; 50%

2. 2.2 inches; 25%

3. 3.45 inches; 25%

CHECKPOINT ✔

4. 2.2 1.8 2.55 3.45 4.9
0 1 2 3 4 5

1.9–2.2: 25%; 2.2–2.55: 25%;
2.55–3.45: 25%; 3.45–4.9: 25%

Exercises

Communicate

1. The quartiles for a data set with 20 values are averages of two adjacent data values, whereas quartiles for data set with 15 values are actual data values.
2. The class had a greater variability in scores on test 1 than test 2. The medians were equal on the two tests.
3. It tells you something about the data's variability, the minimum, median, and maximum, and quartiles. The box covers the middle 50% of the data, and the width of the box is the IQR. A plot allows you to compare similar sets of data.
4. Answers may vary: Possible answer:
Data set 1: 2 5 6 7 8 11 12
Data set 2: 1 5 5 7 9 11 200
The quartiles do not tell you the amount of data values, the maximum and minimum values, or the values between the quartiles.

Portfolio Activity

The ratings used for the sample answers were from the week of February 15–21, 1999.

Show	Rating
E.R.	23.8
Friends	19.0
Frasier	18.8
Jesse	15.9
Veronica's Closet	15.6
60 Minutes	14.9
CBS Sunday Movie	14.3
Touched by an Angel	14.0
ABC Monday Night Movie	12.2
Dateline NBC (Tuesday)	12.2
Storm of the Century	12.0
20/20 (Wednesday)	11.9
Jag	11.6
Everybody Loves Raymond	11.5
Law and Order	11.2
NYPD Blue	10.8
Home Improvement	10.6
Becker	10.5
Drew Carey Show	10.4
Ally McBeal	10.3
Median	**12.1**

The rating points represent the number of households that watched the show. Each point stands for 994,000 households.

Nielsen Ratings

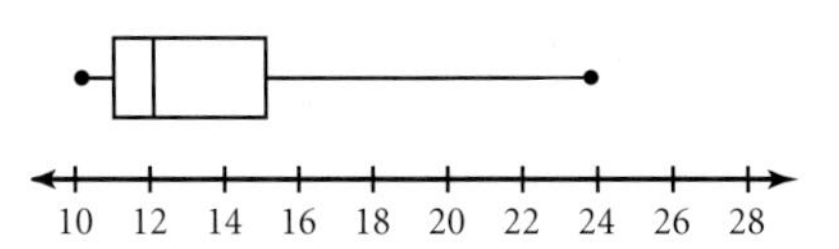

1. The percent of ratings below the median is 50%. The percent of ratings below the lower quartile is 25%. The percent of ratings above the upper quartile is 25%. The percent of ratings in the box is 50%. The percent of ratings in the whiskers is 50%.
2. Answers may vary. Sample answer: The upper whisker is longer than the other whisker. This indicates that the higher ratings are very spread out.
3. Answers may vary. Sample answer: The median is not centered in the box. This indicates that the values are spread out.

Lesson 12.4, pages 792–798

Activity

1. $\sigma \approx 7.07$
2. $\sigma \approx 7.07$; no change
3. $\sigma \approx 7.07k$, where k is a constant; the new standard deviation is the product of the old deviation and the constant, k.
4. $\sigma \approx 3.54$; divide the standard deviation by 2.

CHECKPOINT ✔

5. There is no change when adding or subtracting a constant. The standard deviation is multiplied or divided by the same constant that multiplied or divided the original data.

CHECKPOINT ✔

6. 4 points; 5 points; The standard deviation is one-sixth of the mean.

Exercises

Communicate

1. The mean deviation is calculated by using the absolute value, which is always nonnegative. The standard deviation is the square root of the variance, which is nonnegative because it is calculated by using the square of differences.
2. The standard deviation is the square root of the variance. The standard deviation will be less than the variance for all values of σ^2, except $0 \leq \sigma^2 \leq 1$.
3. The range depends only on two data values, and the interquartile range depends only on the first and third quartiles. The mean deviation and standard deviation take all of the data values into account.

Portfolio Activity

Answers vary according to the data sets selected. Sample answer:

Weather data: The mean is 5551 ÷ 50 = 111.02 days of precipitation. The standard deviation is 1017 ÷ 50 = 20.34 days of precipitation. The mean is quite a bit lower than the highest rating. The standard deviation tells you that the scores vary quite a bit from the mean.

T.V. ratings data: The mean is 271.5 ÷ 20 = 13.575. The standard deviation is 48.425 ÷ 20 ≈ 2.4. The mean is quite a bit lower than the highest rating. The standard deviation tells you that the scores vary quite a bit from the mean.

Lesson 12.5, pages 799–805

Activity

1. $P(S) = \frac{1}{6}, P(F) = \frac{5}{6}$

2. a. $\left(\frac{1}{6}\right)\left(\frac{5}{6}\right)^4$ **b.** $\left(\frac{1}{6}\right)\left(\frac{5}{6}\right)^4$ **c.** $\left(\frac{1}{6}\right)^2\left(\frac{5}{6}\right)^3$ **d.** $\left(\frac{1}{6}\right)^2\left(\frac{5}{6}\right)^3$

CHECKPOINT ✔

3. ${}_5C_1$; ${}_5C_2$

4. a. $P(\text{exactly one success}) = {}_5C_1\left(\frac{1}{6}\right)\left(\frac{5}{6}\right)^4$

b. $P(\text{exactly two successes}) = {}_5C_2\left(\frac{1}{6}\right)^2\left(\frac{5}{6}\right)^3$

CHECKPOINT ✔

5. ${}_5C_3\left(\frac{1}{6}\right)^3\left(\frac{5}{6}\right)^2 \approx 0.32$

Exercises

Communicate

1. The experiment consists of n trials whose outcomes are either successes or failures. The trials are identical and independent with a constant probability, p, of success and a constant probability, $1 - p$, of failure.

2. The formula must account for all of the ways in which r successes out of n trials can occur.

3. $n = 12$, the number of trials (questions)
$r = 9$, the number correct
$p = 0.2$, the probability of getting any one question correct.

4. The distribution is symmetric when $p = 0.5$, is skewed to the left when $p > 0.5$, and is skewed to the right when $p < 0.5$. When there are not enough trials, a variety of distributions (including even-shaped distributions) are possible.

Lesson 12.6, pages 806–813

Activity

1–3. 0.3413; 0.3413

CHECKPOINT ✔

4. 0.6826

5. 0.4772; 0.4772

CHECKPOINT ✔

6. 0.9545

7. 0.4987; 0.4987

CHECKPOINT ✔

8. 0.9973

9. 0.0013; 0.0013

CHECKPOINT ✔

10. 1

Exercises

Communicate

1. A normal curve is defined by the mean, $\bar{x}$, and the standard deviation, σ. The curve is symmetric about the mean, $\bar{x}$. The total area under the curve is 1. The mean, median, and mode are about equal. About 68% of the area is within 1σ of $\bar{x}$. About 95% of the area is within 2σ of $\bar{x}$. About 99% of the area is within 3σ of $\bar{x}$.

2. Find the z-scores for $x_1 = 25$ and $x_2 = 50$. Note that for $x_2 = 50$, $z = 2$ since x_2 is two standard deviations from $\bar{x} = 30$.

$z_1 = \frac{25 - 30}{10} = -0.5$

$z_2 = \frac{50 - 30}{10} = 2$

Use the standard normal curve to find the area between $z_1 = -0.5$ and $z_2 = 2$.

3.

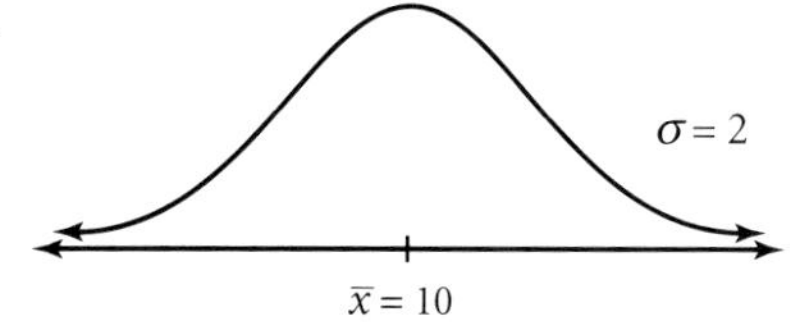

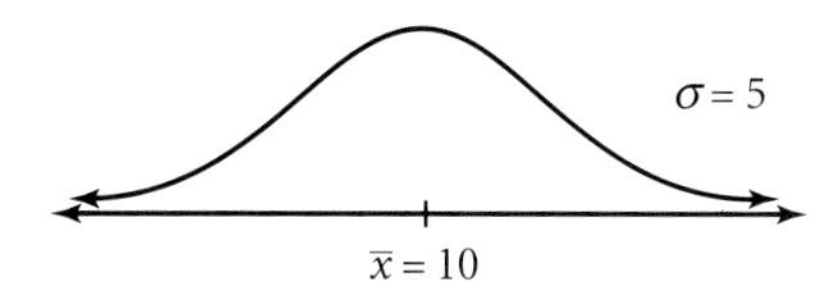

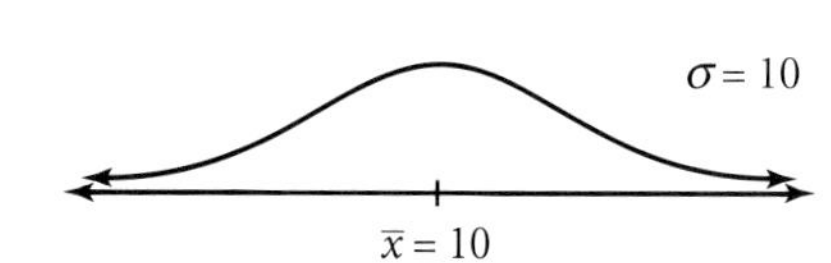

As the standard deviation increases, the curve flattens out.

Chapter 12 Review and Assessment

5.

Stem	Leaf 3|5 = 35
3	5 8
4	5 5 9
5	3 7 8

6.

Stem	Leaf 6|7 = 67
6	7 9
7	3 8 9
8	1 2 7

7.

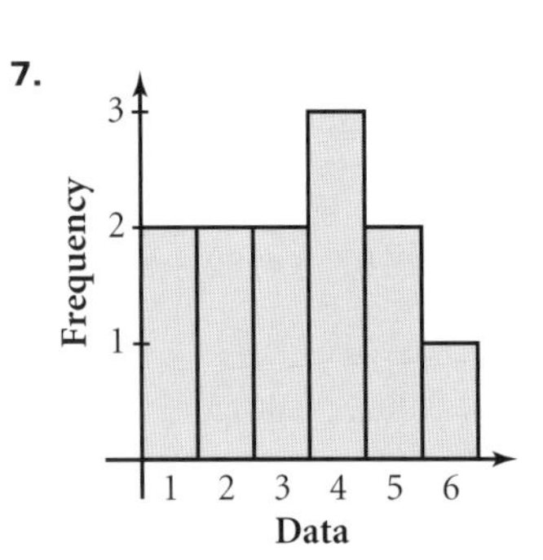

8.

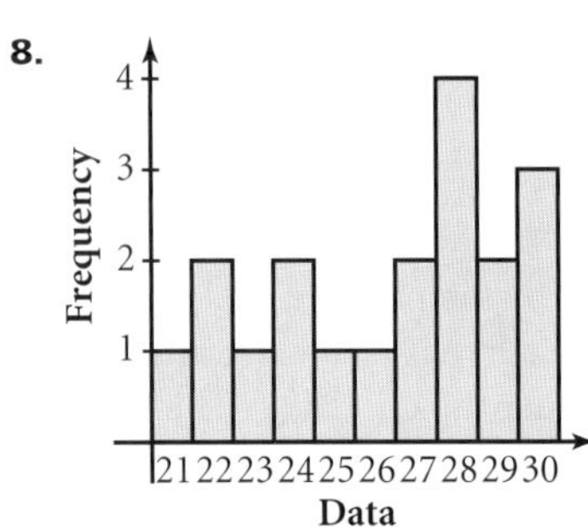

9.

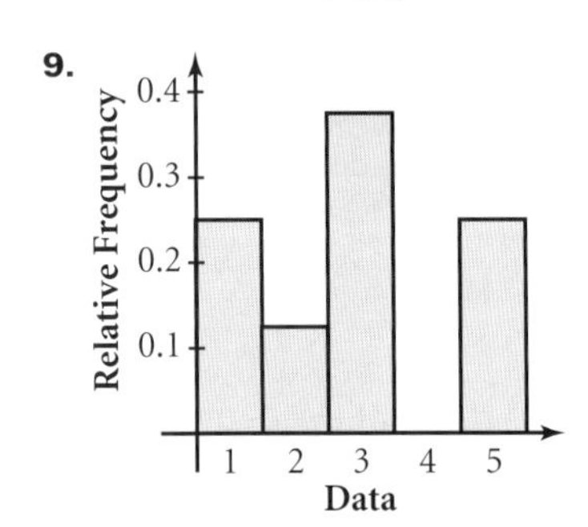

10.

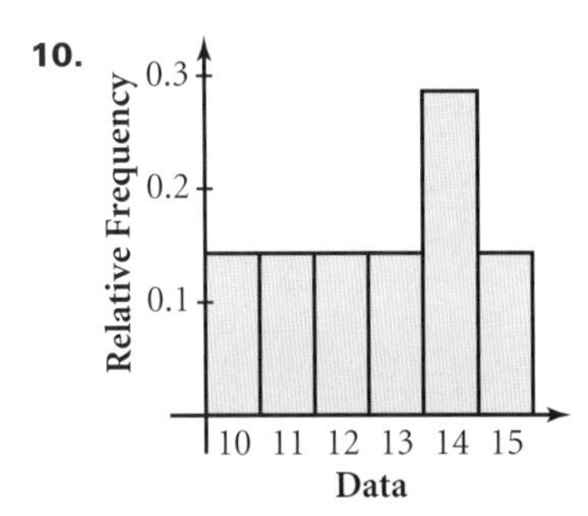

Lesson 13.1, pages 828–835

Activity

1. Answers may vary. Sines, cosines and tangents should be approximately equal.

2. yes; yes; yes

3. All answers should be approximately equal.

CHECKPOINT ✔

4. The ratios for the sine, cosine, and tangent of an acute angle in similar triangles do not depend on the lengths of the sides.

Exercises

Communicate

1. Use the definitions of the trigonometric functions on page 829, where opp. = $BC = 2$, adj. = $AC = 3$, and hyp. = $AB = \sqrt{13}$.

2. Pick a trigonometric function and write the angle in terms of the chosen function and the appropriate sides. Then solve for the angle by using the inverse of the trigonometric function. For example:

$\sin A = \frac{2}{\sqrt{13}}$

$m\angle A = \sin^{-1}\frac{2}{\sqrt{13}} \approx 34°$

3. $\frac{1}{\sin A}$ is 1 divided by the sine of angle A, or $\csc A$. $\sin^{-1} A$ is the angle whose sine is A.

Portfolio Activity

1. $Q(0, 164)$, $R(-164, 0)$, $S(0, -164)$

2.

Rotation (degrees)	Height of P relative to the x-axis (feet)
0	0
90	164
180	0
270	−164
360	0
450	164
540	0
630	−164
720	0
810	164

3.

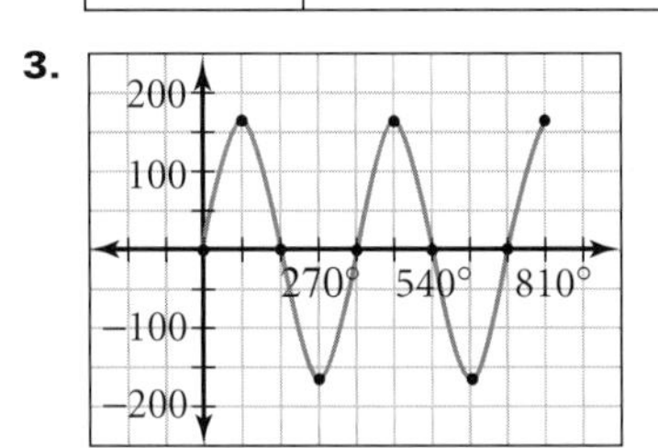

4. The graph does not look linear, quadratic, exponential, or logarithmic.

Lesson 13.2, pages 836–842

Activity

1.

Trig value	Quadrant I	II	III	IV
$\sin\theta$ and $\csc\theta$	+	+	−	−
$\cos\theta$ and $\sec\theta$	+	−	−	+
$\tan\theta$ and $\cot\theta$	+	−	+	−

2. III or IV; II or III; II or IV

CHECKPOINT ✔

3. No; r is always positive.

CHECKPOINT ✔

4. y; x; both x and y

Exercises

Communicate

1. Angles in right triangles must have measures between 0 and 90°. Angles of rotation can be any real number, which includes negative as well as positive values.
2. The signs of the functions may differ. The reference angle and its trigonometric functions are always positive, while an arbitrary angle and its trigonometric functions may be positive or negative, depending on the quadrant it falls in. For example, $\sin 225° \approx -0.7071$ and $\sin 45° \approx 0.7071$.
3. No, you do not need the measure of the angle if you know the x- and y-coordinates of one point on its terminal side in standard position.

Practice and Apply

12.

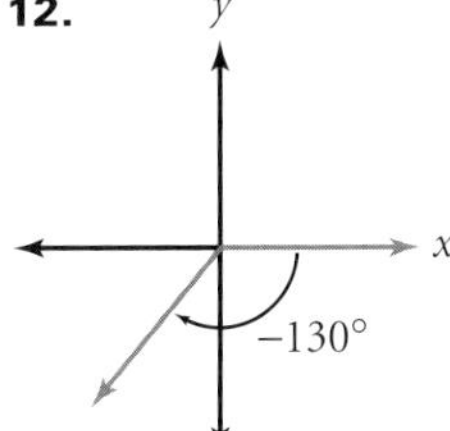

37. $\sin\theta = \frac{1}{\sqrt{17}} = \frac{\sqrt{17}}{17}$ $\quad \csc\theta = \sqrt{17}$

$\cos\theta = -\frac{4}{\sqrt{17}} = -\frac{4\sqrt{17}}{17}$ $\quad \sec\theta = -\frac{\sqrt{17}}{4}$

$\tan\theta = -\frac{1}{4}$ $\quad \cot\theta = -4$

38. $\sin\theta = -\frac{2}{\sqrt{13}} = -\frac{2\sqrt{13}}{13}$ $\quad \csc\theta = -\frac{\sqrt{13}}{2}$

$\cos\theta = -\frac{3}{\sqrt{13}} = -\frac{3\sqrt{13}}{13}$ $\quad \sec\theta = -\frac{\sqrt{13}}{3}$

$\tan\theta = \frac{2}{3}$ $\quad \cot\theta = \frac{3}{2}$

39. $\sin\theta = -\frac{1}{\sqrt{5}} = -\frac{\sqrt{5}}{5}$ $\quad \csc\theta = -\sqrt{5}$

$\cos\theta = \frac{2}{\sqrt{5}} = \frac{2\sqrt{5}}{5}$ $\quad \sec\theta = \frac{\sqrt{5}}{2}$

$\tan\theta = -\frac{1}{2}$ $\quad \cot\theta = -2$

40. $\sin\theta = \frac{4}{5}$ $\quad \csc\theta = \frac{5}{4}$

$\cos\theta = \frac{3}{5}$ $\quad \sec\theta = \frac{5}{3}$

$\tan\theta = \frac{4}{3}$ $\quad \cot\theta = \frac{3}{4}$

41. $\sin\theta = \frac{2\sqrt{29}}{29}$ $\quad \csc\theta = \frac{\sqrt{29}}{2}$

$\cos\theta = \frac{5\sqrt{29}}{29}$ $\quad \sec\theta = \frac{\sqrt{29}}{5}$

$\tan\theta = \frac{2}{5}$ $\quad \cot\theta = \frac{5}{2}$

42. $\sin\theta = \frac{\sqrt{5}}{5}$ $\quad \csc\theta = \sqrt{5}$

$\cos\theta = -\frac{2\sqrt{5}}{5}$ $\quad \sec\theta = -\frac{\sqrt{5}}{2}$

$\tan\theta = -\frac{1}{2}$ $\quad \cot\theta = -2$

43. $\sin\theta = \frac{3\sqrt{13}}{13}$ $\quad \csc\theta = \frac{\sqrt{13}}{3}$

$\cos\theta = -\frac{2\sqrt{13}}{13}$ $\quad \sec\theta = -\frac{\sqrt{13}}{2}$

$\tan\theta = -\frac{3}{2}$ $\quad \cot\theta = -\frac{2}{3}$

44. $\sin\theta = -\frac{\sqrt{3}}{2}$ $\quad \csc\theta = -\frac{2\sqrt{3}}{3}$

$\cos\theta = \frac{1}{2}$ $\quad \sec\theta = 2$

$\tan\theta = -\sqrt{3}$ $\quad \cot\theta = -\frac{\sqrt{3}}{3}$

45. $\sin\theta = -\frac{\sqrt{21}}{21}$ $\quad \csc\theta = -\sqrt{21}$

$\cos\theta = \frac{2\sqrt{105}}{21}$ $\quad \sec\theta = \frac{\sqrt{105}}{10}$

$\tan\theta = -\frac{\sqrt{5}}{10}$ $\quad \cot\theta = -2\sqrt{5}$

46. $\sin\theta = -\frac{3}{5}$ $\quad \csc\theta = -\frac{5}{3}$

$\cos\theta = -\frac{4}{5}$ $\quad \sec\theta = -\frac{5}{4}$

$\tan\theta = \frac{3}{4}$ $\quad \cot\theta = \frac{4}{3}$

47. $\sin\theta = -\frac{8\sqrt{65}}{65}$ $\quad \csc\theta = -\frac{\sqrt{65}}{8}$

$\cos\theta = -\frac{\sqrt{65}}{65}$ $\quad \sec\theta = -\sqrt{65}$

$\tan\theta = 8$ $\quad \cot\theta = \frac{1}{8}$

Look Back

78.

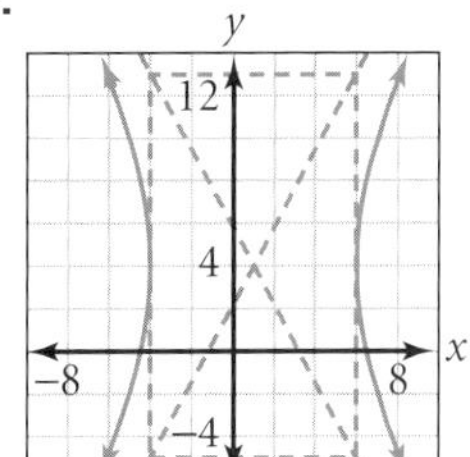

Look Beyond

80. $\sin\theta = \frac{1}{2}$ $\quad \csc\theta = 2$

$\cos\theta = \frac{\sqrt{3}}{2}$ $\quad \sec\theta = \frac{2\sqrt{3}}{3}$

$\tan\theta = \frac{\sqrt{3}}{3}$ $\quad \cot\theta = \sqrt{3}$

81. $\sin\theta = \frac{\sqrt{2}}{2}$ $\quad \csc\theta = \sqrt{2}$

$\cos\theta = \frac{\sqrt{2}}{2}$ $\quad \sec\theta = \sqrt{2}$

$\tan\theta = 1$ $\quad \cot\theta = 1$

82. $\sin\theta = \frac{\sqrt{3}}{2}$ $\quad \csc\theta = \frac{2\sqrt{3}}{3}$

$\cos\theta = \frac{1}{2}$ $\quad \sec\theta = 2$

$\tan\theta = \sqrt{3}$ $\quad \cot\theta = \frac{\sqrt{3}}{3}$

Lesson 13.3, pages 843–850

Activity

1. $c = \sqrt{2}a$

2. $c = \sqrt{3}a$

CHECKPOINT ✔

3. 45-45-90: $1:1:\sqrt{2}$;
30-60-90: $1:\sqrt{3}:2$

Exercises

Communicate

1. It is possible to use a reference angle rather than a calculator when the angle in question is a special angle.

2. The coordinates of point P are $(r \cos \theta, r \sin \theta)$, where r is the radius of the circle centered at the origin that intersects the circle at P. The x-coordinate will be negative because $\cos \theta$ is negative in Quadrant II, while the y-coordinate is positive.

3. Find a coterminal angle with a positive measure that is between 0° and 360°. In this case, the angle is $420° - 360° = 60°$. Since 60° is a special angle, draw a diagram (in the first quadrant) and $\cos 420° = \cos 60°$, which is the adjacent side over the hypotenuse of the triangle, or $\frac{1}{2}$.

Portfolio Activity

1. a. $h = 82$ **b.** $h \approx 116$ **c.** $h \approx 142$

2. a. $a = 426.5$ **b.** $a \approx 460.5$ **c.** $a \approx 486.5$

3. $a = 344.5 + 164 \sin \theta$

4.

Angle (degrees)	Altitude of *P*
0	344.5
30	426.5
45	460.5
60	486.5
90	508.5
120	486.5
135	460.5
150	426.5
180	344.5
210	262.5
225	228.5
240	202.5
270	180.5
300	202.5
315	228.5
330	262.5
360	344.5

5.

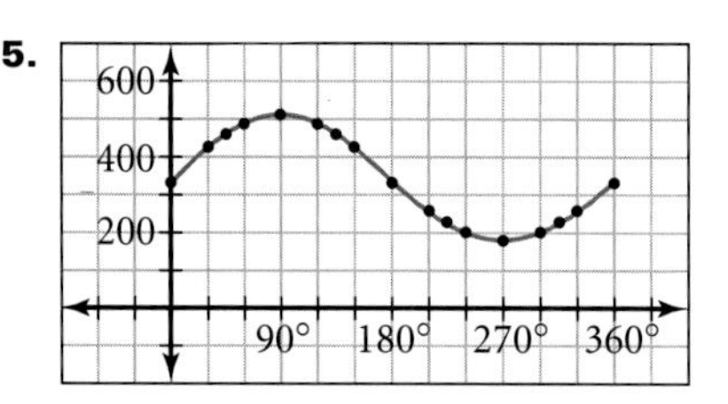

6. Answers will vary. Sample: The two graphs have a similar shape, but the graph of the altitude in Step 5 has no negative y-values. The graph in this activity was easier to sketch because there were more points between each interval of 90°. The second graph has a shorter period.

Lesson 13.4, pages 851–857

Activity

1. Answers may vary. All circles should be a little more than 7.5 cm in diameter.

2. Answers may vary. The graph should approximate a straight line.

3. The slope of the least-squares line should be approximately π.

CHECKPOINT ✔

4. Students should recognize that the slope is π. Answers may vary.

Exercises

Communicate

1. Radian measure reflects the distance along the circumference of the unit circle that is intercepted by a central angle. 2π radians represent a complete rotation. Degree measure is based on arbitrarily assigning a measure of 360° to a full rotation.

2. To convert radians to degrees, multiply by $\frac{180}{\pi}$. To convert degrees to radians, multiply by $\frac{180}{\pi}$.

3. If the radius doubles, the length of the arc intercepted by a central angle will double because $s = r\theta$. If $r' = 2r$, $s' = 2r\theta = 2s$.

4. The linear speed of an object in circular motion is how fast the position of the object changes. The angular speed of an object in circular motion is how fast the angle of rotation of the object changes. Linear speed is distance traveled in a certain amount of time, and angular speed is the change in an angle in a certain amount of time.

Look Back

85.

Data	Frequency	Relative frequency
1	1	$\frac{1}{25} = 0.04$, or 4%
2	1	$\frac{1}{25} = 0.04$, or 4%
3	5	$\frac{5}{25} = 0.20$, or 20%
4	6	$\frac{6}{25} = 0.24$, or 24%
5	6	$\frac{6}{25} = 0.24$, or 24%
6	3	$\frac{3}{25} = 0.12$, or 12%
7	2	$\frac{2}{25} = 0.08$, or 8%
8	1	$\frac{1}{25} = 0.04$, or 4%
Total	25	

86.

Data	Frequency	Relative frequency
65	1	$\frac{1}{17} \approx 0.06 \approx 6\%$
66	3	$\frac{3}{17} \approx 0.18 \approx 18\%$
67	4	$\frac{4}{17} \approx 0.24 \approx 24\%$
68	1	$\frac{1}{17} \approx 0.06 \approx 6\%$
69	1	$\frac{1}{17} \approx 0.06 \approx 6\%$
70	1	$\frac{1}{17} \approx 0.06 \approx 6\%$
71	1	$\frac{1}{17} \approx 0.06 \approx 6\%$
72	2	$\frac{2}{17} \approx 0.12 \approx 12\%$
73	1	$\frac{1}{17} \approx 0.06 \approx 6\%$
74	2	$\frac{2}{17} \approx 0.12 \approx 12\%$
Total	17	

Portfolio Activity

1. 40 seconds; 10 seconds

2.

Angle (degrees)	Altitude of P	Time (seconds)
0	344.5	0
30	426.5	3.3
45	460.5	5
60	486.5	6.7
90	508.5	10
120	486.5	13.3
135	460.5	15
150	426.5	16.7
180	344.5	20
210	262.5	23.3
225	228.5	25
240	202.5	26.7
270	180.5	30
300	202.5	33.3
315	228.5	35
330	262.5	36.7
360	344.5	40

3.

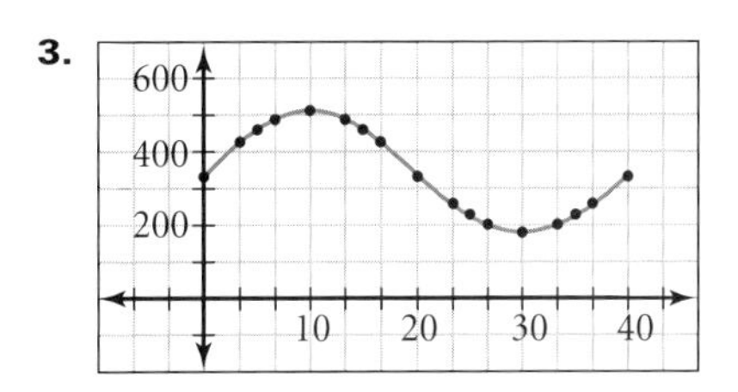

4. The period is 40. The period represents the time in seconds that it takes for one revolution of the ride.

5. linear speed ≈ 25.76 ft/s, or about 17.6 mi/hr

Lesson 13.5, pages 858–866

Activity

CHECKPOINT ✔

1 and 2. Answers may vary. For each value of t, the y-coordinates of each curve are the same.

CHECKPOINT ✔

3. a. 2π

b. The circle will have a radius of 2, so it will be larger. The height of the sine curve will range from –2 to 2.

c. The circle will have a radius of 1, and the period of the sine function will be compressed by $\frac{1}{3}$.

4. For $y_2 = \cos t$:

Steps 1 and 2

Answers may vary. For each value of t, the y-coordinate of the function equals the x-coordinate of the circle.

Step 3

a. 2π

b. The circle will have a radius of 2, so it will be larger. The height of the cosine curve will range from –2 to 2.

c. The circle will have a radius of 1, and the period of the cosine curve will be compressed by $\frac{1}{3}$.

For $y_2 = t$:

Steps 1 and 2

Answers may vary. As the circle approaches the y-axis, the tangent curve approaches its asymptotes.

Step 3

a. π

b. The circle has a radius of 2. The tangent curve increases and decreases more rapidly.

c. The period of the tangent curve will be compressed by $\frac{1}{3}$.

Exercises

Communicate

1. The graphs of $y = \sin\theta$ and $y = \cos\theta$ have the same shape (amplitude and period). The graph of $y = \sin\theta$ is a translation of $y = \cos\theta$ 90° to the right.

2. The tangent function consists of repeated broken pieces, or curves, that each cross the x-axis. The pieces have no bounds, no amplitude, and a period of π.

3. The amplitude of $y = -4\sin\theta$ is larger than that of $y = 3\sin\theta$ because $|-4| > |3|$.

4. Answers may vary. Sample answer:

1. The graph of $y = \sin\theta$ is not broken, but the graph of $y = \tan\theta$ is broken.

2. The graph of $y = \sin\theta$ has a maximum value of $y = 1$ and a minimum value of $y = -1$, but the graph of $y = \tan\theta$ is unlimited in y-values.

3. The period of $y = \sin\theta$ is 2π, but the period of $\tan\theta$ is π.

4. The graph of $y = \sin\theta$ has the same shape as $y = \cos\theta$, but the graph of $y = \tan\theta$ has a different shape.

Guided Skills Practice

7.

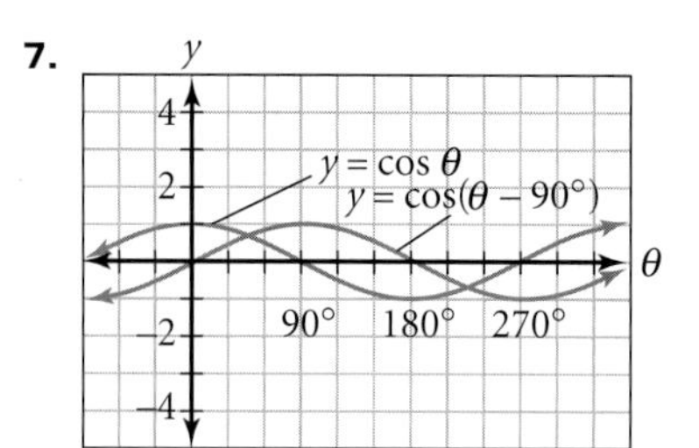

8.

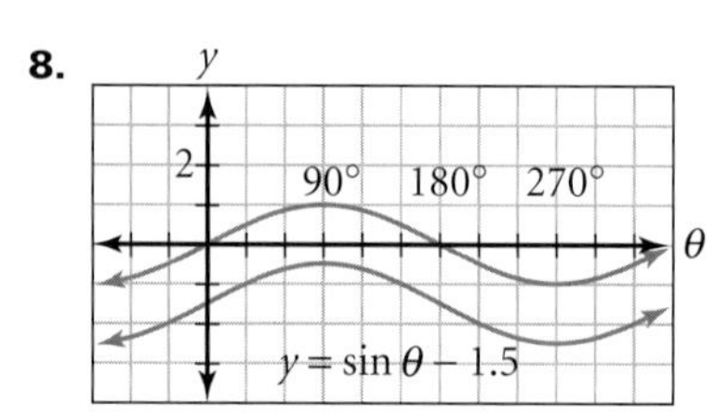

10.

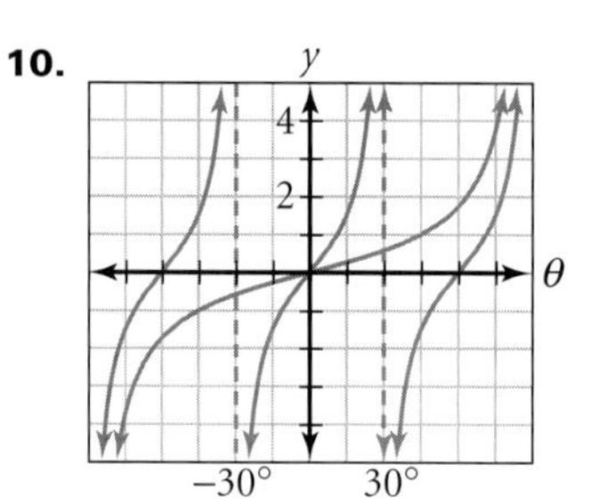

Practice and Apply

34. stretched vertically by a factor of 3 and translated 90° to the left

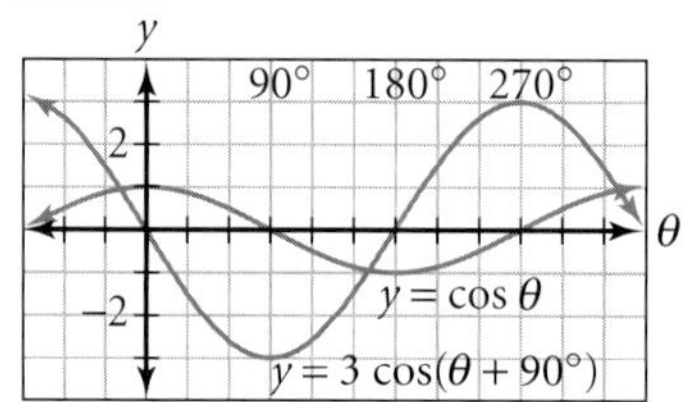

35. stretched vertically by a factor of 2 and translated 90° to the right

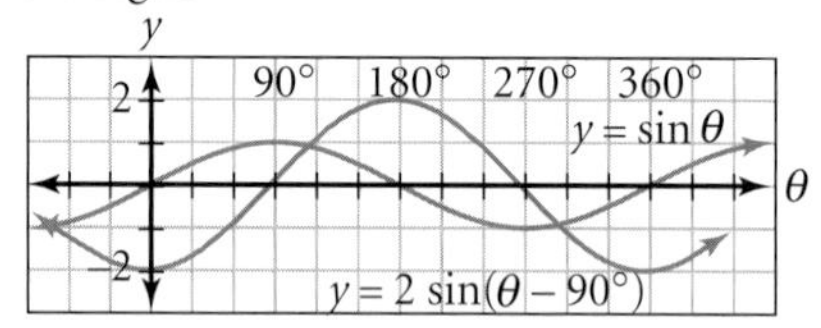

36. reflected across θ-axis, stretched vertically by a factor of 4, and stretched horizontally by a factor of 2

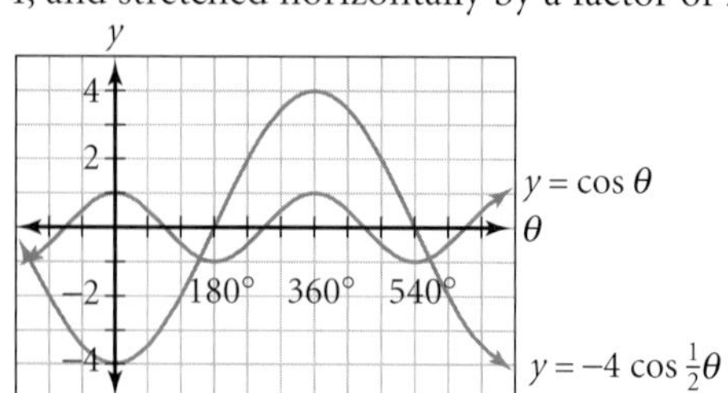

37. reflected across the θ-axis, stretched vertically by a factor of 2, and stretched horizontally by a factor of 3

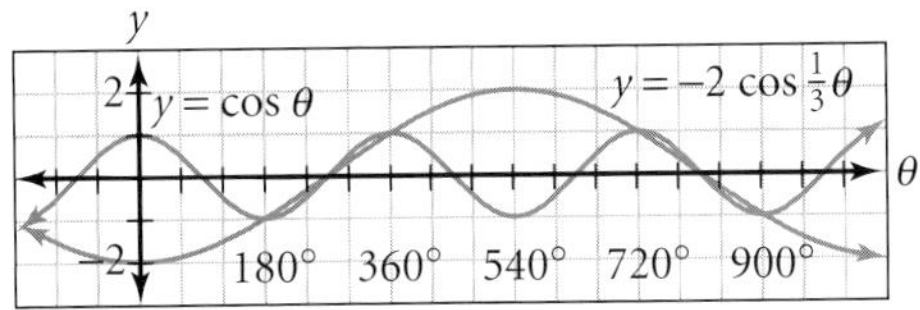

38. compressed vertically by a factor of $\frac{1}{2}$ and horizontally by a factor of $\frac{1}{3}$

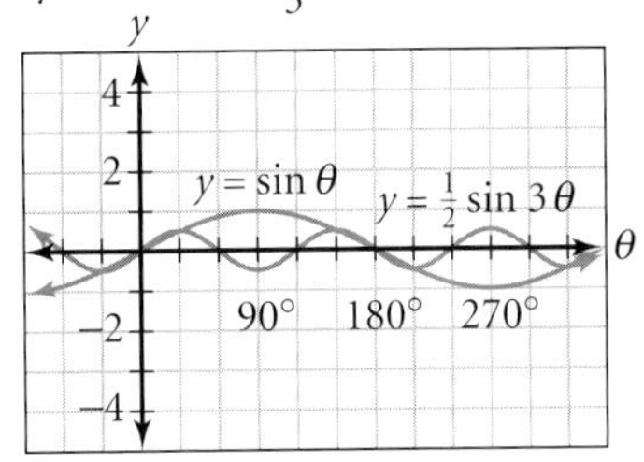

39. compressed vertically by a factor of $\frac{1}{3}$ and horizontally by a factor of $\frac{1}{2}$

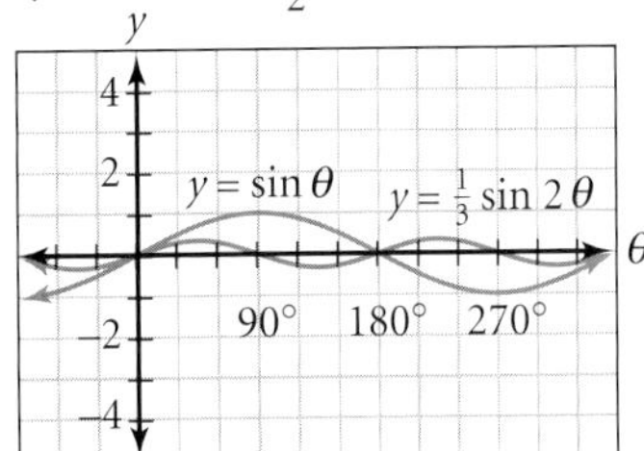

40. stretched vertically by a factor of 3

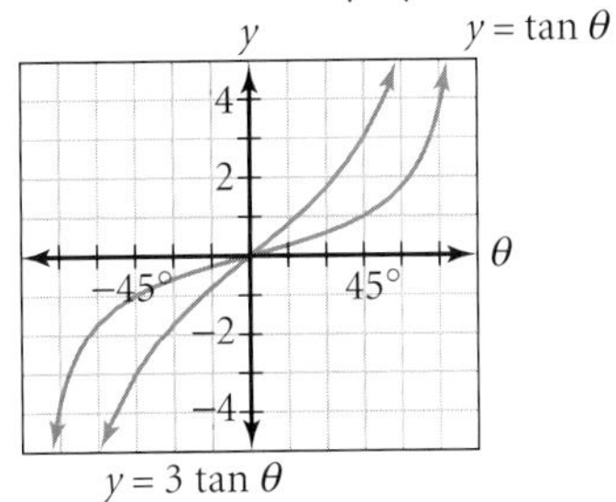

41. stretched vertically by a factor of 2

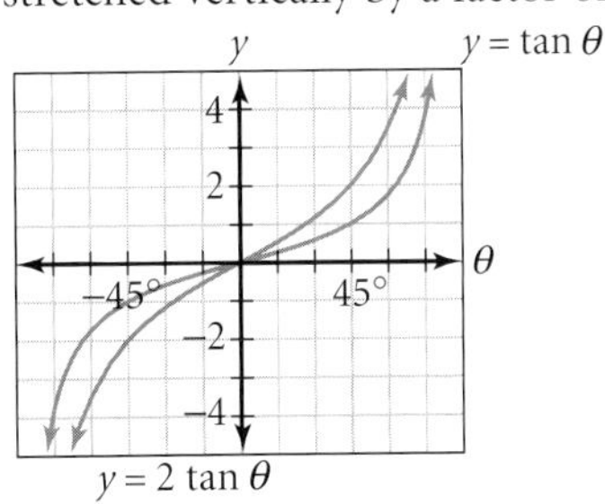

42. translated 3 units up

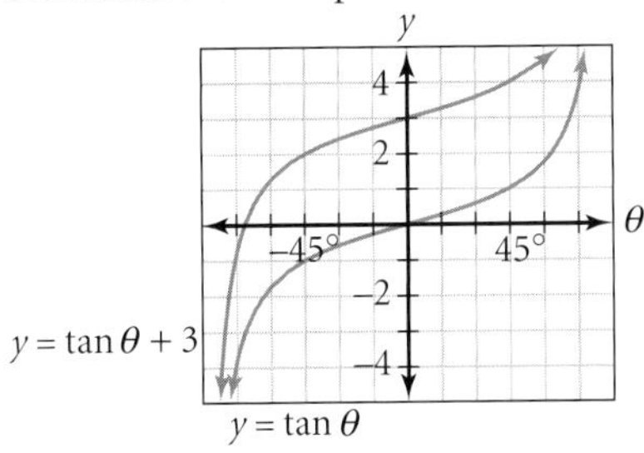

43. translated 2 units down

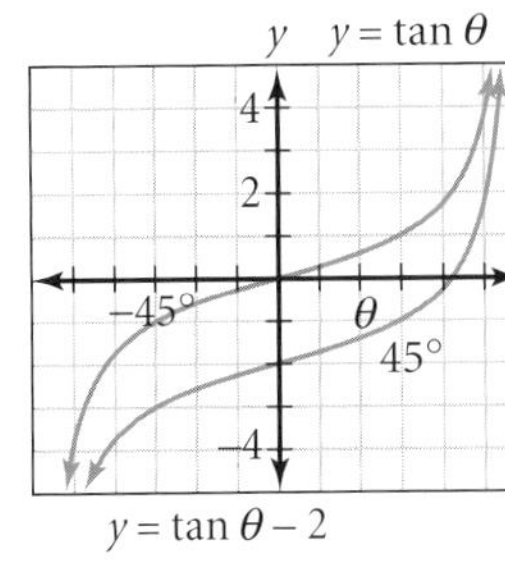

44. compressed horizontally by a factor of $\frac{1}{2}$

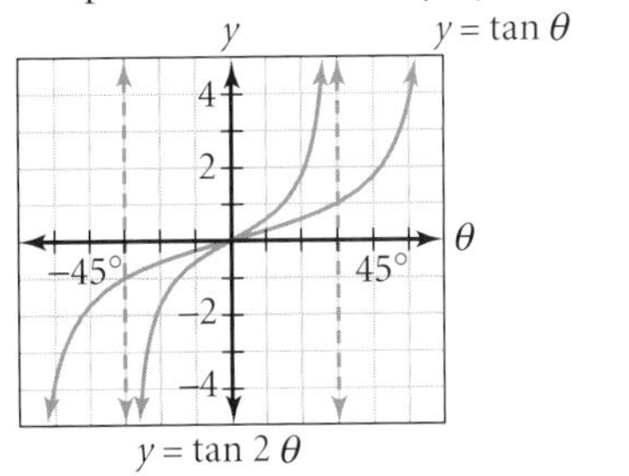

45. compressed horizontally by a factor of $\frac{1}{3}$

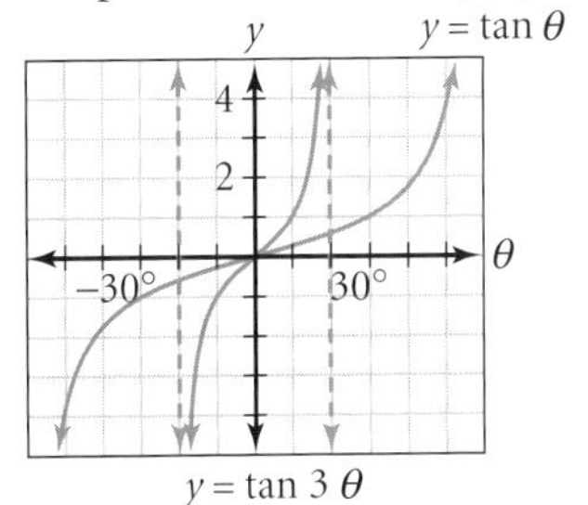

46. stretched vertically by a factor of 2 and horizontally by a factor of 3

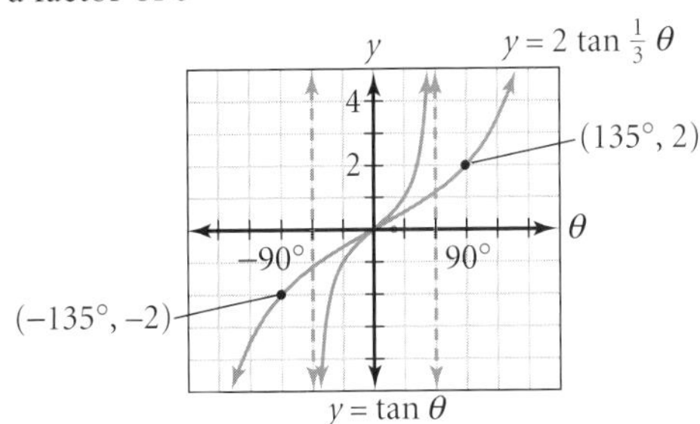

47. stretched vertically by a factor of 3 and horizontally by a factor of 2

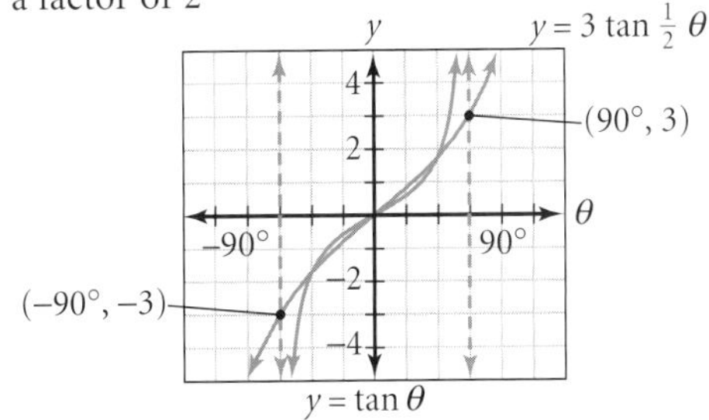

Look Beyond

70. $y = \sec \theta$

θ	y
0°	1
30°	1.155
45°	1.414
60°	2
90°	undefined
120°	−2
135°	−1.414
150°	−1.155
180°	−1
210°	−1.155
240°	2
270°	undefined
300°	−2
330°	1.155

71. $y = \csc \theta$

θ	y
0°	undefined
30°	2
45°	1.414
60°	1.155
90°	1
120°	1.155
135°	1.414
150°	2
180°	undefined
210°	−2
240°	−1.155
270°	−1
300°	−1.155
330°	−2

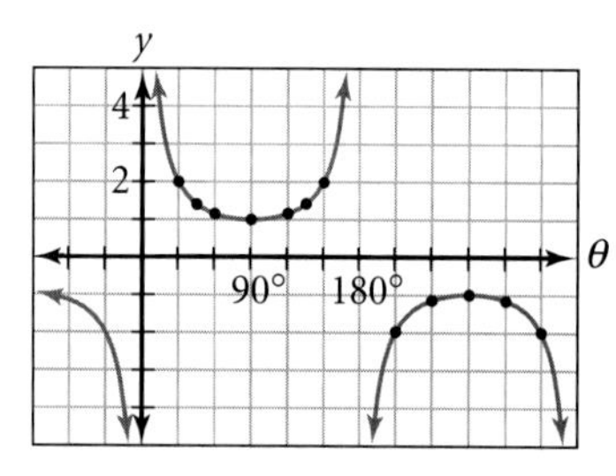

72. $y = \cot \theta$

θ	y
0°	undefined
30°	1.73
45°	1
60°	0.58
90°	0
120°	−0.58
135°	−1
150°	−1.73
180°	undefined

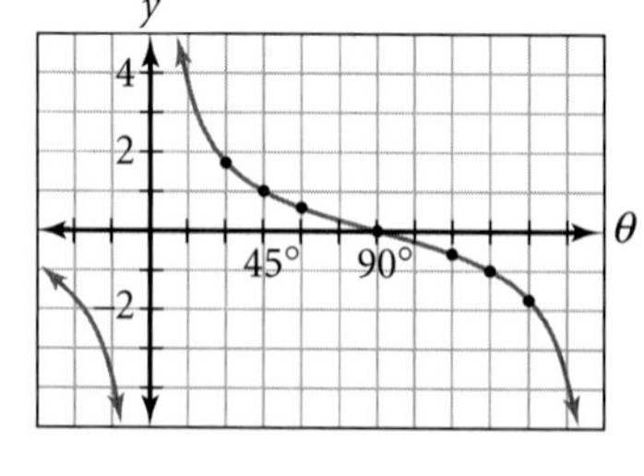

Portfolio Activity

1. amplitude = 164; the amplitude represents the radius of Cosmoclock 21.
2. An equation (using radian measure) is $y = 164\left(\sin \frac{\pi}{20}t\right) + 344.5$.
3. Answers will vary. Sample: The equation fits the data very well since it passes through the points from the table made in Step 2 on page 857.

Lesson 13.6, pages 867–873

Activity

1–3.

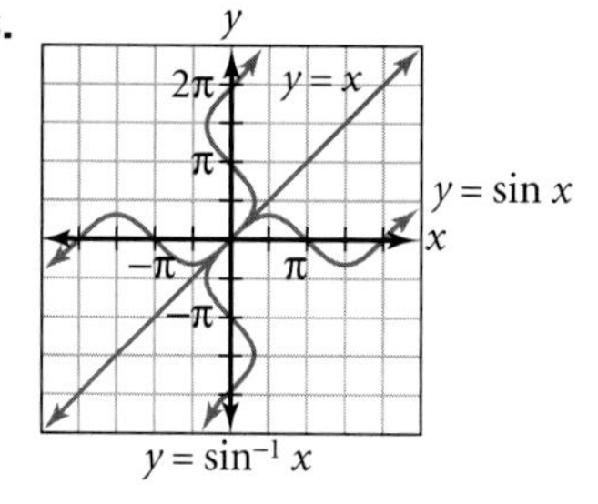

CHECKPOINT ✔

4. The graphs match up with each other; $y = \sin x$ and $y = \sin^{-1} x$ are inverses of each other.

CHECKPOINT ✔

5. $y = \sin^{-1} x$ is the inverse of $y = \sin x$; $y = \sin^{-1} x$ is not a function because some x-values correspond to more than one y-value.

Exercises

Communicate

1. Sin θ, Cos θ, and Tan θ are sin θ, cos θ, and tan θ, respectively, with each domain restricted to the corresponding principle values, and thereby forming one-to-one functions.
2. $\text{Sin}^{-1} x$ is the inverse of Sin x, $\text{Cos}^{-1} x$ is the inverse of Cos x, and $\text{Tan}^{-1} x$ is the inverse of Tan x.
3. $\text{Sin}^{-1} 5$ means "the angle whose sine is 5." The range of the sine function is −1 to 1. $\text{Tan}^{-1} 5$ is defined. The range of the tangent function is all real numbers.

Chapter 13 Project

Activity 1

θ	Distance from x-axis
0°	0
90°	70
180°	0
270°	−70
360°	0
450°	70
540°	0
630°	−70
720°	0
810°	70

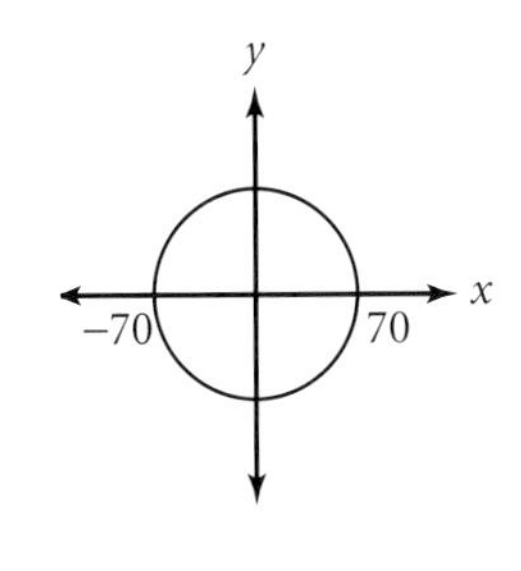

Activity 2

θ	Height (feet)
0°	80
90°	150
180°	80
270°	10
360°	80
450°	150
540°	80
630°	10
720°	80
810°	150

Activity 3

Time (seconds)	Height (feet)
0	80
110	150
220	80
330	10
440	80
550	150
660	80
770	10
880	80
990	150

Lesson 14.1, pages 884–893

Activity

1. 0; side a is not long enough to reach the third side of the triangle.
2. 1; side a is long enough to reach the third side at exactly one point.
3. 2; side a reaches the third side at two different points.
4. 1; side a reaches the third side at exactly one point and is too long to reach the third side a second time.

CHECKPOINT ✔

5. If $a < h$, then 0 triangles can be formed. If $a = h$ or if $a > h$ and $a > b$, then 1 triangle can be formed. If $a > h$ and $a < b$, then 2 triangles can be formed.

Exercises

Communicate

1. First find the missing angle by using the fact that the sum of the angles is 180°. Then apply the law of sines to find the two missing sides.
2. First find the missing angle by using the fact that the sum of the angles is 180°. Then apply the law of sines to find the two missing sides.
3. If one of the sides is less than the height of the triangle formed by the other two sides and the corresponding height, then no triangle is formed.
4. If SSA information is given and the shorter of the given sides is longer than the height of the triangle formed by using the longer given side and given angle, then 2 different triangles may be formed.

Lesson 14.2, pages 894–901

Activity

1 and 2. 2

CHECKPOINT ✔

3. Two solutions are possible because $\sin \theta = x$ has two solutions for any θ on the interval [0, 180).

4 and 5. 1

CHECKPOINT ✔

6. One solution is possible because $\cos \theta = x$ has one solution for any θ on the interval [0, 180).

Exercises

Communicate

1. a, b, c; they represent the lengths of the sides of the triangle.

2. First solve for one angle by using the law of cosines.
$$a^2 = b^2 + c^2 - 2bc \cos A$$
$$\cos A = \frac{a^2 - b^2 - c^2}{-2bc}$$
Then use the law of sines to find another angle. Then find the remaining angle by using the fact that the sum of the 3 angles is 180°.

3. First solve for the remaining side by using the law of cosines, $a^2 = b^2 + c^2 - 2bc \cos A$. Then find another angle by using the laws of sines or cosines. Then find the remaining angle by using the fact that the sum of the 3 angles is 180°.

4. Knowing the angle measures of all three angles in a triangle does not give you enough information to solve the triangle because two similar triangles can have equal angle measures but be different sizes.

Portfolio Activity

1. *Not drawn to scale*

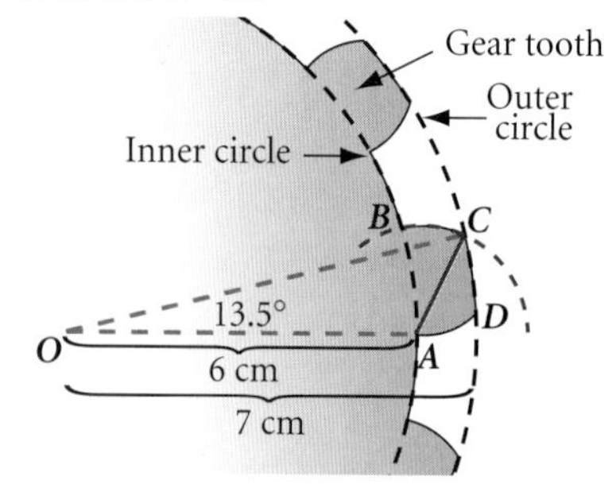

2. $AC \approx 1.8$ cm

Lesson 14.3, pages 902–908

Activity

1 and 2. $y = \sin \theta$

3. $$\begin{aligned} y &= (\csc \theta)(1 - \cos \theta)(1 + \cos \theta) \\ &= \frac{1}{\sin \theta}(1 - \cos^2 \theta) \\ &= \frac{1}{\sin \theta}(\sin^2 \theta) \\ &= \sin \theta \end{aligned}$$

4. $y = \cos \theta$
$$\begin{aligned} y &= \tan \theta(\csc \theta - \tan \theta \cos \theta) \\ &= \frac{\sin \theta}{\cos \theta}\left(\frac{1}{\sin \theta} - \frac{\sin \theta}{\cos \theta}\cos \theta\right) \\ &= \frac{\sin \theta}{\cos \theta}\left(\frac{1}{\sin \theta} - \sin \theta\right) \\ &= \frac{\sin \theta}{\cos \theta}\left(\frac{1 - \sin^2 \theta}{\sin \theta}\right) \\ &= \frac{\sin \theta}{\cos \theta}\left(\frac{\cos^2 \theta}{\sin \theta}\right) \\ y &= \cos \theta \end{aligned}$$

CHECKPOINT ✔

5. When the related function is graphed we can sometimes identify the simplifed expression that is the desired result.

6. Answers may vary.

Exercises

Communicate

1. $\tan \theta = \frac{\sin \theta}{\cos \theta}$

2. Sample answers: Use trigonometric identities to simplify the expression algebraically. Graph the expression to see if it has the same graph as a simpler expression.

3. The trigonometric Pythagorean identity can be simplified into the Pythagorean Theorem by replacing $\sin \theta$ with $\frac{y}{r}$ and $\cos \theta$ with $\frac{x}{r}$.

Practice and Apply

10. $$\begin{aligned} \csc \theta &= \frac{r}{y} \\ &= \frac{1}{\frac{y}{r}} \\ &= \frac{1}{\sin \theta}, \ \sin \theta \neq 0 \end{aligned}$$

11. $$\begin{aligned} \cos^2 \theta &= \left(\frac{x}{r}\right)^2 \\ &= \frac{x^2}{r^2} \\ &= \frac{r^2 - y^2}{r^2} \\ &= \frac{r^2}{r^2} - \frac{y^2}{r^2} \\ &= 1 - \left(\frac{y}{r}\right)^2 \\ &= 1 - \sin^2 \theta \end{aligned}$$

12. $$\begin{aligned} 1 + \cot^2 \theta &= 1 + \left(\frac{x}{y}\right)^2 \\ &= \frac{y^2}{y^2} + \frac{x^2}{y^2} \\ &= \frac{y^2 + x^2}{y^2} \\ &= \frac{r^2}{y^2} \\ &= \left(\frac{r}{y}\right)^2 \\ &= \csc^2 \theta \end{aligned}$$

41. $\frac{\sec\theta}{\cos\theta} = \frac{\frac{1}{\cos\theta}}{\cos\theta}$

$= \frac{1}{\cos^2\theta}$

$= \sec^2\theta$

42. $\frac{\csc\theta}{\sin\theta} = \frac{\frac{1}{\sin\theta}}{\sin\theta}$

$= \frac{1}{\sin^2\theta}$

$= \csc^2\theta$

43. $\frac{\cos\theta}{1-\sin^2\theta} = \frac{\cos\theta}{\cos^2\theta}$

$= \frac{1}{\cos\theta}$

$= \sec\theta$

44. $\frac{\sin\theta}{1-\cos^2\theta} = \frac{\sin\theta}{\sin^2\theta}$

$= \frac{1}{\sin\theta}$

$= \csc\theta$

45. $(\sec\theta)(1-\sin^2\theta) = \left(\frac{1}{\cos\theta}\right)(\cos^2\theta)$

$= \cos\theta$

46. $(\csc\theta)(1-\cos^2\theta) = \left(\frac{1}{\sin\theta}\right)(\sin^2\theta)$

$= \sin\theta$

47. $(\tan\theta)(\csc\theta)(\sec\theta) = \left(\frac{\sin\theta}{\cos\theta}\right)\left(\frac{1}{\sin\theta}\right)\left(\frac{1}{\cos\theta}\right)$

$= \frac{1}{\cos^2\theta}$

$= \sec^2\theta$

48. $(\cot\theta)(\csc\theta)(\sec\theta) = \left(\frac{\cos\theta}{\sin\theta}\right)\left(\frac{1}{\sin\theta}\right)\left(\frac{1}{\cos\theta}\right)$

$= \frac{1}{\sin^2\theta}$

$= \csc^2\theta$

49. $\frac{\sin^2\theta - 2\sin\theta\cos\theta}{\cos^2\theta}$, or $\frac{\sin^2\theta}{\cos^2\theta} - \frac{2\sin\theta}{\cos\theta}$

50. $\tan^2\theta - 2\tan\theta$

55.

$\cos\theta = \pm\sqrt{1-\sin^2\theta}$

$\tan\theta = \pm\frac{\sin\theta}{\sqrt{1-\sin^2\theta}}$

$\csc\theta = \frac{1}{\sin\theta}$

$\sec\theta = \pm\frac{1}{\sqrt{1-\sin^2\theta}}$

$\cot\theta = \pm\frac{\sqrt{1-\sin^2\theta}}{\sin\theta}$

56.

$\sin\theta = \pm\frac{\tan\theta}{\sqrt{\tan^2\theta+1}}$

$\cos\theta = \pm\frac{1}{\tan\theta\sqrt{1+\frac{1}{\tan^2\theta}}}$

$\csc\theta = \pm\frac{\sqrt{\tan^2\theta+1}}{\tan\theta}$

$\sec\theta = \pm\tan\theta\sqrt{1+\frac{1}{\tan^2\theta}}$

$\cot\theta = \frac{1}{\tan\theta}$

57.

$\sin\theta = \pm\frac{1}{\cot\theta\sqrt{1-\frac{1}{\cot^2\theta}}}$

$\cos\theta = \pm\frac{\cot\theta}{\sqrt{1+\cot^2\theta}}$

$\tan\theta = \frac{1}{\cot\theta}$

$\csc\theta = \pm\cot\theta\sqrt{1-\frac{1}{\cot^2\theta}}$

$\sec\theta = \pm\frac{\sqrt{1+\cot^2\theta}}{\cot^2\theta}$

58.

$\sin\theta = \pm\sqrt{1-\frac{1}{\sec^2\theta}}$

$\cos\theta = \frac{1}{\sec\theta}$

$\tan\theta = \pm\sqrt{\sec^2\theta - 1}$

$\csc\theta = \pm\frac{1}{\sqrt{1-\frac{1}{\sec^2\theta}}}$

$\cot\theta = \pm\frac{1}{\sqrt{\sec^2\theta-1}}$

Lesson 14.4, pages 909–916

Activity

1. a. $(\cos\alpha - \cos\beta)^2 + (\sin\alpha - \sin\beta)^2$

$= \cos^2\alpha - 2\cos\alpha\cos\beta + \cos^2\beta + \sin^2\alpha - 2\sin\alpha\sin\beta + \sin^2\beta$

$= (\cos^2\alpha + \sin^2\alpha) + (\cos^2\beta + \sin^2\beta) - 2\cos\alpha\cos\beta - 2\sin\alpha\sin\beta$

$= 1 + 1 - 2\cos\alpha\cos\beta - 2\sin\alpha\sin\beta$

$= 2 - 2\cos\alpha\cos\beta - 2\sin\alpha\sin\beta$

b. $\cos^2\theta + \sin^2\theta = 1$

2. a. $[\cos(\alpha-\beta) - 1]^2 + [\sin(\alpha-\beta) - 0]^2$

$= \cos^2(\alpha-\beta) - 2\cos(\alpha-\beta) + 1 + \sin^2(\alpha-\beta) + 0$

$= [\cos^2(\alpha-\beta) + \sin^2(\alpha-\beta)] + 1 - 2\cos(\alpha-\beta)$

$= 1 + 1 - 2\cos(\alpha-\beta)$

$= 2 - 2\cos(\alpha-\beta)$

b. $\cos^2\theta + \sin^2\theta = 1$

CHECKPOINT ✔

3. yes; $OP = OQ = OR = OS = 1$ by radii of the unit circle, $\triangle POQ \cong \triangle ROS$ by SAS, and $\overline{PQ} \cong \overline{RS}$ by corr. parts; yes; by the principle of powers

CHECKPOINT ✔

4. $2 - 2\cos(\alpha-\beta) = 2 - 2\cos\alpha\cos\beta - 2\sin\alpha\sin\beta$

$-2\cos(\alpha-\beta) = -2\cos\alpha\cos\beta - 2\sin\alpha\sin\beta$

$-2\cos(\alpha-\beta) = -2(\cos\alpha\cos\beta + \sin\alpha\sin\beta)$

$\cos(\alpha-\beta) = \cos\alpha\cos\beta + \sin\alpha\sin\beta$

Exercises

Communicate

1. Answers may vary. Sample answer:

$\cos 75° = \cos(120° - 45°)$

$= \cos 120° \cos 45° + \sin 120° \sin 45°$

2. Answers may vary. Sample answer:

$\sin(-15°)$

$= -\sin 15°$

$= -\sin(45° - 30°)$

$= -(\sin 45° \cos 30° - \cos 45° \sin 30°)$

3. Answers may vary. Sample answer:

$$\begin{bmatrix} \cos 45° & -\sin 45° \\ \sin 45° & \cos 45° \end{bmatrix}\begin{bmatrix} 2 \\ 3 \end{bmatrix} = \begin{bmatrix} \frac{\sqrt{2}}{2} & -\frac{\sqrt{2}}{2} \\ \frac{\sqrt{2}}{2} & \frac{\sqrt{2}}{2} \end{bmatrix}\begin{bmatrix} 2 \\ 3 \end{bmatrix}$$

Practice and Apply

29. Answers may vary. Sample answer:
$\sin(30° + 60°) = 1$
$\sin 30° = \sin 60° = \frac{1 + \sqrt{3}}{2}$
So, $\sin(A + B) \neq \sin A + \sin B$.

30. Answers may vary. Sample answer:
$\cos(30° + 60°) = 0$
$\cos 30° + \cos 60° = \frac{\sqrt{3} + 1}{2}$
So, $\cos(A + B) \neq \cos A + \cos B$.

31. Answers may vary. Sample answer:
$\sin(60° - 30°) = \frac{1}{2}$
$\sin 60° - \sin 30° = \frac{\sqrt{3} - 1}{2}$
So, $\sin(A - B) \neq \sin A - \sin B$.

32. Answers may vary. Sample answer:
$\cos(60° - 30°) = \frac{\sqrt{3}}{2}$
$\cos 60° - \cos 30° = \frac{1 - \sqrt{3}}{2}$
So, $\cos(A - B) \neq \cos A - \cos B$.

45. $\begin{bmatrix} \frac{\sqrt{2}}{2} & -\frac{\sqrt{2}}{2} \\ \frac{\sqrt{2}}{2} & \frac{\sqrt{2}}{2} \end{bmatrix}$

46. $\begin{bmatrix} \frac{1}{2} & -\frac{\sqrt{3}}{2} \\ \frac{\sqrt{3}}{2} & \frac{1}{2} \end{bmatrix}$

47. $\approx \begin{bmatrix} 0.766 & 0.643 \\ -0.643 & 0.766 \end{bmatrix}$

48. $\approx \begin{bmatrix} -0.719 & 0.695 \\ -0.695 & -0.719 \end{bmatrix}$

49. $\begin{bmatrix} -\frac{1}{2} & \frac{\sqrt{3}}{2} \\ -\frac{\sqrt{3}}{2} & -\frac{1}{2} \end{bmatrix}$

50. $\approx \begin{bmatrix} -0.940 & -0.342 \\ 0.342 & -0.940 \end{bmatrix}$

51. $\begin{bmatrix} -\frac{\sqrt{2}}{2} & \frac{\sqrt{2}}{2} \\ -\frac{\sqrt{2}}{2} & -\frac{\sqrt{2}}{2} \end{bmatrix}$

52. $\approx \begin{bmatrix} 0.766 & -0.643 \\ 0.643 & 0.766 \end{bmatrix}$

53.

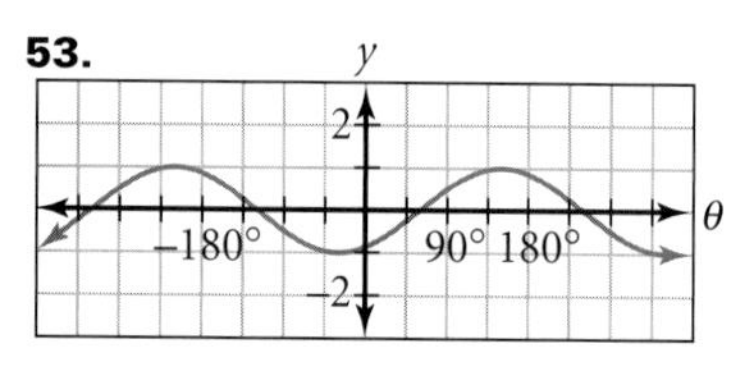

54.

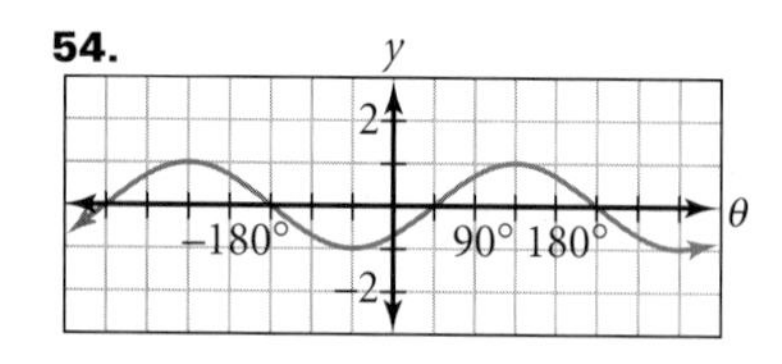

55.

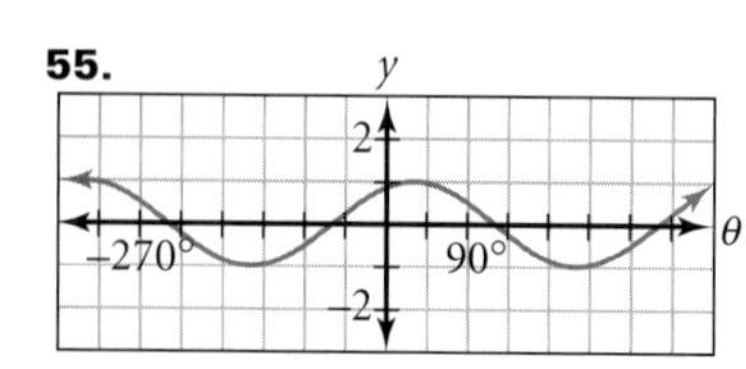

56.

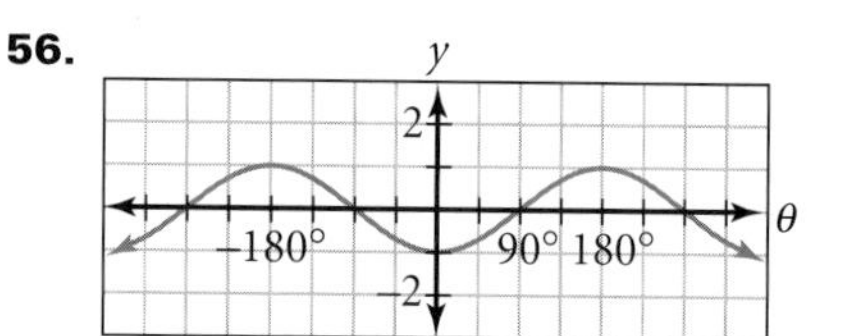

57.

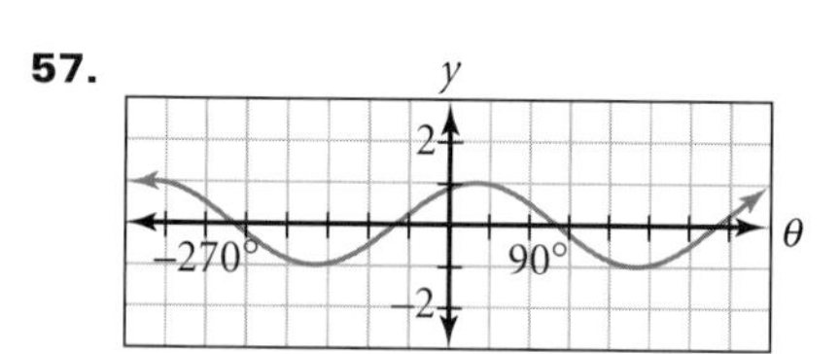

58.

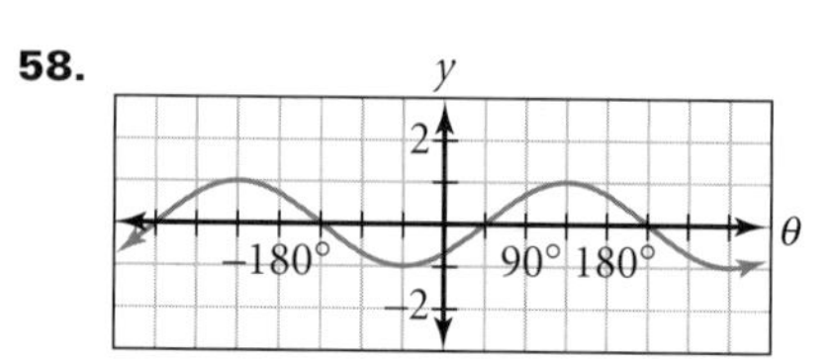

69. $W'(1.50, 2.60)$, $X'(-0.23, 3.60)$, $Y'(1.27, 6.20)$, $Z'(3.00, 5.20)$

70. $W'(-1.50, 2.60)$, $X'(-3.23, 1.60)$, $Y'(-4.73, 4.20)$, $Z'(-3.00, 5.20)$

71. $W'(2.60, -1.50)$, $X'(3.60, 0.23)$, $Y'(6.20, -1.27)$, $Z'(5.20, -3.00)$

72. $W'(-2.60, -0.50)$, $X'(-1.60, -3.23)$, $Y'(-4.20, -4.73)$, $Z'(-5.20, -3.00)$

73. $W'(-2.12, 2.12)$, $X'(-3.54, 0.71)$, $Y'(-5.66, 2.83)$, $Z'(-4.24, 4.24)$

74. $W'(2.60, 1.50)$, $X'(1.60, 3.23)$, $Y'(4.20, 4.73)$, $Z'(5.20, 3.00)$

75. $W'(0, -3)$, $X'(2, -3)$, $Y'(2, -6)$, $Z'(0, -6)$

76. $W'(-1.50, -2.60)$, $X'(0.23, -3.60)$, $Y'(-1.27, -6.20)$, $Z'(-3.00, -5.20)$

78. a.

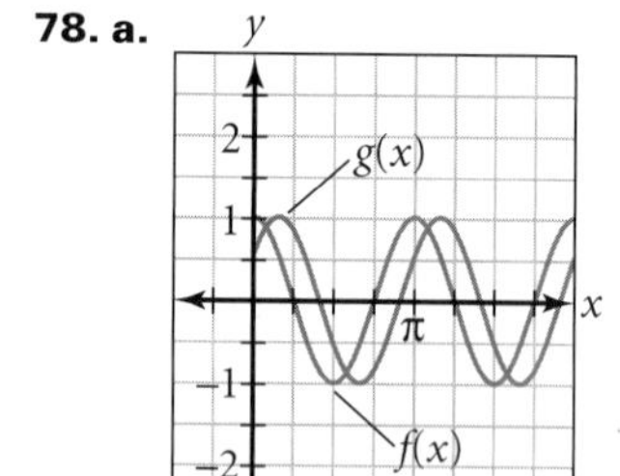

g is a horizontal translation of f one radian to the right.

b.

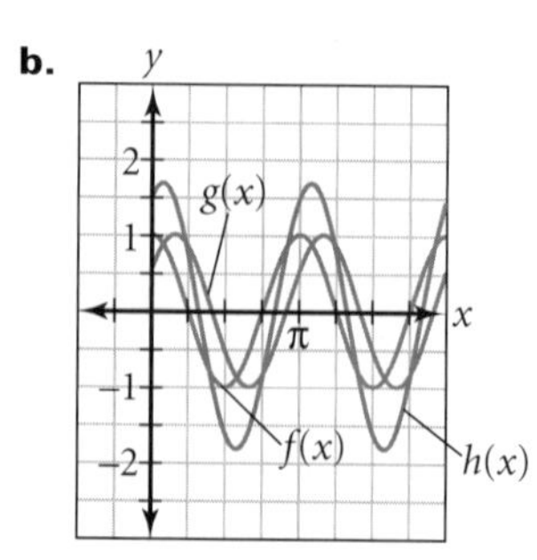

c. They all have a period of π.

d. $h(x) = 2[\cos \frac{1}{2}(4x - 1)]\cos \frac{1}{2}$

Portfolio Activity

1. a. $R_{30°} = \begin{bmatrix} \cos 30° & -\sin 30° \\ \sin 30° & \cos 30° \end{bmatrix}$;

$S = \begin{bmatrix} 8.00 & 7.52 & 9.72 & 9.94 \\ 0.00 & 2.74 & 2.33 & 1.13 \end{bmatrix}$

b. $R \times S = \begin{bmatrix} 6.93 & 5.14 & 7.25 & 8.04 \\ 4.00 & 6.13 & 6.88 & 5.95 \end{bmatrix}$

2. $A'(4.00, 6.93)$, $B'(1.39, 7.88)$, $C'(2.84, 9.58)$, $D'(3.99, 9.17)$

Lesson 14.5, pages 917–921

Exercises

Communicate

1. The ranges of $y = \sin \theta$ and $y = \sin 2\theta$ are all real numbers from −1 to 1, inclusive, whereas the range of $y = 2 \sin \theta$ is −2 to 2.

2. For $0° \leq \theta \leq 360°$, $\sin \frac{\theta}{2}$ is always positive or 0. For $0° \leq \theta \leq 360°$, if θ is in Quadrant I or II, then $\cos \frac{\theta}{2}$ is positive, and if θ is in Quadrant III or IV, then $\cos \frac{\theta}{2}$ is negative.

Lesson 14.6, pages 922–927

Activity

1. 1; 90°

2. 2; 90°, 450°

3.

Interval	$0° \leq \theta < 360°$	$0° \leq \theta < 720°$
Number of solutions	1	2
Solutions	90°	90°, 450°

Interval	$0° \leq \theta < 1080°$	$0° \leq \theta < 1440°$
Number of solutions	3	4
Solutions	90°, 450°, 810°	90°, 450°, 810°, 1170°

$\theta = 90° + n(360°)$ where n is any integer

CHECKPOINT ✔

4. any $a > 1$ or any $a < -1$; sample answer: −2

5. No; if $\sin \theta = a$ has an answer, then that answer plus an integer multiple of 360° will also be an answer.

6. Yes; if $\sin \theta = a$ has a solution, then that solution plus an integer multiple of the solution will also be a solution. Because there are an infinite number of integers, there will be an infinite number of solutions.

Exercises

Communicate

1. Trigonometric equations have a solution for only certain values of the variables, while trigonometric identities are true for all values of the variables.

2. Infinitely many solutions; no solutions; the range of $\sin x$ and $\cos x$ is $[-1, 1]$. So, $\sin x = -1$ infinitely many times, but $\cos x$ never equals 2.

3. Example 1 uses both algebra to solve for $\cos \theta$ and a graph to find the points of intersection. Example 2 uses substitution and algebra, and Example 3 uses trigonometric identities, substitution, and algebra.

INDEX

Definitions of bold face entries can be found in the glossary.

H

I

J

K

L

T

U

V

Z

Credits

PHOTOS

Abbreviated as follows: (t), top, (b), bottom, (l), left, (r), right, (c), center, (bckgd) background.

COVER: Tom Paiva/FPG International. **TABLE OF CONTENTS:** Page T6 (tl), Peter Van Steen/HRW Photo, location courtesy Strait Music Co.; T6 (bl), Mark M. Lawrence/The Stock Market; T7 (tl), (cl), Sam Dudgeon/HRW Photo; T7 (bl), John Langford/HRW Photo; T7 (br), Warren Faidley/International Stock Photo; T8 (bl), Blair Seitz/Photo Researchers; T8 (tl), L.D. Gordon/The Image Bank; T9 (br), UPI/Corbis-Bettmann; T9 (cl), Sam Dudgeon/HRW Photo; T10 (b), VCG/FPG International; T10 (cl), G.A. Plimpton Collection, Rare Book & Manuscript Library, Columbia University; T11 (br), Patrick Cocklin/Tony Stone Images; T11 (l), Telegraph Colour Library/Masterfile; T12 (b), Bill Losh/FPG International; T12 (tl), Miwako Ikeda/International Stock Photo; T13 (b), Andrew Freeman/ SportsChrome-USA; T13 (cl), NASA; T14 (bl), Image Copyright © 2001 PhotoDisc, Inc.; T14 (cl), Mark C. Burnett/Science Source/Photo Researchers; T15 (br), eStudios/HRW Photo; T15 (cl), Superstock; T16 (bl), David Seelig/Allsport; T17 (br), Image Copyright © 2001 PhotoDisc, Inc.; T17 (cl), Ralph H. Wetmore/Tony Stone Images; T18 (b), Stephen Durke/Washington-Artists' Represents; T18 (cl), Mark E. Gibson; T19 (l), Superstock. **PROFESSIONAL ARTICLES:** Page T20(tr), Artbase Inc.; T20(bl), Brett Froomer/ The Image Bank © 2000; T21(tr), Sam Dudgeon/HRW Photo; T22(tr), Artbase Inc.; T22(br), Sam Dudgeon/HRW Photo; T23(tr), John Langford/HRW Photo; T24(tc), Sam Dudgeon/HRW Photo. **CHAPTER ONE:** Page 2 (bc), Superstock; 2 (bl), George Lepp/ Tony Stone Images; 2–3 (bckgd), George Lepp/Tony Stone Images; 2–3 (t), Image Copyright © 2001 PhotoDisc, Inc.; 4 (t), Peter Van Steen/HRW Photo, location courtesy Strait Music Co.; 5 (tl), Ron Chapple/FPG International; 8 (tl), David De Lossy/The Image Bank; 10 (br), 1994 Burke/Triolo Productions/FoodPix; 10 (tl), Uniphoto; 12 (t), John Langford/HRW Photo; 19 (b), Jan Becker; 19 (cl), Ron Tanaka; 20 (tl), Renee Lynn/Photo Researchers; 21 (t), Luis Castaneda/The Image Bank; 25 (br), Michelle Bridwell/HRW Photo; 27 (b), Pascal Rondeau/Tony Stone Images; 29 (t), Uniphoto; 32 (cl), NASA/Science Photo Library/ Photo Researchers; 35 (tl), (tr), Courtesy of the Oriental Institute of the University of Chicago; 36 (tl), Mark M. Lawrence/The Stock Market; 37 (tr), Navaswan/FPG International; 41 (bl), (cl), Photo Courtesy Indianapolis Motor Speedway; 41 (tl), Ron McQueeney/Photo Courtesy Indianapolis Motor Speedway; 43 (tl), UPI/Corbis-Bettmann; 44 (tl), Uniphoto; 45 (tr), Ron Tanaka; 46 (cl), Randal Alhadeff/HRW Photo; 48 (tl), D. Young-Wolff/ PhotoEdit; 50 (b), Aaron Haupt/Photo Researchers; 51 (tr), Roy Morsch/The Stock Market; 52 (cl), 53 (tc), (tr), Maryland Historical Society, Baltimore; 54 (tc), Myrleen Ferguson/PhotoEdit; 56 (tr), Maratea/International Stock Photo; 59 (bl), Margerin Studios Inc./FPG International; 59 (br), Lou Manna/International Stock Photo; 59 (cr), Uniphoto; 61 (c), VCG/FPG International; 67 (br), Mark Gamba/The Stock Market; 68 (br), Uniphoto; 69 (tr), Christian Grzimek/OKAPIA/Photo Researchers; 70 (bl), Image Copyright © 2001 PhotoDisc, Inc.; 70–71 Rob Waymen Photography; 70–71 (c), Ron Tanaka. **CHAPTER TWO:** Page 84 (bl), Paolo Curto/The Image Bank; 84 (c), Janice Travia/Tony Stone Images; 85 (cr), Image Copyright © 2001 PhotoDisc, Inc.; 85 (t), Michael J. Howell/International Stock Photo; 86 (tc), Michael Newman/ PhotoEdit; 86 (tr), John Langford/HRW Photo; 88 (cr), Randal Alhadeff/HRW Photo; 90 (cl), David R. Frazier/Photo Researchers; 91 (br), The Granger Collection; 92 (cr), Lori Adamski Peek/Tony Stone Images; 93 (c), Corbis, photo manipulation by Morgan-Cain & Associates; 94 (tr), 97 (br), Michael Newman/PhotoEdit; 98 (cr), Sam Dudgeon/HRW Photo, prop courtesy Custom Model Products; 100 (tr), Uniphoto; 101 (br), NASA; 102 (t), Uniphoto; 106 (br), Spencer Grant/PhotoEdit; 107 (tl), Randal Alhadeff/HRW Photo; 108 (tr), Rob Waymen Photography; 109 (br), Randal Alhadeff/HRW Photo; 111 (t), Sanford/Agliolo/ International Stock Photo; 114 (tr), Peter Van Steen/HRW Photo, location courtesy Dyer Electronics; 116 (br), Sam Dudgeon/HRW Photo; 116 (tr), Don Smetzer/Tony Stone Images; 124 (truck driver), Dana White/PhotoEdit; 124 (truck), Larry Grant/FPG International; 126 (cl), Sam Dudgeon/HRW Photo; 129 (tr), Robert Brenner/PhotoEdit; 131 (br), Steve Lacey/ Uniphoto, photo manipulation by Morgan-Cain & Associates; 131 (tr), Randal Alhadeff/ HRW Photo; 133 (t), Jeff Kaufman/FPG International; 141 (tl), Alan D. Carey/Vireo; 142 (br), 143 (tr), Corbis; 146 (br), Eric Bouvet/ The Image Bank. **CHAPTER THREE:** Page 154 (b), Don Couch/HRW Photo; 155 (cl), Lucien Clergue/Tony Stone Images; 155 (cr), Bob Daemmrich Photo; 155 (tl), Uniphoto; 155 (tr), David Waldorf/FPG International; 156 (t), Blair Seitz/Photo Researchers; 161 (tl), Judy Unger/FoodPix; 162 (br), Allsport USA/Tony Duffy; 163 (tr), L.D. Gordon/The Image Bank; 164 (inset), (t), Peter Van Steen/HRW Photo, art by Mary Doerr, location courtesy Make-A-Frame; 165 (cl), Randal Alhadeff/HRW Photo; 170 (br), Uniphoto; 170 (tr), Noboru Komine/Photo Researchers; 171 (bl), Peter Van Steen/HRW Photo; 172 (bckgd), Map © by Rand McNally, R.L. #99-S-64; 172 (bl), Photos provided courtesy of Daimler Chrysler; 172 (br), Courtesy Ford Motor Company; 172 (tr), ©1996 GM Corp. Used with permission GM Media Archives; 177 (bl), Peter Van Steen/HRW Photo; 178 (tl), Steve Payne/Uniphoto; 179 (tr), Image Copyright © 2001 PhotoDisc, Inc.; 184 (bc), (bl), Peter Van Steen/HRW Photo; 184 (cr), Ron Tanaka; 185 (l), Ladew Topiary Gardens, Monkton MD, Photography by Runk/Schoenberger from Grant Heilman Photography; 185 (r), Ron Tanaka; 186 (c), Peter Van Steen/HRW Photo; 187(tr), G. Ryan & S. Beyer/Tony Stone Images; 191 (cl), Mark E. Gibson; 192 (br), Randal Alhadeff/HRW Photo; 193 (bl), Image Copyright © 2001 PhotoDisc, Inc.; 195 (tr), George Hall/Check Six; 199 (bl), Tomasso Derosa/Allsport USA; 200 (bl), Rob Waymen Photography; 201 (tl), Lisa Valder/Tony Stone Images; 202 (bl), Runk/ Schoenberger from Grant Heilman; 202 (cl), Peter Beck/The Stock Market; 203 (tiles left to right): (1), Pascal Perret/The Image Bank; (2), Siqui Sanchez/The Image Bank; (3), & (4), John Foxx Images; 203 (bl), Michael Rosenfeld/Tony Stone Images; 203 (tl), Stephen Simpson/FPG International; 206 (bc), (br), Mark E. Gibson. **CHAPTER FOUR:** Page 214 (bl), Scott Barrow/International Stock Photo; 214 (cr), Elliot Smith/International Stock Photo; 214–215 (t), Chip Henderson/Tony Stone Images; 216 (t), Uniphoto; 218 (cr), Peter Van Steen/HRW Photo; 218 (cr), (tr), Randal Alhadeff/HRW Photo, props courtesy of Home Quarters; 222 (bl), 223 (br), Ron Tanaka; 223 (tr), Jan Becker; 225 (tr), Jerry Wachter/Photo Researchers; 227 (cl), Ron Tanaka; 228 (cl), Sam Dudgeon/HRW Photo; 231 (br), Jan Becker; 234 (tr), UPI/Corbis-Bettmann; 237 (c), Courtesy of NSA; 238 (br), Sam Dudgeon/HRW Photo; 242 (br), Randal Alhadeff/HRW Photo; 242 (c), Steve Grohe/ Thinking Machines Corporation 1991; 242 (tr), Image Copyright © 2001 PhotoDisc, Inc.; 243 (cl), Courtesy of NSA; 244 (tr), Bruce Ayres/Tony Stone Images; 245 (cr), Nick Dolding/Tony Stone Images; 251 (tr), 253 (tl), Michelle Bridwell/HRW Photo; 256 (bl), 258 (b), (br), (c), (t), Ron Tanaka; 259 (tr), Wayne Aldridge/International Stock Photo; 261 (br), John Langford/HRW Photo. **CHAPTER FIVE:** Page 272 (b), Superstock; 272 (c), Jim Cummins/FPG International; 272 (cl), Globus Brothers/The Stock Market; 272–273 (t), Image Copyright © 2001 PhotoDisc, Inc.; 274 (t), Clint Clemons/ International Stock Photo; 279 (cr), Jan Becker; 281 (tr), Chuck Mason/ International Stock Photo; 284 (tl), The Bettmann Archive; 284 (tr), G.A. Plimpton Collection, Rare Book & Manuscript Library, Columbia University; 288 (cl), Rob Waymen Photography; 297 (br), Yellow Dog Productions/The Image Bank; 298 (br), Frank Cezus/FPG International; 299 (t), 303 (tr), Mark E. Gibson; 305 (c), David K. Crow/PhotoEdit; 307 (t), Peter Van Steen/ HRW Photo; 312 (tr), Matt Lambert/Tony Stone Images; 313 (br), Frank Cezus/FPG International; 314 (tc), Photos courtesy of Daimler Chrysler; 315 (bl), Science Photo Library/Photo Researchers; 315 (br), Dr. Jeremy Burgess/Science Photo Library/Photo Researchers; 321 (bl), Andy Christiansen/HRW Photo; 322 (t), 325 (tr), Bill Losh/FPG International; 327 (tl), Vincent Graziani/ International Stock Photo; 328 (cl), VCG/FPG International; 330 (t), Robert Waymen Photography; 335 (br), VCG/FPG International; 336 (bl), Randal Alhadeff/HRW Photo; 337 (tr), Peter Van Steen/HRW Photo, location courtesy Austin Shoe Hospital; 338 (basketball), Frank Cezus/FPG International; 338 (cl), 338–339 (b), Image Copyright © 2001 PhotoDisc, Inc.; 339 (br), Corbis; 339 (tl), Image Copyright © 2001 PhotoDisc, Inc. **CHAPTER SIX:** Page 352 (bl), Jerry Jacka; 352 (cl), Patrick Aventurier/Gamma Liaison; 352 (cr), Richard Price/FPG International; 353 (cr), Arizona State Museum/University of Arizona; 353 (t), Mark Newman/International Stock Photo; 354 (cl), Lloyd Sutton/Masterfile; 354 (cr), David Scharf/Peter Arnold Inc.; 355 (cl), Randal Alhadeff/HRW Photo; 356 (tr), Paolo Negri/ Tony Stone Images; 357 (cr), Michelle Bridwell/HRW Photo; 359 (br), Superstock; 360 (br), Telegraph Colour Library/Masterfile; 361 (cr), David Starrett; 362 (t), Richard Ustinich/The Image Bank; 365 (cr), Vincent Graziani/International Stock Photo; 366 (tr), FPG International; 369 (br), David Starrett; 370 (t), Greg Pease/Tony Stone Images; 373 (br), David Starrett; 376 (tl), Stewart Cohen/Tony Stone Images; 376 (tr), Jeff Greenberg/Photo Researchers, Inc.; 377 (cl), (cr), The Granger Collection; 377 (tr), UPI/Corbis-Bettmann; 383 (tr), Doug Martin/Photo Researchers; 384 (cr), David Starrett; 385 (tl), Chris Baker/Tony Stone Images; 385 (tr), David Starrett; 386 (c), (cl), The Bettmann Archive; 389 (cl), David Starrett; 389 (cr), Paul Shambroom/Science Source/Photo Researchers; 390 (c), Patrick Cocklin/Tony Stone Images; 392 (t), Louis Psihoyos/Matrix; 396 (tl), The Granger Collection; 399 (cl), FPG International; 399 (cr), Ed Lallo/Tony Stone Images; 400 (bl), Jefferson National Expansion Memorial/National Park Service; 401 (tr), Alan Nyiri/FPG International; 402 (tl), Allan Seiden/The Image Bank; 402 (tr), Ulf E. Wallin/ The Image Bank; 403 (tr), Dan McCoy/First Light; 409 (br), 410–411 (all), David Starrett; 414 (br), Corbis. **CHAPTER SEVEN:** Page 422 (bl), Thomas Friedmann/ Photo Researchers; 422 (cr), Rohan/Tony Stone Images; 423 (br), Don Couch/HRW Photo; 423 (t), 4 24 (t), Uniphoto; 430 (bl), Mark Joseph/Tony Stone Images; 431 (br), (cl), (cr), Randal Alhadeff/HRW Photo; 432 (t), Bill Losh/FPG International; 437 (tl), Image Copyright © 2001 PhotoDisc, Inc.; 439 (tl), Camille Tokerud/Photo Researchers; 440 (t), Peter Van Steen/HRW Photo; 447 (cl), Randal Alhadeff/HRW Photo; 454 (bl), Sam Dudgeon/HRW Photo; 454 (c), Ron Tanaka; 454 (tr), The Bettmann Archive; 455 (cr), Randal Alhadeff/HRW Photo; 456 (bl), (c), Cedar Point Photos by Dan Feicht; 456 (popcorn), Richard Hutchings/PhotoEdit; 457 (r), Doug Armand/Tony Stone Images; 458 (tr), John Langford/HRW Photo; 462 (cr), Artbase Inc.; 462 (tr), The Bettmann Archive; 464 (bl), Larry Ulrich/Tony Stone Images; 465 (tr), Miwako Ikeda/ International Stock Photo; 466 (cl), 466–467 (b), Randal Alhadeff/HRW Photo. **CHAPTER EIGHT:** Page 478 (bl), Ellen Martorelli/Tony Stone Images; 478 (br), Uniphoto; 479 (t), D. Young-Wolff/PhotoEdit; 479 (t), Photo by Stuart Bowey/Adlibitum from *Discoveries: Great Inventions* © Weldon Owen Pty Ltd; 480 (t), Jerry Wachter/Photo Researchers; 484 (cr), Bob Thomason/Tony Stone Images; 487 (bl), Stan Osolinski/FPG International; 487 (br), Dan Sudia/Photo Researchers; 487 (cl), Mark Joseph/Tony Stone Images; 489 (t), Uniphoto; 495 (cr), Astrid & Hanns-Frieder Michler/Science Photo Library/Photo Researchers; 496 (br), Ken Cavanagh/ Photo Researchers; 498 (tc), Christine Galida/HRW Photo; 498 (tr), 501 (tr), 503 (bl), Randal Alhadeff/HRW Photo; 505 (t), Ed Pritchard/Tony Stone Images; 508 (cl), Leonard Lessio/Peter Arnold Inc.; 510 (cr), Scott Barrow/International Stock Photo; 512 (t), Andrew Freeman/SportsChrome-USA; 518 (br), VGC/FPG International; 519 (br), Brian Bailey/Tony Stone Images; 520 (t), Phil Degginger/Tony Stone Images; 525 (cr), Randal Alhadeff/HRW Photo; 526 (br), Mark Wagner/Tony Stone Images; 527 (c), Tom & Pat Leeson/Photo Researchers; 528 (t), 530 (br), Peter Van Steen/HRW Photo; 535 (tr), Tony Freeman/PhotoEdit; 536 (t), 541 (bl), John Neubauer/PhotoEdit; 543 (tr), Audrey Gibson; 544–545 (from left to right), (1), Chuck Szymanski/ International Stock Photo; (2), E.J. West/Index Stock; (3), Wayne Aldridge/International Stock Photo; (4), Michael Lichter/International Stock Photo; (5), S.I. Swartz/Index Stock; (6), Lester Lefkowitz/The Stock Market; (7), S.I. DeYoung/Index Stock; 550 (br), NASA. **CHAPTER NINE:** Page 560 (bl), David Nunuk/Science Photo Library/Photo Researchers; 560–561 (t), NASA; 560–561 (all), photo manipulation by Uhl Studio Incorporated; 561 (br), David Ducros/Science Photo Library/Photo Researchers; 565 (tl), SIU/Peter Arnold, Inc.; 567 (tr), Bill Stormont/The Stock Market; 570 (t), 573 (tl), Bob Daemmrich Photos; 578 (cr), Don Couch/HRW Photo; 579 (t), Bob Firth/International Stock Photo; 581 (bl), Lonnie Duka/Tony Stone Images; 582 (br), Mark C. Burnett/Science Source/Photo Researchers; 585 (cr), Jeff Zaruba/Tony Stone Images; 586 (cr), John Langford/HRW Photo; 586 (t), David A. Hardy/Science/Photo Library/Photo Researchers; 588 (bl), United States Geological Survey, Flagstaff, Arizona; NASA; 591 (cr), Antonio Rosario/The Image Bank; 593 (cr), NASA/Science Source/Photo Researchers; 594 (cr), Don Couch/HRW Photo; 595 (t), Bryn Campbell/Tony Stone Images; 600 (bl), Sam Dudgeon/HRW Photo; 602 (br), Cliff Hollenbeck/International Stock Photo; 603 (cr), Don Couch/HRW Photo; 604 (br), Brett Froomer/The Image Bank; 605 (tr), Sam Dudgeon/HRW Photo; 606 (t), Alan Oddie/PhotoEdit; 608 (cr), Ron Tanaka, photo manipulation by Jun Park; 614 (t), Image Copyright © 2001 PhotoDisc, Inc.; 615 (br), Image Copyright © 2001 PhotoDisc, Inc.; 618 (br), Art Wolfe/Tony Stone Images. **CHAPTER TEN:** Page 626 (bl), Alison Wright/Photo Researchers; 626 (cr), Ken Hawkins/Uniphoto; 627 (cr), Sam Dudgeon/HRW Photo; 627 (tr), Roslan Rahman/AFP Photo; 628 (t), 629 (br), 630 (tc), 630 (tr), eStudios/HRW Photo; 631 (bl), Matt Bowman/Foodpix; 632 (cl), Roy Morsch/The Stock Market; 633 (tl), eStudios/HRW Photo; 636 (t), Arnold Schönberg Center; 637 (br), Randal Alhadeff/HRW Photo; 638 (br), Steve Satushek/The Image Bank; 639 (br), Ron Tanaka; 640 (cr), Sam

Dudgeon/HRW Photo; 641 (br), Superstock; 644 (br), Christel Rosenfeld/Tony Stone Images; 648 (b), Rudi Von Briel/PhotoEdit; 650 (tr), 651 (b), Copyright 2000, ABC, Inc.; 651 (c), Image Copyright © 2001 PhotoDisc, Inc.; 652 (t), Peter Van Steen/HRW Photo; 654 (cr), Marty Granger/Edge Video Productions/ HRW; 658 (tr), David Young-Wolff/ PhotoEdit; 659 (t), VCG/FPG International; 660 (br), Digital Stock Corp.; 663 (tr), Llewellyn/Uniphoto; 664 (t), Bob Daemmrich Photo; 667 (tr), Don Couch/ HRW Photo; 669 (br), Chip Simons/FPG International; 671 (tl), (tr), Superstock; 672 (cr), Alan Schein/The Stock Market;
673 (tc), Peter Gridley/FPG International; 678 (bl), (tr), Image Copyright © 2001 PhotoDisc, Inc.; 679 (br), Stockman/International Stock Photo. **CHAPTER ELEVEN:** Page 688 (br), Image Copyright © 2001 PhotoDisc, Inc.; 688 (cr), Arizona State Museum/University of Arizona; 689 (t), James Randklev/ Tony Stone Images; 690 (t), Mehau Kulyk/Science Photo Library/Photo Researchers; 693 (c), Runk/Schoenberger from Grant Heilman; 697 (br), Dr. Tony Brain/Science Photo Library/Photo Researchers; 698 (t), Randal Alhadeff/HRW Photo; 699 (t), Courtesy of Rainbird; 703 (tr), Mark Bolster/International Stock Photo; 705 (cr), Stephen Simpson/FPG International; 706 (br), John Langford/HRW Photo; 707 (t), 712 (tr), Don Couch/HRW Photo; 713 (t), John Langford/HRW Photo; 718 (br), Walter Bibikow/FPG International; 720 (t), John Langford/HRW Photo; 726 (cl), Globus, Holway & Lobel/The Stock Market; 727 (tr), Bob Daemmrich/ Uniphoto; 728 (t), *Fractals Everywhere* ©1988 Academic Press; 735 (bl) frame, Image Copyright © 2001 PhotoDisc, Inc.; 735 (t), Stanley Meltzoff; 740 (cr), reprinted with permission of Cambridge University Press, 1959; 741 (t), David Seelig/Allsport; 743 (tl), Courtesy of The Topps Company, Inc.; 746 (bc), (br), Image Copyright © 2001 PhotoDisc, Inc.; 747 (tr), Stephen Simpson/FPG International; 748 (tr), Christine Galida/HRW Photo; 749 (bl), Christine Galida/HRW Photo. **CHAPTER TWELVE:** Page 762 (bl), David Young-Wolff/Tony Stone Images; 762 (c), Mark Scott/FPG International; 762 (tr), Novastock/PhotoEdit; 763 (tl), Janeart/The Image Bank; 764 (t), John Langford/HRW Photo; 765 (br), Robert E. Daemmrich/Tony Stone Images; 766 (bl), E.R. Degginger/Science Source/Photo Researchers; 768 (cl), Billy Hustace/Tony Stone Images; 769 (cl), David De Lossy/The Image Bank; 770 (bl), Sam Dudgeon/HRW Photo; 772 (tr), Loren Santow/Tony Stone Images; 773 (frame), Image Copyright © 2001 PhotoDisc, Inc.; 773 (tr), Rob Lewine/The Stock Market; 774 (tr), Andre Jenny/International Stock Photo; 776 (br), Charles D. Winters/Science Source/Photo Researchers; 778 (tl), Uniphoto; 779 (bl), Robert Ginn/PhotoEdit; 781 (tl), Jeff Hunter/The Image Bank; 781 (tr), John Livzey/Tony Stone Images; 784 (bl), National Baseball Hall of Fame Library & Archive; 784 (br), Milo Stewart Jr./Baseball Hall of Fame Library; 786 (bl), Grandadam/Tony Stone Images; 787 (tl), Michael Newman/PhotoEdit; 788 (tl), UPI/Corbis-Bettmann; 790 (bl), VCG/FPG International; 790 (br), Image Copyright © 2001 PhotoDisc, Inc.; 791 (br), Ron Tanaka, photo manipulation by Jun Park; 792 (tl), John Terence Turner/ FPG International; 792 (tr), David Pollack/The Stock Market; 793 (tr), Superstock; 795 (cl), Peter Pearson/Tony Stone Images; 797 (bl), William Sallaz/The Image Bank; 798 (b), Scott Barrow/International Stock Photo; 799 (t), Yoav Levy/Phototake NYC; 801 (cr), Gabe Palmer/The Stock Market; 801 (tr), Helmut Gritscher/Peter Arnold, Inc.; 803 (cr), Renee Lynn/Science Source/Photo Researchers; 804 (bc), Randal Alhadeff/HRW Photo; 804 (bl), Corbis; 805 (tr), Matt Bradley/Uniphoto; 806 (t), Ralph H. Wetmore/Tony Stone Images; 809 (tr), Robert Essel/The Stock Market; 810 (tr), David Frazier/Tony Stone Images; 811 (cl), Gerhard Gscheidle/Peter Arnold, Inc.; 812 (bl), Sigrid Owen/ International Stock Photo; 812 (cr), Tony Freeman/PhotoEdit; 814 (bl), Scott Barrow/ International Stock Photo; 814 (tc), Image Copyright © 2001 PhotoDisc, Inc.; 814 (tl), Michael Young/HRW Photo; 815 (br), (tr), Image Copyright © 2001 PhotoDisc, Inc. **CHAPTER THIRTEEN:** Page 826 (bl), F. Hidalgo/The Image Bank; 827 (bl), Superstock; 827 (C), Image Copyright © 2001 PhotoDisc, Inc.; 827 (tr), Stephen Studd/Tony Stone Images; 828 (t), S.J. Krasemann/Peter Arnold, Inc.; 833 (cl), Mark E. Gibson; 834 (br), Grant Heilman Photography; 836 (t), Joe McBride/ Tony Stone Images; 841 (tl), Neville Dawson/Check Six; 843 (tc), Tom Tracy/FPG International; 851 (t), Daily Telegraph Colour Library/International Stock Photo; 853 (bl), Roger Tully/Tony Stone Images; 854 (cr), Warren Faidley/International Stock Photo; 855 (tl), Matt Brown/Uniphoto; 858 (tr), D. Roundtree/The Image Bank; 862 (tr), Ron Scherl/Stage Image; 865 (cl), Marc Romanelli/The Image Bank; 872 (cr), Merritt Vincent/PhotoEdit; 874 (br), Peter Pearson/Tony Stone Images; 874–875 (bckgd), UPI/Corbis Bettmann; 875 (tr), Carnegie Library, Pittsburgh. **CHAPTER FOURTEEN:** Page 884 (bc), David Parker/Science Photo Library/Photo Researchers; 884 (br), Spencer Jones/FPG International; 885 (t), Tony Stone Images; 886 (t), Superstock; 892 (cl), VCG/FPG International; 893 (tl), Joe Towers/The Stock Market; 894 (t), Grant V. Faint/The Image Bank; 900 (br), Walter Bibicow/The Image Bank; 901 (tl), Uniphoto; 902 (t), Superstock, photo manipulation by Morgan-Cain & Associates; 905 (cr), Ben Osborne/Tony Stone Images; 906 (c), Image Copyright © 2001 PhotoDisc, Inc.; 907 (br), Superstock; 908 (cl), Francois Gohier/Photo Researchers; 909 (t), Adam Peiperl/The Stock Market; 915 (br), Warren Bolster/Tony Stone Images; 915 (cl), Dennis Hallinan/FPG International; 916 (br), Will Ryan/The Stock Market; 917 (t), Superstock; 920 (frame), Image Copyright © 2001 PhotoDisc, Inc.; 920 (tr), Erich Lessing/Art Resource, NY; 921 (tr), Rich Kane/SportsChrome USA; 922 (tl), Uniphoto; 922 (tr), David Madison/Tony Stone Images; 925 (bl), Henry T. Kaiser/ Uniphoto; 927 (bl), Bill Beatty/Visuals Unlimited; 928 (bl), Corbis;

ILLUSTRATIONS

All technical and line art by Morgan-Cain & Associates. Other art, unless otherwise noted, by Holt, Rinehart & Winston.

Abbreviated as follows: (t) top, (b) bottom, (l) left, (r) right, (c) center.
CHAPTER ONE: Page 3 (cr), Morgan-Cain & Associates; 9 (bl), Stephen Durke/ Washington-Artists' Represents; 22 (bl), Michael Morrow Design; 42 (c), Visual Sense Illustration/Margo Davies Leclair; 43 (t), Ian Phillips; 50 (cr), Leslie Kell; 51 (tl), Leslie Kell; 54 (tr), Stephen Durke/Washington-Artists' Represents; 66 (t), Michael Morrow Design. **CHAPTER TWO:** Page 88 (c), Leslie Kell; 90 (cr), Stephen Durke/Washington-Artists' Represents; 92 (cl), Stephen Durke/Washington-Artists' Represents; 100 (bl), Stephen Durke/Washington-Artists' Represents; 107 (bl), Stephen Durke/Washington-Artists' Represents; 113 (tc), Stephen Durke/Washington-Artists' Represents; 118 (t), Stephen Durke/Washington-Artists' Represents; 123 (c), Stephen Durke/Washington-Artists' Represents; 124 (tr), Leslie Kell. **CHAPTER THREE:** Page 154 (c), Morgan-Cain & Associates; 158 (br), Stephen Durke/Washington-Artists' Represents; 172 (c), Leslie Kell; 182 (tr), Leslie Kell. **CHAPTER FOUR:** Page 214 (c), Morgan-Cain & Associates; 232 (tc), Stephen Durke/Washington-Artists' Represents; 244 (tr), Leslie Kell; 249 (br), Leslie Kell; 250 (tc), Jun Park. **CHAPTER FIVE:** Page 274 (c), Stephen Durke/ Washington-Artists' Represents; 279 (cl), Stephen Durke/Washington-Artists' Represents; 283 (cr), Stephen Durke/ Washington-Artists' Represents; 289 (tr), Stephen Durke/ Washington-Artists' Represents; 290 (t), Uhl Studio Incorporated; 295 (tr), Uhl Studio Incorporated; 307 (cr), Stephen Durke/Washington-Artists' Represents; 312 (cl), Uhl Studio Incorporated; 314 (tc), Stephen Durke/Washington-Artists' Represents. **CHAPTER SIX:** Page 360 (tl), Leslie Kell; 368 (cr), Martha Newbigging; 374 (cr), Stephen Durke/Washington-Artists' Represents; 383 (cr), Stephen Durke/ Washington-Artists' Represents; 391 (cr), Stephen Durke/Washington-Artists' Represents; 398 (br), Nenad Jakesevic; 405 (bl), Stephen Durke/Washington-Artists' Represents; 407 (br), Leslie Kell; 408 (t), Jun Park. **CHAPTER SEVEN:** Page 424 (tr), Stephen Durke/Washington-Artists' Represents; 448 (tr), Uhl Studio Incorporated; 453 (tr), Stephen Durke/Washington-Artists' Represents; 458 (tc), Leslie Kell. **CHAPTER EIGHT:** Page 478 (c), Uhl Studio Incorporated; 484 (tr), Uhl Studio Incorporated; 504 (t), Stephen Durke/Washington-Artists' Represents; 510 (b), Stephen Durke/Washington-Artists' Represents; 533 (tr), Stephen Durke/Washington-Artists' Represents; 544 (c), Leslie Kell. **CHAPTER NINE:** Page 560–1 (all), Uhl Studio Incorporated; 562 (tr), Stephen Durke/Washington-Artists' Represents; 564 (br), Ortelius Design; 577 (bl), Jun Park; 586, (br), Jun Park; 593 (bl), Leslie Kell; 610 (bl), Uhl Studio Incorporated; 613 (t), Stephen Durke/Washington-Artists' Represents. **CHAPTER TEN:** Page 634 (br), Michael Herman; 643 (t), Stephen Durke/ Washington-Artists' Represents; 643 (t), Stephen Durke/Washington-Artists' Represents; 645 (cr), Leslie Kell; 647 (cl), Stephen Durke/Washington-Artists' Represents; 650 (bc), Leslie Kell; 657 (br), Leslie Kell; 662 (br), Stephen Durke/Washington-Artists' Represents; 675 (tr), Leslie Kell; 677 (tr), Leslie Kell. **CHAPTER ELEVEN:** Page 711 (br), Stephen Durke/ Washington-Artists' Represents; 719 (t), Uhl Studio Incorporated; 722 (b), Uhl Studio Incorporated; 725 (br), Michael Herman; 727 (bl), Leslie Kell; 728 (cr), Michael Herman; 744 (bl), Pronk & Associates. **CHAPTER TWELVE:** Page 762 (b), Morgan-Cain & Associates; 763 (tr), Stephen Durke/ Washington-Artists' Represents; 782 (cr), Leslie Kell; 785 (cl), Bernadette Lau. **CHAPTER THIRTEEN:** Page 828 (cl), Morgan-Cain & Associates; 831 (cr), Morgan-Cain & Associates; 831 (bl), Nenad Jakesevic; 833 (tr), Terry Guyer; 834 (cr), Nenad Jakesevic; 835 (cr), Terry Guyer; 837 (tr), Leslie Kell; 849 (cl), Uhl Studio Incorporated; 850 (cr), Terry Guyer; 856 (cr), Uhl Studio Incorporated; 856 (br) David Puckett; 867 (t), Stephen Durke/Washington-Artists' Represents; 872 (bl), Morgan-Cain & Associates; 873 (tr), Terry Guyer; 873 (cr), Stephen Durke/Washington-Artists' Represents. **CHAPTER FOURTEEN:** Page 892 (cr), Morgan-Cain & Associates; 892 (b), Morgan-Cain & Associates; 893 (tr), Morgan-Cain & Associates; 896 (tr), Morgan-Cain & Associates; 901 (br), Leslie Kell; 927 (c), Morgan-Cain & Associates; 932 (all) Morgan-Cain & Associates.

PERMISSIONS

For permission to reprint copyrighted material, grateful acknowledgment is made to the following sources:

The New York Times Company: From "The Quest for True Randomness Finally Appears Successful" by James Gleick from *The New York Times,* April 19, 1988. Copyright © 1988 by The New York Times Company. From "Biggest Division a Giant Leap in Math" by Gina Kolata and "Factoring a 155-Digit Number: The Problem Solved" from *The New York Times,* June 20, 1990. Copyright © 1990 by The New York Times Company.

Sandusky Register, Ohio: From "Coasting to Records" from *Sandusky Register,* July 27, 1989. Copyright © 1989 by Sandusky Register.

St. Louis Mercantile Library Association: Photo caption "St. Louis Can Now Boast of the Nation's Highest Memorial" from "Arch Completed but Fight for Funds to Finish Memorial Still Goes On" from *St. Louis Globe-Democrat,* October 29, 1965. From the collections of the St. Louis Mercantile Library at the University of Missouri, St. Louis.

Time Inc.: From "Time for Some Fuzzy Thinking" by Philip Elmer-Dewitt from *Time,* September 25, 1989. Copyright © 1989 by Time Inc.

Marilyn vos Savant and Parade: From column "Ask Marilyn™" by Marilyn vos Savant from *PARADE,* September 9, 1990. Copyright © 1990 by Parade.